THE COMPACT EDITION OF THE OXFORD ENGLISH DICTIONARY

THE COMPACT EDITION OF THE OXFORD ENGLISH DICTIONARY

COMPLETE TEXT REPRODUCED MICROGRAPHICALLY

VOLUME III

A SUPPLEMENT
TO THE
OXFORD ENGLISH DICTIONARY
VOLUMES I–IV

EDITED BY
R. W. BURCHFIELD

OXFORD
AT THE CLARENDON PRESS
1987

Oxford University Press, Walton Street, Oxford OX2 6DP
Oxford New York Toronto
Delhi Bombay Calcutta Madras Karachi
Petaling Jaya Singapore Hong Kong Tokyo
Nairobi Dar es Salaam Cape Town
Melbourne Auckland
and associated companies in
Beirut Berlin Ibadan Nicosia

Oxford is a trade mark of Oxford University Press

Published in the United States
by Oxford University Press, USA

British Library Cataloguing in Publication Data
The Compact edition of the Oxford English dictionary: complete text reproduced micrographically.
Vol. 3: Supplements 1–4
1. English language—Dictionaries
I. Burchfield, Robert R. W.
423 PE1625
ISBN 0–19–861211–7

Library of Congress Cataloging in Publication Data
The Compact edition of the Oxford English dictionary.
Includes bibliographies.
1. English language—Dictionaries.
I. Burchfield, R. W.
II. Title: Oxford English dictionary.
PE1625.C58 1987 423 87–1592
ISBN 0–19–861211–7 (v. 3)

Printed in the United States of America

This Supplement to the Oxford English Dictionary

is respectfully dedicated to

HER MAJESTY THE QUEEN

by her gracious permission

CONTENTS

PREFACE TO VOLUME I

THE present volume is the first of three which together will replace the 1933 Supplement to the *O.E.D.* It is hoped to publish the remaining two volumes at intervals of not more than three years. The second volume will end at P, and the concluding volume, in addition to words in the range Q to Z, will contain an extensive Bibliography of works cited in the new Supplement.

The vocabulary treated is that which came into use during the publication of the successive sections of the main Dictionary—that is, between 1884, when the first fascicle of the letter A was published, and 1928, when the final section of the Dictionary appeared—together with accessions to the English language in Britain and abroad from 1928 to the present day. Nearly all the material in the 1933 Supplement has been retained here, though in revised form.

Dr. Johnson said in the Preface to his dictionary (1755) that he would admit 'no testimony of living authours':

> My purpose was to admit no testimony of living authours, that I might not be misled by partiality, and that none of my cotemporaries might have reason to complain; nor have I departed from this resolution, but when some performance of uncommon excellence excited my veneration. . . So far have I been from any care to grace my pages with modern decorations, that I have studiously endeavoured to collect examples and authorities from the writers before the restoration, whose works I regard as the wells of English undefiled, as the pure sources of genuine diction.

By such reckoning the terminal date for this new Supplement of modern English words should be certainly not later than the end of the nineteenth century. Sir William Craigie took a similar view in the Preface (1938) to his *Dictionary of American English*:

> The end of the nineteenth century has been selected as a fitting point at which to terminate the admission of new words, however common some of these may have become in recent use. The illustration of those already current before that date, however, is frequently carried into the first quarter of the present century.

A Supplement based on such a premise would, however, have been of restricted value. In the present work we have endeavoured to give shape and historical outline, graced necessarily with 'modern decorations', to a large body of the most recent accessions to the English language. In doing this we have kept constantly before us the opposing concepts of permanence and ephemerality, retaining vocabulary that seemed likely to be of interest now and to future generations, and rejecting only those words, phrases, and senses that seemed transitory or too narrowly restricted in currency.

It is fitting that we should here pay tribute to the Editors of the 1933 Supplement, Dr. C. T. Onions and Sir William Craigie. The replacement of the volume they produced must necessarily mean that copies of it will in future descend, along with other rarely consulted works, into the vaults of the larger libraries. But we feel that neither of these two great scholars would lament this course of events: it is the normal fate of a supplementary volume to vanish into the body of the reference work to which it is for a time annexed, or else to make way for a larger, more recent supplement. The enduring monument to the work of these two great lexicographers, and of their colleagues, remains the *O.E.D.* itself.

The Editor wishes also to record here his personal indebtedness to the late Dr. Onions, who first guided him into the field of lexicography; to Professor Norman Davis and Professor E. G. Stanley, whose assistance on many matters has been deeply appreciated; to numerous transatlantic

friends, including Professor Raven I. McDavid Jr. and Mr. Clarence L. Barnhart; to Dr. G. W. S. Friedrichsen, the principal etymologist of the Oxford dictionaries; to Miss Marghanita Laski (Mrs. John Howard) for her devotion to the Dictionary; to innumerable colleagues in the University of Oxford (especially in St. Peter's College) and the Oxford University Press (especially Mr. C. H. Roberts, Mr. D. M. Davin, and the late Kenneth Sisam) for their friendly advice and helpfulness over a period of many years; and most particularly to his colleagues and assistants on the Dictionary staff itself.

EDITORIAL STAFF

FULL-TIME STAFF

ELIZABETH BROMMER	1959–72	J. P. BARNES	1969–72
A. J. AUGARDE	1960–	SANDRA RAPHAEL	1969–
E. C. DANN	1963–	M. W. GROSE	1969–
VERONICA M. SALUSBURY	1966–	DEIRDRE MCKENNA	1969–
ADRIANA P. ORR	1966–	W. H. C. WATERFIELD	1970–
A. M. HUGHES	1968–	DEBORAH M. COWEN	1970–

PART-TIME STAFF

L. B. FIRNBERG	1962–	PEGGY E. KAY	1967–
ANNE WALLACE-HADRILL	1967, 1971–2	FRANCES M. WILLIAMS	1968–
JELLY K. WILLIAMS	1967–	J. B. SYKES	1971–

Miss Brommer and Miss Salusbury (based in London), Mrs. Orr (in Washington), and Mrs. Kay (in Oxford) were mainly concerned with research (especially for 'first uses') and with the verification of quotations; Mr. Hughes and Mr. Waterfield with scientific terms; and Miss Raphael with plant and animal terms. Mrs. Cowen dealt with terms in Psychology. Mr. Grose was responsible for the collation and assembly of bibliographical information. Miss F. M. Williams assisted with the reading of the proofs. All others named above undertook general editorial work.

Among those who assisted at various stages with editorial work or with bibliographical verification as part of the regular staff were the following: R. C. Goffin (1955–60), E. A. Horsman (1956), Jennifer Dawson (1957–8, 1959), Sally Hilton (1958–60), Joyce M. Hawkins (1958–67), N. van Blerk (1961, Associate Editor of *Woordeboek van die Afrikaanse Taal*), Marjorie Purdon (1962–5), Phyllis Trapp (1964, New Zealand words), Elizabeth Price (1969), and R. J. Dixon (1968–70).

Members of the Editorial Staff received valuable part-time assistance from the following outside helpers: Grace M. Briggs (1959–), Betsy Livingstone (1963–6), N. C. Sainsbury (1963–6, establishment of bibliographical card-index), R. Hall (1963– , philosophical terms), Rita G. Keckeissen (1968–), Dr. J. B. Sykes (1967–71, scientific terms; from 1971 Editor of the *Concise Oxford Dictionary*), and Daphne Gilbert-Carter (1968–9).

Secretarial and Clerical Assistants: Caroline Webb (1958–60), Betty Jennison (1958–60), Jill Cotter (1963–6), Joan Blackler (1966–), Pamela Bendall (1968–), Joyce Harley (1970–), and others. Miss Jennison and Mrs. Cotter also assisted with general editorial work, and Miss Harley with research and the verification of quotations.

CONTRIBUTORS

THIS list contains the names of the principal readers (excluding present members of staff, who have also made substantial contributions) between 1957 and 1971. The main period of reading was between 1958 and 1961 and the material contributed during this period formed the nucleus of the collection of quotations on which the Supplement is based.

A. The following four readers contributed altogether approximately 250,000 quotations:[1]

R. A. Auty	1958–†1967	Miss Marghanita Laski	1959–
W. Kings	1959–71	Mrs. Stefanyja Ross	1959–

B. The following readers supplied between 3,000 and 20,000 quotations each:

D. J. Barr (Canadian sources, etc.)
R. L. Cherry (Thoreau, Mark Twain, etc.)
C. Collier (Australian newspapers)
†P. T. J. Dadley (*Daily Telegraph*)
Mrs. Margaret Gordon (Ornithology, etc.)
Miss Joyce M. Hawkins (P. G. Wodehouse, etc.)
F. M. Henry (Aeronautics)
V. W. Jones (Music, *Punch*, etc.)
Dr. D. Leechman (Canadian sources)
Miss Alison Megroz (*Discovery*, etc.)
G. B. Onions (World War II sources, etc.)
The Revd. H. E. G. Rope[2] (Religion, etc.)
D. Scott (Australasian sources)
Mrs. Gereth M. Spriggs (Antiquarian sources, Arts and Crafts)
F. J. Tidd Pratt (General literature)
†Sir St. Vincent Troubridge (Theatrical sources, etc.)
Miss E. G. Withycombe (Forestry, etc.)

C. Other readers[3] were:

Mrs. J. M. Addenbrooke
N. S. Angus
†Dr. E. de Barry Barnett
Dr. E. H. Bateman
†G. Bennett
Rowland Bowen
Mrs. Ruth C. L. Boxall
P. J. N. Bury
The Revd. R. M. Catling
G. Chowdharay-Best
Mrs. Norah Day
G. W. Dennis
A. H. Douglas
Professor Sir Godfrey Driver
Mrs. Elsie Duncan-Jones
†E. H. Fathers
P. Ferriday
Mrs. A. S. R. Gell
W. Granville
D. Gray
Miss M. Gregory
†G. Griffith
Mrs. Gwynneth Hatton
Dom Sylvester Houédard
Dr. M. D. W. Jeffreys
Emrys Jones
H. L. Jones
Mrs. Jeanne Lindley
†J. P. Lloyd
R. H. T. Mackenzie
W. S. Mackie
J. C. Maxwell
Mrs. Joan Morgan
J. L. Nayler
Mrs. Patricia Norton
Mrs. Elinor Owen
M. B. Parkes
Miss E. Penwarden
W. S. Pierpoint
W. S. Ramson
Mrs. Cherry Record
Miss Alison Redmayne
Miss F. E. Richardson
Miss Louise L. Ross
N. Sabbagha
Miss Ruth C. Salzberger
J. C. Sharp
Professor G. Shepherd
D. Shulman
Mrs. Valentine Sillery
Mrs. B. M. D. Smith
Mrs. Miemie W. Smith
†C. Nelson Stewart
C. P. Swart
Miss Eleanor Symons
D. Halton Thomson
Dr. T. R. Thomson
The Revd. A. F. Thorpe
Mrs. Cynthia Walton
Dr. M. West
E. W. Whittle
J. D. A. Widdowson
H. W. B. Wilson
Dr. L. S. Wittenberg
A. M. Wood

[1] The late Mr. Auty read some 900 books including most of the works of T. S. Eliot, D. H. Lawrence, James Joyce, and W. B. Yeats, and runs of many periodicals (e.g. *Times Literary Supplement* 1930–8, *The Times* 1958–67, *Scrutiny* 1932–8, *Penguin New Writing* (all issues)), works on Linguistics, Cricket, Bridge, etc. Mr. Kings read many scientific books and journals. Miss Laski's reading included, for example, almost all the works of some twentieth-century authors (e.g. G. B. Shaw, Max Beerbohm, Virginia Woolf, Aldous Huxley) and numerous modern crime novels. She read extensively in the literature of the nineteenth century, both novels (e.g. Dickens, C. M. Yonge) and letters (Mrs. Gaskell, George Eliot, G. M. Hopkins). She also read widely in the general field of the domestic arts (old catalogues, books on gardening, cooking, embroidery, etc.) and various modern newspapers and journals (*Guardian*, *Vogue*, etc.). Mrs. Ross made a systematic reading of *Notes and Queries* from 1930 to 1959, *The Listener*, *S.P.E. Tracts*, *American Speech*, and a wide selection of twentieth-century fiction.

[2] At the age of 90 (in 1971) probably the oldest contributor: his name appears among the list of those who contributed to the *O.E.D.* itself.

[3] In the nature of things this list cannot be exhaustive: we hope that no important contributor has been overlooked.

† = deceased.

PROOF-READERS

THE following list contains the names of those who, in addition to the dictionary staff, rendered valuable help by regularly reading the proofs and making suggestions and additions. Proofs of individual items were also submitted to many of those whose names appear in the list of Outside Consultants.

- Professor Audrey R. Duckert, PH.D.
- Dr. B. Foster, M.A., Docteur de l'Université de Paris
- Professor Atcheson L. Hench, A.M., PH.D., LITT.D.
- Miss Marghanita Laski, M.A.
- Mrs. Marguerite Y. Offord, M.A., B.LITT.
- Mrs. Lai-ngau Wong Pauson, PH.B.
- Mrs. Stefanyja Ross, B.A.
- Professor E. G. Stanley, M.A., PH.D.

OUTSIDE CONSULTANTS

THIS list includes the names of some of those who helped us in the formative years immediately after 1957 as well as specialists to whom the editorial staff have turned repeatedly in more recent years for comments and advice. A great many others have given advice on single words and on subjects in which there have been few new terms: they are too numerous to be named here but their assistance is gratefully acknowledged.

- G. C. Ainsworth, PH.D., F.I.BIOL.
- Professor G. E. Blackman, F.R.S.
- Professor A. J. Bliss, M.A., B.LITT., M.R.I.A.
- E. J. Bowen, M.A., D.SC., F.R.S.
- Savile Bradbury, M.A., D.PHIL.
- R. S. Cahn, M.A., D.PHIL.NAT., F.R.I.C.
- Harry Carter, O.B.E., M.A.
- John Carter, C.B.E., M.A.
- Professor Frederic G. Cassidy, PH.D.
- Julio Cortés, PH.D.
- Mrs. Jessie Coulson, B.A. (Editor of the *Little Oxford Dictionary*, etc.)
- G. N. C. Crawford, M.A., M.B., B.S.
- Professor C. D. Darlington, D.SC., F.R.S.
- Professor Norman Davis, F.B.A.
- Professor E. J. Dobson
- Miss Jessie Dobson, B.A., M.SC., A.C.I.S.
- N. S. Doniach, O.B.E., M.A.
- Professor Audrey R. Duckert, PH.D.
- R. D. Eagleson, PH.D., M.A.
- P. G. Embrey, M.A., B.SC.
- D. F. Foxon, M.A.
- R. B. Freeman, M.B.E., M.A.
- G. W. S. Friedrichsen (etymological Editor of several Oxford dictionaries)
- Peter Fryer
- W. K. V. Gale
- P. G. W. Glare, M.A.
- Ives Goddard, PH.D.
- Wilfred Granville
- P. S. Green, B.SC., F.L.S.
- J. S. Gunn, M.A.
- Professor Malcolm Guthrie, PH.D., B.SC., F.B.A.
- R. G. Haggar, R.I., A.R.C.A.
- R. M. Hartwell, M.A., D.PHIL.
- David Hawkes, M.A., D.PHIL.
- R. E. Hawkins, M.A.
- Woodford A. Heflin, M.A., PH.D.
- M. H. Hey, M.A., D.SC.
- A. K. Huggins, M.SC., PH.D.
- Khurshidul Islam, M.A., PH.D.
- Philip Jarrett
- David Kahn, A.B.
- N. R. Ker, PH.D.
- Professor K. Koike
- David Lack, D.SC., F.R.S.
- Douglas Leechman, B.SC., M.A., PH.D.
- Professor Bernard Lennox, M.D., PH.D., F.R.C.P., F.R.C.PATH., M.R.C.P.
- †J. P. Lloyd
- Anthony Loveless, B.SC., PH.D., D.SC.
- Professor Raven I. McDavid, Jr.
- F. H. C. Marriott, M.A., PH.D.
- R. D. Meikle, B.A., LL.B.
- P. A. Mulgan, M.A., B.LITT.
- Iona and Peter Opie, HON.M.A. (Oxon.)
- N. W. Pirie, M.A., F.R.S.
- †Capt. J. L. Pritchard
- Professor R. H. Robins, M.A., D.LIT.
- D. A. Roe, M.A., PH.D.
- †K. S. Sandford
- H. G. A. V. Schenk, M.A., D.PHIL.
- J. S. G. Simmons, M.B.E., M.A.
- Frank H. Smith
- Svante Stubelius, PH.D.
- William C. Sturtevant, PH.D.
- C. H. V. Sutherland, C.B.E., M.A., D.LITT., F.B.A.
- Peter Tamony
- J. B. Trapp, M.A.
- Guenter Treitel, B.C.L., M.A.
- G. W. Turner, M.A.
- J. O. Urmson, M.C., M.A.
- N. van Blerk, M.A.
- Richard L. Venezky, PH.D.
- Professor E. A. Vincent, M.A., M.SC., PH.D., F.R.I.C., F.G.S.
- The Revd. Canon Professor M. F. Wiles, D.D.
- D. R. Woodall, M.A., PH.D.

PREFACE TO VOLUME II

DRYDEN remarks in his *Preface to the Fables* (1700):

> 'Tis with a Poet, as with a Man who designs to build, and is very exact, as he supposes, in casting up the Cost beforehand: But, generally speaking, he is mistaken in his Account, and reckons short of the Expence he first intended: He alters his Mind as the Work proceeds, and will have this or that Convenience more, of which he had not thought when he began. So has it hapned to me; I have built a House, where I intended but a Lodge.

This comment came into my mind when it became apparent that the material in the letters after G could not easily be contained in the two further volumes that were promised when Volume I (A–G) was published in 1972. This second volume of the Supplement ends with N, and there will be two further volumes. The fourth volume will include an extensive Bibliography of works cited in the new Supplement.

The main lines of policy laid down in the first volume are retained in this one, but the material in our quotation files has continued to expand and this expansion is reflected in the vocabulary included in the present volume. It would be difficult to describe every aspect of 'this or that Convenience more' included in the range H–N. Some of the new areas explored are mentioned in papers that I delivered to the Philological Society in 1973[1] and to the Royal Society of Arts in 1975.[2] Others have been dealt with more briefly in several papers on miscellaneous topics.[3] The main conclusions of these papers are, among others, that (i) offensiveness to a particular group, minority or otherwise, is unacceptable as the sole ground for the exclusion of any word or class of words from the *O.E.D.*; (ii) it is therefore desirable to enter new racial and religious terms however opprobrious they may seem to those to whom they are applied and often to those who have to use them, or however controversial the set of beliefs professed by the members of minority sects; (iii) it is also desirable, in order to avoid misunderstanding and consequent hostility, that the somewhat antiquated historical record of words like *Jesuit*, *Jew*, *Negro*, *nigger*, and others already treated in the *O.E.D.* should be brought up to date. These things we have done. Proprietary terms are of more than routine concern to lexicographers and I have endeavoured to establish a policy which safeguards scholarly standards while not doing anything to imperil the proprietary rights of the owners of such terms. It gave me particular pleasure that the United States Trademark Association reprinted my 1973 comments on the subject as part of a regular issue of *The Trademark Reporter*.[4]

For new general vocabulary we have repeatedly and profitably turned to North American sources, including long runs of regional American and Canadian newspapers as well as more traditional sources like the *New York Times* and the *New Yorker*, in addition to the principal publications of the United Kingdom. We have given somewhat more attention in this volume

[1] 'The Treatment of Controversial Vocabulary in the *Oxford English Dictionary*', *Transactions of the Philological Society 1973* (1974), pp. 1–28.

[2] 'The Art of the Lexicographer', *Journal of the Royal Society of Arts*, Vol. CXXIII, May 1975, pp. 349–61.

[3] 'Data Collecting and Research', *Annals of the New York Academy of Sciences* Vol. CCXI (1973), pp. 99–103; 'Some Aspects of the Historical Treatment of Twentieth-Century Vocabulary', *Tavola Rotonda sui Grandi Lessici Storici* (Florence, 3–5 May 1971), Accademia della Crusca, Firenze, 1973, pp. 31–5; (with Valerie Smith) 'Adzuki to Gun: Some Japanese Loanwords in English', *The Rising Generation* (Tokyo) Dec. 1973, pp. 524–6, and Jan. 1974, pp. 593–5; 'Acid to Downer: Some Words for O.E.D.', *Words*, Wai-te-ata Studies in Literature, No. 4 (Jan. 1974) (Wellington, N.Z.); and 'The Prosodic Terminology of Anglo-Saxon Scholars', *Old English Studies in Honour of John C. Pope* (1974), pp. 171–202. See also Sandra Raphael, 'Natural History and the *Oxford English Dictionary*', *Jrnl. Soc. Bibliography Nat. Hist.* Vol. VI (1973), pp. 229–35.

[4] Vol. 65, No. 4, July–August 1975, pp. 291–317.

than in the last to the special vocabulary of the West Indies and, nearer home, of Scotland. The rapid expansion of work in all the sciences has been fully taken into account: anyone interested in the history of scientific words will find much of permanent value in the pages that follow. The terms of the printing industry and the names of plants and animals have continued to yield lexical material of considerable interest. The historical treatment of words again provides many surprises: for example, *minibus* is recorded from 1845, and *mugging*, in its now current sense, turns out to be much older than most people supposed.

Most people, at one time or another, treat words 'as if they are people—beautiful, delinquent, degenerate, regal'.[5] My colleagues and I, who prepared this volume, are no exceptions to the general rule. We do not personally approve of all the words and phrases that are recorded in this dictionary nor necessarily condone their use by others. Nevertheless, in our function as 'marshallers of words',[6] we have set them all down as objectively as possible to form a permanent record of the language of our time, the useful and the neutral, those that are decorous and well-formed, beside those that are controversial, tasteless, or worse.

The late Professor Atcheson L. Hench (University of Virginia, Charlottesville, Virginia) let it be known before his death in 1974 that he wished us to have access to the Hench Collection, a large miscellaneous collection of quotations from Virginian and other newspapers from about 1930 onward. As a result, the *Sun* (Baltimore) and the *Richmond News-Leader*, together with some other daily papers from various cities in the Eastern and Southern United States, appear fairly frequently in entries from the letter M onward. Professor W. R. G. Branford and the staff of the forthcoming *Dictionary of South African English*, especially Mr. John Walker, made valuable contributions to our South African English items, as did also Professor N. G. Sabbagha and Mr. N. van Blerk. Professor K. Koike (especially) and other Japanese scholars have assisted us with the entries for words of Japanese origin, and Dr. L. V. Malakhovski with words from Russian. In 1975 Professor G. A. Wilkes (University of Sydney) allowed us to copy his valuable collection of quotations for Australian colloquialisms and there was time to add some of these to the relevant Australian items in the later letters of this volume. Our indebtedness to G. & C. Merriam Co., described in Volume I, was as great as ever, and I should like to restate our gratitude to Dr. H. B. Woolf and to his successor Dr. F. Mish for their continuing co-operation. Mr. Clarence L. Barnhart and Professor F. G. Cassidy have also made important contributions to this volume by supplying quotations from their dictionary files.

The major libraries in Oxford, London, and Washington, and numerous other libraries in other cities in various parts of the world, continued to give us every possible support as we continued with our work of research and verification. We were able to overcome the difficulties naturally resulting from the dispersal of books and periodicals to new areas distant from the main centres. Special mention should be made of the access to temporary book-stacks allowed to my scientific assistants by the Librarian of the Radcliffe Science Library in Oxford during a period of great upheaval while new sections of the main library were being built.

Sadly not all those who were associated with the Supplement survived to see the publication of this volume. Of members of staff, Miss Elizabeth Brommer died in 1972, a few months before the publication of A–G. Mrs. Joan Blackler (my secretary from 1966 to 1974) and Mrs. Peggy Kay (part-time library researcher from 1967 onward) both died in 1975. The following Contributors or Outside Consultants have also died since Volume I was published in 1972: Professor

[5] A remark (slightly adapted) by the New Zealand writer Janet Frame in *Islands* (Christchurch, N.Z.), Vol. 2, No. 3 (1973), p. 252.

[6] Adopting Joseph Trapp's description of Dryden as 'the best Marshaller of words' as a phrase that comes as near as possible to a description of the perfect historical lexicographer, bearing in mind the *O.E.D.* definition of *marshal*, v. (sense 5) as 'to dispose, arrange or set (things, material or immaterial) in methodical order'.

Sir Godfrey Driver, Mr. W. Granville, Professor A. L. Hench, Dr. M. D. W. Jeffreys, Dr. D. Lack, Mr. J. C. Maxwell, Dr. L. F. Powell, Mrs. Stefanyja Ross, and Miss Phyllis Trapp.

To the list of Contributors in Volume I the name of Dr. D. S. Brewer should be added. Major contributors of quotations in the period 1972–75 included the following: Professor W. S. Avis, D. J. Barr, G. Chowdharay-Best, C. Collier, Professor M. Eccles, R. Hall, T. F. Hoad, Dom Sylvester Houédard, Miss M. Laski, Dr. D. Leechman, Dr. J. Lyman, Professor J. B. McMillan, Mrs. J. M. Marson, Mrs. M. Y. Offord, and D. Shulman. Of these, Miss Laski, Dr. Leechman, Mrs. Offord, Mrs. Marson, and Mr. Chowdharay-Best contributed altogether approximately 70,000 quotations and all the others named supplied between 1,000 and 3,000 quotations each. Smaller, but valuable, sets of quotations were received from numerous others, including the Revd. H. E. G. Rope, R. E. Hawkins, and Mrs. Daphne McColl.

To the list of outside Proof-Readers the names of M. W. Grose, T. F. Hoad, and Dr. Kendon Stubbs should now be added.

The following new Outside Consultants have assisted us in addition to most of those named in Volume I: A. D. Alderson, Professor W. S. Avis, Dr. R. P. Beckinsale, Professor T. Burrow, Sir Alexander Cairncross, Professor Elizabeth Carr, Miss P. Cooray, Dr. S. T. Cowan, M. P. Furmston, B. Greenhill, Professor O. R. Gurney, R. Hall, Professor C. Hart, P. A. Hayward, Dr. R. Hunter, Dr. Russell Jones, Professor J. B. McMillan, Dr. C. I. McMorran, E. Mendelson, Professor G. B. Milner, D. D. Murison, P. H. Nye, Dr. K. P. Oakley, Dr. A. B. Paterson, Professor Dr. I. Poldauf, and N. G. Phillips.

This second volume contains about 13,000 Main Words divided into some 22,000 senses. There are a little under 8,000 defined Combinations within the articles and just over 5,000 undefined Combinations. The illustrative quotations number 125,000.

Finally, the Editor would like to record his personal indebtedness to the following for assistance on many matters: Dr. J. B. Sykes, Deputy Chief Editor of the Oxford English Dictionaries and Editor of the *Concise Oxford Dictionary* (1976), for valuable advice and co-operation at all times; Mr. A. J. Augarde, who has now moved across to the smaller Oxford dictionaries after a long period of service on the *Supplement to the O.E.D.*; the Managers and staff of the branches of the Oxford University Press for their efficiency and encouragement during the Editor's lecture tours of the Far East, the United States, South Africa, and elsewhere in 1972 and in 1974; his colleagues at St. Peter's College in Oxford; and, most particularly, his colleagues and assistants on the Dictionary staff itself, especially for their endurance and perseverance at many times when industrial and economic difficulties had their impact upon the O.E.D. Department as upon every other section of the community.

R.W.B.

Oxford
January 1976

EDITORIAL STAFF

The dates given after the names indicate when each person joined the editorial staff of this dictionary. The letter P precedes the names of those who worked as part-timers.

Senior Assistant Editor: A. J. AUGARDE	1960–76
Assistant Editor (Science): A. M. HUGHES	1968–
Assistant Editor (Natural History): SANDRA RAPHAEL	1969–
Assistant Editor (Bibliography): G. D. HARGREAVES	1973–5

Editorial Assistants:

E. C. DANN	1963–	JEAN H. BUCHANAN	1971–6
VERONICA M. SALUSBURY	1966–	VALERIE SMITH	1972–5
ADRIANA P. ORR	1966–	A. B. BUXTON	1972–5
P†PEGGY E. KAY	1967–75	GILLIAN A. RATHBONE	1973–6
PFRANCES M. WILLIAMS	1968–76	R. E. ALLEN	1974–
W. H. C. WATERFIELD	1970–5	LESLEY S. BURNETT	1974–
DEBORAH M. COWEN	1970–	J. CLAIRE NICHOLLS	1974–
PJOYCE L. HARLEY	1970–	PMARGUERITE Y. OFFORD	1974–

Miss Salusbury and Mrs. Offord (based in London), Mrs. Orr (in Washington), and Miss Buchanan, Miss Harley, Miss Nicholls, and Miss Rathbone were mainly concerned with research (especially for 'first uses') and with the verification of quotations; Mrs. Orr was rejoined by Mrs. Daphne Gilbert-Carter (working part-time) in Washington in 1975. Mr. Waterfield and Mr. Buxton dealt with scientific terms, and Mrs. Cowen with terms in the Social Sciences. Miss Williams assisted with the reading of the proofs. All other Editorial Assistants named above undertook general editorial work.

Among those who assisted at various stages with the editorial work of Volume II as part of the regular staff were the following: PL. B. Firnberg (1962–74), PJelly K. Williams (1967–74), J. P. Barnes (1969–72), M. W. Grose (1969–72), Deirdre McKenna (1969–74), Juliet Field (1973–4), P. E. Davenport (1970–71), Gillian Bradshaw (1972–4), and L. M. Matheson (1974–5).

New members of the Editorial Staff, all of whom joined in 1975 and all of whom assisted with the final stages of Volume II, are D. R. Howlett, J. Paterson, PE. Joan Pusey, Rosemary J. Sansome, W. R. Trumble, and N. S. Wedd.

Members of the Editorial Staff received valuable part-time assistance from the following outside helpers: Grace M. Briggs (1959–) and Rita G. Keckeissen (1968–).

Secretarial and Clerical Assistants: †Joan Blackler (Editor's Secretary, 1966–74), Pamela Bendall (1968–), Kathleen Johnston (1970–), Beta Cotmore (Editor's Secretary, 1974–), and Anne Whear (1975–).

PREFACE TO VOLUME III

IN her *Personal Pleasures* (1935) Rose Macaulay notes that on a blank page at the beginning of her copy of the 1933 *Supplement to the OED* she recorded emendations, corrections, additions, and earlier uses of words.

> To amend so great a work gives me pleasure; I feel myself one of its architects; I am Sir James Murray, Dr. Bradley, Sir William Craigie, Dr. Onions . . .
>
> If there is a drawback to this pure pleasure of doing good to a dictionary, I have not yet found it. Except that, naturally, it takes time.

With the publication of this volume we have now reached the three-quarter mark, proportionately about as far as the point reached by Dr Murray on the *OED* itself before he died in 1915. We are now preparing the fourth and last volume, and it should be ready for publication in 1985. It would not be prudent to start congratulating ourselves yet—no major lexicographical project has been brought to its last word without the final agonies of a marathon runner—but the glittering prize of completion now seems to be within sight at last.

The letter S—the longest one in the alphabet—needed to be divided. Calculations of various kinds were made and in the end it was decided to make the division at *Scythism*, though it is a word of no great account, in order to balance the size of the third and fourth volumes. The fourth and final volume will begin with an entry for the ancient and productive word *sea*.

We have continued to follow the main lines of policy described in the first two volumes, but with changes of emphasis or detail here and there in order to take into account the events of the later part of the past decade, the research interests of scholars in various subjects, and the vicissitudes of the OED Department and of my own life.

During the 1970s the markedly linguistic descriptivism of the post-war years was to some extent brought into question. Infelicities of language, whether in the spoken or the written word, were identified and assailed by a great many people who seemed to believe that the English language itself was in a period of decline. Regular columns largely concerned with verbal error appeared (and many of them continue to appear) in *The Times* (Mr Philip Howard), the *New York Times Sunday Magazine* (Mr William Safire), the *New Statesman* ('This English'), *Encounter* ('In the Margin'), the New Zealand *Listener* (Professor I. A. Gordon), and many other newspapers and journals. The House of Lords devoted a session to the subject on 21 November 1979 and another on 28 January 1981, in the course of which eloquent voices were raised against the use of modish words like *ongoing*, *relevant*, and *viable*.[1]

My own views on these great issues were expressed in several publications[2] and in many lectures and broadcasts—broadly that 'the English language is alive and well, in the right hands'.

> Expressions like *right on* and *hopefully* bring out the worst and the best in men and women. They stand as emblems of social and political divisions within our society. These and other elements lying strewn in the disputed territory of our language are at any given time not numerous but are charged with a significance that goes far beyond the mere linguistic. If you are tempted to fulminate against them, or to feel uneasy about them, bear in mind that the English language has been in the hands of linguistic conservatives and linguistic radicals for more than a thousand years and that, far from bleeding to death from past crudities and past wounds, it can be used with majesty and power, free of all fault, by our greatest writers.[3]

One small legacy of these great debates is that here and there in the present volume I have found

1 The text of these debates is recorded in the relevant issues of Hansard.

2 Especially in *The Quality of Spoken English on BBC Radio* (BBC, 1979), *The Spoken Language as an Art Form: an Autobiographical Approach* (English-Speaking Union, New York, 1981), and *The Spoken Word: a BBC Guide* (BBC, 1981).

3 *The Spoken Language as an Art Form*, p. 17.

myself adding my own opinions about the acceptability of certain words or meanings in educated use. Users of the dictionary may or may not find these editorial comments diverting: they have been added (adapting a statement by John Ray in 1691) 'as oil to preserve the mucilage from inspissation'.

The volume has been prepared in circumstances of great adversity, though not without many mitigating factors. The editorial staff remained more or less constant in number but very different in constitution as some moved away to other posts and others took charge of other projects within the Department. Also, in 1977, we left our 'shabby Victorian villa', 40 Walton Crescent, and moved to 37a St. Giles', a splendid spacious Georgian mansion in central Oxford, with all the disruption brought about by such a move. We lost our printers, William Clowes & Co. Ltd., Beccles, after they had set the letters O and P for this volume. The closing down of their hot-metal department in 1980 delayed the printing of Volume III by six months before new printers were found and the material was transferred to Plymouth or to Oxford. The Department embarked on many other projects, including new editions of nearly all the Oxford dictionaries below the level of the *OED* itself, and completed a number of new dictionaries including the *Oxford Paperback Dictionary* (1979); three dictionaries closely related to it, including the *Oxford American Dictionary* (1980) and the *St. Michael Oxford Dictionary* (1981); the *Oxford Minidictionary* (1981); and several school dictionaries. My governance, at various removes, of these projects inevitably delayed the completion of the third volume of the Supplement. The Oxford University Press found itself locked in internal debates and wrangles about ways and means of surviving in difficult trading conditions. Trading profits turned into trading losses and unpleasantnesses occurred as those responsible for the management of affairs found themselves in inevitable dispute with the unions. The Department could not stand aside and pretend that it existed in an ivory tower of its own choice. The words *picket* and *picketer* are dealt with in this volume: all of us encountered the fact of picketing at intervals while this volume was in preparation. We continued to receive sustained and invaluable assistance from many outside scholars and institutions (see below) but suffered a grievous blow when G. and C. Merriam Co. decided in 1977 that they could no longer help us, as they had in the past, by supplying from their files earlier examples of words than those held in our own files. This volume contains numerous examples of the type '1934 in WEBSTER'[4] because this great American dictionary house felt obliged to cancel an arrangement that I had made with Dr Philip Gove, the Editor of *Webster's Third New International Dictionary*, in 1968.

These adversities have left their mark but the pleasures of historical lexicography remain as a source of endless delight and can be observed in the pages that follow. The burgeoning of the word *O.K.* in its numerous spellings and variations; the engaging curiosities of the letter Q (this with X the letter always dealt with at greatest speed by lexicographers); the words of Yiddish, German, Greek, and Italian origin beginning with *sch-*, a vigorous group if ever there was one; the numerous Chinese loanwords or loan translations—for example, *pipa, putonghua, Little Red Book, running dog*, and *scorched earth*—with the Chinese originals expressed in the revolutionary (and one hopes permanent) Pinyin transliteration system; and the numberless entries with *post-, pre-, pro-*, and *re-* as their first elements. Those who wish to explore the volume further rather than simply use it as a work of occasional reference may derive pleasure (according to taste) under at any rate some of the following assorted entries: *oung*, v. (of an elephant: to drag logs along a stream); *parp* (invented apparently by Enid Blyton); *person*, sb. 2 f (as in *chairperson*); *Pip, Squeak, and Wilfred; pneumonoultramicroscopicsilicovolcanoconiosis* (a factitious word of 45 letters); *Princeton-First Year* (Auden); *puddle-duck* (an earlier use than that in Beatrix Potter's famous work); controversial words like *piccaninny, Sambo*, and *Scientology; scripophily* (one of numerous invented words for various collecting habits); and *scrolloping* (Virginia Woolf).

4 See, for example, the entries for *phrasally, retrain* (verb), *rewire* (noun), *rubbernecker, sales clerk*, and *sales force*.

Some new areas of vocabulary or doctrine that I have explored myself have been dealt with in several papers published elsewhere.[5]

A new feature of this volume is the inclusion of a table showing how we have transliterated foreign scripts—not always, unfortunately, with complete success. I am grateful to Dr J. B. Sykes (in particular), Sir Edward Playfair, Mr T. F. Hoad, and the late Mr. N. A. M. Rankin for very considerable assistance while it was being prepared.

Inevitably and sadly a number of people associated with the Supplement have not survived to see the publication of the third volume. The following contributors or outside consultants have died since Volume II was published in 1976: Professor J. A. W. Bennett (my former tutor), Dr E. J. Bowen, Dr R. S. Cahn, Professor B. Foster, Dr R. A. Hunter, Professor W. S. Mackie, Dr K. P. Oakley, N. A. M. Rankin (from the Dictionary Department itself), the Revd H. E. G. Rope (aged 97), and Nicolaas Van Blerk. Special mention should be made of Dr Douglas Leechman (who died in July 1980) and Professor W. S. Avis (who died in December 1979): there can scarcely be any Canadian item in the Supplement that has not benefited from the work of these two great men. Perhaps the most devastating blow of all came when Mr Gordon Murray, a member of the editorial staff, died in June 1981 at the age of 32.

Major contributors of quotations in the period 1976–80 included the following: †Professor W. S. Avis, G. Charters, G. Chowdharay-Best, C. Collier, Mrs J. Harker, Dom Sylvester Houédard, W. Kings, Miss M. Laski, †Dr D. Leechman, Mrs D. McColl, Mrs J. M. Marson, †E. H. Mart, Mrs M. Y. Offord, Sir Edward Playfair, F. Shapiro, Mrs V. Smith, and Mrs G. M. Spriggs. Of these Miss Laski and Mr Chowdharay-Best contributed approximately 30,000 quotations each, and all the others named supplied between 1,000 and 10,000 quotations each. Indispensable batches of quotations from fiction, including detective fiction, and from works in the whole area of domestic and social life were received from Miss Laski. Our treatment of exotic vocabulary from, for example, the Far East, the Pacific, and the language of politics would have been less thorough without the contributions of Mr Chowdharay-Best, of Asia Minor and the Middle East without those of Sir Edward Playfair, and of American card and board games, and of some other specialized areas, without those of Mr Shapiro.

The outside proof-readers, generously industrious and helpful throughout the preparation of Volume III, were Professor A. R. Duckert, M. W. Grose, T. F. Hoad, Miss Marghanita Laski, Mrs M. Y. Offord, and Professor E. G. Stanley.

The outside consultants to whom we have repeatedly turned while the volume was being prepared were: Dr G. C. Ainsworth, A. J. Augarde, †Professor W. S. Avis, Dr R. P. Beckinsale, Professor A. J. Bliss, Dr S. Bradbury, Dr Jean Branford, †Dr R. S. Cahn, Sir A. K. Cairncross, Professor F. G. Cassidy, Miss Chan Yin-Ling, Miss P. Cooray, Dr J. Cortés, Professor G. N. C. Crawford, Professor S. Deas, P. G. Embrey, D. F. Foxon, R. B. Freeman, W. K. V. Gale, P. G. W. Glare, Dr I. Goddard, R. Hall, R. E. Hawkins, Dr M. H. Hey, M. T. Heydeman, Professor Elizabeth (Carr) Holmes, †Dr R. A. Hunter, Dr D. M. Jackson, P. Jarrett, Dr Russell Jones, Dr N. R. Ker, Dr W. J. Kirwin, Professor K. Koike, Professor J. Leech, †Dr D. Leechman, Professor B. Lennox, Dr A. Loveless, Dr D. J. Mabberley, Professor R. I. McDavid Jr, Dr R. S. McGregor, Professor J. B. McMillan, Dr L. V. Malakhovski, Dr F. H. C. Marriott, R. D. Meikle, D. D. Murison, †Dr K. P. Oakley, I. and P. Opie, Professor C. Rabin, Professor R. H. Robins,

5 'Names of Types of Oil Wells: an Aspect of Short-Term Historical Lexicography', in *Feestbundel F. de Tollenaere* (Leiden, 1977); 'Aspects of Short-Term Historical Lexicography', in *Second Round Table on Historical Lexicography* (delivered in Leiden in 1977, published 1980), pp. 271–86; 'Further Aspects of Short-Term Historical Lexicography', in *James B. McMillan: Essays in Linguistics by his Friends and Colleagues* (University of Alabama Press, 1977), pp. 115–31; 'On that Other Great Dictionary', in *Encounter*, May 1977, pp. 47–50; *The Fowlers: their Achievements in Lexicography and Grammar* (Presidential Address to the English Association, 1979); Preface to a facsimile edition of Samuel Johnson's *Dictionary of the English Language* (London, 1979); 'Dictionaries and Ethnic Sensibilities', in L. Michaels and C. Ricks, *The State of the Language* (University of California Press, 1980), pp. 15–23.

Valuable papers by two of my departmental colleagues were also published in *Exeter Linguistic Studies*, Vol. IV (1979): L. S. Burnett, 'Lexicographical Problems in the Treatment of some Linguistic Terms in a Supplement to the OED'; and S. J. Raphael, 'The Treatment of the Terminology of Natural History in the Oxford English Dictionaries'.

Professor N. G. Sabbagha, R. Scruton, Dr J. B. Sykes, Miss Tao Jie, Professor G. Treitel, G. W. Turner, J. O. Urmson, Professor T. G. Vallance, Dr R. L. Venezky, the Revd Canon Professor M. F. Wiles, and Dr D. R. Woodall. Many others have given us advice and comments on individual entries.

This third volume contains about 18,750 Main Words divided into some 28,000 senses. There are about 8,500 defined Combinations within the articles and some 4,500 undefined Combinations. The illustrative quotations are estimated to number 142,500.

Finally, the Editor would like to record his personal indebtedness to many individual scholars and institutions: Mrs L. S. Burnett and Dr W. R. Trumble, who made signal editorial contributions to the Supplement before they started work on a new edition of the *Shorter Oxford English Dictionary* in 1980; similarly Dr R. E. Allen, who succeeded Dr J. B. Sykes as editor of the *Concise Oxford Dictionary* and *Pocket Oxford Dictionary* in 1981; my other Senior Editors, Mr A. M. Hughes (for Science), Miss Sandra Raphael (for Natural History and Library Research), and, in more recent times, Mr E. S. C. Weiner and Mr J. A. Simpson; the library research staff who have managed to overcome the adversities now strewn in the path of anyone constantly using great libraries, and in particular Miss V. M. Salusbury (until she left in 1977), Miss J. L. Harley (retired 1980), and others who worked in London and in America far from the editorial headquarters in Oxford; Mr J. Paterson for his resolution of many difficult bibliographical problems within the inevitable limits of time; my hosts in Chicago and New York in July 1978 when I startled them and many others by suggesting in a lecture to the American Library Association that American and British English were drifting apart and that one day interpreters would be needed;[6] Liverpool University for their munificence in conferring an Honorary D.Litt. on me in 1978; those who welcomed me in China in May 1979, especially Mr Chen Yuan of the Commercial Press, Peking, and my interpreter, Miss Tao Jie, Peking University; and my hosts in seven cities in the United States in October 1980 when I gave a series of lectures on the English language at the invitation of the English-Speaking Union. Above all, I have continued to build up my indebtedness to those whom I see daily and who have given me superlative support and succour over the last six years, my colleagues and assistants on the Dictionary staff itself.

Oxford R.W.B.
October 1981

[6] The lecture was printed with the title 'The Point of Severance: English in 1776 and Beyond', in *Encounter*, October 1978, pp. 129–33.

EDITORIAL STAFF

The dates given after the names indicate when each person joined the editorial staff of this dictionary. The letter [P] precedes the names of those who worked as part-timers.

Senior Editors (General):	R. E. ALLEN	1974–80
	LESLEY S. BURNETT	1974–80
	J. A. SIMPSON	1976–
	E. S. C. WEINER	1977–
Senior Editor (Science):	A. M. HUGHES	1968–
Senior Editor (Natural History and Library Research):	SANDRA RAPHAEL	1969–
Assistant Editor (Bibliographical Collation):	J. PATERSON	1975–

Editorial Assistants

E. C. DANN	1963–	A. HODGSON	1976–
ADRIANA P. ORR	1966–	YVONNE L. WARBURTON	1976–
DEBORAH D. HONORÉ		JULIA C. SWANNELL	1976–
(*formerly* COWEN)	1970–	D. J. EDMONDS	1977–80
JOYCE L. HARLEY	1970–80	[P]F. D. HAYES	1977–
ROSEMARY J. SANSOME	1975–80	†G. MURRAY	1977–81
N. S. WEDD	1975–81	P. R. HARDIE	1977–80
D. R. HOWLETT	1975–9	ELIZABETH M. KNOWLES	1977–
W. R. TRUMBLE	1975–80	C. F. KEMP	1978–81
EDITH BONNER		ROSAMUND E. MOON	1979–81
(*formerly* ROGERSON)	1976–	AMANDA J. BURRELL	1979–

Members of the editorial staff received valuable part-time assistance from the following outside helpers: Grace M. Briggs (1959–), Rita G. Keckeissen (1968–), Daphne Gilbert-Carter (1975–), and Sally Hinkle (1977–), the first-named in Oxford and the others in New York, Washington, and Boston respectively.

Mr Kemp and Mrs Burrell (both based in London), Mrs Orr (in Washington), and Miss Harley, Miss Warburton, and Miss Knowles (all in Oxford) were mainly concerned with research (especially for 'first uses') and with the verification of quotations. Mr Wedd, Dr Trumble, and Mr Murray dealt with scientific terms, and Mrs Honoré with terms in the Social Sciences. Mr Edmonds assisted with the reading of proofs. All other Editorial Assistants named above undertook general editorial work.

Among those who assisted for relatively short periods with the editorial work of Volume III as part of the regular staff were the following: Veronica M. Salusbury (1966–77), J. Claire Nicholls (1974–7), [P]Marguerite Y. Offord (1974–9), P. J. Broadhead (1977–8), J. S. Wood (1978–80), and Wendy H. Archer (1980–1).

New members of the editorial staff who (from 1980) assisted with the final stages of Volume III were Dr D. B. W. Birk, M. A. Mabe, and Della J. Thompson. Mrs Honoré worked part-time from mid-1980.

Secretarial and clerical assistants: Beta Cotmore (1974–9), Anne Whear (1975–), Katherine A. Shock (1978–9), D. Ann Baker (1978–81), Afra E. Singer (1979–81), and Karin C. E. Vines (1981–).

PREFACE TO VOLUME IV

1. *Volume 4* (OEDS 4)

WHEN Miss Marghanita Laski read the galley-proofs of the entry for *off* in Volume 3 of this Supplement she remarked, 'I am almost completely flummoxed by this, and must ask your indulgence on it'. Professor Audrey Duckert, another of our regular contributors, independently remarked, 'I'm glad I didn't have to write *off* for OED, but I'll never write OED *off*'. *Off* was a complex entry, or set of entries, both in the *OED* and in OEDS 3, and the same is necessarily true of a good many entries in this final volume—those for *un-* (prefix) and *up* (various parts of speech), for example. Our language is a complicated mechanism at the best of times, and sometimes almost frighteningly so when presented in its largest dimension as in the following pages.

But not all of the elements of the English language are complex. Many of the expressions that appear in this final volume merely illustrate the concepts, inventions, and movements of our generation—words of very recent origin (1984) like *yumpie* and *yuppie*, and somewhat older expressions like *self-fulfilling prophecy*, *sputnik*, and *test-tube baby*. We are all in an electronic environment, and the entries for such words as *SNOBOL*, *transputer*, and *wysiwyg* draw attention both to the ingenuity of the world of the green screen and to the manner in which its practitioners embrace the techniques of present-day word-formation.

Every effort has been made to keep up with the language as it developed even while the final volume of the Supplement was being prepared. *The Growing Pains of Adrian Mole* (1984) is quoted under *yoyo*, for example, and there are numerous examples from 1985 sources in the later letters of the alphabet in this volume.

As in earlier volumes no effort has been spared to verify details of the coinage, meaning, or other aspects of each word entered in the Supplement. The Bodleian Library, the British Library, and the Library of Congress, in particular, have probably never experienced such a systematic combing of their resources as has been necessary for the preparation of the four volumes of OEDS. We owe an incalculable debt to the custodians of these libraries, and also to many other specialized libraries in Britain and abroad, for their cooperation. We are also deeply indebted to numerous scholars and men of letters who have over the years assisted us with the definition or the circumstances of origin of expressions they have coined themselves, e.g. *acceptance world* (Anthony Powell), *drogulus* (A. J. Ayer), *dymaxion* (Buckminster Fuller), *hobbit* (J. R. R. Tolkien), *jog* sb.[2] sense 3 (N. F. Mott), *non-event* (I. Gilmour MP), *psephology* (R. B. McCallum), *quark* (M. Gell-Mann), *tracklement* (D. Hartley), and *tribology* (C. G. Hardie). Part of the pleasure and the value of historical lexicography lies in the establishing of the circumstances of a coinage from the coiner himself.

* * *

Sadly the need arises once more to set down the names of people associated with the Supplement who have not survived to see the publication of this final volume: Professor A. J. Bliss, Professor S. Deas, Dr N. R. Ker, John Lyman, Professor Raven I. McDavid Jr., Professor Mitford M. Mathews, Peter Opie, Professor Dr László Országh (Hungary), Professor Dr. I. Poldauf (Czechoslovakia), Professor I. Willis Russell, and Peter Tamony.

Major contributors of quotations in the period 1982–5 included the following: G. Charters (Australia), G. Chowdharay-Best, G. A. Coulson, F. D. Hayes, Miss M. Laski, Sir E. Playfair, and F. R. Shapiro. The product of their work is to be seen on virtually every page of this volume.

The outside proof-readers, who systematically scrutinized sets of galley-proofs and submitted their comments and criticisms during the preparation of Volume IV, were Professor A. R. Duckert, M. W. Grose, T. F. Hoad, Mrs D. D. Honoré, Miss M. Laski, Professor E. G. Stanley, and Mrs H. C. Wright. The volume would have been much the poorer without their expert attention.

The outside consultants to whom we have turned for advice on particular words while the volume was being prepared were: Dr G. E. H. Abraham, Dr G. C. Ainsworth, Professor A. J. Aitken, A. D. Alderson, R. E. Allen, Dr P. W. Atkins, A. J. Augarde, Professor J. R. Baines, †Professor A. J. Bliss, Dr. S. Bradbury, Dr J. Branford, Dr W. H. Brock, Sir A. K. Cairncross, Dr B. G. Campbell, Professor F. G. Cassidy, Dr P. A. Charles, M. J. E. Coode, Dr J. Cortés, Professor G. N. C. Crawford, Mrs U. Dronke, †Professor S. Deas, Dr B. J. Freedman, R. B. Freeman, W. K. V. Gale, P. G. W. Glare, Dr I. Goddard, Dr G. H. Gonnet, P. S. Green, R. Hall, R. E. Hawkins, M. T. Heydeman, Dr D. M. Jackson, P. Jarrett, Dr Russell Jones, Dr D. Julier, Dr W. J. Kirwin, Professor K. Koike, Professor J. D. Latham, Professor J. Leech, Professor B. Lennox, Dr G. Lewis, R. P. W. Lewis, Dr A. Loveless, Dr D. J. Mabberley, Dr R. S. McGregor, Professor J. B. McMillan, Dr T. Magay, Dr L. V. Malakhovski, Dr F. H. C. Marriott, R. D. Meikle, Professor G. Milner, D. D. Murison, Mrs I. Opie, M. B. Parkes, Miss V. Richardson, Professor R. H. Robins, Dr H. M. Rosenberg, Professor J. M. Rosenberg, Professor N. G. Sabbagha, R. Scruton, A. J. Stevens, Dr I. N. Stewart, Dr J. B. Sykes, Associate Professor Tao Jie, Miss D. J. Thompson, Professor G. Treitel, G. W. Turner, J. O. Urmson, Professor T. G. Vallance, Professor R. L. Venezky, Dr M. Weitzman, the Revd Canon Professor M. F. Wiles, and Dr D. Zorc.

This fourth volume contains about 13,500 Main Words divided into some 25,000 senses. There are about 11,000 defined Combinations within the articles and a similar number of undefined Combinations. The illustrative quotations are estimated to number 130,000.

It is appropriate to mention here that the printing of this final volume marks the end of an era in the printing trade. I believe that it may be the last major book to be set up in type by the hot-metal process. The printing house concerned, Latimer Trend of Plymouth, nobly retained its hot-metal department until the entry for *Zyrian* was safely in type.

I should like to take this last opportunity to thank various people: first and foremost, the Delegates of The Oxford University Press, and their senior officers, for allowing this ambitious and very costly project to take its course; all members of my staff since 1957 for their skilled and determined assistance as a seemingly endless procession of problems presented themselves for solution; Miss Marghanita Laski, surely one of the most prolific contributors of illustrative quotations—some 250,000—to any dictionary in history; Clarence L. Barnhart (and more recently his son Robert K. Barnhart) who opened his valuable quotation files to us from the beginning of the preparation of OEDS; and Professor E. G. Stanley, the Rawlinson and Bosworth Professor of Anglo-Saxon in the University of Oxford, a personal friend for nearly forty years, who was among the first to help me build up a team of source-readers for the Supplement and and who has subsequently given close critical attention to sets of galley-proofs in every letter of the alphabet.

2. *A Supplement to the OED* (OEDS)

It is natural when a task of some magnitude has been completed to give some account, however inadequate, of the manner in which the project evolved.[1]

In 1957, when I began, I was both encouraged and worried by the first sentence in the General Explanations of the *OED:* 'The vocabulary of a widely-diffused and highly-cultivated

[1] A fuller account of the evolution of OEDS is given in my article 'The End of an Innings but not the End of the Game', Threlford Memorial Lecture, *The Incorporated Linguist* Volume 23, Number 3, Summer 1984, pp. 114–119.

living language is not a fixed quantity circumscribed by definite limits.' I was also much taken by Dr Murray's employment of the phrase *Lexicon totius Anglicitatis*. 'Limits' and 'totality' plainly suggested that boundaries would need to be set at many stages along the journey.

At the outset I was invited by the Delegates of the Oxford University Press to prepare a new *Supplement to the OED* in a single volume of about 1,275 pages, and to aim at completion within seven years. Both figures seemed reasonable at the time, as far as I can recall. The only significant model before me was that of the *OED* itself. James Murray had been instructed by the Delegates to complete the Dictionary within ten years. But the ten years, I reasoned, had turned into forty-four partly because Murray was a pioneer in the field of historical lexicography,[1] and partly because he and his co-editors had to deal with the language from the time of the first records (eighth century) of English till the late nineteenth century. I had rather less than a century's worth of language to consider.

In the event OEDS has taken nearly twenty-nine years to complete, and one volume has turned into four. By 1957 the language had proliferated at a much greater rate since the beginning of the century than I had at first judged. This dramatic increase was underlined by the publication in 1961 (i.e. not long after I had established my editorial policy and at a point when I thought the main collecting of evidence had been completed) of *Webster's Third New International Dictionary*. The sheer quantity of words in this majestic and influential work made it obvious that I had underestimated the problems confronting me. The whole editorial process, and particularly the extension of the 'definite limits' of inclusion, had to be reconsidered. The mine-shafts into the seams of words needed to be dug much more deeply.

A second circumstance leading to delay was my acceptance of responsibility for the governance of the smaller Oxford dictionaries. In 1957 there were only four Oxford English dictionaries for native speakers apart from the *OED* itself: in chronological order of first publication, the *Concise Oxford Dictionary* (first published in 1911), the *Pocket Oxford Dictionary* (1924), the *Little Oxford Dictionary* (1930), and the *Shorter Oxford English Dictionary* (1933). Now this small flotilla of scattered ships—scattered because in 1957 they had no base to operate from—has turned into an invincible fleet of more than twenty dictionaries and lexical guides, all flying the Oxford flag from their base at 37a St. Giles', Oxford. The editors of all these smaller works, and many of their editorial assistants, first cut their lexicographical teeth on OEDS.

* * *

When I look back at the work of Murray, I cannot but marvel at the permanent value of so much of his editorial policy, and even of his clerical procedures. Like Murray and his editorial colleagues we have worked by hand on dictionary slips, the only difference being that our slips are standardized in size (6 in. × 4 in.). They filed bundles of slips in wooden pigeon-holes after tying them with string or tape. We have placed our slips in an upright manner in strict alphabetical order in the trays of fire-proof cabinets. One procedure of Murray's has been abandoned: as often as not his readers cut up books, including many valuable ones from earlier centuries, in the process of collecting quotations. This would today be regarded as vandalism. We were able to hasten the process of preparing slips containing evidence from much-cited sources (e.g. the technical glossaries of the British Standards Institution), by mass-producing them with typed titles and with the help of a photo-copier and paper guillotine, rather than, as in Murray's day, by the use of hand-setting composition in a printing house. I am sure too that our standards of research and verification of the printed evidence have been consistently higher than those of our predecessors. Victorian standards were lower in such matters; ours is a more pedantic age. The

[1] The *OED* was not in fact the first historical dictionary. Jacob Grimm and Wilhelm Grimm started their *Deutsches Wörterbuch* at an earlier date, and their first volume (*A–Biermolke*) was published in 1854.

resources of even the greatest libraries then were also much more limited than they are now. Many quotations in the *OED* were taken from secondary sources like, for example, C. Pettman's *Africanderisms* (1913) and J. Redding Ware's *Passing English of the Victorian Era* (1909): we have tried always to verify such quotations in the original source. We have also tried always to verify quotations from poems and short stories in their first place of publication and not merely in collected editions.

In matters of editorial policy, except for the abandonment of the once obligatory initial capital in the headwords, we have endeavoured in OEDS to display the entries in the manner of the parent work, even to the extent of retaining Murray's old-fashioned (though very convenient) pronunciation system. If a word has developed new senses we have placed these in their logical place in the numerical sequences first devised nearly a century ago, sometimes needing to make use of strings of asterisks, as explained on p. xxi. We have given the same attention to every department of modern English vocabulary—etymologies, definitions, illustrative examples, combinations, proverbs, idiomatic phrases, and all the rest—as did the lexicographers who preceded us, except that the entries for words entering the language in the twentieth century are more generously illustrated by examples than was judged necessary in the past.

* * *

One unforeseeable circumstance that has had an intangible and yet curiously potent effect on present-day attitudes to historical lexicography is that the period of preparation of OEDS has coincided with the arrival of new linguistic (and especially structuralist) attitudes. What I have elsewhere called 'linguistic burial parties' have appeared, that is scholars with shovels intent on burying the linguistic past and most of the literary past and present. I refer to those who believe that synchronic means 'theoretically sound' and diachronic 'theoretically suspect'. It is theoretically sound, the argument of the synchronicists runs, to construct contrastive sentences or other laboratory-invented examples which draw attention to this or that element of grammar or lexis, and to do *only* that. I profoundly believe that such procedures, leading descriptive scholars never to quote from the written language of even our greatest modern writers, leave one looking at a language with one's eyes partly blindfolded.

I want to dwell on these matters a moment longer because the editorial policy adopted in the four volumes of OEDS was formed in all its essentials between 1957 and 1960, when the new linguistic attitudes were at an embryonic stage. A small measure of autobiography is necessary. Between 1951 and 1957 I had retranscribed the text of the late-twelfth-century set of versified homilies called the *Ormulum*. I had also compiled an *index verborum* to it, and had given lectures to Oxford undergraduates on the language of the *Ormulum* and on the language of other medieval works like the *Peterborough Chronicle*, the *Ancrene Wisse*, and the *Ayenbite of Inwyt*. For such work, indispensable prerequisites included a knowledge of the linguistic monuments of Old English, Old Norse, Gothic, and other Germanic languages, also of Old French, and a professional knowledge of all the elements of comparative Indo-European philology that had a bearing on the vocabulary and grammar of medieval English. I also conducted tutorials and seminars on the language of writers like Spenser, Shakespeare, Milton, Dickens, and many others. In that context the grammaticality of 'Colourless green ideas sleep furiously', and the contrast between *langue* and *parole*, had no relevance at all. Above all else it became clear to me that the entire vocabulary of all the main literary, philosophical, religious, etc., works that had survived from the period since 1100, as well as that part of pre-Conquest English that remained in the language after 1066, had been included in the *OED*. Any omissions were attributable to human frailty not to deliberate design. There were no exclusion zones, no censorings, no blindfoldings, except for the absence of two famous four-letter (sexual) words. Dr Murray, his colleagues,

and his contributors had dredged up the whole of the accessible vocabulary of English (the sexual words apart) and had done their best to record them systematically in the *OED*. I concluded that if the Dictionary had room for the word *thester* as adjective ('dark') and noun ('darkness'), and Orm's *þeossterrleȝȝc* 'darkness', it could, and must, admit the vocabulary of Edith Sitwell and Wystan Auden. Of course the structuralists and other scholars at one or more remove from the primary work of Ferdinand de Saussure could not see this, and they probably never will. It seems that they would prefer to bury Orm's vocabulary along with that of the best writers of the present day. But OEDS, like its parent work, has been hospitable, almost from the beginning, to the special vocabulary, including the once-only uses, of writers like T. S. Eliot, Virginia Woolf, D. H. Lawrence, and others. It must be emphasized that in practice these uses form only a tiny fraction of the vocabulary presented here; in other words the balance of the volumes has not been disturbed by them nor has the publication of the dictionary been delayed by their presence.

Perhaps the main departure from the policy of Murray was my decision to try to locate and list the vocabulary of all English-speaking countries, and not merely that of the United Kingdom. For the most part Murray preferred to fend off overseas words until they had become firmly entrenched in British use. Words more or less restricted to North America, Australia, New Zealand, South Africa, the West Indies, and so on, were treated almost like illegal immigrants. All that has been changed and, as far as possible, equality of attention has been given to the sprawling vocabulary of all English-speaking countries. At a time when the English language seems to be breaking up into innumerable clearly distinguishable varieties, it seemed to me important to abandon Murray's insular policy and go out and find what was happening to the language elsewhere.

* * *

As work on the Supplement proceeded, the number of scientific, technical, and specialized academic words and senses that needed to be included multiplied spectacularly. The astonishing growth in academic and scientific research and in industrial and technological achievements, especially since the 1939–45 war, is plainly reflected in the text of OEDS. The first sputnik was launched in 1957 just as work on OEDS was beginning. As we proceeded with the dictionary, visits to the moon by astronauts and the exploration of outer space by far-travelling rockets became routine features of our age. Nuclear power stations came into being along with quantities of nuclear weapons and other weapons of war, and all the attendant vocabulary. Nobel prize-winners, playwrights, philosophers, and writers of every kind wrote their books and articles. The *New York Times* and other newspapers increased hideously in size, as did the regular issues of learned journals in chemistry, medicine, and all the other academic subjects. New vocabulary reached our language from the wars and revolutions of the twentieth century, and from the considerable extension of leisure travel. A curious by-product of the scholarship of our age is that the metalanguages of linguisticians and philosophers have now reached a point where writers of monographs cannot even reach the starting line without regularly defining exactly what they mean by such ordinary (and certainly not new) expressions as *accent*, *sentence*, *utterance*, and *word*, not to go further afield. Some linguistic scholars can now express themselves only in a manner which is 'as inviting as a tall wall bottle-spiked' (to use Professor Ricks's memorable phrase),[1] others only in tree-diagrams, others again by ritual exercises like distinguishing 'the cat sat on the mat' from 'the mat sat on the cat'. I cannot believe that historical lexicography will crumble or be damaged in any permanent way by these transient schools of thought, but the danger exists.

* * *

[1] *London Review of Books* 6 June 1985, p. 9.

In his famous Dictionary, Dr Johnson expressed his indebtedness to Junius and Skinner in the following manner: 'the only names which I have forborn to quote when I copied their books; not that I might appropriate their labours or usurp their honours, but that I might spare a perpetual repetition by one general acknowledgement'. It is appropriate that I should acknowledge in similar terms our frequent indebtedness to several major works of reference and learned journals throughout the preparation of OEDS: the great regional historical dictionaries, especially *A Dictionary of American English*, *A Dictionary of Americanisms*, and *A Dictionary of Canadianisms*; the slang dictionaries of Eric Partridge; *The Barnhart Dictionary of New English since 1963* and its successor *The Second Barnhart Dictionary of New English*; the multifarious glossaries of the British Standards Institution; innumerable specialized glossaries and textbooks in every subject, as well as the *Encyclopaedia Britannica* and other encyclopaedias and yearbooks; and the learned journals *American Speech* and *Notes and Queries*.

The passage of time has almost made it seem as if the editing of the *OED* and of OEDS has been part of a continuous process. My own work on the final pages of OEDS has also overlapped with the beginning of a new and very ambitious project—the merging in electronic form of the twelve volumes of the *OED* and the four volumes of the Supplement. This imaginative project, the computerization of the *OED*,[1] should ensure that the Oxford tradition—indeed its pre-eminence—in historical lexicography will be maintained into the twenty-first century and beyond.

* * *

It has given me boundless pleasure to 'ascertain the significance' of so many modern English words, and 'to perform all the parts of a faithful lexicographer'. The performance of the task has also taken me to lecture platforms and broadcasting houses in most major countries in the world. In this country I have had the special privilege of advising the BBC on the standards of spoken English presented by them,[2] and the opportunity to discuss the issues with some of the famous broadcasters of our time. I have also participated in many programmes, on both radio and television, in which the English language has been the main topic of discussion or of entertainment, from the studios of Radio 3 and of the external services to the more relaxed ones broadcasting programmes like 'Call My Bluff' and 'Desert Island Discs'. I have tried throughout to insist on the permanent value of the primary canons of my trade: the necessity of recording the indelicate as well as the delicate or neutral works of our century, demotic vocabulary as well as that which is taken to be elegant, words that offend ethnic sensibilities as well as those that cause no offence, overseas English vocabulary as well as that used in the United Kingdom, and the literary language as well as the ordinary printed word.

Now time has moved on. Volumes I and II of OEDS have already been reprinted twice and Volume III once. This Supplement, begun in a small house in a back street in Oxford and finished nearly three decades later in a Georgian mansion in one of Oxford's noblest streets, will surely stand as a lasting testament to the fruitfulness and inventiveness of the language of our age.

In a recent visit to Washington I came across the following statement on a plaque in the Capitol:

> After the departure of British Forces from New York, American Independence was close at hand. George Washington resigned his military commission at the State House in Annapolis before the Continental Congress. 'Having now finished the work assigned to me, I retire from the great theatre of action.' 23 Dec. 1783.

[1] A convenient description of the project by its Editor, E. S. C. Weiner, can be found in the *Journal of English Linguistics* (April 1985), pp. 1–13.

[2] *The Quality of Spoken English on BBC Radio* (BBC, 1979); *The Spoken Word* (BBC, 1981).

With the completion of a task assigned to me in 1957, I now retire from the 'great theatre' of lexicography, and will devote myself in the years ahead to a reconsideration of English grammar.

R.W.B.

Oxford
June 1985

EDITORIAL STAFF

The dates given after the names indicate when each person joined the editorial staff of this dictionary. The letter P precedes the names of those who worked as part-timers.

Senior Editors (General): LESLEY S. BURNETT	1974–80, 1982–4
J. A. SIMPSON	1976–85
E. S. C. WEINER	1977–84
Senior Editor (Science): A. M. HUGHES	1968–85
Senior Editor (Natural History): SANDRA RAPHAEL	1969–83
Assistant Editor (Bibliographical Collation): J. PATERSON	1975–85

Editorial Assistants

E. C. DANN	1963–83	AMANDA J. BURRELL	1979–83
ADRIANA P. ORR	1966–83	DELLA J. THOMPSON	1980–2
DEBORAH D. HONORÉ	1970–80, P1980–3	D. B. W. BIRK	1980–2
EDITH BONNER	1976–83	M. A. MABE	1980–3
A. HODGSON	1976–82	R. C. PALMER	1981–2
YVONNE L. WARBURTON	1976–84	KATHERINE H. EMMS	1981–3
JULIA C. SWANNELL	1976–83	ALANA G. DICKINSON	1981–3
ELIZABETH M. KNOWLES	1977–83	R. C. BEATTY	1981–3

Several of the above assisted with parts of the final stages of work on Volume IV, either part-time or on a free-lance basis.

Members of the editorial staff received valuable part-time assistance from the following library researchers: Grace M. Briggs (1959–85), M. Yvonne Offord (1982–5), Rita G. Keckeissen (1968–85), Daphne Gilbert-Carter (1975–85), Sally Hinkle (1977–85), and G. Chowdharay-Best (1984–5), the first two in Oxford, Mr Chowdharay-Best in London, and the others in New York, Washington, and Boston respectively. Mr F. D. Hayes also continued his work on the reading of sources.

Miss Knowles, Mrs Burrell, and Ms Emms were mainly concerned with research (especially for 'first uses') and with the verification of quotations. Mrs Honoré dealt with terms in the Social Sciences, and Mr Mabe, Mr Beatty, and Mrs Dickinson with scientific terms. All other Editorial Assistants named above undertook general editorial work.

Some other members of the Department and some free-lance readers have assisted with the reading of proofs, especially Dr Margaret A. Cooper, D. J. Edmonds, M. Harrington, Miss Freda J. Thornton, and Dr W. R. Trumble.

Secretarial: Karin C. E. Vines (1981–6).

INTRODUCTION

1. *History of the Project.* In 1933 the *O.E.D.* team, its work done, dispersed, and the two surviving Editors, Dr. C. T. Onions and Sir William Craigie, and their assistants, turned to other matters. The *O.E.D.* library was broken up and quotation slips that had not been used were crated and stored, some later to be dispatched to the United States for use in the preparation of the *Middle English Dictionary* and of the projected dictionary of Early Modern English.[1]

After the conclusion of the 1939–45 war the Delegates of the University Press decided to re-establish a headquarters for the Dictionary and to embark on the preparation of a revised version of the 1933 Supplement to the *O.E.D.* In 1955, as a first step, they invited R. C. Goffin, formerly Deputy Publisher of the Oxford University Press, to establish an office in a private house, No. 40 Walton Crescent, close by the printing-works and the Clarendon Press, and to prepare the way for the editorial staff to arrive. He was joined for a few months by E. A. Horsman, on leave from Durham University. In 1957 the present writer, at that time Lecturer in English Language and Literature at Christ Church, Oxford, accepted the invitation of the Delegates to edit the revised Supplement, and the appointment took effect from 1 July 1957.

There is a passage in the Historical Introduction to the Dictionary describing the 'crowded scene to the eye of the visitor' presented by Sir James Murray's Scriptorium in the garden of his house at 78 Banbury Road:

> If John Baret had been able to look into it, he would have hailed it as another *Alvearie*, with a swarm of workers as busy as those who helped him to compile his own volume.

It was at 78 Banbury Road that Sir James and his staff carried out much of their work in the preparation of the dictionary. Since 1957 No. 40 Walton Crescent has become 'another *Alvearie*'. After some initial disappointments, largely occasioned by the impossibility of finding experienced staff at that time, the preparations for the Supplement proceeded steadily. Useful practical advice was obtained from Dr. Onions, and valuable contacts were made with former members of the dictionary staff who were still alive, among them Dr. L. F. Powell, Dr. G. W. S. Friedrichsen, and Mr. P. T. J. Dadley.

The first phase in the preparation of any dictionary on historical principles is the reading of sources. Since 1957 our readers have extracted about a million and a half quotations from works of all kinds written in the period from 1884 to the present day. The sources included all important literary works (in both prose and verse) of the period, a wide range of scientific books and journals, and large numbers of newspapers and periodicals, ranging from *The Times* to those publications that emanate from the so-called 'underground'. Numerous works containing multiple lexicographical information, for example, articles in *American Speech* and in *Notes and Queries* and the whole of Eric Partridge's *Dictionary of Slang*, were also converted into the form of dictionary slips. Among the material submitted to the Press since 1933 there were three valuable private collections, and these were also added to the quotation files: a set of several thousand quotations assembled from theatrical and other works by the late Sir St. Vincent Troubridge, and a similar number (mostly written on the backs of envelopes or on any scrap of

[1] A description of the work so far undertaken at the University of Michigan on a dictionary of the Early Modern English period is printed in A. Cameron et al., *Computers and Old English Concordances* (Toronto 1970), pp. 94–102. This dictionary and other period and regional dictionaries were first proposed by Sir William Craigie in 1919 ('New Dictionary Schemes Presented to the Philological Society, 4th April, 1919', *Transactions of the Philological Society, 1925–1930* (1931), pp. 6–11).

paper that conveniently lay to hand) from the Revd. H. E. G. Rope and from the late Dr. R. W. Chapman. In 1958 Mr. Clarence L. Barnhart sent to us from his files in New York a set of some 4,500 slips drawn from 1955 issues of *The Times*, *Science News Letter*, and other sources. Some time later Mr. H. W. Orsman presented to us his unique collection of some 12,000 quotations from New Zealand works of the period from the rediscovery of New Zealand by James Cook until about 1950. At a still later stage specialized collections of terms in Archaeology and Forestry respectively were given by Professor C. F. C. Hawkes and Mr. F. C. Ford Robertson. Among the material left behind by the editors of the 1933 Supplement there was also a collection of quotations, numbering about 140,000, from which they had made 'only a restricted selection' (*O.E.D. Suppl.*, Preface), including illustrative examples of words excluded from the dictionary in 1933 because they were not fully established at the time (e.g. *canned* (of music), *usherette*).

Easily the most ambitious reading programme undertaken by any one reader was that of Miss Marghanita Laski. By 1971 her contribution amounted to more than 100,000 quotations, obtained (and copied by hand by Miss Laski herself) from a wide range of sources. Miss Laski described her experiences as a reader in a series of articles in the *Times Literary Supplement*, beginning with the issue of 11 January 1968.

The sifting of sources, the appointment and training of the first members of staff, and other necessary preliminaries were completed by 1964. In that year we turned to the preparation of 'copy' for press and the first instalment, *A–alpha*, was delivered to the University Printer on 27 May 1965. Since then members of the dictionary staff have been engaged simultaneously on two fronts, preparing 'copy' for press from *alpha* onwards, and dealing with the proofs. At a later stage, beginning in May 1970, material in the range E to G was sent to a second printer, Messrs. William Clowes & Sons Ltd., Colchester, leaving the University Press to deal with A to D inclusive.

It will be apparent to users of the Supplement that one result of this method of sending 'copy' to press in instalments as they became ready is that the earlier letters of the Supplement are not quite as up to date as the later ones. For example, it proved possible to add the word *Biafran* to the corrected galleys (though the civil war in Nigeria had not ended when this word was read in the galleys) but not *Anguillan* (to which attention was drawn by events in 1969). Similar considerations apply to numerous other words that readers may hope to find treated in the Supplement. The following table, which shows the dates of dispatch to press of the 'copy' for Volume I of the Supplement, is a useful guide in this connection:

1965	May	A–alpha	1969 Jan.	cruiser–cursus
	Aug.	alpha–antibiosis	Apr.	curtain–dash
	Nov.	antibiotic–end of A		Thereafter in small consignments at regular intervals:
1966	Feb.	B–Benthamism		
	May	benthonic–bond	1969 (remainder)	dash-board–devil
	Sept.	bonded–bucket	1970	devil–end of D
	Oct.	bucket–end of B		E–flathead
1967	Feb.	C–chain		G–get
	May	chain–city	1971	flat-headed–end of F
	Sept.	city–conditioned		get–end of G
1968	Feb.	conditioner–Crimean		
	Dec.	Crimean–cruiser		

The final instalment of the letter G was sent to press in May 1971.

Other important stages[1] in the preparation of Volume I included the publication in the 1958–61 issues of *The Periodical* (the house magazine of the Oxford University Press) of lists of words, with dates attached, for which earlier, or later, or additional quotations were needed;[2] the building up of our reference library of some 7,000 volumes; the appointment of permanent members of staff in London and Washington, thus giving us access to the great reference libraries in those two cities; the establishment of links with language centres (e.g. at the University of Sydney) and with overseas libraries (e.g. in Wellington, New Zealand, where Miss M. Walton and other members of staff of the Turnbull Library have verified local items for us); the appointment to the editorial staff in 1968–9 of some graduates in scientific subjects, a radical departure from the policy adopted by the editors of the main Dictionary;[3] and the creation of a panel of outside consultants, who read and commented on individual items in specialized subjects in galley-proof, and of another group of scholars and writers who read through instalments of galley-proof with a critical eye. Since 1968 we have also received direct and valuable assistance from Dr. Philip B. Gove and his associates at the G. & C. Merriam Co., publishers of *Webster's Third New International Dictionary*, in the form of quotations of earlier date than those in our files for words, such as those in *-ly*, *-ment*, and *-ness*, which elude the most diligent search by research assistants. About one-third of the items of this kind submitted to Dr. Gove were antedated from Merriam-Webster's extensive citation files.

2. *Editorial Policy*. The famous dictum stated in the Introduction to the Dictionary that 'the circle of the English language has a well-defined centre but no discernible circumference' is only partially applicable to the vocabulary contained in this Supplement. The perimeter remains as undefinable as ever. But in any supplementary volumes the domain of new 'common words' known to all English speakers is bound to be somewhat less evident than the scientific, technical, slang, dialectal, and overseas words which have passed into common use in the academic or technical fields, or in the geographical areas, to which they belong. Our aim has been first and foremost to ensure that all 'common words' (and senses) in British written English of the period 1884 to the present day (of those not already treated in the Dictionary) are included. Then, whereas the *O.E.D.* adopted a policy of total literary inclusiveness for the earlier centuries, with the result that all the vocabulary, including *hapax legomena*, of such authors as Chaucer, Gower, and Shakespeare, was included, we have followed a somewhat more limited policy, namely that of liberally representing the vocabulary of such writers as Kipling, Yeats, James Joyce, and Dylan Thomas. The outward signs of the working of this policy may be observed in entries like those for the following words: *apatheia* (a medical word used by Beckett), *athambia* (*hapax legomenon* in Beckett), *Babbitt* (name of a literary 'hero'), *bandersnatch* (a 'Lewis Carroll' word), *bang*, *sb.*[1] 2 (used allusively after T. S. Eliot's line), *barkle*, *v.* (dialectal use in D. H. Lawrence), *baw-ways* (dialectal use in James Joyce), *ectomorph* (anthropometric term adopted by R. Fuller, C. P. Snow, W. H. Auden, etc.), and *elf sb.*[1] 6 (further illustrations in Walter de la Mare, J. R. R. Tolkien, etc., of obvious combinations). Thirdly, we have made bold forays into the written English of regions outside the British Isles, particularly into that of North America, Australia, New Zealand, South Africa, India, and Pakistan. It is already and will remain impracticable for any general

[1] Accounts of the progress of the Supplement may be found in articles by the Editor in *Essays and Studies*, 1961, pp. 35–51, and in the *Oxford Magazine*, 21 Nov. 1969, pp. 68–9. One of our outside consultants, Professor Bernard Lennox, contributed a series of articles entitled 'Reflections of a Lexicographer' to *The Lancet*, beginning with the issue of 8 May 1971.

[2] This necessary exercise brought the detailed needs of the Dictionary to the notice of many people who would otherwise have been unreachable. The response was remarkable and for years afterwards contributions continued to arrive. Among the most devoted of outside helpers reached by this means were Mr. David Shulman (New York) and Mr. D. J. Barr (Almonte, Ontario), who both submitted numerous antedatings and other contributions.

[3] By 1968–9 some scientific words in the earlier pages of the Supplement were already at too advanced a stage of printing for it to be possible to revise them further. Except for these items all scientific words and senses were prepared or revised by the scientific staff of the Supplement in the period between 1968 and the publication of the present volume.

dictionary of English, wherever it is prepared, to absorb all the contents of the great historical dictionaries of regional forms of English that have appeared[1] or are in preparation.[2] In practice we have drawn some items from these dictionaries, and have supplemented them with fresh examples and also with an entirely original vein of words and senses from the same areas. Readers will discover by constant use of the Supplement that the written English of regions like Australia, South Africa, and India has been accorded the kind of treatment that lexicographers of a former generation might have reserved for the English of Britain alone. Fourthly, we have endeavoured to extract from textbooks and journals the central and enduring vocabulary of all major academic subjects, including newish disciplines like Sociology, Linguistics, Computer Science, and the various branches of Anthropology and Psychology. Fifthly, whereas in 1957, when we began our work, no general English-language dictionary contained the more notorious of the sexual words, '*nous avons changé tout cela*', and two ancient words, once considered too gross and vulgar to be given countenance in the decent environment of a dictionary, now appear with full supporting evidence along with a wide range of colloquial and coarse expressions referring to sexual and excretory functions.

In the 1933 Supplement the Editors included a large number of 'Earlier U.S. examples'. These have not normally been retained in the revised Supplement since they have subsequently been absorbed, along with a mass of other material, in the large historical dictionaries of American English. It was also decided to exclude, in the main, pre-1820 antedatings of *O.E.D.* words or senses from general English sources, since the systematic collection of such antedatings could not be undertaken at the present time. Nor have we added later examples to words and senses whose illustration ends in the Dictionary with nineteenth-century examples. In the earlier letters of the alphabet such a policy would have entailed the addition of later-nineteenth-century or of twentieth-century examples for virtually every word and sense listed in the Dictionary. Our policy depends upon the realization by users of the Dictionary that any word or sense not marked '*Obs.*' or '*arch.*' is still part of the current language.

A great many words and senses can be traced to their first appearance in print and we have studiously endeavoured to trace all such 'first uses'. But it should be pointed out that the earliest examples presented here for some other words and senses must necessarily represent merely the first appearance of such words in the printed sources read for the Supplement.

The Main Words, and cross-references, are arranged in a single alphabetical series, as in the Dictionary, but the bold-type head-words are here printed with a lower-case initial letter, except for proper names (in the *O.E.D.* itself all head-words, whatever their status, were given capital initials). The *O.E.D.*'s distinction of Main Words and Subordinate Words has been abandoned and in consequence all head-words appear in the same size and darkness of type. As in the Dictionary, Combinations are normally dealt with under the Main Words which form their first element, and are printed as the concluding section of the article. The asterisk often used in such sections in the Dictionary to draw attention to the word illustrated is here abandoned since the asterisk has other functions in the Supplement (especially preceding cross-references to a word or sense found elsewhere in the Supplement). No substantive changes have been made in the structuring of articles except that an asterisk *, or if necessary a double ** or triple *** asterisk, placed after a sense number indicates a new sense or senses which has to be inserted within the existing

[1] *A Dictionary of American English* (4 vols., 1938–44), edited by W. A. Craigie and J. R. Hulbert; *A Dictionary of Americanisms* (2 vols., 1951), edited by M. M. Mathews; *A Dictionary of Canadianisms* (1967), edited by W. S. Avis and others; and a *Dictionary of Jamaican English* (1967), edited by F. G. Cassidy and R. B. Le Page.

[2] *A Dictionary of the Older Scottish Tongue* (1931–), edited by W. A. Craigie and A. J. Aitken (*A–Mony* published by 1971); the *Scottish National Dictionary* (1931–), edited by W. Grant and D. D. Murison (*A–Selkirk* published); and the *Middle English Dictionary* (1952–), edited by H. Kurath, S. M. Kuhn, and J. Reidy (*A–Lef* published).

numerical sequence because of the custom in the Dictionary of placing the Combinations at the conclusion of each article. To accord with the practice of the Dictionary the ligatures æ and œ have been retained, although the use of these runs counter to the 'house style' of the Oxford University Press.

The system of labelling is unchanged. Thus, for example, the status labels *Obs.* (obsolete), *arch.* (archaic or obsolescent), *colloq.* (colloquial), *dial.* (dialectal), and *slang* are retained notwithstanding the practice in some modern dictionaries of replacing *colloq.* (and sometimes also *slang*) by the label *informal*. Whatever the merits of *informal* it would have been inappropriate to have a different system in the Supplement from that used in the Dictionary itself. The label *N. Amer.* has been used to mean 'recorded in (part(s) of) the United States and Canada'. The Pronunciation Key is in all main particulars unchanged, again in the interests of consistency with the Dictionary. The list of Abbreviations and Signs has been substantially expanded. In the etymologies the characters of all foreign languages except Greek have been transliterated (where necessary) into the roman alphabet.

3. *The Editorial Process*. The various stages involved in the preparation of the material of Volume I for press can be represented diagrammatically:

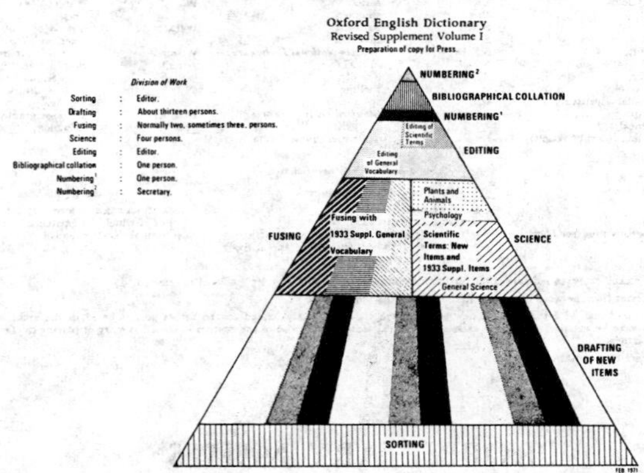

The divisions within the triangle are proportional: they show the amount of 'effort' that was involved at each stage, estimated on the basis of the relative amounts of time expended and the number of people concerned. A brief explanation of the terminology follows.

Sorting: The removal of quotation slips illustrating words and senses that lay outside the terms of reference of the Supplement (pre-1820 antedatings of words and senses already treated in the Dictionary, ephemeral items, etc.), and the rough grouping of quotations into parts of speech and senses.

Drafting of New Items: The preparation of dictionary articles in handwritten form on 6″ × 4″ slips by editorial assistants. Each editorial assistant was expected to draft complete entries, i.e. to ascertain the pronunciation and etymology of each word, to add a definition, and to select and (with the assistance of people specially appointed for the purpose) verify the quotations to be used.

Fusing: The collation of new items submitted by various drafters and the 'fusing' or merging of new items with the words and senses of the standing matter of the 1933 Supplement. Those responsible for 'fusing' also revised the material in the standing matter.

Science: Articles for new scientific words and senses were drafted and scientific items in the standing matter were revised by the scientific staff.

Editing: The bringing together and revision of all material by the Editor.

Bibliographical collation: The process of establishing that the titles of illustrative examples were consistently presented in respect both of the date and spelling, etc., and of the formulaic 'short title' used. For this purpose a very large card-index of verified book-titles, made up of all the works cited in the Supplement, has been compiled and will form the basis of the Bibliography in Volume III.

Numbering: Two stages of numbering of the slips to ensure that the compositors could set the material from the handwritten 'copy'. When the numbering of each instalment was completed the 'copy' was sent to press.

4. *Size of the Supplement*. It is estimated that the Supplement, when completed, will contain some 50,000 Main Words. Volume I contains between 17,000 and 18,000 Main Words divided into some 30,000 senses. There are just under 8,000 defined Combinations within the articles and a similar number of undefined Combinations. The illustrative quotations number 130,000.

R. W. BURCHFIELD

Oxford
October 1971

BIBLIOGRAPHICAL CONVENTIONS

Quotations are normally taken from the earliest available printing of a work; where a later text has been used its date is given in parentheses after the title. Intentional exceptions can be found in often quoted works whose first editions are rare (Joyce's *Ulysses* is simply dated 1922, but quotations are taken from the Random House edition); fuller details will be shown in the Bibliography.

Unverified quotations from secondary sources have an attribution in parentheses after the citation (e.g. Morris, Pettman, etc.), except for quotations from the *Dictionary of American English* and the *Dictionary of Americanisms*, which are silently accepted.

Date. The bold-face date is the date of first printing except for posthumous works. It is sometimes qualified by *c* (*circa*) or *a* (*ante* = before, not later than). The date of delivery of a lecture or production of a play is not normally accepted: the spoken word is dated by its first appearance in print. The date of composition is accepted, however, for dated letters, journals, and the like (of those which have not been written up for publication), but only exceptionally in other cases and only when there is good evidence for the date. Items in collections (e.g. of short stories) which were published earlier elsewhere are given this earlier date when it is known. It should be noted that our criteria are more severe and produce more conservative datings than those of many other reference works.

Author and Title. This part of the citation is used first to identify the user of the quoted word, and secondly to identify the work from which a quotation is taken. Usually there is no conflict, but where there is the identification of the work takes precedence. Monographs are cited either by author and title or by title alone. Corporate authorship is not recognized: the names of institutions, business firms, etc., are not used in place of a personal name, but are added in parentheses at the end of titles. Periodicals and serials are cited by title (usually in abbreviated form); the authors of articles in periodicals are not usually named unless the quotation contains the first use of a word. The information given should be enough to identify the work, but occasionally it is not possible to give an unambiguous indication in the space available; in these cases the Bibliography will give a fuller account. There are also a few quotations which have been accepted from usually reliable readers even though it has not been possible to verify the author and title.

Shortened titles are used throughout: the expanded forms of the titles most often cited in the Supplement will be listed in the Bibliography.

Form of Name. Some frequently quoted authors are cited by surname alone; for most others the title-page of each work is the main authority, with deviations to allow for the standardization of initials where an author's own practice varies or for the purpose of avoiding ambiguity. Pseudonyms are indicated by single quotation marks (except that a few well-known pseudonyms like Geo. Eliot stand without the quotation marks), and authors who have changed their names are cited by the appropriate name for each work. Anonymous works are cited by title, but by author where the authorship has been established and is recognized in common practice. Small capitals usually denote the author of the quotation. The name of an editor of a work, who is not also the author of the actual quotation given, is normally printed in lower case. But quotations from many composite works have been attributed to the editors alone.

SCIENTIFIC TERMS

Lexicographers are now confronted with the problem of treating the vocabularies of subjects that are changing at a rate and on a scale not hitherto known. The complexity of many scientific subjects is such too that it is no longer possible to define all the terms in a manner that is comprehensible to the educated layman. Some indications follow of the policy adopted in this Supplement for the treatment of scientific terms.

Subject Labels. The use of subject labels has been extended somewhat, but relatively broad labels have normally been preferred, e.g. *Geol.* rather than *Stratigraphy*, *Astr.* rather than *Astrophysics*, *Bot.* rather than *Plant Anat.*

First Uses. The first use of each word and sense has been ascertained whenever possible and appears as the first example in the set of illustrative quotations. By 'first use' the compilers of historical dictionaries mean 'the first use traced in a printed source': a word or phrase may have occurred in oral use at an earlier date. If a word was first coined in some other language before being adopted into English, details of the foreign coinage (when traceable) are provided in the etymology. All such foreign coinages have been verified at source since it sometimes happens that the details provided in specialized bibliographies and reference works are inaccurate.

Details of the coinages of plant and animal names are provided in the normal way. When, however, the first use of a term preceded the date accepted as the starting-point for the valid nomenclature of the group involved, a reference to the first valid use is added in the etymology.

Illustrative Quotations. These are not designed as guides to 'further reading' and it is not implied that the works from which they are taken are (or were) necessarily the most authoritative works on the subject, though they often are. The quotations illustrate the sense or senses being treated and amplify the definitions in a way that could not be done as usefully by mere extension of the definitions themselves. Consistency with the principles of the Dictionary obliges us always to include the 'first use' of each word even though it may sometimes appear to present a somewhat archaic or misleading use of the word or sense when judged by later developments.

Words already in the Dictionary. Although with the passage of time the definitions of many basic scientific words in the body of the Dictionary are now out of date or old-fashioned in varying degrees, they have not in general been revised. Such a process would have involved a more radical revision of scientific terminology, necessitating a reconsideration of several centuries of scientific activity, than has been attempted in this Supplement for the general vocabulary.

Systematic Names of Plants and Animals. The inclusion of Latin generic names of plants or animals has depended on the quantity of evidence found for the use of the word in an English context as the name of an individual and not as the name of a genus. Names of groups above generic level have been included only in their anglicized forms, when sufficient evidence for the use of these forms has been traced: thus *dytiscid* has an entry but *Dytiscidæ* has not. Obsolete terms or senses have been included when evidence for their use over a fairly long period was available, e.g. *fossor* (sense 2).

Nomenclature of Plants and Animals. The International Codes of Botanical and Zoological Nomenclature have been taken as guides to taxonomic terms, supplemented by similar codes for more limited groups and authoritative lists of recommended terms and names. An attempt has been made to comply with official changes in the Latin names of plants and animals. The name most recently adopted has been used when a choice was necessary, sometimes with the addition of a common alternative.

The printing conventions now generally accepted for taxonomic literature have been followed, so that only Latin binomials or generic names alone are printed in italics, and specific epithets are not given initial capitals.

LIST OF ABBREVIATIONS, SIGNS, ETC.

Some abbreviations here listed in italics are occasionally, for the sake of clarity, printed in roman type, and vice versa.

Abbreviation	Meaning
a. (in Etym.)	adoption of, adopted from
a (as *a* 1850)	*ante*, 'before', 'not later than'
a.	adjective
abbrev.	abbreviation (of)
abl.	ablative
absol.	absolute, -ly
Abstr.	Abstract(s)
acc.	accusative
ad. (in Etym.)	adaptation of
Add.	Addenda
adj.	adjective
adv.	adverb
advb.	adverbial, -ly
(Advt.),	advertisement
Aeronaut.	in Aeronautics
AF., AFr.	Anglo-French
Afr.	Africa, -n
Agric.	in Agriculture
Alb.	Albanian
Amer.	American
Amer. Ind.	American Indian
Anat.	in Anatomy
Anglo-Ind.	Anglo-Indian
Anglo-Ir.	Anglo-Irish
Anthrop., *Anthropol.*	in Anthropology
Antiq.	in Antiquities
aphet.	aphetic, aphetized
app.	apparently
Arab.	Arabic
Aram.	Aramaic
Arch., *Archit.*	in Architecture
arch.	archaic
Archæol.	in Archæology
Arm.	Armenian
assoc.	association
Astr.	in Astronomy
Astrol.	in Astrology
attrib.	attributive, -ly
Austral.	Australian
A.V.	Authorized Version
bef.	before
Bibliogr.	in Bibliography
Biochem.	in Biochemistry
Biol.	in Biology
Bot.	in Botany
Bulg.	Bulgarian
c (as *c* 1700)	*circa*, 'about'
c. (as 19th c.)	century
Canad.	Canadian
Cat.	Catalan
catachr.	catachrestically
Celt.	Celtic
Cent. Dict.	*Century Dictionary*
Cf., cf.	*confer*, 'compare'
Ch.	Church
Chem.	in Chemistry
Cinemat., *Cinematogr.*	in Cinematography
cl. L.	classical Latin
cogn. w.	cognate with
collect.	collective, -ly
colloq.	colloquial, -ly
comb.	combined, -ing
Comb.	Combinations
Comm.	in Commercial usage
Communic.	in Communications
comp.	compound, composition
compar.	comparative
compl.	complement
Conch.	in Conchology
concr.	concrete, -ly
conj.	conjunction
cons.	consonant
const.	construction, construed with
corresp.	corresponding (to)
cpd.	compound
Cryst.	in Crystallography
Da.	Danish
D.A.	*Dictionary of Americanisms*
D.A.E.	*Dictionary of American English*
dat.	dative
def.	definite, -ition
deriv.	derivative, -ation
dial.	dialect, -al
Dict.	Dictionary; *spec.*, the *Oxford English Dictionary*
dim.	diminutive
D.O.S.T.	*Dictionary of the Older Scottish Tongue*
Du.	Dutch
E.	East
Eccl.	in Ecclesiastical usage
Ecol.	in Ecology
Econ.	in Economics
ed.	edition
E.D.D.	*English Dialect Dictionary*
Educ.	in Education
e.g.	*exempli gratia*, 'for example'
Electr.	in Electricity
ellipt.	elliptical, -ly
Embryol.	in Embryology
e. midl.	east midland (dialect)
Eng.	English
Engin.	in Engineering
Ent.	in Entomology
erron.	erroneous, -ly
esp.	especially
et al.	*et alii*, 'and others'
etc.	et cetera
Ethnol.	in Ethnology
etym.	etymology
euphem.	euphemistically
exc.	except
f. (in Etym.)	formed on
f. (in subordinate entries)	form of
F.	French
fem. (*rarely* f.)	feminine
fig.	figurative, -ly
Finn.	Finnish
fl.	*floruit*, 'flourished'
Fr.	French
freq.	frequent, -ly
Fris.	Frisian
Funk's Stand. Dict.	*Funk and Wagnalls Standard Dictionary*
G.	German
Gael.	Gaelic
Gaz.	Gazette (in names of newspapers)
gen.	genitive
gen.	general, -ly
Geogr.	in Geography
Geol.	in Geology
Geom.	in Geometry
Geomorphol.	in Geomorphology
Ger.	German
Gmc.	Germanic
Goth.	Gothic
Gr.	Greek
Gram.	in Grammar
Heb.	Hebrew
Her.	in Heraldry
Herb.	among herbalists
Hind.	Hindustani
Hist.	in History
hist.	historical
Hort.	in Horticulture
Ibid.	*Ibidem*, 'in the same book or passage'
Icel.	Icelandic
Ichthyol.	in Ichthyology
id.	*idem*, 'the same'
i.e.	*id est*, 'that is'
IE.	Indo-European
imit.	imitative
Immunol.	in Immunology
imp.	imperative
impers.	impersonal
impf.	imperfect
ind.	indicative
indef.	indefinite
inf.	infinitive
infl.	influenced
int.	interjection
intr.	intransitive
Introd.	Introduction
Ir.	Irish
irreg.	irregular, -ly
It.	Italian
J., (J.)	Johnson's *Dictionary* (quoted from)
(Jam.)	Jamieson, *Scottish Dict.*
Jap.	Japanese
joc.	jocular, -ly
l.	line
L.	Latin
lang.	language
Let., *Lett.*	letter, letters
LG.	Low German
lit.	literal, -ly
Lit.	Literary
Lith.	Lithuanian
LXX	Septuagint
Mal.	Malay, Malayan
Manuf.	in Manufacture, -ing
masc. (*rarely* m.)	masculine
Math.	in Mathematics
MDu.	Middle Dutch
ME.	Middle English
Mech.	in Mechanics
Med.	in Medicine
med.L.	medieval Latin
Metaph.	in Metaphysics
Meteorol.	in Meteorology
MHG.	Middle High German
midl.	midland (dialect)
Mil.	in military usage
Min.	in Mineralogy
MLG.	Middle Low German
mod.	modern
mod.L.	modern Latin
(Morris),	E. E. Morris's *Austral English* (quoted from)
Mus.	in Music
Mythol.	in Mythology
N.	North
N. Amer.	North America, -n
N. & Q.	*Notes and Queries*
Nat. Hist.	in Natural History
Naut.	in Nautical language
Neurol.	in Neurology
neut. (*rarely* n.)	neuter
NF., NFr.	Northern French
nom.	nominative
north.	northern (dialect)
Norw.	Norwegian
N.T.	New Testament
Nucl.	Nuclear
Numism.	in Numismatics
N.Z.	New Zealand
obj.	object
obl.	oblique
Obs., *obs.*	obsolete
occas.	occasional, -ly
Oceanogr.	in Oceanography
OE.	Old English (= Anglo-Saxon)
OF., OFr.	Old French
OFris.	Old Frisian
OHG.	Old High German
OIr.	Old Irish
ON.	Old Norse (Old Icelandic)
ONF.	Old Northern French
Ophthalm.	in Ophthalmology
opp.	opposed (to), the opposite (of)
Opt.	in Optics
orig.	origin, -al, -ally
Ornith.	in Ornithology
OS.	Old Saxon
OSl.	Old (Church) Slavonic
O.T.	Old Testament
p.	page
Palæogr.	in Palæography
Palæont.	in Palæontology
pa. pple.	passive or past participle
(Partridge),	E. Partridge's *Dictionary of Slang and Unconventional English* (quoted from)
pass.	passive, -ly
pa. t.	past tense
Path.	in Pathology
perh.	perhaps
Pers.	Persian
pers.	person, -al
Petrogr.	in Petrography
Petrol.	in Petrology
(Pettman),	C. Pettman's *Africanderisms* (quoted from)
pf.	perfect
Pg.	Portuguese
Pharm.	in Pharmacology
Philol.	in Philology
Philos.	in Philosophy
phonet.	phonetic, -ally
Photogr.	in Photography
phr.	phrase
Phys.	in Physics, physical; (*rarely*) in Physiology
Physiol.	in Physiology
pl.	plural; plate
poet.	poetic, -al
Pol.	Polish
Pol.	in Politics
Pol. Econ.	in Political Economy
pop.	popular, -ly
poss.	possessive
ppl. a., ppl. adj.	participial adjective
pple.	participle
Pr.	Provençal
prec.	preceding (word or article)
pred.	predicative
pref.	prefix
pref., Pref.	preface
prep.	preposition
pres.	present
priv.	privative
prob.	probably
pron.	pronoun
pronunc.	pronunciation
prop.	properly
Pros.	in Prosody
Prov.	Provençal
pr. pple.	present participle
Psych., *Psychol.*	in Psychology
Q.	Quarterly (in names of periodicals)
quot(s).	quotation(s)
q.v.	*quod vide*, 'which see'
R.	Royal (in names of periodicals, etc.)
Radiol.	in Radiology
R. C. Ch.	Roman Catholic Church
redupl.	reduplicating
refash.	refashioned, -ing
refl., refl.	reflexive
reg.	regular
rel.	related (to)
repr.	representative, representing
Rhet.	in Rhetoric
Rom.	Roman, Romance, Romanic
Rum.	Rumanian
Russ.	Russian
S.	South
S. Afr.	South Africa, -n
sb.	substantive
sc.	*scilicet*, 'understand' or 'supply'
Sc., *Scot.*	Scotch, Scottish
Sci.	(in) Science, scientific
Sc. Nat. Dict.	*Scottish National Dictionary*
Ser.	series
sing.	singular
Skr.	Sanskrit
Slav.	Slavonic
S.N.D.	*Scottish National Dictionary*
Sociol.	in Sociology
Sp.	Spanish
sp.	spelling
spec.	specific, -ally
(Stanf.),	*Stanford Dictionary of Anglicised Words and Phrases* (quoted from)
subj.	subject, subjunctive
subord. cl.	subordinate clause
subseq.	subsequent, -ly
subst.	substantively
suff.	suffix
superl.	superlative
Suppl.	Supplement
Surg.	in Surgery
s.v.	*sub voce*, 'under the word'
Sw.	Swedish
s.w.	south-western (dialect)
syll.	syllable
Syr.	Syrian
techn.	technical, -ly
Tel.	Telegraph (in names of newspapers)
Telegr.	in Telegraphy
Teleph.	in Telephony
(Th.),	Thornton's *American Glossary* (quoted from)
Theatr.	in the Theatre, theatrical
Theol.	in Theology
Tokh.	Tokharian
tr., transl.	translation (of)
trans.	transitive
transf.	transferred sense
Trig.	in Trigonometry
Turk.	Turkish
Typog., *Typogr.*	in Typography
ult.	ultimate, -ly
unkn.	unknown
U.S.	United States
usu.	usual, -ly
v., vb.	verb
var(r)., vars.	variant(s) of
vbl. sb.	verbal substantive
Vet., *Vet. Sci.*	in Veterinary Science
viz.	*videlicet*, 'namely'
v. str., or *w.*	verb strong, or weak
vulg.	vulgar
W.	Welsh; West
wd.	word
Webster	*Webster's* (*New International*) *Dictionary*
WGmc.	West Germanic
w.midl.	west midland (dialect)
WS.	West Saxon
(Y.),	Yule & Burnell's *Hobson-Jobson* (quoted from)
Zoogeogr.	in Zoogeography
Zool.	in Zoology

Signs and Other Conventions

Before a word or sense

† = obsolete
‖ = not naturalized, alien
¶ = catachrestic and erroneous uses (see Dict., Vol. I, p. xxxi)

In the listing of Forms

1 = before 1100
2 = 12th c. (1100 to 1200)
3 = 13th c. (1200 to 1300), etc.
5–7 = 15th to 17th century. (See General Explanations, Dict., Vol. I, p. xxx)

In the etymologies

* indicates a word or form not actually found, but of which the existence is inferred
:– = normal development of

The printing of a word in SMALL CAPITALS indicates that further information will be found under the word so referred to.

In cross-references * indicates that the word or sense referred to is in the Supplement.

After the number of a sense * and ** (etc.) indicate new senses which are not directly related to the senses so numbered in the main body of the Dictionary, but which have to be inserted within the existing numerical sequence because of the custom in the Dictionary of placing the Combinations at the conclusion of each article.

. . indicates an omitted part of a quotation.

PROPRIETARY NAMES

This Supplement includes some words which are or are asserted to be proprietary names or trade marks. Their inclusion does not imply that they have acquired for legal purposes a non-proprietary or general significance nor any other judgement concerning their legal status. In cases where the editorial staff have established in the records of the Patent Offices of the United Kingdom and of the United States that a word is registered as a proprietary name or trade mark this is indicated, but no judgement concerning the legal status of such words is made or implied thereby.

KEY TO THE PRONUNCIATION

THE pronunciations given are those in use in the educated speech of southern England (the so-called 'Received Standard'), and the keywords given are to be understood as pronounced in such speech.

I. *Consonants and Semi-Consonants*

b, d, f, k, l, m, n, p, t, v, z *have their usual English values*

		(FOREIGN AND NON-SOUTHERN)
g as in *go* (gō^{u}).	þ as in *th*in (þin), ba*th* (bɑþ).	ṅ as in French nasal, e*n*viro*n* (aṅviroṅ).
h ... *h*o! (hō^{u}).	ð ... *th*en (ðen), ba*th*e (bē^{i}ð).	l^{y} ... It. serra*gli*o (serā·l^{y}o).
r ... *r*un (rɒn), te*rr*ier (te·riəɹ).	ʃ ... *sh*op (ʃǫp), di*sh* (diʃ).	n^{y} ... It. si*gn*ore (sin^{y}ō·re).
ɹ ... he*r* (hɜ̄ɹ), fa*r*the*r* (fā·ɹðəɹ).	tʃ ... *ch*op (tʃǫp), di*tch* (ditʃ).	χ ... Ger. a*ch* (aχ), Sc. lo*ch* (lǫχ), Sp. fri*j*oles (frī·χoles).
s ... *s*ee (sī), *s*ucce*ss* (sɒ̆kse·s).	ʒ ... vi*si*on (vi·ʒən), dé*j*euner (deʒō̆ne).	χ^{y} ... Ger. i*ch* (iχ^{y}), Sc. ni*ch*t (niχ^{y}t).
w ... *w*ear (wē^{ə}ɹ).	dʒ ... *j*u*dg*e (dʒɒdʒ).	γ ... North Ger. sa*g*en (zā·γĕn).
hw... *wh*en (hwen).	ŋ ... si*ng*i*ng* (si·ŋiŋ), thi*n*k (þiŋk).	γ^{y} ... Ger. le*g*en, re*g*nen (lē·γ^{y}ĕn, rē·γ^{y}nĕn).
y ... *y*es (yes).	ŋg ... fi*ng*er (fi·ŋgəɹ).	k^{y} ... Afrikaans baardmanne*tj*ie (bā·rtmanək^{y}i).

The reversed r (ɹ) and small 'superior' letters (pe·rĕm^{p}təri) are used to denote elements that may be omitted either by individual speakers or in particular phonetic contexts.

II. *Vowels*

The symbol ¯ placed over a vowel-letter denotes length.

The incidence of main stress is shown by a raised point (·) after the vowel-symbol, and a secondary stress by a double point (:) as in *callithumpian* (kæ:lĭþɒ·mpiăn).

The stressed vowels a, æ, e, i, o, u become obscured with loss of stress, and the indeterminate sounds thus arising, and approximating to the 'neutral' vowel ə, are normally printed ă, æ̆, ĕ, ĭ, ŏ, ŭ.

A break ˌ is used to indicate syllable-division when necessary to avoid ambiguity.

ORDINARY	LONG	OBSCURE
a as in Fr. *à la* mode (a la mod').	ā as in *a*lms (āmz), b*ar* (bāɹ).	ă as in *a*mœb*a* (ămī·bă).
ai ... *aye*=yes (ai), Is*ai*ah (əizai·ă).		
æ ... m*a*n (mæn).		æ̆ ... *a*ccept (æ̆kse·pt), mani*a*c (mē^{i}·niæ̆k).
ɑ ... p*a*ss (pɑs), ch*a*nt (tʃɑnt).		
ɑu ... l*ou*d (lɑud), n*ow* (nɑu).		
ɒ ... c*u*t (kɒt), s*o*n (sɒn).	ɒ̄ ... c*u*rl (kɒ̄ɹl), f*u*r (fɒ̄ɹ).	ɒ̆ ... dat*u*m (dē^{i}·tɒ̆m).
e ... y*e*t (yet), t*e*n (ten).	ē (ē^{ə}) ... th*e*re (ðē^{ə}ɹ), p*ea*r, p*a*re (pē^{ə}ɹ).	ĕ ... mom*e*nt (mō^{u}·mĕnt), sev*e*r*a*l (se·vĕrăl).
‖*e* ... Fr. attach*é* (ataʃe).	*ē* (*ē*^{i}) ... r*ei*n, r*ai*n (rē^{i}n), th*ey* (ðē^{i}).	*ĕ* ... separ*a*te (*adj.*) (se·părĕt).
‖ę ... Fr. ch*e*f (ʃęf).	‖ę̄ ... Fr. f*ai*re (fę̄r').	
ə ... ev*e*r (e·vəɹ), nat*io*n (nē^{i}·ʃən).	ɜ̄ ... f*i*r (fɜ̄ɹ), f*e*rn (fɜ̄ɹn), *ea*rth (ɜ̄ɹþ).	ė ... add*e*d (æ·dėd), *e*state (ėstē^{i}·t).
əi ... *I*, *eye* (əi), b*i*nd (bəind).		
‖*ə* ... Fr. tour d*e* force (tūrd*ə*fors).		
i ... s*i*t (sit), myst*i*c (mistik).	i (i^{ə}) ... b*ie*r (bi^{ə}ɹ), cl*ea*r (kli^{ə}ɹ).	ĭ ... van*i*ty (væ·nĭti).
i ... Psych*e* (səi·k*i*), r*e*act (r*i*ˌæ·kt).	*ī* ... th*ie*f (þ*ī*f), s*ee* (s*ī*).	*ĭ* ... r*e*main (r*ĭ*mē^{i}·n), b*e*lieve (b*ĭ*l*ī*·v).
o ... ach*o*r (ē^{i}·koɹ), m*o*rality (moræ·lĭti).	ō (ō^{ə})... b*oa*r, b*o*re (bō^{ə}ɹ), gl*o*ry (glō^{ə}·ri).	ŏ ... the*o*ry (þ*ī*·ŏri).
oi ... *oi*l (oil), b*oy* (boi).		
o ... her*o* (hi^{ə}·r*o*), z*oo*logy (z*o*ˌǫ·lŏdʒi).	*ō* (*ō*^{u})... s*o*, s*ow* (s*ō*^{u}), s*ou*l (s*ō*^{u}l).	*ŏ* ... vi*o*let (vəi·*ŏ*lėt), par*o*dy (pæ·r*ŏ*di).
ǫ ... wh*a*t (hwǫt), w*a*tch (wǫtʃ).	ǭ ... w*a*lk (wǭk), w*a*rt (wǭɹt).	ǫ̆ ... *au*thority (ǫ̆þǫ·rĭti).
ǫ,ǭ *... g*o*t (gǫt), s*o*ft (sǭft)*.	*ǭ* ... sh*o*rt (ʃ*ǭ*ɹt), th*o*rn (þ*ǭ*ɹn).	*ǫ̆* ... c*o*nnect (k*ǫ̆*ne·kt), amaz*o*n (æ·măz*ǫ̆*n).
‖ö ... Ger. K*ö*ln (köln).	‖ȫ ... Fr. c*œu*r (kȫr).	
‖*ö* ... Fr. p*eu* (p*ö*).	‖*ȫ* ... Ger. G*oe*the (g*ȫ*tĕ), Fr. j*eû*ne (ʒ*ȫ*n).	
u ... f*u*ll (ful), b*oo*k (buk).	ū (ū^{ə})... p*oo*r (pū^{ə}ɹ), m*oo*rish (mū^{ə}·riʃ).	
iu ... d*u*ration (diurē^{i}·ʃən).	iū, ^{i}ū... p*u*re (piū^{ə}ɹ), l*u*re (l^{i}ū^{ə}ɹ).	iŭ, ^{i}ŭ verd*u*re (vɜ̄·ɹdiŭɹ), meas*u*re (me·ʒ^{i}ŭɹ).
u ... unt*o* (ɒ·nt*u*), fr*u*gality (fr*u*-).	*ū* ... tw*o* m*oo*ns (t*ū* m*ū*nz).	*ŭ* ... alt*o*gether (ǭlt*ŭ*ge·ðəɹ).
i*u* ... Matth*ew* (mæ·þi*u*), virt*ue* (vɜ̄·ɹti*u*).	i*ū*, ^{i}*ū*... f*ew* (fi*ū*), l*u*te (l^{i}*ū*t).	i*ŭ* ... circ*u*lar (sɜ̄·ɹki*ŭ*lăɹ).
‖ü ... Ger. M*ü*ller (mü·lĕr).		
‖*ü* ... Fr. d*u*ne (d*ü*n).	‖*ǖ* ... Ger. gr*ü*n (gr*ǖ*n), Fr. j*us* (ʒ*ǖ*).	

^{ə} (see i^{ə}, ē^{ə}, ō^{ə}, ū^{ə}) } see Vol. I of Dict., p.
^{i}, ^{u} (see ē^{i}, ō^{u}) } xxxiv, note 3.
' as in able (ē^{i}·b'l), eaten (ī·t'n) = voice-glide.

* Words such as *soft*, *cloth*, *cross* are often still pronounced with (ǭ) by Southern speakers in England but the pronunciation with ǫ is now more usual.

‖ Only in foreign (or earlier English) words.

TRANSLITERATION OF FOREIGN SCRIPTS

The lists below show the schemes of transliteration used in this Supplement for the most commonly occurring languages that have not adopted the Roman alphabet.

Arabic: (omitted), ب b, ت t, ث ṯ, ج j, ح ḥ, خ ḵ, د d, ذ ḏ, ر r, ز z, س s, ش š, ص ṣ, ض ḍ, ط ṭ, ظ ẓ, ع ʻ, غ ḡ, ف f, ق ḳ, ك k, ل l, م m, ن n, ه h, ة (omitted), و w, ى y; ʼ ʼ; vowels a, i, u, ā, ī, ū.

Chinese: Wade–Giles system without tone-numbers; in Volumes III and IV Pinyin.

Hebrew: א ʼ, ב b, ג g, ד d, ה h, ו w, ז z, ח ḥ, ט ṭ, י y, כ k, ל l, מ m, נ n, ס s, ע ʻ, פ p, צ ṣ, ק q or ḳ, ר r, שׂ ś, שׁ sh or š, ת t;
spirant consonants underlined or with added h; doubled consonant for *daghesh forte*;

vowels a, e, i, o, u; long vowels with macron or circumflex according as written defective or *plene*; shva and reduced vowels superscript or omitted.

Japanese: 'Modified Hepburn' system, British Standard 4812: 1972.

Russian: А а a, Б б b, В в v, Г г g, Д д d, Е е e, Ж ж zh, З з z, И и i, Й й ĭ, К к k, Л л l, М м m, Н н n, О о o, П п p, Р р r, С с s, Т т t, У у u, Ф ф f, Х х kh, Ц ц ts, Ч ч ch, Ш ш sh, Щ щ shch, Ъ ъ ″, Ы ы ȳ, Ь ь ′, Э э é, Ю ю yu, Я я ya.

Sanskrit: अ a, आ ā, इ i, ई ī, उ u, ऊ ū, ऋ ṛ, ॠ ṝ, ऌ ḷ, ए e, ऐ ai, ओ o, औ au, ˙ ṃ, : ḥ, क k, ख kh, ग g, घ gh, ङ ṅ, च c, छ ch, ज j, झ jh, ञ ñ, ट ṭ, ठ ṭh, ड ḍ, ढ ḍh, ण ṇ, त t, थ th, द d, ध dh, न n, प p, फ ph, ब b, भ bh, म m, य y, र r, ल l, व v, श ś, ष ṣ, स s, ह h; post-consonantal vowels -ा -ā, ि- -i, -ी -ī, ु -u, ू -ū, ृ -ṛ, -ॄ -ṝ, ॢ -ḷ, े -e, ै -ai, -ो -o, -ौ -au.

NOTES

Arabic: ْ (sukūn) omitted
ّ (šadda) doubled consonant
Assimilate l of definite article.
Hyphenate article to noun.
Diphthongs aw, ay; nunation an, in, un.
Extra letters in Persian p, ch, zh, g; s, t, z, s̱, ẕ, ż replace ṣ, ṭ, ẓ, ṯ, ḏ, ḍ; vowels include e, o.
Extra letters in Urdu ṭ, ḍ, ṛ.
This is for classical Arabic; colloquial forms may include further letters, e.g. ə at *FELLAGHA.

Hebrew: also for Aramaic and Yiddish.

Japanese: n is assimilated before b, m, p (*kombu*, not *konbu*).

Russian: stress generally marked by acute accent on vowel; stressed ȳ written ý.

Sanskrit: bare stem used (dictionary form); -a is not written in devanagari.
Also for Hindi.

A SUPPLEMENT TO THE OXFORD ENGLISH DICTIONARY

THE OXFORD ENGLISH DICTIONARY

SUPPLEMENT

A. Add: *From A to Z*: see Z 3.

II. 6. Designating a first-class road.

1921 *Autocar* 29 Oct. 829/2 Those roads, however, which have already received numbers are all of the first, or A, category and it is by no means probable that their numbers will suffer any change after once being fixed... Six boundary roads radiating from London are taken as the basis. Road A1 is that from London to Edinburgh; A2 runs from London to Dover; A3 London to Portsmouth.

7. Used of a type of blood.

1927 *Jrnl. Amer. Med. Assoc.* LXXXVIII. 1422/1 Dr. Karl Landsteiner has suggested the substitution of the well known letters O, A, B and AB for the Jansky numbers I, II, III and IV and the Moss numbers IV, II, III and I. The letters will..express the actual constitution of the blood corpuscles with respect to iso-agglutination, as far as it concerns the separation of the groups. 1928 *Jrnl. Exper. Med.* XLVII. 757 They separate the human bloods into four sharply defined groups designated as O, A, B, and AB.

III. (All abbreviations given here with the full stop are frequently used without it.) A, adult, designating films suitable for exhibition to adult audiences; A., a., ampere; A, atom(ic); A, in £A, Australian; Å., Ångström; A.A., A.-A., anti-aircraft; A.A., Automobile Association; A.A.A., Amateur Athletic Association; A.A.F., Auxiliary Air Force; A.A.M., air-to-air missile; A. and R. (see quots.); A.B., Bachelor of Arts (U.S. examples); A.B.C., Aerated Bread Company; A.B.C.A., Army Bureau of Current Affairs, an organization which provided troops with information about current affairs during the 1939–45 war; A.C., A/C, aircraftman; A.C., a.c., alternating current; A.C.C., anodal (or anodic) closure contraction; A.C.S., antireticular cytotoxic serum; A.C.T., Australian Capital Territory; A.C.T.H., *a*drenocortico*t*rop(h)ic *h*ormone; A.C.W., aircraftwoman; A.D.C., aide-de-camp, hence *A.D.C.-ship*, aide-de-campship; A.D.F., automatic direction finder; A.D.G.B., Air Defence of Great Britain; A.D.P., automatic data processing; A.D.(S.), autograph document (signed); A.F., a.f., audio frequency; A.F.C., Air Force Cross; A.F.M., Air Force Medal; A.F.S., Auxiliary Fire Service; A.F.V., armoured fighting vehicle; A.H., *Anno Hegiræ* (see Hegira); A.I., Air Interception; A.I., A.I.D., A.I.H., artificial insemination (by donor, husband); A.I.D., Aeronautical Inspection Directorate (in quot. 1918 Department); A.I.F., Australian Imperial Force; A level, Advanced level (of the General Certificate of Education examination); A.L.(S.), autograph letter (signed); A.M., Albert Medal; A.M.G., Allied Military Government; A.N.(S.), autograph note (signed); A.P., Associated Press; A.R.P. (see *air-raid); A.S.M., air-to-surface missile; A/S.R.S., Air-Sea Rescue Service; A.S.V., air(craft) to surface vessel; A.T.A., Air Transport Auxiliary; A.T.C., Air Traffic Control; A.T.C., Air Training Corps; A.T.C., Automatic Train Control; A.T.S., animal tub-sized; A.T.V., Associated Television; Å.U., A.U., Ångström Unit; A.V.H., initials of Hungarian secret police, also A.V.O.; A.V.M., Air Vice-Marshal; A.W.O.L.. absent without leave (orig. *U.S.*).

1914 *Times* 16 Feb. 6/1 Since the inception of the board [British Board of Film Censors]..627 have been passed for 'public' exhibition with the 'A' certificate. 1935 Ld. Macmillan *Local Govt. Law & Admin.* III. 166 It is the practice of the Board of Film Censors to distinguish between films suitable for universal exhibition ('U' films) and films suitable for exhibition to adult audiences ('A' films) which by implication are unsuitable for children. 1936 *Sunday Express* 13 Dec. 14/1 Cinema Theatres. Academy, Ox.-st. Finnish Epic 'Fredlos' (A). 1889 S. R. Bottone *Electric Bells* ii. 56 Or if we like to use the initials of volts, ampères, and ohms..we may write $\frac{V}{R}$ = A, or $\frac{\text{Volts}}{\text{Ohms}}$ = Ampères. 1937 *Rep. Joint Comm. Chem. Soc. Faraday Soc. etc.* 11 Ampère (in sub-units) = a. 1945 *Daily Mirror* 8 Aug. 1 (*headline*) Jap Radio says Evacuate—'Ware A-Bombs. 1954 *Britannica Bk. of Year* 637/2 *A-test*, a test explosion of an atomic bomb. 1936 *Whitaker's Alman.* 1937 810/2 Total imports, 1935–36..£A104,687,000. 1914 *Times Book of Navy* 145 The abbreviations used in the lists are as follows:..A.A., Anti-air-craft guns. 1917 'Contact' *Airman's Outings* 157 The A.-A. batteries have only worried us to the extent of half a dozen shells. 1943 *N. & Q.* CLXXXIV. 107/2 A.-A. barrage was audible from twenty miles away. 1905 *Autocar* 14 Oct. 443/1 Commonsense precautions taken with regard to villages, and at dangerous points, do not concern the A.A., for therein it is considered lies the duty of the policeman. 1937 *Discovery* July liv/2 (Advt.), Victoria Hotel, A.A. H. & C. Electricity. 1882 *Amateur Athl. Assoc.: Rules*, Laws to be observed at every Athletic Meeting held under the sanction of the A.A.A. 1955 *Times* 20 July 3/5 B. S. Hewson, the A.A.A. mile champion. 1925 *Flight* 15 Jan. 32/2 The designating numbers of the A.A.F. units will start with 600. 1958 *Aero-Space Terms* 1/1 AAM, Air-to-air missile. 1958 *Times* 26 May 7/6 A and R men, dee-jays, pluggers—to use the abbreviated titles by which artists and recording managers, disc jockeys and exploitation men are known. 1959 *TV Times* 14 Aug. 11/1 It all revolves around someone known as an 'A and R man'. The common translation of 'A and R' is Artists and Repertoire—and it is for these two items that the A and R man is responsible. 1773 *Boston News-Let.* 29 Apr. 2/2 Last Friday departed this life..John Alden, A.B., aged 22. 1842 *Knickerbocker* XIX. 429 A man may..write A.B. after his name, and even A.M., and be no great things either. 1895 *Rep. to Harvard Club of Chicago* 2 (D.A.E.), Educational qualifications equivalent to the Harvard A.B. 1894 *Punch* 15 Dec. 285/1, I pass an A.B.C., Where I purchase two or three Cakes and scones. 1941 E. Blunden *Thomas Hardy* 120 Afterwards we went to a Lyons tea-shop, at which he [Hardy] was a little alarmed, being used only to an A.B.C. 1942 *Rep. Comm. on Amenities in Women's Services* (Cmd. 6384) 44 ABCA came into existence in September, 1941, to remedy this ignorance and to mitigate boredom. 1889 E. J. Houston *Dict. Electr. Words* 8 A.C.C.—An abbreviation used in medical electricity for *Anodic Closure Contraction*, or the contraction observed on closing the circuit when the anode is lying over the muscle. 1893 Dunglison *Dict. Med. Sci.* (ed. 21) p. vii, A.C.C., anodal closure contraction. 1938 *Official Year Bk. Australia* 390 A.C.T. 1945 S. J. Baker *Austral. Lang.* x. 187 We use *A.C.T.* or *F.C.T.* to denote the Australian or Federal Capital Territory where Canberra is situated. 1837 Emily Eden *Let.* 28 Oct. in *Up the Country* (1866) I. i. 9 The A.D.C.'s are very apt to assemble over our cabins at night. 1888 Kipling *Plain Tales fr. Hills* 10 The A.-D.-C. in Waiting. 1896 *Punch* 4 Jan. 5/2 Permitted to retain his A.D.C.-ship after promotion from field-rank. 1948 *Shell Aviation News* No. 123, 10/2 When homing on a ground station with the ADF the passage of the aircraft over the station is indicated by a 180° reversal of bearing within a few seconds. 1926 *Flight* 6 May 272/2 The Fighting and Wessex Bombing Areas are under the A.O.C.-in-C., A.D.G.B. 1944 *Times* 1 Mar. 4/5 Sir Archibald Sinclair disclosed for the first time that the organization responsible for the air defence of this country is now known as A.D.G.B. (Air Defence of Great Britain). 1958 *Engineering* 4 Apr. 424/1 Automatic data processing (or ADP) is the modern nervous system of corporate bodies. 1861 *Sotheby & Wilkinson Catal.* 1 July Tenison Mss. 36 A. signifies *Autograph* S. signifies *Signed* L. signifies *Letter* D. signifies *Document*. *Ibid.* 50 *Colbert*, A.L.S. to *Cavalier Bernino*, 1665 —*Perrault* (the architect of the Louvre), A.D.S. with portrait. 1912 F. K. Walter *Abbrev. & Techn. Terms Book Catal.* 1 A.D., Autograph document. A.D.S., Autograph document signed. 1919 *Whitaker's Alman.* 144/1 The Air Force Cross. 1918—A.F.C.—Instituted..for acts of courage or devotion to duty when flying, although not in active operations against the enemy. 1919 *London Gaz.* (Suppl. 5) 19 Dec. 15840/2 Actg. Serjeant Elmo O'Neal Bearden, A.F.M. (South Russia). 1921 S. C. Johnson *Medal Collector* 237 A.F.M.—As for the A.F.C., but the width of the stripes is $\frac{1}{16}$ of an inch...The D.F.M. and A.F.M. ribbon will show at the top corner nearest to the left arm a triangle of white. 1939 *War Weekly* 1 Dec. 166/2 Flying Officer R. C. Graveley, O.B.E., receiving congratulations from members of a Leytonstone A.F.S. station. 1940 *New Statesman* 21 Dec. 642 Most of its best workers are busy on A.R.P., or A.F.S., or evacuation, or some other emergency job. 1939 *War Illustr.* 18 Nov. p. iii/1 Wartime Abbreviations:..A.F.V., Armoured Fighting Vehicle. 1942 *Wardens' Bulletin* (*City of Oxford Civil Defence*) No. 53, 18 June 3 Vehicles. Very dark dull grey with dark brown disruptive design. A.F.V.'s have black cross outlined in white. 1788 Gibbon *Decl. & F.* V. lii. 418 The foundations of Bagdad were laid A.H. 145, A.D. 762. 1940 F. Stark *Winter in Arabia* iv. 25 It [*sc.* the wooden minbar] belonged to the year A.H. 693 (1293 A.D.). 1945 *Electronic Engin.* XVII. 683 With the Battle of Britain by night must be associated the use of A.I. or Air Interception. 1945 *Flight* 21 June 664/2 Later on came airborne interception equipment (A.I.), which comprised a radar set in the aircraft by means of which the target objective could be registered, stalked and attacked. 1918 C. Bright *Telegr., Aeronaut. & War* xix. 291 Steps should be taken..to introduce into the Aeronautical Inspection Department (A.I.D.) more who have had a training that would especially fit them for the work in a technical sense. 1931 *Flight* 16 Jan. 56/2 Those who occupy those front offices such as the A.I.D. and the Meteorological Office staff. 1918 *Official Year Bk. Australia* XI. 1019 Little difficulty was experienced..in..obtaining and training horses for the mounted units of the A.I.F. 1940 *War Illustr.* 16 Feb. 121 The Second A.I.F.—Australian Imperial Force—consisting of men who have volunteered for service either at home or abroad. 1951 *Joint Matric. Board Gen. Cert. Exam. Reg.* 6 If a language is to be specially approved at the A level. 1952 *Jrnl. Educ.* May 220/2 The award of an 'O' level pass on the 'A' level papers. 1851 *Puttick & Simpson Catal.* 29 July, Autograph Letters 1, Letters autograph and signed are thus marked 'A.L.s.' 1918 *Times* 6 Sept. 3/2 It is notified in Army Orders that officers..who have been awarded the Albert Medal are now entitled to add the distinctive letters 'A.M.' after their names. 1912 F. K. Walter *Abbrev. & Techn. Terms Book Catal.* 1 A.N.S. Autograph note signed. 1879 *Chicago Tribune* 4 Mar. 5/4 Now, the A.P. may be a very wicked institution, but a 'Monopoly' it is not. 1947 *Mem. Evidence Subm. to Royal Comm. on Press* 1. 14 The full A.P. service is provided to Press Association and a selection of items is distributed to subscribers in this country. 1958 *Aero-Space Terms* 4/1 ASM, air-to-surface missile. 1944 *U.S. War Dept. Techn. Man., TM 11–467* 364 A.S.V. 1945 *Flight* 23 Aug. 210/1 A centimetric A.S.V., the device carried in Coastal Command aircraft which directed them to the surfaced U-boat. 1949 *Jrnl. R. Aeronaut. Soc.* LIII. 959/2 The inefficiency of A.T.C. (Air Traffic Control) procedures and patterns. 1934 *Railway Mag.* Apr. 290 (*caption*) Diagram of Signum A.T.C. system. 1958 *Engineering* 14 Mar. 336/1 Before the Second World War..in general, the Great Western Railway had A.T.C. and the other British railways did not. *a* 1912 *Spalding & Hodge's Paper Terminol.* ii. 1 Animal Tub-sized, or A.T.S., is a term employed to denote the more costly method of passing the manufactured sheet through a bath or 'tub' of animal size. 1957 *Daily Mail* 4 Sept. 12/8 ATV announce *The £4,000 Question* on September 14. 1953 *N.Y. Times* 29 Mar. § vi. 9/3 The structure of terror in Hungary consists of many overlapping..agencies: It is a pyramid with the *Allam Vedelmi Hasotag* (State Security Authority, or secret police), at the apex. A.V.H., which has 50,000 men and women in its force, is..the party's harsh punitive arm. 1956 *New Statesman* 22 Dec. 818/1 The ..lynching of many persons, A.V.H. men and Party members in particular. 1929 T. E. Lawrence *Let.* 22 Jan. (1938) v. 640, I could do nothing, with the A.V.M., of what I'd hoped. 1956 *Times* 29 Nov. 10/5 The Kadar Government is also concerned..at the treatment being meted out by the population to former members of the A.V.O. (secret police). 1921 *Outing* (U.S.) June 137/1, I was surprised to find one day that unless I left the following morning to rejoin my

regiment I would be an 'a-w-o-l'. 1929 *Amer. Speech* IV. 351 Absence without leave is a military offense designated by the abbreviation awol, usually written without periods and sometimes with small letters, in the pronunciation of which each letter is pronounced. 1949 Wodehouse *Mating Season* x. 98 Nothing sticks the gaff into your chatelaine more than a guest being constantly a.w.o.l. 1957 B. & C. Evans *Dict. Contemp. Amer. Usage* 9/2 awol..in World War I..was pronounced as four letters; in World War II, it was pronounced as a word.

‖**aa²** (ā·ā). *Geol.* [Hawaiian *'a-'ā*.] A rough, scoriaceous lava, one of the two chief forms of lava emitted from volcanoes of the Hawaiian type, the other being *pahoehoe.

1859 R. C. Haskell in *Amer. Jrnl. Sci. & Arts* 2 Ser. XXVIII. 70 We..saw '*pahoihoi*' or solid lava forming, and also '*aa*' or clinkers. 1880 *Encycl. Brit.* XI. 531/2 The *a-a* or spongy lava, which, on account of its extreme roughness and hardness, is carefully avoided by all travellers. 1883 C. E. Dutton in *U.S. Geol. Surv. 4th Ann. Rep.* 95 The second form of the lavas is called by the natives a-a, and its contrast with pahoehoe is about the greatest imaginable. It consists mainly of clinkers sometimes detached, sometimes partially agglutinated together with a bristling array of sharp, jagged, angular fragments. 1914, 1920 [see *block *sb.* 23, *block lava*]. 1944 C. A. Cotton *Volcanoes* iv. 27 Cooling and solidification frequently takes a different course..in lava flows, producing the clinker-like 'aa' lava.

‖**aandblom** (ā·ntblɒm). *S.Afr.* Also 9 **aantblom, avondbloem.** [Afrikaans, f. *aand* evening + *blom* flower; cf. Du. *avondbloem*.] Any of several sweet-scented iridaceous wild flowers, esp. *Hesperantha falcata*. Also dim. **aandblommetjie** (-əkɪ) (earlier often written in Du. form *avondbloemetje*).

1822 Burchell *Trav.* I. 273, I here met with a remarkable species of *Avond-bloem* (*Hesperanthera*). 1870 *Cape Monthly Mag.* Aug. 108 The scent of the 'avond bloemetjies' is fragrant in the air. 1890 A. Martin *Home Life Ostrich Farm* 21 Another of our favourites was the *aantblom*, a kind of ixia. 1915 R. Marloth *Flora S. Afr.* IV. 145 *H. falcata* (*Avond bloemetje*, generally pronounced '*aandbloem*'). 1927 *Chambers's Jrnl.* Mar. 190/1 Wild flowers, from the stately arum lily..to the modest *avondbloem*, which at night perfumed the whole veld with its fragrance. 1927 *Farming S. Afr.* 171 How often does one see..the beautiful 'painted ladies' or 'aandblommetjies' adding scent to the garden in the evening. 1953 V. Buchanan-Gould *Vast Heritage* v. 80 Here the flowers had no perfume except for the rare aandblom.

‖**aasvogel** (ā·sfōᵘgəl). Also **aasvoël.** [S. African Du., f. *aas* carrion (cf. Ees) + *vogel* bird, Fowl *sb.*; Afrikaans *aasvoël*.] A South African vulture, esp. *Gyps kolbii*.

1838 W. C. Harris *Exped. S. Afr.* xix. 186 *note*, *Vultus Fulvus*, and *Vultus Auricularis*: White and Black Aasvogel of the Cape Colonist. 1887 Rider Haggard *Jess* xxvi, They would not be found till the aasvögels had picked them clean. 1903 Kipling *Five Nations* 94 Well the keen *aas-vogels* know it. 1910 J. Buchan *Prester John* iv. 73, I thought it was an aasvogel, but another thought it was a baboon. 1924 R. Campbell *Flaming Terrapin* vi. 90 Great aasvogels, like beetles on a pond, Veered in slow circles. 1939 S. Cloete *Watch for Dawn* xii. 154 'What are you looking for up there?'..'Aasvoëls, brother. Where there is death there are vultures.'

aba², abba² (æ·bă). [Arab. *'abā*.] A sleeveless outer garment, resembling a sack with openings for the head and arms, worn by Arabs.

1811 tr. *Niebuhr's Trav. Arab.* cxxii, in Pinkerton *Voy.* X. 156 A blind tailor at Basra, who earned his bread by making Abbas. 1833 A. Crichton *Hist. Arabia* II. v. 193 A silk gown, over which is thrown a white abba of the finest manufacture of El Hassa. *Ibid.* viii. 350 The common abba is without sleeves, resembling a sack, with openings for the head and arms. 1880 L. Wallace *Ben-Hur* viii. iv. 496 Galileans in his league, carrying short swords under their long abbas. 1934 F. Stark *Valleys Assassins* ii. 74, I enveloped myself in the brown *abba*, tightly pinned under the chin. 1942 M. Home *House of Shade* iii. 54 Over it was an *aba*, or sleeveless coat of fine linen.

abac (ēɪ·bæk). [ad. Fr. *abaque*, ad. L. *abacus*; cf. †Aback.] An alignment chart, a type of nomogram, by means of which the value required is read off by placing a ruler or other straight edge across points representing known values.

F. *abaque* in this sense is used by L. Lalanne, *Description et Usage de l'Abaque* (1845); in an 1846 Eng. tr. of Lalanne, *abaque* is rendered by *abacus*.

1930 R. T. Beatty *Radio Data Charts* 12/1 We have two variable quantities which we multiply together to obtain the third variable, that is, we have to draw scales for three quantities, a, b, c, so that a.b = c... Having thus ascertained the relative sizes of the scales, we can proceed to construct the abac. 1934 *Jrnl. R. Aeronaut. Soc.* XXXVIII. 545 An abac is constructed to give rapid determination of exposure time in reference to time of year, state of sky, colour of subject. 1956 *Electronic Engin.* XXVIII. 533 A rapid method of circuit and transformer design is given in which a series of 'abacs' is used to evaluate the design constants.

abaca. (Examples.)

1818 *Blackw. Mag.* III. 579 Fine timber is to be found in all the islands [Philippines]; also..abacca, a kind of hemp. 1909 *Teachers' Assembly Herald* ii. 43/1 Weaving with the fine threads of abaka. 1925 in J. A. Hammerton *Countries of World* V. 3238/2 The staple abaca, or hemp, the product which first made the name of Manila known all over the world.

aback, *adv.* Add: **4.** *aback of, aback o'*: at the back of, on the back of, behind (cf. *back of*, Back *adv.* 14). *arch.*, *dial.* (see *E.D.D.* and *Sc. Nat. Dict.*), and *U.S.* (where *Obs.* or *dial.*).

1783 W. Fleming *Jrnl. in Kentucky* 8 Jan. in Mereness *Trav. Amer. Col.* (1916) 663 Fern Creek is lost in ponds and low flat land a back of the Falls. 1836 Edward *Hist. Texas* v. 79 Those districts..aback of the older settled ones. 1849 *Gloss. Prov. Words Teesdale* 1 *Aback*, prep. Behind; as, stand aback o' me. 1876 Morris *Sigurd* ii. 150 So he leapeth aback of Greyfell, and rideth the desert bare. 1894 Hall Caine *Manxman* i. xi. 50 Just aback of the lighthouse. 1908 J. Lumsden *Doun i' th' Loudons* 80 Be aff, ye limmer, unman men nae mair, A-back o' granny, is the post for you!

abaction. Delete †, for *Obs.*⁻⁰ read *rare*, and add example.

1949 C. Fry *Lady's not for Burning* 11 He says he has also committed petty larceny, Abaction, peculation and incendiarism.

abalone (æbălōᵘ·ni). *U.S.* Also **abelone, avalone,** etc. [Amer. Sp. *abulón* (f. Monterey Indian *aulun*) in same sense.] A gastropod mollusc of the genus *Haliotis*, used for food; an ear-shell or sea-ear. Also *attrib.*

1850 B. Taylor *Eldorado* I. xvii. 174 The avelone, which is a univalve, found clinging to the sides of rocks, furnishes the finest mother-of-pearl. 1870 *Amer. Naturalist* III. 250 These shells are popularly called Sea-ears...In California the people call them *Abalones*. *Ibid.* 256 Haliotis or Abalone shells. 1882 *Harper's Mag.* Oct. 728 They [*sc.* Chinamen in S. California] prepare..the avallonia meat and avallonia shells for their home market. 1910 *Encycl. Brit.* I. 6 The abalone shell is found..on the southern California coast, and when polished makes a beautiful ornament. 1936 Russell & Yonge *Seas* (ed. 2) xvi. 350 Among the many shells used for decorative purposes we..mention one, known in the Channel Islands as the Ormer and on the Pacific coast of North America as the Abalone.

abandon, *sb.*³ Delete ‖ and add earlier and later examples.

1822 L. Simond *Switzerland* I. 285 There was more *abandon* in what she [*sc.* Mme de Staël] said, than in what she wrote. 1831 *Fraser's Mag.* III. 358/1 It is the heart alone which is surrendered to its 'abandon'. 1922 Joyce *Ulysses* 636 She also was Spanish or half so, types that wouldn't do things by halves, passionate abandon of the south. 1930 W. S. Maugham *Gent. in Parlour* xxxix. 223 There was no abandon in their dance.

abasia (ăbēɪ·siă). *Med.* [mod.L., ad. F. *abasie*, f. A- 14 + Gr. βάσις step: see -ia.] Inability to walk, caused by lack of coordination of the muscles. Hence **aba·sic** *a.* Cf. *astasia.

1890 [see *astasia]. 1891 P. C. Knapp in *Jrnl. Nerv. & Ment. Dis.* XVI. 673 (*title*) Astasia-abasia, with the report of a case of paroxysmal trepidant abasia associated with paralysis agitans. 1909 *Cent. Dict.* Suppl., Abasia, abasic. 1918 A. A. Brill tr. *Freud's Totem & Taboo* 159 A motor inhibition, an abasia or an agoraphobia, becomes perfected and detailed.

abater¹. (Earlier and later examples.)

1609 W. M. *Man in the Moone* sig. C1ᵛ, He [the Tobackonist, i.e. a smoker] is his owne strengths enfeebler..his appetites abater. 1683 Lady R. Russell *Lett.* (1773) x. 14 Time, or necessity, the ordinary abater of all violent passions. 1839 *New Monthly Mag.* LVII. 538 It is not..with the remotest intention of reflecting on the abaters of the dog-cart nuisance.

abat-jour. Add to def.: Also, a device for reflecting light downwards. (Examples.)

1830 in Britton *Dict. Archit.* 1853 J. W. Croker in *Q. Rev.* XCIII. 413 He was confined to a single room..it had one window, closely barred and blinded by an *abat-jour*, which admitted only a small degree of oblique light. 1936 *Burlington Mag.* July 31/2 Two candlesticks behind are formed of leaves; between them is a vertical metal post bearing on the upper part a movable canopy serving as *abat-jour*.

‖**abaton** (æ·bătɒn). *Greek Antiq.* [a. Gr. ἄβατον, neut. sing. of ἄβατος not to be trodden, f. ἀ- priv. + βατός accessible (βαίνειν to walk, go).] An enclosure attached to a temple of Asclepios, where patients slept.

1884 A. C. Merriam in *Amer. Antiquarian* VI. 304 Thereupon sleep fell upon him, and a serpent issuing from the Abaton, healed his toe with his tongue, and when he had done this he glided back again into the Abaton. 1885 *Q. Rev.* Apr. 302 The patients were sleeping in the sacred enclosure or *abaton* of the temple. 1925 *Lancet* 3 Oct. 689/1 In nearly all cases the patient comes to the shrine of Asclepios, sleeps in the 'abaton', has a vision or a dream.

abattoir. Delete ‖ and 'public'. Add earlier and later examples.

1820 M. Starke *Trav. on Continent* i. 5 Among the most prominent improvements..during the late reign are..the five Slaughter-houses, called *Abattoirs*. 1833 *Penny Cycl.* I. 8/2 The abattoirs of Paris are five in number; three being on the right bank of the Seine, and two on the left. 1866 *Cycl. Useful Arts* I. 2/2 Abattoirs have recently been erected in London. 1958 *Times Rev. Industry* Sept. 75/2 Economy installations in Sheffield Corporation abattoir's boiler plant will save the ratepayers £700 a year.

abaya (ăbēɪ·yă). Also **abeih, abbai, abaiya, abeya, abaye,** etc. [Arab. *'abāya*.] = *aba².

1836 Lane *Mod. Egypt.* I. i. 35 A kind of black woollen cloak, called '*abáyeh*. 1855 J. L. Porter *5 Yrs. Damascus* I. iv. 195 The ample folds of an abeih. 1884 S. W. Baker *Heart Africa* iii. 36 (Stanford), A white cashmere abbai. 1922 *Blackw. Mag.* Sept. 358/1 A dazzling yellow 'jibba' covered by a red 'abaiya'. 1942 F. Stark *Lett. from Syria* iv. 172 A man in a black 'abeya galloped past. 1946 A. Koestler *Thieves in Night* ii. iii. 144 Issa came in, his abaye hurriedly thrown over his underclothes.

abba², see *aba².

Abbassid, -ide (ăbæ·sid, æ·băsid, -əid), *a.* and *sb.* Also **Abbasid, -ide.** [f. name of *Abbas* (566–652) + patronymic suff. *-id(e)*.] **A.** *adj.* Of or pertaining to the dynasty (750–1258) of caliphs of Baghdad claiming descent from Abbas, uncle of Mohammed. **B.** *sb.* A member of this dynasty.

1788 Gibbon *Decl. & F.* V. lii. 431 The bloody conflict of the Ommiades and Abbassides. 1872 H. A. Rawes *God in His Works* iii. 54 The Ommiad Caliphate of Damascus and the Abbasside Caliphate of Bagdad began and ended. When the thirty-seven Abbasides began to reign, Zacharias was in the Chair of Peter. 1920 H. G. Wells *Outline Hist.* vi. xxxii. 332/2 Another great Arab family, the Abbas family, the Abbasids..had long been scheming for power. 1923 *Ibid.* vii. xxxiii. 357/2 There was still an Abbasid Caliphate in Bagdad. 1963 *Listener* 14 Mar. 471/1 Under the Abbasid and Umayyad dynasties, Arab and Jew served as transmitters of arts and skills.

abbé. (Earlier and later examples.)

1530 Palsgrave *Lesclarcissement* ii f. lviᵛ, For to the abbe..they say *a lábbe*. 1701 T. Marwood *Diary* in *Cath. Rec. Soc. Publ.* (1909) VII. 100, I walk out to yᵉ Abbé's a League & ½ off. *Ibid.* 108 And [we had] Abbé Villebreuille with us. 1712 Swift *Jrnl. to Stella* 13 Dec. (1948) II. 581, I must see the Abbè Gautier. 1719 Gay *Let.* 8 Sept. in *Lett. Henrietta, C'tess of Suffolk* (1824) I. 34 A French marquis drove an abbé from the table by railing against the vast riches of the church. 1885 Lord R. Gower *Old Diaries* 10 Apr. (1902) 20 Monsieur Floquet..is a grey-haired, abbé-like man. 1955 *Times* 8 July 11/4 A strange, weird, little man, looking something between an actor and an *abbé*.

abbess. Add: **†b.** *transf.* (See quot. 1785.) *Obs.* (See also Farmer *Slang* (1890) I. 3/2.)

1749 J. Cleland *Mem. Woman Pleasure* I. 59 Who should come in but the venerable mother Abbess herself! handed in by a tall, brawny young horse-grenadier. 1770 [see Nun *sb.*¹ 1 c]. 1785 Grose *Dict. Vulg. Tongue*, Abbess, or Lady Abbess, a bawd, the mistress of a brothel. 1793 'Peter Pindar' *Poet. Epistle to the Pope* 33 So an old Abbess for the rattling Rakes, A tempting dish of human nature makes. 1837 W. Kidd *London & its Dangers* 36 The Infernal wretches, who traffic in the souls and bodies of their helpless victims, are called—Lady Abbesses.

Abbevillian (æbᵊvi·liăn), *a.* Also **Abbevillean.** [ad. Fr. *Abbevillien* (H. Breuil 1932, in *Préhistoire* I. 125), f. the place-name *Abbeville* (Somme), France.] = *Chellean *a.* Also *absol.*

[1928 G. B. Brown *Art of Cave Dweller* ii. 36 The so-called early 'drift' period is generally called the 'Chelles' period..though in honour to Boucher de Perthes it might well have been named after Abbeville.] 1934 L. S. B. Leakey *Adam's Ancestors* v. 101 L'Abbé Breuil..has suggested the substitution of the name Abbevillian for Chellean, since Chelles from which the latter name is derived is not a Chellean site. 1937 *Discovery* June 179/2 The Sinanthropus [industry]..belongs to the Lower Pleistocene and is attributed to the Abbevillean in the French palæolithic sequence. 1947 J. & C. Hawkes *Prehist. Brit.* i. 12 Contemporary with the earliest Abbevillian was the Cromer flake culture.

abbey. Add: **4. abbey-counter** (see Counter *sb.*³ 1 a), a medal presented to a pilgrim as a token of his having visited a shrine. (Cf. Pilgrim's *sign*.)

1839–42 C. Knight *Pict. Edition Works Shakspere. Comedies* II. 231/1 Jettons..were commonly used for purposes of calculation, in abbeys and other places... From their being found among the ruins of English abbeys they are usually termed abbey-counters. 1868 Stephens *Runic Mon.* II. 535 There was also a class of Jettons commonly called Abbey-counters, with similar or cognate instructive stamps.

abbeyed (æ·bid), *a.* [f. Abbey + -ed².] Provided with an abbey or abbeys.

1828 [see Mansioned *ppl. a.*]. 1864 A. Jeffrey *Hist. Roxburghshire* IV. i. 56 Bonny Teviotdale, abbeyed, towered and castled. 1864 E. Caswall *May Pageant* iii. 30 Now ancient Shrewsbury appears in sight..Fair-abbey'd formerly ere Faith's decline.

abbozzo (abɒ·tso). Also **abozzo.** [It.] A rough drawing or sketch (for a portrait, etc.); an outline or draft (of a speech, essay, etc.).

1849 [see *alla prima]. 1890 Morley *Diary* 20 Nov. in *Recoll.* (1917) I. 256 A letter from Mr. G. containing an 'abozzo', as he headed it, of what it would be well for us to say. 1905 W. Holman Hunt *Pre-Raphaelitism* I. xiii. 361 Another dashing abozzo, said to be a portrait.

abbreviatory, *a.* Delete *rare*⁻⁰, and add examples.

1935 G. K. Zipf *Psycho-Biol. of Lang.* (1936) iii. 93 These abbreviatory phonetic changes may be 'spontaneous', or accentual, or assimilatory, or dissimilatory. 1943 *Mind* LII. 268 In his notation for proofs, Quine uses ingenious abbreviatory devices.

A.B.C. Add: **4.** *ABC Railway Guide* (example). Also *absol.*

1853 (*title*) A.B.C. or Alphabetical Railway Guide. 1917 W. Pett Ridge *Amazing Years* iii. 41, I found an A.B.C. and selected a train. 1936 A. Christie *ABC Murders* iv. 31 A railway guide, you say. A Bradshaw—*or an ABC*?

abdomen. Add: Also pronunc. (æ·bdɒ̆mĕn), now the more usual.

abdominal. Add: **A.** *adj.* **1.** *abdominal belt.*

1907 H. C. Beeching *Let.* in E. V. Lucas *Post-Bag Diversions* (1934) 25, I have become once more erect and acquisitive. I attribute the change to an abdominal belt. 1909 H. G. Wells *Tono-Bungay* iii. ii. 334 Going about making love indeed!—in abdominal belts!—at his time of life!

B. *sb.* **2.** An abdominal operation.

1932 Kipling *Limits & Renewals* 358 It was the trephining work that had stuck on his mental retina. (Odd! It used to be abdominals with *me*.)

abdominally (æbdɒ·minăli), *adv.* [f. Abdominal *a.* + -ly².] In or with reference to the abdomen.

1889 in *Cent. Dict.* 1897 M. H. Kingsley *Trav. W. Africa* xxvi. 581, I am not thinking of starting a sanatorium for abdominally-afflicted Africans. 1907 *Practitioner* Dec. 845 Nothing definite was ever felt abdominally.

abdomino- (æbdɒ·mino), used as comb. form of Abdomen, as in *abdomino-anterior, -perineal, -posterior* adjs. (see quots.).

1890 Billings *Med. Dict.*, *Abdomino-vesical pouch*, depression in peritoneum as it passes from apex of bladder to the anterior abdominal wall. 1893 Dunglison *Dict. Med. Sci.* (ed. 21), *Abdomino-anterior*, term given to position of fœtus in utero with abdomen presenting anteriorly. *Abdomino-posterior*, term given to position of fœtus with abdomen presenting posteriorly. 1949 *New Gould Med. Dict.* 2/1 *Abdominoperineal*, relating to the abdominal and perineal regions. 1962 *Lancet* 1 Dec. 1172/2 Abdominoperineal excision of the rectum.

abeam, *adv.* Add: Also of aircraft.

1941 J. A. Hammerton *ABC of RAF* 119/1 Abeam (on the Beam). Directly at right angles to the fore and aft line.

Aberdeen (æbəɹdī·n). The name of a city and county in Scotland.

1. *Aberdeen Angus*, one of a breed of polled black beef cattle, originally from Scotland but now widely distributed and exported; also *Angus.

1862 E. Ravenscroft (*title*) The Polled Herd Book containing Pedigrees of Animals of the Polled Aberdeen Angus and Galloway Breeds. 1891 R. Wallace *Rural Econ. Aust. & N.Z.* xxxii. 413 The Aberdeen-Angus Polled cattle are slowly but surely working their way to the front in the Colonies. 1953 E. Smith *English Traditions* 3 Some breeds of cattle..e.g. the Aberdeen Angus, [are reared] for beef.

2. *Aberdeen terrier*, a rough variety of Scotch terrier; also *Aberdeen* ellipt.

1880 H. Dalziel *Brit. Dogs* xxiv. 362 Those selected for prizes..were not *the* old hard-haired Scotch terrier, but a well-known distinct variety yclept the Aberdeen terrier. 1889 *Ibid.* (ed. 2) II. xxiv. 363 Such are the Dandie Dinmont, the Skye, and the Aberdeen Terrier, the latter now merged in the class recognised at our shows as the Scotch Terrier. 1907 B. M. Croker *Company's Servant* xxxiii. 337 Two prim Aberdeens, and a pink-eyed bull terrier. 1922 V. Woolf *Jacob's Room* xiii. 270 The Aberdeen terrier must be exercised.

3. A type of fishing-hook: see quot.

1893 *Outing* (U.S.) XXII. 123/1 Of all the makes and shapes of hooks the 'Aberdeen', of very fine steel-wire, with long, straight shank and a broad, even bend, terminating in a short, barbed point, is the surest shape for large-mouthed trout.

4. Other *attrib.* uses.

1756 A. Butler *Lives Saints* I. 50/1 The Aberdeen Breviary..was printed at Edinburgh, by Walter Chapman, in 1509. 1866 *Illustr. London News* 14 Apr. 366/2 The columns and cornice..are of polished Aberdeen granite. 1904 Goodchild & Tweney *Technol. & Sci. Dict.* 74/2 *Aberdeen Granite*, a grey granite, extremely hard and durable.

Aberdonian (æbəɹdōᵘ·niăn), *a.* and *sb.* [f. med.L. *Aberdonia*: see -ian.] Of or pertaining to, a native of, Aberdeen.

a 1670 Spalding *Troub. Chas. I* (1850) I. 181 To convoy our Abredonianis in thair Kirkcaldie bark. 1773 Boswell *Jrnl. Tour Hebr.* 23 Aug. (1785) 99 That the Aberdonians had not started a single *mawkin* (the Scottish word for hare) for us to pursue. 1791 Newte *Tour Eng. & Scot.* 191 A striking difference between the Moravians and Aberdonians. 1861 *Two Cosmos* iii. iii, The pettifogging Aberdonian had been at first disappointed. 1906 *Westm. Gaz.* 24 Apr. 10/1 A miller on the Don, an Aberdonian.

Abernethy (æ·bəɹnɪ:þi, -ne:þi). [Prob. f. name of John *Abernethy*, surgeon (1764–1831).] A kind of hard biscuit flavoured with caraway seeds.

1836 Dickens *Sk. Boz* II. 107 Eating Abernethy biscuits. 1845 Dodd *Brit. Manuf.* 22 'Abernethy', 'butter', 'fancy', or 'sponge' biscuit. 1860 Jeaffreson *Bk. Doctors* I. 198 The hard biscuits, known as Abernethies. 1886 *Bakers' Guide* 84.

aberrant, *a.* Add: Hence **abe·rrantly** *adv.*

a 1878 Geo. G. Scott *Recoll.* (1879) vii. 291 Skidmore followed my design, but somewhat aberrantly. 1929 *New Statesman* 31 Aug. 614/1 Most unfortunately we have aberrantly accepted a mandate for Palestine.

abessive (æbe·siv), *a.* Chiefly *Finnish Gram.* [f. L. *abesse* to be distant + -ive.] Applied to the case which denotes absence.

1890 C. N. E. Eliot *Finnish Gram.* 132 The comitative and abessive express respectively the presence and absence of an object. 1930 F. R. Blake in *Curme Vol. Ling. Studies* 37 The immaterial adnominal cases are..abessive—coffee *without sugar*.

abettance (ăbe·tăns). = Abetment 1. *rare.*

1829 Scott *Anne of G.* III. v. 128 The affray at La Ferette..had no origin or abettance from us. 1835 *Fraser's Mag.* XI. 347/1 In the Roman Catholic Church of Ireland, the rewards are..for connivance, if not absolute abettance, of the foulest and most atrocious crimes.

‖**ab extra** (æb e·kstră), *phr.* [late L., f. L. *ab* from + *extrā* adv., without, outside.] From outside, from without.

1642 T. Goodwin *Encouragements to Faith* (1650) 7 There are many Demonstrations of his will herein, that may be taken *ab extra* from his [*sc.* God's] Oath. 1652 N. Culverwel *Light of Nat.* xi. 115 The divine understanding never receives the least tincture from an object, no species *ab extra*, but views all things in the pure Crystal of his own essence. 1672–1859 [see *ab intra]. 1864 Mrs. Gaskell *Wives & Daughters* (1866) I. ix. 101 It was a great relief to her to have Mr. Gibson to decide for her... Such decisions *ab extra*, are sometimes a wonderful relief. 1941 W. Temple *Citizen & Churchman* iv. 65 They were no longer to be apart from the world, bringing to it *ab extra* the divine act of redemption.

abhorrible, *a.* Delete † and for '*Obs. rare*⁻¹' read 'Now *rare*'. Add examples.

1822 J. Scott *Life Rev. T. Scott* (ed. 5, 1823) xvi. 531 O Lord, abhor me not—though I be indeed *abhorrible*, and abhor myself! 1849 *Blackw. Mag.* Nov. 646 Hamlet's uncle, Claudius, seems to me to be the most that can be borne of one purely abhorrible. 1862 *Illustr. Times* 17 May 46/3 The relief consists..in being fed upon detestable gruel, and such other abhorrible viands as poor-law guardians may consider economical and proper.

abidingness (ăbəi·diŋnės). [f. Abiding *ppl. a.* + -ness.] Permanence, lastingness.

1857 Newman *Serm. Var. Occas.* ix. 163 The world..discerns and recognizes her abidingness, her unchangeableness. 1860 Pusey *Min. Proph.* 503/2 His [Zechariah's] book opens..contrasting the transitoriness of all human things..with the abidingness of the word of God. 1920 E. I. Watkin in Hess *God & Supernatural* v. 142 The abidingness in another form of the positive value of noble pain.

‖**ab initio** (æb ini·ʃio), *phr.* [L.] From the beginning; hence, as quasi-*adj.* = Initial *a.* 1.

1600 B. Jonson *Ev. Man out of Hum.* Prol., If those lawes..had beene deliuered vs, *ab Initio*; and in their present vertue and perfection. 1767 H. Walpole *Lett.* (1843) Ser. I. let. cv. 344 It ought to have been declared null *ab initio*. 1817 Coleridge *Biog. Lit.* ix. 138 Assuming as a postulate, that both [*sc.* Truth and Being] are ab initio, identical and co-inherent. 1838 J. S. Mill *Diss. & Disc.* (1859) I. 349 He [*sc.* Bentham] begins all his inquiries by supposing nothing to be known on the subject, and reconstructs all philosophy *ab initio*. 1906 *Harmsworth Encycl.* VIII. 6011/2 If such legal right is abused, the act of trespass becomes an *ab initio* one. 1937 Kipling *Something of Myself* viii. 207, I found that when, to save trouble, I 'wrote short' *ab initio* much salt went out of the work. 1943 T. Horsley *Find, Fix & Strike* 21 The Swordfish was used as an *ab-initio* trainer.

‖**ab intra** (æb i·ntră), *phr.* [mod.L.] From within.

1672 T. Jacomb *Romans* v. 186 When 'tis [*sc.* sin] committed with little opposition *ab intra* and in spight of all opposition *ab extra*, I assure you then it hath a great power. 1805 Coleridge *Notebks.* (1962) II. 2444 The difference therefore between Fabrication and Generation becomes clearly indicable/the Form of the latter is ab intra, *evolved*, the other ab extra, *impressed*. 1859 J. Martineau *Ess., Rev. & Addresses* (1891) III. 526 If we neither add anything to our premisses *ab extra*, nor draw anything *ab intra*, that was not comprised in them before, no new thing ever can appear.

abiotic (ēɪbəiɒ·tik), *a.* [f. A- 14 + Biotic *a.*; cf. Gr. ἄβιος without life.] Characterized by the absence of life; inanimate. See also quot. 1950.

1893 in Dunglison *Dict. Med. Sci.* 1950 F. D. Howitt in *Brit. Encycl. Med. Practice* (ed. 2) I. 158 Ultra-violet rays between 1,894 Å and 2,900 Å are called abiotic rays, for they are inimical to life. 1952 *New Biol.* XII. 99 In abiotic times, such molecules, rich in free energy, might survive for long periods. 1961 J. Keosian in *Encycl. Biol. Sciences* 672/2 Abiotic synthesis of organic compounds, first simple then complex, preceded the origin of life on earth.

Abissian, var. *Abyssinian.

abiturient (æbitiūᵊ·riĕnt). [G., ad. mod.L. *abiturient-*, pr. pple. of *abiturire*, desiderative of *abire* to go away.] In Germany, a pupil who is leaving a 'gymnasium' or high-school to enter a university. Also *attrib.*

1863 *Chambers's Encycl.* V. 168/2 The boys attend..till they reach the age of 18, when, after a special examination (the abiturient or maturity examination), they are transferred to the university. 1868 M. Arnold *Sch. & Univ. Cont.* xvii. 179 The *Abiturient*, or leaving boy, must have been two years in *prima*. *Ibid.* 181 These *externi*, as they are called, are not examined along with the *Abiturienten* of the gymnasium. 1917 Jevons in *Times Educ. Suppl.* 7 June 215 Examinations for which abiturient pupils may have to be prepared.

abjoint (æbdʒoi·nt), *v. Biol.* [f. Ab- + Joint *v.*] *trans.* To separate by formation of a joint or partition, as the cells and spores in certain fungi. So **abju·nction**, the process of such separation.

1887 Garnsey tr. *K. Goebel's Outl. Classif. Plants* 83 The non-motile cells are the result of abjunction. *Ibid.*, The extremity of which [sterigmata]..enlarges, and is then abjointed as a gonidium. 1889 E. S. Bastin *College Botany* 313 Conidia..formed from hyphae by the separation of cells in succession from their free ends—a process called abjunction.

Abkhasian (æbkēɪ·ʃɪăn; also -ziăn, -ʒən), *a.* and *sb.* Also **Abkhazian.** [f. *Abkhasia* (see def.) + -an.] **A.** *adj.* Of or pertaining to the U.S.S.R. territory of Abkhasia in the Caucasus. **B.** *sb.* One of the Abkhasian people; also, the language of this people.

1867 *Cornhill Mag.* Oct. 504 Of the early history of the Abkhasian race little is known... Records are wanting among a people who have never committed their vocal sounds to writing; they know that they are Abkhasians, and nothing more. 1876 *Encycl. Brit.* V. 258/2 Petty expeditions against the Circassian and Abkhasian tribes. 1879 *Ibid.* X. 433/1 Uslar's 'Caucasian Family' comprises the following three great divisions: 1. Western Group. Typical races: Tcherkessians and Abkhasians. 1959 E. H. Carr *Socialism in One Country* II. iv. xx. 265 The constitutional anomaly of the region was the Abkhazian republic occupying a small coastal strip on the Black Sea.

ablate, *v.* Delete † *Obs. rare*, and add after def.: *Obs.* in general sense; in modern use as back-formation from Ablation.

1902 *Encycl. Brit.* XXXI. 744 When the cerebral hemispheres have been ablated. 1923 *Times Lit. Suppl.* 16 Aug. 541/1 The low sun..evaporates or 'ablates' the ice continually.

b. *spec.* in *Astronautics* (see quot. 1959²).

1952 *Astrophys. Jrnl.* CXVI. 203 In the meteor work..only the parameter (where Γ is the retardation coefficient and ζ the energy to ablate unit mass of meteor material)..could be determined from the observations. 1959 *Dublin Evening Press* 8 June 5/6 The nose cone is covered with several layers of tough plastic highly resistant to temperature. As the cone falls back through the atmosphere, the layers of plastic are eroded, or 'ablated', off in turn, but they achieve their ultimate object of protecting the cone's main substance. 1959 W. A. Heflin *Aerospace Gloss.* 1/1 *Ablate*, to carry away heat, as generated by friction, from a vital part by arranging for its absorption in a nonvital part, which may melt or vaporize, then fall away taking the heat with it.

Hence **abla·ting** *vbl. sb.* and *ppl. a.*

1959 W. A. Heflin *Aerospace Gloss.* 1/1 For hypersonic reentry bodies, ablating materials must handle temperatures in surrounding air up to 15,000° F. *Ibid.*, *Ablating nose cone*, a nose cone designed to prevent heat transfer to its internal structure by the use of an ablating material.

ablation. Add: **5.** *Astronautics.* (See quot. 1958.)

1951 *Astrophys. Jrnl.* CXIV. 460 More direct methods of determining actual surface losses by ablation should be developed. 1958 in *Amer. Speech* (1960) XXXV. 283 *Ablation*, melting of nose cone materials during reentry of space ships or vehicles into the earth's atmosphere at hypersonic speeds. 1961 *New Scientist* 4 May 241 The blunt end [of the space capsule] acts as an 'ablation shield' for re-entry.

ablatival (æblătəi·văl), *a. Gram.* [f. Ablative + -al.] Pertaining to the ablative case.

1854 Ellicott *Comm. Gal.* v. 1, 81 The usual ablatival explanation, 'quâ nos liberavit'. 1884 *Trans. Amer. Philol. Assoc.* XV. 5 The ablatival uses of the genitive will be shown to be more numerous,..in Sophokles at least, than the grammars would lead one to suppose.

ablative, *a.* Add: **3.** *Astronautics.* Of or pertaining to something which ablates (*ablate *v.* b).

1959 *N.Y. Times* 30 Aug. 42/4 Special ablative material on the capsule's nose would burn away from the heat of friction as the gradually thickening atmosphere slowed the vehicle.

ablaut. Delete ‖ and add earlier and *attrib.* examples.

1849 E. Thomson *Select Monuments* p. xxxi, As if it took its meaning from 'roar'; which in reality is the *ablaut*-form of 'rear'. 1886 *Athenæum* 4 Sept. 302/3 The Gothic word differs in ablaut-grade from its Teutonic cognates. 1923 J. K. Wallenberg *Ayen. of Inwyt* 145 Franck, 'leuk'

derives the Continental forms from a Teut. basis *hlá-* which he considers to stand in ablaut-relation to Du. *lauw*. 1935 *Mod. Lang. Notes* L. 533 The ablaut-patterns of the strong verbs in the ME dialects of the east and central midlands. 1963 in A. Brown & P. Foote *Early Eng. & Norse Studies* 73 On *-hæfi* and *-hæfi* as ablaut variations, see A. Noreen.

able, *a.* Add: **8.** *able-minded, -souled.*

1849 Thoreau *Week on Concord* 81 Let not the apprehension..discourage the cheerful labors of the able-souled man. 1934 G. B. Shaw *On the Rocks* Pref. 151 Every able-bodied and able-minded and able-souled person.

9. Used for the letter 'A' in transmitting messages (first adopted by services signalling units in Dec. 1942). Cf. *ack.

1943 *Signal Training* (*All Arms*) No. 5, p. 7 The letters will be spoken in the following form A able B baker C charlie.

ablegate, *sb.* (Examples.)

[1651 W. MacDonnell *Anglia Liberata* ii. 37 In the beginning of his pretended Answer, nick-names them *Ab-Legats* in the *Diminutive*, and calls them *pretended Ambassadors*, as if they merited not the title of true Legats or Ambassadors.] 1890 W. M. Brady *Anglo-Roman Papers* iii. 239 In 1778, when he [*sc.* Card. Braschi] went as Ablegate to carry the Berretta to the two Cardinals, De la Rochefoucauld and De Rohan. 1927 *Month* Sept. 256 The central interest of the evening was the anticipated arrival of the Apostolic Ablegate, the high ecclesiastic sent by the Pope to bring the red biretta to the new Cardinal.

ablette, var. Ablet.

1885 *Encycl. Brit.* XVIII. 448/2 Spheres of thin glass are filled with a preparation known as 'essence d'orient', made from the silvery scales of the bleak or 'ablette'. 1936 Russell & Yonge *Seas* (ed. 2) xvi. 351 On the Continent the little fresh-water 'ablette' is the main source of pearl essence.

ablow (ăblōᵘ·), *adv.* or *pred. a.* [A *prep.*¹ + Blow *v.*] Blowing.

1893 W. B. Yeats *Poems* (1957) 743 For dreams are the flowers ablow. 1912 Compton Mackenzie *Carnival* xxiv. 268 A black mushroom hat a-blow with rosebuds. 1921 W. de la Mare *Crossings* 12 The tide's aflow, the wind ablow.

ablute (ǣblᶦū·t), *v. colloq.* [Back-formation f. Ablution.] *trans.* and *intr.* To wash (oneself).

1892 *Judy* 9 Mar. 110/2 She went off to..wash her face... I'd previously abluted mine. 1924 *Blackw. Mag.* Nov. 691/1 Postal work would have been sadly disorganised if Father Moti had kept the postal staff abluting. 1959 C. MacInnes *Absolute Beginners* 183, I put on some music and abluted, then made two Nescafés.

abluted, *ppl. a.* Delete † and 'not otherwise found' in etymology (cf. *ablute *v.*), and add later example.

1922 Joyce *Ulysses* 695 He scratched..his partly exposed, wholly abluted skin.

abmigration (æ:bməigrēɪ·ʃən). *Ornith.* [f. Ab- *pref.* + Migration.] (See quots.) Hence **a·bmigrate** *v. intr.*

1923 A. L. Thomson in *Brit. Birds* XVI. 276 Some such term as 'abmigration' might perhaps be used to describe the northward departure in spring, for a new summer area, on the part of birds which had made no corresponding southward journey in the previous autumn. 1929 E. M. Nicholson *Study of Birds* v. 54 Birds which are native to one country, may 'abmigrate', and be found breeding in another, up to thousands of miles away. 1953 *New Biol.* XV. 54 Abmigration occurs when a bird (usually a young one), bred in this country and wintering here, accompanies foreigners which have also wintered here to *their* home countries in the Spring, instead of returning to its own British nesting grounds.

Abney (æ·bni). The surname of Sir William de Wiveleslie *Abney* (1844–1921) in *Abney level* (Level *sb.* 1), a small hand instrument used by surveyors for measuring slopes and angles above the horizon. Also *Abney clinometer*.

1889 G. W. Usill *Pract. Surveying* iii. 33 The Abney Level.—This portable and neat little instrument is a reflecting level and clinometer combined... It was invented by Captain Abney (of photographic renown). 1946 R. J. C. Atkinson *Field Archaeol.* ii. iii. 136 *Abney Clinometer*..The instrument consists of a sighting tube fitted with a pin-hole aperture, [etc.]. 1955 *Oxf. Jun. Encycl.* VIII. 429/1 The Abney level or clinometer, which consists of a sighting tube fixed to a semicircular vertical plate marked in degrees, with a spirit level attached to its index, is..used to measure vertical angles.

abnormalism (æbnɒ·ɹmăliz'm). [f. Abnormal *a.* + -ism.] The condition of being or tendency to be abnormal; also, an abnormal thing. So **abno·rmalist**, one who is characterized by abnormalism (Funk, 1893).

1894 *Thinker* V. 536 If it were so [i.e. that the organism was nothing more than the creature of environment] one would have expected..that abnormalism would be more frequent. 1907 *Daily Chron.* 24 Dec. 5/2 Specialist on Abnormalism.

abnormalize (æbnɒ̄·ɹmăləiz), *v.* [f. Abnormal *a.* + -ize.] *trans.* To render abnormal. Also *absol.*

a 1871 De Morgan *Budget Parad.* (1872) 339 The toe-bone might have been abnormalised by therian..malady. 1890 *Sat. Rev.* 4 Jan. 16/1 The Romantic turned Realist, who tries to avoid Classicality and *das Gemeine*, not merely by individualizing, but by abnormalizing, if we may say so. 1895 *Arena* (Boston) Aug. App. p. vi, The excessive use of alcohol, which abnormalizes (if I may so express it) man's entire being.

abnormous, *a.* Add: Hence **abno·rmously** *adv. rare.*

1878 Trollope *Is he Popenjoy?* II. ii. 26 A brat so abnormously distasteful and abominable.

Abo, abo (æ·bo), *a.* and *sb.* *Australian slang.* [Shortened f. Aboriginal (see note below).] Aboriginal.

The currency of *Abo* was assisted by its use in the Sydney *Bulletin* feature entitled 'Aboriginality' (first numbers 21 Apr. 1888): see quots. [1906–1933].

[1906 *Bulletin* (Sydney) 5 Apr. 14/3 Can any aboliar state definitely how long sheep can do without water? *Ibid.* 2 Aug. 16/4 Have read the 'Abo.'s' on it. *Ibid.* 18 Oct. 17/3 Remarkable the number of 'Abo' writers who have been chased by snakes. 1907 *Ibid.* 17 Jan. 14/2 [He] says Chinese extract the gall-juice of crows for physic. All Abo-shine. 1933 *Ibid.* 1 Nov. 20/4 Would any Aboliar like to enter for the oldest-kangaroo stakes?] 1922 'Te Whare' *Bush Cinema* 38 An 'abo' legend attaches to the great bluff between Bermagni and Tathra. *Ibid.* 91 Fires were lighted, and the 'abos' feasted royally on fish broiled on live coals. 1933 *Bulletin* (Sydney) 25 Jan. 20 The idea of better housing for the abos. 1950 'N. Shute' *Town like Alice* iii. 82 Black boys—black stockmen. Abos.

aboard, *adv.* and *prep.* Add: **A. 1. c.** *all aboard*, the call to warn passengers to get aboard a vessel about to start. Also used (chiefly *U.S.*) in connexion with coaches and trains.

1838 J. C. Neal *Charcoal Sk.* 76 He..gave himself a gentle impulse, crying 'All aboard!' and slid slowly but majestically down. 1871 Barnum *Struggles & Triumphs* 363, I should have expected..to have seen him dressed in a pea jacket, blowing off steam, and crying out 'all aboard that's going'. 1873 J. H. Beadle *Undevel. West* xv. 251 All aboard for Yosemite and the Big Trees. 1903 A. D. McFaul *Ike Glidden* xxxi. 297 He and his bride boarded the train, and the conductor announced, 'All aboard'.

transf. 1878 I. L. Bird *Rocky Mts.* (1879) 148 'Head them [*sc.* cattle] off, boys!' our leader shouted; 'all aboard! hark away!' and..away we all went at a hand-gallop. *c* 1890 McK. Wright in A. E. Woodhouse *N.Z. Farm & Station Verse* (1950) 33 All aboard! all aboard! is the cry They're a ripping lot of shearers in the shed. 1911 W. H. Koebel *Maoriland Bush* viii. 122 Then comes the..signal to commence work [*sc.* shearing]—the stentorian cry of 'All Aboard'. 1928 A. E. Andrews *C. C. Andrews: Recoll. 1829–1922* 169 At Dubuque a non-commissioned officer of the guard would call out 'All aboard', and upstairs we promptly would go.

d. On, or into, a train. Chiefly *U.S.*

1856 M. J. Holmes *L. Rivers* iv. 33 She..told him that 'the trunks..were every one on 'em left!' 'No, they are not..I saw them aboard myself.' 1931 W. G. McAdoo *Crowded Years* i. 7 When the..train..rattled up to the..station..the McAdoo family..climbed aboard.

B. 1. c. In, or into (a train). Chiefly *U.S.*

1855 *Knickerbocker* XLV. 561, I..put myself 'a-board' the six-o'clock Train. 1869 'Mark Twain' *Innoc. Abr.* xii. 79 We..stepped aboard the train.

abode, *sb.*¹ Add: **5.** *abode of love*: see *Agapemone; (*of*) *no fixed abode*: a conventional *phr.* used of a person without a settled habitation.

1852 *Roget's Thes.* 223/1 Abode of love, agapemone. 1931 G. B. Shaw *Let.* 24 Nov. in *To a Young Actress* (1960) 156 If the customer wants to decorate a bedroom in an Abode of Love he (or she) may require something quite different from the keeper of a boarding house in Tunbridge Wells.

1883 *Times* 9 Jan. 4/6 At the Guildhall, yesterday, Charles M'Manes, who said he was a labourer but had no fixed abode, was charged on remand. 1922 Joyce *Ulysses* 447 Henry Flower. No fixed abode. Unlawfully watching and besetting. 1962 *Oxford Mail* 20 Jan. 3/7 Two labourers of no fixed abode, who pleaded 'Guilty' to robbing —— with violence at Stone.

‖**aboideau, aboiteau** (æ·bədōᵘ, -tōᵘ; Fr. abwado, -to). *Canadian.* 9 **ab(b)atteau.** Pl. **-eaux, -s.** [Canadian Fr.] In Nova Scotia and New Brunswick, a sluice-gate in a dike. Hence as *v.* (*rare*), to provide with such a sluice-gate.

1825 *Novascotian* (Halifax) 9 Mar. 86/1 Before half an hour the violence of the water washed away..the complete side of the abatteau. *Ibid.* 16 Mar. 99/2 Heretofore Abbateaus have been constructed in too narrow a base. 1830 Moorsom *Lett. N.S.* 187 The *aboiteau* is provided with a floodgate which opens and closes with the preponderating water. 1876 J. R. Campbell *Hist. Yarmouth Co., N.S.* 5 The abatteau was protected by a long pier running out seaward. 1903 *Bot. Gaz.* Sept. 180 (Cent. D. Suppl.), At first sight it might seem wise to aboideau all rivers at their mouths. 1905 Crosskill *Prince Edward Is.* 85 As a result of such legislation, Aboideaux..have been constructed at the undermentioned places. 1955 in *Jrnl. Canad. Ling. Assoc.* (1958) IV. 9 Bulldozers and all types of modern machinery..are constructing an aboiteau to provide protection to a considerable marsh body.

abolitionist. Add: **c.** Applied to one who advocates the abolition of capital punishment. Used also *attrib.*, esp. of countries, states, which have abolished capital punishment.

1927 E. R. Calvert *Cap. Punishment in 20th Cent.* ii. 23 The abolitionist recognises..the need to protect society against murder, but he realises that the problem cannot be solved by the mere infliction of unpleasant consequences *Ibid.* iii. 48 The relative culture and ethical standards in the abolitionist and non-abolitionist states. 1930 *Times* 3 Apr. 7/7 The cumulative experience of abolitionist countries demonstrated beyond reasonable doubt that the death penalty is not necessarily a deterrent. 1956 *Times* 10 July 6/3 There was no evidence to show that the police in abolitionist countries were any less safe than they were in retentionist countries.

A-bomb: see *A. III.

abominable, *a.* Add: **1. c.** *Abominable Snowman*, name applied to a creature alleged to exist in the Himalayas. (Cf. *yeti.)

1921 *Times* 31 Oct. 10/6 The men were never seen..but footprints were found which were suspected of being those made by these men, who are apparently known to the Tibetans as Meetoh Kangmi, or 'Abominable Snowmen', and small colonies of these people are believed to exist on the slopes of Everest, Chumalhari, and Karola. 1955 *Ann. Reg. 1954* A British party which set out to investigate the Abominable Snowman or Yeti found that legendary creature most elusive and returned with only a few more photographs of footprints.

abord, *sb.* **1.** (Later examples.)

1854 P. G. Patmore *Friends & Acq.* III. xviii. 100 Such were the impressions produced on me by the first *abord* of John Hunt, as I saw him within his prison walls. 1884 *Edin. Rev.* Oct. 557 Her granddaughter has a vivid recollection of Mrs. Merivale in her later days, with a slightly formal and old-fashioned 'abord'.

aborigines. 2. (Additional U.S. and Australian examples of *aborigine*, a form now common esp. in Australia.)

1858 Vielé *Following Drum* 216 The aborigine was inclined to dispute the point. 1876 J. Burroughs *Winter Sunshine* (1883) vi. 133 If the red aborigine ever had his summer of fulness and contentment. 1922 *Times* 8 Aug. 5/4 (*headline*) The Australian Aborigine: His Future Welfare. 1959 *Chambers's Encycl.* I. 804/2 The aborigine's social and spiritual feeling for his own land means that he does not covet the land of other people.

aborning (ăbɒ̄·ɹniŋ), *adv.* and *pred. a.* Chiefly *U.S.* [f. A *prep.*¹ 12 + *borning*, f. Born.] (While) being born or produced.

1934 in Webster. 1943 H. A. Wallace *Century of Common Man* (1944) 87 A future which..will save and share the culture past and now aborning. 1957 D. W. Brogan *French Nation* v. 189 The Franco-Russian alliance was a long time aborning. 1961 F. G. Cassidy *Jamaica Talk* xiv. 297 Some names are obsolete; some died aborning. 1961 *Woman's Own* 16 Dec. 9/1 A home for a child aborning.

abort, *v.* Add: **1. b.** Later U.S. examples, *spec.* in *Aeronautics*.

1962 J. Glenn in *Into Orbit* 41 A bright red light at the very top of this panel is labelled 'Abort'...You could be aborted automatically..or by a command from the ground. 1963 *Amer. Speech* XXXVIII. 118 We lost an engine, so we aborted the mission.

c. *intr.* use of prec. *U.S.*

1946 *Britannica Bk. of Year* 832/1 *Abort*, to fail to complete a mission or flight; said of an aeroplane. 1963 *Amer. Speech* XXXVIII. 118 A specified point on the runway..used as a decision point for aborting. If trouble develops on the take-off roll before go-no-go, it is possible to abort and stop the aircraft on the remaining runway.

abort. Transfer *Obs.* to senses 1, 2 and add: **3.** *Aeronaut.* An abortive flight by an aeroplane, etc.; an aeroplane, etc., that so fails. (See also quot. 1962.) *U.S.*

1958 *Aero-Space Terms*, *Abort 1.* An instance of a rocket missile or vehicle failing to function effectively and not achieving the objective plotted for it. *2.* A rocket missile or vehicle that so fails. 1960 *Times* 18 Oct. 13/6 A thing which detects errors in the interiors of missiles is now known as 'abort sensory equipment'. 1962 *Into Orbit* Glossary 243 *Abort*, an escape; in spacecraft terms an emergency separation of the capsule from the booster—or even the emergency exit of the Astronaut from the capsule. 1963 *Amer. Speech* XXXVIII. 118 The control room had to classify our flight as an air abort.

abo·rtus fe·ver. [L. *abortus* miscarriage.] An infection caused by *Brucella abortus bovinus* or by other species of the genus *Brucella*, and transmitted to man in milk or meat products; undulant fever. Cf. *brucellosis.

1927 *Lancet* 31 Dec. 1414/2 (*title*) Is it Undulant or Abortus Fever in Great Britain? 1932 *Ibid.* 28 May 1152/2 Attempts to induce abortus fever experimentally in volunteers. 1936 *Nature* 21 Nov. 889/1 Well-recognized generic names..the diptheroids..and the organisms of undulant and abortus fevers (*Brucella*).

aboulia, abulia (ăbū·liă, ăbiū·liă). [mod. L., f. Gr. ἀ- priv. + βουλή will; cf. Gr. ἀβουλία thoughtlessness.] Loss of will-power, as a mental disorder. Hence **abou·lic, abu·lic** *a.* (and *sb.*). Also **abouloma·nia** = *aboulia.

1848 Dunglison *Med. Lex.* (ed. 7), *Abulia*, loss of the will, or of volition. 1878 tr. *von Ziemssen's Cycl. Med.* XIV. 542 In some cases this abulia reaches such a degree that the patients..abandon almost every movement, do not leave their bed [etc.]. **1889** *Buck's Handbk. Med. Sci.* VII. 687/2 Desire exists, resolution is formed, but no action, or only feeble action, follows. This would..be a form of *aboulia*. **1890** BILLINGS *Med. Dict.*, *Aboulomania*, a form of insanity characterized by inability to exert the will. **1893** DUNGLISON *Dict. Med. Sci.* (ed. 21), *Abulic*, one who has lost power of will. **1894** GOULD *Dict. Med.*, *Abulic*, characterized by or affected with abulia; of defective will-power. **1899** W. JAMES *Talks to Teachers* xv. 179 A condition of perfect 'abulia', or inability to will or act. **1900** J. ROYCE *Let.* 12 Sept. in R. B. Perry *Tht. & Char. of W. James* (1935) I. 815 That left me..morally tired,—mildly aboulic, so to speak. **1940** *Brit. Jrnl. Psychol.* Jan. 235 When the individual is in a dither of indecision, or when he suffers from temporary *aboulia*.

about. Add:
B. 4. b. *fig.* in reference to mental faculties, etc.: *about one*, at command, in readiness for use.
1622, 1809 [see WIT *sb.* 3 c]. **1747** CHESTERFIELD *Let.* 30 Oct. (1774), Wherever you are, have (as the low, vulgar expression is) your ears and your eyes about you. **1779** *Mirror* 20 Apr. 99 Things were not a bit mended by my wife's sollicitude (who, to do her justice, had all her eyes about her) to correct them. **1873** NEWMAN *Idea Univ.* (ed. 3) vi. 128 He has his eyes ever about him. **1923** *Times Lit. Suppl.* 18 Jan. 33/2 We shall need to have our best wits about us if we are to avoid confusion.

about-face, *v. intr.* and *sb.* orig. and chiefly *U.S.* [Shortening of *right about face* (RIGHT ABOUT B. 1, 2).] = next. (Cf. FACE *v.* 9 b.)
1861 *Harper's Mag.* Aug. 364/2 'About-Face!' was the next command. **1918** FARROW *Dict. Mil. Terms* 1 *About Face*, one of the facings in the school of the soldier, executed by facing to the rear, turning to the right. **1924** *Scribner's Mag.* July 36/1 Morrow got very white—about-faced, and marched out of the room. **1930** *Liberty* 6 Sept. 33/2 That's enough to make us do an about face and give it [*sc.* a car] the gas. **1934** *Baltimore Sun* 20 Sept. 2/1 (*headline*) Backers [of Cotton Act] Do An About Face. **1959** *Times* 20 Feb. 8/2 If he did not mislead the Prime-Minister, his abrupt about-face was certainly disconcerting.

about-turn. [Shortening of *right about turn* (RIGHT ABOUT B. 1, 2).] *About turn!* a military command (cf. TURN *v.* 22 c). Hence as *sb.*, a reversal of position (lit. and fig.), point of view, etc. (cf. TURN *sb.* 8). So as *v. intr.*, to execute an about-turn.
1893 *Infantry Drill* II. 59 The instructor will then give the command About—Turn. By the Right. Quick—March. **1937** J. A. LEE *Civilian into Soldier* 32 This is the army, not the Salvation Army. About Turn! **1942** C. S. LEWIS *Broadc. Talks* I. v. 29 If you're on the wrong road, progress means doing an about-turn and walking back to the right road. **1959** *Times* 3 June 8/1 Congregation to-day executed a graceful about-turn on the question of retaining the classical languages as compulsory subjects for candidates taking Responsions. **1960** *Guardian* 7 July 7/6 The whole party about-turned on the steps.

above, *adv.* and *prep.* Add:
B. 10. Phr. *above oneself*: in a state above the normal; out of hand. Also said of horses when they are overfed and under-exercised, or have not undergone the full training for a race.
1890 KIPLING *From Sea to Sea* (1899) I. 455, I have..seen more decent men 'above' or below themselves with drink, than I care to think about. **1893** *Punch* 23 Sept. 137 Lest the spirits of the travelling tourist there should rise so high that he might run the chance of 'getting a bit above himself', as horsebreakers graphically express it. **1897** *Daily News* 7 Nov. 7/3 Horses run best when they are above themselves. **1937** V. WOOLF *The Years* 282 He was a little flushed, a little above himself.
D. *above-average* adj.; **above-ground,** above the ground; also, not buried, alive (= B. 4; additional examples).
1952 C. P. BLACKER *Eugenics* xi. 306 Children..who make an impression of being promising and above-average in abilities and character. **1960** K. AMIS *New Maps of Hell* (1961) v. 127 The required above-average score for articulateness. **1878** H. TAYLOR *Notes f. Life* in *Wks.* IV. 139 The poetry of those from whom the..be embalmed when the body of their works is no longer aboveground. **1892** *Daily News* 19 Feb. 5/4 In the whole hamlet there were only three above-ground dwellings. **1913** A. G. BRADLEY *Other Days* iii. 90 Marlborians of the sixties are still numerous above-ground. **1940** G. GREENE *Power & Glory* I. iii. 37 A big above-ground tomb.

‖ **ab ovo** (æb ō·vo), *phr.* [a. L. *ab* from + *ōvō*, abl. of *ōvum* egg.] From the (very) beginning.
Cf. Hor. *Sat.* I. iii. 6 *ab ovo usque ad mala* 'from the egg to the apples', alluding to the Roman custom of beginning a meal with eggs and ending it with apples; and Hor. *A.P.* 147 *Nec gemino bellum Troianum orditur ab ovo*, in allusion to the twin egg from which Helen of Troy was born.
a **1586** SIDNEY *Apol. Poet.* (1595) Sig. K1v, If they [dramatic poets] will represent an history, they must not (as Horace saith) beginne *ab ovo*: but they must come to the principall poynt of that one action, which they will represent. **1623** J. CHAMBERLAINE in *Court & Times of Jas. I* (1848) II. 389, I spoke with one yesterday, that on Tuesday heard the king relate the whole story, *ab ovo*, from point to point, with great contentment. **1862** THACKERAY *Philip* I. i. 11 Shall we begin *ab ovo*, sir? *c* **1879** W. JAMES in R. B. Perry *Tht. & Char. of W. J.* (1935) I. 480 Cognition would have to be evolved again almost *ab ovo*. **1927** D. H. LAWRENCE *Mornings in Mexico* 12 Is it possible that we are so absolutely, so innocently, so *ab ovo* ridiculous?

abox, a-box (ăbǫ·ks), *adv. Naut.* [f. A- 2 + BOX *v.*[1] (sense 13).] Applied to the position of the head-yards when only the head-sails are laid aback.
1801 CAPT. FERRIS in *Naval Chron.* VI. 245 With an intent ..to brace the head yards a-box. **1867** [see BRACE *v.*[2]]. **1867** SMYTH *Sailor's Word-bk.* s.v., To lay the head-yards square, or abox. **1922** *Blackw. Mag.* Dec. 803/2 Her sides were flaked with rust, her yards all a-box. **1961** F. H. BURGESS *Dict. Sailing* 7 *Abox*, term applied to yards of a mast that are braced in opposite direction to those on a neighbouring mast; a handy practice to retain a vessel under control.

abradant (ăbrēɪ·dănt), *a.* and *sb.* *U.S.* [f. ABRADE *v.* + -ANT[1].] **A.** *adj.* = ABRADING *ppl. a.* (Cent. Dict. 1889). **B.** *sb.* An abradant substance.
a **1877** KNIGHT *Dict. Mech.* 247/1 *Bath-brick*, a fine silicious material, found in the vicinity of Bath, England, compacted into the form of bricks, and used as an abradant.

abrade, *v.* Add:
3. *intr.* To wear or rub away.
1890 C. DIXON *Stray Feathers* xvi, Birds which assume many nuptial ornaments in spring, gorgets, plumes, and crests, which drop out or abrade as soon as the pairing season has passed.

Abraham-man. Add: **to sham Abraham** (earlier and later examples); also used substantively.
1780 O. GOLDSMITH *Misc. Works* Essay xxiv. p. 154, I shammed Abraham, merely to be idle. *c* **1805** *Astonishing Abraham Newland* 7 I've heard people say, sham Abraham you may, But you mus'nt sham Abraham Newland. **1835** *Court Mag.* vi. 234/1 Now, all this is sheer nonsense—all sham Abraham, pretty one. **1863** C. READE *Hard Cash* II. xix. 284 'Look out,' he cried in some alarm; 'he's shamming Abraham.'

abranchial, *a.* Add: (Earlier example.) Hence **abra·nchialism,** abranchial condition.
1839 TODD *Cycl. Anat.* II. 732/1 The Abranchial Annelida ..are naturally hermaphrodite. **1902** *Encycl. Brit.* XXX. 796/1 The Pterotracheidæ [have] no shell and opisthobranchialism, or even abranchialism in Firoloida.

abrasion. Add: **1.** *abrasion platform*, a flat surface at the edge of the sea produced by the abrading action of the waves.
1922 FLATTELY & WALTON *Biol. Sea-Shore* ii. 36 Lastly we have the more extensive, nearly horizontal plane produced by long-continued wave erosion, and commonly called the abrasion platform. **1937** WOOLDRIDGE & MORGAN *Physical Basis Geogr.* xxi. 322 In other cases it [*sc.* the wave-cut beach-bench] is continued seaward in a flatter surface, the abrasion platform, over which a thin offshore 'veneer' of finer material may be spread.

abrasive, *a.* Add: **b.** *fig.*
1925 T. DREISER *Amer. Trag.* (1926) I. II. xxxiv. 387 His mind was troubled with hard, abrasive thoughts. **1963** EDMUND WILSON in *New Statesman* 8 Feb. 198/3 Abrasive is coming in, in application to literary qualities.
B. *sb.* An abrasive substance or body.
1853 O. BYRNE *Artisan's Handbk.* 17 To polish the tool upon the oil-stone, or other fine abrasive for setting the edge. **1951** *Good Housek. Home Encycl.* 11/1 Abrasives are useful for heavily soiled surfaces, when soap and water or detergents are unsuccessful and some gentle friction is required. **1960** *Jrnl. Iron & Steel Inst.* CXIV. 406/1 A study of bonded abrasives.

abraxas. Add: (Later, incl. *attrib.*, examples.) Also **abrasax.**
1858 R. WALSH *Anc. Coins* (ed. 2) 39 The word Abraxas, sometimes spelled Abrasax, was the great mystery of the Gnostics. **1946** E. WAITE [illegible] 37 He had a large collection of amulets, Gnostic gems and abraxas stones. **1963** [illegible] In certain Egyptian papyri containing Abraxas prayers, we find the prayer coupled with the reminder that the petitioner knows the divine mystic name.

abreaction (æbriæ·kʃən). *Psychiatry.* [f. AB- + REACTION, after G. *abreagierung*.] The liberation by revival and expression of the emotion associated with forgotten or repressed ideas of the event that first caused it. Hence **abrea·ct** *v.*, to eliminate by abreaction; **abrea·ctive** *a.*, of this kind of treatment.
[**1893** J. BREUER & S. FREUD *Studien über Hysterie* iv. 23 Heilung hysterischer Symptome durch..Abreagiren.] **1912** A. A. BRILL tr. *Freud's Sel. Papers Hysteria* (ed. 2) 6 The ab-reaction (abreagiren)..is not the only form of discharge at the disposal of the normal psychic mechanism of the healthy person who has experienced a psychic trauma. The memory of the trauma even where it has not been ab-reacted enters into the great complex of the association. [illegible] **1926** W. MCDOUGALL *Outl. Abnormal Psychol.* 451 The assumption underlying the practice of Abreaction was that the forgotten or repressed 'ideas' were beset (*besetzt*) by a charge of emotional energy or *libido*; and that the process of Abreaction discharges this energy from the system. **1926** *Contemp. Rev.* Aug. 191 It is claimed that they [*sc.* early memories] can be traced down through links of emotion—called 'ab-reaction'—from the conscious mind into the so-called unconscious stratum. **1944** BRENMAN & GILL *Hypnotherapy* (1947) iv. 72 Although the 'abreactive method' has been used extensively in World War II, the alteration in consciousness which is necessary for such a 're-living' has usually been brought about by drugs. **1958** A. HUXLEY *Brave New World Revisited* (1959) 107 The process known in England as 'abreaction therapy', in America as 'narcosynthesis'.

abreption. For '*obs. rare*' read '*obs.*' and add examples.
1686 *Reg. Privy C. Scotland* (1932) XIII. 76 In this case his Majestie was not to be considered in his royal capacity as giving gifts or grants, in which case there is place for abreption and subreption. **1828** *Blackw. Mag.* XXIV. 157/2 This earliest Abreption of this famous Fleece. **1872** MCCLINTOCK & STRONG *Cycl. Bibl. Lit.* IV. 613/2 Such an abreption of bodily functions is calculated to enhance the perceptions of celestial verities.

abrim (ăbri·m), *adv.* or *pred. a.* [f. A *prep.*[1] + BRIM *sb.*[1]] Full to the brim; brimming.
1896 KIPLING *Seven Seas* 114 Weed ye trample underfoot Floods his heart abrim. **1920** A. E. W. MASON *Summons* ii. 16 Partly because of her vivid colouring and because she was abrim with life. **1934** D. L. SAYERS *Nine Tailors* iv. i. 330 Dyke and drain were everywhere abrim. **1951** W. DE LA MARE *Winged Chariot* 55 Abject with misery or with bliss abrim.

abrin (ǣ·brin). *Chem.* [f. mod. L. *Abrus* (see def.) + -IN[1].] A highly poisonous protein contained in the jequirity bean (*Abrus precatorius*).
1884 KLEIN *Micro-org. & Disease* 165 Messrs. Warden and Waddell published in Calcutta during the present year..a large number of observations on the jequirity poison.. They have..proved, that the active principle is a proteid—abrin—closely allied to native albumen. **1897** [see PHYTALBUMOSE].

abristle (ăbri·s'l), *adv.* or *pred. a.* [f. A *prep.*[1] + BRISTLE *v.*[1]] Bristling.
1916 E. POUND *Lustra* 97 A-bristle with antennae to feel roads. **1919** M. BEERBOHM *Seven Men* 98 So a-bristle am I with memories of the meetings I had with its author. **1927** *Chambers's Jrnl.* May 343/2 Stripes sprang to his feet with a growl, every hair abristle.

abroad, *adv.* and *prep.* Add: **A.** *adv.* **5.** (Earlier examples.)
1806 J. BERESFORD *Miseries* I. i. 5 Unless I am quite abroad still; and if so, I will humbly wait, while you clarify my understanding. *a* **1828** J. BERNARD *Retrosp. Stage* (1830) I. ii. 183 The actors appearing to be all abroad when they were at home.
¶ Examples of a comparative form.
1923 [illegible] he had meant to go abroad, and he had gone abroader than anybody less happily disposed than himself had ever dreamed of. **1939** A. THIRKELL *Before Lunch* vii. 190, I never can think why when people say abroad they mean France as a rule, and yet most other places are much abroader than France.
C. *sb.* [cf. *adv.* sense 4 and quot. 1866 in Dict.] Any region outside one's homeland.
1895 K. GRAHAME *Golden Age* 98 She was somewhere over in that beastly abroad. **1925** R. MACAULAY *Casual Commentary* 138 Restaurants and hotels are getting fuller. Abroad is getting fuller. **1934** H. G. WELLS *Exper. Autobiog.* II. 565 'Abroad' was a slightly hostile region of adventure. **1945** N. MITFORD *Pursuit of Love* xv. 114 'Frogs,' he would say, 'are slightly better than Huns or Wops, but abroad is unutterably bloody and foreigners are fiends.' **1958** K. AMIS *I like it Here* iii. 40 Why was abroad accorded exclusively by the Romance-speaking family of nations?

abrupt, *sb.* (Later examples.)
1735 THOMSON *Liberty* I. 314 Whole stately Cities in the dark Abrupt Swallow'd at once. [illegible] When the whole loaded Heaven Descends in Snow, lost in one white Abrupt. **1887** W. C. RUSSELL *Frozen Pirate* I. vi. 73 It was like the face of a cliff, a sheer abrupt.

abrupt, *v.* Delete † and '*Obs. rare*' and add examples.
1819 *Blackw. Mag.* V. 737/2 This gas obtains in greatest abundance in the vicinity of dykes which abrupt the coal. **1940** 'M. INNES' *Journeying Boy* xi. 130 But to abrupt his journey in a strange town..was a procedure..unnecessarily drastic. **1958** *Observer* 29 June 15/6 The film version of 'The Brothers Karamazov'..does not deliberately distort the novel or abrupt it.

abscise, *v.* Delete † and '*Obs. rare*—[1]' and add later examples. Also *Bot.*, *intr.* to fall off, to separate by abscission.
1879 *St. George's Hosp. Rep.* IX. 473 This eye was abscised..and she is now able to admirably manage an artificial substitute. **1909** WEBSTER, *Abscise* (*Bot.*), to separate by abscission. **1940** MEYER & ANDERSON *Plant Physiol.* xxxvi. 651 Leaves, however, are not the only organs or parts of plants which abscise. **1964** S. DUKE-ELDER *Parsons' Diseases of Eye* (ed. 14) xvi. 197 If prolapse of iris has occurred it should usually be abscised.

absciss, *sb.* Add: **2.** *Bot.* **absciss layer,** a layer of cells across the base of the petiole in deciduous leaves.
1893 in *Funk's Stand. Dict.* **1898** H. C. PORTER tr. *Strasburger's Text-bk. Bot.* I. § I. 143 Preparatory to the falling of leaves an absciss layer is formed, by means of which the separation of the leaves from the stem is effected. **1924** HOLMAN & ROBBINS *Textbk. Gen. Bot.* vi. 180 The

conditions which favor leaf fall cause the cells near the lower end of the leaf stalk to become meristematic and to give rise to a zone of delicate, thin-walled cells extending clear across the base of the petiole. This is called the *abscission layer* (ed. 2, 1934, vi. 198 *abscission layer*). **1940** *Bot. Gaz.* CII. 323 (*title*) Effect of growth substances on the abscission layer in leaves of *Coleus*.

absci·ss, *v.* [Back-formation f. ABSCISSION.] *trans.* and *intr.* To cut off; to abscind; to shed. Hence **absci·ssed** *ppl. a.*
1869 G. LAWSON *Dis. & Injuries of Eye* ii. 51 The Staphyloma may be abscissed. *Ibid.* 52 Another way of abscissing the staphyloma is to transfix its base with a Beer's knife. **1918** *Bot. Gaz.* LXVI. 26 The numbers refer to the total number of abscissed petioles at the corresponding dates. *Ibid.* 42 The lower leaves began to abscis. **1958** PRIESTLEY & SCOTT *Introd. Bot.* x. 150 (*caption*) Naturally abscissed branches of oak. *Ibid.* xxxviii. 554 These twigs..may be abscissed after many years of growth.

abscission. Add: **3.** *Bot.* The natural separation of parts following the disorganization of the cells in the abscission layer. Hence *abscission layer* = *absciss layer.
1889 E. S. BASTIN *College Bot.* 403 *Abscision* [sic], a term applied to that mode of the detachment of spores in Fungi which consists in the disorganization of the zone connecting the spores with the hypha. **1916** *Bot. Gaz.* LXI. 225 That this separation of walls actually obtains can be proved by taking rather thick sections of suitable material and..pulling the portions separated by the abscission layer apart on the slide. **1934** [see *ABSCISS *sb.* 2]. **1946** *Nature* 3 Aug. 147/2 Abscission in shot-hole disease of peach. **1950** METCALFE & CHALK *Anat. of Dicotyledons* I. 545 Before flowers or fruits fall off, a definite abscission layer is formed.

‖ **abseil** (æ·bzail). *Mountaineering.* [G. *abseilen*, f. *ab* down, *seil* rope.] The technique of descent of a steep face by means of a doubled rope fixed above the climber. So as *v. intr.*, to use the abseil or 'rope down' technique in descent. Hence **a·bseiling** *vbl. sb.*
1933 G. D. ABRAHAM *Mod. Mountaineering* x. 184 To help in the abseil or descent of steep, almost holdless places. **1941** C. F. KIRKUS *Let's Go Climbing!* v. 76 To abseil you must be unroped. Hang the rope over a belay and see that it hangs down on either side and both ends are resting on the ground. You then have a double rope to slide down. *Ibid.* 78 In the Alps a great deal of abseiling is done. **1954** *Mountaineering* 1954 59 One member of the party ascended the first pitch of Brant, and was then forced to abseil off, for nobody would, or could, follow him. **1958** M. BANKS *Commando Climber* v. 80, I had to untie for the last abseil down to the glacier. **1960** R. C. EVANS *On Climbing* iii. 17 He had been taught to abseil, and the rock over which he had looped his doubled rope had come away.

absent, *a.* and *sb.* Add: †**B.** *sb.* (Later example.)
1702 EDMUND GIBSON *Schedule Review'd* 27 The Absents lose their Right of Voting.
C. **absent-minded** *a.* (examples); **absent voter,** one authorized because of special circumstances to vote (by post, etc.) at a general election though absent at the time from the ordinary place of voting; hence *absent vote*, a vote so authorized (cf. *ABSENTEE 1 b).
1854 THOREAU *Walden* 18 Dreaming and absent-minded all the way. **1890** W. JAMES *Princ. Psychol.* I. iv. 115 Very absent-minded persons in going to their bedroom to dress for dinner have been known to take off one garment after another and finally to get into bed. **1918** *Act* 8 Geo. V c. 64, sched. 1, § 16 The registration officer, if satisfied that there is a probability that the claimant, by reason of the nature of his occupation, service, or employment, may be debarred from voting at a poll at parliamentary elections held during the time the register is in force, shall place the claimant (if registered) on the absent voters list. **1935** *Hansard Commons* Ser. 5. CCLXXXIII. 1335, I do not see any need for imposing as a condition for receiving the benefit of an absent vote the necessity of being habitually engaged in an employment which compels a man to be away from his place of residence for a period of not less than 10 days in each month.

absentee. Add: **1. b.** = *absent voter. Also (chiefly *U.S.*) *absentee ballot, voter, voting*.
1925 *Hansard Commons* Ser. 5. CLXXXIII. 1362 The Territorial Soldier is mentioned [in the Bill] but not the regular soldier, I presume because the regular soldier would come into the class of general absentee. **1932** *N.Y. Times*, 10 Oct. 2/5 Uniform laws governing absentee voting in the United States were urged today by John F. Costello. *Ibid.* 23 Oct. 17/2 (*headline*) Grandmother's 'Duty' to Child Wins Fight for Absentee Vote. **1936** *Daily Oklahoman* (Okla. City) 1 Nov. 12/6 Iowa was the first state to adopt the absentee voter idea. **1957** *N.Z. Offic. Year-bk.* 1957 29 An elector may, however, vote as a 'special voter'... These latter conditions replace the former classes of absentee, postal, and declaration voters.

absenteeism. Add: **b.** *spec.*, in industry, the repeated absence of employees; in schools, of the pupils.
1922 J. D. HACKETT in *Management Engineering* Feb. 1 (*sub-heading*) Absenteeism. **1941** *Punch* 3 Sept. 201 Committee on National Production..we shall do our best to decrease absenteeism during the coming winter. **1957** *Observer* 8 Sept. 1/7 Absenteeism was a factor with such a direct and devastating effect on performance that it could destroy the fruits of the technical advance into which the industry [coalmining] was putting so much effort. **1957** *Oxford Mail* 25 Sept. 1/9 Schools in Oxford City have so far been unaffected, and it was stated today that there is no unusual amount of absenteeism.

absenteeship. (Later example.)
1824 W. COBBETT *Hist. Protestant 'Reformation'* V. § 150 The great cause of the miseries of Ireland, at this moment, is 'absenteeship'.

absinthe. Also 9 absanth. Add: **1. b.** The prairie-sage, sage-brush. *U.S.*
1843 FRÉMONT *Explor. Rocky Mts.* in *Narrative* (1846) 14 The artemisia, absinthe, or prairie sage, as it is variously called. *Ibid.* 56 Absinthe bushes..grew in many thick patches. **1846** SAGE *Scenes Rocky Mts.* iv. 31 Countries abounding with absanth and wild sage. **1849** PARKMAN *Oregon Trail* 146 Multitudes of strange medicinal herbs, more especially the absinth, which covered every declivity.
3. (Earlier example.) Also used of a colour resembling the green of absinthe.
1842 *Blackw. Mag.* LII. 494/2 We took a glass of absinthe to compose our nerves. **1872** *Young Englishwoman* Dec. 644/1 Absinthe, pale green. **1938** J. CARY *Castle Corner* 33 A sunray was turning the rollers from hawthorn to absinthe. **1963** *Harper's Bazaar* May 41 An absinthe suit with a deep draped armhole.

absinthian, *a.* (Later examples.)
1833 W. E. WALL in *Fraser's Mag.* VIII. 659/1 Despair prevented her cup Apsinthian. **1843** T. CARLYLE in *Foreign Q. Rev.* XXXI. 570 Huge temporary Pumpkin, saccharine, absinthian.

absit (æ·bsit). [L. *absit* let (him) be absent, 3rd pers. sing. pres. subj. of *abesse* to be absent.] Permission given to a student to be absent. (Cf. EXEAT.)
1853 DICKENS *Dict. Univ. Cambridge* 3/1 *Absit*, every Undergraduate wishing to leave Cambridge for a whole day, whether including a night or not, must obtain an 'absit' from his Tutor. **1920** 'TWO OF THEM' *Guide to Camb. Univ. Life* 7 Exeats (for one night or more) and absits (for one day) may be obtained from College Tutors on producing a sufficient reason.

‖ **absit omen** (æ·bsit ōᵘ·mĕn). [L., may this omen be absent, i.e. void.] May no ominous significance attach to the words.
1594 JAMES VI OF SCOTLAND *Let.* 5 June in *Lett. Eliz. & Jas. VI* (Camden Soc. 1849) 107 Quhich shall neuer be, except ye constraine me unto it, but absit omen. **1854** THACKERAY *Newcomes* II. xxvi. 292 Absit omen! [illegible]

absolute, *a.* Add: **5.** †**b.** *spec.* Of numbers, parts: complete. (Cf. Plin. *Ep.* 9. 38 *liber numeris omnibus absolutus*, a book complete in all its parts.) *Obs.*
1623 HEMINGE & CONDELL *Pref. Shakesp.*, All the rest, absolute in their numbers, as he conceived them. **1667** MILTON *P.L.* VIII. 421 And through all numbers absolute, though One.
9. Add: *absolute clause, comparative, construction, superlative*.
[illegible]
11. Of a decree or rule: see DECREE *sb.* 4 b, RULE *sb.* 4 a.
1836, 1860 [see NISI]. **1922** JOYCE *Ulysses* 638 Then the decree nisi and the King's Proctor to show cause why and, he failing to quash it, nisi was made absolute.
16. In specific combinations with sbs., as *absolute contraband, drought, initials, majority, zero*, etc. (see the sbs.); also **absolute altitude** *Aeronaut.*, the altitude of an aircraft above the surface of the earth over which it is flying; hence *absolute altimeter, height*; **absolute ceiling** *Aeronaut.* (see quot. 1950; cf. *CEILING *vbl. sb.* 6 b); **absolute humidity** (see quot. 1946); **absolute magnitude** *Astr.*, the magnitude of a star if at a standard distance of 10 parsecs; **absolute music,** self-dependent instrumental music without literary or other extraneous suggestions (opp. *programme music*); hence *absolute musician*; **absolute pitch** *Mus.*, (*a*) a fixed standard of pitch determined by the rate of vibration (see PITCH *sb.*[2] 23); (*b*) used of the ability to recognize or reproduce the pitch of a note; **absolute temperature** *Physics*, temperature calculated from the scale associated with the thermodynamic experiments of Lord Kelvin (1824–1907), and based on an absolute zero (see ZERO 3) of approx. $-273 \cdot 16°$ C.; hence *absolute* (*temperature*) *scale*; **absolute unit** (see quot. 1940); **absolute value** *Math.*, (*a*) of a real number, its value irrespective of its sign—called also *numerical value*, or *modulus*; (*b*) of any complex number $x+iy$, the positive square root of x^2+y^2—called also *modulus* (Webster, 1934) (see MODULUS 2 c).
1931 *Aeronautics* Mar. 65/2 *Absolute altimeter*. The instrument operates through a radio transmitter which..sends out a radio wave so that it is reflected back from the earth... The elapsed time for the circuit of the wave is converted into absolute altitude. **1934** *Aircraft Engin.* June 162/2 'Absolute altitude' [illegible] **1950** *Gloss. Aeronaut. Terms* (*B.S.I.*) 4 *Absolute ceiling*, the height at which the rate of climb of an aircraft would be zero in standard atmosphere under specified conditions. **1936** *Aircraft Engin.* Nov. 298/2 We cannot help wishing it were possible to define an absolute height akin to absolute temperature. **1867** A. BUCHAN *Handy Bk. Meteorol.* 93 It may also be termed the absolute humidity of the atmosphere. **1946** *Humidity of the Air* (*B.S.I.*) 3 *Absolute humidity*, the weight of water vapour present in unit volume of moist air, i.e. grains per cubic foot or grams per cubic metre. **1902** J. C. KAPTEYN in *Publ. Astron. Lab. Groningen* No. 11, 11 We further define the *absolute magnitude* (*M*) of a star..as the magnitude that star would have if it was transferred to a distance from the sun corresponding to a parallax of 0"·1. **1914** A. S. EDDINGTON *Stellar Movements* viii. 170 Absolute magnitudes (magnitude at a distance of 10 parsecs). **1890** G. B. SHAW *London Music 1888–89* (1937) 329 The first sign of a reaction in favor of abstract or 'absolute' music against the great Wagnerian cult of tone poetry and music drama. [illegible] This is the art which Wagner called absolute music. **1946** BACHARACH *Brit. Mus.* iii. 53 All the same *The Planets* are in no sense programme music. One must accept them as seven pieces of 'absolute' music. **1880** *Grove's Dict. Mus.* II. 129/2 Instances of a similar kind from the works even of the most 'absolute' musicians might be multiplied *ad libitum*. **1864** A. J. ELLIS in *Proc. R. Soc.* XIII. 394 A compound tone will be represented by the absolute pitch of its primary and the relative pitches of its partial tones. **1901** E. HUGHES *Mus. Guide* I. 239/2 The vibration-number of a tone also gives it an absolute pitch according to the particular pitch accepted as the standard. **1914** A. EAGLEFIELD HULL *Mod. Harmony* iv. 31 If the possession of the sense of absolute pitch is a *sine quâ non* for the proper reception of such music, then the circle of appreciation at present is narrowed down almost to vanishing-point. **1938** *Oxf. Compan. Mus.* 2/2 Sir Frederick Ouseley..was all his life remarkable for his sense of absolute pitch. At five he was able to remark, 'Only think, papa blows his nose in G'. **1848** W. THOMSON (LORD KELVIN) in *Phil. Mag.* XXXIII. 316 The characteristic property of the scale which I now propose is, that all degrees have the same value; that is, that a unit of heat descending from a body A at the temperature T° of this scale, to a body B at the temperature (T−1)°, would give out the same mechanical effect, whatever be the number T. This may justly be termed an absolute scale, since its characteristic is quite independent of the physical properties of any specific substance. **1852** J. P. JOULE in *Phil. Trans. R. Soc.* I. 67 [illegible] **1873** *First Rep. Brit. Assoc. Comm. Selection Dynam. & Electr. Units* 1 We accordingly recommend the general adoption of the Centimetre, the Gramme and the Second as the three fundamental units; and that they be designated as the 'absolute units' [illegible] **1940** *Chambers's Techn. Dict.* 3/1 *Absolute unit*, a unit which may be defined directly in terms of the fundamental units of length, mass, and time. [illegible] The symbol |x|..indicates the 'numerical' or 'absolute' value of x. **1909** *Cent. Dict. Suppl.* (s.v. *number*), *Absolute value of a complex number*, [illegible] **1930** DURELL & ROBSON *Adv. Trigonom.* 145, *r* is called the modulus, or sometimes the absolute value, of the complex number [illegible] and is denoted by $|x+yi|$ or by mod $(x+yi)$. [illegible] We call the positive number $\sqrt{a^2+b^2} = |a+bi|$ the absolute value or modulus of $a+bi$.

absolutely, *adv.* Add: **11. d.** Also *ellipt.*; *colloq.* (orig. *U.S.*), used as an emphatic affirmative: yes, quite so. (Stressed *absolu·tely*.)
1892 'MARK TWAIN' *Amer. Claimant* xxiv. 154 'Do you mean to say that if he was all right and proper otherwise you'd be indifferent about the earl part of the business?' 'Absolutely.' **1917** A. WAUGH *Loom of Youth* III. i. 154 'But,

sir, was it true to Harrow life?' 'Absolutely; and it's as true to the life of any other Public School.' **1922** JOYCE *Ulysses* 720 Was the narration otherwise unaltered by modifications? Absolutely. **1937** R. STOUT *Red Box* ix. 136 'I trust that we are still brothers-in-arms?—' 'Absolutely. Pals.'
¶ For additional emphasis, *bally* and similar words are sometimes inserted medially in slang use, as *abso-bally-lutely*.
1914 W. L. GEORGE *Making of an Englishman* III. v. 299 Oh, don'tcher care, it's all over, absoballylutely. **1924** C. HAMILTON *Prisoners of Hope* ii. 56 All I ask is that you'll.. let me go on to hit the sheets. I'm absoballylootly all in. **1929** KIPLING *Limits & Renewals* (1932) 367 'Did it cure him?' I asked..'Ab-so-bally-lutely,' said Keede. **1935** E. WEEKLEY *Something about Words* i. 24 This natural tendency to add body and content to words is possibly prehistoric...A crude example of this persisting instinct is offered by the contemporary *abso-bloody-lutely*.

absolutism. Add: **4.** *Philos.* The philosophy of the Absolute (see ABSOLUTE *a.* 13, 14, 15, and ABSOLUTIST *sb.* and *a.* 2).
1878 S. H. HODGSON *Philos. Reflection* I. 121 The same school of objective, or non-Idealist, absolutism. **1884** W. JAMES *Ess. Rad. Empir.* (1912) xii. 279 The one *fundamental* quarrel Empiricism has with Absolutism is over this repudiation by Absolutism of the personal and aesthetic factor in the construction of philosophy. **1890** —— *Princ. Psychol.* I. x. 353 In demanding a more 'real' connection than this obvious and verifiable likeness and continuity, Hume seeks 'the world behind the looking-glass', and gives a striking example of that Absolutism which is the great disease of philosophic Thought.

absolutist. Add: **A.** *sb.* **2. b.** *Philos.* One who maintains that absolute certainty, or some other absolute, is attainable.
1896 W. JAMES *Will to Bel.* (1897) 12 The absolutists in this matter say that we not only can attain to knowing truth, but we can *know when* we have attained to knowing it. *Ibid.* 14 We are all such absolutists by instinct.
B. *adj.* **2.** *Philos.* Of or pertaining to absolutism or absolutists.
1884 W. JAMES *Ess. Rad. Empir.* (1912) xii. 273 Fact holds out blankly, brutally and blindly, against that universal deliquescence of everything into logical relations which the Absolutist Logic demands. **1896** —— *Will to Bel.* (1897) 12 We may talk of the *empiricist* way and of the *absolutist* way of believing in truth. **1936** A. J. AYER *Lang., Truth & Logic* vi. 106 The 'absolutist' view of ethics—that is, the view that statements of value are not controlled by observation..but only by a mysterious 'intellectual intuition'.

absolutistic, *a.* Delete 'rare' and add examples.
1854 *Tait's Mag.* XXI. 352 [It] attempted to reconcile the self-government of the nation with the domination of thirty-four absolutistic princes! **1905** W. JAMES *Meaning of Truth* (1909) iii. 57 It means..a break with absolutistic hopes, when one takes up this inductive view of the conditions of belief. **1940** *Mind* XLIX. 426 Hegel..used his principle of the identity of reason and reality..to defend the idea of the absolutistic state (an idea called, today, 'totalitarianism').
Hence **absoluti·stically** *adv.*
1909 W. JAMES *Pluralistic Universe* 365 An *ipse dixit* of Mr. Bradley's absolutistically tempered 'understanding'.

a:bsolutiza·tion. [cf. next.] The process of making absolute; the action of the verb *ABSOLUTIZE.
1863 D. W. SIMON tr. *Dorner's Person of Christ* Div. II, vol. III. 7 Herein, therefore, is already involved the germ of the absolutization of subjectivity. **1939** *Theology* XXXIX. 454 The Nazi..his dethronement of reason and his absolutization of uncontrolled racial or group impulse. **1951** *Scottish Jrnl. Theol.* IV. 288 The Old Church.. deliberately set itself apart from the rest of Christendom by demanding rigid conformity to its own absolutisations of tradition.

a·bsolutize, *v.* [f. ABSOLUTE *a.* + -IZE.] To make absolute; to convert into an absolute. Hence **a·bsolutizing** *vbl. sb.*
1936 *Mind* XLV. 74 Modes of absolutising what is merely relative. **1952** *Theology* LV. 282 Study of typical answers given in the past can..prevent us from absolutizing our own or any one else's solutions. **1952** *Essays in Crit.* II. 371 Death as the 'absolutizing' of the real experience of sleep. **1958** W. STARK *Sociol. of Knowl.* iii. 118 Philosophers have even tended to absolutize the relative.

absorb, *v.* **5.** (Further examples in *Physics*.)
1899 RUTHERFORD in *Phil. Mag.* XLVII. 123 The α radiation from uranium and its compounds is rapidly absorbed in its passage through gases. **1923** GLAZEBROOK *Dict. Appl. Physics* IV. 582/1 The rays are absorbed according to an exponential law. **1942** STRANATHAN *Particles* xii. 470 The atmosphere..acted merely as an absorbing blanket, absorbing radiation coming from above.

absorbative (æbsǭ·ɹbătiv), *a.* [f. ABSORB *v.* + -ATIVE.] = ABSORPTIVE *a.*
1893 *Longman's Mag.* Mar. 481 He had enjoyed the life with the peculiar appropriative, assimilative, absorbative gust of the typical schoolboy. **1920** GALSWORTHY *In Chancery* I. vii. 66 The two young ones having been supplied with food, the process went on silent and absorbative.

absorbent. A. *adj.* Add: **absorbent cotton** *U.S.*, cotton wool.
1889 in *Cent. Dict.* **1890** BILLINGS *Med. Dict.* I. 7/1 *A*[*bsorbent*] *cotton*, L. gossypium...Cotton freed from adhering impurities, and deprived of oily matter by boiling in a dilute alkaline solution and thorough washing; absorbent and protective; used as a surgical dressing. **1947** *Hygeia* Oct. 767/2 The thermometer should be kept in a small tumbler, one-quarter filled with antiseptic solution, with a little absorbent cotton in the bottom of the glass.

absorber. Add: (Earlier and additional examples in various technical senses.) Cf. SHOCK-*absorber*.
1792 A. YOUNG *Trav. France* 153 He has a pair of scales made at Paris,..an air pump,..an absorber,..a respirator. **1922** GLAZEBROOK *Dict. Appl. Physics* I. 689/2 A low pressure is maintained in the evaporator by causing the evaporated vapour to pass into another vessel, called the absorber, where it comes into contact with cold water in which it becomes dissolved. **1933** *Discovery* Aug. 260/1 The wheel and tyre are sprung in conjunction with absorbers. **1949** *Gloss. Terms Refrigeration* (*B.S.I.*) 4 *Absorber*, the vessel, used in an absorption system, in which the refrigerant vapour is absorbed. **1958** *Chambers's Techn. Dict.* 952/1 *Absorber*, material for capturing neutrons without generating more neutrons, e.g., boron and cadmium, much used for controlling and shielding reactors. **1962** A. NISBETT *Technique Sound Studio* 239 Soft absorbers depend for their action on such things as the friction of air particles in the interstices of the material.

absorbing, *ppl. a.* Add: Examples in *Physics*.
1904 RUTHERFORD *Radio-Activity* iv. 113 Leonard.. has shown that the absorption of cathode rays is nearly proportional to the density of the absorbing matter. **1913** —— *Radioactive Substances* iv. 148 If an absorbing screen is placed in the path of the rays, the velocities of the α particles..should all be diminished by the same amount. **1923** [see *ABSORPTION 4 b].

absorbingly. (Earlier example.)
1836 *Blackw. Mag.* XL. 13/1 Mr. Hillary was at that eventful moment absorbingly engaged with a letter.

absorptiometer. Add: **2.** An apparatus for measuring colour-absorption by means of a photo-electric device.
1939 *Thorpe's Dict. Appl. Chem.* III. 305/2 Moll, Burger and Reichert..have devised a 'spectroscopic absorptiometer' which can be used at any part of the photographable spectrum. **1947** *Nature* 11 Jan. 50/2 Applications of the spectrograph and the photo-electric absorptiometer. **1953** *Electronic Engin.* XXV. 212 The continuous flow self-balancing absorptiometer is an instrument for detecting and giving continuous indication of small changes in colour density of a liquor.
Hence **abso:rptiome·tric** *a.*, of, pertaining to or involving an absorptiometer.
1909 in *Cent. Dict.* Suppl. **1950** *Engineering* 3 Mar. 240/2 The determination of silicon in plain carbon steels..by the absorptiometric method. **1956** *Nature* 24 Mar. 550/2 Conventional absorptiometric methods of gas analysis.

absorption. Add: **4.** *spec.* in (*a*) *Electr.* (See quot. 1904.)
1884 O. HEAVISIDE in *Electrician* 2 Feb. 270/1 If we pass an electric current..across the junction, there will be, by elementary principles, a continuous absorption of energy. *Ibid.* 271/2 The absorption of energy is at the zinc surface where the current goes with the E.M.F. **1904** GOODCHILD & TWENEY *Technol. & Sci. Dict.* 2/1 *Electrical Absorption*, the storing up of a part of the electrical energy of a charged condenser by the dielectric.
(*b*) *Nucl. Physics* and *Radiology*. The reduction in intensity of a beam of radiation during its passage through matter. Also *attrib.*
1899 RUTHERFORD in *Phil. Mag.* XLVII. 128 The absorption coefficient for the α radiation is 1·6, or 160 times as great. **1923** GLAZEBROOK *Dict. Appl. Physics* IV. 582/1 The absorption of β rays is investigated by placing thin sheets of the absorbing material in the path of the rays and measuring the activity through different thicknesses. **1926** R. W. LAWSON tr. *Hevesy & Paneth's Man. Radioactivity* vii. 77 We can detect the increased absorption... Such 'absorption edges' enable us to determine the 'levels' of a series. **1943** *Gloss. Terms Electr. Engin.* (*B.S.I.*) 143 *Absorption-coefficient*, of a homogeneous substance, for X-rays of a given wavelength, the ratio of the linear rate of change of intensity at any point to the intensity at that point. **1947** G. THOMSON *Atom* (ed. 3) xi. 99. Experiments show that these discontinuities in absorption, or Absorption Edges as they are called, are extremely well marked. **1958** *Optima* Mar. 36/1 Neutron absorption converts uranium 238 into element 94—plutonium.
(*c*) *Acoustics*. (See quot. 1900.)
1900 W. C. SABINE in *Amer. Architect* 7 Apr. 4/2 Sound, being energy, once produced in a confined space, will continue until it is either transmitted by the boundary walls, or is transformed into some other kind of energy, generally heat. This process of decay is called absorption. *Ibid.* 12 May 44/1 An attempt to connect the rate of decay..with the absolute co-efficient of absorption of the wall. **1958** *B.S.I. News* Aug. 8/2 Measurement of the absorption coefficient in a reverberation room.
(*d*) *Radio*. Loss of power in the transmission of radio waves. Also *attrib.* Hence also in *Radar*.
1914 *Rep. Brit. Assoc. Adv. Sci. 1913* 607 (*title*) Atmospheric Refraction and Absorption as affecting Transmission in Wireless Telegraphy. **1931** *B.B.C. Year-Bk.* 435/1 Absorption Control. A method of 'controlling' the high-frequency oscillations delivered to the aerial in a wireless telephone transmitter, so that they vary in amplitude at the low frequency of modulation. *Ibid.*, Absorption Wavemeter. A wavemeter consisting of a low-loss oscillatory circuit which is tuned to the transmission to be measured. **1948** TAYLOR & WESTCOTT *Princ. Radar* iii. 26 The related concept of effective absorption area A_a of an aerial.
(*e*) *Refrigeration*. (See quots.)
1922 GLAZEBROOK *Dict. Appl. Physics* I. 689/2 Absorption Machines..act by the absorption of one substance by another, to form a solution or compound, and the subsequent separation of the constituents by the agency of heat. **1949** *Gloss. Terms Refrigeration* (*B.S.I.*) 4 *Absorption system*, a system in which the variation of solubility with temperature or pressure is employed to produce refrigeration.
6. *Comb.* **absorption band,** 'a region of darkness produced in the spectrum of white light which has passed through an absorbing medium' (*Brit. Med. Dict.* 1961) (cf. *spectrum-band*); **absorption factor** *Optics* and *Photometry* (see quot. 1940); **absorption lines,** the dark lines of an absorption spectrum (cf. *spectrum-line*, and *FRAUNHOFER); **absorption spectrum** (see quot. 1940).
1867 Absorption-band [in Dict., sense 4]. **1905** GOODCHILD & TWENEY *Technol. & Sci. Dict.* xi. 698/2 An examination of the transmitted light in the spectroscope will show dark bands in its spectrum corresponding to the wave length of the light absorbed by the substance. The bands are called *absorption bands*, and their nature, position, and extent..serve to identify the substance. **1932** *Gloss. Terms Illum. & Photom.* (*B.S.I.*), Absorption factor. **1940** *Chambers's Techn. Dict.* 3/2 *Absorption factor*, the ratio of the difference between the total luminous flux falling on a surface and the sum of the fluxes transmitted through and reflected from the surface, to the total luminous flux falling on the surface. **1889** *Cent. Dict.* (s.v. *absorption*), (*caption*) Part of Solar Spectrum, showing Absorption-lines. **1928** D. BRUNT *Meteorol.* v. 39 This effect is shown in the spectrum by..a number of narrow dark lines running across the spectrum in definitely fixed positions, marking the absence of the light absorbed. Each element in the absorbing medium produces its own set of dark absorption lines. **1879** HARTLEY & HUNTINGTON in *Proc. R. Soc.* 235 In order to ascertain whether isomeric bodies exhibited similar or identical absorption spectra a series of benzene derivatives was examined. **1940** *Chambers's Tech. Dict.* 4/1 *Absorption spectrum*, the system of absorption bands, or lines, seen when a selectively absorbing substance is placed between a source of white light and a spectroscope. **1950** *Sci. News* XV. 89 Oxyhæmoglobin and oxychlorocruorin have each their characteristic absorption spectrum.

absquatulation (æbskwǫtiulēɪ·ʃən). [f. ABSQUATULATE *v.*: see -ATION.] The action of 'absquatulating' or decamping.
1847 H. N. MOORE *Fitzgerald & Hopkins* 164 Artaxerxes and Euphrosyne, after the absquatulation of their pet daughter with the Irish nobleman incog.,..were completely confounded. *a* **1884** M. PATTISON *Mem.* (1885) vi. 213 M[anuel] Johnson jocularly proposed to write the history of Absquatulation. **1901** 'LINESMAN' *Words by Eyewitness* (1902) 248 The enclosing and utter absquatulation of the commando.

abstentionist. Also *attrib.* (passing into *adj.*).
1927 *Glasgow Herald* 10 Aug. 10/2 The exhilarating risks of a forward, as distinct from a cautious and abstentionist, policy. **1955** *Times* 14 July 6/4 The break-away 'abstentionist' group of the Argentine Conservative..Party has issued a declaration demanding the resignation of President Perón.

abstinence. Add: **3.** *Pol. Econ.* The practice of abstaining from expenditure in order to accumulate capital.
[**1848** MILL *Pol. Econ.* I. I. v. 83 Suppose that every capitalist came to be of opinion that not being more meritorious than a well-conducted labourer, he ought not to fare better; and accordingly laid by, from conscientious motives, the surplus of his profits; or suppose this abstinence not spontaneous, but imposed by law or opinion upon all capitalists.] **1867** J. LAING *Theory of Business* ii. 25 Cost.. is resolvable into prior outlay, on account of labour, and 'abstinence'. **1890** A. MARSHALL *Princ. Economics* IV. vii. I. 289 That sacrifice of present pleasure for the sake of future, which is the chief cause of the accumulation of wealth, has been called abstinence by economists. **1899** J. B. CLARK *Distrib. Wealth* ix. 126 Abstinence is nothing more than electing to take out income in the form of wealth-creating goods, instead of that of pleasure-giving goods.

abstract. Add: **A.** *pple.* and *adj.* **4. d.** In the fine arts, characterized by lack of or freedom from representational qualities.
1915 *Forum* (N.Y.) Dec. 665 This painter no doubt has tried to be significantly abstract. *Ibid.* 670 Dore shows an uninteresting abstract canvas. **1921** A. HUXLEY *Crome Yellow* xii. 116 His work..[is] frightfully abstract now—frightfully abstract and frightfully intellectual. **1929** C. DAY LEWIS *Transitional Poem* I. 16 The intellectual Quixotes of the age Prattling of abstract art. **1948** H. READ *Art Now* 134 In practice we call 'abstract' all works of art which, though they may start from the artist's awareness of an object in the external world, proceed to make a self-consistent and independent aesthetic unity in no sense relying on an objective equivalence. **1948** R. O. DUNLOP *Understanding Pictures* iv. 42 The pure abstract picture in which all representation of objects, all extraneous subject-matter, was finally eliminated and the canvas contained only shapes, spaces, colours.
e. Of other art forms.
1877 G. B. SHAW *How to become Musical Critic* (1960) 14 Detailed programs seem to be a complete mistake. They may impart a certain interest to a composition for those who are incapable of appreciating abstract music. **1890** —— [see *ABSOLUTE *music*]. **1957** *Listener* 19 Dec. 1041/2 In poetry, as in painting, it is perhaps the abstract which is nowadays the easiest to do passably... Mr. Roy Fuller's

competence and self-confidence as an abstract poet in the Auden manner are unquestionable. **1958** *Oxford Mail* 23 Aug. 6/5 The music is thin, the ballet abstract in form and inventive in choreography. **1962** *Times* 15 Jan. 14/5 Of all the common errors to do with ballet, none is now more common, or more erroneous, than that implied by the comparatively recent term 'abstract ballet'.
6. *Comb.* **abstract-concrete** *a.*, concerning both abstract and concrete aspects; **abstract expressionism** (orig. *U.S.*) (cf. sense *A. 4 d) = *ACTION *painting*; hence *abstract expressionist* a. and sb.; **abstract impressionism** (orig. *U.S.*), a form of painting which combines the characteristics of abstract art and impressionism; hence *abstract impressionist* a. and sb.
1874 J. FISKE *Cosmic Philos.* I. I. viii. 215 Molar physics, molecular physics, and chemistry, dealing with abstract laws of motion and force that are gained from experience of concrete phenomena, and appealing at every step to the concrete processes of observation and experiment, may be distinguished as abstract-concrete sciences. **1959** *Listener* 2 July 26/2 The international idioms of pure Abstract, Abstract-Concrete, Tâchiste, Automatist, and all the rest. **1964** *Language* XL. 249 The opposition 'abstract-concrete' as it is customarily used to characterize aphasic speech. **1952** *Amer. Artist* Apr. 26 (*title*) From abstract expressionism to new objectivity. **1957** B. S. MYERS *Art & Civiliz.* 656 The New York school of Abstract Expressionism led by such artists as Jackson Pollock (1912–1956) and Willem de Kooning (b. 1904). **1958** *Observer* 3 Aug. 11/7 William Johnstone's method has an increasing air of abstract-expressionist improvisation. **1957** D. COOPER in *Monet* (Edin. Festival Catal.) 6 We..find the last works of Monet described as 'abstract impressionism', a label which serves the convenient purpose of immediately linking them with that tendency in contemporary painting for which the term 'abstract expressionism' has already been coined. **1958** L. ALLOWAY in *Abstract Impressionism* (Arts Council G.B.) 4/1 The term Abstract Impressionism seems to have been coined by Elaine De Kooning in 1951 at the Arts Club, 8th Street, New York. **1958** *Observer* 15 June 15 Joan Mitchell's ..painting..might indeed be called 'abstract impressionist'. **1958** *Listener* 9 Oct. 570/3 Humphrey Spender comes out as an accomplished abstract impressionist.
B. *sb.* **2.** (Further examples.)
1927 [see *ABSTRACTOR]. **1959** L. M. HARROD *Librar. Gloss.* (ed. 2) 12 *Abstract*. 1. A form of current bibliography in which contributions to periodicals are summarized... When published in periodical form they are known as journals of abstracts. 2. The individual entry. **1962** *Lancet* 19 May 1068/1 Have you ever tried doing abstracts? I once did—for about a year. It was the American articles that caused me the most anguish.
4. A work of art in the abstract style (see sense *A. 4 d).
1950 in WEBSTER Add. **1958** 'N. BLAKE' *Penknife in Heart* ii. 30 The great mainsail curved up into the darkness, like a sculptured abstract. **1959** *House & Garden* July 72 A whole lot of modern designs..included mechanically produced abstracts.

abstractable (æ̆bstræ·ktăb'l), *a.* Also **-ible.** [f. ABSTRACT *v.* + -ABLE.] Capable of being abstracted, in the senses of the verb.
1893 W. JAMES in *Mind* II. 509 There must be *some* things whose resemblance is *not* based on such discernible and abstractable identity. **1943** A. M. FARRER *Finite & Infinite* viii. 94 The only common properly abstractible form is that of mere numerability. **1958** D. J. FURLEY in A. D. Booth et al. *Aspects of Transl.* 64 A poem has meaning in very many different ways..some may be abstractable only at the cost of others.

abstraction. Add: **8.** In the fine arts, the practice or state of freedom from representational qualities; a work of art with these characteristics.
1915 *Forum* (N.Y.) Dec. 662 Sheeler..fears to take the final leap into abstraction, not feeling sufficiently sure of his desires. **1921** A. HUXLEY *Crome Yellow* xii. 117 Soon, he says, there'll be just the blank canvas. That's the logical conclusion. Complete abstraction. **1948** R. O. DUNLOP *Understanding Pictures* iv. 44 Cubism was a half-way house on the road to pure abstraction. **1954** WYNDHAM LEWIS *Demon of Progress in Arts* I. vii. 29 It is usually those of very little talent who furnish the little crowds of people painting empty abstractions.

abstractional. (Earlier and later examples.)
1865 MILL *Comte* 10 Instead of the Theological we should prefer to speak of the Personal..; instead of Metaphysical, the Abstractional or Ontological. **1907** W. JAMES *Meaning of Truth* (1909) vi. 153 Reduced to a bare abstractional scheme. **1942** *Mind* LI. 241 Spinoza's constant use of '*quatenus*' rather than '*qua*' is no mere stylism, but is closely bound up with his rejection of the abstractional view of knowledge.

abstractionism (æ̆bstræ·kʃəniz'm). [f. ABSTRACTION + -ISM.] **1.** The pursuit or cult of abstractions.
1909 W. JAMES *Meaning of Truth* xiv. 284 Abstractionism of the worst sort dogs Mr. Russell in his own trials to tell positively what the word 'truth' means. **1914** F. H. BRADLEY *Ess. Truth & Reality* 143 Prof. James, in *Pragmatism*..insists..on taking the crime in its abstraction as absolutely real. And then he goes on (Hegel would have smiled) to denounce 'abstractionism'.
2. *spec.* The cult or practice of abstract art.
1926 *Lit. Digest* 30 Oct. 28/2 Expressionismus, expressionism..abstractionism, are being tried. **1937** H. READ *Art & Society* ii. 49 Generally a high degree of naturalism excludes a co-existent abstractionism. **1952** *Lit. Guide* Apr. 60/1 Ours is the age of atomic energy..of psycho-analysis and abstractionism in the arts.

abstractionist. Substitute 1844 quot. for 1876 quot. in Dict., and add earlier and later examples. Also as *adj.*
1841 *Kendall's Expositor* I. 373 There are two sorts of Abstractionists in our country...The Virginia Abstractionist is one who recognizes the obligations of oaths, the duty of obeying the Constitution [etc.]...The Kentucky Abstractionist is one who looks chiefly to abstracting money from the people's pockets for the purpose of building up a Nobility. **1844** EMERSON *Ess.* Ser. II. viii. 158 She [*sc.* Nature] punishes abstractionists. **1930** *Time & Tide* 24 May 670 Shall I sum up my criticism of Lady Rhondda's attitude ..by calling it too 'abstractionist'? **1945** A. A. LUCE *Berkeley's Immaterialism* II. 32 Abstractionists are inveterately disposed to think that matter exists.
2. One who practises or advocates the principles of abstract art.
1923 *Internat. Studio* June 229/2 The extremes are held by those who advocate 'innocence of the eye' and the 'pure abstractionists'. **1934** *Burlington Mag.* Aug. 94/1 'The Evolution of Abstract Art' is an intelligent statement by one of the most uncompromising of the younger abstractionists. **1950** A. HUXLEY *Themes & Variations* iv. 184 The pure abstractionist will come forward with a question. Seeing that the non-representational passages in representational works are so expressive, why should anyone bother with representation?

abstractor. Delete † *Obs.* and add: One who makes abstracts; *spec.* as the title of a grade of clerks in the Civil Service. Also *attrib.*
The office of abstractor arose from the requirements of Section 6 of the Births and Deaths Registration Act of 1836, according to which the Registrar General was to prepare 'a general abstract of the number of births, deaths and marriages'. Before 1855 the members of the staff at Somerset House engaged on this work were designated 'task-workers'. From 1890 onwards the term has been applied to a new class of 'assistant clerks' in various departments. The spelling *abstracter* gave place to *abstractor* between 1859 and 1866.
1855 (Aug.) Statistical abstracter [appointed at the General Register Office, Somerset House]. **1897** *Westm. Gaz.* 3 Sept. 1/2 The old writers or copyists are to be done away with; and in their stead an army of 'abstractors' substituted. **1901** *Daily Chron.* 5 Oct. 6/5 Two abstractor clerks in the War Office. **1927** BALBI *Deafness Expl.* title-p., Appointed by the Institution of Electrical Engineers as Abstractor to Science Abstracts in Electro Acoustics. **1964** *Times Lit. Suppl.* 2 Jan. 16/1 (Advt.), [We] require an abstractor..for our new Research and Development Centre at Harlow.

‖ **abstractum** (æbstræ·ktv̆m). *Philos.* Usu. in pl. **-ta.** [mod.L., neut. of L. *abstractus* ABSTRACT *ppl. a.*] = ABSTRACT *sb.* 3.
[**1620** BACON *Instauratio Magna* II. bk. I. § 51, p. 63 Intellectus humanus fertur ad abstracta propter naturam propriam.] **1865** [see ABSTRACT *sb.* 3]. **1865** J. H. STIRLING *Secret of Hegel* II. III. II. 136 Such despairing contemplation is a result of our occupying only the *abstractum* of the Ansichseyn. **1868** N. PORTER *Hum. Intell.* IV. viii. 650 The infinite, etc., may stand for the infinitude, the unconditionedness, the absoluteness of some being—*i.e.*, as an *abstractum* or property of a being. *Ibid.* 651 If they [*sc.* the terms] are used only in the sense of *abstracta*, then the question..is, Can they *be conceived* by the mind? **1932** G. D. HICKS *Berkeley* 159 The *abstracta* in question failed to satisfy this test. **1942** *Mind* LI. 247 The common man would probably hold that the abstracta in all or at any rate in most of these senses are somehow or other 'unreal'.

abstrict (æ̆bstri·kt), *v.* *Biol.* [f. L. *ab* off + *strict-*, ppl. stem of *stringĕre* to bind.] *trans.* and *intr.* To separate by constriction: see next.

abstriction. Add: **2.** *Biol.* Separation by constriction.
1893 TUCKEY *Amphioxus* 41 Part of the embryo..was abstricted by the egg membrane. *Ibid.* 141 An abstriction of the fold formation of the dorsal wall. *Ibid.* 149 They both abstrict completely from the alimentary canal. **1895** OLIVER tr. *Kerner's Nat. Hist. Plants* II. 20 Spores which arise.. by abstriction and abjunction..the effect..is as though the end of the sac had been tied off or abstricted.

abstrusity. (Earlier and later examples.)
1632 H. R[EYNOLDS] *Mythomystes* 42 Those secreter Mysteries, and abstrusities of most high diuinity. **1839** M. TUPPER *Proverb. Philos.* (ed. 3) 285 All will again be for abstrusity. **1925** DREISER *Amer. Trag.* (1926) II. III. xiii. 179 There were criminal lawyers deeply versed in the abstrusities and tricks of the criminal law.

absurd. B. *sb.* Revived as a rendering of Fr. *l'absurde* (Camus).
1954 H. READ *Anarchy & Order* 13 He [Albert Camus] suggested a philosophy of the absurd, and his subsequent work..has been an affirmation of 'absurdism' in politics and ethics, as well as in metaphysics. **1962** *Listener* 13 Dec. 1027/1 The theatre of the absurd, whose master remains Camus.

absu·rdism. [ad. Fr. *absurdisme* (Camus).] (See quots.) Hence **absu·rdist** *a.*
1954 [see *ABSURD B. *sb.*]. **1959** *N. & Q.* CCIV. 380/1 The absurdist dilemma. **1960** *20th Cent.* Apr. 321 The absurdist ethic does not recommend either..virtue or.. crime.

abulia, abulic: see *ABOULIA.

abundance. Add: **5.** In solo whist, a declaration by a player that he will take nine tricks single-handed. Also in Fr. form *abondance*.
1888 WILKS & PARDON *How to Play Solo Whist* 8 *Abondance*. (Abundance.) An independent call to make nine tricks. **1890** R. F. GREEN *Solo Whist* i. 3 *Abondance*, in which case the declaring player engages to make nine out of the thirteen tricks. **1911** *Encycl. Brit.* XXV. 362/1 Abundance in the turn-up suit takes precedence over abundance in other suits. **1961** MERVYN JONES *Potbank* xiii. 53, I went solo and was..criticized for my timidity. 'You could have gone abundance with that lot.'
6. *Ecology*. (See quots.)
1905 F. E. CLEMENTS *Research Methods in Ecol.* 314 *Abundance*, the total number of individuals in an area. **1932** FULLER & CONARD tr. *Braun-Blanquet's Plant Sociol.* iii. 30 Abundance is intended to express the plentifulness (number of individuals) of each species. **1957** P. GREIG-SMITH *Quant. Plant Ecol.* iii. 61 Abundance (defined as mean density within occupied quadrats).
7. *Physics*. (See quots.)
1922 F. W. ASTON *Isotopes* ix. 111 In discussing the nuclear structure of elements the question of their relative abundance in nature is one of great interest. **1955** *Gloss. Terms Radiology* (*B.S.I.*) 13 *Relative abundance*, the proportion of a given isotope in a particular specimen of an element, usually expressed as a percentage. **1962** *Nature* 19 May 621/2 Although the masses of individual nuclides that constitute an element can be measured very accurately, their abundances in an element cannot be determined.

abundancy. Delete †*Obs.* and add later examples.
1893 *Illustr. Lond. News* 7 Jan. 22/2 The abundancy of detail that folk-lore delights in. **1959** *New Scientist* 19 Mar. 625/2 Relative abundancies of barley mutants.

abura (ăbᵘ·ŭ·rā). [Yoruba.] A tree of tropical West Africa, *Mitragyna ciliata*, from which a soft pale wood is obtained; also, the wood of this tree.
1926 *Imp. Inst. Bull.* XXIV. 2 *Abura*..A fairly soft, fine and close-grained wood, of moderate weight and plain appearance, with light reddish-brown heartwood and slightly paler sapwood. *Ibid.* 3 Abura wood should be useful locally for light constructional and other work. **1954** *Archit. Rev.* May 332/2 To relieve the large areas of plastic sheeting internally, there is wall panelling in places in abura and iroba.

a-burton: see *BURTON b.

abusable, *a.* **2.** (Later examples.)
1826 *Edin. Rev.* Sept. 445 What an..abusable power is this to confide to any body of men! **1892** *Sat. Rev.* 4 June 641/1 The venerable but abusable principle of grievance first, Supply afterwards. **1904** G. B. SHAW *Commonsense Munic. Trading* ix. 86 Readjusting the burden of the rates by an obviously abusable method.

abusage. Revived in:
1942 E. PARTRIDGE (*title*) Usage and Abusage. A guide to good English.

abusefully, *adv.* Delete †*Obs.* and add later examples. *dial.* and *colloq.*
1875 W. D. PARISH *Dict. Sussex Dial.* 12 He treated her most abusefully, and threw abroad all her shop-goods. **1914** KIPLING *Divers. of Creatures* (1917) 53 Talkin' most abusefully.

abuzz (ăbv·z), *adv.* or *pred. a.* [f. A *prep.*[1] + BUZZ *sb.*[1] and *v.*[1]] In a buzz; filled with buzzing.
1859 DICKENS *T. Two Cities* III. ix, The court was all astir and a-buzz. **1859** GEO. ELIOT *A. Bede* I. xxi, I hate the sound of women's voices; they're always either a-buzz or a-squeak. **1926** J. G. KERR in *Rep. Brit. Assoc. 1926* 111 The whole air is abuzz with discussions on sex.

Abyssinian (æbisi·niăn), *a.* and *sb.* Also 7 **Abissian,** 7–8 **Abassin,** 7–9 **Abyssin(e).** [f. *Abyssinia*, the name of a country in East Africa (now officially called Ethiopia) + -IAN.] **A.** *adj.* Of or pertaining to Abyssinia, its Christian church, its inhabitants or their language. Also *transf.*
1638 W. CHILLINGWORTH *Relig. Protestants* Answ. 2. 105 This therefore we deny both to your and all other Churches of any one denomination, as the Greek, the Roman, the Abyssine. *a* **1666** J. EVELYN *Diary* 18 Jan. an. 1645 (1955) II. 301 Divers China, Meerian, Samaritan, Abyssin & other Oriental books. **1686** [W. WAKE] *Def. of Exposition of Doctr. Ch. Eng.* 78 The Grecian, Armenian, Abassine Churches..have..differed from the Church of Rome. **1781** GIBBON *Decl. & F.* III. xxxiii. 351 Their names are honourably inscribed in the Roman, the Habyssinian, and the Russian calendar. *a* **1806** WORDSWORTH *Prel.* VI. 592 Como,..a darling bosom'd up In Abyssinian privacy. **1816** COLERIDGE *Kubla Khan* in *Christabel* etc. p. 57 It was an Abyssinian maid And on her dulcimer she play'd. **1928**

E. Sitwell *Five Poems* 16 Rich trees And Abyssinian glooms have fostered these.

b. Abyssinian cat, a breed of domestic cat having long ears and short brown hair ticked with grey; also *ellipt.*; **Abyssinian gold** = Talmi.

1876 W. G. Stables *Cats,* (*caption*) Abissinian..brought from Abissinia at the conclusion of the war. **1893** J. Jennings *Domestic or Fancy Cats* ii. 16 In size, the Abyssinian resembles the self-coloured English cat. **1959** *Chambers's Encycl.* III. 168/2 Abyssinians..are speckled black and grey, differing from tabbies in having no pattern on the body...There is no ground for supposing that this breed originated in the country after which it is named. **1890** Abyssinian gold [see Talmi]. **1933** E. A. Smith *Working Prec. Met.* xx. 381 Copper Zinc Alloys..Abyssinian Gold.

B. *sb.* **a.** An inhabitant of Abyssinia. **b.** A member of the Abyssinian Christian church.

1621 T. Mun *Discourse Trade* 16 The great quantitie of gold & some siluer..yearely brought thither from the Abissians countrie in Ethiopia. **1671** A. Woodhead *Consid. Counc. Trent* xvi. 294 Ethiopians, or Abyssines, agreeing in this Point with the Roman. **1735** S. Johnson tr. *Lobo's Voy. to Abyssinia* Pref. p. ix, He neither exaggerates overmuch the Merits of the Jesuits..nor aggravates the Vices of the Abyssins. **1737** R. Challoner *Cath. Chr. Instr.* viii. 109 'Tis the Practice..of the Cophts or Egyptians, and of the Abassins or Ethiopians, who all use in their Liturgies their ancient Languages. **1753** E. Chambers *Cycl.* Suppl., The Abyssinians are a branch of the Copts, or Jacobites; with whom they agree in admitting only one nature in Jesus Christ, and rejecting the council of Chalcedon. **1845** J. C. Prichard *Nat. Hist. Man* (ed. 2) xi. 103 Hairs of a Negro..and of some Abyssinians..were..viewed both as transparent and opaque bodies. **1936** *Discovery* June 170/1 The Abyssinians—essentially Hamitic in origin—are nowadays much mixed with Semites and negroes. **1957** *Encycl. Brit.* I. 73/1 Although the country is now officially described as Ethiopia, the terms 'Abyssinia' and 'Abyssinians' have been retained as being more properly descriptive of the land and people whether from the ethnological, historical or geographical points of view.

academe. Add: **b.** = Academic B. 2 a.

1938 *Mod. Lang. Rev.* XXXIII. 560 The principal secular cultural influence at work in the seventeenth century is the late, rather jaded Humanism of a bureaucracy, and its bearers are not courtiers, but academes and officials. **1955** *Times Lit. Suppl.* 9 Dec. 733/3 Here..are values put upon things in human life by defence chiefs, airmen, soldiers, security officers, academes, scientists.

academic, *a.* and *sb.* Add: **A.** *adj.* **2. b.** *academic freedom*, the freedom of a teacher to state his opinions openly without censorship, or without the fear of losing his position, etc. (cf. G. *akademische Freiheit*); see also quot. 1963.

1901 *World's Work* July 920/2 Every right-thinking man will stand firmly for academic freedom of thought. **1930** *Jrnl. Abnormal Psychol.* XXV. 156 Academic Freedom... University instructors should have the greatest possible freedom in discussing their opinions with their students. **1963** *Times* 9 Mar. 8/4 The Chancellor of the University of Natal..unveiled a plaque..to commemorate 'the death of academic freedom' through the imposition of racial segregation in South African universities.

4. Not leading to a decision; unpractical; theoretical, formal, or conventional.

1886 *Times* 31 Mar. 7/2 This discussion partook of an academic character, for it was well understood that, whatever the result of the discussions might be, no practical step would be taken in the present Parliament. **1888** H. James in *Scribner's Mag.* IV. 73/1 Mr. Wendover asked her if she liked English society and if it was superior to American..; she thought his question 'academic'—the term she used to see applied in the *Times* to certain speeches in Parliament. **1897** D. G. Hogarth *Philip & Alex. of Macedon* ii. 85 Since the references..to the Olynthian war are in the last degree meagre and vague, and those to Philip merely general, the Olynthiacs would possess for the historian only an academic interest. **1901** C. Eliot in *Foreign Office Confid. Print Ser., E. Africa* LXVIII. 12 ..of Mount Elgon..the frontier should proceed in a straight line..but at present the point seems to be of purely academic interest, as we are not likely to extend our effective Administration to this district. **1929** H. G. Wells *King who was King* vi. § 2. 198 All this discussion, Sirs, is—academic. The war has begun already. **1931** *Economist* 21 Nov. 961/2 The results of the elections can only be of academic interest, as there were no Opposition candidates. **1957** *Times* 19 Nov. 11/2 If Russia's rockets can do what Mr. Khrushchev claims they can the blocking of American ports would scarcely matter.

5. Conforming too rigidly to the principles (in painting, etc.) of an academy; excessively formal.

1889 *Cent. Dict.* (s.v. *academic*), *Figure of academic proportions*, in painting, a figure of a little less than half the natural size, such as it is the custom for pupils to draw from the antique and from life. hence, an *academic figure, composition*, etc., is one which appears conventional or unspontaneous, and smacks of practice-work or adherence to formulas and traditions. **1934** Haskell *Balletomania* xv. 310 It did not take long for the new movement to become solidly and immovably academic. **1941** *Manchester Guard. Wkly.* 17 Jan. 51 As an artist he was never too revolutionary to be easily understood, yet never academic enough to be dull. **1961** *Times* 22 Mar. 16/1 The figure-studies by Puvis are complacent and academic in the worst sense.

B. *sb.* **2. a.** (Later examples.)

1838 *Fraser's Mag.* XVII. 468 He annoyed tutors, proctors, *et hoc genus omne*;..but was, on the whole, a not indecorous young academic. **1955** J. Wain *Interpretations* p. xv, The contributors are mainly either academics—men who draw salaries from Universities—..or..members of the 'literary world'.

academicalism (ækăde·mikăliz'm). [f. Academical *a.* + -ism.] Academical style (in a derogatory sense).

1890 *Athenæum* 14 May 640/1 The execution is marred by conventional coldness and obsolete academicalism.

academician. 1. (Earlier and later examples.)

1748 B. Franklin *Let.* 7 Nov. in *Wks.* (1840) VII. 40 If you have the journal of the French Academicians to Lapland, I should be glad to see it. **1963** *Oxf. Univ. Gaz.* 30 May 1337/1 The Degree of Doctor of Letters, *honoris causa*,..will be conferred upon Academician Mikhail Pavlovich Alexeyev.

academicism. Add: **2.** = *academism 2.

1887 *Century Mag.* Nov. 30 The inroad of academicism and all the subsequent degradation of art. **1905** W. James *Mem. & Stud.* (1911) v. 84 He denounced me for the musty and mouldy and generally ignoble academicism of my character. **1948** Boase in *Jrnl. Warburg & Courtauld Inst.* X. 99 It is dry academicism at its worst, but was admired at the time.

academism. Transfer † *Obs. rare* to sense 1, and add: **2.** The state or quality of being academic (see esp. *academic *a.* 5).

1926 R. Fry *Transformations* 114 The academism of the 'Mannerists', whose ideal consisted in the exaggeration of the manner of Michelangelo and Raphael. **1947** *Horizon* Jan. 19 The atmosphere of academism and late romanticism. **1953** *Ballet Ann.* 1954 63 Thus was born Academic Dancing; and soon the term 'academism' became synonymous with 'routine'.

academy. Add: **7.** *Academy award*: an award of the Academy of Motion Picture Arts and Sciences (Hollywood, U.S.A.) for success in a field connected with cinematographic entertainment; *academy blue* (see quot. 1926).

1941 B. Schulberg *What makes Sammy Run?* x. 190, I know we're going to knock them for a row of Academy Awards. **1950** *Amer. Speech* XXV. 3 *Johnny Belinda*, an Academy award motion picture of 1948. **1926** A. S. Jennings *Paint & Colour Mixing* (ed. 7) xxviii. 301 *Academy Blue*, a mixture of French ultramarine and viridian, ground only in oil and used by artists.

acantho-. Add: **acantho·dian** (-ǫᵘ·d-), earlier also **-ean**, *a.* and *sb.*, (of) a small spiny-finned, shark-like fossil fish of the genus *Acanthodes* found esp. in rocks of the Devonian period.

1852 D. T. Ansted in *Man. Geogr. Sci.* I. xii. 380 The Acanthodians and Dipterians (two families of Ganoids, nearly allied to the Lepidoids). **1861** P. de M. G. Egerton in *Geol. Survey U.K., Memoirs, Organic Remains* Dec. X. 67 The scales..resemble those of the other Acanthodean fishes. **1869** J. Powrie in *Trans. Edin. Geol. Soc.* 286 None of our Old Sandstone Acanthodeans have yet been found of large size. **1894** R. H. Traquair in *Geol. Mag.* Dec. IV. I. 256 We may also, in the series of Acanthodian genera, trace every gradation from the most to the least claviculoid shape. **1937** *Phil. Trans. R. Soc.* B. CCXXVIII. 58 The whole fish is covered with the customary square Acanthodian scales.

‖a capella, a cappella. [It. *cappella* chapel.] = Alla Capella; also see quots.

1876 Stainer & Barrett *Dict. Mus. Terms.* **1901** *Daily News* 3 Jan. 6/3 'The 100th Psalm', which is written for eight voices—a capella style. **1938** *Oxf. Compan. Mus.* 1/1 *A cappella*..This refers to the period up to and including the sixteenth century when church music was written for unaccompanied voices...Hence nowadays often used of choral music as a synonym for 'unaccompanied'. **1940** A. Einstein *Mus. Romantic Era* iv. 36 A succession of delicate and refined *a cappella* composers, such as Mendelssohn and Brahms.

acapnia (ăkæ·pniă, æ-). *Path.* [mod.L., f. Gr. ἄκαπνος without smoke, f. ἀ- priv. + καπνός smoke; see -ia¹. Cf. F. *acapnie*.] Diminution or deficiency of carbon dioxide in the blood.

1908 *Amer. Jrnl. Physiol.* XXI. p. xv, The condition of acapnia lowers the tonus of the peripheral blood vessels. **1911** *Encycl. Brit.* XXIII. 199/1 Professor Angelo Mosso was led..to attribute mountain sickness to lack of carbon dioxide, a condition which he designated by the word 'acapnia'. **1913** *Lancet* 23 Aug. 557/2 Many of the deaths under anæsthetics are due to the condition termed acapnia, that is a depreciation of the carbon dioxide tension in the blood.

acarian (ăkē³·riăn), *a.* [f. Gr. ἄκαρι or mod.L. Acarus + -(i)an.] Pertaining to, caused by, or of the nature of an acarus or mite.

1877 W. T. Fox *Atlas Skin Dis.* 16 The absence of acarian furrows and interdigital vesiculations. **1902** *Encycl. Brit.* XXV. 197 Sheep-scab, a loathsome skin disease due to an acarian parasite.

acaricide. Add: (Later example.)

1947 *Nature* 4 Jan. 32/1 The use of a highly refined petroleum oil for application to orchard trees as an insecticide and acaricide is firmly established as a valuable pest-control treatment with citrus.

Hence **acarici·dal** *a.*

1946 *Ann. Reg.* 1945 349 Only the 'gamma' isomer..is actively insecticidal and possesses in addition acaricidal properties.

acarid (æ·kărid). [ad. mod.L. Acaridæ.] An arachnid of the family *Acaridæ*; a mite. So **acari·dian**.

1875 *Encycl. Brit.* III. 275 Acaridians..are to be found under stones, dead leaves, [etc.]. **1881** A. Leslie tr. *Nordenskiöld's Voy. Vega* I. iii. 147 Arachnids, acarids, and podurids occur most plentifully [in the Arctic regions]. **1910** E. Ray Lankester *Sci. fr. Easy Chair* xiv. 317 Red-spider is a small mite or acarid.

acarine. Delete *Path.*, and add: **B.** *sb.* = *acarid.

1891 *Athenæum* 19 Dec. 837/2 Association..between certain acarines of the family Gamasidæ and certain species of ants.

acaroid, see *accaroid.

acarologist (ækărǫ·lŏdʒist). [f. *acaro-*, used as comb. form of mod.L. Acarus + -ologist.] One who studies or treats of the *Acari*.

1890 *Proc. Zool. Soc.* 416 Almost all the members of the genus [Damæus] have a..globular abdomen, or else one which is discoidal, the latter being considered a separate genus by some Acarologists. **1902** *Ann. & Mag. Nat. Hist.* IX. 311 The sense in which acarologists use the genus *Oribata*.

acarophilous (ækărǫ·filəs), *a. Bot.* [f. as prec. + -philous, after *entomophilous* adj.] Applied to plants that are fertilized by the agency of mites. So **acarophily** (-ǫ·fili), acarophilous character.

1898 *Nature* 3 Nov. 15/1 A number of cases of acarophily among ferns.

acataleptic. Add: **B.** *sb.* An adherent of the doctrine of acatalepsy.

1878 C. P. Krauth *Vocab. Philos. Sci.* 11 All sceptics and Pyrrhonians were called acataleptics.

acathisia, var. *akathisia.

acatholic (ækæ·þǫlik, ē-), *a.* and *sb.* [A- 14.] Non-Catholic. Hence **acatho·licism**.

1809 F. Plowden *Hist. Irel.* I. 120 Not a single writer..foreign or native, catholic or acatholic. *c* **1814** W. Poynter *Let.* in B. Ward *Eve Cath. Emanc.* (1911) II. xxiii. 124 Their call..was surely as valid..as that of..a-Catholic as well as Catholic. **1820** J. Milner *Suppl. Mem. Eng. Cath.* ii. 232 To prevent the virtual choice of a Catholic Bishop by an A-Catholic Ministry. **1891** A. J. Harrison *Prob. Christianity & Scepticism* xiv. 269 If Christians were never enamoured of Acatholic or even Anticatholic interpretations, the Freethinker would have little success. *Ibid.* xv. 272 (*heading*) Acatholicism. **1902** *Encycl. Brit.* XXX. 525/1 The fourth provincial synod of Westminster, which legislated on 'acatholic' universities. **1934** F. Rolfe *Desire & Pursuit of Whole* 142 It is possible..for acatholics to pray with catholics catholically.

Accadian, var. *Akkadian.

accaroid, acaroid (æ·kăroid). Also **a·ccroid**. [Etym. unkn.; appar. not related to Acaroid *a*.] The name given to a resinous gum obtained from various Australian trees, esp. the blackboy or grass-tree, used for preparing varnish, paper size, etc. Usu. *attrib.*, as *accaroid gum, resin*, etc.

1875 *Spons' Encycl.* The blackboy and Botany Bay gum yield red and yellow varieties respectively. **1896** Dunglison *Dict. Med. Sci.* (ed. 12) 987/1 *Xanthorrhœa*, Grass Tree. A genus of trees in Australia...Two resins are obtained from them, one the yellow resin of *Xanthorrhœa* or of New Holland, Botany Bay resin, Acaroid resin or gum. **1933** R. S. Morrell et al. *Synthet. Resins* viii. 307 *Gum Accroides*. This resin possesses various names, acaroid resin, grass tree gum, black boy gum, etc. There are two varieties, red and yellow. The red variety..dissolves in methylated spirit and gives varnishes which possess high..resistance to oil. **1959** *Chambers's Encycl.* II. 84/1 Xanthorrhœa balsams are the accaroid resins from *Xanthorrhœa* species, particularly *X. australis*... The accaroids are varnish resins.

accede, *v.* Add:

3. (Only in pa. pple.) Of an office or dignity: to assign, award.

1818 Coleridge *Friend* I. xii. 134 A most puissant military chieftain of low birth, who will have acceded to him a fellowship with the other Sovereigns of the earth. **1875** A. Trollope *Prime Min.* I. i. 5 And this most precious rank was acceded to him.

accelerando. Add: (Examples.) Also as *sb.*: a gradual increase of speed or a passage where this occurs.

1842 J. F. Warner *Dict. Mus. Terms* 5/1 *Accelerando*, hastening, moving faster and faster, increasing the quickness of the time. **1889** G. B. Shaw *London Music 1888–9* (1937) 110 It [*sc.* an orchestra]..made pauses and ritardandos and accelerandos in the most natural manner. **1938** *Oxf. Compan. Mus.* 816/1 It is difficult to find performances where there are no rallentandos and accelerandos over longer stretches, superposed on the tempo rubato of the shorter ones. **1962** *Listener* 31 May 969/3 Emphasis is placed on the growth and decay of phrases by accelerandi or ritardandi.

accelerate, *v.* Add: **1. b.** To increase the speed of (a railway train, motor-car, motor-engine, etc.); also *absol.* (Cf. sense 3).

1888 *Boy's Own Paper* (Summer No.) 43/2 He made..imperceptible changes in the speed of the train, accelerated it or diminished it so steadily. **1902** Harmsworth et al. *Motors* xv. 328 To accelerate a little when wishing to change speed successfully. **1919** B. H. Davies *Motor Driving* 52 The expansion of cushions of compressed air accelerates them on their downward strokes. **1924** *Motor* 21 Oct. 528/1 Everybody accelerating all they knew to reach the corner first.

c. *Physics.* To impart high velocities to charged particles (as electrons) by electrical or magnetic means.

1931 *Physical Rev.* XXXVIII. 2021 The electric field between tubes is always in a direction to accelerate the ions as they pass from..one tube to..the next.

accelerated, *ppl. a. Physics.* (Additional examples.)

1903 *Proc. Physical Soc.* XVIII. 598 The radiating power of an accelerated electric charge. **1942** Strachan *Particles* vi. 195 On the classical theory of radiation from an accelerated electron it is found..the energy distribution should approach zero asymptotically at short wave lengths.

accelerating, *vbl. sb.* Add: Also *attrib.*, as *accelerating chamber, field, tube* (see *accelerate *v.* 1 c).

1932 *Physical Rev.* XL. 23 The method for the multiple acceleration of ions to high speeds..is so arranged that as an ion travels from the interior of one tube to the interior of the next there is always an accelerating field. *Ibid.* XLII. 901 Charged atoms are accelerated..at each passage through an accelerating chamber. **1942** Strachan *Particles* xi. 431 The accelerating tube, including a source of ions at one end and the target to be bombarded at the other, must have sufficient length to prevent spark-over along the outside of the tube.

acceleration. Add: **1.** Also *spec.* of charged particles (cf. *accelerate *v.* 1 c).

1909 H. A. Lorentz *Theory of Electrons* 120 The radiation from a single electron..is determined by the acceleration. **1932** [see prec.]. **1942** Strachan *Particles* vi. 194 ..negative linear accelerations suffered by the electrons as they strike the target.

b. The process of increasing the speed of a motor-engine or -vehicle; hence, capacity of being accelerated, as an attribute of the vehicle itself.

1901 *Motor-Car World* II. 238/1 Instantaneous acceleration when the clutch is once more engaged. **1926** *Daily News* 18 May 4/5 There are many British cars with fine acceleration.

accelerator. Add: **b.** *Photogr.* A substance used to shorten the duration of development of a negative.

1858 Sutton & Worden *Dict. Photogr.* 3 *Accelerators.* This name is given to any substances used in photographic processes, which have the power to shorten the time of exposure, either in the camera or the printing-frame. *c* **1865** J. Wylde's *Circ. Sci.* I. 157/1 By using the above solution as accelerators, a rich red rose-colour is produced on the plate. **1898** Talbot *Ph. Photogr.* 24 The alkali sets the reducer at work..and the developer may be considered as produced by this accelerator. **1902** Harvey & Perkins *Princ. Optics* xi. 223 The developer usually contains four components: a reducing agent, an accelerator, a preservative, and a restrainer.

c. An apparatus for regulating the speed of the engine in a motor-vehicle, esp. for increasing speed; also *attrib.* as *accelerator pedal*, the pedal that controls the 'throttle'; *accelerator valve* (see quot. 1901).

1901 W. W. Beaumont *Motor Vehicles* viii. 147 The other governor control is known as the accelerator. **1901** *Motor-Car World* Oct. 272/1 A by-pass throttle or 'accelerator' valve, by opening which a full charge of mixture can be admitted at any speed of the engine. **1902** Harmsworth et al. *Motors* vii. 129 If the driver wishes to slow down..he does not necessarily change his gear, but operates the accelerator. **1960** A. H. B. Young *Complete Motorist* ix. 219 The accelerator pedal..has its uses, notably in changing to a higher speed. **1959** *Chambers's Encycl.* IX. 576/2 The driver moves off by depressing the accelerator having first shifted a miniature gear lever into the neutral position. *Ibid.*, The..speeds at which such changes occur are partially dependent upon the use..of the accelerator pedal.

d. A substance used to increase the speed of a chemical or organic process, *spec.* in the vulcanization of rubber or the curing of a plastic.

1908 *Proc. Chem. Soc.* XXIV 132 (*title*) Acids as accelerators in the acetylation of amino-groups. **1912** H. E. Potts *Chem. Rubber Ind.* 52 Calcined magnesia is useful..as an accelerator of vulcanization, particularly for resinous rubbers. **1936** *Amer. Rep.* 1935 III. 56 In controlling the span and speed of life in fruits in store, ethylene is now used as an accelerator of ripening. **1943** R. S. Morrell et al. *Synthet. Resins* (ed. 2) iv. 160 Reduction in curing time of urea-formaldehyde moulding powders can be obtained by using 'latent' accelerators. **1956** *Gloss. Terms Concrete* (*B.S.I.*) 5 *Accelerators*, additives which, when added to concrete during mixing, will appreciably quicken setting and/or hardening.

e. *Physics.* An apparatus which imparts high velocities to free electrons or other atomic particles by electrical or magnetic means, esp. *linear accelerator* (see quot. 1962). Cf. *cyclotron.

1931 *Science* LXXIII. 2030 Focusing occurs along the accelerators and is accomplished quite automatically by the curved electric fields between adjacent cylindrical accelerators. **1935** *Physical Rev.* XLVIII. 495/2 The term 'magnetic resonance accelerator' is suggested...The first word is added to distinguish it from the apparatus of Sloane and Lawrence which can be called a 'linear resonance accelerator'. **1942** Pollard & Davidson *Appl. Nucl. Physics* 55 The simplest type of apparatus utilizing this method is the linear multiple accelerator. **1947** L. W. Alvarez et al. in *Science* CVI. 506/1 (*title*) Initial Performance of a 32-Mev Proton Linear Accelerator. **1953** *Lancet* 17 Oct. 840/1 An 8-million-volt linear accelerator..the first machine of its kind to be built for X-ray therapy. **1958** *Engineering* 7 Feb. 174/3 The accelerator provides a..source of high-energy electrons. **1962** *Gloss. Terms Nucl. Sci.* (*B.S.I.*) 6 Among the most common accelerators..is a linear accelerator, in which the particles move in a straight line, and cyclic, in which they move in a circular or spiral orbit.

accelerograph (ækse·lerǫgrɑf). [f. *accelero-* (see next) + -graph.] An apparatus for recording the succession of pressures developed in a power-chamber by the combustion of a charge; also, an instrument for recording the acceleration of earth-tremors.

1909 in *Cent. Dict. Suppl.* **1950** *Engineering* 10 Feb. 157/1 ... **1956** *Atlantic* 19 Dec. 19/1 A device called the Accelerograph, something like a baby seismograph, is turning this art into a science.

accelerometer (ækselerǫ·mitəɹ). [f. *accelero-* irreg. combining form repr. Accelerate *v.*, etc. + -meter.] An instrument for ascertaining the acceleration of a moving body or for measuring mechanical vibrations.

1904 *Amer. Inventor* 15 July 312 (Cent. Dict. Suppl.), Accelerometer...This instrument consists of two glass vessels connected by a tube, one containing a liquid such as mercury and the other red alcohol. **1910** *Westm. Gaz.* 22 Mar. 5/2 Measuring B.H.P. on the Road, by an accelerometer. **1932** *Discovery* Apr. 114/2 Accelerometers have been carried [in aircraft] to measure the resultant forces experienced in flight. **1946** *Nature* 14 Sept. 365/1 A vibrograph and accelerometer..much used to obtain vibration records on ships at sea.

accent, *sb.* Add: **3.** Without defining word: of a regional English accent (further examples).

1930 H. G. Wells *Autocr. Mr. Parham* II. i. 74 Underbred contradictory people with accents and most preposterous views. **1934** — *Exp. Autobiogr.* II. viii. 522 He spoke with an accent. **1940** D. Abercrombie *Prob. & Princ.* iv. 42 Accent..is a word which, in its popular use, carries a stigma: speaking without an accent is considered preferable to speaking with an accent...The popular, pejorative, use of the word begs an important question by its assumption that an accent is something which is added to, or taken away from..an accepted norm. **1962** *Guardian* They were poor, they had 'accents', the children went to State schools.

8. (Further examples.) Now esp. with *on*: emphasis.

1947 F. Meynell in E. Barker *Char. England* xviii. 389 The design of the components—the type, paper, and binding—will show local accents. Since the difference between good and bad design..may be a question of millimetres in the thickness of a part of a letter..even these slight national accents are of great importance to the bibliophile. **1955** *Times* 5 Nov. 4/3 The accent in exports is likely to shift away from ready-made products to the sulphonates which are necessary for their manufacture. **1958** *Times* 15 July 7/7 Inside the accent is on comfort, with deeply upholstered seats which reduce fatigue to the minimum on long journeys. **1959** *Observer* 29 Mar. 14/8 The accent was on poverty, squalor and ugliness, but the balance was kept with comments on the good old days which never were.

9. *Art.* A touch of colour or light which serves to bring the features of a structure into relief or furnishes a contrast in a scheme of colour.

1849 Ruskin *Seven Lamps Archit.* iii. 79 The Greek workman cared for shadow only as a dark field wherefrom his light figure or design might be intelligibly detached: his attention was concentrated on the one aim at readableness, and clearness of accent. **1888** *Contemp. Rev.* May 712 A few stronger touches, and an accent of light on the neck. **1900** *Westm. Gaz.* 17 Mar. 3/2 A trained eye which discerned at a glance where the accents of a building fell.

10. *attrib.*, as *accent-shift, -shifting*.

1935 D. L. Sayers *Gaudy Night* xviii. 382 How dared he pick up her word 'sleep' and use it four times in as many lines, and each time in a different foot, as though juggling with the accent-shift were child's play? **1940** *Amer. Speech* XV. 200/1 Experimental study of the origin of secondary intonation (accent) in the Slavic languages. Accent-shift has been accompanied by the introduction of strong vowel reduction. **1926** Fowler *Mod. Eng. Usage* 386/1 Words in which accent-shifting is tentative only..'construe' v. (doubtful), co'ntour n.

accented, *ppl. a.* Add: Hence **acce·ntedly** *adv.*

1856 *Titan* July 64 An elderly man..made me a bow; And accentedly said, 'Sir, I have not the pleasure?' *c* **1905** F. Rolfe *N. Crabbe* (1958) xx. 132 He tore off to Crabbe's friend Neddy: thus accentedly putting his fat foot further into everything. **1934** — *Desire & Pursuit of Whole* viii. 70 She was accentedly hearty.

accentor (ækse·ntǫɹ). *Ornith.* [Late L. *accentor* one who sings with another (Isidore), f. L. *ad* to + *cantor* singer, f. *canĕre* to sing.] A genus of passerine singing birds (Bechstein), including the hedge-sparrow or hedge accentor, *A. modularis*; a bird of this genus.

a **1825** [see Hedge 9]. **1890** [see Robin¹ 4 c]. **1909** W. Verner *Life among Wild Birds in Spain* vi. iii. 305 One of the few species met with in these stony wastes was the Alpine Accentor (*Accentor collaris*). **1959** *Edin. Rev.* Jan. 72 The hedge sparrow and golden-crested wren, the former of which is..an accentor, and the latter a regulus. **1954** R. Peterson et al. *Field Guide Birds Brit.* 302 Mountain Accentor. *Prunella montanella*.

b. The golden-crowned thrush or oven-bird, *Siurus auricapillus* of the United States.

1884 Coues *N. Amer. Birds* (ed. 2) 308.

accentualist (ækse·ntiu̯ălist). [f. Accentual *a.* + -ist.] One who advocates a particular theory of accent, *spec.* concerning the interpretation of plainsong (see quot. 1954).

1909 in *Cent. Dict. Suppl.* **1937** *Downside Rev.* LV. 349 Father Robertson allows himself occasionally to enunciate rhythmic principles which run directly counter both to the theory and the practice of the accentualists. **1954** G. Reese *Music in Middle Ages* v. 142 The accentualists..consider the accent the principal—according to some, the only—rhythmical determinant of its melodies. **1954** *Grove's Dict. Mus.* (ed. 5) VI. 810/1 The Accentualists hold that the change from quantity to accent in both Greek and Latin in about the 5th century caused the long and the short to chant to change to notes of equal value and that the principal and secondary accents of particular syllables in particular words..determine the rhythm of that phrase.

accentuator (ækse·ntiu̯ēɪtǫɹ). [agent-n. in L. form f. Accentuate *v.*] One who or that which accentuates; *spec.* (*a*) one who marks words with written accents; (*b*) a device for marking the accent in musical performance.

1876 J. Martin tr. *C. F. Keil's Comm. Ezek.* xxvi. 19–21 I. 382 We should then have to take the clause as independent and affirmative, as the accentuators and the Targum have done. **1923** *Daily Mail* 10 Jan. 2 (Advt.), Autola Player Piano...Automatic Accentuator and Sustaining Pedal.

accept, *v.* Add: **1.** Also, to take or receive with patience or resignation, to tolerate.

1859 Geo. Eliot *Adam Bede* II. xvii. 13 These fellow-mortals, every one, must be accepted as they are. **1862** Wilkie Collins *No Name* I. i. 16 Accept the situation—as the French say. **1934** *Punch* 3 Jan. 6/1 There are not going to be fewer cars, but more, and we must accept them.

b. Of a female animal: to permit (copulation).

1931 Pickard & Crew *Rabbit Breeding* ii. 10 The period of 'season' or 'heat' is the time when a doe accepts the buck. **1958** E. F. Daglish *Dog Owner's Guide* viii. 199 Some bitches will accept the dog earlier, but about the tenth day is usually the ideal.

acceptance. Add: **1. b.** The accepting of copulation by a female animal.

1925 F. A. E. Crew *Animal Genetics* x. 297 The sexual season is divided into..proestrum..œstrum, the phase of desire and acceptance of the male.

6. Also *attrib.*, as *acceptance bank, credit, house*. Cf. *accepting *vbl. sb.* 2.

1923 *Crowell's Dict. Bus. & Finance*, Acceptance Bank. **1927** G. G. Munn *Encycl. Bank. & Finance*, Acceptance Houses. **1948** G. Crowther *Outl. Money* (ed. 2) ii. 70 Usually, the merchant agrees with one of the London banks..to 'accept' the bill for him, and the proceeds can be discounted at a bank or an 'acceptance house' immediately..the bill a 'prime' bill. **1958** *Spectator* 6 June 729/1 Fresh information about the sterling balances, and about acceptance credits, an important method by which international trade is financed in the City of London.

c. *acceptance world*: used by A. Powell in the ordinary *Comm.* sense and also *transf.*, of life 'accepted' as a debt which must eventually be repaid.

1955 A. Powell (*title*) The Acceptance World.

8. The paper or card on which an invitation is accepted.

1873 J. H. Beadle *Undevel. West* xviii. 342 The 'nervous-hysterically Italian hand' resumed its beautiful regularity on pink-tinted 'acceptances'. **1957** R. Carrington *Mermaids & Mastodons* ix. 136 Twenty-one acceptances were received and..the guests arrived at the appointed hour.

9. *Racing.* A horse accepted for a race (freq. in *pl.*)

1923 *Daily Mail* 1 Feb. 9/1 The acceptances..are satisfactory for the 'Lincoln'. **1928** *Ibid.* 25 July 14/5, I expect to find among the acceptances for the Chesterfield Cup at Goodwood. **1946** F. Sargeson *That Summer* 69, I wrapped it in a sheet of newspaper I was keeping because of the acceptances.

accepting, *vbl. sb.* Add: **2.** *attrib.*, as *accepting house*. Cf. prec., sense 6.

1919 R. G. Hawtrey *Currency & Credit* vii. 105 Accepting houses authorise merchants to draw bills, on condition that the necessary funds are provided to meet the bills when due. **1962** *Times* 13 Sept. 11/5 Rothschilds have remained..the only accepting house in the City to have avoided turning itself into a legal company.

acception. For † *Obs.* read *rare* and add later examples.

1. 1921 *Glasgow Herald* 7 Apr. 6/3 Acception of the Premier's proposal followed by its rejection.

3. 1850 E. W. Grinfield *Apol. Septuag.* 179 To use Greek words, in such peculiar acceptions. **1937** *Nature* 20 Nov. 866/2 These initial restrictions on the meaning and acception of experience.

acceptive, *a.* **2.** Delete † *Obs.*, and add:

1883 W. Whitman *Specimen Days* 84 Returning to the naked source-life of us all—to the breast of the great silent savage all-acceptive Mother. **1920** *Edin. Rev.* Jan. 46 Reverently acceptive of every Victorian formula.

Hence **accepti·vity**, the quality or condition of being acceptive.

1920 E. & C. Paul tr. *Baudouin's Suggestion & Autosuggestion* Gloss., *Acceptivity*, the readiness with which the subconscious accepts an idea.

acceptor. Add: **3.** An atom or molecule capable of receiving an electron and so combining with another atom or molecule; *spec.* in a semi-conducting material (see quot. 1950). Cf. *donor 2.

1907 *Chem. Abstr.* 2342 It should be possible to substitute the anode of an electrolytic cell in place of the acceptor in an oxidation process. **1927** N. V. Sidgwick *Electronic Theory of Valency* vii. 116 When a co-ordinate link is formed between two atoms, one of them gives the other a share in two of its own (previously unshared) electrons...We may call the atom which lends the two electrons..the *donor*, and the one which receives them..the *acceptor*. **1940** S. Glasstone *Text-Bk. Physical Chem.* (1941) i. 99 The terms donor and acceptor are often employed to describe the two atoms, or the molecules in which they are present, involved in a dative bond. **1949** *Physical Rev.* LXXV. 865/2 The ionization energy of donors is less than that of acceptors, probably because conduction electrons have a smaller effective mass than holes. **1950** W. Shockley *Electrons & Holes in Semiconductors* (1951) i. 14 Impurities with a valence of five are called 'donor impurities' because they donate an excess electron to the crystal; those with a valence of three are called 'acceptor impurities', since they accept an electron from somewhere else in the crystal to complete the structure of the valence bonds with their neighbors, thus leaving a hole to conduct.

4. An apparatus designed to accept; also *attrib.*, as *acceptor circuit*. Cf. *rejector.

1923 *Harmsworth's Wireless Encycl.* I. 9/2 Any circuit which comprises an inductance and capacity in series, which is in resonance with the frequency applied to it, is said to be an acceptor circuit for that frequency. **1931** *B.B.C. Year-Bk.* 435/1 *Acceptor Circuit*, a tuned oscillatory circuit, having the opposite characteristics of a rejector circuit. **1946** *Electronic Engin.* XVIII. 45 It is better to tune the acceptor well below 25 c/s.

access. Add: **2.** Also *attrib.*, as *access-time*, the time taken to reach 'information' stored in a computer.

1950 W. W. Stifler *High-Speed Computing Devices* xiv. 304 We define the access time of a system as the time required to withdraw a number from storage. **1960** E. Delavenay *Introd. Machine Trans.* ii. 20 Access to data ..is very rapid, the average access time being of the order of a few millionths of a second.

7. a. Also *attrib.*

1943 J. S. Huxley *TVA* 78 (*caption*) Access Road. **1959** John, Duke of Bedford *Silver-Plated Spoon* x. 202 The new access roads were nearly ready. **1962** *Listener* 10 May 800/2 The East German regime's agreement was needed for use of the access routes to the city. *Ibid.* 24 May 902/1 The motor vehicle is demanding completely novel arrangements of buildings and access ways.

acce·ssibleness. [f. Accessible *a.*] Accessibility.

1829 *Examiner* 13 Dec. 793/1 The security against oppression, which his accessibleness gives to his subjects. **1845** J. B. Mozley *Essays* (1878) I. 203 His accessibleness, affability, openness, and all those features of the genuine public character. **1857** W. R. Alger *Genius of America* 15 This..universal accessibleness of honors.

accession. Add: **6. b.** *attrib.*; **accession(s book,** a book in which are entered the accessions to a library: so *accessions list; accession number, stamp*. Hence **acce·ssion** *v. trans.*, to enter in the accessions register of a library (orig. *U.S.*); so **acce·ssioning** *vbl. sb.*

1876 W. F. Poole in *Publ. Libr. U.S.* I. 489 The books must then be entered in the 'accession catalogue'. *Ibid.*, Every work entered has its accession number. **1877** *Library Jrnl.* May 316/1 The accession-book properly kept up is the librarian's official indicator for his whole collection. **1882** Accessions-List 3 Chief older works [Bodl., L.P. Access. lists b. 5]. **1892** G. M. Jones *Salem* (Mass.) *Public Library Rep.* 9 The new books have been promptly accessioned. **1896** *Library Jrnl.* Dec. 129/2 Accessioning, classifying, and cataloguing. **1900** *Library* 1 Mar. 153 The Mitchell Library. Rough Accessions Book. *Ibid.* 154 The books are stamped with an 'accession stamp' on the back of the title-page. **1920** W. C. B. Sayers *Brown's Man. Libr. Econ.* xxxv. 479 Accessioning is done the ordinary way. **1928** *Armstrong Coll. Rep. 1927–8* 73, 5,000 volumes and 3,050 pamphlets have been classified and accessioned but not fully catalogued. **1961** T. Landau *Encycl. Librarianship* (ed. 2) 2/1 The accession number also serves to link the book with the catalogue (or shelf register) and the charging system, and distinguishes between copies of a book when there is more than one.

accessit (ǣkse·sit). [L. *accessit*, 3 sing. pa. t. of *accēdĕre* to approach.] **1.** With reference to French examinations: = Proxime accessit.

1884 *Harper's Mag.* Mar. 597/2 In the competition for the ..prize medal..he gained the *accessit*. **1898** *Daily News* 20 Jan. 8/5 He became a student at the Paris Conservatoire, where, however, he did not take very high honours, gaining only a second accessit in comic opera.

2. A secondary vote given in the election of a Pope: see quots.

1877 *Encycl. Brit.* VI. 240/1 After each scrutiny an 'accessit' takes place; i.e.,..it is open to every voter to declare that he 'accedes' to such or such a candidate. **1882** Schaff *Relig. Encycl.* I. 521 Every morning a ballot is cast, followed in the evening by an 'accessit'.

accessless, *a.* (Later examples.)

a **1750** Aaron Hill *Works* (1753) IV. 296 Th' accessless wilds Of bleak-brow'd mountains. **1852** *Fraser's Mag.* XLV. 452 The corpse of him most dear to me is left, To rot, unwashed, amid accessless rocks.

accessorize (ǣkse·sərəiz), *v.* orig. *U.S.* [f. Accessory *sb.* + -ize.] *trans.* To provide or furnish with an accessory or accessories. Also *intr.* So **acce·ssorizing** *vbl. sb.* (Chiefly in *Advertising*.)

1939 *Amer. Speech* XIV. 316 A hankie to accessorize a costume. **1947** in *Amer. Speech* (1948) XXIII. 69/1 A sign I saw in [a] show-window in Connecticut avenue: Our staff is especially trained to accessorize your home. **1951** *Word Study* Oct. 3/1 Advertising men love to invent words, as in this piece: '..It is a flattering color to almost every man and the easiest to *accessorize*.' **1958** *Vogue* Apr. 142 Cotton in town needs strict accessorizing to avoid any hint of a holiday look. **1961** *Sunday Times* 29 Oct. 37/3 Do accessorise simply, pick plain colours for bag and shoes.

accessory, *a.* and *sb.* Add: **A.** *adj.* **1.** spec. *accessory* (*food*) *factor*.

1912 F. G. Hopkins in *Jrnl. Physiol.* XLIV. 425 (*title*) Feeding Experiments Illustrating the Importance of Accessory Factors in Normal Dietaries. **1919** *Lancet* 23 Aug. 338/1 The kind and degree of accessory factors in the common dietary. *Ibid.*, It would be interesting to discover whether they [*sc.* pickles] fill a gap also by supplying accessory food factors.

b. *accessory nerve*, the eleventh cranial nerve.

1842 Dunglison *Dict. Med. Sci.* (ed. 3) 15/2 Accessory ligament, muscle, nerve, &c. **1858** Gray *Anatomy* 496 The Spinal Accessory Nerve consists of two parts. **1950** R.-M. S. Heffner *Gen. Phonetics* 41 The velum and uvula are controlled, in part at least, by fibers from the accessory, or eleventh cranial nerve.

c. *accessory chromosome* = *sex chromosome* (*sex *sb.*).

1902 C. E. McClung in *Biol. Bull.* III. 43 (*title*) The Accessory Chromosome—Sex Determinant? *Ibid.* 72 My conception of the function exercised by the accessory chromosome is that it is the bearer of..the faculty of producing sex cells.

B. *sb.* **1.** *spec.* (in *pl.*): the smaller articles of (esp. a woman's) dress, as shoes, gloves, etc.; minor fittings or attachments for a motor-car, etc. Occas. in *sing.*

1896 M. Beerbohm *Dandies & Dandies* in *Wks.* 4 The many little golden chains..would have seemed vulgar to Mr. Brummell. For is it not to his fine scorn of accessories that we may trace that first aim of dandyism, the production of the supreme effect through means the least extravagant? **1902** [see *chicken-brooding*, *chicken *sb.*[1] 7]. **1958** *Woman's Own* 5 Nov. 10/1 She was wearing a pastel green suit with white accessories. **1961** *John O' London's* 23 Feb. 212/4 Writing about a woman's 'accessories' (gloves, handbag, umbrella, shoes). **1962** *Observer* 2 Dec. 31/1 *Most important accessories*..Heater..screen-washers..fog-lights ..loose covers..cigar-lighter.

accident, *sb.* Add: **1. d.** *colloq.* An accidental or untimely call of nature.

1899 *Allbutt's Syst. Med.* VIII. 244 The wearing of india-rubber urinals, and other means of avoiding 'accidents'. **1926** *Nation* 9 Jan. 517/2 Then a new child had, as Mabel calls it, 'an accident'. She may have been afraid of asking to go out. **1959** A. Wesker *Roots* II. i. 38 Jimmy Beales give him a real dowsin' down..'cause he had an accident.

6. b. *Textual Criticism.* = *accidental B. *sb.* d.

1942 W. W. Greg *Edit. Probl. Shakes.* p. l, What may conveniently be called the 'accidents' of presentation, namely the spelling, punctuation, and other scribal or typographical details. *Ibid.* p. liii, The only other accident we need consider is line division.

10. *attrib.* and *Comb.*

1866 *Boston Directory* 565/2 The Pioneer Accident Insurance Company in America. **1892** A. Pinero *Magistrate* III, p. 135 Tell Sergeant Lugg to look over the Accident-Book, this morning's Hospital Returns. **1899** *Westm. Gaz.* 2 Mar. 8/3 Last year we paid about 5,000 accident claims. **1899** *Daily Chron.* 10 Aug. 5/7 They were manning the accident boat. **1905** *Ibid.* 19 Sept. 4/4 Life-saving and accident-preventing machinery. **1906** *Ibid.* 15 June 5/5 The accident rate was considered..low. **1918** W. Hutchinson *Doctor in War* (1919) xxii. 231 There seemed to be no even approximately 'accident-proof' type of airman. **1964** *Economist* 4 Jan. 34/2 Motorists..with their accident-potentiality enhanced by alcohol.

b. *accident neurosis*, a neurosis caused or precipitated by an accident; **accident-prone** *a.*, predisposed or likely to cause or attract an accident; also *absol.*; so **accident-proneness**, such predisposition or likelihood.

1901 in Dorland *Med. Dict.* 442/2 Accident neurosis. **1932** J. H. Huddleson *Accidents, Neuroses & Compensation* iv. 59 Accident neuroses make up the most discussed group. **1961** *Listener* 28 Sept. 459/1 The term 'accident neurosis' refers to a disabling complaint of nervous origin: the symptoms are subjective, and there is usually no bodily sign of any emotional disturbance. **1926** Farmer & Chambers *Psychol. Study Accident Rates* 3 'Accident proneness' implies the possession of those qualities which have been found to lead to an undue number of accidents. *Ibid.*, A person can be said to be accident prone without any knowledge of the number of accidents he has sustained. **1951** *Sci. Amer.* Aug. 68/3 A pushed conveyance such as a wheelbarrow..is..less accident-prone than the pulled four-wheel truck. **1952** *Lancet* 28 June 1296/1 Flanders Dunbar made a study of the accident-prone which bears out reasonable expectations. **1954** A. Koestler *Invis. Writing* xv. 167 Their accident-proneness was entirely due to the hazard of a Jewish origin.

accidental. B. *sb.* Add: **d.** *Textual Criticism.* [cf. *adj.* 3.] Applied to any feature that is non-essential to the author's meaning. Cf. *accident *sb.* 6 b.

1942 W. W. Greg *Edit. Probl. Shakes.* p. liv, It is desirable that the preservation of accidentals should be seen in proper perspective. **1959** *N. & Q.* CCIV. 119/2 If no long s was available in his fount, it would surely have been better to have substituted the modern s for this 'accidental' of early script and typography. **1964** *Ibid.* CCIX. 179/1 Such variants as & for *and* throughout suggest that Fischer is more faithful than Clark to accidentals.

accidentally, *adv.* Add: **1. b.** joc. phr., *accidentally on purpose*: with the appearance of an accident although actually on purpose or by pre-arrangement.

1862 Lady Morgan *Mem.* x. 91 Dermody neglected the order—perhaps 'accidentally on purpose'. **1916** R. Frost *Let.* 14 Nov. (1964) 45 My not meeting De la Mare in England was rather accidentally on purpose. **1946** R. Lehmann *Gipsy's Baby* 137 The other one chiming in all wrong every time accidentally on purpose to spoil the effect.

accidented, *ppl. a.* (Additional examples.) (Cf. Accident *sb.* 5.)

1878 R. L. Stevenson *New Arab. Nights* ii. 27 The highest and most accidented of the sandhills immediately adjoined: and from these I could overlook Northmour. **1925** *Blackw. Mag.* May 670/2 Owing to the echoes and the accidented nature of the ground, it was not quite certain where the sounds came from.

‖accidia. [a. med.L. *accidia* (see Accidie).] = Accidie.

1930 *Time & Tide* 28 Mar. 414/2 Half of them suffer from the deadly spiritual disease of accidia. **1947** C. Gray *Contingencies* iii. 89 The combination of medieval *accidia* and modern *weltschmerz*, which we find in his [Liszt's] *Hamlet*.

accidie. Delete † *Obs.* and add later examples.

1891 F. Paget (*title*) The Spirit of Discipline..with an..essay concerning Accidie. **1923** J. M. Murry *Pencillings* 21 The worst attack of accidie..gives way before them. **1936** H. G. Wells *Anat. Frust.* vi. 54 There is nothing before you but sloth and apathy, accidie, which is a lingering suicide.

accidious, *a.* Cf. prec., delete †*Obs.*⁻⁰ and add:

a **1400** *Pore Caitiff* (MS. Harl. 2335) lf. 17, Þe accidious man haþ ydilnesse sleuþe & sleep for his god. **1912** Paget & Crum *Francis Paget* 136 Men who would not go there might..be thought..accidious.

accinge, *v.* Delete † *Obs. rare*⁻¹ and add:

1829 T. L. Peacock *Misfort. Elphin* xiv, This task, to which I have accinged myself, is arduous. **1886** *Oxf. Mag.* 12 May, When Mr. Jesse Collings accinges himself to constructive legislation. **1888** 'Q' *Troy Town* xiv. 166 Peter, instead of adjuring Miss Limpenny to fear no more the heat o' the sun, accinged himself to the practical difficulty.

acclimate, *v.* Add to def.: Now chiefly *U.S.* Add later examples.

1934 F. Scott Fitzgerald *Tender is Night* II. viii. 196 'I'll drop in after dinner,' Dick promised. 'First I must get acclimated.' **1937** S. Cheney *World Hist. Art* (1938) xiii. 400 The Byzantine style, which had been acclimated there during the rule of the Western exarchs.

2. *intr.* = Acclimatize *v.* 2.

1861 Winthrop *C. Dreeme* 174 Until I acclimate to the atmosphere of work. **1864** *Good Words* 228/2 They..acclimated easily, and they prospered.

acclimated, *ppl. a.* (Earlier and later examples.)

1849 in J. R. Commons *Doc. Hist.* (1910) I. 253 The proprietor of several thousand arpents of land..is desirous to dispose of 45 acclimated negroes. **1921** B. Matthews in *S.P.E. Tract* V. 8, I have asked why these thoroughly acclimated French words should not be made to wear our English livery. **1961** *Times* 24 Feb. 3/6 Shivering ceases in the acclimated animal.

acclimation. (Earlier examples.)

1826 Flint *Recoll.* 132 The gradual process of *acclimation*. **1837** T. Bacon *First Imp. Hindostan* I. p. vii, English residents in India imbibe peculiarities in the process of *acclimation*.

accommodate, *v.* Add: **2. b.** *spec.* in *Philol.*

1933 [see *Shin]. **1939** R. W. Chapman *Adjs. fr. Proper Names* 59 *Persien* or *Percien* is in English of the fourteenth century. This in the sixteenth century was accommodated to the regular *-ian*. **1962** E. S. Olszewska in Davis & Wrenn *Eng. & Medieval Stud.* 120 The alliterative formula is

borrowed from a Norse coupling of the two synonyms.. with the second noun accommodated in form to the native cognate.

11. and **12.** Add *absol.* uses.

1881 Mrs. Riddell *Senior Partner* xxxi, A common money-lender willin' to accommodate. **1888** Gunter *Mr. Potter of Texas* xiv, 'Won't you be seated?'..'Certainly! Anything to accommodate!'

accommodation. Add: **1. c.** *spec.* in *Psychol.*

1875 A. Bain *Emotions & Will* (ed. 3) 83 Under a fresh shock this accommodation operates by diminishing the interval of transition. **1895** J. M. Baldwin *Mental Development* vii. 217 Accommodation as it is best to call it in psychology, adaptation in biology. **1901** —— *Dict. Philos. & Psychol.* I. 8/1 As habit is the principle of mental 'conservation of type', so accommodation is the principle of modification of type. **1937** R. M. MacIver *Society* vi. 106 To his complex changeful world man can achieve only a partial adjustment, a compound of conflict and accommodation (by the latter term we mean the process in which the person or the group comes to fit into a given situation and to feel 'at home' within it).

2. b. *Philol.* The action of accommodating or fact of being accommodated (cf. prec., 2 b); adaptation; assimilation.

1871 Kennedy *Pub. Sch. Lat. Gram.* 18 Incomplete Assimilation or Accommodation takes place when the former Consonant is changed, not to the latter, but to one more akin to this in sound. **1925** P. Radin tr. *Vendryès's Language* I. iii. 59 In the first case accommodation takes place, in the second case epenthesis. **1932** Gordon & Onions in *Medium Ævum* I. 127 It is necessary to suppose an English accommodation of the Scandinavian word to account for the forms that the word presents in our dialects.

6. b. *accommodation address*, an address used solely or primarily for convenience of correspondence; freq. one adopted to conceal the whereabouts of the addressee; *accommodation bridge* (see quot. 1954); *accommodation ladder* (earlier examples); *accommodation road*, a road constructed to give access to a property or piece of land not adjoining a public road; *accommodation stage, train*, one stopping at all (or nearly all) the points or stations on the route (*U.S.*).

1894 Littlechild *Reminisc.* vii. 69 One of their number arranged, on some plausible story, for letters to be received for him at an 'accommodation address'—generally a small shop. **1914** 'I. Hay' *Knight on Wheels* I. i. 13 A small greengrocer's shop—an 'accommodation address' of the most ordinary type—whose proprietor admitted that he was in the habit of taking in letters on behalf of some of his customers. **1958** *Economist* 16 Aug. 552/2 A *pied-à-terre* for meetings and an accommodation address and telephone, linked by private line with their main premises. *a* **1809** A. Rees *Cycl.* VI. s.v. *Canal*, For *occupation*, or accommodation bridges..a kind of swing or swivel bridge has.. been adopted. **1954** *Highway Engin. Terms* (*B.S.I.*) 24 *Accommodation bridge*, a bridge serving an accommodation road. **1769** W. Falconer *Dict. Marine* § Z 1 Accommodation-*Ladder*, is a sort of light stair-case, occasionally fixed on the gangway of the admiral..of a fleet. **1839** G. W. M. Reynolds *Pickwick Abroad* i. 7 That faithful domestic descended the accommodation-ladder, and once more stood upon the deck of the vessel. **1881** Accommodation road [see Dict.]. **1909** Kipling *Actions & Reactions* 36 No better than accommodation-roads. **1811** *Columbian Centinel* (Boston) 25 Sept. 3/1 Accomodation stage[-coach]. **1838** *Boston Almanac* 49 Depots on the Providence Rail Road. Accommodation Train. **1843** 'R. Carlton' *New Purchase* 19 So remarkably accommodating were the old-fashioned accommodation stages and stage owners. **1860** O. W. Holmes *Prof. Breakf.-t.* iii. 80 Accomodation train. A good many stops, but will get to the station by and by.

c. *ellipt.* for *accommodation stage, train*. *U.S.*

1829 A. Royall *Pennsylvania* II. 9, I..intended to take the Accommodation in the morning. **1877** 'E. W. Martin' *Hist. Great Riots* 117 The Sharpsville 'accommodation'.. had been lying for two hours without an engine. **1891** C. Roberts *Adrift in America* ii. 33 We went on what is called an 'accommodation', that is, a freight train with a passenger car at the end of it.

7. (Later U.S. examples of *pl.* use.)

1804 G. T. Hopkins *Diary* 24 Feb. in *Maryland Hist. Mag.* (1909) IV. 3 We rode to the house of..a neighbouring farmer, where we found good accomodations. **1931** C. S. Johnson *Negro in Amer. Civ.* xxiv. 338 The wastefulness of exactly dual accommodations [for Negroes and for whites in railway trains].

b. *attrib.* *accommodation house*, a lodging house for travellers (freq. derog.); *accommodation paddock* (Austral. and N.Z.), see quot. 1933; *accommodation unit*, for purposes of official enumeration: a single place of residence.

1823 'Jon Bee' *Slang* 2 *Accommodation-house.* The Reader had better consult 'Fubbs' than we explain the minute difference that exists between these and a 'Dress-house' or a 'Bodikin'. **1857** H. W. Harper *Let. fr. N.Z.* 1 Sept. (1914) 17 There I found what is known as an accommodation house. **1861** Mayhew *London Labour* (Extra vol.) (1862) 249/1 Those who gain their living by keeping accommodation houses..are of course to be placed in the category of the people who are dependant on prostitutes. **1933** *Bulletin* (Sydney) 1 Nov. 21/3 An accommodation-house that has never received its due meed of notoriety. **1933** L. Acland in *Press* (Christchurch, N.Z.) 9 Sept. 15/7 *Accommodation Paddocks*, paddocks kept by publicans and others for the use of travelling stock..unknown, of course, until the roads were fenced in the 'seventies. **1947** C. B. Brereton *No Roll of Drums* xvi. 143, I know your accommodation paddock is eaten out. **1950** D. Maxwell Fyfe in *Hansard Commons* 5th Ser. CDLXXII. 853 One million 'accommodation units' have been built since the war. **1951** *Evening Stand.* 17 Aug. 4/1 The Ministry of Health's classic 'accommodation unit', which Mr. Churchill killed stone-dead by singing the phrase to the tune of Home, Sweet Home.

8. (Earlier examples.)

1790 Hamilton in *Ann. 1st Congress* II. 2058 The accommodations which they might derive in the way of their business, at a low rate. **1803** *Edin. Rev.* II. 102 All accommodation bills are iniquitous. **1807** Scott *Lett.* (1932) I. 411 With respect to accomodations in my opinion we ought to get rid of all that floating balance which.. has hitherto kept us in poverty.

b. *attrib.* as *accommodation note, paper*.

1797 *Deb. Congress* 27 June (1851) 395 Many of such notes were what was called 'accommodation notes'; all acknowledgments of debt, and therefore no proof of wealth. *Ibid.* 28 June 401 Being able to raise money by accommodation notes to pay duties. **1829** Sherwood *Gaz. Georgia* (ed. 2) 75 The Bank to collect the debts due the State, and debtors to be allowed to renew their notes..as persons borrowing money on accommodation paper.

accommodator. Add: **b.** A part-time domestic help. *U.S.*

1948 F. P. Keyes *Dinner at Antoine's* (1949) ii. 32 Tossie was in great demand as an 'accommodator' at débutante parties. **1959** *Encounter* Nov. 43/2 In the evenings, the old 'accommodators' dart about the city, carrying their black uniforms and white aprons in a paper bag. They..go anywhere to cook and serve dinners.

accompaniment. Add: **2.** Hence **accompanime·ntal** *a.*

1917 C. Scott *Philos. of Modernism* ix. 71 There are two ways of treating old airs from an accompanimental point of view. **1959** *Times* 18 May 9/7 There is nothing subordinate..in Schubert's accompanimental patterns.

accomplishment. Add: **5.** *accomplishment quotient* U.S. = *achievement *quotient*.

1924 Bagley & Keith *Introd. Teaching* 192 The accomplishment quotient. **1934**, **1957** [see *achievement 4].

accordion. (Earlier examples.) Add after def.: Now also *piano accordion*, an improved type of accordion, with a piano-keyboard instead of buttons for producing the notes.

1831 *Harmonicon* Mar. 56/1 The Accordion is about nine inches long, four wide, and about the same in depth; not unlike a small tea-caddy in appearance. It is made after the manner of a bellows, and the tone is produced by raising and lowering the upper part, at the same time pressing one of the ivory keys, of which there are five on those commonly used. **1839** *Inventors' Advocate* 17 Aug. 12/2 A daily musical performance on Reisner's Improved Accordion. **1860** Dickens *Uncomm. Trav.* in *All Year Round* 10 Mar. 464/1 It was real talent!..a kind of piano-accordion, played by a young girl. **1938** *Oxf. Compan. Mus.* 788/1 The application of the..piano keyboard (hence the term 'Piano Accordion') was made by Bouton, of Paris, in 1852. **1958** *Listener* 25 Sept. 484/3 The unexpected and ingenious completion of the wind ensemble of the Nonet with a piano-accordion.

b. *attrib.* and *Comb.*, having a series of folds like those of the bellows of an accordion, as *accordion door, pleat, pleating, skirt, wall, window*; *accordion-plaited, -pleated* adjs.

1885 *Catal. Cookery & Food Exhib.* x, The Permanent Accordion Pleating Manufacturing Company. **1888** *Daily News* 25 Oct. 2/6 The sleeves..are made of..nun's-veiling, arranged in accordion pleats. **1895** [see Pleated *ppl. a.*]. **1899** Howells *Ragged Lady* xvii, She had on an accordion skirt. **1905** *Smart Set* Oct. 29/1 A portly front, an accordeon-plaited chin, a thick, oily forehead. **1936** *Times* 18 Jan. 9/5 The skilful use of different planes at the corner accordion windows and fluting on the tower. **1959** 'S. Ransome' *I'll die for You* i. 17 Through the wide-open accordion door.. he saw her. **1959** I. Ross *Image Merchants* (1960) iv. 67 The accordion wall was always open between their two.. offices. **1962** *Times* 22 Jan. 13/3 Accordion-pleated chiffons.

accost, *v.* Add: **7. b.** Of a woman: to solicit in the street for an improper purpose.

1887 *Times* 12 July 11/5 The police-constable said she was in company with a second female, and that they had accosted gentlemen. **1927** *Daily Mail* 29 Sept. 10/1 The Cass case of 40 years ago, in which a young woman of irreproachable character was arrested..and charged..with accosting.

accost, *sb.* (Earlier examples.)

1616 B. Jonson *Cynth. Revels* v. iii, They act their accost severally to the lady that stands forth. **1650** A. B. *Mutatus Polemo* 27 At the first accost there was nothing but dumb shews. **1807** J. Beresford *Miseries* II. xviii. 170 Shooting your affectionate accosts, enquiries, and details, at each other.

accouchement. (Earlier and later examples.)

1803 Eliz. Wynne in *Wynne Diaries* 2 Sept. (1940) III. iv. 89 Mrs. Otway's accouchement which is daily expected. **1955** *Times* 23 Aug. 9/4 Feelings of relief at a safe accouchement.

account, *v.* **4. d.** Add to def.: Also, to beat in a contest.

1928 *Daily Mail* 25 July 14/1 The filly should account for Pure Gem, Falakeh, and company.

account, *sb.* Add: **2. c.** *account day*, on the stock exchange, the last day of the account, on which stock is delivered and paid for and differences are paid; also called *pay day*.

1879 Melsheimer & Laurence *Lond. Stock Exch.* 8 The account days for English and India stocks, &c...are always fixed at least five weeks beforehand. *Ibid.* 18 On the third and last day of the settlement (called the 'account day' or 'pay day') the delivery of securities commences at ten o'clock. **1928** *Morn. Post* 19 Nov., A contract setting out that these 50 shares have been sold for the account November 22—next Thursday, that is—and bought for the following account day, December 6.

4. *on one's own account* (earlier example).

1801 Jane Austen *Let.* 14 Jan. (1952) 109 She..desired me to ask you to purchase for her two bottles of..Lavender Water..provided you should go to the Shop on your own account.

c. *on account; account of*: *ellipt.* for 'on account of (the fact that)'. *slang.*

1936 E. Waugh *Mr. Loveday's Outing* 44 The purser who's different on account he leads a very cynical life. **1936** Wodehouse *Laughing Gas* xi. 117, I was feeling kind of down, on account that tooth of mine was giving me the devil. **1942** *Horizon* July 62 Fred's five foot ten..but I tell him he's still a shrimp, account of I'm so tall. **1948** E. Waugh *Loved One* 52 Take your three days off, Mr. Barlow, only don't expect to be paid for them on account you're thinking up some fancy ideas.

8. d. Delete '*in sporting phr.*' and add earlier and later examples.

1617 F. Moryson *Itinerary* II. II. i. 109 We doe hope to giue her Maiestie a very good account of her Kingdome, and of our selues, vntill wee shall haue cause to sue for more reliefe. **1779** J. Wesley *Jrnl.* 29 Dec. (1786) XIX. 16 We have a musket and a fusee. If you load one, as fast as I discharge the other, I will give a good account of them all. **1809** B. H. Malkin tr. *Lesage's Gil Blas* I. I. ii. 24 Those said eggs of which he had given so good an account. **1883** R. L. Stevenson *Treas. Isl.* xvi. 133 We flattered ourselves we should be able to give a good account of a half-dozen [*sc.* mutineers] at least. **1928** *Daily Mail* 7 Aug. 8/3 They are likely to give a very good account of themselves in the big fight.

9. c. *to hand in one's accounts*, to die. *U.S.*

1873 Aldrich *Marj. Daw* 150 The hotel remains to-day pretty much the same as when Jonathan Bayley handed in his accounts in 1840.

16. phr. *by all accounts*.

1798 Mrs. Inchbald *Lovers' Vows* (ed. 3) II. i, By all accounts the Baroness was very haughty. **1825** J. Neal *Bro. Jonathan* I. i. 10 [She was] the prettiest one, though, 'by all accounts'.

accounting, *vbl. sb.* Add: **2.** Proverbial phr. *there is no accounting for tastes* [cf. L. *de gustibus non est disputandum*].

1823 J. Galt *Entail* xxix. 277 But you know..that there is no accounting for tastes. **1867** A. Trollope *Last Chron. Barset* I. xxxi. 263 He had not the slightest objection to recognizing in Major Grantly a suitor for his cousin's hand... There was..no accounting for tastes. **1903** G. B. Shaw *Man & Superman* III. 101, I don't admire the heavenly temperament..but it takes all sorts to make a universe. There is no accounting for tastes: there are people who like it.

accredit, *v.* Add: **3. b.** To attribute (a thing) *to* a person. *U.S.*

1876 A. Wilder in R. P. Knight *Symb. Lang.* p. xxvii, To the fanatical hordes of Islam..is to be accredited the extinction of the Mystic Orgies of the East. **1900** *N. & Q.* 9th Ser. 22 Dec. 487/1 The introduction of the name [Columbia] as a poetic title for the United States is to be accredited to Dr. Timothy Dwight.

accredited, *ppl. a.* Add: **b.** Used of a grade of milk. (Term disused as official designation since 1954.)

1932 *Min. Agric. & Fish. Pubn.* (*title*) Scheme for the introduction of county registers of accredited milk producers. **1933** *E. Angl. Inst. Agric. Rep. 14th County Clean Milk Comp., 1933* 13 Members of the County Registers of Accredited Milk Producers will be in a very strong position. This Register..was inaugurated in Essex in 1931. **1934** *Milk Marketing Board's Scheme of Accred. Producers* ser. P.B. 3, p. 2 What will the term 'Accredited Milk' stand for? It will be milk from cows that have been clinically tested; from farms where scrupulously clean methods are practised—and itself a product subject to bacteriological tests.

accrescent, *a.* **1.** (Later example.)

1891 *Temple Bar* June 222 Accrescent layers of instruction sandwiched in between patches of narrative.

accrete, *v.* Add: **3. b.** To draw or attract to oneself or itself. Hence **accre·ted** *ppl. a.*

1901 H. B. George *Relat. Geogr. & Hist.* 260 It became the seat of one of the small principalities which happened to accrete other dominions. **1914** G. B. Shaw *Common Sense about War* 11 He, too, accreted fools and knaves, and ended defeated in St. Helena. **1921** *Chambers's Jrnl.* 10 Sept. 648/1 The accreted and reclaimed land.

accretionary (ăkrī·ʃənări), *a.* [f. Accretion + -ary.] Characterized or formed by accretion.

a **1835** J. Phillips *Geol.* in *Encycl. Metrop.* (1845) VI. 674/2 An accretionary rock, formed by the cementation of coralline reliquiæ. **1872** D. Brown *Life John Duncan* 409 The 'real' [body], he says, shrinks at the amputation of a limb, the 'accretionary' part only being cut off. **1932** J. S. Huxley *Prob. Relat. Growth* v. 149 The most familiar examples of organs growing by the accretionary method are shells.

accretive, *a.* (Later examples.)

1852 *Tait's Mag.* XVI. 667 The constitution of the mind is not accretive, but fixed and unalterable. **1889** E. Carpenter *Civilisation* 137 If we take the external view of Variation..modification or race-growth appears as an unconscious or accretive process. **1912** O. Elton *Surv. Eng.*

Lit. I. 249 He likes a complex rather than a merely coordinate or accretive structure.

accroid, var. *ACCAROID.

accrual. Substitute for definition: Accruement; *spec.* in *Law*, = ACCRETION 8 b. Also in *Book-keeping* (see quot. 1949).
1880 [in Dict.]. **1883** ADDISON *Contracts* (ed. 8) II. ii. 430 The subject-matter of the bailment was lost in the lifetime of the devisor, and has not been in the possession of the bailee since the accrual of the title of the devisee. **1913** *Rep. N.Y. & Hudson R. Railroad Co.* 11 The amount of taxes on railroad property..was $462,550.00 larger than the accruals in 1912. **1929** C. I. LEWIS *Mind & World-Order* vi. 194, I can safely predict the accrual of *something* the particular nature of which I cannot now determine. **1949** NORTHCOTT & FORSYTH *Pract. Book-Keeping* vii. 105 An accrual [is]..the proportionate part of a liability that will not fall due for payment until after the Balance Sheet date.
b. *attrib.*, as *accrual basis* (see quot. 1960).
1934 in WEBSTER. **1941** BANGS & HANSELMAN *Accounting for Engineers* II. ix. 135 Natural outgrowths of the fundamental characteristics of keeping business records are two bases of operation, namely, the cash basis and the accrual basis. **1960** NANASSY & SELDEN *Business Dict.* 3 *Accrual basis*, a method of keeping the books of account by which expenses and income are charged to periods to which they are applicable, regardless of when payments for such expenses and income are made.

acculturation (ăkʊltiŭrēⁱ·ʃən). Chiefly *U.S.* [f. AC- *pref.*+CULTURE *sb.*+-ATION.] The adoption and assimilation of an alien culture. Also *transf.* So **accu·ltural** *a.*, involving or produced by acculturation; **accu·lturative** *a.*, involving or producing acculturation; **accu·lture**, cultural elements acquired by acculturation; **accu·lturate, accu·lturize** *vbs. trans.* and *intr.*, to cause to change or to become changed through acculturation.
1880 J. W. POWELL *Study Ind. Lang.* (ed. 2) 46 The force of acculturation under the overwhelming presence of millions has wrought great changes. **1895** *Smithsonian Rep.* 44 The arts and industries of the partially acculturized Papago Indians. **1900** *Rep. Bur. Amer. Ethnol. 1897–8* I. p. xxi, When an invention is accepted and used by others it is accultural. **1904** G. S. HALL *Adolescence* II. 726 There is little acculture [among American Indians]. **1934** WEBSTER, Acculturate, *v.* **1936** *Amer. Anthropologist* XXXVIII. 149 Acculturation comprehends those phenomena which result when groups of individuals having different cultures come into continuous first-hand contact, with subsequent changes in the original culture patterns of either or both groups. *Note*: Under this definition, acculturation is to be distinguished from *culture-change*, of which it is but one aspect, and *assimilation*, which is at times a phase of acculturation. **1950** F. EGGAN *Soc. Org. W. Pueblos* v. 248 Change from a lineage to a 'bilateral' system, under acculturative influences. **1950** *Mind* LIX. 195 A primitive community that is in the process of being acculturated to the West. **1951** R. FIRTH *Elem. Soc. Org.* iii. 81 Terms such as..'acculturation' were introduced to express the way in which new patterns of behaviour or types of relationship were acquired and incorporated into a primitive system. **1959** V. PACKARD *Status Seekers* (1960) xvi. 232 They [*sc.* the private schools] acculturate 'the members of the younger generation..into an upper-class style of life'. **1963** *Time & Tide* Oct. 5/3 Many of the Indians have become a[c]culturised.

accumulated, *ppl. a.* (Earlier example.) Also *accumulated temperature* (see quots.).
1692 C. O'KELLY *Macariæ Excidium* in T. C. Croker *Narr. Contests in Ireland* (1841) 97 His good fortune to have retired before such accumulated misfortunes happened to his country. **1884** *Meteorol. Off. Wkly. Weather Rep.* p. iv, The Accumulated Temperature is expressed in 'Day-degrees',—a Day-degree signifying 1° F. of excess or defect of temperature above or below 42° F. continued for 24 hours, or other number of degrees for an inversely proportional number of hours. **1952** *Jrnl. Ecol.* XL. 376 Most attempts at measuring the effects of temperature on plant growth have involved the summation of daily, weekly or monthly mean temperatures above a certain base-line...Such a use of summed mean temperatures or 'accumulated temperatures' assumes that growth ceases below the base-line employed.

accumulation. Add: **4. b.** *mountain of accumulation*; also *accumulation mountain* (see quot. 1956).
1886 *Sc. Geogr. Mag.* II. 150 Mountains of Accumulation.—Volcanoes may be taken as the type of this class of mountains. **1898** J. GEIKIE *Earth Sculpt.* xvi. 272 Some of these have been piled or heaped up at the surface—they have grown into heights by gradual accumulation, and may therefore be termed *accumulation-mountains*. **1956** J. C. SWAYNE *Conc. Gloss. Geogr. Terms* 9 *Accumulation mountain*, a mountain the growth of which has taken place by accretion from outside, usually due to volcanic, though in some cases to epigenic, action.

accumulator. Add: **3.** (Earlier examples.) In various specific uses: see quots. In a computing machine: that part which collects and compounds the units of 'information' that it receives.
1833 *Brit. Pat.* 6357, I claim generally the use of compressed air as an accumulator of power to be made use of when required. **1856** *Engineer* 23 May 284 The cranes are worked by means of water pressure stored up in a cylinder termed an accumulator. **1878** *Engineering* XXVI. 271 (*title*) Hydraulic Accumulator and Pumps. **1883** GLADSTONE & TRIBE *Chem. Secondary Batteries*, p. ix, It is somewhat unfortunate that they have been called 'accumulators' or 'storage batteries'. **1911** *Encycl. Brit.* XXII. 227/2 A hydraulic accumulator ordinarily consists of a hydraulic cylinder and ram, the ram being loaded with sufficient weight to give the pressure required in the hydraulic mains. **1946** *Electronics* Apr. 310/2 The arithmetic elements include 20 accumulators, 1 multiplier and 1 combination divider and square rooter. **1947** D. R. HARTREE *Calculating Machines* 17 An accumulator..has a number of channels for the reception and transmission of numerical information. **1949** *Gloss. Aeronaut. Terms* (*B.S.I.*) II. 15 *Fuel accumulator*, a device for storing fuel, during a portion of the starting cycle, in order to augment the flow momentarily when a predetermined fuel pressure has been reached. **1949** *Gloss. Terms Refrigeration* (*B.S.I.*) 4 *Accumulator*, a liquid refrigerant container in the low-pressure side of the system.
b. *attrib.* and *Comb.*
1883 *Daily News* 10 Sept. 2/1 This installation is by the International Electric Company, and combines seven series of accumulator stations. **1898** *Engineering Mag.* XVI. 164/1 In Europe..accumulator traction has a decidedly better outlook. **1946** *Electronics* Apr. 310/2 (*caption*) Rear view of two of the accumulator racks. **1946** *Nature* 13 July 54/2 Adequate control of accumulator-charging current.
4. In betting: one who carries forward the amount won on one event and stakes it on a subsequent event; a bet of this kind. Also *attrib.*
1889 BARRÈRE & LELAND *Dict. Slang* I. 14/1 *Accumulator* (racing), a person who backs one horse, and then if it wins results (sometimes including original stakes) goes on to some other horse. **1923** *Turf Guardian Yr. Bk.* 12 In accumulators or commissions subject to contingencies, no further investments can be made after an objection is lodged or a dead heat occurs. **1951** E. RICKMAN *Come Racing* xviii. 182 Doubles, trebles and accumulators are popular among those backers who are particularly attracted by the possibility of winning a substantial sum for a small outlay. **1961** *New Statesman* 15 Sept. 336/1 A professional racing man..felt impelled to have an accumulator bet on every race on the day's card.

accusatrix. Delete † *Obs. rare*⁻¹, and add:
1897 MARY KINGSLEY *W. Africa* 25 Confronted with his accusatrix.

ace. Add: **1. c** Substitute: In lawn tennis, badminton, etc.: an unreturnable stroke, esp. a service that the opponent fails to touch; a point thus scored.
1889 W. M. BROWNLEE *Lawn-Tennis* vi. 158 He was equal to scoring two aces against the fair wind. **1933** *Times* 18 Nov. 5/7 Six of his nine aces were given to him by his opponent failing to bring off his drop shots. **1955** *Times* 28 June 2/7 Against this the luckless Drobny simply could not conjure up the old magic of first service aces.
2. b. [After F. *as*.] In the war of 1914–18, an airman who had brought down ten enemy machines; a crack airman.
1917 *Times* (weekly ed.) 14 Sept. 757 Second Lieutenant Lufbery, the 'ace' of the American Lafayette flying Squadron. **1918** E. SIDGWICK *Jamesie* III. 170 Gabriel is what they call an 'ace' here, a great adventurer of the air. **1921** *Punch* 12 Jan. 26/1 Airman; playwright, Empire-builder,..Ace of all the furious aces, slightly bald D'Annunzio! **1940** *War Illustr.* 12 Apr. 366 To be officially recognized as an air 'Ace' a pilot must have ten definitely established victories to his credit.
c. Chiefly *U.S.* A person outstanding in any activity or occupation; also *attrib.* Also (*U.S. slang*), in *pl.*, anything or anyone outstandingly good.
1919 WODEHOUSE *Damsel in Distress* ii. 35 Put it in your diary, Mac, and write it on your cuff, George Bevan's all right. He's an ace. **1931** G. IRWIN *Amer. Tramp & Underw. Slang* 17 *Aces*, anything or anyone considered to be the best or most desirable. **1932** WODEHOUSE *Hot Water* vi. 113 You're aces, boy. **1935** R. S. WOODWORTH *Psychol.* (ed. 10) x. 245 Any occupation in which master workmen, virtuosos, and 'aces' occur. **1936** *New Yorker* 15 Feb. 12/2 We found ourselves..talking with an ace pulp writer. **1937** *Daily Express* 20 Apr. 23/2 London's ace players [in an orchestra]—improvising hot numbers. **1943** *Amer. Speech* XVIII. 72/1 That broad (female) is aces with me. **1961** *Guardian* 30 Sept. 12/6 The ace byliners found their stories on the back page.
4. ace-high *a.*, valued or esteemed highly (orig. in Poker, used of a hand in which the highest card is an ace). *U.S. colloq.*
[**1878** F. H. HART *Sazerac Lying Club* 154 A discussion on the Russo-Turkish war relieved the tedium of 'ace-high' and single pairs.] **1901** S. E. WHITE *Westerners* xxii. 224 Your bull [terrier] wouldn't be ace high. **1906** 'O. HENRY' *Four Million* (1916) 123 From the very first dose he was ace high and everybody else looked like thirty cents to her. **1948** *Time* 12 Jan. 34/3 In one field, at least, the Russians were still ace-high with Americans last week.

ace (ēⁱs), *v.* [f. prec., 1 c.] *trans.* To score an ace against (an opponent); to gain an ace by playing (the ball).
1923 *Glasgow Herald* 9 July 11 [*Tennis*] His breezy attack, in which the desire to 'ace' his adversary at every stroke was the dominating factor. **1927** *Daily Express* 21 May 9/2 Eight times Tilden aced his service ball.

acedia (ăsī·diă). [L.: see ACEDY.] Sloth, torpor, = ACCIDIE: esp. as a condition leading to listlessness and want of interest in life.
1607 R. PARKER *Schol. Disc. agst. Antichrist* II. 74 The ceremonies..offende the ministers and the Pastors...Many of these are brought to an Acedia by them. **1922** W. R. INGE *Outspoken Ess.* 149 The medieval casuists classified acedia, which is just this temper, among the seven deadly sins. **1927** AUDEN & C. DAY LEWIS in *Oxf. Poetry*, p. v, That acedia and unabashed glorification of the subjective so prominent in the world since the Reformation. **1950** A. HUXLEY *Themes & Var.* 202 To *acedia* and confusion, to nightmare and *angst*, to incomprehension and panic bewilderment.

acentric (ăse·ntrik, ēⁱ-), *a. Biol.* [f. A- 14+ *-CENTRIC.] Of a chromosome: having no centromere. Also as *sb.*
1937 C. D. DARLINGTON in *Nature* 30 Oct. 760/2 When the bivalent..attempts to divide, the 'dicentric' chromatid forms a bridge between the daughter nuclei, and the 'acentric' one incapable of movement, is lost. **1939** STURTEVANT & BEADLE *Introd. Genetics* (1940) viii. 128 Acentric and dicentric chromosomes or chromosome fragments are not usually transmitted from one cell generation to another with any regularity. **1949** DARLINGTON & MATHER *Elem. Genetics* vi. 131 Any single crossing-over within this loop gives two new chromatids, one with two centromeres attached to it, a dicentric, the other with none, an acentric.

acephalic (æsĭfæ·lik), *a.* [f. Gr. ἀκέφαλος (see ACEPHALI)+-IC.] Headless, *lit.* and *fig.* So **acephalia** (æsĭfēⁱ·liă), absence of a head.
1656 BLOUNT *Glossogr.*, *Acephalick*.., without head, title, or beginning. *Ibid.* s.v. *Heresie*, The Acephalick Sects of Barcotabas, [etc.]. **1839–47** *Todd's Cycl. Anat.* III. 718/1 A total defect of the brain is found in that state in which the head is wanting (*Acephalia*)...The acephalic state is very frequent. It is always associated with complete or nearly complete absence of the cranial bones. **1924** *Glasgow Herald* 31 Dec., The acephalic creatures, with eyes in their breasts, of whom..St. Augustine declared he had seen a specimen.

acequia (ăsē·kia). *U.S.* Also **azequia, zequia.** [Sp., ad. Arab. *sāqiah*.] A canal for irrigation; an open drain.
1844 GREGG *Commerce Prairies* I. 152 All the *acequias* for the valley..are conveyed from the main stream. **1857** MAYNE REID *War Trail* v, As the mustang sprang over the zequia. **1859** BARTLETT *Dict. Amer.* (ed. 2) s.v., The irrigating ditches used in Texas and New Mexico are called *Acequias*. ..The word is sometimes spelt *azequia* or *zequia*. **1921** *Chambers's Jrnl.* 14 May 382/2 He takes the ointments and flings them into the acequia—the open drain that still traverses some country houses, where the water dissolves and washes away the contents of the pots. **1958** *Amer. Speech* XXXIII. 106 When the Rocky Mountain Project is completed, it will show where..water ditches are called *irrigation ditches, government canals*, or *acequias* ('sequies').

acetal. 1. (Earlier example.)
1853 URE *Dict. Arts* (ed. 4) I. 1 Acetal, is the subacetate of ether, having for its chemical symbol 3AcO + AcO₃. It is a light colourless ethereous liquid.

acetanilide. Add: Used esp. as an analgesic and antipyretic. (Add further examples.)
1897 *Trans. Amer. Pediatr. Soc.* IX. 41, I have collected several cases of acetanilide poisoning in children where the drug was used externally as an antiseptic. **1956** *Lancet* 24 Mar. 339/1 Acetanilide is a dangerous drug because the body de-acetylates part of it into aniline.

acetate. Add: **2.** A synthetic material in the manufacture of which acetic acid is used, esp. *attrib.*, as *acetate rayon*, rayon made from cellulose acetate (see *CELLULOSE B. c), *acetate silk*, etc. Also **acetyl silk*.
1920 C. SALTER tr. *Georgievics' Textile Fibres* 11 Acetate Silks. **1921** *Jrnl. Soc. Dyers & Colourists* XXXVII. 294 Possibly dyed Acetyl Silk yarn could also be used for weaving with cotton. **1925** *Good Housekeeping* Apr. 142/3 Acetate silk..being made out of..cotton or wood-pulp..with acetic acid. **1926** A. L. WYKES *Working Viscose Silk* 34 Acetate rayons burn more slowly than other rayons. **1932** A. HUXLEY *Brave New World* iii. 58 Her jacket was made of bottle-green acetate cloth. **1940** *Chambers's Techn. Dict.* 6/2 *Acetate film*, positive or negative film consisting of emulsion carried on a base of cellulose acetate. **1964** *Which?* Aug. 252/2 *Acetate*, made from cellulose by chemical treatment of wood-pulp or cotton linters...Used for a wide range of woven fabrics such as satins, taffetas, brocades and moirés for dress and furnishing.

acetazolamide (æ:sĭtăzǫ·lămaid). Also **acetazoleamide** (-zōᵘ·l-). [f. ACET-+AZO-+AMIDE.] A drug used as a diuretic, an anticonvulsant, and in the treatment of glaucoma.
1954 T. H. MAREN et al. in *Bull. Johns Hopkins Hosp.* XCV. 200 Acetazoleamide..is a specific inhibitor of the widely distributed animal enzyme, carbonic anhydrase. **1955** *Lancet* 2 Apr. 706/2 Acetazoleamide has theoretical pharmacological qualifications as an oral diuretic. **1961** *Lancet* 29 July 240/1 When admitted she was having seizures...She was immediately given..750 mg. of primidone, 1 g. of acetazolamide. **1962** *Lancet* 28 Apr. 900/2 In such cases [of acute congestive glaucoma] acetazolamide has proved a useful adjunct to intensive miotic therapy.

a:cetoace·tic, *a.* [f. ACETO-+ACETIC *a.*] **acetoacetic acid**, an unstable acid, present in traces in urine and in increased amount in the urine of diabetic patients; prepared from **acetoacetic ester**: a colourless liquid ester, ethyl acetoacetate.
1903 WALKER & MOTT tr. *Holleman's Org. Chem.* I. 291 Acetoacetic acid..is not of much importance, but its ethyl ester, acetoacetic ester, is a very interesting compound. **1956** *Nature* 24 Mar. 545/1 The formation of large amounts of acetoacetic acid CH_3—CO—CH_2—COOH. **1957** *Technology* Aug. 226/2 Another [chapter] on..synthetic reagents includes..acetoacetic and malonic esters.

acetone. Add: (Earlier and later examples.) Now widely used as an organic solvent and in the preparation of chloroform, etc.
1839 URE *Dict. Arts* 14 *Acetone*, the new chemical name of pyro-acetic spirit. **1922** *Chambers's Encycl.* I. 33/2 Acetone..is a solvent for gums and resins. **1957** V. J. KEHOE *Technique Film & T.V. Make-Up* iii. 31 *Acetone*, solvent used for cleaning spirit gum out of net and gauze hair goods and for various sealers. **1958** *Everyman's Encycl.* (ed. 4) 47/2 Acetone (CH_3·CO·CH_3), or Dimethyl Ketone..is miscible with water, alcohol, and ether in all proportions...It is also used in the preparation of..chloroform, iodoform, and viscose rayon.
attrib. **1928** A. B. CALLOW *Food & Health* ii. 21 In the absence of carbohydrates 'acetone bodies' are produced which give rise to biliousness. **1951** *Gloss. Terms Plastics Industry* (*B.S.I.*) 8 *Acetone resin*, a synthetic resin formed by the reaction of acetone with another compound, such as phenol or formaldehyde. **1961** *Brit. Med. Dict.* 12/2 *Acetone bodies*, aliphatic ketones and hydroxy ketones found in the blood and urine of severe diabetics as the result of the incomplete breakdown of fatty and amino acids, a condition known as ketosis.

acetonuria (æ:sĭtoniū̆ᵊ·riă). *Path.* [f. ACETONE+URIA.] An excess of acetone bodies in the urine; ketonuria.
1894 GOULD *Illustr. Dict. Med.* 26/2. **1907** *Practitioner* Oct. 551 The feature of this disorder..is the occurrence of acetonuria during the attack. **1961** *Lancet* 9 Sept. 566/2 Its hypoglycæmic action in diabetes is associated with an increased incidence of acetonuria.

acetophenetidin (æ:sĭto,fĭne·tidin). *Med.* Also **acetphene·tidin.** [f. ACETO-+PHENETIDIN.] A white crystalline compound used as an antipyretic and analgesic; phenacetin.
1910 *Practitioner* Apr. 552 Kephaldol is a yellowish-white powder..containing the following percentage composition: Citric acid 6 per cent., Salicylic acid 33 per cent., Acet-phenetidine 48 per cent. **1943** *Martindale's Extra Pharmacopœia* (ed. 21) II. 321 Tabellæ Acetophenetidini..contain 90 to 110% of the labelled amounts of acetophenetidin. **1959** *Brit. Pharmaceut. Codex* 547 Phenacetin..*Synonyms*: Acetophenetidin; Phenacetinum.

acetophenone (æ:sĭto,fĭnōᵘ·n). *Chem.* [f. ACETO-+PHEN(YL+-ONE.] Methyl phenyl ketone; HYPNONE; used esp. as a constituent of certain synthetic perfumes.
1871 *Jrnl. Chem. Soc.* XXIV. 258 Ethylbenzene alcohol..is obtained..by hydrogenising acetophenone in alcoholic solution. **1886** [see HYPNONE]. **1924** *Times Trade & Engin. Suppl.* 29 Nov. 243/2 Quite a number of changes have taken place among perfumery chemicals..acetophenone..and nerolin..are all cheaper. **1944** J. GRANT *Hackh's Chem. Dict.* (ed. 3) 8/2 *Acetophenone*,..used medicinally as a hypnotic, and also in perfumes (orange blossoms).

acetyl. Add: *acetyl-cellulose*; *acetyl-salicylic acid*; *acetyl silk*: see *ACETATE 2.
1879 WATTS *Dict. Chem.* VI. (Suppl.). 418 Acetyl-cellulose. **1915** P. E. SPIELMANN tr. *von Richter's Org. Chem.* I. 665 Acetyl Cellulose is formed by the action of glacial acetic acid..and a small quantity of concentrated sulphuric acid..on cellulose. **1897** *Jrnl. Chem. Soc.* LXXII. I. 531 Acetyl salicylic acid..is obtained by adding small quantities of ferric chloride (100 grams) to a mixture of salicylic acid (80 grams) and acetic chloride. **1899** Acetyl-salicylic acid [see *ASPIRIN].

acetylation (ăse:tilēⁱ·ʃən). *Chem.* [f. ACETYL+-ATION.] The introduction of one or more acetyl groups into (a compound) by means of a chemical reaction. Hence **acetylate** (ăse·tilēⁱt), *v.*; **ace·tylated, ace·tylating** *ppl. adjs.*
1895 *Jrnl. Chem. Soc.* LXVII. 447 Acetylation of Cellulose. **1908** *Ann. Rep. Chem. Soc.* V. 83 Many experiments on the..acetylation of amino-groups have also been made. **1909** WEBSTER, Acetylate, *v.* **1927** T. WOODHOUSE *Artificial Silk* iv. 31 The raw material is almost invariably cotton which is acetylated by acetic anhydride in the presence of a catalyst, usually sulphuric acid...The cellulose acetate thus formed is precipitated by the addition of water, and freed from its acetylating solution. **1939** *Nature* 29 July 217/1 Modifications in the acetylation of cellulose for the production of acetate rayon include the use of special solvents. **1946** *Nature* 19 Oct. 553/2 Believing that the usual methods of acetylation give degraded products, Wassermann has attempted to acetylate alginic acid with ketene. **1959** *Times Rev. Industry* Sept. 4/1 Non-cellulosic fibres such as..acetylated cellulose (Tricel).

acetylcholine (æ·sĭtil,kǫ·ləin). *Biochem.* [f. ACETYL+CHOLINE.] The acetyl ester of choline, $C_7H_{17}O_3N$, a chemical secreted at the ends of many nerve-fibres.
1906 *Brit. Med. Jrnl.* 22 Dec. 1789/2 Acetyl-choline, the first of this series, is a substance of extraordinary physiological activity. **1933** *Jrnl. Physiol.* LXXIX. 255 The only choline ester which has hitherto been isolated in a chemically pure state from animal tissues is acetylcholine, which was found to be present in large quantities in the spleens of horses and oxen. **1944** *Ann. Reg. 1943* 353 The electrical and chemical theories of transmission of nerve impulses may be reconciled by the discovery that acetylcholine is an essential link in the generation of the electrical changes recorded during activity.

acetylene (now usu. ăse·tilīn). Add: **b.** *attrib.* in *acetylene gas*; hence *acetylene* (*gas*) *lamp*.
1895 *Nation* 19 Dec. 447/2 Acetylene gas. **1897** *U.S. Patent* in W. E. Gibbs *Lighting by Acetylene* (1898) 139 Acetylene gas lamp. **1900** V. B. LEWES *Acetylene* 466 Acetylene lamps for signalling.
c. *ellipt.*
1915 W. S. MAUGHAM *Of Human Bondage* xlii. 198 The sea of faces, half seen in the glare of acetylene.

acetylenic (ăse:tile·nik, -ī·nik), *a. Chem.* [f. ACETYLEN(E+-IC.] Made from, or involving the use of, acetylene.
1915 *Chem. Abstr.* 3221 (*title*) Preparation of acetylenic nitriles and of 1-halogenated acetylene derivatives. **1946** *Nature* 10 Aug. 205/1 (*title*) Acetylenic Ketones. *Ibid.* 14 Dec. 876/2 The fungistatic activity of a considerable number of ethylenic and acetylenic compounds.

Achæan (ăkī·ăn), *a.* and *sb.* Also **Achaian** (ăkəi·ăn), **Achean.** [f. L. *Achæus*, a. Gr. Ἀχαιός, f. Ἀχαία Achæa.] **A.** *adj.* Of or belonging to Achæa, a name of varying application, in Homeric usage applied to Greece generally, later to a district of the northern Peloponnesus. **B.** *sb.* An inhabitant of Achæa (or Greece).
1567 G. TURBERVILLE tr. *Ovid's Her.* III. 71 In all the Achaian soyle. **1607** TOPSELL *Foure-footed Beastes* 126 The Achaian Harts are said to haue their gall in their tailes. *Ibid.* 315 The Achæans had this degre in high estimat. **1676** HOBBES tr. *Homer's Iliad* II. 28 The rest That in the Army of th' Achæans were. *Ibid.* 33 These were the Leaders of th' Achean forces. **1715** POPE *Iliad* II. 834 The Achaians, Myrmidons, Helleneans bear. **1797** *Encycl. Brit.* I. 63 Achaia was..taken for all those countries that joined in the Achæan league, reduced by the Romans to a province. **1833** *Penny Cycl.* I. 82/1 The history of the Achæans forms an inconsiderable part of the general history of Greece till about B.C. 251. **1925** G. MURRAY *Eumenides* p. viii, That glorified Achaean chieftain who was King of gods and men in the ordinary Homeric tradition. **1956** J. FORSDYKE *Greece before Homer* iv. 83 There is not yet any reason beyond the resemblance in the place-name to connect the people of Ahhiyawa with the Achaians of Greece.

Achæmenian (ækĭme·niăn), *a.* and *sb.* [f. L. *Achæmenius*, f. Gr. Ἀχαιμένης Achæmenes, said to have been the ancestor of Cyrus+-IAN.] Also **Achæmenid** (ăkī·mĕnid), *a.* and *sb.* [-ID³.] **A.** *adj.* Of or pertaining to a member of the dynasty that ruled in ancient Persia from the time of Cyrus the Great (d. 529 B.C.) until the death of Darius in 330 B.C. **B.** *sb.* A member of this dynasty.
1717 G. SEWELL et al. tr. *Ovid's Metam.* I. IV. p. 122 The Vogue of Achæmenian Towns obtain'd. **1885** *Encycl. Brit.* XVIII. 565/1 It is far more likely that Anshan was a place in Persis, the proper family seat of the Achæmenians. **1886** *Ibid.* XXI. 849/1 A residence of the Achæmenian kings. **1900** M. L. MCCLURE tr. *Maspero's Passing of Empires* vii. 736 Compare the tombs of the Achæmenids. **1931** *Times Lit. Suppl.* 8 Jan. 18/1 In the Achaemenid period the soil was rocky and muddy with the leavings of Babylonia and Assyria. **1939** A. J. TOYNBEE *Stud. Hist.* VI. 207 The Achaemenid Great King Artaxerxes Ochus. **1956** —— *Hist. Approach Relig.* iv. 51 This Macedonian Greek founder of an Egyptian successor-state to the Achaemenian Empire.

achætous (ăkī·təs), *a.* [f. A- 14+ Gr. χαίτη hair+-OUS.] Having no setæ.
1896 BENHAM in *Cambr. Nat. Hist.* II. 263 The peristomium is achaetous in the adult.

Achaian, var. *ACHÆAN.

achalasia (ækălēⁱ·ziă). *Path.* [mod.L., f. Gr. ἀ- priv.+χάλασις slackening, relaxation, f. χαλᾶν to relax+-IA¹.] Failure (of a muscle) to relax. Hence **achala·sic** *a.*, of an organ, etc., affected by achalasia.
1914 A. F. HERTZ [= A. F. HURST] in *Q. Jrnl. Med.* VIII. 300 Achalasia of the Cardia (so-called..Cardiospasm)...For the word 'achalasia' I am indebted to Sir Cooper Perry, who coined it for me, as it was obvious that the word 'cardiospasm' and the erroneous idea it conveys would never be discarded unless some less cumbersome expression..was devised. **1927** *Lancet* 19 Mar. 618/2 Achalasia of the cardia is a condition in which the cardiac sphincter does not relax..in the act of deglutition. **1939** *Brit. Med. Jrnl.* 23 Dec. 1225/2 Mega-oesophagus (Cardiospasm)..is now regarded as an achalasia, or failure of the cardio-oesophageal sphincter to relax. **1962** *Lancet* 12 May 1008/2 These receptors are present in both circular and longitudinal muscle from the body of the normal and the achalasic œsophagus.

a-cha·tter. [f. A *prep.*¹ 11+CHATTER *v.*] Chattering.
1828 WILSON in *Blackw. Mag.* XXIV. 277 Morning magpie, a-chatter at skreigh of day. **1876** EGAN tr. *Heine's Atta Troll* 89 Shivering and with teeth a-chatter. **1883** *Harper's Mag.* Jan. 166/1 Eyes blinking and teeth a-chatter.

achene. Substitute for definition: A small dry one-seeded fruit which does not open to liberate the seed. (Additional examples.)
1960 W. B. R. LAIDLAW *Guide Brit. Hardwoods* i. 19 Winged achenes are termed *samaras*, as in ash or elm!... The small nutlets of alder and birch are also achenes. **1961** R. W. BUTCHER *Brit. Flora* I. 21 The achene is an indehiscent dry fruit with the single seed loosely enclosed in a dry, horny pericarp.

Acheulian (ăʃiᵘ·ŭ·liăn), *a. Archæol.* Also **Acheulean.** [ad. F. *acheuléen* (G. de Mortillet 1873, in *Comptes Rendus du Congrès Préhist.* 436), f. the name of Saint-*Acheul* (Somme), France.] Of or belonging to the palæolithic period of Europe succeeding the Chellean, represented by the remains found at Saint-Acheul, near Amiens.
[**1894** DAWSON *Meeting-Place Geol. & Hist.* v. 69 Within the former [*sc.* the palæolithic period] he [Mortillet] believes that it is possible to separate different ages...Respectively the Achulienne, Chellienne, Mousterienne, [etc.].] **1909** in *Cent. Dict.* Suppl. **1921** R. A. S. MACALISTER *Text-Bk. Europ. Archæol.* I. ii. 56 Those [*sc.* terms] of the Lower Palaeolithic are the Pre-Chellean, Chellean, and Acheulean. **1939** *Ann. Reg. 1938* 368 The Swanscombe skull suggests that Acheulean man was not distinguishable morphologically from *Homo sapiens*. **1940** *Oxoniensia* V. 161 Two paleolithic hand-axes of Acheulian type have been discovered.

à cheval: see *CHEVAL.

achievement. Add: **4.** *Psychol.* Chiefly *U.S.* Performance in a standardized test (*achievement test*) or tests. So *achievement age*, *quotient* (abbrev. *A.Q.*): see quot. 1921.
1921 *Univ. Illin. Bur. Educ. Res. Bull.* VI. 5 Medians are the mental age norms, which are used as a basis for translating the point scores into achievement ages. *Ibid.*, Provision is made for comparing a pupil's achievement score..with the norm corresponding to his mental age by dividing his achievement age by the standard score for his mental age. This quotient is called the Achievement Quotient. *Ibid.*, The plan consists of establishing for the achievement tests mental age norms which are used to supplement the usual grade norms. **1934** WARREN *Dict. Psychol.* 1/1 *AQ*, abbrev. for accomplishment (or achievement) quotient. **1957** *Encycl. Brit.* XVIII. 673/2 Tests of proficiencies, also called achievement tests, are intended to measure outcomes of systematic education and training in school or occupation toward a conventionally accepted pattern of skill or knowledge...Several subject tests may be combined into an achievement battery for measuring general school proficiency either in point scores or 'achievement ages' and perhaps 'accomplishment quotients' (AA/CA).

Achilles' heel, see *HEEL *sb.*¹

Achilles' tendon. = TENDON *of Achilles.*
1900 DORLAND *Med. Dict.* 674/2.

achkan (æ·tʃkăn). Also **atchkan.** [Hindi *ackan*.] A long coat worn by men in India.
1911 *Encycl. Brit.* XIV. 419/1 Garments for outdoor wear are the..*achkan* or *sherwāni*...The *achkan*..is buttoned straight down the front. Both *angā* and *achkan* reach to a little below the knee. **1939** M. R. GODDEN *Black Narcissus* v. 52 He was plainly dressed in a dove-coloured achkan and white cotton pantaloons like jodhpurs. **1954** J. MASTERS *Bhowani Junct.* iii. 28 A chuprassi in a scarlet achkan. **1959** *20th Cent.* June 602 An old man in a plain atchkan...This was Jawaharlal Nehru.

achlorhydria (æklōᵊɹhəi·driă, ēⁱ-). *Path.* [mod.L., f. Gr. ἀ- priv.+*chlorhydr*(*ic* (CHLOR-²) +-IA.] Lack of hydrochloric acid in the gastric secretions.
1898 *Allbutt's Syst. Med.* V. 529 The secretion is deficient in hydrochloric acid, in many cases it may be actually wanting (achlorhydria). **1907** *Practitioner* Apr. 558 Both the achlorhydria and the neoplastic process might have a common basis in an alteration of the alkalinity of the tissue fluids. **1930** *Q. Jrnl. Med.* XXIII. 175 Bennett proved that a secretion of hydrochloric acid could be evoked in many cases of apparent achlorhydria.

acholia (ækǫ·liă, ēⁱ-). *Path.* [See ACHOLOUS *a.*; -IA.] A deficiency or an absence of bile. (In quot. 1903 *spec.* = cœliac disease.) Hence **acho·lic** *a.* = ACHOLOUS *a.*
1848 DUNGLISON *Dict. Med. Sci.* (ed. 7) 19/1. **1864** W. AITKEN *Pract. Med.* (ed. 3) II. 813 Occasionally death occurs under symptoms of acholia. **1893** *Funk's Stand. Dict.*, Acholic. **1903** W. B. CHEADLE in *Lancet* 30 May 1497/1 An affection of the digestive tract..is that condition which I have ventured to term 'acholia'...Acholia is specially characterised, as the name implies, by the absence of bile. *Ibid.* 1497/2 Acholic stools occur usually in children under five years of age. **1963** E. D. PALMER *Clinical Gastroenterology* (ed. 2) xvii. 580/2 After the common duct has been obstructed for some time, acholia, or failure of bile production, may result.

acholuria (ækǒlⁱūᵊ·riă). *Path.* [mod.L., f. Gr. ἄχολ-ος lacking bile+-URIA.] Absence of bile pigment in the urine. Hence **acholu·ric** *a.*, characterized by acholuria.
1902 *Lancet* 20 Dec. 1708/2 (*title*) Acholuric icterus. We are indebted entirely to French clinicians for the description of *ictère acholurique*—chronic jaundice of slight degree (subicterus) in which there is no bile in the urine. **1905** GOULD *Dict. New Med. Terms* 23/2cho Aluria. **1910** *Lancet* 22 Jan. 254/2 Four cases of congenital acholuric jaundice..in one family are recorded. **1930** *Q. Jrnl. Med.* XXIII. 222 Panton and Valentine describe one severe case of acholuric family jaundice which did not respond to whole liver.

achondrite (ækǫ·ndrəit). *Min.* [f. A- 14+ *CHONDRITE 2.] A meteorite containing no chondrules.
1904 *Proc. Amer. Philos. Soc.* XLIII. 232 *Achondrites*. Stones poor in Nickel-iron, essentially without round chondrules. **1944** C. PALACHE et al. *Dana's Syst. Min.* I. 121 The stony meteorites without chondri, named *achrondrites*, show porphyritic, ophitic, and granular textures resembling those found in some terrestrial rocks. **1957** *Encycl. Brit.* XV. 338/1 The comparatively few chondrule-free stone meteorites are called achondrites.

achondroplasia (ækǫ:ndroplēⁱ·ziă, ēⁱ-). *Path.* [mod.L., f. Gr. ἄχονδρο-ς without cartilage+πλάσ-ις moulding, conformation+-IA¹.] Abnormal formation of cartilage resulting in a form of dwarfism. Hence **achondropla·siac**, one affected with this condition. Also **achondropla·siac, achondropla·sic, achondropla·stic** (prevailing form) *adjs.*, of, or affected with, achondroplasia.
[**1890** BILLINGS *Med. Dict.* I. 14/2 *Achondroplasie*, (F.). Want of normal cartilage development.] **1893** DUNGLISON *Dict. Med. Sci.* (ed. 21) 10/2 *Achondroplasia*, imperfect formation or development of cartilage. **1897** *Allbutt's Syst. Med.* III. 124 *Achondroplasia*, fœtal cretinism..characterised chiefly by extreme and abnormal shortness of the long bones of the limbs. **1902** *Lancet* 5 Apr. 982/1 Achondroplasic workers in metal are represented on the Egyptian monuments, a circumstance which is interesting in connexion with the lameness and deformity of Hephaistos or Vulcan in classical mythology. **1906** *Lancet* 9 June 1598/1 An achondroplasiac dwarf. **1913** DORLAND *Med. Dict.* (ed. 7) 24/1 *Achondroplastic*, pertaining to or affected with achondroplasia. **1919** [see *DWARFISM]. **1934** *Nature* 5 May 691/1 The dwarfism..is not rachitic but achondroplastic. Five, and probably all six, of the representations of dwarfs at Amarna exhibit the inward turning feet due to talipes. **1951** *Lancet* 30 June 1418/1 (*heading*) Struts for Achondroplasiacs. Have you ever wondered what it would be like to be an achondroplastic dwarf?

achy (ēⁱ·ki), *a.* [f. ACHE *sb.*¹+Y¹.] Full of aches; suffering from continuous or recurring pain.
1875 GEO. ELIOT *Let.* 4 Mar. (1909) 44, I will not write more now, because my head is achy. **1878** DISRAELI in Monypenny & Buckle *Life* (1920) VI. 260, I'm too ill and achy to be out later. **1882** W. HARCOURT in Gardiner *Life* (1923) I. 461, I was rather achy last night. **1926** *Chambers's Jrnl.* 31 July 552/2 His throat was rather achy.

achylia (ăkəi·liă). *Path.* [mod.L., f. as ACHYL(OUS *a.*+-IA¹.] Lack of chyle. Also *achylia gastrica*, absence of chyle from the gastric juices.
1894 GOULD *Illustr. Dict. Med.* 28/1 *Achylia*, absence or deficiency of juice, or of chyle. **1909** *Practitioner* Nov. 735 Iodized solutions of potassium iodide suit the stenotic forms of chronic gastritis and cases of total gastric achylia. **1962** *Lancet* 13 Jan. 107/2 Patients with Bilroth-II resection or other forms of achylia gastrica.

acid. Add: **A.** *adj.* **1. b.** *acid drop*, short for *ACIDULATED *drop*: a sweet made of sugar strongly flavoured with tartaric acid. Also *acid tablet* (formerly *acidulated tablet*).
1836 *Mag. Dom. Econ.* July 28 Acid Drops.—Boil one pound of lump-sugar, one cupful of water, and one table spoonful of vinegar till it snaps like glass. **1889** *Pract. Confectioner* 1 May 5/1 Run through the small acid-drop rollers. **1902** *Boys' Realm* 29 Nov. 388/4 He should carry about him a few acid tablets, and slip one of these in his mouth when the desire for smoking is particularly strong upon him.
Also *fig.*
1869 *Porcupine* 17 July 149/3 *Saturday Review* acid-drops may do for babes—strong men want sharper sauce. **1960** *Times* 27 Oct. 15/2 Her best characters are viewed satirically..but, even so, many of her acid-drops are soft-centred.
2. b. Applied to an intense colour.
1916 H. G. WELLS *Mr. Britling sees it Through* I. v. 175 The dark turf at the wayside..became for a moment an acid green as the glare passed. **1923** D. H. LAWRENCE *Birds, Beasts & Flowers* 151 You acid-blue metallic bird. **1959** *Listener* 29 Jan. 209/1 The acid blues and beautifully placed creamy whites are not to be met with outside Russia.
3. b. Derived from an acid by partial exchange of the replaceable hydrogen, as *acid radical, salt* (in Dict.).
1868 J. P. COOKE *Chem. Philos.* (1870) ix. 83 The metallic atoms are basic radicals, while the non-metallic atoms are acid radicals. **1869** [see ANHYDRIDE]. **1870** [see ANION]. **1947** G. THOMSON *Atom* (ed. 3) xiv. 128 When the salt is formed, water is eliminated and takes with it the oxygen, from the metallic oxide, leaving the metal in combination with the residue of the acid, called the Acid Radical.
c. *acid dye*, a dye used in an acid bath for dyeing textiles, etc.; also one used as a stain in certain kinds of microscopical and histological research. So *acid colour*, the dyestuff itself or the resultant colour.
1888 *Jrnl. Soc. Dyers & Colourists* IV. 107 When wool is boiled in a neutral solution of these acid dyes it merely becomes tinted. **1888** A. SANSONE *Dyeing* vii. 140 There is no end to the shades that can now be produced by means of these acid colours. **1902** *Encycl. Brit.* XXVII. 557/1 *Acid Colours*. These dyestuffs are so called because they dye the animal fibres wool and silk in an acid bath; they do not dye cotton...The colouring matters themselves are of an acid character. **1942** G. BOURNE *Cytol. & Cell Physiol.* i. 2 Acid dyes cannot effectively be used supravitally in this way..to show pre-existing structures; but some of them can most profitably be injected into living animals, certain of whose cells will take them up in a characteristic way.
d. Of, pertaining to, or resulting from an *acid process* of steel manufacture in which the converter or furnace is lined with acid refractory, and acid slag is used for refining. (Cf. BASIC *a.* 2 c.)
1903 H. H. CAMPBELL *Manuf. & Prop. Iron & Steel* 8 The growth of the basic Bessemer practice made it necessary to have a distinguishing name for the old way, and it is therefore called the *acid* process. *Ibid.* 23 If acid steel does not follow exactly the same law as basic steel, then it is certain that they are not the same. **1944** *Jrnl. Iron & Steel Inst.* CL. 177 P, Comparing the 18 basic steels with the 30 acid steels, there is practically no difference in the mechanical-test results.
e. *acid value*, a measure of the free acid content of a substance.
1911 *Jrnl. Soc. Chem. Ind.* 15 Apr. 409/1 Acid value of oils and fats. **1912** THORPE *Dict. Appl. Chem.* (ed. 2) III. 767/2 The acid value indicates the number of milligrams of potassium hydroxide required to saturate the free fatty acids in one gram of an oil or fat. **1931** CARLETON ELLIS *Hydrogenation Org. Subst.* lxi. 853 A method of utilizing the acid values of sodium acid sulphate or bisulphate is recommended by Becquevort and Deguide.
B. *sb.* **b.** *acid test*, the testing for gold by means of aquafortis; *fig.* a crucial test.
1892 G. E. GEE *Jeweller's Assistant* 131 The old-fashioned platinum alloy, in imitation of the bright gold alloys of thirty years ago, can hardly be said to have any golden tinge in its appearance. It certainly withstood the acid test very well, and this will be the only point of resemblance worthy of comparison. **1912** L. J. VANCE *Destroying Angel* (1913) xi. 149 Few professional beauties could have stood, as this woman did, the acid test of that mercilessly brilliant morning. **1918** WOODROW WILSON in *Times* 9 Jan. 8/1 The treatment accorded Russia by her sister nations in the months to come will be the acid test of their good will. **1955** *Times* 27 July 8/6 The acid test, however, would come with the Foreign Ministers' meetings in October, when generalities had to be translated into specific agreements.
2. *to put the acid on* (someone), to exert pressure on (a person) for a loan, a favour, etc. Also *to ply, try, the acid. Austral. slang.*
1906 *Bulletin* (Sydney) 15 Mar., Note 'to put the gas on'—a variant of 'to put the acid on',—the latter familiar slang from the mine-assayer's lexicon. **1935** *Ibid.* 3 Apr. 20/1 He was evidently applying the acid to ascertain if my tombstone was as good as ordered. **1938** J. MOSES *Nine Miles from Gundigai* 10 The barber's shop's a witness-box (They ply the acid there). **1945** BAKER *Austral. Lang.* vi. 124 *To put the hard word* (or *acid*) *on* a girl, to make a request to her for favours. **1947** J. MORRISON in *Coast to Coast 1946* 158 They want to shift the ship at seven. That puts the acid on us.
3. *to come the acid* (cf. A. *adj.* 2): to behave in an 'acid' manner, to be unpleasant or offensive, to speak in a caustic or sarcastic manner. *slang.*
1925 FRASER & GIBBONS *Soldier & Sailor Words* 3 *Coming the acid*, stretching the truth; making oneself unpleasant; trying to pass on a duty; exaggerating one's authority. **1936** J. CURTIS *Gilt Kid* vi. 61 If the queer fellow tried to come any acid he would get hit right on the razzo. **1939** H. HODGE *Cab, Sir?* xviii. 259 Any attempt to 'come the acid', so far from frightening the cabman, will probably result in the cabman's giving him a little fatherly advice. **1953** 'H. CECIL' *Nat. Causes* vii. 95 Why come the old acid? Not even a 'sit down, old man —'.
4. In numerous combs. in *Chem.* (see also A. 3), as *acid-base*; **acid-embossing**, embossing (of glass) by treatment with acid; **acid-fast** [*FAST *a.* 1 h] *a.*, resistant to decolorization by an acid when stained: said esp. of bacteria (hence *acid-fastness*); **acid-forming** *a.*, producing or yielding an acid; **acid-free** *a.*, not containing an acid; **acid-proof** [PROOF *a.* 1 b], *a.*, (*a*) impervious to acid; capable of resisting the deleterious action of acid, (*b*) = *ACID-*fast*.
1917 *Chem. Abstr.* 3050 (*caption*) Observations on acid-base equilibrium in the body. **1925** W. MORSE *Appl. Biochem.* 575 Modern biochemistry considers the mineral substances of the food..as participating in the acid-base metabolism of the body. **1937** MCGRATH & FROST *Glass in Arch. & Decoration* 405 The French artist-technicians, Paule and Max Ingrand, have demonstrated more delightfully than anybody else the possibilities of acid-embossing and sand-blasting. **1904** *Med. Ann.* 578 With regard to examination of sputum, it must be remembered that several observers have reported 'acid-fast' bacilli which were apparently non-tubercular. **1909** *Cent. Dict.* Suppl., Acid-fastness. **1925** C. H. BROWNING *Bacteriol.* viii. 175 Ehrlich showed, however, that tubercle bacilli possessed the remarkable property of acid-fastness, that is to say they withstood the decolorising action of strong acids..when once they had been stained. **1869** H. E. ROSCOE *Elem. Chem.* xvii. 175 Oxides may be divided into (1) Basic oxides; (2) Peroxides; (3) Acid-forming oxides. **1914** *Lancet* 28 Mar. 908/1 The air of New York schools was examined for the acid-forming streptococci. **1930** *Paper Terminol.* (*Spalding & Hodge*) 1 *Acid free*, paper free from any acid or other material likely to have deleterious effects. **1934** R. F. INNES *Causes & Prevention Decay in Leather* 1/2 Modern knowledge demands that leather should not only be 'Acid free' but 'Protected' as well. **1868** G. A. SALA *Notes & Sk. Paris Exhib.* xxvii. 387 They [*sc.* safes] are also stated to be acid proof. **1909** *Installation News* iii. 271 [Conduit] protected by an even *and not too thick* coating of acid proof, moisture resisting, flexible enamel. **1913** DORLAND *Med. Dict.* (ed. 7) 32/2 *Acid-proof*, same as acid-fast. **1937** E. J. LABARRE *Dict. Paper* 95/2 Acid proof paper is generally wood Manilla, Kraft or Rope Wrapping which has been treated to resist acids or acid fumes.

acidæmia (æsidī·miă). *Path.* Also **-emia.** [mod. L., f. L. *acid-um* ACID+Gr. αἷμα blood.] A condition of excessive acidity in the blood.
1906 DORLAND *Med. Dict.* (ed. 4) 24/2 *Acidemia*, abnormal acidity of the blood. **1914** *N.Y. Med. Jrnl.* XCIX. 987/2 (*title*) Alcohol acidemia. **1965** [see *ALKALÆMIA].

acidic, *a.* Add: **3.** Of, pertaining to, or of the nature of, an acid. Also *fig.*
1889 in *Cent. Dict.* **1905** *Harmsworth Encycl.* II. 1162/2 Though neutral to litmus, carbon disulphide exhibits some acidic properties, uniting with alkaline sulphides to form sulpho-carbonates. **1961** *John O'London's* 21 Dec. 678/3 Starkly etched portraits of immigrants in New York heat-wave, doom-laden acidic.

acidity. Add: **b.** *fig.*
1711 P. H. *Two Late Parliam.* 130 He reflected on each mistake with unusual acidity. **1823** *Blackw. Mag.* May 567 The augmented acidity and more naked shallowness of Brougham. **1915** W. S. MAUGHAM *Of Human Bondage* lii. 266 The vicar answered with some acidity.

acidize (æ·sidəiz), *v.* [f. ACID *sb.*+-IZE.] To treat with acid, to acidify; *spec.* to apply acid to an oil (or other) well in order to neutralize the lime.
1909 *Cent. Dict.* Suppl., *Acidize*, to treat with an acid; render acid. **1936** B. M. KINGSTON *Acidizing Handbk.* 10 Since 1932, wells in practically every lime field of the United States have been acidized. **1958** *Times* 5 Nov. 16/7 Since yesterday's statement regarding the Puri well a formation..has been acidized and further flow tests taken.

acidophil, -phile (æsi·dǒfil, -fəil), *a.* and *sb.* [f. ACID *sb.*+-O-+-PHIL(E.] **A.** *adj.* Having an affinity for acids, staining readily with an acid dye. (Cf. OXYPHIL *a.*). **B.** *sb.* A cell or histological element readily stained by an acid dye. Hence **aci:dophi·lic, acido·philous** *adjs.*
1900 DORLAND *Med. Dict.* 21/1 *Acidophil* 1. Easily stained by acid dyes. 2. An element or substance that is readily stained with acid dyes. *Acidophilic, Acidophilous*, readily stained with acid dyes. **1927** HENDERSON & GILLESPIE *Text-bk. Psychiatry* IX. 217 Dunlap was unable to demonstrate the occurrence of acidophile staining in the nuclei of nerve cells in any of his cases. **1932** FULLER & CONARD tr. *Braun-Blanquet's Plant Sociol.* xiii. 311 This concentration [of the soil solution] causes a speedy weeding out of the acidophilous species which are found in the first stages on raw siliceous soils and sands. **1951** G. BOURNE *Cytol. & Cell Physiol.* (ed. 2) ix. 400 Acidophil intranuclear inclusions resembling those produced by viruses have been reported in cells under such conditions that there is no reason to assume any relationship with a virus infection. **1958** *Immunology* I. 13 No carbohydrate-containing substance was demonstrable in the acidophils.

acidophilus (æsidǫ·filǒs), *a.* [mod.L., esp. in *Lactobacillus acidophilus*, a species of bacteria causing fermentation in milk; cf. ACID *sb.*, -PHIL.] *acidophilus milk*, milk fermented by any of several bacteria and used therapeutically.
1921 RETTGER & CHEPLIN *Intestinal Flora* iv. 112 Four strains of *B. acidophilus* have been employed for the production of acidophilus milk. **1929** *Lancet* 25 May 1126/2 Each block..contain[s] some 150 billion bacilli..the number usually present in a pint of comparable acidophilus milk.

acidosis (æsidōᵘ·sis). *Path.* [irreg. f. ACID *sb.*+-OSIS.] An acid condition of the blood such as occurs in diabetes.
1900 E. KLEEN *Diabetes Mellitus* vii. 227 The acid diathesis, the acidosis, causes an increase of ammonia in the blood and the urine in cases of severe diabetes. **1905** HEWLETT tr. *Krehl's Clin. Path.* Index, Acidosis, in diabetic coma. **1913** PEMBREY & RITCHIE *Gen. Path.* 690 The term 'acidosis' is applied to the condition in which abnormal quantities of organic acids fail to be oxidized by the tissues.

acidulated, *a.* Add: *acidulated drop* (*tablet*) = *ACID *drop* (*tablet*).
1836 DICKENS *Astley's* in *Sk. Boz* I. 305 Ma, in the openness of her heart, offered the governess an acidulated drop. **1838** *Mag. Dom. Econ.* Oct. 109 Acidulated drops.—Rasp some orange peel...Add..orange juice,..Dry..then drop. **1851** MAYHEW *Lond. Labour* I. 203/2 Barley-sugar and acidulated drops. **1889** *Pract. Confectioner* 1 May 4/1 Acidulated Raspberry Tablets. **1906** *Daily Chron.* 1 Oct. 6/5 He was not an 'acidulated drop curate'.

acidulation (ăsidiŭlēⁱ·ʃən). [f. ACIDULATE *v.*+-TION.] The action or process of acidulating; the state of being acidulated. Also *fig.*
1849 *Jrnl. R. Agric. Soc.* X. 395 The acidulation of the fermenting sap promotes their growth. **1919** L. STRACHEY *Let.* 27 Nov. in *V. Woolf & Strachey* (1956) 85, I have retired, probably permanently, into this acidulation of ice.

acierate (æ·sierēⁱt), *v.* [f. F. *aciérer* (f. *acier* steel)+-ATE³.] *trans.* To convert into steel. So **acieration** (æsierēⁱ·ʃən) [F. *aciération*], conversion into steel.
1866 *Phil. Trans.* CLVI. 439 The inquiry suggests itself whether acieration would not be promoted by alternation of temperature frequently repeated. **1880** *Encycl. Brit.* XIII. 342/1 This prevents the ready access of carbon and carbon oxide to the covered-up part, and hence hinders or entirely prevents acieration thereat. **1887** *Dublin Rev.* July 55 The beautiful mechanical contrivance of Sir Henry Bessemer by which crude iron..is acierated in half an hour. **1900** S. COLVIN in *Brit. Mus. Return* 51 Proof before the plate was acierated.

acinar (æ·sinăɹ), *a. Anat.* [f. ACIN(US+-AR¹.] Of or pertaining to an acinus.
1936 *Nature* 12 Dec. 1000/2 In an albino strain [of mice], the males treated with œstrone showed a more localized acinar development.

ack (æk) and vars., used for *a* in the oral transliteration of code messages and in telephone communications, as in *ack emma*, for *a.m.* = ante meridiem; air mechanic. See *ACK-ACK, *EMMA. In military use replaced by *able* in Dec. 1942.
1898 *Signalling Instructions* (War Office) 86 The letters T, A, B, M,..will be called toc, ak [1904 *Signalling Regs.* ack], beer, emma. **1918** *Signalling Simplified* 11 Special Names of Letters. (Semaphore and Morse.) A = Ack...Note that, in Signalling, these Special Names must *always* be used, i.e. A is always Ack, M is always Emma, and so on. **1917** 'IAN HAY' *Carrying On* vi. 134 He [the Signaller] salutes the rosy dawn as 'Akk Emma', and eventide as 'Pip Emma'. **1927** D. L. SAYERS *Unnatural Death* III. xxiii. 285 Some damned thing at the Yard, I suppose. At three ack emma! **1930** BROPHY & PARTRIDGE *Songs & Slang, 1914–18* 93 *Ack Emma*, Air Mechanic, in the Royal Air Force. Also a.m. = morning. **1934** V. M. YEATES *Winged Victory* 78 The Ak Emma went off in search of food. *a* **1935** T. E. LAWRENCE *Mint* (1955) xxii. 78 We shorten them [*sc.* our ranks] to LAC, AC I, AC II, and speak of ourselves as 'ack-emmas' (the air mechanic of the Great War) or 'urks'.

ack-ack (æ·k,æ·k), *a.* and *sb.* Also **Ack-Ack.** [Redupl. form of *ACK, repr. *A.A. (see *A III).] **A.** *adj.* Anti-aircraft. **B.** *sb.* Anti-aircraft gun, gunfire, regiment, etc.
1939 *Collier's* 2 Dec. 19/1 One unscheduled move and the machine would have been blown from the air by those long, lean guns the British call Aak Aaks, signalers' code for anti-aircraft. **1940** R. WALKER *Flight to Victory* vii. 49 Engine failure due to ack-ack fire. **1943** E. BONE tr. *W. Wassilewska's Rainbow* vi. 105 'There are more guns under the limes...' 'Ack-ack perhaps?' **1944** *Amer. Speech* XIX. 293 Some fliers who, in their conversation, were careful to distinguish between the enemy's flak and our ack-ack. **1944** 'N. DREW' *Amat. Sailor* ix. 267 Occasionally, for extra special shows, an ack-ack cruiser was in attendance. **1945** *Daily Express* 20 Apr. 4/6 Captain Flack A.A.—Captain Flack, of the Ack-Ack, has been transferred to the Royal Engineers.

ackee, akee (æ·kī). [Native name.] The fruit of the tropical sapindaceous tree *Blighia sapida*; the tree itself.
1794 A. BROUGHTON *Hortus Eastensis* 11 The Akee [from] Africa [presented by] Dr. Tho. Clarke, 1778. **1866** *Treas. Bot. Blighia*..consists of only one species, *B. sapida*, which produces the Akee fruit. **1890** H. T. THOMAS *Untrodden Jamaica* 12 The scarlet blots of the ackee. **1916** *Nature* 1 June 286/1 Dr. Scott..concludes that vomiting sickness, so prevalent in that island [*sc.* Jamaica], is due to poisoning by ackee fruit, Blighia sapida. **1956** J. HEARNE *Stranger at Gate* xix. 151 He chewed the..salty, shredded fish and smooth, yellow gobbets of ackee.

acknowledge, *v.* Add: **2.** *acknowledge the corn* (U.S.): see *CORN *sb.*¹ 7.
5. To show recognition of (see quot. 1881).
1881 OGILVIE (Annandale), *Acknowledge*, to show recognition by some act, as by a bow, nod, smile, lifting the hat, &c., as a mark of friendship or respect; to salute; as, she met him in the street, but barely *acknowledged* him. **1886** 'MAXWELL GRAY' *Silence Dean Maitland* 163 He acknowledged this compliment with a slight bow. **1888** MRS. H. WARD *R. Elsmere* xx, He thought his greeting was acknowledged. *Ibid.* xxvi, Robert smiled slightly, acknowledged the bow, but did not speak.

acmite, var. AKMITE. Now the usual form.
1837–80 [see AKMITE]. **1910** *Encycl. Brit.* I. 149/1 *Acmite*, or Aegirite, a mineral of the pyroxene group...It was first discovered..in the pegmatite veins of the granite..near Kongsberg in Norway, and was named by F. Stromeyer in 1821.

acneiform (æ·kni,ifǫɹm), *a.* Also erron. **acne-form.** [f. ACNE+-(I)FORM.] Of the nature of acne.
1877 W. T. FOX *Atlas Skin Dis.* 25 The acneiform spots do not make their appearance until the disease has been some time in existence. **1884** *Lancet* 31 May 978/2 A Dermatitis taking an acneform character.

acnestis (æknɪ·stis). [mod.L., a. Gr. ἀκνηστις spine, backbone.] That part of the back between the shoulder-blade and the loins which an animal cannot reach to scratch.
1807 in *Edin. Med. & Phys. Dict.* II. Suppl. **1927** *Observer* 3 Apr., That spot known to crossword solvers as the acnestis.

acnode. Substitute for def.: = *conjugate point* (s.v. CONJUGATE *a.* 6 a). Add later example.
1961 C. C. T. BAKER *Dict. Math.* 3 *Acnode*, an isolated point or conjugate point of a curve. A point in whose neighbourhood there is no other point of the curve.

acœlomate (æsī·lomet, æsīlōᵘ·-), *a. Zool.* [f. A- 14+CŒLOMATE *a.*] Having no cœlome or body-cavity; not cœlomate. Also **acœlo·matous** *a.*
1879 tr. *Haeckel's Evol. of Man* II. xxv. 404 Although these acœlomatous Worms have no body-cavity, no blood, no vascular system, they always have a kidney system. **1889** *Cent. Dict.*, Acœlomate. **1921** J. A. THOMSON *Outl. Zool.* (ed. 7) ix. 137 The Cœlentera..form a very large series of Acœlomate Metazoa.

Acol (æ·kǫl). [f. *Acol* Road, Hampstead, London, in a house in which the system was devised.] The name given to a system of bidding in bridge.
1938 COHEN & REESE *Acol Syst. Contract Bridge* 2 The Acol system, deriving its name from the former Acol club where it was developed..first attracted attention in 1937. **1959** *Listener* 11 June 1042/1 Acol has two types of bid for strong hands. **1959** REESE & DORMER *Bridge Player's Dict.* 12 *Acol system*, this system is played by the majority of British tournament players and some of its ideas are reflected in modern American systems such as Stayman and Kaplan-Sheinwold.

acoluthic, var. *AKOLUTHIC *a.*

acone (ăkōᵘ·n, ēⁱ-), *a. Ent.* [f. A- 14+CONE *sb.* 10.] Of the eyes of insects: without a cone.
1888 ROLLESTON & JACKSON *Forms Animal Life* 502 In acone eyes one cell stands in the centre with the six others around, and the visual rods or rhabdomeres are contained one in each cell. **1957** A. D. IMMS *Gen. Textbk. Ent.* (ed. 9) 105 In the acone eyes there is a group of elongate, transparent cone cells but the latter do not secrete any kind of cone whether crystalline or liquid.

acorn. Add: **2. b.** An artificial object resembling an acorn in shape. Also in *Comb.* (see quots.)
1580 T. BAWDEWYN in E. Lodge *Illustr. Brit. Hist.* (1791) II. cliv. 243, I did send yowre Honor..a cup wᵗʰ a cover..two saltes, II acornes. **1795** *Ann. Reg. 1772* (ed. 5) Chron. 85/1 The lightning was attracted by the acorn on the top of the chapel. *a* **1884** KNIGHT *Dict. Mech.* Suppl. 3/2 *Acorn-headed Bolt*, a carriage-bolt with an ornamental head..in shape resembling an acorn. **1935** C. G. BURGE *Compl. Bk. Aviation* 85/1 *Acorn*, a device introduced at the intersection of bracing wires to prevent abrasion. **1935** *Burlington Mag.* LXVII. 150/2 The acorn-bulb [in a drinking-glass]..is exactly matched by a stem in the Thaurin Collection at Rouen. **1943** *Gen* 19 June 42/2 Acorn Tops are screwed on to the ends of brass curtain rods. **1960** H. HAYWARD *Antique Coll.* 9/2 *Acorn clock*, shelf or mantel clock..with the upper portion shaped somewhat like an acorn...*Acorn knop*, a knop or protuberance on the stem of a drinking glass, tooled in the form of an acorn.
3. (Earlier example.)
1769 in W. FALCONER *Dict. Marine.*
5. acorn-sugar = QUERCITE; **acorn tube, valve** *Radio*, a small acorn-shaped valve; **acorn-worm**, a worm-like animal of the class *Enteropneusta*, having an acorn-shaped anterior end to its body.
1899 *Syd. Soc. Lex.*, *Quercite*, the so-called acorn-sugar or oak-sugar. **1934** *Electronics* Sept. 282/1 The 'acorn' tubes which amplify, oscillate, and detect waves as short as 40 centimeters have now reached the stage of practical manufacture. **1937** *Nature* 2 Jan. 34/2 The very small 'acorn' valve for the transmission and reception of telephony on a wave-length of about one metre. **1889** *Cent. Dict.* Acorn-worm. **1955** *Sci. News Let.* 9 Apr. 232/1 Known popularly as the acorn worm, the balanoglossus is found throughout the sea-coasts of the world. **1959** A. HARDY *Fish & Fisheries* v. 116 The most primitive of all chordate animals, the acorn-worm *Balanoglossus* and its relatives.

acosmic (ăkǫ·zmik), *a.* [f. as ACOSM(ISM+-IC.] Of or pertaining to acosmism; not cosmic. So **aco·smical** *a.*
1843 *Blackw. Mag.* LIV. 652/1 This man is an Idealist—or as we would term him,..an Acosmical Idealist; that is, an Idealist who absolutely denies the existence of an independent material world. **1904** G. S. HALL *Adolescence* II. xvi. 537 Some..despair of building up again the world they have lost out of its acosmic elements. **1937** A. HUXLEY *Ends & Means* xiv. 284 Pantheism in either its pancosmic or acosmic form. **1953** R. NIEBUHR *Christ. Realism & Polit. Probl.* (1954) xi. 169 Culture religions are..pantheistic in either the cosmic or the acosmic sense.

acosmist. Add: So **acosmi·stic** *a.*
1866 D. W. SIMON tr. *Dorner's Person of Christ* Div. II, vol. II. 46 He did not make the negative acosmistic theology his final goal. **1899** W. R. INGE *Christ. Mysticism* iv. 138 This judgment followed the appearance of a strongly pantheistic or acosmistic school of mystics.

acoumeter. Add: (Earlier and later examples.) Also, **acoume·tric** *a.*, of or pertaining to acoumetry.
1839 K. GRANT *Hooper's Med. Dict.* (ed. 7) 31 Acoumeter or Acumeter..an instrument invented by Itard for estimating the extent of the sense of hearing. **1890** *Cassell's Fam. Mag.* Feb. 188/1 It is proposed to form phonograms into an acoumetric scale for the examination of the hearing. **1911** S. S. COLVIN *Learning Process* vi. 94 Auditory acuity may also be determined..with elaborate instruments such as Seashore's audiometer or Lehmann's acoumeter.

acoustic. Add: **A.** *adj.* *acoustic impedance*, see *IMPEDANCE; *acoustic mine*, a submarine mine designed to be exploded by sound-waves transmitted under water; *acoustic phonetics*, the study of the sound-waves of speech; so *acoustic-phonetic* adj.
1941 *Hansard Commons* 9 Sept. 5th Ser. CCCLXXIV. 67 [Mr. Churchill] The attack..is now waged continually by the acoustic mine as well as the magnetic. **1941** C. P. R. GRAVES *Life Line* 23 Acoustic mines have been dealt with by our mine-sweepers for months...The sound vibrations of a ship's engines act..in the same way as a finger placed on an ordinary electric bell-push, connecting two electrically alive points and causing current to flow and explode the mine. **1933** L. BLOOMFIELD *Language* v. 75 The phonetician can study either the sound-producing movements of the speaker (*physiological phonetics*) or the resulting sound-waves (*physical* or *acoustic phonetics*). **1948** M. JOOS *Acoustic Phonetics* 5 Acoustic phonetics is now in its infancy...Acoustic phonetics discussion, even when it is carried on by a linguist, must deal with sound as sound. **1960** *Amer. Speech* XXXV. 230 An acoustic-phonetic study of internal open juncture. **1964** *New Statesman* 1 May 705/1 (Advt.), University of Leeds. Applications are invited for the following posts...Lecturer in Acoustic Phonetics.
2. Applied to any device or material designed to lessen sound or noise; sound-absorbent.
1924 *Sci. Amer.* Sept. 165/1 Professor Sabine invented an acoustic tile that is many times as absorbent as the usual masonry surfaces. **1931** STEWART & LINDSAY *Acoustics* vii. 175 The construction of an acoustic filter. **1937** *Discovery* Dec. 388/1 The new..trains have both roof and sides covered with acoustic blankets made of asbestos. **1960** *House & Garden* Aug. 46/3 Acoustic board..1s. 3d. sq. ft. **1961** *Listener* 31 Aug. 310/2 We all know what a difference absorbent acoustic tiles on a ceiling..make to the restful character and quality of a public space.

acoustical, *a.* Add: **1.** *acoustical impedance*, see *IMPEDANCE.
b. = *ACOUSTIC A. 2.
1932 P. E. SABINE *Acoustics & Archit.* ix. 192 The application of one of the more highly absorbent of the acoustical plasters..would have been a natural means of securing the desired reverberation time. *Ibid.* x. 228 A number of acoustical materials were investigated, both as to the viability of bacteria within them and as to the possibilities of adequately disinfecting them.

acoustician. (Earlier and later examples.)
1859 *Nat. Mag.* V. 102/1 He had already been engaged specially, in his capacity of 'acoustician', to remedy some defect of the St. Petersburg Opera-house. **1956** H. WHITEHALL in *Kenyon Rev.* XVIII. 422 The acousticians can measure minute variations in stress, pause and pitch. **1957** *Technology* Oct. 302/1 Acousticians may have largely silenced the famous echo [*sc.* in the Albert Hall].

acoustico-. (Later examples.)
1940 *Chambers's Techn. Dict.* 9/2 *Acousticolateral system*, in Vertebrates, afferent nerve-fibres related to the neuromast organs and to the ear, receptors in aquatic forms of relatively slow vibrations. **1946** S. ULLMANN in *Word* Aug. 118 The phoneme is defined as the smallest acoustico-phonetic unit of speech. **1949** R. JAKOBSON in A. Cohen *Phonemes of English* (1952) ii. 30 Among a multitude of acoustico-motor possibilities there is a restricted number upon which language chooses to set a value. **1951** G. HUMPHREY *Thinking* iii. 87 A subject might begin a reaction series by mentally repeating the letter E when that letter appeared. Gradually this 'acousticokinaesthetic presentation' disappeared. **1957** M. E. BROWN *Physiol. of Fishes* II. ii. 155 Phylogenetically the acoustico-lateralis system of sense organs is as old as the vertebrates. **1960** *Amer. Speech* XXXV. 234 Acoustico-cineradiographic analysis considerations.

acousticon (ăkau·stikǫn). [ad. Gr. ἀκουστικόν neut. sing. of ἀκουστικός ACOUSTIC.] An instrument for helping the deaf to hear.
1900 in DORLAND *Med. Dict.* (22/1). **1920** *Chambers's Jrnl.* 27 Nov. 831/2 An instrument known as 'The Acousticon', which we believe..to be a real boon to those suffering from deafness.

acoustics. Add: **3.** The acoustic properties (of a building, etc.).
1885 H. W. LUCY in *Eng. Illustr. Mag.* Dec. 193/2 Whilst the vast majority of the peers are practically inaudible in the press gallery, I cannot call to mind any individual case in which the public interest materially suffers owing to the faulty acoustics of the chamber. **1895** D. F. TOVEY *Let.* in M. Grierson *D. F. Tovey* (1952) iii. 37 The execrable acoustics of the Sheldonian.

acquaintance. Add: **1. b.** *Philos.* Knowledge of a person, thing, or other entity (e.g. sense-datum, universal) by direct experience of it, as opposed to knowing facts about it. So *knowledge of, by, acquaintance* (opp. *knowledge-about* or *by description*).
1865 GROTE *Expl. Philos.* I. iv. 61 If by knowledge we mean acquaintance or familiarity, kenntniss, then we know the thing in itself. **1885** W. JAMES in *Mind* X. 28 An interminable acquaintance, leading to no knowledge-*about*. **1890** —— *Princ. Psychol.* I. viii. 221 *There are two kinds of knowledge* broadly and practically distinguishable: we may call them respectively *knowledge of acquaintance* and *knowledge-about*. Most languages express the distinction; thus, γνῶναι, εἰδέναι; *noscere, scire*; kennen, wissen; connaitre, savoir. **1905** B. RUSSELL in *Mind* XIV. 479

The distinction between *acquaintance* and *knowledge about* is the distinction between the things we have presentations of, and the things we only reach by means of denoting phrases. **1911** —— in *Proc. Aristot. Soc.* XI. 127 We began by distinguishing two sorts of knowledge of objects, namely, knowledge by *acquaintance* and knowledge by *description*. Of these it is only the former that brings the object itself before the mind. **1954** J. A. C. Brown *Soc. Psychol. Industry* iii. 95 Two kinds of knowledge: 'knowledge-about', based on reflexion and abstract thinking, and 'knowledge-of-*acquaintance*', based on direct experience.

acquiescer (ækwi,e·səɹ). [f. Acquiesce *v.* + -er[1].] One who acquiesces.
a **1800** Mrs. M. Robinson *Mem.* (1801) I. 122, I know too well that I have been sufficiently the victim of events, to become the tacit acquiescer where I have been grossly misrepresented. **1837** Lockhart *Life of Scott* III. x. 329 An undoubting acquiescer in 'the decision of the public'.

acquired, *ppl. a.* Add: In specific senses and phrs.: (*a*) (See quots.)
1842 Dunglison *Dict. Med. Sci.* (ed. 3) 20/1 *Acquired Diseases*, diseases which occur after birth, and which are not dependent upon hereditary predisposition. **1934** *Ann. Reg. 1933* 64 Salaman described what is apparently a clear case of acquired immunity to potato virus in tobacco plants. **1961** Webster *Acquired*: of disease or abnormal states, developed after birth (acquired heart disease)—opposed to *congenital*.
(*b*) *acquired taste*, a taste (for a food or drink) that is gained by experience; also *transf.* applied to any thing or person for which or for whom one has acquired a liking (as distinct from a natural or spontaneous taste).
1858 *Illustr. News of World* 24 July 55/3 A yielding to the acquired taste of tobacco cannot be gratified for any length of time without..affecting the breath. **1881** Ogilvie (Annandale) s.v., Abilities natural and acquired; an acquired taste. **1885** W. S. Gilbert *Mikado* II. 43 He would have loved me in time. I am an acquired taste.
(*c*) *acquired character* [Character *sb.* 8 b], *characteristic*, *Bot.*, *Zool.*, a development in an individual plant or animal occurring during its lifetime through the influence of its environment. (Cf. Lamarckian *a.*)
[**1794** E. Darwin *Zoonomia* I. § xxxix. 503 Many of these acquired forms or properties are transmitted to their posterity.] **1876** E. R. Lankester tr. *Haeckel's Hist. Creation* I. ix. 204 We may first divide all the different phenomena of inheritance into two groups, which we may distinguish as the transmission of inherited characters, and the transmission of acquired characters. **1893** Parker & Rönnfeldt tr. *Weismann's Germ-Plasm* iv. xiii. 392 By acquired characters I mean those which are not preformed in the germ, but which arise only through special influences affecting the body or individual parts of it. **1899** W. Z. Ripley *Races of Europe* (1900) xiv. 382 Whether the short stature of the Jew is a case of an acquired characteristic which has become hereditary, we are content to leave an open question. **1926** J. S. Huxley *Essays Pop. Sci.* II. 22 So-called 'acquired characters', in other words modifications caused by the environment..are not inherited at all, or else to such a slight degree as not to be of any great importance in heredity and evolution. **1945** B. Russell *Hist. West. Philos.* (1946) xxi. 753 If, as Lamarck held,..acquired characteristics were inherited. **1958** *Times* 1 July 11/7 Since acquired characters are not inherited, the supply of heritable variation is indeterminate and blind.

acrawl (ăkrǭ·l), *adv.* or *pred. a.* [f. A *prep.*[1] 11 + Crawl *sb.*[1] and *v.*[1]] Crawling (*with*).
1830 *Blackw. Mag.* XXVIII. 146 Mountain tops..a-crawl with insects, above a few acres of wet! **1834** *Ibid.* XXXV. 1003 Something a-crawl in the ditch. **1909** R. Sabatini *St. Martin's Summer* i. 14 He felt himself the meanest, vilest thing a-crawl upon this sinful earth. **1923** *Chambers's Jrnl.* 29 Dec. 73/1 The slime..seems acrawl with strange forms of life.

acre. Add: **4.** *Irrigation.* **acre-foot**, a unit of volume of water equal to one acre in area and one foot in depth.
1909 in *Cent. Dict.* Suppl. **1958** *New Scientist* 10 Apr. 8/1 It [the Kariba dam] will contain 140 million acre-feet of water.

acreless (ē[i]·kəɹlės), *a.* [f. Acre + -less.] Without acres or landed estates. Also *fig.*
1890 *Temple Bar* June 226 The lady had the bad taste to prefer the acreless to the acred. **1941** L. H. B. Lyon *Tomorrow Revealing* 14 A stealthy and mute fertility still moves The acreless heart.

acridian (ăkri·diăn), *a.* and *sb.* *Ent.* [f. Gr. ἀκριδ-, ἀκρίς locust + -ian; cf. F. *acridien*.] **A.** *adj.* Of or pertaining to an orthopterous insect of the family *Acridiidæ*, comprising certain locusts and grasshoppers. **B.** *sb.* An insect of this family.
1878 *Boston Soc. Nat. Hist. Proc.* XIX. 336 Mr. S. H. Scudder exhibited a number of western Acridians in illustration of one type of dimorphism. **1889** *Cent. Dict.*, Acridian, *a.* **1904** A. P. Morse *Res. N. Amer. Acridiidæ* 7 The Acridian fauna of the southeastern United States. **1930** T. S. Eliot tr. *St. John Perse's Anabasis* 49 My heart gives heed to a family of acridians.

acriflavine (ækriflē[i]·vin). *Med.* [irreg. f. Acri(dine + Flavine.] An orange-red, odourless powder, used as an antiseptic for wounds, cuts, etc.
1917 C. H. Browning et al. in *Brit. Med. Jrnl.* 21 July 71/1 The unsubstituted 3.6-diamino-10-methyl-acridium compound—'flavine', now called 'acriflavine'. **1918** *Lancet* 16 Feb. 257/1 Acriflavine has a very marked bactericidal inhibiting action on streptococci and a less marked on staphylococci. **1943** *Endeavour* Apr. 41/2 It had been shown that acriflavine or proflavine need not kill trout ova which they nevertheless sterilize. **1954** Heilbron & Bunbury *Dict. Org. Comp.* (rev. ed.) I. 28/2 *Acriflavine* (Trypaflavine) $C_{14}H_{15}N_3Cl_2$..commercial product usually contains a proportion of the unmethylated base (hydrochloride).

Acrilan (æ·krilæn). The proprietary name of a synthetic acrylic fibre.
1951 *U.S. Pat. Off., Official Gaz.* 8 May 387/2 The Chemstrand Corporation, Philadelphia. *Acrilan.* For Yarns, Threads, and Yarn and Thread Filaments. **1952** *N.Y. Times* 11 Sept. 52/1 The new synthetics such as Orlon, Dacron, Acrilan and Vicara will find their best use in apparel in blends with the older man-made fibers such as rayon and acetate. **1957** *Observer* 13 Oct. 11/5 Housewives will like the 100 per cent. Acrilan blankets which have newly come into our shops.

acrobacy (æ·krŏbæsi). [cf. Fr. *acrobatie*.] Acrobatics.
1918 *Observer* 17 Nov. 6/4 The 'D.H. 10'..loops the loop and performs any acrobacy which is usually only demanded of a fighting scout. **1923** A. Huxley *Antic Hay* xvi. 230 The *monster*..climbs up the runged back of his chair and stands, by a miraculous feat of acrobacy, on the topmost bar. **1928** A. Haskell *Some Stud. in Ballet* 116 The line is not clearly and scientifically marked between dancing and acrobacy.

acrobatic, *a.* Add: **B.** *sb. pl.* Acrobatic performances or feats. Also *transf.* and *fig.*
1882 G. Macdonald *Weighed & Wanting* II. iv. 28 There was not much popular receptivity for acrobatics in the streets. **1890** *Athenæum* 22 Feb. 239/1 The art and science of what may be called acrobatics have never yet received really adequate treatment. **1915** *Morn. Post* 20 Apr. 7/7 The German railway acrobatics. **1917** W. J. Locke *Red Planet* ix, A mind trained in the acrobatics of a Calvinistic Theology. **1922** *Daily Mail* 4 Dec. 11 Habton is a very fast chaser, but rather given to acrobatics.

acroblast (æ·krŏblɑst). *Biol.* [f. Acro- + -blast.] A body in the spermatid from which arises the acrosome.
An earlier sense 'tissue of the germinal wall' (after G. *akroblast*, J. Kollmann) was short-lived.
1907 H. D. King in *Amer. Jrnl. Anat.* VII. 348 This body is undoubtedly concerned in the formation of the acrosome of the spermatozoön, and therefore I suggest for it the name acroblast as somewhat more appropriate than 'chromatoid Nebenkörper'. **1958** *New Scientist* 13 Nov. 1272/2 The acroblasts..have an onion-like layered membrane structure. **1963** E. V. Cowdry et al. *Special Cytology* (ed. 2) III. 1758 The Golgi bodies and idiozome material have collected at one side of the nucleus to form the acroblast.

acrocentric (ækrose·ntrik), *a.* *Cytology.* [f. Acro- + *-centric.] Of a chromosome: having the centromere close to the end. Hence as *sb.* Cf. *metacentric *a.*, *telocentric *a.*
1945 M. J. D. White *Animal Cytol. & Evol.* ii. 20 A distinction still exists in practice between those [chromosomes] which have the centromere somewhere near the middle and those in which it is very close to the end. The former we shall call *metacentric*, the latter *acrocentric*. **1946** *Nature* 26 Oct. 587/2 The chromosome number of the parthenogenetic females generally amounted to 68, made up of six pairs of metacentric and twenty-eight pairs of acrocentric elements. **1949** I. F. & W. D. Henderson *Dict. Sci. Terms* (ed. 4), *Acrocentric*, a rod-shaped chromosome. **1962** *New Scientist* 22 Nov. 457 The chromosome involved is one of the four small acrocentrics—those having one arm much shorter than the other.

acrochordite, var. *akrochordite.

acrolein. (Earlier example.)
1857 Miller *Elem. Chem.* III. vi. 385 *Acrolein*, or *Acrylic Aldehyd*..a substance which, from its intensely irritating effects upon the mucous membrane.., has received the name of *acrolein*.

acrological (æ:krolǫ·dʒikăl), *a.* [f. as Acrologic *a.* + -al.] = Acrologic *a.* Hence a:crolo·gically *adv.*
1831 *Edin. Rev.* LIII. 383 A task for which his [M. Klaproth] Letters on Acrological Hieroglyphs prove him to be so eminently qualified. **1883** I. Taylor *Alphabet* I. i. 43 The symbol was used 'acrologically', to express simply the initial syllable of the word. **1932** J. Joyce *Let.* 29 Aug. (1957) 324 Short verses for children, or fables or merely acrological rhymes.

acromegaly (ækrome·găli). *Path.* [ad. Fr. *acromégalie* (P. Marie), f. Gr. ἄκρον extremity + μέγας, μεγαλ- great.] A disease characterized by hypertrophy and enlargement of the extremities. Hence **acromegalic** (-mĭgæ·lik) *a.*, pertaining to or of the nature of acromegaly; *sb.*, one affected with acromegaly.
1889 *Brain* July 59 Acromegaly. By Pierre Marie, M.D. **1896** *Godey's Mag.* Feb. 125/1 A..very rare disease,..acromegaly, or the enormous enlargement of the feet, hands, face, and chest. **1902** *Lancet* 27 Sept. 884/2 The historic and folklore record about acromegalic and giant. **1909** *Cent. Dict.* Suppl., Acromegalic, a. and n. **1937** *Jrnl. R. Anthrop. Inst.* LXVII. 161 Acromegalic crania are not infrequently metopic. **1937** W. Inge *Modernism in Lit.* 3 A modernist sculptor will carve figures apparently suffering from elephantiasis or acromegaly. **1953** *Sci. News* XXX. 62 Acromegaly..is a condition caused by an excessive production of growth hormone in the adult. **1964** L. Martin *Clinical Endocrinol.* (ed. 4) i. 29 X-ray of an acromegalic's hand.

acromio-clavicular (ăkrōu·mio,klævi·kiŭlăɹ), *a.* *Anat.* [f. Acromio(n + Clavicular *a.*] (See quot. 1858.)
1858 Gray *Anat.* 158 The *Superior Acromio-Clavicular Ligament* is a broad band of fibres, of a quadrilateral form, which covers the superior part of the articulation, extending between the upper part of the outer end of the clavicle, and the superior part of the acromion...The *Inferior Acromio-Clavicular Ligament*..covers the inferior part of the articulation. **1942** *Lancet* 10 Oct. 424/1 Only one example of congenital dislocation of the acromio-clavicular joint appears to have been recorded.

acronym (æ·krŏnim). orig. *U.S.* [f. Acr(o- + -*onym* after Homonym.] A word formed from the initial letters of other words; hence acrony·mic *a.*
1943 *Amer. N. & Q.* Feb. 167/1 Words made up of the initial letters or syllables of other words..I have seen..called by the name *acronym*. **1947** *Word Study* 6 (*title*) Acronym Talk, or 'Tomorrow's English'. **1950** S. Potter *Our Language* 163 Acronyms or telescoped names like *nabisco* from *National Biscuit Company*. **1954** *Britannica Bk. of Yr. 1954* 638/1 Typical of acronymic coinages, or words based on initials, were..mash (Mobile Army Surgical Hospital).

acroparæsthesia (æ·kro,pærės-, -ĭsþī·siă, -z-). *Path.* [f. Acro- + Paræsthesia.] A disease marked by vasomotor changes in peripheral parts of the body.
1894 *Med. News* LXV. 178/2 (*title*) The Treatment of Acro-Paresthesia (Numbness of the Extremities). **1900** *Lancet* 23 June 1795/2 The affection called 'acroparæsthesia' has..not been described as occurring in this country, though it undoubtedly does occur. **1908** *Practitioner* Aug. 298 Stengel discusses..cases in which numbness, tingling, burning, and other symptoms..known collectively as 'acroparaesthesia', occur in elderly persons who have different degrees of arterio-sclerosis.

acrophobia (ækrofōu·biă). [f. Acro- + Phobia.] Morbid dread of heights. Hence acropho·bic *a.*
1892 D. H. Tuke *Dict. Psychol. Med.* 679 *Acrophobia*, described by Dr. Andrea Verga, means the dread of being in high places. **1899** *Allbutt's Syst. Med.* VII. 886 Both agoraphobia and acrophobia are common forms of imperative ideas. **1959** *New Yorker* 24 Oct. 34/3 We walked over to the winding parapet,..took a shuddering, acrophobic look down the well.

acrophonic (ækrofǫ·nik), *a.* [f. Acrophon(y + -ic.] Pertaining to acrophony.
1909 in *Cent. Dict.* Suppl. **1932** *Times Lit. Suppl.* 7 Jan. 13/1 The Hittite pictographic script which he thought was acrophonic. **1959** A. G. Woodhead *Gr. Inscriptions* 18 The Greek names for the letters of the alphabet are acrophonic, the first letter of the name expressing the sound of the letter to which the name is given.

acroscopic (ækroskǫ·pik), *a.* *Bot.* [f. Gr. ἄκρον apex + -σκοπος viewing + -ic.] Looking, or on the side, towards the apex.
1882 Vines tr. *Sachs' Bot.* 450 In *Azolla*, the leaves of the one row all arise from one cell of the acroscopic part of the segment.

acrosome (æ·krŏsōum). *Zool.* [ad. G. *akrosoma* (M. von Lenhossék 1898, in *Archiv. Mikrosk. Anat.* LI. 282), f. Acro- + -*some* (cf. Gr. σῶμα body).] A cap-like structure forming the anterior part of the head of a spermatozoon. Hence acroso·mal, acroso·mic *adjs.*
1899 J. H. McGregor in *Jrnl. Morphology* XV. Suppl. 91 For this body in mammals Lenhossek proposed the term 'Akrosom', and throughout the present paper I shall call it the *acrosome*. *Ibid.* 93 The attachment of acrosome to nucleus marks the anterior end of the spermatid from the start. **1940** *Chambers's Techn. Dict.* 10/1 Acrosomal. **1951** G. H. Bourne *Cytol. & Cell Physiol.* (ed. 2) vi. 271 There is evidence..that the [Golgi] apparatus is associated with ..the formation of fatty yolk in eggs, and of the acrosome in spermatozoa. **1952** *Amer. Jrnl. Anat.* XC. 169 One to four granules..coalesced, forming one larger structure, the acrosomic granule. **1960** L. Picken *Organization of Cells* vi. 231 The 'acrosomic system' (that is, acrosome, head-cap, and precursors). *Ibid.* 233 The acrosomal vesicle containing the acrosome.

across. Add: **A.** *adv.* **3. b.** *to come across* (*with*): to hand over, contribute (money, information, etc.). *slang* (orig. *U.S.*).
1910 *Sat. Even. Post* 13 Aug. 8/1, I knew pull was required ..but I hadn't learned that I'd have to come across with the price as well. **1915** J. London *Jacket* viii. 62 You might as well come across now and save trouble. **1928** D. Sayers *Lord Peter views Body* 16, I think you ought to come across with the rest of the story. **1938** Wodehouse *Summer Moonshine* ix. 99 He hinted that one had got to come across. **1948** M. Allingham *More Work for Undertaker* xxii. 253 'Did she admit it?' 'Yes...But she wouldn't come across till dawn.'
4. Delete † and read *Obs.* exc. *dial.* Cf. *B. 1 c.

1887 Baring-Gould *Red Spider* I. vii. 110 When folks who look straight before them fall across. *Ibid.*, The two who have got across. **1891** S Weyman *New Rector* ix. 173 Lindo must beware of getting across with him. **1892** *Cornh. Mag.* July 28 Matters were soon across again between the pair. **1897** *Daily News* 4 Mar. 6/1 He is getting across with the farmers now, for he roundly rates them on account of their apathy.
B. *prep.* **1. c.** *to get across*: to annoy, get on the wrong side of (someone). *slang.*
1926 D. Sayers *Clouds of Witness* i. 30 One was always getting across Denver, but it never came to anything. **1960** M. Stewart *My Brother Michael* xiv. 183 He's got across that damned Greek.
2. *across lots* (U.S.): see Lot *sb.* 6 a.
b. Phr. *across the footlights*: from the performers to the audience; hence advb. (by ellipsis) in *to get* or *come across* (*to*), to reach the audience or the public, to make oneself or itself understood or appreciated; similarly *to get it across.* (For *to put it across* see *Put *v.*[1])
1894 G. B. Shaw *Let.* 11 June (1965) 443, I want to see how much of it they succeed in getting across the footlights. **1913** Kipling *Diversity of Creatures* (1917) 190 Tell a fellow now, did I get it across? **1921** *Sat. Westm. Gaz.* 27 Aug. 14/2 Some vitality that may be as far away as you like from lifelikeness..but nevertheless gets across to the reader from the writer. **1923** Silberrad *Lett. Jean Armiter* vi. 148 Sorry—my fault—one fails to get across. **1923** *Westm. Gaz.* 27 Feb., It is very doubtful whether the play would get across the footlights in an ordinary run. *Ibid.* 26 Mar., 'Magda', dressed as a comedy of manners of a hundred years ago, would probably come across more satisfactorily than it does played as..contemporary tragedy. **1927** *Observer* 16 Oct. 15/3 The lower comedy is at present in the making, but Miss B. C., Mr. E. M. and Mr. L. H. are experts at getting it across. **1927** *Daily Tel.* 19 July 12/4 The Earl of Birkenhead..said..he had never succeeded in projecting his personality across the footlights.
c. *across the board* (esp., with hyphens, as *attrib. phr.*). (See quot. 1950.) *U.S.*
1950 Webster Addenda. *Across-the-board*, embracing all classes or categories without exception;—from placing a combination wager on a race horse to win, place, or show, that is, betting 'across the board'; as, an *across-the-board* tax cut. **1958** *Listener* 24 July 116/2 There is a common *cliché* among labour relations specialists in the United States that it is not the across-the-board wage increase ..which is decisive. **1961** *Flight* LXXX. 134/2 There has been no across-the-board reduction since tourist fares were introduced eight years ago. **1964** R. D. Hopper in I. L. Horowitz *New Sociology* xix. 317 Job displacement..is occurring right across the board.

acrost (ăkrǫ·st), *adv.* and *prep.* *U.S. dial.* and *colloq.* [f. Across *adv.* and *prep.* + inorganic -*t*.] **A.** *adv.* Across; from side to side.
1779 W. McKendry *Jrnl.* 6 Sept. in *Proc. Mass. Hist. Soc.* (1886) II. 467 The Lake..is..about 8 miles acrost. **1846** F. M. Whitcher *Widow Bedott Papers* (1856) vii. 77 Kier said he heered her stretch her neck acrost and whisper. **1908** *Dial. Notes* (Boston, Mass.) III. 285.
B. *prep.* Across; *come acrost*, come upon.
1759 in *Essex Inst. Hist. Coll.* (1882) XIX. 145 Ye enemy fird at our men a Crost ye River. **1831** *Louisville* (Ky.) *Advertiser* 17 Oct., I came acrost a feller..fast asleep. **1852** Stowe *Uncle Tom* xxii, A good, round, school-boy hand, that Tom said might be read 'most acrost the room'. **1906** F. Lynde *Quickening* 2 A-smashin' the whisky jug acrost the wagon tire.

acrostical, *a.* Read: Of the nature of, consisting of, or in the form of an acrostic. (Examples.)
1843 J. Holland *Psalmists Brit.* I. 104 The whole Bible is abridged in a sort of scheme of acrostical mnemonics. **1887** Lupton in *W. Smith's Dict. Chr. Biog.* IV. 648/1 The Eighth Book, in acrostichal verse. **1894** *Athenæum* 28 July 128/3 Chaucer's 'A B C', a curious acrostical prayer 'like Psalm cxviii'.

acrostichal, *a.* Add: **2.** *Ent.* [a. G. *acrostichal* (J. Mik *Dipterolog. Untersuchungen*, 1878).] 'Situated in the highest rank or row—used of certain bristles on the mesonotum of muscoid flies' (Webster). Also *ellipt.* as *sb.*
1884 C. R. Osten-Sacken in *Proc. Ent. Soc. Washington* 509 The intermediate pair of rows is represented by two, sometimes one..row of peculiar, minute bristles, which Prof. Mik calls the *acrostichal bristles*. **1925** Imms *Textbk. Ent.* III. 601 Thoracic Bristles..*Acrostichal*: a row along each side of median line. **1939** Webster *Acrostichal n.*, an acrostichal bristle. **1961** J. E. Collin *Empididae* I. 59 Very short biserial acrostichal bristles...There may, however, often be a distinct posthumeral bristle, and all except the acrostichals may be longer and stronger.

a-crow (ăkrōu·), *adv.* or *pred. a.* *poet.* [f. A *prep.*[1] + Crow *v.*] Crowing.
1898 G. Meredith *Poems* I. 141 Straight cocks a-crow. **1919** W. B. Yeats *Two Plays for Dancers* 16 The strong March birds a-crow, Stretch neck and clap the wing.

acrylate. Add: **2.** *acrylate* (*resin*), a synthetic resin made by polymerizing derivatives of acryl.
1935 C. Ellis *Chem. Synthetic Resins* 1461 (*index*) Acrylates. **1936** *Industr. & Engin. Chem.* XXVIII. 271/1 The acrylate and methacrylate polymers are characterized by their excellent adhesion to most surfaces. **1938** *Jrnl. R. Aeronaut. Soc.* XLII. 100 The problem of choosing between cellulose acetate and acrylate resin plastics for aircraft windows. **1942** *Paint Technol.* VII. 205 (*title*) Acrylate Resins in Surface Coatings. **1953** *Encycl. Brit.* XVIII. 411/2 Polymethyl acrylate is softer and rubbery; polybutyl acrylate is softer and more rubbery; polybutyl acrylate is sticky..polyoctyl acrylate..is almost liquid in consistency.

acrylic, *a.* Add: (Earlier example.) Also *acrylic fibre*, *plastic*, *resin*, etc., synthetic substances prepared from acrylic acid or its derivatives. Also *acrylic sb.* (usu. *pl.*).
1855 H. Watts tr. *Gmelin's Handbk. Chem.* IX. 369 Acrylic acid. **1936** *Industr. & Engin. Chem.* XXVIII. 271/2 The acrylic resins show many unique physical, chemical, and mechanical properties. **1939** H. R. Simonds *Industr. Plastics* 69 The important trade names for the acrylic plastics..are: Plexiglas, Lucite, Crystalite, Acryloid. **1941** *Jrnl. R. Aeronaut. Soc.* XLV. 370 Transparent Acrylic Resins for Cockpit Enclosures. **1943** *Jrnl. Amer. Dent. Assoc.* XXIX. 640 Acrylics in Restorative Dentistry. **1945** H. D. Warren in V. E. Yarsley *Plastics Applied* 63/2 For coloured telephones acrylic plastics are used. **1953** S. Holt *Ibid.* 167/1 Acrylic teeth offer advantages in requiring no pin or diatoric anchorage..and do not break in processing or wear. **1947** *Lancet* 22 Mar. 379/2 The acrylic resins..have proved satisfactory for making artificial eyes. **1953** *Economist* 24 Oct. 275 Acrylic fibres (these, broadly speaking, are the fibres that can be given a wool-like 'feel', as distinct from the more silkier texture of nylon or 'terylene'). **1960** *Which?* Jan. 15/1 The acrylics are easy to wash, do not shrink or stretch, and generally need no ironing.

acrylonitrile (ækrilo,nɑi·tril). *Chem.* [f. Acryl(ic *a.* + -o- + Nitrile.] Vinyl cyanide, $CH_2{:}CH.CN$, a colourless liquid used as a copolymer in the manufacture of certain synthetic materials; the nitrile of acrylic acid.
1893 *Jrnl. Chem. Soc.* LXIV. i. 661 β-hydroxypropionitrile..yields acrylonitrile when distilled with phosphoric anhydride. **1947** Mark & Dakrin tr. *Karrer's Org. Chem.* (ed. 3) 180 Acrylonitrile..readily polymerises and therefore serves for the manufacture of plastics and artificial rubber. **1964** N. G. Clark *Mod. Org. Chem.* vi. 121 Acrylonitrile: obtained from acetylene and hydrocyanic acid by passing the mixed vapours at ordinary temperature and pressure into a suspension of cuprous chloride in dilute hydrochloric acid.

act, *sb.* Add: **1. c.** Any operation of the mind, as distinguished from the content or object of that operation. Also *attrib.*, as *act psychology*, psychology regarded as the study of such acts; = *intentionalism.
1694 Locke *Essay* (ed. 2) II. xxi. § 30. p. 134 Desiring and willing are two distinct Acts of the mind. **1890** W. James *Princ. Psychol.* II. xx. 168, I can not feel them by a pure mental act of attention unless they belong to quite distinct parts of the body. **1927** E. B. Titchener *Systematic Psychol.* (1929) iii. 194 The importance of the 'act' in modern psychology derives from the work of Brentano. **1934** H. C. Warren *Dict. Psychol.* 5/1 *Act psychology*, 1. A system of psychology which holds that every psychical phenomenon is characterized by the intentional inherence of an object. 2. A system..in which the data are psychical activities, usually of a subject upon an object. **1936** A. J. Ayer *Lang., Truth & Logic* vii. 188 We do not accept the realist analysis of our sensations in terms of subject, act, and object.
6. c. *act and deed*, that part of a formula used when signing a legal instrument and putting a finger on the seal at the end of the transaction.
1756 D. Garrick *Let.* in C. Oman *D. Garrick* (1958) 177 The act and deed of the wife, in such cases [i.e. business matters], pass for nothing. **1827** Harnewall & Cresswell *Reports* V. 671 [He] produced the parchment, placed it on the table, signed his name, and said, 'I deliver this as my act and deed', putting his finger at the same time on the seal. **1877** W. S. Gilbert *Sorcerer* i. 9 They deliver it—they deliver it As their Act and Deed! **1920** *Encycl. Brit.* I. 156/2 In law it means any instrument in writing, for declaring or justifying the truth of a bargain or transaction, as: 'I deliver this as my act and deed'.
7. (Earlier examples.)
? **1520** *Terens in Englysh* (*Andria*) The furst scene of the furst Act. **1549** Chaloner tr. *Erasmus's Praise of Folie* sig. N iv *verso*, Resteth nowe the fifte acte or parte, wherein it behoueth them to shewe foorth all their cunnyng and profunditie. **1565** Norton & Sackv. *Gorboduc* (*title-p.*), The Tragedie of Gorboduc Whereof three Actes were wrytten by Thomas Norton, and the two laste by Thomas Sackuyle. *Ibid.*, sig. A ii[v] The Order of the dōme shewe before the firste Acte.
†b. An interval or interlude in a play. *Obs.*
1606 J. Marston *Parasitaster* Actum Quintus. Whilest the Act is a playing, Hercules and Tiberio enters, Tiberio climes the tree, and is received above by Dulcimel. **1611** Cotgrave *Dict.*, *Acte*..an Act in a Pause or a Comedie, or Tragedie. **1623** Shakes. *Mids. N.* III. ad fin., They sleepe all the Act. **1653** Middleton & Rowley *Changeling* III. i (*stage-direction*) In the act-time De Flores hides a naked rapier behind a door.
c. One of a series of short performances in a variety programme, circus, etc.
1890 'Biff' Hall *Turnover Club* vi. 63 The usual attraction was 'Professor Etherio, the flying man', who did a rope-walking act. **1912** *Sat. Even. Post* 28 Dec. 19/3 He puts on flesh..Even those managers who..had refused to give engagements to this class of 'act', were soon tumbling over one another. **1933** Priestley *Good Compan.* Back-Stage xviii. 228 Their act was booked for a tour by the African theatres. **1959** *Times* 16 Oct. 12/3 The team had completed their twists, loops and all the tricks of their acts which had been seen in..other air shows.

d. *transf.* An imitation of a theatrical part, a piece of acting; a display of exaggerated behaviour; pretence (of being what one is not); esp. *to put on an act* (colloq.), to show off, to talk for display, to behave insincerely, to act a part.
1928 Barrie *Peter Pan* I in *Plays* 20 We are doing an act; we are playing at being you and father. **1934** Faith Baldwin *Innocent Bystander* (1935) viii. 145 When he spoke of the theatre he wasn't putting on an act. He was himself. **1934** J. O'Hara *Appointment in Samarra* (1935) vii. 275 You put on some kind of an act with Caroline, and..she fell for it. **1939** A. Huxley *After Many a Summer* II. v. 220 It was such a relief not to have to put on that act with Pete for the benefit of uncle Jo. **1946** M. Dickens *Happy Prisoner* viii. 158 This girl's not naturally like that. She's putting on an act. **1953** A. W. Fielding *Stronghold* III. ii. 191 This might have been an act designed to impress us. **1959** *Times* 1 Apr. 8/3 Some men were injured and some were 'putting on an act'.
e. *attrib. act-drop* = Drop *sb.* 16.
1884 (see Execute *v.* 2). **1890** G. B. Shaw in *Star* 28 Apr. 2/3 She made a very rapid impression..by which the audience gave vehement emphasis to after each descent of the act-drop. **1896** [see *Drop sb.* 16]. **1960** *Times* 29 Sept. 16/7 An act-drop is lowered to display a multicoloured abstract design.
8. (Earlier examples.)
1549 Chaloner tr. *Erasmus's Praise of Folie* sig. M iv *verso*, At their Actes and Commencementes ye dooe see them swadled in with so many cappes, coyfes, and furde hoodes as they weare. **1592** T. Nashe *Strange News* I. 271 Acts are but idle wordes, and..Pumps and Pantofles..therefore do do Acts..onelie..to Oxford they trudge..and there are confirmed in the same degree they took at Cambridge. **1607** [R. Parker] *Scholast. Disc. Antichrist* I. 89 For proofe hereof, what need I goe further then to an Vniuersitie Acte, where before a confluence and concourse of people,..a Doctor incipient in Diuinitie publisheth these verses.

act, *v.* Add: **8. b.** Of a play: to be susceptible of being performed (well or otherwise).
1668, **1789** [see Read *v.* 18 a, *v.* 1]. **1821** Byron in Trelawny *Recoll. Shelley & Byron* (1858) 29 My plays won't act..my poesy won't sell. **1936** S. Kaye-Smith *John Galsworthy* 17 Galsworthy's plays have the advantage of acting well—unlike much literary drama.
9. d. (Earlier example.)
1814 Jane Austen *Mansf. Park* III. x. 201 It was somehow or other ascertained, or inferred, or at least acted upon, that they were not at all afraid.
f. *to act up*: to act in an abnormal or unexpected manner; to 'put on an act'; to become unruly, to 'play up'. *colloq.*
1903 A. Adams *Log Cowboy* xvii. 275 The cattle never acted up one night. **1929** D. H. Lawrence *My Skirmish with Jolly Roger* 8 When people act it on, nowadays they are half the time acting up. **1956** J. Hearne *Stranger at Gate* xxxvii. 282 I'm sorry, he said, 'I'm acting up a bit. I feel pretty tight inside.' **1960** *New Statesman* 20 Nov. 786/2 She would have to 'act up' to convince anyone that her danger was in danger.

actability (æktăbi·lĭti). Also irreg. -ibility. [f. Actable *a.* + *-bility.] Of a play: capability of being acted.
1836 *Fraser's Mag.* Apr. 451 Opinions..as to the actibility of certain unacted plays. **1925** *Glasgow Herald* 9 Apr. 4 When Hugh M'Diarmid's Braid Scots play, 'The Purple Patch', appeared,..doubts were expressed as to its actability.

‖**acte gratuit** (akt gratwi). [Fr. (A. Gide).] A gratuitous or inconsequent action performed on impulse.
1933 'Rebecca West' *St. Augustine* iii. 40 It was simply a demonstration against order..; in fact, it was an acte gratuit of the sort that tills M. André Gide with such ecstasy. **1949** N. Blake' *Head of Traveller* I. 25 I'm an Artist. I should only be interested in murder as an *acte gratuit*. **1950** W. H. Auden *Enchafèd Flood* (1951) iii. 96 His criminal *acte gratuit* of shooting the Albatross.

actin (æ·ktin). *Biochem.* [f. Gr. ἀκτίς ray + -in[1].] A protein which with myosin plays an important part in the contraction and relaxation of muscle (see *actomyosin).
1942 A. Szent-Györgyi in *Stud. Inst. Med. Chem. Univ. Szeged* I. 67 Myosin B is a stoichiometric compound of myosin A and another substance. We will call this other substance 'actin'. **1944** *Stud. Musc.* II. 39 This property of actin, that it is capable of existing in two forms, in globular and fibrous form, which can be reversibly transformed into each other, is striking and unique. **1962** *Gray's Anat.* (ed. 33) 40 The thinner filaments believed to be largely composed of actin.

actine. Add: **2.** A 'ray' or radiating part of a sponge-spicule.
1887 Sollas in *Encycl. Brit.* XXII. 416/2 Two actines soldered together by intervening silica.

acting, *vbl. sb.* Add: **5.** *acting manager*, a person responsible for arranging the scenery, costumes and acting of a play (often himself an actor); producer; hence *acting management*.
1733 B. Victor *Mem. Life B. Booth* 7 And the part of Cato was given to Mr. Booth by two of his Masters, who were Acting Managers. **1829** H. Foote *Compan. Theatres* 30 In 1795, a share was purchased by John Grubb, Esq. to whom Mr. Kemble resigned the acting management. **1860**

Players I. 138 He has also had the acting management during the summer months of the Surrey Gardens. **1879** C. E. Pascoe *Dram. List* 237/1 Acting and Stage Manager: George Bartley.
b. Suitable for dramatic performance; *spec.* susceptible of being performed well (cf. *Act *v.* 8 b).
1768 D. Garrick *Let.* 26 Mar. (1831) I. 295 Also five copies of all the acting plays and farces half-bound. **1825** P. Egan *Life of Actor* 203 The School of Adversity..is one of the best acting plays in the whole catalogue of the drama. **1847** J. Boaden *Mem. Mrs. Siddons* II. 244 The Cleopatra of Dryden..did not range among her acting parts. **1869** *Harlequin* 20 June 43 The third act—the finest portion of our acting drama—was consequently inefficient. **1884** A. C. Mowatt *Autobiogr. of Actress* 203, I designed the play wholly as an acting comedy. **1898** L. Merrick *Actor-Manager* xx, It was a finer acting scene for Blanche than for himself. **1901** Clara Morris *Life on Stage* 141 It was an excellent acting part, very sweet and tenderly pathetic in the first act.
c. Applied to versions of plays specially prepared for actors' use (provided with full stage-directions, etc.), as *acting copy*, *edition*, *version*.
1845 J. Cowell *Thirty Years among Players* (1845) I. vii, 19/2, I went home and read Ross and Lennox from the acting copy. **1850** [see Copy *sb.* 6]. **1859** [see Lacy's acting edition of plays. **1855** W. B. Wood *Pers. Recoll. Stage* 165, I could not, however, but regret that he had not adopted the acting version which we formerly used. **1910** W. W. Greg *Shaks. Merry W.* Introd. 31 These four scenes cannot have been altogether omitted in the acting version. **1943** G. Mitchell *Laurels are Poison* xiv. 162 An 'acting copy' of *Richard of Bordeaux*.

acting, *ppl. a.* Add: **3.** As a qualifying adj. (Later examples.)
1836 Marryat *Midsh. Easy* xxv, He..served his time, was acting lieutenant for two years and then somehow or other he bore up for the Church. **1873** *Porcupine* 24 May 124/1 The proud position of acting editor of the *Mail* is a far higher honour than that of any collector of Customs, acting or inactive. **1948** J. Irving *Royal Navalese* 19 *Acting Un-paid*. A rating who is granted permission by his Captain to wear some badge or mark of rank..although, by regulation, he is not entitled to it and cannot therefore be paid accordingly.

actinian (ækti·niăn). *Zool.* [f. Actinia + -an.] A sea-anemone belonging to the genus *Actinia*.
1888 *Athenæum* 30 June 830 A tube-forming actinian (*Cerianthus membranaceus*). **1902** *Fortn. Rev.* June 1012 The actinian..merely responds mechanically to a chemical stimulus.

actinide (æ·ktinɑid). *Chem.* [f. *Actin(ium 2 + -*ide* as in *lanthanide.] One of the elements with an atomic number of 89 (actinium) or higher. Also *attrib.*, as *actinide series*.
1945 G. T. Seaborg in *Chem. & Engin. News* XXIII. 2192/1 This rare-earth-like series begins with actinium in the same sense that the 'lanthanide' series begins with lanthanum. On this basis it might be termed the 'actinide' series. **1950** P. Gaynor *Encycl. Atomic Energy* 14 Recently the discovery of two new actinide elements, *berkelium* and *californium* has been announced. **1958** The actinides and the rare earth elements show a striking similarity of extranuclear electron structure. **1958** R. D. Cowson in G. R. French *Nucl. Handbk.* iv. 521 The second 'rare earth' series starts at actinium and is called the *actinide series* just as the original series starting at lanthanum is termed the *lanthanide series*.

actinium. Add: **2.** *Chem.* A radioactive metallic element found, associated with thorium, in pitchblende. Symbol Ac; atomic weight 227; atomic number 89.
[**1900** A. Debierne in *Compt. Rend.* CXXX. 906.] **1900** *Jrnl. Chem. Soc.* LXXVIII. ii. 350 (*abstr.*) Actinium: A New Radio-active Element. *Ibid.* 351 The rays emitted by actinium produce the same effects as those from radium and polonium. **1903** Rutherford in *Phil. Mag.* XI. 362 In order to obtain a homogeneous source of α rays, the active deposit of actinium was used..This active deposit consists of two products, actinium A and B, the former of which is rayless. **1958** R. D. Cowson in G. R. French *Nucl. Handbk.* iv. 14 It gradually became clear that there existed three naturally-occurring families of radioactive substances, called the *Uranium-Radium Series*, the *Thorium Series* and the *Actinium Series*. *Ibid.* iv. 34 Actinium..can be produced artificially by bombarding radium (Ra^{226}) with neutrons.

actino-. Add: **actinodermatitis**, dermatitis arising from the action of actinic rays; **actinomorphic** *a.* = Actinomorphous *a.*
1906 Dorland *Med. Dict.* (ed. 4) 27/1 *Actinodermatitis*..X-ray dermatitis. **1898** H. C. Porter tr. *Strasburger's Textbk. Bot.* I. i. 16 Such parts of plants as may be divided by each of three or more longitudinal planes into like halves are termed either..Radial, or Actinomorphic.

actinobacillosis (æ:ktino,bæsĭlō[u]·sis). [ad. F. *actinobacillose* (Lignières & Spitz 1902, in *Bull. Soc. Méd. Vét.* XX. 487), f. *Actinobacillus* (see *def.*) + -osis.] A disease affecting cattle and other animals and sometimes man, caused by the bacterium *Actinobacillus lignieresi*.
1903 *Vet. Jrnl.* VII. 110 In actinobacillosis thel ymphatic glands are always involved. **1934** Zinsser & Bayne-Jones *Textbk. Bacteriol.* (ed. 7) lv. 867 Actinobacillosis..is primarily a disease of cattle, occurring chiefly in Argentina, with lesions like those of actinomycosis. **1955** Gaiger & Davies *Vet. Path. & Bact.* (ed. 4) xv. 307 Actinobacillosis is chiefly a disease of cattle; it is occasionally met with in sheep.

actinograph. Add: **b.** *spec.* *Photogr.* An instrument (of which there are various kinds) used for recording the actinic power of the light, to determine the correct time of exposure for a photographic plate. (*Disused.*)
1890 *Anthony's Photogr. Bull.* III. 432 The actinograph [shown in] an ingenious application of the slide rule for the purpose of determining photographic exposures. **1902** *Encycl. Brit.* XXX. 702/1 In Hurter & Driffield's 'Actinograph', the light coefficient is given by a printed card showing the curves for every day in the year.

actinology (æktinǫ·lŏdʒi). [f. Actino- + -logy.] The science of the chemical action of light.
1860 in Mayne *Expos. Lex.* Add.

actinometer. Add: **2.** *Photogr.* (See quots.)
1866 J. W. Swan in *Brit. Jrnl. Photogr.* 16 Mar. 125/2, I name the instrument the 'photographic Actinometer', because it is exclusively designed for regulating the exposure of photographic prints. **1926** Lansley & Chipp *Aims & Methods in Study of Vegetation* 93 The ordinary photographic 'actinometer' or 'exposure meter' is a very useful instrument for roughly measuring the relative light intensities of different habitats. **1958** M. L. Hall et al. *Newnes Compl. Amat. Photogr.* vii. 94 Actinometers..Used for many years, but now obsolete, these measured the light intensity falling on the subject by the time taken by a printing-out paper to darken to match a standard tint.

actinomyces (æ:ktino,mɑi·sīz). [mod.L., f. Gr. ἀκτίς, -ῖν- ray (see Actino-) + μύκης fungus.] The ray-fungus, the presence of which, in cattle, constitutes the disease actinomycosis, forms of which are known as *lumpy jaw* (Lumpy *a.* 1 c) and *wooden tongue* (Wooden *a.* 9). Also *attrib.* Hence **a:ctino-myco·tic** *a.*, resembling, related to, or caused by actinomyces.
1882 *Times* 8 Nov. 5/6 From the peculiar manner in which the fungus grows..it has received the name of Actinomyces, and it is consequently proposed to designate the disease Actinomycosis. **1884** Klein *Micro-org.* xvi. 142 In the centre of the nodules the dense groups of peculiar club-shaped corpuscles—*actinomyces*..Each of these actinomyces-corpuscles appears homogeneous. **1894** *Lancet* 5 May 808/1 The first genuine instance of Actinomycosis in this country. **1900** *Jrnl. Exper. Med.* V. 179 The actinomycotic form of the tubercle bacillus. **1949** H. W. Florey et al. *Antibiotics* I. ii. 33 The lysate was a good antigen and they employed it for the treatment of three cases of actinomycosis.

actinomycetes (æ:ktinomaisĭ·tīz), sometimes anglicized to -mai·sīts). *Biol.* [f. Actino- + -mycetes.] A group of minute organisms of the order Actinomycetales, commonly held to be filamentous bacteria; treated as sing. (of less common occurrence) **actinomycete**.
1916 *Jrnl. Bacteriol.* I. 197 (*title*) Possible Function of Actinomycetes in Soil. **1919** S. A. Waksman in *Jrnl. Bacteriol.* IV. 189 Studies in the metabolism of the actinomycetes. **1935** *Lancet* 2 Nov. 1014/2 (*title*) Pathogenic Actinomycetes. **1947** C. E. Skinner et al. *Henrici's Molds, Yeasts, & the Actinomycetes* (ed. 2) 351 The term actinomycete we write in the lower case to indicate that we do not use it in a taxonomic sense but somewhat in the manner we would use yeast or mold. **1960** *Lancet* 21 July 116/1 The actinomycetes, a heterogeneous group comprising many thousands of species, are found in soil (in which they may represent nearly a quarter of the cultivable micro-organisms), in dust, fresh water, and lake bottoms, in decaying plant and animal residues, and in manure.

actinomycin (æktinomai·sin). *Biochem.* [f. Actino- + Gr. μύκης fungus + -in[1].] An antibiotic substance of the actinomycetes group originally isolated by S. A. Waksman and H. B. Woodruff from certain micro-organisms found in soil.
1940 *Proc. Soc. Exper. Biol.* XLV. 610 The active substance was tentatively designated as 'actinomycin'. It is proposed here to designate the above two preparations as 'actinomycin A' and 'actinomycin B'. **1944** *Ann. Reg. 1943* 359 Actinomycin is bactericidal but lethal to animals. **1960** *Lancet* 20 Apr. 582/1 Other fungus-products—streptomycin, streptothricin, actinomycin, and clavacin.

actinophage (ækti·nofē[i]dʒ). *Biochem.* [f. *Actino(mycetes + *phage.] A virus that attacks actinomycetes.
1947 H. C. Reilly et al. in *Jrnl. Bacteriol.* LIV. 451 The term 'bacteriophage' is usually applied to the virus or phage of bacteria, and the term 'mycophage' to that of fungi. By analogy, the term 'actinophage' may be used to designate the phage of actinomycetes. **1951** *Bacteriol. Proc.* 36 An actinophage has been isolated from cultures and fermenter mashes of *S. aureofaciens*. **1961** *Nature* 18 Feb. 603/2 The possible utility of actinophage susceptibility as a criterion for species differentiation in the genus *Streptomyces* has been the subject of much discussion.

actinophorous, *a.* (Example.)
1863 H. J. Carter in *Ann. Nat. Hist.* XII. 32 The actinophorous rays of..Rhizopods.

actinotherapy (æ:ktinoþe·răpi). [f. Actino- + Therapy.] Treatment of disease by means of light rays. Hence **a:ctinotherape·utic** *a.*, **a:ctinotherapeu·tics** *sb. pl.*
1903 Dorland *Med. Dict.* (ed. 3) 26/1 Actinotherapy. **1906** *Practitioner* June 818 Various methods of treatment other than actinotherapy and X-ray therapy. **1913** Dorland *Med. Dict.* (ed. 7) 35/2 Actinotherapeutics. **1921** E. H. & W. K. Russell (*title*) Ultra-Violet Radiation and Actinotherapy. **1930** *Times* 18 Sept. 14/2 He is Surgeon-in-Charge, Actino-Therapeutic Department, Guy's Hospital. **1947** *Archd. Rev.* XCII. 5 The central corridor, either side of which the X-ray and actinotherapy rooms are disposed. **1958** *Gloss. Terms Electr. Engin.* (*B.S.I.*) 144 *Ultra-violet therapy* (*actino-therapy*), the treatment of diseases by ultra-violet radiation of wavelengths between 4000 Ångström units and 2500 Ångström units.

actino-uranium (æ:ktinoyurē[i]·niŏm). [f. Actino- + Uranium.] The radioactive isotope of uranium of mass 235.
1929 Rutherford in *Nature* 2 Mar. 313/2 The probable mass of this new isotope, which for convenience will be called actino-uranium. **1958** H. J. Gray *Dict. Physics* 14/1 Actino-uranium..is explosive when in sufficiently high concentration, owing to nuclear fission, and was used in the first atomic bombs.

‖**actio in distans** (æ·ktio in di·stæns). Also **actio ad distans.** [L., = 'action upon a thing standing apart'.] The exertion of force by one body upon another separated from it by space, as in some theories of gravitation; action at a distance. Also *fig.*
1846 W. Hamilton in *Reid's Wks.* 852/1 Repulsion..remains, as apparently an *actio in distans*, inconceivable as a possibility [for inclusion among the primary qualities of body]. **1890** W. James *Princ. Psychol.* I. ii. 47 It then becomes less probable that it [sc. hemiopia] is due to an *actio in distans*, probably to the interruption of fibres proceeding from the occipital lobe. **1901** Baldwin *Dict. Philos.* I. 21/2 Intimately associated with gravitation is the question whether it is really an *actio ad distans*—whether it takes place without any intervening medium or other agency. **1960** W. James *Meaning of Truth* x. 221 By a sort of *actio in distans* my statement had taken direct hold of that other fact. **1937** A. H. Murray *Philos. of James Ward* iv. 78 The monads or 'psychoids' are qualitatively distinct and are capable of what in scholastic philosophy was called *actio in distans*.

action, *sb.* Add: **3. b.** Proverb: *actions speak louder than words.*
1845 *Knickerbocker* XXV. 106 He had heard that 'actions speak louder than words', and he acted. **1856** Abraham Lincoln *Coll. Wks.* (1953) II. 352 'Actions speak louder than words' is the maxim. **1906** F. McCullagh *With Cossacks* 178 The gallant foreigner, who could not tell them how he sympathized with them, but whose actions spoke louder than words.
6. d. Used as a film director's word of command.
[**1914** J. B. Rathbun *Motion Picture Making* iii. 68 At the word 'Ready', given by the producer, the camera man starts cranking the machine and the actors stand alert..and an instant after follows the order, 'Act'.] **1923** F. A. Talbot *Moving Pictures* xiv. 218 Ready! Action! Camera!!Go!!!! **1929** *Daily Express* 13 Feb. 10/2 (if there is a) cameraman then the director calls out 'Action' to the actors, then 'Roll 'em' and the cameraman starts the camera.
7. a. *to take action.* (Examples.)
1865 G. W. Lyttelton *Ephemera* I. 174 In the whole.. Session the only 'action taken'—according to the vile modern Yankeeism—by Lord Derby..was to move for a Committee on National Education. **1907** F. G. Lee *Church under Q. Eliz.* I. v. 279 Her Majesty's advisers, therefore, lost no time in taking fresh action consequent upon the publication of this Bull. **1922** Joyce *Ulysses* 309 So Joe starts telling the citizen about..taking action in the matter. **1961** *Times* 12 May 11/1 Did anybody, or a friendly Power, at any time, take action? *Ibid.*, any action, any action, what action was taken?
11. b. *action front!, action rear!*, military commands in artillery regiments to prepare for action against the enemy in front of, behind, the line of guns.
1860 *Man. Artill. Exerc.* 20 Unlimbering or coming into Action...At the word *Action front*, each number repairs to his post as in limbering up. **1875** *Man. Field Artill. Exerc.* vi. 205 Guns..in heavy ground..should be brought into action rear'. **1892** Kipling *Barrack-r. Ballads* 37 But 'e swung 'is 'orses 'andsome when it came to 'Action Front!' **1896** *Field Artill. Training* v. 161 Guns..may be brought into action...by the command 'Action rear', so as to save unnecessary labour and delay. *Ibid.* v. 167 On the command 'Action front' the guns are brought into action and..officers go to their places in action.
16. action committee, group, etc., one chosen to take active steps, esp. in local or national politics; **action current** *Biol.*, the electrical current produced in living tissue during activity; **action painting** (orig. *U.S.*), the theory or practice of the action school in art; a form of abstract art in which the paint

is placed in strokes or splashes on the canvas by the spontaneous or random action of the artist; hence *action painter*, *school*, etc.; **action-photography**, photography representing the subject in action; hence *action-photograph*; also *action picture*, *shot* (*shot *sb.*[1]), etc.; **action potential** *Biol.*, the difference in electrical potential between the excited and unexcited parts of a nerve or muscle; **action song** (see quot. 1938); **action station** (Action *sb.* 10, 11), a position assigned to someone going into action; (*pl.*) the signal to proceed to such a position; **action-time** *Physiol.*, the period between the application of a stimulus and (the resulting) reaction; reaction-time.
1949 *Britannica Bk. of Yr.* 244 Action committee, a Communist committee appointed to purge societies and organizations of all non-Communist elements, as a means of consolidating a communistic revolution. **1952** *Ann. Reg. 1951* 103 [In S. Africa] the 'Commando' started as a small body of ex-servicemen calling itself the War Veterans Action Committee. **1958** *Oxf. Mail* 30 Dec. 1/4 The programme also calls for a scientific action committee to organise research..to evaluate drugs. **1875** *Jrnl. Physiol.* Stud. Univ. Inst. Pt. I. 29 *Action current*. **1913** R. Sanderson in *Phil. Trans. R. Soc.* 1882 CLXXIII. 25 Of the nature of this preliminary disturbance (to which alone the term excitatory variation ought to be applied, it alone being the analogue of the 'action current' of animal physiology) we know nothing. **1934** *Ann. Reg.* 1934 121 Dr. Ewoltsen's journey (in Nigeria). 'The action current is the only known change accompanying excitation in nerve which is competent to stimulate an adjoining nerve. **1955** *Ann. Reg.* 1954 137 The 'Action Group' party in Nigeria. The Action Group, lost ground except for Western Nigeria. **1952** *Observer* 21 Dec. 3/5 Pacific Youth Action Group. **1952** H. Rosenberg in *Art News* 22 (*title*) The American action painters. **1964** *Art* 1 Action painting is painting in the medium of difficulties. **1958** *Times* 24 June 6/4 As an Action painter his contemplative period ends in an eruption of black and scarlet blunt-edged squares. **1958** D. Lewis *Alan Davie* [Exhib. catal.] 1 'Action' painting has ... **1961** Other terms given to it have been 'tachism', 'abstract expressionism', etc., but action painting is the best of these. **1890** *Photogr. Ann.* (Sept.) So-called 'action' photographs..the process. **1961** A photographer ...and..action-photographs. ... **1901** N. V. Lindsay *Art of Moving Picture* ii. 9 Many Action Pictures are indoors. **1935** R. Elbert in *Action Pict.* ..Action-photography. **1913** R. Rogers & F. Morrison ... ***Action potential***...
1929 W. B. Vary *Inferred Animal Physiol.* 187 The action potential..

[illegible]

1936 ..action stations ...

Hence **activation** (æktivē[i]·ʃən), the action of activating; the state of being activated; in specific uses corresp. to those of the vb.; *activation analysis* = *radioactivation analysis.
1906 *Practitioner* Dec. 747 Mixed sera from several animals might be used, in the hope that one might be suitable for activation with human serum. **1919** *Science* L. 268/1 (*title*) Activation theory. **1937** Haldane & Huxley *Animal Biol.* ii. 54 Activation, or the starting-off of the egg on development. **1939** *Engineering* 21 Feb. 273/1 At least one of the molecules concerned must be 'activated' and this 'activation' may well be due to the absorption of radiation. **1955** J. W. McBain *Sorption of Gases by Solids* 19 Methods of activation always involve heating to more or less elevated temperatures. **1940** *Ann. Reg. 1939* 371 Gene mutations were generally regarded as due to individual atomic activations. **1949** *Atomics* Nov. 106/2 Tracers are only rarely radioactive. To overcome this difficulty, the sample may be irradiated either with high energy charged particles or thermal neutrons.. This method is known as activation analysis. **1960** *Nature* CLXXXV. 196 Neutron activation analysis of ancient Roman potsherds.
Also **a·ctivator**, one who (*rare*) or a thing which activates (in various technical senses); a catalyst.
1911 R. C. Punnett *Mendelism* (ed. 3) 45 A ferment which behaves as an activator of the chromogen. **1913** Dorland *Med. Dict.* (ed. 7) 36/1 *Activator*, a substance which renders some other substance active: especially an inorganic substance which combines with an inactive enzyme to render it capable of effecting its proper reaction. **1948** *Electronic Engin.* XX. 225 The inclusion of activators such as tin, copper, antimony..in calcium oxide caused speedy and violent ion-burning. **1958** *Spectator* 4 July 9/3 They chose Maurice Bucknaster, who later became famous as an activator of the French Resistance. **1962** *Lancet* 27 Jan. 193/1 The defect in fibrinolytic activity is associated with 'unavailability' rather than with an absolute deficiency of plasminogen activator. **1963** B. Fozard *Instrumentation Nucl. Reactors* vi. 63 In scintillation phosphors these impurities are known as activators.

actionability (æ:kʃənăbi·lĭti). [f. Actionable *a.* + *-bility.] Liability to action at law.
1883 Moyle *Institut. Justinian* I. 46 Actionability is only one of such causes.

actional, *a.* (Later examples.)
1935 G. O. Curme *Gram. Eng. Lang.* II. 218 The common actional form is employed also as a statal passive.. 'The door *was shut* (statal) at six, but I don't know when it *was shut* (act)'. **1958** Priebsch & Collinson *Germ. Lang.* (ed. 4) i. vi. 206 A dative singular in -*anne* (with doubling caused by *j* on the analogy of neuter actional *ja*-stems like *gibbi* 'chatter').

actioned (æ·kʃənd), *ppl. a.* [f. Action *sb.* 6 b.] Having an action of a specified kind. Usu. preceded by a defining word.
1837 *Times* 20 June 1 (Advt.), Twenty remarkably fine, powerful, young, fresh, good actioned, short-legged Horses, in beautiful condition. **1875** [see Back *sb.*[1] 19 a]. **1928** *Daily Tel.* 30 Jan. 18/7 Quick actioned, weight-shifting Shires, Clydesdales, Suffolks, or Percheron horses. **1949** E. Jenkins *Hyde* 'tired' the two horses she drove herself in a phaeton with two 'handsome actioned' cobs.

activate, *v.* Delete †*Obs.* and add later examples. Hence **a·ctivated** *ppl. a.*, **a·ctivating** *ppl. a.* and *vbl. sb.*
1858 Bennet *Nutrition* ii. 42 Increased muscular vigour ..activates respiration. **1905** *Sat. Westm. Gaz.* 15 July 13 The young English dramatist has very few opportunities of making the hair of a play-goer..activated rely on his activating his digestion. **1926** J. A. Thomson *Man in Light of Evol.* 10 The rarely activated muscles of our rare-trumpet. **1926** *Public Opinion* 22 Oct. 408/3 The plant and its activating power. **1946** H. G. Wells *Babes in Darkling Wood* II. iii. 195 She wanted 'activating'. She had recently acquired that word. **1961** *Always in Stud.* XXV. 26 To motives of self-appointed moralists..Are they activated by concern for public morality? **1964** *Amer. Speech* XXXIX. 271 A flat plate set in the pavement to detect traffic in order to activate traffic lights.
b. *spec.* (*a*) *Chem.*, to render active, reactive, excited (of carbon, molecules, etc.); *activated carbon*, *carbon*, *esp.* charcoal, which has been treated to increase its adsorptive power; (*b*) *Physics*, to make radioactive; (*c*) to aerate (sewage) as a means of purification; *activated sludge*, aerated sewage containing aerobic bacteria.
1902 *Jrnl. Soc. Chem. Industry* XXI. 1102/1 Schobein found that sulphurous acid had a remarkable activating [*activirende*] effect on various oxidising substances, *i.e.*, they were considerably more active in the presence of small quantities of sulphurous acid. **1903** *Electr. World & Engineer* 10 Jan. 86 (C.D. Suppl.), Underground air is not like activated air. **1905** *Sci. Amer. Suppl.* 28 Apr. 22835/1 In these measurements it is necessary to maintain the body to be activated for several hours at a negative potential of several thousand volts. **1907** *Phil. Mag.* Record 3 Aug. 171 The former [ferment, viz. enterokinase] activates the pancreatic juice. **1919** *Jrnl. Industr. & Engin. Chem.* XI. 282 The development of activated carbon as a canister filler was largely the work of two Cleveland organizations. **1921** *Glasgow Herald* 13 Aug. 5 The new process of producing 'activated sludge' by which complete purification of the sewage is achieved. **1921** *Jrnl. Soc. Chem. Industry* XL. (Trans.) 419/1 (*title*) Active Carbon. **1934** *Jrnl. Inst. Sanit. Engin.* Sept. 251 The sewage is settled, treated by the activated sludge process, and passed through the clarification. **1937** A. M. Prentiss *Chemicals in War* xix. 549 The filter [in a gas-mask canister] consists of an oval-shaped perforated sheet-steel container filled with a mixture of 50 per cent activated charcoal and 50 per cent soda-lime. **1938** H. G. W. Lawson tr. *Hevesy & Paneth's Man. Radioactivity* xiv. 189 We find that the wire now shows no trace of radioactive induced after its removal from the Actinium vessel. *Ibid.* x. 119 Most of the elements situated between boron and calcium have been activated under the influence of α-rays. **1956** A. H. Compton *Atomic Quest* 71 Polling the last activated slugs from the Hanford piles.

active, *a.* Add: **4.** *active list* (see List *sb.*[6]); also *transf.*; *active service*, war service in the field, at sea, or in the air.
1858 [see Service[1] 11]. **1838** *Navy List* 20 Sept. 118 Alphabetical List of Masters. Those in Italics are unfit for active Service. **1852** *Ibid.* 20 Dec. 327 The number of captains on the Active List will be reduced to one hundred permanently exceeding 160. **1856** Kipling *Brushwood Boy* 7 'Cleaned' him as he never was cleaned on active service, wiping something off a slate. **1906** *Harmsworth Encycl.* VII. 576/1 No officer on the active list may leave the United Kingdom without special permission. **1907** *Jrnl. Soc. Arts* LV. 1075/1 Of these only the Remington and the Hammond are to-day on the active list.

b. *spec.* (in *Physics*), = *radioactive *a.*
1900 Rutherford in *Phil. Mag.* XLIX. 163 The top layer of paper over the thorium was found to be active on its upper side. **1905** [see *active deposit*]. **1938** H. W. Lawson tr. *Hevesy & Paneth's Man. Radioactivity* (ed. 2) x. 120 Active isotopes that are lighter than the stable types disintegrate with emission of positrons.
8. *Comb.* **active carbon**, **charcoal** = activated carbon, charcoal (see *activate *v.* b); **active current** *Electr.* (see quot. 1926); **active deposit** *Physics* (see quot. 1955); **active mass** *Chem.*, the molecular concentration of a substance (see *mass-action* (*b*) s.v. *mass *sb.*[2])
1918 *Jrnl. Industr. & Engin. Chem.* X. 813/2 If our object is to make an active carbon, none of the mineral matter must be removed before heating. **1921** *Jrnl. Soc. Chem. Industry* XL. (Trans.) 231/1 All the secret process active carbons come upon the market in the form of a fine powder. **1938** *Thorpe's Dict. Appl. Chem.* II. 193/2 Active charcoal is widely used in industry for the recovery of valuable products which exist as attenuated vapours in unsuitable gases. **1924** S. R. Roget *Dict. Electr. Terms* 1 *Active current*, the 'watt' current. **1926** *Gloss. Terms Electr. Engin.* (*B.S.I.*) 30 *Active current*, that component of an alternating current (regarded as a vector quantity) which is in phase with the voltage. **1905** Rutherford in *Phil. Trans. R. Soc.* CCIV. 175 It is speaking generally of the active matter, which causes excited activity, without regard to its constituents, I have used the term 'active deposit'. **1906** —— *Radioactive Transformations* iv. 95 All bodies surrounded by the radium emanation become coated with an invisible active deposit, possessing physical and chemical properties which sharply distinguish it from the emanation. **1955** *Gloss. Terms Radiology* (*B.S.I.*) 8 Active deposit, solid radioactive deposit formed by the decay of a radioactive emanation. **1909** *Cent. Dict. Suppl.*, Active mass. **1910** *Encycl. Brit.* VIII. 711 So that the ratio of the reacting components. **1932** Philbrick & Holmyard *Text Bk. Theoret. & Inorg. Chem.* II. iii. 334 This law was first enunciated by Guldberg and Waage, who stated that *the rate at which a substance reacted was proportional to its active mass*.
b. *sb.* **b.** A person devoted to the active life. (Cf. *adj.* 1 and Contemplative *sb.* 1.)
*c*1380 [see Contemplative *sb.* 1]. *a* **1425** *Cloud of Unknowing* (1944) 51/18 Into this day alle actyves pleynen hem of contemplatyves, as Martha did on Mary. **1671** Phillips *New World of Words*, Actifs, an order of Fryars, that feed on Roots, and wear tawny habits. **1946** E. A. Peers *Fool of Love* vii. 128 It is the active rather than Ramon [Lull] appeals to the present age most widely. **1950** A. Huxley *Themes & Variations* 114 Were there not then, as always, the actives?

activism (æ·ktiviz'm). [f. Active *a.* + -ism.] **1.** A philosophical theory which assumes the objective reality and active existence of everything.
1907 W. R. B. Gibson *Eucken's Philos. of Life* (ed. 2) App. 170 Activism has affinities with Pragmatism, especially on its negative side. **1930** *Ibid.* 172 Activism..is essentially realistic. It assumes the 'objective' validity and 'real' being of entities and relations, as well as the fundamental relational complexes of space, time, number, and change.
2. A doctrine or policy of advocating energetic action.
1920 *Glasgow Herald* 12 Aug. 7 It is a question of repelling the fateful activism of the Entente just as it was necessary in 1916 to stand against German activism. **1933** V. A. Demant *God, Man & Society* iii. 93 The major conflict in modern society between technique to serve life, and the moral philosophy of economic 'activism' which is embodied in economic theory and practice. **1960** *Spectator* 19 Aug. 272 The sizzling flame of activism is visible in both the agricultural and pastoral districts.
Hence **activist** (æ·ktivist), an advocate of activism in either sense; also *attrib.* (passing into *adj.*); **activi·stic** *a.*
1907 W. R. B. Gibson *Eucken's Philos. of Life* (ed. 2) App. 170 Eucken deliberately adopts the activistic label as a distinctive philosophical badge. **1909** *Athenæum* 17 Apr. 469/3 Pragmatism..is tainted with the characteristic activist fallacy of making process an active account for the process of knowledge of the first process which it implies. **1913** E. Underhill *Mystic Way* 35 The positive and activistic mysticism of the West. **1915** *Times* 7 Aug. 7/6 For some, neutrality simply means a passive aloofness. For others, neutrality is active, and these are divided, in the current jargon, into the 'passivists' and the 'activists'. **1949** *Theology* LII. 363 American Christianity has tended traditionally to express itself in an activist form. **1954** Koestler *Invisible Writing* 206 He was not a politician but a propagandist, not a 'theoretician' but an 'activist'.

activity. Add: **1. b.** *spec.* in *Physics*, = *radioactivity; the disintegration rate of a radioactive substance.
1903 Rutherford in *Phil. Mag.* V. 446 The excited activity from radium decays much faster than that from thorium. **1926** R. W. Lawson tr. *Hevesy & Paneth's Man. Radioactivity* xxiv. 174 The a-activity of uranium in equilibrium with all its transformation products is 4.73 times as large as that of the uranium itself. **1958** W. K. Mansfield *Elem. Nucl. Physics* iii. 21 The disintegration rate..is known as the *activity* of the sample...The unit of activity is the curie, which is defined as 3.7×10^{10} disintegrations a second.
4. b. *spec.* (esp. in *pl.*) The parts of a school curriculum devoted to projects carried out by the pupils; also applied *attrib.* to a system or

method of teaching involving such projects. orig. *U.S.*

1923 E. COLLINGS *Exp. with Project Curriculum* vii. 325 Curriculum Principles... A series of concrete and practical activities directed toward some foreseen end. **1934** *Nat. Soc. Study Educ. 33rd Yearbk.*, (*title*) The Activity Movement. **1938** *Amer. Speech* XIII. 204 She suggests that classes organized for 'activities' can get more out of radio than others. **1960** C. W. HARRIS *Encycl. Educ. Research* 852/1 The project and activity methods were commonly used.. in the twenties and thirties.. as a general approach to teaching.

5. *attrib.* and *Comb.*

1905 W. JAMES *Ess. Radic. Empir.* (1912) vi. 185 An activity-process is the form of a whole 'field of consciousness'. **1905** —— *Notebk.* 17 Dec. in R. B. Perry *Tht. & Char. of W. J.* (1935) II. 757 Purely on the plane of the analysis of what experience *is*, the co-ness of content there is but the intellectual preliminary to *activity*-experience. *a* **1910** —— *Some Probl. Philos.* (1911) xiii. 210 Where now is the typical experience [of causation] originally got? Evidently it is got in our own personal activity-situations. **1940** R. S. WOODWORTH *Psychol.* (ed. 12) xi. 364 In one type of experiment, energy output is measured by use of the *activity cage*, which is like a squirrel cage... A mechanical counter shows the number of times the animal's running has turned the wheel round.

actomyosin (æ:ktǫ̈mai·ǫ̈sin). *Biochem.* [irreg. f. Gr. ἀκτίς + MYOSIN.] A complex of actin and myosin.

1942 A. SZENT-GYÖRGYI in *Stud. Inst. Med. Chem. Univ. Szeged* I. 67 The myosin-actin complex will be called 'actomyosin'. **1944** —— *Stud. on Muscle* III. 39 The formation of actomyosin is reversible and the complex readily dissociates under different conditions into its components, actin and myosin. **1957** *Lancet* 21 Dec. 1288/2 The contractile element of muscle is actomyosin.

actor. Add: **5. actor-manager** (earlier example); **actor-proof** *a.*, applied to a play or part in a play of which the excellence is evident apart from the standard of the actor(s). Also, further examples of appos. use, as *actor-management*, *-playwright*, *-producer*. So *actor-manage(r)* vbs. (nonce-wds.).

1907 G. B. SHAW *Let. to G. Barker* 27 Feb. (1956) 75 If I am to be actor-managered out of all my decent leading men.. I shall go and be actor-managed by Alexander. **1894** W. ARCHER *Theatr. 'World' for 1893* 163 As smooth, as adequate, as they could possibly have been in any actor-managed theatre. **1895** G. B. SHAW in W. Archer *Theatr. 'World' of 1894* Pref. p. xviii, In order to bring up the list of real exceptions to the London rule of actor-management to three. **1904** P. FITZGERALD *Garrick Club* viii. 241 What discussions, heated and other, used there to be on this topic of actor-management! **1826** F. REYNOLDS *Life & Times* II. 126, I speak of *Actor-managers* both in town, and in country. **1926** FOWLER *Mod. Eng. Usage* 644/1 The actor-playwright is invariably the better craftsman than the literary man who commences dramatist. **1927** J. P. KENNEDY *Story of Films* viii. 181 There are seventeen actor-producers, men and women producing dramatic films and comedies. **1938** N. MARSHALL in R. D. Charques *Footnotes to Theatre* 110 Gerald du Maurier is another obvious example which can be cited against such a generalisation about actor-producers. **1895** G. B. SHAW *Our Theatres in Nineties* (1931) I. 102 Plays good enough to be comparatively 'actor-proof'. **1923** J. M. MURRY *Pencillings* 206 Molière is infinitely the better dramatist... His plays are as actor-proof as Shakespeare's. **1938** W. S. MAUGHAM *Summing Up* 151 There is no such thing as an actor-proof part.

actorship. (Later examples.)

1773 in *Private Corresp. Garrick* (1831) I. 544 Worthy of your genius, (probably as equal in authorship, as it has been *proved* in actorship, to both the muses of the theatre). **1832** *Rep. Sel. Comm. Dram. Lit.* 25 If novelty were given in the shape of the regular drama, both as respects the authorship and the actorship, if I may use that word. **1840** E. B. BROWNING *Lett.* (1877) I. 17 Only you do more honour to the stage and the actorship than I could do.

actress. Add: **3.** *Comb.* **actress-manageress** (cf. *actor-manager*).

1894 G. B. SHAW in W. Archer *Theatr. 'World' for 1893* Pref. p. xxix, The time is ripe for the advent of the actress-manageress. **1937** *Times* 29 Sept. 8/3 A National Theatre that fulfilled the aspirations of actors and actresses would have to consist of plays every part of which was written for actor-managers or actress-manageresses.

actressy (æ·ktrėsi), *a.* [f. ACTRESS + -Y[1].] Pertaining to or resembling an actress; affectedly theatrical.

1896 G. B. SHAW *Let. to E. Terry* 5 Dec. (1931) 111 You'd feel instantly with her that such a line would be actressy and that the dressing-room was the wrong *scene* for the right *line*. **1911** M. & J. FINDLATER *Penny Monypenny* III. ii. 304 Photographs of beautiful ladies.. dreadful actressy people all teeth and hair. **1958** *Observer* 9 Feb. 11/5 She talked of the theatrical digs she'd known in this part of the world. Her voice was actressy and deep.

actual, *a.* **1.** Delete † *Obs.* and read: *Obs.* exc. in *actual grace*, *actual sin* (see quots.). *R. C. Ch.*

c **1315–1534** [see Dict. †1]. **1577** [see ORIGINAL A. 1 b]. **1638** W. CHILLINGWORTH *Relig. of Protestants* Answ. 2, para. 163, p. 118 Not touching the Virgin Maries freedome from actuall and original sinne. **1742** J. & C. WESLEY *Hymns & Sacred Poems* (1869) II. 338 From actual and from inbred sin Us thou hast wash'd in Thine own blood. **1855** F. W. FABER *Growth in Holiness* xxii, They [*sc.* seven supernatural gifts of the Holy Ghost] are played upon according to the needs of our spiritual life by what are called the actual impulses of the Holy Ghost, and which correspond in their subject-matter to actual grace, standing in the same relation to the habitual gifts as it does to habitual grace. **1859** *Catechism Chr. Doctr.* 18 Q. How many kinds of sin are there? A. Two; original and actual... Q. What is actual sin? A. Every sin which we ourselves commit. **1890** WILHELM & SCANNELL *Man. Cath. Theol.* I. III. II. i. 437 Around this [Habitual] Grace are grouped all other salutary Graces especially 'Actual Grace'. **1957** *Oxf. Dict. Chr. Ch.* 14/2 *Actual sin*, a sin, whether of commission or omission, which is the outcome of a free personal act of the individual will. *Ibid.* 577/1 *Actual grace*, a certain motion of the soul, bestowed by God *ad hoc* for the production of some good act. It may exist in the unbaptized.

actual (æ·ktiuăl), *sb.* [f. the adj.] *pl.* Actual qualities, actualities (opposed to *ideals*).

1541, 1587 [see ACTUAL *a.* 3]. *a* **1902** S. BUTLER *Way of All Flesh* (1903) vii. 31 He may have had an ill-defined sense of ideals that were not his actuals. **1926** D. WATSON *Church at Work* x. 119 Christian ideals must be made actuals.

actualist (æ·ktiuălist). [f. ACTUAL *a.* 3.] One who aims at actuality or realism. Hence **actuali·stic** *a.*

a **1866** J. GROTE *Moral Ideals* (1876) xv. 375 To which the *actualist* ever answers, The moral world.. is *given* by human constitution and circumstance. **1887** *Harper's Mag.* Jan. 324/1 In his first essay in the field of fiction he turns out an actualist, whose first wish seems to be truth to his facts and the meaning of them. **1893** *Funk's Stand. Dict.*, Actualistic. **1921** HANNAY & COLLINGWOOD tr. *Ruggiero's Mod. Philos.* 201 Over against this actualistic concept of life,.. we find maintained.. an absolutely intellectualistic conception. **1934** *Essays & Stud.* XIX. 145 An actualistic drawing-room play.

actuality. Add: **4. b.** *spec.* in *Cinemat.*, *Television* (see quot. 1941). Also *attrib.*

1929 H. G. WELLS *King who was King* I. § 1. 10 The films began with 'actualities', the record of more or less formal current events. **1941** *B.B.C. Gloss. Broadc. Terms* 3 *Actuality*, presentation of real persons and things to give a picture of contemporary life in a particular aspect; documentary. **1944** L. MACNEICE *Columbus* 12 The radio dramatist.. must select his actuality material with great discrimination. **1963** *Listener* 3 Oct. 501/2 Television is a medium far more successful at documentary or 'actuality' than at fiction.

actuarially (æktiuē°·riăli), *adv.* [f. ACTUARIAL *a.* + -LY[2].] In relation to actuarial principles; on an actuarial basis.

1884 *Athenæum* 12 July 39/3 That.. every society hereafter formed has its actuarially certified table. **1886** *N. Amer. Rev.* Sept. 233 The trade-unions of England are, actuarially speaking, bankrupt. **1905** *Daily News* 12 Apr. 9/3 These schemes.. were found.. to be actuarially sound.

actuary. Add: **2. b.** *spec.*, one whose office it is to manage the deposits in a savings bank.

1816 E. CHRISTIAN *Plan for County Provident Bank* 19 The Actuary, in the name of the Institution, is to receive such Deposits under £20 as may be offered, which he shall enter into his Deposit Book. **1817** J. BOWLES *Provident Institutions called Savings' Banks* (ed. 3) 45 The actuary shall keep all the accounts of the institution in a regular set of books, and be responsible for the accuracy of every individual account. **1919** W. THOMSON *Dict. Banking* (ed. 2) 13/2 *Actuary*, the chief official in a savings bank.

actuate, *v.* **6.** Delete † *Obs.* and add later examples.

1920 *Chambers's Jrnl.* 25 Dec. 55/1 The [human] mass.. came round with an almost dismaying swiftness. 'How could such a bulk actuate with such rapidity?' Clement thought. **1924** O. LODGE *Making of Man* v. 113 He is beginning to learn.. that the portion of consciousness now actuating and made manifest in his brain is but a small part of the whole.

Hence **a·ctuator**, one who or a thing which actuates.

1889 in *Cent. Dict.* **1890** C. MERCIER *Sanity & Insanity* xii. 299 The higher nerve regions are the actuators of conduct. **1920** L. BAIRSTOW *Appl. Aerodynamics* vi. 281 W. E. Froude introduced into airscrew theory the idea of an actuator. **1931** *Electronics* Sept. 107/2 The electrostatic actuator consists of a grill or slotted plate which is slipped into the front opening of a condenser microphone. **1956** *Spaceflight* I. 24/1 Pivoting motors.. to provide pitch and yaw control, movement being obtained by electro-hydraulic actuators.

‖ **actus purus** (æ·ktǫ̈s piū°·rǫ̈s). *Philos.* Also **purus actus**. [med. L., 'pure act', tr. Gr. ἐνέργεια.] Actuality unmixed with potentiality; pure act (ACT *sb.* 3). Also, a mental act uncontaminated by elements of sensory awareness (see quots. 1890).

[*a* **1274** AQUINAS *Sum. Theol.* I. 14. 2 Cum igitur Deus nihil potentialitatis habeat, sed sit actus purus.] **1707–8** BERKELEY *Commonpl. Bk.* (1944) No. 701 The substance of Spirit we do not know.. it being purus actus. *Ibid.* No. 828 The Will is purus actus or rather pure Spirit. **1878** S. H. HODGSON *Philos. of Reflection* II. 12 Pure energy, *actus purus*, without anything merely potential. **1890** W. JAMES *Princ. Psychol.* I. ix. 245 They are known, these relations,.. by an *actus purus* of Thought, Intellect, or Reason. *Ibid.* xii. 474 The conceptualists.. invent.. as the vehicle of the knowledge of universals, an *actus purus intellectûs*; or an Ego, whose function is treated as quasi-miraculous... The nominalists.. dislike *actus puros*.

acushla (ăku·ʃlă). *Anglo-Ir.* [f. Ir. *a* O + *cuisle* vein, pulse (of the heart); cf. Ir. *cuisle mo chroidhe*, my heart's pulse, my darling.] Dear heart; darling. (Used as a term of address.)

1842 S. LOVER *Handy Andy* vi. 61 'No, *a cushla*,' whispered the aunt. **1865** DION. BOUCICAULT *Colleen Bawn* I. i. 11 Come to your own Eily, that has not seen you for two long days. Come, acushla agrah machree. **1899** SOMERVILLE & 'ROSS' *Exper. Irish R.M.* v. 112 'Come here to me, acushla,' says I to him. **1936** 'F. O'CONNOR' *Bones of Contention* 208 Cross me, acushla, and I'll shift my tent.

acute, *a.* Add: **2.** (Examples of *fig.* sense.) Severe; crucial.

1932 'E. PRICE' *Enter—Jane* I. 11 Jane Turpin's financial position.. might well be described as acute. **1941** *Spectator* 12 Dec. 547/1 For months past a state of acute tension has existed in the Far East. **1942** *Punch* 4 Feb. 83/2 Housing Shortage Grows Acute.

acyclic, *a.* Add: **2.** *Dynamics*, etc. That does not move in circles.

1873 MAXWELL *Electr. & Magn.* I. 137 As the negative region continues to expand till it fills all space, it loses every degree of cyclosis it has acquired, and becomes at last acyclic. **1902** *Encycl. Brit.* XXVII. 570/2 The system now behaves, as regards the co-ordinates $q_1, q_2 \ldots q_m$, exactly like the acyclic type there contemplated. **1963** *New Scientist* 7 Feb. 294 A commercial-scale acyclic generator has been constructed and put to work.

3. *Chem.* Applied to an organic compound that contains no 'cycle' or ring of atoms.

1909 *Chem. Abstr.* 2145 (*title*) Cyclization of acyclic diketones. **1913** BLOXAM & LEWIS *Chem.* 544 There are two great divisions of organic compounds (*a*) The acyclic, open-chain, fatty or aliphatic.. series; (*b*) the cyclic or closed-chain series.

acyl (æ·sil, ē[i]·səil). *Chem.* [G. *acyl* (C. Liebermann 1888, in *Ber. d. deutschen chem. Gesellschaft* XXI. 3372); cf. AC(ID, -YL.] Name used to designate an organic radical derived from an acid, as acetyl, benzoyl, etc.; an acid radical; also *attrib.* Hence **a·cylate** *v.*, to introduce an acyl radical into (a compound); **a:cyla·tion**.

1899 *Jrnl. Chem. Soc.* LXXVI. I. 553. **1901** *Ibid.* LXXX. I. 118 Transformation of O-Acyl Derivatives.. into the Isomeric C-Acyl Derivatives. *Ibid.* 187 Wandering of Acyl Groups. **1907** *Chem. Abstr.* 1290 Indanthrene is not acylated by benzoic anhydride. **1908** *Ibid.* 2088 The reactions of phenylhydrazine and its acyl derivatives.. were studied. **1910** *Ibid.* 177 Acylation of Amines and Phenols. **1910** *Encycl. Brit.* III. 756/1 With formic or acetic acids.. the para-semidines give acyl products possessing no basic character. **1949** H. W. FLOREY et al. *Antibiotics* xxi. 807 A free NH group, whose presence could be demonstrated by acylation. **1959** *Times* 6 Mar. 13/7 Various penicillins can be regarded as acyl derivatives.

ad (æd), colloq. abbrev. of ADVERTISEMENT and ADVERTISING. Also *attrib.* and *Comb.*, as *ad-man*, *ADMASS.

1841 W. M. THACKERAY in *Britannia* 1 May 284/1, I'll have my books properly reviewed; or else, I'll withdraw my ads. **1852** *Household Words* V. 5/2 We know that the really interesting 'ads.' are in the body of the paper. **1902** HOWELLS *Lit. & Life* 268 Ad is a loathly little word, but we must come to it. It's as legitimate as lunch. **1909** *Collier's* 22 May 15/2 So in a sense, the ad-man is a public entertainer. **1919** [see *WANT *sb.*[1]]. **1922** JOYCE *Ulysses* 158 Best paper by long chalks for a small ad. **1933** *Scrutiny* I. 400 Like all successful ad-men he has come to believe quite uncritically in what he sells. **1942** M. McCARTHY *Company she Keeps* (1943) v. 132 He was the Average Thinking Man.. that.. ad-writers try to frighten. **1957** *Observer* 10 Nov. 15/4 That side of modern life.. which bears the finger-smears of the ad. man. **1958** *Ibid.* 28 Sept. 21/7 The heroine.. is straight out of the ad. pages in the shiny feminine magazines. **1959** *Spectator* 19 June 875/2, I cannot change my opinion.. that 'admags' in their present form are contrary to the intentions of the framers of the TV Act.

-ad, *suffix* invented by J. Barclay in *A New Anatomical Nomenclature*, 1803, in the sense of 'towards' (the part denoted by the main element of the word), as *CAUDAD, *CEPHALAD, DEXTRAD, DORSAD, LATERAD, NEURAD.

adagial, *a.* (Later examples.)

1869 *Contemp. Rev.* XII. 219 This constant resort of Theocritus to adagial expressions. **1913** BARING-GOULD & FISHER *Lives of Brit. Saints* IV. 368 The poem is of an adagial or moral character.

adagio, *sb.* Add: **2.** A dance or ballet movement in adagio time. Also *attrib.*

1830 R. BARTON tr. *Blasis's Code of Terpsichore* II. i. 56 Can any thing be more ludicrous than to see a thick-set dancer.. gravely figure off in a slow and mournful *adagio*? **1922** BEAUCLERK & EVRENOV tr. *Svetlov's Karsavina* 16 No dancer could find a better partner in *pas de deux* or *adagio*. **1929** M. LIEF *Hangover* 179 That new adagio team I imported from Vienna. **1946** WODEHOUSE *Joy in Morning* iii. 18 He spun round with a sort of guilty bound, like an adagio dancer. **1959** *Sunday Times* 29 Mar. 11/5 This *adagio* should never be devoid of character when it is part of the full-length ballet.

adalin (æ·dălin). *Chem.* Also **adaline**. [a. G. *adalin* (1910 *Medizinische Klinik* 20 Nov. 1859/1).] (See quots.)

1911 *Jrnl. Chem. Soc.* C. II. 1120 Adaline is a sedative producing light, but lasting, sleep. **1912** *Ibid.* CII. I. 244 Adaline (α-bromo-α-ethylbutyrylcarbamide).

Adam[1]. Add: **1. b.** Phr. *not to know* (a person) *from Adam*: not to recognize him; (*as*) *old as Adam*: primevally old. Also, *since Adam was a boy*, etc.

1784 *London Sessions* Feb. 400/1 Some man stopped me, I do not know him from Adam. **1795** T. WILKINSON *Wand. Patentee* IV. 129 He was so great a stranger I should not have known him from our father Adam. **1839** *Spirit of Times* 26 Oct. 397/1 As great races.. as have ever been run since Adam was a yearling. **1840** DICKENS *Old C. Shop* xxxviii, He called to see my Governor this morning,.. and beyond that I don't know him from Adam. **1867** 'COLONIST' *Life's Work Austral.* 82 Though old as Adam, love is still the theme that interests all hearts in all countries. **1900** BUCHAN *Half-Hearted* xx, I found people I didn't know from Adam drinking the old toasts. **1918** MULFORD *Man fr. Bar-20* ii. 25 You hunt up that pen you've had since Adam was a boy.

Adam[2] (æ·dăm). The surname of the brothers Robert Adam (1728–92) and James Adam (1730–94) used *attrib.* or *ellipt.* (at first usu. in pl.) to designate buildings, furniture, etc., designed by them.

1872 LADY C. SCHREIBER *Jrnl.* (1911) I. 463 We got three very good 'Adam' pedestals. **1898** *Lady's Realm* July 389/1 Adams rooms. **1900** *Jrnl. Soc. Arts* XLVIII. 374/2 The 'Adams' is the most delicate and refined of all styles founded on the classic. **1903** *Connoisseur* Mar. 21 Adam, and other Furniture. **1914** H. A. VACHELL *Quinneys'* iv. § 1, The sweetest table, genuine Adam. **1918** J. ALFRED GOTCH *Eng. Home* ix. 280 Robert was the most gifted, and it is his work which gave rise to the well-known 'Adam' style. **1920** GALSWORTHY *In Chancery* I. xii. 99 The fine reading-room was decorated in the Adam style. **1926** —— *Silver Spoon* I. ii, A blend between Adam and Louis Quinze. **1926** *Times* 31 Mar. 28 Beautiful Adam Residence.

Adam-and-Eve. The Biblical names of the first man and woman.

1. *dial.* A name applied locally to various plants, roots, etc. (See *E.D.D.*)

1789 B. RUSH *Let.* 8 June in T. J. Pettigrew *Mem. J. C. Lettsom* (1817) II. 439 Some of the country people call it *Adam-and-Eve*. **1808** *Monthly Mag.* 1 Dec. 421/2 *Adam and Eve*, the male and female handed orchis. **1959** *Times* 30 Dec. 8/7 A woman will carry a piece of Adam-and-Eve root in a little bag round her neck. It's really *Aplectum hyemale*, useful in bronchial troubles.

2. As *vb.*: rhyming slang for 'believe'.

1925 FRASER & GIBBONS *Soldier & Sailor Words* 3 Could you Adam and Eve it. **1956** E. A. THORNE *Baby & Battleship* II. 89 A *baby*! Would you Adam-and-Eve it!

adamant. Add: **B.** *adj.* Unshakeable, inflexible, esp. *to be adamant*, stubbornly to refuse compliance with requests. (The point at which the sb. use passed into adj. is not determinable.)

1936 in *Webster Collegiate Dict.* **1943** A. CHRISTIE *Moving Finger* vii. § 1. 77 Both Joanna and I tried to make her change her mind, but she was quite adamant. **1948** 'N. SHUTE' *No Highway* v. 135 C.A.T.O. were adamant that they would not carry Mr. Honey back across the Atlantic in one of their aircraft. **1960** H. E. BATES *Aspidistra in Babylon* 157 In her adamant, challenging, desperate fashion she seized my arm.

Hence **a·damantly** *adv.*

1961 *Sunday Times* 2 July 32/1 'When it's a girl her mother should be the one interviewed,' he stated adamantly.

adamantinoma (æ·dămæntinōᵘ·mă). *Path.* [mod.L., f. ADAMANTIN(E *a.* + *-OMA.] (See quot. 1926.) Cf. ODONTOME.

1922 W. G. CARNATHAN in *Jrnl. Tenn. Med. Assoc.* 1921–22 XIV. 408 (*title*) A study of adamantinoma, with report of a case. **1926** R. J. E. SCOTT *Gould's Med. Dict.* 31/1 *Adamantinoma*, an epithelial tumor resembling in structure the enamel organ of a developing tooth. **1931** *Nomencl. Dis.* (*Min. of Health*) (ed. 6) 84 *Adamantinoma*, fibrotic disease of the jaw. **1952** *Lancet* 1 Nov. 862/2 (*title*) The Multilocular Cystic Tumour or Adamantinoma of the Jaw.

adamellite (ædăme·ləit). *Min.* [ad. G. *adamellit* (A. Cathrein 1890, in *N. Jahrb. Mineral.* I. 74) f. Mt. *Adamello*: see -ITE[1].] A kind of granite from Mt. Adamello in North Italy.

1896 J. F. KEMP *Handbk. Rocks* 121 *Adamellite*, a name proposed by Cathrein as a substitute for tonalite. **1910** *Encycl. Brit.* VIII. 289/2 Among varietal designations given to rocks of the diorite family are.. 'adamellite' for the quartz-mica-diorite or tonalite of Monte Adamello (Alps).

adamsite[2] (æ·dămzəit). [f. the name of Roger *Adams* (b. 1889), an American chemist: see -ITE[1].] An arsenical compound for use in chemical warfare, etc.

1923 *Nature* 17 Nov. 742/2 The most remarkable results were obtained with an arsenic compound, phenarsazine, known in poison-gas warfare under the names 'D.M.' and 'Adamsite'. **1937** A. M. PRENTISS *Chemicals in War* III. xiii. 275 *Adamsite* (DM), used for cloud attack primarily against an enemy's forward elements to cause temporary casualties. **1961** *Guardian* 17 Mar. 1/7 A gas called Adamsite ..similar but superior to the German gases.. intended to make troops vomit and to tear off gas masks.

Adam's needle. Add: **1.** (Earlier example.) Also *Adam's needle-and-thread*.

1760 J. LEE *Introd. Bot.* 294/1 Adam's Needle, *Yucca*. **1891** *Cent. Dict.* [s.v. *Yucca*], From their sharp-pointed leaves with threads hanging from their edges, *Y. filamentosa* and *Y. aloifolia* are known as *Adam's needle and thread* and as *Eve's thread*. **1938** B. DAMON *Grandma* 73 A clump of yuccas, 'Adam's needle-and-thread'.

Adansonia. Add: (Earlier example.)

1797 T. HOLCROFT tr. *Stolberg's Travels* II. xciv. 485 The famous African tree, which is called Barbab, and described by.. Adanson, a French Botanist, after whom it has been named *Adansonia*.

2. *Paper-making.* (See quot. 1937.)

1888 CROSS & BEVAN *Text-Bk. Paper-Making* vi. 104 Jute and Adansonia are largely used for papers where strength is of more importance than appearance. **1937** E. J. LABARRE *Dict. Paper* 3/1 *Adansonia*, inner bark of *Adansonia digitata* or Baobab.. used to a limited extent for wrapping papers.

adapt, *v.* Add: **2. b.** To construct or produce by adaptation *from*.

1805 J. WILD (*title*) Dramas Adapted (from the Original French) to the English Stage. **1849** [see sense 2 in Dict.]. **1852** C. READE (*title*) The Lost Husband. A drama.. written and adapted from the French. **1911** (*title*) The Concise Oxford Dictionary of Current English adapted by H. W. Fowler and F. G. Fowler.. from The Oxford Dictionary.

3. *intr.* To undergo modification so as to fit for a new use, etc. Const. *to*. Also *absol.*

1956 M. BRYAN *Intent to Kill* vi. 67 In our country, the rich have no sense of responsibility. I wonder how they will adapt to the future. **1962** *Listener* 3 May 762/1 Birds certainly adapt to the urban community, and particularly well to the suburban community. *Ibid.* 19 July 84/1 There is an absolute lack of imagination, or failure to adapt, a refusal to face the need for change.

adaptation. Add: **2. b.** spec. in *Opt.* The adjustment of the eye to variations in the intensity or colour of light. Also = ACCOMMODATION 1 b.

1881 in *Syd. Soc. Lex.* **1920** *Jrnl. Gen. Phys.* II. 499 The phenomenon of retinal adaptation is one of the most familiar facts of sensory physiology. *Ibid.* 516 During the dark adaptation of the human eye, its visual threshold decreases to a small fraction of its original value in the light. **1950** L. C. THOMSON in *Brit. Jrnl. Ophthalmology* Mar. 145 The sensory receptor cells in the retina.. play a part in visual adaptation. **1960** R. A. WEALE *Eye & its Function* vii. 110 Chromatic adaptation experiments, in which the eye is exposed continuously to a coloured background of moderate luminance.

5. *Biol.* Organic modification by which an organism or species becomes adapted to its environment.

1859 DARWIN in *Jrnl. Linn. Soc.* Zool. III. 50 The most vigorous and healthy males, implying perfect adaptation, must generally gain the victory in their contests [for the females]. **1875** *Encycl. Brit.* I. 145/2 Adaptation.. is usually restricted.. to imply such modifications as arise during the life of an individual, when an external charge directly generates some change of function and structure. **1897** H. F. OSBORN in *Science* 15 Oct., Ontogenetic adaptation.. enables animals and plants to survive very critical changes in their environment. **1904** H. E. CRAMPTON in *Biometrika* III. 114 A rigid.. organization, incapable.. of structural alterations as the result of 'functional adaptation'. **1923** J. S. HUXLEY *Ess. of Biologist* i. 13 If the degree of adaptation has not increased during evolution, then it is clear that progress does not consist in increase in adaptation.

adapter, -or. Delete 'rarely' and add: **3.** *Electr. Engin.* (See quots.)

1907 T. O'C. SLOANE *Stand. Electr. Dict.* (ed. 12) 11 *Adapter*, a screw coupling to engage with a different sized screw on each end; one of the uses is to connect incandescent lamps to gas-fixtures. **1913** *E. E. Becker & Co. Illustr. Catal. Sci. App.* 539 (*caption*) Plug adaptor, for tapping current off any ordinary bayonet-catch lamp-holder. **1943** *Gloss. Terms Electr. Engin.* (*B.S.I.*) 123 *Socket-outlet adaptor*, an accessory for insertion into a socket-outlet and containing metal contacts to which may be fitted one or more plugs for the purpose of connecting to the supply portable lighting fittings or other current-using appliances.

adaptive, *a.* Add: *adaptive radiation*: H. F. Osborn's term for the evolutionary process which produces divergent forms in different environments (see quots. and cf. SPECIALIZATION b).

1902 H. F. OSBORN in *Bull. Amer. Mus. Nat. Hist.* XVI. 92 They represent an adaptive radiation for different local habitat, different modes of feeding, fighting, locomotion, etc. **1902** —— in *Amer. Naturalist* May 353 One of the essential features of divergent evolution.. has been termed by the writer 'adaptive radiation'. This term seems to express most clearly the idea of differentiation of habit in several directions from a primitive type. **1927** HALDANE & HUXLEY *Animal Biol.* xi. 240 This adaptation to different modes of life, while.. we call it *specialization* when we are thinking only of one species of animal, is called *adaptive radiation* when we are thinking of the group as a whole.

adaptometer (ædæptǫ·mitəɪ). [f. ADAPT(ATION + -OMETER.] (See quots.)

1917 A. DUANE *Fuchs's Textbk. Ophthalm.* (ed. 5) ii. 127 A better instrument than Förster's is the *adaptometer* of Nagel. **1928** R. J. E. SCOTT *Gould's Med. Dict.* (ed. 2) 31/2 *Adaptometer*, an instrument for measuring the time taken in retinal adaptation. **1934** H. C. WARREN *Dict. Psychol.* 6/1 *Adaptometer*, any device for measuring the course or degree of sensory adaptation, in terms of fall or rise of threshold or sensitivity. **1942** *Brit. Jrnl. Psychol.* July 4 Certain hysterical and pathological eye conditions undoubtedly affect adaptometer results.

adaxial (ǣdæ·ksiăl), *a.* *Bot.* [f. L. *ad* to, toward + *axi-s* axle + -AL.] Toward the axis or central line. Cf. ABAXIAL *a.*

1900 B. D. JACKSON *Gloss. Bot. Terms* 5/1 *Adaxial*, the side or face next the axis, ventral. **1903** *Bot. Gaz.* XXXVI. 104 The ovaries [in Casuarina] are flattened laterally, in contrast to the adaxial flattening of the wings in Pinus.

adazzle (ădæ·z'l), *adv.* and *pred. a.* [f. A *prep.*[1] + DAZZLE.] In a dazzle; dazzling.

1832 *Blackw. Mag.* Feb. 281 Every line shall be a line of light—all a-dazzle with appropriate words. **1877** G. M. HOPKINS *Poems* (1918) 30 With swift, slow; sweet, sour; adazzle, dim. **1923** W. DEEPING *Secret Sanct.* xxix. 302 Snow.. lay on the hedges.. and all the hillsides were a-dazzle with it as the sun rose.

‖ **ad captandum (vulgus)** (æd kæptæ·ndǫ̈m (vǫ̈·lgǫ̈s)), *a.* and *adv. phr.* [L. *ad* for + *captandum* neuter of the gerundive of *captāre* to entice, allure.] (Calculated) to take the fancy of (the crowd).

1762 SMOLLETT *Launcelot Greaves* I. x. 201 These paltry tricks, *ad captandum vulgus*, can have no effect but on ideots. **1837** DICKENS *Pickw.* x. 95 Such an *ad captandum* argument, as the offer of half a guinea. **1882** W. JAMES *Let.* 8 Jan. in R. B. Perry *Tht. & Char. of W. J.* (1935) I. 737 Arguments which I cared little about.. were used merely on account of their availability and *ad captandum* power. **1926** R. FRY *Transformations* 112 How considerable an artist lies hidden behind that *ad captandum* sentimentality. **1931** —— in W. Rose *Outl. Mod. Knowl.* 961 But vulgar and *ad captandum* as his art was, he [*sc.* Caravaggio] had a touch of genius.

add, *v.* Add: **1. a.** *spec.* in *Horse-racing* (cf. *ADDED *ppl. a.* c).

1816 *Racing Cal.* 1815 XLIII. 15 The Old Stakes of 10 gs. each, with 20 gs. added. **1848** *Sporting Life* 5 Aug. 277/1 By subscriptions of 20 sovs. each.. with 100 added from the racing fund. **1934** *Times* 18 Oct. 5/4 The Cesarewitch Stakes, a handicap, of 25 sovs. each, with 1,000 sovs. added.

4. Examples of *add up*. Also *absol.*

1754 D. FENNING *Brit. Youth's Instructor* (ed. 2) 12 When you have two or three rows of figures.. add up the first row, [etc.]. *Ibid.* 25 To add up well.. is a great Step towards Multiplication.

5. *intr.* in phr. *to add up*: **a.** To make the desired, expected, or correct total.

1850 DICKENS *Dav. Copp.* xli, The figures made her cry. They wouldn't add up, she said. **1864** *Good Words* 316/2 On one occasion, it struck me that a series of figures just given by the traffic-manager of a railway company, would not add up. **1885** *Sat. Rev.* 21 Feb. 243 Of the sixteen articles of diet enumerated, the percentages of five only will 'add up'. **1893** *Chambers's Jrnl.* 26 Aug. 532/1 Account books that would not add up right.

b. *to add up to*, to amount to, to signify, to come to mean. Also *absol.*, *to add up*, to make sense, to be significant or consistent.

1933 S. HOWARD *Alien Corn* I. 35 What is it that adds up to? A trio! **1936** *Time* 10 Feb. 23/1 The ship, plus the service, plus the passenger list, plus the Caribbean—it all adds up to the best vacation I ever had. **1938** E. S. GARDNER *Case Shoplifter's Shoe* (1939) ix. 158 All right.. I wasn't home last night. So what does that add up to? **1942** *Daily Progress* (Charlottesville, Va.) 27 Apr. 8/4 How does it [*sc.* the evidence] all add up? **1945** M. ALLINGHAM *Coroner's Pidgin* xi. 88 I've tried to make it mean something else, but it doesn't add up any other way. **1945** A. J. P. TAYLOR *Course German Hist.* 8 There were and, I dare say, are many millions of well-meaning kindly Germans, but what have they added up to? **1958** *Oxf. Mag.* 6 Mar. 342/1 As a contribution to the philosophy of religion, this volume of essays does not add up to much.

added, *ppl. a.* Add: **b.** *added sixth*: a major sixth added to a triad; the chord thus obtained.

1876 in STAINER & BARRETT *Dict. Mus. Terms*. **1931** G. JACOB *Orchestral Technique* viii. 76 All 'added 6ths' cannot be arranged as above. **1959** D. COOKE *Lang. Mus.* ii. 65 The Delius.. reminds us that the 'lushness' of his music is often due to his use of the chord of the 'added sixth'.

c. *added money*, esp. in *Horse-racing*, money added (by a racing association, etc.) to the stakes (cf. STAKE *sb.*[2] 3).

1880 *Illustr. Sporting & Dramat. News* 4 Dec. 278/3 The withdrawal of the subsidy augurs badly for the annual grant of £2,000 usually given towards the added money for the Grand Prix. **1886** *Kennel Club Calendar & Stud Book 1885* XIII. 95 Pointer Puppy Stakes.. first £50 and £10 added money. **1923** *Empire Rev.* June 631 Racecourse executives are perforce more generous in the matter of added money than was the case in the pre-war days.

addendum. Add: **b.** *Mech.* A term applied to certain dimensions of toothed gearing (see quots.).

1841 R. WILLIS *Princ. Mechanism* 98 Now this addition to the radius of the pitch circle [of a gear] is called by clockmakers the *addendum*, which term I shall, for convenience, employ. **1873** RANKINE & BAMBER *Mech. Text Bk.* II. ii. 81 The radius of the pitch circle of a circular wheel is called the *geometrical radius*; that of a circle touching the crests of the teeth is called the *real radius*; and the difference between these radii, the *addendum*. **1954** *Gloss. Terms*

Toothed Gearing (B.S.I.) 11 *Addendum*. 1. Of a helical or spur gear. a. *Reference addendum*, the radial distance from the reference cylinder to the tip cylinder. b. *Working addendum*, the radial distance from the pitch cylinder to the tip cylinder. 2. Of a wormwheel. Half the difference between the diameter of the pitch circle and the throat diameter. 3. Of a bevel gear. The distance from the pitch circle to the tip cone, measured on the back cone.

adder[3]. Add: **2.** An adding-machine.

1890 *N.Y. Herald* Jan. (Advt.), The Adder is so called because really too simple to be styled a 'machine'.

3. A unit in an electronic computer (see quot. 1962).

1948 A. SVOBODA *Computing Mechanisms* ii. 37 The bar linkage adder.. consists of essentially the same parts as the linkage of Fig. 2–11. **1954** *Electronic Engin.* Sept. 376/2 The sample instruction.. asks the machine to take the number.. and to add this to the contents of the accumulator... In the last 8 digits [of the instruction] is the binary code which operates the circuit to perform the required function, the adder. **1962** *Gloss. Terms Autom. Data Processing* (*B.S.I.*) 100 *Adder*, a unit with two or more input variables and one output variable which is equal to the sum, or a weighted sum, of the input variables.

adder[4]. **5.** Further examples in *fig.* or *poet.* uses.

1824 SCOTT *St. Ronan* III. viii. 209 What was it the old, adder-tongued murderer said to Clara Mowbray? **1874** SWINBURNE *Bothwell* II. xiii. 182 What could you bring.. but adder-stings? **1896** A. E. HOUSMAN *Shropshire Lad* xxvi, The adders' tongues. **1908** WILDE *Ballad Reading Gaol* 7 The gallows-tree With its adder-bitten root. **1926** E. SITWELL *Wooden Pegasus* 21 Adder-flames flare and spout From his lips. **1946** L. B. LYON *Rough Walk Home* 29 Crushing the adder-coloured dung. **1946** E. SITWELL *Fanf. Eliz.* iii. 18 Adder-voices shrilling and hissing.

addict (æ·dikt), *sb.* [f. ADDICT *v.*] One who is addicted to the habitual and excessive use of a drug; chiefly with qualifying sb., as *drug*, *morphia addict*. Also *transf.*

1909 OSCAR JENNINGS *Morphia Habit* vi. 78 As shown by *post-mortem* examinations in morphia addicts. **1920** *Glasgow Herald* 18 Oct. 9 The Chicago Brickwell Institute for drug addicts. **1920** *Outward Bound* Oct. 38/1 The morphia addict is a doomed man. **1924** *Westm. Gaz.* 19 Dec., People who.. get into the habit of going to the cinema.. for drugs to induce sleep, and often end up by becoming opium, morphine, or heroin addicts. **1936** *Daily Exp.* 27 Jan., Even many working men are night club addicts.

addiment (æ·dimėnt). *Biochem.* [ad. L. *addimentum, f. *addere* to ADD; see -MENT.] = *COMPLEMENT *sb.* 5 i.

1901 *Lancet* 19 Oct. 1029/1 If an animal be.. given two M.L.D. [minimum lethal dose] and two serum equivalents it.. dies from the infection. This has been explained as due to a deficiency of addiment in the animal concerned. **1903** *Jrnl. Hygiene* Jan. 52 The compound ferment (complement, addiment) upon this theory.

adding, *vbl. sb.* Add: **b.** *attrib.*: **adding-machine**, an instrument for the mechanical adding up of numbers.

1874 in KNIGHT *Dict. Mech.* **1911** H. S. HARRISON *Queed* viii. 102 He was as definite as an adding machine, as practical as a cash register. **1929** *Times* (weekly ed.) 7 Feb. 147/4 Clerks recording quickly on adding machines.

Addison (æ·disǫn). *Path.* The name of an English physician, Thomas Addison (1793–1860), used to designate the condition resulting from defective functioning of the suprarenal glands, characterized by progressive anæmia, debility, and a brown pigmentation of the skin. Also *Addison's anæmia*, *Addisonian anæmia*, pernicious anæmia. Hence **A·ddisonism**, Addison's disease. Cf. BRONZED *ppl. a.* 4.

[**1856** TROUSSEAU in *Bull. Acad. Med.* XXI. 1036 Une maladie bizarre, à laquelle M. le docteur Addison.. a donné le nom de *peau bronzée*, de *maladie bronzée*, et à laquelle je crois devoir donner désormais le nom de *maladie d'Addison*.] **1856** *Med. Times & Gaz.* 30 Aug. 219/1 Lecturing on.. the Bronze Disease, shown by Addison to be so frequently connected with disease of the supra-renal capsule, M. Trousseau suggests that it should be denominated 'Addison's Disease'. **1875** J. S. BRISTOWE *Theory & Pract. Med.* IV. 582 Addison's disease occurs much more frequently in males than in females. **1906** *Lancet* 4 Aug. 288/1 Bozzet has classified.. a few other diseases showing pigmentation under the title of Addisonism. **1910** *Practitioner* Jan. 126 The appearance presented by the tongue described by Dr. W. Hunter as specific for Addisonian anæmia. **1933** *Discovery* Mar. 83/1 Pernicious anæmia, or as it is more generally known, Addisonian anæmia, was first described by Addison, a physician at Guy's, in 1856. **1959** *Sci. News* XV. 132 The progress of Addison's disease can be checked by administering suitable extracts of the adrenal glands of animals.

addition, *sb.* **4.** Delete † *Obs.* and add later examples.

1802 *Hull Advertiser* 13 Nov. 4/5 A List of Certificates issued.. pursuant to the Acts of Parliament granting a Duty on such Certificates. Names. Additions. Places of Abode. **1936** R. W. CHAPMAN *Names* 282 The Lord President of the Council and the Financial Secretary to the Treasury are not often so named; nor.. does one hardly remember who are the owners of those additions.

7. *Chem.* *attrib.*, as *addition compound*, *product*, one formed by the direct addition of one substance to another without change of valency. (Cf. *ADDITIVE *a.* b.) Also *addition agent* (see quot. 1940).

1879 *Jrnl. Chem. Soc.* XXXVI. 51 The addition-products of cinnamic acid.. with the view of ascertaining the state of these bodies on exposure to the air. **1888** ROSCOE & SCHORLEMMER *Treat. Chem.* III. iv. 390 Addition Products of Phthalic Acid. **1921** T. M. LOWRY *Inorg. Chem.* II. xxiii. 375 *Addition and substitution of Chlorine.*.. In general, it may be said that chlorine forms two types of derivatives, namely, (*a*) addition-compounds.. (*b*) substitution derivatives. **1940** *Chem. Abstr.* 2073 In this solution, the electrolyte, contains.. according to numerous patents, various other substances known as 'addition agents' either to start the plate or to improve the 'throwing power' of the bath. **1946** *Chambers's Techn. Dict.* 12/2 *Addition agent*, a substance added to the electrolyte in an electro-deposition process in order to improve the character of the deposit formed.

additive, *a.* Add: **b.** *Chem.* Applied to a product, process, etc., characterized by addition (see prec., sense *7).

1875 *Jrnl. Chem. Soc.* XXVI. 1006 Aromatic additive compounds. **1875** BLOXAM *Chem.* (ed. 3) 84 The carbonates may be expressed either by additive formulæ, showing the bases which combine with carbonic acid to produce them, or by substitutive formulæ. **1876** *Jrnl. Chem. Soc.* XXIX. I. 338 The Laws which regulate Direct Additive Reactions. **1890** W. H. PERKIN & F. S. KIPPING *Org. Chem.* 101 Compounds.. that are produced by the addition of an atom, or atoms, to the molecules of a compound.. are named additive compounds. **1908** *Jrnl. Chem. Soc.* XCIV. I. 719 The oxidation of amines is initially an additive process.

c. Of a process of colour reproduction in which the primary colours are superimposed one upon another.

1906 E. J. WALL tr. *König's Natural-Color Photogr.* II. 71 The additive methods of three-colour photography by optical synthesis. **1934** HARDY & PERRIN *Princ. Optics* xv. 314 Colors combined in this manner obey the laws governing the addition of stimuli. This is therefore called the *additive method* of combining colors. **1935** *Discovery* July 199/1 We can reproduce colour either by adding primary coloured lights to black (by projection on to a screen in a dark room), or by *subtracting* primary coloured lights from a white light.. containing all colours.. Processes can thus be classified as additive or subtractive.

d. Examples of other technical uses.

1933 E. N. GOWAN *Mod. Statistics* iv. 67 The Additive Nature of χ^2. A useful property of χ^2 is that several values can be added together. **1939** *Brit. Jrnl. Psychol.* July 14 If mild mental impairment were inherited by multiple Mendelian additive factors, the sibs should regress towards the norm. **1946** C. E. WEATHERBURN *First Course Math. Statist.* 177 (*heading*) Additivity of chi-square. **1950** P. B. MEDAWAR in *Listener* 27 Dec. 1068/1 The word 'additive' refers to a particular pattern of co-operation or interaction between genes. An additive pattern of interaction implies (amongst other things) that there will be no such thing as hybrid vigour in respect of intelligence.

Hence as *sb.*, something that is added; esp. (in various technical uses) a substance added (to a mixture, alloy, etc.) in order to impart specific qualities to the resulting product.

1945 *Electronic Engin.* XVII. 516/1 The use of the monomer styrene.. as an additive to the solution of the polymer in solvent. **1948** *Economist* 31 July 192/2 The supply of 'additives' for petroleum products. **1954** A. NISA *Morphology* (ed. 2) vi. 103 Such bound forms are.. noncitics—additives, replacives, subtractives. **1959** *Times* 27 Apr. (*Special Suppl.*), Rubber [ed.] p. viii/3 In road building it is a valuable additive to road surfacing materials.

additivity (æditi·viti). [f. ADDITIVE *a.* + -ITY.] The state or condition of being additive.

1908 *Jrnl. Chem. Soc.* XCIV. II. 937 Connexion between Residual Affinity and Additivity. **1946** *Nature* 28 Sept. 471/2 Kraus's assumption on the additivity of the spectra has been questionable. **1964** HEISENBERG in *Cambridge Rev.* 24 Oct. 51/2 It is only on account of the long range and of the additivity of the forces from different sources that a reasonable 'classical limit' of the photon field operator exists.

additory, *a.* Delete † *Obs.* and add:

1805 [see DETRACTORY *a.*]. **1897** *Amer. Jrnl. Philol.* XVIII. 27 The three distinct values of *etiam*.. the temporal ('still', with negative 'yet'), the additory ('also') and the intensive ('even').

address, *v.* Add: **8. f.** *U.S.* To force (a judge) out (of office) by a petition to the executive. (Cf. ADDRESS *sb.* 11 b.)

1822 *Missouri Intell.* 2 July 3/2 If any of the judges have corruptly discharged their duties, impeach them. If they are incompetent, address them out. **1857** R. H. COLLINS *Kentucky* I. 37 David Ballengall, an assistant judge.. [was] 'addressed' out of office, because a Scotchman unnaturalized.

address, *sb.* Add: **7. b.** Place of residence, house.

1888 GUNTER *Mr. Potter of Texas* xix, He doesn't know where his son is to be found.., otherwise he would drive to his address at once. **1936** G. B. SHAW *Millionairess* III. 178 First I must go round to the address Tim has given me and arrange that we send them our stuff direct. **1937** *Punch* 29 Dec. 718/1 He has returned.. to the same old address in South-West London that has sheltered him so long. **1938** ... in West End London...

fig. **1934** G. B. SHAW *Too True to be Good* II. 68 What you think is love and interest, and all that, is not real love at all: three quarters of it is only their absurd address.

c. *Computers.* (See quot. 1956.)

1949 WILKES et al. *Prep. Programs Electronic Digital Computer* I. 10 The address specified in this particular order is then increased by one. **1955** BOOTH & BOOTH *Automatic Digital Calculators* ii. 12 It is, therefore, Faster, Faster 11 In order to tell the machine to place a word in a given storage location it is necessary to specify the number, or 'address', of the location. **1956** BERKELEY & WAINWRIGHT *Computers* 334/2 Address, digital computers, A label, name, or number identifying a register, a location, or a device where information is stored.

11. b. A formal request, directed to the executive by both branches of the legislature, requesting the removal of a judge from office. *U.S.*

1822 *Missouri Intell.* 2 July 3/2 If any of the judges.. must be removed, even without cause, still let it be done by address. **1882** H. ADAMS *J. Randolph* 131 The Constitutions of England, of Massachusetts, of Pennsylvania, authorized the removal of judges by address.

12. *attrib.* (in sense 7) *address-book*, *-card*.

1877 LADY C. SCHREIBER *Jrnl.* (1911) II. 42 Several mentioned in the address book 'n'existent pas'. **1854** *Poultry Chron.* I. 354/2 So as to have all the purchasers' address-cards quite ready by the close of the show.

addressee. Substitute for def.: One who is addressed; *spec.* the person to whom a letter, packet, etc., is addressed. (Earlier and later examples.)

1810 R. PETERS *Let.* 25 Nov. in W. Jay *Life J. Jay* (1833) II. 332 Nothing must go with a pamphlet but the mere direction, under the pains and penalties of rousing the correspondent or addressee in all the shape of a penny letter. **1832** BENTHAM *Essay on Lang.* in *Wks.* (1843) VIII. 328/1 By a third addressee, the sign is presented to the addressee. **1884** *N. & Q.* 12 July 22/1 The strong presumption that this offers in favour of this youthful nobleman as the addressee of the sonnets.

addressing, *vbl. sb.* Add: **b.** *attrib.*: **addressing-machine**, a machine for printing addresses on newspaper-wrappers, etc.

1866 *U.S. Pat.* 47,741 (*title*) Addressing-Machine. A Machine for Printing the Directions on Newspapers or Wrappers. **1955** *Vocab. bibliothec.* (Unesco) 160/1 Addressing machine; addressograph.

addressograph (ădre·sǫgraf). [f. ADDRESS *sb.* + -O + -GRAPH.] An addressing-machine for printing addresses. (Proprietary term.)

1902 *Modern Business* Sept., Index to Advertisers p. iii/2, **1924** *Public Opinion* 11 Jan. 37/1 The addresses have that dull, uninteresting look that an addressograph gives them. **1958** *Observer* 7 Dec. 3/6 Barbara, an addressograph operator, showed her companions how the Government has already progressed by way of the addressograph plate to the punched card.

adduct (æ·dǫkt), *sb.* *Chem.* [ad. G. *addukt* (Diels and Alder 1931, in *Ann. d. Chem.* 1911), f. ADD(ITION *sb.* + PRODUCT *sb.* [ad. (G. *produkt*).] An addition product (see *ADDITION *sb.* 7).

1944 *Brit. Pat.* 552,644 3 Equal parts of the cyclopentadiene adduct of maleic anhydride and blown linseed oil.. were reacted together. **1952** *Chem. Abstr.* 9953 Urea and thiourea adducts, diisotriphenyl adducts, cyclodextrin adducts.. are treated.

Adélie. Also **Adelie.** The name of Adélie Land in the Antarctic, used *attrib.* or *ellipt.* of a kind of penguin (*Pygoscelis adeliæ*) found there.

1907 E. A. WILSON in *Nat. Antarctic Exped.* 1902–4 Nat. Hist. II. ii. 26 We had Adélie Penguins in the water all around us. *Ibid.* 39 The numbers of adult Adélie rapidly increased. **1959** SHACKLETON *Heart of Antarctic* I. 2 During the day we had seen adélie and ringed penguins. **1930** J. S. HUXLEY *Bird-Watching* vi. 96 Adélie penguins lay their eggs in a little depression bordered by stones. *Ibid.* An Adélie rookery in the Antarctic. **1959** *New Biol.* XXIX. 107 The truly Antarctic birds are the Emperor and Adélie penguins.

adelo- (ădĕl·o, ædĭlo), comb. form of Gr. ἄδηλος not manifest or evident, unseen, in **adelocodo·nic** *a.* (see quot.); **adelomo·rphic**, **-mo·rphous** *adjs.*, applied to the central cells of the peptic glands.

1871 G. J. ALLMAN *Gymnobl. Hydroids* 30 The gonophore is always borne as a bud.. It may be referred to one or other of two principal types, based respectively on the greater or less approach to the completely formed medusa. The peculiar condition by which one of these types is characterised may be termed phanerocodonic, while that which distinguishes the other may be designated as adelocodonic. **1891** W. D. HALLIBURTON *Text-bk. Chem. Physiol.* xxx. 655 These cells [of the cardiac glands] were called principal cells by Heidenhain, adelomorphous cells by Rollett, and central cells on account of their position. **1875** A. GAMGEE tr. *Hermann's Elem. Hum. Physiol.* III. 99 'Hauptzellen' (Heidenhain), or 'adelomorphous cells' (Rollett).

Adelphi (ădĕ·lfi). The name of a group of buildings in London between the Strand and the Thames, laid out by the four brothers, James, John, Robert, and William Adam (see *ADAM[2]) and hence called *Adelphi* (Gr. ἀδελφοί brothers); the name of the theatre in the vicinity of these buildings, at which a certain type of melodrama was prevalent *c* 1882–1900, and so allusively.

1804 *Opera* 17 Mar. 432/1 Those who expected that the advent of a new Adelphi dramatist would be marked by a new development in Adelphi drama, were deceived. **1889** *Sat. Rev.* 9 Nov. 521/1 The 'Adelphi' as opposed to the Drury Lane type of drama has recently died out in the West End. **1903** CHESTERTON *R. Browning* iii. 45 So... which is now called an Adelphi melodrama. **1928** —— *Generally Speaking* 231 His trouble cannot have been as deep as hell and as shallow as an Adelphi melodrama.

adelphogamy (ædelfǫ·gămi). *Zool.* [f. Gr. ἀδελφός brother + γάμος marriage, after MONOGAMY.] The mating of brothers and sisters, as in certain ants.

1906 W. M. WHEELER tr. *R.A.F. de Réaumur's Nat. Hist. Ants* 46 Adelphogamy, or the mating between males and virgin females of the same formicary undoubtedly occurs. **1948** C. OGDEN tr. *Forel's Soc. World of Ants* II. 406 Marriage takes place.. between newly-matured brothers and sisters sprung from the same mother... Hence this is a clear case of adelphogamy.

adenine (æ·dĕnin). *Chem.* Also **-in**. [ad. G. *adenin* (A. Kossel 1885, in *Ber. d. deutschen chem. Gesellschaft* XVIII. 79), f. Gr. ἀδήν gland + -INE[5].] A crystallizable base, $C_5H_5N_5$, found in certain glands and animal and vegetable tissues.

1885 *Jrnl. Chem. Soc.* XLVIII. II. 566 The author [sc. A. Kossel] has isolated from the pancreatic gland of the ox, a new base, $C_5H_5N_5$, to which he gives the name adenine. **1887** A. M. BROWN *Anim. Alkaloids* II. i. 75 Adenine, $C_5H_5N_5$ is in large transparent crystals, contains three molecules of water of crystallization. **1931** [see *ADENOSINE]. **1937** *Thorpe's Dict. Appl. Chem.* I. 140 Adenine.. discovered in the pancreatic gland and spleen of the ox, occurs in all cell nuclei. **1955** *Chem. & Ind.* 3 The most stable fact about adenine.. that the cell substance in glands and animal tissues rich in cells... It has been extracted from tea leaves.. from beet juice, [etc.]. **1964** *New Biol.* XXXI. 34 The base is the biologically most important nucleotide is adenine.

adenitis (ædĕnai·tis). *Path.* [mod. L., f. Gr. ἀδήν gland + -ITIS.] Inflammation of a gland.

1848 DUNGLISON *Med. Lex.* (ed. 7). **1853** ERICHSEN *Sci. & Art Surg.* xxxiii. 424 Inflammation of the lymphatic glands, or adenitis. **1879** *St. George's Hosp. Rep.* IX. 674 Erysipelas and adenitis in 6, 1 of whom died. **1949** H. W. FLOREY et al. *Antibiotics* I. 1 no. The best results were obtained in osteitis of the small joints, adenitis, [etc.].

adenocarcinoma (æ:dĕno,kā.ɪsinōᵘ·mă). *Path.* [f. ADENO- + CARCINOMA.] A malignant adenoma.

1889 in *Cent. Dict.* **1900** *Jrnl. Amer. Med. Assoc.* 31 Mar. 775/1 *Adenocarcinoma*.. Nine tumors show the general plan of the gland. **1907** *Practitioner* May 618 The tumour.. was an adeno-carcinoma. **1962** *Lancet* 13 Jan. 85/2 The much higher proportion of adenocarcinomas in the Birmingham series.

adenohypophysis (æ:dĕno,həipǫ·fisis). [mod. L. (Berblinger, 1932), f. ADENO- + HYPOPHYSIS.] The glandular or anterior part of the hypophysis.

1935 DORLAND & MILLER *Med. Dict.* (ed. 17) 48/1 *Adenohypophysis*, the anterior or glandular portion of the hypophysis as distinguished from the neurohypophysis. **1952** [see *ADRENOCORTICOTROPHIN]. **1962** *Lancet* 2 June 1177/1 Results of experiments in laboratory animals suggest that the long-term administration of steroid preparations containing œstrogen may have a harmful effect on the adenohypophysis.

adenoid, *a.* Add: (Earlier examples.)

1839 K. GRANT *Hooper's Med. Dict.* (ed. 7) 35/2 *Adenoides*, adenoid, (glandiform; resembling a gland. **1865** W. TURNER *Paget's Surg. Pathol.* 541 The name 'adenoid tumours' is sometimes applied to these tumours of the mammary gland.

B. *sb. pl.* Adenoid growths or vegetations.

1891 *Medical Annual* 341 Admitting that usually adenoids are of slow development during childhood. **1901** CHEYNE & BURGHARD *Man. Surg. Treatm.* v. 365 The removal of the adenoids is more easily accomplished during the earlier and deeper stage of the anæsthesia. **1912** OSLER & McCRAE *Princ. Med.* (ed. 8) 468 'Adenoids' have become recognized as the most important cause of mouth-breathing in childhood.

adenoidal, *a.* Add: **b.** Having the voice or appearance of one suffering from adenoids. Hence **adenoi·dally** *adv.*

1926 E. F. HURST *Humoresque* 195 Whole rows of ladies, with the slightly open-mouthed, adenoidal expression of vivacious romance, sat forward in their chairs. **1933** 'J. SUTHERLAND' *Circle of Stars* xvi. 167 The adenoidal young daily maid had gone. **1933** ST. JOHN ERVINE *Theatre in my Time* 131 The mouth of a.. disturbed adenoidal voice. **1942** L. A. G. STRONG *Bay* 9 The Father began to scold, the son to expostulate adenoidally.

adenoidectomy (æ:dėnoide·ktǫmi). *Surg.* [f. *ADENOID(S *sb. pl.* + Gr. ἐκτομή cutting out.] Excision of the adenoids.

1906 in DORLAND *Med. Dict.* (ed. 4). **1936** *Lancet* 28 Nov. 1274/1 (*caption*) A Ring Knife for Adenoidectomy. **1962** *Nature* 13 Jan. 2.. With tonsillectomies and adenoidectomies accounting for over half of the surgical cases.

adenoidy (æ·dėnoidi). *Path.* [f. *ADENOID *a.* + -Y.]

1946 K. TYNAN *Bull Sea* in *Harper's* 2 in. [..] impossible, adenoidy, Cockney subalterns. **1932** J. B. PRIESTLEY *Dangerous Corner* I, But which was the husband and who the thin and adenoidy one?

adenoma (ædĕnōᵘ·mă). [mod.L., f. Gr. ἀδήν gland: see *-OMA.] A benign tumour with the structure or appearance of a gland or originating in a gland (cf. *ADENOCARCINOMA). Hence **adenomatous** (-ō°·mătǝs) *a.*, of the nature of an adenoma, glandular.

1869 W. TURNER *Paget's Surg. Path.* (ed. 3) xxviii. 558 Glandular tumour, Adenoid Tumour, or Adenoma. **1870** T. HOLMES *Syst. Surg.* (ed. 2) IV. 578 Adenomatous or Glandular Growths. **1899** *Allbutt's Syst. Med.* VIII. 737 Adenoma of the ear-gland. **1907** *Practitioner* Sept. 336 The more difficult ones to deal with were the adenomatous ones, where a secondary growth was widely disseminated, and a whole lobe or more had to be excised, owing to the impossibility of removing the adenomata singly. **1964** L. MARTIN *Clinical Endocrinol.* (ed. 4) iii. 85 Solitary hypersecreting adenoma may be single or multiple.

adenosine (ădĕ·nosin). *Biochem.* Also **adenosin**. [f. G. *adenosin* (P. A. Levene and W. A. Jacobs 1909, in *Ber. d. deutschen chem. Ges.* XLII. 2703), blend of *ADENINE and *RIB(OSE.] A nucleoside ($C_{10}H_{13}N_5O_4$) found chiefly in striated muscle tissue and derived from ribose nucleic acid.

1909 *Chem. Abstr.* 2563 *d*-Ribose is also produced by the hydrolysis of adenosin. **1931** *Jrnl. Biol. Chem.* IX. 171 The other nucleoside (adenosine) is composed of *d*-ribose and adenine, and upon acid hydrolysis gives directly pentose (ribose) and adenine. **1939** *Proc. Physiol.* LXVII. p. xxxvi, The activity of adenosine liberated from the pirate is of the same order as the substance isolated from heart muscle. **1954** *Lancet* 22 May 1077/2 Hereditary spherocytosis (H.S.)... In some H.S. bloods both the metabolic defect and the fragility were reversible with mannose, glucose, or adenosine.

b. *Comb.*, as *adenosine monophosphate* (abbrev. *A.M.P.*), *diphosphate* (*A.D.P.*); and esp. *adenosine triphosphate* (*A.T.P.*), a derivative of adenosine important as a source of energy for the contraction of muscles and for other biochemical processes.

1911 *Jrnl. Biol. Chem.* LX. 180 This is a demonstration of adenosine-deaminase in the practical absence of guanosine-deaminase. **1929** *Science* LXX. 381/1 The adenosine phosphoric acid isolated about two years ago from voluntary muscle is not identical with that obtained from yeast nucleic acid. **1938** *Chem. Abstr.* 6275 Reduction of Co by the substrate is favored by adenosine triphosphate. **1939** *Nature* 18 Feb. 281 Di-adenosine tetraphosphate.. consists of one mol. adenylic acid and 1 mol. adenosine triphosphate. **1939** M. STEPHENSON *Bacterial Metabolism* (ed. 2) iv. 80 Under the influence of the enzymes present in yeast extracts of dried yeast.. A.T.P. can be dephosphorylated in two stages.. yielding first ADP... **1946** *Nature* 23 Mar. 353 The adenosine diphosphate and adenosine triphosphate were equally as effective as the triphosphate. **1955** *Sci. Amer.* May 44/2 The young man.. is the cytoplasm of fruit cells.. take up inorganic phosphate and add it to adenosine monophosphate (AMP) to form adenosine diphosphate (ADP). **1960** *New Biol.* XXXI. 34 The nucleotide itself is called adenosine monophosphoric acid (AMP).

adenovirus (æ·dĕno,vaiᵊrǝs). *Path.* [f. ADENO(ID *a.* + VIRUS.] Any of certain viruses causing respiratory diseases, first found in adenoid tissue of man. (Called 'adenoid degeneration agent' by W. P. Rowe and others in *Proc. Soc. Exp. Biol.* (1953) LXXXIV. 573.)

1956 ENDERS et al. in *Science* CXXIV. 119/2 Agreement was reached on the term *adenovirus* group, which suggests a characteristic involvement of lymphadenoid tissue, as well as the tissue of the first reported isolation. **1957** *Times* 21 Nov. 15/4 During the past years American workers have isolated an increasing number of viruses from the throats of individuals with what are generally known as upper respiratory infections—that is, infection of the nose, throat, pharynx, and larynx. At least 14 such viruses have been now isolated, and to these has been attached the name of the title of adenoviruses. **1964** *Lancet* 25 Sept. 718/1 Our results have shown that the H.A of adenoviruses was a specific, though highly complex, phenomenon.

adenyl (æ·dėnil). *Chem.* [f. *ADEN(INE + -YL.] A radical contained in adenine. Hence **adeny·lic** *a.*, esp. *adenylic acid*, a nucleotide found in red blood corpuscles, muscle, etc., consisting of adenine, ribose, and phosphoric acid.

1894 GOULD *Illustr. Dict. Med.*, Adenyl. **1894** *Jrnl. Chem. Soc.* LXVI. I. 156 The nucleic acid prepared from the thyroid gland of calves yields adenine on boiling with water, and is therefore called adenylic acid. **1929** *Biochem. Jrnl.* XXIII. 54 Muscle study showed that the chemical changes accompanying contraction are a complicated chain of linked reactions involving creatine, adenylic, and phosphoric acid, and carbohydrates and the complexity grows. **1946** *Nature* 23 Nov. 746/2 Here also the hydrogenation would be accompanied by a hydrolysis of the coenzyme molecule followed by a transfer of the phosphate of the adenylic acid to other phosphate acceptors.

adequate, *a.* Add: **2. b.** Without const.: equal to the occasion, competent to deal with the situation. So **adequacy**.

1924 ANNE D. SEDGWICK *Little French Girl* II. viii, The deliberate adequacy with which those Vervier advanced to meet the situation. *Ibid.* ix, Alix, in Mamma's place, would have been adequate. **1930** ... Adequacy.

c. In slightly derogatory sense: hardly sufficient or acceptable; barely reaching a minimum standard.

1900 M. BEERBOHM in *Sat. Rev.* 24 Feb. 234/1 It [the production of a play] may simply what the dramatic critics call 'adequate', meaning 'inadequate'. **1948** *Times* 11 Feb. 3/1 The standard rapidly sinks to a level which is, at best, adequate but at worst incompetent.

adermin (ădɜ·ɪmin). Also **adermine.** [G. R. Kuhn and G. Wendt 1938, in *Ber. d. deutschen chem. Ges.* LXXI. 1534), f. A- [14] + Gr. δέρμα skin + -IN [1].] Pyridoxin; vitamin B_6.

1938 *Chem. Abstr.* 3423 The substance.. is designated as adermin. **1939** *Chem. Abstr.* 505 It is converted by acetic anhydride into.. acetyladermin... **1939** A. B. DAVIES in *Nature* ... pyridoxine... 3-hydroxy-4,5-bis(hydroxymethyl)-2-methylpyridine. **1939** GYÖRGY & ECKARDT in *Nature* CXLIV. 512/2 Vitamin B_6 adermin or pyridoxin... The term *pyridoxin* appears to be preferable.

adespota (ădĕ·spǫtă), *sb. pl.* *Bibliography*. [neut. pl. of Gr. ἀδέσποτος without owner, f. ἀ- priv. + δεσπότης master, DESPOT.] Literary works not attributed to (or claimed by) an author.

Orig. used as a title of collections of anonymous Greek poetry.

1867 *Times* 25 Oct. 10 A bibliographical paper on Burns's adespota; verses fugitive, unauthorized, or apocryphal. **1904** A. W. POLLARD in *Library* 2nd Ser. V. 20 The cards were then sorted into order according to countries, towns and presses, with a large section of adespota. **1927** F. B. KERNOW *Printers' & Publ. Devices Introd.* p. 1, To lump together a large number of books of every date and style in a class of Adespota.

adessive (ădĕ·siv), *a.* *Gram.* [f. L. *adesse* to be present + -IVE.] Denoting the case used (in Finnish, etc.) to express position in or presence at a place.

1860 *Trans. Philol. Soc.* 1857 38 Cases are genitive, dative and the like all the world over.. But besides these they are also caritive, adessive and the like has yet to be investigated. **1890** C. N. E. ELIOT *Finnish Gram.* 131 The first member (inessive, adessive, essive) denotes originally rest in a position. **1890** GATSCHET *Klamath Indians* II. 1. 481 *Adessive* case. **1936** A. H. WHITNEY *Teach Yourself Finnish* (1959) 89 The Adessive Case has the ending -lla.

‖ **à deux** (a,dǒ), *adv.* and *a.* [F., for two, between two.] Of, for, or pertaining to, two people.

1876 STAINER & BARRETT *Dict. Mus. Terms* 183/2 *A deux*, for two voices or instruments. **1886** E. BLOOM *Dict. Coll. Cupid* II. iv. 85 Some keen happiness à deux; some two happy souls together. **1891** E. W. HENRY... *Whist* 19 ... had been out, but planned a luncheon à deux. **1928** D. H. LAWRENCE *Woman who rode Away* 83 This dictating business, this ten hours a day intercourse, à deux. **1959** *Listener* 22 Jan. 157/2 Strindberg regards marriage as a ghastly solitary confinement à deux.

adhesion. Add: **1.** Also *spec.* in *Physical Chem.* (see also quots.).

1853 MAYNE *Expos. Lex.* 19/1 *Adhesion*, term for that property by which certain bodies attract others; or for the power by which their particles adhere to each other. **1876** *Encycl. Brit.* VI. 361 The forces which are concerned in these phenomena are those which act between small particles of matter, and are sometimes called cohesive forces of cohesion, and those which act between portions of matter of different kinds, which are called forces of adhesion. **1948** GLASSTONE *Physical Chem.* (ed. 2) vii. 483 This work is a measure of the energy required to separate the solid from the liquid and is called the work of adhesion between solid and liquid.

c. *Mech.* The grip (of a wheel on a track, etc.) produced by friction, or the friction itself. Also *attrib.*

1862 W. J. M. RANKINE *Man. Civ. Engin.* III. i. 639 The 'adhesion', as it is called, or force which prevents the driving wheels from slipping on the rails. **1889** *Cent. Dict.*, *Adhesion-car*, a railroad-car provided with means for increasing the adhesive or tractive power beyond that due to the weight of the car. **1922** *Brit. Jrnl.* XXXIII. 167/2 Adhesion lines.. may be used on gradients of not more than one in twenty. **1958** *Engineering* 14 Mar. 342/1 The high cost of these locomotives is apparently justified from the good adhesion characteristics.

2. *to give in one's adhesion* (connexion).

1861 TROLLOPE *Tales of All Countries* Ser. 2 (1863) 295 His first idea had been that she was too clever... would be delightful. He had blushed as he had given in his adhesion.

adhesive, *a.* Add: **1.** *adhesive tape.*
1934 in WEBSTER. **1951** *Good Housek. Home Encycl.* 28/1 Seal the tin with adhesive tape.

B. *sb.* [The *adj.* used *absol.*] An adhesive substance.
1912 *Sci. Amer.* 17 Aug. 144/3 Gluten, gum arabic, dextrine, are also..used, though their use is mainly that of simple adhesives. **1958** *Observer* 11 May 10/3 There is no universal adhesive. If the manufacturer claims that the adhesive is good for leather, it probably is. It's no good using it for a teacup handle.

b. *Philately.* An adhesive postage stamp (opp. one impressed on a card or wrapper, etc.).
1881 *Stamp-Collector's Ann.* 9 The 5, 20, and 50 pf. adhesives have now followed the example of the 10 pf. **1893** H. EWEN (*title*) Priced Catalogue of the Used Postage Adhesives of Great Britain.

‖**ad hoc** (æd hǫk). [L., lit. 'to this'.] For this purpose, to this end; for the particular purpose in hand or in view.
1659 R. BAXTER *Key for Catholics* II. iv. 451 *Ad hoc* the Magistrate is the only Judge what is sound doctrine. **1809** *Edin. Rev.* Jan. 433 The conscripts are..examined..by a special commission, created *ad hoc* by the prefect. **1875** W. R. GREG *Misc. Ess.* (1882) vi. 147 A sum not far off two millions per annum will have to be provided *ad hoc* by the Chancellor of the Exchequer.

b. *attrib.* or as *adj.* Devoted, appointed, etc., to or for some particular purpose.
[**1853** GREVILLE *Mem.* (1887) III. I. ii. 51 There are already symptoms of a possible combination *ad hoc.*] **1879** in E. Yates *Time* I. 3 The special matter that brought about the *ad hoc* departure from the Lawrentian policy. **1900** *Daily News* 18 Apr. 7/1 The discussion of the constitution of the educational authorities was sensational, inasmuch as it led to the Conference declaring for the ad hoc principle. **1904** *Fabian News* Aug. 29/1 A report..on the total abolition of *ad hoc* bodies was read. **1928** ABP. CANTERBURY in *Daily News* 8 Feb. 10/1 A growing sense of the usefulness of Reservation of the ad hoc kind. **1955** *Times* 14 May 11/1 He was still employed by the B.B.C. on an *ad hoc* basis as a programme producer.

c. Hence (nonce-wds.): **ad hoc** *v.*, to use *ad hoc* measures or contrivances, to improvise; so **adhoc(k)ing** *vbl. sb.*; **ad hoc-ery,** the use of such measures; **ad-hoc-ness,** the nature of, or devotion to, *ad hoc* principles or practice.
1930 A. FLEXNER *Universities* II. 71 '*Ad-hoc-ness*', if..I may be permitted to invent an indefensible word, is not confined to courses or curricula. **1936** *Mind* XLV. 249 There is a suspicion of '*ad-hoc*-ness' about the 'explanation'. **1960** C. RAY *Merry Eng.* 136 The Civil Servant who complained..about 'all this adhocking'. **1960** *20th Cent.* Dec. 588 The Children's Department, generous-hearted, chaotic, brilliantly ad-hocing. **1961** *Economist* 14 Oct. 124/1 Britain thrives on 'anomalies' and 'ad hoc-eries'.

ad hominem. (Earlier examples.)
1599 R. PARSONS *Temp. Ward-Word* vi. 79 This is an argument..which logicians call, *ad hominem.* **1633** W. AMES *Fresh Suit* I. x. 105 Some arguments, and answers are *ad hominem*, that is, they respect the thing in question, not simply, but as it commeth from such a man.

adhortative (ædhǭ·ɹtătiv), *a.* [f. ADHORTAT(ION + -IVE.] Expressing adhortation or exhortation: applied to the verbal mood which expresses an exhortation.
1845 STODDART *Gram.* in *Encycl. Metrop.* I. 52/2 Some writers have distinguished from the imperative, the precative, the deprecative, the permissive, the adhortative, [etc.]. **1934** PRIEBSCH & COLLINSON *Germ. Lang.* vi. 316 The adhortative *gehen wir.* **1963** VISSER *Hist. Syntax Eng. Lang.* I. iii. 190 In the Pres. D. suggestive or adhortative questions in which *to be* is construed with *to do* (as in: 'Why *don't* you *be* reasonable?').

adiabatic, *a.* Substitute for def.: Impassable to heat; involving neither loss nor acquisition of heat. Add later example.
1933 *Flight* 2 Feb. 99/2 Closed, or adiabatic, cockpits, wherein the pressure could be kept at 430 mm.

adiaphanous, *a.* Delete 'rare⁻⁰' and add examples.
1658 T.C. in S. Austin *Naps upon Parnassus* sig. B3, Upon the Gurmundizing Quagmires and most Adiaphanous Bogs, of the Author's obnubilated Roundelayes. **1826** KIRBY & SPENCE *Introd. Entomol.* IV. xlvi. 285 *Adiaphanous*..which does not transmit the light at all. **1833** J. WATERWORTH tr. *Véron's Rule Cath. Faith* 77 Nor will our adversaries ever be able to prove..that actions, which are prescribed by human laws, continue..indifferent or adiaphanous. **1855** J. SCOFFERN in *Orr's Circle Sci.* IV. 103 We have transcalent and non-transcalent substances..corresponding with the expressions *diaphanous* and *a-diaphanous.*

adiate (æ·diĕ¹t), *v. Roman-Dutch Law.* [app. irreg. f. L. *adīre* to approach + -ATE³.] *trans.* To accept (an inheritance) as heir under a will; in South Africa, to accept as beneficiary under a will. Hence **adiation** (ædiĕ¹·ʃən).
1829 in J. W. Knapp *Rep. Cases Privy Council* (1831) I. 111 Adiation (*aditio in hereditatem*) is a question more of intention than action. **1845** C. HERBERT tr. *Grotius' Introd. Dutch Jurispr.* II. xx. 143 An instituted heir, who adiates freely the inheritance, may deduct therefrom a fourth part. *Ibid.* xxi. 147 Repudiation must take place after the inheritance falls in by death and before adiation. **1896** JUTA *Selection of Leading Cases* II. 111 If the survivor has adiated and accepted benefits under the will. **1915** R. W. LEE *Roman-Dutch Law* IV. i. 286 If he [*sc.* the 'extraneus heres'] accepted or acted as heir, he was said to 'adiate' the inheritance (*adire hereditatem*), and from that moment was in the position of a universal successor.

adipocellulose (æ:dipose·liulōᵘs). [f. ADIPO(SE + CELLULOSE.] A compound of cellulose and suberin, as in cork tissue.
1888 MORLEY & MUIR *Watt's Dict. Chem.* I. 721/1 This view of its [cork's] constitution..is summed up in the group term Adipocellulose, by which it is proposed to designate them. **1895** C. F. CROSS et al. *Cellulose* II. 228 We..adopt the terms Suberose and Cutose for the compound adipocelluloses. **1900** JACKSON *Gloss. Bot. Terms* 6/1 *Adipocelluloses*, a group of bodies which constitute the cuticular tissues of leaves and fruits. **1937** E. J. LABARRE *Dict. Paper* 96/1 *Adipo-cellulose*, a compound so termed because on reduction it yields acids analogous to those obtained by the reduction of fats and cork tissue. The group embraces the cuticular tissue of such plants as straw, esparto and cotton.

adiposis (ædipōᵘ·sis). *Path.* [mod.L., f. L. *adip-*, *adeps* fat: see -OSIS.] Obesity or fatness of the body; fatty degeneration (of an organ).
1842 DUNGLISON *Med. Lex.* (ed. 3).

aditus (æ·ditŏs). *Zool.* [L., lit. 'approach', f. *adīre*, f. *ad* to + *īre* to go.] An incurrent canal in a sponge. Hence **adital** (æ·dităl) *a.*
1887 SOLLAS in *Encycl. Brit.* XXII. 415/1 The prosopyles..may..be prolonged into..a *prosodus* or *aditus*...By the extension of the prosodal or adital canals..a still higher differentiation is reached.

Adj. (ædʒ), colloq. abbrev. of ADJUTANT *sb.* 2.
1930 R. BLAKER *Medal without Bar* iv. 29 The Adj. [wants you]. He sent the office Corporal along in a great hurry. **1959** 'D. BUCKINGHAM' *Wind Tunnel* v. 37, I signed a 'blood chit' in the Adj.'s office, absolving everyone from blame if I broke my neck.

adjectival, *a.* Add: **b.** = *ADJECTIVE *sb.* 1 b.
1910 H. G. WELLS *New Machiavelli* (1911) I. ii. 31 My mother would never learn not to attempt to break him of swearing..refusing to assist him to the adjectival towel he sought. **1932** D. L. SAYERS *Have his Carcase* viii. 92 The decease of a damned dago, hr'rm, in an adjectival four-by-three watering-place like Wilvercombe. **1959** G. MITCHELL *Man who grew Tomatoes* xiii. 167 Beresford told him to take his adjectival charity elsewhere.

adjectivally (ædʒĕktəi·văli), *adv.* [f. ADJECTIVAL *a.* + -LY².] In an adjectival manner, as an adjective; = ADJECTIVELY *adv.*
1867 F. W. FARRAR *Greek Syntax* Introd. § 38 The fact that substantives are frequently used adjectivally. **1928** E. G. R. WATERS *St. Brendan* p. cxcvi, The tonic forms of the possessive pronouns are frequently used adjectivally.

adjective, *sb.* Add: **1. b.** Euphemistically substituted for an expletive adjective; usu. *attrib.*
1851 DICKENS in *Househ. Words* 14 June 270/1, I won't, says Bark, have no adjective police and adjective strangers in my adjective premises! I won't, by adjective and substantive! [**1888** KIPLING *Soldiers Three* (ed. 3, 1889) 66 They..slept until it was cool enough to go out with their 'towny', whose vocabulary contained less than six hundred words, and the Adjective.] **1894** *Idler* Feb. 102 To know where the adjective blazes they are going. **1900** E. WELLS *Chestnuts* 29 Now..we must have some (adjective) fun.

adjective, *v.* Add: **2.** To furnish with an adjective. Also *intr.* (*colloq.*) to use adjectives. So **a·djectived** *a.* or *ppl. a.*, qualified by an adjective or adjectives.
1804 *Med. & Physical Jrnl.* XII. 335 *Vaccine*, French, is from Latin:..Milk is by Pliny adjectived with the word, *lac vaccinum.* **1871** EARLE *Philol. Eng. Tongue* 341 Clough took the liberty of thus adjectiving Lord Macaulay..'I have only detected one error myself, but it is a very Macaulayesque one'. **1892** 'H. S. MERRIMAN' *From one Generation to Another* xi, He paid for his pleasure in..the adjectived items [i.e. 'ripping', 'topping'] of hospitality. **1920** *Sunday at Home* Apr. 423/2 In her place I think I should have 'adjectived' a good deal more.

adjectively, *adv.* Add: **b.** (Cf. *ADJECTIVE *sb.* 1 b.)
1918 *Pilot* (Boston, U.S.) 9 Feb. 4/7 The effect of zero weather..on the public..is adjectively bad.

adjectivism (æ·dʒĕktivi:z'm). [f. ADJECTIVE *sb.* + -ISM.] The (excessive) use of adjectives.
1890 *Sat. Rev.* 5 Apr. 423/1 All our isms—Romanticism, Naturalism, Socialism, Æstheticism, Undogmaticism, Adjectivism. **1904** G. S. HALL *Adolescence* II. xvi. 467 Adjectivism, adverbism, and nounism, or marked disposition to multiply one or more of the above classes of words.

adjectivity (ædʒĕkti·vĭti). [f. ADJECTIVE *sb.* + -ITY.] Addiction to the free use of adjectives.
1889 *Sat. Rev.* 5 Jan. 22/1 The adjectivity of his description. **1894** *Athenæum* 14 Apr. 469/1 Mrs. Ward..has checked the reckless fluency of her 'adjectivity'.

adjectivize (æ·dʒĕktivəi:z), *v.* [f. ADJECTIVE *sb.* + -IZE.] To make into an adjective. So **a·djectivi:zed** *ppl. a.*
1901 *Publ. Mod. Lang. Assoc. Amer.* XVI. 142 The completely adjectivized participle. **1949** E. A. NIDA *Morphology* (ed. 2) iv. 100 Derivational morphemes may..adjectivize nouns, *truthful.*

adjoin, *v.* Add: **7.** *Math.* (See quot. 1903 and cf. *ADJUNCTION 3 a.)
1903 L. E. DICKSON *Theory Algebraic Equations* vi. 62 In the language of Galois..we derive the domain R¹ = (1, i) from the included domain R = (1) by *adjoining* the quantity i to the domain R. **1904** [see *ADJUNCTION 3 a].

adjourner (ădʒɔ̄·ɹnəɹ). [f. ADJOURN *v.* + -ER¹.] One who adjourns or is in favour of an adjournment.
1893 *Columbus* (Ohio) *Even. Disp.* 24 Aug., The weakening among Senators..upon which..the confidence of the early adjourners is based. **1893** *Westm. Gaz.* 25 Nov. 7/2 There were only 44 adjourners, while 142 members preferred to proceed with the..business.

adjudicator. (Earlier example.)
1835 *New Monthly Mag.* XLV. 20 It is the best adjudicator we can have, for rather than condemn rashly, it acquits both parties.

adjunct. Add: **A.** *adj.* Also spec. *adjunct professor*, (in some institutions) a university or college teacher ranking immediately below a professor. *U.S.*
1826 *Catal. Univ. Cambridge (Mass.)* 6 John W. Webster, M.D., Adjunct Erving Professor of Chemistry. **1840** J. QUINCY *Harvard Univ.* II. 305 In 1808, John Collins Warren, M.D., and in 1809, John Gorham, M.D., were appointed Adjunct Professors. **1931** W. G. MCADOO *Crowded Years* ii. 24 My father was invited to Knoxville as Adjunct Professor of History and English in the University of Tennessee.

B. *sb.* **2.** Also *spec.* = *adjunct professor. *U.S.*
1876 D. C. GILMAN *University Probl.* (1898) 29 Promoting them because of their merit to successive posts, as scholars, fellows, assistants, adjuncts, professors, and university professors.

5. b. In Jespersen's terminology, a word or group of words of the second rank of importance in a phrase or sentence. Cf. *PRIMARY *sb.*, *SUBJUNCT.
1914 O. JESPERSEN *Mod. Eng. Gram.* II. i. 2 In the combination *extremely hot weather*..*hot*, which defines *weather*, is a secondary word or an adjunct. **1924** — *Philos. Gram.* vii. 97 It will be useful to have the special names *adjunct* for a secondary word in a junction, and *adnex* for a secondary word in a nexus. *Ibid.* 98 Other examples of substantives as adjuncts are *women* writers, a *queen* bee, *boy* messengers, and (why not?) *Captain* Smith. **1934** M. CALLAWAY in *Language* X. 366 As used by Jespersen, 'adjunct' covers both 'attribute' and 'appositive' as used by Sweet. **1935** W. F. LEOPOLD in *Jrnl. Eng. & Germ. Philol.* XXXIV. 415 The 'principals', 'adjuncts', and 'subjuncts' of his [Jespersen's] *Modern English Grammar* have now given way to the simpler and more mechanical terms 'primary', 'secondary', and 'tertiary'.

adjunction. Add: **3. a.** *Math.* [ad. F. *adjonction* (Galois 1846).] The relation holding between sets when without overlapping one another they are so 'joined' or continuous as to form another complete set; also, the process of putting them into this relation.
1896 *Bull. Amer. Math. Soc.* Dec. 103 The formation of an algebraic 'domain' and..the nature of the process of 'adjunction' introduced by Galois. **1904** F. CAJORI *Theory of Equations* xiii. 135 This process of obtaining the domain $\Omega(a)$ from Ω is called *adjunction*. We say that we *adjoin* a to Ω and obtain $\Omega(a)$. **1947** COURANT & ROBBINS *What is Math.?* (ed. 4) iii. § 2. 132 It is assumed that $\sqrt{k}$ is a new number not lying in *f*, since otherwise the process of adjunction of $\sqrt{k}$ would not lead to anything new.

4. *Logic.* The operation consisting in the joint assertion in a single formula of two previously asserted formulæ (see quot. 1932). Now often called *conjunction.*
1932 LEWIS & LANGFORD *Symbolic Logic* vi. § 1. 126 *Adjunction.* Any two expressions which have been separately asserted may be jointly asserted. That is, if *p* has been asserted, and *q* has been asserted, then *pq* may be asserted. **1962** KNEALE *Devel. Logic* ix. § 4. 550 In *Principia Mathematica*, which allows inference from *P* and $P \supset Q$ to *Q*, it is possible to dispense with Lewis's rule of adjunction because $p \supset [q \supset (p.q)]$ is a theorem.

adjust, *v.* Add: **2. b.** To adapt oneself *to*; to get used to. Also *absol.*
1924 J. J. B. MORGAN *Psychol. of Unadjusted School Child* III. viii. 121 It may be that the child will not adjust and will later develop a more serious form of dissociation. **1938** E. BOWEN *Death of Heart* II. iv. 244 'He is not really; he's in my brother's office.' 'Well, after all,' said Evelyn, adjusting to this. **1955** *Amer. Dialect Soc.* XXIV. 14 The inability of the criminally-inclined to 'adjust' to normal society. **1962** *Listener* 9 Aug. 207/1 He may try to adjust by staying with people of the same group as his family. **1962** *Amer. N. & Q.* Sept. 4/1 He adjusted rapidly to the rough life of the open range.

3. b. *refl.* Const. *to.*
1859 GEO. ELIOT *Lifted Veil* in *Blackw. Mag.* July 39/2 Our impulses, our spiritual activities, no more adjust themselves to the idea of their future nullity. **1888** MRS. H. WARD *R. Elsmere* II. xvii. 85 Lady Charlotte's eye-glass, having adjusted itself for a moment to the distant figure of

the Rector..turned back towards the squire. **1943** J. S. HUXLEY *Evol. Ethics* iii. 22 The individual..can adjust himself..to the ethical standards of his society.

adjustable, *a.* Add: *adjustable pitch*: freq. *attrib.* (with hyphen), applied to an aeroplane propeller the blade angle of which can be adjusted when the engine is at rest (opp. *variable pitch). Also *absol.*, a pitch setting capable of such adjustment.
1900 *Car* 5 May 451/1 Look to your propeller which..has an adjustable pitch. **1918** *Flight* 15 Feb. 156/1 The airscrews are of the three-bladed adjustable-pitch type. **1949** *Gloss. Aeronaut. Terms* (B.S.I.) II. 19 *Adjustable-pitch propeller*, a propeller, the blades of which can be adjusted to any required pitch when the propeller is at rest.

adjusted, *ppl. a.* Add: **4.** Adapted to one's environment (often preceded by defining word, as *well adjusted*). Cf. *ADJUSTMENT 1.
1924 J. J. B. MORGAN *Psychol. of Unadjusted School Child* I. i. 4 It is the function of the parent and teacher to encourage the successfully adjusted children. **1940** R. S. WOODWORTH *Psychol.* (ed. 12) xvii. 590 If an individual is participating intelligently and wholeheartedly in what is going on in his environment, we have no hesitation in calling him well adjusted. **1957** D. KARP *Leave me Alone* vi. 97 Don't talk to me about being 'adjusted'.

adjuster. Substitute for def.: One who adjusts, settles or regulates. Spec. in *Insurance*, a person who assesses liabilities and claims arising out of loss, damage, etc.; *average adjuster*: see AVERAGE *sb.*² 4 b. (Add earlier and later examples.)
1673 E. HICKERINGILL *Gregory* 252 Designed for a Confession of Faith, the Adjuster of Controversies. **1733** *London Gaz.* 16–19 May 3/1 Your Majesty..the steady Arbiter and great Adjuster of the Affairs of Europe. **1830** *Examiner* 31 Jan. 62/1 If the adjusters have the best of the bargain, it lowers the credit of the insurers. **1954** R. WESCOTT *While Rome Burns* 143 This play had been written..by a young insurance adjuster.

b. Something that adjusts (in various technical uses).
1884 KNIGHT *Dict. Mech.* Suppl. 6/1 *Adjuster (Surgical)*, an instrument for bringing into coaptation the parts in case of ruptured perineum. **1909** *Westm. Gaz.* 6 Apr. 4/1 The steering-gear, too, is fitted with an adjuster at the base of the steering column.

adjustive (ădʒɒ·stiv), *a.* [f. ADJUST *v.* + -IVE.] Tending to adjust, concerned with adjustment.
1883 ROMANES *Mental Evol. in Anim.* I. 17 Adjustive movements due to reflex action, and adjustive movements accompanied by mental perception. **Ibid.** 18 Adjustive action. **1935** *Brit. Jrnl. Psychol.* July 109 Only 'Adjustive' and 'Receptive' processes are dealt with. **1959** *Sci. Prog.* XLVII. 50 Thirdly there are the incoming stimuli which inform the pilot all the time of the discrepancy between the performance and his intention, and which also convey back to him results of his adjustive movements.

adjustment. Add: **1.** Further examples, esp. in contexts of emotional adaptation.
1912 A. A. BRILL tr. *Freud's Sel. Papers on Hysteria* i. 8 Those psychic traumas which are not rectified by reaction are also prevented from adjustment by associative elaboration. **1922** R. S. WOODWORTH *Psychol.* iv. 72 Much used..the 'adjustment' of the individual to his environment. **1937** *Brit. Jrnl. Psychol.* 20 If we ask the question, 'What is to be the adjustment..the individual to an adjustable machine which can be set for one or another set of work. **1957** P. HALMOS *Towards Measure of Man* i. 4 ..so Modern psychologists and psychiatrists distinguish two main stages in the process of adjustment. The instinctual development of the infant..and the treatment he receives from his environment..combine in a primary process of adjustment. ..A secondary process of adjustment..goes on from about the fifth year of life till its end. **1959** *Times Lit. Suppl.* 16 Jan. 29/2 Three of them emerge to face life again; the fourth..refuses to make the effort of adjustment.

5. *attrib.* in *adjustment award, committee, levy, method.*
1907 C. H. JUDD tr. *Wundt's Outl. Psychol.* iii. 257 The *adjustment-methods.* Among these, we have..the 'method of minimal change', and then as a special form of this to suit the case of adjustment until equality is reached, the 'method of average error'. **1909** KIPLING *Traffics & Discov.* 251 The Adjustment Committee—the umpires of the Military Areas. **1920** *Act 10 Geo. V*, c. 4 § 2 If the profits..exceed the sum apportioned to that undertaking..the excess shall be payable to the Controller by the owner of the undertaking and shall be recoverable as a debt due to the Crown, and the amount so payable is in this Act referred to as adjustment levy. *Ibid.*, Any sum so payable [by the Controller] is in this Act referred to as an adjustment award.

adjustor (ădʒɒ·stǫɹ). *Anat.* and *Zool.* [f. ADJUST *v.*, after L. agent-nouns in *-or*, as *retractor*.] Name for certain muscles in Brachiopoda; see quot. 1895. Also *attrib.* in *adjustor muscle.*
1895 SHIPLEY in *Cambr. Nat. Hist.* III. 477 There are three pairs of adjustor muscles..called respectively the central..., external..., and posterior..adjustors, whose action adjusts the shells when all contract together, and brings about a certain sliding movement of the shells on one another when they act independently. **1923** *Glasgow Herald* 11 Aug. 4 The more adjustors a creature has, the higher its capacity for effective behaviour.

adjutancy. 1. (Earlier examples.)
1775 R. ALDEN in H. P. Johnston *N. Hale* (1901) 157, I would accept of a lieutenancy but should prefer an adjutancy. **1796** JANE AUSTEN *Let.* 9 Jan. (1952) 3 He has got a scheme..about getting a lieutenancy and adjutancy in the 86th.

adjutor¹. Delete †*Obs.* and add later examples.
1670 J. FLAMSTEED in Rigaud *Corr. Sci. Men* (1841) II. 96 My friend, who is my sole adjutor, knows your hand. **1874** CARLYLE *Let.* 27 Apr. (1904) II. 296 Taït and some of his adjutors can set to work next year. **1883** *Nat. Observer* 15 Apr. 540/2 That gentleman, as became an adjutor of the greatest living thing in petticoats, was equal to the occasion.

adjuvancy (æ·dʒuvănsi). [f. ADJUVANT *a.*: see -ANCY.] Assistance, help.
1884 W. STAMLAND *Songs after Sunset* 54 Whose designing flattery Bought my adjuvancy with foul intent. **1896** *Durham Univ. Jrnl.* XII. 14 The students..endeavoured to secure his adjuvancy in their theological studies.

Adlerian (ædlī*ə*·riăn), *a.* [See -IAN.] Of, pertaining to, or characteristic of the psychologist Alfred *Adler* (1870–1937) or his teaching. Also *sb.*, a follower or adherent of Adler.
1925 *Times Lit. Suppl.* 1 Mar. 151/2 This latest medical pamphlet contains three papers in which Adlerian psychology is applied to various phases of child life. **1952** H. G. WELLS *You can't be too Careful* v. i. 237 A certain section of the mixture of peoples called the Jews..suffers from an Adlerian assertiveness. *Ibid.* vi. i. 269 A natural Adlerian revolt against the overbearing religious dogmatisms of the Middle Ages. **1959** *Times Lit. Suppl.* 1 Aug. 457/4 They were thus given to..a mixed audience..of all 'schools' of dynamic psychology: Freudians, Adlerians, eclectics and a few Jungians. **1966** *Listener* 13 Jan. 71/2 The Adlerian principle of compensation through power.

ad lib., abbrev. of AD LIBITUM *adv. phr.*
1811 in T. BUSBY *Dict. Mus.* (ed. 3). **1894** STEVENSON & OSBOURNE *Ebb-Tide* I. v. 66, I'll have fizz *ad lib.*, or it won't wash. **1900** in A. E. T. WATSON *Young Sportsman* 27 The catalogue may be extended *ad lib.* for those who require their spring-balance weighing-machine, vaseline, and small bottle of colourless paraffin. **1932** *Times Lit. Suppl.* 24 Sept. 719/1 Now he could have all he wanted for his two marbles —living material *ad lib.*, hundreds of test-tubes for his culture in place of dozens. **1960** N. DEL MAR R. *Strauss* I. vi. 183 He never gave them [saxophones] any solo work, marked them *ad lib.*, and constantly cut them in later performances.

B. *adj.* (**ad-lib**). Extemporized, improvised, spontaneous, unrehearsed. orig. *U.S.* Also as *sb.*, something spoken or played extempore (chiefly *U.S.*).
1925 *Amer. Speech* I. 36/1 'Can the ad lib!' which means, politely, 'Will you be good enough to hush!' **1935** *Peabody Bulletin* (Baltimore) Dec. 42/2 Licks: Phrases used in solo choruses that are..catchy. They occur in 'ad lib' solos or hot choruses. **1958** *Daily Tel.* 11 June 15 That was an ill-considered ad-lib remark. **1958** G. LASCELLES in P. Gammond *Decca Bk. of Jazz* viii. 107 With his early big band..he retained a reasonable ad-lib feeling.

ad-lib (æ:d,li·b), *v.* orig. *U.S.* [f. prec.] *trans.* and *intr.* To speak extempore; to announce without script, improvise (words, etc.) esp. in the course of a stage or broadcast performance. Hence **ad-li·bbed** *ppl. a.*, improvised; **ad-li·bber**, one who ad-libs; **ad-li·bbing** *vbl. sb.* and *ppl. a.*
1919 F. HURST *Humoresque* 165 'Easy money, friends,' Miss Hoag would ad lib, to the line-up outside her railing. **1926** WHITEMAN & MCBRIDE *Jazz* iii. 73, I remember one day asking one of my English musicians, 'Can you ad lib?'. 'Certainly,' answered the man, rather nettled, 'I can ad lib anything.' **1927** *Sat. Even. Post* 5 Mar. 54/3 The text of musical plays..is much more fluid..and considerable 'ad libbing,' as impromptu speeches are called, is indulged in. **1929** *Bookman* (U.S.) Mar. 149/2 'Ad-libbing' is improvising. 'When Jones got lost back stage I had to ad-lib all over the place until he came on.' **1936** *Harper's Mag.* 4 Apr. 573/1 The preacher or orator whose message is canned beforehand never reaches the emotional heights of persuasion that his 'ad libbing' brother does on a rival rostrum. **1958** *Scrutiny* VIII. 216 It seems much more likely from the extensive *ad-libbing* (to borrow a word from the Americans) Mr. Roberts allows himself that Holme's positions were useful to him in working out problems. **1953** *Listener* 13 Nov. 825/1 This..was by no means the end of the ad-libbed script. **1955** KOESTLER *Trail Dinosaur* 46 'Remember Fred Karno'..66 in the sketch he 'ad-libbed' a bit, to use Fred Karno's phrase. **1960** *Amer. Speech* XV. 65 Fadiman is a catholic ad libber. **1966** C. K. *Wills Power Elite* i. 15 The easy ad-libbed gave the celebrity 'spontaneously' echoes. **1958** I. BROWN *Words in Our Time* 15 The adlibber in the theatre is one who introduces what are called 'gags'. **1958** 'N. SHUTE' *Rainbow & Rose* iii. 95 The whole squadron in the stalls chi-biking at us and Judy ad-libbing back at them across the footlights.

ad libitum, *adv. phr.* Add: Earlier and later examples of general and musical uses.
1660 PHILANDEMUS *Art of Survey* iv. vi. 38 These may be contrived in Parallelograms, Squares, Circles..composed and tricked *ad libitum.* **1671** LOCKE *Hum. Underst. (Draft B)* (1931) § 50 p. 139 Imperfect ideas, and consequently words, *ad libitum* by every particular man made use of for signs of those ideas. **1724** in *Short Explic. For. Wds. Mus. Bks.* 7. **1823** *Harmonicon* Mar. I. 40/2 Air, with accompaniments, ad libitum, for the flute, by T. A. Rawlings. **1845** J. E. WARNER *Lessons in Music* 200 Sometimes the movement is expressly committed to the taste of the performer by placing the term *ad libitum*..over the passage. **1848** LYTTON *Harold* I. I. ii. 38 To marry wives *ad libitum.* **1922** JOYCE *Ulysses* 691 *Love's Old Sweet Song*..open at the last page with the final indications *ad libitum*, *forte*, [etc.].

1951 W. APEL *Harvard Dict. Mus.* 16 *Ad libitum*, an indication which gives the performer the liberty to (1) vary from strict tempo..(2) to include or omit the part of some voice or instrument..(3) to include a cadenza according to his own invention.

Hence as *adj.*
1769 GRAY *Jrnl.* 10 Oct. in *Corresp.* (1935) III. 1103 Many neat buildings of white stone, but a little disorderly in their position ad libitum like Kendal. **1801** T. BUSBY *Dict. Mus.* s.v., We speak of an *Ad Libitum* piano; or, an *Ad Libitum* cadenza. **1821** *Edin. Rev.* XXXV. 343 Armed with the *ad libitum* reserve of fact-hardness. **1823** *Harmonicon* June I. 83/1 Clementi has arranged for the piano-forte, with ad libitum accompaniments, many of the best symphonies of Haydn and Mozart.

‖**ad litem** (æd loi·tem). *Law.* [L.] For a lawsuit or action (applied esp. to a guardian appointed for such a purpose).
1768 W. BLACKSTONE *Commentaries on Laws of England* III. xvii. 427 The court of exchequer can only appoint a guardian *ad litem*, to manage the defence of the infant if a suit be commenced against him. **1838** [see GUARDIAN 2 b]. **1883** LORD COLERIDGE in *Law Rep. Queen's Bench Div.* XI. 252 Guardians ad litem are relieved from the necessity of answering interrogatories. **1959** JOWITT *Dict. Eng. Law* I. 884/1 A guardian *ad litem* is appointed by the court or other proceeding on behalf of an infant, or a lunatic or idiot not so found..who is defendant or a respondent to a proceeding in the court. **1961** *Lancet* 12 Aug. 366/1 The husband, by the Official Solicitor as his guardian *ad litem*, who denied the allegations.

ad-man: see *AD.

admass (æ·dmæs). Also **ad-mass.** [f. *AD advertisement + MASS *sb.*²] (See quot. 1955.) That section of the community which is easily influenced by mass methods of publicity and entertainment; also applied to the advertising, etc., processes or agents themselves.
1955 J. B. PRIESTLEY & J. HAWKES *Journey down Rainbow* iii. 51 *Admass.* This is my name for the whole system of an increasing productivity, plus inflation, plus a rising standard of material living, plus high-pressure advertising and salesmanship, plus mass communications, plus cultural democracy and the creation of the mass mind, the mass man. **1957** *Times Lit. Suppl.* 4 Oct. 593/2 The broadening cultural pattern in the age of ad-mass. **1957** *Listener* 10 Oct. 583/2 Ordinary, intelligent people..who do not wish to be caught up in the Admass culture, yet whose education has not provided them with a complete means to resist it. **1959** N. MARSH *Singing in Shrouds* v. 97 The entire Commercial TV admass.

admin., colloq. abbrev. of ADMINISTRATION, ADMINISTRATIVE *a.*, or ADMINISTRATOR. Freq. *attrib.*, as *admin. block, building.* Also *ellipt.*, administration building, administrative order, etc.
1942 T. RATTIGAN *Flare Path* ii. ii, Probably challenge me to a duel or something. Rapiers at dawn behind the Admin Block. *Ibid.* 67, I'd have a fine time as a Squadron Leader admin. at a training school. **1951** A. C. CLARKE *Sands of Mars* viii. 98 There was nothing very exciting about Admin. It might have been any office building. *Ibid.* x. 119 She worked in the Admin. building. **1955** J. CHRISTOPHER *Year of Comet* I. 17 You are under San Diego for admin but any reports you make go direct to Geneva. **1958** *Spectator* 15 Aug. 219/1 All this is in part an admin. problem. **1961** W. BUCHAN *Helen All Alone* 26 A mass of practical details—sheer 'admin'.

administratively, *adv.* (Earlier example.)
1837 *Athenæum* 18 Nov. 842/2 To render Hungary, if not completely, at least administratively, independent of Austria.

Admiralty. Add: **3.** The title *Lords Commissioners of Admiralty* was dropped in 1964.
4. b. In full *Admiralty coal*, a type of large steam coal supplied to the Royal Navy under contract.
1914 *Coal & Iron* 11 May 709/2 For best Admiralties up to 20s. is still being quoted. **1915** *Ibid.* 4 Jan. 337/2 Best Admiralty large steams are not dealt in to any extent, the Naval authorities still controlling supplies. **1928** *Daily Mail* 9 Aug. 12/1 The Ferndale Collieries, which are among those producing the finest Admiralty coals in the world.

8. *Rhet.* Command of the seas (esp. in phr. *the price of admiralty*).
1893 KIPLING 'Song of the English' in *Eng. Illustr. Mag.* May 535 If blood be the price of admiralty Good God, we ha' paid in full! **1923** E. K. CHATTERTON *Auxiliary Patrol* viii. 121 By the end of the following year three more minesweepers..and three more motor boats had been lost; that, of course, loss of life. Such was the inevitable price of Admiralty.

admiration. Add: **2. b.** Phr. *to admiration* [cf. F. *à merveille*], so as to elicit admiration; in an admirable or excellent manner. Slightly *arch.*
1633 W. AMES *Fresh Suit* Pref., This book..hath made all the world to admiration. **1680** EVELYN *Diary* May an. 1646 (1955) II. 501 They are curious in Straw-worke among the Nunns, even to admiration. **1681** H. MORE *Expos. Proph. Dan.* App. I. 259 You shall find this part of the Prophecy fulfilled to admiration. **1793** [in Dict., sense 2]. **1821** C. LAMB *Old Benchers* in *Lond. Mag.* Sept. 282/2 [He] moulded heads in clay or plaister of Paris to admiration, by the dint of natural genius merely. **1930** V. WOOLF *Beau Brummell* 5 He..tied his cravat to admiration.

admire, *v.* Add: **1. d.** To like, be desirous (*to* do something). *U.S.*
a **1770** J. MECOM *Lett. to B. Franklin* (1859) 194, I should admire to come and see and hear all about every thing. **1816** PICKERING *Vocab.* s.v., *To admire*, to like very much, to be very fond of. This verb is much used in New England in expressions like the following: I should admire to go to such a place; I should admire to have such a thing, &c. **1839** MARRYAT *Diary Amer.* II. 223 'Have you ever been at Paris?' 'No; but I should admire to go.' **1869** [in Dict.]. **1876** B. HARTE *Gabriel Conroy* IV. i, 'Why didn't you come into the parlour?' she said...'I didn't admire to to-night,' returned Gabriel. **1939** J. STEINBECK *Grapes of Wrath* xx. 348 'Well, I'd admire to git a hand,' said the young man.

‖**ad misericordiam** (æd mi:zerikǭ·ɹdiăm). [L.] Of an appeal, argument, etc.: to mercy, to pity.
1824 *Edin. Rev.* XLI. 55 The fallacy of those arguments *ad misericordiam* on which the agriculturists now principally rest their claims to protection. **1863** THACKERAY *Round. Papers* 73 No day passes but that argument *ad misericordiam* is used. **1885** *Manchester Exam.* 27 Feb. 5/3 He now made an *ad misericordiam* appeal for an extension of that time, on the ground of his ignorance of the practice. **1929** E. MARJORIBANKS *Marshall Hall* x. 359 Later, in his final speech, he was making an appeal *ad misericordiam* for his client.

admission. Add: **1. b** *attrib.*, as *admission fee, money, ticket.*
1667 [in Dict., sense 1]. **1779** *Mirror* 12 June, I shall..present you with a dozen admission tickets. **1825** *Morning Chron.* 2 June 3/2 To give him a general passport to the Theatre, without paying any admission money. **1829** H. FOOTE *Compan. Theatres* 32 The lowering of the admission-money to the boxes. **1842** *Knickerbocker* XX. 498 Certain persons..being stationed at the gates to exact..admission fees. **1888** GUNTER *Mr. Potter of Texas* xx, No one demands an admission fee.

c. *ellipt.* for *admission fee, ticket.*
1792 *Observer* 4 Mar. 4/2 A mere admission for thirteen years sold the other day for fifty-eight guineas. **1802** G. F. COOKE in W. Dunlap *Mem. of G.F.C.* (1813) I. 216 Dispatched a note containing an admission for two on Monday evening to Covent-Garden theatre. **1833** I. F. ALDRIDGE *Let.* in *N. & Q.* (1872) 4th ser. X. 211/1, I enclose an admission for the printers, and one for yourself.

admittance. Add: **5.** *Electr.* The reciprocal of impedance, measured in mhos.
1887 O. HEAVISIDE in *Phil. Mag.* XXIV. 482 It is naturally suggested to call *J* the 'admittance' of the combination. But it is not to be anticipated that this will meet with so favourable a reception as impedance, which term is now considerably used. **1931** *B.B.C. Year-Bk.* 435/1 *Admittance*, the admittance of a circuit is the reciprocal of its impedance or apparent resistance. **1949** *Electronic Engin.* Apr. 145 A method is presented for finding the input admittance of an amplifier.

admitted, *ppl. a.* Add: **2. b.** *admitted clerk*, a clerk qualified by admission to the roll of solicitors kept by the Law Society.
1906 A. BENNETT *Whom God hath joined* i. 6 He was an 'admitted' clerk in a solicitor's office. That is to say, he had acquired the right to practise for himself as a solicitor, but he did not practise for himself. **1961** *Times* 9 Feb. 2/6 Wanted, Admitted or Unadmitted man..experienced in conveyancing and trust accounts.

‖**ad nauseam** (æd nǭ·siæm). [L., = 'to sickness'.] To a sickening extent, so as to excite disgust.
[**1616** T. ADAMS *Sacrifice of Thankefulnesse* 34 Wee haue heard this often enough. *Ad nauseam usque.*] **1647** J. TRAPP *Comm. upon Four Evang.* Matt. vi. 7, p. 196 Doe not iterate or inculcate the same things odiously & *ad nauseam.* [**1693** N. MATHER in J. Owen *Two Discourses concerning Holy Spirit* sig. A3, They are not filled..with novel and uncouth Terms foreign to the Things of God, as the manner of some Writers is *ad nauseam usque.*] **1814** *Edin. Rev.* Apr. 73 He had already spoken *ad nauseam* on this very subject. **1907** W. DE MORGAN *Alice-for-Short* xxv, His frequent use of this expression compels repetition *ad nauseam.* **1955** *Bull. Atomic Sci.* Jan. 15/1 The cliché that fear is altogether and always uncreative is being repeated ad nauseam.

adnex (æ·dneks). *Gram.* [f. AD- + *NEX(US.] In Jespersen's terminology, a word (usu. a verb) or group of words of secondary importance in a 'nexus' or predicative relation.
1924 [see *ADJUNCT B. 5 b]. **1924** O. JESPERSEN *Philos. Gram.* vii. 100 Finite forms of verbs can only stand as secondary words (adnexes), never either as primaries or as tertiaries. **1933** — *Syst. Gram.* 17 *The dog runs*, nexus: *runs*..is adnex to *dog*. **1933** A. R. RADCLIFFE-BROWN *Andaman Islanders* (rev. ed.) 504 In nexus the primary precedes the adnex, or in other words the nominative precedes the verb.

adnexed (ædne·kst), *ppl. a. Bot.* [f. L. *adnex-us*, pa. pple. of *adnectĕre* (see ANNEX *v.*) + -ED¹.] Applied to the gills of certain fungi that reach but do not join the stem.
1836 BERKELEY *Fungi* II. 107 Gills adnexed. **1891** W. G. SMITH *Berkeley's Outl. Brit. Fungology* Suppl. 8 Gills adnexed, shining white. **1947** F. A. & F. T. WOLF *Fungi* vii. 360 (*caption*) Diagrams representing the manner in which gills of mushrooms may be attached to the stipe...Free, Adnate, Decurrent,..Adnexed.

adobe. Add: (Earlier examples.) Also (all *obs.*) **adaube, adobey, adoby,** etc.
1748 *Earthquake Peru* (ed. 2) iii. 268 Adobes, that is, large Bricks, about two Feet long. **1845** GREEN *Texian Exped.* viii. 91 The guard..occupied a small adoby house. **1850** G. HINES *Voyage* 188 Two or three small buildings are enclosed in an adobey wall.

2. A house made of adobe. *U.S.*
1821 DEWEES *Lett. from Texas* (1852) 21 The remainder of the buildings are adobes. **1881** *Amer. Naturalist* XV. 25 The adobe at one moment seemed near, and the next very far off. **1898** F. REMINGTON *Crooked Trails* 25 A little broken adobe.

3. *Geol.* The loess-like earth or clay from which adobe bricks can be made; a deposit of this material; (see quot. 1889). Also a 'cement' made from this clay.
1856 G. H. DERBY *Phoenixiana* (1859) xix. 133 We have..Indians employed..in mixing adobe for the type moulds. **1869** BROWNE *Adv. Apache Country* 118 This concrete, or adobe, was cast in large blocks, several feet square. **1889** I. C. RUSSELL in *Geol. Mag.* July 291 A peculiar calcareous clay which is used..for the manufacture of sun-dried bricks, known by the Spanish name 'adobe'. The earth from which these bricks are made is also designated by the same name. We have therefore adopted it..as a convenient term by which to designate the fine subaërial accumulations in general, exclusive of eolian sands. **1944** A. HOLMES *Princ. Physical Geol.* II. xiii. 269 In the semi-arid regions of the western States and in the Mississippi valley there are thick deposits of *adobe* which correspond in all essentials to the loess of Europe and Asia.

4. *attrib.* (See also examples under sense 1.)
1841 T. J. FARNHAM *Great Western Prairies* 136 We spent the 2d and 3d most agreeably with Mr. Walker in his hospitable adobie castle. **1845** J. C. FRÉMONT *Exp. to Oregon* 245 We gave a shout at the appearance on a little bluff of a neatly built *adobe* house with glass windows. **1895** *Outing* (U.S.) Aug. 355/1 The centenarian bells that hang from buckskin thongs in the adobe towers. **1897** *Ibid.* Feb. 457/1 A mile or so of broad adobe country road. **1951** R. BRADBURY *Illustr. Man* (1952) 123 Newer and better fireworks..banged against adobe café walls.

adolesce (ædǒle·s), *v.* [Back-formation f. ADOLESCENT.] *intr.* To become adolescent or pass through adolescence.
1909 H. G. WELLS *Tono-Bungay* II. iv. 214 He does his silly utmost to prevent our reading and seeing the one thing, the one sort of discussion we find..supremely interesting. So we don't adolesce; we blunder up to sex. **1934** G. B. SHAW *On the Rocks* 166 The training of the scholar and the sportsman may split and diverge as they adolesce; but they must start from a common training and a common morality as children. **1938** L. MACNEICE *I Crossed the Minch* ix. 128 In September I shall be thirty. How much longer..shall [I] take over adolescing?

Adonis. **1.** Substitute for def.: A beautiful or handsome young man. (Add earlier and later examples.)
1622 MABBE tr. *Aleman's Guzman d'Alf.* II. ii. 21 My Master..made me another *Adonis*, in the neatnesse and gallantry of my cloathes, and delicacie of Perfumes. **1624** MASSINGER *Parl. Love* II. ii, A leper,..in respect of thee, Appears a young Adonis. **1768** TUCKER *Lt. Nat.* II. I. xxiii. 225 How it would divert our ladies below to hear two such Adonises talking so sweetly of our reciprocal passion! *a* **1800** COWPER *On Female Inconstancy*, She who call'd thee once her pretty one, And her Adonis, now inquires thy name. **1888** GUNTER *Mr. Potter of Texas* viii, George! in a month this chap 'll be an Adonis.

3. For 'Clifton Blue' read 'Clifden Blue'.

adopt, *v.* Add: **2. b.** Of a local authority: to take over (from private ownership, etc.) responsibility for a road, etc.
1862 *Act* 25 & 26 *Vict.* c. 61 §45 It shall and may be lawful for the Council of every such Borough in England and Wales..to adopt all or any of such Parish Roads and Highways as the Council shall in its Discretion consider advisable. **1907** *Justice of Peace (Reports)* LXXI. 564/3 In 1904 the widening and laying out of Stubbington Avenue as a road forty feet wide for the whole of its length was completed, but it had never been adopted by any resolution of the council of the plaintiffs. **1958** *Times* 22 Mar. 7/6 A frontager who considers that they [*sc.* roads] should have been adopted.

8. To approve, to confirm (accounts, reports, etc.).
1906 GALSWORTHY *Man of Property* 178, I propose then that the report and accounts be adopted. **1958** *Oxford Mail* 16 Aug. 8/7 The best balance sheet the club has ever had was unanimously adopted at the annual general meeting of the Oxford Club League.

adopted, *ppl. a.* Add: **1. b.** *spec.* of a road (cf. *ADOPT *v.* 2 b).
1938 *Times* 31 May (Building Soc. No.) ix/3 Roads are either adopted or unadopted. **1951** E. BARKER *Princ. Soc. & Polit. Theory* II. iii. 50 An 'adopted' road is incorporated into the road system of a town.

adoptee (ădǫptī·). [f. ADOPT *v.* + -EE¹.] An adopted person.
1892 *Sat. Rev.* 6 Aug. 179/1 That odd provision of French law which permits adoption—in case the adoptee has saved the life of the adopter. **1962** *Guardian* 16 Nov. 8/7 The..psychiatric figures for adoptees.

adoption. Add: **2. d.** Of a road: see *ADOPT *v.* 2 b.
1890 *Act* 53 & 54 *Vict.* c. 59 §41 Adoption of private streets...The urban authority..may..declare the whole of such street or part of a street to be a highway repairable by the inhabitants at large.

3. Of accounts, reports, etc.: approval, confirmation (cf. *ADOPT *v.* 8).

1881 *Evening News* 26 July 3/6 The chairman in moving the adoption of the report said it was with great satisfaction that he informed the shareholders of the favourable traffic receipts of 1880.

adorant (ădōᵊ·rănt), *a. poet.* [f. ADORE *v.* + -ANT¹.] = ADORING *ppl. a.*
1819 KEATS *Fall of Hyperion* (1905) I. 283 'Shade of Memory'—Cried I, with act adorant at her feet. **1893** *19th Cent.* Nov. 842, I..make petition on adorant knee.

adorative, *a.* Delete '† *Obs.*' Add to def.: also, adoring. Add later examples.
1855 J. C. JEAFFRESON in *Fraser's Mag.* LI. 559/2 Such is the boy's state of mind, adorative and confident. **1861** *Temple Bar* III. 284, I always quitted the college amidst a commotion of adorative assurances. **1885** J. C. JEAFFRESON *Real Shelley* II. ix. 271 A tribute of adorative homage.

adoze. (Earlier example.)
1849 *Blackw. Mag.* LXVI. 23 Lying on banks a-dose or poetising.

adradial, *a.* Substitute for def.: Situated near or beside a ray. Add: **B.** *sb.* An adradial organ.
1888 ROLLESTON & JACKSON *Anim. Life* 717 The adradials open one into each of the eight meridional or ctenophoral vessels. **1892** J. A. THOMSON *Outl. Zool.* x. 134 Tentacles—first four corresponding to the angles of the mouth (perradials), and then other four (interradials) between these, and then eight intervening adradials.

adread (ădre·d), *ppl. a.*, var. ADRAD *ppl. a.*¹ *arch.*
1580, 1855 [see ADRAD *ppl. a.*¹]. **1887** T. MARTIN tr. *Schiller's Ring of Polycrates* in *Blackw. Mag.* Nov. 684 Back drew the monarch, all a-dread.

a-dream, *adv.* Add: (Earlier and later examples.) Also *pred. a. poet.*
1830 *Blackw. Mag.* Feb. 276/2 That sets the imagination dimly a-dream of mermaids. **1909** W. J. LOCKE *Septimus* ii. 23 He stared into vacancy, his pale blue eyes adream. **1923** D. H. LAWRENCE *Birds, Beasts & Flowers* 115 And slowly, very slowly, as if thrice adream, Proceeded to draw his slow length curving round. **1933** W. DE LA MARE *Fleeting* 68 Adam and Eve lay adream.

adrectal (ædre·ktăl), *a. Zool.* [f. AD- + RECTAL *a.*] Situated at or near the rectum; *spec.* of or pertaining to the purpuriparous glands of certain molluscs.
1883 *Encycl. Brit.* XVI. 648/2 The presence of glandular plication of the surface of the mantle-flap..and an adrectal gland (purple-gland).

‖**ad referendum** (æd refĕre·ndɒm). [mod.L., = 'for reference'.] In diplomatic use, a phr. describing the acceptance of proposals by representatives subject to the assent of their principals.
1781 J. ADAMS in *Wks.* (1852) VII. 488 They will take the proposition *ad referendum* immediately. **1787** *Gentl. Mag.* 1015/2 Congress have taken this generous offer of his *ad referendum.* **1815** WELLINGTON *Dispatches* (1838) XII. 287 The agreement was read to the whole, and taken *ad referendum* by the Russian and Prussian Ministers. **1906** HARDY *Dynasts* II. v. i. 258 Prince Eugène will..make the formal offer in his name...Which I can but receive *ad referendum.*

‖**ad rem** (æd rem). [L.] (Pertaining or pertinent) to the matter or subject in hand; to the purpose.
1608 MIDDLETON *Fam. Love* v. H 4, *Ad rem, ad rem*, master Poppin: leaue your allegories,..and to the point. **1621** BURTON *Anat. Mel.* To Rdr. 40 To speake *ad rem*, who is free from passion? **1680** J. HOWE *Let. to Person of Qual.* 23 What I can find in his Sermon hath any aspect or design that way is either *ad rem*, or *ad hominem.* **1865** RUSKIN in *Daily Tel.* 7 Sept. (Cent. Dict.), Your statements of practical difficulty are..more *ad rem* than my mere assertions of principle. **1905** *Spectator* 4 Feb. 180/2 It is more *ad rem* to consider whether a satisfactory answer to Newman's question is to be found in the second book on our list. **1940** *Mind* XLIX. 174 It can hardly be regarded as *ad rem* in our present discussion that is directed to *human* knowledge.

adrenal (ădrī·năl), *a.* and *sb.* [f. AD- I + L. *rēnes* kidneys: cf. RENAL *a.* and *sb.*] **A.** *adj.* = SUPRARENAL *a.*; *adrenal cortical hormone*, adrenocorticotrophic hormone.
1875 HUXLEY & MARTIN *Elem. Biol.* xiii. 173 The Adrenal glands are yellow bodies imbedded in the ventral face of the kidney. **1921** L. BERMAN *Glands regulating Personality* vii. 150 Christina may be adrenal cortex centred and so masculinoid. **1927** *Lancet* 8 Oct. 745/2 (*title*) Virilism due to an Adrenal Cortical Hypernephroma. **1934** *Ann. Reg. 1933* 60 Harris devised a chemical test for vitamin C, which is probably ascorbic acid and a specific constituent of the adrenal cortex. **1946** *Lancet* 10 Aug. 203/1 As the nitrogen output rises..so the urinary content of adrenal cortical hormone reaches relatively enormous levels. **1950** *Sci. News* XV. 132 The adrenal gland..is readily distinguishable into two parts, the rind or cortex, and the core or medulla.

B. *sb. pl.* Suprarenals.
1882 *Trans. Path. Soc.* XXXIII. 341 Adrenals. **1883** *Encycl. Brit.* XV. 365/1 The 'suprarenal bodies' or 'adrenals'. **1927** HALDANE & HUXLEY *Anim. Biol.* i. 18 The adrenals, the parathyroids, the pancreas, the reproductive organs, and probably other organs also produce internal secretions.

adrenalectomy (ădrī:năl,e·ktŏmi). *Surg.* [f. prec. + Gr. ἐκτομή cutting out.] Excision of the adrenal glands. So **adre:nale·ctomized** *ppl. a.*
1910 *Lippincott's New Med. Dict.* 22/2 *Adrenalectomy*, excision of suprarenal body. **1918** *Amer. Jrnl. Physiol.* 1 May 103 The following protocols are samples of our negative experiments on adrenalectomized rabbits. **1936** *Nature* 18 July 124/2 A high potassium intake is detrimental to the life of adrenalectomized animals. **1946** *Lancet* 3 Aug. 179/2 Adrenalectomy has not proved a satisfactory method of treating patients in whom prostatic cancer shows recrudescence after relief through castration.

adrenaline (ădre·nălin, ădrī·n-). Also **-in.** [f. *ADRENAL + -INE⁵.] A hormone secreted by the adrenal glands and affecting circulation, muscular action, etc.; also (**A-**), this substance extracted from the adrenal glands of animals or prepared synthetically and used for medicinal purposes (U.S. trade name); cf. *EPINEPHRINE.
See quot. 1901 for the discovery of the substance and the invention of the name, which have, however, been claimed also for Dr. Norton L. Wilson.
1901 *Amer. Jrnl. Physiol.* V. 457 The most important contribution to our knowledge of the active principle of the suprarenal gland..is from Dr. Jokichi Takamine who has isolated the blood-pressure-raising principle of the gland in a stable and pure crystalline form...To this body..he has given the name 'Adrenalin'. **1902** *Jrnl. Chem. Soc.* LXXXII. I. 68 It is claimed that suprarenin is identical with Takamine's adrenalin. **1907** OSLER & MCCRAE *Syst. Med.* I. 758 Experimentally..adrenalin..has a powerful influence on carbohydrate metabolism. **1920** *Discovery* Mar. 88/2 Adrenaline prescribed for Addison's disease. **1922** J. C. SQUIRE in *Public Opinion* 3 Mar. 202/1 The toad secretes too much adrenalin. **1962** *Lancet* 29 Dec. 1337/1 When adrenaline is infused intravenously the vein used stands out as a white cord.

adrenalism (ædrī·năliz'm). [f. *ADRENAL + -ISM.] The state arising from incorrect functioning of the adrenal glands; suprarenalism.
1922 G. B. SHAW in S. & B. Webb *Eng. Prisons under Local Govt.* p. lxiii, The stigmata of adrenalism.

adrenergic (ædrĕnɜ̄·ɹdʒik), *a.* [f. *ADREN(ALINE + Gr. ἔργ-ον work + -IC.] Of the sympathetic nerve-fibres: liberating adrenalin or a substance resembling adrenalin; also, stimulated by adrenalin.
1934 H. H. DALE in *Jrnl. Physiol.* LXXX. 11 P, We seem to need words which will briefly indicate action by two kinds of chemical transmission, due in the one case to some substance like adrenaline, in the other case to a substance like acetylcholine, so that we may distinguish between chemical function and anatomical origin. I suggest the words 'adrenergic' and 'cholinergic' respectively. **1935** *Ann. Reg. 1934* II. 53 There was increased recognition of the cholinergic mechanism as having the more general application in the functions of the nervous system and of the adrenergic mechanism as being a more specialised and a more recent development. **1955** *Sci. News Let.* 19 Mar. 179/2 Adrenergic chemicals inhibit, or stop, transmission of nerve impulses across the nerve junctions called synapses. Cholinergic chemicals excite, or give the go signal, to such impulses.

adrenin (ădrī·nin). Also **-ine.** [f. *ADREN(AL + -IN¹.] = *ADRENALINE.
1908 E. A. SCHÄFER in *Brit. Med. Jrnl.* 30 May 1281 It would..be well to substitute the term 'adrenin' for the proprietary name 'adrenalin'. **1910** *Year-Bk. Pharmacy* 177 (*heading*) Adrenine as an Emergency Treatment in Cases of Non-corrosive Poisoning. **1932** *Discovery* Oct. 339/1 Adrenin, the internal secretion of that ductless gland the suprarenal capsule.

adreno- (ădrī·no, ădre·no), comb. form of *ADRENAL *a.* and *sb.*, and *ADRENALINE, as in **adre·nochrome** [G. *adrenochrom* (*Chem. Zentralbl.* (1909) II. 1584).], a red oxidation product of adrenalin; **adre:no-co·rtical** *a.*, of or pertaining to the adrenal cortex or the hormone (adrenalin) which it secretes; **adre:noge·nital** *a.*, of or pertaining to the adrenal glands and the genitals; **adre:noly·tic** *a.*, inhibiting the effect of adrenalin or the action of adrenergic nerves; also *sb.*; **adre:notro·p(h)ic** *a.*, = *ADRENOCORTICOTROP(H)IC *a.*
1913 DORLAND *Med. Dict.* (ed. 7) 39/2 *Adrenochrome*, a sulphur compound of the suprarenal gland for the internal treatment of skin diseases. **1962** *Lancet* 27 Jan. 201/1 Practically none, or none at all, of the infused adrenaline could have been metabolised via adrenochrome or its products in either schizophrenic or normal subjects. **1936** H. D. ROLLESTON *Endocrine Organs in Health & Disease* iii. 64 Corticotropic..or Adreno-cortical Stimulating Hormone. **1941** *Lancet* 8 Mar. 313/2 A case of adreno-cortical syndrome. **1949** *Vitamins & Hormones* VII. 294 Both the sex hormones and the adrenocortical hormones are involved in the regulation of protein, carbohydrate, and electrolyte metabolism. **1931** *Brit. Med. Jrnl.* 2 May 746/1 The accentuation of the male secondary sexual characters..may be most aptly described under the heading of 'Virilism', rather than..as the 'Adreno-genital syndrome'. **1947** *Proc. Soc. Exp. Biol. & Med.* LXVI. 152/2 Dibenamine..is primarily a powerful adrenolytic, but a weak sympathicolytic agent. **1948** *Ibid.* LXVII. 298 Several sympathicolytic substances have mainly adrenolytic properties. **1932** *Lancet* 12 Aug. 347/1 (*title*) The adrenotropic hormone of the anterior pituitary lobe. **1938** *Encycl. Brit. Bk. of Yr.* 227/2 The adrenotrophic, thyrotrophic, and gonadotrophic hormones.

adrenocorticotrop(h)ic (ădrī·no,kǭ:ɹtikotrǫ·fik, -ǫ·pik), *a.* [f. *ADRENO- + *CORTICO- + TROPHIC *a.*] Stimulating or controlling the adrenal cortex; esp. *adrenocorticotrop(h)ic hormone* (abbrev. *A.C.T.H.*), a hormone produced by the pituitary gland which influences the action of the adrenal cortex.
1936 *Proc. Soc. Exp. Biol. & Med.* XXXV. 281 (*title*) Effect of Adrenocorticotropic Extracts on Accessory Reproductive Organs of Castrate Rats. **1937** *Ibid.* 649 Preparation and biological assay of adrenocorticotropic hormone. **1938** J. D. REESE & H. D. MOON in *Anatomical Rec.* LXX. 543 The..increase in the adrenal of normal rats following administration of the adrenocorticotrophic hormone of the anterior hypophysis is well known. **1944** *Anatomical Rec.* LXXXIX. 163 Effect of adrenocorticotropic hormone (ACTH) on the anterior pituitary of the adrenalectomized young male rat. **1950** *Lancet* 7 Jan. 11/1 A.C.T.H. seems to be the most promising tool medical investigators have today. **1962** *Lancet* 28 Apr. 888/1 Acute failure was precipitated by the adrenocorticotrophic stimulus of isoniazid.

Hence **adre:nocorticotro·p(h)in** [-IN¹] = **adrenocorticotrop(h)ic hormone.*
1952 *Proc. Soc. Exp. Biol. & Med.* LXXIX. 252 (*title*) Cytologic Changes in Rat Adenohypophysis Following Administration of Adrenocorticotrophin or Cortisone. **1959** *New Biol.* XXVIII. 131 Some tumours of the pituitary gland secrete abnormally large amounts of a hormone, adrenocorticotrophin, known as ACTH for short, which is required for the normal maintenance and function of the adrenal cortex.

‖**adret** (adre). *Geogr.* [F. (orig. dial.), f. *à* to + OFr. *dre(i)t*: see ADROIT *a.*] A (mountain) slope which faces the sun. Cf. *UBAC.
1931 *Geogr. Rev.* XXI. 424 The valley has an east-west trend and so a well defined *adret* (sunny side) and *ubac* (shady side). **1936** R. PEATTIE *Mountain Geogr.* iv. 90 An exception to the law of exposure is found in the existence of sunny slopes on an *ubac*. These slopes are the 'secondary *adrets*'. **1937** A. GARNETT *Insolation & Relief* i. 4 The familiar contrasts of land utilisation on *adret* and *ubac* slopes.

adrip. (Earlier example.)
1830 *Blackw. Mag.* July 142 A head of hair all adrip like that of Neptune.

adscribe, *v.* (Further example.)
1864 G. M. HOPKINS *Let.* 14 Aug. (1938) 64, I am adscribed third into Alfred Erskine and Bond.

adscript. Add: **B.** *sb.* **3.** A comment or note added to a manuscript; esp. one which in error becomes incorporated in the text.
1889 W. G. RUTHERFORD *Thucyd. IV* p. xxxiv, In place of this word [*scholium*]..I would suggest another..if we anglicise the Latin *adscriptum* on the analogy of *postscript* and *rescript* we get exactly what we want. *Ibid.* p. xl, Adscripts combined with glosses may modify the whole structure of a sentence. **1898** F. BLASS *Philol. of Gospels* 161 The record of its original place must have been preserved..by some adscript..*Here comes in the passage on the adulteress.* **1929** E. LOBEL in *Classical Q.* XXIII. 78 A very large number of marginal or interlinear adscripts have made their way into the text of Aᵉ.

adsorb (ædsǭ·ɹb), *v.* [Back-formation from ADSORPTION.] *trans.* To collect by adsorption. Hence **adso·rbent,** an adsorbing substance; **adso·rbing** *ppl. a.*; **adso·rbate** [-ATE⁴], the material which is adsorbed; an adsorbed substance.
1882 *Nature* XXVI. 139 Continuing his researches on 'adsorption', or condensation of gases on surfaces of solids, Herr Kayser..has studied the influence of the adsorbing material. **1906** *Bio-Chem. Jrnl.* I. 485 As to other factors which influence the adsorption v. Bemmelen points out the following:—(1) the adsorbing substance, (2) the solvent, (3) the substance to be adsorbed. **1923** *Discovery* Sept. 231/2 Impurities deposited on or adsorbed by the crystals. **1928** W. A. CASPARI *Struct. & Prop. Matter* ii. 37 Some charcoals will adsorb hundreds of times their bulk of gas. *Ibid.* 38 Animal and vegetable fibres..are tolerably good adsorbents, owing to the large surfaces presented by their internal structure. **1934** *Discovery* July 198/2 Clay behaves like an insoluble weak acid, in that it can react with, and adsorb the actions of electrolytes in the soil. **1938** R. W. LAWSON tr. *Hevesy & Paneth's Man. Radioactivity* (ed. 2) xvii. 163 A cation will be adsorbed..when it forms with the anion of the adsorbing salt a compound, the solubility of which in the solvent is small. **1938** *Nature* 26 Nov. 946/1 The first International Standard for Vitamin B, which consisted of an adsorbate of the antineuritic vitamin. **1952** *New Biol.* XII. 102 When only monomolecular layers of the crystal are interacting with the adsorbate, more symmetric crystals can exercise such actions, if adsorption takes place on an asymmetric crystal face.

adsorption. Add def.: The process by which specific gases, liquids or substances in solution adhere to the exposed surfaces of materials, usually solids, with which they are in contact. Add further examples.
1904 R. A. LEHFELDT tr. *Nernst's Theor. Chem.* 129 Adsorption.—Charcoal shaken with an iodine solution or placed in an atmosphere of iodine vapour condenses appreciable amounts of iodine on its surface; this is known as 'adsorption'. **1912** E. HATSCHEK *Introd. Physics & Chem. Colloids* (1913) i. 5 One other property of colloids..is their capacity for taking dissolved substances out of solution and retaining them...This phenomenon..is now generally called 'Adsorption'...Specially striking is the power of 'selective adsorption'..possessed by many substances. **1928** W. A. CASPARI *Struct. & Prop. Matter* ii. 37 Adsorption from liquids is of the highest technical importance.

b. *attrib.* and *Comb.*, as *adsorption force*; **adsorption compound,** a compound formed between an adsorbing surface and the substance adsorbed; **adsorption isotherm,** a curve plotted to show the relative proportion of adsorption and concentration in a fluid or solution at a constant temperature.
1908 *Chem. Abstr.* 1571 (*title*) Colloids and their Adsorption Compounds. **1913** *Ibid.* 438 Negative adsorption isotherms. **1918** A. M. WILLIAMS in *Proc. R. Soc. Edin.* XXXIX. 48 (*title*) Adsorption isotherm at low concentrations. **1920** C. F. CROSS & E. J. BEVAN *Paper-Making* (ed. 5) i. 65 The lignocelluloses are adsorption compounds of lignones and cellulose. **1938** R. W. LAWSON tr. *Hevesy & Paneth's Man. Radioactivity* (ed. 2) xv. 147 Non-specific adsorption forces do not play a part in this process.

Hence **adso·rptional** *a.*, pertaining to adsorption.
1928 W. A. CASPARI *Struct. & Prop. Matter* ii. 35 Adsorptional effects are by no means limited to the liquid state.

‖**adsum** (æ·dsɒm). [L.] 'I am here', as an answer in a roll-call, etc.
1593 SHAKES. *2 Henry VI* I. iv. 26 *Spirit.* 'Ad sum.' **1854** THACKERAY *Newcomes* II. xlii. 373 He lifted up his head a little, and quickly said 'Adsum!' and fell back. It was the word we used at school, when names were called. *c* **1909** D. H. LAWRENCE *Collier's Friday Night* (1934) ii. 56 Ne'er mind; I'll say 'adsum' every time. *Recording Angel*; 'Ernest Lambert.'—'Adsum!' **1915** J. E. G. HADATH *Sheepy Wilson* i. 2 Before ever he had uttered his first *Adsum* or slipped timidly for the first time into form.

adult, *sb.* Add: **1. b.** spec. *adult education*, the further education of those over ordinary school age (as in the universities), but commonly used of that provided by local educational authorities, etc.
1851 J. W. HUDSON (*title*) The history of adult education. **1922** G. M. TREVELYAN *Brit. Hist. in 19th Cent.* 164 The case of adult education received its first stimulus from the Industrial Revolution in the desire of mechanics for general scientific knowledge, and the willingness of the more intelligent part of the middle class to help to supply their demand. **1942** *Spectator* 6 Mar. 244/3 The Cambridgeshire Technical Institute..is the centre of adult education, both humane and vocational, in Cambridge and district. **1958** *Whitaker's Almanack* 509 Adult Education is carried on in the United Kingdom by universities and university colleges..local education authorities..and by a wide variety of voluntary organizations.

adulterously, *adv.* (Later example.)
1922 JOYCE *Ulysses* 718 The matrimonial violator..had not been outraged by the adulterous violator of the adulterously violated.

adurol (ædiūᵊ·rǫl, æ·diurǫl). [G.] A haloid substitution product of hydroquinone, used as a photographic developer.
1899 *Brit. Jrnl. Photogr.* 3 Mar. 139/1 Adurol is a new photographic developer, said to be obtained from hydroquinone by a patented process. **1912** G. MARTIN *Industr. Chem., Org.* 659 'Adurol' developer consists of the stock solutions: (A) 10 g. adurol + 80 g. cryst. sod. sulphite + 500 c.c. H_2O; (B) 60 g. K_2CO_3 + 500 c.c. H_2O.

advance, *v.* Add: **2. b.** Of a colour: to appear to be nearer to the eye than other colours in the same plane; to stand out.
1876 S. R. KOEHLER tr. *von Bezold's Theory of Color* v. 197 Another point to which attention must be called concerns the advancing and retiring qualities of certain colors...Warm hues advance while the cold retire; if the brightness is not equal, the light colors advance while the dark retire. **1908** J. A. H. HATT *Colorist* vii. 41 Beside the distinction of luminous and sober in colors, we also have the attributes of warm and cold, also advancing and retiring. **1958** *Listener* 24 July 131/2 This is well known to artists, who call red an advancing, and blue a receding colour.

advance, *sb.* **3.** Add to def.: Esp. in *pl.*, amorous overtures or approaches.
1817 COLERIDGE *Sibylline Leaves* 137 True, I woo'd her..but she Met my advances with empassion'd pride. **1891** G. MOORE *Impressions & Opin.* 138 The Duke would make some absent-minded advances, which in an absent-minded way would be repelled. **1898** G. B. SHAW *Philanderer* II. 108 No woman writes such a letter to a man unless he has made advances to her. **1959** M. CUMBERLAND *Murmurs in Rue Morgue* xi. 71 Men don't make 'advances' any longer. Modern man propositions a girl. **1959** D. EDEN *Sleeping Bride* ii. 10 Teaching English to..French children was one thing, but having to fight off advances from their father was another.

V. *Comb.* and *attrib.* Add: *advance agent, announcement, notice, publicity*; **advance copy,** a copy of a book sent out in advance of publication; **advance growth** (see quots.); **advance(d) guard** (further example); also *fig.*, = *AVANT-GARDE 2; **advance note** (see quot. 1886).
1882 SWEET & KNOX *Texas Siftings* 38 Do you think I am the advance agent of a variety show? **1897** *Congress. Rec.* Mar. 177/1 That grand advance agent of prosperity,

William McKinley. **1885** *Art Annual* I (Advt.), Advance, Announcements from the Prospectus for 1885/86. **1899** *Academy* 25 Nov. 591 Mr. Donnelly conceived the request to be for an advance copy. **1903** E. Almack *Eikon Basilike* Pref. p. iv, The present edition has been set up from an advance copy' of the first edition. **1891** W. Schlich *Man. Forestry* II. ii. 155 In almost every mature wood groups of young growth are found, which have sprung up here and there before the regeneration cuttings have been commenced; such young growth is called 'advance growth'. **1953** *Brit. Commonw. For. Terminol.* I. 11 *Advance growth*, young trees which have established themselves in openings in the forest, or under the forest cover, before regeneration fellings are begun. **1758** in *Essex Inst. Coll.* XVIII. 113 The Advance Guard..have cleared off the Trees and built Breastworks. **1898** G. B. Shaw *Our Theatres in Nineties* (1932) III. 309 That is why Mr Pinero, as a critic of the advanced guard in modern life, is unendurable to me. **1931** C. St. John *E. Terry & B. Shaw* 297 His refusal..made him specially obnoxious to the advance guard of the drama. **1959** *Times* 20 Feb. 11/6 Poplar became the advanced guard in London Labour. **1959** *Times Lit. Suppl.* 6 Nov. p. xxix/1 Advance guard Americans are convinced..that ..Abstract Expressionism radiates the world over. **1845** *Act* 8 & 9 *Vict.* c. 116 § 7 The Owner..of any Merchant Ship ..shall not pay or advance, nor give..an Advance Note for any Part of the Wages of any Seaman [etc.]. **1886** *Encycl. Brit.* XXI. 606/1 Advance notes—that is, documents promising the future payment of money on account of a seaman's wages conditionally on his going to sea. **1900** T. E. Pemberton *Kendals* 281 If this much-advertised play were as original, as artistic, as great as the advance notices have painted it. **1932** D. L. Sayers *Have his Carcase* iv. 56 Good advance publicity means sales.

advanced, *ppl. a.* **2.** Add to def.: sometimes *spec.* of women, esp. those advocating or favouring women's rights or equality with men.

1871 *N.Y. Tribune* 2 Feb. (De Vere), The shortsightedness of the advanced female to the interest of her own cause. **1876** C. M. Yonge *Womankind* xxii. 177 The advanced school [of brides] are said to prefer a civil marriage. **1895** S. Hale *Lett.* (1919) 288 She is a lion-hunter *enragée*, advanced female, views, everything. **1896** M. Corelli *Mighty Atom* ii. 24 Two or three ugly 'advanced' young women who have brought their bicycles and go tearing about the country all day.

b. Of study: on a higher level than the elementary; of a degree: (usu.) superior to a bachelor's degree; *spec.* in some universities: *advanced student*, one who, because of qualifications obtained at a former place of study, is admitted to a more advanced course than would normally be available to him; so *advanced status*, etc.

1790 *Laws of Harvard Coll.* 7 No person shall be admitted to an advanced standing, unless..he shall be found qualified. **1871** L. H. Bagg *At Yale* 689 Whether an 'advanced student' comes from a private tutor or from another college, or drops from a higher class at Yale, makes little difference. **1892** *Brigham Young Acad. Circular* 12 Requests for examination for advanced credit must be made to the faculty in writing. **1916** *Oxf. Univ. Gaz.* 22 Nov. 138/2 In a Congregation to be held on Tuesday, January 30, 1917.. the form of Statute creating a new status of Advanced Student..will be promulgated. **1952** M. McCarthy *Groves of Academe* (1953) vi. 113 We don't insist on the Ph.D. or even the Master's; in fact, we regard advanced degrees as a liability. **1963** *Oxf. Univ. Gaz.* 20 June 1484/2 Permitted to transfer from B.Litt. Status to Advanced Status.

advection (ædve·kʃən). *Meteorol.* [ad. L. *advectiōn-em*, n. of action from *adveh-ĕre* to carry to; f. *ad* to + *veh-ĕre* to carry.] The transfer of heat by the horizontal movement of the air. Also *attrib.*, as *advection fog*. Hence **adve·ctive** *a.*, of, or caused by, advection. Cf. Convection.

1910 *Rep. Brit. Assoc. Advancem. Sci. 1909* 103 Interchange of air in the upper region would be mainly by advection and the two [upper and lower] regions might be appropriately named advective and convective regions. **1941** *Jrnl. R. Aeronaut. Soc.* XLV. 82 The non-adiabatic cooling may be due to two causes, viz., the radiative cooling of the underlying surface, and the advective cooling that results when warm air streams over a colder surface. We may therefore speak of radiation fogs, advection fogs and upslope fogs. **1956** J. C. Swayne *Conc. Gloss. Geogr. Terms* 9 *Advection fog*, a fog formed when relatively warm, moist air passes over colder land or water, e.g. the sea-fogs of the Grand Banks of Newfoundland.

advenient, *a.* Delete †*Obs.* and add later examples.

1923 *Times Lit. Suppl.* 13 Sept. 596/2 There is in knowledge both an advenient physical influence from the thing and a projicient psychical reference to the thing. **1936** *Nature* 28 Mar. 522/1 Colour as experienced is..an indication..of the presence of specific chemical changes in the retina and the choroid under advenient electromagnetic influence.

Adventist (æ·dvĕntist). Also **adventist.** [f. Advent + -ist.] A member of any of the various religious sects holding millenarian views. Also *attrib.* Hence **a·dventism**, the principles or tenets of adventists.

Second Adventist, orig. the fuller designation of the followers of William Miller (died 1849), who promoted the doctrine that the second coming of Christ and the end of the world were near at hand; a Millerite.

1843 *Signs of the Times* 15 Nov. 109 in F. D. Nichol *Midnight Cry* (1944) x. 156 The English Adventists. **1844** in *Cist Cincinnati Misc.* (1845) I. 29 As the day approaches ..the faithful adventists continue nearly the whole time, day and night at the Tabernacle. **1874** Wellcome *Hist. Second Advent Message* 592 (D. A. E.), Judaism is being taught. If brethren do not mean to teach it let them tell us so and not teach this under the cloak of Adventism. **1876** [see Seventh-day 2 b]. **1878** in G. L. Prentiss *Life & Lett. E. Prentiss* (1882) 504 Neither Mr. Prentiss or myself have ever had any sympathy with Second Adventists. **1883** *Encycl. Brit.* XVI. 320/2 At present the number of Millerites or Adventists is estimated at from 15,000 to 20,000. **1895** J. R. Harris *Union with God* xi. 188 The existence of an adventist movement in Jerusalem. **1927** Flint & Tait tr. *Fülöp-Miller's Mind & Face of Bolshevism* 78 The sects with a more rationalist tinge, the 'Adventists' and the 'New Adventists'.

adventitial, *a.* Delete †*Obs.* and add:

1901 *Jrnl. Exper. Med.* VI. 69 The adventitial lymphatic sheath is in most cases distended.

adventive, *a.* and *sb.* Transfer '*Obs.*' to senses in Dict., and add:

A. *adj.* **2.** *Bot.* Of a plant: growing spontaneously but not native to the locality in which it appears; imperfectly naturalized.

1867 Gray *Man. Bot.* (ed. 5) 16, I have classified our introduced plants..into two sorts, the thoroughly *naturalized*, and the *adventive*;..the second, those which are only locally spontaneous. **1888** F. A. Lees *Flora of W. Yorks.* 313 It is luxuriant as to size and adventive. **1919** I. Hayward & G. C. Druce (*title*) The adventive flora of Tweedside. **1936** *Nature* 11 Apr. 600/2 A survey of the weed flora of the northern part of the United States shows that a large proportion of the adventive weeds are from Europe.

B. *sb.* **2.** An adventive plant.

1888 F. A. Lees *Flora of W. Yorks.* 313 Campanula Trachelium L...by the tram-road from Manor Coal-pits, Sheffield...An adventive here, clearly.

adventure, *sb.* Add: **5. b.** An instance of adventurism (see *adventurism 2); applied disparagingly to any act or policy considered to be dangerous, e.g. as likely to involve the country concerned in war. (Russ. *avantyura* is used in a similar sense.)

1932 [see *adventurism 2]. **1957** *New Statesman* 18 May 630/1 Mr. Macmillan..argued..that the Suez adventure in no way influenced Egypt's attitude to the negotiations. **1958** *Listener* 30 Oct. 682/2 [citing Moscow radio] The intensification of the policy of adventure and provocation of People's China, and the drawing up of plans for a new adventure in the Taiwan Straits area. *Ibid.* 27 Nov. 864/2 Outside interference, from East or West, delays such a solution, by diverting Arab attention to political problems and adventures, and thus impeding the emergence of constructive Arab statesmanship.

10. adventure playground, a playground where children are provided with miscellaneous equipment, often waste material, from which they may contrive their own amusement (see quot. 1953[1]).

1953 Lady Allen of Hurtwood *Adventure Playgrounds* (Nat. Playing Fields Assoc.) 3 How does an Adventure Playground differ from the usual playground? There is no asphalt, no see-saws, swings or slides, except those created by the children themselves out of waste material freely available on the site. **1953** *Playgrounds for Blocks of Flats* (Nat. Playing Fields Assoc.) 20 The employment of a supervisor, play leader or attendant is not considered necessary, except in the case of 'adventure' (creative or building) playgrounds. **1960** *Times* 25 Mar. 24/2 A steam-roller.. is for the children's 'adventure playground'.

adventure, *v.* Add: **5. b.** *trans.* To venture to say or utter.

1881 C. E. L. Riddell *Sen. Partner* II. i. 7 'I've been looking up my songs, Mr. McCullagh,' added the eldest daughter..'And we have been practising reels,' adventured Miss Vanderton. **1898** *Daily News* 19 Oct. 3/1 He adventured the opinion that 'some members opposite' were 'unaccustomed to the amenities of debate'. **1900** L. B. Walford *One of Ourselves* xiv, 'Did he tell you about us?' she adventured, cautiously.

adventureship (ædve·ntʃəɹʃip). [f. Adventure *sb.* + -ship.] Adventurous practice.

1847 J. S. Mill *Lett.* (1910) I. 131 The inordinate impulse given to..the general spirit of adventureship. **1879** Morley *Burke* ii. 33 An unpleasant taint of speculation and financial adventureship hung at one time about the whole connexion.

adventurism. Add: **2.** The tendency in a government or politician to take risks in the management of national affairs, esp. in foreign policy (cf. Russ. *avantyurizm*).

1932 H. Nicolson *Public Faces* i. 17 Only three months before they had ousted the Churchill Government on a charge of adventurism. And here..was a weapon of adventure such as no British Government had ever possessed before. **1934** —— *Curzon: Last Phase* 52 The best analogy of the British imperial system is that of an old-fashioned and conservative banking house..an identical tradition, an identical caution, the same suspicion of adventurism. **1954** *Ann. Reg. 1953* 294 The two 'deviations' were 'passivity'..and 'commandism' or 'precipitate haste and adventurism'. **1957** *Observer* 3 Nov. 1/1 Of the three official accusations against Marshal Zhukov, that of 'adventurism' appears to be based on no evidence whatever.

adventurist (ædve·ntiŭrist), *sb.* [f. Adventure *sb.* + -ist.] One inclined to adventurism (various senses). Also *adj.*

1920 *Public Opinion* 16 Apr. 372/2 Our own military adventurists have been unable to stampede the Government. **1943** *Amer. Speech* XVIII. 310 A mid-western newspaper headline telling of the death of Ignatius Timothy Lincoln in Shanghai a few years ago read 'Ignatius Lincoln, *Adventurist*, Dies'. **1958** I. Brown *Words in our Time* 16 Marshal Zhukov..was formally accused of being adventurist and personalist. **1962** *Guardian* 28 Dec. 8/2 In accusing Mr. Kruschev's Cuban policy of having been both 'adventurist' and 'capitulationist' the Peking 'People's Daily' has scored a fair point. **1963** *Ann. Reg. 1962* 214 Vigorously condemning Albanian 'adventurists and dogmatists'.

adverbial, *sb.* (Later examples.)

1924 H. E. Palmer *Gram. Spoken Eng.* II. 234 The term *Adverbials* is a convenient designation of adverbs and their equivalents, viz. Group-Adverbs, Adverbial Phrases and Adverbial Clauses. **1960** I. L. Gordon *Seafarer* 44 *Þurh* with an abstract noun is a frequent method of expressing the adverbial of manner or state. **1962** J. Söderlind in *Contrib. Eng. Syntax* (Gothenburg Stud.) 100 Thus the names of days are used as adverbials without a preposition.

advert (æ·dvɜ̄ɹt), colloq. abbrev. of Advertisement. Cf. *ad.

1860 J. Blackwood *Let.* 6 Jan. in *Lett. George Eliot* (1954) III. 244, I do not mean this for announcement in the advert. but that we may consider [etc.]. **1954** J. B. Priestley *Magicians* vi. 120, I don't know what you're going to put in the adverts—the usual bilge, I suppose. **1955** 'R. Gordon' *Dr. at Large* iii. 41 'Those vacancies are already filled.' 'But the advert. only came out yesterday!'

advertisement. Add: **5.** Also *fig.*

1952 A. Wilson *Hemlock* III. i. 216 Ours has not perhaps been the ideal advertisement for such a union.

6. = Advertising *vbl. sb.* 2.

1902 *Encycl. Brit.* XXV. 94/2 Advertisement, or advertising, as the process of purchasing publicity is now more commonly called, is [etc.]. **1909** H. G. Wells *Tono-Bungay* II. iii. 194 Advertisement has revolutionized trade and industry. **1922** Joyce *Ulysses* 667 The infinite possibilities hitherto unexploited of the modern art of advertisement.

7. *attrib.* and *Comb.* (in senses 5 and 6), as *advertisement agent, board, curtain, -manager, -sticker.*

1897 G. B. Shaw *Our Theatres in Nineties* (1932) III. 27 But there is a clown, who acts extensively as an advertisement agent. **1933** H. G. Wells *Bulpington* iii. 81 Mr. Parkinson was an advertisement agent. **1880** *Harper's Mag.* Dec. 72 The advertisement boards were freshly covered with clean notices. **1951** *Oxf. Compan. Theatre* 13/1 *Advertisement curtain*, an outer curtain or act-drop used mostly in smaller theatres, covered with advertisements of local shops and manufacturers. **1902** A. Bennett *Truth abt. Author* x, in *The Academy* 21 June 634/2, I could play Blucher at the Waterloo of the advertisement-manager. **1778** S. Foote *Capuchin* II. 114 You became advertisement-sticker to lottery-offices, auctioneers..and mountebank-doctors.

advertising, *vbl. sb.* and *ppl. a.* **2.** (Further examples.)

1779 W. A. Smyth (*title*) The Publican's Guide..by which every retailer will be enabled to..sell on terms equal to..the generality of advertising merchants. **1816** Jane Austen *Emma* II. xvii. 324 There are advertising offices, and..by applying to them I should have no doubt of very soon meeting with something that would do. **1832** F. Trollope *Dom. Manners* I. iv. 49 We went to the office of an advertising agent. **1843** *Edin. Rev.* Feb. 2 The advertising system. **1850** *Hunt's Merch. Mag.* XXIII. 580 Mr. V. B. Palmer keeps what he terms the '*American Newspaper Advertising Agency*'. **1863** *Chicago Post* 29 Sept. (D.A.), Last night the Varieties bloomed forth a set of magnificent new scenes and a beautiful advertising curtain. **1913** *Maclean's Mag.* May 135/1 We live in the Advertising Age. **1936** *Discovery* Mar. 85/1 Advertising agencies are supplied with source material for advertisements of plant products. **1962** *Rep. Comm. Broadc. 1960* 82 The [Independent Television] Authority told us that they were moving towards a reform of advertising magazines rather than to their disappearance.

advisee (ædvəizī·). [f. Advise *v.* + -ee[1].] The person advised.

1824 *New Monthly Mag.* XI. 12 The adviser reminded the advisee of those happy days. **1856** T. Carlyle *Let.* 25 Apr. (1923) 297 'Nobody follows advice,' they say; which means withal, 'Advice never hits the case; the case is not known to any Adviser, but only to the Advisee.' **1952** M. McCarthy *Groves of Academe* (1953) iv. 70 To have advisees in the reading courses double up in the tutorial hour.

advisory, *a.* (Earlier U.S. examples.)

1778 *Result Convention at Ipswich, Essex* (Mass.) 45 We think therefore that the members of that court ought never to be advisory to any officer in the state. **1789** Morse *Amer. Geogr.* (1792) 170 The churches claim no jurisdiction over each other, and the power of ecclesiastical councils is only advisory.

‖ **advocaat** (ædvokā·t). Also **advokaat.** [Du., shortened f. *advocatenborrel* egg-and-brandy liqueur, f. *advocaat* Advocate *sb.* + *borrel* drink.] A Dutch liqueur made with eggs, sugar and brandy; also, a glass of this.

1935 in A. L. Simon *Wines & Liqueurs from A to Z*. **1938** M. Allingham *Fashion in Shrouds* xxii. 420 There's some gin and French and a little Advocaat..in the cupboard. **1951** P. Branch *Lion in Cellar* vi. 78 Three gins, a cherry brandy, and an Advocaat. **1962** P. Helm *Death has Thousand Entrances* ii. 41 We drank *advokaat*, sweet and thick.

‖ **advocatus diaboli** (ædvokē¹·təs dəi,æ·bŏləi). [mod.L.; cf. Advocate *sb.*] 'Devil's advocate' (Devil *sb.* 25 b).

1842 H. C. Robinson *Let.* 29 Dec. (1927) I. 473 You are aware that here I am considered as a sort of Advocatus Diaboli. **1883** *Guardian* 21 Mar. 412/1 The mere *advocatus diaboli* who is content to damage an opponent. **1929** *Music & Lett.* X. 194 It is a pleasant thing to play *advocatus diaboli* with the knowledge that one is going to allow oneself to cross over to the angels' bench later.

adynamical (ædinæ·mikăl), *a.* [f. as Adynamic *a.* + -al.] Not dynamical.

1900 *Jrnl. Inst. Electr. Engineers* Apr. 396 The properties of electric and magnetic force are explicable upon dynamical principles; so far there is no known necessity for seeking for adynamical properties in the ether.

æcial, *a.*: see *æcium.

æcidial (isi·diăl), *a. Bot.* [f. Æcidi(um + -al.] Of, pertaining to, or resembling an æcidium.

1880 *Smithsonian Rep.* 324 A monograph..by Von Thümen contains an account of the æcidial forms attacking *Coniferæ*. **1891** *Athenæum* 23 May 671/1 The extraordinary abundance and wide distribution of the teleutosporic stage as compared with the comparative scarcity of the æcidial stage. **1927** Gwynne-Vaughan & Barnes *Struct. & Dev. Fungi* 268 Centrally, where the pseudoperidium arches over the aecidial contents, it is derived from the cells first cut off in each row.

æcidiospore (isi·diospōᵊɹ). *Bot.* [f. Æcidi(um + -o + Spore.] A spore formed in an æcidium.

1880 S. H. Vines tr. *Prantl's Elem. Textbk. Bot.* IV. 131 Some Uredineæ are known in which æcidiospores produce æcidia. **1922** *Nature* 8 Apr. 462/2 Its recurrence is explained only through fresh infection by æcidiospores produced on Barberry.

æcium (ī·siŏm). *Bot.* [mod.L., f. Gr. αἰκία injury.] U.S. term for Æcidium. Hence **æ·cial** *a.* = *æcidial *a.*; **æ·ciospore** = *æcidiospore.

1905 J. C. Arthur in *Bot. Gaz.* XXXIX. 221 For the sorus of the first spore-stage, usually..called aecidium, [etc.]..I propose *aecium*..derivatives *aecial, aeciospore*, etc. **1907** —— in *Jrnl. Mycology* XIII. 203 Cultures of Uredineae..aecial stage. **1929** —— et al. *Plant Rusts* i. 6 Aeciospores are binucleate and germinate..by producing a hypha. *Ibid.* 11 The aecia are the essential sori produced by the gametophytic mycelium in macrocyclic rusts...The cupulate form of aecium, generally known as the cluster-cup or aecidium. **1937** *Nature* 8 May 800/2 The uredial and telial stages have regularly occurred on the gramineous hosts, and the pycnial..and æcial stages on the common barberry. **1957** *Encycl. Brit.* IX. 934/1 The aeciospores are borne in fructifications (sori) known as aecia, which are frequently cup-shaped receptacles..containing chains of spores, which arise from a primordial layer of cells at the base of the aecidium.

ægis. Add: **2.** Examples of *under the ægis* (*of*). Now freq. in senses 'auspices, control, etc.'.

1910 *Encycl. Brit.* III. 936/2 Under the aegis of the Billiard Association a tacit understanding was arrived at that the position must be broken up, should it occur. **1958** P. Gammond *Duke Ellington* I. 18 They make their valuable individual contributions, but under the Ellington aegis they find themselves constantly enriched musically. **1963** *B.S.I. News* May 14/2 These basic criteria and recognized methods of assessment, drawn up under the aegis of BSI.

ægithognathous (īdʒiþǫ·gnăþəs), *a. Zool.* [f. Gr. αἴγιθος, name of an unknown bird + γνάθος jaw.] Having the formation of palate characteristic of the family *Ægithognathæ* (perching birds, woodpeckers, swifts): see quot. 1894. Hence **ægitho·gnathism**, the condition of being ægithognathous.

1875 W. K. Parker in *Encycl. Brit.* III. 699/1 All the *Coracomorphæ* have the ægithognathous palate. **1884** Coues *N. Amer. Birds* (ed. 2) 172 Ægithognathism..is exhibited almost unexceptionally by the great group of Passerine birds. **1894** R. B. Sharpe *Handbk. Birds Gt. Brit.* I. 1 The palate is said to be 'ægithognathous', or 'Passerine', when the *vomer* is broadened and blunt, or truncated, at the anterior end, and is not connected with the *maxillo-palatines*, which, consequently, are widely separated from each other.

ægophonic, *a.* (Examples.)

1855 G. H. Barlow *Pract. Med.* xii. 257 The more ordinary tremulous bronchophony, or, as some have termed it, ægophonic bronchophony. **1876** J. S. Bristowe *Theory & Pract. Med.* II. iii. 386 The ægophonic sound..is often distinctly feebler than the normal voice-resonance.

ægophony. (Earlier and later examples.)

1834 J. Forbes tr. *Laennec's Treat. Dis. Chest* (ed. 4) I. iii. 27 Perfect pectoriloquy..is changed..into a simple resonance..such as to be with difficulty discriminated from ægophony and bronchophony. **1866** [see Bronchophony]. **1876** J. S. Bristowe *Theory & Pract. Med.* II. iii 386 In association with pectoriloquy, bronchophony, or ægophony there can generally be detected a distinct whiff of tubular quality.

Æluroid (il¹ū°·roid), *a.* (*sb.*) *Zool.* [f. mod.L. *Æluroidea* neut. pl., f. Gr. αἴλουρος cat: see -oid.] Belonging to, or having the characters of, the division *Æluroidea* of Carnivora, comprising the feline and allied families; as *sb.*, an animal of this division.

1869 *Proc. Zool. Soc.* 22 Cryptoprocta is a member of the Æluroid group. *Ibid.* 27 In the presence of a short cæcum ..Hyæna conforms with the Æluroids.

aeneolithic (ē,ī:nīoli·þik), *a.* [f. L. *aēneus* of copper or bronze + Gr. λίθος stone + -ic.] Of or pertaining to the period of the neolithic age in which copper was used together with flint implements. See also *eneolithic.

1901 Sergi *Mediterr. Race* xii. 240 In Italy this period is termed *æneolithic*, that is to say, the period of copper and polished stone together. **1928** H. C. Dawson *Age of Gods* iii. 51 The new cultures which entered Europe in neolithic and aeneolithic times followed two main lines of diffusion. **1937** *Proc. Prehist. Soc.* III. 357 Iberian art of the aeneolithic period influenced Ireland.

Æolian, *a.* Add: **1.** (Earlier example.)

1589 Puttenham *Arte Eng. Poesie* II. x. 70 The Dorien.. the Phrigien..the Lydien, and..the Eolien.

2. (Earlier examples.)

1605 Sylvester tr. *Du Bartas's Deuine Weekes & Workes* [Week] II. [Day] ii. p. 399 Th' Æolian Crowd obayes his mighty call, The surly surges of the waters fall. **1693** Dryden *Juvenal* x. 293 And Eurus never such hard usage found In his Eolian Prisons underground.

b. spec. *Geol.*, of formations produced or deposited by the action of wind.

1853 R. J. Nelson in *Q. Jrnl. Geol. Soc.* IX. 201 They [the Bermudas] may be placed..as a peculiar post-tertiary formation of a composite character..Neptunian below and 'Æolian' above. *Ibid.* 207 (*caption*) Horizontal section of 'Æolian' rock. **1879** [in Dict.]. **1889** I. C. Russell in *Geol. Mag.* July 289 The subaërial deposits now accumulating in the arid portion of the United States may be divided into four classes: 1, Eolian Sands; 2, Talus Slopes; 3, Alluvial Cones; and 4, Calcareous Clays. **1910** *Encycl. Brit.* XI. 656/1 Some [rocks]..are accumulated by the drifting action of wind upon loose materials, and are known as 'aeolian' formations. Familiar instances of such wind-formed deposits are the sand dunes along many parts of the sea coast. **1935** *Nature* 1 June 909/1 It seems safe to conclude that the bulk of the ridge is relatively late Pleistocene in age, and that it is essentially æolian in origin.

Æolianly (i,ōᵘ·liănli), *adv.* [f. Æolian *a.* + -ly[2].] With an Æolian sound; with a sound as of an Æolian harp.

1849 Symington *Harebell Chimes* 129 Plaint melody Sung by the mermaids of the wave, Æolianly. **1886** M. F. Tupper *My Life as Author* 393 Moaning Æolianly as it went.

æolienne, occas. var. of *éolienne.

Æolism (ī·ŏliz'm). [f. Æol(ian *a.* or Æol(ic *a.* + -ism.] A style or idiom characteristic of or restricted to the Æolic dialect of Greek.

1847 Grote *Hist. Greece* II. xxix. 117 He composed in the Laconian dialect—a variety of the Doric with some intermixture of Æolisms. **1884** *Amer. Jrnl. Philol.* V. 521 First must be eliminated from the so-called Aeolisms all phenomena which, so far from deserving the name of Aeolisms, do not so much as occur in Aeolic. **1950** H. L. Lorimer *Homer & Monum.* viii. 459 All 'Aeolisms'..in these two categories can be equally well explained as examples of primitive Ionic preserved by the exigencies of metre.

æon. Add: **1. b.** *attrib.* and *Comb.*

1916 O. Sitwell in E. & O. Sitwell *20th Cent. Harlequinade* 25 From far within his æon-battered brain Well up those wanton wistful images. **1923** *Blackw. Mag.* July 61/3 The aeon-long passage of water a-down the rock has worn its surface to a glassy smoothness. **1938** W. de la Mare *Memory* 29 A storm-cock shrilled its aeon-old refrain. **1948** E. Sitwell *Notebk. W. Shakes.* vi. 53 In one of the most terrible aeon-moments of the play.

æonic (i,ǫ·nik), *a.* [f. Æon + -ic.] Age-long, lasting an æon.

1883 *Harper's Mag.* Sept. 622/1 Such a period of æonic sleep may have been requisite for the evolution. **1924** H. E. Fosdick *Mod. Use Bible* 251 A world where all moral living is only an accidental episode in an æonic physical process. **1932** *Discovery* Apr. 108/1 They [palæontologists] make the very dry bones live, reconstructing ancient scenes of the aeonic drama.

Æpyornis (īp¹,ǭ·ɹnis, e-). [mod.L. (I. G. St.-Hilaire 1850, in *Ann. Sci. Nat.* (*Zool.*) XIV. 209), f. Gr. αἰπύς high + ὄρνις bird.] A genus of extinct gigantic struthious birds known from remains discovered in Madagascar; a bird of this genus.

1851 *Ann. & Mag. Nat. Hist.* VII. 163 The gigantic bird of Madagascar...We shall give to this genus the name of *Æpyornis*. **1883** *Encycl. Brit.* XV. 171/1 Madagascar was the home of..the extinct *Æpyornis*, whose eggs, found in a sub-fossil state, are the largest known. **1895** [see Dromornis]. **1927** Haldane & Huxley *Anim. Biol.* xii. 277 Largest flightless birds (Moa, Aepyornis, large Ostrich). **1957** R. Carrington *Mermaids & Mastodons* v. 92 Several species of *Aepyornis* are now recognized, the largest still being *Aepyornis maximus*. The mounted skeleton of this bird..has a height of just under ten feet. **1959** *Chambers's Encycl.* VIII. 779/2 There are also fossil remains of large, extinct, bird-like animals not found elsewhere, such as the Aepyornis, remnants of whose eggs are strewn over the beaches of the extreme south.

aerate, *v.* Add: **1.** (Earlier examples.)

1799 W. Nicol *Pract. Planter* iv. 121 If we admit that fallowing and aerating land is of advantage. **1851** *Fraser's Mag.* XLIII. 633/1 Mopping the dew from their brows, or aerating their persons in the breeze.

Also *fig.*

1958 *Listener* 23 Oct. 665/2 Not a breath of this was allowed to aerate this particular talk.

3. (Example.)

1905 *Harmsworth Encycl.* I. 82/1 A more recent introduction is the sparklet. This is a bottle with a hollow removable screw top, for holding a soft steel capsule containing liquid carbon dioxide sufficient to aerate the bottle of water.

aerated, *ppl. a.* Add: **1.** (Earlier example.) Also *aerated concrete* (see quot. 1956).

1800 S. Mitchill *Let.* 3 Dec. in T. J. Pettigrew *Mem. of Lettsom* (1817) III. 218 An opinion, held by several eminent men, that aërated pus was of an acid quality. **1947** *Brit. Standard 1364* (*title*) Aerated Concrete Building Blocks. **1956** *Gloss. Terms Concrete* (*B.S.I.*) 13 *Aerated concrete*, lightweight concrete in which the lightness is obtained by the formation of bubbles of air or gas in the plastic mix which are retained on setting and hardening.

aeration. 4. (Examples.)

1874 Knight *Dict. Mech.* I. 19/1 The aëration of sparkling champagne and Catawba is produced by adding a small amount of white sugar to the wine in bottling, the slight fermentation eliminating alcohol therefrom and liberating carbonic acid gas. **1905** *Harmsworth Encycl.* I. 81/3 Various additions are made, either to the bottle before filling, or to the liquid in bulk before aeration.

aerator. Add: *spec.* **a.** A contrivance for fumigating grain. **b.** An apparatus for forcing air or carbon dioxide into liquids. **c.** An apparatus for aerating turf.

1874 in Knight *Dict. Mech.* **1891** *Sci. Amer.* 4 Apr. 218/1 Aerator..a portable device having a receiver near the top of a standard, the receiver having numerous small perforations, while lower on the standards are pans..for the aeration of warm and fresh milk. **1892** *Jrnl. Soc. Chem. Industry* 30 Nov. 896/1 Improvements in Æreators for Treating Liquids. **1950** *N.Z. Jrnl. Agric.* Dec. 538/1 If rain has fallen on the swath, it will be advisable to lift the swath with an aerator so that it is brought up on the top of the stubble, where it dries out rapidly. **1958** *House & Garden* Feb. 90/1 Proper surface aeration is most important for good turf and when your lawn is small this aerator is the tool to do it.

aerial, *a.* Add: Now generally with pronunciation (ē°·riăl) except in poetry when the metre calls for four syllables.

5. *esp.* with reference to locomotion in the air by means of aircraft (*aerial navigation, ship, transport*, etc.); conducted by aircraft (*aerial attack, photography, top-dressing* (N.Z.), *warfare*, etc.); dropped from an aircraft (*aerial bomb, mine, torpedo*, etc.); in other uses relating to aircraft or aviation, as *aerial camera, corridor, screw*, etc. Some of the collocations (listed for convenience of reference in alphabetical order) are falling into disuse in favour of the corresponding expressions with *air-* as first element (see *air *sb.*[1] B).

1915 *Readers' Guide to Periodical Lit. 1910–14* III. 19/2 Aerial advertising company. **1919** H. G. Anderson *Med. & Surg. Aspects of Aviation* i. 4 The French conceived the idea of having aerial ambulances to convey quickly the wounded. **1919** N. J. Gill *Aerial Arm* ix. 162 Fleets and armies...These ancient services [have] to be reinforced by the new aerial arm. **1908** H. G. Wells *War in Air* viii. § 4 The drachenflieger appeared as little flecks on either wing of this aerial Armada. **1909** A. L. Rotch *Conquest of Air* v. 176 In view of the possibility of this method of aerial attack on land or sea it is unfortunate that..the United States was alone among the other first-class powers to agree to prohibit..the discharge of projectiles and explosives from balloons. **1881** W. D. Hay *300 Years Hence* x. 246 It is the first aërial battle—may we not say it is the last, too? **1784** *Morning Herald* 18 Mar. 2/3 The aerial boat. **1836** *New Monthly Mag.* Sept. 60 Mr. Southey's aerial boat. **1919** N. J. Gill *Aerial Arm* vii. 136 Freedom..is blighted by the curse of aerial bombs. *Ibid.* 135 It is..conceded that there is no form of frightfulness so trying to the nerves as aerial bombing. **1945** *Ann. Reg. 1944* 21 The aerial bombing of Germany reached a new pitch of intensity. **1910** *Flight* II. 96/2 An Aerial 'Bus for Pau...The airship is to carry eight passengers besides the crew. **1919** H. Shaw *Text-bk. Aeronaut.* xxii. 249 The modern aerial camera is a modified form of the original 'press' camera used in the early days of aerial work. **1785** in S. Stubelius *Balloon* (1960) 54 Aerial car [*sc.* basket of a balloon]. **1957** *Everybody's Weekly* 16 Feb. 10 Whether you want an aerial car or bus..the amazing helicopter will do the job. **1887** tr. *J. Verne's Clipper of Clouds* xxii. 231 An aerial combat was beginning in which there were none of the chances of safety as in a sea-fight. **1921** *Flight* XIII. 293/1 The aerial corridor for machines entering or leaving France..has now been enlarged. **1884** *Cassell's Family Mag.* 764/2 On arriving at Villebon the aërial craft was steered gradually round. **1910** *Flight* II. 655/1 The movement of aerial craft. **1912** *Flight* IV. 471/1 On June 8th will take place, starting from London Aerodrome, Hendon, the first aerial Derby. **1893** *Ann. Rep. Aëronaut. Soc.* 72 Mr. Horatio Phillips has, for the first time in aërial dynamics, conveyed a weight of 400 lbs., with the aid of superposed narrow surfaces. **1784** *Universal Mag.* LXXIV. 18 A full account of the late wonderful Aërial Excursions. **1900** *Science* XII. 798/1 What..will become of national frontiers when the aërial fleets can cross them with impunity? **1922** *Flight* XIV. 218/2 New ones [aeroplanes] which should bring down the rate of aerial freight to the value of those now charged in France for first-class railway passengers. **1919** N. J. Gill *Aerial Arm* v. 100 For purposes of self-defence, one or more passengers are carried as aerial gunners in addition to the pilot. **1918** F. H. Colvin *Aircraft Mech. Handbk.* xxi. 311 School of Aerial Gunnery. **1918** E. S. Farrow *Dict. Mil. Terms* 9 *Aërial Lighthouses*, aerial beacons to guide aviators at night through the atmospheric ocean. **1906** *Sci. Amer.* 21 Apr. 327/2 Compared with any other means of transportation, the aerial line seems miraculously safe. **1920** *Discovery* Mar. 80/1 It is probable that kite balloons will be used as landmarks for the main aerial lines over the world. **1909** *Flight* I. 801/1 Baron v. Roenne gives some interesting particulars regarding a proposed aerial liner. **1843** in S. Stubelius *Balloon* (1960) 56 The Project of Aërial Locomotion refuted...The aerial locomotive will *then* go up, and..*not till then*. **1784** *Morning Herald* 16 Feb. 3/4 The aerial navigators..mounted in the gallery of the balloon..the cords, which held the aerial machine, were cut. **1897** Baden-Powell in *United Service Mag.* Apr. 46 There are two distinct schools of inventors of aërial machines, the subject of navigable balloons being very different from that of flying machines proper. **1905** *Aeronaut. Jrnl.* Oct. 57 (*title*) The Vertical Screw Aërial Machine. **1911** *Daily Mail* 23 Aug. 3/6 A contract has also been entered into..for conveyance of the aerial mail from London to Windsor. **1921** H. E. Porter *Aerial Observation* viii. 342 Aërial maps of forests have been made. **1908** *Trans. Amer. Soc. Mech. Engin.* XXX. 681 If, then, a nation can submerge a mine for the destruction of ships from underneath the water, why can it not drop an aërial mine upon a ship from above? **1938** *Jrnl. R. Aeronaut. Soc.* XLII. 618 A new type of aerial mine has been invented...These mines are suspended under captive balloons. **1903** *Aeronaut. Jrnl.* 8 When we can order round our aerial motor to take us straight to our destination. **1804** G. Cayley *Note-book* (1933) 80, I am well convinced that Aerial Navigation will form a most prominent feature in the progress of civilization. **1922** Aerial navigation [see *navigation]. **1784, 1825** Aerial navigator [see *navigator]. **1951** *Times Wkly.* 31 Jan. 18/5 New British Aerial Navigator. The Flight Log. A new British automatic system of aerial navigation which will ease the complicated task of the pilot is going into quantity production...The equipment..automatically shows the pilot his position on a map. **1783** in *W. H. Robinson's* (Newcastle-on-Tyne) *Catal. no. 14* (1926) 49 [*Aeronautical Cartoon*]. The Montgolfier, A first Rate of the French Aerial Navy. **1909** *Sci. Amer.* 12 June 443/3 London newspapers comment bitterly on the fact that Germany spends nearly eighty times as much as Great Britain for the creation of an aerial navy. **1914** Hamel & Turner *Flying* xvi. 280 The aerial observation officer must be highly trained... Nothing appears yet to have been done towards making staff officers capable of taking aerial observations. **1917** 'Contact' *Airman's Outings* 189 The planning of bomb raids and concerted aerial offensives. **1897** *Scribner's Mag.* XXII. 618/1 The first aërial photograph taken in America..was on a wet plate from a balloon over the city of Boston in 1862. **1919** H. Shaw *Text-bk. Aeronaut.* xxii. 252 Only experts can be expected to read all the information that may be contained in an aerial photograph. **1897** *Scribner's Mag.* XXII. 617/2 M. Nadar, of Paris, was one of the earliest experimenters in aërial photography. **1937** *Discovery* Oct. 306/1 In recent years aerial photography has played an important part in revealing the location of abandoned cities, for the artificial configuration of the ground appears in the air photograph and reveals the work of man. **1868** *Ann. Rep. Aëronaut. Soc.* 37 The measure of his aërial planes. **1911** *World of Stamps* Oct. 6/1 The first aerial post..was that instituted in besieged Paris in 1870. **1908** H. G. Wells *War in Air* xi. § 2 The second aerial power in Europe at this time was France. **1884** *Knowledge* VI. 230/1 The accompanying engraving represents an aërial propeller recently patented. **1910** *Flight* II. 867/1 A 'skimmer' with an aerial propeller. **1915** *Lancet* 12 June 1249/2 The dangers of an aerial raid. **1914** *R.F.C. Training Man.* II. 22 Aerial reconnaissance..may be considered under three heads: strategical, tactical and protective. **1908** *Sci. Amer.* 3 Oct. 218/1 The aerial scout [i.e. an aeroplane], rising high into the air, would command a vast field of observation. **1856** *Brit. Patent* 2993 [The] first impetus..is easily sustained and increased by turning the aërial screw. **1867** *Ann. Rep. Aëronaut. Soc.* 31 The French Aerial Screw..was now exhibited. **1892** O. Chanute in *Railroad & Engin. Jrnl.* Mar. 133/2 A proposed aerial screw machine. **1784** *Morning Herald* 18 Mar. 2/3 Mr. Blanchard had..given notice of an ærial ship in which he was to take flight through the air. **1834** *Times* 15 Aug. 3/2 It was intended, or given out so in Paris, that on the 15th instant (this day) the aerial ship of Count de Lennox was to set out on the voyage to England. **1865** *Mech. Mag.* XIV. 64/1 Mr. Low, another American aëronaut, has constructed what he terms an aërial ship. **1917** 'Contact' *Airman's Outings* 180 Throughout the Somme Push we were able to maintain that aerial superiority without which a great offensive cannot succeed. **1919** *Aeronautics* 2 Oct. 322 (*heading*) Antarctic Aerial Survey. **1921** *Ibid.* 28 Apr. 304/1 The Government of India has issued a paper which deals with some experiments in aerial surveying which have been made. **1958** *Oxf. Mail* 27 June 1/2 Two United States.. aerial tankers landed at Brize Norton today after breaking the New York–London record. **1946** *N.Z. Jrnl. Agric.* LXXIII. 193/2 The extensive areas of copper-deficient peat; land in New Zealand provided a suitable opportunity to try out aerial topdressing under circumstances reasonably favourable to success. **1959** A. H. McLintock *Descr. Atlas N.Z.* p. xxi, Throughout this hill country aerial topdressing has made spectacular progress in improving the carrying capacity of the land. **1896** *Invention* 6 June 356/2 Mr. Rich ..has invented an aerial torpedo capable of dealing death and destruction to..an ordinary sized county. **1938** *News Review* 25 Aug. 24/1 Three weeks ago a Japanese aerial torpedo fell within 50 feet of him at Hankow, splashing him with mud. **1908** H. G. Wells *War in Air* viii. § 2 They were declared to be aerial torpedo-boats, and the aeronaut was supposed to swoop close to his antagonist and cast his bombs as he whirled past. **1881** W. D. Hay *300 Years Hence* ix. 202 So important was aerial traffic become. **1910** *Times* 21 Mar. 6/6 Aerial traffic is not to circulate at the height of less than 50 mètres over buildings and enclosed property, nor are aerial vehicles to stop over enclosed property at any height below 500 mètres. **1917** 'Contact' *Airman's Outings* 27 When you visit the Continent after the war,..travel by the Franco-British service of aerial transport. **1784** *Universal Mag.* LXXIV. 20/2 The sensations of the two aërial travellers. *Ibid.* LXXV. 264/1 The gallery of one of these aerial vehicles. **1809** G. Cayley in W. Nicholson *Jrnl. Nat. Philos.* Nov. 167 In such proportion may aerial vehicles be loaded with inactive matter. **1784** *Morning Herald* 1 Dec. 2/4 The first ascent of the aerial vessel. **1838** T. M. Mason *Aeronautica* 325 The first of these [restrictions] regards the form of the aerial vessel. **1909** *Times* 27 Sept. 8/2 Breakage of the propeller blade is a serious danger in every form of aerial vessel. **1908** H. G. Wells *War in Air* viii. § 2 The early battles of the aerial war were no doubt determined by attempts to realise the old naval maxim, to ascertain the position of the enemy's fleet and to destroy it. **1895** *Knowledge* XVIII. 276/1 Aerial warfare..likely to ensue when aerial navigation becomes an accomplished fact. **1784** *Morning Herald* 1 Dec. 2/4 The Devonshire Aerial Yacht. **1920** Aerial yacht [see *air *sb.*[1] B. III. 5].

6. b. *aerial cableway, ropeway*: see quots.; *aerial railway, tramway*, a track consisting of overhead wires, cables, or rails supporting carriages, usually driven by electricity; *aerial wire*, a wire or wires (by extension also a rod, etc.) supported in the air for radiating or receiving electromagnetic waves; an antenna; hence applied to things connected with this, as *aerial array, circuit, switch, system*, etc. Also *sb.*, short for *aerial wire*.

1874 Knight *Dict. Mech.*, *Aerial railway*, an attempt to govern the balloon or aërostat by guiding rails or wires stretched between posts. **1889** *Cent. Dict.*, *Aerial railway*, a name sometimes applied to systems of transportation by cars suspended from a rail or rope above them. **1903** *Buddhism* I. 157 A gorgeous car..ran on an aerial ropeway to the top of the scaffolding from the Pagoda platform. **1904** *Sci. Amer.* Suppl. LVII. 23438/3 (*title*) Aerial tramways as an economic means of transportation. **1910** *Encycl. Brit.* VII. 62/2 The aerial cableway is a development of the ropeway, and is a conveyor capable of hoisting and dumping at any desired point. **1957** in *Amer. Speech* (1963) XXXVIII. 204 *Aerial tramway*, usually two enclosed cabins which counter balance each other and transport a group of skiers at once. **1959** *Chambers's Encycl.* IX. 196/1 *Aerial Ropeways* are used for transporting material in bulk over long distances and over hilly ground or valley...They consist of a series of towers carrying suspension wires to which are attached carriers for the material.

1899 Marconi in *Jrnl. Inst. Electr. Engin.* XXVIII. 274 A vertical conductor *W*, which I will call the aërial conductor. *Ibid.* 289 The aërial wire comes through the framework of a skylight. **1902** [see *antenna 5]. **1906** A. F. Collins *Man. Wireless Telegr.* 208 Aerial, a word much used instead of the longer term *aerial wire*. *Ibid.*, Aerial switch, a switch used to throw the aerial wire into connection with the spark-gap and out of connection with the detector, and vice versa. **1908** *Westm. Gaz.* 8 Dec. 9/4 The four aerials connected with the mast cover about an acre and a half. **1913** *Yr. Bk. Wireless Telegr.* 415 *Aerial Circuit.*—Starts at the free or insulated end of the aerial and ends with the connection to earth. **1930** *Times* 25 Mar. 11/3 There appears to be no reason for supposing that an effective aerial array would need masts over 180 ft. in height. **1930** *B.B.C. Yr.-Bk. 1931* 407 One 70-foot mast supports an umbrella aerial system serving the D.F. equipment. **1937** *Discovery* Feb. 43/2 A standard di-pole G.E.C. receiving aerial for television.

aerialist (ē°·riălist). [f. Aerial *a.* + -ist.]

1. = Pillarist 1. *rare.*

1846 W. Trollope *Jortin's Eccl. Hist.* II. xxxvi. 108 [St. Simeon Stylites'] example was imitated by other aerialists of great nerve and little brains.

2. A performer on the high wire or trapeze.

1905 *Barnum & Bailey's Program* 15 A cluster of most remarkable champion aerialists. **1950** *Oxf. Jun. Encycl.* IX. 2/1 An 'aerialist' is an acrobat who performs on the tight-rope or trapeze, up in the dome of the circus tent. **1961** *Guardian* 22 May 5/1 The aerialists are so high and perilous that their bland composure fools no one.

aerially, *adv.* (Earlier example.)

1821 *Examiner* 1 Oct. 618/1 A light handkerchief..floats aërially from her neck.

aerify, *v.* Delete *rare*−⁰ and add examples.

1. *fig.* **1821** *New Monthly Mag.* II. 362 The first symptoms of that spirit of insubordination..were probably viewed with equal contempt by the living..and by their aerified predecessors. **1891** 'L. Malet' *Wages of Sin* II. v. iv. 224 The aerified pleasure-seeking little world of Tullingworth.

2. 1858 *Jrnl. R. Agric. Soc.* XIX. 82 Lungs which aërify their blood.

aero- (ē°·rŏ, now rarely ē·ərŏ). Add: †**aero-elastic** (in form *aereo-elastic*), *a.* (see quot.); **ae:ro-e·mbolism** *Path.* (see quot. 1939); **ae·rogel** *Chem.*, a gel in which the liquid has been replaced by air or gas; **ae:ro-ge·nerator**, an electric generator operated by wind; **ae·rograph** = *air-*brush*; **ae:ro-mecha·nics**, the branch of mechanics which treats of the equilibrium and motion of air and other gases, and of bodies sustained in them; **ae·ro-oti·tis me·dia** *Med.*, inflammation of the middle ear caused by a change in atmospheric pressure; cf. *aviator's ear* (*aviator 2 b); **aerophagia** (-fē¹·dʒiă) *Path.*, also **aero·phagy** [ad. F. *aérophagie* (M. L. Bouveret 1891, in *Rev. de Méd.* XI. 148): see -phagy], the swallowing of air; **aeropla·nkton** *Biol.*, a collective name for all the forms of minute organic life drifting in the air; also *attrib.*; **ae:rothera-peu·tics** (*Syd. Soc. Lex.*, 1881), **-the·rapy** = Pneumatotherapeutics; **aero·tropism** *Bot.* [ad. G. *aërotropismus* (H. Molisch 1885, in *Wiener Akad. Sitzungsber.* XC. I. 137), f. Gr. τροπή turning (τρέπειν to turn): see -ism], the property, exhibited esp. by the growing roots of plants, of bending or turning towards a source of air; hence **aerotro·pic** *a.*

1747 R. James *Pharmacopœia Univ.* II. iii. 146 [Temperating medicines] operate by an expansive and aereo-elastic Quality, such as that which is inherent to Nitre. **1939** H. G. Armstrong *Aviation Med.* xxi. 342 Aeroembolism may be defined as the disease produced by a rapid decrease of pressure below 1 atmosphere, such as may occur in aircraft flights to high altitude, and which is marked by the formation of nitrogen bubbles in the body tissues and fluids. **1923** W. E. Gibbs *Clouds & Smokes* vi. 112 Frequently a low specific gravity is due to the particles possessing the very open, spongy structure of an 'aerogel'. **1923** R. Whytlaw-Gray et al. in *Proc. R. Soc.* A. CII. 613 It seems likely that our thicker deposits of ZnO and CdO possess a conformation closely analogous to many gels, and the term 'aerogel' suggested by Prof. Donnan is not inappropriate. **1951** *Sci. News Let.* 1 Sept. 144/1 The modern napalm bomb..contains a silica aerogel. **1946** P. H. Thomas (*title*) The Wind Power Aerogenerator. *Ibid.* (Foreword), This study was made to determine the character of structures suitable for aerogenerators devoted to utility power supply. **1898** *Brit. Jrnl. Photogr.* 29 Apr. 274/1 Enlargements finished with the aerograph in water colours. **1939** *Archit. Rev.* LXXXV. 101 (*caption*) Detail of a panel by Sigmund Pollitzer, on polished black glass, using a fine aerograph-type sand-gun nozzle which enables the sandblaster to direct the blast accurately and, by reducing the pressure, obtain gradations of tone. **1909** *Cent. Dict.* Suppl., Aeromechanics. **1938** *Jrnl. R. Aeronaut. Soc.* XLII. 394 A very old and fundamental principle of aeromechanics. **1937** Armstrong & Heim in *Jrnl. Amer. Med. Assoc.* 7 Aug. 419/1 In the United States the term 'aviator's or aviation ear' has begun to appear in the literature, while in Germany the terms 'barotrauma' and 'tonetrauma' have been suggested. The former are obviously unsuitable and the latter may be criticized as not being descriptive of the disease. We therefore suggest 'aero-otitis media'..as a suitable descriptive term. **1949** *Lancet* 14 May 826/1 Clinical observations on aero-otitis media in aviators, compressed-air workers, and subjects in low-pressure-chamber tests. **1901** *Lancet* 30 Mar. 953/1 In certain cases aerophagia takes a grave form: it produces vomiting, aggravates the dyspepsia, and leads to emaciation. *Ibid.* 953/2 In horses aerophagia is a recognized disorder and is known to horse-dealers in France as 'tic à l'air'. **1901** Dorland *Med. Dict.* (ed. 2) 26/1 Aerophagy. **1932** Fuller & Conard tr. *Braun-Blanquet's Plant Sociol.* iv. 76 The quantitative investigation of the edaphon and aeroplankton is still in its infancy, and one must be content [with]..an approximate idea of the great wealth of species of air and soil floras. **1938** *Nature* 7 May 828/1 The possible importance of the aeroplankton yeasts as virus carriers is not, of course, limited to the case of foot-and-mouth disease. **1894** C. T. Williams (*title*) Aero-Therapeutics, or the Treatment of Lung Diseases by Climate. **1876** *Sci. Amer.* 29 July 63/1 Aerotherapy..Air..is forced by steam power and kept at a pressure somewhat above that of the open atmosphere. Dr. Carlo Forlanini is the discoverer and advocate of this treatment. **1918** *Lancet* 30 Nov. 757/1 His..suggestions.. about aerotherapy, hydrotherapy, poultices. **1889** *Cent. Dict.*, Aërotropism. **1898** H. C. Porter tr. *Strasburger's Text-Bk. Bot.* I. II. 263 Thermotropism..Rheotropism..and Aerotropism, a form of chemotropism, are additional phenomena, which have been distinguished as arising from the special action of external stimuli. *Ibid.* 281 The pollen grain then grows out into a tube which is acted upon by chemotropic (including hydrotropic and aerotropic) influences. **1919** F. O. Bower *Bot. Living Plant* viii. 128 Such plants are described as 'pot-bound'. This condition is due to the fact that the roots grow towards a source of free oxygen...Such a response is styled aerotropism, and roots curving towards the source are positively aerotropic.

b. With reference to aircraft, aviation, travel or conveyance by aircraft, or to operations conducted by aircraft; also in various names of aircraft or their parts: as †*aero-biplane, -bus, -car, -cycle, -engine, -meteorograph*, †*-motion, -photography* (hence *-photographic* adj.), †*-screw*, †*-surface*, †*-taxi*, †*-traffic*; *aero-chemical, -marine, -medical, -naval* adjs.; also **aero club**, a club for the pursuit and promotion of aviation; also †*aero meet*, †*race*; †**aero-curve**, an occasional substitute for *aeroplane (senses 1 and 2 b); †**aerodone** (-dōᵘn), **-done·tics** (see quots.); **ae·rodyne** (-dəin) [back-formation from Aerodynamic *a.*], a generic term for heavier-than-air aircraft; †**aeromotive** [F. *aéromotive*, after *locomotive*], a heavier-than-air aircraft; also *attrib.*; **ae·ro-mo·tor** †(*a*) a heavier-than-air aircraft; (*b*) an aircraft motor (Funk, 1928) (chiefly *U.S.*); †**aeronat** [ad. F. *aéronat* (1889 *Aéronaute* 158), f. L. *natāre* to swim]: see quot. 1908.

1874 *Ann. Rep. Aëronaut. Soc.*, The Aero-bi-plane or First Steps to Flight. **1906** *Daily Chron.* 11 Dec. 6/6 'Aerobus' might be preferred [to 'airvan']. **1913** *Daily Mail* 4 Oct. 6 The Grahame-White aerobus established a new passenger-carrying record by taking up nine passengers. **1960** *Aeroplane* XCIX. 850/1 To date, the transport helicopter has been most successful in airline service as a special purpose vehicle ('aerobus') transferring passengers between airports in the Chicago and New York areas. **1965** *Guardian* 10 Feb. 1/1 The Minister..announced that discussions were in progress for 'an aerobus' to provide cheap travel for large numbers. **1910** *Flight* 13 Aug. 634/2 Airmen..are either Aeronauts or Aviators, according as the Aerocar that they control is an Airship or an Aeroplane. **1926** *Chambers's Jrnl.* 14 Aug. 581/2 Aero-cars may take off from here. **1937** *Daily Herald* 16 Apr. 10/3 The second stage of an aero-chemical attack will be the passing down of fluid poisons stored, under pressure or not, in containers carried by aircraft. **1900** *Science* XII. 799/1 Through the cooperation of the Aëro Club, balloon races were organized. **1894** J. Means in *Aeronaut. Ann. 1895* 157 The design here given calls for aeroplanes as being more easily made than aerocurves modelled after the wings of birds. **1897** O. Chanute in *Aeronaut. Ann.* 41 The first machine which was repaired..was the aerocurve, with three superposed fixed surfaces. **1901** *Flying* Dec. 17 The twin-screw propelled 'aerocurve machine'. **1902** F. Walker *Aërial Navig.* 115 Aeroplanes and Aerocurves. When these are made of fabric, yacht duck may be employed. **1901** *Sci. Amer.* 2 Mar. 130/3 Mortureux; aerocycle with four wings. **1957** *Britannica Bk. of Yr. 1956* 512/1 *Aerocycle*, a small helicopter for use by combat infantrymen. **1907** F. W. Lanchester *Aerodynamics* 393 *Aerodone*, from the Greek ἀεροδόνητος, lit. *tossed in mid air*; *soaring*. To denote a gliding or soaring model or machine; in particular, any gliding or soaring appliance destitute of propelling apparatus or auxiliary parts; in contradistinction to *aerodrome*. *Ibid.*, *Aerodonetics*, the science specially involved in problems connected with the stability or equilibrium of an aerodone or aerodrome, or of birds in flight, and with the phenomenon of soaring. **1908** *Daily Report* 21 Nov. 2/4 'Aerodone' denotes a motorless flying machine, a glider. **1906** W. Turnbull in *Sci. Amer.* 211/1 Aerodyne. **1907** — in *Physical Rev.* XXIV. 285, I use the word 'aërodyne' in preference to 'flying-machine', to denote an aëroplane-supported machine, driven by mechanical power through the air. **1934** *Shell Aviation News* No. 32. 4 There are specialists in engines, heavier-than-air craft (aerodynes), [etc.]. **1913** *Aeroplane* 6 Feb. 124/2 'Aero', which Mr. O'Gorman suggested as an abbreviation for aeroplane, is useful as a prefix, for instance in aero-show, aero-engine, and so forth. **1963** *Listener* 17 Jan. 124/1 Rolls-Royce are to put 16,000 employees in their aero-engine division on short time. **1917** H. Woodhouse *Textbk. Naval Aeronautics* 100 (*caption*) The 125 horse-power Aeromarine hydro-aeroplane. **1937** *Jrnl. R. Aëronaut. Soc.* XLI. 634 Aeromedical research (esp. the chemistry of breathing). **1962** *Flight Internat.* LXXXI. 262/2 The 'aeromedical observer', a doctor, monitors the astronaut's physical condition. **1910** *Boston Daily Globe* 4 Sept. 21/1 Aero meet. **1945** C. P. Lent *Rocket Research* 74/1 Friez-type aero-meteorographs. **1878** *Ann. Rep. Aeronaut. Soc.* 44 In no case in which man has succeeded in fast locomotion by machines has he rigidly copied nature, and in every case he has beaten it. In aëromotion man will do the same. **1881** W. D. Hay *300 Yrs. Hence* ix. 214 Health..has been marvellously improved by the universal custom of aëromotion. **1865** C. H. Turnor *Astra Castra* vii. 256 The balloon can only therefore be..the prologue; the true piece is the aeromotive, which supersedes the balloon. **1868** *Aëronaut. Soc., Catal. First Exhib.* 11 Model of an Aeromotive Engine. **1902** F. Walker *Aërial Navig.* p. viii, The construction of dirigible balloons, aërostats, aëroplanes, and aëromotors to be..illustrated by various types already made. *Ibid.* p. ix, The aëromotor or air-ship will always be of great value. **1935** 'J. Guthrie' *Little Country* vi. 139, I am working on the boat..a hundred horse-power Curtis Vth aero-motor in her. **1903** *Aeronaut. Jrnl.* 41 A steerable balloon [is called] an aëronat. **1908** *Daily Report* 21 Nov. 2/4 'Aeronat' is a dirigible, motor-driven balloon or airship. **1959** *Economist* 21 Feb. 663/1 United States aero-naval strength in the Formosa Strait. **1918** H. Woodhouse *Textbk. Mil. Aeronautics* 93/2 Detailed description of the British aerophotographic organization will be of great assistance. **1918** E. S. Farrow *Dict. Mil. Terms* 9 *Aërophotography*, the art of photographing from aeroplanes or airships. **1939** *Geogr. Jrnl.* XCIII. 331 It..we ever undertake contouring of the Highlands of Scotland..the best way to do it will be by means of aerophotography and a plotting machine. **1912** *World Alman.* (N.Y.) 432 Cross-country Aero Races European Circuit Race. **1902** *Flying* June 135 Studying this matter of aerocurves and aeroscrews. **1894** J. D. Fullerton in *Proc. Internat. Conf. Aerial Navig.* 241 Sustainer, or aerosurface, which rests on the air. **1902** *Aeronaut. World* (U.S.) 1 Oct. 65/2 Hofman's Flying Machine..is furnished with three large square flat aero-surfaces. **1904** J. Fullerton *Rep. Aerial Navig.* 1 Aero-surfaces, or Flying Machines proper. **1909** *Flight* 2 Jan. 14 In the aeroplane we have something new, something that is not standardised down, and commercialised so far as to have aero-taxis and aero-buses. **1906** *Lancet* 15 Dec. 1674/1 (*title*) Aero-traffic.

aerobatics (-bæ·tiks), *sb. pl.* [After *acrobatics*.] Feats of expert aviation, performed esp. for display. Hence **aeroba·tic** *a.*; **ae·robat**, one who performs aerobatics; hence as *v. intr.*, to perform aerobatics; also *trans.*

1917 'Contact' *Airman's Outings* 128 Watching the aerobatics and sham fights of the pool pupils. **1918** H. Barber *Aerobatics* II. 58 There are many who..will never shine as aerobatic pilots. **1929** *N. & Q.* CLVII. 359/2 The word *aerobat*..is now used for an airman who does 'stunts'. **1930** *Flight* XXII. 659/1 Mr. George Murray aerobated in Capt. Broad's special Moth. **1943** T. Horsley *Find, Fix & Strike* 43 The pupil who 'shows off'..or who can never take an aircraft up without aerobating it. **1955** *Times* 20 Aug. 2/5 The merit of aerobatics is that everything depends on the skill of the pilot. **1961** C. B. Smith *Testing Time* iii. 44 Gustav Hamel, the brilliant aerobat..was summoned by the King to Windsor to demonstrate his looping of the loop.

aerobe (ē·ərōᵘb, ē°·rōᵘb). *Biol.* [ad. F. *aérobie* (Pasteur, see *anaerobe), f. Gr. ἀήρ air + βίος life.] One of a group (*Aerobia*) of micro-organisms which live on free oxygen derived from the air. So **aero·bian, aerobic** (-ǫ·bik), **aero·bious** *adjs.*, living on the oxygen of the air; of the nature of or pertaining to aerobes; **aero·bically** *adv.*; **a:erobio·sis**, life sustained by the oxygen of the air; **a:erobio·tic** (-ǫ·tik) *a.*, pertaining to or characterized by aerobiosis, aerobic.

1879 Faulkner & Robb tr. *Pasteur's Ferment.* 210 The first aërobian ferment. *Ibid.*, An aërobious ferment. **1884** [see *anaerobic *a.*]. **1885** Vines in *Encycl. Brit.* XIX. 51/2 In aerobiotic plants the normal processes of destructive metabolism..may be replaced for a..time by those abnormal processes of which fermentation is the outward expression. **1886** *Buck's Handbk. Med. Sci.* III. 64/1 Those [micro-organisms] which thrive only with free oxygen; these he calls *aerobes*. **1887** A. M. Brown *Anim. Alkaloids* 117 Four-fifths of our tissues live aerobically; and..the remaining fifth part..lives anaerobically, that is, after the fashion of putrid ferment. **1896** *Allbutt's Syst. Med.* I. 513 Obligatory aerobes, which must be supplied with oxygen. **1900** *Jrnl. Soc. Arts* XLVIII. 387/2 A larger number of bacteria were..facultatively aerobic. **1933** *Biol. Abstr.* 1929 III. 2332/3 Bacteriophagy and aerobiosis. **1949** H. W. Florey et al. *Antibiotics* I. II. ii. 76 One incubation may be aerobic and the other anaerobic. **1951** J. P. Marbarger *Space Medicine* 36 The most important energy source of the organisms, the biological oxidation or aerobic respiration, is a process during which the body's nutritive substances.. are slowly combusted into water and carbon dioxide by oxygen. **1959** *Times Lit. Suppl.* 27 Mar. 183/2 He will know ..the meaning of an aerobiosis, the action of sulphate-producing bacteria, [etc.].

ae:robio·logy. [f. Aero- + Biology.] The study of airborne micro-organisms or spores and their distribution, esp. as agents of infection. Hence **ae:robiolo·gical** *a.*; **ae:robio·logist**.

1937 *Biol. Abstr.* XI. 2348/2 Aerobiology. **1942** F. R. Moulton *Aerobiology* p. iii, Although aerobiology has developed as a specialized field of investigation only within the past ten or fifteen years, it had its origin in the pioneer experiments of Spallanzani, in 1776, and in the work of Pasteur, Tyndall and others who used the methods of the aerobiologist..in combating the theory of spontaneous generation of life. *Ibid.* 1/2 Agriculture and medicine appear to be the general fields in which there is greatest present need for aerobiological investigations. **1951** *Nature* 27 Oct. 721/2 (*caption*) Exposures on aerobiological flight IV of 1947. **1957** H. A. Hyde in *Nature* 4 May 890/1 Aerobiology may be defined as the study of the aerial transport of plants and animals or viable parts thereof, including flying animals (such as insects) in so far as passive movement is a major factor in their dispersal.

aerodrome (ē°·rŏdrōᵘm). [In sense 1, ad. Gr. ἀεροδρόμος *a.*, running through or traversing the air; in sense 2, f. Aero- + Gr. δρόμος course, racecourse (cf. *hippodrome*).]

†**1.** S. P. Langley's name for an aeroplane: = *aeroplane 2 b. Also *attrib. Obs.*

1891 S. P. Langley *Exper. Aerodynamics* 49 An actual working aerodrome model with its motor. **1896** A. G. Bell in *Smithsonian Rep.* 6 Witnessing the successful flight of some of these aerodromes. **1907** [see *aerodone* s.v. *aero- b]. **1908** G. H. Bryan in *Nature* 29 Oct. 668 Mr. Farman mounted with M. Delagrange on the latter's aërodrome, which flew a considerable distance with a heavy load.

2. †**a.** A place where a balloon or flying-machine is housed; a hangar [Fr. *aérodrome*]. *Obs. rare.*

1902 *Westm. Gaz.* 29 Jan. 6/2 He..soared above the Prince's castle to the aerodrome. **1921** M. Corelli *Secret Power* viii. 79 The building..was easily recognisable as a huge aerodrome. *Ibid.* 80 The lady of many millions had commanded an air-ship to be built..with an aerodrome for its safe keeping and anchorage.

b. (The current sense.) A large tract of open, level ground, together with the runways, hangars and other installations, for the operation of aircraft. Also *attrib.*

1909 F. Lanchester in *Flight* 2 Jan. 13/1, I regret to see that the misuse of the word 'aerodrome' is receiving support in your columns...I suppose because a hippodrome is a big open space for horses, you think that an aerodrome should be a big open space for flying machines. **1909** *Flight* 20 Feb. 104/1 *Aerodrome*, a field or place set apart for the practice of flight. **1911** Grahame-White & Harper *Aeroplane* iv. 124 A circle had been whitewashed on the aerodrome..to act as a mark in which the aviators were to descend. **1922** *Encycl. Brit.* XXX. 48/1 The London terminal aerodrome at Croydon, Sur., may be taken as typical of a modern air-port for commercial traffic. **1934** *Jrnl. R. Aeronaut. Soc.* XXXVIII. 725 There was erected the aerodrome beacon. **1959** *Chambers's Encycl.* I. 95/2 Siting requirements for aerodromes often conflict, as when a city lying in a valley or close to a mountain range must be served.

†**aerodro·mic,** *a. Obs.* [f. prec. + -ic.] Pertaining to flying-machines; †**aerodro·mics**, the art of constructing and using flying-machines.

1891 S. P. Langley *Exper. Aerodynamics* 5 The yet inchoate art of constructing suitable mechanisms for guiding heavy bodies through the air..which art..I will provisionally call aerodromics. **1894** Chase & Kirchner (*title*) The Coming Railroad: The Chase-Kirchner Aerodromic System of Transportation. **1896** A. G. Bell in *Smithsonian Rep.* 6 Investigations connected with aerodromic problems.

aerodynamics. Add: Pronunc. now usu. (ē°:ro,dəinæ·miks). (Additional examples.) Hence **aerodynamic** *a.* [cf. G. *aerodynamische* (1835)], pertaining to aerodynamics [cf. Dynamics 1], esp. to the effects produced on aircraft or other solid bodies by the air through which they pass, or the effects produced in air by the motion of solid bodies through it.

In quot. 1891, the art of flying through the air by some mechanism, the use of flying-machines, aviation.

1891 S. P. Langley (*title*) Experiments in Aerodynamics. **1908** W. H. Story tr. *Hildebrandt's Airships* ix. 90 Aerodynamic airships. **1920** *Advis. Comm. Aeronautics, Rep. & Mem.* No. 651, p. 5 Aerodynamic balance of the aileron. **1922** *Edin. Rev.* Oct. 214 The improvements in range and aerodynamic and engine efficiency of aircraft. **1923** Glazebrook *Dict. Appl. Physics* V. 199/2 *The Main Aerodynamic Balance*..This balance was designed so that it could be used to measure forces about three axes perpendicular to one another and the moment about a vertical axis. **1935** *Jrnl. R. Aeronaut. Soc.* XXXIX. 826 With regard to the use of sponsons or stubs, he had understood Mr. Coombes to say that one could reckon on some aerodynamic lift from the stub. **1936** *Aircraft Engin.* Sept. 241/2 The idealized case is that of a monoplane in which the aerodynamic centres of the wing and tail plane lie in one straight line. **1949** *Jrnl. R. Aeronaut. Soc.* LIII. 659/2 It provided new incentives for understanding stability and control..aerodynamic heating..and similar general problems. **1950** *Gloss. Aeronaut. Terms* (*B.S.I.*) I. 28 *Aerodynamic balance*, a balance designed for measuring aerodynamic forces or moments. **1950** *Sci. News* XV. 83 Rockets are powerful research tools...Their value in aerodynamics lies in their ability to propel full-sized or model aircraft..at supersonic speeds in order that the unknown aerodynamic forces in play at these speeds can be measured. **1962** *Gloss. Aeronaut. Terms* (*B.S.I.*) I *Aerodynamic centre*, the point about which the rate of change of pitching moment with incidence is zero. *Ibid.* 5 *Aerodynamic balance*, the degree to which the hinge moment of a control surface is reduced by balancing.

So **ae:rodyna·mical** *a.*, **-ally** *adv.*; **ae:rodyna·micist**, one skilled in aerodynamics.

1908 *Aeronaut. Jrnl.* 84 The design and perfection of true aërodynamical flying ships of the future. **1921** *Glasgow Herald* 8 Sept. 10/2 Department for aerodynamical research. **1928** C. F. S. Gamble *North Sea Air Station* x. 147 The Wight seaplane..was of similar type to the Short, but its aerodynamical qualities were inferior to its prototype. *Ibid.* xiii. 210 Aerodynamically, it was of interest in being fitted with an air-brake in the form of adjustable flaps. **1939** *Jrnl. R. Aeronaut. Soc.* XLIII. 138 It is their duty to co-ordinate the efforts of the draughtsmen,..designers and aerodynamicists. **1960** [see next].

aeroelasticity (ē°:ro,ilǣsti·siti). [f. Aero- + Elasticity, cf. *photo-elasticity*.] **a.** A branch of mechanics dealing with the effects of aerodynamic forces on elastic bodies, esp. the non-rigid structures of aircraft; **b.** the quality of being liable to distortion by aerodynamic forces. Hence **ae:roela·stic** *a.*; **ae:roela·stics** *sb. pl.*, aeroelasticity (sense a); **ae:roelasti·cian**, one skilled in aeroelasticity.

1935 *Aircraft Engin.* Oct. 261/2 This branch of study, which might appropriately be labelled 'aero-elasticity', is steadily increasing in importance. **1936** *Aeronaut. Res. Comm., Techn. Rep. 1934–35* II. 748 It was found to permit of a very convenient treatment of the aerodynamics of the reversal of aileron control problem, and applications in other 'aero-elastic' problems may be possible. **1938** *Aircraft Engin.* Mar. 73/2 With increasing speeds and the consequent continued expansion of the field of aeroelasticity, stiffness and test methods are being applied..to wings. **1946** *Jrnl. R. Aeronaut. Soc.* L. 613 The developments in aeroelastic science during the past ten years. **1946** *Nature* 21 Dec. 897/1 The problems of aero-elasticity are..complicated by the fact that the wings are swept back. **1947** *Jrnl. R. Aeronaut. Soc.* LI. 418/2 (*caption*) The aeroelastician's viewpoint showing mechanical vibration as outlying territory of the aeroelastic domain. **1960** *Times* 24 Oct. 2/6 Aerodynamicists..are required for the following sections:- Aeroelastics, Stability and Control, [etc.]. **1962** *New Scientist* 8 Nov. 344/1 (Advt.), *Senior Aeroelastician*. This post is suitable for a graduate..who has acquired a number of years experience working on aeroelastic problems of aircraft or missiles.

ae·rofoil [f. *aero- b + Foil *sb.*[1]] A wing, aileron, tailplane or other lifting surface of an aircraft; any surface designed on similar principles; also *attrib.*, and applied to a type of ship's propeller (see quot. 1948).

1907 F. W. Lanchester *Aerodynamics* v. 179 The author does not employ the term *aeroplane* outside its correct signification, that is to say, to denote other than a true or *plane* aeroplane; the misuse of the word being avoided by the introduction of the word aerofoil, to denote a supporting member, or organ of sustentation of undefined form. **1914** *Aeronaut. Jrnl.* XVIII. 315 *Aerofoil*, a structure, analogous to the wing or tail of a bird, designed to obtain a reaction from the air approximately at right angles to the direction of its motion. **1920** *Conquest* I. 437/1 The effects of air-flow upon various aerofoil surfaces. **1931** *Discovery* May 147/2 The general use of what is now known as 'aerofoil' blade propellers. **1948** R. de Kerchove *Internat. Maritime Dict.* (1958) 5/1 *Aerofoil propeller*, a propeller designed in accord with aerodynamic principles based on the circulation theory instead of as a true screw...The margin of efficiency of the aerofoil propeller over the true screw is about 10 to 12%. **1950** *Engineering* 24 Mar. 333/3 The propellers [of a launch] are of the standard..type in which vertical aerofoil-section blades are fitted. **1959** *Chambers's Encycl.* XIV. 612/1 In 1923 aerofoils were first fitted to the leading edges of normal windmill sails by Dekker in the Netherlands. **1961** *Spectator* 14 July 53 Part of the aerofoil lift develops as the forward speed on the cushion increases.

ae·rogram. [f. AERO- + -GRAM.] **1.** A message sent 'through the air', i.e. by radio.
1899 *Daily News* 20 Nov. 7/1 'Wanted, a new name for wireless telegraphy', Miss Collett, Hyde-park-mansions, suggests..'Aerogram'. **1901** *Westm. Gaz.* 20 Dec. 6/2 (*heading*) Halfpenny-a-word aerograms.
2. An X-ray photograph of an organ injected with air.
1935 DORLAND & MILLER *Med. Dict.* (ed. 17) 53/1 *Aerogram*, a roentgenogram of an organ after it has been injected with air.

aerographer. (U.S. examples.)
1889 in *Cent. Dict.* **1945** *Jane's Fighting Ships 1943–44* 432/1 (*caption*) Aerographer [in U.S. Navy].

aerography. (Later examples.)
1918 E. S. FARROW *Dict. Mil. Terms* 9 *Aërography*, the study of the structure of the atmosphere and charting the physical conditions at all levels. The work is done principally by kites, and balloons. **1920** *Nature* 17 June 479/2 The substitution of the new name aerography for the older meteorology has not changed the leopard's spots.

aerology. Add: (Earlier example.) Also *spec.* [app. ad. G. *aerologie* (W. Köpper, 1906)], the branch of meteorology which treats of the phenomena of the upper air. Hence **ae:rolo·gical** *a.*
1736 BAILEY *Dict. Britannicum* (ed. 2) Pref., *Aerology*, a Treatise or Philosophical Discourse of the Air. **1912** *Aëronaut. Jrnl.* July 166 The training he had received from his distinguished father..made any facts regarding aerology which he [C. H. Ley] might contribute well worthy of..consideration. **1917** *Geogr. Rev.* III. 397 A..movement is under way..to establish aërological stations for observers. **1932** *Discovery* Aug. 271/2 Three sets of meteorological and aerological observations were made at very different stations. **1932** *Flight* XXIV. 974/1 Aerology (meteorology) seems sound, but perhaps too brief.

aeronaut. Add: **1.** Also used of any air pilot.
1809 G. CAYLEY in W. Nicholson *Jrnl. Nat. Philos.* XXIV. 173 Let there be a convenient seat for the aeronaut. **1908** H. G. WELLS *War in Air* i. § 3 They flew in machines heavier than air. But they smashed...The breeze upset them..a passing thought in the mind of the aeronaut upset them.
Hence *transf.* and *fig.*
1790 [in Dict.]. **1840** DICKENS *Old C. Shop* xxxiii. 277 The historian takes the friendly reader by the hand, and springing with him into the air..alights with him upon the pavement of Bevis Marks. The intrepid aeronauts alight before a small dark house.

aeronautic, *a.* Add: (Earlier and later examples.)
1784 *Hibernian Mag.* Sept. 489/2, I shall..attempt a description of his [*sc.* Richard Crosbie's] Aeronautic Chariot. **1908** H. G. WELLS *War in Air* iv. § 1 The Emperor forgave him and placed him in control of the new aeronautic arm of the German forces. *Ibid.* xi. § 2 None of these countries had prepared for aeronautic warfare on the magnificent scale of the Germans. **1918** E. S. FARROW *Dict. Mil. Terms* 9 *Aëronautic Maps*, maps showing the contours and configuration of the land as closely as possible to the way it looks to the aviator from the air.
Hence **ae:ronau·tica** *sb. pl.* [mod.L., see AERONAUTICS], matters or facts of aeronautics.
1753 [see AERONAUTICS]. **1838** T. M. MASON (*title*) Aeronautica. **1883** *Catal. Pat. Off. Libr.* II. 4 (*title*) Aëronautica Illustrata. A complete Cabinet of Aërial Ascents and Descents from the earliest period.

aeronautical, *a.* (Additional examples.)
1910 in A. E. Berriman '*Flight*' *Manual* p. vi (Advt.), Aeronautical Engineers. **1916** F. W. LANCHESTER *Aircraft in Warfare* xix. 202 The question of the future of the Aeronautical Arm is not purely the concern of the Army and Navy..it is essentially an affair of the Nation. **1957** *Technology* Mar. 8/1 Early this century the design and building of flying machines was being undertaken by pioneers with no experience as aeronautical engineers. **1959** *Chambers's Encycl.* V. 191/1 *Aeronautical Engineering*. This branch is concerned with the design and construction of both heavier-than-air and lighter-than-air types of aircraft.

aeronautics. (Earlier example.)
1824 *Encycl. Brit.* Suppl. I. 65/2 We prefer, as more correct and appropriate [than *aerostation*], the word Aeronautics, now generally adopted to express aërial navigation.

aerophane. (Earlier and later examples.)
1829 *Ladies' Pocket Mag.* I. 35 Several dresses of colored satin have long sleeves of white *crepe Aerophane, a la Mameluke*. **1830** *New Monthly Mag.* XXIX. 17 Aerophane and *couleur de soupir*..float before me. **1915** *Observer* 11 July 1/3 (Advt.), White Hat, in *aerophane*, small rolled brim underlined black taffeta.

ae·rophone (-fōᵘn). [f. AERO- + *-PHONE.] **1.** A device invented by Edison for amplifying sound. (Disused.)
1878 *Telegraphic Jrnl.* 15 May 200/2 A company of English merchants have offered Mr. Edison £60,000 if he can successfully apply his aerophone or talking fog-horn to 'the local telegraph wires' in London. **1894** W. JERROLD *Electricians & Marvels* vi. 120 The 'aerophone' a great voice of two hundred and fifty times the capacity of the human lungs, which is designed to be used between lighthouses or lightships and vessels at sea.
2. Any musical instrument in which the sound is produced by a vibrating column of air.
1937 *Times Lit. Suppl.* 17 Apr. 288/2 Those [instruments] which employ..a column of air (aerophones). **1959** *Collins Music Encycl.* 333/2 *Aerophones*, woodwinds, brass, and instruments using a free reed. **1960** *Times* 18 Mar. 4/6 The instrument is a thin plate from the inner side of the birch bark, technically described as an aerophone with a tongue but without a soundboard.

ae·rophore (-fō³). Also **aerophor.** [f. AERO- + -PHORE. Cf. Fr. *aérophore*.] A form of respirator containing a quantity of air.
1877 *Encycl. Brit.* VI. 73/2 Apparatus, originating in France, known as aerophores, which enable the miner to carry sufficient fresh air for his own respiration, and to keep a lamp alight for a short time in a totally irrespirable atmosphere. **a 1884** KNIGHT *Dict. Mech.* Suppl. 8/2 The aërophore devised by Herr Schultz, captain of the fire brigade at Aschaffenburg, Bavaria, depends upon the regeneration of the exhaled air, the oxygen being reproduced as it is consumed. **1931** *Discovery* Apr. 128/1 The Aerophor apparatus [for use in mines] is probably the most interesting type of all on account of the fact that it makes use of liquid air.

aerophysics (ē³:rofi·ziks). [f. AERO- + PHYSICS.] The physics of the atmosphere; *spec.* the branch of physics concerned with the movement of solid bodies, as guided missiles, space vehicles, etc., through the air.
1897 A. MCADIE *Equipment & Work of Aero-Physical Observatory* 3 A laboratory..devoted to investigation and research in aero-physics. **1948** *Jrnl. Brit. Interplanetary Soc.* VII. 177 This building..is equipped generally with facilities for all types of basic and applied research in aero-physics, electronics, [etc.]. **1962** *Times* 16 Mar. 2/5 Advanced fields of aircraft instrumentation and aerophysics.

aeroplane. Read: (ē³·rŏplē¹n, *now rarely* ē·aro-), *sb.* [In sense 1, f. AERO- + PLANE *sb.*³; in sense 2, ad. F. *aéroplane* (1855), f. Gr. ἀερο- (see AERO-) + F. *plan* plane (see PLANE *v.*³).]
†1. A plane (or slightly curved) light framework or 'surface' forming part of a flying-machine, and serving to sustain it in the air. *Obs.* (Later called simply *plane*, also *wing*.)
1866 WENHAM in *Ann. Rep. Aëronaut. Soc.* 33 In the flying mechanism of beetles..when the wings..are opened, they are checked by a stop, which sets them at a fixed angle. It is probable that these serve as 'aeroplanes', for carrying the weight of the insect. *Ibid.* 37 A thin steel tie-band, served as the foundation of the superposed aeroplanes. **1869** [in Dict.]. **1901** F. WALKER *Aerial Navig.* 117 Air-ships as combinations of aerostat, aeroplane, and propelling apparatus. **1902** G. BACON *Balloons* 111 What are called 'aeroplanes'—large flat surfaces, light but rigid, inclined at a suitable angle to the horizon.
2. **†a.** An airship provided with planes. *Obs.*
1884 *Pall Mall Gaz.* 28 Aug. 4/1 As soon as the Aero-Plane has been seen floating to and fro over the city of San Francisco, steered at pleasure this way and that, and carrying a number of passengers.
b. (The current sense.) A heavier-than-air flying-machine supported by such planes or wings and mechanically driven.
The use in quot. 1873¹ appears to derive from sense 1 and to owe nothing to Fr. *aéroplane* (Stubelius *Airship* 239).
1873 D. S. BROWN in *Ann. Rep. Aëronaut. Soc.* 17, I think [impetus] will be more requisite with respect to the aero-plane than any other vehicle. *Ibid.* 19 Mr. Bennett introduced an Aeroplane invented by a Frenchman, to be worked by a screw by native power derived from elastic springs. **1874** *Ibid.* 69 Flight has thus been accomplished on three different principles, and the practicability of a flying-machine proved. M. Penaud..thinks the aeroplane to be the only practicable machine. **1892** H. S. MAXIM in *Century Mag.* Oct. 957/2 Ascertaining how much power was required to perform flight with a screw-driven aeroplane. **1897** LANGLEY *Memoir Mech. Flight* (1911) I. 8, 7 His [*sc.* A. Penaud's] aeroplane is a toy in size, with a small propeller whose blades are made of two feathers..and whose motive power is a twisted strand of rubber. **1910** R. FERRIS *How it Flies* 453 *Aeroplane*, the type of flying machine which is supported in the air by a spread of surfaces or planes formerly flat and therefore truly 'planes' but of late are curved. **1920** *Blackw. Mag.* July 74/1 The perfected aeroplane is the obvious instrument to suppress war. **1934** ORVILLE WRIGHT *Let.* 10 Jan. in *Papers W. & O. Wright* (1953) II. 1162 The helicopter type of aeroplane offers several seemingly insurmountable difficulties. **1943** [see *AIRCRAFT]. **1951** *Oxf. Jun. Encycl.* IV. 4/1 The power propelling the aeroplane may be of a number of types—the ordinary piston-engine (internal combustion) driving a propeller (air-screw), the gas-turbine driving a propeller, or the gas-turbine 'pure jet'.
¶ In form indicating the once common *dial.* or *vulg.* but now *obs.* pronunc. (-ī-).
1876 *Pop. Sci. Monthly* Feb. 454 The aeroplanes have propelling surfaces which are nearly plane and slightly inclined to the horizon. **1913** S. LEWIS *Trail of Hawk* xxii. 203 It's a new areoplane (that's the way he pronounced it), and that dingus in front is a whirling motor. **1925** *Aeronautics* 13 Oct. 256/2, I have never heard anyone pronounce aeroplane in any other way than 'airoplane', with perhaps an occasional 'arioplane' or even 'hairyoplane'. **1935** 'J. GUTHRIE' *Little Country* xxix. 411 Half a dozen veterans of the forgotten Maori wars...What with these 'aereoplanes', war, like everything else, was not what it had been!
3. *attrib.* and *Comb.*
1872 *Ann. Rep. Aëronaut. Soc.* 15 At an angle of ten degrees, about one man power would be sufficient to drive an aeroplane machine twenty miles an hour. **1896** H. S. MAXIM in *Aëronaut. Ann., Nat. & Artif. Flight*, 18 The next machine..was on the kite or aeroplane system. **1902** *Aëronaut. World* (U.S.) I. Oct. 58/1 This bird-like aeroplane machine. **1914** *Lancet* 4 July 55/1 A special aeroplane constructed to accommodate, besides the aviator, a medical man and also a concealed litter on which the wounded man can be placed for transport...This aeroplane ambulance [etc.]. **1920** *Proc. Air Conference, London* 96 Movement by sea is a slow business unless aeroplane carriers are available. **1923** KIPLING *Irish Guards in Gt. War* I. 50 'Aeroplane duty' was another invention of those early days. A Company was told off daily to look out for aeroplanes. **1932** H. NICOLSON *Public Faces* xi. 301 She was an aeroplane carrier of His Majesty's Navy. **1958** *Spectator* 10 Jan. 47/1 HQ Tank Corps arranged with the RFC to have daily aeroplane photographs taken of the front over which tanks would advance.
Hence **ae·roplane** *v. intr.*, to fly or glide like an aeroplane; to fly or travel in an aeroplane; **ae·roplaning** *vbl. sb.*; **ae·roplanist**, one who flies an aeroplane. Now *rare*.
1906 *Daily Mail* 16 Nov. 7/5 The first successful aeroplanist in Great Britain will win..as much money as the Soap Trust has already lost. **1907** *Ibid.* 19 Feb. 7/7 M. Santos Dumont..felt that for some years to come aeroplaning would remain a sport. **1911** G. B. SHAW *Blanco Posnet* Pref. 367 Motoring has its risks; aeroplaning has its risks. **1912** *Let. to G. Harker* 1 May (1956) 182 We went to the aeroplanists' sheds again...Flying just about to begin as usual. **1914** — *Misalliance* I. never drink tea. *Tarleton*. Bad thing to aeroplane on, I should imagine. **1927** *Observer* 7 Aug. 24 We cannot ignore the fact that aeroplaning is beginning to progress as a pastime, chiefly because of traffic conditions on our roads. **1934** *Times Educ. Suppl.* 17 Feb. 59/2 Model aeroplanists [wend their way] to Hampstead Heath.

aerosol (ē³·rŏsǫl). [f. AERO- + SOL(UTION: see SOL *sb.*⁶).] **1.** A system of colloidal particles dispersed in the air or in a gas, as mist, fog, etc.
1923 R. WHYTLAW-GRAY et al. in *Proc. R. Soc.* A. CII. 600 Aerosol is a convenient term to denote a system of particles of ultra-microscopic size dispersed in a gas, suggested to us by Prof. Donnan. **1937** *Nature* 31 July 201/1 Systems in which the dispersion medium is a gas are usually known as aerosols. **1943** *Lancet* 25 Feb. 443/2 The introduction by Trillat (1938) [*Bull. Acad. Méd. Paris*, CXIX. 64] of the conception of 'aerosols' for the purpose of air-sterilisation. **1949** H. W. FLOREY et al. *Antibiotics* II. xv. 663 The inhalation of penicillin as an aerosol.
2. A substance packed under pressure in a container with a spraying device, esp. *attrib.*, as *aerosol bomb, pack, spray*, etc. Hence, a container of this kind.
1944 *Reader's Digest* May 45/2 Today every soldier in mosquito country is armed with an 'aerosol bomb'—a specially designed container about twice the size of a hand grenade. **1957** *Times* 25 Nov. 11/4 Special foam shaving cream in aerosol container, price 12s. 6d. **1957** *Listener* 5 Dec. 963/1 Aerosol Paint-sprays: Most people by now are familiar with the aerosol pack. **1958** *Vogue* July 82/1 For the beach an oily preparation is best, and these packed in aerosols are likely to be more freely and frequently applied. **1958** *Woman's Own* 10 Sept. 22/1 Does your cosmetic drawer contain either a lipbrush, hair-lacquer, scent atomiser, or an aerosol spray?

ae·rospace. *orig. U.S.* [SPACE *sb.*¹ 8.] The earth's atmosphere and outer space; also *attrib.*
1958 W. A. HEFLIN *Aero-Space Terms* p. iii, This glossary of aero-space terms is published to give guidance in a specialized vocabulary that deals with space weapons, space vehicles, and the physical laws that govern them. **1960** *John o' London's* 7 Apr. 395/1 The Air Force now is operating in an age of 'aero-space'. **1964** *Flight Internat.* LXXXII. 113 The concept of an 'aerospace plane' that takes off from and lands on conventional airfields yet is capable of orbital flight and effective manoeuvres in space is one which is actively being considered at present. **1965** *Guardian* 6 June 13/1 The change from an aircraft industry to an aerospace industry is one which has not yet been made in Britain.

aerosphere. (Example.)
1912 C. M. DOUGHTY *Cliffs* 121 And signals winged, of mens sollicitudes, From the large compass of the aerosphere Continually were received.

aerostat. 1. Delete quot. 1865 and substitute:
1895 *Knowledge* 2 Dec. 276/2 Suppose an aerostat came overhead on warlike deeds intent, carrying dynamite missiles.
b. A generic term for lighter-than-air aircraft.
1906 *Sci. Amer.* 8 Sept. 175/1 'Aerostat'..applies merely to the ordinary balloon. **1909** C. C. TURNER *Aerial Navig. To-day* 299 *Aerostat*, balloons of any form, as distinct from heavier-than-air aerial machines. **1919** W. B. FARADAY *Gloss. Aeronaut. Terms* 15 *Aerostat*, a generic term for any type of aircraft which derives its lift chiefly from aerostatic forces. **1950** *Gloss. Aeronaut. Terms* (B.S.I.) I. 7 *Lighter-than-air aircraft* (*Aerostat*: not the preferred term).
2. Delete (Mod. Fr...etc.) and substitute: (In quots. 1870, 1871, repr. Fr. *aérostier*.) (Add earlier examples.)
1788 R. BAGE *James Wallace* (1824) 392/1 Simon the magician, the first aerostat upon record. **1806** M. EDGEWORTH *Leonora* II. lxv. 109 To mount with a friend in a balloon, amid crowds of spectators, who..applaud the courage of the aerostats.

aerostatic, *a.* **2.** (Earlier and later examples.)
1783 (*title*) The Air Balloon: Or a Treatise on the Aerostatic Globe lately invented by Mons. Montgolfier. **1865** C. H. TURNOR *Astra Castra* vii. 158 The balloon..will be employed in various works of *aerostatic photography*, for which I was the first to take patents in France and abroad seven years ago. **1957** *Observer* 17 Nov. 17/3 The author has a highly developed taste for aerostatic necrology.

aerostatical, *a.* (Later example.)
1861 P. D'AMÉCOURT *Brit. Pat.* 1929, My aerostatical apparatus (which I intend denominating aeronef or helicopter).

aerostatics. (Earlier and later examples.)
1784 *Morning Herald* 18 Mar. 1/4 The laws of Aerostatics, Pneumatics, and Hydrostatics. **1887** tr. *J. Verne's Clipper of Clouds* ii. 15 They were not engineers by profession, but simply amateurs of all that appertained to aerostatics.

aerostation. Substitute for etym.: [a. Fr. *aérostation*, after *navigation*: see AEROSTAT and -ATION.]
2. Add: (Earlier example.) Now applied only to the operation of lighter-than-air aircraft.
1784 *Morning Herald* 16 Sept. 3/2 Aerostation. The brothers Robert have advertised their experiment with their balloon. **1950** *Gloss. Aeronaut. Terms* (*B.S.I.*) I. 6 *Aërostation*, the operation of lighter-than-air aircraft.

ae:rothermodyna·mic, *a.* [f. AERO- + THERMODYNAMIC *a.*] Pertaining to the thermodynamic effects of air and other gases; *aerothermodynamic duct*, see *ATHODYD; hence **ae:rothe:rmodyna·mics** *sb. pl.*; **ae:rothe:rmodyna·micist**, one skilled in aerothermodynamics.
1945 *Westinghouse Engineer* Mar. 51/1 The..name of 'athodyd' (from Aero-THermoDYnamic Duct). **1946** *Jrnl. R. Aeronaut. Soc.* L. 446/2 The 'engine'..is merely a duct in which aerodynamic and thermal changes occur, and it has been described as an aero-thermodynamic-duct. **1951** D. O. DOMMASCH et al. *Airplane Aerodynamics* ii. 26 The necessity of considering thermodynamic effects in high-speed aerodynamics has led to the use of the term *aerothermodynamics*, which identifies work in which thermal effects are considered. **1958** *Times Rev. Industry* May 38/3 A high-speed induced-flow transonic and supersonic [wind-] tunnel..has been installed in a new 'aerothermodynamic' laboratory. **1960** *Times* 10 Feb. 3/2 Aerothermodynamicists.

Aeschylean (ēskilī·ăn), *a.* [f. L. *Aeschylus*, ad. Gr. Αἰσχύλος (see def.) + -AN.] Of, pertaining to, or characteristic of Aeschylus, the Athenian tragic poet (525–456 B.C.), or his works, style, etc.
1844 J. F. BOYES *Illustr. Tragedies Æschylus & Sophocles* i. p. xxxvii, His style and expressions are so bold, as often to border on the Æschylean. **1850** KINGSLEY *Alton Locke* I. ix. 138 The Eschylean grandeur, the terrible rhythmic lilt. **1871** SWINBURNE *Songs bef. Sunrise* 178 With wind-notes as of eagles Æschylean. **1921** *Spectator* 12 Mar. 333/1 An Aeschylean trilogy, the three connected plays, lasted most of a day. **1958** *Ibid.* 22 Aug. 257/2 The last, Aeschylean novel.

Æsopic (īsǫ·pik), *a.* Also (now *U.S.*) **Esopic.** [ad. late L. *Æsōpicus*, f. *Æsōpus* ad. Gr. Αἴσωπος.] Of, pertaining to, or characteristic of Æsop, a semi-legendary Greek fabulist of the sixth century B.C. So **Æsopian, -ean** (īsō·piăn), *a.* [f. late L. *Æsōpius*: see -IAN]. Hence **Æsopism** (ī·sǫpiz'm), an Æsopic characteristic.
1728 CHAMBERS *Cycl.* s.v. *Fable*, Esopic Fables. **1831** CARLYLE in *Foreign Q. Rev.* VIII. 321 The old prevalence of the Didactic, especially of the Æsopic, is every where manifest. **1869** TOZER *Highl. Turkey* II. 266 The Æsopic and similar fables of later date. **1875** *Encycl. Brit.* I. 212/1 The Latin writers of Æsopean fables. **1889** *Cent. Dict.* s.v., A table in the Æsopian style. **1905** J. M. ROBERTSON *Did Shakes. write Titus Andron.* about the Æsopism... **1957** K. A. WITTFOGEL *Oriental Despotism* ix. 420 Originally Lenin used an 'Aesopian' (slave) language to speak to those oppressed by the government.

æsthesia (īsþī·siă, -ziă). [mod.L., ad. Gr. αἴσθησις (see ÆSTHESIS).] Capacity for feeling and sensation; = ÆSTHESIS.
1879 in R. HUNTER et al. *Encycl. Dict.* I. 87/3 *Æsthesia*, perception by the senses, feeling. **1931** *Times Lit. Suppl.* 8 Jan. 17/3 A general æsthesia akin to that of a creature in which the special sense organs are not yet differentiated.

æsthete. (Additional examples.) Cf. *ÆSTHETIC *a.* 4 and *sb.* 3.
1882 W. HAMILTON *Æsthetic Movement* p. vii, The Æsthetes are they who pride themselves upon having found out what is the really beautiful in nature and art, their faculties and tastes being educated up to the point necessary for the full appreciation of such qualities. **1914** C. MACKENZIE *Sinister Street* II. III. i. 516 My dear chap, he was absolutely barred. M' tutor used to like him, but ..I don't mind telling you, he's really an æsthete.

æsthetic. Add: **A.** *adj.* **2.** (Earlier example.)
1821 COLERIDGE in *Blackw. Mag.* X. 254, I wish I could find a more familiar word than æsthetic, for works of taste and criticism.
4. *spec.* Of or pertaining to a late nineteenth-century movement in England of artists and writers who advocated a doctrine of 'art for art's sake'.
1868 W. PATER *Æsthetic Poetry* in *Appreciations* (1889) 213 The 'æsthetic' poetry is neither a mere reproduction of Greek or medieval poetry, nor only an idealisation of modern life and sentiment. **a 1882** D. G. ROSSETTI *St. Agnes* in *Coll. Wks.* (1886) I. 410 The journal of the worthy poet-critic.. was much too æsthetic to permit itself many readers. **1882** W. HAMILTON *Æsthetic Movement* 31 The leaders of the Æsthetic School in poetry have been styled fleshly poets, delighting in somewhat sensually-suggestive descriptions of the passions. **1950** E. H. GOMBRICH *Story of Art* xxv. 402 Whistler became a leading figure in the so-called 'aesthetic movement' which tried to make out that artistic sensibility is the only thing in life worth taking seriously.
B. *sb.* **1.** Delete '† *Obs.*' and insert '*Hist.*'
1798 WILLICH *Elem. Crit. Philos.* 139 Aesthetic commonly signifies the Critique of Taste, but with Kant, the science containing the rules of sensation. **1825** CARLYLE *Schiller* III. 174 The only department [of transcendentalism] to which he attached himself with his ordinary zeal was that which relates to the principles of the imitative arts, with their moral influences, and which in the Kantean nomenclature has been designated by the term *Æsthetics*, or the doctrine of sentiments and emotions. **1875** *Encycl. Brit.* I. 212/1 Kant..under the title Transcendental Æsthetic, treats of the *a priori* principles of all sensuous knowledge.
2. b. (Earlier example.)
1822 *New Monthly Mag.* IV. 149 He accordingly applied himself diligently to study the spirit of classical Tragedy, and the principles of *Æsthetic*.
3. = ÆSTHETE; an adherent of the æsthetic movement (see above *A. 4).
1883 L. TROUBRIDGE *Life amongst Troubridges* (1966) 164 The great Oscar Wilde..is grown enormously fat,..not at all the aesthetic he used to look. **1894** *Cosmopolitan* XVII. 122 The æsthetics..who proclaim the infinite superiority of art to nature. **1946** *English Studies* XXVII. 49 It is not unsympathetic to the Aesthetics, for it seeks to understand them.

æsthetical, *a.* (Earlier example.)
1798 WILLICH *Elem. Crit. Philos.* 104 Those judgments.. are called aesthetical, which relate to the Beautiful and the Sublime.

æsthetically, *adv.* Substitute for first example:
1827 DE QUINCEY *Murder* in *Blackw. Mag.* Feb. 200/2 It may also be treated æsthetically, as the Germans call it, that is, in relation to good taste.

æsthetician. (Earlier example.)
1829 *Virginia Lit. Museum* I. 199/1 There is no question, which has given rise to more frequent disputation amongst Æstheticians..than that of the origin of beauty.

æstheticism. (Additional example.)
1882 W. HAMILTON *Æsthetic Movement* 32 It is at the Lyceum Theatre that æstheticism in all its beauty can best be seen. The present revival of 'Romeo and Juliet' may be pronounced to be the most exquisite rendering of that sweet poem.

‖ **afanc** (avaŋk). Also 9 **avanc, avangc, addanc** (aðaŋk). [Welsh *afanc* beaver, f. med. Welsh *adanc, afanc*:— Celt. **abankos*, f. **ab-* water (whence, from the base **abona*, Welsh *afon*, Bret. *aven* river; Ir. *ab, abann* river), repr. by med. Breton *avancq* 'bièvre, espèce de castor', Ir. *abac* beaver, dwarf. Cf. L. *amnis* (f. **ab-nis*) stream, river, of Italo-Celtic origin.]
It seems likely that the original sense in Welsh and Irish was 'river animal' which by extension came also to mean 'water sprite'. For the semantic development of Ir. *abac*, an aquatic creature, beaver, dwarf, cf. the history of LEPRECHAUN (O.Ir. *luchorpáin* small bodies, applied in the 8th c. to water-sprites).
An aquatic monster in Celtic mythology.
[**1781** T. PENNANT *Tour in Wales* II. 134 A deep, wide, and still water, called *Llyn yr Afangc*, or *The Beavers Pool*.] **1834** G. ROBERTS tr. *Theophilus Evans's View Primit. Ages* (c 1864) I. v. 129 As to the *Afanc*, it is the common opinion that he was a kind of large water-dog, with a broad tail called a Beaver. **1838** C. GUEST tr. *Mabinogion* I. 341 The Addanc of the Lake slays them once every day. *Ibid.* 381 In the Triads, mention is made of the Addanc, or Avanc of the Lake, as an aquatic monster. **1859** *Cambrian Jrnl.* VI. 146 It was a woman who enticed the avangc out of..the lake. **1923** OGDEN & RICHARDS *Meaning of Meaning* iii. 149 Before the appearance of an image, say, of an afanc, something can be observed to occur which is often misleadingly described as 'an intention of imagining' an afanc. **1960** *Listener* 29 Sept. 512/2, I hope no one revives the *afanc* which used to emerge from a lake near my house in Cardiganshire to devour cattle.

afara (ăfā·rā). [Yoruba.] A tall West African tree (*Terminalia superba* or *T. scutifera*) yielding a straight-grained wood; = *LIMBA. Also, its timber.
1920 A. H. UNWIN *West Afr. Forests* ix. 371 *Terminalia scutifera*. Shingle Wood. Afara (Yoruba). It attains a height of about 200 feet and a girth of about 16 feet. **1934** A. L. HOWARD *Timbers of World* (ed. 2) 9 *Afara*, reported by the Imperial Institute as a wood very similar to oak in colour, and possessing a straight grain. **1955** *Nomencl. Commerc. Timb.* (*B.S.I.*) 87 The plain light-coloured wood is called light afara, light limba, limba clair or limba blanc; the figured heartwood dark afara, dark limba, limba noir or limba bariolé...White afara is a name used in Nigeria for the tree as distinct from the timber.

afebrile (ē¹fī·brǝil, æ-, -fe-), *a.* [f. A- 14 + FEBRILE *a.*] Unaccompanied by fever; not feverish.
1875 F. DELAFIELD tr. *von Ziemssen's Cycl. Med.* I. 124 The cases of febrile and afebrile abdominal catarrh. **1901** *Practitioner* Mar. 303 The afebrile cases of lobar pneumonia. **1952** M. E. FLOREY *Clin. Application Antibiotics* I. i. 14, 4 patients treated with penicillin for 7½ days became afebrile in 72 hours.

affair. Add: **3.** (Further examples.)
1837 DICKENS *Pickw.* ii. 20 'I want your assistance..in an affair of honour,' said Mr. Winkle. **1922** JOYCE *Ulysses* 230 The annual dinner you know. Boiled shirt affair. **1933** N. COWARD *Design for Living* I. 19 We could carry on a backstairs affair for weeks without saying a word about it.
4. Examples of use without a preceding adj.
1866 WENHAM in *Ann. Rep. Aëronaut. Soc.* 37 The affair [a glider] falling over sideways, broke up the right-hand set of webs. **1903** H. G. WELLS *12 Stories* 11 He did not make the affair large enough to carry a man...The first flight of this first practicable flying machine took place over some fields near Burford Bridge. **1905** *Smart Set* Sept. 127/1 There were holes for the eyes and strings that tied the affair and held it in place.

‖ **affaire** (afɛ̄·r). [Fr.] = AFFAIR 3; *esp.* (often in form *affaire de* (or *du*) *cœur*, 'an affair of the heart', a love affair, an amour, an amatory episode.
1809 *Q. Rev.* II. 349 The connection with Egeria resolves itself, of course into an *affaire du cœur*. **1819** T. HOPE *Anastasius* I. viii. 174 The only thing he could have liked was an *affaire de cœur* with the favourite Sultana. **1845** GEO. ELIOT *Let.* 6 Apr. (1954) I. 185 My unfortunate 'affaire' did not become one till.. **1908** 'IAN HAY' *Right Stuff* ii. 28 In either case the affaire terminates then and there. **1928** A. WAUGH *For many Waters* v. 208 That life of parties and affaires that had seemed so infinitely desirable in the early twenties. **1940** M. DICKENS *Mariana* iv. 93 The affaire Grady was not something to add a spice to school life, and did not affect in any way her love for him. **1956** A. WILSON *Anglo-Saxon Att.* ii. i. 260 But you and she had an *affaire de cœur*! **1958** — *Middle Age of Mrs. Eliot* ii. 271 It would be so awful to have one of those office affaires that they have in the women's mags.

‖ **affairé** (afɛ̄re). [Fr.] Busy, involved.
1901 'LINESMAN' *Words by Eyewitness* xiii. 266 The great Public, *affairé* and astute as it is. **1927** *Blackw. Mag.* Nov. 600/1 The attractive *affairé* young lady..appropriated him for her own mysterious ends. **1928** D. H. LAWRENCE *Let.* 7 Jan. (1962) II. 1032 Well, don't be too *affairé*. **1954** E. JENKINS *Tortoise & Hare* xvi. 214, I hear Hunter is very much *affairé* with his little typist.

affect, *sb.* Restrict †*Obs.* to senses in Dict. and add: **1. c.** *Psychol.* [G. *affekt*.] (See quots.)
1891 J. M. BALDWIN *Handbk. Psychol.* II. 314 Affects..are the feelings antecedents of involuntary movements; as motives, including affects, are the inner antecedents of acts of will. **1894** W. JAMES *Coll. Ess. & Rev.* (1920) 358 We may also feel a general seizure of excitement, which Wundt, Lehmann, and other German writers call an *Affect*, and which is what I have all along meant by an emotion. **1923** *Wkly. Westm. Gaz.* 24 Mar. 18/1 Their psychic lives are overfull of complexes, levels and affects. **1926** W. MCDOUGALL *Outl. Abnormal Psychol.* 26 The terms 'affect' and 'affective' denote the emotional-conative aspect of all mental activity.
attrib. and *Comb.*
1934 H. C. WARREN *Dict. Psychol.* 7/1 *Affect psychoses*, psychoses which are especially characterized by disturbances in the emotional life. **1943** *Horizon* VIII. 271 The personality, rich in affect-life. **1944** *Mind* LIII. 180 The dream-process is an affect-regulative mechanism. **1949** A. KOESTLER *Insight & Outlook* v. 68 The affect-amplifying emergency-mechanisms of the sympathico-adrenal system became gradually superfluous. **1951** C. KLUCKHOHN in Parsons & Shils *Towards Gen. Theory Action* iv. ii. 390 Affect-laden customs or traditions. **1958** *Listener* 17 July 95/1 These areas of the brain..have come to be known as the 'pleasure' centres and 'pain' centres. Perhaps it would be best to call them the 'affect' systems, one which seems to be particularly related to positive affect (or pleasantness) and the other to negative affect (or unpleasantness).

affectability (ăfe:ktăbi·lĭti). [f. next + -ILITY.] = AFFECTIBILITY.
1894 *Q. Rev.* Oct. 395 The coloured lens of 'affectability' through which they look at most things, makes them strikingly unfit to discourse in the abstract of justice. **1908** D. F. HARRIS *Functional Inertia of Living Matter* i. 1 Affectability can, then, be defined as the power, tendency, capacity, or disposition to be affected by a stimulus, i.e., to *exhibit response*. **1919** W. MCDOUGALL *Introd. Soc. Psychol.* (ed. 14) 449 The most fickle and shallow temper results from the opposite conjunction, namely, high affectability with low intensity and persistence. **1961** *Brit. Med. Dict.* 51/2 *Affectability*, the quality or state of being responsive to a stimulus.

affectable (ăfe·ktăb'l), *a.* = AFFECTIBLE *a.*
1764 J. ERSKINE *Law of Scotl.* (ed. 3) 57 Neither is the right assignable by the minister, or affectable with his debts. **1766** KAMES *Decis. Crt. Sess. 1752–68* 279 The proper effects of the office-bearer will not be affectable by such a diligence. **1920** E. WALKER in F. S. Marvin *Rec. Dev. Europ. Thought* xi. 275 Neither honour nor artistic personality is affectable by external considerations.

affection, *sb.* Add: **1. b.** *Celtic Philol.* Mutation or umlaut of a vowel under the influence of a following sound.
1911 J. MORRIS JONES in *Encycl. Brit.* XXVIII. 269/1 Short vowels have been affected by vowels in succeeding syllables. These 'affections' of vowels are as follows:—

(*a*) I-affection, caused by *i* in a lost termination.. (*β*) A-affection.. (*γ*) Penultimate affection. **1913** —— *Welsh Gram.* 89 Affection is of two kinds in Welsh: 1. ultimate, when it takes place in the syllable which is now the last, having been brought about by a sound in a lost termination; 2. non-ultimate, when the affecting sound being generally preserved in the ultima. **1937** LEWIS & PEDERSEN *Conc. Comp. Celt. Gram.* 3 Changes due to vowel affection' in Ir. (> *i*),.. in W. (> *y, ei*),..in Co. *y* is found written *e* where there is no affection. **1953** K. JACKSON *Lang. & Hist. Early Brit.* 578 It would be better..to date *ā*-affection in the first half or middle of the fifth century.
12. (Earlier example.)
1588 W. KEMPE *Educ. Children* sig. G 3ʳ, The Rhetoricall pronounciation and gesture fit for every word, sentence, and affection.

affectionless, *a.* (Later (non-*poet.*) example.)
1959 B. WOOTTON *Social Sci. & Social Path.* iii. 81 Bowlby's celebrated study of affectionless thieves.

affective, *a.* Add: **7. b.** *Psychol.* Of, pertaining to, or characterized by feelings or affects (see *AFFECT *sb.* 1 c).
1891 J. M. BALDWIN *Handbk. Psychol.* II. xiii. 313 Affective Nature of All Stimuli to Movement... Stimuli..are all phenomena of feeling. **1897** tr. *T. Ribot's Psychol. of Emotions* 1 In all affective manifestations there are two elements: the motor states or impulses, which are primary; the agreeable or painful states, which are secondary. *Ibid.* xi. 153 Others recall the circumstances *plus* the revived condition of feeling. It is these who have the true 'affective memory'. **1912** A. A. BRILL tr. *Freud's Sel. Papers on Hysteria* (ed. 2) i. 7 Some important memories..on their return acted with the full affective force of new experiences. **1922** *Brit. Jrnl. Psychol.* (*Gen. Sect.*) Oct. 121 Love and hate ..are built of emotional stuff—they are *affective* phenomena. **1926** [see *AFFECT *sb.* 1 c]. **1950** D. RIESMAN in *Psychiatry* XIII. 1/2 His [Freud's] own deep affective involvement in an idea.
Comb. **1895** *Amer. Jrnl. Psychol.* VII. 81 *Gemüthsvorgang*, affective or affective-conative process. **1921** D. H. LAWRENCE *Psychoanalysis* i. 22 The great affective-passional functions and emotions. **1925** I. A. RICHARDS *Princ. Lit. Crit.* xi. 91 The affective-volitional aspect of mental activity. **1947** M. M. LEWIS *Lang. in Soc.* i. 20 British psychologists.. have suggested the term 'orectic' as an equivalent of 'affective-conative'.
c. *affective fallacy* (see quots.).
1948 W. K. WIMSATT & M. C. BEARDSLEY in *Poetry* Dec. 155 Affective fallacy...a confusion between the poem and its *results* (what it *is* and what it *does*)... The affective fallacy is coupled with the intentional fallacy.., the former being a confusion between the poem and its results, the latter a confusion between the poem and its origins. Examples of the affective fallacy range from Plato's feeding and watering of the passions, Aristotle's counter-theory of catharsis, and the Longinian 'transport' of the audience. **1959** *Times Lit. Suppl.* 20 Feb. 97/1 The Affective Fallacy, for Mr. Wimsatt..is the fallacy of the *frisson*, of the excited response to the isolated single line; or, more broadly, of the admirer of Dylan Thomas, say, who says: 'I don't understand a word of it, but how wonderful!'

affectivity (æfekti·viti). *Psychol.* [f. *AFFECTIV(E *a.* 7b + -ITY.] Emotional susceptibility.
1907 *Brain* XXX. 160 The extent of this expectation curve rises in normal individuals, depending upon their varying degree of affectivity. **1917** C. R. PAYNE tr. *Pfister's Psychoanalytic Method* xi. 303 In it [i.e. autistic thinking] the affectivity predominates. **1924** A. A. BRILL tr. *Bleuler's Textbk. Psychiatry* i. 32 Under the term affectivity we comprise the affects, the emotions, and the feelings of pleasure and displeasure. **1952** V. GOLLANCZ *My dear Timothy* xx. 360 Nursing is, for mother and child, one long delightful and highly charged game, in which the easy warm affectivity of a lifetime is set up.

affenpinscher (æ·fĕnpi:nʃǝɹ). [G., f. *affe* monkey, APE + *pinscher* terrier.] A small breed of dog, related to the Brussels griffon.
1903 W. D. DRURY *Brit. Dogs* (ed. 3) lxii. 629 The Affenpinscher is an alert, intelligent little dog of some 7 lb. to 8 lb. in weight. It has a round skull well covered with stiff hair... The colour is different shades of red, as well as grey and yellowish: while there is often a black mask. **1922** J. MAXTEE *Popular Toy-Dogs* II. 47 In Germany there is the Affenpinscher, which unquestionably is one of the chief constituents that were used in producing the Griffon Bruxellois.

affettuoso. Add: (Earlier and later examples.) For 'slow tune' read 'slow time'.
1724 *Short Explic. For. Words in Mus. Bks.* 7 *Affetto*, or *con affetto*, or *affettuoso*, by which Words is signified, that the Musick must be performed in a very moving, tender, or affecting Manner, and therefore not too fast, but rather slow. **1766** C. ANSTEY *New Bath Guide* (ed. 2) II. ix. 70 *Song*. 'Tis this that makes my Chloe's lips Ambrosial sweets distil; *Affettuoso*. **1813** 'H. HORNEM' [= BYRON] *Waltz* 4 A..tune, that reminded me of the 'Black Joke', only more 'affettuoso'. **1962** *Times* 2 Feb. 13/3 Ornate and affettuoso composer.

‖ **affiche** (æfi·ʃ, Fr. afiʃ). [F., f. *afficher*:— L. type **affīgicāre* (see AFFICHE *v.*).] A paper containing a notice to be affixed to a wall, etc.; a placard, poster.
[**1757** M. POSTLETHWAYT tr. *J. & P. L. Savary's Univ. Dict. Trade* (ed. 2) I. 23/2 *Affiche*, so the French call those bills, or advertisements, which are pasted up in public places.] **1818** MOORE *Fudge Fam. Paris* viii. 87 Then we stare into shops—read the evening's *affiches*. **1819** H. BUSK *Vestriad* v. 241 Soon those who spell the grand affiche peruse. **1833** T. HAMILTON *Men & Mann. Amer.* (1843) I. 11, When the sphere of my intelligence became enlarged with regard to this *affiche* [*sc.* placard on a wall]. **1884** J. SHARMAN *Cursory Hist. Swearing* i. 6 The usual notice-board..covered with a trellis-work of crimson tape for the purpose of retaining the various *affiches*. **1920** *Punch* 31 Mar. 249/1 A couple of lurid *affiches* which declared that 'Exhampton Is So Exhilarating'.

affiliate. B. *sb.* Add to def.: as an affiliated organization, company, etc.
1930 *Economist* 4 Oct. 608/1 The establishment of German bank 'affiliates' has also helped considerably the development of the Amsterdam bill market. **1931** G. T. CARTINHOUR *Branch, Group & Chain Banking* v. 74 Control of a number of banks may be exercised directly or indirectly by a particular bank through a security affiliate as well as a holding company. **1953** *Economist* 18 July 189/1 Two affiliates of the internationally owned Iraq Petroleum Company.

affiliation. Add: **2. b.** An affiliated part of an organization.
1818 *Ann. Reg. 1817* 22 Busily..at work, establishing branches and affiliations. **1922** *Daily Mail* 14 Nov. 4 (Advt.), The London Joint City and Midland Bank or its affiliations, the Belfast Banking Company, Ireland, and the Clydesdale Bank, Scotland.
c. Association, connection, esp. in politics. *U.S.*
1852 *Congress. Globe* 15 Mar., App. 323/3 Certain merchants with whom he has affiliations in New Mexico. **1862** *Ibid.* Jan. 589/2, I am here almost without any affiliation in political sentiment. **1893** *Congress. Rec.* Feb. 2301/1 The black man..is being educated, and can see where his political affiliation can best be allied. **1904** ROOSEVELT in *N.Y. Times* 23 Mar. 2, I have not the slightest idea what your political affiliations are.

affine, *sb.* and *a.* Delete † *Obs.* and add:
A. *sb.* (Later examples.)
1893 *Spectator* 6 May 592/1 Because they [a son and his father's goddaughter] are in some sense close spiritual affines. **1950** M. WILSON in Radcliffe-Brown & Forde *African Syst. Kinship* 124 The crux of Nyakyusa ideas of marriage: relations between *affines* (abako) are ideally permanent—a divorce should never occur; a dead husband should be replaced by his heir, a dead wife by her younger sister or brother's daughter.
B. *adj.* (Earlier and later examples.)
1650 W. CHARLETON tr. *J. B. van Helmont's Ternary of Paradoxes* sig. f 4, Whatever soundeth but analogous or affine, that doth Reason positively judge, consonant and homogeneous to verity. **1883** *Academy* 13 Oct. 240/1 [The statement is] free from that acrimonious spirit in which writers of a creed more affine to that of the Church of England frequently indulge when criticising her traditions. **1927** C. C. MARTINDALE *Relig. of World* 67 Man with one part of himself was affine to the rest of creation, and with another, was affine to God.
2. *Math.* Preserving finiteness (see quots.).
[**1748** L. EULER *Introd. in Analysin Infinitorum* II. xviii. 239 Quia Curvæ hoc modo ortæ inter se quandam Affinitatem tenent, has Curvas *affines* vocabimus.] **1918** VEBLEN & YOUNG *Projective Geom.* II. iii. 72 Any projective collineation transforming a Euclidean plane into itself is said to be affine; the group of all such collineations is called the affine group, and the corresponding geometry the affine geometry. **1923** A. S. EDDINGTON *Math. Theory of Relativity* vii. 214 If a displacement AB is equivalent to CD, then AC is equivalent to BD. This is the necessary condition for what is called affine geometry. **1923** P. FIELD *Projective Geom.* 21 An Affine Transformation..is a perspective having the centre of perspective at infinity.

affirmance. 1. and **2.** (Later U.S. examples.)
1885 *Century Mag.* XXIX. 730 The affirmance of the judgment of the court below by the General Term. **1888** BRYCE *Amer. Commw.* I. 505 A majority of the Supreme court seems to have placed upon this ground..its affirmance of that competence of Congress to declare paper money a legal tender for debts.

affirmative (ăfǝ̄·ɹmătiv), *v.* [f. the sb.] *trans.* To agree to (a suggestion or remark).
1861 G. MEREDITH *E. Harrington* III. xii. 203 Andrew again affirmatived his senior's remarks. **1871** —— *H. Richmond* II. xvii. 234 He..affirmatived her motion to ring the bell for the servants.

affix, *sb.* Add: **4.** Of a dog's name (see quots.).
1893 *Kennel Club Rules* 2 A name which has been duly registered..cannot be again accepted for registration of a dog of the same breed, without the addition of a distinguishing number, prefix or affix. **1954** C. L. B. HUBBARD *Compl. Dog Breeders' Man.* xii. 123 Affixes consist of names attached to the names of dogs, either in front of the dogs' names (prefix) or behind (suffix) in order to identify the dogs with a particular breeder or kennel.

affixal (æ·fiksăl), *a.* *Gram.* [f. AFFIX *sb.* + -AL.] Pertaining to an affix.
1873 [see IMPERATIVAL *a.*]. **1953** C. E. BAZELL *Linguistic Form* 18 Basic, affixal and inflectional morphemes.

affli·ctedly, *adv.* In an afflicted manner, distressfully.
1888 DOUGHTY *Trav. Arabia Deserta* II. 169 The stranger answered him afflictedly, 'Eigh me.'

affluent, *a.* and *sb.* **A.** *adj.* **4.** Freq. in phr. *affluent society*.
1958 J. K. GALBRAITH (*title*) The affluent society. **1958** *Listener* 25 Sept. 449/2 Many economists have lifted their eyes..to the coming age of abundance; even more sociologists have inveighed vaguely against the affluent society. **1959** *Observer* 1 Nov. 1/5 Mr. James Griffiths..attributed Labour's election defeat to its being the party of change at a time when Britain had entered the 'affluent society'.

afforestable (æfǫ·rēstăb'l), *a.* [f. AFFOREST *v.* + -ABLE.] Capable of being afforested.
1928 *Britain's Industr. Future* (*Liberal Ind. Inquiry*) Index 489 Estimate of afforestable land. **1958** *Times* 24 Jan. 15/3 The planting of the total afforestable area with pines is nearing completion.

afformative (ăfǭ·ɹmătiv), *a.* (*sb.*) *Philol.* [f. AF- + FORMATIVE *a.*] Affixed as a formative element; said of a letter, syllable, etc. As *sb.*, an afformative particle; a suffix (esp. in Semitic languages). Cf. PREFORMATIVE.
1821 M. STUART *Heb. Gram.* 139 The Imperative has no Praeformatives like the Future, but only *Afformatives*, or Suffix-formatives. **1858** *Trans. Philol. Soc.* I. xii. 179 (*title*) On constant affixes and afformatives in the Hebrew language. **1864** E. B. PUSEY *Daniel* i. 47 The inflection of a passive participle by the afformatives of the preterite. **1880** B. DAVIES tr. *F. W. W. Gesenius's Heb. Gram.* 114 The characteristic *Pa'thăch* of the second syllable becomes *Sh'wâ* before an afformative beginning with a vowel. **1942** *Antiquity* XVI. 325 The afformative of the perfect tense. **1952** *Archivum Linguisticum* IV. 61 Substantives involving both preformative *ma-* and afformative *-ān* > *-ōn* are not unknown in North-west Semitic.

affranchisement. (Examples.)
1799 *Ann. Reg. 1792* (Dodsley) Hist. Eur. 81/1 The decree for the affranchisement of the mulattoes. **1887** *Harper's Mag.* Mar. 581/1 Affranchisement, independence, freedom, secured for others, had anew their echo in the hearts of the Russian people. **1931** *Times Lit. Suppl.* 8 Jan. 24/4 Mr. Woodward's reading of modern history as a gradual 'affranchisement'.

affricate (æ·frikĕt), *sb.* *Phonetics.* [ad. L. *affricātus*, pa. pple. of *affricāre*, f. *ad* to + *fricāre* to rub.] A close combination of an explosive consonant or 'stop' with an immediately following fricative or spirant of corresponding position, as in Ger. *pf, z* (= ts). Also called **affri·cative.**
1880 SAYCE *Introd. Sci. Lang.* I. 270 Where a spirant or fricative is immediately preceded by an explosive, a double sound or affricative is the result (*e.g.* German *pf*, Armenian *t'f*). **1891** *New Eng. Dict.* s.v. CH (consonantal digraph), The combination CH..was introduced [into Roman spelling] to represent the Greek aspirate or affricate X. **1895** P. GILES *Man. Compar. Philol.* 70 Another series of sounds which must be..distinguished from spirants and aspirates is the affricates. **1905** L. A. MAGNUS *Respublica* 74 Here we might safely say final *c* is preserved as a sibilant or affricate. **1950** D. JONES *Phoneme* p. xii, The affricate tʃ (Eng. *ch*).

affricate (æ·frikē¹t), *v.* Restrict † *Obs.*⁻⁰ to sense in Dict. and add: **2.** To convert into an affricate (see prec.). Also *intr.* So **a·ffricated** *ppl. a.*
1891 *New Eng. Dict.* s.v. CH, *Ch*..was introduced initially, in Upper German, for the affricated sound of *c* (*k*) as *chamara* (kχa·mara). **1902** E. W. SCRIPTURE *Elem. Exper. Phonetics* xxi. 307 The 'constant diphthongs' ts tš.. are affricates, that is, occlusives with fricative releases... The former seems to be an affricated t, the latter an affricated r. **1946** PRIEBSCH & COLLINSON *Germ. Lang.* (ed. 2) II. i. 118 He suspects that Ostrogoths affricated medial *t* between A.D. 553 and 580. **1964** *Language* XL. 26 The sibilant affricated in close juncture with /l/ but did not voice.

affrication. Restrict † *Obs. rare*⁻¹ to sense in Dict. and add: **2.** The conversion (of a sound) into an affricate.
1934 *Trans. Philol. Soc.* 87 A pure tenuis passes necessarily through the stages of aspiration and affrication. **1934** PRIEBSCH & COLLINSON *Germ. Lang.* I. ii. 32 So-called assibilation or palatal affrication of *k* and *g* before front vowels. **1953** *Archivum Linguisticum* V. 69 The choice of the same diacritic mark to indicate affrication and velarization is to invite ambiguity.

Afghan (æ·fgæn). Also 9 **Afghaun.** [Name of the people of Afghanistan, a country lying north of West Pakistan.] **1.** A blanket or wrap of knitted or crocheted wool. Also *attrib.*
1833 CARLYLE *Sartor Res.* I. v, in *Fraser's Mag.* VIII. 670/1 Afghaun shawls. **1868** ELIZ. PRENTISS *Let.* 13 Dec. in G. L. Prentiss *Life & Lett. E. P.* (1882) 260 Mr. P. has come up-stairs rolled up in your afghan. **1887** STOCKTON *Hundredth Man* xxxii. 375 Miss Burns was crocheting an afghan. **1947** J. STEINBECK *Wayward Bus* xiv. 195 The bedspread, now, a giant afghan she had knitted herself in little squares.
2. A coarse-woven rug or carpet. Also *attrib.*
1877 F. H. BURNETT *Theo* iv. 89 It was an Afghan Miss Elizabeth was making now. **1901** J. K. MUMFORD *Oriental Rugs* xii. 236 Afghan carpets. **1920** T. E. LAWRENCE *Let.* 16 Feb. (1938) 299 As for the rugs, please take any that seem worthy to you. There were two Afghans in the Arab Bureau. **1960** *News Chron.* 12 Sept. 6/4 Good quality Afghans, say 12 ft. × 9 ft., can cost anything from £110 to £250.
3. In full *Afghan* (*grey*)*hound*. A swift hunting dog of the Near East resembling a shaggy greyhound. Cf. *BARUKHZY.
1895 [see *BARUKHZY]. **1905** H. DE BYLANDT *Dogs of All Nations* I. 767 (*caption*) Ideal Afghan Greyhound. **1911** *Greyhound Stud Book* 228 The President read a letter.. admitting to the 'Appendix' of the Stud Book an imported Afghan greyhound. **1925** E. M. AITKEN *Pets* i. 28 For the purpose of classifying dogs on its registers the Kennel Club has two main divisions, sporting breeds and non-sporting breeds. These again are subdivided, the miscellaneous sporting kinds being Afghan Hounds, Basset Hounds, Beagles [etc.]. **1950** *Oxf. Jun. Encycl.* IX. 191 Carvings on the walls of the Balkh caves of Afghanistan of about 2200 B.C. show that already an Afghan Hound, almost the same as the modern Afghan, was established in that country.

afghani (æfgā·ni). [Pashto.] The principal monetary unit of Afghanistan, divided into 100 puls.
1927 *Statesman's Year-Bk.* 643 The following new currency was introduced in March 1926: Gold coins: the *amani*, equivalent to Rs. 20 Afghani, and the half-*amani* and the 20 *pooli*: silver coins, the Afghani rupee..and the half-Afghani. **1933** M. ALI *Progressive Afghanistan* xv. 186 Students, in addition to their monthly stipends of 50 Afghanis each, get free meals. **1934** AHMAD & AZIZ *Afghanistan* xiv. 125 The unit is an *Afghani* weighing ten grammes of silver, 900 fine, which is sub-divided into 100 *puls*. **1963** *Whitaker's Almanack* 821/1 The free rates..in June, 1960, were approximately..106 Afghanis = £1.

afibrinogenæmia (ē¹fai:brino͵dʒenī·miă, æ-). *Path.* [mod.L., f. A- 14 + FIBRINOGEN + Gr. αἷμα blood: see -IA.] A condition resulting from the lack of or diminution in the amount of fibrinogen in the blood.
1943 *Biochem. Jrnl.* XXXVII. p. xvi, Studies on the Plasma Fibrinogen and other Protein Fractions in a Case of Afibrinogenaemia.

‖ **aficionado** (afiþiona·ðo; anglicized as ăfisi̇̄ŏnā·dō). Pl. **-os.** [Sp. = amateur, f. pa. pple. of *aficionar* to become fond of, f. *afición* AFFECTION.] A devotee of bull-fighting; by extension an ardent follower of any hobby or activity.
1845 R. FORD *Handbk. Travellers in Spain* I. ii. 178 This sham fight is despised by the *torero* and *aficionado*, who aspire only to be in at the death. **1902** W. D. HOWELLS *Lit. & Life* iii. 58 The last [bull] was uncommonly fierce, and when his hindquarters came off or out, his forequarters charged joyously among the aficionados on the prisoners' side. **1957** *Times* 12 Oct. 7/6 The bull-fight is the most Spanish of spectacles... Some old *aficionados* will go so far as to say that it is dying.
1882 C. G. LELAND *Gypsies* 25 The *aficionados*, or Romany ryes, by whom I mean those scholars who are fond of studying life and language from the people themselves. **1928** F. O. LINDLEY *Diplomat off Duty* iv. 64 All amateurs, or to use a much more expressive Spanish word, *aficionados*, of bathing agree that the full flavour of the pastime is only tasted in beautiful surroundings. **1948** J. STEINBECK *Russian Jrnl.* (1949) iii. 37 A little swing band was led by Ed Gilmore, who is a swing *aficionado*. **1957** *Technology* Apr. 70/3 The *aficionados* of science fiction and golf. **1959** J. WAIN *Trav. Woman* 41, I didn't tell you I had a son who was an *aficionado* of railways, did I?

‖ **afikoman, aphi-** (æfikōᵘ·măn). Also **afikomen, afikuman,** etc. [Heb. *aphīqōmān*, f. Gr. ἐπικώμιον revel (G. Dalman).] Near the beginning of the Jewish Passover service, a piece broken from the second of the three cakes of unleavened bread and put aside to be eaten at the end of the meal.
1891 M. FRIEDLÄNDER *Jewish Rel.* II. 382 The part laid aside is called *afikuman*. **1892** I. ZANGWILL *Childr. Ghetto* II. I. xxv. 208 The *Afikuman* [ed. 1901, *Afikoman*], or *Motso* specially laid aside for the final morsel. **1905** *Westm. Gaz.* 20 Apr. 9/3 The Afikomen is always the centre of a little ceremony associated with..Jewish holidays, the giving of gifts. **1960** *Jewish Chron.* 8 Apr. 33/2 Grandpa split the middle matzo and hid half of it away as the *Afikoman*.

aflare (ăflē³·ɹ), *adv.* and *pred. a.* [f. A *prep.*¹ + FLARE.] Flaring; spread out, expanded; also, blazing, glowing. esp. *fig.*
1908 T. HARDY *Dynasts* III. I. vii. 40 Large pieces of canvas aflare sail away on the gale like balloons. **1924** R. CAMPBELL *Flaming Terrapin* iii. 46 Beauty ran all aflare Through nerve and bone. *Ibid.* 52 Wolfing huge coals with iron jaws aflare. **1959** *Times* 24 Mar. 15/3 When her interest was engaged the broad, strong features would be aflare with excitement.

aflash (ăflæ·ʃ), *adv.* and *pred. a.* [f. A *prep.*¹ + FLASH.] Flashing; in a flash.
1876 G. M. HOPKINS *Poems* (1918) 85 And crush-silk poppies aflash. **1885** W. B. YEATS in *Dublin Univ. Rev.* July 136/2 Your eyes are all a-flash.

‖ **à fond** (afoń), *adv.* [Fr., lit. 'to the bottom'.] Thoroughly, fully.
1813 WELLINGTON *Disp.* 9 May (1838) X. 366 That which ought to be done is to examine a subject of this kind *à fond*. **1928** A. CHRISTIE *Myst. Blue Train* xv. 124 The Comte de la Roche knows one subject *à fond*: Women.

a fortiori. (Earlier and later examples.)
1606 R. PARSONS *Answer to Coke* iii. 52 Yet cannot bee either *tacitè* or *à fortiore*..take vnto him, all the power.. which the said President and fellowes haue. **1712** SHAFTESBURY *Plastics* in *Second Characters* (1914) 158 From hence argument *a fortiori*: How indecent is obscenity? **1827** *Blackw. Mag.* Feb. 207/2 One might, *a fortiori*, count on his being murdered. **1961** *New Scientist* 16 Mar. 688/1 Anyone who, being a mathematician or a scientist, loses his sight after leaving school—or, *a fortiori*, in mid-career—cannot easily escape feelings of deep dismay.

afoul. For '*adv.* prop. *phr.*' read '*adv.* and *pred. a.* Orig. *Naut.*'
Add to def.: in collision; fouled. Const. *of.* Phr. *to fall, run afoul of*: to fall, run foul of (see FALL *v.* 86 and RUN *v.* 10 c). Now chiefly *U.S.*
1824 *Blackw. Mag.* Oct. 416/1 We see no reason for encouraging anybody in running afoul of other people's countries. **1840** DANA *Bef. Mast* (1854) xv. 76 After paying out chain, we swung clear, but our anchors were no doubt afoul of hers. **1841** TOTTEN *Naval Text-bk.* 328 A vessel ran a-foul of us. **1893** *Scribner's Mag.* June 793/2 He sometimes falls afoul of the night lines baited for eels. **1937** F. SCOTT FITZGERALD *Let.* July (1964) 16, I ran afoul of a bastard named de Sano. **1958** C. FERGUSON *Naked to Enemies* I. iii. 29 He stood pledged to aid them if they..fell afoul of the law on any personal matter. **1961** B. VAWTER *Consc. of Israel* ix. 244 The modern neutralist or other espouser of unpopular causes..can hardly fail to run afoul of the jingoistic society in which he lives.

afraid, *ppl. a.* Add: **1.** (Later examples.)
1864 BROWNING *Dramatis Personæ* 77, Trust God: see all, nor be afraid! **1872** J. DOOLITTLE *Chinese Vocab.* II. 684/2 Those that came were not afraid: those that were afraid did not come. **1915** T. S. ELIOT in *Poetry* VI. 132 And I have seen the eternal Footman hold my coat, and snicker, And in short, I was afraid.
2. c. Add to def.: *I am* (or *I'm*) *afraid*: often used *colloq.* with little or no implication of fear or danger, in the sense of 'I regret to say; I regretfully or apologetically admit, report, etc.; I suspect; I am inclined to think'. Const. *that*, or simple clause.
1596 SHAKES. *Tam. Shrew* v. ii. 89, I am affraid sir, doe what you can Yours will not be entreated. **a 1678** H. SCOUGAL *Importance & Difficulty of Ministerial Function* in *Wks.* (1765) 240 This..was the humour of some in his days; and I am afraid the case is not much better in ours. **1709** ADDISON *Let.* 14 Dec. (1941) 197, I am afraid if this matter comes on it will be necessary to have Copys of the Office Books. **1740** GRAY *Let.* 16 July (1900) I. 76 Disagreeable enough (as most necessities are) but, I am afraid, unavoidable. **1813** JANE AUSTEN *Pride & Prej.* I. x. 104, I am afraid you do not like your pen. Let me mend it for you. **1853** MRS. GASKELL *Cranford* iii. 47, I did many a thing she did not like, I'm afraid—and now she's gone! **1911** F. SWINNERTON *Casement* vi. 209 'I'm afraid,' said he, rather stiffly, 'that I don't know anything about his habits.' **1959** *Observer* 14 June 22/6 It would be less kind, but true, I am afraid, to find in this book a quite invincible taste for the mediocre.

Aframerican: see *AFRO-.

Afric (æ·frik), *a.* and *sb.* *arch.* or *poet.* Also **Af(f)rick,** 6 **Africke.** [ad. L. *Africus* African.]
A. *adj.* Of or pertaining to Africa; African.
B. *sb.* A native of Africa; an African Negro. *rare.*
1590 SPENSER *F.Q.* III. iii, Beyond the Africk Ismael. **1591** J. HARINGTON tr. *Ariosto's Orlando Fur.* I. i, The Moores transported all their might On Affrick seas the force of France to breake. **a 1592** GREENE *Friar Bacon* (1594) sig. F3ᵛ l. 1382 The Africke Dates *mirabiles* of Spaine. **1667** MILTON *P.L.* I. 585 Or whom Biserta sent from Afric shore. **1733** SWIFT *On Poetry* 12 So Geographers in Afric-Maps With Savage-Pictures fill their Gaps. **1876** *N. American Rev.* CXXIII. 446 Then will the Afric indeed have changed his skin and the leopard his spots. **1897** L. JOHNSON *In Ireland* 98 Make kind to him the Afric sun, The Afric stars and moon. **1956** J. MASEFIELD in *Times* 27 Jan. 9 So be the Africk visit of our Queen.

African (æ·frikăn), *sb.* and *a.* Also 6–7 **Af(f)ricane,** (6 **Aph-**). [OE. (only pl.) *Africanas*, ad. L. *Africānae*, f. *Africa*, sb. use of fem. (sc. *terra* land) of *Africus*, f. *Afri* (sg. *Afer*) ancient people of N. Africa; see -AN.]
A. *sb.* A native or inhabitant of Africa; an African Negro.
In local usage applied to the particular Negro (negroid) race or races of a territory in Africa, as (*S. Afr.*) the Bantu people of South Africa.
c 888 K. ÆLFRED *Boeth.* xvi. § 2 Þa he feaht wið Africanas. **c 1205** *Layamon's Brut* (1847) 25379 Mid him com moni Aufrican. *Ibid.* 27501 Of Ethipe wes þe an, þe oðer wes an Aufrican. **13..** *K. Alisaunder* (1952) 2021 And lete armen pine Affricanes. **1564** HARDING *Answ. Jewel* 61ᵛ, He being an Aphricane borne, and writing to Aphricanes. **1567** T. STAPLETON *Counterblast* Pref., I reporte me to the Africans, who falling from the vnitie of the Romaine See..became in time Infidelles. **1671** A. WOODHEAD *Consid. Council of Trent* ii. 20 Which was the chief matter stood upon by the Africans against Pope Bonifacius. **1687** W. D. tr. *B. Le Bovier de Fontanelle's Discourse Plurality of Worlds* iii. 50 All Faces in general are wrought according to one and the same model, but those of two great Nations, of the Europeans, for example, and Africans seem to be fram'd to two particular models. **1780** A. BUTLER *Lives Saints* IX. 172 The concurring suffrages of sixteen ancient and worthy bishops (two of whom were Africans). **1806** *Gleanings in Africa* xvi. 121 The case of the unhappy African is..entirely different. The unoffending negro, in the forests and morasses of Africa, never so much as meditated hostility against Europe. **1849** CARLYLE in *Fraser's Mag.* XL. 672/2 Our West-Indian policy..of keeping down the labour-market in those islands by importing new Africans. **1952** L. MARQUARD *Peoples & Pol. of S. Afr.* i. 1 Today..8,500,000 Africans form part of the population of South Africa.
b. *spec.* Applied to a white resident of Africa.
1815 A. PLUMPTRE tr. *Lichtenstein's Trav. S. Afr.* II. iv. xxxv. 96 Like all other settlers here, they [the French settlers] are become entirely Africans. **1923** O. SCHREINER *Thoughts on S. Afr.* ii. 77 The young English African who has never been in Europe may boast that South Africa is the finest country on earth. **1934** 'N. GILES' *Ridge of White Waters* I. xiii. 150 As an African, with Dutch blood in me, I am suspected of Republican sympathies. *Ibid.* xiv. 163 Why does the Government give the vote to Hollanders and leave good Africans out in the cold? **1953** P. H. ABRAHAMS *Return to Goli* v. 179, I am convinced that of all the whites in the plural societies of Africa only the descendants of the Trekkers have made that deeply subjective..transition in their relations with the African earth that has made pure Africans of them.
B. *adj.* **a.** Of or pertaining to Africa (the continent, or the ancient Roman province of that name); occurring in some specific names of plants, trees, etc. (see quots.). **b.** Belonging to or characteristic of the Negro peoples of Africa.
1564 HARDING *Answ. Jewel* f. 89ᵛ, This reconciliation.. of the Affricane churches to the catholike church. **1594** HOOKER *Eccl. Pol.* III. 128 The Africane bishopes in the Councell of Carthage. **1597, 1611,** etc. [see MARIGOLD 1 b]. **1607** TOPSELL *Four-f. Beasts* 99 The Affrican Camels, are much more woorth then the Asian. *Ibid.* 249 A medicine made of an African Sparrow. **1624** [SCOTT] *2nd Pt. Vox Populi* 23 That (insolent and african pride) of restrayning him from that liberall..converse..with the Lady Maria Infanta. **1646** 'THOS. CARRE' *Occas. Disc.* 147 The Epist. of the Africane Bishops to Pope Celestine. **1706** PHILLIPS (ed. Kersey), *Ulpicum*..African Garlick. **1782** LATHAM *Gen. Syn. Birds* I. II. 532 African C[uckow]. **1861** BENTLEY *Man. Bot.* 645 *Oldfieldia africana*, yields..African Oak or African Teak. *Ibid.* 677 *Sanseviera zeylanica*..[produces] African Hemp or Bowstring Hemp. **1869** ALLIES *Form. Christendom* II. 277 Tertullian adds the witness of the African church to that of the Asiatic and Gallic churches in Irenæus. **1913** C. PETTMAN *Africanderisms* 24 *African walnut* (*Schotia brachypetala, Sond*). A tree bearing handsome, bright scarlet flowers. The wood is said to be much like walnut, but closer in the grain, and takes a splendid polish. **1921** W. BULLOCK *Timbers for Woodwork* 24 *African Walnut*..is a production from the west coast of Africa, being obtained from the French Congo, the Cameroons and the Nigerian territories...The colour of this timber is the only feature in which it compares with walnut, being quite unlike in every other respect. **1941** F. L. MULFORD *House Plants* 25 The so-called African-violet (*Saintpaulia ionantha*) is a blue-flowered, constant-blooming plant under house culture. **1958** *Economist* 13 Dec. 24/1 The idea of the 'African personality', about which African leaders talk so grandiosely.

Africana (æfrikā·nă, -ē¹·nă). [f. *Africa* + ANA *suff.*] Books, documents, or the like, relating to objects peculiar to, or connected with, Africa, in particular Southern Africa, especially those of value or interest to collectors.
1908 (*title of catalogue, Davis & Sons, Durban*) Africana: a list of work [*sic*] dealing with South Africa. **1943** *Africana Notes & News* Dec. 4 Africana is a word of many different meanings; to one it means books and other printed or manuscript material, to another it means objects other than books. For *Africana Notes and News* it is used in its widest sense, covering prints, maps,..books, pamphlets,..furniture, weapons, pictures and by-gones of all kinds—it is restricted only from a geographical point of view; it denotes not the whole of Africa but only Southern Africa. **1949** L. G. GREEN *Land of Afternoon* vi. 78 Greatest of all Africana collectors was Sidney Mendelssohn. **1952** E. H. BURROWS *Overberg Outspan* v. 122 A rare item of *africana* of a dozen or so pages.

Africa·nder, Afrika·nder. Also **Africaander, Afrikaander.** [see *AFRIKANER.]
1. = *AFRIKANER 1. *obsolescent.*
1822 BURCHELL *Trav.* I. 21 All those who are born in the colony speak that language [*sc.* Dutch], and call themselves Africaanders, whether of Dutch, German, or French origin. **1834** *C. Gd. Hope Lit. Gaz.* IV. 103 (Pettman), The number of matches that have taken place between the fair Africanders. **1882** *De Patriot* (Cape Colony) in *Encycl. Brit.* (1902) XXVI. 568/2 The Afrikander Bond has for its object the establishment of a South African Nationality by spreading a true love for what is really our fatherland. **1884** *Q. Rev.* July 150 The Africanders would hoist their own flag. *Ibid.*, An Africander republic. **1899** W. J. KNOX LITTLE *Sk. & Stud. S. Afr.* I. iv. 92 The Dutch Afrikander is wanting in this quality. The English Afrikander too suffers. **1900** A. H. KEANE *Boer States* viii. 161 Any African-born white person, whether of Dutch, English, or German origin, is an Afrikander in the social, if not in the political, sense of the term. **1905** [see *Afrikanderize* v. below]. **1939** J. S. MARAIS *Cape Coloured* v. 163 The majority of the slaves were 'Afrikaanders', that is to say Colonial-born. **1957** R. CAMPBELL *Portugal* 104 The word Afrikander is often applied by Englishmen to the human inhabitants of South Africa, who are Afrikaners without the 'd'.
2. A South African breed of cattle or sheep. Also *attrib.*
1852 *D'Urban Observer* 9 Jan. 4/1 There are..three or four recognized breeds now common in the country, viz., the *Fatherland*, the *Africanda*, and the *Zulu* and *Bastard Zulu*. *Ibid.* 4/2 Zulu cows crossed by Fatherland and Africander bulls. **1868** J. CHAPMAN *Trav. Int. S. Afr.* I. viii. 174 Rather higher than an Africander ox, with immense horns. **1874** F. OATES *Jrnl.* 6 Sept. in C. G. Oates *Matabele Land* (1881) iii. 48 Lee has just sold twelve red oxen—Africanders, with white faces. **1955** J. H. WELLINGTON *S. Afr.* II. v. 69 Non-woolled sheep..such as the Black-head Persian, the Africander, [etc.].
3. *Bot.* = *AFRIKANER 3.
1870 *Cape Monthly Mag.* Oct. 225 Yellow sorrel, pink 'Africanders', and the most lovely Magenta coloured wild 'figs'. **1915** R. MARLOTH *Flora of S. Afr.* IV. 157 G. [= *Gladiolus*] *grandis*..Often called the 'large brown Afrikander'. **1959** *Cape Argus* 15 Aug. 8/8 The blue afrikander (*gladiolus recurvus*..) is scenting the sandveld.
4. Afrikaans. *rare.*

1886 G. A. Farini *Through Kalahari* xxvi. 434 A little nigger..said in Afrikander, 'That is Mr. Scott's..house.' 1902 G. M. G. Hunt (*title*) A Handy Vocabulary, English-Afrikander, Afrikander-English.

Hence **Africa·nderdom** = **Afrikanerdom*; **Africa·nderism** = **Afrikanerism*; **Africa·nderize** *v.* = **Afrikanerize* v.

1884 *Pall Mall Gaz.* 9 Oct. 2/2 Shall we throw in our lot with Afrikanderism, abjuring our nationality for evermore? 1891 *Sat. Rev.* 17 Jan. 59/2 The..apology for Afrikanderism which Sir Gordon Sprigg included in his speech on Imperial Federation. 1893 *Standard* 21 Apr. 6/2 The sympathy of Africanderdom. 1899 *Daily News* 16 Nov. 4/5 If South Africa was to be saved to the Empire, it would be by Afrikanderdom, and not by Downing-streetism. 1905 Milner *Let.* 14 Apr. in C. Headlam *Milner Papers* (1933) II. 552 A separate Afrikander nation and State, comprising, no doubt, men of other races, who are ready to be 'afrikanderized'. 1909 *State* Dec. 701 If an English boy learns Dutch he is apt to acquire what are popularly called Dutchisms or Africanderisms.

Africanism (æ·frikăniz'm). [f. *African + -ism.] An African mode of speech or idiom. Also, African qualities or characteristics in the aggregate.

1641 Milton *Reform.* i. 38 He that cannot understand the sober..stile of the Scriptures, will be ten times more puzzl'd with the knotty Africanisms..of the Fathers. 1836 *New Monthly Mag.* XLVII. 152, I have spent some days in a town where every thing is pure Africanism. 1851 Trench *Exp. Serm. on Mt.* (ed. 2) Introd. ii. 27 The harsh Africanisms of Tertullian and Arnobius. 1882 B. F. Westcott in Smith & Wace *Dict. Chr. Biogr.* (1887) IV. 139/2 The principles which he [*sc.* Origen] affirmed..are fitted to correct the Africanism which, since the time of Augustine, has dominated Western theology. 1884 G. W. Cable *Creoles of Louisiana* xxxiii. 260 He [*sc.* the rich Creole] dropped..the Africanisms of his black nurse.

b. The policy which advocates that the indigenous inhabitants should have political control in Africa; African nationalism.

1957 Hailey *African Survey* v. 251 It seems advisable on this occasion to give prominence to the use of the term 'Africanism' rather than 'nationalism'. 1959 *Cape Times* 14 Apr. 8/6 Africanism can be accepted as a solution in one sense, viz., that Africa in future must belong to the Africans, that is to those peoples who have chosen Africa as their home.

Africanist (æ·frikănist), *sb.* (and *a.*). [f. *African + -ist.] An expert or specialist in African affairs, culture, etc.

1895 *19th Cent.* XXXVIII. 455 (*heading*) Africanists in Council. 1926 E. W. Smith *Chr. Mission in Afr.* vii. 56 A representative gathering attended by the leading Africanists of many nationalities. 1932 W. L. Graff *Language* xi. 433 Some Africanists make of the latter a distinct West African group.

b. An adherent of Africanism (sense b); an African nationalist.

1958 *Cape Times* 22 Dec. 12/7 What is the origin of the Africanists, the extreme Black nationalist group which recently broke away from the African National Congress? 1960 *Times* 12 Feb. 13/5 The Africanist would..like to know to what extent rule of Europeans in the West Indies ever resulted in the creation of economic privilege of the kind built up by European minorities in Africa. 1963 Fenner Brockway *African Socialism* iii. 47 Touré is..the classical Africanist, unwilling to be the tool of any external Power or bloc of Powers, rejecting alike Western capitalism, European social democracy and Soviet communism.

Hence as *adj.* Also **Africani·stic** *a.*

1958 *Spectator* 13 June 760/2 The African National Congress and other Africanist forces. 1959 *Cape Times* 28 Apr. 10/8 A Pan-Africanism, which is in a narrow and violent sense Africanistic. 1960 *Daily Tel.* 22 Aug. 8/2 Maybe Mr. Krushchev will find himself in the awkward position of being more 'Africanist' and more racialist than some of the independent African leaders themselves.

Africanize (æ·frikănəiz), *v.* [f. *African + -ize.] To give an African character to; to make African; to subject to the influence or domination of African Negroes. Hence **A:fricaniza·tion**; **A·fricanized** *ppl. a.*; **A·fricanizing** *vbl. sb.*

1853 Jas. Buchanan *Let.* 12 Nov. in J. F. Rhodes *Hist. U.S.* (1893) II. 26 A violent..article in the Washington *Union* charging them with an intrigue with Spain to 'Africanize' Cuba. 1856 S. Cartwright in J. F. Claiborne *Life of Quitman* (1860) II. 230 The Clayton-Bulwer Treaty, the preposterous claims..and the Africanization of tropical America. 1865 *Cincinnati Commercial* 4 July 1/2 A Yankee voice with Africanized accent. 1884 *N. Amer. Rev.* Nov. 429 When the Africanizing and ruin of the South becomes a clearly seen danger. 1905 *Tablet* 21 Oct. 649/2 They have become thoroughly Africanised, speak only the Ethiopian language. 1954 (*title*) A Statement on the Programme of the Africanisation of the Public Service (Gold Coast Govt., Accra). 1958 *Economist* 13 Dec. (African Suppl.) 7/2 The fate of Africa in the next decade, therefore, depends upon economic advance catching up with political advance in the 'Africanised' north and west. 1960 *Listener* 29 Sept. 498/2 The 'Africanization' to which so many firms have had to bow, by promoting their messengers and office boys into managing directors and retaining their Europeans merely as 'advisers'.

Africanoid (æ·frikănoid), *a.* [f. *African + -oid.] Resembling the African types of mankind.

1899 Ripley *Races Eur.* 397 A long-headed member of the Africanoid races. 1921 *19th Cent.* May 884 The character-making quality did not come from Asianoid or Africanoid races, it was supplied by the Teuton.

Africanthropus (æ:frikănþrōᵘ·pŏs, -kæ·nþropŏs). *Anthropology.* [mod.L., f. *Africa* + Gr. ἄνθρωπος man.] The name given to a type of primitive hominid of the Pleistocene in Africa, known from remains found near Lake Njarasa (or Eyasi) in Tanzania. In full *Africanthropus njarasensis*.

[1939 H. Weinert in *Zeitschr. f. Morphologie & Anthropologie* XXXVIII. 252 Africanthropus njarasensis.] 1946 L. S. B. Leakey in *Jrnl. East Afr. Nat. Hist. Soc.* XIX. 43 The Eyasi Skull..has now been studied in detail by Dr. Weinert, who has created a new genus *Africanthropus* for it. 1948 A. L. Kroeber *Anthropol.* iii. 90 *Africanthropus njarasensis* is a discovery, made by Kohl-Larsen in 1935, whose definitive description is one of many delayed by World War II.

Afridi (ăfrī·di). Also **Afreedi.** [Pashto.] A Pathan people inhabiting the mountainous region between Pakistan and Afghanistan; a member of this people. Also *attrib.*

1815 M. Elphinstone *Kingdom of Caubul* iii. i. 356 The Khyberees consist of three independent tribes,..the Afreedees, Shainwaurees, and Oorookzyes. 1875 *Encycl. Brit.* I. 227/2 The Momands, Afridis, Vasiris, &c. 1908 Mrs. H. Ward *D. Mallory* i. ii. 27 In the English House of Commons, there were men..who held an Afridi life dearer than an English one. 1923 Kipling *Land & Sea Tales* 4 The little hillsman of the North-east Indian Frontier, Afreedi, Pathan, Biluch. 1932 —— *Limits & Renewals* 209 He was a hard-bitten Afridi from the Khaibar hills.

Afrikaans (æfrikā·ns), *sb.* and *a.* Also (rarely) **Africaans.** [= Du. *Afrikaansch* (now written *Afrikaans*): see *African and -ish.] **A.** *sb.* The modified form of Dutch spoken in South Africa. Formerly also called *Cape *Dutch*, *South African Dutch*, the Taal.

[1900 A. H. Keane *Boer States* p. xix, *Taal*, Cape Dutch, called by the Netherlanders *Afrikaansch*.] 1908 *East London Dispatch* 20 Oct. 4 (Pettman), I have always regarded (high) Dutch as my mother tongue and Africaans (low Dutch) as a hodge-pot sort of a language. 1921 *Glasgow Herald* 21 July 4 The dream of the young Dutch Nationalist is of a great Dutch South African Republic, the language of which would be Afrikaans. 1925 *Times* 25 Mar. 13/2 Afrikaans, the South African form of the Dutch language,..had been introduced in schools and churches in South Africa, and there were now proposals that it should become the official language of the Union, side by side with English. 1952 [see *Afrikaner 1].

Comb. 1946 *Mind* LV. 45 In South Africa among the Afrikaans-speaking community ('the Dutch').

B. *adj.* Of, pertaining to, or designating the Afrikaans language or the Afrikaans-speaking people (cf. *Afrikaner 1).

1923 J. Reynolds (*title*) Maskew Miller's Afrikaans Exercises for Secondary Schools. 1927 *Off. Year Bk. of S. Afr.* 16 At present German origin seems acceptable in the case of only a few Afrikaans words. *Ibid.*, It has been calculated that the proportion of German blood in the Afrikaans colonists up to the year 1806 was about 27 per cent. *Ibid.* 22 The Afrikaans language..enjoys complete official recognition to-day. 1935 *Times* 8 Nov. 13/3 The Afrikaner Broederbond, originally a laudable society interested in Afrikaans culture ..but now a secret society aiming for an independent Afrikaans form of government. 1948 *Cape Times* 25 Aug. 3 The Cape Town Afrikaanse Sakekamer (Afrikaans Chamber of Commerce).

Afrikaner (æfrikā·nər). Also **Afrikaaner.** [Afrikaans, earlier (Cape) Du. *Afrikaander*, f. *Afrikaan* sb., African + *-d*)*er*, pers. suff. modelled on termination of *Hollander* Dutchman. Cf. *Africander.]

1. An Afrikaans-speaking white South African, esp. one of Dutch descent; also occas. applied to any white citizen of South Africa. Also *attrib.* (Replacing *Africander.)

1824 Burchell *Trav.* II. 619/2 (*index*) Africaanders, or Afrikaäners. 1905 R. Fenton *Peculiar People* p. vi, As General Piet Joubert said to me one day: 'We are Afrikaaners, not Dutchmen.' 1926 S. G. Millin *S. Africans* vi. ii. 157 The South African of Dutch descent whom one loosely and inappropriately speaks of as a Dutchman, but who prefers to define himself as an Afrikaner. 1930 *Times Lit. Suppl.* 26 June 525/2 The Afrikaner attitude towards the British Empire. 1937 W. K. Hancock *Survey Brit. Commonw. Affairs* I. 269 By struggle the French Canadians and the Afrikaners won from the British respect for their individual ways of life. 1952 L. Marquard *Peoples & Policies S. Afr.* iii. 65 In this book Afrikaner is used to mean a South African of European extraction whose mother-tongue is Afrikaans. 1958 *Observer* 20 Apr. 4/5 In the South African general election the Nationalists..have set the Afrikaner nation upon a dominating rock.

2. = *Africander 2.

1918 *Off. Year Bk. of S. Afr.* 386 There was created a distinct type [of cattle] which came to be known as the *Afrikaner*. 1939 *Nature* 11 Nov. 819/1 Zootechnical studies on Afrikaner, imported, and grade cattle.

3. *Bot.* An iridaceous South African flower belonging to any of various species of *Gladiolus* or *Homoglossum*.

1801 J. Barrow *Trav. S. Afr.* i. 25 The *Gladiolus*, which is here called *Africaner*, is uncommonly beautiful with its tall waving spike of striped flowers, and has also a fragrant smell. 1950 Adamson & Salter *Flora Cape Penin.* 263 *H.* [= *Homoglossum*] *Watsonium*..Red Afrikaner. 1959 *Cape Times* 7 Sept. 2/6 Of the gladioli, the brown Afrikaners attracted most attention.

Hence **Afrika·nerdom**, the Afrikaner people or nation in South Africa; Afrikaners collectively; Afrikaner nationalism; **Afrika·nerism**, (*a*) the policy, ideals or aspirations of Afrikaners; Afrikaner nationalism; (*b*) an Afrikaans word or idiom used in South African English; **Afrika·nerize** *v.*, to bring under the influence or rule of Afrikaners; to make like an Afrikaner; so **Afrika·nerizing** *vbl. sb.* and *ppl. a.*, **Afrika:neriza·tion.**

1926 S. G. Millin *S. Africans* vi. i. 156 There was a spirit of Afrikanerdom abroad. 1934 A. J. Barnouw *Lang. & Race Probl. S. Afr.* 26 This language [*sc.* Afrikaans] is to the Nationalists the hall mark of their Afrikanerism. 1935 *Cape Times* 8 Nov. 8/4 Wherever these..ministers of religion use the words 'Afrikaner' and 'Afrikanerdom' they mean 'Afrikaans-speaking Afrikaner' and 'Afrikaans-speaking Afrikanerdom'. 1942 'B. Knight' *Sun Climbs* iii. xxv. 229 He is lost in his dreams of Afrikanerism. 1947 *Forum* 5 Apr. 5/2 The Afrikanerising of the cities..had gone ahead by leaps and bounds. 1952 E. H. Burrows *Overberg Outspan* vii. 190 The afrikanerizing influence she exerted on her husband. 1952 B. Davidson *Rep. S. Afr.* iii. ii. 158 The 'Afrikanerization' of public life and education 'in the Christian National sense'. 1955 Saron & Hotz *Jews in S. Afr.* xi. 207 Of those who fought on the Boer side, some were Jewish burghers who..had become thoroughly Afrikanerized. 1958 A. Paton *Hope for S. Afr.* vii. 52 As in the past Afrikaners were subject to a process of anglicization, so now English-speaking people are subject to a process of Afrikanerization. 1959 *Cape Argus* 21 Nov. 11/3 Among the words which fall in the category of Afrikanerisms, now in common use in the English language in this country, are bobotie, gousblom. 1960 *20th Cent.* July 63 The Nationalists sought to Afrikanerize the white trades unions.

Afro- (æ·fro), used as comb. form of L. *Afer*, *Afr-* African, as in *Afro-American* (also *Aframerican*) adj. and sb., (a person) of African descent born in America (spec. the United States); *Afro-Aryan* adj., African and Aryan; *Afro-Asian* adj., of or pertaining to both Africa and Asia; also sb.; similarly also, *Afro-Caribbean*, *-Cuban*, *-European*, *-Spanish*, adjs. and sbs.

1853 *Voice of Fugitive* (Windsor, Ont.) 21 June 2/4 In our opinion, the true policy of the Afro-American race..is to emigrate to Canada, the West Indies. 1890 *Advance* 23 Jan. 61/2 To encourage all State and local leagues..in obtaining for the Afro-American an equal chance. *Ibid.* 80/1 The Afro-American Convention in Chicago. 1898 *Westm. Gaz.* 31 May 3/1 She is a New Orleans Creole, her mother being an Afro-American, and her father a Louisiana Frenchman. 1910 H. H. Johnston *Negro in New World* 390 In music the Aframerican..may achieve triumphs. 1934 C. Lambert *Music Ho!* iii. 201 By jazz..I mean the whole movement roughly designated as such, and not merely that section of it known as Afro-American. 1939 W. Hobson *Amer. Jazz Mus.* 29 Afro-Americans have been the chief rhythmic originators in the forty-year spread of both ragtime and jazz. 1944 H. L. Mencken in *Amer. Speech* XIX. 161 When the New York *Times* announced in an editorial on March 7, 1930, that it would capitalize the word *Negro* thereafter, there were loud hosannahs from the Aframerican intelligentsia. 1914 *Lancet* 4 Apr. 966/2 An Afro-Aryan child aged 3 years, the offspring of a male African negro and a female Cingalese. 1955 *Newsweek* 17 Jan. 14/1 The critical issue at the 30-nation Afro-Asian conference here in April..will be a behind-scenes struggle for power between India's Nehru and Red China's Chou En-lai. 1956 *Ann. Reg. 1955* 165 The Afro-Asian conference..at Bandung..was attended by Ministers of twenty-nine states, from Liberia to Japan, but not including Soviet Asia. 1958 *Observer* 6 July 12/6 More immediate fears tend to overshadow the long-term political value of wooing the Afro-Asians. 1958 *Oxf. Mail* 14 Feb. 9/6 Lessons in Afro-Caribbean dancing..for..members of the Oxford University Ballet Club. 1959 *Encounter* Dec. 53/1 In the bad old days, when..the Afro-Caribbeans had little but humiliation. 1949 L. Feather *Inside Be-bop* vi. 41 George Russell..penned an Afro-Cuban drums suite. 1956 M. W. Stearns *Story of Jazz* (1957) xix. 252 Perhaps the most stable pattern..was established by Machito and his Afro-Cubans. 1895 A. H. Keane *Ethnology* xiv. 409 The original Aryan type..resembled that of the Afro-European as represented by the Mauritanian Berbers. 1959 *Listener* 5 Nov. 791/2 He has the same right to talk of the land in which he has grown up as any other Afro-European. 1946 *Jazz Writings* 3/2 Spanish American rhythms in the blues (such as the rhythm of the Afro-Spanish 'habanera', created in Havana, Cuba).

afrormosia (æfrǭ³ɹmōᵘ·ziă). [G. (H. Harms in Engler and Prantl *Pflanzenfamilien* (1906) Suppl. III. 158), f. *Afro- + mod.L. *Ormosia* (1811), genus of trees.] A North and West African tree of the genus *Afrormosia* (fam. *Leguminosæ*), esp. *A. laxiflora* and *A. elata*; also, its timber.

1920 A. H. Unwin *W. African Forests* ix. 271 Nigerian Timber Trees..*Afrormosia elata*. African Satinwood, Yellow Satinwood...The timber planes well..it is very hard and withstands the attacks of white ants. 1925 H. V. Lely *Useful Trees N. Nigeria* 11 *Afrormosia* Laxiflora..a small or medium-sized tree very common in bush or tree savannah, averaging 30 feet in height. The wood..very dark brown..is used for axe and hoe handles. 1956 *Archit. Rev.* CXX. 126/1 The furniture and joinery are of afrormosia (a wood very similar to teak in colour and quality), except for the small chairs which are black, with rush seats. 1960 *Times* 7 Mar. 3/3 A consumers' paradise of transistor radio sets, electric food mixers, and afrormosia coffee tables.

aft, *adv.* Add: **2. a. b.** Also of an aircraft.

1948 'N. Shute' *No Highway* ix. 242 We climbed up into the fuselage and went aft through the luggage bay.

4. (Further examples.)

1896 *Strand Mag.* XII. 323/1 At the fore-end and at the aft-end of the vessel. 1915 'Bartimeus' *Tall Ship* i. 9 He..was precipitated through his cabin door across the aft-deck. 1948 'N. Shute' *No Highway* i. 27 The aft windows of the cabin. 1961 *Engineering* 20 Jan. 123/1 The aft-engine concept pioneered..in the Caravelle..has..been adopted by Boeing for its short to medium range 727.

after, *prep.* Add: **1. b.** *after you* (and similar, incl. extended, expressions): a formula used in yielding precedence; *after you with*, colloq. request for the next turn at (something).

1650 R. Heath *Clarastella* Epigrams 33 Oh! after him is manners. *a* 1652 [see Manner *sb.*¹ 7a]. 1721 J. Kelly *Sc. Proverbs* 42 After you is good Manners. Spoken when our Betters offer to serve us first. 1738 Swift *Polite Conv.* ii. 139 Oh! Madam; after you is good Manners. 1899 R. Whiteing *No. 5 John St.* xiv. 139, I remember the fine-company style of Tildor's tea-party, 'After you's manners', whenever we passed the plate. 1927 W. E. Collinson *Contemp. Eng.* 31 The card-expressions now most prevalent in a figurative application are drawn in the main from bridge e.g...*after you, partner.* 1935 Isherwood *Mr. Norris* ix. 136 'After you.' 'No, please.' 1955 E. Pound *Classic Anthol.* iii. 152 Taught 'em to bow and stand aside, Say: after you, and: if you please.

4. b. *to be after*: also, to attend to, keep watch upon, 'see to'.

1856 C. M. Yonge *Daisy Chain* i. vii. 67 You are a little bit of a sloven, and..some one must be always after you.

6. In expressions of the type *day after day*, *man after man*.

c 1175, 1866 [in Dict.]. 1611, 1830 [see Year 7a]. 1631, 1881 [see Time *sb.* 31]. 1674 R. Strange *S. Thomas Cantilupe* xxiii. 293 And soe strophe after strophe till the hymne was ended. 1798 Coleridge *Anc. Mar.* 115 Day after day, day after day, We stuck, nor breath nor motion. 1868 Dilke *Greater Britain* I. i. iii. 34 Time after time I heard the complaint, 'The Yankees treat us shamefully, I reckon.' 1887 [see Hour 1a]. 1893 Beatrice Harraden *Ships that Pass* i. vii, 'It seems so little to ask,' she cried to herself time after time.

8. *after hours*: also, after the regular hours of opening (of a public house, etc.). Also (both senses) *attrib.* and in form *after-hour*.

1929 *Punch* 23 Jan. 90/2 A few austere theatricals drinking coffee and orange-juice because it was after hours. 1930 *Times Educ. Suppl.* 5 Apr. p. iv/4 These after-hour labours, done in a man's own time. 1947 in R. de Toledano *Frontiers of Jazz* xvi. 175 His unflagging interest in after-hours music. 1957 *Economist* 19 Oct. 256/2 It informed broker clients that..after hours facilities would in future be confined to 'small routine business'.

b. Past (a certain hour). Now chiefly *dial.* and *U.S.*

1732 B. Lynde *Diary* (1880) 24, a.m. ½ after 5 I went with son's horse. 1774 P. V. Fithian *Jrnl.* (1900) 271, I..rode thence to Westmoreland Court House ten Miles by half after six. 1775 in *Essex Inst. Hist. Coll.* XLVIII. 52 We was preaded [i.e. paraded] about half after two in the morning. 1809 Mar. Edgeworth *Mme de Fleury* i, in *Tales Fashionable Life* II. 167 It was now half after four. 1905 *N.Y. Even. Post* 27 Jan. 3 About half after twelve the roof of the building fell in with a crash. 1961 Arthur Miller *Misfits* i. 13 'Young man? You have the time?'..'It's twenty after nine.' '*After!*' Isabelle comes farther out on the porch and calls up to a second-floor window: 'Dear girl? It's twenty after!'

10. (Further example.)

1846 Tennyson in *Punch* X. 106/1 Surely, after all, The noblest answer unto such Is kindly silence when they brawl.

12. c. (Later example.)

1601 Holland tr. *Pliny's Hist. World* vii. lvi. 188 Anacharsis the Scythian, or after some, Hyperbios the Corinthian, invented the cast of turning the roundell or globe.

after, *a.* Add: **3. a.** *after ages*: see also After- 9 in Dict. and Suppl.

III. Add to def.: Also, the rear or latter part. (Earlier and later examples.)

1650 B. *Discolliminium* 49 He would..turne so nimbly that he could not see which was his 'fore, which his after. 1903 W. S. Blunt *7 Golden Odes* 39 Who knows to-morrow, who the after of days, the years we see not? 1937 P. J. McCann *St. Benedict* 10 We can place his life and work in the sequence of a continuous development, and estimate it with reference to a before and an after.

after, colloq. abbrev. of Afternoon.

1890–1934 in Wentworth *Amer. Dial. Dict.* (1944) 9/1. 1906 E. Dyson *Fact'ry 'Ands* viii. 104 Iv yeh don't do yer fair share iv yacker this after, I'll punt ther slacks off yer. 1934 J. O'Hara *Appt. in Samarra* (1935) ii. 50 Tonight, or this after', when Ed showed up at the Apollo, he probably would be in a bad humour. 1939 J. Mulgan *Man Alone* xviii. 235 Boss wants us to get the hay in up top this after.

after-. Add: **I. 1.** Further *attrib.* examples (very common in 20th c. use), as *after-sales*; *after-shave* (also *-shaving* adj.); also *ellipt.*: after-shave lotion. See also *after-war.

1895 *Daily News* 14 Sept. 5/7 The after-lunch drive was through more lovely country. 1905 E. Wharton *House of Mirth* (1906) ii. ix. 430 The noisy after-theatre supper. 1906 *Daily Chron.* 4 Jan. 6/7 A severe line is drawn by the 'after-season sale' between the masculine and feminine shop. 1914 J. Collings *Colon. Rur. Brit.* I. vi. 113 The remainder of the children whose after-school career was traced went into industrial or commercial occupations. 1922 'R. Crompton' *More William* iv. 69 'We di'n't ought to have set off before dinner,' said the squire with after-the-event wisdom. 1939 G. Greene *Confid. Agent* i. ii. 72 The after-office rush was over. 1943 L. B. Lyon *Evening in Stepney* 18 Be small, be mute, you after-midnight tears. 1945 *Amer. Speech* XX. 165 The after-shaving lotion may also leave your face feeling 'softer and smoother'. 1946 *Ibid.* XXI. 169 Merchandisers have emphasized terms [of lotions] used in virile, mostly expensive, sports:..Field and Stream After Shave [etc.]. *Ibid.*, Ascot After Shave Lotion. 1955 *Times* 13 May 6/5 British manufacturers of aircraft and aero-engines are devoting special attention to 'after-sales' service. 1958 P. Mortimer *Daddy's Gone A-Hunting* v. 27 He still managed to look suave and tidy, to emanate the bitter-sweet smell of money and after-shave. 1959 I. & P. Opie *Lore & Lang. Schoolch.* xviii. 377 Their favourite after-dark games. 1961 *Guardian* 19 Jan. 9/7 A navy chiffon after-six dress. 1962 *Ibid.* 3 Jan. 4/1 An after-ski poncho. 1963 *Amer. Speech* XXXVIII. 203 The popularity of skiing and the after-ski atmosphere. *Ibid.* 205 After-ski boot. 1964 F. Bowers *Bibliogr. & Textual Criticism* ii. iii. 56 This bibliographical after-the-event interpretation of the Folio error.

4. b. (Further examples.)

1694 Motteux tr. *Rabelais* iv. xxii. 93 Hall your after-misen bowlins.—Hawl, Hawl, Hawl. 1769 Falconer *Dict. Marine* s.v. *After*, The After-Sails usually comprehend all those which are extended on the mizen-mast, and on the stays between the mizen and main-masts. They are opposed to the head-sails. 1813 Southey *Nelson* I. i. 28 The Glasgow..was in flames, the steward having set fire to her while stealing rum out of the after-hold. *Ibid.* iii. 124 He ordered..the driver and after-sails to be brailed up and shivered. 1851 Melville *Moby Dick* II. i. 1, Archy..whose post was near the after-hatches. 1883 *Man. Seamanship for Boys' Training Ships* 175 To reeve an after-guy. 1897 G. du Maurier *Martian* vii. 318 He made the ladies as comfortable as he could on the after-deck. 1898 Kipling in *Morning Post* 11 Nov. 5/1 An Admiral..goes up on the after-bridge. 1933 J. Masefield *Bird of Dawning* 209 An open locker against the after-bulkhead caught his eye. *Ibid.* 211 I've got to..get down into the after-hold.

6. *after-echo*, *-gust*.

1918 D. H. Lawrence *New Poems* 16 The after-echo of fear. 1925 O. Jespersen *Mankind, Nation & Individual* ix. 169 Fear of the naked word, an after-echo of the view held by savage tribes. 1876 Swinburne *Lett.* (1960) III. 239, I cannot write of it now without feeling bitterly an aftergust or afterglow of that enthusiasm. 1885 G. Saintsbury *Marlborough* iv. 54 That aftergust of the plot which blew off the head of Sir John Fenwick.

7. (Further examples.)

1596 Lodge *Marg. Amer.* sig. F1, To make your after-good in deede more savourie. 1890 W. James *Princ. Psychol.* I. x. 393 In the highest phase..there is no after-memory whatever until the next trance comes. 1924 S. C. Chew (*title*) Byron in England. His fame and after-fame. 1933 *Mind* XLII. 359 The after-influence of each philosopher should be dealt with. 1937 J. M. Murry *Necessity of Pacifism* vii. 116 The life and teaching and death and after-influence of Jesus of Nazareth.

9. *after-age* (earlier and later examples).

1608 Chapman *Byron's Trag.* i. i, And of his worth, let after ages say. 1632 Massinger & Field *Fatall Dowry* iii. sig. H1, Something I must do mine owne wrath to asswage, And note my friendship to an after-age. 1646 in J. W. Draper *Cent. Broadside Elegies* (1928) No. 17, p. 37 His fame to after-ages shall Sound out in praise. 1955 *Essays & Studies* VIII. 16 Compositors are far more likely than authors to allow after-ages to see their spelling idiosyncrasies.

II. a·fter-beat *Mus.*, (*a*) a note or tone falling on an upbeat; (*b*) *spec.* the last two notes of a trill; **after-blow** (later example); also *Metallurgy*, a continued period of the blow, after decarbonization is complete, in the basic Bessemer process (cf. Blow *sb.*² 4); **a·fter-Christ**, a second Christ; **a·fter-Chri:stian** *a.*, having ceased to be Christian; also *sb.*; hence **a:fter-Christia·nity**; **a·fter-cure**, convalescence or further treatment taken after a period of treatment; **after-damp** (earlier example); **a·fter-discha:rge**, a discharge (esp. of neural impulses) after the initial cause has been removed; **after-leech** (earlier example); **a·fter-ri:pening** (see quots.); **after-sight**: delete † and add later examples; **a·fter-sound**, a subsequent sound; an echo; cf. After-image; **after-view**: delete † and add later examples.

1908 R. Dunstan *Cycl. Dict. Mus.* 18/1 *After-beat*, last two notes of a Trill. 1927 *Melody Maker* June 597/3 The rhythmic section should support with a 'straight' but well-accented 'after-beat' rhythm. *Ibid.* Sept. 925/2 The stick taps out the second and fourth (or 'after') beats on the cymbal, thus accentuating these after-beats. 1881 *Encycl. Brit.* XIII. 346/1 The elimination of phosphorus..could be very largely effected..this action chiefly taking place during the 'after blow'. 1891 Phillips & Procháska tr. *Wedding's Basic Bessemer Process* iv. 104 It is useless to pour off the slag before the after-blow. 1899 H. S. Bowden *Relig. of Shakespeare* v. 232 He begs that his dismissal..may be at once..and not come as an after-blow to destroy his only hope. 1910 H. P. Tiemann *Iron & Steel* 17 The period before the drop of the flame is called the fore blow, the latter one the after blow. 1881 G. M. Hopkins *Sermons & Dev. Writ.* (1959) 100 The Holy Ghost makes of every Christian another Christ, an AfterChrist. 1886 C. S. Devas *Stud. Fam. Life* iii. 274 We cannot alter the past, or be as though England..had never been a Christian country...We must of necessity be either Christian or After-Christian. *Ibid.*, Popular writers of After-Christian France. 1906 —— *Key to World's Progress* i. 57 Further details of After-Christianity ..are here unnecessary. 1911 *Month* Mar. 282 St. Paul's description of the Fore-Christians of his day applies equally to the After-Christians of ours. 1901 W. James *Let.* 6 Aug. in R. B. Perry *Tht. & Char. of W. J.* (1935) II. 199 We leave here on Saturday..and take the after-cure in the Vosges. 1860 *Mining Gloss.* (ed. 2) 48 *After-damp*, destructive gas (carbonic acid) remaining in the workings after an explosion of fire-damp. 1932 Dorland & Miller *Med. Dict.* (ed. 16) 51/2 *After-discharge*, a response to stimulation in a sensory nerve which persists after the stimulus has ceased. 1941 *Brit. Jrnl. Psychol.* July 74 After short light exposures..this after-discharge is followed by a short period of complete darkness. 1949 A. Koestler *Insight & Outlook* x. 148 Compare the short refractory period or afterdischarge of nerves. 1769 W. Falconer *Univ. Dict. Marine* Kk. 1 The foremost perpendicular or sloping edge is called the *fore* leech, and the hindmost the *after* leech. 1867 *Iowa Agric. Soc. Rep.* (1868) 188 To make the wine: gather the fruit with the stems on...Leave for three or four days in a cellar.. thus causing after-ripening. 1872 *Vermont Bd. Agric. Rep.* 72 Shortly after, begins after-ripening, a chemical change, whereby the starch, abundant in the unripe or green fruit, is transformed into sugar. 1935 *Forestry* IX. 30 The need of the embryo for some process of development or 'after-ripening' after the seed has been shed. 1953 *Brit. Commonw. Forest Terminol.* i. 11 *After-ripening*, biochemical or physical changes occurring in seeds..and fruits after harvesting when ripe in the ordinary way. *a* 1878 Whyte Melville *In Lena Delta* (1885) iv. 50 Aftersight informed us of much that our foresight had overlooked. 1942 T. S. Eliot *Little Gidding* ii. 11 Speech impelled us To purify the dialect of the tribe And urge the mind to aftersight and foresight. 1909 *Cent. Dict.* Suppl., *After-sound*, a subjective sensation of sound which remains after the sound itself has ceased. 1957 L. Durrell *Bitter Lemons* 52 The beadle crashed at the church bell..and then left the silence to echo round us in wing-beats of aftersound. 1800 Coleridge tr. *Schiller's Death of Wallenst.* pref., The feelings that arise from an afterview of the original. 1951 L. MacNeice tr. *Goethe's Faust* ii. ii. 204 The afterview of that sorrowful fearful night.

a·fter-bu:rner. [f. next.] An auxiliary burner fitted to the exhaust-pipe of a turbo-jet engine to increase its thrust.

1947 *Air Reserve Gazette* Oct. 405/1 An after burner is a thrust augmentor which is, in effect, the turning of the jet pipe into an auxiliary ram-jet engine. 1948 *Sci. News Let.* 22 May 323/3 The afterburner being installed is a cylindrical device eight feet long which is attached on the exhaust nozzle of the Westinghouse turbo-jet engine which powers this plane. 1955 *Times* 30 Aug. 11/3 Afterburner, or reheat, systems provide a means of augmenting the thrust of jet engines by burning additional fuel in the jet pipe.

a·fter-bu:rning, *vbl. sb.* [cf. G. *nachbrennen*; After- 9.] **1.** In internal combustion engines, that diminishing combustion which follows the fuller force of the first ignition.

1887 J. A. Ewing in *Encycl. Brit.* XXII. 525/1 The process of combustion..is essentially gradual; when ignition takes place it begins rapidly, but it continues to go on at a diminishing rate throughout the stroke. That part which takes place after the maximum pressure is passed is the phenomenon of after-burning. 1931 A. W. Judge *Automobile & Aircraft Engines* (ed. 2) i. 59 The part of combustion which occurs after the point of maximum pressure is termed the 'after burning' portion.

2. Combustion in the after-burner of a jet-propelled aircraft.

1946 *Aeroplane Spotter* 19 Oct. 244/1 For the supersonic range of speeds, what is known as 'after-burning' was to be employed for generating even more thrust. The 'after-burning' consisted simply of fuel jets disposed annularly around the tail pipe or nozzle much in the same way as planned for the Athodyd or ram-jet. 1949 *Flight* 8 Sept. 285/1 Exhaust reheat, or 'afterburning', is the name given to the process of burning fuel in the exhaust pipe of a jet-propulsion unit.

a·fter-care. [f. After- 9 + Care *sb.*]

1. Lit. = later (i.e. after-coming) concern. Cf. After *a.* 3.

1762 Fielding *J. Wild* i. iv, in *Wks.* II. 242 An accident ..provided Mr. Wild a better tutor than any after care or expence could have furnished him with. ['after' om. in ed. 1, 1743 and in 1754 ed.] 1855 R. Browning *Men & Women* I. 229 Nor bring a moment's trouble on success With after-care to justify the same?

2. *spec.* In medical and social contexts: care or attention bestowed after the conclusion of a course of treatment, etc.; freq. *attrib.*

1854 *Poultry Chron.* I. 285/2 No after-care can then counterbalance the effects of former negligence. 1894 *Daily News* 12 Jan. 5/5 The After-Care Association facilitates the readmission of poor female convalescents from lunatic asylums into social life. 1921 *Act 11 Geo. V* c. 12 § 2 Arrangements..for the after-care of persons who have suffered from tuberculosis. 1927 *Daily Tel.* 8 Feb. 13/7 Changes which are contemplated in connection with St. Dunstan's Institute..will not affect the work of after-care. *Ibid.* 3 May 14/1 Some after-care committee should protect young people from jobs which led nowhere. 1952 *Times* 21 Nov. 3/5 Problems of supervision and after-care of offenders released from prison and other institutions were being discussed. 1962 *Lancet* 27 Jan. 219/2 The delivery and aftercare should be the responsibility of the same general-practitioner/midwife team which supervised the antenatal care.

after-chrome (ɑ·ftəɪkrōᵘm), *a.* [f. After- 9 + Chrome.] Of, pertaining to, or designating a process of textile dyeing in which the material, after being dyed or printed, is

treated with a chromium compound. Also as *v.*; so *after-chroming* vbl. sb.

1905 L. Cassella *Dyeing of Wool* 63 For shading the already afterchromed dyeings the same Anthracene Colours ..are usually applied. *Ibid.* 139 After boiling..the material is then afterchromed in the usual way. *Ibid.*, The usual quantity of bichrome need be used for afterchroming. 1922 *Encycl. Brit.* XXX. 869/2 Colouring matters of this type are known in the trade as 'after-chrome' colours. 1946 *Nature* 21 Dec. 920/2 Laboratory methods of dyeing after-chrome blacks. 1963 A. J. Hall *Textile Sci.* iv. 184 In the second (*after-chrome*) method, the wool is first dyed with the chrome mordant dye and is afterwards treated with the bichromate.

aftercomer. Delete †*Obs.* and add later examples.

1853 D. Rock *Ch. of Fathers* III. ii. xi. 55 That one church which Christ..left to be taught and governed by his apostles and their after-comers in the apostleship. 1879 G. M. Hopkins *Poems* (1918) 40 After-comers cannot guess the beauty been. 1958 E. Blunden *War Poets 1914–18* ii. 19 When after-comers walk through some well-tended war cemetery.

aftercoming, *ppl. a.* Delete †*Obs.* and add later example.

1961 *Lancet* 22 July 189/2 Fœtal distress, the aftercoming head, and face presentation.

a·ftercoo:ler. [After- 9.] An apparatus for cooling air discharged from a compressor.

1903 W. C. Popplewell *Compressed Air* iii. 62 In some cases this intercooler is used as an 'after-cooler', through which the air is allowed to pass after leaving the high-pressure cylinder of the compressor. 1916 D. Penman *Compressed Air Practice in Mining* vi. 95 Where the capacity of the receiver is insufficient to allow the air to cool down..after-coolers are often placed in the pipe-line somewhere between the receiver and the compressor. 1944 *Jrnl. R. Aeronaut. Soc.* XLVIII. (Abstr.) 673 Inter-coolers may be either of the air to air or liquid to air type and can be positioned either between the two stages of the supercharger (intercooler proper) or between the second stage of the engine (so-called aftercoolers). 1948 D. A. Wrangham *Heat Engines* (ed. 2) v. 94 With the object of removing moisture, coolers are sometimes fitted after the last stage, and for this reason are called *After coolers*.

after-course. Add: **1.** (Later example.)

1629 Parkinson *Parad.* iii. xxi. 594 The most excellent sorts of Peares, serve..to make an after-course for their masters table.

fig. 1749 J. Cleland *Mem. Woman of Pleasure* I. 197 He gave me an after-course of pleasure, in a natural burst of tender gratitude and joy.

2. (Further example.) Delete 'Properly two words'.

1922 Joyce *Ulysses* 138 That small act..determined the whole aftercourse of both our lives.

aftercrop. Add: (Further examples.) Also *fig.*

1740 Cibber *Apol. Life* xiv. 268 This unexpected After-crop of Cato, largely supplied to us, those Deficiencies. 1789 Cowper *Let.* 12 Aug. (1824) II. 195 We are now gathering from our meadows, not hay, but muck..that the after-crop may have leave to grow. 1831 W. Howitt *Bk. of Seasons* 338 Sowing wheat upon the fallows, also after crops of tares, clover, early peas.

a·fter-days, *sb. pl.* [After- 9 and After *a.* 3 a.] Later or subsequent days. Less commonly *sing.*, a subsequent day or period.

1635 Quarles *Emblemes* ii. xiii. 113 But something whispers in my dying ear, There is an After-day; which day I feare. 1700 S. Wesley *Ep. to Friend conc. Poetry* line 451 Like old Ennius he design'd What After-days have polish'd and refin'd. 1795 Southey *Joan of Arc* i. 173 Happy those Who in the after-days shall live. 1814 Wordsworth *Excursion* i. 153 In the after day Of boyhood. 1828 Scott *F. M. Perth* xxi, That celebrated Lindsay, Earl of Crawford, who, in his afterdays, was known by the epithet of the Tiger-Earl. *a* 1878 Caswall *Tale of Tintern* (1907) iv. v. 49 And He thine offering will repay Most fully in an after day! 1878 [see After *a.* 3 a]. 1904 Hardy *Dynasts* i. i. iii. 43 To retrospective eyes of afterdays.

a·fter-death. [After- 1.] An existence that follows death; a future life.

1899 W. S. Blunt *Satan Absolved* 40 In the long after-death Ye shall be burned with fire. 1907 *Folk-Lore* June 164 That belief in an after-death or life of souls. 1929 B. Jarrett *Hist. Europe* x. 546 An after-death punishment for evil.

a·fter-effe:ct. [After- 7.] A delayed effect; an effect following after an interval.

1817 [see After- 7]. 1927 A. H. McNeile *Introd. N.T.* vii. 212 Christians at the capital were still feeling the after-effects of Nero's mad outburst. 1940 'G. Orwell' *Inside Whale* 67 At the time this causes rebellion and vomiting, but it may have different after-effects in later life. 1949 J. F. Embree in M. Fortes *Social Structure* 223 The after-effects of Military Government..live on long after the period of occupation. 1956 A. H. Compton *Atomic Quest* v. 304 The second aftereffect is that from the..radioactive materials from the bomb itself.

afterglow. Add: **b.** *Physics.* The phosphorescent light persisting in a gas or on the screen of a cathode-ray tube after the cessation of the electric current. Also *attrib.*

1893 J. J. Thomson *Rec. Research Electr.* ii. 185 A phosphorescent glow, which often lasts for several seconds after the discharge has ceased...All the gases I examined which do polymerize have shown the after-glow. 1934 *Nature* 28 July 140/1 At certain adjustments of pressure, velocity, and voltage, the [nitrogen] afterglow shows many of the colours of the aurora. 1943 *Electronic Engin.* XVI. 196/2 A certain amount of flicker..can be largely overcome by the use of a cathode-ray tube with a long afterglow screen. 1945 *Ibid.* XVII. 384/1 The screen of the tube has..an afterglow in total darkness of approximately 10 seconds. 1951 E. W. Anderson *Princ. Air Navigation* x. 176 The face of the [radar] screen is painted with a fluorescent 'afterglow' material so that the impression of the trace persists for a second or so.

after-guard. (Earlier and later examples.)

1801 *Knight & Mason* IV. v. 60 Darby Drumconda, one of the after-guard. 1927 J. Sampson *7 Seas Shanty Bk.* 43 If sung in proper time, it is not looked upon with favour by the afterguard.

after-image. Add: (Earlier and later examples.) Also *transf.*

1874 J. Sully *Sensation & Intuition* iii. 41 Plateau sought to determine the time during which this after-image (Nachbild) continues in decreasing intensity. 1889 G. B. Shaw in *Eng. Illustr. Mag.* Oct. 53 The after-image of the lyric drama witnessed is deeply engraved in the memory, aural and visual. 1890 W. James *Princ. Psychol.* I. xv. 635 Peculiarities in an after-image, left by an object on the eye. 1962 *Listener* 12 Apr. 647/1 Though they [*sc.* abstract paintings] are concerned with *gestalt* effects, and with after-images, they are not out to batter one's eyes into submission.

a·fter-life. [After- 9.] **1.** A subsequent or future life.

a 1593 Marlowe *Hero & Leander* (1598) (Ded.) sig. A 2, The impression of the man, that hath beene deare unto us, living an after life in our memorie, there putteth us in minde of farther obsequies. 1615 Hoby *Curry-combe* iii. 149 Saduces..denied the Resurrection, and by consequence any after-life. 1925 F. W. Norwood in J. Marchant *Life after Death* 70 Concerning Jesus and the After Life. 1925 D. Thomas *Ibid.* 104 Young children whose minds were unversed in speculations about the After Life.

2. The later period of one's life.

a 1678 H. Scougal *Life of God* (1726) 204 The lessons which afflictions teach us, are then most advantageous when we learn them betimes, that we have the use of them in the conduct of our after lives. *a* 1805 Wordsworth *Notebook Y* in *Prelude* (1926) 556 Untutor'd minds stop here, and after-life Leads them no further. 1817 Scott *Harold* vi. xix, Eivir! since thou for many a day Hast follow'd Harold's wayward way, It is but meet that in the line Of after-life I follow thine. 1833 [see After- 9]. 1837 Lockhart *Scott* I. 323 Archibald Constable, in after life one of the most eminent of British publishers. 1878 W. S. Jevons *Pol. Econ.* Pref. 6 To instil..notions on subjects with which all must in after-life be practically conversant. 1923 J. M. Murry *Pencillings* 77 He gives a picture of the after-life of the stockinged Fauntleroys.

af·ter-light. [After- 9.] An afterglow (also *fig.*); also, the light of what is known afterwards, hindsight.

1894 W. J. Locke *At Gate of Samaria* xvi. 194 She..noticed a look upon Thornton's face, —the after-light, as it were, of a sneer, before the features had time to reset. 1923 H. E. G. Rope *City of Grail* 47 The afterlight of sunset in those summers long gone by. 1940 W. Empson *Gathering Storm* 32 When this leaves the green afterlight of day. 1950 W. S. Churchill *2nd World War* III. xxxiv. 585 These three documents, with which, in the afterlight..I am content.

afternoon. Add: **1. b.** *pl.* as *adv.*: in the afternoon; of an afternoon. Cf. Nights *adv.*

1896 *Vermont Agric. Rep.* XV. 36, I prefer to gather sap afternoons. 1911 *Rep. Lab. & Soc. Conditions in Germany* III. 31 The boys..attended school from 8 o'clock till 11 in the forenoon, and from 2 to 4 afternoons.

3. (Further examples.)

1748 Richardson *Clarissa* I. xvii. 110, I will go down.. and excuse your attendance at afternoon tea. 1754 Fielding *Voy. Lisbon* (1755) 104 Our ladies..drank their afternoon tea at an alehouse. 1882 [see Tea *sb.* 4 a]. 1903 A. Bennett *Leonora* iii. 76 He was admiring the spacious room and herself in her beautiful afternoon dress. 1927 E. Glyn *'It'* xii. 111 She had put on her most demure garment—an afternoon frock of crimson crêpe. 1931 A. Christie *Sittaford Mystery* i. 10 Two women in afternoon frocks rose to greet the staunch old warrior.

b. *attrib.* or quasi-*adj.*, in sense 'procrastinating, slow, lazy', as *afternoon farmer*. *dial.* (see *E.D.D.*).

1743 W. Ellis *Mod. Husb.* June iii. 30 Others, where the Crop is thin, will not hough at all; but these are your Afternoon Farmers, as we call them. 1889 *Standard* 28 Nov. 2/1 The rain and snow..have come too soon for a few 'afternoon farmers', who have not yet put in all their wheat. 1894 *N. & Q.* V. 153/2 In West Middlesex the expression 'an afternoon farmer' is frequently used in talking of a farmer who is behind hand in his work.

Hence (*rare*) **afternoo·ny** *a.*

1885 T. H. Huxley *Let.* 23 Feb. in L. Huxley *Life of T. H. H.* (1900) II. vi. 96 There is something idle and afternoony about the air which whittles away one's resolution.

afternoo·ner. Used in comb., as in *Saturday-afternooner*, one who has a holiday on Saturday afternoons.

1906 *Daily Chron.* 28 July 7/6 Thus giving early-closers and Saturday-afternooners a chance. 1927 *Daily Tel.* 7 June 3/1 If the amendment which I suggest is wrong for the Saturday afternooners, it is wrong for cricket.

afterpiece. 1. (Earlier example.)

1779 T. Holcroft *Let.* in *Memoirs* (1816) III. 251 We have a new afterpiece of Mr. Sheridan's coming out this evening, 'The Critic'.

after-reckoning. (Earlier and later examples.)

1623 Webster *Devils Law-case* iv. ii. 566, I could never away with after reckonings. 1829 Southey *Oliver Newman* (1845) 34 Sure that no after-reckoning will arise, Of shame, or sorrow.

afters (ɑ·ftəɪz), *sb. pl. colloq.* (formerly only *dial.* and *vulg.*). [f. After *a.* or *adv.*] The course which follows the main course of a meal. Cf. *seconds.

1909 J. R. Ware *Passing Eng.* 3/1 *Afters* (Devon), sweets—pies and puddings. 'Bring in the afters' is a common satirical remark in poor Devonshire houses, especially when there are no 'afters' to follow. Also used in Scotland, *e.g.*, 'Hey mon, a dinner, an' nae afters!' 1919 *Athenæum* 29 Aug. 822/2 'Afters' is in no sense an army word..beyond the fact that the mass of the army is composed of the working classes...Used in its equivalent sense to sweets, pudding, entremets or dessert, it may not show much imagination. 1940 *Manchester Guardian Weekly* 11 Oct. 255 The meat course costs from fourpence to sixpence, and the 'afters', as Londoners call puddings, from twopence to threepence. 1940 'N. Shute' *Landfall* x. 256 'What's he got for afters?' 'Plummy duff.' 1953 Scott & Fisher *1,000 Geese* x. 120 We ate our corned beef and cheese, crisp-bread and biscuits, with dates for afters.

a·fter-sensa:tion. [After- 6.] = After-image (but less commonly used of visual impressions).

1867 J. Marshall *Outl. Physiol.* I. 431 After-sensations.. are noticed in regard to all the senses. 1874 G. H. Lewes *Problems* I. 149 We may consider the gradations of Sensation, After-sensation, Imagination, and Hallucination. 1893 Dunglison *Dict. Med. Sci.* (ed. 21) 28/2 *After-sensation*, continuance of impression after withdrawal of stimulating cause. 1909 E. B. Titchener *Text-bk. Psychol.* i. § 40. 152 Followed by an after-sensation of cold. 1952 D. J. O'Connor *John Locke* iv. 81 It cannot be said that a red after-sensation, for example, is a property of the self who perceives it in the same sense as the red colour of a cherry is a property of the cherry.

a·fter-shaft. *Ornith.* [tr. mod.L. *hyporachis.*] = Hyporachis.

1867 W. S. Dallas tr. *Nitzsch's Pterylography* I. i. 8 The aftershaft originates from the underside of the feather beneath the umbiliciform pit. 1871 T. H. Huxley *Man. Anat. Vertebr. Anim.* vi. 274 In very many birds each quill bears two vexilla; the second, called *aftershaft* (*hyporachis*), being attached on the under side of the first close to the superior umbilicus. 1884 E. Coues *Key N. Amer. Birds* (ed. 2) 84 The after-shaft, when well developed, is like a duplicate in miniature of the main feather, from the stem of which it springs, at junction of calamus with rhachis, close to the umbilicus.

a·fter-shock. [After- 6.] A lesser shock coming after the main shock of an earthquake.

1894 *Seismolog. Jrnl. Japan* III. 72 All the after-shocks have been weaker than the respective initial earthquakes. 1903 R. D. Oldham in *Mem. Geol. Survey India* XXXV. ii. 2 In the early months of the aftershocks, there appeared to be a distinct increase in the frequency of earthquakes. 1946 *Nature* 30 Nov. 784/2 The first, at 22 hr. 26·3 min. G.M.T., was an aftershock of the destructive Dominican Republic earthquake of August 4 off Samana Peninsula.

after-supper. Add to def.: Also, = Rere-supper (quots. 1590 and 1637 in Dict.). *Obs.* Add further *attrib.* examples.

1852 Dickens *Bleak Ho.* xxxii. 312 His unpremeditated after-supper stroll. 1879 F. W. Robinson *Coward Conscience* I. ii. vii. 254 An evening party, in the full vigour of its after-supper dances.

a·fter-time. [After- 9.] A later or future time; posterity. Freq. in *pl.* (See also After *a.* 3 a.)

1597 Shakes. *2 Hen. IV* iv. ii. 52 You are too shallow (*Hastings*)..To sound the bottome of the after-Times. ?1603 S. Daniel *Def. Ryme* (1904) II. 380 When after-times shall make a quest of inquirie, to examine the best of this Age. 1664 S. Butler *Hudibras* ii. iii. 170 As Withers in Immortal Rime Has registerd, to after-time. 1816 Shelley *Alastor* 266 To remember their strange light in many a dream Of after-times. 1842 Tennyson *Morte d'Arthur* 35 Wheresoever I am sung or told In aftertime, this also shall be known. 1858 Bagehot *Coll. Wks.* (1965) II. 48 It is for the slow critic of after-times to piece together their teaching. 1902 W. James *Var. Relig. Exper.* xvi. 381 As a rule they [mystical states] carry with them a curious sense of authority for after-time.

a·fter-trea:tment. [After- 7.] Later or subsequent treatment.

1831 W. Youatt *Horse* xii. 227 No after treatment will be necessary, except that the animal should be sheltered from intense heat. 1910 *Practitioner* July 29 Some of the more recently published cases of recovery in this branch of surgery have shown an advance in the after-treatment. 1925 B. Beetham in E. F. Norton *Fight for Everest, 1924* iii. 326 The after-treatment of the [photographic] negatives is often somewhat of a nightmare. 1946 *Nature* 5 Oct. 475/1 The strength of casein fibre could be increased by after-treatment with basic zinc-chloride and formaldehyde. 1962 J. T. Marsh *Self-Smoothing Fabrics* xviii. 296 The use of sodium copper tartrate as an after-treatment. 1964 S. Duke-Elder *Parsons' Diseases of Eye* (ed. 14) xxvii. 427 The most important point about after-treatment is keeping the pupil well dilated.

a·fterwale. [f. After- 4 a + Wale *sb.*¹ 7.] A back wale on a horse's collar.

1889 in *Cent. Dict.* 1968 [see Wale *sb.*¹ 7]. 1946 N. Wymer *Eng. Country Crafts* v. 48 He will cut out his leather afterwales, damp them, and sew them to the body with white leather laces.

a·fter-war. [After- 1.] The period after a war; post-war. Used esp. *attrib.* or quasi-*adj.*

1919 A. A. Milne *First Plays* 113 The most suitable career for a young man in after-war conditions. 1931 *Times Lit. Suppl.* 17 Sept. 694/3 The numerous shady railroad flotations of the after-war period. 1939 'G. Orwell' *Coming up for Air* iii. i. 184 It isn't the war that matters, it's the after-war. The world we're going down into.

afterward, *adv.* and *prep.* Add: **D.** *sb.* A later or subsequent time; the after-life. *rare.*

1906 K. Trask *Night & Morning* 27 Yea, all the Afterward—beyond the grave—Could have no terrors.

afterwards, *adv.* Add: Also as sb. The future; the future life, the after-life.

1901 'Linesman' *Words by an Eyewitness* (1902) 343 The little spark must keep the great cold world warm until the Afterwards. 1902 E. Glyn *Refl. Ambrosine* 285 An English girl would have a blank prospect in front of her for the afterwards. 1922 D. H. Lawrence *England, my England* (1924) 47 To..mingle and commingle with the one darkness, without afterwards or forwards. *a* 1930 —— *Last Poems* (1932) 288 About the afterwards As a matter of fact, we know nothing.

a·fterword. [f. After- 9, after *foreword.*] A passage added at the end of a book, etc., as an epilogue or the like.

1890 *Caxton's Eneydos* (E.E.T.S.) p. xx, Afterwords by F. J. Furnivall. 1900 Mead *Fragm. Faith* 605 (*heading*) Afterword. 1911 J. Gwynn (*title*) The Present Position of Protestantism in Ireland and an Afterword. 1939 *Mind* XLVIII. 433 The after-word that he thought it essential to print.

a·fterwork. [After- 9.] Later work; work done afterwards.

1907 L. Binyon in *Sat. Rev.* 30 Mar. 389/2 Six leaves of a sketchbook joined together, just as they were, with no afterwork. 1958 M. L. Hall *Newnes Compl. Amat. Photogr.* 1 There is the man who considers that the picture is created the moment it is taken, to whom 'afterwork' is the work of the devil, and there is the man who spends joyful evenings printing in clouds and retouching out mistakes.

a·fter-world. [After- 6.] **1.** A future generation; posterity.

1596, 1649 [see After- II]. 1913 G. Murray *Euripides & his Age* ii. 50 The judgement of the afterworld upon them will depend on the side we take in a never-ending battle.

2. A world after death.

1615 Hoby *Curry-combe* iv. 181 He speaks of the state of th' elect in the after-world. 1887 H. S. Holland *Christ or Ecclesiastes* 128 That great after-world for which..this age-long story of man is but a preparation. 1949 D. L. Sayers *Dante's Divine Comedy* 17 The positions they occupy in the Three Kingdoms of the After-world.

a·fter-years, *sb. pl.* [After- 9.] The years that come, came, or will come after; the later years (of a man's life, etc.). Rarely *sing.*, a later year.

1814 Wordsworth *Excursion* viii. p. 378 Sottish vice.. To which in after years he may be rouzed. 1837 Lockhart *Scott* I. 341 Had the subject been taken up in after years, we might have had another Marmion. 1858 W. T. Matson *Poems* 459 Love..sagely musing o'er Fate's riddle-book, The secret coins of many an after-year. 1869 Tennyson *Coming of Arthur* 157 In one great annal-book, where after-years Will learn the secret of our Arthur's birth. 1902 Gairdner *Eng. Ch. 16th C.* iv. 45 In spite of all the frightful demoralization of his after years he retained both characters to the very end.

aftward, *adv.* Add examples. Also in form **aftwards,** and as *adj.*

1627 John Smith *Sea Grammar* ii. 5 The..forward or aftward beames. *Ibid.* 7 The ships quarter is from the maine mast aftward. *Ibid.* v. 19 The vse of those staies are to keepe the Masts from falling aftwards. 1887 W. Morris tr. *Odyss.* I. xii. 410 And the mast withal fell aftward.

afzelia (æfzī·liă). *Bot.* [mod.L., f. the name of Adam *Afzelius* (1750–1837), Swedish botanist.] A genus of trees of tropical Africa and Asia yielding a marketable timber; a tree of this genus; also, its timber.

1797 J. E. Smith in *Trans. Linn. Soc.* IV. 221 Twenty New Genera of Plants..Afzelia. *Ord. Nat.* Leguminosæ.. Locus *Africa æquinoctialis*..Nomen dedi in inventoris honorem, celeberrimi *Adami Afzelii.* 1874 Lindley & Moore *Treas. Bot.* II. 1258/1 *Afzelia*,..with..large flowers borne in short panicled racemes at the end of the branches. 1955 *Nomencl. Commerc. Timbers* (*B.S.I.*) 10 The trade name *afzelia* is proposed for all species of this genus. In practice the West African species are usually grouped together as a single commercial timber. The East African species is usually marketed separately. 1959 *House & Garden* July 41, 3-seater settee, afzelia wood frame.

‖**agacerie** (agasəri). [Fr.] Allurement, coquetry.

1809 *Q. Rev.* II. 182 One class of readers may perhaps be amused by the *agaceries* of the *filles de chambre*. 1818 Mrs. Opie *New Tales* III. 95 Till her mother gave her..a very significant frown, her *agaceries* were addressed to me. 1848 A. H. Clough *Bothie* ii. 13 These, I think, no less than other agaceries, cloy one. 1853 Mrs. Gaskell *Ruth* I. iii. 66 He had seen others..with many more *agaceries* calculated to set off the effect of their charms. 1883 Mrs. Oliphant *Sheridan* v. 174 That very transparent *agacerie* by which foolish men are sometimes attracted in the lower ranks of life.

again. A.. *adv.* **4. a.** Add further example (in request for a repetition of what has previously been said). Cf. *come again* (see *come *v.* 51 d).

1853 Dickens *Bleak Ho.* lvii. 549 'You know Mr. Skimpole!' said I. 'What do you call him again?' returned Mr. Bucket.

against, *prep.* Add: **12. c.** Esp. in phr. *against the government*, opposed to the established view, rulers, etc. (cf. *agin *prep.*).

1934 H. G. Wells *Exper. Autobiogr.* II. viii. 604 The Bastables are an anarchistic lot. Her soul was against the government all the time. 1962 *Sunday Express* 25 Feb. 6/5 Milton and Dante were usually against the Government.

‖**agal** (agā·l). Also †**ageil, aggal, agghal.** [Arab. *'iqāl* bond, rope.] A fillet to keep the keffiyeh in position.

1855 J. L. Porter *Five Years in Damascus* II. xi. 87 Instead of the ample white turban, he wore the *kefiyeh* and *ageil* of the Arabs. 1920 *Blackw. Mag.* May 594/2 All three attired in costly silk robes and kefias and aggales (the band that secures the kefia or turban). 1934 *Times Lit. Suppl.* 22 Feb. 115/2 The nomadic Beduin who wear the keffiyeh and agghal. 1951 A. Christie *They came to Baghdad* v. The dignity of the Arab dress, the inevitable *Keffiyah* of black and white held in place by the black silk *agal*.

agamid (æ·gămid). *Zool.* [f. mod.L. *Agamidæ*: see Agama, -id³.] A lizard of the family Agamidæ. Also as *adj.*

1889 in *Cent. Dict.* 1934 *Geogr. Jrnl.* Nov. 393 The agamid lizard, *Phrynocephalus theobaldi*. 1958 C. H. Pope *Reptiles round the World* 175 Africa has a few monitors and some agamids. 1959 *New Biol.* XXX. 79 Three related families of lizards, the iguanids, agamids, and chamaeleons.

agammaglobulinæmia (ēⁱgæ·mă,glǫ·biŭlinī·miă). *Path.* [mod.L., f. A- 14 + **gamma globulin* + Gr. αἷμα blood.] Lack of gamma globulin in the blood.

1952 O. C. Bruton in *Pediatrics* IX. 722 *Agammaglobulinemia*. The complete absence of gamma globulin in human serum with a normal total protein..does not appear to have as yet been reported in the literature. 1954 *Lancet* 2 Oct. 671/2 *Agammaglobulinæmia*. A syndrome has recently been described in America in which recurrent severe infections are associated with a virtual absence of γ-globulin from the serum, the levels of the other plasma-protein fractions remaining within normal limits. 1961 *Lancet* 29 July 246/1 In congenital agammaglobulinæmia, where the incidence of a rheumatoid-like condition is high, family studies have revealed a hereditary factor.

agamoid, *a.* Add: **B.** *sb.* An agamoid lizard.

1882 *Encycl. Brit.* XIV. 736/2 *Calotes* is another genus of agamoids peculiar to the East Indies. 1886 *Ibid.* XX. 469/2 The bulk of the Lacertilian fauna is composed of Skinks, Geckos, Agamoids, and *Varanidæ*.

agapanthus (ægăpæ·nþŏs). [mod.L., f. Gr. ἀγάπη love + ἄνθος flower.] A genus of South African liliaceous plants, having large umbels of blue, violet or white flowers on a stout scape, of which the bright blue and white species are commonly cultivated for ornament; also, a plant of this genus.

1789 W. Aiton *Hortus Kewensis* I. 414 Agapanthus.. African blue Lily. Nat. of the Cape of Good Hope. Cult. 1692, in the Royal Garden at Hampton-court. 1817 *Botanical Cabinet* I. 42 Agapanthus Minor..a native of the Cape of Good Hope...It differs from umbellatus both in leaves and flowers. 1874 G. M. Hopkins *Jrnl.* 2 Oct. (1937) 214 A splendid thick-stemmed carnation-coloured lily called valotta..in the greenhouse next to an agapanthus. 1879 W. Cory *Lett. & Jrnls.* (1897) 448 He made flowering plants huddle round the trunks of trees, agapanthus under plane. 1887 Rider Haggard *Jess* i. 1 The agapanthus which is so familiar to us in English greenhouses. 1962 *New Statesman* 5 Jan. 28/2 There are one or two places in Cornwall where a much more beautiful plant than mesembryanthemum is colonising uncultivated land by seeding itself freely: agapanthus.

agape. Add: (Earlier example.)

1607 R. Parker *Scholast. Disc. agst. Antichrist* i. ii. 70 The Christians had their *Agapæ* at communions.

2. Now used commonly in its simpler N.T. sense of Christian love (of God or Christ or fellow Christians: see Dict. s.v. Charity); freq. contrasted with Eros, earthly or sexual love.

1856 H. G. M. Lascelles *Compensation* I. x. 88 Mr. Grant was..animated by real charity, or love, the 'Agape' —so rare to find—of which she had read and admired the true meaning in her Greek testament. 1932 A. G. Hebert tr. *A. Nygren's Agape & Eros* i. i. 23 Christian love is essentially Agape. *Ibid.* 32 The idea of Agape is not merely a fundamental idea of Christianity, but the fundamental idea *par excellence*. The idea of Agape is a new creation of Christianity. 1938 J. Burnaby *Amor Dei* i. 15 From the fact that the world into which Christianity brought the Gospel of divine Agape, of God's self-giving love to men, already knew that thirst for the Divine which Platonism called Eros, it did naturally result that when Christians spoke of love they did not always mean the same thing. 1950 W. H. Auden *Enchafèd Flood* (1951) iii. 100 He exhibits Christian forgiveness and Christian *agape* without the slightest effort. 1953 R. Niebuhr *Chr. Realism & Polit. Probl.* (1954) ix. 132 The *agape* form of love in the New Testament fails to be appreciated particularly in two of its facets: (*a*) the equality of the 'two loves'..(*b*)..the notion of sacrificial love. 1955 P. Tillich *New Being* (1956) i. vi. 47 Calculating love is not love at all. Jesus did not raise the question about how much *eros* and how much *agape*, how much human passion and how much understanding was motivating the woman.

Agapemone (ægăpī·mōnĭ). [irreg. f. Gr. ἀγάπη love + μονή dwelling, abode (μένειν to stay, remain).] Proper name of an association of men and women established at Spaxton in Somerset by the Rev. Henry James Prince; a similar establishment conducted by his successor, the Rev. John Hugh Smyth-Pigott, at Clapton, London. Also *gen.*, an establishment of this kind, an abode-of-love; esp. with unfavourable implication. Hence **Agapemonian** (-mōᵘ·niăn), **Agape·monite** *adjs.* and *sbs.*

1850 *Daily News* 22 Mar., The Agapemonites.—It is said that there is a general split amongst this deluded sect at their abode near Bridgewater. 1851 *Illustr. London News* 29 Mar. 253/3 The Agapemone, or the Abode of Love, is the residence of a religious body, which calls itself the Family of Love. 1854 *Edin. Rev.* Apr. 377 The Agapemone of Bridgwater is full of crazy fanatics. 1859 *Sat. Rev.* 30 Apr. 527/2 A carriage-full of Brothers and Sisters in Love singing the Agapemonite Psalter. *Ibid.*, Any of the subsequent Agapemonian extravagances. 1860 Dickens *Uncommerc. Trav.* in *All Yr. Round* 29 Sept. 590/1 The happy nature of my retirement is most sweetly expressed in its being the abode of Love. It is, as it were, an inexpensive Agapemone. 1888 J. D. Hooker *Let.* 22 Aug. in L. Huxley *Life & Lett. J. D. H.* (1918) II. 317 The moment you allow of 'promiscuous intercourse' it is all up and the thing degenerates into an agapemone. 1893 *Funk's Stand. Dict.*, Agapemonian, *n.* 1899 *Daily News* 10 Jan. 3/3 About four years ago the deceased [Henry James Prince] assisted in the opening of..the first church of the Agapemonians. 1908 *Daily Chron.* 24 Sept. 1/1 The Agapemonites are extremely reticent about their peculiar religion. 1908 *Times* 10 Nov. 12/4 The date..on which a party of Cambridge undergraduates were going to raid the 'Agapemone' at Spaxton, and tar and feather the Rev. Smyth Pigott. 1951 M. Kennedy *Lucy Carmichael* ii. i. 86 The Staff, the Students, etc., all have to have an *agapemone* in the assembly hall, and sing carols.

agar (ēⁱ·gaɹ). Short for next. Also *attrib.*

1889 *Jrnl. Chem. Soc.* LVI. 817 Diffusion in Agar Jelly. 1892 *Pall Mall Gaz.* 3 Aug. 7/2 Cultures were made in broth, gelatine, and agar. 1909 *Practitioner* Nov. 596 An ordinary 2 per cent. agar medium. 1946 *Nature* 31 Aug. 293/2 A circular disk was cut from the fruit-body..and placed on a nutrient agar plate of typhoid bacillus culture.

agar-agar (ēⁱ·gaɹ,ēⁱ·gaɹ). Also occas. **agal-agal.** [Malay.] Any of certain East-Indian seaweeds, esp. the Ceylon moss *Gracilaria lichenoides*, from which a gelatinous substance is extracted and used in China for soups and the manufacture of transparent silk and paper, and in bacteriology as a solidifying agent in culture media.

1813 W. Milburn *Orient. Comm.* II. xxiii. 304 Agal Agal is a species of sea-weed, in which some trade is carried on by the Chinese. 1820 J. Crawfurd *Ind. Archipelago* III. ix. ii. 181 The articles of the return cargo [to China]..embrace.. *agar-a-gar*, or sea-weed, *tripang*, or sea-slug. 1863 Watts *Dict. Chem.* I. 61 Agar-agar or Bengal Isinglass. 1886 Crookshank *Bacteriology* 65 Agar-agar has the advantage of remaining solid up to a temperature of about 45°. 1896 *Lancet* 28 Mar. 835/2 Löffler's serum agar-agar. 1929 W. Deeping *Roper's Row* xvi. § i. 169 A culture of germs on a plate of agar-agar.

Agassiz trawl (æ·găsiz trǭl). [f. the name of Alexander E. *Agassiz* (1835–1910), American naturalist + Trawl *sb.*] A type of beam-trawl.

1910 *Encycl. Brit.* VIII. 573/1 (*caption*) Agassiz or Blake Trawl. From Alexander E. Agassiz's *Three Cruises of the 'Blake'*. 1936 Russell & Yonge *Seas* (ed. 2) 262 A small net known as the Agassiz trawl..has the advantage that whichever way it may fall on the sea bottom it can still fish effectively. 1959 H. Barnes *Oceanogr. & Marine Biol.* i. 32 Separate problems arise in the sampling of plankton very close to the bottom...In one method, a net is mounted on an Agassiz trawl frame.

agate, *sb.* Add: **1. b.** A marble made of glass, etc., resembling agate. Also *attrib.*

1843 [see Taw *sb.*²]. 1921 *Glasgow Herald* 26 Nov. 6 The marbles we played with..were called 'nicks',..'agates' (black, blood and milk). 1934 *Amer. Speech* IX. 75/1 *Agates* or *Aggies*, marbles made of agate and usually used as shooters. 1952 J. Steinbeck *East of Eden* xxxvi. 365 Cal was able to develop his marble game and set about gathering in all the chalkies and immies, glassies and agates.

4. (Earlier example.) Also *attrib.* (see quots.).

1838 U.S. Mag. & Democratic Rev. I. 61 Light faced Book and Job Printing Types..Diamond, Pearl..Agate. 1884 Chicago News Let. 5 Apr., Commercial advertising 20 cents per line Agate measure. 1956 F. H. Collins Authors' & Printers' Dict. (ed. 10) 8/2 Agate line (U.S.), measure advertising space, 1/14 in. deep and one column wide.

6. *agate-eyed* adj.; **agate-shell** (example): also called **agate-snail**; **agate-ware**, a kind of pottery coloured to resemble agate; also, enamelled iron or steel ware for household utensils; also *attrib.*

1925 C. Day Lewis Beechen Vigil 30 Panic thereafter Came agate-eyed, gibbering, past the gate. 1884 G. W. Tryon Struct. Conch. III. 59 Achatina, Etym. Agate-shell..Mostly African. 1889 Cent. Dict., Agate-snail. 1901 E. Step Shell Life xix. 379 The Agate Snail (Cæcilianella acicula), though only about one-fifth of an inch in length, is interesting on several grounds. 1857 J. Marryat Pottery & Porcelain (ed. 2) viii. 154 Agate and jasper ware. 1865 Jewitt Wedgwoods Index, Agate-ware vases. 1879 Cassell's Techn. Educ. I. 367 During the reigns of Anne and George I, an improved ware was made of sand and pipe-clay, coloured with oxide of copper and manganese, forming the well-known 'agate-ware' and 'tortoiseshell-ware'.

agba (æ·gbă). [Yoruba.] A West African tree (*Gossweilerodendron balsamiferum*); also, its timber.

1920 A. H. Unwin W. Afric. Forests & Forestry ix. 304 Pterogopodium, Agba, Pink Mahogany... It is a very large forest tree, with clear bole of 90 feet. 1934 A. L. Howard Timbers of World (ed. 2) 9 Agba, reported by the Imperial Institute as a good, useful wood of very serviceable width. 1952 Archit. Rev. CXII. 408/3 The desk for Unilevers, designed by Charles Kenrick, in agba, with a black bean top and reeded hardboard case.

age, *sb.* Add: **1. c.** *age and area*: designating a theory that the area occupied by a culture, language, animal species, etc., is a measure of its antiquity.

1915 J. C. Willis in Phil. Trans. R. Soc. CCVI. 337 Genera will obviously tend to follow the age and area rule more closely than species. 1922 —— Age & Area vi. 62, I called this hypothesis, that on the average the area occupied by species in a country depended upon their age within that country, by the convenient jingle of 'Age and Area'. 1922 J. Small in J. C. Willis Age & Area xiii. 126 Inequalities more or less cancel out when the genera are taken in groups of ten or more as specified for the Age and Area hypothesis.

d. Expostulatory phr. *be* (less commonly *act*) *your age*: behave as becomes your years, *i.e.* in a responsible manner; don't be childish. orig. *U.S.*

1925 New Yorker 26 Sept. 18/2 (caption) Be Your Age. 1931 E. Linklater Juan in America ii. i. 63 Aw, be your age! 1932 Amer. Speech June 328 Act your age, 'don't be childish'; 'stop the foolishness'. 1933 Punch 11 Jan. 29/3 Son. Aw shucks! Doancher know nuttin'? Cummawn, be yerr age, Paw. 1948 'N. Shute' No Highway iii. 70 Do you think the Inspection would have let this aircraft fly if there was any danger of that sort of thing? Be your age. 1951 S. Kaye-Smith Mrs. Gailey ix. 180 Rosamund..spoke irritably. 'Oh, be your age!'

e. *of an age*: of the same age. Const. *with*.

1934 H. G. Wells Exper. Autobiogr. I. iii. 105, I got more mental stimulus from some of my school-fellows who were of an age with me. Ibid. II. viii. 627 We were both about of an age; to be exact he was six months younger than I.

6. b. *Cards.* The 'eldest hand' in the game of poker.

1882 C. Welsh Poker; how to play it 47 Before the dealer begins to deal the cards, the player next to his left, who is called the ante-man, or age, must deposit in the pool an ante not exceeding one-half the limit previously agreed upon. 1889 Guerndale Poker Bk. v. 33 It would be C's place to bet first, he being to the left of the Age.

10. b. Add to def.: Also in *pl.*

1889 W. S. Gilbert Gondoliers ii. 32 As at home we've been remaining—We've not seen you both for ages. 1922 Joyce Ulysses 742, I suppose he died of galloping drink ages ago.

III. *age-class, -determination, -distribution, -fellow, -grade* (hence *-grading* vbl. sb.), *-limit, -range, -scale, -set, -war; age-ago, -cold, -dimmed, -encamped, -enfeebled* (example), *-established, -gnarled, -moulded, -new, -proof, -weary, -worn* (earlier and later examples), adjs.; **a·ge-group,** a number of persons or things classed together as of similar age; hence *age-grouping* vbl. sb.; **a·ge-hardening** *Metallurgy*, (of certain alloys) the process or result of *ageing.

1923 Kipling Irish Guards in Gt. War I. 325 That age-ago retreat from Mons. 1905 Terms Forestry & Logging 6 Age class, all trees in a stand whose ages are within given limits. 1920 R. H. Lowie Primitive Society xi. 302 Schurtz..is as certain of the uniform priority of age-classes when compared with clubs or secret organizations as Morgan is of the necessary priority of matrilineal descent. 1929 Age-Class [see *age-fellow]. 1950 W. de la Mare Inward Comp. 86 Cliffs of age-cold stone. 1926 R. W. Lawson tr. Hevesy & Paneth's Man. Radioactivity xxvi. 216 Age determination from the Helium content. 1832 W. C. Bryant Poems 57, I shall see the day..with an age-dimmed eye. 1909 Cent. Dict. Suppl., Age-distribution. 1934 Planning I. xxvii. 5 The death-rate like the birth-rate, has fallen fast in recent years, and here again the age-distribution of the population must be considered as well as the crude rate. 1913 Kipling Songs from Books 157 Age-encamped Oblivion Tenteth every light that shone!

1807 J. Barlow Columbiad v. 531 Their maim'd, their sick, their age-enfeebled sires. 1925 R. Graves Welchman's Hose 49 Age-established brooks run dry. 1929 N. & Q. Anthrop. (ed. 5) ii. 56 Those Age-Fellows who have been initiated together may be looked upon as an Age-Set or Age-Class. 1954 J. G. Peristiany in Instit. Prim. Soc. iv. 44 To-day the council of elders, which includes..age-fellows of the wrong-doer, puts a collective curse on his head. 1933 W. de la Mare Fleeting 101 The age-gnarled thorn. 1906 N. W. Thomas Kinship Organisations i. 2 The other kind of association, to which the name age-grades is applied, is composed of a series of grades, through which..each man passes in succession, until he attains the highest. a 1942 B. Malinowski Sci. Theory Culture (1944) v. 50 Studying an Australian tribe, we would have to follow the small family groups..the age-grades, and totemic clans. 1948 K. Davis Human Soc. (1959) iv. 107 Reliance upon age-grading is very prominent in African societies. 1950 C. F. Hockett in Language XXVI. 449 (title) Age-grading and linguistic continuity. 1904 Gen. Rep. Census Eng. & Wales 1901 147 The following Table, which gives the proportions of blind per million living at the earlier age-groups, shows [etc.]. 1930 Times Educ. Suppl. 7 June 257/2 The public elementary school age-group 10 to 11. 1936 J. T. Jenkins Fishes Brit. Is. (ed. 2) 177 Probably the best method of determining the growth and age-groups..is by measuring large numbers of individuals caught together on the same ground. 1937 Proc. Prehist. Soc. III. 182 To divide the sites into three age-groups merely by observing their heights above modern sea-level. a 1942 B. Malinowski Sci. Theory Culture (1944) vi. 57 More frequently the organization according to sex is related to..age groupings or age-grades. 1921 Hanson & Gayler in Jrnl. Inst. Metals XXVI. 345 The extent of the age-hardening which takes place is roughly proportional to the amount of magnesium silicide in solution at the moment of quenching. 1932 Discovery May 145/1 After quenching in water from about 475°–500° C. its [duralumin's] strength increases with time. This is known as 'age-hardening' which is now recognized as an extremely important phenomenon, not confined to this type of alloy. 1898 Strand Mag. XV. 331/2 It is doubtful whether the inexorable age limit will not preclude his inclusion in the next Conservative Ministry. 1917 Aberdeen Univ. Rev. June 259 The raising of the upper age limit [for examinees] to twenty or twenty one. 1925 D. H. Lawrence St. Mawr 77 The rocks..heavy with age-moulded roundnesses. 1938 R. Graves Coll. Poems 176 Time was my chronicler, my deeds age-new. 1928 Daily Express 7 Nov. 4 Annette Kellermann..demonstrates..that she is practically age-proof. 1929 Nat. Soc. for Study of Educ. Yearbk. XIV. 695 The chief concern of adults should be to make sure that..the age-range is kept fairly narrow. 1908 W. McDougall Introd. Soc. Psychol. iv. 109 The time of ripening of any instinct..is liable to be shifted forwards or backwards in the age-scale during the course of racial evolution. 1929 Age-set [see *age-fellow]. 1940 E. E. Evans-Pritchard Nuer vi. 255 The age-set system is a further exemplification of the segmentary principle. 1954 J. G. Peristiany in Instit. Prim. Soc. iv. 40 The initiation rituals..provide him with an age-set; that is, with a group of age-mates who remain his social co-evals through life. 1932 Wyndham Lewis Doom of Youth iv. i. 201 The 'Age-War' is really a Father-and-Children-war. 1895 W. B. Yeats Poems 147 And demons have lifted The age-weary eyelids from the eyes that of old Turned gods to stone. 1836 Todd Cycl. Anat. I. 805/1 The infirm and age-worn patients of Salpetrière. 1933 W. de la Mare Fleeting 135 An image of age-worn stone.

age, *v.* Add:

3. *Calico-printing.* To fix the mordants and printed colours in (cloth, etc.) by the process of ageing. Also *intr.*, to undergo this process. So **aged** *ppl. a.*

1849 [see *ageing vbl. sb. 2]. 1862 C. O'Neill Dict. Calico Printing 8/1 The difference of appearance will be..in favour of the aged or exposed part. 1890 W. J. Gordon Foundry 177 The calico..has to be dried and aged. 1910 E. Knecht et al. Man. Dyeing (ed. 2) II. 649 The dyeing of aged blacks. 1912 Knecht & Fothergill Textile Printing 138 If they [sc. the goods] are simply printed in aluminium mordants, one day may be quite sufficient to fully 'age' them. Ibid. 141 It is preferable to let them 'age' for a day or two in pile before dyeing.

4. *trans.* To mature by keeping in storage, by exposing to the air, etc.

1852 Swindells & Nicholson Brit. Pat. 390 1 For oxydating metallic solutions, and for ageing and raising various colouring matters. 1854 W. E. Staite Brit. Pat. 468 2 Madder which, technically speaking, has not been 'aged'.

5. To calculate the age of (a horse).

1887 M. H. Hayes Soundness & Age of Horses vi. 94 If a colonial animal in, say, September showed the condition of mouth just described, we should age him as five years old.

6. *intr.* Of iron, the iron core of an electrical transformer: to suffer a continuously increased loss in hysteretic quality.

1896 [implied in *ageing vbl. sb. 4]. 1899 S. R. Roget in Proc. R. Soc. 23 Jan. 154 Brands of transformer steel, which are practically 'non-ageing'. 1902 Encycl. Brit. XXVIII. 121/1 Brands of steel are now obtainable which do not age in this manner.

aged, *ppl. a.* Add: **1. a.** *aged parent*, applied joc. to a parent (whether elderly or not).

1861 Dickens Gt. Expect. II. vi. 92 You don't object to an aged parent, I hope? 1934 A. Ransome Coot Club iii. 40 'Tell her we won't be late. Macaroni cheese to-night. Specially for you, A.P.' That was Starboard talking to her Aged Parent.

3. *spec.* of a horse.

1869 F. Fitzwygram Horses & Stables lviii. 563 A moderately fresh aged horse is..more useful..than a young untried horse. 1882 [in Dict.]. 1951 E. Rickman Come Racing with Me ii. 17 Any horse or mare above six years is 'aged'.

ageing, *vbl. sb.* Add:

2. *Calico-printing.* The action or process of fixing the colours and mordants by exposing the printed goods to the action of a warm, moist atmosphere or by running them through hot steam (see quot. 1961). Also *attrib.*

1849 J. Thom Brit. Pat. 12,610 4 The ageing of goods by..passing them through a chamber containing aqueous vapour. 1850 S. Pincoffs Brit. Pat. 13,080 5 Improvements in the ageing process..consist in..the mode of introducing into the ageing room..a heated and moistened atmosphere. 1859 Ibid., The ageing stove. 1862 C. O'Neill Dict. Calico Printing 8/2 Ageing liquor. 1881 Instr. Census Clerks (1885) 69 Ageing Machine Minder. 1882 Jrnl. Soc. Chem. Industry I. 188 To prevent condensation,..the walls of the aging [sic] house are double. 1961 Blackshaw & Brightman Dict. Dyeing & Textile Printing 6 Ageing, originally a process in which printed fabric was exposed to a hot moist atmosphere. At the present time the term is almost exclusively applied to treatment of printed fabric in moist steam in absence of air.

3. The action or process of maturing pottery clay, wine, logwood, tobacco, etc. Also *attrib.*

1860 Tomlinson Arts & Manuf. 2nd Ser. Pottery 31 The mixture of clay and flint..ought to be kept..in order that the materials may become more intimately united...During this ageing, as it is called, a fermentation takes place. 1874 Knight Dict. Mech., Ageing (Wine and Liquors). Ibid., Wine-ageing Apparatus. 1910 E. Knecht et al. Man. Dyeing (ed. 2) I. 331 Ageing or Maturing of Logwood. 1935 Discovery Nov. 340/2 The ageing of tobacco is dependent on the action of micro-organisms.

4. The continually increasing hysteresis of iron under alternating magnetization, or when subjected to great heat.

1896 D. C. & J. P. Jackson Alternating Currents 539 Ageing of Transformer Cores. Ibid., The ageing seems to have the greatest effect upon poor qualities of iron.

b. In various other techn. uses (see quots.).

1902 Encycl. Brit. XXVIII. 90/1 Ageing of lamps. 1916 Jrnl. Soc. Chem. Industry XXXV. 873/2 After varying periods of ageing..both [rubber] compounds are much over-vulcanised at 4½ hours. 1930 Engineering 19 Sept. 358/1 Artificial ageing of Duralumin. 1930 Field & Weill Electro-Plating iv. 50 After some use a solution works better than an entirely new one. This is called 'ageing'. 1937 Jrnl. Inst. Civil Engin. VII. 115 There are records of the ageing of at least one pipe from which the local growth-rate [of internal roughness] can be found. 1945 Ann. Reg. 1944 386 The addition of small quantities of finely powdered copper..has the added advantage of increasing the resistance [of synthetic rubber] to ageing. 1958 Engineering 28 Mar. 410/3 The heat-treatment furnace is designed for solution treatment and artificial ageing, as well as annealing. 1959 B.S.I. News Nov. 21 Tests [of copper conductors] include..heat shock, heat ageing, [etc.].

agelessness (ē¹·dʒlėsnės). [f. Ageless a. + -ness.] The quality of being ageless.

1905 W. Watson Poems II. 137 And so the people of this land possess, Age after age, unaltering agelessness. 1906 A. C. Benson From a College Window xviii, The mighty unresting Heart, to whose vastness and agelessness the whole mass of these flying and glowing suns are but as a handful of dust. 1927 Observer 1 May 6 He had a rare gift with boys and young men, whom he could draw out by sheer agelessness.

age-mate. Delete † *Obs.* and add later examples.

1920 Q. Rev. July 171 Age-mates are associated at initiation. 1951 N. & Q. Anthrop. (ed. 6) ii. 68 Without definite age-sets there may be social and economic ties between age-mates. 1954 [see *age-set]. 1965 New Society 6 May 11/1 These children saw their environment as more threatening than did their agemates.

agency. Add: **6.** The office of an Indian agent, or the establishment forming the headquarters of one. *U.S.*

1707 S. Carolina Statutes at Large (1837) II. 314, I will neither directly nor indirectly trade with any Indian..during the time of my agency. 1824 Publ. Stat. U.S.A. (1856) IV. 25 It shall be the duty of each Indian agent to reside and keep his agency within, or near the territory claimed by the tribe or tribes of Indians for which he may be agent. 1878 J. H. Beadle Western Wilds xiii. 198 We concluded we had better see the Creeks at home, and started afoot for the Agency. 1895 C. King Fort Frayne xvi. 228 Eleven o'clock came and..no further authority from the agency.

attrib. 1873 J. H. Beadle Undevel. West xxv. 527 The last grain in the Agency storehouse was issued to them on the 14th. 1901 S. E. White Westerners viii. 57 Rain-in-the-Face was at once an agency Indian and a reckless man.

agenda. Pl. of Agend (sense 3), treated as a singular (a use now increasingly found but avoided by careful writers).

1907 N.U.T. Conference Agenda (Oxford) 30 This Conference protests against the action of the Executive in printing Supplementary Agendas for Conference. 1928 Daily Express 13 Jan. 1/1 Further conferences are to follow with..the Trade Union Council to tabulate agendas and settle difficulties. 1928 Observer 1 July 30/5 It sometimes happens that an agenda promises sensations. 1957 E. Hyams Into Dream ii. ii. 101 It's a short agenda, by the way, only two items. 1963 Times Rev. Industry June 3/1 Union agendas are increasingly devoted to industrial and economic issues.

b. *attrib.*, as **agenda-paper,** the paper containing the agenda of a meeting.

1887 Westmor. Gaz. 10 Dec. 2/5 (Stanf.), The next business stated on the agenda paper was to sign a petition [etc.]. 1905 Westm. Gaz. 24 July 2/2 The Czar..and the Kaiser..are meeting to-day...We are not given the agenda-paper of their conversation.

agendum (ădʒe·ndŏm). [See Agend.] = Agend 3.

1898 Westm. Gaz. 23 Mar. 1/2 A prearranged agendum and precise rules of debate. 1920 Daily Tel. 25 May 12/4 To cause the closing of the session before its agendum had been fully dealt with.

agene (ē¹·dʒīn). [Trade-name in U.S.A., f. Age v. or sb. + -ene.] Nitrogen trichloride used in bread-making for improving, stabilizing, and artificially ageing the flour. Hence **ageniza·tion**, the addition of agene to flour; **a·genized** *ppl. a.*

1932 Cereal Chem. IX. 360 When nitrogen-trichloride, commercially known as 'Agene', is applied to flour..its aging effect is..to be noticed. 1937 Hackh & Grant Chem. Dict. (1938) 26/1 Agene process, the bleaching of flour with nitrogen trichloride. 1947 Lancet 23 Aug. 284/1 Acceptability of bread as an article of food is largely increased by the use of an 'improver'...The one in most extensive use in this country, in America, and in Western Europe is nitrogen trichloride: known commercially as 'agene'. Ibid., Experiments which show that 'canine hysteria' can be produced by feeding dogs on agenised flour. 1949 Lancet 22 Jan. 143/2 In 1946, Sir Edward Mellanby showed conclusively that the agenisation of flour rendered it highly toxic to dogs. 1954 Times 22 Dec. 3/5 (heading) Use of Agene to be Discontinued.

agent, *sb.* Add:

1. b. *Telepathy.* The person who originates the impression received by the percipient.

1882 Proc. Soc. Psychical Research I. 119 In Thought-transference..both parties (whom, for convenience' sake, we will call the Agent and the Percipient) are supposed to be in a normal state. 1886 [see Percipient sb. b]. 1961 W. H. Salter Zoar xi. 149 Spontaneous cases [of telepathy] do occasionally occur in which no such connection between apparent agent and apparent percipient can be traced.

4. b. (In full *Indian agent.*) An official appointed to represent the government in dealing with an Indian people. *U.S.*

1707 S. Carolina Statutes at Large (1837) II. 311 Thomas Nairne..is..appointed the agent to reside among the Indians. 1816 Register of Officers & Agents U.S. 62 Indian agents have been allowed from 2 to 8 rations per day in addition to their annual compensation. 1886 Capt. Bell Report in Nation (1888) 15 Mar. 211/1 There can be but one head to an Indian agency, and the agent should be that head, if discipline is to be maintained. 1901 S. E. White Westerners xxxv. 317 He could not recall all the story he had told the Indian agent.

c. Ellipt. for *road-agent* (see Road sb. 12). *U.S.*

1876 Weekly Calaveras Chron. (Mokelumne Hill, Calif.) 29 July 3/1 The driver finally succeeded in satisfying the 'agent' that no express box was carried by San Andreas. a 1904 S. E. White Blazed Trail Stories 152 Nex' time I drives stage some of these yere agents massacrees me from behind a bush.

d. A secret agent; a spy.

1932 W. H. Auden Orators iii. 108 The agent clutching his side collapsed at our feet, 'Sorry! They got me!' 1946 Ann. Reg. 1945 230 Switzerland..had been full of German agents. 1956 A. H. Compton Atomic Quest ii. 117 Relaying secret information to Russian agents.

agent, *v.* Delete †*Obs.* and add later examples. So **a·genting** *vbl. sb.*, the work or duties of an agent.

1751 [see Extractor 2 b]. 1924 Galsworthy White Monkey ii. i, A man called Smith..who's done most of the agenting for the German business. 1951 L. Z. Hobson Celebrity (1953) vi. 84 The deal had been 'agented' by the author's own brother.

agentive (ē¹dʒe·ntiv), *a. Gram.* [f. Agent sb. + -ive.] Of or pertaining to a noun or suffix that indicates an agent or agency; also applied to certain verbal cases in some languages.

1903 Amer. Anthropologist Jan.–Mar. 26 Syntactical Cases and Appositions [in native languages of California]. Agentive, Subjunctive, Objective, Possessive. 1921 E. Sapir Lang. v. 87 The word farmer has an 'agentive' suffix -er that performs the function of indicating the one that carries out a given activity, in this case that of farming. 1941 E. A. Speiser Introd. Hurrian v. 211 Even the agentive verb is always in included position unless accompanied by the proper associative element. 1964 J. Vachek in D. Abercrombie et al. Daniel Jones 194 His suggestion that the morpheme /-ə/ forming the comparatives contrasts with the 'agentive' /-ə/.

‖ **agent provocateur** (aʒaṅ provokatœr). [Fr., = provocative agent.] An agent employed to induce or incite a suspected person or group to commit an incriminating act.

1877 W. de Horsey Let. 8 Jan. in A. Ponsonby Henry Ponsonby (1942) xiv. 324 You may think that I am looking through very coloured spectacles when I attribute..the Bulgarian atrocities to Russian intrigue—that Russian 'agents provocateurs' prepared the Servian rebellion. 1896 Le Queux Secret Service viii. 169 It was hinted to an agent provocateur that your death would be gratifying. 1923 Daily Mail 1 Mar. 8 Constantinople had long been the Mecca of the professional spy and agent provocateur.

age-old (ē¹·dʒ,ōuld), *a.* [f. Age sb. + Old a.] Having existed for an age or ages; very old.

1910 Sat. Westm. Gaz. 22 Jan. 6/2 When, with self-knowledge sated, The age-old hand upon its treasure rests. 1937 C. M. Arensberg Irish Countryman v. 174 The countryman will make his last resort to age-old folk-belief. 1939 D. Cecil Young Melbourne viii. 239 Frederic and Emily..found all their age-old hostility towards her melting away.

ager (ē¹·dʒəɹ). [f. Age v. + -er¹.] **a.** An ageing apparatus or chamber. Also *attrib.*, as *ager man, minder.* **b.** = ager man. **c.** An inspector of electric lamps.

1907 Cycl. Textile Work VI. 126 The steam ager consists of a long chamber containing two series of rolls..over which the cloth can be passed up and down. 1912 Knecht & Fothergill Textile Printing 143 Mather & Platt's Rapid Steam 'Ager'..was brought out in 1879. 1921 Dict. Occup. Terms (1927) § 308 Ager, flash ager; examines electric incandescent lamps, before and after cap is fitted on. Ibid. 398 Ager, ager man, ager minder, age minder, ageing machine minder..minds ageing machine.

ageusia (ăgiū·siă). *Med.* [mod.L., f. Gr. ἀ- priv. + γεῦσις sense of taste.] = Ageustia.

1848 in Dunglison Med. Lex. (ed. 7) 30/2. 1899 Allbutt's Syst. Med. VI. 797 Ageusia, therefore, implies either an extension of the neuritis into the Fallopian canal, or, according to some investigators, a concomitant lesion of the fifth nerve, which sometimes shows itself in diminished sensation over the paralysed side of the face.

agey (ē¹·dʒi), *a. arch.*, var. Agy *a.*

1547 W. Salesbury Dict. in Engl. & Welshe sig. G4v, Hoedloc, agey. 1672 Locke Let. in B. Rand Locke & Clarke (1927) 78 Those are earthly pleasures for clodpate mortals, and we agey men contemn them. 1852 P. J. Bailey Festus (ed. 5) 132 Thee, agey world, thee, universal Heaven.

agglutinate, *v.* Add:

2. b. *Bacteriology.* To cause agglutination or coalescence of (bacteria or red blood-corpuscles). Also *absol.* or *intr.*, to undergo agglutination. Hence **agglu·tinating** *ppl. a.*; **agglu·tinable, agglu·tinative** *adjs.*; **agglutinabi·lity,** the quality or property of being agglutinable; **agglutina·tion**; **agglu·tinator**; **agglu·tinin** (also -inine), an agent that causes agglutination; **agglu·tinogen**, a substance present in bacteria or blood cells, which stimulates the formation of agglutinins; hence **agglutino·genous** *a.*; **agglu·tinoid**, an agglutinin that has lost its agglutinophoric group, but retains the haptophoric group for the cell; **agglu·tinophore,** a molecular complex of the agglutinins to which their agglutinating property is due; hence **agglutinopho·ric** *a.*

1896 Lancet 19 Sept. 806/2 The 'agglutinines' found in the serum of immunised animals. Ibid., The use of the agglutinative action of human serum for the diagnosis of enteric fever. 1898 Brit. Med. Jrnl. 3 Sept. 589/1 Agglutinating or sedimentary properties of serums..their power of causing clumping (agglutination). Ibid. 592/2 A. S. F. Grünbaum..pointed out that even inanimate substances..might be agglutinated by various serums. 1900 Jrnl. Chem. Soc. LXXVIII. ii. 560 No satisfactory theory [to explain agglutination]..has yet been offered, although Bordet's, that an agglutinating agent (agglutinine) acting upon an agglutinable substance..is regarded as most rational. 1901 Jrnl. Exper. Med. V. 361 Any given race [of bacilli] does not necessarily produce the same quantities of the different constituents at different times, and hence the variations of agglutinability, virulence, etc. 1902 Jrnl. Exper. Med. 17 Mar. 289 Ricin, a strong agglutinator. 1903 Jrnl. R. Microsc. Soc. Feb. 78 The agglutinative and agglutinogenous functions are subject to the greatest variations. 1903 Lancet 4 Apr. 946/1 The existence of the agglutinins was one of the earliest results of modern investigations into immunity. Ibid. 946/2 The bacilli may lose their agglutinability,..viz., their agglutinable substance loses its functional atom-group. Ibid., A higher dilution would sometimes agglutinate when a lower one would not. Ibid., Agglutinins may become converted into agglutinoids. 1904 Jrnl. Med. Research Oct. 314 A. Joos..describes two other complementary agglutinable substances of the bacilli themselves, which he calls α and β agglutinogens. 1915 Lancet 13 Nov. 1086/2 The agglutination test..gave a positive result in practically every case of enteric fever. 1947 Lancet 1 Feb. 193/1 The term 'universal donor' was introduced when it was thought there were only two corpuscular agglutinogens. 1947 Sci. News IV. 49 Blood grouping has become more and more complex with the discovery of new specific agglutinable (clump-forming) substances. 1948 Ibid. VI. 98 When blood of a wrong (incompatible) group is given in a blood transfusion, the unsuitable red cells are clotted together and broken by agglutinins in the patient's circulation. 1950 Ibid. XV. 106 In no case did the serum agglutinate the red cells which came from the same blood sample.

aggrade (ăgrē¹·d), *v. Geol.* [f. Ag- + Grade, after *degrade*.] *trans.* To fill up (a river bed, valley, etc.) with detritus. Also *intr.*, to build up by aggradation. (The opposite of Degrade v. 6.) Hence **aggra·ded, aggra·ding** *ppl. adjs.*; also **aggrada·tion,** the process of raising a surface by the deposition of detritus; whence **aggrada·tional** *a.*

1898 T. C. Chamberlin in Jrnl. Geol. VI. 524 The degradation of the one furnishes the material for the aggradation of the other. 1902 Ibid. X. 758 The alternate depositional and aggradational work..done by the Missouri river. Ibid. 759 The channel at present is in a slightly aggraded and apparently still aggrading stage. Ibid. 760 The aggradation stage has..been inaugurated by the detour of the river...The present aggrading washes have made a little bottom in the lower twenty rods of the valley. 1904 Chamberlin & Salisbury Geol. (1905) I. i. 2 The deposition of material, whether on the land or in the sea, is aggradation. Ibid. 178 Streams carrying glacial drainage are usually aggrading streams. Ibid. 184 The stream in flood aggrades its plain, and degrades its channel. 1937 Wooldridge & Morgan Physical Basis Geogr. xii. 173 Flood alluviation, like aggradation by gravel, is closely linked with the development of meanders. 1946 F. E. Zeuner Dating the Past v. 130 In its course near the sea..a river aggrades during the mild phases and erodes during the cool phases.

aggravator. Add: †**2.** *slang* (also in corrupt form (*h*)*aggerawator*). A greased lock of hair (see quot. 1860 in Dict.). *Obs.*

1835 Dickens in Bell's Life in London 4 Oct. 1/1 His hair carefully twisted into the outer corner of each eye, till it formed a variety of that description of semi-curls, usually known as 'haggerawators'. 1859 F. Fowler Southern Lights 38 The ladies..are addicted to..strained hair, embellished with two or three C's—aggravators they call 'em—running over the temple. 1861 Temple Bar I. 226 The broad bull neck, and the 'aggerawator' curl.

aggregable, *a.* Delete ?*Obs.* and add: That may be aggregated *with* (other property).

1910 Ld. Halsbury Laws Eng. XIII. § 253. 204 Property accruing to a deceased person's estate after his death..is aggregated with the other aggregable property passing on the death of such person. 1924 Westm. Gaz. 12 Feb., The Parliamentary estates were aggregable with the marriage settlement funds. 1927 Daily Tel. 12 Apr. 9/1 Property which is [so] settled..is not aggregable when passing on the death of the life tenant with the other property passing on that death.

aggregate, *ppl. a.* and *sb.* Add: **A.** *ppl. a.*

9. *aggregate recoil*: the ejection, from the surface of a radioactive sample, of atoms additional to those which recoil on disintegrating (*B.S.I. Gloss. Terms Nucl. Sci.* 1962).

1919 R. W. Lawson in Nature 13 Feb. 464/2 To the recoil of a compact cluster of atoms of the active matter when one of the atoms contained in it disintegrates with an ejection of an α-particle..I recently gave the name of 'aggregate recoil'. 1926 —— tr. Hevesy & Paneth's Man. Radioactivity vi. 61 Aggregate recoil phenomena can also be observed with preparations in which the polonium was not deposited electrolytically.

B. *sb.* **4.** (Earlier examples.)

1795 Kirwan Elem. Min. (ed. 2) I. 338 Masses of different aggregates inhering or adhering to each other. Ibid. 370 Derivatives..differ from aggregates in this, that the associated ingredients are not visibly distinct.

5. Substitute for def.: Gravel, sand, slag or the like added to a binding agent to form concrete, tarmacadam, etc. (Further examples.)

1930 Engineering 19 Dec. 764/3 The importance of mineral aggregates for concrete. 1933 Archit. Rev. LXXIII. 217/1 The solid concrete balustrade has had the aggregate exposed. 1949 P. C. Carman Chem. Const. of Engin. Mat. xvi. 464 By mixing cement with sand or 'fine aggregate' and broken rock or 'coarse aggregate'..the resulting concrete is..stronger than cement itself. 1958 Daily Mail 16 July 7/2 Coated roadstone—known as 'tarmac'—which is a mixture of tar or bitumen with aggregates of natural stone or..slags.

6. *Metallurgy.* (See quots.)

1935 A. Sauveur Metallogr. of Iron & Steel (ed. 4) i. 8 When an alloy contains more than one of these phases, it is generally referred to as an aggregate. 1958 A. D. Merriman Dict. Metall. 3/1 Aggregate..In reference to metals and alloys, the term is applied to mechanical mixtures of two or more phases. Quenched steel, for example, is an aggregate of three phases: solid solution of carbon in gamma-iron, alpha-iron and iron carbide.

aggregation. Add: **1. b.** *Ecology.* The act or process of organisms coming together to form a group; a group so formed; = *association 12. (See also quot. 1927.)

1905 F. E. Clements Research Methods in Ecology 314. 1912 J. S. Huxley Indiv. in Anim. Kingdom iv. 110 In the making of Volvox, community-life—mere aggregation—came first, division of labour last. 1927 Haldane & Huxley Animal Biol. xi. 235 Aggregation is the joining together of a number of separate units to form a super-unit, as when coral polyps unite to form a colony. 1929 Weaver & Clements Plant Ecol. i. 3 The individuals come to be grouped, as a result of propagation, a process termed aggregation.

aggress, *v.* Add: Hence **aggre·ssed** *ppl. a.*; also as *sb.*

1801 Ann. Reg. 1799 ii. 285/2 The aggressed party shall have been properly indemnified for the damages suffered. 1890 J. Middlemass Two False Moves III. iii. 37 Are you the aggressor or the aggressed? 1930 Observer 7 Sept. 8/7 The aggressed State would raise loans..but the service of the loans would be guaranteed by the signatory States.

aggressin (ăgre·sin). *Med.* Also **-ine.** [G.; cf. Aggress v. and -in¹.] Bail's term (*Archiv f. Hygiene* (1905) LII. 340) for a hypothetical substance which increases the toxic effect of bacteria in the body.

1906 Johns Hopkins Hospital Bull. Dec. 406/2 The sterile fluid seeming..to produce conditions favorable for the growth and reproduction of the bacteria..Prof. Bail called it aggressine at the suggestion of Prof. Kruse. 1907 Chem. Abstr. 2804 The aggressin exudates of Bail can be separated in two parts by means of dialysis. 1913 Dorland Med. Dict. (ed. 7) 42/2 Aggressin, a substance supposed to exist in the body of an infected animal and to hasten or make aggressive the action of the bacilli.

aggression. Add: **2.** (Later examples.)

1955 Bull. Atomic Sci. Mar. 97/1 This is clearly the future

reserved for all those who are unable or unwilling to resist Communist aggression or infiltration. 1957 R. N. Carew Hunt Guide Communist Jargon 1 Aggression can be predicated only of imperialist Powers, which is in line with the distinction made in communist theory between 'just' and 'unjust' wars. 1959 Observer 19 Apr. 1/4 A 21-nation United Nation General Assembly Committee has voted to delay for three years an attempt to define aggression.

3. *Psychol.* Hostile or destructive tendency or behaviour, held to arise from repressed feelings of inferiority, frustration or guilt. (Except in textbooks not clearly distinguishable from senses 1 or 2.)

1912 A. A. Brill tr. Freud's Sel. Papers on Hysteria vii. 160 We no longer deal here with sexual passivity but with pleasurably accomplished aggressions. 1917 Glueck & Lind tr. Adler's Neurotic Constitution (1921) v. 156 The neurotic succeeds in..this new line so skillfully as to manage to set up an aggression which enables him to dominate and torture others. 1918 A. A. Brill tr. Freud's 3 Contribs. to Theory of Sex 22 The sexuality of most men shows a taint of aggression. 1932 S. Herbert Unconscious in Life & Art 153 For the repression rouses the innate tendency to aggression against those who interfere with the original libidinous impulses. 1943 J. S. Huxley Evol. Ethics iii. 22 Victims of a distorted and hyperactive conscience, charged with aggression which seeks an outlet in attacking in others all those tendencies of which itself bears the unconscious guilty burden.

aggressive, *a.* Add: **2. b.** *Psychol.* Of, pertaining to, or characterized by, aggression (sense 3 above).

1913 A. A. Brill tr. Freud's Interpr. of Dreams iv. 134 In the sexual make-up of many people there is a masochistic component, which has arisen through the conversion of the aggressive, sadistic component into its opposite. 1917 Glueck & Lind tr. Adler's Neurotic Constitution (1921) v. 157 Forms of the aggressive impulse become accentuated through the feeling of inferiority. 1949 Koestler Insight & Outlook v. 56 A component..of aggressive-defensive self-assertion, has been recognized in laughter. 1950 B. Wootton Test. Soc. Sci. ii. 8 Aggressive behaviour means physical actions like letting off bombs, or, on a milder scale, giving vent to cross words or looks.

c. Self-assertive, pushful; energetic, enterprising. Chiefly *U.S.* and *Canadian colloq.*

1930 Vancouver newspaper (Advt.), Aggressive clothing salesman with ambition, to manage a large..store. Good salary..for the right man. 1956 Winnipeg Free Press 19 Jan. 34/1 We require 2 salesmen!..Only aggressive men need apply. 1959 H. E. Salisbury Shook-up Generation xi. 154 'Aggressive group work'..means that instead of sitting in offices..the Youth Board goes out on to the street, finds the youngsters who are in trouble..and begins to work with them right in the neighbourhood. 1966 Which? Aug. 268/1 Co-operative Societies are giving private industry an example of aggressive retailing.

aggressively, *adv.* (Earlier and later examples.)

1800 Ann. Reg. 1795 (Dodsley) Hist. 145/1 Unwillingly to act aggressively, in so dubious a case, he proposed a negotiation for peace. 1847 De Quincey in Tait's Edin. Mag. XIV. 680/2 Johnson..relied sturdily upon his natural powers for carrying him aggressively through all conversational occasions. 1963 Publishers' Weekly 23 Sept. 34/1 An aggressively modern bookstore has been opened in Oxford.

aggressiveness. (Later example.)

1943 J. S. Huxley Evol. Ethics viii. 68 It appears more than probable that, for instance, certain psycho-physical types are predisposed to callous and tireless aggressiveness.

aggressor. Add: Also *attrib.*, as *aggressor nation, state.*

1768 [in Dict.]. 1930 S. O. Levinson in E.R.A. Seligman Encycl. Soc. Sci. (1935) I. 485/2 In both the American Revision and the Protocol an attempt was made to base the definition of an 'aggressor nation' on refusal of arbitration. 1940 King George VI in Times Weekly 27 Nov. 11/1 My peoples and My Allies are united in their resolve to continue the fight against the aggressor nations until freedom is made secure. 1941 Spectator 5 Sept. 225/1 Great Britain, left to face the aggressor States alone..was keeping the enemy at bay.

aggrievedly (ăgrī·vėdli), *adv.* [f. Aggrieved *ppl. a.* + -ly².] In an aggrieved manner.

1883 R. Broughton Belinda xiii, 'I offered to teach it to Sarah,' he says aggrievedly. 1886 R. A. King Shadowed Life vi, He growled as aggrievedly as though it was his wife. 1906 Daily Chron. 3 Feb. 4/4 Mr. Manktelow expressed frank incredulity and they went aggrievedly.

agila (æ·gilă). Also 6 **aguila,** 7 **aguala,** 8 **agala,** 9 **agilla.** [ad. Sp., Pg. *aguila* in *palo de aguila, pao d'aguila* (= wood of a.): see Eagle-wood and Agalloch.] A resin or resinous wood: = Agalloch, Calambac. Also *agila-wood.*

1588 Aguila [see Cayolac]. 1699 Dampier Voy. II. i. 8 Pepper, Lignum Aloes, and Aguala Wood. 1727 A. Hamilton E. Indies II. 193 It [sc. the Siam Coast] produces much Agala and Sapan-woods. 1846 Lindley Veg. Kingd. 579 Aloes-wood, Agila-wood, or Eagle-wood. 1871 Agilla [see Calambac]. 1881 Spons' Encycl. Industrial Arts 1523 Agar, Agila, Akyaw, Calambak, Eagle-wood, Kayugaru, or Lignum-aloes.

agin (ăgi·n), *prep.* Widespread dialectal var. Again *prep.* (= against) and Against *prep.*, often used jocularly and esp. in phr. *agin the government*, taken to represent the typical Irishman's or countryman's attitude in politics.

1823 E. Moor Suffolk Words 5 'A struv agin um as long as 'a could. 1841 Lever C. O'Malley iii, 'Have ye a great spite agin him?' 'I have,' said I fiercely. 1847 Sporting Life 23 Oct. 114/1 A pound to a hay-seed agin' the bay. 1878 W. S. Gilbert H.M.S. Pinafore ii. 26 I'm unpleasant to look at, and my name's agin me. 1888 Nation (N.Y.) 2 Aug. 81/3 It was as natural for them to support the party in power..as it was for the 'Pat' of the anecdote to reply, when asked on landing which party he belonged to: 'I'm ag'in the gover'mint.' 1904 Athenæum 5 Nov. 615 He was by nature 'agin' the Government'. 1920 Galsworthy In Chancery ii. xii, Possession, vested rights; and anyone 'agin' 'em—outcast. 'Thank Heaven!' he thought, 'I always felt "agin" 'em, anyway!' 1920 Punch 21 July 53/1 To compel the poor thing [sc. the motor car] to crawl is 'agin natur''. 1960 Guardian 4 Nov. 14/3 The Kennedy crowds..are noticeably..people who are agin the Government. 1961 New Statesman 28 Apr. 656/3 Ashington has one other grievance agin the government. 'Why don't they encourage light industry to come yere?' 1963 Listener 17 Jan. 120/2 The real breeder, the chap with green fingers for dealing with living things, knows that artificial insemination is 'agin nature'.

agistment. **1.** (Further examples.)

1885 Sheffield Tel. 30 June (Advt.), Agistment. Good pasture for cattle and horses. 1955 Times 16 July 11/5 A number of cattle had to be sent away on agistment.

agitatingly (æ·dʒitē¹·tiŋli), *adv.* [f. Agitating *ppl. a.* + -ly².] In such a manner as to cause agitation.

1819 Blackw. Mag. VI. 160 That such a mind should have known miserable thoughts so well as to be enabled thus agitatingly to paint them. 1887 Pictorial World 28 Apr. 410 He had been deeply, not to say agitatingly, impressed.

agitation. Add: **8.** (Further example.) Also, *spec.* of agitation on behalf of Communism.

1923 E. Rice Adding Machine iii. 46 Too damn much agitation, that's at the bottom of it... Foreign agitators, that's what it is. 1946 L. Harry Gould Marxist Glossary 11 Agitation, the act of rousing the masses to political action around some particular social injustice. 1948 J. Towster Pol. Power in U.S.S.R. x. 195 For the duration of the campaign, agitation points (Agitpunkty) are set up in the election precincts. 1957 R. N. Carew Hunt Guide to Communist Jargon 3 Ozhegov's Dictionary of the Russian Language defines agitation as 'Oral and written activity among the broad masses which aims at inculcating certain ideas and slogans for their political education and for attracting them to the solution of the more important social and political tasks'. 1958 Economist 1 Nov. 427/1 But in the Soviet Union, the Department of Propaganda and Agitation is one of the seven sections of the central committee of the Communist party.

agitator. Add: **2.** (Further examples.) After the Bolshevik Revolution freq. applied *spec.* to Communist agitators.

a 1734 R. North Examen (1740) i. iii. 195 The visible Agitators of all the Seditions and Troubles of King Charles the Second's Reign. 1852 Emily Tennyson in Rev. Eng. Stud. (1964) XV. 402 Do not speak of him as an 'agitator'... The word has come to have so evil a meaning, a sort of hysterical lady meaning if nothing worse. 1920 Independent 31 Jan. 161/1 Bolshevik Representative Klishke—'Soviet Russia will not allow itself to be used as a dumping ground for agitators from America.' 1931 New Statesman 4 July 4/1 Mr. Winston Churchill, the notorious British agitator (we adopt the phraseology of the Morning Post when describing M. Bukharin and other distinguished Russians now in this country) has now decided that disarmament is impossible, because of the menace of Russia. 1934 T. S. Eliot Rock i. 31 On the fore-stage, an agitator is addressing a tattered crowd. 1938 Auden & Isherwood On the Frontier i. ii. p. 46 Col. Hussek. Tcha! Another lightning strike at the Docks!..Mrs. Vrodny. I'm sure it's only due to Westland agitators. 1959 Observer 29 Mar. 11/4 Mrs. Nora Jefferey said at the British Communist Party congress..that the Communist aim should be: Every member a propagandist and agitator.

3. Add to def.: In various techn. uses. (Further examples.)

1809 J. Dickinson Brit. Pat. 3191, Machinery for Making and Cutting Paper. Ibid. 7, B is an agitator consisting of a number of arms..and this being turned..keeps the stuff in motion. 1839 Ure Dict. Arts 72 The agitator is then suspended to a spring R, and..the operator gives an alternating rapid movement, which agitates the solution. 1937 Times 13 Apr. xiii/4 The paint on arrival from the manufacturers is subjected to what is termed 'agitating' or mixing. When it enters the containers it meets two more agitators, one at the top and one at the bottom of the container. 1958 Ibid. 2 June ix/5 Gas is used to boil the clothes and an electric agitator helps with the washing. 1958 Listener 19 June 1006/2 Mechanical equipment such as valves, pumps, centrifuges and agitators in the radioactive section.

Agit-prop, agit-prop (æ·dʒit,prǫ:p, æ·g-). [f. Russ. *agitprόp*, f. *agit(átsiya* agitation + *prop(agánda* propaganda.] A department of the Central Committee of the Russian Communist Party responsible, with its local branches, for 'agitation and propaganda' on behalf of Communism; its activities. Also, a person engaged in agitprop. Also *transf.*

1934 N. & Q. CLXVI. 73/1 The A[g]itprop, the central organ for propaganda and agitation, has sent word round to writers, newspapers and publishers, that there is to be an organisation for mass-laughter. 1935 Time 17 June 38/1 Far more serious, far more earnest is the Depression-born movement of workers' theatres which are currently putting on 'agit-prop' (agitational propaganda) plays in 300 U.S. cities. 1936 Times Lit. Suppl. 22 Feb. 146/3 Primitive 'agitprop' and mass-chant plays. 1949 'G. Orwell' Nineteen Eighty-Four 307 The tendency to use abbreviations of this kind was most marked in totalitarian countries and totalitarian organizations. Examples were such words as Nazi, Gestapo, Comintern, Inprecorr, Agitprop. 1950 A. Koestler in God that Failed 51 The job of Agit.-Prop. fell to me soon after I had joined the cell. 1952 G. A. Duncan in R. S. Sayers Banking in Brit. Commonw. viii. 314 They [sc. the banks] have made no effort to counter the flood of often malicious..but sometimes pointed agitprop. 1952 Economist 1 Mar. 508/2 Any businessman who goes to Moscow in the belief that he will be able to strike an effective blow for anything he believes in and the Communists do not is simply inviting the 'Agitprop' experts to make a monkey of him. 1959 Spectator 6 Nov. 629/2 The whole tone [of the play] is ten times heavier and cornier than any of the agitprop from the old Unity Theatre.

agley·, *adv. Sc.* Also **aglee.** [f. A *prep.*¹ + Glee, Gley *v.*] Asquint, askew, awry.

1785 Burns To a Mouse in Poems (1786) 140 The best laid schemes o' Mice an' Men, Gang aft agley. 1813 E. Picken Misc. Poems I. 67 We haena mense like cruel man; Yet tho' he's paukier far than we, Whatreck, he gangs as aft aglee. 1880 [see Go v. 4 d]. 1887 R. L. Stevenson Underwoods xx, Or lads that tak a keek a-glee At sonsie lasses. 1929 C. Day Lewis Transit. Poem ii. 23 These are they who built My house and never a stone of it laid agley.

aglist (ăgli·st), *adv.* Also **agli·sten.** [f. A *prep.*¹ + Glist *v.*] Glistening.

1858 Mrs. Oliphant Laird of Norlaw lxvi, All aglist with early morning dews and sunshine. 1891 Murray's Mag. Aug. 291 The garden all aglisten 'twixt two showers. 1916 E. Blunden Harbingers 32 Carriage-roofs aglisten with rain.

aglucone: see *aglycone.

agly, *adv.* Delete entry, and see *agley *adv.*

aglycone (ăglai·kōun). *Chem.* [ad. G. *aglykon* (A. Windaus et al. 1925, in *Zeitschr. f. Physiol. Chemie* CXLV. 38), f. A- 14 + Glyco- + -o)ne.] The portion of a glycoside which remains when the sugar component is removed; called an **aglu·cone** when the sugar involved is glucose.

1925 Jacobs & Hoffmann in Proc. Soc. Exp. Biol. & Med. XXIII. 214 The digitalis glucosides, digitoxin, and gitoxin..contain different although possibly related aglucones, respectively digitoxigenin and gitoxigenin. 1938 A. T. Cameron Textbk. Biochem. (ed. 5) iv. 94 The 'aglycone' (non-carbohydrate portion) of the glucoside. Ibid. v. 122 These aglycones..possess chemical structures which relate them to the sterols. 1941 Ann. Reg. 1940 344 The aglucone fraction of the cardiac glucosides..[has] the same skeleton as cholesterol. 1964 Guthrie & Honeyman Chem. Carbohydrates (ed. 2) viii. 58 The non-sugar portion [of a glycoside], known as the aglycone (aglucone when the substance is a glucoside), is an alcohol or a phenol.

agnomination. **2.** (Later examples.)

1682 Keach & Delaune Philologica Sacra ii. ii. 3 in τροποσχηματολογια II, Agnomination, or Likeness of Words..is when by the Change of one Letter or Word, the Signification thereof is also changed. 1962 G. K. Hunter John Lyly iv. 244 The highly patterned prose..is not there simply because Lyly was suffering from a bad dose of agnomination or isocolon.

agnosia (ægnōu·siă). *Path.* [mod.L., a. Gr. ἀγνωσία ignorance (ἀ- priv., γνῶσις knowledge: see Gnosis).] Freud's term (*Zur Auffassung der Aphasien*, 1891) for loss of perception (sense 7).

1900 Dorland Med. Dict. 27/2 Agnosia, loss of the perceptive power; loss of the power to recognize persons or things seen. 1940 Hinsie & Shatzky Psychiatric Dict. 16/1 In psychiatry the term agnosia is employed to indicate the loss or disuse of knowledge of objects, when the knowledge appears to have been altered by emotional circumstances. 1953 E. Stengel tr. Freud's On Aphasia 78 For disturbances in the recognition of objects, which Finkelnburg called asymbolia, I should like to propose the term 'agnosia'. 1961 Brit. Med. Dict. 57/1 Agnosia is found in relation with the senses: auditory agnosia, gustatory agnosia, olfactory agnosia, optic agnosia, and tactile agnosia.

agnostical (ægnǫ·stikăl), *a.* [f. Agnostic + -al.] = Agnostic *a.*

1884 Punch 20 Sept. 135 The Agnostical or Nothingarian creed. 1886 Sat. Rev. 25 Dec. 849 Our Agnostical friend passed to the paradise of Mahomet. 1907 Daily Chron. 23 Mar. 6/4 Anthropomorphism is the most agnostical and atheistical habit of thought.

agogic (ăgǫ·dʒik), *a.* [f. Gr. ἀγωγός leading, guiding, f. ἄγειν to lead + -ic.]

†**1.** Of or pertaining to modelling in wax.

1662 Evelyn Sculptura i. i. 7 Neither the Paradigmatic, Agogic, or any of the Plastic [Arts], can Genuinely..be call'd Sculpture.

2. *Mus.* [G. *agogik* (Riemann 1884).] Applied to a kind of accent consisting in a lengthening of the time-value of the note. Also **ago·gical** *a.* So **ago·gics,** the use of agogic accents.

1893 J. S. Shedlock tr. Riemann's Dict. Mus. (1899) 13/1 Agogic accent is the name given by H. Riemann..to the slight prolongation of the note-value, in rhythms, which are in conflict with the species of time. Ibid., Agogics..relates to the small modifications of tempo..which are necessary to genuine expression. 1921 Mus. Assoc. Proc. 1919–20 23 Treatises..on Aesthetics, Expression, Agogics, Phrasing. 1922 S. Grew Art of Player-Piano 62 'Agogic restraint' is the term found by German pedagogues to describe this detail of rubato. Ibid. 63 Clashings of discordant notes may be..intensified by an 'agogical' pressure. 1962 Times 15 Jan. 14/4 There are subtleties of agogic accent.

agon. Add: **2.** A verbal contest or dispute between two characters in a Greek play. Also *transf.*

1887 Amer. Jrnl. Philol. VIII. 197 The participants in an Agon are the Choros.., the two antagonists, the judge, and the clown. 1914 F. M. Cornford Origin Attic Comedy i. 2 What is now generally called the Agon, a fierce 'contest' between the representatives of two parties or principles, which are in effect the hero and villain of the whole piece. 1918 R. C. Flickinger Greek Theater & its Drama 41 The agon, a 'dramatized debate' or verbal duel between two actors, each supported by a semi-chorus. 1932 T. S. Eliot Sweeney Agonistes 23 Fragment of an Agon.

agonal (æ·gŏnăl), *a.* [f. Gr. ἀγών (see Agon) + -al.] **1.** Of or pertaining to an Agon.

1770 tr. Mme. Fiquet du Boccage's Lett. II. 38 Sea-fights are now exhibited in the agonal circus called the square of Navonne. 1838 Fraser's Mag. Nov. 516 Come to us Our saviour, and the patron of our games: And with thee come our Agonal deities all.

2. [f. Agony.] Of, pertaining to, or occurring during agony, esp. the death agony.

1901 Jrnl. Exper. Med. Jan. 344 A condition which, if not due solely to agonal contraction, might at least favor the production of diverticula in the upper part. 1961 Lancet 29 July 266/1 Antemortem autolysis may occur when the agonal period is drawn out.

agonist. Delete † and add: **2.** A person engaged in a contest or struggle; a protagonist. (For the *spec.* sense in quot. 1914 cf. *agon 2.)

1914 F. M. Cornford Origin Attic Comedy v. 71 Three, or sometimes four, rôles are involved in the Agon...First there are the two Adversaries (as we shall call them). For the sake of convenience, we shall distinguish them as the 'Agonist' and the 'Antagonist'. The Agonist is the hero, who is attacked, is put on his defence, and comes off victorious. 1921 Glasgow Herald 4 Aug. 5/3 He knows too well the respective roles of agonist and spectator of life. 1933 E. K. Chambers Eng. Folk-Play 23 The culminating point of the Drama is of course the Combat. It will be convenient to call the champion who falls the Agonist and his vanquisher the Antagonist. 1934 Punch 14 Feb. 195/1 Since this is a novel and not an economic treatise, the high lights are naturally focussed on particular agonists.

3. One who advertises in an 'agony column' (see Agony 1 a).

1915 Chambers's Jrnl. 6 Feb. 149/1 Yet somebody must respond, or a number of the 'agonists' would require to retire..from the business. 1934 I. Brown in Essays of Year 1933–4 p. xx, He even scours the advertisements, for the Agonists of The Times are often helpful.

agonizing, *ppl. a.* Add: **1. b.** *agonizing reappraisal*, a reassessment of a policy, position, etc., painfully forced on one by a radical change of circumstances, or by a realization of what the existing circumstances really are. (Chiefly a political and journalistic catch-phrase.)

1953 J. F. Dulles in N.Y. Times 15 Dec. 14/3 If..the European Defence Community should not be effective; if France and Germany remain apart...That would compel an agonizing reappraisal of basic United States policy. 1958 Economist 18 Oct. 222/2 New Zealand..is being forced into an agonizing reappraisal of its domestic policies. 1958 Star 9 Dec., As if in response to new directions from an agonising reappraisal in MCC's room at lunch, the scoring spurted as Cowdrey twice swung Benaud to the leg fence.

agony. Add: **1. a.** *agony column* (earlier and later examples).

1863 Fun 3 Oct. 23/2 Our own agony column. 1930 Wyndham Lewis Let. 30 July (1963) 190 The agony-column of the Times has echoed the rage of people who considered themselves attacked in the Apes.

4. b. *transf.* and *fig.*

1835–40 etc. [see Pile v.¹ 2 b]. 1863 Geo. Eliot Let. 23 Oct. (1956) IV. 111 We shall soon be in the agonies of moving. 1924 R. Campbell Flaming Terrapin ii. 25 The mountains frown, Locked in their tetanous agonies of stone. 1932 W. B. Yeats Words for Music 2 Dying into a dance, An agony of trance, An agony of flame that cannot singe a sleeve.

agora. Add: Pl. **agoræ.** (Earlier and later examples.) Also *transf.*

1598 Hakluyt Voyages I. 489 The Emperor himselfe, who hath no other seat of Empire but an Agora, or towne of wood, that moueth with him whithersoeuer hee goeth. 1862 E. Falkener Ephesus i. iv. 63 The Greeks, in these hot climates, loved to have water in some form in the centre of their agoræ. 1886 W. T. Stead in Contemp. Rev. May 654 The telegraph and the printing-press have converted Great Britain into a vast agora, or assembly of the whole community. 1941 Auden New Year Let. iii. 53 The agora of work and news Where each one has the right to choose His trade, his corner and his way.

Comb. 1948 L. MacNeice Holes in Sky 66 Ancient Athens Was a sparrow-chatter of agora-gibes.

agoraphobic (æ:gorăfōu·bik), *a.* (*sb.*) [f. Agoraphobi(a + -ic.] Of or pertaining to agoraphobia; (one who is) suffering from or affected with agoraphobia. Also *transf.*

1884 Lancet 27 Dec. 1141/1 The giddiness which accompanies his agoraphobic attacks. 1898 Lancet 19 Nov. 1322/2 All the time that I was an 'agoraphobic' I attended church regularly with my family, but we always sat close to the door. There I was safe. 1936 A. Strachey tr. Freud's Inhib., Sympt. & Anx. vii. 89 An agoraphobic patient may be able to walk in the street provided he is accompanied..by someone he knows and trusts. 1958 Times 4 Feb. 6/4 Chardin comes at once to mind as an agoraphobic figure quite unconcerned with anything outside his own four walls.

agraffe. Add: **1.** (Earlier example.)

a 1666 Evelyn Diary c. 16 Nov., an. 1643 (1955) II. 88 The agraffe of his royal mantle.

2. Various technical uses: see quots.

1883 Lond. & Prov. Music Trades Rev. Mar. 13/2 Erard [in 1808]..substituting for a long, pinned wooden bridge as many little brass bridges as there were notes. The strings passing through holes bored through the little bridges, called agraffes, or studs, turned upwards towards the wrest-pin. 1900 Dorland Med. Dict. 27/2 Agraffe, an instrument for keeping together the edges of the wound in operation for harelip. 1918 E. S. Farrow Dict. Mil. Terms 11 Agraffe,..the coupling pin in artillery.

agrammatism (ăgræ·mătiz'm). *Path.* [f. Gr. ἀγράμματ-ος illiterate (see Agrammatist) + -ism.] A form of aphasia marked by an inability to form sentences grammatically.

[1881 Syd. Soc. Lex. I, Agrammatismus, inability to form a grammatical sentence.] 1888 F. P. Foster Illustr. Encycl. Med. Dict. I. 117/1 Agrammatism. 1909 A. Pick in Rev. Neurol. & Psychiatry VII. 757 When the question came under consideration, more than ten years ago, it took the form of inquiring whether we could assume for agrammatism as one of the higher mental functions a circumscribed localisation. 1956 Jakobson & Halle Fundamentals of Lang. ii. iv. 72 A typical feature of agrammatism is the abolition of inflection.

agranulocyte (ăgræ·niŭlosəit). [f. A- *pref.* 14 + *granulocyte.] A leucocyte without cytoplasmic granules (see also quot. 1928). So **agra:nulocy·tic** *a.*, of or pertaining to such a leucocyte; **agra:nulocyto·sis** [mod. L., f. G. *agranulozytosen* (W. Schultz 1922, in *Deutsche Med. Wochenschr.* XLVIII. 1496)], an agranulocytic condition.

1927 D'A. Prendergast in Canad. Med. Assoc. Jrnl. XVII. 446 (title) A Case of Agranulocytic Angina. 1928 R. J. Scott Gould's Med. Dict. (ed. 2) 43/2 Agranulocytosis, absence, or great decrease in the number of granulocytes. 1928 Jrnl. Amer. Med. Assoc. XCI. 1718/2 This case resembles..the case of agranulocytic angina first reported by Dr. Werner Schultz in 1922. 1932 Dorland & Miller Med. Dict. (ed. 16) 55/2 Agranulocyte. 1946 Nature 24 Aug. 270/1 Despite the severe leucopenia and agranulocytosis which developed in the rats on the purified diet, the rate of growth of the nymphs was not affected. 1954 Sci. News Let. 16 Jan. 36/3 The granulocyte cell is most commonly involved in myeloid leukemia and in the fighting of infections. The agranulocyte cell originates in the lymph glands and carries off waste materials.

agrapha (æ·grăfă), *sb. pl.* The sing. form **agraphon** is rare. [Gr., neut. pl. of ἄγραφος unwritten.] The collective name given to sayings attributed to Jesus but not recorded in the canonical Gospels.

[1889 A. Resch (title of book written in German) Agrapha.] 1890 Church Q. Rev. XXXI. 6 Resch has been successful in calling attention to some 'agrapha' which his predecessors had not noticed. 1900 B. Jackson 25 Agrapha 8 J. G. Körner (†1785) of Leipzig, De sermonibus Christi ἀγράφοις, [was] the first..to use the term Agrapha of the extra-canonical Sayings of the Lord. 1910 Encycl. Brit. I. 382 Agrapha (i.e. 'unwritten'), the name given to certain utterances ascribed, with some degree of certainty, to Jesus, which have been preserved in documents other than the Gospels. 1920 H. G. E. White Sayings of Jesus from Oxyrhynchus p. xxx, Harris pointed out that agrapha are quoted by St. Paul, Clement of Rome and Polycarp. 1950 Sc. Jrnl. Theol. III. 300 In particular he lays the axe to the word mysterion as not being a genuine word of Jesus, in spite of the fact that an agraphon survives through Clement of Alexandria.

agraphy (æ·grăfi). Anglicization of Agraphia.

a 1901 Myers Hum. Pers. (1903) I. 65 Elements of agraphy, of word-blindness, of word-deafness appear.

agree, *v.* Add: **4.** (Further example.)

1928 Times 15 Aug. 7/5 The actual figures of profits were agreed between the accountants.

5. Delete †*Obs.* and add later examples.

1928 Britain's Industr. Future (Liberal Ind. Inquiry) 140 These councils should have the power to agree factory rules. 1959 Bookseller 13 June 1982/1 The Russians have agreed a wide list of categories. 1963 Listener 23 May 877/3 Miss Laski's letter..shows once more the difficulty of agreeing a definition of mysticism.

12. a. *I couldn't agree* (*with you*) *more*: I am in complete agreement (with you).

1942 J. B. Priestley Black-out in Gretley viii. 185 'I couldn't agree with you more,' he said, grinning. 1953 —— Try it Again in Best One-Act Plays 1952–3 109 Kramer. Whatever I do can hardly make things worse. Helen. I couldn't agree more. 1960 L. Cooper Accomplices ii. i. 77 You think it's a nasty cold-blooded business..? I couldn't agree more.

agreeable, *a.* Add: **6. c.** *to do* (or †*make*) *the agreeable*, to make oneself pleasant, show courteous attentions. (Cf. Fr. *faire l'agréable.*)

1825 H. Wilson Memoirs I. 52 It fatigues me to death to be eternally making the agreeable to a set of men who might be all buried, and nobody would miss them. 1834 J. K. Townsend Narrative 31 Mar. in R. G. Thwaites Early Western Travels (1905) XXI. i. 127, I endeavored to do the agreeable to the fair ones. 1851 J. J. Hooper Widow Rugby's Husb. 168 The Colonel does the agreeable to strangers.

agreed, *ppl. a.* Add: **5. b.** *agreed syllabus* (see quots.).

1944 Act 7 & 8 Geo. VI, c. 31 § 114 (1) 'Agreed syllabus' means..an agreed syllabus of religious instruction. 1945 Guide Educ. Syst. Eng. & Wales (Min. Educ.) 57 Agreed syllabus, syllabus governing undenominational religious instruction in county and controlled schools, and drawn up or adopted for each area by a conference representing the religious denominations, the teachers and the L.E.A.

‖ **agrément** (agremaṅ). Also (*pl.*) **agrémens.** [F.: see Agreement.] **1.** *pl.* = Agreement 9.

1711 Spectator No. 28, 2 Apr., I had ghessed by the little Agréemens upon his Sign, that he was a Frenchman. 1749 Chesterfield Let. 14 Nov. (1774) I. 489 You must..captivate them by the agrémens, and charms of conversation. 1841 Geo. Eliot Lett. (1954) I. 96 All the agrémens it [sc. honeysuckle] brings us. 1848 Thackeray Van. Fair lxii. 566 We..talked..about the agrémens of the place. It was very agreeable for the English. 1863 'Ouida' Held in Bondage I. vii. 153 Lansquenet, racing, Coralies, champagne, and all one's other habitual agréments. 1904 W. James Let. 1 Jan. in R. B. Perry Tht. & Char. of W. J. (1935) II. 201 America does not offer the agréments to a tourist which almost any part of Europe offers.

2. *Mus. pl.* Grace-notes; embellishments.

1789 Burney Hist. Mus. III. i. 87/1 Extraneous episodes, or fashionable divisions, which being the agrémens, or trimmings, of the times, become antiquated. 1871 J. Hiles Dict. Mus. Terms 9 Agréments, graces, embellishments, ornaments. 1879 Grove Dict. Mus. I. 44/1 The agrémens used in modern music..are the acciacatura, appoggiatura, arpeggio, mordent, nachschlag, shake or trill, slide, and turn.

3. The approval given by the government of a country to a diplomatic representative of another country.

1917 E. Satow Guide to Diplomatic Practice I. xiv. 193 The generally existing diplomatic practice to ask, previously to any nomination of a foreign minister, the consent (agrément) of the Government to which he is to be accredited. 1939 H. Nicolson Diplomacy x. 236 It is customary..to sound a foreign government privately before making a formal application for an agrément. 1955 Times 2 May 10/3 The present Burmese Ambassador to India has been accorded an agrément by the Queen as Ambassador to the United Kingdom.

agrestal (ăgre·stăl), *a.* [f. L. *agrest-is* (see Agrest) + -al¹.] = Agrestial *a.* Also as *sb.*

1858 [see Agrestial a.]. 1863 J. G. Baker N. Yorkshire xiv. 183 To designate the different kinds of locality we may employ a series of adjectives such as sylvestral, pratal, pascual, ericetal, uliginal, agrestal. 1926 G. C. Druce in J. J. Walker Nat. Hist. Oxf. Distr. 116 Naturally these Agrestals are not the same from one century to another. Ibid. 117 Closely allied to the Agrestal are the Ruderal species.

agrestial, *a.* (Further examples.)

1842 Fraser's Mag. Sept. 298/1 A vast accession of richly productive agrestial territory. 1855 Jrnl. R. Agric. Soc. XVI. 584 A large farmer..seems to have been afflicted by..a sense of agrestial invasion.

agribusiness (æ·gribiznės). orig. *U.S.* [f. Agri(culture + Business.] (See quot. 1955).

1955 Harvard Business School Bull. Autumn 41 'Agribusiness'—a term coined to define the many diverse enterprises which produce, process, and distribute farm products or which provide supporting services. 1960 Farmer & Stockbreeder 29 Mar. 78/2 The broiler industry..had to go in for 'vertical integration'...It is not farming, it is agribusiness. 1962 Times 3 Dec. Agric. Suppl. i/5 So long as agriculture is carried on by farmers and not by so-called 'agribusiness'.

agricultural, *a.* Add: **b.** *slang.* Applied to a clumsy stroke in cricket.

1937 Partridge Dict. Slang 186/1 (s.v. cow-shot), A more clumsy shot, made by a standing batsman, is termed an agricultural one. 1955 Times 25 July 3/4 Keith..took an agricultural swing at Wardle and was bowled.

c. *agricultural ant*: a species of ant, such as the *Pogonomyrmex barbatus* of Texas, that clears the vicinity of its nest of verdure or herbage except for that on which it feeds.

1868 Amer. Naturalist II. 157 Such structures are remarkable..reminding us of the intelligence shown by the Agricultural Ant of Texas. 1882 [see Harvesting ppl. a.].

agriculturalist. (Earlier example.)

1802 Ann. Reg. 1801 Chron. 21/1 The company exceeded five hundred agriculturalists.

agriculturally (ægrikɒ·ltiŭrăli), *adv.* [f. AGRICULTURAL *a.*+-LY².] With regard or reference to agriculture.
1821 *Examiner* 50/1 Why say a word about it—why not pass it agriculturally, in prudent silence? **1883** HOLME LEE *Loving & Serving* I. xi. 211 His land is poor agriculturally. **1885** *Standard* 2 Sept. 2/4 The Board have not lost sight of the dealing with the sewage agriculturally.

agrimensorial (æ:grimensǭ·riăl), *a.* [f. L. *agrimensor* land-surveyor+-IAL.] Of or pertaining to land surveying.
1878 H. C. COOTE *Romans of Brit.* 67 The incision was made in the form of a cross. This was the agrimensorial '*antica et postica*'. *Ibid.* 83 An inscribed agrimensorial stone found at Drumburgh. **1892** C. MCLEAN ANDREWS *Old Eng. Manor* 33 Mr. Coote..believed that all Britain was laid out on the exact lines of the Roman agrimensorial system. **1958** *Antiquity* XXXII. 26 Studies agrimensorial and tenurial.

agrimotor (æ·grimōᵘtǫɹ). [f. L. *agri-*, *ager* land+MOTOR *sb.*] A motor tractor for agricultural work.
1917 *Town Topics* 10 Feb., The immediate future of agrimotors in this country. **1920** *Country Life* 10 Jan. p. lxi (Advt.), The Crawley Agrimotor Equal to 8 men and 16 horses.

agriology (ægriǫ·lŏdʒi). [f. Gr. ἄγριος wild, savage: see -OLOGY.] The comparative study of the history and customs of savage or uncivilized peoples. Hence **agriolo·gical** *a.*
1878 *Fraser's Mag.* XVII. 730 Trying the law..of euphony and prosody in face of the agriology of the day. *Ibid.* 733 Mr. Sayce..threatens to make agriological scalps the only wear in glottology. **1886** *Jrnl. Educ.* 1 Apr. 152/1 The new lights thrown on the early stages of society by..what may be called Agriology.

‖ **agrogorod** (ægrogō³·rŏd). Pl. **-a.** [Russ., f. *agro-* as in AGRONOMIC *a.* + *górod* town.] A group of amalgamated collective farms (kolkhozes) forming an administrative unit; a 'rural city'. Also, by partial translation, **agro-city, -town.**
1951 *Soviet Stud.* Oct. III. 158 Under the leadership of the Politburo member Nikita Khrushchev, a campaign was begun for a great enlargement of the individual kolkhozy by 'voluntary' mergers, by corresponding consolidation of villages into what were proudly called 'agro-cities'. **1951** *Sun* (Baltimore) 20 Mar. 12/2 What Stalin has now launched is a program of combining Russia's 250,000-odd collective farms into not more than 100,000 giant farm enterprises, to be called agro-towns or agricultural towns. **1952** *Ann. Reg. 1951* 198 Khrushchev himself had dismissed contemptuously the earlier grandiose dreams of '*agrogoroda*', or rural cities, and had substituted the idea of '*kolkhoz* settlements'. **1959** E. CRANKSHAW *Khrushchev's Russia* 83 He [*sc.* Khrushchev] had had his wild ideas, like the premature scheme for *agrogoroda*; but he had had his good schemes too.

agrology (ăgrǫ·lŏdʒi). [f. Gr. ἀγρό-ς field, land+-LOGY; cf. F. *agrologie*.] **a.** The science of soils. **b.** *spec.* In Canada, 'professional agriculture' (see quot. 1946). So **agro·logist,** an expert or specialist in agrology; a 'professional agriculturist'.
1916 B. D. JACKSON *Gloss. Bot. Terms* (ed. 3) 415/1 *Agrology*, the science of soils, and their support of special vegetation. **1946** *Statutes Prov. Saskatchewan* LXVIII. 652 'Agrologist' means any person registered as an agrologist under the provisions of this Act...'Practising agrology' and similar forms of expression mean..experimenting with, or advising on the application of scientific principles and practices relating to the cultivation, production, improvement..of agricultural plants. **1962** *Careers in Prof. Agric.* (*Canadian Imp. Bank of Commerce*) 1 The need for more professional agriculturists (agrologists) in Canada.

agronome. (Earlier and later examples.)
1838 *Monthly Chron.* II. 212 The rule adopted by the South American agronomes. **1945** H. G. WELLS *Happy Turning* vi. 22 The agronomes—there are no farmers in Dreamland—come along and tell us, 'We can produce all the food.'

agronomic, *a.* (Further examples.)
1891 *Times* 28 Sept. 13/5 Agronomic stations have been created for the purpose of enlightening agriculturalists. **1957** *Times* 2 July Agric. Suppl. i/2 The improved agronomic techniques..have..been taken up and tested by Norfolk farmers.

agterskot (a·χtəɹskǫt). *S. Afr.* [Afrikaans, f. *agter* after+*voor*)*skot* advance payment, f. *voorskiet* to advance (money).] The final payment for a crop by a farmers' co-operative society or similar body to members, consisting of the difference between the total amount due for the season and the advance-payment (*VOORSKOT). Also *transf.*
1944 *Cape Argus* 18 May 7/6 More than 1,000 persons went to Rustenburg..from many parts of the Transvaal to receive back pay (agterskot) for tobacco sold to the Magaliesberg Co-operative Tobacco Planters Association. **1950** *Ibid.* 8 Sept. 8/6 The country is reaping an 'agterskot' in reverse from devaluation. **1958** *Cape Times* 12 Dec. 2/8 An *agterskot* amounting to £31,000 is being paid to lucern seed farmers in Oudtshoorn district.

aguacate (agwakā·te). [Sp.: see AVOCADO.] The alligator pear or avocado. Also *attrib.*
1897 *Blackw. Mag.* Nov. 686/1 Tall mangoes and aguacates. **1926** *Chambers's Jrnl.* 23 Jan. 126/2 Banana peelings, aguacate skins.

aguardiente (agwardie·nte). [Sp., = brandy, f. *agua* water+*ardiente* ARDENT.] A coarse kind of brandy made in Spain and Portugal. Also applied to any distilled spirituous liquor; in south-western U.S., native whisky.
[**1818** *New-Eng. Palladium* (Boston, Mass.) 28 Sept. 3/2 Isaac McLellan & Co...have for sale..100 pipes Spanish Rum or Auquedent.] **1824** W. BULLOCK *6 Months' Res. in Mex.* xvii. 212 Shops for the sale of native and Spanish brandy, (aguardiente). **1828, 1854** [see MESCAL 1]. **1899** ST. BARBE *Mod. Spain* 67 Much of the fun of the fair consists in sipping *aguardiente*, a strong liqueur. **1926** D. H. LAWRENCE *Plumed Serp.* iv. 80 The hateful sugar-cane brandy, *aguardiente*.

ah (ā), *v.* [f. AH *int.*] *intr.* To say 'ah' (as an expression of surprise, wonder, etc.); also, = *ER *v.*
1897 'MARK TWAIN' *Foll. Equator* in *Writ.* (1900) V. 111 We..went oh-ing and ah-ing in admiration. **1924** D. H. LAWRENCE *Indians & Entertainment* in *Adelphi* Nov. 500 The men sing in unison..for hours..it goes on..gurgling, aah-h-h-ing! of the male voices. **1931** [see *ER]. **1934** 'REBECCA WEST' *Mod. Rake's Progr.* 113 These visitors from the next civilisation..will walk about oh-ing and ah-ing. **1960** C. RAY *Merry Eng.* 128 The spectators ooh! and ah! and roar their delight.

ahem, *int.* Add to def.: Also, to express disapproval by a factitious clearing of the throat. (Further examples.) Hence as *sb.*
1848 G. E. JEWSBURY *Let.* 12 Aug. (1892) 252 George Sand corresponds with Miss —, and calls her the 'sister of her soul'. Ahem! **1928** R. CAMPBELL *Wayzgoose* ii. 58 Then having seen his error, [he] paled with fear And coughed—Ahem, we'll leave the matter here! **1929** D. H. LAWRENCE *Pansies* 147 They were just a bloody collective fraud, That was what their *Ahem!* meant.

ahem (ăhe·m), *v.* [f. AHEM *int.*] *intr.* To exclaim 'ahem!'; also *trans.*, to pass off with the exclamation 'ahem!'.
1839 *Fraser's Mag.* XIX. 125 He immediately 'a-hems' away his jocularity. **1876** MEREDITH *Beauch. Career* III. ix. 166 Tuckham brushed his hand over his mouth and ahemed. **1891** —— *One of our Conq.* I. xiii. 240 He..pulled the waistcoat, and swelled it, ahemming.

‖ **ahimsa** (āhi·msā). *Hindu Philos.* [Skr., f. *a* without+*himsa* injury.] The doctrine of non-violence or non-killing.
1875 MONIER WILLIAMS *Indian Wisdom* x. 249, I am told ..that, notwithstanding the strict rules of *a-hiṇsā*, the 'Society for Prevention of Cruelty to Animals' might find work to do in some parts of India. **1884** H. JACOBI *Gaina Sutrâs* in *Sacred Bks. East* XXII. p. xxi, The stress which is laid on the ahimsâ or not killing of living beings. **1913** J. N. FARQUHAR *Crown of Hinduism* vii. 263 The idea at the basis of ahimsa is that all life is sacred, and that no holy man can take life. **1915** M. K. GANDHI *Speeches & Writings* (ed. 3, 1922) 236 Our religion is based upon *ahimsa*, which in its active form is nothing but Love. **1936** A. HUXLEY *Eyeless in Gaza* xi. 150 How foolish of Satan to tempt a, by definition, *ahimsa*-practising Messiah with fame, dominion, ambition. **1962** *Economist* 27 Oct. 330/2 *Ahimsa* is not the only thread in Indian political tradition.

a-histo·ric (æ-, ē¹-). [A- 14+HISTORIC *a.*] Not historic; lacking an historical background.
1937 WYNDHAM LEWIS *Let.* 21 Nov. (1963) 246 My mind is *ahistoric*, I would welcome the clean sweep. **1941** L. B. NAMIER in *19th Cent.* Nov. 276 The Jews who, by reducing themselves to the level of a nondescript, a-historic group, cast away the dignity of a nation. **1952** AUDEN *Nones* 58 Their a-historic Antipathy forever gripes All ages and somatic types.

a-histo·rical, *a.* [f. A- 14+HISTORICAL *a.*] Regardless of, or indifferent to, the historical aspect; not historical.
1957 J. C. MCKINNEY in Becker & Boskoff *Mod. Sociol. Theory* vii. 228 American sociology has been generally ahistorical in its approach to the study of society. It has instead concerned itself with the realm of 'contemporary' events. **1960** H. READ *Forms of Things Unknown* ix. 146 It will be one of my purposes..to maintain that the negation of the historical present is not to be confused with an apathetic nihilism..that on the contrary, this a-historical art..is the only positive evidence of renewal..in the visual arts of our time. **1962** EVANS-PRITCHARD *Essays Soc. Anthrop.* iii. 46 Durkheim, though perhaps not anti-historical, had been ahistorical, at any rate in the sense that his developmental studies were in the field of evolutionary typology rather than of history proper.

a-hold, *adv. phr.* Restrict †*Obs.* to sense in Dict. and add: **2.** *a-hold of*: holding (something). Also in *to catch, lay, take*, etc., *a-hold of, on, upon*: see HOLD *sb.*¹ 2. Also **a-holt** (see HOLT²). *colloq.* or *dial.*
1872 E. EGGLESTON *End of World* xi. 77 You gripped a-holt of the truth. **1879** *Scribner's Monthly* May 17/1 With one bee a-hold of your collar..and another a-hold of each arm. *a* **1881** LANIER *Poems* (1892) 17, I will heartily lay me a-hold on the greatness of God. **1887** MORRIS *Odyss.* x. 264 He caught ahold upon me. **1925** E. HEMINGWAY *In our Time* (1926) v. 79 Nick dropped his wrist. 'Listen,' Ad Francis said. 'Take ahold again.' **1959** C. MACINNES *Absolute Beginners* 107 Some of the birds tried to get aholt of me.

A-horizon: see *HORIZON *sb.*

ahorse (ăhǭ·ɹs), *adv.* [f. A *prep.*¹+HORSE *sb.*; cf. ME. *on hors* (HORSE *sb.* 15).] = A-HORSEBACK.
1860 W. H. RUSSELL *Diary India* II. 126 Young ladies and gentlemen..a-horse and a-foot. **1925** *Chambers's Jrnl.* Xmas No. 854/2 When he and I encountered, ahorse on a road, we pulled up our horses and conversed.

ahu·nger, *adv.* or *pred. a.* [f. A- *pref.*+HUNGER; cf. ANHUNGERED.] Hungry; hungering.
c **1450** *Mirk's Festial* 127/10 Þe pore..aboden.., sor ahongyr. **1830** *Blackw. Mag.* Mar. 511/1 The maw of the public, ahunger and athirst for stories of peril. **1922** JOYCE *Ulysses* 261 Lenehan, small eyes ahunger on her humming.

a-hungry, *a.* Delete †*Obs.* and add earlier and later examples.
c **1460** *Towneley Myst.* iii. 499 The ravyn is a hungrye All way. **1832** *Fraser's Mag.* Nov. 527 The baby in the cradle is for ever a-hungry, or a-thirsty. **1852** *Tait's Mag.* XIX. 228 The Whigs..were a-thirst and a-hungry for office. **1933** C. MILLER *Lamb in Bosom* (1934) viii. 91 Jasper found that he was not a-hungry.

Aich('s) metal. [Named after the patentee Johann *Aich* (3 Feb. 1860).] An alloy of copper, zinc, and iron, used in gun-making.
1867 BLOXAM *Chem.* 342 Aich (or Gedge's) metal. **1875** HUNT & RUDLER *Ure's Dict. Arts* (ed. 7), *Aich metal.* **1958** MERRIMAN *Dict. Metallurgy* 3/1 *Aich's Metal*, a brass of the 60/40 type that has good casting properties.

aid, *sb.* Add: **1. b.** *in aid of*, in support of (a cause or charity). Hence, *fig.* and *colloq.* (presumably having its origin in the freq. use of the phr. in appealing for the public support of a cause), about, concerned with; esp. in phr., often disparaging, *what's this* (or *that!*) *in aid of?*, what is the meaning or purpose of this?, what is this all about?
1837 *Playbill* in M. Morley *Old Marylebone Theatre* (1960) 20 A Benefit will take place in Aid of the Funds of the New Alms Houses. **1860** S. S. HENNELL (*title*) Thoughts in aid of faith. **1881** W. S. GILBERT *Patience* I. 19 In aid—in aid of a deserving charity, I've put myself up to be raffled for! **1915** *Times* 22 Oct. 11/3 Queen Alexandra..was present at the Empire Theatre *matinée* in Aid of the British Red Cross Society. **1918** *Punch* 20 Nov. 332 (*caption*) Oh Mother,.. they've given us a whole holiday to-day in aid of the war. *a* **1935** T. E. LAWRENCE *Mint* (1936) 127 The hut lights were on and he had brought me a tin of tea and a hot sausage roll. 'Scram up!' he called...'What's all this in aid of?' I asked, stupidly. **1935** MARSH & JELLETT *Nursing-Home Murder* xv. 231 'That's your disillusioned expression, Fox,' said Alleyn. 'What's it in aid of?' **1942** 'BLAKE' *We Rendezvous at Ten* ii. 41 The Group Captain called down the table to Roger: 'Find out what that's in aid of, Roger, will you?' **1949** E. BOWEN *Heat of Day* xvii. 315 What you *were* in aid of..often was a mystery to me. **1956** 'M. INNES' *Old Hall, New Hall* viii. 70 He couldn't quite make out what Olivia's questions and speculations were in aid of.
2. b. *to call* or *crave in aid*, properly a legal phrase, also in a loose transf. use.
1927 *Observer* 8 May 16/2 Imagination craves the wireless in aid. **1928** *Ibid.* 1 July 13/4 Many [bishops]..would call in aid, as justifying their action, the use of the *Jus Liturgicum* inherent in their office.
3. (Later example in *Horsemanship.*)
1953 G. BROOKE *Introd. Riding* i. 16 During the period that the novice is riding his first mount, he should learn the aids (correct and combined applications of his hands and legs).
b. Freq. with defining word, as *approach, artificial, hearing, homing, legal, radio-navigational, visual aid*: see these words (in Suppl.).
1924 *Lancet* 31 May 1140/2 A new acoustic aid for the deaf. **1955** *Oxf. Jun. Encycl.* XI. 128/2 Hearing may be greatly improved by the use of..electrical aids which amplify sounds.
c. *spec.* Material help given by one country to another. Also *attrib.* and *Comb.* Cf. *MARSHALL.
1940 *Economist* 5 Oct. 421/1 The United States' aid to Britain would be rendered ineffective. **1946** *Ann. Reg. 1945* 100 The difficulties of procuring American aid for Britain on acceptable terms. **1951** *Ann. Reg. 1950* 337 The U.S. aid conventions with the Associated States [of Indo-China]. **1958** *Spectator* 17 Jan. 65/3 Congress would like to buy missiles with foreign-aid money. **1964** *Listener* 16 Apr. 614/1 Since the Soviet Union and..China have joined in the game of competitive aid-giving the Western Powers, it is argued, cannot afford to drop out.
7. b. *U.S.* = AIDE. See also *AID-DE-CAMP.
1780 S. HOLTEN in *Essex Inst. Coll.* (1920) LVI. 94 One of General Lincoln's aids is arrived with the accounts of the surrender of Charlestown. **1832** J. P. KENNEDY *Swallow Barn* I. xix. 190 Ned and myself formed part of his retinue, like a pair of aids somewhat behind the commander-in-chief. **1907** *Chicago Tribune* 8 May 2 Gen. A. W. Greely.. arrived with his..aids.
8. aid-post, a post at which wounded soldiers receive first medical attention.
1916 'BOYD CABLE' *Action Front* 49 To walk..to the nearest aid-post and hospital. *a* **1917** E. A. MACKINTOSH *War, the Liberator* (1918) 149 The Aid Post was like a shambles with blood and wounded men.

aid-de-camp, occas. (chiefly *U.S.*) var. AIDE-DE-CAMP.
1690 R. DAVIES *Jrnl.* 4 Aug. (1857) 134 Scravenmore's aid-de-camp came to us again. **1732** [see AIDE-DE-CAMP]. **1762** C. BRIETZCKE *Diary* 9 Feb. in *N. & Q.* (1959) CCIV. 136/2 Jack spoke to Mr. Percival in the Boxes who was aid de camp to Hodgson. **1776** *Jrnl. Cont. Congress* V. 418 Resolved, That the aids de camp of the commander in chief rank as lieutenant colonels. **1878** J. H. BEADLE *Western Wilds* xxxiii. 532 He..was made full captain and aid-de-camp of General McClellan.

aide. (Earlier and later examples.)
1777 J. M. LINCOLN *Papers R. Lincoln* (1904) 11 They.. fired on the flag and killed an ade. **1826** COOPER *Mohicans* xxxiii, Attended by the aide of Montcalm with his guard. **1952** *Manch. Guard. Weekly* 12 June 3/2 General MacArthur..is on inactive status, but he receives..the services of three military aides.

aided, *ppl. a.* Add: **b.** *spec.* Receiving financial aid; esp. (of a school) assisted by monetary grants from the government. So *grant-aided, state-aided*, adjs.
1882 [see STATE *sb.* 40 b]. **1943** *Educ. Bill, Explan. Memorandum* (Cmd. 6492) 5 Aided schools: The managers of these schools will continue to appoint their own teachers and have the teachers' salaries and other maintenance charges paid by the local education authority.

‖ **aide-mémoire** (ē¹·dmemwāɹ, ędmemwār). [F., f. *aider* to help+*mémoire* MEMORY.] An aid to the memory; esp. in diplomatic use, a memorandum (cf. MEMOIR 2).
1846 G. G. LEWIS et al. (*title*) Aide-Mémoire to the Military Sciences. **1855** H. LAXTON (*title*) Examples of Building Construction, intended as an Aide-Memoire for the Professional Man. **1885** *Athenæum* 1 Aug. 151/1 *Catalogue Illustré du Salon*..is more than a very useful *aide mémoire* of the great collection. **1923** *Westm. Gaz.* 3 July, No written document on the subject of the British questionnaires, even in the way of an aide memoire, will be handed to the Foreign Secretary. **1941** AUDEN *New Year Let.* I. 312 This aide-mémoire on what they say. **1955** *Times* 1 Aug. 5/6 The United States has sent to Bulgaria a strongly worded aide-mémoire, transmitted through the Swiss Government, protesting against the 'brutal attack'. **1957** *BBC Handbk.* 23 An agreement reached in 1947 between the BBC, the Government, and the Opposition, and embodied in an *Aide-Mémoire*.

aileron (ē¹·lĕrǫ̆n). [a. F. *aileron*, dim. of *aile* wing.] One of the hinged flaps on the trailing edge of a wing of an aeroplane for maintaining or restoring its balance when flying.
1909 *Aero* 25 May 1 The ailerons or small planes between the main surfaces are used instead of wing-flexing for balancing. **1916** *Air* I. 11 Elevator surface and aileron surfaces which are constructed of steel frames covered with linen. **1937** *Jrnl. R. Aeronaut. Soc.* XLI. 191 Partial flap (slotted or split type) combined with drooping ailerons. **1953** *Sci. News* XXIX. 99 The ailerons on an aircraft are usually situated on the trailing edges of the wings and extend approximately halfway along the wing from the tips.

ailurophobia (əil¹ŭ³rofō³·biă). [mod. L., f. Gr. αἴλουρος cat+-PHOBIA.] Morbid fear of cats. Hence **ailu·rophobe**, one affected with ailurophobia; **ailuropho·bic** *a.*
1905 S. W. MITCHELL in *Amer. Med.* IX. 851 (*title*) Of ailurophobia and the power to be conscious of the cat as near, when unseen and unheard. **1905** A. LANG in *Morning Post* 16 June 4/3 Female ailurophobes are just as capable of not knowing that a cat is in the room..as of saying that it is there when it is not. *Ibid.*, Finding a lady, rather ailurophobic, in a low dress at dinner Tippoo suddenly leaped up and alighted on her neck. He was never so friendly with non-ailurophobes. **1920** C. VAN VECHTEN *Tiger in House* (1921) ii. 30 Cats only give affection where it is deserved, except..when they annoy an ailurophobe with their attentions.

aim, *v.* Add: **5. b.** With infinitive: to intend; to attempt. Formerly chiefly *dial.* and *U.S.*, now *colloq.*
1602 MARSTON *Antonios Revenge* v. i. sig. I 2ʳ, Alberto drawes out his dagger, Maria her knife, ayming to menace the Duke. **1649** [in Dict.]. **1665** in *Rhode Isl. Col. Rec.* II. 120 And this the Court hath done, aimeing alsoe therein to save the towne. **1785** HUTTON *Bran New Wark* (1879) 24 Aiming to hev a good conscience. **1878–** in *Eng. Dial. Dict.* **1891** HARDY *Tess* (1892) xxv. 204 Aiming to arrive about the breakfast hour. **1909** J. BIGELOW *Retrosp.* I. iii. 57 A[n]..article in which I aimed to compare and contrast the duties of Roman lawyers with those of our own time and country. **1940** F. VAN WYCK MASON *Stars on Sea* II. viii. 283 A seaman called through tattooed hands, 'We ain't Spanishers. We don't aim to harm you.'
6. b. *to aim off* (in *Rifle-Shooting*), to alter the aim to allow for the wind. So *aiming-off* vbl. sb., *aim-off* sb. Also *transf.*
1918 E. S. FARROW *Dict. Mil. Terms* 12 *Aiming-off*, altering the point of aim laterally, so as to give deflection to the rifle-barrel without using the wind gauge. **1931** *Small Arms Training* (*War Off.*) V. v. 70 Aiming off for wind.., using wind-gauge to give effect of wind. **1932** J. A. BARLOW *Elements Rifle Shooting* iv. 44 We have to learn to aim at some spot other than the one which we want to hit...We will leave for the moment the business of judging the amount of aim-off required and concentrate on the problem of how to aim off. **1957** *Spaceflight* I. 64/2 One of the most difficult problems in the launching of a guided missile, that of aim-off, is entirely eliminated, as the rocket flies continuously within the guiding beam, always endeavouring to intercept its target.
7. (Further examples.) Also *transf.*
1776 M. ANGELO *Juv. Sports* (ed. 2) ix. 81 If you stand too far from it [the wicket], you may be knocked out by the bowler, before you can recover your bat after aiming a stroke. **1927** *Punch* 4 May 480/2 The maid appeared again, aimed me at the consulting-room and loosed me off. **1962** *Which?* Oct. (Car Suppl.) 139/2 Offside headlamp [was] aimed high.

aiming, *vbl. sb.* Add: *aiming point*, in *Gunnery* (see quots.).
1910 *Encycl. Brit.* I. 692/1 In war, the target, even if visible, is often indistinct, and in this case..an 'aiming point'..a conspicuous point quite apart and distinct from the target, has to be employed. **1918** E. S. FARROW *Dict. Mil. Terms* 12 *Aiming Point*, in gunnery, a stationary object..upon which the panoramic sight is directed after the proper deflection is set off.

ain't² (ē¹nt). *dial.* and *vulg.* var. *hain't*, have not.
1845 SIMMS *Wigwam & Cabin* Ser. I. 7 But you ain't said, ..who was your Carolina gineral. **1875** in *Eng. Dial. Dict.* s.v. *Have* v. I. ii. **1880** TOURGÉE *Fool's Err.* II. x. 473, I ain't done nothing. **1887** MARY E. WILKINS *Humble Romance* 3 You've been crying, ain't you? **1910** W. M. RAINE *B. O'Connor* iii. 37 We ain't got one chance in a hundred, Jim. **1917** MATHEWSON *Sec. Base Sloan* xiv. 195 More than that I ain't got the right to say.

Ainu (əi·nu). Also 9 **Aino.** [Ainu, lit. 'man'.] The name of a Caucasoid people in Japan and the U.S.S.R. and of the language spoken by them. Also *attrib.*
1819 tr. *Golovnin's Recoll. Japan* v. 242 Their manners shew that the Ainu and Kuriles are one people. **1843** J. C. PRICHARD *Nat. Hist. Man* xxi. 226 (*heading*) The Ainos, or Kurilians. **1864** J. HUNT tr. *Carl Vogt's Lect. Man* v. 127 We find a small, nearly extinct tribe of the Kuril islands,—the Ainos,—whose body is so completely covered with shaggy hair, that it gave rise to the Japanese tradition, that the Aino mothers suckled young bears, which gradually became men. **1880** I. L. BIRD *Unbeaten Tracks in Japan* II. 37 He spoke both Aino and Japanese in the low musical tone which I find is characteristic of Aino speech. **1880** D. P. PENHALLOW in *Sci. Amer.* Suppl. 4 Sept. 3884/2 As an article of clothing, for which use alone it appears to be utilized, the Aino cloth has several good qualities. *a* **1884** KNIGHT *Dict. Mech.* Suppl. 12/2 *Aino cloth*, a cloth made by the Ainos..from the divided fibers of the elm, beaten so as to obtain bast layers, which are split and woven. **1893** A. H. S. LANDOR (*title*) Alone with the Hairy Ainu. **1931** H. G. WELLS *Work, Wealth & Happiness of Mankind* (1932) xiii. 666 The existence of really primitive and undeveloped peoples in small numbers, like the white hairy Ainu or the pygmies. **1932** W. L. GRAFF *Language & Languages* xi. 411 Ainu, spoken by some 20,000 people in the southern part of the island of Sakhalin.

air, *sb.*¹ Add:
1. c. The air considered as a medium for the transmission of radio waves; colloq. = *RADIO *sb.* 2, esp. in phr. *on the air*, (being) broadcast by radio transmission; so *off the air*.
1927 *Observer* 11 Dec. 16 The only New York church which is 'on the air'. **1928** *Daily Express* 13 Apr. 11/1 They will speak into the microphone as usual, but before being put 'on the air' their voice modulations will be turned upside down. **1940** *N. & Q.* CLXXIX. 66/1 On and off the air. **1955** *Times* 9 May 10/5 Radio and, particularly, television will play a large part in the election, and the campaign of the air has already begun. **1956** A. H. COMPTON *Atomic Quest* 255 The Hiroshima radio went off the air. **1964** *Evening Standard* 4 Feb. 16/2 Every policeman on the beat may be 'on the air' in a few years. Experiments with pocket radios..have established that they..can be of immense value to the constable on patrol.
attrib. **1928** *Sunday Dispatch* 16 Dec. 12/2 *Air Hogs*.. Certain people..have been interrupting wireless transmission by crude signals from a private station. **1944** *Amer. Speech* XIX. 49 The New York agencies..spread *fifth column* over the pages of every subscribing newspaper in the country and onto the air waves as well. **1955** T. H. PEAR *Eng. Soc. Differences* xi. 246 The listening public's approval or disapproval of the 'air-time' given to each.
3. a. Also, considered as a medium for the operation of aircraft; a collective term for aircraft or aerial power; esp. in *Comb.*, as *air arm, cover, offensive, warfare* (see below B. III. 2). So *by air*, by aeroplane.
1917 I. D. FISHER *Let.* 11 July in R. H. S. Bacon *Life Ld. F.* (1929) II. xxi. 303 The air is going to win the war. **1919** *Daily Tel.* 17 Feb. 6/4 (*headline*) Egypt to the Cape by Air. *Ibid.* 19 Feb. 12/6 (*headline*) Heroes of the Air. *Ibid.* 13/4 (*headline*) Who Owns the Air? **1944** *Picture Post* 9 Sept. 8 We'd got the air, we'd got the guns. **1945** *Hutchinson's Pict. Hist. War* 27 Sept.–13 Mar. 54 Our land-based air was very thin indeed.
c. *in the air.* *fig.* 1. *a.* (Earlier example); *b.* (Earlier example); colloq. phr. (*up*) *in the air*, of persons: in doubt, uncertain; of ideas or theories: speculative, hypothetical; *c.* (slang) *up in the air*: excited, as in anger (orig. *U.S.*). 2. *Mil.* (Earlier and later examples.)
1875 A. W. WARD *Hist. Eng. Dram. Lit.* I. iv. 325 The appreciation of Shakspere and the dramatic art perceptible in both these great writers was, as the phrase is, in the air,—in the air, *i.e.*, breathed by those who stood on the height of European culture.
1752 H. WALPOLE *Let.* 28 Oct. (1903) III. 124 Don't look upon this paragraph as a thing in the air. **1910** GALSWORTHY *Justice* IV, Keep him in the air; I don't want to see him yet. ..Keep him hankering. **1933** D. L. SAYERS *Murder must Advertise* viii. 140 You might have let somebody know. I was left rather up in the air this morning. **1940** in Harrisson & Madge *War begins at Home* v. 107, I didn't hear anything for a long time. They sort of leave you in the air. **1943** W. TEMPLE *Let.* 8 June (1963) 80 The faith he professes in that book is very much up in the air and devoid both of practical and philosophical attachments. **1956** C. WILSON *Outsider* ii. 44 The reader is left feeling oddly 'up in the air' about it all. No happy finale, no dramatic tying up of loose ends. **1956** *Essays in Criticism* VI. 193 By modern standards the article is a bit up in the air.
1906 *N.Y. Even. Post* 13 Jan. 4 Representatives..have.. 'gone up in the air' because they could not 'land' their men. **1928** E. WALLACE *Again Sanders* ii. 49 Abiboo, who is a strict Mussulman, got up in the air because Bones suggested he might have been once a guinea-pig. **1930** *Punch* 21 May 577/3 Why the Prime Minister should have 'gone up in the air', as they say, because it appeared in print that Gandhi was about to be arrested..was not revealed.
1865 M. ARNOLD *Ess. Crit.* 261 No intelligent man can read the *Tractatus Theologico-Politicus*..without feeling that, as a speculative work, it is, to use a French military expression, *in the air*; that, in a certain sense, it is in want of a base and in want of supports. **1923** KIPLING *Irish Guards in Gt. War* I. p. x, There was hardly an operation in which platoons..brigades, or divisions were not left with one or both flanks in the air.
d. *to give* (a person) *the air*: to dismiss (him); to reject. Also with *get*. *U.S. slang.*
1900 G. ADE *More Fables in Slang* 85 (*title*) The Fable of why Essie's Tall Friend got the Fresh Air. **1924** P. MARKS *Plastic Age* 202 'How about my studies?' Hugh retorted. 'I suppose you want me to give them the air.' **1934** WODEHOUSE *Thank You, Jeeves* x. 135 Surely you don't intend to give the poor blighter the permanent air on account of a trifling lovers' tiff? **1949** R. GRAVES *Seven Days in New Crete* xvii. 207, I couldn't change her views.. nor could she convert me to hers, even when she threatened to give me the air.
e. *to give* (a ball) *air*: in *Cricket*, to deliver (a ball) with a high trajectory. In *Football*, etc., to keep the ball constantly in movement.
1920 E. R. WILSON in P. F. Warner *Cricket* 88 Slow bowlers are right to 'give the ball air' to a nervous or slow-footed batsman. **1929** *Daily Express* 7 Nov. 18/7 The ball was given plenty of 'air', the pace of the passing and the accuracy of handling a greasy ball reflecting the greatest credit on every one concerned. **1932** *Ibid.* 20 Sept. 10/2 He is doing what every good footballer seeks to do when the play becomes too close. He is opening up the game by 'giving the ball fresh air'.
15. c. *to put on airs*: to assume an unjustified air of superiority.
1781 [see PUT *v.*¹ 46 d]. **1832** *Deb. Congress* 30 Jan. 203, I am aware that, at times, States have attempted to put on airs, and set up their own against federal opinions. **1860** O. W. HOLMES *Prof. Breakf.-t.* v. 93 None of them like too well to be told of it, but it must be sounded in their ears whenever they put on airs. **1952** T. WILLIAMS *Summer & Smoke* II. i, It is understandable that she might be accused of 'putting on airs' and of being 'affected'.
B. air- in *Comb.*
I. (Additional combinations.)
1. *air-crisping, -entraining; air-cleaner, -cleanser.*
1865 G. M. HOPKINS *Poems* (1948) 37 Let me be to Thee as the circling bird, Or bat with tender and air-crisping wings. **1926** *Lancet* 26 June 1292/1 The 'Deodos' Air Cleanser. The purpose of this apparatus is to purify and medicate the air of rooms and buildings by means of a vapour. **1929** *Times* 2 Nov. 4/7 The carburettor is to hand, and there is a useful air-cleaner. **1956** *Gloss. Terms Concrete* (*B.S.I.*) 7 *Air-entraining agent*, an admixture to Portland cement or to concrete which causes a small quantity of air to be incorporated..in the concrete during mixing. **1962** *Which?* Oct. (Car Suppl.) 139/1 Stones [were] found in air cleaner.
2. *air-freighted, -swept.*
1851 H. MELVILLE *Moby Dick* III. xxiv. 159 Rolled in the sea like an air-freighted demi-john. **1853** M. ARNOLD *Scholar-Gipsy* in *Poems* 202 And air-swept lindens yield Their scent. **1901** *Guide to Felixstowe* (*Ward, Lock*) 2 It is an air-swept place, this sunny Felixstowe.
3. *air-pale, -sweet, -thin, -white.*
1918 E. SITWELL *Clowns' Houses* 11 Each in an air-white crinoline. **1920** —— *Wooden Pegasus* 73 In air-pale waves. **1938** W. DE LA MARE *Memory* 67 A quiet, air-sweet October day. **1942** E. SITWELL *Street Songs* 31 The air-pale petals of the foam seem flowers. **1948** —— *Notebk. W. Shakes.* xii. 141 There are echoes..some more air-thin than the sound of which they are a memory.
4. *air-built* (additional example).
1888 G. M. HOPKINS *Poems* (1918) 68 Cloud-puffball, torn tufts, tossed pillows flaunt forth, then chevy on an air-Built thoroughfare.
5. *air-breath, -bubble* (earlier example), *-eddy, -supply.*
1756 F. HOME *Exper. Bleaching* II. iv. 76 A few hours after it has been there, air-bubbles arise, the liquor swells, and a thick scum is formed. **1851** H. MELVILLE *Moby Dick* III. i. 21 The air-eddy made by the sudden tossing of a pair of broad flukes. **1877** F. SCHUMANN *Man. Heating & Ventilation* 18 *Air Supply*. The following formulæ will demonstrate the necessity of a greater supply of pure air. **1885** W. B. YEATS in *Dublin Univ. Rev.* July 136/1 For there came an air-breath cool. **1909** H. G. WELLS *Tono-Bungay* III. iv. 407 That cold side that gives you the air-eddy I was beginning to know passing well. **1959** E. F. LINSSEN *Beetles* I. 112 The larvae of *Haliplus*..must renew their air-supply at the surface of the water since they breathe through spiracles.
6. *air-world.*
1888 G. M. HOPKINS *Poems* (1918) 89 How the boys..Are earthworld, airworld, waterworld thorough hurled. **1906**

Westm. Gaz. 8 Sept. 13/1 We imagine ourselves stopping in just that way to chat with a friend in the highways of the air-world.
7. *air-bottle, -furnace* (earlier example).
1732 in *Cal. State Pap., Amer. & W. Indies* 22 Feb. (1939) 230 To lett us know whether air furnaces are allow'd of, because at one of the works there is one built. **1918** *Jane's Pocket Aeronaut. Dict.* 8 *Air-bottle*, container for compressed air used for starting big engines. **1941** W. S. CHURCHILL *Speech* 25 June in *Secret Session Speeches* (1946) 28 By..somewhat increasing the compressed air-bottle which drives them, they [*sc.* U-boats] were able to fire volleys of torpedoes.
II. air-bag, a bag inflated with air; see also quot. 1836; **air-bath,** the protracted exposure of the body to the free action of the air as a form of medical treatment (cf. SUN-*bath*); **air-bell** [BELL *sb.*³], a small bubble of air; *spec.* one formed in a photographic developer, etc., and appearing as a spot on a plate, film, or paper; **air-blast,** a blast of air; *spec.* in various technical uses (see quots.); **air-bloomery** (earlier example); **air-brake** (example); **airbrasive** *a.* and *sb.* [(A)BRASIVE *a.*] *Dentistry* (see quots.); **air-break,** (*a*) [BREAK *sb.*¹ 5] *Cricket*, a 'twist' or deviation in the air, of the ball when bowled; (*b*) [*BREAK *sb.*¹ 17 b] *Electr. Engin.*, *attrib.* (see quot. 1910); **air-breathing** *a.*, applied to a jet-engine requiring the intake of air for the combustion of its fuel; **air-brush,** a device for spraying colour over a surface by means of compressed air; **air-burst,** the bursting of a shell or bomb in the air; hence as *v. intr.*; **air-castle** (earlier example); **air-channel,** a channel for the passage of air, in various structures; **air-compressor,** a machine for compressing air; **air-cooled** *a.*, cooled by means of a current of air; so **air-cool** *v. trans.*, **air-cooling** *sb.* and *a.*; **air-cooler,** an apparatus or appliance for reducing the temperature; **air-core,** esp. *attrib.* (so *-cored* adj.) *Electr.*, applied to a type of transformer or coil in which the central core consists of an air-filled space instead of a magnetic material; **air-course** = *AIRWAY 1; **air-crossing,** a passage or arched way to carry one air-passage over another in a mine; **air-cushion,** (*b*) a cushion (CUSHION *sb.* 2 a) of air; used esp. *attrib.* of a type of craft or vehicle buoyed up by a cushion of air; **air-drain** (example); hence *air-drained* adj.; **air-drainage** (see quots.); **air-dried** *a.*, dried by the action of the air; so *air-dry* v. trans.; **air-driven** *a.*, actuated by means of compressed air; **air-dry** *a.*, dry to such a degree that on exposure to the air no further moisture is given off; **air embolism** *Path.*, an embolism caused by an air-bubble in the blood-stream; **air-extractor** (see quot.); **air-filter,** an apparatus for extracting extraneous particles, germs, etc., from air; **air-flow,** the flow of air, *spec.* that encountered by the surface of an aircraft in flight or by a motor-car in motion; **air-freshener,** a substance or device for freshening the air (of a room); **air-gap,** a gap or hole through which air passes; *Electr.*, the air-filled space in a magnetic or electric circuit, as between the poles of a magnet or the terminals of an electrostatic machine; **air-gas,** a mixture of air and a vaporous hydrocarbon mixture (e.g. petroleum), used esp. as an illuminant; **air-gauge,** an instrument for measuring and indicating the pressure of air or gases; **air-glow** [GLOW *sb.* 2]: see quots.; **air-hardening** *a.* *Metallurgy*, applied to a metal that can be cooled in air; so *air-harden* v. trans., **air-hardened** *ppl. a.*; **air-heater** (see quots.); **air-intake,** an inlet or duct for air; **air-jacket,** (*b*) a jacket (JACKET *sb.* 2) in which air or gas is circulated to diminish loss of heat from the enclosed vessel; so *air-jacketed* adj.; **air-layering** [cf. LAYER *v.* 1] *Hort.* (see quot. 1934); cf. CIRCUMPOSITION; **air-machine,** in a mine, a contrivance by which pure air is forced into ill-ventilated parts; **air-mass** *Meteorol.*, a body of air of uniform temperature and humidity; **air-mattress,** one inflated with air; **air-pad,** a pad inflated with air; **air-passage,** (*a*) a passage through which air travels, e.g. the nasal passages, bronchial tubes, etc.; (*b*) *Bot.* the large intercellular space in the stems and leaves of some plants; **air-pit,** a ventilating shaft in a mine; **air-port** [PORT *sb.*³] (examples); also, an aperture for admitting air in a gas-burner; **air-pressure,** atmospheric pressure; pressure of air; **air-quake** (later examples); **air-receiver** [RECEIVER¹ 5], *spec.* a vessel for the storage of compressed air (*Funk's Stand. Dict.* 1928); **air-resistance,** the resistance of air to a moving body; **air rifle,** one actuated by the force of compressed air; **air scoop,** a scoop for diverting the wind (see quots.); **air-seasoned** *a.*, of timber = *air-dried*; also **air-season** *v. trans.* and *intr.*, **air-seasoning** *vbl. sb.*; **air-space,** a space for the use or passage of air, e.g. for respiration, insulation, etc. (see quots.); **air speed,** the velocity of moving air or wind; **air-stream,** (*a*) a current of air, *spec.* in *Meteorol.*; (*b*) = *air-flow*; **air-tube** (earlier and later examples of sense B. I. 7); (*b*) the inner tube of a pneumatic tyre; **air-tunnel** = *wind-tunnel*; **air-twist,** a spiral used for decorative effect in the stem of a wine-glass; so *air-twisted* adj.; **air-volcano,** an eruptive orifice from which volumes of gas are discharged with mud and stones; **air-washer** (see quot. 1949); **air-wave,** an atmospheric wave as of compression, rarefaction, or progression; **air wheel,** a balloon-tyre (see quots.).
Also in various names of instruments or apparatus actuated by the elastic force of compressed or heated air (often = 'pneumatic'), as *air-cylinder, -drill, -locomotive*. Also *AIR-LIFT 1.
1836 Air-bag [see AIR- I. 7]. *a* **1877** KNIGHT *Dict. Mech.* I. 769/1 An india-rubber air-bag. *a* **1884** *Ibid.* Suppl. 12/2 Air bags for raising sunken ships. **1948** 'N. SHUTE' *No Highway* vii. 175 There's no equipment here to lift an aircraft of this size. We've got no air bags. **1791** BOSWELL *Johnson* (1887) III. 168 He..walked in his room naked, with the window open, which he called taking an air bath. **1885** *Buck's Handbk. Med. Sci.* I. 467 It is often desirable not to employ too much water, but to expose the body freely, giving an air bath. **1959** *Encounter* Oct. 48/2 Seven minutes air-bath in the shade; four minutes in the sun; then a shower. **1815** Air-bell [see BELL *sb.*³]. **1889** *Anthony's Photogr. Bull.* II. 143 Carbon Printing in Winter...The difficulty I experienced in avoiding air bells. **1945** T. H. SAVORY *Spiders Brit. Isles* (ed. 2) 53 As winter approaches the spider hibernates. Either it closes the mouth of its air-bell or it finds an empty shell..and fills it with air. **1962** *Gloss. Terms Glass Industry* (*B.S.I.*) 38 *Air bell*, a bubble of irregular shape formed..in the manufacture of optical glass. **1889** *Cent. Dict.*, Air-blast. **1902** *Encycl. Brit.* XXXIII. 422/1 In..air-blast transformers, apertures are left in the core by means of which the cooling air can reach the interior portions. **1946** H. P. YOUNG *Electr. Power Syst. Control* (ed. 2) vii. 194 Air-blast breakers can be classified into three main types depending upon the manner in which the compressed air is directed at the arc. **1845** *North Brit. Rev.* IV. 128 An air-bloomery..was dependent, for its blast, upon the varying currents of air that played around the hill on which it was placed. **1872** *Rep. Comm. Patents 1871* I. 253 Westinghouse, George, Jr.,..Valve device for steam-power air-brake couplings. **1945** R. B. BLACK in *Jrnl. Amer. Dental Assoc.* XXXII. 956/2 The airbrasive process employs for its action a very fine—almost pinpoint—stream of compressed air into which a suitable finely divided abrasive agent has been introduced. **1953** I. GLICKMAN *Clin. Periodontology* xxxix. 665 *Airbrasive* which consists of fine abrasive powder (Dolemite in a stream of carbon dioxide) is used for removing surface deposits from the teeth. **1900** *Cricket* 29 Mar. 41/3 There is no necessity to mention Noble's air-breaks any more. **1910** *Hawkins's Electr. Dict.* 6/2 *Air Break Switch*, a type of switch designed to break the circuit in the open air or in an enclosed air space, as distinguished from an oil break switch. **1958** *B.S.I. News* Aug. 16 Heavy-duty composite units of air-break switches and fuses for voltages not exceeding 660V. **1956** in G. Merrill *Guided Missile Design, Operations Res.* III. i. 374 Pulsejet engines will initiate pressure waves of considerable energy, and the relatively long time it takes to warm-up an air breathing engine forces consideration of the effect of heat in the exhaust gases. **1964** *Economist* 7 Mar. 891/2 Manned aircraft powered by air-breathing jets. **1889** *Cent. Dict.*, Air-brush. **1901** *Brit. Jrnl. Photogr.* 1 Nov. 696 The ærograph [*read* aero-] is probably better known to the majority of photographers as the air-brush. **1916** 'B. CABLE' *Doing their Bit* iv. 58 The quick and even painting of the shells by air-brush spray. **1934** H. HILER *Notes Technique Painting* iii. 241 It may then be varnished..preferably with an air brush. **1917** 'DIXHUIT' *Artillery Experience* v. 62 Air-bursts of shrapnel are conspicuous. **1946** *Jrnl. R. Aeronaut. Soc.* L. 486/1 This particular rocket..air burst over Sweden. **1950** in *Effects of Atomic Weapons* (*Los Alamos Scient. Lab.*) ii. 30 The brownish or peachlike tint of the cloud which has been reported, particularly in the Bikini 'Able' airburst, is apparently due to nitrogen dioxide. **1795** T. WILKINSON *Wand. Patentee* I. 22 By attempting a visionary comparison, which has just now struck my air-castle imagination. **1840** C. HOWARD *Farming at Ridgemont* 140 A tunnel is formed by placing a wooden pipe..exactly over the centre of the air-channel. **1927** HALDANE & HUXLEY *Animal Biol.* iv. 96 These contain air-channels.. which run within the bodies of elongated cells. **1874** Air-compressor [see COMPRESSOR g]. **1892** P. BENJAMIN *Mod. Mechanism* 17 The Norfolk Compound Air-Compressor. **1908** *Westm. Gaz.* 21 July 4/2 An..ingenious air-compressor, specially designed for use on motor-vehicles. **1899** *Motor-Car World* I. 59/1 An air-cooled Aster motor of 2½ h.p. **1909** *Westm. Gaz.* 1 Apr. 4/1 The seven cylinders of the Gnome..when they are revolving at a high speed..will be very efficiently air-cooled. **1914** E. A. POWELL *Fighting in Flanders* iii. 73 The Lewis gun..is air-cooled. **1935** *Economist* 18 May 1141/1 A number of these aircraft are fitted with Rolls Royce engines, but the Gloster 'Gauntlet' ..is equipped with an air-cooled engine. **1962** *Which?* Jan. (Car Suppl.) 4/1 Air-cooled: relies on fan-driven air, not water-filled radiator. *a* **1875** KNIGHT *Dict. Mech.* I. 34/2 Shaler's air-cooler. **1865** N. S. SHALER *U.S. Pat.* 47,991 (*title*) Air Cooling Apparatus. **1909** *Westm. Gaz.* 25 Mar. 2/2 The designer has adopted the expedient of revolving the cylinders en bloc around fixed cranks, whereby he is enabled to successfully adopt air-cooling. **1909** *Ibid.* 9 Sept. 4/3 The sparking plugs, which are provided with air-cooling ribs. **1894** *Phil. Mag.* XXXVII. 405 The air-core transformer used for the experiment consisted of two coils wound one inside the other. **1906** A. RUSSELL *Alternating Curr.* II. viii. 231 The ideal air core transformer, that is, the air core transformer the resistance of the primary coil of which is zero. **1953** AMOS & BIRKINSHAW *Television Engin.* I. x. 265 (*caption*) Air-cored coils.., iron-cored.., Ferrite-cored. **1882** *Imp. Dict.*, Air-course. **1937** *Times* 6 Feb. 8/5 Work on 14's face should have been stopped at least until a return air-course, far removed from the intake and leading directly to the main return, had been made. *a* **1884** KNIGHT *Dict. Mech.* Suppl. 15/1 *Air-crossing* (Mining), an arch built over a horse-way or other road, with a passage or airway above it. **1911** *Act* 1 & 2 *Geo. V*, c. 50 § 42 (3) All air-crossings shall..be so constructed as not to be liable to be destroyed in the event of an explosion. **1960** *Aeroplane* XCIX. 770/1 This new craft has, in fact, been designed to enable operators to obtain practical experience with air-cushion craft 'in the field'. **1962** *New Scientist* 19 Apr. 79/2 The air cushion of the VA-3 is derived from peripheral and intersecting slots. **1965** *Guardian* 5 Jan. 3/3 Air cushion vehicles—or hovercraft, if you prefer the term. **1843** *Jrnl. R. Agric. Soc.* IV. 357 An air drain round the building is, in damp situations, highly useful. **1848** *Ibid.* IX. 341 No indications of wetness appeared on the two air-drained pieces. **1944** *Geogr. Jrnl.* June 252 Valleys and basins are more frosty, depending again on the degree to which the relief and other factors permit air drainage from extensive uplands inland. **1948** WHITE & RENNER *Human Geogr.* IV. xx. 340/2 At night the air, which is cold and therefore heavy, drains down the slopes settling in the lower elevations. Hence danger of frost is least at the head of the piedmont plain and greatest at its foot. This..is known as air drainage. **1889** *Cent. Dict.*, Air-dried. **1891** W. SCHLICH *Man. Forestry* II. iv. 248 The Ash yields an excellent timber, hard and heavy, specific gravity when air-dried = ·75. **1908** *Chambers's Jrnl.* July 543/1 Peat which can be air-dried to such an extent that only some 25 per cent of moisture is retained. *a* **1912** *Paper Terminol.* (*Spalding & Hodge*) II. 1 *Airdried*..is applied to hand-made and exceptionally good machine-made writings and brown papers, when dried slowly by exposure to a uniform temperature. **1897** *Daily News* 1 Nov. 7/1 The air-driven hydraulic pump. **1856** *Jrnl. R. Agric. Soc.* XVII. I. 194, I..then allowed it to become air-dry, by keeping it for some days in a safe place, in a heated room. **1949** *Gloss. Terms Timber* (*B.S.I.*) 13 In Great Britain the moisture content of air-dry timber may range between 14..and 23 per cent according to the season of the year and the species of timber concerned. **1890** BILLINGS *Med. Dict.* I. 32/2 *Air embolism*, the presence of free atmospheric air within the vascular system during life in sufficient quantity to give rise to symptoms of obstruction. **1905** *Lancet* 9 Dec. 1738/1 (*title*) Death of a Diver from Air Embolism. **1936** *Archit. Rev.* LXXX. p. lviii/1 The removal of smells from kitchens, of steam from bathrooms and of smoke from smokerooms is well worth while and there are a number of air-extractor devices on the market that deserve the consideration of the architect. **1861** J. STENHOUSE (*title*) The successful application of Charcoal Air-Filters to the ventilation and disinfection of Sewers. *a* **1884** KNIGHT *Dict. Mech.* Suppl. 15/2 *Air Filter*, a protective ventilator consisting of a cloth interwoven with thin brass wire to act as a filter for the air. **1927** *Daily Tel.* 10 Feb., Motor manufacturers are urged to provide air-filters on all motor vehicles. **1911** R. M. PIERCE *Dict. Aviation* 16 *Air-flow*, the flow or movement of the air. **1915** *Aeronautics* 17 Nov. 327 Let us first see wherein mainly the behaviour of the air flow and its resistance differ from that prevailing in the case of a flat plane. **1935** *Discovery* Oct. 309/1 The occasional opening of doors is quite inadequate for regulating the air flow [in a film studio]. **1940** E. C. SHEPHERD *Britain's Air Power* 18 The bomber began to take on a load of ice it had not expected—ice on the edge of the wing to spoil its shape and interfere with the air-flow. **1961** P. STREVENS in *Papers in Lang. & Lang. Teaching* (1965) xi. 134 If the breath-stream is forced to pass through a narrow constriction, the air-flow becomes turbulent. **1962** D. SLAYTON in *Into Orbit* 22 When you are re-entering the atmosphere..you must get the capsule into a position where its blunt end is pointed straight down into the airflow. **1949** *Good Housekeeping* (N.Y.) Nov. 135/1 Have you ever used an air freshener—a special product that camouflages unpleasant odors with clean countrylike scents? **1960** *Guardian* 22 Feb. 6/7 An air-freshener in aerosol form. **1962** *Which?* Mar. 90/1 There are a great many air fresheners on the market, which claim to 'dispel', 'kill', 'neutralise' or 'suppress' unwanted smells. **1848** MRS. GASKELL *Mary Barton* xvi, Air-gaps were to be seen in their garments. **1899** R. ROUTLEDGE *Discov. & Inv. 19th C.* (ed. 13) 541 A stout wire interrupted by an air gap in its centre provided with small brass balls. **1902** *How to make Things* 3/2 A miniature flash of lightning breaks through the insulating air-gap between the balls or oscillators. **1942** C. A. COTTON *Geomorphology* (ed. 3) vi. 73 The former gorge, or water gap, through this stratum is now no longer traversed by a stream, and becomes an 'air gap'. **1873** *Pract. Mag.* II. 399/1 Air Gas Machine..an improved apparatus for forcing air in uniform quantities into a carbureter. **1879** *Encycl. Brit.* X. 101/2 This air-gas is now largely used both in America and Europe for lighting mansions, churches, factories, and small rural districts. **1909** *Chambers's Jrnl.* June 411/1 The application of what is generically termed 'air-gas' to domestic uses is one of far-reaching possibilities. **1841** *Civil Engin. & Arch. Jrnl.* IV. 13/1 The..instruments employed..to determine the pressure of the steam,..namely, the barometer-gauge, the air-gauge, etc. **1951** J. G. VAETH *200 Miles Up* ii. 26 *Air glow*, which is a term applied to the light of the night sky (excluding starlight and moonlight), lends itself to spectroscopic examination and has been found to contain light emissions characteristic of nitrogen, oxygen, and sodium. **1958** *Sci. News* XLVIII. 12 One effect of the normal Sun on the atmosphere..is the production of the faint emission from the atmosphere at night which is called the night airglow. **1914** H. BREARLEY *Case-Hardening of Steel* vi. 72 The surface of the air-hardened steel is less hard

than that of water or oil quenched steel. **1930** *Engineering* 23 May 680/3 Steel A was air-hardened from 950 deg. C after soaking for 20 minutes. **1906** E. R. MARKHAM *Amer. Steel Worker* (ed. 2) 278 The steel hardens when..exposed to the air. It is styled 'Air Hardening Steel', more generally known, however, as Self-Hardening Steel. *a* **1875** KNIGHT *Dict. Mech.* I. 49/1 *Air-heater*, a stove or furnace so arranged as to heat a current of passing air, for warmth or ventilating purposes. **1944** *Gloss. Terms Gas Ind.* (*B.S.I.*) 34 *Air heater*, an appliance designed to heat spaces by the forced circulation of large volumes of warmed air. **1918** *Jane's Pocket Aeronaut. Dict.* 8 *Air intake pipe*, a pipe fitted to the carburetter or induction system through which only air is drawn. **1922** *Encycl. Brit.* XXX. 26/2 (*caption*) Air Intake [in a wind tunnel]. **1931** T. E. LAWRENCE *Let.* 14 July (1938) 729 Scoop-tubes like air intakes thrust through the floor amidships. **1958** *Times* 19 June 6/3 He had been sucked into the air intake of a jet engine. **1909** *Westm. Gaz.* 4 May 4/2 An engine having specially designed air-jackets. **1936** *Techn. Rep. Aeronaut. Res. Comm. 1934–35* I. 19 A wind tunnel investigation has been made of an air-jacketed engine. **1900** L. H. BAILEY *Cycl. Amer. Hort.* II. 894/2 In a conservatory, merely a ball of sphagnum bound around the branch with twine will serve an equally good purpose...This kind of propagation is known as air-layering. **1934** WEBSTER, *Air layering*, a form of propagation, employed with certain plants whose branches cannot be brought to the ground for layering, in which a portion of a branch or stem, sometimes girdled, is kept covered, as by wrapping with moist soil, moss, or the like, until it forms roots and may be detached from the parent and planted...The process is specified as *pot layering* when the rooting medium is enclosed, as in a pot or box, and as *marcottage*, or *Chinese layering*, when the ball of earth or moss is merely tied about the stem. **1957** M. FREE *Plant Propag.* vi. 183 In air-layering a suitable shoot is selected, the stem is wounded by removal of a cylinder of bark..where it is desired that roots should form. **1855** LEIFCHILD *Cornwall Mines* 282 The underground boys work the air-machines. **1893** F. WALDO *Mod. Meteorol.* iv. 318 If a swiftly moving air mass moves into a quiet mass of air, then the resistance is considerable. **1942** W. G. KENDREW *Weather* xii. 65 *Air mass* is the term applied to a part of the atmosphere large enough to play an appreciable part for a period of at least some hours in the meteorology of any region. **1949** M. MEAD *Male & Female* xi. 240 A hunting-trip with an air-mattress. **1876** *Trans. Clinical Soc.* IX. 23 An air-pad was applied to the tumour. **1836** Air-passage [see AIR- I. 7]. **1878** RANKINE *Steam Engine* (ed. 9) 459 Air Passages—Blowing Apparatus—Chimney. **1911** *Encycl. Brit.* XXIII. 199/2 A foreign body in the air-passages may be impacted above the vocal cords. **1709** T. ROBINSON *Nat. Hist. Westmld.* v. 30 If the Miners should not open their Air-Pits and keep their Thurling-Ways clear. **1839** URE *Dict. Arts* 969 These air-pits do not in general exceed 7 feet in diameter. **1788** A. FALCONBRIDGE *Acct. Slave Trade* 24 Most of the ships..are provided, between the decks, with five or six air-ports on each side of the ship. **1867** SMYTH *Sailor's Word-bk.* 28 *Air-ports*, large scuttles in ships' bows for the admission of air, when the other ports are down. **1944** *Gloss. Terms Gas Ind.* (*B.S.I.*) 33 *Air port*, the aperture in an aerated burner adjacent to the injector through which primary air is admitted to the mixing tube. *a* **1875** KNIGHT *Dict. Mech.* I. 50/2 Gruber's air-pressure filter. **1946** *Nature* 9 Nov. 674/2 Using air-flow through a capillary to regulate the air-pressure difference. **1965** W. S. ALLEN *Vox Latina* 7 *Ceteris paribus*, stressed sounds produce greater intensity of air-pressure. **1891** *Daily News* 13 Oct. 5/4 General Dyrenforth's experiments in rain-making by means of explosions, or what he calls 'terrific airquakes', have not convinced his scientific opponent. **1950** D. GASCOYNE *Vagrant* 28 Till all night's spark-sprayed dome is stunned with quick air-quakes of gold. **1951** *Gloss. Terms Plastics Ind.* (*B.S.I.*) 34 *Air receiver*, (*a*) an accumulator or storage vessel charged with compressed air or inert gas used in a high pressure hydraulic system as an alternative to a weight-loaded accumulator; (*b*) an accumulator or storage vessel used as a reservoir in pneumatic systems. **1959** *B.S.I. News* Apr. 3 The welding of air receivers is likely to be separately discussed by a group of experts on welding and may become the basis for welding of pressure vessels as a whole. **1901** *Sci. Amer.* 9 Feb. 82/2 Light has recently been thrown upon the question of air resistance of railway trains. **1908** *Aeronautics* Mar. p. xviii, Trials of a cellular aeroplane ..have demonstrated that weight is a less important factor than air-resistance. **1936** *Discovery* Feb. 40/1 To diminish air-resistance by the streamlining of both engine and train. **1902** *Sears, Roebuck Catal.* (ed. 112) 298/2 Quackenbush Improved Nickel Plated Air Rifle..$4.35. **1958** *Daily Tel.* 30 June 15/8 A girl was injured by an air rifle. **1919** W. B. FARADAY *Gloss. Aeronaut. Terms* 55 *Air Scoop*, a projecting cowl, which, by using the dynamic pressure of the relative wind or slip stream, serves to maintain air pressure in the interior of the envelope. **1920** *Flight* XII. 663/2 In Italy the British method of supplying air to the ballonets through airscoops fitted in the slipstream of the propellers is never used. **1929** *Daily Express* 1 Jan. 6/3 The air-scoops projecting from every porthole in a vain attempt to manufacture a breeze with the ship's motion, fail to fulfil their functions. **1917** J. B. WAGNER *Seasoning of Wood* x. 151 The wood is allowed to air-season for several months to a year. *Ibid.* 154 The present methods of air-seasoning in use have been determined by long experience. **1919** H. S. BETTS *Timber* v. 150 A kiln is used also when partially air-seasoned or even fully air-seasoned material is to be dried further. **1930** *Forestry* IV. 36 In air-seasoning both the temperature and humidity of the available air are dependent on local climatic conditions. **1889** *Cent. Dict.*, Air-space. **1893** DUNGLISON *Dict. Med. Sci.* (ed. 21) 32/1 *Air space*, space filled with air from rupture or other injury to air-cells. **1900** *Lancet* 11 Aug. 458/1 That this conference approves that the standard of air space for dwellers in cities and large towns be raised to 500 cubic feet for every adult and to 250 cubic feet for each child under 10 years of age. **1936** *Discovery* June 197/2 It appears that the dielectric constant is lower in the case of those organs having much fatty tissue and air-spaces such as the bones and lungs. **1957** *Archit. Rev.* CXXI. 213 Rust-proof rails swing out, cabinets are double cased (airspace insulation) in zinc coated steel. **1922** *Encycl. Brit.* XXX. 27/1 The larger Eiffel tunnel gives an air speed of 40 metres per second. **1869** Air-stream [see AIR- I. 5]. **1913** J. C. HUNSAKER tr. *Eiffel's Resistance of Air & Aviation* 239 Stability was verified by suspending the model [aeroplane] in the air stream upon a horizontal axis. **1922** *Encycl. Brit.* XXX. 27/1 The experimental section of an Eiffel type wind tunnel consists of an air stream as it crosses an open room from wall to wall. **1958** *Times* 23 Sept. 11/4 His pilgrimage coincides with the fair example of a wet, westerly airstream. **1961** *New Scientist* 16 Mar. 672/1 Such a cooling mechanism [for rockets] is superior..because a liquid is simply swept away in the air-stream. **1961** L. F. BROSNAHAN *Sounds of Language* i. 2 The conversion of some of the kinetic energy of this airstream into acoustic energy. **1826** KIRBY & SPENCE *Entomol.* III. xxx. 154 The two prolegs, which M. Latreille thinks are air-tubes. **1877** *Engineering* 16 Nov. 381/3 The air-tube of a diver's dress. **1894** *Work* 315/2 Repairing Air-tube of 1892 Dunlop. **1953** J. S. HUXLEY *Evolution in Action* iii. 75 [The insects] breathe by means of air-tubes, which convey oxygen direct to the tissues. **1933** *Jane's Fighting Ships* 26 A ½ inch scale model was prepared for testing in the air-tunnel at the National Physical Laboratory. **1897** A. HARTSHORNE *Old Eng. Glasses* 58 The beaded stems, out of which the air twists were derived, continued to be made in Holland. *Ibid.*, Air-twisted stems of various kinds. **1903** *Burlington Mag.* III. 63/1 The secret of the construction of two of the classes—namely, the brilliant, and the combined opaque and air-twist—seems to have been lost. **1916** J. S. LEWIS *Old Glass* 62 The air-twist probably began with a 'tear'. **1879** GEIKIE in *Encycl. Brit.* X. 250 Certain remarkable orifices of eruption..to which the names of mud-volcanoes, salses, air-volcanoes, and macalubas have been applied. **1905** *Lancet* 25 Feb. 507/2 The Stellite Air Deodoriser ..is an effectual air-washer and as such it may obviously have numerous hygienic applications. **1949** *Gloss. Terms Refrig.* (*B.S.I.*) 4 *Air washer*, a water-spray system or other device for cleaning air, capable of serving also as a cooler, humidifier, or dehumidifier. **1879** W. JAMES in *Jrnl. Speculative Philos.* XIII. 85 Notwithstanding the brilliant conjectures of the last few years which assign different acoustic end-organs to different rates of air-wave, we are still greatly in the dark about the subject. **1881** Air-wave [see AIR- I. 5]. **1895** H. LAMB *Hydrodynamics* Index, Air-waves, effect of viscosity on. **1930** *Flight* XXII. 404/1 The Air Wheel..is revolutionary in design, and, as its name implies, the cover has the dual function of a tyre and wheel, being a full balloon cover mounted directly on a hub attached to the 'plane undercarriage. **1932** *Times* 29 Feb. 17/5 A new tire for small cars has been introduced...It is known as the air wheel and is based on aeronautical experience.

III. Of or pertaining to aircraft.
1. In numerous combinations (tending to supersede *AERIAL) relating to locomotion in the air by means of aircraft, as *air drill, navigation, pageant, service* [SERVICE¹ 32], *traffic*; carried or conducted by aircraft, as *air parcel, photography* (also *-photo*(*graph*)), *post, survey, tour* (so *-tourist*); *air-flying, -launched, -portable* (so *-portability*), *-sailing* adjs.; also, **air-bridge,** a link between points provided by air transport; **air carrier** = *AIRCRAFT *carrier*; **air circus** [cf. *CIRCUS 2 d], a squadron of aeroplanes; an air display, an air pageant; **air Derby,** see *DERBY 1 d; **air edition,** see *AIRMAIL b; **air ferry,** an aircraft or system of aircraft for the conveyance of passengers and goods; **airfoil** *U.S.* = *AEROFOIL; **air freight,** freight conveyed by air; also as *v. trans.*; hence *air-freighting* vbl. sb.; **air fuelling,** the refuelling of one aircraft by another in flight; **airgraph** (also **Airgraph**), a form of airmail registered by Kodak Ltd., in which the correspondent's letter is photographed on a reduced scale; a letter so transmitted; **air hog,** cf. *road hog* (ROAD *sb.* 12 and *ROAD HOG); **air letter,** a letter conveyed by air, esp. one written on a folding form of special design; **air-mark** *v. trans.* (see quot. 1929); hence *air marker, -marking*; **air-mast,** a mast to which airships are moored; **air mile,** a nautical mile used as a measure of distance flown by aircraft; hence *air mileage*; **air-minded** [*MINDED *ppl. a.*] *a.*, interested in or enthusiastic for the use and development of aircraft; so *air-mindedness*; **air miss** (see quots.); **air plot** (see quot. 1951); **air-pocket,** a local condition of atmosphere, as a down current or sudden change of wind velocity, which causes an aircraft to lose height suddenly; **air position,** position in the air, (*a*) for tactical purposes [see POSITION *sb.* 7 c]; (*b*) that an aircraft would have reached if the flight had been in motionless air; also *attrib.*; **airscape,** a view taken from the air (cf. SCAPE *sb.*³); **air-sea rescue,** applied to a branch of the Royal Air Force, whose task is to rescue airmen and passengers from the sea, and to such operations; **air sense,** cf. *road sense* (*ROAD *sb.*); **air-sick** *a.* [after SEA-SICK *a.*], sick from the motion of an aircraft; **air-sickness** (later example: see sense II in Dict.); **air space, airspace,** the air considered as a medium for the operation of aircraft; **air speed,** the velocity of an aircraft (or of anything flying, e.g. a bird) in relation to the air through which it is moving; also *attrib.*; **air terminal,** (*a*) the terminal point of an air-line, also called *air terminus*; (*b*) the town office of an air-line, equipped for the reception of passengers; **air-to-air** *a.*, from one aircraft to another; so *air-to-ground, air-to-surface* adjs.
1939 *Baltimore Sun* 17 Apr. 9/1 The New Zealand service will constitute the air line's second 'air bridge' of the Pacific. **1948** *Newsweek* 9 Aug. 27/1 The Berlin 'air bridge' —as the Germans call it—claimed its first American victims on July 9. **1920** *Proc. Air Conf., London* 99 Air carriers were designed and commissioned towards the end of the war. **1932** *Flight* XXIV. 1227/1 His 'Air Circus' carried 250,000 passengers. **1933** *Aeroplane* 18 Oct. 690/1 The accident arose from a collision..during an air circus. **1940** D. WHEATLEY *Faked Passports* viii. 94 After the Armistice he [Goering] was ordered to surrender the planes of his famous air-circus to the Americans. **1932** *Flight* XXIV. 584/2 A squadron of 'Furies' was given just 15 minutes in which to show off their air drill. **1916** *Aerial Age Weekly* 11 Sept. 793 (*heading*) Air Ferry over Great South Bay. **1932** *Flight* XXIV. 933/1 The daily air ferry services between Shoreham, Portsmouth and Ryde. **1912** C. M. DOUGHTY *Clouds* 119 Their air-flying enemies. **1922** *N.A.C.A.* (*U.S.*) *Rep. Nomencl. Aeronaut.* 621 Airfoil. **1930** *Flight* 29 Aug. 972/2 The addition of an airfoil fuselage. **1929** *Aerial A.B.C.* Feb.–Apr. 19 The conditions for air freight services apply to all goods which are accepted by an air traffic company. *Ibid.* May–July 41 Luggage can also be forwarded as air freight. **1959** W. D. PEREIRA *North Flight* ix. 143 Make sure that box is air-freighted to-night. **1947** *Aircraft Engin.* Jan. 27/1 Now..that civil air freighting is upon us..the problems involved in the efficient handling of freight must be tackled. **1937** *Jrnl. R. Aeronaut. Soc.* XLI. 285 Air fuelling offers another alternative, but the large aerodrome ..seems the simplest of all methods of increasing the economy of air transport operation. **1941** *Engineer* 2 May 296 To meet the need for cheapening and expediting homeward postal communication from the British Forces in the Middle East, the Post Office is introducing an airgraph service. **1941** *Sphere* 6 Dec. p. i, Airgraph letters should be written in black ink. **1945** *Comment from Italy* (*Three Arts Club*) 41 For weeks now there had been little for him,—just an occasional Airgraph,—nothing more. **1955** H. & A. GERNSHEIM *Hist. Photogr.* 254 The airgraph service which operated between 1941 and 1945, in which by modern microfilm methods myriads of messages were flown between families in England and the fighting services in the four corners of the world. **1909** *Westm. Gaz.* 9 Feb. 4/1 Pointing out how the flying-machine is likely to violate every international law and rudely trespass on every private right and privilege, characterising the intrepid navigators as air-hogs and human vultures. **1949** A. R. WEYL *Guided Missiles* 108 The most powerful air-launched missile of its kind. **1951** *Jrnl. Brit. Interplanetary Soc.* X. 217 The 'Skyrocket'..adopted the technique of air-launching from a B.29. **1920** *Flight* XII. 781/2 Threepenny air-letter postage between London and Amsterdam, with the prospect of a similar charge to Paris, is getting a little nearer sanity. **1951** *Oxf. Jun. Encycl.* IV. 17/1 An air letter is written on a special form supplied by the Post Office. This is made of thin paper and impressed with a sixpenny stamp. **1929** *Times* 12 Mar. 12/2 Thousands of cities and towns throughout the States have been 'air-marked' by civic and trade associations. During last year one oil company alone painted names on 4,200 stations...These markings, together with a standardized system of indicating obstructions, such as high tension cables..have proved a very valuable aid to air pilotage. **1948** *Shell Aviation News* No. 117, 5/1 An extensive air-marking programme along Skyway One is now under way. *Ibid.* No. 122, 4/2 The Civil Aeronautics Administration is sponsoring a programme for air marking cities, towns and villages throughout the country. An air marker is a sign on rooftop or ground, visible from the air, which enables a pilot to orient himself when lost. **1927** *Daily Tel.* 1 Nov. 10/6 The selection of a site on the south side of the St. Lawrence for the erection of an Imperial airmast. **1919** *Sphere* 6 Dec. p. viii/1 Fifteen hundred air miles at 107 m.p.h. **1945** *Yorks. Post* 19 Apr. 1/1 Nulde, 20 air miles east of Amsterdam. **1948** *Jrnl. Inst. Navig.* I. 63 The aircraft's fuel consumption..must be balanced against savings in air mileage. **1951** *Gloss. Aeronaut. Terms* (*B.S.I.*) III. 10 *Air-mileage unit*, an instrument which derives continuously and automatically the air distance flown, and feeds this function into other automatic instruments. **1928** *Daily Express* 20 June 8/3 At last, I believe, people are becoming 'air-minded'. **1927** *Glasgow Herald* 2 Nov. 13 The expansion of aviation systems and the spread of a sense of 'airmindedness'. **1930** *Flight* 3 Jan. 4 That great wave of airmindedness that followed Lindbergh's transatlantic flight. **1960** *Guardian* 10 Nov. 1/5 The new Minister did not explain..the exact technical significance of an 'air-miss'..(The Air Ministry says it is 'rather like a near miss on the ground'). **1962** *UK 'Air Pilot'* p. RAC 35 Whenever a pilot considers that his aircraft may have been endangered by the proximity of another aircraft during flight..he should make an airmiss report. **1871** *Ann. Rep. Aëronaut. Soc.* 62 The minds of many thinking men have been, during the present century, turned to this interesting subject of air navigation. **1927** *Air* Dec. 55/2 The first Birmingham Air Pageant. **1928** *Aerial A.B.C.* Apr. 20 Air parcels may be posted at any District or Branch Post Office. **1919** *Geogr. Jrnl.* LIII. 330 (*heading*) Air Photography in Archaeology. *Ibid.*, Had I not been in possession of these air-photographs the city would probably have been merely shown by meaningless low mounds. **1920** *Flight* XII. 233/1 The achievements of air photography during the War were very remarkable. **1923** *Geogr. Jrnl.* LVII. 359 Here two air-photos will certainly reveal the course of undiscovered Roman roads. *Ibid.* 363 The field archæologist has much to gain in future from an alliance with the air-photographer, particularly in England. **1959** *N. & Q.* CCIV. 1/1 From air photographs Dr. St. Joseph was able to show the size and shape of Roman fields in the Fenlands. **1942** D. C. T. BENNETT *Compl. Air Navig.* (ed. 4) v. 172 The Air Plot Method is similar in principle..but instead of flying on one constant course only, a number of courses may be followed. **1951** *Gloss. Aeronaut. Terms* (*B.S.I.*) III. 6 *Air plot*, a continuous plot of true heading steered and air distances flown. **1913** C. GRAHAME-WHITE *Aviation* v. iv. 155 A lessening of pressure—or what is more familiarly

known as an 'air pocket'. **1933** *Boys' Mag.* XLVII. 24/2 We shall probably bump a bit, owing to air-pockets. **1959** *Times* 16 Jan. 10/5 The British idea was to develop an air-portable gun for both roles. **1959** *Star* 19 Feb. 9/5 Mr. Christopher Soames, Secretary for War, coined a new watchword today for Britain's all-regular Army of the future—air portability. **1917** F. A. Collins *Air Man* vi. 140 The English had not chosen their battlefield, or rather air-position, and thus fought at a disadvantage. **1937** D. C. T. Bennett *Compl. Air Navig.* v. 154 The difference of the air position so obtained in relation to the ground position (i.e. the departure point) is the wind effect for the total time. **1945** *Sci. Amer.* June 349/1 One of the recently revealed secrets..is the 'air position indicator', an instrument that..gives continuous readings of latitude and longitude plus a continuous record of nautical air miles flown. **1911** *Daily Mail* 11 Sept. 3/4 (*heading*) First Air-Post. *Ibid.*, An air post cannot be expected as yet to behave with the same clockwork regularity as an earth post. **1919** *Liverpool Jrnl. Commerce* 20 Nov. 6/4 Within the past few months regular air post services have sprung into being. **1897** *Aeronaut. Ann.* 92 The..care needed in making changes in an air-sailing machine. **1921** *Flight* XIII. 193 (*caption*) Winter in Switzerland: An airscape of the popular resort, Davos. **1941** *Flight* XXXIX. 361/1 The various rescue services..have been co-ordinated under one central control known as the Directorate of Air/Sea Rescue Services. **1941** J. A. Hammerton *ABC of RAF* (ed. 2) 74 Abbreviations of Titles and Terms employed by the R.A.F...A/SRS. Air Sea Rescue Service. **1942** *Aeroplane* 13 Nov. 562/3 A Supermarine Walrus of the Air-Sea Rescue Service alighted on the sea in the middle of a German minefield. **1958** *Times* 10 July 15/2 The manufacture..of..inflatable liferafts and other air-sea rescue aids. **1919** *Conquest* Dec. 65/1 The successful execution of aerial acrobatics involves the possession..of that indefinable quality which, for want of a better word, we will call 'air-sense'. **1919** *Sphere* 1 Nov. p. x/3 Outside the R.A.F. there were no records of a daily air service to guide the Avro company in organising such an undertaking. **1922** *Daily Mail* 8 Dec. 12 An 1,800 miles air service from Copenhagen to Brindisi is being planned. **1785** F. Farley *Bristol Jrnl.* 14 May in *N. & Q.* (1938) CLXXV. 79/1 Air sick. **1873** Air-sick [see *airman]. **1908** H. G. Wells *War in Air* vii. § 1 Even the air-sick men flushed and spoke. **1958** G. Greene *Our Man in Havana* III. ii. 123, I suppose he's feeling air-sick again. **1908** H. G. Wells *War in Air* vii. § 2 For a time he was not a human being, he was a case of air-sickness. **1911** R. Wallace in Grahame-White & Harper *Aeroplane* xiii. 289 A State should have full dominion in the air space above its territory. **1959** *Listener* 19 Mar. 512/1 Persia protests to U.S.S.R. that Russian aircraft have violated her air space eighty-one times in past three months. **1961** *New Scientist* 20 Apr. 103/2 It is generally agreed that national sovereignty in the so-called 'airspace' is limited to some level above the Earth's surface. **1910** R. Ferris *How it Flies* 453 Air-speed—the speed of aircraft as related to the air in which they are moving; as distinguished from land-speed. **1912** *Aeronautics* Dec. 391/1 It is possible with an air-speed indicator to read..how far one is above this danger point. **1937** D. C. T. Bennett *Compl. Air Navig.* v. 165 The following abbreviations are acceptable for use in Navigation Logs..A/S = Air Speed. **1941** E. C. Shepherd *Military Aeroplane* 17 Height and forward speed..are shown on a sensitive altimeter and on an airspeed indicator. **1942** *Gen* 15 July 50/1 Watch that airspeed! Eleven hundred and you're levelling out. **1918** *Times* 6 Dec. 12/2 (*headline*) An Air Survey..Surveying the country by means of aerial photographs. **1924** O. G. S. Crawford (*title*) Air Survey and Archæology. **1925** *Flight* XVII. 735 (*title*) Air surveying. **1933** *Discovery* Feb. 57/1 Air surveying is used extensively by the United States Geological Survey for the preparation of topographic maps. **1921** *Aircraft Yr. Bk.* 79 The principal communities which are situated along this air route should create thoroughly modern air terminals. **1935** C. G. Grey in N. Tangye *Air is Our Concern* i. 10 Though Hounslow Heath was actually the first London Air Terminal, it was given up because it was on the wrong side of London. **1956** *Times* 2 Feb. 5/1 A temporary air terminal will be erected on the platform to take those services now being handled at B.E.A.'s Waterloo air terminal. **1919** *Sphere* 10 May 108/1 The air terminus for London is Hounslow. **1941** *Flight* XL. 48 f/2 Drogue targets for air-to-air gunnery training. **1955** *Sci. News Let.* 26 Mar. 197/3 The Air Force has unveiled its newest guided missile..described as the 'only air-to-air missile with a "brain" of its own'. **1957** *Times* 22 Aug. 6/6 Air-to-air guided missiles on wing-tip launchers. **1942** *Hutchinson's Pict. Hist. War* 10 June–1 Sept. 135 (*caption*) A..high-wing monoplane..has proved useful for reconnaissance and air-to-ground co-operation by the British. **1958** C. C. Adams et al. *Space Flight* 51 Next to nothing is known about air-to-air or air-to-surface rockets, though a few surface-to-air missiles are known. **1923** *Daily Mail* 29 Jan. 13 (*heading*) Air tours at a penny a mile. **1929** *Punch* 20 Mar. 326/3 The Minister is satisfied that before many months we air-tourists will be taking our twelve-day flips to Kenya. **1912** H. E. Richards *Sovereignty over Air* 4 We must consider the principle on which the relations of States are to be conducted with regard to air traffic. **1933** *Flight* XXV. 524/2 (*heading*) Air Traffic Control. **1951** *Oxf. Jun. Encycl.* IV. 381/2 The advent of the fast, all-weather aircraft, and the demand for frequent and regular services, have made it essential to establish strict rules for air-traffic control. **1929** *Lancet* 12 Jan. 105/2 The facilities offered by air transport to patients travelling abroad. **1946** in *Amer. Speech* (1948) XXIII. 76 Air transportability. **1879** P. Brannon (*title*) The air-boat for arcustatic air-travel. **1923** G. Collins *Valley of Eyes Unseen* 326 The great strides recently made in the art of air-travel. **1963** *New Yorker* 8 June 96 Showcase is air-travel light. **1951** L. MacNeice tr. *Goethe's Faust* II. Act II. p. 205 She departs. Enter, above, the air-travellers. **1908** *Westm. Gaz.* 1 July 12/1 Thirteen persons who made a successful air-trip from the Champ de Mars. **1959** *Elizabethan* Apr. 5/3 The chance of an air-trip overseas.

2. Pertaining to the air as a sphere of offensive or defensive operations, as *air alert, attack, -bombing* (also *-bomber*), *defence, observation, offensive, reconnaissance, strike, support, supremacy, warfare*; used of a bomb, missile, etc., discharged from an aeroplane, as *air bomb, torpedo*; also, **air control,** control of an area by means of air power (in quot. 1915 *spec.* to give the correct range for artillery fire); **air cover,** protection by aircraft during a military operation; **air-drop** (chiefly *U.S.*), the landing of troops or supplies by parachute; also as *v. trans.* & *intr.*; **air-head,** cf. *beach-head*; **air mine** (see quots.); **air power,** power of defensive and offensive action dependent upon a supply of aircraft, missiles, etc. (cf. Sea-power 2); **air umbrella,** a force of aircraft used to give air protection to a military operation; **air warden** (see *air-raid).

1941 *Times Weekly* 5 Feb. 2/3 In spite of air alerts and privations, the population have not lost their courage. **1959** *Times* 18 May 7/2 The reason why Strategic Air Command does not maintain an air alert, with aircraft carrying nuclear weapons in the air twenty-four hours a day. **1914** *Sphere* 26 Dec. 318/1 The possible air attack over London. **1915** Grahame-White & Harper *Aircraft in Gt. War* v. ii. 172 Air attacks on cities have appeared despicable largely because they are so new. **1941** N. Macmillan *Air Strategy* xiv. 109 A true conception of the object of the air blockade. **1914** *Sci. Amer.* 15 Aug. 113/2 (*heading*) The Air Bomb. **1915** Grahame-White & Harper *Aircraft in Gt. War* v. ii. 172 The Germans..were prepared to use every instrument..drifting mines, air bombs. *a* **1930** D. H. Lawrence *Last Poems* (1932) 54 And most murderous of all devices Are poison gases and air-bombs Refinements of evil. **1929** F. P. Gibbons *Red Napoleon* (1930) viii. 198 American field guns..became the target of air bombers and low flying combat 'planes. **1934** *Flight* XXVI. 141/2 The idea of air bombing, which General Groves and..his supporters are trying to popularise. **1915** Grahame-White & Harper *Aircraft in Gt. War* vi. ii. 271 It was certainly unfortunate for the Germans that, as their air control for artillery grew less effective, that of the Allies should have begun to reach its full efficiency. **1930** *Flight* 3 Jan. 1 These are the air control of Iraq and the Air Defences of Great Britain. **1942** *Hutchinson's Pict. Hist. War* 10 June–1 Sept. 245 Our bombers and fighters thrust fiercely into the attack, affording continuous and effective air cover to our attacking forces. **1916** *Sphere* 26 Feb. 207/1 (*heading*) The Problem of Air Defence. **1922** *Encycl. Brit.* XXX. 87/1 Air defence..deals with the arrangements which deny to enemy aircraft access to vulnerable points. **1949** *Britannica Bk. of Year* 687/1 *Airdrop*, that which is dropped from an airplane, such as supplies. **1950** *Baltimore Sun* 29 Apr. (edition B*) 1/7 D-day of the exercise was marked by the biggest of peacetime airdrops. **1958** *N.Z. News* 1 July 4/1 An airdrop of prefabricated sections for ten bivouacs..was made by the New Zealand Forest Service. **1951** A. M. Ball *Compounding & Hyphenation* 23/2 Air-drop, *v.* **1955** *Time* 3 Oct. 29/2 The U.S. International Cooperation Administration this month began air-dropping 1,000 tons of rice to the 100,000 peasants who inhabit the region. **1966** H. Harrison *Plague fr. Space* iii. 33 We had airdropped in during the night. **1941** N. Macmillan *Air Strategy* xv. 128 Long range air escorts..to protect the fleets of bombers. **1784** Air-fight [see Air- I. 4]. **1908** H. G. Wells *War in Air* ix. § 3 The devastation and ruins of the greatest air fight in the world. **1944** *Amer. N. & Q.* Sept. 84/2 *Airhead*, the counterpart, in air action, of *beachhead*; used by Major Eliot F. Noyes..speaking before the Soaring Society of America, Aug. 5, 1944, in referring to the behind-German-lines base established by Allied gliders during the Normandy invasion. **1945** *Times* 1 Mar. 5/6 Two large 'air-heads' each with two transport strips and one light plane strip were built between Kalemyo and the river. **1944** *Times* 18 Mar. 4/7 A big-scale exercise with paratroops, air-landing units, [etc.]..was in progress. **1914** *Sci. Amer.* 15 Aug. 114/1 The aerial mine is inferior to the sea-mine not only in its vulnerability to currents, but also in its visibility. Against these handicaps the air-mine can oppose only its cheapness and lightness. **1939** *War Weekly* 202/3 A German sketch visualising the use of air mines against aeroplanes. The sketch shows a swarm of hydrogen-filled balloons released during an air-raid. Each balloon has hanging from it a chain with a mine attached. **1943–4** *Hutchinson's Pict. Hist. War* 27 Oct.–11 Apr. 26 Among the vast armada that was sent up against the invading force of bombers was a special squadron which towed air mines...The effect..was to catch the American aircraft in the tow ropes or destroy them by the blast from the mines. **1923** Kipling *Irish Guards in Gt. War* I. 30 Artillery fire, directed by air observation. **1944** *Times* 5 Jan. 3/1 Some of the credit for the flexibility and accuracy with which our superior weight of guns is used must go to our 'air o.p.' (observation post) squadrons...I have seen the little 'air o.p.' circling over the enemy lines. **1915** Grahame-White & Harper *Aircraft in Gt. War* vii. i. 295 (*heading*) Strategy of an Air Offensive. **1944** *Ann. Reg. 1943* 29 The British air offensive against Germany..set up a new record. **1923** Kipling *Irish Guards in Gt. War* I. 57 The days of the merciless air-patrols had yet to come. **1908** H. G. Wells *War in Air* iii. § 5 The immense aeronautic park that had been established..to give Germany..the air power and the Empire of the world. **1909** in F. T. Jane *All World's Air-Ships* 327/3 'Air power' can hardly be more than one of many factors in deciding the issue of future wars. **1940** *Economist* 11 May 851/2 The superiority of air power over sea power. **1915** Grahame-White & Harper *Aircraft in Gt. War* i. ii. 25 An air reconnaissance would have told him that Blücher..was actually marching north. *Ibid.* ii. iv. 47 (*heading*) Air-Scouting and Tactics. **1945** *Times* 3 May 3/3 Destroyers of the East Indies Fleet bombarded airfields..and followed this up with an air strike. **1944** T. H. Wisdom *Triumph over Tunisia* xvii. 143 The function of what [Air Marshal] Coningham now called the Air Striking Force. **1935** Burge *Compl. Bk. Aviation* 553/2 A deliberate attempt was made by both sides to gain air superiority. **1941** *Aeronautics* Oct. 49/2 Fortunately the Royal Navy has never completely neglected the importance of air support to naval forces. **1916** *Sphere* 29 Jan. 109 (*heading*) The latest German attempt to challenge British air supremacy. **1940** E. C. Shepherd *Britain's Air Power* 7 On an Army front air supremacy is essential to success in these days. **1874** 'Mark Twain' & C. D. Warner *Gilded Age* I. xviii. 233 Colonel Sellers..was the inventor of the famous air-torpedo. **1916** *Illustr. War News* 8 Mar. 6/1 (*caption*) A French air-torpedo caught in a tree over a German trench. **1941** *Hutchinson's Pict. Hist. War* 14 May–8 July 138 They slink along from port to port under the protection of their air umbrella. **1915** Grahame-White & Harper *Aircraft in Gt. War* v. xiv. 252 The axiom that 'might is right' may apply very forcibly to the air wars of the future. **1916** *Fortnightly Rev.* XCIX. 1062 Air warfare on the scale indicated..opens up possibilities in the way of air raids for landing considerable bodies of men.

3. Of that branch of a country's armed forces which fights in the air, as *air arm* [cf. *fleet *sb.*[1] 1 d], *armada, cavalry, fleet, service*; also, **air force,** a military or naval force organized for conducting operations in the air; that part of the military forces of a country (in Great Britain, the Royal Air Force) which consists of officers and men with aircraft and other necessary equipment; so in titles of officers, as *air commodore, marshal* [see *marshal *sb.*], *officer*; of non-commissioned ranks, as *air-gunner* (hence *-gunnery*), *mechanic*; also *air council, ministry* (see quot. 1959), *-scout, staff*; **air-bomber,** a bomb-aimer; **Air Training Corps** (abbrev. A.T.C.), an organization for the training of cadets for the Royal Air Force; **Air Transport Auxiliary** (see quot.). For *air crew, pilot* see *4; see also *airman.

1917 *Flying* 19 Sept. 129/2 Why not remove the 'air arm' at once from 'the naval and military control'? **1940** E. C. Shepherd *Britain's Air Power* 7 The Navy has its own air arm designed to work with the ships of the Fleet. **1917** *Flying* 31 Oct. 225/2 The disaster which befell the German air armada. **1911** *Times* 25 Feb. 7/3 The Balloon School is being reorganized and will be transformed into an Air Battalion. **1943** *Hutchinson's Pict. Hist. War* 17 Feb.–11 May 101 We now group pilots, navigators and air bombers together on entry into the Service. **1944** *Times* 8 July 2/2 As the air-bomber takes his aim he [etc.]. **1917** 'Contact' *Airman's Outings* 203 We shall see a great extension of ground attacks by air cavalry. **1965** *Observer* 11 July 2/5 The United States is about to reinforce its troops in Vietnam with a new high-powered 'air-cavalry' division. **1919** *Flight* XI. 1044 His Majesty..has approved of new titles for the commissioned ranks of the Royal Air Force..Air Commodore. **1920** *Ibid.* XII. 113/1 The chair will be taken at 8 p.m. by Air Commodore E. M. Maitland. **1917** *Act 7 & 8 Geo. V* c. 51 An Act to make provision for the establishment, administration, and discipline of an Air Force, the establishment of an Air Council, and for purposes connected therewith. **1908** H. G. Wells *War in Air* iv. § 3 The German airfleet. **1946** A. Lee *German Air Force* ii. 19 The Luftwaffe was organized territorially into Air Fleets (Luft-flotten). There were four immediately before the war. **1917** Air force [see *air council*]. **1918** *Flight* 6 June 605/1 'The Air Force Cross', to be awarded to officers and warrant officers for acts of courage. *Ibid.*, 'The Air Force Medal', to be awarded to non-commissioned officers and men for acts of courage. **1920** *Act 10 & 11 Geo. V* c. 76 § 11 (2) Where possession is reasonably required for naval, military, or air force purposes. **1959** *Chambers's Encycl.* I. 185/2 The view that an air force should be free to operate independently or that it should have equal status to the land and sea forces is predominantly an Anglo-American conception. **1928** *R.A.F. Regulations Amendm. List* July 3 Airmen selected for employment as air gunners will be required initially to qualify at a short course of air gunnery. **1944** *Times* 11 Apr. 4/4 Air gunners reported they shot down 43 enemy fighters. **1917** *Flying* 18 July 480/3 Second-Lieutenant Fletcher warmly commended the gallant conduct of First Air Mechanic Merritt. **1928** C. F. S. Gamble *North Sea Air Station* i. 39 The term 'air mechanic' is in use, though no such naval rating really exists. By air mechanic is meant a man who has been through a course of training either at the Central Flying School or at Eastchurch. **1916** *Flight* VIII. 112/1 (*heading*) An Air Ministry at last. **1959** *Chambers's Encycl.* I. 197/2 *Air Ministry* is the department responsible in the United Kingdom for the organization and direction, under the minister of defence, of the Royal Air Force. **1920** *Act 10 Geo. V* c. 7 § 11 (3) The expression 'air officer' means any officer above the rank of group captain. **1963** *Times* 16 Apr. 12/3 Air Vice-Marshal T. N. Coslett has been appointed Air Officer Commanding-in-Chief, R.A.F. Maintenance Command. **1939** *War Illustr.* 11 Nov. 282/3 Officers of Air Rank. **1914** *Times* 24 June 4/1 The Royal Naval Air Service..will form part of the Military Branch of the Royal Navy...A certain number [of officers] will..be selected to fill the higher posts in the Air Service. **1911** R. M. Pierce *Dict. Aviation* 19 *Air-scout*, a scout who operates in the air; an aerial observer. **1914** *Illustr. War News* 19 Aug. 43 (*caption*) The value of the air-scout: military entrenching viewed by an airman from a height of about 1000 feet. **1922** *Times* 17 Jan. 11/5 He thought him an officer very likely to become one day Chief of the Air Staff. **1940** Air Staff [see *fleet *sb.*[1] 1 d]. **1941** *Times* 10 Jan. 2/3 It is proposed to establish an Air Training Corps to provide pre-entry training for candidates for air crew and technical duties. **1941** *Flight* XXXIX. 57/2 The birth of the A.T.C. **1939** *Ibid.* XXXVI. 373/2 The units have been strengthened by..groups of pilots from..Air Transport Auxiliary. A.T.A. was originally formed by British Airways for..assisting the regular airline people to maintain communications during and after the expected full-scale bombing attacks on this country. **1919** Air Vice-Marshal [see *marshal *sb.*].

4. Of persons engaged in the flying, operation, or maintenance of aircraft, as †*air-boy, crew, -girl, navigator* (in quot. 1834 transf.), *pilot*, †*sailer* (*-or*), *stewardess*; also, **air hostess,** a stewardess in a passenger aircraft; *airman; **airwoman** (see *airman). See also *3.

1873 *Cassell's Mag.* VIII. 134/1 We saw two air-boys leaning over the side of the car. **1921** *Flight* XIII. 477/1 The two University air crews are staying there. **1940** *Times Weekly* 27 Nov. 6 It is one of the great merits of the R.A.F. curriculum..that it turns out air crews of which every member can at a pinch take over the work of any other. **1955** *Times* 21 July 4/6 While finding that there was nothing in the aircrew's tour of duty to cause undue fatigue, the report urges that B.O.A.C. should consider some limitation of hours of duty of an aircrew at an airport. **1928** *Daily Express* 20 June 1/3 All first impressions vanished...The boyish airgirl [*sc.* Miss Earheart] became a feminine woman. **1945** P. A. Larkin *North Ship* 14 The Polish airgirl in the corner seat. **1934** *Baltimore Sun* 6 Feb. 22/1 The air hostess was the overnight guest of Captain and Mrs. W. O. Schrum. **1936** *N.Z. Herald* 24 Mar., A knowledge of nursing is essential for an air hostess..for the experience of handling people. **1939** *Flight* 14 Dec. 490/1 The K.L.M. has found it wiser to employ stewards rather than Air Hostesses on the London line. **1958** *Times Lit. Suppl.* 17 Oct. 598/4 The unanimity with which air hostesses give as their reason for choosing this profession the desire to meet people and see distant places. **1834** Carlyle *Sart. Res.* II. v. in *Fraser's Mag.* IX. 306/1 A hapless Air-navigator, plunging, amid torn parachutes, sand-bags, and confused wreck, fast enough, into the jaws of the Devil! **1915** *Sphere* 20 Feb. 198/1 A well developed system of meteorological reports can be of such help to the air navigator. **1913** *Stamp Collecting* 27 Sept. 27/2 The provisional air pilot was arguing with the Republican officials. **1918** E. Wallace *Tam o' the Scouts* 211 (*title*) Aircraft by 'An Air Pilot'. **1923** J. W. Simpson *Ess. & Mem.* 169 The confident courage that inspires air-pilots. **1834** Carlyle *Sart. Res.* II. v. in *Fraser's Mag.* IX. 306/1 The thunderstruck Air-sailor is not wanting to himself in this dread hour. **1897** *Aeronaut. Ann., Scientific value of flying models*, The air-sailer who..adds the thrust of a screw to the forces he is accustomed to deal with. **1908** H. G. Wells *War in Air* vii. § 5 Then Bert..had his first experience of the work of an air-sailor. **1936** *Punch* 9 Dec. 646/1 To Chloe, an 'Air Stewardess'. My Chloe rides the heavens in a roaring silver hull, She serves up morning coffee over Basle and Istanbul.

5. In names of various types of aircraft, as *air ambulance* (cf. *ambulance *aeroplane*), *-bomber* [see *2], *-bus, -car, freighter, liner,* †*-machine,* †*-sailer* (*-or*), *scout, taxi, vehicle, yacht*; **air-boat,** (*a*) a lighter-than-air aircraft; (*b*) = *flying *boat*. Also, *aircraft, *air-plane 2, *airship, *air-vessel 3.

1921 *Aeronautics* 2 June 394/1 The first air ambulance..is painted aluminium with a large Red Cross painted on the fuselage and beneath the wings. **1933** *Lancet* 16 Dec. 1381/2 Air ambulance detachments under the British Red Cross. *Ibid.* 15 July 160/2 An air-ambulance service for the conveyance of urgent..cases to the hospitals and infirmaries in Glasgow. **1870** tr. *F. Marion's Wonderful Balloon Ascents* III. iv. 218 The air-boat of M. Pline seems to us one of the best ideas; but the working of it presents many difficulties. **1876** C. B. Mansfield *Aerial Navig.* II. xiii. 436 The action of rowing an air-boat must be much simpler than the same exercise on water. **1913** *Britannica Year-bk.* 343 The second class [of seaplane], variously termed 'flying-boat' and 'air-boat', consists essentially of a long boat-shaped hull, wherein the passengers' seats are contained, and on which the planes are built up. **1926** *Glasgow Herald* 18 Aug. 9 New British all-metal air-boats. **1910** *Times* 4 May 11/6 Probably when there are air-buses we shall call their drivers airmen. **1960** *Aeroplane* XCVIII. 468/1 A subsonic short- to medium-stage high passenger-density aircraft, for operation at low fares. This we call the Air-Bus. **1829** Air-car [see *airworthy *a.*]. **1911** Grahame-White & Harper *Aeroplane* ii. 41 Further developments, in passenger-carrying, are expected during 1911, when 'air-cars', carrying four and six occupants as their regular equipment, will be introduced. **1962** *Flight Internat.* LXXXII. Suppl. 6/3 The air car [i.e. a hovercraft] is capable of operating over land, water, sand, swamps, snow, or thin ice, with equal ease. **1930** *Pop. Sci. Monthly* Dec. 55/2 (*caption*) One of the air freighters put in service on a recently opened Pacific coast line. **1908** *Daily Mail* 25 May 7/6 The cost of working the air-liner was represented as small. **1955** *Times* 11 July 9/3 Twenty people died in the airliner, which crashed on..a scheduled flight. **1783** *Morning Herald* 17 Sept. 3/1 In a few days I shall have finished an air-machine, which will ascend, descend, or describe at pleasure a horizontal line. **1881** W. D. Hay *300 Years Hence* xii. 326 Tenders on the opposite side of the great air-machine begin to load. **1910** *Times* 23 Aug. 7/5 People look for the coming of the day when air machines are to become a practical means of regular locomotion. **1897** *Aeronaut. Ann.* No. 3, p. 2 The development of the motorless air-sailer. **1923** *Daily Mail* 17 Apr. 8 The engineless air-sailor. **1913** Grahame-White & Harper *With Airmen* xi. 276 The air scouts whirl over the enemy's troops at the rate of sixty miles an hour. **1914** *Sphere* 3 Oct. p. ii, One of these dismantled air scouts. **1920** *Flight* XII. 459/2 The chief concern of the many Americans who attended Mr. Handley Page's recent lectures on aviation in the United States appeared to be to discover when air taxis would be possible. **1927** *Observer* 7 Aug. 11/3 It was an ordinary air-taxi flight from Brooklands to Ascot. **1963** *Economist* 14 Dec. 1125/3 Charter and air-taxi flights in the area. **1902** *Aeronaut. Jrnl.* July 51/1 Some accomplishment on the part of an air vehicle. **1898** *Ibid.* July 54/2 The millionaire who indulges in an air yacht. **1920** *Flight* XII. 865/1 A converted..flying boat, fitted up as an 'aerial yacht'...This air yacht—elegantly furnished with two cabins seating 10 passengers..was officially launched..on June 22.

6. In names of parts of aircraft, or of apparatus used in aircraft or for the navigation of aircraft, as *air chart, frame* [*frame *sb.* 11 h], *log, map, propeller* (so *prop* colloq.), *sextant*; also, **air-brake,** a movable plane or flap on the wing of an aeroplane, that can be lowered to decrease its speed; **airscrew,** a power-driven screw for producing pull or thrust by rotation in the air; a propeller.

1914 *Aeronaut. Jrnl.* July 228 Air brakes..must not tend to produce any upsetting effect on the machine. **1928** C. F. S. Gamble *North Sea Air Station* xiii. 210 Fitted with an air-brake in the form of adjustable flaps in the trailing edge of the lower plane adjacent to the fuselage. **1920** *Flight* XII. 854/2 These air charts, which are constructed on Mercator's projection, measure approximately 20 ins. by 18 ins. **1951** *Oxf. Jun. Encycl.* IV. 290/1 An air map presents an accurate picture of the ground below, illustrating..all conspicuous geographical features...An air chart, drawn to a much smaller scale and without the elaborate detail of the map, is designed simply to enable the navigator to plot his position during the journey. **1931** *Aircraft Engin.* Jan. 9/2 Enclose both engines..in a high speed air frame. **1957** *Technology* July 187/1 A turbo-jet engine which rose straight into the air by itself with no airframe at all [the 'flying bedstead']. **1928** V. E. Clark *Elements Aviation* 138 *Air log*, an instrument for measuring the linear travel of an aircraft relative to the air. **1943** 'T. Dudley-Gordon' *Coastal Command at War* xii. 115 One navigator..always takes a portable typewriter with him on a raid...So I thought that I should do my air logs on the typewriter. **1913** Grahame-White & Harper *With Airmen* viii. 204 It has been decided that..certain districts should be marked out on air-maps, and that aeroplanes should not be allowed to fly over them. **1951** Air map [see *air chart*]. **1935** T. E. Lawrence *Let.* 5 Apr. (1938) 867 You can push an air-prop pitch up to great steepness, so long as the revs are not extravagant. **1910** R. Ferris *How it Flies* x. 208 The form of the air-propeller has passed through a long and varied development. [**1784** tr. *J. P. Blanchard's Jrnl.* 6 The fly, acting on the air as a screw, appeared to me the most suitable and efficacious mode which an aeronaute can adopt to advance in a calm.] **1894** *Proc. Internat. Conf. Aerial Navig., Chicago* 265 For aeroplanes driven by screw propellers..there must always be two air screws..rotating in opposite directions. **1914** *Aeronaut. Jrnl.* XVIII. 315 *Airscrew*, used as a generic term to include both a propellor and a tractor screw. **1916** M. A. S. Riach *Air-Screws* ii. 14 (*heading*) The forces acting on an air-screw blade. **1951** *Oxf. Jun. Encycl.* IV. 8/2 The gas-turbine is widely used in aircraft to turn an ordinary propeller or airscrew...For long, fast journeys the airscrew-turbine is much used. **1922** *Flight* XIV. 754/3 They can now supply..revolution indicators..air sextants, [etc.].

7. Of land or buildings used for the operation or maintenance of aircraft, as *air base, park* (chiefly *U.S.*), *shed, station*; also, **ai·r-drome** *U.S.* = *aerodrome 2 b; **ai·r-strip,** a strip of land prepared for the taking off and landing of aircraft, often for temporary use. Also *airfield, *airport.

1919 *Athenæum* 23 May 360/2 'Air-base', 'aircraft',..'air mechanic', [etc.]..are now everyday terms. **1938** *Flight* XXXIV. 424 e/2 The cost of the Shannon air base..will be close on half a million pounds. **1917** E. N. Fales *Learning to Fly* v. 97 The airdrome..is used exclusively for flying, and may be as large as a mile square. **1943** Airdrome [see *airfield]. **1929** *Daily Tel.* 22 Apr. 2/5 Ten air parks..and sixty landing grounds will be provided. **1944** *Amer. Speech* XIX. 304 'Airparks', the ATS said, would be small landing fields in or near communities for the use of private flyers. **1915** *Whitaker's Almanack 1916* 464/2 French airmen raided the German air-sheds at Freiburg. **1911** *Aeronautics* Apr. 13/2 Starting and Landing Stations..A cumbersome expression...'Air Stations' have been suggested as alternatives. **1914** *Whitaker's Almanack 1915* 774 *Fort George* (Cromarty Firth).—British Naval Air Station. **1923** *Daily Mail* 17 July 10 Ocean Air-stations. **1942** *Newsweek* 7 Dec. 27/3 Then..further airstrips for landing the transport planes were built by the troops as they went along the jungle trails. **1944** *Times* 18 Mar. 3/2 The American forces..immediately began a drive for the airstrip. **1956** W. Slim *Defeat into Victory* xii. 251 An airstrip which served as an emergency landing ground on the Hump route.

8. Of routes or courses taken through the air by anything flying, esp. by aircraft, as *air lane, road, route*; also, **air corridor,** a route to which aircraft are restricted, esp. one over a foreign country. Also *airway 2

1922 *Flight* XIV. 34/1 (*heading*) Abolition of Air 'Corridors'. The regulations which have hitherto been in force relating to the 'corridors' by which aircraft might enter and leave the U.K. have now been abolished. **1948** *Daily Mail* 22 Apr. 1/3 The R.A.F. have introduced air corridors from which Russian and Eastern European planes must not stray as they fly over the British zone of Germany. **1911** R. M. Pierce *Dict. Aviation* 16 *Air-lane*, a lane or road thru the air. **1958** *Listener* 13 Feb. 269/1 Do not think..that the airlines fly as the crow flies...They fly along prescribed air lanes..they meander and zig-zag. **1909** *Westm. Gaz.* 30 Aug. 2/3 Already we hear of rules of the air-road. **1926** Kipling *Debits & Credits* 359 He Who bids the wild-swans' host still maintain their flight on Air-roads over islands lost. **1911** *Technical World Mag.* Sept. 117/2 (*title*) Marking out air routes. **1935** *Economist* 26 Oct. 816/1 The Company will continue to be the Goverment's chosen instrument for the operation and development of Empire air routes.

air (ē^ə^ɹ), *sb.*[2] *Sc.* Also **aer, aire, ayr(e, er.** [ON. *eyrr*; cf. Norw. *ør, øyr* sandbank, gravel-bank.] A gravelly beach. (See *Sc. Nat. Dict.* s.v. air *n.*[4])

a **1795** G. Low *Tour thro' Ork. & Schet.* (1879) 11 A house on the Aire..bears the empty name of the Fish-house. **1809** A. Edmondston *State Zetl. Is.* I. iii. 140 Most of the extensive beaches on the coast are called *airs*; as *Stour-air, Whale-air, Bou-air*. **1868** D. Gorrie *Summers & Winters in Orkneys* ix. 365 This *aith, ayre,* or spit of land. **1933** *Geogr. Jrnl.* LXXXI. 505 These 'green fish' were bought by the merchants and dried on stony ayres. **1936** *Nature* 19 Sept. 512/2 The most interesting find..was at Braewick..where storm water had breached an 'ayr' or storm beach.

air-balloon. Delete def. in Dict. and substitute:

†**1.** = Balloon *sb.*[1] 3. *Obs.*

1753 [in Dict.].

2. A globose bag filled with gas so as to ascend in the air. Cf. Balloon *sb.*[1] 6.

1783 [see Balloon *sb.*[1] 6]. **1783** *Morning Chron.* 8 Sept. 3/4 The first air-balloon he made was filled with fumous particles. **1784, 1789, 1829** [in Dict.]. **1907** *Daily Mail* 9 July 7/4 In July 1901 M. Santos Dumont flew from St. Cloud round the Eiffel Tower and back in his steerable air balloon.

3. An inflatable toy balloon.

1895 Croker *Village Tales* 155 The child was jumping for joy, and had a green air-balloon in his hand. **1908** H. G. Wells *War in Air* vi. § 2 Small children's air-balloons of the latest model attached to string became a serious check to the pedestrian in Central Park. **1944** O. Sitwell *Autobiogr.* (1945) I. II. vi. 227 Fashionable beauties, with psyches that resembled air-balloons, inflated, light and highly coloured.

air-balloonist. Substitute for quot.:

1817 Kirby & Spence *Entomol.* II. xxiii. 346 The aërial excursions of our insect air-balloonist.

ai·r-borne, *a.* Carried through the air (see Air- I. 2); (of aircraft) having left the ground; in flight; (of troops) carried by aircraft.

1641, 1880 [see Air- I. 2]. **1909** C. M. Doughty *Cliffs* 57 We've seaborne, airborne and now subsea fleets. **1920** *Flight* XII. 48/1 An air-borne letter delivered by hand in Paris for 2s. 6d. **1937** B. H. L. Hart *Europe in Arms* iii. 30 The most striking features of the Army are its development of tank and air-borne units. **1941** *War Fortnightly* 20 June 1876/1 It [Crete] was the first occasion on which a major operation had been undertaken by air-borne troops, without the aid of sea-borne or any other troops. **1943** *Aeronautics* Feb. 48/1 An Airborne Division of the British Army.

air-chamber. Add: **1. b.** A chamber filled with air in a boat, airship, etc., to provide or assist buoyancy.

1881 W. D. Hay *300 Years Hence* vii. 133 In the upper part [of the boat] was the entrance and air-chamber. **1882** *Encycl. Brit.* XIV. 570/2 The buoyancy of the institution's lifeboat..is secured chiefly by means of a watertight deck..and two large air-chambers, one in the bow, the other in the stern. **1908** H. G. Wells *War in Air* viii. § 1 The airship was remarkably simple to construct: given the air-chamber material, the engines, [etc.]..it was really not more complicated..than an ordinary wooden boat had been..before.

ai·r-condi:tioning, *vbl. sb.* [cf. *condition *v.* 9.] The process of cleaning air and controlling its temperature and humidity before it enters a room, building, etc., and in certain manufacturing processes. Hence **ai·r-condi:tion** *v. trans.*; **ai·r-condi:tioned** *ppl. a.*; **ai·r-condi:tioner,** an apparatus for conditioning the air (of a room or building).

1909 S. W. Cramer *Useful Information for Cotton Mfrs.* (ed. 2) IV. 1395, I finally hit upon the compound word 'Air Conditioning'..suggested by the use of the term 'conditioning' in the treatment of yarn and cloth. *Ibid.* 1411 Well-known 'Air Conditioners' of both individual and central station types. **1930** *Engineering* 11 July 34/2 Air-conditioning is dealt with fully, with the methods for washing, cleaning, humidifying, cooling and drying the air. **1930** *Discovery* Sept. 317/2 Tobacco leaf..is extraordinarily sensitive to changes of atmospheric humidity, and this has led in recent years to much development in the matter of air-conditioning. **1933** *Archit. Rev.* LXXIII. 101 Its thirty-five stories of air-conditioned, sound-insulated..offices. **1935** Burge *Complete Bk. Aviation* 111/2 Air conditioners have been designed to supply fresh air, and to maintain a comfortable cabin temperature while a machine is on the ground. **1937** *Times* 22 Sept. 10/1 The trustees of the National Gallery will discuss a scheme to air-condition the gallery. **1938** *Encycl. Brit. Bk. of Year* 28/2 The built-in central plant system provides the best performance at the lowest cost wherever a number of rooms, or where one large room is to be air-conditioned...Quite recently unit air-conditioners of improved design and of relatively low cost have been produced. **1939** *Nature* 6 May 769/2 The large-scale air-conditioning and refrigeration installed in many South African gold mines enable gold companies to explore much greater depths. **1958** *Engineering* 14 Mar. 352/1 Originally, air conditioning was not introduced for reasons of comfort but was developed for the controlled processing of materials. **1958** *Times Lit. Suppl.* 28 Mar. 176/1 An admirably equipped strong room for manuscripts with its own air-conditioning and humidifying plant.

aircraft (ē^ə^·ɹkrɑft). [f. *air- III + Craft *sb.* 9.] Flying-machines collectively; a flying-machine.

Since the 1930s commonly restricted to denote an aeroplane (as distinct from a balloon or airship) or aeroplanes collectively. The nineteenth-century examples refer to balloons and airships.

1850 J. Wise *System of Aeronautics* xvi. 102 The air-craft has but one medium, the water-craft has two. **1876** C. B. Mansfield *Aerial Navig.* 2 Air-craft seems an appropriate term for the whole apparatus, including boat and gas vessel, or car and balloon. *Ibid.* II. viii. 323 The inventors of air-crafts. **1903** *Aeronaut. Jrnl.* VII. 81/1 His world-famed aircrafts. **1909** *Daily Chron.* 26 Feb. 1/2 The vast commercial possibilities that the manufacture and world-wide use of air craft offer. **1910** Rotch in *Epitome Aeronaut. Ann.*, [Suppose] an aircraft to possess the very moderate speed of 9 metres per second. **1910** *Daily Mail* 27 May 6/1 The three types of aircraft—the balloon, the airship, and the aeroplane. **1912** F. T. Jane *All the World's Air-craft* III. 7/1 It has..been deemed advisable gradually to change the title of this annual to *All the World's Air-craft*, in order to avoid all risk of misunderstanding as to its scope—'air-craft' being apparently the only generic which cannot be associated with a single type. **1915** *Whitaker's Almanack 1916* 465/1 Sixteen German aircraft attempted to fly over the English Channel. **1933** *Aeroplane* 19 July 124/2 The Air Council will do well to make a note of the fact that the word 'aircraft', which they allow their subordinates to use in the limited sense of the word 'aeroplane', covers 23 classes of air vehicles. **1936** *Discovery* Aug. 238/1 Like all mapping cameras it is installed in the aircraft on a mount. **1942** R. Hillary *Last Enemy* ii. 34 The proper handling of an aircraft. **1943** W. S. Churchill *Second World War* (1952) V. 566 Will you please make the following terminology effective in all British official correspondence: For 'aeroplane' the word 'aircraft' should be used...It is a good thing to have a rule and stick to it. **1944** H. St. G. Saunders *Per Ardua* ix. 122 The constant coast-wise patrols carried out by three kinds of aircraft, land planes, seaplanes, and airships. **1945** *Yorks. Post* 19 Apr. 1/2 Nearly 1,000 aircraft of R.A.F. Bomber Command. **1955** *Times* 9 May 10/5 An aircraft of the Queen's Flight. **1960** C. H. Gibbs-Smith *Aeroplane* I. xv. 119 The helicopter..soon became an indispensable type of aircraft.

b. *attrib.*, as *aircraft hand* [Hand *sb.* 8], abbrev. A.C.H.; **aircraft carrier,** a ship that carries and serves as a base for aircraft; so *aircraft-carrying* adj.

1912 *Aeronautics* Nov. 365/1, 4 machines are under construction at the Royal Aircraft Factory. **1919** *Times* 30 Dec. 4/6 The aircraft carrier Hermes..is to be towed from the Tyne to Devonport. **1921** *Airman* Nov. 113/1 [In the] aircraft of the future..we shall find such people as cooks, fire-fighters..and aircrafthands; these latter, of course, to be employed as guards, and to do odd jobs which will be numerous no doubt on this type of aircraft. **1922** T. E. Lawrence *Let.* 1 Oct. (1938) 363, I can't ask the corporal how an aircraft hand addresses an air-vice-marshall. **1925** —— *Let.* 7 Oct. (1938) 484 Five are fitters, five are riggers, five are A.C.H.'s. **1925** *Jane's All World's Aircraft* A. 10a/1 The following are the aircraft-carrying ships owned by the British Navy. **1940** E. C. Shepherd *Britain's Air Power* 12 The latest aircraft carriers can carry many aeroplanes. A total of 60 would not be an over-estimate, though the exact complement of any such floating aerodrome must not be given. **1942** O. J. Lissitzyn *Internat. Air Transport* 430 (*heading*) Route miles and aircraft miles flown. **1957** *Technology* Mar. 8/1 In the short space of fifty years the aircraft industry has come a long way.

Hence **ai·rcraftman** (abbrev. A.C., A/C), the lowest non-commissioned rank in the Royal Air Force. Also (in non-official use) **ai·rcraftsman.**

1920 *Glasgow Herald* 2 Aug. 6 The demand for experienced pilots and aircraftsmen. **1921** *King's Regs. R.A.F.* 8 The ranks of Warrant Officers, Non-Commissioned Officers, and men of the Royal Air Force..are as follows..*Non-Commissioned Officers.* Flight Sergeant. Sergeant. Corporal. *Men in the Ranks.* Leading Aircraftman. Aircraftman. **1922** T. E. Lawrence *Let.* 6 Nov. (1938) 376 Address now No. 352087, A.C. 2 Ross. **1924** *Lancet* 19 Jan. 164/2 Examined the olfactory condition of 100 air craftsmen, and found only one who could not distinguish petrol, orange, or peppermint among the test solutions offered to him. **1943** C. H. Ward-Jackson *It's a Piece of Cake* 10 *A/C Plonk*, Aircraftman 2nd Class.

So **ai·rcraftwoman** (abbrev. A.C.W.), the lowest rank in the women's air service.

1939 *Times* 11 Sept. 5/6 Airwomen..will receive 2s. 2d. a day on entering, 2s. 8d. a day when mustered as aircraftwomen first class. **1941** J. Hammerton *ABC of RAF* 53/1 Aircraftwomen 2nd class. **1942** *Ibid.* (ed. 2) 58 *A.C.W.*, Aircraftwoman. **1943** K. B. Beauman *Wings on her Shoulders* v. 57 A plotter, A. C. W. Cooper, told me about the attack on North Weald.

Airedale (ē^ə^·ɹdē^i^l). The name of a district in the West Riding of Yorkshire; hence short for *Airedale terrier*, one of a breed of large rough-haired dogs.

1880 Dalziel *Brit. Dogs* 377 The Airedale or Bingley Terrier. **1889** *Ibid.* (ed. 2) II. 385 The name Terrier, applied to the Airedale, is admittedly a stretching of the term beyond its original meaning. **1916** F. M. Jowett *Airedale Terrier* 26 A dog brimful of true Airedale character and type.

airer. (Later examples.)

1884 *Manchester Examiner* 10 Dec. 3/6 Mrs. Jackson is no rabid airer of grievances. **1920** *Cornhill Mag.* Sept. 3/3 The 'airers' of wounded soldiers were dreaming of jaunts without hospital blue. **1959** *Which?* Aug. 94/1 Airer Driers..all work on the same basic principle. Washing is spread out on horizontal rods and is dried by warm air from an electric heater.

airfield (ē^ə^·ɹfīld). [f. *air- III + Field *sb.*] An area of land where aircraft are accommodated and maintained and may take off or land.

1935 *Aero Digest* May 13/3 Officials..have a fundamentally clear conception of how airfield projects ought to be handled. **1936** *Baltimore Sun* 18 May 7/8 All airfields in Virginia, with one or two exceptions, are below the standard sizes recommended by the Department of Commerce. **1940** B. Ward *Russian Foreign Policy* 30 The Baltic States..'acquiesced' in the cession of islands, the leasing of harbours and airfields. **1943** W. S. Churchill *Second World War* (1952) V. 566 For 'aerodrome' either 'airfield' or 'airport' [should be used]. The expression 'airdrome' should not be used by us.

airing, *vbl. sb.* Add: **1. b.** *airing cupboard*, a cupboard for airing linen and clothing.

1917 A. Waugh *Loom of Youth* I. vii. 82 The School House changing-room...In the far corner there was an airing cupboard. **1958** *Times* 2 June ix/3 An immersion heater..Small and compact, it is particularly suitable for fitting in an airing cupboard.

4. [Air *v.* 5.] Display, exposure to public notice.

1870 G. Meredith *H. Richmond* (1871) I. x. 166, I really felt that I was justified in giving my irritability an airing. **1884** *Eng. Illustr. Mag.* I. 440 It was little more than scientific gossip, and the occasional airing of certain theories. **1965** *Listener* 25 Nov. 874/3 An informative and lively series which could certainly stand another airing.

airish, *a.* Restrict †*Obs.* to sense 1 and add:

1. (Later example.)

1612 T. James *Iesuits Downefall* 61 He was not a meere man; but some Fairies brat, or begotten by an Incubus, or aerish spirit, vpon the bodie of a base woman.

2. (Later U.S. examples.)

1882 J. H. Beadle *Western Wilds* xxxviii. 613 Going westward on any line one will find the winters growing dryer, also more 'airish'. **1885** 'C. E. Craddock' *Prophet Gt. Smoky Mts.* 267 It air toler'ble airish in the fog.

airless, *a.* Add: **2.** *airless injection* [*injection]: see quot. 1940.

1930 *Engineering* 11 July 39 (*heading*) Airless-injection oil engine at the Royal Agricultural Show. **1940** *Chambers's Techn. Dict.* 19/2 *Airless injection*, the injection of liquid fuel into the cylinder of an oil engine by a high-pressure fuel pump, so dispensing with the compressed air necessary in the early Diesel engines.

airlessness (ē^ə^·ɹlėsnės). [f. Airless *a.* + -ness.] The condition of being airless.

1831 J. Wilson *Unimore* in *Blackw. Mag.* XXX. 144 Idle all at once her sails Hang in the airlessness. **1857** Dickens *Dorrit* II. xxiii, The airlessness and closeness of the house. **1885** 'Lucas Malet' *Col. Enderby's Wife* II. iv, The shut and darkened windows produced an effect of airlessness. **1913** E. F. Benson *Thorley Weir* i. 5 The baking airlessness of town.

ai·r-lift. [Lift *sb.*[2]] **1.** [*air- II.] A pumping device operated by compressed air.

1893 *Patent Specif.* No. 22372 My process..I term the 'air lift' process. **1902** *Encycl. Brit.* XXXIII. 776/1 The object attained by the air-lift is precisely the same as that attained by putting a pump some distance down a bore-hole. **1917** *Proc. Inst. Mech. Engin.* 628 In an air-lift pump in operation the air-bubbles are rising through a mixture of variable density.

2. [*air- III.] Transportation of supplies or troops by air, esp. during a state of emergency. Also as *v. trans.*, hence **ai·rlifted** *ppl. a.*, **ai·rlifting** *vbl. sb.*

1945 *Life* 17 Dec. 112/2 General Ho Ying-chin used the American air lift to pass four crack American-equipped..armies over the heads of the Communists into Shanghai. **1948** *News Chron.* 20 Sept. 1/1 This is the first British plane to crash on the air lift, which began in June. **1949** *Baltimore Sun* 5 Jan. 7/1 The Soviet-licensed newspaper *National Zeitung* today accused British and American officers of shunting more than 8,000 tons of airlifted coal into the black market of fuel-short Berlin. *Ibid.* 2 July 1/1 It may lead to the airlifting of other highly perishable fruits and vegetables. **1952** in *Amer. Speech* (1953) XXVIII. 50 Thus whirlybirds airlifted almost 11,000 troops to safety. **1957** *Ann. Reg. 1956* 77 The infantry battalion which had been air-lifted from Calgary to Halifax in readiness for transfer to Egypt. **1958** *Oxf. Mail* 19 July 1/2 The United States today began a massive oil airlift to save Jordan from oil starvation. **1960** *Guardian* 20 July 1/4 The United States had sent support in the airlifting of United Nations troops.

ai·r-line. 1. (In Dict. s.v. Air- II.) Add earlier and later examples. Chiefly *U.S.*

1813 J. Quincy in *Deb. Congress* 1 Jan. 544 They will not rigidly observe any air-lines or water-lines in enforcing their necessary levies. **1829** J. F. Cooper *Wish-ton-wish* ii. 27 This clearing, which by an air line might have been half a mile from the place where his horse had stopped. **1904** *Chicago Tribune* 1 Aug. 4 The judge held that distance was to be measured by air line or 'as the crow flies', not as the main traveled road leads. **1919** *Sphere* 22 Mar. 254/2 Flying routes will soon show a tendency to become fixed, just as birds fly along 'air lines' from point to point.

attrib. **1863** *Congress. Globe* Feb. 813/1 This is a proposition to construct an air-line railroad between Washington and New York. **1895** *Outing* Dec. 214/2 Pursuing an even air-line route across the roughest country.

b. In *nonce attrib.* use: Sent through the air.

The reference is to letters dropping through the air as a theosophical manifestation in India.

1888 Kipling *In Black & White* 66 The Religion never seemed to get much beyond its first manifestations; though it added an air-line postal *dak*, and orchestral effects.

2. [*air- III.] A line of aircraft for public service.

1914 *Argus* (Melbourne) 16 July 13 The Defence flying school at Point Cook has been inaccessible..except by air line. **1937** *Discovery* May 163/2 Few people realise how vast is the network of air lines which now links up the United States with Central and South America. **1956** *Times* 2 Feb. 5/1 Work will begin this month on the..platform above District Line railway tracks close to Cromwell Road, Earls Court, for future use as London's main air line terminal. **1958** [see *air- III. 8].

ai·r-lock. [Air- I.7.] **a.** = Lock *sb.*[2] 10; also, a similar chamber in a space-craft, etc.

1857 *Brit. Almanac* 99 Each cylinder..is filled with compressed air, by which it is kept free from water, and by means of chambers at the top furnished with doors or valves, on the principle of the canal lock, and called 'air-locks'. **1877** *Encycl. Brit.* VI. 63/1 A cylinder of wrought iron, within which a tubular chamber, provided with doors above and below, known as an air-lock. **1926** *Blackw. Mag.* Sept. 322/1 A diver could lie inside a great steel cylinder undergoing compression and have tea passed in to him through an airlock. **1951** A. C. Clarke *Sands of Mars* ii. 12 Help him through the airlock when the tender couples up.

b. [Air- I. 5.] A stoppage of the flow of liquid in a pump or pipe by a bubble of air. So **air-locking.**

1909 in Webster. **1920** *Flight* XII. 219/1 The avoidance of air-locks in pipe systems. **1927** D. L. Sayers *Unnatural Death* xi. 132 'Blew through the filler-cap,' said his lordship with a grin. 'Air-lock in the feed, old son, that's all.' **1936** *Aircraft Engin.* Nov. 321/1 Airlocking was studied in the laboratory by observing the behaviour of a fuel-air system in glass pipes.

ai·rmail. [Mail *sb.*[3]] Mail conveyed by air; a service for conveying letters, parcels, etc., by air; also *attrib.* Hence as *v. trans.*, to send by air.

1913 *Stamp Collecting* 13 Dec. 210/1 There have been many other German air mail flights. **1920** *Lancet* 11 Sept. 581/2 A Royal air mail service from London to Holland was started on July 6th. **1922** *Glasgow Herald* 4 Apr. 9 The despatch of air-mail letters..between London and Paris. **1928** *Post Office Guide* July 55 A special Blue Air Mail label should be affixed to the top left hand corner of every Air Mail packet. **1932** S. Gibbons *Cold Comfort Farm* xx. 272 The papers arrived by air-mail at noon. **1942** Wyndham Lewis *Let.* 8 July (1963) 327 This is going to be air-mailed, and a carbon of it will go by ship. **1950** *Lancet* 8 July 70/1 Air-mail envelopes with coloured borders.

b. Applied to a type of thin paper intended for dispatch by airmail; also applied to an edition of a newspaper specially printed for conveyance by air (also called *air edition*).

1935 *Brit. Paper* Summer 13/1 A new air mail paper..a tub-sized air-dried paper for air mail purposes. **1944** *Times Weekly* 9 Aug. 7/2 *The Times* now publishes an Air Edition daily. This consists of..copies of full size printed on India paper. **1948** *Hansard Commons* CDXLVI. 994 Only two of our newspapers have airmail editions. **1958** *Times* 14 June 8/7 Among the exhibits here is a copy of the Air Mail edition of *The Times*, printed on paper made from Sunn Hemp..and rag. **1959** *Engineering* 27 Feb. 262/3 [The machine] can handle a wide range of stock, from air-mail to board.

airman (ē^ə^·ɹmæn). [f. *air- III + Man *sb.*[1], after *seaman*.] One who is engaged in the flying or operation of aircraft, esp. as a pilot or a member of an air crew; *spec.* an enlisted person in the Royal Air Force.

1873 *Cassell's Mag.* VIII. 135/2 The airmen were running..and the balloon was pitching so much as almost to make Bob feel a little air-sick. **1910** *Daily Mail* 4 June 4/2 A flight of over forty miles above the sea, during which the airman is in continual danger. **1918** *Times* 26 Aug. 3/4 A light blue uniform has been approved for Warrant Officers, N.C.O.'s, and Airmen of the Royal Air Force. **1920** *Act 10 & 11 Geo. V* c. 30 § 41 (5) The expression 'airman' means a man of the regular air force. **1924** *King's Regs. R.A.F.* p. vii, Airman, or Airmen. These words, wherever they occur, will be held to include a warrant officer, a N.C.O., an aircraftman, and a boy. **1927** T. E. Lawrence *Let.* 2 Dec. (1938) 553 Airmen mustn't fly machines:—that is a privilege of officers; and R.A.F. officers are very unlike R.A.F. airmen. **1934** *Lancet* 30 June 1377/1 The term 'airman' in the R.A.F. is used to designate 'other ranks' as distinguished from officers. It does not, per se, imply flying capacity.

So **ai·rwoman.**

1911 *Chambers's Jrnl.* 6 May 364/1 Crowds were collecting..to see the first English airwoman make her widely advertised attempt on the height-record. **1941** *Hutchinson's Pict. Hist. War* 22 Jan.–18 Mar. 229 Airwomen had made an unqualified success of every trade in which they had taken the place of airmen.

airmanship. Add in def. after 'balloon': or other aircraft. (Later examples.)

1908 *Daily Chron.* 15 Aug. 4/6 'Airmanship' is the newest addition to the language. **1910** *Daily Mail* 4 June 4/2 When airmanship was in the elementary stage. **1955** *Times* 12 July 8/7 The emphasis in training will be on developing officers rather than airmen, though the course will include airmanship.

air-pipe. (Later examples.)

1889 F. R. Stockton *Gt. War Syndicate* vi. 84 If the air-pipes..could be rendered useless the crew [of the submarine vessel] must inevitably be smothered. **1899** W. S. Churchill *River War* I. i. 2 The Soudan is joined to Egypt by the Nile, as a diver is connected with the surface by his air-pipe.

air-plane (ē^ə^·ɹplē^i^n). [f. Air- + Plane *sb.*[3]]

1. The air regarded as a horizontal plane. *nonce-use.*

1874 *Belgravia* IV. 168 Not all the rudders and flappers..can ever enable the aeronaut to navigate his machine *horizontally*—to one hand or the other of the air-plane on which he is sailing.

2. [Alteration of Aeroplane, after *air- III.] Also **airplane.** †**a.** = *aeroplane 1; **b.** = *aeroplane 2 b (esp. *U.S.*). Also *attrib.*

a. 1896 J. Challis in *Invention* 13 June 380/2 The combined use of the screw and aeroplane (why not call it air plane) principles.

b. 1907 *Westm. Gaz.* 19 July 4/2 It is this ease of going against the current, with no motive force in evidence, that is..the despair of the aeronauts with their air-planes. **1916** Buchan *Nelson's Hist. War* XIV. 48 Airplane reconnaissance. **1917** *N.A.C.A.* (*U.S.*) *Rep. Nomencl. Aeronaut.* 31 *Airplane*..This term is commonly used in a more restricted sense to refer to airplanes fitted with landing gear suited to operation from the land. If the landing gear is suited to operation from the water, the term 'Seaplane' is used. **1918** *King's Regs. R.A.F.* § 1024 A, The supply of patent fire-extinguishers is to be limited to the following services:—Motor boats. Motor vehicles. Airplanes. Seaplanes. **1928** A. Lloyd James *Broadc. English* I. 21 *Aeroplane*..the [B.B.C. Advisory] Committee [on Spoken English] advises the use of the word *airplane*. **1930** H. G. Wells *Autocr. Mr. Parham* IV. i. 279 The airplane carrier *Courageous*. **1937** *Daily Express* 2 Feb. 3/4 The islanders clamber into the airplane as though they were boarding a bus. **1956** W. A. Heflin *U.S. Air Force Dict.* 33/1 'Airplane' is preferred to 'aeroplane' in American usage, having received official sanction in Army publications as early as 1918.

airport (ē^ə^·ɹpō^ə^ɹt). [Port *sb.*[1], in transf. use of sense 2.] An aerodrome, esp. one with a customs-house, to which aeroplanes resort to load and unload, and at which passengers embark or disembark. (For *air-port* in an unconnected use see Air- II in Dict. and Suppl.)

In quot. 1921 [after Seaport], a landing-place for a seaplane.

1919 *Aerial Age Weekly* 14 Apr. 235/1 There is being established at Atlantic City the first 'air port' ever established, the purposes of which are..to provide a municipal aviation field,..to supply an air port for trans-Atlantic liners, whether of the seaplane, land aeroplane or dirigible balloon type. **1921** *Aeronautics* 19 May 351/2 The flight..was made in..two hours, the machine landing above Westminster Bridge...It was the first occasion on which the Thames had been used as an air port for a machine from abroad. **1924** *Lancet* 9 Feb. 309/2 The vigilance practised by sanitary authorities at our seaports will require to be exercised in even greater degree at the great airports of the future. **1926** *Glasgow Herald* 13 Nov. 4 The scene [at Croydon] is characteristic of the airports of all the big cities. **1933** *Jrnl. R. Aeronaut. Soc.* XXXVII. 22 Another important feature of night lighting is the airport beacon, usually of the rotating or flashing type, and mounted upon the terminal building. **1943** [see *airfield]. **1946** *Daily Tel.* 26 July 6/3 The course began at the helicopter airport at Bridgeport.

air-pump. Add to def.: In a steam-engine.

1802 Trevithick & Vivian *Brit. Pat.* 2599 In the steam engines constructed and applied according to our said Invention..we use a new method of condensing by an injection above the bucket of the air pump. **1875** *Encycl. Brit.* I. 429/2 *Air-pump*, in steam-engines, is the pump which draws the condensed steam, along with the air which is always mixed with it, and also the condensing water..away from the condenser, and discharges it into the hot well.

ai·r-raid. [Raid *sb.*] A raiding attack by aircraft upon an enemy. Also *attrib.*, as *air-raid alarm, precautions* (abbrev. A.R.P.), *shelter, warden* (also *air warden*), *warning*.

1914 *Whitaker's Almanack 1915* 823/1 British air raids on Cologne and Dusseldorf. **1916** [see *air warfare*, *air- III. 2]. **1916** Lanchester *Aircraft in Warfare* 190 Air-raids on Great Britain by Zeppelin do not pay. **1917** *Flying* 1 Aug. 18/2 The Home Secretary stated..that air-raid warnings would be given by signal rockets. **1919** 'I. Hay' *Last Million* p. xi, Above all, we hope to see the air-raid shelters gone. **1935** N. Hammer (*title*) A Catechism of Air Raid Precautions. **1935** *Lancet* 27 Apr. 1018/1 Mr. Mander asked the Home Secretary the precise nature of the work to be carried out by the Home Office Air-raids Precautions Department at 5, Princes-street, Westminster. **1936** *Ibid.* 19 Dec. 1465/2 Each subdivision being required to make its own plans for police and fire brigade services, rescue, [etc.]..and air-raid wardens. **1937** *Ibid.* 2 Oct. 812/2 A.R.P. These sinister initials are being made more and more familiar by a spate of books on air-raid precautions. **1938** *Times Weekly* 27 Jan. 8/2 The appointment by local authorities in Great Britain of voluntary air wardens and fire-fighters. **1938** *Lancet* 9 July 90/2 The current theory is that there will be an air-raid warning, after which the population will go to shelters..and stay there till the all-clear signal. **1940** *Times Weekly* 7 Aug. 18 This was the first time that his Majesty had had the experience of an air-raid alarm sounding while carrying out a tour of inspection.

airship (ē^ə^·ɹʃip). Formerly **air-ship.** [f. *air- III + Ship *sb.*[1]; cf. G. *luftschiff*.] A dirigible motor-driven balloon, *spec.* one of an elongated cigar-shaped form having the gas-bags enclosed inside a rigid structure; also, esp. in the U.S., applied generally to other types of aircraft.

1819 in J. Milbank *First Cent. Flight in Amer.* (1943) v. 71 To ascend first in a balloon of the common construction, and afterwards to carry into operation his principles for navigating airships. **1838** T. M. Mason *Aeronautica* 327 Count Lennox's air-ship [*sc.* a balloon]. **1891** O. Chanute *Aerial Navig.* 7 It was not until 1852 that Henri Giffard..laid down the foundation for eventual success by ascending with a spindle-shaped air ship driven by a steam-engine. **1894** —— *Progr. Flying Machines* 66 In 1885 Mr. Foster patented an air ship consisting of two screws. **1900** [see Zeppelin]. **1910** C. C. Turner in *Jrnl. R. Soc. Arts* LVIII. 156/1 The common word airship can be applied to all vessels that travel in the air, but it is gradually becoming restricted to the dirigible balloon. **1927** —— *Old Flying Days* xxv. 344 To this day the Americans call aeroplanes 'airships'. **1950** *Gloss. Aeronaut. Terms* (*B.S.I.*) I. 46 *Airship*, a power-driven lighter-than-air aircraft.

b. *fig.*

1829 Carlyle in *Foreign Rev.* III. 449 Over all which Chamouni-needles and Staubbach-Falls, the great *Persi-fleur* skims along in this his little poetical air-ship. **1833** —— *Sart. Res.* I. xi. in *Fraser's Mag.* VIII. 682/1 What vacant, high-sailing air-ships are these, and whither will they sail with us?

Hence †**ai·rshipman.**

1904 *Pall Mall Mag.* Jan. 12/1 We air-shipmen are steam-boat captains and not sailing yachtsmen.

air-tight (ē^ə^·ɹtəit), *sb. U.S.* [f. the adj.]

1. An air-tight stove.

1844 'Jonathan Slick' *High Life N.Y.* II. xxxi. 227 Speakin' of stoves, Par, I got..what they call an air-tight. **1909** J. C. Lincoln *Keziah Coffin* i. 12 The stovepipe was attached to the 'air-tight' in the dining room.

2. An article of food put up in an air-tight vessel.

1897 A. H. Lewis *Wolfville* 330 What's air-tights?..Air-tights is can peaches, can tomatters, an' sim'lar bluffs. **1907** S. E. White *Arizona Nights* I. xvi. 219 On top of a few incidental pounds of *chile con*, baked beans, soda biscuits, 'air tights', and other delicacies.

Hence **ai·r-ti:ghtness.**

1852 D. Booth *Art of Brewing* (ed. 2) 65 Pay the strictest attention to the air-tightness of the vats.

air-vessel. Add:

1. *fig.*

1857 Geo. Eliot *Amos Barton* ii in *Blackw. Mag.* LXXXI. 7/1 We are poor plants buoyed up by the air-vessels of our own conceit.

2. (Earlier example.)

1744 J. T. Desaguliers *Course Exper. Philos.* II. xii. 510 A large Copper Air-Vessel, which receives the Water forced into it by the Action of 2 pumps.

†**3.** [*air- III.] A flying-machine. *Obs.*

1821–2 Byron in Medwin *Conversat. Ld. B.* (1824) I. 199, I suppose we shall soon travel by air-vessels; make air instead of sea-voyages. **1916** *Sphere* 9 Dec. 183 The strafing of Germany's zeppelin fleet: a pictorial presentation of the air vessels brought down and wrecked.

†**4.** The gas-bag of a balloon. *Obs.*

1870 *Sci. Amer.* 8 Jan. 33/2 An air vessel of 100 feet diameter, two thirds filled with coal gas..would be all sufficient for a practicing machine.

airward, *adv.* (Later examples.)

1880 G. Macdonald *Book of Strife* 247 All winged things came from the waters first; Airward still many a one from the water springs. **1937** W. de la Mare *This Year, Next Year*, With whirr of wing Will airward spring.

ai·rway. Also **air-way.** Transfer from Air- II and read:

1. A passage for air, esp. one for ventilation in a mine.

1851 [see Intake *sb.* 4]. **1880** [see Air- II]. **1908** *Daily Chron.* 7 Mar. 5/5 Free the return air-way from noxious gases.

b. A passage for air into the lungs; also, *spec.* a device to keep this passage open.

1908 *Lancet* 15 Feb. 491/1 Should there be much jaw spasm at the moment when it is desired to introduce the 'air-way' it may be necessary to separate the teeth by means of a Mason's gag. **1911** *Ibid.* 11 Nov. 1335/2 Insisting on the routine use in every administration of an anaesthetic of establishing an oral airway by means of a mouth-prop and tongue-clip. **1962** *Ibid.* 28 Apr. 879/2 To protect the patient against obstruction of the airway, endotracheal intubation with a cuffed tube is highly desirable during abdominal surgery.

2. A route through the air, esp. one regularly followed by aircraft from airport to airport.

1873 *Punch* 1 Feb. 44 [This book] professes to give..account of the..customs of..one of the planets...Are their railways, or airways, or whatever their means of locomotion may be called, as well managed as our own? **1908** *Westm. Gaz.* 3 Oct. 3/2 An impression of 1920...Extract from Passenger Handbook of the Great Eastern Airway Company for June. **1911** L. Blériot in Grahame-White & Harper *Aeroplane* 218 The Atlantic will, also, beyond doubt have its regular 'airway'. **1937** *Discovery* May 163/2 A message was sent to an airways agent. **1958** *Economist* 1 Nov. 434/2 Outside the airways..the need is for international agreement on a standard form of navigation.

b. airway beacon (see quot. 1940).

1937 *Aeronaut. Res. Comm. Rep. & Mem.* No. 1793 p. 1 Light signals used in aviation..on or near the airway, airway beacons. **1940** *Chambers's Techn. Dict.* 19/2 *Airway beacon*, a powerful light (often flashing a morse sign), for the guidance of aircraft.

3. A radio channel (cf. *air *sb.*[1] 1 c). *U.S.*

1934 in M. Weseen *Dict. Amer. Slang* xii. 165. **1946** *Baltimore Sun* 10 Oct. 18/8 By that time a radio broadcaster had appeared with a portable microphone but Ted had nothing for the airways, even after most of the other players had taken their turns at the 'mike'.

airworthy (ē^ə^·ɹwɒ̄ɹði), *a.* Also **air-worthy.** [f. *air- III + -worthy, after Seaworthy *a.*] Of aircraft: in a fit condition for travelling through the air. Hence **ai·rwor:thiness.**

1829 *Mech. Mag.* XI. 181 The airiner has time, in all cases (assuming that the air-car is air-worthy) to concert proper measures for his safe descent. **1864** M. Reid *Cliff-Climbers* lii. 342 Whether their aerial ship would prove herself air-worthy. **1909** *Westm. Gaz.* 11 May 5/2 That their vessel would be airworthy when completed. **1909** H. G. Wells *Tono-Bungay* IV. i, I had satisfied myself..of the real air-worthiness of Lord Roberts β [*sc.* a dirigible balloon]. **1920** *Flight* XII. 1182/2 Ensuring that machines are maintained in an airworthy condition after the issue of the airworthiness certificate. **1955** *Times* 23 May 6/2 The results..will be reviewed by the Air Registration Board, and their bearing on airworthiness certification will be considered.

airy, *a.* Add: **9.** *Comb.*

1879 G. M. Hopkins *Poems* (1918) 46 The gold-wisp, airy-grey Eye, all in fellowship.

airy-fairy (ē^ə^·ri,fē^ə^·ri), *a. colloq.* [After Tennyson's 'airy, fairy Lilian' (*Lilian*, 1830).]

1. Delicate or light as a fairy. Also as *sb.*

1869 W. S. Gilbert *Bab Ballads, Only a Dancing Girl* iii, No airy fairy she, As she hangs in arsenic green, From a highly impossible tree, In a highly impossible scene. **1898** *Westm. Gaz.* 1 Oct. 8/1 The low-necked airy-fairy ladies' ball gown. **1907** *Ibid.* 30 Mar. 16/1 To find partners for the airy-fairy creatures.

2. Fanciful (in disparaging sense).

1920 D. H. Lawrence *Lost Girl* iv. 55 He had already an airy-fairy kind of knowledge of the whole affair. **1957** F. Hoyle *Black Cloud* iv. 85, I am concerned with facts not with motives, suspicions, and airy-fairy nothingness. **1966** *Listener* 8 Sept. 336/2 Favoured by..some men of the market place, whose ideas I believe to be airy-fairy.

aisle. Add: **2.** Delete quot. 1807 and add earlier examples.

1789 E. Darwin *Loves of Plants* iv. 9 Long ailes of Oaks. **1818** Keats *Endymion* IV. 977 Through the dark pillars of those sylvan aisles.

4. b. *broad aisle* (U.S.): see *broad aisle.

5. A passage-way in a building (esp. a theatre, cinema, etc.), a train, etc. orig. *north. dial.* and *U.S.*

1755 in J. N. Scott *Bailey's Dict.*, *Isle*..a long passage in a church or public building. **1827** *Western Monthly Rev.* I. 73 The long aisles of all the stories [of a factory] to the fourth loft. **1842** Fanny Butler in *Bentley's Misc.* XII. 2 The seats..are placed down the whole length of the vehicle, one behind the other, leaving a species of aisle in the middle for the uneasy..to fidget up and down. **1851** J. J. Hooper *Widow Rugby's Husb.* 103, I have seen him..charge..into one door of the court-house, dash furiously along the aisle [etc.]. **1873** *Sat. Rev.* 22 Nov. 662/2 The Deputy-Sheriff placed his prisoners in the smoking-car of the train..The aisle was packed. **1880** L. Wallace *Ben-Hur* v. xiii, As the four stout servants carried the merchant in his chair up the aisle [in the circus], curiosity was much excited. **1890** *N. & Q.* 19 July 53/1, I have heard the space between the counters of a shop called 'the aisle' in Liverpool. **1903** A. I. Bacheller *Darrel of Blessed Isles* xiv. 148 Small boys would be chasing each other up and down aisles [of the school]. **1909** *Daily Chron.* 16 Feb. 4/7 [In America] all gangways and narrow paths whether in theatres, shops, or omnibuses, are 'aisles'. **1921** Wodehouse *Jill the Reckless* xviii. 260 The audience began to move up the aisles. **1961** R. Graves *More Poems* 33 Bring the charge-nurse scuttling down the aisle With morphia-needle levelled. **1965** G. Melly *Owning Up* vi. 61 Pat would..scurry down the coach aisle.

b. Colloq. phr. *to have, lay, send* (people) (*rolling*) *in the aisles*: to make (an audience) laugh uncontrollably; to be a great theatrical success. Also *transf.*

1940 Wodehouse *Quick Service* xii. 136, I made the speech of a lifetime. I had them tearing up the seats and rolling in the aisles. **1943** D. W. Brogan *Eng. People* vii. 202 This trick had the population, white and coloured, rolling in the aisles. **1954** N. Coward *Future Indef.* I. 17 This, to use a theatrical phrase, had them in the aisles! In fact, two of my inquisitors laughed until they cried. **1959** *Sunday Express* 11 Oct. 6/5 A book that sends my English friends rolling in the aisles. **1959** *Times* 14 Dec. 13/4 We looked forward to a school play which would really lay them in the aisles.

6. A double row of wheat-sheaves set up to dry. *local.*

1794 T. Davis *Agric. Wilts.* 76 The general custom of Wiltshire, is, to set up the sheafs in double rows,..and the sheaves so set up are called an aile. **1839** 'M. Gray' *Last Sentence* II. III. i. 206 Paler gold of piled sheaves 'in aisle' on upland slopes. **1904** *Daily Mail* 10 Sept. 3/7 In the Isle of Wight, what is locally described as an 'aisle' of corn standing in a field..was struck by lightning.

aitch (ē^i^tʃ). Name of the letter H; cf. Ache *sb.*[3] Also *Comb.*, as *aitch-dropping, -free* adjs. Hence **ai·tchless** *a.*, that does not aspirate his *h*'s.

1887 H. Baumann *Londinismen* p. xvi, Avowing himself 'An Aitch Dropper'. **1892, 1894** [see H 1 a, 2]. **1900** G. Swift 75 The aitch-less 'Arry. *Ibid.* 115 A bounder of the aitch-dropping type. **1903** Shaw *Man & Superman* II. 50 This man takes more trouble to drop his aitches than ever his father did to pick them up. **1907** *Daily Chron.* 9 July 3/3 Humanity, with a capital aitch. **1925** *Contemp. Rev.* June 746 The aitch-free accent. **1937** 'G. Orwell' *Road to Wigan Pier* xiii. 257 Even the aitchless millionaire, though sometimes he..learns a B.B.C. accent, seldom succeeds in disguising himself as completely as he would like to.

ajostle (ădʒɒ·s'l), *adv.* or *pred. a.* [f. A *prep.*[1] + Jostle.] Jostling; in a jostle.

1893 *Nat. Observer* 22 July 246/2 The quiet distance is all a-jostle and a-quake. **1908** A. S. M. Hutchinson *Once aboard Lugger* VI. v. 323 George put a hand to his head. This young man's senses were ajostle and awhirl.

ajoupa (ădʒū·pă). [Fr., repr. the Creole name.] In the West Indies, a hut or wigwam built on piles and covered with leaves or branches.

1871 Kingsley *At Last* I. viii. 298 A tribe of Chaymas built their palm-leaf ajoupas upon the very spot where the lake now lies.

akalat (ăkæ·lăt). [Native (Bulu) word.] A native name of babblers of the genus *Illadopsis* in West Africa, and of chats of the genus *Sheppardia* in East Africa.

1930 G. L. BATES *Handbk. Birds W. Afr.* 379 Malacocincla fulvescens fulvescens. Brown Akālat *Bulu* Akalat. *Ibid.* 381 Malacocincla cleaveri batesi. Blackcap Akalat. ..The Blackcap Akalat is similar in habits to the others. **1953** D. A. BANNERMAN *Birds of West & Equat. Afr.* II. 847 In the White-breasted Akalat (*I. rufipennis*) the throat and middle line of the belly are white; in the Blackcap Akalat the under parts are even whiter.

Akan (ā·kăn), *a.* and *sb.* Forms: 7 **Arcanys,** 8 **Acanni, Acanny, -s,** 9– **Akan.** [Native name.] The name of a group of Negro peoples inhabiting Ghana and neighbouring regions of West Africa, and of their group of languages.

1694 T. PHILLIPS in *Coll. of Voyages & Travels* (1732) VI. 224/2 The Arcanys, who are the best traders to our ships and castles, and have the purest gold, are an inland people. **1705** tr. *W. Bosman's New Descr. Guinea* vi. 77 Next..we come..to the description of Acanny, whose Inhabitants..were famed for great Traders...And that which they vended was always so pure and fine, that to this day the best Gold is called by the Negroes, Acanni Sica, or Acanny Gold. **1897** J. M. SARBAH *Fanti Customary Laws* I. 2 The Akan language is nevertheless the parent language—the language of diplomacy and courtiers. *Ibid.* 3 The words 'Akan' (Akanfu) arose probably from the way the Mfantsifu referred to those who remained at Takieman. The word Akan to our mind means a remnant. **1923** R. S. RATTRAY *Ashanti* ix. 113 The Brong are, in my opinion, undoubtedly a branch of the Akan stock, to which the Ashanti and the Fanti belong.

akasa, akasha (ākā·ʃā). *Hindu Philos.* [Skr. *ākāśá-* ether, atmosphere.] One of the five elements: ether (see quot. 1858[1]). Hence **aka·sic, aka·shic** *adjs.*, of, pertaining to, or existing in the akasa.

1858 H. T. COLEBROOKE *Relig. & Philos. Hindus* vi. 154 Five elements, produced from the five elementary particles or rudiments...A diffused, etherial fluid (*ácása*), occupying space: it has the property of audibleness, being the vehicle of sound, derived from the sonorous rudiment or etherial atom. *Ibid.* x. 253 The Bauddhas do not recognise a fifth element, *ácása*, nor any substance so designated. **1885** *1st Rep. Comm. Soc. Psychical Res.* 30 The akāsic substratum of ponderable things. **1917** 'RAMACHARAKA' *14 Lessons in Yogi Philos.* vi. 105 These akasic records contain the 'memory' of all that has passed. **1938** M. FIRTH tr. *F. Werfel's Hearken unto Voice* i. 17 In some incomprehensible way..*akāsha* contains in each of its particles, simultaneously and pervading the whole of space, all the phenomena and happenings of the cosmos. **1962** F. BANKS *Frontiers of Revel.* v. 66 An 'akashic record', that showed me..the truth of the assertion that 'the universal ether is the recording angel of the Christian terminology'.

akathisia, acathisia (ækăþi·siă). *Path.* [mod.L., ad. Czech. *akathisie* (L. Haškovec in *Sborn. Klin. v Praze* (1901–2) III. 193), Fr. *akathisie*, f. Gr. ἀ- priv. (A- 14) + καθίσις sitting: see -IA.] Inability to sit; morbid fear of sitting. Also used *joc.* (quot. 1938).

1904 *Jrnl. Nerv. & Ment. Dis.* XXXI. 195 *Akasthesia* [sic]. The symptom which the author has called by this name consists in the development of peculiar involuntary movements whenever the patients attempt to seat themselves. **1938** S. BECKETT *Murphy* vii. 119 It was true that Cooper never sat, his acathisia was deep-seated and of long standing. **1953** E. PODOLSKY *Encycl. Aberrations* 3/2 The inability to sit down or the dread to sit down is known as acathisia. **1961** *Lancet* 29 July 267/2 Akathisia.

ake (æ·ki, Maori a·ke). Also **ake-ake** (a·keˌa·ke, anglicized ækiæ·k); 9 **aki, aki-aki, haki.** [Maori.] The native name in New Zealand for for the small hardwood tree *Dodonæa viscosa* and trees of the genus *Olearia*, as *O. traversii.*

1835 W. YATE *Account of N.Z.* (ed. 2) ii. 47 Aki, called the *Lignum vitæ* of New Zealand. **1840** J. S. POLACK *Manners N. Zealanders* II. 261 The *Akki* a species of *lignum vitæ*, when young is much used for boat-timbers. **1844** *Barnicoat's Jrnl.* (MS.) 188 New Zealand woods, totara, haki, mairi. **1847** *Annals of Diocese of N.Z.* (*S.P.C.K.*) 239 Akeake (Metrosideros buxifolia), very hard and heavy, fit for cabinet work. **1851** MRS. R. WILSON *New Zealand* 43 The ake (Dodonæa spatulata) and towai..are almost equal, in point of colour, to rosewood. **1851** E. WARD *Jrnl.* 3 June (1951) 189 The largest of the akiaki trees. **1851** H. R. RICHMOND *Let.* 15 June in *Richmond-Atkinson Papers* (1960) I. ii. 97 The hakihaki for the handles of axes. **1879** J. HECTOR *Handbk. N.Z.* 94 Ake.—A small tree, 6–12 feet high. Wood very hard, variegated black and white; used for Native clubs. **1882** W. D. HAY *Brighter Britain!* II. 195 The Ake-ake..gives a handsome wood for cabinet work. **1917** *Chambers's Jrnl.* 6 Jan. 90/2 The dense scrub, consisting of silver-pine, ake-ake, and alpine vegetation. **1928** W. BAUCKE *Where White Man Treads* (ed. 2) 246 The Chathams grew only one durable timber, the ake-ake, which, because of its insufficient bulk, was seldom without heart-rot, shakes, and other blemishes.

akee, var. *ACKEE.

a-kimbo. Add: Now usu. hyphenless. Also *transf.* and *fig.* (see quots.), and as *adj.*

1922 JOYCE *Ulysses* 516 The Fan (folded akimbo against her waist). **1943** I. BROWN *Just Another Word* 24 'She got terribly akimbo' became a species of Mayfair slang for what was earlier called 'high horse'. I have also heard it used by stage people for over-acting. 'So and so was a bit akimbo to-night.' **1959** *New Yorker* 5 Dec. 146 He tended to match all of Coleman's near-atonal plunges with akimbo melodic lines of his own.

Akita (ăkī·tă). [Name of a district in northern Japan.] A medium-sized dog of Japanese breed.

1928 F. T. BARTON *Kennel Encycl.* 8/1 *Akita.* The title of this breed came from the name of the district, in the northern part of Japan. **1945** C. L. B. HUBBARD *Observer Bk. Dogs* 17 *Akita*..is a typical Spitz, with broad-pointed skull, stiff 'fur' on the back but softer hair elsewhere, and a curled..and bushy tail. **1948** —— *Dogs in Britain* xvii. 170 The Shika Inu is popularly called the Akita and is the only member of the breed to have been imported into Britain.

Akkadian (ăkē[i]·diăn), *a.* and *sb.* Also Accadian (see Dict. s.v.). [f. *Akkad, Accad*, name of a city (prob. to be identified with *Agade*) founded by Sargon I, and of the northern part of ancient Babylonia: see -IAN. Cf. F. *accadien*, G. *akkadisch.*] **A.** *adj.* **1.** Formerly, of a dialect related to Sumerian (see note s.v. SUMERIAN *a.*). **2.** Of or belonging to an eastern Semitic language of northern Babylonia, known from cuneiform inscriptions, or to the people of this region or their culture. **B.** *sb.* **1.** The Akkadian language. **2.** An inhabitant of Akkad or northern Babylonia.

c **1855** E. HINCKS (*title*) On the Relation between the newly-discovered Accadian Language and the Indo-European, Semitic, and Egyptian Languages. **1874** SAYCE in *Trans. Soc. Biblical Archæol.* III. 468 Elamu..is but a translation of the old Accadian name Susiana. *Ibid.* 484 In both Elamite and Susian, as well as in Accadian, the genitive relation may be expressed by simple position. **1875, 1878** [see SUMERIAN *a.* and *sb.*] **1884** SAYCE *Fresh Light fr. Anc. Mon.* ii. 24 The Accadians had been the inventors of the pictorial hieroglyphics..afterwards developed into the cuneiform..system of writing. **1921** G. A. F. KNIGHT *Nile & Jordan* iii. 31 The still earlier non-Semitic Akkadian civilization which the dynastic Babylonians dethroned. *Ibid.*, Eridu..means in Akkadian 'the city of the good (god)'. **1948** D. DIRINGER *Alphabet* I. i. 49 In the long development of the cuneiform writing of the Mesopotamian Semites, we can distinguish in particular six periods: (1) The Early Accadian period and Ur III, roughly from the middle of the twenty-fifth century B.C. to the middle of the twenty-second century B.C. **1958** A. TOYNBEE *East to West* liii. 160 The Akkadians themselves had acquired the vast irrigated oasis in the waist of Mesopotamia where the two rivers all but meet.

akoluthic (ækŏlū·þik), *a.* Also **acoluthic.** [f. Gr. ἀκόλουθος following + -IC.] Following, subsequent; *spec.* R. Semon's term (G. *akoluthe* in *Die Mneme* (ed. 2) 1908, p. 18) designating the phase following the initial excitation of a sensation.

1921 B. RUSSELL *Analysis of Mind* ix. 175 At the beginning of a stimulus we have a sensation; then a gradual transition; and at the end an image. Sensations while they are fading are called 'akoluthic' sensations. **1923** B. DUFFY tr. *Semon's Mnemic Psychol.* vi. 136 To describe these after-effects which are immediately connected with synchronous ones I have chosen the term 'acoluthic'. *Ibid.* 137 In the acoluthic phase excitation, of which the intensity begins to diminish rapidly when the stimulus ceases,..results nevertheless in sensations which we describe as acoluthic.

akrochordite (ækrokǭ·ɹdəit). *Min.* Also **acro-.** [ad. Sw. *akrochordit* (G. Flink 1922), f. Gr. ἀκροχορδών wart: see -ITE[1].] Hydrated basic arsenate of manganese and magnesium found in small spherical aggregates.

1922 G. FLINK in *Geol. Fören. Förh.* XLIV. 776 Summary. *Akrochordite*,..the individual grouped in wartlike aggregations. Found in..Långban mines. **1932** W. E. FORD *E. S. Dana's Textbk. Min.* (ed. 4) V. 720 *Akrochordite*..Monoclinic. In spherical aggregates of minute, nearly parallel crystals. **1955** M. H. HEY *Index Min. Species* (ed. 2) 255 *Akrochordite.* $Mn_4Mg(AsO_4)_2(OH)_4.5H_2O$...Syn. Acrochordite.

akvavit, var. *AQUAVIT.

-al, *suff.*[2] *Chem.* The first syllable of ALDEHYDE and ALCOHOL, used to form the names of substances which are aldehydes or derived from alcohol; e.g. *bromal, butyral, chloral, ethal.* Also as a general termination for pharmaceutical products, as *barbital, *hormonal (sb.).

à la, *phr.* Add: (Earlier and later examples.) Also *à la débandade*: in confusion or disorder; *à la fourchette*: see *fork-breakfast* (FORK *sb.* 16); also, eaten informally; *à la Russe*: in Russian fashion; *spec.* designating a meal for which the table is dressed with flowers and dessert, while the courses are served from other tables or from another room. See also *À LA CARTE, *À LA PAGE.

1589 PUTTENHAM *Eng. Poes.* III. xxv. 250 The breech a *la Françoise.* a **1666** EVELYN *Diary* 19 Nov. an. 1644 (1955) II. 259 A deepe basso-relievo a l'antique. **1739** R. West *Let.* 21 June in Walpole *Lett.* (1857) I. 22 We supped à l'Angloise. Imprimis, we had buttock of beef. **1812** [see FORK *sb.* 16]. *a* **1823** D. WORDSWORTH *Second Tour Scotl.* in *Jrnls.* (1941) II. viii. 353 Sheltered in the Cabin, and took a breakfast *a la fourchette*—not at the gentry end. *a* **1828** —— *Tour Continent* in *Ibid.* II. 20 An elegant gauze mob-cap with flowers, and robe *à la française.* **1828** SCOTT *Jrnl.* 3 Mar. (1890) II. 135, I do better *à la débandade* than I could with rules of regular study. **1828** *Souvenir* II. 95 (Stanford), The collar..is of fancy spotted silk, tied à-la-Russe, and fastened at the back of the neck. **1851** *London at Table* II. 41 Serving a dinner *à la Russe.* **1863** *Blackw. Mag.* Mar. 305/2 The modern service *à la Russe*, adopted in all good houses, has struck a decisive blow at the old English heavy dinners. **1874** J. H. WALSH *Dom. Econ.* IX. ii. 701/1 A table set out *à la Russe* for a party of eighteen. **1919** G. B. SHAW *Heartbreak Ho.* p. xx, Police raid *à la Russe.* **1926** D. H. LAWRENCE *Plumed Serp.* vii. 123 Some of the organdie frocks had green legs and green feet, some had legs *à la nature.* **1939** A. TOYNBEE *Stud. Hist.* IV. 440 A pair of mounted men-at-arms..were eventually to drive the legionary off the field *à la débandade.* **1957** V. NABOKOV *Pnin* 157 A supper *à la fourchette.*

b. *Cookery.* In the names of particular dishes, denoting the manner in which they are prepared; as *à la broche, brochette*: (cooked) on a spit or skewer; *à la meunière*: see *MEUNIÈRE.

1653 I. D. G. tr. *De la Varenne's French Cook* 58 Joint of Mutton after the Kingly manner (a la royale). **1723** J. NOTT *Cook's & Confectioner's Dict.* No. 498 Beef a la Braise. *Ibid.* No. 59c To dress a Carp a la daube..take the Carp, fill the body..with this Farce, set it a stewing..pour on it your Ragoo and serve it up. **1747** H. GLASSE *Art of Cookery* ii. 20 Beef *à la Daub. Ibid.* 24 Leg of Mutton *à la Royale. Ibid.* 44 *To à la Daube* Pigeons. **1853** E. K. KANE *U.S. Grinn. Exped.* xxxiv. 309 He made me this morning an idea of white bear's liver, *à la brochette.* **1882** [in Dict.]. **1906** MRS. BEETON *Housek. Managem.* lxii. 1652 *À la Broche*, roasted in front of the fire on a spit or skewer. **1959** R. POSTGATE *Good Food Guide* 31 What it can claim real distinction for is its cooking à la broche; chicken, duck or game roasted on the spit should be chosen.

c. With English sbs. and names.

1808 JANE AUSTEN *Let.* 15 June (1932) 189 Yesterday passed quite *à la* Godmersham. **1814** —— *Mansf. Park* II. x. 218 Her happiness on this occasion was very much a-la-mortal, finely chequered. **1881** [in Dict.]. **1926** GALSWORTHY *Silver Spoon* II. x. 191 As for Foggartism, they didn't—*à la* 'Evening Sun'—pooh-pooh it.

alabamine (ælăbæ·min). (*Disused.*) [f. *Alabama*, name of a State in U.S.A. + -INE[5].] The name originally proposed for *ASTATINE. Cf. also *ANGLO-HELVETIUM.

1932 F. ALLISON et al. in *Jrnl. Amer. Chem. Soc.* LIV. 616 We suggest the name Alabamine and symbol Am for element 85. **1936** *Nature* 13 June 969/2 The problems of illinium, masurium, virginium, and alabamine.

alabastron (æ:lăbɒ·strŏn). [a. Gr. ἀλάβαστρον globular perfume jar.] = ALABASTER 2.

1848 G. DENNIS *Cities & Cemeteries of Etruria* I. p. c, An *alabastron* of stone from Chiusi..having a hole in the crown for pouring out the ointment or perfume. **1928** D. H. LAWRENCE *Etruscan Places* (1932) iii. 79 A servant..presumably offering the *alabastron*, or ointment-jar. **1940** *Antiquity* XIV. 245 Among other shapes (of pots) the alabastron is prominent.

‖ **à la carte** (alakā·rt). [See A LA and CARTE[1] 5.] 'By the card (or menu)': said of meals which are ordered at a hotel or restaurant in separate items, each having its specified price. (Distinguished from a TABLE D'HÔTE meal, which is served at a fixed inclusive price.) Also *transf.*

1826 SHERER *Ramble in Germany* 252 He will find comfortable apartments, civil attendance, excellent fare, *à la carte*, at any hour. **1914** 'I. HAY' *Lighter Side School Life* vii. 189 If you want to have your son educated *à la carte*, you must get a private tutor for him. **1933** MRS. C. S. PEEL *Life's Enchanted Cup* xvi. 211 A restaurant for the staff, where meals were served *à la carte.* **1944** RUNYON (*title*) Runyon à la carte. **1966** *Automobile Assoc. Members Handbk.* 1966/67 30 A minimum charge for the à la carte menu.

alairy (ălē[ə]·ri), *adv. dial.* [Cf. ALIRY *adv.* in Dict. and Suppl.] (See quots.)

The expression is doubtless much older but printed evidence is lacking.

1916 'N. DOUGLAS' *London Street Games* 5 There are other ball-games, such as..One-Two-Three-And-A-Lairy (I wish I knew what a-lairy meant). **1958** *Medium Ævum* XXVII. 113 The game often played by little girls, in Northern England at least, where the player bounces a ball on the ground, catching it with one hand, keeping the score by chanting 'One, two, three, alairy..', and accompanying each 'alairy' with a circular swing of the leg so as to describe a loop around but not obstruct the rising ball. **1959** I. & P. OPIE *Lore & Lang. Schoolchildren* vii. 108 One, two, three a-lairy, My ball's down the airie.

alalonga (ælălǫ·ŋgă). Also **-lunga.** [mod.L., a. local It. *alalonga*, f. *ala* fin + *longa* (It. *lunga*) long.] An albacore, *Thunnus (germo) alalonga.*

1854 C. D. BADHAM *Prose Halieut.* xi. 201 The trembling thunnies, pelamyds, and alalongas, which covered the bottom of the net. *Ibid.* 203 Alalongas, whose long pectorals had been draggled in the mire.

alamanda, alamander, erron. varr. *ALLAMANDA.

alamandine (ælămæ·ndin). [ad. med.L. *alamandina*, altered f. L. *alabandina* (*gemma*), f. *Alabanda* a city of Caria.] A kind of garnet. Cf. ALMANDINE *sb.*

1895 J. W. ANDERSON *Prospector's Handbk.* (ed. 6) 96 Garnet (including carbuncle, alamandine, pyrope, &c.). **1921** W. W. WATTS in *Trans. Scott. Ecclesiol. Soc.* 152 A ring with an alamandine (a garnet-coloured stone).

alameda. (Earlier examples.)

1797 *Encycl. Brit.* IV. 10/1 The public walk, or Alameda, is pleasant in the evening. **1807** R. SOUTHEY *Lett. fr. England* iii. 26 Every town has not its Norney [as in Exeter] as with us its *alameda.*

Alan[2] (æ·lăn). Pl. **Alani** (ălā·ni, ălē[i]·ni), **Alans.** [ad. L. *Alānus, -i.*] A member of an ancient Scythian people, first encountered near the Caspian Sea. Also *attrib.* or as *adj.* Hence **Ala·nic** *a.*, of or pertaining to the Alani.

a **1450** tr. (prose) *Vegetius' De Re Militari* 19 b For þow among gothus and alanus & hunus horsmen vsen to ben armed. **1598** HAKLUYT tr. *J. de Plano Carpini's Tartars* in *Princ. Navigations, Voiages* I. 66 On the South side it hath the Alani. **1601** P. HOLLAND tr. *Pliny's Nat. Hist.* IV. xii. p. 83 The bastard and degenerate Scythians..and anon the Alani. **1614** A. GORGES tr. *Lucan* x. p. 443 He whom the fierce Alanis might, Nor Scythians rage could ought affright. **1776** GIBBON *Decl. & F.* I. xii. 329 He had negotiated with the Alani, a Scythian people, who pitched their tents in the neighbourhood of the lake Mœotis. **1875** *Encycl. Brit.* I. 442/1 The Alani were frequently in conflict with the Roman power. **1913** E. H. MINNS *Scythians & Greeks* iv. 37 Klaproth first proved in 1822 that the Ossetes are the same as the Caucasian Alans. *Ibid.* xviii. 555 The Alan or Tauric tongue. *Ibid.* [The name] is almost certainly Iranian, but we cannot take this as throwing light on the Tauri for it is more probably Alan. **1930** C. A. MACARTNEY *Magyars* 2 They [*sc.* the Magyars] moved southward, and were in contact with Alanic and other Caucasian elements. **1932** C. H. DAWSON *Making of Europe* v. 87 Reinforced by ..Sarmatian Alans from across the Danube. *Ibid.*, Gothic and Alan contingents, serving under their own leaders, became the mainstay of the Roman armies.

alanna (ălæ·nă). *Anglo-Ir.* Also **alan(n)ah.** [Ir. *a leanbh* my child!] My child! Used as a form of address or as a term of endearment.

1839 C. J. LEVER *Confessions H. Lorrequer* iii. 29 Shutting one eye knowingly, with an air of great secrecy, [he] whispered out, 'Miss Betty—Miss Betty, alanah!' **1841** 'H. LORREQUER' [C. J. LEVER] *Ch. O'Malley* I. iii. 16 'Do you want to break your neck entirely?' 'No Brackely, not mine.' 'Whose then, alannah?' **1847** A. TROLLOPE *Macdermots* I. viii. 176 Who else, should I main, alanna; sure isn't he your own beau? *Ibid.* xii. 307 Every one that entered, said, 'Well, Mary,' or, 'Well, alanna, how's yourself?' **1907** G. B. SHAW *John Bull's Other Island* III. 56 Come on, alanna, an make the paste for the pie. **1922** JOYCE *Ulysses* 580 Ireland's sweetheart, the king of Spain's daughter, alanna. **1948** T. H. WHITE *Elephant & Kangaroo* xxiii. 183 Jump in, alannah.

‖ **à la page** (alapaʒ). [Fr., lit. 'at the page'; *être à la page*, to be up to date.] Up to date, up to the minute.

1936 A. HUXLEY *Eyeless in Gaza* ii. 12 His awful suspicions that people are beginning to find him a bore, no longer *à la page.* **1948** *Mind* LVII. 522 These vagaries were in a sense useful, because they put Campanella more *à la page* than ever. **1961** *Times* 14 Feb. 6/7 In medieval times..ladies 'thought it indispensable' to carry small pocket mirrors kept in shallow circular lidded boxes (which sounds à-la-page today). **1963** *Sunday Express* 6 Oct. 19/2 To be really *à la page*, which is how the girls in Paris are at the moment, invest also in a small chunky handbag.

alarm, *sb.* Add: **4.** and **11.** *alarums* (or *alarms*) *and excursions*, a stage-direction occurring in slightly varying forms in Shakes. *Hen. VI* and *Rich. III* (e.g. 3 *Hen. VI*, v. ii. *init.*); hence used playfully by recent writers for: skirmishing, confused fighting or onsets, sudden divagations, etc.

1891 [see *CRASH *sb.*[1] 5]. **1895** K. GRAHAME *Golden Age* 29 Alarums and Excursions. **1907** W. RALEIGH *Shakesp.* iv. 102 The whole First Act of *Coriolanus* is..full of alarums and excursions and hand-to-hand fighting. [**1910** G. K. CHESTERTON (*title*) Alarms and Discursions.] **1922** BARRIE *Courage* 9, I want you to hold..That to gain courage is what you come to St. Andrews for. With some alarums and excursions into college life.

7. Delete 'Usually *alarum*'. Also *ellipt.* = *alarm-clock.*

1910 A. BENNETT *Clayhanger* III. vi. 366 The ticking parcel drew the discreet attention of the doctor...'It's only an alarm,' said Edwin.

12. Esp. in phr. *alarm and* (orig. *or*) *despondency.*

1879 *Act* 42 & 43 *Vict.* c. 33 § 5 Every person subject to military law who..by word of mouth or in writing spreads reports calculated to create unnecessary alarm or despondency..shall..be liable to suffer penal servitude. **1950** V. PENIAKOFF *Private Army* II. v. 128 A message came on the wireless for me. It said: '*Spread alarm and despondency*'...The date was, I think, May 18th, 1942. **1957** 'N. SHUTE' *On the Beach* viii. 249 They don't want to create alarm and despondency until they've got to.

b. In modified sense: apprehension; uneasiness as to consequences.

1833 *Reg. Deb. Congress* 26 Feb. 1781 This General Assembly views with alarm..the proposition..for abandoning..the principle of protection. **1834** *Ibid.* 18 Feb. 604 They regard with alarm the late measures of the President. **1937** I. GERSHWIN *Foggy Day*, I viewed the morning with alarm, The British Museum had lost its charm.

13. *alarm call, cry* = ALARM-*note*; **alarm-bird,** a name applied to various birds, as the kookaburra, the wattled peewit (*Lobivanellus lobatus*) of Australia, the African touracou (*Schizorhis concolor*), etc.; **alarm-clock** (examples); **alarm-cord,** the cord of an alarm-bell, *esp.* (*U.S.*) the communication-cord on a railway train; **alarm-gun** (examples); **alarm-watch** (later example).

1848 J. GOULD *Birds of Austral.* VI. pl. 9 *Lobivanellus lobatus*..has obtained the name of the Alarm Bird from its rising in the air, flying round and screaming at the approach of an intruder. **1940** V. POHL *Bushveld Adv.* vii. 155 The popular belief that the 'Go-away' or 'Alarm birds' scare off game. **1938** *Brit. Birds* XXXII. 93 The alarm call was a single note repeated several times before the bird took to wing. **1697** A Larum Clock [see LARUM *sb.* 3]. **1751** *Boston News-Let.* 10 Oct., Eight day and small alarm clocks. **1835** J. TODD *Student's Manual* 69 The students in Yale..have generally the alarm-clock. **1961** C. MCCULLERS *Clock without Hands* iv. 90 He..would not let himself daydream in the morning after the alarm clock went off. **1872** E. EGGLESTON *End of World* v. 40 In behind the donjon chimney he pulled an alarm cord. **1920** H. ELIOT HOWARD *Territory in Bird Life* iv. 139 The wary Redshank, poised on flickering wings, forgets its mournful alarm cry. **1928** D. H. LAWRENCE *Lady Chatt.* x. 134 The mother hen's wild alarm-cries. [**1757** *General Orders* 53 (D.A.E.), After the larem gun is fired.] **1760** S. NILES *Indian Wars in Mass. Hist. Soc. Coll.* (1861) ser. IV. V. 433 Alarm-guns were fired at St. George's Fort. **1826** G. R. GLEIG *Campaigns Brit. Army at Washington* (ed. 2) xii. 165 Wherever a light-house or signal station was erected, alarm guns were fired. *a* **1875** E. H. KNIGHT *Dict. Mech.* I. 57/2 *Alarm-watch*, an instrument not necessarily a time-piece, with going works, and adapted to..sound an alarm after a specific interval of time.

alarmist. Add: **a.** (Earlier example.)

1793 BURKE *Corr.* 23 Aug. (1844) IV. 135 We must continue to be vigorous *alarmists.*

b. (Earlier example.)

1800 *Aurora* (Philadelphia) 28 Nov. (Thornton), The little alarmist Jacobin doctor found he had mistaken his man.

alarm-post. (Earlier example.)

1694 *Regulation for Garrison Duty* in J. Muller *Syst. Camp-Discipline* (1757) 1 Each Regiment to have an Alarm-post to repair to, in case of Fire, or any other extraordinary Alarm.

Alaska (ălæ·skă). The name of a State in the north-west of the United States, used *attrib.* to designate things connected with it in origin: as *Alaska cedar, pine*, trees indigenous to western N. America; *Alaska sable, seal*, used in the fur trade as spurious names for skunk and other pelts.

1884 SARGENT *Rep. Forests* 7 The most valuable species of the northern Coast Forest [is]..the Alaska cedar (*Chamæcyparis* [*nootkatensis*]). **1897** *Boston Even. Transcript* 11 Sept. 24/3 Skunk skins are one of the biggest items in the fur market. They go under the attractive name of 'Alaska sable'. **1897** SUDWORTH *Arborescent Flora* 45 *Tsuga mertensiana*..Alaska Pine [of] Northwestern lumbermen. **1910** *Encycl. Brit.* I. 475/1 The yellow or Alaska cedar, a very hard and durable wood of fine grain and pleasant odour. **1921** A. C. LAUT *Fur Trade of America* iv. 43 Skunk as skunk simply wouldn't sell; so skunk became 'Alaska sable'. *Ibid.* iii. 33 Plucked otter is sold dyed for Alaska seal.

b. baked Alaska, a dessert consisting of a centre of sponge cake and ice cream with a light covering of rapidly cooked meringue.

1909 FARMER *Boston Cook Book* 448 Baked Alaska... Make meringue of eggs and sugar..., cover a board with white paper, lay on sponge cake, turn ice cream on cake.., cover with meringue, and spread smoothly. Place on oven grate and brown quickly in hot oven. **1954** *Menu* (*M/S Stella Polaris*) 13 Dec., Tenderloin Steak with Vegetables—Baked Alaska.

Alaskan (ălæ·skăn), *a.* and *sb.* [f. prec. + -AN.] **A.** *adj.* Of or pertaining to Alaska (see prec.) and its inhabitants.

1868 F. WHYMPER *Travel in Alaska* xxii. 255 The 'medicine man' of the Tchukchis (and also of the North Alaskan peoples, who use the same term) was..represented in Greenland. **1890** FRAZER *Golden Bough* II. 116 Alaskan hunters preserve the bones of sables and beavers out of reach of the dogs for a year. **1921** *Contemp. Rev.* Feb. 286 It seems certain that the Alaskan races..are of the same ultimate stock as ourselves. **1936** *Discovery* July 200/2 Archaeological evidence is conclusively in favour of an Alaskan origin [for the Eskimo].

B. *sb.* A native or inhabitant of Alaska.

1871 *Lakeside Monthly* July 16 (*title*) The Alaskans at home. **1881** C. M. WILLARD *Life in Alaska* (1884) 16 It is the custom of the Alaskans to compel the murderer to stay beside the corpse until it is finally disposed of. **1924** *Chambers's Jrnl.* 19 Apr. 325/1 The largest of all..as the Alaskans call it, the 'tyee' salmon, sometimes weighs over one hundred pounds.

alastrim (ălæ·strim). [Pg., f. *alastrar* to spread.] A contagious eruptive disease resembling smallpox.

1913 DORLAND *Med. Dict.* (ed. 7), *Alastrim*, same as *amaas.* **1922** *Times* 19 May 10/4 Last week no fewer than 38 cases of smallpox were notified in England and Wales... One of these cases was notified as 'alastrim'. [**1926** *Encycl. Brit.* Suppl. III. 559/2 An outbreak [of mild smallpox] in South Africa was described by de Korté in 1904, under the name of Amaas or Kaffir-pox, another by Ribas in Brazil in 1910, under the name of Alastrim.]

alba (æ·lbă). [Pr., f. Rom. *alba (cf. Sp. *alba* dawn), f. L. *albus* white.] A dawn-song of the Provençal poets. Cf. AUBADE.

1821 *New Monthly Mag.* I. 284 The *Albas* and the *Serenas* of the Provençal poets were the stanzas, which were sung by them at the break or close of day, in honour of their mistresses. [**1934** A. JEANROY *La Poésie des Troubadours* II. vii. 292 Réveillés, à l'aurore, par le cri du guetteur, deux amants qui viennent de passer la nuit ensemble se séparent en maudissant le jour qui vient trop tôt; tel est le thème, non moins invariable que celui de la pastourelle, d'un genre dont le nom est emprunté au mot *alba*, qui figure parfois au début de la pièce, et régulièrement à la fin de chaque couplet, où il forme refrain.] **1936** C. S. LEWIS *Alleg. Love* iv. 196 Before them lies the morning and the beautiful antiphonal *alba.*

Albanian (ælbē[i]·niăn), *a.*[1] and *sb.*[1] *Hist.* [f. med.L. *Albania* Scotland (Ir. *Alba*, gen. *Alban*) + -AN.] **A.** *adj.* Of or pertaining to Scotland. **B.** *sb.* A Scot. So **Albanic** (ælbæ·nik) *a.* [cf. Ir. *Albanach*, Gael. *Albannech*], Scottish.

1561 NORTON & SACKV. *Gorboduc* v. ii. 137 The proude attemptes of this Albanian prince. **1779** A. BUTLER *Lives of Saints* (ed. 2) I. 446 St. Gildas the Albanian, or the Scot. **1865** T. MᶜLAUCHLAN *Early Scott. Ch.* xvi. 208 The great convention at Drumceat in Ulster, where the question of Albanic independence was fully discussed.

Albanian (ælbē[i]·niăn), *a.*[2] and *sb.*[2] [f. *Albania* (see below) + -AN.]

Albania is the med.L. and general mod. name of the country, which is called *Shqipnija* by the inhabitants, who call themselves *Shqipetars*; in med. Gr. Ἀλβανία, with variants Ἀλβανητία, Ἀρβανητία, the inhabitants being called Ἀλβάνοι, Ἀλ-, Ἀρβανῆτες (in Turkish *ARNAUT).]

A. *adj.* Of or pertaining to Albania, a country in the western part of the Balkan peninsula, its inhabitants or their language.

1596 Z. I. tr. *J. de Lavardin's Hist. Scanderbeg, King of Albanie* IX. 381 The Albanian souldiers enclining to the peace which had beene so longe demanded. **1788** GIBBON *Decl. & F.* VII. lxviii. 466 In the Albanian war, he [*sc.* Mahomet II] studied the lessons..of his father. **1813** *Q. Rev.* Oct. 283 The Hungarian and the Albanian languages have some traits of resemblance to each other. **1869** TOZER *Highl. Turkey* I. 186 The Mahometans are mostly Albanian. **1905** M. E. DURHAM *Burden of Balkans* x. 227 A young Albanian officer. **1920** — in *Contemp. Rev.* Aug. 210 The long-forbidden Albanian language was printed.

B. *sb.* **1.** A native or inhabitant of this country.

1596 Z. I. tr. *J. de Lavardin's Hist. Scanderbeg, King of Albanie* x. 387 Mahomet..might at this time haue bene beaten downe..if Italy & the other Prouinces of Christendom would haue..harkened to the..admonitions of the Hungarians, the Greekes, and the Albanians. **1788** GIBBON *Decl. & F.* VI. lxvii. 457 Under his [*sc.* Scanderbeg's] conduct, the Albanians were invincible. **1813** *Q. Rev.* Oct. 284 The Albanians speak a language of which a considerable portion is Greek, Latin, German, Sclavonian or Turkish. **1920** *Contemp. Rev.* Aug. 210 Freed from Turkish prohibition,..the cult of the national language became almost a religion to the Albanians.

2. The language of this people (the descendant of Ancient Illyrian).

1813 *Q. Rev.* Oct. 283 The Finnish is said to have some coincidence with the Greek, the Hungarian with the Finnish, and the Albanian with all its neighbours. **1905** M. E. DURHAM *Burden of Balkans* xii. 298 Books printed in Albanian by the Church press at Skodra. **1920** — in *Contemp. Rev.* Aug. 210 Up till 1908 the foreign schools, where alone Albanian could be taught as a written language, had for political purposes frequently changed the spelling of their school books. *Ibid.* 212 He [*sc.* Essad Pasha] could only understand Turkish and Albanian.

Albanian (ælbē[i]·niăn), *a.*[3] and *sb.*[3] *Hist.* [L. *Albānia.*] **A.** *adj.* Of or pertaining to the ancient province of Albania on the Caspian Sea.

1607 TOPSELL *Four-f. Beasts* 146 These Dogges growe to an exceeding greate stature, and the next vnto them are the Albanian Dogs. **1788** GIBBON *Decl. & F.* IV. xl. 110 The name of Caspian or Albanian gates, is properly applied to Derbend. **1853** *Encycl. Brit.* II. 444/1 The whole country, formerly called Albania, now goes under the names of Daghistan, Schirwan, and Leghistan...The ancient historians take notice of the Albanian men as tall, strong-bodied, and..of a very graceful appearance. **1948** D. DIRINGER *Alphabet* II. v. 327 A potsherd from Old Ganja may contain an Albanian inscription.

B. *sb.* **1.** A native or inhabitant of this province.

1579 NORTH *Plutarch* 695 The Albanians lye towards the East, and Mare Caspium. **1781** GIBBON *Decl. & F.* II. xix. 153 The king of the Albanians, who led his independent tribes from the shores of the Caspian. **1948** D. DIRINGER *Alphabet* II. v. 327 Many ancient and modern savants dare to connect the Albanians of the Balkans with the Caucasian Albanians.

2. The language of this people.

1948 D. DIRINGER *Alphabet* II. v. 327 According to some scholars, Caucasian Albanian still survives in the Udi language.

albarello (ælbăre·lo). Also **alberello.** [It. *alberello* pot, phial, app. dim. of *albero, albaro* (silver) poplar, f. late L. *albarus* white poplar, f. L. *albus* white.] A Majolica jar used esp. as a container for drugs.

1873 C. D. E. FORTNUM *Maiolica* p. lxxv, The *albarello*.., or drug pot, universally known under that name, is made of different sizes and always of one piece. **1893** *Funk's Stand. Dict.*, Alberello. **1904** H. WALLIS (*title*) Italian Ceramic Art: The Albarello, a Study in early Renaissance Maiolica. **1925** B. RACKHAM in R. Fry et al. *Chinese Art* 17 The jar is no less pleasing. In shape it approaches the *albarello* of the Near East and Italy. **1950** *Ashmolean Mus. Dutch & Flemish Still-life Pictures* 173 A blue pottery *albarello* with foliage. **1959** G. SAVAGE *Antique Coll. Handbk.* 71 It was also fashionable for pharmacies to be gaily decorated with painted drug-jars, the most popular being the tall, waisted jar known as the *albarello*, which was originally derived from Persia, by way of Moorish Spain.

albedo. Add: **1.** Further examples: applied to the proportion of light reflected from various surfaces. Also *attrib.*

1936 *Brit. Jrnl. Psychol.* Jan. 313 Experiments on albedo perception. **1936** *Nature* 11 July 70/1 The albedo of different types of soil and vegetation. **1937** *Jrnl. R. Aeronaut. Soc.* XLI. 410 The albedo of clouds and the earth's surface.

2. A white structure, tissue, or material (Webster 1934); *spec.* the white pith of the inner peel of citrus fruits.

1949 J. B. S. BRAVERMAN *Citrus Products* iii. 77 Penetrating further into the peel of citrus fruit, one comes to the white, spongy, parenchymatous layer (mesocarp), generally known as the albedo.

3. *Nuclear Physics.* (See quots.)

1949 W. E. SIRI *Isotopic Tracers* v. 134 The effect of a reflector..is expressed in terms of its albedo γ, defined as the ratio of the number of neutrons that flow back in across the boundary to the number that flow out. **1962** *Gloss. Terms Nucl. Sci.* (*B.S.I.*) 8 *Albedo*, of a given object, the fraction of radiation (e.g. slow neutrons) reflected from it (Symbol: β).

albergo (ælbǭ·ɹgo). Pl. **albergi.** Occas. anglicized **alberg.** [It.: cf. AUBERGE.] An inn (in Italy).

1617 MORYSON *Itin.* I. 154 Three houses like Colledges, called *Albergi*, for those that make long stay in the Citie. **1673** RAY *Journ. Low C.* 303 The *Alberghi* or Halls of the eight several Nations..of the Order. **1808** 'HELEN ST. VICTOR' *Ruins of Rigonda* I. 137 By your naming payment, I should suppose that you had mistaken my poor Cottage for an Albergo, some distance from this. **1889** 'F. PIGOT' *Strangest Journ.* 149, I was thus able..to rest at the country albergs. **1926** *Spectator* 20 Feb. 318/2 Ostia..consisted of a few forlorn houses, a modest *albergo*, and a wilderness of sand.

Albert. Add: **1.** (Earlier and later examples.)

1861 *Daily Rev.* (Edinburgh) 16 Dec. 4/5 (Advt.), Rings, Eardrops, Albert and Guard Chains. **1865** DICKENS *Let.* 22 Apr. (1880) II. 228 What's an Albert chain? **1922** JOYCE *Ulysses* 431 Wearing gent's sterling silver waterbury keyless watch and double curb Albert with seal attached.

2. Albert Medal, a medal awarded since 1866 for 'gallantry in saving life at sea or on land'. Abbrev. A.M. (see *A III).

1866 *Times* 16 June 7/4 The Queen has conferred the decoration of the Albert Medal on Samuel Popplestone. **1902** *Encycl. Brit.* XXXI. 340/1 The Albert Medal..consisting of two classes.

3. A size of writing paper.

1859 *Stationers' Hand-bk. & Guide Paper Trade* (ed. 2) II. 71 *Albert*, a fancy note paper, of any sort, 3½ by 6 inches, produced from foolscap 8vo. **1876** *Stationer's Price & Cost Book* 3 *Albert Note*..is the 8vo of Fcap., 3½ by 6 inches, and it may be had at any price, according to the quality of paper from which it is cut. *c* **1912** *Paper Terminol.* (*Spalding & Hodge*) II. 1 *Albert*, a standard size of note paper. *Ibid.* 18 Albert, about 3½ × 6. **1954** *Ibid.* 11 *Albert*, a standard size of..notepaper 6 × 4 in.

Alberti bass (ælbǭ·ɹti bē[i]s). *Mus.* [f. the name of Domenico *Alberti* (*c* 1710–1740), an Italian musician + BASS *sb.*[5]] A style of bass accompaniment consisting of broken chords or arpeggios.

1876 STAINER & BARRETT *Dict. Mus. Terms* 20/1. **1954** *Grove's Dict. Mus.* (ed. 5) 93/2 It is not very probable that he [Domenico Alberti] actually invented the 'Alberti Bass', ..but he certainly brought it into undue prominence in his sonatas. **1957** *Encycl. Brit.* I. 527/1 An extremely familiar, but nowadays little-esteemed, formula of arpeggio accompaniment..known as the Alberti bass.

Albian (æ·lbiăn), *a.* and *sb. Geol.* [ad. Fr. *Albien* (1842, A. D'Orbigny, *Paléont. franç. terrains crétacés*), f. *Alba* (now *Aube*), a department of France + -IAN.] Epithet of a division of the Cretaceous formation in Europe (see quot. 1910); belonging to or found in this, as a fossil.

1847 in R. Chambers *Vestiges* (ed. 6) xii. 163 [citing translated statement of F. J. Pictet] Is it probable that the albian fauna had been completely annihilated, and then.. replaced by a fauna altogether new, and so similar to it? **1910** *Encycl. Brit.* I. 503/1 In England it is usual to regard the Albian stage as equivalent to the Upper Greensand *plus* Gault, that is, to the 'Selbornian' of Jukes-Browne.

Albigenses (æ:lbidʒe·nsīz), *sb. pl.* [med.L. (12th c.), f. *Albiga*, L. name of Albi, a city in south-west France.] The name of a sect of Catharist heretics of the south of France in the 12th and 13th centuries.

1625 PURCHAS *Pilgrims* II. VIII. vi. 1254 By exciting Princes and People to roote them out..thus Innocent the third oppressed the Albigenses. **1646** [see TABORITE]. **1692** P. ALLIX (*title*) Remarks upon The Ecclesiastical History of the Antient Churches of the Albigenses. **1757** A. BUTLER *Lives of Saints* III. 397 The abominations of the heresy of the Albigenses. **1832** [in title below]. **1910** *Encycl. Brit.* I. 505/2.

Hence **Albige·nsian** *a.* and *sb.*, **Albige·nsic** *a.*

1604 R. PARSONS *Treat. Three Convers. Engl.* II. Ep. ded., Manie different sectaryes also of opposite opinions,..as Waldensians, Albigensians,..Thaborits..& Lutherans. **1828** G. S. FABER *Sacr. Cal. Prophecy* III. 28 The Albigensic church of the south of France. **1832** S. R. MAITLAND *Albigenses & Waldenses* 14 Fugitive Waldensians (not Albigensians) driven out of Lyons. *Ibid.* 17 Either the Albigensian, or Waldensian, sect. **1922** H. J. WARNER (*title*) The Albigensian Heresy.

albinistic (ælbini·stik), *a.* [f. ALBINISM: see -ISTIC.] Of, pertaining to, or affected with albinism.

1880 in WEBSTER *Suppl.* **1912** *Brit. Mus. Return* 115 Skull of an albinistic Pekinese Spaniel.

albinotic (ælbinǫ·tik), *a.* [f. ALBINO + -OTIC.] Of, pertaining to, or affected with albinism: = prec.

1872 COUES *N. Amer. Birds* 237 Albinotic or melanotic, and other abnormally colored specimens.

Albion (æ·lbiŏn). *poet.* or *rhet.* [OE. *Albion*, f. L. *Albion, -ŏnis* (Pliny), Gr. Ἀλουίων (Ptolemy):—Celtic *Albio, gen. *Albionis, whence Ir.-Gael. *Alba*, gen. *Alban* Scotland (cf. med.L. *Albania*: see *ALBANIAN *a.*[1]); usu. referred to *albho- (L. *albus*) white, the allusion being to the white cliffs of Britain.] Great Britain. Phr. *perfidious Albion*, rendering F. *la perfide Albion*, a rhetorical expression for 'England', with reference to her alleged treacherous policy towards foreigners.

The phr. 'la perfide Albion' is said to have been first used by the Marquis de Ximénès (1726–1817) (*N. & Q.* (1932) CLXII. 107/2).

c **900** tr. *Bæda's Hist.* (1890) I. I. 24 Breoton is garsecges ealond, ðæt wæs iu geara Albion haten. *c* **1205** *Layamon's Brut* (1847) 1243 Albion hatte þat lond. **1387** TREVISA *Higden* (Rolls) II. 5 Firste þis ilond hiȝte Albion, as it were þe white lond. *c* **1399** CHAUCER *Purse* l. 22 in *Wks.* (1894) I. 406 O conquerour of Brutes Albioun. *a* **1592** GREENE *Fr. Bacon* (1594) sig. E2ᵛ, As if that Edward gaue me Englands right, And richt me with the Albion diadem. **1593** SHAKES. 2 *Hen. VI* I. iii. 48 Is this the Gournement of Britaines Ile? And this the Royaltie of Albions King? **1605** —— *Lear* III. ii. 91 Then shal the Realme of Albion, come to great confusion. [**1653** J. BOSSUET *Œuvres* (1816) XI. 469 L'Angleterre, ah! la perfide Angleterre, que le rempart de ses mers rendoit inaccessible aux Romains.] **1713** POPE *Windsor-For.* 5 When Albion sends her eager Sons to War. **1757** GRAY *Progr. Poesy* II. st. iii in *Odes* 9 When Latium had her lofty spirit lost, They sought, oh Albion! next thy sea-encircled coast. **1841** W. M. THACKERAY in *Fraser's Mag.* June 711/2 Ferocious yells of hatred against perfidious Albion were uttered by the liberal French press. [**1846** R. FORD *Gatherings fr. Spain* iv. 37 If there be a thing which '*La perfide Albion*', 'a nation of shopkeepers', dislikes,..it is a bankrupt.] **1850** WORDSW. *Prel.* (1926) x. 239 Since I had seen the surge Beat against Albion's shore. **1903** A. MCNEILL *Egregious English* 11 The French dislike of perfidious Albion may be reckoned to a great extent an intermittent matter. **1941** H. G. WELLS *You can't be too Careful* III. viii. 146 There I was—a lovely crossing—saying Adieu to the white cliffs of Albion. **1958** *Observer* 18 May 9/4 He [*sc.* General de Gaulle] felt it to be essential..that he should maintain a proud and haughty demeanour towards 'perfidious Albion'.

albite. Add: Hence **albitiza·tion** (see quot. 1940).

1909 *Geol. Mag.* VI. 250 (*title*) Albitization of Basic Plagioclase Felspars. **1940** *Chambers's Techn. Dict.* 20/2 *Albitisation*, in igneous rocks, the process by which a soda-lime feldspar (plagioclase) is replaced by albite (soda-feldspar). **1959** J. J. REED in G. J. Williams *Econ. Geol. N.Z.* (1965) x. 156/2 The hardness of the argillites is due largely to albitization.

albitite (æ·lbitəit). *Min.* [f. ALBIT(E + -ITE.[1]] A form of albite (see quot. 1940). Hence **albiti·tic** *a.*

1896 H. W. TURNER in *Amer. Geologist* XVII. 380 Some of these dikes are composed of coarsely granular albite only. Such a rock may be called an albitite. **1940** *Chambers's Techn. Dict.* 20/2 *Albitite*, a rare type of soda-syenite consisting almost entirely of albite, with a small content of coloured silicates. **1955** BROWN & DEY *India's Min. Wealth* (ed. 3) xvii. 636 Intermediate in composition are the albititic jadeitites which grade into the albitites—granulitic rocks composed almost exclusively of untwinned albite.

album[1]. Add: **4.** Also *attrib.*

1936 C. J. WEST *Bibliog. Pulp & Paper Making* 612 Changes in the composition of thick album paper and oiled paper. **1937** E. J. LABARRE *Dict. Paper* 97/1 Album papers and boards are cover papers, or double thick pasted board..intended for photographic albums.

5. *attrib.*, passing into *adj.* Applied to verses of a sort or quality suitable for inclusion in an album (often with disparaging implication).

1918 V. WOOLF *Writer's Diary* 7 Aug. (1953) 3 I'm much impressed by the extreme badness of B's [Byron's] poetry... Why did they think this album stuff the finest fire of poetry? **1953** C. DAY LEWIS in *Proc. Brit. Acad.* XXXVII. 163 Album verse that ranges between the inept and the perfunctory.

6. A long-playing gramophone record or a set of such records.

1957 *Gramophone* Apr. 427/2 That he was too little appreciated in his lifetime makes this memorial album (another record is to follow) even more..important. **1963** *Oxford Mail* 28 Jan. 6/6 Many jazz album covers have little connection with the contents. **1967** *Times* 16 Dec. (Advt.), A four record album with illustrated colour brochure.

albumen. Add: **4.** *attrib.* Designating a photographic paper or plate coated with albumen; also denoting a process using such a paper or plate.

1856 R. HOWLETT *Methods Printing Photogr. Pict.* 7 The Albumen Process. The most simple process for obtaining positive prints is that upon albumenized paper. **1867** SUTTON & DAWSON *Dict. Photogr.* 126 This consists in subjecting sensitised albumen-paper to the fumes of ammonia. **1889** *Internat. Ann. Anthony's Photogr. Bull.* II. 403 The albumen paper asserts its old place in the photographic galleries. *Ibid.*, The clearness..of the albumen pictures.

albuminuria. Add def.: The presence of albumen in the urine. (Add earlier example.) Hence **albuminu·ric** *a.*

1842 DUNGLISON *Dict. Med. Sci.* (ed. 3) 30/1 *Albuminuria*, the condition of the urine in which it contains albumen. **1879** *St. George's Hosp. Rep.* IX. 500 One male, aged 40, was the subject of long-standing albuminuric disease. **1908** *Practitioner* Jan. 15 Examination of the fundus, revealing albuminuric retinitis.

albumose (æ·lbiumō[u]s). *Chem.* [f. ALBUM(IN + -OSE[2].] An intermediate digestion-product of albuminous matter, passing into peptone; a species of proteose.

1884 *Jrnl. Chem. Soc.* XLVI. 1389 A new method of preparation..whereby four different forms of albumose are separated. **1889** G. M'GOWAN tr. *Bernthsen's Org. Chem.* 516 Albuminous matters undergo change when acted upon by the juices of the stomach at a temperature of 30–40°, pepsin converting them in the first instance into Anti- and Hemialbumoses, both of which then go into peptone; trypsin likewise gives rise to the two above albumoses. **1891** [see PTOMAINE]. **1946** *Nature* 20 July 87/1 The albumoses and allied compounds, constituting about 14 per cent of the material [*sc.* meat extract], although supplying first-class protein, are of small quantitative significance in the diet.

alcapton, alcaptonuria: see *ALKAPTON.

alcheringa, (æltʃəri·ŋgă). Also **a·lchera, a·ltjira, A-.** [Austral. Aboriginal, lit. = 'dream-time'.] In the mythology of some Australian Aboriginal peoples, the 'golden age' when the first ancestors were created.

1899 SPENCER & GILLEN *Nat. Tribes Central Australia* ix. 324 'It was so in the Alcheringa,' was considered as perfectly satisfactory by way of explanation. **1910** *Encycl. Brit.* X. 162/2 The Arunta believe that the souls of the primal semi-bestial ancestors of the Alcheringa or 'dream time' are perpetually reincarnated. **1939** S. DELL tr. *Jung's Integr. Personality* (1941) iii. 92 The palæolithic Australians ..identify themselves..with their ancestors of the so-called alcheringa period. **1959** S. H. COURTIER *Death in Dream Time* iv. 43 Alchera..is the name given by Central Australian tribes to the time when the world was being made. *Ibid.*, The natives..who had learned the white man's tongue found the word *dream* to come nearest to what they meant by *Alchera.* **1965** W. E. H. STANNER in R. & C. Berndt *Aborig. Man in Austral.* 214 A full understanding of the Aboriginal view of life and the world requires a careful study of the whole body of doctrine about the Dream Time (*altjira, bugari*).

Alcmanian (ælkmē[i]·niăn), *a.* Also **Alcma·nic.** [f. L. *Alcmānius*, f. *Alcmān*, Gr. Ἀλκμάν, the name of a Greek lyric poet of the 7th c. B.C.] Used to designate a dactylic tetrameter, or a distich consisting of a dactylic hexameter followed by this verse. Also *sb.*, an Alcmanian verse or distich. (See also quot. 1853.)

1850 P. SMITH in W. Smith *Dict. Gr. & Rom. Biogr.* I. 107/1 The shorter dactylic lines into which Alcman broke up the Homeric hexameter...The Cretic hexameter was named Alcmanic, from his [Alcman] being its inventor. **1853** *Encycl. Brit.* II. 453/1 *Alcmanian*, an ancient lyric kind of verse, consisting of two dactyles and two trochees: as *Virgini|bus pue|risque| canto.* **1879** J. W. WHITE tr. *J. H. H. Schmidt's Rhythmic & Metric* 95 The Alcmanian group, consisting of a dactylic hexameter followed by a dactylic tetrameter. **1962** D. S. RAVEN *Greek Metre* 84 *Alcmanic*, a name given to the dactylic tetrameter —⏑⏑ —⏑⏑ —⏑⏑ (–).., and, in Latin, to its combination by Horace with the hexameter.

alcogel (æ·lkŏdʒel). *Chem.* [f. ALCO(HOL + GEL(ATIN: see *GEL.] A gelatinous precipitate from a colloidal solution in alcohol.

1864 [see *ALCOSOL].

alcohol. Add: **5. b.** In full *alcohol fuel*, a fuel used in internal combustion engines, guided missiles, etc.

1901 *Sci. Amer.* LXXXIV. 344/3 (*title*) Alcohol as fuel for motor carriages. *Ibid.*, The champions of the alcohol motor scored another triumph in the Paris-Roubaix races. **1904** GOODCHILD & TWENEY *Technol. & Sci. Dict.* 11/1 *Alcohol fuel*..In France and Germany cheap alcohol, made from potatoes or beetroot, is utilised to a considerable extent in place of petrol. **1935** *Jrnl. R. Aeronaut. Soc.* XXXIX. 470 These engines are fitted with positive superchargers and run on exceptionally high compression ratios on alcohol fuel. **1940** *Chambers's Techn. Dict.* 20/2 *Alcohol fuel*, volatile liquid-fuel consisting wholly, or partly, of alcohol, able to withstand high-compression ratios without detonation. **1958** *Aero-Space Terms* 3/1 *Alcohol*, ethyl alcohol (C_2H_5OH) or methyl alcohol (CH_3OH), used with liquid oxygen as a bipropellant. *Ibid.*, Ethyl alcohol and liquid oxygen were used in the German V-2.

alcoholic, *a.* Add: **4.** Addicted to alcohol (cf. sense *B. 2 below).

1957 *Times Lit. Suppl.* 8 Nov. 670/5 The girl is virtually imprisoned by the neurotic, psychotic and (it is apparent by this time) alcoholic novelist.

B. *sb.* **1.** (Example.)

1870 *Food Jrnl.* 1 Apr. 141 Any restriction on the sale of alcoholics.

2. One who is addicted to excessive consumption of alcoholic drinks, a drink addict.

1891 G. T. KEMP in *Q. Jrnl. Inebriety* Jan. (Funk), Chronic alcoholics. **1907** *Daily Chron.* 4 Sept. 3/1 There is..a time coming when the alcoholic will be a rarity. **1909** *Westm. Gaz.* 25 Feb. 8/1 Warning him that deceased was a 'chronic alcoholic', and must have a bottle of port a day.

alcoholist (æ·lkŏhǫlist). [See -IST. Cf. ALCOHOLISM.] One who is addicted to, or advocates, drinking alcoholic drinks.

1888 *Forum* (N.Y.) Sept. 103 Of 250 chronic alcoholists nearly 90% had fatty degeneration of the liver. **1894** *Pop. Sci. Monthly* Nov. 99 A moderate alcoholist. **1920** W. J. LOCKE *House of Baltazar* iii, Old Jack Bonnithorne, the champion alcoholist of the moorland.

alcoholizer (æ·lkŏhǫləi:zəɹ). [See -IZER.] = prec.

1890 'R. BOLDREWOOD' *Miner's Right* xxiv, I pity those alcoholisers...And they are often such Bayards in their sane periods.

alcosol (æ·lkŏsǫl). *Chem.* [f. ALCO(HOL + SOL(UTION: see *SOL *sb.*[6]] A colloidal solution in alcohol.

1864 T. GRAHAM in *Proc. R. Soc.* XIII. 337 If I may be allowed to distinguish the liquid and gelatinous hydrates of silicic acid by the irregularly formed terms of hydrosol and hydrogel of silicic acid, the two corresponding alcoholic bodies now introduced may be named the alcosol and alcogel of silicic acid. **1922** A. W. STEWART *Physico-Chem. Themes* ix. 160.

alder, *sb.*[1] Add: **4. b. alder-fly** (also ellipt. **alder**) = ORL-*fly.*

1828 H. DAVY *Salmonia* 24 The fly you see is called by fishermen the alder fly, and is generally in large quantities before the May fly. **1859** C. KINGSLEY *Glaucus* (ed. 4) 195 The water-crickets..change into..the dark chocolate 'Alder' (Sialis lutaria). **1875** [see ORL-*fly*]. **1902** *Encycl. Brit.* XXV. 447/2 Among the flies which are general favourites with dry-fly fishers are the..alder..and the May-fly. **1928** *Daily Express* 11 Aug. 4/2 On Dartmoor streams..I found a small alder and a black gnat as effective as anything.

alderman. Add: **5.** *slang.* **a.** (See quots.)

1782 G. PARKER *Hum. Sketches* 31 Nick often eat a roast fowl and sausage with me, which in cant, is called an Alderman double slang'd. **1859** LORD W. LENNOX *Pict. Sporting Life* I. vi. 169 A turkey, garnished with sausages —technically termed, an alderman hung in chains.

†**b.** = JEMMY *sb.* 6. *Obs.*

1872 DIPROSE *London & London Life* xxi. 159 The 'alderman'..is a 'head bar', which would open any safe. The smaller bars were called 'citizens'. **1883** *Standard* 14 May 5/8 A complete set of safebreaking tools had been.. left behind, including wedges, an 'alderman', a jemmy.

Aldermaston (ǭ·ldəɹmɑstən). The name of a village in Berkshire, site of the Atomic Weapons Research Establishment, used *attrib.* and *ellipt.* of demonstrations and protest marches by 'anti-bomb' demonstrators, as *Aldermaston march, marcher.*

1958 *Observer* 2 Nov. 6/1 There are plenty of Aldermaston badges, with their semaphore symbol of ND—nuclear disarmament. **1958** *Times Lit. Suppl.* 28 Nov. 689/4, I am not a Communist, a pacifist or an Aldermaston marcher. **1960** *New Left Rev.* Nov.–Dec. 5/2 Next year's Aldermaston will really be a *fighting* Aldermaston. **1962** *Guardian* 19 Oct. 3/2 Neutralist and pacifist movements abroad..call their marches 'Aldermastons'.

Alderney (ǭ·ldəɹni). [The name of one of the Channel Islands.] Properly, the designation of the cattle bred in Alderney; popularly used *attrib.* or *ellipt.* as a general name for the cattle (mostly light or dark fawn-coloured) of the Channel Islands, including those of Jersey and Guernsey; also, an animal of this breed.

1771 SMOLLETT *H. Clinker* I. 2 Let Morgan's widow have the Alderney cow. **1772** F. BOSCAWEN *Let.* 11 July in C. F. Aspinall-Oglander *Admiral's Widow* (1942) v. 34 We..have a great deal of garden, two fields and an Alderney cow. **1816** JANE AUSTEN *Emma* I. iv. 50 Eight cows, two of them Alderneys. **1834** YOUATT *Cattle* 267 The Alderney, considering its voracious appetite..yields very little milk. **1853** MRS. GASKELL *Cranford* i. 7 An old lady had an Alderney cow, which she looked upon as a daughter. **1854** DICKENS *Hard Times* II. x, She didn't give any milk, ma'am; she gave bruises. She was a regular Alderney at that. **1875** in J. Coleman *Cattle Gt. Brit.* 139 The Channel Island breed of cattle, popularly known in this country as 'Alderneys', consists of two classes...The Guernsey is the larger of the two, usually of a light fawn colour...The Jersey class is smaller. **1910** *Encycl. Brit.* V. 539/2 (*Cattle*) The term Alderney is obsolete, the cattle of Alderney being mainly a type of the Guernsey breed. **1918** GALSWORTHY *Five Tales* v. 292 Listening to the starlings and skylarks, and the Alderney cows chewing the cud.

alderwoman. Add **1.** (Earlier example.)

1557 in W. H. Turner *Sel. Rec. Oxford* (1880) 264 Mʳᵉˢ. Brydgman shall go nexte unto the Alderwomen.

2. A woman holding the position of an alderman.

1768 MORANT *Hist. Essex* II. 400/2 A Gild of women of our Lady's Lights, to which belonged an Alderwoman, and two wardens. **1895** [see MAYORESS 2]. **1900** *Daily News* 24 May 3/1 The London Government Act of last year.. prohibited women serving as alderwomen or councillors on borough Councils.

Aldine (æ·ldəin, ǭ·ldəin), *a.* and *sb.* Read: [ad. mod.L. *Aldinus, -a* (sc. *editio*), f. *Aldus*, latinized form of *Aldo* (see below).] The designation of the editions of Greek and Latin classics (including many *principes*) printed or published at Venice by Aldo Manuzio (the more familiar name of Teobaldo Manucci, 1450–1515), and his family (*c* 1490–1597); also applied to certain styles of display types. As *sb.*, an Aldine book or edition.

The device characteristic of Aldine books is a figure of a dolphin on an anchor.

1802 DIBDIN *Rare Eds. Classics* Introd. 27 A most beautiful copy of this Aldine edition. **1827** — *Ibid.* (ed. 4) II. 342 The Aldine Demosthenes of 1504. *Ibid.* 343 This edition..departs from the Aldine in some places. **1837** [in Dict.]. **1862** BURTON *Bk. Hunter* 101 Old editions of the classics in vellum bindings—Stephenses or Aldines. **1880** A. LANG *XXII Ballades in Blue China* 24 In rich men's shelves they take their ease—Aldines, Bodonis, Elzevirs! **1883** *Pall Mall Gaz.* 18 Sept. 3/2 An Aldine dolphin. **1895** [see DOLPHIN 4].

Aldis[1] (ǭ·ldis). *Aldis lens*: see quot. 1911.

1904 J. T. TAYLOR *Optics Photogr.* xxx. 204 This optician [*sc.* Hugh Lancelot Aldis] applied for another patent, which was granted in February, 1902. The construction is known as the Aldis lens. **1911** *Encycl. Brit.* XXI. 511/1 In 1902 H. L. Aldis issued the 'Aldis lens', *f*/6, a doublet composed of a cemented meniscus in front and single double-convex back lens.

Aldis[2] (ǭ·ldis). The name of Arthur Cyril Webb *Aldis*, used *attrib.* or *ellipt.* in the proprietary names of certain of his inventions, as *Aldis lamp* (for signalling in Morse code), *Aldis (unit) sight* (for artillery, aircraft, etc.).

1917 *R.N.A.S. Anti-Submarine Rep. No.* 3 4 The Aldis lamp used in surface craft, trawlers, etc., for communication with aircraft has given satisfactory results. **1918** J. M. GRIDER *War Birds: Diary* 11 July (1927) 231 His altimeter and Aldis sight were both hit. **1928** C. F. S. GAMBLE *North Sea Air Station* xv. 280, I at once reported the incident by means of Aldis lamp to H.M.S. *Dryad. Ibid.* xvi. 295 In some instances the ring was mounted in a telescope of unit magnification, one of the best examples of which was the Aldis sight. **1944** 'N. SHUTE' *Pastoral* viii. 190 Would you like me to make him a signal on the Aldis, sir, and tell him to bale out? **1962** *Oxf. Illustr. Dict.* 18/2 *Aldis lamp*, signalling lamp, used esp. in navy and air force, in which Morse signs are transmitted by rotating a mirror at whose focus the light is located.

aldo- (ældo), before a vowel **ald-,** used as combining form of ALDEHYDE in the names of chemical compounds as *aldo-hexose, -keten, -pentose*; **aldoxim(e,** the oxime of an aldehyde.

1883 *Jrnl. Chem. Soc.* XLIV. 1104 Lactones are not converted into aldoximes by the action of hydroxylamine. **1906** SUDBOROUGH *Bernthsen's Org. Chem.* 308 The relationship of the aldohexoses to the aldopentoses. **1907** J. B. COHEN *Org. Chem. Adv. Stud.* I. 295 An aldehyde sugar containing six atoms of carbon would be termed an aldohexose. *Ibid.* 310 Aldo-Pentose. **1912** F. G. POPE *Mod. Research Org. Chem.* 276 Two classes of ketens have been recognized, namely, the keto-ketens..and the aldo-ketens.

aldolase (æ·ldŏlē[i]z). *Biochem.* [a. G. *aldolase* (O. Meyerhof et al. 1936, in *Biochem. Zeitschr.* CCLXXXVI. 301): see ALDOL and *-ASE.] An enzyme present in muscle extract.

1940 *Chem. Abstr.* 7947 The zymohexase of Meyerhof and Lohmann..and aldolase are identical. **1941** DORLAND & MILLER *Med. Dict.* 69/1 *Aldolase*, an enzyme in muscle extract which causes aldol condensations between phosphodihydroxyacetone and aldehydes to produce ketophosphoric acid. **1964** G. H. HAGGIS *Introd. Molecular Biol.* ii. 34 Samples of aldolase and glyceraldehyde-3-phosphate dehydrogenase (enzymes involved in glycolysis) were obtained and crystallized.

aldose (æ·ldō[u]s). *Chem.* [f. ALD(EHYDE + -OSE[2].] An aldehyde sugar.

1894 *Jrnl. Chem. Soc.* LXVI. II. 489 (*title*) Reaction for aldehydes: differentiation of aldoses and ketoses. **1900** W. H. HOWELL *Amer. Text-Bk. Physiol.* (ed. 2) I. 561 Bodies..possessing either the constitution of an aldehyde-alcohol,—CH(OH)CHO, called aldoses, or of a ketone-alcohol. **1902** J. J. R. MACLEOD in A. P. Beddard et al. *Pract. Physiol.* II. i. 161 Chemically, monosaccharides are either aldehydes or ketones, the former are called aldoses, the latter ketoses.

aldosterone (æ:ldoste·rōᵘn, -stī[ə]·rōᵘn). *Biochem.* [f. *ALDO- + *STER(OL + -ONE.] A steroid hormone isolated from the adrenal gland. Hence **a:ldoste·ronism,** a disease characterized by excessive amounts of aldosterone in the blood and urine.

1954 *Lancet* 12 June 1226/2 Because of the presence of an aldehyde group at the 18 position on the steroid ring the name 'aldosterone' has been proposed (in place of 'electrocortin'). **1955** *Ibid.* 4 June 1167/1 Many cases of potassium-losing nephritis..were really cases of primary aldosteronism. **1962** *Ibid.* 1 Dec. 1125/2 When aldosterone antagonists became readily available patients were included only when diuresis was complete.

aldrin (æ·ldrin). [f. the name of Kurt *Alder* (1902–1958), a German chemist + -IN[1].] A white crystalline insecticide.

1949 *U.S. Dept. Agric. Release Interdepart. Comm. on Pest Control*, (*title*) Aldrin, a coined name for an insecticidal product. **1953** *Ann. Reg. 1952* 412 A new American insecticide, aldrin, was used..to supplement other methods of [locust] control. **1955** D. E. H. FREAR *Chem. of Pesticides* (ed. 3) 62 *Aldrin*, the first published tests on the material were made by Kearns, Weinman, and Decker early in 1949. This compound..was later assigned the coined name *aldrin*, chosen because it was made by the Diels-Alder reaction.

aleatico (æliˌæ·tiko). [It.] A sweet Italian red wine.

1821 SHELLEY *Let.* 5 Aug. (1964) II. 313 If you take any wine let it be Aleatico. **1824** ALEX. HENDERSON *Hist. Wines* vi. 237 The *Aleatico*, or red muscadine..is produced in the highest perfection at Montepulciano. **1921** A. HUXLEY *Crome Yellow* xx. 221 The carminative virtues..of Marsala, of Aleatico, of stout, of gin.

aleatoric (ē[i]liˌătǭ[ə]·rik), *a.* [f. L. *āleātōr-ius* (see ALEATORY *a.*) + -IC.] Dependent on uncertain contingencies; done at random.

1961 *Times* 28 July 13/5 Indeterminate, or aleatoric music. **1962** *Publishers' Weekly* 12 Mar. 19 A new art form, known as 'aleatoric', meaning subject to elements of chance, is making headway among the younger generation of poets in Europe. Its leading exponent..Maxwell Volker..creates poetry by cutting out single words or long strips from newspapers and magazines which he haphazardly pastes together to create aleatoric verse.

alecithal (ăle·siþăl), *a. Embryol.* [f. A- 14 + Gr. λέκιθος yolk: see -AL.] Of an ovum: with little or no food-yolk.

1880 BALFOUR *Comp. Embryol.* I. iii. 90 It is convenient to distinguish the ova which segment uniformly by some term; and I should propose for this the term alecithal, as implying that they are without food-yolk, or that what little food-yolk there is, is distributed uniformly. [*Note*] For this term as well as for the terms telolecithal and centrolecithal I am indebted to Mr. Lankester.

Aleck (in *smart Aleck*): see *SMART *a.*

Alemannic (ælĕmæ·nik), *a.* and *sb.* Also formerly **All-, -anic.** [ad. late L. *Alemannicus*, f. *Alemanni* pl., Gr. Ἀλαμανοί, ad. Germ. *Alamanniz, prob. f. ALL + MAN *sb.*[1] and so denoting a wide alliance of peoples.] The name of a confederation of Germanic tribes occupying the territory between the Rhine, the Main, and the Danube; also, of the dialects of Old High German spoken by them or the modern representatives of these in Alsace, Switzerland, and S. W. Germany. Also **Alema·nnian** *a.* and *sb.*; **Alema·nnish** [after G. *Alemannisch(e)*, *a.* and *sb.*

1776 GIBBON *Decl. & F.* I. xi. 310 Aurelian had chosen these veteran troops..whose valour had been severely tried in the Alemannic war. **1797** *Encycl. Brit.* I. 469/2 Allemannic language. **1813** *Q. Rev.* 258 German, Himina. Alemannish, Himil. Classical German, Himmel. **1814** J. JAMIESON (*title*) Hermes Scythicus: or, the radical affinities of the Greek and Latin languages to the Gothic: illustrated from the Moeso-Gothic, Anglo-Saxon, Francic, Alemannic. **1842** *Penny Cycl.* XXIV. 268/1 Suabian branch [of modern High German]...*Allemannic*, commonly so called in the south-west corner of the Black Forest. **1879** *Encycl. Brit.* X. 517/2 For Bavarian we may quote some old glossaries, for Alemannian the interlinerary versions of the Benedictine Rule. *Ibid.* 519/2 Alemannian is best characterized by its rigidly keeping its original vowel qualities. **1888** WRIGHT *O.H.G. Primer* 1 Upper German, spoken in the highlands of South Germany, and consisting of the Alemanic and Bavarian dialects. **1934** PRIEBSCH & COLLINSON *German Lang.* II. vii. 326 The Alemannic Group includes Swabian. **1951** E. S. DUCKETT *Alcuin* iii. 98 The new Queen, Liutgard, was of Alemannian origin.

[Delete ALEMBICATE *v.* and substitute:]

alembicated (ăle·mbikē[i]tĕd), *ppl. a.* [f. L. type *alembicatus + -ED[1]. Cf. F. *alambiqué.*]

Of ideas, expression, etc.: over-refined, over-subtilized (as if by passing through an alembic).
1786 MRS. PIOZZI *Anecd. Johnson* 197 His mind was like a warm climate, which brings every thing to perfection suddenly and vigorously, not like the alembicated productions of artificial fire, which always betray the difficulty of bringing them forth when their size is disproportionate to their flavour. 1818 LADY MORGAN *Fl. Macarthy* I. i. 8 Theories of alembicated refinement. 1823 *New Monthly Mag.* VII. 194 Alembicated systems for explaining conduct, and reconciling contradictions of character, are rarely satisfactory. 1836 [in Dict.]. 1913 W. RALEIGH in *Eng. Rev.* May 228 The prose style of Boccaccio..is not a simple style—rather it is curious and alembicated.

alembication (ăle:mbikēⁱ·ʃən). [f. as prec. + -ATION.]
1. Over-subtlety or over-refinement of expression.
1893 STEVENSON *Vailima Lett.* 2 June (1895) 271 This sort of trouble..produces nothing when done but alembication and the far-fetched.
2. Refinement; concentration; = DISTILLATION 3 b, 4 c.
1936 *Proc. Brit. Acad.* XXI. 195 The masterpieces of this process of alembication are the views of Petworth Park, painted in about 1830. 1951 O. DOWNES in *N.Y. Times* 24 Mar. 9/3 The reading [of Verdi's 'Requiem'] was the alembication of a lifetime's thought and experience.

alert, *adv., a.,* and *sb.* Add: **B.** *adj. Comb.,* as *alert-looking* adj.
1926 F. W. CROFTS *Insp. French & Cheyne Myst.* xix. 270 A tall, alert-looking young man entered the room.
C. *sb.* **1. b.** *spec.* A signal given by means of a siren or hooter to indicate that an air attack is imminent; an air-raid alarm or warning; also, the state of preparedness so produced or the period during which this alarm is in effect. (Used esp. in the war of 1939–45.)
1940 W. S. CHURCHILL in *Hansard Commons* 5 Sept. ser. v. CCCLXV. 46 It was felt that the red warning should be taken merely as an alert. *Ibid.*, We immediately resumed our work under the conditions of alert. 1940 *Flight* 26 Sept. 243/2 London had three 'Alert' warnings in the morning of Monday, September 16, each lasting about half an hour. 1941 [see *AIR- III. 2]. 1949 E. BOWEN *Heat of Day* viii. 151 There were no alerts just at present, not so much as a yellow.

alert, *v.* Delete *rare*⁻¹ and add later examples. Also **ale·rted** *ppl. a.*; **ale·rting** *vbl. sb.* and *ppl. a.* (Cf. *ALERT *sb.* 1 b.)
1941 *Baltimore Sun* 19 Dec. 14/7 Directors of civilian defense air-raid control centers will alone be responsible for..alerting their..districts. 1943 *Ibid.* 12 Jan. 5/3 An alerted zone. 1948 *Cape Times Week-end Mag.* 17 Jan. 2/3 All bank tellers and others handling large sums have now been alerted. 1950 *Evening News* 11 Feb. 1/1 The Shrewsbury Flood Warden Service was alerted to-day. 1954 A. W. FIELDING *Hide & Seek* ix. 114 We spent the whole night with our senses more alerted than the situation.. demanded. 1955 *Financ. Times* 10 Sept. 4/2 The advantages of alerting the public to economic realities. 1958 *Times Lit. Suppl.* 9 May 256/4 Something of such a writer's concentrated quality gives even his most expansive verse an alerting brevity.

aleukæmic (ælⁱukī·mik, ēⁱ-), *a. Path.* Also -emic. [f. A- 14 + *LEUKÆMIC *a.*] Not characterized by a high white-cell count; esp. in *aleukæmic leukæmia.*
1904 A. STENGEL in *Amer. Jrnl. Med. Sci.* CXXVIII. 524 Some authorities regarded the latter [*sc.* splenic anæmia] as an aleukæmic variety of splenic leukæmia. 1949 *Blakiston's New Gould Med. Dict.* 562 *Aleukemic leukemia*, a form of leukemia in which the total leukocyte count is normal, despite leukemic changes in the tissues and qualitative changes in the blood. 1962 *Lancet* 13 Jan. 64/2 When splenomegaly occurred alone the initial diagnosis was often obscure, and 3 cases had been diagnosed as aleukæmic leukæmia.

Aleut (ălⁱū·t, (U.S.) æ·li,ūt). [Origin unknown, perh. from a native word.] **a.** A native of the Aleutian Islands. Also *attrib.* **b.** The language of this people, distantly related to Eskimo.
1780 W. COXE *Acc. Russ. Disc. betw. Asia & Amer.* 263 The inhabitants of Unalashka..now begin to call themselves by the general name of Aleyut, given them by the Russians. 1881 *Encycl. Brit.* XII. 826/1 Of the Aleuts, whose collective name is 'Ungungun', or 'People', there are two divisions. *Ibid.* 826/2 The Thlinkeet language seems to be completely isolated, showing nothing beyond the faintest verbal resemblance to the Aleut. 1890 *Independent* (N.Y.) 9 Oct. 1/2 The word *Aleut*, pronounced *al-lee-yoot*, is of uncertain origin. *Ibid.*, The Aleuts themselves do not recognize the word, their own name for their tribe being Oonangan. 1936 *Discovery* Nov. 351/2 It is not certain that the skull found..is that of an Aleut. 1956 J. WHATMOUGH *Lang.* iii. 41 Aleut was reduced to picture writing even in the last century. 1962 *Listener* 25 Oct. 690/1 The anatomical knowledge of the Aleut Eskimo.

Aleutian (ălⁱū·ʃən), *a.* and *sb.* [f. prec. + -IAN.] **A.** *adj.* Of or pertaining to the Aleutian Islands off the western coast of Alaska. **B.** *sb.* A native of the Aleutian Islands.
1780 W. COXE *Acc. Russ. Disc. betw. Asia & Amer.* (map facing title-p.), Aleütian Islands. 1814 tr. *G. H. von Langsdorff's Voy. & Trav.* II. ii. 48 The Aleutians are not addicted to smoking, but are passionately fond of snuff. 1870 *Alaska Times* (Sitka) 7 May 2/3 On last Monday afternoon war was declared between the Sitka Si-washes and the native Aleutians. 1902 *Encycl. Brit.* XXV. 242/2 The Aleutian district derives its name from traditions of islands beyond the Oliutorsk Cape of Kamchatka. 1911 FRAZER *Golden Bough* (ed. 3) I. I. iii. 123 An Aleutian hunter of sea-otters. 1937 *Times Lit. Suppl.* 4 Dec. 923/1 A fair crossing of the Aleutian ledge.

ale-wife². The form *aloofe* in quot. 1678 in Dict. is app. a misprint, as all other 17th-c. instances have *ale-wife, -wives,* e.g.:
1633 in *New Plymouth Col. Rec.* (1855) I. 17 Whereas God, by his providence, hath cast the fish called alewiues or herrings in the middest of the place. 1634 in *Mass. Col. Rec.* (1853) I. 114 [He] is to sell the Alewyves hee takes there at 5 s. the thousand. 1652 J. ELIOT *Let.* 28 Feb. in *Strength out of Weakness* 5 Where the Fish we call Alewives come, there we built a Bridge. 1670 S. CLARKE *Acc. Plantations* 37 Big-bellied Alewives, Mackrils richly clad With Rainbow colours.

Alexander (ælĕgza·ndəɹ), *sb.*² A kind of cocktail.
1930 H. CRADDOCK *Savoy Cocktail Book* 1. 18 Alexander Cocktail, ⅓ Dry Gin, ⅓ Crème de Cacao, ⅓ Sweet Cream. 1951 N. BALCHIN *Way through the Wood* vii. 98 The sort of places where one always drinks Alexanders. 1958 A. L. SIMON *Dict. Wines* 55/1 *Alexander*, (1) ⅓ gin; ⅓ Crème de Cacao; ⅓ cream. Frappé. (2) ⅔ jigger Rye Whisky; ⅓ jigger Bénédictine; twist orange peel on top.

Alexandra (ælĕgza·ndrā). **1.** The name of *Alexandra* (1844–1925), wife of King Edward VII, used *attrib.* to designate a manner of walking affected by fashionable society in imitation of her limp when she was Princess of Wales. So **Alexa·ndrian** *a.*³ (*temporary.*)
1870 G. M. HOPKINS *Further Lett.* 1 Mar. (1956) 110 He is going to make use of the 'Alexandrian step' when he can walk, that is the step wh. suddenly became the thing at court when the Princess of Wales got lame with a sore knee. 1876 *Chambers's Jrnl.* 15 Jan. 35/1 The Grecian bend and the Alexandra limp—both positive and practical imitations of physical affliction. 1880 C. M. YONGE *Autobiogr. P. Applecheeks* in *Bye-words* 304 Her Alexandra limp, and all her most unnatural airs.
2. In full *Alexandra fly*: an artificial fly used by anglers.
1880 F. FRANCIS *Bk. Angling* (ed. 5) viii. 286 A fly with a silver twist body and a bunch of peacock herl for the wing, which they [*sc.* trout] have a great fancy for, the fly minnow, known as the Alexandra in fact. 1882 D. FOSTER *Scientific Angler* xii. 199 Gaudily-dressed lake flies..are good. But the combination more suitable..is the dressing known as the 'Alexandra fly'. This is as large as a full-sized Sewin fly..the shank carefully wrapped with broad silver tinsel, to represent the body. 1926 *Chambers's Jrnl.* XVI. 601/1, I have an Alexandra, a confection of peacock feather and tinsel.

Alexandrian (ælĕgza·ndriăn), *a.*² and *sb.*² [f. L. *Alexandria*, Gr. Ἀλεξάνδρεια, the capital of ancient Egypt in classical times: see -AN.]
1. Of, pertaining to, or characteristic of, Alexandria. As *sb.*, an inhabitant of Alexandria.
1584 GREENE *Gwydonius* 63 But if yᵉ Alexandrians obtained yᵉ conquest, yᵉ Duke of Metelyne should peacably depart the countrie. 1606 SHAKES. *Ant. & Cl.* II. vii. 102 This is not yet an Alexandrian Feast. 1607 TOPSELL *Four-f. Beasts* 627 The Babilonians and the Alexandrians loued diuersity of colours in their garments. 1611 BIBLE *Acts* vi. 9 Certaine of the Synagogue..of the Libertines, and Cyrenians, and Alexandrians, and of them of Cilicia, and of Asia, disputing with Steuen. 1671 A. WOODHEAD *Consid. Council Trent* vii. 104 Arius, an Alexandrian Presbyter. 1775 ASH *Dict.*, *Alexandrian*, a native of Alexandria. 1875 *Encycl. Brit.* I. 494/1 The famous lighthouse..cost 800 talents, which, if Alexandrian, is equivalent to £248,000. 1949 *Oxf. Classical Dict.* 35/2 The Alexandrian citizenship was allowed to continue and carried with it certain privileges.
2. Belonging or akin to the schools of philosophy in Alexandria, esp. those which produced Philo Judaeus, the Neo-Platonists Ammonius, Plotinus, Iamblichus, and the Christian fathers Clement of Alexandria and Origen. As *sb.*, a member or follower of one or other of these schools.
1753 E. CHAMBERS *Cycl. Suppl.* s.v., The chief Alexandrian philosophers, were Ammonius, Plotinus, Origen, Porphyry, [etc.]. 1809 E. JERNINGHAM (*title*) The Alexandrian School; or, A Narrative of the First Christian Professors in Alexandria. —— *Ibid.* 18 The Alexandrians at length silenced the adverse party. 1840 *Q. Rev.* June 67 What in Plato was a religious philosophy, became, in the hands of the Alexandrians, a philosophical religion. *Ibid.*, John Smith, Cudworth, Norris, and More, were Alexandrian, not Athenian Platonists. 1858 DONALDSON *Contin. C. O. Müller's Lit. Anc. Greece* III. liii. 172 Gnosticism originated in the Orientalized Platonism of the Alexandrians. 1883 FARRAR *Ep. Hebrews* Introd. 35 The most marked feature of the Epistle to the Hebrews is its Alexandrian character. *Ibid.* 36 It is Alexandrian in its method of dealing with Scripture.
3. Belonging or akin to the school of Greek literature, esp. poetry, which flourished at Alexandria under the Ptolemies, and is regarded as a 'silver age'; derivative, imitative, artificial, addicted to recondite learning. As *sb.*, a member or imitator of this school, or a writer who shows himself to be a kindred spirit.
1840 *Q. Rev.* June 110 The great mass of Alexandrian literature has perished by its own acknowledged worthlessness. 1858 DONALDSON *Contin. C. O. Müller's Lit. Anc. Greece* II. xlv. 449 Of all the writers of the Alexandrian school, the bucolic poets have enjoyed the most universal and permanent popularity. 1877 SELLAR *Virgil* 42 Virgil.. has used the Alexandrians more freely than any other Greek writers, with the exception of Homer. 1887 MAHAFFY *Gr. Life & Thought* xi. 235 Antimachus of Colophon.. certainly was great enough to set up a model which had fatal effects on the Alexandrians. 1904 T. R. GLOVER *Stud. Virgil* 68 Ovid adds to his Alexandrian learning an air of humour which gives it quite a new complexion. 1958 *Times Lit. Suppl.* 7 Feb. 76/1 An Alexandrian style that abounds in hidden allusions, a version..of the academic rococo.
Hence **Alexa·ndrianism**, the method and doctrine of the philosophical and Christian schools, or the style and characteristics of the literary school, mentioned above.
1860 WESTCOTT *Introd. Study Gosp.* i. 75 The society of the Therapeutæ was indeed the practical corollary of Alexandrianism. 1878 J. C. COLLINS *Tourneur's Plays* II. 223 Lycophosed...This is a good instance of Tourneur's Alexandrianism. 1883 FARRAR *Ep. Hebrews* Introd. 38 The Alexandrianism of the Epistle appears most clearly in the constant parallels which it furnishes to the writings of Philo. 1960 H. READ *Forms of Things Unknown* II. vi. 97 Alexandrianism, which is another name for cultural decadence.

Alexandrine, *a.*² and *sb.*² Add examples = *ALEXANDRIAN *a.*² and *sb.*² 2 and 3.
1877 SELLAR *Virgil* 42 Yet even in him [*sc.* Horace] the influence of the Alexandrine tone is apparent, especially in his treatment of the subjects taken from the Greek mythology. 1893 C. T. CRUTTWELL *Lit. Hist. Early Christ.* II. 442 The Alexandrine Fathers. 1904 T. R. GLOVER *Stud. Virgil* 165 From Euripides and the Alexandrines the love motive found its way into Latin poetry. 1923 J. B. BURY *Hellen. Age* i. 3 The Latin poets..owed more to..the Alexandrines.
Hence **Alexa·ndrinism** = *ALEXANDRIANISM.
1880 CHEYNE *Proph. Isaiah* I. 238 Hence the Greek writer of Wisdom need not be credited with Alexandrinism. 1901 J. C. COLLINS *Ephem. Crit.* 336 He [*sc.* Catullus] has not, indeed, escaped the taint of Alexandrinism.

alexia (ăle·ksiă). *Path.* [mod.L., badly f. A- 14 + Gr. λέξις speech (λέγειν to speak, confused with L. *legere* to read), after AGRAPHIA.] Inability to see words or to read; word-blindness.
1878 G. B. SHATTUCK tr. *von Ziemssen's Cycl. Med.* XIV. 774 It is..by no means easy to say how much of the alexia and paralexia is to be ascribed to the hemiopia. 1904 TITCHENER tr. *Wundt's Physiol. Psychol.* I. 305. 1964 in de Reuck & O'Connor *Disorders of Language* 215 Alexia occurs significantly more often in left-handed people with left-sided lesions than in right-handed people with similar lesions.

alexin (ăle·ksin). *Biochem.* Also **alexine.** [a. G. *alexin* (H. Buchner 1891, in *Münchener Med. Wochenschr.* XXXVIII. 437/1, in form *alexine*), f. Gr. ἀλέξειν to ward off + -IN¹.] A name for a class of substances found in blood-serum, having the capacity of destroying bacteria, etc., and thus guarding against infection; *spec.* = *ADDIMENT, *COMPLEMENT *sb.* 5 i.
1892 *Brit. Med. Jrnl.* II. 728/2 Proteid substances capable of killing bacteria, and now known as alexins, can be obtained from the bodies of various animals. 1896 *Allbutt's Syst. Med.* I. 568 Buchner..suggested that the destructive power of serum is due to certain proteid substances, to which, unfortunately, he gave the name of alexins. 1901 *Jrnl. Chem. Soc.* LXXX. II. 256 Alexin is found in larger quantity in the polynuclear leucocytes than in normal blood serum of dogs and rabbits. 1904 [see THERMOLABILE *a.*]. 1949 H. W. FLOREY et al. *Antibiotics* I. i. 62 The bacilli so treated formed an excellent antigen for the fixation of alexine.

alfa (æ·lfă), var. HALFA.
1857 [see HALFA]. 1883 *Nature* XXVII. 299/2 English journals, many of which are printed on paper made from alfa, a plant cultivated in those remote regions [*sc.* Algeria], and manufactured in England. 1935 S. BECKETT *Echo's Bones* (note *ad fin.*), 250 [*sc.* copies] on Alfa paper numbered 1 to 250. 1937 E. J. LABARRE *Dict. Paper* 141/2 Esparto.. also termed Alfa.

alfilaria (ælfilēᵊ·riă). Also -eria. [Mexican Sp., f. Sp. *alfiler* pin, from the shape of the carpels.] (See quots.)
1868 *Overland Monthly* Aug. 180/1 Burr clover,.. alfalfa, bunch grass and alfilaria represent the general pasture of the mountains. 1888 FARMER *Americanisms*, Alfilaria (*Erodium cicutarium*). Also known as Storksbill, Pin-grass, Filaree, &c. A valuable forage plant of the dry regions from Colorado and New Mexico to Southern California. 1897 B. HARRADEN *Hilda Strafford* 115 Carpets of the little pink blossom of the alfilaria, the first spring flower.
Also **alfileri·lla.**
1889 in *Cent. Dict.* 1897 *Outing* (U.S.) XXIX. 551/2 The sleek Herefords lifted their white faces from the ferny circlets of alfilerilla.

Alfredian (ælfrī·diăn), *a.* Also Ælfredian. [f. *Alfred* + -IAN.] Characteristic of Alfred, King of the West Saxons (871–899), or of his writings.
1826 J. J. CONYBEARE *Illustr. Anglo-Saxon Poetry* 258 The style of these Ælfredian versions is distinguished from that of the Cædmonian school by its great simplicity of diction. 1871 H. SWEET *Alfred's Past. Care* p. vi, I have given a short sketch of the characteristics of Alfredian English as distinguished from those of the later period. 1899 C. PLUMMER *Two Saxon Chrons. Parallel* II. p. cxiv, The Alfredian Chronicle up to 892, itself compiled from earlier materials under Alfred's supervision. 1958 *Times Lit. Suppl.* 2 Jan. 14/3 The proverb in Alfredian English. 1933 C. L. WRENN in *Trans. Philol. Soc.* 74 We may then consider how far our material will enable us to deduce Alfred's norm, and then inquire more fully into Sweet's own 'normalized Early West Saxon'—or *Alfredian*, as we have seen that it should strictly be termed. 1935 A. H. SMITH *Parker Chron. 832–900* 8 The 'Ælfredian' Chronicle was circulated throughout the kingdom.

alfresco, *adv.* **1.** Delete † *Obs.*, and add later examples.
1866 *Athenæum* 6 Mar. 333/1 The prehistoric artist worked al fresco, executing patterns or figures. 1939 *Burlington Mag.* Sept. 108/1 The painting of the flesh—which is completely al fresco.
2. (Earlier *attrib.* examples.)
1811 L. M. HAWKINS *Countess* I. ii. 32 A little lad who had reported an *alfresco* orchestra as consisting of two horns and a hautboy. 1816 JANE AUSTEN *Emma* III. vi. 92 Mr. Woodhouse was safely conveyed in his carriage..to partake of this al-fresco party.

Algerian (ældʒiə·riăn), *a.* and *sb.* [f. *Algeria* + -AN.] **A.** *adj.* Of or pertaining to Algeria or Algiers, in northern Africa. Also *ellipt.*, Algerian wine. *Algerian onyx*: see quot. 1904.
1874 [R. L. PLAYFAIR] *Handbk. Trav. Algeria* 4 Algerian hotels are similar to the French country hotels. *Ibid.* 112/2 The mosque..owes its chief interest to its 8 columns of Algerian onyx. 1904 GOODCHILD & TWENEY *Technol. & Sci. Dict.* 101/2 *Oriental alabaster*, a stalagmitic variety of calcite; of white or brownish concentric laminæ. Used as an ornamental stone under the name Algerian onyx. 1912 'R. DEVEREUX' *Alg. Algeria* ix. 165 In 1906 Algerian wine was introduced into the English market. 1953 R. FULLER *2nd Curtain* i. 10 The waiter came with a bottle of Algerian wine. *Ibid.* 11 Garner shielded his face with his glass of Algerian. 1958 *Times* 16 May 6/1 The Algerian insurrection has now spread to..Corsica.
B. *sb.* An inhabitant or native of Algeria or Algiers.
1625 PURCHAS *Pilgrimes* II. IX. xii. 1564 Opinions foolish and Deuilish in Algier. The Algerians..the foolish conceits of other Mahumetans..that Fooles and Dwarfes are Saints. 1914 M. D. STOTT *Real Algeria* ix. 4 The Algerian regards his own intelligence and aptitude in every respect superior to that of the Frenchman.

Algerine (ældʒĕrī·n), *sb.* and *a.* [f. *Alger-ia* + -INE¹.] **A.** *sb.* **1.** An inhabitant or native of Algeria or Algiers; esp. a pirate from Algiers.
1657 J. SOMERS 12 May in *Cal. Clarendon State Papers* (1876) III. 287 Their whole fleet should not have been taken by the Algerines. 1666 PEPYS *Diary* 26 Jan. (1896) VIII. 211 The Algerines having blown up the fort. 1728 J. MORGAN *Hist. Algiers* I. 433 Hassan Basha, with his Algerines, determined to attack that small Peninsula. 1785 JEFFERSON *Writ.* (1894) IV. 25 The Algerines..have taken two of our vessels. 1819 SALAMÉ *Narr. Exped. Algiers* 30 At a few minutes before three, the Algerines, from the Eastern battery, fired the first shot at the Impregnable. 1870 THORNBURY *Old Stories* 259 The Algerines..against the English audacity. 1890 S. LANE-POOLE *Barbary Corsairs* 291 The French..beating back the Algerines as they advanced.
†**2.** *transf.* One who acts like a pirate. Hence **Algeri·nism.** *U.S. Obs.*
1841 FOOTE *Texas & Texans* 83 A transaction which will..call down retributive vengeance upon these American Algerines. 1844 *Congress. Globe* 11 Mar. 360/1 The gentleman from Rhode Island had talked of 'ruffianism' in that State, and of 'Algerines'; but if the proposition he made to this House was not a specimen of 'Algerinism', he apprehended it was not to be found in Rhode Island. 1862 WINTHROP *John Brent* (1883) vii. 57 He's one er them Algerines what don't know a dark line..till it begins to make motions.
3. A popular name for one of the cetaceans; a dolphin or porpoise. Also as *adj.*
1849 H. MELVILLE *Mardi* I. xiii. 63 Of a somewhat similar species [to the Black Fish] were the Algerines; so called, probably, from their corsair propensities; waylaying peaceful fish on the high seas. 1851 — *Moby Dick* I. xxxii. 228 *Algerine Porpoise.* A pirate. Very savage. He is only found, I think, in the Pacific. 1911 *Cent. Dict.*, *Algerine*, a grampus: because common off the coast of Algiers.
B. *adj.* Of or belonging to Algiers or Algeria; characteristic of Algerine pirates.
1657 R. CLERKE 31 Mar. in *Cal. Clarendon State Papers* (1876) III. 265 The money sent by the Turks to hire the Algerine fleet has been lost at sea. 1682 in L. Hertslet *Treaties* (1820) I. 63 In like manner no Algerine passenger being on board any ship or vessel in enmity with the said King of Great Britain shall be in any way molested. 1728 J. MORGAN *Hist. Algiers* I. 348 The Algerine Turks. 1772 H. WALPOLE *Let. to W. Cole* 15 Dec. (1818) 84 The plunder of your prints by that Algerine hog. 1818 FEARON *Sketches* 150 He [*sc.* the captain] is an American, tall, determined, and with an eye that flashes with Algerine cruelty. 1819 SALAMÉ *Narr. Exped. Algiers* 175 It is well known that the Algerine government are very rich, by their plundering at sea. 1864 MRS. GASKELL *French Life* i, in *Fraser's Mag.* Apr. 438/2 The window-curtains and portières are made of handsome dark Algerine stripe. 1870 THORNBURY *Old Stories* 250 An Algerine frigate. 1873 *Handbk. Trav. Algeria* (John Murray) 12 The audacity of the Algerine pirates at this time was unparalleled. 1925 E., O. & S. SITWELL *Poor Young People* 3 Wears an Algerine turban.

algesimeter (ældʒesi·mĕtəɹ). [f. Gr. ἄλγησις sense of pain + -METER.] = *ALGOMETER. Hence **algesime·tric** *a.*
1896 *Index-Catal. Libr. Surg.-Gen. U.S.* I. 252/1 Algesimeter. 1909 *Cent. Dict. Suppl.* 1909 *Amer. Jrnl. Psychol.* xii. 343 An algesimeter, an instrument for exerting a pressure of known intensity on the surface of the skin. 1953 *Brit. Jrnl. Psychol.* Nov. 284 Keys..found a very much wider threshold scatter for his algesimetric results.

algicide (ældʒisəid). Also (*rare*) **a·lgacide.** [f. ALGA + -CIDE 1.] That which kills algæ; *spec.* a preparation used for destroying algæ.
1904 *Science* 9 Dec. 805/1 (*title*) The use of Copper Sulphate as an Algicide in the Treatment of Water Supplies. 1909 *Cent. Dict. Suppl.*, Algicide. 1950 *Endeavour* IX. 143/1 Algæ may cause serious trouble in waterworks, in small fish-ponds, and even on the soil of commercial gardens and on house walls, so that the need for algicides is obvious.

algid, *a.* Add: Also **a·lgide.** *spec.* in *algid cholera*, a form of cholera, esp. Asiatic cholera, which is marked by copious watery alvine discharges, etc.
1847 E. A. PARKES *Res. Path. & Treatm. Asiatic or Algide Cholera* i. 4, I have ventured to employ the term 'Algide Cholera' as a synonym for the 'Cholera gravior' of Orton and others. 1873 F. T. ROBERTS *Theor. Pract. Med.* 685 Epidemic, Asiatic, Algide, or Malignant Cholera. *Ibid.* 690 Stage of Collapse, Algide Stage. 1889 *Buck's Handbk. Med. Sci.* VIII. 610 Algid dysentery. 1901 DORLAND *Med. Dict.* (ed. 2) 16/2.

algin (æ·ldʒin). *Chem.* Also **a·lgine.** [f. ALGA + -IN¹.] A nitrogenous substance, resembling gelatin, obtainable from certain algæ. So **a·lgic** or **algi·nic** acid; **a·lginate**, a salt of alginic acid.
1883 E. C. C. STANFORD in *Chem. News* XLVII. 255/2 This fluid contains a unique substance..to which, from its source [*sc.* sea-weed], I have given the name algin. 1885 *Jrnl. Soc. Chem. Ind.* IV. 520/1 Insoluble algin, or alginic acid, has a sp. gr. of 1·5, and closely resembles horn. 1888 *Chambers's Encycl.* I. 160 The cell-walls of our common brown seaweeds..contain a substance called algin. 1901 *Manuf. Leather* xxvii. 505 Algine dissolved in water in the proportion of 1 in 100. 1932 Constituents of the cell walls of the *Fucoideae* are..algin, the calcium salt of alginic acid. 1937 *Thorpe's Dict. Appl. Chem.* I. 201 Commercial algin or sodium alginate. 1940 J. B. SPEAKMAN in *Textile Manufacturer* LXVI. 464/2 Since sodium alginate dissolves in water to give solutions of high viscosity, whereas calcium alginate is insoluble, filaments are readily obtained. 1946 *Lancet* 24 Aug. 279/1 Calcium alginate filaments can be prepared as an absorbable hæmostatic gauze for application to oozing surfaces... Alginates..are relatively cheap, and can be sterilised by heat. 1955 *Times* 14 July 9/3 The establishment in Scotland, during the Second World War by the Ministry of Supply, of factories for the manufacture of salts of alginic acid extracted from the brown seaweeds.

Algol¹ (æ·lgɒl). *Astr.* [ad. Arab. *al ghūl* (see GHOUL).] The β star of the constellation Perseus, of variable brightness.
1390 GOWER *Conf.* III. 130 The tacchid sterre..Is hote Algol the clere rede. 1706 [see MEDUSA'S HEAD]. 1855 *Proc. Amer. Jrnl.* IV. 371 Algol, or β Persei, unquestionably belongs to the most remarkable of all the variable stars. 1880 E. C. PICKERING in *Proc. Amer. Acad. Arts & Sciences* XVI. 1 (*title*) Dimensions of the Fixed Stars, with special reference to Binaries and Variables of the Algol Type. 1909 [see NOVA 2]. ...'wink' lasting for several hours.

Algol² (æ·lgɒl). [f. *ALGO(RITHMIC *a.* + L(AN-GUAGE *sb.*] An international algebraic language for use in programming computers.
1959 PERLIS & SAMELSON in *Numerische Mathematik* I. 41 (*title*) Report on the Algorithmic Language ALGOL. 1960 [see *ALGORITHMIC *a.*]. 1961 *Times* 3 Oct. (Computer Suppl.) p. x/7 Algol..is used widely as a communication language.

algolagnia (ælgolæ·gniă). *Path.* [mod.L., f. G. *algolagnie* (A. von Schrenck-Notzing, *Sugg.-Therap.* (1892) ii. vii. 125), f. Gr. ἄλγος pain + λαγνεία lust.] A form of sexual perversion (see quots.). Hence **algolagni·stic** *a.*
1901 DORLAND *Med. Dict.* (ed. 2) 21/1 *Algolagnia*, abnormal and distorted activity of sexual impulse toward persons of opposite sex, including sadism, masochism. 1908 M. E. PAUL tr. *I. Bloch's Sexual Life of our Time* xxi. 558 De Sade..collected almost all the facts..regarding the algolagnistic phenomena in ethnology. 1924 E. & C. PAUL tr. *L. C. Baudouin's Psycho-Analysis & Aesthetics* 101 The word 'algolagnia' was coined to denote the inner unity of a tendency embracing two instincts, the instinct of suffering and that of making others suffer, the pathological forms of these instincts being known as 'masochism' and 'sadism' respectively. 1940 HINSIE & SHATZKY *Psychiatric Dict.* 20/1 Sadism thus came to be known as active algolagnia, while masochism was called passive algolagnia. 1949 H. HARE *Swinburne* vi. 126 This active sadism is expressed over and over again throughout *Poems and Ballads*... The algolagnia is carried to the last extreme.

algometer (ælgɒ·mitəɹ). [f. Gr. ἄλγος pain: see -METER.] An instrument for measuring degrees of sensitiveness to pain. So **algometric** (ælgome·trik) *a.*, pertaining to such measurement; **algometry** (-ɒ·metri), the use of the algometer.
1880 *Index-Catal. Libr. Surg.-Gen. U.S.* I. 190/1 (*heading*) Algometry. 1890 in BILLINGS *Med. Dict.* 1892 *Fortn. Rev.* May 355 By experiments made with the electric algometer it is clearly shown that woman is inferior to man [in general sensibility, including sensitiveness to pain]. 1897 E. W. SCRIPTURE *New Psychol.* 303 The pressure algometer consists..of a strong spring by means of which a rubber disc or point is pressed against the surface to be tested. 1904 G. S. HALL *Adolescence* II. 4 Other algometric tests..have led to somewhat different results.

Algo·nkian, *a.* and *sb.* [Var. of next.]
1. See next.
2. *Geol.* An epithet designating a period between the Archæan and the Palæozoic; proterozoic. As *sb.*, this period or system of rocks (in the U.S.A.).
These rocks are a feature of the region of Lake Superior, a territory of the Algonquian Indians.
1890 J. W. POWELL in *10th Rep. U.S. Geol. Survey* I. 20 At a recent conference of geologists..it was decided to make but one period of the Algonkian, and the name 'Algonkian' was chosen to designate that period. 1893 — in *14th Ann. Rep. U.S. Geol. Survey* I. 100 The chronologic series from the Algonkian up to the Cretaceous. 1893 GEIKIE *Text-bk. Geol.* vi. i. § 1. 713 Another enormous succession of rocks comprised under the general name of 'Algonkian', but consisting of several distinct formations, separated from each other by unconformabilities. 1925 J. JOLY *Surface-Hist.* v. 80 These are probably of Algonkian age; that is, they are some of the earliest rocks in which life has been recognized.

Algonquian, -kian (ælgɒ·ŋkiăn, -ŋkw-), *a.* and *sb.* [f. *ALGONQUIN, by alteration of ending: see -IAN.] **A.** *adj.* Of or belonging to a large group of N.-American Indians which includes the Algonquin tribe proper. **B.** *sb.* **1.** A member of this group. **2.** Any one of the languages or dialects spoken by the people of this group.
1885 J. W. POWELL in *Bur. Amer. Ethnol. Rep.* VII. 47 The area formerly occupied by the Algonquian family was more extensive than that of any other linguistic stock in North America. 1891 J. C. PILLING (*title*) Bibliography of the Algonquian Languages. *Ibid.*, Pref. p. iii, The Algonquian speaking peoples. 1900 *Lowdon's Races of Man* 576 The original home of the Algonquians was the region around Hudson's Bay. 1907 F. W. HODGE *Handbk. N. Amer. Ind.* I. 40/1 The Central Algonquians are the Cree, Ottawa.. 1919 R. H. LOWIE *Hist. Ethnol. Theory* vi. 63 He recognized the criteria of..the 'Omaha' system and indicated its occurrence among the Algonkian..family. *Ibid.* vii. 84 He [*sc.* Tylor] definitely ascribes the latter view only to Algonkians, Fijians, and Karens.

Algonquin, -kin (ælgɒ·ŋkin), *sb.* and *a.* [a. F. *Algonquin*, perh. contracted f. *Algoumequin* (17th c.). In Micmac *algoomeaking* or *-making* means 'at the place of spearing fish and eels'.] **A.** *sb.* **1.** An Indian of a people encountered in the districts of Ottawa and Quebec; also used as collect. sing., this people. **2.** More widely = *ALGONQUIAN *sb.* **3.** The language of this people. **B.** *adj.* Pertaining to any of the above.
1625 PURCHAS *Pilgrims* III. 1750 The Estechemins, Algoumequins, and Mountainers. 1667 TRACY in *N.Y. Documents* (1855) III. 150 Our authority now have hindred the Algonquins from making warre upon them [*sc.* the Dutch]. 1698 L. HENNEPIN *New Discov.* Contin. xxiv. 95 The Algonkians. 1705 BEVERLEY *Virginia* III. 24 They have a sort of General Language..what Lahontan calls the Algonkine. 1705 J. HARRIS *Navigantium* II. 909/2 [We] made towards them, crying out in the Iroquese and Algonquin Languages. 1765 *Exped. Henry Bouquet* (1868) 153 Nipissins, Algonquins, living towards the heads of the Ottawa river. 1778 T. HUTCHINS *Topogr. Descr. Virginia* etc. 67 Algonkins. 1845 H. R. SCHOOLCRAFT *Oneóta* 171 The Algonquin tribes. 1851 — *Ind. Tribes U.S.* I. 306 Their language is pure Algonquin. 1865 PARKMAN *Pioneers of France* in *Wks.* (1899) I. xii. 383 This neighborhood was the seat of the principal Indian population of the river [Ottawa], ancestors of the modern Ottawas..Usually called Algonquins, or Algonquins, by Champlain and other early writers,—a name now often used to designate not a single tribe, but a large family of cognate tribes. 1867 — *Jesuits N. Amer.* in *Wks.* (1899) III. ii. 4 Tribes speaking various dialects of the Algonquin. 1884 C. G. LELAND (*title*) The Algonquin Legends of New England, or, Myths and Folk Lore of the Micmac, Passamaquoddy, and Penobscot Tribes. 1949 J. BRODRICK *Proc. Saints* 199 She composed..a prayer book in Algonquin.

algophobia (ælgofōᵘ·biă). *Path.* [mod.L., f. Gr. ἄλγος pain + -PHOBIA.] Morbid fear of pain.
1897 in R. W. GREENE et al. *Lippincott's Med. Dict.* 33/2.

algorism. Add: **2.** = *ALGORITHM 2.
1960 [see *ALGORITHM 2].

algorithm. Read: **1.** See ALGORISM.
2. *Math.* A process, or set of rules, usually one expressed in algebraic notation, now used esp. in computing, machine translation and linguistics.
1938 HARDY & WRIGHT *Introd. Theory of Numbers* x. 135 The system of equations..is known as Euclid's algorithm. 1960 E. DELAVENAY *Introd. Machine Transl.* 129 *Algorithm* or *algorism*.., used by computer programmers to designate the numerical or algebraic notations which express a given sequence of computer operations, define a programme or routine conceived to solve a given type of problem. 1964 F. L. WESTWATER *Electronic Computers* ix. 146 An Algorithm is a set of rules for performing a calculation. 1966 OWEN & ROSS tr. *Revzin's Models of Lang.* ii. 22 A..more convenient way of arranging the phonemes is suggested. It is given by an instruction (an 'algorithm') consisting of six points.

algorithmic (ælgŏri·ðmik), *a.* [f. ALGORITHM + -IC.] Expressed as or using an algorithm or algorithms. Cf. *ALGOL² and prec. 2.
1881 J. VENN *Symbolic Logic* iv. 98 'Symbolic', as I understand it, being almost exactly the equivalent of 'algorithmic'. 1940 *Mind* XLIX. 249 He thinks of the subject [*sc.* the calculus] not merely as an algorithmic method. 1959 [see *ALGOL²]. 1960 *Communic. Assoc. Comput. Mach.* III. 300 The algorithmic language has three different kinds of representation—reference, hardware, and publication. 1964 *Language* XL. 167 Further syllabic studies for algorithmic prediction of English parts of speech.

algraphy (æ·lgrăfi). [ad. G. *algraphie* (J. Scholz 1896, in *Papier Zeitschr.* XXVIII. I. 1450), f. AL(UMINIUM: see -GRAPHY; cf. LITHOGRAPHY.] The art or process of printing from aluminium plates. Hence **algra·phic** *a.*
Scholz did not use the term in the orig. patent of 1893 (Germ. Patent D.R.P. 72470).
1897 *Studio* Sept. 244/2 An alternative device which would give equal facilities to the artist, and would preserve the technical character of lithography, while it would remove the necessity for depending upon the use of lithographic stone itself,..a process..discovered by Mr. Scholz of Mayence, who has patented his invention under the name of 'Algraphy'. 1898 *Daily Chron.* 14 Oct. 3/1 The algraphic plate is a substitute for the ordinary lithographic stone. 1914 E. H. RICHTER *Prints* 10 Plates of metal are often substituted for stone (zincography, algraphy).

ali- (æli), in *Anat.* combining form of L. *āla* wing (as in *aliform, aliped, alisphenoid,* etc.) denoting 'pertaining to the "wings" or lateral expansions' of certain parts, as **alie·thmoid** *a.*, pertaining to the lateral expansions of the ethmoid bone of certain birds; also **aliethmoi·dal** *a.*; **alina·sal** *a.*, pertaining to the lateral parts of the nostrils; **alise·ptal** *a.*, pertaining to a cartilaginous partition in the nasal passage of the embryo of a bird; also as *sbs.*
1869 W. K. PARKER in *Philos. Trans. R. Soc.* CLIX. 759 The rudiments of the 'aliethmoidal' and 'aliseptal' cartilages of the nasal labyrinth. *Ibid.* 800 The roots of the cartilaginous aliethmoid. *Ibid.* 780 The alinasal fold. *Ibid.* 798 The alinasals give rise to the nasal turbinal.

alibi. Add: **B.** *sb.* (Earlier and later examples.)
1743 FIELDING *J. Wild* in *Miscellanies* III. IV. iii. 303 A single *Alibi* would have saved them. 1939 T. S. ELIOT *Old Possum's Pract. Cats* 34 He always has an alibi, and one or two to spare: At whatever time the deed took place—Macavity wasn't there!
b. *colloq.* In weakened sense: an excuse, a pretext; a plea of innocence; a person providing an excuse, etc. orig. *U.S.*
1912 *Collier's* 20 Apr. 15/2 'Getting your alibi ready?' asked Zeider. 1922 W. T. TILDEN *Lawn Tennis* i. 1 Don't offer alibis for losing. 1936 W. HOLTBY *South Riding* v. § 3. 297 Pip's devotion gave her, she considered, a complete alibi in all charges of frustration and virginity. 1949 'C. HARE' *When Wind Blows* xiv. 181 Tom and Maureen are my alibis. 1951 L. P. HARTLEY *Fellow Devils* xxi. 216 Low spirits make you seem complaining..I have an alibi because I'm going to have a baby.
Hence **a·libi** *v. trans.*, to clear by an excuse; to provide an alibi. orig. *U.S.* Also (in *U.S.*) *intr.* Both *colloq.*
1909 P. ARMSTRONG *Alias Jimmy Valentine* (MS.) III, I'm going to alibi Doyle until he'll think he's lost his eye sight. *Ibid.*, You cant get away from the scar on your left wrist if you alibied yourself into hell. 1917 *Collier's* 13 Oct. 16/1, I ain't trying to alibi, it was a solid bone play. 1926 J. BLACK *You can't Win* i. 3, I am not lugging in the fact.. to alibi myself away from anything. *Ibid.* xx. 318, I could say I was looking for his room and he would alibi for me. 1930 'E. QUEEN' *French Powder Myst.* xxxi. 257 There's a sagacious chauffeur to alibi one. 1958 J. CANNAN *And be a Villain* iv. 111 Aunt Primrose..hadn't opportunity. She's alibi-ed by Mrs. Fitch.

Alice (æ·lis). **1.** The name of the heroine of two books by 'Lewis Carroll' (C. L. Dodgson, 1832–98), 'Alice's Adventures in Wonderland' (1865) and 'Through the Looking-Glass' (1871), used allusively with reference to these books, their heroine or her fantastic adventures. So *Alice-in-Wonderland* used *attrib.*; *Alice band*: a type of hair-band, as worn by Alice in Tenniel's illustrations to 'Through the Looking-Glass'.
1925 D. H. LAWRENCE *Let.* 6 Oct. (1962) II. 857 England is the most fantastic Alice-in-Wonderland country. 1931 *Observer* 19 Apr. 22/2 They [*sc.* trout] flourish and increase in size..at an Alice-in-Wonderland rate. 1935 E. BOWEN *House in Paris* I. ii. 34 'Oh, don't be such a baby!' said Henrietta..with her most Alice-ish air. 1940 *N. & Q.* CLXXIX. 395/1 Tweedledee in the first of the Alice books. 1945 E. BOWEN *Demon Lover* 170 Weeks of exile from any hairdresser had driven Miss Bates to the Alice-in-Wonderland style. 1955 'C. BROWN' *Lost Girls* xiii. 141 Holding it [*sc.* her hair] in place with an Alice band. 1959 *Sunday Express* 14 June 1/4 Princess Anne, wearing a dress of flowered silk and an Alice band over her fair hair. 1961 *Guardian* 21 Mar. 2/4 A £1,500 council house is costing my city over £6,000. This is an Alice in Wonderland situation.
2. The name of *Alice* Roosevelt Longworth, daughter of the American President Theodore Roosevelt, used *attrib.* to designate a light greenish-blue colour. orig. *U.S.*
1921 *Collier's* 19 Feb., Mood change carefree to gay record-creation causing such change Alice blue gown. 1946 T. JONES *Skinny Angel* 149 She sat straight-spined on the edge of her chair as she said this, a lovely figure in Alice blue. 1959 *Woman's Own* 16 May 12/3 An Alice blue dress.

alicyclic (ælisi·klik, -səi·k-), *a. Chem.* [ad. G. *alicyclisch* (E. Bamberger 1889, in *Ber. d. Deut. Chem. Ges.* XXII. 767), f. *ALI(PHATIC *a.* + *CYCLIC *a.* 7.] Combining the properties of aliphatic and cyclic compounds (see *CYCLIC *a.* 7).
1891 *Jrnl. Chem. Soc.* LX. II. 1097 Alicyclic Homology. 1900 E. F. SMITH tr. *von Richter's Org. Chem.* II. 18 Aliphatic-cyclic or alicyclic saturated and unsaturated compounds. 1907 J. B. COHEN *Org. Chem. Adv. Stud.* I. 553.

alien, *a.* and *sb.* Add: **B. 3. b.** *transf.* A word from one language used but not naturalized in another language.
1884 *New Eng. Dict.* p. ix, *Aliens* are names of foreign objects, titles, etc., which we require often to use, and for which we have no native equivalents. 1926 FOWLER *Mod. Eng. Usage* 193/2 Only faddists will engage in alien-hunting and insist on finding native substitutes for *tête-à-tête, agent provocateur,* [etc.]. 1933 *Shorter Oxf. Eng. Dict.* p. vii, *Natives* are words of Old English origin, *denizens* are borrowings from foreign languages which have acquired full English citizenship, *aliens* are words that retain their foreign appearance and to some extent their foreign sound.

alienation. Add: **1. b.** spec. *alienation of affection(s)*: see quot. 1961. *U.S.*
[1861 *Rep. Cases Wisconsin Supr. Court 1860* XI. 430 The evidence offered by the defendant, for the purpose of showing that the affections of the wife had been previously alienated..should have been admitted.] 1867 in N. Howard *Practice Rep. Supreme Court N.Y.* XXXII. 145 Does such alienation of affection—such refusal to recognize and receive the plaintiff as her husband, and to live with him as his wife—..constitute a cause of action, when caused as charged in the complaint? 1922 *Dominion Law Rep.* LXVI. 144 An affection that was merely flickering faintly with life might be finally killed by the act of adultery. Would even this not be an alienation of affection? 1933 in S. N. Grant-Bailey *Lush on the Law of Husband & Wife* (ed. 4) i. 36 It would seem that the English temperament is sufficiently sagacious to make the importation into the English law of the alienation of affection as a ground for action so remote as to be a negligible danger. 1949 M. MEAD *Male & Female* xv. 299 Alienation-of-affection cases between two men, which assume that the woman is a gently pliant lily, ring just as false. 1961 WEBSTER, *Alienation of affection*, the transfer of a person's affection from someone who has certain rights or claims to such affection to a third person who is held to be the instigator or cause of the transfer.
c. *Marxism.* (See quot. 1962.) (Marx's term *Entäusserung*, which he used in 'Zur Judenfrage', *Deutsch-Französische Jahrbücher*, 1844, has been rendered by 'self-alienation' and 'dehumanization' as well as by 'alienation'.)
1926 H. J. STENNING tr. *Marx's Sel. Essays* 95 After Christianity had completed the alienation of man from himself..Judaism [could] attain to general domination and turn the alienated individual..into alienable and saleable objects. 1958 *Listener* 7 Aug. 194/1 Marx, or at any rate the early Marx, has used a concept, Hegelian in origin, which Berdyaev found immensely fruitful for his own discussion of the idea of objectivity: the concept of 'alienation'. Men turn or are turned into impoverished things, dependent on power outside themselves. 1962 E. KAMENKA *Ethical Foundations Marxism* 12 The philosophico-ethical conceptions that underlie the younger Marx. Chief among these conceptions is that of 'alienation': the notion that in modern capitalistic society man is estranged or alienated from what are properly his functions and creations and that instead of controlling them he is controlled by them.
d. *Theatr.* In full *alienation effect.* [tr. G. *verfremdungseffekt* (Brecht, 'Verfremdungseffekte in der chinesischen Schauspielkunst', 1937, in *Schriften z. Theater*, 1957).] An effect sought by the German playwright Brecht and followers, aimed at the destruction of many of the conventions of theatrical illusion.
[1948 E. R. BENTLEY *Brecht's Private Life of Master Race* 91 The meaning of the device is, in a word, *Verfremdung*. The audience is put at a distance from the events related, is prevented from identifying itself with any character because each actor is all the time shifting roles.] 1949 —— in *Theatre Arts* Jan. 38/2 It is Brecht's contention that.. we need a kind of acting..that will set the action before us rather than involve us in the action... The German word which Brecht has made up to describe the distancing or estranging of the action is Verfremdung, here translated as 'alienation'. Any device which promotes such alienation is called an A-effect. 1956 K. TYNAN in *Observer* 2 Sept. 10/2 The famous 'alienation-effect' was originally intended to counter balance the extravagant rhetoric of German classical acting. 1956 S. WANAMAKER in *Internat. Theatre Ann.* 125 His [Brecht's] principal theory: objectivity of the spectator, distancing, alienation of the audience's feeling and involvement. 1962 *Listener* 29 Nov. 932/2 This method of description seems almost like a parallel to Brecht's 'alienation effect'. We watch, we judge, but we do not participate.
6. *alienation coefficient* or *coefficient of alienation* in *Statistics*: a ratio expressing the degree of lack of correlation of two variables.
1919 T. L. KELLEY in *Jrnl. Appl. Psychol.* III. 61 Just as ·484 is the coefficient of correlation between intelligence and vocational choice, so may ·875 be called the coefficient of alienation between the same two things. 1923 —— *Statist. Method* xi. 289 We have called $k_{0·12}$ the multiple alienation coefficient...We will define $k_{01·2}$ as the partial alienation coefficient. 1936 J. P. GUILFORD *Psychometric Methods* III. xi. 362 The expression $\surd(1-r^2)$ is known as the *coefficient of alienation*, and it measures the absence of relationship just as *r* measures its presence.

alienize (ēⁱ·liĕnəiz), *v.* [See -IZE.] *trans.* To make alien, reduce to the status of an alien (in quot. 1841 rendering W. *alltudio*). Hence **a·lienized** *ppl. a.*, made foreign.
1841 *Anc. Laws & Inst. Wales* II. 333 Her brothers alienized her children when they gave her to an alltud. 1860 MEREDITH *Evan Harrington* iv, This extremely alienised idea of the nature of a member of the Parliament of Great Britain.

alienness (ēⁱ·liĕn,nes). [f. ALIEN *a.* + -NESS.] The fact or quality of being alien.
1929 A. HUXLEY *Do what you Will* 33 Absolute alienness which no amount of Esperanto and international government..will ever..completely abolish. 1939 A. J. TOYNBEE *Study Hist.* V. 363 If alienness were really a hindrance and not a help to the spread of a 'higher religion'.

alight, *v.*¹ **6.** (Examples in *Aeronaut.*)
1786 J. JEFFRIES *Narr. 2 Aerial Voyages* 16 After alighting for a moment..M. Blanchard threw out the remaining part of our sand ballast. 1958 *Times* 17 Oct. 3/4 To see and feel the aircraft obeying the dictates of unseen devices..as we were about to alight along the centre-line of the runway.

alighting, *vbl. sb.*¹ Add: **b.** *Aeronaut.* = *LANDING *vbl. sb.* 1 d. Also *attrib.*
1909 V. LOUGHEED *Vehicles of Air* 466 *Alighting gear*, the under mechanism of an aeroplane, used to cushion its descent and to bring it to a stop as it reaches the ground. 1914 *Aeronaut. Jrnl.* XVIII. 315 *Carriage*, that part of the aircraft..intended..to absorb the shock of alighting. 1927 *Flight* 3 Feb. 54/1 Compulsory alightings must be made at the end of each seven laps.
c. **alighting-board**: a board placed in or near a hen-house, bee-hive, or the like upon which hens, bees, etc., can alight.
1855 *Poultry Chron.* III. 61/2 It [*sc.* a bee] has measured every inch of the alighting-board, and now it flies off. 1926 *Chambers's Jrnl.* 12 June 435/1 Distraught bees run about on the alighting-board. 1950 *New Zealand Jrnl. Agric.* Nov. 446 The hinged alighting board [of the hen house] can be raised at night.

align, *v.* Now the usual spelling (cf. ALINE *v.*). Add: **4.** *Politics.* To bring into line with a particular tradition, policy, group, or power. Also *refl.*
1934 WEBSTER s.v., To align nations against warfare. 1942 *Ann. Reg. 1941* 279 President Vargas..now showed himself..anxious to align his country with the democracies. 1955 *Treatm. Brit. P.O.W.'s in Korea* (H.M.S.O.) 2 If a prisoner refused to be educated..he was voluntarily aligning himself with the forces of reactionary capitalism.

alignment. Now the usual spelling (cf. ALINEMENT). Add: **1. b.** In various techn. uses (see quots.).
1892 *Perry & Co. Typewriter Prospectus* 2 Alignment is the first quality to be considered in purchasing a typewriter. 1904 GOODCHILD & TWENEY *Technol. & Sci. Dict.* 11/1 *Alignment*, the correct placing in line of the parts of a machine or apparatus—*e.g.* the mandrel and loose centre of a lathe, or the front and back wheels of a bicycle. 1940 *Gloss. Highway Engin. Terms* (*B.S.I.*) 8 *Alignment*, the precise direction given to the centre line of a highway in plan. 1955 *Gloss. Acoust. Terms* (*B.S.I.*) 29 *Magnetic head alignment*, the adjustment of the magnetic gap in relation to the magnetic medium.
5. *Politics.* The process or result of aligning (see *ALIGN *v.* 4); the grouping or agreement of parties or powers. Also *transf.*
1933 *Planning* 25 Apr. 2 Its approach is not partisan, but cuts across existing political alignments. 1940 G. F. HUDSON *Turkey, Greece & E. Medit.* 31 With Britain and France arrayed against Germany, and Italy neutral, the alinements in Western Europe are in the main those of the autumn of 1914. 1955 D. DAVIE *Articulate Energy* xiii. 154 The alignment of forces, for and against authentic syntax in poetry, is particularly clear. 1957 *Times* 11 May 6/3 The preservation of Jordan last month has bought time, and may have shifted the formal power of alignments of the Middle East.

6. *Comb.* **alignment chart** = *NOMOGRAM; hence *alignment diagram.*
1910 J. B. PEDDLE *Constr. Graphical Charts* ii. 15 A type of chart which has received considerable attention of late..is that known as the alinement chart. *Ibid.* iii. 31 So far the alinement charts as described have only taken account of three variables. This is not a necessary limitation and we will next consider a case in which the number of variables is four. 1915 *Trans. Amer. Soc. Civil Engin.* LXXVIII. 1382 Nomographic solutions for formulas... The alignment diagram has all the advantages over diagrams plotted by rectangular co-ordinates. 1918 J. LIPKA *Graphical & Mech. Computation* iii. 44 The fundamental principle involved in the construction of nomographic or alignment charts consists in the representation of an equation connecting three variables, $f(u, v, w) = 0$, by means of three scales along three curves (or straight lines) in such a manner that a straight line cuts the three scales in values of *u, v,* and *w* satisfying the equation.

‖ **aligoté** (aligote). [F.] A white grape used for wine-making in Burgundy; the wine made from this grape.
1912 P. ALLEN *Burgundy* xvii. 247 The best wines of all are called Aligote, and the second-best—still very good—are called Pinot. 1924 H. W. ALLEN *Wines of France* iii. 167 The best ordinary white wines are made in Burgundy from the *aligoté* grape. *Ibid.*, The *aligoté* wine..keeps well. 1958 *Spectator* 13 June 782/3 The travelling qualities of aligote—cheapest and least fragile of white burgundies.

alike, *a.* For '†(Formerly in various other const.) *Obs.*' read 'In various other const.; now *rare.*' Add further examples.
1867 MORRIS *Jason* IX. 170 A golden glittering sun That seemed well-nigh alike the heavenly one. 1920 D. H. LAWRENCE *Lost Girl* vi. 112 Oh, but he was always alike. 1938 E. BOWEN *Death of Heart* III. i. 339 Thomas and Portia turned their alike profiles in the direction from which the breeze came.

alikeness. Delete †*Obs.* and add later examples.
1920 *Times* 25 Sept. 11/6 This alikeness is not confined to the features and clothing of Americans, but it is noticeable in their thought and conversation. The whole tendency is towards levelling. 1924 *Public Opinion* 6 Dec. 572/1 The fundamental alikeness of God and man. 1960 E. BOWEN *Time in Rome* ii. 65 There is an alikeness between all and any nights in the same place.

aline: see *ALIGN *v.*

alinement: see *ALIGNMENT.

alio-relative (æ:lio,re·lătiv), *sb. Philos.* [f. L. *ali-us* other + -o + RELATIVE *a.* and *sb.*] Peirce's term for 'a relation which no term can have to itself' (see quot. 1873). Hence as *adj.* = *IRREFLEXIVE *a.*
1873 C. S. PEIRCE in *Mem. Amer. Acad. Arts & Sci.* IX. II. 369 Simple relatives are divisible into those which contain elements of the form (A:A) and those which do not. The former express relations such as a thing may have to itself, the latter (as cousin of —, hater of —) relations which nothing can have to itself. The former may be termed *self-relatives*, the latter *alio-relatives*. 1920 B. RUSSELL *Introd. Math. Philos.* (ed. 2) iv. 33 It often happens that a relation is an aliorelative without being asymmetrical. 1934 *Mind* XLIII. 224 This is also guaranteed by my definition.. provided R [*sc.* a relation] is assumed alio-relative.

aliphatic (ælifæ·tik), *a. Chem.* [f. Gr. ἀλειφατ-, ἄλειφαρ unguent, fat + -IC]. Fatty; epithet of organic compounds having an open-chain structure.
1889 M'GOWAN tr. *Bernthsen's Org. Chem.* vi. 146 The saturated alcohols yielding the saturated monobasic fatty acids, or 'acids of the aliphatic series' as they are termed. 1895 F. P. MÖLLER *Cod-liver Oil & Chem.* 7 The aliphatic series. *Ibid.* 83 The aliphatic hydrocarbons. 1958 *Sci. News* XLIX. 44 In general, dimethyl silicone fluids are readily soluble in aliphatic and aromatic hydrocarbons.

aliquot. Add: **B.** *sb.* (Further examples.) (See also quot. 1955.)
1955 *Gloss. Terms Radiology* (*B.S.I.*) 65 *Aliquot*, a small sample of radioactive material assayed in order to determine the radioactivity of the whole. 1962 *Lancet* 6 Jan. 53/2 Deep-frozen aliquots were sent to the departments where the methods were evolved.

aliry, *adv.* Add to etym.: [Perh. f. OE. *lima lyre* loss of (the use of) limbs (E. J. Dobson in *Eng. & Germ. Stud.* (1948) I. 60): then quot. 1362 in Dict. would mean 'made their legs lame, acted as if paralysed'. Connection with mod. dial. *ALAIRY *adv.* is also possible.]

aliturgic (ælitə̄·ɹdʒik), *a. Eccl.* [f. A- 14 + LITURGIC *a.* Cf. F. *aliturgique* and Gr. ἀλειτούργητος.] Of a day: on which the liturgy is not celebrated. So **alitu·rgical** *a.*
1872 O. SHIPLEY *Gloss. Eccl. Terms*, *Aliturgical Days*, those days when the holy sacrifice is not offered. 1898 *Church Times* 7 Apr., Those who desired to communicate on that [*sc.* Good Friday] and other aliturgic days. 1912 A. FORTESCUE *Mass* 186 In the Milanese rite all Fridays in Lent are still strictly aliturgical days.

alive, *adv.* or *pred. a.* Add: Also occas. as *adj.* in *attrib.* position (senses 1 and 5).
1938 E. BOWEN *Death of Heart* III. i. 329 Portia said, in a hardly alive voice: 'I thought you said you had finished everything.' 1959 D. EDEN *Sleeping Bride* xix. 150 Her intensely alive face..her eager response. 1961 K. NORWAY *Waterfront Hosp.* iv. 67 Group Captain Hurst is the most alive man I know.
5. (*all*) *alive,* (*alive*), *oh!* (from the cry of fish sellers): very much alive and active. *alive and kicking*: see KICKING *ppl. a.*
c 1709 *Squire Bickerstaff Detected* 7 Can any Man of common Sense think it..not much beneath the Dignity of a Philosopher, to stand bawling before his own Dore—Alive! Alive! Hoa! 1825 T. CREEVEY *Let.* 16 Mar. in *Creevey Papers* (1963) xii. 205 Our *York* is all alive O! He dined at Sefton's this day week as gay as a lark. 1854 J. R. PLANCHÉ *Once upon a Time* I. iv. 16 *Sub.* Girls, where's your mother? *Chloe.* Coming at a wish. *Sub.* She is — and all alive O! like your fish. 1859 [see O *int.* 3]. 1884 J. YORKSTON *Cockles & Mussels*, As she wheel'd her wheel barrow Through streets broad and narrow, Crying 'cockles and mussels alive, alive o!' 1941 'N. BLAKE' *Abom. Snowman* xix. 215 He and Andrew have skipped together, both alive-oh. 1956 'M. INNES' *Old Hall, New Hall* I. x. 91 Joscelyn.. was suddenly struck all of a heap by Edward's all alive-oh Grecian girl.
b. *transf.* Of things, in specific uses. (See quots.)
1845 S. JUDD *Margaret* II. i. 186 Her Mother 'stirs it off' and a due quantity of the 'quick' and 'alive' crystal sweet is the result. 1892 GUNTER *Miss Dividends* xvi. 234 The locomotives..are moving about slowly, with a view to keeping themselves what is technically called 'alive'—that is their steam up, sufficient to give them power of motion. 1937 *Times* 13 Apr. p. xxv/2 The spare parts department.. carries approximately 40,000 different components, all of which are kept 'alive' for a period of no less than five years. 1943 'T. DUDLEY-GORDON' *Coastal Command* ii. 20 It [*sc.* a torpedo] becomes 'alive' only after running through the water for a certain distance, so that the arming vane rotates a certain number of times.
c. *Electr.* Charged with electric current; connected to a source of electricity. Cf. LIVE *a.* 4.
1898 E. J. HOUSTON *Dict. Electr. Words* (ed. 4) 674/2 *'Alive'*, (1) a name sometimes given to a live wire or circuit. (2) an active wire or circuit. 1903 [see *LIVEN *v.*]. 1958 *Which?* I. II. 6/1 If the casing becomes alive and you touch it while in contact with the bath, you may get electrocuted.
6. (Earlier U.S. example.)
1789 MORSE *Amer. Geogr.* 205 The markets are alive with them [*sc.* fish].

aliveness (ăləi·vnĕs). [f. ALIVE *a.* + -NESS.] The quality of being alive or full of vigour. Also, sensitiveness *to* (one's surroundings, etc.).
*a*1866 J. GROTE *Moral Ideals* (1876) 322 This would be the real fountain of youth, supposing..there were energy to maintain the perpetual aliveness. 1866 MRS. CARLYLE in T. W. Reid *Life Ld. Houghton* (1891) II. xv. 146 A woman ..going about in society alive and well. To be sure, both the aliveness and the wellness 'may be strongly doubted', as they say in Edinburgh. 1892 *Independent* (U.S.) 7 Oct. 35/1 Aliveness, if we may use such a phrase, to the spirit of the age. 1921 J. WOOD in L. S. Hunter *John Hunter* ii. 18 He had all Dawson's aliveness to the world in which he lived.

alizarin. Add: Also **alizarine.** Add to def.: Also *attrib.*, designating dyestuffs derived from, or similar in action to, alizarin; the colours of these dyes.
1876 *Jrnl. Chem. Soc.* II. 234 The contamination of the printed cotton with iron is thus prevented, and only the pure alumina lake, that is to say, the pure alizarin-red, remains upon the cotton. 1885 C. T. DAVIS *Manuf. Leather* xliii. 735 Alizarine red (a pale flesh color) is produced by rubbing the cleansed and trodden skins with a solution of alizarine or extract of madder in weak soda lye and rinsing in water. 1901 *Kynoch Jrnl.* II. 36/2 The use of picric acid as a dye has been superseded by the use of alizarine and naphtol yellows. 1954 *Archit. Rev.* CXV. 286/2 It makes a particularly good textured background for pictures, and is manufactured in six colours: natural, light corn, tobacco, strawberry, alizarin green and egyptian green. 1962 *Listener* 11 Jan. 76/1 A fortress of almost vertical rocks.. burnt umber and alizarin red.

alkalæmia (æ:lkălī·miă). *Path.* [mod.L., f. ALKALI + Gr. αἷμα blood: see -IA¹.] A condition of increased alkalinity resulting from a disturbance of the acid-base equilibrium of the blood.
1922 *Lancet* 9 Sept. 560/2 (*title*) Periodic Alkalæmia with Alkalosis in the Adult. *Ibid.* 561/2 When the value of the ratio is decreased the blood reaction becomes more alkaline than normal—that is, a condition of *alkalæmia* results. 1924 *Q. Jrnl. Med.* 405 (*title*) Disturbance of the Acid-Base Equilibrium of the Blood to the Alkaline Side: Alkalæmia. *Ibid.*, In this paper only examples of alkalæmia due to alkali excess are presented, but the alkalæmia due to carbon dioxide deficit also has clinical importance. 1961 *Brit. Med. Dict.* 65/2 *Alkalæmia* a state in which there is an increase in the hydrogen-ion concentration in the blood without changes in the bicarbonate content of the blood. 1965 G. H. BELL et al. *Textbk. Physiol. & Biochem.* (ed. 6) xxxiii. 667 Raised pH, that is alkalæmia, or lowered pH, acidæmia.

alkali. Add: **3. b.** Native alkali (i.e. various salts) existing in excess in the soil of certain areas in the Western States; hence, a region abounding in alkali. *U.S.*
1848 E. BRYANT *California* viii. 124, I found the liquid bitter with salt and alkali. 1869 S. BOWLES *Our New West* xiv. 278 In looking out..on the starry heavens,..one almost forgets alkali. 1872 'MARK TWAIN' *Sketches* 119 That awful five days' journey, through alkali, sagebrush, peril of body. 1904 S. E. WHITE *Blazed Trail Stories* x. 181 They had ridden solitary over the limitless alkali of the Arizona plains.
attrib. 1848 W. CLAYTON *Latter-Day 'Saints' Emigr. Guide* i, Emigrants have lost many of their teams in the neighbourhood of the *Alkali lakes.* 1869 S. BOWLES *Our New West* xiv. 275 The alkali dust, dry with a season's sun, ..was thick and constant. *Ibid.* 276 Bare alkali plains stretch out..for miles. 1871 SCHELE DE VERE *Americanisms* 177 The Alkali Flats are now crossed by the Pacific Railroad. 1878 J. H. BEADLE *Western Wilds* xxiv. 388, I am convinced there is little to see but..alkali flats and sandhills. 1886 *Boston* (Mass.) *Herald* 16 July, It is only an 'alkali sink'—a natural well, filled with a paste as yielding as water.
7. **alkali-cellulose**, a compound of cellulose and an alkali.
1901 CROSS & BEVAN *Res. on Cellulose* 31 Hydrolytic changes should occur in the cellulose molecule when kept for prolonged periods as alkali-cellulose.

alkalimetric, *a.* (Earlier examples.)
1844 *Mech. Mag.* XL. 335/2 Instead of using 'beads for preparing the alkalimetric and acidimetric test liquors.. hydrometers may..be employed. 1856 W. A. MILLER *Elem. Chem.* II. xi. 739 The 35 divisions of alkalimetric acid used in the experiment.

alkaline, *a.* Add: **3.** Of soils or areas: charged or permeated with alkali. *U.S.*
1850 L. SAWYER *Way Sketches* (1926) 109 Great care should be taken to avoid the alkaline waters found along the route and the animals should never be picketed out upon the low alkaline bottoms. 1869 S. BOWLES *Our New West* xiv. 277 It would seem as if these alkaline valleys of the Great Interior Basin were too cold. 1870 *Amer. Naturalist* IV. 29 A desert section proper and one more particularly pertaining to the alkaline flats.

alkalosis (ælkălōᵘ·sis). *Path.* [mod.L., f. ALKAL(I + -OSIS.] A condition of excessive alkalinity in the body-tissues and blood. Cf. *ACIDOSIS.
1912 *Chem. Abstr.* 642 The meat intoxication is due to a condition of alkalosis, the alkalosis probably being brought about by 'NH₃ salts'. 1930 *Q. Jrnl. Med.* XXIII. 393 'Acidosis' is herein defined as a reduction in the bicarbonate content of the blood plasma..and 'alkalosis' as an increase above normal in this substance. 1965 J. R. ROBINSON *Acid-Base Regulation* (ed. 2) xiii. 31 It therefore seems preferable to restrict the terms 'acidosis' and 'alkalosis' to conditions in which the total concentration of buffer base is less than or greater than the normal.

alkane (æ·lkēⁱn). *Chem.* [f. *ALK(YL + -ANE 2 b.] Any member of the paraffin series of hydrocarbons (see PARAFFIN *sb.* 3).
1899 E. F. SMITH tr. *von Richter's Org. Chem.* (ed. 3) I. i. 79 Saturated or limit hydrocarbons, paraffins, alkanes. 1949 F. S. & F. B. KIPPING *Perkin & Kipping's Org. Chem.* I. iv. 63 The names of all the hydrocarbons of this series (also called the *alkanes*) end in *ane*.

alkapton (ælkæ·ptɒn). *Med.* Also **alcapton.** [a. G. (C. Boedeker 1861, in *Ann. d. Chemie u. Pharm.* CXVII. 101), f. AL(KALI + Gr. κάπτον, neut. pres. pple. of κάπτειν to swallow greedily.] A reducing substance which causes urine, left standing, to turn dark through oxidation. Also *attrib.* Hence **alkaptonu·ria** [-URIA], the disease in which this substance is present; **alkaptonu·ric** *a.* and *sb.*
1888 *Jrnl. Chem. Soc.* LIV. 1121 Alcaptonuria... A certain kind of human urine darkens on the addition of alkalis. Bödeker..isolated from such urine a substance to which he gave the name 'alcapton'. 1899 GARROD in *Jrnl. Physiol.* XXIII. 512 The urine of an alkaptonuric individual. *Ibid.*, Alkapton urines. *Ibid.*, Some cases of alkaptonuria. 1905 —— & HELE *Ibid.* XXXIII. 198 An adult male alkaptonuric.

alkargen. (Earlier example.)
1843 J. S. MILL *Logic* III. ix. § 1. I. 482 The substance called alkargen, discovered by Bunsen,..has not the slightest injurious action upon the organism.

alkene (æ·lkīn). *Chem.* [f. *ALK(YL + -ENE.] Any member of the olefine series of hydrocarbons.
1899 [see OLEFINE]. 1944 L. F. & M. FIESER *Org. Chem.* iii. 53 The Geneva system provides the convenient generic term alkene, which designates the hydrocarbons of the series comprising ethylene and its homologues. 1964 N. G. CLARK *Mod. Org. Chem.* vi. 87 Thus the olefins are systematically known as the 'alkenes'.

alkine (æ·lkəin). *Chem.* [ad. G. *alkin* (A. Ladenburg 1881, in *Ber. d. Deut. Chem. Ges.* XIV. 2126) f. ALC(OHOL + -INE⁵.] †**1.**

ALKYD

Ladenburg's name for a tertiary base containing an alcoholic group; also called *alkamine*. *Obs.*

1882 *Jrnl. Chem. Soc.* XLII. 165 *Alkines*...*Piperpropylalkine*..dissolves in water with evolution of heat. **1886** E. F. SMITH tr. *von Richter's Org. Chem.* 264 Oxy-alkyl bases may be obtained from the allyl amines...The bases obtained from the secondary amines are *alkamines* or *alkines*. **1904** *Jrnl. Chem. Soc.* LXXXVI. I. 685 Alkines combine with the haloids of organic alkyloxy- or acetoxy-acids...The products closely resemble the difficultly prepared, unsubstituted alkine esters in their physiological action.

2. Also **a·lkyne** (now the usual form). Any member of the acetylene or ethyne series of hydrocarbons, with the general formula C_nH_{2n-2}, containing a triple bond.

1909 in *Cent. Dict.* Suppl. **1938** *Jrnl. Org. Chem.* II. 1 In 1913 Lebeau and Picon reported that 1-alkynes were produced in good yield by the action of certain alkyl iodides on sodium acetylide in liquid ammonia. **1944** L. F. & M. FIESER *Org. Chem.* iv. 75 The properties of some of the members of the alkyne series.

alkyd (æ·lkid). [f. *ALKY(L + ACI)D.] In full *alkyd resin*. Any of a group of synthetic resins (see quot. 1929). Also *attrib.* and *Comb.*

1929 *Chem. Abstr.* 2308 Alkyd resins include all complexes resulting primarily from the interaction of a polyhydric alc[ohol] and polybasic acid. **1943** E. A. BEVAN in R. S. Morrell et al. *Synthetic Resins & Allied Plastics* (ed. 2) xi. 281 The alkyd resins, sometimes known as 'glyptals', are formed by the condensation of polybasic acids with polyhydric alcohols. **1943** *Jrnl. R. Aeronaut. Soc.* XIV. Abstr. 480 Alkyd type finishes are relatively new. **1958** *Woman* 22 Feb. 14/2 Choose an alkyd-based enamel.

alkyl (æ·lkil). *Chem.* [f. G. *alk*(*ohol* ALCOHOL + -YL.] In full *alkyl radical*. Any of the series of radicals derived from paraffin hydrocarbons by the removal of a hydrogen atom, as methyl (CH_3), ethyl (C_2H_5), etc. So *attrib.* or as *adj.*, derived from, or related to, the paraffin series of hydrocarbons. Formerly also called *alphyl*. Cf. *ARYL.

1879 *Jrnl. Chem. Soc.* XXXVI. 522 (*title*) Boiling points of the ethereal salts of hydroxy- and alkyloxy-acids. **1882** *Ibid.* XLII. 831 (*title*) Synthesis of so-called alkyl disulphoxides. **1883** *Ibid.* XLIV. 919 Action of the Alkyl-derivatives of the Halogen-substituted Fatty Acids on Aniline. *Ibid.* 276 The alkyl salts of the $C_nH_{2n}O_3$ series. **1889** M'GOWAN tr. *Bernthsen's Org. Chem.* 37 The halogen alkyls, $C_nH_{2n+1}X$. **1906** M'GOWAN & SUDBOROUGH *Bernthsen's Org. Chem.* 22 The monovalent residues, C_nH_{2n+1} (methyl, ethyl, etc.), which form the radicals of the monovalent alcohols, $C_nH_{2n+1}OH$, are frequently termed *alkyls*, or *alphyls*, while the divalent residues, C_nH_{2n}, are known as *alkylenes*. **1953** *Sci. News* XXX. 91 One of the detergents..is an alkyl sulphate, with an average chain-length of about eight to ten carbon atoms. **1957** *Technology* July 168/4 Detergents usually contain a surface active agent, alkylbenzenesulphonate, which influences surface adsorption and, incidentally, causes the familiar foaming (of doubtful utility). **1964** N. G. CLARK *Mod. Org. Chem.* viii. 125 These compounds may also be regarded as alkyl-substituted water molecules.

Hence **a·lkylate** *v. trans.*, to introduce an alkyl radical into (a compound); so **a·lkylated** *ppl. a.*, **a·lkylating** *ppl. a.* and *vbl. sb.*, **alkyla·tion**; **a·lkylate** *sb.*, a product of alkylation; **a·lkylene**, an olefine of the ethylene series.

1889 M'GOWAN tr. *Bernthsen's Org. Chem.* 46 Olefines or Hydrocarbons of the Ethylene Series (Alkylenes): C_nH_{2n}. *Ibid.* 272 Alkylated ureas. **1900** *Jrnl. Chem. Soc.* LXXVIII. I. 619 Dimethyl Sulphate as an Alkylating Agent. *Ibid.* LXXVII. I. 729 Alkylation by means of Dry Silver Oxide and Alkyl Halides. **1903** *Amer. Chem. Jrnl.* Apr. 293 The acylhalogenalkyl(alphyl)amines, by treatment with zinc ethyl, are simply reduced and not alkylated. **1904** F. P. FOSTER *Appleton's Med. Dict.* 86/1 *Alkylate*, a compound derived from a monatomic alcohol by the substitution of a metal for the hydroxylic hydrogen. **1958** *New Scientist* 8 May 23/1 There are alkylation processes at Stanley in Cheshire, and at Shellhaven, the latter for making those alkylates used in detergents. **1961** *Lancet* 29 July 250/1 Chemotherapy with various alkylating agents. **1964** N. G. CLARK *Mod. Org. Chem.* vi. 98 In the presence of strong acids (concentrated sulphuric acid or hydrofluoric acid) or aluminium chloride, $AlCl_3$, olefins react with tertiary paraffins, whereby the alkyl group corresponding to the olefin is attached to the paraffin; the paraffin is said to undergo 'alkylation'.

alkyne, see *ALKINE 2.

all. Add: **A.** *adj.* **6.** (Additional examples.) Also, as much as, altogether, quite; *for all of* (cf. FOR *prep.* 26 b) *U.S.*, as far as concerns (a person or thing).

1903 T. B. ALDRICH *Ponkapog Papers* (1904) 169 All of Herrick's geese were swans. **1932** J. H. FREDERICK *Devel. Amer. Commerce* vii. 77 There were insufficient American vessels to handle all of the traffic.

1829 R. C. SANDS *Writings* (1834) II. 57 Miss Cross, who was all of six feet high. **1854** M. J. HOLMES *Tempest & Sunshine* xix. 124 No;—he may have her and go to the old boy for all of Josh. **1883** 'MARK TWAIN' *Life on Miss.* xviii. 220 It must have been all of fifteen minutes..of dull, homesick silence. **1911** H. S. HARRISON *Queed* xvii. 209 Thus they parted, almost precipitately, and, for all of him, might never have met again in this world. **1944** 'G. GRAHAM' *Earth & High Heaven* I. iii. 65 He was all of thirty-three, solitary and unsure of himself.

7. b. In scoring at games, denoting that both sides have made the stated score; e.g. *love all* = neither side has scored.

1742 HOYLE *Short Treatise Whist* xi. 25 Suppose the Game to be Nine all. **1837** T. BACON *First Impr. fr. Nat. in Hindostan* I. 252 In the fourth 20, they were even; 67 all. **1878** *Laws of Lawn-Tennis* 12 If both players win five games, the score is called games-all. **1883** *Nat. Hist. Jrnl. & School Reporter* VII. 155 Up to within ten minutes when time was called the score was one goal all. **1898** Love all [see LOVE *sb.* 10 b]. **1965** *Listener* 17 June 915/2 Game all. Dealer West.

8. b. all but (also with hyphen), used *adject.*, almost complete or entire; in ellipt. use: almost; also as *sb.*

1810 M. L. WEEMS *Let.* in *Wks.* (1929) III. 14 Doctor Fendall..sold 60 doll[ar]s all but. **1816** TUCKEY *Narr. Exped. R. Zaire* (1818) i. 18 Negro washerwomen, whose state of *all but* nudity. **1831** [in Dict.]. **1881** GILBERT *Patience* II, *Col.* (apologetically). I'm afraid we're not quite right. *Ang.* Not supremely, perhaps, but, oh, so all-but! Oh, Saphir, are they not quite too all-but? **1914** 'I. HAY' *Knight on Wheels* xix. 192 Most of them are Impossibles, but there are a good many All-Buts. **1920** 'W. S. PALMER' *Christianity & Christ* 43 Until the great all-but-men brought forth true men. **1935** W. EMPSON *Poems* 3 Our all-but freedom.

f. In phr. after an oath or obscenity, as *damn all*, nothing at all. Cf. *BUGGER *sb.* 2 e.

1922 JOYCE *Ulysses* 417 Proud possessor of damnall. **1926** D. L. SAYERS *Clouds of Witness* x. 187, I'll tell you my story as shortly as I can, and you'll see I know damn all about it. **1930** BROPHY & PARTRIDGE *Songs & Slang* 114 *Damn all*, nothing. Used to give emphasis...A bowdlerization of a foul expression. **1939** J. MULGAN *Man Alone* iii. 31 There's —— all grows up here.

9. b. at all: used without a negative and affirmatively (later examples). *Irish*, dial., and *U.S. local*.

1894 CAINE *Manxman* II. 56 Is the woman mad at all? **1895** J. BARLOW *Strangers at Lisconnel* 262 And what at all have you got there? **1916, 1932** in H. WENTWORTH *Amer. Dial. Dict.* (1944) 34/2. **1945** *Amer. Speech* XX. 15 This affirmative use, although it has disappeared from speech at the formal level, lives on in Irish dialect and in colloquial speech in certain parts of America, especially after a superlative, as in the sentence 'We had the best time at all'.

C. *adv.* **2. c.** In idiomatic hyperbolic use, esp. in phr. *to go all* ——.

1932 S. GIBBONS *Cold Comf. Farm* v. 71 She will..go all arty-and-crafty. **1934** A. HUXLEY *Beyond Mexique Bay* 254 We will assume that the Indians have gone all hygienic. What are the results? **1935** *Punch* 16 Oct. 448/2 If Mr. Morley does go all Rider Haggard for a few pages towards the end of the book the circumstance is scarcely to be deplored. **1947** M. GILBERT *Close Quarters* iii. 50, 'I have questioned Canon Beech-Thompson,' replied Pollock shortly. 'And got an imperial raspberry?' said Halliday... 'I suppose he went all Crockford at once.'

5. b. all one (later example).

1930 C. WILLIAMS *War in Heaven* iii. 35, I gather it's all one to you whether we take it or leave it?

c. all for [FOR *prep.* 7 a], entirely in favour of, on the side of. *colloq.*

1864 A. TROLLOPE *Can you forgive Her?* xix. 150, I am all for Mr. Cheesacre. I can't abide anything that's poor. **1934** D. L. SAYERS *Nine Tailors* II. ii. 105, I'm all for a treasure-hunt. **1945** J. B. PRIESTLEY *Three Men in New Suits* iii. 38 'You've got to..see that the country's run properly.' 'I'm all for that,' said Herbert.

7. (Additional examples.)

1921 JOYCE *Lett.* (1957) 160 The last word (human all-too-human) is left to Penelope. **1932** D. GASCOYNE *Roman Balcony* 75 On farms, on all-too-barren fields. **1952** AUDEN *Nones* 23 Far from tall Restituta's all-too-watchful eye.

E. all- in *comb.* The sections that follow contain additional illustrative examples (esp. of the 20th century) for combinations with *all-* as first element.

3. *all-destroyer* (example), *-knower*.

1890 W. JAMES *Princ. Psychol.* I. x. 362 These declarations on Kant's part of the utter barrenness of the consciousness of the pure self, and of the consequent impossibility of any deductive or 'rational' psychology, are what, more than anything else, earned for him the title of the 'all-destroyer'. **1902** —— *Var. Relig. Exper.* vi. 138 Only by an All-knower can we finally be judged. **1909** —— *Pluralistic Universe* iii. 126 Again, the absolute is always represented idealistically, as the all-knower.

5. In *attrib.* phrases: made wholly of (a substance), as *all-aluminium*, *-metal*, *-steel*, *-wool* (earlier example; also *fig.*); containing or consisting of something exclusively, as *all-fire*, *-freight* (hence *all-freighter*), *-grass*, *-male*, *-star* [STAR *sb.*[1] 5]; *all-sea*, cf. *all-rail*. (Cf. also *IV below.)

1868 *Chambers's Encycl.* X. 268/1 Many of the names used in the all-wool class are retained in this, with the addition of the word 'union', as union merino. **1876** G. M. HOPKINS *Wr. Deutschland* (1918) st. 23 To bathe in his fall-gold mercies, to breathe in his all-fire glances. **1889** *Daily Even. Bulletin* (San Francisco) 13 July 1/6 Manager Harris' team is not now so much of an 'all star aggregation' now as it was. **1908** A. W. TILBY *Eng. People Overseas* I. ii. 32 It now became evident that the all-sea route was really the safest. **1912** MRS. P. CAMPBELL *Let. to G. B. Shaw* (1952) 25 An 'all star' show is fit only for Kings and Queens! **1912** W. RALEIGH *Let.* 12 Jun. (1926) II. 379 He..uses only all-wool five-star Romany when he orders a drink. **1927** W. E. COLLINSON *Contemp. Engl.* 89 All-steel houses. **1927** *Melody Maker* Aug. 785 (Advt.), One of the greatest 'all-star' records ever issued. **1930** *Engineering* 7 Feb. 159/3 A new ambulance aeroplane has recently been delivered...It is an all-metal machine. **1933** *Ann. Reg. 1932* II. 47 One of the most expensive of the year's [cinema] productions was 'Grand Hotel', which had a showy all-star cast. **1933** *Discovery* Oct. 325/2 An all-aluminium train was exhibited for the first time at the Chicago World's Fair this year. **1941** G. STAPLEDON in H. J. Massingham *Engl. & Farmer* vii. 148 To the holders of these all-grass farms I would address myself. **1960** *Times* 14 Nov. 13/6 British companies that fly regular long haul all-freight flights. **1961** *Ann. Reg. 1960* 429 This play had an all-male cast. **1961** *Guardian* 24 Jan. 14/1 As an all-freighter in airline use it is unique.

6. *all-beauteous* (examples), *-fair* (example), *-important* (later example), *-powerful* (earlier and later examples), *-surfy*. Hence as *sb.*, *all-fair*, *-wise*. Also with the implication 'entirely, exclusively', as *all-coloured*, *-red* (sense *13 below), *-white*.

1692 E. SCARISBRICK *Life Lady Warner* (ed. 2) v. i. 294 Hereby to encrease her confidence in his All-Powerful assistance. **1785** T. DWIGHT *Conq. Canäan* II. 29 Flowers all-beauteous. **1847** EMERSON *Poems* 15 This monument of my despair Build I to the All-Good, All-Fair. **1851** BUCKLEY tr. *Homer's Il.* 60 She beheld the all-beauteous neck of the goddess. *Ibid.* 91 His step-mother, all-fair Eëribæa. **a1889** G. M. HOPKINS *Poems* (1918) 71 Thundering all-surfy seas. **1899** *Daily News* 28 Oct. 7/6 An all-white framework of symmetrical petals. **1929** D. H. LAWRENCE *Pansies* 30 The All-wise has tired us of wisdom. **1937** *Burlington Mag.* Nov. 211/2 The meaning of certain all-important words. **1939** *Ibid.* Sept. 124/1 Rogier's all-powerful personality. **1940** 'G. ORWELL' *Inside Whale* 47 Something resembling the colonial attitude..exists..in all-white communities. **1958** *Times* 13 Aug. 5/4 Whose last musical, also with an all-coloured cast, was *Carmen Jones*. **1960** *Guardian* 21 Mar. 1/4 Negroes arrested after refusing to leave two all-white public libraries, were each released on bail.

¶ *all-inclusive*, *-pervasive* (example).

1885 W. JAMES *Coll. Ess. & Rev.* (1920) 282 Dr. Royce's novel reason for believing that all that is has the foundations of its being laid in an infinite all-inclusive Mind. **1910** —— in *Hibbert Jrnl.* VIII. 758 There is no complete generalisation,..no all-pervasive unity. **1944** H. A. HODGES *W. Dilthey* iv. 54 There is no all-inclusive purpose which could constitute the meaning of life. **1965** *Engl. Studies* Feb. 2 His suspicion is all-inclusive.

b. (Cf. *IV below.) Representative of the whole, as distinguished from part, of an area or country, as *all-American* (hence as *sb.*), *-Russian*.

1888 *Outing* (U.S.) Nov. 166/2 The All-American team.. is composed of men picked from the ranks of the representative ball teams of America. **1920** *Ibid.* Nov. 84/3 The little cripple sometime All-American. **1920** B. RUSSELL *Pract. & Theory Bolshevism* v. 76 The All-Russian Soviet, which is constitutionally the supreme body. **1959** *Times Lit. Suppl.* 14 Aug. 469/4 All-American footballer.

7. *advb.* with pr. pple. *all-absorbing*, *-accepting*, *-containing*, *-destroying* (example), *-embracing* (examples), *-encompassing*, *-enveloping*, *-pervading* (earlier example), *-providing* (example), *-satisfying* (example), *-seeing* (earlier example).

1610 G. FLETCHER *Christs Victorie* xxiv, He heau'ns al-seeing eye, He earths great Prophet. **1807** J. BARLOW *Columbiad* x. 483 Her cloudless ken, her all pervading soul Illume, sublime and harmonize the whole. **1834** J. S. MILL in *Monthly Repos.* VIII. 387 The universal and all-absorbing struggle to be or to appear rich. **1847** J. D. MORELL *Hist. View Speculative Philos.* (ed. 2) I. 15 Researches..so wide and all-embracing in their extent. **1848** A. BRONTË *Tenant Wildf. Hall* II. xviii. 365, I cannot live here, and be for ever silent on the all-absorbing subject of my thoughts and wishes. **1848** J. R. LOWELL *Oak* 42 in *Poet. Wks.* (1873) 76/2 Lord! all thy works are lessons,—each contains Some emblem of man's all-containing soul. **1866** G. M. HOPKINS *Poems* (1948) 147 My all-accepting fixed eye. **1880** W. JAMES *Will to Believe* (1897) 237 The vague Asiatic profession of belief in an all-enveloping fate. **1892** W. B. YEATS *C'tess Kathleen* Pref. 8 No dramatic method elastic and all-containing enough. **1902** W. JAMES *Var. Relig. Exper.* vi. 142 The jealousy of the gods, the nemesis that follows too much happiness, the all-encompassing death. **1907** —— *Meaning of Truth* (1909) vi. 157 We get nearer and nearer to realities, we approximate more and more to the all-satisfying limit. **1912** W. DE LA MARE *Listeners* 82 This all-encompassing hush. **1930** 'R. CROMPTON' *William's Happy Days* vi. 129 Lounging at his ease—the gracious, generous all-providing male. **1932** W. B. YEATS *Words for Music* 14 All-destroying sword-blade, still Carried by the wandering fool. **1935** *Burlington Mag.* Nov. 235/1 Art, which, like human nature, should be all-embracing in its scope. **1939** R. CAMPBELL *Flowering Rifle* II. 66 The former all-accepting soar above To triumph over death and die for love. **1940** L. MACNEICE *Last Ditch* 9 Night came down upon the bogland With all-enveloping wings.

8. *advb.* with pa. pple. *all-accomplished* (example), *-armed* (later example), *-seen*, *-starred*.

1756 TOLDERVY *Hist. Two Orphans* IV. xxxiii. 184 Where the modest, and all-accomplished landlady, Mrs. Rogers, was..invited to sit down. **1842** TENNYSON *Sir Galahad* in *Poems* II. 178 All-arm'd I ride, whate'er betide Until I find the holy Grail. **1865** G. M. HOPKINS *Poems* (1948) 142 And I must have the centre in my heart To spread the compass on the all-starr'd sky. **1917** D. H. LAWRENCE *Look! We have come Through!* 101 There, sure in sinless being, All-seen, and then all-seeing..We might have lain.

10. *sb.* **a.** *all-alikeness*, *-aloneness*, *-powerfulness* (later example), *-whiteness*; **b.** *all-inclusiveness*, *-sufficiency* (earlier example); **c.** *all-absorbingness*, *-pervadingness* (earlier example).

a. **a 1897** W. JAMES in R. B. Perry *Thought & Char. W. James* (1935) I. 482 Evolution is a change from a

nobowish untalkaboutable all-alikeness to a somehowish.. not-all-alikeness. **1921** D. H. LAWRENCE *Sea & Sardinia* v. 164 Save us from proletarian homogeneity and khaki all-alikeness. **1924** *Glasgow Herald* 17 July 9 London's vastness..[is] yet capable of baffling him and impressing him with a feeling of strangeness and 'all-alone-ness'. **1927** E. GRUNDY *Happy Pianist* vi. 32 C major [scale] is technically the most difficult, on account of its 'all-whiteness'. **1956** D. GASCOYNE *Night Thoughts* 18 Cave-night which every night His all-aloneness drives him back into. **1957** *New Biol.* XXII. 57 These disturbing cases, which test to the limit our faith in the all-powerfulness of Natural Selection.

b. **1619** J. SEMPILL *Sacrilege Sacredly Handled* vii. 79 Hee vseth Tithing, as a chiefe argument to confirme them in the Al-sufficiencie and Eternitie of Christs Priest-hood. **1882** W. JAMES *Let.* 8 Jan. in R. B. Perry *Thought & Char. W. James* (1935) I. 738 What are they but limitations to the all-inclusiveness of any single being? **1938** R. H. LOWIE *Hist. Ethnol. Theory* ix. 131 We are struck by the all-inclusiveness of his interests.

c. **1849** J. S. MILL *Lett.* (1910) I. 146 The limitation may be in our faculties, and in the allpervadingness, to us, of a contrary experience. **1945** *Scrutiny* XIII. 131 The all-absorbingness of the purpose.

12. *adj.* in *-ed*. *all-aged* (earlier and later examples), *-coloured* (later example), *-sided* (hence *all-sidedness*).

1876 *Coursing Calendar* 51 The Craven Cup, for all-aged bitches, contained some already known to fame. **1880** RUSKIN *Bible of Amiens* i. 3 All-coloured velvets, pearl-iridescent colombettes. **1881** W. JAMES in *Unitarian Rev.* XVI. 407 May you still be the champions of mental completeness and all-sidedness. **1905** *Terms used in Forestry* 15 *Many-aged*, a forest through all parts of which many different classes of trees tend to distribute themselves. When all age classes are thus distributed, the forest is all-aged. **1933** *Times* 2 June 15/6 We are sick of the weak-kneed, allsidedness of people who see everything and do nothing.

III. 13. Special Combs.: **all-ages**, horses of all ages entered for a race; **all-along** *adv. Bookbinding* (see quot. 1960); **all-court** [COURT *sb.*[1] 4] *a.*, in *Lawn Tennis* applied to a game played both from the base-line and from near the net; **all-electric** *a.*, using only electrical power; **all-or-none**, (*a*) used esp. *attrib.* of the nature of a response of a muscle or nerve fibre, etc., to a stimulus (see quots.); (*b*) = *ALL-OR-NOTHING 2; **all-red** *a.*, used to indicate a telegraph-line, a trading route, etc., lying throughout in British territory (from the practice of colouring British and Imperial territory red in our maps); **all risks**, applied *attrib.* to a type of comprehensive insurance; **all-sliming** *vbl. sb.* and *ppl. a.* (*Mining*), applied to the process of reducing all the ore to slime (see SLIME *sb.* 4) in the extraction of gold; so *all-slimed* adj.; **all-sorts** (cf. SORT *sb.*[2] 12 b), as *sb.* (see quots.) and *attrib.*; *spec.* of confectionery (in full *liquorice all-sorts*), a mixture of black-and-white or brightly coloured sweets containing liquorice; **all-through** *a.*, of a journey: performed without a change of conveyance; **all-time** *a.*, (*a*) taking one's full time; (*b*) of a record level or figure, etc., for all time up to the present (orig. *U.S.*); **all-up (weight)**, the total (weight) of the machine, crew, passengers and cargo of an aircraft when in the air; **all-welded** *a.*, having all the joins welded (opp. *riveted*, etc.); **all-wing** *a.*, of aircraft (see quot. 1956).

1876 *Coursing Calendar* 323 The all-ages were not a good lot, but they contained one or two who had already distinguished themselves this season. **1880** J. W. ZAEHNSDORF *Art Bookbinding* 167 When a volume is sewed, and the thread passes from kettle-stitch to kettle-stitch, or from end to end in each sheet, it is said to be sewed 'all-along'. **1960** G. A. GLAISTER *Gloss. of Book* 5/2 *All along*, the method of hand sewing the sections of a book on cords or tapes when the thread goes from kettle-stitch to kettle-stitch inside the fold of each section. Also used to describe a machine-sewn book in which each section has the full number of stitches. **1927** *Observer* 5 June 21 His main weakness is a preference for the base-line rather than the all-court game. **1920** *Electr. Rev.* 27 Feb. 274/2 On Friday, last week..a Press visit to the 'All-Electric House' exhibited.. at the Ideal Home Exhibition, took place. **1924** S. R. ROGET *Dict. Electr. Terms* 4/2 *All-electric Signalling*, automatic or other railway signalling in which the signals are actuated as well as controlled electrically. **1960** AUDEN *Homage to Clio* 25 This all-electric room Where ghosts would feel uneasy. **1900** J. BURDON-SANDERSON in Sharpey-Schafer *Text-bk. Physiol.* II. 449 The characteristic relation of response to stimulus known as the 'all or none' principle, according to which the vigour of the response evoked by a stimulus adequate to produce an effect, is independent of the strength of the stimulus. **1912** E. H. STARLING *Princ. Human Physiol.* v. 230 The rule of 'all or none'..was first enunciated for heart muscle. **1920** W. H. R. RIVERS *Instinct & Unconscious* vi. 45 There is [in these reactions] an absence of graduation...They tend to occur in their full strength. This form of reaction is known in physiology as the 'all-or-none' reaction. **1928** J. T. MACCURDY *Comm. Princ. Psychol. & Physiol.* 168 Adrian..found that, in the isolated fibre, the strength of the impulse, once it was set up, remained constant and could not be increased by augmenting the strength of the stimulus. This is known as the all-or-none law of the nerve impulse. **1932** J. S. HUXLEY *Probl. Rel. Growth* VI. iv. 178 The 'all-or-none' law of the action of the sex-hormones on comb-growth. **1934** *Discovery* June 165/1 The mechanism of repression is an all-or-none mechanism. **1940** AUDEN *Another Time* 70 Are her fond responses All-or-none reactions? **1895** G. HUTCHISON in *N.Z. Hansard* XCI. 854 Arguments..in favour of the 'All Red' line. **1923** *Daily Mail* 12 Mar. 9 A huge 'All-Red' aeroplane, spanning the Empire in such a series of great, long-distance night-and-day flights as shall obviate any necessity for alighting on foreign soil. **1911** *Law Times Rep.* CV. 413/2 The insurance is to be a complete insurance against all risks..an insurance which would be known as an all risks insurance. *Ibid.* 412/2 Lloyd's 'all risks' policy was inadequate. **1962** D. TENCH *Law for Consumers* vi. 74 It is a mistake to imagine that a comprehensive policy covers the consumer against every conceivable happening... Even an 'All Risks' policy does not cover you against all risks. **1934** *Brit. S. Afr. Ann. 1934–35* 103/2 Two or three-stage grinding..reduces all the ore to an all-slimed product. **1920** A. H. FAY *Gloss. Mining & Mineral Industry* 26/2 *All-sliming*, crushing all the ore in a mill to so fine a state of subdivision that only a small percentage will fail to pass through a 200-mesh screen. **1933** *Brit. S. Afr. Ann. 1933–34* 97/2 The extended 'all-sliming' process plant came into full commission in November, 1929. **1955** *Times* 20 May 19/1 The reduction plant has been converted to operate on an all-sliming basis, and it is expected that this modification will result in small improvements in reduction costs and gold residue values. **1823** 'JON BEE' *Slang* 3 *All sorts*, or All Nations, spirits compounded of all the drippers in a cellar, and the pewter save-all on a gin-shop counter. **1931** W. HOLTBY *Poor Caroline* iii. 76 Pink sugared hazel nuts, and Liquorice All Sorts. **1937** W. DE LA MARE *This Year, Next Year*, An all-sorts shop. **1961** *20th Cent.* Feb. 99 The architectural all-sorts of the actor's metropolitan Mecca. **1912** MRS. HUGH FRASER *Further Remin.* I. 10 After an all-through journey from Rome, our train crawled into the Gare de Lyons. **1930** H. G. WELLS *Autocr. Mr. Parham* I. i. 4 An all-time job with a garrulous advertisement contractor. **1933** *Sat. Even. Post* 10 June 61 Brings cost of power to new all-time low. **1939** 'N. BLAKE' *Smiler with Knife* ii. 29 An all-time high in adenoidal growth. **1959** N. MARSH *Singing in Shrouds* vi. 110 An all-time-low in inhibitions and an all-time-high in what it takes. **1959** *Observer* 29 Mar. 3/3 The market in industrial equities had soared by the end of 1958 to an all-time peak. **1933** *Jrnl. R. Aeronaut. Soc.* XXXVII. 774 We have been able to build a flying boat with an all-up weight of 56 tons. **1958** *Economist* 25 Oct. 349/3 New York airport, for one, has too short a runway to permit it to take off at full all-up weight. **1936** *Ibid.* 18 Jan. 150/2 The first all-welded spiral guide-frames gasholder to be built in this country is now in course of erection at Newhaven. **1931** *Flight* 25 Sept. 970a/1 The ideal all-wing type of airplane would incorporate only wing formation, utilising all power output for lifting purposes. **1956** W. A. HEFLIN *U.S. Air Force Dict.* 42 *All-wing*, of an airplane: Designed and built as a single wing, without conventional fuselage or empennage.

IV. In modern (esp. 20th-century) use, *all* with a sb. forms an attrib. phr.: **a.** with names of countries, states, etc., denoting that the whole is concerned; **b.** with words denoting a period of time: lasting or functioning throughout that period, as *all-day*, *-night* (hence *all-nighter*: see NIGHTER); **c.** (with the sb. either in sing. or pl.) denoting that all the persons or things expressed by the sb. are concerned or included. (Cf. also senses E. 5, 6 b above.)

a. **1853** 'C. BEDE' *Verdant Green* I. xi. 102 Verdant's score was always on the *lucus a non lucendo* principle of derivation..and he felt that he should never rival a Mynn or be a Parr with any one of the 'All England' players. **1878** (*title*) Laws of Lawn-Tennis, as adopted by The Marylebone Cricket Club, and The All England Croquet and Lawn-Tennis Club. **1904** *Independent* (U.S.) 27 Oct. 951/1 The selection of 'All-America' teams seems to have become a mania. **1906** *Dublin Rev.* Oct. 332 We are forced onwards to the necessity of an all-world State if we would escape an all-world anarchy. **1928** *Manch. Guardian Weekly* 28 Sept. 242/3 The all-India leaders. **1937** *Ann. Reg. 1936* II. 133 The All-Africa Convention. **1952** C. P. BLACKER *Eugenics* viii. 160 In 1926..there took place..the first All-Union [i.e. in U.S.S.R.] census of population.

b. **1861** [see ALL-NIGHT-MAN]. **a1870** in *Dict. Amer. Eng.*, An all-day horse. **1888** A. C. GUNTER *Mr. Potter of Texas* xx, The..Press Club..just beginning to assume its usual all-night gayety and brilliancy. **1895** [see NIGHTER]. **1901** *Dialect Notes* II. 134 *All-day sucker*, a piece of candy on small stick to be disposed of by sucking for some time. **1906** 'O. HENRY' *Four Million* 215 An all-night lunch Counter. **1930** B. MAINE *Rondo* I. ii. 38 There are some who profess to find excitement in encountering..all-nighters on the Embankment. **1939** J. STEINBECK *Grapes of Wrath* xxvi. 330 Ever' place you look is restaurants...Lookit that all-nighter there. **1955** A. HUXLEY *Genius & Goddess* 32 You know those enormous lollipops on sticks that children lick at all day long? Well, that's what her fear was—an all-day sucker. **1959** E. K. WENLOCK *Kitchin's Road Transport Law* 8/2 A cheap 'all-day' ticket. **1967** *Melody Maker* 28 Jan. 11 'It's an all-nighter,' he revealed. 'And I want you to play three 45 minute spots.'

c. **1909** *Westm. Gaz.* 8 Nov. 7/1 The National All-Grades Movement of the Amalgamated Society of Railway Servants. **1928** *Times* 3 July 16/6 The new all-purpose Fairey Napier aeroplane. **1928** *Manch. Guardian Weekly* 23 Nov. 401/3 The all-parties Constitution. *Ibid.* 26 Oct. 328/2 Six-cylinder Thornycroft all-weather body cars. **1929** *Punch's Almanack 1930* 4 Nov. p. xxxv, The All-Mains set draws its power from the house electrical supply. **1930** *Times Educ. Suppl.* 8 Mar. 107/4 With the present system of parochial 'all-age' schools it had not been possible to provide for the..needs of the pupils. **1933** *Jrnl. R. Aeronaut. Soc.* XXXVII. 2 In Europe our aerodromes, being surfaced with excellent turf, are available for landing in every direction. In other words, they are 'all way' fields and the envy of the Americans. **1937** *Discovery* Feb. 44/2 An alternative [television] set incorporating an 8-valve all-wave sound receiver. **1939** D. CECIL *Young Melbourne* v. 114 A petition was brought forward in the House of Lords asking for an all-party Government. **1955** W. W. DENLINGER *Complete Boston* I. 159 The dog..was many times Best Dog in Show at all-breed shows. **1955** *Times* 8 July 5/1 The Gloster Javelin delta winged all-weather fighter, which is soon to go into service with the R.A.F. **1960** *Farmer & Stockbreeder* 1 Mar. 142/1 An all-purpose mounted implement capable of drastically reducing the time and labour spent in preparing seed beds.

allactite (ælæ·ktəit). *Min.* [ad. Sw. *allaktit* (A. Sjögren 1884, in *Geol. Fören. Förhandl.* VII. 109), f. Gr. stem ἀλλακτ-, ἀλλάσσειν to change: see -ITE[1].] A native arsenate of manganese.

1892 DANA *Syst. Min.* (ed. 6) 800 *Allactite*...Monoclinic. ..Numerous vicinal planes in the prismatic zone. **1920** *Brit. Museum Return* 145 Allactite from Sweden.

allamanda, Alla- (ælămæ·ndă). Also erron. **alamanda, alamander.** [mod.L. (Linnæus 1771), f. the name of the Swiss scientist J. N. S. *Allamand*.] A plant of the genus so named of apocynaceous trees of tropical America, cultivated for their large funnel-shaped yellow flowers.

1796 CURTIS in *Bot. Mag.* X. pl. 338 Willow-leaved Allamanda. **1879** W. CORY *Lett. & Jrnls.* 7 Oct. (1897) 449 An alamanda with big petals of a very lovely yellow. **1918** C. W. BEEBE *Jungle Peace* iv. 83 The sheets of bougainvillea blossoms, of yellow allamandas..brought joy to the eye and the nostril. **1957** P. MANSFIELD *Final Exposure* xiv. 213 The wind had been..beating the branches of the alamander against the wall and plucking the pale flowers from the stems.

Comb. **1918** W. H. HUDSON *Far Away* xxii. 294 The large alamanda-like flower of a purest divine yellow.

‖ **alla prima** (a·lla prī·mă, prəi·-), *advb. phr.* passing into *sb.* [It., lit. 'at once'.] The direct method of painting without underpainting.

1849 M. P. MERRIFIELD *Orig. Treatises Arts of Painting* I. vi. p. cxxxiv, Paolo Veronese painted generally *alla prima* with more body than Titian..so that the finished picture was little more than the abbozzo; that is, he painted up his picture at once. **1961** M. LEVY *Studio Dict. Art Terms* 12 *Alla prima*, direct painting. The completion of a painting by a single application of pigments, in contrast to a painting which is completed in stages by the application of successive layers of pigments.

‖ **allargando** (alaɹga·ndo). *Mus.* Pl. **allargandi.** [It., pres. pple. of *allargare* to broaden.] A musical direction indicating that the time is to be made slower and slower and the tone fuller; also a passage of music in this style.

1893 J. S. SHEDLOCK tr. *Riemann's Dict. Mus.* 19/1 *Allargando*, becoming broader (slower); specially used in place of *ritardando* (*rallentando*), where the tone is to be increased. **1959** D. COOKE *Lang. Mus.* v. 256 The first three notes..are accented and held back (*allargando*): there is the maximum of rhetorical emphasis here. **1961** *Times* 28 Jan. 3/7 A tendency to break the flow of the music by introducing *ritardandi*, *allargandi*.

all-around, *a. U.S.* = ALL-ROUND *a.*

1883 [see AROUND *adv.* 4]. **1886** *Harper's Mag.* June 18/2 This arrangement gives a clear sweep..enabling the forward gun to cover an all-around fire. **1904** *Brooklyn Daily Standard Union* 7 July 6 Our army in the Philippines seems to be composed of 'all-around' men. **1904** *Forum* Oct. 257 The most comprehensive and in many ways the best all-around American city school exhibit. **1920** *Harvey's Weekly* 5 June 4/1 Senator Knox is the best-equipped all-around international statesman.

Hence **all-arou·ndness.** *U.S.*

1888 *Voice* (N.Y.) 6 Sept., The all-aroundness of our chieftain's character. **1929** D. H. LAWRENCE *Paintings* sig. E 1ᵛ, But intuition needs all-aroundness, and instinct needs insideness.

allative (æ·lătiv), *a. Gram.* [f. L. *allāt-* ppl. stem of *afferre* to bring to + -IVE.] Denoting the case used in some languages, e.g. Finnish, to express motion to or toward.

1860 *Trans. Philol. Soc. 1857* 34 The declension of the personal pronouns [in the Tushi Language] is as follows.. Singular..Affective..Allative. **1924** O. JESPERSEN *Philos. Gram.* xiii. 182 Finnish has no dative proper, but the 'allative' which expresses motion to or into the neighbourhood of, often corresponds to the Aryan dative. **1928** *Mod. Lang. Rev.* Apr. 137 Thus, he might say, in Finnish the indirect object is in the allative case. **1939** L. H. GRAY *Found. Lang.* vii. 194 The Eskimo *allative*, denoting that toward which an action tends. **1951** *Archivum Linguisticum* III. 57 An allative-dative type of case suffix.

all clear, *phr.*, also used as *sb.* A signal giving information that there is no danger; *spec.* the signal that hostile aircraft have left the neighbourhood ('raiders passed'). Also *attrib.* and *transf.*

1902 J. CONRAD *Youth* 9 There was a moment of confusion, yelling, and running about. Steam roared. Then somebody was heard saying, 'All clear, sir.' **1917** *Times* 18 Oct. 6/3 It has been decided that the 'All clear' signal after air raids shall in future be given by bugle calls. **1922** WODEHOUSE *Girl on Boat* xvii. 282 Webster had promised to come and knock an all-clear signal on the door. **1923**

—— *Inimit. Jeeves* v. 55 The effect she had on me whenever she appeared was to make me want to slide into a cellar and lie low till they blew the All-Clear. **1936** *Economist* 15 Feb. 347/2 The 'all-clear' for armament expansion on a great scale is being given. **1939** T. S. ELIOT *Old Possum's Practical Cats* 40 The [railway] signal goes 'All Clear!' **1939** *War Weekly* 3 Nov. 41/2 The alarm was at 2.30 and the 'all clear' half an hour later. **1942** *Hutchinson's Pict. Hist. War*, 18 Mar.–9 June 9 (*caption*) After the 'All Clear'... Malta after a 'raiders passed' signal had been given, showing the people about to resume their daily occupations. **1945** E. WAUGH *Brideshead Rev.* I. iii. 70, I felt a sense of liberation and peace such as I was to know years later when after a night of unrest, the syrens sounded the 'All Clear'.

allegiant, *a.* Add: *sb.* One who owes or renders allegiance; a subject.

1886 *N. Amer. Rev.* CXLII. 125 Strangers shall have the same personal rights as the allegiants. **1888** TOURGÉE *Lett. to King* v. 67 In the one sense it [*sc.* the word 'citizen'] is used to distinguish the American people from the allegiants of a foreign power.

allegoric, *a.* For quot. 1859 read:

1827 DE QUINCEY in *Blackw. Mag.* XXI. 14 [as in Dict.].

allegorism (æ·ligŏriz'm). Delete †*Obs.* and for def. read: The use of allegory; the allegorical method of interpreting Scripture. (Cf. ALLEGORIST.)

In quot. 1567 in Dict., which should be attributed to Harding, not Jewel, 'allegorisme' is an error for 'algorisme'.

1889 *Sat. Rev.* 18 May 613/1 [His views] rest upon what the early Fathers called Allegorism—that is, on the spiritual, not the literal, interpretation of our Lord's words. Nor is Allegorism 'Gnostic'. **1901** *Jewish Encycl.* I. *Allegorical interpretation* [of Scripture]. Expositors of this system may be called allegorists; the system itself, allegorism. **1919** P. H. OSMOND *Myst. Poets Engl. Ch.* 350 No doubt there is a rather slippery descent from this type of mysticism, through symbolism, to mere allegorism.

allegorization. For quot. 1859 read:

1827 DE QUINCEY in *Blackw. Mag.* XXI. 14 [as in Dict.].

allegretto. (Earlier and later examples.)

1740 in J. GRASSINEAU *Mus. Dict.* 3. **a 1817** JANE AUSTEN *Lesley Castle* in *Love & Freindship* (1922) 70, I had..constantly hollowed whenever she played, *Bravo, Bravissimo*, ..*allegretto*,..and *Poco presto* with many other such outlandish words, all of them..expressive of my admiration. **1833** [see ANDANTINO *adj.*]. **1877** G. B. SHAW *How to become Mus. Critic* (1960) 13 The symphony..was very fairly played, especially the *allegretto*. **1966** *Listener* 3 Mar. 329/2 A supremely concentrated account of the allegretto.

allegro, *a.*, *adv.*, and *sb.* Add: **B, C.** *Mus.* (Earlier and later examples.)

1683 PURCELL *Sonnata's III Parts* Pref., *Allegro*, and *vivace*, a very brisk, swift, or fast movement. **1752** [see PRESTO *a.*[1]]. **1754** LD. CHESTERFIELD in *World* 14 Nov. 588 His adagio's, his allegro's..and his jiggs. **1876** G. B. SHAW *How to become Mus. Critic* (1960) 10 The first portion is an *allegro agitato*. **1947** A. EINSTEIN *Mus. Romantic Era* xi. 125 With no other composer than Mendelssohn does there appear so frequently in the Allegro movements the indication *con fuoco* or *appassionato*. **1963** AUDEN *Dyer's Hand* 487 The tempo indication is *allegro assai*...The duet.. is sung *allegro assai*.

B. 2. *Philol.* Applied to a word or phrase conventionally shortened in actual speech.

1939 L. H. GRAY *Found. Lang.* iii. 65 Only words of very common use have these double forms (*lento-* and *allegro-*forms). **1964** R. H. ROBINS *Gen. Linguistics* viii. 347 The reduction or abbreviation of the forms of frequently used words ('allegro forms' like *howd'yedo*, and *yes'm* for *yes madam*).

allele (æ·līl). *Biol.* Also **allel.** [ad. G. *allel*, abbrev. of next.] = *ALLELOMORPH. Hence **alle·lic** *a.*; **alle·lism** = *ALLELOMORPHISM.

1931 J. A. JENKINS in *Univ. Calif. Publ. Agric. Sci.* VI. 390 The most probable assumption is that..there are many different combinations of alleles. **1934** *Genetics* July 310 (*title*) Alleles of the Mallard Plumage Pattern. **1935** *Science* 12 July 37/2 Allelic..allelism. **1939** *Nature* 11 Nov. (Suppl.) 822/1 D. Ferriman found that oxycephaly and acrocephalosyndactyly are dominant conditions, probably allelomorphic, and that a third allele may be acholuric jaundice. **1940** HINSIE & SHATZKY *Psychiatric Dict.* 21/1 Each gene is an allele of, or is allelic to, the other...Brown and blue eyes in man are alleles of each other, as is red and white color in snapdragons. **1953** J. S. HUXLEY *Evol. in Action* i. 26 Each particular kind of gene can exist in a number of slightly different forms or 'allels', each allel exerting a slightly different effect in development. **1956** C. H. WADDINGTON in A. Pryce-Jones *New Outl. Mod. Knowl.* 232 Geneticists speak of the whole family of genes.. as occupying a 'locus'. The different members of the family are the 'alleles' of the locus.

allelomorph (æli·lomǭɹf). *Biol.* [f. Gr. ἀλληλ- one another + -o + μορφή form.] One of several alternative forms of the same gene, occupying the same relative positions in homologous chromosomes. Hence **alle:lomo·rphic** *a.*, pertaining to or of the nature of allelomorphs; **alle:lomo·rphism**, the existence, transmission, or inheritance of allelomorphs.

1902 BATESON & SAUNDERS in *Rep. Evol. Com. R. Soc.* I. 159 [If] two similar gametes meet, their offspring will be no more likely to show the other allelomorph than if no cross had ever taken place. *Ibid.* 127 The strictly allelomorphic or Mendelian distribution of characters among the gametes. *Ibid.* 142 It does not appear as yet that simple allelomorphism occurs between any two colours, of which neither is xanthic or albino. **1906** R. H. LOCK *Variation, Hered. & Evol.* vii. 184 The dominant and recessive allelomorph seem to represent respectively the presence and absence of something. *Ibid.* x. 265 Allelomorphic characters. **1914** T. H. MORGAN in *Amer. Naturalist* XLVIII. 502 (*title*) The theoretical distinction between multiple allelomorphs and close linkage. *Ibid.*, Dexter pointed out that the mode of treatment that Nabours followed..is the procedure of multiple allelomorphism. **1949** DARLINGTON & MATHER *Elements Genetics* v. 114 In addition to such simple allelomorphs, however, particular loci in scores of species show whole series of alternatives, multiple allelomorphs as they are called. **1957** E. B. FORD *Butterflies* ix. 188 Three, four, or more forms of the same gene have on occasion been encountered, so giving rise to 'multiple allelomorphs'.

allemande, *sb.* Add: **3.** *Cookery.* **allemande sauce** [F. *sauce allemande*, also used], a rich velouté sauce thickened with egg yolks. Also *ellipt.*

[**1827** 'M. DODS' *Cook & Housewife's Man.* (ed. 2) 329 *Sauce à l'Allemand, or German Sauce*..is extensively used for dressed meat-dishes.] **1863** MRS. BEETON *Bk. Househ. Managem.* x. 244 Sauce Allemande, or German Sauce. **1868** M. JEWRY *Warne's Model Cookery & Housek. Bk.* 54/1 *Allemande*, reduced white velouté sauce, thickened with cream and yolks of eggs, seasoned with nutmeg and lemon-juice. **1951** *Good Housek. Home Encycl.* 332/1 *Allemande Sauce*, a rich version of velouté sauce.

allemande, *v.* [f. the sb.] **1.** *trans.* To move (a person) as in an allemande.

1821 GALT *Annals of Parish* xxxviii. 308 He..allemanded her along in a manner that should not have been seen.. out of a king's court.

2. *intr.* To dance an allemande. Also *fig.*

1830 *Fraser's Mag.* I. 298/1 With what grace might he *allemand* in a quadrille. **1835** *Nautical Mag.* July 416 A girl in the same set, who was tying her shoe-string when she ought to have been allemanding. **1890** BARING-GOULD *Arminell* xlviii, The love-making below stairs is..full of restraints and shynesses, of setting to partners, and allemanding about them.

allergic (ǣlə·ɹdʒik), *a.* [f. next + -IC.] Pertaining to or characterized by allergy.

1911 C. E. VON PIRQUET in *Arch. Internal Med.* VII. 265 The allergic condition does not occur immediately after the first injection. *Ibid.* 428 The practical method of diagnosis by the allergic reactions will be extended to other diseases. **1919** *Jrnl. Amer. Med. Assoc.* LXXIII. 759/2 Allergic action of drugs. **1925** W. W. DUKE *Allergy, Asthma* 246 A large proportion of patients with asthma, 'allergic coryza', erythema,..and even allergic shock, are actually sensitive specifically and solely to the action of a physical agent, such as light, heat, cold, or mechanical irritation. **1926** SOLLMANN *Pharmacol.* (ed. 3) 86 Allergic hypersusceptibility is a special type of idiosyncrasy in which the patient reacts to special substances. **1958** *Immunology* I. 22 Allergic reactions have occurred in human patients after treatment with ACTH.

b. *fig.* Sensitive (esp. antipathetic) *to*. *colloq.*

1937 *New Yorker* 23 Jan. 20 (*caption*) He's allergic to Philadelphia scrapple. **1942** *Punch* 26 Aug. 169/3 Colonels have a curious effect on me. Quite frankly, I am allergic to them. **1958** *Observer* 5 Oct. 19/4 Allergic as I am to opera on television..I cannot be relied on for an objective opinion of the B.B.C.'s two-hour presentation.

allergy (æ·ləɹdʒi). *Path.* [ad. G. *allergie* (von Pirquet 1906, in *Münchener Mediz. Wochenschr.* LIII. 1457/2), f. Gr. ἄλλος other, different + ἐργεια (known only in comp. ἐνέργεια. see ENERGY), used for 'reactivity': see -Y[3].] The altered degree of susceptibility of a body produced by a sensitizing dosage of or exposure to some foreign material; more widely, hypersensitivity to the action of some particular foreign material, as certain foods, pollens, micro-organisms, etc.

1911 C. E. VON PIRQUET in *Arch. Internal Med.* VII. 260 We might rightly use the word 'allergy'..as a clinical conception. **1919** *Jrnl. Amer. Med. Assoc.* LXXIII. 759/2 Coca's classification of hypersensitiveness into 'anaphylaxis' and 'allergy' will serve to clarify this whole field...Anaphylaxis is an antigen antibody reaction, artificially induced by immunologic processes. Allergy is used to express the natural hypersensitiveness of the individual not produced by immunologic processes. **1925** W. W. DUKE (*title*) Allergy, Asthma, Hay Fever, Urticaria and Allied Manifestations of Reaction. *Ibid.* 20 Wells and others contend that 'allergy' is a broad useful term, for under its heading can be described a group of peculiar phenomena of altered reactivity the underlying cause of which is as yet unproved. **1959** *Which?* Sept. 116/2 The person who is specially sensitive to some article of food—for instance, shellfish—may make himself very ill after only a mouthful. This is *true* allergy...True allergy or specific sensitivity to synthetic detergents is rare.

b. *fig.* A feeling of antipathy to some thing or person.

1944 AUDEN *For Time Being* (1945) 118 Before the Diet of Sugar he was using razor blades And excited soon after with an allergy to maidenheads. **1951** KOESTLER *Age of Longing* 186 The European public..had developed a violent allergy to all kinds of rationing, saving and public-spirited exhortation.

Hence **a·llergen** [after *ANTIGEN], a substance producing allergy; **allergenic** (ælə·ɹdʒe·nik) *a.* Also **a·llergist**, one who specializes in allergic phenomena.

1912 *Jrnl. Pharm. & Exp. Therap.* III. 232 The same amount of allergen given in lesser concentration elicits a more intense reaction. **1913** DORLAND *Med. Dict.* (ed. 7), Allergen..allergenic. **1923** *Westm. Gaz.* 16 Feb. 8/6 The majority of cases of spasmodic asthma were due to some allergen. **1934** WEBSTER, Allergist. **1942** in F. R. Moulton *Aerobiology* 7 Relatively little is known about air-borne bacteria as possible allergens. **1945** *Lancet* 22 Dec. 822/1 (*heading*) Alarums of an Allergist. **1957** *New Scientist* 7 Nov. 39/1 These agents [*sc.* antibodies] are called allergens when they produce allergic reactions like hay fever.

alleviate, *v.* Add: **2.** Also *absol.*

1888 MRS. H. WARD *R. Elsmere* xli, The constant effort to serve and to alleviate.

alleviative, *a.* (Examples.)

1823 *Examiner* 24 Aug. 552/3 Any measure alleviative of distress. **1931** *Birmingham Post* 13 May 7/5 He was admitted to hospital for an alleviative operation.

alley. Add: **3. b.** A back-lane running parallel with a main street. *U.S.*

1729 in *Baltimore Town Rec.* (1905) 10 The commissioners.. shall cause the same Sixty Acres to be..divided into convenient Streets, Lanes, and Allies, as near as may be into Sixty equal lots. **1747** *Ibid.* 22 To Survey the Same and lay it out into Lotts with convenient Streets and Alleys. **1817** S. R. BROWN *Western Gaz.* 90 There are three streets,.. besides lanes and alleys. *Ibid.* 101 Each block of lots has the advantage of two 16 feet alleys. **1835** J. MARTIN *Gazetteer Virginia* 139 Fire plugs are connected with the distributing pipes at every intersection of the alleys with 2nd and 3rd streets. **1890** J. A. RIIS *How other Half Lives* (1891) 21 A notorious Fourth Ward alley.

c. *to be up a person's alley*: to be up a person's street (see *STREET *sb.*). *slang.*

1931 M. E. GILMAN *Sob Sister* v. 65 It's about time a good murder broke, and this one is right up your alley. **1936** D. CARNEGIE *How to win Friends* (1938) IV. viii. 247 Bridge will be in a cinch for you. It is right up your alley. **1941** AUDEN *New Year Let.* II. p. 37 All vague idealistic art That coddles the uneasy heart, Is up his alley. **1954** R. BISSELL *High Water* (1955) iii. 32 Right up your alley with Donald Duck and Mickey Mouse and all them other uplifting characters you are always studying up on. **1965** *New Statesman* 9 Apr. 583/2 Its slogans and chirpy recommendations are right up her spiritual alley.

10. alley cat (chiefly *U.S.*), a cat that frequents alleys, a stray cat. Also (*U.S. slang*) *transf.* (see esp. quot. 1942).

1904 *Atlantic Monthly* Mar. 369/1 If you were just an alley-cat you wouldn't even get the chloroform. **1914** E. POUND in *Blast* I. 50 For her laughter frightens even the street hawker And the alley cat dies of a migraine. **1916** DON MARQUIS in *Evening Sun* (N.Y.) 1 Sept. 10/4 Ours is the zest of the alley cat. **1926** S. LEWIS *Mantrap* iv. 44 Thinking up a way of insulting that mangy alley cat! **1941** *Time* 16 June 85/1 [list of words not to be used in film scripts] Alley cat (applied to a woman). **1942** BERREY & VAN DEN BARK *Amer. Thes. Slang* §418.2 *Ragamuffin*..alley cat. *Ibid.* §439.2 *Slut*, alley cat. *Ibid.* §507.2 *Prostitute*, alley cat. **1946** M. DICKENS *Happy Prisoner* viii. 149 They're as quarrelsome as a couple of alley cats.

alley-way (æ·liwēⁱ). orig. *U.S.* Also **alloway** (æ·lawēⁱ). [f. ALLEY + WAY *sb.*[1]] A narrow passage or avenue, e.g. between houses or between rows of cabins in a ship.

1788 P. M. FRENEAU *Misc. Wks.* 223 The article stipulated..that the alley-way should be sufficient for the passing and repassing of the plaintiff. **1854** *Harper's Mag.* IX. 849/2, I was taken to the Auburn state-prison. And as I walked along the concealed alley-ways,..I bethought me of my theft of fruit. **1888** *Murray's Mag.* III. 238 The covered way, or 'alloway', on the port side, which led from the quarter-deck to the fore part of the ship. **1891** C. ROBERTS *Adrift Amer.* 231 Alley-way between the after-deck house and the bulwarks. **1894** *Outing* (U.S.) XXIV. 286/2 The main street of this town was as dirty as the filthiest alley-way found in any large city of America. **1920** *Blackw. Mag.* Jan. 108/2 A local marshman as a guide, or you will soon find yourself lost in the maze of alleyways through the giant reeds. **1924** WODEHOUSE *Bill the Conqueror* xi. 193 Her decks and alleyways were crowded with voyagers and those who had come to see those voyagers off.

all-father. c. (Earlier and later examples.)

1826 KIRBY & SPENCE *Entomol.* III. xxxv. 694 The wisdom and skill of the Allfather. **1850** C. KINGSLEY *Alton Locke* II. xv. 225 Had we not an All-Father?..Did He not love us, too, even as we loved each other? **1945** AUDEN *Coll. Poetry* 69 Only the All-Father Can change the cast or give them Easier lines to say.

all-fired, *a.* Add: Hence as *adv.*

1837 *Yale Lit. Mag.* II. 149 Star's an all-fired good ox. **1861** HUGHES *Tom Brown at Oxford* III. vii. 130, I knows I be so all-fired jealous, I can't abear to hear o' her talkin', let alone writin' to ——. **1935** M. M. ATWATER *Murder in Midsummer* x. 96 Tell him to get all-fired busy on it.

all-firedly, *adv.* (Earlier and later examples.)

1833 A. GREENE *Life & Adv. Dr. Duckworth* II. 176 He was seldom downright drunk; but was often all-firedly sprung. **1924** H. DE SÉLINCOURT *Cricket Match* vi. 192 I'm most all-firedly sorry about it.

alligator[2]. Add: **3. alligator gar, snapper, terrapin, turtle** (see defs. s.v. GAR, TERRAPIN, etc.).

1821 in *Texas Hist. Q.* VII. 300 Found another Karanqua encamp at which was..Alligator heads and the skins of Alligator Gars. **1843** [see GAR *sb.*]. **1944** *Reader's Digest*

July 110 A huge alligator gar rips their nets to pieces, destroying the labor of weeks. **1884** GOODE *Nat. Hist. Aquat. Anim.* 153 An Alligator Snapper of perhaps forty pounds weight. **1835** W. G. SIMMS *Partisan* 317 Three enormous terrapins of that doubtful brood which the vulgar in the southern country describe as the alligator terrapin. **1888** Alligator terrapin [see SNAPPER *sb.*[1] 6 c]. **1842** J. E. DE KAY *Zool. N.Y.* III. 8 The Snapping Turtle..is one of our largest turtles...In other sections, it is known under the names of Logger-head, Alligator Turtle and Couta. **1885** J. S. KINGSLEY *Stand. Nat. Hist.* III. 452 The elongated tail of this animal..has..given rise to the popular name, 'alligator-turtle'.

b. Used *attrib.* of a device or apparatus, as clips, forceps, etc., resembling in form or action the jaws of an alligator (cf. sense 2 in Dict.).

a 1884 KNIGHT *Dict. Mech.* Suppl. 22/1 Alligator Forceps. **1940** *Chambers's Techn. Dict.* 23/1 *Alligator wrench*, a tool with fixed serrated jaws, used for twisting and screwing pipes into position. **1952** *Brit. Jrnl. Psychol.* XLIII. 86 Alligator clips with platinum electrodes soldered to the jaws were attached to the rat's ears.

a·lligatoring, *vbl. sb.* orig. *U.S.* The cracking and retraction of paint, varnish, etc., caused by contraction.

1911 *Engin. News* (N.Y.) 27 July 121 Many of the paints which lack any evidence of cracking, checking, or alligatoring. **1953** *Glossary Paint Terms* (*B.S.I.*) 11 *Crocodiling or alligatoring*, a drastic type of crazing producing a pattern resembling the hide of a crocodile.

all-in. Also **all in.** **1.** As predicative adj. **a.** Exhausted. *colloq.*

1903 A. D. MCFAUL *Ike Glidden* xxii. 201 The horse was holding steady up to his clip, but it could be easily seen that he was 'all in'. **1904** ADE *Breaking into Society* 53 'I'm all in,' said the Wreck. **1916** 'B. M. BOWER' *Phantom Herd* xiv. 244 You better find him right there in the blizzard—hurt maybe—anyway just about all in. **1952** M. LASKI *Village* iii. 56 You look all in...Been doing too much,' that's what it is.

b. Completely or wholeheartedly involved.

a 1910 'O. HENRY' *Strictly Business* (1917) xx. 221 'On the dead level,' said Cork, holding her close with one arm, 'when it comes to you, I'm all in.' **1925** W. DEEPING *Sorrell & Son* viii. 76, I am all in on this adventure. Either we touch port—or we founder. **1933** H. G. WELLS *Shape of Things to Come* III § 1. 263 It was becoming impossible to retain self-respect, to be happy within oneself, unless one was 'all in' upon that one sound objective.

2. As attrib. phr.: Inclusive of all; in *Wrestling*, without restrictions. Also *absol.*

1890 *Daily News* 17 Feb. 3/6 The all-in [Billiards] tournament at the Westminster Aquarium. **1900** N. GOULD *Settling Day* xvi. 135, 'I'll take that wager.' 'All in, run or not?' **1908** *Westm. Gaz.* 6 June 9/3 The 'all-in' system—that is to say, railway and hotel accommodation combined. **1913** J. E. G. HADATH *Schoolboy Grit* viii. 88 A fight is just a fight: Catch-as-catch-can, All-in, and Best-your-enemy-anyhow! **1924** F. W. THOMAS *Rain & Shine* 125 Our Double-Decked Combination Endowment and All-in Policy. **1927** *Daily Tel.* 7 Mar. 2 The exceptionally high all-in cost of 1s. 1½d. per lb. **1927** *Manch. Guardian Weekly* Oct. 329/2 A notable appeal for disarmament and all-in arbitration. **1929** *Times* 8 Jan. 14/1 The two-part tariffs comprise a domestic 'all-in' rate and a business 'all-in' rate. The charge per unit for electricity for..lighting, heating, cooking, and domestic and business appliances is 1d. per unit. **1934** E. J. HARRISON *Wrestling* ii. 16 Alike, therefore, in the Cumberland and Westmorland, Græco-Roman, Catch-as-Catch-Can, All-in, or even the Japanese *sumo* styles of the art, lightweights do not usually contend against heavyweights. *Ibid.* v. 68 Any aspirant for mat honours..will not seek to explore and master the mysteries of All-in until he has gained a good working knowledge of orthodox Catch-as-Catch-Can. **1944** 'G. ORWELL' in *Horizon* X. 242 To the extent that all-in wrestling is worse than boxing.

allo- (æ·lo), combining form of Gr. ἄλλος other, different, as in ALLOTHEISM, *ALLOCHIRIA, etc. Also in various technical and scientific subjects: **1.** *Chem.* Used to distinguish one of two organic isomers, as *allocinnamic*, *-maleic*, *-mucic* (acids).

1893 *Jrnl. Chem. Soc.* LXIV. I. 513 Allocinnamic acid. **1894** *Ibid.* LXVI. I. 173 Allofurfuracrylic Acid and the Formation of Allo-Acids. *Ibid.*, The allo-form of cinnamylacrylic acid. **1949** S. & L. M. MIALL *New Dict. Chem.* (ed. 2) 21/1 With sterols and related substances the prefix allo- should indicate that rings A and B are in the trans position to each other, but allocholesterol is an exception.

2. *Biol.* In comp. with *-ploid*, as *ALLOPOLYPLOID, *ALLOTETRAPLOID: having sets of chromosomes derived from different species by hybridization (opp. *AUTO- c).

3. In *Philology*: see *ALLOGRAPH[2], *ALLOMORPH[2], *ALLOPHONE.

allobarbitone (æ:lobā·ɹbitōᵘn). [f. *ALLO- + *BARBITONE.] A white crystalline compound, diallylbarbituric acid ($C_{10}H_{12}N_2O_3$), used as a sedative and hypnotic.

1934 *Lancet* 22 Dec. 1404/2 The claim is made..for allobarbitone B.P.C...that it alleviates pain. **1953** *Faber Med. Dict.* 18/2 *Allobarbitone*, a hypnotic similar to barbitone, more rapidly excreted and of less prolonged action.

allocation. Add: **1. d.** *spec.* The allotment of available materials, provisions, etc., by the government or other authority.

1940 *Economist* 24 Aug. 237/1 The Select Committee recommends that allocation—that is, the rationing of materials—should be centralised and applied to every material that is scarce enough to form a bottleneck in the process of production. **1951** E. A. G. ROBINSON in D. N. Chester *Lessons Brit. War Econ.* iii. 57 As the war developed the need for critical allocation of resources increased progressively. **1951** R. PARES *Ibid.* x. 154 The allocation and use of scarce resources.

e. A portion allocated in this way.

1940 *Times* 6 Feb. 5/3 Allocations will be based on the actual sales made by the butchers. **1947** J. HAYWARD *Prose Lit. since 1939* 19 Commercial publishers..had some cause to resent the intrusion of a rival with an unlimited allocation of paper. **1947** *Ann. Reg. 1946* 30 [The Minister of Food] has been successful in getting an allocation of 1,550,000 tons [of rice] for India.

allochiria (ælokəiᵊ·riă). *Path.* Also **allocheiria.** [mod. L., f. *ALLO- + Gr. χείρ hand: see -IA[1].] A condition in which a sensation is referred to a point other than the true one (esp. the corresponding point on the other side of the body).

1881 H. OBERSTEINER in *Brain* IV. 153 The patient is not clear..as to which side of the body has been touched. I would term the phenomenon 'Sensory Allochiria', or shortly 'Allochiria' or confusion of sides. **1907** E. JONES in *Brain* XXX. 525 Under the name allochiria two fundamentally different conditions have hitherto been confused. A patient's mistake in determining the side of the stimulus may be (i) part of a general defect in localisation..or (ii) a specific defect independent of any error in localisation. **1910** *Lancet* 1 Jan. 44/2 The difficult subject of allochiria. By this term neurologists have commonly understood the false reference of a point stimulated to a corresponding spot on the opposite side of the body. **1943** I. S. WECHSLER *Textbk. Clinical Neurol.* (ed. 5) 44 In certain cerebral lesions..a patient sometimes refers the sensation to a corresponding spot on the opposite limb (*allocheiria*).

So **allochi·ral, -chei·ral** *adj.* (*a*) relating to the other hand; *spec.* = HETEROCHIRAL *a.* (Funk 1893); so *allochi·rally* adv.; (*b*) pertaining to or exhibiting allochiria.

1889 W. THOMSON *Math. & Physical Papers* (1890) III. 410 Two men of exactly equal and similar external figures would be allochirally similar if one holds out his right hand and the other his left. **1900** DORLAND *Med. Dict.* 32/1 *Allochiral*, exhibiting reversed symmetry; having the relation of the right hand to the left. **1911** T. L. STEDMAN *Med. Dict.* 28/2 *Allochiral*, relating to or suffering from allochiria.

allochthon (ælǫ·kþŏn). *Geol.* [G., or back-formation f. next.] An allochthonous rock formation; opp. *AUTOCHTHON 4.

1942 M. P. BILLINGS *Struct. Geol.* x. 181 Rocks..have traveled many miles from their original place of deposition and are said to be *allochthonous*..these rocks are sometimes called the *allochthon*. **1957** *Amer. Geol. Inst., Gloss. Geol.* 7/2 *Allochthon*, rocks that have been moved a long distance from their original place of deposition by some tectonic process, generally related to overthrusting or recumbent folding, or perhaps gravity sliding. Used in contrast to autochthon. **1961** J. CHALLINOR *Dict. Geol.* 15/2 The 'allochthon' (allochthonous folds and thrust masses) is the far-travelled ground.

allochthonous (ælǫ·kþŏnəs), *a. Geol.* [f. G. *allochthon* (K. W. von Gümbel 1888, in *Grundzüge d. Geologie* I. II. v. 615), f. *ALLO- + χθών, χθονός earth, soil: see -OUS.] Applied to organic deposits and rock formations: consisting of or formed from transported material; not formed *in situ* (opp. *AUTOCHTHONOUS *a.* c).

1911 E. A. N. ARBER *Nat. Hist. Coal* v. 110 We now turn to the alternative hypothesis of the Drift theory or Allochthonous formation of coal as Gümbel termed it...The material, which formed the mother substance of coal seams, was, on this view, always drifted..and transported from a distance. **1916** C. C. FORSAITH in *Bot. Gaz.* LXII. 33 Allochthonous peat (that type of peat which has been deposited by a gradual accumulation of floated, drifted, and wind-blown vegetable material in permanent and more or less quiet bodies of open water). **1932** J. A. STEERS *Unstable Earth* iii. 114 Nappes characteristic of the true geosyncline which have often been moved far from their place of origin, and to which the term allochthonous is applied. **1939** W. H. TWENHOFEL *Princ. Sedimentation* v. 162 Before fossil plants may be used to reconstruct a physical environment it must first be determined whether the plant material is autochthonous, that is, whether it grew where it is found; or allochthonous, that is, whether it was transported to where it is found. **1942, 1961** [see prec.].

allo-erotism (æ:lo,e·rŏtiz'm). *Psychol.* Also **-ero·ticism.** [ad. G. *alloerotismus* (Freud 1899, *Let.* to W. Fliess 9 Dec. in *Aus d. Anfängen d. Psychoanalyse* (1950) 324). f. *ALLO- + EROTISM, EROTICISM.] Erotism aroused by another person (opp. *AUTO-EROTISM). So **allo-ero·tic** *a.* [cf. G. *alloerotisch* (Freud 1899).]

1921 J. C. FLÜGEL *Psycho-anal. Stud. of Family* ii. 14 (To use a convenient term of Ferenczi's) allo-erotic aspects of the mind. **1934** H. C. WARREN *Dict. Psychol.* 10/1 Alloeroticism, allo-erotism. **1936** *Brit. Jrnl. Psychol.* XXVI. 284 The contrast between object-love, or allo-erotic libido, and self-love, narcissistic libido. **1954** MOSBACHER & STRACHEY tr. *Freud's Let.* 9 *Dec. 1899 to W. Fliess* in *Origins of Psycho-Analysis* 304 The lowest of the sexual strata is auto-erotism...This is superseded by allo-erotism (homo- and hetero-). *Ibid.*, Hysteria..is allo-erotic; the main highway it follows is identification with the beloved person.

allogenic (ælodʒe·nik), *a.* [f. Gr. ἀλλογενή-ς (see ALLOGENEITY) + -IC.]

1. *Geol.* (See quot. 1888.)

1888 F. H. HATCH in J. J. H. Teall *Brit. Petrogr.* 423 *Allogenic* (*Allothigenic*), applied to such minerals as are of more ancient origin than the rocks which they compose. **1893** [see *AUTHIGENIC *a*]. **1937** WOOLDRIDGE & MORGAN *Physical Basis Geogr.* xix. 277 Many..of these streams are 'allogenic', deriving much of their water from terrains beyond the Chalk outcrops.

2. *Ecology.* Caused by external factors, opp. *AUTOGENIC *a.* (see also quot. 1962).

1931 W. B. MCDOUGALL *Plant Ecol.* (ed. 2) xix. 259 Successions that are brought about by the reactions of the plants themselves are sometimes called autogenic successions while those caused by climatic or physiographic factors are called allogenic successions. **1934** *Geogr. Jrnl.* LXXXIII. 499 The further stages in the succession may be brought about by the operation of external factors alone, and when this is so the succession is termed an allogenic succession. **1962** H. HANSON *Dict. Ecol.* 14 *Allogenic succession*, the kind of succession in which one kind of community replaces another because of a change in the environment which was not produced by the plants themselves, e.g., decrease in soil moisture by improved drainage.

allogenous, *a.*: see ALLOGENEOUS *a.* (Later examples.)

1920 *Glasgow Herald* 6 May 6 The Russia of the Soviets is still the old Muscovite tyranny which denies to allogenous races the right to live. **1950** *Jrnl. Theol. Stud.* I. 102 The body from Mary, with its *centrum vitae*, afforded to God the Word an allogenous manward *prosopon*.

allograph[2] (æ·lograf). *Philol.* [f. *ALLO- 3 + *GRAPH(EME.] **a.** A particular form of a letter of an alphabet. **b.** One of a number of letters or letter-combinations representing a phoneme; thus *f* (in *friend*) and *gh* (in *cough*) are two of the allographs representing the phoneme /f/. Hence **allogra·phic** *a.*

1951 R. A. HALL JUN. in *Archivum Linguisticum* III. 117 We now have *allomorph* and *morpheme*, *alloseme* and *sememe*, *allograph* and *grapheme*. **1956** A. MCINTOSH in *Trans. Philol. Soc.* 43 At other times it [*sc.* the word 'letter'] is used of the particular allographic form a grapheme may have in a given context, e.g. when we speak of 'the *s* used in final position in Greek'; again it may be used of a single instance of an allograph, as when we say 'that's a badly formed letter'. **1962** *Amer. Speech* XXXVII. 229 A proposal to describe graphemes in terms of phonemic model, i.e., grapheme: phoneme::graph: phone:: allograph: allophone. **1963** *Language* XXXIX. 234 We might imagine that an Arcadian, struck by the fact that the Laconians sometimes spelled with ⟨wo⟩ what was to him simply /o:/, introduced ⟨wo⟩ here, considering it merely an initial allograph of ⟨o⟩.

allometry (ǣlǫ·mētri). *Biol.* [f. *ALLO- + -METRY.] (See quot. 1936.) Hence **allome·tric** *a.*, pertaining to or exhibiting allometry.

1936 HUXLEY & TEISSIER in *Nature* 9 May 780/2 To denote growth of a part at a different rate from that of a body as a whole or of a standard, we propose the term *allometry*. *Ibid.*, *Positive* and *negative allometry* denote respectively growth-rates of the part above and below that of the standard. **1938** *Nature* 3 Sept. 437/2 Limbs must exhibit positive heterogony, or allometry..during fœtal development. **1940** G. R. DE BEER *Embryos & Ancestors* iv. 25 The relative growth-rates of the allometric organ and of the body remain constant during long periods. **1941, 1942, 1949** [see *ALLOMORPHOSIS 2]. **1959** *Chambers's Encycl.* I. 276/1 The huge claws of the adult lobster and the armament of the stag are thus evidently the outcome of a strongly positive allometric growth.

allomorph[1] (æ·lomǭɹf). *Min.* [f. *ALLO- + Gr. μορφή form.] A distinct crystalline form without change of chemical composition. Hence **allomo·rphic** *a.*[1], **allomo·rphism.**

1866 KING & ROWNEY in *Q. Jrnl. Geol. Soc.* XXII. I. 187 All these varieties are considered to belong to one axial system—the monoclinate; but being obviously of other forms than one, and as they are all chemically alike, we propose to call them *allomorphs*. Like aragonite, calcite is also subject to allomorphism. *Ibid.* 193 It is our present impression that flocculite is merely an allomorph of serpentine. *Ibid.*, There is an allomorphic variety of serpentine known as chrysolite.

allomorph[2] (æ·lomǭɹf). *Philol.* [f. *ALLO- 3 + *MORPH(EME.] A morphemic alternant; one of two or more morphs making up a morpheme. Hence **allomo·rphic** *a.*[2]

1948 E. A. NIDA in *Language* XXIV. 420 Morphemic alternants can conveniently be called allomorphs. Accordingly, allomorphs are related to morphemes as allophones are related to phonemes. **1949** —— *Morphology* (ed. 2) ii. 21 Allomorphic differences of tone may be described in terms of different tonal surroundings. **1951** TRAGER & SMITH *Outl. Eng. Struct.* II. 60 Suffixes, especially inflectional ones, usually have many allomorphs. **1953** C. E. BAZELL *Ling. Form* i. 8 Allomorphs are not to be regarded as segments, since this way of treating them gives rise to the impression that they are not minimal units. **1962** E. F. HADEN et al. *Resonance-Theory Linguistics* iii. 30 The forms..*showed*,..*handed*,..are..analyzable into stem + suffix:../ʃow/+/d/, /hænd/+/id/...Each of these endings signals the same grammatical meaning, namely, 'past', and they can all be classed together in the morpheme {-D} (or {past})...The several allomorphs of this morpheme ({M_1}) are statable as sets of phonemes, but more accurately as sets of allophones.

allomorphosis (ælomǭ·ɹfōsis, -mǫɹfōᵘ·sis). *Biol.* [f. *ALLO- + MORPHOSIS.] **1.** = METAMORPHOSIS (Mayne *Expos. Lex.* Add. 1860).

2. Evolutionary allometry: a term used for comparison of different phylogenetic groups (see quots.).

1941 J. S. HUXLEY et al. in *Nature* 23 Aug. 225/1 We wish to suggest that *allometry* be the covering term, and *allomorphosis* the term for phylogenetic comparison. This indicates that all these differences..concern differing rates of growth in the individual, and..different (morphological or chemical) patterns brought about..in the completed individual of different groups. *Ibid.*, *Allomorphosis*, the relation of parts of organisms at some definite age to wholes or parts also at some definite age, but of different groups (races, varieties, species, genera). **1942** J. S. HUXLEY *Evolution* ix. 529 For comparison of relative proportions in different types it will..be best to use *allomorphosis*..keeping allometry as a general and inclusive term. **1949** *Blakiston's New Gould Med. Dict.* 42/1 *Allomorphosis*, a type of allometry which deals with the relation of a part to the whole, or to another part, in a series of organisms..as the relation of jaw length to skull length of the adult in a series of breeds of dogs.

allopatric (ælopæ·trik), *a. Biol.* [f. *ALLO- + Gr. πάτρα fatherland: see -IC.] Applied to organisms that occupy different geographical areas. Opp. SYMPATRIC. Hence **allopa·try,** the occurrence of allopatric forms.

1942 E. MAYR *Systematics & Origin of Species* vii. 149 Two forms (or species) are *allopatric*, if they do not occur together, that is if they exclude each other geographically. **1953** —— et al. *Methods & Princ. Syst. Zool.* ii. 29 Closely related allopatric forms are usually subspecies of a polytypic species. *Ibid.* v. 101 The word *allopatric* is essentially an antonym of *sympatric* and means therefore geographical distribution without geographical overlap. There are five kinds of allopatry that may be encountered by the taxonomist. **1958** *Canadian Jrnl. Zool.* XXXVI. 148 Reproductive isolation is more likely to be present between closely related sympatric species than between closely related allopatric species. **1960** *Evolution* XIV. 82/1 The best estimate of the course of geographic speciation comes from analysis of disjunct, allopatric populations.

allophone (æ·lofōᵘn). *Philol.* [f. *ALLO- 3 + *PHONE(ME.] Any of the variants making up a single phoneme.

1938 B. L. WHORF *Lang., Thought, & Reality* (1956) 126 Allophones or positional variants. **1941** TRAGER & BLOCH in *Language* XVII. 223 Sound-types as members of a phonemic class are called allophones. **1949** C. E. BAZELL in *Archivum Linguisticum* I. 4 Two comparisons are familiar, the one between the allophones of a phoneme and the alternants of a morpheme, the other between the allophones and the semantic variants of the morphemic *Gesamtbedeutung*. **1957** S. POTTER *Mod. Linguistics* ii. 39 A phoneme is a closely coherent group or bundle of sounds consisting of one more frequent phone, or chief member, together with other related phones or allophones which take its place in particular contexts. **1964** R. H. ROBINS *Gen. Linguistics* iv. 131 Members of phonemes are often called *phones* or *allophones*, and it is common practice to write phone symbols between square brackets and phoneme symbols..between oblique brackets. Thus [t] and [tʰ] are allophones of the /t/ phoneme.

Hence **allopho·nic** *a.*, of or pertaining to an allophone; **allopho·nically** *adv.*

1938 B. L. WHORF *Lang., Thought, & Reality* (1956) 126 Allophonic constellation. **1953** L. F. BROSNAHAN *Some O.E. Sound Changes* 82 By themselves these inducing phones are unable to induce anything more than those allophonic modifications of the affected vowels which are present in normal speech. **1956** D. JONES *Outl. Eng. Phonetics* (ed. 8) x. 49 The words [*sc.* F. *loup, boucle*] may be transcribed allophonically..as *lu, bukḷ*, ḷ representing a voiceless *l*.

allopolyploidy (æ:lopǫ·liploidi). *Biol.* [ad. G. *allopolyploidie* (Kihara and Ono 1927, in *Zeitschr. für Zell. und Mikr. Anat.* IV. 480), f. *ALLO- 2 + *POLYPLOIDY.] The state or occurrence of a polyploid having its sets of chromosomes derived from different species by means of hybridization.

1928 DARLINGTON in *Jrnl. Genetics* XIX. 243 A useful distinction has been made amongst polyploids..according as their gametic complements are constituted by the reduplication of similar series (auto-polyploidy), that is, by the doubling of the chromosome number in a theoretically pure line, or by the combination of dissimilar series (allo-polyploidy), that is, by doubling in a hybrid. **1930** E. MÜNTZING in *Hereditas* XIII. 293 As to the nature of the chromosome increase the case may be one of auto-polyploidy (doubling of the same genome) or allo-polyploidy (addition of different genomes). **1952** *New Biol.* XIII. 46 Among the newer crops which have arisen by allopolyploidy in recent years, the most notable are the various berries related to blackberry and raspberry of the genus *Rubus*.

Hence **a:llopo·lyploid,** a polyploid of this type. Also as *adj.*

1928 DARLINGTON in *Jrnl. Genetics* XIX. 243 It is impossible to draw any but an arbitrary line across the innumerable gradations between a pure type, as pure as in-breeding can produce, and a hybrid;..it is thus impossible to distinguish in absolute terms between an allo-polyploid and an auto-polyploid. **1938** E. B. FORD *Study Heredity* iv. 95 The polyploids..belong to the group called autopolyploids, in which all the sets of chromosomes are derived from the same species. More striking..are the allopolyploids, which result from chromosome doubling after hybridization. **1953** *Rep. 13th Internat. Hort. Congress 1952* 66 Allopolyploid or amphidiploid groups [of plants]..should be treated as species.

all-or-nothing, all or nothing. 1. Applied *attrib.* to a piece of mechanism in a repeating watch (see quot. 1940); also used *absol.*

1765 *Ann. Reg.* 1764 79/1 The all or nothing piece [in a watch]. **1843** *Penny Cycl.* XXVII. 108/1 (*Watch, Repeating*) The whole..carried by the all-or-nothing piece *TR*. *Ibid.* 108/2 When the quarter-rack is brought back to its original position..the part *m* will have passed between the end *R* of the all-or-nothing, which in its passage *m* will have pressed outwards. **1940** *Chambers's Techn. Dict.* 22/2 *All-or-nothing piece*, a piece of the mechanism of a repeating watch which either allows the striking of the hours and quarters or entirely prevents it.

2. As phr., indicating that a principle, policy, etc., must be accepted without qualification or not at all; also *attrib.*

1862 EMERSON *Jrnl.* Jan. (1914) IX. 361 My estimate of my own mental means and resources is all or nothing; in happy hours, life..infinitely rich, and sterile at others. **1882** W. JAMES *Will to Believe* (1897) 292 The silly hegelian all-or-nothing insatiateness. **1891** W. WILSON tr. *Ibsen's Brand* v. 290 *The Figure.* All these fell and pallid spectres can be laid with *three words*. *Brand.* Speak them! *The Figure.* All or nothing! **1936** J. C. POWYS *Maiden Castle* (1937) vi. 222 They don't need your damned all-or-nothing truths! **1950** B. WOOTTON *Test. Social Sci.* iii. 49 Questions which (like nearly all the significant issues in life) cannot but be matters of more or less are constantly answered in terms of yes or no, all or nothing. *Ibid.* vi. 132 The peculiarity of religious systems is that this threat to security has a peculiar all-or-nothing character. **1957** R. N. CAREW HUNT *Guide to Communist Jargon* p. xiv, The 'all-or-nothing' nature of communist thinking, which leads to the rejection as 'opportunism' of whatever stops short of the absolute goal of revolution. **1960** in L. Pincus *Marriage* i. 21 An all-or-nothing response in which the infant feels itself and its world to be entirely good, or entirely bad.

3. = *all-or-none* (*a*): see *ALL III. 13.

1922 *Encycl. Brit.* XXXII. 104/2 In the case of the heart muscle, this fact has long been known and was given the name of the 'law of all-or-nothing'..This law has now been shown to hold for voluntary muscle and for nerve. **1932** *Times Lit. Suppl.* 11 Feb. 93/2 The 'all-or-nothing' law of nervous conduction. **1964** J. Z. YOUNG *Model of Brain* ii. 23 We now know that nerve fibres are able to propagate all-or-nothing signals with a high safety factor and without decrement.

allotetraploid (æ:lote·träploid), *a.* and *sb. Biol.* [f. *ALLO- 2 + *TETRAPLOID.] = *AMPHIDIPLOID *a.* and *sb.*

1930 W. J. C. LAWRENCE in *Genetica* XII. 272 Hybrid tetraploid (allo-tetraploid). *Ibid.* 277 The most probable origin of an allotetraploid would seem to be hybridisation of two self-incompatible diploid species. **1938** E. B. FORD *Study Heredity* x. 225 The allotetraploids..may be quite fertile forms, though sterile with the two species from which they are derived. *Ibid.* 239 An allotetraploid plant is as completely isolated from its parent species as if it had been transported to a desert island. **1957** M. ABERCROMBIE et al. *Dict. Biol.* 14 *Allotetraploid* (*amphidiploid*), allopolyploid which arises when an ordinary hybrid between two different species, containing a set of chromosomes from each parent, doubles its chromosome number. An ordinary hybrid is usually sterile because its chromosomes cannot pair during meiosis. But if it becomes an allotetraploid it solves this difficulty because each chromosome then has a homologue with which it can pair...Artificial preparation of allotetraploids has great scope in the production of new agricultural and horticultural varieties.

Hence **allote·traploidy,** the state or occurrence of such a tetraploid.

1942 J. S. HUXLEY *Evolution* iii. 87 Allotetraploidy is almost confined to plants, because of the favourable conditions provided by their vegetative growth for the rare chromosome-doubling.

allothigenic (ælǫ:þidʒe·nik), *a. Geol.* [ad. G. *allothigene* (E. Kalkowsky 1880, in *Neues Jahrb. f. Min.* 4), f. Gr. ἄλλοθι elsewhere: see *-GENIC.] See quot.; opp. *AUTHIGENIC *a.*

1888 [see *ALLOGENIC *a.* 1].

allotment. Add: **1. a.** *spec.* The payment of part of a seaman's wages to a nominated person; also, the amount so paid; esp. *attrib.* in *allotment note*, a note authorizing such an allotment.

1867 SMYTH *Sailor's Word-Bk.* 32 *Allotment*, a part of the pay apportioned monthly to the wives, children, mothers, or destitute fathers of the warrant and petty officers, seamen, and marines. **1894** *Act* 57 & 58 *Vict.* c. 60 § 141 (2), The seaman may require that a stipulation be inserted in the agreement for the allotment by means of an allotment note, of any part (not exceeding one half) of his wages in favour either of a near relative or of a savings bank. **1953** 'R. GORDON' *Doctor at Sea* vi. 82 Long voyages and young wives don't mix. You leave the allotment of your pay and if you don't get a letter at every port you wonder what's up.

4. In modern use: see quot. 1959.

1917 *Times* 14 Aug. 3/3 Holders of municipal allotments..claim for loss by malicious damage, the hearts of 700 cabbages having been cut out. **1922** *Encycl. Brit.* XXX. 81/2 The powers conferred upon the [Food Production] Department by D.O.R.A., which were delegated to town and urban district councils, enabled them to take possession of any unoccupied land for the purpose of letting it as allotments. **1944** *Amer. Speech* XIX. 294 In Britain the playing of tennis and golf has been succeeded by the week-end pastime of working in one's 'allotment', which is a kind of 'Victory Garden'. **1959** *Chambers's Encycl.* I. 276/2 *Allotment*, a small plot of land let for cultivation, sometimes by private landowners but usually by local authorities for individuals or allotment associations. They were originally provided for country labourers by their overlords, but now are found in and near towns.

6. *allotment-garden* (earlier and later examples), *-gardening, -holder* (examples); **allotment letter,** also *letter of allotment* (see quot. 1882 and cf. sense 1 a in Dict., quot. 1882).

1863 DICKENS *Uncommercial Traveller* in *All the Year Round* 1 Aug. 542/1 Certain allotment-gardens by the roadside. **1959** *Chambers's Encycl.* I. 276/2 There is in Britain a legal distinction between allotments and allotment gardens (Allotments Act, 1922) important as regards determination of tenancy and compensation. **1944** AUDEN *For Time Being* (1945) 113 Allotment gardening has become popular. **1848** J. S. MILL *Pol. Econ.* I. ix. § 4. 1. 180 The labour which the peasant, or even the allotment holder, gladly undergoes. **1920** Allotment holder [see next]. **1882** R. BITHELL *Counting-House Dict.* 169 *Letter of Allotment*, a letter issued in answer to a letter of application for a portion of a public loan, or for shares in a commercial undertaking. **1909** A. E. DAVIES *Money, Stock & Share Markets* xii. 98 Once the allotment letter has been posted it is impossible for the applicant to withdraw his application. **1955** *Times* 15 July 13/1 The allotment letters were being posted by the Bank of England last evening.

allotmenteer (ălǫ:tmĕntiᵊ·ɹ). [f. ALLOTMENT + -EER.] One who holds or rents an allotment (of land).

1920 *Evening Times* (Glasgow) 1 Oct. 4 Allotment holders..have had fine weather for the ingathering of their crop, and the average of quality—as shown at the recent exhibition arranged by allotmenteers..—is very high. **1959** *Times* 21 Apr. 15/5 As an 'allotmenteer' of over 30 years' standing, I must challenge the suggestion..that an allotment cannot be a thing of beauty.

allotriomorphic (ælǫ:triomǭ·ɹfik), *a. Min.* [ad. G. *allotriomorph* (H. Rosenbusch *Mikrosk. Physiog. d. Massigen Gest.* (1886) i. 11), f. Gr. ἀλλότριος belonging to another (ἄλλος other) + μορφή form: see -IC.] = XENOMORPHIC *a.* (opp. *idiomorphic*). Hence **allo:triomo·rphically** *adv.* Cf. *ANHEDRAL *a.*

1887 *Geol. Mag.* Mar. 123 The form of an individual mineral in a rock may..be..dependent on causes other than the molecular forces proper to the mineral (*allotriomorphic*). **1888** [see XENOMORPHIC *a.*]. **1888** W. S. BAYLEY in *Amer. Naturalist* Mar. 209 When none of the constituents are..[idiomorphically] developed the structure is *allotriomorphically granular*. **1922** C. H. DESCH *Metallography* (ed. 3) ix. 179 Each grain [of metal] is, to use the mineralogical term, an 'allotriomorphic' crystal, the outline of which is determined by the presence of neighbouring grains. **1958** A. D. MERRIMAN *Dict. Metallurgy* 5/1 *Allotriomorphic*, a term applied to crystals which have taken their shape from their surroundings and have not been allowed to develop the contours prescribed by their crystalline form.

allotrope (æ·lotrōᵘp). [Back-formation f. ALLOTROPY.] An allotropic form of a substance.

1889 *Cent. Dict.*, *Allotrope*, one of the forms in which an element having the property of allotropy exists: thus, the diamond is an allotrope of carbon. **1942** J. N. GREENWOOD *Gloss. Metallographic Terms* (ed. 2) 6 The importance of allotropes in connection with metals and alloys is that they constitute different phases, and increase the possibilities of changing the mechanical properties by heat treatment.

all out, *advb. phr.* Restrict †*Obs.* to senses 2 and 3 in Dict. and add:

1. (Later examples.)

1835 M. DOYLE *Common Sense for Common People* 18 *Bench.*..Were you drunk? *Tom.* Not all out, I could keep my feet on the floor. **1852** W. R. WILDE *Irish Pop. Superstitions* iv. 121 In the islands of the extreme west, except from sheer old age, or some very ostensible cause, no-one is ever believed to 'die all out'.

4. Using or involving all one's (or its) strength or resources; 'fully extended'; at top speed.

1895 *Windsor Mag.* I. 120/1 There is no fun in going all-out. **1919** *Punch* 19 Mar. 216/2 The car..bolted down-hill all out. **1925** E. F. NORTON *Fight for Everest 1924* 143 Irvine..was willing..to 'go all out', as he put it, in an utmost effort to reach the top. **1928** GALSWORTHY *Swan Song* II. vi. 149 He..made up his mind to go 'all out' for his Uncle Hilary's slum-conversion scheme. **1938** E. WAUGH *Scoop* iii. 60 One screw [*sc.* of an aeroplane] swinging slow, one spinning faster, one totally invisible, roaring all-out. **1940** *War Illustr.* 16 Feb. 110/1 Germany wants oil..but if she is to go all out for victory she must have at least 10,000,000 tons a year.

Hence as *adj. phr.*

1908 *Westm. Gaz.* 29 Sept. 4/2 To put a vehicle to an all-out test. **1955** A. L. ROWSE *Expansion Eliz. England* vii. 261 Not committing herself to an all-out war in Europe. **1959** M. M. KAYE *House of Shade* iii. 39 An all-out bid for control.

all over, *advb. phr.* Add: **1.** (Later examples.)

1815 H. BROUGHAM *Let.* 14 July in T. Creevey *Corr.* (1903) I. xi. 243 Such an exertion..I already *ache all over* with it. **1852** E. TWISLETON *Let.* 1 July (1928) i. 10 Both were thoroughly well-bred and polite—lady and gentleman all over. **1855** THACKERAY *Let.* 28 Apr. (1946) III. 443 His house was all over pictures, drawings, casts, statues, and medals. **1903** E. BUTLER *Lett. fr. Holy Land* 42 He and Isaac are all over pistols and weapons of various sorts. **1906** *Harper's Mag.* Oct. 764 They'd been hunting all over for her. **1922** JOYCE *Ulysses* 761, I often felt I wanted to kiss him all over.

b. As *sb.* Everywhere. Chiefly *U.S.*

1904 *Post Express* (Rochester, N.Y.) 12 Sept. 3 News Flashes from All Over. **1911** *N.Y. Even. Post* 12 Jan. 16 They came from all over, and showed it. They were..from every section of the country. **1956** D. JACOBSON *Dance in Sun* I. i. 7 There were cars, we saw from the registration plates, from all over.

c. *transf.*, *to be all over* (a person): to display great warmth or affection towards (one); *to be all over oneself* (see quot. 1925). *colloq.*

1912 E. PUGH *Harry the Cockney* x. 102 The worst of women is..they never leave you alone...They're all over you. **1923** S. KAYE-SMITH *End House Alard* I. 23, I wouldn't come yesterday. I thought your family would be all over you, and I didn't like ——. **1925** FRASER & GIBBONS *Soldier & Sailor Words* 5 *All over oneself*, extremely pleased. Over confident; e.g.:'He's all over himself because he's got leave.' 'He used to be all right, but now he's promoted he's all over himself.' **1931** A. CHRISTIE *Sittaford Mystery* v. 43 'Were they friendly?' 'The lady was..All over him, as you might say.'

d. (See *OVER *adv.* 2 b.)

2. (Examples.)

1765 C. BRIETZCKE *Diary* 16 June in *N. & Q.* (1963) CCVIII. 307/1 Walkd in the Park with Mr. Stanhope who seems to think it is all over with Us. **1805** NELSON in Southey *Life* (1813) II. ix. 260, I am a dead man, Hardy..I am going fast:—it will be all over with me soon. **1845** CARLYLE *Cromwell's Lett. & Speeches* I. i. 134 Sir Thomas [Steward] makes his will in this same month of January, leaving Oliver [Cromwell] his principal heir; and on the 30th it was all over. **1857** RUSKIN *Elem. Drawing* iii. 216 If you once begin to..try this way and that with your colour, it is all over with it and with you. **1863** 'OUIDA' *Held in Bondage* I. x. 223 We were profoundly thankful when it was all over and done with. **1963** 'J. LE CARRÉ' *Spy who came in fr. Cold* xi. 101 And then I asked him, 'Is this goodbye?'—whether it was all over.

all-over, *adj. phr.* **1.** See ALL OVER *adv. phr.* 3.

2. Covering every part, esp. of ornamental patterns or designs (see quot. 1893).

1859 C. TOMLINSON *Illustr. Useful Arts* 39/2 The patterns of floor-cloths should not be elaborate...What are called *all-over* patterns..have the best effect. **1893** *Jrnl. Soc. Arts* XLI. 372/2 A common cry now is, that what is termed an 'all-over' treatment is best; that is, the areas of the details of the design should be nearly all equal, so that no feature should obtrude itself in the design. **1916** HAMLIN *Hist. Ornament* i. 5 In 'all-over' patterns the units are arranged along two or more intersecting systems of lines so as to cover a broad surface. **1932** *Times Lit. Suppl.* 7 July 493/1 Hazlitt's 'Conversations of Northcote'..may be met with in at least six 'original' bindings, ranging from all-over boards with label, through half-cloth and full-cloth. **1960** B. ROBERTSON *Jackson Pollock* 45 The significance of Pollock's late, all-over style, is not only aesthetic.

Also as *sb.* (See quots.)

1838 *Penny Cycl.* XI. 220/1 The amalgam for this description of work [i.e. the gilding of buttons] is brought to a much stiffer consistence than that which is used for 'all-overs'. **1916** HAMLIN *Hist. Ornament* i. 5 Continuous 'all-overs' forming a mesh of two sets of intersecting lines are called quarries. **1957** M. B. PICKEN *Fashion Dict.* 6/1 *All-over*,..fabric completely covered with a design, such as lace.

3. = *OVER-ALL *adj. phr.*

1933 *Irish Press* 27 Jan. 6/2 The Irish people have answered. .. They have given Fianna Fáil its all-over majority. **1959** *N. & Q.* CCIV. 45/2 His all-over conclusion is that [etc.].

all-overish, *a.* Add: (Examples.)

1832 *Blackw. Mag.* XXXII. 647 He said he was all-overish, and at last he began to strain and bock. **1833** *Sk. & Eccentr. D. Crockett* 52, I wish I may be shot if I know how I felt; but I tell you what, it made me feel quite all-overish. **1851** JANE WELSH CARLYLE *Let.* 5 Mar. (1924) 347, I still feel sick and sore and miserably *all-overish*. **1929** *Daily News* 17 July 11/2 He said he had a sore throat and felt 'All-overish'.

2. Ubiquitous; over the whole extent.

1853 MARCUS DODS *Let.* 18 July (1910) 37 There was an all-overish sort of freshness about your last [*sc.* letter]. **1920** D. H. LAWRENCE *Let.* 15 Mar. (1932) 501 There are a good many English people, but fewer than [in] Capri, and not so all-overish.

3. (Cf. *ALL-OVER *adj. phr.* 2.)

1885 LEWIS F. DAY in *Art Jrnl.* 45/1 The besetting danger of mere ornament is that it is so apt to be monotonous and all-over-ish.

all-overishness. (Earlier examples.)

1820 M. WILMOT *Let.* 3 Mar. (1935) 53 Then poor nurse had her share [of influenza]—Mlle hers and Nanny hers—and I had an alloverishness. **1832** *Blackw. Mag.* XXXII. 647 So great was my all-overishness.

a:ll-o·verness. [f. *ALL-OVER *adj. phr.* + -NESS.] **1.** = ALL-OVERISHNESS.

1820 in *Amer. Speech* (1965) XL. 127 I'm seized with an alloverness, I faint! I die! **1828** *Blackw. Mag.* XXIV. 191, I could almost fain be sick—not *very* sick, but have a gentle all-overness—a tranquil debility. **1840** *Fraser's Mag.* XXI. 474/1, I felt attacked by the malady best designated by the term all-overness.

2. The quality of being 'all-over' (see *ALL-OVER *adj. phr.* 2).

1953 *Archit. Rev.* CXIV. 194/1 There is continual interplay..between the flat, tawny-coloured all-overness and the coming out of the objects. **1955** P. HERON *Changing Forms of Art* 122 His form is very powerful; but it is distinct from the form of Renoir in that it is developed entirely in terms of this 'alloverness'.

allow, *v.* Add: **7.** (Earlier U.S. examples.)

Also used in aphetic form *'low.*

1825 NEAL *Bro. Jonathan* I. ii. 28 Her large eyes would sparkle—so the men 'allowed'—like the mischief. *a* **1861** WINTHROP *John Brent* (1883) iii. 26, I allowed from seeing you handle that thar hoss, that you had got your hand in on women. **1871** G. A. TOWNSEND *Mormon Trials* 13 Well, in the first place, he allowed he was doing his religious duties, and he allowed that he had got to live with some one else.

8. c. Also with ellipse of inf.: To permit to go or come *in, out,* etc.

1864 TROLLOPE *Small House at Allington* II. xxv. 260 He could not have been allowed again into my drawing-room. **1882** BESANT *All Sorts* xiv, Then he allows the Conservatives..back again, and gives them another show. **1911** *Rep. Labour & Social Conditions Germany* III. 76 The miners who were in the company were allowed to some parts. **1915** RUPERT BROOKE *Let.* 12 Mar. in *Coll. Poems* (1918) p. cxlii, We were allowed ashore from 5 to midnight. **1924** D. GARNETT *Man in Zoo* 19 He was allowed out every evening after closing-time.

12. b. *refl.*

1751 JOHNSON *Rambler* 13 Apr., He, therefore, that allows himself to be dissatisfied while he can perceive any error or defect, must refer his hopes of ease to some other period of existence. **1874** HARDY *Far fr. Madding Crowd* II. ii. 27 She contemplated it firmly, allowing herself, nevertheless,..to dwell upon the happy life she would have enjoyed had Troy been Boldwood. **1888** MRS. H. WARD *R. Elsmere* xvii, Robert, meanwhile..had been allowing himself a little deliberate study of Mr. Wendover. *Ibid.* xliii, Madame de Netteville allowed herself plenty of jests with her intimates. **1896** 'H. S. MERRIMAN' *Sowers* vii, He would not allow himself the luxury of being the first arrival.

allowable, *a.* Add: **4.** *spec.* of the restricted production of oil (see quots.). *U.S.* and *Canada.* Hence as *sb.*

1940 M. ALBERTSON in E. DeGolyer *Elem. Petroleum Ind.* xii. 287 Many methods of allocation of allowable oil to producers are in use. *Ibid.*, When oil-producing capacity became sufficient..restriction of production was enforced. Allowables were established for wells, leases, pools, and states. **1949** M. MUSKAT *Physical Princ. Oil Production* x. 454 Total production from the state is fixed..and this total is subdivided or prorated among the fields in the state. These prorated field 'allowables' have generally been far below..production capacities. **1960** *Guardian* 9 Sept. 10/1 Their owners..keep up the price of oil by a quota system which restricts production to so many days' output (called 'allowables') in each month.

allowance, *sb.* Add: **7. a.** In military use, esp. in *pl.* the sums of money (distinct from the *pay*) paid for various purposes or services. *family allowance*, see *FAMILY *sb.* 11.

1794 *Certain Rules & Orders to be Observed by the Corps of Fencible Infantry* (*War Office*) 15 Allowances to Officers. We are graciously pleased to grant the following allowances, *viz.* To every Captain of a Company .. Fifty-six Pounds Ten Shillings *per Annum*. *Ibid.* 19 Every..Non-Commissioned Officer, and Private Man shall receive, in Aid of the Expence of his Necessaries, an Annual Allowance of 13s. 2½d. **1795** (*title*) Copy of Warrant for establishing a consolidated Allowance at a daily rate, for soldiers of cavalry and infantry. *Ibid.* 18 The several Allowances called Bread Money..the annual Allowance to each Man in the Infantry for Alteration of Clothing. **1837** T. BACON *First Impr. fr. Nat. in Hindostan* I. 252 Upon pay and allowances not exceeding £20 a month. **1853** [see CLOTHING *vbl. sb.* 5 b]. **1888** KIPLING *Plain Tales fr. Hills, False Dawn* 40 Pay and allowances of nearly fourteen hundred rupees a month. **1949** F. MACLEAN *Eastern Approaches* II. ii. 190 They had drawn special equipment..and special pay and allowances.

8. b. A deduction from the weight a race-horse is required to carry.

1823 'JON BEE' *Slang* 3 *Allowances* (turf) Mares and geldings running against horses are allowed weight (usually 3 lbs. each); also, if coming of untried parents..fillies *always* carry less than colts. **1881** *Encycl. Brit.* XIII. 202/2 The competitors carried the same weights, with the exception of a slight allowance [to fillies] for sex. **1955** *Times* 10 Sept. 3/1 By this success Robinson lost the right to claim an apprentice's allowance.

c. *Mech.* (See quots. 1940, 1958.)

1903 *Trans. Amer. Soc. Mech. Engin.* XXIV. 1173 Before making any changes in our present allowances I wrote to a number of engine builders. **1940** *Chambers's Techn. Dict.* 23/2 *Allowance*, a difference in dimensions prescribed in order to allow of various qualities of fit between mating pieces. **1958** *Van Nostrand's Sci. Encycl.* (ed. 3) 887/1 Allowance is an intentional difference in the size of mating parts. Tolerance is the permissible variation in the size of a part.

alloway: see *ALLEY-WAY.

alloy, *sb.* Add: Now commonly with pronunc. (æ·loi). **I. 5.** Also *attrib.*, as *alloy steel.*

1902 *Encycl. Brit.* XXIX. 572/1 Alloy Steels and Cast Irons are those which owe their properties chiefly to the presence of one or more elements other than carbon. **1922** *Ibid.* XXXI. 927/2 Certain alloy steels..can be rendered (or kept) completely 'austenitic'..by quenching.

all right. (See RIGHT *a.* 15, *adv.* 13 c in Dict. and Suppl.) Hence **1.** sb. use of phr., esp. in *bit of all right*: see *BIT *sb.*² 4 h.

2. *adj. phr.* Used to indicate approval. *colloq.*

1953 'S. RANSOME' *Hear No Evil* (1954) vii. 71 Was it an all-right evening? **1962** M. PROCTER *Body to Spare* xxv. 186 He seemed an all-right bloke to me.

all round, all-round. Add: **A.** (Earlier and later examples.)

1728 E. CHAMBERS *Cycl.* s.v. *Foundry*, They first loosen them a little all round, with a small cutting Instrument. **1861** *Baily's Monthly Mag.* July 140 A stronger Eleven 'all round' never came from Cambridge. **1959** H. P. TRITTON *Time means Tucker* 7 Work was scarce and wages low, and conditions all-round were tough.

C. (Earlier and later examples.)

1867 *Routledge's Every Boy's Ann.* Aug. 15 A particularly good all round cricketer. **1895** W. JAMES *Coll. Ess. & Rev.* (1920) 403 We find the all-round men like Washington, Cavour, and Gladstone. **1958** *New Statesman* 6 Dec. 802/3 An excellent all-round performance by the Guildford Repertory Company.

Hence **a:ll-rou·ndness,** the quality of being all-round; all-round nature or quality (cf. *ALL-AROUNDNESS).

1914 W. E. BEET *Mediaeval Papacy* ii. 43 The most striking feature of Gregory's character and work was its all-roundness. **1950** A. HUXLEY *Themes & Variations* 47 Doubting the inevitability and all-roundness of progress, Biran began to feel very dubious about that Future to which the Liberals looked forward with such sanguine hope.

all-rounder. (Earlier and later examples.)

1857 J. E. RITCHIE *Night Side of London* 146 One or two awful young swells with excruciating all-rounders were present. **1875** *Contemp. Rev.* XXV. 741 An all-rounder in attainments .. 'who had run through the circle of the sciences, and understood them all equally well'. **1938** *Times* 7 Jan. 13/6 Nor, even if the organization were bigger, would it be easy to clap the staff into their pigeon-holes, when so many jobs call for the all-rounder. **1955** *Times* 12 May 14/4 Mr. Gilbert Jessop. A Great Cricketing All-Rounder. **1959** *Listener* 5 Mar. 402/1 The academically successful student is an all-rounder. He plays games and he takes a good part in general college life.

allurance. Delete †*Obs.* and add later examples.

1845 'BON GAULTIER' *Bk. Ballads* 118 That the liquor bath allurance, Well I understand. **1949** O. NASH *Versus* 4 He use no lotions For allurance.

allure, *sb.*¹ Delete †*Obs.* and add later examples.

1901 *Westm. Gaz.* 2 Jan. 1/3 Biskra..Doré like are its allures. **1924** J. FARNOL *Loring Mystery* (1925) xix. 126 He seemed to find a strange allure in this forbidding prospect.

allusion. 4. Add to def.: A passing or incidental reference (cf. ALLUDE *v.* 5). Also *attrib.* in *allusion book*, a collection of references to a writer or his works.

1874 C. M. INGLEBY *Shaksp. Allusion-Books* I. p. i, A Section of our Reprints appears under the title of *Allusion-Books*. By this term we intend to cover not only those books which afford some *allusion*, or indirect reference, to Shakspere or to a work of his, but also those which directly deal with either: i.e. which mention him by name as the author of such and such a play, or as a poet worthy of praise or of blame. **1939** H. MACDONALD *Dryden Bibliogr.* Pref. p. viii, This book is therefore an allusion book as well as a bibliography.

alluvial, *a.* Add: Applied to various formations, as *alluvial cone, fan,* etc.: see the sbs.

B. *sb.* An alluvial deposit; alluvium; *spec.* 'the common term in Australia and New Zealand for gold-bearing alluvial soil' (Morris).

1866 *Harper's Mag.* Sept. 544/1 Nearly two thousand acres of the rich alluvial at the junction of the Ohio and Great Miami rivers. **1888** 'R. BOLDREWOOD' *Robbery under Arms* III. xviii. 270 The whole of the alluvial will be taken up, and the Terrible Hollow..will re-echo with the sound of pick and shovel. **1890** *Pall Mall Gaz.* 16 Dec. 2/1 There is..every indication of mineral wealth, but it has not been sufficiently proved yet for one to say whether there is good alluvial or not. **1944** J. A. LEE *Shining with Shiner* (1963) 60 Men were out shovelling at and turning over the alluvials.

alluviation (ălⁱū:viēⁱ·ʃən). [f. ALLUVI(UM + -ATION.] The process of depositing alluvium.

1847 C. LANMAN *Summer in Wilderness* xxii. 132 They have been striped with various colors by mineral alluviations. **1877** G. K. GILBERT *Rep. Geol. Henry Mts.* 139 The changes by planation and alluviation. **1937** WOOLDRIDGE & MORGAN *Physical Basis Geogr.* xii. 173 Flood alluviation, like aggradation by gravel, is closely linked with the development of meanders.

allyl. Add: **2.** *allyl isothiocyanate* (C_3H_5NCS), *allyl-thio-urea* (see THIOSINAMINE); **allyl plastic, resin,** a plastic or synthetic resin made by polymerization of an allyl compound.

1884 ROSCOE & SCHORLEMMER *Treat. Chem.* III. 394 *Allyl-thio-urea*, $CS.N_2H_3(C_3H_5)$, was obtained in 1834... Will..proposed for it the name of *thio-sinamine* in place of that which had been hitherto used, namely mustard-oil ammonia. **1891** THORPE *Dict. Appl. Chem.* II. 626/2 The *volatile oil*, consisting of allyl isothiocyanate is formed from the brown mustard on addition of water. **1895** *Proc. Chem. Soc.* 216 Iodine..acts similarly to bromine on allyl-thiourea. **1907** WALKER & MOTT tr. *Holleman's Text-bk. Org. Chem.* (ed. 2) I. 323 The isothiocyanic esters are also called *mustard-oils* after allyl isothiocyanate, to which the odour and taste of mustard-seeds are due. **1911** *Encycl. Brit.* XXVIIe 794/2 The most important derivative [of thiourea] pharmacologically is allyl-thiourea. **1943** *Mod. Plastics* Nov. 96/2 The capacity for forming or drawing the new allyl plastics into curved windows is restricted..because of the absence of plastic flow. *Ibid.* June 102 (*caption*) Physical Properties of Cast Allyl Resins and Laminated Allyl Resins. *Ibid.* Nov. 96/1 The manner in which the Columbia allyl resins differ from..commercial plastics can be clarified..by indicating why the development of the new materials has appeared desirable. **1959** *Chambers's Encycl.* X. 776/2 *The Allyl Plastics.* The allyl resins are among the newest plastic materials, and are based on allyl alcohol produced in the cracking of oil.

allylic, *a.* (Earlier example.)

1857 W. A. MILLER *Elem. Chem.* III. vi. § 2. 386 Hofmann and Cahours have succeeded in discovering the allylic..alcohol.., a compound which stands in the same relation to acrolein that ethylic alcohol does to ordinary aldehyd.

Ally Sloper (æ·li slōᵘ·pəɹ). The name of a character in a series of humorous publications, having a prominent nose and receding forehead and noted for his dishonest or bungling practices; hence used allusively.

1873 (*title*) Some Playful Episodes in the Career of Ally Sloper..pictorially portrayed by Marie Duval, and explained..by Judy's Office Boy [*i.e.* C. H. Ross]. **1877** C. H. Ross (*title*) Ally Sloper's Book of Beauty. **1884** (*title*) Ally Sloper's Half Holiday. **1901** *Ann. Brit. Sch. Athens* VI. 126 The young man [in the carnival] contents himself with..a roughly-made domino, or, thanks to the steadily increasing influx of Western culture..an 'Ally Sloper' mask. **1922** JOYCE *Ulysses* 495 Prognathic with receding forehead and Ally Sloper nose. **1925** FRASER & GIBBONS *Soldier & Sailor Words* 170 *Ally Sloper's Cavalry*, the Royal Army Service Corps. A humorous perversion of the initials of the Army Service Corps.

almacantar. Add examples with spelling *almucantar* (now the usual form). Also used in *sing.*

1863 W. CHAUVENET *Man. Spherical & Pract. Astron.* I. i. 19 Small circles parallel to the horizon are called *almucantars*. **1923** D. CLARK *Plane & Geodetic Surv.* II. ii. 83 The almucantar, or small circle of equal altitudes.

2. A telescope fitted with horizontal wires and mounted on a float resting on mercury, used for observing the rising and setting of stars.

Transfer quot. 1880 to this sense, and add:

1901 *Proc. Durham Philos. Soc.* II. 6 In the Almucantar we claim that the telescope turns with faultless truth about its theoretical axis of rotation. **1920** *Nature* CV. 329/1 The almucantar, in which horizontality is secured by the device of floating the whole in a mercury bath.

almighty, *a.* Add: **2.** For 1837 example read:

1836 W. IRVING in *New Yorker* 12 Nov. 115 The almighty dollar [etc., as in Dict.].

(Further examples of *almighty dollar* and similar phrases.)

1842 DICKENS *Amer. Notes* I. iii. 63 The almighty dollar sinks into something comparatively insignificant. **1891** F. W. BAIN *Antichrist* ii. 104 A mere conglomeration of fortune hunters, worshipping the Almighty Dollar. **1947** *Time* 16 June 33/1 There is a limit to the sacrifices some Britons would make for the sake of the almighty greenback.

almond. Add: **3.** (Examples.)

1855, 1892 [see SUGARED *ppl. a.* 1 d].

5. d. *attrib.* Applied to eyes shaped like an almond, esp. of certain Asian peoples. So *almond-eyed* adj.

1870 'MARK TWAIN' *Wks.* (1900) XIX. 145 An 'unsuspecting, almond-eyed son of Confucius'. **1891** WILDE *Critic as Artist* in *Intentions* 179 The almond-eyed sage of the Yellow River. **1915** W. S. MAUGHAM *Of Human Bondage* xxii. 94 He [a Chinaman] laughed..and his almond eyes almost closed as he did so. **1935** *Discovery* Aug. 244/1 Mongol is but a generic term, associated with almond eyes and high cheek bones.

e. *attrib.*, of a face.

1878 WILDE *Ravenna* 8 The almond-face which Giotto drew so well.

9. (Earlier examples.)

1735 J. MOORE *Columbarium* 39 But amongst all [the colours of Tumblers], there is a Mixture of three Colours, vulgarly call'd an Almond, perhaps from the Quantity of Almond colour'd Feathers that are found in the Hackle: Others call it an Ermine. **1765** (*title*) A treatise on domestic pigeons..to which is added, a most ample description of that celebrated and beautiful pigeon called the Almond Tumbler. **1854** *Poultry Chron.* I. 359/2 Marks in the Almond Tumbler. ..Should it be an almond, it will most likely have a great deal of black about it.

10. *almond-scented* adj. (example); **almond-comfit** = ALMOND 3; **almond-peach** (example); **almond rock** (earlier example).

1853 MRS. GASKELL *Cranford* xv. 299 If a little child came in to ask for an ounce of almond-comfits..she always added one more. **1629** J. PARKINSON *Parad.* III. xv. 582 The Almond Peach, so called, because the kernell of the stone is sweete, like the Almond, and the fruit also somewhat pointed like the Almond in the huske. **1859** SALA *Tw. round Clock* 56 The relative merits of almond-rock and candied horehound. **1881** WILDE *By Arno* in *Poems* 162 The almond-scented vale.

almoner¹. Add: **4.** A hospital official who has duties concerning patients' welfare (also formerly concerning patients' payments).

1896 M. STEWART in *Mag. Lond. School Med. for Women* May 146 The appointment of an almoner is an experimental measure which has been adopted conjointly by the [Royal Free] Hospital and the Charity Organisation Society. **1922** *Encycl. Brit.* XXXI. 384/2 Many hospitals had adopted the almoner system, that is, specially appointed hospital employees who approach the individual patient who comes to hospital, explain the needs of the hospital, and invite the patient to contribute according to his ability. **1964** *Times* 10 Oct., The almoner has died, at least in name,..giving way to a new title, Medical Social Worker.

Alnaschar (ælnæ·ʃəɹ). The name of a beggar in the 'Arabian Nights' who destroys his means of livelihood because he indulges in visions of riches and grandeur; applied allusively to a person given to such illusions. Also *attrib.* Hence **Alna·scharism; Alna·scharize** *v. intr.*

1712 ADDISON *Spect.* No. 535 13 Nov., Alnaschar was entirely swallowed up in this chimerical Vision. **1812** M. EDGEWORTH *Tales Fash. Life* IV. i. 13 With maternal *Alnascharism*, she had, in her reveries, thrown back her head with disdain, as she repulsed the family advances of some wealthy, but low-born heiress. **1840** *New Monthly Mag.* LIX. 349 Higgins had been Almascharizing [*sic*]—building *chateaux en Espagne*. **1840** THACKERAY *Shabby Genteel Story* vi. in *Fraser's Mag.* Aug. 236/2 'I'll..make designs hallusive of my passion for her.' And so our pictorial Alnaschar dreamed and dreamed. **1849** —— *Pendennis* I. xxxiii. 329 What an Alnaschar I am because I have made five pounds by my poems. **1853** MRS. GASKELL *Ruth* I. ii. 55 She..was busy with Alnaschar visions of wide expenditure. **1896** G. B. SHAW in *Contemp. Rev.* Feb. 205 Those delicious Alnaschar dreams.

alocasia (ælokēⁱ·ziă, -kēⁱ·ʒⁱă). *Bot.* [Said to be an alteration of COLOCASIA.] A plant of a genus from tropical Asia cultivated for its foliage, closely allied to COLOCASIA.

1860 *Bot. Mag.* LXXXVI. t. 5190 Alocasia Metallica. Bronze-leaved Alocasia. **1882** *Garden* 4 Feb. 86/3 Alocasias..should be potted before growth begins. **1884** *De Candolle's Orig. Cultivated Pl.* 75 Apé, or Large-rooted Alocasia—*Alocasia macrorrhiza*.

alogical (ēⁱlǫ·dʒikăl), *a.* [f. A- 14 + LOGICAL *a.* Cf. F. *alogique*.] Non-logical; not based upon reason or formed by an act of judgement; opposed to logic. Also *absol.*, that which is alogical. Hence **alo·gically** *adv.*; **a:logica·lity.**

1694 MOTTEUX *Rabelais* v. xviii, Foreign Civilians..have uncivily drawn alogical and unreasonable Consequences from it [*tiré à consequence alogicque, c'est à dire desraisonnable*]. **1881** G. S. HALL *Asp. German Culture* 43 There is an immanent teleology in his [Julius Bahnsen's] universe; but it is not merely alogical, but anti-logical, and even anti-causal. **1907** E. B. BAX *Roots of Reality* 19 Undoubtedly the alogical is..a primary element in all experience. *Ibid.*, In ordinary consciousness..the ultimate elements of a reality or thing are an alogical feeling or sensation, and a logical form or category. *Ibid.* 161 In every process of consciousness a contradiction lies embedded, based on the antithetic character of its two ultimate elements, the mark of which we have found to consist respectively in alogicality and logicality. *Ibid.* 173 The aim of art..is to express the unity and harmony of experience..in the world of immediate feeling—in a word, *alogically*. **1908** J. M. BALDWIN *Thought & Things* II. 386 The Logical and the A-logical. **1909** W. JAMES *Let.* in R. B. Perry *Tht. & Char. W. J.* (1935) II. 680, I am a-logical, if not illogical, and glad to be so. **1925** J. E. TURNER *Theory Direct Realism* 25 It is not surprising..that this attitude..should be regarded as instinctive non-rational, and alogical. **1957** P. LAFITTE *Person in Psychol.* xiv. 205 What was described as alogical thinking was quite logical thinking in a different frame of reference.

alogism. Restrict '†*Obs. rare*⁻¹' to sense in Dict. and add: **2.** The fact or quality of being alogical.

1911 J. M. BALDWIN *Thought & Things* III. iv. 47 The 'alogism' of Bergson, which James cites as nearer his own, seems, however,..in its outcome, nearer to that of Bradley, since it is a mode of higher intuition or immediacy to which Bergson makes appeal. **1935** R. B. PERRY *Tht. & Char. W. James* II. 681 An almost morbid alogism, or antipathy to the mode of thinking which employs definitions, symbols, and trains of inference.

along, *adv.* Add: **1. b.** Some way on (in the progress of time). *U.S.*

1870 'MARK TWAIN' *More Distinction* in *Sketches* (1872) 282 In the one case, you start out with a friend along about eleven o'clock. **1883** —— *Life Mississippi* xxviii, Far along in the day, we saw one steamboat. **1886** *Harper's Mag.* Oct. 808/1 He come to the house 'long in the first part of the evenin'. **1897** 'MARK TWAIN' *Following Equator* xxi. 290 He was along toward fifty. **1902** *N.Y. Tribune* 26 Apr. 82 The afternoon was well along by this time.

c. *along back*: at or for some time in the past. *U.S.*

1851 N. KINGSLEY *Diary* (1914) 165 Worked as usual to-day, took out 50 ounces and 4 dollars, which gains on our days along back. **1880** *Harper's Mag.* Dec. 85/1 She's had an easy time along back, but she's seen the last on't.

d. *right along*: continuously; without interruption. *U.S.*

1856 *Mich. Agric. Soc. Trans.* VII. 806 His corn grew right along, for it could not help it. **1936** WODEHOUSE *Laughing Gas* xiv. 150 'And you come from England?' 'Yes.' 'London?' 'Yes.' 'Lived there right along, I guess?'

2. (Earlier examples of *to get along.*)

1830 S. SMITH *Sel. Lett. J. Downing* (1834) 34, I wish you'd write me..whether you think I could get along with the business [of Governorship]. **1837** [see GET *v.* 53 b].

b. *to be along*, to come to a place, to call, to arrive.

1831 MRS. HOLLEY *Texas* (1833) 21 The captain..sent word that he would be along for us about sun-set. **1892** 'MARK TWAIN' *Amer. Claimant* xvi. 162 They'll be along as soon as it's done. **1951** 'J. WYNDHAM' *Day of Triffids* i, Any time now they'd be along with pneumatic drills. **1959** N. KNEALE *Quatermass Exper.* IV. 114 The official pathologists' report should be along in half an hour.

aloof. A. 7, B. Add later *adj.* examples. Also in sense detached, unsymphathetic'.

1872 GEO. ELIOT *Middlem.* III. vi. lviii. 287 What that cleverness was—what was the shape into which it had run as into a close network aloof and independent. **1893** A. J. BALFOUR in *Daily News* 27 Apr. 6/7 Democracy..perhaps required the cold and aloof reasoning of a statesman like Lord Derby. **1913** *Stage Year Book 1913* 19 The imported 'yes' and 'no' of an aloof and distant Censor. **1934** H. NICOLSON *Curzon: Last Phase* 8 His public manner..created the legend of a man, conceited, reactionary, unbending and aloof. **1948** LD. D. CECIL *2 Quiet Lives* I. 14 She..hid her true self behind the shield of an aloof formality.

aloofly (ălū·fli), *a.* and *adv.* [f. ALOOF *a.* + -LY¹, ².] **A.** *adj.* Characterized by aloofness; 'distant', unsympathetic. *rare.* **B.** *adv.* So as to keep, or as if keeping, aloof.

1901 F. CAMPBELL *Love* 118 His cold justice and aloofly Christianity. **1921** *Chambers's Jrnl.* Mar. 203/1 A tall moon rode aloofly across the sky. **1924** *Blackw. Mag.* Oct. 514/1 This aloofly displeased person now came and spoke to us. **1963** *New Statesman* 24 May 806/3 In case I sound aloofly virtuous.

alopecoid (ælopī·koid), *a.* and *sb. Zool.* [f. Gr. ἀλώπηξ fox: see -OID.] Belonging to the division of the genus *Canis* of which the fox is the type; vulpine. As *sb.*, an animal of this division.

1880 [see THOOID *a.* (*sb.*)].

‖ **à l'outrance:** see OUTRANCE.

1837 J. F. COOPER *Recoll. Europe* I. ix. 310 There was a famous quarrel, *à l'outrance*, about it. **1860** *Once a Week* 20 Oct. 476/2 Francis II. will be called upon to make his choice between casting in his lot with the defenders *à l'outrance* of Gaeta, or making his escape by sea.

Alp¹, alp. Add: **1.** (Examples of the sense 'mountain pasture'.)

1827 M. WILMOT *More Lett.* (1935) 276 An Alpe..is a cottage and little establishment to which a shepherdess conducts a certain number of flocks and herds in the spring. **1857** B'NESS VON TAUTPHŒUS *Quits* I. xvi. 253 Is this Peissenberg what you call an alp or alm..is it one of those pasture-grounds on the mountains where..the people send their cattle in summer? **1922** W. G. KENDREW *Climates of Cont.* xi. 54 Between the forests and the cap of perpetual snow is a zone of alp, pasture land with drought-resisting bushes.

alpargata (alpaɹgā·tă). Also 9 **alpargate.** [Sp.] A type of sandal (see quot. 1897).

A form in earlier use was *alparca*: see Stanford Dict. Anglicised Words.

1827 J. P. HAMILTON *Trav. Int. Prov. of Columbia* I. 217 They had not a pair of shoes and stockings in the whole battalion; the officers wearing alpargátes. **1845** R. FORD *Handbk. Trav. Spain* I. i. 35 The *caleseros* on the eastern coast wear..the ancient Roman sandals, made of the *esparto* rush, with hempen soles, '*alpargatas*'. *Ibid.* II. xii. 926 In dry weather they prefer the sandal *Alpargata*, which, however, will not stand much wet. **1897** GADOW *In Northern Spain* iv. 63 My companions wore *alpargatas*, namely, canvas shoes with thick soles of plaited hemp or grass. **1909** BELLOC *Pyrenees* v. 169 Remember that, in Alpargatas, you will always end the day with wet feet. **1920** *Glasgow Herald* 6 Mar. 3 Spinning, and twisting heavy jute yarns for alpargatas. **1940** HEMINGWAY *For whom Bell Tolls* i. 12, I wonder if he has a pair of boots or if he rides in those *alpargatas*.

alpeen (æ·lpīn). *Anglo-Ir.* Also **-ine.** [Ir. *ailpín.*] A cudgel, stout-headed stick.

1828 BANIM *Anglo-Irish* II. 188 The..alpeens of the primitive Irish. **1844** THACKERAY *Box of Novels* in *Wks.* (1900) XIII. 411 Here comes Rory O'More thundering down with his big alpeen. **1847** TROLLOPE *Macdermots* xiii, Thady had an alpine in his hand, and was preparing to strike a blow at the Captain.

alpenglow (æ·lpĕnglōᵘ). [Partial tr. G. *alpenglühen*, less freq. *alpenglut*, f. *alpen*, gen. pl. of *alp* + *glühen, glut* fire, flame, glow (see GLEED *sb.*).] The rosy light of the setting or rising sun seen on high mountains.

1871 TYNDALL *Fragm. Sci.* 282 On August 23, 1869, the evening Alpenglow was very fine. **1880** *Scribner's Monthly* XX. 346/2 The whole picture stood revealed in the full flush of the alpenglow.

alpenhorn (æ·lpĕnhǭɹn). Also **alphorn.** [Both G., f. *alpen* (see prec.) + HORN *sb.*] (See quot. 1879.) Also *attrib.*

1864 [see ALP¹ 4]. **1879** GROVE *Dict. Mus.* I. 56 *Alpenhorn*, or *Alphorn*, an instrument with a cupped mouthpiece, of wood and bark, used by the mountaineers in Switzerland. **1895** *Westm. Gaz.* 13 Aug. 2/3 The time-honoured Alp-horn melodies.

alpenrose (æ·lpĕnrōᵘz). *Bot.* [G., f. as prec. + ROSE *sb.*] = *Rose of the Alps* (see ROSE *sb.* 3 b).

1914 *Gardening Illustr.* 24 Jan. 49/2 R. hirsutum had disappeared while R. ferrugineum was abundant...They are the two varieties of the well-known Swiss Alpenrose, which looks handsome in large sheets of colour. **1924** 'L. MALET' *Dogs of Want* i. 17 Bilberries and dwarf thickets of juniper and *alpen-rose* grew in the shelter of outcrops of rock and scattered boulders. **1960** *Times* 5 Mar. 9/1 Alpenroses, gentians, and other alpine plants may often be found.

alpha. Add: In technical, as in general, contexts frequently written α.

3. d. *Metallurgy.* (*a*) Applied to the first of a series of allotropic forms of a metal, as *alpha iron*. (*b*) Applied to a solid solution in a range of alloys, as *alpha brass*, the first of a series of alloys of copper and brass: that in which there is the highest proportion of copper.

1885 *Jrnl. Iron & Steel Inst.* II. 618 The author [Osmond] believes..that there are two..varieties of iron, α and β. The crystalline iron α is obtained by tempering at a red heat, followed by a slow cooling. This is transformed into the β variety (which the structure of tempered steel shows to be amorphous) either by a permanent alteration at a low temperature, or by rapid cooling from a red heat..in the presence of carbon, or some other body (manganese, tungsten) exercising the same influence on the..steel. **1902** *Encycl. Brit.* XXIX. 572/2 We must recognize the further complication in the constitution of iron, due to its having at least three distinct allotropic modifications, α, β, and γ, each corresponding to a distinct range of temperature. α iron is the weak, ductile, magnetic variety.., β iron is the non-magnetic variety, [etc.]. **1914** W. ROSENHAIN *Physical Metallurgy* viii. 167 Iron is capable of existing in at least three diverse or 'allotropic' conditions. These are generally known as the γ, β and α forms of iron. **1957** *Encycl. Brit.* XII. 669/2 Pure iron or ferrite at room temperature consists of tiny crystals of so-called alpha iron, crystallized in body centred arrangement.

1904 HEYCOCK & NEVILLE in *Phil. Trans. R. Soc.* A. CCII. 9 Crystals of five different types, which we designate α, β, γ, δ, η... The α crystals are solid solutions, apparently isomorphous with pure copper. They may contain any percentage of tin not greater than 9 per cent. The β crystals are also solid solutions, perhaps isomorphous with the α. **1919** *Jrnl. Inst. Metals* XXII. 370 (*heading*) Thermal expansion of Alpha and Beta Brass between 0° and 600° C. **1923** GLAZEBROOK *Dict. Appl. Physics* V. 223/1 As the α solution becomes more and more saturated the tensile strength of the alloys increases...The αβ alloys cannot be worked cold, but are always rolled, or extruded, hot.

e. *Physics. alpha radiation*, the first of three types of radiation emitted by radioactive substances, and consisting of positively charged particles. So *alpha particle* (identified with the helium nucleus: see quot. 1908), *ray.*

1899, 1902 [see *BETA 2 f]. **1903** RUTHERFORD in *Phil. Mag.* V. 184 The α rays are complex, and probably consist of particles projected with velocities lying between certain limits. **1903** —— in *Nature* 20 Aug. 366/2 The determination of the mass of the α body..supports the view that the α particle is in reality helium. **1908** —— & GEIGER in *Proc. R. Soc.* A. LXXXI. 172 We may conclude that an α-particle is a helium atom, or, to be more precise, the α-particle, after it has lost its positive charge, is a helium atom. **1925** F. SODDY in *Evol. Light Mod. Knowl.* x. 371 The radiant α- and β-particles expelled by the radio-elements. **1938** R. LAWSON tr. *Hevesy & Paneth's Man. Radioactivity* (ed. 2) ix. 94 The α-particle..a kind of 'molecule' of the 'nuclear atoms' possessing especial stability, and arising from the union of 2 protons and 2 neutrons.

Hence ellipt. for *alpha particles* in *alpha decay, emission, emitter; alpha-emitting* adj.

1936 *Physical Rev.* L. 977 The radioactive α-decay is discussed as a many-body problem. *Ibid.*, The probability of α-emission consists therefore of two parts. **1937** *Ibid.* LI. 818/2 Possible evidence for the existence of a new alpha-emitting isotope of uranium. **1949** W. E. SIRI *Isotopic Tracers & Nucl. Radiations* iv. 95 It is assumed that the nuclear radius is nearly constant for all alpha emitters. **1962** *Newnes Conc. Encycl. Nucl. Energy* 25/1 Spontaneous α-emission takes place with many of the nuclei heavier than lead, each species decaying with its characteristic half-life.

f. *alpha rhythm, waves*, the normal rhythmic activity of the brain recorded in an electro-encephalogram, having a frequency of eight to thirteen cycles per second.

[**1930** H. BERGER in *Jrnl. f. Psychologie & Neurol.* XL. 162 Der Kürze halber werde ich im folgenden die Welle erster Ordnung als Alphawellen = α-W., die Wellen zweiter Ordnung als Betawellen = β-W., bezeichnen.] **1935** [see *BETA 2 h]. **1936** *Archives Neurol. & Psychiatry* XXXVI. 1215 Most prominent is a rhythm of approximately 10 a second..variously known as the 'Berger rhythm', the 'alpha waves' or the '10 cycle rhythm'. *Ibid.* 1220 There is a group of persons in whom the alpha rhythm is rarely visible in the spontaneous record, although it may usually be found for a second or two shortly after the subject has closed his eyes. **1941** P. BARD *Macleod's Physiol. in Mod. Med.* xix. 238 The most prominent rhythm is one which ranges from 8 to 13 and averages 10 waves a second. This is sometimes called the 'Berger rhythm', but has more generally come to be known as the *alpha rhythm*.

4. An examiner's first-class mark. Also *transf.*

1902 S. T. in *Oxford Mag.* 22 Jan., And what I deemed an α rules (Like markets) flat as β+. **1907** A. D. GODLEY *Ibid.* 23 Jan., One who to all Experience gave An Alpha or Epsilon. **1923** *Times Lit. Suppl.* 25 Oct., The critic (who can hardly help feeling on this occasion like an examiner with prize competition papers before him) must give the *alpha* here to Mr. Sichel. **1929** *Ibid.* 2 May, Nor can we give an alpha mark to [etc.]. **1962** C. S. LEWIS *Let.* 10 Aug. (1966) 305 Yes, I..have read Gombrich and give him alpha with as many plusses as you please.

alphabet, *sb.* Add: **1. c. alphabet book,** a book for teaching the alphabet; **alphabet soup,** a clear soup containing pieces of paste or biscuit shaped like letters of the alphabet.

1922 JOYCE *Ulysses* 49 One of the alphabet books you were going to write. **1907** *Black Cat* June 15 Alphabet soup—that thin, clear soup, with little noodle or cracker letters in it.

4. Delete †*Obs.* and add later example.

1923 G. B. SHAW in *Nation* 10 Feb. 715/1 The soldiers would have to be inoculated with a whole alphabet of microbes.

alphabetiform (ælfăbe·tifǭɹm), *a.* [See -FORM.] Resembling the letters of an alphabet.

1901 G. SERGI *Mediterranean Race* xiv. 296 The alphabetiform signs in megalithic inscriptions. **1908** A. J. EVANS in R. R. Marett *Anthropol. & Classics* 12 Linear signs which ..present a truly alphabetiform character.

alphabetization (æ:lfăbĕtaizēⁱ·ʃən). [f. ALPHABETIZE *v.* + -ATION.] The process of arranging words in alphabetic order; the result of this, an alphabetic series or list.

1889 H. B. WHEATLEY *How to Catalogue* 203 In the alphabetization of a catalogue. **1902** *Academy* 19 July 76/2 The index to the completed work, covering under one alphabetization the ninth edition and the new volumes.

alphameric (ælfăme·rik), *a.* Contraction of *ALPHA-NUMERIC *a.*

1956 BERKELEY & WAINWRIGHT *Computers* II. i. 29 This is..*alphameric coding*, a system of coding or abbreviation such that information for the machine to handle may be reported not only in digits but also in letters.

alpha-numeric (æ:lfă,niume·rik), *a.* Also **alpha-numerical.** [f. ALPHA(BET *sb.* + NUMERIC(AL *a.*] Consisting of or employing both letters and numerals; applied esp. to a system of coding in computers.

1955 BRADSHAW & NEWMAN in M. J. Dooher *Electronic Data Processing in Industry* i. 18 Typical methods of classifying, as to both type of classification and examples, are: (*a*) alphanumeric codes, commonly used for customer accounts where it is desirable to have the ease of numeric filing coupled with the advantages of alphabetic reference. **1959** *Brit. Standard 3174* (*title*) Alpha-numeric punching codes for data processing cards. **1961** *Times* 3 Oct. (Computer Suppl.) p. viii (Advt.), It has a built-in parity check on all alpha-numeric input information. **1962** *B.S.I. News* Apr. 15/1 Mixed alpha-numerical systems .. caused many mistakes in transcription.

Alpine, *a.* Add: **2.** Designating a physical or racial type (see quots.).

1899 W. Z. RIPLEY *Races of Europe* (1900) iv. 72 The great central highland seemed indeed to constitute a veritable focus of this peculiar physical type...This geographical characterization of the broad-headed variety entitled it, in our opinion, to be called the Alpine type, in distinction from the two others above mentioned [*sc.* the Teutonic and the Mediterranean]. **1921** E. SAPIR *Language* x. 225 The mass of the German-speaking population..belong to..the shorter, darker-complexioned, short-headed Alpine race... The distribution of these 'Alpine' populations corresponds in part to that of the old continental 'Celts'. **1934** R. BENEDICT *Patterns of Culture* (1935) i. 11 The so-called race line..is held to divide the people of Baden from those of Alsace, though in bodily form they alike belong to the Alpine sub-race.

3. *Comb.* **Alpine anemone,** a small blue flower indigenous to mountain areas in N.E. America and Europe, now widely introduced elsewhere; **Alpine club,** one for alpinists; **Alpine fir,** a tall tree (*Abies lasiocarpa*), indigenous to the mountains of western North America; **Alpine rose**: see ROSE *sb.* 3 b and cf. *ALPENROSE.

1892 C. M. YONGE *Old Woman's Outlook* 49 The garden shows their blue brother, the Alpine anemone. **1859** Alpine Club [see CLUB *sb.* 14]. **1864** TROLLOPE *Can you forgive Her?* I. v. 33 What fine fellows those Alpine club men think themselves. **1900** W. ROBINSON *Eng. Flower Garden* (ed. 8) 407/1 *A. Lasiocarpa* (Alpine Fir)—A beautiful spire-like tree 150 feet high with white bark, and very small cones, purple, 2 to 3 inches long, and red male flowers.. Alaska, B. Columbia.

B. *sb.* (Also with small initial.) An alpine plant, or one that grows on high ground; also, such a plant grown in rock-gardens, etc. So *alpine frame, garden, house.*

1830 B. MAUND *Bot. Gard.* III. No. 234 *Onosma taurica*. A plant of it may also be kept in a pot amongst the alpines. **1841** MRS. LOUDON *Ladies' Comp. Fl. Gard.* (1846) 257 Of such plants as Saxifrages or other Alpines. **1881** *Encycl. Brit.* XII. 250/1 Many of the little alpines may be brought into the front line planted between suitable pieces of stone, or they may be..placed on an artificial rockery. *Ibid.* 290/1 Give abundance of air to the greenhouse, conservatory, and alpine frame in mild weather. **1907** R. FARRER *My Rock-garden* 26 *Arenaria gothica* justly heads the list of our alpines. **1925** E. F. NORTON *Fight for Everest 1924* viii. 162 As we descended, the lichens, starveling grasses and dwarf alpines rapidly succeeded each other and in turn gave way to larger growths. **1933** *Jrnl. R. Hort. Soc.* LVIII. 223 With the development of the alpine house..it would be possible to grow numbers of them. **1951** *Dict. Gardening* (*R. Hort. Soc.*) (ed. 2) I. 86/1 Alpine Garden.

2. A racial type. (Cf. *adj.* *2.)

1916 MADISON GRANT *Passing of Great Race* (1917) iv. 120 The Alpines are of stocky build and moderately short of stature. **1922** A. J. TOYNBEE *Western Question in Greece & Turkey* vi. 113 The Central Asian 'Mongoloid' type is found in Anatolia, but it is rare, and the great mass of the Turkish-speaking peasantry are 'Alpines'. **1935** HUXLEY & HADDON *We Europeans* v. 152 The Alpines (Eurasiatics), who are of medium build, rather dark, and broad-headed.

alpinism (æ·lpəiniz'm, -pin-). [ad. F. *alpinisme*: see ALPINE *a.* and -ISM.] Climbing of the Alps or of high mountains.

1884 *Standard* 5 July, Alpenism is, in this country at least, a relaxation, not a pursuit more serious than the nature of it demands. **1888** *Ibid.* 9 July 5/4 For these members of the upper crust of Alpinism, Switzerland is ceasing to have its old attractions. **1892** *Ibid.* 24 May 5/2 The Geographical Society has..admitted 'Alpinism' to be geography. **1960** KOESTLER *Lotus & Robot* II. vii. 190 Sporting activities such as skiing or alpinism, which are of recent origin.

alright, a frequent spelling of *all right.*

1893 *Durham Univ. Jrnl.* Nov. 186, I think I shall pass alright. **1897** *Westm. Gaz.* 16 Dec. 9/3 Witness said, 'Alright, come along.' **1924** H. W. FOWLER in *S.P.E. Tract* XVIII. 1 (*subtitle*) Open Court on 'Alright'. **1925** MARQUESS CURZON in Marq. of Zetland *Life* (1928) III. 378, I am sure I shall get through alright. **1926** H. W. FOWLER *Mod. Eng. Usage* 16/1 There are no such forms as *all-right, allright*, or *alright*, though even the last, if seldom allowed by the compositors to appear in print, is often seen..in MS.

Alsatian, *a.* and *sb.* Add: **B. 2.** The registered Kennel Club name for the German Shepherd Dog (*deutscher Schäferhund*), formerly known as the *Alsatian wolf-dog* (occas. *wolf-hound*).

The name *Alsatian* was adopted in order to avoid the associations of *German*. The dog does not belong to Alsace, nor is there a wolf strain in its composition.

1917 A. CROXTON SMITH in *Ladies' Field* 12 May 416/3 The French or Alsatian sheepdogs, which are now becoming familiarised to us. **1922** R. LEIGHTON *Compl. Bk. Dog* 119 The dogs lately introduced into Great Britain as the Alsatian Wolfdog and into the United States as the German sheepdog. *Ibid.* 120 The Alsatian was known in England before the war. **1923** [see WOLF-HOUND]. **1926** D. BROCKWELL (*title*) The Alsatian. *Ibid.* 25 The so-called Police Dog, or German Shepherd Dog,..variously known as the Alsatian Wolf Dog, Belgian Police Dog, and French Police Dog. **1948** C. L. B. HUBBARD *Dogs in Britain* xviii. 197 The breed [German Shepherd Dog], on its re-importation during 1918–1919, was named the Alsatian Wolfdog.

also, *adv.* Add: **A. 3. b.** In phr. *also ran*, applied to horses in a race which do not get a 'place'; also *fig.* of any unsuccessful competitor or inferior person or thing.

1896 ADE *Artie* 21 They ain't even in the 'also rans'. **1904** *Enquirer* (Cincinnati) 6 Feb. 13/3 (*heading*) George B. Cox—He Heads The List Of Also Rans. **1908** G. H. LORIMER *Jack Spurlock* xi. 269 Sporty old three-bottle ancestors, with a genius for throwing deuces and picking also-rans. **1921** F. KLICKMANN *Trail of Ragged Robin* 73 Bella's name [in a list of scholarship winners] was not even among those who 'also ran'. **1924** B. PAIN in *Nash's Mag.* June 7 He is so hard that he puts the nether millstone among the also-ran. **1926** *Daily Gaz.* (Karachi) 11 Oct. 5 Karachi Autumn Meeting. Also ran: Teddy...Also ran: Sir Visto. **1934** *Bulletin* (Sydney) 2 May 25/2 As a football tipster, he is in the also-ran class. **1946** R. KNOX *Epistles & Gospels* 73 It is not enough that he should have put his name down as a candidate in the race for salvation; he may yet be found, if he is not careful, among the also-rans, the ἀδόκιμοι, the people who are 'counted out'. **1949** WODEHOUSE *Mating Season* xxiv. 209 Dobbs..was more laboured in his movements and to an eye like mine, trained in the watching of point-to-point races, had all the look of an also-ran.

alstrœmeria (ælstrœmĭᵊ·riă). [mod. L., f. the name of Claude *Alstrœmer*, Swedish naturalist (1736–96): see -IA¹.] A plant of the genus of tropical American amaryllidaceous plants so named, grown in hothouses for their flowers.

1791 *Bot. Mag.* IV. 125 Striped-Flower'd Alstrœmeria... We treat it, and successfully, as a stone plant. **1833** *Penny Cycl.* I. 447/2 Gaily-marked flowers called Alstrœmerias. **1882** *Garden* 30 Sept. 292/3 Alstrœmerias are real gems, and should be in every garden. **1961** *Amateur Gardening* 2 Dec. 4 Alstrœmerias .. with their azalea-like colours and flower heads, are good middle-border plants.

Altaian, *a.* and *sb.* For def. read: = *ALTAIC.

Altaic, *a.* Read: *a.* and *sb.* **A.** *adj.* Of or pertaining to the Altai Mountains, or to a family of languages comprising Turkish, Mongol, and Tungus. **B.** *sb.* The Altaic family of languages.

1832 W. MACGILLIVRAY *Trav. Humboldt* xxviii. 415 Between the Altaic range and Teen-shan are Zungaria and the basin of the Ele. **1850** [in Dict.]. **1887** C. R. CONDER (*title*) Altaic hieroglyphs and Hittite inscriptions. **1905** *Dublin Rev.* Jan. 190 The Finns..belong rather to the Altaic group. **1933** BLOOMFIELD *Lang.* iv. 68 The Turkish (Turco-Tartar or Altaic) family of languages covers a vast main area, from Asia Minor, conquered, at the end of the Middle Ages, by the Ottoman Turks...These languages..

are spoken by some 39 millions of people: Turkish, Tartar, Kirgiz, Uzbeg, Azerbaijani are the more familiar language-names. **1954** Pei & Gaynor *Dict. Linguistics* 10 *Altaic*, a sub-family (also called Turco-Tartaric or Turkish) of the Ural-Altaic family of languages. It consists of three main branches: Turkic, Mongol and Manchu or Tungus... Some linguists consider Mongol and Manchu distinct sub-families of the Ural-Altaic family, and consider only the members of the Turkic group as the true Altaic languages; hence the alternative designation Turkish for the Altaic sub-family. **1959** *Chambers's Encycl.* I. 303/1 At one time the Altaic family was considered to be related to the Uralian languages, but this view is no longer held.

altar. Add: **A. 6.** Each of the steps or ledges up the sloping sides of a graving-dock. [Suggested by *altar-step*: see B. II.]

1840 *Civil Engin. & Arch. Jrnl.* III. 273 The object of these altars is for the convenience of placing the shores against the hull of a vessel at any height, and for resting the ends of spars for staging. **1885** Vernon-Harcourt *Harbours & Docks* I. 457 The sides of a graving dock..are constructed with steps, or altars, for receiving the timber props which support the vessel in an upright position.

B. I. 4. *altar-taper.*

1860 G. M. Hopkins *Escorial* in *Poems* (1930) 129 The Altar-tapers flar'd in gusts.

II. altar-card, any of a set of three cards placed on the altar (one in the middle, and one at either end), containing certain portions of the Eucharistic prayers to assist the priest's memory; **altar-tomb** (earlier example).

1849 F. Grant *Let.* 4 July in 'G. Ramsay' *Thomas Grant* (1874) v. 85 A subscription for publishing on stone a set of illuminated altar cards. **1884** Addis & Arnold *Cath. Dict.* s.v. *Altar*, Under the crucifix there ought to be an altar-card, with certain prayers which the priest cannot read from the Missal without inconvenience. **1906** *Church Times* 30 June 842/1 A 'wooden altar card' of the 18th century. It is a thick wooden panel, about 2 ft. by 18 in., painted to imitate a book, bearing the words, 'The Lord Jesus after He was betrayed,' etc., from the prayer of Consecration in the English liturgy; it stood on the altar of Redbourn Church, Herts, until about 1850. **1739** F. Blomefield *Hist. Norfolk* I. 237 Under this, is an Altar-Tomb cover'd with a Black Marble.

alter *v.* Add: **1. b.** *transf.* To geld or spay (an animal). *U.S.* and *Austral.*

1821 T. B. Hazard *Nailer Tom's Diary* (1930) 555/2 Worner Knowles oltered my four Boar Pigs. **1849** *Austral. Pastoralists' Rev.* 15 Aug. 295 For this reason bulls were rarely altered (castrated) till they were four or five years old.

alteration. Add: **1.** (Later example.)

1922 Joyce *Ulysses* 467 To alteration one pair trousers eleven shillings.

5. *attrib.*: **alteration hand,** one who alters or remakes ready-made clothes to suit customers' requirements, or renovates old or second-hand garments.

1901 *Daily Chron.* 9 Sept. 9/4 Mantles.—Wanted a good alteration hand. **1961** *Evening Standard* 21 Aug. 17/6 Alteration Hand (female)..C. & A. Modes Ltd.

alter ego (ǭ·ltəɹ ī·go, æ·ltəɹ e·go). [L. (Cicero), *alter* another, *ego* I. Cf. Gr. ἄλλος ἐγώ, ἕτερος ἐγώ.] A second self: an intimate and trusted friend; a confidential agent or representative. Hence **a:lter-e·goism,** altruism; **a:lter-egoi·stic** *a.*, altruistic.

1537 R. Layton *Let.* 4 June in *Lett. Suppress. Monast.* (1843) 156 Ye muste have suche as ye may trust evyn as well as your owne self, wiche muste be unto yowe as *alter ego*. **1622** Mabbe tr. *Aleman's Guzman d'Alf.* I. ii. 24 She would tell him, that I was his *alter ego*, that he and I were one. **1650** Trapp *Comm. Gen.* ii. 18 One..that may be to him as an *Alter-ego*, a second-self. **1652** N. Culverwel *Lt. Nature* 10 We use to call a friend *Alter ego*. **1872** Geo. Eliot *Middlem.* v. li. 148 These people might not take that high view of you which I have always taken, as an *alter ego*, a right hand. **1880** Meredith *Trag. Comedians* I. v. 93 The pleasure she had of the sensational comparison was in an alteregoistic home she found in him, that allowed of her gathering a picked self-knowledge. **1886** *Law Times Rep.* LIV. 856/1 He who makes the contract agrees to the condition that it shall not be binding on the person whose *alter ego* or representative he is if he has made any misrepresentation, or has been guilty of any concealment. **1901** M. F. Libby in *Amer. Jrnl. Psychol.* XII. 470 The social affections in Shaftesbury generally mean the alteregoistic affections. *Ibid.* 485 His tendency to see patriotic and cosmic affections as an expansion of the narrower forms of alteregoism, as shown in love, family, and party relations. **1926** D. H. Lawrence *Plumed Serpent* i. 10 'Isn't that fun!' 'No,' said Kate, her little *alter ego* speaking out for once. **1936** *Mind* XLV. 72 The wider forms of alter-egoism (so-called altruism, which is really *égoïsme à deux, à trois*, etc.). **1939** A. J. Toynbee *Study Hist.* VI. 44 The One True God whose *alter ego* Allāh was now proclaimed to be.

fig. **1856** S. Dobell *Eng. in time of War* 80 Methinks the fruit But alter ego of the root.

alternance (ǭ·l-, æ·ltəɹnăns). [Fr.: cf. Alternate *a.* and *sb.* and -ance.] Alternation, variance. Chiefly in *Philol.*

J. Baudouin de Courtenay *Versuch einer Theorie phonetischer Alternationem* (1895) uses the German word *Alternation* in the sense illustr. by quots. 1934, 1935.

1921 *Times Lit. Suppl.* 20 Oct. 676/2 There are..two tendencies or groups of tendencies which alternately dominate literature... His book..is the most coherent essay we have seen to attach this alternance to a psychological reality. **1934** J. R. Firth in *Le Maître phonétique* XLVI. 45 Kruszewski, however, extended the term phoneme to include sound alternances associated with changes of morphological categories. **1935** —— in *English Studies* XVII. 11 Vowel alternance is also a very important morphological instrument in the strong conjugation of verbs. There are thirty vowel alternances for our babies to learn. **1957** C. L. Wrenn in West & Kimber *Deskb. Correct English* p. vi, Doubt-provoking alternances in spelling or in grammatical usage. *Ibid.*, The alternance of spellings and usages of all kinds has struck me very forcibly.

alternant. Add: **B.** *sb.* **2.** *Logic.* An alternative proposition or term; one of the components in an alternation (see *alternation 7 b); = *disjunct *sb.*

1892 W. E. Johnson in *Mind* I. 237 The alternants of the alternative synthesis. **1912** *Mind* XXI. 163 The argument..proves that there are *alternants* of universals like isosceles and scalene with respect to triangularity. **1921** W. E. Johnson *Logic* I. iii. 31 In the alternative function 'Either *p* or *q*' *p* and *q* are alternants. *Ibid.* 36 It may very well be the case that the alternants in an alternative proposition are..'exclusive' to one another. **1957** J. G. Brennan *Handbk. Logic* vi. 95 If at least one of the alternants must be true, then if one of them is false, the other is true.

3. *Philol.* The name given to any one of the possible variants found within any particular feature of speech or language. Cf. *alternance.

1926 L. Bloomfield in *Language* II. 160 Absence of sound may be a phonetic or formal alternant. **1942** Bloch & Trager *Outl. Ling. Analysis* 57 We call the three forms [-ez, -z, -s] alternants of the same morpheme. **1949** C. E. Bazell in *Archivum Linguisticum* I. 2 There are variant expressions of the morpheme (the so-called morpheme-alternants) which may encounter phonemes of distinct morphemes in so far as they have no phonemes in common.

alternate, *a.* and *sb.* Add: **A.** *adj.* **5. b.** *Electr.* *alternate current* = alternating current (see next).

1858 Wheatstone *Brit. Pat.* 1241 8 No contrivance is..required to invert or to stop out the alternate currents. **1877** Knight *Dict. Mech.* I. 78/1 Alternate currents of opposite character are generated in each set of bobbins, the polarity being changed at the moment of polar passage. **1878** *Design & Work* 23 Feb. 134/2 In each revolution, there will have been induced 16 alternate currents.

8. = Alternative 3. *U.S.*

1961 in Webster. **1962** *Amer. Speech* XXXVII. 109 Only the *PI&E* records [sic], and this as an alternate end to four words.

B. *sb.* **1.** (Further examples.) Now chiefly *U.S.*

1915 L. M. Phillipps *Form of Colour* iv. 70 On the completion of Santa Sophia an alternate confronted the Byzantine architects. **1952** *Times* 17 Oct. 7/5 Their official reasons for proposing an extension of Gatwick were that it was required..as an 'alternate' to London Airport. **1962** J. Glenn in *Into Orbit* 37 In the interests of safety all the major systems had to be studded with alternates or stand-by components.

2. One who is appointed to act in place of a delegate who is unable to be present; a substitute. *U.S.*

1848 *N.Y. Weekly Tribune* 26 Feb. 4/1 *Resolved*, That the Chair appoint a Committee..to report to this Convention thirty-six delegates to the National Convention; also an alternate to each delegate. **1888** Bryce *Amer. Commw.* II. iii. lxix. 542 To every delegate there is added a person called his 'alternate',..to replace him in case he cannot be present..; if from any cause the delegate is absent, the alternate steps into his shoes. **1895** *Denver Times* 5 Mar. 2/7 Each precinct is entitled to delegates and alternates as follows.

3. A person who alternates with another in the occupation, or performance of the duties, of an office.

Cf. F. *alternat* (used also in Eng. context), the arrangement according to which rotation of office is maintained among persons of equal rank, etc.

1898 *Westm. Gaz.* 21 Apr. 4/3 When sitting at our Board as an alternate in London for Mr. Rhodes. **1908** *Ibid.* 21 July 5/2 In 1903 he was appointed alternate to the Chief of the Admiralty Staff.

alternating, *ppl. a.* Add: **d.** *alternating current* (abbrev. A.C., a.c.): an electric current which reverses its direction at regular intervals. So *alternating machine*, etc.

1839 *Ann. Electr., Magn., & Chem.* III. 389 The alternating currents, from the semi-revolutions of the armatures, are converted into a current of the same direction, by the application of my pole changer. **1869** *Brit. Pat.* No. 3196 A foot motion..is employed to generate a rapid succession of alternating currents. **1879** G. Prescott *Sp. Telephone* xiv. 491 Alternating magneto machine. **1884** F. Krohn tr. *Glaser de Cew's Magn.- & Dyn.-Electr. Mach.* 249 The Ferranti alternating current generator. **1888** E. Atkinson tr. *Mascart & Joubert's Electr. & Magn.* II. 289 Use of the Electrodynamometer with Alternating Currents. **1903** *Electr. Times* 31 Dec. 962/1 (*heading*) The Ferranti A.C. Meter. **1910** *Hawkins' Electr. Dict.* 2/1, a.c., abbreviation for *alternating current*. **1922** Glazebrook *Dict. Appl. Physics* II. 30/2 The potentiometer is then thrown on to the A.C. supply. **1931** *Times Trade & Engin. Suppl.* 24 Jan. p. iv/3 The Mooltan has British Thomson-Houston a.c. generators.

alternation. Add: **7.** *Logic.* The truth-function that has the value 'true' whenever at least one of its components is true, or that has the value 'false' only when every component is false; the function that is usually symbolized by 'v' (L. *vel*) and corresponds to the inclusive sense of the word 'or' that is sometimes rendered by 'and/or'.

1874 W. S. Jevons *Princ. Science: Treat. Logic* I. v. 81 We require a sign of the alternative or disjunctive relation, equivalent to one meaning at least of the little conjunction *or* so frequently used in common language. ... In my first logical Essay I..adopted the common sign ·|·; but this sign should not be employed unless there exists exact analogy between mathematical addition and logical alternation. **1890** E. E. C. Jones *Elem. Logic* 110 Mill and Jevons..insist upon the non-exclusiveness of alternation. **1940** Quine *Math. Logic* i. 12 Alternation—composition of statements by means of the connective 'or'. **1952** A. N. Prior *Formal Logic* i. 8 Writers who use 'disjunction' in this way generally call the non-exclusive 'Either *p* or *q*', the 'alternative' function or 'alternation' of its arguments... The alternation of two propositions is also often called their logical sum.

b. A compound statement or formula formed by joining two or more statements or formulas by the connective symbol 'v' or the word 'or'.

1894 J. N. Keynes *Formal Logic* (ed. 3) ii. ix. 231 To deny an alternation is the same thing as to affirm a conjunction. **1950** Quine *Methods Logic* (1952) i. 19 'Either' and 'or' may be used to mark the boundaries of the first component of an alternation.

alternative, *a.* Add: **6.** *alternative vote*, a system of voting in which the voter places the names of the candidates in the order in which he supports them. Cf. *preference voting.

1910 *Rep. R. Comm. Elect. Systems* (Cd. 5163) 3 *The Alternative Vote.* Here the voter is invited to arrange the candidates in the order of his choice by placing the figures 1, 2, 3..against their names. **1926** Hoag & Hallett *Proport. Represent.* x. 483 (*heading*) The Alternative Vote (The Single Transferable Vote as a Majority System). *Ibid.*, Such a preferential ballot is provided under the single transferable vote used as a majority system, known in Australia simply as 'preferential voting', in Great Britain and Canada as the 'alternative vote'.

alternatively, *adv.* Add: **1.** Now freq. = as or by way of an alternative.

1884 *Law Rep. Queen's Bench* XIII. 674 The defendants claimed 30,000 *l.* in respect of their counter-claim, and alternatively 30,000 *l.* damages. **1912** *Times* (weekly ed.) 4 Oct. 794 Mr. Paul Taylor imposed the full penalty of £5, or, alternatively, sentenced the prisoner to one month's imprisonment. **1921** *Act* 11 & 12 *Geo. V* c. 13 § 1 Liable..to a fine not exceeding [25 *l.*], or alternatively or in addition thereto to be imprisoned..for a term not exceeding three months. **1927** *Sunday Express* 8 May 4 What you do..is to secure a photograph of the hall... Alternatively, you may make an exact drawing of the hall.

alternator (ǭ·ltəɹnē[i]təɹ). *Electr.* [f. Alternate *v.* + -or.] A dynamo giving an alternating current.

1892 *Sci. Amer.* 16 Apr. 246/1 (*heading*) 250-Kilowatt 'Mordey-Victoria' Alternator. *Ibid.*, The electromotive force of the alternator varies only about five per cent, even if the full load is thrown on or off. **1893** G. Kapp *Dynamos* 10 The alternator, in which mechanical energy of rotation is converted into the energy of an alternating current. **1919** R. Stanley *Wireless Telegr.* (ed. 2) I. 86 An alternator is a machine in which the difference of potential, or voltage, induced has not a constant value, as in an ordinary direct current generator, but rises and falls and reverses in direction many times per second. **1962** *Newnes Conc. Encycl. Electr. Engin.* 337/2 The turbo-alternator is essentially a high-speed construction..for coupling to steam or gas turbines..the salient-pole alternator is suited to..lower speeds and may therefore be driven by water turbines or internal-combustion engines.

alterne (ǭ·ltəɹn, æl-). *Ecology.* [Cf. F. *alterne* adj. alternate and Altern *a.*] A contrast, often sharply defined, shown by vegetation in adjoining areas; an area exhibiting such a contrast.

1916 F. E. Clements *Plant Succession* 115 Significance of alternation. Alternation is the consequence of disturbed or incomplete zonation. Such areas produce *alternes*, which it now seems can always be related to more primary zones. **1920** —— *Plant Indicators* 73 Communities show general structural features, such as zones, alternes, layers... Alternes are due to the interruption of zonation through any cause whatsoever, but they are especially typical where disturbed..areas are found. **1926** Tansley & Chipp *Study of Vegetation* iv. 53 When there is no such regular spatial or temporal change, different communities *alternate* with one another, and the different areas occupied by these may be called *alternes* in contrast to zones. **1929** Weaver & Clements *Plant Ecology* 7 Where conditions change abruptly instead of gradually and zonation is disturbed or incomplete, vegetation exhibits *alternes*.

althorn (æ·lthǭɹn). *Mus.* [G.: cf. Alt[2].] A high-pitched instrument of the saxhorn family; an alto or tenor saxhorn.

1859 C. Mandel *Treat. Instrumentation Milit. Bands* 38 The Tenor Horn or Baryton (Alt-Horn in B Flat). **1879** Grove *Dict. Mus.* I. 57/2 *Althorn*, an instrument of the Saxhorn family... It is exclusively used in military music. **1939** A. Carse *Mus. Wind Instr.* xx. 296 Tenor Horn, E Flat Saxhorn... The earliest of these various alto or tenor instruments was probably the German *Althorn* or *Altkornett* which appeared in Berlin round about 1830. *Ibid.* 299 Instruments in C or B flat have also been called 'alt-horns'.

altimeter. Add: Also with pronunc. (æ·ltimītəɹ).

2. A form of aneroid barometer which indicates the altitude reached, esp. in aviation. Also *attrib.* See also *radio altimeter.

1918 'Boyd Cable' *Air Men o' War* 179 His altimeter showed him to be a bare couple of hundred feet up. **1923** *Blackw. Mag.* July 101/1 His altimeter needle crawled down from nine to eight thousand feet. **1950** *Gloss. Aeronaut. Terms* (*B.S.I.*) 20 *Altimeter calibrator*, an apparatus for measuring the instrument errors of an altimeter. *Contacting altimeter*, an instrument in which electrical contacts are made or broken at a predetermined height. *Ibid.* 41 *Recording altimeter*, an instrument by which variation in height is recorded against time.

altimetric (æltime·trik), *a.* [f. Altimetry + -ic[1].] = Altimetrical *a.*; relating to or concerning altimetry. Spec. *altimetric frequency curve*: in morphological analysis, a graph constructed to show the distribution of areas of a certain height, or of the highest or lowest points, in a given area.

1900 *Geogr. Jrnl.* Oct. 472 He proposed to carry a chain of altimetric observations to Kara-koshun and Chaklik. **1935** H. Baulig *Changing Sea Level* 42 (*caption*) Altimetric curves (from frequency of spot heights) for peninsular Brittany, [etc.]. **1937** Wooldridge & Morgan *Physical Basis Geogr.* xvii. 266 Altimetric frequency curves... If the map is divided into small uniform squares, we may estimate the elevation of the highest point in each square, or use simply the spot heights shown on the map, providing that these, in general, mark elevated points. A curve may then be constructed showing altitudinal distribution of such points over the region. **1948** *Jrnl. Inst. Navig.* I. 57 It would be unlikely that altimetric determinations of drift over the Atlantic Sea would be reliable at levels below 5000 feet. **1952** Monkhouse & Wilkinson *Maps & Diagrams* ii. 89 Two altimetric frequency curves of Cornwall and Devon. **1958** F. E. Zeuner *Dating the Past* (ed. 4) vii. 225 Resting on an irregular rock-floor..a beach conglomerate is found which, on altimetric grounds, can be correlated with the Late Monastirian phase of the Last Interglacial.

altiplanation (æ:ltiplănē[i]·ʃən). *Geol.* [f. Alti- + *planation.] (See quot. 1916.)

1916 H. M. Eakin *Yukon-Koyukuk Region* 78 The author has suggested the term 'altiplanation' to designate a special phase of solifluction that, under certain conditions, expresses itself in terrace-like forms and flattened summits and passes that are essentially accumulations of loose rock materials.

altitude. Add: **8.** (Later example.)

1849 *Lett. Miss Riversdale* III. 7 You are getting into your English altitudes.

9. *attrib.*, as *altitude table*; **altitude chamber** *Aeronaut.*, a chamber in which the air pressure, temperature, etc., can be regulated to simulate conditions at different altitudes; **altitude control** (see quots.); **altitude sickness,** sickness brought on by ascent to a high altitude.

1935 *Jrnl. R. Aeronaut. Soc.* XXXIX. 909 A full-scale engine (R. R. Condor IIIA) working in an altitude chamber. **1921** R. M. Pierce *Dict. Aviation* 23 *Altitude-control*, the controlling apparatus by which the altitude of an airship is regulated. **1919** W. B. Faraday *Gloss. Aeronaut. Terms* 11 *Altitude control*, a device fitted to a carburettor or other part of induction system..to obtain a correct mixture of the fuel gas at high levels. **1934** *Flight* 5 Feb. 111/2 Limitation of the movement of the altitude control in aircraft might well be discontinued. **1920** W. H. Wilmer *Aviation Med.* [illegible] Altitude sickness..robs the pilot of the opportunity of recognizing that he is in danger. **1907** F. Ball (*title*) Altitude tables..designed for the determination of the true bearing..of the sun..without logarithmic computation.

altitudinous (æltitiū·dinəs), *a.* [f. L. *altitūdin-*, *altitūdō* Altitude.] Used affectedly for: high, lofty.

1868 H. T. Tuckerman *Collector* 55 The confined and altitudinous cells into which so many of the complacent victims of these potentates are stowed. **1890** *Cornhill Mag.* Feb. 146 The keen air of this altitudinous city exhilarated him. **1921** *Public Opinion* 15 June 57/3 Enveloped in the mists of his altitudinous thoughts.

altivolant, *a.* Delete †*Obs.* and add later examples.

1833 W. E. Wall in *Fraser's Mag.* VIII. 659 Bore him altivolant from Salem's towers. **1960** *Aeroplane* XCIX. 59/1 Another altivolant classification was for altitude sustained by an aircraft for 15–25 km. or go min.

alto, *sb.*[2] and *a.* Add: **A.** *sb.* **6.** *ellipt.* for *alto saxophone*. Also *attrib.* Hence **a·ltoist, a·lto-man,** an alto saxophonist.

1876 Stainer & Barrett *Dict. Mus. Terms* 385/2 *Saxophones*..are six in number, the high, soprano, alto, tenor, baritone and bass. **1927** *Melody Maker* June 553/3 Lionel Clapper..trots out the 'hottest' extemporisations on his alto. **1928** *Ibid.* Feb. 184/2 The 'hot' alto chorus by Breed. **1949** L. Feather *Inside Be-Bop* ii. 12 Charlie Parker offers inspired alto solos. **1952** B. Ulanov *Hist. Jazz in Amer.* (1958) xxi. 276 Alto-men of course—they all imitated Bird. **1956** M. Stearns *Story of Jazz* (1957) xviii. 242 Altoist Julian 'Cannonball' Adderley from Florida.

B. *adj.* Also, applied *gen.* to the second highest member of a family of musical instruments: high, tenor.

Whether *alto* in such contexts is an adjective or a combining form (i.e. Alto- 1) is not certainly determinable, and the presence or absence of a hyphen is an uncertain guide. Many of the examples given below could be placed with equal justification s.v. Alto- 1.

1802, 1856 [see Alto- 1]. 1939 [see *althorn].

b. *spec.* **alto saxophone,** the second highest member of the saxophone family, usually pitched in E flat. Abbrev. **alto sax.** Hence (often written with hyphens) **alto saxist, saxophonist,** one who plays the alto saxophone.

1869 C. Mandel *Mandel's Syst. Mus.* xvi. 68 There are various kinds of Saxophones. The smallest..is in B flat; the next, or Alto Saxophone, is in E flat. **1889** Grove *Dict. Mus.* IV. 780/2 A. Adam gives an effective solo to the E♭ Alto Saxophone in his opera 'Hamlet'. **1927** *Melody Maker* Aug. 785/2 The best features are the alto saxophone in the third chorus and the 'hot' fiddle in the first half of the last chorus. **1955** Keepnews & Grauer *Pict. Hist. Jazz* iii. 40 Cobb himself could play trumpet, clarinet, alto sax and banjo. **1958** S. Traill in P. Gammond *Decca Bk. of Jazz* vi. 82 Alto-saxophonist Boyce Brown and pianist Floyd Bean both play in real Chicago style. A. Morgan *Ibid.* xii. 443 The resultant band..contained..alto-saxist Lee Konitz.

c. alto-horn, (*a*) = *althorn; (*b*) *U.S.*, an alto saxophone.

1934 Webster, *Alto horn*, the althorn. **1940** C. Sachs *Hist. Mus. Instrum.* 429 The *alto horn*, called *Althornett*..is coiled either in the shape of a trumpet, or upright as a tuba, or circular as a horn. Its usual pitch is E♭ or F. **1946** R. Blesh *Shining Trumpets* (1949) vii. 161 To some extent, the alto horns and piccolos dropped out of the march band during this period. **1957** W. C. Handy *Father of Blues* v. 64, I had bought a tenor sax for myself, but W. N. P. Spiller, our alto horn player, appropriated it mostly for his own use.

alto-cumulus (æ:lto,kiū·miŭlŏs). *Meteorol.* [mod.L. *alto-*, f. L. *altus* high + Cumulus.] A cloud-formation made up of rounded masses similar to cumulus clouds but at a higher altitude; a cloud of this kind. Similarly **alto-stra·tus,** a cloud-formation consisting of a more or less uniform continuous layer or veil resembling stratus or cirro-stratus clouds but lying in the middle-cloud region. So **alto(-) cloud,** an alto-cumulus or alto-stratus cloud; also *alto* ellipt.

1894 *Nature* 8 Feb. 344/2 Clouds having altitudes from 3000 to 6000 metres...Alto-cumulus...Alto-stratus. **1898** *Jrnl. Sch. Geogr.* (U.S.) Oct. 297 The development of alto-cumulus and heavy cumulus clouds over the land, while out to sea only small cumuli were visible. **1905** Clayden *Cloud Studies* 59 From cirro-cumulus and cirro-stratus we pass through almost insensible gradations to the denser forms classed together in the alto group. *Ibid.* 62 The simplest alto cloud is alto-stratus.

altogether. Add: **B.** *adv.* **4.** In all, in total amount.

1797 H. Cox *Jrnl. Resid. Burmhan Emp.* (1821) 93 We were in the palace tent altogether about an hour and a half. **1811** Jane Austen *Sense & Sens.* I. ii. 24 Altogether, they will have five hundred a-year amongst them. **1871** S. T. Hall *Morning Studies* iii. xii. 158 Debt amounting altogether to not much more, perhaps, than twenty pounds. **1928** H. W. Freeman *Joseph & His Brethren* xxx. 259 Altogether they did not spend as much on the whole meal as some of his other customers on drink alone.

5. On the whole, taking everything into account.

a1817 Jane Austen *Persuasion* (1818) IV. xi. 256 Though we could have wished it different, yet altogether we did not think it fair to stand out any longer. **1888** W. R. Inge *Society in Rome* 44 Altogether, Roman slavery at this time contrasts favourably in many ways with the negro slavery of Christian nations. **1927** W. E. Collinson *Contemp. Eng.* 43 Altogether I cannot think of any modern writer who has exercised so far-reaching an influence on our every-day speech.

C. *sb.* **b.** *the altogether* (colloq.): the nude.

1894 Du Maurier *Trilby* I. 185, I have sat for the 'altogether' to several other people. **1908** *Daily Chron.* 16 Apr. 5/7 Mme. Sarah Bernhardt frankly says she sees nothing wrong in the 'altogether'. **1947** N. Balchin *Lord, I was Afraid* 52 Should I get a kick out of just seeing a girl in the altogether?

attrib. **1896** *Punch* 25 Jan. 45/2 O, Röntgen..Your worse than 'altogether' state Of portraiture we bar *in toto*!

c. *pl.* A set of tights for the whole body.

1927 *Observer* 24 July 13/2 Sokolova impersonated Death in scarlet altogethers.

¶ There is a common tendency to write *altogether* where *all together* is logically preferable. Cf. sense A. 2 in Dict.

1765 Mrs. Glasse *Art of Cookery* p. iv, Put all the ingredients together again,..strain it off well,..and give it a boil altogether. **1837** T. Bacon *First Impr. Nat. in Hindostan* I. 243 Of infinitely greater importance to the creation than the sun, moon and stars altogether. **1861** Trollope *Orley F.* I. xxix. 225 In that field the dogs were now running, altogether, so that a sheet might have covered them. **1880** Grove *Dict. Mus.* II. 574/2 The pipes of the early organs are said to have sounded at first altogether. **1930** E. Raymond *Jesting Army* I. iii. 45 A medical officer..and the whole of his Sick Parade ran altogether.

alto-stratus: see *alto-cumulus.

altricial (ăltri·ʃ'ăl), *a.* *Ornith.* [f. mod.L. *Altrices* division of birds, f. L. *altric-*, *altrix*, fem. of *altor* nourisher (*alĕre* to nourish) + -al.] = *nidicolous *a.*; opp. Præcocial *a.*, q.v.

1872 Coues *N. Amer. Birds* 224 Pigeons are altricial, and monogamous. **1884** *Ibid.* (ed. 2) ii. i. 33 § 8. Altricial birds such as are reared by the parents in the nest. **1885** *Athenæum* 1 Aug. 146/2 The altricial herons. **1902** [see Præcocial *a.*].

alula (æ·liūlă). Pl. **alulæ.** [mod.L. dim. of *āla* wing.]

1. *Ornith.* The bastard wing of a bird (see Bastard *a.* 5 c).

1772 [see Bastard *a.* 5 c]. **1959** Van Tyne & Berger *Fund. Ornith.* iii. 82 The group of feathers borne by the 'thumb' or 'pollex' is called the alula. *Ibid.* 83 Alula quills should be numbered from the innermost to the outermost.

2. *Ent.* A small scale-like appendage at the base of each wing of many *Diptera*, above the halteres. Also, a similar appendage beneath the elytron in some water-beetles.

1817 Kirby & Spence *Entomol.* II. 359 (*Diptera*) Their winglets (*Alulæ*). **1877** *Encycl. Brit.* VI. 127 In certain water beetles (Dytiscidæ) a pair of *alulæ*, or winglets, are developed at the inner angle of the elytra. **1899** D. Sharp in *Camb. Nat. Hist.* VI. vii. 447 On the hind margin of the wing, near the base, there is often a more or less free lobe: the alula.

alum, *sb.* Add: **5.** *alum-house*; **alum-mine,** raw material from which alum is obtained.

1612 W. Strachey *Travaile into Virginia* (1849) I. i. 33 We doe allready heare the Indians talke both of allum mines and of copper. **1669** *New Cabl. Weekly* 25 The Liquor is conveyed to the Alum-house. **1830** G. Colman *Random Rec.* I. vi. 187 Most readers will pardon me for not taking them into the Alum-House, to explain the several methods of crystallization &c.

alum, *v.* Delete '*rare*' and add earlier and later examples.

a1430 in *Middle Eng. Dict.* **1598** Florio *Worlde of Wordes* 14/3 *Allumare*, to alume with alume. Before they can be died into any light colour. **1735** J. Barrow *Dict. Polygraphicum* II. s.v. *silk*, How to alum the boiled silk. **1877** [illegible] **1889** *Internat. Ann. Anthony's Photogr. Bull.* II. 108 The plate should be alumed before and after the operation.

alumed, *ppl. a.* (Earlier and later examples.)

a1425 in *Middle Eng. Dict.* **1893** *Pall Mall Gaz.* 30 Jan. 7/3 Three hundred alumed skins. **1896** *Kodak News* Aug. 29/1 During development, fixing and aluming, it showed no sign of blistering... Using an alumed fixing bath..improved matters.

alumina. Add: **b.** *attrib.*

1908 *Chem. Abstr.* 735 Process of producing alumina solutions compounded with plant extracts. **1909** *Cent. Dict. Suppl.*, *Alumina cream*, freshly precipitated aluminium hydrate held in suspension in water. **1958** R. Mayer *Artist's Handbk.* ii. 40 *Alumina hydrate*, aluminium hydroxide, artificially produced. A white, fluffy, light-weight powder which becomes virtually colourless and transparent when it is ground in oil. **1963** *Lancet* 5 Jan. 13/1 The management of the isotope on an alumina column has proved quite easy and the expense not unreasonable.

aluming, *vbl. sb.* (Later examples.)

1862 C. O'Neill *Dict. Calico Printing & Dyeing* 15/1 By aluming the silk is considered to take the dye better. **1896** [see *alumed *ppl. a.*].

aluminium. Add: **b.** (Additional examples.) Also *aluminium brass*, an alloy of aluminium and brass; *aluminium foil*, paper-thin sheet aluminium, used as wrapping material, etc.; *aluminium hydroxide*, a compound used in medicine and industry.

1866 Roscoe *Elem. Chem.* xx. 181 Aluminium chloride, Al_2Cl_6, is a volatile yellow solid body... it is used in the manufacture of the metal. *Ibid.*, The soluble aluminium sulphate, $Al_2(SO_4)_3$, is prepared..for the use of the dyer by decomposing clay by acting upon it with sulphuric acid. **1871** *Ibid.* (ed. 3) xvii. 188 The hydroxides..may be represented as six molecules of water in which half of the hydrogen is replaced by a hexad group, thus: Aluminium hydroxide..(or Hydrate of Alumina). **1875** Knight *Dict. Mech.* I. 78/1 There are also aluminium silver, aluminium tin, aluminium bronze..for certain parts, such as journals of engines..it has proved itself superior to all other metals. **1892** *Photography* 4 Aug. 11. 71 Aluminium foil is easily ignited and gives no smoke. **1907** G. S. Newth *Inorg. Chem.* (ed. 12) iii. viii. 618 Aluminium hydroxide unites with many soluble organic colouring matters, and precipitates them from solution as lakes. **1909** *Cent. Dict. Suppl.*, *Aluminium brass*. **1934** *Discovery* June 162/1 It [*sc.* Alucol, a patent medicine] was aluminium hydroxide, and was given in doses of 1 grm. from three to six times a day. **1958** A. D. Merriman *Dict. Metall.* 8/1 *Aluminium Bronze*, this is the name conferred on a series of copper-aluminium alloys which may or may not also contain additions of iron, manganese, nickel, tin and zinc... It has been used for making imitation gold cigarette cases and trinkets. **1959** *Engineering* 20 Feb. 250/1 The heat transmission properties of aluminium brass are good, and the material is remarkably resistant to corrosion. **1962** *House & Garden* Jan. 54/2, 2 pieces of aluminium foil to cover breasts of chicken. **1964** N. G. Clark *Mod. Org. Chem.* 546 Aluminium chloride, $AlCl_3$, is a pale yellow, fuming, hygroscopic solid, which hydrolyses in contact with water.

aluminize, *v.* Add: **2.** *trans.* To spray or coat with aluminium. Hence **alu:miniza·tion; alu·minized** *ppl. a.*; **alu·minizing** *vbl. sb.*

1934 *Sci. Amer.* Feb. 81/3 With aluminized mirrors and quartz prisms and lenses, the stellar spectrum between 3400 and 3000 angstroms should now be easily observable. *Ibid.*, The construction of an aluminizing apparatus big enough to hold a good-sized mirror..and provided with the necessary devices for..volatizing the aluminium, involves considerable time and cost. **1938** *Times* 2 June 4/5 The cost of aluminizing the mirror has been found to be prohibitive, and the observatory will have to be content with a silvered mirror. **1940** *Jrnl. Iron & Steel Inst.* CXLII. 114 A This process is usually known as 'aluminising', and the coating may be said to consist of three layers. **1946** W. Lewis *Thin Films & Surfaces* vii. 60 Another advantage of aluminized mirrors as compared with silvered mirrors..is that aluminium films do not scatter light. *Ibid.* 61 Aluminization improves the reflectivity of speculum metal gratings. **1957** Amos & Birkinshaw *Television Engin.* I. 301 Aluminised Cathode-ray Tube Screen. **1960** *Metallurgia* Jan. 15/1 The coatings were of the hot-dipped aluminium, aluminised (aluminium sprayed and subjected to a diffusion heat treatment), and chromised types.

alumino·graphy. Fuller form of *algraphy.

1909 in Webster.

alumino-silicate (ăl[1]ū:mino,si·likēt). [f. Alumino- + Silicate.] A natural compound which occurs in many minerals, esp. feldspars, forming the basis of all clays and used industrially as a constituent of strengthened glass. Also *attrib.*

1907 *Chem. Abstr.* 2547 The action of free humus acids on alumino-silicates is identical with that of water. **1934** W. H. Taylor in *Proc. R. Soc.* A. CXLV. 80 (*title*) Nature and Properties of Aluminosilicate Framework Structures. **1935** *Chem. Abstr.* 7550 The clays should therefore be considered aluminosilicates. **1946** *Electronic Engin.* XVIII. 299 Experiments with new aluminosilicate glasses..are giving interesting results. **1962** *Aeroplane* 29 Mar. 43/2 Wind-screens for supersonic transports incorporating alumino-silicate..are to be produced by Triplex.

aluminothermy (ăl[1]ū:minoþɔ̄·ɹmi). [ad. G. *aluminothermie* (H. Goldschmidt 1900, in *Zeitschr. Ang. Chem.* 919).] A process by means of which high temperatures are produced by the combination of aluminium and oxygen. So **alu:minothe·rmic** *a.*, of or pertaining to aluminothermy; **alu:minothe·rmics,** a collective name for processes involving aluminothermy.

[**1900** *Engineering Mag.* Aug. 755/2 An important new industry, to which the name 'Aluminothermie' has been given.] **1904** H. Goldschmidt in *Electro-chem. Ind.* II. 145 Production of pure metals free from carbon by the aluminothermic method. **1904** *Railroad Gaz.* I. 111 (*title*) Alumino-thermics and its use for welding and metallurgical purposes. **1909** *Cent. Dict.* Suppl., Aluminothermy. **1916** *Chambers's Jrnl.* Apr. 270/1 In the preparation of the cobalt for combination with the steel, what is known as the alumino-thermic process is practised. **1958** *Chambers's Techn. Dict.* 955/1 Aluminothermic process..also known as the thermite process, is used especially for the oxides of metals which are reduced with difficulty. **1958** A. D. Merriman *Dict. Metallurgy* 8/2 Aluminothermy..is used for welding certain articles *in situ*.

aluminous, *a.* Add: **2.** *aluminous cement.*

1930 *Engineering* 14 Feb. 238/1 The net effect..is..to avoid the use of either salt [sodium chloride or calcium chloride] in the preparation of reinforced concrete or with aluminous cements. **1940** *Chambers's Techn. Dict.* 26/1 The aluminous cements..possess rapid-hardening qualities.

aluminum (ălū·minəm). The usual form in the U.S. of Aluminium.

alumna (ălɒ·mnă). Pl. **alumnæ.** [L., fem. of *alumnus*.] A female graduate or former student of a school, college, or university. Chiefly *U.S.*

1882 M. Harland *Eve's Daughters* 177 The Alumnæ and Alumni of Oberlin. **1896** *Century Mag.* LI. 798/1 The average salary of the alumna teacher would be below rather than above $1000 a year. **1939** A. Huxley *After Many a Summer* vi. 69 Successive generations of the College's Alumni and Alumnæ.

alumnus. Add to def.: Also, a graduate or former student. (Examples of U.S. use, esp. in *pl.*)

1696 Sewall *Diary* 12 Oct. (1878) I. 435 Lt. Govr...promised his Interposition for them, as become such an Alumnus to such an Alma Mater. **1823** J. & R. C. Morse *Traveller's Guide* 320 The number of alumni, that is, the number who have been educated at each college since its establishment. **1843** Hopkins in B. H. Hall *College Words* (1851) 7 So far as I know, the Society of the Alumni of Williams College was the first association of the kind in this country... It was formed September 5th, 1821. **1890** *Harper's Mag.* Apr. 799/1 The associated alumni of a college organized into a club preserve..the old feeling of comradeship.

attrib. **1843** Hopkins in B. H. Hall *College Words* (1851) 8 Last year, for the first time, the voice of an Alumnus orator was heard at Harvard. **1851** *Ibid.*, An Alumni Society was formed at Columbia College in the year 1829. **1895** *Century Mag.* Sept. 794/2 How often at an alumni banquet is intellectual supremacy in college life praised? **1906** *Springfield* (Mass.) *Weekly Republ.* 28 June 10 Tuesday was alumni day at Yale, when hundreds of old graduates..gathered in alumni hall.

alundum (ălɒ·ndəm). [f. Alu(minium + Cor)undum.] The proprietary name of a hard material produced by fusing alumina in an electric furnace, used chiefly as an abrasive agent.

U.S. Patent by C. B. Jacobs, 1900, but the word *alundum* was not used in it.

1904 *Sci. Amer.* Suppl. 24 Sept. 24018/2 *Alundum*, the trade name for artificial corundum, is an abrasive made by a process due to C. B. Jacobs and others. **1905** *Trans. Faraday Soc.* I. 301 Artificial corundum, or, as the Americans call it, 'alundum', also deserved notice..as, in addition to being highly refractory, it was of extreme hardness. **1906** *Westm. Gaz.* 27 Dec. 2/1 The advent of the electric furnace has brought about a revolution by the introduction of alundum. **1918** *Nature* 4 July 350/2 The fusion of bauxite gives another abrasive, alundum, which has practically superseded other materials for the grinding of steel.

alure. Delete †*Obs.* and add examples.

1878 M'Vittie *Chr. Ch. Cath.* 63 Round north and south transepts and on to the alure of choir. **1919** *Proc. Soc. Antiq. Scot.* LIII. 38 Its base in the ditch must have been visible from the allure on the curtain.

alurk, *adv.*, prop. *phr.* Restrict †*Obs.* to sense in Dict. and add: **2.** Lurking. *rare.*

1895 F. Thompson *Sister-Songs* 20 And whitest witchery, a-lurk in that Authentic cestus of two girdling arms.

alveolar. Add: **A.** *adj.* **1. b.** Pertaining to or resembling an alveolus in a membrane, air-cell of the lungs, etc. Cf. *alveolus senses d, e, f.

1848 *Quain's Anat.* (ed. 5) I. p. cclxxxi, This peculiar character of the mucous membrane, which might be called 'alveolar', is seen very distinctly in the gall-bladder;..still more minute alveolar recesses with intervening ridges may be discovered..on the mucous membrane of the stomach. **1882** *Ibid.* (ed. 9) 228 The flattened cells which compose the basement-membrane may send delicate lamellar processes between the alveolar cells. **1885** E. A. Schaefer *Essent. Histology* xxv. 123 Section of injected lung, including several contiguous alveoli... Between the capillaries is seen the homogeneous alveolar wall with nuclei of connective-tissue corpuscles. **1889** *Cent. Dict.* s.v., *Alveolar passages*, the passages into which the respiratory bronchial tubes enlarge. **1927** Haldane & Huxley *Anim. Biol.* vii. 153 This, which is called the alveolar air, can be obtained at the end of a deep breath out.

B. *sb.* **2.** *Phonetics.* An alveolar sound or letter. Cf. *apico-alveolar sb.

1895 P. Giles *Man. Compar. Philol.* i. v. 68 The sounds called dentals—*t, d, th, dh,* where *th* represents not the sound in *then* or *thin* but *t* followed by breath—are in English pronunciation not dentals but alveolars, being produced by the pressure of the tongue against the roots of the teeth. **1910** *Encycl. Brit.* VII. 725/1 The English *d*..in phonetic terminology is called an alveolar. In the languages of India..both true dentals and alveolars are found. **1961** R. B. Long *Sentence & its Parts* xix. 430 This obstruction [in the flow of air] can occur..in the front of the mouth above the teeth, as for the alveolars.

Hence **alveola·rity,** the quality of being alveolar; an instance of this.

1952 A. Cohen *Phonemes of English* ii. 36 If now we put 'normal' [t] against [p] we find: alveolarity v. labiality. **1964** R. H. Robins *Gen. Linguistics* iv. 155 The nine separate phonemes..are maintained as distinctive units by six features: plosion, voice, nasality, bilabiality, alveolarity, and velarity.

alveolo-palatal (ælvi·ōlo,pæ·lătăl), *a.* *Phonetics.* [f. Alveolo- + Palatal *a.*] Designating sounds formed by placing the tongue against the alveolar arch and the palate; of or relating to this position.

1942 A. H. Marckwardt *Introd. Eng. Lang.* i. 33 The tongue-tip is fairly well back on the alveolar ridge, and the blade of the tongue..against the extreme front portion of the hard palate; this forms the lengthened trough or channel, *alveolopalatal* in position, through which the breath stream must escape. *Ibid.* 40 The shift of the tongue from the alveolar to palatal position has resulted in the establishment of the alveolopalatal affricate. **1950** D. Jones *Phoneme* p. xiii, Alveolo-palatal fricatives (Polish *ś, ź*).

alveolus. Add: **d.** A small depression on the mucous membrane of the stomach.

1848 *Quain's Anat.* (ed. 5) II. 1025 The gastric mucous membrane..is seen to be marked..with little depressions or cells named *alveoli*. **1858** Gray *Anat.* 605 The entire surface of the mucous membrane [of the stomach] is covered with small shallow depressions or alveoli.

e. An air-cell of the lungs.

1859 A. T. H. Waters in *Proc. R. Soc.* X. 17 Structure..of the human lung... Here and there circular orifices exist, leading to smaller air-sacs, sometimes only to a small group of 'air-cells', or *alveoli*. **1873** T. H. Green *Introd. Path.* (ed. 2) xxxvii. 295 Minute spots of a more decidedly yellow colour..which are very common in capillary bronchitis, are merely alveoli containing mucus and purulent matter which has been drawn in by the act of inspiration. **1909** *Practitioner* Dec. 859 The alveoli themselves are distended with a mixture of gases.

f. An acinus of a compound gland.

1867 *Quain's Anat.* (ed. 7) I. p. clxxxvii, [The trabeculæ] are mostly lamellar in form, and divide the space into small compartments, *alveoli*, from $\frac{1}{50}$ to $\frac{1}{30}$ of an inch wide. **1880** E. Klein *Atlas of Histology* xxiv. 192 The alveoli of all salivary glands..are surrounded by a dense network of capillary blood-vessels. **1920** *Gray's Anat.* (ed. 21) 1087 Each lobule [in compound racemose glands] consists of the ramification of a single duct, the branches ending in dilated ends or alveoli. *Ibid.*, (*caption*) Surface view of alveolus. **1953** *Faber Med. Dict.*, 7/1 *Acinus*, one of the minute berry-like beginnings of the excretory ducts of a racemose gland. Syn. alveolus.

always, *adv.* Add: **3.** Delete †*Obs.* or *dial.*, substitute the following def. and add examples: In any or every circumstance; whatever the circumstances; whatever happens, whatever one may do or say; in any event, anyhow.

[*c* 1460, 1488] **1600** [see Provided *ppl. a.* 5, 5b]. **1778** F. Burney *Evelina* xxxvii, You will always make my respects to the hospitable family to which we are so much obliged. **1846** Dickens *Dombey* viii. 78 Never so distressed as by the company of children—Florence alone excepted, always. **1872** Geo. Eliot *Middlem.* iii. xxiii. 5 Fred had always (at that time) his father's pocket as a last resource. **1888** Mrs. H. Ward *R. Elsmere* xxx, Always supposing there were no risk in the matter. **1910** *Punch* 9 Feb. 104/3 It don't look as if I'm goin' to 'ave a job this afternoon. 'Owever, no matter. There's always the work'us.

4. *Comb.* Qualifying an adj. or ppl. adj. used attributively.

1855 Whitman *Leaves of Grass* 27 Sea of unshovelled and always-ready graves! **1902** *Academy* 16 Aug. 183/1 The spirit who is not weighed down with the cares of an always-ageing body. **1952** C. Day Lewis tr. *Virgil's Aen.* vi. p. 134 Giving no rest to the always-replenished vitals. **1955** P. Larkin *Less Deceived* 39 Bargains, suffering, and love, Not this always-planned salute.

alypin (ăləi·pin). *Med.* Also **-ine.** [G. (E. Impens 1905, in *Deut. Med. Wochenschrift* XXXI. 1154/2), f. Gr. ἄλυπος painless + -in[1].] A proprietary term for a glycerin derivative, used as a local anæsthetic.

1905 *Lancet* 29 July 321/2 A new anæsthetic compound termed 'alypin'. **1908** *Practitioner* Jan. 143 A 5 per cent. solution of cocaine and alypine. **1917** *Calif. State Jrnl. Med.* XV. 268 (*title*) Two cases of poisoning from the use of alypin in the urethra. **1928** *New Eng. Jrnl. Med.* CXCIX. 267 (*title*) Deaths from alypin poisoning.

Alzheimer's disease (æ·ltshəiməɹz dizī·z). [f. the name of Alois *Alzheimer* (1864–1915), German neurologist.] A grave disorder of the brain which manifests itself in premature senility.

1912 S. C. Fuller in *Jrnl. Nervous & Mental Dis.* XXXIX. 440 (*title*) Alzheimer's Disease (*senium præcox*): the report of a case and review of published cases. **1930** *Lancet* 19 July 138/2 Two cases of Alzheimer's disease... In one typical case, that of a woman aged 51, there was a gradual failure of mental power. **1952** *Ibid.* 27 Sept. 635/1 The characteristic changes of Alzheimer's disease could be designated..disseminated paramyloidosis of the brain. **1955** H. H. Merritt *Textbk. Neurology* vi. 417 The microscopic changes in Alzheimer's disease are somewhat similar to those which occur with senility. There is a diffuse loss of cells in all layers of the cortex, secondary gliosis, argentophile plaques (Alzheimer plaques) and neurofibrillar degeneration.

A.M., a.m. Add: **1.** (Further examples.) Also (*colloq.*) in sense of 'morning'. Cf. *ante meridiem.

1776 A. R. Robbins *Jrnl.* (1850) 6 Exercised and walked around with the officers in A.M. **1839** G. W. M. Reynolds *Pickwick Abroad* i. 3 At about ten o'clock A.M., [etc.]. **1889** *Cent. Dict.* s.v., I arrived here this A.M...that is, this morning or forenoon. **1925** E. F. Norton *Fight for Everest 1924* i. iv. 81 Overcast and warm, light snow in early a.m. **1955** *Times* 9 May 5/5 At 2.50 a.m...the stolen car approached.

Amadis (æ·mădis). Also **amadis.** [Name of the hero of a romance of chivalry, and title of one of Quinault's operas (1684).] A close-fitting sleeve buttoned at the wrist, which became fashionable from the costume of Amadis as worn in the opera of the name. Also, an embroidered edging on a bodice.

1835 *Court Mag.* VI. p. xxi/1 An attempt has been made to bring up again the Amadis sleeve. **1898** Lady Mary Loyd tr. *O. Uzanne's Fashion in Paris* ii. 38 Their bodices..were known as 'canezous', a scalloped embroidery called 'amadis' was carried round the edges and wrists.

amalgamating, *ppl. a.* Add: **b.** Of a language: inflexional; characterized by the use of inflexions to express the grammatical relations of words.

1877 J. Peile *Philology* (ed. 2) 51 But you must not suppose that any one language is so absolutely 'isolating', 'agglutinative', or 'amalgamating', as to exclude all traces of the other methods. *Ibid.* 54 The second great group of amalgamating languages is called Indo-European. **1908** T. G. Tucker *Introd. Nat. Hist. Lang.* 87 Languages commonly spoken of as 'inflexional', sometimes as 'organic'. We shall prefer to call them 'amalgamating'. These modify the sense and relation of words by variations of the terminating elements.

amalgamation. 2. (Further U.S. examples.)

1837 *Baltimore Com. Transcript* 8 June 2/1 (Th. Suppl.), *Amalgamation.* A black man and a white woman were lately brought before the Police Court in Boston charged with unlawfully marrying. **1905** *N.Y. Even. Post* 11 Oct. 4 If the white race are permeated with race consciousness, there is no danger of amalgamation.

Amapondo (amapǭ·ndo). Also **Amaponda.** = *Pondo; the tribe of the Pondos. Also *attrib.*

1824 Brownlee *Let.* in G. Thompson *Trav. & Adv. in S. Afr.* (1827) I. xviii. 209 A tribe called Amaponda, who live on the coast to the eastward of the Tambookies. **1837** F. Owen *Diary* (1926) 15 Faku, the chief of the Amapondas. **1838** *Ibid.* 120 The Amaponda country—Faku's tribe—beyond the Umzimvubu between Port Natal and Caffraria. **1871** C. M. Yonge *Pioneers & Founders* x. 258 The next tribes, the Amapondas, were scrupulously honest. **1876**, **1884** [see *Pondo].

amarantite (æmăræ·ntəit). *Min.* [ad. G. *amarantit* (A. Frenzel 1888, in *Min. und Petrogr. Mittheilungen* IX. 398), f. Amarant(h: see -ite[1].] A hydrous ferric sulphate of brownish-red colour (see quot. 1938).

1890 *Amer. Jrnl. Sci.* XL. 199 *Amarantite.* The crystallization is triclinic. *Ibid.* 201 Amarantite occurs associated with, and sometimes imbedded in a finely fibrous orange-colored mineral, probably sideronatrite. **1938** [see *hohmannite]. **1938** M. C. Bandy in *Amer. Mineralogist* XXIII. 745 *Amarantite group*... The members of this group are brownish red basic hydrous iron sulphates, in one of which, amarantite, dehydration data indicate that the water is given off in three stages... The formulae are, accordingly, written to indicate these various kinds of water.

Amarna (ămā·ɹnă). In full *Tell el-Amarna*, the modern name of Akhetaten, the site of a city in ancient Egypt built by Amenophis IV in the 14th century B.C. on the east bank of the Nile near Mallawi; used *attrib.* to designate remains found in this city, esp. certain tablets discovered in 1887 (see quots.).

1888 E. A. W. Budge in *Proc. Soc. Biblical Archæol.* X. 554 On the Tell el-Amarna tablets Amenophis III is addressed by his prenomen. **1893** C. R. Conder *Tell Amarna Tablets* p. xi, The Hebrews..appear for the first time in monumental history in the Tell Amarna letters. **1901** J. Hutchison tr. *Niebuhr's Tell el Amarna Period* ii. 8 Ai, called Haya in the Amarna letters, received golden honours. **1923** E. S. Bristowe *Oldest Lett. in World* i. 7 With the Amarna Tablets we come up against the mystery of the Hebrew race. **1960** K. M. Kenyon *Archaeol. in Holy Land* viii. 206 The Amarna letters..provide contemporary evidence for the break-up of the Asiatic empire re-established by Thotmes III. **1962** J. Gray *Archaeol. & O.T. World* iv. 85 The Amarna Tablets, the correspondence of the native chiefs and Egyptian deputies in Palestine and Syria with the suzerain power in the reigns of Amenhotep III (1411–1375 B.C.) and his son the 'heretic' Akhnaten (1375–1358 B.C.). **1962** *Listener* 19 July 106/3 The curious lack of confidence which Amarna art sometimes displays.

amateur. Add: **2.** (Earlier example.) See also quot. 1861.

1786 *European Mag.* Dec. 421/1 Dr. Percival..writes on philosophical subjects as an amateur rather than as a master. **1861** B. Hemyng in Mayhew *London Lab.* Extra vol. (1862) 221/2 This class [of prostitutes] have been called the 'amateurs', to contradistinguish them from the professionals, who devote themselves to it entirely as a profession.

b. (Earlier and later examples.)

1805 *Wynne Diaries* 8 May (1940) III. vii. 166 The Amateur performers were Mrs. W. Jerningham on the Harp, myself on the Piano. **1818** E. Blaquière tr. *Pananti's Narr. Res. Algiers* xiv. 266 There are, also, many amateur performers, but these always practice at home. **1822** De Quincey *Confess.* 7 The number of *amateur* opium-eaters (as I may term them) was, at this time, immense. **a1855** C. Brontë *Emma* in *Cornhill Mag.* (1860) Apr. 496 Any secret quest was to his taste; perhaps there was something of the amateur detective in him. **1953** B. Gordon-Cumming *Gentle Rain* 160 I've appointed myself amateur detective and am trying to discover a few things.

3. (Earlier and later examples.)

a1828 J. Bernard *Retrosp. of Stage* (1830) I. iv. 78 A party of ladies and gentlemen, who were going to get up an amateur play at Poole. *Ibid.* II. iii. 73 This was..the best piece of amateur acting I ever saw. **1873** C. M. Yonge *Pillars of House* III. xxv. 42 He dabbled in everything that was *not* his proper occupation—concerts, amateur theatricals, periodical literature. **1892** *Pall Mall Gaz.* 27 Feb. 6/2 The judge: Was this an amateur company?—Yes; they took money out of it.—The judge: Oh, then, I don't call that amateur. **1962** 'S. Nash' *Killed by Scandal* i. 21 Young men who had wasted their time on amateur theatricals.

b. Used disparagingly. Cf. sense 2.

1814 M. Edgeworth *Patronage* I. vi. 183 Sir Amyas talked a great deal of amateur nonsense. **1903** Kipling *Five Nations* 194 'Ow we're sugared about by the old men ('Eavy-sterned amateur old men!) That 'amper an' 'inder an' scold men. **1950** T. S. Eliot *Cocktail Party* I. i. 37 *Peter.* I met her here, about a year ago. *Edward.* At one of Lavinia's amateur Thursdays? *Peter.* A Thursday. Why do you say amateur? *Edward.* Lavinia's attempts at starting a salon.

amateurship. (Earlier example.)

1827 De Quincey *On Murder* in *Blackw. Mag.* XXI. 209/2 Wearied with the frigid pleasures..of mere amateurship, he had quitted England.

amatol (æ·mătǫl). [irreg. f. Am(monium + *-tol-* in *trinitrotoluene.] A high explosive consisting of a mixture of trinitrotoluene (T.N.T.) and ammonium nitrate.

1918 E. Colver *High Explosives* 252 In England these mixtures [trinitrotoluene and ammonium nitrate] are designated amatol. **1922** H. F. Moulton *Life Ld. Moulton* vii. 191 Finally in February 1915 Lord Kitchener..gave his casting vote on the side of big production, and A. 6 were told that they might proceed with the production of amatol.

amatorious, *a.* Delete †*Obs.* and add to def.: Inclined to love, amorous. Hence **amato·riousness.**

1746 Francis tr. *Horace, Ep.* I. iii. 31 Sure to gain, for amatorious Lays, The Wreaths of Ivy, with unenvied Praise. **1887** *Longman's Mag.* Nov. 108 A fine balance between domestic and 'amatorious' interest on one hand, and romance on the other. **1893** *National Observer* 1 Apr. 489/1 Girls and boys..jostle and jest at one another with a certain violence of amatoriousness. *Ibid.* 23 Sept. 482/1 The divagations of amatorious poets with fragile women. **1897** W. E. Henley in *Poetry R. Burns* IV. 249 It was natural and honourable in a young man of this lusty and amatorious habit to look round for a wife.

amaurotic, *a.* Add: **b.** Applied to an extreme type of hereditary imbecility, with the symptoms of amaurosis.

1896 B. Sachs in *N.Y. Med. Jrnl.* LXIII. 703/1 It will be better..to find a clinical designation, and I would propose the name *amaurotic family idiocy*. **1900** *Pediatrics* X. 9 Sachs has suggested and applied the name 'amaurotic family idiocy' to the condition. In view of the blindness that invariably follows, the term 'amaurotic' is distinctive and correct. **1937** I. H. Page *Chem. of Brain* iii. 93 (*heading*) Tay-Sachs amaurotic familial idiocy.

amazingness (ămē[i]·ziŋnės). [f. Amazing *ppl. a.* + -ness.] The state or condition of being amazing.

1860 Pusey *Min. Proph.* 562/1 The surpassing amazingness of it [*sc.* God's goodness] in the work of our redemption.

Amazonic (æmăzǫ·nik), *a.* Also **amazonic.** [f. L. *Amāzonic-us* (see Amazonical *a.*).] = Amazonian A. 1.

1889 *Cornhill Mag.* Aug. 179 Amazonic war, the war of female relatives armed to the teeth. **1920** D. H. Lawrence *Lost Girl* iii. 45 Her curious Amazonic power left her again. **1927** *Times* 13 Feb. 9 Miss Davies was neither a scholar nor a student. Nor was there anything amazonic about her, as there undoubtedly was about her friend, George Eliot.

Amazonism (æ·măzǫni:z'm). [f. Amazon + -ism.] Amazonian character or condition; a condition in which women have the supremacy. Also **Amazo·nianism.**

1873 E. J. M. Collins *Transmigr.* II. xv. 247 A world in which women were analytical..would result in absolute Amazonism. **1903** L. F. Ward *Pure Sociol.* xiv. 338 It might be supposed that woman would prove the dominant sex in primitive hordes... The..most striking form of evidence pointing this way consists in a class of facts that may be roughly grouped under the general head of *amazonism*. **1909** W. J. Locke *Septimus* ii, She had done with men... In that she prided herself on her Amazonianism.

ambari (æmbā·ri). Also **-ee.** [Urdu *ambārā, ambārī*.] The fibre of an Indian plant, *Hibiscus cannabinus*, used for making ropes and coarse cloth; brown Indian hemp (also *ambari hemp*); the plant itself.

1855 in *Imp. Dict.* Suppl. **1873** H. Drury *Useful Plants of India* (ed. 2) 243 Hibiscus cannabinus..Ambaree... The bark of this species is full of strong fibres which the inhabitants of the Malabar coast prepare and make into cordage. **1887** Moloney *Forestry W. Afr.* 282 The plant is largely grown in Western India, both as a pot-herb and for its fibre, known as 'Ambari', which much resembles jute. **1910** *Encycl. Brit.* XIII. 263/1 Deccan or Ambari hemp,..an Indian and East Indian malvaceous plant.

ambassador. Add: **5.** *ambassador-at-large* (U.S.), an ambassador appointed to perform special duties, and not accredited to any one government or sovereign. Also *transf.*

1908 *Busy Man's Mag.* Sept. 122/1 If he ever does get his deserts, he will be designated as ambassador-at-large for the Canadian Pacific Railway. **1933** *U.S. Naval Inst. Proc.* May 740/2 Conversations in London..between Premier MacDonald and Norman H. Davis, President Roosevelt's ambassador-at-large.

ambassado·rially, *adv.* [f. Ambassadorial *a.* + -ly[2].] According to the capacity or manner of an ambassador; as an ambassador.

1889 *Clergyman's Mag.* V. 165 St. Paul, directing an offender to be restored, says, 'I forgive the offence in the person of Jesus Christ'—ambassadorially. **1960** *Times* 27 Aug. 3/2 She was ambassadorially non-committal about tomorrow evening.

ambatch (æ·mbætʃ). Also **ambach, ambash.** [app. of Ethiopic origin.] A leguminous tree or shrub, *Æschynomene elaphroxylon*, of tropical Africa, with very light spongy wood; also called *pith-tree*.

1864 J. A. Grant *Walk across Africa* p. xv, Ambadj; native name for the pith-tree. **1884** J. Colborne *With Hicks Pasha in Soudan* 100 Here, too, we came across the famous ambatch.., one of the most extraordinary growths of African vegetation. **1887** [see *pith-tree*, Pith *sb.* 8]. **1889** H. M. Stanley in *Daily News* 4 Dec. 3/1 A..sand bank overgrown with sedge and ambatch. **1935** C. S. Forester *Afr. Queen* ix. 169 On either bank now appeared broad fringes of reeds—papyrus and ambash. **1957** Bannerman *Birds Brit. Isles* VI. 91 Not until the nose of the canoe was pushed into the ambatch bushes did the brooding birds leave their nests.

amber, *sb.*[1] Add: **3.** *spirit of amber* (earlier example).

a1700 [see Spirit *sb.* 22].

9. *Theatr.* An amber-coloured spotlight. Also *attrib.*

1913 A. Bennett *Regent* ix. 260 The curtain rose...'What about that amber, Cosmo?' Mr. Marriner cried. **1921** G. B. Shaw in *Times Lit. Suppl.* 17 Mar. 178/2 Take your ambers out of your number one batten. **1933** P. Godfrey *Back-Stage* vii. 90 'What's in your perches?' 'Ambers, sir.' *Ibid.* i. 18 The amber circuits in No. 1 batten.

B. *adj.* **b.** Designating the intermediate cautionary light in road traffic signals, between red (= stop) and green (= go). Also, as *sb.*, the amber-coloured light itself; hence *fig.*, an indication of approaching change or danger.

1929 *Min. Transport Roads Dept. Mem.* No. 297 The signal indications should be given in the following order: (1) Red, (2) Red and Amber together, (3) Green, (4) Amber... The purpose of the Amber is to give warning to drivers of vehicles of an impending change from Red to Green or Green to Red. **1933** *Traffic Signs (Size, Col. and Type) Prov. Regs.* § 27 (a) Three lights shall be used facing the stream of traffic which the signal is intended to control, one red, one amber and one green. **1937** Auden & MacNeice *Lett. fr. Iceland* 32 And always need a noise, the radio or the city, Traffic and changing lights, crashing the amber. **1956** *Ann. Reg. 1955* 421 The increase from 3 to 3½ per cent...was described as an 'amber light' of general warning. **1961** A. Wilson *Old Men at Zoo* iii. 136 In view of Godmanchester's remarks, I regard this as the amber warning. I'm afraid you must make it a must.

C. 1. c. *amber-solid*; **e.** *amber-foaming, -sanded, -tinged, -tinted, -toned.*

1881 Wilde *Panthea* in *Poems* 178 The hot and amber-foaming must. **1939** A. E. Housman *Coll. Poems* 192 Oh, the pearl seas are yonder, The amber-sanded shore. **1949** S. Spender *Edge of Being* 15 The sense felt behind darkened walls, An amber-solid world. **1866** G. M. Hopkins *Jrnl.* 3 May (1959) 134 Swallows..shewing their amber-tinged breasts. **1895** *Daily News* 20 Dec. 2/3 Allusion was also made to his amber-tinted hair. **1878** A. Nesbitt *Catal. Glass Vessels S. Kens. Mus.* 120 Amber-toned glass.

2. amber-plum, a variety of yellow plum (earlier example); **amber shell, snail** = Succinea.

1629 Parkinson *Parad.* iii. xiii. 578 The Amber plum is a round plum, as yellow on the outside almost as yellow waxe. **1835** W. Kirby *Creat. Anim.* I. ix. 291 The amber shells, at least one species, is [*sic*] stated to swim occasionally on the surface of the water. **1858** W. Baird *Cycl. Nat. Sci.* 528/1 *Succinea*, Amber Snail.

a·mber-fish. [Amber *sb.*[1]] A brightly coloured fish of the genus *Seriola*, found esp. in warm parts of the Atlantic.

1674 Josselyn *Two Voy.* 107 The Sea-bream, Dorado, or Amber-fish, they follow ships..and are good meat. **1706** Phillips *New World of Words* (ed. 6), *Dorado*..a Fish otherwise call'd the Sea-bream; or Amber-fish, the Head of which in the Water is Green, and the Body as Yellow as Gold. **1799** A. Ellicott in C. V. Mathews *Life & Lett.* (1908) 186 A great abundance and variety of fish..such as..Amber-fish. **1888** Goode *Amer. Fishes* 232 The Amber-fish, *Seriola carolinensis*, is quite common off the West Florida coast. **1897** *Outing* (U.S.) XXIX. 330/2 Not inferior to the kingfish for sport is the amber fish, or 'amberjack'.

a·mberjack. Also **amber jack.** [Amber *sb.*[1]] A species of amber-fish, esp. *Seriola dumerili*.

1893 *Funk's Stand. Dict.*, *Amber-jack*, an amber-fish (*Seriola lalandi*). **1897** [see prec.]. **1960** *Catal. Names Fishes Medit.* (Gen. Fish. Council Medit.) 161 Seriola dumerili... Amber jack.

amberoid, var. *ambroid.

ambi- (æmbi), repr. L. *ambi-* both, on both sides (*ambo* both), in various (chiefly scientific) terms (see words in *ambi-* in Dict. and Suppl.). **ambiciliate** (-si·liėt), *a.* *Ichth.*, having the scales on both sides of the body minutely toothed along the edges; **ambicolorate** (-kǫ·lōrėt, -kɒ·l-), *a.* *Ichth.*, applied to flatfishes abnormally coloured on both sides instead of having the under side white; so **ambicolora·tion.**

1894 *Proc. Zool. Soc.* 439 Ambicolorate fish appear to be always what one may call 'ambiciliate' also. *Ibid.* 435 Why Cyclopean examples should be ambicolorate. *Ibid.* (*heading*) 432 On an Adult Specimen of the Common Sole..with Symmetrical Eyes, with a Discussion of its bearing on Ambicoloration.

ambience (æ·mbiėns). Also ‖**ambiance** (aṅbiaṅs). [f. Ambient *a.*: see -ence; cf. F. *ambiance*.] Environment, surroundings; atmosphere.

The Fr. form *ambiance* is used in *Art* for the arrangement of accessories to support the main effect of a piece.

1889 *Harper's Mag.* Sept. 500/2 The form which we discern in the dreamy ambience is of supreme elegance. **1902** W. Watson *Ode on Coronation of King Edward VII* 5 Slowly in the ambience of this crown Have many crowns been gathered. **1923** R. H. Myers *Mod. Music* iv. 47 No other composer has ever reproduced in music with such complete success the very perfume and ambiance of a literary text. **1944** *Burlington Mag.* June 156/1 But the present picture was never meant to be microscopically dissected thus, for it is..an impression, a single figure in its *ambiance*, which is vaguely suggested as reflections in a mirror. **1952** *Ballet Ann.* VI. 25 The costumes and sets.. have such a suggestion of space that they give the Sadler's Wells stage the *ambiance* of Covent Garden. **1957** *London Mag.* Jan. 52 For some writers the urban ambience may provide just the kind of stimulus they need. **1961** *Listener* 5 Oct. 527/2 The Zoo provides a colourful ambience for this Administrative Novel [*sc.* Angus Wilson's 'Old Men at the Zoo']. **1965** *N. & Q.* CCX. 15/1 The way in which the poet by the use of the traditional vocabulary gives the impression that he was introducing his heroine into a Germanic ambience.

ambiens (æ·mbienz). *Ornith.* [pres. pple. of L. *ambīre* to go round, surround, f. *ambi-* around + *īre* to go.] A muscle in the thigh of certain birds, so called from the way in which it winds in passing from the hip to the foot.

1873 A. H. Garrod in *Proc. Zool. Soc.* 630 The ambiens and the accessory femoro-caudal are absent. **1884** Coues *N. Amer. Birds* (ed. 2) 193 The ambiens arises from the pelvis about the acetabulum, and passes along the inner side of the thigh.

ambient, *a.* **3.** (Later examples.)

1928 E. A. Wilcox *Electric Heating* vi. 128 Tank temperatures are constantly maintained at 100° F. above surrounding (or *ambient*) temperatures. **1958** *Engineering* 28 Mar. 393/3 The air pressure within the dome is maintained at the not uncomfortable figure of 2 lb per sq in above the ambient pressure.

‖ **ambiente** (æmbie·nte). [It. and Sp., f. L. *ambient-em*: see Ambient *a.* and *sb.*] = *Ambience.

1926 D. H. Lawrence *Plumed Serp.* xii. 197 He was utterly still..soft and unroused, within his own *ambiente*. **1927** —— *Lett.* (1962) II. 988 So with the mind. One's *ambiente* matters awfully. **1965** *House & Garden* Dec. 37/1 Within the entertaining ambiente of this decoratively practical kitchen/dining-room. **1966** M. Steen *Looking Glass* iv. 75, I couldn't afford it, but I liked the *ambiente*.

ambiguity. Add: **3. b.** *spec.* in *Literary Criticism* (see quots.).

1930 W. Empson *Seven Types of Ambiguity* i. 1 An ambiguity, in ordinary speech, means something very pronounced, and as a rule witty and deceitful. I propose to use the word in an extended sense, and shall think relevant to my subject any consequence of language, however slight, which adds some nuance to the direct statement of prose. [ed. 3, 1953: I..shall think relevant to my subject any verbal nuance, however slight, which gives room for alternative reactions to the same piece of language.] **1962** W. Nowottny *Lang. Poets Use* vii. 146 The term 'ambiguity' now has wide currency as a means of referring to diverse ways in which the language of poetry exhibits a charge of multiple implications and fits itself to contain within the form of discourse aspects of human experience whose difference or distance from one another might seem such as not easily to permit their coherent assembly in linguistic form. *Ibid.* 172 'Ambiguity' in its current critical sense of the manysidedness of language.

ambili·ngual, *a.* and *sb.* [*Ambi-, after *bilingual.*] (See quots.) Hence **ambili·ngualism**, the condition of being an ambilingual.

1959 J. C. Catford in Quirk & Smith *Teaching of English* vi. 164 In everyday speech the word 'bilingual' generally refers to a person who has virtually *equal* command of two or more languages. If a special term is required for such persons of equal linguistic skill (which is very difficult to measure) I should prefer to call them 'ambilinguals'. Ambilinguals are relatively rare. **1964** M. A. K. Halliday et al. *Ling. Sciences* iv. 78 The endpoint where a speaker has complete mastery of two languages and makes use of both in all uses to which he puts either. Such a speaker is an 'ambilingual'. True ambilingual speakers are rare. *Ibid.*, Even those who approach or attain true ambilingualism are still usually unable to translate without instruction.

ambitioned, *ppl. a.* (Earlier and later examples.)

1651 R. Boyle *Parthenissa* (1655) I. i. 5 My performing it has not onely produc'd the ambition'd effect of making you perceive your fault. **1824** *New Monthly Mag.* XI. 555 A long-ambitioned and long-promised addition to her summer finery. **1873** C. M. Yonge *Pillars of House* III. xxix. 137 'I should not have thought it a woman's work.' This, the most ambitioned praise a woman can receive.

ambitiously, *adv.* **2.** (Example.)

1822 *New Monthly Mag.* V. 332 Scenic description.. laboriously and ambitiously worked up.

ambitty (æmbi·ti), *a. Glass-making.* Also **ambetti, ambetty, ambitti.** [app. ad. F. *ambité*, of obscure origin.] Designating glass which becomes devitrified while it is being worked. Also *absol.*

1883 H. Chance *Crown & Sheet Glass* in H. J. Powell et al. *Glass-Making* 104 Glass made with sulphate of soda is less liable to devitrify or as it is termed, become 'ambitty' in the pot during the time of working. **1885** F. Miller *Glass-painting* 52 A beautiful make of glass is anbetti [*sic*]. **1885** *Spon's Mechanic's Own Book* 630 'Ambilti' [*sic*] (single and double) is a sheet glass, originally of Italian manufacture. **1902** Suffling *Art of Glass Painting* 63 Ambetty sheet. **1953** *Glass for Glazing* (*B.S.I.*) 20 *Ambetti, antique venetian*, [etc.]..translucent 'antique' glasses used chiefly for leaded lights and decorative purposes.

caught sight of a muffled figure in an ambulance that stopped for water for its thirsty mules.

ambulando: see *Solvitur ambulando.

ambulant, *a.* Add: **1.** (Earlier and later examples.)

1619 Bacon *Let.* 20 Nov. in Spedding *Lett. & Life* (1874) VII. 61 Sir Edward Coke was at Friday's hearing, but in his night-cap; and complained to me he was ambulant and not current. **1885** G. Meredith *Diana* II. ii. 58 The zealous Irishman might be trusted to become an ambulant advertizer.

†**3.** *Path.* and *Med.* **a.** Of a disease: shifting from one part of the body to another; = Wandering *ppl. a.* 2 g. *Obs.*

1879 *St. George's Hosp. Rep.* IX. 703 Such inquirers.. must bear in mind the existence of ambulant fever. **1881** A. B. Ball in *von Ziemssen's Cycl. Med.* Suppl. 682 Ambulant œdema.

b. Of a disease: allowing the patient to walk about, not confining him to bed; also of medical treatment in which the patient is allowed or ordered to walk about, and of a patient who is able to walk.

1913 Dorland *Med. Dict.*, *Ambulant, Ambulatory*.. Walking or able to walk; not confining the patient to bed. **1927** *Daily Tel.* 31 May 15/5 Additional provision for what was called ambulant treatment of those suffering from lupus. **1958** *Hosp. O. & M. Service Reports* I. 11 Patients who attend from a distance (and who have no private transport) are dependent either on public transport or, if non-ambulant, on the ambulance service.

ambulatory, *a.* Add: **5.** *Path.* and *Med.* = prec. sense 3.

1857 Dunglison *Med. Lex.* s.v., A morbid affection is said to be 'ambulatory'..when it skips from one part to another. **1882** Quain *Dict. Med.* I. 38/1 *Ambulatory*, a term given to typhoid fever, showing that the patient is able to walk about during the attack. **1903** *Westm. Gaz.* 21 Feb. 6/1 That the cause of death was ambulatory typhoid. **1947** L. K. Ferguson *Surg. Ambulatory Patient* (ed. 2) p. ix, Surgery of the ambulatory patient is the surgery performed more often by the younger men and general practitioners. *Ibid.* i. 1 (*heading*) A survey of the field of ambulatory surgery.

ambusher (æ·mbuʃəɹ). [f. Ambush *v.* + -er[1].] One who makes an ambush.

1893 in *Funk's Stand. Dict.* **1920** *Glasgow Herald* 20 Nov. 7 The ambushers, he said, were all dressed in khaki.

amebiasis, var. *Amœbiasis.

‖ **âme damnée** (âm,dane). [Fr., lit. 'damned soul'.] A devoted adherent; a tool.

1823 Scott *Pev. Peak* IV. xii. 288 He is the *ame damnée* of every one about my court—the scape-goat, who is to carry away all their iniquities. **1830** C. C. F. Greville *Mem.* (1874) II. xiii. 96 He [*sc.* Sefton] is the *âme damnée* of Lord Grey, and defends everything of course. **1879** M. E. Braddon *Vixen* III. vii. 185 Their *âmes damnées*, the men who hold their hods and mix their mortar. **1917** Asquith *Let.* 13 Nov. (1933) I. 46 We are having Col. House, the Head of the American 'Mission', to lunch: he is President Wilson's own particular *âme damnée*. **1938** S. Beckett *Murphy* iv. 54 Neary..boarded the first train for Dublin, accompanied by his *âme damnée* and man-of-all work, Cooper.

ameloblast (âme·loblæst). *Zool.* [f. En)amel *sb.* + -o + -blast.] A columnar cell which secretes enamel in the teeth of the higher vertebrates.

1882 J. L. Williams in *Dental Cosmos* XXIV. 229 Prof. Eames, of St. Louis, introduced the term ameloblast..as an appropriate one for the inner layer, or true enamel cells. **1886** W. X. Sudduth in W. F. Litch *Amer. Syst. Dentistry* I. iii. 641 A slight layer of enamel has been formed... Between this layer of enamel and the layer of ameloblasts a space is noticed. **1942** E. S. Horning in G. Bourne *Cytology & Cell Physiol.* vi. 178 The cells destined to become ameloblasts in the developing enamel organ were found to contain relatively large amounts of calcium and magnesium.

amen, *int.* Add: **A. 2. b.** *attrib.* **amen corner, seat** *U.S.*, that part of a meeting-house occupied by persons who assist the preacher with occasional and irregular responses; also *transf.*; **Amen glass**, an eighteenth-century drinking-glass with part of the Jacobite version of 'God Save the King', concluding with the word 'Amen', engraved upon the bowl.

1860 *Harper's Mag.* Jan. 279/2 The Rev. Judson Noth, a local Methodist preacher,..was one of the best 'scotchers' that occupied the 'Amen Corner'. **1868** *All Year Round* 31 Oct. 490/1 Sunday found them, judge and lawyers, seated in the 'amen corner'. **1877** Habberton *Jericho Road* xiv. 128 In an 'amen' seat sat an old half-breed. **1884** *Congress. Rec.* 24 Apr. 3207/1 When commiserated upon the fact that he was compelled to go to what is commonly known here as the amen corner, [he] frankly said that any seat in the Senate was better than none. **1894** *Ibid.* Jan. 1502/2 One of those saintly Republican monopolists who sit in the 'Amen corner' of protected privilege. **1904** Harben *Georgians* vii. 67 [They] were in their places in the 'amen corner', at the right of the crude pulpit.
[**1897** A. Hartshorne *Old Eng. Glasses* xxiv. 349 This glass also has on the bowl the first two verses of the song, and the crowned cypher with *amen* under it.] **1924** J. Bles *Rare Eng. Glasses XVII & XVIII Cent.* 98 This interesting goblet is engraved in diamond point with the crowned cypher JR, forward and reversed with the figure 8..in the same manner as the majority of 'Amen' glasses. **1926** G. R. Francis *Old Eng. Drinking Glasses* xvi. 167 Of this type are most of the 'Amen' glasses, which were all similar in the inscription; some ten or twelve are now known to exist. **1936** *Burlington Mag.* May xvii/1 An *Amen* Jacobite glass, with the full Jacobite National Anthem inscribed upon it, had realised as much as £300 last year.

amenability. 1. (Earlier U.S. example.)

1789 J. Madison *Writ.* (1904) V. 373 A unity in each has been resolved on, and an amenability to the President alone, as well as to the Senate by way of impeachment.

amenable, *a.* **4.** (Earlier examples.)

1828 Mill *Autobiogr.* (1924) 298 Make them amenable to the general tribunal of the public at large. **1843** —— *Logic* I. ii. i. 216 Such of them [*sc.* assertions]..as, not being amenable to direct consciousness or intuition, are appropriate subjects of proof.

amendment. 4. (Earlier and later examples.)

1303 R. Brunne *Handl. Synne* (1901) 9225 Ne never hadde þey amendement..at any corseynt, But at þe vyrgyne Seynt Edyght. **1789** R. F. Greville *Diaries* 17 Feb. (1930) 228 The King for some time past, had been in a State of Amendment &..he was this day in a State of convalescence. **1814** Jane Austen *Mansf. Park* III. xvi. 260 Tom's amendment was alarmingly slow.

amenity. Add: **1. a, 3. b.** Also with pronunc. (âmī·nĭti). In mod. use (freq. in pl.) applied to the more 'human' and pleasurable environmental aspects of a house, factory, town, etc., as distinguished from the features of the house, etc., considered in or by itself. Also *concr.* (usu. in sing.), a particular advantageous or convenient feature of this kind. Also *attrib.* Also, **amenity bed** (see quots.).

1908 *Royal Comm. Care Feeble-Minded, Min. Evid.* II. 63/1 *Social Amenities.* The experience we have gained emphasises the desirability of organised recreation. **1928** *Britain's Industr. Future* (*Liberal Ind. Inquiry*) iv. xxiv. § 9. 336 Amenity woodland definitely uneconomic. **1929** *Oxford Times* 8 Feb. 13/4 The payment of £88 for the purchase of the land; the payment of £250 as compensation for the loss of amenities and disturbance of existing garden and grounds. **1936** *Times* 2 Apr. 10/3 Repairable cottages of amenity value..could be acquired and sympathetically repaired by the local authorities. **1951** *Brit. Med. Jrnl.* 13 Oct. Suppl. 146/2 Amenity beds, for which the maintenance charge is almost negligible and medical service is given under the National Health Service Act. **1951** B. J. Collins *Devel. Plans Explained* 42 *Amenity*, the quality which makes a desirable residence desirable, a favoured locality favoured, or enchanting views in all directions enchanting. **1952** *Lancet* 2 Aug. 229/1 No privileges can be bought within the service (except the amenity bed, for which a relatively small charge has been made). **1957** *Times* 12 Dec. 18/4 Arrangements were made to provide an amenity centre for the labour on Effingham and Seventh Mile Estates, the centre comprising a clubroom with cinema and a playing field. **1958** *Times* 1 July 1/3 There are, of course, many holdings below 20 acres, especially those that are part-time or amenity holdings where the earning of an income from the holding is not of great importance. **1958** *Listener* 11 Sept. 368/2 Where the people themselves want a new amenity—a school, a meeting house, a road to link up with the outside world, [etc.]. **1964** G. L. Cohen *What's Wrong with Hospitals?* i. 23 'Amenity beds'..were designated under the Act for patients who want more privacy and will pay extra for it.

amenorrhœal, *a.* (Earlier example.)

1833 J. Forbes et al. *Cycl. Pract. Med.* I. 70/2 A torpid or amenorrhœal condition of the uterus.

‖ **a mensa et thoro** (ēi me·nsă et tōə·ro). *Law.* [L., lit. 'from table and bed'.] In the older English law (before 1857) = *judicial separation* (Separation 3): see Divorce *sb.*

1600 [see Separation 3]. **1628** Coke *Littleton* v. § 36. 32 When the husband and wife are diuorced *à vinculo matrimonij*, as in case of precontract, consanguinitie, affinitie, &c. and not *à mensa & thoro* onely as for adulterie. *a* **1683** J. Owen *Sermons* (1721) 572 This divorce *a mensa & thoro* only is no true divorce, but a mere fiction of a divorce. **1857** [see Separation 3]. **1957** O. R. McGregor *Divorce in Engl.* i. 3 The ecclesiastical court would only pass sentence of divorce *a mensa et thoro* (a divorce from bed and board), which had the effect of a modern judicial separation.

ament[2] (ē·ment, āme·nt). [ad. L. *āment-, āmens*: see Amenty.] A person congenitally deficient in mind or intellect; a born idiot or imbecile.

1894 in Gould *Dict. Med.* **1912** *Ch. Q. Rev.* LXXIII. 326 We will classify them all (idiots, imbeciles, or feeble-minded) under the name 'Ament', meaning people without mind in contrast to the class of Dement, which we will assume to mean all those who have been sane, but have lost their mind. **1935** W. de la Mare *Early One Morning* 552 Aments..begin to talk late, have certain special defects, are sluggish of thought. **1952** A. F. & R. F. Tredgold *Mental Deficiency* (ed. 8) iv. 61 Aments vary very greatly in the degree of their defect.

amental, *a.*[2] Delete *nonce-wd.* and add to def.: Also, non-mental. Add further example.

1938 S. Beckett *Murphy* xi. 247 An amental pattern as precise as any of those that governed his chess.

amentia (ēi·me·nʃiă, ame·-). [L., see Amenty.] **1.** Mental deficiency; feeble-mindedness, either congenital or resulting from damage to the brain in early childhood.

1398 Trevisa *Barth. De P.R.* (1535) vii. p. lxxxvii, Amentia and madness is all one, as Plato sayeth. **1800** tr. *W. Cullen's Nosology* II. iv. 131, I have brought Amnesia and Amentia under one genus,..because they are in general conjoined. **1879** [see Amenty]. **1887** *Buck's Handbk. Med. Sci.* IV. 88/2 The word [*sc.* idiocy] was used until Esquirol first clearly applied it..in connection with both amentia and the terminal stage of chronic insanity. **1914** A. F. Tredgold *Mental Deficiency* (*Amentia*) (ed. 2) xiii. 226 Secondary Amentia and its Clinical Varieties. Toxic, Inflammatory and Vascular Amentia. *Ibid.* 228 The fact that the onset of these cases is so often attended with convulsions causes them to be frequently designated 'epileptic' or 'eclampsic' amentia.

2. *Psychiatry.* A particular psychotic state (see quot.).

1924 J. Riviere et al. tr. *Freud's Coll. Papers* II. 252 In.. amentia, the acute hallucinatory confusion which is perhaps the most extreme and striking form of psychosis, the outer world is..not perceived.

American, *a.* and *sb.* Add:

A. 1. Also, of or pertaining to its inhabitants.

1885 *Century Mag.* Apr. 953/2 To use an expression made popular, we believe, by General Hawley some years ago.., dynamiting is '*not the American way*'! **1937** Hemingway *To Have & Have Not* iii. xvi. 232 The Colt or Smith and Wesson..so well designed to end the American dream when it becomes a nightmare. **1960** *Observer* 17 Jan. 20/6 The spread of personal prosperity in America has led away from, rather than towards, the American Dream. *Ibid.* 20/7 The American Dream, the reasonable expectations of Americans, are by tradition that all men shall be equal. **1961** M. McCarthy *On the Contrary* (1962) I. 40 Apologists for the American Way of Life find themselves condoning injustices.

b. *American language* (usu. with *the*), (i) a language of American Indians: (ii) American English (see sense *3). Also *American tongue.*

1643 J. Langley in *R. Williams' Key into Lang. of Amer.* 200, I have read over these thirty Chapters of the American Language, to me wholly unknowne. **1689** I. Mather *Brief Relation of State of New Engl.* 16 In an Indian Town..was an Englishman, who being skilful in the American Language, Preached the Gospel to them in their own Tongue. **1789** Webster *Dissertations Eng. Lang.* i. 22 Numerous local causes..will introduce new words into the American tongue. **1800** in Mencken *Amer. Lang.* (1936) I. i. 11 (*title*) On the Scheme of an American Language. *Ibid.*, Grammars and dictionaries should be compiled by natives of the country, not of the British or English, but of the *American* tongue. **1839** [in Dict.]. **1936** Mencken *Amer. Lang.* (ed. 4) I. iii. 23 This occasional tolerance for things American was never extended to the American language.

2. c. *U.S. spec.* (See quot. *a* 1861.)

1837 *Diplom. Corr. Texas* (1908) I. 187 A large number of fine American horses..which there is no doubt had been stolen from citizens of Texas. **1846** E. Bryant *What I saw in Calif.* (1849) vi. 37 Such [Indians] as rode ponies were desirous of swapping them for the American horses of the emigrants. *a* **1861** Winthrop *John Brent* (1862) ii. 14 He was an American horse,—so they distinguish in California one brought from the old States. **1878** J. H. Beadle *Western Wilds* xvi. 253, I rode a good-sized American horse.

3. Special Combs. **American bar** [Bar *sb.*[1] 28], the name given (outside the U.S.) to a bar serving refreshments in allegedly American style; **American blight** = Apple-*aphis*; **American cheese,** cheese of the Cheddar type, made in the U.S.; **American cloth,** (*a*) = *Americani; (*b*) also **American oilcloth**, see Cloth *sb.* 9 c and add examples; **American English,** the form of English used by the inhabitants of the U.S.; **American leather** (see Leather *sb.* 1); **American organ,** see Organ *sb.*[1] 3 c and add earlier example; **American plan,** 'the system of charging an inclusive price for room and board in a hotel' (see D.A.E.), contrasted with **European plan*; **American sheeting** = *Americani; **American supper, tea,** a social function for raising funds to which the guests contribute by bringing or buying food and drink.

1862 American bar [see *cobbler 3]. **1869** B. J. Spedding *Ino* 25 Trio. Air, 'American Bar'. **1886** L. P. Richardson *Dark City* ii. 20 If the English bar is a curiosity to an American, the American bar is still more so. **1913** G. W. Hills *John Bull Ltd.* 229 [Many English hotels] proudly bear aloft the sign of relief yclept 'American Bar'; but sign and beverages are alike delusions. **1815** Kirby & Spence *Introd. Entomol.* I. vi. 196 The greatest enemy of this tree, and which has been known in this country [England] only about twenty years, is the apple-aphis, called by some *Coccus*, and by others the *American blight*. **1882** American blight [see Blight *sb.* 2 b]. **1804** *Guardian of Freedom* (Frankfort, Ky.) 10 Mar. 2/1 (D.A.), Cheese, American per lb. 18. **1860** Dickens *Uncomm. Traveller* vi, in *All Year Round* 24 Mar. 515/1 Your waiter..is carrying in seventeen pounds of American cheese. **1879** American cheese [see Cheddar]. **1860** J. A. Grant in *Blackw. Mag.* (1865) XCVII. 107 The body sewed up in an American cloth. **1889** J. C. Willoughby *East Africa & its Big Game* App. ii. *Eng.-Swaheli Vocab.*, American cloth, *Amerikano.* **1896** C. T. C. James *Yoke of Freedom* 85 Not a single ring of stickiness was to be found upon the American-cloth table-cover. **1896** H. G. Wells *Wheels of Chance* iv, A neat packet of American cloth behind the saddle contained his change of raiment. **1904**

AMOLE

amole (ămō·le). Also **ammole.** [Mexican Sp.] The root or bulb of any one of several plants found in Mexico and California, used as a detergent; also any of such plants, esp. *Chlorogalum pomeridianum*, called also *soap-plant* (see Soap *sb.* 6 b). Also *attrib.*

1831 Beechey *Voyage* II. 43 A very useful root called in that country [California] *amoles.* **1845** Frémont *Exped.* (1846) 269 Great quantities of *ammole* (soap plant), the leaves of which are used in California for making..mats for saddle-cloths. **1884** *Encycl. Brit.* XVII. 401/2 *Yucca filamentosa*, commonly called amole or soap-weed. **1886** *Proc. U.S. Nat. Mus.* VIII. 518 Lechuguilla is the most important of the soap or 'amole' plants of Southwestern Texas and Northern Mexico. **1927** *Chambers's Jrnl.* 227/1 Soap grows for him..in the bulbs of the curious Amole. **1928** D. H. Lawrence *Woman Who Rode Away* 96 They..washed her all over with water and the amole infusion.

amontillado (amǫntilā·do, -tilā·do). [Sp.; f. *Montilla*, a town in Spain + *-ado* -ate².] Formerly, a wine of the sherry type produced in Montilla; now, a matured sherry in which the 'flor' has developed.

1825 Henderson in *Q. Jrnl. Sci. & Arts* XVIII. 130 The driest species of Sherry is the Amontillado. **1833** C. Redding *Mod. Wines* 190. *a* **1845** Poe *Cask Amont.* in *Wks.* (1864) I. 347, I was silly enough to pay the full Amontillado price without consulting you in the matter. **1886** Ruskin *Præterita* II. ix. 325 A certain quantity of the drier Amontillado, from the hill districts of Montilla. **1961** *Times* 16 Jan. 13/5 There are three main types of sherry..*finos*, which are dry, light, and pale; *amontillados*, which are softer, fuller, and less dry; and *olorosos*, which are sweet, rich dessert wines.

b. *attrib.* in fig. sense.

1862 Thackeray *Philip* xvii, 'By the housekeeper, do you mean Mrs. Baynes?' I ask, in my *amontillado* manner. **1921** *Spectator* 22 Jan. 109/1 We will take as our next example of what we might call Mr. Max Beerbohm's Amontillado style 'Hosts and Guests'.

amoral (ēⁱ·mǫ·răl), *a.* Delete *nonce-wd.* and add further examples. So **amo·ralism, amo·ralist, amora·lity.**

1892 S. Weir Mitchell in *Century Mag.* July 343/2 You are amoral, not immoral. **1910** *Westm. Gaz.* 19 Mar. 3/2 He argued..that Nature was a-moral, ethically neutral. **1915** A. S. Neill *Dominie's Log* v. 56, I have no morals, I am an a-moralist, or should it be a non-moralist? **1917** Kipling *Divers. Creatures* 176 He's the Absolutely Amoral Soul. I've never met one yet. **1920** *Glasgow Herald* 27 Nov. 6 Dada..leads to amoralism. **1923** *Observer* 10 June 8/3 Stupidity, brutality, and general amorality. **1926** *British Weekly* 16 Sept. 487/4 That great sceptic and amoralist [Anatole France]. **1927** R. A. Taylor *Leonardo* II. ii. 103 The bright amoral virtue of courage. **1962** J. Gray *Archaeol. & O.T. World* v. 112 The fertility-cult of Canaan may have been quite amoral.

amorce (ămǭ·ɪs). ? *Obs.* [ad. F. *amorce* (OF. *amorse*) bait, lure, priming, f. OF. *amordre*, f. *à* + *mordre* to bite.] A charge of fine-grained powder for priming a small fire-arm; a cap for a toy pistol.

1802 C. James *Milit. Dict.*, *Amorce*, an old military word for fine-grained powder, such as is sometimes used for the priming of great guns, mortars or howitzers; as also for small-arms, on account of its rapid inflammation. A port fire, or quick match. **1883** *B'ham Weekly Post* 15 Dec. 7/5 Summoned for having in his possession a small quantity of manufactured amorces, he not having a license for the sale of explosives...These toy pistol caps..were made of a very dangerous explosive. **1889** *Standard* 2 Dec. 4/8 To restrain Mr. Cadwell from making amorces for toy pistols.

amorism (æ·mŏriz'm). [f. as Amorist + -ism.] The disposition or practice of amorists; amorous sentiment or intrigue.

1897 *Star* 30 Mar. 1/7 We were in the land of romantic amorism. **1903** *Athenæum* 17 Jan. 77/1 Half old-world Spanish, half topsy-turvy Oriental in its fatalism and passionate amorism.

amorist. Add: **2.** One who treats of love; a writer of amatory literature. Also *attrib.*

1641 Milton *Reason Ch.-Govt.* II. 41 A work not..like that which flows..from the pen of some vulgar Amorist, or the trencher fury of a riming parasite. **1824** *Blackw. Mag.* XVI. 191 Our most eminent amorist..Tom Moore. **1882** Palgrave in Grosart *Spenser's Wks.* IV. p. lx, Amourist literature. **1905** *Athenæum* 1 Apr. 390/3 The poet..is imagined as a mild and amiable amorist. **1909** Jusserand *Lit. Hist. Eng. People* III. 468 The..amourist writers of Elizabethan times.

amosite (æ·mŏsəit). *Min.* [f. *Amosa*, formed from the initial letters of *A*sbestos *M*ines *o*f *S*outh *A*frica + -ite¹.] A form of asbestos found only in certain areas of South Africa. Also *attrib.*

1918 A. L. Hall *Asbestos in S. Africa* 12 *Amosite*, chemically characterized by high percentage of iron with variable amounts of aluminium, magnesium, and calcium. *Ibid.* 13 Amosite..the special variety of monoclinic iron amphibole from the Lydenburg and Pietersburg Districts. **1928** *Observer* 1 July 3 Four years later [in 1895] the chrysolite deposits of the Carolina district began to be exploited, followed by the discovery of amosite fibre in the Transvaal. **1961** *Engineering* 2 June 770 Marinite is a homogenous pressed lime-silica board having an amosite asbestos fibre content.

‖**amour courtois** (amu̇r kurtwa). [F.] Courtly love (see *courtly *a.* 2 b).

1907 Chambers & Sidgwick *Early Eng. Lyrics* 262 This poetry [*chanson d'amour* or *chanson courtois*]..has practically but one theme, that of the *amour courtois.* **1950** N. Coghill in *Ess. & Stud.* III. 10 Its sentiments are those of *amour courtois.* **1963** *Medium Ævum* XXXII. 59 The alleged resemblances between Andreas Capellanus' 'system' and the actual poetry of *amour courtois* break down at every point.

amourette. 1. (Earlier examples.)

1825 Scott *Betrothed* II. xi. 187 This comes of meddling with men's *amourettes.* **1826** Disraeli *Viv. Grey* I. viii. 53 Master Vivian entered into all these amourettes in very beautiful style.

amour-propre. Add: Freq. without hyphen. (Earlier and later examples.)

1775 I. Moore *Let.* 12 June in Duke of Argyll *Int. Soc. Lett. of 18th Cent.* (1910) II. 390 He is not found of the Company of his Superiours either in Rank or Understanding. The first put him under Restraint, and the others offend his *amour propre.* **1955** *Times* 27 July 9/2 Mr. Malcolm MacDonald..has returned to the scene to counsel caution and soothe wounded amour-propre.

amp., abbreviation of Ampere.

1886 O. Heaviside in *Electrician* 29 Jan. 227/2 But ampère shortened to am or amp is abominable. **1907** *Install. News* July 8/2 Suction gas plant and 200 amp. dynamo. *Ibid.*, A private plant supplying 100 amps.

Ampelopsis (æmpilǫ·psis). *Bot.* [mod. L., f. Gr. ἄμπελος vine + ὄψις appearance.] A genus of climbing plants allied to the vine; (with small initial) a plant of this genus, as one of the various species of Virginia creeper.

1807 J. Donn *Hort. Cantab. or Catal. of Plants* (ed. 4) 49 Ampelopsis. heart-leaved. N. America. **1829** Loudon *Encycl. Plants* 176 Ampelopsis..is commonly employed for covering old walls, for which the rapidity of its growth renders it very suitable. **1843** *Florist's Jrnl.* 1842 III. 148 The honeysuckle, the jasmine, the ampelopsis,..may be made to contribute largely to the general appearance of a garden. **1866** [see *pepper-vine*, Pepper *sb.* 5]. **1900** *Echo* 25 Sept. 1/3 The crimsoning leafage of ampelopsis. **1914** Compton Mackenzie *Sinister Street* III. vii. 646 The teatray gothic of Balliol, and Trinity with its municipal ampelopsis. **1961** *Amateur Gardening* 30 Sept. Suppl. 3/2 Most of the ampelopsis are hardy climbers related to the Virginia Creepers.

amperage (æmpēᵊ·redʒ, æ·mpērėdʒ). *Electr.* [f. Ampere + -age, after *voltage.*] The strength of an electric current expressed in amperes. Also *fig.*

1894 *Work* 17 Mar. 139/2 The longer the length of wire you use on any armature, the higher will be the voltage, but the lower the amperage. **1924** Mégroz *Walter de la Mare* 105 The living poet's 'Sunk Lyonnesse' in which each word has a far higher amperage of power. **1941** *Illustr. London News* 18 Feb. 191/1 Ordinary X-ray pictures are taken by long exposure to rays generated by high voltage and low amperage current.

ampere. Add: Now usually written without accent. Def.: see quot. 1963. **ampere-hour**, the quantity of electricity equal to a current of one ampere flowing for one hour; abbrev. Ah; **ampere-turn**, a unit of magnetomotive force, expressed as the product of the number of turns in a coil and the current in amperes; abbrev. A.T.

1884 J. T. Sprague *Electricity* xii. 526 To equally magnetize different sized cores, the 'ampere-turns' must be in the proportion of the square root of the cube of the diameters. **1885** *Electrician* 27 Nov. 49/1 Secondary cells..giving 280 ampere hours. **1912** G. Kapp *Electricity* vi. 166 The excitation produced by a coil may be conveniently expressed by the product of amperes and turns, or 'ampere-turns'. **1919** R. Stanley *Wireless Telegr.* I. 452 A proper check..of the number of ampere-hours' charge and discharge. **1963** Jerrard & McNeill *Dict. Sci. Units* 17 The 9th Meeting of the International Weights and Measures Congress in 1948 defined the ampere as the intensity of the constant current which, when maintained in two parallel straight conductors of infinite length and of negligible cross section placed one metre apart in a vacuum, produced between them a force equal to 2 × 10⁻⁷ M.K.S. units of force per metre length. This unit, known as the absolute ampere, replaced the international ampere which had been defined in 1908.

Amperian (æmpēᵊ·riăn), *a.* Pertaining to the French physicist André-Marie *Ampère* (1775–1836) or his theory of electromagnetic currents.

1866 E. Atkinson *Ganot's Elem. Treat. Physics* (ed. 2) x. iv. 674 At the north pole of a magnet the direction of the Ampèrian currents is opposite that of the hands of a watch, and at the south pole the direction is the same. **1879** G. Prescott *Sp. Telephone* xiv. 496 The Amperian currents in the inducing magnet.

amperometer. Add: Hence **amperome·tric** *a.*, esp. *amperometric titration*, titration by electromagnetic methods.

1941 Kolthoff & Lingane *Polarography* viii. xxxiii. 448 As it is the current which is measured during the titration, we prefer to call this type of titration 'amperometric titration'. **1960** *Jrnl. Iron & Steel Inst.* CXCV. 374/1 Amperometric titrations may be conducted with an indicator electrode in combination with a reference electrode, or by means of two indicator electrodes.

amphetamine (æmfe·tāməin, -in). [f. *a*lpha-*met*hyl-*phe*ne*t*hyl*amine.*] A synthetic drug which stimulates the heart and respiration, constricts blood-vessels, and induces sleeplessness.

1938 *Amer. Jrnl. Psychiatry* XCV. 371 (*title*) The effective use of phenobarbital and benzedrine sulfate (amphetamine sulfate) in the treatment of epilepsy. **1939** W. Bloomberg in *New Engl. Jrnl. Med.* 26 Jan. 129/1 Amphetamine (Benzedrine) sulfate was introduced into therapeutics in 1935 by Prinzmetal and the author as an effective agent in the prevention of symptoms in narcolepsy. **1946** *Lancet* 14 Dec. 865/1 Amphetamine is a powerful stimulant of the central nervous system, an action which it owes to its ability to increase cerebral respiration. **1955** *Sci. News Let.* 9 Apr. 235/3 Amphetamine, popularly known as an ingredient of 'pep pills', has been reported one of the drugs used in brain washing by the Communists. **1959** *Times* 29 May 14/6 The use of amphetamine drugs in sport. **1962** *Lancet* 15 Dec. 1282/2 Hypertensive patients receiving amphetamine in addition to reduce obesity.

amphi- (æmfi). Add:

amphidiarthrosis (æ:mfidəi,aɪþrōᵘ·sis) *Anat.*, a form of articulation (see quots.); **amphidromic** (-drǫ·mik), *a.*² [Gr. ἀμφίδρομος running both ways] (see quots.); **amphikaryotic** (-kæriǫ·tik), *a. Biol.* [Gr. κάρυον nut, kernel], of a nucleus, having two haploid sets of chromosomes; **amphimixis** (æmfimi·ksis) *Biol.* [mod.L. (A. Weismann, 1891), f. Gr. μίξις mingling: cf. ἀμφιμιγνύναι to mix up thoroughly] (see quot. 1949); **Amphineura** (-niūᵊ·rā), *sb. pl. Zool.* [mod.L. (H. von Jhering, 1876, in *Jahrb. d. deut. Malacozool. Ges.* III. 128): cf. Gr. νεῦρον nerve], a group of bilaterally symmetrical molluscs containing the chitons and related forms; **a·mphiodont** (-odǫnt), *a. Ent.* [Gr. ὀδούς, ὀδοντ- tooth], applied to a form of the mandibles in stag-beetles intermediate between the priodont and teleodont; **amphipneustic** (-(p)niū·stik), *a. Zool.* [f. as Amphipneust + -ic], having breathing-pores at both ends of the body, as certain dipterous larvæ; **a:mphiprosty·lar** *a.*, of or pertaining to an amphiprostyle; **amphirhine** (æ·mfirəin), *a. Zool.* [ad. mod.L. *Amphirhinus*, neut. pl. *Amphirhina*, f. Gr. ῥίν- nose], belonging to the *Amphirhina*, a division (comprising the large majority) of skulled vertebrates, having the nasal orifice double; two-nostrilled; opp. Monorhine *a.*

1842 Dunglison *Dict. Med. Sci.* (ed. 3), *Amphidiarthrosis*, a name given by Winslow to the temporo-maxillary articulation, because, according to that anatomist, it partakes both of ginglymus and arthrodia. **1890** Billings *Med. Dict.*, *Amphidiarthrosis*, joint having characteristics of two classes, both hinge and gliding, as articulation of lower jaw. **1909** *Cent. Dict.* Suppl., Amphidromic. **1938** *Nature* 11 June 1067/1 The effect of uniting the two waves which will have the higher part of their crests on opposite shores is to produce a set of amphidromic points. These are points at which there is no tidal rise and fall of the water level. **1947** *Sci. News* IV. 92 In the northern hemisphere the tidal oscillation will appear to rotate in an anticlockwise direction round such a point, and cotidal lines, joining points which have high water at the same time, will radiate from it. The point is called an amphidromic point. *Ibid.*, The rotation of the tide-producing forces round the sea also operates so as to produce an amphidromic system. **1909** J. W. Jenkinson *Experim. Embryol.* 267 One blastomere has a male and a female nucleus, and therefore 2 *n* chromosomes (amphikaryotic), while the other has only a female (thelykaryotic). **1893** Parker & Rönnfeldt tr. *Weismann's Germ-Plasm* 20 Amphimixis..consists in the mingling of two individuals or of their germs, and owing to its constant connection with reproduction in multicellular organisms it is usually spoken of as 'sexual reproduction'...Amongst unicellular organisms..amphimixis is widely spread..in the form of conjugation. **1909** Sorley *Interpr. Evolution* 23 It throws off certain cells which have the power of reproducing organisms like itself—this result being dependent in all the higher organisms upon amphimixis. **1913** Amphimixis [see *apomixis]. **1949** Darlington & Mather *Elements Genetics* 377 *Amphimixis*, reproduction by the fusion of two gametes in fertilization. As opposed to Apomixis. **1889** *Cent. Dict.*, Amphineura. **1906** G. Bourne tr. *Pelseneer's Mollusca* ii. 40 When *Chaetoderma* and *Neomenia* were investigated from an anatomical point of view, von Jhering united them..in a division of 'Worms', which he called Amphineura. **1922** *Chambers's Encycl.* I. 238/2 Amphineura are primitive gasteropods. **1945** Step & Wells *Shell Life* 181 The orders [of Gastropoda] are only four, but several..have been divided into sub-orders. The first of these orders (Amphineura) consists of the Mail-shells. **1883** Amphiodont [see Priodont *a.*]. **1932** J. S. Huxley *Probl. Rel. Growth* vii. 209 A specimen is classified as Amphiodont if a gap is present which is considered larger than the normal gap between two teeth, Prionodont if it is considered not to exceed this size. **1891** H. M. & M. Bernard tr. *A. Lang's Compar. Anat.* I. 482 This amphipneustic tracheal system is found in many parasitic or half-parasitic *Diptera* larvæ. **1899** *Camb. Nat. Hist.* VI. 450 Other larvæ have a pair of stigmata placed at the termination of the body, and another pair near the anterior extremity..; these larvæ are said to be 'amphipneustic'. **1957** Richards & Davies *Imms's Testbk. Ent.* (ed. 9) I. 134 *Amphipneustic*.—Only the prothoracic and the posterior abdominal spiracles are open. This type is a common one among larval Diptera. **1875** *Encycl. Brit.* II. 459 A temple with a portico at each end is said to be amphiprostylar. **1877** T. H. Huxley *Anat. Invert.* 147 The other amphirhine fishes.

amphiaster (æ·mfi,æ·stəɪ). [f. Amphi- + *aster 4.] **1.** *Biol.* [Fr. (H. Fol 1877, in *Arch. des Sci. Phys. et Nat.* Apr. 441).] A spindle-shaped formation in a developing ovum, with radiations at each end, thus resembling two star-shaped figures conjoined. Hence **amphiastral** *a.*, pertaining to or involving an amphiaster.

1885 *Q. Jrnl. Microsc. Sci.* XXV. 131 In Echinoderms the germinal vesicle and spot are not directly but only partially fragmented into the 'Amphiaster de rebut' or 'directive spindle'. **1901** G. N. Calkins *Protozoa* 81 A central or 'astral' granule which in some cases has been seen to divide, and to form an amphiaster. **1925** E. B. Wilson *Cell* (ed. 3) ii. 156 As far as the history of the chromosomes is concerned the anastral types of mitosis..do not differ in any important way from the amphiastral. **1952** G. H. Bourne et al. *Cytol. & Cell Physiol.* (ed. 2) iv. 179 The importance of..amphiaster formation in mitosis and cell division.

2. *Zool.* A sponge-spicule with rays at each end.

1887 Sollas in *Encycl. Brit.* XXII. 417/2 The sigmaspire becoming spined produces the spiraster;..this..by simultaneous concentration of its spines into a whorl at each end becomes the amphiaster. **1934** L. W. Sharp *Introd. Cytology* (ed. 3) xi. 155 In animal cells the amphiastral figure commonly develops essentially as follows.

amphibian. Add: **A.** *adj.* **3.** Of, pertaining to, or designating an amphibian (sense *B. 3).

1920 *Glasgow Herald* 26 June 7 A service employing flying boats or 'Amphibian' machines of a sea-type of service at sea and land aircraft. **1932** R. N. Liptrot in *Handbk. Aeronaut.* (R. Aeronaut. Soc.) ii. 55 Where amphibian gear is provided add 4 per cent. of the flying weight. **1935** H. G. Wells *Things to Come* v. 33 In the foreground a smooth-flowing river, or lake, that reflects the scene—suddenly the mirror is broken as enormous amphibian tanks crawl up out of the water. **1939** *War Illustr.* 21 Oct. 186 An interesting development made by the Soviet army is the amphibian tank.

B. *sb.* **2.** *fig.* A person having two modes of existence or a double character.

1902 *Daily Chron.* 5 May 4/3 These quiet, undistinguished amphibians of Sunwich. **1903** *Ibid.* 7 Jan 4/4 Such is the natural logic of the amphibian.

3. A seaplane, tank or other vehicle able to operate both on land and on water.

1920 *Glasgow Herald* 2 Aug. 6 The land tests for seaplanes (amphibians). **1923** *Sci. Amer.* May 306/1 A Man-Made Amphibian..a successful boat that would also travel on its bottom, or on its own wheels rather, upon dry land. **1933** *Jane's Fighting Ships 1933* 507/3 Aircraft in operation, 11 seaplanes, 2 UO seaplanes, 1 Viking Flying Boat, 3 Douglas amphibians. **1937** L. Hart *Europe in Arms* iii. 31 There are great numbers of two-men light tanks which can 'swim' rivers and are based on the Carden-Lloyd amphibian. **1957** R. Watson-Watt *3 Steps* 350 Tests in a *Walrus* amphibian on the slipway showed a useful surface-to-surface range.

amphibious, *a.* Add: **2. b.** Of a vehicle, aeroplane, etc.: = *amphibian A. 3; of a military operation: involving both land and sea forces; of soldiers: trained or used for amphibious warfare.

1915 *Illustr. War News* 7 Apr. 17 A new and ingenious amphibious automobile..constructed for the Austrian Army..the invention combines..features both of a boat and of a motor-car. **1928** H. Rowan-Robinson *Artillery* vi. 46 Even if amphibious tanks can be built, rivers will be normally crossed at the bridges. **1931** *Illustr. London News* 31 Oct. 675/2 The Carden-Loyd Amphibious Tank has all the fighting qualities of the latest light Tank, and, in addition, can navigate deep water as easily as it can cross the roughest country. **1941** *New Statesman* 15 Feb. 151/2 There remains the possibility of 'amphibious' warfare. **1943** W. S. Churchill in *Hansard Commons* ser. v. CCCXC. 563 Amphibious operations of peculiar complexity and hazard. **1945** *Ann. Reg. 1944* 56 'Ducks', amphibious lorries which were equally at home on land and in the water. **1945** *Daily Tel.* 17 May 5/6 Vice-Adml. Barbey, Commander of Amphibious Forces in the South-West Pacific.

amphibrachic (-bræ·kik), *a. Prosody.* [f. Amphibrach + -ic.] Consisting of amphibrachs.

1822 *Blackw. Mag.* XI. 441 The amphibrachic verses of French poetry. **1874** *Chambers's Encycl.* VI. 428/2 In the opening of Byron's *Bride of Abydos*..each of the three lines is in a different metre, the first dactylic, the second amphibrachic, the third anapæstic.

amphidiploid (æ:mfi,di·ploid), *a.* and *sb. Biol.* [f. Amphi- + *diploid *a.*] Having a diploid set of chromosomes derived from each of its parents; double diploid; = *allotetraploid *a.* Also as *sb.*, a hybrid of this kind. Hence **a:mphidi·ploidy,** the state of being amphidiploid.

1930 *Univ. Calif. Publ. Agric. Sci.* VI. 84 An amphidiploid *C. capillaris*–*C. dioscoridis* hybrid obtained by M. Nawashin..was smaller. *Ibid.*, Amphidiploids..involve two species of the same genus. **1932** N. I. Vavilov in *Proc. VI Internat. Congress Genetics* I. 341 The phenomenon of amphidiploidy of sterile hybrids proved to be rather frequent... Several cases of amphidiploidy were produced in hybrids of wheat and Aegilops. **1936** *Nature* 23 May 874/1 An amphidiploid wheat. The plant..had 42 chromosomes, and the evidence indicates that it was an amphidiploid produced by the parthenogenetic development of an egg... The amphidiploids were fertile.

amphitheatre. 4. (Earlier example.)

1860 G. A. Sala *Twice round Clock* 253 The dwellers in the high-up amphitheatre or gallery.

ample, *a.* **1. a.** Add to def.: Of persons: of large proportions; esp. in euphemistic use, of stout or well-covered women.

1860 Whyte-Melville *Mkt. Harb.* 60 Mrs. Dove, an ample lady, with the remains of considerable beauty. **1900** 'Sarah Grand' *Babs* (1901) iii, She was already more ample than a woman of thirty-eight need be in active life. **1959** *Times* 23 Feb. 12/3 Drawings of..blowsily ample girls undressing.

Amplidyne (æ·mplidəin). *Electr.* Also **a-.** [Trade-name (General Electric Co., U.S.A.), f. Ampli(fier + -dyne as in *metadyne.] A direct-current generator in which the output is controlled by small changes in the input.

1940 *Electronics* Apr. 34 Dr. E. F. W. Alexanderson..of the General Electric Company revealed a new type of generator known as the Amplidyne... The Amplidyne generator responds almost instantaneously to changes in power input. **1943** F. Felix in *Jrnl. Amer. Soc. Nav. Engin.* LV. 775 One of the most important developments of recent years is the amplidyne, a product of GE engineering. **1958** *Guided Missiles* vii. 388/1 Power amplification is obtained by electronic means or by using a dynamotor-type amplifier such as a motor-generator set or an amplidyne.

amplification. Add: **6.** *Electr.* The action of amplifying (see *amplify *v.* 2 b). Also *attrib.*

1915 *Year-Bk. Wireless Telegr.* 665 Amplification (Magnification). The ratio of the useful effect obtained by the employment of the amplifier to the useful effect obtained without that instrument. **1919** J. A. Fleming *Thermionic Valve* vii. 249 Method..for Measuring the Amplification Factor. **1920** H. J. Van der Bijl *Thermionic Vacuum Tube* vii. 160 The constant μ..can therefore be referred to as the amplification constant. **1935** *Discovery* Sept. 278/2 Special valves are being developed suitable for direct signal amplification at the ultra-high frequencies. **1942** *Electronic Engin.* XV. 127/3 Existing formulæ for the amplification factor of the three electrode valve..assume..that the distance between the grid and anode is large. **1943** *Gloss. Terms Telecomm.* (*B.S.I.*) 32 *Amplification factor*, the voltage factor of the anode and the control electrode, the anode current remaining unchanged. **1962** *Newnes Encycl. Electr. Engin.* 825/2 It [*sc.* the transistor] can be used..to give current or voltage amplification as desired.

amplified, *ppl. a.* Add: **3.** *Electr.* (See *amplify *v.* 2 b.)

1922 Glazebrook *Dict. Appl. Physics* II. 856/1 Energy coming in on either [telephone] line will in part go into R, which..sends out amplified energy to both lines.

amplifier. Add: **4.** *Electr.* An appliance or circuit which amplifies. (See *amplify *v.* 2 b.)

1914 *Electrician* 12 June 402/2 After developing the Audion as detector, amplifier and oscillator, Dr. Lee de Forest has recently described..the small-sized 3½ volt amplifier bulb. **1918** in J. A. Fleming *Thermionic Valve* (1919) 274 Where the de Forest valve acts as an amplifier it is outside any claim on the Fleming patent of 1904. **1933** *Boys' Mag.* XLVII. 108/1 The complete amplifier could be housed in any average existing gramophone, or could be converted, with the addition of a tuned circuit, into a radio receiver. **1941** *B.B.C. Gloss. Broadc. Terms* 3 *Amplifier*, apparatus used to increase the volume of a programme output by means of energy drawn from an external source, such as the supply mains.

amplify, *v.* **2.** Delete †*Obs.* and add: **b.** *Electr.* To increase the strength of (an electrical current, signal, etc.).

1915 *Proc. Inst. Radio Engin.* III. 282 By connecting the pliotron as an amplifier, as shown in Figure 10, the high frequency currents received from the grid may be amplified from 100- to 600-fold. **1928** *Daily Chron.* 9 Aug. 4/1 The Archbishop's voice was heard, amplified loudly, but indistinctly, telling us of the sacredness of Ypres. **1931** *B.B.C. Year Bk.* 279 On arrival at the Brookmans Park control room, the music is amplified to the correct strength and is then passed to the sub-modulators in the main transmitters. **1953** Amos & Birkinshaw *Television Engineering* I. i. 17 In the present BBC television system vision and sound signals are transmitted over separate links but it is common practice for the aerial and early stages of receivers to accept and amplify vision and sound transmissions together. **1955** *Oxf. Jun. Encycl.* VIII. 470/1 The transistor can be used in the same way as a thermionic triode valve to amplify weak electric currents.

amplifying, *ppl. a.* Add: **2.** *Electr.* (See *amplify *v.* 2 b.)

1920 *Chambers's Jrnl.* X. 274/1 The Boche listening apparatus or amplifying telephone. **1922** Glazebrook *Dict. Appl. Physics* II. 888/2 Amplifying valves differ from those used for transmitting purposes in more than mere dimensions and power.

amplitude. Add: **6. c.** *Electr.* The maximum departure of the value of an alternating current or wave from the average value. Also *attrib.*, as *amplitude distortion.*

1895 S. P. Thompson *Elementary Less. Electr. & Magn.* x. 487 The impressed electromotive force follows a sine law..where D is the maximum value or amplitude attained by E. **1931** *B.B.C. Year-Bk.* 436/1 *Amplitude Distortion*, in electrical apparatus, the variation in response at different amplitudes with an input of constant frequency. **1945** *Electronic Engin.* XVII. 640 Either the amplitude or the frequency of the sweep may be increased. **1953** Amos & Birkinshaw *Television Engin.* I. i. 18 All the components of the vision signal must be reproduced at the correct amplitude to avoid distortion in the reproduced image.

d. *Electr.* *amplitude modulation*, modulation of a wave by variation of its amplitude; also, the system using such modulation (abbrev. A.M.). Cf. *frequency modulation.

1922 [see *frequency modulation]. **1932** F. E. Terman *Radio Engin.* x. 357 In all the commonly used systems of radio communication the intelligence is transmitted by varying the amplitude of the radiated waves...Communication carried on in this way is said to take place by means of amplitude modulation. **1942** *Electronic Engin.* XIV. 650 The action of the limiter largely prevents the frequency distortion effect present with amplitude modulation. **1944** *Ibid.* XVII. 58 In an A.M. transmitter... **1962** A. Nisbett *Technique Sound Studio* 240 *Amplitude modulation* (AM), a method whereby the information in an audio signal is carried on the much higher frequency of a radio wave. The envelope of the amplitude of the radio wave in successive cycles is equivalent to the wave form of the initial sound. Historically, AM is the method which was used first and such transmissions now crowd the short, medium and long wave bands...For high quality transmission AM has largely given way to FM (frequency modulation).

amplitudinous (æmplitiū·dinəs), *a.* [f. L. *amplitūdin-em* (see Amplitude) + -ous.] Ample, capacious.

1904 *Daily Chron.* 9 Sept. 4/7 What monsters there be; Huge, amplitudinous, endless, and ravenous, Serpents that people the sea. *a* **1913** F. Rolfe *Desire & Pursuit* (1934) v. 38 His gesture was as amplitudinous as his drawl and deep guffaw. **1930** *Times Lit. Suppl.* 3 July 547/1 The well-known statuettes of amplitudinous female forms.

Hence **amplitu·dinously** *adv.*, amply.

1921 Joyce *Lett.* (1957) 164 Struggling with the acidities of Ithaca..to prepare for the final amplitudinously curvilinear episode Penelope.

ampoule (æ·mpūl), var. of Ampul. Also **ampule. a.** A small sealed (glass) vessel used for storing sterilized materials prepared for injection. **b.** A similar vessel or phial containing other materials.

1644 [see Dict. 2]. [**1886** S. Limousin in *Archives de Pharmacie* I. 145 (*title*) Ampoules hypodermiques; nouveau mode de préparation des solutions pour les injections hypodermiques.] **1907** *Chemist & Druggist* 1 June 843/2 The method of storing hypodermic and other liquids in closed glass bulbs or ampoules has long been in vogue in France. **1909** *Ibid.* 30 Jan. 169/1 M. Limousin..introduced the ampoule as a convenient method of preserving hypodermics. **1909** *Lancet* 8 May 1365/2 An ampoule filler. **1922** *Nature* CIX. 252/1 An apparatus for the routine standardisation of ampoules containing radium compounds and emanation by the γ-ray ionisation method was installed recently at the Institut Curie. **1930** *Discovery* Sept. 290/2 [A] device for detecting carbon monoxide [in mines]. This consists of an easily crushed cotton-covered ampoule filled with a solution which changes colour when exposed to air containing carbon monoxide. **1937** *Nature* 27 Feb. 380/1 Anahaemin..is issued as a solution in ampoules and rubber-capped vials. **1945** *Times* 7 Aug. 10/2 Special ampoules for time fuses. **1947** J. Steinbeck *Wayward Bus* xix. 239 A little ampule [containing poison] to wear around his neck. **1959** *Observer* 3 May 4/6 Messengers scurried in every direction to retrieve the condemned ampoules [of polio vaccine].

amputee (æmpiuti·). [f. Amput(ate *v.* + -ee¹.] One who has lost a limb or other part of the body by amputation.

1910 *St. Barts. Hosp. Jrnl.* XVII. 90/1 Please put the patient *both* to bed, and then, perhaps, we'll see Which is the amputated part and which the amputee. **1939** *Lancet* 14 Oct. 837/2 In place of endless gadgets we now have a few standard 'set ups' round which limbs to suit any particular amputee can be constructed. **1945** *N.Y. Times* 28 Sept. 23/6 A permanent lodge where vacationing members of the Bilateral Leg Amputee Club of America may spend their vacations. **1958** *Oxford Mail* 11 Feb. 3/2 Loose lino, chairs and rugs were all snares for the unwary amputee.

amuletic, *a.* Delete †*Obs. rare* and add examples.

1742 [see Prophylactic *a.*]. **1855** E. Rich in Smedley et al. *Occult Sciences* 359 The amuletic medicine which cured disease. **1932** *Times Lit. Suppl.* 17 Mar. 198/2 Although referred to as 'seals'..their purpose may be amuletic.

amusedly (ămiū·zėdli), *adv.* [f. Amused *ppl. a.* + -ly².] As being amused; with amusement.

1844 *Fraser's Mag.* XXX. 63/1 He looked amusedly at the baggage. **1864** Meredith *Emilia in England* xv, Leaning back and contemplating him amusedly. **1901** W. J. Locke *Usurper* iii. 29 He wandered amusedly around the baccarat tables. **1955** T. W. Thomas in *Ess. in Crit.* V. 78 The phrase is..Lou's but reported to hold her amusedly at a distance.

amusement. Add: **7.** Frequent in *Comb.* in senses 5 and 6, as *amusement-lover* (*-loving*), *-mad*, *-seeker* (*-seeking*); also *amusement arcade, centre, hall, park.*

1870 D. J. Kirwan *Palace & Hovel* (1963) xvii. 154 The cheap amusement halls of London are of the very lowest kind to be found anywhere. **1898** *Daily News* 8 Sept. 5/1 It may surprise old amusement lovers to learn that one of the original troupe of Christy Minstrels..is still alive. **1904** *Westm. Gaz.* 31 Dec. 16/3 The average amusement-seeker has a limited amount to spend. **1906** *Bungalow* Dec. 4/2 The Tivoli..caters for the amusement-loving people. **1906** 'O. Henry' *Four Million* (1916) xxv. 252 [He] hied himself to play pennies in the slot machines at the Amusement Arcade. **1907** *Westm. Gaz.* 10 May 12/1 London in particular..is 'amusement mad'. **1908** *Daily Chron.* 9 Dec. 4/7 He would secure to the amusement-seeking public a far more genuine and substantial benefit. **1909** *N.Y. Even. Post* 13 June 4 This is, or was to have been, the year of the amusement park. **1923** R. D. Paine *Comrades Roll. Ocean* ix. 161 It had been picked up empty in one of the amusement parks outside the city. **1924** *The Studio* (1925) Autumn no., 76 (Advt.), Advertising in the Amusement Park at Wembley. **1935** 'J. Guthrie' *Little Country* xxi. 330 Ernest..clung to the sides of his boat and wondered why they called the place amusement park. **1936** *Discovery* Sept. 267/1 All the resources of amusement that a great amusement-centre [*sc.* Blackpool] can offer. **1937** *Ibid.* June 816/2 Amusement halls, a club, swimming baths, parks and other comforts. **1947** *Min. Town & Country Planning Bull. Selected Appeal Decisions* I. 11 (*heading*) Amusement arcade in holiday resort. *Ibid.*, An amusement caterer purchased three small single storey lock-up shops. *Ibid.*, He opened two of the shops in succession, as an amusement arcade. **1958** *Times* 23 Aug. 8/6 A lady who also runs an amusement arcade and whose eyes flicker in time with the lights of her pin-ball machine.

amusing, *ppl. a.* Add: **3.** Also, as a vogue-word in trivial applications. Also *absol.*

1925 A. Huxley *Along Road* III. 200 The great invention of more recent years has been the 'amusing'...All bad art, whose badness is a positive..quality..may be said to be amusing. **1929** D. L. Moore *Pandora* ix. 168 'Amusing'..is at the moment the first favourite..'an amusing pair of shoes', 'amusing little sandwiches', 'an amusing kind of lipstick'. **1934** C. Lambert *Music Ho!* II. 79 The scrapbook taste which is considered so modern and 'amusing' when applied to interior decoration. **1939** O. Lancaster *Homes Sweet Homes* 62 Steel-engravings and wax fruit enjoy a come-back on an 'amusing' basis. **1960** *Times* 15 Sept. 15/3 The cult of the 'amusing'. This lethal word has been applied during the past 30 years to one art after another. From ballet to interior decoration, from the Parisian music of 1920 to *objets trouvés*, clever people in search of high fashion have turned amusing.

amusingly, *adv.* (Earlier example.)

1812 *Examiner* 3 May 282/2 Our minds should be..impressed with the immorality of the habit, otherwise so amusingly depicted.

amusingness. (Earlier and later examples.)

1823 *Blackw. Mag.* XIV. 641/2 Touching the reality (as well as the amusingness) of spectral appearances, I protest..against being put down as a scoffer. **1886** R. Broughton *Dr. Cupid* I. iii. 26 Thoughts..that augur but ill for the amusingness of his dinner. **1959** *Times* 11 Sept. 16/5 Our entertainment is made to depend on..the general amusingness of the dialogue.

amylase (æ·milēˈs, -z). *Biochem.* [f. Amyl² + *-ase.] An enzyme or organic catalyst which disintegrates starch, present in animals and plants; = Diastase.

1893 *Jrnl. Chem. Soc.* LXIV. II. 587 The dried, as well as the fresh, organism contains an enzyme 'amylase', similar to the diastase of malt. **1899** J. R. Green *Soluble Ferments* ii. 15 (*heading*) Diastase (Amylase, Ptyalin). **1910** *Practitioner* June 823 The hydrolysis is brought about by means of enzymes, such as ptyalin, amylopsin or amylase, lactase and glucase.

amyloid, *a.* and *sb.* Add: For etym. read: [G. (Vogel and Schleiden 1839, in *Ann. Phys. & Chem.* XLVI. 327), f. L. *amylum* (see Amyl¹) + -oid.] **A.** *adj.* **2.** *Path.* Applied to a form of degeneration of various organs, or to the albuminoid substance (formerly supposed to be akin to starch) produced in this; = Lardaceous *a.*; (transfer the quots. in B. 3 to this sense and add earlier and additional examples).

1859 J. Paget *Let.* 1 June in *Mem. & Lett.* (1901) 224, I have only a very imperfect knowledge..of what has been done..on amyloid degeneration. **1873** [see Lardaceous *a.*]. **1881** [see Cholesterin].

B. *sb.* For senses 1 and 2 substitute:

1. A substance akin to starch; any member of the group of carbohydrates including starch and related substances, as cellulose (with quots. 1872, 1873).

For sense 3 substitute:

2. *Chem.* The substance formed in amyloid degeneration, also called Lardacein.

amyloidosis (æ:miloidōᵘ·sis). *Path.* [mod.L., f. Amyloid *a.* and *sb.* + -osis.] Amyloid degeneration.

1900 in Gould *Pocket Med. Dict.* (ed. 4). **1961** *Lancet* 9 Sept. 599/1 Amyloidosis in typhoid fever.

amylolysis (æmilǫ·lisis). *Biochem.* [f. Amylo- + *lysis; cf. Amylolytic *a.*] The conversion of starch into soluble products by the action of enzymes.

1890 Billings *Med. Dict.* **1909** *Practitioner* Dec. 818 The contents [of the stomach] showed deficient amylolysis.

amylopectin (æ:milope·ktin). *Chem.* [ad. F. *amylopectine* (Maquenne and Roux 1905, in *Comptes Rendus* CXL. 1305), f. Amylo- + Pectin.] A mucilaginous constituent of starch.

1905 *Jrnl. Chem. Soc.* LXXXVIII. I. 511 The second constituent [of starch] is amylopectin. **1957** *Sci. News* XLV. 83 A branched-chain molecule such as the amylopectin component of starch.

amyloplast (æ·milǫplɑst). *Bot.* [ad. F. *amyloplaste* (L. Errera *L'Épiplasme des Ascomycètes* (1882) x. 74), f. Amylo- + -plast.] A colourless granule in a plant-cell, around which a starch-grain is formed.

1886 S. H. Vines *Lect. Physiol. Plants* x. 180 Schimper has observed that the formation of starch-grains is commonly effected, in parts of plants not exposed to light, by certain specialised portions of the protoplasm which are termed starch-forming corpuscles or amyloplasts. **1955** *New Biol.* XIX. 90 The amyloplasts, the specific sites of starch formation.

amylopsin (æmilǫ·psin). *Biochem.* Also **-ine.** [f. Amylo-, after *pepsin*: cf. Steapsin.] The amylolytic ferment of the pancreatic juice.

1881 *Syd. Soc. Lex.* s.v., *Amylopsine*, a name given by Defresne to that ferment of the pancreatic juice which converts starch into sugar. **1886** *Buck's Handbk. Med. Sci.* II. 452/1 A third ferment, *amylopsin*, acts on starch as ptyalin does. **1894** *Jrnl. Chem. Soc.* LXVI. II. 103 The production of ferments by the pancreatic cell is successive; it is an operation of two acts, of which the first is the formation of amylopsin, the second of trypsin. **1937** *Thorpe's Dict. Appl. Chem.* I. 690/1 The proteolytic enzyme (trypsin), and the amylolytic enzyme (amylopsin).

amyrin (æ·mirin). Also **-ine.** [ad. F. *amyrine* (S. Baup 1851, in *Jrnl. de Pharm.* XX. 323), Amyr(is + -in¹.] A resin obtained from a Mexican species of *Amyris* (*A. elemifera*).

1876 *Jrnl. Chem. Soc.* II. 423 Amyrin is found in elemi in the form of microscopic prisms. **1889** *Chambers's Encycl.* IV. 288/1 When treated with cold alcohol, it [*sc.* Elemi] partly dissolves, leaving about 20 per cent. of a white resinous substance called Amyrine.

Amytal (æ·mităl). Also **amytal.** Trade-name of a white crystalline powder used as a sedative.

1926 *Jrnl. Pharmacol.* XXVI. 379 Isoamyl ethyl barbituric acid (Amytal). **1928** *Chem. Abstr.* 460 (*caption*) The effect of insulin on the metabolism of dogs under amytal anesthesia. **1930** *Trade Marks Jrnl.* 26 Mar. 484/1 *Amytal*, a medicated preparation in tablet form for human use as a sedative and hypnotic. Eli Lilly and Company. **1930** *Lancet* 14 June 1302/1 Amytal is the trade-name of iso-amylethyl barbituric acid, and it has now been employed as an anæsthetic in a fair number of cases. *Ibid.*, It appears that amytal, if used without risk, is rather a hypnotic than a true anæsthetic. **1938** J. Steinbeck *Long Valley* 118 Here's some sodium amytal. One of these capsules will calm him down.

anabatic, *a.* Add: **2.** *Meteorol.* Of a wind: caused by the (local) upward motion of warmed air. Cf. *katabatic *a.*

1918 *Meteorol. Gloss.* 29 *Anabatic*, referring to the upward motion of air due to convection. **1919** Faraday *Gloss. Aeronaut. Terms* 31 A local wind is called anabatic if it is caused by the convection of heated air, as, for example, the breeze that blows up valleys when the sun warms the ground.

anabiosis (æ:năbəi,ōᵘ·sis). *Biol.* [mod.L., ad. Gr. ἀναβίωσις, f. ἀναβιόειν: see Anabiotic *a.*] A coming to life again; revival, resuscitation; see also quot. 1963.

1890 Billings *Med. Dict.*, *Anabiosis*, the power which certain organisms possess of regaining vital activity after being dried and heated. **1913** J. G. Frazer *Belief in Immortality* I. III. 85 John Hunter, supported by his experiments on *anabiosis*, hoped to prolong the life of man indefinitely by alternate freezing and thawing. **1926** *Chambers's Jrnl.* 1 May 349/2 From latent life there can be a protoplasmic resurrection, learnedly called 'Anabiosis'. **1963** *Times* 28 Feb. 8/2 Spacemen might one day be put in a state of anabiosis—in which the body functions are greatly slowed down and food is not necessary—for long space journeys, and then revived as they neared their destination.

anabolic (ænăbǫ·lik), *a. Biol.* [f. Gr. ἀναβολή (*lit.* a throwing up), ascent, etc. + -ic.] Pertaining to, involving, or exhibiting anabolism; constructively metabolic: opp. to Katabolic.

1876 [see Katabolic *a.*]. **1885** *Encycl. Brit.* XIX. 13/2 An upward series of changes (*anabolic* changes). **1889** Geddes & Thompson *Evol. Sex* ii. 26 The males live at a loss, are more katabolic...The females..live at a profit, are more anabolic—constructive processes predominating in their life, whence indeed the capacity of bearing offspring. *a* **1901** F. W. H. Myers *Hum. Pers.* (1903) II. 514 The katabolic as well as the anabolic forces, the output as well as the intake of the bodily frame, are amenable..to subliminal control. **1962** *Lancet* 12 May 986/2 The second, or anabolic, phase in the newborn child also resembles the anabolic state after injury.

anabolism (ănæ·bŏliz'm). *Biol.* [f. as prec. + -ism.] The 'ascending' process in metabolism, in which simpler substances, as nutritive matter, are transformed into more complex ones, and thus built up into the living structure of the organism; constructive metabolism: opp. to Katabolism.

1886 W. H. Gaskell in *Jrnl. Physiol.* VII. 46 In other words, metabolism includes the two opposite processes of destruction and construction, or as they may be called of katabolism and anabolism. **1889** Geddes & Thomson *Evol. Sex* x. 122 These upbuilding, constructive, synthetic processes are summed up in the phrase anabolism. **1897** Willis *Flower. Pl.* I. 206 Metabolism..may be divided into anabolism, the building up, and katabolism, the breaking down, of complex materials. **1959** *Chambers's Encycl.* II. 318/1 Growth, and all other processes involving chemical synthesis, are processes in which energy is consumed, and all such are grouped together under the title of anabolism. *Ibid.* IX. 309/1 Processes which lead to an increase in chemical complexity making up anabolism. *Ibid.*, Only the energy of their food can be put to service for anabolism, muscular movement and so on.

anabranch (æ·năbrɑnˈʃ). *Physical Geogr.* [f. Ana(stomosing *ppl. a.* + Branch *sb.*] Esp. in Australia: a branch stream which turns out of a river and re-enters it lower down.

1834 J. R. Jackson in *Jrnl. R. Geogr. Soc.* IV. 79 Thus, such branches of a river as after separation re-unite, I would term anastomosing-branches; or, if a word might be coined, ana-branches. **1847** Leichhardt *Jrnl.* ii. 35 The river..divided into ana-branches which..made the whole valley a maze of channels. **1849** Sturt *Central Austr.* I. 93 To ascertain how high the back-waters of the Murray had gone up the Ana-branch of the Darling. **1907** H. Lawson *Send round the Hat* in *Prose Wks.* (1948) 649 He rode round the outside track and came in on to the river just below where the anabranch joins it.

anachronism. Add: **2.** Also *transf.* of persons.

1871 [in Dict.]. **1899** B. Harraden *Fowler* I. vii, 'Sentiment,' she repeated. 'It is absurd to try and hustle sentiment off the scenes.'..'You are always an anachronism,' he said, quietly. **1952** M. McCarthy *Groves of Academe* iii. 37 She herself was a smoldering anachronism, a throwback to one of those ardent young women of the Sixties, Turgenev's heroines.

‖**anaclasis** (ănæ·klăsis). *Pros.* Also **-klasis.** Pl. **-ases.** [mod.L., a. Gr. ἀνάκλασις bending back (see Anaclastic *a.*).] In Ionic verse: an interchange of the final long syllable of the first metron with the opening short syllable of the second.

1784 J. B. Seale *Anal. Gr. Metres* 25 In the intermediate places a second Pæon is occasionally joined to a second or third Epitrite, so that the two Feet together are equal in time to two Ionic Feet. This is called an Ανακλασις. **1830** J. Seager tr. *Hermann's Doctrine of Metres* xxxviii. 98 This method is termed ἀνάκλασις, and the verses themselves ἀνακλώμενοι, because the change in the numbers is not made in one Ionic foot, but in two, the end of the one, and the beginning of the other being changed. **1900** [in def. of Galliambic *a.*]. **1901** T. D. Goodell *Chapt. Gr. Metric* v. 144 Not only were ionic kola extended to eighteen lines, with or without anaklasis, but also the plain iambic and trochaic. **1938** G. D. Thomson tr. *Aeschylus: Oresteia* II. 327 A rearrangement of the opening syllables (anaclasis). **1949** *Oxf. Classical Dict.* 567/1 The view..that the anacreontic is derived from the ionic dimeter by the interchange of the final long of the first metron with the opening short of the second ('anaclasis').

anaclastic. Add: **A.** *adj.* **3.** *Pros.* Of, pertaining to, or involving anaclasis.

1889 in *Cent. Dict.* **1912** J. W. White *Verse Gr. Comedy* viii. 185 Normal and anaclastic metres may correspond in strophe and antistrophe. **1949** *Oxf. Classical Dict.* 566/2 The *hypodochmius*, or 'anaclastic' dochmius, inverts the first two elements, – ⏑ – ⏑ –. **1957** W. Beare *Latin Verse & Europ. Song* vii. 83 Ionic Verse (a minore ⏑ ⏑ – –, a maiore – – ⏑ ⏑)..is..highly emotional, particularly in its 'anaclastic' or 'broken' form (⏑ ⏑ – ⏑ – ⏑ – –).

anaclinal (ænăkləi·năl), *a. Geol.* [f. Gr. ἀνά back + κλίν-ειν to bend + -al.] Of a valley, river, etc.: descending in a direction opposite to the dip of the underlying rocks.

1875 J. W. Powell *Explor. Colorado River* xi. 160 Transverse valleys..three varieties are noticed..(*c*) anaclinal, valleys that run against the dip of the beds. **1941** C. A. Cotton *Landscape* iv. 25 Those flowing in the reverse direction are 'anti-dip' or 'anaclinal' streams.

anaclitic (ænăkli·tik), *a. Psychol.* [f. Gr. ἀνάκλιτ-ος for reclining (ἀνακλίνειν to lean back, recline) + -ic.] Orig. in phr. *anaclitic type* (tr. G. *Anlehnungstypus* (Freud), lit. 'leaning-on type'), a person whose choice of a 'love object' is governed by the dependence of the libido on another instinct, e.g. hunger; also in extended use, characterized by dependence on another or others (see quots.).

1922 J. Riviere tr. *Freud's Introd. Lect. Psycho-Analysis* III. xxvi. 356 The anaclitic type (*Anlehnungstypus*) in which those persons who become prized on account of the satisfactions they rendered to the primal needs of life are chosen as objects by the Libido also. **1922** J. Strachey tr. *Freud's Group Psychol.* vii. 60 The boy has begun to develop a true object-cathexis towards his mother according to the anaclitic type. **1930** W. Healy et al. *Struct. & Meaning of Psychoanalysis* II. 106 These attachments are likely in the beginning to be strongly anaclitic in nature; the child will turn first to those who assist it in its helplessness and gratify its self-preservative needs. **1934** H. C. Warren *Dict. Psychol.* 12/1 *Anaclitic object-choice*, taking one's earliest attachments (to mother or nurse) as a model for the selection of the first love object. Contr. w. *narcissistic object-choice.*

anacrustic (ænăkrʊ·stik), *a. Pros.* [ad. Gr. ἀνακρουστικ-ός, f. ἀνακρούειν: see Anacrusis.] Characterized by anacrusis.

1878 G. M. Hopkins *Let. to Bridges* 30 May (1935) 53 The rhythm is anacrustic or, as I should call it, 'encountering'.

anæmia. (Earlier example; the 1807 example cited in the *Stanford Dict. of Anglicized Words* is not certainly the same word.)

[**1761** J. Lieutaud *Précis de la Médecine Pratique* (1765) I. I. 81 L'Anemie. Cette maladie (anemia) qu'on peut regarder comme l'épuisement des vaisseaux sanguins.] **1824** *Trans. Med. Chir. Soc. Edin.* 202 A disease, under the title of Anæmia, has been described by Becker.

anæmic, *a.* Add: **1. b.** *transf.* and *fig.* Lacking in vigour, strength, or spirit. *colloq.*

1898 *Daily News* 16 Sept. 4/5 The Americans..are not anæmic. They are overflowing with vigour and enterprise. **1899** *Ibid.* 21 Oct. 7/7 Some of the tweeds are made in rather anæmic, sickly tones. **1906** *Westm. Gaz.* 1 Dec. 20/3 [He] must indeed have made an anæmic diamond declaration. **1941** N. Coward *Australia Visited* iv. 28 Far too much anæmic nonsense was put forward by the young intelligentsia.

anaerobe (ănē·ərōᵘb, ănēᵊ·rōᵘb). *Biol.* [ad. F. *anaérobie* (Pasteur 1863, in *Comptes Rendus* LVI. 1192), f. An- 10 + *aerobe.] A micro-organism of the group *Anaerobia*, which can live without free oxygen. So **anaero·bian, anaerobic** (-ǫ·bik), **anaero·bious** *adjs.*, of the nature of or pertaining to anaerobes; capable of living without free oxygen; **anaero·bically** *adv.*; **ana:erobio·sis,** life in a medium devoid of free oxygen; **ana:erobiotic** (-ǫ·tik) *a.*, pertaining to or characterized by anaerobiosis, anaerobic.

1879 Faulkner & Robb tr. *Pasteur's Ferment.* iv. 116 We may divide living beings into two classes, aërobian..and anaërobian. **1884** Klein *Micro-Org.* vi. 34 Some bacteria require free access of oxygen, and are called aërobic (Pasteur); others grow without free oxygen, and are anaërobic (Pasteur). **1884** W. Stirling tr. *L. Landois' Physiol.* I. iv. 374 This fungus [*sc.* Bacillus butyricus]..is a true anaërobe, and grows only in the absence of O[xygen]. **1885** Vines in *Encycl. Brit.* XIX. 51/2 It is just the anaerobiotic plants which are most highly endowed with the property of exciting fermentation. **1887** Anaerobically [see *aerobe]. **1891** Linsley tr. *Fraenkel's Bacteriol.* 115 That the anaërobia in particular distinguish themselves by generating gases is already known. **1894** *Pop. Sci. Monthly* June 278 With special reference to anaerobiosis and gas production among bacteria. **1914** G. B. Shaw *Misalliance* 15 There was a regular terror of a countess with an anaerobic system [of drainage]. **1931** *Discovery* Sept. 303/2 The distinction, seldom made, between anaerobic respiration and fermentation is pointed out. **1949** H. W. Florey *Antibiotics* I. I. i. 32 He isolated two anaerobes, one of which,..had apparently no direct effect in inhibiting the growth of any of a considerable number of organisms. **1958** *New Biol.* XXV. 87 Where the lagoons are not too deep, i.e. where the surface area is sufficiently large in relation to the volume of sewage, it is possible to dissolve, by diffusion at the surface, sufficient oxygen from the air to prevent anaerobiosis and consequent odour. **1959** J. Clegg *Freshwater Life* (ed. 2) 68 The decomposition of the organic remains by anaerobic bacteria results in the formation of sulphuretted hydrogen. **1961** *Lancet* 22 July 210/2 The cells..are proliferating under virtually anaerobic conditions.

anæsthesia. (Now fully naturalized; pronunc. usu. -þī·ziă.) (Additional example.) Also, **local anæsthesia,** anæsthesia of a limited area of the body; opp. to **general anæsthesia,** anæsthesia of the whole body. Cf. *basal anæsthesia*, *basal *a.* 3.

1846 O. W. Holmes *Let. to W. T. G. Morton* 21 Nov. in E. Warren *Letheon* (ed. 2, 1847) 79 The state should, I think, be called 'Anæsthesia'. This signifies insensibility... The adjective will be 'Anæsthetic'. Thus we might say the state of Anæsthesia, or the anæsthetic state. **1848** *Monthly Jrnl. Med. Sci.* July 48 He could produce thus *local* anæsthesia in the worm. **1867** *Times* 14 Mar. 12/6 He [*sc.* Dr. Richardson] next demonstrated the principle and practice of the local method by ether spray...In veterinary surgery this local anæsthesia is, Dr. Richardson said, applicable to all operations. **1900** *Lancet* 21 Apr. 1125/2 (*title*) The production of local anæsthesia in the ear. **1910** *Encycl. Brit.* I. 909/1 *Local Anaesthesia*..The discovery of methods by which the insensibility may be confined to the area of operation and the loss of consciousness avoided...It is only lately that it has been successfully applied to the severer [operations]. It is very doubtful whether local anaesthesia will ever replace general in the latter class. **1936** *Discovery* Nov. 357/2 One who has had some personal experience of anaesthesia, both general and local.

anæsthesiology (ænesþīziǫ·lŏdʒi, ænis-.) [f. Anæsthesi(a + -ology.] The study and practice of anæsthesia and anæsthetics. So **anæsthesio·logist,** a person versed in anæsthesiology.

1914 *Stedman's Med. Dict.* 46/1 Anesthesiology. **1947** Dorland & Miller *Med. Dict.* 101/1 Anesthesiologist. **1955** *Sci. News Let.* 28 May 351/2 Resident anesthesiologists can quickly learn the techniques so as to reinforce the hypnosis daily. **1959** *Times Suppl. Rubber Industry* 27 Apr. x/6 Anaesthesiology, a branch of the medical sciences of ever-increasing importance.

anæsthetic. Add: **A.** *adj.* **1.** (Earlier example.)

1846 [see *anæsthesia].

B. *sb.* **local anæsthetic,** a substance which by application or injection induces local anæsthesia; opp. to **general anæsthetic,** a substance which induces general or total anæsthesia. Cf. *basal anæsthetic*, *basal *a.* 3.

1851 J. Arnott *Neuralgic Affections* 20 It is very natural that the disappointment from the exaggerated statements..should indispose the surgeon to put trust in any local anæsthetic, without such corroborative evidence. **1876** Dunglison *Dict. Med. Sci.* 47/2 The ethers, rhigolene and other agents, when applied to a part in the form of spray, by their evaporation benumb it; and thus act as local anæsthetics. **1879** *Syd. Soc. Lex.* s.v., General anæsthetics are commonly employed in the form of vapour. **1910** *Encycl. Brit.* I. 909/1 The earliest local anaesthetic was cold, produced by a mixture of ice and salt. In place of this cumbersome method, the skin is now frozen by means of a fine spray of ether or ethyl chloride directed upon it. **1955** *Oxf. Jun. Encycl.* XI. 4/2 A patient to whom a general anaesthetic is administered loses consciousness, while a local anaesthetic affects only the area of operation, the patient remaining fully conscious. Cocaine..the most commonly used local anaesthetic, was introduced in 1879.

anagenesis (ænă,dʒe·nėsis). *Biol.* [mod.L., f. Gr. ἀνά up + γένεσις origin, Genesis.] Progressive or 'upward' evolution of species (opp. *catagenesis). So **anagenetic** (æ:nă,dʒėne·tik) *a.*

1889 A. Hyatt *Genesis of Arietidae* in *Mem. Museum Comp. Zool.* XVI. II. 71 (*title*) Anagenesis, or the Genesis of Progressive Characteristics. *Ibid.* 74 The law of succession in anagenesis, therefore, is that progressive species..were the direct descendants of progressive varieties or forms. **1893** E. D. Cope in *Monist* III. 637 The process of evolution may be either progressive (Anagenesis) or retrogressive (Catagenesis). **1896** —— *Organ. Evol.* ix. 475, I have termed these classes the Anagenetic, which are exclusively vital... The anagenetic class tends to upward progress in the organic sense. **1953** J. S. Huxley *Evolution in Action* iii. 75 Bernhard Rensch, the German biologist..separated evolutionary processes into cladogenesis, or branching evolution..and anagenesis, or upward evolution.

anaglyph. Add: **2.** *Photogr.* A composite stereoscopic picture printed in superimposed complementary colours.

1897 E. J. Wall *Dict. Photogr.* (ed. 7) 30 *Anaglyph*, a means of producing stereoscopic effect due to MM. Louis Ducos du Hauron and D'Almeida. *Ibid.* 31 The principle of the anaglyph is applicable to the stereoscopic projection with the magic lantern. **1910** *Lancet* 5 Nov. 1366/2 The method of anaglyphs—i.e. the projection of the right stereoscopic image in red and the left in green, the composite image being viewed through spectacles furnished with red and green glasses. **1925** *English Mechanics* 13 Mar. 123/1 The anaglyph is a composite picture consisting of the two pictures forming a stereoscopic pair, one being printed in red, and the other in green; the two pictures are not quite superimposed, but a small amount of lateral displacement is given to one. This composite 'anaglyph' is viewed through a mask, containing two ocular apertures, the left one being covered with a red transparent screen and the right with a green one. Each eye, therefore, sees only the picture taken with the corresponding lens of the stereo camera, with the result that the two views are merged and stereoscopic relief obtained.

anaglypta (ænăgli·ptă). [a. L. *anaglypta* work in bas-relief; cf. Anaglyptic *a.*] A special type of thick embossed wallpaper. Also *attrib.*

1887 *Trade Marks Jrnl.* 12 Oct. 1142 Anaglypta..Thomas John Palmer..Decorative Material in the Nature of a Covering, sold in the Piece, for Application to Walls, Ceilings, or other Surfaces. **1905** *Daily Chron.* 9 June 9/7 (Advt.), Paperhanger wants work, high-class or commons, anaglypta, lincrusta, &c., a speciality. **1920** *Punch* 30 June 505/1 You can stare at the anaglypta ceiling. **1934** *Times* 15 Feb. 17/5 Has he not, paperhanger and decorator too, hung lincrusta and anaglypta, as well as other papers difficult to hang?

anagogic. Add: **A.** *adj.* **2.** *Psychol.* (See quots.)

1917 S. E. Jelliffe tr. *Silberer's Probl. Mysticism & Symbolism* III. i. 241 Anagogic interpretation..is a form of functional interpretation. *Ibid.* 242 The true functional phenomenon, as I have so far described it, pictures the actual psychic state or process; the anagogic image appears on the contrary to point to a state or process that is to be experienced in the future. **1934** H. C. Warren *Dict. Psychol.* 12/1 *Anagogic*, a term applied by Silberer and Jung to the moral, spiritual, allegorical, or uplifting trends of the unconscious. **1940** Hinsie & Shatzky *Psychiatric Dict.* 29/1 Silberer believes that dreams are susceptible of two different interpretations. One is..[the] infantile sexual interpretation. The other is related to the spiritual, idealistic, non-sexual forces of the unconscious; hence, the expression anagogic interpretation. The latter is of the kind stressed by Jung. **1954** R. F. C. Hull tr. *Jung's Coll. Wks.* XVI. 9 The synthetic or anagogic view..asserts that certain parts of the personality which are capable of development are in an infantile state.

anal, *a.* Add: **1. b.** *anal eroticism* (or *erotism*) [cf. G. *analerotik* (Freud 1908, in *Psychiatrisch-Neurologische Wochenschr.* IX. 465)], erotic gratification from stimulation in the anal region; so *anal-erotic* adj., of, or pertaining to, anal eroticism; *anal-erotic* sb., a person who seeks anal-erotic gratification.

1913 E. Jones in *Papers on Psycho-Analysis* (1918) xxxi. 540 (*title*) Hate and Anal Erotism in the Obsessional Neurosis. **1917** C. R. Payne tr. *Pfister's Psychoanalytic Method* I. x. 201 The young man..was analerotic to a high degree. *Ibid.*, His father..is still deeper in analeroticism than he. **1924** *Freud's Coll. Papers* II. iv. 47 Anal erotism is..useless for sexual aims. *Ibid.* xvi. 169 A wish for a child, in which..an anal-erotic and a genital impulse..coincide. **1931** *Amer. Jrnl. Orthopsychiatry* Oct. 515 Eventually poor Pons fell victim to gall-bladder disease, an occurrence of some significance in an anal erotic.

anallantoidian (æ:nælǣntoi·diăn), *a.* and *sb. Zool.* [f. An- 10 + Allantoidian.] **A.** *adj.* Having no allantois in the embryo, as the lower vertebrates. **B.** *sb.* An animal thus characterized. (Opp. to Allantoidian.)

1861 R. T. Hulme tr. *Moquin-Tandon's Elem. Med Zool.* II. I. 62 Anallantoidians—Batrachia, Pisces, Myelairia.

analog. U.S. variant of Analogue.

analogate (ănæ·lŏgĕt). [f. Analog(ous *a.* + -ate¹.] A thing, concept, etc., shown to be analogous; an analogue.

1884 *Dublin Rev.* Apr. 307 *Being* cannot be predicated univocally of God and the creature, but by analogy of attribution of the second class, wherein God is the primary analogate. **1949** E. L. Mascall *Exist. & Analogy* v. 100 The way in which it has being depends in the last resort upon its relation to the self-existent Being which is the prime analogate of all.

analogue. Add: **3.** *analogue* (U.S. *analog*) *computer*, a computer which operates with numbers represented by some physically measurable quantity, such as weight, length, voltage, etc. (cf. *digital *a.* 4). Also, *analogue device, machine*, etc. Hence *analogue computing*, computing by this process.

1946 D. R. Hartree in *Nature* 12 Oct. 500/1 The American usage is 'analogue' and 'digital' machines. **1947** *Electronic Engin.* XIX. 178/1 Electrical analogue computing. **1948** G. R. Stibitz *Function Unit Theory of Counting Computers* i. 2 The ordinary desk calculator is a digital machine. If the numbers are not broken down into digits, but are represented by physical quantities proportional to them the computer is called 'analog'. A slide rule is an 'analog' device. **1948** *Electronics* July 116 The REAC (Reeves Electronic Analog Computer) can solve such problems, and more complex ones. **1949** *Jrnl. R. Aeronaut. Soc.* LIII. 628/1 Electrical and electronic analogue computers have recently been developed on a commercial scale, and have been found useful in problems of aeroplane dynamics, heat transfer, and so on. *Ibid.*, The analogue machine is not strictly a computing machine, but rather a direct physical analogue of the system being studied. **1952** J. Diebold *Automation* ii. 25 Analogue computers, especially the electrical ones, are very useful for the construction of models, or analogues, of industrial processes and systems. **1960** E. Delavenay *Introd. Machine Transl.* 130 The analog computer..simulates the problems it is asked to solve, the digital computer..works out numerical solutions to problems, by calculations made with and on digits.

analphabet, *a.* Delete †*Obs. rare*⁻¹ and insert *rare.* Add later example.

1961 C. Winston *Hours Together* (1962) vii. 140 She had..been..barely able to trace her name on those infrequent documents..which entered their analphabet domain.

B. *sb.* Also **analphabete.** [after F. *analphabète*, G. *analphabet*, etc.] One who is totally illiterate or unable to read.

1881 *Encycl. Brit.* XIII. 460 As late as..1861..[in Italy] in a population of 21,777,331 there were..16,999,701 'analphabetes', or persons..absolutely unable to read. **1914** *N.Y. Times Current History* 12 Dec. 79 There are no analphabets to be found among them. **1938** S. Beckett *Murphy* x. 205 The skill is really extraordinary with which analphabetes..circumvent their dread of verbal commitments.

analphabetic, *a.* Delete *rare*⁻¹ and add later example.

1940 E. Pound *Cantos* lv. 49 Tartar Yuen ruled as protector Cut down taxes, analphabetic.

2. *Phonetics.* Of or pertaining to a system of phonetic transcription whereby sounds are represented not by single letters or single signs, but by composite symbols made up of a number of individual signs each representing a feature of the sound.

1889 Jespersen *Articulations of Speech Sounds Repr. Analphabetic Symbols* i. 8 The next step will be an ultra-alphabetic or *analphabetic* system of writing, symbolizing not sounds, but elements of sounds. **1937** —— *Analytic Syntax* i. 13 In phonetics my own Antalphabetic (formerly called Analphabetic) system provides analogous means for symbolizing the formation of speech-sounds. **1967** D. ABERCROMBIE *Elem. Gen. Phonetics* 112 Analphabetic notations.. represent each segment by a composite symbol made up of a number of signs put together.

B. *sb.* = *ANALPHABET *sb. rare.*

1947 I. A. RICHARDS in *Ess. & Stud.* XXXII. 8 Two-thirds of us on this planet are..analphabetics.

analysability (æ:nălǝizăbi·lĭti). [f. ANALYSABLE *a.* + -ITY.] The quality of being analysable.

1925 A. N. WHITEHEAD *Sci. & Mod. World* (1926) x. 232 The complexity of an eternal object means its analysability into a relationship of component eternal objects. **1932** A. H. GARDINER *Theory of Speech & Lang.* iv. § 58. 216 The analysability of utterances into subject + predicate cannot be made the touchstone of the sentence. **1965** N. CHOMSKY *Aspects of Theory of Syntax* i. 56 If the structural analyses that define transformations are restricted to Boolean conditions on Analyzability..it will be impossible to formulate many 'structure-dependent' operations as transformations.

analysand (ănæ:lizæ·nd). *Psycho-analysis.* [f. ANALYSE *v.* + *-AND².] The subject of, or patient in, psycho-analysis; one whose psycho-analysis is being attempted.

1933 H. CRICHTON-MILLER *Psycho-anal.* i. 105 'The transference'. By this is denoted the relation of the analysand to the analyst. **1949** *Brit. Jrnl. Psychol.* Sept. 8 The risks of failure might plunge both analyst and analysand into very deep waters. **1953** R. F. C. HULL tr. *Jung's Coll. Wks.* VII. 136 Every analysand starts by unconsciously misusing his newly won knowledge in the interests of his abnormal, neurotic attitude. **1962** *Listener* 29 Mar. 568/2 Psycho-analysis is primarily historical and reductive, deducing the analysand's present situation from his past history.

analysandum (ănæ:lizæ·ndǝm). *Philos.* [f. ANALYSE *v.* + *-andum* (L. neut. gerundive termination, 'thing fit to' undergo the action of the verb: see *-AND²).] The proposition or concept to be analysed; in a philosophical analysis, the expression to be clarified, or that which is signified by the expression to be clarified.

1907 S. H. HODGSON in *Proc. Arist. Soc.* VII. 117 But in whichever way we define apperception..we have by no means surmounted the difficulty of distinguishing, in that universal panorama which is our *analysandum*, what is due to conation from what is due to perception. **1932** *Proc. Arist. Soc.* XXXIII. 77 The kind of analysis that is possible ..depends upon the kind of combination, or complex, which the analysandum is. **1944** [see next]. **1956** J. O. URMSON *Philos. Analysis* iv. 53 To say '*p* is equivalent to *q*' where *p* is the analysis and *q* the analysandum is to utter a tautology if true. **1960** in —— *Concise Encycl. Western Philos.* 18/1 An analysis..is a sort of definition, a kind of equation with the puzzling expression, the *analysandum*, on the left-hand side and the new expression, sometimes called the *analysis*, sometimes the *analysans*, on the right.

analysans (ănæ:lizæ·nz). *Philos.* [f. ANALYSE *v.* + *-ans* (L. pres. pple. termination).] The clarifying expression, in a philosophical analysis; the concept signified by such an expression.

1944 *Mind* LIII. 73 He suspects that the statement of an analysis must, in some sense, be about the *expressions* used for the analysandum and the analysans, as well as about these concepts. **1960** [see *ANALYSANDUM].

analyse, *v.* Add: **3. b.** Short for *PSYCHO-ANALYSE *v.*

1909 A. A. BRILL tr. *Freud's Sel. Pap. Hysteria* (1912) 122 In those patients whom I have analyzed there existed psychic health until..there appeared an experience.. that the person decided to forget. **1919** M. K. BRADBY *Psycho-analysis* x. 126, I have had no nightmare..since I was analysed. **1921** ROSE MACAULAY *Dangerous Ages* v. § 4, I think you'd be awfully wise to get analysed. **1958** *Times Lit. Suppl.* 28 Mar. 173/4 (*heading*) The Analysts Analysed.

4. b. *to analyse out*: to discover or isolate by a process of analysis. Hence, *colloq.* to work out the elements of (a situation): see quot. 1952.

1890 W. JAMES *Princ. Psychol.* I. xiii. 503 If any single quality or constituent..of such an object, have previously been known by us isolatedly,..then that constituent..may be analysed out from the total impression. **1952** M. McCARTHY *Groves of Academe* (1953) ix. 169 'It's a dead give-away, Domna,' he expatiated. 'Analyse it out for yourself.' **1954** *Brit. Jrnl. Psychol.* May 146 It is possible to isolate *g* [*sc.* factor in intelligence], and progressively to analyse out the verbal, spatial and other content factors. **1958** *Spectator* 30 May 692/1 The author of our study would need to look into these and analyse out some common factors.

8. *Mus.* To provide an analysis of (*ANALYSIS 10).

1885 G. B. SHAW *How to become Musical Critic* (1960) 110 When the work analyzed is a familiar one..the analysis.. is a stereotyped reprint. **1890** *Ibid.* 189 A calm confidence in his power..to 'analyze' the last movement of the Jupiter Symphony. **1935** D. F. TOVEY *Ess. Mus. Analysis* I. 1 If the defects of the works analysed are too notorious to be ignored, he must [etc.].

9. *Philos.* To subject to logical or philosophical analysis (see *ANALYSIS 9).

1910, 1958 [see *ANALYSIS 9 b].

analysis. Add: **1. c.** *in the last* (or *final*) *analysis* [after F. *en dernière analyse*], when reduced to its fundamental elements; at the conclusion of the investigation or examination involved; all things duly considered and weighed.

[**1791** J. MACKINTOSH *Vindiciæ Gallicæ* i. 89 Corporate property is here as sacred as individual, because in the ultimate analysis it is the same.] **1844** E. A. POE in *Chambers's Jrnl.* 30 Nov. 345/2 Now this mode of reasoning in the schoolboy, whom his fellows termed 'lucky',—what, in its last analysis, is it? **1877** *Independent* 29 Mar. 13/2 In the last analysis there will be a painless universe! **1885** *Harper's Mag.* Mar. 648/1 The loveliest doll in the last analysis is merely sawdust. **1902** W. H. FITCHETT *Let.* 18 Oct. in D. P. Hughes *Life H. P. Hughes* (1904) xix. 535, I have always felt that, in the last analysis, the question of union was a religious one. **1918** B. RUSSELL *Logic & Knowl.* (1956) 194 Analysis..always depends, in the last analysis, upon direct acquaintance with the objects which are the meanings of certain simple symbols. **1957** L. F. BROSNAHAN *Genes & Phonemes* 11 The total experience incorporated in a language, in the vocabulary, morphology and syntax, must, in the last analysis, have been gained through the senses. **1963** V. NABOKOV *Gift* iii. 188 In the final analysis all girls aspire to be beauties.

2. *Cricket.* (Earlier and later examples.)

1854 F. W. LILLYWHITE *Guide to Cricketers* 22 The analysis of the Bowling is given of all those matches possessing the most interest. **1904** WARNER *How we recovered Ashes* iii. 57 His [*sc.* Rhodes'] analysis—two wickets for seventy-eight runs—is misleading, as analyses very often are.

7. (Later examples.)

1902 E. T. WHITTAKER (*title*) A Course of Modern Analysis. *Ibid.* Pref., The first half of this book contains an account of those methods and processes of higher mathematical analysis, which seem to be of greatest importance at the present time; [illegible] the properties of infinite series and complex integrals, and their applications to the analytical expression of functions. **1959** G. & R. C. JAMES *Math. Dict.* 10/2 *Analysis*, that part of mathematics which uses, for the most part, algebraic and calculus methods—as distinguished from such subjects as synthetic geometry, number theory, and group theory.

9. *Philos.* **a.** The resolution into their elements, as a philosophical procedure, of complex things, facts, propositions, and concepts. Freq. with defining word, as *logical* (cf. also sense b), *philosophical analysis.*

1828 J. S. MILL in *Westm. Rev.* IX. 144 To perform the logical analysis of an argument, in the manner pointed out by the doctrine of the syllogism, is not the best means of discovering whether it contain a flaw. *Ibid.* 171 The philosophical analysis of Predication, the explanation of what is the immediate object of belief when we assent to a proposition, [illegible] **1883** F. H. BRADLEY *Logic* I. ii. 93 It is wholly unjustifiable to take up a complex, to do any work we please upon it by analysis, and then simply [illegible] these results of our analysis [illegible] *Stud.* (1911) vii. 169 Some psychologists are fascinated by the dissecting out, whether by logical analysis or by brass instruments, of whatever elements mental processes may be [in living action]. **1903** B. RUSSELL *Princ. Math.* iv. 42 The correctness of our philosophical analysis of a proposition may..be usefully checked by..assigning the meaning of each word in the sentence expressing the proposition. *Ibid.* vii. 83 A less complete analysis of propositions into subject and assertion has also been considered. *Ibid.* liii. 466 A distinction is made..between conceptual analysis and real division into parts. **1918** —— *Logic & Knowl.* (1956) 192 The analysis of apparently complex things..can be reduced..to the analysis of facts which are apparently about those things.

b. (Freq. in recent Philos.) The procedure or the result of finding an expression exactly equivalent to a given word, phrase, or sentence, for the purposes of clarification.

At first as analysis was intended simply to reveal the logical structure of the analysandum, but later was often intended as leading to, or justifying, the metaphysical claim that an element eliminated by the analysis was not real, or was not fundamental: then sometimes called *reductive analysis*, etc. (see quot. 1956).

1910 WHITEHEAD & RUSSELL *Principia Mathematica* iii. 69 In all such cases, the proposition must be capable of being so analysed that what was the grammatical subject shall have disappeared. Thus when we say 'the round square does not exist', we may, in a first attempt at such analysis, substitute 'it is false that there is an object *x* which is both round and square'. **1912** RUSSELL *Probl. Philos.* v. 91 The fundamental principle in the analysis of propositions containing descriptions is this: Every proposition which we can understand must be composed wholly of constituents with which we are acquainted. **1949** *Mind* LVIII. 46 A later and more radical form of reductive analysis is characterised by the complete repudiation of 'consciousness' and the replacement of mental 'acts' as the constituents of mind by events which are common to both mental and physical complexes. **1956** J. O. URMSON *Philos. Analysis* iii. 39 The elimination of logical constructions..is then another sort of analysis. It was variously called new-level (as opposed to same-level), or philosophical (as opposed to logical), or directional, or reductive, analysis. Different names were patronized by different analysts. **1958** G. J. WARNOCK *Eng. Philos. since 1900* vi. 27 Moore involved both himself and others in difficulties resulting from the unquestioned assumption that any analysis must be of a standard pattern. It was always to consist in providing a verbal paraphrase of what was to be analysed, in the form of a longer, more explicit, but strictly synonymous phrase or sentence. **1960** [see *ANALYSANDUM].

c. *philosophical analysis*: as a branch of Philosophy, that conducted by means of, or understood as consisting in, analysis (senses a or b); philosophy concerned with the clarification of existing concepts and knowledge, by a method of reformulation, rather than with conceptual revision or addition to knowledge; *esp.* the analytic movement associated with Bertrand Russell, Moore, Wittgenstein and others.

1843 MILL *Logic* I. II. ii. 238 Philosophical analysis confirms the indication of common sense, that the function of names is but that of enabling us to *remember* and to *communicate* our thoughts. **1887** S. H. HODGSON *Let.* 8 Apr. in R. B. Perry *Tht. & Char. W. James* (1935) I. 642 What is 'the mind'?..I suspect it is a mere *façon de parler*.. which has a basis neither in psychological construction nor in philosophical analysis. **1936** AYER *Lang., Truth & Logic* ii. 62 The possibility of philosophical analysis is independent of any empirical assumptions. **1943** *Mind* LII. 275 The subject [*sc.* mathematical logic]..may claim an important place in the corpus of philosophical analysis. **1956** J. O. URMSON (*title*) Philosophical Analysis: Its development between the two world wars.

10. *Mus.* A critical description of a musical work, designed to make clear its structure.

1885 [see *ANALYSE *v.* 8]. **1935** D. F. TOVEY *Ess. Mus. Analysis* I. 1 Some of the analyses..were written on occasions where there was no opportunity for musical quotations.

11. *Psychol.* The mental process of discrimination, by separate attention, of the separable elements in a totality of simultaneous sensory impressions, or in any other complex experience.

1890 W. JAMES *Princ. Psychol.* I. xiii. 502 Our first way of looking at a reality is often to suppose it simple, but later we may learn to perceive it as compound. This new way of knowing the same reality may conveniently be called by the name of *Analysis*. It is manifestly one of the most incessantly performed of all our mental processes. **1938** R. S. WOODWORTH *Exper. Psychol.* xxv. 645 To judge the lines and angles of a figure requires analysis which is difficult because the observer is engrossed in the appearance of the figure as a whole.

12. *Psychiatry.* Short for *PSYCHO-ANALYSIS.

1907 *Brain* XXX. 179 In hysteria with very little trouble the complex may be revealed by analysis, and with a good prospect of therapeutic advantage. **1912** A. A. BRILL tr. *Freud's Sel. Papers on Hysteria* (ed. 2) iv. 76 It is very difficult to examine a case of neurosis before it has been subjected to a thorough analysis. **1913** *Lancet* 10 May 1345/2 Analysis might record some latent content having nothing to do with hunting and would perhaps show Dr. Brown that he no more fears hunting in dreams than he does awake. **1916** C. E. LONG tr. *Jung's Analytical Psychol.* vii. 208 An attempt has..been made to compare analysis with the reasoning method of Dubois, which is in itself a rational process. This comparison does not..hold good, for the psychoanalyst strictly avoids argument and persuasion with his patients. **1958** *Spectator* 1 Aug. 160/1 He [*sc.* Freud] was always attracted by telepathy (so much so that he alarmed his English followers, who feared that it would give their enemies an excuse to sneer at analysis as occultism).

analyst. Add: **4.** *Mus.* One who carries out analysis (*ANALYSIS 10) of a musical work.

1885 G. B. SHAW *How to become Musical Critic* (1960) 109 The false estimates into which the analyst has been led by the necessity of judging the score by eye instead of by ear. **1893** [see STREPITOUS *a.*].

5. *Psychiatry.* Short for *PSYCHO-ANALYST.

1914 EDER & MOLTZER tr. *Jung's Theory Psychoanalysis* in *Psychoanal. Rev.* I. 425 Transference to, and dependence upon the analyst could be considered as a sufficient end, with a definite therapeutic effect, if the analyst were in every respect a great personality, capable..to guide the patients. **1920** *Internat. Jrnl. Psycho-anal.* I. 128 The ideal situation for analysis is when someone, otherwise master of himself, is suffering from an inner conflict which he is unable to resolve alone, so that he brings his trouble to the analyst and begs for his help. **1952** V. GOLLANCZ *My Dear Timothy* 55 In moneyed circles with a tinge of culture people talk of 'my analyst' as glibly as our grandmothers used to talk of 'my grocer'.

6. *Philos.* A practitioner of philosophical analysis (*ANALYSIS 9 c); an adherent of the analytic movement in philosophy.

1936 AYER *Lang., Truth & Logic* ii. 52 The majority of those who are commonly supposed to have been great philosophers were primarily not metaphysicians but analysts. **1941** *Mind* L. 166 Metaphysical propositions are indeed nonsense, as the Analysts say, but only if taken literally; and it is unfortunate that many of the Analysts have been only too literal-minded. **1945** *Mind* LIV. 194 Philosophical analysts can do either of two things; they may undertake what one could call case studies or they may engage in the construction of systems.

analytic. Add: **A.** *adj* **1.** spec. in *Math.* (Later examples; cf. *ANALYSIS 7.)

1902 E. T. WHITTAKER *Course Mod. Analysis* iii. 47 If $f(z)$ is an analytic function, regular at all points in the interior of a contour, then $\int f(z)dz = 0$, where the integration is taken round the contour. **1938** HARDY & WRIGHT *Theory of Numbers* p. v, Thus Chs. XII–XV belong to the 'algebraic' theory of numbers,..Ch. XXII to the 'analytic' theories. **1959** G. & R. C. JAMES *Math. Dict.* 11/1 (*heading*) Analytic continuation of an analytic function of a complex variable.

Also, further examples in philosophical contexts (cf. *ANALYSIS 9 a, b, c).

1865 MILL *Exam. Hamilton* xx. 405 Mr. Mansel expressly limits the province of Logic to analytic judgments—to such as are merely identical. **1870** —— in *Fortnightly Rev.* VIII.

123 Taine..in the case of the axioms of geometry,..classes them among 'analytic propositions'—that is, truths latently included in the ideas which are the subject of them, to be proved by evolving them out of the ideas. **1936** E. NAGEL in *Jrnl. Philos.* XXXIII. 5 (*title*) Impressions and Appraisals of Analytic Philosophy in Europe. *Ibid.* 9 Analytic philosophy is ethically neutral formally. **1946** AYER *Lang., Truth & Logic* (ed. 2) vi. 78 A proposition is analytic when its validity depends solely on the definitions of the symbols it contains, and synthetic when its validity is determined by the facts of experience. *Ibid.*, Analytic propositions are devoid of factual content. **1946** *Mind* LV. 339, I do not think that either Prof. Moore or most of the other adherents of Contemporary British Analytic Philosophy are ready to draw the consequences from this fact. **1948** *Mind* LVII. 292 Analytic philosophy consists at least partly, in replacing a concept, or a set of concepts, by another concept, or set of concepts.

c. *analytic psychology*, the branch or school of psychology mainly based on the work of Locke, which endeavours to analyse ideas and trace them to their origins.

a **1854** MILL *Draft Autobiog.* (1961) 76 [= 1873 *Autobiog.* iii. 68] Under my father's direction, my studies were carried into the higher branches of analytic psychology. **1889** W. JAMES *Coll. Ess. & Rev.* (1920) 311 The widely prevalent notion that analytic psychology has proved the space-perceptions of the eye to be but reproduced experiences of touch and locomotion.

d. *Mus. analytic programme*, a concert programme containing analyses (*ANALYSIS 10) of the works performed.

1885 G. B. SHAW *How to become Musical Critic* (1960) 77 The analytic program..costs an additional shilling.

3. Short for *PSYCHO-ANALYTIC *a.*

1912 A. A. BRILL tr. *Freud's Sel. Papers on Hysteria* (ed. 2) vii. 178 The analytic therapy..concerns itself with the genesis of the morbid symptoms. **1959** *Times Lit. Suppl.* 13 Mar. 148/2 Certain analytic terms (which they prefer to avoid) such as 'acting-out'.

analytical, *a.* Add: **1. c.** *Math.* Applied to geometry treated by means of algebra, in the Cartesian representation of curves and surfaces by equations.

1826 H. P. HAMILTON *Princ. Analytical Geom.* 2 Algebra may be applied to investigate the Theorems, and to resolve the Problems of Geometry...Application of these elementary branches of Mathematics forms the Science of Analytical Geometry. **1845** *Encycl. Metrop.* I. 709 This branch of the subject is usually distinguished by the name of Algebraic, or Analytical, Geometry. **1881** J. M. DYER (*title*) Exercises in Analytical Geometry.

d. *Math.* Pertaining to *ANALYSIS (sense 7).

1902 [see *ANALYSIS 7]. **1902** E. T. WHITTAKER *Course Mod. Analysis* iii. 46 The definition of functionality must now be translated into analytical language.

2. b. = *ANALYTIC *a.* 1.

1854 J. COULTHARD tr. *C. W. von Humboldt's Sphere & Duties of Govt.* viii. 105 It is the analytical philosopher alone..who is able to arrive at his results through the calm, but cold processes of reason. **1865** MILL *Exam. Hamilton* xviii. 354 This doctrine ignores the famous distinction.. between Analytical and Synthetical judgments. Analytical judgments are supposed to unfold the contents of a concept. **1945** *Mind* LIV. 194 My concern is exclusively with analytical philosophy. *Ibid.*, Both the casuist and the systematist, if they are analytical philosophers, are interested in the individual clarifications that are peculiar to this kind of philosophising. **1956** J. HOLLOWAY in A. Pryce-Jones *New Outl. Mod. Knowl.* 30 The philosophy of analysis thus tends to have two somewhat different branches: the analysis of *a priori* or analytical propositions..as represented by what Russell and Wittgenstein had to say about formal logic and mathematics; and the analysis of propositions which are expressed in words, which can occur in the ordinary layman's thinking...On the whole, modern analytical philosophy has developed most in this second field.

c. *analytical psychology* = *ANALYTIC *psychology*; also applied to the psychology of C. G. Jung, as distinguished from that of Freud (psycho-analysis), esp. the practical procedure of psychological analysis laid down by Jung.

1835 MILL *Diss. & Disc.* (1859) I. 114 Of Locke's Essay, the beginning and foundation of the modern analytical psychology, we cannot speak but with the deepest reverence. **1916** C. E. LONG tr. (*title*) C. G. Jung. Collected Papers on Analytical Psychology. **1945** H. CRICHTON-MILLER *Psycho-analysis* (ed. 2) i. 13 If we respect the history of terms, no system which diverges from the Freudian can correctly be called psycho-analysis. For purposes of differentiation Jung uses the term Analytical psychology (which happens to be a plagiarism from Stout). **1962** A. M. DRY *Psychol. of Jung* p. xi, The International Association for Analytical Psychology has been formed.

d. *analytical programme* = *ANALYTIC *programme*.

1885 G. B. SHAW *How to become Musical Critic* (1960) 57 A careful analytical program.

3. *analytical chemistry*, the branch of chemistry concerned with analysis (ANALYSIS 3).

1879 A. J. BERNAYS (*title*) Skeleton notes on analytical chemistry, for students in medicine. **1938** R. W. LAWSON tr. *Hevesy & Paneth's Man. Radioactivity* (ed. 2) xxii. 220 The..large domain of the analytical chemistry of the radio-elements.

analytically, *adv.* **1.** Add to def.: By the analytical method; by the methods of analytical geometry.

1879 *Encycl. Brit.* X. 408/1 It would be possible, analytically, or by the method of coordinates, to develop the truths of geometry in a systematic course.

analyticity (ænăliti·siti). *Philos.* [f. ANALYTIC *a.* + -ITY.] The property, in propositions or statements, of being analytic.

1939 *Mind* XLVIII. 76 Moreover, on the basis of the analyticity of (a) alone absolutely nothing can be known about a consequence-relation involving '*p* ⊃ *q*, *p*' as a first member. **1953** W. V. O. QUINE *From a Logical Point of View* ii. 22 Kant's intent, evident more from the use he makes of the notion of analyticity than from his definition of it, can be restated.

analytico-, combining form of Gr. ἀναλυτικό-ς analytic, prefixed to an adj. to denote **a.** 'pertaining to analytical ...' as *analytico-chemical*; **b.** 'analytical and ...', as *analytico-synthetic*.

1920 T. P. NUNN *Education* xiii. 173 Thus the function of the nervous system is never purely integrative nor purely analytic, but always analytico-synthetic. **1938** R. W. LAWSON tr. *Hevesy & Paneth's Man. Radioactivity* (ed. 2) xxii. 230 This analytico-chemical difference. **1961** T. LANDAU *Encycl. Librarianship* (ed. 2) 81/2 Attempts have been made..to build up classification schemes inductively from certain fundamental concepts which may be combined in various ways to form a synthesized concept for a whole book. Then groups of books in these analytico-synthetic schemes may be arranged into classes.

anamesite (ænæ·mĕsait). *Min.* [ad. G. *anamesit* (K. C. von Leonhard *Die Basalt-Gebilde* (1832) I. 150), f. Gr. ἀνάμεσος intermediate + -ITE¹.] A fine-grained variety of basalt.

1876 F. H. LAWRENCE tr. *B. von Cotta's Rocks Classified* iii. 135 Basaltic Rocks..a fine-grained dark grey to black mass, in which we are unable to distinguish between labradorite and augite. This fine-grained variety has been specially named by von Leonhard as *Anamesite*. *Ibid.* 136 The different stages of compactness of texture may be typified by the three names Dobrite, Anamesite, Basalt. **1879** RUTLEY *Stud. Rocks* xiii. 253 Those [basalts] in which the constituents are too small to be recognised without a magnifying power, but in which a crystalline texture is yet clearly discernible, are styled anamesites. **1937** A. JOHANNSEN *Petrography* III. 291 The name *anamesite* was given by von Leonhard to those rocks of the basalt family which, in texture, stand between the dense basalts and the coarse dolerites...It is seldom used at the present time.

anamnesis. Add: **b.** *Liturgiology.* That part of the Eucharistic canon in which the sacrifice of Christ is recalled and pleaded.

1894 H. LUCAS in *Dublin Rev.* CXIV. 123 The Anamnesis, a prayer commencing with the words *Μεμνημένοι οὖν*, and answering more closely to the Roman *Unde et memores* than to any Gallican *Post Secreta* or Mozarabic *Post Pridie.* **1912** A. FORTESCUE *Mass* ii. 103 The next prayer (*Post pridie* or *Post mysterium*) contains the Anamnesis and Epiklesis of the Holy Ghost. **1945** G. DIX *Shape of Liturgy* vii. 205 It was precisely the institution-narrative which would need amplifying and the *anamnesis* section which would have to be supplied.

anamorphic (ænămǭ·ɹfik), *a.* [f. ANAMORPH(ISM + -IC.] **1.** *Geol.* Applied to a zone characterized by anamorphism (see *ANAMORPHISM 3).

1904, 1914 [see *ANAMORPHISM 3].

2. *Cinemat.* (See quot. 1960.)

1954 *New Internat. Yr. Bk. 1953* 326/1 The anamorphic lens used in CinemaScope was invented by Professor Henri Chretian, a French scientist who has been working on the lens for the past twenty-five years. **1958** *Engineering* 21 Mar. 384/1 Anamorphic lenses are also available to provide wide-screen pictures in the home. **1960** O. SKILBECK *Film & TV Working Terms* 9 *Anamorphic*, type of lens which 'squeezes' the width of an Angle, compared with its height, in a camera and does the converse in a Projector.

anamorphism. Add: **3.** *Geol.* Metamorphism below oxidation level resulting in the formation of complex minerals from simpler ones.

1904 VAN HISE *Treat. Metamorphism* iv. 169 The lower zone may therefore properly be called the zone of anamorphism, or anamorphic zone. *Ibid.* 170 The zone of anamorphism may be defined as the zone in which alterations of rocks result in the production of complex compounds from simple ones. **1914** RIES & WATSON *Engin. Geol.* iii. 204 *Anamorphic zone.* The zone of anamorphism corresponds to the zone of flowage, in which there is great pressure in all directions. It is a zone of *reconstruction*, and is especially characterized by silication involving decarbonation, dehydration, and deoxidation.

anamorphoscope (ænămǭ·ɹfǭskōᵘp). [f. ANAMORPHO(SIS + -SCOPE.] A device, usually in the form of a vertical cylindrical mirror, made so as to give a correct image of an anamorphosis (see ANAMORPHOSIS 1).

a **1884** in KNIGHT *Dict. Mech.* Suppl. **1932** *Times Educ. Suppl.* 12 Nov. p. iv/1 The exhibits in the historical section will include the 'Anamorphoscope', a set of 12 designs showing distorted figures in perfect formation in a cylindrical mirror (date about 1635).

Ananias (ænănai·ăs). Name of a man who, 'with Sapphira his wife, sold a possession and kept back part of the price' (Acts v. 1, 2); used allusively for a liar.

1876 in Sperber & Trittschuh *Amer. Polit. Terms* (1962) 15/1 It is a matchless product of Uriah Heep 'harmony', or of Ananias 'harmony'. **1890** *Harper's Mag.* Apr. 795/1 Since Locke's time newspaper Ananiases have not been infrequent. **1948** 'J. TEY' *Franchise Affair* xxiii. 266 The report [of the trial] was printed under a heading which read: Ananias Also Ran.

anaphase (æ·năfēⁱz). *Biol.* [a. G. *anaphase* (E. Strasburger 1884, in *Arch. f. mikrosk. Anat.* XXIII. 260), f. ANA- + PHASE.] The later stage in mitosis, between metaphase and telophase, during which the daughter chromosomes move apart towards the opposite poles.

1887 W. HILLHOUSE tr. *E. Strasburger's Bot.* xxxii. 363 The further phases of the separation of the sister-segments, which appertain to the receding phases of division, the anaphases. **1901** CALKINS *Protozoa* 263 Late anaphase. **1946** *Nature* 17 Aug. 239/1 The birefringence disappears at fertilization and reappears during the anaphase of the first mitosis.

anaphora. Add: **1. b.** *Gram.* The use of a word which refers to, or is a substitute for, a preceding word or group of words.

1933 BLOOMFIELD *Language* xv. 266 The word *one*..replaces a with anaphora of the noun with no other modifier is present (Here are some apples; take one). **1964** M. A. K. HALLIDAY et al. *Linguistic Sciences* viii. 148 In English these [*sc.* non-structural features] include grammatical anaphora, grammatical substitution and lexical anaphora; the first is reference back by personal pronouns and by deictics such as 'the', 'this' and 'his'; the second is the use of 'do' and 'one' in the verbal and nominal groups, as in 'I might do' and 'a big one'; the third is the repetition of a lexical item, or occurrence of a second item from one lexical set. **1964** R. H. ROBINS *Gen. Linguistics* iv. 248 Membership of these gender classes also governs the lexical relationship of anaphora or back reference which may hold across sentence boundaries. *Boy, girl,* and *snake* are referred to by *he, she,* and *it*, respectively.

2. [late Gr., f. Gr. ἀναφορά offering.] *Liturgiology.* That part of the Eucharistic service which includes the consecration, oblation, and communion; in the Western Church, the canon of the mass. Hence **ana·phoral** *a.*

1744 *Anc. Liturgy Ch. Jerus.* Pref. p. iii, All that Part, both of this and the other ancient Liturgies, which precedes the Anaphora, is a latter Addition to the Service of the Church. *Ibid.* p. v, The proper Anaphora, or Eucharistical Service, *viz.* from the Sursum Corda..to the Ite in Pace. **1832** W. PALMER *Orig. Liturg.* I. 27 Omitting..any comparison of the introductions of these two liturgies, I will compare their *Anaphoræ*, or solemn offices. **1859** NEALE *Liturgies of S. Mark* p. xi, The Anaphoral portion has these four divisions: The great Eucharistic Prayer: The Consecration: The Intercession for quick and dead: and The Communion. **1930** A. G. HEBERT tr. *Brilioth's Euch. Faith* ii. 20 The following versicle and response, in the anaphora of Hippolytus. **1938** *Theology* XXXVII. 265 Early anaphoral traditions of Syria and Egypt.

anaphoric (ænăfǫ·rik), *a.* (and *sb.*). [f. ANAPHORA + -IC.] Of, pertaining to, or constituting anaphora (sense *1 b); referring to or standing for a preceding word or group of words. Hence as *sb.*, an anaphoric word.

1914 JESPERSEN *Mod. Eng. Gram.* II. I. x. 247, I propose to apply the word *anaphoric* to *one* (or any other word) if it refers to some word already mentioned, while I say *independent* if there is no such contextual reference. **1933** BLOOMFIELD *Language* xv. 251 Anaphoric substitutes..say only that the particular form which is being replaced (the antecedent) has just been mentioned. **1934** PRIEBSCH & COLLINSON *German Lang.* vi. 315 Anaphoric expressions (e.g. *the former, he, this*) placed ahead of new matter. **1957** R. W. ZANDVOORT *Handbk. Eng. Gram.* III. vii. 177 Anaphoric *one* after a definite article or an adjective is usually called a *prop-word*. **1960** S. STUBELIUS *Balloon* 28 While *machine* was from the outset the standard anaphoric for the aeroplane in the press, *apparatus* was only occasionally used.

anaphorical (ænăfǫ·rikăl), *a.* [f. *ANAPHORIC *a.* + -AL.] = prec. Hence **anapho·rically** *adv.*

1914 JESPERSEN *Mod. Eng. Gram.* II. I. x. 247 *The little one* is used anaphorically if it means 'the little flower' or whatever it is that has just been mentioned. **1924** —— *Philos. Gram.* xviii. 237 When the prop-word *one* is anaphorical (i.e. refers to a word mentioned already). **1933** BLOOMFIELD *Language* xv. 251 In English, finite verb expressions are anaphorically replaced by forms of *do, does, did*, as in *Bill will misbehave just as John did.* **1947** H. JACOB *Planned Aux. Lang.* iv. 82 The adjective is used anaphorically with the plural -s.

anaphylaxis (ænăfilæ·ksis). *Path.* [mod.L., ad. F. *anaphylaxie* (Portier and Richet 1902, in *Compt. Rend. Soc. de Biol.* 170), f. ANA- + Gr. φύλαξις watching, guarding. Cf. PROPHYLAXIS.] Extreme sensitivity of tissues to the reintroduction of an antigen. Also *transf.* So **anaphyla·ctic** *a.*, esp. in *anaphylactic shock*, a rapid physical reaction to such reintroduction.

1907 *Jrnl. Med. Research* XVI. 143 (*heading*) On Serum Anaphylaxis in the Guinea-Pig. Animals may react to certain toxic or foreign substances..by an increased susceptibility or anaphylaxis. *Ibid.* 174 It is important to consider certain general relations which exist between the

anaphylactic serum and the affected organism... Sera of animals other than the horse have an anaphylactic power against the guinea-pig. **1909** *Arch. Int. Med.* III. 519 Anaphylaxis.., also called hypersusceptibility, supersensitiveness, is a condition of unusual or exaggerated susceptibility of the organism to foreign substances. **1910** *Lancet* 9 July 84/1 Deep ether narcosis..modifies anaphylactic shock. **1913** J. M. BLIGH tr. *Richet's Anaphylaxis* i. 1 Anaphylaxis is the opposite condition to protection (phylaxis). **1925** C. H. BROWNING *Bacteriology* x. 232 A similar phenomenon is that of anaphylaxis in which a parenteral injection of foreign serum, which is quite harmless to a normal individual, may cause serious symptoms in one who has received a previous injection of serum; this result follows only when both injections consist of serum of the *same* foreign species. **1939** *Nature* 6 May 767/2 Anaphylactic phenomena in the rhesus monkey after intravenous injection of horse serum. *Ibid.* 768/1 Fatal anaphylaxis usually occurred within six minutes of the injection of egg-white. **1953** HINSIE & SHATZKY *Psychiatric Dict.* (ed. 2) 571/1 In psychic anaphylaxis..the reaction is..a response to the sensitizing agent rather than to the activating agent. **1955** *Sci. News Let.* 9 Apr. 234/3 These severe sting reactions, and similar severe reactions to bites of biting insects, are what scientists call anaphylactic shocks. **1959** *Listener* 24 Dec. 1115/2 Anaphylactic shock, allergy, and hypersensitivity are all aberrations or miscarriages of the immunological process.

anaptyxis (ænăpti·ksis). *Phonetics.* [mod.L., a. Gr. ἀνάπτυξις unfolding.] (See quot. 1895¹.) So **anapty·ctic, -ical** *adjs.*, pertaining to anaptyxis.

1885 *Athenæum* 18 July 76/1 The irregular appearance of the anaptyctic vowel. **1895** P. GILES *Man. Compar. Philol.* 169 Anaptyxis..the development of a vowel between two consonants. **1895** *Athenæum* 14 Sept. 347/2 ['Cyclar' is] formed from 'cycle' on the analogy of 'circular' from 'circle' (more correctly from *circulus*). The irregular anaptyxis cannot be defended on historical grounds. **1953** *Archivum Linguisticum* V. 83 The introduction of an anaptyctic vowel next to *H* is a logical step where contiguous consonants in zero grade forms, including *H*, impede the speaker. **1955** *Sci. Amer.* Aug. 79/3 The insertion of an extra vowel in the body of a word, giving rise to an additional syllable, is called 'anaptyxis'. Note athaletic, ellum, fillum, siggunel,..and 'he ran thataway'.

anarch. A. Add: **2.** An advocate of anarchy, an anarchist.

1884 W. CORY *Lett. & Jrnls.* (1897) 508 We occupied the Delta in 1882 to prevent anarchs from hindering trade on the Canal. **1885** R. L. & F. STEVENSON *Dynamiter* 194, I recognise in you the marks of an accomplished Anarch. **1966** *New Statesman* 26 Aug. 299/2 The self-righteousness of a sour young anarch.

anarchial, *a.* (Earlier and additional examples.) For '1789' in Dict. read '1823 (ed. 2)'.

1710 'G. LOVEWHIG' *Last Will & Test. of C-h of E-d* 9 All other like Republican, Anarchial, and Factious Innovators. **1831** *Blackw. Mag.* XXIX. 49 That anarchial and revolutionary spirit which had shewn itself. **1848** *Tait's Mag.* XV. 479 Thence two civilizations, two distinct societies: one Moldo-Wallachian and aristocratic; the other Illyrian and Mahometan; in Bosnia, anarchial; in Bulgaria, uncertain; in Servia, democratic.

anarchic, *a.* Add: **b.** Pertaining to *ANARCHY 1 b. So **a·narchist**.

1850 [see *ANARCHY 1 b]. **1889** W. DONISTHORPE *Individualism* 256 Under a truly anarchic system. **1889** S. WEBB in Shaw *Fabian Ess. Socialism* 44 The complete freedom preached by..the scientific Anarchists of to-day.

anarchism. Delete *rare* and add later examples.

1893 G. B. SHAW (*title*) The Impossibilities of Anarchism. **1913** J. A. ESTEY *Revolutionary Syndicalism* v. 128 The 'veritable abyss', which separates the ideal of Anarchism (i.e. philosophic Anarchism) from the ideal of Syndicalism. **1962** *New Statesman* 7 Sept. 287/1 His method is to examine the 'Family Tree' of anarchism, with its roots in Lao-Tse, Zeno and the Essenes.

anarchistic (ænaɹki·stik), *a.* [f. ANARCHIST + -IC: see -ISTIC.] Belonging to, characteristic of, or adhering to anarchists or anarchism. Hence **anarchi·stically** *adv.*

1884 *Standard* 5 Dec. 5/5 The seizure of Revolutionary and Anarchistic prints. **1900** H. G. WELLS *Love & Mr. Lewisham* xxiii, That Honesty is essentially an anarchistic and disintegrating force in society. **1932** *Times Lit. Suppl.* 29 Dec. 990/3 They live at the expense of the society they anarchistically reject.

anarcho-syndicalism (ænā:ɹko ˌ si·ndikăliz'm). [comb. form. of ANARCHY + SYNDICALISM.] = SYNDICALISM. So **ana:rcho-sy·ndicalist** *sb.*, = SYNDICALIST; also as *adj.*

[**1913** J. A. ESTEY *Revolutionary Syndicalism* i. 31 Jaurès..was throwing them back into Anarchist-Syndicalism.] **1934** in WEBSTER. **1937** A. KOESTLER *Spanish Test.* I. ii. 49 The party and Trades Union of the Anarchists and Anarcho-syndicalists. **1938** *Ann. Reg. 1937* 241 Anarcho-syndicalist opposition was forced underground. **1940** H. READ *Philos. Anarchism* iv. 28 Whatever may be the merits and demerits of the anarcho-syndicalist system, it can and does work. **1949** J. S. SCHAPIRO *Liberalism* xiv. 362 Proudhon has been exalted as the father of anarchosyndicalism. **1955** H. HODGKINSON *Doubletalk* 12 Anarcho-syndicalism, or direct action by anarchist factory workers, is equally anathema. *Ibid.*, The anarcho-syndicalists carry their treacherous activities under the banner of the 'protection of the rights of the individual and his free development'.

anarchy. Add: **1. b.** A theoretical social state in which there is no governing person or body of persons, but each individual has absolute liberty (without implication of disorder).

1850 *Eclectic Rev.* XCI. 167 Confessions of an Anarchist... Proudhon proceeds..that 'all men are equal and free. Society is, therefore, by nature and destination, autonomic ..there is no government'...We see nothing a-head that warrants us in supposing that man is about to be regenerated; and, for the present, must pronounce *anarchy* to be a delightful dream! **1884** RAE *Contemp. Socialism* vii. 281 This idea of a 'genial anarchy'..has always been the favourite social remedy of the Russian revolutionary party. **1889** W. DONISTHORPE *Individualism* 282 Scientific anarchy is..the end towards which society is moving. **1892** *Daily News* 27 Apr. 5/8 Anarchy means the placing in common of all this world's riches to allow each to consume according to his needs. Anarchy is a great family where each will be protected by all and will take whatever he requires.

2. (Later examples.)

1959 *Daily Tel.* 23 Feb. 10/5 The spirit of anarchy today current in the visual arts. *Ibid.*, A form of emotional anarchy even more destructive of talent than the slovenly disregard of technique.

anarthria (ænā·ɹþriă). [mod.L., ad. G. *anarthrie* (see below), f. Gr. ἀναρθρία want of vigour (ἄναρθρος without strength, f. ἀν- (AN- 10) + ἄρθρον joint): see -IA¹.] Defective articulation in speech. Hence **ana·rthric** *a.*, of or pertaining to anarthria (*Cent. Dict.*, 1889).

First used in G. by E. Leyden 1867, in *Berl. klin. Wochenschr.* 25 Feb. 78/1, adopting a suggestion of Dr. Tobias.

1881 *Syd. Soc. Lex.*, *Anarthria*, disjointed speech, an impairment of the articulation usually dependent upon bulbar paralysis. **1897** H. T. PERSHING in T. L. Stedman *20th Cent. Practice* X. ix. 779 Stammering, syllable-stumbling, and scanning...The word anarthria may be used to designate any of these defects. **1915** C. B. CRAIG in *Jrnl. Amer. Med. Assoc.* LXIV. 51/2 Where this area is the site of disease, the resultant symptom is designated by Marie [1906 *La Semaine Médicale* 23 May 243/3] anarthria, and is the simple motor aphasia of the older writers. **1926** *Contemp. Rev.* CXXIX. 96 We have seen a number of hopelessly incurable general paralytics mumbling unceasingly through their anarthria—'Every day and in every way, I am getting better.' **1961** *Brit. Med. Dict.* 92/1 *Anarthria*, the loss of the power of articulating words, the condition being caused by a local lesion.

anastate (æ·năstēⁱt). *Biol.* (*Disused.*) [f. ANA- + Gr. στατός placed.] A substance formed in the process of anabolism in a living organism: opp. to KATASTATE.

1885 M. FOSTER in *Encycl. Brit.* XIX. 19 The substances or mesostates appearing in the former [*sc.* the anabolic series of processes] we may speak of as anastates, those of the latter we may call katastates. *Ibid.*, In the animal-cell the initial anastates seem..generally more complex than the final katastates. **1889** GEDDES & THOMSON *Evol. Sex* vii. 88.

anastigmat (ænăsti·gmæt). *Photogr.* [a. G. *anastigmat* (Miethe), back-formation from *anastigmatisch* adj.: see next.] An anastigmatic lens or system of lenses.

1890 *Brit. Jrnl. Photogr.* 19 Sept. 604/2 Doublet Anastigmat, No. 102, by Carl Zeiss Jena. **1894** *Amer. Ann. Photogr.* 100 The Anastigmat is the most rapid lens I have tried. **1902** *Westm. Gaz.* 23 June 8/2 All scientists have declared it impossible to construct a true Anastigmat lens without the aid of the Jena glass. **1908** *Ibid.* 6 June 14/2 The best lens for copying purposes is..one of the flat field anastigmats. **1957** AMOS & BIRKINSHAW *Television Engineering* I. ix. 187 For lens combinations it [*sc.* elimination of curvature of field] can be achieved, if the lenses are separated, by correct choice of lens constants. A lens combination thus corrected is termed an anastigmat.

anastigmatic (ænăstigmæ·tik), *a.* [f. AN- 10 + ASTIGMATIC *a.*; cf. G. *anastigmatisch* and STIGMATIC *a.* 9.] Not astigmatic; free from astigmatism: applied to a compound lens so constructed as to correct the astigmatic aberration.

1890 P. RUDOLPH *Brit. Pat. 6028* 1 It is..necessary to choose the ratios of the single lenses..in such a manner that the opposed astigmatic differences are of the same magnitude, in view of attaining compensation of this astigmatism, or anastigmatic correction of the entire objective. **1893** *Year-Bk. Photogr.* 10A (Advt.), Zeiss' Patent Anastigmatic Lenses..are the result of calculations with the new Jena Glass. **1897** *Outing* (U.S.) XXX. 340/2 The lens must be one of the almost faultless anastigmatic type. **1901** *Brit. Jrnl. Photogr.* 22 Nov. 744/1 The anastigmatic flatness of field. **1902** [see STIGMATIC *a.* 9]. **1953** *Electronic Engin.* XXV. 138/1 Anastigmatic lenses are employed in this unit so that a high degree of optical accuracy is obtained.

anathematical. Transfer †*Obs.* to sense B in Dict. and add examples of adj.:

1882 *Li-quor Christmas Ann.* 1. 27/1 To profane my lips with an anathematical expression. **1927** *Scots Observer* 22 Jan. 15/1 The tenets of militarism were fathered upon Carlyle and his name made anathematical.

anatine (æ·nătǝin), *a.* and *sb.* [ad. L. *anatinus*, f. *anat-*, *anas* duck: see -INE¹.] **A.** *adj.* Of or pertaining to, resembling or characteristic of, a duck. **B.** *sb.* A bird of the duck family.

1862 H. G. ADAMS *Wild Fl., Birds, & Insects* 274 Those [birds] of the Anatine group. **1875** *Encycl. Brit.* III. 712/2 The Anserines and Anatines. **1893** NEWTON *Dict. Birds* 543 Their [*sc.* the Mergansers'] structure does not much depart from the Anatine or rather Fuliguline type.

Anatolian (ænătōᵘ·liăn), *a.* and *sb.* Also 6 **Natolian.** [f. *Anatolia* (cf. Gr. ἀνατολή east), Asia Minor + -AN.] **A.** *adj.* Of or pertaining to Anatolia or its inhabitants. **B.** *sb.* An inhabitant of Anatolia.

1590 MARLOWE *2nd Pt. Tamburl.* I. i, Natolians, Sorians, blacke Egyptians. *Ibid.* ii, The Natolian king. *Ibid.* II. ii, With full Natolian bowles Of Greekish wine. **1679** [see ORTHODOX A. 4]. **1788** [see THEME *sb.* 7]. **1888** *Encycl. Brit.* XXIII. 655/2 The cause of national education is seldom forgotten in the legacies of patriotic Anatolian Greeks. **1926** *Spectator* 17 Apr. 693/2 Groups of Anatolians whom I found assembled at the various stations. **1934** *Times Lit. Suppl.* 10 May 331/2 Curzon had prepared a memorandum.. arguing that Turkey..must be maintained in her Anatolian homeland. **1958** *Listener* 18 Dec. 1034/2 It would be unwise to assume that the Mycenaean Greek was exactly like his Anatolian contemporary at Hattusas.

Anatolic (ænătǫ·lik), *a.* [f. Gr. ἀνατολικός eastern: see prec. and -IC.] = *ANATOLIAN *a.*

1853 G. FINLAY *Hist. Byzantine Empire* I. i. 14 Seven great themes are particularly prominent in Asia Minor, Optimaton, Opsician, the Thracesian, the Anatolic, [etc.].. and the Armeniac. **1886** *Amer. Jrnl. Archæol.* II. 124 A vast change had come over the Anatolic Theme: great parts of it, including the eastern and southern and much of the central regions, had been occupied by the Seljuk Turks. **1934** A. J. TOYNBEE *Study Hist.* II. D. iii. 79 In the central and north-eastern parts of the Anatolian Plateau or, in the administrative terminology of the day, in the Anatolic and Armeniac army corps districts..of the East Roman Empire.

anatriæne (æ·nătrǝiˌīːn). *Zool.* [f. Gr. ἀνά up, back + TRIÆNE.] In sponges, a triæne with recurved prongs.

1887 [see PROTRIÆNE].

anatropy (ănæ·trǒpi). *Bot.* [f. Gr. type *ἀνατροπία, f. *ἀνάτροπος: see ANATROPOUS *a.*] The condition of being anatropous.

1848 [see ORTHOTROPY].

anautotomic (ænˌǭtotǫ·mik), *a.* *Geom.* [f. AN- 10 + AUTOTOMIC *a.*] Of a curve: not intersecting itself; having no multiple point: opp. to AUTOTOMIC *a.*

1901 A. B. BASSET *Cubic & Quartic Curves* Pref. p. vii, I have..introduced the words *autotomic* and *anautotomic* to designate curves which respectively do and do not possess multiple points.

anauxite (ǣnǭ·ksǝit). *Min.* [ad. G. *anauxit* (A. Breithaupt 1838, in *Jrnl. f. Prakt. Chem.* XV. 325), f. Gr. ἀναυξής not increasing, so named 'because the mineral does not swell up before the blowpipe' (Chester): see -ITE¹.] A pearly-white translucent variety of CIMOLITE.

1850 J. D. DANA *Syst. Min.* (ed. 3) 288 Anauxite, *Breithaupt*..pearly; greenish-white; translucent. **1883** *Encycl. Brit.* XVI. 424/2 Anauxite..Granular..Translucent, pearly.

ancestor, *sb.* Add: **1. b.** (Earlier and later examples.)

1854 *Wesleyan Methodist Mag.* X. 623 Ancestor-worship ..has linked and attached itself most powerfully to the heart of every Chinese. **1883** MAINE *Early Law & Custom* iii. 55 Ancestor-worship, the worship of father, grandfather, and great-grandfather, has among the Hindus a most elaborate liturgy and ritual. **1928** H. C. DAWSON *Age of Gods* iii. 47 The churingas or 'ancestor stones' of the modern Australian natives. **1936** C. DAY LEWIS in Day Lewis & Stebbing *Imag. & Thinking* 11 The profound desire for a feeling of continuity, to which some of us young poets have applied the expression 'ancestor-worship'. **1957** V. W. TURNER *Schism & Cont. in Afr. Soc.* p. xxi, The misfortunes of life.. are attributed to the punitive action of ancestor spirits.

c. *Law.* A person who precedes another in the course of inheritance, and from whom an inheritance is derived, whether in the direct line of descent or not: correlative to *heir*. *collateral ancestor*: see COLLATERAL *a.* 4.

1628 COKE *Littleton* 380 b, If lands had beene giuen to the husband and wife and their heires, and the husband had made a Feoffement to another, to whom a Collaterall Ancester of the wife had released and died. **1651, 1767** [see HEIR *sb.* 1]. **1768, 1809** [see ANCESTRAL *a.* 1 b.] **1959** JOWITT *Dict. Eng. Law* I. 116/2 Under the law as it stood before the Law of Property Act, 1925, an ancestor meant any person from whom real property was inherited.

Hence **a·ncestor** *v. trans.*, to be the ancestor of; also *transf.*; **a·ncestored** (-ǝɹd) *ppl. a.*, having an ancestor or ancestors (of a specified kind); **a·ncestorship,** the position of an ancestor.

1776 H. WALPOLE *Let.* 9 Sept. (1904) IX. 412 Few men are so well ancestored in so short a compass of time. **1853** *Tait's Mag.* XX. 604 The thorough-bred, orientally ancestored horse. **1883** MAX MÜLLER *India, What can it teach Us?* vii. 239 Ancestorship as a natural ingredient of religion among all savage nations. **1921** *Times Lit. Suppl.* 10 Feb. 92/2 Their younger brother ancestored the well-known family of Howard-Vyse, of Stoke Place, Slough. **1940** BRYANT & AIKEN *Psychol. of English* iv. 33 The *Ursprache* which ancestored our own English.

ancestrally (ænse·străli), *adv.* [f. ANCESTRAL *a.* + -LY².] By or in respect of ancestry.

1882 G. ALLEN *Colin Clout's Calendar* xvii. 97 Ancestrally, yellow-rattle is a near relation of the pretty little blue veronicas. **1886** *Athenæum* 6 Mar. 328/3 Whether the vertebrate eye..will turn out..to be ancestrally derived from a number of modified ancestral gills. **1923** R. FROST *New Hamp.* (1924) 50 With something of the baby grip Acquired ancestrally in just such trees.

anchor, *sb.*¹ Add: **3.** (Further techn. examples.)

1906 *Westm. Gaz.* 29 Aug. 2/1 The tram lines that corrugate most freely are those which are laid on the concrete foundation without anchors. **1951** *Gloss. Terms Plastics (B.S.I.)* 44 *Anchor.* In injection moulding. An undercut extension to the feed. It is usually located on an ejector pin to facilitate the removal of the stalk from the mould.

b. *Billiards.* A stroke in which the two object-balls are kept close to or against the cushion so that a series of cannons can be made without disturbing their position; in full *anchor cannon, shot, stroke.* Also *anchor baulk-line*: one of four lines drawn on a billiard table to form spaces in which the number of cannons allowed is restricted; hence *anchor-space*.

1901 *World of Billiards* 25 Sept. 460/2 When the two object balls are at rest in any one of the eight spaces, one or both of the object balls must be driven out... The same condition applies to the anchor. **1904** S. A. MUSSABINI *Billiards Expounded* i. 20 The greatest thing known to close-cannon play is the 'anchor stroke'. **1907** *Daily Chron.* 15 Mar. 8/5 T. Reece..went on with his 'anchor' cannon until he had reached an item of 1,825. **1907** T. REECE in *Ibid.* 27 Mar. 5/5 To the ordinary, everyday billiard player..I would not recommend a cultivation of this 'anchor'. **1910** *Encycl. Brit.* III. 939/1 The 'anchor baulk-lines' form a tiny compartment, 6 in. by 3, and are drawn at the end of a baulk-line where it touches the rail and so divides the compartment into two squares. Only one shot is allowed in this 'anchor-space', unless a ball be driven out of it. **1911** C. ROBERTS *Compl. Billiard Player* xx. 146 (*caption*) Position of balls in the famous anchor cannon, from which T. Reece made 499,135 in 1907. **1922** *Westm. Gaz.* 16 Oct., The push-stroke and the anchor-cannon were eliminated when the need for their elimination became obvious.

c. *Athletics.* The end member of a tug-of-war team, who secures the rope by looping it round his body; = *anchor-man (below).

1909 T. A. COOK et al. *4th Olympiad* 92 Tug of War.. The Americans were magnificent athletes, but were not aware how to tie an anchor or how to place their men. **1911** *Encycl. Brit.* XXVII. 365/1 When a tug-of-war takes place out of doors the men, or at least the 'anchors', are allowed to dig holes in the ground for their feet.

d. *Athletics.* The person who runs the last section of a relay race; also *attrib.*

1934 in WEBSTER. **1958** *Observer* 27 July 22/2 On the anchor leg, M. C. Spence..finished first by ten yards for his team. *Ibid.* 3 Aug. 16/1 M. Spence..held off Britain's J. E. Salisbury on the anchor stage to win by a long yard. **1961** *Sunday Times* 16 July 20/7 He must surely be our best prospect as an anchor runner for the relay against America next week.

e. *pl.* Brakes. *slang.*

1936 *Daily Herald* 5 Aug. 8/4 List of busmen's slang phrases..Anchors (Brakes). **1965** PRIESTLEY & WISDOM *Good Driving* vii. 55 There is more to it..than just putting on the brakes—or, to use the colourful language of the sporting motorist, 'clapping on the anchors'.

7. anchor bolt (see quots.); **anchor escapement** = *recoil escapement*; **anchor-ice** (earlier U.S. example); **anchor-line,** a line attached to or serving as an anchor (see also quot. 1909); also *transf.* and *fig.*; **anchor-man**: restrict † to sense in Dict. and add *transf.* and *fig.* examples, as the person at the end of a group tugging a rope, roped together, etc.; also, a compère of a radio or television programme; **anchor-ring,** (*b*) *Geom.*, = ANNULUS 2.

a **1875** KNIGHT *Dict. Mech.* I. 96 *Anchor bolt* (Machinery), one having an expanded shank to prevent its drawing out. **1957** *Gloss. Terms for Stone used in Building (B.S.I.)* 21 *Anchor bolt*, a T-shaped bolt for attaching fascia and similar stones to a supporting R.S.J. **1854** E. B. DENISON in *Encycl. Brit.* VII. 8/1 The 'anchor' escapement..appears to have been invented by the celebrated Dr. Hooke as early as the year 1656. **1815** *Niles' Reg.* IX. 201/1 On the same day the anchor-ice began to run a little. **1906** *Westm. Gaz.* 29 Sept. 13/3 The first thread spun [by a spider] is secured to the ground as an anchor-line. **1909** *Cent. Dict.* Suppl., *Anchor-line*, a line attached to a small buoy and to one fluke of an anchor: used in towing a raft of logs and to free the anchor when fast to rocks or snags. (U.S.) **1910** BELLOC *Verses* 47, I stand at home and slip the anchor-line. **1911** *Encycl. Brit.* XXVII. 365/1 Some rules allow the 'anchor-men', who hold the ends of the rope, to fasten it to their persons. **1955** E. HILLARY *High Adv.* ix. 173 Ang Temba went first..; then Tom Bourdillon; and finally Charles Evans as anchor-man. **1957** L. P. HARTLEY *Hireling* i. 9 The driver..[had] been the anchorman in many a tug-of-war. **1958** *Spectator* 29 Aug. 278/2 Mr. Williams is almost always there on this kind of occasion as a tubby anchor-man. **1958** *Observer* 28 Dec. 3/1 Its [*sc.* a television programme's] remarkable compère or anchor-man, Cliff Michelmore. **1863** FROST & WOLSTENHOLME *Solid Geom.* xiv. 213 An anchor ring, supposed to be generated by the revolution of a circle about an axis in its plane not intersecting the circle.

anchorage¹. Add: **3. b.** *spec.* in *Dentistry.*

1912 *Catal. Dental Manuf. Co. Ltd.* G. 71 Sufficient anchorage must be made to withstand the force of mastication, and to provide for firm retention of the inlay within the cavity. **1917** *Recalled to Life* Sept. p. xx, They have platinum anchorages baked in the teeth and large, strong pins soldered to the anchorages after baking of the porcelain is complete.

anchoret, -ite. Add: *anchorite* is now the prevalent form.

1. b. (Later example.)

1929 C. DAY LEWIS *Transitional Poem* II. 29 Then life's pistons..Begin to tickle the most anchorite ear.

anchor-hold². *Hist.* [f. ANCHOR *sb.*² + HOLD *sb.*¹] An anchorite 'hold', abode, or retreat; the cell of an anchorite; = ANCHORAGE².

1631 WEEVER *Anc. Funeral Mon.* 150 Their solitarie little cells..carrie still the name of..Anchor-holds. *a* **1666** WOOD *City of Oxford* (1889) I. 356 The Anchorhold of S. Giles Church. **1802** FOSBROKE *Brit. Mon.* (1843) 372 The *Destina* (for so these anchor-holds or stalls, affixed to larger buildings were called), occupied by Dunstan soon after he became a Monk. **1922** *Times* 22 Apr. 9/4 Both the church and its 'anchorhold', or anchorite's cell, are more than once mentioned in documents of the college [*sc.* Merton].

anchorless, *a.* Delete *rare* and add earlier and later examples.

1832 *Fraser's Mag.* VI. 739/2 With all my anchorless thoughts adrift. **1853** C. BRONTË *Villette* I. vi. 94 My homeless, anchorless, unsupported mind. **1914** R. BROOKE *Let.* 7 Mar. in *Coll. Poems* (1929) p. cxii, I really do feel a little anchorless. I shall be glad to be back among you all, and tied to somewhere in England. **1945** P. A. LARKIN *North Ship* 17 Then would the moon go raving, The moon, the anchorless Moon go swerving Down at the earth.

anchor-stock. Add: (Earlier examples.)

c **1302** *Pipe Roll 30 Ed. I. lm. 2 b* In vj Anchoris, vj ankerstokes. **1346–1688** [see STOCK *sb.*¹ 24 a].

b. Delete def. of *anchor-stock fashion*, substitute 'see quot. 1867', and add example.

1867 SMYTH *Sailor's Word-Bk.* 40 *Anchor-stock fashion*, the method of placing the butt of one wale-plank nearly over the middle of the other; and the planks being broadest in the middle, and tapered to the ends, they resemble an anchor-stock.

anchovy. Add: **2.** *anchovy essence, paste*; **anchovy butter,** a paste made of anchovies mixed with butter, used as a filling for sandwiches, savoury biscuits, etc.; **anchovy-toast** (earlier example).

1845 E. ACTON *Mod. Cookery* (ed. 2) iv. 127 Anchovy Butter. (*Excellent.*) **1959** *Listener* 13 Aug. 263/1 Mashed liver sausage, or anchovy butter, or any other favourite paste. **1846** 'A LADY' *Jewish Man. Cookery* ii. 35 Pound it [*sc.* fish]..with a couple of anchovies, or a little anchovy essence. **1856** DICKENS *Out of Season* in *Househ. Words* 28 June 555/2 Anchovy Paste..and the whole stock of luxurious helps to appetite. *c* **1771** S. FOOTE *Maid of Bath* I. 11 Get me an anchovy toast.

anchovy-pear. (Earlier example.)

1696 [see *river pear*, RIVER *sb.*¹ 6].

‖**ancien régime** (ansiaṅ reʒim). [Fr., lit. 'old rule'.] See RÉGIME 2 b.

1794 MORRIS in *Amer. State Papers* (1832) *For. Relat.* I. 404 If once that terror were, by superior force, to receive a counter direction, the *Ancien Régime* or any other régime, would, I think, be submitted to without the slightest struggle. **1839** F. A. KEMBLE *Jrnl. of Residence Georgia Plant.* (1863) 82 Two little filthy children, however, seemed to be still under the *ancien régime* of non-ablution. **1856** C. M. YONGE *Daisy Chain* II. xviii. 539 An old French Marquis..a dear old man, quite of the *ancien régime*. **1864** MRS. GASKELL *French Life* i, in *Fraser's Mag.* Apr. 436/2 She received the homage of the ladies and gentlemen of the ancien régime. **1944** A. L. ROWSE *Eng. Spirit* ii. 25 The oldest and greatest monarchy in Europe, the French *ancien régime*. **1961** *Guardian* 7 Mar. 7/6 The *ancien* (pre-Nasser) *régime* Egyptian politicians.

ancient, *a.* and *sb.*¹ Add: **A. 3. b.** *ancient history*: see HISTORY *sb.* 3.

4. b. *ancient lights*: see LIGHT *sb.* 13.

c. *ancient monument*: a monument made or set up long ago; *spec.*, a monument or other edifice scheduled as being of historical, architectural, or archæological interest and protected by Act of Parliament from damage or destruction. Also *attrib.* and *transf.*

1593 (*title*) Discription..of all the Ancient Monuments, Rites, and Customes..within the Monastical Church of Durham (1842, Surtees Soc.). **1873** *House of Commons Papers, Reports from Committees* XI. 54 [= 950] A Memorial from the Committee for the Conservation of Ancient Monuments. **1877** *Encycl. Brit.* VII. 308/1 His [*sc.* Döbrentei's] great work is the *Ancient Monuments of the Magyar Language* (*Régi Magyar Nyelvemlékek*). **1880** C. P. KAINS-JACKSON *Our Ancient Monuments* 1 We give below a list of the ancient monuments of the United Kingdom as scheduled in the bill repeatedly introduced into the House of Commons by Sir John Lubbock, Bart. [etc.]. *Ibid.* 6 Then it [*sc.* Wayland Smith's Forge] became a relic simply, a 'monument' in the sense of the framers of the Ancient Monuments Bill. **1911** *Encycl. Brit.* XVIII. 797/1 There are four acts: the Ancient Monuments Protection Acts of 1882, 1900 and 1910, and the Ancient Monuments Protection (Ireland) Act 1892. **1958** *Listener* 23 Jan. 150/2 The Ancient Monuments Department of the Ministry of Works has set out to preserve what remains.

B. 2. (Later examples.)

1833 ROBERT GRANT in *Bickersteth's Christian Psalmody* 17 Our Shield and Defender—the Ancient of Days Pavilioned in splendour, and girded with praise.

b. *transf.* A very old person or thing. *jocular.*

1935 T. E. LAWRENCE *Let.* 5 Apr. (1938) 867 I've only ridden the ancient-of-days twice this year. **1937** KIPLING *Something of Myself* i. 10, I was shown an Ancient of Days who, I was told, was the Provost of Oriel.

ancientry. Add: **1.** (Later examples.)

1913 KIPLING *Songs from Books* 21 Witness hereby the ancientry Of Oak and Ash and Thorn! *a* **1954** F. BRETT YOUNG *Wistanslow* (1956) i. 7 Not merely venerable to me in its ancientry but also impressive as a work of art.

5. (Later example.)

1904 H. JAMES *Golden Bowl* I. I. vi. 111 Small florid ancientries, ornaments, pendants, lockets.

Ancona¹ (ænkōᵘ·nă). [Name of a town in Italy.] A breed of domestic fowl characterized by black and white mottled plumage and leaden-grey shanks.

1853 WINGFIELD & JOHNSON *Poultry Bk.* 110 The *Ancona* is a first cousin to the Minorca, its sole point of difference being a mottled or splashed plumage, black and white, in about equal proportions. **1873** L. WRIGHT *Bk. Poultry* 357 We entertain scarcely any doubt that the origin of Anconas is to be found in accidental 'sports' of this colour from crossing Black and White Minorcas. **1960** *Farmer & Stock-breeder* 8 Mar. 135/1 One of my Ancona bantams lays eggs with blood spots.

‖**ancona²** (ænkōᵘ·nă). Pl. **ancone.** [It., = med.L. (Stat. Guild of Painters at Venice, A.D. 1271): etym. uncertain; perh. an alteration of Gr. εἰκόνα, acc. of εἰκών image, ICON.] An altar-piece, esp. one consisting of a group of pictures or painted panels connected by architectural structure.

1874 R. H. BUSK *Tirol* 146 A priceless work of Alb. Durer, an '*Ancona*', showing forth in its various compartments the history of the Passion. **1885** *Athenæum* 19 Sept. 377/2 The Van Eycks' 'Adoration of the Lamb'..[is] a true representative of the Italian *ancona* or group of pictures included in a single altarpiece. **1887** *Ibid.* 20 Aug. 248/3 The work is neither more nor less than an ancona, which is different in kind as well as in scale from a triptych. 'Altar-piece' would be a correct term here, if the author fears to use 'ancona'. **1903** *Burlington Mag.* I. 309/2 This altarpiece now hangs on the north wall of the choir of the Collegiata. It is a Gothic *ancona* in which four scenes are represented. **1906** H. R. F. BROWN tr. *Molmenti's Venice, Mid. Ages* II. xii. 123 It is in these *ancone*, sacred images painted on wood, that we find the origin of Venetian painting.

attrib. **1900** *Speaker* 28 Apr. 116/1 The ancona form of composition..was characteristic of the old Venetian school.

and, *conj.*¹ Add: **B. 3. c.** *and/or*: a formula denoting that the items joined by it can be taken either together or as alternatives.

1855 *Law Jrnl. Reports* XXIV. II. Exchequ. 199/2 The parties were to 'load a full and complete cargo of sugar, molasses, and/or other lawful produce'..the words 'and' and 'or' being introduced into the charter-party. **1895** POLLOCK & MAITLAND *Hist. Eng. Law* I. I. v. 152 In medieval Latin *vel* will often stand for *and*..Often it is like the *and/or* of our mercantile documents. **1916** H. BARBER *Aeroplane Speaks* ii. 85 The jamming of the rudder and/or elevator. **1929** *Penrose's Ann.* XXXI. 99 A good proportion of cotton and/or linen in the furnish of a paper. **1959** *Camb. Rev.* 2 May 454/2 The Press has rather plumped for the scholar as writer, and/or as bibliophile. **1960** E. BOWEN *Time in Rome* iii. 82 The young set-apart creature, waiting at home for her fifteenth birthday and/or the next vacancy in the Atrium.

4. (Further examples.) Cf. *GOOD *adv.* d.

1575 R. LANEHAM *Let.* in *Leisure Hour* (1884) 631/1, I am..jolly and dry of a mornings. **1846** [see NICE *a.* 15 d]. **1887** T. DARLINGTON *Folk-Speech of S. Cheshire* 109 'Fine an' vexed' = exceedingly vexed.

8. b. (Later examples.)

1896 A. AUSTIN *Eng. Darling* III. i. 63 Face a head gust and it will steady you. **1933** D. L. SAYERS *Murder must Advertise* iv. 72 Spray with Sanfect and you're safe.

9. Add at end of def.: 'or phrase'. (Further examples.)

1818–1884 [see MISTAKE *sb.* 2 c]. **1853** [see WONDER *sb.* 6 e]. **1890** F. GOLDIE *Ven. Ed. Arrowsmith* (C.T.S.) 2 Robert Arrowsmith's father..was often thrown into gaol—and we know what gaols were in those days. **1914** S. A. HIRSCH in A. G. Little *R. Bacon* v. 128 Another large portion of the Greek Grammar is taken up by Bacon's treatment of accentuation and prosody (pp. 95–144), and no wonder! **1930** G. K. CHESTERTON *Resurr. Rome* v. 202 The French would certainly have recovered the stolen French provinces whenever they could; and quite right too.

10. For quot. 1780 read:

1780 MRS. THRALE *Let.* 10 June (1788) II. 150 Do go to his house, and thank him.

C. ¶ Used *subst.* (Earlier and later examples.)

1513, 1613 [see IF *sb.*]. **1823** tr. *J. Campan's Mem. M. Antoinette* (ed. 2) I. x. 259 Five different requests—such an office, *or* such a mark of distinction, *or*..and so on... The *ors* were changed into *ands.*

-and, *suffix*[2]. A formative element representing the termination *-andus, -a, -um* of the gerundive of Latin verbs in *-āre*. Examples of words in *-and* are *analysand, confirmand, educand, graduand, multiplicand, operand, ordinand*. The meaning of these words is passive, thus *ordinand* 'person to be ordained'. This element has never been a living suffix, having no separate existence apart from the Latin gerundive form from which it is derived. The gerundial endings are sometimes retained in their Latin (neut.) form (with pl. *-a*), as in *avizandum, memorandum, notandum*.

Andalusian (ændălū·ʃiăn, -ziăn), *a.* and *sb.* Forms: 7–9 **Andalucian, Andaluzian.** [f. *Andalusia* (see below) + -AN.] **A.** *adj.* Of or pertaining to Andalusia, a southern province of Spain, or its inhabitants or speech. **B.** *sb.* **a.** A native or inhabitant of Andalusia. **b.** The variety of Castilian spoken in Andalusia. **c.** A Mediterranean breed of domestic fowl, rabbit, etc.

1612 SHELTON *Quix.* (1619) I. ii. 12 The Oast thought he had called him a Castellano, or Constable,..whereas he was indeede an Andaluzian. **1615** *Ibid.* (1617) II. i. 10 A famous Andaluzian Poet wept, and sung her teares: and another famous and rare Poet of Castile her beauty. **1740** P. PINEDA *New Spanish Dict.* (s.v. *Andaluz*), The Castilians, who have no good opinion of the Andaluzians. **1823** BYRON *Don Juan* XII. lxxv. p. 42 She cannot step as does an..Andalusian girl from mass returning. *Ibid.* XIII. xxiii. p. 66 And Juan, like a true-born Andalusian, Could back a horse. **1839** G. DENNIS *Summer in Andalucia* II. xvi. 385 This failing of the Andalucians betrays itself in many ways. **1854** *Poultry Chron.* I. 101 Several so-called Andalusians, bred by a Minorca cock out of a Maltese hen...The Andalusian fowls were introduced to the notice of amateurs by Mr. Taylor, of Shepherd's Bush. **1862** G. BORROW *Wild Wales* III. xxiv. 275 The place upon the whole put me very much in mind of an Andalusian village overhung by its sierra. **1868** DARWIN *Var. Anim. & Plants under Domestication* I. iv. 105 Various allied sub-breeds [of rabbit] are reared on the Continent, such as the so-called Andalusian, which is said to have a large head with a round forehead, and to attain a greater size than any other kind. **1879** *Cassell's Fam. Mag.* Apr. 274/2 The experiment was once tried of fattening a youthful Andalusian [rabbit] entirely on clover hay. **1882** CAULFEILD & SAWARD *Dict. Needlework* 6/2 Andalusian Wool..is also called Victoria Wool, and is a fine soft warm make of woollen thread or yarn...It is the same wool as the Shetland, but is thicker. **1887** *Encycl. Brit.* XXII. 351/1 The word 'dialect' is still more appropriately applied to Andalusian than either to Asturian or Navarrese-Aragonese. **1900** J. D. M. FORD in *Studies & Notes in Philol. & Lit.* VII. 1 Andalusian and other dialect peculiarities are only incidentally concerned. **1911** —— *Old Sp. Readings* p. vii, Andalusian does not differ radically enough from Castilian to oblige us to regard it as a separate dialect. **1920** *Discovery* July 200/2 This form of fowl is bluish-black with black lacing, and is called the Blue Andalusian. **1924** KIPLING *Debits & Credits* (1926) 229 The fifth bull rushed out—an unthinking black Andalusian. **1927** *Blackw. Mag.* Sept. 313/1 A sky of Andalusian azure. **1936** W. J. ENTWISTLE *Sp. Lang.* vi. 220 Since 1600..there has been no 'correct' Andalusian, and the dialect has gathered to itself all current vulgarisms.

Andaman (æ·ndămæ̆n), *a.* and *sb.* [f. *Andaman* Islands in the Bay of Bengal.] **A.** *adj.* Of or belonging to the Andaman Islands, or pertaining to their inhabitants or language, esp. in *Andaman islander*. **B.** *sb.* The language of the Andaman people.

1848 J. CRAWFURD in *Jrnl. Indian Archipelago* II. 206 On comparing the native portion of the language of Wageou with that of the Sámang, and the words of the Andaman, no resemblance can be found between them. **1860** MAYNE REID *Odd People* 412 We trace in the Andaman islander the true physiognomy of a negro. *a* **1873** MILL *Ess. Relig.* (1874) 193 It is to suppose that God could not, in the first instance, create anything better than a Bosjeman or an Andaman islander. **1936** J. R. KANTOR *Obj. Psychol. Gram.* xv. 208 An animate and inanimate division is also found in Andaman, the language of the Andaman Islands in the Bay of Bengal. **1952** *Jrnl. R. Anthrop. Inst.* LXXI. 17/1 An analysis of the non-totemic Andaman Islanders. **1955** *Nomencl. Comm. Timbers* (*B.S.I.*) 66 *Pterocarpus dalbergioides*, Andaman padauk; Andaman redwood (U.S.A.).

Andamanese (æ:ndămănī·z). [f. as prec. + -ESE.] An aboriginal of the Andaman Islands; also the language of the Andaman people; also *attrib.* or as *adj.* Hence **A:ndamane·sian** *a.* Also **Andama·ner,** an Andaman islander.

1727 A. HAMILTON *New Acc. E. Ind.* II. xxxviii. 68 The Andemaners have no Notions of a Deity. **1833** *Penny Cycl.* I. 513/2 The Andamaners eat rats, guanas, and snakes. **1862** *Andamanese Vocab. & Phraseol.* 1 Those words marked thus * are strongly suspected to be Hindostanee, or words from some other language picked up by the Andamanese. **1875** *Jrnl. Anthrop. Inst.* IV. 464 The larger shed was filled with 110 Andamanese of all ages. **1887** M. V. PORTMAN (*title*) A Manual of the Andamanese Languages. Part I.—Dictionary, English-Andamanese. **1907** *Indian Antiquary* XXXVI. 220 For the purpose of an adequate presentation of Andamanese..I gradually framed a Theory of Universal Grammar. **1922** *Blackw. Mag.* June 787/2 A keen-sighted Andamanese saw two men on a little hill. **1932** W. L. GRAFF *Language & Languages* xi. 416 The Andamanesian dialects spoken by the natives of the Andaman Islands.

a **1942** B. MALINOWSKI *Scientific Theory of Culture* (1944) vi. 61 Among the..Andamanese, we can not speak of the political organization of the tribe, since this does not exist.

anderoon (ændĕrū·n). Also **anderun.** [Pers. *enderūn* prop. internal, interior, f. *ender* within.] The apartments of a (Persian) harem; also, any inner room of a house.

1840 SAVILE in *New Monthly Mag.* LVIII. 189 In the anderoon..sat Khodadad, in sweet converse with his beloved Semira. **1842** THACKERAY *Sultan Stork* (1887) 7 The ladies they kept so secretly in the anderoon. **1900** *Westm. Gaz.* 20 July 2/1 The household gathering outside the *anderoon* till the Shah emerges. **1934** F. STARK *Valleys of Assassins* iii. 213 An inner *anderun* where the women slept.

Anderson (æ·ndəɹsən). *Anderson shelter*, a small prefabricated air-raid shelter devised by Mr. (later Sir) William Paterson, a Scottish engineer, and adopted while Sir John Anderson was Home Secretary (1939–40). Also *ellipt.*

1939 *New Statesman* 3 June 860/2 Goats sheltered from high explosive in Anderson shelters were claimed to be quite unhurt. **1939** *War Illustr.* 29 Dec. 535 An Anderson shelter erected in a kitchen because there is no garden space available. **1940** *New Statesman* 9 Nov. 465/2 Where's my rabbits?..Kept 'em in the Anderson. **1944** *Modern Reading* x. 87 'Cold?' Dad asked her. 'No. Like toast. That's the best of Andersons, you can get warm in them.' **1952** *Oxf. Jun. Encycl.* X. 89/1 Anderson shelters were small, curved, steel huts, partially buried in people's gardens and covered with 2 or 3 feet of earth to protect them from the effects of explosion.

Andine (æ·ndəin), *a.* [f. *Andes* + -INE[1].] = ANDEAN *a.*

1900 *Daily Chron.* 23 Aug. 3/4 The explorations carried out..for the Argentine Government in the Andine region. **1924** *Chambers's Jrnl.* Mar. 188/2 A vast ancient crater of the Andine range.

Andorran (æ̆ndǫ·răn), *a.* and *sb.* Also 9 **Andorrian.** [f. *Andorr(a*, a small independent state in the eastern Pyrenees, and its capital: see -AN, -IAN.] **A.** *adj.* Of or pertaining to Andorra. **B.** *sb.* A native or inhabitant of Andorra.

1837 J. E. MURRAY *Summer in Pyrénées* (ed. 2) I. v. 99 The leader of our little band was a considerable proprietor in the valley, who was well known to the Andorrians. *Ibid.* 117 This Andorrian piscator..was a labourer. **1856** *Chambers's Jrnl.* 12 Apr. 235/2 It was not difficult to recognise in the cavalier an Andorran of the upper class. **1865** M. EYRE *Over the Pyrenees* iv. 118 The Andorrans are passionately fond of dancing. *Ibid.* 123 An Andorran peasant..always carries arms. **1906** H. BELLOC *Hills & Sea* vi. 36 In the further Andorran valley. **1909** —— *Pyrenees* v. 268 The Andorrans have all the vices and virtues of democracy clearly apparent. **1967** *Chambers's Encycl.* I. 420/1 Each of the two suzerains..has a court..for Andorran affairs.

‖ **andouillet.** Add pronunc. (an̈dūie). Also **andouillette** (an̈dūiet). Delete ?*Obs.* and add earlier and later examples.

1611 J. DONES *Paneg. Verses* in T. Coryat *Crudities* sig. f5v [Rabelais] Whose *Papagants, Andouilets*, and that traine Should be such matter for a Pope to curse. **1877** *Kettner's Book of Table* 30 We know little of the andouillette in England. **1906** I. BEETON *Househ. Managem.* xxiv. 737 Andouilettes of game. **1958** R. GODDEN *Greengage Summer* ix. 106 Andouillette is a kind of sausage. **1965** *House & Garden* Jan. 60/1 *Andouillette*. This sausage, similar to andouille but fresh, is usually cooked slowly in white wine or champagne and served with crisp potatoes and watercress.

andradite. (Examples.)

1868 DANA *Syst. Min.* (ed. 5) 268 Named *Andradite* by the author after the Portuguese mineralogist d'Andrada, who described and named the first of the included subvarieties, Allochroite. The included kinds vary so widely in color and other respects that no one of the names in use will serve for the group. **1903** F. RUTLEY *Mineralogy* (ed. 13) II. 114 The Iron-lime garnets are included under the name Andradite. **1914** *Brit. Mus. Return* 227 Garnet (andradite) from China.

Andrew. Add: **4.** *Naval slang.* Also *Andrew Millar* or *Miller*. †**a.** A ship, esp. of war. Also *Andrew Millar's lugger*. *Obs.*

[**1596** s.v. DOCK *v.*[2] 2.] **1812** VAUX *Flash Lang.*, *Andrew Miller's lugger*, a king's ship or vessel. **1864** HOTTEN *Slang Dict.* 67 *Andrew Millar*, a ship of war.

b. The Royal Navy; also (obs.), a government department or authority.

1867 SMYTH *Sailor's Word-Bk.* 40 *Andrew*, or *Andrew Millar*, a cant name..for government and government authorities. **1916** 'TAFFRAIL' *Carry On!* 29 Terms..heard every day in 'Andrew', as the bluejacket calls the Navy. **1955** G. FREEMAN *Liberty Man* I. iv. 62 That's 'ow it is in the Andrew...That's what we call the navy. *Ibid.*, A press man called Andrew Miller..forced so many blokes to join, that it got known as Andrew's navy. Then they shortened it to the Andrew.

Andrewsite (æ·ndruzəit). *Min.* [f. the name of Thomas *Andrews* + -ITE[1].] A bluish-green hydrous phosphate of iron and copper.

1871 *Chem. News* XXIV. 99 A mineral recently found in Cornwall..has been analysed in the Museum Laboratory, and Professor Maskelyne named it Andrewsite, in honour of the distinguished President of the Chemical Section of the British Association, Dr. Andrews, of Belfast.

andro- (before a vowel **andr-**), repr. Gr. ἀνδρο-, combining form of ἀνήρ man, male (see words in ANDRO- in Dict.). **androcentric** (ændrose·ntrik) *a.*, having man, or the male, as its centre; so **androcentri·city**; **androcracy** (ændrǫ·krăsi) [-CRACY], the rule of man or the male, male supremacy; **androcratic** (-kræ·tik) *a.*, pertaining to or involving androcracy; **andromonœcious** (-mǫnī·ʃiəs), *a. Bot.* [MONŒCIOUS *a.*], having male and hermaphrodite flowers on the same plant; **andromonœcism** (-mǫnī·siz'm) *Bot.*, the condition of being andromonœcious.

1903 L. F. WARD *Pure Sociol.* xiv. 292 The androcentric theory is the view that the male sex is primary and the female secondary in the organic scheme, that all things center, as it were, about the male. **1959** *Guardian* 6 Nov. 6/5 The Fathers of the Church accepted from their cultural environment the androcentric standpoint. **1954** *Theology* LVII. 326 Do we see anything of this androcentricity in the Christian ministry? **1903** L. F. WARD *Pure Sociol.* xiv. 376 The stage of gynæcocracy was succeeded by the stage of androcracy, and the subjection of woman was rendered complete. **1893** *Athenæum* 7 Oct. 494/1 Marital relations among tribes in the enjoyment of an androcratic government being generally far more satisfactory. **1903** L. F. WARD *Pure Sociol.* xiv. 399 The androcratic régime, during which woman had no voice in the selecting process. **1877** Andromonœcious [see *ANDRODIŒCIOUS *a.*]. **1888** G. HENSLOW *Origin Floral Struct.* 227 Andromonœcism signifies that the same plant bears both male and hermaphrodite flowers.

androconium (ændrokōᵘ·niəm). *Zool.* [mod.L., f. Gr. ἀνδρο- male + κονία dust.] Usu. in pl. *androconia*, scales on the wings of certain male Lepidoptera from which the attractive scent of the male is diffused; = PLUMULE 3 b. Hence **androco·nial** *a.*

1877 S. H. SCUDDER in *Proc. Amer. Acad. Arts & Sci.* XII. 157 These peculiar scales [on a male butterfly], or androconia, as they may be called in reference to their masculine nature, were first noticed by Bernard Deschamps... Deschamps called them plumules. **1914** T. B. FLETCHER *Some Indian Insects* v. 49 The flocculent yellow androconial hairs which line the interior of the pocket and which doubtless emit a smell attractive to the female. **1937** *Nature* 13 Feb. 265/2 The androconial scale and its development in the genus *Erebia*. **1957** E. B. FORD *Butterflies* (ed. 3) v. 96 The attractive scent of the male is scattered by the androconia, usually situated on the upper side of the fore-wings.

androdiœcious, *a.* In etym., for 'διά apart' read 'δι- (DI-[2]) twice'. Add earlier example. Hence **androdiœ·cism,** the state of being androdiœcious.

1877 C. DARWIN *Different Forms Flowers* i. 13 There are plants which produce hermaphrodite and male flowers on the same individual, for instance, some species of Galium, Veratrum, &c.; and these might be called andro-monœcious. If there exist plants, the individuals of which consist of hermaphrodites and males, these might be distinguished as andro-diœcious. **1888** G. HENSLOW *Origin Floral Struct.* xxiv. 277 Androdiœcism signifies that the same species has both male and hermaphrodite plants.

androgen (æ·ndrodʒĕn). *Biol.* [f. *ANDRO- + -GEN.] A substance, as a male sex hormone, capable of developing and maintaining certain male sexual characteristics (cf. *ŒSTROGEN). Hence **androge·nic** *a.*

1936 *Jrnl. Amer. Med. Assoc.* 18 July 212/2 The Nomenclature of 'Male Hormones'..*Androgen* to designate substances possessing masculinizing activity. **1939** I. F. & W. D. HENDERSON *Dict. Sci. Terms* (ed. 3) 16/2 *Androgenic*, stimulating male characters; masculinising; *appl.* hormones; *appl.* tissue capable of elaborating an androgenic hormone. **1939** *Lancet* 22 Apr. 948/1, 50 mg. of testosterone is equivalent to 3500 I.U. of androgenic activity. **1946** *Nature* 24 Aug. 276/2 The significance of androgen and œstrogen excretion in the urine in relation to ageing. **1962** *Lancet* 1 Dec. 1166/1 The action of the œstrogens is far less specifically related to precise molecular structures than is that of the androgens or corticoids.

androgenesis (ændrodʒe·nėsis). *Biol.* [mod. L. f. *ANDRO- + GENESIS; cf. G. *androgenese* (M. Verworn 1891, in *Pflüger's Archiv ges. Physiol.* LI. I. 81).] (See quot. 1925.) Hence **androgene·tic** *a.*, pertaining to androgenesis.

1916 B. D. JACKSON *Gloss. Bot. Terms* (ed. 3) 21/1 *Androgenesis*, the growth of an individual from a male cell. **1918** C. PACKARD in *Biol. Bull.* XXXV. 50 The inference is that these embryos have developed under the influence of the sperm nucleus and are therefore androgenetic. **1925** E. B. WILSON *Cell* (ed. 3) v. 464 Androgenesis. By this term we may designate the activation of the egg by the sperm followed by development without the participation of the egg nucleus. *Ibid.* 465 The nuclei of the embryo are solely of paternal origin, *i.e.*, descendants of the sperm-nucleus alone. This process is designated as 'androgenetic'.

androgyne (æ·ndrodʒəin), *a.* [f. the sb.] = ANDROGYNOUS *a.* Also *absol.*

1848 *Tait's Mag.* XV. 703/2 The planets being androgyne, like plants, *copulent avec eux-memes* and with the other planets. **1958** J. G. BENNETT *Subud* viii. 164 The force of sex..reunites the separated parts to produce the androgyne fourth gradation of the human essence. **1962** *Listener* 8 Mar. 415/2 The symbolists' fascination with the unnatural in Byzantium—in a taste for the androgyne and the perverse.

androgynous, *a.* **4.** (Earlier example.)

1760 J. LEE *Introd. Bot.* I. xxi. 64 Androgynous, Male and Female, such as upon the same Root bear both male and female Flowers.

android. Delete *rare* and add recent examples from Science Fiction. Also *attrib.*

1951 C. SIMAK *Time & Again* (1956) i. 1 Human gossip as well as android and robot gossip. **1958** *Spectator* 19 Sept. 379/1 Today SF must be more than a blood-and-sex day-dream spattered with words like androids (robots made of flesh and bone).

andromeda. 2. (Earlier and later examples.)

1760 [see ROSEMARY 3]. **1765** J. BARTRAM *Jrnl.* 30 Dec. (1769) 8 The..dwarf-myrtle, andromeda..and other evergreens. **1960** *Oxf. Bk. Wild Flowers* 118/2 Bog Rosemary or Marsh Andromeda.

Andromedid (ændrǫ·mīdid). *Astr.* [f. ANDROMEDA 1 + *-ID[2]: cf. LEONID, PERSEID.] = ANDROMED(E.

1898 *Pall Mall Gaz.* 21 Nov. 2/3 The Leonids are remarkably swift; the Andromedids strikingly slow. **1904** *Westm. Gaz.* 17 Nov. 12/1 The Andromedids form part of the débris of Biela's Comet.

andromedotoxin (ændrǫ:mīdotǫ·ksin). [a. G. (P. C. Plugge 1882, in *Pharm. Weekblad* 1 Oct.), f. ANDROMEDA 2 + TOXIN.] A poisonous crystalline substance found in various ericaceous plants, esp. the genus *Andromeda*.

1883 *Jrnl. Chem. Soc.* XLIV. 349 The poisonous principle of *Andromeda japonica* has been examined with similar results by P. C. Plugge, who calls it Andromedotoxin. **1902** *Westm. Gaz.* 16 July 2/2 A chicken fed for four days on increasing doses of the andromedotoxin from this plant [sc. *Kalmia latifolia*]. **1956** *New Gould Med. Dict.* (ed. 2) 70/1 Andromedotoxin,..has potent hypotensive action.

androsterone (æ̆ndrǫ·stĕrǫn, æ:ndrostī·-rōᵘn). *Biochem.* [f. *ANDRO- + *ster(ol* as in *CHOLESTEROL + -ONE.] A male sex hormone.

1934 *Brit. Chem. Abstr.* A. 1221/2 (*title*) Synthesis of the testicular hormone (androsterone and stereoisomerides thereof). **1935** *Ann. Reg. 1934* 53 The conversion by Ruzicka of cholesterol into the male sex hormone..gave promise that the three sex hormones (androsterone, progestin, and oestrin) will be available for clinical use in chemically pure crystalline form. **1935** *Lancet* 13 July 77/1 Solutions of the crystalline male hormone androsterone injected into rats will maintain normal prostate and seminal vesicles after adult castration.

anecdotal, *a.* Add: **b.** In extended use in *Art* (cf. *ANECDOTE *sb.* 2 c).

1933 C. H. C. BAKER *Brit. Painting* xxii. 219 Forbes' *Fish Sale*—a vivid out-of-door piece, is his least anecdotal picture. **1937** S. SITWELL *Narr. Pictures* i. 3 These [paintings] are as full of anecdotal detail as any painting by Hogarth.

anecdotalism (ænėkdōᵘ·tăliz'm). [f. ANECDOTAL *a.* + -ISM.] A propensity for telling anecdotes. Also *spec.* in *Art*.

1901 E. H. FOWLER *World & Winstow* xiii. 320 'Admiral Kingston tells a great many very prosy stories,' continued Major Trayne. 'Anecdotalism is his besetting sin.' **1928** *Sunday Express* 6 May 12/2 The Royal Academy ..is full of vitality, experiment, and imaginative creativeness. There is hardly a trace of sickly sentimentalism and sugary anecdotalism. **1961** W. H. SALTER *Zoar* iii. 28 Defects such as these..brought on them the stigma of 'anecdotalism'.

So **anecdo·talist,** a person given to or adept in telling anecdotes.

1911 *Daily News* 25 Aug. 7/1 Maupassant as well as Mr. Kipling was included by Mr. Moore among the anecdotalists. **1960** *Spectator* 12 Feb. 228 He is a superb anecdotalist, endowed with vast self-confidence and the gift of imagery ten times the size of life.

anecdotard (ænėkdōᵘ·tăɹd). *rare.* [Jocular blend of ANECDOTE and DOTARD *sb.*] A dotard given to recounting anecdotes. Cf. ANECDOTAGE 2.

1894 *World* 14 Mar. 24/2 The anecdotard..is busying himself in smoking-rooms and other centres of 'the latest intelligence'. **1937** KIPLING *Something of Myself* v. 145 Americans are too much anecdotards; the French too much orators for this light-handed game.

anecdote, *sb.* Add: **2. c.** spec. in *Art*, used of a painting, etc., that depicts a small incident.

1933 C. H. C. BAKER *Brit. Painting* xviii. 188 His [*sc.* Mulready's] *Last In*..is typical of his concessions to anecdote, at the expense of design and unity of rhythmic control. **1937** S. SITWELL *Narr. Pictures* ii. 28 Truth to life..he [*sc.* Hogarth] achieved..by every detail in his anecdote, by the aptness of every small incident that he depicted. *Ibid.* 29 The utmost detail of anecdote has been lavished upon every incident depicted.

anecdote (æ·nėkdōᵘt), *v. rare.* [f. the sb.] **a.** *intr.* To tell anecdotes. **b.** *trans.* To tell anecdotes to; to entertain with anecdotes; also, to make (a person) the subject of an anecdote or anecdotes.

1786 H. MORE *Let.* in W. Roberts *Mem. of H.M.* (1834) II. ii. 34, I left Mrs. Boscawen to anecdote with them. **1867** [see ANECDOTED *ppl. a.*]. **1900** *Academy* 28 Apr. 347/2 His wish not to be interviewed, anecdoted, or otherwise disturbed.

anecdotic, *a.* Add: **1.** (Earlier example.)

1786 'P. PINDAR' *Poet. Ep. to Boswell* 18 Amidst the anecdotic mine, Thou labour'st hard to bid thy Hero shine.

b. spec. in *Art*.

1933 C. H. C. BAKER *Brit. Painting* xxi. 208 We notice.. a steady revulsion from anecdotic 'period' pictures. The pity is that so habitual had become the yearning to tell a story that the anecdotic spirit persisted even in the modern realistic genre of the '70's.

anecdotically, *adv.* (Earlier example.)

1836 *New Monthly Mag.* XLVII. 416 They may be thrown in *anecdotically*, or conundrum-wise.

anechoic (ænekōᵘ·ik), *a.* [AN- 10 + ECHO *sb.* + -IC.] Free from echo.

1948 *Electronics* Nov. 106/1 The usual method of obtaining the sound field by the use of free-field or anechoic chambers was felt to be too expensive. **1959** *Engineering* CLXXXVIII. 65 (*title*) Anechoic chamber for noise research.

anemophily (ænīmǫ·fĭli). *Bot.* [f. Gr. ἄνεμος wind + -PHILY.] Pollination with the assistance of the wind.

1883 D'ARCY W. THOMPSON tr. *H. Müller's Fert. Flowers* IV. 591 In a few cases reversion to anemophily has taken place. **1896** G. HENSLOW *How to study Wild Flowers* 105 The flowers are purple..with the usual tufted stigma characteristic of anemophily.

anemotropism (ăne:mo,trǫ·piz'm). [f. Gr. ἄνεμος wind + TROPISM.] Tropism in response to wind.

1899 W. M. WHEELER in *Archiv f. Entwicklungsmechanik* VIII. 373 This peculiarity [in the flight of Diptera]..is an orientation of the body with respect to the wind. As it appears to be a true tropism I shall call it anemotropism. **1938** J. R. CARPENTER *Ecol. Gloss.* 21 *Anemotropism*, tropic response of organisms to wind and air currents.

anencephalic, *a.* (Later example.)

1960 *New Scientist* 4 Aug. 335/2 Some anencephalic infants do survive briefly after delivery.

anencephaly (ænėnse·făli). *Physiol.* Also (formerly) in mod.L. form **anencephalia.** [ad. mod.L. *anencephalia*; cf. ANENCEPHALOUS *a.*] The condition of being anencephalous.

1832 tr. G. BRESCHET in *Meckel's Man. Anatomy* II. 495 The following terms have been proposed..anencephalia, absence of the encephalon. **1835–36** TODD *Cycl. Anat.* I. 744/1 *False acephalia* and *anencephalia*. **1889** in *Cent. Dict.* **1954** *New Biol.* XVII. 115 Anencephaly is a fatal defect rather common in man, in which the skull and skin over the top and back of the head are missing and so is the major part of the brain. **1960** *New Scientist* 4 Aug. 335/2 Among such [congenital] defects is anencephaly, the absence of a large part of the brain.

an-end, *adv. phr.* Add: **5.** In the direction of the length; directly ahead. Chiefly *Naut.*

1601 DEACON & WALKER *Spirits & Diuels* 68 You must not be haled hedlong an end with an inueterate opinion. **1769** FALCONER *Dict. Marine* (1789), *Avoir vent de bout*, to have the wind right an-end, or a-head. **1801** ADML. HYDE PARKER *Let.* 15 Apr. in *Ann. Reg. 1801* (1802) Chron. 82/1 They were riding with two cables an end. **1867** SMYTH *Sailor's Word-bk.* s.v., To strike a spar or plank *an-end* is to drive it in the direction of its length. *Ibid.*, *Every rope an end*, the order to coil down the running rigging, or braces or bowlines, after tacking, or other evolution. Also, the order, when about to perform an evolution, to see that every rope is clear for running.

aneroidograph (ænēroi·dograf). [f. ANEROID *a.* and *sb.* + -O: see -GRAPH.] An instrument for recording the reading of an aneroid barometer (cf. BAROGRAPH).

1918 *Meteorological Gloss.* (*Met. Office*) 30 *Aneroidograph*, a self-recording aneroid. An aneroid-barometer provided with mechanism for recording the variations of pressure of the atmosphere.

aneuploid (æ·niūploid), *a. Biol.* [a. G. (G. Täckholm 1922, in *Acta Horti Berg.* VII. 234), f. AN- 10 + *EUPLOID *a.*] Not euploid. Hence as *sb.* So **a·neuploi:dy** [G. *aneuploidie* (Täckholm)], the condition of being aneuploid.

1931 S. H. YARNELL in *Genetics* XVI. 455 (*title*) A study of certain polyploid and aneuploid forms in Fragaria. *Ibid.* 464 The chromosome number falls within the polyploid series and..its appearance is that of an aneuploid type. **1932** C. D. DARLINGTON *Rec. Advances in Cytology* 494 *Aneuploid*, having an uneven multiple of the basic number of chromosomes through purely numerical aberration—therefore an unbalanced polyploid. **1934** WEBSTER Aneuploidy. **1937** *Nature* 28 Aug. 368/2 Three aneuploid species [of the grass *Poa*] have chromosome numbers suggestive of a 9*n* origin. **1939** *Ibid.* 14 Oct. 649/1 The various peculiar arrangements seen in meiosis..in all sorts of species hybrids,..aneuploids, and auto- and allo-polyploids. **1946** *Ibid.* 12 Oct. 520/1 Reduced fertility in autotetraploids depends primarily on irregular chromosome distribution in meiosis, leading to the formation of aneuploid micro- and macrospores and consequently to pollen abortion. **1956** *Ibid.* 25 Feb. 376/2 Males with average litter sizes of nine or more were not examined cytologically. This may result in the loss of small translocations which do not cause lethal aneuploidy of the zygote.

aneurin, -ine (ăniūᵊ·rin). *Biochem.* [f. A(NTI- + POLY)NEUR(ITIS + *VITAM)IN.] Vitamin B_1; = *THIAMIN(E.

1935 B. C. P. JANSEN in *Nature* 16 Feb. 267/2, I propose to call the present vitamin B_1 in the future aneurin...Dr. Donath and I were the first to obtain this vitamin in a crystalline state. **1945** *New Biol.* I. 16 Aneurine (the anti-beri-beri vitamin, B_1)..and riboflavine..are present in appreciable quantities. **1946** *Nature* 31 Aug. 306/1 A conditioned pyridoxine deficiency in rats receiving a diet of high aneurin content.

anfractuous, *a.* Add: **1.** (Later example.)

1953 C. DAY LEWIS *Ital. Visit* iv. 49 A gorge of a street, anfractuous, narrow.

2. [After F. *anfractueux* craggy.] Rugged, craggy.

1920 T. S. ELIOT *Sweeney Erect* in *Ara Vus Prec* 22 Paint me the bold anfractuous rocks Faced by the snarling and yelping seas.

3. *slang* (*jocular*). Fractious, irritable.

1932 KIPLING *Limits & Renewals* 140 If they've been hoicked out of bed, *ad hoc*, they're apt to be anfractuous.

angareb, -eeb, -ep (æ·ŋgărēⁱb). Also **-er-, -ib.** [Native name.] A stretcher or light bedstead used by the Arabs, and in Egypt and the Sudan.

1867 S. W. BAKER *Nile Trib. Abyssinia* viii. 182 The angareps, or native bedsteads,..are simple frameworks upon legs, covered with a network of raw hide. **1885** *Daily News* 3 July 5/4 Camels..heavily laden with angerebs transversely placed and resting on the flank upon a huge box. **1900** CONAN DOYLE *Green Flag* 329 He lay upon his angareeb still debating it. **1925** *Blackw. Mag.* Sept. 424/2 The harassed native Bimbashi sought his angerib.

angary (æ·ŋgări). [ad. F. *angarie*, ad. L. *angaria*: see ANGARIATE *v.*] In full *right of angary* (F. *droit d'angarie*): the right of a belligerent to use and destroy, if necessary, the property of neutrals.

1880 HALL *Internat. Law* 655 The most recent cases of the exercise of the right of angary occurred during the Franco-German War of 1870–71. **1902** *Encycl. Brit.* XXXI. 129.

angeio-, see ANGIO-.

angekok (æ·ŋgėkǫk). [Eskimo.] An Eskimo sorcerer or medicine-man.

1767 J. GAMBOLD tr. *Crantz's Hist. Greenland* I. 206 When a dearth befalls them [*sc.* Greenlanders] at sea, an Angekok must undertake a journey thither. **1819** *Edin. Rev.* XXXI. 345 Their conjurors or angekoks are merely a superior order of jugglers. **1837** [see CONJURE *v.* 5 a]. **1842** MOFFAT *Mission. Labours S. Africa* 305 The angekoks of the Greenlanders..and the greegrees of Western Africa. **1895** KIPLING *2nd Jungle Bk.* 147 The *angekok*, the sorcerer, frightened them into the most delightful fits.

angel, *sb.* Add: **A. 7.** *angels on horseback* (see quot. 1900).

1888 MRS. BEETON *Bk. Househ. Managem.* § 2078 Angels on Horseback. (Fr. — Anges à Cheval). **1900** 'SARAH GRAND' *Babs* (1901) xv, Angels on Horseback, now—those delicious little morsels of oysters rolled in bacon, and served on crisp toast.

8. *slang.* A financial backer of an enterprise, esp. one who supports a theatrical production. orig. *U.S.*

1891 J. MAITLAND *Amer. Slang Dict.* 18 *Angel*..One who possesses the means and inclination to 'stand treat'. **1900** G. ADE *More Fables* 190 There was no more Capital coming from the Angels. **1921** WODEHOUSE *Jill the Reckless* xi. 161 Ike hasn't any of his own money in the thing...The angel is the long fellow you see jumping around. **1948** *Times Lit. Suppl.* 17 Apr. 218/5 In the United States there is more money, more paper..the magazines are fatter, the 'angels' are richer. **1953** *Economist* 28 Mar. 853/1 That increasingly rare being, the 'angel' who will risk his money in a theatrical venture.

9. *R.A.F. slang.* Height; *spec.* a height of 1,000 feet. Usu. in *pl.*

1943 P. BRENNAN et al. *Spitfires over Malta* i. 18 'Gain your angels quickly...' We continued to climb. *Ibid.* iii. 71 We climbed into sun, Woody advising us to get as much angels as possible. **1943** HUNT & PRINGLE *Service Slang* 11 '20 M.E.s at Angels one owe' means '20 Messerschmitts at 10,000 ft.' **1946** J. IRVING *Royal Navalese* 23 *Angels*, a measurement of one thousand feet in height, used in the air combat code.

10. An 'unexplained' mark on a radar screen.

1947 W. B. GOULD in *Proc. Inst. Radio Engineers* XXXV. 1105/1 Radar equipment..has given fairly consistent unexplainable echoes at altitudes between approximately 300 and 3000 yards. For want of a better term, these echoes have been dubbed 'Angels' by Signal Corps personnel. **1958** *Listener* 30 Oct. 691/1 New radar sets of much higher power, on which the display was sometimes covered with small echoes, called 'angels' by the operators. *Ibid.*, A Swiss biologist, working with British radar equipment at Zurich airport, proved that 'angels' were the echoes from small birds on migration. **1962** *New Scientist* 12 Apr. 23/3 A substantial proportion of angel echoes are indeed attributable to birds.

B. 1. Selected additional examples illustrating mod. poet. usage. **a.** *angel-vampire, -warder.* **c.** *angel-guarded* (example), *-infested,*

-loosened. **d.** *angel-wise* (example). **e.** *angel-evening, -event.*

1929 E. BLUNDEN *Near & Far* 41 The tender amaranthine domes Of angel-evenings. *Ibid.* 49 The bright Angel-event of sunset's fresh creation. **1917** D. H. LAWRENCE *Look! We have come Through!* 101 But we storm the angel-guarded Gates of the long-discarded Garden. **1940** G. BARKER *Lament & Triumph* 31 The Seven Seas with their angel-infested crests. **1938** W. DE LA MARE *Memory* 26 The noon sun's angel-loosened archery. **1936** W. H. AUDEN *Look, Stranger!* 41 The white angel-vampires flit. **1879** G. M. HOPKINS *Poems* (1918) 44 Frowning and forefending angel-warder. **1918** W. DE LA MARE *Motley* 54 The Shape who hoofs applause..Hoots—angel-wise—'the Cause!'

2. angel-cake (orig. *U.S.*), a variety of sponge-cake; **angel-face,** used, esp. as a term of address and freq. ironically, for a person with an 'angelic' or innocent face (cf. quot. 1833 in Dict., sense B. 1; see also quot. 1925); **angel('s)-food** (**-cake**), (orig. *U.S.*), angel-cake; **angel-kind** (later example); **angel skin** [tr. F. *peau d'ange*], a fabric with a smooth waxy face; **angel sleeve,** a long loose sleeve.

1886 *Good Housek.* (N.Y.) 10 July 127/2, I always use the pan sold as an 'angel cake pan'. **1904** *N.Y. Times* 13 June 8 To have angel cake would be sacrilegious. **1905** *N.Y. Even. Post* 4 Aug. 7 Angel cake, sponge cake, and ice-cream cake have conspired to relegate the seed cake to practical oblivion. **1909** J. MASEFIELD *Tragedy of Nan* II. p. 28 None but angel-cakes 'd be fit eating for you, Miss Nan. **1956** 'N. SHUTE' *Beyond Black Stump* vii. 201 He would usually take with him an enormous peach pie or an angel cake. **1913** WODEHOUSE *Little Nugget* I. i. 14 'Ogden, darling..stay by me, angel-face.' 'Oh, shush!' muttered angel-face. **1925** FRASER & GIBBONS *Soldier & Sailor Words* 6 *Angel face*, Air Force slang for any young, or boyish looking, Probationary Flight Officer. **1932** E. WALLACE *When Gangs Came* xxi. 181 I've finished talking about it, angel-face. **1881** MRS. OWENS *Cook Bk.* 161 Angel's food. In other words, White Sponge Cake. **1920** SINCLAIR LEWIS *Main St.* vii. 88 They distributed..stuffed olives, potato salad, and angel's-food cake. **1951** *Good Housek. Home Encycl.* 335/1 *Angel Cake* (*Angel Food Cake*), an extremely light, feathery cake of the sponge type. **1874** R. MORRIS tr. *Blickling Homilies* vii. 92 All angel-kind shall look through the aperture on mankind. **1935** *Times* 13 Nov. 15/5 The short jacket is of white angel skin with blue sleeves. **1951** F. STARK *Beyond Euphrates* 186 A white evening gown of a satin they call angel-skin. **1862** CHESNUT *Diary* (1905) 204 She saw them coming in angel sleeves, displaying all their white arms. **1887** E. B. CUSTER *Tenting on Plains* (1889) v. 174 The sting was inflicted..in the far back days of 'angel sleeves', which fell away from the arm to the shoulder. **1906** *Daily Chron.* 19 Sept. 5/6 Capes of fur with large 'angel' sleeves.

angel (ēⁱ·ndʒĕl), *v.* [f. the sb.] **1.** *trans.* To finance or back (an enterprise, esp. a theatrical production). *slang* (chiefly *U.S.*).

1929 M. LIEF *Hangover* 235 He's trying to get me to angel one of his plays. **1948** LAIT & MORTIMER *New York: Confid.* (1951) xxi. 194 A revue angeled by gangster dough. **1949** *Newsweek* 16 May 60/2 Last week..Aunt Anita agreed to angel a new Manhattan morning tabloid.

2. *intr.* To gain height. Cf. *ANGEL *sb.* 9. *R.A.F. slang.*

1941 *Reader's Digest* Feb. 54 The boys of Britain's R.A.F. have developed a language all their own. A fighter pilot is told to 'scramble', instead of take off; then he 'angels upward'.

angel-fish. Add to def.: Also applied to various other fish, such as *Pterophyllum scalare* and some species of the family Chætodontidæ or Ephippidæ (mainly those in N. and Central American Atlantic waters).

1873 T. GILL *Catal. Fishes East Coast N. Amer.* 24 *Chætodontidae*..angel-fish (*Berm.*) West Indies; occasional northwards. *Ibid.* 29 *Ephippidae*..Moon-fish; angel-fish (South Carolina). **1888** GOODE *Amer. Fishes* xviii. 146 The Moon-fish, *Chætodipterus faber*, is..abundant... In the northern parts of Mexico it is called 'Spade-fish'; from Florida to Charleston the Angel-fish. **1905** D. S. JORDAN *Stud. Fishes* II. xxiii. 404 The angel-fish or isabelita (Holacanthus ciliaris), orange-red, sky-blue, and golden, as though gaudily painted, is the best known species [of the family Chætodontidæ]. **1931** J. R. NORMAN *Hist. Fishes* ii. 16 The beautiful Angel-fish (*Pterophyllum*) of the rivers of South America, a familiar object in aquaria, has a very much compressed and almost circular body. **1962** *Listener* 22 Nov. 852/2 My black-and-silver lace angel fish was a most appealing wide-eyed beauty.

angelicize, *v.* Delete *rare*[-1] and add earlier and later examples. Add to def.: Also *intr.*, to act in an angelic manner.

1825 *Blackw. Mag.* XVII. 52/1 The better half..will be for ever on the wing..not merely coquetting but *angelicising* with men—floating and flying literally. **1838** *Fraser's Mag.* Feb. 148/2 His notions of gentlemanly conduct so completely *angelicised* the mode of its exhibition. **1902** *Academy* 8 Feb. 149/1 He [Thackeray] had to 'angelicise' the old gentleman at the club. **1948** W. DE LA MARE *Chardin* 24/1 His children, never belittled, prettified or angelicized.

angelology. (Earlier example.)

1846 GEO. ELIOT tr. *Strauss's Life of Jesus* I. 106 The manner in which this interpreter [*sc.* Olshausen]..treats of angelology and demonology.

angelophany. (Earlier example.)

1846 GEO. ELIOT tr. *Strauss's Life of Jesus* I. 67 The Theophany and Angelophany of the Old and New Testament.

angelship. Delete *nonce-wd.* and insert def.: The condition of being an angel; a mock title for an angel.

(Further examples.)

1802–12 [see MOUSESHIP]. **1907** J. H. MCCARTHY *Needles & Pins* i. 8 Two human beings..are transmuted from their humanity into an angelship of intimate bliss.

anger, *sb.* Add: **2.** As a literary nonce-use (quot. 1937). Later with overt or implicit reference to J. Osborne's play *Look Back in Anger* (first performed 1956). Cf. *ANGRY *a.* 3 c.

1937 H. G. WELLS *Brynhild* vii. 100 It brings my Anger back. I am an Angry Man...Almost professionally. You don't know my books? **1957** J. HOLLOWAY in *Hudson Rev.* Autumn 424 Notes on the 'School of Anger'. *Ibid.* 426 The anger of Sassoon's war poem was not, of course, the kind..of Amis, Wain, Osborne, etc. **1958** *Spectator* 4 July 17/3 Students here have become sceptical about protest campaigns, petitions and the Anger Cult.

Angevin (æ·ndʒėvin), *a.* and *sb.* [a. F. *Angevin* :—med.L. *Andegavin-us*, f. *Andegavum* Angers, capital of Anjou (*Andegavia*).]. **A.** *adj.* Of or pertaining to Anjou, a former province of France, or any sovereign, government, etc., derived thence; *spec.* in *Eng. Hist.*, belonging to or characteristic of the Plantagenet kings (beginning with Henry II) descended from Geoffrey, count of Anjou, and Matilda, daughter of Henry I; pertaining or relating to their descendants, the period of history which they cover, etc. **B.** *sb.* A native of Anjou; an Angevin ruler.

1653 T. URQUHART *Pantagruel* II. xxvi. 170 One hundred and fifty thousand whores..whereof some are Amazons, some Lionnoises,..Angevines, [etc.]. **1737** MRS. THOMSON tr. *M. Baudier's Hist. Marg. Anjou* 4 Henry..complained of the Angevin's Escape. **1769** G. LYTTELTON *Hist. Hen. II* (ed. 3) I. 162 The Angevin family had been long very powerful and illustrious. **1834** *Penny Cycl.* II. 35/1 The Angevin prince invaded Italy with an army of 30,000 men. **1874** GREEN *Short Hist.* ii. § 7. 94 To understand the history of England under its Angevin rulers, we must first know something of the Angevins themselves. **1887** KATE NORGATE *Eng. under Angevin Kings* I. 3 Those wonderful Angevin counts who..grew into a sovereign house. **1887** *Encycl. Brit.* XXII. 27/2 The Angevin conquest of Sicily. **1924** J. S. C. BRIDGE *Hist. France* II. 173 Charles had vindicated his Angevin claims. **1935** T. S. ELIOT *Murder in Cath.* I. 32 Let the Angevin Destroy himself, fighting in Anjou.

angico (andʒī·ko). [Pg.] A Brazilian name applied to the gum, etc., of the tropical S. American tree *Piptadenia rigida*, which yields a hard durable timber, and an astringent bark and a gum both used medicinally.

1867 SIMMONDS *Dict. Trade* Suppl., *Angico Gum*. **1897** WILLIS *Flowering Pl.* II. 298 P[iptadenia] rigida..yields Angico gum used like gum-arabic.

anginal (æ·ndʒinăl, ændʒəi·năl), *a.* [f. ANGINA + -AL.] Pertaining to angina, i.e. quinsy, or *angina pectoris*; = ANGINOUS. Also **anginic** (ændʒi·nik) *a.* So **angi·niform, a·nginoid** *adjs.*, resembling *angina pectoris*.

1811 LETTSOM in Pettigrew *Life* (1817) III. 4 The anginal stridula may make its onset without that exudation in the throat. **1882** J. D. HOOKER in L. Huxley *Life* (1918) II. 260, I have had a ten days' bout of my Anginic pains. **1889** *Cent. Dict.*, Anginoid. **1891** *Lancet* 14 Feb. 371/2 A series of anginoid attacks. *Ibid.* 371/1 Such cases might be termed 'anginous' or 'anginal'. **1906** *Westm. Gaz.* 27 Sept. 4/2 Tea..gave me an awful pain, almost anginiform. **1910** W. JAMES *Let.* 25 May (1920) II. 337, I have a dilatation of the aorta, which causes anginoid pain of a bad kind whenever I make any exertion.

angiocardiography (æ:ndʒi,o,kăɹdiǫ·grăfi). [f. ANGIO- + CARDIOGRAPHY.] X-ray examination of the thoracic vessels and the heart after the intravenous injection of a substance opaque to X-rays. Hence **a:ngioca·rdiogram,** a cardiogram taken during and immediately after such an injection.

1938 *Lancet* 24 Dec. 1476/2 Method of radiography of the heart and great vessels after the intravenous injection of opaque fluid. Angiocardiography is mainly applicable to congenital malformations in children, and can be used after death where a full post-mortem examination is refused. **1948** *Brit. Jrnl. Radiology* XXI. 381/1 The normal angiocardiogram. After injection of 20 c.c. of contrast agent local venous pressure is rapidly raised. **1951** *Brit. Encycl. Med. Practice* VI. 235 The size and shape of the various chambers of the heart and the great vessels arising from it may be studied during life by means of angiocardiography. A radio-opaque substance is injected rapidly intravenously and X-ray films are taken at frequent intervals to enable the course of the dye to be followed through the heart chambers and lungs.

angiocarpous, *a.* Add to def.: Also, having the apothecium enclosed within the thallus, as certain lichens.

1851 W. A. LEIGHTON (*title*) The British Species of Angiocarpous Lichens. [So also quot. 1875 in Dict.]

angiogram (æ·ndʒi,ogræm). [f. ANGIO- + -GRAM.] The tracing made by an angiograph.

1933 A. EGAS MONIZ in *Lancet* 18 Nov. 1147/1 The speed of circulation is very different in the brain as compared with other parts of the head...We have ascertained this fact by studying the arteriograms and phlebograms of diverse cases, but confirmation was obtained by a series of angiograms, taken from second to second, after the injection of thorotrast in the common carotid and the external and internal carotids.

angiography. 2. Add to def.: Recently extended to denote the examination of the blood vessels by radiography. Hence **angiogra·phic** *a.*, of or pertaining to angiography.

[**1933** EGAS MONIZ & ALVES in *Rev. Neurologique* XL. I. 375 Dans le domaine radiologue nous obtenons non seulement l'artériographie du cerveau mais aussi les angiographies du passage du thorotrast par les capillaires et par les veines du cerveau.] **1933** *Lancet* 18 Nov. 1157/2 By taking a series of radiograms at intervals of one second, it has been found possible to display the cerebral veins and sinuses as well as arteries, and Moniz has therefore changed the name of the procedure from arteriography to angiography. **1950** P. ALMEIDA LIMA *Cerebral Angiography* vii. 156 The localization of aneurysms in the arterial region is almost always clear and well defined in the angiographic images. *Ibid.* x. 212 The angiographic method lends itself to application in physiological studies of the cerebral circulation. **1955** *Gloss. Terms Radiology* (*B.S.I.*) 56 *Angiography*, the radiological examination of blood vessels following direct injection of a contrast medium. **1956** *Acta Radiologica* XLVI. 480 (*title*) Angiographic Appearances in Transtentorial Herniation of the Brain.

angioma (ændʒi,ōᵘ·mă). *Path.* Pl. **-ata.** [mod.L., f. Gr. ἀγγεῖον vessel + *-OMA.] A tumour produced by dilatation or new formation of blood-vessels. Hence **angio·matous** *a.*

1871 T. H. GREEN *Introd. Path.* xxiv. 186 The angiomata, or vascular tumours, are tumours consisting of blood-vessels held together by..connective tissue. **1897** *Trans. Amer. Pediatric Soc.* IX. 154 Angiomatous tumours. **1962** *Lancet* 1 Dec. 1136/2 A man of 53 had an angioma of the spinal cord and a neurogenic bladder.

angioneurosis (æ:ndʒi,o,niurōᵘ·sis). [ad. G. *angioneurose* (H. Quincke 1882, in *Monatshefte f. Prakt. Dermatologie* I. 130), f. ANGIO- + NEUROSIS.] Neurosis of the blood vessels. So **a:ngioneuro·tic** *a.*, characterized by angioneurosis.

1887 *Lond. Med. Rec.* 33/1 (*heading*) Strübing on Acute (Angioneurotic) Œdema. **1889** *Cent. Dict.*, Angioneurosis. **1908** *Practitioner* LXXX. 824 Colcott Fox, Quincke, and Walker put urticaria as an angioneurosis, while Pepper.. calls angioneurotic œdema, 'giant urticaria', with symptoms of heat, swelling, itching. **1964** S. DUKE-ELDER *Parsons' Dis. Eye* (ed. 14) xxxi. 490 An intermittent and acute œdematous condition due to unstable vasomotor reactions, frequently determined on an allergic basis, is relatively common in the lids (angioneurotic œdema).

angle, *sb.*[2] Add: **1.** ¶ *angle of approach, attack, entry, incidence* (earlier and later examples), *trail.*

1626 etc. [see INCIDENCE 4]. **1908** *Westm. Gaz.* 30 May 7/3 The angle at which the plane is inclined to the direction of motion, the technical term for that angle being the angle of attack. **1910** R. FERRIS *How it Flies* xx. 454 *Angle of Entry*, the angle made by the tangent to the curve of the aeroplane surface at its forward edge, with the direction, or line, of travel. *Ibid.*, *Angle of Incidence*, the angle made by the chord of the arc of a curved 'plane', or by the line of a flat plane, with the line of travel. *Ibid.*, *Angle of Trail*, the angle made by the tangent to the rear edge of a curved plane with the line of travel. **1932** H. NICOLSON *Public Faces* II. 29 Even the angles were so difficult: the angles of approach: the actual moment of initiation: the excruciating angle of continuance. **1948** *Jrnl. R. Aeronaut. Soc.* LII. 441/2 Approaches at a constant angle could be made by means of an 'angle of approach light', which was a light having three coloured sectors arranged one above the other, so that the pilot saw a red light if the aircraft was too low, a green light if the aircraft was approaching at the correct angle, and a yellow light if the aircraft was too high. **1962** *Gloss. Aeronaut. Terms* (*B.S.I.*) IV. 1 *Angle of incidence, angle of attack*, the angle between the chord of a wing or the reference line in a body and the direction of the undisturbed flow in the absence of sideslip. *Ibid.* v. 4 *Rigging angle of incidence*, the angle between the chord of the main or tail plane and the horizontal when the aeroplane is in the rigging position.

c. The point or direction from which one views or approaches an object, circumstance, event, subject of inquiry, etc.; standpoint; hence (loosely) = ASPECT 9; (*slang*) scheme, illegal method. Freq. with defining word. Cf. *SLANT *sb.*[1]

1872 GEO. ELIOT *Middlem.* III. xxiii. 13 Tacit expectations of what would be done for him by Uncle Featherstone determined the angle at which most people viewed Fred Vincy in Middlemarch. **1922** H. WALPOLE in S. Lewis *Babbitt* p. viii, Mr. Lewis turns the figure round and allows us to view it from every possible angle. **1928** A. WAUGH *Nor Many Waters* (1930) vi. 241 The account of the cricket from an Australian angle. **1935** *Proc. Prehist. Soc.* I. 10 The study and appreciation of a culture from this angle imposes fresh obligations upon the archaeologist. **1936** *Punch* 14 Oct. 430/1 Let us approach the subject from a mathematical or statistical angle. **1942** *World Rev.* May 6/1 The right type of advertising man might be better as an adviser on the

propaganda angle. **1942** A. CHRISTIE *Body in Library* xviii. 157 The selection of the poor child, Pamela, the approach to her from the film angle. **1944** 'N. SHUTE' *Pastoral* v. 116 The old stagers..the men who knew all the angles, who had great experience. **1958** S. ELLIN *Eighth Circle* (1959) II. ix. 103 His angle was to sell off tapes and pictures to the dirt magazines. **1958** *Spectator* 22 Aug. 259/3 This collection of essays..feels obliged to use such a selling angle to persuade America that a favourable attitude to colonialism can be held without giving hostages to Marxist criticism.

d. *Squash Rackets.* In full *angle shot*: a shot struck at the side wall so that it rebounds directly to the front wall without touching the floor.

1926 C. ARNOLD *Squash Racquets* ii. 12 The angle shot can be made from either court, forehand or backhand. **1933** *Times* 4 Dec. 6/6 In the fourth game Elsmie set an even faster pace by means of cleverly played angle shots. **1934** *Times* 13 Jan. 4/5 He played the angles cleverly and showed a delicate touch when using the drop.

e. *Photogr.* In full *camera angle*: the direction or viewpoint from which a photograph is taken. So *angle shot*, esp. in *Cinemat.*, a shot taken with the camera at an angle to the horizontal.

1928 *Amer. Speech* III. 352 Directors are obsessed with camera angles. **1937** *Mod. Encycl. Photogr.* I. 70/2 The angle shot is now extensively employed in order to give emphasis and dramatic force. **1959** HALAS & MANVELL *Technique Film Animation* 336 For variety of continuity a different angle is often used in long shots, medium shots and close-ups of the same subject.

8. b. *angle-tower.* Also **angle-brick** (see quot. *a* 1884); **angledozer, angle-dozer** (æ·ŋg'l,dōᵘ·zəɹ) [cf. *BULLDOZER], a type of bulldozer on which the blade can be set at an angle; **anglepoise, angle-poise** [Proprietary term] *a.*, designating a type of swivelled reading-lamp; also *ellipt.*; **angle shades** [cf. SHADE *sb.* 3 c], collectors' name for a species of moth, *Phlogophora meticulosa*.

a **1884** KNIGHT *Dict. Mech.* Suppl. 36/2 *Angle-brick*, bricks specially formed to enable other than square angles to be turned. **1937** *Discovery* Sept. 289/2 Angle-bricks and half-sized bricks have been found. **1940** *Gloss. Highway Engin. Terms* (*B.S.I.*) 54 *Bulldozer, Angledozer*, a power operated machine of the grader type employed for spreading and levelling by pushing loose excavated material. **1942** *Times* 9 Oct. 2/2 There are..machines for levelling—motor propelled scrapers—tractors, dumpers, angle-dozers and bulldozers. **1958** J. S. SCOTT *Dict. Civ. Engin.* 8 *Angle dozer*, a bulldozer with the mouldboard set at an angle so that it pushes earth sideways instead of straight ahead. **1940** GRAVES & HODGE *Long Week-end* xx. 350 Angle-poise reading-lamps that would swing and bend in any desired direction. **1949** *Trade Marks Jrnl.* 22 June 548/2 Angle-poise..Electric lamps embodying adjustable brackets or supports. Herbert Terry & Sons Limited..Redditch; Manufacturers—5th February, 1947. **1961** I. MURDOCH *Severed Head* vi. 52 He switched off the centre lights and turned on a single anglepoise lamp on the work table which he swung round towards the pedestal. *Ibid.* 56 Alexander..had turned the anglepoise back to shine upon his unfinished head. **1843** J. DUNCAN in W. Jardine *Naturalist's Libr.* XL. 235 (*heading*) The Angle Shades. *Phlogophora Meticulosa*..Angle-Shades Moth. **1938** C. A. HALL *Pocket-Bk. Brit. Butterflies* 86 The Angle-Shades..is on the wing in June and again in the autumn. **1961** R. SOUTH *Moths Brit. Isles* I. 294 (*heading*) The Small Angle Shades...The pale reniform mark on the outer edge of the blackish central area is the prominent feature of this pinkish or purplish brown moth. *Ibid.* 295 The Angle Shades... In this position [*sc.* with the wings folded] the moth is very like a crumpled decaying leaf. **1911** T. E. LAWRENCE *Let.* 8 June (1938) 110 No one knows how the outwork at Gaillard was breached: I fancy..that the angle tower was mined over the filled up moat. **1958** *Listener* 9 Oct. 558/1 Telephone wires, aerials, angle towers, and sub-stations.

angle, *v.*[2] Add: **2.** Also *intr.* = *refl.*

1863 W. H. GOODE *Outp. Zion* II. xvii. 381 Thence angling across the country, we..entered the Lane road. **1883** *Century Mag.* Oct. 923/2 He..once more alters his direction, and so twists off, 'angling' across the meadow. **1953** A. UPFIELD *Murder must Wait* xx. 177 She wondered why he didn't walk direct to the..boulders, why he angled this way and that.

3. *intr.* To turn or move at an angle, diagonally, or obliquely; to lie in an oblique direction.

1741 *Boston Rec.* 274 They have computed the charge of fixing piers..set angling cross the channel. **1835** *Fraser's Mag.* XI. 39 The circuitous route you are obliged to take—angling off at an infinite variety of points. **1869** *Rep. U.S. Comm. Agric. 1868* 258 About half the saplings may be laid along outside one stake, then inside of the next, and thence angling across to the other row. **1881** MAYNE REID *Free Lances* II. xxxii. 116 [The road] angles abruptly to the right. **1897** *Times* 5 Feb. 14/5 The Majestic..broke her starboard quarter rope, which caused her to angle across the entrance and to become jammed. **1942** R. CHANDLER *High Window* (1943) xxix. 189 Two davenports angled across the corners of the room and there was one gold chair.

b. *trans.* To strike, put, or drive at an angle.

1920 *Westm. Gaz.* 16 Oct. 2/2 She attracted the returns to her by cleverly angling the ball. **1953** A. UPFIELD *Murder must Wait* xxv. 222 Bony had walked like a white man, angling his feet at twenty-five minutes to five.

4. *trans.* To present (a story, description, subject of inquiry, etc.), esp. in journalism, so as to suit a particular point of view. Cf. *ANGLE *sb.*[2] 1 c.

1937 *Harper's Mag.* Dec. 54/2 The good communist editor to-day turns a news story..over to his copy-writer and says: 'Class-angle that, Jim.' **1951** M. DICKENS *My Turn to make Tea* ii. 13 You..almost never see the proprietor, although you feel his presence, because you have to angle your writing his way. **1959** C. MACINNES *Absol. Beginners* 132 The leader columns are angled at the more intelligent portions of the population.

angleberry. (Later examples.)

1844 H. STEPHENS *Bk. Farm* II. 156 Warts and angleberries are not uncommon excrescences upon cattle. **1914** P. MACGILL *Children of Dead End* vi. 32 One of the horses was spavined and the other was covered with angle-berries.

angled, *ppl. a.* Add: **2.** *angled deck*, a flight deck on an aircraft carrier on which the landing path is inclined to the ship's fore-and-aft axis.

1952 *Flight* 21 Nov. 636/2 The angled deck has the effect of increasing the operational length of the carrier by 40 per cent... Preliminary trials..have been conducted by the Royal Navy and the U.S. Navy—where it is known somewhat misleadingly as the 'canted deck'. **1955** *Ann. Reg. 1954* 12 Three aircraft carriers with angled flight decks would be completed..within the year.

5. Of a story, description, question, etc.: presented so as to suit a particular point of view; distorted, biased (see *ANGLE *v.*[2] 4).

1958 *Manchester Guardian* 29 Jan. 6/5 'Pravda' carried 370 words on Mr. Macmillan's reply... It contained some direct quotation but in the main it was angled comment. **1958** *Times* 10 Oct. 8/3 A great deal of unwholesome, often erotically angled, American rubbish. **1960** J. FINGLETON *4 Chukkas* i. 1 The journalist..has to look for what is known as the 'angled' story.

angler[1]**. 1. b.** Read: *fig.*; *spec.* in *Thieves' Cant*: see quot. 1673. *Obs.* Also in extended use (see quot. 1903) in U.S.

1567 T. HARMAN *Caveat* 16 These hokers or Angglears be peryllous and most wicked knaues. **1612** DEKKER *O per se O* sig. N4v, Hee is an Angler for Duds, who hath a Ferme in the Nab of his Filch. **1673** R. HEAD *Canting Academy* 68 Anglers are so called, because they have a Rod or Stick with an Iron hook at the end of it, with which they Angle in the night at Windows or any other place in which they cannot convey their hand. **1823** 'JON BEE' *Dict. Slang* 4 *Anglers*, thieves who with a hook at the end of a mop-stick drag to them the ends of cloth which may lie exposed, and so pull out entire pieces. **1930** CLAPIN *New Dict. Americanisms* 19 *Angler*, in thieves' slang, a street prowler, generally belonging to a gang of petty thieves, and who is always on the lookout for opportunities to commit small larcenies.

2. Add to def.: In full *angler-fish*. Also applied to other pediculate fishes having a free dorsal spine, as in the family Antennariidæ.

1653 [see SEA-ANGLER]. **1845** [see TOAD-FISH]. **1884** GOODE *Nat. Hist. Aquatic Anim.* 173 Marbled Angler, *Pterophryne histrio*. **1889** *Cent. Dict.*, Angler-fish. **1902** *Westm. Gaz.* 28 Jan. 1/3 An enormous specimen of the rare and curious angler-fish (*Lophius piscatorius*) known as the 'Fishing Frog'. **1951** R. CAMPBELL *Light on dark Horse* xix. 274 The fishing frog or angler-fish.

Anglic (æ·ŋglik), *sb.* [f. ANGL(O- + -IC.] A simplified form of English spelling devised by the Swedish philologist R. E. Zachrisson (1880–1937), and intended for use as an international auxiliary language.

1930 ZACHRISSON *Anglic: A New Agreed Simplified English Spelling* 12 Anglic, which is based on an analysis of all English words in general use, supplies the demand for an agreed method of simplified spelling without adding new letters to the alphabet, and at the same time keeps the continuity with the conventional English orthography. **1935** A. LLOYD JAMES *Broadc. Word* iii. 125 The result is the system known as Anglic, which, based on the foundations laid by Professor Zachrisson, of Uppsala, is what English spelling will probably be when the English-speaking world decides that the moment has come for spelling reform. **1937** J. R. FIRTH *Tongues of Men* vi. 79 Anglic simplifies the spelling, Basic English the vocabulary.

Anglicanism. (Earlier example.)

1838 [see *ANGLO-CATHOLICISM].

A·nglicanize, *v.* [f. ANGLICAN *a.*] *trans.* **1.** To Anglicize.

1904 L. O. BRASTOW *Repres. Mod. Preachers* 50 Although in remote lineage he may have been Scotch, he had become thoroughly Anglicanized.

2. To make Anglican (in doctrine, character, etc.).

1919 N. HILL *Story Sc. Church* vii. 143 Events..were destined to interrupt for some years James' policy of seeking to strengthen his position by Anglicanising the Scottish Church. **1962** *Observer* 20 May 31/3 Many of our cathedrals, having duly 'Anglicanised' the chancel more and more through three centuries, have destroyed the *raison d'être* of the basilican plan.

‖ **Anglice** (æ·ŋglisi), *adv.* Also **anglice.** [med.L., adv. corresp. to *anglicus*: see ANGLIC *a.*] In English; in plain English.

1602 [see WINDFALL 1]. **1665** R. HEAD *Eng. Rogue* XXXIX. 139 An old Comrade..had lately *heav'd a Booth, Anglice* broken open a Shop. **1718** J. OZELL tr. *Pitton de Tournefort's Voy. Levant* I. i. 36 Here we lay at the Sign of the Moon and seven Stars (*anglicè* in the open Air). **1760** STERNE *Tr. Shandy* I. xx. 132 Baptism shall..be administered..by injection—*par le moyen d'une petite Canulle*,—Anglicè *a squirt*. **1828** H. STEUART *Planter's Guide* (ed. 2) 465 The workmen objected to the execution of three spits deep, at the same price per Fall (anglicè pole), as had been paid for two spits. **1848** H. W. HAYGARTH *Bush Life Austral.* xii. 135 A set of the veriest 'Russians' (Anglicè, wild things) *he* ever had anything to do with. **1908** 'IAN HAY' *Right Stuff* iii. 44, I handed him a quarter's salary in advance, gave him two days' holiday wherein to 'make his arrangements'—*Anglicè*, to replenish his wardrobe.

Anglicism. 2. Delete '†*Obs.*' Add to def.: English character, the quality of being English; also, imitation of or support for what is English.

1781 JOHNSON *Lives Eng. Poets* II. 86 If his language had been less idiomatical, it might have lost somewhat of its genuine Anglicism. **1810** JEFFERSON *Writ.* XII. 373 The Anglicism of 1808..is a longing for a King, and an English King rather than any other. **1818** M. BOUCHER *Let.* 1 Oct. in Lady Morgan *Memoirs* (1862) 146 Frenchmen..have adopted the opposite mania of Anglicism, down to the most trifling fashions. **1827** LAMB in W. Hone *Table Bk.* I. 302/1 Christianism; and true hearty Anglicism of feelings, shaping that Christianism; shine throughout his beautiful writings. **1831** *Blackw. Mag.* XXIX. 519 Among other allegorical emblems of nations, as the representative and express image of Anglicism, he drew a naked man. **1871** LOWELL *Study Windows* iii. 62 She [*sc.* England] has a conviction that whatever good there is in us is wholly English, when the truth is that we are worth nothing except so far as we have disinfected ourselves of Anglicism. **1876** *Dublin Rev.* July 103 Mr. Carlyle, whom nobody will suspect either of narrow Anglicism or of unreasoning antipathy to other nations.

Anglicist (æ·ŋglisist). [f. ANGLICISM: see -IST.] **1.** An advocate or favourer of Anglicism, in any sense; in quots., one who advocated the use of English in Indian schools.

1867 P. M. MEHTA *Speeches & Writings* (1905) 6 Two parties, who may be conveniently denominated as the Vernacularists and the Anglicists. **1878** G. SMITH *Life John Wilson* xvi. 529 Dr. Duff, Macaulay, and the Anglicists under Lord William Bentinck. **1892** *Athenæum* 24 Sept. 411/3 The controversy between the 'Orientalists' and the 'Anglicists', which evoked Macaulay's famous minute.

2. (See quot. 1930.)

1930 K. MALONE in *English Jrnl.* XIX. 642 In various papers I have used *Anglicist* in the technical sense 'worker in Anglistics', and this practice of mine seems to have given the term a certain currency. **1959** *Brno Studies* I. 23, *ĭ* (termed '*i impurum*' by the older generations of Anglicists).

Anglicize, *v.* **1.** (Earlier example.)

1710 SEWALL *Letter-book* (1886) I. 401 The best thing we can do for our Indians is to Anglicize them.

Anglicized, *ppl. a.* (Earlier example.)

1774 PENNANT *Tour Scotl.* (ed. 3) 69 John of Sterling, with his army of Anglicised Scots, sat down before it.

Anglist (æ·ŋglist). [a. G. *Anglist*, f. L. *Anglus* English (see ANGLE *sb.*[3]): see -IST.] A student of English, or scholar versed in English language or literature.

1888 *Jrnl. Educ.* Jan. 32 The 'Anglist' is warned that he has yet stricter duties to fulfil than the 'Germanist'. Most Anglists, it appears, in Germany conceive that they have done their duty if they have studied thoroughly Gothic, German, and English. **1930** K. MALONE in *English Jrnl.* XIX. 640 On the Continent our subject is known as *Anglistics* ..and a worker in it is called an *Anglist*.

Hence **Angli·stics** *sb. pl.*, the study of English language or literature.

1930 [see *ANGLICIST]. **1936** *Amer. Speech* XI. 101/1 *ELH*, the only American journal devoted exclusively to *anglistics*. **1947** *Eng. Studies* XXVIII. 129 English studies or Anglistics (the usual Continental term) is a discipline which deals with the speech and writings of the whole Anglo-Saxon world.

Anglo-. Add: **1. a. Anglo-Jewish** (example); so *Anglo-Jewry*; **Anglo-Norse** *a.*, pertaining to the Norse in England during the Anglo-Saxon period, esp. to their language; also as *sb.*; **Anglo-Scandinavian** *a.*, pertaining to England and Scandinavia, esp. to the Norsemen in England in the Anglo-Saxon period; also as *sb.*

1871 *Times* 7 July 5/2 An influential meeting..for the purpose of constituting the Anglo-Jewish Association of the Universal Israelitish Alliance. **1910** *Westm. Gaz.* 9 Apr. 5/1 It may..be accepted that only books worthy of preservation are paid this compliment in Anglo-Jewry. **1961** *New Left Rev.* Jan.–Feb. 60/2 An official Anglo-Jewry, represented by such institutions as the Chief Rabbi. **1927** E. V. GORDON *Introd. O. Norse* 306 The change of Norse *w* to bilabial *v* had not taken place in Anglo-Norse. *Ibid.* 304 The Anglo-Norse form borrowed in English..shows no trace of the fronting of *g*. **1936** P. THORSON (*title*) Anglo-Norse Studies. An Inquiry into the Scandinavian Elements in the Modern English Dialects. **1908** W. G. COLLINGWOOD *Scandinavian Britain* ii. 167 Edward's reign was disturbed throughout by a struggle between the Anglo-Scandinavians and the Franco-Scandinavians. **1924** MAWER & STENTON *Introd. Surv. Eng. Place-Names* ix. 183 The time has therefore not yet come for any general estimate of the character of Anglo-Scandinavian personal nomenclature in the phases which preceded its final disappearance.

2. *Anglo-Boer, -Soviet.*

1900 *Westm. Gaz.* 4 May 11/3 A large proportion of the money hoarded up, as a consequence of the Anglo-Boer War, will be set into circulation so soon as peace shall be declared. **1961** *Times* 18 May 18/4 Events that led up to the Anglo-Boer war. **1924** *Glasgow Herald* 12 July 9/3 If an agreement which is at present being considered by the Anglo-Soviet Conference is finally approved it will have the effect of benefiting the Soviet Government. **1939** E. H. CARR *Propaganda in Internat. Politics* 19 The Anglo-Soviet Trade Agreement of March 1921. **1943** *Lancet* 26 June 818/2 The Anglo-Soviet Medical Council.

A:nglo-Ame·rican, *sb.* and *a.* [f. ANGLO- + AMERICAN.] **A.** *sb.* **a.** An American of English origin.

[**1738** *Remarks on Trial of J. P. Zenger* i. 16 (*Signature of letter*), I am yours, &c. Anglo-Americanus.] **1787** S. S. SMITH *Ess. Complexion* (1788) 194 The Anglo-Americans on the frontiers of the states, who acquire their sustenance principally by hunting. **1792** J. MORSE *Amer. Geogr.* (ed. 2) 63 The greater part, however, are descended from the English; and for the sake of distinction, are called Anglo-Americans. **1856** SIMMS *Charlemont* 19 That sleepless discontent of temper, which, perhaps,..is the moral failing in the character of the Anglo-American. **1948** *Seventeen* June 4/4 We had a Chinese boy cheerleader, a Mexican girl cheerleader, and three Anglo-Americans.

b. In contrast to the non-English races in, or on the borders of, the United States.

1828 C. SEALSFIELD *Americans as they Are* xi. 132 The Anglo-Americans, however, treat their slaves throughout better than the French. **1842** *McDonogh Papers* (1898) 64 The Anglo-Americans, a term by which all Americans and strangers generally are called by the natives of Louisiana of French descent. **1858** *Texas Almanac* 114 The Anglo-Americans of Texas were threatened with subjection to military despotism.

B. *adj.* **a.** Of or pertaining to Americans of English origin.

1797 PINCKNEY in *State P.* (1819) III. 140, I should have made some observation on being termed Anglo-American, but, on inquiry, I found it was customary to call all my countrymen so, to distinguish us from the inhabitants of St. Domingo, and the other French West India Islands. **1841** H. S. FOOTE *Texas & Texans* I. 91 The heroic achievements of these three hundred Anglo-American soldiers. **1851** C. CIST *Cincinnati* 34 Destined to render the Anglo-American race paramount throughout this great continent. **1880** G. W. CABLE *Grandissimes* liii. 402 A comparison of.. Anglo-American and Franco-American conventionalities.

b. Of or belonging to both England and America.

1812 H. MARSHALL *Hist. Kentucky* 81 The Anglo-American army, then on its march to attack the French and Indians, posted in Fort Duquesne. **1843** *Knickerbocker* XXII. 90 The 'American Book Circular' recently put forth by Mr. Geo. P. Putnam, of the Anglo-American House of Wiley and Putnam. **1898** W. JAMES *Coll. Ess. & Rev.* (1920) 436 The great Anglo-American alliance against the world, of which we nowadays hear so much. **1933** *Discovery* Dec. 359/1 An Anglo-American agreement has now been reached regarding a transatlantic air service.

Hence **A:nglo-Ame·ricanism.**

1803 *Christian Observer* II. Sept. 554 Occasional vulgarisms (possibly Anglo-Americanisms). **1841** H. S. FOOTE *Texas & Texans* I. 110 In the land [*sc.* England] whence what we now call Anglo-Americanism has derived its origin.

Anglo-Catholic. Add: **A.** *adj.* (Earlier examples.)

1838 C. SEAGER (*title*) The Daily Service of the Anglo-Catholic Church, adapted to family or private worship. **1838** C. M. LE BAS in *Brit. Crit.* XXIV. 83 Both of them appeal to the Homilies and Formularies of the Anglo-Catholic Church. **1838** W. PALMER *Treatise on Church of Christ* I. p. vi, Many of the ancient errors against which the masters of Anglo-catholic theology contended in the sixteenth and seventeenth centuries, have been permitted to sink into oblivion.

B. *sb.* **2.** (Earlier example.)

1842 P. GELL *Sermon Preached at Visitation of Archdeacon of Derby* 33 The Anglo-Catholics consider it essential to be ordained by bishops receiving their appointment in regular succession from the apostles.

Anglo-Catholicism. (Earlier example.)

1838 J. H. NEWMAN in *Brit. Crit.* XXIV. 61 The heroine ..after going through the phases of Protestantism..seeks for something deeper and truer in Anglicanism, or, as Mr. Palmer more correctly speaks in his recent work, Anglo-Catholicism.

Anglocentric (æ:ŋglo͵se·ntrik), *a.* [f. ANGLO- + CENTRIC *a.*] Centring on England; centred in England. Hence **A:nglocentri·city.**

1886 *Ann. Rep. U.S. Treas.* p. ix, An Anglo-centric monetary system. **1898** L. A. TOLLEMACHE *Talks with Gladstone* 114 A belief in the anthropomorphic and anthropocentric—I had almost said Anglo-centric—government of the physical world. **1960** *Times Lit. Suppl.* 7 Oct. 642/2 The excuses for such Anglocentric history are now wearing thin. *Ibid.* 642/4 A similar Anglocentricity of outlook..colours Mr. Watson's treatment..of the French Revolution. **1962** *Economist* 26 May 776/2 The Commonwealth has moved further and further away from being Anglocentric.

A:nglo-Fri·sian, *sb.* and *a.* Also 9 **Anglo-Friesic.** [f. ANGLO- + FRISIAN *a.* and *sb.*] **A.** *adj.* **1.** Pertaining to both English and Frisian. **2.** In the use of H. Sweet and some later writers: pertaining to a subdivision of the West Germanic branch of the Germanic group of languages, the hypothetical parent language of Anglo-Saxon and Old Frisian. **B.** *sb.* In senses of the adj.

1836 HALBERTSMA in J. Bosworth *Orig. Eng., Gmc. & Scand. Lang.* 75 Low-Saxon has all the appearance of German grafted on an Anglo-Friesic tree. The words are Anglo-Friesic with German vowels, as if the Friesians, in adopting the German, retained the consonants of the old language. **1877** H. SWEET in *Trans. Philol. Soc. 1875–6* 562 The language spoken by these tribes before the migration may be called Anglo-Frisian, and its characteristics may be ascertained with considerable certainty from a comparison of the oldest English and Frisian. **1888** —— *Hist. Eng. Sounds* 85 Within Low German English and Frisian again form a special group 'Anglo-Frisian'. **1907** H. M. CHADWICK *Orig. Eng. Nation* 99 But have we any justification for believing that a language of Anglo-Frisian type was spoken beyond the sea to the north? **1913** P. SIPMA *Phonol. & Gram. Mod. W. Frisian* 2 From time immemorial English and Frisian have had in common a certain number of peculiarities in their system of vowels and consonants: these must have been proper to the original Anglo-Frisian language. **1948** *Trans. Philol. Soc. 1947* 14 This view [in *T.P.S. 1939*]..attached over much weight to literary Old Saxon, a language highly divergent from Old Frisian, and assumed that we must postulate a stage in which there was an Anglo-Frisian unity rather closer than the loose unity of Frisian, English, and Saxon, i.e. non-High German West Germanic. **1958** A. S. C. ROSS *Etymology* ii. 98 In Anglo-Frisian, initial and medial *k* before a front vowel > [ts] [tš].

A:nglo-Ga·llic, *a.* [f. ANGLO- + GALLIC *a.*[1]] Pertaining, relating, or common to both England and France; *spec.* in *Numismatics* (see quot. 1962). Hence *Anglo-Ga·llicized* adj.; *Anglo-Ga·llicism*, a French word or phrase adopted into English.

1757 A. C. DUCAREL (*title*) A series of above 200 Anglo-Gallic..Coins of the antient Kings of England. **1821** *New Monthly Mag.* II. 310 Take then your new-fangled Anglo-gallicism *trait*. **1901** J. E. MOON (*title*) Catalogue of Anglo-Saxon, English and Anglo-Gallic Coins. **1926** FOWLER *Mod. Eng. Usage* 476/2 *Purposive*..the Latin suffix *-ive* is unsuited to the delatinized & anglogallicized *pur-*. **1960** L. D. STAMP *Brit. Struct. & Scenery* (ed. 5) xii. 141 A partly enclosed sea, which has been called the Anglo-Franco-Belgian Basin or the Anglo-Gallic Basin, covered the south-east of England, the north-east of France and the greater part of Belgium. **1962** R. A. G. CARSON *Coins* 249 The..term 'Anglo-Gallic' has been generally adopted to describe the coinage in the name of English rulers for the territories in France which they at different times controlled.

A:nglo-Helve·tium. Earlier name for *ASTATINE. Cf. also *ALABAMINE. (*Disused.*)

1942 *Nature* CL. 768 As a tribute to the scientific work of our two countries, we propose to name the element 85 'anglohelvetium'.

Anglo-Indian (æ:ŋglo͵i·ndiăn), *a.* and *sb.* [f. ANGLO- 1 b + INDIAN.] **A.** *adj.* Of, pertaining to, or characteristic of India under British rule, or the English in India. **B.** *sb.* **a.** A person of British birth resident, or once resident, in India. **b.** A Eurasian of India.

1826 J. MALCOLM *Polit. Hist. India 1784–1823* II. xi. 248 The mixed population of Calcutta, Madras and Bombay, made up of European Half Castes, or Anglo-Indians. **1842** *Ainsworth's Mag.* II. 63 He had been at Massorie, that sanatorium of Anglo-Indians. **1845** E. ACTON *Mod. Cookery* (ed. 2) xiv. 288 We think..the proportion of onion and garlic by one half too much for any but well seasoned Anglo-Indian palates. **1847** *Howitt's Jrnl.* 30 Jan. 67/2 The transmission of the Anglo-Indian mails. **1861** SWINHOE *N. China Camp.* 153 The Chinese *Tsaou*, called Bier by the Anglo-Indians, is a somewhat cylindrically shaped fruit. **1907** *Westm. Gaz.* 11 Dec. 2/1 Calcutta..merits the epithet of Anglo-Indian better than anything else in India. **1929** *Cowley Evangelist* June 134 A congregation of Indian, Anglo-Indian and European people, all happily joining together. **1929** *Church Times* 14 June 726/4 The Archdeacon of Madras said that the Anglo-Indian, and particularly the lower class of Anglo-Indian, was the crux of the whole question. **1934** *S.P.E. Tract* XLI. 21 The term 'Anglo-Indian' used to be applied to people of British birth who had lived long in India. In 1911 the Government of India decided to substitute 'Anglo-Indian' for 'Eurasian' as the official term for those of mixed descent. **1941** O'MALLEY *Mod. India & West* xv. 552 Anglo-Indian literature is really a subject in itself.

A:nglo-I·rish, *a.* and *sb.* [f. ANGLO-.] **A.** *adj.* Of, pertaining to, or descended from both the English and the Irish.

1839 T. C. CROKER *Pop. Songs of Ireland* p. vi, The Anglo-Irish settlers degenerated. **1905** *Daily Chron.* 8 Feb. 3/3 This Gaelic world..did immeasurably more for poetry than the Anglo-Irish spheres.

B. *sb.* **a.** Collectively, persons of English descent born or resident in Ireland, or descendants of mixed English and Irish parentage. **b.** The English language as spoken or written in Ireland.

1792, 1834 [see ANGLO- 1 b]. **1927** J. J. HOGAN *Eng. Lang. in Ireland* 16 The Poems..present a full mirror of medieval Anglo-Irish as used about 1300, the time of its greatest extension. **1940** L. MACNEICE *Last Ditch* 4 And the mist on the Wicklow hills Is close..As the Irish to the Anglo-Irish.

A:nglo-I·sraelite. One who holds that the English-speaking peoples represent the 'lost' tribes of Israel. Also *attrib.*, as *Anglo-Israelite theory, theorist.* So **A:nglo-I·sraelitism**; also **A:nglo-I·sraelism.**

1875 'CLERICUS' *Anglo-Israel Theory Refuted* 8 The other errors of Anglo-Israelite advocates (while holding some truth) is their ignoring any of these acts of God but *one*. **1876** J. C. MCCLELLAN (*title*) Anglo-Israelism: its pernicious Nature fully exposed. **1884** R. ROBERTS (*title*) Anglo-Israelitism refuted. **1886** *Brit. & For. Evang. Rev.* Jan. 65 The Anglo-Israelite Theory. *Ibid.* 72 The Anglo-Israelite theorists hold that they [*sc.* the Ten Tribes] were 'lost' first, and then their punishment and humiliation was changed to a brilliant future. **1897** *Westm. Gaz.* 22 Nov. 3/3 Anglo-Israelitism, which, judged on an historical theory, is simply a foolish craze, has yet no little importance as a sentiment. **1963** *Times* 20 Aug. 9/6 Busy advocates of.. Anglo-Israelism.

A:nglomani·acal, *a.* [f. ANGLOMANIAC.] Of the nature of Anglomania.

1893 *Outing* (U.S.) XXII. 115/1 Bunker's utmost anglomaniacal yearnings had never caused him to be taken for a Briton.

A:nglo-No·rman, *a.* and *sb.* [f. ANGLO- + NORMAN *sb.*[1] and *a.*] **A.** *adj.* Of or pertaining to the Normans in England after the Norman Conquest, or to their descendants, or to the variety of Norman French spoken in England after the Conquest.

1767 A. C. DUCAREL (*title*) Anglo-Norman Antiquities considered, in a Tour through part of Normandy. **1767** LD. LYTTELTON *Hist. Hen. II* I. I. 158 It is from his reign we must date the first regular settlement of the Anglo-Norman constitution. **1801** G. ELLIS in *Spec. Early Eng. Poets* I. ii. 37 The Anglo-Norman jargon was only employed in the commercial intercourse between the conquerors and the conquered. **1811, 1874** [see ANGLO- 1 a]. **1847** F. MADDEN *Laȝ. Brut* I. p. xii, The Anglo-Norman metrical chronicle of the *Brut*. **1865** W. WHITE *Eastern England* I. xvii. 235 An Anglo-Norman church with its round arches, arcades, and zig-zag mouldings. **1923** J. VISING (*title*) Anglo-Norman Language and Literature. **1960** E. G. STANLEY *Owl & Nightingale* 34 It is..worth noting that similar use of proverbs is made in the Anglo-Norman debate, Chardry's *Petit Plet*.

B. Also as *sb.*

1735 M. SHELTON tr. *Wotton's Sh. View of Hickes's Treasury* 41 Hickes observes, that all the Charters which shine with guilt Crosses, and painted Images, (or Figures,).. were spurious, and forged by Anglo-Normans long afterwards. **1758** J. RAYNER tr. *Anc. Dial. concerning Exchequer* I. 33 That division of the country in the time of the Anglo-Saxons, was called among the English *shire*, which was afterwards called county by the Anglo-Normans. **1847** F. MADDEN *Laȝ. Brut* III. 453 *Stal*, in A[nglo]-Norman, *estal*. **1953** A. J. BLISS in *Archivum Linguisticum* V. 22 (*title*) Vowel-quantity in Middle English Borrowings from Anglo-Norman.

So **A:nglo-Norma·nic** *a.* and *sb.*

1707 FLEETWOOD *Chron. Prec.* iii. 36 Remember that the 120 are *Anglo-Saxonic* Shillings, and the 50 are Anglo-Normanic ones. **1796** [see ANGLO- 1 a]. **1845** STODDART in *Encycl. Metrop.* I. 73/1 The modern English Language is founded on the Anglo-Normannic, of which the two earliest specimens referred to by Hickes are the Life of St. Margaret and the Description of Cokaygne.

Anglophile (æ·ŋglofəil), *a.* and *sb.* Also **-phil.** [ad. F. *anglophile*, f. ANGLO- + -PHIL, -PHILE.] **A.** *adj.* Friendly to England or to what is English. **B.** *sb.* One who is friendly to England. Hence **Anglophi·lia, A·nglophilism,** friendliness to England; **Anglophi·lic** *a.*, Anglophile.

1867 *Contemp. Rev.* IV. 88 The *Revue des deux Mondes*, a thorough 'Anglo-phile' periodical. **1883** *Chambers's Jrnl.* 13 Jan. 18/1 This telephone..was an Anglophile, and would only respond to the honoured name of Faraday. **1892** *Athenæum* 26 Mar. 400/3 When prudence dictated assistance to the Dutch, the Huguenots, or the 'Anglophile' party in Scotland. **1896** *Westm. Gaz.* 2 Oct. 2/1 To show how Anglophilia and Anglophobia counteract each other. *Ibid.* 5 Dec. 6/3 The New York *Evening Sun*..has frequently attacked Mr. Bayard for what it regards as his extreme Anglophilism. **1920** ROBB *Thre Prestis of Peblis* (S.T.S.) Introd. p. xxxv, The unpopularity he thus incurred as an anglophile. **1950** *Amer. Speech* XXV. 90 The large Anglophilic group in the city.

Anglophobe. Add: *attrib.* or as *adj.*

1901 *Daily Chron.* 6 Aug. 3/1 The revival of Anglophobe reaction was further strengthened by the Parliamentary activity of Clericalism.

Anglophobia. (Earlier U.S. example.)

1793 JEFFERSON *Writ.* (1895) VI. 250 We are going on here in the same spirit still. The Anglophobia has seized violently on three members of our Council.

Anglophobiac (æŋglofōᵘ·biæk). [f. ANGLOPHOBIA after *mania, maniac.*] = ANGLOPHOBE. Also *attrib.* or as *adj.* So **Anglopho·bian** *a.*

1893 *N. Amer. Rev.* Aug. 170 The work of an Anglophobiac who labors to widen..the schism. **1894** *Pop. Sci. Monthly* XLV. 476 The Anglophobiac American who proposed cutting a canal through Yucatan. **1896** *Daily News* 11 Jan. 7/5 If..scurrilous little newspapers of Paris indulge in Anglophobian diatribes. **1902** *Daily Chron.* 26 May 5/1 The New York 'Sun', a violently Anglophobiac paper.

Anglophobic, *a.* Delete *rare* and add further examples.

1894 *Westm. Gaz.* 1 June 3/1 Our old Anglophobic friend, M. Deloncle. **1907** *Ibid.* 6 May 11/1 The Anglophobic outbursts of the German Press. **1958** *Listener* 13 Nov. 786/2 On our side, there is a counterpart Anglophobic sentiment.

Anglophobist. Now normally stressed on the third syllable. (Further examples.)

1854 *Blackw. Mag.* LXXV. 13/1 In all matters relating to the East he is an Anglophobist (to coin a word for the occasion). **1904** *Westm. Gaz.* 24 Sept. 12/1 François Garnier, a bitter Anglophobist.

A:nglo-Ro·man, *a.* [f. ANGLO-.] **a.** English Roman Catholic. Hence **A:nglo-Ro·manism,**

English Romanism. **b.** Of or pertaining to England and Rome.

c **1840** A. W. PUGIN in E. S. Purcell *Life & Lett. A. Phillipps de Lisle* (1900) II. 218 The Oratory..fitted up.. with a sort of Anglo-Roman altar. **1866** 'A LAYMAN' (*title*) Anglo-Romanism unveiled; or, Canon Oakeley and Dr. Newman at issue with the Catholic and Roman Church, and with one another. **1891** GLADSTONE in *Westm. Gaz.* (1898) 20 May 8/1, I share the dissatisfaction which many feel at the attitude in England of the *Anglo*-Roman body. **1895** *Ibid.* 12 Jan. 8/2 An essential part of that history [of the sixteenth and seventeenth centuries] is what I may term Anglo-Roman. **1913** C. G. BAYNE (*title*) Anglo-Roman Relations, 1558–1565.

Anglo-Saxon, *sb.* and *a.* Add: II. **B.** *absol.* (Later example.)

1955 QUIRK & WRENN *O.E. Gram.* 1 In the eighteenth and nineteenth centuries the term *Anglo-Saxon*..was the commonest name for the language; but, although still sometimes used by scholars it has gradually been replaced in the last hundred years by the more scientific term *Old English*.

III. **A, B.** (Earlier U.S. and later examples.)

Quot. 1871 should perhaps be placed in sense II. B.

1832 R. CHOATE in *Deb. Congress* 13 June 3515 The whole circle of the..arts, trades, and branches of manufacture, which characterize the..industry of the Anglo-Saxon race of men. **1846** *Spirit of Times* (N.Y.) 6 June 177/3 The Anglo-Saxon 'never can acknowledge the corn' to the cross of negro and Indian. **1871** 'L. CARROLL' *Through Looking-Glass* vii, He's an Anglo-Saxon Messenger—and those are Anglo-Saxon attitudes. **1875** W. JAMES *Coll. Ess. & Rev.* (1920) 16 But the thing which to our Anglo-Saxon mind seems so outlandish is that crowds of dapper fellows, revelling in animal spirits and conscious strength, should enroll themselves in cold blood as his [*sc.* Schopenhauer's] permanent apostles. **1888** KIPLING in *Lett. of Marque* (1891) xvi. 119 A snowy-bearded chowkidar..threw himself into Anglo-Saxon attitudes. **1904** CONRAD *Nostromo* II. vi. 180 It is part of the truth of things which hurts the—what do you call them?—the Anglo-Saxon's susceptibilities. **1924** R. GRAVES *Mock Beggar Hall* 63 Yet commonsense, the Anglo-Saxon flair Seems weakest on its vaunted practical side. **1956** ANGUS WILSON (*title*) Anglo-Saxon Attitudes.

IV. Used for 'the English language'. '(Of) the English language (of the modern period)' *U.S.*; freq. with the implication 'plain, unvarnished, forthright'. *colloq.*

A. *adj.*

1859 'J. DOWNING' *Thirty Years out of Senate* 10 The best and truest exposition of the peculiar Yankee dialect of the Anglo-Saxon language that there is extant. **1863** C. LYELL *Antiquity of Man* xxiii. 466 Among the [Germans of Pennsylvania]..I found the newspapers full of terms half English and half German, and many an Anglo-Saxon word which had assumed a Teutonic dress, as 'fencen', to fence, instead of umzäunen. **1927** in *Amer. Speech* (1928) III. 376 Several Laborites were suspended in the House of Commons..to the accompaniment of..the hurling of bald Anglo-Saxon epithets traditionally classed as unparliamentary. **1927** *Sat. Rev. Lit.* 23 Apr. 772/4 All nine of the tabooed Anglo-Saxon monosyllables. **1958** *Spectator* 31 Jan. 133/2 The Bishop was reported in reputable newspapers as having said (and in more Anglo-Saxon terms) that he had been reliably informed of the truth of this fact.

B. *sb.*

1866 J. C. GREGG *Life in Army* xv. 137 Occasionally a word of honest, hearty Anglo-Saxon, or a 'bit of the brogue', to remind you that you are not in Naples, but in New Orleans. **1872** H. A. WISE *Seven Decades of Union* 141 He [*sc.* Senator Leigh of Virginia] was a purist in his Anglo-Saxon. **1917** in *Amer. Speech* (1929) IV. 271, I like your stilted style best Jack. When you descend to the Anglo Saxon you get too much in dead earnest. **1926** *Amer. Speech* I. 265/1 Specimens of the jargon daily spoken by witnesses believing they talk pure Anglo-Saxon. **1927** *Yale Rev.* Jan. 414 Tell me what you forget and I will tell you what you are, says the psycho-analyst. But I can do this, too, and in plain Anglo-Saxon. The man who insists on telling me what he forgets is a fool. **1947** K. MALONE in *Word Study* Oct. 2/2 In current speech *Anglo-Saxon* often means plain English. In this use, the word has *Latin* for antonym.

†**Anglo-Saxo·nic,** *a. Obs.* [f. ANGLO-.] = SAXONIC *a.* 2. Also as *sb.*

1672 PENN *Truth Vin.* 53 Besides, several other Languages, amongst others a Gothick and Anglo-Saxonick Testament. **1686** PARR *Life of Usher* 101 His great activity..to advance the Restauration of our old Northern Antiquities, which lay buried in the Gothick, Anglo-Saxonick, and other the like obsolete Languages. **1739** J. LEWIS *Hist. Transl. Bible into Eng.* (ed. 2) i. 7 In his Observations on the Anglo-Saxonic Version of the Gospels. *Ibid.*, To note what Books of the Old Testament he translated into Anglo-Saxonic. **1796** *Ann. Reg. 1767* (ed. 5) 150/1 The heavy Gothic by Sir C. Wren is distinguished as Anglo-Saxonic, the lighter as Saracenic.

A:nglo-Sa·xonize, *v.* [f. ANGLO-SAXON + -IZE.] *trans.* To make Anglo-Saxon.

1883 MARQUIS OF LORNE in *Contemp. Rev.* Nov. 641 That great Anglo-Saxonizing amalgamation mill, the United States, would soon efface their [*sc.* French Canadians'] language. **1898** *Westm. Gaz.* 14 July 2/1 These silly efforts to Anglo-Saxonise the United States.

Anglo-Sa·xony. *rare.* [f. ANGLO-SAXON + -Y[3].] A rhetorical expression for 'Great Britain and the United States'.

1905 *Daily Chron.* 13 Mar. 3/7 Miss Annie Russell is artistically eager to render his 'Jinny the Carrier' all over Anglo-Saxony. **1937** WYNDHAM LEWIS *Blasting & Bombardiering* v. iv. 272 T. S. Eliot—need I say the premier poet of Anglo-Saxony?

A:nglo-verna·cular, *a.* [f. ANGLO-.] Pertaining to or consisting of English and an Indian language.

1877 CALDWELL *Evangel. Work Tinnevelly* 2 Our S.P.G. Anglo-Vernacular School. **1888** KIPLING *Plain Tales fr. Hills* 269 He wound up with a six-shot Anglo-Vernacular oath. **1915** (*title*) A List of Educational Books..for Schools in India, Burma and Ceylon..(Vernacular, Anglo-Vernacular and English).

angola. Add: **2.** *Angola cat.* Also *ellipt.* (See ANGORA 1.)

1851 BORROW *Lavengro* i, The cheerful hum of the kettle and the purring of the immense angola. **1854** MRS. GASKELL *Let.* ?15 May (1966) 285 We are going to have a little kitten ..with long hair...It is called an Angola or a Persian cat.

angon (æ·ŋgǫn). [med.L., ad. Gr. ἄγγων.] (See quot. 1893.)

1875 *Encycl. Brit.* II. 555/1 Iron head of Angon from a grave at Darmstadt. **1893** *Athenæum* 9 Sept. 361/3 Among the most curious of the weapons our ancestors used was the angon, a sort of long spear with a double barb—one tongue of which was longer than the other and flame-shaped. **1908** *Victoria Hist. Kent* I. 377 One of the rare iron weapons usually called angons. **1940** *Antiquity* XIV. 14 Through one of the drop-handles of the outer bowl the shafts of three iron angons had been thrust.

Angora, angora. Add: **1.** (Earlier example.) Also applied to a variety of rabbit having fine white fur and pink eyes; also ellipt. as *sb.*, a rabbit of this variety, its fur or a fabric made from it.

1819 T. HOPE *Anast.* II. i. 5 Of the things themselves whose appellations he had learnt, he seemed to have no more idea than the huge Angora cat which sat purring by his side. **1849** tr. *Wyss's Swiss Family Robinson* Ser. II. xvii. 152 Fritz has two magnificent Angora rabbits in his pouch. **1908** J. A. THOMSON *Heredity* x. § 4. 365 Hurst paired white Angora rabbits..with 'Belgian hare' rabbits. **1927** E. V. KNOX *Awful Occasions* 58 The vexed topic of dandelions as a diet for half bred Angoras. **1960** *Which?* Jan. 19/2 *Angora*, hair from the angora rabbit. Very fine and soft. Angora sheds fibres easily. *Ibid.* Feb. 39/2 *Angora*..a fabric made from Angora rabbits' hair with the addition of up to 5 per cent wool to facilitate spinning.

Angrian, *a.* [f. *Angria* (see below) + -AN.] Of or pertaining to Angria, an imaginary African empire invented by the Brontë children (Charlotte, Branwell, Emily, and Anne) in stories that they composed. Also as *sb.*

1834 C. BRONTË *Twelve Adv.* (1925) 168 The wife of Northangerland, the prima donna of the Angrian Court. **1834** P. B. BRONTË in *Misc. Unpub. Writ.* (1936) I. 435 Sound the loud Trumpet o'er Afric's bright sea; Zamorna hath triumphed: the Angrians are free. **1936** *Times Lit. Suppl.* 3 Oct. 784/3 These Angrian games are the most curiously exciting ever played by children. **1954** K. TILLOTSON *Novels of Eighteen-Forties* I. 97 Recollected from Charlotte's early reading, through her own Angrian romances.

angry, *a.* Add: **3. c.** *spec.* Dissatisfied with and outspoken against the prevailing state of affairs, current beliefs, etc.; esp. in phr. *angry young man* (abbrev. A.Y.M.).

The expression 'angry young man' and variants of it became commonly used, esp. by journalists, after the production of J. Osborne's play *Look Back in Anger* (first performed 1956). The phrase did not occur in the play but was applied to Osborne by G. Fearon, a press reporter (see quot. 2 Oct. 1957), thence used particularly of young writers, usually of provincial and lower middle-class or working-class origin, who denounced or satirized the 'Establishment' (*q.v.) and the abuses of the time; later applied by extension to any person, group, etc., in Britain and elsewhere who considered the times to be out of joint. For the formation, cf. 'bright young thing' (*BRIGHT *a.* 7 b).

(*a*) (General and nonce uses.)

1937 [see *ANGER *sb.* 2]. **1941** 'R. WEST' *Black Lamb* (1942) I. 157 Their [*sc.* the Dalmatians'] instinct is to brace themselves against any central authority as if it were their enemy. The angry young men run about shouting. **1951** LESLIE PAUL (*title*) Angry Young Man. **1954** J. B. PRIESTLEY *Magicians* vi. 132 Too much resentment, too much cheap cynicism. And he's expecting too much, in the wrong way. He's the contemporary Angry Little Man.

(*b*) With overt or implicit reference to Osborne's *Look Back in Anger*; also extended uses.

1956 *Reporter* (N.Y.) 18 Oct. 33/1 The angry young man is a stock character in literature. **1957** G. FEARON in *Daily Tel.* 2 Oct. 8/7, I had read John Osborne's play. When I met the author I ventured to prophesy that his generation would praise his play while mine would, in general, dislike it...'If this happens,' I told him, 'you would become known as the Angry Young Man.' In fact, we decided then and there that henceforth he was to be known as that. **1957** *Times Lit. Suppl.* 4 Oct. 593/1 The 'angry young men' of England (who refuse to write grammatically and syntactically in order to flaunt their proletarian artistry). *Ibid.* 8 Nov. 674/1 *Declaration* is a volume of essays in which four of the 'angry' movement attempt to formalize their beliefs. **1958** *Times* 1 Feb. 7/5 The angry young man who feels that life is short, and that he must make his mark early by carping at established ideas and institutions. **1958** *Spectator* 11 July 67/2 A vigorous but implausible book about an AYM, *strictly* lower-class. **1958** *Listener* 4 Dec. 961/2 A fine, hard-bitten performance of an Angry Old Man. **1959** *Times* 1 Jan. 9/3 Mr. Forster..has been angry but never Angry.

angry (æ·ŋgri), *sb.* [f. the adj.] An 'angry' person (see *ANGRY *a.* 3 c); an 'angry young man'.

1957 *Times Lit. Suppl.* 15 Nov. 689/3 Writers, such as Mrs. Lessing, who have never been called Angries at all. **1958** *Observer* 14 Sept. 4/5 The 'beat generation' is beginning to acquire the same kind of dubious place in American culture as the Young Angries in Britain. **1959** *Bookseller* 7 Mar. 1133/1 The 'beats' are represented by Anatole Broyard..and Carl Solomon; the 'angries' by John Wain, Colin Wilson, John Osborne, [etc.].

‖**angst** (æŋst). [G.] Anxiety, anguish, neurotic fear; guilt, remorse.

[**1849** GEO. ELIOT *Let.* 5 Aug. (1954) I. 293 'Die Angst' she says often brings on a pain at her heart. **1922** C. J. M. HUBBACK tr. *Freud's Beyond Pleasure Principle* ii. 9 Apprehension (*Angst*) denotes a certain condition as of expectation of danger..even though it be an unknown one. **1941** *Philosophy* XVI. 260 To Heidegger *Angst*, dread, is the fear of metaphysical insecurity.] **1944** 'PALINURUS' *Unquiet Grave* 22 Angst may take the form of remorse about the past, guilt about the present, anxiety about the future. **1950** A. HUXLEY *Themes & Variations* 202 To *acedia* and confusion, to nightmare and *angst*, to incomprehension and panic bewilderment. **1956** C. P. SNOW *Homecomings* viii. 65 Discussing other people whose lives were riven by angst—it domesticated her wretchedness a little to have that label to pin on. **1964** *New Statesman* 10 Apr. 574/1 Telly-angst is a natural corollary of architectural news-angst generally.

Comb., as *angst-forming, -ridden, -wrought* adjs.

1944 'PALINURUS' *Unquiet Grave* 43 There need be nothing angst-forming about the sexual act. **1958** *Times* 14 Feb. 6/1 (Advt.), The..contrast drawn between those years and our *Angst*-ridden era. **1958** *Observer* 27 Apr. 15/3 Petrushka is a tied-up angst-wrought neurotic.

Ångström (unit) (ǭŋström). Also **angstrom** (æ·ŋström). [The name of A. J. *Ångström* (1814–74), a Swedish physicist.] A hundred-millionth of a centimetre (10^{-8} cm.), used in measuring the wavelengths of light, X-rays, etc. (also called *absolute Ångström*); the same unit used as a measure of the thickness of thin film. Abbrev. Å, Å.U.; also A.U.

The *International Ångström* (I.Å.) was defined in 1907 in terms of the wavelength of cadmium which in standard conditions is 6438·4696 I.Å. When the metre was defined in terms of the wavelength of krypton in 1960 the Ångström became exactly equal to 10^{-8} cm.

[**1887** *Phil. Mag.* XXIII. 258 As all observers so far seem to accept the measures of Ångström, I have decided that a table of my results would be of value.] **1897** *Rep. Brit. Assoc. Advancem. Sci.* 556 No lines were found to shift more than a fraction..of an Ångström unit. **1906** *Trans. Internat. Union Solar Research* I. VIII. 230 *Resolutions concerning standards of Wave-length.* The wave-length of a suitable spectroscopic line shall be taken as the primary standard of wave-length. The number which defines the wave-length of this line shall be fixed permanently and thereby define the unit in which all wave-lengths are to be measured. This unit shall differ as little as possible from 10^{-10} metres and be called the Ångström. **1911** KAYE & LABY *Phys. & Chem. Constants* 9, 1 Å.U. = 10^{-10} metre. **1916** ALLEN & MOORE *Text-Bk. Pract. Physics* III. vii. 300 Ångström Units (A.U.). **1921** *Discovery* Sept. 226 An X-ray of wave length equal to two-tenths of an Ångström Unit (100-millionth of a centimetre). **1933** S. W. COLE *Pract. Phys. Chem.* (ed. 9) x. 255 Ångström units (Å or A.U.). **1943** A. E. MIRSKY in Nord & Werkman *Advances in Enzymology* III. 18 Histone spreads to give a film 7–9 Å thick. **1951** *Engineering* 18 May 589/1 Dr. J. S. Courtney-Pratt has used interferometric methods to study the uniformity and thickness of thin films...He was able to measure the thickness correct to the nearest Ångstrom. **1957** *Technology* July 168/4 Those [fluorescing agents] emitting a visible blue light in the 4,000–4,400 angstrom range are commercially useful in detergents. **1957** H. BARRELL in *Nature* 21 Dec. 1388/1 The value λ_R = 6438·4696 A. was adopted as the spectroscopic reference standard in 1907 and thereby became the means of defining the international angstrom—the unit that has since served for all spectroscopic measurements of wave-length.

anguipede, -ped (æ·ŋgwipīd, -ped), *a.* [ad. L. *anguipēs* (*-ped-*), f. *anguis* serpent + *pēs* foot.] Having feet or legs in the form of serpents, serpent-footed: an epithet of certain giants of ancient mythology.

1883 A. S. MURRAY *Gr. Sculpt.* II. 305 A winged anguipede giant. **1888** A. H. SMITH *Catal. Engr. Gems Brit. Mus.* 100 Athenè..treading down anguipede Giant.

angular, *a.* Add: **2. a.** (Further examples.)

1835–6 Supra-angular [see SUPRA- 1 b]. **1836** W. BUCKLAND *Geol. & Min.* I. xiv. 176 Head of the Crocodile. In the lower jaw, *u*, marks the dental bone; *v*, the angular bone; *x*, superangular. **1866** R. OWEN *Anat. Vertebrates* I. ii. 141 The mandibular arch..the 'articular' piece; that beneath it, which develops the angle of the jaw, when this projects, is the 'angular' piece; the piece above..is the 'surangular'.

c. (Further examples.)

1819 PLAYFAIR *Nat. Phil.* (ed. 3) I. 67 The angular velocity with which the bodies will begin to revolve. **1835** M. SOMERVILLE *Connex. Physical Sciences* (ed. 2) 455 The angular velocity of the earth is at the rate of 180° in twelve hours. **1837** [see DISTANCE *sb.* 5 e]. **1858** SUTTON & WORDEN *Dict. Photogr.* 21 The angular aperture of a lens is the angle which its diameter subtends at its principal focus. **1870** [see MOMENTUM 4]. **1873** J. N. LOCKYER *Elem. Lessons in Astr.* viii. 292 The mean angular diameter of the Moon is 31′ 8″·8. *Ibid.*, Knowing the real and also the

apparent angular diameter, we can at once determine the distance. **1883** *Encycl. Brit.* XV. 692/1 The angular accelerations about these axes are equal at that instant. **1885** RANKINE & MILLAR *Man. Appl. Mech.* (ed. 11) v. ii. 505 Angular momenta are compounded and resolved like forces, each angular momentum being represented by a line whose length is proportional to the magnitude of the angular momentum. **1895** P. G. TAIT *Dynamics* 254 Thus the linear acceleration of each of the masses is equal to *a* times the angular acceleration of the pulley. **1921** *Discovery* Sept. 237/1 Professor Eddington..made an estimate of the probable angular diameters of some of the brighter stars. **1929** RATCLIFFE *Physical Princ. Wireless* i. 3 The angular frequency will often be referred to as the frequency of the oscillation. **1940** *Chambers's Techn. Dict.* 35/2 *Angular aperture*, the ratio of the working diameter to the focal-length of a lens, i.e. reciprocal of the *f*-number. *Ibid.*, *Angular frequency*, the frequency of a steady recurring phenomenon expressed in radians per second. *Ibid.* 554/1 *Angular momentum* is the product of the moment of inertia and the angular velocity of a body. **1949** W. E. SIRI *Isotopic Tracers & Nucl. Radiations* i. 12 In any nuclear reaction, spin and angular momentum must be conserved as well as mass and energy. *Ibid.* xiii. 375 The angular distribution of beta particles from a point source of small mass is isotopic. **1952** *Sci. News* XXIII. 41 Many nuclei behave as though they have angular momentum—that is to say, part of the system is either spinning on its own axis or revolving around the centre in an orbit, or both.

5. For ANGLE *sb.*[2] 8 read ANGLE *sb.*[2] 7.

angular (æ·ŋgiŭlaɹ), **angulare** (æŋgiŭlā·ri), *sb. Zool.* [ad. mod.L. *angulāre*, neut. of *angulāris* adj. (see ANGULAR *a.*).] In some vertebrates, the angular bone of the lower jaw.

1846 R. OWEN *Lect. Comp. Anat. Vertebrate Animals* v. 113 It sends upwards a pointed coronoid process to which ..the masticatory muscles are attached; one short square plate downwards, to join with the angular. **1896** [see SUPRA- 1 b]. **1905** D. S. JORDAN *Study Fishes* xxxiv. 606 Each half of the lower jaw consists of..the articular, angular, dentary, and splenial (coronoid). Most of these bones are armed with teeth. **1928** G. R. DE BEER *Vertebrate Zool.* v. 71 The quadrate articulates with a bone of the lower jaw called the articular...The ventro-posterior part of the lower jaw is formed by the angular. *Ibid.* xx. 298 The angular becomes converted into the tympanic bulla..and the supra-angular is represented by the processus Folii.

anguloa (æŋgiulōᵘ·ă). *Bot.* [mod.L. (1794), named by Ruiz and Pavon, *Fl. Peruv.*, in honour of Francisco de *Angulo*, a Spanish naturalist.] A member of a genus of South American plants so called, belonging to the family Orchidaceæ.

1836 *Mag. Bot. & Gardening* III. 29/2 Anguloa Superba. Superb Anguloa..is found growing on trees in Peru, sucking their sap by insinuating its roots beneath the bark...It should be planted in pots filled with porous stones. **1882** *Garden* 11 Mar. 168/2 Anguloas of all the different species ought to be potted. **1962** *Amateur Gardening* 7 Apr. 6/1 The anguloas, or Cradle orchids, are so named from the movable lip which bears a fanciful resemblance to a rocking cradle.

Angus (æ·ŋgɒs). The name of a county in Scotland, applied *attrib.* to a breed of cattle. Also *absol.* Cf. *ABERDEEN *Angus*.

1842 D. LOW *Breeds Dom. Animals* I. II. 13 The Polled Angus breed. *Ibid.* 14 The Angus is a good breed. **1844** [see *DEVON]. **1882** MACDONALD & SINCLAIR *Hist. Polled Aberdeen or Angus Cattle* viii. 136 No stock but polled Angus was at Kinnaird until about 1834. **1956** J. HEARNE *Stranger at Gate* iv. 31 This was the problem..how to keep beef on the mixed Angus and shorthorn stock in a climate that never had any spring or autumn.

Angustura. The word *Angostura* as a brand of aromatic bitters is a registered trade mark. It derives its name from the fact that it was originally made in the town of Angostura, but it is not made from the bark exported from Angostura.

‖**anhedonia** (ænhidǫ·niă). *Psychiatry.* [ad. F. *anhédonie* (Ribot, 1896), mod.L., f. Gr. ἀν- priv. + ἡδονή pleasure.] Inability to feel pleasure.

1897 tr. *T. A. Ribot's Psychol. of Emotions* iii. 53 *Anhedonia* (if I may coin a counter-designation to analgesia) has been very little studied..but there are cases of an insensibility relating to pleasure. **1902** W. JAMES *Var. Relig. Exper.* vi. 145 One can distinguish many kinds of pathological depression. Sometimes it is mere passive joylessness and dreariness, discouragement, dejection, lack of taste and zest and spring. Professor Ribot has proposed the name *anhedonia* to designate this condition. *Ibid.* 146 Prolonged seasickness will..produce a temporary condition of anhedonia. **1960** HINSIE & CAMPBELL *Psychiatric Dict.* (ed. 3) 44/1 *Anhedonia*..is seen often in schizophrenic patients.

anhedral (ænhī·drăl, -he·drăl), *a.* [f. AN- 10 + HEDRAL *a.*] Applied to crystals that are not bounded by plane faces; also called *ALLOTRIOMORPHIC, *xenomorphic*. So **anhe·dron,** a crystal of this kind.

1896 L. V. PIRSSON in *Amer. Geologist* XVII. 94 Since the term *crystal*, when strictly used, means a body possessing not only a certain internal molecular structure, with definite physical properties, but more especially an outward symmetrical form with plane faces, it is evident that there is no good term for the rounded or formless masses in which minerals occur in rocks. Therefore, after consultation with Prof. E. S. Dana, the term *anhedron* (meaning without planes) is proposed for these formless masses which possess the internal structure of crystals; and such minerals may also be spoken of as having an anhedral development. **1898** *Amer. Jrnl. Sci.* V. 361 The rock..shows numerous tabular phenocrysts..and a few anhedra of augite. **1913** G. A. J. COLE *Outl. Min.* ii. 20 The individual crystal in such cases is sometimes said to be *anhedral*, because it has no good bounding planes, in opposition to the well-crystallised or *euhedral* forms. **1950** HARTSHORNE & STUART *Crystals & Polarising Microscope* (ed. 2) iii. 44 Crystals which are bounded by plane faces are described as *idiomorphic* or *euhedral*, and irregular ones as *allotriomorphic* or *anhedral*. **1961** J. CHALLINOR *Dict. Geol.* 7/2 'Anhedral', 'euhedral', and 'subhedral', in mineralogy, are probably to be preferred to 'allotriomorphic', 'idiomorphic', and 'hypidiomorphic' as avoiding any confusion between mineralogical terms and petrographical-textural terms.

anhima (æ·nhimă). Also **aniuma.** [Pg., f. Tupi.] The kamichi or horned screamer (*Palamedea cornuta*).

1773 [see SCREAMER 3 a]. **1774** GOLDSM. *Hist. Earth* V. v. v. 389 A bird..called the Anhima,..a native of Brasil. **1869–73** Aniuma [see SCREAMER 3 a].

anhinga (ænhi·ŋgă). [Tupi.] Any bird of the genus *Anhinga*, esp. the American snake-bird, *A. anhinga*.

1769 T. PENNANT *Indian Zool.* xii. 13/1 The black-bellied Anhinga. We give it this epithet, to distinguish it from an American species with a silvery belly. This kind is found in Ceylon and Java..neck extremely long; the bill strait, long, and sharp-pointed. **1816** TUCKEY *Narr. Exped. River Zaire* (1818) 82 Mr. Cranch shot some birds amongst which..an anhinga. **1838** [see SNAKE-BIRD 1]. **1872** COUES *N. Amer. Birds* 332 Few other birds, such as cormorants and anhingas, resemble the *Pygopodes* in this respect.

anhypostasia (æ:nhəipostēⁱ·siă). *Theol.* Also **anhypo·stasis.** [mod.L., f. Gr. ἀνυποστασία, ἀνυπόστασις unsubstantiality, f. ἀν- priv. + ὑπόστασις substance, substantial existence (see HYPOSTASIS 5).] Lack of a substantial or personal existence. So **anhyposta·tic, -sta·tical** *adjs.*, having no independent or personal existence; describing the human nature of Christ which had no existence apart from the hypostatic union.

1862 D. W. SIMON tr. *Dorner's Person of Christ* (Div. II.) II. 92 Maintaining that Luther regarded the humanity of Jesus as the mere means of the manifestation of the person of the Logos, and taught most clearly that humanity was selfless or anhypostatical. **1863** *Ibid.* (Div. II.) III. 300 He describes the Logos..as the primal personality.., he holds the significance of the doctrine of the anhypostasis of humanity to be, that God is the essence of humanity. **1877** SCHAFF in Smith & Wace *Dict. Chr. Biogr.* I. 495/1 The anhypostasia, impersonality, or, to speak more accurately, the enhypostasia, of the human nature of Christ; for anhypostasia is a purely negative term, and presupposes a fictitious abstraction, since the human nature of Christ did not exist at all before the act of the incarnation, and could therefore be neither personal nor impersonal. **1941** *Jrnl. Theol. Stud.* XLII. 123 He sees no absurdity in Cyrilline anhypostatic manhood, with its corollary, the 'passionless suffering' of the Word. **1946** E. L. MASCALL *Christ, Christian & Church* 8 According to [the doctrine of *enhypostasia*], the humanity of Christ is neither *hypostatic* (that is, possessing a human person) nor is it *anhypostatic* (that is, without a person altogether), but it is *enhypostatic* (that is, it is constituted in the person of the divine Word). **1954** *Sc. Jrnl. Theol.* VII. 249 By *anhypostasia* classical Christology asserted that in the *assumptio carnis* the human nature of Christ had no independent *per se* subsistence apart from the event of the Incarnation, apart from the hypostatic union.

ani (ā·ni). [Sp. *aní* or Pg. *ani*, f. Tupi.] A bird of the genus *Crotophaga* (family Cuculidæ), of which several species are found in the warmer parts of America and in the West Indies.

1829 E. GRIFFITH et al. *Cuvier's Anim. Kingd.* VII. 476 The Ani, Crotophaga L. Are known by their thick bill. **1893** NEWTON *Dict. Birds* 191 There are other eggs, as those of the Anis, *Crotophaga*, the Grebes, *Podicipedidæ*, [etc.] which are more or less covered with a cretaceous film. **1918** W. BEEBE *Jungle Peace* (1919) iii. 45 Heavy-billed anis *whaleeped* and fluttered clumsily ahead of us. **1964** A. L. THOMSON *New Dict. Birds* 170/2 The anis are medium-sized non-parasitic cuckoos with almost black plumage, long and nearly square-ended tails, and extremely heavy hooked bills.

aniconic (æn͵əikǫ·nik), *a. Gr. Antiq.* Also **anik-.** [f. AN- 10 + ICONIC *a.*] Applied to simple material symbols of a deity, as a pillar or block, not shaped into an image of human form; also to the worship connected with these. Hence **aniconism** (ǣnəi·kǒniz'm), the use of, or worship connected with, such symbols.

1892 C. WALDSTEIN *Excav. Heraion Argos* I. 19 Clement of Alexandria..tells us that the aniconic σανίς was superseded by the iconic image, which he calls βρέτας. **1893** *Athenæum* 1 July 38/2 The various stages of development of her [*sc.* Hera's] *agalmata*: the rudest of all, the anikonic. ..A..pillar which may have been the actual anikonic image of the goddess. **1907** *Academy* 24 Aug. 818/1 Aniconism does not necessarily imply..anthropomorphism. **1912** L. R. FARNELL *Higher Aspects Gr. Relig.* i. 4 Certain aniconic sacred things that we may call fetishes—the hewn stock or pillar, the meteorite, the axe. **1956** I. A. RICHMOND in R. L. S. Bruce-Mitford *Rec. Archæol. Excav. in Brit.* 67 As often in religions which have sprung from Asia, there are iconic and aniconic traditions...The Zend-Avesta is a production of the second.

anima (æ·nimă). *Psychol.* [L., 'mind, soul'.] Jung's term for the inner part of the personality or character, as opposed to the *persona* or outer part; also, the feminine component of a male personality. Cf. *ANIMUS 2.

1923 H. G. BAYNES tr. *Jung's Psychol. Types* xi. 593, I term the outer attitude, or outer character, the *persona*; the inner attitude I term the *anima*, or *soul*. *Ibid.* 595 If, therefore, we speak of the *anima* of a man, we must logically speak of the *animus* of a woman. **1926** W. MCDOUGALL *Outl. Abnormal Psychol.* ix. 198 The *Anima* is not to be regarded as the whole of the Collective Unconscious; it is only a selection from it. **1943** *Horizon* VIII. 262 A seductive mixture which fascinated the *anima* in Kierkegaard. **1962** A. M. DRY *Psychol. of Jung* iv. 96 In women the phenomena corresponding to the moods of the anima-ridden man are 'opinions'. **1962** R. MANHEIM tr. *J. Jacobi's Psychol. of C. G. Jung* (ed. 6) iii. 112 Typical anima figures in literature are Helen of Troy.., Beatrice in the *Divine Comedy*, [etc.].

animadverter. Delete *arch.* **1.** (Later examples.)

1792 SCOTT *Let.* 10 Sept. (1932) I. 21 He will have many advisers and animadverters upon the naughtiness of his ways. **1958** *Times Lit. Suppl.* 13 June 322/3 The most censorious Republican (or Democratic) animadverter on the 'mess in Washington'.

animal. Add: **A. 1. b.** The living body or soft fleshy part of a mollusc, crustacean, etc., as distinguished from its shell or other hard part.

1834 MCMURTRIE tr. *Cuvier's Anim. Kingd.* 248 These Mollusca are arranged in several families according to the form of their shell, which appears to bear a constant relation to that of the animal. **1868** DANA *Syst. Min.* (ed. 5) 677 Carbonaceous matters..derived from the animals of the shells, corals, etc. out of which the limestones were.. made.

3. b. With *the*. The animal nature in man: cf. BEAST *sb.* 1 c.

1809 C. SIMEON *Memoirs* (1847) 272 Less mixture of the *animal* I never expect to see in this world. **1907** H. A. VACHELL *Her Son* i, The animal in this girl was about to spring upon her. **1919** M. K. BRADBY *Psychoanalysis* 231 His fleshly desires were strong, and he was unmerciful to the animal in himself.

6. *colloq.* A person, thing; esp. in phr. (*there is*) *no such animal*.

1922 E. F. MURPHY *Black Candle* II. xxii. 322, I would like to ask these same 'old-timers' 'how many square shooting addicts have you found in your experience?' I can hear them roar and say 'There is no such animal'. **1963** *Camb. Rev.* 27 Apr. 386/2 Teachers must remember that they are dealing with learners, which is another animal altogether. **1963** *Times Rev. Industry* May 85/1 Computer makers would therefore have us believe that there is no such animal as a typical programmer.

B. 4. (Later examples.)

1923 G. SANTAYANA (*title*) Scepticism and Animal Faith. **1947** *Mind* LVI. 336 We have..no reason for believing any of these inferences; they are all a matter of custom or habit, or, if one prefers more recent terminology, of 'animal faith'.

b. Characteristic of or resembling (that of) a lower animal. Also *Comb.*, as *animal-bodied* adj.

1922 D. H. LAWRENCE *England, my England* (1924) 216 The wild, bare, animal shoulders. **1924** HICHENS *After the Verdict* III. xv, Fine-souled and animal-bodied men.

5. (Further examples; cf. *animal pole* below and VEGETATIVE *a.* 1 d.)

1843 T. W. JONES in *Brit. & Foreign Med. Rev.* XVI. 547 In the rabbit's ovum..two layers of the blastodermic vesicle can be demonstrated; which Bischoff therefore from this time calls *serous* or *animal* and *mucous* or *vegetative*. *Ibid.* 552 The whole embryo is as yet composed only of the thickened central part of the animal layer; the vegetative layer lies quite smooth at its under surface. **1902** E. B. WILSON *Cell in Devel. & Inheritance* (ed. 2) viii. 379 The smaller cells of the upper hemisphere [*sc.* of the ovum] represent the 'animal layer', outer germ-layer or ectoblast from which arise the epidermis, the nervous system, and the sense-organs. This fact..led to the designation of the two poles as *animal* and *vegetative*.

C. 1. *animal-lover, -name, -ornament*; **animal black,** that formed by the carbonization of animal substance (cf. BONE-*black*, IVORY-BLACK); **animal electricity** (earlier example); **animal flower** (earlier example); **animal food** (examples); **animal grab** [GRAB *sb.*[2] 5], a card game similar to 'snap'; **animal heat** (examples); **animal magnetist** (earlier example); **animal pole** *Embryology* (see quots. and cf. POLE *sb.*[2] 7); **animal psychology,** the study of the behaviour of animals; hence *animal psychologist*; **animal size** [SIZE *sb.*[2]], a size made from gelatine; **animal (tub)-sized, tub-sizing** (see quot. 1937 and TUB *sb.* 10).

a **1875** KNIGHT *Dict. Mech.* I. 106/2 *Animal black*, carbonaceous matter obtained by the calcination of bones in close vessels. **1957** E. POUND tr. *Rimbaud* 15 As factories of suet and animal-Black spread out the whiff and flavour from Grenelle. **1793** R. FOWLER (*title*) Experiments and Observations relative to the Influence lately discovered by M. Galvani, and commonly called Animal Electricity. **1767**

Animal flower [see ACTINIA]. **1749** FIELDING *Tom Jones* I. I. i. 3 The several Species of animal and vegetable Food. **1819** KEATS *Let.* 26 (?) Oct. (1958) II. 225, I have left off animal food. **1912** E. M. DELL *Way of Eagle* liii. 350, I can play Animal Grab as well as anybody. **1941** J. CARY *House of Children* x. 38 The unexpected playmate who..forgot some pressing duty in order to play..animal grab, even more noisily than we. **1779** A. CRAWFORD *Experiments & Obs. on Animal Heat* iv. 81 Animal heat depends, indirectly, upon a change which the blood undergoes in the course of the circulation. **1874** ROSCOE *Elem. Chem.* 441 The whole of the animal heat is derived from the combustion of the materials of the body. **1910** F. E. WHITE (*title*) The animal lover's birthday book. **1928** R. CAMPBELL *Wayzgoose* ii. 43 The usual animal-lover's sloppiness which is popular everywhere. **1792** *Looker-On* No. 20, 15 May 155 A great number of animal magnetists were among this crowd of philosophers. **1931** C. L'E. EWEN *Hist. Surnames* xiii. 333 The Anglo-Saxons commonly bestowed animal-names upon their children. **1937** *Burlington Mag.* Feb. 99/1 Intricate geometrically conceived animal-ornament. **1887** C. O. WHITMAN in *Jrnl. Morphol.* I. 108 The next step consists in the formation of four ectoblastic micromeres which eventually present the figure of a quarter-foil at the animal pole. *Ibid.* 111 The macromeres..take no further part in the cleavage, if we except the budding off of ectoblastic micromeres at the animal pole. **1890** BILLINGS *Med. Dict.* I. 69/2 *A*[*nimal*] *pole*, pole of ovum at which there is least yolk, and where the polar globules are extruded; where, also, subsequent segmentation is most rapid. **1961** *Brit. Med. Dict.* 1134/2 *Animal pole*, that end of the early ovum which shows.. greatest proliferation and metabolic activity, and which eventually forms the head end of the embryo. **1894** CREIGHTON & TITCHENER tr. *Wundt's Lects. Human & Animal Psychol.* xxiii. 342 The inclination of animal psychologists to see the intellectual achievements of animals in the most brilliant light. *a* **1942** B. MALINOWSKI *Sci. Theory Cult.* (1944) ix. 89 The concept of drive is better omitted from any analysis of human behavior, unless..we understand that we have to use it differently from the animal psychologists or physiologists. **1881** G. J. ROMANES *Animal Intelligence* p. vii, If it is remembered that my object in these pages is the mapping out of animal psychology for the purposes of a subsequent synthesis, I may fairly.. receive credit for a sound scientific intention. **1887** Animal size [see *tub-size*, TUB *sb.* 10]. *a* **1912** Animal tub-sized [see A.T.S., *A III]. **1937** LABARRE *Dict. Paper Terms* 98/1 *Animal tub-sizing*, abbr. A.T.S., same as tub-sizing. **1940** *Chambers's Techn. Dict.* 36/1 *Animal-sized*, paper which has been hardened by passing the sheet through a bath of gelatine. More costly than engine-sized.

animalist. Add: **4.** An artist who makes figures of animals; an animal-painter or -sculptor; also, a writer of stories of animals.

1886 *Century Mag.* Feb. 484/1 [Barye] brought envy and malice on his head through the erection..in the Tuileries gardens of his colossal bronze lion and serpent. It was then the sneer of 'animalist' began. **1890** *Universal Rev.* 15 July 412 The animalists greatly repeat themselves. **1909** *Daily Chron.* 19 Jan. 4/4 The whole crowd of animalists, from Æsop to Mr. Jack London.

Hence **animali·stic** *a.*, of or pertaining to animalism.

1877 L. STEPHEN in *Cornhill Mag.* Apr. 427 One may see in him [*sc.* Kingsley] the type of character which, under other conditions, produces the 'diabolical' or rather the animalistic school of art and literature. **1955** *Bull. Atomic Sci.* Mar. 76/3 Should we regret having to abandon many poetic illusions of the vast, pantheistic or animalistic interpretation of the world?

animalization. Add: **1.** (Earlier example.)

1767 J. PRIESTLEY *Hist. & Pres. State Electr.* IV. 489 The processes of calcination, vegetation, animalization.

b. The process of animalizing vegetable fibre.

1862 O'NEILL *Calico Printing & Dyeing* 14 The present view of animalisation is, that it is not possible to animalise a fabric in any other way than by actually depositing upon it the animal matter in question.

4. The action of making into or representing as an animal. (Cf. ANIMALIZE *v.* 1.)

1886 HUXLEY in *19th Cent.* Apr. 493 In the theology of both the Babylonians and the Egyptians there is abundant evidence..of..the deification of animals, and the converse animalisation of Gods.

animalize, *v.* **1.** Delete † *Obs.* and add:

1835 [see ANIMALIZED *ppl. a.* 2]. **1886** [implied in *ANIMALIZATION 4]. **1889** *Cent. Dict.* s.v., The Egyptians animalized their deities.

2. b. To convert (vegetable fibre) into a substance resembling animal fibre.

1862 [see prec. 1 b]. *a* **1875** KNIGHT *Dict. Mech.* I. 106/2 *Animalizing fiber*, the process of conferring upon vegetable fiber the physical characteristics of animal fiber.

‖**anima mundi** (æ:nimă mɒ·ndəi). [med.L. (Abelard), = 'soul of the world'; app. formed to render Gr. ψυχὴ τοῦ κόσμου. Cf. WORLD-*soul, -spirit*, *WELTGEIST.] The soul of the world; a power or spirit supposed to be diffused throughout the material universe, organizing and giving form to the whole and to all its parts, and regularizing the motions and alterations of the parts.

1678 CUDWORTH *Intell. Syst.* I. iv. 561 All the Perfection of a Mundane Soul, may perhaps be attributed to God in some sense, and be [*printed* he] called, Quasi Anima Mundi, As it were the Soul thereof. **1698** J. LOCKE *3rd Let. to Stillingfleet* in *Wks.* (1823) IV. 363 There might be added other senses, wherein the word φύσις may be found, made use of by the Greeks..; as particularly Aristotle, if I mistake not, uses it for a plastic power, or a kind of *anima mundi*, presiding over the material world, and producing the order and regularity of motions, formations, and generations in it. *a* **1748** WATTS *Improvement of Mind* II. v. § 2 in *Wks.* (1753) V. 340 Take a Platonist, who believes an *anima mundi*, a universal soul of the world to pervade all bodies, to act in and by them according to their nature, and indeed to give them their nature and their special powers. **1890** W. JAMES *Princ. Psychol.* I. x. 346, I find the notion of some sort of *anima mundi* thinking in all of us to be a more promising hypothesis. **1948** *Mind* LVII. 240 Especially the belief in an *anima mundi* and in the comprehensive analogy between man and the universe are present in Herbert's discussion. **1957** G. BOAS *Dominant Themes of Mod. Philos.* viii. 280 This was not the first time that a correlation between the *anima mundi* and the Holy Spirit had been made; indeed Abelard had been condemned at the Council of Sens in 1140 for identifying the two. **1958** *Spectator* 17 Jan. 80/3 His [*sc.* Wordsworth's] belief in an *anima mundi* was encouraged by ideas taken from Cambridge authorities such as Newton himself.

animate, *ppl. a.* Add: **4. b.** *Gram.* Denoting living beings.

1902 *Amer. Anthropologist* Jan.–Mar. 27 The distinction between animate and inanimate gender is still preserved in both Penobscot and Abenaki. **1939** S. C. BOYANUS *Spoken Russian* ii. 5 The Accusative of masculine animate nouns is identical with the Genitive in the singular and plural.

animated, *ppl. a.* Add: **1. c.** (Additional example.)

1813 *Theatr. Inquisitor* II. 184 Correct and animated pictures of existing manners..are always gratifying.

d. Of or pertaining to cinematographic animation or 'moving pictures', esp. to cinema cartoons.

[**1895** *Cassell's Fam. Mag.* Mar. 320/1 The handle is turned..so as to move the series of photographs rapidly past the eyes, and give the impression of a single animated portrait.] **1897** C. M. HEPWORTH *Animated Photogr.* xii. 93 For animated photograph work it [*sc.* the 'Stigmatic' lens] can be made with an aperture of *f*/2. *Ibid.* xiii. 94 On taking animated photographs. **1903** *Westm. Gaz.* 28 Aug. 8/2 The America Cup race will be shown by animated pictures. **1912** [see next]. **1915** *Harper's Wkly.* 11 Dec. 574/3 Even cartoons began to come in—'animated' cartoons, as they are called. **1921** A. C. LESCARBOURA *Cinema Handbk.* (1922) xiv. 435 The basis of animated cartoons, animated models or so-called mechanigraphs, animated sculpture and so on is the fact that one frame can be exposed at a time with the usual motion picture camera. **1934** *Discovery* Feb. 50/2 The animated picture is that of a continuously repeated cycle of effects. **1935** R. MACAULAY *Pers. Pleasures* 135 There is a news reel on;..troops walk past with that jerky gait peculiar to animated photography. **1960** *Times* 11 Feb. 3/4 Up to about 1950 the animated film all over the world was dominated by the influence of Disney.

animation. Add: **8.** *Cinemat.* The production of 'moving pictures'; the technique by means of which movement is given, on film, to a series of drawings (*esp.* for an animated cartoon). (See quot. 1959.)

1912 F. A. TALBOT *Moving Pictures* i. 7 What we describe as animated photography is not animation at all. All that happens is that a long string of snap-shot photographs..are passed at rapid speed before the eye. **1919** A. C. LESCARBOURA *Behind Motion-Picture Screen* xvi. 302 The animation of a picture calls for a large number of separate drawings. **1958** *Observer* 9 Mar. 8/5 Commercial television has brought a boom in animation, with comic men and goofy animals bouncing out from everywhere. **1959** HALAS & MANVELL *Technique Film Animation* 336 *Animation*, the art of giving apparent movement to inanimate objects. The word is also used for the sequence of drawings made to create the movement, and for the movement itself when seen on the screen.

animatism (æ·nimătiz'm). [f. ANIMAT(E *a.* + -ISM.] The ascription of psychic qualities to inanimate as well as animate objects. So **animati·stic** *a.*

1900 R. R. MARETT in *Folk-Lore* June 171 It is Animism in the loose sense of some writers, or, as I propose to call it, *Animatism*. **1910** *Encycl. Brit.* II. 53/1 The term [animism] is often extended to include pantheism or animatism, the doctrine that a great part..of the inanimate kingdom, as well as all animated beings, are endowed with reason.. identical with that of man. *Ibid.*, One portion of the savage explanation of nature may have been originally animistic, another part animatistic. **1940** HINSIE & SHATZKY *Psychiatric Dict.* 33/2 Animatism..is vividly examplified..in schizophrenic subjects, who often personify the whole of the inanimate world, as if the latter possessed the same mental qualities as human beings do. **1941** R. R. MARETT *Jerseyman at Oxford* xi, It reminds me of an occasion long ago when Andrew Lang and Count Goblet d'Alviella were engaged in public and printed controversy over the meaning of my term 'animatism'—one of which, by the way, I am rather ashamed as a verbal impropriety though it has happened to become current. **1952** *Theology* LV. 75 The evidence as to the chronological sequence in which animatism precedes animism is conflicting.

†**animatograph** (ænimæ·tograf). *Obs.* [Hybrid f. L. *animātus* ANIMATE *a.* + -GRAPH.] An early name for the cinematograph; also, a cinematographic camera. Also *fig.* in reference to literary description. Hence †**animatographic** (-græ·fik) *a.*

1896 *Daily News* 26 Mar. 3/6 At a neighbouring house we have the 'cinematographe', and by way of varying the title Mr. Paul calls his pictures the 'Animatographe'. **1897** *Westm. Gaz.* 5 Feb. 7/1 Animatographic Records for British Museum...Photographs taken for the purpose of the animatograph, or theatregraph, as it is variously called. **1897** *Ibid.* 25 Sept. 2/3 The animatographic fight between Fitzsimmons and Corbett. **1898** *Field* 12 Feb. 213/2, I noticed..a man with an animatograph, and, as there was a good fence to start with, and I know one spill, the photos should not lack animation. **1902** *Westm. Gaz.* 3 Oct. 6/2 Animatograph scenes of life in the Royal Navy.

animator. Add: **2.** *Cinemat.* An artist who produces the original drawings for an animated film; sometimes applied to other persons concerned with the preparation of an animated film.

1919 A. C. LESCARBOURA *Behind Motion-Picture Screen* xvi. 306 After the master artist or animator has indicated the changes from one drawing to the next, his assistants work out the drawings. **1936** A. JENKINSON *America came my Way* xvii. 142 The 'originals' drawn by the animators are traced on to celluloid by the tracers.

animistic, *a.* Add: Hence **animi·stically** *adv.*, in an animistic manner.

1884 *Brit. Q. Rev.* LXXIX. 439 The doctrine of transubstantiation in the Romish Church..is the same thing, only that the new element is animistically imparted to the food, instead of being taken from it. **1942** *Mind* LI. 3 Speaking animistically, but obliged by the prestige of science to speak jokingly, 'My car goes when she chooses.'

animus. Add: (Earlier and later examples.)

1820 *Ann. Reg. 1819* 74/2 The original design..was demonstrative of the *animus* of the projectors. **1953** L. EDEL *Henry James, Untried Years* IV. 195 Henry James expressed..the starch that congealed the blood of some New Englanders, with an often ill-concealed..animus.

2. *Psychol.* Jung's term for the masculine component of a female personality. Cf. *ANIMA.

1923 [see *ANIMA]. **1943** *Horizon* VIII. 262 The dominating *animus* peeping through the light-heartedness of the young girl. **1943** H. READ *Educ. through Art* iv. 95 According to Jung, the conscious aspect of the individual's personality..is balanced..by a contra-sexual counterpart—that individual's animus (the male counterpart in the case of a woman) or anima (the female counterpart in the case of a man). **1962** J. JACOBI *Psychol. of C. G. Jung* (ed. 6) iii. 111 The animus-possessed woman, opinionated and argumentative, the female know-it-all, who reacts in a masculine way and not instinctively.

anion. Add modern pronunc. (æ·nəiǫ̆n) and substitute for def.: An ion carrying a negative charge which moves towards the anode (positive electrode) during electrolysis. Opposed to *cation*. (Earlier example.) Hence **anion exchange,** ion exchange involving anions; also *attrib.*; so **anion exchanger,** a substance capable of anion exchange. Also **anion-active** *a.*, having an active anion.

1834 W. WHEWELL *Let.* 5 May in I. Todhunter *W. Whewell* (1876) II. 182 If you take *anode* and *cathode*, I would propose for the two elements resulting from *electrolysis* the terms *anion* and *cation*,..and for the two together you might use the term *ions*. **1931** S. MATTSON in *Soil Science* XXXI. 325 Cation exchange predominates at high pH values whereas anion exchange predominates at low pH values. **1941** *Industr. & Engin. Chem.* XXXIII. 1271/2 The high-capacity anion-exchange resin was Amberlite IR-4. *Ibid.*, Table I indicates that a practical equilibrium adsorption value is attained with both anion-exchangers after 6-hour contact with either hydrochloric or sulfuric acid. **1946** *Nature* 26 Oct. 585/2 It is well known that the addition of divalent metallic ions greatly enhances the surface-active properties of anion-active detergents. **1963** A. J. HALL *Textile Sci.* vi. 296 Surface active agents or surfactants can be usefully classified into three main groups—*anion-active, non-ionic*, and *cation-active*. *Ibid.* 297 All the above surface active types are *anion-active*, that is, the effective hydrophobic component is present in the anion when the substance is dissolved in water and ionises into anions and cations.

Hence **anio·nic** *adj.*, (*a*) of or pertaining to anions; (*b*) = **anion-active*.

1922 *Chem. Abstr.* 3417 Evidence of salt properties is further verified by identity of negative charges with the no. predicted for the periodic group in which the anionic metal is found. **1928** *Jrnl. Chem. Soc.* II. 906 It would..be expected that the tendency to migration of different potentially mobile anions (X) contained in systems otherwise identical would follow the order of anionic stability. **1948** *Q. Rev.* (*Chem. Soc.*) II. 312 Although there are many examples of systems in which cationic exchange is known to take place, there are relatively few reports of the anionic exchange. These are almost all in the field of inorganic chemistry. The term 'anionic exchanger' is often applied, however, to materials which absorb acid molecules from aqueous solution. **1950** *Thorpe's Dict. Appl. Chem.* X. 800/1 The anion is the active part of the molecule, hence the term anionic agents. The best detergents are in this class. **1962** *Lancet* 26 May 1109/2 The drug slowly exchanges with cations (or anions in the case of anionic drugs) in the lumen of the alimentary tract and is then free to be absorbed.

anionoid (æ·niənoid, ānəi·ǫ̆noid), *a. Chem.* [f. ANION + -OID.] Resembling an anion (see quots.).

1925 A. LAPWORTH in *Proc. Manch. Lit. & Philos. Soc.* LXIX. p. xix, Some ions..are shown to have some properties in common and are termed 'cationoid'...[Some] have other properties in common and are termed 'anionoid'. *Ibid.*, The 'anionoid' properties of ethylenic hydrocarbons are held to be closely related to Thomson's observation that methyl is frequently observed in vacuum tubes with a positive charge, but never with a negative charge. **1936** *Discovery* May 162/2 The function of anionoid and kationoid reagents in ionic reactions. **1943** H. GILMAN et al. *Org.*

Chem. (ed. 2) II. xxv. 1859 The terms cationoid and anionoid..refer to the characteristic acceptor or donor activity of reactive cations or anions. 1953 C. K. Ingold *Struct. & Mech. in Org. Chem.* v. 203 [Lapworth] likened all reagents either to anions or to cations: an 'anionoid' reagent was one which behaved analogously to an anion, or, as Robinson subsequently expressed it, to an 'active' anion; and a 'cationoid' reagent was one which behaved like a cation, or an active cation. 1958 Packer & Vaughan *Org. Chem.* ii. 57, B is said to be electron-donating, anionoid (anion-like) or nucleophilic.

aniono·tropy. *Chem.* [f. Anion + -o + Gr. -τροπία turning.] Tautomerism characterized by the movement of an anion. So **anionotro·pic** *a.*
1928 Burton & Ingold in *Jrnl. Chem. Soc.* i. 904 It can be seen in a general way that opposite conditions will prevail in mobile-anion tautomerism (anionotropy). *Ibid.* 905 The effect of groups in promoting anionotropic change. 1941 *Trans. Faraday Soc.* XXXVII. 721 (*heading*) Mechanism and Kinetics of Anionotropic Change. 1942 *Nature* 31 Jan. 128/2 Anionotropic change is, in some respects, similar to prototropic change... In anionotropy we are concerned with the movement of negative groups, and there is evidence that the interconversion may proceed through the intermediate formation of a positive mesomeric ion and a complementary negative ion.

aniridia (ænəiᵊri·diă, -iri·diă). *Path.* [mod. L., f. Gr. ἀν- (An- 10) + ἰριδ-, ἶρις Iris + -ia.] Traumatic or congenital absence of the iris. Cf. Irideremia.
1860 in Mayne *Expos. Lex.* Suppl. 1885 *Ophthalmic Rev.* IV. 247 *Congenital Aniridia.* Mr. Lang showed a mother and son in both of whom the iris on both sides was totally deficient. 1961 *Lancet* 22 July 170/2 Traumatic subtotal aniridia in which 60% of the iris is missing. *Ibid.* 171/1 This technique [for implanting an artificial iris] might be applicable not only to cases of *traumatic* aniridia but also to..cases of *congenital* aniridia.

aniseed. Add: **1. b. aniseed ball,** a round, hard sweet flavoured with aniseed.
1924 'R. Crompton' *William—the Fourth* viii. 125 'Have an aniseed ball?' said Ginger.

aniso-. Add: Also (before a vowel) **anis-. aniseikonia** (ænəi:səikōᵘ·niă) *Path.* [Gr. εἰκον-, εἰκών image], a defect of vision marked by the presence of unequal images in each eye; so **aniseikonic** (-kǫ·nik) *a.*, characterized by aniseikonia; **anisocoria** (-kǫ·riă) *Path.* [Gr. κόρη pupil], inequality in the size of the pupils of the eye; **anisocytosis** (-səitōᵘ·sis) *Path.* [Gr. κύτος receptacle], abnormal variation in the size of cells, esp. of the red blood corpuscles; **aniso·gamete** *Biol.* [cf. Isogamete], either of two unequal uniting gametes; so **aniso·gamous** *a.*, characterized by **aniso·gamy,** the union of two unequal gametes in reproduction.
1934 *Dartmouth Coll. Dept. Res. Physiol. Optics*, (*title*) Iseikonic Lenses; lenses to correct Aniseikonia. 1935 *Amer. Jrnl. Ophthalmol.* XVIII. 1014 (*title*) Aniseikonia—A Factor in the Functioning of Vision. *Ibid.* 1018/2 Aniseikonic patients who come to the clinic are being tested..after their aniseikonia has been measured and corrected on the ophthalmo-eikonometer. 1944 *Electronic Engin.* XVII. 189 The aniseikonic effect is introduced by using optical systems of different focal lengths. 1961 *Lancet* 22 July 167/2 The patients are..examined on the space eikonometer..which is an ingenious instrument enabling one to test for aniseikonia. 1902 *Lancet* 12 July 69/2 The anisocoria met with in some cases of thoracic aneurysm might be sufficiently explained by unilateral alterations in blood-pressure. 1910 *Practitioner* Apr. 420 Wall and Ainley Walker..have come to the conclusion that the most common cause of anisocoria is unequal blood pressure in the ophthalmic arteries. 1904 Stedman *Dunglison's Dict. Med. Sci.* (ed. 23) 64/1 *Anisocytosis*, irregularity in size of associated cells, especially of red blood cells. 1956 *Nature* 17 Mar. 524/1 During the first two weeks of cortisone treatment, the degree of polychromasia and anisocytosis seen in the normal blood picture became increased. 1891 Hartog in *Nature* 17 Sept. 484/1 *Anisogamy*: the union of two gametes differing chiefly in size. 1891 [see Isogamy]. 1901 G. N. Calkins *Protozoa* 221 No sharp line, however, can be drawn between conjugation in isogamous and anisogamous forms.

aniuma: see *anhima.

ankaramite (æŋkărēⁱ·məit). [a. F. *ankaramite* (Lacroix 1916, in *Comptes Rendus* CLXIII. 182), f. *Ankaramy* in Madagascar: see -ite¹.] A melanocratic basaltic rock (see quots.).
1926 G. W. Tyrrell *Princ. Petrol.* vi. 129 Some basaltic types..need to be distinguished from basalts proper... When augite is the predominant mafic mineral the term *ankaramite* (Ankaramy, Madagascar) is applied. 1933 R. A. Daly *Igneous Rocks & Depth of Earth* xvi. 398 The ankaramites appear to be analogous mixtures of approximately basaltic liquid with concentrated crystals of pyroxene. 1939 *Geogr. Jrnl.* XCIV. 134 Other basic lavas of which ankaramite is one notable type. 1946 *Nature* 3 Aug. 172/1 The commonest rock-type is trachybasalt, accompanied by variations in the ultra-basic direction through olivine-rich trachybasalt and ankaramite to limburgite.

ankh (æŋk). Also **ank.** [Egyptian, = life, soul.] A figure resembling a cross, with a loop or ring forming a handle instead of the upper arm: used in ancient Egyptian art as a symbol of life. Also called *crux ansata*. Cf. Tau 2 b.
1888 A. H. Smith *Catal. Engr. Gems Brit. Mus.* 49 In front, Uraeus; beneath, ānχ and Uraei. 1895 Elworthy *Evil Eye* 279. 1896 T. Wilson *Swastika* 766 The *Crux Ansata*..according to Egyptian mythology, was Ankh, the emblem of Ka, the spiritual double of man. 1910 A. Churchward *Signs & Symb. Primord. Man* xix. 409 The cross on the church of Amba Derho is a form of the Ank-cross. 1961 S. Lloyd *Art Anc. Near East* iv. 125 (*caption*) In his hand he carries two *ankh*, symbols of the resurrection.

ankle, *sb.* Add: **3. ankle-boot,** (*a*) a boot reaching to or a little above the ankle; (*b*) a covering for a horse's ankle, used as a protection; **ankle-jerk,** a reflex movement of the ankle-joint produced by tapping the Achilles tendon; **ankle-length** *a.*, of a garment, that reaches down as far as the ankles; **ankle-sock,** a sock reaching to just above the ankle; **ankle-tie** = *ankle-straps*.
1840 *New Monthly Mag.* LVIII. 505 He has no stockings under his ankle-boots. 1917 G. Frankau *City of Fear* 16 The ankle-boots and the puttees, caked stiff with the Flanders mud. 1936 *Discovery* July 228/2 Many parachutists wear special ankle boots. 1958 J. Hislop *From Start to Finish* 166 *Ankle-boots*, felt or cloth pads made to go round a horse's legs below the knee and fastened with straps. 1888 A. L. Ranney *Lect. Nerv. Dis.* ii. 173 The *ankle-jerk*. If the muscles of the tendo-Achilles be put upon the stretch by flexion of the foot, a blow upon that tendon will cause a similar extension of the foot. 1962 *Lancet* 1 Dec. 1133/2 Ankle-jerks were absent and knee-jerks much reduced. 1903 *Daily Chron.* 25 July 8/4 Ankle-length ones [skirts] have gained votaries across the Channel. 1950 H. McCloy *Through Glass* (1951) ix. 89 An ankle-length housecoat. 1936 W. Holtby *S. Riding* iv. vi. 262 Her young body, partially covered by pink brassière, trunks, slippers and white ankle-socks. 1962 V. Nabokov *Pale Fire* 123 In those days growing boys of high-born families wore on festive occasions..sleeveless jerseys, white ankle-socks with black buckle shoes, [etc.]. 1873 E. S. Phelps *Trotty's Wedding Tour* i. 3 Her little ankle-ties swung tormentingly and carelessly to and fro against the wood-pile. 1923 Kate D. Wiggin *Gard. Memory* 5 Children with brief legs, white stockings, and ankle-ties.

ankle (æ·ŋk'l), *v.* [f. the sb.] **1.** *intr.* To use the ankles to good effect in cycling (see quot. 1961).
1896 *C.T.C. Monthly Gaz.* July 323/1 How to learn to ankle is a question of some difficulty. 1961 F. C. Avis *Sportsman's Gloss.* 146/1 *Ankle*, to turn the pedal with the foot so that this pushes downwards with the forward stroke, and claws the pedal upwards from the rear, the ankle forming a kind of axis upon which the foot turns.
2. *intr.* To walk, go. *slang.*
1926 Maines & Grant *Wise-crack Dict.* 5/1 *Ankled by*, went by and not riding. 1930 'I. Hay' & Wodehouse *Baa, Baa, Black Sheep* ii. 49 Let's ankle out of this. 1932 Wodehouse *Hot Water* xv. 243 Ankling into the hospital and eating my grapes with that woman's kisses hot upon your lips.
b. *trans.* To walk in. *poet.*
1949 Dylan Thomas in *Botteghe Oscure* IV. 399 The heron, ankling the scaly lowlands of the waves.

anklet. (Earlier example.)
1819 Shelley *Cyclops* in *Poet. Wks.* (1904) 795 She was bewitched to see the many-coloured anklets.

ankus (æ·ŋkǝs). Also **ankush.** [Hindi *ankus*, Pers. *anguzh*, f. Skr. *ankuça*.] In India, an elephant-goad. Also *attrib.*
1886 Jacob & Hendley *Jeypore Enamels* 11 The pointed end of the goad..is termed *Ankas* or *Ankush*...It forms part of the *khillat* or 'dress of honour' given by the Maharajah of Jeypore. 1895 Kipling *2nd Jungle Bk.* 128 It was a two-foot ankus, or elephant-goad—something like a small boat-hook. 1951 J. Masters *Nightrunners* vii. 87 Each mahout..brandished his ankus. *Ibid.* 89 Oblivious of the ankus hook driven through her ear.

ankylosaurus (æ:ŋkilosǭᵊ·rǝs). *Palæont.* [mod.L., f. Gr. ἀγκύλο-ς crooked + σαῦρος lizard.] A cretaceous dinosaur of the group including the genus so called.
1908 B. Brown in *Bull. Amer. Mus. Nat. Hist.* XXIV. 190 *Stereocephalus*..is probably ancestral to *Ankylosaurus*. 1934 *Discovery* July 210/1 The tank-like *Ankylosaurus*, 'the most ponderous animated citadel the world has ever seen', had only to squat tight to defy even *Tyrannosaurus*. 1962 E. Neaverson *Stratigr. Palaeont.* 572 The nodosaurs *Ankylosaurus*, *Dyoplosaurus*..are likewise recorded from both areas.

ankylostomiasis (æ:ŋkilostomǝi·ăsis). *Path.* Also **anchylo-.** [mod.L., f. *Ankylostoma* (see below) + -asis.] A disease, characterized chiefly by iron-deficiency anæmia, caused by a nematode worm (*Ankylostoma duodenale*, or some similar species) parasitic in the intestines; also called *hook-worm disease, tunnel-disease*.
1887 J. D. MacDonald tr. in *Papers Legislative Council Ceylon 1886* No. lvii. 19 By the term anchylostomiasis, we designate the various pathological conditions which arise from the presence of the parasite anchylostoma duodenale in the human intestines. 1889 *Brit. Med. Jrnl.* 21 Sept. 656/2 Sudden death from ankylostomiasis. 1897 *Allbutt's Syst. Med.* II. 1043 The essential symptoms of ankylostomiasis are those of a progressive anæmia associated with symptoms of intestinal catarrh. 1931 J. Arthur Thomson in W. Rose *Outl. Mod. Knowl.* 219 Hookworm disease, or ankylostomiasis.

anlage (a·nlāgǝ). Pl. **anlagen, anlages.** Also with anglicized pronunc. (æ·nlēⁱdʒ). [G., = foundation, basis, f. *anlegen* to establish, f. *an-* on + *legen* Lay *v.*¹] The rudimentary basis of an organ or organism; in *Embryology*, the first accumulation of cells recognizable as the beginning of a part or organ.
1892 Minot *Human Embryol.* 432 Comparative anatomy renders it probable that every vertebra had ribs primitively, and most of them have still in the human embryo the anlages of ribs. 1907 Driesch *Sci. & Philos. Organism* (1908) I. 81 You will get a complete larva only from that part [of the gastrula] which bears the 'Anlage' of the endoderm. *Ibid.* 112 A leaf may be formed out of the Anlage of a scale, if all the leaves are cut off. 1962 *Lancet* 26 May 1129/1 This does not exclude its potential teratogenous effect in man, where the critical phases of determination of the various organ anlages are probably concentrated in distinct periods.

anlaut (a·nlaut). *Philol.* [G., f. *an* on + *laut* sound.] The initial sound of a word. So **a·nlauting** *ppl. a.*, serving as anlaut, initial.
1881 in *Imp. Dict.* 1884 Einenkel *St. Kath.* p. xvii, [MS.] Z frequently softens the anlauting *f* to *u*, that is, *v*. Whenever this *u* is no more than a peculiarity of Z's,..I gave it up for the *f*, which C always has as 'anlaut'. 1892 G. Dunn in *Classical Rev.* Feb. 2/2 This representation of γ by ζ only occurs with certainty as an anlaut. 1933 *Trans. Philol. Soc. 1931–2* 28 Many philologists..distinguish between (1) absolute *Anlaut*.., (2) relative *Anlaut*... The first is the position of a sound at the beginning..of a phrase, the second is its position at the beginning..of a word within a phrase.

anna. Delete portion of definition in brackets. Add later example.
1959 *Chambers's Encycl.* VII. 448/2 On 1 April 1957 the decimal system came into use [in India] and the rupee was divided into 100 *naye paise* (new paisas) instead of into 16 annas as formerly.

annalistically (ænăli·stikăli), *adv.* [f. Annalistic *a.*: see -ically.] In annalistic order; by way of annals.
1881 Freeman in Stephens *Life & Lett.* (1895) II. 222 The years 1088–1093 are easy to tell, almost annalistically. 1894 *Academy* 24 Feb. 164/2 Chronologically and annalistically arranged.

Annamese (ænǎmī·z), *a.* and *sb.* Also 9 **Anamese.** [f. *Annam* (see below) + -ese.] **A.** *adj.* Of or pertaining to Annam, a province of Vietnam, or its inhabitants. **B.** *sb.* **a.** An inhabitant of Annam. **b.** The language of Annam.
[1808 *Asiatick Researches* X. 267 The Anam language has neither genders, numbers, nor cases.] 1826 J. Conder *Modern Traveller: Birmah* [etc.] 322 Adrian..was interred..with all the pomp and ceremonies prescribed by the Anamese religion. *Ibid.* 332 The Anamese employ several sounds..which are incapable of being pronounced by a Chinese. 1862 R. G. Latham *Elem. Compar. Philol.* ix. 61 The collective name for them is Anam, or Annam; whence we get the adjectives Anamese or Anamitic. 1867 W. D. Whitney *Lang. & Study of Lang.* ix. 336 The principal nations of Farther India are the Annamese or Cochin-Chinese, the Siamese, and the Burmese... Annamese culture is of Chinese origin. 1875 —— *Life & Growth of Lang.* xii. 239 The various languages of Farther India—as the Annamese or Cochin-Chinese, the Siamese, and the Burmese. 1910 *Encycl. Brit.* II. 62/1 The Annamese language is composed of monosyllables. 1910 G. M. Vassal *On & Off Duty in Annam* viii. 134 The birth of a son..confers a celestial benediction on an Annamese. 1931 H. G. Wells *Work, Wealth & Happiness of Mankind* (1932) vii. 278 It is a description of brutal compulsion and unhappiness, inflicted in this case upon Annamese victims. 1932 W. L. Graff *Lang. & Languages* xi. 417 Annamese, spoken along the eastern coast of Tonkin. 1933 H. G. Wells *Bulpington* ix. 349 All these poor devils from the ends of the earth, Polynesians, Annamese, Coolies, Gurkhas.

Annamite (æ·nǎmǝit), *a.* and *sb.* Also 9 **Anamite.** [f. as prec. + -ite.] = *Annamese *a.* and *sb.* Hence (now rare) **Annami·tic** *a.* [cf. mod.L. *Annamiticum* (A. De Rhodes, *Dictionarium Annamiticum* 1651)].
1838 P. S. du Ponceau *Diss. Nature & Char. Chinese Syst. Writing* i. vii. 87 Of one of them only, a dictionary and a short grammar exists in Europe. It is the Anamitic, or, as some write it, Annamitic, the language of the country of Anam. 1847 *Jrnl. Indian Archipelago* I. 112 The Anamite language is monosyllabic. *a* 1861 H. Mouhot *Trav. Central Parts Indo-China* (1864) I. vii. 205 My Annamite was with me, and filled the office of interpreter. 1862 [see prec.]. 1877 *Trans. Philol. Soc. 1875–6* 83 Cambojan..has loan-words from Malay, Pali, Annamite, and Mon. *Ibid.* 84 Every town has at least four names, being known under a different combination of syllables by the Siamese, Annamites, Cambojans and Savage peoples. 1910 G. M. Vassal *On & Off Duty in Annam* 1 The 'Annamitic Chain' expands in North Tongking, traverses Annam, where the highest

peaks..are to be found. 1921 E. Sapir *Language* 205 Chinese has flooded the vocabularies of Corean, Japanese, and Annamite for centuries, but has received nothing in return. 1936 *Times Lit. Suppl.* 14 Nov. 915/4 The presence of black Senegalese and yellow Annamite troops.

annatto, annotta, annotto, varr. Anatta.

annexational (ænĕksēⁱ·ʃənăl), *a.* [f. Annexation + -al.] Of, pertaining to, or relating to annexation. So **annexa·tionism,** advocacy of, or a policy aiming at, annexation; **annexative** (ăne·ksătiv) *a.*, disposed to annex territory, given to annexation.
1850 Carlyle *Latter-day Pamph.* (1872) iv. 127 A regular statistic of Annexationism. 1869 *Nation* 8 Apr. 267 They..are availing themselves of the strong annexational fever which now rages. 1918 A. Gray tr. *R. Grelling's The Crime* II. i. 24 That a belligerent state..should..make known annexational intentions. 1921 *Contemp. Rev.* Dec. 723 A nation..insatiably aggressive and annexative. 1940 *Manch. Guardian Weekly* 19 Apr. 312/3 There assembled..in the Prussian Parliament on July 7 1915..about fifty persons who were concerned to oppose this mischievous annexationism.

annex(e, *sb.* Usual pronunc. now (æ·neks).

annexive (ăne·ksiv), *a.* *Gram.* [f. L. *annex-* (see Annex *v.*) + -ive.] Expressing annexation: = Conjunctive *a.* 3 b.
1853 Edwards & Taylor tr. *Kühner's Gram. Gr. Lang.* § 321 A copulative coördinate sentence is either annexive or enhansive. 1913 J. M. Jones *Welsh Gram.* 440 Conjunctions are..i. annexive..ii. disjunctive.

annexure (ăne·ksiŭɹ, -ʃiŭɹ). [f. L. *annex-* (see Annex *v.*) + -ure.] Something annexed: = Annex *sb.*, Annexment 1.
1878 *Fraser's Mag.* XVIII. 85 A large and profitable annexure to our vast possessions on the Indian continent. 1902 *Westm. Gaz.* 7 July 5/4 The annexures to the petition. 1904 *Blue-bk. Papers Rel. to Thibet* 15, 17.

annihilation. Add: **3.** *Physics.* (See quot. 1962.)
1938 R. W. Lawson tr. *Hevesy & Paneth's Man. Radio-activity* (ed. 2) vii. 72 Processes in which the total energy of annihilation is emitted in a single quantum of energy 10^6e.V. 1942 J. D. Stranathan *Particles Mod. Physics* 374 (*heading*) Creation and Annihilation of Electron-Positron Pairs. 1962 *Gloss. Terms Nucl. Sci.* (*B.S.I.*) 9 *Annihilation*, a collision between a particle and its anti-particle..in which they both disappear, their energy being converted into annihilation radiation.

Anno Domini. Add: **b.** *jocular colloq.* as *sb.* Advanced or advancing age. Also *attrib.*
1885 F. Gale *Hon. R. Grimston* xvi. 284 Whenever he felt less able to do things than formerly, he used to say he was afraid 'Anno Domini' was the cause. 1892 F. Pollock *Leading Cases* 93 Yet nothing mortal may deny The march of *Anno Domini* Not e'en the Senior Fellow. 1900 *Daily News* 3 Feb. 2/5, I suffer from an incurable complaint—the complaint of Anno Domini. 1906 E. V. Lucas *Fireside & Sunshine* 186 When the time came for A. to take the bat he was unable to do so. *Anno Domini* asserted itself. 1921 *Times* 31 Mar., My resignation is due to what we call the *anno Domini* clause, which has come into operation.

announce, *v.* Add: **1. c.** To make (a person) known *as* so-and-so; often *refl.*
1837 E. Howard *Old Commodore* II. xix. 184 Announcing himself as a magistrate made a considerable impression on the seamen. 1856 De Quincey *Confess.* in *Wks.* V. 150, I announced myself as a passenger 'booked' for that night's mail. 1888 Mrs. H. Ward *R. Elsmere* ii, He announced himself as safely installed at Oxford. 1891 Hardy *Group of Noble Dames* 125 She could not live without announcing herself to him as his mother. *Ibid.* 158 She..requested him to allow her to..announce him as having died of malignant ague.

†**announce,** *sb.* *Obs.* [f. Announce *v.*; cf. Annonce.] **1.** = Announcement.
1787 J. Nichols in Welsted *Wks.* p. xxvi, This friendly announce is somewhat premature. 1818 Lady Morgan *Fl. Macarthy* III. iii. 106 The announce of the judges' carriages..induced the whole party to rise. 1827 *Blackw. Mag.* XXI. 210/1 The baker..seemed to have been struck by catalepsy at my first announce.
2. announce bill, a poster advertising a theatrical performance or the like.
1824 R. Humphreys in J. Decastro *Memoirs* 88 Here follows his announce bill for that night. 1866 M. Mackintosh *Stage Reminiscences* vi. 69 She [*sc.* Madame Vestris] then drove round to Fairbrother's, in Bow-street (the great playbill printer of those days), and ordered her 'announce' bills.

announcer. Add: **b.** *Broadcasting.* The person who announces the subjects of a programme or the items of current news, etc.
1922 *Radio Broadcast* Dec. 138 You have heard the announcer say: 'This is XYZ, the Blankety Blank station at So-and-So.' 1923 *Radio Times* 28 Sept. 19 (Announcer) 'The Stock Exchange was very lively to-day.' 1929 G. N. Clark in *S.P.E. Tract* XXXIII. 415 The professional announcers are not so dangerous as they seem. They belong to a class, very numerous under modern conditions, of what may be called persons of far-reaching insignificance. *Ibid.* 419 Coherence which is given..by the force of an individual personality... Those of us who value [this kind] of coherence regard with distaste the pure speech of the announcers. 1957 Auden & Kallman *Magic Flute* i. p. 57 For interruption is what we expect, Since that new god, the Paid Announcer, rose.

annual. Add: **A. 1. c.** *annual ring* = Ring *sb.*¹ 7 c.
1879 in R. Hunter et al. *Encycl. Dict.* 1902 *Encycl. Brit.* XXV. 417/1 Annual rings. 1905 *Sci. Amer.* Suppl. 25 Mar. 24433/1 A partial or entire separation of two consecutive annual rings. 1928 *Forestry* II. 127 The term 'annual ring', originally used in describing the structure of European trees, was later expanded to include the zones of growth discernible..in the trunks of tropical trees.
B. 3. b. Add to def.: Less frequently applied to an animal. Also *hardy annual* (lit. and fig.): see Hardy *a.* 4 b. (Further examples.)
1685 Evelyn *Diary* 6 Aug. (1955) IV. 462 The Apothecaries Garden of simples at Chelsey: where there..[are] many rare annuals. 1767 Abercrombie *Ev. Man his own Gardener* 49 To prepare for sowing some of the more curious forms of annuals. 1834 *Paxton's Mag. Bot.* I. 18 Treatment of Hardy Annuals. 1883 Day *Indian Fish* 31 The various modes in which the reproduction of these fishes is carried on... Whether the parents are monogamous, polygamous, or are annuals dying after the reproductive process has been accomplished. 1905 *Westm. Gaz.* 20 June 3/2 Presumably a time will come when even such a wonderfully hardy annual [as Sarah Bernhardt] will fail to reappear.

annulism (æ·niŭliz'm). [f. Annul(us + -ism.] Annulate structure.
1841 [see Radiism].

annunciation. Add: **4.** *attrib.*: Annunciation lily, a Madonna lily such as is depicted in pictures of the Annunciation.
1877 [see sense 2 in Dict.]. 1899 *Daily News* 4 Feb. 6/3 The dress is white velvet, embroidered straight up the front with long sprays of Annunciation lilies. 1907 *Daily Chron.* 2 May 4/5 The large drawing room,..decorated with tall Annunciation lilies..and exquisite roses.

annunciator. Add: **b.** (Earlier example.)
1853 A. Bunn *Old Eng. & New Eng.* 39 There is an appendage to this office..called an 'Annunciator', invented, we believe, by Jackson, of New York city, whereby all the bell-pulls of the house are brought within one focus.
c. Various techn. uses (see quots.).
1886 *Marine Engin.* 1 Feb. 287/2 The Baird Annunciators..indicate in the pilot house or elsewhere..whether the engines of a ship are in motion and if so in what direction they are moving. 1904 *Daily Chron.* 25 Mar. 3/6 The annunciator is an electrical apparatus by means of which the name of the member [of Parliament] speaking is printed immediately he rises to his feet on a broad paper band which runs horizontally. 1905 *Ibid.* 7 Apr. 4/6 The 'Annunciator'..announces to the smoking-room and libraries of the House of Commons the name of the Member addressing the Legislative Chamber. 1936 *Gloss. Terms Railway Signalling* (*B.S.I.*) 6 *Annunciator*, a device used to indicate audibly the passage of a train at a given point.

annunciatory, *a.* Delete *rare*⁻⁰ and add earlier and later examples.
1834 W. Roberts *Memoirs Hannah More* III. xi. 161 That one of the most illustrious females..should have her tenderest feelings thus barbarously sported with..in annunciatory advertisements. 1885 *Chambers's Jrnl.* 19 Sept. 606/2 Before sundown their [*sc.* the mosquitoes'] annunciatory hum began.

‖**annus mirabilis** (æ·nǝs məirēⁱ·bilis, mirā·bilis). [mod.L., lit. 'wonderful year'.] A remarkable or auspicious year.
1667 Dryden (*title*) Annus Mirabilis: the year of wonders, 1666. *a* 1700 Evelyn *Diary* an. 1660 (1955) III. 239 Annus Mirabilis. 1767 Ld. Chesterfield *Let.* 1 June (1774) II. 518 This has been every where an *annus mirabilis* for bad weather. 1885 *Good Words* Jan. 63/1 The years of evil fame which followed the *annus mirabilis* of 1815. 1940 *19th Cent.* Feb. 157 In 1848, the *annus mirabilis* of European history, a movement arose which shook the core of Europe. 1959 *Listener* 13 Aug. 251/1 By then he [*sc.* Tennyson] was successful and famous, his *annus mirabilis* of 1850 already three years behind him.

anoa (ănōᵘ·ă). [Native name.] An animal of the genus of the same name; a small wild ox of the Celebes.
1845 Lord Derby 6 Nov. in J. E. Gray *Gleanings Menagerie Knowsley* (1850) ii. 47, I have arranged our exchange for the *Anoa* and *Burchell's Zebra*. 1875 *Encycl. Brit.* II. 695/2 The singular ruminant, Anoa. 1888 *Athenæum* 1 Dec. 740/1 Prof. J. B. Steere [read a paper] on the 'tamaron', a bovine animal found in the island of Mindoro, Philippines, which he believed to be allied to the *anoa* of Celebes. 1930 *Times Educ. Suppl.* 27 Sept. p. iv/2 The Anoa, or dwarf buffalo, is confined to the island of Celebes.

anoci-association (ănōᵘ·si-). *Surg.* [f. A- 14 + L. *noc-ēre* to hurt + -i- + Association.] Minimizing of operational shock by means of an anæsthetic routine introduced by the American surgeon G. W. Crile (1864–1943) in 1908.
1911 G. W. Crile in *Jrnl. Amer. Med. Assoc.* LVII. 1814/1 The operation was done..under..anesthesia..completely protecting the brain from all harmful associations—a state best designated by a new word, viz., *anoci-association*. 1913 *Lancet* 23 Aug. 557/1 Anoci-association involved the cutting out of all conditions inimical to brain areas. 1961 *Brit. Med. Dict.* 106/2 *Anoci-association*... The patient is kept free from fear by management and narcotics, remains in ignorance of the time of the operation, and is anaesthetized in such a way that no adaptive response is excited. The field of operation is completely blocked by local anaesthetics so that traumatic impulses do not reach the brain [etc.].

anodal (æ·nodăl), *a.* [f. Anode + -al.] Pertaining to the anode; *anodal closure contraction*: see A.C.C. (s.v. *A III).
1882 *Athenæum* 8 July 50/3 The character (anodal or kathodal) of the electric charge. 1886 Buck *Handbk. Med. Sci.* II. 651/1 The second, anodal closure contraction, occurs when the anode is applied to the nerve or muscle.

anode. Add: (Earlier example.)
1834 W. Whewell *Let.* 25 Apr. in Todhunter *W. W.* (1876) II. 179, I have considered the two terms you [Faraday] want to substitute for *eisode* and *exode*, and recommend instead of them *anode* and *cathode*.
c. *attrib.* and *Comb.*, as *anode bend, circuit, current, surface, tap*; **anode mud,** the insoluble residue which forms on the anode during electrolytic refining.
1928 *Morn. Post* 26 Jan. 13/4 The best all-round results are often obtained when the rectifier is of the anode bend type. 1940 *Chambers's Techn. Dict.* 37/2 *Anode bend*, the more or less abrupt curve in the anode-current versus grid-voltage characteristic of a triode, which occurs at small values of anode current. 1919 *Radio Rev.* Nov. 54 There are..two sources of potential, the one in the anode circuit being of fixed value, that in the grid circuit being variable. 1920 *Ibid.* Aug. 545 The values of the anode-current..are then given. 1963 B. Fozard *Instrumentation Nucl. Reactors* xi. 129 The production of positive ions in the residual gas due to collisions between anode-current electrons and molecules. 1912 J. W. Mellor *Mod. Inorg. Chem.* xxi. 380 The anode is enclosed in filter cloth bags to facilitate the collection of the 'anode-mud'. Considerable amounts of silver and gold are obtained from the slimes or 'anode mud'. 1890 G. Gore *Art Electrolytic Separation Metals* ii. G. 188 The amount of anode surface per vat in different works therefore varies from 24 to 322 square feet. 1919 R. Stanley *Textbk. Wireless Telegr.* v. 91 It will suffice to say here that an anode tap may be usefully employed to increase the oscillating power and oscillating aerial current. 1929 J. A. Ratcliffe *Physical Princ. Wireless* iii. 48 The anode lead..should be connected to the aerial inductance by way of a movable contact known as the Anode Tap.

anodic (ænǫ·dik), *a.* Read: [f. Gr. ἄνοδος way up (f. ἀνά up + ὁδός way) or directly f. Anode + -ic.] **1.** *Electr.* = *anodal *a.*
1837 W. Whewell *Let.* 14 Oct. in S. P. Thompson *M. Faraday* (1898) iv. 164 As for positive and negative, I do not see why *cathodic* and *anodic* should not be used. 1844 Noad *Lect. Electricity* (ed. 2) 200 The anodic division..will finally become colourless, owing to the separation of chlorine. 1889 [see *A.C.C. s.v. *A III]. 1890 Billings *Med. Dict.* s.v., *A*[*nodic*]*-closure-contraction*,..contraction of a muscle when circuit is closed by placing the anode over the motive-point. *A.-opening-contraction*,..contraction when circuit is opened by removal of anode from over motive-point. 1943 *Electronic Engin.* XVI. 170 The operations involved are:..rinsing, anodising, modifying of the anodic film and plating.
b. *anodic bath*, the electrolyte in the anodic oxidation process; *anodic oxidation, process, treatment*, anodizing (see *anodize *v.*).
1924 Bengough & Stuart *Brit. Pat.* 223,995 The invention consists in subjecting a surface of aluminium or aluminium alloy to anodic treatment in an electrolytic bath. 1924 U. R. Evans *Corrosion of Metals* xi. 185 In the anodic process the recesses always receive adequate covering—no matter where the cathodes are placed. 1926 *Engineering* 27 Aug. 274/1 The film produced by anodic oxidation in a bath containing a chromate, bichromate, or—best of all—chromic acid, protects the metal..against corrosion. 1932 *Jrnl. R. Aeronaut. Soc.* Jan. 15 Chromic acid from the anodic bath..might cause trouble. *Ibid.* 11 In modern anodic oxidation plants stainless steel cathodes are frequently used.
2. *Physiol.* Of nerve force: proceeding towards a nerve-centre; afferent. (Cf. Cathodic *a.* 1.)
1852 M. Hall *Diastaltic Nerv. Syst.* 32 The influence which has induced this effect must have been anodic and cathodic in one and the same lumbar nerve.
3. *Bot.* (See quot. 1882 s.v. Cathodic *a.* 2.)
Hence **ano·dically** *adv.*
1905 J. McCabe tr. *Haeckel's Wond. Life* 101 Most of the flagellate infusoria..are anodically sensitive or positively galvanotactic. 1932 *Jrnl. R. Aeronaut. Soc.* Jan. 18 The modern tendency was to make tanks of either duralumin or aluminium, and the Air Ministry required that these should be anodically treated.

anodize (æ·nǒdəiz), *v.* [f. Anod(e + -ize.] To cover aluminium or its alloys with a protective layer of aluminium oxide by means of electrolysis. Hence **a·nodizing** *vbl. sb.*, **a·nodized** *ppl. a.*, covering, covered, with such a layer; **anodiza·tion,** the process of anodizing; **a·nodizer,** apparatus used for anodizing.
1931 *S.A.E. Jrnl.* XXIX. 148 Methods of protecting parts from exposure to corrosive influences are considered..in heat-treating, nitriding, anodising. 1933 *Archit. Rev.* LXXIV. 228/1 The wood doors are..fitted with anodized aluminium..handles. 1936 *Nature* 24 Oct. 708/1 The production of adherent oxide coatings on aluminium and its alloys by anodic treatment (anodizing). 1947 *Electronic Engin.* XIX. 170 Other controlling factors are indicated such as the concentration of the electrolyte and the nature of the alloy to be anodised. 1948 'N. Shute' *No Highway* iv. 109 Scratches on the paint and anodising [of the tailplane]. 1952 *Electronic Engin.* XXIV. 282 In carrying out anodization by the sulphuric acid process, the acid strength used industrially is about 36 per cent by weight. 1958 *Times Rev. Industry* Feb. 68/1 Recently one London factory brought an automatic anodizer into operation.

anodynous, *a.* (Later U.S. example.)
1904 'O. Henry' *Cabbages & Kings* v. 89 The prestige of drifting music on moonlit waters gave it an anodynous charm.

anoetic, *a.* Add: **2.** *Psychol.* Relating to or characterized by **anoesis** (æno,ī·sis), a hypothetical state of consciousness in which there is sensation but no thought.
1896 G. P. Stout *Anal. Psychol.* I. 51 Objective reference supervening on purely anoetic experience would be a completely new psychical fact. 1902 *Encycl. Brit.* XXXII. 63 Such a consciousness has been happily named anoetic. Whether or no it actually exists is another matter... But relative anoesis suffices here. 1916 W. McDougall *Introd. Social Psychol.* (ed. 10) Suppl. i. 360 Anoetic sentience; a mere feeling or sentience involving no objective reference.

anointment. For †*Obs.* read *arch.* and add:
1868 Milman *S. Paul's* 40 The dying yearned in vain for anointment with the blessed oil. 1887 Hardy *Woodlanders* I. iv. 59 As if your knee-jints were greased with very saint's anointment.

anolyte (æ·nǒləit). [f. Ano(de + Electro)lyte.] That part of the electrolyte which adjoins the anode. Cf. *catholyte.
1890 G. Gore *Art Electrolytic Separation Metals* ii. E. 79 In the cases of electrolysis of two liquids, separated by a porous partition, it is sometimes convenient to call the liquid containing the anode the anolyte, and that containing the cathode the catholyte. 1903 W. McA. Johnson in *Electrochem. Industry* I. 454 Whenever a current passes from the anode to the anolyte, or from the catholyte to the cathode, then chemical change ensues. 1949 F. A. Lowenheim in A. E. Knowlton *Stand. Handbk. for Electr. Engin.* (ed. 8) XX. 1926 The solution itself is the electrolyte; that surrounding the cathode is the catholyte, and that surrounding the anode the anolyte.

anomaloscope (ănǫ·mălǫskōᵘp). *Optics.* [ad. G. *anomaloskop*, f. Anomalo- + -scope.] An apparatus invented by W. A. Nagel (1870–1911) for measuring abnormality in colour vision.
Apparatus of this kind was first announced by Nagel in 1898 but he did not at first call it an 'Anomaloskop'.
1923 L. C. Martin *Colour* x. 150 Nagel's anomaloscope is a simple instrument similar to a direct-vision spectroscope, in which a yellow may be matched by a mixture of red and green. The brightness of the yellow and the ratio of red to green are indicated by the readings on two micrometer heads. The numerical divergences from the normal readings are taken as arbitrary measures of the 'anomaly'. 1929 W. D. Wright *Re-det. Trichr. Mixture Data* 27 So-called anomalous matches made on the Nagel anomaloscope. 1962 *Lancet* 15 Dec. 1269/1 Anomaloscope examination..reveals deviations from normal.

anomalous, *a.* Add: **2. d.** *anomalous dispersion*: the dispersion of light in the vicinity of an absorption band, where the refractive index changes rapidly with wavelength, being abnormally high on one side of the band and abnormally low on the other.
1883 R. T. Glazebrook *Phys. Optics* x. 277 In ordinary dispersing media the refractive index increases as the wave-length and the period of vibration decrease;..in media in which anomalous dispersion takes place this is not the case, for it is no longer true that waves of short period have a greater refractive index than waves of long period. 1899 A. S. Percival *Optics* viii. 179 Anomalous dispersion is best displayed in substances that have strongly marked absorption bands. 1904 J. Walker *Analytical Theory of Light* xvii. 337 There is an intimate connection between anomalous dispersion and the absorptive power of a substance. 1936 *Discovery* June 198/1 Such molecular rearrangements as anomalous dispersion and dipole phenomena.

anomalure (ănǫ·mălⁱūᵊɹ). *Zool.* [ad. mod.L. *Anomalūrus*, f. Gr. ἀνώμαλος Anomalous *a.* + οὐρά tail.] An animal of the African genus *Anomalurus* of rodents resembling the flying squirrels, and having projecting scales on the tail which serve for assistance in climbing; a scale-tailed squirrel.
1876 *Proc. Zool. Soc. 1875* 88 In external appearance the Anomalures very closely resemble the larger Flying Squirrels. *a* 1886 *Cassell's Nat. Hist.* III. 96 Whilst in the Flying Squirrels this spur springs from the wrist.., in the Anomalures it projects from the elbow. 1920 *Brit. Museum Return* 102 An Anomalure..from Sierra Leone.

anomaly. Add: **5.** *Meteorol.* (See quots.)
1853 E. J. Sabine tr. *Dove's Distribution of Heat* 20 We require..to exhibit the relation of the actual temperature of each place to the mean or normal temperature of its geographical latitude. I call the difference between the actual and normal temperature the 'thermic anomaly'. 1922 W. G. Kendrew *Clim. Cont.* i. i. 3 The 'anomaly of temperature' for that place, a positive anomaly if the place is warmer than the mean, a negative anomaly if it is colder.
b. *Geogr.* A local departure from the normal pull of gravity.
1924 H. Jeffreys *Earth* 121 This anomaly is always negative. In other words, the gravity on a mountain top is less than elsewhere. 1944 A. Holmes *Princ. Physical Geol.* xviii. 404 This band of what are called 'negative anomalies of gravity' implies that there is a corresponding deficiency of density in the materials of the crust beneath.

anomie, var. *anomy.

Anomœan (ænomī·ăn), *a.* and *sb.* *Theol.* Also **Anomean.** [f. mod.L. *Anomœus*, ad. Gr. ἀνόμοιος unlike, dissimilar (f. ἀν- priv. + ὅμοιος like, similar) + -an.] Belonging to, or a member of, an extreme sect of Arians who held that the Father and the Son are unlike in essence: = Heteroousian: opp. to *Homœan.
[1526 R. Whytford *Martiloge* sig. k1ʳ, Yᵉ heretykes called..anomeyes.] 1683 Cave *Ecclesiastici* 153 The Anomæan Heresie. *Ibid.* 224, The Head of the Homoiousian, against the Anomæans or Heterousian Sect. 1780 A. Butler *Lives Saints* V. 31 The Anomæans or rank Arians. 1840 Milman *Hist. Chr.* III. iii. v. 47 The party of the Anomeans triumphed, while Aetius, its author, was sent into banishment. 1874 J. C. Robertson *Hist. Chr. Ch.* ii. vi. 63 The perpetual virginity of the Saviour's mother was denied by the anomœan Eunomius. 1882–3 [see Heteroousian].

anomy (æ·nomi). Restrict †*Obs.* to sense in Dict. and add: **2.** Also commonly in French form **anomie.** [F. (Durkheim *Suicide*, 1897).] Absence of accepted social standards or values; the state or condition of an individual or society lacking such standards.
1933 E. Mayo *Human Probl.* vi. 130 We are facing a condition of *anomie*, of planlessness in living, which is becoming characteristic both of individual lives and of communities. 1938 R. K. Merton in *Amer. Sociol. Rev.* III. 674 The integration of the society becomes tenuous and anomie ensues. 1951 Spaulding & Simpson tr. *Durkheim's Suicide* ii. v. 253 The state of de-regulation or anomy is..further heightened by passions being less disciplined, precisely when they need more disciplining. 1959 *Guardian* 18 Sept. 8/5 An emphasis on order and stability (as opposed to anarchy or *anomie*). 1959 B. Wootton et al. *Soc. Sci. & Soc. Path.* ii. 69 Sociologists have thought it worth while to coin a special term—'anomie'—to describe the unorthodox social values, norms and attitudes to which 'underprivileged' children may be conditioned.
So **ano·mic** [ad. F. *anomique* (Durkheim, 1897)] *a.*, of, pertaining to, or characterized by such a state (see quots.).
1950 D. Riesman et al. *Lonely Crowd* iii. xiv. 287 My use of anomic..covers a wider range than Durkheim's metaphor: it is virtually synonymous with 'maladjusted'. 1951 Spaulding & Simpson tr. *Durkheim's Suicide* ii. v. 258 The third sort of suicide..results from man's activity's lacking regulation and his consequent sufferings. By virtue of its origin we shall assign this last variety the name of *anomic* suicide. 1961 B. R. Wilson *Sects & Society* 8 It is the leisure time of the individual in which his anomic circumstances will be most acutely felt.

anon (ănǫ·n). Also **anon.** (with full point), **Anon.** Abbrev. of Anonymous *a.* Hence as *sb.*, a person (esp. a writer or composer) whose name is unknown or not given.
1736 Pope *Works* IV. 235 A compleat Key to the What d'ye call it. Anon. By Griffin a Player, supervis'd by Mr. Th—. 1815 *Gentl. Mag.* LXXXV. ii. 539/2 Besides the great composers..the Author has drawn his materials from the following: Anon, Baillot, Borri, [etc.]. 1902 J. M. Barrie *Little White Bird* v. 55 She had sworn to hunt Mr. Anon down. 1947 *Punch* 5 Feb. 135 To-day I brought off a first-class scoop, An interview with Anon., The most prolific poet of us all. 1951 V. H. Galbraith *Hist. Research in Med. Eng.* 10 The medieval historians are dim figures. The vast majority would be classed in modern anthologies as 'anon'.

anonyma (ănǫ·nimă). [mod.L., fem. of L. *anōnymus* (see Anonymous *a.*).] †**1.** A woman of the demi-monde. *Obs.*
1864 Sala *Quite Alone* I. i. 2 Bah! there are so many Anonymas now-a-days. 1889 *Modern Society* 13 July 852 His mother-in-law mistakes his buxom laundress for a fair anonyma.
2. *Anat.* The innominate artery.
1871 T. H. Huxley *Anat. Vert.* viii. 380 The arteries arise from the arch of the aorta, as in Man, by an *anonyma*, a left carotid and left subclavian.

anonymity. (Earlier example.)
1820 *Blackw. Mag.* Oct. 98/2 An attempt was made to fasten anonymity on Blackwood's Magazine.

anonymuncule. (Earlier example.)
1867 Swinburne *Let.* 25 Dec. (1962) VI. 263, I have always found that these 'anonymuncules' vanish or collapse as soon as one attempts to set foot on them.

anopheles (ănǫ·filīz). *Ent.* [mod.L. (Meigen *Syst. Beschr. Eur. Zweifl. Ins.* (1818) I. 10), a. Gr. ἀνωφελής unprofitable, useless.] A mosquito of the genus *Anopheles*, which conveys the parasite of malaria. Also *attrib.* Hence **ano·pheline** *a.* and *sb.*
1899 *Daily News* 28 Sept. 3/4 We could kill most of the anopheles grubs here in a few hours with kerosene oil. 1902 *Encycl. Brit.* XXIX. 498/2 The parasite has now been found to live in the body of the Anopheles. 1915 G. B. Shaw *Doctors' Delusions* (1931) 133 Inoculation with anopheles vaccine. 1915 *Indian Jrnl. Med. Res.* July 180 The exact identity of Arabian anophelines has been for some time a matter of interest. 1920 J. Ritchie *Animal Life Scot.* 511 Mosquitoes,..I mean Anopheline, or..'spot-winged' Gnats. 1929 *Times* 10 Jan. 13/6 The anopheline mosquitoes, which spread malaria, breed chiefly in marshes and streams.

anopisthograph (æ:n,opi·sþogrɑf), *a.* [f. Gr. ἀν- (An- 10) + ὀπισθόγραφος Opisthograph.] Having no writing (or printing) on the back; inscribed only on one side. Also **a:nopisthogra·phic, -ical** *adjs.*; hence **a:nopisthogra·phically** *adv.*
1871 H. Bradshaw in *Coll. Pap.* (1889) xiv. 263 If a fragment is found printed only on one side it has hitherto been described as 'a remarkably interesting specimen of anopisthographic typography'. 1887 Bigmore tr. *Bouchot's Printed Bk.* i. 15 Gutenberg..desired..that the leaves of his books should not be anopist[h]ograph, or printed only on one side. 1887 *Book-Lore* VI. 159 Two leaves were..pasted together, back to back, forming what is termed an anopist[h]ographical book. 1893 E. G. Duff *Early Printed Bks.* 13 There must have been some reason for printing these books on one side only..or, as it is called, anopisthographically.

anopsia (ǣnǫ·psiă), var. Anopsy. See also quot. 1842.
1842 Dunglison *Dict. Med. Sci.* (ed. 3) 49/2 *Anopsia*, a case of monstrosity in which the eye and orbit are wanting. 1880 [see Anopsy]. 1937 *Brit. Jrnl. Psychol.* Apr. 360 These people have developed a functional anopsia with regard to conditions obtaining in other parts or groups of their social environment.

anorak (æ·nǒræk). Also ‖**anoraq.** [Greenland Eskimo.] A weatherproof jacket of skin or cloth, with hood attached, worn by Eskimos; a similar garment in countries other than Greenland.
1924 [see *muckluck]. 1934 I. W. Hutchison *North to Rime-Ringed Sun* xvii. 191, I displayed..the gay beaded 'anorak' and scarlet embroidered trousers and boots of the Greenland Eskimo woman. 1936 *Discovery* Mar. 91/2 The [Greenland] bride was resplendent in her handsome national dress of scarlet boots, sealskin trousers, and *anoraq* with magnificent bead collar. 1937 E. A. M. Wedderburn *Alpine Climbing* ii. 17 A wind-proof 'Anorak' or glacier smock, with a hood and without zip-fasteners..is a life-saver in foul weather. 1950 tr. *Mountaineering Handbk.* ii. 24 In rain an Anorak worn over the shoulders is an excellent protection. 1952 Morin & J. A. Smith tr. *M. Herzog's Annapurna* iii. 50 We put on our nylon anoraks..which were both snow- and wind-proof. 1959 *Vogue* Oct. 180 Swiss ski clothes..poplin anorak with attached hood.

anore·xia. Now the usual form of Anorexy. Also, **anorexia nervosa,** a condition marked by emaciation, etc., in which loss of appetite results from severe emotional disturbance.
1650 [see Anorexy]. 1873 W. W. Gull *Let.* 30 Apr. in *Trans. Clinical Soc.* (1874) VII. 26 The case appears to be an extreme instance of what I have proposed to call 'Apepsia hysterica', or 'Anorexia nervosa'. 1874 *Ibid.* 25, I used the term *Apepsia hysterica* but..*Anorexia* would be more correct. The want of appetite is, I believe, due to a morbid mental state... We might call the state hysterical... I prefer, however, the more general term 'nervosa'. 1939 *Psychosomatic Med.* I. 335/1 The syndrome of anorexia nervosa. 1963 P. H. Johnson *Night & Silence* xxiv. 173 He tried to eat it, but was taken by a terrible anorexia.

anormal, *a.* Delete †*Obs.* and add further examples. In recent use (ē¹,nǭ·rmăl), as antonym of 'normal' when the associations of 'abnormal', *i.e.* 'unhealthy', 'unnatural', would be inappropriate.
1936 *Mind* XLV. 504 An important distinction between 'normal' and 'anormal' series is defined. 1951 *Essays & Stud.* IV. 127 The collocations are unique and personal, that is to say, a-normal.

anorthosite (ænǭ·ɹþosəit). *Min.* [f. F. *anorthose* + -ite¹.] A granular igneous rock composed mainly of a soda-lime feldspar. Hence **anorthosi·tic** *a.*, of the nature of, or containing, anorthosite.
1863 T. S. Hunt in *Amer. Jrnl. Sci.* 2nd Ser. XXXVI. 222 A series of strata characterized by a great development of anorthosites. *Ibid.* 224 These anorthosite rocks are often compact, but more frequently granitoid in structure. 1925 J. Joly *Surface-Hist. Earth* iii. 47 These rocks..present to our view many varieties of structure and mineral composition; e.g. in the granites, gneisses, anorthosites, diorites. 1939 *Geogr. Jrnl.* XCIV. 104 A diorite..so altered by movement and pressure that it has acquired a foliated structure. They want to call it 'anorthositic gneiss'.

another. 1. b. Add to def.: *you're* (or *you are*) *another!*, a colloq. phrase properly used in retorting a charge upon the person who makes it (cf. Tu quoque); hence humorously as a meaningless or vaguely contemptuous retort.
1553 [see Dict.]. 1569 Preston *Cambyses* (1898) 786 And thou calst me knave, thou art another! 1749 [see Dict.]. 1836 Dickens *Pickw.* xv, 'Sir,' said Mr. Tupman, 'you're a fellow.' 'Sir,' said Mr. Pickwick, 'you're another.'

1882 [see Dict.]. 1887 Lowell *Democracy* 10, I find little..to edify me in these international bandyings of 'You're another'.
3. *another place*: a traditional phrase used by members of parliament to designate the other House, i.e. the House of Lords or the House of Commons.
1789 [see Place *sb.* 5 d]. 1883 [see Dict.]. 1908 *Westm. Gaz.* 27 July 1/3 Lord Lamington, was a busy member of the House of Commons..before he went to 'another place' on his father's death in 1890. 1927 *Observer* 10 July 15/7 Usually M.P.'s speak dispassionately of 'another place', but Mr. Baldwin the other day made it 'the other end of the passage', and Mr. Snowden 'the other end of the corridor'. 1960 *Guardian* 24 Nov. 18/6 Has a sovereign the ultimate right to unseat a member of Parliament and transfer him, against his will, to 'another place'?

anoxæmia (ænǫksī·miă). *Path.* [mod.L., f. Gr. ἀν- (An- 10) + Ox(ygen + Gr. αἷμα blood + -ia¹. Cf. F. *anoxémie*.] A deficiency of oxygen in the blood.
1890 Billings *Med. Dict.* I. 74/1 *Anoxæmia* or *Anoxyhæmia*. 1919 *Proc. Physiol. Soc.* 25 Jan. p. lxiv, While..the accelerated output of adrenalin must be supposed to participate in the effect, the rise in blood sugar must be in the main due to a direct action of the anoxæmia upon the liver. 1920 [see *anoxic *a.*]. 1922 *Encycl. Brit.* XXXI. 902/2 The chief sign of anoxaemia is cyanosis. 1931, 1939 [see *anoxia].

anoxia (ǣnǫ·ksiă). *Path.* [mod.L., f. Gr. ἀν- (An- 10) + Ox(ygen + -ia¹.] Deficiency of oxygen in the tissues (see quot. 1931).
1931 Peters & van Slyke *Quant. Clin. Chem.* I. 577 Anoxia literally means oxygen lack. We shall use the term to cover *any condition which retards oxidation processes in the tissues*... The less specific word anoxia seems preferable as a general term to cover the condition defined, 'anoxemia' being limited to..oxygen deficit in the blood. 1939 *Nature* 15 Apr. 619/2 'Pressor episodes', that is to say, a narrowing of the small vessels with resulting deprivation of oxygen (*anoxia* or *anoxæmia*). 1957 *New Scientist* 20 June 16/1 In the thin upper atmosphere, lack of oxygen may result in a progressively intoxicated state... *Anoxia*, the aviator's most insidious and lethal enemy, is perhaps the most difficult to detect and counter.
Hence **ano·xic** *a.*, characterized by or causing anoxia.
1920 J. Barcroft in *Lancet* 4 Sept. 487/1 The anoxic type of anoxæmia is the most serious. 1925 —— *Resp. Function Blood* (ed. 2) i. xii. 155 The same amount of exercise considerably reduces the oxygen content of the arterial blood in an animal which is already anoxic. 1955 *Sci. Amer.* Dec. 60/3 But equally we do not ignore the well-acclimated man accustomed to the low pressure of the mountain heights, who surely represents the maximum of adaptation to anoxic conditions.

ansate (æ·nsēⁱt), *a.* [ad. L. *ansātus*, f. *ansa* handle.] = Ansated *ppl. a.* Also in L. (fem.) form in *ansata cross* for *crux ansata* (= *ankh).
1891 Kinns *Graven in Rock* viii. 292 Iris and Horus are each holding the ansata cross or emblem of life. 1891 *Athenæum* 31 Oct. 591/1 Centurial stone, ansate, 17 in. by 8 in.

ansated, *ppl. a.* Delete †*Obs.*⁻⁰ and add:
1785 *European Mag.* VIII. 131 The Tau, or ansated Cross of the Egyptians. 1798 *Proc. Soc. Antiquaries* IV. 2 An ansated patera. 1891 *Athenæum* 16 May 644/1 [The stone has] well-cut letters..in an ansated panel.

ansation (ænsēⁱ·ʃən). [f. L. *ans*(*a* handle + -ation, after *ansate *a.*] The provision of handles.
1859 *Jrnl. Brit. Archæol. Assoc.* XV. 69 There is rarely any attempt at ansation; the nearest approach to handles being heavy perforated knobs placed a little beneath the mouth.

‖**Anschauung** (a·nʃau,uŋ). [G. = 'looking at', mode of view, f. G. *anschauen* to look at (MHG. *aneschouwen*); cf. Show *v.*] **1.** Sense-perception; *spec.* in Kantian philosophy = Intuition 5 c (see note in Dict.).
[*a* 1856 Hamilton *Logic* (1860) I. vii. 127 We are..in want of a general term to express what is common to the presentations of Perception, and the representations of Phantasy, that is, their individuality and immediacy. The Germans express this by the term *Anschauung*, which can only be translated by *intuition*,..which literally means *a looking at*.] 1865 Mill *Exam. Ham.* xvii. 327 By intuitions Mr. Mansel means the Anschauungen of Kant, or what Mr. Mansel himself otherwise calls Presentations of Sense. *Ibid.* xviii. 352 It is the Anschauungen, the intuitions, the presentations of experience, which we in this case compare and judge. [1889 E. Caird *Crit. Philos. of Kant* p. xi, *Anschauung* I have generally translated by 'Perception', rarely by 'Intuition', as the term Intuition seems in English to carry with it associations which are misleading. Sometimes I have used 'Pure Perception' where the context seemed to require it.] 1901 Baldwin *Dict. Philos.* I. 569/1 High authorities have of late questioned the equivalence of intuition and the Kantian Anschauung.
2. An outlook, attitude, or point of view.
1907 G. B. Shaw *Major Barbara* Pref. 148 The Shavian *Anschauung* was already unequivocally declared. 1922 *Internat. Jrnl. Psycho-anal.* III. 377 In many respects however they..are merely 'points of view' (*Anschauungen*).

‖**Anschluss** (æ·nʃlus, a·n-). [G., 'addition, annexation, union', f. *anschliessen* to join, annex.] Annexation or union, *spec.* of Austria to Germany (either the actual union in 1938 or as proposed before that date).
1924 *Ann. Reg. 1923* 192 This..could be explained by the decline of interest in the *Anschluss* idea, due to the calamitous condition of Germany. 1929 *Social Sci. Abstr.* 3496 *Anschluss* would be disastrous to Austria and to the peace of Europe. 1939 J. N. L. Baker *Atlas of War* 5/1 The Anschluss with Austria, proposed periodically since 1917 and forbidden by the treaty, was accomplished by force in March 1938. 1955 *Times* 9 May 11/2 A united Germany must not use an *anschluss* to extend her power into the Balkans and eastern Europe. 1958 *Economist* 18 Oct. 219/1 Ability to opt to join the Commonwealth might give certain states a valuable policy counter, when other forms of Anschluss seem to be in the offing.
Also **a·nschluss** (rare) *v. trans.*, to annex, join.
1945 in *Amer. Speech* (1948) XXIII. 69/1 Hitler's next big coup was to anschluss Austria. 1945 *News Rev.* 29 Mar. 5/3 What would anschlussed Austria do?

Anselmian (ænse·lmiăn), *a.* [f. St. *Anselm* + -ian.] Of or pertaining to St. Anselm (1033–1109), Archbishop of Canterbury, the scholastic philosopher, and esp. to his ontological argument for the existence of God.
1884 W. James in *Mind* IX. 283 Only the Anselmian proof can keep Fact out of philosophy. *a* 1910 —— *Some Probl. Philos.* (1911) iii. 43 The anselmian or ontological proof of God's existence, sometimes called the cartesian proof, criticised by Saint Thomas, rejected by Kant, re-defended by Hegel. 1943 E. L. Mascall *He who Is* viii. 98 What we cannot imagine we can nevertheless conceive, as we saw in discussing the Anselmian definition of God.

Anselmic (ænse·lmik), *a.* [f. as prec. + -ic.] = *Anselmian *a.*
1872 G. S. Morris tr. *Ueberweg's Hist. Philos.* I. ii. 379 The character of the Anselmic philosophy was not that of the whole period. 1932 *Times Lit. Suppl.* 18 Feb. 106/3 The objective or Anselmic theory, sometimes called the Satisfaction Theory. 1939 J. Baillie *Our Knowledge of God* iv. 151 Something that closely resembled the Anselmic proof which St. Thomas had condemned.

‖**an sich** (an zi·χʸ), *phr.* [G., = 'in itself'. Orig. taken from Ger. philos.] In itself; in the abstract; not in relation to anything else. Cf. *Ding an sich.
1846 J. D. Morell *Hist. View Speculative Philos.* II. v. 142 An existence may be viewed in relation to itself, or in relation to the things around it; it may be existence *an sich* or existence *für andere*. 1909 W. James *Pluralistic Universe* v. 215 Others, as Kant for example, have denied intellectualism's pretension to define reality *an sich* or in its absolute capacity. 1928 T. S. Eliot in E. Pound *Sel. Poems* p. xvii, Not the matter *an sich*, which is unknowable, but the matter as we know it. 1956 A. Huxley *Adonis & Alphabet* 197 Much more widespread than the love of truth is the appetite for marvels, the love of the Phony *an sich*, in itself and for its own sweet sake.

answer, *sb.* Add:
4. b. *answer-back*: a rejoinder or repartee; also *fig.*
1924 J. A. Thomson *Sci. Old & New* xvi. 90 First there is the gall—an answer-back which the plant makes to the irritation which follows when the gall-midge lays an egg in the soft tissue. 1925 —— *Sci. & Relig.* vi. § 9. 194 The struggle for existence is a formula covering all the answers-back that organisms make to environing difficulties.
6. b. *to know all the answers*, to be fully knowledgeable or expert; to be worldly-wise or experienced. (Cf. similar colloq. phrases s.v. Know *v.* 15.)
1933 G. Laven *Rough Stuff* p. ix, *To know all the questions and answers*, to know your way about. To be clever in things relating to crime and law-breaking. 1935 G. & S. Lorimer *Heart Specialist* v. 128 'You may be one of these carefully reared dames,' he said with a trace of doubt in his voice, 'but you sure know all the answers.' 1937 C. Boothe *The Women* in *Famous Plays 1937* 622 *Miriam.* Getting wise, aren't you? *Mary.* Know all the answers. 1940 J. Thurber *Fables for Our Time* xii. 30 It is better to ask some of the questions than to know all the answers. 1955 A. L. Rowse *Expansion Eliz. Eng.* x. 408 The positive old lady in the garden, who knew all the answers and could not be told anything, had not ceased to be a marvellous politician. 1956 A. Wilson *Anglo-Saxon Attitudes* ii. iii. 367 She'd been a glamour girl, but she knew all the answers.
9. answer print *Cinemat.* (see quot. 1940).
1940 *Chambers's Techn. Dict.*, *Answer print*, the first print from the edited negative, shown to the producers of the sound-film for final approval before release. 1959 Halas & Manvell *Technique Film Anim.* xix. 234 The studio manager..records deadline dates for line-tests, rough cuts and answer print screenings.

answer, *v.* Add: **12. e.** *spec.* in Horse-racing, *to answer* (*the question*): (of a horse) to respond to a call made by the jockey. (Cf. *to ask the question*, *ask *v.* 2 b.)
1894 H. Custance *Riding Recoll.* vi. 88, I..asked 'King Lud' the question. He answered in the most generous manner possible,..and won. 1894 *Idler* June 545 The certain winner of the Derby—if he is able to answer the question I am going to put to him.
14. Also, *to answer back*.
1904 Stanislaus Joyce *Dublin Diary* (1962) 52 She is obstinate and inclined to answer back a great deal. 1926 *Sat. Rev.* 20 Mar. 376/1 In dialogue the characters do not answer so much as answer back. 1929 H. G. Wells *King who was King* ii. § 4. 66 The King receives him coldly and admonishes him with evident severity... Against all etiquette he answers back. 1937 F. Stark *Baghdad Sketches* 172 One who, after many years of patient trampling upon, suddenly answers back. 1948 *Daily Herald* 7 Aug. 2/2 This sort of person is compensating himself for not being allowed to Answer Back as a child.

answerability (ɑ:nsərăbi·liti). [f. Answerable *a.* + -ity.] = Answerableness 2.
1929 *Another Milestone* (Brit. & For. Bible Soc.) 18 The answerability of the Christian West for such a situation. 1949 *Sc. Jrnl. Theol.* II. 280 Terms like Responsibility and its Anglo-Saxon equivalent answerability, have become household words on the lips of those who have contrived to keep abreast of recent theological thought.

ant. Add: **4.** *ant-colony, -community, -man*; **ant-bird,** any bird of the ant-eating family Formicariidæ, ant-thrush; **ant-cow,** a popular name for the aphides kept and tended by ants for the sake of the sweet fluid that they extract from them; **ant-eggs, ants' eggs,** properly, the eggs of ants; popularly applied to their larvæ or pupæ; **ant-orchis,** a terrestrial Australasian orchid of the species *Chiloglottis gunnii* (Morris *Austral English*, 1898); **ants' wood** = *saffron plum* (Saffron B. b); **ant-wren,** W. Swainson's name for a South American ant-eating bird of the genus *Formicivora*, included by P. L. Sclater in the subfamily Formicariinæ in his division of the family Formicariidæ, the ant-thrushes.
1858 Sclater in *Proc. Zool. Soc.* 13 Apr. 202 Synopsis of the American Ant-birds (Formicariidæ). 1897 J. H. Comstock *Insect Life* viii. 276 The workers are by far the most interesting portion of the ant colony. 1909 H. C. McCook (*title*) Ant communities and how they are governed. 1924 J. A. Thomson *Sci. Old & New* xii. 67 The success of the ant-community depends on a semi-repression of the workers. 1875 *Encycl. Brit.* II. 98/1 The Coccidæ in America take the place of the European aphides as ant-cows. 1901 H. G. Wells *First Men in Moon* xxii. 272 These ant-like beings, these ant-men. 1949 E. Sitwell *Canticle of Rose* 254 But I saw the little Ant-men as they ran. 1666 *Phil. Trans.* No. 23. 426 Upon opening of these Banks, I observe first a white substance, which..looks like the scatterings of fine white Sugar or Salt... This same substance..I finde in the Ants themselves, which I take to be the true Ants Eggs. *Ibid.* 427 In the Morning they bring up those of their Young (that are vulgarly call'd Ants Eggs) towards the top of the Bank. 1834 *Penny Cycl.* II. 61/1 To collect the cocoons (popularly and erroneously called ants'-eggs) in quantity as food for nightingales and larks. 1884 Ants' Wood [see Saffron B. b]. 1825 W. Swainson in *Zool. Jrnl.* (1826) II. 146 The *Formicivoræ*, or Ant-Wrens, are all of them very small.

antagonism. Add: **1. d.** *Biochem.* Inhibition of or interference with the action of a substance or organism (as a salt, microbe, mould, etc.) by another substance or organism.
1874 W. Roberts in *Phil. Trans. R. Soc.* CLXIV. 466 There was an antagonism between the growth of certain races of Bacteria and certain other races of *Bacteria*. 1938 *Ann. Reg. 1937* 349 Much research was done on synergism and antagonism of vitamins. 1949 H. W. Florey *Antibiotics* II. viii. xv. 636 When a mixture of micro-organisms is grown together, certain bacteria or moulds will inhibit the growth of other bacteria or moulds. This phenomenon is known as 'bacterial antagonism'.

antagonist. Add: **6.** *Biochem.* A substance or organism that inhibits or interferes with the action of another substance or organism. Cf. *antagonism 1 d. Freq. with defining word.
1889 Shoemaker & Aulde *Treat. Materia Med., Pharmacol.* I. i. 122 After the ingestion of a poison its physiological antagonist might be administered with the expectation that the effects of the first poison might be counteracted by the second. *Ibid.* 123 Morphine and atropine are well-known physiological antagonists. 1961 *Lancet* 12 Aug. 371/1 The adverse effects commonly produced by cytotoxic and cytostatic agents such as..folic-acid antagonists. *Ibid.* 2 Sept. 512/2 Only one other patient with periodic paralysis has been treated with an aldosterone antagonist.

antalphabetic (æ:nt,ælfăbe·tik), *a.* *Phonetics.* [f. Ant- + Alphabetic *a.*] = *analphabetic *a.* 2.
1933 Jespersen *Linguistica* Index 454/1 Antalphabetic symbols. 1937 [see *analphabetic *a.* 2].

antapex (ænt,ēⁱ·peks). *Astr.* Also **anti-apex.** [f. Ant- + Apex.] The point on the celestial sphere, situated in the constellation Columba, away from which the sun is moving; the point opposite to the apex of the solar way (cf. *solar apex* s.v. Solar *a.* 7).
1890 A. M. Clerke *Syst. Stars* xxiii. 320 There would be the semblance of a general retreat from the 'apex' or solar *point de mire*, coupled with a thronging-in from all sides towards the opposite point, or 'anti-apex'. 1906 J. C. Kapteyn in *Rep. Brit. Assoc. 1905* 258 If..we determine these lines of symmetry for several points of the sky and prolong them, they must all intersect in two points, which are no other than the apex and the antapex. 1911 *Encycl. Brit.* XXV. 791/1 The annual parallactic motion is equal to four times the parallax, for a star lying in a direction 90° from the solar apex; for stars nearer the apex or antapex it

is foreshortened. **1923** *R. Astr. Soc., Monthly Notices* LXXXIII. 465 Two quantities are of fundamental importance—the angular distance of a star from the apex or antapex of the solar motion, and the position angle of the star relative to the apex or ant-apex. **1928** W. M. Smart *Sun, Stars & Universe* x. 171 The sun..is moving towards the solar apex with a speed of 12 miles per second, and therefore our group of stars will apear to us to move as a whole towards the solar antapex.

ant-bear. Add: (Earlier example.)

1555 Eden tr. *P. Martyr of Angleria's Hystorie of West Indies* in *Decades* fol. 189 There is also on the firme lande an other beaste cauled *Orso Formigaro*, that is, the Ante beare.

2. The English name for the S. African aardvark. Also *attrib.*

1894 *Daily News* 8 Nov. 6/1 A poor little boy, who had tried to hide himself in an ant-bear hole. **1910** *Encycl. Brit.* I. 3/1 The typical form, *Orycteropus capensis*, or Cape antbear from South Africa. **1951** A. Roberts *Mammals S. Afr.* (1954) 239 Order Tubulidentata. Family Orycteropodidæ: Antbears. The family comprises the Antbears of the Old World, or Aardvarks (in Afrikaans now spelt Erdvarks), which are survivors of an ancient type of animal.

ante (æ·nti), *sb.*[2] Orig. and chiefly *U.S.* Also erron. **anti.** [a. L. *ante* before.] In *Poker*, a stake put up by a player (usually, the eldest hand) before the draw; similarly in other card games. Also *attrib.* in *ante-man* = *age *sb.* 6 b.

1838 *Victims of Gaming* 86 The one next to the dealer [at brag] puts up into the pool..any sum he may choose, unless the amount has been..fixed by the company. This sum is called the ante. **1844** J. Cowell *Thirty Years* 94 The dealer makes the game, or value of the beginning bet, and called the anti. **1853** J. G. Baldwin *Flush Times Alab.* 8 A negro *ante* and twenty on the call, was moderate playing. **1882** *Poker* 7 To begin the pool, the player next to the dealer, on his left, must put up money, which is called an 'Ante'. *Ibid.* 8 It is best generally for the ante-man to make good and go in.

b. *transf.* An amount paid in advance; price, subscription, means. Phr. *to raise the ante*, to procure money or means.

1890 *Harper's Mag.* Feb. 428/2, I raised the ante, and sold three hundred papers at ten cents each. **1934** *Bulletin* (Sydney) 1 Aug. 36/1 Weekly premiums are paid, and after a conviction a member pays a double ante, because his fine will be heavier next time. **1959** *Economist* 16 May 555/2 Raising the Ante... The problem for the News of the World was how it was to find the cash to make the purchase. **1964** *Wanganui Chronicle* 25 Mar. 1/9 Few people in Wanganui and district had the ante to go into the fishing industry.

ante (æ·nti), *v.* Orig. and chiefly *U.S.* Also erron. **anti.** [f. prec.] *trans.* To put (an ante); also *transf.*, to bet, stake; to pay *off, up*; to hand over, surrender. Also *absol.*

1846 J. J. Hooper *Adv. Simon Suggs* x. 129 Ante up! ante up, boys—friends I mean—don't back out! *Ibid.* xii. 144 Exsept..500 dollers I anteed off amongst the boys of a night, I couldn't git off a sent. **1854** in Thornton *Amer. Gloss.* (1912) 971 Playin' at billiards an' monte Till they've nary red cent to ante. **1857** *Knickerbocker* Jan. 43, I did hear that you anted off 1000 shares in trade for Texas lands. **1859** Bartlett *Dict. Amer.* (ed. 2), To *Anti*, to risk; to venture a bet; as, 'I'll anti all I'm worth on that'. This term is derived from the game of poker. **1888** Farmer *Americanisms*, To *Ante up* is to pay, as well as to wager. **1889** R. Guerndale *Poker Bk.* ii. 13 The player after the dealer must Ante first, before the draw. *Ibid.* vii. 42 B Antes one, and the cards are dealt by A. **1896** H. A. Vachell *Rom. Judge Ketchum* III. viii. 267 Death..and I hev bin pardners many a time, an' when he passes the word, I'll ante up with a smile if I kin. **1900** H. Lawson *On Track* 157 The man that doesn't ante gets the best of this world. **1904** 'G. B. Lancaster' *Sons o' Men* vii. 98 Look here; you anti them [missing sheep] up, at once. **1919** W. H. Downing *Digger Dialects* 8 *Ante up*, to surrender anything. **1938** Wodehouse *Summer Moonshine* viii. 88 You pay as you enter. Can you ante up? **1945** D. Wheatley *Man who missed War* i. 13 The income tax payer would have to ante-up quite a bit more in the pound. **1948** V. Palmer *Golconda* v. 33 'Ante up for your union ticket..,' went on Mahony.

ante-. B. *adjs.* **1. anteconsonantal,** preceding a consonant; **antecubital** *Anat.*, pertaining to the inner surface of the forearm.

1909 Webster, Anteconsonantal. **1927** *Mod. Philology* Nov. 226 OHG *za-* is the anteconsonantal form of *zar-*. **1937** C. E. Bazell in *Jrnl. Eng. & Ger. Philol.* Jan. 6 Part of a plural system wherein *j* represented before a vowel the ante-consonantal *i* of the other cases. **1893** *Funk's Standard Dict.*, Antecubital. **1908** *Practitioner* Sept. 486 The injection made into one of the large veins of the ante-cubital region. **1962** *Lancet* 29 Dec. 1372/2 In other cases an antecubital vein was used.

ante-bellum (æ·nti be·lǒm). [Lat. phr., 'before the war', used *attrib.* or as *adj.*] Previous to the war, i.e. *spec.* the American civil war (1861–5), the S. African war (1899–1902), or either of the wars of 1914–18 and 1939–45.

1862 M. B. Chesnut *Diary* 14 June (1905) 188 Her face was as placid and unmoved as in antebellum days. **1879** Tourgée *Fool's Errand* xv. 82 Evidently thinking that his connection with this ante bellum barbarity had somehow increased his importance. **1882** *Rep. Ho. Repr. Prec. Met. U.S.* 306 One of the productive mines of antebellum days. **1905** *Westm. Gaz.* 4 Jan. 1/3 The ante-bellum negotiations. **1924** *Contemp. Rev.* Apr. 438 Deaths of infants from overlaying..were also only one-third of the ante-bellum number. **1955** *Times* 28 May 7/5 In the particular case of our basic raw material, paper, the *ante-bellum* price for certain grades has actually trebled itself.

antebrachial, *a.*, more correct f. Antibrachial *a.* So **antebrachium** (anti-) *Biol.* [med.L.], in mammalia, the part of the arm or fore-limb from the wrist to the elbow; the forearm.

1877 [see Brachium]. **1880** [see Antibrachial *a.*].

antecede, *v.* Delete *arch.* and add to **1**:

1902 H. Spencer *Facts & Comments* 204 Recognizing the properties of Space..as anteceding all creation..and all evolution. **1914** C. A. Mercier *Astrol. in Med.* 1 Its origin is so remote as to antecede all written records.

a·ntechamber, *v.* *intr.* and *trans.* To wait or wait for in or as in an antechamber: cf. Antechambering *vbl. sb.*

1891 *Pall Mall Gaz.* 2 Oct. 1/3 All intriguing Paris was antechambering him. **1900** W. A. Ellis *Wagner* I. II. ix. 291 My poor dog..was antechambering in the street in wait for his more fortunate master, allowed to antechamber among men.

antecian, var. Antœcian.

A:nte-Commu·nion. In full *Ante-Communion Service* [Ante- B. 2], a popular term for that part of the Communion Service in the Book of Common Prayer which extends from the beginning of the office to 'the end of the general Prayer "For the whole state of Christ's Church militant here in earth"' and is appointed to be said 'upon the Sundays and other Holy-days (if there be no Communion)'.

According to the American rite, the Ante-Communion ends with the Gospel.

1827 *Gentl. Mag.* XCVII. II. 487 There was no offertory, and no prayer for the church militant. I am not aware by what authority this part of the Ante-Communion Service is now so commonly omitted on Sundays. **1871** F. G. Lee *Dict. Ritual Terms, Ante-Communion.* (1) An Anglican term used to designate that portion of the Liturgy or Communion service which, commencing with the Introit, or the Lord's Prayer, closes with the end of the Nicene Creed. (2) This term is also used for the introductory part of the Eucharistic office, when it only, and nothing further, is intended to be used. **1899** *Westm. Gaz.* 2 Jan. 4/2 Those of you who feel that it is not necessary to hear the ante-Communion service twice on Sunday.

‖**ante meridiem** (æ·nti mėri·diem), *phr.* [Lat. phr., 'before midday'.] Before midday; applied to the hours between midnight and the following noon; abbreviated a.m. or a.m. (q.v. in Dict. and Suppl.).

1563 Dee *Diary* 28 Sept. (1842) 2 Mr. John Ask ante meridiem, by York six myle on this syde. **1647** Lilly *Chr. Astrol.* iv. 41 If your hour of the day be in the morning, or as we say *Ante Meridiem*, or before noon.

antenatal, *a.* Add: **2.** Of, pertaining to, or concerned with the health and well-being of women during pregnancy.

1917 *Maternity & Child Welfare* (Local Govt. Board Rep.) p. xi, At Bradford there is an ante-natal clinic at the Maternity Hospital. *Ibid.* 19 There is a municipal ante-natal centre. **1930** *Lancet* 20 Dec. 1359/2 Many local authorities have already provided facilities for antenatal supervision.

antenna. Add: **1.** (Earlier example.)

1646 Sir T. Browne *Pseud. Ep.* III. xviii. 153 Insects that have *antennæ*, or long hornes to feele out their way, as Butter-flies and Locusts.

2. (Further examples.)

1918 E. Pound *Pavannes & Divisions* 43 My soul's antennæ are prey to such perturbations. **1959** *Listener* 17 Dec. 1082/1 This is where an author with sound learning, a seeing eye, and sensitive 'antennae' can be of great assistance.

4. A dorsal sense-organ in rotifers.

1886 A. G. Bourne in *Encycl. Brit.* XXI. 5/2 A structure found in many Rotifers, and variously known as the 'calcar', 'siphon', 'tentaculum', or 'antenna'.

5. *Radio.* An aerial wire or other device for radiating or receiving radio waves, an aerial: see *aerial *a.* 6 b.

1902 J. A. Fleming in *Encycl. Brit.* XXXIII. 230/2 The great improvement introduced by Marconi was the employment of this vertical air-wire, aerial, antenna, or elevated conductor. **1904** *Physical Rev.* Sept. 197 In 1898 he [*sc.* Marconi] saw the importance of the direct grounding of the receiving antenna. **1915** A. E. Seelig tr. *Zenneck's Wireless Telegr.* 150 Every radio station has an open oscillator, the 'antenna', that part of the antenna which is suspended in the air being called the 'aerial'. **1922** Glazebrook *Dict. Appl. Physics* II. 1044/1 Electric waves..received on the antennae of wireless stations. **1960** *Daily Tel.* 22 Feb. 14 The radar antenna shown above..has an 84-ft. 'dish'. **1961** *Word Study* Apr. 3/1 The biological scientist will insist that the creatures have *antennae*; an electronics technician writes about the *antennas* of a microwave installation. **1962** *Observer* 10 June 20/1 A huge 340-ton horn antenna at Andover, Maine, will beam signals at the satellite.

b. *attrib.* and *Comb.*

1906 J. A. Fleming *Princ. Electr. Wave Telegr.* ix. 556 This is dissipated..as heat in the antenna and antenna circuit. **1916** —— *Radiotelegr.* (ed. 3) 155 A single mast..having two sprits attached to it by means of which an antenna wire is upheld in the form of a vertical rectangle. **1922** Glazebrook *Dict. Appl. Physics* II. 1038/1 The condenser C is intended to represent the antenna capacity, and R..the same power as is actually occasioned by the antenna resistance and radiation. **1940** *Chambers's Techn. Dict.* 39/1 *Antenna array*, a group of two or more antennae spacially arranged to have particular directional radiating and/or receiving properties. *Ibid.* 39/2 *Antenna system*, the whole of the equipment of a radio transmitter or receiver associated with the antenna-to-earth circuit. **1946** *Electronic Engin.* XVIII. 20 An automatic V.H.F. direction finder using a fixed..antenna system. **1963** *Ann. Reg. 1962* 447 The U.S.S.R. built the highest tower in Europe..a height of 1,383 feet with its antenna mast.

6. A 'feeler' attached to a naval mine. Also *attrib.*

1933 *Jane's Fighting Ships 1933* 461/1 Body of mine, spherical, 3 feet diameter. Form of detonation uncertain; said to consist of antennae fitted with..magnetic pistol... Antennae said to consist of thin conducting wires and magnetic pistols. **1947** *10 Eventful Years* IV. 632 *Antenna mine*, a large naval mine..exploding on contact with a metal object.

ante-partum (æ·nti,pā·ɹtǒm), *a.* [Lat. phr., 'before birth', used *attrib.* or as *adj.*] (Occurring) before birth, esp. of things affecting the mother immediately before or during the early stages of labour.

1908 *Practitioner* Mar. 358 Crédé's method..cannot prevent ante-partum infection. **1961** *Times* 26 Apr. 19/5 Antepartum haemorrhage.

ante-post (æ·nti,pōᵘst), *a.* [f. Ante- + Post *sb.*[1] 2.] Of betting (see quot. 1902).

1902 *Encycl. Brit.* XXVI. 236/1 Betting is of two kinds: 'post', when wagering does not begin until the numbers of the runners are hoisted on the board; and 'ante-post', when wagering opens weeks or months before the event. **1914** *Daily Express* 19 Sept. 5/2 If there had been any important ante-post betting..many would have burnt their fingers badly.

‖**ante rem** (æ·nti rem). *Philos.* [med.L. (Albertus Magnus), 'before the thing'.] Used as *phr.* and also *attrib.* Prior to the existence of a particular or physical thing; *spec.* of the philosophical theory (a form of Realism 1) which holds that the universal is logically prior to, i.e. capable of existing independently of, the particular; also, of the universal so regarded.

c **1873** W. James in R. B. Perry *Tht. & Char. of W. J.* (1935) I. 499 Our notion of a future time with its material content forms a sort of matrix *ante rem* into which in its time the *res* fits. **1904** —— *Coll. Ess. & Rev.* (1920) 446 Like empiricism, it is individualistic and phenomenalistic; it places truth *in rebus*, and not *ante rem*. **1927** C. R. S. Harris *Duns Scotus* I. vi. 215 The distinction of the threefold aspect of the universal *ante rem, in re*, and *post rem* forms the starting-point of the epistemological theory of Scotus no less than that of Albert or Aquinas. **1953** H. H. Price *Thinking & Exper.* ix. 266 Nothing in this world is a perfect instance of circularity, according to the *Ante Rem* theory. *Ibid.* x. 302 One might deny that there are subsistent and extra-mental universals, whether of the *Ante rem* or *In re* variety.

antergic (æntɜ̄·ɹdʒik), *a.* *Physiol.* [f. Ant- *pref.* + Gr. ἔργον work + -ic, after *synergic*.] Acting in opposition to the action of another part, as a muscle: = Antagonistic *a.* 2: opp. to Synergic *a.*

1892 W. R. Gowers *Man. Dis. Nerv. Syst.* I. 7 This antergic contraction is essential for the exertion of force.

anteriad (æntiᵊ·riæd), *adv.* [f. Anterior + *-ad.] Towards the anterior part or surface of the body.

1903 *Trans. Amer. Microsc. Soc.* Nov. 66 (Cent. D. Suppl.), The rudiment of the gall-bladder which in the previous stage..opens dorsad..in the present stage..opens anteriad.

anterior, *a.* Add: **1. b.** *Anat.*, *Bot.*, and *Zool.* Situated in the front or nearer the head, fore-part, etc.: opp. to Posterior A. 3. In *Bot.* also, below, inferior. In *Human Anat.* now usually (because of the erect posture of man) = Ventral *a.*; similarly for other animals having an upright posture.

1733 G. Douglas tr. *Winslow's Anat.* I. 75 The Clavicle is divided into a Body or middle Part, and two Extremities, one anterior, inferior, and internal..; the other posterior, superior, and external. **1826** Kirby & Spence *Entomol.* IV. 335 *Anterior*, the fore or upper wings. *Ibid.* 339 *Anterior* or *Exterior*, the outer margin of the wing, or that from the body. **1829** Loudon *Encycl. Plants* 1094 *Anterior*, growing in front of some other thing. **1853** Darlington *Flora Cestrica* (ed. 3) Gloss., *Anterior*, in front, or below,—as that part of a flower next to the bract, or farthest from the axis of inflorescence. **1875** W. Turner *Hum. Anat.* I. v. 213 The anterior cerebral vesicle bends downwards from the middle vesicle. **1880** A. Gray *Bot. Text-bk.* (ed. 6) I. 160 The portion of the flower which faces the subtending bract is the anterior. **1886** Buck *Handbk. Med. Sci.* II. 507/2 The anterior oblique dislocation of Bigelow. **1893** H. Morris *Hum. Anat.* 1127 In front of the sterno-mastoid is the

anterior triangle. **1897** D. H. Scott *Struct. Bot.* (ed. 4) I. 23 The two outer sepals are so placed that one lies next the axis (*posterior*), and the other remote from it (*anterior*). **1937** *Discovery* Jan. 16/2 The cortex of the hemispheres, the so called anterior brain.

c. *Phonetics.* = Palatal A. 2.

1902 E. W. Scripture *Elem. Exper. Phonetics* xxiii. 316 The *n* contact involves an anterior occlusion. **1925** P. Radin tr. *J. Vendryès's Lang.* I. i. 21 The vowels known as anterior or palatal. **1933** L. Bloomfield *Lang.* vi. 99 One distinguishes, usually, between *anterior* or *palatal* position and *posterior* or *velar* position.

antevert, *v.* **1.** Delete †*Obs.* and add later example.

1914 *N. Amer. Rev.* Sept. 387 The Cossacks..rode down the sidewalks..in an effort to antevert a meeting of the students.

anthelmi·nthic (-þik), more correct form of Anthelmintic.

1684 [see Anthelmintic A]. **1714** Fuller *Pharmacopœia* (ed. 2) 76 The Decoction called Sacrum..is a most noble..Anthelminthic. *Ibid.* 203 Give Anthelminthics..to make the Worms weak and languishing. **1871** Kingsley *At Last* xiii, Their bitter anthelminthic oil.

anthemion (ænþɪ·miǫn). Pl. **-mia.** [a. Gr. ἀνθέμιον flower.] = Honeysuckle 4. Also *attrib.*

1865 J. K. Colling *Art Foliage Sculpture & Decoration* I. iii. 25 Variations upon the Anthemion ornament may be seen upon Etruscan vases. **1899** R. Glazier *Man. Hist. Orn.* 17 The anthemion..is derived from the traditional lotus and bud of Egypt, Assyria, and India. **1912** R. Fry in H. G. Wells et al. *Socialism & Great State* 265 The chairs into the wooden seats of which some tremendous mechanical force has deeply impressed a large distorted anthemion. **1955** R. Fastnedge *Eng. Furn. Styles* 279 *Anthemion ornament*, of Greek derivation..popularized in England during the second half of the eighteenth century.

anthilly (æ·nt,hi:li), *a.* [f. Ant-hill + -y[1].] Full of or abounding in ant-hills.

1796 W. H. Marshall *Rural Econ. W. Eng.* II. 212 Rough anthilly land. **1892** *Field* 19 Nov. 766/2 Anthilly fields of grass.

anthocyanidin (æ:nþosai,æ·nidin). [a. G. *anthocyanidin* (R. Willstätter and A. E. Everest 1913, in *Ann. d. Chemie* CCCCI. 205); cf. next and *cyanidin.] A plant pigment formed by the hydrolysis of an anthocyanin.

1914 *Chem. Abstr.* 335 The anthocyanidin reaction..consists in extg. the fresh plant material. **1914** *Proc. R. Soc.* B. LXXXVII. 444 In no case could any trace of anthocyanidin be found. **1934** *Ann. Reg. 1933* 61 The widespread distribution in plants of a new class of colourless anthocyanidins was established.

anthocyanin(e (æ:nþosai·ānin, -nəin). Delete ?*Obs.* and add later examples. Also **anthocy·an.** Substitute for def.: One of the blue, violet, or red pigments in plants. Also *attrib.*

1894 F. W. Oliver et al. tr. *Kerner von Marilaun's Nat. Hist.* I. 288 The under side of the leaf is coloured violet by a pigment called anthocyanin. **1902** *Encycl. Brit.* XXXI. 763/2 The red pigment anthocyan, which is found very commonly in young developing shoots, petioles, and midribs. **1913** W. Bateson *Mendel's Princ. Hered.* I. iii. 38 The addition of various other factors produces anthocyan reds. **1918** A. W. Stewart *Rec. Adv. Org. Chem.* (ed. 3) 209 The anthocyanins. **1924** *Empire Review* May 583/1 The anthocyanin pigments belong to the great class of the glucosides. **1956** *Sci. News* XL. 42 The red and yellow pigments, the anthocyanins and the flavones, were all studied in Java.

anthodium (ænþōᵘ·diǒm). *Bot.* Pl. **-ia.** [mod.L., f. Gr. ἀνθώδης (see Anthoid *a.*) + *-ium*.] A name for the flower-head or *capitulum* in Compositæ.

1858 in A. Gray *Introd. Struct. & Syst. Bot.* 523/2. **1861** Bentley *Man. Bot.* 203 *The Capitulum, Anthodium, or Head.*—This kind of inflorescence constitutes the Compound Flower of Linnæus.

anthologer (ænþǫ·lǒdʒəɹ). [f. Anthology + -er[1].] = Anthologist.

1906 *Westm. Gaz.* 13 Oct. 11/3 These are far fewer than the examples..that have not yet found the anthologer. **1953** *Sat. Rev.* 7 Mar. 13/2 The anthologers set about their task with imagination and taste.

anthologize (ænþǫ·lǒdʒəiz), *v.* [f. Anthology + -ize.] **a.** *trans.* To make an anthology of or from; to use in an anthology. **b.** *intr.* To make an anthology or anthologies; to compile as, or in the form of, an anthology; (of a large work) to yield suitable pieces for an anthology. Chiefly in *pa. pple.*, *vbl. sb.*, or *gerund*.

1892 *Sat. Rev.* 9 Apr. 426/2 The anthologized poet. **1892** *Daily News* 12 July 4/8 Where poets dead and gone are concerned,..the practice of selecting and anthologising is comparatively innocuous. **1923** *Times Lit. Suppl.* 4 Jan. 9/3 Many writers of verse are familiar to the public through an anthologized example. **1924** *Ibid.* 13 Nov. 724/2 The thirty-four lines..deserve to be rescued from Professor Macaulay's grim black covers and anthologized. **1928** *Manch. Guardian Weekly* 17 Aug. 133/2 The anthologising offences of Palgrave, Sir Arthur Quiller-Couch and Mr. J. C. Squire. **1959** *20th Cent.* June 612 One can get an idea of Huxley's own range..from a little book..anthologized from his works by his widow. **1965** *New Statesman* 24 Dec. 1006/3 *Private Eye* magazine does not anthologise well.

anthology. Add: **2.** Also *transf.*, esp. of paintings, songs, etc., and other art forms.

1961 *Listener* 21 Dec. 1068/1 The quays..bore an anthology of western European coastal traffic: Bilbao, Stockholm, Hamburg, Glasgow. **1965** *Ibid.* 11 Nov. 760/1 Fry put on his second and even more demanding anthology of post-Impressionism. **1967** *Ibid.* 3 Aug. 129 Some of the LPs that Sinatra began bringing out in the mid-Fifties..are virtually anthologies of pop songs from the previous 20 years.

6. *anthology-piece.*

1935 *Scrutiny* IV. 168 A favourite anthology-piece *When the lamp is shattered.* **1965** *Listener* 2 Sept. 350/2 The..'Battle of San Romano' by Uccello..is an anthology piece so familiar from countless small reproductions that its gigantic scale when one sees it again is breathtaking.

Anthony Eden (æ:ntǒni ī·dĕn). [English politician, 1897–1977.] A black Homburg hat of the type often worn by Sir Anthony Eden (later Lord Avon). Also in shortened form **Eden.**

1940 Graves & Hodge *Long Week-End* xxi. 376 Anthony Eden, the Foreign Secretary, had reintroduced the black Homburg hat, known as the 'Eden' in Savile Row. **1956** D. Davin *Sullen Bell* II. iv. 128 Dark suit and Anthony Eden hat during the week, tweed jacket and flannels at the week-end. **1958** *Spectator* 31 Jan. 130/1 How did the soft black Homburg, with a bound, turned-up edge, come to be called an 'Anthony Eden'? **1959** S. Gibbons *Pink Front Door* xv. 180 He was just off somewhere—white scarf, black Anthony Eden and all.

anthophilous (ænþǫ·filəs), *a.* *Ent.* [f. Gr. ἄνθος flower + -philous.] Loving or frequenting flowers, as an insect (correlative to Entomophilous *a.*); *spec.* belonging to the division *Anthophila* of hymenopterous insects, comprising the bees.

1883 D'Arcy W. Thompson tr. *H. Muller's Fert. Flowers* 33 The larvæ of beetles which are anthophilous.

anthracene. Add: anthracene oil (see quot. 1940).

1874 *Jrnl. Chem. Soc.* XXVII. 853 Red or anthracene oils and creasote oils. **1920** *Conquest* May 323/1 A final fraction above 270° C., when the anthracene oil or green oil is obtained. **1940** *Chambers's Techn. Dict.* 40/1 *Anthracene oil*, a coal-tar fraction boiling above 270° C., consisting of anthracene, phenanthrene, chrysene, carbazole, and other hydrocarbon oils.

anthracitization (æ:nþrăsəitəizē̆·ʃən). [f. Anthracite + -ization.] The process of becoming changed from bituminous coal into anthracite.

1903 *Progr. Geol. Surv. U.K. 1902*, 49 An investigation of the anthracitisation of the coals. *Ibid.* 190 The anthracitisation is more rapid in a north and south direction.

anthracnose (ænþræ·knōᵘs). [a. F. *anthracnose* (Fabre and Dunal 1853, in *Bull. Soc. Centrale d'Agric. de l'Hérault* XL), f. Anthrax + Gr. νόσος disease.] A fungal disease of plants, esp. of vines and beans, characterized by dark spots or lesions.

1886 *Rep. Commissioner U.S. Dept. Agric.* (Mycology) 112 Anthracnose, *Sphaceloma ampelium* DeBy...a comparatively new disease in this country... In Europe it has been known for many years..as 'Charbon', 'Brenner', [etc.]. **1888** [see *charbon 3]. **1938** *Nature* 12 Nov. 881/2 The fungus [*Colletotrichum lagenarium*] causes leaf spotting or anthracnose upon a wide range of varieties. **1950** *N.Z. Jrnl. Agric.* Oct. 325/1 Black spot or anthracnose appears on early spring growth [on grape-vines] and the most serious damage usually occurs before blossom time.

anthracosis (ænþrăkōᵘ·sis). *Med.* [mod.L., f. Gr. ἀνθρακ-, ἄνθραξ coal + -osis.] Deposition of carbon particles in the lung; also, a severer form of such deposition, e.g. coal miner's pneumoconiosis. Hence **anthracotic** (-ǫ·tik) *a.*, pertaining to or affected with anthracosis.

1838 T. Stratton in *Edin. Med. & Surg. Jrnl.* XLIX. 490 Black infiltration of the whole lungs..has been called the black lung of coal miners, and may more shortly be defined *anthracosis.* **1848** Dunglison *Med. Lex.* (ed. 7), Anthracosis. **1875** C. H. Jones & Sieveking *Path. Anat.* (ed. 2) 492 'Colliers' Phthisis', or as it has been called, anthrakosis. **1900** *Jrnl. Exper. Med.* V. 156 The cut section of the lungs, which were markedly anthracotic, was smooth. **1912** Adami & McCrae *Pathol.* 325 Section of an anthracotic lung. **1930** *Engineering* 25 Apr. 536/2 French experts consider that anthracosis, caused by coal-dust, is a form of silicosis.

anthrone (æ·nþrōᵘn). *Chem.* [f. Anthra- + Ket)one.] A colourless crystalline ketone, $C_{14}H_{10}O$.

1913 *Chem. Abstr.* 1872 Anthracene..Derivatives of Anthrone. **1921** E. de B. Barnett *Anthracene* 96 Anthrone itself was first obtained by Liebermann. **1956** *Nature* 7 Jan. 38/1 A slightly modified anthrone method..for the quantitative estimation of carbohydrate in chromatographic fractions.

anthropic, *a.* Add to def.: Concerned with or relating to human beings; in *Geol.* applied to the period of the deposits in which human remains are found.

1884 Harrison in *19th Cent.* Mar. 505 The conclusion that the future of religion is to be, not only..anthropomorphic—but frankly anthropic. **1884** Blackmore *Sir Thomas Upmore* xiii, My dear little anthropic nautilus, I can do nothing. **1893** J. W. Dawson *Salient Points* 465 The age of which we have been writing the history, is that which has been fitly named the Anthropic.

anthropocentra·lity. *rare.* [f. Anthropo- + Centrality.] = *anthropocentricism.

1934 *Theology* XXVIII. 267 The humanist ethic..leads us to the blind alley of anthropocentrality.

anthropoce·ntrically, *adv.* [f. Anthropocentric *a.* + -al + -ly[2].] In an anthropocentric manner or way.

1953 *New Biol.* XIV. 16 Anthropocentrically minded observers have suggested [etc.]. **1963** *Times* 28 May 13/5 The second is that urban minds stop regarding agriculture anthropocentrically as if it always represented a human efficiency problem.

anthropocentricism, anthropocentrism (-se·ntrisiz'm, -se·ntriz'm). [f. Anthropocentric *a.* + -ism.] An anthropocentric view or doctrine.

1909 *Cent. Dict.* Suppl., Anthropocentricism. **1909** Webster, *Anthropocentrism*, the assumption that man is the center of all things. **1912** J. H. Moore *Ethics & Educ.* xvii. 141 There is a doctrine called Anthropocentricism... According to this theory, man is the centre of the universe. **1922** *Q. Rev.* July 97 The last shreds of anthropocentrism have been worn away. **1936** V. A. Demant *Chr. Polity* v. 103 To confine theological seriousness to God's redemptive work in man..is a disguised form of the very anthropocentricism which this doctrine intends to oppose. **1941** A. Huxley *Grey Eminence* iii. 79 Theocentrism produces better ethical results than anthropocentrism and moralism. **1956** *Sc. Jrnl. Theol.* IX. 72 The anthropocentricism and moralism of Hebrew theology.

anthropogenic (-dʒe·nik), *a.* **1.** [f. Anthropogeny + -ic.] Of or pertaining to anthropogeny.

1889 in *Cent. Dict.* **1952** Gerth & Martindale tr. *Weber's Anc. Jud.* ix. 227 The cosmogonic and anthropogenic myths are of secondary importance in Yahwistic religiosity.

2. [f. Anthropo- + *-genic.] Having its origin in the activities of man.

1923 A. G. Tansley *Pract. Plant Ecol.* iv. 48 Where he [*sc.* man] has introduced a more or less permanent modifying factor or set of factors, we have biotic (anthropogenic) climaxes or some stage of development towards them. **1939** —— *Brit. Islands & Vegetation* vi. 129 All the factors mentioned—felling, mowing, grazing, and fire—are sometimes loosely termed 'biotic factors', though they would be more properly called *anthropogenic factors*, and the climaxes they produce *anthropogenic climaxes*. Among the anthropogenic factors the grazing factor is biotic in the strict sense. **1947** *Nature* 11 Jan. 45/1 The virgin forest..and even the forest nursery are viewed as possible systems of this kind maintained either by natural or anthropogenic forces. **1960** N. Polunin *Introd. Plant Geogr.* vii. 200 Leaving aside the so-called 'anthropogenic relics' whose areas have become drastically reduced through the activities of Man.

anthropogeography (ænþrōᵘ:po,dʒi,ǫ·grăfi). [f. Anthropo- + Geography, after G. *anthropogeographie* (F. Ratzel 1882).] That department of geography which treats of the relations of the earth to mankind as its inhabitants. So **anthropogeographer** (-ǫ·grăfəɹ), one versed in anthropogeography; **anthropogeographic** (-ogræ·fik), **-ical** *adjs.*, pertaining to anthropogeography.

1652 *Hermeticall Banquet* 120 The new Anthropogeographicall Map. **1895** *Geogr. Jrnl.* Oct. 375 Biogeography..has three sections—phytogeography..zoogeography..and anthropogeography, or the geography of men. *a* **1899** Brinton *Basis Soc. Relat.* (1902) iv. 181 Even the most determined of the 'anthropo-geographers' will not deny that the power over the mind which they attribute to geographical features diminishes in proportion as culture increases. **1899** *Geogr. Jrnl.* Feb. 171 Anthropogeography is a convenient term under which to include all those aspects of geography that deal with the relations of humanity..to the earth... 'Applied Geography' might be taken as an alternative term. **1939** *Jrnl. R. Anthrop. Inst.* 45 Anthropogeographical studies in Greenland. **1940** S. W. Boggs *Internat. Boundaries* ii. 27 The term 'boundary' as applied to any of the anthropogeographic types.

anthropogony (ænþropǫ·gǒni). [ad. Gr. ἀνθρωπογονία.] **a.** The origin of man. **b.** The investigation or an account of this.

a **1871** Mansel *Gnostic Heresies* (1875) 36 A scheme of cosmogony and anthropogony, running parallel to each other, man being regarded as the microcosm, or image in miniature of the world. **1874** J. W. Watson & M. J. Evans tr. *Van Oosterzee's Chr. Dogmatics* I. 361 A comparison with other Eastern Anthropogonies gives a result most favourable to the Mosaic.

anthropoid. Add: **A. b.** Shaped like a man.

1912 T. E. Lawrence *Home Lett.* (1954) 228 It [a sarcophagus] was white marble..anthropoid but Greek-featured. *Ibid.* 235 Anthropoid means human 'shaped': the sarcophagus is like the sort of mummy-coffin that has a face carved on it. **1920** *Brit. Mus. Return* 43 All the rectangular and anthropoid coffins..have been incorporated.

c. Of a human being: of ape-like form or character.

1930 *English Jrnl.* XIX. 608 Mr. Mencken watched with alert eyes the simian antics of the anthropoid rabble at the Dayton farce. **1939** C. K. Allen *Law in the Making* (ed. 3) 56 Some incalculably remote age when the life of anthropoid men was 'nasty, brutish, and short'.

anthropologist. (Earlier example.)

1798 Willich *Elem. Crit. Philos.* 22 Plattner, that excellent Anthropologist,..has employed rational scepticism against the Kantian System.

anthropologize (ænþropǫ·lǒdʒəiz), *v.* *colloq.* [f. Anthropology + -ize, after *philosophize*, etc.] **a.** *trans.* To explain, treat, or study anthropologically. **b.** *intr.* To pursue anthropology.

1939 'M. Innes' *Stop Press* II. v. 285 Modish enough to play at anthropologising an unknown culture. **1941** R. R. Marett *Jerseyman* xii. 178 To anthropologize in the field would have attracted me greatly. **1957** *Times Lit. Suppl.* 1 Nov. 656/3 But the more serious side of Mr. Speirs's anthropologizing is to be found, for instance, in his deep sense of the adjustment of medieval life..to the rhythms of the year.

anthropometer (-ǫ·mɪtəɹ). [See -meter.] **a.** One who studies or practises anthropometry. *rare.* **b.** An anthropometrical instrument.

1883 *Ann. Rep. Smithsonian Inst. 1881* 499 Man is an animal..exhibiting in his adult form those characteristics which engage the attention of the anatomist, the physiologist, and the anthropometer. **1898** A. C. Haddon *Study of Man* xvi. 446 The rod of the Anthropometer should be held vertically in front of the face of the subject. **1951** *Notes & Queries on Anthropol.* (ed. 6) I. 12 In the field the anthropometer may be used, taking care to keep the instrument vertical.

So **anthropo·metrist** = sense a above.

1883 *Pall Mall Gaz.* 3 Oct. 2 Surely the anthropometrists will do harm if they encourage the craze for tallness. **1904** G. S. Hall *Adolescence* I. 19 Anthropometrists think growth in height..antagonistic to growth in girth. **1956** *Antiquity* XXX. 123 Not all of the anthropometrists..could remember exactly how they had executed each of the measurements.

anthropomorph (ænþrōᵘ·po-, æ·nþrōᵘpomǭɹf). [ad. Gr. ἀνθρωπόμορφος: see Anthropomorphous *a.*] A representation of the human form in art.

In quot. 1894 used with sarcastic application.

1894 *Daily News* 26 Nov. 6/5 Your fashionable and self-adoring man or 'anthropomorph'. **1895** A. C. Haddon *Evol. Art* 185 New Zealand is one of the places where anthropomorphs abound. **1913** J. Rendel Harris *Boanerges* ii. 14 There was an ornithomorph, and..several theriomorphs, before the anthropomorph.

anthropomorphic, *a.* Add: **2.** Having or representing a human form: = Anthropomorphous *a.*

1886 [see Zoomorphic *a.* 2 b]. **1905** A. S. Griffith tr. *Capart's Prim. Art Egypt* iii. 59 The designs..borrowed from animals (zoomorphic designs), from the human figure (anthropomorphic), and occasionally from manufactured objects (skeuomorphic).

anthropomorphiza·tion. [f. Anthropomorphize *v.* + -ation.] = Anthropomorphism 1 b.

1880 W. James *Coll. Ess. & Rev.* (1920) 216 All such discussions rest on an anthropomorphization of outward force. **1963** *Punch* 2 Jan. 7/3 My..conviction..that anthropomorphisation, attributing to animals the characteristics of human beings, was a lot of rot.

anthropomorphological, *a.* Delete *rare*⁻⁰ and add example.

1863 W. James *Let.* 13 Sept. (1920) I. 51, I send a photograph of Gen. Sickles... It is a part of a great anthropomorphological collection which I am going to make.

anthroponymy (ænþropǫ·nīmi). [f. Anthropo- + Gr. ὠνυμία, f. ὄνομα name: cf. Toponymy.] Personal names as a subject of study. So **anthro·ponym,** a person's name.

1937 J. Orr tr. *I. Iordan's Introd. Romance Ling.* iii. 222 A special study of toponymy and anthroponymy. **1951** (*title*) Third International Congress of Toponymy and Anthroponymy..1949. Programme [etc.]. **1958** D. Tait in Middleton & Tait *Tribes without Rulers* 170 Kitiak is therefore a toponym... Others are of the form Udzado or Gbiedo, these are anthroponyms.

anthropopsychism (ænþrōᵘ:posəi·kiz'm). [f. Gr. ἄνθρωπος man + ψυχή soul + -ism.] The ascription of mental faculties or characteristics like those of man to the Divine Being or the agencies at work in nature. So **anthro:popsy·chic,** *a.*, **anthro:popsy·chically** *adv.*

1884 Duke of Argyll *Unity of Nature* v. 168 It is not the Form of Man that is in question. It is the Mind and Spirit of Man—his Reason, his Intelligence, and his Will... The question is of a..fundamental analogy..between the Mind which is in us and the Mind which is in Nature. The true etymological expression for this idea..would be, not Anthropomorphism, but Anthropopsychism. *Ibid.* viii. 289 Professor Tyndall himself cannot describe this System without using the most intensely anthropopsychic language: 'The continued effort of animated nature is to improve its conditions and raise itself to a loftier level.' **1884** *Edin. Rev.* Apr. 514 The Duke admits that much of the language which is anthropopsychically used, is of a metaphorical character. **1898** *Literary Guide* 1 Jan. 10, I desire to abstain from any thoughts about the Ultimate Reality which suggest anthropomorphism or anthropopsychism.

anthroposophy. Add: **2.** A movement inaugurated by Rudolf Steiner (1861–1925) to develop the faculty of cognition and the realization of spiritual reality. So **anthropo·so·phical** *a.*, of or pertaining to anthroposophy; **anthropo·sophist,** an adherent of anthroposophy.

1914 tr. *Steiner's Spiritual Sci.* 26 This building will be for the use of the Anthroposophical Society. **1916** H. Collison *New Impulse* 19 The anthroposophist claims the ability to distinguish between subjective and objective phenomena. *Ibid.*, The object of anthroposophy is to enable man to work consciously on every part of his system, and be in full command of his own faculties. **1922** G. Kaufmann *Fruits of Anthroposophy* i. 4 While Natural Science is..*Anthropology*—the Spiritual Science..is *Anthroposophy.* *Ibid.* 6 The Anthroposophist comes forward with his claim. *Ibid.* ii. 39 The period of Rudolf Steiner's life-work when his teaching of Anthroposophical Science was to begin. **1927** tr. *Steiner's Anthroposophical Leading Thoughts* 11 Anthroposophy is a path of knowledge, to guide the Spiritual in the human being to the Spiritual in the universe. **1933** E. Bowen-Wedgwood tr. *Steiner's Anthroposophic Movement* 51 Robert Zimmermann wrote his book *Anthroposophy* [*Anthroposophie im Umriss*, 1882]. And from this *Anthroposophy* I afterwards took the name.

anthurium (ænþiū·riǒm). [mod.L., f. Gr. ἀνθ-ος flower + οὐρά tail.] A member of the large genus so called of tropical American perennial plants of the family Araceæ.

1839 in R. Sweet *Hort. Brit.* (ed. 3) 633. **1884** [see *tail-flower*, Tail *sb.*[1] 14]. **1891** *Daily News* 14 May 6/2 A superb Goodyear bouquet of scarlet anthuriums and green wheatears. **1961** *Amat. Gardening* 30 Sept. Suppl. 3/3 There are also some very beautiful foliage plants among the anthuriums.

anti (æ·nti), *a.* and *sb.* [Anti-[1] used as a word.] **A.** *adj.* or quasi-*adj.* Against or antagonistic to some person or thing.

1857 Mundy *Antipodes* (ed. 4) xix. 209 The 'Anti' journals joined with them in coarse personalities. **1939** R. C. K. Ensor *Who Hitler Is* 6 The circumstances of Vienna made both schools anti-Semite; those of Austria made them both anti-Slav, and in particular anti-Czech. The 'anti' feelings were intense. **1948** 'J. Tey' *Franchise Affair* xii. 129 'Some letters on the correspondence page.'..'All anti, I suppose.' **1953** R. Lehmann *Echoing Grove* 299 'You were violently anti all that, weren't you?' 'You make me sound like a Hyde Park tub-thumper.'

B. *sb.* One who is against or antagonistic to some person or thing; *spec.* (U.S.) used as abbrev. of *anti-federalist.

1788 *Columbian* July 414/1 It was agreed to raise the constitution that the anti's had burnt. **1801** *Spirit of Farmers' Museum* 56 There Feds shall cease to charge the Antis With making Frenchmen rule brave yankees. **1826** M. Edgeworth *Thoughts on Bores* in *Janus* 96, I invite every true friend of literature and of good conversation, *blues* and *antis*, to contribute their assistance. **1889** *Pall Mall Gaz.* 27 June 1 The 'antis' have no option but to take it up. **1963** *Times* 17 May 15/4 Our times are fertile of protestation; anything from new roads or buildings to the methods of national defence calls forth the campaigning of the 'antis'.

anti, erron. var. *ante *sb.*[2], *v.*

anti-[1]. Add: **B. I.** As a living formative, this prefix has been extraordinarily productive in the 20th century: additional illustrative examples (esp. of the 20th century) are provided below as well as some new sense-divisions.

2. b. *anti-metaphysics, -philosophy, -romance, -science.*

1940 R. G. Collingwood *Essay on Metaphysics* viii. 81 In the second part I pass from metaphysics to anti-metaphysics, by which I mean a kind of thought that regards metaphysics as a delusion and an impediment to the progress of knowledge, and demands its abolition. **1818** Coleridge *Philos. Lect.* (1949) ii. 107 Materialism..is an anti-philosophy arising out of a thorough coldness of the moral feeling. **1877** F. H. Laing *Ld. Bacon's 'Philosophy' Examined* xiv. 123 The Baconian view..might rather be called an *anti-philosophy* than a philosophy. *Ibid.* 125 The negative anti-philosophy of Bacon. **1653** J. Davies tr. *C. Sorel's The Extravagant Shepherd*, (*continuation of title*) The anti-romance: Or, the history of the shepherd Lysis. **1911** G. B. Shaw *Doctor's Dilemma* p. xli, The medical profession is coming more and more to represent, not science, but a desperate and embittered anti-science.

c. In recent use, of a form of art or literature seemingly opposed to the basic conventions or traditions of the form in question or to the form itself. Cf. *anti-hero.

1935 D. Gascoyne *Short Survey Surrealism* ii. 35 The third review to appear was *Littérature* (which really meant Anti-literature). **1939** Dylan Thomas *Let.* 11 Sept. (1966) 236, I like the idea of Miller's anti-literature. [**1948** J.-P. Sartre in N. Sarraute *Portrait d'un Inconnu* (1956) 7 Un des traits les plus singuliers de notre époque littéraire c'est l'apparition, çà et là, d'œuvres vivaces et toutes négatives qu'on pourrait nommer des anti-romans.] **1955** J. Bazaine in A. C. Ritchie *New Decade* 12 The nature of painting is abstraction... It is not only 'anti-painting' that reminds us, in terms all too obvious, that a work of art exists only in the degree to which it denies itself, exceeds itself, blasphemes. **1958** *Archit. Rev.* CXXIV. 1/1 His [*sc.* Debord's] celebrated anti-film *Hurlements en Faveur de Sade* is chiefly notable for the fact that the uproar it creates is designed and predictable. **1958** M. Jolas tr. *J.-P. Sartre's Pref. to N. Sarraute's Portr. of Man Unknown* (1959) p. vii, These anti-novels maintain the appearance..of the ordinary novel... But..their aim is to make use of the novel in order to challenge the novel. **1958** *New Statesman* 25 Jan. 101/1 One of the cleverer heirs of that nineteenth-century tradition of playmaking against which M. Ionesco's anti-theatre is presumably directed is M. André Roussin. **1960** *Listener* 25 Feb. 368/1 Rag-time..flourished about the same time as Dada. It distorted and 'ragged' traditional melodies and rhythms into something like an anti-music. **1963** *Times* 11 Feb. 5/2 One begins to apply these doctrines to the anti-play, to Mr. Ionesco and to the other writers who seem to be..in alliance with him.

Hence, of those who practise such a form of art or literature.

1935 D. Gascoyne *Short Survey Surrealism* ii. 30 [Picabia] was known to the others as the Anti-painter, just as Tzara was known as the Anti-philosopher. **1956** *Art News* Jan. 13/4 To be an anti-painter, as Appel clearly would like to be, one must first be a painter. **1959** *Listener* 5 Nov. 764/1 The anti-artists are those who in their work have attempted to deny or break with every conceivable canon of style, taste, or convention that may have been established by the practice of artists in the past. **1962** *Ibid.* 8 Mar. 406/1 The new French school of 'anti-novelists'.

d. *anti-particle*: an elementary particle of the same mass as a given particle but having an opposite electrical charge, or (of an uncharged particle) differing in the direction of its magnetic moment (see also quot. 1962), etc. So *anti-neutrino, -neutron, -nucleon, -proton* (postulated earlier and called 'negative proton' but first discovered in 1955); also *anti-matter*, matter composed of anti-particles.

1931 Dirac in *Proc. R. Soc.* A. CXXXIII. 61 We may call such a particle an anti-electron. **1934** *Chem. Abstr.* 2263 (*title*) Theoretical remarks on the..symmetry between particles and antiparticles. *Ibid.* 5327 (*title*) Tensorial fields accompanying the Dirac electron: neutrino and antineutrino. **1942** *Ibid.* 1235 Such a theory..demands the existence of antiprotons and antineutrons. **1946** *Nature* 24 Aug. 280/2 Since Dirac's treatment was able to predict the existence of the positron it is to be expected that anti-nucleons, produced by removing nucleons from negative energy states, also exist. **1953** *Sci. News Let.* 14 Mar. 169/1 The existence or creation of anti-matter, or the negative analogue of the proton..is theoretically possible. **1955** *N.Y. Times* 23 Oct. E9/6 (*heading*) Discovery of the Anti-Proton Ends a Long Search. **1958** *Chambers's Techn. Dict.* (ed. 3) 956/2 *Antineutron*, recently discovered particle which can mutually annihilate a neutron, with the evolution of vast energy. **1959** *New Scientist* 5 Nov. 854/1 The main ingredients of anti-matter—antiprotons, antineutrons and anti-electrons—can be set in the tables alongside the protons, neutrons and electrons of which ordinary matter is composed. **1962** *Newnes Conc. Encycl. Nuclear Energy* 33/2 *Anti-particle*, that particle (known or hypothetical) whose interaction with a given particle results in their mutual annihilation. Examples of particle-antiparticle pairs are electron-positron, proton-antiproton, and so on. *Ibid.* 521/1 The antineutrino..accompanies electron emission, whereas the neutrino is associated with positron emission.

II. 3. a. Formed on adjs. derived from proper names of persons, parties, groups, or nations, as *anti-Bolshevik, -Comintern, -English* (earlier and later examples), *-European, -German, -Humian* (or *-Humean*), *-Jewish, -Judaic, -Nazi, -Negro, -Russian, -Soviet, -Stratfordian, -Wagnerian.*

1919 *Spectator* 20 Sept. 360/1 Whether we should do good or harm by trying to help the anti-Bolshevik Russians to a limited extent. **1936** *Times* 30 Nov. 12/4 (*Report of speech by H. Hess*). He [*sc.* Hitler] had lessened this grave, permanent danger to the nation by an act of such world-political significance as the anti-Comintern alliance concluded between Germany and Japan. **1808** Jane Austen *Let.* 1 Oct. (1952) 212 The Man describes well, but is horribly anti-english. **1960** *Spectator* 30 Sept. 469 A republic would reduce slightly the anti-English feelings of the Afrikaners. **1948** B. G. M. Sundkler *Bantu Prophets S. Afr.* iii. 66 He was convinced that certain ministers of the African Methodist Episcopal Church were engaged in political anti-European propaganda. **1900** *Daily News* 7 Sept. 5/2 The anti-German nature of the recent official Russian enunciations. **1879** W. James in *Mind* IV. 330 In the second volume of Lewes's *Problems* we find this anti-Humean view that the effect is the 'procession' of the cause. **1933** *Mind* XLII. 140 The more thoroughgoing and more famous anti-Humian arguments of Kant. **1817** M. Edgeworth *Harrington* iii. 57 When Jacob appeared in the school room, the antijewish party gathered round him. **1956** A. H. Compton *Atomic Quest* 208 The anti-Jewish feeling that was being fomented here by the Nazis. **1885** Lightfoot *Apostolic Fathers* II. I. 371 Marcion was markedly Anti-judaic. **1933** *Time* 20 Nov. 20/3 Last week an irate anti-Nazi raiding party entered the cemetery. **1862** Greeley in J. A. Logan *Great Conspiracy* 432 The Rebels are everywhere using the late Anti-Negro riots in the North..to convince the Slaves that they have nothing to hope from a Union success. **1947** M. M. Lewis *Lang. in Society* x. 217 Books..have been..used to intensify anti-Negro

attitudes. **1901** *Westm. Gaz.* 31 Oct. 2/1 Anti-Russian political meetings. **1940** B. Ward *Russian Foreign Policy* 26 In the early days of 1939..Hitler himself still seemed determined to pursue his anti-Russian policy. **1920** *Daily Herald* 22 Mar. 1/3 (*heading*) Smashing Reply to Anti-Soviet Propaganda. **1939** *Ann. Reg. 1938* 182 To bring Poland into the anti-Soviet coalition. **1948** *N. & Q.* CXCIII. 111/2 We have enjoyed the three latest numbers of *Baconiana* and an anti-Stratfordian brochure (from the same source): 'Was Shakespeare educated?' **1895** G. B. Shaw *Sanity of Art* (1908) 33 The old anti-Wagnerian confusion.

b. and **c.** Formed on adjs. belonging to or formed on common nouns of every description, or when *anti-* simply reverses the adj., as *anti-æsthetic, -art, -artistic, -authoritarian, -bacterial, -bourgeois, -capitalistic, -cholinergic, -colonial, -corrosive* (earlier and later examples), *-decadent, -depressant, -dogmatic(al), -everything, -formal, -historical, -human, -humanitarian, -infective, -logical* (earlier example), *-malarial, -materialistic, -mentalistic, -metaphysical, -microbial, -microbic, -military, -nationalistic, -naturalistic, -neuritic, -nuclear* (= opposed to nuclear weapons), *-philanthropic, -philosophic* (earlier example), *-philosophical, -physical, -psychological, -red* (see *red *a.*), *-revolutionary* (earlier example), *-rheumatic, -scientific, -sensational, -sensual, -sexual, -snobbish, -thyroid, -totalitarian, -tubercular -tuberculous, -utilitarian* (earlier example), *-viral, -western, -white.*

1902 G. K. Chesterton *Twelve Types* 171 Savonarola could not have been fundamentally anti-æsthetic, since he had such friends as Michael Angelo, Botticelli, and Luca della Robbia. **1942** Wyndham Lewis *Let.* 15 Sept. (1963) 337 Or is it just racial—that in Anglosaxon countries the Catholics like everybody else are a little anti-art? **1896** G. B. Shaw *Let.* 20 Feb. (1965) 604 Your nature is radically anti-artistic. **1947** M. Lowry *Let.* 6 May (1967) 145, I take it you do not of course mean your own anti-artistic emotion of self-pity. **1937** A. Koestler *Spanish Testament* I. ii. 48 As a logical consequence of their anti-authoritarian attitude, Anarchists proclaim their belief in the policy of direct action. **1897** *Lippincott's Med. Dict.* 67/2 Antibacterial. **1903** *Med. Record* 28 Mar. 511/1 The leucocytes have a special antibacterial action. **1947** *New Biol.* II. 77 In the case of such a substance as penicillin at this stage of the research its only known property was its antibacterial effect. **1923** J. M. Murry *Pencillings* 211 He had..a point of view which would have enabled him to see the comic side of Flaubert's anti-bourgeois *emportement.* **1887** S. Moore & Aveling tr. *Marx's Capital* II. xxv. 798 How, then, to heal the anti-capitalistic cancer of the colonies? **1950** Webster Add., Anticholinergic. **1962** *New Scientist* 26 Apr. 143/1 The effectiveness of the new anticholinergic drugs used in treating duodenal ulcers. **1952** *Atlantic Monthly* Aug. 10/2 Anticolonial sentiments are strong in many South American republics. **1824** Col. Hawker *Instr. Yng. Sportsmen* (ed. 3) 469 A specification of a new 'Anti-corrosive' percussion powder. **1947** *New Biol.* III. 128 First it receives two or more coats of an anti-corrosive or protective paint, which prevents corrosion of the hull. **1916** H. G. Wells *Mr. Britling sees it Through* iii. 75 'It's got its anti-decadent side,' said Mr. Direck. **1962** *Lancet* 29 Dec. 1342/2 It is well known that antidepressant drugs are of most use in the mild-to-moderate degrees of depression. **1859** Newman *Lect. & Ess. Univ. Subj.* iii. 75 In illustration and defence of the anti-dogmatic principle in political and social matters. **1940** *Mind* XLIX. 423 Progressive, anti-dogmatic science is critical—criticism is its very life. **1846** J. D. Morell *Hist. View Specul. Philos.* I. iii. 255 Simon Foucner..revived the spirit of the new academy, and with its anti-dogmatical principles, firmly opposed the views of Descartes and Malebranche. **1898** Rider Haggard *Doctor Therne* v. 92 Mr. Strong was indeed anti-everything. **1848** Bailey *Festus* (ed. 3) 111 The antiformal spirit wants no word Whereby to mark its union with the soul. **1934** C. Lambert *Music Ho!* v. iv. 316 While romanticism is not specifically anti-formal it is specifically anti-symphonic. **1937** J. Orr tr. *I. Iordan's Introd. Romance Linguistics* iv. 336 The Dutch scholar, C. de Boer..whose approach..is so definitely anti-historical. **1854** Geo. Eliot tr. *Feuerbach's Essence Chr.* p. xxxv, A superhuman, *i.e.*, anti-human, anti-natural religion. **1965** *English Studies* Feb. 73 [The New Criticism]..is therefore both anti-historical and anti-human. **1945** Popper *Open Society* I. vi. 76 His [*sc.* Plato's] political demands are purely totalitarian and anti-humanitarian. **1899** *Daily News* 1 Dec. 3/4 The anti-infective inoculation..protecting the mucous membrane against the invasion of the microbe. **1933** *Discovery* May 158/1 Vitamin A has been called the 'anti-xerophthalmic' and 'anti-infective' vitamin. **1821** *Blackw. Mag.* IX. 397 Its mode of reasoning is the most impudent and antilogical that can be conceived. **1893** *Funk's Stand. Dict.*, Antimalarial. **1932** *Lancet* 9 Apr. 796/1 Atebrin, a new antimalarial preparation. **1882** W. James *Let.* 8 Jan. in R. B. Perry *Thought & Char. W. J.* (1935) I. 737, I wanted to shew that physiology could send out an anti-materialistic blast as well as she had been supposed to emit materialistic ones. **1942** A. Koestler in *Horizon* V. 391 Perhaps the common denominator we are looking for can best be described as an 'anti-materialistic nostalgia'. **1952** *Archivum Linguisticum* IV. 67 It may be doubted whether linguists and logicians, even those free from any anti-mentalistic bias, will be satisfied. **1877** F. H. Laing *Ld. Bacon's 'Philosophy' Examined* vii. 74 Bacon..by his anti-metaphysical, *i.e.*, anti-philosophical efforts, brought the whole study..into thorough disrepute. **1935** *Mind* XLIV. 108 The prevailing positivistic and anti-metaphysical philosophies of science. **1910** *Lippincott's New Med. Dict.* 67/1 *Antimicrobic, -bial*, antagonistic to microbes. **1949** H. W. Florey et al. *Antibiotics* I. p. v, In this book plant products with antimicrobial properties have been included. **1901** Dorland *Med. Dict.* (ed. 2) 53/2 *Antimicrobic*, checking the growth of microbes. **1850** R. Hort *Horse Guards* 100 For the advocates and friends of the service to successfully combat the anti-military feeling abroad. **1920** *19th Cent.* Aug. 204 Antinationalistic class war. **1902** W. James *Var. Relig. Exper.* xvi. 422 The mystic range of consciousness..is anti-naturalistic. **1881** *Syd. Soc. Lex.* I, *Antineuritic*, term applied to remedies that prevent inflammation in nerves. **1934** *Times Lit. Suppl.* 22 Feb. 130/1 The potency of preparations containing the antineuritic vitamin B_1 has been tested. **1958** *Listener* 18 Dec. 1029/2 Switzerland who had forbidden an anti-nuclear conference to be held on her soil. **1850** J. S. Mill in *Fraser's Mag.* XLI. 26/1, I must first set my anti-philanthropic opponent right. **1923** D. H. Lawrence *Ladybird* 13 So, her reckless, anti-philanthropic passion could find no outlet. **1818** Coleridge *Philos. Lect.* (1949) 72 The scheme..as delivered by Hesiod, is an anti-philosophic Atheism. **1877** Anti-philosophical [see *anti-metaphysical* above]. *a* **1603** T. Cartwright *Confut. Rhem. N.T.* (1618) 511 We reiect not your Transubstantiation, because it is Metaphysicall, or aboue nature..but we refuse it as Antiphisicall, or against nature. **1916** D. H. Lawrence *Twilight in Italy* iii. 124 Hamlet is far more even than Orestes..a mental creature, anti-physical and anti-sensual. **1890** W. James *Princ. Psychol.* I. v. 141 The conception of consciousness as a purely cognitive form of being..is thoroughly anti-psychological. **1933** *Mind* XLII. 262 Joad is uncompromisingly anti-psychological. **1927** D. L. Sayers *Unnatural Death* xix. 216 The 'Daily Yell' wrote anti-Red leaders and discovered a plot. **1946** 'G. Orwell' *Crit. Essays* vii. 122 He stays in Spain long enough to pick up a few anti-red atrocity stories. **1828** J. S. Mill in *Westm. Rev.* IX. 295 Placing him under the protection of the anti-revolutionary general Bouillé. **1817** Anti-rheumatic [see *antibilious *a.*]. **1886** W. M. Rossetti *Pref.* to *Coll. Wks. D. G. Rossetti* I. p. xxi, Superstitious in grain, and anti-scientific to the marrow. **1927** A. Huxley *Proper Stud.* 19 The prophets of the democratic-humanitarian religion have at all times, from the eighteenth century down to the present day, denounced the upholders of Christian orthodoxy as anti-scientific. **1846** J. D. Morell *Hist. View Specul. Philos.* II. v. 18 The tendency of the Scottish philosophy..was clearly and decidedly anti-sensational. **1916** Anti-sensual [see *anti-physical* above]. **1890** W. James *Princ. Psychol.* II. xxiv. 437 What might be called the *anti-sexual instinct*, the instinct of personal isolation, the actual repulsiveness to us of the idea of intimate contact with most of the persons we meet, especially those of our own sex. **1929** B. Russell *Marriage & Morals* iv. 34 So far we have been considering pro-sexual elements in religion; anti-sexual elements, however, existed side by side. **1890** W. James *Princ. Psychol.* I. x. 293 Stiffen ourselves as we will by appealing to anti-snobbish first principles, we cannot escape an emotion..of respect and dread. **1908** H. Mackenzie in Allbutt & Rolleston *Syst. Med.* (ed. 2) IV. I. 382 An antithyroid serum, prepared by Merck from the blood of thyroidless sheep, has also been employed. **1962** *Lancet* 22 Dec. 1317/1 She was treated with antithyroid drugs, and later by subtotal thyroidectomy. **1941** 'G. Orwell' in *Horizon* IV. 134 An attempted definition of fundamental human rights, of anti-totalitarian tendency. **1946** *Nature* 30 Nov. 783/2 [France] has a special anti-tubercular service for university students. *Ibid.* 14 Dec. 864/2 This compound..has very poor anti-tubercular power. **1904** T. L. Stedman *Dunglison's Dict. Med. Sci.* (ed. 23) 74/2 *Antituberculotic* or *antituberculous*, designed to arrest the progress of tuberculosis. **1958** *Times* 15 Oct. 19/1 Anti-tuberculous drugs. **1861** J. S. Mill in *Fraser's Mag.* LXIV. 666/1 This is admitted even by anti-utilitarian moralists. **1934** *Ann. Reg. 1933* II. 63 The development, in horse and baboon serum, of anti-viral properties promises a means of immunising populations to yellow fever. **1949** I. Deutscher *Stalin* vi. 209 His 'anti-western' tone was vague enough not to cause any objection. **1906** *Westm. Gaz.* 2 Jan. 3/1 The contention that yellow labour is anti-white labour he characterised as a 'foolish and atrocious falsehood'. **1935** G. Gorer *Africa Dances* Itin. iv. 277 There is..anti-white feeling; at Accra I had..the experience of being hated by passers-by.

¶ (2) Used as substantives, as *anti-bacterial, -depressant, -diuretic, -malarial, -perspirant.* Also *antibiotic, -coagulant, -convulsant, -oxidant, -pruritic.

1944 *Lancet* 6 May 605/2 Such an influence is to be expected if natural antibiotics are involved in the interplay between plant and environment, and certain natural anti-bacterials appear to play such a part. **1962** *Ibid.* 29 Dec. 1341/2 The original assessment as to response to anti-depressants was judged almost entirely on the basis of double-blind trials. **1941** *Dorland's Med. Dict.* (ed. 19) 118/2 *Antidiuretic*, a drug which checks urinary secretion. **1935** *Discovery* Oct. 291/1 Two remarkable synthetic anti-malarials, plasmochin and atebrin. **1957** *Housewife* Sept. 88/3 Two deodorants and anti-perspirants make news.

4. Attributive phrases, consisting of *anti-* governing a sb. (i) Against or opposed to, as *anti-art, -betting, -Blimp, -business, -dumping* (see *dump *v.*[1] 2 c), *-Establishment, -litter, -noise, -reform* (earlier example), *-scrape, -segregation, -sex, -slavery* (earlier example), *-treat, -union* (= opposed to union; opposed to trade-unionism), *-war.*

1932 O. Sitwell *Dickens* 16 The genuine anti-art bias which has possessed most English people since the triumph of Cromwell. **1905** *Daily Chron.* 13 July 5/5 (*headline*) Anti-Betting Campaign. **1941** *Horizon* III. 220 It is a magnificent anti-Blimp pamphlet. **1938** *New Statesman* 15 Jan. 100/1 It [*sc.* the Wall Street market] had been depressed by the anti-business speeches of the President and his lieutenants. [**1934** Webster, Antidumping.] **1948** *Ann. Reg. 1947* 198 In the cases where the exports of one country were assisted by dumping..importing countries might impose anti-dumping duties. **1958** *New Statesman* 12 July 37/2 An independent, basically 'anti-Establishment' tabloid with a deliberately mass appeal. **1917** *Greater New York* 19 Nov. 11/3 The Anti-Litter League. **1908** *Westm. Gaz.* 1 Aug. 3/2 New York has started an anti-noise crusade. **1961** *B.S.I. News* Sept. 23 (*title*) Anti-noise regulations. **1835** J. S. Mill in *Lond. Rev.* I. 255 They..will fight the battle of half-reform with anti-reform artillery. **1877** W. Morris in J. W. Mackail *Life W. Morris* (1899) I. xi. 346 You have not yet joined our Anti-Scrape Society. **1958** *New Statesman* 28 June 839/2 Churches, trade unions and professional organisations are being torn apart by the refusal of their members in the South to follow national anti-segregation policies. **1936** *Discovery* Nov. 360/2 Headmasters with an anti-sex bias. **1820** *Niles' Reg.* XIX. 127 An attempt..to get up, what its projectors call, an 'anti-slavery' ticket. **1909** *Daily Chron.* 24 July 1/5 He has just turned them [*sc.* the saloons] into..'anti-treat' places, where no one will be allowed to get intoxicated or to treat a friend. **1813** *Niles' Reg.* V. 3/2 They may expose the British anti-union demagogues. **1905** *Daily Chron.* 24 Apr. 4 The anti-Union feeling in Norway grew. **1878** A. Pinkerton *Strikers, Communists* (1884) 206 (D.A.), The officers of the road secured a volunteer anti-union engineer and fireman to move the cars. **1938** *Ann. Reg. 1937* 281 The 'independent' steel companies stuck to their anti-union policies. **1856** S. G. Goodrich *Recoll. Lifetime* (1857) II. 51 The democrats were overjoyed that Colonel S. took pains to show his hatred and contempt for the anti-war party. **1945** H. Read *Coat Many Colours* xv. 73 The spate of anti-war literature.

(ii) Commonly used in combinations denoting an agent, device, product, etc., that inhibits, limits, or counteracts a condition, effect, etc., as *anti-crease, -dazzle, -drag, -fallout, -flash, -foam, -fouling, -glare, -halation, -ice, -knock, -roll, -shrink, -skid, -spin.* Also *anti-fat, *anti-freeze. For combinations of this type used as sbs. (also *attrib.*) in *Biochem.*, etc., see sense *7 b.

1938 *Encycl. Brit. Bk. of Yr.* 636/2 New permanent finish substances, including anti-crease treatment with synthetic resin materials. **1920** *Flight* XII. 408/1 A large assemblage of cars of all types was brought together..to test a variety of anti-dazzle devices, both British and foreign. **1963** *Times* 7 June 8/4 A two-position anti-dazzle driving mirror. **1918** H. J. Stephens *Gloss. Aeronaut. Words & Phr.* 40 Wires intended mainly to resist forces in the opposite direction to the drag are sometimes called 'anti-drag wires'. **1962** *Guardian* 9 July 14/3 Anti-fallout pills..are thought to provide a resistance to radioactive iodine in the event of fallout. **1922** *Encycl. Brit.* XXX. 127/2 In order to reduce the flash on discharge of a gun *anti-flash charges* have been under experiment. **1960** *Times* 10 June 7/7 The wings painted in white anti-flash paint. **1934** Webster, Antifoam. **1949** H. W. Florey et al. *Antibiotics* xvii. 695 The synthetic media were much more sensitive to antifoam reagents than corn steep liquor-lactose media. **1913** V. B. Lewes *Oil Fuel* iv. 91 Volatile spirit used in making up anti-fouling compositions. **1930** *Punch* 19 Feb. p. ix, Anti-glare headlamps. **1905** *Westm. Gaz.* 8 Apr. 6/3 Orthochromatic non-curling and anti-halation films. **1959** W. S. Sharps *Dict. Cinematogr.* 77/1 *Antihalation layer*, the layer interposed between the emulsion and base of film in order to prevent excessive halation. **1935** *Aircraft Engin.* Nov. 279/3 The application of anti-ice dopes..is..hardly a practical solution of the problem. **1921** *Flight* XIII. 412/2 Benzole and similar anti-knock fuels are, however, permitted. **1938** *Encycl. Brit. Bk. of Yr.* 426/1 As a rule, British cars prefer using a harder spring to obtain anti-roll characteristics. **1959** *Times* 16 Jan. 8/1 Her [*sc.* the liner's] anti-roll stabilisers ensure the utmost comfort at sea. **1946** *Nature* 19 Oct. 554/2 The unshrinkable finish is produced by means which differ from those suggested for the majority of other anti-shrink reagents. **1904** *Westm. Gaz.* 8 Dec. 10/2 A couple of chain bolts broken by catching against an anti-skid appliance had to be replaced. **1941** *Aeroplane Spotter* 29 May 195/3 Should the aeroplane develop a flat spin from which it cannot be pulled out the pilot releases the anti-spin parachute.

¶ Used as sbs. (see also senses 7 and *7 b).

1929 *Encycl. Brit.* V. 369 This substance [*sc.* lead tetra-ethyl] is of importance as an 'anti-knock' in petrol or gasoline. **1937** *Times* 13 Apr. x/3 About a year ago the regulations to deal with the lighting of road vehicles, and in particular with 'anti-dazzle', were issued. **1958** *Engineering* 21 Feb. 253/1 The development of suitable anti-foams to increase the utilisation of coil area of existing plant.

(iii) Employed or used for defence against enemy forces, weapons, etc., as *anti-air-raid, -airship, -gas, -missile, -personnel* (of bombs, mines, etc., designed to kill or injure human beings), *-satellite, -submarine, -tank, -torpedo* (see Dict., sense 4), *-U-boat.* Also *anti-aircraft.

1934 J. Hilton *Good-bye, Mr. Chips* xiv. 95 The anti-air-raid blinds that had to be fitted on all the windows. **1939** *Ann. Reg. 1938* 184 Militia formations for anti-air-raid and coast defence. **1909** *Times* 13 July 13/4 Another 'Anti-airship' gun mounted on a motor car is exhibited. **1915** *Illustr. War News* 9 June 40/1 (*caption*) The anti-gas protective respirators which they are seen wearing. **1940** *Hutchinson's Pict. Hist. War* 7 Aug.–1 Oct. 25 Wearing respirators and anti-gas clothing, men of the Royal Marines engaged in anti-gas training on the West Coast. **1956** Anti-missile [see *missile *sb.*]. **1962** *Listener* 29 Mar. 546/2 The first-fruits of an anti-missile defence will be to deflect a small proportion of an attack. **1939** *Jrnl. R. Aeronaut. Soc.* XLIII. 909 Anti-personnel bombs must explode as soon as they touch the ground. **1945** *Lancet* 14 July 44/1 Anti-personnel mines are of many types but can be divided into three main groups—the 'S' mine, 'Schu', and the light Teller anti-tank mine adapted for anti-personnel use. **1961** *New Scientist* 20 July 140/3 To avoid interception and destruction by anti-satellite missiles, *Midas III* carries a number of small vernier rockets that enable its course to be changed. **1914** Domville-Fife *Submarines, Mines & Torpedoes* ix, (*heading*) Anti-Submarine Tactics. **1942** *Ann. Reg. 1941* 272 The anti-submarine net guarding the harbour was open. **1916** W. S. Churchill *Memorandum* 9 Nov. in *World Crisis, 1916–18* (1927) II. 564 Let me point out the need of establishing without delay an Anti-Tank Committee ... We ought to have a complete anti-tank outfit by the spring. **1918** Farrow *Dict. Mil. Terms* 26 *Anti-tank Gun*, a gun especially designed and prepared to meet the onrush of the tank or mobile fort. **1944** *Return to Attack* (Army

Board, N.Z.) 19/2 Then the staccato bark of the anti-tank guns could be heard as they were turned on the escaping vehicles. **1881** Anti-torpedo [see Dict., sense 4]. **1919** L. R. Freeman *Sea-Hounds* ix. 198 Probably some sort of patrol or anti-U-boat worker, for a guess, perhaps, a 'Q'. **1939** *War Illustr.* 2 Dec. 374/2 Our Navy is stronger. Our anti-U-boat forces are three times as numerous.

III. Substantives uniform with, or formed on, adjs. and attrib. phrases. **5** and **6.** *anti-authoritarian, -Blimp, -Bolshevik, -Bolshevist, colonial(ist), -Comintern, -everything, -Fascist, -formalist, -historicist, -humanist, -imperialist, -intellectual, -Judaic, -mentalist, -metaphysician, -militarist, -Nazi, -revisionist, -revolutionist* (earlier example), *-sensationalist, -Soviet, -Stratfordian, -suffragist, -ugly* (= one opposed to ugliness in architecture, etc.), *-vaccinist, -vivisectionist, -Wagnerite.*

1910 *Encycl. Brit.* I. 916/2 After the names of 'Federalists' and 'Anti-authoritarians' had been used for some time by these federations the name of 'Anarchists'..prevailed. **1941** 'G. Orwell' *Lion & Unicorn* I. 49 The English intellectuals..[were] purely *negative* creatures, mere anti-Blimps. **1918** *New Statesman* 28 Dec. 250/1 Even the most enthusiastic of Russian anti-Bolsheviks would be satisfied. **1921** G. B. Shaw *Ruskin's Politics* 32 Look at all that has been done, not only by Bolshevists, but by anti-Bolshevists, by ourselves, and by all the belligerents! **1956** *Newsweek* 2 Apr. 41/1 Russia's new 'reasonableness' appeals to neutralist nations, to hotheaded anticolonials and, dangerously, to our allies. **1960** *Times* 27 June 13/7 Anti-colonialists say that the western countries dug themselves in commercially even in Siam. **1939** *Times* 26 Aug. 11/2 It [*sc.* the Russo-German agreement] has already produced a vigorous protest from Japan, where it is held that the pass has been sold by the leader of the anti-Cominterns. **1848** O. W. Holmes in *Poems* (1851) 253 Lean, hungry, savage, anti-everythings. **1926** *Sat. Rev.* 28 Aug. 228/1 Up to October, 1920, I have found the names of thirty-three anti-Fascists who were killed by the Fascists. **1851** J. B. Eastwick *Dry Leaves* (ed. 3) xi. 186 All mere outward show!—says Anti-formalist, who wears his hat in church, and talks during divine service louder than the minister. **1936** *Mind* XLV. 272 Leśniewski, an anti-formalist who has constructed a complete system of the foundations of mathematics. **1937** J. Orr tr. *I. Iordan's Introd. Romance Ling.* iv. 299 Such comparatists and anti-historicists as Gilliéron and Vossler. **1904** W. James in *Mind* XIII. 466 This is just another of those objections by which the anti-humanists show their own comparatively slack hold on the realities of the situation. **1898** *Westm. Gaz.* 17 Dec. 1/3 The antithesis between imperialists and anti-imperialists. **1937** V. Gollancz in 'G. Orwell' *Road to Wigan Pier* p. xviii, Mr. Orwell..is at one and the same time an extreme intellectual and a violent anti-intellectual. **1916** J. R. Harris & Burch *Testimonies* I. 49 It is a very favourite quotation with the earlier anti-Judaics. **1948** L. Spitzer *Ling. & Lit. Hist.* i. 15 The antimentalists who would suppress all expressions of opposition to their theories. **1896** *Academy* 1 Aug. 77/1 We godless ones and anti-metaphysicians. **1941** *Mind* L. 187 The reactionary anti-metaphysician is primarily a reactionary in science and only secondarily a reactionary in metaphysics. **1905** *Daily Chron.* 27 Dec. 1/7 The trial of twenty-eight anti-militarists was begun yesterday in Paris. **1938** *Encycl. Brit. Bk. of Yr.* 193/2 With Hitler's coming into power in Germany in 1933 ..the conflict between the Nazis and the anti-Nazis in Danzig increased. **1898** *Westm. Gaz.* 26 Oct. 7 In view of probable trouble between revisionists and anti-revisionists. **1828** J. S. Mill in *Westm. Rev.* IX. 272 The marquis de Ferrières, .. a decided royalist and anti-revolutionist. **1890** W. James *Princ. Psychol.* II. xx. 252 A thorough-going anti-sensationalist. **1940** *Tablet* 4 May 419/2 That is why..a sop was conveniently thrown to the anti-Soviets over the mysterious affair of the Soviet Embassy's plain language telegram. **1940** *N. & Q.* CLXXIX. 181/1 Anti-Stratfordians either do not make themselves acquainted with the facts recently discovered, or, knowing them, give them no place in their arguments concerning the authorship of the plays. **1897** *Westm. Gaz.* 7 July 1/2 We might have supposed that some satanic anti-suffragist had contrived this thing. **1959** *Times* 3 June 8/6 Anti-Uglies at Peacehaven. **1822** *Blackw. Mag.* XII. 786 [He] treated with utter scorn..every hint of the anti-vaccinist. **1908** *Practitioner* LXXXI. 479 Hostile agitation of the anti-vivisectionists. **1889** G. B. Shaw in *Star* 29 Nov. 2/4 He is by no means an anti-Wagnerite.

¶ Formations containing a second element ending in *-ist*, used *attrib.* or as *adj.*, as *anti-capitalist, -colonialist, -Communist, -Fascist, -historicist, -humanist, -imperialist, -intellectualist, -materialist, -militarist, -revisionist, -sensationalist, -substantialist, -vaccinist.* Also *ANTISOCIALIST.

1902 *Encycl. Brit.* XXVIII. 492/1 It was not a phase of the anti-capitalist movement. **1949** I. Deutscher *Stalin* iii. 56 A 'bourgeois' (that is anti-feudal, but not anti-capitalist) revolution. **1955** *Times* 5 July 9/5 The Americans, who proclaim their anti-colonialist spirit, are nevertheless paving the way for the British. **1958** *Ann. Reg. 1957* 179 Anti-colonialist speeches were made by a number of delegates and..resolutions were approved in which Western imperialism was condemned in all its forms. **1934** Webster, Anti-communist. **1938** *Encycl. Brit. Bk. of Yr.* 171/2 In 1936 an 'Anti-Communist Pact' was formed by Germany and Japan, and in Nov. 1937 this was joined by Italy. **1927** M. W. Davis *Pol. Handbk. Europe* 57 National Union Party: An anti-Fascist organization. **1938** *Ann. Reg. 1937* 238 The loss of Málaga, due in no small measure..to the lack of cohesion between the anti-Fascist elements in the city. **1945** Popper *Open Society* 201 From the anti-historicist point of view, the question [whether the state originates through a conquest] is of no great importance. **1930** H. Crane *Let.* 8 June (1965) 352 The anti-humanist symposium. **1908** T. Roosevelt *Let.* 4 Mar. (1919) 222 This ought to be read before all the..anti-imperialist societies of the present day. **1937** E. Snow *Red Star over China* i. 20 Were the [Chinese] Reds really anti-imperialist? **1952** C. P. Blacker *Eugenics* 276 The democratic anti-imperialist front. **1907** W. James *Pragmatism* ii. 54 All these..are anti-intellectualist tendencies. **1890** W. B. Yeats *Let. to K. Tynan* 28 Feb. (1953) 111 An article on Blake and his anti-materialist Art. **1905** *Daily Chron.* 27 Dec. 5/6 Anti-militarist placards. **1898** *Westm. Gaz.* 26 Oct. 1/2 No such deadly blow could have been aimed at the Government by any open opponent of revision as this..open avowal of anti-revisionist faith. **1931** *Times Lit. Suppl.* 22 Jan. 52/2 The Church, which was anti-revisionist. **1964** *Ann. Reg. 1963* 209 In the Communist Parties..pro-Chinese members broke away to form anti-revisionist groups. **1890** W. James *Princ. Psychol.* II. xvii. 9 Such passages as the following abound in anti-sensationalist literature: 'Sense is a kind of dull, confused, and stupid perception obtruded upon the soul from without.' **1882** —— in *Princeton Rev.* July 62 Anti-substantialist writers strangely overlook this function in the doctrine of substance. **1945** *Mind* LIV. 214, I do, in fact, like all positivists, hold views..greatly influenced by the anti-substantialist analyses of Berkeley and Hume. **1898** *Daily News* 20 Apr. 4/7 The anti-vaccinist agitation.

7. b. Names of material agents that counteract or inhibit the effect of another substance; in *Biochem.*, spec. a substance that is an antibody (to the antigen to which it corresponds). Examples: *anti-auxin, -cholinesterase, -coagulin, -enzyme, -hormone, -metabolite, -prothrombin, -streptolysin, -thrombin, -trypsin, -virus, -vitamin.* Also *ANTITOXIN; cf. *ANTIBODY. Hence the names of specific agglutinins present in blood serum, as *anti-A, -B, -globulin, -Rh(esus)*; also *ANTIHISTAMINE, *ANTI-SERUM. (See also sense *4 ¶.)

1929 O. C. W. Prausnitz *Standardisation Therapeutic Sera* 51 The designation [of diagnostic sera] should also indicate the blood corpuscle groups with which they reacted, thus: 'Test serum A (anti-B)' and 'Test serum B (anti-A)'. **1932** L. W. H. Bertie tr. *L. Lattes's Individuality of Blood* i. 17 The corresponding agglutinins are denoted by 'anti-A' and 'anti-B' respectively. **1961** *Brit. Med. Dict.* 55/1 *Anti-A*, or *α, agglutinin*, the agglutinin in the serum corresponding to the agglutinogen A with which it will combine, causing clumping and haemolysis of the erythrocytes containing the latter. ..*Anti-B*, or *β, agglutinin*. **1955** *Sci. News Let.* 4 June 361/3 They can prevent the tar-extract effect on plants by adding a recently-discovered plant hormone, antiauxin, to the tar at the time of application. The antiauxin is a crystalline substance extracted from the leaves of plants which are in flower. **1951** *Dorland's Med. Dict.* (ed. 22) 106/2 *Anti-cholinesterase*, a substance which inhibits the action of cholinesterase. **1905** Gould *Dict. New Med. Terms* 80/1 *Anticoagulin*, a substance formed in the body antagonistic in its action to that of a coagulin. **1952** *New Biol.* XIII. 86 The saliva [of the bed-bug] contains an anticoagulin which prevents the blood from clotting and so facilitates its passage through the narrow feeding channel. **1903** *Lancet* 4 Apr. 946/1 The human stomach wall contains an anti-enzyme to its own ferment. **1909** *Cent. Dict.* Suppl., Anti-globulin. **1940** *Biol. Rev.* XV. 27 The antihormones are specific in so far as, for instance, antigonadotrophic hormone does not antagonize the action of the thyrotrophic hormone and vice versa. **1949** I. F. & W. D. Henderson *Dict. Sci. Terms* (ed. 4) 26/2 *Antihormones*, substances which prevent the effect of hormones; chalones. **1947** *Dorland's Med. Dict.* (ed. 21) 120/2 *Antimetabolite*, a closely similar but inactive compound which tends to replace an essential metabolite. **1956** *Nature* 14 Jan. 59/2 The abortifacient action of certain anti-metabolites which are related to nucleic acid metabolism. **1930** *Chem. Abstr.* 422 (*title*) Anti-complementary action of antiprothrombin. **1936** Stedman *Med. Dict.* (ed. 13) 71/2 Antistreptolysin. **1949** *New Gould Med. Dict.* 79/2 *Antistreptolysin*, an antibody which operates against the hemotoxin of hemolytic streptococci. **1909** *Cent. Dict.* Suppl., Antithrombin. **1910** *Encycl. Brit.* IV. 82/2 The body has been found to possess the power of making a substance, antithrombin, which can combine with thrombin forming a substance which is quite inactive as far as clotting is concerned. **1905** *Jrnl. Physiol.* XXXII. 394 The mixture of trypsin and antitrypsin is kept at first at a low temperature. **1913** Dorland *Med. Dict.* (ed. 7), *Antitrypsin*, an antibody of blood-serum having an inhibitive action on trypsin. **1927** H. Plotz tr. *A. Besredka's Local Immunization* 53 The other [substance] is atoxic, thermostable..and antagonizes the former substance. For convenience, we will call this substance, 'antivirus'. **1949** H. W. Florey *Antibiotics* I. i. 45 Artificial immunity could be conferred locally on tissues by the use of what he termed an 'antivirus', which was prepared from old broth cultures of bacteria. **1942** *Chem. Abstr.* 5503 It is suggested that heated fats contain an antivitamin A of undet[ermine]d nature. **1948** *Sci. News* VII. 56 Many substances are now known which antagonise the action of certain of the vitamins.. These 'anti-vitamins' may occur in natural foodstuffs.

attrib. **1934** F. E. D'Amour et al. in *Proc. Soc. Exp. Biol.* XXXII. 192 It seemed to us desirable to investigate the possibility of anti-hormone production against estrin. **1941** Landsteiner & Wiener in *Jrnl. Exp. Med.* LXXIV. 310 The serum of these patients contained anti-Rh iso-agglutinins. *Ibid.* 309 Certain anti-*rhesus* immune sera contain agglutinins specific for the human agglutinogen M. **1948** *Sci. News* VII. 67 A true anti-vitamin effect is shown by the fact that feeding nicotinic acid to the mice overcomes the deleterious action of the 3-acetyl-pyridine. **1951** *Dorland's Med. Dict.* (ed. 23) 1184/1 *Rh antiserums* (*anti-Rh serums*), human serums containing Rh antibodies. .. Rh_0 is also the original Rh factor discovered by Landsteiner and Wiener with the aid of immune anti-rhesus serum produced in animals. **1951** *Proc. Soc. Exp. Biol.* LXXVI. 427/1 Phosphorus-containing anticholinesterase agents are currently in use in this country as agricultural insecticides. **1956** *Nature* 18 Feb. 329/2 Antiglobulin sera obtained by our technique contain antibodies against globulin denatured on the surface of red cells as a result of antigen-antibody reaction. **1958** *Times* 5 July 10/7 A group of patients who have been operated on for cancer of the bladder are now being given this anti-enzyme substance, in the hope that it may reduce the chance of new tumours developing.

c. *anti-horse, -human*, etc., applied to, or pertaining to, an anti-serum prepared from the blood serum of the animal specified.

1906 *Practitioner* Dec. 754 A powerful anti-human serum may be injected intravascularly in man without necessarily producing the serum disease. **1909** W. D'Este Emery *Immunity & Specific Therapy* xi. 317 The symptoms are present when not the slightest trace of antihorse precipitin is demonstrable. **1946** *Nature* 5 Oct. 486/2 It was decided to find out if the test would give positive results only with rabbit anti-human serum, and not with rabbit anti-sera prepared with the serum of various animals. **1964** M. Hynes *Med. Bacteriol.* (ed. 8) vii. 75 The horse-serum compound described above evokes the production of some anti-horse-serum antibodies in the rabbit.

8. Abstract substantives in *-ism*, as *anti-authoritarianism, -Bolshevism, -capitalism, -colonialism, -Communism, -Fascism, -foreignism, -Germanism, -historicism, -imperialism, -intellectualism, -militarism, -nationalism, -nominalism, -pragmatism, -rationalism, -revolutionism, -sensationalism, -Wagnerism, -westernism.* Also *ANTI-AMERICANISM, *ANTI-CLERICALISM.

1939 I. Berlin *K. Marx* ix. 203 The Swiss, Italian and Belgian sections [of the First International], bred on the anti-authoritarianism of Proudhon and Bakunin. **1919** G. B. Shaw *Heartbreak House* p. xliv, By..calling the process anti-Bolshevism. **1934** Webster, Anticapitalism. **1940** B. Ward *Russ. Foreign Policy* 4 The old fear of encirclement reappeared in the disguise of anti-capitalism. **1955** *Ann. Reg. 1954* 97 This cloud was the advent of the 'quit Africa' policy of anti-colonialism. **1927** *New Statesman* 21 May 174/2 There seems to be an attempt to create a big issue of Communism versus anti-Communism. **1923** *Time* 19 May 11/1 Anti-Fascism. **1958** *Listener* 23 Oct. 656/2 The anti-fascism of the nineteen-thirties. **1926** *British Weekly* 18 Nov. 203/2 The bloodshed of May 30 in Shanghai carried anti-foreignism to fever heat throughout the country. **1910** *Daily Chron.* 9 Apr. 1/3 For Great Britain anti-Germanism is not merely a matter of ill-temper..nor even of dogma, but a view of world affairs which has grown up on historical and religious grounds. **1965** *New Statesman* 27 Aug. 281/2 Anti-Germanism in Britain is largely a prejudice of the middle-aged, the middle class and the intellectuals. **1943** *Jrnl. Hist. Ideas* IV. 431 The anti-historicism of Descartes and Malebranche expresses itself also in the fact that they failed to realize another essential aspect of history, namely, its social and cumulative character. **1899** *Daily News* 30 May 5/5 They champion anti-Imperialism, free silver, and the rest of the doctrines dear to the Populists. **1909** Anti-intellectualism [see *FIDEISM]. **1921** M. Ginsberg *Psychol. of Society* iii. 35 This anti-intellectualism is really open to much the same sort of objections as the 'intellectualism' which it attacks. **1906** J. Joyce *Let.* 19 Aug. (1966) II. 151 He was ridiculing..antimilitarism. **1907** *Daily Chron.* 16 Sept. 4/6 Socialists are entitled to say that Socialism is not anti-nationalism, anti-militarism in M. Hervé's sense, nor anti-patriotism. **1906** *Ibid.* 29 Sept. 5/7 The danger..was the growth of a spirit of what he could only call anti-nationalism. **1897** W. James *Let.* 22 Dec. in R. B. Perry *Thought & Char. W. James* (1935) II. 419 Anti-nominalism, categories [etc.]. **1960** *Encounter* Mar. 41/1 The two lines of anti-nominalism meet..in French existentialism. **1907** W. James in *Jrnl. Philos.* IV. 465 If that be my friend Pratt's definition of a pragmatist, I can only concur with his antipragmatism. **1909** A. O. Lovejoy *Let.* 27 Aug. in R. B. Perry *Thought & Char. W. James* (1935) II. 595, I mean the extreme of anti-rationalism to which the book gives expression. **1936** *Mind* XLV. 272 By anti-rationalism is meant the rejection of any form of mystical intuition. **1930** Anti-revolutionism [see *anti-Westernism* below]. **1909** W. James *Plur. Universe* vii. 277 To say nothing of your traditional Oxford devotion to Aristotle and Plato, the leaven of T. H. Green probably works still too strongly here for his anti-sensationalism to be outgrown quickly. **1895** G. B. Shaw *Sanity of Art* (1908) 33 The huge blunder of anti-Wagnerism. **1930** L. Fischer *Soviets in World Affairs* II. xvii. 535 The elements of his *credo* were monarchy, anti-revolutionism and anti-Westernism. **1959** *Economist* 14 Feb. 569/1 They have exploited Afro-Asian anti-westernism.

IV. With reduplication of the prefix. Also *anti-anti* used *absol.*

1896 W. E. Gladstone in *Daily News* 29 May 6 Of course, I am strongly *anti*-anti-Semitism. **1900** M. Beerbohm *Around Theatres* (1924) I. 182, I gave them [i.e. my reasons] merely because they will explain equally why a play ought not to end on a high note, and will thus confute the anti-anticlimax party. **1951** R. Campbell *Light on Dark Horse* xiv. 200, I was dumbfounded by this anti-anticlimax, if I may coin such a word. **1953** *Time* 7 Dec. 24 A well-worn anti-anti-Communist technique; he denounced..Brownell's handling of the White case as 'McCarthyism'. **1961** *New Statesman* 25 Aug. 234/2 It is not Communism that has been preoccupying Senators of the far right lately, but something called 'anti-anti-Communism', a barbarism coined by the intellectual wing of the McCarthy movement. **1962** *New Scientist* 15 Mar. 607/1 The weapons..include..anti-missile and anti-satellite missiles, 'anti-antis' and decoys. *Ibid.* 22 Mar. 677 (*caption*) Satellite bombers..'antis' and 'anti-antis'. *Ibid.* 677/1 There is every prospect of a major arms race in space, with the development of a variety of 'anti' and 'anti-anti' devices.

anti-aircraft (æ:nti,ɛ̄ᵊ·ɹkrɑft), *a.* [Anti-[1] 4.] Used for defence against hostile aircraft; esp. *anti-aircraft gun*, one for firing shells at a very high elevation. Hence *ellipt.* or as *sb.*, anti-aircraft defences, regiment, etc. Abbrev. **A.A.* (see *A III); see also *ACK-ACK.

1914 *Scotsman* 25 Sept. 5/4 An anti-aircraft gun of the Third Army Corps. **1919** *Athenæum* 23 May 360/2 The anti-aircraft force. **1939** *Times Weekly* 15 Feb. 8 Two Regular regiments of anti-aircraft artillery were being brought to London to give additional protection to the city. **1947** L. Hastings *Dragons are Extra* ix. 197, I had been passed A1 by the doctors in London and found myself back in uniform in anti-aircraft. **1959** *New Statesman* 3 Jan. 6/3 The Nike-Zeus anti-aircraft missile system. **1959** *Listener* 24 Dec. 1108/2 The anti-aircraft did its stuff.

anti-American (ænti,ăme·rikăn), *a.* [f. Anti-[1] 3 a + American.] Hostile to the interests of the United States, opposed to Americans. Hence **anti-Ame·ricanism**, a spirit of hostility towards Americans.

1773 in *Amer. Speech* (1945) XX. 269 The anti-American doctrine of the parliament's right in all cases. **1838** in J. S. Buckingham *America* (1841) I. ix. 173 Now our young men are rogues and fops, with extravagant anti-American notions. **1844** P. Hone *Diary* (1889) II. 238 Most of them seem to have escaped the foppery of foreign manners and the bad taste of anti-Americanism. **1932** H. Nicolson *Public Faces* xi. 301 Not only did he regard it [*sc.* the communication] as fresh in tone, but he regarded it as anti-American in substance..not merely against the Monroe Doctrine, but against the freedom of the seas. **1958** *Spectator* 1 Aug. 167/1 It is not anti-Americanism to feel that now, more than ever, we need to cultivate our own idiom and qualities. **1963** *Guardian* 21 Feb. 16/7 The Labour Party—which is anti-American without being pro-European.

antibilious, *a.* (Earlier example.)

1817 Jane Austen *Sanditon* (1954) ii. 373 The Sea air & Sea Bathing..were nearly infallible..; They were..anti-bilious & anti-rheumatic.

antibiosis (æ:ntibəiōᵘ·sis, -bi-). *Biol.* [mod.L., ad. F. *antibiose* (P. Vuillemin 1890, in *Compt. Rend. Assoc. Franç. Avanc. Sci. 1889* II. 525), f. Anti-[1] 2 + Sym)biosis.] A condition of antagonism between organisms, esp. micro-organisms (opp. *symbiosis*).

1899 H. M. Ward in *Ann. Bot.* XIII. 555 Extremes.. where one of the two associated organisms is injuring the other, as exemplified by many parasites... This state of affairs has been termed *Antibiosis*. **1949** H. W. Florey *Antibiotics* I. i. 73 Although much work has been done and considerable interest has been taken in antibiosis during the last 60 years it is only recently that it has reached its present intensity.

antibiotic, *a.* Restrict *rare* to sense in Dict. and add: **2.** [ad. F. *antibiotique* (P. Vuillemin 1890, in *Compt. Rend. Assoc. Franç. Avanc. Sci. 1889* II. 526).] Injurious to or destructive of living matter, esp. micro-organisms. So **a:ntibio·tically** *adv.*

1894 *Phil. Trans.* B. CLXXXV. 312 Production of a bactericidal, or at least antibiotic, substance. **1942** S. A. Waksman et al. in *Soil Science* LIV. 295 The most active antagonistic cultures were studied in detail, and several antibiotic substances were isolated. **1945** S. A. Waksman *Microbial Antagonisms & Antibiotic Substances* 271 *Antibiotic*, inhibiting the growth or the metabolic activities of bacteria and other microorganisms by a chemical substance of microbial origin. **1946** *Nature* 6 July 24/2 Antibiotically active cell-free watery extracts could be prepared from cultures on Dorset's egg medium.

3. Hence as *sb.*, an antibiotic substance: one of a class of substances produced by living organisms and capable of destroying or inhibiting the growth of micro-organisms; *spec.* any of these substances used for therapeutic purposes. Also used of synthetic organic compounds having similar properties.

1944 *Lancet* 18 Mar. 375/2 (*title*) The Mould Antibiotics. **1949** H. W. Florey et al. *Antibiotics* II. xlvii. 1438 The antibiotics comprise substances with diverse chemical structures and biological activities. They range in their action from those which inhibit the growth of certain strains of bacteria in a highly selective manner to those which are relatively toxic to all living cells. **1958** *Listener* 16 Oct. 620/1 The discovery of 'interferon', a sort of virus antibiotic.

Also in *Comb.*

1956 *Nature* 25 Feb. 368/1 The search for new antibiotic-producing organisms, particularly streptomyces, from Malayan soil continues. **1961** *Times* 17 Mar. 5/4 Of the antibiotic-resistant staphylococci..90 per cent were isolated from infants born in the hospital.

antibody (æ·ntibǫdi). *Physiol.* [tr. G. *antikörper*: see Anti-[1] 2 a.] Any of the proteins, naturally present in the body or produced in response to the introduction of an antigen, which react with specific antigens.

1901 Hektoen & Riesman *Pathol.* 231 Substances which appear during spontaneous or artificial infection or intoxication are known as antibodies (*Antikörper*) and antitoxins. **1921** R. A. Freeman in *Edin. Rev.* July 33 Those antibodies by which the existence of aggregates of the lower organisms is brought to an end. **1955** *Sci. News Let.* 8 Jan. 19/2 Antibodies are the substances formed in the body to fight off invading disease germs..a different kind of antibody being formed for each different kind of disease germ.

Also *fig.*

1921 R. A. Freeman *Social Decay* 287 The Social Antibody is not Mechanism itself but the domination by it of human life and human activities. **1947** J. Hayward *Prose Lit. since 1939* 47 The writer himself creating the antibodies to protect his integrity.

a:nti-Bri·tish (stress variable), *a.* [f. Anti-[1] 3 a + British.] Hostile to the interests of Britain, opposed to the British. Hence **anti-Britisher**, one who is antagonistic to the British or British interests; **anti-Britishism**, a spirit of hostility towards the British.

a **1845** Syd. Smith *Irish Ch.* in *Wks.* (1859) II. 334/1 Such a piece of anti-British villany. **1898** *Daily News* 4 Oct. 4/5 Germany has withdrawn her anti-British support from the Transvaal. **1902** *Daily Chron.* 10 Mar. 7/5 The result is that Dutch Reformed ministers are, as a rule, anti-British. *Ibid.* 15 Sept. 5/3 Commandant Wolmarans,.. an extremely bitter anti-Britisher. **1909** H. G. Wells *Tono-Bungay* III. iv, The captain helped him to express his own malignant Anti-Britishism. **1928** *Observer* 18 Mar. 17/4 There is no anti-Britishism here, but I am puzzled by the objection to our being pro-American.

anticathode (æ:nti,kæ·þōᵘd). *Electr.* [Anti-[1] 2.] The metal plate opposite the cathode in an X-ray tube, upon which the cathode rays impinge and produce X-rays.

1907 A. F. Hertz in *Practitioner* Apr. 524 A pencil, which points directly towards the anticathode, is fastened behind the fluorescent screen. **1913** *Physical Rev.* II. 416 The anticathode..consists of a single piece of wrought tungsten, having at the end facing the cathode a diameter of 1·9 cm. **1923** Glazebrook *Dict. Applied Physics* IV. 595/2 The electrons (cathode rays)..are directed on a heavy metal anticathode or target. **1951** J. R. Partington *Gen. & Inorg. Chem.* (ed. 2) viii. 191 When cathode rays strike a material target, which in the X-ray tube consists of a metal..plate called an anti-cathode, this emits a radiation which passes outside the tube.

anti-Catholic, *a.* and *sb.* (Earlier examples.)

1665 N. French *Let.* 19 Sept. in *Hist. Wks.* (1846) I. 136 A principal Leader in the Anti-catholick Ormonian Faction. **1780** H. Walpole *Let.* 6 Feb. (1904) XI. 125 Lord George Gordon..dresses..like the first Methodists; though I take the modern ones to be no anti-Catholics.

anticer (æ·ntəisəɹ). [Contraction of *ANTI-ICER.] = *ANTI-ICER.

1935 B. Lockspeiser in *Aircraft Engineering* Nov. 280/3 The anticer has been tested under ice forming conditions on ..aeroplanes. **1941** *Aeroplane Spotter* 23 Jan. 30/3 *Anticer*, a device for preventing the formation of ice on aircraft, usually in the form of a liquid paste. Should not be confused with De-icers, which are devices for removing ice when it has formed.

anti·cipatingly, *adv.* [f. Anticipating *ppl. a.* + -ly[2].] With anticipation.

1851 H. Melville *Moby Dick* II. xxi. 158 That weapon must..be anticipatingly tossed out of the boat. **1888** Walt Whitman in *Century Mag.* (1911) Dec. 250/1 His whole face would light up anticipatingly as he spoke. **1904** *Daily Chron.* 19 Mar. 7/5 The listeners..tittered anticipatingly.

anticle·rical, *a.* and *sb.* [Anti-[1] 3 b.] Opposed to clericalism. Hence as *sb.*, a person who is opposed to clericalism.

1845 S. Austin tr. *Ranke's Hist. Ref.* II. iv. v. 517 Here too, the inclinations of the council and citizens had received an anti-clerical bias from a great number of circumstances. **1881** *Times* 16 May 7/4 An anti-Clerical Congress..closed to-day with a speech by M. Louis Blanc. *Ibid.* 19 July 5/1 When the anti-Clericals—for it would be absurd to call them Liberals—fail to obtain their provocation in one direction [etc.]. **1883** [see Anti-[1] 3 b]. **1906** J. E. C. Bodley *Church in France* 51 The anti-Clericals were a minute and feeble minority. **1937** R. H. Lowie *Hist. Ethnol. Theory* xi. 193 Anticlerical critics have suggested that the field work undertaken under Father Schmidt's auspices has been unduly coloured by Catholic or personal prejudices.

Hence **anti-cle·ricalism**.

1886 *Times* 31 Mar. 5/2 These measures may be regarded as the swing of the pendulum from clericalism to anti-clericalism. **1898** J. E. C. Bodley *France* I. 141 Free-thinkers contravene the basis of their own profession in erecting anti-clericalism into a dogma.

anticlimactic (æ:ntikləimæ·ktik), *a.* [f. Anticlimax, after Climactic *a.*] Of the nature of an anticlimax. So **a:nticlima·ctically** *adv.*

1898 *Westm. Gaz.* 28 May 2/3 That sounds a little anti-climactic. **1915** W. J. Locke *Jaffery* xix, I urged, somewhat anticlima[c]tically after my impassioned harangue [etc.]. **1917** McKenna *Sonia* i. 37 He began valiantly enough, and then anticlimactically as he caught sight of me, 'What d'you want?' **1941** J. Ransom *New Criticism* i. 100 When the artist works over the practical failures in human life, and the ambiguous anticlimactic half-successes.

anticlinal, *a.* **A. 1.** Substitute for def.: Applied to a line or axis from which strata slope down or dip in opposite directions; also said of the fold or bend in such strata, or of a ridge so formed. Opposed to Synclinal. (Later examples.)

1875 J. W. Powell *Explor. Colorado River* xi. 160 Longitudinal valleys..three varieties (*a*) anticlinal, valleys which follow anticlinal axes. **1929** H. G. Busk *Earth Flexures* ii. 7 An 'anticlinal bend' is any relatively sharp flexure in an anticlinal sense, that is, where the beds of the one limb dip gently towards the apex, and the beds of the other dip relatively steeply away from it.

3. *Bot.* Applied to those cell-walls at a growing-point which have a direction at right angles to the surface: opp. to Periclinal *a.* 2. Also as *sb.*, an anticlinal wall or plane.

1882 Vines tr. *Sachs's Bot.* 951 The planes of the walls in a growing-point are classified thus: *a. Periclinal*... *b. Anticlinal*, those which intersect the surface and the periclinal walls at right angles... If the outline..of the growing-point is a parabola..the anticlinals being the orthogonal trajectories of the periclinals, constitute a system of confocal parabolas. **1960** I. F. & W. D. Henderson *Dict. Sci. Terms* (ed. 7) 33/2.

Hence **anticli·nally** *adv.*

1855 J. Phillips *Man. Geol.* ii. 45 Thus in fig. 16 the strata are synclinally and anticlinally bent, but the cleavage is vertical, or nearly so. **1954** *New Biol.* XVI. 119 In the outer layer known as the tunica the cells divide anticlinally, that is by walls at right angles to the surface. **1965** G. J. Williams *Econ. Geol. N.Z.* xix. 349/2 The Ahaura basin at that time was but a shallow shelf or even anticlinally warped.

anticlinorium (æ:ntiklinōᵊ·riŭm). *Geol.* [mod.L., f. as Anticlinal + -orium.] **1.** (See quot.)

1874 J. D. Dana *Man. Geol.* (ed. 2) IV. 752 An upward bend of the crust, or geanticlinal, is of itself an elevation; and such an elevation is an *anticlinorium*.

2. (See quot. 1893.) (The only current sense.)

1893 *13th Ann. Rep. U.S. Geol. Surv.* II. 220 It often happens that the result of the combination of many anticlines and synclines is to form a complex structure, which, regarded as a whole, is either synclinal or anticlinal. The former is called a synclinorium, the latter an anticlinorium. **1920** J. Geikie *Struct. & Field Geol.* (ed. 4) ix. 142 When a broad zone has bulged up under lateral pressure..we may have one great arch composed of numerous subordinate wrinkles or minor folds and flexures. A complex arch of this kind is termed an Anticlinorium. **1938** *Geogr. Jrnl.* XCI. 281 The whole San Carlos range is an arcuate anticlinorium trending east–west with convexity to the south.

Hence **a:nticlino·rial** *a.*, of or pertaining to an anticlinorium.

1940 *Geogr. Jrnl.* XCV. 332 On the west Burma is separated from India by a complex anticlinorial fold of Alpine-Himalayan age.

anti-clockwise (æ:nti,klǫ·kwəiz), *a.* and *adv.* [f. Anti-[1] 3 c + Clock *sb.* + -wise.] = Counter-clockwise.

1898 G. Wherry in *Lancet* 1 Jan. 24/1 Mathematicians often use the expression 'clockwise' or 'anti-clock-wise' to indicate the way of a spiral coil. **1914** Domville-Fife *Submarines, Mines & Torpedoes* 163 A tiny three or four-cylinder engine which operates twin screws, moving 'clockwise' and 'anti-clockwise'. **1917** 'Contact' *Airman's Outings* 153 A little left rudder again puts the needle into an anticlockwise motion. **1927** *Daily Express* 23 Sept. 3/3 A clockwork motor that is wound up anti-clockwise.

anti-coagulant (æ:nti,ko,æ·giŭlănt), *a.* and *sb.* [Anti-[1] 3 b.] **A.** *adj.* That retards or prevents coagulation, esp. of the blood. **B.** *sb.* Any substance that is antagonistic to coagulation.

1905 Gould *Dict. New Med. Terms* 80/1 *Anticoagulant*, adj. and sb. **1909** *Jrnl. Physiol.* XXXVIII. 497 The large quantity of anti-ferment present in hirudin indicates that it also plays a part in the anti-coagulant action. **1920** *Discovery* Apr. 105/2 This effect is not spoilt by the admixture of citrate, though this is used as an anti-coagulant outside the body. **1958** *Observer* 2 Mar. 13/3 He takes anti-coagulants constantly. **1965** *New Scientist* 22 Apr. 245/2 The greatest advance in effective rodenticides was the development of the anti-coagulants such as dicoumarin.

anti-coi·ncidence. *Physics.* [f. Anti-[1] 4 + *coincidence 7.] Applied *attrib.* to a device, or to a system employing a device, which rejects or discriminates against measurements of events which occur simultaneously or are separated by a predetermined time, used esp. in experiments with cosmic rays and in nuclear physics. Also, the indication of the occurrence of an ionizing particle in such a device.

1939 G. Herzog in *Physical Rev.* LV. 1266/1 To investigate the nature of cosmic-ray particles a special anti-coincidence amplifying set was used to trip the cloud chamber only by those cosmic-ray particles whose further path lies below a given range. **1940** —— *Ibid.* LVII. 67/2 Anticoincidences can be used to suppress the recording of the discharges in a coincidence circuit when one additional G-M counter is fired simultaneously. **1948** L. Janossy *Cosmic Rays* iv. 40 If the background of coincidences, caused by the ionizing rays, could be eliminated, the arrangement would be vastly more sensitive for the detection of any non-ionizing rays. The solution of this problem was achieved by the important technique of *anti-coincidences*. *Ibid.*, By a suitable electrical arrangement the coincidence counters are made *insensitive* whenever the anti-coincidence counters register a particle. **1953** A. D. & K. H. V. Booth *Autom. Digital Calculators* vi. 42 The circuitry involves only standard coincidence and anti-coincidence units.

anti-convulsant (æ:nti,kǫnvɒ·lsănt), *a.* and *sb.* [Anti-[1] 3 b.] **A.** *adj.* That retards or prevents convulsions (*Lippincott's New Med. Dict.*, 1910). **B.** *sb.* A substance that is antagonistic to convulsions.

1943 *Lancet* 6 Nov. 578/2 Petit mal or slow-wave activity is less amenable to phenobarbitone and other anticonvulsants than are major fits. **1946** *Nature* 19 Oct. 541/2 Although the patient was under the influence of large doses of anticonvulsant drugs and was almost free from spontaneous attacks. **1966** *Lancet* 24 Dec. 1371/2 Phenobarbitone sodium and intramuscular paraldehyde in optimum doses, were used as anticonvulsants.

anticryptic (æntikri·ptik), *a. Zool.* [f. Anti-[1] 3 c + *cryptic *a.* 3.] Applied to markings or coloration serving for concealment to the disadvantage of some other animal: distinguished from Procryptic *a.*

1890 E. B. Poulton *Colours Anim.* xvii. 336 Protective and Aggressive Resemblances are classed as Cryptic Colours (Procryptic and Anticryptic). *Ibid.* 337 Pseudepisematic colours..are special instances of Anticryptic colours. **1960** I. F. & W. D. Henderson *Dict. Sci. Terms* (ed. 7) 33/2.

anti-de·mocrat. [Anti-[1] 5.] One who is opposed to democracy or to the principles of the Democratic party in the United States (see Democrat *a.* 2)

1802 (*title*) The Republican; or, Anti-Democrat [cited in D.A.]. **1855** Bristed in *Fraser's Mag.* LII. 521 Many of the Southern anti-Democrats..carried them [*sc.* some doctrines] to greater length than the Democrats themselves. *a* **1937** J. L. Stocks *Reason & Intuition* (1939) vii. 112 The imposture of collective wisdom, of which anti-democrats talk. **1945** Popper *Open Society* I. x. 161 He [*sc.* Thucydides] probably did not support the treacherous acts of the extreme anti-democrats.

So **a:nti-democra·tic** *a.*, **-demo·cracy.**

1837 *U.S. Mag. & Democr. Rev.* Oct. 10 That extensive anti-democratic corruption of sentiment in some portions of our people. **1842** J. S. Mill *Let.* 6 Apr. in *Wks.* (1963) XIII. 514 Some of the opinions are likely to be considered ultra-liberal, although .. rather anti-democratic. **1881** Antidemocratic [see Anti-[1] 3 b]. **1909** G. K. Chesterton *Orthodoxy* ix. 281 These supernatural things are never denied except on the basis of anti-democracy or of materialist dogmatism. **1939** *War Illustr.* 9 Dec. 416/3 Fascism remains anti-Communist, but it also remains obstinately anti-democratic.

anti-dip (ænti,di·p). *Physical Geogr.* [cf. Dip *sb.* 5.] Used *attrib.* (see quot. 1961).

1900 *Proc. Cotteswold Club* XIII. 178 It may be premised that dip streams are termed consequents..and anti-dip streams, obsequents. **1961** L. D. Stamp *Gloss. Geogr. Terms* 27/1 *Anti-dip stream*, a stream flowing in approximately the opposite direction to the regional dip.

antidoron (æntidōᵊ·rǫn). *Eastern Church.* [Gr. ἀντίδωρον return-gift, f. ἀντι- Anti-[1] + δῶρον gift.] = Eulogia b, c.

1850 Neale *Holy East. Ch.* I. 525 Before the people are dismissed, the Priest gives them the *antidoron*; i.e. a part of the bread blessed for the prothesis. **1892** A. J. Maclean & W. H. Browne *Catholicos of East* 209 The Syrian deacon.. may also make the holy loaves for consecration and for the *Antidoron*. **1957** *Oxf. Dict. Chr. Ch.* 1005/1 Among the E. Orthodox..the so-called 'antidoron'..i.e. what remains of the loaves from which the Eucharistic Bread is cut, is held to share in the liturgical offering, and is distributed as a consolation to those unable to receive Holy Communion.

anti-fat (æ·nti,fæt), *a.* and *sb.* [Anti-[1] 4, 7.] **A.** *adj.* That counteracts the formation, or reduces the amount, of fat. **B.** *sb.* (*Disused.*) A substance used to counteract obesity.

1886 *Chambers's Jrnl.* III. 46/1 He must feed on either locusts or grasshoppers, which are both supposed to be anti-fat diet. **1887** Bentley *Man. Bot.* (ed. 5) 768 This Alga is the essential constituent in the nostrum termed Anti-Fat. **1909** *Brit. Med. Assoc., Secret Remedies* ix. 92 Anti-Fat..acts solely upon the food in the stomach.

anti-fe·d, -fe·deral, -fe·deralist. *U.S.* [Anti-[1] 5.] An opponent of federalism in the early years of American independence. Also **anti-fe·deral** *a.*, **-fe·deralism.**

1787 *Independent Gazetteer* (Philad.) 28 Sept., This anti-federalist should reflect that his name may yet be known and himself branded with infamy as an enemy to the happiness of the United States. **1788** R. King in *Life & Corr.* (1894) I. 327 Of three thousand votes given..not more than two hundred were in favor of the antifederal Ticket. **1788** C. Gore in R. King *Life & Corr.* (1894) I. 348 Non-residence in one candidate..who was abhorred for his anti-federalism. **1788** *Maryland Jrnl.* 3 June (Th.), The famous Dr. Spring asked a lady on which side she was, fed, or antifed. **1800** *Aurora* (Philad.) 28 Nov. (Th.), By anti-federalist, the same is meant as by jacobin,..and the like. **1804** Fessenden *Orig. Poems* (1806) 55 And swears no anti-federal noddy Has half a soul to bless his body. **1805** —— *Democr. Unveiled* I. 13 As Torries many of you vex'd us As Antifederals then perplex'd us. *Ibid.* 112 Supported by the factious heads Of ever restless anti-feds.

antiferromagnetic (æ:nti,feromægne·tik), *a.* and *sb.* [ad. G. *antiferromagnetisch* (L. Hulthèn 1936, in *Proc. Kon. Akad. Amsterdam* XXXIX. 190), f. Anti-[1] 3 + Ferromagnetic *a.*] **A.** *adj.* Characterized by oppositely directed (or anti-parallel) alignment of electron spin. **B.** *sb.* A magnetic substance with these properties.

L. Néel 1932, writing in French in *Ann. Phys.* XVIII. 5f., introduced the idea of antiferromagnetism but did not use the words *antiferromagnétique, -magnétisme*.

1936 *Sci. Abstr.* A. XXXIX. 548 Cases where interchange action occurs, but with a positive coefficient, are indicated as antiferromagnetic. **1938** *Physical Rev.* LIV. 82/1 Manganese probably is an antiferromagnetic substance. **1949** *Ibid.* LXXVI. 1257/1 At absolute zero all of the atoms on one lattice have their electronic magnetic moments aligned in the same direction and all of the atoms on the other lattice have their moments antiparallel to the first. Above the Curie temperature the thermal energy is sufficient to overcome the tendency of the atoms to lock antiparallel. ..Materials exhibiting the characteristics described above have been designated 'antiferromagnetic'. **1955** *Sci. News* XXXVII. 36 In ferro-, antiferro-, or ferri-magnetics the alignment of atomic magnets is ordered and periodic.

Hence **a:ntiferroma·gnetism**, the quality of being antiferromagnetic; the phenomena exhibited by antiferromagnetic bodies.

1938 F. Bitter in *Physical Rev.* LIV. 79 The proposed equations of state..may be regarded as an attempt to describe antiferromagnetism (i.e., the case in which oppositely directed spins on adjacent atoms have the lowest energy). **1961** *Encycl. Dict. Physics* I. 222/1 The state of complete antiferromagnetism (antiparallel alignment of spin vectors) will be attained only at absolute zero.

anti-fogma·tic. *U.S.* [f. Anti-[1] 7 + *fogmatic*, jocular formation on Fog *sb.*] An alcoholic liquor taken to counteract the effects of damp or wet.

1789 *Massachusetts Spy* 12 Nov. 4/2 Rum. Its great utility in preserving the planters from the effects of the damp and unwholsome air of the morning, has given it the medical name of an Antifogmatick. **1824** in Horry & Weems *Life Marion* (1833) ix. 77 Now suppose you take a glass of peach..they say it is good of a rainy morning..a mighty antifogmatic. **1852** *As good as a Comedy* (Philad.) 134 (Th.), Tom Nettles [was] mixing a couple of rosy anti-fogmatics.

anti-freeze (æ·nti,friz), *a.* and *sb.* [Anti-[1] 4, 7.] **A.** *adj.* That reduces the tendency (of a liquid) to freeze. **B.** *sb.* A chemical agent added to water to lower its freezing point.

1913 A. L. Clough *Dict. Automobile Terms* 13 Anti-freeze solution. **1935** *Amer. Speech* X. 122/1 The material is..limited by excluding the names of all products such as antifreezes which have neither a paraffin nor an asphalt base. **1937** D. Aldis *Time at her Heels* i. 10 They kept the Ford out in the street in front of the house. It worked all right if they remembered to get it filled with anti-freeze. **1959** *Which?* Dec. 174 Most of the anti-freeze compounds sold in Britain are based on ethylene-glycol because methanol, though cheaper, has disadvantages.

anti-g (æ:nti,dʒī·), *a.* [Anti-[1] 2 a + G III. b.] Short for *ANTI-GRAVITY, esp. in *anti-g suit*, = *G-SUIT.

1945 *Flight* XLVII. 36/1 (*heading*) Anti-g suit for fighter pilots. A simple but effective means of protecting the fighter pilot from the effects of 'positive g' during violent manœuvres is now in general use with U.S.A.A.F. fighter pilots. **1961** *Ibid.* LXXX. 50/2 Ducts from the main air systems lead to..the pilot's anti-g trousers and air ventilated suit.

antigen (æ·ntidʒen). *Biochem.* [G. (E. Friedberger and C. Moreschi 1905, in *Centralbl. f. Bakteriol.* XXXIX. 456), f. *ANTI(BODY + -GEN.] A foreign substance which, when introduced into a living organism, stimulates the production of an antibody.

1908 Park & Bolduan in Nuttall & Graham-Smith *Bacteriol. of Diphtheria* 628, v. Pirquet and Schick attribute the immediate or hastened reaction and indeed the serum disease as a whole to an interaction between the antigen and the antibody. **1911** *Jrnl. Chem. Soc.* C. II. 812 When an optimum dose of antigen (goat's or rat's blood) is injected intravenously into a dog. **1955** *Sci. News Let.* 16 Apr. 253/1 The test reaction is based on the body's ability to react to and destroy bacteria and foreign protein invaders. These foreign agents are called antigens, while the body's own destructive agents are referred to as antibodies. **1959** *Chambers's Encycl.* VII. 398/1 Most antigens are proteins or contain substantial amounts of protein, and they have a high molecular weight.

Hence **antige·nic** *a.*, of the nature of an antigen, **antige·nically** *adv.* So **a:ntigeni·city**, the condition or property of being antigenic.

1913 *Arch. Internal Med.* XI. 85 The 'antigenic' function of all the extracts resides solely in the lipoid fraction. **1916** Olitsky & Bernstein in *Jrnl. Infectious Diseases* XIX. 259 We have shown that organisms grown on serum media incorporate..those serum elements which possess serum antigenic properties. **1946** *Nature* 27 July 120/1 The hæmagglutinins are complexes of a phospholipid (probably lecithin) with an antigenically specific virus constituent. *Ibid.* 17 Aug. 239/1 The loss in antigenicity during preparation of the toxoid (measured by flocculation) was about half that usually found.

anti-gra·vity. [Anti-[1] 2 a.] A force opposed to that of gravity. Chiefly *attrib.* Also in shortened form (*colloq.*) **anti-grav.** Cf. *ANTI-G.

1945 in *Amer. Speech* (1946) XXI. 142/1 The anti-gravity pneumatic pants apply pressure to the pilot's abdomen and legs, preventing the blood from pooling in the lower extremities. **1946** A. C. Clarke in *Across Sea of Stars* (1959) 155 The three explorers stepped out of the airlock and adjusted the antigravity fields of their suits. **1949** *Ibid.* 58 Antigravity.. the ultimate secret that man had never learned. **1953** C. L. Harness *Flight into Yesterday* (1964) ii. 32 He had found a way to beat the tremendous solar gravity by..an anti-grav mechanism.

anti-hero (æ·nti,hīᵊ:ro). [Anti-[1] 1 b.] One who is the opposite or reverse of a hero; esp. a chief character in a poem, play, or story who is totally unlike a conventional hero. So **a·nti-her:oine.**

1714 [see Anti-[1] 1 b]. **1897** W. P. Ker *Epic & Romance* iv. 346 The ineffectual sorrows and good intentions of the anti-hero Fromont. **1907** F. W. Chandler *Lit. Roguery* I. ii. 68 A work of the Eulenspiegel type..its anti-heroes remain less roguish than Till. *Ibid.* iv. 148 Other women figured as anti-heroines in less literary tracts. **1958** *Listener* 14 Aug. 248/3 Anti-heroes and unpleasant young men are now so morbidly the fashion. **1959** *Sunday Times* 27 Dec. 6/4 It was the decade of..the new, shambling, oafish anti-heroes, flotsam of the Welfare State—Lucky Jim and his first cousin Joe Lampton. **1962** *Times* 26 Jan. 16/5 Miss Wendy Baldwin presented a drab, vulgar anti-heroine.

Hence **anti-hero·ic** *a.*

1876 [see Anti-[1] 3 c]. **1959** *Times Lit. Suppl.* 20 Nov. 670/4 *Ulysses* was and remains the first great masterpiece of anti-heroic literature.

a:ntihi·stamine, *a.* and *sb.* [Anti-[1] 7 b.] **A.** *adj.* That counteracts the effect of histamine. **B.** *sb.* A synthetic drug having this property, used esp. in the treatment of certain allergic conditions. So **a:nti-histami·nic** *a.* and *sb.*

1933 A. C. H. Yen & Hsi-Chun Chang in *Proc. Soc. Exper. Biol.* XXXI. 338 The sensitivity to histamine in guinea pig and man may be accounted for by the absence of antihistamine activity in their blood. **1942** *Jrnl. Immunology* XLIV. 219 In a series of studies on the antihistamine activity of a number of derivatives of phenoxyethylamine.. (1571 F) prevented certain histamine effects. **1947** *Lancet* 17 May 678/1 (*title*) The Anti-histamine Drugs. **1950** *Chem. & Engin. News* XXVIII. 846/1 Opinion is divided on the use of antihistaminics as cold cures. **1955** *Sci. News Let.* 16 July 40/2 Many of the itch remedies used today..such as anti-histaminics..may depend for their results on a partial blocking of the proteinases, or itch enzymes. **1957** *N.Z. Listener* 26 July 15/1 The antihistamines, discovered to have a quietening effect and to act specifically against motion sickness. **1964** S. Duke-Elder *Parsons' Diseases of Eye* (ed. 14) xxii. 313 Treatment should consist of the administration of anti-histaminic or diuretic drugs in the early stages.

anti-icer (ænti,əi·səɹ). [f. Anti-[1] 7 + Ice *sb.* + -er[1].] A device for preventing the formation of ice on an aeroplane. Cf. *ANTICER.

1935 *Aircraft Engineering* Nov. 281/2 Figures 7 and 8 are reproductions..illustrating the action of the anti-icer under ice forming conditions. *Ibid.*, The fin to which an anti-icer is fitted is also free from ice.

Antilegomena (æ:ntilĕgǫ·mĕnă), *sb. pl.* [Gr. ἀντιλεγόμενα (Eus. Caes.), neut. pl. of pres. pple. pass. of ἀντιλέγειν to gainsay, pass. to be disputed.] (See quot. 1957.)

[**1847** Crusé tr. *Eusebius's Eccl. Hist.* (ed. 4) 247 He has given us abridged accounts of all the canonical Scriptures, not even omitting those that are disputed (The Antilegomenoi).] **1875** *Encycl. Brit.* II. 128/2 The following is a catalogue of the *Antilegomena*. **1957** *Oxf. Dict. Chr. Ch.* 62/1 *Antilegomena*, the word is used by Eusebius of Caesarea.. of those Scriptural books of which the claim to be considered a part of the NT canon was disputed. He subdivided them into (1) those 'generally recognized' (γνώριμοι), viz. Jas., Jude, 2 Pet., 2 and 3 Jn.; and (2) the 'spurious' (νόθοι), viz. the 'Acts of Paul', the 'Shepherd of Hermas', the 'Apoc. of Peter', the 'Ep. of Barnabas', the 'Didache', and (perhaps) the 'Rev. of St. John'.

anti-life (æ·nti,ləif), *a.* and *sb.* [Anti-[1] 2 b, 4 + Life *sb.*] **A.** *adj.* Opposed to living fully and harmoniously with the natural order. **B.** *sb.* An alleged contrary principle to full and natural living; an attitude, or course of action, in accordance with this principle.

1926 D. H. Lawrence *Plumed Serpent* vi. 112 But he, too, was widdershins, unwinding the sensations of disintegration and anti-life. **1929** —— *Let.* 12 June in F. Lawrence *Not I, But the Wind* (1935) 252 The real principle of Evil is not anti-Christ or anti-Jehovah, but anti-life. I agree with you, in a sense, that I am with the anti-Christ. Only I am not anti-life. **1938** 'N. Blake' *Beast must Die* III. xii. 214 Somewhere deep within her, thought Nigel, there is a stone core of lifelessness, an anti-life principle. **1956** C. Wilson *Outsider* iii. 47 The atmosphere of the Existentialist Outsider is unpleasant to breathe. There is something nauseating, anti-life, about it.

Antillean (æ:ntili·ăn, ænti·liăn), *a.* [f. *Antilles* + -an.] Of or belonging to the Antilles, a group of islands in the West Indies, or their inhabitants.

1876 A. R. Wallace *Geogr. Distribution of Animals* II. xiv. 60 (*heading*) The West Indian Islands, or Antillean Sub-region. *Ibid.* 62 On the western group of islands..the chief peculiarities of Antillean zoology are developed. **1888** A. Agassiz in *Bull. Mus. Compar. Zool.* XIV. 116 The deep soundings..do not lend much support to the theory of an Antillean continent. **1902** *Encycl. Brit.* XXV. 361/1 Central America, the West Indies, and various submarine ridges by which they are connected with one another and with the mainland to the west..all belong together in what has been termed the Antillean mountain system. **1961** L. D. Stamp *Gloss Geogr. Terms* 532 *Stratigraphical Terms*..Antillean. Miocene orogenesis of West Indies.

antilog, abbreviation of Antilogarithm.

1910 G. C. Douglas (*title*) Douglas' Two Colour Logs & Anti-Logs. **1927** A. G. Cracknell (*title*) Elementary Practical Mathematics. With Tables of Logs and Antilogs. **1959** G. & R. C. James *Math. Dict.* 16/2 *Antilogarithm*, the number whose logarithm is the given number; *e.g.* antilog $_{10}2$ = 100.

antilogism (ænti·lŏdʒiz'm). *Logic.* [f. Gr. ἀντιλογία contradiction + -ism; the existence of late Gr. ἀντιλογισμός 'countercharge' was unkn. to the coiner.] An inconsistent triad, or set of three propositions which cannot all be true together, obtained by taking the premisses of any valid syllogism together with the negation of the conclusion. Hence **anti:logi·stic** *a.*, **anti:logi·stically** *adv.*

The term, though implicit in earlier writers (see quot. 1962), was first used by Mrs. C. Ladd-Franklin.

1902 Mrs. C. Ladd-Franklin in *Baldwin's Dict. Philos. & Psychol.* II. 633/1 The three propositions taken together constitute an inconsistency, or an incompatibility, or, as it may perhaps be called, to distinguish it from the syllogism, an antilogism. **1922** W. E. Johnson *Logic* II. iv. 78 An antilogism may be defined as a formal disjunction of two, three, or more propositions, each of which is entertained hypothetically. *Ibid.* 80 From this single antilogistic dictum we construct the dicta for the first three figures of syllogism. **1962** W. & M. Kneale *Devel. of Logic* iv. 278 Burleigh..goes on to remark that Aristotle's procedure of indirect reduction depends on the principle that the premisses and the negation of the conclusion of any valid syllogism form together an inconsistent triad (or antilogism as it has sometimes been called).

antimacassar. Add: **b.** *transf.* Applied *attrib.* to that which is typical of the period (chiefly the 19th cent.) when antimacassars were in general use.

1913 *Eng. Rev.* Feb. 482 The Libraries..are largely swayed by the letters they receive, generally from old ladies (worthy, antimacassar people), who regard art with pious misgivings and..spend part of their leisure in denouncing it. **1938** *Scrutiny* VII. 397 Antimacassar evangelicanism and Sunday-school sanctimoniousness.

Hence **a:ntimaca·ssared** [-ed[2]] *a.*, covered or adorned with an antimacassar. Also *transf.* (see *b).

1907 A. Bennett *Grim Smile of Five Towns* 56 Seated in the antimacassared arm-chair. **1914** *Times* 13 May 11/5 The antimacassared ease of early Victorian times may have been too much of a good thing, but that was better than the excesses of the modern athletic girl. **1920** D. H. Lawrence *Lost Girl* vii. 137 A horse-hair antimacassar-ed sofa. **1928** *Daily Express* 31 July 8/3 The antimacassared chairs, the horse-hair sofa, and the picture of Queen Victoria's jubilee.

Anti-Ma·son. *U.S.* [Anti-[1] 5 b.] One who is opposed to Freemasonry, used esp. (in the U.S.) of a member of the Anti-Masonic (political) Party.

1828 in T. Weed *Autobiogr.* (1883) xxvi. 307 If under these multiplied difficulties the Anti-Masons incline to bestow their votes upon Mr. Southwick. **1838** *Congress. Globe* Apr., App. 275/1 Counterfeit Democrats, National Republicans, Antimasons and Abolitionists. **1878** J. H. Beadle *Western Wilds* xii. 186 They parted and held their own, and they daresn't an anti-mason show hisself.

So **Anti-Maso·nic** *a.*; **Anti-Ma·sonry.**

1826 (*title*) The Anti-Masonic Enquirer. **1827** in T. W. Barnes *Mem. T. Weed* (1883) 31 The subject of my communication was anti-Masonry. **1828–30** H. D. Ward (*title*) The Anti-Masonic Review and Magazine. **1880** H. O'Rielly (*title*) American Political-Antimasonry. **1899** A. H. Quinn *Pennsylv. Stories* 59 His grandfather..had been one of the local leaders of the Anti-Masonic party.

anti-natural (æ:nti,næ·tiŭrăl, -tʃərăl), *a.* [f. Anti-[1] 3 c + Natural *a.*] Contrary or antagonistic to nature; the reverse or opposite of natural.

a **1603** T. Cartwright *Confut. Rhem. N.T.* (1618) 228 Your Diuinity in the Supper is Anti-naturall..that is..contrary to nature. **1728** 'Martinus Scriblerus' *Art of Sinking* v. 19 This happy and antinatural way of thinking. **1897** *Daily News* 24 Nov. 9/5 Such wealth is represented as not only 'anti-social', but 'anti-natural'. **1947** C. S. Lewis *Miracles* xiv. 144 The anti-Natural or pessimistic religions.. such as Buddhism or higher Hinduism, tell us that Nature is evil and illusory. **1957** *Sc. Jrnl. Theol.* X. 230 Intellectually he [*sc.* modern man] often thinks of the supernatural as the preter- or anti-natural.

anting (æ·ntiŋ), *vbl. sb.* [f. Ant + -ing[1], after G. *einemsen* (E. Stresemann 1935, in *Ornith. Monatsber.* XLIII. 138).] The action of birds in rubbing on their plumage ants or other insects that secrete acrid juices; also, a similar action of birds with various other objects (see quot. 1959). Hence (as back-formation) **ant** *v. trans.* and *intr.*, to act in this way.

1936 *Jrnl. Bombay Nat. Hist. Soc.* XXXVIII. 631 Dr. Stresemann suggests the use of a special term for this 'rubbing in' process..which may be translated into English ..as 'anting',—e.g. a bird ants itself or its feathers, even when objects other than ants..are used in the process. **1944** *Ibis* July 404 Some birds..practise 'anting' more or less consistently, while others of related species do not 'ant' at all. **1959** *Observer* 1 Mar. 19/4 Starlings and rooks will 'ant', without ants, on smoking chimney-pots. Tame birds will ant with matches or cigarette ends.

anting-anting (a·ntiŋ,a·ntiŋ). [Tagalog.] A supposed supernatural influence having the power of protecting its possessor from harm; also, an amulet or charm having the same qualities.

1890 Foreman *Philippine Isl.* 129 The most ignorant classes..believe that certain persons are possessed of a diabolical influence called *anting-anting*, which preserves them from all harm. **1898** *Daily News* 28 July 2/2 Supernatural qualities of anting-anting. **1900** F. H. Sawyer *Inhab. Philippines* 215 The famous Tulisanes or bandits, thoroughly believe in the power of the *Antin-Antin* or amulet to render them invulnerable to bullets.

antinode (æ·ntinōᵘd). [f. Anti-[1] 2 + Node 6.] A point or line of maximum disturbance in a vibrating body, a region of maximum amplitude between nodes: for *spec.* applications see quots.

1882 J. D. Everett *Vibratory Motion & Sound* v. 51 Comparing together the four positions of the string.. points A, C, and A[1] remain permanently at rest, and the points B and D midway between them undergo the largest displacement... The points of permanent rest, A, C, and A[1] are called nodes and the points of maximum displacement, B and D, antinodes. **1889** *Cent. Dict.*, *Antinode*, a point of a vibrating string where the amplitude of vibration is greatest. **1910** *Hawkins' Electr. Dict.* 14/2 *Anti-nodes*, the points in wave motion which vibrate through the greatest amplitude, half-way between the points at rest, called *nodes*. **1926** *Gloss. Terms Electr. Engin.* (B.S.I.) i. 199 *Antinode*, in a system which has a non-uniform distribution of R.M.S. current or voltage, any point at which the R.M.S. value is at its maximum is called an antinode of current or of voltage respectively. **1930** *Engineering* 29 Aug. 255/2 Measurement of wavelength by noting successive current antinodes along the wires. **1959** *Chambers's Encycl.* VIII. 554/1 These points are called nodes. At the intermediate points..the ether particles oscillate with maximum amplitude; these points are called antinodes.

Antiochene (æntəi·ŏkīn), *a.* and *sb.* [ad. L. *Antiochēnus*, f. *Antiochia* Antioch.] **A.** *adj.* Of or pertaining to Antioch in Syria, esp. to the school of theology represented chiefly by the church at Antioch in the 4th and 5th centuries. **B.** *sb.* An adherent of this school.

1845 Newman *Devel. Chr. Doctr.* i. 72 The Antiochenes and other heretics sometimes were Arians, sometimes Sabellians. **1884** *Encycl. Brit.* XVII. 355/2 The antagonism between the Alexandrian and Antiochene schools of theology. *Ibid.* 356/2 The Antiochenes continued to maintain for a considerable time an attitude of antagonism towards Cyril and his creed. **1894** Rivington *Prim. Ch.* 192 To understand..the Antiochene troubles of that century aright, it is necessary to bear in mind that St. Meletius entered upon his episcopate at Antioch under false pretences. **1939** *Burlington Mag.* Aug. 44/1 Authentic monuments of Antiochene art. **1939** P. Hughes *Pop. Hist. Ch.* ii. 36 The Antiochenes, remaining obdurately aloof, were excommunicated too.

Also **Antiochian** (æntiǫ·kiăn), *a.* and *sb.*, in the same senses.

1840 E. Cox tr. *Döllinger's Hist. Ch.* I. ii. § 7. 176 The Antiochian fathers. **1867** C. M. Yonge *Pupils S. John* xviii. 289 The Antiochians were satirizing his bearded figure. **1927** J. H. Willey *Early Ch. Portraits* v. 107 Hence came the famous skill of the Antiochians in coining names.

anti-oxidant (æ:nti,ǫ·ksidănt). [Anti-[1] 3 b.] An agent which inhibits oxidation. Also *attrib.* or *adj.*, that inhibits oxidation.

1934 *Chem. Abstr.* XXVIII. 7504 The use of antioxidants in the stabilization of cracked gasolenes. **1943** *Jrnl. Biol. Chem.* CXLVII. 516 In all experiments reported..on the effect of heat on antioxidant activity, the treatment involved moist heat under pressure. **1958** *Economist* 20 Dec. 1051/1 Antioxidants protect fats, oils and foods containing these products against development of off-flavour, unpleasant odour, or discolouration.

anti-papal (ænti,pē·păl), *a.* [Anti-[1] 3.] Opposed to popery and the pope.

1639 [see Anti-[1] 3 b]. **1649** Milton *Eikon.* xxvii. 214 He..charges strictly his Son after him to persevere in that Anti-Papal Scisme. **1894** J. C. Hedley *Retreat* xxi. 257 The anti-Catholic, anti-Papal,..and anti-spiritual literature. **1902** Gairdner *Eng. Ch. 16th Cent.* viii. 141 Some anti-papal measures.

antiparallel, *a.* Add: **3.** *Physics.* Parallel but moving or directed in opposite directions: used chiefly of vectors. Also as *adv.*

1927 C. G. Darwin in *Proc. R. Soc.* A. CXVI. 230 When the electron *wave* is analysed by a Stern-Gerlach experiment, the associated *rays* (or particles) are polarised antiparallel. **1938** F. Bitter in *Physical Rev.* LIV. 81/2 The case in which the internal forces are such as to make interacting spins antiparallel to each other. **1949, 1961** [see *ANTIFERROMAGNETIC, -MAGNETISM].

antipasto (æntipɑ·sto, ‖antipasto). [It., f. as Antepast.] Hors d'œuvre. Also *fig.*

1934 in Webster. **1939** C. Morley *K. Foyle* xxx. 300, I don't mean just what we got from the antipasto at Enrico's place. **1952** S. J. Perelman *Ill-Tempered Clavichord* (1953) 102 The gambol in the cupola..is merely antipasto for a real shindig at the Praying Ghats, the sacred stairs fronting the Ganges. **1961** I. T. Ross *Requiem for Schoolgirl* vi. 102, I demanded a large *antipasto* and a bottle of Valpolicella.

antipatharian (æ:ntipăþēᵊ·riăn), *a.* (*sb.*) *Zool.* [f. mod.L. *Antipatharia* (f. *Antipathes*, name of the typical genus) + -an.] Belonging to the order *Antipatharia* of corals (the black corals); as *sb.*, a coral of this order. So **antipathid** (ænti·păþid), a coral belonging to the Antipathidæ, a family of this order.

1890 *Athenæum* 17 May 644/1 Prof. F. Jeffrey Bell [read a paper on] the antipatharian corals..containing..an account of a very remarkable antipathid from the neighbourhood of..Mauritius. **1899** *Nat. Science* Jan. 90 Large numbers of hydroids, antipathids, and the crinoid *Antedon phalangium* were dredged. **1924** *Glasgow Herald* 1 Nov. 4 The black corals or antipatharians.

antiphlogistin (æ:ntiflǫdʒi·stin). Also -ine. [f. Antiphlogistic + -in[1].] A proprietary form of kaolin poultice used as a convenient form of treatment for the relief of inflammatory lesions.

1901 Dorland *Med. Dict.* (ed. 2). **1910** W. Bennett in *Practitioner* June 750 The induction of hyperaemia is certainly sometimes beneficial and is most easily effected by the use of some such material as Antiphlogistine. **1965** G. McInnes *Road to Gundagai* xi. 189 Some were in favour of..antiphlogistine to 'draw the boil out'.

antipodal, *a.* Add: **3.** *spec.* in *Bot.* Applied to cells at the base of the embryo sac, formed by division of the nucleus. Hence as *sb.*, a cell of this kind.

1876 *Encycl. Brit.* IV. 146/1 In some cases at the base of the embryo-sac a few cells are formed, which have been termed antipodal cells. **1898** *Nat. Science* June 375 The variety in the place of origin of the embryos from egg-cells, synergids, antipodal cells, or nucellus. **1946** *Nature* 10 Aug. 204/2 After fertilization, the antipodals also behave normally.

Antipodean, *a.* Add: **B.** *sb.* In the senses of the adj.

1640 R. Brome *Antipodes* II. iv. sig. E3[v], *Per.* And what are those? *Doct.* All Antipodeans. **1890** *Athenæum* 22 Feb. 239/2 The 'antipodeans'—the jugglers who work with their legs. **1890** A. P. Morton tr. *H. Le Roux's Acrobats & Mountebanks* ix. 222 He..rarely leaves the 'carpet'; he is a juggler or an antipodean. *Ibid.* 225 The crural muscles attain an extraordinary expansion and strength in the antipodeans. **1914** C. Mackenzie *Sinister St.* II. IV. iv. 942 They were moral antipodeans to the magistrate or the legislator or the social reformer. **1921** *Glasgow Herald* 21 Jan. 8 The Englishmen..are probably slightly weaker in Australia than the best side the Antipodeans can put on the field.

antipodes. Add: **6.** *Chem.* (with pronunc. æ·ntipōᵘdz, *sing.* **antipode**). An enantiomorphic compound.

1918 J. B. Cohen *Org. Chem. Adv. Stud.* (ed. 2) II. iii. 169 The two complementary active forms are variously termed *active components, antipodes*, or *enantiomorphs*. **1947** A. J. Mee tr. *Karrer's Org. Chem.* (ed. 3) iv. 96 Two such isomerides, which differ neither in chemical nor in general physical properties, but which rotate the plane of polarization of light by the same amount, one to the right, the other to the left, are called *antipodes*, or *enantiomorphic* forms.

antipruritic, *a.* Add: Hence as *sb.*, a substance tending to relieve itching.

1894 in Gould *Illustr. Dict. Med.* **1920** J. M. H. Macleod *Dis. Skin* vii. 119 Anti-pruritics allay irritation and itching, and act either directly on the sensory nerve terminations, or indirectly by evaporation or by astringing the blood-vessels.

antipyrin (æntipəiᵊ·rin). Also -ine. [f. Anti-pyr(etic + -in[1].] The commercial name of a benzene derivative, $C_{11}H_{12}N_2O$, used as an antipyretic: discovered in 1884 by L. Knorr, professor at Jena.

1884 *Jrnl. Chem. Soc.* XLVI. 1378 By distillation with zinc-dust, antipyrine yields benzene. **1893** [see Tolypyrin]. **1902** *Encycl. Brit.* XXXIII. 271/1 Antipyrin belongs rather to the furfurol group.

antique, *a.* Add: **8.** Applied to old furniture, pictures, china, and other articles of virtu, esp. as sought for and collected by amateurs; *absol.* with *the*, antique articles collectively. Also *sb.*, such an article; freq. *attrib.* as in *antique dealer, furnisher, shop.*

In practice 'old' with reference to antiques is frequently interpreted to mean 'more than 100 years old'.

1771 J. Wedgwood *Let.* 13 Jan. (1965) 102 The first will enable us to make Tablets and figures &c without cracking, and the third to make *real Antiques* you know. **1822** *P.O. London Directory* 17 Antique furniture and ornamental china dealer. **1851** (*title*) Antiquarian Gleanings in the North of England, being examples of antique furniture, plate,..etc. drawn and etched by William B. Scott. **1868** H. T. Tuckerman *Collector* 99 When an ancient dame ..drew one [*sc.* a miniature] of her husband from an antique cabinet. **1877** *G. Shrimpton's Oxford Direct.* 33 Walford & Son, jewellers, and dealers in antique china. **1884** *Eng. Illustr. Mag.* Dec. 210/1 Brass and copper vessels..many of them shapely as antiques. **1888** (*title*) Antique and Modern Point Lace. **1904** *Daily Chron.* 29 Jan. 8/1 A story told in the 'antique' shops shows the cunning of those concerned in the business. **1908** R. & E. Shackleton (*title*) The quest of the antique, being some personal experiences in the finding of old furniture. **1920** *19th Cent.* July 166 Morrison was staying with a little antique-dealer in the West of England.

9. Applied to a type of paper (see quot. *a* 1912).

1890 A. Watt *Art of Paper-Making* xiv. 157 Some descriptions of paper, as 'antique' and 'old style', for example, are surfaced with good cardboard instead of copper or zinc plates. *a* **1912** *Paper Terminol.* (*Spalding & Hodge*) ii. 1 *Antique*, a term originally applied to machine, made papers made in imitation of old handmade printings. It denoted colour and finish. It is now used to describe any good bulking paper with a rough surface. 1958 Times Lit. Suppl. 3 Oct. 567/4 To combine two qualities of paper in the same book, that is to print the text on antique paper and the plates on coated paper.

antique, v. Add: **2.** trans. To give an antique appearance to (furniture, etc.) by artificial means. Hence **anti·quing** vbl. sb., **anti·qued** ppl. a.

1923 Daily Mail 1 Mar. 8 Having completed the article, he proceeds to do the 'antiquing'. 1934 E. Bowen Cat Jumps 212 The flat sprang into sight: 'art' distemper, six of the World's Best Pictures, the dark shame angles of antiqued oak. 1959 F. Astaire Steps in Time i. 3 Mr. Shipp claims I am aging him the way you antique furniture, at the rate of several years per week.

antiquity. Add: **4. c.** The early ages of the Christian era; the early centuries of the Church; more explicitly *Christian antiquity*.

1564 Harding Answ. Jewel 173 To see antiquitie for proufe hereof: Let him reade [etc.]. 1574 Bristow Brief Treat. Diverse Plain Waies (1599) 54 All Antiquity is full of such practise. 1753 Challoner Cath. Chr. Instr. 77 This Custom..is as ancient as Christianity, as appears from the most certain Monuments of Antiquity. 1850 Newman Difficulties of Anglicans ii. 34 He would..have given up the Establishment, rather than have rejected antiquity. 1860 A. P. de Lisle in E. Purcell Life (1900) I. x. 185 Christian Antiquity.

anti-rabic (ænti,ræ·bik), a. [irreg. f. Anti-¹ 3 b + Rabies + -ic.] Having the property of counteracting the virus of rabies; curing, or relating to the cure of, rabies. Also **anti-rabietic** (rēⁱbi,e·tik) a., **anti-rabific** (rēⁱbi·fik), a.; **anti-rabies** used *attrib.*

1887 Science IX. 186 The Russian antirabic inoculation institution [in Odessa]. 1888 Nature 22 Nov. 73/2 The anti-rabietic treatment of M. Pasteur. 1894 Ibid. 8 Mar. 437/2 The establishment in the chief town of each province of an antirabific laboratory. 1919 Times Weekly 25 Apr. 404 The treatment of patients with specific anti-rabic material. 1938 Brit. Jrnl. Exp. Path. XIX. 378 The successful cultivation of rabies virus in tissue cultures..has raised the possibility of simplifying antirabic immunization by the use of such cultures. 1961 Times 23 Dec. 9/2 He was the first human being to receive anti-rabies inoculation.

a:nti-re·sonance. [Anti-¹ 2 a + Resonance.] The opposite of resonance in various senses (see quots.).

1923 Bell System Techn. Jrnl. II. 12 Series resonance and shunt anti-resonance coincide if both are included in an internal transmitting band. 1936 Gloss. Acoustical Terms (B.S.I.) 7 Anti-resonance, a condition resulting from a combination of the mass and stiffness reactances of a system in which a response to a sinusoidal stimulus of constant magnitude reaches a minimum at a particular frequency. 1948 M. Joos Acoustic Phonetics 96 Anti-resonance, suppression of a narrow band of frequencies.

antireticular (æ:nti,rĭti·kiŭlǎɹ), a. Med. [Anti-¹ 3 b.] Acting against the reticulum cells of the reticulo-endothelial system; spec. in *antireticular cytotoxic serum* (abbrev. A.C.S.), a cytotoxic serum prepared from animal and human spleen and bone marrow, claimed to have a beneficial effect on the reticulo-endothelial system.

1942 tr. B. E. Linberg in A. A. Bogomolets Therapeutic Action of the Antireticular Cytotoxic Serum 'ACS' 154 Everywhere where a weakening of the physiological system of the connective tissue takes place it can be activized by means of ACS. 1943 Brit. Med. Jrnl. 14 Aug. 203/1 The favourable effect of antireticular cytotoxic serum on the course of infectious diseases of the nervous system.

a·nti-self. [Anti-¹ 1 b + Self C. 3.] (See quots.)

1917 W. B. Yeats Wild Swans at Coole 25 And look most like me, being indeed my double, And prove..The most unlike, being my anti-self. 1922 — Trembl. Veil I. xviii. 58, I know very little about myself and much less of that anti-self. 1931 E. Wilson Axel's Castle ii. 45 The other self, the anti-self or the antithetical self. 1946 Essays & Stud. XXXI. 38 The Mask, the anti-self, the concrete image which is the opposite of the conscious normal Self. 1949 Wellek & Warren Theory of Lit. vii. 72 A work of art.. may be the 'mask', the 'anti-self' behind which his real person is hiding.

anti-Semitism (ænti,se·mitiz'm). [f. Anti-¹ 8 + Semitism.] Theory, action, or practice directed against the Jews. Hence **anti-Se·mite**, one who is hostile or opposed to the Jews; **anti-Semi·tic** a.

1881 [see Anti-¹ 3 a, 5]. 1882 [see Anti-¹ 8]. 1935 Economist 24 Aug. 366/1 The Nazi Party stalwarts..have all been leading an anti-Semitic, anti-Catholic, anti-Protestant..crusade. 1941 J. S. Huxley Uniqueness of Man ii. 50 Germanic nationalism on the one hand and anti-Semitism on the other.

antisepsis. Delete ‖ and 'surgical' (in def.). Now the current word, superseding *antisepticism.

antise·pticism (-siz'm). Now *rare*. [f. Antiseptic + -ism.] The process or principles of the prevention or treatment of sepsis by antiseptic means. So **antise·ptize** v. trans., to treat antiseptically; to disinfect by means of an antiseptic.

1877 R. Barwell in Brit. Med. Jrnl. 28 Apr. 506/1, I shall gather together the results of a number of major operations, and show you what effect antisepticism has had upon my practice. 1888 M. Mackenzie Illness Fredk. the Noble 109 The Professor of Surgery, in the matter of antisepticism, attended to the letter rather than to the spirit. 1910 Contemp. Rev. Mar. 402 The bottle and its contents must be antiseptized at a heat of 123 degs. centigrade.

anti-serum, antiserum (ænti,sɪə·rəm). Biochem. Pl. -sera. [Anti-¹ 7 b.] **a.** A contraction of *antitoxic serum* (see *Antitoxic a.). **b.** A serum containing a high level of antibodies, esp. one that can be used in the treatment of disease.

1901 Nuttall in Daily Chron. 28 Dec. 3/7 No other bloods excepting those of monkeys give a reaction to the anti-serum for human blood. 1910 Encycl. Brit. III. 176/2 So far as bacterial immunity is concerned, the anti-serum exerts its action either on the toxin or on the bacterium itself; that is, its action is either antitoxic or anti-bacterial. 1948 Endeavour VII. 49 The assumption that most antibody molecules have two regions able to combine with antigen..accounts for some of the properties of antisera. 1966 New Scientist 2 June 587/2 Blood serum obtained from the treated animal is known as antiserum and the antibodies in it will combine specifically with the virus.

antisocial, a. Add: **2.** (Earlier examples.) Also, spec. in *Sociol.*, pertaining to a class of persons or actions devoid of or antagonistic to normal social instincts or practices.

1802 James Mackintosh in Memoirs (1835) I. iv. 176 A collection of all the rebellious, antisocial, blasphemous books..published during the Revolution. 1844 Dublin Rev. Mar. 31 The dark, malignant, atrocious, and utterly anti-social character, which the Republican party in its contest with the new government has exhibited. 1889 S. Olivier in G. B. Shaw Fabian Ess. Socialism 109 Some kinds of anti-social action are so unreasonable, that we brand them as insane. 1946 Guinness Princ. Social. iv. 77 The anti-social or criminal [class], in which the consciousness of kind is approaching extinction. 1904 Westm. Gaz. 7 Dec. 2/3 To insist on not paying when asked by the conductor is at once inconvenient and anti-social.

Hence as sb., one who exhibits such tendencies.

1945 C. Burney Dungeon Democracy i. 24 The 'antisocials'—that is..those who had been locked up for antisocial activities.

antisocialism (æntisǝᵘ·ʃǝliz'm). [f. Antisocial a. + -ism.] **1.** Opposition to socialism.

1867 Catholic World Apr. 1/1 A rampant, and apparently victorious, socialism, or more properly, anti-socialism, threatened the destruction of society itself. 1897 J. B. Bury Gibbon's Decl. & F. II. App. 543 [It] was converted into a charge of odium generis humani.. symptoms of anti-socialism and other characteristics of Christianity.

2. Antagonism to socialism.

1905 Daily Chron. 26 June 5/4 Mr. Deakin declined on behalf of the Liberal Protectionists to accept Anti-Socialism as the issue at the next General Election.

antisocialist. Delete rare⁻¹ and add: **2.** One who is antagonistic to socialism. Also *attrib.* or as adj.

1862 J. S. Mill Pol. Econ. (ed. 5) II. iv. vii. 357 The vitality of these associations..enabled about twenty of them to survive..the anti-socialist reaction. 1914 J. Collings Coloniz. Rur. Brit. I. xv. 157 The well-to-do generally are alarmed, and in formal Property Defence Associations, Anti-Socialist Leagues, [etc.].

antisociality (æ:ntisǝᵘʃiæ·liti). [f. Antisocial a. + -ity.] The quality or condition of being antisocial: **a.** aversion to social intercourse; **b.** antagonism to the laws of ordered society.

1818 T. L. Peacock Nightmare Abbey vii. 100 That amiable discontent and antisociality, which you reprobate in our present partial-window literature. 1902 Amer. Jrnl. Psychol. Oct. 586 The anti-sociality of so many delinquents is due to the arrest which so often takes place at the end of childhood. 1952 B. Wootton et al. Soc. Sci. & Path. x. 303 Investigators have made attempts..to restrict their category of 'delinquent' at least to those whose anti-sociality is both persistent and pronounced.

antiso·cially, adv. [f. Antisocial a. + -ly².] In an antisocial manner.

1908 H. G. Wells First & Last Things III. § 29. 215 Sexual relations..tend to..set up a peculiar emotional partnership. It is a partnership that kept secret may work as anti-socially as a secret business partnership. 1943 J. L. Brierly English Law 39 If individual rights are in danger of being used anti-socially, it is for Parliament to re-define them in terms which will prevent their abusive use.

anti-spadix (ænti,spēⁱ·diks). Zool. [Anti-¹ 2.] A group of four tentacles situated opposite to the spadix in the male nautilus.

1883 E. R. Lankester in Encycl. Brit. XVI. 674/2 These four tentacles may be called the 'anti-spadix'. 1888 Rolleston & Jackson Anim. Life 457. 1960 I. F. & W. D. Henderson Dict. Sci. Terms (ed. 7) 34/2.

antispasmodic, sb. (Later examples.)

1868 A. B. Garrod Mat. Med. 393 The direct antispasmodics appear to give tone to the spinal cord and other parts of the nervous system. 1955 Sci. News Let. 2 Apr. 212 The Jamestown weed..whose alkaloids have been used in the manufacture of such sedatives and antispasmodics as atropine, hyoscyamine and scopolamine.

anti-static (ænti,stæ·tik), a. [Anti-¹ 4.] **1.** Radio. Designed to minimize 'static' (= *atmospherics).

1938 Archit. Rev. LXXXIII. 284 Antistatic aerials are provided for all flats for broadcast reception and separate aerials and wiring for television.

2. In various other specific applications: counteracting the effect of static electricity.

1952 Jrnl. Textile Inst. XLIII. P 212 The resistance of nylon can be reduced by as much as 10,000 times by treatment with a suitable anti-static agent. 1954 Ibid. XLV. P 293 The principle of the anti-static lubricant is that it should provide a conducting coat on each filament. 1956 W. J. Roff Fibres, Plastics & Rubbers iii. 312 Surface resistivity..reduced by the presence of mould growth, and by application of 'anti-static' dressings. 1968 Oxf. Mail 19 Nov. (Suppl.) 14/4 A solution of anti-static fluid is used to moisten the pad, which then collects the dust and removes the static.

antistrophe. Add: **4.** Cryst. (See quot., and cf. Metastrophe 2.)

1895 N. Story-Maskelyne Crystallogr. v. § 80. 99 A solid figure..is [symmetrical] to a plane of symmetry when corresponding points equidistant from the plane would lie on any line drawn perpendicularly to the plane. Where the solid figure presents symmetry to only a single plane (and not to a centre also) the corresponding portions of its surface..are to each other as either would be to its own image if seen reflected by the plane of symmetry as by a mirror. Def.—Such a correspondence of form will be termed antistrophe, and such figures will be said to be antistrophic to each other. 1899 W. J. Lewis Crystallogr. 21.

anti-sun (æ·ntisʌn). [Anti-¹.] The point in the sky diametrically opposite the sun; esp. a point opposite the sun in azimuth, of the same altitude as the sun.

1902 Encycl. Brit. XXX. 705/2 Babinet located a neutral point or zone about as far from the anti-sun as was Arago's from the sun itself.

a:ntisymme·trical, a. Math. Also **a:ntisymme·tric**. [Anti-¹ 3 c.] The reverse or opposite of Symmetrical a. in various senses (see quots.). Hence **anti-sy·mmetry**, the property of being antisymmetrical.

1913 L. Silberstein Vectorial Mechanics vi. 96 The decomposition of the general operator into a symmetrical and an antisymmetrical part (the last being the so-called antisymmetrical part) can be effected in but one way. 1923 J. Rice Relativity vi. 126 If its components satisfy the relations $F_{\mu\nu} = -F_{\nu\mu}$ it [sc. the tensor] is called 'anti-symmetric'. Ibid. 127 So the anti-symmetry is preserved after transformation. 1926 P. A. M. Dirac in Proc. R. Soc. A. CXII. 669 If there is interaction between the electrons, there will still be symmetrical and antisymmetrical eigenfunctions..An antisymmetrical eigenfunction vanishes identically when two of the electrons are in the same orbit. 1939 Mind XLVIII. 113 The law of Fermi-Dirac for antisymmetric wave-functions. 1948 E. A. Milne Vectorial Mechanics iii. 40 A tensor T is said to be anti-symmetrical if its components in any triad satisfy the relation $T_{\beta\alpha} = -T_{\alpha\beta}$. 1950 E. Schrödinger Space-Time Structure ii. 16 Envisage a covariant antisymmetric tensor of the fourth rank T_{iklm}. By antisymmetric we mean that an exchange of any two subscripts should merely produce a change of sign in the component. 1962 E. Bruce Inform. Transform. Rational Gram. vii. 136 Where R(x, y) and R(y, x) always imply identity of x and y the relation is called antisymmetric.

antitheist. (Earlier examples.)

1847 T. Chalmers Nat. Theol. (1849) I. ii. 59 He is not an antitheist. 1855 Geo. Eliot Essays (1884) 171 An antitheist—that is, one who deliberately..opposed and hated God.

antitoxic (æntitǫ·ksik), a. (sb.). [Anti-¹ 3 b.] Having the quality of counteracting the effect of a toxin: of the nature of an antitoxin. Also as sb.

1890 Billings Med. Dict., Antitoxics, antidotes. 1894 Daily News 1 Dec. 3/4 The anti-toxic serum treatment of diphtheria. 1894 Liberal 24 Nov. 41/2 Bacterial products being gifted with an antitoxic power. 1905 G. A. Reid Princ. Heredity x. 124 'Passive' immunity which results from the injection of antitoxic sera. 1946 Nature 19 Oct. 557/1 Thus electrophoresis can serve as control of the various factors..that govern the purification of antitoxic sera by means of enzyme digestion.

antitoxin (æntitǫ·ksin). Also *erron.* **-ine.** [Anti-¹ 7 b.] A substance which has the property of counteracting the effect of a toxin; one of the antibodies capable of neutralizing toxins. Also *attrib.*

1892 Pop. Sci. Monthly Sept. 629 Acquired immunity depends upon the formation of an antitoxine in the body of the immune animal. 1893 Fortn. Rev. Jan. 115 Antitoxin was used, and resulted in perfect recovery. 1895 Pop. Sci. Monthly Sept. 715 The principles..employed in the antitoxine treatment of diphtheria. 1904 [see Toxin a, b]. 1941 Nature 26 Apr. 515/2 Antitoxin is stored in a sterile condition ..in glass containers.

anti-trust (æ:nti,trʌ·st), a. U.S. [Anti-¹ 4.] Opposed to trusts (Trust sb. 7 b) or similar monopolistic combinations.

1890 Congress. Record 21 Mar. 2465/2 Will the senator inform me upon what ground the Missouri anti-trust bill was declared unconstitutional in his own State? 1938 Amer. Rev. 1937 177 A return to the old American policy of enforcing the 'anti-trust' laws forbidding all combinations 'in restraint of trade'. 1962 Times 9 Jan. 11/2 Under the anti-trust arrangements mergers can be disallowed.

antitype. Add: (Earlier example.)

1613 Sc. Ch. Rev. Gospel vii. 55 If the order of Priesthood there spoken be an eternall order, and yet be such an order as to whom the tythes are due, then be the Priest, either Melchisedech the type, or Christ the antitype, yet the same thing is still enjoined.

2. One of the opposite or contrary kind.

1926 G. K. Chesterton Incredul. Father Brown v. 146 An antitype; a sort of extreme exception that proves the rule. 1937 'M. Innes' Hamlet, Revenge! i. ii. 39 He looked lazily and amiably round his guests; the very anti-type..of King Claudius of Denmark.

antivenin (æntive·nin). Also **-ine, -venene** (-vini·n). [f. Anti-¹ 7 b + Venin sb.] Any antitoxin used as an antidote to the venom of serpents. Also as adj.

1895 T. R. Fraser in Brit. Med. Jrnl. 17 Aug. 416/2 (title) The treatment of snake poisoning with antivenene..derived from animals protected against serpents' venom. 1902 Encycl. Brit. XXXIII. 271/2 The serum is found to act as an anti-venin. 1904 Brit. Med. Jrnl. 10 Sept. 574 Polyvalent antivenenes. 1909 Westm. Gaz. 17 Aug. 6/3 The antivenine treatment. 1912 Adami & McCrae Pathol. 168 By the repeated injection of minute quantities of venom into lower animals, antitoxins—or antivenins..can be obtained for all the animal poisons. 1959 Times Lit. Suppl. 3 Apr. 191/3 The main quarry is snakes..and his aim to bring them in alive—so that they can be milked of their venom to make antivenene.

antler-moth. A noctuid moth (*Charæas* or *Cerapteryx graminis*), the larva of which is destructive in meadow lands. Also in shortened form **antler.**

1775 M. Harris Eng. Lepidoptera 9 Phalena. English Names..Antler. 1832 T. Brown Bk. Butterfl. & M. I. 75 The Antler Moth..devours a considerable variety of grasses. 1897 Daily News 13 Sept. 8/2 The grub of the antler moth..has been known to commit such havoc that..the meadows looked as if a fire had passed over them. 1961 R. South Moths Brit. Isles (ed. 4) Ser. I. 196 Antler Moth..has the fore wings greyish brown or reddish brown, sometimes tinged with ochreous in the paler parts.

Antonian (æntǝᵘ·niǎn), a. and sb. [f. L. *Antoni(us* Anthony + -an.] **A.** adj. Of or pertaining to St. Anthony of Egypt or St. Anthony of Padua (1195–1231).

1904 tr. J. Rigauld's Life of St. Anthony of Padua Introd. 11 A short account of the sources of Antonian history. Ibid. 12 The Antonian library of Padua. 1911 C. M. Antony St. Antony of Padua p. xiii, For the present..Antonian studies are at a standstill. 1919 E. C. Butler Benedictine Monachism i. 12 The monachism that derived from St Anthony was eremitical in character. This type, which I shall call the Antonian, prevailed throughout northern Egypt. 1919 C. C. Martindale Upon God's Holy Hills i. 12 Violent bodily penances, rarely heard of in Antonian literature.

B. sb. Any of various communities claiming the patronage of, or continuity with the community of, St. Anthony of Egypt (c 251–356); = Antonine.

1907 Cath. Encycl. I. 20/2 Some few houses..of the Antonians..are also under the direction of Abbots. 1932 Pax Oct. 161 There are also 771 monks who are neither priests nor deacons..divided as follows: Maronite Antonians..Chaldean Antonians.

Antonine (æ·ntǝnǝin), sb. and a. Also 6 **Antonyne**, 7 **Antonin.** [f. L. *Antonin(us* (see Antonian a.), + -ine¹, or as variant of Antonian.] **A.** sb. **1.** A disciple or follower of St. Anthony of Egypt (c 251–356); = Antonian.

1550 Image of Ipocrysy 276, 260 in Skelton Poetical Wks. (1843) II. 440/1 Some be Paulines, Some be Antonynes, Some be Bernardines. 1898 H. Thurston Life St. Hugh iv. iii. 467 The Order of Antonines sprang into existence in 1090.

2. pl. The Roman emperors Antoninus Pius and Marcus Aurelius Antoninus who reigned in 138–161 and 161–180 respectively.

1686 G. Burnet Some Lett. iv. 209 Burying began in the times of the Antonins. 1770 H. Walpole Let. 6 May (1904) VII. 380 If we have the arts of the Antonines,—we have the fustian also. 1776 Gibbon Decl. & F. I. i. 8 The general system of Augustus was equally adopted and uniformly pursued by Hadrian and by the two Antonines. 1836 Wiseman Lect. Princ. Doctr. Cath. Ch. I. i. 13 The study of philosophy..under the patronage of the Antonines..was become very prevalent. Ibid. II. xvi. 223 Prudent and philosophical men, like the Antonines. 1959 Encycl. Brit. XIX. 504/1 The internal tranquillity and the good government which have made the age of the Antonines famous.

B. adj. **1.** Of or pertaining to St. Anthony of Egypt or St. Anthony of Padua (1195–1231); = *Antonian a.

1898 H. Thurston Life St. Hugh iv. iii. 469 The poor sufferers, and the Antonine monks who tended them. 1902 E. Guest tr. Lepitre's St. Antony of Padua i. 3 The Antonine biographers of our own times. 1908 J. J. Walsh Popes & Science (1912) x. 276 The Antonine Congregation of Vienna, which was especially devoted to the care of patients suffering from the 'holy fire'.

2. Of or pertaining to the Antonines (sense A. 2 above) or their rule. *Antonine Wall*: a Roman frontier-wall between the Forth and the Clyde, built for Antoninus Pius in A.D. 142.

1776 H. Folkes in Archæologia I. 117 (title) On the Trajan and Antonine Pillars at Rome. 1875 Encycl. Brit. II. 142/2 More than 1900 years Antonine's wall continued to be the boundary of the Roman territories in Britain.. 1899 (title) The Antonine Wall Report, being an Account of Excavations, etc., made under the Direction of the Glasgow Archæological Society during 1890–93. 1900 Daily News 9 July 6/3 Another historical work dealing with the Flavian and Antonine periods. 1937 Discovery July 208/1 Antonine coins.

A·ntony over. U.S. (See quot. 1872.)

1872 Schele de Vere Americanisms 579 Antony over, a game of ball played by two parties of boys, on opposite sides of a schoolhouse, over which the ball is thrown. Used in Pennsylvania. 1899 Amer. Dial. Notes..Horse xi. 118 Why he and the alligator moved the dresser out from the wall and began to play 'ant'ny over' with my eye.

antral, a. Delete rare⁻⁰ and add examples.

1893 R. F. M. Thomson in Jrnl. Laryng. & Otol. LXIX. 829 (title) Air-Embolism following Antral Lavage. 1962 Lancet 8 Dec. 1200/1 There was an overflow of acid on to the antral area.

Antwerp (æ·ntwɜːp). The Belgian city of that name, used *attrib.* in *Antwerp edge, edging stitch*, an embroidery stitch used for decorating and finishing edges and hems; *Antwerp lace*, a coarse-textured lace (see quot. 1960); *Antwerp pigeon*, a variety of homing or carrier pigeon (ellipt. in quots. 1839 and 1876). Also *Antwerp blue, brown, red* (see quots.).

1832 Field Chromatography xi. 112 Antwerp Blue is a lighter-coloured and somewhat brighter Prussian blue. Ibid. xii. 120 Antwerp Brown is a preparation of asphaltum ground in strong drying oil. 1839 New Sporting Mag. June 378 The Antwerps are a later introduction into our country..I believe little was known of them before the famous Antwerp match in July 1830. 1865 Mrs. F. B. Palliser Hist. Lace vii. 116 Antwerp lace would have disappeared from the scene had it not been for the attachment evinced by the old people for one pattern,..generally known by the name of 'pot lace'. 1876 Fulton Ill. Pigeons 238 A few years ago, when Antwerp fanciers were few and far between. Ibid. 281 The bird now known as the Antwerp Carrier, Voyageur, or Homing Pigeon. 1882 Caulfeild & Saward Dict. Needlework 272 Antwerp edge, a needle point edging to braid on cordonnet. 1885 [see *Harlem]. 1885 J. Scott Taylor Field's Chromatography. Modernized 187 Antwerp Red, a variety of Red Ochre. 1904 Encycl. Amer. XIV. 343 Two main types of the Belgian homer have been distinguished as the Antwerp and the Liege varieties, the former being larger but less graceful in form than the latter. 1934 M. Thomas Dict. Embroidery Stitches 138 Knot stitch, also known as Antwerp Edging Stitch, and Knotted Blanket Stitch. 1940 R. Mayer Artist's Handbk. (1951) ii. 47 Antwerp blue, a pale variety of Prussian blue made by reducing pure Prussian blue with 75% of an inert pigment, usually alumina hydrate. 1960 H. Hayward Antique Collecting 15/2 Antwerp lace, pillow lace, strong and heavy looking, the patterns outlined in a cordonnet of thick untwisted thread and made at the same time as the fond chant ground.

anuran (ǎnūᵊ·rǎn, -iŭə-), a. Zool. [f. mod.L. *Anura* (J. Hogg 1839, in Mag. Nat. Hist. 270) (f. A- 14 + Gr. οὐρά tail) + -an.] Of or pertaining to the Anura, an order of tailless amphibians. Hence as sb.

1900 Amer. Naturalist XXXIV. 689 The segmentation of several other anuran eggs that contain much yolk has not as yet been described. 1907 E. J. Bles in J. G. Kerr Wk. of J. S. Budgett xvi. 445 (title) Notes on Anuran Development. 1926 J. S. Huxley Essays Pop. Sci. 293 The different behaviour of the Anuran tadpole's limbs. 1932 — Probl. Rel. Growth vi. 181 Even the normal growth of the Anuran tadpole's limbs is slightly heterogonic.

anuria. Now the usual form (see Anury). (Earlier and later examples.)

1838 R. Willis Urinary Diseases I. ii. 28 The title under which this suspension or abolition of the secreting office of the kidney is usually spoken of in this century, is Ischuria renalis; but as ischuria means an impeded discharge of urine, I prefer adopting the title Anuria. 1894 Lancet 15 Dec. 1426/2 He mentioned the case of a man, now alive and well, who had twice had complete anuria for ten days. 1922 in F. W. Price Textbk. Practice Med. xv. 1097 Anuria, suppression, as opposed to retention of urine.

anvil, sb. Add: **3. c.** In full *anvil cloud* (see quots.).

1894 W. C. Ley Cloudland v. 80 It is not difficult to understand the nature of the environment in which the Anvil cloud has its birth. 1903 S. Barber Cloud World 108 Anvil Cloud. The 'Anvil' shape of a cloud is generally regarded as one of the best intimations of an approaching gale. 1920 G. A. Clarke Clouds iii. 39 One special feature of cumulo-nimbus is the development above the domed portion of a mass of condensation which has a shape very closely resembling an anvil, if judged by its profile appearance, but which is really a tabular flat-topped mass, rounded in plan and widest at the top. The edges of this 'anvil' are frayed out into the fibrous form associated with the cirrus clouds.

4. *anvil-ding.*

1876 G. M. Hopkins Wr. Deutschland (1918) st. 10 With an anvil-ding And with fire in him forge thy will.

anxiety. Add: **4.** Psychiatry. A morbid state of mind characterized by unjustified or excessive anxiety, which may be generalized or attached to particular situations. Freq. *attrib.* and *Comb.*, as *anxiety-producing, -ridden* adjs.; **anxiety complex** (cf. *Complex sb. 3); **anxiety hysteria**, a form of anxiety neurosis (see quot. 1923); **anxiety neurosis** [tr. G. *angstneurose* (Freud 1895, in Neurolog. Zentralbl. XIV. 55)], **anxiety state**, names technically applied to such a condition of anxiety.

1904 G. S. Hall Adolescence I. iv. 285 The anxiety neurosis was relatively more common in women than in men. 1909 A. A. Brill tr. Freud's Sel. Papers on Hysteria vi. 134, I call this symptom-complex 'anxiety neurosis' (Angstneurose) because the sum of its components can be grouped around the main symptom of anxiety. Ibid. vi. 136 A quantum of freely floating anxiety which controls the choice of ideas by expectation. 1912 Ibid. (ed. 2) xii. 210 We must slightly modify our procedure in anxiety-hysteria (phobias). 1913 Lancet 26 Apr. 1164/1 Anxiety dreams and anxiety neuroses represent a breaking down of this compromise. 1923 R. Gables tr. Stekel's Conditions of Nervous Anxiety I. ii. 20 Freud..proposed to distinguish two sorts of anxiety neuroses: one with a pure somatic basis, Freud's genuine anxiety neurosis, and the other on a psychical basis, which he terms 'anxiety-hysteria'. 1926 W. McDougall Outl. Abnormal Psychol. xiv. 269 The first factor in the production of the anxiety state is the bringing into activity of the instinct of flight. 1931 M. C. D'Arcy Nature of Belief ii. 39 Another cause for failure is..what has been called an anxiety state. Those who suffer from this are unable to make up their minds: they have had their balance upset by some shock to their old convictions, and as a consequence they oscillate to and fro. 1942 A. L. Rowse Cornish Childhood 109 It undoubtedly produced an anxiety-complex.. 1958 New Biol. XXVII. 74 The taboo word which is presented subthreshold..is classified as an anxiety-producing word. 1960 C. Day Lewis Buried Day iv. 64, I went to school..self-important, artificial, anxiety-ridden.

5. Phr. *Age of Anxiety*: the title of W. H. Auden's poem applied as a catch-phrase to any period characterized by anxiety or danger.

1947 W. H. Auden (title) The Age of Anxiety. 1953 Economist 7 Nov. 413/1 The main change is to be found between the nineteenth and the twentieth century outlooks..A survey which ignores it..fails to achieve a complete relevance to the current Age of Anxiety. 1958 Times 11 Nov. 4/4 He [sc. Jackson Pollock] was very much an American product of the age of anxiety.

anxious, a. Add: **2.** *anxious bench* or *seat* (additional examples; see quot. 1839). U.S.

1832 F. Trollope Dom. Manners Amer. I. vii. 111 The poor creatures..seated themselves on the 'anxious benches'. 1835 Reed & Matheson Visit to Amer. Churches I. 13, I was speedily led to conclude that..[the preacher] was about to try the anxious seat. 1839 Marryat Diary Amer. II. xi. 184 In front of the pulpit there was a space railed off, and strewed with straw, which I was told was the anxious seat, and on which sat those who were touched by their consciences or the discourse of the preacher. 1888 J. Kirkland McVeys 19 'Seekers' were sought for and urged forward to the 'anxious seat' or 'mourners bench' by zealous friends.

b. *on the anxious bench* or *seat* (fig.), in a state of anxiety. U.S.

1839 Knickerbocker XIII. 345 He did look as if he had been on 'the anxious seat', as he used to say, when things puzzled him. 1862 Mrs. Stowe Pearl Orr's Isl. II. ii. 159 What a life you did lead me in them days! I'm sure you kept me on the anxious seats a pretty middlin' spell. 1906 N.Y. Even. Post 23 Nov. 1 The entire diplomatic corps at Havana is..on the 'anxious bench'. 1910 W. M. Raine B. O'Connor 25 That extra hour and a half crashes our escape, and we weren't on the anxious seat any without it.

any, a. and pron. Add: **1. e.** *any old*: any..whatever. *any old how*: see *Anyhow. (Cf. *Old a.) *slang* (orig. U.S.).

1896 Ade Artie xviii. 171 Any old farmer..could buy up him and a hundred more like him. 1911 R. W. Chambers Common Law ii. 63 'Would you like to have a chance to study?'..'Study? What?' 'Sculpture—any old thing!' 1916 'B. Cable' Doing their Bit ii. 23 Mails take any old time to do their journey. 1918 W. J. Locke Rough Road v, Mate, Bill, Joe—any old name. 1940 War Illustr. 12 Jan. p. ii/2 It is just one man's reactions to circumstances of the moment and his thoughts on 'any old thing'. 1958 B. Hamilton Too Much Water iv. 78 His steward..just shoves some fruit in his cabin, any old time.

2. b. A large or considerable (number, quantity, etc.). *colloq.*

1861 O. W. Norton Army Lett. (1903) 26 In the woods near us we found any quantity of grapes and chinquapins. 1862 Ibid. 50 We cut down any number of the poles. 1876 Coursing Calendar 125 Irish Nell led any number of lengths.

4. b. Colloq. phr. *not having any* (with verb 'to be'): to want no part in something; to turn down a proposition or to reject an overture of friendship. Also, more positively, to refuse to tolerate a situation.

1902 J. Milne Epistles of Atkins ix. 166 A Boer cries to the trenches of the besieged, 'Have you any whiskey?' The cry back is 'Yes', and an invitation to call for it; but no, 'he isn't having any'. 1918 Kipling Land & Sea Tales

(1923) vii. 116 They tried to get into touch with the natives. ..But the natives weren't havin' any. They took to the bush. 1923 D. H. Lawrence Studies in Classic Amer. Lit. vii. 133 Hester urges Dimmesdale to go away with her, to a new country, to a new life. He isn't having any. 1943 J. B. Priestley Daylight on Saturday xxii. 177 They wanted me to be in it. But I wasn't having any, thanks. 1955 A. L. Rowse Expansion Eliz. England v. 171 Lady Mary Hastings was thought of for promotion to the bed of Ivan the Terrible. She was not having any.

7. *any more*: see More adv. 4 a.

b. At all. (Without comparative adj. or adv.) *dial.* (occas. *slang*) and *U.S.*

1735 [see Gentle v. 2 b]. 1780 S. Holten Jrnl. in Essex Inst. Coll. (1920) LVI. 96, I have not traveled any this day on account of my horses. 1817 Analectic Mag. (Philad.) IX. 437 (Th.), If our readers are any like ourselves, we think they cannot help laughing. 1869 'Mark Twain' Innoc. Abroad iv. 45 It is a good tune—you can't improve it any. 1886 R. E. G. Cole Gloss. Words S.W. Linc. 7 He's not worked any sin' June. She can't sit up any. 1890 Kipling in Harper's Wkly. 22 Nov. 911/1 You don't want being made more drunk any. 1911 H. P. Fairchild Greek Immigr. to U.S. 101 Costa was not used to springs, and he did not mind this any. 1937 A. Christie Death on Nile xxv. 245 We're used to responsibility. Doesn't worry us any. 1958 Punch 17 Sept. 386/1 This couldn't have helped the Yogi any.

8. c. *any one's* (or *anyone's*) *guess*, etc.: see *Anybody 3.

1958 N. F. Simpson Resounding Tinkle I. ii, in Observer Plays 240 How close we're getting to the original tonight is anyone's guess.

anybody, sb. or pron. Add: **2. a.** (Further examples.)

1866 Trollope Last Chron. v. 34 Everybody, who was anybody, knew that Mr. Walker was convinced of the man's guilt. 1961 Harper's Bazaar Dec. 47/2 Everybody who is anybody in the business world will be seeking the latest status symbol.

3. In colloq. phrs., as *anybody's game, match, race*, designating a contest in which the competitors are so evenly matched that either side (any competitor) could win; also *fig.*; *anybody's guess*, an unpredictable matter, a question to which no one knows the answer.

1840 Spirit of Times 4 Jan. 523/2 It was anybody's race yet! 1853 F. Gale Public Sch. Matches 58 Sixty-nine runs and six wickets down; anybody's match, by jingo! 1865 J. Pycroft Cricketana vii. 152 Last year's match had been left unfinished, just in that interesting state in which it is called 'anybody's game'. 1898 Forum Jan. 576 In Greater New York, it was what is called 'anybody's race', till close upon the day of election. 1938 Time 21 Nov. 70/2 What this type of angry, incoherent prose will prove is anybody's guess. 1955 E. Bowen World of Love x. 179 Anybody's game, she had thought...Though which of them, dead man and living girl, had been the player, and which the played-with? 1958 Times 27 Sept. 9/4 How many less serious accidents there were, is, therefore, anybody's guess.

anyhow, adv. Add: **1.** Examples of sense 'however imperfectly', in random fashion, unmethodically. Also (*slang*) with insertion of *old*: *any old how* (cf. *Any a. and pron. 1 e).

1828 [in Dict.]. 1853 Mrs. Gaskell Ruth II. iii. 64 Just try..to think of all the odd jobs as to be done well..not just slurred over anyhow. 1896 Pop. Sci. Monthly Feb. 538 They are apt at first to be thrown in anyhow. 1933 Punch 23 Aug. 216/1 Scenes like a splash of confetti Hurled any old how. 1937 M. Hillis Orchids on your Budget (1938) v. 79 The kind of woman who gets herself up any-old-how.. deserves just what she gets. 1949 F. Sargeson I Saw in my Dream xiv. 191 He was all dressed up in his Sunday best.. but his hair was any old how.

2. b. *any old how*, in any case, at any rate. *slang.*

1924 P. Creswick Beaten Path xxxiii. 183 Oh, likes! Yes, she likes me. But liking's nothing. Well, any old how, I had to tell you. 1958 F. Norman Bang to Rights II. 34 Any old how..he managed to get behind him.

anyplace (e·niplēⁱs), adv. *U.S. colloq.* [f. Any + Place sb., after Anywhere adv.] Anywhere.

1934 in Webster. 1936 Punch 25 Mar. 340/3 It's perfectly good American grammar. Why, you even say 'I can't find it any place'. 1948 J. Steinbeck Russ. Jrnl. (1949) v. 111 The tycoon..makes it impossible for the correspondent to publish it anyplace else.

anything. Add: **1. b.** In various phrases: *anything but*, by no means, not at all, the very reverse (of); (*he didn't do*) *anything else*, (*U.S. colloq.*) phr. denoting a strong affirmation; *anything goes*: see *Go v. 19 d; *if anything*: see *If; *like anything*: see Like adv. 1 b; *too..for anything*, extremely, excessively. *colloq.* (cf. Too adv. 2 b.). See also *Happen v. 1 a.

1805–6 Wordsworth Prelude (1926) 376 Grief call it not, 'twas anything but that. a1832 F. Trollope Notebks. in Dom. Manners Amer. (1960) 428 Too hot for anything. Too bad for anything. a1859 in Bartlett Dict. Amer. 10 Loco Foco. Didn't Gen. Cass get mad at Hull's cowardice, and break his sword? Whig. He didn't do anything else. 1874 Hardy Far from Madding Crowd I. xxix. 320 His being higher in learning and birth than the ruck of soldiers is anything but a help to him. 1897 Westm. Gaz. 1 Oct. 2/1 The anything-but-particular denominationalists. 1909 Dial Notes III. iv. 'He didn't do anything else', meaning he did just that. 1925 Chesterton Tales of Long Bow viii. 281 'Really,' she said, laughing, 'you are too ridiculous for anything.' 1933 Times Lit. Suppl. 16 Feb. 106/2 Richard Roe, the posthumous anything-but-hero. 1944 S. E. Hicks Beach A. & M. Verena i. 12 'It sounds as if you were not altogether a Christian young gentleman.' 'Anything but.' 1960 News Chron. 27 July 4/2 The aspirin age needed its drugs largely because the fair sex tried so hard to look anything but.

anywhere, adv. Add: **b.** Used with *from ...to*, to indicate limits of variation. *U.S.*

1897 Outing XXIX. 471/1 The tarpon will be anywhere from fifty to three hundred feet away when the boat is ready to follow him. 1909 'O. Henry' Options (1916) 13 I'll guarantee an increase of anywhere from ten thousand to a hundred thousand a year.

Anzac (æ·nzæk). A word made up from the initials of *Australian and New Zealand Army Corps*, and used colloq. for a member of that corps, or to designate any Australian or N.Z. serviceman. Also *attrib.*, as *Anzac Day*, the occasion (or anniversary) of the landing of the corps in the Gallipoli Peninsula on 25 April, 1915.

1915 Sphere 16 Oct. 55/3 The term, Anzac, which recent events on the Gallipoli Peninsula have rendered so prominent, is derived as follows [etc.]. 1916 W. R. Birdwood in C. E. W. Bean Anzac Book p. ix, It may be of interest..to hear the origin of the word 'Anzac'. When I took over the command of the Australian and New Zealand Army Corps in Egypt a year ago, I was asked to select a telegraphic code address for my Army Corps, and then adopted the word 'Anzac'. 1916 Daily Mail 1 Nov. 4/4 'Anzac'..is a word that bids fair to be reckoned among the immortals. 1920 Chambers's Jrnl. June 374/2 The thing had been started by one of the Anzacs venturing the modest opinion that if Britain had had a million Australian troops, they..would be..in Berlin. 1920 Glasgow Herald 26 Apr. 9 Anzac Day. 1921 Q. Rev. CCXXXVI. July 18 This..is..the surest guide to an understanding of things Australian—that the Anzac, the 'Digger' in his best and worst qualities alike, is a fair type of his fellow countrymen. 1941 Times Weekly 23 Apr. 5 The Australian and New Zealand divisions fighting in Greece are to be known as the Anzac Corps. 1944 J. H. Fullarton Troop Target xiv. 104 'What was wrong with being an Anzac?' demanded Peter Doe, whose father had died at Anzac Cove. 'Nothing,' said Robin. 'Except that to nine people out of ten Anzac means Australian.'

Anzus (æ·nzɒs). Also **ANZUS.** [f. the initials of *Australia, New Zealand, United States*.] The combination of Australia, New Zealand, and the United States for the security of the Pacific, usu. *attrib.* in *Anzus Alliance, Council, Pact*, etc.

1952 Sydney Morning Her. 18 July 3/4 The U.S. Secretary of State, Mr. Dean Acheson, yesterday suggested that the Pacific Council be called the 'Anzus Council'. Ibid. 9 Aug. 2/1 (headline) Anzus Alliance 'gets off to a good start'. 1952 Times 16 Oct. 5/6 Britain's Exclusion from A.N.Z.U.S. 1952 N.Y. Times 15 Dec. 1/6 The Alliance.. known as ANZUS, was devised to safeguard the Antipodes against a resurgence of Japanese militarism. 1959 Times 27 Oct. 12/5 Senior diplomatic and military officials of the United States, Australia, and New Zealand met..to-day for the annual review of the Anzus defence pact.

aoristically, adv. (Later examples.)

1860 G. P. Marsh Lect. Eng. Lang. xiv. 300 In most languages..even the forms grammatically expressive of time are, in general propositions, employed aoristically, or without any reference to time. 1880 Expositor 381 There is no need to suppose these aorists used aoristically (as they would be in classic Greek).

aortitis (e,ɒɹtəi·tis). Path. [f. Aort(a + -itis.] Inflammation of the aorta.

1842 in Dunglison Dict. Med. Sci. (ed. 3). 1906 Practitioner Nov. 581 Other causes of systolic aortic murmurs are aortitis and atheroma.

aoudad (ǎ·udæd). Also **audad, udad.** [Fr. form of native name.] A species of wild sheep, *Ovis* or *Ammotragus tragelaphus* (often domesticated), found in N. Africa.

1861 Proc. Zool. Soc. 234 Additions to the Menagerie.. Aoudad, Ovis tragelaphus. 1864 J. Ormsby Autumn Rambles 70 The maned moufflon of the Atlas,..the Aoudad of modern naturalists. 1886 Encycl. Brit. XXI. 785/1 The.. goat-like aoudad, O. tragelaphus. 1902 Ibid. XXXIII. 483/2 The magnificent udad or Barbary sheep.

aoul (ǎ·ul). Also **aoull, aul.** [Eastern Turkish *aul*.] A Caucasian or Tartar village or encampment.

1828 J. B. Fraser Kuzzilbash I. iv. 48 Men, women, and children, poured tumultuously from the nearer aouls. 1877 D. M. Wallace Russia II. xxi. 39 The aoul consisted of about twenty tents..scattered about in sporadic fashion. 1884 Leisure Hour 461/1 One or two of the mounted young men are sent from the aul..to select a suitable spot for a new encampment. 1960 Times 27 July 11/7 The villagers coming down from their auls in the hills.

Apache (ǎpæ·tʃi). Also 9 **Appache.** [Mexican Sp. (see quot. 1907).] **1.** A people of Athapascan Indians in New Mexico and Arizona; a member of this people; also their language. Also *attrib.*

1745 H. Moll's Atlas Minor Plate 46 (D.A.E.), Apaches. 1797 Encycl. Brit. I. 553/1 ..the Apaches, the Hiaquese, ..are at least as tall as the tallest Europeans. 1834 A. Pike Prose Sk. 126 He..overtook four Apaches. Ibid. 138 A filthy, ragged fellow with.. Apache moccasins. 1907 F. W. Hodge Amer. Indians I. 63 Apache (probably from ápachu, 'enemy', the Zuni name for the Navaho, who were designated 'Apaches de Nabaju' by the early Spaniards in New Mexico)...They were first mentioned as Apaches by Oñate in 1598. 1921 E. Sapir Lang. iv. 71 Such languages as Navaho, Apache, Hupa, [etc.].

2. (ǎpæ·ʃ, ‖apaʃ) Also **apache.** [Fr., f. prec.] A ruffian of a type infesting Paris; also *gen.* a man of ruffianly behaviour.

1902 Westm. Gaz. 22 Oct 8/1 The leader of the band of roughs in Paris known as the 'Apaches'. 1909 Times 9 Feb. 4/4 Those apaches with which Brussels is haunted. 1920 Times Lit. Suppl. 10 June 363/2 Something a little more Bacchic than the calculated extravagances of the drawing-room apache would occasionally relieve the atmosphere. 1933 'G. Orwell' Down & Out iii. 20 He wore side whiskers, which are the mark either of an apache or an intellectual.

b. *attrib.*: **apache dance**, a violent dance in which the partners are dressed as apaches; so *apache dancer.*

1914 G. W. Young From Trenches i. 12 The apache element..was soon brought into order. 1918 H. G. Wells Joan & Peter xi. 443 Huntley wanted to teach Joan an Apache dance. 1922 Joyce Ulysses 469 In workman's corduroy overalls..and apache cap. 1928 Punch 30 May 602 Friend (referring to 'Apache' hat). I like that one, dear. You see, it's noticeable without being really fierce. 1945 H. L. Mencken Amer. Lang. Suppl. I. 186 During the [1914–18] war or soon afterward apache-dancers began to appear in the United States. 1962 J. D. Salinger Franny & Zooey 57 It isn't enough to treat her with the doting brutality of an apache dancer towards his partner.

aparejo (æpǎrē·ho). *U.S.* Also **app-.** [Sp. *aparejo* preparation, harness, tackle.] A pack-saddle.

1844 J. Gregg Commerce Prairies I. 180 It is necessary too for the aparejo to be firmly bound on to prevent its slipping and chafing the mule's back. 1845 T. J. Green Texian Exped. xiv. 228 An aparejo, a kind of pack-saddle upon which is packed all kinds of produce. 1895 C. King Fort Frayne ii. 22 Ten days rations were set aside in readiness to be packed on the apparejos the moment word should come. 1904 Omaha Daily Bee 6 July 4 The aparejo..as a valuable method of holding the cargo and protecting the animal's body from injuries, has never been improved upon.

apartheid (ǎpǎ·ɹthēⁱt). [Afrikaans, lit. 'separateness', f. Du. *apart* (ad. F. *à part* Apart adv.) + *heid* -hood.] Name given in South Africa to the segregation of the inhabitants of European descent from the non-European (Coloured or mixed, Bantu, Indian, etc.); applied also to any similar movement elsewhere; also, to other forms of racial separation (social, educational, etc.). Also *fig.* and *attrib.*

[1929 J. C. Du Plessis in Die N. G. Kerk in die O.V.S. en die Naturelle-Vraagstuk 22 In hierdie grondbegrip van Sendingwerk en nie in rassevoorordeel nie, moet die verklaring gesoek word vir die gees van apartheid wat ons gedragslyn nog altoos gekenmerk het.] 1947 Cape Times 24 Oct. 7/7 Mr. Hofmeyr said apartheid could not be reconciled with a policy of progress and prosperity for South Africa. 1948 Ibid. 12 Aug. 1/1 Mr. P. O. Sauer..will explain the application of the apartheid policy on the railways. Ibid. 13 Aug. 8 It is always easy to discern the immediate benefits or comforts conferred on the apartheid-minded Europeans, but impossible to discern the benefits conferred on the non-Europeans. 1949 Ibid. 18 July 9/3 Apartheid is to be introduced at the Kimberley Post Office as soon as necessary structural alterations can be made. Separate counters will be provided for European and non-European customers. 1949 Manch. Guardian 13 July 4/6 Thus Dr. Malan's policy of 'Apartheid' for the non-Europeans, which is only the Dutch word for Field Marshal Smuts's policy of 'segregation', which in turn is only a pretty word for repression, is achieving a position of 'Apartheid', in the literal sense of isolation, for the nation as a whole. 1950 Hansard Commons CCCCLXXVI. 2020 It does not really justify making a sort of political apartheid as the basis of one's foreign policy. 1953 J. Packer Apes & Ivory ii. 17 This residential and social apartheid is not artificial. It is in the very nature of life in South Africa. What is new in apartheid is the Immorality Act which forbids intimacy between White and Brown. 1955 Times 5 July 6/3 The Archbishop of Canterbury, Dr. Fisher, drew a parallel yesterday between the political apartheid which he had seen in South Africa, separating the nation, and ecclesiastical apartheid which prevented unity among the churches. 1958 Times Lit. Suppl. 21 Nov. 674/5 The tristichs deprived of their rhyming nexus suggest only a metrical apartheid. 1959 Times 28 Feb. 7/3 Some system of apartheid in Central Africa would result. 1959 News Chron. 13 Aug. 4/1 Without going to extreme lengths of apartheid, it should still be..possible to allow those who smoke to do so..on a bus top, reserving the lower deck to those who find the habit revolting. 1961 Times 15 Mar. 14/2 The South African Broadcasting Corporation said the word apartheid would now not be used except in direct quotation...It would use the word 'self-development' to describe the Government's race policies. 1963 Listener 25 Apr. 699/1 It was Sir Charles Snow who first put about the idea of cultural apartheid.

apartment. Add: **1. b.** A set of rooms forming one dwelling-place in a building containing a number of these. Chiefly *N. Amer.* (Corresponding to *flat* in British use.)

1874 Scribner's Monthly VIII. 63/2 The apartments, some of six, others of ten rooms, including kitchen and servants' rooms, are designed for strictly independent house-keeping. Ibid., Each of the family apartments includes a small antechamber or private entrance..a parlor,..a kitchen, three

bed-rooms, and a bath-room. 1890 Harper's Mag. Jan. 327/1 Mr. and Mrs. Delancy Robinson reside in a cosy flat, or 'apartment', as they prefer to call it, in New York city. 1903 N.Y. Even. Post 12 Sept., The chief distinction between a flat and an apartment, according to the accepted definition, is that the apartment has an elevator. 1935 Holt Marvell These Foolish Things 5 A tinkling piano in the next apartment.

c. *attrib.*: **apartment house**, a building divided into residential suites of rooms (chiefly *N. Amer.*); so *apartment block, building, hotel.*

1874 Scribner's Monthly VIII. 63/1 The pioneer apartment houses in New York are the well-known Stuyvesant Buildings, the first..having been opened in the fall of 1870. 1883 Chicago Tribune 4 May 9/3 To Rent..Flat in elegant apartment building, 7 rooms. 1909 Eaton & Underhill Runaway Place 238 The vast apartment hotels along the Park front. 1932 A. Huxley Brave New World v. 88 Henry's forty-story apartment house in Westminster. 1955 Jrnl. Canadian Ling. Assoc. Oct. 5 The classes into which Canadian English words commonly fall..(11) Everyday words formed into combinations unfamiliar abroad, such as apartment block. 1956 D. Gascoyne Night Thoughts 24 Through streets of dwelling-houses and apartment-blocks.

apatelite (ǎpæ·tēləit). Min. [ad. F. *apatélite* (A. Meillet 1842, in Revue Scientifique XI. 255), f. Gr. ἀπατηλός illusive, deceptive + -ite¹.] A hydrous ferric sulphate, found in yellow nodules in clay.

1844 Dana Syst. Min. (ed. 2) 617 Apatelite, a sulphate of the peroxyd of iron from Meudon and Auteuil, peculiar in containing but little water. 1883 Encycl. Brit. XVI. 401/2 Apatelite, reniform-earthy, yellow, from Auteuil near Paris.

apatetic (æpǎte·tik), a. Zool. [ad Gr. ἀπατητικός fallacious, f. ἀπατᾶν to deceive: see -ic.] Applied to markings or coloration deceptively resembling those of another species or of the environment.

1890 Poulton Colours Anim. xvii. 338 Apatetic Colours. 1926 Contemp. Rev. Sept. 369 Data affecting Epigamic colours have not been collected and examined to anything like the extent of those which concern Apatetic and Sematic colours.

apatheia (æpǎþi·ǎ). Also **apa·thia.** [ad. Gr. ἀπάθεια: see Apathy.] = Apathy 1.

[1887 Encycl. Brit. XXII. 567 This remarkable development of Stoic principles leads to the demand for the entire suppression of the affections (ἀπάθεια).] 1893 T. H. Huxley Evol. & Ethics 28 That 'apatheia' in which desire, though it may still be felt, is powerless to move the will. 1955 C. B. Cox in Essays & Studies VIII. 77 The Stoic doctrine of apatheia, the principled refusal to experience more emotion than is forced upon one. 1956 S. Beckett Godot I. 42 A personal God..Who from the heights of divine apathia.. loves us dearly.

ape-man. Anthropol. One of a hypothetical genus of mammals supposed by Haeckel to have been intermediate in character and development between the apes and man. Cf. *missing link* (b) s.v. Missing ppl. a. 4. Also in extended use (quot. 1914).

1879 tr. Haeckel's Evol. Man II. 182 The ape-men, or Alali, were therefore probably already in existence toward the close of the Tertiary Epoch, during the Pliocene Period. 1895 Westm. Gaz. 17 Dec. 8/2 Professor Virchow..held that Dubois's discoveries did not at all prove the existence of an antediluvian ape-man. 1914 E. R. Burroughs Tarzan of Apes (1917) xxvi. 240 The ape-man threw the warm carcass of Numa across his shoulders and took to the trees once more.

aperçu. Add: (Earlier and later examples.) Also, a revealing glimpse; an insight.

[1821 Hazlitt Table-T. I. xv. 362 To give the modern reader un petit aperçu of the tone of literary conversation about five or six and twenty years ago.] 1828 Disraeli Voyage Capt. Popanilla iv. 33 [He] flew off to an airy aperçu of the French Revolution. 1844 J. S. Mill in Edin. Rev. LXXIX. 22 Every one of the chapters we have cited is full of interesting aperçus. 1957 Times Lit. Suppl. 27 Dec. 783/2 Almost always in the course of each essay there are aperçus of great penetration. 1963 Med. Ævum XXXII. 152 The notes..are reminiscent, in their expansive relevance and out-of-the-way aperçus, of the best nineteenth-century editing.

aperiodic (ēⁱpīˀri,ǫ·dik), a. [f. Gr. ἀ- priv. (A- 14) + Periodic a.¹] Not periodic; without regular recurrence. In various techn. senses (see quots.); spec. of a galvanometer, without periodic vibrations, 'dead-beat'.

1879 Encycl. Brit. X. 50/2 An intermediate stage called the aperiodic state. 1884 Krohn tr. Glaser de Cew's Magn.- & Dyn.-Electr. Mach. 228 The commutator am-meter is an aperiodic galvanometer. 1894 W. L. Dallas in Indian Meteorol. Mem. VI. 2 The annual periodic changes and the aperiodic changes [are] both very slight. 1930 Engineering 31 Jan. 129/3 Another new instrument shown was an aperiodic balance, in which air damping is provided in order to bring the beam to rest rapidly. 1943 Gloss. Terms Telecomm. (B.S.I.) 68 Aperiodic aerial, an aerial designed to operate over a substantial range of frequencies. 1952 E. Hyams Soil & Civilization i. 7 The man's motions abrupt, aperiodic and angular.

aperitif (‖aperitif, ǎpe·ritīf). Also ‖ **apéritif.** [F. *apéritif*:—L. *aperitīvus*, f. *aperīre* to open.] An alcoholic drink taken, before a meal, to stimulate the appetite.

1894 Idler Mar. 187 They..sipped their aperitifs tranquilly. 1905 Daily Tel. 22 Feb. 9/2 A teacup-full of whisky which was handed to him by the cook as an aperitif. 1929 Graphic 12 Jan. 54 A memento of Prince George's visit to Bermuda...He is having an after-the-bathe apéritif. 1959 Listener 30 July 191/1 Apéritif ports and sherrys can be served very lightly chilled, too.

fig. 1920 Sat. Rev. 21 Aug. 154/2 But the young artists.. mistook the apéritif for the dinner. 1937 Burlington Mag. June p. xxvi/1 A magnificent scene by Jacob van Ruisdael.. and a pair of huge game and fish compositions..are apéritifs for any Gargantua.

aperture. Add: **4.** *aperture number, ratio* (see quot. 1953).

1889 W. A. Watts in Year-bk. Photogr. 1889 91 The desirability of ascertaining the aperture ratio (f/n) of each stop employed is universally admitted. 1953 Amos & Birkinshaw Telev. Engin. I. 166 The aperture number or aperture ratio, which is defined as the ratio of the focal length of the system to the diameter of the entrance pupil.

6. The opening in the sight of a rifle. Usu. *attrib.*, as *aperture sight.*

a1884 Knight Dict. Mech. Suppl. 43/1 Aperture-sight, another name for the open bead-sight. 1913 A. G. Fulton Notes on Rifle Shooting 8 In choosing an aperture backsight, a pattern which is perfectly rigid throughout..should be chosen...One can hardly be better than the latest No. 9c B.S.A. aperture sight. Ibid. 9 The size of aperture depends ..on the individual. 1915 F. H. Lawrence Let. 26 Apr. in Home Lett. T. E. Lawrence (1954) 708 If Pearson wants my rifle sell it him. The aperture back sight wants to be screwed down tight.

a·pex, v. [f. Apex sb.] *trans.* To form with an apex or pointed top; to raise to a point or tip.

1905 Holman Hunt Pre-Raph. I. 87 Should the several parts of the composition be always apexed in pyramids? 1924 Scribner's Mag. Jan. 56/1 Phil apexed his fine even eyebrows in the direction of Mary.

apfelstrudel: see *apple strudel*, *Apple sb. B. II.

aphakia (ǎfēⁱ·kiǎ). Path. [mod.L., f. Gr. ἀ- priv. (A- 14) + φακ-ός lentil + -ia¹.] Absence of the crystalline lens of the eye. Hence **apha·kic** a. and sb.

1864 W. D. Moore tr. F. C. Donders' On Anomalies of Accommodation & Refraction ii. 84 Of the disturbances in the lenticular system the condition of total absence of the lens, which I have termed aphakia, comes almost exclusively under observation. 1874 Dunglison Dict. Med. Sci. 69/2 Aphakia, an anomalous state of refraction caused by the absence of the crystalline lens, as after operations for cataract. 1889 Cent. Dict., Aphakic, adj. 1894 Pop. Sci. Monthly XLIV. 559 Dazzled and dazed the scientific mind is at present like the aphakic, suddenly brought to see, but not recognizing or knowing what he sees. 1964 S. Duke-Elder Parsons' Dis. Eye (ed. 14) viii. 92 The retinal image of the aphakic eye is about a quarter larger than the emmetropic retinal image.

aphasia. Substitute for def.: Loss of speech, partial or total, or loss of power to understand written or spoken language, as a result of disorder of the cerebral speech centres. Add further examples.

1890 W. James Princ. Psychol. I. ii. 54 The condition in question is word-deafness, and the disease is auditory aphasia. 1907 Practitioner Oct. 544 Pierre Marie..recognises three terms:—(1) Aphasia of Wernicke, in which intelligence is impaired, but speech possible. (2) Aphasia of Broca, in which intelligence is impaired, and no speech possible. (3) Anarthria. 1926 H. Head Aphasia I. I. vi. 91 Complete sensory aphasia invariably leads to some intellectual want of power to manipulate written speech.

aphasiac. Delete rare⁻¹ and add later example.

1929 W. J. Locke Ancestor Jorico i. 14 The ordinary aphasiac could be got at through his ears.

B. adj. = Aphasic a.; also *transf.* (*jocular*), incapable of 'speaking' or sounding. *rare.*

1892 Stevenson & Osbourne Wrecker xxi. 326 The old familiar room,..with..the aphasiac piano. 1906 Dorland Med. Dict. (ed. 4), Aphasic, Aphasiac, pertaining to or affected with aphasia.

aphasic, a. Add to A: Of, characteristic of, or characterized by aphasia.

1874 Brit. Med. Jrnl. 2 May 575/1 Hemiplegia is, as a rule, noticed at some stage of the progress of aphasic disease. 1901 Lancet 26 Jan. 276 Slight aphasic symptoms. 1906 Westm. Gaz. 2 June 9/1 A curious form of aphasic amnesia.

aphesis. (Examples.) Add to def.: Now also used in the sense of Aphæresis.

1880 J. A. H. Murray in Trans. Philol. Soc. 175 The Editor can think of nothing better than to call the phenomenon Aphesis..and the resulting forms Aphetic forms. 1930 A. Western in Gram. Misc. Jespersen 135, I do not quite see the difference between aphesis and aphæresis, but use the former term as the shorter and therefore more convenient of the two. 1932 W. L. Graff Lang. vi. 234 A loss at the beginning is called aphaeresis or apbesis..bishop < Lat. episcopus, knife and write in which k and w were formerly sounded.

aphetic, a.² (Examples.)

1880 [see prec.]. 1884 New Eng. Dict. s.v. A-, Many words with a- in one or other of the preceding senses have aphetic forms with the a- lost, as adown down, amid mid. 1905 Eng. Dial. Dict., Tankerous, an aphetic form of 'cantankerous'. 1959 N. & Q. CCIV. 434/2 Serued..may be merely an aphetic form of OF. deservir.

aphetically, adv.² (Examples.)

1884 New Eng. Dict. s.v. Across, Across..also aphetically cross. 1914 Ibid. s.v. Spend, The ME. spende..may also have been formed aphetically from OF. despendre.

aphetize, v. (Examples.)

1884 New Eng. Dict. s.v. Acute, In the sense of sharp in business, shrewd, it [acute] is familiarly aphetized..to 'cute. 1900 B. A. P. Van Dam Shakespeare ii. 23 The very frequent use of aphetised forms was specially characteristic of the negligent parlance of every-day life.

aphid (ēⁱ·fid, æ·fid), anglicized form of Aphis (from the pl. *aphides*).

1884 Browning Ferishtah's Fancies xii. 117 The aphis feeds, nor finds his leaf Untenable, because..Lightning strikes sere a moss-patch close beside, Where certain other aphids live. 1925 R. W. G. Kingston in E. F. Norton Fight for Everest 285 Aphids were numerous at 15,000 feet.

aphikoman, var. *afikoman.

aphonia (ǎfōᵘ·niǎ). See Aphony.

aphoristical, a. Delete Obs.⁻⁰ and add example.

1846 Geo. Eliot tr. Strauss's Life of Jesus II. II. vi. § 76. 102 The dictum accords with either context, and from its aphoristical conciseness would be likely to recur.

aphotic (ǎfōᵘ·tik), a. [f. A- 14 + *Photic a.] Not reached by sunlight (cf. *photic a.).

1903 W. R. Fisher tr. Schimper's Plant-Geogr. III. v. i. 782 The aphotic or dark region, in which only non-assimilating organisms can exist. 1913 J. Murray Ocean vii. 137 The Aphotic or Deep-Sea Zone extends from the lower limit of the photic zone down to the bottom of the greatest 'deeps'. 1928 E. Neaverson Stratigraph. Palaeontol. iv. 97 Two general light-regions recognised by oceanographers are based on the distribution of plants: the photic region..and the aphotic region, where the light is insufficient for this process [sc. photosynthesis] to be carried on.

aphototropic (ǎfōᵘ:totrǫ·pik), a. Bot. [f. A- 14 + Phototropic a.] Bending or turning away from the light (cf. Apheliotropic a.). So **aphoto·tropism.**

1903 Nature 9 July 237/2 At the moment of hatching, Convoluta is aphototropic. 1916 B. D. Jackson Gloss. Bot. Terms (ed. 3) 28/2 Aphototropism, turning away from light.

aphrodisiacal, a. Delete †Obs. and add later examples.

1869 J. Davenport Aphrodisiacs iii. 111 An account of certain aphrodisiacal charms practised by women of his time. 1918 A. Bennett Pretty Lady xxiii. 151 The rag-time music..bathed everyone at the supper tables in a mysterious aphrodisiacal fluid. 1959 New Biol. XXX. 47 Few Europeans are familiar with that great delicacy of Chinese cuisine, bird's nest soup (or its alleged aphrodisiacal properties).

apical, a. Add: **2.** Phonetics. Pertaining to articulation with, or sounds made by, the tip of the tongue. Hence as sb., an apical sound.

1899 Rippmann Elem. Phonet. 71 The part of the tongue which helps to form the narrowing is the front rim. The front rim of the tongue: apical formation. 1902 E. W. Scripture Elem. Experimental Phonet. xxi. 296 An articulation of the extreme point [of the tongue] is often termed 'apical'. 1935 G. K. Zipf Psycho-Biol. of Lang. (1936) iii. 78 The apical trill or 'rolled r' of Italian. 1964 E. Palmer tr. Martinet's Elem. Gen. Linguistics ii. 49 It [a consonant] is apical if it is pronounced with the tip of the tongue (apex) like the [t] of touche. Ibid. iii. 64 'Apicals' are always stops in French.

apico- (æ·piko, ēⁱ·-), comb. form of Apical a., as in *apico-basal, -dental, -palatal* adjs., *apico-alveolar* adj. and sb.

1926 J. S. Huxley Ess. Pop. Sci. xviii. 256 In the frog's eggs at the onset of gastrulation there exist two main gradients—the apico-basal from animal to vegetative pole, ..and the dorso-ventral. 1958 Archivum Linguisticum X. 101 A direct change from apico-dental [θ] to labio-dental [f]. 1962 Amer. Speech XXXVII. 166 Under terminal contour before apico-alveolars. Ibid. 169 It is normally a lengthened monophthong in Ocracoke speech..except under a terminal contour before an apico-alveolar consonant. 1964 Language XL. 58, q, gq, and nq (voiceless, voiced, and nasal apicopalatal clicks).

apicultural (ēⁱpi,kʌ·ltiŭrǎl, -tʃər-), a. [f. Apiculture + -al.] Of or pertaining to apiculture.

1884 J. Phin Dict. Pract. Apiculture p. iii, In matters of general apicultural practice we have depended largely upon our own experience. 1914 Kipling Diversity of Creatures (1917) 392 Apicultural exhibitions. 1961 Bee Research Assoc. Handbk. 7 'Bee World' is recognized as the leading international apicultural journal.

apiculturist (ēⁱpi,kʌ·ltiŭrist, -tʃər-). [f. Apiculture + -ist.] One who practises apiculture, a bee-keeper.

1883 (title) The American Apiculturist. 1884 W. H. Harris Honey-Bee 271 Apiculturists, like agriculturists, are subject to many and great alternations of hope and fear.

APIOID

apioid (ē¹·pi,oid, æ·p-). *Geom.* [f. Gr. ἄπιον pear: see -OID.] **a.** A species of plane curve, being that one of a pair of Cartesian ovals which is within the other. **b.** A species of solid of revolution, being the form assumed by a rotating liquid when the velocity of rotation exceeds a certain amount. Hence **apioi·dal** *a.*, pertaining to or having the form of an apioid.
1898 G. W. Myers in *Astrophys. Jrnl.* VIII. 163 The distance of centers does not materially differ from the sum of the radii of the components [of the star U Pegasi], suggesting the probable existence of the 'apioidal' form of Poincaré. **1905** A. M. Clerke *Mod. Cosmog.* vi. 104 An 'apioid', or pear-shaped body, replaces the antecedent ellipsoid.

aplanat (æ·plănæt). [a. G. *aplanat* (A. Steinheil, 1866); cf. Aplanatic *a.*] An aplanatic lens. Also *attrib.*
1890 *Anthony's Photogr. Bull.* III. 129 After Steinheil invented his applanat [*sic*], Dallmeyer copied the same and called it Rapid Rectilinear. **1895** J. A. Hodges *Photographic Lenses* viii. 44 Prior to 1866, when Steinheil introduced his rapid aplanat working at f/7, all doublet lenses required comparatively small apertures in order to obtain sufficient definition and flatness of field. **1901** *Brit. Jrnl. Photogr.* 1 Nov. 695/1 These aplanats consist of strong refracting flint glasses, whose quotient and colour diversion do not much differ. **1951** *Engineering* 9 Mar. 297/2 The two-mirror aplanat telescopes.

aplanogamete (æ·plănogămī:t). *Biol.* [f. Gr. ἀ- priv. (A- 14) + πλάνος wandering (Plano-²) + Gamete.] A non-ciliated stationary gamete or conjugating cell, as distinguished from a Planogamete.
1886 [see Gamete].

aplasia (ăplē¹·ziă). [mod.L., f. Gr. ἀ- priv. (A- 14) + πλάσις formation (πλάσσειν to form, mould): see -IA¹.] Congenital irregularity or absence of an organ or structure (see Aplastic *a.*).
1885 in Dict. s.v. Aplastic *a.* **1908** *Practitioner* Sept. 460 Aplasia of the muscularis of the last sections of the large gut. **1962** *Lancet* 6 Jan. 45/2 Aplasia of the gallbladder and of the appendix.

aplastic, *a.* (Later example.)
1952 M. E. Florey *Clin. Applications Antibiotics* xxi. 567 In aplastic anaemia and leukaemia penicillin in large doses can usually control infection.

a-ple·nty, aplenty, *adv.* or *pred. a.* orig. U.S. [f. A *prep.*¹ + Plenty B. I. 2 c.] In plenty; in abundance.
1830 J. F. Cooper *Water Witch* III. ii. 68 A sailor's blessing on you—fair winds and a plenty. **1841** [see Plenty B. I. 2 c]. **1876** 'Mark Twain' *Tom Sawyer* vi. 59 Sho, there's ticks a-plenty. **1910** Mulford *Hopalong Cassidy* xi. 72 There's water a-plenty there. **1917** J. C. McCorquodale *In Divers Moods* 30 And [I] dreamed wild dreams aplenty Of how I should Do what I could. **1945** *Scrutiny* XIII. 99 There is in passing ambiguity aplenty.

aplite (æ·plait), var. Haplite.
1879 [see Haplite]. **1923** *Nature* CXII. 117/2 The well-known aplite from Meldon in Devonshire. **1965** G. J. Williams *Econ. Geol. N.Z.* xii. 181/2 Pegmatites, and possibly aplites, are also represented and several large quartz lodes.

aplustre (ăplʊ·stri). [L. (pl. *aplustria*), ad. Gr. ἄφλαστον.] The curved and ornamented stern of an ancient Greek or Roman ship.
1705 Addison *Italy* 344 The one holds a Sword in her Hand to represent the Iliad,..as the other has an *Aplustre* to represent the Odyssy, or Voyage of Ulysses. **1842** W. Smith *Dict. Gk. & Rom. Antiq.* 58/2 A bird..which perches on the aplustre of the ship Argo. **1880** Wallace *Ben-Hur* 141 The aplustre of the galley. **1929** *Times* 1 Feb. 7/5 Their [Phoenicians'] earliest coins..show a long rakish vessel with ..a handsome aplustre astern.

apnœic (æ̆pnī·ik), *a. Path.* [f. Apnœa + -ic.] Characterized by or suffering from apnœa.
1883 *Science* I. 524/2 Rabbits in which the vagi are intact are made apnoeic by free artificial respiration. **1961** *Lancet* 9 Sept. 573/2 The patient suddenly became unresponsive and apnœic. **1963** *Ibid.* 5 Jan. 46/2 The anæsthetist accepts unconscious, apnœic patients..as part of his everyday routine.

apocalyptic, *a.* **2.** Further examples in sense 'revelatory, prophetic'.
1950 *Economist* 14 Jan. 57/2 The apocalyptic promises of Communism can be met by the sober reality of rising standards of living. **1955** *Times* 18 July 7/5 It would be foolish, even dangerous to work oneself up into a frenzy of apocalyptic fervour.
B. 2. Apocalyptic teaching, philosophy, or literature.
1898 R. H. Charles in Hastings *Dict. Bible* I. 109/2 Prophecy and Apocalyptic..both claim to be a communication through the Divine Spirit of the character and will and purposes of God. **1940** *Scrutiny* IX. 286 A periodical which has previously discussed the problems of the relation between culture and the coming social-economic revolution mainly in terms of the crudest Marxian apocalyptic. **1957** *Oxf. Dict. Chr. Ch.* 67/2 In the NT, the element of Apocalyptic appears in various places.

apocalypticism (ăpǫ:kăli·ptisiz'm). [f. Apocalyptic *a.* + -ism.] An apocalyptic doctrine or belief, esp. one based on an expectation of the imminent end of the present world order.
1884 *Encycl. Brit.* XVII. 842/2 No one has dealt such deadly blows to Chiliasm and Christian apocalypticism as Origen. **1948** J. L. Adams tr. *Tillich's Protestant Era* ii. 23 Later apocalypticism emphasizes those miraculous elements, thus breaking through all limits of time and space. **1952** *Sc. Jrnl. Theol.* V. 120 Apocalypticism threatened to paralyse the life of the Church.

apocalyptism (ăpǫkăli·ptiz'm). [f. Apocalypt(ic *a.* + -ism.] = prec.
1889 *Edin. Rev.* Jan. 77 The second book of Esdras represents a literary development of Hebrew Apocalyptism. **1915** W. R. Inge in F. J. Foakes-Jackson *Faith & War* v. 103 A belief, corresponding to the apocalyptism of the Jews, was widespread that the 'Kingdom of Saturn'..was immediately to appear. **1930** —— *Chr. Ethics* i. 19 Apocalyptism has left traces in almost all the books of the New Testament.

apocalyptist. Delete *rare* and add earlier example. Also, the writer of any apocalyptic work; the writer of a commentary on the Apocalypse.
1835 *Fraser's Mag.* XI. 332/2 Does the apocalyptist represent the Angel of the Covenant with a Rainbow on his head? **1889** *Cent. Dict.*, *Apocalyptist*, an interpreter of the Apocalypse. **1922** W. R. Inge *Outspoken Ess.* (2nd Ser.) ii. 101 The idea of 'the Kingdom of God'..in the writings of the apocalyptists. **1957** *Oxf. Dict. Chr. Ch.* 67/1 The Apocalyptists were pre-eminently writers, directing their attention towards the end of things and to the destiny of the world in general.

apocatastasis. Also **apokatastasis. 1.** Add to def.: *spec.* in *Theol.* (see quot. 1957).
1867 R. E. Wallis tr. *Delitzsch's Syst. Biblical Psychol.* vii. 552 No doctrine..contradicts the Holy Scripture in a more unwarrantable manner than that of the so-called Apokatastasis. **1956** *Sc. Jrnl. Theol.* IX. 242 All of this is not to infer a doctrine of *apokatastasis.* **1957** *Oxf. Dict. Chr. Ch.* 67/2 *Apocatastasis*, the Greek name..for the doctrine that ultimately all free moral creatures—angels, men, and devils—will share in the grace of salvation.

apocentre (æ·posentəɹ). *Astr.* [f. Apo- + Centre *sb.*] The point in the eccentric orbit of a body at which it is most distant from the body or point around which it moves.
1901 Newcomb *Stars* 160 In some cases there may be two pericentres and two apocentres to the apparent orbit. **1902** *Science* 7 Feb. 221 The apparently needless introduction of new terms in place of the familiar old ones, such as the logically inappropriate apocenter, pericenter, for apastron, periastron, in connection with double star orbits.

apochromatic (æ:pokromæ·tik), *a.* (*sb.*) [f. Gr. ἀπό from, after Achromatic *a.*] Epithet of an improved form of achromatic lens invented by Abbe. Also as *sb.*, an apochromatic lens; in *Photogr.* further shortened to **a·pochroma:t.** So **apochro·matism,** apochromatic condition or quality.
1887 *Jrnl. R. Microsc. Soc.* 2nd Ser. VII. 23 In the new objectives..the elimination of these errors realizes an achromatism of higher order than has hitherto been attained. The objectives of this system may be therefore distinguished from achromatic lenses in the old sense of the word by the term apochromatism, and may be called apochromatic objectives. **1890** *Anthony's Photogr. Bull.* III. 318 The apochromatic objectives give absolutely the same photographic image as that seen upon the screen. *Ibid.* 319 In 1884, before the apochromatics were brought out. **1901** *Photogr. Jrnl.* 31 July 324 Remarkably perfect microscope objectives—the Abbe apochromats. **1939** W. Clark *Photogr. by Infrared* xvi. 342 The 'apochromatic' lenses..Apochromats of low aperture are..sometimes used for photoengraving and colour photographic cameras.

apocrine (æ·pǒkrin), *a. Histol.* [f. Apo- + Gr. κρίν-ειν to separate.] (See quot. 1961.)
1926 H. Homma in *Bull. Johns Hopkins Hosp.* XXXVIII. 365 (*title*) On apocrine sweat glands in white and Negro men and women. **1940** Macleod & Muende *Handbk. Path. Skin* xxiv. 252 Instead of opening on the surface of the skin like the small sweat-glands, the apocrine glands open into the hair-follicle between the funnel and the mouth of the sebaceous gland. **1955** *Sci. News Let.* 1 Oct. 213/2 The distinctive apocrine odor of the axilla (underarms). **1961** *Brit. Med. Dict.* 123/1 *Apocrine.* 1. A term descriptive of a gland cell which loses part of its protoplasmic substance when it is secreting. 2. Applied to sweat glands which differ from ordinary (eccrine) sweat glands in that they occur only in hairy regions.

apodal (æ·pǒdăl). [f. mod.L. *Apoda* (f. Gr. ἀποδ-, ἄπους footless) + -al.] A batrachian belonging to the order Apoda, a group of the Cæcilians.
1856 *Encycl. Brit.* XII. 229/1 Sub-Order III.—Ribbon Apodals. **1902** *Ibid.* XXV. 383 To say nothing of the scales, present in many genera of Apodals and absent in all Caudates.

apo-enzyme (æ:po,e·nzəim). *Chem.* [a. F. *apoenzyme*, f. Apo- + *Enzyme 2.] (See quot. 1961.)
[**1936** *Union Internat. de Chimie Compt. Rend.* 43 Le complexe total sera appelé *holoenzyme* et le résidu, après séparation de ses activateurs, sera appelé *apoenzyme*.] **1936** *Chem. Abstr.* 6397 A comprehensive treatise discussing.. character and action of coenzymes and apoenzymes. **1944** *Jrnl. Pharm. & Exp. Therap.* LXXXI. 246 This enzyme like substance appears to attack the apoenzyme or protein part of the oxidative enzyme system. **1950** [see *co-enzyme]. **1961** *Brit. Med. Dict.* 123/2 *Apo-enzyme*, the protein carrier, or enzyme proper (e.g. carboxylase) which, with a co-enzyme (co-carboxylase), carries out an enzymic reaction.

apogamy (ăpǫ·gămi). *Bot.* [f. Gr. ἀπό Apo- + γάμος marriage (cf. Gamo-).] Absence of sexual reproduction; asexual reproduction, agamogenesis; *spec.* in ferns and other cryptogams, production of the perfect plant directly from a bud on the prothallus instead of by the usual sexual process. So **apogamic** (æpǫgæ·mik), **apogamous** (ăpǫ·gămǝs) *adjs.*, characterized by or of the nature of apogamy; agamogenetic (*spec.* in the way described above); hence **apo·gamously** *adv.* (Cf. *Apospory, etc.)
1878 Vines in *Jrnl. Bot.* VII. 360 A transition from an oophore to a sporophore without the intervention of sexual reproductive organs. Apogamy (De Bary). *Ibid.* 361 We may speak of this plant as being 'aposporous', using a word which is symmetrical with the term 'apogamous', applied by De Bary to those Ferns in whose life-history no process of sexual reproduction occurs. **1883** *Ann. Rep. Smithsonian Inst. 1881* 403 Forms where oogonia are found without male pollinodia..considered as representing a distinct apogamous species. **1886** *Jrnl. R. Microsc. Soc.* 2nd Ser. VI. 298 The author could not detect any act of impregnation [in a parasite on the olive], and believes that reproduction is apogamic. **1886** Vines in *Encycl. Brit.* XX. 431/2 By the suppression either of the sexually produced spore or of the asexually produced spore; the former is an instance of apogamy, the latter of apospory. *Ibid.* 431/1 [Spores] formed..without a sexual process—in a word, apogamously. **1921** *Ann. Bot.* XXXV. 184 We see..in the true apogamy of other species, a further manifestation of the powers of heterosis..ending in apomictical reproduction.

apogee. Add: **4.** The point in the trajectory of a missile, rocket, or the like at which it is at its greatest distance from the earth.
1958 in *Aero-Space Terms.* **1961** *Flight* LXXX. 756/1 When the satellite reaches the 22,300 mile apogee of the trajectory..the solid-propellant apogee motor will be used to inject the satellite into a circular, near-synchronous orbit. **1962** J. Glenn in *Into Orbit* 6 The apogee or highest point of the capsule's orbit was over eight times that altitude.

apohyal (æpohəi·ăl). [f. Apo- + Hy(oid + -al.] (*a*) *Ornith.*, the ceratobranchial bone. (*b*) *Ichth.*, the basihyal bone.
1860 in Mayne *Expos. Lex.* Suppl. **1884** Coues *Key N. Amer. Birds* (ed. 2) 207 Ceratobranchials proper, commonly called apo-hyals.

apokatastasis, var. Apocatastasis (in Dict. and Suppl.).

‖**apo koinou, apo-koinou** (æ·po koi·nɑu), *advb. phr. Gram.* [Gr. ἀπὸ κοινοῦ in common.] Applied to a construction consisting of two clauses which have a word or phrase in common (see quots.).
1892 Kellner *Hist. Outl. Eng. Syntax* 62 The whole construction may be apprehended as a sentence *with one subject and two predicates.* This is the so-called construction ἀπὸ κοινοῦ. **1927** Jespersen *Mod. Eng. Gram.* III. vii. 133 Our constructions are often explained on the so-called *apo-koinou* principle, according to which something is expressed only once instead of twice, but in such a way that the hearer connects it both with what precedes and with what follows. *Ibid.* 134 In the later language, too, we have many sentences with contact-clauses in which the *apo-koinou* explanation cannot be applied on account of the word-order, as the common element does not stand in the middle. **1963** Visser *Hist. Syntax Eng. Lang.* I. i. 11 (*heading*) In apo koinou constructions. Utterances of the type 'I have an uncle is a myghty erle' consist of two syntactical units with one element in common (in this example 'an uncle') which functions as the subject in the second unit. *Ibid.*, The apo koinou construction with 'understood' subject already occurred in Old English, although instances are not frequent and mainly restricted to combinations the second part of which contained the verb 'hatan'.

apolau·sticism (-siz'm). [f. Apolaustic *a.* + -ism.] Devotion to enjoyment.
1883 *Sat. Rev.* 15 Dec. 763 He combined the lessons of economy with apolausticism. **1894** 'X.L.' *Aut Diabolus aut Nihil* 6 He was indeed only fervent in his apolausticism.

apolitical (ē¹poli·tikăl), *a.* [f. A- 14 + Political *a.*] Detached from, not interested in or concerned with, political issues or activities.
1952 M. McCarthy *Groves of Academe* (1953) iii. 49 You are not political. You are a-political. **1958** *Times Lit. Suppl.* 28 Nov. 689/4 By an apolitical attitude I meant the attitude of the heroes of *Hurry On Down* or *Lucky Jim*, for instance; the attitude which Mr. Priestley once described as one of 'opting out'. **1960** *Guardian* 12 Apr. 8/3 Comparatively a-political young people may be quite willing to join a Young Conservative club for the sake of its social activities.

Apollina·rianism. The doctrine of the Apollinarians.
1877 P. Schaff *Creeds Gk. & Lat. Churches* II. iii. 63 The addition of a new Creed is justified by the subsequent Christological heresies (Apollinarianism, Nestorianism, and Eutychianism). **1921** *Contemp. Rev.* Nov. 623 The Dean of Carlisle warns theologians against the peril of Apollinarianism or Monotheism. **1946** E. L. Mascall *Christ* i. 11 The heresies of Apollinarianism and Nestorianism.

Apollinaris (ăpǫlinē³·ris). Short for *Apollinaris water*, an effervescent mineral water produced at Apollinarisburg near Bonn in Germany, and used as a beverage.
1875 R. Browning *Inn Album* i. 9 Dinner, Apollinaris, —what they please. **1878** *Fun* 26 Jan. (Hoppe), Sweet champagne and Apollinaris—Sham and Polly, as it is slangily called. **1897** Flandrau *Harvard Episodes* 335 Mrs. Lauriston brought some apollinaris from her table. **1914** J. Joyce *Dubliners* 114 Weathers said he would take a small Irish and Apollinaris.

Apolline (ăpǫ·ləin, -lin), *a.* [ad. L. *Apollineus*, f. *Apollin-*, Apollo.] Pertaining to or resembling Apollo: = Apollonian *a.* 1. So **Apollinic, -inian** (æpǫ̆li·nik, -i·niăn) *adjs.* in same sense.
1605 P. Erondell *French Garden* x, It was called of the ancient heathen, the Apolline Stone. **1884** *Encycl. Brit.* XVII. 808/2 Apolline oracles, such as the Delphic. **1886** *Ibid.* XX. 360 Apollinic (Delphic) religion. **1923** J. M. Murry *Pencillings* 220 Elderly spinsters made the fortune of a lady-novelist who..invariably represented one of their kind as the beloved of an ardent, Apolline youth. **1924** E. & C. Paul tr. *Baudouin's Psycho-analysis & Aesthetics* vi. 255 Dionysian art is being converted into apollinian art. **1963** *Times* 19 Apr. 15/5 The most characteristic features of Apollo and the Apollinian movement.

Apollonian, *a.* Add: **B.** *sb.* A worshipper or follower of Apollo; one having the characteristics of Apollo (opp. Dionysian).
1925 Chesterton *Everl. Man.* II. v. 274 There would still be intelligent Apollonians apparently worshipping the sun-god. **1936** G. B. Shaw *Millionairess* Pref. 120 The Teutons and Latins, the Apollonians and Dionysians.

‖**apologia** (æpǒlōᵘ·dʒiă). [L., see Apology *sb.*] = Apology *sb.* 1; esp. a written defence or justification of the opinions or conduct of a writer, speaker, etc.
The currency of the word is largely due to J. H. Newman's *Apologia pro Vita Sua*, 1864.
1784 J. Newton (*title*) Apologia. Four letters to a Minister of an independant Church: by a Minister of the Church of England. **1876** [see Apology *sb.* 1]. **1883** *Sat. Rev.* 10 Nov. 613/1 The Duke [of Argyll] has put his own version of the story on record. This apologia is a pamphlet, entitled *Crofts and Farms in the Hebrides.* **1903** *Westm. Gaz.* 20 Jan. 8/2 They may be taken as his 'Apologia'—though not in any sense an apology—for the achievements of his official career. **1927** A. H. McNeile *Introd. N.T.* ii. 12 His [St. Mark's] Gospel is not an *apologia* to Jews but an *apologia* to the world of the truth of Christianity.

apomict (æ·pomikt). *Biol.* [Back-formation from next.] An organism produced or reproducing by apomixis. Also *attrib.*
1938 J. R. Carpenter *Ecol. Gloss.* 24 *Apomict population*, a parthenogenetically produced population. **1939** *Nature* 11 Feb. 252/2 Apomicts should receive distinct treatment. **1953** *Rep. 13th Internat. Hort. Congress 1952* I. 54 The *apomict* (abbreviated as *ap.*) is a plant reproducing by means of seed, the embryos of which are produced without fertilization.

apomictic (æpomi·ktik), *a. Biol.* Also **apomi·ctical.** [f. Apo- + Gr. μικτός mixed + -ic; see next.] Pertaining to or produced by apomixis; reproducing without sexual fusion. Hence **apomi·ctically** *adv.*, by means of apomixis.
1913 *Amer. Naturalist* XLVII. 283 Parthenogenesis, the apomictic origin of a sporophyte from an egg. **1920** *Trans. Nat. Hist. Soc. Northumberland & Durham* V. 278 We have on the one hand the tetraploid *Pimpinellifoliae* microgenes pollinated normally, and on the other the tetraploid *Villosae* apomictical... It is a reasonable assumption to make that the apomictical roses are derived from the sexual types. **1921** *Ann. Bot.* XXXV. 183 Seedlings showing hybrid characters have originated from flowers, pollinated with foreign pollen, growing on the same bush as those castrated, and producing seed apomictically. **1939** *Nature* 4 Mar. 383/1 Embryos sometimes arise apomictically from unfertilized eggs. *Ibid.* 22 Apr. 684/1 Reproduction may be.. entirely non-sexual, apomictic.

apomixis (æpomi·ksis). *Biol.* [mod.L. (H. Winkler 1908, in *Progressus Rei Bot.* II. 303), f. Apo- + Gr. μῖξις mingling.] Reproduction of organisms without fertilization (see quot. 1932). (Opp. *amphimixis*; see *Amphi-.)
1913 *Amer. Naturalist* XLVII. 282 Winkler divides all reproductive phenomena into three divisions, namely: Amphimixis, Pseudomixis, and Apomixis. **1920** *Trans. Nat. Hist. Soc. Northumberland & Durham* V. 277 We are compelled to take up the view that in the *Rosae* apogamy, or whatever form of apomixis they present, is a phenomenon originating in hybridity. **1932** C. D. Darlington *Rec. Adv. Cytol.* xv. 416 Apomixis may be defined (following Winkler, 1908) as a system of reproduction having the external character of sexual reproduction but omitting one or both of its essential internal processes.

apomorphine. (Examples.)
1888 *Brit. Med. Jrnl.* 14 Jan. 110/2 (*heading*) Poisoning by Nitre: Treatment with Apomorphine. **1891** *Daily News* 6 Oct. 2/7 One of the apomorphine tablets were missing. *Ibid.*, The deceased..had recourse to the apomorphine in order to produce vomiting. **1966** *Listener* 17 Mar. 401/2 The hero of William Burroughs's latest novel is apomorphine, a vaccine, prepared by boiling morphine in hydrochloric acid, whose function it is, through its action on the back brain, to regulate human metabolism and cure junk-addiction.

apophatic (æpofæ·tik), *a. Theol.* [ad. Gr. ἀποφατικός negative (see G. W. H. Lampe *Patristic Greek Lexicon* s.v.).] Applied to knowledge of God obtained by way of negation. Hence **apopha·ticism,** an apophatic approach to knowledge of God.
1869 D. W. Simon tr. *Dorner's Doctrine of Person of Christ* Div. II. vol. I. 427 The distinction between the communicable and the incommunicable in God (in which, probably, we may trace the influence of the cataphatic and apophatic theology). **1938** G. Reavey tr. *Berdyaev's Solitude & Society* I. ii. 33 *Apophatic* knowledge..knowledge in the process of discarding all notions and determinations. **1956** V. White *God the Unknown* I. ii. 19 The Greeks called it *apophatic* theology—'denying' theology. St. Thomas calls it the *via remotionis* or the *via negativa*: the negative way of removing from our statements about God all that he is not. **1957** tr. *V. Lossky's Myst. Theol.* ii. 38 Apophaticism..is, above all, an attitude of mind which refuses to form concepts about God. **1961** H. Armstrong in I. T. Ramsey *Prosp. Metaphys.* vi. 104 Negative or apophatic theology..certainly does not lead to complete ignorance.

apophony (ăpǫ·fǒni). *Philol.* [ad. F. *apophonie*, f. Apo- + Gr. φωνή sound.] Variation in vowel quality in the formation of grammatically related words, as in Eng. *give, gave*, G. *sprechen, sprach*. (Also called *ablaut* and *vowel gradation*.)
1883 I. Sydow *German Convers.-Gram.* 253 The Germans call this..Ablaut, that may be rendered by apophony. **1894** V. Henry *Compar. Gram. Eng. & Ger.* 358 Mod. German, though keeping the apophony, obscures it by borrowing the metaphony from the subjunctive.

apophysis. Add: **3.** *Geol.* A branch from the main mass of an intrusive igneous rock.
1888 F. H. Hatch in J. J. H. Teall *Brit. Petrography* 424 *Apophysis*, a vein or branch from the main mass (boss or dyke) of an igneous rock. **1893** A. Geikie *Geol.* (ed. 3) IV. vii. 580 All over the world it is common for eruptive bosses of this rock to have a fringe of intrusive veins (*Apophyses*). **1925** B. N. Odell in E. F. Norton *Fight for Everest* 293 Thoroughly metamorphosed and crystalline limestone resting on the schorl granite, which sent off apophyses into it.

apopyle (æ·popəil). *Zool.* [f. Gr. ἀπό Apo- + πύλη gate.] An aperture by which an endodermal chamber in a sponge communicates with the central cavity or paragaster.
1887 W. J. Sollas in *Encycl. Brit.* XXII. 414/1 A comparatively large aperture, which we may term for distinction an apopyle. **1888** [see Paragaster].

aporetic, *a.* Delete †*Obs. rare*⁻¹ and add later examples.
1935 A. C. Bouquet tr. *Przywara's Polarity* iii. 104 In its 'aporetic' character the entire subsequent development of the true teaching of St. Thomas is displayed. **1949** *Mind* LVIII. 247 In the 'aporetic dialogues' [of Plato]..Essence never appears.

aporia. Add: **2.** A perplexing difficulty.
[**1888** *Athenæum* 18 Aug. 219/3 No quibble was too sophistical, no ἀπορία too transparent, for him to think it worth examination.] **1893** W. Clarke Robinson tr. *Ten Brink's Hist. Eng. Lit.* II. 80 The solution of many an *aporia*, as attempted by the idealistic thinker. **1902** *Daily Chron.* 12 Dec. 3/4 Mr. Kidd does not seem to us to surmount this *aporia* very successfully.

aposematic (æposimæ·tik), *a. Zool.* [f. Gr. ἀπό Apo- + σηματ-, σῆμα sign (cf. Sematic *a.*).] Applied to colours, markings, or other attributes serving to warn or alarm, and thus to repel the attacks of enemies. So **aposeme** (æ·posīm), an aposematic marking, etc.
1890 Poulton *Colours of Animals* xvii. 337 Pseudaposematic colours..deceptively resemble Aposematic colours. **1902** *Encycl. Brit.* XXVII. 147 We often see the combination of Cryptic and Sematic methods, the animal being concealed until disturbed, when it instantly assumes an Aposematic attitude. **1920** F. A. Dixey in *Rep. Brit. Assoc. Advancem. Sci. 1919* 202 Adopting Professor Poulton's terminology, we may say that..one form may possess..the Aposemes belonging to two distinct models. *Ibid.* 204 The ..most conspicuous of these common aposemes or danger-signals belong to the under surface. **1940** *Nature* 3 Aug. 144/2 Aposematic or warning colours [in animals].

apospory (ăpǫ·spǒri). *Bot.* [f. Apo- + Gr. σπόρος seed + -y³.] Absence or suppression of spores where they are normally produced, as in certain ferns, mosses, and other cryptogams in which the sexual organism (prothallus in ferns, etc., or perfect plant in mosses) is developed directly from the sporangium or from the leaf instead of from a spore. So **apo·sporous** *a.*, characterized by or of the nature of apospory; hence **apo·sporously** *adv.* (Cf. *Apogamy.)
1878 Vines in *Jrnl. Bot.* VII. 361 To indicate..that no spores are..produced..by the sporophore of *Chara*, we may speak of this plant as being 'aposporous'. **1884** *Jrnl. Linn. Soc., Bot.* XXI. 360 (*title*) On Apospory in Ferns. **1886** [see *Apogamy]. **1886** *Encycl. Brit.* XX. 431/2 In the aposporous Ferns and Mosses and in the Characeæ the oophore is developed as a bud from the sporophore. **1892** *Athenæum* 12 Nov. 667/3 Prothalli developed aposporously.

a posteriori. Add: (Earlier and later examples.)
1624 Francis White *Replie to Fisher* sig. C4ᵛ, Your other argument..is, *à posteriori*, from an example of the..French King, Henry the fourth, to whom you wish his Maiestie to bee a parrallell. **1647** H. Hammond *Power of Keyes* sig. A3ᵛ, This, I conceive, is not by me magisterially dictated, but already demonstrated *à posteriori*, by the experience which the few last moneths have yeelded us. **1873** H. Spencer *Study Sociol.* vii. 174 Accounts of existing uncivilized races.. show us *à posteriori*, what we might infer with certainty *à priori*.
2. *Facetious.* From behind, on the back, on the buttocks. Cf. Posterior B. 2.
1762 Smollett *Launc. Greaves* ix. 200 One of them clapped a furze-bush under the tail of Gilbert, who, feeling himself thus stimulated *a posteriori*, kicked and plunged and capered. **1837** Carlyle *Fr. Rev.* II. III. v. 182 Accelerated..by smitings, twitchings,—spurnings, *a posteriori*, not to be named. **1861** *Temple Bar* Nov. 534 A golden cross sewn on *à posteriori*.

apostle. Add: **III. apostle-bird** (also *apostle*), a name applied to various birds in Australia (see quots.).
1901 J. A. North *Nests & Eggs of Birds* (Austral. Mus., Sydney, Spec. Catal. 1) I. 18 From their habit of associating in flocks, they are known nearly all over New South Wales, and the south-eastern portions of Queensland, as the 'Twelve Apostles', a name also shared in some parts with that of 'Happy Family' for *Pomatostomus Temporalis.* **1934** Webster, *Apostle bird*, the gray jumper. **1945** S. J. Baker *Austral. Lang.* xii. 211 The Grey-crowned Babbler is known also as the *apostle-bird.* **1962** *New Scientist* 15 Mar. 606/1 They are members of the babbler or apostle-bird family (Timaliidae). **1964** A. Landsborough Thomson *New Dict. Birds* 442/2 The Apostlebird *Struthidea cinerea*.. is..about 13 inches in length..of grey plumage..and has a short, stout bill. It gets its name from its habit of going about in parties of about twelve. The bird lives in the drier parts of eastern Australia.

apostolate. Add: **b.** A society or sodality of persons having as their object the propagation of a method or rule of faith, life, or conduct.
1897 *Westm. Gaz.* 1 Oct. 4/2 The little settlement is an apostolate of mercy. **1905** *Tablet* 14 Oct. 607/2 The mere mention of that oxymoron the 'lay-apostolate' causes among Catholics of the elder generation a wise shaking of heads. **1911** *Catholic Encycl.* XII. 107 Ven. Vincent Mary Pallotti..gave to his society the name of 'Catholic Apostolate', afterwards changed by Pius IX to the 'Pious Society of Missions'.

apothecial (æpoþī·ʃiăl), *a.* [f. Apothecium + -al.] Of or pertaining to an apothecium.
1882 J. M. Crombie in *Encycl. Brit.* XIV. 559 Apothecial reactions for the most part take place either externally on the epithecium or internally on the hymenial gelatin.

apotropaic (æ:potropē¹·ik), *a.* [f. Gr. ἀποτρόπαιος averting evil (f. ἀποτρέπειν to turn away, avert) + -ic.] Having or reputed to have the power of averting evil influence or ill luck. Hence **apotropa·ically** *adv.*
1883 *Encycl. Brit.* XV. 570/1 The sacrifice of the 'October horse' in the Campus Martius..had also a naturalistic and apotropaic character. **1904** W. M. Ramsay in Hastings *Dict. Bible* V. 115/1 The..employment of a bull's head on.. sarcophagi..evidently..had at first an apotropaic purpose. **1918** L. Strachey *Eminent Victorians* 230 The same doctrine led him [*sc.* Gordon]..to append, in brackets, the apotropaic initials D.V. after every statement in his letters implying futurity. **1945** *Proc. Prehist. Soc.* XI. 55 In the centre, an apotropaic ornament, a severed head between two volutes. **1956** W. H. Auden *Old Man's Road*, Apotropaically scowling, a tinker Shuffles past.

a-pout (ăpɑu·t), *adv.* [f. A *prep.*¹ + Pout.] Pouting.
1886 J. Ashby-Sterry *Lazy Minstrel* 150 Ah! sweet are those eloquent lips a-pout. **1893** F. Thompson *Poems* 74 (*The Poppy*) With mouth wide a-pout for a sultry kiss. **1922** Joyce *Ulysses* 268 Richie cocked his lips apout.

apozymase (æ:pozəi·mē¹s). *Chem.* [G. (Neuberg & Gottschalk 1925, in *Biochem. Zeitschr.* CLXI. 248), f. Apo- + Zymase.] The residue of zymase after separation of cozymase.
1926 *Chem. Abstr.* 923 (*title*) Apozymase and cozymase: phosphorylation. **1930** J. B. S. Haldane *Enzymes* 136 Zymase (yeast press-juice) can be separated into a colloidal, thermolabile fraction (apozymase), and a thermostable dialysable organic fraction (cozymase)... An amount of cozymase produces..activation of a given apozymase. **1946** *Nature* 21 Dec. 900/2 In the inactivation of cozymase by apozymase..the reaction velocity was little affected by variation in cozymase concentration.

Appalachian (æpălē¹·tʃiăn), *a. Appalachian revolution* [*Revolution], the formation of the Appalachian Mountains in eastern N. America.
1856 J. D. Dana *Amer. Geol. Hist.* 16 The deposits, with small exceptions, were a single unbroken record, until this Appalachian revolution. **1915** Pirsson & Schuchert *Text-bk. Geol.* II. xli. 749 During the Appalachian Revolution much of eastern North America was again thrown into pronounced folds.

appalling, *ppl.a.* Add: (Earlier example.)
1817 Shelley *Laon & Cythna* I. xi. p. 6 Within the sphere of that appalling fray.
2. *colloq.* In weakened sense: distasteful, 'shocking'. Cf. Dreadful *a.* 3.
1919 *Punch* 7 May 357 What appalling rot! **1935** *Discovery* Aug. 222/2 The work involved is truly appalling.

appallingly, *adv.* Add: (Earlier example.)
1825 *Examiner* 11 Apr. 235/1 The prisoner was appallingly altered.
2. In weakened sense: exceedingly; very. *colloq.*
1937 in Partridge *Dict. Slang.* **1946** R. Lehmann *Gipsy's Baby* 123 We got appallingly hungry working so late.

Appaloosa (æpălū·să). Also (now mostly disused) **Apaloochy, Apalousey, Appaloosie, Appalucy.** [ad. *Opelousa*, place-name in Louisiana, or f. *Palouse* River, Idaho, U.S.A.: see *D.A.*] A breed of horse with white hair and dark patches of colour on the body, much used in former times by Indian tribes of western N. America.
[**1849** in *Amer. Speech* (1944) XIX. 69 Nach diesen kommt das sogenannte Kreolen-pferdchen, auch Opelousas poney genannt, ein ausgezeichnetes Damenpferd.] **1924** Davies *Skyline Trail* 49 They find death in a dramatic flare: Trying to ride the apaloochy mare. **1947** DeVoto *Across Wide Missouri* 77 The Nez Perces had learned selective breeding..and had developed a distinctive stock called the Pelouse horse, the 'Appaloosa' of a later date. **1950** in *Amer. Speech* XXV. 310 By 1890 the name had become Apalousey, but today's accepted spelling is Appaloosa. **1963** *Times* 7 Mar. 6/6 Experts believe the Appaloosa to be the oldest breed of horse in the world... They belong to China.

‖**apparat** (æpără·t). [Russ., a. Ger. *apparat* apparatus, instrument, f. L. *apparātus* Apparatus.] The party machine of the Communist party in Russia, etc. Also *attrib.*
1950 A. Koestler *God that Failed* I. i. 46, I found Communist apparat-work much less efficient than its scared opponents presume. **1952** —— *Arrow in Blue* xxviii. 262 This is true not only of members of the 'Apparat' but of militant Communists in general. **1955** H. Hodgkinson *Doubletalk* 14 *Apparat*, the administrative apparatus, usually of State or Party; the 'Party machine'. **1960** *Encounter* Mar. 69/1 The dead-weight of a clumsy bureaucratic *Apparat* having been disposed of, a fresh team of leaders..succeeded in tackling the reconstruction of the intellectual formulation of Social Democratic policy [in E. Germany].

‖**apparatchik** (æpără·tʃik). Pl. **apparatchiki, apparatchiks.** [Russ.] A member of the *Apparat; also, a Communist agent or spy.
1941 Koestler *Scum of Earth* 122 The dark silhouette of the Tchekist, the 'Aparat-chik', or G.P.U. agent had replaced the once bright and lively symbols of the struggle for a happier world. **1950** —— *God that Failed* I. i. 46 A full-fledged 'apparatchik' (the homely euphemism used in the Party for agents and spies). **1957** *Observer* 29 Dec. 6/7 As the year ended, the domination of Khrushchev and the ideologists and *apparachiki* of his machine seemed complete. **1960** *Encounter* Mar. 41/1 The *apparatchiks* of Stalinism. **1963** *Camb. Rev.* 16 Feb. 277/1 The party bureaucrat, or *apparatchik*, is distinguished from the ordinary party member by his professional attachment to the party, for as a rule he devotes himself exclusively to party activity.

apparatus. Add: **3. c.** In full *critical apparatus*: = next. (Further examples).
1868 *Jrnl. Philol.* I. No. 2. 66 His collation of the poem, with a description of the MS., a careful apparatus, and an interesting explanatory commentary. **1876** *Hermathena* II. 389 Professor W. Ramsay compiled his text very carefully, with apparently all the requisite data and critical apparatus.
4. = *Apparat; any (Communist) organization.
1935 tr. in S. & B. Webb *Soviet Communism* I. App. vii. 369 Resolution of the Central Executive Committee..to do away with the shortcomings of organisation in the work and structure of the apparatus of the Narkomsovkhos of the USSR. **1946** *N.Y. Times* 7 July 1/6 For the Soviet government 'apparatus', as the Russians use the word, is a political machine. **1949** *Manch. Guardian Weekly* 7 July 14/4 To get a common friend to join a Communist apparatus. **1950** *Time* 13 Feb. 13/2 He..had turned over the films to the Soviet spy apparatus. **1959** E. H. Carr *Socialism in One Country* II. xix. 199 The rising power of the secretariat led to the emergence of a new and prominent feature in the party and Soviet landscape—the 'apparatus' and the 'apparatus-man', the body of obscure and anonymous officials who were the cogs of the..party machine.

‖**apparatus criticus** (æpărē¹·tǝs kri·tikǝs). [mod.L.: see Apparatus and Critical *a.*] A collection of palæographical and critical matter accompanying an edition of a text.
1865 Conington *Vergili Opera* I. p. x, The publication of Ribbeck's *apparatus criticus* has made a new recension necessary. **1867** *Athenæum* 8 June 758/1 In the notes are found all the various readings of MSS., as given by preceding editors, followed by the testimonies of ancient writers in support of them; the whole forming as complete a *criticus apparatus* as could be desired. **1887** *Classical Rev.* Mar. 17/1 The *apparatus criticus* is still simple and the new editor's personality is rarely obtruded on the reader. **1928** *Observer* 4 Mar. 7/3 The second half of the volume comprises the Geographical Text..with a complete *apparatus criticus*.

apparent, *a.* Add: **8.** *Comb.* **apparent magnitude** [Magnitude 3], the magnitude of a celestial body classified according to its apparent brightness, opp. **absolute magnitude*; **apparent (solar) time**: see quot. 1940.
1875 *Encycl. Brit.* II. 822/1 The gathering of stars of the leading orders of *apparent* magnitude in the galactic zone shows that stars of many orders of *real* size and brightness are there gathered together. **1902** [see **absolute magnitude*]. **1940** *Chambers's Techn. Dict.* 522/1 *Apparent magnitude* is the measure of the brightness on Pogson's logarithmic scale, in which each step of one whole magnitude represents a light ratio of 2·512, and this increases numerically with decreasing brightness. **1694** J. Smith *Horolog. Disquisitions* 30 Now the Clock being still naturally inclined to lose, I therefore..set him again..so he is again too fast for the sun 3' 45" (which is the most I suffer him to differ from the apparent Time). **1706, 1764, 1834** [see Time *sb.* 27]. **1922** H. Spencer Jones *Gen. Astr.* 48 If the sun is used as a reference body the time so determined is called apparent solar time. **1940** *Chambers's Techn. Dict.* 45/1 *Apparent solar time*, the hour angle, at any moment, of the *true*, or *apparent*, *sun* as distinguished from the *mean sun*... Sundials read apparent solar time. **1942** F. Debenham *Astrographics* (ed. 2) 20 The shadow of the real sun will only give us the irregular real sun time (called *Apparent Time*), so we have to add or subtract the Equation of Time to get Mean Time.

apparentation (ăpē³rĕntē¹·ʃǝn, ăpær-). [f. *Apparent(ed *ppl. a.* + -ation; cf. F. *apparentage* alliance, connexion (by marriage).] The relation between an earlier society and a later one that is apparented to it.
1934 A. J. Toynbee *Study of Hist.* I. i. iv. 44 The 'affiliations' and 'apparentations' between one society and another resemble the relations between parent and child. **1939** *Ibid.* IV. 79 The society that passes away through incorporation into another society, and not through apparentation and dissolution, preserves..some continuity in its material fabric. **1946** R. G. Collingwood *Idea of Hist.* IV. i. 160 One of these categories is *affiliation* and its correlative *apparentation*.

apparented (ăpē³·rǝntĕd), *ppl. a.* [ad. F. *apparenté* related, akin.] Of societies: related, as an earlier society to a later one that is derived from it.
1934 A. J. Toynbee *Study of Hist.* I. i. i. 52 Unidentified societies which may be 'apparented' or otherwise related to any of the societies which..our operations have brought to light. **1952** E. Hyams *Soil & Civilization* pl. III (*caption*), Potato and maize, on the basis of which Incarial and apparented civilizations were built, are frequent ceramic motifs.

apparitional. Add: **2.** Belonging to the sphere of mere appearances or phenomena.
1899 A. C. Lyall *Asiatic Studies* Ser. II. 89 The Hindu in his conviction of the illusory nature of all phenomena.. will not contest the authenticity, in an..apparitional sense, of historic religions.

appeal, *v.* Add: **6. c.** *Cricket.* To call upon an umpire for his decision; to make an appeal (see *Appeal *sb.* 4 b).
1744 *Laws* [of Cricket] in *New Dict. Arts & Sci.* (1755) IV. 3460/1 [The umpires] are not to order any man out, unless appealed to by one of the players. **1830** M. R. Mitford *Our Village* IV. 26 They..finally appealed to the umpires as to the fairness of the play. **1897** K. S. Ranjitsinhji *Jubilee Bk. Cricket* vi. 254 When a catch at the wicket is appealed for, an umpire should..take into consideration both sound and sight.
9. Also, to be attractive or pleasing *to* (a person).
1881 Mrs. J. H. Riddell *Senior Partner* III. iii. 70 Mr. McCullagh, to whom Mr. Mostin's general assertion appealed with the force of experience. **1898** Saintsbury *Short Hist. Eng. Lit.* VIII. iii. 524 Its poetical and romantic attractions.. appeal even to a person so little poetical as Hobbes. **1928** C. Singer *From Magic to Science* i. 16 Such a point of view appealed greatly to the Middle Ages.
b. To 'make an appeal'; to be attractive.
1907 *Smart Set* Mar. 23/1 The speciousness of Betty's words appealed.
10. (Further examples.)
1828 Webster s.v., We say the cause was appealed before or after trial. **1900** *Westm. Gaz.* 22 Aug. 2/2 Possibly the case will be appealed. **1932** E. Wilson *Devil take Hindmost* xvii. 192 The defense will appeal the case to the Supreme Court. **1963** *Publisher's Weekly* 2 Sept. 45/1 Curtis has announced that it will appeal the verdict.

appeal, *sb.* Add: **3.** *Court of Appeal* (earlier example). Also *Court of Appeals* (U.S.).
1777 *Journals Cont. Congress* (1907) VIII. 607 The propriety of establishing a court of appeals. **1849** Dickens *Let.* 2 Feb. in W. C. Macready *Diaries* (1912) II. 417 The indispensable necessity there is for a public and solemn Court of Appeal in all criminal cases. **1855** Fernando Wood in X. D. MacLeod *Biogr.* (1856), There is..a Court of Appeals, to which to apply against the subjection.
4. b. *spec.* in *Cricket.* A call made to an umpire, by any player, for his decision on any point arising during a match, esp. whether a batsman is 'out' or not.
1844 F. W. Lillywhite *Hand-Bk. Cricket* 20 After the umpire has given you out, further appeal is useless. **1882** [see sense 4 in Dict.]. **1908** W. E. W. Collins *Country Cricketer's Diary* iv. 71 'How's that?' came the second appeal, as the wicket-keeper, ball in hand, pulled up a stump.
6. b. *fig.* Attractive influence or power; esp. in phr. *to make an appeal* (*to*) = *Appeal *v.* 9.
1916 H. Walpole *Dark Forest* II. i. 209 The appeal of her ignorance and strength and credulity—ah! she won our hearts simply whenever she pleased. **1920** Rose Macaulay *Potterism* I. i. § 4 Mrs. Potter was rather sadly aware that she made no appeal to the twins. *Ibid.* I. iv. § 1 His was the sort of beauty which..makes so strong an appeal to the senses of the sex other than that of the possessor. **1925** W. Deeping *Sorrell & Son* xxiii. § 3 The enticements that might be expected to make an appeal to a very young man. *Ibid.* xxvii. § 1 She was standing close to him,..and Kit was conscious of the sudden shock of her appeal. **1926** *Grand Mag.* Aug. 587/2 The sport of manhunting had lost its appeal.

appealing, *vbl. sb.* Add to def.: Also, the action of making an appeal; attractive influence or power.
1876 G. M. Hopkins *Wr. Deutschland* (1918) st. 27 The appealing of the Passion is tenderer in prayer apart. **1889** *Times* 19 June 15/3 A moderate sum, if we consider..the arbitrating, the judging, the appealing that are thrown in for this modest shilling.

appealing, *ppl. a.* Further examples merging into sense 'attractive'.
1891 Hardy *Tess* II. xxxvi. 234 Corporeal presence is sometimes less appealing than corporeal absence. **1898** *Daily News* 1 Mar. 6/1 Froment..is..the most appealing character in the book. **1922** Joyce *Ulysses* 358 The old love was waiting..with blue appealing eyes. **1932** E. Bowen *To North* xx. 220 We find her a very dear person: there is something appealing about her.

appearance. Add: **12. b.** *to save appearances*: see also Save *v.* 12; *to keep up appearances* (earlier examples).
1760 C. Churchill *Night* (ed. 3) 14 Keep up appearances, there lies the test, The world will give thee credit for the rest. **1836** [see Keep *v.* 57 e].

appearing, *ppl. a.* Add: **2. c.** With defining word preceding, as *best-*, *fine-appearing*. *U.S.*
1839 W. D. Williamson in *Mass. Hist. Soc. Coll.* (1846) 3rd Ser. IX. 95 All these..rendered him the best appearing Indian ever seen in this quarter. **1879** Howells *Lady of Aroostook* xviii. 215 'She is very fine-appearing,' said Lydia. Staniford smiled at the countrified phrase. **1897** *Outing* (U.S.) XXX. 352/1 First prize for best-appearing club at Long Branch.

appease, *v.* Add: **1. a.** (Later example.)
1920 W. S. Churchill *Let.* 24 Mar. in *World Crisis: Aftermath* (1929) xvii. 378, I should be prepared to make peace with Soviet Russia on the best terms available to appease the general situation.
2. c. *Politics.* In derog. sense (cf. sense 2 a, quot. 1750, in Dict.), used esp. of the British Prime Minister's efforts from 1937 to 1939 to placate, and so stave off the threatened aggression of, the Axis powers: to engage in a policy of appeasement (see next).
1939 *Ann. Reg. 1938* 104 So far were they from trying to 'appease' the Dictators that they might rather be described as 'facing up' to them. **1940** *War Illustr.* 16 Feb. 106/3 If at any time..there are signs of a renewed desire for appeasement, let the fate of the Polish people remind us of the power we wish to appease. **1940** *New Statesman* 21 Dec. 650/1 He took a different view of the Nazis, whom he thought we could successfully appease. **1940** *New Republic* 23 Dec. 852/2 While England is appeasing Franco in this strictly limited sense, Franco is certainly not appeasing world opinion.

appeasement. Add: **4.** Freely used in political contexts in the 20th century, and since 1938 often used disparagingly with allusion to the attempts at conciliation by concession made by Mr. Neville Chamberlain, the British Prime Minister, before the outbreak of war with Germany in 1939; by extension, any such policy of pacification by concession to an enemy.
1919 *Gen. Smuts' Messages to Empire: Problem of Peace* 14 In our policy of European settlement the appeasement of Germany..becomes one of cardinal importance. **1920** W. S. Churchill *Let.* 24 Mar. in *World Crisis: Aftermath* (1929) xvii. 378 Here again I counsel prudence and appeasement. Try to secure a really representative Turkish governing authority, and come to terms with it. **1929** J. M. Keynes in *Nation & Athenæum* 9 Mar. 782/2 Apart from Russia, Mr. Churchill appears, in a degree to which public opinion has done much less than justice, as an ardent and persistent advocate of the policy of appeasement—appeasement in Germany, in Ireland, in Turkey. **1934** Ld. Lothian *Let.* in *Times* 4 May 15/5 A limitation of armaments by political appeasement. **1936** Anthony Eden in *Hansard Commons* 5th Ser. CCCX. 1446, I assure the House that it is the appeasement of Europe as a whole that we have constantly before us. **1937** W. K. Hancock *Survey Brit. Commonw. Affairs* I. 262 Equality and Appeasement, 1926–1936. **1938** *Encycl. Brit. Bk. of Year 1938* 194/1 Economic appeasement must precede any world-wide political appeasement. **1938** *Times* 3 Oct. 13/2 The policy of international appeasement must of course be pressed forward... There must be appeasement not only of the strong but of the weak... With the policy of appeasement must go the policy of preparation—preparation not so much for war as against war. **1939** *Ann. Reg. 1938* 10 One of the new Foreign Minister's first steps was to extend to Germany the methods of appeasement—as the Prime Minister was fond of calling them—which were now being tried with Italy. **1939** *New Statesman* 29 July 165/1 First, provided that there is a Russian pact, proposals that now smell of appeasement in the most dangerous sense at once become proper and, indeed, the only possible policy.
transf. **1940** *Mind* XLIX. 327 Thus objectivity or qualified spatio-temporality is the '*Lebensraum*' provided for the 'appeasement' of the finite *actus-potentia.* **1949** F. Maclean *Eastern Approaches* III. iv. 351 Clearly appeasement formed no part of her nature.

appeaser. Add: **2.** *Politics.* One who supports a policy of appeasement (see *Appease *v.* 2 c, *Appeasement 4).
1940 *Economist* 12 Oct. 456/1 Obviously, the United States is not going to play the appeaser's game. **1940** in *Amer. Speech* (1941) XVI. 145 Wouldn't it be better to make our own country impregnable..? People who say so are being called 'fifth columnists' and 'appeasers'.

appendectomy (æ:pende·ktǒmi). *Surg. U.S.* [f. Append(ix *sb.* + Gr. ἐκτομή cutting out.] = *Appendicectomy.
1894–5 D. Mason in *N. Orleans Med. & Surg. Jrnl.* XXII. 570 (*title*) Appendectomy. **1903** *N.Y. Med. Jrnl.* 4 July 5/2 The methods of appendectomy can be divided into two classes, those with a stump and those without a stump. **1961** J. Heller *Catch-22* (1962) xxxiii. 347 A nude, ridiculous man with a blushing appendectomy scar appeared in the doorway.

appendical, *a.* Add: **b.** Of or pertaining to the vermiform appendix. *rare.*
1909 *Practitioner* Nov. 646 A typical history of appendical perforation.

appendicectomy (ăpe:ndise·ktǒmi). *Surg.* [f. L. *appendic-*, stem of Appendix *sb.* + Gr. ἐκτομή cutting out.] Excision of the vermiform appendix of the cæcum.
1894 T. Myles in *Med. Press & Circular* LVIII. 637/2 (*title*) Appendicectomy, with notes of two cases. **1905** *Nursing Times* 23 Dec. 666/2 Appendicectomy: After Results... The late after-results of removal of the appendix.

appendicitis (ăpe:ndisəi·tis). *Path.* [mod.L., f. as prec. + -itis.] Inflammation of the vermiform appendix of the cæcum.
1886 R. H. Fitz in *Amer. Jrnl. Med. Sci.* Oct. 323 As a circumscribed peritonitis is simply one event..in the history of inflammation of the appendix, it seems preferable to use the term appendicitis to express the primary condition. **1887** Buck *Handbk. Med. Sci.* IV. 185/2 The symptoms of appendicitis, pure and simple, are probably very slight, and not easily recognizable. **1905** Kelly & Hurdon *Verm. App.* 382 Some cases of appendicitis occasioned by influenza.

appendix, *sb.* **3. a.** Add after def.: *spec.*, short for *vermiform appendix* (of the cæcum): see Vermiform *a.* 3 a.
1886 [see prec.]. **1902** *Scotsman* 2 July 7/3 This operation, removal of the appendix. **1905** Kelly & Hurdon *Verm. App.* 2 The first recorded case of disease of the appendix is the classical one of Mestivier, reported in 1759. **1905** H. G. Wells *Modern Utopia* App. 377, I had seen the ancestral cæcum shrink to that disease nest, the appendix of to-day.

apperceive, *v.* Transfer †*Obs.* to sense in Dict. and add:
2. *Psychol.* To be or become conscious of perceiving; to comprehend (something perceived) by a mental act which unites and assimilates the perception to a mass of ideas already possessed; to have apperception of: see next.
1876 J. Sully in *Mind* Jan. 41 Where two impressions are simultaneously apperceived, it is because they are such as can be brought under one complex impression as parts of the whole. **1892** —— *Human Mind* I. 163 The new presentative element is said..to be apperceived or assimilated by a pre-existing cluster of ideas or an ideal mass. **1894** *Academy* 7 July 6/2 A thousand people, observes Lange, may read Virgil; but every one will apperceive him differently.
Hence **apperceiving** *ppl. a.* and *vbl. sb.*; *apperceiving mass* [cf. J. F. Herbart's expression (1825) 'appercipirende Vorstellungsmasse'] = *apperception mass* (see next, 3 b, and cf. **apperceptive mass* and *Mass *sb.*²); whence **apperceivingly** *adv.*
1890 [see next, 3 b]. **1893** C. De Garmo et al. tr. *Lange's Apperception* (1896) 101 When ideal norms are apperceivingly active in the field of knowledge..true culture is attained. **1890** W. James *Princ. Psychol.* II. xix. 109 A child who hitherto has seen none but four-cornered tables apperceives a round one as a table; but by this the apperceiving mass ('table') is enriched. **1914** R. M. Jones *Spir. Reformers in 16th & 17th Cent.* xiii. 263 The representative of the old system..condemning a dawning movement which with his apperceiving material he could not understand.

apperception. Add: **3.** *Psychol.* The action or fact of becoming conscious by subsequent reflection of a perception already experienced; any act or process by which the mind unites and assimilates a particular idea (esp. one newly presented) to a larger set or mass of ideas (already possessed), so as to comprehend it as part of the whole: see quots.
1876 J. Sully in *Mind* Jan. 36 The entrance of a presentation into the internal field of view is termed a Perception; its entrance into the point of view an Apperception. **1887** J. Dewey *Psychol.* 89 Apperception is the relating activity which combines the various sensuous elements presented to the mind at one time into a whole, and which unites these wholes, recurring at successive times, into a continuous mental life, thereby making psychical life intelligent. **1893** C. De Garmo et al. tr. *Lange's Apperception* (1896) 28 Apperception is the subsumption of a notion, usually newly given and more or less individual, under a predicate which is more complete..and..usually older and more familiar. Apperception does not always follow perception immediately, for years sometimes intervene between the learning of a fact and its comprehension. **1923** H. G. Baynes tr. *Jung's Psychol. Types* xi. 524 Apperception is a psychic process by which a new content is articulated to similar already-existing contents in such a way as to be understood, apprehended, or clear.
b. *attrib.*, as *apperception mass, process, product.* (H. Steinthal (1871) used the expressions 'Massen Apperception' and 'Massen Apperceptions-Process'.)
1890 W. James *Princ. Psychol.* II. xix. 109 An apperception-product arises: the knowledge of the perceived being as a horse. *Ibid.*, Apperception-processes can perfectly well occur in which the new observation transforms or enriches the apperceiving group of ideas. **1896** W. J. Eckoff tr. *J. F. Herbart's ABC of Sense-Perception* 88 Conscience..we have seen to be a complex apperception mass aboriginally compounded of innumerable presentations. **1933** *Brit. Jrnl. Psychol.* July 24 Subjects having a marked perseveration of the ideas and 'apperception masses' previously stimulated would presumably be more likely..to overlook words wrongly spelt.
Hence **apperce·ptionism** *Psychol.*, the explanation and systematization of the process of apperception; **apperce·ptionist,** one who holds or affects the doctrine of apperceptionism; **apperceptioni·stic** *a.*, of, pertaining to, or characterized by apperceptionism.
1903 H. Münsterberg *Harvard Psychol. Stud.* I. 644 Without returning to apperceptionism we can overcome the one-sidedness of associationism if full use is made of the means which the world of phenomena offers to theory. *Ibid.* 653 Apperceptionistic psychology. **1904** *Jrnl. Philos., Psychol. & Sci. Methods* 18 Aug. 466 (Cent. D. Suppl.), The idealist's view is that of the 'apperceptionists'.

apperceptive (æpəɹse·ptiv), *a.* [f. prec., after Perceptive *a.*] Pertaining to or involving apperception; also = next; *apperceptive mass* = *apperception mass* (see prec.).
1884 E. Montgomery in *Mind* July 381 It is after all nothing but our own apperceptive faculties, potentially idealised, that are made to serve for the consciousness of a universal subject. **1886** F. A. Caspari tr. *P. Radestock's Habit & Education* 45 Blending or apperceptive synthesis, that is, a union of impressions following one another, in which the latter exist in the new conception produced by their union. **1891** G. F. Stout in *Mind* XVI. 31 Apperceptive processes differ for the most part from those which result in the formation of a percept. *Ibid.* 37 An apperceptive system is, apart from fatigue, more prompt to apperceive the more recently it has been brought into play. **1892** *Monist* Jan. 306 Under normal conditions the necessary alternation of different apperceptive masses produces a corresponding variation in the conditions of belief. **1952** W. J. H. Sprott *Social Psychol.* x. 211 All these psychological elements..form the 'apperceptive mass', which receives new material, and the matrix out of which changes emerge.

appercipient (æpəɹsi·piĕnt), *a.* [f. as prec., after Percipient *a.*] Having or exercising the faculty of apperception.
1906 H. H. Joachim *Nature of Truth* iii. 93 Every judgement..is informed, conditioned, and to some extent constituted by the appercipient character of the mind which makes it. **1922** S. Grew *Art of Player-Piano* 161 Only the musician may be entirely appercipient as regards the subtle beauty of poetry.

appetiteless (æ·pitəit,lės), *a.* [f. Appetite *sb.* + -less.] Having no appetite; marked by want of appetite.
1853 Surtees *Sponge's Sp. Tour* lx, Having finished a poor appetiteless breakfast. **1890** *Temple Bar* July 303 He sits appetiteless over his solitary breakfast.

appetitive, *a.* Add: **1.** *spec.* in *Biol.* (see quots.).
1951 W. H. Thorpe in *Bull. Animal Behaviour* Mar. 37/1 Appetitive Behaviour = The variable introductory phase of an instinctive behaviour pattern or sequence. **1958** *New Biol.* XXVII. 74 Sometimes an animal moves over a large area before performing a specific action... This variable introductory activity in a behaviour sequence is called appetitive behaviour.

applanate (æ·plănē¹t), *a. Bot.* [ad. mod.L. *applānātus*, pa. pple. of *applānāre*, f. *ap-* = *ad* to + late L. *plānāre* to level, f. *plānus* Plane.] Of a flattened or horizontally expanded form.
1887 W. Phillips *Brit. Discomycetes* 2 Pileus sessile and applanate. *Ibid.* 101 *Peziza depressa*..from its applanate growth. **1950** *Proc. Prehist. Soc.* XVI. 123 Macroscopically, the Flixton specimens showed the applanate, then ungulate, shape typical of the species.

apple, *sb.* Add: **1. b.** Phr. (*as*) *sure as God made little apples*, and similar phrases.
1874 M. Clarke *His Natural Life* (1875) III. xv. 261 I'll tie you up and give you fifty for yourself, as sure as God made little apples. **1912** Mulford & Clay *Buck Peters* xxii. 198 It's Buck as sure as little apples Kesicks. **1926** J. Black *You can't Win* (1927) ix. 121 Sure as God made little apples I'll see that you get ten days. **1942** M. Lasswell *Suds in Eye* ix. 112 I'm gonna learn to read sure as God made little apples.
c. (Later examples.)
1785 [see Pear *sb.* 2]. **1940** E. Step *Wayside & Woodland Trees* 40 The Wild Apple has not the pyramidal form of the Wild Pear.
7. a. (Later example.)
1827 *Blackw. Mag.* XXII. 374/1 Dull people turn up.. the apples of their eyes on beholding Prose by a Poet.
b. (Later examples.)
1930 R. Campbell *Poems* 11 Live and die The apple, nay the onion, of his eye? **1941** F. Gruber *Hungry Dog* (1950) x. 81 He may have been the apple of *your* eye, but to me he was only a cinder.
8. *apple(s) and pears*: rhyming slang for 'stairs'; also (*ellipt.*) *apples*.
1857 'Ducange Anglicus' *Vulg. Tongue* 1, Apple and Pears, stairs. **1909** Ware *Passing Eng.* 9/1 Bill an' Jack's gone up apples. **1914** C. Mackenzie *Sinister Street* II. IV. ix. 1100, I soon shoved him down the Apples-and-pears. **1962** J. G. Bennett *Witness* xviii. 218 One of the removal men asked him if a sofa was to go 'up the apples'.
B. I. 2. *apple-bright, -cheeked* (examples), *-faced* (earlier example), *-green* (earlier and later examples), *-rotten, -scented* (example), *-yellow.*
1930 E. Sitwell *Coll. Poems* 90 Their apple-bright and ruddy flesh. **1864** Tennyson *Enoch Arden*, etc. 158 A bevy of Eroses apple-cheek'd. **1921** W. de la Mare *Crossings* 86 A shy, fat, apple-cheeked child. **1837** Dickens *Pickw.* II. xlii. 457 A little white-headed apple-faced tipstaff. **1648** Hexham *Groot Woorden-Boeck*, *Appel-groen*, Apple-greene. **1797** Heideloff *Gallery of Fashion* Nov. in R. W. Chapman *Sense & Sensibility* (1933) 387 Hungarian robe of apple-green satin. **1917** D. H. Lawrence *Look! We have come Through!* 54 The dawn was apple-green. **1905** 'Q' *Shining Ferry* III. xxvi. 271 The schooner might be apple-rotten. **1923** W. de la Mare *Riddle* 2 The cool apple-scented pantry. **1953** Bannerman *Birds Brit. Isles* I. 180 Young birds..are more grey-green, that colour replacing the brighter apple-yellow of the adult.
b. apple head, a type of rounded skull found in certain small dogs; so *apple-domed, -headed* adjs.
1883 W. G. Stables *Our Friend the Dog* vii. 59 *Apple-headed*, the roundness of the top of the skull. **1922** R. Leighton *Compl. Bk. of Dog* v. xx. 290 These small dogs usually have 'apple-heads'. **1948** C. L. B. Hubbard *Dogs in Britain* xix. 216 The head [of the Chihuahua] is well rounded and apple-domed. **1959** *Observer* 1 Feb. 11/3 'Apple' heads (that is a thick, domed skull with snipey fore-face)..are particularly deplored [in toy poodles].
3. a. *apple-bloom* (example), *-juice* (earlier and later examples), *-seed, -stem, -time*; **b.** *apple-orchard* (earlier U.S. example); **c.** *apple-dumpling, -pasty, -pudding*: earlier examples.
1949 Blunden *After Bombing* 50 As new apple-bloom May be by hailstones ravaged. **1711** W. Byrd *Secret Diary* 18 Aug. (1941) 391 At dinner I ate some apple dumpling. **1708** E. Cook *Sot-weed Factor* 19 There with good punch and apple juice We spent our hours without abuse. **1961** *Harrods Food News* 11/1 Canadian Apple Juice, natural. **1721** *New-Eng. Courant* 14–21 Aug. 2/2 There was a larger Apple Orchard at that Place. **1728** E. Smith *Compleat Housewife* (ed. 2) 108 Apple-Pasties to Fry. **1710** W. Byrd *Secret Diary* 24 Aug. (1941) 222, I ate some apple pudding for dinner. **1699** J. Lord *Let.* 21 Feb. in *Mass. Hist. Soc. Coll.* (1861) 4th Ser. V. 306 Apple-seeds, sown by us since we came, came up in January. **1930** T. S. Eliot *Ash-Wed.* 19 Spitting from the mouth the withered apple-seed. **1907** Yeats *Deirdre* 10 Praise the blossoming apple-stem. **1865** Swinburne *Chastelard* IV. i. 120 And now the flower, and deadly fruit will come With apple-time in autumn.
II. apple-bee, (*a*) *dial.* a wasp; (*b*) *U.S.* (see Bee *sb.*¹ 4); **apple-borer** *U.S.*, an insect attacking apple-trees; **apple-box, -gum,** names for species of *Eucalyptus*; **apple-brandy** (earlier U.S. example); **apple-bug,** a water-beetle of the family Gyrinidæ, which exudes a milky liquid having an odour of apples; **apple-butter** (earlier and later examples); **apple charlotte**: see *Charlotte; **apple-dowdy** chiefly *U.S.*, a kind of apple-pie made in a deep dish (cf. Pandowdy); **apple-jack** (earlier and later U.S.

examples); **apple-oil**, a synthetic chemical used to imitate the odour of apples in confectionery; **apple-peru** *U.S.*, the thorn-apple; **apple-polishing** *vbl. sb.* (*U.S. slang*), currying favour; toadying; so **apple-polisher**, a toady; **apple-sauce** (see sense B. I. 3 c in Dict.), (*a*) lit. (earlier example); (*b*) *transf.* nonsense, absurdity; insincere flattery (*U.S. slang*); **apple-slump** *U.S.* (see quot. 1872); **apple snow**: see Snow *sb.*¹ 4 a; **apple strudel** [partial tr. of G. *apfelstrudel* (also used), f. *apfel* apple + *strudel* flaky pastry], a baked sweet consisting of a spiced mixture of apples rolled in flaky pastry; **apple-woman** (earlier example).

1808 Monthly Mag. XXVI. 421/2 Apple-bee, a wasp. C[ornwall]. 1912 C. Mackenzie Carnival xxxix. 397 It was vain for Thomas to assure her that apple-bees did not sting without provocation. 1859 Trans. Ill. Agric. Soc. 1857–8 III. 344 The most destructive of these..is that known as the apple borer. 1890 Melbourne Argus 9 Aug. 4/6 An ironstone hill..with apple-box and ironbark dotted about. 1944 F. D. Davison in Coast to Coast 238 The creek made a horseshoe bend under its bower of apple-box-trees. c 1780 in Maryland Hist. Mag. (1907) II. 256, [I] accepted 13 gals. of peach brandy in satisfaction of the damage... He cheated me with apple brandy. 1832 J. P. Kennedy Swallow B. I. xii. 129 The apple-bugs (as school-boys call that glossy black insect which frequents the summer pools, and is distinguished for the perfume of the apple) danced in busy myriads over the surface of the still water. 1869 Rep. U.S. Commissioner Agric. 1868 80 The fifth family, gyrinidae, comprises those oval water-beetles usually known by the name of 'whirligigs' or apple-bugs. c 1774 Crèvecœur Sk. 18th-Cent. Amer. (1925) 105 We often make apple-butter. 1870 Congress. Globe Apr. 2685/1 Apple-butter is a substitute for butter; it is spread upon bread and eaten in like manner. 1923 W. Nutting Massachusetts 241 Did ever a dish of apple dowdy go to the spot like that? 1952 M. Laski Village vii. 114 To make an apple-dowdy in the kitchen. 1847 Leichhardt Jrnl. viii. 264 Another Eucalyptus [E. Stuartiana]..with smooth upper trunk and cordate ovate leaves..; we called it the Apple-gum. 1963 W. S. Ramson in Austral. Quart. XXXV. Sept. 53 Some have been named after their supposed resemblance to the foliage or timber of European trees, like apple gum, [etc.]. 1816 'Old Scene Painter' Emigrant's Guide 30 A partial distillation is also made from apples..called Apple-Jack. 1932 E. Wilson Devil take Hindmost i. 1 The old cider mills are still barrelling applejack and hard cider. 1857 Apple oil [see Valerianate]. 1867 Bloxam Chem. 553 The valerianate of amyle..which has the flavour of apples,..is known as apple-oil. 1784 Cutler in Mem. Amer. Academy (1785) I. 419 Apple-peru..Common by the waysides. August. 1850 Hawthorne Scarlet L. (1851) i. 60 A grass-plot, much overgrown with burdock, pig-weed, apple-peru, and such unsightly vegetation. 1928 Amer. Speech III. 218 Apple polisher. 1947 E. A. McCourt Music at Close 116 The apple-polishers in the front row laughed with forced heartiness. 1935 A. G. Kennedy Current Eng. ii. 28 The college boy with his pet phrases such as apple-polishing, currying favor with a professor. 1739 E. Smith Compleat Housewife (ed. 9) 104 Boil them as you do Apple-sauce. 1921 Collier's 1 Jan. 18/4 That's all apple sauce! 1924 Wodehouse Bill the Conqueror xii. 210 It sounds to me a good deal like apple sauce. Seems like there ain't no sense in it. 1926 S.P.E. Tract XXIV. 119 Applesauce (noun or interjection). One of the latest pieces of slang in this country [sc. U.S.A.]. It has two quite distinct meanings, (1) nonsense! and (2) flattery. It is commonly used as a term of jocular contempt in reply to effusive but unjustifiable flattery. 'He a great author? Apple-sauce!' 1934 J. O'Hara Appt. in Samarra (1935) ii. 45, 'I just didn't want to spoil your evening, that's all.' 'Applesauce,' said Irma. 1831 H. J. Finn Amer. Comic Annual 140 The pumpkin pies and apple slump..were smoking on the table. 1872 Schele de Vere Americanisms 415 Apple-Slump is..a favorite New England dish, consisting of apples and molasses baked within a bread-pie in an iron pot. 1884 E. E. Hale Christmas in Narragansett i. 11 Guessed that they had done justice to..Polly's apple-slump. 1863 Mrs. Beeton Bk. Househ. Managem. xxix. 703 Apple Snow,..10 good-sized apples, the whites of 10 eggs, the rind of 1 lemon, ½ lb. of pounded sugar. a 1887 Apple snow [see Snow sb.¹ 4 a]. 1936 Lucas & Hume Au Petit Cordon Bleu 148 Apfelstrudel. 1963 Punch 2 Jan. 27 What I lose on the nursery slopes I gain on the apfelstrudel. 1923 Hemingway In our Time (1926) xi. 164 We have some apple strudel if you want it. 1935 M. Lane Faith, Hope, No Charity iv. 108 Ada's friend from the Jewish caterers..came round with a four-pound apple strudel. 1741 Pope & Arbuthnot Mem. M. Scriblerus vi. 46 Yonder are two Apple-women scolding.

apple, *v.* Add: **1. b.** *intr.* Of turnips: to swell into globular shape.
1712 J. Morton Nat. Hist. Northants. ix. 483 Unless the Soil has some mixture of Sand the Turnips do not apple, as they call it: that is, do not bottom well. 1731 Miller Gard. Dict. s.v. Rapa, If the Autumn should not prove very mild, [the Turnips] will not have time to Apple before Winter. 1796 [see Dict.].

apple-cart. [Apple B. I. 3 b.] Add: Phr. *to upset the* (or *someone's*) *apple-cart*, humorously used for 'to cause an upset'; *esp.* to upset a person's plans, to ruin the undertaking. Also in similar expressions.
1788 J. Belknap in Mass. Hist. Soc. Coll. (1882) 5th Ser. III. 17 S. Adams had almost overset the apple-cart by intruding an amendment of his own fabrication on the morning of the day of ratification [of the Constitution]. 1796 Grose Dict. Vulgar Tongue (ed. 3) s.v., Down with his apple-cart; knock or throw him down. 1834 Carruthers Kentuckian in N.Y. I. 101 Smash my apple-cart, if there wasn't more crying..than I've seen at many an honest man's funeral. 1848 in Hodder Life Shaftesbury (1887) xvii. 396 If the Prince goes on like this, why he'll upset our apple-cart. 1871 G. P. R. Pulman Rustic Sketches (ed. 3) Gloss. 77 'Don't upsit th' apple-cart.' That is—be careful you do not let fall anything carried. 1883 Pall Mall Gaz. 26 Oct., If the Control had done more it might have upset the apple-cart altogether. 1897 Review of Rev. Aug. 114 Somebody will lose patience, and then over will go the apple-cart. 1955 Sci. Amer. Aug. 68/2 Names such as ammoniated tincture of valerian can safely be revealed to the patient without upsetting the psychological applecart.

apple-pie. Add: **apple-pie bed** (examples); **apple-pie order** (earlier and later examples); also *ellipt.*
1781 J. Woodforde Diary (1924) I. 302 Had but an indifferent night of Sleep, Mrs. Davie and Nancy made me up an Apple Pye Bed last night. 1840 New Monthly Mag. LVIII. 246 He..began to fancy that the bed was too small for him, when..little Oxtowne..told him..'it was only an apple-pie'. 1917 A. Waugh Loom of Youth iv. vii. 308 His dormitory made him apple-pie beds.
1780 T. Pasley Private Sea Jrnls. 12 May (1931) 87 Their Persons Clean and in apple-Pie order on Sundays. 1898 W. Raleigh Let. 13 Sept. (1926) I. 208 We had an artist in the back room..so I could not leave it as apple-pie as I should have liked. 1904 J. London Sea-Wolf xi. 88 The boat-pullers and steerers have..put their boats in apple-pie order.

Appleton layer (æ·p'ltən lēⁱ·ər). *Physics.* [f. the name of Edward V. Appleton (1892–1965), British physicist + Layer.] The upper, or F, stratum of the ionosphere above the Heaviside (or E) layer.
Appleton himself used the term 'F layer' for this stratum (*Proc. Internat. Scientific Radio Union*, Washington Meeting, 1927).
1932 Techn. Tables & Gloss. (B.B.C.) 44/1 Appleton Layer, an ionised layer of the upper atmosphere thought to exist above the Kennelly–Heaviside layer. It has been suggested that it is this layer which is responsible for the reflection of short waves which are attenuated but not reflected in their passage through the Kennelly–Heaviside layer. The height of the Appleton layer above the earth has been measured as approximately 230 kilometres. 1943 Electronic Engin. XV. 342 Presumably the Heaviside, Appleton..etc. layers in the upper atmosphere also give rise to harmonics of radio waves.

apple-tree. Add: **2.** In Australia applied to various indigenous trees, esp. to a species of Eucalyptus (= *apple-gum*: see *apple sb. B. II), and to another myrtaceous tree, *Angophora subvelutina*.
1801 in Hist. Rec. Australia (1915) III. 414 The Timber at the back blue-gum and apple trees. 1820 J. Oxley Jrnl. Exped. N.S. Wales 187 That species of eucalyptus which is vulgarly called the apple tree. 1885 Mrs. C. Praed Head Station I. iv. 60 A giant eucalyptus of the kind called 'appletree'. 1885 Spons' Mechanics' Own Bk. 127 Apple (Australian) (Angophora subvelutina). The so-called apple-tree of Queensland yields planks 20–30 in. in diameter.

appliable, *a.* Restrict †*Obs.* to senses 1 and 3, and add later examples for sense 2.
1855 Jrnl. R. Agric. Soc. XVI. ii. 563 A practical and appliable knowledge. 1874 Contemp. Rev. XXIV. 731 Abundance..of wealth..appliable to other purposes.

appliance. Add: **3. b.** *spec.* A fire-engine.
1899 Daily News 10 May 7/2 The driver..and one of the firemen on the 'manual' were injured, and the appliance sustained some damage. 1958 Listener 14 Aug. 247/3 The fire-engines ('appliances' to the Service) tore off into the night.

application. Add: **11.** *Comb.*: **application money**, the sum of money paid when applying for the allotment of shares.
1900 Westm. Gaz. 12 Mar. 9/1 The lists may be closed earlier than Thursday..to avoid the unnecessary locking up of application money. 1907 Ibid. 9 Mar. 15/1 A special form of application is provided, under which the application money will have to be paid as in the case of new subscriptions.

applicator. Delete *rare* and add further examples. Also *attrib.*
1909 Westm. Gaz. 29 Jan. 1/2 Mr. Hartigan..had invented a radium applicator. 1959 Woman's Own 20 June 52/1 (Advt.), The only home perm lotion in a tube with its own easy-to-use applicator-tip. 1962 Listener 10 May 831/3 This can be spread with a synthetic sponge applicator.

applied, *ppl. a.* Add: **2.** (Additional examples.)
1912 A. Brazil New Girl at St. Chad's x. 152 One of her boys has turned out so clever that he has been sent to the Technical School to study 'applied arts'. 1928 Forestry II. 133 It is probably far wiser for the applied scientist to credit insects with acumen than to attribute to them nothing but forced movements or tropisms. 1933 Burlington Mag. Apr. p. xvi/2 The second collection contains works of fine and applied art from a castle in Upper Hesse, Germany. 1936 Discovery Apr. 129/1 The study of practical psychology or applied psychology. 1958 Year's Wk. Eng. Stud. 1956 37/1 His book will reward study by all interested in what now tends to be called 'applied linguistics'.
3. *Dressmaking.* Laid on as appliqué.
1880 E. Glaisher Needlework iv. 37 Velvet does not bear applied work well. 1882 Caulfeild & Saward Dict. Needlework s.v. Appliqué, The various applied pieces are laid in position one at a time, and secured by being sewn down round their edges. 1901 Westm. Gaz. 25 July 3/2 A lace applied collar. 1910 Ibid. 29 Jan. 15/2 Applied velvet.

apply, *v.* Add: **1. c.** *Dressmaking.* To lay on as appliqué; to trim or ornament with appliqué.
1880 E. Glaisher Needlework v. 49 The heraldic figures..are applied in white cloth. 1882 Caulfeild & Saward Dict. Needlework 12/1 The foundation was generally muslin, and the net applied or let in. 1901 Westm. Gaz. 14 Mar. 4/2 Everything is applied with insertions or cut-out flowers and figures of lace.

appointive, *a.* For *rare* read *U.S.* For def. read: **1.** Dependent on appointment; that is filled by appointment; holding one's place by appointment.
1881 [see Dict.]. 1886 N. Amer. Rev. Aug. 203 Whether the entire judiciary should be appointive or elective. 1889 Nation (N.Y.) 4 July 1/1 Mr. Geer being an appointive officer of the Custom-house. 1914 T. C. Smith Wars Eng. & Amer. iii. 54 The Quebec Act altered the government of the province by the creation of an appointive council.
2. 'Of or pertaining to appointment; appointing; as, the appointive power of the President' (*Cent. Dict.*).

appointment. Add: **4.** Also *attrib.*, as *appointment book*.
1864 Dickens Mut. Fr. I. viii. 65 Would you take a seat..while I look over our Appointment Book? 1950 T. S. Eliot Cocktail Party ii. 91 The Nurse-Secretary enters, with Appointment Book.
8. Esp. in phr. *by appointment*, by or as by royal warrant.
1864 Dickens Mut. Fr. I. v. 33 He..settled it with himself..that he was errand-goer by appointment to the house at the corner. 1962 Guardian 2 Dec. 12/1 A list of Royal Warrant holders..shows that only two firms are now permitted to style themselves 'By Appointment to the late Queen Alexandra'.

apport, *sb.* Transfer †*Obs.* to senses 1 and 2, and add: **3.** The production of material objects, supposedly by occult means, at a spiritualistic séance; also, an object so produced. Usu. in *pl.*
1894 A. Lang Cock Lane 53 Objects are brought from places many miles distant, and tossed on the table. These are technically termed apports. 1926 Conan Doyle Hist. Spiritualism I. xiv. 327 A number of fresh flowers and fruits, still wet, fell upon the table—a phenomenon of apports.

apportioner. Delete †*Obs.*⁻⁰ and add:
1884 Browning Ferishtah's Fancies ix. 83 The apportioner of every lot of ground. 1893 St. James' Budget 23 June 6/3 The apportioners of the Chantrey Bequest. 1905 C. Kernahan Visions 259 The..dishonest Apportioner of life's good and evil.

appraisive (ăprēⁱ·ziv), *a.* [f. Appraise *v.* + -ive.] Of language, etc.: used in making, or of the type used in making, appraisals or valuations. Hence **apprai·sively** *adv.*
1934 in Webster. 1946 C. W. Morris Signs, Lang. & Behav. v. 129 The work itself does not, as fiction, appraisively signify. Ibid. 136 Poetry may serve as an example of discourse which is primarily appraisive-valuative. Ibid. 138 Appraisive-incitive discourse may be illustrated by the language of morality. Ibid. 140 'Criticism' (or 'evaluation') serves to illustrate appraisive-systemic discourse. 1956 J. Whatmough Lang. vi. 88 The merely designative use of language has in each case become appraisive. 1967 Nabokov Speak, Memory (ed. 2) ii. 40 Her grim father would..give the heaviest racket an appraisive shake.

appreciation. Add: **2. b.** An expression (in speech or writing) of one's estimate of something: often implying a favourable estimate (cf. 4).
1858 M. Pattison Essays (1889) II. xix. 344 In the last page he has written, in his small and fine hand, a short appreciation of the book and its author. 1889 Pater (title) Appreciations, with an essay on style. 1907 J. A. Hammerton (title) Stevensoniana: an anecdotal life and appreciation of R. L. Stevenson.

appreciative, *a.* (Earlier and later examples.)
a 1698 J. Fraser Mem. (1738) vii. 239 Tho' in my sensitive Faculty I find not these Impressions of Joy and Fear, yet do I find them in my estimative appretiative Faculty. 1943 H. Read Polit. of Unpolit. xii. 134 Developing the creative and appreciative exercise of the aesthetic impulse in the child.

appressorium (æpresōᵊ·riŏm). *Bot.* Pl. **-ia.** [mod.L., f. L. *appress-*, ppl. stem of *apprimĕre*, f. *ad* Ad- + *premĕre* to press: see -orium.] The organ by which certain fungi attach themselves to their hosts.
1897 W. G. Smith tr. Tubeuf's Diseases Plants Cryptogr. Parasites i. i. 10 The mycelium retains its hold by adhesion-discs or appressoria, and from certain parts of these a fine thread-like process is given off. 1902 Encycl. Brit. XXVIII. 555/1 Appressoria are also formed by some parasitic Fungi, as a minute flattening of the top of a very short branch..or the swollen end of any hypha which comes in contact with..the host. 1937 Gwynne-Vaughan & Barnes Struct. & Devel. Fungi (ed. 2) 187 An external disc or appressorium, from which,..the haustorium proper arises and pushes into the epidermal cell.

appro., appro (æ·pro). Abbrev. of *approbation or *approval, in the phrase *on* (or *upon*) *appro*. Also *attrib.*
1874 Hotten Slang Dict. 74 Most of the extensive show of chains, watches, and trinkets in a shop window is obtained 'on appro', i.e., 'on sale or return'. 1901 Punch 26 June 467/1 You ordered it 'upon appro'. 1906 Daily Chron. 20 June 6/6 We have..purchases on 'appro.'. 1908 Westm. Gaz. 10 Jan. 2/1 The goods on appro. we left to Lavinia's maid. 1910 Ibid. 10 Feb. 8/1 The head salesman..said it was not an 'appro' transaction. 1950 Wodehouse Nothing Serious 40 She took him on again—on appro., as it were. The idea was that if he proved himself steady and serious, those wedding-bells would ring out.

approach, *v.* Add: **4. b.** Also, to make an overture or proposal to; to seek for an interview; to make advances to (a person) with a view to influencing his actions; to attempt to influence or bribe.
1857 Lawrence (Kansas) Republican 30 July 2 An editor of this place had approached him..offering inducements to him to become an organ under it. 1893 Congress. Rec. Sept. 1874/1 Nearly every bit of everything that is said about public men being corrupted or approached is false. 1929 Daily News 10 Apr. 7/3 When the 'Daily News' approached the Admiralty on the subject an official said [etc.].
12. *Golf. intr.* To play the approach stroke. Hence **approa·ching** *vbl. sb.* Also **approa·cher.**
1887 W. G. Simpson Golf viii. 140 When the player's ball is within less than a driver shot of the hole, approaching commences. Ibid. 146 Approachers of equal skill indulge in all the variations [of grip]. 1898 'House' on Sport I (Advt.), Auchterlonie's special registered approaching cleek kept in stock. 1903 Westm. Gaz. 9 Jan. 2/1 Mr. Laidlay used to approach better than any other man I have ever seen play.
13. *Aeronaut.* Of an aircraft or pilot, to make an approach (see *approach *sb.* 13).
1927 Aviation XXII. 1080/2 Upon approaching an airdrome at night, the pilot is interested in the limits of the landing area. 1950 Gloss. Aeronaut. Terms (B.S.I.) i. 8 Approach, to manœuvre an aircraft into position relative to the landing area for flattening-out and alighting.

approach, *sb.* Add: **3. b.** Advances made to a person for the purpose of improperly influencing his actions. *U.S.*
1893 Congress. Rec. Sept. 1874/1 The idea that..[Mr. Hooper] was subject to approach is ridiculous.
11. *Golf.* The play or stroke by which a player hits his ball on to the putting-green or approaches the hole; chiefly *attrib.*, as *approach play, shot, stroke*.
1879 Encycl. Brit. X. 765/2 Having got within some moderate distance of it [sc. a hole], he proceeds to make his 'approach shot'. Ibid., The 'approach' and the 'putting' are by far the most difficult..parts of the game. 1887 W. G. Simpson Golf viii. 155 There are four clubs used for ordinary approach work—the putter, the cleek, the iron, and the mashy. 1903 Westm. Gaz. 10 Mar. 2/3 Plenty of opportunities still will be afforded for approach strokes. Ibid., Approach play is not to be considered in laying out a hole. 1919 Wodehouse Damsel in Distress i. 23 Does it [sc. love] make you slice your approach-shots? Ibid. xv. 173 His approach-putting has to be seen to be believed.
12. *Bridge.* Denoting a type of bidding (see quots.). Usu. *attrib.*, as *approach bid, bidding; approach(-forcing) system*, etc.
1929 Bridge World Nov. 14/1 The Approach System at Auction, also introduced by the Culbertsons, practically revolutionized bidding methods. 1930 E. Culbertson Contract Bridge Blue Book vi. 89 Whenever a hand contains a biddable suit..the suit and not the no-trump should be preferred. (The Approach principle.) 1930 A. E. M. Foster Contract Bridge for All i. 7 You read of approach bids. Ibid. 11 The idea in approach bidding is..to decide on the final declaration after the reaction of partner and opponents has been discovered. 1931 J. H. Reford Contract up-to-date i. 34 The system advocated in this book embodies..the approach-forcing system.
13. *Aeronaut.* The final stage in flight before landing; also, the air space through which the approach is made. Also *attrib.*
1930 Airway Age XI. 381/1 In addition to..green and white boundary and approach lights, each hangar is equipped with obstacle and flood lights. 1931 Flight 6 Feb. 121/2 An approach which ended in the aircraft reaching the after limits of the deck too low..resulted in the undercarriage being severely damaged. 1931 P. V. H. Weems Air Navigation ix. 129 Landing field approach lights are green lights in the boundary lighting circuit at the end of runways. 1933 Bur. of Standards Jrnl. of Research XI. 485 The compass bearing is followed until after hearing the signals from the approach marker beacon. 1940 War Illustr. 19 Jan. 620 With wheels and landing flaps lowered the pilot makes his approach. 1944 Jrnl. R. Aeronaut. Soc. XLVIII. 285 The best approach speed, flaps and undercarriage down, is about 95 m.p.h. 1948 Daily Tel. 23 June 4/6 Radar navigational and approach aids could be standardized by international companies. 1951 Gloss. Aeronaut. Terms (B.S.I.) iii. 22 Approach, the air space over an approach area. 1958 Times 24 Oct. 5/3 (caption) Testing..the new precision approach radar (talk-down) equipment installed in the Approach control room at London Airport.
14. *attrib.*, as *approach island* (see *island *sb.* 2 c), *road* (cf. *access 7 a), *span*.
1958 Times 24 June 5/5 A motor cycle..came into collision with the kerb of an approach island. 1833 Loudon Encycl. Archit. 463 The approach-road to the house. 1962 Listener 7 June 989/1 The control of large motor-coaches on the very narrow approach-roads to the moor. 1924 Times Trade & Engin. Suppl. 29 Nov. 252/3 The bridge..will be carried on five main piers and built in four arch and two approach spans.

approaching, *ppl. a.* Add: **4.** *Bridge.* (See quots. and cf. *approach *sb.* 12.)
1926 Auction Bridge Mag. Sept. 174/1 A call may be tentative, perhaps to be followed by a second one. In America they call this the approaching bid. 1927 M. C. Work Contract Bridge (1928) v. 131 Approaching bid, one designed to guide partner toward the most advantageous declaration.

approbation. Add: **3. b.** *on approbation*: on approval (see *approval b).
1880 L. Higgin Handbk. Embroidery 101 Designs on paper are not supplied under any circumstances nor can work be sent out on approbation. 1901 Strand Mag. XXI. June p. i/2 (Advt.), The London Shoe Co. Ltd. Goods sent on approbation. 1920 Conquest Jan. p. viii, Please send these books on Approbation.

approfound (æprofau·nd), *v.* [ad. F. *approfondir*, f. *ap-* Ad- + *profond* Profound *a.*] *trans.* To go deeply into, to search the depths of (a subject of study). (A Gallicism.)
1885 R. G. White Stud. Shakespeare 35 Subtle theories as to Shakespeare's purpose in this play have been set forth by critics who engage in the task of approfounding him. 1897 Sat. Rev. 8 May 520 These seven pages of M. Mallarmé would take longer to approfound than all the back numbers of 'Cosmopolis' put together.

approval. Add: **b.** *on approval*: phr. in commercial use denoting that goods sent to a customer are submitted for his examination only, without obligation to purchase (if they are returned undamaged). Also without prep. and *attrib.*
1870 St. James's Mag. IV. 598 They had evidently brought most of the things 'on approval'. 1877 Design & Work 15 Dec. 667/2 Watch..English lever..approval. Ibid. 22 Dec. 690/3 Sheets of stamps sent on approval. 1880 Bazaar, Exchange & Mart Suppl. 2 Jan. 1/1 Breech-loaders. Great Bargains..Approval. Deposit. Ibid. 5/2 Mulready envelope and 380 foreign stamps. Sent on approval. 1891 Stamp Collecting June 21 Is there no way of suppressing the Approval-sheet Swindler? 1902 Connoisseur Jan. (Advt.), Stamps... We are sending out Approval Sheets and Books. 1906 Drapers' Record 27 Oct. 190 These goods cannot be sent on approval.

approved, *ppl. a.* Add: **5. approved school**, a place of training for boys or girls who have been found guilty of offences or exposed to moral danger.
1932 Hansard Commons CCLXI. 1179 The question of the approved schools, as they will in future be called—the schools which are known to the public at present as reformatory and industrial schools. 1933 Act 23 Geo. V c. 12 § 107 (1), 'Approved school' means a school approved by the Secretary of State under section seventy-nine of this Act. 1938 Times 7 Dec. 10/3 The effect of the change of name from 'Reformatory' and 'Industrial' to 'Approved Schools' cannot be exaggerated. No youth is ashamed of having been in an approved school. 1956 A. Wilson Anglo-Saxon Attitudes ii. i. 228 Larrie's an orphan, an institution boy who's been in a lot of trouble, he's had three convictions for petty thieving and he's been to an Approved School.
6. approved society, a Friendly Society (see Friendly *a.* 8) legally empowered to administer benefits under the National Insurance Act of 1911 and subsequent Acts down to that of 1946.
1911 Act 1 & 2 Geo. V c. 55 § 23 (1) Any society..which complies with the requirements of this Act..may be approved by the Insurance Commissioners, and, if so approved, shall be an approved society for the purposes of this Part of this Act. 1952 Oxf. Jun. Encycl. X. 171/1 Under the 1911 National Health Act certain registered friendly societies, known as 'approved societies', were used by the Government as part of the health scheme to pay out as 'sick benefit' sums of money provided by the Government... But after the National Insurance Act of 1946 this use of friendly societies by the Government was discontinued.

apraxia (ăpræ·ksiă). *Path.* [mod.L., ad. Ger. *apraxie* (H. Steinthal *Einleit. in d. Psychologie* (1871) 458), a. Gr. ἀπραξία inaction.] Inability to perform purposeful movements; loss of ability to 'do'.
1888 M. A. Starr in Med. Record XXXIV. 497 (title) Apraxia and Aphasia; their varieties, and the methods..for their detection. 1922 R. S. Woodworth Psychology iii. 57 Injury to the 'super-motor centers' causes loss of skilled movement, and produces the condition of 'apraxia', in which the subject, though knowing what he wants to do, and though still able to move his limbs, simply cannot get the combination for the skilled act that he has in mind. 1961 Lancet 12 Aug. 363/2 The nature of the fundamental deficits underlying constructional apraxia..has not been elucidated.

‖**après coup** (aprę ku), *advb. phr.* [Fr., lit. 'after stroke'.] As an afterthought; after the event.
1871 Geo. Eliot Let. 27 Feb. (1955) V. 136 These letters, which must speak better for him than the finest eloquence après coup. 1887 Athenæum 18 June 796/3 The name has probably been affixed to the book après coup, so to speak, and to allure the unwary reader. 1890 W. James Princ. Psychol. II. xxviii. 630 Our abstract and general discoveries usually come to us as lucky fancies; and it is only après coup that we find that they correspond to some reality.

‖**après-ski** (aprę skī). [Fr.] The time when skiing is over for the day; = *after-ski* (*after- I. 1). Freq. *attrib.*, of clothes worn, entertainment, etc., at such a time.
1954 R. Martin Your Ski Holiday iv. 35 The 'après-ski' outfit..is based on the outdoor version. Ibid. viii. 62 There is always a queue of Britons to read it [sc. an English newspaper], during the hours of 'après-ski'. 1959 Guardian 27 Nov. 8/4 The allurement, for après-ski, of a fluffy wool cuddle-skirt. 1961 Sunday Express 1 Jan. 11/1 The ski-suit above..[is] for the indoor après-ski party.

aprick (ăpri·k), *adv.* or *pred. a. poet.* [f. A *prep.*¹ + Prick *v.*] Erect, pricked or pricking up.
1856 Mrs. Browning Aurora Leigh vi. 237 Watching gnats a-prick upon a pond. 1898 O. Seaman In Cap & Bells 15 (At the sign of the cock) Amazon spurs aprick at heel. 1903 Academy 21 Feb. 186/2 The earth was quick with green a-prick.

apricot. Add: **1. b.** *transf.* The pinkish yellow colour of an apricot. Also *Comb.*, as *apricot-coloured, -tinted* adjs.
1906 Daily Chron. 19 Apr. 6/1 'Juliet', a large apricot-tinted rose. 1907 Galsworthy Country House ii. ii. 119 The moon, tinted apricot and figured like a coin, hung above the cedar-trees. 1907 Westm. Gaz. 24 June 3/1 A portrait of a little girl in an apricot-coloured frock and pink ribbons. 1923 Daily Mail 28 Feb. 1 The newest Paris shades of Coral, Lilac,..Apricot. 1933 D. Parker After such Pleasures (1934) 12 Her apricot satin chaise-longue.
3. apricot sickness *S. Afr.* [lit. tr. Afrikaans *Appelkoossiekte*, so named because it often appears at the beginning of summer when apricots ripen], a form of dysentery, said to be caused by a bacillus.
1945 Cape Times 27 Jan. 11/3 Apricot sickness is most troublesome when the entire gastro-intestinal tract is affected; that is to say, when there is vomiting as well as diarrhoea.

April. Add: **1. b.** (Additional examples.)
1912 W. de la Mare Listeners 42 Under the April-grey calm waste of the skies. 1917 T. S. Eliot Prufrock 20 These April sunsets, that somehow recall My buried life. 1940 Blunden Poems 1930–1940 252 Bright-tressed, ready-smiling, April-eyed. 1953 E. Sitwell Gardens & Astronomers 37 Upon your wood-wild April-soft long hair.
3. *April-fool-day* (example): now commonly *April Fool's Day*; also *april-fool v.*, *-ing vbl. sb.*
1832 Carlyle in Fraser's Mag. V. 260 Let the foolish April-fool-day pass by. 1863 H. A. London in Amer. Speech (1951) XXVI. 181/1, [I] April fooled some fellows. 1903 Encycl. Brit. XXXV. 45/2 April Fool's Day. 1919 E. H. Jones Road to En-dor (ed. 2, 1920) xviii. 185 We set out..in true April-fooling spirit. Ibid. 186 The old British custom of April-fooling.

Aprilian (ēⁱpri·liăn), *a.* [f. April + -ian.] Of or characteristic of April.
1902 Daily Chron. 24 Mar. 8/1 Rain of a spitefully Aprilian character. 1904 Bliss Carman Songs Northern Garden (1905) 10 Fear not the mighty instinct, The great Aprilian Creed; The House of Spring is open And furnished for thy need.

apriorism. Add to def.: Also, the philosophical doctrine of *a priori* or innate ideas (see A priori 3). So **aprio·rist**, one who holds this doctrine; also *loosely*, one given to *a priori* reasoning; also *attrib.* or as *adj.*; hence **apriori·stic** *a.*, pertaining to apriorism or apriorists.
1873 Lewes Probl. Life & Mind Ser. 1. I. 412 This will be disputed by the a priorists. 1874 Watson & Evans tr. Van Oosterzee's Christian Dogmatics I. 141 No authority..must..be conceded to such an aprioristic criticism. Ibid. II. 596 However little..inclined to an abstract a-priorism. 1889 Athenæum 2 Feb. 152/3 The problem of external perception has a unique character among the controversies that divide the empiricists and the apriorists. 1889 G. B. Shaw Fabian Ess. Socialism 177 The apriorist notion that among free competitors wealth must go to the industrious. 1891 Monist I. 635 Empiricism is wrong because it can at best show the temporal succession of two phenomena, and apriorism is wrong because a priori knowledge lies in the subject alone and not in the object. 1914 Teixeira de Mattos tr. Maeterlinck's Unknown Guest iii. 42 The 'apriorists', who hold that the idea of time is innate. 1930 Philosophy V. 449 This Experimental Empiricism or Operational Apriorism is obviously attractive. 1934 T. S. Eliot Eliz. Essays 173 His statement is too apriorist to be quite trustworthy. 1958 Times Lit. Suppl. 17 Jan. 25/4 The vast advances of science..have set up strong, though often unformulated, resistances against aprioristic ethics or philosophies.

apron, *sb.* Add: **3.** (Earlier examples.)
1790 Pennsylv. Packet 22 Apr. 2/1 A new Sulkey, with a top and apron. 1835 Dickens in Evening Chron. 21 July 1/3 Cabs, with trunks and band-boxes between the drivers' legs and outside the apron.
4. b, c. (Earlier examples.)
1633 T. James Voyage 75 The Aprons of our Gunnes. 1711 W. Sutherland Ship Builder's Asst. 25 Raising the Stem and false Stem (or Apron) together.
g. An endless belt or band for conveying material of any kind, or a receptacle on such a belt. Also *apron-conveyor*.
1835 Ure Philos. Manuf. iii. 155 His business is to weigh the wool, and spread it in definite quantities on a travelling apron, which feeds the first pair of rollers. a 1875 Knight Dict. Mech. I. 126/1 Apron, a board or leather which conducts material over an opening, as, the grain in a separator, the ore in a buddle or frame, etc. 1909 Cent. Dict. Suppl. s.v. Conveyer, An apron-conveyer consists of endless chains upon which wood or metal slats are placed close together, forming a continuous traveling platform.
h. A protective covering (on a bank, etc.) against the action of a river or the sea.
a 1884 Knight Dict. Mech. Suppl., Apron, a protecting surface of logs and brush..to protect or to form revetment for river sides.
i. *Geol.* An outspread alluvial deposit, esp. one at the extremity of a glacier or the mouth of a river.
1889 N. S. Shaler in U.S. Geol. Survey 9th Ann. Rep. 548 In front of the kames normally comes the morainal apron...Traces of this frontal apron exist along the southern shore of Cape Ann. 1955 Sci. Amer. Mar. 82/2 The rivers of the continents have been building aprons into the sea with their wash of silt, clay, sand and gravel.
j. *Theatr.* (See quot. 1903.) Also *apron stage*.
1903 A. B. Walkley Dramatic Criticism 108 The 'apron' is the technical name for the stage-area in front of the curtain. In the Elizabethan theatre it jutted right out among the public, who surrounded it on three sides. This 'apron' slowly shrank..till at last in our day it has altogether disappeared. 1933 P. Godfrey Back-Stage i. 20 Turns are being played on the half-stage or upon the apron. 1950 G. B. Shaw in New Statesman 6 May 511/2 The Elizabethan stage..with..an apron stage projecting into the auditorium.
k. A defence against hostile aircraft consisting of a series of wires suspended from a cable to which captive balloons are attached.
1917, etc. Balloon apron [see *balloon *sb.*¹ 10 c]. 1919 C. C. Turner Struggle in Air 1914–18 ix. 134 Then the 'apron' was devised, to create a sufficiently dangerous and invisible obstruction which caused raiders to maintain a certain minimum altitude.
l. *Mil.* (See quot. 1918.) Also, a strip of barbed-wire entanglements.
1918 E. S. Farrow Dict. Mil. Terms, Apron, in fortification, that portion of the superior slope of a parapet or the interior slope of a pit designed to protect the slopes against blast. 1926 F. M. Ford Man could Stand Up ii. i. 77 There was their own apron—a perfect village!—of wire over which he looked.

m. A flat, usually paved, surface in front of a building, hangar, etc., *esp.* an area on an aerodrome prepared for the easy handling, (un)loading, etc., of aircraft on the ground.
1925 March's Thesaurus Dict. App. 1110/1 Apron, an open working surface in front of a hangar. 1933 Jrnl. R. Aeronaut. Soc. XXXVII. 6 Where the concrete or asphalt apron in front of the buildings meets the aerodrome. 1954 C. Armstrong Better to eat You iii. 24 Opposite a three-car garage there was a wide paved apron. Ibid. vi. 63 He..ran across..the parking apron. 1955 Daily Tel. 28 Jan. 6/5 The waving base is on the 'airside' as distinct from the 'landside' of the building, and thus gives a clear view of the aircraft leaving the 'terminal apron'.
n. A wooden platform trailed aft by the mother ship of a flying boat as a landing platform or for smoothing rough water. (Now disused.)
1933 Discovery Dec. 366/1 To make the call for refuelling the Westfalen is equipped with an apron which can be let out over the stern and trailed slightly below the surface of the water. The seaplanes will thus have a fairly smooth area of water on which they can alight if the sea is rough. 1941 Flight 24 Apr. 294/1 These ships made use of 'aprons' trailed in the water aft of the vessel...The advantage of the apron was that it reduced to a few minutes the time which the flying boat had to spend on the open sea.
o. *Photogr.* (See quot. 1940.)
1935 News Chron. Amat. Photogr. v. 89 Next tuck the film end underneath the flexible apron so that the clip is holding both the apron and the end of the film. 1940 F. J. Mortimer Wall's Dict. Photogr. (ed. 15) Apron, band of celluloid or other material used to separate successive turns of film when rolled up for development in a tank.
5. b. (See quot. 1876.)
1855 Fraser's Mag. LI. 267 These [crabs] were lying apron upwards on the ground. 1876 Field & Forest II. 73 This mass is very conspicuous even in the rapidly swimming crab, and causes the abdominal flap (called apron by fishermen) to be opened almost to its fullest extent.
c. = Tablier 3. Also called *Hottentot apron, pudendal apron*.
1909 in Cent. Dict. Suppl. 1956 Winick Dict. Anthropol. (1957) 33 There is some controversy as to whether the Hottentot apron is a genetic trait or an intentional deformation.
6. apron-cloth (see sense 3); **apron-piece** (example); **apron wall** *Build.* (see quot. 1934).
1857 D. H. Strother Virginia Illustr. ii. 139 The apron-cloth was drawn up over their legs and with a..crack of the whip they started into the storm. 1859 J. Gwilt Encycl. Archit. (ed. 4) 894 Apron, or Pitching Piece, an horizontal piece of timber, in wooden double-flighted stairs, for supporting the carriage pieces or rough strings and joistings in the half spaces or landings. The apron pieces should be firmly wedged into the wall. 1934 Webster, Apron wall, that portion of the enclosing wall of a building between a window-head or lintel of one story and the sill of the window in the story above or, sometimes, only that portion between the floor line and the window sill of the same story. 1952 Archit. Rev. CXI. 262 A new system of wall construction (which can be briefly described as 3-ply apron walls formed in-situ).

aproned, *ppl. a.* (Later examples not in Comb.)
a 1916 H. James Ivory Tower (1917) iii. i. 152 Only to collapse again like aproned puppets on removal of pressure from the squeak. 1922 Joyce Ulysses 121 He walked on.., passing an old man, bowed, spectacled, aproned.

apron-string. Add: (Additional examples.) Also *transf.* Also **apron-string** *v. trans. rare.*
1678 Ray Prov. (ed. 2) 226 To hold by the Apron-strings, i.e. in right of his wife. 1698 J. Fryer New Acc. E. India & Persia iii. iii. 115 They..would subjugate us, as they do all others that are harness'd with the Apron-strings of Trade. 1825 Eng. Life I. 165 A man of your inches ought to be above making such an apron-string booby of yourself. 1848 A. Brontë Wildfell Hall I. iii. 44 Even at his age, he ought not to be always tied to his mother's apron string. 1912 A. S. M. Hutchinson Happy Warrior iv. ii. 196 That he should submit to be thus chained, thus apron-stringed! 1960 Times 26 Apr. 16/2 The drama..points insistently, in the last two acts, away from Rimsky-Korsakov's apron-strings.

apropos. Add: Also ‖ à propos.
D. Phr. *à propos de bottes* [Fr., lit. = 'with regard to boots'], without serious motive, without rhyme or reason.
1757 Chesterfield Let. 23 Sept. (1774) II. 378 A propos de bottes, for I am told he always wears his; was his Royal Highness very gracious to you, or not? 1845 Q. Rev. LXXVI. 533 The first introduces that prince—very much àpropos de bottes—for the purpose of denying that he had any party. 1925 A. Huxley Those Barren Leaves i. i. 9 She would remember an ancient floater—just like that, à propos de bottes. 1934 'G. Orwell' Burm. Days (1935) xviii. 225 This was àpropos de bottes, but the subject needed no introduction.

aprosexia (æprose·ksiă). *Med.* [mod.L., a. Du. *aprosexia* (Guye 1887, in *Nederl. Tijdschr. v. Geneesk.* XXIII. 381), a. Gr. ἀπροσεξία, f. ἀ- priv. (A- 14) + προσέχειν to turn (the attention): see -ia¹.] Abnormal inability to concentrate one's attention.
1889 A. A. G. Guye in Brit. Med. Jrnl. 28 Sept. 709/2 (title) On aprosexia, being the inability to fix the attention and other allied troubles in the cerebral functions caused by nasal disorders. Ibid. 710/1 What I call aprosexia is the inability to fix the attention on any definite more or less abstract subject. 1894 W. James in Psychol. Rev. 528 His aprosexia is complete, and he is incapable of interest in anything whatever. 1901 Baldwin Dict. Philos. & Psychol. I. 88/1 Aprosexia is also used in a more general sense for the inability to fix the attention; when due to neurasthenia, it would be neurasthenic aprosexia; when due to nasal obstruction, nasal aprosexia. 1948 Brit. Jrnl. Psychol. Dec. 119 The abnormalities of attention are classified as 'aprosexia', 'hyperprosexia', and 'paraprosexia'.

aprosopia (æprosōᵘ·piă). *Teratology.* [mod.L., f. Gr. ἀ- priv. (A- 14) + πρόσωπ-ον face + -ia¹.] Absence or imperfect development of the face.
1842 in R. Dunglison Dict. Med. Sci. (ed. 3) 61/2. 1922 Joyce Ulysses 403 Aprosopia due to a congestion.

apsidiole (æpsi·diǫl). *Archit.* Also **absidiole.** [ad. F. *absidiole* a small apse.] A small or secondary apse.
1889 in Cent. Dict. 1905 F. Bond Goth. Arch. (1906) ix. 170 The semicircular apses and absidoles [sic] of Romanesque become polygonal in Gothic. 1951 Archit. Rev. CX. 338/3 The ambulatory with an apsidiole half buried in the ground enabling pilgrims to see the relics in the crypt.

apt, *a.* Add: **4. c.** Inclined, disposed (in a single instance).
1677 R. Cary Palæol. Chron. ii. ii. i. iv. 195, I am apt to think, that..Vashti is meant. 1706 Hearne Rem. & Collect. (1885) I. 297, I am apt to think he has not consulted Books enough upon this occasion. 1899 E. E. Hale Lowell 126, I am apt to think that this modest man was the first person ..to recognize [etc.].

‖**apterium** (æptiə·riŏm). *Ornith.* Pl. **-ia.** [mod.L. (Nitzsch 1833), f. Gr. ἀ- priv. (A- 14) + πτερόν feather.] Each of the featherless spaces on the skin of a bird intervening between the feathered tracts or pterylæ.
1867 W. S. Dallas tr. Nitzsch's Pterylography 16 To these feathered bands I give the name of Feather-tracts or Contour-feather-tracts (pterylæ, Federn-fluren), and to the naked bands, or those which are not beset with contour-feathers, that of featherless spaces (apteria, Federn-raine). 1894 [see Pteryla].

apterygial (æptĕri·dʒiăl), *a.* *Ichthyol.* [f. Gr. ἀ- priv. (A- 14) + Pteryg- + -(i)al.] Destitute of fins, finless.
1902 Nature 25 Sept. 526 We may distinguish the Cyclostomes as apterygial and epicraniate.

Aptian (æ·ptiăn), *a.* (*sb.*) *Geol.* [f. place-name *Apt* (Vaucluse, France) + -ian.] Epithet of a division of the Lower Cretaceous formation in France and Belgium; belonging to or found in this, as a fossil.
1864 J. D. Dana Man. Geol. (ed. 2) iii. iii. 480 The subdivisions of the Cretaceous are variously named in different parts of Europe. Lower Cretaceous.—Superior Neocomian of D'Orbigny (the Wealden being the inferior); also his Aptian. Ibid. 776/2 Aptian group. 1885 A. Geikie Geol. (ed. 2) 831 In northern France the Aptian stage is chiefly clay. 1921 Brit. Mus. Return 151 Aptian fossils from Whale's Head.

aptitude. Add: **4.** *Comb.*, as **aptitude test** *U.S.*, a test designed to determine a person's capacity in any given skill or field of knowledge.
1923 Amer. Inst. Criminal Law Jrnl. XIV. 376 An aptitude test for policemen. 1926 Sci. Amer. Feb. 93/3 The aptitude test simply consists in making telegraph signals..and then testing the memory of the men. 1952 M. McCarthy Groves of Academe (1953) iv. 53 By the use of aptitude tests, psychological questionnaires, even blood-sampling and cranial measurements, he hoped to discover a method of gauging student-potential.

‖**aptychus** (æ·ptikŏs). *Palæont.* Pl. **aptychi.** [mod.L., f. Gr. ἀ- priv. (A- 14) + πτυχή fold, layer.] A calcareous plate or pair of plates found in the terminal chamber of some ammonites.
1877 Huxley Anat. Inv. Anim. 536 The Aptychi..occupy the middle of the posterior wall or terminal chamber of the Ammonite, and have their bases towards its mouth. 1893 H. Woods Palæont., Invert. vii. 176 In the last chamber of some Ammonites and a few other genera, a pair of calcareous or horny plates, known as the Aptychus, are occasionally found.

Apulian (ăpiū·liăn), *a.* and *sb.* [f. L. *Apulia*, *Appulia* + -an.] **A.** *adj.* Of or pertaining to the ancient province or the modern geographical district of Apulia in southern Italy, or its inhabitants. **B.** *sb.* An inhabitant of Apulia.
1607 Topsell Four-f. Beasts 187 There was such another charme or incantation among the Apuleians. Ibid. 625 All Christendome yealdeth praise and price vnto it next after the Apulian and Tarentinian wooll. 1709 S. Dunster tr. Horace's Sat. ii. i. p. 127 Or rather, with intent, to keep the Lucanians and Apulians in awe. 1757 J. Dyer Fleece ii. 102 They..swell their fleeces, equal to the worth Of cloath'd Apulian. 1820 Shelley Cenci iii. i. p. 45 That lonely rock, Petrella, in the Apulian Appennines. 1889 Cent. Dict., Apulian pottery, a name given to the Italo-Greek pottery found in Apulia and southeastern Italy generally, especially to the vases with red figures on a lustrous black ground, some of the most important examples of which are from this region. 1932 Times Lit. Suppl. 21 Jan. 37/2 The Apulian officials and organizers spreading throughout Frederick's Italian territories. 1955 S. Runciman Eastern Schism ii. 38 In 1020 an Apulian called Melo raised a revolt in southern Italy against Byzantium.

aqua. Pronunc. now usu. (æ·kwă).

aqua² (æ·kwă). [Abbrev. of Aquamarine.] A light greenish blue colour.
1936 New Yorker 8 Feb. 56 (Advt.), Gold, navy, grey, rustic, blue, rose, aqua. 1959 Woman's Own 23 May 52/3 She admired the plate which was aqua.

aqua- (æ·kwă). L. *aqua* 'water' used as a combining form or quasi-*adj.*, esp. in expressions referring to aquatic entertainment; **aquacade** (æ·kwăkēⁱd) *U.S.* [*-cade], an elaborate display of aquatic events and turns.
1887 Gentl. Mag. June 549 When the 'Théâtre Nautique' first opened its doors the bill presented..a three act aqua-drama of Chinese life, entitled 'Kao-Kang'. 1930 Birmingham Post 28 Feb. 8 A daring new aquatic sport introduced at Winter Haven, Florida, U.S.A.—riding in an 'aqua-glider' fitted with an aeroplane wing and attached to a speed boat. 1935 A. P. Herbert What a Word! v. 164 Every week some new thing is flung into the news...Cinema, vitamin...—and, I regret to say, television and aquadrome—here are only a few. 1936 London Transport, Country Walks, 1st Ser., A big lake set with wooded eyots, the Aquadrome is really an old ballast pit. 1937 Daily Progress (Charlottesville, Va.) 16 Nov. 6/5 Billy Rose, the little man who creates big shows, is pictured with Eleanor Holm Jarrett, his star of the Cleveland aquacade. 1949 R. Graves Seven Days in New Crete xx. 248 Several naked girls slipped into the pool and swam languidly around like a Hollywood aquacade. 1952 Granville Dict. Theatr. Terms 17 Aqua show.. Aquatic turns are performed in tanks on the stage or in suitable settings according to the size of the theatre. 1961 Daily Mail 12 Sept. 13/2 Mr. B. is running an 'aqua-camp'. ..He has bought a 16 ft. cabin cruiser..for use by the campers, who bring canoes and diving equipment.

a·quacu:lture, occas. var. Aquiculture. Also, = *hydroponics.
1887 Athenæum 21 May 667/1. 1929 W. F. Gericke in Amer. Jrnl. Bot. XVI. 862 (title) Aquaculture. A means of crop-production..growth of plants in water. 1962 Daily Tel. 4 Apr. 12/3 The systematic exploitation of the sea—marine husbandry or aqua-culture.

aquafer, var. *aquifer.

aquake (ăkwēⁱ·k), *adv.* or *pred. a. poet.* [f. A *prep.*¹ + Quake *v.*] In a quaking condition; quaking.
1875 Morris Æneids iii. 616 My mates aquake with dread. 1883 Meredith Poems of Joy of Earth 54 The valley aquake with the tread Of an iron-resounding hoof.

aqualung (æ·kwălʌŋ). [f. *aqua- + Lung.] A portable diving apparatus consisting of containers of compressed air strapped on the back which feed air automatically through a valve and mouthpiece to the diver as he requires it.

Also *attrib.* Hence **a·qualunger** (-ŋ-), a user of the aqualung; **a·qualung** *v. intr.*, to use this equipment; **a·qualunging** *vbl. sb.*
1950 Sci. News Let. 11 Nov. 306/3 The underwater photographer is equipped with an 'Aqualung', an automatic compressed air, self-contained diving unit. 1952 J. Y. Cousteau in Nat. Geogr. Mag. Oct. 434 Moving in a dense element, Aqualungers must learn to be languid. 1953 Cousteau & Dumas Silent World i. 1 In June 1943..a new and promising device..an automatic compressed-air diving lung conceived by Émile Gagnan and myself...No children ever opened a Christmas present with more excitement than we did when we unpacked the first 'aqualung'. Ibid. 96 (caption) Auguste Piccard, inventor of the Bathyscaphe, the undersea dirigible, tries aqualunging at Dakar. 1959 Elizabethan June 21/1 Fishermen are still bringing up old cannon and bits of bronze from the great galleons of the Spanish Armada, some awaiting excavation by aqualung divers. 1961 I. Fleming Thunderball xviii. 199 You oughtn't to aqualung by yourself.

aquamanile (æ:kwămānai·li, -ī·li). Also **aquæmanale.** [late L., ad. (rare) L. *aquaemanālis* basin for washing the hands, f. *aquae* (gen. sing. of *aqua* water) + *mānāle* ewer.] **a.** A water vessel or ewer, freq. made in the form of an animal or bird, etc. **b.** *Eccl.* A basin used for washing the hands at the celebration of Mass.
1875 S. Cheetham in Smith & Cheetham Dict. Chr. Antiquities I. 134/1 Aquamanile, the bason used for the washing of the hands of the celebrant in the liturgy. The aquamanile with the urceus are the bason and ewer of the sacred ceremony. 1889 Cent. Dict., Aquæmanale. 1910 R. Fry in Burlington Mag. Aug. 289/2 A late Sassanid aquamanile in the form of a bird. 1911 Encycl. Brit. XV. 544/2 In the 13th and 14th centuries a special type of metal ewer takes the form of animals, men on horseback, &c.; these are generally known as aquamaniles. 1939 Oxoniensia IV. 113 Sherd from a cylindrical cup, or perhaps from the body of an aquamanile, T. ⅛ in. 1956 Daily Tel. 5 Dec. 8/3 A bronze equestrian aquamanile..was used as a water vessel for church purposes. The water pours out from the aperture in the horse's forehead. 1957 Mankowitz & Haggar Conc. Encycl. Eng. Pott. & Porc. 8/1 Aquamanile, a ewer for handwashing in medieval times, usually in the form of a mounted figure or animal. 1962 Victoria & Albert Mus. Internat. Art Treasures Exhib. 74/2 A Dinanderie Aquamanile, of circular turret form,..supported on three human legs and feet.

aquanaut (æ·kwănǭt). [f. *aqua- + Gr. ναυτ-ικός, f. ναύτης sailor, ναῦς ship; cf. *astronaut.] An underwater 'explorer' or swimmer.
1881 W. D. Hay 300 Years Hence vii. 132 The small apparatus fixed to the back of every aquatic helmet, and..called in the early days 'the aquanaut's lungs'. 1964 Discovery Oct. 9/3 The experiences of these 'Aquanauts' vividly underlined the need for new underwater tools. 1965 Observer 28 Mar. 13/1 'Thermal protection'..is..a serious problem for 'aquanauts'. 1966 Daily Tel. 31 Oct. 9/8 Each 'aquanaut'..has small measured weights sewn into the arms and legs of his special diving suit.

aquaplane (æ·kwăplēⁱn). orig. *U.S.* [f. *aqua- + Plane *sb.*³] A board which rides on the surface of the water when towed with its rider by a speedboat. Hence *v. intr.*, to ride standing on such a board. Hence also **a·quaplaner**, a rider of an aquaplane; **a·quaplaning** *vbl. sb.*, the sport of riding an aquaplane; also [cf. *plane *v.*²], 'gliding' or skidding out of control on water (quots. 1961, 1963).
1914 Outing (U.S.) May 143/2 The wonderful sensation of shooting through the air and skimming the surface of the water may be enjoyed..by the use of the waterboard or aquaplane. Ibid. 147/2 With these cautions strictly heeded, the sport of aquaplaning at once becomes as thoroughly safe as it is wonderfully exciting. 1915 Harper's Weekly 9 Oct. 347/1 When the aquaplane made its first public appearance two years ago, the machine was a long, narrow board which could be used, without guide ropes, behind boats whose speed did not exceed 15 miles an hour. It was about 12 feet long, a foot or so wide, and was marked to show where to stand at various speeds. 1928 Daily Express 14 July 4/5 There will be..a women's aquaplane race and various other 'speedbug' activities. Ibid. 24 Aug. 15/2 Lady Cunard..is often to be seen admiring the prowess of aqua-planers or shooting out to the islands in a rapid motor-boat. 1941 F. E. Baume I lived these Years viii. 183 Lady Catherine Manley..aquaplaned behind a launch from Boulogne to Folkestone. 1961 Aeroplane CI. 34/1 The phenomenon of 'aqua-planing', in which an aircraft can ride down a wet runway supported on a film of water with the effectiveness of the wheel brakes reduced virtually to zero. 1963 Guardian 18 Dec. 3/5 The phenomenon known as 'aqua-planing', when a car tyre leaves the surface of a wet road... There is a speed at which the tyre can no longer disperse the water, so that it 'aquaplanes'.

aquarist (æ·kwărist, *U.S.* ăkwēᵊ·rist). Also formerly **aqua·riist** (*U.S.*). [f. Aquar(ium + -ist.] One who keeps an aquarium.
1893 Funk's Stand. Dict., Aquariist, the manager or keeper of an aquarium. 1920 Webster, Aquariist or aquarist. 1924 (title) The Amateur Aquarist. 1958 Times Lit. Suppl. 10 Jan. 23/5 The object of this book is to enable the aquarist to identify most of the tropical fishes available for culture in this country.

aquarium. Add: Also *attrib.* and *Comb.*
1875 Encycl. Brit. II. 217/2 An impetus to the popular aquarium movement. 1900 E. Glyn Visits of Eliz. 27 A regular aquarium specimen of turbot sat on its dish opposite him. 1931 A. Huxley Cicadas 44 Oh, how aquarium-still, how brooding-warm This paradise! 1941 Blunden T. Hardy ix. 201, I must lay by A Pair of Blue Eyes as one of the works of Hardy which have at best aquarium value.

aquatinted (æ·kwă,ti:ntėd), *ppl. a.* [f. Aquatint *v.* + -ed¹.] Etched in, or as in, aquatint.
1880 Hardy Trumpet-Major II. xxviii. 271 The rain decreased, and the lovers went on..aqua-tinted by the weak moon and mist. 1931 Blunden Votive Tablets 186 External nature, aquatinted with remarkable sense of solitude. 1958 Observer 10 Aug. 11/3 A strangely aqua-tinted telerecording.

aquavit (ækwăvī·t). Also **akvavit** (akvavī·t). [ad. Norw., Sw., Da. *akvavit* Aqua-vitæ.] A colourless or yellowish alcoholic spirit distilled from potatoes or other starch-containing plants; the 'schnapps' of Scandinavia.
1890 T. Bennett Handbk. Travellers in Norway (ed. 26) 36 Those who suffer from cardialgy (heartburn) and indigestion are recommended to take the Norwegian 'Akvavit'—a kind of brandy flavoured with carraway seeds, which must be drunk without being mixed with water. 1896 Wine List (T. W. Stapleton & Co.) Dec., Aquavit (Norwegian)—54/-. 1920 G. Saintsbury Notes on Cellar-Book viii. 130 The Norwegian 'Aquavit', which used to flow so freely. 1936 New Yorker 7 Mar. 74/4 A Swedish Aquavit which is remarkably soft and smooth. 1964 Punch 15 Jan. 92/2 Fortified by a full potion of akvavit.

aqueo-. Add: **aqueo-glacial** *a. Geol.*, formed or acted upon by glacial ice or water; **aqueo-igneous**: substitute for def. 'of, pertaining to, or resulting from the joint action of heat and water', and add examples.
1892 W. M. Davis in Proc. Boston Soc. Nat. Hist. XXV. 321 Aqueo-glacial gravels are as inconspicuous as deposits of till..but they are seen in the Catskill trench. 1880 J. D. Dana Man. Geol. (ed. 3) IV. v. 744 The view that the fusion of lavas is due to the combined action of moisture and heat, or is aqueo-igneous, was early presented by Scrope. 1920 Discovery Apr. 110/1 The earth's surface was originally in a state of aqueo-igneous fusion.

aqueous, *a.* Add: **4.** as *sb.* Short for *aqueous humour*: see 1 b.
1879 St. George's Hospital Rep. IX. 481 During one week she had the aqueous tapped daily.

aquiculture. Add: (Later example.) Cf. *aquaculture.
1947 Endeavour VI. 7 Aquiculture, or the cultivation of aquatic life, has been conducted on a very limited scale compared with agriculture.

aquifer (æ·kwifəɹ). *Geol.* Also **aquafer** (irreg. form). [f. L. *aqui-*, combining form of *aqua* water + *-fer* bearing (f. *ferre* to bear).] A water-bearing or aquiferous stratum.
1901 Science 22 Nov. 794/1 The artesian system shows four or five aquifers, or water-bearing strata, more or less completely separated from one another. 1939 Nature 12 Aug. 274/2 The limestones are important aquifers for artesian and subartesian water. 1962 Engineering 9 Mar. 332 The whole system of aquifer storage can be likened to a U-gauge with one open limb and a short closed limb. 1966 Listener 1 Sept. 309/2 There are only two methods of [water] storage. in surface reservoirs and in the aquifers from which we pump a part of our supplies.

aquilegia (ækwilī·dʒ'ă). *Bot.* [mod.L., f. as Aquilege.] A plant of the genus of this name consisting of several species of ranunculaceous plants, the flowers of which are pentamerous with spurred petals; = Columbine *sb.*² 1.
The genus-name *Aquilegia* is in Lyte 1578, Gerarde 1597, etc.
1871 W. Robinson Hardy Flowers i. ix. 31 The larger type of alpine and herbaceous plants, beginning with such as the Aquilegias, and rising to the finer Phloxes. 1920 Chambers's Jrnl. May 326/2 Old-world flowers..as delphinium and antirrhinum and aquilegia. 1931 H. H. Thomas Amateur's Week-end Gard. Book 100 The Columbine or Aquilegia is one of the most delightful of early summer flowers. 1934 G. A. Phillips Aristocrats of Flower Border viii. 109 Aquilegias are mid-border plants.

Arab, *sb.* **1.** Read: One of the Semitic race inhabiting Saudi Arabia and neighbouring countries. Add later examples.
1893 F. Adams New Egypt 65 All, or nearly all, retain a sense of their superiority to the 'Arabs' (under this term, which is quite inaccurate, of course, except in so far as it alludes to the speech, it is usual to include all the Egyptians). 1902 [see *ethnic *sb.* 2]. 1936 Discovery June 172/2 In Africa the term 'Arab' is commonly applied to any people professing Islam, however much Negro, Hamitic, or other foreign blood may run in their veins.
2. (Earlier examples.)
a 1666 Evelyn Diary 25 Oct. an. 1644 (1955) II. 195 A stable of incomparable Horses of all Countries, Arabs, Turks, Barbs, [etc.]. 1844 Disraeli Coningsby iii. iii. 274 'A fine Arab, the finest in the world!' said the Duke, who was very fond of horses.

araban (æ·răbæn). *Chem.* [a. G. *araban*, f. Arab(ic *a.* + -an.] = Arabin.
1892 Jrnl. Soc. Chem. Industry XI. 931/1 The Pentosans (Wood-Gum, Xylan, and Araban) of Lignified Fibre. 1938 Thorpe's Dict. Appl. Chem. II. 287/1 The pentoses do not occur as such in plants, but in the form of condensation products of high molecular weight, termed 'pentosans' (araban or xylan). 1943 Ibid. VI. 155/2 Araban..yields arabinose, galactose and glycuronic acid on hydrolysis.

Arabdom (æ·răbdəm). [f. Arab *sb.* + -dom.] Arabs collectively; the state or condition of being an Arab.
1949 International Affairs 475 The fate of Arabdom. 1957 Listener 24 Oct. 663/1 Distinguished Englishmen who were drawn to Arabdom by a kind of elective affinity. Ibid., Jordan east of the river was somehow different from all Arabdom. 1958 [see *Arabism 2].

arabesque, *sb.* Add: **4.** *Ballet.* A pose in which the dancer stands on one foot with one arm extended in front and the other arm and leg extended behind.
1830 R. Barton tr. C. Blasis' Code of Terpsichore ii. v. 74 Nothing can be more agreeable to the eye than those charming positions which we call arabesques, and which we have derived from antique basso relievos, from a few fragments of Greek paintings, and from the paintings in fresco at the Vatican, executed after the beautiful designs of Raphael. 1911 J. E. C. Flitch Mod. Dancing iii. 42 One of her [sc. Marie Taglioni's] most wonderful attitudes was an arabesque which gave her the appearance of actually flying. 1928 A. L. Haskell Some Stud. in Ballet 151 Everything in it depends on line, absolute precision of movement and the purity of the arabesque.
5. *Mus.* (See quot. 1879.) Also *transf.*
1864 Cramer, Beale & Wood's New Eds. Piano Forte Works Misc. Ser., No. 4 (title) Arabesques, by R. Schumann. 1879 Grove's Dict. Mus. I. 80/2 Arabesque..(1) The title has been given..by Schumann to one of his pianoforte pieces.., which is written in a form bearing some analogy to that of the rondo, and it has been since occasionally used by other writers for the piano. (2) The word 'Arabesque' is sometimes used by writers on music to express the ornamentation of a theme. 1924 A. D. Sedgwick Little French Girl iii. vii, Listening to a blackbird that fluted golden arabesques.

arabe·sque, *v.* [f. Arabesque *sb.*] *trans.* To ornament in arabesque. Hence **arabe·squing** *ppl. a.*
(Delete Arabesqued *ppl. a.*, which is the pa. pple. of the vb., and incorporate the quots.)
1858 [see Dict.]. 1861 Sala in Temple Bar I. 306 The same embroideress who arabesqued the hems of her under-skirts pinked the shrouds. 1867 H. Latham Black & White 21 A cell in which a weaver had arabesqued the walls. 1877 [see Dict.]. 1908 Saintsbury in Camb. Hist. Eng. Lit. II. 191 A sort of vignetting or arabesquing fringe and atmosphere of exaggeration and fantasy.

Arabian, *a.* and *sb.* Add: **A.** *adj.* *Arabian horse* = Arab *sb.* 2.
1737 S. Berington Mem. G. di Lucca 47 The Arabian Horses are the best in the World, tho' not very large. 1807 'Cervantes Hogg' Rising Sun III. xvi. 166 The prince,..mounted on an Arabian horse as fleet as the wind, followed with the utmost eagerness.
B. *sb.* **b.** = *Arabian horse.
1770 H. Brooke Fool of Qual. V. 299 Her six spotted Arabians. 1792 A. Young Trav. in France 16 There are all kinds of horses, but chiefly Arabian, Turkish and English. Three years ago four Arabians were imported. 1824 M. R. Mitford Our Village I. 101 That gay, gallant boy, on the gallant white Arabian. 1966 R. Stout Death of Doxy (1967) ix. 94 She liked horses and had four Arabians.

Ara·bianize, *v.* [f. Arabian *a.* + -ize.] *trans.* To make Arabian, give an Arabian character to; to assimilate to the Arabian language.
1893 F. Adams New Egypt 9 The fiercely protracted effort of Islam to Arabianise Egypt.

arabica (ăræ·bikă). Used *ellipt.* for the tree *Coffea arabica* (see Coffee 3) or for coffee obtained from this tree. Also *attrib.*
1922 W. H. Ukers All about Coffee xx. 201 Strong, chilly winds and intensely hot sunlight are foes of coffee trees, especially of the arabica variety. Ibid. 215 Of this area, 110,903 acres were planted with robusta, 15,314 acres with arabica. 1959 Times 22 Sept. 11/6 The Southern Cameroons, where crops of both arabica and robusta are grown. 1964 Economist 28 Mar. 1198/1 The high standards of cultivation required for Arabica coffee.

arability (ærăbi·liti). [f. Arable *a.*: see *-bility.] Capability of being used as arable land.
1879 Athenæum 28 June 817/2 The term [sc. hide] being ..variable according to the arability..of the land. 1963 Spectator 22 Feb. 224 The arability of its [Venezuela's] llanos.

arabis. Add: A plant of the genus.
1870 W. Robinson Alpine Flowers i. 48 The dwarf green Iberises,..Arabises, [etc.]. 1876 Rhoda Broughton Joan v, Milk-white arabis haunted by the drowsy booming bees. 1905 Westm. Gaz. 10 June 16/1 The white blossoms of the double arabis.

Arabism. Delete ? *Obs.* and add examples.
1898 Sayce Early Israel iii. 109 Hebrew had retained a few 'Arabisms', a few traces of its ancient contact with Arabic-speaking tribes. 1958 Archivum Linguisticum X. 26 The Arabism bellota, absent from Portuguese.

2. The state or condition of being an Arab; the influence of Arab culture. Also, Arab nationalism, sympathy with Arab movements of self-assertion.

1874 Draper *Hist. Conflict Relig. & Sci.* iii. 99 Arabism, which had done so much for the intellectual advancement of the world, came to an end when the Turks and the Berbers attained to power. **1952** *Middle East Jrnl.* VI. 473/2 Arabism, now the creed of a large number of people, is not the only political philosophy in vogue in the Arab world. **1958** *Times* 16 July 11/6 *'Aruba*, 'arabism' or 'arabdom', the word which is on the lips of every politician and the pen of every journalist, means the fact of being an Arab. The positive elements of arabism are not so much racial as a sense of common history, and, much more potent, the existence of a common language.

Arabize (æ·răbəiz), *v.* [f. Arab + -ize.] *trans.* To make Arab; to give an Arab or Arabic character to. Hence **A·rabized** (-əizd), **A·rabizing** *ppl. adjs.*, **Arabiza·tion.**

1883 Cust *Mod. Lang. Africa* I. viii. 83 The Arabizing process [in Africa] has taken place in various ways. **1884** *Science* 12 Dec. 531/1 These Arabs of the Sudan are not true Arabs, but to a great extent merely Arabized negroes. **1898** Sayce *Early Israel* iii. 114 The dialect of Edom agreed with Hebrew in those Arabising peculiarities. **1920** *19th Cent.* Aug. 229 The civil administration of the country should be Arabised to a greater extent. **1931** C. G. Seligman in W. Rose *Outl. Mod. Knowl.* 466 The prestige of the dominant religion..accounts for such arabisation as has taken place. **1937** *Times* 20 Oct. 15/6 They [sc. the Arab leaders] regard..the Zionist Jew, who..refuses to be Arabized, as an obstacle to that unity.

arachidic, *a.*: see Arachis.

arachidonic (ăræ:kidǫ·nik), *a. Chem.* [f. Arachid(ic *a.* + *-onic.] *arachidonic acid*: an unsaturated fatty acid found in animal fats.

1913 J. Lewkowitsch *Chem. Technol. of Oils* I. iii. 211 Arachidonic Acid, $C_{20}H_{32}O_2$. The existence of an acid of this composition is inferred from the formation of octo-bromo-arachidic acid... As Hartley has not named this acid the author suggests the term *arachidonic acid*. **1924** L. G. Wesson in *Jrnl. Biol. Chem.* LX. 183 Arachidonic acid, so named by Lewkowitsch, is the tetra unsaturated, normal, aliphatic C_{20} acid, which is present in brain tissue. **1964** J. A. Lovern in H. Barnes *Oceanogr. & Marine Biol.* II. 187 The mammal can convert linoleic acid..into the essential arachidonic acid (eicosa-5,8,11,14-tetraenoic acid), without which it cannot survive.

arachin (æ·răkin). [f. Arach(is + -in[1].] A globulin found in arachis oil, groundnuts, etc.

1905 G. M. Gould *Dict. New Med. Terms* 88/2 *Arachin*, $C_{20}H_{40}O_2$, a glycerid of arachic acid. It occurs as the chief constituent of Rambutan tallow obtained from the seeds of *Nephelium lappaceum*. **1913** J. Lewkowitsch *Chem. Technol. of Oils* I. i. 27 *Triarachin, Arachin*, $C_3H_5(O.C_{20}H_{39}O)_3$, (prepared by *Berthelot* from diarachin and arachidic acid) is ..soluble... Arachin occurs in arachis oil, and in smaller quantities in rape oil and butter fat. **1946** *Nature* 5 Oct. 475/1 Hitherto..it has merely been possible to convert globular proteins, such as casein, arachin and egg albumin, into the fibrous form.

Arachis, arachidic. Add: (Examples.) arachis oil, peanut oil.

1853 Mayne *Expos. Lex.* 80/2 Arachis. **1866** W. Odling *Lect. Anim. Chem.* 31 Monatomic Fatty Acid Series.. $C_{20}H_{40}O_2$ Arachidic. **1875** Gamgee tr. *Hermann's Physiol.* 13 The series of the fatty acids at present known includes.. arachidic acid. **1876** *Encycl. Brit.* IV. 151/2 In some Leguminosæ, as Arachis, Cathartocarpus Fistula, and the Tamarind, the fruit must be considered a legume, although it does not dehisce. **1889** *Cent. Dict.* s.v., *Arachis-oil*, the oil expressed from the seeds of *Arachis hypogæa*, the fine limpid nut-oil of commerce. **1900** Arachis oil [see *peanut]. **1916** *Chambers's Jrnl.* May 313/1 The arachis nut (popularly known as the 'monkey nut').

arachnitis: see *arachnoiditis.

arachnoid, *a.* **2.** (Earlier example.)

1789 *Loiterer* 26 Sept. No. 35, p. 9 The dura and the pia Mater, and the arachnoid coat, entirely consisted of the Wings of Moths.

arachnoidi·tis. *Path.* [f. Arachnoid + -itis.] Inflammation of the arachnoid membrane. Also in less correct form **arachni·tis**.

1827 J. Forbes tr. *Laennec's Dis. Chest* (ed. 2) i. i. 71, I have witnessed the supervention of peritonitis, severe dysentery, and arachnitis, to fluxes suppressed by the use of hot wine and spices. **1857** Dunglison *Dict. Med. Sci.* (ed. 15), *Arachnitis*, more properly *Arachnoiditis*. **1910** *Practitioner* July 48 Sometimes the effusion is localised and secondary to atrophy of the convolutions, and sometimes it may be general and due to serous arachnoiditis (Meningitis serosa externa). **1961** *Lancet* 22 July 178/1 Spinal arachnoiditis, and weakness in the legs were noted.

Aragonese (ærăgǫnī·z), *a.* and *sb.* Also 9 **Arragonese**. [ad. Sp. *aragonés*, f. *Aragón* name of a region and former kingdom of north-eastern Spain; see -ese.] **A.** *adj.* Of or pertaining to Aragon, or its inhabitants, or their speech. **B.** *sb.* An inhabitant of Aragon; the dialect of Spanish spoken in Aragon.

1513 R. Fox *Let.* 16 May (1929) 63 The treuxe..is..proclamed bothe in Spaigne and France bytwyxt the Frenche and Aragonesse Kyngis. **1823** C. Butler *Contin. Lives Saints* App. p. xvi, Charles was acknowledged by his Arragonese subjects. **1823** T. Ross tr. *Bouterwek's Hist. Span. Lit.* I. 7 The geographical boundaries, which..separated the Portuguese from the Castilians, and the latter from the Arragonese. **1843** Borrow *Bible in Spain* I. xiii. 266, I went to the secretary, whose name was Oliban, an Aragonese...'Then His Excellency cannot give you permission,' said the Aragonese secretary. **1932** W. L. Graff *Lang. & Langs.* x. 377 Spanish group, with the Castilian, Andalusian, Aragonese..dialects. **1939** S. de Madariaga *Columbus* xxv. 309 The future Catalina..let her Aragonese lover know of the mines.

Arahat: see *Arhat.

Arakanese (ærăkănī·z), *a.* and *sb.* Also 9 **Ar(r)acanese.** [f. *Arakan* (see def.) + -ese.] **A.** *adj.* Of or pertaining to Arakan, a district on the west coast of Burma, or its inhabitants. **B.** *sb.* **1.** An inhabitant of Arakan. **2.** The Burmese dialect spoken by the people of Arakan.

1820 A. Judson *Diary* 20 Feb. in F. Wayland *Mem. A. Judson* (1853) I. 211 Arracanese, who speak a language similar to the Burman. *Ibid.* 26 Feb. 213 Collect the Arracanese converts. **1882** in E. Forchhammer *Notes Lang. & Dial. Burma* (1884) 6 The Burmese-Chin language-family comprises the Arakanese, Tavoy (?), Burmese, Chin. **1910** *Encycl. Brit.* II. 315/2 The Arakanese are of Burmese origin..and they have a dialect and customs of their own. **1934** 'G. Orwell' *Burmese Days* i. 2 One of those vivid Arakanese longyis..which the Burmese wear on informal occasions.

aralia (ărē̆ⁱ·liă). [mod.L., of uncertain origin.] A genus of trees, shrubs, and herbs, the type of the family *Araliaceæ*; also, a plant of this genus.

1756 P. Browne *Civ. & Nat. Hist. Jamaica* ii. ii. 189 Aralia..The *Galapee*, or Angelica Tree. **1829** J. C. Loudon *Encycl. Plants* 230 *Aralia*. Aralia. *Araliaceæ*. **1897** [see Ginseng 2]. **1908** *Westm. Gaz.* 25 Nov. 5/1 A tall aralia growing in a corner of the well-lighted entrance-hall of the hotel. **1914** *Chambers's Jrnl.* Mar. 154/1 The fertilisers may be given in weak doses to foliage plants such as aralias. **1942** L. Bennett (*Jamaica*) *Dial. Verses* 25, I was gwine to sen' pure orellia bush, As yuh wanted so-soh greens. **1953** *West Indian Med. Jrnl.* II. 237 Aralia guilfoylei..Aralia. The leaves are used in Jamaica to prepare tea for colds. It is common in gardens as a hedge plant and is not indigenous.

araliaceous (ărēⁱliēⁱ·ʃəs), *a. Bot.* [f. mod.L. *Araliaceæ*, f. *Aralia*, the typical genus: see -aceous.] Belonging to the family *Araliaceæ*, comprising various herbs, shrubs, and trees, including the ivy.

1866 Brande & Cox *Dict. Sci. & Art* II. 780/1 *Panax*, a name applied to some plants of the Araliaceous order. **1884** *Spectator* 24 May 685 Virtues..attributed to the araliaceous plant Jinseng.

Aramaicism (ærămē·isiz'm). [f. Aramaic + -ism.] An Aramaic idiom; = Aramaism.

1898 Blass *Philology of the Gospels* xi. 194 In the [first twelve chapters of the Acts] Aramaicisms abound.

Aramaize (ærămē·əiz), *v.* [f. as prec. + -ize.] *trans.* To render Aramaic, imbue with Aramaisms.

1868 Lightfoot *Ep. to Philippians* 147/2 In Aramaised Greek.

Aran (æ·răn), *a.* Of, pertaining to, or characteristic of the Aran Islands, off the west coast of Ireland; *spec.* designating a type of patterned knitwear. So **A·raner**, a native or inhabitant of the Aran Islands.

1902 *Irish Rosary* Jan. 21/2 The Araners of to-day..are deeply religious. **1919** W. B. Yeats *Two Plays for Dancers* 2 He seems an Aran fisher. *Ibid.* 3 Until an Aran coracle puts in At Muckanish. **1962** *Times* 1 Jan. 13/1 A great deal of 'Aran knitwear' is now sold that never saw the three limestone islands. **1963** *Guardian* 30 Oct. 6/5 Aran patterns, fisherman's ribs, cables, and lace stitches.

Aranda (ăræ·ndă). Also **Aranta, Arunta,** †**Aralta.** [Native name.] **A.** *sb.* **1.** An aboriginal people of central Australia; a member of this people. **2.** The language of this people. **B.** *adj.* Pertaining to this people or their language.

1891 W. H. Willshire *Aborigines of Central Australia* iv. 16 The 'Aralta' draw peculiar designs on flat upright rocks. **1896** W. B. Spencer *Horn Sci. Exped. Central Australia* I. i. 39 These natives belong to the Arunta tribe, which occupies a large tract of land stretching from the Macumba Creek in the south to about seventy miles north of Alice Springs. **1910** *Encycl. Brit.* I. 947/1 The Aruntas.. believe that local spirits of trees..enter women as they pass by their haunts. **1927** Spencer & Gillen (*title*) The Arunta: a study of a Stone Age people. **1951** R. Firth *Elem. Soc. Organiz.* i. 3 The Aranda of Central Australia practise sub-incision. **1956** J. Whatmough *Lang.* ii. 33 Aranta in Australia, which has practically no formally distinctive 'parts of speech'. **1959** S. H. Courtier *Death in Dream Time* iii. 29 The Aranda tribes and..their language, ways, customs, ceremonies and myths.

araneidal (ærănī·idăl), *a. Zool.* = Araneidan *a.* So **arane·idiform** *a.* = Araneiform *a.*

1826 Kirby & Spence *Entomol.* III. xxx. 171 The *Corydalina*..is Chilopodiform, but with a tendency to the Araneidiform Type. **1895** *Naturalist* Jan. 29 The Araneidal Fauna of the northern counties of England.

araneology (ărē̆ⁱni,ǫ·lŏdʒi). [f. L. *arānea* spider + -(o)logy.] The department of zoology relating to spiders. Hence **ara:neolo·gical** *a.*, belonging to araneology; **araneo·logist,** one versed in araneology.

1798 *Monthly Mag.* Jan. 53/2 Araneology...The Araneological Calendar. **1875** *Encycl. Brit.* II. 296/1 An eminent araneologist. **1884** *Science* 4 July 24/1 Facts..new to the field of American araneology.

Arapaima (ærăpai·mă). [South American native name.] A genus of fishes of the family Osteoglossidæ, remarkable for their size; (with small initial) a fish of this genus.

1840 R. H. Schomburgk *Brit. Guiana* 39 The Arapaima or Pirarucu (*Sudis Gigas*), and..the Lau-lau, are from ten to twelve feet long. **1896** Lydekker *Roy. Nat. Hist.* V. 478 The true arapaima (*Arapaima gigas*) of the larger rivers of Brazil and the Guianas..occupies the proud position of being the largest fresh-water bony fish. **1908** C. F. Holder *Big Game at Sea* xix. 288 The *arapaima*..the game fish of South American waters—a monster that attained a length of twelve feet and a weight of twelve hundred pounds. **1955** *Times* 11 May 20/1 Capture of an arapaima fish at Karanambo, Rupununi.

araponga (ærăpǫ·ngă). Also **arapunga.** [F., f. Pg.] The campanero or bell-bird of South America.

The name *arapunga/araponga* was established in R. P. Lesson's *Traité d'Ornithologie* (1831).

1824 H. E. Lloyd tr. *Spix & Martius' Trav. in Brazil* I. ii. ii. 247 Above all these strange voices, the metallic tones of the uraponga sound from the tops of the highest trees, resembling the strokes of the hammer on the anvil, which.. according to the position of the songster, fill the wanderer with astonishment. **1881** *Imp. Dict.*, Arapunga. **1910** *Encycl. Brit.* IV. 443/2 Species having strident voices and peculiar unmusical calls, like..the *araponga* (*Chasmorhynchus nudicollis*). **1951** J. C. Fennessy *Sonnet in Bottle* iii. ii. 62 An araponga, a tall bird with a screeching note.

Arawak (æ·răwæk, -ăk), *sb.* and *a.* Also †**Arrowauk,** †**Arowack** and variants. **A.** *sb.* **1.** A group of Indian peoples in South America; a member of one of these peoples. **2.** The language of these peoples. **B.** *adj.* Of or pertaining to the Arawaks or their language. Hence **Arawa·kan** *a.* and *sb.*

1769 E. Bancroft *Nat. Hist. Guiana* iii. 253 The Arrowauks..are a friendly, hospitable people. *Ibid.* 304 An Indian ..of the Arrowauk tribe. **1851** R. G. Latham *Ethnol. Brit. Colonies* vi. 259 The Arawaks are our nearest neighbours, and, consequently, the most Europeanized. **1868** W. H. Brett *Indian Tribes of Guiana* vi. 117 These names respectively signify in Arawâk, 'the resorts of the Ituri', [etc.]. **1910** *Encycl. Brit.* II. 322/2 The Arawaks have given their name to a linguistic stock of South America, the Arawakan, which includes many once powerful tribes. The Arawakans were once numerous. **1932** W. L. Graff *Lang. & Langs.* xi. 431 Arawakan continued to be used only by the females and Carib by the males. **1937** R. H. Lowie *Hist. Ethnol. Theory* viii. 89 The discovery of unsuspected Arawak and Carib tribes in the interior of Brazil. **1965** *Canad. Jrnl. Linguistics* Spring 101 Arawakan and other Caribbean area groups.

arbiter. **4.** ‖*arbiter elegantiarum, elegantiæ* [L., lit. 'judge of elegance': Petronius Arbiter was the *elegantiæ arbiter* of Nero's court (Tacitus *Ann.* xvi. 18)], a judge of matters of taste, an authority on etiquette.

1818 Lady Morgan *Fl. Macarthy* II. iii. 175 He looked up to Lord Frederick Eversham, as the *arbiter elegantiarum* of that system. **1841** Craik & MacFarlane *Pict. Hist. Eng. Geo. III* I. 651/1 Derrick..succeeded Nash as *arbiter elegantiarum* at Bath. **1933** Balmer & Wylie *When Worlds Collide* v. 49 A connoisseur of life and living—an *arbiter elegantiae*. **1957** R. N. Carew Hunt *Guide Communist Jargon* xxi. 76 Zhdanov was appointed Stalin's *arbiter elegantiarum* in the late 'forties.

arbitrable, *a.* Delete †*Obs.* and add later U.S. examples.

1896 *Columbus* (O.) *Even. Dispatch* 18 July 1/4 Secretary Olney points out in support of his amendments that they make all disputes prima facie, arbitrable. **1915** T. Roosevelt *America & World War* xi. 237 They would not be made arbitrable, any more than an individual's right to life and limb is made arbitrable.

arbitrage. **3.** (Earlier example.)

1875 *Encycl. Brit.* II. 311/2 The great Government loans are..the natural subject-matter of arbitrage.

arbitrage (ā·ɹbitrāʒ), *v.* [f. the sb.] *intr.* To engage in arbitrage (see Arbitrage 3). Hence **a·rbitraging** *vbl. sb.*

1900 S. A. Nelson *ABC of Wall Street* 126 *Arbitraging*, trading in two markets in order to profit by the difference in prices. **1923** J. M. Keynes *Tract on Monetary Reform* iii. 129 The surcharge representing the profit of a bank for arbitraging between spot and forward transactions may much exceed the moderate figure indicated above. **1953** *Economist* 16 May 461/1 Possible for authorized banks in these countries to arbitrage between the six currencies concerned.

arbitrageur (āɹbitraʒō̆r). *Comm.* Also in anglicized form **a·rbitrager.** [Fr.] (See quot. 1875.)

1870 O. Haupt *London Arbitrageur* p. iv, To furnish the experienced *arbitrageur* with a number of valuable hints. **1875** *Encycl. Brit.* II. 311/1 Arbitrage properly known as such, is the business of an arbitrageur, who is almost always a member of a Stock Exchange or 'Bourse', and his arbitrations..are..in Government and other stocks and shares. **1923** J. M. Keynes *Tract on Monetary Reform* iii. 128 If in these conditions the purchasers of forward dollars, other than arbitragers, exceed sellers of forward dollars, then this excess of demand for forward dollars can be met by arbitragers. **1948** G. Crowther *Outl. Money* (ed. 2) vii. 219 These rates are all in equilibrium, there are no divergences out of which an arbitrageur can make a profit.

arbitral, *a.* **1.** (Later U.S. example.)

1911 *Amer. Year Book 1910* 103 The most significant event of the past year in the development of international arbitration has been the virtual establishment..of the New Court of Arbitral Justice.

arbitrary, *a.* Add: **5.** *Printing. arbitrary character*: a character used to supplement the letters and accents which constitute an ordinary fount of type. Also as *sb.*

1829 [see Character *sb.* 3 a]. **1890** (*title*) Clarendon Press Inventory of Accents, Arbitrary Characters, &c. compiled April 1890. **1900** H. Hart *Notes Century Typogr.* 139, I thought it unnecessary to 'set' the matrices for all the arbitrary characters. **1928** *Periodical* 15 Feb. 17 The variety of type used, the many languages involved, and the multiplication of 'arbitraries' have demanded technical knowledge and minute accuracy to an extent probably unequalled in any other work.

arbitra·tional, *a.* [See -al.] Pertaining to, involving, or effected by arbitration.

a **1889** A. Hayward *Ethics of Peace* (Cent. D.), Arbitrational settlement of the Alabama claims. **1939** *Mind* XLVIII. 161 Doing nothing to enforce a standard of behaviour but merely giving arbitrational decisions between conflicting parties.

arbitrationist (āɹbitrēⁱ·ʃənist). [f. Arbitration + -ist.] One who is in favour of arbitration.

1884 *Pall Mall Gaz.* 24 June 3/1 Fine doctrine this for the arbitrationists! **1902** *Q. Rev.* Oct. 667 As an arbitrationist, he has been a hero with the emotional Radicals.

arbitrative (ā·ɹbitrēⁱtiv), *a.* [f. Arbitrate *v.* + -ive.] Having power to arbitrate; done by arbitration.

1831 Waddington *Hist. Church* I. ii. xiii. 221 The arbitrative authority of the Primitive Bishops was tolerated ..by the Pagan Emperors. **1875** R. J. Hinton *Eng. Radical Leaders* 117 He urged arbitrative tribunals as one of the better modes of treatment.

Arbor Day (ā·ɹbəɹ dēⁱ). orig. *U.S.* [L. *arbor* tree. Cf. Arbor[2].] A day set apart by law, orig. in the state of Nebraska, afterwards observed throughout the U.S. and adopted in Australia, Canada, and New Zealand, to be yearly observed for the planting of trees.

1872 in *Encycl. Brit.* (1902) XXXI. 112/2 Resolved that Wednesday, the 10th day of April 1872, be..set apart..for tree-planting in the state of Nebraska, and the State Board of Agriculture hereby name it Arbor Day. **1888** *Daily Inter-Ocean* 14 Mar. (Farmer), Governor Oglesby has designated April 13 as Arbor Day in Illinois. **1892** *Congress. Rec.* June 5404/2 In every quarter of the country 'arbor days' are days named by law, and also by custom, for planting forest trees to make lumber for the generations yet to come. **1902** *Encycl. Brit.* XXXII. 731/2 In order to encourage tree-planting [in S. Australia], a yearly school holiday devoted to this purpose, and known as Arbor Day, was established in 1886. **1967** *N.Z. Listener* 28 July, In New Zealand..Arbor Day is celebrated on the first Wednesday in August.

arboricidal (āːɹbŏrisəi·dăl), *a.* [f. L. *arbor* tree + -cid(e 2 + -al.] Given to cutting down trees; tree-felling.

1866 Blackmore *Cradock Nowell* xii, Condemnation of this arboricidal age. **1877** E. H. Yates *Celebrities at Home* I. 50 The arboricidal tastes of the Master of Hawarden.

arboricide (aɹbōᵊ·risəid). [See prec.] The wanton destruction of trees.

1899 H. G. Graham *Social Life of Scotl. 18th Cent.* I. v. 199 This crime of arboricide was distressingly frequent. **1930** *London Mercury* Mar. 386 The charge of arboricide Mr. Lansbury was able to meet completely.

arboricultural, *a.* (Earlier example.)

1828 Steuart *Planter's Guide* 355 A proposal for the establishment of an Arboricultural Society in Scotland.. would be universally approved.

arboriculture. (Earlier example.)

1828 Steuart *Planter's Guide* i. 5 Arboriculture..will at length share the same distinction.

arborification (āɹbōᵊ:rifikēⁱ·ʃən). [f. L. *arbor* tree + -fication.] A tree-like or branching formation.

1809 J. Graham in *Philos. Mag.* XXXIII. 192 No description..can convey any idea either of the beauty or curiosity of these several icy arborifications. **1936** C. S. Lewis *Alleg. Love* vii. 304 The shield of Arthur is borrowed from the shield of Atlant,..Fradubio's arborification from Astulph's.

arborize, *v.* Add: *intr.* To make tree-like formations.

1907 *Practitioner* June 860 Some of the fibres from the posterior nerve-root pass directly into the posterior grey cornu, and arborise around cells at its base.

arborvirus (āɹbŏ̄ɹvəiə·rɒs). Also **arbovirus.** [f. Ar(thropod + Bor(ne *ppl. a.* + Virus.] A virus borne and transmitted by an arthropod. Also *attrib.*

1957 *Trans. N.Y. Acad. Sci.* XIX. 219 All known Arbor viruses are pathogenic for Swiss mice. **1959** C. E. Gordon Smith in *Brit. Med. Bull.* XV. 235/1 'Arborvirus'..is a convenient abbreviation but must not be thought of as having Linnean rank. **1962** *Times* 26 June 3/1 (Advt.), The immunochemistry of viruses, especially those of the pox-virus and arborvirus groups. **1965** C. Andrewes *Common Cold* ix. 77 The virus families..are:..(vi) Arboviruses.

arbutus. Add: **2.** *U.S.* With the pronunciation (aɹbiū̆·tɒs), applied to a North American ericaceous plant, *Epigæa repens*, called also *trailing arbutus, ground laurel*, and May-flower, and prized as a harbinger of spring.

1785 H. Marshall *Arbustrum Amer.* 42 *Epigæa repens*. Trailing arbutus. This grows naturally upon northern hills, or mountains, with trailing shrubby stalks. **1869** J. G. Fuller *Uncle John's Flower Gatherers* 14 It comes forth with softly perfumed garments, and we call it Trailing Arbutus. **1888** Whittier *Writ.* II. 35 The trailing arbutus, or mayflower, grows abundantly in the vicinity of Plymouth, and was the first flower that greeted the Pilgrims after their fearful winter.

arc, *sb.* **5.** Add further *attrib.* examples.

1894 *Phil. Trans. R. Soc.* A. CLXXXV. 983 The photographic arc spectrum of electrolytic iron. **1900** *Jrnl. Soc. Arts* XLVIII. 820/1 The multiple arc furnace..has lately become much used, because it enables the area of arc heating to be greatly extended. **1935** *Discovery* Sept. 260/2 The appearance in the arc-spectrum of the 'raies ultimes' of copper at λ 3274 and λ 3248 completed the identification of the pigment. **1940** *Chambers's Techn. Dict.* 47/1 *Arc furnace*, an electric furnace in which the heat is produced by means of an electric arc between carbon electrodes, or between a carbon electrode and the furnace charge. *Ibid.* 47/2 *Arc spectrum*, a spectrum originating in the non-ionised atoms of an element; usually capable of being excited by the application of a comparatively low stimulus, such as the electric arc. **1946** *Nature* 2 Nov. 631/1 Faulty power lines which are maintained in operation through the use of arc-suppression coils. **1956** *Ibid.* 28 Jan. 157/2 Resistance-type (calrod) heaters embedded in the cast-copper charge oven and arc-chamber support block.

arc (āɹk), *v.* Inflected **arced, arcing** (with *c* pronounced k). [f. Arc *sb.*] **1.** *intr.* To form an electric arc. So **a·rcing** *vbl. sb.*

1893 *Westm. Gaz.* 20 June 6/3 In 'arcing' or jumping from one conductor to another the current damaged the [gas] pipes. **1905** *Daily Chron.* 19 Apr. 6/5 The contact or arcing between the telephone cables and the conductor rail. **1908** *Installation News* II. 19/2 The fuse is provided with a phosphor bronze arcing tongue. **1910** *Ibid.* IV. 63/1 Fires ..due to the melting of soft compo gas-pipe, by arcing. **1943** *Gloss. Terms Electr. Engineering* (*B.S.I.*) 59 *Arcing contacts*, contacts which open after and close before the main contacts to protect the latter from injury by an arc.

2. To move or fly in an arc.

1954 J. Christopher *22nd Cent.* 116 Bouncing round in three complete circuits before arcing down to land. **1961** *Guardian* 4 May 4/2 The rocket arc-ed majestically into the cloudless sky.

‖**Arcades ambo** (ā·ɹkădīz æ·mbo). [L. phr. (Virgil *Ecl.* vii. 4), lit. 'both Arcadians', i.e. both pastoral poets or musicians.] Two persons of the same tastes, profession, or character (often derogatory).

1821 Byron *Juan* iv. xciii. p. 117 Each pull'd different ways with many an oath, 'Arcades ambo', *id est*—blackguards both. **1882** T. Mozley *Remin.* lxxxiii. II. 92 [Denison and Neate] were *Arcades ambo*. They could talk and chaff about anything...They were both good scholars, rather above the Oxford run. **1888** Rider Haggard *Colonel Quaritch* xvii, 'Well, I'm a lawyer too, and a pretty sharp one—*arcades ambo*,' said Johnnie with a coarse laugh.

arcanist (aɹkēⁱ·nist). [f. Arcan(um + -ist.] Used *spec.* of a person who has knowledge of a secret process of manufacture, esp. of the manufacture of porcelain.

1905 R. L. Hobson *Catal. Coll. Eng. Porc. Brit. Mus.* p. xv, The royal houses vied with each other in securing the services of the arcanists from Meissen. **1939** *Burlington Mag.* Oct. 172/2 The death of the arcanist and manager..brought to an end..the first period of the [porcelain] factory. **1959** *Times* 10 Oct. 9/3 We also had our wandering arcanists.

arch, *sb.* Add:

4. b. *Anat.* One of the arches formed by the tarsal and metatarsal bones of the foot; *fallen arch*, one that has flattened. Also *attrib.*, as *arch support*, a device worn in the shoe to provide support for the arch of the foot.

1858 G. M. Humphry *Treat. Human Skeleton* 492 The plantar arch. The foot is..made in the shape of an arch. The summit of the arch is at the top of the astragalus. The hinder limb of the arch is formed..by the os calcis; and the anterior limb is formed by the other tarsal and the metatarsal bones. *Ibid.* 495 The arch yields..at the joint between the astragalus and the navicular bone, the person becomes 'flat-footed'. *Ibid.*, The well-formed foot presents other arches besides the one chief plantar arch which we have been considering. **1875** *St. George's Hosp. Rep.* VII. 211 Cases in which the tarsal arch has given way, but has not been obliterated. **1939** M. L. Spring Rice *Working-Class Wives* v. 109 She has had varicose veins..for which the doctor has advised elastic stockings and arch supports. **1945** B. Macdonald *Egg & I* (1947) i. 16 She toed out and had trouble with her arches.

7. Chiefly *pl.* Collectors' name for certain species of moths: see quots.

1766 M. Harris *Aurelian* 64 *Red Arches*. This Moth is generally taken by beating the boughs of the oak..the latter end of June. **1832** J. Rennie *Butterfl. & Moths* 169 The Rufous Arch (*S[emasia] rufana*, Stephens). **1869** E. Newman *Brit. Moths* 407 The Gray Arches (*Aplecta nebulosa*)...The costal margin of the fore wings is very slightly arched, the margin very slightly scalloped. *Ibid.* 408 The Silvery Arches (*Aplecta tincta*)...The fore wings are slightly arched beyond the middle of the costal margin. **1921** *Conquest* Sept. 496/2 The Dark Arches (*Xylophasia monoglypha*).

arch (āɹtʃ), *v.*[2] [Back-formation f. Archer, Archery.] *intr.* To practise archery. So **a·rching** *vbl. sb.* and *ppl. a.*

1648 A. Ross *Myst. Poet.* (ed. 2) viii. 179 Apollo's Musick, and skill in arching, were more pleasing to him. **1871** W. Cory *Let.* 25 Aug. in *Extr. Lett. & Jrnls.* (1897) 275 We arch—we don't often hit the target..but we know how to bend a bow. **1950** C. Fry *Venus Obs.* 39 All the arching duchesses.

arch-, *prefix.* Add: **2.** Further examples: *arch-criminal* (example), *-metaphysician, -mistress, -rationalist, -scandalmonger, -sceptic, -scoundrel* (example), *-snake, -tempter* (example).

1938 R. G. Collingwood *Princ. Art* v. 85 Hair-raising fiction concerned with arch-criminals, gunmen, and sinister foreigners. **1905** W. James *Let.* 24 Aug. (1920) II. 232 When you write your treatise against philosophy, you will be classed as the arch-metaphysician. **1930** R. Campbell *Poems* 18 Your muse..Arch-mistress of the slowly crawling theme. *a* **1910** W. James *Some Probl. Philos.* (1911) ii. 35 Plato, the arch-rationalist, explained the details of nature by their participation in 'ideas'. **1920** D. H. Lawrence *Let.* 4 Jan. (1962) I. 606 She is staying with an arch-scandalmonger. **1936** *Mind* XLV. 336 Their specific intellectual relationship to the arch-sceptic himself [*sc.* Hume]. **1896** *Westm. Gaz.* 21 May 2/1 There is no knowing how many gullible young women this arch-scoundrel might not have married and fleeced. **1881** G. M. Hopkins *Sermons & Dev. Wr.* (1959) 199 So that if the Devil is symbolised as a snake he must be an archsnake and a dragon. **1916** Joyce *Portr. Artist Young Man* iii. 135 Eve yielded to the wiles of the arch tempter.

4. *arch-mediocrity* (example).

1844 Disraeli *Coningsby* I. ii. i. 155 The Arch-Mediocrity who presided, rather than ruled, over this Cabinet of Mediocrities.

a·rcha:bbey. [Arch- 4.] The head abbey of a Benedictine congregation. So **a·rchabbot,** also called *abbot-general*.

1881 B. Weldon *Engl. Congreg. St. Benedict* p. ix, The Arch-Abbey of Monte Cassino. **1889** *Cent. Dict.*, Archabbot. **1897** Bede Camm *Benedictine Martyr* iv. 82 In the Order of St. Benedict..a confederacy of Abbeys and Priories, ..united together under one President (sometimes called Abbot-General or Arch-Abbot). **1912** *Cath. Encycl.* XV. 648/2 The location where St. Vincent Archabbey, College, and Seminary stand to-day. **1923** Seton *Memories of Many Years* 264 We had dinner at different tables, in the gallery. Arch-Abbot Krug..presided at the principal one.

Archæan, *a.* Substitute for def.: Of or pertaining to crystalline rocks of Pre-Cambrian age (see quots.). Add earlier and additional examples. Also *ellipt.* as *sb.*

1872 Dana in *Amer. Jrnl. Sci.* Apr. 253, I propose to use for the Azoic era and its rocks the general term *Archæan*. **1903** Geikie *Textbk. Geol.* (ed. 4) II. vi. i. 861 Pre-Cambrian, also called Archæan. **1944** A. Holmes *Princ. Physical Geol.* vii. 104 The term Archæan refers to the oldest Pre-Cambrian crystalline rocks of a given region. **1944** Rastall in *Geol. Mag.* LXXXI. 159 There is now a marked tendency in international geology to draw a clear distinction between the ancient highly metamorphic continental blocks and the overlying..sedimentary but unfossiliferous groups. These may be conveniently called the Archaean and Algonkian respectively.

archæocyte (ā·ɹki,ŏsəit). *Zool.* Also **archeocyte.** [f. Archæo + -cyte.] A wandering amœboid cell, esp. in a sponge: see quots.

1887 W. J. Sollas in *Encycl. Brit.* XXII. 419/2 Amœboid wandering cells or *archæocytes* are scattered through the matrix of the collenchyme. **1900** E. A. Minchin in E. R. Lankester *Treat. Zool.* II. iii. 58 The archaeocytes may be considered from two points of view: first, as wandering cells, or *amoebocytes*; secondly, as reproductive cells, or *tokocytes*. **1951** E. N. Willmer in G. Bourne *Cytology* (ed. 2) xi. 463 Two archeocytes are seen at the posterior end.

archæologize (āɹki,ǫ·lŏdʒəiz), *v.* [f. Archæolog(y + -ize.] **1.** *trans.* To treat or explain archæologically. *rare.*

1884 Lytton in *19th Cent.* Dec. 886 The attempt to archæologise the Shakespearean drama is one of the stupidest pedantries of this age of prigs.

2. *intr.* To study or practise archæology; to play the archæologist. Hence **archæo·logizing** *vbl. sb.*

1885 *Harper's Mag.* Mar. 654/1 To archæologize in that delightful air..is one good way of enjoying existence. **1914** T. E. Lawrence *Let.* 28 Feb. (1938) 165 He forbade Newcombe to map, and me to photograph or archaeologise. I photographed what I could, I archaeologised everywhere. **1954** G. Boas in *Essays & Studies* VII. 35 Besides archæologizing round Oxford he practised revolver shooting for fun. **1958** E. L. Mascall *Recov. Unity* vi. 134 Avoiding both the archaeologising which would make the Church's liturgical life irrelevant..and also the modernism.

archæologue (ā·ɹki,ŏlǫg). *arch.* Also **archæolog.** [ad. F. *archéologue* archæologist, f. Gr. ἀρχαιολόγος telling of ancient times (cf. Archæology).] An archæologist or antiquarian.

1839 Wiseman in *Dublin Rev.* Jan. 23 The learned Cav. Borghesi..consulted in his retreat by the first archeologues of Germany, for his extraordinary sagacity in antiquarian difficulties. **1863** Dickens *Let.* 29 July (1880) II. 205 If you were coming to the archæologs at Rochester. **1876** *Nation* 7 Dec. 342/2 Even an American archæologue or art-student might have preferred to keep all the Cesnola discoveries together in one museum.

archaic, *a.* Add: **a.** *spec.* in *Archæol.*, designating an early or formative period of artistic style or culture.

1846 [see Dict.]. **1902** E. A. W. Budge *Hist. Egypt* II. i. 1 With the ending of the IIIrd Dynasty we close our chapter on the archaic period of Egyptian civilization. **1928** H. C. Dawson *Age of Gods* xiii. 289 The decline of the Archaic Culture and the invasions of the warrior peoples at the beginning of the second millennium. **1961** S. Lloyd *Art Anc. Near East* ii. 64 The first two Egyptian dynasties are usually referred to as the 'Archaic period'.

archaicist (aɹkē·isist). [f. Archaic *a.* + -ist.] = Archaist b.

1896 M. Beerbohm *Works* 20 The kilt is now confined entirely to certain of the soldiery and to a small cult of Scotch Archaicists. **1957** H. Read *Tenth Muse* iii. 32 Gauguin, far from being a modernist of his time, was a traditionalist and even an archaicist.

archaism. Add: **2.** Also, an archaic feature in script.

1954 N. Denholm-Young *Handwr. Engl. & Wales* 42 A sufficient number of archaisms survived into the thirteenth and fourteenth centuries...The most obvious of such features are (i) the retention of majuscule R. **1960** N. R. Ker *Eng. MSS. Cent. after Norman Conquest* 22 He might have some difficulty in interpreting two archaisms of the script, the special *ra* and *rt* ligatures.

archaistically (āɹke,i·stikăli), *adv.* [f. Archaistic *a.*: see -ically.] In an archaistic way; as an archaism.

1891 S. R. Driver *Introd. Lit. O.T.* vi. 327 'Ephraim' must in this case be used emblematically or archaistically. **1965** W. S. Allen *Vox Latina* ii. 60 An early disyllabic form..is sometimes preserved or archaistically revived in Plautus.

archaize, *v.* Add: (Examples.) Also **a·rchaizer**, one who uses archaisms, an archaist.

1882 *Encycl. Brit.* XIV. 332 It may be remembered that Varro was himself something of an archaizer. **1906** *Academy* 1 Dec. 543/1 Poets in uncritical times do not archaise. **1921** Glover *Jesus in Exper. Men* viii. 134 The rather fabulous 'Age of Faith' is not for us, however much we archaize. **1933** *Burlington Mag.* Dec. 287/1 The artist obviously has consciously archaized it. **1939** A. J. Toynbee *Stud. Hist.* VI. v. 67 Modern Ottoman Turkish—as it was until the other day, when the archaizers took it in hand.

Archangel[1]. Add: **2. c.** A book-name for *Archangelica officinalis* and allied plants, as *Angelica sylvestris* (wild archangel): = Angelica 1.

1855 A. Pratt *Flower. Pl.* III. 54 *Angelica sylvestris* (Wild Archangel). **1884** W. Miller *Dict. Names Plants*, Archangel, *Archangelica officinalis*.

Archangel[2] (ā·ɹkēⁱndʒĕl). Name of a district of European Russia and its chief town, used *attrib.* in *Archangel mat*, a bast mat used by horticulturists as a protective covering for plants, garden frames, etc.

1854 *F. & A. Dickson's* (of Chester) *Catal.*, Mats, large Archangel. **1909** *Daily Chron.* 9 Mar. 7/5 The plants being covered..with what are known as Archangel mats. **1954** A. G. L. Hellyer *Encycl. Garden Work* 15/1 Archangel mats..were much used in gardens for covering frames or tender plants in winter. Nowadays they are seldom seen.

a:rchconfrate·rnity. *R.C. Ch.* [Arch- 1.] 'A confraternity empowered to aggregate or affiliate other confraternities of the same nature, and to impart to them its indulgences and privileges' (*Cath. Encycl.*).

1636 R. Mason (*title*) A Manuell of the Arch-Confraternitie of the Cord of the Passion. **1661** Blount *Glossogr.* (ed. 2) s.v. *Rosary*, Such Romanists who are of the Arch-confraternity of the Rosary, instituted by St. Dominick. **1844** *Dublin Rev.* Mar. 11 The wonderful blessings, which have followed in the train of this religious association, have induced his present Holiness to..raise it to the title of an arch-confraternity. **1910** Gasquet *Order of Visitation* ii. 57 The Archconfraternity of the Guard of Honour of the Sacred Heart.

archebiosis (āɹkibəi,ōᵘ·sis). *Biol.* [mod.L., f. Gr. ἀρχή beginning + βίος life + -osis (cf. βίωσις way of life).] H. C. Bastian's term for 'spontaneous generation'; = Abiogenesis.

1872 Bastian *Beginnings of Life* i. v. 232. **1872** Darwin *Let.* 28 Aug. in F. Darwin *Life & Lett.* (1887) III. 169, I should like to live to see Archebiosis proved true. **1874** Fiske *Cosmic Philos.* I. ii. viii. 430 Archebiosis, or the origination of living matter in accordance with natural laws, must have occurred at some epoch of the past.

archenteron (aɹke·ntĕrǫn). *Zool.* [mod.L., f. Gr. ἀρχή beginning + ἔντερον intestine.] The primitive intestinal or alimentary cavity of a gastrula. (Cf. Metenteron, Perienteron.) Hence **archenteric** (-ente·rik) *a.*, pertaining to the archenteron.

1877 [see Metenteron]. **1881** E. R. Lankester in *Encycl. Brit.* XII. 548/2 The archenteron or primitive digestive space. *Ibid.*, The closure or shutting off of the axial from the periaxial portion of the archenteric space. **1962** D. Nichols *Echinoderms* x. 120 An inpushing at one end forms the beginning of the larval gut, or *archenteron*.

archeocyte, var. *Archæocyte.

archespore, archispore (ā·ɹki-, ā·ɹkispōᵊɹ). *Bot.* and *Zool.* Also **archesporium.** [ad. mod.L. *archesporium* (K. Goebel 1880, in *Bot. Zeitg.* XXXVIII. 546), f. Gr. ἀρχε-, ἀρχι- Archi- + Spore.] A protoplasmic body from which spores or similar reproductive bodies are developed. Hence **archespo·rial** *a.*, pertaining to or of the nature of an archespore.

1882 Vines tr. *Sachs's Bot.* 403 By repeated divisions the archesporial cell produces the spore-mother-cells. **1901** G. N. Calkins *Protozoa* v. 151 The entire organism takes part in the formation of archispores (or sporoblasts), each archispore gives rise to spores, and each spore to sporozoites. **1902** Oliver tr. *Kerner's Nat. Hist. Plants* II. 95 Within [the anther], large cells..form what is known as the *archesporium*. *Ibid.* 96 The archesporial cells divide, giving rise to the pollen-mother-cells.

archetypal, *a.* Add: **2.** *spec.* In the psychology of C. G. Jung: of, pertaining to, concerned with, or constituting an archetype (see *Archetype 2 c). Freely used in *Literary Criticism*, esp. of motifs which recur in mythologies, fairy tales, etc., e.g. the Great Mother, the Wise Man, the Enchanted Prince, and by extension of any pervasive symbolic representation.

1923 H. G. Baynes tr. *Jung's Psychol. Types* 277 The latent primordial image of the goddess, which is in fact the archetypal soul-image. **1926** W. McDougall *Outl. Abnormal Psychol.* 203 Jung would regard all the main features of this dream as instances of archetypal thinking thrown up from 'the Collective Unconscious'. **1934** M. Bodkin (*title*) Archetypal Patterns in Poetry. **1945** Koestler *Yogi & Commissar* iii. i. 122 Jung showed that certain archaic or archetypal images and beliefs are the collective property of our race. **1948** H. Read *Art Now* (ed. 4) v. 110 If we can accept the hypothesis of the collective unconscious as formulated by Jung, it is even possible that an artist like Picasso is able to reveal those archetypal images which are its characteristic content. **1950** *Brit. Jrnl. Psychol.* June 236 Archetypal images and motives are particularly prevalent in decisive phases of life. **1957** N. Frye *Anat. of Crit.* ii. 99 Archetypal criticism is primarily concerned with literature as a social fact.

archetype. Add: **2. c.** In the psychology of C. G. Jung: a pervasive idea, image, or symbol that forms part of the collective unconscious. For the use of the term in *Literary Criticism* see *Archetypal *a.* 2.

1919 Jung in *Brit. Jrnl. Psychol.* X. 22 A factor determining the uniformity and regularity of our apprehension.. I term the archetype, the primordial image. **1923** H. G. Baynes tr. *Jung's Psychol. Types* 475 Since earliest times, the inborn manner of *acting* has been called *instinct*, and for this manner of psychic apprehension of the object I have proposed the term *archetype*...This term embraces the same idea as is contained in 'primordial image'...The archetype is a symbolical formula, which always begins to function whenever there are no conscious ideas present. *Ibid.* 507 These archetypes, whose innermost nature is inaccessible to experience, represent the precipitate of psychic functioning of the whole ancestral line. **1957** N. Frye *Anat. of Crit.* ii. 99, I mean by an archetype a symbol which connects one poem with another. **1962** A. M. Dry *Psychol. of Jung* iv. 92 For the most part it is the archetypes, not the instincts, with which Jung is concerned.

archetypical, *a.* Delete †*Obs. rare* and add earlier and later examples. Also *absol.* Hence **archetypica·lity**, archetypical character.

1737 B. Martin *Bibliotheca Techn.* x. 200 Perfect or Adequate Ideas represent their archetypical objects, compleatly and perfectly. **1868** D. W. Simon tr. *Dorner's Hist. Person Christ* (Div. 2) III. 203 Supposing the only thing necessary to redemption to be a consciousness of the idea of the archetypical. *Ibid.* 204 A productivity of His archetypicality lies in His royal plenipotence and deed. **1949** *Commentary* VIII. 41/1 [Joel Chandler] Harris, the archetypical Southerner, sought the Negro's love, and pretended he had received it.

Archibald (ā·ɹtʃibǫ̆ld). *slang.* [f. the proper name.] = *Archie.

1915 *Sphere* 20 Mar. 304/1 They laugh at the 'Archibalds' which fling destruction at them whenever they come within range. **1916** H. G. Wells *Mr. Britling* ii. iv. 334 Our anti-aircraft guns were having a go at it. Then, as suddenly, Archibald stopped. **1917** G. B. Shaw in *New Republic* 6 Jan. 271/2 He must have been flying at a speed of nearly a hundred miles an hour, perhaps 12,000 feet up (to avoid Archibalds). **1917** G. S. Gordon *Let.* 12 Mar. (1943) 72 Our Archibalds are peppering a Boche aeroplane.

archibenthal (āɹkibe·nþăl), *a.* [f. mod.L. *archibenthos*, f. Archi- + βένθος depth + -al.] Belonging to or inhabiting the **archibe·nthos**, or depths of the primitive (palæozoic) ocean.

1904 *Science* 7 Oct. 463 The archibenthal species..have a greater range than those restricted to..the shallow waters of the coast. **1912** J. Murray & Hjort *Depths of the Ocean* viii. 459 The archibenthal area (that is to say, the steep continental slopes).

archiblast (ā·ɹkiblæst). *Embryol.* [f. Archi- + -blast.] The formative yolk in an ovum, which constitutes the germ; also applied to the epiblast. (In both uses distinguished from Parablast 2, q.v.) Hence **archibla·stic** *a.*

1876 [see Archi- 2]. **1876** [see Parablast 2]. **1885** W. Stirling tr. *Landois' Human Physiol.* II. 1128 He calls these structures parablastic, in opposition to the archiblastic, which belong to the three layers of the embryo.

archicerebrum (āɹkise·rĭbrɒm). *Zool.* [mod. L., f. Archi- 2 + Cerebrum.] The primitive brain of an insect or other arthropod.

1881 [see *syncerebrum*, s.v. Syn-]. **1882** [see Archi- 2].

Archie (ā·ɹtʃi). *slang.* [abbrev. of *Archibald: see quot. 1922.] An anti-aircraft gun, orig. applied to those used by the Germans in the war of 1914–18. Hence **Archie** *v. trans.*, to fire at with an anti-aircraft gun.

Occas. used in the war of 1939–45 (see *ack-ack).

1915 H. Rosher *Let.* 19 June in *In R.N.A.S.* (1916) 116 There are some beastly Archies..which come unpleasantly near first shot. **1917** 'Contact' *Airman's Outings* vi. 159 Only somebody who has been Archied from Pluspres can realise what it means to fly right over the stronghold at four thousand feet. **1918** 'Wings' *Over German Lines* 30 The German gunners..are putting up a barrage of 'Archie' shells. **1920** *Blackw. Mag.* Dec. 757/1 So soon as they crossed the lines, they were heavily 'archied'. **1922** Raleigh *War in Air* I. 343 The anti-aircraft guns got their name of 'Archies' from a light-hearted British pilot, who when he was fired at in the air quoted a popular music-hall refrain—'Archibald, certainly not!' **1939** *News Review* 30 Nov. 14/3 On each occasion fighters and heavy 'Archie' barrages drove the Nazis off. **1940** N. Monks *Squadrons Up!* i. 29 Fly over the 'archie' (anti-aircraft) batteries.

archiepiscopacy. Add: **a.** (Further example.)

1678 Godolphin *Repert. Canon.* 4 Touching the Original of Episcopacy and Archiepiscopacy in Britain.

b. Delete †(*obs.*) and add later examples. Also, an archbishop, archiepiscopal persons collectively or individually.

1848 Ullathorne in B. Ward *Seq. Cath. Emanc.* (1915) II. xxvii. 220 The leading impression appeared to be that your Lordship wished the Archiepiscopacy, and sought for it. **1892** J. A. Farrer *Books Condemned* iv. 98 Dering appears only really to have aimed at the abolition of Laud's archiepiscopacy. **1901** M. Beerbohm *Variety of Things* (1928) 207 Royalty had opened it. Archiepiscopacy had blessed it.

archiepi·scopally, *adv.* [f. Archiepiscopal *a.* + -ly[2].] In an archiepiscopal way; in the manner of an archbishop.

1839 *Fraser's Mag.* XX. 122 So fond is Mr. Philip of episcopacy, that at page 248 he archiepiscopally consecrates John Bunyan, 'Bishop Bunyan'. **1884** Tennyson *Becket* iii. iii, *Herbert*. And how did Roger of York comport himself? *Walter Map*. As magnificently and archiepiscopally as our Thomas would have done. **1886** F. C. Philips *Jack & Three Jills* II. x. 174 His demeanour was archiepiscopally grave.

Archimedean, *a.* Add: *Archimedean drill* (see quot. 1940).

1889 in *Cent. Dict.* **1914** *Lancet* 4 July 35/2, I have found the all-metal archimedean drill..much more satisfactory than other mechanical drills for making the screw-holes in plating fractures. **1940** *Chambers's Techn. Dict.* 48/2 *Archimedean drill*, a drill in which to-and-fro axial movement of a nut on a helix causes an alternating rotary motion of the bit.

archipallium (āɹkipæ·liɒm). *Anat.* [mod.L., f. Archi- 2 + Pallium.] A portion of the brain of earlier development than the neopallium, comprising the olfactory area of the cerebral cortex. Hence **archipa·llial** *a.*, of or pertaining to the archipallium.

[**1904** L. Edinger *Vorlesungen Nerv. Zentralorgane* (ed. 7) I. xvii. 288.] **1908** H. E. Santee *Anat. Brain & Spinal Cord* (ed. 4) iii. 98 Rhinencephalon (archipallium). **1921** Tilney & Riley *Form & Functions Central Nervous System* xxxvi. 657 The mammalian rhinencephalon..consists of a primordial (sometimes called peripheral) and an archipallial (often referred to as central) portion. **1933** *Proc. Nat. Acad. Sci.* XIX. 7 This cortex..consisting of one thin layer of cells which are arranged in three chief sheets or areas—medially the archipallium (hippocampal cortex), laterally the palaeopallium..and between these the dorsal or 'general' cortex.

archipelagoed (āɹkipe·lăgōᵘd), *pa. pple.* Interspersed (*with*...) like an archipelago.

1880 'Mark Twain' *Tramp Abroad* xlix, Gravy, archipelagoed with mushrooms. **1906** W. D. Howells *Delightful Eng. Towns* 91 The tender blue sky, thickly archipelagoed with whity-brown clouds.

archiphoneme (ā·ɹkifōᵘnīm). *Linguistics.* [ad. F. *archiphonème* (R. Jakobson 1929, in *Trav. Cercle Ling. de Prague* II. 8), f. Archi- + Phoneme.] A phonological unit comprising the totality of distinguishable features common to two or more phonemes. Hence **archiphone·mic** *a.*

1937 A. Tanakadate in *Mélanges de Linguistique offerts à J. van Ginneken* 360 Taguti finds in films of the Japanese sonants wave elements corresponding to their surds, which are their archiphonemes. **1952** A. Cohen *Phonemes of English* ii. 35 In the phenomenon known as neutralization of opposition he [*sc.* Troubetzkoy] works with the notion of archiphoneme which is: 'die Gesamtheit der distinktiven Eigenschaften, die zwei Phonemen gemeinsam sind.' **1956** *Archivum Linguisticum* VIII. 115 The suspension of phonematic opposition which is often called 'archiphonemic'. **1964** E. Palmer tr. *A. Martinet's Elem. Gen. Linguistics* iii. 69 If the phoneme is defined as the sum of the relevant features, the archiphoneme is the sum of the relevant features common to two or more phonemes which alone present them all.

archipresbyterate (ā:ɹkiprezbi·tərĕt). [ad. med.L. *archipresbyterātus*.] The office of archpriest; the order of archpriests; the term of office of an archpriest.

1915 J. H. Pollen in *Month* Nov. 495 Mary..thought.. that Persons had made the Pope erect the Archipresbyterate in England. **1921** T. J. Campbell *Jesuits 1534–1921* v. 153 It is clear that the Jesuits are not responsible for the establishment of an archipresbyterate instead of an episcopate to rule England. **1936** *Downside Rev.* LIV. 504 A letter written in the interregnum between the Archipresbyterates of Birkhead and William Harrison.

archispore: see *Archespore.

a·rchitect, *v.* [f. the sb.] To design (a building). Also *transf.* and *fig.* Hence **a·rchitected** *ppl. a.*, designed by an architect; **a·rchitecting** *vbl. sb.* and *ppl. a.*

1818 Keats *Let.* 23 July (1931) I. 219 This was architected thus By the great Oceanus. [But see Architecture *v.* in Dict.] **1890** *Harper's Mag.* Apr. 809/2 We would not give being the author of one of Mr. Aldrich's beautiful sonnets to be the author of many 'Wyndham Towers', however skilfully architected. **1912** Rose Macaulay *Views & Vag.* viii. 153, I have no sort of interest in the architecting or building trades. **1913** Raleigh *Some Authors* (1923) 3 He has come out of the prison-house of theological system, nobly and grimly architected. **1923** *Public Opinion* 29 June 622/3 A..vague notion that a building ought to be architected.

architectonic, *a.* **1.** (Later examples.)

1931 E. Wenham *Domestic Silver* ii. 8 The various ornamental bands such as the egg and tongue, the ovolo, acanthus leaves and other foliation,..were all of architectonic origin. **1936** *Burlington Mag.* Apr. 191/1 The general effect produced by Boulle is somewhat plastic and architectonic.

architectural, *a.* Add: *spec.* of furniture or other household objects: resembling architecture in style or ornament. Also *transf.* and *fig.* Cf. Architecture *sb.* 5.

1876 C. M. Yonge *Womankind* xxx. 268 Architectural furniture..seems to me simply fitted for puppets. **1941** Auden *New Year Let.* 14 Repair the antique silence the insects broke In an architectural passion. **1959** D. Cooke *Lang. Mus.* i. 2 Medieval music was largely architectural in conception. **1960** H. Hayward *Antique Coll.* 17/2 *Architectural clock*, a clock in which the hood, in long-case, and the top in mantel clocks, is in the style of a classical pediment.

architecture, *sb.* **5.** (Further examples.)

1907 F. R. Sabin in Morris & McMurrich *Human Anat.* (ed. 4) v. 490 (*heading*) The Architecture of the Heart. **1936** *Discovery* Nov. 363/1 The architecture of molecules. **1959** D. Cooke *Lang. Mus.* i. 1 We speak of the 'architecture' of a symphony, and call architecture, in its turn, 'frozen music'. **1962** 'C. E. Maine' *Darkest of Nights* ii. 32 He's our protein chemistry expert and he knows a great deal about the architecture of the Hueste virus.

architecture, *v.* (Later examples.)

1893 *Strand Mag.* VI. 268/1 The house..was architectured by John Belcher from plans by its owner. **1939** Auden & Isherwood *Journey to War* 120 The slope has been architectured into terraces.

archival, *a.* Add pronunc. (āɹkəi·văl) and examples.

1903 *Daily Chron.* 5 Mar. 3/2 His photographs and his archival researches. **1962** *Unesco Bull. for Libraries* XVI. 9 Archival material is usually recorded for preference on roll microfilm.

archive. Add: **4.** *attrib.* and *Comb.*

1937 *Discovery* Oct. 323/1 The promotion of archive science. **1959** *Chambers's Encycl.* I. 570/1 Archive keeping is essential for a civilized community.

arc-lamp. [See Arc 5.] A lamp in which the light is produced by an electric arc. Also *attrib.*

1882 [see Arc 5]. **1906** *Daily Chron.* 30 Nov. 7/3 The arc-lamp standards were torn up and twisted like straw. **1911** *Encycl. Brit.* XVI. 659/2 F. P. E. Carré in France in 1876 began to manufacture arc lamp carbons of high quality from coke, lampblack and syrup. **1932** C. Morgan *Fountain* i. 10 A steep bank already equipped with arc-lamps, sentry boxes and barbed wire.

arc-light. [Arc 5.] = prec. So **arc-lighting.**

1880 *Wesleyan-Methodist Mag.* Dec. 951/2 Electric lighting by incandescence was just as simple as arc lighting was difficult. **1881** *Cassell's Family Mag.* Aug. 571/2 The arc lights, in which the electricity leaps across a gap of air from the point of one carbon rod to another. **1897** *Daily News* 29 Apr. 3/5 Repairs of the arc-lighting mains. **1933** C. Day Lewis *Magnetic Mountain* 34 Switch on the arc-lights.

‖**arco** (ā·ɹko). *Mus.* [It.] The bow; used as a direction in string music when the bow is to be resumed after a *pizzicato* passage. Hence as *adj.* and *adv.*, (played) with the bow.

1740 Grassineau *Mus. Dict.* 6 *Arco*, a bow or fiddle stick. *Stromenti d'Arco*, instruments played with a bow. **1806** T. Busby *Dict. Mus.* (ed. 2), *Arco*..is frequently used in violin music, in opposition to the term *pizzicato*, to denote that the bow is again to be used, instead of applying the fingers to the strings. **1931** G. Jacob *Orch. Technique* ii. 17 Note the effect of the *pizzicato* chord..and that the directions 'arco, unis.', have not been omitted from the 1st violin part on its re-entrance. **1955** L. Feather *Encycl. Jazz* (1956) 65 Present day bassists have a legitimate technique and sound in both pizzicato and arco (bow) work. **1958** J. Aldam in P. Gammond *Duke Ellington* iii. 207 His technique, *arco* or *pizzicato*, is impressive.

arcosolium (āɹkosōᵘ·liɒm). Pl. **-ia.** [med.L., f. *arcus* bow, arch + *solium* seat, sarcophagus.] An arched cell or niche, vaulted in semi-circular form, serving as a tomb in the Roman catacombs.

1876 [see *table-tomb*, Table *sb.* 22]. **1905** *19th Cent.* Nov. 785 Both the 'table-tomb' and the 'arcosolium' are as a rule confined to those many crypts..which opened out of the various galleries.

Arctalian (āɹktēⁱ·liăn), *a. Zoogeography.* [f. mod.L. *Arctalia* (f. Gr. ἄρκτος: cf. Arctic + ἅλς sea) + -an.] Belonging to or designating the marine region called *Arctalia*, comprising the seas of the northern hemisphere as far south as the isocryme of 44° Fahr.

1885 T. Gill in *Proc. Biol. Soc. Washington 1882–4* II. 33 A primary combination of the marine faunas is most natural under the categories of Tropicalian, Arctalian, and Notalian. *Ibid.* 33 The Arctalian realm..for the sake of definition, might be confined to the seas of the northern hemisphere, limited southward by the course of floating ice..but..the isocryme of 44° seems to be a more natural approximate limit.

Arctic, *a.* and *sb.* Add: **A.** *adj.* **1. c.** In Special Combs., as *Arctic fox*, a small fox of the arctic regions (*Vulpes lagopus*); *A. hare*, the polar hare; *A. hysteria* (see quot. 1924); *A.* (*sea*) *smoke* (Smoke *sb.* 3 b): see quots.; *A. stone age*, an early Stone Age culture of Scandinavia; *A. willow*, a low shrub (*Salix arctica*) with pale foliage.

1772 *Phil. Trans. R. Soc.* LXII. 370 Arctic Fox,..*Canis Lagopus*, Linn. Severn River. A most beautiful specimen in its snowy winter furr. **1840** C. H. Smith *Nat. Hist. Dogs* II. 237 The Arctic fox is smaller than the common, measuring only one foot eleven inches. **1863** Arctic fox [see Blue *a.* 12]. **1910** *Encycl. Brit.* V. 371/2 The Arctic fox..of which there is a blue and a white phase, has the tail very full and bushy. **1842** *Trans. Lit. & Hist. Soc. Quebec* IV. 133 The animals frequenting this country, are..the Arctic Hare, a variety of a very large size, [etc.]. **1910** *Encycl. Brit.* XII. 546/1 The other land mammals, the polar bear, the polar fox, the Arctic hare..are perfectly circumpolar forms. **1924** S. Novakovsy in *Ecology* V. 113 Travellers passing through the snowy deserts of extreme northeastern Siberia have more than once come across a nervous disorder, now most commonly known as Arctic Hysteria. **1940** Baughman *Aviation Dict.* 30/1 Arctic smoke. **1941** *Jrnl. R. Aeronaut. Soc.* XLV. 79 Fogs are sometimes observed when cold air streams over a water surface the temperature of which is very much higher than the air temperature. These fogs are known as Steam Fogs or Arctic Sea Smoke. **1945** R. W. Mudge *Meteorol. for Pilots* vi. 106 Land and river steam fog, at times known as *arctic sea smoke* or *convection fog*, is the result of cold air flowing over a much warmer water surface. [**1909** A. W. Brøgger in *Videnskabs-Selskabets Skrifter II, Hist.-filos. Klasse* (*title*) Den arktiske stenalder i Norge.] **1926** R. A. Smith *Guide Antiq. Stone Age* (*Brit. Mus.*) (ed. 3) 159 The Arctic Stone Age..is represented by slate lance- and arrow-heads, celts..and pottery [etc.]. **1937** E. V. Gordon tr. *Shetelig & Falk's Scand. Archaeol.* vi. 73 A conspicuous and attractive feature of the Arctic stone age in the north of Scandinavia is a distinctive naturalistic art..in rock-carvings and rock-paintings in Norway and Sweden. **1911** *Encycl. Brit.* XIX. 833/1 A carpet of mosses allows the arctic willow (*Salix polaris*) to develop.

3. Applied by Blytt to the earliest of the successive periods of vegetation in Scandinavia after the glacial period.

1876 A. Blytt *Immigr. Norwegian Flora* 67 We must presume that the arctic flora was here before all the others; that the subarctic came next; that the subboreal and the subatlantic..came later than the boreal and the atlantic. *Ibid.*, Arctic plants should be found on the present shores which were submerged during the arctic time.

B. *sb.* **2.** *pl.* Thick waterproof over-shoes for winter wear; in full *arctic boots*. *U.S.*

1867 *Daily Territorial Enterprise* (Virginia, Nev.) 1 Mar. 1/1 The 'arctic' boots are taking the place of rubber over-shoes,..and are more serviceable. **1878** 'Mark Twain' in *Atlantic Monthly* XLI. 327 He shook the snow of his native city from his arctics. **1883** M. Arnold *Lett.* (1895) II. 239, I have bought a pair of *arctics*, the lined waterproof boots which everybody here [in Boston] wears in winter.

Arctoid (ā·ɹktoid), *a.* (*sb.*) *Zool.* Also **arctoid.** [f. mod.L. *Arctoidea* neut. pl., f. Gr. ἄρκτος bear (see Arctic *a.*): see -oid.] Resembling a bear; *spec.* belonging to, or having the characters of, the division *Arctoidea* of Carnivora, comprising the bears and allied animals; as *sb.*, an animal of this division. Also **Arctoi·dean** *a.*

1869 W. H. Flower in *Proc. Zool. Soc.* 14 Jan. 24 The reproductive organs [of the dogs] belong neither to the Arctoid nor to the Æluroid type. *Ibid.* 30 An arboreal, prehensile-tailed, omnivorous, Procyonine Arctoid. **1885** *Athenæum* 3 Jan. 20/3 The view that the Pinnipedia were evolved from some arctoid, probably ursine, form of land Carnivora. *Ibid.* 2 May 570/1 The arctoidean carnivorous mammals. **1923** *Glasgow Herald* 27 Mar. 6/2 The polecat..has nothing to do with any cat...It is Arctoid, not Feline.

arcus senilis (ā·ɹkɒs sinəi·lis). *Path.* [L., lit. 'senile bow'.] A narrow opaque band encircling the cornea, common in old age.

1795 J. Ware *Remarks Ophthalmy* (ed. 3) App. 25 In elderly persons an opacity..appears round the whole circumference of the Cornea...This change in the structure of the Cornea has been described by authors under the name of Gerontoxon, vel Arcus Senilis. *Ibid.*, After the Arcus Senilis had been some time formed, a slight Ophthalmy ensued. **1920** A. E. W. Mason *Summons* i. 8 The pupils of his pale eyes were ringed with so pronounced an *arcus senilis* that they commanded the attention like a disfigurement. **1942** *Lancet* 5 Sept. 287/2 Arcus senilis and arteriosclerosis both arise in elderly people..but..it is safe to say that the possession of an arcus senilis does not mean its owner has vascular disease.

arc welding: see Arc *sb.* 5 and *Electric *a.* 2 b. Hence *arc welder*.

1890 *Sci. Amer.* Suppl. XXX. 12271 (*caption*) Coffin's Arc Welding Machine. *Ibid.* 12272/1 Mr. Coffin even proposes to make a very minute arc welder for jewelers' use. **1923** *Jrnl. Inst. Electr. Engineers* LXI. 253/1 Arc welding was originally tried for repair work. **1926** *Gloss. Terms Electr. Engin.* (*B.S.I.*) i. 212 *Arc Welding*, the welding of metals by means of heat generated by passing a current between the metals and an electrode of carbon or other material.

Ardri (ā·ɹd,ri). *Anglo-Irish.* Also **-righ.** [f. Ir. *árd-* chief- + *ríog* king.] A head king. Hence **A·rdriship.**

1881 *Imp. Dict.*, Ardrigh, Ardriagh. **1889** *Athenæum* 17 Aug. 215/2 The usurpation of the ardrigh-ship by Brian Boru. **1896** *Westm. Gaz.* 29 Dec. 3/1 To make the Ardriship (or head kingship) a reality. **1922** *Edin. Rev.* Apr. 243 Ireland will now welcome any native Government strong enough..to give her peace, whether Free State, or Republic, or an Ard-ri chosen from one of her royal races to rule once more from the hill of Tara.

ardurous, *a.* For *rare*⁻¹ read *rare* and add further *poet.* examples.

a **1770** Chatterton *Poems* (1777) 25 And glowe ardurous onn the Castle steeres. **1775** T. Barlow *Let.* in F. Burney *Early Diary* (1889) II. 63 Stop, oh ardurous Pen, and presume not. **1918** P. Maubyn *Wartime Ballad* 20 She felt ardurous in her breast The rising fires of love.

area. Add: **2. b.** *area-door, -gate* (examples), *-head, -railing* (earlier and later examples), *-sneak* (earlier examples), *-sneaking* (so *-sneak* v.), *-step*; **area-way** = Area 2; also *U.S.*, an area serving as a passage-way.

1864 Dickens *Mut. Fr.* i. v. 34 A piece of fat black water-pipe which trailed itself over the area-door. **1913** C. Mackenzie *Sinister St.* I. i. iv. 62 The area door slammed. **1841** Dickens *Barn. Rudge* xxxv, Area-gates is left open. **1917** T. S. Eliot *Prufrock* 31, I am aware of the damp souls of housemaids Sprouting despondently at area gates. **1841** Dickens *Barn. Rudge* viii, To retail at the area-head above pennyworths of broth. **1836** —— in *Bell's Life in London* 17 Jan. 1/1 The hungry wayfarer..plods wearily by the area railings. **1946** R. Campbell *Talking Bronco* 18 Like a cane That's rattled down an area-railing. **1812** J. H. Vaux *Flash Dict.*, *Area sneak*, or *Area slum*, the practice of slipping unperceived down the areas of private houses, and robbing the lower apartments of plate or other articles. **1839** Dickens *Nickleby* lix, Why wasn't I a thief, swindler, housebreaker, area-sneak? **1861** Mayhew *Lond. Labour* Extra vol. (1862) 311/2 Most of the lodgers were out prowling over the various districts of the metropolis, some picking pockets, others area-sneaking. **1906** *Daily Chron.* 14 May 6/3 To prevent burglary and area-sneaking. **1834** Dickens *Sk. Boz* (1836) I. 147 The area and the area-steps..were..clean and bright. **1899** M. P. Breen *Thirty Years N.Y. Politics* 731 Surplus rain..made its way into area-ways. **1907** *Chicago Even. Post* 4 May 3 The building is connected with the main hospital by a covered areaway. **1931** D. Runyon *Guys & Dolls* (1932) iii. 56 Then Louie gets the back door open and takes it on the lam through an areaway.

4. a. Also in many specific uses: (i) with adjs., as *depressed, derestricted area*, etc.; (ii) with sbs., as *defence, sterling area*, etc.: see under the first elements. Also *attrib.* and *Comb.*, as *area board* (Board *sb.* 8 b); **area bombardment, bombing**, the bombing of an extended area rather than of a strictly limited target; similarly *area shoot* (see quot. 1925).

1920 *Act* 10 & 11 *Geo. V.* c. 50. § 7 Where a district is co-extensive with an area, the district committee shall perform the functions of the area board as well as of the district committee. **1925** Fraser & Gibbons *Soldier & Sailor Words* 8 *An area shoot*, a widespread bombardment over a whole district to make it untenable by the enemy. **1941** *Western Evening Herald* (Plymouth) 11 June 3/4 The city is divided into areas, each under the control of an area commander. **1942** *Aeronautics* Oct. 33/3 Area bombing was probably the correct type of bombing to adopt against the German ships in Brest. **1944** *Hansard Lords* 5th Ser. CXXX. 739 Do the Government understand the full force of what area bombardment is doing and is destroying now? *Ibid.* 741 The first outstanding raid of area bombing was..in the spring of 1942. **1962** *BBC Handbk.* 37 There are regional and area news bulletins..and other events which are carried locally. *Ibid.* 70 Most VHF transmitters cover an area smaller than a BBC region and can be used for what is called area broadcasting.

5. (Additional examples.)

1955 D. Davie *Articulate Energy* vii. 87 The images serve to indicate roughly the area of experience that the poet is dealing with. **1957** G. Faber *Jowett* v. 85 The feeling which dictated his unequivocal choice of words..was strongly rooted in a still hateful area of memory.

arena. Add: **5.** Applied *attrib.* to a style of play production in which the stage is so positioned in the auditorium that it is surrounded by the audience, who thus, as in the Greek *theatron*, see the players 'in the round'.

This style was introduced in November 1932 by Glenn Hughes, director of the Washington School of Drama. His first productions were known as 'Penthouse' from the place of their performance; the word *arena* came later as the technique was adopted elsewhere in America and abroad.

1944 N. Felton in *National Theatre Conference Bull.* (U.S.) VI. 17 (*title*) Arena Theatre. Method for Producing a Play. **1948** *Theatre Arts* June–July 58 Called variously central staging, theatre-in-the-round, or arena theatre, the new form has suddenly made any large unencumbered room a possible stage. **1949** *Here & Now* (N.Z.) Oct. 14/2 From the point of view of a touring company the advantages of an arena production are fairly obvious. **1955** *Times* 11 May 7/2 The auditorium..is in the shape of a semi-circle, but may be changed into a circle for what is called the 'arena stage', when the movable sides of the proscenium opening are taken away and the curve of the cyclorama is extended to make a wall behind three additional rows of seats on the stage itself.

Arenig (ărē̆ⁱ·nig). The name of a mountain in Merionethshire, Wales, applied *attrib.*, esp. to a series of rocks of the Lower Ordovician age. Hence **Areni·gian** *a.*, of or pertaining to this series.

1854 A. Sedgwick in *Phil. Mag.* VIII. 362 Festiniog group (Middle Cambrian)..Arenig slates and porphyries, &c. **1865** Lyell *Elem. Geol.* xxvii. 564 This Arenig group may therefore be conveniently regarded as the base of the great Silurian system. **1879** *Encycl. Brit.* X. 331/2 *Arenig or Stiper Stone Group.* These rocks consist of dark slates, shales, flags, and bands of sandstone. **1929** Evans & Stubblefield *Handbk. Geol. Gt. Brit.* iii. 59 The lowest part of the Ordovician has been also called the Arenig or Arenigian Series.

areopagus. (Later examples.)

a **1670** Hacket *Scrinia Reserata* (1693) II. § 159. 169 And therefore, my H. Lordships, here I have fixt my Areopagus, and dernier Resort, being not like to make any further Appeal. **1900** *Daily News* 28 Apr. 5/3 A permanent areopagus has been created, with fixed arbitrators, secretaries, &c. **1919** tr. *Von Tirpitz's Mem.* I. xvi. 249 The European 'areopagus' (as Bethmann called it), proposed by Grey, ought not to have been refused. **1920** *Q. Rev.* Apr. 475 The Commission of Reparations, whose duty it should be to become a..Rhadamanthine Areopagus.

aretalogy (ærĕtæ·lŏdʒi). [ad. Gr. ἀρεταλογία, f. ἀρετή excellence, wondrous deed, miracle: see -logy.] A narrative of the miracles performed by a god or semi-divine hero. So **aretalogical** (ærĕtălǫ·dʒikăl) *a.*

1887 W. Cory *Lett. & Jrnls.* (1897) 525 The aretalogical succession is that traced back to the Scipio and Regulus of Cicero, Livy, and Horace. **1912** J. S. Phillimore tr. *Philostr. Apollon.* I. p. xiii, A life, or rather an edifying *Aretalogy*, of Pythagoras. **1925** W. R. Halliday *Pagan Background* vi. 185 The aretalogy, which narrates the miraculous acts of some thaumaturge.

Aretine, var. *Arretine *a.*

'arf, arf (ɑ̈ɪf). Slang corruption of Half *sb.*, *a.* or *adv.*

1854 *Punch* XXVI. 82/1 *First Polite Native.* Who's 'im, Bill? *Second ditto.* A stranger! *First ditto.* 'Eave 'arf a brick at 'im. **1889** Farmer *Americanisms* 201/2 The rough's traditional 'arf brick. **1903** Shaw *Man & Superman* II. 70 Here! Mister! arf a mo! steady on! **1915** 'Bartimeus' *Tall Ship* 69 Not 'arf. **1934** T. S. Eliot *Rock* I. 12 'Arf a mo', 'arf a mo'. It's lucky for you two as you've got someone what's done a bit o' lookin' into things to keep you in line.

argal (ɑ̈·ɪgăl), *sb.* Short form of Argali.

1904 R. G. Thwaites *Jrnls. Lewis & Clark Exped.* I. 169 The Rocky Mountain sheep or argal (*Ovis montana*). **1928** Childe *Most Anc. East* ii. 44 The third variety of Old World sheep, the argal, lives to the east of the urial.

Argand. Add: **2.** Surname of Jean Robert *Argand*, a French mathematician (1768–1822), applied to a diagram used for the graphical representation of a complex number.

1908 J. H. Jeans *Math. Theory Electr. & Magnet.* viii. 258 The representation of a complex quantity in a plane in this way is known as an Argand diagram. **1959** G. & R. C. James *Math. Dict.* 20/1 *Argand diagram*, two perpendicular axes on one of which real numbers are represented and on the other pure imaginaries, thus providing a frame of reference for graphing complex numbers.

argel (ɑ̈·ɪdʒĕl). *Med.* [ad. Arab. *ḥarjil* (Sharaf *Dict. Med.*).] The leaves of the African asclepiadaceous plant *Solenostemma argel*, formerly used to adulterate senna; also, the plant itself.

1803 *Phil. Mag.* XV. 58 The plant arguel, called senna of Mecca, placed in the class of the cynanchum by C. Nectoux, who has given an ample description of it. **1818** A. T. Thomson *Lond. Disp.* (ed. 2) 93 The leaf of argel is an inch or 14 lines long, while that of senna never exceeds nine lines. **1874** Garrod & Baxter *Mat. Med.* 242 This addition to senna is important, as the argel is supposed to gripe and nauseate. **1920** H. G. Greenish *Mat. Med.* (ed. 3) 34 Argel leaves, *Solenostemma Argel*, Hayne;..formerly regularly mixed with the senna but now of rare occurrence.

Argentine (ɑ̈·ɪdʒĕntəin), *a.*² and *sb.*³ [ad. Sp. *Argentina*: see def.]

A. *adj.* Of or belonging to Argentina, the federal republic (*República* or *Confederación Argentina*) which occupies the greater part of the southern end of South America, and is named from the Rio de la Plata (Sp. *plata* silver).

1830 *Ann. Reg. 1829* xii. 243/1 As to the confederation which formed the Argentine republic,..the different provinces had their armies in the field, and were fighting equally bitterly against each other. **1875** *Encycl. Brit.* II. 488/2 Along the Argentine slopes of the Andes. **1891** T. Child *Spanish-Amer. Republics* 280 The Argentine love of showy novelties. **1917** G. Ross *Argentina & Uruguay* 46 According to Argentine Law, all children born on Argentine soil are *ipso facto* Argentines. **1920** *Cornhill Mag.* Sept. 334 No two Argentine horses are of the same mind for more than a few seconds at a time.

b. *Argentine ant*, a small brown South American ant, *Iridomyrmex humilis*, which has been introduced into and become a household and orchard pest in parts of the United States, South Africa, and elsewhere.

1908 C. W. Woodworth in *California Agric. Exper. Station Circular* No. 38, p. 1 (*title*) The Argentine Ant in California. **1948** F. Wightman *Wind is Free* (1949) vii. 105 Like the Argentine ant in South Africa, the white ant appeared suddenly on the island and now is attacking everything.

B. *sb.* **1.** A native or inhabitant of Argentina. Also, a horse or pony of a breed peculiar to Argentina.

1829 *Ann. Reg. 1828* Chron. 432/1 An intrepid chief, with a handful of Argentines, has recovered our old possessions. **1860** *Chambers's Encycl.* I. 388/2 The Banda Oriental has been a bone of contention between the Brazilians and the Argentines. **1891** T. Child *Spanish-Amer. Republics* 275 In all that concerns civilization, the Argentines look up to the French. **1893** C. E. Akers *Argentine Sketches* 42 In very rare cases is an Argentine found doing anything that requires severe physical exertion. **1901** Kipling *Five Nations* (1903) 163 Atop of a sore-backed Argentine.

2. *The Argentine*, Argentina.

1891 T. Child *Spanish-Amer. Republics* 276 In the Argentine and in Uruguay. *Ibid.* 279 The auctioneer..is a great personage in the Argentine.

Hence **Argentinian** (ăɪdʒĕnti·niăn) *a.*, belonging to the Argentine; also *sb.* = **Argentino** (ăɪdʒĕntī·no), a native of the Argentine.

1918 C. B. Jordan tr. *V. Blasco Ibáñez's Four Horsemen of the Apocalypse* (1919) I. i, I am an Argentinian citizen. *Ibid.*, She was speaking alone to the Argentinian. **1920** *Cornhill Mag.* Sept. 335 The work of the Texans was neat and expeditious, that of the Argentiños clumsy and slow.

arghan (ɑ̈·ɪgæn). [Etym. unknown.] The fibre of a South American plant, later cultivated in Malaya, which can be spun into a strong fine yarn. Also *attrib.*

1922 *Conquest* Oct. 485 Arghan, the new British textile. **1924** J. S. M. Ward *Textile Fibres* 52–3. **1927** *Chambers's Jrnl.* 236 The Arghan plant is Sir Henry [Wickham's] second gift to the Inde.

arginase (ɑ̈·ɪdʒinēⁱz, -s). *Chem.* [G. (Kossel and Dakin 1904, in *Z. Physiol. Chem.* XLI. 322): see next and *-ase.] An enzyme capable of hydrolysing arginine into ornithine and urea.

1904 *Nature* 16 June 160/2 The enzyme has been named 'arginase', and is the first representative of the class of urea-forming enzymes capable of being isolated and of acting outside the body. **1931** *Lancet* 9 May 1021/2 Eldbacher and Merz have recently shown that 'all malignant mammalian and human tumours..are distinguished from normal tissue by their arginase content'. **1953** *New Biol.* XV. 90 The enzyme arginase..is present in large amounts in all the body tissues [of the dogfish] except blood and brain.

arginine (ɑ̈·ɪdʒinəin-in, -in). *Chem.* Also **-in.** [G. *arginin* (E. Schulze and E. Steiger 1886, in *Chem. Berichte* XIX. 1177), of uncertain derivation.] An amino-acid, $C_5H_{13}N_4COOH$, present in many animal proteins and certain vegetable tissues.

1886 *Jrnl. Chem. Soc.* L. 725 The nitrate of the base, to which the authors give the name of *arginine*.., crystallises out on cooling in groups of slender needles. **1919** *Nature* 20 Nov. 322/2 The importance of the diamino-acids lysin, histidin, and arginin must be recognised. Their presence has been shown to be necessary for growth. **1930** *Biochem. Jrnl.* XXIV. 1181 In the case of tumour tissue there is experimental evidence that arginine induces an increase in the rate of growth. **1949** *New Biol.* VII. 77 The basic proteins..are relatively rich in the amino-acid arginine.

Argive (ɑ̈·ɪgəiv), *a.* and *sb.* [ad. L. *Argīvus* (Gr. Ἀργεῖος) pertaining to Argos.]

A. *adj.* Of or belonging to the ancient city of Argos or the territory of Argolis. Hence used in Homer and later classical writers as = Grecian, Greek.

1598 Chapman *Iliad* I. Argt., Apollos Priest to th' Argiue Fleete doth bring Gifts for his daughter. *Ibid.* 2 Till in my Royall Argiue court, her bewties strow my bed. **1676** Hobbes *Iliad* I. 1 The two Atrides..Who of the Argive Army were the best. **1858** Birch *Anc. Pottery* II. 107 The Lacedæmonian, Teian, Chian, and Argive cups were also esteemed. **1875** Morris *Æneids* II. 254 And now the Argive host comes forth. **1925** G. Murray *Eumenides* 15 Myself, mine Argive people, and my home Shall without war be hers.

B. *sb.* A native of Argos or of Argolis; *gen.* a Greek.

c **1530** T. Elyot tr. *Plutarch's Educ. Children* sig. E*, Polynyces the yonger brother, hauynge in his ayde the Argyues. **1579** S. Gosson *School of Abuse* fol. 7*, The Lacedæmonians instructed by Tyrtæus verse, The Argiues by the melody of Telefilla. **1676** Hobbes *Iliad* XVII. 267 Just so Patroclus body tugged they, Trojans to Troy, and Argives to the Fleet. **1716** Pope *Iliad* VIII. 436 Long since had Hector stain'd these Fields with Gore, Stretch'd by some Argive on his native Shore. **1875** Morris *Æneids* I. 40 And Pallas, might not she Burn up the Argive fleet and sink the Argives in the sea..? **1922** Foakes-Jackson & Lake *Beginnings Chr.* II. 12 The ivory or other material..supplied by the Elians, Athenians, or Argives.

a·rgle-ba·rgle, *sb.* [f. *argle-bargle* vb.: see Argle *v.*] Disputatious argument, bandying of words, wrangling.

1872 A. J. Cupples *Tappy's Chicks* 252 During these days of 'argle bargle', as our smith's wife called it. *a* **1881** Carlyle in W. A. Knight *Retrospects* (1904) 15, I have for a long time given up the argle-bargle of metaphysics. **1927** *Observer* 11 Dec. 15/2 Can they..stand up to a good and sufficient argle-bargle that lasts for the best part of three hours?

So **a·rgy-ba·rgy (argue-bargue)** *v. intr.* and *sb.* (orig. *Sc.*).

1887 *Jamieson's Sc. Dict.* Suppl. s.v. *Argewe*, The terms *argie-bargie, argie*, and *bargie*, are applied to such contentions. **1888** Barrie *Auld Licht Idylls* 35 I'se nae time to argy-bargy wi' ye. **1905** H. G. Wells *Kipps* I. ii, An occasional argey-bargey that sprang up between Carshot and Buggins at dinner. **1906** W. De Morgan *Joseph Vance* xxiii, He argue-bargues with you like a winkle that won't come out of its shell. **1922** *Blackw. Mag.* July 65/1 Do not argy-bargy with such scoundrels. **1948** J. B. Priestley *Linden Tree* II. i, 'Avin' a proper argy-bargy in 'ere, aren't you? Losing your tempers too.

argon (ɑ̈·ɪgǫn). *Chem.* [mod.L., f. Gr. ἀργόν, neuter of ἀργός idle, inactive, f. ἀ- priv. + ἔργον work.] A colourless odourless gas occurring in very small quantity (less than 1 per cent.) in the air, used in 'gas-filled' electric light bulbs, etc. Discovered in 1894 and named from its chemical inertness. Symbol Ar, atomic number 18, atomic weight 39·944.

1894 *Daily News* 28 Dec. 3/2. **1895** Ld. Rayleigh & W. Ramsay in *Phil. Trans.* CLXXXVI. A. 187 Argon, a New Constituent of the Atmosphere. *Ibid.* 234 The gas deserves the name 'argon', for it is a most astonishingly indifferent body, inasmuch as it is unattacked by elements of very opposite character. **1913** *Bloxam's Chem.* (ed. 10) 293. **1958** *Engineering* 7 Feb. 187/1 It is necessary to melt titanium either in the presence of a suitable inert gas, such as argon, or else in a vacuum.

Argonaut. Add: **1. b.** *transf.* Also with small initial.

1807 W. Irving *Salmagundi* 20 Mar. 109 A celebrated Roman knight..who..became a great favourite of prince Madoc, and accompanied that famous Argonaut in the voyage which ended in the discovery of this continent. **1893**

K. A. Sanborn *S. California* 1 The *fin-de-siècle* Argonaut, in Pullman trains, flees the cold and grip.

c. An adventurer who went to California in search of gold soon after its discovery there in 1848. *U.S. Obs. exc. Hist.*

1848 in C. C. Cutler *Greyhounds of the Sea* (1930) xiv. 134 The ship will be commanded by a man of experience, and if the adventurers of the *Everett* do not bring home their full share of the golden fleece, we have overestimated the qualities of the Argonauts. **1948** P. Johnston *Lost & Living Cities of California Gold Rush* 28/1 How those argonauts who first beheld this contorted country must have longed for its cool river beds.

arguably (ɑ̈·ɪgiu̇ăbli), *adv.* [f. Arguable *a.* + -ly².] As may be shown by argument or made a matter of argument.

1890 *Sat. Rev.* 22 Feb. 216/1 His policy, if sometimes arguably mistaken, was almost always a..generous policy. **1920** *Q. Rev.* Apr. 494 The real for good and moral righteousness is arguably more purely Jewish in its origin than [etc.]. **1959** *Times* 18 Dec. 5/3 Mozart's sinfonia concertante for violin and viola, arguably the greatest of the genre.

argufier (ɑ̈·ɪgiu̇fəi·əɪ). [f. Argufy *v.*] One who argufies or is given to arguing.

1805 Fessenden *Democr. Unveiled* vi. 203 His honour might have pass'd..For quite a decent country Squire, And no bad Jury—argufier. **1871** Jowett *Plato* III. 494 We know that they are tremendous arguers, and are able to impart their own skill to others. **1880** W. Clark Russell *Sailor's Sweetheart* i, I have noticed that your people who are pretty well agreed are always the fiercest arguers.

argument. Add: **3. c.** *argumentum e* (or *ex*) *silentio*, an argument from silence: used of a conclusion based on lack of contrary evidence.

1934 Toynbee *Stud. Hist.* II. 274 This *argumentum ex silentio* does not, of course, go very far. **1939** L. H. Gray *Foundations of Lang.* iv. 99 We perceive that the absence of a common expression in the Indo-European period for a given concept or thing by no means necessarily implies the non-existence of that concept or thing at that period. The *argumentum e silentio* is notoriously fragile. **1962** *Listener* 6 Sept. 364/1 Doesn't Dr. Needham give the Chinese the benefit of a doubt, sometimes, with an *argumentum e silentio*?

argy-bargy, etc.: see under *Argle-bargle.

argyle, var. Argyll.

1797 M. Underwood *Treat. Dis. Children* III. 74 The pot is formed in the shape of an Argyle, or gravy-pot. **1922** *Weekly Dispatch* 31 Dec. 6 Especially quaint were the argyles, in which there was a receptacle for hot water so that the gravy might not be cooled down by the use of a cold ladle. **1931** E. Wenham *Domestic Silver* v. 67 When, in the eighteenth century, it became fashionable to carve the joint on the table, a now almost forgotten gravy-holder, known as an argyle, was popular.

Argyll Robertson (ɑɪgəi·l rǫ·bəɪtsǝn). The name of a Scottish physician (1837–1909) applied *attrib.* to a pupil of the eye which fails to contract in response to light, but contracts in accommodation to near distance.

1881 J. H. Jackson in *Trans. Ophthalmol. Soc. 1880* I. 149 The common condition described is what is called the Argyll Robertson pupil; the pupil does not act to light and does act during accommodation. **1908** *Practitioner* Jan. 11 The existence of a sluggish reaction to light, or the Argyll-Robertson pupil inactive to light while active to accommodation, would be certain evidence in favour of general paralysis. **1951** *Encycl. Brit.* XVI. 853/2 In 1869 Argyll Robertson described this sign that the eye-pupil (in case of locomotor ataxia) is inactive to light but acts upon accommodation in the great majority of cases.

argyrodite (ɑɪdʒi·rodəit). *Min.* [ad. Ger. *argyrodit* (A. Weisbach 1886, in *N. Jahrb. f. Mineralogie* II. 67), f. Gr. ἀργυρώδης rich in silver (ἄργυρος silver): see -ite¹.] A mineral (Ag_8GeS_6) containing silver, sulphur, and germanium.

1886 *Athenæum* 13 Mar. 364/2 Prof. Clemens Winkler..describes a new element—to which he has given the name of 'Germanium'—in a mineral named argyrodite. **1902** *Encycl. Brit.* II. 488/2 Argyrodite..is a silver sulphogermanate..and crystallizes in the cubic system. **1959** *Times Rev. Industry* Sept. 24/2 Ores like argyrodite, containing silver and germanium sulphides,..are not mined in appreciable quantities.

argyrol (ɑ̈·dʒirǫl). *Pharmacy.* [Proprietary term, f. Gr. ἄργυρος silver + -ol.] Vitellin of silver, a dark brown powder of which the aqueous solution is used as a local antiseptic.

1908 *Offic. Gaz. U.S. Pat. Off.* Apr. 1423/1 Barnes & Hille..Argyrol. Description of Goods—Antiseptics. **1910** *Practitioner* Mar. 397 The ducts..should be swabbed out on each of these days with a solution of argyrol. **1956** *Blakiston's New Gould Med. Dict.* 106/1 *Argyrol*, proprietary preparation somewhat similar to mild silver protein; used as a nonirritating antiseptic in infections of the mucous membranes. **1960** *Times Lit. Suppl.* 28 Oct. 698/4 Argyrol is an antiseptic drug used in treating infections of the eye and throat.

Arhat, Arahat (ɑ̈·rhæt, æ·rəhæt). Also with small initial letter. [a. Skr. *arhat* adj., deserving. *Arahat* is the Pali form.] A Buddhist saint of the highest rank. Hence **A·r(a)hatship**, the state of an Arhat.

[**1850** R. S. Hardy *Eastern Monachism* xxii. 282 To say that any one has 'seen nirwána', is to say that he has become a rahat..There are some persons who obtain the rahatship instantaneously.] **1870** F. M. Müller tr. *Buddha's Dhammapada* in *Buddhaghosa's Parables* p. cx, 'Arhat', the venerable, the perfect, who has reached the highest stage that can be reached, and from which Nirvána is perceived. *Ibid.* p. lxxxi, Ridding themselves of lust by the wisdom which Arhat-ship confers. **1877** T. W. R. Davids *Buddhism* 84 The Buddhist Arahats or saints. **1882** *Encycl. Brit.* XIV. 226/1 The central point of primitive Buddhism was the doctrine of 'Arahatship'. **1912** H. G. Rawlinson *Bactria* vii. 111 The Siamese tradition of Nagasena's attainment to Arhatship. **1956** Toynbee *Hist. Approach Relig.* vi. 84 The Arhat, as the Hinayanian philosopher is called in the Buddhist terminology.

-arian (ɛ̄ᵊ·riăn) *suffix*, based on L. *-ārius* -ary¹ with the addition of -an, used to form adjectives or corresponding sbs. The earliest formations of a general character are *disciplinarian* (sb.) of the late 16th century, *agrarian, antiquarian* (sb.), *corpuscularian, proletarian*, and *veterinarian* (sb.) of the 17th century; of the numeral adjs. *quinquagenarian* (1569) and *septuagenarian* (1715) are the earliest recorded. The commonest use of the suffix is in terms denoting religious or moral tenets, as *Millenarian, Predestinarian, sectarian, Sublapsarian, Supralapsarian, Trinitarian, Unitarian* of the 17th century, *Tractarian* (1824), on the analogy of which were formed *humanitarian, necessitarian, utilitarian*, the jocular *anythingarian* and *nothingarian* of the 18th century, and the nonce-word *strictarian* (after *sectarian*).

1838 Southey *Devil's Walk* in *Poet. Wks.* III. 94 My Utilitarians, My all sorts of -inians And all sorts of -arians My all sorts of -ists. **1867** F. H. Ludlow *Little Brother* 125 I believe there are strictarians who think it too gay.

ariboflavinosis (ɑ̈rəi·boflēⁱvinǫᵘ·sis). *Path.* [mod.L., f. A- 14 + *Riboflavin: see -osis.] Deficiency of riboflavin; the condition resulting from such a deficiency.

1939 J. W. Oden et al. in *Public Health Rep.* (U.S.) LIV. 792 The ease with which these cases were found..leads us to believe that ariboflavinosis is, in all probability, a common dietary deficiency disease in the southern United States. **1942** *Amer. Rev.* 1941 344 Important advances in vitamin research included..descriptions..of symptoms of human ariboflavinosis.

aridify (ări·difəi), *v.* [f. Arid *a.* + -ify.] *trans.* To make arid. Hence **ari·dified** *ppl. a.*

1920 *Glasgow Herald* 5 Mar. 8 As in Charles Darwin's case, prolonged and continuous absorption in hard thinking may aridify the emotions and destroy the appreciation of genuine poetry. **1930** W. de la Mare *Desert Islands* 149 Delusions aridify arguments. **1937** .. in *Proc. Brit. Acad.* 1936 XXI. 249 Only a puritan aridified by a sense of duty could finally condemn it [sc. rhyme].

aridly (æ·ridli), *adv.* [f. Arid *a.* + -ly².] In an arid manner; with aridity.

1883 R. Broughton *Belinda* II. iii. iv. 228 'I think,' he says aridly, 'that we had better keep to the subject in hand'. **1906** Harry Dymond ii. vi. 201 Were I as coarse a wife As I am limned in English caricature..You could not speak more aridly.

Arimasp (æ·rimæsp). [ad. L. *Arimaspi* pl., Gr. Ἀριμασποί, said to mean in Scythian 'one-eyed'.] One of the Arimaspi, a mythical race of one-eyed men in northern Europe, occupied in endeavouring to wrest gold from the griffins (γρῦπες) who guarded it. They are sometimes figured on Greek vases, fighting with griffins. Also *attrib.* or as *adj.*

1579 [see Griffin¹ 1]. **1600** tr. *de Torquemada's Sp. Mandeville* .. In the Arimaspes, being a people with one eie. **1872** C. W. King *Antique Gems & Rings* Index 471 Arimasp and Gryphon. **1889** *Cent. Dict.* s.v., Figures of Arimasps occur sometimes in Greek art, represented in Oriental dress and fighting griffins. **1905** H. B. Walters *Anc. Pottery* II. 148 In one instance an Arimasp woman is seen shooting at a Gryphon of curious type. **1923** E. H. Minns *Scythians & Greeks* iv. 56 An Arimasp such as we find on the calathos from the Great Bliznitsa.

Arimaspian (ærimæ·spiăn). [f. L. *Arimaspi* (see prec.) + -ian.] = prec.

1667 Milton *P. L.* II. 945 As when a Gryfon through the Wilderness..Pursues the Arimaspian, who by stelth Had from his wakeful custody purloind The guarded Gold. **1774** J. Bryant *Anc. Mythol.* I. 380 This may have been one reason..why the Cyclopians and Arimaspians are represented with one eye. **1827** *Blackw. Mag.* XXI. 780/1 Mermaid or salamander..miser or arimaspian. **1913** E. H. Minns *Scythians & Greeks* v. 113 The representations of Arimaspians and griffins in art. *Ibid.* xi. 391 The centre group, in which there are two griffins to one Arimaspian.

arise, *v.* Add: **18.** *arising out of*: used, with loose construction, to introduce a circumstance, action, proposal, etc., arising out of an event, statement, etc.

1922 Joyce *Ulysses* 510 Arising out of the question of my honourable friend..may I ask [etc.]? **1928** *Daily News* 8 Dec. 9/3 Arising out of the Goddard case, [a man] was summoned..for threatening..an ex-Constable.

arising, *vbl. sb.* Add: **2.** *pl.* In concrete use: applied to materials that 'arise' as secondary or waste products of an operation (see quots.). Spec. in naval use: the remains of consumable stores.

1923 'Bartimeus' *Seaways* ix. 131 'Arisings' are an important item of Naval store accounts. They represent what is left over. For instance, the 'arisings' of a candle is a puddle of wax, which is the property of the State. **1955** *Metal Age* Feb. 1 (Advt.), Cobalt..Tungsten, Vanadium. Buyers of stocks and continuous arisings. **1957** *Times* 20 Nov. 4/3 The London board..invited tenders for their 'arisings of scrap metal' over the ensuing 12 months. **1958** *Optima* Mar. 40/2 The final waste from the separation process consists of mixed fission products with a specific activity of up to 600 curies per gram..These arisings..must be stored without possibility of escape for many decades.

aristo (ări·sto). *colloq.* [Fr., shortened form of *aristocrate*.] An aristocrat.

1864 M. J. Higgins *Ess.* (1875) 172 In 1795, the aged Honorius died..in consequence of the..imprisonment to which he had been subjected as an aristo. **1888** L. R. Gower *Rec. & Remin.* (1903) 418 Charles Villiers..said how snubbed Dizzy used to be in old days by the aristos whom he got to serve him in his later life. *Ibid.* 113 Now that aristo-Liberals vote often with Conservatives. **1958** *Times Lit. Suppl.* 31 Jan. 51/3 Jacobins think about nothing but Liberty..aristos about nothing but adultery?

Aristophanic (æ:ristofæ·nik), *a.* and *sb.* [ad. L. *Aristophanicus*, Gr. Ἀριστοφανικός.]

A. *adj.* Of or pertaining to Aristophanes, the Athenian comic dramatist, or his plays; used by, occurring in, resembling that of, or characteristic of Aristophanes.

1827 Carlyle *Germ. Rom.* I. 6 He had laughed with his whole heart, in a true Aristophanic vein. **1827** P. W. Buckham *Theatre of Greeks* (ed. 2) 109 The expressions which the Aristophanic Euripides employs in assailing Æschylus. **1835** T. Mitchell *Acharn. of Aristoph.* 148 The word .. occurs in the other Aristophanic writings. **1873** Symonds *Gk. Poets* viii. 245 If we are to seek for an approximation to the spirit of Aristophanic humour, we shall find it perhaps in Rabelais.

B. *sb.* A logaœdic tripody beginning with a dactyl (–⏑⏑–⏑–⏑); a first Pherecratic.

1874 J. M. Marshall *Horace Od.* I. p. xxxiv, Sapphic (the greater). Alternately an Aristophanic and a greater Sapphic.

Aristotle's Lantern: see Lantern *sb.* 6 b.

arithmetization (ări:þmĕtaizē·ʃǝn). [f. Arithmetize *v.* + -ation.] The action, or the result, of arithmetizing, in various senses.

1902 *Encycl. Brit.* XXVIII. 54 The Theory of Functions of Real Variables..has been remodelled in the process of complete arithmetization. **1903** B. Russell *Princ. Math.* xxxii. 259 What is called the arithmetization of mathematics has shown that all the problems presented..are already present in pure arithmetic. **1930** W. James Sedgwick *Philos.* (1911) xi. 173 The 'new infinite' and the 'number-continuum' are outgrowths of a general attempt to accomplish what has been called the 'arithmetization' of quantity. **1939** *Mind* XLIV. 502 An arithmetization of syntax, based upon the method of Gödel, in which numbers are assigned to simple signs.

arithmetize, *v.* Restrict †*Obs.* to sense 1 and add: *trans.* **2.** *Math.* **a.** To reduce (any other part of mathematics) to arithmetic, by defining mathematical entities and operations in terms of the natural numbers and their properties. Hence *gen.*, to express in a numerical form.

1892 H. B. Fine in *Bull. N.Y. Math. Soc.* I. 173 It is not merely that the purely arithmetical problems growing out of algebra were attractive to him [sc. Kronecker]—he 'arithmetized' algebra itself. **1937** *Mind* XLVI. 311 A time in which there is genuine change or process can scarcely be 'arithmetized' or 'logicised' in its full meaning. **1948** Brodie & Coleman tr. *L. Chwistek's Limits of Sci.* p. xxx, Weierstrass and Kronecker..maintained that..it is logically possible to arithmetize all portions of mathematics. **1951** *Ess. in Criticism* I. 5 There has been a strong tendency in modern times to arithmetize criticism.

b. To apply the metamathematical method invented by Kurt Gödel of assigning a special number to each part of an utterance in order to formulate mathematical proofs about its properties.

1937 A. Smeaton tr. *R. Carnap's Logical Syntax Lang.* § 19. 57 The definitions and sentences of syntax arithmetized in this way do not differ fundamentally from the other definitions and sentences of arithmetic. **1940** *Mind* XLIX. 240 The method of arithmetising axiom systems, used to such effect in Gödel's work, is fully explained.

Hence **ari·thmetizing** *vbl. sb.*, the action of the verb.

1896 *Bull. Amer. Math. Soc.* II. 243 The arithmetizing of mathematics began originally..by ousting space intuition. **1919** B. Russell *Introd. Math. Philos.* i. 4 Pythagoras..was the discoverer of the most serious obstacle in the way of what is called the 'arithmetising' of mathematics.

arithmomania (ări:þmomēⁱ·niă). *Psychiatry.* [f. Gr. ἀριθμό-ς number + Mania. Cf. Fr. *arithmomanie*.] A compulsive desire to count objects and make calculations. So **arithmoma·niac**, one who suffers from arithmomania.

1890 W. S. Colman in *Lancet* 5 July 14/2 In France cases have been described as 'arithmomaniacs', epileptics who are usually of weak intellect and occupy themselves habitually with simple arithmetical calculations. **1892** D. H. Tuke *Dict. Psychol. Med.* 678 He eventually experienced the like trouble in regard to..the number of articles (books, dishes, &c.) falling under his observation (*arithmomania*). **1895** tr. *M. Nordau's Degeneration* III. i. 242 Erotomania (love madness), arithmomania (madness of numbers), oniomania (madness for buying), etc. **1924** J. Riviere et al. tr. *Freud's Coll. Papers* I. vii. 132 Obsession of arithmomania. A woman became obliged to count the boards in the floor, the steps in the staircase. **1934** S. Beckett *More Pricks* iv. 89 The voice of the arithmomaniac was heard.

arizonite (ærizǭᵘ·nəit). *Min.* [f. *Arizona*, a state of U.S.A. + -ite¹.] A titanate of iron found in Arizona.

1909 C. Palmer in *Amer. Jrnl. Sci.* XXVIII. 355 The crystallographic determinations strengthen the view that this titanate of iron cannot be assigned to any known species, but is entirely new. I propose to name it *Arizonite*. **1947** H. B. Graves *Mineral Key* 136 *Arizonite.* Hardness, 5.5.. Color, steel gray.

Arkansas (ā·ɪkǎnsǭ, also (outside U.S.) aɪkæ·nzǎs). [Name of a state of U.S.A.]

1. A fine-grain variety of novaculite found in Arkansas, U.S.A. Also, a whetstone made of this. In full *Arkansas hone, stone* (see quots.).

1869 G. A. Rogers in *Eng. Mechanic* 22 Oct. 125 A sharp-edged Arkansas can be rubbed on the outside. **1875** *Carpentry & Join.* 34 We often find it a good plan to rub down a tool on a bit of Arkansas of coarse quality. **1885** *Spon's Mechanic's Own Bk.* 242 Oilstones.—These are of several kinds, the best known being the Charnley Forest, Turkey, Arkansas, and Washita brands. **1889** *Cent. Dict.* s.v. *stone*, *Arkansas stone*, a fine-grain whetstone found in Arkansas, and used to sharpen surgical and dental instruments. **1910** *Encycl. Brit.* XIII. 653/1 Of Arkansas stones there are two varieties..The finer kind, known as Arkansas hone, is obtained in small pieces at the hot springs.

2. *Arkansas toothpick*: see Toothpick 4.

arm, *sb.*¹ Add: **2. a.** *a child* or *baby in arms* (examples).

1819 Keats *Let.* 24 Sept. (1958) II. 215 A child in a[r]ms was passing by his chair toward the mother, in the nurses a[r]ms. **1848** Thackeray *Van. Fair* lxvii. 618 You are no more fit to live in the world than a baby in arms. **1937** E. Garnett *Family from One End St.* viii. 146 All the children went half price, the younger ones, with luck, 'in arms'.

b. (*at*) *arm's length*: also *Comm.*, designating a sale or transaction that is in accordance with market values; *to make a long arm* (earlier example); see also Long *a.*¹ 1 c; *as long as one's arm* (examples).

1846 J. F. Cooper *Redskins* (1888) xii. 198 Leases as long as my arm, I calcerlate? **1857** H. Melville *Confidence-Man* vi. 39 A fellow with a face as long as my arm. **1860** Trollope *Framley P.* (1861) II. 9 God be with the good old days when I could..make a long arm for a hot potato whenever the exigencies of my plate required it. **1938** M. Allingham *Fashion in Shrouds* xx. 370 Jock has a record as long as your arm. **1961** Webster, *Arm's length*, the condition or fact that the parties to a transaction or negotiation are independent and that one does not dominate the other (sale at arm's length) (arm's length bargaining). **1962** *Taxation of Short-Term Gains* (Cmnd. 1710) 9 Where an asset changes hands as a result of a bargain which is not at arm's length (i.e. generally speaking, at an artificial price) it is normally deemed to pass for its then market value. **1962** *Economist* 8 Dec. 1040/1 Some [companies]..are..'arm's length buyers' of imported crude wherever—and as cheap as—they can get it.

c. Slang phr. *under the arm*: inferior, poor, bad.

1937 in Partridge *Dict. Slang* 17/1. **1958** F. Norman *Bang to Rights* 31, I read no matter how bad the book and some are right under the arm, stand on me. **1963** *New Statesman* 18 Oct. 537/1 All that's under the arm (i.e. no good).

III. *arm-badge, -twister*; **arm-band**, a band worn round the arm; **arm-cylinder**, a cylindrical ornament worn on the arm; **arm-guard**, (*a*) in *Boxing*, defence with the arm; (*b*) a protective covering for the arm; **arm-lock**, a close hold by the arm in wrestling and judo; also *v. trans.* (*nonce-wd.*); **arm-rest**, a device constructed for the arm to rest upon; **arm-ring**, an ornamental ring worn on the arm; an armlet.

1931 A. P. Herbert *Derby Day* III, The policeman's helmet and arm-badge are on the table as he is off duty. **1797** B. Hawkins *Let.* 26 Nov. (1916) 253 The goods he wants are..3 pair arm bands, 3 pair wrist bands. **1846** T. L. McKenney *Memoirs* I. 178, I opened a box and took out a pair of silver arm-bands. **1906** *Yorks. N. & Q.* III. 101/1 A fragment of a jet arm-band was found on the same horizon. **1922** D. H. Low tr. *Ballads of Marko Kraljević* 135 On his feet two fetters, On his arms two armbands. **1952** *Manch. Guardian Weekly* 21 Feb. 5/1 After them came the Sheffield's officers, in white uniforms with black armbands. **1937** *Antiquity* XI. 114 The axes, dagger, and arm-cylinder of the Danish hoard. **1957** Childe *Dawn Europ. Civ.* (ed. 6) v. 69 From Asia came an arm-cylinder of twisted silver wire (like a gold one from Troy II) found in a grave on Levkas. **1889** Michell *Boxing* 148–9 Figure ix, Left arm guard. *Ibid.* 156 Right arm guard. **1899** *Daily News* 20 June 3/5 Two massive gold arm-guards, set with..diamonds. **1905** A. M. Binstead *Mop Fair* vi. 102 Whose simple faith in all mankind has been mercilessly arm-locked and thrown clear off the mat. **1905** F. R. Toombs *How to Wrestle* (1906) 61 (*caption*) The Near Leg Hold and Arm Lock. **1956** K. Tomiki *Judo* 127 Kansetsu-waza..*ude-hishiji* (armlock) and *ude-garami* (entangled armlock). **1889** *Cent. Dict.*, Arm-rest. **1898** Sloane *Stand. Elect. Dict.* App., *Arm rest*, a device for lessening the fatigue of holding a telephone receiver to the ear. **1908** *Daily Chron.* 12 Mar. 5/6 The deceased's head was..over the mouth of the barrel, which was leaning against the arm-rest. **1962** *Which?* (Suppl.) July 91/2 This measurement excludes the permanent centre armrest in the two sports cars. **1903** W. B. Yeats *In Seven Woods* 49 I'll give you gifts, but I'll have something too An arm-ring or the like. **1933** E. E. Evans-Pritchard *Ess. Social Anthrop.* (1962) vii. 143 Azande..made a pact by drinking water from a gourd which contained an iron arm-ring. **1938** *New Statesman* 8 Jan. 58/2 A great arm-twister, and the sort of man who ought never to have been born. **1967** K. Giles *Death in Diamonds* ix. 164 A brute and a bully, an arm-twister and a threatener.

arm, *sb.*² **9.** Add to def.: Also used of other branches of the armed forces, esp. the **air force** (often called *the fourth arm*). See *Fleet *sb.*¹ 1 d.

1908 H. G. Wells *War in Air* iv. 106 The Emperor..placed him in control of the new aeronautic arm of the German forces. **1914** *Engineering* 4 Sept. 312/2 Aircraft in Warfare: the Dawn of the Fourth Arm. By F. W. Lanchester. **1916** [see *Aeronautical *a.*]. **1917** [see *Air *sb.*¹ B. III. 3]. **1919** [see *aerial arm]. **1940** *Ann. Reg. 1939* 62 This was the first occasion on which the Civil Defence Forces had been recognized as a Fourth Arm.

V. *arms-control, -dump* (*Dump *sb.*⁴ 1 c), *-smuggler, -store, town*; **arms race**, competition between unfriendly nations or other groups in the accumulation and development of weapons.

1961 *Economist* 14 Jan. 109/2 Their preferred term, 'arms control', embraces not only disarmament by controlled agreement but also such measures as the banning of nuclear tests. **1939** 'N. Blake' *Smiler with Knife* xi. 168 She knew there was an arms-dump beneath Major Keston's house. **1936** *Hansard Commons* 5th Ser. CCCIX. 1842 This House cannot agree to a policy which in fact seeks security in national armaments alone and intensifies the ruinous arms race between the nations, inevitably leading to war. **1937** *Daily Express* 20 Apr. 2/7 Arms Race Worry...All were worried at the armament race. **1938** Auden & Isherwood *On Frontier* II. ii. 78 The arms race is good for another five years at least. **1964** *Ann. Reg. 1963* 138 In a speech of 10 June..Mr. Kennedy called for a halt to the arms race. **1937** Koestler *Spanish Test.* 24 A spy or an arms smuggler or a foreign agent. **1899** *Westm. Gaz.* 3 Mar. 5/1 A number of artillerymen were removing a quantity of ammunition from one arms-store to another. **1944** *Ourselves in Wartime* 95 The attack on Coventry was the forerunner of similar attempts to smash the arms towns.

armada. Delete ‖ and substitute pronunc. (aɪmā·dă).

Armageddon (āɪmăge·dǫn). [See Rev. xvi. 16 (A.V.)] The place of the last decisive battle at the Day of Judgement; hence used allusively for any 'final' conflict on a great scale. Also *attrib.*

1811 Shelley *Let.* 12 Jan. (1964) I. 45 Do we not now see Superstition decaying..except where Faber..and several others of the Armageddon-Heroes maintain their posts. **1886** Suffolk & Craven *Racing* 247 As long as we have racing we shall have betting—that ceaseless war between layers and backers will still be waged...At present we see no sign of a final Armageddon. **1896** Kipling *England's Answer* in *Poems* (1919) I. 237 In the day of Armageddon, or the last great fight of all. **1910** *Encycl. Brit.* II. 561/1 From the application of the word Armageddon to the great battle of the End of Time comes the use of the phrase 'an Armageddon' to express any great slaughter or final conflict. **1917** F. M. Ford *Let.* 5 Jan. (1965) 83, I am sure you could not have done a better 'bit' during Armageddon. **1928** W. Deeping *Old Pybus* ii. § 2 Mr. Pybus had been able to speak of the war as Armageddon without cribbing an obvious bleat from the popular press.

Armagnac (aɪmæ·nyæk). A superior brandy made in the district of France formerly called Armagnac (department of Gers).

1850 A. K. Johnston *Dict. Geogr.* 539/2 Gers, a dep. in the S.W. of France...Wine is produced..and a good deal of it is converted into Armagnac brandy. **1910** *Encycl. Brit.* XI. 904/2 The remainder [of the wine produced in the department of Gers] is chiefly manufactured into brandy, known by the name of Armagnac, second only to Cognac in reputation. **1920** Saintsbury *Notes on Cellar-bk.* 119 You may drink Armagnac after Burgundy now and they won't quarrel.

armament. Add: **5.** *attrib.*

1914 W. Raleigh *Let.* 5 Dec. (1926) II. 407 One has to think of public gains, the private gains being not very obvious except to..armament syndicates. **1919** G. B. Shaw *Peace Conference Hints* vii. 107 Hegemonies are impossible, and attempts at them certain to end in armament races and finally in war. **1936** *Hansard Commons* 5th Ser. CCCIX. 1851 They were engaged in the same armaments race, piling up more and more armaments.

armamentarium (ā:ɪmămentē̆ᵊ·riǝm). [L., lit. 'arsenal, armoury'.] The equipment of medicines, instruments, and appliances used by a medical man. Also *transf.* and *fig.*

1874 [see Armamentary]. **1906** V. Horsley in S. Paget *Life* (1919) 236 A whole armamentarium of drugs. **1935** *Nature* 7 Sept. 385/1 The pituitary body provides the brain with an armamentarium of hormones, which are secreted in several ways.

armature. Add: **7. b.** A framework of wire, wood, etc., round which a sculptor builds a clay or plaster model.

1903 E. Lanteri *Modelling* II. 137 (*caption*) Armature in Twisted Copper Wire. **1947** J. C. Rich *Materials & Methods of Sculpture* ii. 29 Armatures constructed of flexible metal wire, pipe or tubing are very often employed as artificial 'skeletons' or supporting structures over which the 'flesh' or clay is formed and shaped. *Ibid.* 35 There are several types of armatures available commercially for use as supports in modeling heads, busts, and figures.

8. *Electr.* [from 6 a.] A core of laminated iron wound round by coils of insulated copper wire in which an electrical current is generated when it rotates in a magnetic field (as in a dynamo), or which, when a current is passed through the wire, provides the motive power (in a motor). Also applied to the moving member in any electromagnetic device, as a relay, electric bell, etc.

1835 W. Sturgeon in *London & Edin. Phil. Mag.* Sept. 233, I produce electric shocks..by revolving coils of wire (having an iron axis or armature) in front of the poles of a horse-shoe magnet. **1902** *Encycl. Brit.* XXVII. 575/1 Since these inductors are very commonly mounted on an iron structure, which may be likened to the keeper or 'armature' of a magnet rotating between its poles, the term 'armature' has been extended to cover not only the iron core, but also the wires on it, and is often applied to the copper conductors themselves even when there is no iron core. **1922** Glazebrook *Dict. Appl. Physics* II. 198/1 Alternators are made in almost all cases with the armature stationary, and the field revolving. **1931** *Boys' Mag.* XLV. 157/1 The aerial wires are connected to opposite sides of the buzzer interrupter, one to the fixed contact and the other to the pillar carrying the vibrating armature.

b. *attrib.* and *Comb.*

1884 F. Krohn tr. *Glaser de Cew's Magn.- & Dyn.-Electric Mach.* 34 The armature-bobbins revolve as near the magnetic poles as possible. *Ibid.* 247 The employment of only half as many inducing magnets as of armature magnets. **1902** J. Black *Illustr. Carp. & Build. Techn. Man.* II. vii. 79 The winding of the armature conductors inside insulating tubes. **1902** *Encycl. Brit.* XXVII. 580/1 The separate small sections of the armature winding. **1908** *Daily Chron.* 24 Aug. 9/6 Armature winder wanted. **1910** *Encycl. Brit.* VIII. 768/1 The iron armature core must be laminated. **1934** *B.B.C. Year-Bk.* 172 An alarm..operates in the event of the occurrence of a leakage between either armature winding and the armature iron.

arm-chair, armchair. Add: *attrib.*, in the home; hence domesticated, comfortable; often applied to persons who confine themselves or are addicted to home-made views or criticism of matters in which they take no active part, or of which they have no first-hand knowledge, as *armchair critic, politician, travel(ler)*.

1858 *Househ. Words* 10 July 82/1 Gentler heads and hands have been at work there since, or that county would not rejoice in its line of 'arm-chair cobs'. **1886** *Times Register Events 1885* cxxxii, Mr. Chamberlain..met the expostulations..of his moderate allies with sneers at..'the arm-chair politicians'. **1895** W. James *Let.* 12 Oct. in R. B. Perry *Tht. & Char. W. J.* (1935) II. 9 As an arm-chair professor, I frankly admit my great inferiority as a laboratory-teacher and investigator. **1896** W. H. S. Aubrey *Stock Exch. Investm.* i. 5 The arm-chair critic of politics, war, literature, or finance. **1902** *Lond. Q. Rev.* Jan. 70 Even to the armchair traveller the perusal of this Handbook would..suffice to attest the reality of the transformation of the..province. **1914** *Eng. Rev.* Sept. 259 Arm-chair strategists. **1914** E. A. Powell *Fighting in Flanders* vi. 133 After..the arm-chair historians have settled down to the task of writing a connected account of the campaign. **1935** *Economist* 19 Jan. 128/1 The book is perhaps a little too encyclopaedic for armchair reading, but it will be very valuable as..a source of reference. **1936** *Discovery* Aug. 259/2 Armchair travel becomes easier and more pleasant every day. **1952** C. Day Lewis tr. *Virgil's Aeneid* XI. 246 We don't want big talk from an armchair critic.

armed, *ppl. a.*¹ Add: **1.** *armed camp*, used of any territory regarded as fully armed ready for war; *armed neutrality* (earlier and later examples).

1861 M. B. Chesnut *Diary* 15 Aug. (1949) 111 This appalling picture of New York as an armed camp. **1898** *Daily News* 6 Oct. 5/7 The armed camp, which we call the continent. **1780** J. Adams *Let.* 2 May in F. Wharton *Revol. Diplom. Corr.* (1889) III. 646 The invitation of the Empress of Russia to accede to an armed neutrality. **1827** [see Neutrality 1]. **1906** W. James *Coll. Ess. & Rev.* (1920) 462 Pragmatism, according to Papini, is thus only a collection of attitudes and methods, and its chief characteristic is its armed neutrality in the midst of doctrines.

Armeniac, *a.* Delete †*Obs.* and add recent examples.

1853, 1934 [see *Anatolic *a.*].

Armenian. Add: **B.** *sb.* **c.** The Indo-European language of Armenia.

1718 [see Turkish *sb.* 1]. **1817** Byron *Let.* 2 Jan. (1846) 336/1 Padre Paschal, with some little help from me..is also proceeding in a MS. Grammar for the English acquisition of Armenian, which will be printed also. **1847** E. Riggs (*title*) A vocabulary of words used in modern Armenian. **1910** *Encycl. Brit.* II. 571/2 Armenian lost its genders long before the year A.D. 400.

Armeno-, used as comb. form of *Armenia* or *Armenian* (L. *Armenius*, Gr. Ἀρμένιος).

1898 P. Geddes in *Ethical World* 12 Mar. 167/1 The Turcophobes and Turcophiles, the Hellenophobes and Hellenophiles, Armenophobes and Armenophiles of recent journalism. **1898** *Daily News* 25 Aug. 4/7 The Armenophile movement in England. **1905** *Westm. Gaz.* 15 Nov. 2/1 The Armenophobia of Russian officialdom.

Armenoid (ā·ɪmĕnoid), *a.* [G. *armenoid* (F. von Luschan 1892, in *Archiv f. Anthrop.* (*Correspondenz-Bl.*) Oct. 98/2); cf. Armenian and -oid.] Of or pertaining to a type of the Alpine race of ancient western Asia. Also *absol.* and as *sb.*

1899 A. H. Keane *Man, Past & Present* xii. 457 Armenoid, akin to von Luschan's pre-Semitic of Asia Minor. *Ibid.* xiii. 514 E. Chantre constitutes in Western Asia an Armenoid group. **1911** F. von Luschan in *Jrnl. R. Anthrop. Inst.* XLI. 244 When I first upheld in 1892, in my paper on the anthropological position of the Jews, the homogeneous character of these groups, I called them 'Armenoids'. **1934** F. Stark *Valleys of Assassins* ii. 193 This is an Armenoid type of head, with marked flattening of the occiput. **1935** Huxley & Haddon *We Europeans* iv. 122 The so-called 'Jewish' nose is really Armenoid. *Ibid.* vi. 178 The Armenoid..type is widely distributed in the near East. This ethnic type is of medium stature, stocky and fleshy.

armistice. Add: **2.** (Earlier example.)

1814 L. M. Hawkins *Rosanne* II. xxxv. 131 There was an armistice between father and daughter.

3. Armistice Day, the day, 11 Nov. 1918, on which the armistice was concluded which brought the war of 1914–18 to an end; also, any anniversary of that day. Combined, since the war of 1939–45, with Remembrance Day.

1919 *Times* 7 Nov. 12/1 The Armistice-day service at St. Paul's Cathedral will be the office of Holy Communion. *Ibid.* 12 Nov. 15/1 The first anniversary of Armistice Day was celebrated throughout the Empire yesterday. **1946** *Times* 8 Oct. 4/6 There will be general cooperation in the observance of Remembrance Day as there has been in previous years in the observance of Armistice Day. *Ibid.* 8 Nov. 2/1 Remembrance Day..takes the place in the calendar that till 1938 was filled by Armistice Day; but instead of November 11 it will be the previous Sunday, except when November 11 or 12 is a Sunday.

armorial, *a.* and *sb.* Add: **A. 2.** Of porcelain, etc., bearing heraldic arms.

1907 F. A. Crisp (*title*) Armorial China. A Catalogue of Chinese Porcelain with Coats of Arms in the Possession of Frederick Arthur Crisp. **1925** A. Tudor-Craig *Armorial Porc. 18th Cent.* iii. 10 Errors in many of the coats of arms on armorial china. **1933** *Burlington Mag.* June p. xvii/1 The fine display at the recent British Antique Dealers' exhibition of armorial porcelain.

armorially (aɪmǭᵊ·riăli), *adv.* [f. Armorial *a.* + -ly².] In an armorial manner; in a way that pertains to arms or armour.

1861 *Temple Bar* Aug. 134 With big iron boots encasing their legs armorially. **1887** *N. & Q.* 7th Ser. IV. 54/2 Armorially speaking, I find no trace of Cornish Gunns in Burke's 'General Armory'.

Armorican (aɪmǫ·rikăn), *a.* [f. Armorica + -an.] **1.** See Armorica. **2.** *Geol.* Of or pertaining to mountain-building movements which occurred towards the end of the Palæozoic era, or the mountains then formed, traceable from southern Ireland across southern Britain and Brittany to central France. Hence **Armo·ricanoid** *a.*, resembling such a movement.

1906 H. B. C. Sollas tr. *Suess's Face of Earth* II. 93 This is the great pre-Permian range of western Europe. The traces of its interior and presumably most elevated zone lie in Brittany and the Vendée; for this reason we give these fragments the general name of the *Armorican chain*. **1913** C. Lapworth *Birmingham Country* 43 E. W...is suggestive of the..'Armorican Movement' of Northern France and the southern part of the British Isles; but to avoid all implications whatsoever of geological *age*, this trend in the Midlands can only be safely referred to as *Armoricanoid*. **1960** L. D. Stamp *Britain's Struct.* (ed. 5) xii. 123 The Carboniferous Period was brought to a close by the great Armorican earth movements. *Ibid.* 124 The Armorican mountains and valleys determined the main pattern of the geography which was destined to persist throughout the Mesozoic.

armour, *sb.* **6. b.** Substitute for def.: The steel or other metallic protective sheathing of a warship, military fortification, vehicle, or aircraft. Add earlier examples.

1855 W. M. Gwin (*title*) Report [of] the Committee on Naval Affairs, who were instructed to inquire into the expediency of using submarine armors in the United States navy. (D.A.E.) **1861** in *Offic. Rec. Union & Confed. Navies* (U.S.) 1st Ser. IV. 222 (D.A.E.), This addition would nearly treble the strength of the armor.

c. Armoured vehicles collectively.

1944 *Return to Attack* (Army Board, N.Z.) 18/1 Effective enemy 88-millimetre guns, which moved with the tanks, had done much damage to our armour. **1945** *News Chron.* 18 Apr. 1/5 British armour..has pushed to within 25 miles of Hamburg.

11. armour glass, toughened glass (cf. *Armour-plate 2); **armour-piercing** *a.*, (of a shell, bomb, gun, etc.) designed to pierce the armour-plating of ships, tanks, etc.; **armour-shelf**, a wide shelf extending below the water-line of an armoured ship, supporting the edge of the armour plating.

1932 *Jrnl. Soc. Glass Technol.* XVI. 478 Armour Glass..similar to that made by Messrs. Pilkington Bros. in this country. **1897** *Daily News* 19 July 9/5 Their 12-inch armour-piercing shot. **1922** *Encycl. Brit.* XXX. 122/1 Armour-piercing projectiles for maximum penetration. **1940** R. W. B. Clarke *Britain's Blockade* 17 Armour plate of high resistance power and armour-piercing bombs of great penetrating power. **1868** E. J. Reed *Shipbuilding in Iron & Steel* xx. 460 The armour shelf having been completed for a portion of the length amidships, the working of the skin-plating behind armour and of the longitudinal girders is commenced. **1890** W. J. Gordon *Foundry* 35 The next, in frame up to her armour-shelf, is an iron-clad of 10,000 tons, building for Her Majesty's Navy.

armour-clad, *ppl. a.* Add: (Earlier examples.) Also *transf.*

1862 *Cassell's Illustr. Fam. Paper* 11 Jan. 104/2 The armour-clad vessels of the South. **1863** [see Ironclad *a.* 1 *note*]. **1901** *Westm. Gaz.* 11 Apr. 10/2 Miolania, a large armour-clad aquatic tortoise. **1903** Lydekker *Mostly Mammals* 308 Armour-clad whales. **1930** *Engineering* 25 Apr. 557/1 The struggle between cubicle and armour-clad [switch] gear is still going on.

armoured, *ppl. a.* Add: **2.** Also, of military vehicles, aircraft, etc., and of forces supplied with armoured vehicles or warfare carried on by their use.

1900 J. Ralph *Towards Pretoria* vii. 115 The destruction of a British armoured train. **1911** *Sci. Amer.* 22 Apr. 410/3 Armoured cars in the Mexican Revolution...An ordinary box car..is sheathed inside with quarter-inch steel plates. **1914** *Illustr. War News* 21 Oct. 16/1 Armoured aeroplanes carrying machine-guns. *Ibid.* 28 Oct. 32 An armoured car outside the Headquarters Hotel at Ostend. **1944** *Return to Attack* (Army Board, N.Z.) 15/1 His bogged transport and armoured fighting vehicles were vulnerable to air attack. **1945** *News Chron.* 18 Apr. 1/5 The advance of the 7th Armoured Division and the 4th Armoured Brigade..was cut in two places. **1947** R. W. Cooper *Nuremberg Trial* viii. 166 General Guderian, Germany's foremost expert on armoured warfare.

b. Of cables, hoses or the like: encased in armouring.

1898 E. J. Houston *Dict. Electr. Words* (ed. 4) 67/1 *Cable, Armored*, an electric cable provided, in addition to its insulating coating, with a protective coating or sheathing, generally made of metal tubing or wire. **1930** *Engineering* 7 Mar. 324/3 The suction pipe of the pump is an armoured hose. **1967** *Gloss. Mining Terms* (*B.S.I.*) VIII. 5 *Armoured flexible conveyor*, a heavy duty chain conveyor designed to carry a coalcutter or power loader on the framework and capable of being continually snaked forward, section by section, as the face advances.

3. Of glass: toughened. (Cf. *Armour-plate 2.)

1935 *Economist* 7 Dec. 1140/2 Laminated or sandwich glass [and]..toughened or armoured glass..are termed safety glass..used for [car] windows and windscreens. **1949** *Archit. Rev.* CV. 181 An armoured glass table the length of the wall below the window provides a large working space.

armouring (ā·ɪmǝriŋ), *vbl. sb.* [f. Armour *v.* + -ing¹.] An external covering, usually of steel wire or bands, to give protection to electric cables or the like.

1924 in S. R. Roget *Dict. Electr. Terms* 11/2. **1930** *Engineering* 14 Feb. 223/2 The cables are clamped in cone clamps for securing the sheath or armouring. **1940** *Chambers's Techn. Dict.* 51/2 Armouring causes eddy current loss in single-core cables.

armour-plate. Add: (Earlier and later examples.)

1860 J. Clare in *Evening Standard* 30 Oct. 3/5, I am the originator of placing timber between..the internal metal and fastening and the external armour-plating. **1863** A. Wynter *Subtle Brains* xxii. 235 The Admiralty have lately taken to manufacture their own armour-plates. **1908** H. G. Wells *War in Air* I. 11 Some confusion of ideas about armour plates. **1940** [see *armour-piercing a.].

2. The trade name of a kind of toughened plate glass; = *armour(ed) glass.

1931 in *Trade Marks Jrnl.* 25 Nov. 1560/1. **1932** *Jrnl. Soc. Glass Technol.* XVI. 78 Now comes a further development, namely, the glass described by Messrs. Pilkington Bros. as 'Armour plate' or 'toughened glass'. **1933** *Ibid.* XVII. 149 'Armourplate' is the trade name given to plate glass which ..has acquired a degree of strength far in excess of that with which the original plate glass was endowed, and the property, if broken, of shattering into small and comparatively harmless fragments. **1941** *Archit. Rev.* LXXXIX. 97 The balustrade is formed of 'armour-plate' glass.

armour-plated, *a.* Add: (Earlier example.) Also *fig.* Insensitive to attack; callous.

1863 *Illustr. London News* 13 June 658/1 The frames of armour-plated ships. **1894** *Cassell's Fam. Mag.* Dec. 69/2 Hard armour-plated gambling. **1906** *Westm. Gaz.* 17 July 10/1 His armour-plated indifference. **1909** *Ibid.* 13 Aug. 3/1 His incisive humour, which got between the joints of his armour-plated victims.

armure (ā·ɪmiuᵊɪ). [Fr.] A woollen or silk fabric, or a mixture of the two, with a twilled or ribbed surface.

1880 in Webster Suppl. **1882** Caulfeild & Saward *Dict. Needlework*, *Armure*, this is a silk textile; plain, striped, ribbed, or with a small design. Sometimes it is made of wool and silk. There is also Satin Armure and Armure Bosphore. **1894** *Daily News* 20 Jan. 5/7 Rich cream-coloured silk of the sort known as 'armure'. **1923** G. G. Denny *Fabrics* I. 16 *Armure*..so called because the fabric has a small ridgy or pebbly pattern which suggests chain armor.

army. Add: **III. army ant,** a popular name for certain predatory ants (esp. of the genera *Dorylus* and *Eciton*) which move about in swarms (also called *driver ants, visiting ants*, etc.); **army co-operation**, co-operation of the air force with the army; freq. *attrib.*; **army-list** (later examples); **Army Service Corps** (in 1918 renamed *Royal Army Service Corps*), that part of the army establishment which is concerned with commissariat and transport; previously called the Commissariat Staff Corps; abbrev. (R.)A.S.C.; **army-worm** (substitute:), *U.S.* the larva of any of various destructive moths.

1874 T. Belt *Naturalist in Nicaragua* ii. 17 Ecitons, or Foraging Ants...In Nicaragua they are generally called 'Army Ants'. **1961** P. Gray *Encycl. Biol. Sciences* 41/1 Ant colonies may contain..several millions, as in the African *Anomma* army ants. **1935** C. G. Burge *Compl. Bk. Aviation* 532/2 Co-operation with the army is performed by army co-operation aircraft. **1940** E. C. Shepherd *Britain's Air Power* 26 They are called Army Co-operation aeroplanes and they work invariably with and for the Army. **1857** Mrs. Gaskell *Let.* 5 June (1966) 451 Will you send me an *East Indian* Army-List, if there is such a thing? **1935** G. Greene *Basement Room* 152 He would look up his record in the old Army Lists when he got home. **1871** *Hansard* CCVI. 961 The Army Service Corps is divided into three branches—supply, transport, and stores, with officers and men attached. **1900** S. L. Norris *S. Afr. War* iii. 57 To Cape Town there went from England..two companies A.S.C. **1916** 'Boyd Cable' *Action Front* 124 Army Service Corps supply points. **1930** J. Fortescue (*title*) The Royal Army Service Corps. A History of Transport and Supply in the British Army. **1819** D. Thomas *Travels Western Country* 155 The *army worm*..has ravaged the meadows. **1852, 1865** [see Dict.]. **1961** R. South *Moths Brit. Isles* (ed. 4) 1st Ser. 205 The White speck or American Wainscot (*Leucania Unipuncta*)..is known in America..as the 'Army Worm'.

Arnaut (aɪnɑu·t). Also **Arnaout**, etc. [See *Albanian *a.*² and *sb.*²] An Albanian, *spec.* one serving in the Turkish army. Also *attrib.*

1717 Lady M. W. Montagu *Let.* 1 Apr. (1965) I. 319 The Arnounts..are natives of Arnountlick, the Ancient Macedonia,..being the best militia in the Turkish Empire. **1802** C. James *Mil. Dict.*, *Arnauts*, Turkish light cavalry. **1833** *Penny Cycl.* I. 258/1 The population of Albania has always been of a warlike character...At present, under the denomination of Arnauts, they rank among the flower of the Ottoman army. **1863** Paton *Hist. Egypt. Revol.* II. xxiii. 256 Instances occurred of Arnaut soldiers getting richly dressed prostitutes, adorned with jewels, into their houses. **1910** *Encycl. Brit.* IX. 107/1 The Arnaut (or Albanian) soldiers in the service of Khosrev tumultuously demanded their pay. **1922** D. H. Low tr. *Ballads of Marko Kraljević* 129 A wild Arnaut woman bore me.

Arnoldian (aɪnǭᵘ·ldiăn), *a.* and *sb.* [f. Thomas *Arnold* (1795–1842), Headmaster of Rugby School, or his son Matthew *Arnold* (1822–88), poet and critic, + -ian.] **A.** *adj.* Belonging to or characteristic of either Thomas Arnold or his son Matthew Arnold. **B.** *sb.* An admirer or follower of either Arnold.

1888 Mrs. H. Ward *R. Elsmere* I. I. v. 112 The 'earnestness' of an Arnoldian Rugby. **1902** *Daily Chron.* 24 Dec. 3/6 The Arnoldian tradition had almost died out...When it was announced that the new Headmastership was to be given to the Liberal Principal of Kneller Hall, everybody was gratified, even the boys, and the old school resumed its Arnoldian glory. **1904** *Ibid.* 19 Mar. 3/1 In only one respect is it likely to disappoint Arnoldians seriously. **1953** *Essays in Criticism* III. 2 It was in that conviction..that its founders gave this journal its Arnoldian title. **1957** *Times Lit. Suppl.* 1 Nov. 658/4 The reference will produce the refutation of some possible charge of want of Arnoldian seriousness. **1966** *Punch* 5 Jan. 31/2 Dennis..sees in him..something Arnoldian and headmasterly.

A·rnoldism. [f. as prec. + -ism.] Doctrine, theory, or practice formed after the precepts and example of either Thomas Arnold or his son Matthew.

c 1845 T. Keble in G. Battiscombe *John Keble* (1963) xiii. 253 Mr. Ward seems to me so strongly tinctured with Arnoldism that it shows itself in all his writings, that particular delight..which T. A. had in speaking provokingly of people and things whom one respected. 1858 J. Cairns *Let.* 20 May in A. R. MacEwen *Life & Lett.* (1895) xvi. 433 It is Arnoldism in its aphelion, and I fear will not come back to the sun. 1888 Gladstone *Let.* in *Westm. Gaz.* (1910) 2 Apr. 1/3 [Mrs. Humphry Ward]..is a fruit, I think, of what must be called Arnoldism. 1892 *Nation* 28 Apr. 322/2 The predilections we have indicated and the slavish Arnoldism of the essays, except when dealing with the Hebrews, disable the writer in the higher field [of criticism]. 1934 R. Campbell *Broken Record* vii. 153 Like Arnoldism-of-Rugby Russellism has this in common with it—that it presupposes the child to be a villain and that only a super-amount of 'treatment' can save him.

aroba, var. Araba.

aromal (ărō^u·măl), *a.* [f. Aroma + -al.] Of or pertaining to, concerned with, or involving, aroma or aromas.

1848 *Tait's Mag.* XV. 704 Man, being in a proper state, *presto*, the aromal spherical bath will start forth and do its work. *Ibid.*, That the earth clogs the sun, in default of aromal discharges. *Ibid.*, Still, though dead, she has her use, that of 'mummy, or aromal loadstone'. 1851 Whittier *My Summer with Dr. Singletary* ii, in *Writ.* (1889) V. 206 Nature's healing ministrations came to me through all my senses. I felt the medicinal virtues of her sights, and sounds, and aromal breezes. 1888 A. S. Wilson *Lyric of Hopeless Love* § l. p. 168 Love is..an aromal fire That sweeps the spirit to dismiss All but its own desire.

around, *adv.* Add: **4.** Perhaps orig. *U.K.* (cf. quot. 1816). Now coming back into British use under U.S. influence. (Further examples.)

1816 Jane Austen *Emma* I. x. 187 Emma..was beginning to think how she might draw back a little more, when they both looked around, and she was obliged to join them. 1936 Wodehouse *Laughing Gas* xxvi. 269 'I want to know why you haven't tied him up.'..'We was aiming to get around to it later.' 1952 *Manch. Guardian Weekly* 9 Oct. 7/3 They tended to get rubbed out before he got around to it.

5. a, b. (Earlier and later examples.) For *shop, sleep around*, see the verbs.

1776 A. R. Robbins *Jrnl.* (1850) 6 Exercised and walked around with the officers in A.M. 1834 *Congressional Globe* 6 Feb. 492 In looking around, the President had a right to select an officer who would honestly..discharge his duty. 1848 Bartlett *Dict. Amer.* 170 *To hang around*, to loiter about. 1869 S. Bowles *Our New West* viii. 170 We generally 'boarded around'. 1870 in Schele de Vere *Americanisms* (1872) 435 She..is witty,..and must be pleasant to have around. 1926 J. Black *You can't Win* iii. 16 When a new boy showed up, I was the one to show him around. 1927 E. Wallace *Hand of Power* xliv. 215, I shall be just eating around. 1948 Gore Vidal *City & Pillar* (1949) ii. ix. 195 Well, I'll see you around, Cy. 1965 J. Fleming *Nothing is Number* ii. v. 87 She hasn't been seen since the beginning of last term...All I know is that she hasn't been around.

c. Going about; astir; out of bed. *U.S.*

1849 N. Kingsley *Diary* (1914) 58 Our Captain is out around to day as it is so pleasant. 1884 *Lisbon* (Dakota) *Star* 18 July, He is now able to be around, but has not yet fully recovered. 1887 F. Francis Jr. *Saddle & Mocassin* 61 Mr. Maroney ain't long gone to bed...I guess he'll be around at midday. 1908 S. E. White *Riverman* xvi. 142, I wasn't here until nine o'clock. I thought, of course, you'd be around.

d. Phr. *to have been around*: to have gained much experience in the world. *colloq.* (orig. *U.S.*).

1927 Hemingway *Fiesta* vii. 72 We've all been around. I dare say Jake here has seen as much as you have. 1938 'G. Graham' *Swiss Sonata* vi. 229 You do have a very unnerving effect on me sometimes, although I'm four years older and I've been around, as Ted would say.

B. *prep.* **4. b.** Of time, amount, etc.: about, sometime near. *U.S.*

1888 *N.Y. Mercury* (Farmer), Presuming he was born around three o'clock in the afternoon, he is under Leo and the Sun. 1920 *Daily Tel.* 3 Apr. 10/6 (N.Y. Lett.), S. Motor Company shares, which usually sell around $100 each. 1931 W. G. McAdoo *Crowded Years* x. 158 The convention adjourned around four o'clock.

arousable (ărau·zăb'l), *a.* [f. Arouse *v.* + -able.] That can be aroused.

1890 W. James *Princ. Psychol.* II. xix. 122 The particular process which currents from the sense-organs arouse would seem under normal circumstances to be arousable in no other way. 1894 *Blackw. Mag.* July 36 The Tuaregs are a wild people of a kind easily arousable to religious fanaticism.

arpeggiated (ɑɹpe·dʒie^i tĕd), *ppl. a. Mus.* [f. Arpeggio + -ate[3] + -ed.] Of a chord or series of chords: played or sung in arpeggio.

1901 *Westm. Gaz.* 31 May 2/1 Arpeggiated accompaniments of a rather old-fashioned type. 1922 S. Grew *Art of Player-Piano* 96 The initial note preceded by an arpeggiated chord. 1924 P. A. Scholes *1st Bk. Gramophone Record* 21 A melody..smoother in motion, and hence well contrasted with the preceding, arpeggiated, theme.

arpeggiation (ăɹpedʒiē^i·ʃən). *Mus.* [f. Arpeggio + -ation.] Playing or singing in arpeggio.

1889 in *Cent. Dict.*

||**arpeggione** (arpeddʒ^i o·ni). [G., f. It. *arpeggio*.] A stringed musical instrument of the early 19th century (see quots.).

1879 Grove *Dict. Mus.* I. 89/1 *Arpeggione*, or Guitar Violoncello, a stringed instrument, played with a bow, which was invented by G. Staufer, of Vienna, in 1823. 1959 *Chambers's Encycl.* XIV. 328/1 The arpeggione, now completely forgotten, derives some importance from the fact that no less a composer than Franz Schubert wrote a sonata for arpeggione and pianoforte.

arrange, *v.* Add: **3.** (Earlier example.) Also *intr.*

1808 *Monthly Pantheon* I. 85/1 An immense expense..the reward of the composer, who selected and arranged the music. 1831 F. Reynolds *Playwright's Adv.* iv. 64 A musician, who, by teaching, arranging, and composing, acquired an income of above one thousand pounds 'per annum'.

arrangeable (ărē^i·ndʒăb'l), *a.* [f. Arrange *v.* + -able.] That can be arranged.

a 1832 Bentham *Deontol.* (1834) I. i. vii. 88 Sanctions are arrangeable according to their nature, or according to their sources. 1858 Carlyle *Fredk. Gt.* I. v. iii. 565 Let these be as in the Treaty of Utrecht; arrangeable in the lump.

arra·nging, *ppl. a.* [f. Arrange *v.* + -ing[2].] Of a debtor: that makes an arrangement with his creditors.

1920 *Act* 10 & 11 *Geo. V* c. 30 § 26 In the distribution of the property of a bankrupt or arranging debtor.

array, *sb.* Add: **6. b.** *Math.* An arrangement of quantities or symbols in rows and columns; = Matrix 6.

1880 R. F. Scott *Theory of Determinants* i. 2 These when written one under the other in rows form a rectangular array...In the theory of determinants we have frequently to deal with several such arrays. 1902 *Encycl. Brit.* XXV. 277/2 The square array being termed the *matrix* of the determinant. 1908 G. H. Hardy *Pure Math.* 301 Arrange all the possible products of pairs u_m v_n in the form of a doubly infinite array

$u_0 v_0\ u_1 v_0\ u_2 v_0 \ldots$
$u_0 v_1\ u_1 v_1\ u_2 v_1 \ldots$
$u_0 v_2\ u_1 v_2\ u_2 v_2 \ldots$

1943 Margenau & Murphy *Math. Physics & Chem.* x. 288 A collection of real or complex quantities is called an array if it can be displayed in an orderly table of rows and columns.

c. In *Statistics*, an arrangement of correlated phenomena or data.

1896 K. Pearson in *Phil. Trans. R. Soc.* A. CLXXXVII. 260 Let A and B be two correlated organs (variables or measurable characteristics) in the same or different individuals, and let the sub-group of organs B, corresponding to a sub-group of A.., be extracted. Let the first of these sub-groups be termed an *array*, and the second a *type*. 1938 L. J. Holman *Simplified Statistics* iii. 24 This arrangement, whether of men or any other items in order of size or other value, is called by statisticians an array. 1957 Kendall & Buckland *Dict. Statist. Terms* 11 *Array*,..More usually, the term denotes some special arrangement of the observations, *e.g.* in order of magnitude.

12. An assembly of directional radio aerials. See *Aerial *a.* 6 b, *Beam *sb.*[1] 24.

1919 G. A. Campbell *Coll. Papers* (1937) 280 Antenna Arrays. 1929 *Bell Syst. Techn. Jrnl.* VIII. 309 Comparisons of calculated and observed directional diagrams of..wave-antenna arrays are..discussed. 1930 [see *aerial *a.* 6 b]. 1938 *P.O. Electr. Engin. Jrnl.* XXXI. 94/1 Array type antennæ..are usually employed for ultra-short wave links. ..Arrays consisting of up to 24 half-wave elements..may be erected. 1940 [see *antenna 5 b]. 1965 *New Scientist* 16 Sept. 676 A laser beam can correlate signals received by an extremely complex radar array.

arrayment. Delete ?*Obs.* and add to def.: Also, the act or fact of arraying or being arrayed.

1875 Morris *Æneids* ii. 450 Other some..The nether doorways of the place in close arrayment hedge. *Ibid.* xii. 442 Then in arrayment close Antheus and Mnestheus rush to war. 1892 *Guardian* 11 May 706/2 Whose light arrayment is a mist of green. 1907 *Daily Chron.* 8 Nov. 4/6 The same arrayment of civic dignitaries in scarlet and fur.

arrear, *sb.* Add: **7. b.** Also used as a *sing.* *U.S.*

1902 *Rep. Librarian Congress* 30 They constitute a large arrears, which should be dealt with speedily.

arrester. Add: **1. c.** (Also *arrestor*.) A device in chimneys, etc., as an electrical precipitator, for preventing smoke or fire; also *spark-arrester*. Cf. also Lightning 3 e.

1838, 1879 [see Spark *sb.*[1] 7]. 1881 [in Dict.]. 1922 W. Schlich *Man. Forestry* (ed. 4) I. 190 Measures necessary to prevent forest fires must be taken, such as..the provision of efficient spark arresters. 1950 *Hansard Commons* CCCCLXXVIII. 271/1 The British Electricity Authority are in process of..installing what are known as [smoke] arresters.

d. A contrivance for bringing to rest aircraft alighting on an aircraft carrier; usu. *attrib.*, as *arrester gear, hook, wire*. See also *arresting *ppl. a.* b.

1926 *Flight* XVIII. 696/2 It is in connection with deck landing that the arrester gear is most likely to be of service. 1927 V. W. Pagé *Mod. Aircraft* (1928) xvii. 714 (*heading*) Catapults and Airplane Arresters. 1937 *Aeroplane* 9 June 691/2 About the same time the Fleet Air Arm gave up arrester-wires and took to trusting to wheel-brakes on bare decks. 1940 *Flight* 15 Feb. 147/1 A deck arrester hook is fitted. 1954 P. K. Kemp *Fleet Air Arm* vii. 97 A more satisfactory type of arrester gear was eventually fitted ['Courageous' and 'Glorious'] consisting of transverse wires on the after part of the deck which were picked up by a trailing hook on the aircraft.

arresting, *ppl. a.* Add: **b.** *Aeronaut.* That stops or checks aircraft. Cf. *Arrester 1 d.

1936 W. M. Burgess *Warships To-day* iii. 48 A touch on the arresting wires..and slowly the aeroplane comes to rest. 1959 *Gloss. Aeronaut. Terms* (B.S.I.) iii. 25 *Arresting gear*, a device, usually installed in aircraft carriers, for bringing an aircraft to rest.

arrestive, *a.* **1.** (Earlier example.)

1834 T. Cowan *Art of Poetic Painting* 3 Strength in expression, and arrestive power in all forms of discourse and composition, consist..in the employment of descriptive and illustrative imagery.

Arretine (æ·rĕtain), *a.* Also **Aretine**. [ad. L. *Ar(r)etinus*, f. *Ar(r)etium* Arretium (see def.): see -ine[1].] Of or pertaining to Arretium (modern Arezzo), an ancient city of central Italy; *spec.* designating fine red pottery made at Arretium and elsewhere from *c.* 100 B.C. until the Flavian period; = *terra sigillata.

1782 J. Elphinston tr. *Martial* p. 426 (*title*) Aretine (or Arretian) Ware. 1858 S. Birch *Hist. Anc. Pottery* II. 342 The two collections of Arretine vases at Arezzo. *Ibid.* 343 Patina of Aretine Ware. 1863 W. Chaffers *Marks on Pottery* 12 The pattern round the top of the Aretine vases is evidently the ovolo, or egg and arrow decoration. 1880 *Encycl. Brit.* XIII. 496/1 Intrusions which may be called Emilian have also been noted..in the district where the Arno and the Tiber take their rise (Arretine dialects). 1954 *Proc. Prehist. Soc.* XX. 8 The Arretine stamps found at Silchester belong to a number..which do not figure in the extensive series from Camulodunum.

arrhythmia (ări·þmiă). *Path.* [mod.L., f. Gr. ἀρρυθμία: see Arrhythmy.] Want of rhythm or regularity; *spec.* of the pulse.

1888 F. P. Foster *Illustr. Med. Dict.* I. 391/1 *Arrhythmia*. 1907 *Practitioner* LXXIX. 492 Note the cardiac arrhythmia. *Ibid.* 494 Various types of arrhythmia of the heart's action. 1961 *Lancet* 22 July 185/1 The sixth patient had complained only of occasional arrhythmia. 1966 *Ibid.* 24 Dec. 1389/1 Propranolol..has been used in the treatment of cardiac arrhythmias.

arrhythmic, *a.* (Earlier and later examples.)

1853 *Fraser's Mag.* XLVIII. 469/1 The fitting rhythm will never be found, the subject itself being arrhythmic. 1958 E. Cooke *Theory of Music* v. 53 There is nothing so wearisome as listening to music which is arrhythmic. 1963 R. P. Dales *Annelids* vi. 130 Only arrhythmic activity can be observed if the ganglia of the intact part of the cord are subsequently touched.

arride, *v.* **2.** For ?*Obs.* read *arch.* Add later examples.

1863 *Temple Bar* IV. 468 The malt consumer is mightily consoled and arrided by the notion. 1884 R. Browning *Ferishtah* x. 87 That verse of mine..much arrides myself. 1906 *Academy* 7 Oct. 2027/1 That which would have most arrided Elia in the note on the poet's description of Hell. 1937 Kipling *Something of Myself* iv. 92 My normal output seemed to have the gift of arriding *per se* the very people I most disliked.

||**arrière-pensée** (aryęr paṅse). [Fr. = 'behind-thought'.] A concealed thought or intention.

1824 Lady Morgan *Salvator Rosa* I. vi. 300 Such drear and fearful aspects of nature, mingled with such views of society, conceal an 'arrière pensée'. 1884 *Eng. Illustr. Mag.* Dec. 246 He knew that I had guessed his *arrière-pensée*. 1900 H. B. Forman *Compl. Wks. Keats* I. 106 The phrase..is probably merely inactivity, without the required *arrière-pensée*.

||**arriero** (ari,ē^ə·ro). [Sp.] A muleteer.

1826 Head *Rough Notes* 185 As a mulish-looking sort of man who used to terrify all the arrieros and peons who passed. 1832 W. Irving *Alhambra* I. 14 The most valuable part of our luggage had been consigned to the arriero. 1845 [see Muleteer]. 1923 *Blackw. Mag.* Sept. 339/1 We were probably reckoned as rather less than the arriero who owned two lean donkeys.

arris. 2. arris rail (earlier and later examples).

1833 Loudon *Encycl. Archit.* 458 The calf-pens to be parted off with oak posts and arris rails (rails presenting two surfaces to the eye, which two surfaces unite in forming an edge or arris between them). 1962 *Listener* 23 Aug. 299/1 Even if you cannot treat all the timber [of the fence] in this way, you may at least be able to treat the ends of the arris rails. 1963 *Gloss. Terms Timber* (B.S.I.) 48 *Arris rail*, a rail, mainly used in fencing, formed by cutting a square section across its diagonal.

arrival. Add: **3.** Also *attrib.*, esp. in connection with the place (platform) or time at which a train arrives at a station, or an aeroplane at an airport.

1838 F. W. Simms *Public Wks. Gt. Brit.* 2 On the opposite side an arrival stage or platform is erected. 1886 *Encycl. Brit.* XX. 235/2 The two general arrival platforms, one on each side of the cab road, are 721 feet long by 12½ wide.

1897 *Westm. Gaz.* 31 Dec. 3/3 It is the arrival time, the departure time having closed several minutes later. 1903 *Ibid.* 25 Mar. 6/3 A new wide roadway for carriages..will be used by arrival passengers. 1967 P. Lorraine *W. T. L. One to Curtis* xii. 116 He..told the taxi-driver to take him straight past the main entrance and deposit him at the Arrival departure. 1968 *Listener* 20 June 806/1 Gate Number 9 at Heathrow's arrival lounge.

arrive, *v.* Add: **10.** Of a person: to be successful, establish one's position or reputation. [After F. *arriver*.]

1889 E. C. Dowson *Let.* 3 Feb. (1967) 32 Imagine a man of low origin, extremely strong, cynical & determined to 'arrive'. 1897 E. Knox *New Egypt* 197 He is a younger man, and has, in the large sense of the word, only arrived comparatively recently. 1934 [see *arriviste]. 1936 *English Studies* XVIII. 53 The book was Herrick's greatest success...With *Together* Herrick arrived.

arrived, *ppl. a.* Also **arrivéd**. [ad. F. *arrivé*.] That has reached a position of success or distinction. Also *absol.*

1896 *Godey's Mag.* Feb. 133/2 On the outskirts of the Monceau quarter, which is peopled with 'arrived' artists. 1899 *Westm. Gaz.* 3 July 3/1 Shall Elizabeth come of that arrived race? 1907 *Daily Chron.* 4 Nov. 3/3 Mrs. Atherton is one of the arrived. 1960 *20th Cent.* Nov. 437 The arrived poets of the post-war generation.

||**arriviste** (arivist). [Fr., f. *arriver* to arrive + *-iste* -ist.] One who is bent on 'arriving', *i.e.* on making a good position for himself in the world; a pushing or ambitious person, a self-seeker. So as *adj.* Hence **arrivisme**, the attitude or behaviour of an arriviste. Also occas. in anglicized form **arrivist**.

1901 Gabrielle Gissing *Let.* 12 July in R. A. Gettmann *Geo. Gissing & H. G. Wells* (1961) 189 What we call an 'arriviste'. 1926 Sir W. Raleigh *Let.* 10 June (1926) II. 331 Haldane is arriviste and worldly—not in the least perceptive. 1934 *National Rev.* Dec. 479 The latter are still arrivistes who have so far failed to arrive. 1935 *Contemp. Rev.* Aug. 174 He..was an arriviste..an arriviste naked and unashamed. 1934 C. Lambert *Music Ho!* iii. 64 Both Stravinsky and Chirico stand outside the more unscrupulous and arriviste work of their disciples. 1936 'C. Brahms' *Footnotes to Ballet* ii. 51 But 'arriviste' though the theme might be, it still found time for..the dancing. 1936 M. Franklin *All that Swagger* xxi. 199 Arrivism was his creed. 1944 A. L. Rowse *Eng. Spirit* xii. 92 Its *arrivisme* can be exaggerated: people did not rise, under Elizabeth, from the bottom of the social scale.

arrow, *sb.* Add: **II.** *caustic arrow* (Surg.): see quot. 1890.

1879 *St. George's Hosp. Rep.* IX. 384 A tumour around the anus was removed by the insertion of caustic arrows. 1890 Billings *Med. Dict.*, *Caustic arrows*, small, pointed cylinders or cones made of chloride-of-zinc paste or other caustics, intended for introduction into the substance of morbid growths.

III. broad arrow. 10. b. The figure of an arrow-head having the point upwards, with which the clothes of convicts were formerly marked. Also *allusively*. Hence **broad-arrowed** *a.*, marked with this.

1865 'Ouida' *Strathmore* [illegible] a lifer. 1886 F. Hopkinson *Valley Tasmanian Trees* 50 The tiny double puncture which is the viper's broad-arrow of death. 1896 *Daily News* 29 July 7/4 Dr. Jameson..was..garbed in the prison shoes and broad-arrowed suit. 1901 *Daily Chron.* 17 June 5/6 His broad-arrow trousers. 1914 *N. & Q.* 11th Ser. IX. 482/2 The Prison Commissioners..write that..the broad arrow..has been used in Convict Prisons and Hulks for more than 84 years, and is also used in Australia. 1934 A. Lee *Child. of Poor* ..that shuffling parade of men in broad-arrowed clothes.

IV. 1. a. *arrow-straightener*; **c.** *arrow-leaved* (earlier example), *-pointed*, *-shaped* (example); **d.** *arrow-line*, *-point* (example).

1812 Withering *Bot. Arrangement* (ed. 5) III. 972 Arrow-leaved Spleenwort. 1902 Belloc *Path to Rome* 246 My unwavering arrow-line to Rome. 1952 J. Fisher *Fulmar* xvi. 248 Fulmars were seen..prospecting by a direct arrow-line. 1779 Adam *Anat. Ind.* 231 One of those arrow-point spines. 1766 *Complete Farmer* s.v. *spinage*, The prickly spinage with arrow-pointed leaves. 1926 M. Leinster..specks of..arrow-shaped leaf. 1919 H. G. Wells *Outl. Hist.*..arrow-straightener. 1935 *Discovery* July..human figures,..

2. *arrow-head* (*U.S.* = Arrow-head 4); **arrow-proof** *a.*, proof against arrows; **arrow-slit** (earlier example); **arrow-weed**, (*a*) = Arrow-head 4; (*b*) a western American composite shrub, *Pluchea sericea*; **arrow-worm**, any worm of the class Chætognatha of small transparent marine worms.

1880 *Harper's Mag.* June 70 The frog pond with lush growth of arrow leaves and pickerel weed. 1931 Clute *Common Names Plants* 22 The starchy tubers of the arrow-leaf (*Sagittaria latifolia*). 1831 *Tatler* 1 Nov. 417/2 Our author and his party..were obliged to provide themselves with arms and with arrow-proof dresses. 1906 *Daily Chron.* 13 Nov. 6/4 The horsemen..upon horses dressed in thick arrow-proof coats. 1851 Melville *Moby Dick* III. xvi. 117 From their arrow-slit in the skull, the priests perceived me. 1846 J. W. Abert in Emory *Mil. Reconn.* 434 Some brackish pools..bordered with the..arrow weed (*sagittaria sagittifolia*). 1876 *Field & Forest* II. 55 These Mexican jumping seeds,..are derived from a plant called arrow weed, or *Yerba de flecha*, and *Colliguaja* by the Mexicans. 1889 *Cent. Dict.*, Arrow-worm. 1952 J. Fisher *Fulmar* xviii. 428 The arrow-worm *Sagitta*..was found..in fulmars in Spitsbergen.

arrow, *v.* Add: **4.** *intr.* To move swiftly through space, like an arrow in its flight; to dart. Also *to arrow one's way*.

1827 *Blackw. Mag.* XXII. 446 About an hour ago did we ..see that identical salmon..arrowing up the Tay. 1866 Whyte Melville *Cerise* xliv, Streams..where the otter lurked and vanished, where the noble salmon himself came arrowing up triumphant from the sea. 1905 W. Holman Hunt *Pre-Raphaelitism* I. 71 Here the kingfisher arrowed his way. 1927 *Chambers's Jrnl.* 4/1 She [*sc.* a filly] arrowed over her hurdles like the 'class' young lady that she was. 1936 R. Lehmann *Weather* (1951) i. v. 72 A figure..came arrowing down the last flight in one straight skim.

5. *trans.* To send forth like an arrow. *poet.*

1892 Tennyson *Death Œnone* [etc.] 39 Shadow-maker, shadow-slayer, arrowing light from clime to clime.

arrowed, *a.* Add: **c.** Pierced with arrows. **d.** Wearing the broad arrow.

1908 E. M. Sneyd-Kynnersley *H.M.I.* xxv. 296 The visitor to Continental picture galleries catches sight of S. Sebastian—'arrowed, but unharrowed'. 1930 R. Campbell *Adamastor* 70 Arrowed like convicts.

arrow-wood. Add to def.: Also applied to other shrubs having straight tough shoots, as *Euonymus, Pluchea sericea*, etc. (Further examples.)

1709 Lawson *New Voy. Carolina* 100 Arrow-wood, growing on the banks, is used, by the Indians, for arrows and gun-sticks. 1717 J. Petiver *Petiveriana* III. 221 Arrow-wood. Used by the Indians for Arrows and Gun-sticks. 1819 McMurtrie *Sk. Louisville* 217 *Euonymus carolinensis*, Indian Arrow Wood. 1850 *New Eng. Farmer* II. 60 The first [dogwood]..is the Arrowwood.

arroyo. (Earlier examples. The erron. form *arroya* also occurs.)

[1807 *Deb. Congress* 1st Sess. I. App. 571/1 The country east of the Sabine to the Arroyo Hondo.] 1845 Frémont *Exped. Rocky Mts.* 252 We discovered..groves of oak trees on a dry arroyo. 1846 Sage *Scenes Rocky Mts.* xix, The banks of this *arroyo* are very steep and high.

arsanilic (ăɹsăni·lik), *a. Chem.* [f. Ars(enic + Anilic *a.*; cf. G. *arsanilsäure* (Ehrlich and Bertheim 1907, in *Chem. Zentralbl.* II. 898).] Of or derived from arsenic acid and aniline, esp. *arsanilic acid*, $C_6H_8O_3NAs$ (= $NH_2.C_6H_4.AsO(OH)_2$), a compound used in the preparation of arsenicals.

1907 *Chem. Abstr.* 2715 The 'arsenic anilide' is really *p-aminophenylarsinic acid* or *arsanilic acid*, as the author [*sc.* Ehrlich & Bertheim] terms his compound. 1937 *Thorpe's Dict. Appl. Chem.* I. 486/2, *p-Arsanilic acid*..was first obtained by Béchamp by heating a mixture of aniline and arsenic acid at 190°–200° with stirring. 1959 *Times* (Agric. Suppl.) 7 Dec. p. vii/1, Arsanilic acid..used both as the free acid and as the sodium salt..to improve feathering and pigmentation [in poultry].

arse, *sb.* Add: **1. b.** *arse over tip*, head over heels.

1922 Joyce *Ulysses* 461 Arse over tip. Hundred shillings to five. 1932 W. S. Maugham *Narrow Corner* xi. 73 I'm pretty nimble on me feet, but I nearly come arse over tip two or three times.

3. *arse-end*; also spec. *arse-end Charlie* (see quot. 1942); *arse-hole* (later examples); **arse-licking** *vbl. sb.* and *ppl. a.*, toadying.

1880 Arse-end [in Dict., sense 2]. 1942 R. Hillary *Last Enemy* 132, I was shot down acting as Arse-end Charlie to a Squadron of Hurricanes. Arse-end Charlie is the man who weaves backwards and forwards above and behind the Squadron to protect them from attack from the rear. 1958 P. Scott *Mark of Warrior* i. 52 We *are* out of date. We get the arse-end of any new equipment that's going. 1930 E. Pound *XXX Cantos* xiv. 61 Faces smeared on their rumps...Addressing crowds through their arse-holes. 1948 *Landfall* II. 178 It's absolute comfort from arse-hole to breakfast-table. 1950 Dylan Thomas *Let.* 17 July (1966) 350 This arsehole of the universe..this..fond sad Wales. 1912 D. H. Lawrence in F. Lawrence *Memoirs* (1961) 189 [Written in D. H. L.'s hand over a flattering letter] Arse-licking. 1958 P. Scott *Mark of Warrior* i. 30, I can't go up and say Were you my brother's C.O.?..it'd look like arse-licking. 1960 F. Raphael *Limits of Love* i. ix. 116 The new fighting Yid still squeals like the old arse-licking kind.

arse, *v.* Delete †*Obs.* and add: *to arse about, around*, to 'mess around', fool about. (Cf. *Ass *v.* 2.) *slang.*

1922 Joyce *Ulysses* 307 Arsing around from one pub to another. 1944 'N. Shute' *Pastoral* ii. 22 Up in London you arse around and go to the local. 1947 N. Streatfeild *Grass in Piccadilly* 148 Mrs. Dill would understand. She would never arse about yattering all over the place. 1960 A. Wesker *I'm Talking about Jerusalem* i, Don't arse around Ronnie, the men want their tea.

arsenical, *a.* Add: **B.** *sb.* A substance containing arsenic.

1909 in *Cent. Dict.* Suppl. 1943 [see *atoxyl]. 1949 H. W. Florey et al. *Antibiotics* II. xxxvi. 1199 It seems clear that doses of arsenicals and penicillin..were effective in experimental animals when combined. 1955 *Sci. News Let.* 16 July 40/2 Many of the itch remedies used today..such as ..arsenicals, may depend for their results on a partial blocking of the..itch enzymes.

arsine[1]. Now usu. (ăɹsi·n). (Examples.)

1876 *Encycl. Brit.* V. 541/1 Arsine and stibine..are formed whenever hydrogen is evolved in presence of an arsenic or antimony compound. 1910 *Ibid.* II. 652/1 Arsenic *trihydride* (arsine or arseniuretted hydrogen), AsH_3, is formed by decomposing zinc arsenide with dilute sulphuric acid. *Ibid.*, The arsines and arsine chlorides are liquids of overpowering smell, and..exert an extremely irritating action on the mucous membrane. 1922 F. W. Aston *Isotopes* vi. 77 The gases phosphine PH_3 and arsine AsH_3 were used in the experiments on these elements.

arsine[2] (ăɹsi·n), var. Arsheen.

1932 B. Kincead tr. *M. Ilin's What Time is It?* 107 The meter counted out the versts, sagens, and arsines.

arsinoïtherium (ă:ɹsino,iþī^ə·riɒm). *Palæont.* [mod.L., f. Gr. Ἀρσινόη, an ancient Egyptian city of the Fayum, so named, after his second wife Arsinoë II, by Ptolemy II Philadelphus (308–246 B.C.), who developed it as the metropolis of that region + θηρίον, dim. of θήρ wild beast: cf. Therio-.] An animal of an extinct genus of mammals, as large as the rhinoceros, with two large and two small horns.

1902 H. J. L. Beadnell *Prel. Note Arsinoitherium Zitteli* 3 Discovery of Eocene mammalian and reptilian remains made last year by the Geological Survey of Egypt...The most important of these is a large, heavily built, ungulate, about the size of a rhinoceros, and for which the writer proposes the generic name *Arsinoitherium*, from Queen Arsinoe, after whom the Fayum was called in Ptolemic times. 1904 *Daily Chron.* 4 Jan. 9/1 The four-horned arsinoïtherium. 1930 *Discovery* Nov. 386/2 The arsinoitherium, a huge beast the size of a rhinoceros, was among the most fantastic mammals which have ever existed.

arsonist. Delete *rare*[−1] and add later examples.

1958 M. Pugh *Wilderness of Monkeys* xii. 149 A born arsonist should never try to be a fireman. 1962 *Economist* 3 Nov. 452/2 They attacked Unip as a party of thugs, arsonists and intimidators.

arsphenamine (ɑɹsfe·nămain, -in). [f. Ars(enic + Phen- + Amine.] A synthetic compound of arsenic, $C_{12}H_{12}O_2N_2As_2 2HCl$, used in the treatment of spirillic diseases, as syphilis and yaws. Cf. *Salvarsan.

1917 *Jrnl. Amer. Med. Assoc.* LXIX. 1989/2 The Federal Trade Commission has entered orders for licenses to three firms to manufacture and sell arsphenamine, the product heretofore known under the trade name of salvarsan, patent rights to which have been held by German subjects. 1935 *Lancet* 13 July 74/2 A sensitivity to the arsphenamines may develop gradually. 1944 *Ibid.* 11 Mar. 336/2 A similar process may be at work in bone-marrow dyscrasias following arsphenamine therapy. 1950 G. M. Findlay *Recent Adv. Chemotherapy* (ed. 3) I. i. 8 The first of these arsenobenzene derivatives, arsphenamine..contained both amino- and hydroxy-groups in the benzene nucleus.

arsy-versy, *adv.* (Later examples.)

1881 Duffield *Don Quixote* II. 360 You do not go enchanted, but only with your head turned arsie varsie. 1957 S. Beckett *All that Fall* 29 Like Dante's damned, with their faces arsy-versy.

art, *sb.* Add: **5.** *art for art's sake* (further examples). Hence in many allusive phrases (see quots.).

[1836 V. Cousin *Cours de Philosophie 1818* 224 Il faut de la religion pour la religion, de la morale pour la morale, comme de l'art pour l'art.] 1920 B. Russell *Pract. & Theory Bolshevism* iv. 48 There it stands, this old art, the purest monument to the nullity of the art-for-art's-sake doctrine. 1925 A. Huxley *Those Barren Leaves* ii. i. 86, I was not an art-for-arter. 1928 —— *Point Counter Point* xvi. 291 We're frankly missionaries, not an art for art concern. 1937 'G. Orwell' *Road to Wigan Pier* xii. 243 Our leading writers, who a dozen years ago were art for art's saking for all they were worth..are now taking a definite political standpoint. 1942 *Burlington Mag.* May 115/1 The first exponents of 'art for art' did not, as do their descendants, uphold the claims of the senses abstractly and in isolation. 1948 J. W. Aldridge in *Penguin New Writing* XXXV. 115 *Ulysses* represents the extreme of the art-for-art's-sake doctrine. 1950 E. C. Pettet in *Essays & Studies* III. 45 The advocate of the 'practical' Shakespeare, allied on this occasion with the art-for-art's-saker, says [etc.].

7. *attrib.*

1912 W. Owen *Let.* 12 June (1967) 141, I definitely abandon the *thought* of Divinity Training *till* at least an Arts Degree is won. 1946 *Universities Quarterly* I. 52 The Arts faculties ought to include sufficient knowledge of general science to provide a general appreciation of science and the scientific method as applied to the problems of daily life. 1960 *Guardian* 20 Apr. 8/3 It should be possible to make scientists literate and arts men 'numerate'. 1967 J. Philip et al. *Best of Granta* 104 *Granta's* characteristic role in the 'sixties has been as an avant-garde arts magazine.

17. arts and crafts: the arts of decorative design and handicraft; *spec.* work done by or under the auspices of the Arts and Crafts Exhibition Society, founded in London in 1888,

or similar later work; also *attrib.* Hence *colloq.*) **art-and-crafty, arts-and-craftsy** *adjs.*, pertaining to or characteristic of the arts and crafts or of the 'arts and crafts' movement, esp. its more pretentious side. Cf. *arty-and-crafty *a.*

1888 *Times* 29 Sept. 6/1 The Arts and Crafts Exhibition at the New Gallery..may best be described as an exhibition with a purpose. 1894 *Studio* 48/1 The impression..on the mind of a person who had previously had little experience of collective 'Arts and Crafts' may not be without practical value. 1899 J. W. Mackail *Life W. Morris* II. xviii. 200 The newly-formed association was at first known by the name of the Combined Arts. The name of the Arts and Crafts was the invention of Mr. Cobden-Sanderson. He was also..responsible for another of the new departures made in the first Arts and Crafts Exhibition. 1902 [see *arty-and-crafty *a.*]. 1909 *Westm. Gaz.* 24 Dec. 2/2 An art and crafty tea-table. 1939 O. Lancaster *Homes Sweet Homes* 44 The fervent mediaevalism..found its final expression in the Arts-and-Crafts movement. 1943 G. Greene *Min. Fear* i. i. 8 Short blunt fingers prickly with big art-and-crafty rings. 1957 R. Campbell *Portugal* vi. 202 The arts-and-craftsy, self-conscious supervision, patronage, and spurious jollity given to morris dancing..in England.

V. *Comb. art-activity, -appreciation, -collecting* vbl. sb., *-collection, -collector, -connoisseur, -correspondent, -critic* (earlier and later examples; hence *art-critical* adj., *-critically* adv., *-criticism*), *-dealer, -instinct, -lover* (hence *-loving* ppl. a.), *-magazine, -monger, -product* (example), *-sale, -school* (examples), *-student, -style, -teaching* vbl. sb., *world*. Also **art centre** (see Centre *sb.* 6 a); **art director**, one who is responsible for the décor, properties, scene-painting, etc., in a theatre or in cinematographic films; **art editor**, one who is responsible for the illustrations or the section devoted to the arts in a book, magazine, etc.; hence *art-edit* v. trans. (*rare*); **art-form** [cf. G. *kunstform*], (*a*) an established form taken by a work of art, e.g. a dialogue, novel, sonata, sonnet, triptych, madrigal; (*b*) a theme or motif constituting a traditional subject of works of art, e.g. the Madonna and Child; (*c*) a medium of artistic expression; **art gallery** [Gallery *sb.* 6], a building or portion of a building devoted to the exhibition of works of art; formerly also *art museum*; **art history** [cf. G. *kunstgeschichte*], the history of art, esp. as an academic study; hence *art-historian, art-historical* adj.; **art object**, an object of artistic value; = *objet d'art; **art paper**, paper coated on one or both sides with china clay or the like to give a smooth surface; coated paper (see *coated *ppl. a.* 3); **art square**, a patterned square of carpet woven in a single piece; **art union** (further examples); in Australia and New Zealand, a lottery with cash prizes.

1923 D. H. Lawrence *Stud. Classic Amer. Lit.* vi. 93 The rhythm of American art-activity is dual. 1896 *Peterson Mag.* VI. 225/2 Art-appreciation, like art-creation, is a slow evolution. 1937 *Burlington Mag.* June 310/1 General hints on art-appreciation for the ordinary public. 1908 A. Bennett *Buried Alive* vii. 176 London, the acknowledged art-centre of the world. 1967 *Listener* 20 July 76/2 One of those exclusive civic centres or art centres in the South Kensington or South Bank tradition. 1902 *Encycl. Brit.* XXV. 683/2 They acted as a most healthy stimulus to art collecting. *Ibid.*, The first really important art collection to come under the hammer. 1936 *Burlington Mag.* July 46/2 The Veronese art-collector Bendetto Maffei. 1904 W. James *Let.* 30 June (1920) II. 206 The bulk of 'Modern Painters' and the other artistic writings..have made us take him [*sc.* Ruskin] primarily as an art-connoisseur and critic. 1882 Wilde *Let.* 19 Feb. (1962) 96, I would undertake to be your art-correspondent for London and Paris—two articles a month. 1865 Rossetti *Let.* 15 Nov. (1965) II. 580 The Art-critic's original *dicta*. 1866 *Argosy* (Midsummer) 61, I should certainly have liked to consult our great modern art-critic before making so daring a statement. 1944 H. Treece *Herbert Read* 53 Read has always been more artist than art critic. 1936 *Burlington Mag.* June 303/1 Extensive art-historical and art-critical work. 1880 Swinburne *Let.* 17 May (1960) IV. 143 A sample of the 'first manner' (to speak art-critically) of the poem. 1867 *Fine Arts Q. Rev.* II. 174 We must recognize an immense advance in the tone and character of art-criticism, especially in..leading journals. 1891 Wilde *Intentions* 258 It is only in art-criticism, and through it, that we can apprehend the Platonic theory of ideas. 1935 *Burlington Mag.* Nov. 202/1 The merciless scrutiny of modern art-criticism. 1934 A. Woollcott *While Rome Burns* 11 The young art-dealer was not precisely what would have been called pro-Ally. 1933 *Archit. Rev.* LXXIII. 127 The newest and most satisfactory examples of decoration in the world of the movie art-director. 1933 P. Godfrey *Back-Stage* xiii. 160 The first art director to do so. 1923 T. E. Lawrence *Let.* 13 Dec. (1938) 443 Hogarth will literary-edit the proofs for me: & Kennington art-edit the blocks. 1877 *Harper's Mag.* Dec. 53/2 The day editor [puts]..news relating to art in the hands of the art editor. 1941 *Oxoniensia* VI. 93 The art-editors of the Commission are not entirely aware why a good photograph is good, or a bad one not good. 1868 G. M. Hopkins *Notebks.* (1937) 97 An intellectual attraction for very sharp and pure dialectic or, in other matter, hard and telling art-forms. 1887 *Magazine of Art* 135/2 Some such accidental juxtaposition of decorative bird-woman and galleys..suggested the *art-form* or myth of Odysseus and the Sirens. 1894 *World* 21 Feb. 23/1 This type of musical farce is not an elevating or intellectual art-form. 1928 H. Read *Phases of Eng. Poetry* i. 11 Anglo-Saxon poetry is already a highly developed art-form. 1929 H. G. Wells *King who was King* i. 8 (*heading*) The Film, the Art Form of the Future. 1952 D. Talbot Rice *Eng. Art 871–1100* v. 132 Before the crosses were finally eclipsed as an art-form, a very wide repertory of different types of decoration was to appear on their shafts and heads. [1845 Disraeli *Sybil* III. v. i. 14 Something of the splendour or the rarities of the metropolis; its public buildings, museums, and galleries of art. 1860 *Birmingham Council Proc.* 15 May 195 The scheme..should comprise a central Reference Library.., a Museum and Gallery of Art. 1863 *Ibid.* 5 May 198 The Fine Art Gallery now being erected in connection with the Central Free Library.] 1865 *Ibid.* 28 Nov. 37 An offer..to deposit in the Art Gallery several valuable pictures belonging to the Society. 1885 M. Davitt *Prison Diary* II. xxiv. 46 Every public library in all towns of say 5000 inhabitants should have in connection with it a museum and an art gallery. 1946 *Ann. Reg. 1945* 333 This year of victory saw the gradual resumption of their normal functions by most of our public art galleries and museums. 1890 Wilde *Critic as Artist* in *Wks.* (1948) 958 The Greeks..wrote essays on art, and produced their art-historians. 1907 *Daily Chron.* 5 June 6/4 In its way the most thorough piece of art-historian work that was ever produced. 1957 *Times Lit. Suppl.* 27 Dec. 790/1 The writer is an archaeologist and historian of art, not an 'art-historian', so those interested in this unfamiliar aspect of Greek art are better referred not to the text but to the plates. 1933 R. Fry *Art-History as Acad. Study* 11 The whole tendency of their [*sc.* the Germans'] art-historical studies has been to regard works of art almost entirely from a chronological point of view..without reference to their aesthetic significance. 1874 *Temple Bar* XLII. 204 Incidentally may here be mentioned, though not strictly within the limits of art history, the wonderful power which was exercised this year by a portrait of the period. 1876 *Mind* I. 477 Hasty inductions drawn from a narrow area of art-history are erected into general principles. 1927 R. H. Wilensky *Mod. Movement in Art* i. 51 The genius is so rare in art history. 1891 Wilde *Pict. Dorian Gray* iv, She has not merely art, consummate art-instinct, in her, but she has personality also. 1874 *Daily Tel.* 4 May 5/6 This is..what artists and art-lovers will thank him for. 1861 Trollope *Tales of all Countries* Ser. ii (1863) 52 The haunts in Rome which are best loved by art-loving strangers. 1934 *Burlington Mag.* Oct. 146/2 The English might have a reputable art magazine. 1928 A. Huxley *Point Counter Point* v. 71, I envy you art-mongers your success. 1856 *Art-Jrnl.* 2nd Ser. II. 93/1 If the money spent in the Art-museums of France was taken into consideration, it would be found that the assistance.. was on a..liberal scale. 1857 *Ibid.* 239/1 As a national Art-museum it [*sc.* the South Kensington Museum]..is a success. 1904 *Westm. Gaz.* 20 Oct. 1/2 Everyone..remarked how very flat the picture market has been, compared with that of 'art objects'. 1913 T. E. Lawrence *Let.* 16 Oct. (1954) 269 Arab glass..is the rarest art object in the world. 1962 W. Nowottny *Lang. Poets Use* vi. 137 Different ways of looking at a house..as an art-object. 1905 *Jrnl. Soc. Chem. Industry* XXIV. 771/2 The unpleasantness and fatigue caused by the reflection of light from the surface of high-glazed art papers. 1958 *Times Lit. Suppl.* 3 Oct. 567/4 To print both illustrations and text on coated paper (that is, so-called art paper). 1904 W. James in *Atlantic Monthly* July 102/2 All his [*sc.* Spencer's] dealings with the art-products of mankind. 1882 W. F. Poole *Index Period. Lit.* (ed. 3) 64/1 Art and Art Sales in England, in the 18th Century. 1902 *Encycl. Brit.* XXV. 684/1 The greatest art sale in the annals of Great Britain. 1866 *Once a Week* 3 Feb. 134/1 The Universities do not teach art, the Art-schools do not teach anything else. 1935 *Discovery* Jan. 17/2 Groups drawn from art schools. 1902 *Encycl. Brit.* XXVI. 605/1 The products of these two processes are well known under the trade-names of 'Parquet Carpets' and 'Art Squares'. 1849 *Art-Jrnl.* XI. 107/3 In ordinary times there resides at Paris a numerous body of artists and Art-students. 1934 R. Benedict *Patterns Cult.* (1935) iii. 48 What has happened in the great art-styles happens also in cultures as a whole. 1857 Ruskin *Pol. Econ. Art* ii. 101 The most singular concentration of art-teaching and art-treasure. 1839 Dickens *Let.* 18 Feb. (1965) I. 509 All good fortune to the Art Union. 1849 *Sydney Morning Herald* 27 Nov. 1/4 The undersigned guarantees the..sums of £50 and £40 to the drawers of the 1st and 2nd Prizes, in his Art Union. 1851 Rossetti *Let.* 30 Aug. (1965) I. 103 A Notice about an Art-Union print. 1931 *Dominion* (Wellington) 26 Dec. 13/7 Here we have art unions freely sanctioned..for every conceivable object from sports of every kind to first aid. 1948 D. Ballantyne *Cunninghams* 148 But it would be a long, long time before she got a new dress—unless her ship came home and she won an art union. 1966 *Courier-Mail* (Brisbane) 5 May 3/11 A Scottish couple..yesterday won first prize in a Queensland consultation, the Scarborough art union. 1890 *Atlantic Monthly* Dec. 753/1 January and February, 1890, saw the culmination of a new movement in the art world of Paris.

18. a. Designed to produce an artistic effect, as *art furniture, needlework, pottery*, etc.

1868 *Building News* 25 Dec. 869/3 (*heading*) Messrs. Walford and Donkin's art furniture. 1870 Art-furniture [see sense V in Dict.]. 1879 M. E. Braddon *Vixen* I. xvii. 327 Your last piece of art needlework. 1880 L. Higgin *Handbk. Embroidery* 98 The School was founded in 1872... It was first established, under the title of School of Art-Needlework, in Sloane Street; but in 1875 was removed to the present premises in the Exhibition Road. 1881 C. C. Harrison *Woman's Handiwork* 47 Canton flannel,..a soft downy fabric,..comes in all the 'art' shades. 1885 C. M. Yonge *Two Sides of Shield* I. v. 75 'Don't you love art needlework?' 'Maude Sefton has been working Goosey Goosey Gander on a toilet-cover.' 1887 G. B. Shaw *Don Giovanni Explains* in *Wks.* (1932) VI. 99 Some exquisitely fine fabric in an 'art shade' of Indian red. 1887 *Trade Marks Jrnl.* 9 Feb., 'Liberty' Art Fabrics. 1893 Yonge & Coleridge *Strolling Players* x. 77 What she called 'an art-frock' in Liberty silk. 1894 Mrs. H. Ward *Marcella* i. vii. 61 Marcella wore 'art serges' and velveteens. 1895 *British Warehouseman* Feb. 17/2 Chintzes, Art Blinds, Window Hollands,..Art Serges. *Ibid.* 38/2 A new..shade-card, comprising all the newest art tints. 1897 *Daily News* 23 Mar. 7/1 Great art-pottery establishments..are busy in the preparation of vases and other articles. 1900 J. K. Jerome *Three Men on Bummel* viii. 171 A yard or so of art muslin. 1930 W. S. Maugham *Cakes & Ale* xii. 151 Curtains of art serge and a bilious green.

b. [cf. G. *kunstlied, -musik*.] Produced by an artist, composed with conscious artistry: said of poetry and music, opp. to *popular* or *folk*, as *art ballad, music, song*.

1890 A. B. Bach (*title*) The Art Ballad. Loewe and Schubert. *Ibid.* 19 Schubert was the creator of the art song, Loewe the creator of the art ballad. 1934 Webster, Art music. 1940 [see *artify *v.*]. 1950 M. J. C. Hodgart *Ballads* iii. 49 Ballad music, like other folk music, sounds strange to anyone familiar only with 'art' music. *Ibid.* 57 Folksong is at variance with modern 'art' song, in which the practice is to make the musical stresses correspond to the speech-stresses. 1959 L. Bernstein *Joy of Music* (1960) 172 The songs of these shows are closer to art songs than they are to Tin Pan Alley.

c. Applied *spec.* to a theatre, cinema, etc., specializing in consciously artistic productions, opp. to *commercial, popular*, etc.; hence *art film*.

1929 S. Cheney *Theatre* xxiii. 520 Georg Fuchs of the Munich Art Theatre. 1932 W. Rothenstein *Men & Memories* II. xix. 154 Rumours of Craig's success at the Moscow Art Theatre reached London. 1933 P. Godfrey *Back-Stage* xiii. 160 The studio or art theatre exists..to prevent dramatic art from being wiped out by the commercially minded. 1944 L. MacNeice *Columbus* 15 The radio play..is competing with the Soviet art-cinema rather than with Hollywood. 1959 *Encounter* XIII. 52 The most the art-house circuit can do is $600,000. 1960 *Guardian* 15 Dec. 6/3 French art film makers..spin romantic webs around works of art. 1962 *Listener* 8 Mar. 448/2 Films as sheer entertainment..are slowly being ousted by these art-films. 1967 *Guardian* 5 Aug. 7/8 The Latin Quarter is rich in art cinemas.

VI. ||**art** (ar), the French equivalent, occurring in certain phrases used in English contexts, as:

a. art autre [lit. 'other art'] = *Tachism(e.

1959 *Vogue* June 115 *Art autre* makes painting a truly visual art. 1963 *Times* 17 May 18/4 Official taste remains complacently becalmed at action-painting and *art autre*.

b. art brut [lit. 'raw art'], primitive or unsophisticated art.

1955 A. C. Ritchie *New Decade* 21 Jean Dubuffet.. intensified interest in *l'Art brut*, the work of prisoners, mediums, the insane and other non-professionals. 1960 *Times* 10 Oct. 7/2 A certain aestheticism, under the obligatory roughnesses of *art brut*, is discernible.

c. art mobilier [lit. 'portable art'], a term applied to prehistoric decorated or carved objects.

1946 *Proc. Prehist. Soc.* XII. 153 Engraved plaques of clay-slate from the Bann diatomite indicate..that the *art mobilier* of the Iberian Peninsula has, to some extent, its counterpart in Ireland also. 1959 J. D. Clark *Prehist. S. Afr.* x. 259 The famous *art mobilier* which is found together with the implements and other occupation material.

d. art moderne, modern art.

1934 D. Parker *After Such Pleasures* 97 There was no tallying the gifts of Charvet handkerchiefs, *art moderne* ash-trays, etc. 1937 L. Bromfield *Rains Came* i. xxiii. 115 Atrocious bits of *art moderne*.

e. art nouveau (ăɹ nūvō^u) [lit. 'new art'] (often with capital initials), a style of art developed in the last decade of the 19th century, characterized by the free use of ornament based on organic or foliate forms and by its flowing (i.e. non-geometrical) lines and curves. (Called 'Jugendstil' in Germany.)

[1899 *Studio* XVII. 44 Jewellery..executed at Mr. Bing's establishment 'l'Art Nouveau'.] 1901 *Times* 15 July 12/5 It is much to be regretted that the authorities of South Kensington have introduced into the Museum specimens of the work styled, 'L'Art nouveau'. 1908 G. B. Shaw *Lett. to G. Barker* (1956) 137 A model cemetery with Art Nouveau tombstones. 1909 J. Thorp *Æsthetic Conversion* (*Heal & Son*) 10 The *art nouveau*, with its meandering tulips and inconsequent squirms and dots. 1928 J. Buchan *Runagates Club* 103 A new plate of gun-metal and oxidised silver, lettered in the best style of *art nouveau*. 1939 O. Lancaster *Homes Sweet Homes* 54 A recurrent passion for tortuous curves and sinuous lines..which..finds expression in..the flamboyant Gothic of the later Middle Ages, mid-eighteenth century rococo and, most deplorably, in *Art Nouveau*. 1939 A. Thirkell *Before Lunch* iv, An oxidized silver stand representing the Three Graces in Art Nouveau style.

art, *v.*[2] Restrict *Obs.* to senses in Dict. and add: **5.** *to art up*: to make arty; to decorate in an arty fashion. *colloq.*

1929 *Amer. Mag.* Jan. 126/3 The knowledge that spinning wheels are perfectly good antique stuff..had been impinged upon..by the desire to art up and decorate indiscriminately. *Ibid.* 127/2 The general arting up they have done, with regard to cottages, tea rooms, gift shops..must have had its effect. 1940 A. Thirkell *Cheerfulness* vi. 104 The necks ..wanted arting up a bit, as the frocks were so very plain in cut. 1959 *Motif* 3 Sept. 16/1 Attempts to art up the superstructure, to trick it out with forms that have no reasonable development.

artefac. = next. (Disused.)

1906 Chamberlin & Salisbury *Geol.* III. 502 Following European precedent, the earlier students classed the

rougher artefacts as paleolithic...The better fashioned artefacs were classed as neolithic. 1911 *Encycl. Brit.* XXI. 836/1 The difficulty of employing *artefacs* of stone as chronological indicators.

artefact (ā·ɹtifækt), *sb.* and *a.* Also **arti-**. [f. L. *arte*, abl. of *ars* art + *factum*, neut. pa. pple. of *facere* to make. (Cf. Sp., Pg. *artefacto*, It. *artefatto*, adj. and sb.)] **A.** *sb.* Anything made by human art and workmanship; an artificial product. In *Archæol.* applied to the rude products of aboriginal workmanship as distinguished from natural remains.

1821 Coleridge in *Blackw. Mag.* X. 256 The conception of all these, as realized in one and the same artéfact, may be fairly entitled, the Ideal of an Ink-stand. a 1834 [see Artifact]. 1890 D. G. Brinton *Races & Peoples* ii. 75 This is shown..by the presence of artefacts and shells from the Pacific in old graves on the Atlantic coast. 1922 *Class. Q.* XVI. 24 The shadows seem to be real till their originals are exposed as the paltry artefacts they are. 1925 *Times Lit. Suppl.* 13 Aug. 529/1 The distribution of artifacts. 1927 G. Murray *Class. Tradition* 243 Poetry..is an 'artifact'—I mean, it is a thing made.

b. In technical and medical use, a product or effect that is not present in the natural state (of an organism, etc.) but occurs during or as a result of investigation or is brought about by some extraneous agency.

1908 *Practitioner* LXXXI. 383, I must quote a case of dermatitis artefacta, which was hardly hysterical...The alternatives for diagnosis..were (1) Some local pyodermic infection. (2) A trophic lesion (?). (3) An 'artefact'. 1940 *Chambers's Techn. Dict.* 52/2 *Artefact* (Zoology), any apparent structure which does not represent part of the actual specimen, but is due to faulty preparation. 1943 *Electronic Engin.* XVI. 189/3 The reverse kick [on the cathode-ray oscilloscope]..was not to be expected on theoretical grounds...This artefact is probably unimportant. 1946 *Brit. Jrnl. Psychol.* Sept. 22 The recognition of artefacts, such as those due to circulatory or respiratory changes.. comes with experience. 1947 *Electronic Engin.* XIX. 82/2 An amplifier which will be free from the annoying 'artefacts' which are so often due to self-generated input stage noise. 1961 *Brit. Med. Dict.* 138/2 *Artefact*. In histology, a misleading appearance in a preparation caused by some form of contamination, or by physical or chemical changes induced by manipulation or the reagents employed in making the preparation. In electro-encephalography, any wave that has its origin elsewhere than in the brain. In dermatology, a self-induced lesion.

c. *transf.*

1934 Toynbee *Stud. Hist.* III. iii. 156 It is a mere accident..that the material tools which Man has made for himself should have a greater capacity to survive..than Man's psychic artif[a]cts. 1949 *Foreign Affairs* XXVII. 384 The police power of a government cannot be a pure political artifact.

B. *adj.* Made by human art and workmanship. *rare.*

1909 J. A. Stewart *Plato's Doctr. Ideas* 179 The rêverie-image of an object natural or artefact.

artel (aɹte·l). [Russ. *artél'*.] In Russia, an association of craftsmen or other workers for work in common. Also *attrib.* in *artelman* [partial tr. of Russ. *artél'shchik*].

1884 J. Rae *Contemp. Socialism* vii. 280 The *artel* or labourers' association and the commune. 1892 *Daily News* 22 Jan. 5/2 The only other kind of trade organization existing in the Empire is known as the 'Artel', and is peculiar to Russia. 1895 *Chambers's Jrnl.* 9 Nov. 705/2 The authority of an artelman is unlimited. 1900 L. Maude tr. *Tolstoy's Resurrection* 226 We are all masons, and belong to the same *artel*. 1921 S. Graham *Europe—Whither Bound?* vi. 89 An *artel* of Rostof engineers. 1955 H. Hodgkinson *Doubletalk* 27 A brigade, or *artel*, chosen without regard for family connections, undertake particular functions—ploughing, reaping, processing, milking, etc.—as and where required.

arterial, *a.* Add: **2.** Applied esp. to main roads or lines of transport or communication. Also as *sb.*, an arterial road.

1865 *Chr. Misc. & Family Visitor* Ser. ii. XI. 154/1 The termini of the different trunk or arterial railways. 1867 [in Dict.]. 1898 *Daily News* 17 Nov. 4/4 He urged that wide arterial improvements of the streets were needed. 1920 *Act* 10 & 11 *Geo. V* c. 57 § 2 Land required for or in connection with the construction of any arterial road. 1932 H. Nicolson *Public Faces* xii. 310 Jane Campbell..sped along the Kent arterial.

arteriogram (aɹtī^ə·riogræm). [f. Arterio- + -gram.] †**1.** = Sphygmogram. *Obs.*

1885 Stirling tr. *Landois's Text-bk. Human Physiol.* I. ii. § 67. 134 In every pulse-curve—sphygmogram or arteriogram—we can distinguish the ascending part of the curve.

2. A picture obtained by arteriography (sense *b).

1929 *Amer. Jrnl. Roent.* XXII. 80 New technique of arterial encephalography...Cases are described in which the arteriograms were of more or less diagnostic value. 1944 *Lancet* 26 Aug., The third arteriogram in the series..shows the outline of the popliteal artery.

arteriography. Add to def.: † = Sphygmography. *Obs.* Also, earlier example.

1842 Dunglison *Dict. Med. Sci.* (ed. 3) 70/2 *Arteriography*, a description of the arteries.

b. Examination of the arteries by means of radiology after injection of a radio-opaque material. Hence **arteriogra·phic** *a.*, of or pertaining to arteriography.

1929 D. M. Greig *Edin. Med. Jrnl.* XXXVI. 632 Arteriography in the living subject. The injection of arteries by.. shadow-casting material and subsequent radiography..on the living subject is recent and as yet restricted to few operators...Moniz, Pinto and Lima, who have carried out seventy cerebral arteriographies without..untoward symptom. 1944 *Lancet* 26 Aug. 271/1 It was necessary to determine the exact size, shape and anatomical location of the sac. Arteriography is the only certain method of deciding this preoperatively. 1955 *Gloss. Terms Radiology* (*B.S.I.*) 56 *Arteriography*, the radiological examination of arteries following direct injection of a contrast medium. 1962 *Lancet* 8 Dec. 1225/1 Based on arteriographic criteria, 7 (45%) of the 15 arterial clots were lysed after approximately 60 hours' intensive therapy.

arteriolar (aɹtī^ə riō^u·lăɹ), *a.* [f. Arteriol(e + -ar[1].] Of, belonging to, or of the nature of, an arteriole.

1941 in Dorland *Med. Dict.* 1961 *Lancet* 9 Sept. 573/2 Examination of the fundi showed bilateral retinal arteriolar narrowing.

arteriosclerosis (aɹtī^ə:riosklī^ə rō^u·sis). *Path.* [f. Arterio- + Sclerosis.] Abnormal thickening and hardening of the walls of the arteries, occurring chiefly in old age.

1886 D. Macalister tr. *Ziegler's Text-bk. Path. Anat. & Pathogenesis* ii. 21 Gull and Sutton were the first to call special attention to the arterial changes associated with contracted kidney...However..they did not sufficiently distinguish between primary arteriosclerosis and the secondary form. 1901 *Practitioner* Mar. 315 The patient may be handicapped in his struggle with the fever by arterio-sclerosis. 1948 tr. *A. Brodal's Neurol. Anat.* 409 Among conditions which in some cases have benefited from sympathetic denervation may be mentioned..arteriosclerosis.

Hence **arte:riosclero·tic** *a.*, of, pertaining to, or affected by arteriosclerosis; as *sb.*, an arteriosclerotic person.

1886 D. Macalister tr. *Ziegler's Text-Bk. Path. Anat. & Pathogenesis* ii. 20 The affection..is best described by the term arteriosclerotic atrophy. The renal arteries and their branches in aged persons are very frequently the seat of sclerotic changes. 1907 W. James *Let.* 2 Jan. in R. B. Perry *Tht. & Char. W. J.* (1935) II. 452 All is well with the James family—all but an increase of my arterio sclerotic symptoms in the last six months. 1909 *Practitioner* Nov. 616 The arterio-sclerotic kidney. 1961 *Lancet* 5 Aug. 306/2 Arteriosclerotic brain disease. *Ibid.* 26 Aug. 492/2 A short bruit can be related to kinking and lengthening of the iliac arteries in arteriosclerotics.

arteriovenous (aɹtī^ə:riovī·nəs), *a.* [f. Arterio- + Venous *a.*] Of, pertaining to, or affecting an artery and a vein.

1880 [see Arterio-]. 1890 Billings *Med. Dict.* I. 109/1 *Arterio-venous aneurism*, an aneurism opening into a vein. 1946 *Lancet* 20 July 85/2 The signs of an arteriovenous fistula were found in his right femoral region. 1961 *Ibid.* 29 July 250/1 Systolic or continuous murmurs can be heard in the upper abdomen in patients with.. arteriovenous shunts.

artery, *sb.* Add: **4.** (Earlier and later examples.)

1850 Glisan *Jrnl. Army Life* (1874) 27 Those great arteries of commerce—the railroads. 1924 Galsworthy *White Monkey* iii. vii. 258 But along the main artery at the end the traffic streamed and rattled.

b. A major river in a river-system. *U.S.*

1805 Jefferson in E. O. Rowland *Life W. Dunbar* (1930) 177 We shall delineate with correctness the great arteries of this great country. 1827 J. F. Cooper *Prairie* xxiv, The mighty arteries of the Missouri and Mississippi.

artesian, *a.* Add: By extension applied to water obtainable by artesian boring.

1893 *Pop. Sci. Monthly* XLII. 607 In the West many people discredit the possibility of artesian water in many favorable localities, because of the absence of consolidated strata. 1897 *Westm. Gaz.* 21 Apr. 8/2 A very large part of the interior of Western Australia is wholly unfitted for the purposes of human habitation. Its appearance may ultimately be changed by the discovery of artesian water.

arthralgia (aɹþræ·ldʒ^i ă). *Path.* [mod.L., f. Gr. ἄρθρον joint + ἄλγος pain.] Pain in a joint. Hence **arthra·lgic** *a.*

1848 Dunglison *Dict. Med. Sci.* (ed. 7), *Arthralgia*, arthrodynia, gout. 1878 E. S. Wood tr. *von Ziemssen's Cycl. Pract. Med.* XVII. 566 An attack of colic or arthralgia [in lead-poisoning]. 1890 Billings *Med. Dict.*, *Arthralgic*.., pertaining to arthralgia.

arthro-. Add: **arthrodesis** (aɹþrǫ·disis) *Surg.* [ad. G. *arthrodese* (E. Albert 1878, *Lehrb. d. Chirurgie* (1883) II. 505), f. Gr. δέσις binding together (δέειν to bind)], an operation to produce ankylosis; **arthrography** (example); (*b*) examination of the joints by radiography; **arthro·meter** [see -meter], an instrument (patented in 1919 by W. W. P. Falconer) for measuring the movement-range of joints; hence **arthro·metry**, the science or study of such measurements; **arthrome·trical** *a.*, of or pertaining to arthrometry; **a·rthroplasty** [see -plasty], the construction of an artificial joint; hence **arthropla·stic** *a.*; **arthrotomy** (aɹþrǫ·tŏmi) *Surg.* [see -tomy; cf. Fr. *arthrotomie*], incision of a joint.

1901 *Amer. Jrnl. Med. Sci.* CXXII. 597 *Arthrodesis.* This operation is performed with the aim of establishing a firm right angular ankylosis at the ankle-joint, thus allowing the patient to dispense with apparatus. 1903 Tubby & Jones *Surgery of Paralyses* i. iii. 89 Complete paralysis of the muscles around the ankle-joint calls invariably for arthrodesis. 1952 M. E. Florey *Clin. Appl. Antibiotics* I. vii. 212 An operation for arthrodesis of the knee. 1857 Dunglison *Dict. Med. Sci.* 96/2 *Arthrography*, a description of the joints. 1937 *Lancet* 8 May 1119/1 Arthrography also explains the unsatisfactory clinical results sometimes obtained in cases where the radiographic report on the reduction [of a dislocation] has been encouraging. 1949 G. Claessen *Diagnostic Radiol.* ii. 71 The semilunar cartilages of the knee joint do not show on the straight X-ray film except when calcified, but can be demonstrated after the injection of a contrast solution into the joint. This is known as arthrography. 1918 *Lancet* 9 Nov. 632/2 The illustration depicts a new device for measuring the angles of movements of joints...This 'arthrometer' is simple in construction and easily adjusted to the limbs. 1918 W. W. Falconer (*title*) Arthrometry or the Measurement of the Movements of Joints. *Ibid.* 27 In connection with joint-movement mensuration I have..prepared a complete series of 'Arthrometrical Charts'. 1913 Dorland *Med. Dict.* (ed. 7), Arthroplastic. 1918 *Reveille* Aug. 67 There were very few stiff joints in which arthroplastic operations were urgently needed. 1890 Billings *Med. Dict.*, Arthroplasty. 1962 *Lancet* 19 May 1036/1 In the Ullevål series, 16 patients had a primary arthroplasty with Thompson's prosthesis. 1888 F. P. Foster *Med. Dict.* I. 438/1 Arthrotomy. 1907 *Practitioner* Apr. 575 Robin..reiterated his condemnation of arthrotomy, except in the case of suppuration supervening in the articulation. 1961 *Lancet* 30 Sept. 745/1 In the infected hips..early arthrotomy should be undertaken.

Arthropoda, arthropod. (Later examples.) Hence **arthropod-borne** *a.*, of a pathogenic organism carried by an arthropod. Cf. *arborvirus.

1922 *Encycl. Brit.* XXXI. 896/1 Such arthropoda.. being specific 'nurses' or intermediary hosts of the parasite actually causing the disease, are known as 'carriers' or 'vectors'. 1924 *Glasgow Herald* 27 Sept. 4/2 A crowd of joint-footed Arthropod animals..that grip firmly with their claws. 1956 *Nature* 25 Feb. 367/2 A comprehensive..study of acute febrile illnesses..was undertaken to..determine the part that arthropod-borne viruses might play in their causation. 1962 *Lancet* 29 Dec. 1363/2 One suggestion is that this may be a virus tumour dependent on some arthropod vector. 1968 *Times* 14 Nov. 8/8 Insects belong to the large group of animals known as arthropods.

Arthurian (aɹþiū^ə·riăn), *a.* and *sb.* [See -ian.] **A.** *adj.* Of or pertaining to the legendary British king Arthur, his court, and his knights; also, resembling these or the romances in which they figure.

1853 J. H. Burton *Hist. Scot. 1689–1748* I. 174 If any reality could be extracted from the Arthurian histories. 1869 Stuart-Glennie *Arthurian Localities* 5 The Arthurian Cycle. 1875 A. W. Ward *Eng. Dram. Lit.* I. 121 The Arthurian legend. *Ibid.*, The Arthurian cycle of legend. 1899 A. Nutt *Celtic & Mediæval Romance* 14 The later works of the Charlemagne cycle are in detail, tone and spirit often as 'Arthurian' as any purely Breton romance. Huon and Ogier are Arthurian heroes who have strayed by accident to the Court of Charlemagne. 1942 *Burlington Mag.* Feb. 45/1 William Morris and his ideal of the Arthurian unity of the four countries.

B. *sb.* A follower of King Arthur, a knight of the Round Table; one who studies the Arthurian legend.

1907 J. W. H. Atkins in *Camb. Hist. Eng. Lit.* I. xiv. 312 The noble Arthurian [*sc.* Gawain]..grants the life of the defiant Golagros. 1957 G. Ashe *King Arthur's Avalon* vi. 221 Giraldus..was no gullible Arthurian. 1966 *English Studies* XLVII. 145 Many Arthurians..will feel that the Celtic elements in Geoffrey have been somewhat neglected.

Arthurianism (aɹþiū^ə·riăniz'm). [f. prec. + -ism.] A form of doctrine, theory, or practice based upon the actions and beliefs of the legendary King Arthur.

1908 A. Noyes *Wm. Morris* ii. 22 One true Church of Arthurianism. 1932 *Times Lit. Suppl.* 21 July 521/2 The Arthurianism of the sixteenth century comes to little more than a topic for pageantry.

article, *sb.* Add: **14.** (Earlier example.)

1796 Grose *Dict. Vulg. Tongue* (ed. 3), *Articles*, breeches: coat, waistcoat, and articles.

b. Applied to a person: often derog. Formerly, in the U.S., used of a slave considered as an 'article' of merchandise. *slang.*

1811 *Lexicon Balatronicum*, *Article*, a wench. A prime article. A handsome girl. She's a prime article (*Whip slang*), she's a devilish good piece, a hell of a goer. 1837 H. Martineau *Society* II. 325 The creditors..answered that these young ladies [his 'quadroon' nieces] were 'a first-rate article'. 1844 Dickens *Mart. Chuz.* xxvi. 320 You're a nice article, to turn sulky on first coming home! 1852 Stowe *Uncle Tom* i, There's an article now! You might make your fortune on that ar gal in Orleans, any day. 1857 Trollope *Three Clerks* III. ii. 41 She's the very article for such a man as Peppermint. **1863** L. M. Alcott *Hospital Sketches* v. 80 Here was a genuine article—no, not the genuine article at all, we must go to Africa for that—but the sort of creatures generations of slavery have made them. **1928** E. Sutton tr. *A. Londres's Road to Buenos Ayres* iii. 20 Some 'articles' are..women from seventeen to twenty years old. **1957** M. K. Joseph *I'll Soldier No More* (1958) x. Listen, you sloppy article, who was on guard from twelve to two last night?

c. Colloq. euphemism for chamber-pot.

1922 Joyce *Ulysses* 715 Pitcher and night article (on the floor, separate). **1939** *Statesman* (Calcutta) 24 July, A mad-man..crowned his amusement this morning by bringing an article which he..sat on. **1958** J. Cannan *And be a Villain* iii. 62 How could he be so rude, she asked, when he said 'pot' instead of 'bedroom article'.

d. With defining word: applied to something judged to be authentic of its kind, not an imitation or counterfeit, esp. (*the*) *genuine article.* Cf. *the real thing* (Real a.[2] 4 c).

[**1863** (see sense *b).] **1864** C. M. Yonge *Trial* II. iv. 82, I think that's so! Genuine article—no mistake. **1923** A. Bennett *Regent* v. vi. 161 Mr. Rollo Wrissell belonged to one of the seven great families which once governed England. Edward Henry breathed to himself, 'This is the genuine article'. **1919** D. Ashford *Young Visiters* (1951) v. 34 His mother was a decent family called Hyssops of the Glen so you see he is not so bad and desirous of being the correct article. **1922** Joyce *Ulysses* 643 Education (the genuine article).

articulable (ɑːtɪ·kiŭlăb'l), *a.* [f. Articulate *v.* + -able.] That can be articulated. Also *absol.*

1833 *Fraser's Mag.* VII. 706 Word-worshipping even the articulable clothing wherein the clear and ethereal harmony of Goethe is invested. **1897** W. James *Lett.* 28 Apr. (1920) II. 58 Life and mysticism exceed the articulable, and if there is a One, it must remain only mystically expressed. **1916** F. von Hügel *German Soul* 92 How impossible in practice is the maintenance of international obligations of any articulable kind.

articulacy (ɑːtɪ·kiŭlăsi). [f. Articul(ate *a.* + -acy.] The quality or state of being articulate; articulateness.

1934 in Webster. **1943** E. M. Almedingen *Frossia* ix. 313 She was angry, and anger denied her articulacy. **1960** J. B. Broadbent *Some Graver Subject* ii. 70 Satan's mobility, his articulacy, and muscularity..make him the most vital character in *Paradise Lost*.

articular, *a.* **2.** *Gram.* Add to def.: Having an article prepos

1906 Moulton *Gram. N. T. Greek* I. 70 There is a very marked increase in the use of the articular nominative in address. **1926** Plater & White *Gram. Vulgate* 40 The Gerund with *in* is frequently used to render the Greek Articular Infinitive with preposition.

articulated, *ppl. a.* Add: **1.** Also *fig.*

1886 A. Weir *Hist. Basis Mod. Europe* iii. 122 The pettifogging lawyers who swarmed around the minutely articulated social structure.

2. (Further examples.)

1950 *Gloss. Aeronaut. Terms (B.S.I.)* 1. 44 *Articulated blade*, a blade connected to the rotor head by one or more hinges or pivots. **1958** *Daily Mail* 19 Aug. 4/8 It should be fairly easy to devise some sort of articulated joint [in the steering column of a car] which would cause the column to fold up in a crash.

b. Of a road vehicle: consisting of elements joined in a flexible arrangement, as *articulated lorry*. Abbrev. *artic* (slang). Also, of a locomotive, train, bus, etc., one fitted with a bogie or bogies, or other mechanical device making for flexibility.

1923 *Sci. Amer.* Jan. 12/3 Attention has been drawn to the articulated train, in which the abutting ends of the passenger cars are carried upon a common truck. **1930** *Engineering* 30 May 715/3 (*heading*) Articulated Steam Rail Coaches for the Egyptian State Railways. **1932** *Statutory Rules & Orders* 1931 1158 'Articulated vehicle' means a heavy motor car, with a trailer drawn thereby. **1936** *Discovery* Nov. 356/2 High-speed articulated trains of three coaches. **1951** *Landfall* V. 106 Running those two big articulated cars for him. **1958** *Times* 12 Apr. 7/6 Joe had a brand-new articulated truck, or artic (where the driving cab is separately joined to the lorry).

articulative (ɑːtɪ·kiŭlătiv), *a.* [f. Articulat(e *v.* + -ive.] Of or pertaining to articulation.

1849 A. M. Bell *Princ. Speech* xi. 11. 155 The analogy between the articulative actions of R and S is generally of much service in bringing the tongue to the position for the latter element. **1877** W. R. Alger *Life E. Forrest* I. 68 The ponderous guttural and sweetness of articulative energy.

articulator. Add: **3.** An apparatus used in prosthetic dentistry to obtain correct articulation of the teeth.

1851 *Amer. Jrnl. Dental Sci.* Jan. 181 Cut a V shaped groove down the back of the cast, and oil it well, so that the articulator will not stick to it; then mix some plaster and drop it into the impression...The articulator should be about half an inch thick. **1953** *Faber Med. Dict.* 47/2 *Articulator*, instrument for supporting the casts of the jaws while arranging artificial teeth on the plates. **1963** G. F. Kantorowicz et al. *Inlays, Crowns & Bridges* vii. 74 A model of all teeth of the opposing jaw will allow assessment of the occlusion without the use of an articulator.

4. A movable vocal organ (as the tongue). (See quot. 1942.)

1942 Bloch & Trager *Outl. Ling. Anal.* 13 The vocal organs which can be moved more or less freely and can thus be made to assume a variety of positions; and *points of articulation*, fixed points or areas lying above the articulators, which these may touch or approach. **1943** K. L. Pike *Phonetics* vii. 120 At the time in the production of some sound when any movable part of the vocal apparatus causes any *stricture*..it becomes an *articulator* unless it is a closure performing the function of an initiator. **1946** J. Murray in *Q. Jrnl. Speech* XXXII. 203/1 The second interpretation of 'explosive' pertains exclusively to plosives which characterize the syllabic thrust following closures of the glottis, the articulators and glottis.

articulatory, *a.* (Later examples.)

1890 W. James *Princ. Psychol.* II. xviii. 64 In persons whose auditory imagination is weak, the articulatory image seems to constitute the whole material for verbal thought. **1935** G. K. Zipf *Psycho-Biology of Lang.* (1936) v. 230 Conceptual words relate our speech to our experience, articulatory words relate our conceptual words to one another. **1953** C. E. Bazell *Linguistic Form* iv. 46 Some plausible articulatory-acoustic interpretation can be given of the distributional phoneme. **1955** H. A. Gleason *Introd. Descriptive Linguistics* xiv. 187 Articulatory phonetics..is concerned with the study of sounds usable in speech in terms of the mechanisms of their production by the human vocal apparatus. **1964** B. Honikman in D. Abercrombie et al. *Daniel Jones* 73 By articulatory setting is meant the disposition of the parts of the speech mechanism and their composite action. **1964** R. H. Robins *Gen. Linguistics* iii. 84 It [sc. phonetics] can be studied primarily as the activity of the speaker in terms of the articulatory organs and processes involved; this is called articulatory phonetics.

Hence **articula·torily** *adv.*, in relation to any matter of vocal articulation.

1952 A. Cohen *Phonemes of English* ii. 23 Two sounds of a language, related either acoustically or articulatorily, which never appear in the same phonetic context, are..

artifact: see *Artefact.

artificer. Add: **2.** (Earlier and later U.S. examples.)

1758 S. Hamilton *Lett. to Washington* (1899) II. 321 The 2d. Company of Artificers of the 2d. Regiment. **1861** *U.S. Army Regulations* 520 Every volunteer non-commissioned officer, ..and artificer, who enters the service.

b. A mechanic in the Royal Navy. Also *attrib.* and *Comb.*

1884 *Encycl. Brit.* XVII. 294/1 The engine-room artificers, leading stoker, and stokers are under his [sc. the engineer's] immediate control. **1899** [see *Tiffy*]. **1902** *Monthly Rev.* Aug. 92 The Admiralty have conferred the charge of machinery in small vessels on fifty-six artificer-engineers. **1957** *H.M. Royal Forces Navy Openings for Boys* (*H.M.S.O.*) (ed. 2) 22 Aircraft Artificers. An apprentice in this branch is trained to undertake the most difficult maintenance and repair work on Naval aircraft. *Ibid.*, 27 Radio electrical artificers..deal with wireless and radar equipment.

artificial, *a.* Add: **1. b.** *artificial light* (earlier examples).

1662 Fuller *Worthies* (1662) 191 Darkness..made artificial Lights to appear with the more Solemnity. **1828** F. Warren *Electro-Magnetism* 10 Sir H. Davy..was enabled to produce the most intense artificial light ever seen.

c. *spec.* (a) In *Physics*, applied to the disintegration of an element or particle (under bombardment, etc.), to radioactivity that is induced (as opposed to that which occurs naturally), to isotopes produced in this way, and to the radioactivity of such isotopes; (b) popularly applied to objects, real or imaginary, put into orbit by man, e.g. *artificial moon, planet, satellite.*

1920 Rutherford in *Proc. R. Soc.* A. XCVII. 398 The region of mass 3 arises in the artificial disintegration of lighter atoms like oxygen and nitrogen. **1926** R. W. Lawson tr. *Hevesy & Paneth's Man. Radioactivity* xxvii. 228 Recent years have seen the initiation of the artificial disintegration of the elements, an achievement first successfully carried out, and continued, by Rutherford in 1919. **1934** Joliot & Curie in *Nature* 10 Feb. 201 (*heading*) Artificial production of a new kind of radio-element. *Ibid.*, 201/2 These experiments give the first chemical proof of artificial transmutation. **1938** R. W. Lawson tr. *Hevesy & Paneth's Man. Radioactivity* (ed. 2) vii. 73 Neutrons disrupt the silicon nucleus and call forth an artificial radioactivity. *Ibid.* x. 123 It is possible to produce a number of artificial radio-elements in a variety of ways.

1880 P. Greg *Across Zodiac* I. iii. 51 The weeks I spent in the solitude of this artificial planet..[in] a voyage through space. **1946** H. Harper *Dawn Space Age* 68 Pirquet has worked out plans for an artificial island, or rather an 'artificial moon', stationed out in space..as a permanent fuelling-point for space-craft setting out on interplanetary voyages. **1963** *Listener* 14 Feb. 287/1 Mariner has now entered a permanent orbit round the Sun, and has joined the increasing company of dwarf 'artificial planets'.

3. b. *spec.* in *Contract Bridge* (see quots.).

1930 E. Culbertson *Contract Bridge Blue Bk.* xx. 267 'Strength showing' optional or artificial bids..are a polite warning to the enemy not to fall into the trap. **1950** J. Culbertson *Contract Bridge for Beginners* (rev. ed.) xv. 162 'Artificial' bidding. When the partner bids, for example, five hearts to show that he has two aces, his bid has nothing whatsoever to do with the hearts in his hand. **1961** *Listener* 28 Sept. 485/2 South's One Diamond was artificial.

5. *artificial aid*, an aid (e.g. crampons, pitons) used in climbing; hence *artificial climbing*, climbing with the assistance of such aids; *artificial fertilizer, manure*, a chemical fertilizer or manure; *artificial horizon* (earlier example); also, a gyroscopic instrument used in an aircraft to indicate the attitude of the aircraft in relation to the horizon; *artificial insemination*, injection of semen into the uterus (of an animal) by other than the natural means (abbrev. *A.I.*); so *A.I.D.*, *A.I.H.*, artificial insemination (of a woman) with a donor's, the husband's, semen (see also *donor d); *artificial kidney*, an apparatus that performs the functions of a human kidney in place of a diseased or injured organ; *artificial language*, an invented language, esp. one designed for international use; *artificial lung*, an apparatus which supplements the supply of oxygen available to the lungs or induces respiration; cf. *iron lung*; *artificial respiration*, a manual or mechanical procedure designed to restore the natural function of breathing when this has been suspended; *artificial rubber*, a manufactured plastic substance resembling rubber; *artificial silk*, see *silk*; *artificial sunlight*, a name given to light radiated from an ultra-violet ray lamp; see also *sun-ray.

1934 E. R. Blanchet in S. Spencer *Mountaineering* v. 107 *Artificial Aids*.—On rocks these are met with more and more frequently. The rope is no longer used merely for safety but also as a means of climbing and may be looked upon as an artificial aid. **1966** C. Evans *On Climbing* iii. 53 'Artificial' climbing, on sections otherwise unclimbable, is the province of the advanced climber. **1907** *Roy. Soc. S. Australia Trans.* Index 1877–1906 37 Artificial fertilisers. **1936** *Economist* 11 Jan. 71/2 The increase in imports is mainly due to larger imports of artificial fertilisers. **1943** J. S. Huxley *TVA* 116 The artificial fertiliser industry. **1801** C. Gould *Brit. Patent* 2539 Artificial horizon, to be attached to and used with the quadrant or sextant for taking altitudes on land or water. **1912** *Engineering* 1 Mar. 285/3 The artificial horizon..brought out..as an attachment to sextants especially suitable for observations from balloons. **1920** *Flight* XII. 1158/1 In common with all artificial horizons when used in aircraft, however, the level is subject to error due to the acceleration of the machine acting on the bubble. **1897** *Proc. R. Soc.* LXI. 52 (*heading*) The artificial insemination of mammals. **1923** H. M. Box *Trial M. Stopes* (1967) 105, I am referring to that paragraph about artificial insemination of women with seed taken from other men. **1936** *Discovery* Oct. 328/2 We might..by 'artificial insemination' produce a race containing as many 'Lenins or Darwins' as we pleased. **1945** *Brit. Med. Jrnl.* 13 Jan. 40/1 We ourselves have used, and in certain respects developed, the technique of artificial insemination (A.I.) during the past five years. *Ibid.* 40/2 (*heading*) Male sterility and Other Indications for Artificial Insemination with Donated Semen (A.I.D.). *Ibid.* 43/1 In a Jewish marriage..artificial insemination of the wife with the husband's semen..

[…] *Ibid.* 43/1 In a series of 50 successive cases A.I.H. was carried out because of low invasive power of the semen. **1947** *Lancet* 19 Apr. 523/1 The signed statement of consent, usually obtained by the doctor from both the husband and wife before insemination with donor's semen (A.I.D.) is performed, is likely to be written largely. **1957** C. L. Russell in *Artificial Human Insemination* vii. 56 A.I.D. involves adultery, whatever the motives, circumstances, or consequences. **1949** 'G. Orwell' *Nineteen Eighty-Four* I. 68 All children were to be begotten by artificial insemination (*artsem*, it was called in Newspeak). **1959** *Observer* 15 Jan. 11/5 Intended originally for the small farmer, A.I. has proved of increasing interest also to the larger farmer and the pedigree breeder. **1913** *Brit. Med. Jrnl.* 16 Aug. 387/1 Salicylic acid injected into a vein was excreted in some cases rather more rapidly by this artificial kidney than by the normal secretory channels. **1949** *Jrnl. Pharmacol. & Exper. Therapeutics* Jan. 276 The apparatus constitutes what has been called an artificial kidney in the sense that it allows the escape of the diffusible constituents of the blood. **1961** *Lancet* 1 Sept. 551/2 Representatives of several artificial-kidney units..expressed the opinion that clotting in the artificial kidney was commoner when machines of large surface-area were used. **1864** F. M. Müller *Lect. Sci. Language* 2 ser. xi. 61 And though an artificial language might be much more perfect, more regular, more easy to learn, than any of the spoken tongues of man. **1885** W. F. Farrar *Chapters on Lang.* iii. 37 All attempts to frame an artificial language have been a failure. **1919** G. Willis *Philos. of Speech* iv. 63 An artificial language, such as Esperanto, can never acquire any literary value, although it may possibly provide a useful vehicle of scientific or philosophic thought. **1870** *Century Mag.* Nov. 29/2 This result is obtained by means of an *artificial lung*..The diver carries this apparatus on his back. **1947** *Sci. Digest* Sept. 74 (*caption*) Oxygenated blood, external artificial lung. **1872** (*title*) The artificial manure, its chemical and England. **1882** *Encycl. Brit.* XIV. 568/1 The artificial manures which contain the essential..soluble phosphates, nitrogenous artificial manures. **1932** A. Huxley *Mortal Coils* 31 His land was farmed in the best modern way—silos and artificial manures and continuous cropping and all that. **1852** *New Orleans Med. & Surg. Jrnl.* IX. 209 One thing that saved the child's life—the Artificial Respiration. **1854** W. Marcet in *Med. Times & Gaz.* 22 Apr. 401 An account of a new instrument for performing artificial respiration. **1922** *Man. Seamanship* (*H.M.S.O.*) I. 371 To effect artificial respiration put yourself astride or on one side of the

patient's body, in a kneeling position, facing his head. **1911** C. Christy *African Rubber Ind.* xvii. 232 There are hundreds of materials called rubber substitutes..but these are not artificial rubbers. **1930** *Engineering* 18 Apr. 525/2 Artificial rubber gave no spot diagram. **1942** *Electronic Engin.* Mar. 668/1 Polyethylene and polyisobutylene (artificial rubber) are almost as good. **1927** *Lancet* 29 Oct. 947/2 London clinic for artificial sunlight treatment. **1928** *Daily Express* 27 June 3/6 The Committee on Artificial Sunlight in Industrial Hygiene.

B. *spec.* Artificial manure.

1860 in Thirsk & Imray *Suff. Farming 19th Cent.* (1958) 57 Artificial consumed. **1937** W. Deeping *Doomsday* xix. 2 Plenty of stock..saved you from having to spend too much on artificials.

b. An artificial flower.

1840 H. Cockton *Valentine Vox* xxvi. 196 A bouquet of variegated artificials on one side. **1848** Mrs. Gaskell *Mary Barton* I. i. 8 Esther, I see what you'll end at with your artificials, and your fly-away veils.

c. An artificial bait.

1949 A. Wanless *Fly Fisherman's Alphabet* 53 There are not so many flies which trout feed on and the artificials are definite imitations of them. **1959** *Times* 7 Feb. 9/3 There have been many occasions when the lure or bait caught on..spun artificials.

artify (ɑ̄·ɹtifai), *v. colloq.* [f. Art *sb.* + -ify.] To bring art into; to make 'arty'. Hence **a·rtifying** *vbl. sb.*

1905 *Westm. Gaz.* 27 Oct. 11/4 Man's work..is 'to artify the world, and to be ever spiritualizing it towards the beauty of holiness'. **1940** *Scrutiny* VIII. 398, I am not in favour of a rapprochement between art-music and commerce because..the artifying of 'low' music has been demonstrably almost spent. **1940** J. Betjeman in *Archit. Rev.* LXXXVIII. 162 Certain brewers..say they give the people's idea of art, where the brewer has not arty-fied them or destroyed them for some wretched trade-mark. **1962** *Times* 28 Nov. 15/4 What can happen when an attempt is made to 'artify' folk elements.

artillery. Add: **8. artillery wheel,** a heavily-built dished wheel used on gun-carriages; a similar wheel on motor vehicles.

1904 Rhys Jenkins *Motor Cars* 210 The wheels [of the Daimler car] are of wood, artillery pattern, fitted with pneumatic tyres. **1902** *Car* 15 Oct. 252/2 The Lanchester car was not strong enough to have artillery wheels, and so was fitted with wire suspension wheels. **1904** A. B. F. Young *Complete Motorist* 200 The wheels [of the Baby Peugeot car] are of the artillery type.

artiness: see *Arty *a.*

artisa·nship. [See -ship.] The work and activity of an artisan or of artisans collectively.

1847 Carlyle *Germ. Romance* IV. 14 Intellectual artisanship..is less desirable than intellectual manhood. **1852** Chalmers in *Fraser's Mag.* III. 60 The toils of busy artisanship. **1832** Carlyle *Crit. & Misc. Ess.* (1840) IV. 102 Work out thy Artisanship in the spirit of an Artist. **1884** *N. Brit. Rev.* I. 139 Its [sc. the Byzantine Empire's] Industry and commerce. **1907** W. de Morgan *Alice-for-Short* xiv. The sudden substitution of inspiration for mere artisanship.

artist. Add: **10.** Usu. preceded by a defining word: a person, 'chap', 'fellow'; also, one devoted to or unusually proficient in something (reprehensible). *U.S., Austral.*, and *N.Z. slang.*

1890 Farmer *Slang* I. 75/1 *Artist* (American thieves'), an adroit rogue; a skilful gamester. **1919** Downing *Digger Dial.* 9 One-star artist—a second lieutenant. **1932** Runyon in Wentworth & Flexner *Dict. Amer. Slang* (1960) 468/2 Chiv artist. **1934** *Bulletin* (Sydney) 22 Aug. 49/2 Syd..classified him as a species of booze-artist. **1940** *Ibid.* 3 Jan. 35/4 Dingbatting to a one-pip artist may be a bit more tree and easy. **1941** Baker *Dict. Austral. Slang* 6 *Artist*, a fellow, a bloke. An expert, a specialist; also one who indulges in excesses, e.g. 'bilge a.', 'booze a.', 'bull a.'. **1947** D. M. Davin *Roads from Home* iii. 49 A real artist for the booze, isn't he?

11. *attrib.* and *Comb.*

1894 G. B. Shaw *Music in London* 1890–4 (1932) III. 177, I still call for the artist-craftsman to give us once more a flute that is a flute, and a trumpet that is a trumpet. **1902** *Westm. Gaz.* 27 Jan. 2/1 The artist-author has studied his subject from every point of view. **1908** *Athenæum* 25 July 108/2 The monotonous laments of the artist-hero Tikipu. **1929** S. Cheney *Theatre* xxiii. 517 The development of artist-directors stimulated interest in the search for 'form' as an attribute of stage art. **1930** O. Lancaster *Homes Sweet Homes* 62 An 'artist designed' table of unusual design. **1959** Halas & Manvell *Technique Film Animation* 13 A few individual artist-producers who have managed to combine a characteristic style of work with adequate business organization.

artiste. (Earlier and later examples.)

1823 *Drama, or Pocket Theatr. Mag.* IV. 355 He [sc. Charles Vestris] is clearly the least distinguished artiste to the whole race of Vestris. **1871** R. H. Horne in *Gentl. Mag.* Sept. 473 At his house were met artists and artistes of various kinds, all 'on their promotion' and students in literature, poetry, and science. **1909** Wodehouse *Swoop* II. viii. 101 They stood talking and blocking the gangway, as etiquette demands that a successful artiste shall. **1955** *Times* 19 Aug. 4/1 The artistes—they lose their *e* on notice boards inside the reception office—..were absent today.

artistly, *adv.* (Earlier examples.)

1754 J. Cleland *Let.* 31 July in *Garrick's Private Corr.* (1831) I. 56 The plots of two excellent plays..being wove artistly into one. **1811** Simond *Jrnl. of Tour in Gt. Brit.* 8 June (1815) II. 219 Tables, sofas, and chairs, were artistly *deranged* about the fire-places.

artotype (ā·ɹtotaip). [irreg. f. Art *sb.* + -o + -type.] A form of *collotype process*, the collotype process. Hence, the print or impression produced by this process.

1879 *Photogr. News* 9 May 222/1 In America..the introduction of a method of photo-lithographic printing, named *Artotype*. [has] given a sudden impetus to..this mode of mechanical printing. *Ibid.* 222/1 The 'Artotype' process owes its invention to Europe. It is..the most recent..form of collographic printing patented by Herr Obernetter. **1893** Cabal. *Orig. & Early Editions of English Writers from Langland to Wither* (Grolier Club) p. xi, List of facsimiles. *Artotypes. Ibid.* T. Landau *Encycl. Librarianship* (ed. 2) 26/1 *Artotype*, a photo-engraved illustration made by one of the gelatine processes.

arts-master. Restrict †*Obs.* to sense 2 in Dict. and add further example of *art master* in sense 1.

1930 *Encycl. Brit.* II. 703/1 The Society of Art Masters..is an association of teachers of art schools.

artsy (ā·ɹtsi), *a. colloq.* [f. *arts (and crafts)* + -y.] **artsy-(and-)craftsy:** = *arty-(and-)crafty* (see *Arty-and-crafty *a.* and *Art *sb.* 17).

1902 *Temple Bar* CXXVI. 63 The furniture is of the 'artsy-craftsy' type. **1942** Coward *Blithe Spirit* II. iii. 85 It's no use trying to defend Ruth's taste to me—it's thoroughly artsy crafty. **1942** J. B. Priestley *Black-out* iii. 27 That kind of woman frequently opened just that kind of artsy-craftsy shop.

art work. [f. Art *sb.* + Work *sb.*] **1.** A work of art. (Cf. Art *sb.* 14). Cf. G. *Kunstwerk*. Also, = Art *sb.* 6.

1877 Ruskin *St. Mark's Rest* ii. v. 60 The entire body of noble art-work belonged to this time. **1889** G. B. Shaw *How to become Mus. Critic* (1960) 168 They have a right..to expect..the faithful and earnest representation of any art work which appeals seriously to them. **1896** —— *Ibid.* 249 [Wagner's music-drama] The mightiest art work our century has produced. **1903** A. Bennett *Truth about Author* iv. 57 They dreamed of great art-works, lovely compositions, impassioned song. **1941** L. MacNeice *Poetry of Yeats* i. 7 Even before the artist has started his art-work proper he is..re-acting emotionally to his subject. **1953** S. Spender *Creative Elem.* iii. 64 The status first serves as an outside object, an art work.

2. The illustrative or decorative material in printed matter, as distinct from the text.

1926 C. Morley *Thunder on Left* ix. 122 He spoke of the Elevated Railroad's limited appropriation for promotion..asked what was her usual price for art work. **1959** 'S. Ransome' *I'll Die* ix. 102 She stopped at a certain illustration..You said, 'I see you're admiring my art work.'

arty (ā·ɹti), *a. colloq.* [f. Art *sb.* + -y[1].] A jocular epithet for furniture, decoration, etc., of artistic pretensions; also applied to persons who wish to be regarded as artistic in taste, dress, etc.

1901 *Academy* 16 Mar. 121 The *Kensington* is its title; it is broad in the page, handsomely printed, and decidedly arty. **1910** *Daily Chron.* 5 Apr. 9/5 The house filled with badly made 'arty', not artistic, furniture. **1921** Galsworthy *Six Short Plays* 119 Pretty?..Arty? **1925** *Sat. Rev.* 7 Nov. 534/1, I do not like the way..in her 'arty' window ..taking place. **1927** *Daily Express* 30 Apr. 8/2 Artists and 'arty' persons—between the two there is a great gulf fixed. **1930** *Observer* 30 Nov. 16/2 They were arty-minded, art-promoting, and in love with the colonies. **1930** 'G. Orwell' *Coming Up* iv. 20 They were arty-looking houses, another of those sham-Tudor colonies.

Hence **a·rtiness.**

1901 *Academy* 12 Oct. 337 The infected age of artiness. **1907** *Daily Chron.* 17 Apr. 6/6 Go something beyond the 'artiness' of the day. **1928** *Observer* 22 July 9/2 The same disarming artiness on the cover page.

arty (ā·ɹti), slang abbrev. of Artillery *sb.*

1942 *Gen* 15 June 30/2 Their arty drove us off. He hit, exploded and burned. **1944** L. Glassop *We were the Rats* iii. xxvii. 154 Hundreds of Jerries..moved their arty up to within rifle range of our forward posts.

arty-and-crafty (ā·ɹti,ænd,krɑ·fti), *a. colloq.* Also (commonly) **arty-crafty.** [f. *arts and crafts* + -y[1].] Characteristic of the work done by or under the auspices of the 'arts and crafts' movement (see *Art *sb.* 17); jocularly applied to furniture, etc., of specially artistic style but not conspicuously useful or comfortable; also used of their makers; thus, pretentiously or quaintly artistic. Also *absol.* Hence **arty-crafty** *sb.*, an 'arty-crafty' person; **arty-(and-)craftiness**, 'arty-crafty' quality or characteristics. (Cf. *art-and-crafty*, etc., s.v. *Art *sb.* 17.)

1902 *Daily Chron.* 28 July 3/4 Canon Barnett..spoke of a coming Exhibition of Arts and Crafts, which, if not too 'Arty-and-Crafty', should be excellent. **1920** A. Huxley *Limbo* 30, I liked..the way you went for the Arty-Crafties. **1925** —— *Along Road* iii. 200 What I may call 'arty-craftiness' or 'peasantry' is a Tolstoyan derivation from the quaint. **1925** *Sat. Rev.* 7 Nov. 524/1 The 'arty and crafty' furniture that is uncomfortable in use,..the model house in which no man can be at ease. **1927** *Observer* 18 Sept. 15/4 Country lore gathered chiefly from the works of arty-and-crafty peasants from Belsize Park. **1930** E. Wallace *Lady of Ascot* xiii. 136 What on earth can you find in Wolverhampton that can appeal to your arty-crafty mind? **1932** S. Gibbons *Cold Comfort Farm* v. 71 She will only go and keep a tea-room in Brighton and go all arty-and-crafty about the feet and waist. **1932** *Daily Express* 30 June 5/5 Provided the 'bun' is not placed low on the nape of the neck, suggesting a mixture of 'arty-crafty' and demure. **1937** 'G. Orwell' *Road to Wigan Pier* xii. 232 Deliberately to revert to primitive methods..would be a piece of dilettantism, of pretty-pretty arty and craftiness. **1945** L. A. G. Strong *Tongue in Your Head* 47 The 'poetry voice' removes it from the sphere of art..to the arty-crafty sphere. **1962** *John O'London's* 29 Nov. 508/3 The general public, as distinct from the arty-crafties.

Arundel (æ·rəndĕl). The name of the *Arundel* Society (1848–97) for promoting artistic knowledge, named in memory of Thomas Howard, 2nd earl of Arundel (1586–1646), used *attrib.* to designate prints, engravings, etc., issued by this Society.

1872 Mrs. D. A. Goddard (*title*) The Arundel Society pictures. **1879** C. M. Yonge *Magnum Bonum* xxxv. 529 He found Esther telling her little sister..histories of Arundel Society engravings. **1914** C. Mackenzie *Sinister St.* II. iv. iv. 949 Corridors hung with Arundel prints and faded photographs of cathedrals. **1937** A. Thirkell *Summer Half* ii. 60 An Arundel print of the Martyrdom of St Ursula.

Arunta, var. *Aranda.

arvo (ā·ɹvo). *Austral. slang.* [repr. voiced pronunc. of *af-* of *afternoon* + *-o[2].] Afternoon.

1933 *Bulletin* (Sydney) 6 Dec. 42/1 Alone on the bench this Sunday arvo. **1952** J. Cleary *Sundowners* iii. 154 That how you spend your Sunday arvos, Rupe? **1957** 'N. Culotta' *They're a Weird Mob* (1958) iii. 34 I'll try an' drop back this arvo.

ary-, shortened form of Aryteno-.

1879 *St. George's Hosp. Rep.* IX. 162 The right ary-epiglottidean fold. **1897** *Trans. Amer. Pediatric Soc.* IX. 180 A union of the aryepiglottic folds. **1931** G. O. Russell *Speech & Voice* iii. xvi. 176 The aryepiglottal..muscle. **1964** J. C. Catford in D. Abercrombie et al. *Daniel Jones* 34 Contraction of the ary-epiglottic folds.

Aryan. Add: **A.** *adj.* (Earlier example.)

1839 *Penny Cycl.* XIII. 309/1 The Medo-Persic, or Arian branch; at the head of which stands the Zend.

b. *spec.* Of or pertaining to the ancient Aryan people.

The idea current in the 19th cent. of an Aryan race corresponding to a definite Aryan language was taken up by nationalistic historical and romantic writers. It was given especial currency by de Gobineau, who linked it with the theory of the essential inferiority of certain races. The term 'Aryan race' was later revived and used for purposes of political propaganda in Nazi Germany (see 2).

1851 [see Celt[1] 2]. [**1855** M. A. de Gobineau *Essai sur l'inégalité des races humaines* IV. 350 La race germanique était pourvue de toute l'énergie de la variété ariane. Il le fallait pour qu'elle pût remplir le rôle auquel elle était appelée.] **1858, 1872** [in Dict.]. **1911** J. Lees tr. *H. S. Chamberlain's Foundations 19th Cent.* I. iv. 266 Anthropologists, ethnographers and even historians, theologians, philologists and legal authorities find the idea 'Aryan' more and more indispensable...Though it were proved that there never was an Aryan race in the past, yet we desire that in the future there may be one. **1916** Madison Grant *Passing of Great Race* (1917) v. 62 The name 'Aryan race' must also be frankly discarded as a term of racial significance. **1939** J. S. Huxley '*Race' in Europe* 24 Biologically it is almost as illegitimate to speak of a 'Jewish race' as of an 'Aryan race'.

2. Under the Nazi régime (1933–45) applied to the inhabitants of Germany of non-Jewish extraction.

1932 L. Golding *Let. to Hitler* 25 What you and many of your followers imagine is the Arian manner. **1933** W. Norman Brown (*title*) The Swastika, a Study of the Nazi Claims of its Aryan Origin. **1934** *Ann. Reg. 1933* II. 170 Germany:..During April the so-called Aryan Decrees were introduced. **1940** *War Illustr.* 5 Jan. 574/1 The Founder of the Christian faith, he [sc. Rosenberg] asserts, was of 'Aryan' not Jewish descent. **1943** *Ann. Reg. 1942* 183 The town..was cleared of its 'Aryan' population and made a centre, or rather a concentration camp, for all Jews.

B. *sb.* **2.** *spec.* under the Nazi régime (cf. sense *A. 2).

1933 tr. *Hitler's Mein Kampf* in *Times* 25 July 15/6 The exact opposite of the Aryan is the Jew. **1933** *Education* 1 Sept. 170/2 The basic idea of the new law is that non-Aryans, that is to say mainly Jews, shall be admitted to Universities and to the higher and middle schools only in accordance with their ratio of the total population of Germany. **1940** *War Illustr.* 16 Feb. 107 Pitiable in the extreme is the plight of the Jews in Poland under Nazi rule..On their backs they must wear a triangle of yellow cloth to show that their faith is abominable to the Nazi, and they must walk in the gutter, for the pavement is for Aryans only.

Aryanism (ē̆ə·riăniz'm). [f. Aryan + -ism.] **a.** Aryan principles or method of administration; belief in the historical existence of an Aryan race. **b.** A theory asserting the cultural and racial superiority of those of 'Aryan' descent. (Cf. *Aryan A. 1 b, 2.)

1888 S. C. B. Pitman tr. *Renan's Hist. Israel* I. i.7 Strict

monogamy was the law of primitive Aryanism. **1905** *Harmsworth Encycl.* I. 403/1 The advocacy of the opposite theory is termed Aryanism—that Europe owes its culture and civilization to the Aryan, a tall, blond dolichocephalic race represented by the modern Germans, Scandinavians, and English. **1934** *Times Educ. Suppl.* 6 Jan. 1/2 The brutalities of the new régime..and the solemn absurdities of 'Aryanism' shocked outside nations.

Aryanization (ē̆ə·riănəizē̆ɪ·ʃən). [f. Aryanize *v.* + -ation.] The act of Aryanizing or fact of being Aryanized.

1889 I. Taylor *Aryans* iv. § 2. 212 The Aryanization of Europe doubtless resembled that of India. **1904** *Athenæum* 24 Dec. 871/3 This Aryanization of inferior races.

b. *spec.* Of the Nazi régime (cf. *Aryanize *v.* b).

1939 *Sun* (Baltimore) 16 Feb. 1/4 A decree was issued today supplementing the December 5, 1938, order for the 'Aryanization' of German economic life, and providing for the payment of seventy per cent. of the 'Aryanization' profits to the German treasury. **1941** *Manchester Guardian Weekly* 31 Jan. 82 In the provinces young Nazis have invented their own technique of 'Aryanisation'...They make raids on Jewish families and demand contributions, with the threat that refusal to comply will mean the concentration camp.

Aryanize, *v.* Add: **b.** Under the Nazi régime, to make 'Aryan', to confer 'Aryan' status upon (a 'non-Aryan', e.g. a Jew); also, to dispossess a 'non-Aryan' of property, civil rights, etc.

1935 L. L. Snyder *From Bismarck to Hitler* viii. 81 Only in their outer forms will Japan and other 'bearers of civilization' remain Asiatic; inwardly they will become Aryanized. **1939** B. R. Epstein tr. *M. L. Berges's Cold Pogrom* 215 A half Jew might become 'Aryanized' if he had been a soldier. **1941** *Manchester Guardian Weekly* 31 Jan. 82 Plundering of Jewish property goes on, and the majority of Jewish shops in Prague have already become 'Aryanised'. **1952** Koestler *Arrow in Blue* 253 Though the Ullsteins were Jews, they tried to Aryanize the firm by degrees.

aryballos (æribæ·lǫs). *Greek Antiq.* [ad. Gr. ἀρύβαλλος bag, purse, oil-flask.] A flask with a globular body and narrow neck used as a container for oil or unguent.

1848 G. Dennis *Cities & Cemeteries Etruria* p. xcix, Ointment and perfume vases. .. The *aryballos* receives its name from its resemblance to a purse. **1858** S. Birch *Hist. Anc. Pott.* II. ii. ix, The *aryballos* was a vase always described as like a purse...It..was carried with the strigil to the bath. **1882** J. F. Molloy *It is no Wonder* II. x. 296 An aryballos, signed by Xenophantus, and magnificently ornamented by a wonderful crowd of figures in relief. **1935** Richter & Milne *Shapes & Names Ath. Vases* 16 *Aryballos.* Oil bottle with narrow neck, commonly used by athletes at the bath. **1946** *Antiquity* XX. 53 The second, a flat-bottomed aryballos is an early middle Corinthian piece.

aryl (æ·ril, also commonly æ·rəil). *Chem.* [ad. G. *arryl* (D. Vorländer 1899, in *Jrnl. Prakt. Chemie* LIX. 247), now *aryl*, f. Ar(omatic + -yl.] A general term for an organic radical derived from an aromatic hydrocarbon by the removal of one hydrogen atom. So **aryla·tion,** the introduction of an aryl group into a substance.

1906 M'Gowan & Sudborough tr. *A. Bernthsen's Text-bk. Org. Chem.* xxi. 365 The purely aliphatic alkyl radicals are termed alphyl groups, and the aromatic, aryl. **1918** J. C. Cain *Manuf. Intermed. Prod. for Dyes* 4 Arylation is usually effected by means of the corresponding amine. **1955** Brown & Dey *India's Mineral Wealth* (ed. 3) x. 391 Some of the more stable silicone esters, the aryl and higher alkyl silicates, are said to show promise as liquid media for heat-exchange and hydraulic systems. **1958** *Oxf. Univ. Gaz.* 23 Apr. 881 A quantitative study of the arylation of naphthalene by free radical reactions. **1959** *Times* 17 Mar. 4/3 Alkyl aryl sulphonates.

as. Add: **B. II. 8.** *As you were!* (in Drill): additional examples. Also *transf. colloq.*

1626 B. Jonson *Staple of Newes* in *Wks.* (1938) VI. 358 When my Muster-Master..cries, Faces about to the right hand, the left, Now, as you were. **1680** Radcliffe in *Rochester's Poems* 133 And be Godly a while ho, and then as you were. **1864** Hotten *Slang Dict.* 68 'As you were', a military phrase in drilling; used in a Slang sense to one who is going on too fast in his assertions, and wants recalling to moderation. **1871** J. W. T. *Questions & Answers on Company Drill* 28 The men [are] frequently to be brought back to the 'ready' by the command 'as you were'. **1925** Fraser & Gibbons *Soldier & Sailor Words* 11 *As you were*: The ordinary military word of command, used colloquially by way of acknowledging a mistake in anything said, e.g., 'I saw Smith—as you were—I mean Brown.' **1935** *Man. of Ceremonial* (*War Office*) ii. 14 Numbers 1 and 4 of the front rank will then prove. On the command 'As you were', they will drop their arms to the side.

11. b. *spec.* in theatrical use.

1847 *Semi-Wkly. News* (Fredericksburg, Va.) 7 Oct. 1/4 His robust health..gave him great advantage where unusual muscular strength is necessary, as Damon, Matamor, the Gladiator, etc. **1954** *Times* 18 Oct. 2/3 *Toad of Toad Hall*, with Mr. Leo McKern as the irrepressible Toad. *Ibid.* 2/4 Mr. Victor Mature..as an ambitious officer of low birth,..Miss Jean Simmons as an honest serving wench.

VII. 33. c. *as from*, in formal dating: from, after. Also, in the *U.S.*, *as of.*

1916 *Ann. Reg. 1915* 50 The British Government declared war against Bulgaria as from 10 p.m. on October 15 and the French Government as from 6 a.m. on October 16. **1918** *Act* 8 *Geo. V*, c. 5 § 1 (1), The foregoing provision shall..have effect as from a date to be specified in the Order. **1943** *Oxfordshire County Council Quart. Meeting* 10 Feb., In these circumstances your Sub-Committee recommend that ..the temporary addition to the basic salary of the Coroner for the Northern District be reduced from £70 to £45 as from the 1st of April, 1943.

1900 'Mark Twain' *Let.* 26 Jan. in *Howells Lett.* (1960) II. 716 Various samples & side-lights which bring the [human] race down to date, & expose it as of yesterday. **1955** D. W. Maurer in *Publ. Amer. Dialect Soc.* XXIV. 9 For as of 1955 well over one half of the crimes against property in the United States were committed by youths under twenty-five. **1956** A. H. Compton *Atomic Quest* 311 As of now there is no international authority whose strength comes near to being sufficient. **1957** D. Karp *Leave me Alone* vi. 96 I'm resigning from the committee as of now.

C. 1. *I thought as much* (examples).

1593 Shaks. *2 Hen. VI* ii. i. 15 *Glost.* My Lord, 'tis but a base ignoble minde, That mounts no higher then a Bird can sore. *Card.* I thought as much, hee would be aboue the Clouds. **1778** F. Burney *Evelina* I. xxi. 150 'Only see, here's Miss!' cried the brother. 'Well, I declare I thought as much!'

asbestos. The only current form is *asbestos.* Add: (Further examples.) *Comb.*: **asbestos cement**, a mixture of asbestos and cement moulded into sheets, pipes, etc., and used as a building and insulating material; **asbestos cloth**, a cloth woven from asbestos fibres; **asbestos curtain**, (in a theatre) a safety curtain made of asbestos and other fire-proof material.

1762 G. Gordon *Let.* 21 June in *N. & Q.* (1906) 10th Ser. V. 208/1 Manner of making asbestos paper. **1857** R. Hunt *Guide Mus. Pract. Geol.* 242 Asbestus [ed. 2, 1859, asbestos] cloth..may be exposed to the fire without being consumed. **1890** R. H. Jones *Asbestos* viii. 140 Asbestos cement is also very largely used for covering boilers, steam-pipes, hot-blast furnaces, and stills. *Ibid.* ix. 153 M. Henry Irving..laid special stress on the necessity of shutting off the stage from the auditorium by means of an asbestos curtain. **1920** *Blackw. Mag.* Mar. 336/2 The great asbestos-lagged exhaust from the oil-engines. **1933** *Archit. Rev.* LXXIV. facing p. 47 (*caption*) The variety of colour in the different plywood, plaster board and asbestos cement coverings placed on the steel pavilions. **1944** *Coast to Coast 1943* 165 The walls were asbestos-lined. **1964** N. Freeling *Double-Barrel* iii. 99 She..moved the asbestos mats..and clanked the lid back on the pot.

asbestosis (æzbestōᵘ·sis). *Path.* [mod.L., f. Asbestos + -osis.] A disease of the lungs caused by inhaling particles of asbestos. Also *pulmonary asbestosis.*

1927 W. E. Cooke in *Brit. Med. Jrnl.* 3 Dec. 1024/1 (*title*) Pulmonary Asbestosis. *Ibid.* 1027/2 From what I have seen clinically of pulmonary asbestosis it resembles silicosis of the lungs in the marked shortness of breath on slight exertion. **1930** *Lancet* 1 Mar. 445/2 Although asbestos and asbestos products have been used in industry for a long time, it is only during the last few years that pulmonary asbestosis has been recognised as a serious industrial disease.

ascariasis (æːskărəi·ăsis). *Path.* [f. Gr. ἀσκαρίς intestinal worm + -asis.] Infestation, esp. of the gastro-intestinal canal, with *Ascaris lumbricoides*; the disease resulting from this infestation.

1888 in F. P. Foster *Med. Dict.* I. 448/2. **1904** *Brit. Med. Jrnl.* 29 Oct. 1170/2 (*title*) A case of ascariasis with 'enteric' temperatures. **1923** *Nature* 19 May 683/2 An epidemic of ascariasis on a skunk-farm.

ascaridol(e (äskæ·ridǫl). *Chem.* [f. *ascarid* (see Ascarides *sb. pl.*) + -ol 3.] A terpene peroxide, $C_{10}H_{16}O_2$, the active principle of chenopodium oil, used in the treatment of hook-worm disease.

1908 *Schimmel & Co. Semi-Ann. Rep.* Apr. 114 The body, which we will call ascaridol on account of its pronounced action against ascarides, absorbs..2 atoms bromine. **1911** *Jrnl. Amer. Chem. Soc.* XXXIII. 1405 Schimmel & Co., in 1908, reported an extensive investigation of the oil... They separated a liquid substance, designated by them 'ascaridol'...Ascaridole afforded on analysis the formula $C_{10}H_{16}O_2$. **1947** *Lancet* 7 June 799/1 Ascaridole has undoubtedly been responsible for occasional fatalities, but.. none of the intestinal anthelmintics is free from such a risk.

ascaris (æ·skăris). *Zool.* [mod.L. (Linnæus, 1758), ad. Gr. ἀσκαρίς intestinal worm: see Ascarides.] The sing. of Ascarides; a parasitic nematode worm belonging to the genus so called.

1797 tr. Gmelin *Linnaeus' Syst. Nat. Hist.* XIII. 486 Intestinal Worms..within other animals. Ascaris—Body round, tapering each way; head with three vesicles. **1807** Morris & Kendrick *Edin. Med. Dict.* s.v. *Ascaris*, The body of the ascaris is cylindrical, filiform, and tapers at both ends. **1927** Haldane & Huxley *Animal Biol.* ii. 59 The number of chromosomes is always the same for a given race of animals or plants..in the Mexican salamander 28; in Ascaris 4. **1962** *Lancet* 29 Dec. 1360/2 Only 17% were found to be free from intestinal parasites, of which trichuris (in 72%) and ascaris (23%) were common.

ascender. Add: **c.** In *Printing* and *Palæography*, an ascending stroke; a stroke which extends above the body of a letter.

1934 in Webster. **1938** A. H. Smith in *London Mediæval Stud.* I. 204 After *neah* there is a long ascender which may be part of *h* or *l* or *b*. **1955** C. E. Wright *Bald's Leechbook* 24 The hook at the top of the ascender of b.

ascensional, *a.* Add: **2. b.** *ascensional screw*: a helical propeller that gives a lifting motion to an airship or flying-machine.

1901 *Westm. Gaz.* 15 Aug. 6/1 With my ascensional screws ..I am indifferent to wind, because I can rise and descend in search of a suitable belt of air. **1903** *Daily Chron.* 27 Sept. 7/4 A new flying machine...It is the invention of Mr. H. A. Chubb, and it rises by means of ascensional screws or helices.

Ascensiontide (äse·nʃəntəid). [f. Ascension 2 + Tide *sb.* 6.] The period of ten days from Ascension Day to Whitsun Eve. Also *attrib.*

1871 (*title*) Ascensiontide. Thoughts on the Ascension Day Collect. **1898** *Westm. Gaz.* 19 May 10/3 The Ascensiontide feast. **1900** D. C. Tovey *Lett. T. Gray* I. 90 These three spent the festival of Ascensiontide 1741, in Venice together.

ascesis (ăsī·sis). Also **askesis** (æskī·sis). [ad. Gr. ἄσκησις exercise, training, f. ἀσκεῖν to exercise.] The practice of self-discipline.

1873 Pater *Renaissance* p. xii, The charm of *ascesis*, of the austere and serious girding of the loins in youth. **1880** —— *Greek Studies* (1895) 267 The sanity of soul and body.. the perfecting of both by reasonable exercise or *ascêsis*. **1890** E. Johnson *Rise Christendom* 107 In the conduct of life they establish a strict ascesis..as a means of a closer communion with the Divine. **1924** J. Kelman *Prophets of Yesterday* i. 24 The Greek idea of *askesis* was but the habit of the athlete. **1944** Auden *For Time Being* (1945) 84 The vision That plain men can predict through an Ascesis of their senses.

ascetism (äse·tiz'm). [f. Gr. ἀσκητής (see Ascetic) + -ism.] = Asceticism.

1850 W. Howitt *Year-bk. of Country* iv. 106 Lent, with all its ascetism and abstinence, and gloom, is over. **1870** *Contemp. Rev.* XIII. 391 The melodramatic ascetism upon which such great store was then set. **1889** *Macm. Mag.* Jan. 236/1 There is certainly not a tinge in Shakespeare of sympathy with Catholic ascetism.

asciferous (ăsi·fĕrəs), *a.* [f. mod.L. Ascus: see -ferous.] = Ascigerous.

1882 *Encycl. Brit.* XIV. 557 There is a parallelism between the fructification of lichens and the asciferous section of fungi.

asco-. Add: **ascocarp** (æ·skokāɹp) [Gr. καρπός fruit], the fruit body of an ascomycetous fungus; **ascogenous** (äskǫ·dʒĭnəs), *a.* [Gr. γενής producing + -ous], that produces asci; **ascophore** (æ·skofōᵊɹ), an ascophorous hypha or filament.

1887 Garnsey & Balfour tr. *De Bary's Compar. Morphol. Fungi* 492/1 *Ascocarp*, in Ascomycetes: sporocarp producing asci and ascospores; its three kinds are apothecium or discocarp, perithecium or pyrenocarp, and cleistocarp. **1907** W. R. Fisher *Schlich's Man. Forestry* IV. iii. ii. 465 Black spermagonia subsequently appear before winter, but their spores do not germinate, as ascocarps do not develop till the second year after infection. **1910** *Encycl. Brit.* XI. 340/2 The other divisions of the Ascomycetes..do not bear the asci free on the mycelium but enclosed in definite fruit bodies or ascocarps. **1940** *Chambers's Techn. Dict.* 54/2 *Ascocarp*, the fructification of the Ascomycetes, containing asci and ascospores; it may be a cleistocarp, an apothecium, or a perithecium. **1882** Vines *Sachs's Bot.* 239 The ascogenous filaments and the asci which proceed from them. **1959** *Chambers's Encycl.* VI. 118/1 The male and female nuclei do not fuse in the ascogonium but become associated in pairs and pass into the ascogenous hyphae. **1877** Bennett tr. *Thomé's Text-bk. Bot.* 282 The filaments which bear the asci (*ascophores*).

Ascomycetes (æ:skomaisī·tīz). *Bot.* [mod.L., f. Asco- + Gr. μυκητ-, μύκης mushroom.] A class of fungi (see Ascomycetous *a.*); in *sing.* **ascomycete** (æ:skomai·sīt, -maisī·t), a fungus belonging to this class.

1857 M. J. Berkeley *Introd. Cryptogamic Bot.* 269 Fungales..*Ascomycetes*.—Asci formed from the fertile cells of an hymenium. *Ibid.* 270 *Ascomycetes, Berk.*..The essential character of this important division consists in the development of definite or indefinite sporidia within certain of the external cells of the hymenium called asci. **1875** *Q. Jrnl. Microsc. Sci.* XV. 144 Reess..terms the mother-cells *asci*, the daughter-cells *ascospores*, and the *Saccharomyces* an *Ascomycete* in the broadest sense of the term. **1922** H. Gwynne-Vaughan *Fungi* ii. 34 The Ascomycetes include over 15,000 species, all of which, excepting only the yeasts, possess a well-developed mycelium of richly-branded and septate hyphae. **1935** *Ann. Reg. 1934* 55 Fungal sexuality received much attention and it looks as if some ascomycetes may have one and others two nuclear fusions. **1953** J. Ramsbottom *Mushr. & Toadst.* iii. 18 Ascomycetes, which include cup-fungi, morels, ergot, candlesnuff fungus, and yeasts.

ascorbic (ăskǫ·ɹbik), *a. Chem.* [f. A- 14 + Scorb(ut)ic *a.*] *ascorbic acid*, vitamin C, the anti-scorbutic vitamin.

1933 Szent-Györgyi & Haworth in *Nature* 7 Jan. 24/2 'Hexuronic Acid' (Ascorbic Acid)...We wish to ascribe the name ascorbic acid to the crystalline substance $C_6H_8O_6$. **1935** *Nature* 14 Sept. 439/1 Vitamin C (ascorbic acid)..is found..to accelerate the growth and increase the length of the shoots of germinating wheat.

Ascot (æ·skǫt). Name of a village near Windsor in Berkshire, used *ellipt.* for a fashionable race-meeting held at Ascot Heath in June; freq. *attrib.*, applied esp. to hats, dresses, etc., designed for or suitable for wearing in the Royal Enclosure at Ascot. *Ascot tie* (see quot. 1957); also, *U.S.*, simply *Ascot.*

[**1711** *London Gaz.* 12–16 July *Advertisements.* Her Majesty's Plate of 100 Guineas will be run for round the new Heat on Ascott Common near Windsor, on Tuesday the 7th of August next, by any Horse, Mare or Gelding, being no more than six Years old.] **1814** *Sporting Mag.* June 128/2 The visit to Ascot of the Prince Regent, Emperor of Russia, and King of Prussia, with the other illustrious Foreigners..during the late Meeting, was too distinguished an honour paid to an English race-course, to be unnoticed in our pages. **1824** *Ibid.* June 131/1 Although the weather was not propitious, Ascot races never went off with more eclat. *Ibid.* 132/2 It would be well for himself and his subjects if there were more Ascots than one in the year. **1828** *Ibid.* June 202/2 Ascot still remains a pattern to all race courses throughout the kingdom. *Ibid.* 205/1 The Ascot Cup, value 100 sovs. **1867** 'Ouida' *Under Two Flags* I. ii. 29 Cottage for Ascot week, yachting with the R.V.Y. Club. **1899** R. Whiteing *No. 5 John St.* xiv. 148 Sinclair is trying to form a party..to run over to the Grand Prix... 'What about Ascot?' cries a protesting voice. **1907** B. M. Croker *Company's Servant* xxix. 300 Ladies in Ascot frocks, and men with field glasses. **1908** 'O. Henry' *Gentle Grafter* in *Wks.* (1928) 218 He had on a white flannel suit.. and a pink ascot tie. **1912** C. Mackenzie *Carnival* xxxiv. 352 His Ascot tie of red satin. **1957** M. B. Picken *Fashion Dict.* 10/1 *Ascot.* 1. Broad neckscarf, usually double. Tied so that ends come horizontally on each side of knot, then cross diagonally. Worn by men and fastened with scarf-pin. 2. Double scarf that is informally looped under the chin. Worn by men and women. **1959** *Listener* 19 Feb. 344/1 He loyally goes through the motions, military and social (Bank Picquet, Tower Guard, débutantes balls, Ascot, Lord's), that are required of him. **1967** *Guardian* 24 June 12/8 The flowery, floppy Ascot hat.

ascription. Add: **4. b.** *spec.* The formula used by a preacher at the end of a sermon in which he ascribes praise to God ('Now to God the Father..be ascribed all praise..').

1899 Dearmer *Parson's Handbk.* 201 It has become customary to conclude the sermon with an ascription.

Asdic (æ·zdik). [f. the initials of *A*llied *s*ubmarine *d*etection *i*nvestigation *c*ommittee.] An echo-sounding device for the detection of submarines. Cf. *sonar.

1939 *War Illustr.* 29 Dec. 538/2 Asdic..mentioned by Mr. Churchill in one of his speeches..[is] a type of secret apparatus now used by the Navy. **1940** *Nature* 3 Aug. 171/1 On May 21, Dr. R. W. Boyle received the Flavelle Medal of the Society..in recognition of his researches, particularly on the development of 'Asdics' and his work on ultrasonics. **1943** *Penguin New Writing* XVIII. 27 Taff is an Asdic rating. They work with headphones and they pick up echoes underwater and can get the bearing and distance of the echo and often they can tell if it's a sub or not. **1952** J. V. Noel *Naval Terms* 13 *Asdic*, British echo-ranging sound gear, equivalent to our [sc. American] Sonar. **1958** *Spectator* 8 Aug. 196/2 asdic, the device which sends out pulses of sound through the sea and records the echo where these are reflected back from a submarine.

-ase, *Chem.*, a suffix taken from the ending of Diastase, and used in naming enzymes, as *lactase, maltase, pectase*, etc.

aseismic (ăsəi·zmik, ē̆ɪ-), *a.* [f. A- 14 + Seismic *a.*] Resistant to the destructive effects of earthquakes.

1884 *Nature* 24 Jan. 290 A house that my experience suggests as being aseismic. **1908** C. G. Knott *Physics of Earthquake Phenomena* iv. 50 D. and T. Stevenson, the well-known lighthouse engineers, made the method the foundation principle of their 'aseismic joint'. **1939** *Archit. Rev.* LXXXV. 223 A third structural problem was introduced by the necessity of aseismic design.

asepsis (äse·psis). *Med.* [f. A- 14 + Sepsis.] Absence of micro-organisms likely to cause infection; methods or treatment that, by the use of sterilized instruments, etc., are aimed at preventing septic infection.

1892 F. J. Thornbury in *N.Y. Med. Jrnl.* LVI. 741/1 Asepsis permanent and reliable..may be accomplished in a simple manner by sterilizing the dressings by heat applied..in the form of steam. **1900** *Daily News* 7 May 3/2 The conditions as regards asepsis were by no means ideal. **1919** W. T. Grenfell *Labrador Doctor* (1920) iv. 70 Removing a leg at the thigh..without any attempt at or idea of asepsis. **1964** M. Hynes *Med. Bacteriol.* (ed. 8) iii. 26 The foundation of modern surgery is asepsis—the creation and preservation of a bacterium-free environment.

aseptically (äse·ptikăli), *adv.* [f. Aseptic *a.*: see -ically.] So as to be free from sepsis. Also *transf.*

1901 *Jrnl. Path. & Bacteriol.* VII. 260 All apparatus was.. carefully sterilised, and the operation wounds healed aseptically. **1940** *Mind* XLIX. 385 Every philosophical question answers itself if asked *aseptically* and fully enough.

asepticize (äse·ptisəiz), *v.* [f. Aseptic *a.* + -ize.] *trans.* To render aseptic, sterilize.

1892 *N.Y. Med. Jrnl.* LVI. 755/1 A new combination sterilizer for asepticising simultaneously water, instruments, and dressings. *Ibid.* 755/2 The dressings, after having been asepticized, are removed. **1907** *Practitioner* June 803 Bottles and forceps were asepticised by being dipped in carbolic lotion.

asexual, *a.* Add: **b.** In general contexts: without sexuality.

1896 L. Eckenstein *Woman under Monasticism* ix. 307 The high estimation of virgin purity..was advocated by the leaders of thought..and..the asexual existence..was extolled as virtue in itself. **1903** *Daily Mail* 10 Sept. 2/7 All doctors will tell you that, athletic or not, women are more asexual than men. **1928** D. H. Lawrence *Let.* ? 17 Aug. (1962) 1077, I feel I've shot it [sc. the book] like a bomb against all their false sex and hypocrisy—as my Florentine doctor said, against all their asexual sexuality.

asexualization (äse:ksiuăləizē̆ɪ·ʃən, -ɪzē̆ɪ·ʃən). *Surg.* [f. A- 14 + Sexualiz(e *v.* + -ation.] The action or process of sterilizing, or rendering sexually impotent, an animal or human being.

1895 E. Stuver in *Ohio Med. Jrnl.* VI. 193 (*title*) Asexualization for the limitation of disease, and the prevention and punishment of crime. **1909** *Boston Med. & Surg. Jrnl.* 21 Oct. 584/2 (*title*) Asexualization as a Remedial Measure in the Relief of Certain Forms of Mental, Moral and Physical Degeneration. **1961** *Brit. Med. Dict.* 144/1 *Asexualization*, the process of sterilization..in the male by castration or vasectomy; in the female by removal of ovaries or ligation of the uterine tubes.

ash, *sb.*[2] Add: **1. e.** *ash*(*es*) *of roses*: a greyish-pink colour.

1872 *Young Englishwoman* Nov. 599/1 *Marmotte*, a dark ashes-of-roses shade. **1893** *Daily News* 17 May 3/4 The soft greyish pink known as 'ashes of roses'. **1901** Kipling *Five Nations* (1903) 113 Opal and ash-of-roses, Cinnamon, umber, and dun. **1927** *Daily Express* 14 Mar. 5 Pomegranate, mulberry red, mushroom, and ashes of roses, a colour that looks like blue ash with a touch of flame, are among the fashion reds.

4. b. *the ashes*: in *Cricket*, a term originating in a mock obituary notice published in the *Sporting Times* 2 Sept. 1882, after the sensational victory of Australia at the Oval on Aug. 29 of that year, announcing the cremation of the dead body of English Cricket and the taking of the ashes to Australia; hence, the losing or winning of the rubber in the series of test matches played periodically between the chosen representatives of English and Australian cricket is frequently described as the loss, or the recovery (if previously lost) or retaining (if previously held) of the ashes.

[**1882** *Sporting Times* 2 Sept., In Affectionate Remembrance of English Cricket Which died at the Oval on 29th August, 1882. Deeply lamented by a large circle of sorrowing friends and acquaintances. R.I.P. N.B.—The body will be cremated and the ashes taken to Australia.] **1903** Warner in *Westm. Gaz.* 8 Dec. 2/1 If we fail to bring home 'the ashes' it will certainly not be for want of trying. **1921** *Glasgow Herald* 21 Jan. 8 We must accept the transference of the 'Ashes' philosophically.

6. b. Ash-colour.

1876 *Field & Forest* II. 41 Admitting that *tephrocotis* 'has the least ash on the head', how can this fact be attributed to climatological influences?

8. a. *ash-brown, -grey* (later examples). Also, in instrumental relation, *ash-smeared.*

1921 *Glasgow Herald* 25 June 5/7 The spotted fly-catchers ..are ash-brown above with a central dark line on the head feathers. **1959** *Times* 8 Sept. 12/7 On the log wall was an ashbrown felt. **1889** Yeats *Wanderings of Oisin* 125 An ash-grey feather. **1904** W. de la Mare *Henry Brocken* 88 The cadent wail of the ash-grey birds. **1901** Kipling *Kim* i. 4 Then there were holy men, ash-smeared *faquirs* by their brick shrines. **1927** D. H. Lawrence *Mornings in Mexico* 160 Some of these ash-smeared men held armfuls of snakes.

b. ash-barrel (chiefly *U.S.*), a barrel for holding ashes; also occas. *ashes barrel*; **ash-blond(e** *a.*, applied to hair of a light blond colour; as *sb.*, the colour itself; a person with hair of this colour; **ash-box,** a receptacle for ashes, (*a*) a pan beneath a fire-grate, (*b*) a dust-bin; **ash-can** (chiefly *U.S.*) = *ash-bin*; also *fig.*; in *U.S. Services' slang*, a depth charge; **ash-drawer,** (*b*) a drawer for ashes beneath a fire-grate; **ash-grate,** a grate that fits over the ash-hole; **ash-hoist,** a hoist for removing ashes from the ash-pit of an engine house, the stoke-hold of a vessel, etc.; **ash-hopper** *U.S.*, a lye cask, resembling a hopper in a mill; **ash-man,** a collector and remover of ashes; **ash-pan,** also, earlier term for *ash-tray*; **ash-receiver** *U.S.*, an ash-tray; **ash-riddling,** the northern custom of riddling ashes on the hearth on St. Mark's Eve as a method of divination; **ash-shoot,** a shaft through which the ashes are shot, or are raised from the stokehold to the deck of a ship; **ash-tray,** a small bowl or other receptacle for tobacco-ash.

1846 D. Corcoran *Pickings from Portfolio* 61 They were ..knocking over the ashes barrels, shying stones at the lamps. **1870** 'F. Fern' *Ginger-Snaps* 55 Garbage-heaps and ash-barrels before the door of poverty. **1905** Kipling *Actions & Reactions* (1909) 6 That wilderness which is reached from an ash-barrel of a station called Charing Cross. **1947** E. H. Paul *Linden on Saugus Branch* 136 The Protestant deacons..did not fail to note..the empty bottles in the ash barrels. **1903** *Daily Chron.* 9 Mar. 3/1 My hair is a beautiful ash-blonde. **1926** *Bulletin & Scots Pictorial* 29 June 10/3 A mass of ash-blond hair was braided and coiled around her head. **1947** E. Hyams *Wm. Medium* x. 197 They were ash-blondes, with heavenly skins and large blue eyes. **1833** Loudon *Encycl. Archit.* § 605 The bottoms of these chambers should have an ash-box fitted into them. **1847** *Rep. Comm. Patents 1846* (*U.S.*) 261 The chamber of combustion and its grate and ash-box. **1899** W. C. Morrow *Bohemian Paris* 103 The rows of heaped ash-cans that lined the way. **1919** *World's Work* Oct. 604/2 The depth charge looked like the innocent domestic ash can, and that was the name by which it soon became popularly known. **1920** *Chambers's Jrnl.* Mar. 151/2 One of them upsets the ash can. **1944** Auden *For Time Being* (1945) 71 In a wet vacancy among the ash cans. **1959** G. Jenkins *Twist of Sand* iv. 68 'I give it five minutes before the ash-cans come.' Waiting for a depth-charge attack is probably as bad as the attack itself. **1833** Loudon *Encycl. Archit.* § 599 With the fire-places near the hearth, with front bars beveled inwards, and with an ash-drawer, the fender may be made very low. **1920** *Chambers's Jrnl.* Apr. 264/2 An ash-drawer is arranged under the grate with the hot chamber below it. **1833** Loudon *Encycl. Archit.* § 605 The ash-grate, being taken out, and its contents thrown in the dusthole, might be replaced, and the chamber again filled with coals. **1837** *Mag. Domestic Econ.* III. 111 There should be an ash-grate and pit made under the fire-place, that the ashes may drop through the ash-grate into the pit, and leave the cinders over it. **1920** *Blackw. Mag.* Sept. 303/2 The noise of the ash-hoist greeted his ear. **1809** *Trans. Amer. Philos. Soc.* VI. 240 'Cubic salts'..thrown upon the ash-hoppers..are supposed to assist in precipitating the lime. **1885** 'C. E. Craddock' *Prophet Gt. Smoky Mts.* i. 18 Some fifteen or twenty hounds that suddenly materialized from the ash-hopper. **1873** Aldrich *Marj. Daw* 142 He has fought the ashman's boy, the grocer's boy [etc.]. **1959** E. Fenwick *Long Way Down* iii. 22 The ashman's visits: depressing, but routine. **1857** E. FitzGerald *Let.* 3 Oct. (1889) I. 265 The last Cheroot he had tried lay three quarters smoked in its little China ash-pan. **1898** G. B. Shaw *Mrs. Warren's Profession* iv. p. 216 A double writing-table..with a cigar box, ash pans. **1930** D. Parker *Laments for Living* 62 The ash-receivers, of Japanese manufacture, were in the form of grotesque heads..given.. mouths stretched into great gapes, into which those who had the heart for it might flick their ashes. **1825** Brockett *North Country Words* s.v. *Ass, Ass-riddlin*, the riddling or sifting of the ashes on the hearth, on the eve of St. Mark. The superstitious notion is, that, should any of the family die within the year, the shoe will be impressed on the ashes. **1893** K. Simpson *Jeanie* 240 Ash-riddling is an old custom which I like to keep up. **1889** *Cent. Dict.*, Ash-shoot. **1898** Kipling *Fleet in Being* 5, I heard Swinburne laying down the law to his juniors by the ash-shoot. **1920** *Blackw. Mag.* Sept. 303/2 He..watched the fireman unhook his first bucket of ashes from the hoist and carry them to the ash-shoot. **1887** Kipling *Plain Tales fr. Hills* (1890) 116 He had tipped a bagful of his powder into the big silver ash-tray. **1926** M. Sutherland *One o' the Herd* v, A lacquered Chinese table with..a box of cigarettes and an ashtray upon it.

ash, *v.*[1] (Later example.)

1940 F. Kitchen *Brother to Ox* i. 4 'Ashing a lad's behind' was the recognized form of punishment.

Ashanti (ăʃæ·nti). Also 8 **Assanti,** etc., 9– (*now rare*) **Ashantee.** [Native name.] One of the Akan peoples of West Africa; a member of this people. Also, their language. Hence **Asha·ntian** *a.*

1705 tr. *Bosman's Descr. Coast of Guinea* vi. 76 The Plunder after this Victory took up the Asianties fifteen days time...This messenger of ours, who is now in the Asiantean Camp. **1721** J. Atkins in *New Coll. Voyages & Travels* (1745) II. iii. iv. 454/1 It seems, the Santis, or Assantis..had lately..driven them from their Habitations. **1824** J. Dupuis *Jrnl. Residence in Ashantee* i. 21 A party of Ashantees..suddenly approached. **1831** H. I. Rickett (*title*) Narrative of the Ashantee War. **1873** J. Marshall *Let.* 9 Nov. in W. R. Brownlow *Memoir J. M.* (1890) ii. 19 His party had fired on some Ashantis two days before. **1880** *Ibid.* iii. 51 Since the Ashanti War, communication is becoming more constant..with the interior. **1874** J. G. Christaller et al. (*title*) A Dictionary, English, Tshi-Asante-Akra. **1875** —— *Gram. Tshi* p. xv, As a common name of the 'Asante and Fante Language' we have put down the name Tshi. **1880** G. A. Sala *Amer. Revisited* (1882) I. xvi. 227 [The features] of his companion were of almost brutish Ashantian or Dahomian ruggedness. **1962** *Listener* 25 Jan. 156/1 There should be no reference to Fantis, Ashantis, Ewes, Gas, Dagombas, 'strangers', and so forth, but we should all call ourselves Ghanaians.

ash-cake. *U.S.* [Ash *sb.*[2] 8 b.] A cake baked in or under the ashes of a fire.

1809 Horry & Marion *Life F. Marion* 47 A rasher of fat bacon from the coals, with a good stout lump of an ash cake. **1844** *Congress. Globe* App. 631/3 At this dinner, a large ash-cake was baked containing about three bushels of corn meal. **1888** 'C. E. Craddock' *Keedon Bluffs* 25 Her father..deftly constructed an ash-cake.

Asherah (aʃī·ra). Pl. **Asherahs, Asherim** (-rim). [Heb. *Ashērāh.*] A wooden post, pillar, or trunk of a tree used as the symbol of the goddess Ashera, occurring near the altar in Canaanitish high places devoted to the worship of Baal. Also the goddess herself, associated with Baal in Syrian, Phœnician,

ASHERY

and Hebrew heathen worship. (Cf. GROVE 2 a.)
1863 G. GROVE in W. Smith *Dict. Bible* I. 120/2 Asherah is the name of the image or symbol of the goddess. This symbol seems in all cases to have been of wood, and the most probable etymology of the term (..*to be straight*..) indicates that it was formed of the straight stem of a tree, whether living or set up for the purpose. **1884** *Bible* (R.V.) Exod. xxxiv. 13 Ye shall cut down their Asherim. *Note.* Probably the wooden symbols of a goddess Asherah. *Ibid.*, Deut. xvi. 21 Thou shalt not plant thee an Asherah of any kind of tree beside the altar of the Lord thy God. **1886** CONDER *Syrian Stone-Lore* v. 189 The 'hangings for the grove', or robe for the Asherah—the sacred tree erected even in the Jerusalem Temple in Josiah's time. **1912** H. F. HAMILTON *People of God* I. II. iv. 55 Poles, which served as substitutes for trees, called *asherim*. **1925** W. EWING *J. E. H. Thomson* xviii. 259 The altars, the pillars, and *Asherahs* were not destroyed.

ashery. (Examples.)
1831 A. SHERBURNE *Memoirs* viii. 185, I much wished to set up an ashery. **1837** *Fraser's Mag.* XVI. 690 Manufactories for reducing common ashes into pots and pearls are sometimes erected on a pretty large scale; in such cases, the individual to whom the 'ashery' belongs buys up all the ashes in the district. **1884** L. F. & R. L. ALLEN *New Amer. Farm Bk.* 62 Spent lye of the asheries, is the liquid which remains after the combination of the lye and grease in manufacturing soap.

a-shimmer (ăʃi·məɹ), *advb. phr.* [A *prep.*[1] + SHIMMER.] Shimmering.
1926 *Chambers's Jrnl.* June 340/1 The marsh around them a-shimmer with heat. **1930** R. CAMPBELL *Adamastor* 80 His myriad spectacles a-shimmer Confront the lighted halls.

a·shing, *vbl. sb.* [f. ASH *v.*[2] + -ING[1].] **1.** Sprinkling or strewing with ashes.
1842 J. BUEL *Farmer's Companion* (ed. 3) 206 Draining improves the quality of the herbage, and marling, liming, or ashing increases the quantity.
2. Reducing to ashes, *spec.* for the purpose of analysis.
1937 HACKH & GRANT *Chem. Dict.* (1938) 88/2 *Ashing*, the process of burning organic matter, especially during analysis. **1946** *Nature* 20 July 96/1 Appreciable losses occur during ashing, owing to the volatility of inorganic components and adsorption to the surface of the ashing vessel. *Ibid.* 28 Dec. 946/2 Even low-temperature ashing of filter samples may alter the crystal structure.

Ashkenazim (æʃkĕnā·zim), *sb. pl.* [mod. Heb., f. *Ashkenaz*, the name of a son of Gomer (Gen. x. 3, 1 Chron. i. 6), son of Japheth, son of Noah, typifying a race of people identified with the Ascanians of Phrygia, and, in medieval times, with the Germans.] Jews of middle and northern Europe as distinguished from Sephardim or Jews of Spain and Portugal. Hence **Ashkena·zic** *a.*, of or belonging to the Ashkenazim.
1839 R. M. MCCHEYNE *Let.* 23 July in *Familiar Lett.* (1848) 109 One of the Ashkenazim..invited us secretly to his house. **1842** BONAR & MCCHEYNE *Narr. Mission Jews in 1839* iv. 330 There are no rabbies properly speaking among the Ashkenazim. **1892** ZANGWILL *Childr. Ghetto* I. 4 All the Ashkenazic tribes lived very much like a happy family. *Ibid.* 38 Spanish Jews look down on the later imported Ashkenazim, embracing both Poles and Dutchmen in their impartial contempt. **1914** *East & West* XII. 154 The true Zionists are mostly Ashkenazim Jews from all Europe.

ashlar (æ·ʃləɹ), *v.* Also **ashler.** [f. ASHLAR *sb.*] *trans.* To face with ashlar; cf. ASHLAR *sb.* 3.
1836 *Mirror of Lit.* XXVIII. 147/1 It has been generally done by 'ashlering' them..with thin blocks or slabs of the same stone. **1895** HARDY *Jude the Obscure* v. vii, Sometimes setting the parapet of a town-hall, sometimes ashlaring an hotel at Sandbourne.

ashlering, var. ASHLARING *vbl. sb.*
1757 [see ASHLARING *vbl. sb.* 2]. **1858** *Skyring's Builders' Prices* 91 To ashlering take plain work to face, bed, and joint. **1940** *Chambers's Techn. Dict.* 54/2 *Ashlering*, the vertical timbers or quarterings, 2½ to 3 feet long, fixed in attics between floor-joists and rafters as supports for a partition wall, to cut off the sharp angle under the lower end of the rafter.

ash-plant. [ASH *sb.*[1]] A sapling of the ash tree, used as a walking-stick, whip, goad, etc.
1850 'H. HIEOVER' *Pract. Horsemanship* 180 Sit tight, and lay your ash plant well into his ears. **1852** MUNDY *Our Antipodes* III. i. 26 He..trudges away..supported by his son on one hand and an ash-plant in the other. **1918** P. MACGILL *Glenmornan* vi. 140 He hit a bullock near him with his ash-plant. **1935** S. SPENDER *Destructive Element* 82 Stephen..is only recognizable at all by being made inseparable from his ashplant.
Hence as *v.*, to flog with an ash-plant (cf. ASH *v.*[1]).
1923 G. B. SHAW *Doctors' Delusions* (1932) 346 Nothing will persuade me that any boy..really likes being kicked and ashplanted into doing the drudgery of a professional footballer.

ashram (ā·ʃrăm). [ad. Sk. *asrama* hermitage, f. *ā* near to, towards + *śrama* exertion, earnest endeavour.] In India, a place of religious retreat, sanctuary, or hermitage. Hence **a·shramite,** an occupant of an ashram.
1917 W. W. PEARSON *Shantiniketan* 17 There was a farewell ceremony according to the ancient Hindu custom when a guest leaves an ashram for the outer world. *Ibid.* 47 An ashram, or religious retreat, where those in search of peace might have an opportunity for quiet and meditation. **1926** *Contemp. Rev.* Nov. 616 The traditional ashram is built of wattle and mud, and its roofs are of leaves. **1933** *Daily Express* 31 July 1/1 Gandhi has decided to vacate his ashram (seminary) at Sabarmati and start the mysterious 'fresh sacred mission' on Tuesday. This will begin with a march of thirty-two of the ashramites. **1959** *Times* 10 July 7/7 The buildings in the *ashrams* at Tiruppatur and Tenkasi are so strikingly Dravidian in effect that they could almost be mistaken for Hindu temples.

Asian, *a.* and *sb.* Add: **b.** In recent official use superseding *Asiatic* because of the alleged depreciatory implication of the latter term.
1930 *Kenya Legislative Council Debates* 4 Apr. 69 The first thing I notice about the motion is the word 'Asian'. I have tried to look it up, but I can find no dictionary which gives it—it is usually 'Asiatic'. **1930** *Official Gaz.* (*Kenya*) 9 Dec. 2825 (*heading*) Asian Clerical Staff Confirmation Examination. **1953** *Times Lit. Suppl.* 6 Feb., *Asian Review.* Volume XLIX, Number 177. January, 1953...After 68 years of existence the widely circulated quarterly hitherto known as the *Asiatic Review* makes a slight but important change of designation. The reason is to be found in the sensitiveness of the cultured classes of Asia about nomenclature...The term 'Asiatic' has come to be regarded with disfavour by those to whom it applied, and they feel entitled to be brought into line with usage in regard to Europeans, Americans and Australians. **1953** *Encounter* Oct. 74/1 Through European schools on Asian soil, and by study in Europe and America, young Asians entered the cultural world of the 20th century.
2. *Asian 'flu*, the popular name of a kind of influenza caused by a virus first identified in Hong Kong in 1957. Also *Asian influenza*.
1957 *Economist* 31 Aug. 670/2 The type of virus officially called A/Singapore/1/57, and popularly known as..Asian flu. **1958** *Punch* 1 Jan. 68/3 Three days later..all of us, including the cat, were struck down with Asian flu. **1958** *Listener* 19 June 1000/2 When Asian influenza strikes the population, we no longer appeal on our knees for help.

Asianic (ē[i]ʃiæ·nik), *a.* [f. ASIAN *a.* + -IC.]
1. Of or pertaining to Asia Minor.
1883 I. TAYLOR *Alphabet* II. 116 A syllabic writing,..which prevailed throughout the whole of Asia Minor, and which has been designated by Professor Sayce as the Asianic syllabary. **1927** W. RAMSAY (title) Asianic Elements in Greek Civilisation. **1928** C. DAWSON *Age of Gods* xiii. 289 The bulk of the population belonged to the native Asianic stock.
2. Pertaining to or characterized by the florid and inflated literary style characteristic of the Asiatic Greeks in the three centuries preceding the Christian era.
1920 *Q. Rev.* July 42 His elaborate Asianic style.

Asianize (ē[i]·ʃănəiz), *v.* [f. ASIAN *a.* + -IZE.] *trans.* To make Asian in character, habits, customs, etc. Hence **A·sianized** *ppl. a.* So **A:sianiza·tion,** the process of becoming Asian.
1893 F. ADAMS *New Egypt* 50 Europe has indeed annexed this outpost of Asianised Africa [Cairo]. **1921** *19th Cent.* May 884 A humanity asianised in sentiment. *Ibid.* 888 The Arab and Hindu..who have undergone asianisation in the continent of their adoption.

Asiatic, *a.* Add: *Asiatic cholera* (see CHOLERA 3).
1831 T. W. CHEVALIER (*title*) On Asiatic Cholera. **1832** *Deb. Congress* 30 June 3833/1 A deliverance from the impending judgment..of the Asiatic cholera.

aside, *adv.* Add: **9.** Of speech: earlier example.
1801 M. EDGEWORTH *Belinda* II. xiv. 72 'You won't blow us to Clary,' added he *aside* to her ladyship.
10. aside from (= Brit. 'apart from'). **a.** Besides, in addition to; without reckoning or including. *U.S.*
1818 TICKNOR in *Life, Lett., & Jrnls.* (1876) I. 206 Aside from this, the mere show is more magnificent than can be seen at any other court in Europe. **1847** L. COLLINS *Kentucky* 507 The college..possesses revenues, aside from tuition, sufficient to maintain the faculty. **1905** *Forum* Apr. 598 The city, aside from being a great industrial..centre, enjoys ..a high moral record.
b. Except for. *U.S.*
1861 in *Maryland Hist. Mag.* (1910) V. 303 Aside from the upheavals made by our engineers,..I don't think I have ever seen a more dreary region. **1902** H. E. BOURNE *Teaching Hist. & Civics* 303 Aside from the wars in Europe..the history of the colonies..is uneventful. **1931** H. F. PRINGLE *Theodore Roosevelt* I. iii. 35 Aside from boxing, Roosevelt's.. activities at Harvard are of slight importance.

as if, *phr. Philos.* [Cf. As 9 b; often used to represent G. *als ob* (freq. in Kant and taken up by Vaihinger as a key expression).] Introducing a supposition, or way of conceiving some entity or situation, that is not to be taken literally, but yields some insight or convenience in metaphysics. Also *absol.* and *attrib.* Hence **as-i·fness.**
1892 J. ROYCE *Spirit Mod. Philos.* iv. 113 You *act as if God were* your..visible companion, as if the moral law.. were spoken in your ear.., as if the unsearchable God..were as familiar to you as your daily walk..This is to do what Kant calls postulating God's existence. **1900** —— *World & Indiv.* I. v. 206 This *as if*, or *as it were*, becomes to some thinkers, a sort of ultimate category. **1921** HANNAY & COLLINGWOOD tr. *Ruggiero's Mod. Philos.* ii. 61 A champion has now arisen in Vaihinger to erect the 'as if' into a philosophical method! **1924** C. K. OGDEN tr. H. Vaihinger (*title*) The Philosophy of 'As If'. **1937** G. W. ALLPORT *Personality* (1938) xi. 289 There is no real objection to this 'as if' epistemology provided it is applied equally to all the data of psychology. **1937** *Mind* XLVI. 114 The 'as-if' concepts, like 'wave', 'corpuscle', etc., which we borrow from macroscopic perception in order to interpret the nature of phenomena too small to be perceived in their own right. **1940** H. H. PRICE *Hume's Theory of Ext. World* v. 154 There are different degrees of as-ifness. This is acknowledged in ordinary speech, where we find such phrases as 'to some extent as if', 'rather as if', 'very much as if', 'exactly as if'.

asigmatic (ē[i]sigmæ·tik), *a.* [f. A- 14 + SIGMATIC *a.*] Not sigmatic, formed without sigma.
1893 F. W. WALKER in *Classical Rev.* July 292/1 The one so called asigmatic tense that is not digammated is generally admitted to have dropped a sigma. **1967** *Language* XLIII. 628 An asigmatic aorist of *kheō*.

asile. (*Poet.* example.)
1925 HARDY *Human Shows* 45 When an inner court outspreads As 'twere History's own asile.

asilid (æ·silid), *a. Ent.* [ad. mod.L. *Asilidæ*, f. *asilus*: see next and -ID[3].] Belonging to the Asilidæ, a family of dipterous insects comprising the hornet-flies.
1904 *Athenæum* 24 Dec. 881/1 [Prof. E. B. Poulton] exhibited a photograph..showing the Xylocopid model and its Asilid mimic. **1951** COLYER & HAMMOND *Flies Brit. Isles* viii. 118 There is some evidence for the view that the instantaneous collapse of Asilid attack is due to the injection of poison. **1961** *New Scientist* 8 June 589/3 When feeding the [blind] flies are also exposed to predatory wasps, asilid flies and dragon flies.

asilus (ăsəi·ləs). *Ent.* [L. *asilus* gadfly.] A member of a genus of flies belonging to the order Diptera, family Asilidæ; a hornet-fly, hawk-fly, or robber-fly.
1752 J. HILL *Hist. Anim.* 32 The great, smooth, black, and red Indian Asilus. **1866** BLACKMORE *Cradock Nowell* xxv, They discuss the asilus and the oestrum.

ask, *v.* Add: **2. b.** *to ask* (a horse) *the question*: to call upon him for a special effort.
1894 H. CUSTANCE *Riding Recoll.* vi. 88 Until the last ten strides, when I really asked 'King Lud' the question.
4. c. *I ask you*, exclamatory phr. indicating disgust or asseveration.
1855 DICKENS *Dorrit* I. ii. 12 'Now, I ask you,' said Mr. Meagles...'I ask you simply, as between man and man,..DID you ever hear of such damned nonsense as putting Pet in quarantine?' **1898** G. B. SHAW *Arms & Man* III in *Plays Pleasant & Unpl.* II. 73 Now, I ask you, would a woman who took the affair seriously have sent me this? **1902** CONRAD *Youth* 44 Now, I ask you, can anybody stand this kind of thing? **1932** *Punch* 18 May 536, I ask you—not a taxi in sight! **1937** C. DAY LEWIS *Starting Point* I. iii. 41 'My God, look at these people! Just look at them—I ask you,' he exclaimed.
d. colloq. phr. *if you ask me*: in my opinion.
[**1856** DICKENS *Dorrit* II. ix. 394 And if you ask me for my advice, it is that you set off tomorrow.] **1910** GALSWORTHY *Justice* II, If you ask me, I don't think he was quite compos when he did it. **1930** J. B. PRIESTLEY *Angel Pavement* ii. 65 Girls *are* a bit silly, if you ask me, and it's a good job for the men they are. **1932** *Punch* 14 Dec. 654/2 If you ask me, a little of that sort of thing would brighten up the trade wonderfully.
5. d. colloq. phr. *ask me another*: I do not know (the answer to your question). Also *ask me, ask me a harder*, etc.
1910 A. BENNETT *Clayhanger* III. ii. 344 'Why's he wearing his best clothes?' Clara demanded...'Ask me another!' said Edwin. **1912** A. BRAZIL *New Girl at St. Chad's* ix. 139 'Then who was it?' Honor shook her head. 'Ask me a harder!' she said briefly. **1915** MRS. H. WARD *Eltham House* xii. 212 'Tell me about Madge Whitton. There are all sorts of rumours..'..Lewson shrugged his shoulders—'Ask me another, Duchess.' **1928** E. WALLACE *Gunner* ii. 24, I hope you are wearing warm undies. Why are undies indelicate and sable coats ladylike? Ask me. It's one of the mysteries. **1933** I. COMPTON-BURNETT *More Women* xiii. 202 'Devoted?' said Josephine, raising her brows. 'Ask me another. I am not in a position to give you an account of their feelings.'
16. b. phr. *to ask for trouble* (or the like), also (as slang or colloq. substitute) *to ask for it*, to act in such a way as to bring trouble upon oneself, to give provocation.
1902 *Captain* VII. 549/2 The fly-away contingent were what is spoken of as 'asking for' bent spines and injured hearts. **1909** E. P. OPPENHEIM *Jeanne of Marshes* II. xiv. 274 'Whatever happens to him,' Forrest said, 'he's asking for it.' **1909** *Westm. Gaz.* 16 Sept. 12/2 There are vainglorious players in this world who, to use a vulgarism, 'ask for it', and deserve it when they get it. **1915** *Times 'Red Cross Story Book* 71 Certainly we were 'asking for it', as my ..offspring did not fail to remark. **1916** 'BOYD CABLE' *Action Front* 165 Silly fools...What do they want to hoist that huge Red Cross flag up there for, where any airman can see it? Fairly asking for it, I call it. **1925** E. F. NORTON *Fight for Everest, 1924* 343 It is asking for trouble to give out any money except the daily ration allowance. **1946** 'M. INNES' *From London Far* I. v. 42 'The damned scoundrels!'..The girl was philosophical. 'I asked for it, all right.' **1960** H. PINTER *Room* I, I don't know how they live down there. It's asking for trouble.

21. b. *colloq.*, with adverbs: *to ask back*, to reciprocate an invitation; *to ask down*, to invite (someone) to come and stay in the country; *to ask out*, to invite to something, esp. to an entertainment.
1834 DICKENS in *Monthly Mag.* Feb. 152 If I see him.. tomorrow, perhaps I'll ask him down. **1864** H. C. ADAMS *White Brunswickers* x. 167 Billy moved heaven and earth to get asked out on the same days. **1887** KIPLING *Plain Tales fr. Hills* (1890) 199 Now and again he was asked out to dinner. **1888** MCCARTHY & PRAED *Ladies' Gallery* I. xii. 289, I was asked out in a kind of way. **1922** 'R. CROMPTON' *More William* xiv. 231 But if he asks you to his you must ask him back. **1934** H. G. WELLS *Exper. Autobiogr.* II. viii. 532 People, often strange people, were beginning to ask us out.

ask, *sb.*[1] Delete †*Obs.*, and add later examples.
1781 T. TWINING *Let.* 8 Dec. in *Recreat. & Stud.* (1882) 108, I am not so unreasonable as to desire you to..answer all my asks. **1886** 'CAVENDISH' *Whist* 127 When your three comes down in the next round, it is not an ask for trumps.

askance, *adv.* **3.** (Later examples.)
1914 T. HARDY *Satires of Circ.* 32 Down there they are dubious and askance. **1938** E. BOWEN *Death of Heart* I. iii. 56 'Well, what, Matchett?' Anna said touchily. 'Young people like to wear what is usual.' Anna had been askance. *Ibid.* I. iv. 76 Looking at all this in here with a wild askance shrinking eye.

Askar (æ·skăɹ). [Arabic, *'askar* army.] A native Moroccan infantryman. Also as *pl.*
[**1902** *Encycl. Brit.* XXX. 840/1 The infantry (Askárla) are mostly rough, ill-trained levies.] **1918** FARROW *Dict. Mil. Terms*, *Askar*, in the Moroccan army, a native infantryman or foot soldier. **1925** *Blackw. Mag.* Nov. 619/1 An askar of the Riffi army. *Ibid.* 619/2 At our approach, bearded askar come to the salute.

Askari (æskă·ri, æ·skări), *sb. pl.* Also **-is.** [Arabic *'askarī* soldier.] Native soldiers of East Africa. Also as *sing.*
1889 LD. LUGARD *Diaries* (1959) I. 53, I have now 4 Headmen, 10 Askari, 96 porters..and 9 Persians. **1906** *Macm. Mag.* Oct. 929 Before him stood a..native, who acted as interpreter and clerk of the court, and a dozen *askari* whose onerous duty it was to preserve order. **1910** ROOSEVELT *African Game Trails* i. 18 The askaris were uniformed, each in a red fez, a blue blouse, and white knickerbockers. **1927** *Chambers's Jrnl.* 694/2 An askari or native constable took it in turns with two others to guard the market.

asked (ɑskt), *ppl. a.* [pa. pple. of ASK *v.*] Mentioned as a request. Also *asked-for*.
1900 *Westm. Gaz.* 5 Feb. 2/2 The asked-for area. **1901** *Ibid.* 9 July 4/3 The asked price. **1936** [see *ASKING *ppl. a.* b].

askeletal (ē[i]ske·lităl), *a.* [f. A- 14 + SKELETAL *a.*] Having no skeleton.
1904 J. MCCABE tr. *Haeckel's Wonders of Life* x. 288 The askeletal vermalia.

asker[1]. **1.** (Later examples.)
1886 J. J. WRIGHT *Little Asker* viii. 187 For every boy and girl..is only a little asker, made of such mind and spirit as can live, and learn. **1925** CHESTERTON *Everlasting Man* I. viii. 183 The sages had degenerated..into hired rhetoricians or askers of riddles.

askesis, var. *ASCESIS.

asking, *vbl. sb.* Add: **8.** *asking price* (*rate*), the price (or rate) asked for or set by the seller. Cf. ASK *v.* 9 b.
1755 MRS. C. CHARKE *Life* 151, I flew..to all the Brokers in Town, to buy my Houshold Furniture, gave the asking Price for every thing I bought. **1852** *Harper's Mag.* V. 534/1 The merchants..have two prices—an 'asking price' and a 'taking price'. **1924** *Times Trade & Engin. Suppl.* 29 Nov. 242/4 This is higher than recent f.o.b. sales for this autumn's delivery, but about £2 a standard lower than the 'asking' prices at the commencement. **1930** *Economist* 10 May 1051/1 The asking rate on 90-day bankers' bills has been cut to 2⅞ per cent.

asking, *ppl. a.* Add: **b.** *asking bid* (see quot. 1936).
1936 E. CULBERTSON *Contr. Bridge* 15 *Asking Bid*, an artificial bid requesting specific information in the asked suit. **1959** REESE & DORMER *Bridge Player's Dict.* 16 Asking-bids are a conventional way of locating controls for slam purposes.

Aslib (æ·zlib). Also **ASLIB, A.S.L.I.B.** [f. the initials.] The Association of Special Libraries and Information Bureaux, an organization set up in 1926 to facilitate the coordination of information, etc. Also *attrib.*
1926 *Assoc. Spec. Libr. & Inform. Bur.* (*Prospectus*) 2 History of the Development of A.S.L.I.B...On March 29th, 1926, a meeting was held at the Institution of Mechanical Engineers at which the new Association was formally inaugurated. **1928** (*title*) The Aslib Directory. **1937** *Discovery* June 192/2 It is obtainable by 'Aslib' members. **1945** *Lancet* 29 Dec. 856/1 The ASLIB Microfilm Service. **1951** *Engineering* 27 Apr. 503/3 Aslib..move into new premises.

a-smoke (ăsmōu·k), *advb. phr.* or *pred. a.* [f. A *prep.*[1] + SMOKE.] Smoking.
1827 *Blackw. Mag.* XXII. 554 Cottages all a-smoke. **1875** MORRIS *Æneids* VIII. 106 With yet warm blood the altars were a-smoke. **1904** B'NESS VON HUTTEN *Pam* 190 Its chimneys were all asmoke.

Asmonean, var. *HASMONEAN.

asocial (ē[i]sōu·ʃăl), *a.* [f. A- 14 + SOCIAL *a.*] Not social; antagonistic to society or social order; (*colloq.*) inconsiderate of or hostile to other people. Cf. ANTISOCIAL *a.*
1883 MAUDSLEY *Body & Will* III. i. 241 New products of an asocial or antisocial kind are formed in the retrograde metamorphosis of the human kind. **1920** W. H. R. RIVERS *Instinct & the Unconscious* xix. 156 Some channel which leads in an asocial or antisocial direction. **1932** *Daily Express* 20 Sept. 9/4 One cannot foresee to what end their asocial behaviour or maladjustments will ultimately lead. **1952** M. MCCARTHY *Groves of Academe* (1953) ii. 17 This ugly, a-social man, at home and suddenly garrulous in their midst.
Hence as *sb.*, an asocial person. *rare.*
1945 KOESTLER *Yogi & Commissar* III. iv. 249 Most asocials have some such sort of jealously-guarded private philosophy. **1961** R. KEE *Refugee World* iv. 41 A proportion of 'a-socials' there—people who *wouldn't* live anywhere else rather than couldn't.

asp. See *ASS *sb.* 3*.

asparagus. Add: **2.** *asparagus-bed* (earlier example); also *transf.* (sometimes *ellipt.*), an anti-tank obstacle (*slang*); *asparagus-tip*; **asparagus-bean** *U.S.*, a tropical American bean (*Dolichos sesquipedalis*); **asparagus pea,** the Goa bean.
1856 COZZENS *Sparrowgr. Papers* vii. 85 The asparagus bean, a sort of long-winded esculent, inclined to be prolific in strings. **1733** ELLIS *Chiltern & Vale Farming* xxxv. 257, I remember..the case of an Asparagus Bed. **1939** *War Illustr.* 16 Dec. 426 Anti-tank 'asparagus', which consists of railway lines set in concrete at an angle of 45 degrees pointing towards the enemy lines and in some cases with high-explosive caps. **1943** HUNT & PRINGLE *Service Slang* 12 *Asparagus bed*, a form of anti-tank obstacle, designed to slow up the A[rmoured] F[ighting] V[ehicle] while it comes under fire. **1859** A. VAN BUREN *Sojourn in South* 155 The Asparagus Pea, with a small round pod that grows from a foot to three in length, makes a choice dish at table. **1958** *Times* 1 Feb. 9/6 An oddity among the peas is the so-called 'asparagus pea', which makes a spreading plant and produces flanged, rectangular pods. **1932** H. SIMPSON *Boomerang* x. 264, I remember particularly asparagus-tips rolled up in brown-bread. **1965** J. WAINWRIGHT *Death in Sleeping City* I. 44 The appetising meal of braised cutlets, chipped potatoes and asparagus tips.

aspartic, *a.* Add: *aspartic acid* (earlier example).
1836 BRANDE *Man. Chem.* (ed. 4) 1042 When asparagin is long boiled with hydrated oxide of lead, magnesia, or other bases, it is resolved into ammonia, and into a new acid, called the aspartic acid.

aspect, *sb.* Add: **5. b.** *aspect ratio*, (*a*) *Aeronaut.*, the ratio of the span to the mean chord of an aerofoil; also, the ratio of the square of the span to the total area of the aerofoil; (*b*) of a television or cinematographic picture: the ratio of the width to the height.
1907 F. W. LANCHESTER *Aerodynamics* vi. 208 The normal pressure is a *continuous function* of the aspect ratio of the plane. **1910** R. W. A. BREWER *Art of Aviation* xvii. 232 Aspect ratio is the ratio of the length of span to that of chord. **1930** *Flight* 21 Feb. 234/1 As regards the aerodynamic efficiency of the Do.X, the aspect ratio (if one may be so old-fashioned as to use this expression) is low. **1936** *Electronics* June 28/1 The formula used is $f = \frac{1}{2}$ (no. lines) × aspect ratio × (no. pictures per second), where the aspect (width to height) ratio is 4/3. **1957** AMOS & BIRKINSHAW *Television Engin.* I. i. 29 For a given number of lines and aspect ratio the bandwidth occupied by the components of the video signal is directly proportional to the picture frequency.
9. b. *Gram.* In Russian and other Slavonic languages, a verbal category of which the function is to express action or being in respect of its inception, duration, or completion, etc.; by extension applied to such forms in other languages.
The earlier term in Slavonic grammars was 'branch'.
1853 C. P. REIFF *Eng.-Russ. Gram.* I. 86 The aspects have not all the same number of tenses; the imperfect aspect is used in all the three tenses; the perfect is employed in the preterit and future, while the iterative is met with only in the preterit. **1884** J. NESTOR-SCHNURMANN *Russ. Man.* 97 The variations in form of the same action are expressed by what is called in Russian branches or aspects..of a verb... There are four branches, viz.: the Indefinite, the Perfect, the Semelfactive.., and the Iterative...A Fifth Aspect might be added, viz.: the Inchoative. **1921** E. SAPIR *Lang.* v. 114 Aspect is expressed in English by all kinds of idiomatic turns. **1924** JESPERSEN *Philos. Gram.* xx. 286 It is generally assumed that our Aryan languages had at first no real forms in their verbs for tense-distinctions, but denoted various aspects, perfective, imperfective, punctual, durative, inceptive, or others.
14. *Ecology.* The characteristic seasonal appearance or constitution of a plant community.
1905 F. E. CLEMENTS *Research Meth. Ecology* iv. 296 The seasonal changes of a formation, which are called aspects, are indicated by changes in composition or structure, which ordinarily correspond to the three seasons, spring, summer, and autumn. **1916** —— *Plant Succession* vii. 132 In boreal and alpine regions the number of aspects is often but two, vernal and æstival, and the societies correspond. **1926** TANSLEY & CHIPP *Stud. Vegetation* ii. 17 The word 'aspect' is used as a technical term for the seasonal phases of vegetation. Thus we speak of the *prevernal, vernal, æstival, autumnal* and *hiemal* aspects of British vegetation. Societies dominated by species vegetating at particular seasons are often called *aspect societies*. **1952** P. W. RICHARDS *Tropical Rain Forest* viii. 191 The Tropical Rain forest..has no marked seasonal 'aspects' and no 'resting' period.
15. In *Signalling*: an indication given by means of a light (see quot. 1936).
1926 D. R. LAMB *Modern Railway Operation* xiii. 152 The colour light signal is now generally adopted in connection with power installations. The type which has so far found favour in Great Britain is that giving three aspects, green, orange and red, signifying, respectively, 'all right', 'caution' and 'danger'. **1936** *Gloss. Terms Railway Signalling* (*B.S.I.*) 47 *Signal aspects*, a term used to describe light indications of signals as opposed to indications given by semaphore arms.

aspectual, *a.* Restrict †*Obs. rare*[-1] to sense in Dict. and add: **2.** *Gram.* Of or pertaining to an aspect or aspects (see *ASPECT *sb.* 9 b).
1950 in WEBSTER Addenda. **1959** R. QUIRK in *Teaching of English* i. 33 We can even make suggestions about a kind of aspectual meaning with this verb by reason of the analogy of our frequentative series of English verbs like *sparkle, gabble, wriggle*. **1963** J. LYONS *Structural Semantics* vi. 117 The aspectual schemes set out..for the 'action'—'event'—'state' verbs.

aspergill. (Examples.)
1864 [see ASPERGILLUM]. **1899** *Blackw. Mag.* Aug. 157/2 It is the alliance of the sword and the aspergill. **1922** JOYCE *Ulysses* 333 Eyes on a dish, wax candles, aspergills, unicorns.

aspergillic (æspəɹdʒi·lik), *a.* [f. ASPERGILL(US + -IC.] *aspergillic acid*, an acid derived from *Aspergillus flavus* and used as an antibiotic.
1943 E. C. WHITE & J. G. HILL in *Jrnl. Bacteriol.* XLV. 433 The present paper presents details of the isolation of an active crystalline substance from liquid cultures of our strain of *Aspergillus flavus*...We propose to call our substance 'aspergillic acid'. **1949** FLOREY et al. *Antibiotics* I. vii. 315 Aspergillic acid ($C_{12}H_{20}O_2N_2$)..crystallizes in pale yellow rods. *Ibid.* 318 Sanders..found that aspergillic acid..inhibited the growth of nearly all of a large number of fungi pathogenic to man.

aspergillin (æspəɹdʒi·lin). *Biol.* [ad. F. *aspergilline* (G. Linossier 1891, in *Comptes Rendus* CXII. 489), f. ASPERGILL(US + -IN[1].]
1. A black pigment found in the spores of various species of *Aspergillus*.
1891 *Jrnl. Chem. Soc.* LX. 751 The spores of *Aspergillus niger*, when treated with very dilute ammonia, yield a dark-coloured solution from which a slight excess of hydrochloric acid throws down *aspergillin* as a black, bulky, flocculent precipitate. **1949** FLOREY et al. *Antibiotics* I. vii. 348 Since the name 'aspergillin' was originally proposed by Linossier (1891) for the black water-insoluble pigment of the spores of *A*[*spergillus*] *niger*, it would appear desirable to avoid applying it to other substances.
2. A name given to various antibiotic substances obtained from different species of *Aspergillus*.
1943 *Jrnl. Pharmacol.* LXXVIII. 164 In an abstract of this paper..our antibacterial substance is referred to as *aspergillin*. Due to the possible confusion with White's aspergillic acid..we thought it advisable to change the name of our antibacterial substance to *flavicin*. **1944** N. F. STANLEY in *Austral. Jrnl. Sci.* VI. 151/2 (*title*) Aspergillin, a Stable Antibacterial Substance of High Potency Produced by a Species of Aspergillus...Aspergillin is more soluble in organic solvents than in water.

aspergillosis (æspəɹdʒilōu·sis). *Path.* [mod.L., ad. F. *aspergillose* (A. Lucet 1896, in *Bull. Soc. Cent. Méd. Vét.* XIV. 593), f. ASPERGILL(US + -OSIS.] Infection with the fungus *Aspergillus*, most often found in the respiratory organs of birds and mammals including man.
[**1897** A. LUCET in *Vet. Jrnl.* XLV. 227 Some of which [*sc.* subjects] even (for instance, the birds), in a good many cases contract *spontaneous Aspergillose*.] **1898** *Med. Chron.* IX. 390 The basis of the following investigation was a case of aspergillosis observed by the author in March, 1896. **1902** *Encycl. Brit.* XXXI. 534/2 Aspergillosis, or pigeon-breeders' disease, is the result of the infection with the Aspergillus fumigatus. **1955** GAIGER & DAVIES *Vet. Path. & Bacteriol.* xxi. 414 *Aspergillosis*,..popularly known as brooder pneumonia. **1965** *Economist* 20 Feb. 780/2 He [*sc.* a penguin]..was found to have contracted the disease known as aspergillosis. Aspergillosis is a fungus which spreads through the lungs and is not unknown in humans.

asperging (ăspɔ̄·ɹdʒiŋ), *vbl. sb.* [f. ASPERGE *v.* + -ING[1].] The action of sprinkling (with holy water). Also *attrib.*
1865 TYLOR *Early Hist. Mankind* ix. 258 The priest stands with an asperging-brush in his hand, with which he sprinkles them with holy water.

aspermia (ăspɔ̄·ɹmiă, ē[i]-). [mod.L., f. A- 14 + Gr. σπέρμα seed: see -IA[1].] The lack of, or inability to ejaculate, semen; absence of spermatozoa from the semen as ejaculated. Hence **aspe·rmic** *a.*, of or pertaining to aspermia, lacking sperm.
1853 DUNGLISON *Dict. Med. Sci.* (ed. 12) 110/1 Aspermia. **1925** *Glasgow Herald* 5 Dec. 4 An egg-cell which normally requires to be fertilised by a sperm-cell may be launched on a voyage of aspermic development. **1946** *Nature* 23 Nov. 729/2 The groups [of semen samples] cover a very wide range from complete aspermia to slight oligospermia. **1956** *Ibid.* 28 Jan. 190/1 (*heading*) Zinc in aspermic human semen.

aspersory (ăspɔ̄·ɹsəri), *sb.* [ad. med.L. ASPERSORIUM.] A holy-water sprinkler; an aspersorium or aspergillum.
1881 F. E. WARREN *Liturgy Celtic Ch.* 116 It rather resembles an aspersory than a pastoral staff. **1897** *Pall Mall Mag.* Mar. 367 A holy-water stoup with an aspersory stood at the feet [of a corpse].

aspheric (ăsfe·rik, ē[i]-), *a.* [f. A- 14 + SPHERIC *a.*] Not spherical. Hence **aspheri·city,** want of sphericity.
1923 W. R. RAYTON in *Jrnl. Optical Soc. Amer.* VII. 197 Osculating paraboloids are, in general, much closer approximations to the required aspheric surfaces than are the osculating spherical surfaces. **1932** HARDY & PERRIN *Princ. Optics* xvi. 347 Any surface having other than a plane or spherical form is said to be *aspheric*...Aspheric surfaces are used to reduce the spherical aberration. **1944** *Jrnl. Optical Soc. Amer.* May 279/2 The asphericity of its plates is about four times greater than the asphericity of a parabolic mirror of the same diameter. **1948** *Electronic Engin.* XX. 314 An aspheric plastic objective lens..focuses infra-red radiation from a distant beacon on to the photocathode.

aspherical (ăsfe·rikăl, ē[i]-), *a.* [f. A- 14 + SPHERICAL *a.*] = prec.
1922 A. H. LEVY tr. *M. von Rohr's Eyes & Spectacles* ii. 75 We consider an aspherical surface [of a lens] to be formed by adding material to a spherical surface. **1923** GLAZEBROOK *Dict. Appl. Physics* IV. 160/1 *Aspherical lens*, a type of spectacle lens designed to correct the aberration introduced when the eye looks towards the edge of the lens instead of to the centre.

asphyxiant (æsfi·ksiănt), *a.* and *sb.* [f. ASPHYXIA + -ANT.] **A.** *adj.* Causing asphyxia. **B.** *sb.* Any chemical substance that causes asphyxia.
1854 *Englishwoman in Russia* 302 Long-range guns and asphyxiant balls. **1888** *Chambers's Encycl.* I. 501 The term asphyxiants is applied to any gases having a suffocating.. effect on the human system. **1921** *19th Cent.* July 34 If in the Great War we had refused to use asphyxiant gas.

aspidistra (æspidi·stră). [mod.L., f. Gr. ἀσπιδ-, ἀσπίς shield + *-istra*, after *tupistra*.] A plant of the genus so called, belonging to the family Liliaceæ and native to China and Japan; frequently grown as a pot-plant, and often regarded as a symbol of dull middle-class respectability.
1822 *Bot. Register* 628 *Aspidistra lurida*. Dingy-flowered Aspidistra. **1852** G. W. JOHNSON *Cott. Gard. Dict.* s.v., The mushroom-shaped stigma by which Aspidistræ are characterised. **1920** 'O. DOUGLAS' *Penny Plain* iii, An aspidistra in a pot completed the table decorations. **1925** W. DEEPING *Sorrell & Son* IV. i, Carrying out the aspidistras and washing them in the yard. **1936** 'G. ORWELL' (*title*) Keep the aspidistra flying.
¶ Illiterate forms were formerly frequent.
1895 *Daily News* 4 Dec. 6/7 The broad leaves of the aspidestria. **1899** *Westm. Gaz.*, They are just moderate-sized palms, ferns, and aspidestra.
Hence **aspidi·stral** *a.*, of or pertaining to an aspidistra; abounding in aspidistras. Also *transf.*
1936 'G. ORWELL' *Keep Aspidistra Flying* iii. 73 He was a made man—or, by Smilesian, aspidistral standards, *un*made. *Ibid.* viii. 198 Why chew leathery beef in the aspidistral dining-room when he had ten quid in pocket? *Ibid.* x. 270 Flaxman's wife had forgiven him, and he was back at Peckham, in aspidistral bliss. **1939** R. CAMPBELL *Flowering Rifle* I. 13 The weed of Life that grows where air is hot With 'Meetings' for its aspidistral pot.

Aspinall (æ·spinǫl), *v. trans.* To paint (articles of household furniture, esp. when old and shabby) with Aspinall's Enamel Paint. Hence **A·spinalling** *vbl. sb.*
1889 *Jrnl. Educ.* 1 Aug. 388/1 Three low wicker chairs 'Aspinalled' in dark green. **1891** M. E. MANN *Winter's Tale* II. I. xi. 25 The ..stool, which Erica herself had aspinalled in a spare hour. **1891** (*title*) Guide to Aspinalling. By A Lady. **1908** A. BENNETT *Buried Alive* vii. 165 What a smell of paint!..You surely haven't been aspinalling that bath-room chair?

aspirating (æ·spire[i]tiŋ), *ppl. a.* [f. ASPIRATE *v.* + -ING[2].] That operates by aspiration or suction.
a 1884 KNIGHT *Dict. Mech.* Suppl. 51/2 *Aspirating Filter*, one in which the action is expedited by the withdrawal of air from beneath the filtering material. *Ibid.*, *Aspirating Winnowing Machine*, one which draws air through the grain instead of blowing it. **1895** *Catal. Surg. Instruments* (Arnold & Sons), Aspirating Syringe, Aspirating Guarded Needle. **1902** *Times* 25 Sept. 3/4 The operation was performed with an aspirating needle.

aspiration. Add: **7.** The action or process of drawing in, out, or through by suction; *esp.* the drawing out (of fluids, gases, etc.) by means of an aspirator.
1842 *Civil Engin. & Archit. Jrnl.* V. 203/2 The fluid was absorbed throughout all the pores of the [felled] tree, by a process which is termed 'aspiration'. **1879** *Encycl. Brit.* IX. 344/2 Apparatus, in which the principle of aspiration, or drawing currents of air through the grain, is now extensively employed. **1881** *Trans. Obstetr. Soc. Lond.* XXII. 57 Aspiration was frequently resorted to, with the view of arresting the growth of the fœtus. **1950** *Engineering* CLXIX. 31/1 'Aspiration', or separation by means of ascending air currents, is much used in [flour] milling.
attrib. **1902** C. N. & A. M. WILLIAMSON *Lightning Conductor* 18 The 'aspiration pipe'..had worked loose.

aspirational (æ:spirē[i]·ʃənăl), *a.* [f. ASPIRATION + -AL.] Belonging to or characterized by aspiration.
1887 H. R. HAWEIS *Christ & Christianity* I. iii. 83 Its sense of mystery feeble, and consequently its lift and aspirational power almost *nil*. **1916** A. M. RIHBANY *Syrian Christ* (1919) II. ii. 66 As pious in his imprecations and curses as he is in his aspirational prayer. **1967** *Listener* 30 Nov. 731/2 Everybody banging away at that final aspirational aria.

aspirin (æ·spirin). *Chem.* [G. (C. Witthauer 1899, in *Die Heilkunde* Apr.), shortened form of *Acetylirte Spirsäure* (i.e. acetylated spiræic acid) + -IN[1].] A white crystalline compound, acetylsalicylic acid, used esp. as an analgesic and antipyretic; with *an* and *pl.*, a dose of this in tablet form. Also *attrib.*
1899 *Jrnl. Chem. Soc.* LXXVI. II. 605 Physiological Action of Aspirin (Acetylsalicylic Acid). **1901** *Ibid.* LXXX. II. 408 Aspirin (o-acetoxybenzoic acid) does not increase the flow of urine, but somewhat raises the total output of solids. **1922** *Blackw. Mag.* Apr. 460/1 He was always at hand with tea and brandy and aspirin. **1923** MRS. A. SIDGWICK *None-go-by* xxix. 243, I think that frontal neuralgia is worse. You can give me an aspirin if you like. **1924** *Blackw. Mag.* June 801/2 Aspirin tablets. **1937** 'G. ORWELL' *Road to Wigan Pier* xii. 224 Every aspirin-eater in the outer suburbs. **1954** *Numbers* July 4 Flash of wet overcoats; artificial neon-lit expressions. Aspirin age. World gone sour. **1960** [see *ANYTHING 1 b].
fig. **1930** *Time & Tide* 30 Aug. 1085/1 This aspirin remedy could continue only so long as the price of wheat and wool was high. **1957** M. SWAN *Brit. Guiana* I. i. 21 The living standards would eventually improve without the necessity for this aspirin from the Government.

asportation. (Later examples.)
1889 F. W. MAITLAND *Coll. Papers* (1911) II. 173 The imaginary defendant is charged..with the asportation of all that is asportable. **1890** *Harper's Mag.* Oct. 702/1 If the silver..could be deposited..in bars too heavy for asportation.

Asquithian (æskwi·þiăn), *a.* and *sb.* [f. the name of H. H. *Asquith* (afterwards Earl of Oxford and Asquith), prime minister of Great Britain 1908–16 + -IAN.] **A.** *adj.* Pertaining to, resembling, or supporting Asquith as leader of the Liberal party or, later, as leader of a group of the divided party. **B.** *sb.* A supporter of Asquith.
1910 *Spectator* 2 Apr. 530/1 The Asquithian Peers will, we venture to say, prove the most intractable of 'backwoodsmen'. **1928** *Daily Express* 24 Apr. 3/3 Montague..belonged to the central Asquithian group in the Cabinet,..and as an Asquithian he preferred to support McKenna rather than Lloyd George. **1928** H. FYFE *Brit. Liberal Party* viii. 229 Liberals were still at loggerheads. Asquithians and Lloyd Georgians fought against each other. **1960** *Spectator* 14 Oct. 553 An old-fashioned Asquithian Liberal.

ass, *sb.* Add: **3.*** *Paper Manuf.* (The relationship of the form *asp*, also used, is unexplained.) A donkey-rest (see DONKEY 3 b).
a 1875 KNIGHT *Dict. Mech.* I. 170/2 *Ass*, a post in the bridge of a pulp-vat to lay the mold upon while the water drains from it. Used in the hand-made paper work. **1927** *World's Paper Trade Rev.* 24 June 2002/2 *Ass*, the wooden strut against which the vatman places the mould for a moment to drain, shaped like a bow. **1947** D. HUNTER *Papermaking* (ed. 2) vi. 178 The coucher placed the mould ..against the inclined drainage-horn, or 'asp', which allowed the surplus water to drain back into the vat. *Ibid.* xv. 437 The coucher..leans the mould against the asp or horn at the proper angle.
4. b. ass-cart, a cart drawn by an ass.
1821 *Blackw. Mag.* Apr. 412/2 They met the country people coming in—some in horse-carts, others in ass-carts. **1902** YEATS *Where there is Nothing* (1903) II, A little kennel of straw under the ass-cart.

ass, vulgar and dial. sp. and pronunc. of ARSE. Add: U.S. examples (Webster 1961 'often considered vulgar'), and later U.K. examples.
1930 J. DOS PASSOS *42nd Parallel* I. 100 My ass to habeas corpus. **1934** J. O'HARA *Appointment in Samarra* (U.S. ed., 1953) iv. 119 You give me a pain in the ass. **1949** 'N. BLAKE' *Head of Traveller* ii. 37 Put it to rights in two shakes of a cat's ass, you would. **1959** W. GOLDING *Free Fall* iv. 96 You sit on your fat ass in your 'ouse all the week. **1967** *Observer* 8 Oct. 31/2 All our trousers are designed to fit round the ass and not sag at the crutch.

ass, *v.* Add: **2.** Now freq. in (orig. schoolboys') *slang*: to fool *about*. (Cf. *ARSE *v.*)
1899 KIPLING *Stalky* vii. 196 Don't mind learnin' my drill, but I'm not goin' to ass about the country with a toy Snider. **1932** A. J. WORRALL *Eng. Idioms for Foreign Students* 1 Those boys are not working; they are just assing about. *Ibid.*, Don't ass about with that valve, you'll break it.

assagai, *v.* (Earlier examples.)
1836 *Editor Grahamstown Jrnl.'s Nar.* 185 (Pettman), When only a few yards from the village..he was assegaied. **1863** W. C. BALDWIN *Afr. Hunting* iii. 71 A Kaffir who had threatened to assagai one of Walmsley's Kaffirs.

Assamese (æsămī·z), *a.* and *sb.* [f. the name of the Indian State *Assam* + -ESE.] **A.** *adj.* Of or pertaining to Assam or its inhabitants. **B.** *sb.* **1.** A native of Assam. **2.** The language of Assam.
1826 J. CONDER *Mod. Traveller* XI. 245 The Assamese came out of their hiding-places. *Ibid.* 246 The poverty of their native soil, and the fertility of the plains of Assam, induced the Sing-fos to settle in the plains, which they cultivated by means of Assamese captives. **1836** *Jrnl. Asiatic Soc.* V. 195 A great deal of opium is grown by the Miris, which they barter for grain with the Assamese. **1837** *Ibid.* VI. 18 Their [*sc.* the priests'] vernacular and common dialect, as well as that of the people, is *Assámese*. **1875** *Encycl. Brit.* II. 718/1 Aurangzeb..despatched a considerable force for the..invasion of the Assamese territory. *Ibid.* 721/1 The vernacular Assamese possesses a close affinity to Bengali. **1961** *Times* 18 May 12/4 The hill leaders were firmly opposed to Assamese as the official language of the state.

assassin. Add: **4.** *assassin-like* (later example); **assassin bug,** a predaceous insect of the family Reduviidæ.
1895 J. H. & A. B. COMSTOCK *Man. Stud. Insects* xiv. 137 Family Reduviidæ...There are many bugs which destroy their fellows, but the members of this family are so preeminently predaceous that we call them the Assassin-bugs. **1937** *Discovery* Dec. 368/2 A..cheerful brute occurs in North America, where it is known as the 'big bed bug' and the assassin bug. **a 1846** B. R. HAYDON *Autobiogr.* (1927) vii. 104 On this principle I have acted in not making the assassins so assassin-like as perhaps they were.

assassinate, *v.* Add: **3.** (Later examples.) (Cf. MURDER *v.* 2.)
1920 WODEHOUSE *Damsel in Distress* ix. 115 The same feeling which a composer with an over-sensitive ear would suffer on hearing his pet opus assassinated by a schoolgirl. **1962** *Guardian* 15 Nov. 3/1 Helping the Prime Minister in his political battle to assassinate Mr. George Brown.

assault, *sb.* Add: **8.** Applied *attrib.* to equipment or troops used in making an assault, as *assault boat, craft, ship, troops*; **assault course,** a course for training in assault; also *assault practice*.
1915 *Bayonet Fighting* (*H.M.S.O.*) 1 To obtain efficiency with the bayonet, the men..should frequently carry out the Final Assault Practice. **1916** *Bayonet Training* vi. 26 The 'points' will also be practised..on dummies placed, as a preparation for the Final Assault Course, in positions of increasing difficulty. **1941** *War Illustr.* 18 Apr. 398 Collapsible boats of wood and canvas, known as assault boats, much larger than the rubber boats, are used when a number of troops are to be carried across water. **1943** *Combined Operations* ii. 16 (*caption*) Assault Course. In the battle schools, British troops are trained under realistic conditions. *Ibid.* 17 Landing craft are carried by infantry landing ships, originally known as assault ships. **1943** *Hutchinson's Pict. Hist. War* July 247/2 As the assault craft headed towards the beaches, everything was very quiet. **1945** *Daily Tel.* 12 June 1/1 Assault troops of the Australian Ninth Division ..quickly established beachheads against almost no resistance. **1959** *Times* 8 Sept. 4/2 Rotodyne vertical take-off and landing aircraft..will be able to carry 70 fully equipped assault troops.

assaultive (ăsǫ·ltiv), *a.* [f. ASSAULT *v.* + -IVE.] Liable or wishing to commit an assault.
1955 *Sci. News Let.* 2 July 4/1 She became assaultive, wanted to kill herself and had delusions of persecution. **1961** E. HUNTER *Mothers & Daughters* II. 213 The moment the jacket was removed..she became assaultive.

assay, *sb.* Add: **6. b.** Esp. in *biological assay* (= *bio-assay), the determination of the strength of a substance by means of a test on an organism, usu. in comparison with the effect of a standard preparation.
1922 BURN & DALE *Rep. Biol. Stand.* I. 50 We do not assume that it will be necessary to make a new batch of the Standard for the assay of every batch of the commercial extract. **1927** BURN & ELLIS in *Pharmaceut. Jrnl.* 9 Apr. 384/1 (*title*) The Biological Assay of the Specific Alkaloid of Ergot. **1928** H. H. DALE in J. H. Burn *Methods Biol. Assay* p. x, The most serious difficulty in biological assay is that due to the variability of the living reagent. **1961** *Lancet* 22 July 214/1 The daily urinary loss of vitamin B_{12}.. was measured in three patients by microbiological assay.
VI. *assay office.*
1773 *Act* 13 *Geo. III*, c. 52 § 4 Such Silver Vessel, Plate, or manufactured Silver shall be marked..with the Mark of the Company within whose Assay Office such Plate shall be assayed and marked. **1851** *San Francisco Herald* 1 Feb. 2/1 Arrangements for opening the United States Assay

Office in this city, are being rapidly completed. **1853** in Mrs. C. Clacy *Lady's Visit to Gold Diggings* xv. 233 The quantity of gold taken to the Assay-office, during four consecutive weeks, amounting to less than four thousand ounces. **1869** 'MARK TWAIN' *Innoc. Abr.* xviii. 179 The cargoes of 'crude bullion' of the assay offices of Nevada.

assay, *v.* Add: **4. c.** To show (a certain yield) by assay; to yield on assay. Also *absol.*
1882 *Rep. Prec. Metals U.S.* 305 Assay as high as $100 to the ton. **1892** *Graphic* 26 Mar. 387/1 A 'pocket'..which assays a good many ounces to the ton. **1927** *Sunday Times* 13 Feb. 2 Ore,..assaying as high as 7·3 dwts. (31s.).
d. To test the strength of a substance by means of a test on an organism (see *ASSAY *sb.* 6 b); so *assayed* ppl. a.
1922 BURN & DALE *Rep. Biol. Stand.* I. 51 A comparison of each batch with its accurately assayed predecessor should suffice. **1928** *Lancet* 20 Oct. 820/1 Drugs..can be assayed with a sufficient degree of accuracy by biological methods. **1962** *Ibid.* 6 Jan. 24/1 Tissue-culture methods have been used for some time to assay the effect of drugs on normal and neoplastic cells.

assemblage. Add: **2.** In various *techn.* uses: the joining, putting together of parts (of a machine); a collection (e.g. of artefacts); a work of art consisting of miscellaneous objects fastened together.
a 1875 KNIGHT *Dict. Mech.* I. 171/1 This system of interchangeability and assemblage..is one of the most beautiful triumphs of modern mechanism. **1958** *Listener* 6 Nov. 752/2 The success of the total effect was due to the cutting and assemblage of the recordings under the composer's care. **1959** J. D. CLARK *Prehist. S. Afr.* ii. 55 The preservation of more complete faunal assemblages. **1961** *N.Y. Times* 4 Oct. 42/2 An 'assemblage' is a work of art made by fastening together cut or torn pieces of paper, clippings from newspapers, photographs, bits of cloth, fragments of wood, metal,..shells or stones, or even objects such as knives and forks,..automobile fenders, steel boilers, and stuffed birds and animals. **1963** *Listener* 7 Feb. 254/2 His [*sc.* John Latham's] assemblages do not strike me as random, when considered as arrangements of forms and planes.

assemble, *v.*[1] Add: **2. b.** To put together (the separately manufactured parts of a composite machine or mechanical appliance); also with the machine as obj. Also **asse·mbled** *ppl. a.*, **asse·mbling** *vbl. sb.* and *ppl. a.* Also in extended use.
1852 *Harper's Mag.* V. 158/1 When the several parts are all finished, the operation of putting them together so as to make up the musket from them complete, is called 'assembling the musket'. **1865** *Mech. Mag.* 31 Mar. 200/2 When all these parts are assembled together. **1888** *Sun* 21 Mar. (Farmer), The steel forgings have been made and turned over to our ordnance officers to assemble into guns. **1894** *Harper's Mag.* July 252/1 The assembling, as the process of shrinking on the various parts which go to make up the modern cannon is called. *Ibid.* 256/2 The partly assembled gun. **1907** *Westm. Gaz.* 9 Nov. 14/2 The vertically assembled differential case is being gradually superseded by the horizontally assembled case. **1909** *Ibid.* 28 Apr. 2/2 Structural steel is imported punched, riveted, and assembled to be set up. *Ibid.* 16 Sept. 5/1 If all the parts are not there when the time comes to assemble the chassis. **1923** *Ibid.* 15 Sept., Assembling and packing cycle bells. **1926** *Time* 14 June 34/2 In 1924 he profited $25,000,000 by shipping parts to district assembling plants. **1961** K. REISZ *Technique Film Editing* (ed. 9) 278 *Assemble*, to carry out the first process in film editing, namely, to collect together the required shots and join them in provisional order, thus producing a rough cut.

‖**assemblée** (asãble). [Fr.; see ASSEMBLY.] Var. ASSEMBLY (esp. sense 7).
1712 SWIFT *Jrnl. to Stella* Let. xxxviii (1948) II. 460 Was last night at Lady Betty Germain's assemblée. **1787** MATY tr. *Riesbeck's Trav. Germ.* II. xxxi. 47 [In Hungary] every town with four or five houses in it, has its *assemblées*, and *redoutes*. **1938** *Times Lit. Suppl.* 5 Mar. 153/1 Mr. Tourtellot is prepared to join the ranks of the admiring *assemblée* and to gaze at her with uncritical eyes.

assembler. Add: **3.** One who assembles a machine or its parts.
1898 *Cricket & Football Field* 1 Oct. 3 'Assemblers'—men who buy [cycle] frames..then put in whatever fittings they or their customers may be inclined to use. **1908** *Daily Chron.* 21 Feb. 10/6 Arc lamp assemblers. **1909** *Ibid.* 21 Aug. 6/6 One cannot get such a cycle ready made, but a good assembler can build it up.
4. The mechanism of a linotype machine which assembles the matrices and space bands to form a line of type. Also *attrib.*
1902 *Encycl. Brit.* XXXIII. 523/2 They are set up in proper order in the assembler block. **1905** C. T. JACOBI *Printers' Handbk.* (ed. 3) 93 When the line is accidentally overset..it is necessary to remove some of the matrices from the assembler. **1908** J. S. THOMPSON *Mechanism of Linotype* (ed. 3) ii. 15 The assembler shute spring..should be curved so as to tend to throw the bottom of the matrix towards the assembler wheel. **1925** J. R. ROGERS *Linotype Instruct. Book* 1 Beginning with the touch of a finger upon a Keybutton..a matrix is released and falls by gravity upon a constantly running belt which delivers the matrix into the 'assembler'.

assembly. Add: **1. b.** The assembling of troops; freq. *attrib.* Cf. sense III in Dict.
1917 H. W. YOXALL *Jrnl.* 18 Sept. in *Fashion of Life* (1966) iv. 37 Went with..two runners to peg out our assembly area...The marking-out of the assembly position was not difficult. **1918** E. S. FARROW *Dict. Mil. Terms* 42 The assembly position must be far enough to the rear to enable [etc.]. **1919** *King's Royal Rifle Corps Chron.* 1916 62 The Battalion..proceeded to the assembly trenches. **1923** KIPLING *Irish Guards* I. 324 The Battalion moved nearer their assembly-areas.
c. The action or method of assembling a machine or composite article; the parts so assembled.
1914 *Engineering Mag.* XLVII. 6 The boards travel.. down the line, growing in completeness as they move, each 'team' working simultaneously on opposite sides of the board, adding some step to the assembly. *Ibid.* 882/2 Axles to be placed in chassis assembly at the Highland Park shops are lifted up off the chain-line. **1922** *Autocar* 10 Nov. 983 Arrol-Johnston gear assembly, showing how the whole can be removed without disturbing the box itself...Rear axle assembly. **1935** *Discovery* June 183/2 The assembly of an ordinary electric-lampholder consisting of eighteen parts. **1937** *Ibid.* June 187/1 The glass envelope housing the electrode assembly. **1948** 'N. SHUTE' *No Highway* ii. 43 The port tailplane and elevator, the port landing wheel assembly. **1962** A. NISBETT *Technique Sound Studio* 247 The whole assembly can be very light in weight and may be cheap to produce.
Also *attrib.*, as *assembly-belt* (also *fig.*), *-plant, -room, -shop*; **assembly line** (orig. *U.S.*), a group of machines and workers concerned with the progressive assembly of some product; also *attrib.* and *fig.*
1938 AUDEN & ISHERWOOD *On Frontier* 17 The assembly-belt is like an army on the move. **1945** SINCLAIR LEWIS *C. Timberlane* (1946) i. 13 He had a wife unremittingly productive of babies, for whose assembly-belt production he felt only accidentally responsible. **1914** *Engineering Mag.* XLVII. 858/2 The study of the finishing and assembling of front-axle components shows how labor-costs may be.. reduced..by the use of sliding assembly lines, chain-driven for the final assembling, but having the partial assemblies moved by hand. **1926** *Sci. Amer.* July 41/1 The illustration shows the chassis on the assembly line. **1943** S. C. MENEFEE *Assignment* (1944) I. v. 103 The assembly-line method of putting bombers together has great possibilities. **1945** *Reader's Digest* Oct. 82/2 The syndicate developed an assembly-line method of producing beef. **1955** *20th Cent.* June 548 From the top of a school assembly line they are carried by their parents or the State to a conveyor belt that runs through the spiry city [*sc.* Oxford] for three years or so. **1931** *Economist* 28 Feb. 432/1 The project to build an assembly plant at the new port of Gdynia has not been carried any further. **1897** *Outing* (U.S.) XXX. 279/2 They are then polished, nickelplated,..and sent to the assembly-room to take their places in the wheel. **1914** *Auto-motor Jrnl.* 21 Mar. 362/1 It is only because every part is dead true before it reaches the assembly shop.
d. The assembling of parts of a film or sound recording. Also *attrib.*
1949 W. H. OFFENHAUSER *16-Mm. Sound Motion Pict.* x. 328 Editing and assembly are loose and broad terms customarily used to describe the processes that occur between the original records and the release prints. **1960** O. SKILBECK *Film & T.V. Terms* 12 *Assembly*, rushes of a film joined together in script scene order, with spare takes and number boards eliminated; but not, as yet, edited. **1962** A. NISBETT *Technique Sound Studio* 246 *Copy editing*, the copying of selected extracts from recorded material into sequence on a main programme assembly tape.
IV. assembly-ground, a place where birds assemble esp. for mating; **assembly house** (*a*) = *sembly-house* (SEMBLY 4); (*b*) a house in which assemblies (see sense 7 in Dict.) were held; **assembly-man** (earlier U.S. example); **assembly-place,** a place in which an assembly is held; *spec.* = *assembly-ground*.
1906 E. SELOUS in *Zoologist* X. 419 There were, too, some good examples of the deportment of Ruffs on the assembly-ground. **1920** H. ELIOT HOWARD *Territory in Bird Life* v. 173 In the first place, there are the assembly grounds to which the birds repair season after season; and then, on the assembly grounds, there are the territories. **1502** in *Oxf. Univ. Statutes* (1888) 217 In Domo vocata *le Assemblie-House* in Universitate prædicta. **1749** B. M. CAREW *Apol. for Life* 28 He saw a fine House, and demanding whose it was, they told him it was the Assembly-House [in Maryland]. **1762** O. GOLDSMITH *Life R. Nash* 46 People of fashion [at Bath] make public breakfasts at the assembly-houses. **1837** HALIBURTON *Clockm.* Ser. I. xxxii. 347 He sot up for an Assembly-man. **1936** *Discovery* Sept. 263/1 The last occasion in which a new assembly-place in this country was chosen was in 1919. **1941** J. S. HUXLEY *Uniqueness of Man* ix. 198 Assembly-places [of birds] for mating.

assented (ăse·ntĕd), *ppl. a.* [f. ASSENT *v.* + -ED[1].] Applied to bonds or stocks deposited under an agreement by which the owners assent to some proposed change affecting their amount, nature, or status. (Webster 1961.)
1907 M. ROLLINS *Money & Investments* 11 *Assented Stocks* (*or Bonds*), in the event of a corporation passing through a reorganization where the security holders are requested to give 'assent' to a certain plan..some banking house..is usually selected to receive the securities. **1957** *Economist* 7 Sept. 866/1 Operators therefore went largely for the most readily available stocks—Dawes assented and the Young assented. **1959** *Ibid.* 16 May 659/1 Most of the bonds are quoted in an 'assented' form, but those British subjects who have held on to 'non-assented' bonds for surtax purposes will have to sell to foreign residents to avoid tax on the accrued interest that becomes payable once assent to the debt settlement is given.

assertion. Add: **6.** *Comb.* **assertion-sign** *Logic*, the sign introduced by G. Frege in 1879 to indicate that the signs following it express a proposition which is asserted to be a true judgement; the same sign used in related senses; also in extended use of other signs considered equivalent in function.
[**1903** B. RUSSELL *Princ. Math.* § 477 The sign of judgment (*Urtheilstrich*) does not combine with other signs to denote an object.] **1906** —— in *Amer. Jrnl. Math.* XXVIII. 160, I have adopted from him [*sc.* Frege] the assertion-sign. *Ibid.* 161 The sign '⊢' is called the assertion-sign; it may be read 'it is true that' (although philosophically this is not what it means). **1910** WHITEHEAD & RUSSELL *Principia Math.* i. 8 The sign '⊢', called the 'assertion-sign', means that what follows is asserted. It is required for distinguishing a complete proposition, which we assert, from any subordinate propositions contained in it but not asserted. **1922** tr. *Wittgenstein's Tractatus* 4. 442, Frege's assertion sign '⊢' is logically altogether meaningless. **1947** H. REICHENBACH *Elem. Symbolic Logic* vii. 336 In written language the assertion sign is supplied by the period at the end of a sentence, meaning: the writer asserts the sentence. **1953** G. E. M. ANSCOMBE tr. *Wittgenstein's Philos. Investig.* I. § 22 Frege's assertion sign marks the *beginning of the sentence*. Thus its function is like that of the full-stop.

asserto·ric, *a. Logic.* [f. as ASSERTORIAL *a.* + -IC.] = ASSERTORIAL *a.* Also as *sb.*, a proposition of this type.
1889 in *Cent. Dict.* **1905** *Athenæum* 8 Apr. 439/2 One kind of propositions, viz. assertorics. **1929** N. K. SMITH tr. *Kant's Crit. Pure Reason* 110 The assertoric proposition deals with logical reality or truth.

assertorically, *adv.* (Earlier example.)
1838 F. HAYWOOD tr. *Kant's Crit. Pure Reason* 233 A possible intuition, whereby objects can be given to us out of the field of sensibility, and the understanding used assertorically beyond the same.

assessee (ăsesī·). [f. ASSESS *v.* + -EE[1].] One whose property or income is assessed.
1726 *Laws of Sewers* 194 The Assessees of the Lands held by the Adventurers under the Trustees. **1959** *Economist* 7 Mar. 883/2 The expenditure of wives and minor children will be aggregated with that of the assessee even where their incomes are separate.

assession. Add: **2.** In the Duchy of Cornwall, the action of assessing and letting to rent the lord's demesnes, which was done at a court held for the purpose; also attrib. in *assession-court, -roll*. Also **asse·ssioning** *vbl. sb.* and *ppl. a.*; **assessionable** (ăse·ʃənăb'l) *a.*, occas. **asse·ssional** *a.*, applied to seventeen manors of the Duchy in which the lands were let by courts of assession.
1820 J. SCARLETT in G. Concanen *Rep. Rowe* v. *Brenton* (1830) App. 77 There is no custom stated for the free tenants to attend the assession-court, only the conventionaries. **1820** BROUGHAM *Ibid.*, They have done enough to differ the present case, of tendering the assession-roll, from the circumstances under which it was tendered before. **1828** BARNEWALL & CRESSWELL *Rep. King's Bench* VIII. 740 The seventeen manors hereafter mentioned as assessionable manors. *Ibid.* 750 A roll called an Assession Roll, which purported to be an account of the acts done by certain assessors in the 7 Edw. 3., under a commission to them by John Earl of Cornwall. **1830** G. CONCANEN *Rep. Rowe* v. *Brenton* (title), The right to minerals in the assessional lands of the duchy of Cornwall. *Ibid.* Introd. 23 The mode of letting the lands of the assessional manors by the assessioning commissioners. *Ibid.*, The assessioning of the same manors occurred at intervals of less than seven years. *Ibid.* 36 The conventionary tenants of the seventeen assessionable manors of the Duchy. **1839** *Penny Cycl.* XIV. 390/1 The courts..were called assessions, or courts of assession. The course usually was to let the land until the next assession. **1848** *Act* 11 & 12 *Vict.* c. 83 (*title*) An Act to confirm the Awards of Assessionable Manors Commissioners.

assibilation. (Earlier and later examples.)
1850 *Proc. Philol. Soc.* IV. 262 The assibilation of the final guttural is a matter far too important to be discussed incidentally. **1933** JESPERSEN *Essentials Eng. Gram.* v. 57 Assibilation..is the name of the development of new [ʃ, ʒ, tʃ, dʒ] from combinations with [j].

assification (æ:sifikē[i]·ʃən). [n. of action, f. as ASSIFY: see -FICATION.] The action of making an ass of (a person); asinine act.
1823 *Blackw. Mag.* XIV. 576/2 The last assification I shall notice..would settle the business in the most scrupulous court *de Lunatico inquirendo*. **1879** C. M. YONGE *Magnum Bonum* III. xxxii. 682, I ruined you all..with that assification.

assign, *v.* **4. b.** (Earlier example.)
1827 P. CUNNINGHAM *Two Years in N.S. Wales* II. xxviii. 188 Convict servants are now assigned on application being made for them in a printed form to the land board.

assignation. Add: **10.** *attrib.* (sense 7), *assignation house* (U.S.), a brothel.
1870 OLIVE LOGAN *Bef. Footlights* 538 Denounced the National theatre as the vilest of 'assignation houses'. **1943** R. OTTLEY *'New World A-Coming'* 28 Don't come.. bothering me with any more protests about assignation houses until you can bring concrete evidence of such houses.

assignment. Add: **13.** A task assigned to one; a commission or appointment. orig. *U.S.*
c 1848 GARLAND in W. C. Church *Ulysses Grant* (1897) iii.

45 Lt. Grant can best serve his country..under this assignment. **1897** *Scribner's Mag.* Aug. 232/2 The reporters.. were waiting to be sent off on their first assignments before getting breakfast. **1898** *Daily News* 4 Jan. 3/1 Here a man goes out on an assignment and if the person sought does not wish to express his views, he comes back empty-handed. **1910** C. E. Mulford *Hopalong Cassidy* xvii. 99 At first his assignment had pleased, but as hour after hour passed with growing weariness, he chafed more and more. **1949** F. Maclean *Eastern Approaches* iii. vi. 373 His next assignment [as an officer helping partisans] was in enemy-occupied Albania. **1964** G. L. Cohen *What's Wrong with Hospitals?* i. 17 The nursing profession has therefore perfected its own technique of fragmentation: 'task assignment'. This enables one patient's needs to be split up among many nurses. **1967** *Oxford Mag.* 10 Feb. 205/2, I set them [sc. Canadian students] an exercise and they bring me an assignment.

||**assimilado** (ăsimilā·do). [Pg., pa. pple. of *assimilar* to assimilate.] Used as *sb.* to designate an African in Portuguese East and West Africa who has been admitted to Portuguese citizenship. Also *attrib.*
1953 *N.Y. Times* 15 Jan. 3/1 In Portuguese East Africa, he said, a native wishing to become a member of the European group had to be officially recognized as an 'assimilado'. **1953** *Wall St. Jrnl.* 24 July 4/5 It's even possible for a Mozambique native to become a full Portuguese citizen. About 4,000 Africans have reached this 'assimilado' status. **1958** *Spectator* 30 May 679/3 The Portuguese *assimilado* system. **1961** *Economist* 6 May 550/1 Out of a black population of 4½ million, Angola has 30,000 *assimilados* who enjoy the basic rights denied to their fellows.

assimilate, *pple.* and *sb.* Restrict †*Obs.* to senses in Dict. and add: **B. 2.** Something which has been assimilated (see Assimilate *v.* 7).
1935 *Ann. Reg. 1934* 56 Work on the transport of assimilates and other dissolved substances. **1936** *Nature* 14 Nov. 851/2 The movement of assimilate in seedling tomato plants.

assimilate, *v.* Add: **1. d.** *Philol.* To render (a sound) accordant, or less discordant (*to* another sound in the same or a contiguous word). Also *intr.*
1854 *Proc. Philol. Soc.* V. 200 In our own language..it is to be expected that some traces of the law of assimilated vowels should appear. **1871** H. J. Roby *Gram. Latin Lang.* I. i. viii. 48 Before *s*, *d* is assimilated or falls away. **1946** E. A. Nida *Morphology* ii. 43 Nasal consonants assimilate regressively according to the point of articulation of the following consonant.

assimilation. Add: **1. b.** *Philol.* The action of assimilating or fact of being assimilated: see prec.
1850 *Proc. Philol. Soc.* IV. 89 The law for the assimilation of vowels..will account for the introduction of an *o* in *biodh-mur*,..before the *u* of the final syllable. **1871** B. H. Kennedy *Public Sch. Lat. Gram.* 18 Complete Assimilation occurs, when, of two meeting Consonants, the former becomes the same as the latter. **1885** Cook tr. *Sievers's O. E. Gram.* § 86. 38 A partial assimilation of the basic vowel to the following sound. **1936** *Language* XII. 246 An assimilation is produced by the replacement of some phoneme or phonemes by other phoneme or phonemes shortly to be uttered.
6. *Psychol.* The process whereby the individual acquires new ideas, by interpreting presented ideas and experiences in relation to the existing contents of his mind. Used with some manner of qualification or specification by various writers.
1855 [in Dict., sense 3]. **1873** G. H. Lewes *Problems of Life & Mind* 190 Since interpretation means mental assimilation, the significance of the phenomena must depend upon the pre-perceptions and pre-conceptions which they arouse. **1896** G. F. Stout *Analyt. Psychol.* II. 118 Assimilation there must always be, inasmuch as the existence of a given experience coincides with the re-excitement of some preformed disposition. **1897** C. H. Judd tr. *Wundt's Outlines Psychol.* 227 Assimilations, or associations between the elements of like compounds. *Ibid.* 228 Assimilations are a form of association that is continually met with, especially in the case of intensive and spacial ideas. **1923** W. McDougall *Outl. Psychol.* 397 By some psychologists who followed Locke's way of 'ideas', yet saw that 'ideas' cannot be generated by association alone..assimilation was made the fundamental mode of growth.
7. *Geol.* The absorption of extraneous matter by an igneous magma.
1903 R. A. Daly in *Amer. Jrnl. Sci.* XV. 270 The 'marginal assimilation' theory of plutonic intrusion..a hypothesis of slow caustic action by magmas that have advanced into the overlying earth-crust by their own energetic solvent action. **1909** A. Harker *Nat. Hist. Igneous Rocks* iii. 83 On the *assimilation hypothesis*, still supported by some French geologists, an igneous rock-magma is supposed to be capable of melting and incorporating freely the solid rocks which it encounters.

assimilationist (ăsi:milēɪ·ʃənist). [f. Assimilation + -ist.] One who advocates the integration of different races, cultural groups, etc.
1928 L. Wirth *Ghetto* p. ix, The ghetto becomes the physical symbol for that sort of moral isolation which the 'assimilationists'..are seeking to break down. **1930** *Times Lit. Suppl.* 23 Jan. 51/3 General Smuts is not an assimilationist. The whites must not impose their civilization on the natives. **1942** L. B. Namier *Conflicts* 132 The 'emancipation' movement of the 'assimilationists' arose in the individualist era; it knew only Jews, but no Jewish nation.
b. *attrib.* or *quasi-adj.*
1934 W. Steed *Hitler* iv. 126 The younger Jews.. revolted inwardly against assimilation, and denounced assimilationist doctrine.

assimilative, *a.* Add: **1. b.** *spec.* in *Philol.* (cf. *Assimilation 1 b).
1936 *Language* XII. 246 In consonantal clusters it is common to find that the posterior consonant exerts an assimilative influence upon the preceding. **1946** E. A. Nida *Morphology* ii. 29 If assimilative changes are quite extensive in some pattern of the language, the related sets are frequently described as 'vocalic harmony'.

assimilatory, *a.* (Example in *Philol.*)
1921 E. Sapir *Language* viii. 186 This assimilatory change was regular, i.e., every accented long *o* followed by an *i* in the following syllable automatically developed to long *ö*.

Assisian (ăsi·siăn, -·ziăn), *a.* and *sb.* Also **Assisan.** [See -an.] **A.** *adj.* Of or pertaining to the town of Assisi in central Italy; *spec.* of St. Francis of Assisi. **B.** *sb.* An inhabitant of Assisi. *The Assisian*, St. Francis.
1870 Mrs. Oliphant *Francis of Assisi* i. 18 This absolute penury must have struck the sick fancy of the young Assisan [sc. St. Francis]. *Ibid.* iv. 66 The Assisan bishop, who knew something of the temper of the lords of Umbria. *Ibid.* xvii. 281 As anxious as the Assisans that the headquarters of the Order at the Portiuncula should be the future shrine of the saint. **1893** F. Thompson *Poems* 72 The Assisian, who kept plighted faith to three, to Song, to Sanctitude, and Poverty. **1905** *Athenæum* 30 Sept. 431/1 She—like the Assisian..exhibited on her own body the Stigmata. *a* **1907** F. Thompson *St. Ignatius Loyola* (1909) iii. 76 Another story recalls the Assisian Saint. **1923** G. K. Chesterton *St. Francis* iii. 52 The Papal cause..aroused enthusiasm among a number of young Assisians. *Ibid.* 53 The legal authority of the Assisian magistrates.

assist, *sb.* Delete †*Obs. rare*⁻¹ and add: Now only *U.S.* (Earlier and later examples.)
1597 G. Markham *Devoreux* f. 15ʳ, st. 84, I hop'd, by her assist, this fourth Henrie. **1923** *Daily Mail* 5 May 8 The supporting bid, the assist,..and 'the switch' assume a new value. **1958** B. Ulanov *Hist. Jazz* vii. 73 He usually played second cornet back of Oliver's lead, an alternately delicate assist and blasting support. **1957** W. H. Whyte *Organ. Man* viii. 107 An assist from others can help a lot. **1966** *New Yorker* 4 June 18 (Advt.), Credit Hart Schaffner & Marx tailors with a skilful assist.
b. *Baseball.* 'The act of a player who handles the ball in assisting to put an opposing player out; also the credit given by the scorer to a player so assisting' (*Dict. Americanisms*).
1877 *Constit. Nat. League Baseball* 40 An assist should be given to each player who handles the ball in a run-out or other play of this kind. **1896** Knowles & Morton *Baseball* 101 The fielder who handles the ball in sufficient time to aid in retiring a base-runner is credited with an assist. **1917** C. Mathewson *Sec. Base Sloan* xiv. 187 Five strike-outs, three assists and no errors was considered a fine record.

assist, *v.* Add: **6. a.** Frequently with adv. or advb. phr. denoting that in which the assistance is given.
1837 E. Howard *Old Commodore* I. v. 186 Assist the chaplain down the poop-ladder. **1843** Lever *Jack Hinton* xxv, He assisted me on with my great-coat. *Ibid.* xxvii, As I assisted her from the carriage, I could not but mark the flashing brilliancy of her eye. **1855** —— *Sir Jasper Carew* xxix, He assisted me off with my coat. **1864** Mrs. Gaskell *Wives & Dau.* (1866) I. v. 47 'May I help you to potatoes?' or, as Mr. Wynne would persevere in saying, 'May I assist you to potatoes?' **1881** Mrs. J. H. Riddell *Senior Partner* II. xi. 228 'It's no use my trying to put in a friendly oar', said Mr. McCullagh, assisted perhaps to this last figure of speech by the sight of an outrigger spinning down the stream. **1924** Rose Macaulay *Orphan Island* xi, Miss Smith..was assisted from her hammock.
7. c. Of the dealer's partner in the game of euchre: to order the adoption of the suit of the card turned up as trump.
1878 *Encycl. Brit.* VIII. 654/2 If the first hand passes, the second may say 'I assist', which means that the dealer (his partner) is to take up the trump. *Ibid.* 655/1 A player can declare to play alone when he or his partner orders up, or when his partner assists.

assistance. Add: **3. b.** *ellipt.* for *National Assistance* (see *national *a.*).
1956 A. Wilson *Anglo-Saxon Att.* ii. i. 244 Ten quid's not going to put us on Assistance. **1959** *New Statesman* 1 Aug. 125/2 The forthcoming increases will be a temptation to the 'casual' labourer who has a large family to shift from earnings to assistance.

assistancy (ăsi·stănsi). Also 7–8 **assistency.** [f. Assistant *sb.*: see -ancy.] The position of an assistant.
1608 T. Fitzherbert *Let.* 23 Aug. (1948) 29 You have confirmed Mʳ Singleton in his Assistency. **1757** A. Butler *Lives Saints* III. i. 328/1 The general..has five assistants nominated by the general congregation, who prepare all matters to his hands, each for the province of his Assistency. **1790** W. Maclay *Jrnl.* 3 May (1890) vii. 255 There is a prospect of Tench Coxe succeeding Duer in the assistancy of the Treasury. **1884** *Lancet* 27 Dec. (Advt.), Wanted an Assistancy, or to manage a Branch. **1909** *Practitioner* Nov. 723 Thirty years ago the average medical man took an assistancy directly he was qualified. *Ibid.*, Qualified men..are ..not available for assistancy work.

assistant. Add: **A.** *adj.* **3.** *assistant curate*, see Curate 2; *assistant master, mistress* (in a school); *assistant professor* (chiefly U.S.); *assistant stage-manager.*
1851 C. Cist *Cincinnati* 304 In September..he received the appointment of assistant professor of mathematics. **1872** *Schoolmaster* 6 Jan. 4/1 Wanted..an Assistant Master, Ex P.T., for Canal-street Wesleyan Schools, Derby. *Ibid.*, Wanted, in a Mixed School, an ex P.T., as Assistant Mistress. **1898** L. Merrick *Actor-Manager* iii. 38 He was to be described on the playbills as 'Assistant Stage-manager'. **1902** A. C. Benson *Schoolmaster* 29 As to corporal punishment, the doubtful privilege of dispensing it is, at my own School, not conceded to the assistant-masters. **1905** [see *associate *ppl. a.* 1]. **1952** M. McCarthy *Groves of Academe* (1953) iv. 74 All these, on the instructorial or assistant-professor level, constituted the bulk of Jocelyn's faculty.
B. *sb.* **3. c.** A shop assistant (see *shop *sb.*).
1853 Mrs. Gaskell *Ruth* I. iv. 110 Her eldest son..was an assistant in a draper's shop. **1908** *Rep. Truck Committee* (Cmd. 4442–4) 132/2 The smartest assistant is the one who can sell to customers worthless goods.

assisted, *ppl. a.* (Examples.) *spec.* Applied to subsidized passages for emigrants.
1853 *14th Gen. Rep. Colonial Land & Emigration Commissioners* (1854) 139 Remittances may be made for either of the following objects:—*Assisted passages.*—To assist the nominees in paying the expense of their passage and out-fit. *Free passages.* **1856** *16th Gen. Rep. Emigration Commissioners* 9 Emigration from the United Kingdom, which we shall divide into 'unassisted', or that of which the expense is defrayed out of private or local funds, and 'assisted', or that which is carried on by this Board with funds derived from the Colonies (at present the Australian only) to which the emigrants are sent. **1888** C. M. Yonge *Our New Mistress* xvi. 148 A number of papers about Australia, assuring me of all the advantages of emigration, promising me..an assisted passage. **1937** *Jrnl. R. Aeronaut. Soc.* XLI. 275 An aircraft with the assisted take-off, and therefore not requiring the maximum power output of the engine. **1945** *Guide Educ. System* (Min. Educ. Pamph. No. 2) 57 *Assisted School*, school not maintained by, but receiving some financial assistance from, an L.E.A. **1954** P. K. Kemp *Fleet Air Arm* xx. 216 Brown..successfully took off in about half the length of the flight deck. For this he had an assisted take-off with rockets. **1946** *Encycl. Brit.* XV. 470/1 Assisted passage schemes have been arranged which considerably reduce the cost of travel to the settler.

assize, *sb.* Add: **12. b.** *attrib.* *assize-ball, -week* (later example); **assize sermon,** a sermon preached at the holding of assizes.
1832 F. Trollope *Dom. Manners* xiv. 215 The dancing was not quite like..what we see at an assize or race-ball in a country town. **1853** Mrs. Gaskell *Ruth* I. i. 13 The assize-balls had been discontinued. **1699** Sewall *Letter-book* (1886) I. 216 Mr. Noyes..preached an excellent Assize Sermon. **1859** Reade *Love me Little* iii, Shall you go to the assize sermon? **1845** *Ainsworth's Mag.* VII. 503 Some years ago..I acted in Lancaster during the assize week.

associate, *ppl. a.* and *sb.* Add: **A.** *ppl. a.* **1.** (Further examples.)
1822 *N. Amer. Rev.* XIV. 51 M. Stuart, Associate Professor of Sacred Literature in the Theological Seminary at Andover. **1896** J. L. Ford *Lit. Shop* (ed. 3) vi. 62 Mr. Johnson..in his capacity of associate editor of the *Century Magazine.* **1900** *Daily News* 14 Nov. 6/3 Associate-editor of the 'Engineer'. **1905** *N.Y. Even. Post* 25 Dec., The additional class is that of associate professors. All assistant professors under the old order were made 'associate'. **1931** H. F. Pringle *T. Roosevelt* ii. xii. 432 The *Outlook* office where the ex-President was an associate editor.
B. *sb.* **7.** *Psychol.* An idea, or other mental content, connected with another by any of the forms of association.
1700 [see sense 6 in Dict., quot. erron. dated 1690]. **1880** W. James *Coll. Ess. & Rev.* (1920) 207 In the trance-subject's mind any simple suggestion will be both believed and acted on, because none of its usual associates are awakened. **1931** *Brit. Jrnl. Psychol.* XXI. 279 The effect of partial elimination of associates on order of precedence in respect of memory.

associated, *ppl. a.* Add: **1.** (Additional examples.) *Associated Press* (abbrev. *A.P.*), an association of American newspapers.
1849 *New Orleans Picayune* 9 May 2/2 The Associated press..occupy the telegraph many hours. **1879** [see *A III]. **1948** *Daily Ardmoreite* (Okla.) 25 Apr. 11/1 A poll conducted by the Associated Press. **1955** *Times* 29 June 7/5 Details of the programmes which will be transmitted by the Associated Broadcasting Company when commercial television begins in September were announced. **1958** *Daily Tel.* 30 June 13/4 Associated-Rediffusion, the main ITV company for London.

associating, *ppl. a.* Add: *associating fibres* = *association fibres* (see next, sense 9).
1885 *Harper's Mag.* Mar. 640/1 One set serves to connect the cells of different areas of the cortex (the 'associating fibres').

association. Add: **5. b.** *Chem.* The aggregation of molecules to form a loosely-bound complex.
1895 *Bull. Philos. Soc. Washington* XII. 158 They behave as simple oxide molecules, capable of arranging themselves in different associations according to physical circumstances. **1904** *Amer. Jrnl. Sci.* XVII. 427 Molecular weights of liquids, with a few words about association. **1940** *Chambers's Techn. Dict.* 553/1 *Molecular association*, the relatively loose binding together of the molecules of a liquid or vapour in groups of two or more.

7. Examples of recent usage; phr. *association of ideas* (examples).
1700 Locke *Hum. Und.* (ed. 4) ii. xxxiii. 221 Of the Association of Ideas. **1759** J. Adams in *Wks.* (1850) II. 68 The principle in nature is imitation, association of ideas, and contracting habits. **1890** W. James *Princ. Psychol.* I. xiv. 555 Association occurs as amply between impressions of different senses as between homogeneous sensations. **1894** G. T. Ladd *Psychol.* xiii. 264 The very limited nature of the application of the so-called laws of the association of ideas to the entire mental life. **1905** E. B. Titchener *Exper. Psychol.* II. i. 192 We show him a word; he is to react when the word has suggested something, no matter what. The word *sea* may arouse the idea of land or water or ships or some particular sea or some particular incident at sea,—anything it likes. Associations of this sort are termed, technically, *free* associations. **1938** R. S. Woodworth *Exper. Psychol.* xv. 340 In free association the laws of association are supposed to have full sway. **1958** Hayward & Harari tr. *Pasternak's Dr. Zhivago* ii. ix. 262 Through an unaccountable association of ideas started by the sight of the real town outside the window..Yury remembered the distant panorama of the town.
b. Psychol. *laws of association*, see quot. 1897²; *mediate association*, association by unconscious or unnoticed intermediaries; *simultaneous, successive association*, forms of association of ideas in which the process of connexion is simultaneous or falls into two stages. Also *attrib.*, as *association philosophy, psychology, test, theory, time.*
1820 T. Brown *Lect. Phil. Human Mind* II. xl. 346 The other supposition..ascribes our trains of ideas to associations previous to the suggestion itself,—to laws of association in short, in the sense in which that phrase is distinguishable from laws of suggestion. **1833** J. S. Mill in *Monthly Repos.* VII. 663 The association-philosophy as taught by Hartley. *a* **1856** W. Hamilton *Lect. Metaph. & Logic* (1860) IV. xxx. 122 Our Cognitions, Feelings, and Desires are connected together by what are called the *Laws of Association*. **1864** J. S. Mill *Let.* 3 Apr. in D. Duncan *Life & Lett. H. Spencer* (1908) 115 You and Bain..have succeeded in affiliating the conscious operations of mind to the primary unconscious organic actions of the nerves, thus filling up the most serious lacuna..in the association psychology. **1890** W. James *Princ. Psychol.* I. xiv. 558 The difference [between the apperception-time and the reaction-time], called by Wundt the *association-time*, amounted, in the same four persons, to 706, 723, 752, and 874 thousandths of a second respectively. **1897** C. H. Judd tr. *Wundt's Outlines Psychol.* 13 Intellectualistic psychology has in the course of its development separated into two trends. ..The logical theory... The association-theory. *Ibid.* 225 The following forms were discriminated: association by similarity and contrast, and association by simultaneity and succession. These class-concepts gained by a logical dichotomic process were dignified with the name 'laws of association'. **1924** J. Riviere et al. tr. *Freud's Coll. Papers* II. 13 Association test. **1938** [see sense *7]. **1959** *Listener* 29 Oct. 722/2 He [sc. Jung] used what were called association tests..The subject of the experiment was given a number of stimulus words and asked to react with another word or phrase to each.
9. *Physiol.* Used *attrib.*, as *association area, centre, field, link, path, sphere*, of those portions of the cortex of the brain which connect the sensory and motor areas, and are supposed to be concerned with ideation, etc.; *association fibres* (in *Funk's Stand. Dict.* 1900), nerve fibres connecting different areas of the brain cortex, as distinguished from the commissural fibres; so *association organ, system.*
1880 H. C. Bastian *Brain as Organ of Mind* xxiii. 452 The connecting, or, as Meynert terms them, the 'association system' of fibres of the Brain. **1890** W. James *Princ. Psychol.* II. xviii. 75 During waking hours every centre communicates with others by association-paths. **1901** Allchin *Man. Med.* III. Physiol., Introd. 31 A portion of the 'association' field of the cortex. **1902** *Encycl. Brit.* XXXI. 742/1 The areas of intervening cortex, arriving at structural completion later than the..sense-spheres, are called by some association-spheres. **1902** W. James *Var. Relig. Exper.* xvi. 427 Other alienists..have explained 'paranoiac' conditions by a laming of the association-organ. **1904** J. McCabe tr. *Haeckel's Wonders of Life* i. 13 In 1894 Flechsig showed that there are four central sense-regions.. in the gray cortex of the brain, and four thought-centres ('association-centres', or phroneta). **1909** *Cent. Dict.* Suppl., Association area. **1932** *Brit. Jrnl. Psychol.* XXIII. 22 These lesions [of brain tissue] represented interference with association links. **1952** *Sci. News* XXIII. 63 Histologically, we recognize in the so-called association areas [of the brain] the structures responsible for linking sensory stimuli with motor response.
10. A personal connection or link; esp. *attrib.* in *association book, copy*, a volume showing some mark of personal connection with the author or a former owner (of note).
1882 *Gentl. Mag.* CCLII. 92 Speaking of books with an association reminds us of that most destructive craze of the present day, the collection of book-plates. **1901** *Munsey's Mag.* Oct. 80/1 His remarkable collection of 'Association Books'. **1912** *Times* (weekly ed.) 7 June 444 The collection is chiefly remarkable for what are termed in America 'association books'. **1914** *W. M. Murphy's Catal. Bks.* 19 Jan. 5 Association Copy. **1918** *Times* 21 Feb. 3/3 It is very rich..in what Americans call 'association books', such as the copy of 'Vanity Fair' which Thackeray sent to Charlotte Brontë. **1928** *N. & Q.* 12 May 341/2 Presented by Sir Walter Scott to Lydia White in 1808—double association of good interest.
11. Applied to the game of football played according to the rules of the Football Association formed in 1863, as distinguished from the Rugby game. (Cf. *soccer*, Socker.)
1867 *Routledge's Handbk. Football* 53 Football Association Rules. **1873** *Football Annual* 16 To play *with the feet* is the main object of Association Football. Hands should not, and must not be used. **1880** [in Dict., sense 2]. **1885** Shearman & Vincent *Football* 30 In the Association game no collaring, and therefore no running with the ball, is allowed. *Ibid.* 45 Before the days of the Rugby Union and Association rules. *Ibid.* 53 The ball..is several ounces heavier than an Association ball. **1920** K. R. G. Hunt (*title*) Association football.
12. *Ecol.* A group of associated plants within a formation (see *formation 5 b).
[**1807** Humboldt & Bonpland *Essai sur la Géographie des Plantes* 13 La Géographie des plantes..c'est cette science qui considère les végétaux sous les rapports de leur association locale dans les différens climats.] **1900** B. D. Jackson *Gloss. Bot. Terms* 25/2 Plant Associations. **1909** Groom & Balfour tr. *Warming's Œcology of Plants* xxxv. 145 An association is a community of *definite floristic composition* within a formation. **1911** A. G. Tansley *Types Brit. Veg.* 10 Thus each of the types of vegetation, woodland, scrub and grassland, within a given formation, is a plant-association. **1916** F. E. Clements *Plant Succession* vi. 128 The association as usually understood becomes what is here termed the consociation, in so far as it is a climax community. This is the association with a single dominant. **1918** G. E. Nichols in *Trans. Connecticut Acad.* XXII. 275 In any unit area where more than one association is represented, the associations, taken collectively, constitute an association complex.

associational, *a.* (Later examples.)
1899 W. James *Talks* xii. 119 The laws of memory..are incidents of our associational constitution. **1951** R. Firth *Elem. Social Organization* ii. 43 The associational standard is applied to the way in which it affects social relations.

associationism. (Earlier and later examples.)
1875 *Encycl. Brit.* II. 732/2 To account for the fact of synthesis in cognition, in express opposition to associationism. **1890** W. James *Princ. Psychol.* I. x. 338 All the individual thoughts and feelings which have succeeded each other 'up to date' are represented by ordinary Associationism as in some inscrutable way 'integrating' or gumming themselves together.

associationist. Add: Also *attrib.* or as *adj.* = next.
1876 *Mind* I. 322 It is only with a very small part..of what is distinctive in the teaching of the associationist philosophers that we are at present concerned. **1880** W. James *Coll. Ess. & Rev.* (1920) 209 The English associationist school..had also represented choice and decision as nothing but the resultant of different ideas failing to neutralize each other exactly. **1881** —— *Will to Believe* (1897) 128 The strict associationist school..under the domination of Mill, Bain, and Spencer dominated us but yesterday. **1886** A. Weir *Hist. Basis Mod. Europe* xii. 500 The associationist psychology. **1890** W. James *Princ. Psychol.* I. x. 353 The chain of distinct existences into which Hume thus chopped up our 'stream' was adopted by all of his successors as a complete inventory of the facts. The associationist Philosophy was founded. **1951** E. E. Evans-Pritchard *Social Anthropology* iii. 44 When anthropologists even as recent as Tylor and Frazer looked to psychology for aid it was to associationist psychology that they looked.

associationistic (ăsōᵘ:ʃieɪʃəni·stik), *a.* [f. Associationist + -ic.] Of or relating to associationism or associationists.
1901 W. James *Mem. & Stud.* (1911) vii. 148 All psychology, whether animistic or associationistic, was written on classic-academic lines. **1935** *Brit. Jrnl. Psychol.* XXV. 277 We are here making use of the associationistic 'meaning theory'.

associative, *a.* Add: **2.** *Psychol.* Of, pertaining to, characterized by, or subserving, the association of ideas. (Cf. Association 7.)
1812 Coleridge in *Omniana* II. 13 The fancy, or the aggregative and associative power. **1864** J. S. Mill *Lett.* (1910) II. 7 He [sc. Spencer] has a great mastery over the obscurer applications of the associative principle. *a* **1873** —— *Three Ess. Relig.* (1874) 199 When analysed to the bottom on the principles of the Associative Psychology, the brain..is, like matter itself, merely a set of human sensations either actual or inferred as possible. **1923** J. S. Huxley *Ess. of Biologist* i. 24 Through associative memory, present behaviour is modified by past experience. **1957** *New Biol.* XXIV. 123 Associative learning—a term which covers both classical conditioned reflexes and learning by trial and error.

associes (ăsōᵘ·sɪ̆iz). *Ecol.* [mod.L., irreg. f. L. *associāre*: see Associate *ppl. a.* (cf. quot. 1926).] (See quot. 1916.)
1916 F. E. Clements *Plant Succession* vi. 136 The associes is the developmental equivalent of the association... It is composed of two or more consocies, *i.e.*, developmental consociations, just as the association consists of two or more consociations. Like the association, it is based upon life-form, floristic composition, and habitat, but differs from it in as much as all of these are undergoing constant or recurrent developmental changes. **1926** Tansley & Chipp *Stud. Vegetation* ii. 19 The three higher ranks..we distinguish from the corresponding climax communities by using the suffix *-es*, thus *associes, consocies, socies.* **1964** V. J. Chapman *Coastal Vegetation* i. 11 The *associes, consocies* and *socies* are their equivalents in a sere (succession).

assonance. Add: **2. b.** In extended use: = *half-rhyme; the correspondence or rhyming of one word with another in the final (sometimes also the initial) consonant, but not in the vowel. Also applied by philologists, in studying rhyming pairs of words (i.e. with identical vowel), to final consonants of such similarity of articulation as to be acceptable, with poetic licence, in a rhyming position.
1917 R. Graves *Let.* c. 22 Dec. in W. Owen *Lett.* (1967) 595 You [sc. Wilfred Owen] have found a new method and must work it yourself—those assonances instead of rhymes are fine. **1920** E. Blunden in *Athenæum* 10 Dec. 807/1 The discovery of final assonances in place of rhyme may mark a new age in poetry. **1934** C. Day Lewis *Hope for Poetry* iii. 17 His [sc. Wilfred Owen's] one innovation is the constant use of the alliterative assonance as an end rhyme—(mystery, mastery; killed, cold). *Ibid.* x. 72 Owen's alliterative assonance. **1948** G. L. Brook *Harley Lyrics* 20 Assonance sometimes takes the place of rhyme; the most frequent examples are of *m* : *n* and *ng* : *nd* (e.g. *tyme* : *pyne*). **1960** E. G. Stanley *Owl & Nightingale* 112 An example..of assonance on dentals. **1960** D. S. R. Welland *Wilfred Owen* vi. 116 Not only in its use of the 'mean/moan/men' assonance does this passage anticipate 'Insensibility'.

assonant, *sb.* (Earlier and later examples.)
1834 *Penny Cycl.* II. 510/1 In English, *hardy, manly*, and *carry*, would be assonants. **1845** R. Ford *Handbk. Trav. Spain* i. 191 *Amor* and *razon* are assonants. **1904** Brandin & Hartog *Bk. Fr. Pros.* i. iv. 47 *Bergère* and *pimprenelle* are assonants..the 'Symbolistes' have reintroduced assonance into poetry.

assonantal, *a.* (Earlier and later examples.)
1852 *Fraser's Mag.* XLV. 651 Assonantal rhymes..are of not unfrequent occurrence. **1925** D. Corkery *Hidden Ireland* vii. 179 They had, perforce, to use the simpler assonantal metres. **1938** R. Graves *Coll. Poems* p. xv, Internal assonantal rhyme (*heather, wither*).

assonate, *v.* (Later examples.)
1952 R. Campbell *Lorca* 40 We could assonate the word 'filbert' with it [sc. 'winter']. **1966** Bennett & Smithers *Early ME. Verse & Prose* 291, [rð] and [rd] do not commonly assonate in ME texts.

assortative (ăsǭ·ɹtătiv), *a.* [f. Assort *v.* + -ative.] That assorts; *assortative mating*, non-random mating, mating on the basis of the possession by the partners of similar characteristics, circumstances, etc.
1897 K. Pearson in *Phil. Trans. R. Soc.* 1896 A. CLXXXVII. 258 Preference of individuals with an organ or characteristic of given size for mates with the same or another organ or characteristic of a size, the average of which differs from the..population average. This type of sexual selection which may be spoken of as assortative mating is measured mathematically by the coefficient of correlation between the two organs or characteristics in mated pairs. **1902** *Encycl. Brit.* XXXIII. 637/1 Assortative mating exists when individuals which mate are not paired at random, but a definite correlation is established between the characters of one mate and those of the other. **1955** *New Biol.* XVIII. 37 In a few animal species negative assortative mating, that is to say a tendency of like to mate with unlike, has been observed. Positive assortative mating occurs in flowering plants which are fertilized by insects.

assorted, *ppl. a.* Add: Also, comprising various sorts.
1897 G. B. Shaw *Theatres in Nineties* (1932) III. 246 Dickens..made desperate efforts to take his assorted heroines quite seriously by resolutely turning off the fun. **1937** D. L. Sayers *Busman's Honeymoon* xv. 302 'Will you have some of these little biscuits? Dear me, what a remarkable variety!'..'They come assorted in boxes.'

asso·rter. *U.S.* [f. Assort *v.* + -er¹.] One who assorts.
1897 *Scribner's Mag.* Jan. 17/2 The assorter, who distributes the bundles [of goods] into lots to go to the different parts of the city. **1915** W. Holt *Beacon for Blind* vii. 70 He was anxious to bring deaf and dumb assorters into the Post Office.

assortment. Add: **2.** *spec.* in *Forestry* (see quots.). Also *attrib.*
1896 W. R. Fisher *Schlich's Man. Forestry* V. i. v. 248 The various pieces into which a tree may be converted by the woodcutter are termed rough assortments of timber. **1953** *Brit. Commonw. Forest Terminol.* i. 13 *Assortment table*, a volume table in which the volumes are differentiated according to size classes or other criteria affecting utilization.

assuasive, *a.* (Later example.)
1854 Dickens *Hard Times* iii. iii. 282 His so quiet and assuasive father-in-law.

as-su·chness. [f. *as such* (Such 38 c) + -ness.] Absolute existence or possession of qualities, independently of all other things whatever; the character of a thing when viewed as it is in itself, regardless of anything else.
1909 W. James *Pluralistic Universe* ii. 47 It [sc. Bradley's Absolute] is us, and all other appearances, but none of us *as such*, for in it we are all 'transmuted', and its own as-suchness is of another denomination altogether. **1962** *Times Lit. Suppl.* 4 May 311/3 Mr. Middleton..has a beautiful love and respect also just for the as-suchness of things: the sniffing man in the restaurant.

assuetude (æ·switiud). Delete †*Obs. rare*⁻¹ and add later examples.
1830 Marryat *King's Own*, xxxvi, To whose..beauty, from assuetude, he had..been blind. **1873** *Contemp. Rev.* XXII. 699 The petrified assuetudes and porcelain effeminacies of the Chinese. **1890** *Daily News* 15 Nov. 5/6 At the same time..the effect of assuetude may become apparent.

assumingly (ăsiū·miŋli), *adv.* [f. Assuming *ppl. a.* + -ly².] In an assuming manner, presumptuously.
1839 J. P. Smith *Scripture & Geol.* 215 What I may, not assumingly, call the whole body of geologists. **1927** *Scots Observer* 12 Mar. 3/5, I ken him fine commented the other assumingly.

assumptionist (ăsʌ·mᵖʃənist). [f. Assumption + -ist.]
1. One who bases his arguments on assumption.
1891 G. A. Denison *Let.* 9 Mar. in *Fifty Years at East Brent* (1902) 344 He is a remarkable assumptionist—he is no logician.
2. *R.C. Ch.* A member of the congregation entitled Augustinians of the Assumption, 'which had its origin in the College of the Assumption, established at Nîmes, in France, in 1843' (*Cath. Encycl.*).
1898 *Westm. Gaz.* 30 Sept. 6/2 An article in the *Croix* by its editor, the Assumptionist Father Bailly. **1901** *Daily News* 7 Jan. 4/7 The Assumptionists, who have played such an infamous part in French politics the last three years.

assuring, *ppl. a.* Add: **2.** That takes out a policy of assurance.
1891 *Charity Organis. Rev.* VII. 262 Each assuring member secured a small payment at death.

Assyro- (ăsi·ro), comb. form of Assyrian *a.*, as **Assyro-Babylonian** *a.*, of or pertaining to both Assyria and Babylonia; *spec.* designating an ancient Semitic language of Mesopotamia; hence as *sb.*, = *Akkadian *sb.* 1.
1835 [see *Medo-]. **1888** A. H. Sayce *Hittites* ii. 22 A large collection of clay tablets..in the Assyro-Babylonian language. **1932** W. L. Graff *Lang.* xi. 400 The only East Semitic language known is Akkadian, which was spoken in Ancient Babylonia and Assyria and is therefore sometimes called Assyro-Babylonian. **1952** O. R. Gurney *Hittites* iii. 82 The Hittite and Assyro-Cappadocian texts.

astable (ā-, ēɪstēɪ·b'l), *a.* [f. A- 14 + Stable *a.*] Not stable. In *Electr.* (see quot. 1960).
1951 *Electronic Engin.* XXIII. 481/2 An astable multivibrator is synchronized by line frequency pulses at a multiple of line frequency. **1952** *Proc. Inst. Radio Engin.* XL. 1531/2 The operation of this circuit may be monostable, bistable, or astable (oscillatory), depending on the nature of the emitter load. **1953** *Electronic Engin.* XXV. 239 A form of astable pulse generator..has been used by several workers. **1960** Cooke & Markus *Electronics & Nucl. Dict.* 26/2 *Astable circuit*, a circuit that alternates automatically and continuously between two unstable states. *Ibid.*, *Astable multivibrator*, a multivibrator in which each tube alternately conducts and is cut off.

astasia (ăstēɪ·siă). *Med.* [mod.L., a. Gr. ἀστασία unsteadiness, f. ἄστατος unstable, never standing still (ἀ- priv. + στατός standing): see -ia¹.] †**a.** Dysphoria. (Dunglison 1842.) *Obs.* **b.** [ad. Fr. *astasie.*] Inability to stand, as a hysterical disorder. So *astasia-abasia*, inability to stand or walk. Cf. *abasia.
[**1888** P. Blocq in *Archives de Neurologie* XV. 24 Sur une Affection caractérisée par de l'Astasie et de l'Abasie.] **1890** *N.Y. Med. Jrnl.* LII. 612/2 Those symptoms, then, of difficulty in standing..and ataxia of movement for the act of walking, but not for other muscular acts, corresponded.. to the condition described by Blocq under the title *astasia* and *abasia.* **1891** [see *abasia]. **1960** Hinsie & Campbell *Psychiatric Dict.* (ed. 3) 74/1 Although astasia-abasia is not confined to any single psychiatric group, it is probably most common in conversion hysteria.

astatic, *a.* (Earlier and later examples.)
1827 J. Cumming *Man. Electro Dynamics* i. 13 The moveable conductors..are astatic; that is, they are so constructed as not to be affected by the magnetic influence of the earth. **1940** *Chambers's Techn. Dict.* 56/1 *Astatic galvanometer*, a moving-magnet galvanometer in which the magnets form an astatic system. **1962** T. Axson in G. A. T. Burdett *Automatic Control Handbk.* vi. § 10 This fundamental characteristic of not tending to any one position has given rise to the name 'Astatic' voltage relay.

astatine (æ·stătin, -īn). *Chem.* [f. Gr. ἄστατ-os unstable (see *astasia) + -ine⁵.] A radioactive element of short half-life, which is rare in nature but can be made artificially; the heaviest element of the halogen group. (Formerly called *alabamine and *Anglo-Helvetium.) Symbol At, atomic number 85, principal isotope 211.
1947 D. R. Carson et al. in *Nature* 4 Jan. 24/1 We propose to call element 85 'astatine' from the Greek ἄστατος, unstable. Astatine is in fact the only halogen without stable isotopes. The corresponding chemical symbol proposed is 'At'. **1950** F. Gaynor *Conc. Encycl. Atomic Physics* 21 *Astatine*, a very rare radioactive element..first announced by Allison and Murphy under the name *alabamine* in 1929; also referred to formerly as eka iodine. *Ibid.*, The name astatine was officially adopted by International Union of Chemistry and announced by the American Chemical Society on September 23, 1949. **1950** N. V. Sidgwick *Chem. Elements* II. 1260 *Element No. 85* (*Astatine*). There seems to be no doubt now that this element has been obtained as artificial product of nuclear bombardment.

astatki (æstæ·tki). [ad. Russ. *ostátki* (pronounced asta·tki), pl. of *ostátok* remainder.] The waste product of the distillation of Russian petroleum atomized with steam and made combustible for use as fuel. Also *attrib.*
1885 *Jrnl. Soc. Chem. Ind.* IV. 78/1 Petroleum residuum or astatki is the only fuel employed in distilling petroleum at Baku. **1902** *Encycl. Brit.* XXXI. 646/1 Large quantities of refuse were produced—known by the Russian name of *astatki.* *Ibid.* 646/2 Practically all the steam power in South Russia..is now raised from *astatki* fuel.

Astbury (æ·stbə̆ri). The surname of John *Astbury* (1688–1743), used (esp. *attrib.*) to designate a type of Staffordshire pottery; also in *Astbury-Whieldon* (see quot. 1929).
[**1874** W. Chaffers *Collect. Hand Bk. Marks Pottery* 141 Pottery. Staffordshire..Astbury.] **1904** A. H. Church *Eng. Earthenware* (ed. 2) 55 A two-handled loving cup with design in white relief on a red body the whole glazed with a yellow lead glaze was sold..under the designation of Astbury ware. **1929** H. Read *Staffs. Pottery Figures* (Plate 17, *caption*), The term 'Astbury type' is used for those figures in which the decoration depends mostly on a use of coloured clays; the term 'Astbury-Whieldon type' when this technique is combined with decoration in coloured glazes. **1962** *V. & A. Mus. Internat. Art Treasures Exhib.* 62/2 Two Astbury figures in creamware splashed with manganese.

aster. Add: **4.** *Cytology.* **a.** A star-shaped achromatinic structure surrounding the centrosome of a cell during mitosis. **b.** The star-shaped grouping of the chromosomes during mitosis.
1879 E. Klein in *Q. Jrnl. Microscop. Soc.* XIX. 414 We pass..to large nuclei..in which the deeply-stained fibrils are arranged like a single aster ('Monaster'), apparently terminating freely at the periphery, but connected into a central network... Next, we trace these [fibrils] into a nuclei without a membrane in which the fibrils are similar in appearance to the preceding ones, but arranged as a double aster ('Dyaster'). **1888** Rolleston & Jackson *Anim. Life* p. xxv, A star or aster with a pronucleus as a centre. **1909** J. W. Jenkinson *Exper. Embryol.* 107 Each sperm forms its own aster, and these combine with one another to form various irregular mitotic figures (triasters, tetrasters, and so on). **1920** L. Doncaster *Introd. Study Cytology* iii. 31 The centrosome with its system of rays is called an *aster*, and the two asters with the sheaf of fibres connecting them are the achromatic or mitotic spindle. **1963** *New Scientist* 7 Feb. 305 When an egg is fertilised, two structures called asters appear in the cell and a spindle forms between them.

astereognosis (ăste:riognōᵘ·sis, -stɪ̆ᵊ:rio-). *Med.* [mod.L., f. A- 14 + *stereognosis.] Inability to recognize the shapes and special relationships of objects, *esp.* as a symptom of disorder of the central nervous system. Also **astereogno·sia.**
1900 in Dorland *Med. Dict.* 82/2. **1912** *Proc. R. Soc. Med.* V. (Neurol.) 150 (*title*) Case of Asteriognosis, probably due to a Lesion of the Posterior Columns in the Cervical Region. **1941** *Brit. Jrnl. Psychol.* XXXI. 334 This case of astereognosia,..offers no support whatever to the assumption of a psycho-physical process conveying more than mere sensory qualities.

asterisked (æ·stĕriskt), *ppl. a.* [f. Asterisk *v.* + -ed.] Marked with an asterisk; printed with asterisks replacing omitted letters or words.
1897 *Westm. Gaz.* 20 July 2/2 An asterisked speech [in 'Hansard']. **1911** C. Mackenzie *Passionate Elopement* xx. 183 All the world was deeply engrossed in reading about his asterisked self and his asterisked neighbour.

asteroid, *a.* and *sb.* Add: **A.** *adj.* **2.** *Zool.* Of, belonging to, or characteristic of the Asteroidea.
1888 Rolleston & Jackson *Anim. Life* 5 The Asteroid larva closely resembles the Holothurioid. **1900** F. A. Bather et al. *Echinoderma* 3 No sharp line can be drawn between Asteroid and Ophiuroid structure.
B. *sb.* **4.** *Zool.* A star-fish of the class Asteroidea.
1841 E. Forbes *Brit. Starfishes* i. 3 The Crinoids as analogues of the Polypes are lower than the Asteroids. **1888** Rolleston & Jackson *Anim. Life* 563 The apical system of plates is well defined in the young Asteroid. **1900** F. A. Bather et al. *Echinoderma* 14 The Asteroids were probably the last group to branch off from the fixed Echinoderms. **1921** *Chambers's Jrnl.* XI. 502/1 The mauve of these beautiful asteroids blends perfectly with the soft gray sands of the North Pacific.

asthenia. (Earlier and later examples.)
1802 *Med. & Physical Jrnl.* VII. 246 Nervous diseases from direct asthenia. **1933** *Brit. Jrnl. Psychol.* July 22 Perseveration due simply to mental asthenia and lack of spontaneity. **1938** S. Beckett *Murphy* iv. 49 That long hank of Apollonian asthenia.

asthenic, *a.* Add: **b.** Applied to a type of physique (see quot. 1937).
1925 W. J. H. Sprott tr. *Kretschmer's Physique & Character* i. ii. 19 Three ever-recurring principal types of physique have emerged..which we will call 'asthenic', 'athletic', and 'pyknic'. **1927** Henderson & Gillespie *Text-bk. Psychiatry* vii. 118 The thin, visceroptotic 'asthenic' type. **1937** [see *athletic *a.* 3].
B. *sb.* One who is asthenic.
1893 in *Funk's Stand. Dict.* **1925** E. & C. Paul tr. *Janet's Psychol. Healing* i. iii. 389 Deschamps finds it necessary to admit that, under the influence of strong emotion, an asthenic can perform actions which are beyond his powers at ordinary times. **1925** W. J. H. Sprott tr. *Kretschmer's Physique & Character* i. ii. 21 The essential characteristic of the type of the male asthenic is..a deficiency in thickness combined with an average unlessened length.

asthenolith (æsþe·noliþ). *Geol.* [f. as next + -lith.] The material in the asthenosphere.
1929 B. Willis in *Bull. Geol. Soc. Amer.* XL. 311 An asthenolith may be defined as a body of magma formed in consequence of a rise of temperature in the asthenosphere because of concentration of heat energy.

asthenosphere (æsþe·nosfīᵊɹ). *Geol.* [f. Gr. ἀσθεν-ής weak (f. Gr. ἀ- priv. + σθένος strength) + -o + Sphere *sb.*] (See quot. 1914.)
1914 J. Barrell in *Jrnl. Geol.* XXII. 659 The theory of isostasy shows that below the lithosphere there exists in contradistinction a thick earth-shell marked by a capacity to yield readily to long-enduring strains of limited magnitude... Its comparative weakness..is its distinctive feature. It may then be called the sphere of weakness—the *asthenosphere.* **1915** *Ibid.* XXIII. 43 The weakest part of the asthenosphere is of the order of one-hundredth of the maximum strength of the lithosphere. **1922** *Encycl. Brit.* XXXI. 214/1 A yielding but not necessarily molten layer, Barrell's asthenosphere..which lies some 80 to 100 m... below the general surface of the geoid. **1966** *New Scientist* 10 Mar. 635/1 The seat of crustal upheavals may be the so-called 'asthenosphere' or low-velocity layer in the Earth's upper mantle.

asthma. Add: **2.** *attrib.* **asthma herb** *Austral.* (see quot. 1887).
1887 Moloney *Forestry W. Afr.* 411 Australian Asthma-Herb (*Euphorbia pilulifera*, L.). **1889** J. H. Maiden *Useful Native Plants Australia* 183 Queensland Asthma Herb.

asthmatic, *a.* **3.** *fig.* (Earlier example.)
1806 J. Beresford *Miseries* I. x. 242 An asthmatic pair of bellows.

||**asthore** (æsþǭ·ɹ). *Anglo-Ir.* [Irish, f. *a* O + *stór* treasure.] My treasure; (my) darling.
1894 Somerville & Ross *Real Charlotte* III. xli. 132 Come in out o' the rain, asthore. *a* **1896** J. Todhunter in E. A. Sharp *Lyra Celtica* (1896) 170 Sure it's our complaint that's on us, asthore, this day. **1897** Ross & Somerville *Silver Fox* v. 64 'Oh, asthoreen,' wailed his mother... 'Take the cup o' tay, asthore, don't be talkin' that way.' **1906** A. P. Graves in C. V. Stanford *Nat. Song Bk.* 156 I've found my bonny babe a nest On Slumber Tree: I'll rock you there to rosy rest, Astore Machree!

astigmatic, *a.* Add: **2.** Of an artificial lens: correcting astigmatism.
1881 *Ophthalmic Rev.* Nov. (Advt.), Every description of Astigmatic, Cylindrical and Sphero-Cylindrical Spectacle Lenses. **1958** *Listener* 6 Nov. 728/2 This distortion is immediately removed if his pictures are photographed through a one-degree astigmatic lens at an axis of fifteen degrees off the vertical.

astigmatism. Add: (Earlier example.)
1849 G. B. Airy in *Trans. Camb. Philos. Soc.* VIII. 361 Twenty years ago, I had the honour of submitting to this Society a statement of the effects of a mal-formation in my own left eye. The nature of the effect was this: that the rays of light coming from a luminous point and falling upon the whole surface of the pupil do not converge to a point at any position within the eye, but converge in such a manner as to pass through two lines at right angles to each other, (a geometrical phenomenon, to which the term *astigmatism* was very happily affixed by the present Master of Trinity College [Whewell]).
b. *Optics.* A similar defect in a lens.
1859 Parkinson *Optics* iv. 48 When a pencil of rays is reflected or refracted at the surface of a medium, the reflected or refracted rays will not, in general, pass accurately through one point. This peculiarity is sometimes called astigmatism. **1892** J. T. Taylor *Optics Photogr.* xxiii, Astigmatism is a serious fault for a lens to possess in any marked degree.
2. *fig.*
1934 G. Bottomley in *Ess. & Stud.* XIX. 142 This mental astigmatism..was not noticed while the fires of great genius flamed high.

astilbe (ăsti·lbi). *Bot.* [mod.L. (F. Hamilton (*olim* Buchanan), 1825), f. Gr. ἀ- priv. (A- 14) + στιλβή, fem. of στιλβός glittering (στίλβειν to glitter).] A plant of a genus of perennials of the family Saxifragaceæ with terminal clusters of small white flowers.
1843 A. Gray in *London Jrnl. Bot.* II. 123 Three species belong to a single and highly natural genus, for which the name of *Astilbe* must be retained. *Ibid.* 124 The stamens of the original *Astilbe* are probably sometimes double the number of the sepals. **1863** —— *Man. Bot. N. United States* 142 *Astilbe*, false goatsbeard. **1956** *Dict. Gardening* (R. Hort. *Soc.*) (ed. 2) I. 215/1 The Astilbes grow well in rich garden soil if it is moist in summer. **1959** *Listener* 14 May 866/3 Aruncus and astilbe..and many primulas are suitable.

astonish, *v.* Add: **4.** Also *absol.*
1904 L. T. Meade *Love Triumphant* iii. i, There might come to you a knowledge which would astonish and terrify.

astragal. Add: **2. b.** (See quot. 1940.)
1858 *Skyring's Builders' Prices* 33 Astragal sashes. **1940** *Chambers's Techn. Dict.* 56/2 *Astragal*, a specially shaped bar used for connecting together glazing bars or sheets of glass in a window. *Ibid.*, *Astragal tool*, a special tool..used in wood-turning for turning beads and astragals.

astragalomancy. (Earlier and later examples.)
1640 E. Chilmead tr. *Ferrand's 'Ερωτομανία* xxiii. 178, I shall omit to speak here of Astragalomancy, that was done with Huckle bones. **1853** *Encycl. Brit.* III. 780/2 Astragalomancy..was practised in a temple of Hercules in Achaia. **1962** C. L. Wrenn in Davis & Wrenn *Eng. & Med. Studies* 312 Pauly-Wissowa..also treats of astragalomancy.

astrakhan. Add: **b.** A kind of cloth (see quots.) used chiefly as an edging or trimming for garments. Also (both senses) *attrib.*
1887 *Cassell's Fam. Mag.* Oct. 698/1 The curled cloths or woollen Astrachans used for trimmings and for entire jackets. **1898** *Daily News* 23 Mar. 5/6 Mr. Gladstone.. closely muffled up in a brown astrachan-lined overcoat. **1904** *Tailor & Cutter* 4 Aug. 479 Astrachan: A cloth made of worsted material, with a long and closely-curled pile, in imitation of the fur of that name. **1958** Hayward & Harari tr. *Pasternak's Dr. Zhivago* i. vi. 183 Her astrakhan cape hung open over the..quaking layers of her double chin. **1965** *Which?* Mar. 96/2 Most astrakhan cloth is imitation fur, usually made from wool and mohair. For coats, hats, trimmings.

astral, *a.* (and *sb.*). Add: **2.** *astral lamp* (earlier U.S. example).
1831 *Encycl. Amer.* VII. 398/1 In the astral and sinumbral lamps..the oil is contained in a large horizontal ring.
3. *Theosophy.* Pertaining to or consisting of a supersensible substance considered to be next above the sensible world in refinement and held to pervade all space. So *astral body*, the ethereal counterpart or shadow of a human or animal body.
[**1691** R. Kirk *Secret Commonw.* (1815) i. 9 That what the Low-countrey Scotts calls a Wreath..is only exuvious Fumes of the Man approaching Death, exhal'd and congeal'd..and called astral Bodies, agitated as Wild-fire with Wind.] **1877** Helena P. Blavatsky *Isis Unveiled* in *Secr. Doctrine* (1888) II. 74 When those circulations—which Eliphas Levi calls 'currents of the astral light'—in the universal ether..take place in harmony with the divine spirit, our earth..enjoys a fertile period. **1880** Anna Kingsford in Maitland *A. Kingsford* (1913) I. 401 In man the astral fluid becomes transformed into human life at the moment of conception. *Ibid.*, All they whose bodies have decomposed leave, or have left, their shadow in the astral space. **1881** Sinnett *Occult World* 162 Even our astral bodies, pure ether, are but illusions of matter so long as they retain their terrestrial outline.
4. *Cytology.* Of or pertaining to an aster.
1896 E. B. Wilson *Cell* ii. 74 The contractile elements are formed by certain of the astral rays which grow into the nucleus. **1901** Calkins *Protozoa* iii. 82 In others they [*sc.* the rays] are focussed in a central or 'astral' granule (*Gymnosphæra*, *Actinophrys*, *Sphærastrum*, etc.), which in some cases has been seen to divide like a centrosome and to form an amphiaster. *Ibid.* viii. 278 In some cases this structure resembles the astral system of Metazoa. **1910** *Encycl. Brit.* VII. 714/2 The remaining radiations at the two poles of the spindle are the 'astral rays'.
B. *sb.* **1.** (Earlier U.S. example.)
1838 *Knickerbocker* XII. 57 As she drew the flowers on the centre-table more under the light of the astral.
2. An astral body.
1880 Anna Kingsford in Maitland *A. Kingsford* (1913) I. 400 The Astral is not an entity, for it cannot reproduce itself. It is an imprint only, a shadow, a reflect, an echo. **1888** Helena P. Blavatsky *Secr. Doctrine* (ed. 2) I. 639 Our more intimate astral, or inner man.

astrally, *adv.* Delete *rare*⁻¹ and add: In, or through the powers of, the astral body.
1886 *Sat. Rev.* 4 Dec. 751 Whether the Thibetan Adepts go spooking astrally through the world. *c* **1900** W. B. Yeats *Lett.* (1954) 344 Our thaumaturgists..called her up astrally, and told her to leave him.

astream (ăstrī·m), *adv.* and *pred. a.* [f. A *prep.*¹ + Stream.] In a stream; that streams, streaming, flowing.
1763 C. Smart *Song to David* st. 85, Glorious the northern lights astream. **1925** G. Murray tr. *Eumenides* 51 Joy be a-stream in your ways, as the fire that bloweth A-stream from beacon and brand. **1928** Blunden *Undertones of War* 299 Their rosy scarves are spied astream. **1929** C. K. Scott-Moncrieff tr. *J.-R. Bloch's —— & Co.* (1930) i. iv. 43 Two or three marauding brigs..arrived one after another, their backs astream with water. **1930** R. Campbell *Adamastor* 79 Round which, astream through flowering vales.. The great Zambezis wreathe their coils.

astride, *adv.* and *prep.* Add: **C.** *adj.* Of a seat on horseback: belonging or proper to one riding astride.
1889 in *Cent. Dict.* **1907** *Daily Chron.* 25 Oct. 3/5 Makers of riding habits are going to make a special effort to show that the 'astride' seat can be made elegant. **1930** S. G. Goldschmidt *Fellowship of Horse* ix. 132 Some say that astride riding is safer. **1931** *Times Lit. Suppl.* 26 Nov. 944/4 The Dianas of our modern horse-shows will be mildly surprised to learn..that the astride seat is impossible for women.

astro-. Add:
astrobio·logy, a branch of biology concerned with the discovery or study of life on the celestial bodies; (*astro-biology* (ad. F. *astrobiologie* (R. Bertholet)) and *astro-biological* also occur in *Mind* (1934) XLIII. 269 and *Mind* (1937) XLVI. 116 applied to a pseudo-metaphysical interpretation of certain ancient beliefs of the Chaldeans); **astrobo·tany** [ad. Russ. *astrobotanika* (Tikhov 1945)], the study of plant organisms on the celestial bodies; hence **astrobo·tanist**; **astroco·mpass,** a form of compass for determining the direction of true north relative to the stars; used for correcting the errors of magnetic or gyroscopic compasses; **a:stro-dyna·mics** (see quot. 1955); **astro-fix,** the process or an act of determining the position of an aircraft by observation of the stars; **astro-hatch** = *astrodome; **astrophile,** delete † and add later examples (also *fig.*); **astropla·nkton,** living material, postulated by Liebig, Kelvin and others but not yet observed, drifting in space; **astro-shot** (*colloq.*), **astro-sight,** an observation of the stars for the purpose of determining the position of an aircraft in flight.
1955 *Sci. News Let.* 13 Aug. 107/2 Among operational sciences and skills..would be..astro-biology. **1958** C. C. Adams et al. *Space Flight* x. 240 The International Mars Committee..surveyed astrobiology, spectroscopic evidence of vegetation on Mars [etc.]. **1952** *Sci. News Let.* 5 Jan. 2 There is much attention [in the Soviet Union] to what is called astrobotany. **1956** *Newsweek* 22 Oct. 26/2 Gabriel Tikhov, Soviet 'astrobotanist', reports confirmation of his theory that higher forms of vegetation, similar to trees, exist on Mars. **1951** *Gloss. Aeronaut. Terms* (*B.S.I.*) iii. 10 *Astro-compass*, a non-magnetic instrument which gives the direction of true north relative to a celestial body. **1955** *Sci. News Let.* 13 Aug. 107/1 Astrodynamics, dealing with 'the dynamic flight behavior of space vehicles'. It would include the performance kinetics and dynamics, stability and control of spacecraft, similar to aerodynamics in aeronautics. **1941** J. A. Hammerton *ABC of RAF* 111 If the aircraft is over cloud and with clear sky above, the navigator may use the stars and get an 'astro' fix. **1944** 'N. Shute' *Pastoral* iv. 97 He said 'Navigator to Captain. Can you get her up above this for an astro fix?' **1941** A. O. Pollard *Bombers over Reich* x. 140 White frost..on the astrohatch. **1944** *Times* 18 Jan. 6/3 Over the target area one engine was put out of action and the astro-hatch shattered. **1903** *Daily Chron.* 6 June 5/2 Dr. Anderson..refuses..the name of astronomer... As he makes his observations with the naked eye from his own study window, with merely a pocket telescope..he prefers the name astrophil. **1936** L. Pearsall Smith in *S.P.E. Tract* XLVI. 204 The fever of perfection is not catching; and if it be foolish for these astrophils to hitch their wagons (in Emerson's phrase) to this remotely glittering star, surely they cannot reasonably be supposed to inflict any serious damage on the solar system. **1954** J. B. S. Haldane in *New Biol.* XVI. 25 One of the earliest parties to land on the moon should be able to look for astroplankton, that is to say spores and the like, in dust from an area of the moon which is never exposed to sunlight. **1946** V. Tempest *Near Sun* i. 15 An Astro-shot is a check on navigation that is made by the use of a sextant and the stars. **1943** G. L. Cheshire *Bomber Pilot* i. 10 To check..by visual pin-point if possible: if not, by taking drifts or Astro sights.

astroblast (æ·stroblɑst). *Anat.* [f. Astro- + -blast.] A primitive cell that develops into an astrocyte.
1901 in Dorland *Med. Dict.* (ed. 2). **1940** Gladstone & Wakeley *Pineal Organ* xxv. 382 Section through the spinal cord of a..chick embryo..*b.* supportive spongioblast, *c.* astroblast, or displaced epithelial cells.

astrocyte (æ·strosəit). *Anat.* [f. Astro- + -cyte.] **a.** A star-shaped cell of the neuroglia tissue in the central nervous system. **b.** A star-shaped bone corpuscle (1893 in *Funk's Stand. Dict.*).
1898 E. L. Billstein tr. *P. Stöhr's Text-Bk. Histology* (ed. 2) v. 170 The neuroglia principally consists of nucleated elements, the glia-cells... There are two kinds of glia-cells, ependymal cells and astrocytes. **1934** E. A. Sharpey-Schafer *Histology* (ed. 13) xviii. 191 Protoplasmic cells..radiating processes spring from the cytoplasm; hence the names astrocyte and spider-cell, by which these neuroglia-cells are also known. **1945** W. E. Le Gros Clark *Tissues of Body* (ed. 2) xiii. 367 Scattered throughout the grey and white matter of the central nervous system are stellate cells... They are divided into two categories, protoplasmic astrocytes..and fibrous astrocytes.

astrocytoma (æ:stro͵səitōᵘ·mă). *Path.* Pl. **-mata, -mas.** [mod.L., f. prec. + *-oma.] A tumour consisting wholly or largely of astrocytes.
1930 D. S. Russell & H. Cairns in *Jrnl. Path. & Bacteriol.* XXXIII. 389 The tumour described is evidently a fibrillary astrocytoma. **1961** *Lancet* 16 Sept. 657/1 Well-differentiated astrocytomas. **1962** *Ibid.* 6 Jan. 25/2 Several astrocytomata have also been treated with derivatives of melphalan and with chlorambucil.

astrodome (æ·strodōᵘm). *Aeronaut.* [f. Astro- + Dome *sb.*] A transparent dome on the top of the fuselage of an aircraft from within which astronomical observations can be made.
1941 *Aeropl. Spotter* 16 Jan. 23 (*caption*) Astro Dome. **1942** *We speak from the Air* xi. 34 As second pilot I was in the astro-dome keeping a look-out all round. **1943** *Times* 11 Mar. 2/5 The astrodome was shot away; and the bomb doors would not close.

astrographic (æstrogræ·fik), *a.* [f. Astrography + -ic, after *photographic.*] Of, pertaining to, used in, or produced by astrography.
1893 *Westm. Gaz.* 5 June 6/3 With the astrographic equatorial 722 photographic plates..were taken. **1895** *Athenæum* 8 June 743/1 The shutter of the astrographic dome was torn off and blown into the courtyard. **1895** H. H. Turner in *Oxf. Univ. Gaz.* 11 June 588/1 The De la Rue Astrographic Telescope is in good order and has been in.. use throughout the year. **1903** —— *Astrographic Chart* 13 The real work of the Astrographic Chart consists in measuring the positions of the stars, which will tell us ultimately their motions. **1946** *Nature* 17 Aug. 221/2 A better system for astrographic work is the Schmidt camera, in which the aberrations of a spherical mirror are removed by a figured plate at its centre of curvature.

astroid (æ·stroid), *a.* and *sb.* [f. Gr. ἀστροειδής star-like, f. ἀστήρ star: see -oid.] **A.** *adj.* Star-shaped, star-like (cf. Asteroid *a.*).
1909 in Webster.
B. *sb.* *Geom.* A hypocycloid with four cusps.
1897 H. Lamb *Infinit. Calculus* ix. § 138 The 'four-cusped hypocycloid'..is sometimes called the 'astroid'.

astronaut (æ·stronǭt). [f. Astro- + Gr. ναύτ-ης sailor, after Aeronaut.]
1. Name given to a space ship. *rare*⁻¹.
1880 P. Greg *Across the Zodiac* I. 27 In shape my Astronaut somewhat resembled the form of an antique Dutch East-Indiaman.
2. One who travels in space, i.e. beyond the earth's atmosphere; a student or devotee of astronautics. Cf. *cosmonaut.
1929 *Jrnl. Brit. Astr. Assoc.* June 331 That first obstacle encountered by the would-be 'Astronaut', viz., terrestrial gravitation. **1954** *N.Y. Times* 4 Apr. E9/7 The escape velocity from the earth is 25,000 miles an hour, yet astronauts talk glibly of achieving it, though they are fully aware of the heat that will be generated. **1957** P. Moore *Sci. & Fiction* xvii. 171 The astronauts taking off for the planet Hesikos remain standing upright. **1961** *Times* 6 May 8/3 President Kennedy spoke to Commander Alan Shepard by radiotelephone a few minutes after the astronaut was delivered by helicopter to the deck of the aircraft carrier Lake Champlain. **1962** J. Glenn in *Into Orbit* 4 We Astronauts have a mission... We are helping to break the bonds that have kept the human race pinned to the earth. **1966** *Electronics* 31 Oct. 39 Should a single goal—such as a manned landing on Mars in the 1980's—be chosen for the nation's space program once the Apollo astronauts have landed on the moon? **1971** *International Herald Tribune* 3 June 1 American astronauts and Soviet cosmonauts today swapped anecdotes about their experience in space exploration.

astronautics (æstronǭ·tiks). [ad. F. *astronautique* (coined by J. H. Rosny aîné (pseud. for J. H. H. Boëx-Borel) (1856–1940) in 1927; see *N. & Q.* (1960) CCV. 312 ff. and *Amer. Speech* (1961) XXXVI. 169 ff.): see -ic 2; cf. Aeronautics.] The art or science of locomotion outside the earth's atmosphere. Hence **astronau·tical** *a.*, pertaining to or concerned with astronautics or astronauts.
1929 *Jrnl. Brit. Astr. Assoc.* June 332 Prof. Oberth..has just been awarded the £80 prize offered for the most successful solution of the 'astronautical' question. **1929** *Mech. Engin.* (*A.S.M.E.*) Nov. 864/2 (*title*) Astronautics. **1931** *Internat. Cycl. Aviation Biogr.* s.v. Robert Esnault-Pelterie: This volume [*sc.* R. Esnault-Pelterie's *L'Astronautique* (Paris, 1930)] affords a most thorough groundwork upon which the future development of the science of 'Astronautics' will doubtless depend. **1932** *Times Lit. Suppl.* 16 June 439/3 France has a Committee of Astronautics. **1952** *Jrnl. Brit. Interplanetary Soc.* May 126 The term 'astronautical engineer' cannot yet have the same significance as 'aeronautical engineer'..although one day we feel sure it will have. **1960** *Times* 21 Mar. 13/6 There is need for an international law of astronautics.

astronavigation (æ:stronævigēɪ·ʃən). [f. Astro- + Navigation.] Determination of the position and course of aircraft by means of observation of the stars; applied also to the possible or actual use of this means in vehicles travelling in space. Also **a:strona·vigator,** an air pilot, or automatic air-piloting machine, using astronavigation.
1942 W. Simpson *One of our Pilots is Safe* ii. 46 We had neither the sextant nor the detailed knowledge necessary for astro-navigation. **1949** *Aeronautics* Jan. 52/1 When the weapon enters the area of enemy held territory..an automatic astro-navigator will take over. **1958** *Daily Mail* 10 Nov. 4/3 The huge 'astro-navigation trainer', a planetarium sort of thing which taught the space navigator to determine his position by measuring the position of the earth and the moon with respect to the immovable background of the fixed stars.

astronomical, *a.* Add: **1. b.** Of figures, distances, etc.: immense, similar in magnitude to those used in astronomy.

1899 *Daily News* 7 Oct. 8/2 He..excused his delay on the ground that in stock-taking..they discovered that they omitted one credit line of thirty thousand pounds. Such familiarity with astronomical finance made Mr. Rylands somewhat irritable. **1934** G. B. SHAW *Too True to be Good* Pref. p. 7 The odds against a poor person becoming a millionaire are of astronomical magnitude. **1947** *Evening News* 23 Apr. 2/1 Britain 'owes' astronomical sums in war-debts to India and Egypt. **1953** E. HYAMS *Gentian Violet* i. 9 The value of stage, film, broadcasting and other rights was astronomical.

3. *Comb.*, as *astronomical clock*: a clock which keeps sidereal time; *astronomical telescope*: a telescope designed for astronomical use, commonly one not giving an upright image (opp. *terrestrial telescope*); *astronomical triangle*: 'a triangle on the celestial sphere whose vertices are the pole, the zenith, and the observed body' (Webster 1961); = **celestial triangle*; *astronomical unit*: the mean distance between the earth and the sun (approx. 93 million miles), used as a unit for measuring distances within the solar system.

1856 D. LARDNER *Handbk. Astr.* I. vi. 136 The rate of the astronomical clock is so regulated that [etc.]. **1904** GOODCHILD & TWENEY *Technol. & Sci. Dict.* 31/1 *Astronomical clock*, a clock keeping sidereal time. It indicates 0 h. 0 m. 0 s. when the first point of Aries crosses the meridian. **1882** *Encycl. Brit.* XIV. 594/2 We can now understand the working of the ordinary *astronomical telescope*... The object glass furnishes an inverted but real image of a distant body, *within our reach*. **1917** A. H. HOLT *Man. Field Astr.* ii. 10 This spherical triangle..is so much used in field astronomy that it is called the astronomical triangle. *Ibid.* 11 Each part of the astronomical triangle, with the exception of the angle of the star, may be expressed in terms of the observer's position on the earth's surface (latitude) or the co-ordinates of the star. **1909** *Cent. Dict.* Suppl. s.v. *Unit*, Astronomical unit. **1963** JERRARD & MCNEILL *Dict. Sci. Units* 20 *Astronomical unit*..approximately equal to the mean distance between the sun and the earth ($1{\cdot}496 \times 10^{11}$m)...Radar determinations carried out since 1960 indicate the astronomical unit will soon be known to an accuracy better than 1 in 10^6.

a:strophoto·graphy. [f. ASTRO- + PHOTOGRAPHY.] The photography of celestial bodies. So **a:strophotogra·phic, -gra·phical** *adjs.*, of, pertaining to, used in, or produced by astrophotography. Also **a:stropho·tograph** *sb.*

1858 SUTTON & WORDEN *Dict. Photogr.* 28 *Astrophotography*, a convenient name for the application of Photography to the delineation of the planets and constellations. **1889** WINTERHALTER (*title*) The International Astrophotographic Congress. **1892** *Pall Mall Gaz.* 9 Aug. 2/1 The astrophotographic telescope. *Ibid.*, The new astrophotographical telescope is..controlled by an electric current. **1903** A. M. CLERKE *Problems in Astrophysics* 5 Astrophotography is an art, and has a technique of its own. **1938** *Times Lit. Suppl.* 3 Dec. vi/2 A vision of the exposed night sky—shown in over 30 astrophotographs.

astrophysical, *a.* (Later examples: see ASTRO- and next.)

1956 *Nature* 18 Feb. 299/2 The explanation lies in the advance of radio technology as well as in astrophysical phenomena at radio wave-lengths. **1958** *New Scientist* 20 Nov. 1299/2 The Soviet Union's most powerful telescope, the 48-inch reflector at the Crimean Astrophysical Observatory.

astrophysics (æstrofi·ziks). [f. ASTRO- + PHYSICS.] That branch of astronomy which treats of the physical or chemical properties of the celestial bodies. Hence **astrophy·sicist**, a student of astronomical physics.

1889 D. DUNKIN *Midnight Sky* 20..a subject for the investigations of the astro-physicist, the examination of the luminous spectra of the heavenly bodies has proved a remarkably fruitful one. **1890** *Sat. Rev.* 9 Aug. 176/1 The new science of 'astrophysics'. **1901** NEWCOMB *Stars* i. 10 The astronomer, or astrophysicist as he now calls himself. **1933** *Discovery* May 165/1 Some time ago it seemed to be the belief of the astrophysicists that the age of the stars and of the galactic system must be much greater than that of the earth. **1938** R. W. LAWSON tr. *Hevesy & Paneth's Man. Radioactivity* (ed. 2) xxvi. 289 The results of their observations [of cosmic radiation] are no less important in astrophysics than in atomic research. **1946** *Nature* 23 Nov. 764/2 A modern laboratory for spectroscopic research—a most important branch in connexion with future developments in astrophysics. **1962** *Sci. Survey* V. 70 Astrophysicists—those who apply physics to astronomical problems—have provided a coherent account of the life history of the stars.

astropyle (æ·stropail). *Zool.* [f. Gr. ἄστρον star + πύλη gate.] A tubular aperture or funnel-like membranous projection found in some radiolarians.

1887 E. HAECKEL *Radiolaria* in *Rep. Challenger* XVIII. 1. p. lv, This osculum is closed by a radiate cover (astropyle or operculum radiatum). **1888** ROLLESTON & JACKSON *Anim. Life* 876 Or again there is a main oral aperture or astropyle....The astropyle consists of a tubular proboscis, rising from the centre of a radially striated disc or operculum. **1901** CALKINS *Protozoa* iii. 70 The Cannopylea, in which the membrane around the pores is drawn out into funnel-like projections termed astropyles.

astrosphere (æ·strosfiᵊɹ). *Biol.* [ad. G. *astrosphäre* (H. Fol 1891, in *Anatomischer Anz.* VI. 273), f. ASTRO- + SPHERE.] **a.** The central portion of the aster exclusive of the astral rays; the centrosphere. **b.** The whole aster exclusive of the centrosome.

1896 E. B. WILSON *Cell in Development & Inheritance* Gloss. 333 Astrosphere..1. The central mass of the aster, exclusive of the rays, in which the centrosome lies..(Fol, 1891, Strasburger, 1892). 2. The entire aster exclusive of the centrosome..(Boveri, 1895). **1901** CALKINS *Protozoa* 278 This structure resembles the astral system of Metazoa, in consisting of an outer spherical mass with radiating processes (astrosphere).

Asturian (ăstiūᵊ·riăn), *a.* and *sb.* [ad. Sp. *asturiano*; cf. L. *Astur, -uris* Asturian.] **A.** *adj.* Of or pertaining to Asturias, an ancient province of northern Spain (now Oviedo), or its inhabitants; *spec.* designating a culture, remains of which have been found in Asturias (see quots.). **B.** *sb.* **1.** A native or inhabitant of Asturias. **2.** The Castilian dialect spoken by natives of Asturias.

1612 T. SHELTON tr. *Cervantes' Don Quixote* (1896) I. III. ii. 132 There likewise served in the Inne an Asturian Wench. *Ibid.* 136 He stretched forth his armes to receive his beautifull Damzell the Asturian. **1834** *Penny Cycl.* II. 539/1 The Asturian farmers also plant..scarlet-runners together with the Indian corn. *Ibid.* 540/1 The Asturians speak the Castilian language. **1875** *Encycl. Brit.* II. 824/2 There is a special clan among the Asturians called the Baqueros. **1887** *Ibid.* XXII. 350/2 *Castilian Dialects*..*Asturian*—The Asturian idiom, called by the natives *bable*, is differentiated from the Castilian by the following characters [etc.]. **1924** C. D. MATTHEW tr. *H. Obermaier's Fossil Man in Spain* x. 349 Protoneolithic phases in Europe.... The only such phase known in Spain is the Asturian, which..is post-Azilian and pre-Neolithic in age. *Ibid.*, The cultural phase which the present writer has styled the 'Asturian'. **1959** *Chambers's Encycl.* V. 452/2 On the north coast of Spain are found the middens of the Asturians who hunted land game and collected shell-fish on the shore during late mesolithic times.

a-swarm, *advb. phr.* (Earlier and later examples.)

1830 *Blackw. Mag.* XXVII. 947 A shower of natural leaf-born flies a-swarm in the air. **1840** BROWNING *Sordello* I. 147 Tutti Santi, think, a-swarm With Ghibellins, and yet he took no harm! **1890** J. A. SYMONDS *Essays* i. 166 The atmosphere..was aswarm with them.

a-swirl (ăswɔ̄·ɹl), *adv.* or *pred. a.* [f. A- *pref.* 2 + SWIRL.] In a swirl, swirling.

1909 H. G. WELLS *Tono-Bungay* IV. iii. 490 The water all a-swirl with the wash of shipping. **1915** C. C. MARTINDALE *In God's Army* 125 Despite the..mists that set the imagination a-swirl. **1924** R. CAMPBELL *Flaming Terrapin* vi. 34 The breeze..passed by with golden locks aswirl.

asyllabic (æ-, ēⁱsilæ·bik), *a.* [f. A- 14 + SYLLABIC *a.*] Not syllabic; not constituting a syllable; = ASYLLABICAL *a.*

1827 S. LEE *Hebr. Gram.* ix. 169 This letter, when Paragogic, is..asyllabic, and is mostly found between two nouns in the state of construction. **1953** C. E. BAZELL *Ling. Form* v. 50 A single phoneme (e.g. Latin *i*) may occur in syllabic or in asyllabic function.

asylum (ăsai·lŏm), *v. rare.* [f. the sb.] *trans.* To give protection to; to place in an asylum. Also *refl.*

1794 J. COURTENAY *Pres. State Manners, Arts, & Politics of France & Italy* 43 Th' assassin *asylums* himself in the Church, And we see him in every fine portico lurch. **1843** *Times* 8 Mar. 5/3 Do they wish to spill blood—they have only to play a few pranks—get asylum'd a month and a day. **1866** J. B. ROSE tr. *Ovid's Fasti* ii. 139 Crime thou asylum'd, crime expell'st be.

asymbolia (æsimbōᵘ·liă). *Path.* [ad. G. *asymbolie* (C. Finkelnburg 1870, in *Berl. klin. Wochenschrift* 19 Sept. 461), f. A- 14 + SYMBOL *sb.*²: see -IA¹.] Inability to understand or use words or signs: a form of aphasia.

1876 DUNGLISON *Dict. Med. Sci.* **1922** *Jrnl. Abnormal Psychol.* VI. 218 When the nature of these objects (symbolia) was tested, grave errors were committed by the patient and she was absolutely unable to name any of them...Asymbolia was therefore complete.

asymmetric, *a.* Add: **b.** *Chem.* Exhibiting, pertaining to, or affected by asymmetry (sense 2 b below); = *stereo-isomeric*. Spec., *asymmetric synthesis* (see quots.). Cf. **optical isomerism*.

1875 *Jrnl. Chem. Soc.* XXVIII. 862 The author [sc. J. H. van't Hoff] proceeds to show that combinations which possess an atom of carbon combined with four different univalent groups, and which he terms an asymmetric atom, present anomalies with respect to isomerism. **1881** [in *Dict.*]. **1885** *Encycl. Brit.* XIX. 314/2 A carbon atom combined with four different atoms or compound radicals may therefore be called an asymmetric carbon atom. **1902** *Ibid.* XXVI. 722/1 The number of modifications of an asymmetric compound—which is not itself symmetric—is 2^n, where *n* is the number of such carbon atoms. **1904** *Jrnl. Physical Chem.* VIII. 528 By asymmetric synthesis is meant the synthesis of a substance containing an asymmetric carbon under such conditions that there shall be formed an excess either of the dextrorotatory or of the levorotatory compound. **1949** *Thorpe's Dict. Appl. Chem.* IX. 117/2 A molecule containing one asymmetric carbon atom possesses two isomeric structures, which are mirror images of each other. *Ibid.* 118/1 Asymmetric syntheses involving photochemical reactions under the influence of circularly polarised light have been successfully attempted. **1962** P. J. & B. DURRANT *Adv. Inorg. Chem.* viii. 237 A rotating molecule may..be regarded as a spinning top, and on this analogy a polyatomic molecule may be placed in one of the four following classes: A. Symmetric tops. B. Spherical tops. C. Linear molecules. D. Asymmetric tops.

asymmetry. Add: **2.** (Later example.)

1948 GLASSTONE *Physical Chem.* xii. 904 The influence on the velocity of the ion is known as the relaxation effect, or sometimes as the asymmetry effect, because it arises from the lack of symmetry in the electrical atmosphere of a moving ion.

b. *Chem.* Lack of symmetry in the spatial arrangement of atoms or groups in a molecule; = *stereo-isomerism*.

1875 *Jrnl. Chem. Soc.* XXVIII. 862 He [sc. J. H. van't Hoff] has deduced the following rules...Derivatives of optically active combinations lose their rotatory power when the asymmetry of their carbon-atoms disappears. **1885** *Encycl. Brit.* XIX. 314/1 What peculiarity of constitution can give a molecule this helicoidal asymmetry? **1902** *Ibid.* XXVI. 721/1 The doctrine of asymmetry may be extended to elements other than carbon. **1959** *Chambers's Encycl.* III. 357/1 When two of the four attached groups are similar, as in propionic acid,..the asymmetry vanishes together with the possibility of optical isomerism.

asymptomatic (ăsimptomæ·tik), *a. Path.* [f. A- 14 + SYMPTOMATIC *a.*] Without symptoms; producing or exhibiting no symptoms.

1932 in DORLAND *Med. Dict.* (ed. 16). **1962** *Lancet* 13 Jan. 63/1 Splenectomy has not been indicated to control these asymptomatic blood changes.

asymptotically, *adv.* (Later examples.)

1879 J. TYNDALL *Frag. Sci.* (ed. 6) II. xv. 418 The theory is not a thing complete from the first, but a thing which grows, as it were asymptotically, towards certainty. **1886** A. B. BUCKLEY *Moral. Element in Gospels* i. 41 A moral ideal which in the natural order of things can never be more than asymptotically approximated. **1959** E. PULGRAM *Introd. Spectogr. Speech* viii. 63 The resonance curve of a damped resonator..approaches asymptotically, but never reaches, a zero level of power.

asynapsis (æ-, ēⁱsinæ·psis). *Biol.* [f. A- 14 + SYNAPSIS.] Absence of synapsis; failure of chromosomes to pair in meiosis. Hence **asyna·ptic** *a.*

1930 G. W. BEADLE in *Cornell Univ. Agr. Exp. Sta. Mem.* CXXIX. (*title*) Genetical and Cytological Studies of Mendelian Asynapsis in *Zea mays*. **1937** C. D. DARLINGTON *Rec. Adv. Cytol.* (ed. 2) vii. 291 In 'asynaptic' maize which has as a rule no pairing of chromosomes at meiosis, the progeny nevertheless show the results of normal crossing-over. **1949** DARLINGTON & MATHER *Elem. Genetics* v. 110 A so-called asynaptic gene often..produces a general reduction in the crossing-over of all the chromosome pairs when homozygous.

asynchronous, *a.* (Later examples.)

1912 G. KAPP *Electricity* viii. 219 To the late Professor Ferraris of Turin belongs the merit of having discovered a principle of alternating current working by which the motor may be started by the alternating current itself without bringing it first up to the speed of synchronism. Motors of this kind are called 'asynchronous' or 'non-synchronous' motors. **1940** *Chambers's Techn. Dict.* 58/2 *Non-synchronous motor*..an a.c. motor which does not run at synchronous speed.. Also called *asynchronous motor*. **1962** *Gloss. Terms Autom. Data Processing (B.S.I.)* 9 *Asynchronous working*, the performance of a sequence of operations, such that each operation starts as a result of a signal that the previous operation has been completed or that the equipment required for the next operation is now available.

asyndetic, *a.* Add: Hence **asyndetically** *adv.*

1928 in *Funk's Standard Dict.* **1959** M. SCHLAUCH *Eng. Lang.* in *Mod. Times* xi. 57 The smoothly, or the adding adjectival relative clauses asyndetically to the main one.

asynergy (ăsi·nəɹdʒi). Also **asyne·rgia**. *Path.* [f. A- 14 + SYNERGY.] Lack of co-ordination of parts or organs normally acting in harmony.

1866 MAYNE *Expos. Lex.* 1496/2 *Asynergia*. **1909** DORLAND *Med. Dict.* **1950** *Practitioner* Dec. 851 The laryngitis of singers..is made evident..by muscular troubles (vocal asynergy). **1934** H. C. WARREN *Dict. Psychol.* 22/2 *Asynergia*, a term introduced in 1899 by J. Babinski to describe the underlying symptom of cerebellar deficit, i.e. an inability to carry out complex motor acts involving harmonious cooperation of separate muscle groups.

at, *prep.* Add: **1. b.** Used with the cardinal points of the compass, as *at the East(ward)*, to indicate parts of the country. *U.S.*

1636 JOHN WINTHROP in *Proc. Mass. Hist. Soc.* Coll. (1863) 4th Ser. VI. 515 If Mr. Mayhew hath bought the provisions at the east. **1646** JOHN WINTHROP *Jrnl.* (1908) II. 263 In Sept. 2 in R. C. Winthrop *Life & Lett. J. Winthrop* (1867) II. xviii. 357 Some hurt was done here..much fish and salt lost at eastward. **1692** in *Essex Inst. Hist. Coll.* XLII. 143 Marke How that is now dead, who dyed at the Eastward. **1782** *S. C. Lit. Frnl.* Ibid. I. 123 My company being at the Sotherd, the money was drawn for them for 2 months. **1851** HAWTHORNE *Ho. Sev. Gables* xiii. 210 A still unsettled claim to a very large extent of territory at the eastward. **1883** J. QUINCY *Figures of Past* (1884) 345 Characteristic of slave-holders when upon their good behavior at the North.

c. *Naut.* Indicating the quarter of the wind. *U.S.*

1635 R. MATHER *Jrnl.* (1850) 18 Afore noone the wind waxed strong at north. **1732** FRANKLIN *Poor Rich. Alm.* 1733 10 Clouds and winds at southwest. **1780** WM. HEATH *Let.* 30 July in *Mass. Hist. Soc. Coll.* (1905) 7th Ser. V. 93 The wind which now blows at east. **1848** J. F. COOPER *Bee-hunter* II. xiv. 203 The wind stood at the westward.

d. Used superfluously after *where*. *U.S.* and *Brit. dial.* (see *E.D.D.*).

1859 BARTLETT *Dict. Amer.* (ed. 2), *At* is often used superfluously in the South and West, as in the question 'Where is he at?' **1899** A. NICHOLAS *Idyl of Wabash* 34 Where does he live at? **1903** *N.Y. Sun* 8 Nov. 6 The business world wants rest. It wants to know where it is at. **1911** E. FERBER *Dawn O'Hara* xx. 294 This is where I get off at. **1914** G. ATHERTON *Perch of Devil* i. 8 She..disliked..not knowing where she was at.

3. c. (Later examples.) Cf. also *Sc. Nat. Dict.*, *at* (sense 1).

1842 FITZGERALD *Let.* 31 Mar. (1889) I. 94 Alfred [Tennyson] is busy preparing a new volume for the press: full of doubts, troubles &c. The reviewers will doubtless be at him. **1899** E. WHARTON *Greater Inclin.* iv. § 3. 111 All his people are at him, you see—well, I know their little game! Trying to get him away from me.

13. c. *esp.* with verbs of speaking, with implication of indirect attack. See also TALK *v.* 3 d.

1711 Preaching at [in Dict.]. **1863** COWDEN CLARKE *Shakes. Char.* v. 133 The latter..always make her speak at her husband. **1863** TROLLOPE *Rachel R.* II. iv. 78 Had he been then present, she would have risen up and spoken at him, as she had never spoken before.

15. d. By (auction or sale; retail or wholesale). Orig. *U.S.* Cf. AUCTION *sb.* 2.

1726 *Boston News-Let.* 3 Mar., Valuable books, many more than a thousand, to be sold at auction. **1825** NEAL *Bro. Jonathan* I. 12 The education, which they had been laying in, at wholesale, during the summer season. **1860** [see AUCTION *sb.* 2]. **1900** DRANNAN *On Plains & Mtn.* 485 As soon as we arrived at San Francisco we commenced selling our horses at private sale. **1932** GAYDEN *Leader-ship* 135 They got the land at $2 an acre and immediately offered it at auction. **1967** PHILIP WILSON (*title*) Art at Auction. The Year at Sotheby's and Parke-Bernet, 1966–67.

35. *at all*: see ALL *a.* 9 b; *at that*: see THAT *dem. pron.* 5 c.

40. *at about*, at approximately.

1843 G. BORROW *Bible in Sp.* II. iii. 56 At about seven o'clock in the evening we reached Aldea Gallega. **a 1882** TROLLOPE *Autobiogr.* (1883) I. ix. 214, I have been paid at about that rate. **1915** V. WOOLF *Voyage Out* iii. 37 At about that hour he reappeared. **1929** D. H. LAWRENCE *Paintings* 24 Bit, At about the time of the Elizabethans. **1945** E. WAUGH *Brideshead Rev.* II. v. 272 My divorce case..was due to be heard at about the same time.

At, AT, A.T.: see *A.T.S. (as separate entry).

-at, *suffix.* **1.** The original form of *-ATE¹, surviving in a number of words, as *commissariat* (1), *concordat, diplomat, format, quadrat, secretariat* (also *-ale*). Most of these are loan-words from French.

2. Representing the 3rd pers. sing. pres. tense ending of Latin verbs of the first conjugation, as *ægrotat, habitat*.

atabrine, var. *ATEBRIN.

atactic, *a.* Add: **2.** *Path.* Of or pertaining to or affected with ataxia.

1880 W. JAMES *Feeling of Effort* 20 in *Anniv. Mem. Boston Soc. Nat. Hist.*, The special case of the limb being completely anæsthetic, as well as atactic. **1956** JACOBSON & HALLE *Fundamentals of Lang.* II. iv. 74 In one variety of aphasia,..sometimes..labeled 'atactic', the word is the sole linguistic unity preserved.... All other sound-sequences are either alien and inscrutable to him or he merges them into familiar words by disregarding their phonemic aberrations.

3. Characterized by or exhibiting irregularity in the spatial arrangement of parts in a molecule. Cf. *ISOTACTIC *a.*, *SYNDYOTACTIC *a.*

1957 *Chem. Abstr.* 15599 [Abstract of art. by G. Natta et al.] Polypropylenes of a type now defined as 'atactic'. **1957** *Technology* July 176/2 The side groups are situated quite at random along the chain; this type of combination is termed 'atactic'. atactic polypropylene is a rubbery material freezing below –30 deg. C.

ataman, var. HETMAN.

1835 *Court Mag.* VI. 82/1 'We may not do this,' replied one of the Cossacks, 'without the consent of our Ataman.' **1920** *Glasgow Herald* 18 Aug. 7 The Ataman of the Don Cossacks. **1924** *Blackw. Mag.* Feb. 171/2 The Ataman had been waiting a whole day for the last of the attacking party to come in. **1950** H. H. CARR *Bolshevik Revol.* I. 294 At the head of each community was an elected ataman.

atamasco lily (ætămæ·sko). Also **'atta-musco.** [N. Amer. Ind.] A plant, *Zephyranthes atamasco*, of the south-eastern U.S., bearing a single lily-like flower.

1629 PARKINSON *Parad.* 87 *Narcissus Virginensis*. The Virginia Daffodil... The Indians in Virginia do call it Attamusco. **1743** CATESBY *Carolina* I. App. 112 The Atamusco Lilly is a native of Virginia and Carolina. **1760** J. LEE *Introd. Bot.* App. 305 Atamasco Lily, *Amaryllis*. **1813** MUHLENBERG *Catal. Plants* 34 Atamasco lily; Car. Georg. Pen.; Fl. Jun. **1956** *Dict. Gardening (R. Hort. Soc.)* (ed. 2) IV. 2307 Atamasco Lily...perianth 2½ to 4 in. long, white or white flushed pink...United States, from Missouri and Virginia to Florida.

atap, var. *ATTAP.

ataractic, -axic (ætăræ·ktik, -æ·ksik), *a.* [f. Gr. ἀτάρακτος without disturbance, calm + -IC; cf. ATARAXY.] **A.** Calm, serene.

1941 H. MILLER *Colossus of Maroussi* i. 90 Mycenæ..reared in anthropophagous luxury, reptilian, ataraxic, stunning and stunned.

B. *Med.* Of drugs: inducing calmness, tranquillizing. Hence as *sb.*, a drug of this kind.

1955 H. D. FABING in *Neurology* V. 327/1 The Epicureans were eagerly in search of the term 'ataraxia' which meant *freedom from confusion, peace of mind*... It is proposed, therefore, that drugs of this type be designated *ataraxics*, and that the adjectival form, *ataractic*, be used to describe this therapeutic property in drugs. *Ibid.* 328/2 Proposal is made to adopt the generic term, *ataraxics*, for pharmacological agents such as chlorpromazine, rauwolfia compounds, Frenquel, and others, which bring about ataraxy, or freedom from confusion. *Ibid.*, 'Why can't some new term be devised to screen potential new ataractic drugs?' **1956** *Newsweek* 21 May 68/1 The other tranquillizers..are all known as ataraxics. **1957** *Times* 11 Mar. 6/6 The rapidly increasing use of drugs described as 'tranquillisers' and 'ataraxics' has become a cause of concern in many countries. **1957** *New Scientist* 12 Dec. 26/1 So-called ataractic drugs—that can cure the 'incurable' mental case.

atavistic, *a.* Add: (Later examples.)

1915 W. S. MAUGHAM *Of Human Bondage* xxvi. 108 Some atavistic inheritance of the cave-dweller. **1922** JOYCE *Ulysses* 676 The specific reappearance of atavistic delinquency. **1932** E. WAUGH *Black Mischief* v. 168 Was it some atavistic sense of a caste, an instinct of superiority, that held him aloof?

Hence **atavi·stically** *adv.*

1884 *N. Amer. Rev.* Sept. 255 The ancient types crop out atavistically. **1897** E. P. EVANS *Evol. Ethics* i. 33 The lower classes..reflect atavistically the ideas and passions of primitive man. **1926** *Blackw. Mag.* Apr. 446/2 Some of them bolted atavistically up the nearest tree.

atcha, atchoo, var. *ATISHOO.

atchcan, atchkan, var. *ACHKAN.

atebrin (æ·tebrin). Also (chiefly U.S.) **atabrine.** (Both are proprietary terms.) [ad. -ATE¹ 1 c + BRIN(E).] A synthetic antimalarial drug, quinacrine dihydrochloride; also called *MEPACRINE.

The word is recorded from 1933 but was not applied to the anti-malarial drug discovered by the German chemists H. Mauss and F. Mietzsch in 1930 (when it was known by the laboratory name of 'Erion') until 1932. See **1933** *Klinische Wochenschrift* 19 Aug. 1276–8.

[**1933** *Trade Marks Jrnl.* 12 Mar. 390 *Atebrin*. Chemical Substances Prepared for Use in Medicine and Pharmacy... The Bayer Comp., Limited, Manchester.] **1932** *Lancet* 16 Apr. 826/1 Atebrin, originally called *erion*, has been produced recently by the makers of plasmoquine. **1933** *Discovery* Apr. 118/1 Atebrin, made by the chemists Mietzsch and Mauss.. in some experiments..has been more successful than quinine in curing and preventing relapses (of malaria). **1935** F. STARK *Let.* 17 Mar. in *Coast of Incense* (1953) i. 91 Decided it must be malaria—took masses of atebrin. **1943** *Times* 1 Sept. 4/3 The sick had few medical supplies and no quinine or atebrin. **1948** *Chem. Abstr.* 7577-d [index 4075] Atabrine. **1957** *New Yorker* 19 June 75/1 The only malaria suppressive available for the Buna campaign was quinine; even quinine was not on hand. **1961** J. HELLER *Catch-22* (1962) xxviii. 291 I'm sorry about making such a fuss about those Atabrine tablets on the way over. If you want to catch malaria, I guess it's your business, isn't it?

atelectatic (æ:tilektæ·tik), *a. Path.* [f. ATELECTASIS.] Characterized by atelectasis. Also **atelectic** *a.*

1875 CARNEGIE tr. *Hermann's Human Physiol.* 159 The lungs left to themselves contain no air: they are atelectic [sic], like the lungs of the fœtus before it has 'breathed'. **1880** *Flint Princ. Med.* (ed. 5) 187 Occlusion of the bronchi immediately connected with the atelectatic spots can be demonstrated.

ateleiosis, ateliosis (ătelīᵊᵘ·sis, ătelai-, ǽtĭl-). [f. A- 14 + TELEIOSIS.] Dwarfism; defective or arrested development. Hence **ateleio·tic, -liotic** *a.* Also as *sb.*, a person suffering from ateleiosis.

1904 HASTINGS GILFORD in *Medico-Chirurg. Trans.* LXXXV. 306 This delay of growth and development is so evidently the main feature, that I have suggested that the disease should receive a name which emphasizes this fact. I have proposed that it should receive the name of *Ateleiosis*. *Ibid.* 345 It is probable that ateleiotic dwarfs of the second class may be distinguished from all other dwarfs by their physiognomy alone. **1920** *Clinical Jrnl.* 7 Dec. 140/1 A sexual ateleiotic who began a family of children who, except one who was an asexual ateleiotic, were normal. **1923** *Proc. R. Soc. Med.* VI. 2 Clin. 137 (*title*) Ateleiosis in a boy, aged 45. **1932** J. S. HUXLEY *Probl. Rel. Growth* iv. 131 In ateliotic dwarfs, the proportions of the limbs to the trunk and of the limb-segments to each other are not affected.

Atellan, *a.* and *sb.* (Earlier and later examples.)

1600 P. HOLLAND tr. *Livy* vii. 251 The Actours in the Atellane interludes. **1875** A. W. WARD *Hist. Dram. Lit.* I. i. 10 The ribald jests of Atellanes and mimes. **1938** R. GRAVES *Coll. Poems* 166 And what Atellan orgies of the soul Were celebrated then among the rocks They testify themselves in books That rouse them.

||a tempo (a te·mpo). *Mus.* [It., lit. 'in time'.] A direction to perform a passage, etc., in the tempo indicated, as *a tempo rubato, giusto*, etc. Also, a direction to return to the previous tempo.

1740 GRASSINEAU *Mus. Dict.* 6 *Atempo giusto*, signifies to sing or play in an equal, true and just time. **1834** *Penny Cycl.* II. 549/2 *A tempo*, in music, signifies, that after any change in motion, by retardation or acceleration, the original movement is to be restored. **1886** R. DUNSTAN *Man. Mus.* 111. 70 Terms of Pace, or Speed, *A tempo*, in strict time (used after *Accel.*, *Rall.*, or *Rit.*).

Aten (ā·tən). Also **Aton.** [ad. Egyptian *itn.*] One of the names of the sun in ancient Egypt; the name by which the sun or solar disc was worshipped particularly during the reign of Amenophis IV (Akhnaten) in the 14th century B.C. Hence **A·tenism**, worship of the Aten.

[**1841** G. WILKINSON *Manners & Customs Anc. Egyptians* ser. II. I. xiii. 297 The name Aten or Atin is found to call to mind Atin, or Atys, the Phrygian Sun.] **1877** *Encycl. Brit.* VII. 738/1 Amenophis IV..introduced a new religion, the worship of Aten, the solar disk. **1906** J. H. BREASTED *Hist. Egypt* xviii. 360 Already under Amenhotep III an old name for the material sun, 'Aton', had come into..use, where the name of the sun god might have been expected... Under the name of Aton, then, Amenhotep IV introduced the worship of the supreme god. **1925** J. W. JACK *Date of Exodus* i. 20 There is no connection between the Atenism of that king [sc. Akhnaten] and the religion of Israel. **1933** *Times Lit. Suppl.* 22 June 419/2 The priest of the Aten 'Panhesy'..which may be said to cover a little more than half-a-century down to 1346 B.C. **1935** W. EMPSON *Poems* 12 Stand, well-chased Phoebus..Atos of maggots of reflected glitter. **1939** A. TOYNBEE *Study Hist.* VI. 328 (*heading*) Atonism. **1961** A. GARDINER *Egypt of Pharaohs* ii. 222 A curious addition states that the Mnevis-bull of Heliopolis should likewise be buried in the Aten's city, another sign how dependent the new Atenism was upon one of the oldest of Egypt's religious cults.

A tent. *U.S.* [f. the letter A, from its shape.] A tent with sides sloping downwards from a ridge pole.

1863 GRAY & ROPES *War Lett.* (1927) 187 Beside them was a lot of negro laborers (not even soldiers) in brand new A tents. **1888** *Century Mag.* Jan. 447/1 Three well used A tents, or as we call them 'wedge' or common tents. **1904** ELIZ. ROBINS *Magnetic North* xvi. 284 Down in the desolate hollow a ragged A tent stood away from the prevailing wind. **1923** *Chelsea Camping Out* 432 An 'A' tent is almost as easy to pitch, especially if it is hung on a ridge rope tied between two trees.

Aterian (ătīᵊ·riăn), *a.* [ad. F. *atérien* (M. Reygasse 1922, in *Rec. de Notices & Mém. Soc. Arch. Constantine* LIII. 171), f. Bir el Ater in Algeria + -IAN.] Of or pertaining to a form of Mousterian culture found in northern Africa. Also *ellipt.* as *sb.*

1925 G. G. MACCURDY *Human Origins* I. 289 The Aterian is a specialized Mousterian found in Mauretania.. **1929** M. C. BURKITT *Old Stone Age* 243 Another important..industry is the Aterian. **1931** *Man* XXXI. 83 (*caption*) A specialized Mousterian industry with 'Aterian' affinities. **1941** F. E. ZEUNER *Prehist. Archaeol.* N.W. *Afr.* iv. 35 This site shows that the Aterian not only grew out of the archaic Mousterian, but that it followed the archaic Mousterian very closely, with only a comparatively short interval between. **1946** G. CATON-THOMPSON in *Jrnl. R. Anthrop. Inst.* LXXVI. 87 The Aterian industry, which appears to be centred in North-West Africa, has been described as a Mousterian with the addition of the tanged point. This is an understatement. It habitually includes other artefacts equally unfamiliar or rare in the Mousterian.

||à terre (a tɛ̄r), *adv.* and *adj. phr.* [Fr.] On the ground; *spec.* in *Ballet.* Also *fig.* (cf. *TERRE-A-TERRE).

1922 BEAUMONT & IDZIKOWSKI *Man. Class. Theat. Dancing* 19 When the entire base of the foot touches the ground, the foot is said to be *à terre*. If we speak of..the *third* position, etc. it is understood that the position is *à terre*, that is, with both feet *flat on the ground*. **1926** D. H. LAWRENCE *Plumed Serp.* ix. 149 Without aim or purpose, they lived absolutely *à terre*, in the dark, volcanic earth. **1951** *Ballet Ann.* V. 138/2 One dancer indeed is so indifferent to the ballet-dancing demands that the *corps de ballet* should be on *pointe*, she is firmly *à terre*. **1961** *Times* 13 Apr. 16/7 She has a lightness and effortless purity, à terre and in elevation.

Atestine (ăte·stin, -ain), *a.* and *sb.* [ad. L. *Atestinus* of Ateste + -INE¹.] **A.** *adj.* Of or pertaining to the ancient city of Ateste in north-eastern Italy (now Este); *spec.* designating the pre-Roman culture of Ateste. **B.** *sb.* An inhabitant of Ateste.

1924 D. RANDALL-MACIVER *Villanovans & Early Etruscans* ii. 4 Immediately north of the Po were the Atestines,.. To avoid race-names we have chosen this word in preference to 'Euganean'. **1931** *Antiquity* V. 386 In the later tombs we have figured bronzes of Atestine style presumably of early 5th century age. **1957** *Encycl. Brit.* VIII. 733/2 The close cousinship of the Bolognese Villanovans and the Atestines is proved by the complete identity not only of the burial rite, cremation, but of the forms and details of their graves. *Ibid.* 734/1 The third period, especially the first half of the 5th century B.C., marks the zenith of Atestine art.

Athabascan, var. *ATHAPASCAN.

athambia (ăþæ·mbiă). *rare*⁻¹. [a. Gr. ἀθαμβία imperturbability.] Imperturbability.

1956 S. BECKETT *Godot* i. 42 From the heights of divine apathia divine athambia divine aphasia loves us.

Athanasian, *a.* and *sb.* (Later examples.)

1781 GIBBON *Decl. & F.* III. xxxvii. 557 He [sc. Gundamund] recalled the bishops, and restored the freedom of Athanasian worship. **1853** C. LAMB *Let.* of Elia to R. Southey] in *London Mag.* VIII. 402/1 T. N. T...a little tainted with Socinianism, and—, a sturdy old Athanasian. **1850** NEWMAN *Difficulties Anglicans* i. 3 The same popular voice..may..dispense with the Athanasian Creed altogether. *Ibid.* i. 24 It will be obvious to allege that..provided we hold fast this 'scriptural fact', it matters not whether we be Athanasians, Sabellians, Tritheists, or Socinians.

Athapascan, -paskan (æþăpæ·skăn), *a.* and *sb.* Also **-bascan, -baskan, †-pasca.** [f. Cree *Athap-askaw*, lit. 'grass or reeds here and there' (Webster) + -AN.] **A.** *adj.* Of or pertaining to a widely spread people of North American Indians or their language.

[**1776** *Cumberland House Jrnl.* 27 June (Hudson Bay Rec. Soc., 1951) I. 60 he says that he supposed there were an hundred Canoes of them, the chiefest part A Thapuskow Indians.] **1846** J. SCOULER in *Edin. New Philos. Jrnl.* XLI. 171 An inspection of the vocabularies of the languages spoken on the north-west coast, will aid us in defining the limits of the Athabascan family. **1871** L. H. MORGAN *Ancient Soc.* i. i. 10 It leaves in the Upper States of Sewarty the Athapascan tribes of the Hudson's Bay Territory. **1915** R. H. LOWIE in *Amer. Anthrop.* XVII. 597 In the Athapascan..or..all the Athabaskan, or all the Southwestern systems. **1933** BLOOMFIELD *Language* iv. 72 The Athabascan family covers all but the coastal fringe of northwestern Canada [etc.].

B. *sb.* **1.** A member of this people.

1846 J. SCOULER in *Edin. New Philos. Jrnl.* XLI. 170 To the west of the Rocky Mountains, the Athabascans, under the names of Tacullies or Carriers, occupy the country called New Caledonia. **1851** R. G. LATHAM *Ethnol. Brit. Colonies* vi. 257 To separate, not only Caribs from Algonkins, or Peruvians from Athabascans, but Peruvians from Caribs [etc.]. **1871** L. H. MORGAN *Syst. of Consanguinity* v. 231 The Athapascans depend for subsistence upon fish and game. **1877** *Encycl. Brit.* VI. 448/1 The still more important myth of the north-west Athapascas. **1910** *Ibid.* 842/2 The Athapascan..covered all north-western Canada with his open and portable birch-bark canoe. **1938** R. H. LOWIE *Hist. Ethnol. Theory* vii. 121 The Canadian Athabaskans are introduced as 'vigorous, but poorly endowed'.

2. The language of this people.

1889 in *Cent. Dict.* (s.v. *Athabascan*). **1932** A. HUXLEY *Brave New World* vi. 120 Extinct languages, such as Zuñi and Spanish and Athapascan. **1933** *Publ. Mod. Lang. Assoc.* XLVIII. 620 In Athabascan..the idea 'to carry' is expressed by different verbs according to whether the load is light or heavy. **1965** *Canad. Jrnl. Linguistics* Spring 78 Languages of sure affiliation, e.g. Sarcee (Athapaskan).

athematic (æ-, ēⁱþimæ·tik), *a.* [f. A- 14 + THEMATIC *a.*] **1.** *Gram.* Of verb-forms: having suffixes attached immediately to the verb-stem without a connecting (thematic) vowel; also used of nouns (see quot. 1959). Hence used of languages which have such verb-forms.

1894 W. M. LINDSAY *Lat. Lang.* viii. 454 In Latin almost every athematic verb becomes thematic in 1 Sg. Pres. Ind. **1959** A. CAMPBELL *O.E. Gram.* xi. 255 Place-names of Celtic origin seem frequently to adopt the forms of the athematic nouns in OE, having gen. sg. in *-e*, otherwise no endings, e.g. a. and d.s. *Cent* Kent.

2. *Mus.* Lacking, or not composed of, deliberate themes.

1935 *Musical Times* Oct. 942/1 The three main forces of the universe..are each symbolised, not by themes—because the music [of Hábas] is athematic—but by the character of the thematic material and the orchestration. **1959** *Listener* 22 Oct. 704/2 The symphony is described by the composer [sc. Gerhard] as 'athematic' (the literal antonym of 'thematic'), from which it is obvious that each part, while indispensable to the whole, is meaningless when divorced from its context. **1960** *20th Cent.* Nov. 460 Webern..adopted Schönberg's serial technique to organize this athematic music.

Athenæum. Add: **2. a.** (Earlier and later U.S. examples.)

1807 *Monthly Anthol.* May 226 The Trustees with their associates are made a body corporate by the title of the Proprietors of the Boston Athenæum. **1847** HOWE *Ohio* 390 The atheneum was commenced as a library company. **1882** E. K. GODFREY *Nantucket* 13 The Athenaeum and other literary societies.

b. Add to def.: esp. the Athenæum Club in London. (Earlier and later examples.)

1799 (*title*) Laws and Regulations of the Athenæum in Liverpool. **1807** SOUTHEY *Lett. from England* II. xl. 224 The history of their [sc. Liverpool's] Athenæum is a striking instance of their spirit—by this name they call a public library, with a reading-room for the newspapers and other journals. **1834** H. C. ROBINSON *Diary* 14 Mar. (1967) 131 When the alarm of the cholera took place two years ago he gravely proposed turning the Athenæum into a cholera hospital. **1845** *Ainsworth's Mag.* VII. 424 It would almost seem that the Athenæum had a private quarrel with Mr. Pettigrew. **1887** W. S. GILBERT *Ruddigore* (1890) II. p. 36 Who found Athenæum and local museums. **1936** N. & Q. CLXX. 386/1 Thackeray..was more of the Athenæum type.

c. (Examples.)

1835 DICKENS *Let.* 21 Jan. (1965) I. 54, I can't get an Athenæum, and Literary Gazette—so not even a penny Magazine. **1848** H. C. ROBINSON *Diary* 5 Aug. (1967) 248 There is in the *Athenæum* a severe article as respects Talfourd's style.

Athenian (ăþī·niăn), *a.* and *sb.* Also **6 Attenian.** [ad. L. *Athēniensis* a. and sb. (whence OE. *Athēniense sb. pl.*, the Athenians), f. Gr. Ἀθῆναι Athens + -IAN; cf. Gr. Ἀθηναῖος Athenian.] **A.** *adj.* Of or pertaining to Athens, the leading city of ancient Greece, now the capital of Greece. Cf. ATTIC *a.* and *sb.*¹ Also *transf.* and *fig.*

1590 SPARKES *Myds. N.* I. i. 12 Stirre vp the Athenian youth to merriments. **1603** MILTON *Death of Fair Infant* st. 1, 5 in *Poems* 17 Aquilo his charioter By boistrous rape th' Athenian damsel got. **1791** BOSWELL *Johnson* I. 32 For an Athenian blockhead is the worst of all blockheads. **a 1822** SHELLEY *Homer's Hymn to Minerva* 2 in *Post. Wks.* (1907) 694 Athenian Pallas! tameless, chaste, and wise. **1840** E. BULWER *Let.* in C. H. Shattuck *Bulwer & Macready* (1958) 157 The Scotch of Macbeth & but hardly be fetched by our ears, or more learned than I am in that Athenian tongue. **1851** R. W. BROWNE *Hist. Class. Lit.* II. xiv. 211 The Greek orator, or rather,..the Athenian orator, for oratory flourished only in Athens, composed, his speech in private, before he delivered it. **1933** D. P. POCOCK tr. *Durkheim's Sociol. & Philos.* iii. 57 The Roman or the Athenian ideals were closely related to the particular organizations of these two cities.

B. *sb.* **a.** A native or inhabitant of Athens; also (see quot. 1598) the speech of this people.

[**c 893** ÆLFRED tr. *Orosius* 78/22 Sona swa Athenienses wisten..] **a 1400** SKELTON tr. *Diod. Siculus* I. 40 The Athenyensis contrine this oppysson.] **1598** TIDDALE *Adv.* (etc.) ii All the Athenians an straungers wyche were there gave them selves to nothing els but other to tell or to heare newe tydynges. **1530** *Cranmer's Bible Acts* xvii. 21 All the Athenians & straungers which were there. **1590** T. NICOLLS (*title*) The hystory writtone by Thucidides the Athenyan. **1590** SPARKES *Mids. N.* IV. ii. 32 If I tell you, I am no true Athenian. **1840** R. BRATHWAIT *Barnabees Jrnl.* sig. B5 To Oxford came I...Each thing ther's the Muses Minion, Queenes College-Hors speakes pure Athenian. **1849** JOHN TAYLOR *Let. sent to London* 13, I would have gone to a fellow that conjures with a paire of sheares and a Sieve, sure that grave Athenian would have told me all. **1788** GIBBON *Decl. & F.* VI. lxii. 255 The Athenians are still distinguished by the subtlety and acuteness of their understandings. **1877** L. H. MORGAN *Ancient Soc.* II. xiv. 349 Bachoven has collected and discussed the evidence of female authority..among the..Athenians.

athermic (ăþɔ̄·ɹmik), *a.* [ad. F. *athermique*, f. A- 14 + Gr. θέρμη heat + -IC.] **a.** That is not pervious to heat or heat-rays. Cf. ADIATHERMIC *a.* **b.** Without fever or rise of temperature (Dorland 1900). **c.** (See quot. 1909.)

1862 *Catal. Internat. Exhib.* II. XIII. 16 Melloni's apparatus for illustrating the athermic and diathermic character of minerals. **1909** WEBSTER, *Athermic*, heatless; as, an athermic motor.

atherosclerosis (æ:þēro,sklīᵊrōᵘ·sis). *Path.* [ad. G. *atherosklerose* (F. Marchand 1904, in *Verh. d. Kongr. f. Innere Med.* XXI. 58), f. Gr. ἀθήρη = ἀθάρη (see ATHEROMA) + -O + SCLEROSIS.] A form of arteriosclerosis with atheromatous degeneration of blood-vessel walls; = ATHEROMA b. Hence **a:therosclero·tic** *a.* and *sb.*

1910 *Jrnl. Amer. Med. Assoc.* LV. 546/1 (*title*) Atherosclerosis in Youth of 17 with Pulmonary Tuberculosis and Chronic Nephritis. [By] M. Simon. **1914** *Trans. Assoc. Amer. Physicians* XXIX. 513 Our knowledge of atherosclerotic processes in dogs is scant. **1958** *New Scientist* 25 Dec. 1582/3 Among the atherosclerotics, only three did not benefit. **1962** *Lancet* 28 Apr. 889/2 Most aneurysms of the abdominal aorta are probably a result of the atherosclerotic and degenerative changes in the media of the vessel wall. **1964** *Listener* 20 Feb. 311/2 Atherosclerosis..is known to be associated with certain biochemical abnormalities in the blood—raised levels of serum cholesterol and other components.

athetesis (æþiti·sis). [a. Gr. ἀθέτησις, f. ἀθετεῖν (see next).] In textual criticism: the rejection (of a passage) as spurious.

1887 J. S. REID in *Classical Rev.* June 135/2 He is judiciously conservative and takes no notice of such extravagant *atheteses* as those put forward by Lütjohann in a paper on the *Cato Maior*. **1888** LEAF *Iliad* II. 435 Aristarchos' athetesis of 614–7 has been generally accepted, but..on inadequate grounds. **1962** WACE & STUBBINGS *Compan. to Homer* 222 There are two grades of rejection: complete omission, and athetesis which denotes the part of suspicion which would lead a modern editor to consider the use of square brackets.

athetize (æ·þitaiz), *v.* [f. Gr. ἄθετος set aside + -IZE: formed to render Gr. ἀθετεῖν to set aside, reject as spurious.] *trans.* To set aside (a passage) as spurious. Hence **athetized** *ppl. a.*

1885 JEBB in *Jrnl. Hellenic Stud.* VI. 320 The solution is to athetize ll. 35–41 (of the *Iliad*). **1888** LEAF *Iliad* II. 435 The meaning is thoroughly Homeric, and the number of lines which have been suspected on undeniably valid grounds since the time of Aristarchos, who athetized no less than fourteen. **1889** *Athenæum* 26 Oct. 553/1 Dr. Leaf..controverts with success the athetizers and subjectors of Homeric verses. **1962** WACE & STUBBINGS *Compan. to Homer* 223 He used the *obelus* as Zenodotus had done, and he added other signs, especially the ἀντίσιγμα..to mark groups of athetized lines.

athetosis (æþitōᵘ·sis). *Path.* [f. Gr. ἄθετος without position or place + -OSIS.] A form of cerebral palsy, chiefly affecting children, characterized by continuous slow involuntary movements of the extremities. Hence **athetoid** (æ·þitoid), **athetotic** (æþitǫ·tik) *adjs.*, pertaining to or characteristic of athetosis.

1871 W. A. HAMMOND *Treat. Dis. Nervous Syst.* 654 Under the name of athetosis, I propose to describe an affection..which is mainly characterized by an inability to retain the fingers and toes in any position. **1875** *Proc. Amer. Neurol. Assoc.* I. 195 (*title*) A case of athetoid affection. **1882** *Edin. Med. Jrnl.* XXVII. 975 (*title*) Athetosis and athetoid movements in the insane. **1897** *Trans. Amer. Pediatric Soc.* IX. 158 In the analysis of 151 cases of cerebral paralysis in children, he found..athetosis six times. **1948** DAWSON WILLIAMS *Med. Dis. Inf.* 362 The general character of athetotic movements. **1950** *Practitioner* July 69 The right side of the face was then noticed to be paralysed.. **1951** ... [illegible] ... **1962** *Lancet* 26 May 1131/2 Although the athetosis ceased within fifteen minutes, respiration did not drop to 16 for ninety minutes.

athlete. Add: **4.** *athlete's foot*, a popular name for tinea or ringworm of the foot.

1928 *Lit. Digest* 22 Dec. 16/1 Athlete's foot..from which more than ten million persons in the United States are now suffering. **1942** *Lancet* 18 July 75/1 After what we have been told about the danger of phenol and its derivatives no-one ever here is likely to begin trials of a remedy for 'athlete's foot' (tinea interdigitalis), advocated in America, which consists of equal parts of phenol and camphor.

athletic, *a.* and *sb.* Add: **A.** *adj.* **3.** In Kretschmer's system, designating a type of physique characterized by well-developed muscles (see quots.); = *mesomorphic*.

1925 W. J. H. SPROTT tr. *Kretschmer's Physique & Character* ii. 14 The male athletic type is recognized by the strong development of the skeleton, the musculature and also the skin. **1937** R. H. THOULESS *Gen. & Social Psychol.* (ed. 2) vi. 105 Kretschmer distinguished three bodily types: the *asthenic* (large muscle and bones), the *athletic* or *leptosomatic* (lean, flat-chested, and narrow-shouldered), and the *pyknic* (with tendency to rounded contours of face and body).

athleticize (æþle·tisaiz), *v.* [f. ATHLETIC *a.* + -IZE.] *trans.* To make athletic.

1860 *Godey's Mag.* Apr. 447/1 Are we to be so athleticised that we will disdain all forgeries? **1897** *Eclectic Mag.* Oct. 523 France, superficially, has become Anglicised, athleticised.

Athoan (æþōᵘ·ăn), *a.* [f. Gr. Ἄθωος of Athos + -AN.] Of or pertaining to Athos, the most easterly of the three peninsulas of Chalcidice in Macedonia, or to the 'Holy Mountain' at the eastern end of this peninsula, or to the monasteries of Athos (cf. *ATHONITE).

1869 H. F. TOZER *Highl. Turkey* I. v. 119 The new iconostase..is composed partly of Tenian and partly of Athoan marble. **1939** A. TOYNBEE *Study Hist.* IV. 361 An artist whose art would appear to be the antithesis of the rigid canon of the Athoan iconists.

athodyd (æ·þodid). *Aeronaut.* [f. *Aero-* + *thermodynamic duct.*] A jet engine deriving its propulsive power from the combustion of fuel in a duct or tube into which air is admitted and compressed as a result of forward speed. Now usu. called *ram-jet.

1945 *Westinghouse Engineer* Mar. 51/1 The continuous-firing duct engine is relatively little known but is credited to Lorin of France in 1913. It has been under development principally in England under the..name of 'athodyd'. **1946** *Air Reserve Gaz.* July 7/1 The athodyd engines at the tail-plane tips are highly efficient at speed. **1950** *Jrnl. R. Aeronaut. Soc.* Apr. 217/1 The principle of the ram-jet or athodyd has been understood for a number of years.

Athole brose. [*Athole, Atholl*, a district in Perthshire.] See BROSE b and *Sc. Nat. Dict.*

at home. Add: **A.** *advb. phr.* **4.** *Cribbage.* See *HOME *sb.*¹

-athon: a combining form, barbarously extracted f. *MAR)ATHON, used occas. in the U.S. (*talkathon, walkathon*), rarely in Britain, to form words denoting something carried on for an abnormal length of time. (*Amer. Speech* (1934) IX. 76, 317/2.)

Athonite (æ·þonait), *a.* [f. L. **Athōnis* of Athos, f. *Athōs*, f. Gr. Ἄθως + -ITE¹.] Of or pertaining to the monasteries of Athos. Also as *sb.*

1887 J. A. RILEY *Athos* xi. 163 St. Athanasius the Athonite was a Georgian by nation, who came from Trebizonde to Mount Athos about the year 950, and founded the Lavra in 963 or 964. **1935** *Times Lit. Suppl.* 10 Oct. 621/3 Frescoes so characteristic of Athonite Churches. **1957** *Oxf. Dict. Chr. Ch.* 101/2 A curious rule of the Athonite monks forbids women, or even female animals, to set foot on the peninsula. **1963** *Sunday Times* (Colour Mag.) 24 Mar. 19 A..representative of the Athonite police force..scrutinized my letter of ecclesiastical recommendation.

athrepsia (æþre·psiă). *Path.* [ad. Fr. *athrepsie* (Parrot 1874, in *Progrès Méd.* 24 Oct. 637/1), f. A- 14 + Gr. θρέψ-ις nourishing + -IA¹.] Malnutrition, esp. in infants; marasmus.

1885 R. & F. BARNES *Syst. Obstetric Med.* II. iii. 122 *Athrepsia* is a frequent morbid state in the new-born. **1905** *Med. Ann.* 405 There would appear to be such a condition as 'athrepsia', or marasmus which is not dependent upon any discoverable source.

athrocyte (æ·þrosait). *Cytology.* [f. Gr. ἀθρόος crowded, collected + -CYTE.] A cell having the property of absorbing and retaining solid particles. Hence **athrocy·tic** *a.*; **a:throcyto·sis**, the capacity for such absorption.

1935 *Chem. Abstr.* 2631 The athrocytic function among the Hirudinea... Athrocytic cells (athrocytes) are found in renal or celomic organs, and serve to maintain the constancy of comps. of the interior fluids of the animal... Athrocytosis is the absorption by living cells of submicroscopic electronegative particles. **1948** *Lancet* 24 July 150/1 In blackwater fever..some engulfed the pigment and athrocytic apparatus. **1951** G. BOURNE *Cytol. & Cell Physiol.* (ed. 2) xi. 463 The proximal convoluted tubule cells have the capacity for storing vital dyes and other particulate matter (athrocytosis).

atishoo, atchoo (ăti·ʃū), *int.* and *sb.* A representation of the characteristic noises accompanying a sneeze.

[**1873** R. BROUGHTON *Nancy* iii. 41, I sneeze loudly and irrepressibly. Atcha! Atcha!] **1878** *Punch* 26 Jan. 30/4 A cough tears your lungs, but a sneeze tears you through.. A'ch—cooooh—h! A'—tshh—h!] **1878** *Hood's Comic Ann.* 54 I've got such a cold. Er-tchiou—er-tchiou! **1914** W. KEENE *Kaulcal Childs' Games* I. 112 Ezekiel sneezed. It was a convulsive 'atchoo'. **1920** *Punch* 10 Nov. 383 There, that's all right. A-a-a-tishoo! **1960** *Times* 8 Jan. 11/4 A sneeze may still mean the onset of a pestilential cold in the head and the thought prevents our taking a proper pleasure in our atishoo.

Atkins (æ·tkinz). Also **Mister Atkins.** [See THOMAS 3.] = *Tommy Atkins* (s.v. TOMMY 7).

1892 KIPLING *Barrack-r. Ballads* 6 Oh it's Tommy this, an' Tommy that, an' 'Tommy, go away'; But it's 'Thank you, Mister Atkins', when the band begins to play. **1893** *Daily News* 15 Nov. 4/3 Our citizen soldiers have been unduly agitated... Pathetic pictures have been drawn in the imagination of our sturdy civilian 'Atkinses' of wives left weeping. **1901** 'LINESMAN' *Words by Eyewitness* I. 3 Private Atkins..is from the masses himself. **1928** E. WEEKLEY *Eng. Lang.* vii. 47 *Padre*, father, adopted in the sense of priest by the natives of India and then acquired from the natives by Mr. Atkins.

Atlantic, *a.* Add: **1. c.** Of or pertaining to countries bordering on the Atlantic Ocean, in recent times esp. with reference to the political alliances of these countries.

1776 H. WALPOLE *Let.* 17 Apr. (1904) IX. 349, I now submit to recall my thoughts to America....The army that was to overrun the Atlantic continent, is not half set out yet. **1931** H. G. WELLS *Work, Wealth & Happiness* (1932) viii. 305 We of the Atlantic world are too disposed to be ungrateful to the vast experiments Communism has made. **1944** W. LIPPMANN *U.S. War Aims* 80 In addition to the United States, the United Kingdom, and France..the Atlantic Community includes..[the South American countries],..Australia, Belgium,..Canada, [etc.]. **1960** R. L. C. FITZGIBBON *Kissing had to Stop* xii. 220 She [sc. Britain] could become..the junior partner of an omnipotent Atlantic bloc. **1961** *Listener* 5 Oct. 493/1 Granted that Americans are not interested in Atlantic union, the emotional value to them of European union is enormous.

d. Crossing the Atlantic Ocean.

1839 *S. Lit. Messenger* V. 5/2 The packet owners have carried the Atlantic mail..for twenty years. **1858** *Harper's Mag.* Oct. 700/2 We had learned articles proving that the Atlantic cable could never succeed under the existing conditions. **1895** KIPLING *Devil & Deep Sea* in *Day's Work* (1898) 141 Her crew signed and signed again with the regularity of Atlantic liner boatswains.

4. Applied by Blytt to one of the successive periods of vegetation in Scandinavia after the glacial period, and later by others to the climate of other areas.

1876 [see *ARCTIC *a.* 3]. **1935** *Discovery* July 198/1 The following climatic phase of the Postglacial period, the 'Atlantic Period', was considerably damper than the Boreal. **1959** J. D. CLARK *Prehist. S. Afr.* vii. 169 The Makalian Wet Phase probably equates in time with the warm Atlantic stage in Northern Europe, between *c.* 5,500 B.C. and 2,500 B.C. **1960** B. W. SPARKS *Geomorphology* ix. 216 Godwin has distinguished the following phases in the development of the Fens...*The pre-Boreal period*..approximately 8300 to 7600 B.C. *The Boreal period*. From about 7600 to 5500 B.C. *The Atlantic period*. The period 5500 to 3000 B.C. was one of increasing dampness and extensive peat formation. **1964** G. MANLEY in WATSON & SISSONS *Brit. Isles: Syst. Geogr.* 162 This event [sc. the breaking of the last land bridge between Britain and the continent] is conveniently associated with the onset of the 'Full-Atlantic' conditions under which the mild winters that resulted from the broadening of the North and Irish Seas.

5. Phrases: *Atlantic Alliance* = **Atlantic Pact*; also the countries concerned in the Atlantic Pact; *Atlantic Charter*, a declaration of eight common principles in international relations, drawn up by the British Prime Minister, Winston Churchill, and President Franklin D. Roosevelt, on behalf of the British Empire and the U.S.A., at their meeting in the Western Atlantic in August 1941; *Atlantic Pact*, an agreement made in 1949 to provide for the defence of countries with seaboards on the North Atlantic (cf. *N.A.T.O.); *Atlantic States*, those of the United States situated on the Atlantic coast; *Atlantic Wall*, the line of fortifications constructed by the Germans to defend the Atlantic coast of Europe in the war of 1939–45.

1958 *New Statesman* 15 Feb. 185/3 Does the Atlantic alliance come first—or the principles it is pledged to uphold? **1941** *Hutchinson's Pict. Hist. War* 9 July–30 Sept. 150/1 The President of the United States and the British representative in what is aptly called the Atlantic Charter, have jointly pledged their countries to the final destruction of the Nazi tyranny. **1958** *New Statesman* 1 Feb. 130/3 The verbal flavour of the chimay, the ruthless persecution of the unfortunate—there have been in common with Buchenwald than with the Atlantic Charter. **1949** *Times* 19 Mar. 5/1 The Atlantic Pact is a much more than an American guarantee of Europe; it is a co-operative venture which will call for equal efforts. **1780** *Deb. Congress U.S.* (1834) I. 1155 [The] want of an uniformity in the Atlantic States for the purpose of encouraging their agriculture. **1833** J. P. KENNEDY *Swallow Barn* I. xiv. 133 Old Nick..is falling into the sere and yellow leaf, especially in the Atlantic states. **1961** KURATH & MCDAVID (*title*) The Pronunciation of English in the Atlantic States. **1944** *Hutchinson's Pict. Hist. War* 12 Apr.–26 Sept. 189/1 Their fast transports of Hitler's so-called Atlantic Wall were breached.

Atlantosaurus (ætlæ:ntosǭ·rǫs). *Palæont.* [mod.L., f. Gr. Ἀτλαν-, ἄτλας (see ATLAS *sb.*¹) + -O- + σαῦρος lizard.] An extinct genus of huge dinosaurian reptiles of the sub-order Sauropoda; also, a reptile of this genus.

1877 O. C. MARSH in *Amer. Jrnl. Sci.* XIV. 514 The gigantic *Titanosaurus*, Atlantosaurus, from the upper Jurassic.... It is the representative of a distinct family, which may be called *Atlantosauridae*. In the type genus, *Atlantosaurus*, one of the most important characters is the pneumaticity of the vertebræ. **1882** *Cliftonian* Nov. 189 We have seen what gigantic representatives of the race (the Atlantosaurus, and the rest) existed in those old times. **1896** J. FISKE *Excurs. Evolut.* i. 12 Professor Marsh has lately discovered the atlantosaurus of Colorado, nearly one hundred feet in length and thirty feet in height—the largest land animal as yet known. **1893** *Graphic* 11 Feb. 134/1 A view of the thigh-bone of the atlantosaurus.

atlatl (æ·t'læt'l). [Amer. Indian (Nahuatl) *atlatl* spear-thrower.] A throwing-stick used by American Indians and Eskimos.

1871 E. B. TYLOR *Prim. Cult.* I. ii. 60 The highest people known to have used the spear-thrower proper are the Aztecs. Its existence among them is vouched for..by its name 'atlatl'. **1910** F. W. HODGE *Handbk. Amer. Indians* II. 746/1 *Throwing stick*. This implement, called also throwing board, dart sling, and atlatl, is an apparatus for hurling a lance, spear, or harpoon at birds and aquatic animals. **1933** A. A. MORRIS *Digging in Southwest* iv. 45 The bow and arrow had not yet come into use, but instead game and enemies were reduced by long darts hurled with an *atlatl*.

||atman (ā·tmæ̆n). *Hindu Philos.* [Skr. *ātmán* essence, the highest personal principle of life; cf. OE. *ǣðm, ēðm* breath, ETHEM.] The self or soul; the supreme principle of life in the universe.

1785 C. WILKINS tr. *Bhagvat* xiii. 105 He who beholdeth all his actions performed by *Prăkrĕĕtĕĕ*, nature, at the same time perceiveth that the *Ātmā* or soul is inactive in them. **1859** J. R. BALLANTYNE *Christianity contr. w. Hindu Philos.* p. xxiii, The soul (*ātman*) is spoken of as an entirely different entity from the mind (*manas*). **1867** F. M. MÜLLER *Chips* I. iii. 70 For *ātman*, originally breath or spirit, comes to mean Self and Self alone. **1933** J. BAILLIE *Life Everl.* v. 139 The soul of man is *atman*, and *atman* is Brahma, the one supreme reality. **1945** A. HUXLEY *Time must have Stop* xxx. 275 'Devotion can be defined as the search for the reality of one's own Atman.' And the Atman, of course, is the spiritual principle in us, which is identical with the Absolute.

atmosphere, *sb.* Add: **4.** Prevailing psychological climate; pervading tone or mood; characteristic mental or moral environment; fascinating or beguiling associations or effects. (Examples in *Dict.* are of the type 'atmosphere of—'.)

1817 COLERIDGE *Biog. Lit.* I. iv. 84 The original gift of spreading the tone, the *atmosphere*, and with it the depth and height of the ideal world around forms, incidents. **1854** W. C. ROSCOE in *Prospective Rev.* X. 398 [Shakespeare] leaves his meaning to rest in great measure on the atmosphere that hangs about his language, rather than on its dictionary meaning and grammatical construction. **1869** M. ARNOLD in *Cornhill Mag.* Nov. 600 Bring in new changes for the time a man's spiritual atmosphere. **1884** [see EFFECT *sb.* 3 b]. **1885** 'VERNON LEE' *Euphorion* I. 27 Their intellectual atmosphere was as clear as our own. **a 1902** S. BUTLER *Way of all Flesh* (1903) xi. 27 Genial mental atmosphere. **1922** G. SANTAYANA *Soliloq. in England* ix. 30 What governs the Englishman is his inner atmosphere, the weather in his soul. **1923** H. G. BAYNES tr. *Jung's Psychol. Types* v. 230 The religion of the last two thousand years has..thereby, created an atmosphere which remains wholly uninfluenced by any intellectual disavowal. **1923** WODEHOUSE *Inimitable Jeeves* xi. 130, I never know, when I'm telling a story, whether to cut the thing down to bare facts or whether to shove in a lot of what they call atmosphere. **1934** L. SIEVEKING *Stuff of Radio* i. 19 Together the music of the orchestra, the aeroplane and sea sounds, and the dialogue of the three men, created genuine 'atmosphere', evoked emotion. **1948** F. J. R. SUTHERLAND *Pref. Eighteenth Cent. Poetry* i. 31 We tend to..overtake the intellectual atmosphere to a sort of compressed hemisphere.

b. *spec.* Applied to the background sounds that evoke a particular mood, impression, setting, etc., in a broadcast programme, etc. Also *attrib.*

1941 *B.B.C. Gloss. Broadc. Terms* 4 *Atmosphere*, sounds forming the background incidental to an event such as a race meeting, procession, etc. Hence *atmosphere microphone*, microphone specially placed to pick up such sounds. **1958** K. REISZ *Technique Film Editing* (ed. 9) ii. 186 The sound editor can do little more than choose a piece of accompanying atmosphere music from his library. **1962** A. NISBETT *Technique Sound Studio* ii. 41 A record of court-room 'atmosphere' completed the picture by providing an occasional cough or shuffling noise. *Ibid.* viii. 141 At this point, in order to maintain the fullest continuity, we can use a crossfade which keeps the atmosphere running throughout the pause.

a·tmosphered, *ppl. a.* [f. ATMOSPHERE *v.*] Having or provided with atmosphere (sense *4).

1920 *Times Lit. Suppl.* 12 Feb. 103/2 The scenes..are no less sharp and clear and 'atmosphered'. **1962** *Listener* 1 Feb. 224/3 The amorphous and exquisitely atmosphered landscapes.

atmospheric, *a.* Add: **2.** (Earlier and later examples.)

1789 A. BENNET *New Exper. Electr.* 103 It may be proper to mention such theory of atmospheric electricity as appears to me consonant to the general operations of nature. **1934** *Discovery* Mar. 68/2 A report on fog and atmospheric pollution. **1947** *Archit. Rev.* CI. 149/1 Heavy atmospheric pollution occurs in most of the larger towns in this country, with consequent detriment to health and environment.

4. Evoking or designed to evoke an atmosphere (sense *4).

1908 [see *ATTRACT *v.*] **1927** *Melody Maker* Sept. 942 (Advt.), Modern atmospheric music by the World's most eminent writers. **1946** *Ann. Reg.* 1945 343 An atmospheric story, 'The Seventh Veil', broke all box-office records. **1963** *Punch* 16 Feb. 261/1 Visually beautiful, atmospheric little tragedy that could have been spun in Chekhov.

atmospherics (ætmǫsfe·riks), *sb. pl.* [pl. of ATMOSPHERIC *a.*, after *acoustics*, etc.: see -IC 2.] Atmospheric disturbances of electrical origin causing interference with communication in wireless telegraphy, television, etc. Also occas. in *sing.*

1905 A. Y. FORREST in J. Erskine-Murray *Handbk. Wireless Telegr.* (1907) x. 181 Upon changing over to the Fessenden interference preventer, the terrific atmospheric was cut down to such an extent that I was enabled to receive the report. **1913** *Year-bk. Wireless Telegr.* 334 No signal nor atmospheric can put it out of action. **1915** *Wireless World* Apr. 55/1 To protect the condenser from strong 'atmospherics', a spark gap should be connected to the two plates of the condenser. **1919** R. STANLEY *Text-bk. Wireless Telegr.* I. 173 Irregular noises are produced in the receiver telephones which seriously interfere with the reception of the regular signals; they are due to 'Atmospherics', or 'strays', or Xs. **1934** *Discovery* Mar. 73/2 The Air Ministry chose this wavelength because it is only in this region that uninterrupted communication, free from all possibility of interference and atmospherics can be depended upon at all times. **1947** CROWTHER & WHIDDINGTON *Sci. at War* i. 15 The atmospheric is a very brief phenomenon. **1947** *Sci. News* IV. 58 It is an accepted fact that 'atmospherics' are produced also by a radiation which comes to us direct from the centre of the Galaxy.

atocha (ătō·tʃă). *U.S.* [Sp.] Esparto. Also *attrib.*, as *atocha grass, plant.*

1869 *Rep. U.S. Commissioner Agric. 1868* 262 There are two classes of this plant, the 'atocha', properly so called, and the coarse or 'bastard' atocha... The atocha grass, which is called esparto, is not cut like ordinary grass, but is pulled up from its socket. *Ibid.* 264 It is at about this elevation [sc. 3,500 feet] where the snow usually commences, that the atocha plant ceases to grow.

atoke (æ·tōᵘk). [G. (E. Ehlers *Die Borstenwürmer* (1868) II. 453), ad. Gr. ἄτοκος, f. ἀ- priv. (A- 14) + τόκος birth.] The sexless part of certain polychætous worms. Hence **atokal**

(æ·tŏkăl), **atokous** (æ·tŏkəs) *adjs.*, non-sexual; producing only asexual progeny.

1896 *Cambr. Nat. Hist.* II. 277 Ehlers employed the term 'epitokous', whilst he called the 'Nereid' phase 'atokous', under the impression that the worm did not become mature in this condition. **1903** J. S. Kingsley tr. *Hertwig's Zool.* 311 The atoke..forms chains of dimorphic individuals which later separate. **1904** *Jrnl. R. Microsc. Soc.* Apr. 183 The total length [of the marine worm palolo] averages 400 mm., about one-fourth of which is in the anterior atokal part. **1916** [see *EPITOKOUS a.].

atole (ătōᵘ·lɪ). *U.S.* [Amer.-Sp., f. Nahuatl *atolli*.] A kind of corn or other meal; gruel or porridge made of this.

1716 in *S.W. Hist. Q.* XXXI. 56 They make large pots in which to keep water, make *atole*, and to preserve other things. **1798** tr. *Peyrouse's Voyage* 86 A breakfast of barley meal..It is boiled in water; the Indians give this food the name of *atole*. **1844** J. Gregg *Commerce of Prairies* I. 153 A sort of thin mush, called *atole*, made of Indian meal, is another article of diet. **1946** R. Peattie *Pacific Coast Ranges* 41 All partook of a morning meal of atole.

atom, *sb.* Add: **2.** and **3.** Now generally held to consist of a positively charged nucleus, in which is concentrated most of the mass of the atom, and round which orbit negatively charged electrons.

5. b. *logical atom*: one of the essential and indivisible elements into which some philosophers hold that statements can be analysed.

1873 C. S. Peirce in *Mem. Amer. Acad. Arts & Sci.* IX. II. 343 The logical atom, or term not capable of logical division, must be one of which every predicate may be universally affirmed or denied... A logical atom, then, like a point in space, would involve for its precise determination an endless process. **1918** [see *ATOMISM 1 b]. **1958** G. J. Warnock *Eng. Philos. since 1900* v. 54 Russell's world of indefinitely numerous, independent logical atoms is the metaphysical opposite of Bradley's Absolute.

c. *to smash to atoms* (examples). Cf. also *BLOW v.¹ 24 a.

1851 [see SMASH v.¹ 2]. **1905** A. Conan Doyle *Return of Sherlock H.* 226 His second bust..had been smashed to atoms where it stood.

IV. *attrib.* and *Comb.* **a.** Loosely used *attrib.* for *atomic*, as *atom age*, *bomb* (so *atom-bomb* v., *-bombing* vbl. sb.), *scientist*, *spy*. **b.** Also, used with the sense of *atomic energy*, as *atom-driven*, *-powered*, and of *atomic weapons*, as *atom-free*. **c.** Objective, as *atom-splitting*. **d.** Similative, as *atom-like*. Also **atom-smasher** *colloq.*, an apparatus for accelerating charged particles; so *atom-smashing* vbl. sb.

1946 *Daily Tel.* 24 Sept. 3/7 A pre-vision of interplanetary travel in the atom age. **1945** *Times* 7 Aug. 4/1 An impenetrable cloud of dust and smoke had covered the target area after the atom bomb had been dropped at Hiroshima. **1945** *Evening News* 13 Aug. 1 (*headline*) Ultimatum to Japs: Accept by 6 p.m.—or be atom-bombed. **1946** *Britannica Bk. of Year* 832/1 Atom bombing. **1948** *John o' London's Weekly* 15 Oct. 491/2 The story concerns a group of survivors in the ruins of atom-bombed Los Angeles. **1953** C. Day Lewis *Ital. Visit* ii. 30 Gleefully brush past atom-bomb cauliflowers [sc. cloud-shapes]. **1950** *Jane's Fighting Ships 1950–51* Add. 5/2 Plans for an atom-driven submarine..are reported. **1958** *Listener* 18 Dec. 1026/2 The creation of an atom-free zone in central Europe. **1890** W. James *Princ. Psychol.* I. x. 348 The enjoyment of the atom-like simplicity of their substance *in sæcula sæculorum* would not to most people seem a consummation devoutly to be wished. **1953** *Encounter* Oct. 59/2 An American atom physicist. **1949** *Air Trails* 21 May 21 Atom-powered bombers! How soon? How big? What shape? **1953** *Ann. Reg. 1952* 403 On 14 January the first official announcement was made in the United States about the production of an atom-powered aircraft carrier. **1945** *Sci. News Let.* 17 Nov. 312/3 Atom scientists federate to help Congress. **1937** *Lit. Dig.* 12 June 17/3 The Westinghouse atom-smasher will look like a gigantic aluminum pear. **1932** *Ibid.* 28 May 26/1 Atom-smashing has been accomplished before, and it seems to be still only a laboratory wonder. **1942** *Electronic Engin.* Sept. 166/1 Several types of accelerator have been developed for providing high velocity particles with which to bombard the nuclei of atoms—a practice popularly known as 'atom smashing'. **1960** *New Statesman* 2 Jan. 8/1 With the aid of their big atom-smashing machines, other Americans manufactured fragments of anti-matter. **1939** *Discovery* Apr. 181 (*caption*) A New Type of Atom-Splitting. **1946** *War Illustr.* 4 Jan. 572 (*caption*) Japanese Atom-Splitting Device. **1959** *Listener* 19 Mar. 514/2 The ideological atom spies.

atomic, *a.* and *sb.* Add:

A. *adj.* **1.** *atomic heat*: see HEAT *sb.* 2 e; *atomic mass* (= *atomic weight*): the mass of an atom, usually expressed in *atomic mass units* (in one scale, a unit equal to $\frac{1}{16}$ of the mass of an oxygen atom, or *spec.* of the oxygen isotope ^{16}O; since 1960, when it was formally adopted by the International Union of Pure and Applied Physics, equal to $\frac{1}{12}$ of the mass of the principal isotope of carbon, ^{12}C), abbrev. *a.m.u.*, *amu*; *atomic number*: (of a chemical element) the number of unit positive charges present in the nucleus of its atom, being the physical property which determines the position of the element in the periodic table; symbol Z; *atomic structure* (see quots.); *atomic weight*: (after the adoption of oxygen as a standard) the ratio between the weight of one atom of the element and $\frac{1}{16}$ of the weight of an atom of oxygen; (also add earlier and later examples, and see WEIGHT *sb.*¹ 10 d, and *atomic mass, above).

1898 S. W. Holman *Matter, Energy, Force & Work* 153 The terms *atomic masses*, *combining masses*, etc., are likewise used instead of 'atomic weights', [etc.]. **1946** *Electronic Engin.* May 153/1 This discovery accounted for the large discrepancy between the actual atomic mass of neon (20.2) and the..view that such masses should be whole numbers. **1955** *Gloss. Terms Radiology* (B.S.I.) 18 Atomic mass unit..Abbreviation: amu. **1958** Mansfield *Elem. Nucl. Physics* i. 10 The atomic mass M, not to be confused with the molecular weight, is the mass of an isotope in atomic mass units. Numerically the atomic mass and mass number differ by less than 1 part in 100. In fact, the mass number is obviously the atomic mass rounded off to an integer. **1962** *Nature* 19 May 621/1 Some confusion of meaning..could be avoided if 'atomic mass' were reserved for the species of matter now known as nuclides and 'atomic weight' were used only in its traditional connotation, that is, for elements. **1963** Jerrard & McNeill *Dict. Sci. Units* 20 The masses of atoms and molecules are generally given in atomic mass units. These units are based on a scale in which the mass of the carbon isotope C_6^{12} is taken to be 12. **1821** T. Thomson in *Ann. Phil.* n.s. I. 5 The following table exhibits the numbers assigned to the atom of these bodies. ..They approach very nearly to the other atomic numbers contained in the table. **1913** H. G. J. Moseley in *Phil. Mag.* 6th Ser. XXVI. 1028 Table I. N atomic number. *Ibid.* 1031 We are therefore led by experiment to the view that N is the same as the number of the place occupied by the element in the periodic system. This atomic number is then for H 1. **1958** Mansfield *Elem. Nucl. Physics* i. 4 The *atomic number* Z, is the number of protons in the nucleus and, since the atom as a whole is electrically neutral, this also equals the number of the electrons. **1897** W. F. Magie tr. C. *Christiansen's Elem. Theor. Physics* i. 48 (*heading*) On the Molecular and Atomic Structure of Bodies. **1938** R. W. Lawson tr. *Hevesy & Paneth's Man. Radioactivity* (ed. 2) viii. 79 The theory of atomic structure has cleared up many points on the origin and nature of the individual types of rays. **1820** T. Thomson in *Ann. Phil.* XVI. 329 The greater number of chemical writers in this country have adopted the atomic weights assigned by Wollaston. **1821** *Ibid.* n.s. I. 3 We found reason to conclude that the atomic weight of every body is a multiple of the weight of an atom of hydrogen. **1827** A. Fyfe *Elem. Chem.* II. 493 The atomic weights are given both according to the oxygen and hydrogen scale. **1942** J. D. Stranathan *Particles* v. 169 Results obtained with the new mass spectrograph..made it certain that atomic weights are in general not quite whole numbers; but they are very close to whole numbers. *Ibid.* 182 It is rather fortunate that 99·76 % of all O is of atomic weight 16. **1962** *Nature* 19 May 621/2 The chemist..needs to know the average mass of the mixture of nuclidic species which constitute the element. This is the quantity that has been called the atomic weight of the element.

2. a. *Philosophy*. In modern philosophy: unanalysable, irreducible, ultimate, essential; also, of a sentence: without conjunctions or other connective words.

1912 L. Wittgenstein *Let.* (to Russell) in *Notebks. 1914–16* (1961) 120, I believe that our problems can be traced down to the *atomic* propositions. **1918** B. Russell in *Monist* 523 An atomic proposition is one which does mention actual particulars, not merely describe them but actually name them. **1922** tr. *Wittgenstein's Tract. Log.-Phil.* 31 An atomic fact is a combination of objects (entities, things). **1929** Wittgenstein in *Knowledge, Exper. & Realism* (Aristotelian Soc. Suppl., Vol. IX) 163 The propositions which represent this ultimate connexion of terms I call, after B. Russell, atomic propositions. **1933** *Mind* XLII. 38 Similar to the species of geometry, we might have in logic 'a logic of atomic propositions' and 'a logic of molecular propositions'. **1948** B. Russell *Human Knowl.* ii. ix. 145 We give the name 'atomic sentence' to one not containing logical words. **1956** G. Ryle in Ayer et al. *Revol. Philos.* 10 The analysis of compound propositions into their simple elements, the conjunctionless or 'atomic' propositions.

β. Physics and extended uses. (In 2 c, d, e the early quots. refer to the theory of atomic structure and to various theoretical applications.) Cf. *ATOM IV.

c. Applied to research and researchers in atomic energy, structure, etc. (e.g. *atomic physics, physicist, scientist*); to apparatus (e.g. *atomic furnace, pile* (see *PILE *sb.*³), *reactor*) used to disintegrate atomic nuclei; and to ships, industrial plant, etc., deriving power or driving-force from the harnessing of atomic energy (*atomic-powered* ppl. a.).

1882 J. Tyndall in *Knowledge* II. 371/1 (*title*) A problem in atomic physics. **1914** H. G. Wells in *Century Mag.* LXXXVII. 343/2 The new atomic aëroplane became indeed a mania. *Ibid.* 571/1 The swift aëroplane, with its atomic engine as noiseless as a dancing sunbeam. **1925** F. J. Reynolds *Marvels of 1924* 24 The specialty of Dr. Robert A. Millikan..is atomic physics and he has done as much as any other American in this branch. **1933** *Discovery* May 154/2 The quantum of action..taken together with the existence of the elementary particles, forms the foundation of atomic physics. **1945** W. Davis (*title*) Atomic Power Plants of the Future. **1945** *War Illustr.* 9 Nov. 439/1 (*caption*) An atomic-powered locomotive;..an atomic power-house;..a streamlined atomic-powered liner. **1946** *Sci. News Let.* 23 Mar. 187 Atomic scientist warns against misinterpretation. **1949** *Jrnl. R. Aeronaut. Soc.* Oct. 1013 A British atomic physicist. **1949** *Atomics* Aug. 3 This work is particularly aimed at building an atomic power plant that will..provide steam. **1950** *Amer. Speech* XXV. 24 The surprising thing about *pile* and *reactor* is that only a few alternatives have ever been used for them: *nuclear furnace, atomic furnace, atomic-energy machine*. **1950** *Sat. Even. Post* 22 July 26 (*headline*) We're betting our shirts on the atomic submarine. **1952** *Jane's Fighting Ships 1952–53* p. vii/1 The second atomic submarine will be of the same general design as the nuclear-powered *Nautilus*. *Ibid.*, Her [sc. the submarine's] power plant will use a different kind of atomic reactor. **1953** *Ann. Reg. 1952* 403 In June President Truman was present at the laying of the keel of the first atomic-powered submarine *Nautilus*. **1955** *Times* 4 July 5/1 The heat from the Atomic Energy Authority's six new reactors (atomic furnaces). **1958** *Listener* 11 Dec. 992/1 Fourth atomic reactor at Calder Hall comes into operation. **1959** *Times* 16 Sept. 10/2 The Soviet atomic icebreaker *Lenin* left Leningrad to-day on her maiden voyage into the Baltic. **1962** *Ibid.* 4 Apr. 15/1 There is every promise that the Sizewell nuclear power station..will be the first atomic installation to produce electricity as cheaply as the most modern conventional power station.

d. *atomic energy*: the energy released by the fission of the atomic nuclei of certain heavy elements such as uranium 235 or plutonium or by the fusion of light nuclei; also *attrib.*

1906 *Nature* 9 Aug. 357/2 Nevertheless, there is a sense in which it may be said that we are profiting by atomic energy. **1914** H. G. Wells *World set Free* i. 40 Holsten..was destined to see atomic energy dominating every other source of power. **1921** *Flight* XIII. 299/2 This is indeed the beginning of the liberation of atomic energy. **1924** *Sci. Amer.* Aug. 120/1 Atomic energy is the phrase of the hour. **1946** A. Boyd *U.N. Handbk.* v. 72 The member states had all nominated their representatives to the Atomic Energy Commission by May 1946. **1953** *Economist* 14 Nov. 505/1 The White Paper sets out to describe the kind of organisation to which the Government has decided to entrust the development of atomic energy.

e. Of weapons: deriving their destructive power from the partial conversion of such energy, as *atomic bomb* (hence as vb., to attack with this type of bomb, *-bombing* vbl. sb.; *atomic bomber*, an aircraft designed to carry atomic bombs), *device*, *shell*, etc. Cf. *HYDROGEN *bomb*. Also, of, pertaining to, possessing, or employing atomic weapons, as *atomic club*, *control*, *warfare*.

1914 H. G. Wells *World set Free* ii. 96 The three atomic bombs, the new bombs that would continue to explode indefinitely. **1917** S. Strunsky in *Yale Rev.* Jan. 295 When you can drop just one atomic bomb and wipe out Paris or Berlin, war will have become monstrous and impossible. **1925** *Punch* 11 Feb. 152/2 When, like the bursting of atomic bombs, Cats call to cats and Toms miaul to Toms. **1932** H. Nicolson *Public Faces* xii. 325 We must now assume that a single atomic bomb is capable of destroying all matter within a circumference of seventy to eighty miles from the point of explosion. **1944** G. B. Shaw *Everyb. Polit. What's What* xxxii. 286 Neither Vril nor Prospero's magic nor flying islands have ever existed, nor has the atomic bomb yet been invented. **1945** *Times* 7 Aug. 4/1 President Truman announced..yesterday that the first atomic bomb had been dropped 16 hours before by an American aircraft on Hiroshima. **1945** *Reader's Dig.* Dec. 2/2 The atomic-bombers of nation A may be on their way to bomb nation B while their own homeland is being turned into a crematorium. **1945** *War Illustr.* 28 Sept. 345 (*title*) I Visited the First City to be Atomic Bombed. *Ibid.* 9 Nov. 438/1 The results of the atomic bombing of Hiroshima and Nagasaki. **1946** *Ibid.* 4 Jan. 572/1 The manufacture and storage of atomic weapons. *Ibid.* 18 Jan. 604/3 Our Royal Navy is rushing plans to revolutionize the fleet to resist atomic warfare. **1946** *Jane's Fighting Ships 1944–45* p. iv/2 The possibility of ships mounting guns that fire atomic shells, to say nothing of the potentialities of the atomic torpedo. **1948** *Ann. Reg. 1947* 250 No progress was made in the discussions on the problem of atomic control. **1954** *Ann. Reg. 1953* 376 The main spring test was made on 17 March with an 'atomic device' having about three-quarters of the power of the bombs dropped on Hiroshima and Nagasaki, but containing less fissionable material. **1955** *Bull. Atomic Sci.* Mar. 79/2 Atomic weapons do not belong to a dream world. **1955** Koestler *Trail of Dinosaur* 238 The only deterrent against atomic aggression is an atomic stockpile. **1957** *Times Lit. Suppl.* 1 Nov. 654/4 Can atomic attack be circumscribed? **1958** *Times* 1 July 10/4 The French Ministers, as had been expected, last night emphasized their wish to see France become a member of the 'atomic club'. **1958** *Economist* 25 Oct. 297/2 If the Chinese had missiles with atomic warheads opposite Formosa now.

f. In journalistic and colloquial use: of the age, era, etc., marked by the various applications of atomic energy.

1945 *War Illustr.* 9 Nov. 439 Wonders of the Atomic Age in the not-too-distant future. **1946** A. Boyd *U.N. Handbk.* i. 25 It is already clear that the arrival of the 'atomic age' has brought in its train the possibility of outbursts of destruction on an incomparably greater scale [than in the war of 1939–45]. **1950** *Amer. Speech* XXV. 27 The 'Atomic Era' will depend, both as a new term and a new thought, upon the use the generality of us give it. **1959** *New Statesman* 19 Dec. 874/2 What are you to do about all the new long atomic-age words?

g. Forming phraseological *combs.* with *sbs.*, as (senses 1, 2) **atomic beam** (see quot. 1962); **atomic clock**, an instrument which attains extreme accuracy by measuring time in terms of the vibration-rate of molecules of ammonia or cæsium atoms; **atomic hydrogen**, hydrogen dissociated into atoms; so **atomic-hydrogen (arc) welding**, a form of arc-welding in which hydrogen is dissociated into atoms

by passing through an electric arc and then recombines, thus supplying intense heat; hence *atomic-hydrogen torch*, etc.; (senses *2 c, d, e) **atomic power**.

1928 *Physical Rev.* XXXI. 646 The increase of the divergence of the atomic beam upon reflection at the crystal surface may obviously be due to any of several causes. **1962** *Gloss. Terms Nucl. Sci.* (B.S.I.) 10 *Atomic beam*, gas atoms emerging from a small aperture into a high vacuum and collimated by one or more additional apertures so as to form a narrow beam. **1938** *Jrnl. Optical Soc. Amer.* July 215 (*title*) Experimental Study of the Rate of a Moving Atomic Clock. **1958** *Times* 14 Oct. 5/3 The caesium atomic clock, which is on exhibition at the British clock and watch industry's display in the Goldsmiths' Hall, London. **1962** E. Bruton *Dict. Clocks & Watches* 15 The most accurate atomic clock or frequency standard ever made employs as its 'pendulum' the vibration of the caesium atom, which is at 9,192,631,770 cycles a second. **1915** I. Langmuir in *Jrnl. Amer. Chem. Soc.* XXXVII. 428 (*heading*) Diffusion of Atomic Hydrogen away from the Wire. **1930** H. Chatley *Princ. Rocket Propulsion* 5 Esnault-Pelterie..discusses the possibility of using atomic hydrogen as the propellant but little is known of this highly active material. **1933** *Welding Industry* June 141 Atomic-Hydrogen arc welding, which originated in America..is rapidly finding favour. **1914** H. G. Wells in *Century Mag.* LXXXVII. 704/2 The year of crisis that followed the release of atomic power. **1945** *Daily Tel.* 7 Aug. 1/1 Congress would be asked to investigate how atomic power might be used to maintain the future peace. **1953** *Economist* 14 Nov. 507/2 The argument is also gaining ground that a shortage of coal will compel Britain to turn to atomic power for marginal supplies of electricity, regardless of cost.

atomically, *adv.* (Later examples in other senses of the adj.)

1888 *Athenæum* 18 Aug. 214/2 Ourselves atomically constituted in body, soul..and spirit..we are in constant rapport with external influences of similar kinds. **1955** *Bull. Atomic Sci.* May 191/3 The example..cited, amphibious landings in the face of an atomically armed foe, presents no greater difficulties than defense of a shore line against an atomically armed invader.

atomicity. Add: **1. b.** The number of atoms in the molecule of an element (Webster 1900).

2. In modern philosophy: capacity for being reduced to or analysed into atomic propositions or other elements; cf. *ATOMIC A. 2 a, *ATOMISM 1 b.

1929 A. N. Whitehead *Process & Reality* iii. i. 333 A prehension, considered genetically, can never free itself from the incurable atomicity of the actual entity to which it belongs. **1940** B. Russell *Inq. Meaning & Truth* xii. 169 Can we construct an adequate language in which the principle of atomicity holds? **1959** —— *My Philos. Devel.* x. 118 The principle of atomicity is stated by Wittgenstein in the following terms: 'Every statement about complexes can be analysed into a statement about their constituent parts, and into those propositions which completely describe the complexes' (*Tractatus*, 2.0201).

atomism. Add: **1. b.** In modern philosophy: the theory that all statements, propositions, situations, etc., are composed of mutually independent, simple, primary, and irreducible elements; the elucidation and study of these elements; spec. *logical atomism*.

1914 B. Russell *Sci. Method in Philos.* 16 The philosophy which I wish to advocate may be called logical atomism or absolute pluralism, because while maintaining that there are many things, it denies that there is a whole composed of those things. **1918** —— in *Monist* 497 The reason that I call my doctrine *logical atomism* is because the atoms that I wish to arrive at as the..last residue in analysis are logical atoms and not physical atoms. **1956** D. F. Pears in Ayer et al. *Revol. Philos.* 49 You see how the theory of logical atomism develops. You begin with statements, subject them to analysis and find that they are built up out of parts.

3. *Psychol.* (See quot. 1934.)

1883 F. H. Bradley *Logic* II. ii. i. 276 The philosophy of Experience is psychological Atomism. **1934** H. C. Warren *Dict. Psychol.* 23/1 *Psychological atomism*, the theory of mind which assumes that experiences are composed of elementary psychic units or atoms.

atomist. Add: **1. b.** An adherent or student of logical atomism. Cf. *ATOMISM 1 b.

1941 *Mind* L. 166 They may themselves choose to be called Logical Atomists or Logical Positivists, or may repudiate all such titles. **1956** P. F. Strawson in Ayer et al. *Revol. Philos.* 97 Atomists and Positivists alike accepted the skeleton language of the new mathematical logic.

atomistic, *a.* Add: **1. b.** Of or pertaining to logical atomism. See *ATOMISM 1 b.

1918 B. Russell in *Monist* 496 The logic which I shall advocate is atomistic, as opposed to the monistic logic of the people who..follow Hegel. When I say that my logic is atomistic, I mean that I share the common-sense belief that there are many separate things.

3. Of or pertaining to psychological atomism. See *ATOMISM 3.

1883 *Encycl. Brit.* XX. 60/2 Such a statement is liable to all the objections already urged against what we may call atomistic psychology. *a*1910 W. James *Some Probl. Philos.* (1911) iv. 87 To such an atomistic plurality the associationists reduce our mental life. **1940** R. S. Woodworth *Psychol.* (ed. 12) ii. 38 An atomistic psychology attempts to explain any total activity by analysing it into its elements.

atomistically, *adv.* (Later examples, *spec.* in *Psychol.*)

1924 A. S. Pringle-Pattison *Locke's Human Und.* p. xx, This start with sensations, atomistically conceived as simple or particular ideas..is a presupposition of the older psychology, common to Locke, Berkeley, and Hume. **1935** K. Koffka *Princ. Gestalt Psychol.* vii. 271 Atomistically considered, there is no disparate point on the right side.

atomization. Add: (Earlier example.)

1866 H. Beigel *On Inhalation* v. 49 The minuteness of atomization.

2. *transf.* The process of reducing or dividing into small units; fragmentation.

1935 *Archit. Rev.* LXXVIII. 86/2 The characteristically capitalist atomization of society into isolated economic units. **1953** A. K. C. Ottaway *Educ. & Soc.* v. 98 The break-down or 'atomization' of the process into one operation per worker leads to a lack of interest. **1954** Koestler *Invis. Writing* xxii. 244 There are various answers to this—atomisation, dispersion, the sorry end of all independent Communist splinter-groups.

atomize, *v.* Add: **2. b.** To reduce (a liquid) to very small particles or to the condition of spray.

1865 *Lancet* 25 Feb. 203/2 He believed that atomising the fluids was unnecessary, on the ground that they could be introduced in the simplest manner by ordinary inhalation. **1902** *Encycl. Brit.* XXXI. 646/2 Utilizing the waste product [of petroleum] as fuel by spraying or atomizing it with steam.

3. To damage or destroy with an atomic weapon. *colloq.*

1945 *Daily Mirror* 11 Aug. 1/3 Nagasaki, the second..Japanese city to be 'atomised'. **1961** A. Wilson *Old Men at Zoo* ii. 90 Poor old Rackham's looking quite down in the mouth now that we aren't all going to be atomized into eternity.

atomized, *ppl. a.* Restrict †*Obs. rare*⁻¹ to sense in Dict. and add: **2.** Reduced to a fine spray; also, (of a solid) reduced to minute particles.

1865 Morell Mackenzie in *Lancet* 25 Feb. 202/2 (*heading*) On the treatment of chronic disease of the lungs by the inhalation of atomised liquids. **1870** T. Holmes *Syst. Surg.* (ed. 2) IV. 530 The inhalation of 'atomised fluids', or spray. **1926** *Punch* 26 May p. ix (Advt.), An electro-vaporiser secures a perfectly atomised gas for starting. **1958** *Gloss. Terms Powders* (B.S.I.) 5 *Atomized powder*, (a) powder produced by the dispersion of molten metal or other material by spraying under conditions such that the material breaks down to powder, (b) powder produced by disintegration of a material through internal rupture caused by rapid change of external pressure.

3. Damaged or destroyed by an atomic weapon. *colloq.*

1950 G. Greene *Third Man* x. 95 Tainted, like the soil of an atomised town.

4. *transf.* Disunited, dissociated; having lost social unity.

1942 J. S. Huxley in *Polit. Q.* XIII. 387 A dingy, discontented, and atomized black proletariat. **1945** Koestler *Yogi & Commissar* iii. iii. 215 Europe east of the Alps is atomized and in a state of political, economical, ideological chaos. **1949** I. Deutscher *Stalin* iv. 98 But the atomized mass of Moslem labourers did not lend itself easily to propaganda or organization.

atomizer. (Earlier and later examples.)

1865 Morell Mackenzie in *Lancet* 25 Feb. 203/1 The author observed that his own atomiser is very simple and can be used very easily. The liquid is driven from a fine glass pipe on to a projection in a bell-shaped tube, by the descent of a piston. **1904** [see next]. **1930** *Engineering* 3 Jan. 3/1 The quantity of fuel supplied to each atomiser is regulated on the by-pass principle.

atomizing, *ppl. a.* Add: **2.** Reducing to a fine spray.

1867 B. Durham in *Chicago Med. Jrnl.* XXIV. 245 (*title*) Atomizing instruments. **1894** *Jrnl. Amer. Med. Assoc.* XXIII. 973 Topical treatment of the air passages, with exhibition of a new atomizing vaporizer. **1904** Goodchild & Tweney *Technol. & Sci. Dict.* i. 33/1 Atomiser or atomising carburetter. **1951** *Good Housek. Home Encycl.* 204/1 An atomising jet is needed, and the paraffin oil is burnt as a vapour.

Aton, var. *ATEN.

atonal (ătōᵘ·năl, æ-, ēⁱ-), *a.* *Mus.* [f. A- 14 + TONAL *a.*] Applied to a style of composition in which there is no conscious reference to any scale or tonic. So **ato·nalism** *sb.*, **ato·nalist** *sb.* and *a.*, **atona·lity** *sb.*, **ato·nally** *adv.*

1922 A. E. Hull in *Musical Opinion* Oct. 48/1, I have been working for two years at a system of non-tonal harmony, which I had long been unable to christen. Now, after visiting no less than seven foreign countries I not only find that the thing is widely known as Atonality, but [etc.]. *Ibid.* 48/3 Keyboard chord-writing as well as linear, tonal as well as Atonal. **1923** *Mus. Assoc. Proc. 1922–3* 67 We find the principle of polytonality or atonality superseding the old key system. **1928** *Daily Express* 25 June 10/6 Eugene Goossens himself said recently that 'modern music composition has come to an impasse in its trend towards laboratory atonalism'. **1929** P. C. Buck *Hist. Mus.* iii. 34 Nor as a means of destroying the key-scale as the atonalists of today are using it. **1930** *Mus. Assoc. Proc.* 97 To write atonally in a harmonic form at present is to produce a thought 'heterodyne'. **1936** *Scrutiny* V. 153 Atonalism would seem to be an adequate medium for the expression of certain precious, morbid, esoteric sensibilities. *Ibid.* The fact that the works of the most relentless of the atonalists, Webern, rarely last more than a minute or so..is surely significant. **1952** B. Ulanov *Hist. Jazz Amer.* (1958) xxiii. 330 He can teach almost anything..from counterpoint to atonalist formations. **1958** *Times Lit. Suppl.* 4 July 372/3 What began as atonality is now called dodecaphony..or..serialism. **1958** *Listener* 18 Dec. 1051/1 Schönberg who was also writing atonally in such a work as the *Five Pieces* for orchestra. **1963** *Ibid.* 14 Feb. 313/3 Luigi Dallapiccola, an atonalist who has remained faithful to his country's abiding concern for melody.

at one, *advb. phr.* **6. at-oneness.** Delete *rare* and add further examples.

1960 C. Day Lewis *Buried Day* iv. 81 The strange at-one-ness between myself and the boat. **1962** *Listener* 25 Oct. 659/1 A sense of at-oneness with the universe.

atonicity (ætǫni·sĭti). [f. ATONIC *a.* + -ITY.] The fact or quality of being atonic (in various senses).

1900 Dorland *Med. Dict.* 85/2 *Atonicity*, atonic quality. **1930** L. Bloomfield in *Curme Vol. Ling. Stud.* 54 This usage does not reflect old atonicity.

atopy (æ·tǫpi). *Med.* [ad. Gr. ἀτοπία unusualness, f. ἄτοπος unusual (ἀ- priv. + τόπος place).] A form of hypersensitivity in which acute reactions occur, on exposure to the antigen, in some special organ or tissue. So **a·topic**, **ato·pic** *a.*, (*a*) out of place (Webster 1909); (*b*) *Med.* pertaining, relating to, or characterized by atopy.

1923 A. F. Coca in *Jrnl. Immunol.* VIII. 166 The group of abnormal hypersensitiveness includes..those idiosyncrasies that are controlled by the dominant gen...This latter sub-group evidently needs a special term by which it may be conveniently designated, and this need is satisfactorily met with the word *atopy*, which was..suggested by Professor Edward D. Perry. *Ibid.* 167 (*heading*) Atopic hypersensitiveness. **1925** Coca & Grove *Ibid.* X. 448 It will be convenient to introduce here a term with which to designate the specifically reacting substances in the serum of atopic individuals...We would like to suggest the non-committal term 'atopic reagin'. **1948** J. H. Black *Vaughan's Practice of Allergy* vi. 37 Atopy will be used in this volume..to indicate that group of allergic phenomena in which a reagin mechanism can be demonstrated...Atopic eczema can thus be differentiated from contact eczema. **1964** S. Duke-Elder *Parsons' Dis. Eye* (ed. 14) xix. 275 An Atopic Cataract which tends to develop rapidly, appears frequently in sufferers from severe and widespread skin diseases—atopic eczema, [etc.].

-ator (ēⁱtəɪ), *suffix.* See -OR 2 b, c.

‖**à tort et à travers** (a,tǫr,e,a,travẽr), *advb. phr.* [Fr., lit. 'wrongly and across'.] At random, haphazardly; without fixed principle.

1749 Ld. Chesterfield *Let.* 24 Oct. (1774) I. 479 Pray speak it [sc. Italian] in company, right or wrong, *à tort ou à travers*. **1778** F. Burney *Diary* 25 July (1842) 21 They were..afraid of praising *à tort et à travers*, as their opinions are liable to be quoted. **1834** J. S. Mill in *Monthly Repos.* VIII. 391 That kind of philosophic pedantry, which, when it has got hold of a few truths which it conceives to be a test of superiority over the vulgar, applies them *à tort et à travers*. **1856** C. M. Yonge *Daisy Chain* I. i. 47 She found herself going *à tort et à travers* all the morning. **1921** G. B. Shaw in *Times Lit. Suppl.* 17 Mar. 178/2 It bristles with mad hyphens *à tort et à travers*.

atoxyl (ătǫ·ksil). *Chem.* [G. (1902), as trade name regd. by Ver. Chem. Werke, Charlottenburg, f. A- 14 + TOX(IC *a.* + -YL.] An organic arsenical compound (esp. as formerly used hypodermically in skin diseases).

Equivalent names include arsamin and soamin.

1906 *Daily Chron.* 18 Sept. 3/4 The treatment of the patients was based on the simultaneous use of atoxyl and strychnine. *Ibid.* 21 Dec. 5/3 Professor Koch decided to employ atoxyl injections of half a gramme. **1911** *Allbutt's Syst. Med.* IX. 104 Optic atrophy caused by arsenical injections, in the form of atoxyl. **1920** J. M. H. MacLeod *Diseases of Skin* 111 Arsacetin and orsudan have the advantage over atoxyl of being more stable and capable of being boiled without decomposition. **1943** *Endeavour* Apr. 41/1 The organic arsenicals, atoxyl, salvarsan, neo-salvarsan. **1957** A. H. Douthwaite *Hale-White's Materia Medica* (ed. 30) 383 Pentavalent Arsenicals. *Sodii Aminoarsonas* (B.P.C.). Synonym, Atoxyl. **1964** S. Duke-Elder *Parsons' Dis. Eye* (ed. 14) xxiii. 352 Arsenic is specially liable to cause optic atrophy, usually total, when administered in the form of pentavalent compounds such as atoxyl or soamin.

atraumatic (æ-, ēⁱtrǫmæ·tik), *a.* [f. A- 14 + TRAUMATIC *a.*] Not causing trauma or injury; applied *spec.* to surgical instruments or techniques that minimize injurious effects.

1934 in Webster. **1950** K. W. Starr in *Brit. Surg. Practice* VIII. 140 *Atraumatic sutures*. In plastic procedures on nerves, blood-vessels, tendons, [etc.]..the finest atraumatic needle (8–20 millimetres) is required. **1957** A. V. Partipilo *Surg. Technique* (ed. 6) i. 33 An atraumatic technique avoids pinching, pulling, crushing, twisting, and tearing of the tissues.

atresia. Add: (Earlier example.)

1807 in Morris & Kendrick *Edin. Med. Dict.*

2. Disappearance by degeneration, as of the follicles in the mammalian ovary.

1903 *Jrnl. R. Microsc. Soc.* Aug. 485 In the ferret, ovulation does not occur in the absence of coition, without which the follicles undergo atresia. **1962** *New Scientist* 7 June 506/3 The others [sc. egg cells] are eliminated from the ovary at different stages of development by a process of spontaneous degeneration known as atresia.

Hence **atre·sic**, **atre·tic** *adjs.*, pertaining to or resulting from atresia; imperforate.

1897 *Lippincott's Med. Dict.* 101/1 *Atresic*, characterised by atresia. **1903** *Nature* 20 Aug. 384/2 Other atresic follicles are reduced to fibrous tissue or remain cystic. **1903** *Jrnl. R. Microsc. Soc.* Aug. 485 The atretic follicle differs from the developing corpus luteum in the absence of any discharge to the exterior. **1962** *New Scientist* 7 June 506/3 No distinction was made between normal and atretic egg cells.

atriopore (ēⁱ·triopǭᵊɪ). *Zool.* [f. ATRIUM + Gr. πόρος passage, PORE *sb.*¹] The posterior opening of the atrium or cavity in the body of the lancelet.

1892 *Athenæum* 13 Aug. 227/2 There are no buccal cirri, and, so far as can be made out, no atriopore [in a certain specimen of *Amphioxus*]. **1896** Kirkaldy & Pollard *Boas's Text Bk. Zool.* 355.

atrioventricular (ēⁱ·trio,ventri·kiŭlăɪ), *a.* *Anat.* [f. *atrio-*, comb. f. ATRIUM + VENTRICULAR *a.*] Of or belonging to the atrial and ventricular cavities of the heart.

1879 *St. George's Hosp. Rep.* IX. 93 The heart showed..very slight thickening of the atrio-ventricular valves. **1962** *Lancet* 12 May 990/2 In a few experiments a sling of rubber tubing was placed in the atrioventricular groove.

atrium. 2. a. Substitute for def.: Either of the two upper cavities (*left* and *right atrium*) of the heart into which the veins pour the blood.

atrochal (æ·trokăl), *a.* *Zool.* [f. A- 14 + TROCHAL *a.*] Having no definite rings of cilia.

1880 F. M. Balfour *Comp. Embryol.* I. 274 The atrochal forms are to be regarded as larvæ which never pass beyond the primitive stage of uniform ciliation. *Ibid.*, The atrochal larvæ are not common.

atrocity. Add: **3.** (Examples from the period of the war of 1914–18.)

1915 *Sphere* 22 May 197 The British report on German atrocities in Belgium. **1918** W. Owen *Let.* 25 Oct. (1967) 589, I have found in all these villages no evidence of German atrocities. The girls here were treated with perfect respect.

5. *attrib.* and *Comb.*, as *atrocity-monger* (so *-mongering* vbl. sb. and ppl. a.), *propaganda*, *story*.

1896 *Westm. Gaz.* 18 Feb. 1/2 The massacres were a tale, either grossly exaggerated or altogether invented by atrocity-mongering journals. **1897** *Ibid.* 28 Aug. 2/3 We should be very cautious about accepting these atrocity stories. **1899** *Ibid.* 18 Oct. 3/3 In the words of General Colley, is not all this atrocity-mongering calculated to 'make our soldiers either cowards or butchers'? **1905** A. Bennett *Tales of Five Towns* I. 94 You see roundabouts, swings,..atrocity booths, quack dentists. **1914** E. A. Powell *Fighting in Flanders* v. 129 Let them hear our side of this atrocity business. **1930** G. B. Shaw *What I really wrote about War* p. ix, The atrocity mongers who are using Belgium as a stick to beat Germany. **1937** Koestler *Spanish Testament* iv. 84 We know how much harm the preposterous atrocity propaganda engaged in by both sides caused during the Great War.

‖**à trois** (a,trwa), *phr.* [Fr.] Arranged for or shared by three persons; in a group of three. Cf. *MÉNAGE À TROIS.

1881 L. Troubridge *Jrnl.* 31 Dec. in J. Hope-Nicholson *Life amongst Troubridges* (1966) xii. 158 A cheery dinner *à trois* and a quietly nice evening. **1897** G. B. Shaw *Let.* in J. Dunbar *Mrs. G. B. S.* (1963) x. 153, I bar the seaside *à trois*. **1962** *Observer* 18 Feb. 27/2 He lives *à trois* with a worried, irritating wife and a niece whom he loves like an incestuous father. **1966** L. Davidson *Long Way to Shiloh* v. 72 We were sitting taking lemon tea *à trois*.

atrophying (æ·trŏfi,iŋ), *ppl. a.* [f. ATROPHY *v.* + -ING².] That atrophies; spec. *atrophying cirrhosis*, cirrhosis characterized by shrinkage and shrivelling of the organ.

1886 *Brit. Med. Jrnl.* 2 Oct. 650/2 Atrophying cirrhosis was diagnosed.

atropine. Add: Also **atropin.** (Earlier and later examples.)

1836 R. D. & T. Thomson *Records of General Sci.* IV. 149 Atropin may be obtained in a crystalline state by dissolving it in the smallest possible quantity of boiling water. **1908** *Practitioner* Mar. 364 Atropin should be instilled into the eye to dilate the pupil. **1927** Haldane & Huxley *Animal Biol.* ix. 198 Stimulating substances, such as atropin, caffein, etc.

b. *attrib.* and *Comb.*

1879 *St. George's Hosp. Rep.* IX. 477, I prescribed atropine drops. **1920** *Chambers's Jrnl.* 11 Sept. 654/1 In England atropine-poisoning is mostly accidental.

atropinization (ătrǫ:pinəizēⁱ·ʃən). [f. ATROPIN(E + -IZATION.] The condition of being affected by atropine; the process of atropism.

1880 W. James in *Annivers. Mem. of Boston Soc. Nat. Hist.* 14 Slight atropinization of one eye might cause such strong accommodative innervation. **1949** H. W. Florey et al. *Antibiotics* I. vii. 303 After atropinization the blood pressure was raised by from 20 to 30 mm. Hg. **1964** S. Duke-Elder *Parsons' Dis. Eye* (ed. 14) xviii. 244 The most effective treatment is to intensify atropinization in order to allay the inflammatory congestion.

A.T.S., abbrev. of *Auxiliary Territorial Service* for women (1938–48). Often with pronunc. (æts), whence the colloq. sing. form **At, AT, A.T.**, a member of the A.T.S. Cf. *AUXILIARY *a.* 1 b.

1939 *War Illustr.* 18 Nov. p. iii/1 Wartime Abbreviations ..A.T.S.—Auxiliary Territorial Service. **1940** *New Statesman* 13 Apr. 486 It looks as though, while democratising the army, we are aristocratising the A.T.S. **1941** *Ibid.* 6 Dec. 469 You would not become an adventuress In the ranks of adventurous A.T.S., Where brave girls cook for the Sergeants' Mess And the batwoman busily bats. **1942** *Ann. Reg. 1941* 90 The Wrens, the Waafs, and the Ats...The two former had no difficulty in filling their ranks from volunteers. The Ats, however,..were not so popular. **1944** *Lancet* 15 Apr. 516/1 The ATS with one accord Say 'Ma'am' when they address her, And though this is an ugly sound It does not much distress her. *Ibid.* 517/1 (*heading*) Rehabilitation for ATS. **1958** J. Betjeman *Coll. Poems* 94 As beefy ATS Without their hats Come shooting through the bridge.

atta (æ·tā). *Anglo-Indian.* [Punjabi *aṭṭa*.] Wheaten flour or meal.

1860 W. H. Russell *Diary India* II. 206 Each man with his viaticum of atta in skin-bags tied over his hips. **1920** *Blackw. Mag.* Oct. 445/2 Sacks of atta for the troops. **1937** H. W. Tilman *Ascent of Nanda Devi* iv. 31 Coolie food..could be bought locally; that is rice, atta (wheat flour) and satu.

attaboy (æ·tăboi), *int.* *slang* (chiefly *U.S.*). Also **at-a-boy**, **ata boy**. [Said to represent careless pronunc. of *that's the boy!* (*BOY *sb.*¹ 2 c).] An exclamation expressive of encouragement or admiration. Hence **attagirl**, etc., as *nonce-wds.*

1909 *Amer. Mag.* May 40/2 Back of Chance's war cries, 'At-a-boy', or 'Now ye're pitching', may be hidden a whole command to his team. **1917** C. Mathewson *Sec. Base Sloan* xxii. 298 'Ata boy!' called the Damascus catcher. **1924** P. Marks *Plastic Age* iv. 26 Suddenly one of the girls..caught the bag deftly...'Ray! Ray! Atta girl! Hot dog! Ray, ray!' **1925** H. L. Foster *Trop. Tramp Tourists* 101 The marines rose from their chairs to encourage the new performer: 'Attaboy, soldier! Attaboy! Shake 'em doggies!' **1927** *Blackw. Mag.* July 4/2 Cries, also, of 'Atta-boy!' leave no doubt that the greeting is not entirely British. **1929** Wodehouse *Summer Lightning* xv. 286 'Give me two minutes to get the car out and five to make the trip and I'll be with you.' ''At-a-boy!' said Millicent. ''At-a-baby!' said Hugo. **1932** James Joyce in *New Statesman* 27 Feb. 261/1 Attagirl! **1936** Auden & Isherwood *Ascent of F6* ii. iii. 101 Chin up! Kiss me! Atta Boy! Dance till dawn among the ruins of a burning Troy! **1946** H. Croome *Faithless Mirror* x. 102 I'm celebrating. Atta girl! **1958** J. Betjeman *Coll. Poems* 153 Look at that little mite with *Attaboy* Printed across her paper sailor hat.

attach, *v.* Add: **6. b.** *Mil.* and *Naval.* To allocate for service to a particular unit: chiefly *pass.*

1802 C. James *Mil. Dict.* s.v., Officers and non-commissioned officers are said to be attached to the..army, regiment, [etc.]..with which they are appointed to act. **1909** *Regs. for Mobilization* 8 The equipment..which he would take with him when transferred or attached to another unit.

attaché case (ătæ·ʃēⁱ kēⁱs, ătæ·ʃikēⁱs). A small rectangular case (orig. such as those used by attachés) for carrying papers, documents, and the like.

1904 *Army & Navy Co-op. Soc. Price List*, 'Attaché' cases. Green and brown leather, double-action lock. **1918** H. Walpole *Green Mirror* ii. i. 146 The two young men in perfect attire and attaché cases. **1946** M. Dickens *Happy Prisoner* vi. 106 Mary Brewer burst in at the door, carrying her little fibre attaché case.

attachéship. (Earlier examples.)

1834 *Tait's Mag.* I. 440/1 Attachéship is, in fact, too onerous a calling for any man to adopt. *Ibid.* 440/2 The Honourable Arthur is promoted to *paid* attachéship.

attachment. Add: **10.** *Mil.* and *Naval.* The fact or condition of being attached to a particular unit.

1904 *Regs. for Mobilization* (Provisional) i. 12 Units which do not exist as such in peace are completed in officers by special appointments, and as regards other ranks by reservists, attachments from other corps, and transfers from existing units. **1914** *Daily Express* 5 Oct. 5/1 Each man wearing an armlet indicating his attachment to these special siege batteries. **1955** *Times* 5 Aug. 7/1 Numbers of Egyptian officers and men have recently been serving on attachment with the British forces in the Zone.

attack, *v.* Add: **4.** (Earlier example.)

1812 Shelley *Let.* 17 Dec. (1964) I. 216 Mrs. Shelley is attacking Latin with considerable resolution.

7. *Mus.* *intr.* and *trans.* (See quot. and cf. *ATTACK *sb.* 7.)

1835 *Court Mag.* VI. 264/1 The instruments do not *attack* properly—that is to say, they do not come in simultaneously with sufficient precision to form a sharp, crisp chord, as if proceeding from a single instrument. **1967** *Oxford Mail* 27 Nov. 6/1 Those boys from the Chapel Royal and St. Paul's..attack their leads with the confidence of professionals.

attack, *sb.* Add: **2. b.** Lacrosse. *the attack*: the 'attack fields' collectively; *attack field*: see quot. 1892.

1885 E. T. Sachs *Lacrosse for Beginners* 43 As matches are played, the ball falls more often to the defence than to the attack. *Ibid.* 52 The Attack Fields. These two players play rather forward of centre. **1892** G. A. Hutchison *Outdoor Games* 507 The players stand all down the field, from goal-keeper to goal-keeper, a defence man watching one of the opposite attack. *Ibid.* 508 The three men nearest the opponents' goal (they are called the 'home' first, second, and third, first being nearest the goal) should practice throwing at goal...The two players next nearest are called the 'attack fields'. **1902** *Westm. Gaz.* 28 Apr. 7/3 His place in the attack field.

7. *Mus.* [after It. *attacca*.] The action or manner of beginning a piece, passage, or phrase, in respect of precision and clarity. Also, more generally in the arts, brilliance of style, decisive rendering. Cf. sense 4.

1871 *Monthly Pkt.* Oct. 401 Their splendid 'attack', no dropping in one after another...They knew how to look at their music and the conductor both at once. **1879** Grove *Dict. Mus.* I. 100 *Attack*, a technical expression for decision and spirit in beginning a phrase or passage. **1891** *Durham Univ. Jrnl.* 11 Nov. 216 There was a certain want of precision in the attack. **1905** *Westm. Gaz.* 12 Dec. 3/2 Mr. Runcie has fancy, verve, and what artists call 'attack'. **1906** B. Stoker *Personal Rem. H. Irving* II. li. 84 In this play Irving was very decided as to the 'attack'. He had often talked with me about the proper note to strike at the beginning of the play. **1938** A. L. Haskell *Ballet* vii. 198 Riabouchinska is not a purely classical dancer, lacking the necessary hardness and attack. **1956** I. Deakin *At Ballet* 206 *Attack*: In ballet it has the meaning of the deliberation behind the performance of the various steps. **1961** G. Millerson *Technique Telev. Production* iii. 45 The face would merge into the background—the picture would lack 'attack'—appear flat. **1962** *Listener* 13 Sept. 409/2 One cannot imagine a performance more remarkable for attack, verve, clarity of articulation, and gradations of tone colour.

8. *attrib.* and *Comb.*, as (sense 1) *attack-aircraft-carrier, -bomber, -division, -formation, -order, -practice*; (sense 5) *attack-rate*.

1899 *Daily News* 6 Mar. 8/5 Small-pox breaks out, and..the attack rate is..limited to exactly 10 per cent. in each class. *Ibid.* 27 Dec. 8/4 We..marched four or five miles in regular attack formation. **1901** *Westm. Gaz.* 8 July 7/3 One officer, two non-coms., and twenty men, who march eleven miles and finish at Stickledown Ranges with the attack practice. **1923** Kipling *Irish Guards* II. 179 The Companies dressed in attack-order. **1932** Auden *Orators* ii. p. 76 1st Army: 15 attack divisions, 2 ordinary divisions. **1952** *Amer. Speech* XXVII. 4 Search planes, dive bomber, attack bomber. **1957** *Jane's Fighting Ships 1957–58* 407/3 It is the intention to build six nuclear powered, attack aircraft carriers. **1961** *Lancet* 26 Aug. 473/2 Another outbreak of respiratory-syncytial-virus infection, again with a peak attack-rate in infants under 7 months.

attain, *sb.* Delete †*Obs.* and add later example.

1925 Hardy *Human Shows* 141 A painter of high attain.

attap (æ·tăp). Also **adap, atap.** [Native word.] The name used in Malayo-Javanese regions for any palm-fronds used in thatching, *esp.* those of the palm *Nipa fruticans*; hence, a thatch made of these. Also *attrib.*

[**1672** Baldæus *Ceylon* 164 Het Huys is van Leem gemaakt, ende gedekt met *Atap*, ofte bladeren von Palmeer-boomen.] **1817** Raffles *Java* I. iv. 166 In the maritime districts *átap*, or thatch, is made almost exclusively from the leaves of the *nípa* or *búyu*. *Ibid.*, The leaves of the *gébang*..are too large and brittle to form durable *átap*. **1886** *Jrnl. Anthrop. Inst.* XV. 293 The roof is thatched with the common Brettam attaps in the same way as Malay houses. **1897** *Blackw. Mag.* Nov. 637/2 Overhead lizards ran in the attap thatch. **1924** *Chambers's Jrnl.* 72/1 The atap-thatched, pile-perched native hutches. **1928** *Daily Express* 13 Mar. 12/2 The coolie women of Bangkok, who cook their food..over an open fire in front of their 'atap' houses. **1951** 'N. Shute' *Round Bend* iii. 89 The little atap village was just by the strip.

attempt, *sb.* Add: **2. b.** A concrete result of an attempt.

1871 L. W. M. Lockhart *Fair to See* xxiii, His first attempt [sc. a letter] ran thus.

attention. Add: **5.** (Earlier example.)

1792 *Rules & Regs. for Formations of H.M.'s Forces* 6 Upon the word *Attention*, no one shall have materially lost his dressing in the line.

b. *to stand* (*at, to*) *attention*: to stand in the military attitude assumed at the word of command 'Attention!' Also *to draw oneself up, spring*, etc., *to attention*.

1859 Whyte Melville *Holmby House* (1860) xviii. 276 He ..stood gaunt and dripping at 'attention'. **1866** —— *Cerise* (ed. 3) I. i. 11 Like a soldier who springs to 'attention'. **1879** H. Hartigan *Stray Leaves* ser. ii. 146 He drew himself up to attention, and performed the lance exercise. *Ibid.* 181 The old man was standing at attention. **1892** Kipling & Balestier *Naulahka* 259 A trooper..stood to attention at the horse's head. **1895** *Cornhill Mag.* Dec. 633 He..saluted, and stood at attention. **1899** *Allbutt's Syst. Med.* VIII. 8 Make the patient 'stand attention'.

6. *attrib.* and *Comb.*

1898 Titchener *Primer Psychol.* v. 88 The attention-wave rises and falls at short intervals: attention fluctuates. **1901** —— *Exper. Psychol.* I. ii. viii. 197 The attention period (the time-interval from maximum to maximum of sensation). *Ibid.* 198 A continuous attention-strain. **1902** W. James *Let.* 20 Apr. (1920) II. 164 It is an uplifting thought that truth is to be told at last in a radical and attention-compelling manner. **1934** *Jrnl. Social Psychol.* V. 313 Results which present conflicting records of the attention-spans of the subjects observed. **1949** *Mind* LVIII. 125 The book is written with grace and humour, and in an attention-holding way. **1964** L. S. Hultzén in D. Abercrombie et al. *Daniel Jones* 90 Attention-calling modifications tend to occur.

attentional (ăte·nʃənăl), *a.* [f. ATTENTION + -AL.] Of or pertaining to attention as a psychological concept.

1896 Titchener *Outl. Psychol.* vi. 132 The passive attention of the animal or the child is the first stage of attentional development. **1901** —— *Exper. Psychol.* I. ii. 206 The experiments on attentional time-displacement form one of the most interesting..chapters of experimental psychology.

attentive, *a.* Add: **1. b.** Of or pertaining to attention as a psychological concept.

1890 Baldwin *Handbk. Psychol.* v. 69 A strong effort of attentive thought. **1890** W. James *Princ. Psychol.* I. xi. 420 All forms of attentive effort would be exercised at once. **1892** J. Sully *Human Mind* I. iv. 75 The region of clear consciousness or of attentive consciousness. *Ibid.* ix. 301 The process of attentive integration is really more complex than is here supposed. **1898** Titchener *Primer Psychol.* v. 84 The attentive state. **1901** —— *Exper. Psychol.* I. ii. viii. 186 There are serious differences of opinion concerning the nature of the attentive consciousness. **1934** *Jrnl. Social Psychol.* V. 314 An investigation of the attentive activity of preschool children.

attentiveness. (Later examples.)

1820 *Indicator* 11 Oct. (1822) 6 Their..spirits carry them away from a proper attentiveness to others. **1921** *Contemp. Rev.* Sept. 366 A wider degree of attentiveness involving eye, ear, and the gregarious instinct.

attenuate, *v.* Add: **5.** *Electr.* To introduce attenuation; in *pass.*, to be subjected to attenuation. Cf. *ATTENUATION 4.

1886 Lord Rayleigh in *Phil. Mag.* 5th Ser. XXII. 490 If we had the means of observing the passage of signals at various points of a long cable, we should find them not merely retarded..as we recede from the sending end, but also attenuated. **1892** Heaviside *Electr. Papers* II. 133 The act of reflection attenuates. *Ibid.* 346 During transmission along the circuit, the vibrations are attenuated. **1959** *Chambers's Encycl.* VII. 696/2 The lower part of the Heaviside layer is of particular importance..because it is in this region that radio waves used for long-distance communication are attenuated.

attenuation. Add: **2.** *spec.* in brewing and distilling.

1882 W. T. Brannt tr. *Thausing's Beer* 707 The decrease in density [of the beer-worts] is called attenuation. **1956** *New Biol.* XXI. 17 Beers..which are satisfactory in respect of attenuation.

3. *spec.* of a disease, or of the pathogenicity of a micro-organism. Also *attrib.*

1882 [in Dict.]. **1944** C. D. Darlington in *Nature* 5 Aug. 167/1 In the attenuation process, the nucleus is mutafacient with respect to the virus. **1964** M. Hynes *Med. Bacteriol.* (ed. 8) vi. 62 The virulence of a given organism can often be reduced by physical means such as drying or heat. The reduction of virulence is known as attenuation.

4. *Electr.* The decrease in amplitude of an electrical signal or current. Also *attrib.*

1887 Heaviside in *Electrician* 24 June 143/2 The idea of attenuation, expressed in a more roundabout manner in terms such as diminution of amplitude, and so forth, is nothing new; the *word* 'attenuation' I found Lord Rayleigh use, and at once adopted it myself. **1931** *B.B.C. Year-Bk.* 436/2 *Attenuation Factor*, a factor indicating the rate of reduction in amplitude of an ether wave as the distance from the point of origin increases. **1931** *Wireless World* 2 Sept. 228/1 The attenuation of the higher modulation frequencies. **1935** *Discovery* Sept. 278/2 The cut-off or attenuation of television signals, due to intervening obstructions.

attenuator (ăte·niu,ēⁱtəɪ). *Electr.* [f. ATTENUATE *v.* + -OR.] A device which introduces attenuation.

1924 *Wireless World* 3 Sept. 643 The mechanical filter, or as it has sometimes been termed..the mechanical attenuator. **1931** *B.B.C. Year-Bk.* 436/2 *Attenuator*, an arrangement of calibrated resistances to introduce loss into a circuit. **1963** B. Fozard *Instrumentation Nucl. Reactors* x. 122 Variation of the over-all gain is effected by the use of switched resistance attenuator networks.

attenuity (ætĕniū·iti). [f. ATTENUATE after *tenuity*.] = TENUITY 1.

1830 *Blackw. Mag.* XXVIII. 387 Not so much as the taperest wine-glass wire-woven into almost invisible attenuity. **1861** *Temple Bar* III. 437 No cotton-yarn of such attenuity as this can be..woven. **1888** *Echo* 22 June 1/4 Drawing his thin willowy form..to its full attenuity.

attest, *v.* Add: **5.** Also, to administer the oath of allegiance to a military recruit (see quot. 1812 in Dict.); used esp. in connection with the 'Derby Scheme' of 1915. Also *intr.*, to enrol oneself as ready for military service when called up. Hence **attested** *ppl. a.*

1915 Ld. Derby in *Times* 20 Oct. 10/1 They would be medically examined, and, if found fit, attested there and then. **1917** J. H. Worrall *Tribunal Hand-bk.* 128 Police magistrates who..will not..see that the attested man before them is shorn of justice, because, unfortunately for him, he attested. **1922** *Encycl. Brit.* XXX. 212/2 In Oct. 1915, the 'Derby Scheme' or 'Group System' was initiated by Lord Derby...Between Oct. 25 and the middle of Dec. 2,000,000 men were attested under this system.

attested, *ppl. a.* Add: **b.** Of cattle or milk: approved by authority as free from disease.

1934 *Hansard Commons* CCXC. 1131 Our programme providing..for the payment of premiums to attested herds should..bring us perhaps 1,000 herds in the first year.

attic, *sb.*² Add: **4.** *Anat.* The upper part of the tympanum of the ear.

1889 J. Leidy *Elem. Treat. Human Anat.* (ed. 2) xvi. 893 The attic of the tympanum is a pyramidal cavity above the atrium with which it communicates by a horizontal, fore and aft oval aperture. **1891** *Med. Ann.* 159 Pathological Changes in the External 'Attic' of the Tympanic Cavity. **1900** Dorland *Med. Dict.* 87/1 *Attic*, the part of the tympanum that is situated above the atrium.

Atticist. Add: Hence **Attici·stic** *a.*

1919 H. J. Cadbury *Style of Luke* i. ii. 38 The gulf between New Testament Greek in general and Attic or Atticistic Greek is..being exaggerated. **1934** A. Toynbee *Study Hist.* I. i. B. ii. 25 This phase of Hellenic history is commonly called 'the Hellenistic Age', but 'the Atticistic Age' is the proper name for it.

attingent, *a.* and *sb.* Restrict †*Obs.* to *sb.* and add later example.

1895 M. H. N. Story-Maskelyne *Crystallogr.* ii. § 15. 16 To each octant there will correspond three adjacent octants, which have each one axial plane in common with the original octant; three attingent octants, in contact with it only along an axis.

attitude. Add: **2. c.** *Aeronaut.* (See quots.) Also *attrib.*

1910 R. Ferris *How it Flies* 455 *Attitude*, the position of a plane as related to the line of its travel. **1914** *R. Engineers' Jrnl.* Nov. 311 *Attitude*, an aeroplane's or wing's position relative to the direction of motion through the air. **1953** *Flight* 17 July 78/2 The automatic pilot..applies control movements proportional to the attitude changes. **1962** J. Glenn in *Into Orbit* 209 The capsule's attitude would have to be near-perfect when the rockets fired, or the angle of re-entry would be affected.

d. *Dancing.* A posture or disposition of the body; spec. a form of arabesque (see quot. 1957).

1721 J. Weaver *Lectures upon Dancing* 137 Dancing is an elegant, and regular Movement, harmonically composed of beautiful Attitudes, and contrasted graceful Postures of the Body. *Ibid.* 145 *Attitude* is a Posture, or graceful Disposition of the Body, in Standing; Sitting; or Lying. **1830** R. Barton tr. *Blasis' Code of Terpsichore* ii. v. 74 That particular position technically termed attitude is the most elegant, but at the same time the most difficult which dancing comprises. **1911** [see *ARABESQUE *sb.* 4]. **1922** Beaumont & Idzikowski *Man. Class. Theatr. Dancing* 27 There are an infinite number of *attitudes*, so that they depend on the taste of the professor or *choregraphe*. **1957** G. B. L. Wilson *Dict. Ballet* 32 *Attitude*, position derived from the statue of Mercury by Giovanni da Bologna.

4. b. = *attitude of mind* (ATTITUDE 4). Cf. senses 2 b, 3.

*a*1873 Mill *Three Ess. Relig.* (1874) 126 Along with this change in the moral attitude of thoughtful unbelievers towards the religious ideas of mankind, a corresponding difference has manifested itself in their intellectual attitude. **1909** Titchener *Lectures Exper. Psychol. Thought-Processes* iii. 112 Attitude, the background of meaning or reference against which a mental process is seen, may [etc.]. **1922** H. E. Palmer *Eng. Intonation* p. viii, We all recognize immediately..each of the attitudes associated with the tones. **1937** G. W. Allport *Personality* (1938) xi. 294 Both *attitude* and *trait* are indispensable concepts...Ordinarily *attitude* should be employed when the disposition is bound to an object or value, that is.., when it is aroused by a well-defined class of stimuli, and when the individual feels toward these stimuli a definite attraction or repulsion. **1941** *Punch* 3 Sept. 203/1 The attitude of ordinary people..towards mathematics..may be summed up as the same as their attitude to the police force. **1948** Krech & Crutchfield *Theory & Probl. Social Psychol.* v. 151 Many psychologists regard the study of attitudes as the central problem of social psychology. **1958** G. J. Warnock *Eng. Philos. since 1900* 169 A marked capacity for abstract thought is compatible with an 'attitude to life' entirely ordinary, or even dull.

5. *Lit. Criticism. Spec.* in the use of I. A. Richards, 'the non-overt impulse to action involved in the poetic experience of the reader'.

1925 I. A. Richards *Princ. Lit. Crit.* xv. 112 These imaginal and incipient activities or tendencies to action, I shall call attitudes. *Ibid.* xvi. 132 For it is the attitudes evoked which are the all-important part of any experience.

6. *attrib.* and *Comb.*, as *attitude measurement, research, scale, study, test, theory*; *attitude-taking* vbl. sb.

1904 W. James *Mem. & Stud.* (1911) vi. 140 Spencer's 'Ethics' is a most vital and original piece of attitude-taking in the world of ideals. **1929** Thurstone & Chave *Measurement of Attitude* p. xii, The total allocation of an individual to a position on an attitude scale is an abstraction. *Ibid.* i. 1 (*heading*) Theory of attitude measurement. **1934** *Jrnl. Social Psychol.* V. 387 The attitude test was administered again. **1935** H. C. Warren *Dict. Psychol.* 24/1 *Attitude scale*, a scale for measuring degrees of attitude upon a particular question. **1940** *Mind* XLIX. 228 In the Introduction Ross divides attempted definitions of ethical terms into 'attitude-theories' and 'consequence-theories'. **1941** *Jrnl. Social Psychol.* XIII. 429 (*heading*) On the use of certain qualitative methods of attitude research. **1958** *Listener* 28 Aug. 308/3 All types of attitude-studies could contribute something. **1960** *Language & Speech* III. 223 Attitude measurement..seemed a promising technique. *Ibid.*, Osgood's semantic differential..was the attitude-measuring technique used in the experiment. **1960** *Times Rev. Industry* Jan. 54/1 Attitude and motivation research assesses what qualities of the product give it its appeal to some consumers and not to others.

attitudinal, *a.* (Later examples, *spec.* in *Psychol.*)

1909 Titchener *Lectures Exper. Psychol. Thought-Processes* v. 181 The different *Aufgaben* come to consciousness, in part, as different feels of the whole body...I find ..that these attitudinal feels are touched off in all sorts of ways. *Ibid.* 248 We may not..infer that the attitudes and the attitudinal constituents of the thoughts..are vestigial derivatives of visual imagery. **1935** H. C. Warren *Dict. Psychol.* 24/1 *Attitudinal* (or *attitude*) *test*, a phrase applied somewhat broadly to numerous tests of non-intellectual mental traits. **1941** *Jrnl. Social Psychol.* XIII. 455 Inconsistency, both of the attitudinal and merely verbal sort, appeared often enough to demand some explanation. **1960** *Language & Speech* III. 223 (*heading*) Attitudinal meanings conveyed by intonation contours. **1962** A. C. Gimson *Introd. Pronunc. Eng.* x. 253 (*heading*) The Attitudinal Function of Intonation.

attitudinarian. Add: (Later examples.)

1831 J. Boaden *Life Mrs. Jordan* I. xiii. 310 Dancing was certainly more strictly a science, and either the *jumper* or the *attitudinarian* undervalued by the masters of the art. **1870** *Contemp. Rev.* XV. 281 The attitudinarians of the Ritualistic Churches.

2. *attrib.* or quasi-*adj.*

1823 *Examiner* 9 June 379/2 When an artist intends only to give some dashes of luxuriant colour and hints of attitudinarian character,..we will not think of wishing for the presence of higher qualities.

attitudinizer. (Earlier examples.)

1824 *Blackw. Mag.* XV. 635 My first flame was a flaunting, airy, artificial attitudinizer. **1833** J. S. Mill in *Monthly Repos.* VII. 68 [= 1859 quot. in Dict.].

attract, *v.* Add: **5.** To pilfer or steal. *slang.*

1891 Kipling *Light that Failed* vii. 135 'Do they [sc. the students] still steal colours at lunch-time?' 'Not steal. Attract is the word... I'm good—I only attract ultramarine; but there are students who'd attract flake-white.' **1933** 'E. Cambridge' *Hostages to Fortune* 24 He 'attracted' some timber and built a boat house.

attractant (ătræ·ktănt). [f. ATTRACT *v.* + -ANT¹.] That which attracts; a substance used to attract.

1926 *Jrnl. Econ. Entomology* XIX. 546 It soon became evident that if the fundamentals of 'attractants' and repellents were to be studied, special apparatus for measuring or recording the responses of insects to these substances must be developed. **1953** *Brit. Commonw. Forest Terminol.* i. 14 *Attractant*, a substance attractive to insects, rodents, etc., used in population surveys or in control traps or baits. **1955** *Sci. Amer.* Aug. 76/3 It is neither an attractant nor a repellent to unconditioned salmon. **1959** *Times* 9 Jan. 13/2 The attractive scents of female butterflies are produced by scattered gland cells...Other examples are the sexual attractants of cockroaches.

attraction. Add: **9.** (Earlier examples.)

1829 *Harlequin* 20 June 43 These performances, though possessing much novelty, did not prove sterling attractions. **1832** *Rep. Sel. Committee Dram. Lit.* 45 You may draw as beautiful a picture, but not so as to produce that sort of scenic effect which is the great attraction.

10. attraction sphere = *CENTROSPHERE (*a*).

1896 E. B. Wilson *Cell in Developm. & Inher.* 36 The centrosome..lies outside, though near, the nucleus, in the cyto-reticulum, surrounded by a granular, reticular, or radiating area of the latter known as the *attraction-sphere* or *centrosphere*. *Ibid.* 334 *Attraction-sphere*, the central mass of the aster from which the rays proceed.

attributal (æ·tribiutăl, ătri·b-), *a.* [f. ATTRIBUTE *sb.* + -AL.] Of the nature of an attribute.

1894 *Standard* 30 July 6/6 Why do not Welshmen take as surnames..the attributal adjectives by which they are known? William Thomas, distinguished by Tynawern (the house in the wood) would always be identified as William Tynawern. **1902** *Jrnl. Amer. Folk-lore* Jan.–Mar. 14 It is oftentimes difficult to discover their identities, since they bear many attributal or descriptive names.

attributeless (æ·tribiutlės), *a.* [f. ATTRIBUTE *sb.* + -LESS.] Without attributes.

1894 *Thinker* VI. 448 The Vedântin is more drawn by the fascination of that attributeless immensity. **1924** *Expositor* Feb. 107 Esoterically he holds that nothing is real but the attributeless Brahman.

attrition. Add: **2. b.** *Mil.* The wearing down of the enemy's strength and morale by unremitting harassment, esp. in phr. *war of attrition*.

1914 *Sphere* 21 Nov. 181/1 This is a war of attrition, in which each side tries to wear down the other. **1915** Kitchener *Memorandum* in Lloyd George *War Mem. D. Lloyd George* (1933) I. xii. 435 The end of the War must come through one of the two following causes: (1) by a decisive victory..or (2) by attrition. **1918** E. S. Farrow *Dict. Mil. Terms* 45 *Attrition*, in a military sense, the act of wearing away the enemy's strength, increasing his mortality list, and lowering his morale. **1919** Haig *Desp.* 21 Mar. (1919) 326 The rapid collapse of Germany's military powers..would not have taken place but for that period of ceaseless attrition. **1927** W. S. Churchill *World Crisis 1916–18* I. ii. 45 The only method of waging war on the Western Front was by wearing down the enemy by 'killing Germans in a war of attrition'. **1958** *Listener* 13 Nov. 791/3 Nor did Montgomery, unfairly scornful though he is of generalship in the first world war, disdain tactics of attrition at times.

fig.

1930 *Daily Express* 30 July 3/7 Fine weather at the Oval may mean an endurance test—perhaps a full week of slow batting and attack by process of attrition.

attritional (ătri·ʃənăl), *a.* [f. Attrition + -al.] Characterized by attrition.

1847–9 *Todd's Cycl. Anat.* IV. 530/1 The preliminary breaking up of structure which appears to be chiefly physical or attritional in the normal cartilage. **1960** *Times* 20 Feb. 10/2 McMorris..delved into the archives of slow-scoring... Attritional cricket is his speciality.

attuition (ætiu,i·ʃən). *Psychol.* [f. L. *ad* (see Ad-)+Tuition after *intuition*.] A hypothetical apprehension higher in order than mere animal sensation and lower than human perception. Hence **attui·tional, attu·itive** adjs. So **attuent** (ǣtiū·ĕnt), *a.* [L. *tuēnt-*, pres. pple. of *tuēri* to look at], that has the function of, or is characterized by, attuition; **attu·it,** something of which one becomes conscious by attuition; **attuite** (ǣtiū·it) *v. trans.*, to become conscious of (an object) by attuition; **attu·itively** *adv.*, by attuition.

1884 'Scotus Novanticus' *Metaph. Nova et Vetusta* 6 This word (Attuition) is here coined, in order to indicate that state of Consciousness which lies between Sensation-proper and Perception-proper. *Ibid.* 11, I may now define *Attuition* to be the reflex co-ordination of the elements or units of Sensation into an image or synopsis: it is a synthesis *in* and *for* the Conscious Subject. *Ibid.*, The range and character of attuitional intelligence. *Ibid.* 29 The Will then has now affirmed the attuited totality in perception as a one totality. *Ibid.* 49 The Attuent consciousness at the moment of receiving the 'impression' of Extension, locates it as outside itself. *Ibid.* 143 An animal is attuitively conscious of the *à posteriori* categories, but it cannot categorize, because it cannot affirm. They are present to it as sense-attuits. **1888** J. Martineau *Study Relig.* I. ii. i. 180 In the higher quadrupeds..a state is reached which may be called Attuition, marked by discrimination of particular objects from each other in space..but without corresponding discrimination of them from the attuent subject. *Ibid.* 184 How I should feel if I were..reduced to the 'attuent' condition, I find it impossible to judge. **1896** W. Caldwell *Schopenhauer's System* iii. 157 Our consciousness of other things—*i.e.*, our *attuitive* knowledge. **1902** *Encycl. Brit.* XXX. 677/2 Supposing..that a man, or a dog, through association 'attuites' sequence and invariableness of succession.

attune, *v.* Add: †**4.** *Wireless Telegr.* To tune in. *Obs.*

1899 R. Routledge *Discov. 19th C.* (ed. 13) 546 A Geissler tube, when its circuit is properly attuned, can be lighted up by the magneto-electric disturbances propagated without material contacts, and this itself would constitute a method of signalling to a distance. **1901** *Westm. Gaz.* 23 Dec. 5/1 The message..would reach every attuned recorder within that distance. **1902** *Ibid.* 30 Apr. 12/2 Inventions covering the attuning of transmitters and receivers.

‖**atua** (a·tu,a, ătū·ă). Also **akua.** Earlier forms: 8 **eatua** (erron.), 9 **hotooa.** [Maori, Raratongan, Tahitian, Samoan *atua*, Tongan *'otua*, Hawaiian *akua*.] A Polynesian name for a supernatural being, god, or demon.

1769 Cook *Jrnl.* 28 June (1955) I. 113 Near the great Morie were many large Altars..on [which] were exposed meat for Eatua or God. **1773** *Ibid.* 17 Sept. (1961) II. 234, I then asked if they sacrificed men to the Eatua. **1773** J. Hawkesworth *Acct. Voy. Disc. S. Hemisphere* II. i. xv. 165 It was a representation of Mauwe, one of their Eatuas, or gods of the second class. **1817** J. Martin *Account of Natives of Tonga Islands* II. xviii. 103 Hotooas, gods, or superior beings, who have the power of dispensing good and evil to mankind. **1832** A. Earle *Narr. Resid. N.Z.* (1966) 178 The natives say, 'It is Atua, the Great Spirit, coming into them, and eating up their inside.' **1851** J. C. Richmond in Scholefield *Richmond–Atkinson Papers* I. 103 Though he may have some faith in the atonement as a charm to exorcise 'atuas'..his Xtianity does not go a great deal further. **1884** Lady Martin *Our Maoris* ii. 35 He..told us that an Atua Maori,—that is to say, a man who professes to be possessed by a spirit,—had been disturbing the neighbourhood. **1902** E. Clodd *T. H. Huxley* iv. 160 The 'atuas' include gods good and evil, home and foreign, as well as the souls of men. **1928** W. Baucke *Where White Man Treads* (ed. 2) 211 Each tohunga diagnoses after his own skill, and in the ripeness of his wisdom applies what his atua has revealed to him.

atwitch (ătwi·tʃ), *adv.* or *pred. a.* [f. A *prep.*[1] +Twitch *v.*[1]] Twitching; jerking.

1910 J. Farnol *Broad Highway* II. v. 219 That hoary head all a-twitch with eagerness. **1957** J. Frame *Owls do Cry* xi. 46 The four poplars..atwitch with trickle of air.

atypical (æ-, ēⁱti·pikăl), *a.* [f. A- 14+Typical *a.*] Not typical; not conformable to the ordinary type. Hence **atypica·lity,** the quality of not being typical; **aty·pically** *adv.*, in an atypical manner.

1885 C. H. Fagge *Princ. Med.* I. 100 Carcinomata can be..distinguished from other epithelial growths by their being 'atypical'. **1916** *Times* 25 Oct. 7/3 The cases up to this time were mostly mild and atypical. **1917** C. R. Payne tr. *Pfister's Psychoanalytic Method* i. xi. 292, I have also found the serpent atypically as allusion to the pretendedly poisonous tongue of the wife. **1918** *Sat. Westm. Gaz.* 13 Apr. 11/1 She,..something atypically for a woman, appears to be more interested in places than people. **1935** *Proc. Prehistoric Soc.* I. 152 The stations in Crimea and Transcaucasia..have yielded exclusively rather atypical flake implements. **1957** *Archivum Linguisticum* IX. 83 The atypicality of -o in feminines. **1959** B. Wootton *Soc. Sci. & Soc. Path.* v. 161 Criminologists have interested themselves more in the heavy incidence of crime upon the young than in the relatively atypical late entrants. **1964** S. Duke-Elder *Parsons' Dis. Eye* (ed. 14) xiv. 143 Three atypically large viruses of ophthalmic interest.

aubergine. Add: **b.** A purple colour resembling that of the fruit. Also *attrib.* and as *adj.* *aubergine purple*, a shade of purple found in Oriental porcelain.

1895 *Windsor Mag.* I. 109/1 An aubergine silk lining. **1898** *Daily News* 29 Sept. 3/4 Cashmere in a tone of Burgundy that is sometimes called aubergine, from the deep reddish mauve of the vegetable. **1903** *Daily Chron.* 31 Oct. 8/4 Aubergine..is a tint with some claret and a touch of violet in it. **1906** S. W. Bushell *Chinese Art* II. 23 The deep purple, or *aubergine* (*ch'ieh tzŭ*), of the Chun-chou wares. **1909** *Daily Chron.* 29 Jan. 4/5 Aubergine coloured cloth. **1927** *Daily Tel.* 5 Apr. 13/3 Examples of the Ming dynasty..vases, jardinières, bowls,..an aubergine fish-bowl.

aubrietia (ǭbrī·ʃ¹ă). Also very freq. in erron. form **aubretia.** [mod.L., f. the name of Claude *Aubriet* (1668–1743), after whom it was named by Adanson in 1763: see -ia[1].] The common name of a member of the genus *Aubrieta* of spring-flowering dwarf perennial plants belonging to the family Cruciferæ, of trailing habit, bearing flowers of various shades of purple, grown as a rock plant or garden edging.

'Aubrieta is the original spelling of the generic name but the form Aubrietia has been so generally used that it has come into common speech and we retain that form as the common name' (R. Hort. Soc. *Dict. Gardening*, 1956, I. 222/1).

1829 in Loudon *Encycl. Plants* (1836) 544. **1870** W. Robinson *Alpine Flowers* i. 48 Look at what we could do with the dwarf green Iberises, Helianthemums, Aubrietias, Arabises, [etc.]. **1883** F. M. Peard *Contrad.* I. 227 Lilac aubretias clamber up the meadows behind Titian's house. **1911** 'A. Marshall' *Eldest Son* xxx, The station-master's arabis and aubrietia were making a fine show.

Aubusson (obüsoṅ). [Name of a manufacturing town, dept. Creuse, France.] Tapestry made at Aubusson, *esp.* a carpet made of this, more explicitly *Aubusson carpet*.

[**1774** H. Walpole *Descr. Strawberry-Hill* 102 The bed is of tapestry of Aubusson, festoons of flowers on a white ground, lined with crimson silk.] **1851** *Illustr. Exhibitor* p. xxx, Tapestry, carpets, [etc.]. ..Castel, E., France, Aubusson carpet. **1866** *Argosy* Jan. 142 The elegant Aubusson carpet. **1900** E. Glyn *Visits Eliz.* (1906) 110 The room ..is hung with aubusson. **1927** J. B. Priestley *Adam in Moonshine* vi. 101 Wasn't that the dreadful hairy, smelly one [*sc.* a Russian conspirator] who spoilt your Aubusson? **1961** *Connoisseur* Dec. p. lii, Fine Aubusson Carpets and Rugs.

‖**au courant** (o kuraṅ). [Fr.] Acquainted with what is going on; aware of current developments: usu. const. *with* or *of*.

1762 Gibbon *Jrnl.* 13 Oct. (1929) 162 Having neglected my Journal from the 14th of September, I was forced to bestow some days to put myself *au courant*. **1830** J. S. Mill *Let.* 20 Aug. in *Wks.* (1963) XII. 59 They are quite eager to place me *au courant* of all their proceedings. **1885** *Law Q. Rev.* Apr. 138 To keep themselves fairly *au courant* with what was being decided in the various Courts. **1928** *Daily Tel.* 23 Oct. 11/3 It would have been wiser if the United States had been kept au courant of the negotiations from the very outset. **1962** *Listener* 8 Mar. 427/2 He also believed that Chambers 'has a set of spies here in Rome'—to be *au courant* what Robert is doing.

auction, *sb.* Add: **2. b.** = *auction bridge* (*Bridge *sb.*[2] b). Also *attrib.*

1908 W. Dalton *Auction Bridge* p. iv, If you are accustomed to play Bridge for 6*d.* points, or £2 10*s.* per 100, play Auction for £1 per 100. **1918** A. Bennett *Pretty Lady* xxviii. 196 Auction afterwards? **1929** M. C. Work *Complete Contract Bridge* p. xi, Slowly but surely auction players began to realize that the new game had attractions. **1930** J. B. Priestley *Angel Pavement* xi. 580 They were now playing their third rubber of auction.

c. *Auction Bridge.* The act or process of bidding; a bid made in this way.

1908 'Auctioneer' *Auction Bridge for Three or Four Players* 31 The player who wins the auction takes the dummy as his partner. **1927** L. Hattersley *Contract & Auction Bridge Clarified* (ed. 2) i. 20 The Auction begins with the first declaration after each deal and continues until the final pass. **1958** *Listener* 2 Oct. 541/2 North bid Three Clubs and East became the declarer in Six Diamonds after this auction: *South* No bid [etc.].

d. Phr. *all over the auction*, everywhere. *Austral. slang.*

1930 K. S. Prichard *Haxby's Circus* v. 63 'Did I step on it, Doc?' Will asked. 'Step on it? You waltzed all over the bloomin' auction.' **1960** 'N. Shute' *Trustee from Toolroom* i. 11 You'd be surprised at the number of letters that there are—all over the auction.

5. *auction-room, -sale*; **auction bridge, pool:** see *Bridge *sb.*[2] b, *Pool *sb.*[2]

1767 J. Wedgwood *Let.* 31 May (1965) 55 Everybody would be apt to stroll into an Auction room. **1775** John Andrews *Lett.* (1866) 81 The soldiers..took quarters..in Gould's auction room or store. **1791** Boswell *Johnson* 8 Apr. 1775 A certain celebrated actor was just fit to stand at the door of an auction-room with a long pole, and cry Pray gentlemen, walk in'. **1936** V. A. Demant *Christian Polity* ii. 36 The auction-room flavour of our secular moralities. **1820** *Deb. Congress U.S.* I. 367 The evils arising from auction sales. **1888** [see Sale *sb.*[1] 1 b]. **1898** *Daily News* 3 Mar. 3/3 Auction sale of rights of patronage was absolutely prohibited.

auctorial (ǭktō·riăl), *a.* [f. L. *auctor* Author *sb.*: see -orial.] Of or pertaining to an author. (Cf. Authorial *a.*)

1821 *Examiner* 300/1 Pricked to it..by auctorial jealousy. **1900** M. H. Spielmann *Ruskin* 65 What is his reflection on his own auctorial life? **1926** A. Brent Smith *Studies & Caprices* 153 Auctorial bias. **1949** Wellek & Warren *Theory of Lit.* iii. 27 Eliot makes the judgment of responsibility depend on both auctorial intention and historic effect.

aucuba. Add: **2.** *aucuba mosaic* [*Mosaic *a.*[1]], a mosaic disease which attacks the leaves of solanaceous plants.

1922 H. M. Quanjer in *Rep. Internat. Potato Conf.* 136 On account of its resemblance to the variegation of Aucuba japonica, I call it Aucuba-mosaic [of potato]. **1936** *Nature* 21 Mar. 500/2 It was not..possible to distinguish serologically the viruses of tobacco mosaic, aucuba mosaic and tomato streak.

audad, var. *Aoudad.

Audenesque (ǭdĕne·sk), *a.* [f. the name of W. H. *Auden* (b. 1907), + -esque.] Resembling in matter, style, or quality the works of Wystan Hugh Auden, poet and critic.

1940 *Scrutiny* IX. 291 Facetious quasi-ballads in the Audenesque mode. **1944** *Horizon* IX. 209 The point of this little Audenesque folly is that history has revealed to us the volcano under the Victorian world.

‖**au désespoir** (o dezęspwār). [Fr.] In despair.

1766 C. Anstey *New Bath Guide* i. 9 My Spirits flag, my Life and Fire Is mortify'd *au Desespoir*. **1768** Sterne *Sent. Journ.* I. 142 Monsieur his master was *au desespoire* for her re-establishment from the fatigues of her journey. **1801** *Wynne Diaries* 6 Sept. (1940) III. iii. 62 Mrs. Otway has been a widow ten days...She does not seem au desespoir. **1859** Trollope *Bertrams* II. v. 109 Some time since she was giving advice that it should be broken off, and now she was *au désespoir* because that result had been reached.

Audi (odi). *Archæol.* Applied *attrib.* to remains of the lower Aurignacian period resembling those found at l'Abri *Audi*, a rock-shelter near Les Eyzies, Dordogne, France, and to the culture they represent. Also *absol.*

1921 M. C. Burkitt *Prehist.* iv. 72 In the Audi tool, the pointed knife edge is produced by a flake struck off the flint, this edge being sometimes trimmed. **1927** Peake & Fleure *Hunters & Artists* iv. 42 The characteristic Audi tool is found with considerable frequency throughout most Lower Aurignacian deposits. **1932** *Antiquity* VI. 192 In France the Aurignacian culture is divisible into five stages: the Audi, the Châtelperron, [etc.].

audience. Add: **7. c.** *transf.* Listeners to radio programmes or viewers of television.

1928 *B.B.C. Handbk.* 1929 259 The audience for broadcast entertainment has already far outstripped in size any other audience in the world. **1936** *B.B.C. Ann.* 85/2 The audience for the daily broadcasts to schools constitutes another special section of the public. **1952** *Ann. Reg. 1951* 400 Television's *For the Children*..won an avid audience.

d. *attrib.* and *Comb.*, as *audience participation*, sharing by an audience in a broadcast programme, etc.; *audience-rating*, assessment of the audience of a radio or television programme; *audience research* (cf. *listener research*), see quot. 1951; hence *audience-researcher*.

1812 *Dramatic Censor for 1811* 99 The hall, or audience part of the House, to comprise the segment of a circle. **1940** E. McGill *Radio Directing* x. 201 On audience-participation broadcasts the script is no more than a guide-post. *Ibid.* 203 An audience-participation program should never be built around a person who is not a ready improviser. **1940** *Q. Jrnl. Speech* Feb. 134 Experimental productions may teach the playwright..the importance of audience-reactions in the revision of a script. **1948** *Penguin Music Mag.* Feb. 52 Intelligent audience-participation is more and more possible. **1950** L. A. G. Strong *Which I Never* ii. 48, I was thinking less of intrinsic quality, of skill, than of what I believe is termed audience appeal. Box-office. **1950** *Times* 6 Sept. 2/5 An analysis by the B.B.C. Audience Research Department of the social grades of listeners. **1951** *B.B.C. Year Book* 144 The BBC maintains an Audience Research Department to advise it on the

habits, tastes, and opinions both of listeners and of viewers. **1955** Koestler *Trail of Dinosaur* 91 The radio performances of a Bach cantata and of a sobbing crooner are compared on the same scale of audience-rating. **1959** *Observer* 8 Feb. 18/3 I.T.V...is now estimated by the audience-researchers to have an average daily audience or viewing public of 5,250,000.

audile (ǭ·dəil), *a.* and *sb.* [Irreg. f. L. *audīre* to hear + -ile.] **A.** *adj.* Pertaining to or received through the auditory nerves. Of a person: of or pertaining to an audile.

1897 *19th Cent.* Aug. 229 The phenomena..may be dichotomised as (1) audile, (2) visual. The audile subdivide into (1) Footsteps. (2) Voices..(3) Raps..(4) and (5) Noises.. (6) A detonating noise. **1909** *Daily Chron.* 22 Feb. 4/7 Unless you are a microcephalous idiot, you are either Audile, Motile, or Visile. **1919** E. Barker in H. G. Wells *Outl. Hist.* xv. 86/2 Homer..is audile, not visual. **1956** H. Read *Art of Sculpture* iv. 71 Thus there are visual types, tactile types, and audile types.

B. *sb.* A person in whom auditory images are predominant over motile and visual presentations.

1886 *Mind* July 415 M. Paulhan, an audile, declares..he can represent the auditory images of *i* and *u* while the motor presentation of *a* is being presented. **1917** J. Adams *Student's Guide* 23 Some prefer to learn through the eye, others like to learn through the ear, still others through the sense of touch. The first kind are called *visuals*, the second *audiles*, the third *tactiles*.

audio- (ǭ·dio). [Combining form, f. L. *audī-re* to hear + -o; cf. Audiometer.] **I.** **audio frequency, audio-frequency,** a frequency capable of being perceived aurally; abbrev. **A.F., a.f.**; **audioge·nic** *a.*, caused by sounds of high frequency, *spec.* of a seizure so induced; **au·diogram** [see -gram], the diagram traced by an audiometer; **audio·logy,** the science or study of hearing; hence **audio·logist,** one who specializes in audiology; **au·diophile** [see -phil], a devotee of high-fidelity reproduction of sound (chiefly *U.S.*); **au:diospe·ctrograph,** an instrument for analysing sound; a diagram traced by this instrument; **au:diospectro·meter** (see quot. 1957); **audio-visual** *a.*, pertaining to both hearing and vision, esp. of mechanical aids to teaching (see quot. 1959).

1913 *Proc. Inst. Radio Engin.* I. 102 The audio-frequency produced is equal to the difference in the fundamental oscillation frequencies. **1916** *Wireless World* Aug. 344 (*title*) The Design of the Audio Frequency Circuit of Quenched Spark Transmitters. **1918** W. H. Eccles *Wireless Telegr.* (ed. 2) 487 A.F. Abbreviation for 'Audio Frequency'. **1919** E. W. Stone *Elements of Radiotelegr.* 17 Frequencies from 25 to 10,000 cycles per second are termed audio frequencies. **1936** *Electronics* Jan. 44/2 Resistance coupled a–f amplifiers. **1941** *B.B.C. Gloss. Broadc. Terms* 4 Audio-Frequency: Rate of oscillation corresponding to that of sound audible to the normal human ear (i.e. within the range of about sixteen cycles per second to about fifteen thousand cycles per second). **1943** *Electronic Engin.* XVI. 69 No single curve can be taken as universally representative of the distribution of speech energy throughout the audio frequency band. **1941** C. T. Morgan & H. Waldman in *Jrnl. Compar. Psychol.* XXXI. 1 (*title*) 'Conflict' and Audiogenic Seizures. **1959** *Chambers's Encycl.* XI. 334/1 These so-called audiogenic seizures..perhaps have more in common with epileptic convulsions than with neurotic manifestations as ordinarily understood. **1930** *Archives of Otolaryngology* XII. 760 An audiogram is a graphic representation of hearing capacity. **1945** *Electronic Engin.* XVII. 451 Audiograms, i.e. plots of hearing acuity against frequency, taken from a large number of deaf people. **1947** Hallowell Davis *Hearing & Deafness* i. 4 'Audiology', meaning the science of hearing seems to be a useful name for this field. *Ibid.* [S. R. Silverman] xiv. 363 In consultation with the clinical audiologist (a person trained in meeting problems of deafness) he [*sc.* the otologist] is prepared to suggest whatever measures are indicated. **1952** *Lancet* 15 Nov. 967/1 In clinical audiology the normal unit of intensity is the decibel. **1951** *High-Fidelity* I. 4 (*title*) Audio-Phile's Bookshelf. **1953** *Electronic Engin.* XXV. 306 An 'audiophile' whose desire is to achieve high quality reproduction. **1954** *Ohio Jrnl. Sci.* LIV. 298 Audio-spectrographs of the song of *N. ensiger*. *Ibid.* 303 A tape recording of the song of *Neoconocephalus ensiger* (Harris) was analyzed by means of an audio-spectrograph. **1962** *Science Survey* XVII. 278 The sounds must be recorded because only then can they be examined by the oscillogram and audio-spectrograph. **1952** Benson & Hirsh in *Jrnl. Acoust. Soc. Amer.* XXIV. 453 (*title*) A simplified audiospectrometer. **1957** *New Scientist* 9 May 26/2 The panoramic audiospectrometer shows the frequencies present in a sound at any given instant... A second type of audiospectrometer called the 'Sonagraph' graphs the variation in frequency with time for a part of the sound. **1937** Townsend & Stewart (*title*) Audio-visual aids for teachers. **1959** L. M. Harrod *Libr. Gloss.* (ed. 2) 26 *Audio-visual aids*, material such as gramophone records, tape recordings and various visual aids used as an adjunct to teaching.

II. Independent use of prec. as a quasi-sb. used *attrib.*: of or pertaining to frequencies within the range of audibility; relating to the reproduction, transmission, or reception of sound. Also **audio typist,** one who types directly from material previously recorded (as on magnetic tape); hence **audio typing** *vbl. sb.* and *ppl. a.*

[**1924** Gibson & Cole *Wireless of To-day* 11 *Audio*, within the limits of audition.] **1930** *Proc. Inst. Radio Engin.* XVIII. 159 If..the radio tubes obtain their grid and plate voltage from the same points supplying the first audio tube [etc.]. **1935** *Discovery* Sept. 277/2 They are providing ever better products and service to enable the listening public to get more enjoyment from the 'audio' programmes ..and will be ready to cater for those who wish..to see such 'video' items as may become available. **1937** *Ibid.* Nov. 331 (*caption*) Audio Transmitter. **1943** *Gloss. Terms Telecommunic.* (*B.S.I.*) 3 *Audio range*, the range of frequencies audible to the normal human ear. **1948** *Audio Visual Guide* XIV. 7 (*title*) A brief survey of currently available audio equipment. **1958** *Observer* 20 Apr. 10/5 A closer understanding between audio engineers and musicians. **1959** *Times* 3 Sept. 14/1 The audio-typing pool. *Ibid.*, Most kinds [of tape-recording machines] enable the employer to dictate letters into a hand microphone... With the help of another machine..an 'audio-typist' completes the process. **1960** *Times* 14 Jan. 19/4 Two audio channels—each with a bass 'woofer' and treble 'tweeter' to cover the extremely wide compass of stereo sound. **1962** H. E. Beecheno *Introd. Bus. Stud.* viii. 75 By audio-typing methods he can dictate material when he is free. **1966** *Daily Tel.* 19 Aug. 20/2 Nearly all studiously avoid the term 'typing pool'. Instead, 'audio room' or 'the typing service unit' is preferred.

Occas. used *absol.*

1940 *Broadcasting* 1 June 32 In addition to *video*, television employs *audio* (= sound signals).

audiometer. (Further examples.) For def. read: An instrument for measuring the sensitivity of the ear to sounds of different frequencies.

1879 B. W. Richardson in *Proc. R. Soc.* XXIX. 65 Professor Hughes' new Instrument for the Measurement of Hearing; the Audiometer. *Ibid.* 70 The world of medicine.. is under a deep debt of gratitude to Professor Hughes for his simple and beautiful instrument, which I have christened the audimeter, or less correctly but more euphoniously, the audiometer. **1919** *Trans. Amer. Otol. Soc.* XV. 24 The pitch range audiometer as we have it today, measures the tonal range from 30 double vibrations to 10,000 double vibrations; that is, the practical range of hearing. **1959** *Chambers's Encycl.* VI. 721/1 Audiometers are of two types: one tests hearing for speech and the other acuity for pure-tones of the kind produced by tuning-forks.

Hence **audio·metry,** the testing and measurement of the sense of hearing; **au:diome·tric** *a.*, of or pertaining to audiometry; **au:diometri·cian, audio·metrist,** one who specializes in audiometry.

1889 *Cent. Dict.*, Audiometry, audiometric. **1924** *Volta Rev.* Jan. 10 (*title*) Audiometric Measurements and Their Uses. **1931** *Lancet* 4 July 20/2 A method of audiometry by tuning-forks. **1948** *Hansard Commons* CCCCXLV. W.A. 221 To recruit trained audiometrists to ensure the suitable allocation of the aural aids. **1952** *Lancet* 15 Nov. 967/2 Any worker who was to be exposed continuously to a noise of more than 90 decibels' intensity should have his hearing tested by modern audiometric methods before starting work and at intervals afterwards. **1963** *Hark!* IV. 70 They were helped by the loan of equipment for free hearing tests, with two audiometricians in attendance throughout the displays.

audion (ǭ·diǫn). *Radio.* [Coined in 1906 by C. D. Babcock, assistant to the inventor, Lee de Forest; f. L. *audire* to hear+*-on*.] A former trade name for a three-electrode thermionic valve, used as a detector and an amplifier. (*Disused.*)

1911 Lee de Forest *U.S. Pat.* 995,126 1/2, O represents an oscillation detector..herein shown as an audion. **1914** E. H. Armstrong in *Electrical World* 12 Dec. 1149 (*title*) Operating Features of the Audion. **1916** *Chambers's Jrnl.* 63/1 The audion, or wireless lamp, which plays such a part in wireless telephony between Arlington, Hawaii, California, Paris and other distant parts.

audition, *sb.* Add: **1. b.** A trial hearing or performance of an actor, singer, etc., seeking employment.

1881 *Scribner's Monthly* May 122/1 The director of the Académie de Musique..fixed a day for her *audition* at the theatre. **1908** *Evening News* 18 June 3/3 When she was nineteen she was given an 'audition' at the Santa Cecilia Conservatoire. **1926** *Westm. Gaz.* 30 Jan., The plaintiff gave auditions to several girl saxophonists to fill the vacancy. **1933** P. Godfrey *Back-Stage* xv. 190 The chorus-girl gets her jobs by attending auditions.

audition (ǭdi·ʃən), *v.* [f. the sb.] **1.** *trans.* To give an audition to (an applicant); to test by means of an audition.

1935 *Punch* 18 Sept. 325/3 'Players who wish to be auditioned..' *B.B.C. Advertisement.* **1958** *Sunday Times* 26 Jan. 20/4 Here the sixty-eight-year-old playwright auditions Jacqueline Foster for a part. **1959** *Times* 14 Sept. 5/1 When I auditioned English dancers for the London production.

2. *intr.* To undergo an audition; to be tested by means of an audition. orig. *U.S.*

1937 *Variety* 17 Mar. (*heading*) Ice-Skater auditions in Rockefeller Plaza to ag[en]cy 14 floors up. **1938** *Amer. Speech* XIII. 194 Candidates for radio work at first are given an audition; later they simply *audition*. **1955** Begley & MacCrae (*title*) Auditioning for TV: How to prepare for success as a television actor.

auditor. Add: **4. b.** (See quot.)

1919 Moore-Anderson *Sir Robert Anderson* i. 4 Of his University life [at Trinity Coll., Dublin] he..cherished pleasant memories..associated with the College Historical Society, of which he became Auditor, a position corresponding to that of President of the Union at Oxford or Cambridge.

auditorium. Add: **1.** (Earlier and later examples, of public buildings.) Also (*U.S.*) applied to the entire building.

1854 W. Irving *Let.* 22 Nov. in P. M. Irving *Life & Lett.* (1864) IV. 158, I mingled in the crowd, and heard Bancroft's erudite address from the 'auditorium'. **1928** F. Hurst *President is Born* iv. 35 Bek..arranged his one-man [water-colour] show in the Auditorium of the Tallahassee High School. **1929** H. G. Wells *King who was King* viii. 248 The lights in the auditorium [of the cinema] go up. **1954** *Manch. Guardian Weekly* 18 Feb. 8/4 Joe's first stop was Charleston, West Virginia, where..he drew a good crowd of 2,800 to an auditorium that had seats for 3,517.

b. *attrib.*

1898 J. Hollingshead *Gaiety Chron.* ii. 50 The room belonged to the *auditorium* side of the curtain. **1908** *Westm. Gaz.* 7 July 10/1 The design of the Auditorium building is in the Renaissance style of architecture.

auditory, *a.* Add: **1. b.** in *Comb.*

1936 J. Kantor *Objective Psychol. Gram.* iii. 32 Comparative grammar deals with auditory-vocal mass phenomena abstracted from the linguistic adjustments of the persons *using* such languages. **1964** M. A. K. Halliday et al. *Linguistic Sciences* iii. 59 The term 'consonant'..is..an auditory-articulatory label.

Audubon (ǭ·dŭbǫn). The name of J. J. *Audubon* (1785–1851), American ornithologist, used *attrib.* or in the possessive case of certain birds.

1837 Townsend in *Jrnl. Acad. Nat. Sci. Phila.* VII. 191 Audubon's Warbler..inhabits the forests of the Columbia river. **1909** Webster s.v. *shearwater*, The Manx shearwater..and the Audubon's shearwater..common about the West Indies, Florida coasts, etc., are small species. **1933** *Discovery* Feb. 62/2 The petrel known as Aud[u]bon's shearwater is sometimes confused with the capped petrel, 'Le Diablotin', owing to the fact that both birds have the same colloquial name in the different islands [*sc.* Dominica and Grenada].

au fait, *advb. phr.* Add: (Earlier and later examples.) Also const. *to*, or (most freq. in recent use) *with*.

*a*1743 Ld. Hervey *Mem.* (1952) v. 221 Your Majesty being *au fait* of all transactions..for these last ten years. **1821** *New Monthly Mag.* II. 220 Yet, to be *au fait* at nonsense, is no easy matter. **1849** C. Brontë *Shirley* I. vi. 157 For the benefit of those who are not 'au fait' to the mysteries of the 'Jew-basket' and 'Missionary-basket',..these 'meubles' are willow-repositories. **1906** *Nature* 13 Sept. 485/1 The author is..not *au fait* with any up-to-date work. **1916** T. E. Lawrence *Let.* 22 Dec. (1938) 213, I hope..to run up first to Feisal and so put you au fait with his intentions. **1936** *Punch* 3 June 624/2 Now Daniel would have been perfectly *au fait* with such a situation.

‖**aufgabe** (au·fgabə). [G.] A task, an assignment; used esp. in *Psychol.* Cf. *Task *sb.*

1902 W. James *Var. Relig. Exper.* xvi. 389 The notion is thoroughly characteristic of the mystical level, and the *Aufgabe* of making it articulate was surely set to Hegel's intellect by mystical feeling. **1909** —— *Pluralistic Universe* vi. 238 This amount is the datum or *gabe* which reality feeds out to our intellectual faculty; but our intellect makes of it a task or *aufgabe*. **1917** M. A. May in *Arch. Psychol.* xxxix. 2 *Aufgabe*. There are two gross meanings: (1) The problem set by the experimenter, the instructions given, and (2) the problem understood and actualized by the observer or subject.

‖**Aufklärung** (au·fklęruŋ). [G., 'enlightenment'.] Enlightenment (sense 2), illuminism; the name given to a European intellectual movement in the 18th c. laying claim to extraordinary intellectual illumination and enlightenment. So **Au·fklärer** *pl.*, the members of this movement.

1801 H. Crabb Robinson in E. J. Morley *Crabb Robinson in Germany 1800–1805* (1929) 49 Till the *Aufklärer* or Enlighteners—contrived to draw to themselves the larger body. **1842** J. S. Mill *Let.* 22 Aug. in *Wks.* (1963) XIII. 543, I hope to hear from yourself somewhat more about the Berlinische Aufklärung from personal knowledge. **1856** [see Illuminism]. **1865** J. H. Stirling *Secret of Hegel* I. p. xix, Those remnants of the *Aufklärung*, of Eighteenth Century Illumination, which still exist among us. *Ibid.* p. xxii, The one principle of the Aufklärung, its single outcome—the Right of Private Judgment. **1933** J. Baillie *Life Everlasting* (1934) viii. 259 Humanism had its real fruitage in the *Aufklärung*.

‖**au fond** (o foṅ), *advb. phr.* [F.] At bottom, basically, essentially.

1782 Lady Derby *Let.* 1 Jan. in Duke of Argyll *Intimate Society Lett. of 18th C.* (1910) I. 280 Prince Hen: says he [*sc.* Prince Ernest of Prussia] is charming, but *au fond* he is french. **1847** Geo. Eliot *Let.* 10 May (1954) I. 235 But *au fond*, dear Mary, I have no impiety in my mind at this moment. **1890** G. B. Shaw in *World* 19 Nov. 23/2 If you know your Don Giovanni *au fond*. **1933** M. Lutyens *Forthcoming Marriages* 182 *Au fond* every man is a hunter, a cave-man.

‖**augen** (au·gən). [pl. of G. *auge* Eye *sb.*[1]] Applied to a variety of gneissic rock containing eye-shaped masses of feldspar or quartz; porphyritic gneiss; esp. *augen-gneiss*.

1885 A. Geikie *Text-bk. Geol.* (ed. 2) vii. 132 Porphyritic gneiss or Augengneiss, in which large eye-like kernels of orthoclase are dispersed. **1910** *Encycl. Brit.* XII. 149/2 When large felspars, of rounded or elliptical form, are visible in the gneiss, it is said to have augen structure. *Ibid.* 352/2

The larger crystals are converted into lenticular or elliptical 'augen'. **1925** N. E. Odell in E. F. Norton *Fight for Everest 1924* iii. iii. 297 It would certainly appear that the augen variety [of gneisses] were the younger and intrusive facies. **1940** C. M. Rice *Dict. Geol. Terms* 27/2 *Augen-gneiss*, a general term for gneissose rocks,..containing 'eyes', i.e., phacoidal or lenticular crystals, or aggregates, which simulate the porphyritic crystals of igneous rocks.

auger. Add: **2*.** 'A large spiral bit used to mix a material and force it through a die (as in a brickmaking machine or a meat grinder); the rotating helical member of a screw conveyor' (Webster 1961). Hence as *v. trans.*, to convey by an auger.

1934 Webster, *Auger*, a rotating screwlike device for advancing chaff in some forms of spiral conveyors; a large spiral bit used to mix a material and force it through a die, as in a brick or a sausage machine. **1960** *Farmer & Stockbreeder* 5 Jan. 74/3 Oats are shovelled or augered into the feed mixer until a flashing light indicates the feed mixing drum is full. *Ibid.* 8 Mar. 42/2 Regulated auger feed from food hoppers to mangers. **1963** *Gloss. Terms Agric. Machinery* (*B.S.I.*) 82 *Auger conveyor*, a grain conveyor in which material is moved by the screwing action of a helix inside a tube or trough.

Auger (ōᵘ·ʒe). [Name of P. V. *Auger*, French physicist (b. 1899).] Used *attrib.* in *Auger effect, electron* (see quot. 1962).

[**1930** G. Wentzel in *Physik. Zeitschr.* XXXI. 1007/2 Den wir erläutern wollen..am Auger-Effekt.] **1931** *Chem. Abstr.* XXV. 3911 This reviews quantum processes such as.. the Auger effect. **1949** M. G. E. Cosyns et al. in *Proc. Physical Soc.* LXII. 803 The criterion..is that slow electrons should be admitted as Auger electrons only if the beginning of the track can be certainly identified within 1 micron of the end of the meson. **1962** *Newnes Concise Encycl. Nuclear Energy* 53/1 *Auger effect*, the non-radiative transition of an atom from an excited electronic energy state to a lower state with the emission of an electron... The electron ejected in the Auger effect is known as an Auger electron.

augmented, *ppl. a.* **2. b.** Read: Of an interval: greater by a chromatic semitone than a perfect, or than a major, interval of the same name: opp. to *diminished*. (Earlier and later examples.)

1825 [see Fifth *sb.* 2 a]. **1947** A. Einstein *Mus. in Rom. Era* ix. 100 The Sanctus, which leaves the major key of F by way of an augmented fifth chord. **1954** *Grove's Dict. Mus.* (ed. 5) IV. 522/1 An augmented interval, so limited, is a chromatic interval regarded as a perfect or a major diatonic interval extended by adding to it a chromatic semitone.

augmentor (ǭgme·ntǫɪ). [f. Augment *v.*+-or.] **1.** var. Augmenter.

2. *Physiol.* Applied to a nerve or nerve cell by the stimulation of which the cardiac contractions are increased. Also *attrib.*

1902 *Nature* 6 Nov. 3 The anabolic inhibitory and the katabolic augmentor parts. **1907** *Practitioner* Nov. 692 The imperfectly oxygenated blood has to be sent through the heart and other muscles oftener per minute than healthy blood; the augmentor nerve is, therefore, called into action.

3. A substance which increases the action of an auxetic substance although unable itself to initiate cell-division.

1915 Stedman *Med. Dict.* (ed. 3) 90/2 *Augmentor*, a hypothetical substance supposed to increase the action of an auxetic or a kinetic. **1918** J. A. Thomson in *New Statesman* 26 Jan. 400/1 The dying away in autumn and winter produces substances ('auxetics') which later on promote.. an increasing quantity of certain other substances ('augmentors') which give more power to the elbow of the first.

‖**au grand sérieux** (o graṅ seriö), *advb. phr.* [F.] In all seriousness: const. to treat, take a matter, a person, etc., *au grand sérieux*. Cf. *Au sérieux.

1849 Thackeray *Pendennis* I. xvi. 139 Pen..took the matter *au grand sérieux*, with the happy conceit and gravity of youth. **1865** Mrs. Gaskell *Wives & Daughters* (1866) II. v. 48 You mustn't go and take me *au grand sérieux*. **1930** *Times Lit. Suppl.* 9 Oct. 800/2 The author..is being treated *au grand sérieux*..to an extent which he does not quite merit. **1936** C. S. Lewis *Alleg. Love* iv. 163 They all took Chaucer's love poetry *au grand sérieux*.

au gratin: see *Gratin.

Augustal, *a.* For †*Obs.* read *Hist.* and add: *Augustal Prefect* (L. *præfectus Augustalis*), the title of the prefect of Egypt.

1731 *Hist. Litteraria* II. 57 The great power that was lodged in the Augustal Praefect of Egypt. **1907** W. G. Holmes *Age of Justinian & Theodora* II. vii. 475 The Augustal Praefect resigned the control of all Egypt for that of Alexandria and the adjacent country.

Augustan, *a.* Add: **2.** (Earlier and later examples.) Also, applied *spec.* to 17th and 18th cent. English literature; and *absol.*

1712 J. Oldmixon *Refl. on Dr. Swift's Let.* 19 King Charles the Second's Reign, which probably may be the Augustan Age of English Poetry. *a*1745 Swift *Thoughts on Var. Subjects* in *Wks.* (1897) I. 284 Charles the Second's reign..is reckoned, though very absurdly, our Augustan age. **1759** Goldsmith *Bee* 24 Nov. No. 8. p. 235 (*title*) An Account of the Augustan Age of England. **1772** H. Walpole *Let.* 21 July (1904) VIII. 184 What a figure will this our Augustan age make; Garrick's prologues, epilogues, and verses, Sir W[illiam] Chambers's *Gardening*, Dr. Nowel's sermon. **1861** M. Arnold *On Transl. Homer* i. 29 Chapman translates his object into Elizabethan, as Pope translates it into the Augustan of Queen Anne. **1916** G. Saintsbury (*title*) The Peace of the Augustans. *Ibid.* 389 To see the beauties, to hear the music, and to taste the sweetness or the tartness, the bitter and the salt, of Augustan poetry. **1952** I. Jack (*title*) Augustan satire. Intention and idiom in English poetry 1660–1750.

3. (Earlier example.)

1565 Shacklock tr. *S. Hosius's Hatchet of Heresies* 71 *margin*, In his Apologie of the Augustane confession.

Augustanism (ǭgv·stăniz'm). [f. Augustan *a.*+-ism.] The condition of being Augustan.

1903 T. Watts-Dunton in *Chambers's Cycl. Eng. Lit.* III. 5/1 [Gray's] chief poem, the famous elegy, furnishes a striking proof of the poet's slavery to Augustanism. **1904** *Athenæum* 16 Jan. 73/1 The period of Augustanism in English literature—that age of acceptance which began after Milton and ended with Gray and Collins. **1955** J. Lawlor in *Ess. & Stud.* VIII. 61 This is the real achievement of Augustanism..the imposition of order, the self-possession of the artist.

Auguste (au·gust). [Adopted in Fr. spelling *auguste*, f. G. *August* (*der dumme August* circus clown).] A type of circus clown (see quots.).

The 'explanation' of the origin of the word in quot. 1935 is one of several unsubstantiated stories. Robert, *Dict. de la Langue Franç.*, regards Fr. *auguste* as a noun that has arisen by antonymy from the adj. *auguste* 'august, majestic', but evidence is lacking.

1910 W. J. Locke *Simon the Jester* iv. 42 Auguste..is the generic name of the clown in the French Hippodrome. **1935** *Observer* 15 Dec. 13/3 The Auguste is the man with the battered bowler hat, his jacket on back to front..who always gets the worst of it... The name is supposed to have come about this way: A century ago an English clown, playing in Germany, had the novel idea..of going into the circus with a battered hat and his clothes all wrong. The audience began shouting out, 'August! August!' (The German phrase, 'dumme August' is equivalent to our word 'zaney' or 'addle-pate'.) And the name has stuck to this day. **1950** *Oxf. Jun. Encycl.* IX. 139/1 In the late 19th century a new type of clown appeared in the circus. He was called an 'auguste', and wore ill-fitting ordinary clothes..and a large bulbous red nose.

Augustean, *a.* For *rare*⁻¹ read *rare* and add further example.

1690 [Atterbury] *Pref. to Second Part of Mr. Waller's Poems* sig. A4, I question whether in Charles the Second's Reign, English did not come to its full perfection; and whether it has not had its Augustean Age, as well as the Latin.

auk. Substitute for def.: Any bird of the family *Alcidæ* of diving birds, predominantly black, white, or grey in colour, inhabiting mainly the colder parts of the northern oceans and characterized by short wings, tail, and legs, and webbed feet. The auks include the guillemot, puffin, razor-bill, little auk, and the extinct and flightless great auk. (Earlier and later examples.)

1674 [see Razor-bill 1 a]. **1802, 1894** [see Gare-fowl].

auklet (ǭ·klėt). [f. Auk+-let.] Applied to any of various species of small auk.

1886 R. Ridgway *Man. N. Amer. Birds* 13 Simorhynchus pusillus (Pall.) Least Auklet. **1901** R. C. Lehmann *Anni Fugaces* 135 The last Great Auk..left..Two orphaned Auklets in their shell. **1956** R. Peterson et al. *Field Guide Birds Brit. & Europe* (ed. 2) 297 Paroquet auklet..Crested auklet.

aul, var. *Aoul.

Aularian, *a.* Add: Pertaining to or characteristic of a hall in a collegiate university, esp. in respect of its power of self-administration.

*c*1849 *Stranger's Guide Oxford* (ed. 6) 92 As the foundation of this establishment marks a new era in the academical annals of this University, it is not without reason..that.. the name of *New College* should still attach to this erection. Before this period, however, the Aularian system was generally prevalent. **1895** Rashdall *Universities Europe Mid. Ages* II. 624 Aularian Statutes at Oxford. *Ibid.* 626 Even these Oxford Statutes which had for their very object to assert the authority of the University and of the Principals contain traces of the old Aularian self-government.

auld, *a.* (Examples of *auld lang-syne*.)

1692 'Jacob Curate' *Scotch Presb. Eloq.* iii. 101 Now billy Jonah, wilt thou go to Nineveh for ald lang syne [*marginal note*, old kindness]. **1721** Ramsay *Poems* 72 [*title*] The Kind Reception. *To the Tune of* Auld lang syne. **1788** [see Langsyne]. **1827** Scott *Two Drovers* i. in *Chron. of Canongate* I. xiii. 303, I am Hugh Morrison from Glenae, come of the Manly Morrisons of auld langsyne. **1896** G. B. Shaw in *Our Theatres in Nineties* (1931) II. 140 That smile may have meant sentimental memories of auld lang syne. **1901** W. Laidlaw *Poetry & P.* 58 On auld lang syne I'll stand and ponder.

‖**au mieux** (o myö), *advb. phr.* [Fr., lit. 'at the best'.] On the best of terms *with* (someone); on very intimate terms.

1848 Thackeray *Van. Fair* li. 450 Both attachés of the embassy..declared that they were *au mieux* with the charming Madame Ravdonn. **1849** —— *Pendennis* I. xxxviii. 374, I thought you used to be *au mieux* in that quarter. **1953** E. M. Forster *Hill of Devi* 145 The Maharajah, out of sheer mischievousness, accused him of being *au mieux* with a lady.

‖**au naturel** (o natürel), *advb. phr.* [F.] In the natural state; cooked plainly; uncooked; undressed.

1817 *Blackw. Mag.* Nov. 192/2 They in general speak with great contempt of our national partiality for roast beef and potatoes 'au naturel'. **1827** Disraeli *Vivian Grey* v. xv. 303 To say the truth, I think it [*sc.* the make-up] is *au naturel*. **1860** *Leisure Hour* 22 Nov. 186/2 With good round paunch *au naturel*. **1862** Thackeray *Philip* I. xvi. 300, I.. forget whether it was a cold dagger *au naturel*, or a dish of hot coals *à la Romaine*, of which they partook. **1905** Mrs. H. Ward *Marriage W. Ashe* x. 181 You would have preferred ankles *au naturel*?

Aunjetitz (au·nyətits). The German form of the name of a village near Prague, Czechoslovakia, applied attrib. to the remains found there of an important Early Bronze Age culture and to the culture itself.

1925 V. G. Childe *Dawn Europ. Civilization* xii. 191 The first original bronze age civilization of Central Europe is called, after the great cemetery south of Prague, the Aunjetitz culture. *Ibid.* 191 Typical Aunjetitz mug. **1949** *Proc. Prehist. Soc.* XV. 99 The first effective European adoption of distinctive bronze-pin types..took place in the Early Bronze Age of Central Europe, primarily in the Aunjetitz culture with its centre in Bohemia.

aunt. Add: **1. b.** Also used *dial.* (see E.D.D.) as 'a term of familiarity or respect applied to elderly women, not necessarily implying relationship'. Also *transf.* Cf. *Auntie b.

Universal aunt is taken from the name of a bureau in London undertaking such services, e.g. conducted tours, errands, as might be rendered by a maiden aunt. Cf. Aunthood.

1793 [see Uncle *sb.* 2 b]. **1921** *Star* 28 Feb. 4/5 (headline) Professional Aunts on Hire. **1921** *Times* 15 Oct. 13/5 This sort of universal aunthood to the whole neighbourhood... It has been left to our own day to adopt her [*sc.* Miss Mulock's] idea in the more direct and concrete form of 'Universal Aunt'. **1928** H. Williamson *Pathway* xvii. 376 'I..thought I would look in, Aunt Connie.' 'Please don't call me Aunt Connie. I'm sorry to be so explicit.' **1944** H. G. Wells *'42 to '44* 144 Jane Austen is one of my dearest aunts. **1959** *Observer* 11 Jan. 15/1 There is an element of knight-errantry in the investigations with which *Shopper's Guide* and *Which?* continue to plough their way through the labyrinth of consumers' goods...We, the public, are at last being championed by a battalion of scientific aunts. **1962** 'J. Le Carré' *Murder of Quality* ii. 21 She..became..their guide, friend and universal aunt.

d. *U.S.* (See quots.)

1835 Longstreet *Georgia Scenes* 110 'There they are, Aunt Glory.' *Footnote*: 'Aunt' and 'Mauma'..are terms of respect commonly used by children to aged negroes. **1869** *Atlantic Monthly* Oct. 479/2 A pleasant, industrious 'Aunt Sally', a mulatto. **1904** Harben *Georgians* 106 A negro woman, Aunt Amanda,..passed in and out.

4. Aunt Sally. b. A nickname for a wicket-keeper in cricket. *colloq.*

1898 G. Giffen *With Bat & Ball* 239 Practice may improve an 'Aunt Sally'..but unless he has natural genius.. a lad is not likely to become a star wicket-keeper. **1927** *Observer* 29 May 28/1 A 'keeper'..who combines batsmanship with all the 'Aunt Sally's' excellencies.

c. *fig.* An object of unreasonable or prejudiced attack.

1898 G. B. Shaw in *Sat. Rev.* 30 Apr. 592/2 This comes of an author making no serious attempt to get to the point of view of the character he professes to have dramatised—of simply conspiring with the stupid section of the pit to make an Aunt Sally of it. **1905** [see *Ibsenistic *a.*]. **1958** *Oxf. Mag.* 1 May 398/2 Palpable gross hits at a favourite Aunt Sally, modern psychology.

5. *My (sainted) aunt!* and similar phrases, as trivial exclamations.

1888 *Boy's Own Paper* Summer No. 35/2 'My aunt!' exclaimed Guy, with a start. **1888** Kipling *Story of Gadsbys* (1889) 66 Prince Kraft a stable-boy—Oh, my Aunt! **1919** J. Buchan *Mr. Standfast* viii. 168 My holy aunt! The General disguised as Charlie Chaplin! **1921** Wodehouse *Jill Reckless* 301 'My sainted aunt!' he said slowly. **1928** A. Huxley *Point Counter Point* x. 163 Rampion..turned up his eyes. 'Oh, my sacred aunt!' he said.

6. Special collocations: *Aunt Edna*, used of a typical theatre-goer of conservative taste; *Aunt Emma*, used in croquet of a typically unenterprising player (or play); *Aunt Fanny*, in various slang phrases expressing negation or disbelief.

1953 T. Rattigan *Coll. Plays* II. p. xii, Let us invent a character, a nice, respectable, middle-class, middle-aged, maiden lady, with time on her hands and the money to help her pass it...Let us call her Aunt Edna...Now Aunt Edna does not appreciate Kafka...She is, in short, a hopeless lowbrow. **1958** N. F. Simpson *Resounding Tinkle* in *Observer Plays* 241 The author..leans forward..to make simultaneous overtures of sumptuous impropriety to every Aunt Edna in the house. **1960** E. P. C. Cotter *Tackle Croquet this Way* 66 Whatever happens don't become an Aunt Emma player. **1963** *Croquet* Aug. 3/1 Aunt Emma is banished for ever. **1967** *Ibid.* Aug./Sept. 13/2 He played too much 'Aunt Emma'. **1945** M. Dickens *Thurs. Afternoons* i. 69 She's got no more idea how to run this house than my Aunt Fanny. **1946** J. Irving *Royal Navalese* 24 Tell that to my old Aunt Fanny. **1954** 'G. Carr' *Death under Snowdon* v. 54 'Agree my Aunt Fanny,' retorted the other loudly.

auntie, aunty. Add: **b.** In U.S. applied esp. to a Negress (earlier examples). Also used in Britain, formerly only *dial.* (see E.D.D.), as a term of familiarity or respect applied to an elderly woman. Now increasingly used (*a*) in some social classes by a younger person of an unrelated older family friend, (*b*) *transf.* of an institution, etc., considered to be of conservative style or approach; *spec.* the B.B.C. Cf. *Aunt 1 b.

1835 J. H. Ingraham *South-West* II. 241 Nor are planters indifferent to the comfort of their gray-headed slaves... They always address them in a mild and pleasant manner as 'Uncle' or 'Aunty'. **1852** *Knickerbocker* Oct. 326 So long as the race of good old colored 'aunties' do the cooking. **1865** G. W. Nichols *Story Gt. March* 132 [To Negro woman] 'What was it that struck you, aunty?' **1937** Partridge *Dict. Slang* 20/2 *Auntie*..like *uncle*, used by children for a friend of the house. **1938** E. Bowen *Death of Heart* iii. vi. 439 'Rightie-o, auntie,' said the driver. **1959** *Guardian* 14 Oct. 6/2 A part-time teacher..paid her children's 'auntie' £2 a week to take them for afternoon walks. **1960** Willmott & Young *Family & Class in Lond. Suburb* i. 4 The East End is a different class altogether—people there call you Dad or Auntie. **1968** *Woman* 27 Jan. 18/3, I had read..about Mollie Dundas's unit for autistics at this hospital. Mollie was in urgent need of voluntary 'aunties', substitute mother figures, to work with her in treating them. *Ibid.* 3 Feb. 15/2 At this clinic I had been an 'auntie' ..for some months. **1931** A. Huxley *Music at Night* 134 The *Times* finds it profitable to employ someone..to talk to us every morning about our dear old Culture-Aunties and Uncles. **1958** J. Cannan *And be a Villain* i. 31, I saw about Uncle Edmund in auntie *Times*. **1959** J. Braine *Vodi* xxi. 231 Honestly, that woman's magazine stuff, just the sort of advice these damned aunties give. **1962** *Listener* 22 Mar. 529/1 The BBC needs to be braver and sometimes is. So let there be a faint hurrah as Auntie goes over the top.

‖**au pair** (o pę̄r, ōᵘ pē̄ɹ). [F., on equality.] Applied to an arrangement between two parties by which mutual domestic services are rendered formerly without consideration of money payment; *esp.* of a young girl learning the language of a foreign country while rendering certain services in return for hospitality. Also *attrib.* Hence as *sb.*, a person who is 'au pair'.

1897 *Girl's Own Paper* 16 Oct. 63/1 An arrangement.. frequently made is for an English girl to enter a French, German or Swiss school and teach her own language in return for joining the usual classes. This is called being *au pair*. **1928** *Sunday Express* 22 July 15 To have a German lady on 'au pair' terms in my house to teach German to my children in return for education in English. **1960** *Oxford Times* 1 Jan. 2/2 *Au pair* to help with young children. **1962** *Times* 22 May 13 An agency for placing *au pair* girls. **1967** *Guardian* 3 July 5/5 The au pair's Italian boy friend.

‖**au pied de la lettre** (o pye də la lętr'), *phr.* [Fr., lit. 'to the foot of the letter'.] Exactly, down to the last detail; literally.

1782 H. Walpole *Let.* 10 Nov. (1904) XII. 368 The Romans..loved to be obeyed *au pied de la lettre*. **1854** Mrs. Gaskell *North & South* (1855) II. vi. 78 Some one had told you that stolen fruit tasted sweetest, which you took au pied de la lettre, and off you went a-robbing. **1863** *Blackw. Mag.* Feb. 192/1 She does not always intend that we should believe her *au pied de la lettre*. **1936** A. Huxley *Eyeless in Gaza* xiii. 173 A purely figurative and metaphorical sanctity taken *au pied de la lettre*.

aura. Add: **2. b.** (Later example.) Also, a distinctive impression of character or aspect.

1901 *Harper's Mag.* XLI. 808/2 Tragedy is as old as story-writing itself—the *aura* thereof being the very breath of every great drama since Eden. **1921** *Glasgow Herald* 25 Aug. 5 The genteel aura of the upper circle. **1959** *Economist* 27 June 1164/2 No such charges were brought against Mr Strauss, though his Wall Street aura rankles with progressive Democrats.

c. A supposed subtle emanation from and enveloping living persons and things, viewed by mystics as consisting of the essence of the individual, serving as the medium for the operation of mesmeric and similar influences. So **au·ral** *a.*[3]

1870 P. B. Randolph *Seership!* (1884) 77 Peculiar substances can be charged with the efflux or aura of the human being. **1918** Violet Tweedale *Veiled Woman* vii. 83 [His] personality happens to be very distasteful to me. Our auras don't blend. **1920** Rose Macaulay *Potterism* iii. i, Unless..the immortal soul wraps itself about in some aural vapour that takes the form it wore on earth.

auramine (ǭ·rămin, -in). *Chem.* [G. (proprietary, 1884), f. L. *aur-um* gold + Amine.] A synthetic yellow dyestuff of the diphenylmethane series.

1884 *Jrnl. Chem. Soc.* XLVI. 1450 Auramine..is the first artificial yellow dye which can be fixed on the vegetable fibre with tannic acid like aniline dyes... Auramine also becomes fixed on wool, giving colours of great purity. **1888** *Jrnl. Chem. Soc.* LIV. 156 Commercial auramine is the hydrochloride of a base obtained by the action of ammonium chloride on tetramethyldiamidobenzophenone. **1920** Cross & Bevan *Textbk. Paper-Making* (ed. 5) ix. 269 Auramine Yellow, very largely employed in paper dyeing, stands near but not in this group, being a derivative of diphenylmethane. **1963** *Lancet* 5 Jan. 24/2 It is analogous to the detection of isolated tubercle bacilli in sputum after their staining with an auramine-rhodamine dye combination.

aurantia (ǫræ·nʃiă). [L. *aurantia*: see Orange *sb.*[1]] An orange-yellow dye colour; see also quot. 1940.

1877 *Jrnl. Chem. Soc.* XXI. 310 Aurantia...This dyestuff ..produces splendid orange tints on wool and silk. **1892** Woodbury *Encycl. Photogr.* 62 *Aurantia*, a beautiful orange dye, the ammonia salt of an acid. It is used as a sensitiser in orthochromatic photography. *Ibid.*, To make an aurantia screen for orthochromatic photographic purposes. **1940** *Thorpe's Dict. Appl. Chem.* (ed. 4) IV. 14/2 *Aurantia*..is a dyestuff, but is of more importance as an explosive.

aureate, *a.* **2.** Delete †*Obs.* and add later examples; *spec.* designating or characteristic of a highly ornamental literary style or diction (see quots.).

1819 T. Campbell *Spec. Brit. Poets* I. ii. 93 The prevailing fault of English diction, in the fifteenth century, is redundant ornament, and an affectation of anglicising Latin words. In this pedantry and use of '*aureate terms*', the Scottish versifiers went even beyond their brethren of the south. **1908** G. G. Smith in *Camb. Hist. Eng. Lit.* II. iv. 93 The chief effort was to transform the simpler word and phrase into 'aureate' mannerism, to 'illumine' the vernacular. **1919** J. C. Mendenhall *Aureate Terms* i. 7 Such long and supposedly elegant words have been dubbed 'aureate terms', because..they represent a kind of verbal gilding of literary style. The phrase may be traced back.. in the sense of long Latinical words of learned aspect, used to express a comparatively simple idea. **1936** C. S. Lewis *Allegory of Love* vi. 252 This peculiar brightness..is the final cause of the whole aureate style.

aureation (ǭri,ēⁱ·ʃən). [f. Aureate *a.* + -tion.] The condition of being aureate in literary style.

1908 G. Saintsbury in *Camb. Hist. Eng. Lit.* II. viii. 219 Undue aureation and undue beggarliness being equally avoided. **1936** C. S. Lewis *Allegory of Love* vi. 251 The aureation of his allegories.

aureole, *sb.* Add: **6.** *Geol.* The belt of metamorphosed rocks surrounding an igneous intrusion.

[**1884** A. J. Jukes-Browne *Student's Handbk. Physical Geol.* ii. xii. 442 By French writers the term *aureole* has been introduced to designate the concentric zone of metamorphosed rock which surrounds an intrusive mass of igneous rock.] **1896** J. F. Kemp *Handbk. Rocks* 126 *Aureole*, the area surrounding an igneous intrusion that is affected by contact metamorphism. **1905** J. Geikie *Structural & Field Geol.* xv. 214 The zone or aureole of altered rocks surrounding a large batholith of granite may be a mile or more in width.

aureole (ǭ·ri,ōᵘl), *v.* [f. Aureole *sb.*] *trans.* To encircle with or as with an aureole or halo. Chiefly in *pa. pple.*

1888 *Macm. Mag.* July 188 Their forms are intertwined with rainbows and aureoled with light. **1896** Mary Cowden-Clarke *My Long Life* 13 His seraph-like face,..aureoled by its golden hair. **1905** C. Kernahan *Visions* 240 The cloud-rack which aureoled the moon. **1933** L. A. G. Strong *Sea Wall* 258 The slanting rays of the sun aureoled it with mellow gold.

aureomycin (ǭ:riōməi·sin). [f. L. *aure-us* golden + Gr. μύκ-ης fungus + -in[1].] Name of an antibiotic substance, chlortetracycline, derived from the mould *Streptomyces aureofaciens*, or produced artificially, and used in medical treatment esp. of lung and rickettsial diseases.

1948 B. M. Duggar in *Ann. N.Y. Acad. Sci.* LI. 177 We are concerned with a species of the established *Streptomyces*, for which the name *Streptomyces aureofaciens* is being proposed... The antibiotic (aureomycin) from this species of mold will be evaluated. **1948** *Lancet* 16 Oct. 618/1 Aureomycin, derived from a strain of *Streptomyces aureofaciens* and effective against numerous gram-positive and gram-negative organisms..has been obtained as a yellow crystalline hydrochloride. **1949** H. W. Florey et al. *Antibiotics* II. 1533 The action of aureomycin was thought to be bacteriostatic rather than bactericidal. **1949** *Britannica Bk. of Yr.* 687/1 *Aureomycin*,..a golden-coloured antibiotic related to streptomycin, and effective in some diseases where streptomycin and penicillin fail.

‖**au reste** (o rest), *phr.* [F.] As for the rest.

1619 J. Chamberlain *Let.* 26 June in T. Birch *Court & Times Jas. I* (1848) II. 176 *Au reste*, he is fallen to his old diet. **1752** H. Walpole *Let.* 6 June (1903) III. 97 *Au reste*, he is just where he was. **1825** H. Wilson *Memoirs* I. i. 22 'Au reste,' continued Julia, 'some day, perhaps soon, you shall know all about me.' **1853** C. Brontë *Villette* I. xi. 193 Au reste it was only the most temporary expedient in the world.

aureus (ǭ·ri,ŭs). Pl. aurei (-əi). [L., substantival use (sc. *nummus*) of *aureus* golden, f. *aurum* gold.] A Roman gold coin of the late republic and empire, of the value of 25 silver denarii.

1609 Holland *Amm. Marcell.* xx. iv. 149 He..promised unto them all throughout five *aurei* apeece. *a*1666 Evelyn *Diary c* 7 May an. 1645 (1955) II. 398 As to Coynes & Medails..10 Asses make the Roman Denarius,..10 Denarius's an Aureus. **1895** G. Allen *The Woman who Did* xv. 162 His profile was clear-cut, like Trajan's on an aureus. **1940** *Oxoniensia* V. 142 The donation of..an *aureus* of Hadrian, and a *siliqua* of Gratian. **1959** *Chambers's Encycl.* X. 127/2 Constantine gave it a last reform, introducing an aureus of gold of 72 to the pound which lasted to the end of the empire.

‖**au revoir** (orəvwār). [F., = lit. 'to the seeing again', i.e. in anticipation of seeing (meeting) you again.] An expression implying farewell for the present; hence as *sb.*, a farewell of this kind. Also occurs in slang abbrevs. *aurev.*, *au 'voir*, and as a malapropism in the form *au reservoir*.

1694 N. H. *Ladies Dict.* 15/1 He must kiss his Landlady .., and so parting, says *a Revoir*, Madam, till I see you again. **1761** Chesterf. *Let. to his Son* 6 Dec. (1774) 455 *Au revoir*, as Sir Fopling says, and God bless you. **1801** Maria Edgeworth *Belinda* II. xv. 117 Instead of adieu, I shall only say—*Au revoir!* **1853** 'C. Bede' *Adv. Verdant Green* x. 93 I'm going now in search of an appetite, and I should advise you to take a turn round the Parks and do the same. *Au reservoir!* **1905** *Daily Mail* 5 Aug. 6/1 The King and Queen, after an au revoir of ceremonious courtesy, left for their own yacht. **1921** T. E. Lawrence *Let.* 1 Oct. (1938) 334 This life goes on till February 28 next year. Au 'voir. **1922** Joyce *Ulysses* 418 Au reservoir, Mossoo. Tanks you. **1924** Galsworthy *White Monkey* i. viii. 59 Oh! Well! You'll bob in anyway. Aurev! **1948** S. G. Spaeth *Hist. Pop. Mus. in Amer.* (1960) vi. 221 That lilting waltz..appeared in 1892, and in the following year Harry Kennedy made his lasting gift to barber-shop quartets with *Say Au Revoir but Not Goodbye*.

auric (ǭ·rik), *a.*[2] [f. Aura+-ic.] Of or pertaining to an aura.

1889 in *Cent. Dict.* **1910** *Westm. Gaz.* 13 Apr. 5/2 An auric veil. **1921** Galsworthy *To Let* i. vi. 62 Why! Look at their photographs of auric presences. **1956** R. M. Lester *Towards Hereafter* iii. 39 Every one of us has auric radiations.

Aurignac, name used *attrib.* (See next.)

1863 C. Lyell *Geol. Evidences Antiquity of Man* x. 192 The Aurignac cave adds no new species to the list of extinct quadrupeds. **1875** *Encycl. Brit.* II. 336 [The bones of] other extinct mammals, alongside of human remains and works of art, in the famous Aurignac caves of the Pyrenees. **1939** R. Campbell *Flowering Rifle* ii. 61 The Aurignac man could spare us better bones. **1946** F. E. Zeuner *Dating Past* ix. 299 The foyer..contains Aurignac bone points.

Aurignacian (ǭrinyēⁱ·ʃ¹ăn), *a.* and *sb.* [ad. F. *Aurignacien* (H. Breuil 1906, in *Comptes Rendus Congr. Int. d'Anthrop.* 329), f. the place-name *Aurignac* (France)+-ian.] **A.** *adj.* Of or pertaining to the Aurignac cave of the Pyrenees; belonging to the *Aurignac era* or *period*, that indicated by the remains and works of art found in the cave. Also *absol.*

1914 *Q. Jrnl. Geol. Soc.* LXX. p. xcviii, A bed was found, which yielded the incised drawings..as well as numerous mammalian remains and flint-implements; and this is regarded as of Aurignacian age. Immediately below the last-mentioned bed a deposit of sand..was penetrated.. and this deposit, also referred to the Aurignacian, was found to contain an enormous number of bones. **1920** J. Ritchie *Influence Man Anim. Life Scot.* vi. iii. 344 The Aurignacian painting of two Reindeer fronting each other. **1920** *Q. Rev.* Oct. 377 The course of Art, from the Aurignacian era (circa 35,000 B.C.) till to-day. **1926** *Times Lit. Suppl.* 2 Sept. 574/3 During the greater part of the upper palæolithic age the culture of Europe was what is termed Aurignacian.

B. *sb.* A man or woman of this period.

1915 R. Lankester *Diversions of Naturalist* xxx. 280 Bushman-like Aurignacians. **1963** E. S. Wood *Field Guide Archæol.* ii. ii. 102 Aurignacians lived in Paviland Cave, Gower (Glamorgan).

aurora. Add: **5. b.** *aurora borealis*: transf.

1818 'Thos. Brown, the Elder' *Brighton* I. 131 She was eclipsed by an *Aurora Borealis* in the matrimonial sphere. **1922** Joyce *Ulysses* 469 The aurora borealis of the torchlight procession leaps.

auscultate, *v.* (Earlier and later examples.)

1862 Thackeray *Philip* III. vi. 122 This practitioner studied, shampooed, auscultated Tregarvan. **1892** Stevenson & Osbourne *Wrecker* xv. 233 It was therefore necessary..to auscultate what remained [of the ship], like a doctor sounding for a lung disease. **1908** *Practitioner* Feb. 272 The child is first auscultated in the upright position.

‖**au sérieux** (o seriö), *advb. phr.* [F.] Seriously (cf. Seriously *adv.*[2] 1 b). Cf. *Au grand sérieux.

1836 J. S. Mill *Lett.* (1910) I. 104 What you write would be taken *au sérieux* and not as a mere play of intellect and fancy. **1888** W. James *Let.* 29 Mar. in R. B. Perry *Tht. & Char. W.J.* (1935) I. 705 To me the argument seems irresistible, so long as we take the relation of really *intending* an object, *au sérieux*. **1962** *Times* 16 Feb. 15/1 D'Oyly Carte fanatics who are believed to take all Savoyardery *au sérieux*.

‖**auslaut** (au·slaut). *Philol.* [G., f. *aus-* denoting termination+*laut* sound.] The final sound of a syllable or word. Cf. *Anlaut, *Inlaut.

1881 in *Imp. Dict.* **1892** *Classical Rev.* Feb. 3/2 It is further probable that the form *-endus* for *-vendus* arose primarily in verbal roots with vocalic auslaut. **1936** M. Schlauch in *Science & Society* I. 34 Purely vocalic *Auslaut* is to be associated with a developed patriarchal-totemistic culture of hunting.

Ausonian (ǫsōᵘ·niăn), *a.* [f. L. *Ausonia* Lower Italy, poet. Italy, Gr. Αὐσονία poet. Italy, f. Αὔσων, son of Ulysses, who was fabled to have settled there.] Of or pertaining to Ausonia or to the Ausonians, the primitive inhabitants of middle and lower Italy; hence, Italian. Also *sb.*, a native of Ausonia, an Italian.

1607 Topsell *Four-footed Beasts* 481 The Masilian and Ausonian shepherds were..afraid of this lion. **1667** Milton *P.L.* I. 739 And in Ausonian land Men call'd him Mulciber. **1775** Ash *Dict.* Suppl., *Ausonian*, sb. **1821** Byron *Proph. Dante* II. 131 p. 233 The Ausonian soil. **1826** E. H. Dacre *Morus* 207 All that in this beautiful world is fair and lovely, mountains, woods, rivers, and Ausonian skys. **1850** Tennyson *Pal. Art* in *Poems* (ed. 6) 115 Or..stay'd the Ausonian king to hear Of wisdom and of law. **1876** Disraeli *Let. to Lady Chesterfield* 31 Mar. in Buckle *Life* (1920) V. xii. 472 The spring of Ausonian lands. **1882** *Encycl. Brit.* XIV. 344/2 The Auruncans, or Ausonians as they were termed by Greek writers.

Aussie (ǫ·si, in Australia ǫ·zi), *sb.* and *a. colloq.* [Hypocoristic, f. Australian *sb.* and *a.*: see -ie.] **1.** (An) Australian.

1917 *Harefield Park Boomerang* 16 Mar. 3 Send us a barrel of 'Aussy's' shyppe. **1918** (*title*) Aussie. The Australian Soldiers' Magazine. *Ibid.* 2/1 A certain Aussie Unit. **1919** *Athenæum* 8 Aug. 727/2 [The soldier] gave nicknames to the Overseas troops, as 'Aussies', 'Diggers', or 'Dinkums' for Australians. **1931** I. L. Idriess *Lasseter's Last Ride* i. 7 Fred Colson, a cheerful Aussie bushman. **1957** S. Hope *Diggers' Paradise* 78 Most Aussies, contrary to popular belief, are town-dwellers.

2. Australia.

1917 *Harefield Park Boomerang* 18 Apr. 9/2 That a passage to 'Aussie' [sic] they'll get, Is considered an 'odds-on' bet. **1918** *Australian in Weymouth* 3 July 8/2 So soon returns to Aussie with his 'better half'. **1928** W. Smith *Jean of Tsinuck Country* xxii. 215 You meet a few bums from Aussie. **1956** F. B. Vickers in *Coast to Coast* 72 You're in Aussie now. Not your country.

austempering (ǫste·mpəriŋ), *vbl. sb. Metallurgy.* [f. *aus(tenite[2] + tempering *vbl. sb.*] A quenching process used in the manufacture of steel or cast iron (see quot. 1949).

1937 E. S. Davenport in *Steel* 19 Mar. 42/1 The heat treating process known as 'austempering' consists essentially of heating steel to an appropriate temperature above the critical range to render it austenitic. **1949** E. T. Rolfe *Dict. Metallurgy* 17 *Austempering*, a patented heat treatment involving the controlled hot-quenching of steel (or cast iron), which has previously been heated above the transformation range, so as to give rise to austenite.

Austenian (ǫsti·niăn), *a.* and *sb.* **A.** *adj.* Of or pertaining to Jane Austen, novelist, 1775–1817, or her writings. **B.** *sb.* = *Austenite[1]. So **Austenish** (ǫ·stěniʃ) *a.*, characteristic of Jane Austen's work; **Austenite**[1] (ǫ·stěnəit), an admirer of Jane Austen's writings.

1898 J. Jacobs *Jane Austen's Emma* Introd. p. vii, If 'Emma' is not the most striking of Jane Austen's works, it is the most Austenish. **1901** *Westm. Gaz.* 11 Dec. 3/1 The Austenian topography seems to me tolerably complete. **1905** *Daily Chron.* 16 May 3/7 With Mr. Perugini Austenites will have a quarrel. **1928** *Observer* 22 Jan. 6 The claim that this unknown book is at least better than 'Sense and Sensibility' is calculated to prejudice Austenians against it from the outset. **1963** *Times* 7 Mar. 15/5 To the perfect Austenite Mr. Liddell's method of approach will be a delight and a provocation.

austenite[2] (ǫ·stěnəit). *Metallurgy.* [a. F. *austénite* (Osmond 1896, in *Bull. Soc. d'Encourag. de l'Industrie* 738), f. the name of Sir William C. Roberts-Austen (1843–1902): see -ite[1]. (See quot. 1910).] Hence **austeni·tic** *a.*, containing austenite.

1902 [see Troostite 2]. **1905** *Nature* 18 May 69/1 The polyhedral or 'austenitic' type of structure has never been obtained alone in a pure carbon steel... The austenitic structure appears to be that of the nose of the tool in actual use. **1910** *Encycl. Brit.* XIV. 804/2 Austenite is the name of the solid solution of an iron carbide in allotropic γ-iron. *Ibid.*, On cooling into region 6 or 8 austenite should normally split up into ferrite and cementite, after passing through the successive stages of martensite, troostite and sorbite. **1922** *Nature* 24 June 818/1 The gamma-iron lattice of austenitic steels is enlarged by dissolved carbon. **1937** [see *Austempering *vbl. sb.*]. **1947** *Nature* 11 Jan. 50/1 Austenitic irons have found a number of industrial uses. **1966** *Economist* 1 Oct. 80/1 Austenitic steels and S-bends.

austerity. Add: **4. b.** Applied *attrib.*, esp. during the war of 1939–45, to clothes, food, etc., in which non-essentials were reduced to a minimum as a war-time measure of economy. Also *absol.*

1942 *Times Weekly* 2 Dec. 14/2 A General Limitation Order—..which suggests that the United States have got quite a way on the road to austerity. **1942** *Times of India* 31 Dec. 8/4 The first of a fleet of 'austerity' buses has just been completed in Britain.. Fittings are kept to essentials and anything approaching luxury has been..cut out. **1944** *Times* 16 Feb. 8/3 Mr. Dalton..said that austerity clothing was not unsaleable. On the contrary, many men evidently thought that at 30 coupons austerity suits were a good bargain. **1951** R. Knox *Stimuli* iv. 9 There is a real humility in imitating a god who was born in a utility nursery, and laid in an austerity cradle to match it.

Austinian (ǫsti·niăn), *a.* Of or pertaining to John Austin (1790–1859) and his theory of government. Hence **Austinianism**.

1875 H. J. S. Maine *Lect. Early Hist. Institutions* xii. 362 The Austinian conception of Sovereignty has been reached through mentally uniting all forms of government in a group by conceiving them as stripped of every attribute except coercive force. **1882** *Encycl. Brit.* XIV. 357/2 The Austinian definition of law as a command creating rights and duties. **1917** H. J. Laski *Stud. Probl. Sovereignty* ii. 66 It goes back to that passionate Erastianism of Luther which was the only answer he could make to the Austinianism of Rome. **1921** *Contemp. Rev.* Dec. 858 The unified, mechanical, Austinian State.

Austral, *a.* Add: **2.** Of or pertaining to Australia or Australasia.

1823 W. C. Wentworth *Australasia* 21 Grant that yet an Austral Milton's song..flow deep and rich along;—An Austral Shakspeare rise. **1835** *Penny Cycl.* III. 122/2 The Austral negroes may be considered as still living in the lowest state of civilization. **1855** W. Howitt *Land, Labour, & Gold* I. 43 Every servant in this Austral Utopia thinks himself a gentleman. **1898** E. E. Morris (*title*) Austral English. A Dictionary of Australasian Words, Phrases, and Usages. **1904** *Westm. Gaz.* 30 Mar. 7/2 The Austral Club at 41, Norfolk House, St. James' Court, Buckingham Gate. **1921** (*title*) The Austral Avian Record. A scientific journal devoted primarily to the study of the Australian avifauna. **1967** S. L. Tucker *Profane Skage* 279 [In 1793] Austral English was barely born.

Australasian, *a.* and *sb.* (Later examples.)

1838 *Dublin Rev.* July 176 New Zealand is included in the vast diocese of our Australasian Bishop. **1937** D. Cowie *N.Z. from Within* 153 [New Zealand] refuses to allow the term..Australasian to be used in her hearing. **1940** *Chambers's Techn. Dict.* 60/1 *Australasian region*, one of the primary faunal regions into which the surface of the globe is divided; includes Australia, New Guinea, Tasmania, New Zealand, and the islands south and east of Wallace's line.

Australasiatic (ǫstrălēiʃi,æ·tik), *a.* [f. *Australasia* (see Australian + Asiatic *a.*] Consisting of or characterized by a mixture of Australian and Asiatic elements. Also *absol.* (? Obs.)

1819 *New Whig Guide* in *Blackw. Mag.* V. 96 Graf-calee, which is a kind of Australasiatic Delos. **1890** *Cornhill Mag.* July 15/1 But 'tis neither Cockney nor Yankee, but a nasal blend of both:—in a word, it was Australasiatic of the worst description.

Australian, *sb.* and *a.* Add: **A.** *sb.* **2. b.** *New Australian*, a recent immigrant resident in Australia, esp. one from Europe.

1926 J. Doone (*title*) Timely Tips for New Australians. **1952** Mitchell in *Chambers's Shorter Eng. Dict.* Suppl., *New Australian*, an immigrant, usually a European; first applied to displaced persons from the Baltic countries. **1959** *N.Z. Listener* 9 Jan. 3/1 The hero of the story, Nino Culotta, is a New Australian—or what is now known as a 'New Aussie'.

3. *absol.* uses of the adj. (contextual sense editorially inserted in square brackets).

1890 E. Ward *Jrnl.* 3 Oct. (1951) 40 Australians [ships] are not common enough to on the high seas. **1896** N. Newnham-Davis *Three Men & God* 123 The additional felicity of refreshing an old bucking Australian [horse] belonging to the Master. **1901** *Westm. Gaz.* 4 June 12/1 That well-known collector..had a collection of Australians [stamps] alone worth £10,000. **1930** *Economist* 8 Nov. 860/2 The improving tendency of Australians [shares]. **1945** C. L. B. Hubbard *Observer's Bk. Dogs* 21 Sydney Silky Dogs are frequently crossed with Australians [Australian terriers].

B. *adj.* **b.** In specific combinations with sbs., as **A. ballot** (see quot.); **A. bluebell creeper**, an evergreen vine of the genus *Sollya* with blue flowers; **A. cattle-dog** (see quots.); **A. cranberry**: see Cranberry 2; **A. crawl**: see *Crawl *sb.*[1] 5; **A. currant**: see Currant 3; **A. honeysuckle**, any of several plants of the genus *Banksia* (see Honeysuckle 3 a); **A. kelpie**: see *Kelpie[2]; **A. lady-bird** (see quots.); **A. mahogany**, any of several eucalypts having hard red timber (Webster, 1909); **A. rules**, regulations governing a football game having certain characteristics of both association and rugby football, with eighteen players on each side; also, the game itself; **A. terrier**, a type of terrier bred in Australia, resembling the Skye and Yorkshire terriers.

1888 *Nation* (N.Y.) 2 Aug. 91/2 By introducing the secret 'Australian ballot' in Congressional elections..the use of bribery in the choice of Congressmen might be discouraged to some extent. **1903** A. B. Hart *Actual Government* 74 The so-called Australian ballot system, under which all the candidates appear upon one ballot, prepared and distributed by the state, and the voter indicates on the ballot his choice of candidates. **1896** T. W. Sanders *Encycl. Gardening* (ed. 2) 368 *Sollya* (Australian Bluebell Creeper). **1916** H. L. G. van Wijk *Dict. Plantnames* II. 81/2 Australian blue bell creeper. **1926** *Australian Encycl.* II. 453/1 A blending of the native bred dog..and the smooth-haired Scotch sheep-dog...Such is the Australian cattle-dog. **1945** C. L. B. Hubbard *Observer's Bk. Dogs* 20 Australian Cattle Dog. Australian Heeler. This is an Australian manufactured breed much resembling a miniature Alsatian. **1881** *Encycl. Brit.* XII. 140/2 Australian or heath honeysuckle is the Australian *Banksia serrata*. **1898** Morris *Austral Eng.* 18/2 *Banksia*,..the so-called *Australian Honeysuckle*. **1902** *Encycl. Brit.* XXVII. 636/2 One of the most notable examples of the use of insect allies is the case of the Australian Lady Bird, *Adalia cardinalis*. **1952** Mitchell in *Chambers's Shorter Eng. Dict.* Suppl., *Australian ladybird*, a small blue and orange beetle introduced into California in 1888 to combat the cottony-cushion scale. **1948** T. Corkhill *Gloss. Wood* 19 Australian Mahogany. *Dysoxylon fraseranum*. Also called Australian Rosewood. Reddish colour and fragrant. Characteristics of red cedar. Very durable and stable. Used for general purposes, furniture, aircraft. **1925** *Australian Encycl.* I. 479/1 Of matches under Australian (then known as Victorian) rules seven were won and 12 were lost. **1933** *Bulletin* (Sydney) 22 Mar. 30 The Australian rules football carnival in Sydney next August. **1958** *Australian Encycl.* IV. 134/2 Generally, it [*sc.* Australian football] is known as the Australian Game, National Football, or Australian Rules. **1909** Webster, Australian terrier. **1910** *Encycl. Brit.* VIII. 378 (*caption*) Australian Terrier. **1928** F. T. Barton *Kennel Encycl.* 26 The so-called Australian terriers are nothing more or less than inferior specimens of the Yorkshire terriers. **1966** *Weekly News* (N.Z.) 1 June 40 Australian terrier..a low-set, compact, active little dog which may be silver-grey with tan markings, or sandy.

Australiana (ǫstrē¹:liā·nă). [f. *Australia* + Ana *suff.*] Things relating to or characteristic of Australia.

1855 *Putnam's Monthly Mag.* V. 598 Australiana. The Campbell Town Election. **1886** R. Henty (*title*) Australiana or my early life. By..first white native of the first settlement of Victoria, Australia. **1941** Baker *Dict. Austral. Slang* 6 *Australiana*, Australian cultural or artistic life; the record of the country's culture and history. **1958** *Times* 7 Aug. 9/7 There is a growing interest in Australiana. **1962** *Listener* 4 Oct. 510/2 The September number of *The London Magazine*..consists of Australiana. **1967** S. H. Courtier *See who's Dying* iii. 36 He collects books, original editions, early Australiana and that sort of thing.

Australianism (ǫstrē¹·liăniz'm). [f. Australian + -ism.] An idiom or mode of expression peculiar to Australian English; an attitude of mind characteristic of Australia or an Australian. So **Australa·sianism**.

1891 *Daily News* 26 June 5/2 He appears to think that our speech is in danger of being invaded by Australianisms. **1891** *Harper's Mag.* July 215/1 The coming degradation of the English language by the invasion of Australasianisms. **1905** *Daily Chron.* 22 Nov. 6/7 There are 'Australianisms' enough to make a dictionary an essential for the proper understanding of an Antipodean journal. **1911** E. M. Clowes *On Wallaby* vii. 179 One place in Melbourne..where the Frenchman, the German,..and Russian throw off the garment of Australianism, and eat, look, and speak like men of their own country. **1965** G. McInnes *Road to Gundagai* xvi. 281 They would prove their Australianism by rallying to England's defence.

Australianize (ǫstrē¹·liănəiz), *v.* [f. Australian + -ize.] *trans.* To naturalize as an Australian; to make Australian in habits, customs, etc. So **Austra·lianized** *ppl. a.*

1883 *St. James's Gaz.* 10 May, Are the latter wronged..in having to become for instance, 'Australianized'? **1888** D. B. W. Sladen *Australian Ballads* 280 Even in his own particular line of 'Australianised Calverley'. **1908** H. Jones in Hetherington *Life & Lett.* (1924) 213 She is, I think, the wife of an old Scot who has been Australianized.

Australianness (ǫstrē¹·liăn,nės). [f. as prec. + -ness.] The condition of being Australian or having qualities peculiar to Australia or its inhabitants.

1954 *Landfall* VIII. 27 An essential Australianness which is apt to recede when too deliberately pursued. **1960** *Encounter* May 28 The British migrant..must not..attack the country itself or its Australianness.

australite (ǭ·străləit). *Min.* [ad. G. *australit* (F. E. Suess 1901, in *Jahrb. d. K.K. Geol. Reichsanstalt 1900* L. 194), f. Austral *a.* + -ite[1].] A form of tektite found in Australia and neighbouring countries. Cf. *billitonite, *tektite.

1909 *Geol. Mag.* VI. 411 Occurrences of the related 'obsidianites' (referred to by Professor Weinschenk as Billitonite and Australite) in the Malay Peninsula. **1916** *Bull. U.S. Nat. Mus.* XCIV. p. x (*caption*) Australites and an obsidian button from Australia. **1936** *Geogr. Jrnl.* LXXXVIII. 287 Tektites under various names are found in many parts of the world, but those which are peculiar to Australia, designated australites, appear to be more numerous as well as much more distinctive than any other class.

Australoid, *a.* See *s.v.* Australian and add later examples. Also *sb.*

1910 *Encycl. Brit.* II. 748/2 It is also probable that the Australoid family extends into south Arabia and Egypt. **1913** H. Johnston *Pioneers in Australasia* ii. 49 The ancestors of the Tasmanians and Australoids were driven forth into the forests of Africa and southern Asia...The Australoids..became the native race of the Australian continent.

Australopithecus (ǫstrē¹:lopi·piθekəs). [mod.L., f. L. *austrālis* southern + Gr. πίθηκος ape.] In full *Australopithecus africanus*, a form of anthropoid ape known from fossil remains discovered at Taungs in Bechuanaland in 1924 (cf. *Taungs). Hence **Austra:lopi·thecine** *a.* and *sb.*, pl. -s; also in mod.L. form *Australopithecinæ*.

1925 R. A. Dart in *Nature* 7 Feb. 195, I propose that the first known species of the group be designated *Australopithecus africanus*. **1937** *Ann. Reg. 1936* 32 A Pleistocene skull of Australopithecus type..was found near Johannesburg. **1949** *Jrnl. R. Anthrop. Inst.* LXX. 23 The diminutive gap which separates the Australopithecine group proper from human status. **1953** *N.Y. Times* 3 Jan. 16/7 A small and aberrant..specimen of Paranthropus, one of the South African apemen or Australopithecines.

Australorp (ǫstrălǫrp). [f. Austral(ian + Orp(ington).] (See quot. 1929.)

1922 *Daily Mail* 9 Dec. 14 (Advt.), Australorps imported. Australian Black Orpingtons. World's Record Layers—1,750 eggs 6 birds, 12 months. **1929** E. Brown *Poultry Breeding* i. x. 139 *Australorps*. The fowls to which this name is given are practically the same as the type of the original Black Orpington...When brought over to Europe..the name here employed was applied to..distinguish them from those which had been so markedly changed. **1965** *Sunday Mail Mag.* (Brisbane) 3 Oct. 14/1 Australorp. The bird is of medium size, has a white skin, a single comb, and beautiful black plumage.

Austrasian (ǫstrē¹·ʃiăn), *a.* and *sb.* [f. med.L. *Austrasia*, *Ostrasia* + -an.] **A.** *adj.* Of or pertaining to the Germanic part of the Frankish empire east of the Rhine. **B.** *sb.* A native or inhabitant of Austrasia.

1781 Gibbon *Decl. & F.* III. xxxviii. 598 The remote province was separated from his Austrasian dominions, by the intermediate kingdoms of Soissons, Paris, and Orleans. **1833** S. A. Dunham *Hist. Eur. Middle Ages* II. i. 13 He restored Brunehild to the Austrasians. **1874** Stubbs *Const. Hist.* I. v. 9 The Austrasian domination was more purely Germanic than the Neustrian which it superseded. [illegible] her army stood as unshakable as Austrasian Franks.

Austrian (ǫ·striăn, ǫ-), *a.*[2] and *sb.* [f. *Austria* + -an.] **A.** *adj.* Of or pertaining to Austria, a country in central Europe. **B.** *sb.* A native or inhabitant of Austria.

1620 *Two very lamentable Relations* [ad fin.], The true nearer Austrian Evangelical Committees and Ambassadours. **1660** A. Brock *Compend. Hist. Turks* 356 Turacious led Ferdinands main Battel, strengthened on one side with the Styrians, on the other with the Austrian Horse. *Ibid.*, The Stirians were worsted by the Transilvanians, and Boskai's left wing being mostly unexpert Souldiers, were likewise overthrown by the Austrians. **1701** John Savage *Knolles's Turkish Hist.* II. 114 In the year 1680, the Austrian Troops entered Hungary. **1741** *Defence of Rights of House of Austria* 27 To this Purpose the Austrians had..offered large Sums of Money. **1807** W. Coxe *Hist. House of Austria* I. i. 145 The Imperial investiture established the indivisibility of the Austrian territories. *Ibid.* [illegible] The elector, at the head of..10,000 Austrians, and 6,000 Polish cavalry, marched towards Holstein. **1876** [see Slavonian *sb.* 1]. **1932** Cecily Hamilton *Mod. Austria* i. 2 More than one frontier of the Austrian state is a barrier between those who may one day be enemies. *Ibid.* ix. 126 The Italian march halted, arrested by no more than 30,000 Austrians.

Austric (ǫ·strik), *a.* [ad. G. *austrisch* (W. Schmidt *Die Mon-Khmer-Völker* (1906) 70), f. L. *Austr-* (see Austrian *a.*) + -ic.] Designating or pertaining to a family of languages comprising the Austro-Asiatic and Austronesian families.

1927 [see *Austro-[1]]. **1933** Bloomfield *Language* iv. 71 Some scholars believe both the Munda and the Mon-Khmer families to be related to the Malayo-Polynesian family (forming the so-called Austric family of languages).

Austro-[1] (ǫ·stro), combining form of *Austrian.

1834 J. H. Stocqueler *Fifteen Months' Pilgrimage* II. iv. 98 It marks to the traveller proceeding northward the boundary of Austro-Papal influence. **1851** *Fraser's Mag.* Feb. 141/2 The Austro-Russian ascendancy over Germany. **1866** *Macm. Mag.* XIV. 387 The state of public opinion which preceded the Austro-Prussian war. **1868** *Amer. Ann. Cycl.* 57/2 The whole Austrian monarchy is now officially called the Austro-Hungarian empire. **1910** *Westm. Gaz.* 9 Apr. 4/1 The links between Italy and her ally can hardly be strengthened without provoking a fresh outburst of Austrophobia in Italy, or weakened without making the Austrophobes rather more dangerous. **1920** *Q. Rev.* July 204 The endeavours of far-seeing Austro-German patriots. **1954** Koestler *Invisible Writing* xxiii. 354 The so-called teachings of the so-called Austro-Marxist school.

Austro-[2] (ǭ·strǫ), combining form of Austral *a.* and of Australian.

1869 A. R. Wallace *Malay Archipelago* i. 14 We can draw a line among the islands, which shall so divide them that one half shall truly belong to Asia, while the other shall no less certainly be allied to Australia. I term these respectively the Indo-Malayan, and the Austro-Malayan divisions of the Archipelago. **1899** A. H. Keane *Man Past & Present* v. 145 Sumbawa and Celebes..are..included in the Austro-Malayan zoological and botanical regions. **1927** G. A. Grierson *Ling. Survey India* I. i. ii. 32 This 'Austric Family'..he [*sc.* W. Schmidt] divided into two sub-families, the 'Austro-Nesian' and the 'Austro-Asiatic'. **1936** *Discovery* Jan. 20/1 The Austro-Malayan sub-region. **1961** [illegible] The Mon-Khmer branch of the great linguistic family known as Austroasiatic.

Austronesian (ǫstrŏnī·ʃiăn), *a.* [f. G. *austronesisch* (W. Schmidt 1899, in *Mitteil. Anthrop. Ges. Wien* XXIX. 248/1), f. L. *austrālis* southern + -o + Gr. νῆσος island: see -ian.] = Malayo-Polynesian *a.*

1903 *Amer. Anthropol.* V. 164/1 Father Schmidt..accepts that part of his [*sc.* Müller's] theory which sees in the Melanesians a mixed race sprung from the Papuan aborigines and Austronesian (Malayo-Polynesian) immigrants. **1927** [see prec.]. **1958** *Times* 6 Nov. 7/2 Bennett..considers that beyond doubt the cradle of Austronesian tongues lies 'somewhere in south-east Asia' while Malaya itself is the homeland of the Austronesian family of languages.

autacoid (ǫ·tăkoid), *a.* and *sb.* [f. Gr. αὐτ-ός self + ἄκ-ος remedy + -oid.] (Of, pertaining to, or designating) a drug-like substance produced by the organs of internal secretion.

1914 E. A. Schäfer in *Internat. Congr. Med. 1913* ii. ii. 21 A convenient Greek derivative to denote this quality is 'acoid'..and to further denote that these principles are formed within and by the body, i.e. are natural, we may employ the prefix 'aut-'. We then get the complete expression 'autacoid substance' to denote any drug-like substance produced in one organ which can be extracted from the internally secreting organs. **1928** *Daily Express* 6 July 6 These are chemical substances of a drug-like character distributed by the circulation of the blood, to act where they are needed upon various organs and tissues. They are best referred to as 'autacoids'...Among the best known of the autacoids are adrenalin, pituitrin, secretin, and thyroxin.

autarchic (ǫtā·ɹkik), *a.*[1] [f. Autarchy[1] + -ic.] Absolute, despotic.

1893 in *Funk's Standard Dict.* **1936** [see *autarkic *a.*].

autarchic, *a.*[2], var. *autarkic *a.*

autarchy[1]. **1.** (Earlier and later examples.)

1665 N. N. *Medela Medicinæ* sig. A5 Epist. ded., Who are so confident, as to attempt an Autarchie in Physick, and long for an opportunity to oppress and trample upon the noblest sort of Philosophers and Physicians. **1838** F. Lieber *Man. Pol. Ethics* I. ii. xii. 411, I call autarchy that state in which public power, whole and entire, unmitigated and unmodified, rests..in the hands of a monarch, or the people, or a particular class.

autarchy[2]: see *autarky.

autarkic (ǫtā·ɹkik), *a.* Also (irreg.) **autarchic**. [f. next (cf. Autarchy[1]) + -ic.] Of, pertaining to, or characterized by autarky; (economically) self-sufficient.

1883 L. F. Ward *Dynamic Sociology* I. vii. 466 Stages in the inception, development, and progress of social aggregation..1, the solitary, or autarchic, stage. **1936** *Economist* 11 Jan. 74/1 The world must retrace its..steps towards arbitrary autarkic divisions. **1936** R. F. Harrod in *Times* 1 Dec. 12/3 Since then autarkic States have generally to be autarkic also, genuine confusion may be caused, unless the spellings are kept distinct. **1939** A. G. B. Fisher *Econ. Self-Suff.* 22 Any attempt to carve the world's surface into suitable large-scale autarkic blocs must..mean serious conflict. **1958** *Economist* 26 July 269 The..to make the sterling area into more of an autarkic block.

autarky (ǫ·tāɹki). [See Autarchy[2].] = Autarchy[2], self-sufficiency. (Earlier and later examples.)

1617 Samuel Ward *Balme from Gilead* 18 The Autarchie and self-sufficiency of God. **1635** H. Valentine *Foure Sea-Sermons* 10 It may as well stand upon its bottome, and boast of its Autarkie, and selfe sufficience. **1657** J. Trapp *Commentary Ezra, Neh., Psalms* 669 These have their autarky, or self-sufficiency; such as godliness never without. **1951** Hoggart *Auden* v. 150 He may deny the necessity for any metaphysic, and retreat into a cold intellectual autarky. **1957** T. S. Eliot *On Poetry* i. 23 A general autarky in culture simply will not work: the hope of perpetuating the culture of any country lies in communication with others.

b. *spec.* (A policy of) economic self-sufficiency in a political unit.

1934 *Sun* (Baltimore) 11 Apr. 10/3 Internationalism..would at least check the present steady drive toward..autarchy. [illegible] autarky, political isolationism and Fascism. **1939** J. Hope Simpson *Refugee Question* 5 The doctrine of 'autarky'..has materially affected normal international trade.

Hence **au·tarkist** *sb.*, one who advocates or practises autarky. Also *attrib.*

1938 *New Statesman* 15 Jan. 75/1 The only chance of inducing the autarkist countries to open their frontiers to the free flow of goods and money. **1939** A. G. B. Fisher *Econ. Self-Suff.* 14 The autarkist is not greatly interested in higher standards of living.

autecology (ǭtīkǫ·lŏdʒi) [ad. G. *autökologie* (Schröter, 1898), f. *aut(o-[1] + *ecology.] The study of the ecology of an individual plant or species, opp. *synecology. Hence **au:teco·lo·gical** *a.*

1910 Flahault & Schröter *Phytogeogr. Nomencl.* (IIIrd Internat. Bot. Congr.) 24 Ecology thus includes the study of the habitat conditions and the adaptation-phenomena of the single species (Autecology). **1926** H. H. Allen in Tansley & Chipp *Aims & Methods Stud. Veg.* xxii. 370 The high-mountain vegetation provides, further, much valuable material for autecological work. **1932** Fuller & Conard tr. *Braun-Blanquet's Plant Sociol.* v. 81 The chief interests of most ecologists have centred in autecology, the life relations of the individual plant.

authentic, *a.* Add: **9.** Also, composed in an authentic mode. *authentic cadence*: that form of perfect cadence in which the (major) chord of the dominant immediately precedes that of the tonic. Opp. to Plagal *a.*

1806 Callcott *Mus. Gram.* 160 The Dominant..derives its name from the ancient Church Tones, in which it was the Fifth in the Authentic, and the Octave in the Plagal Scales, but always a Fifth above the final or modern Tonic. **1873** [see Dict.]. **1879** Grove *Dict. Mus.* I. 105/2 'Ein feste Burg' and 'Eisenach' are examples of 'authentic' melodies, and the Old 100th and Hanover of 'plagal' ones. **1880** *Ibid.* II. 18/1 When first employed in polyphonic music, the Authentic scale was usually transposed.

authenticated, *ppl. a.* (Earlier example.)

1846 Geo. Eliot tr. *Strauss's Life of Jesus* I. i. iv. § 32. 203 It is an authenticated fact, that the circumcision..did not take place..under Herod.

authigenic (ǫþidʒe·nik), *a. Geol.* [ad. G. *authigene* (E. Kalkowsky 1880, in *Neues Jahrb. f. Mineralogie* 4), f. Gr. αὐθιγενής born on the spot, native + -ic.] Originating where found (see quots.). Cf. *allothigenic *a.*

1888 F. H. Hatch in J. J. H. Teall *Brit. Petrogr.* (Gloss.), *Authigenic*, a term applied by Kalkowsky to a mineral constituent, to denote that it came into existence with, or after, the rock containing it. **1890** C. R. Van Hise in *Bull. Geol. Soc. Amer.* I. 231 It has been seen that much if not all of the mica is authigenic. **1893** A. Geikie *Text-bk. Geol.* (ed. 3) 65 Such crystals, which are obviously more ancient than those forming the general mass of the rock, have been called *allogenic*, while those which belong to the time of formation of the rock, or to some subsequent change within the rock, are known as *authigenic*.

author, *sb.* Add: **6.** Frequent in appos. use.

*a***1806** D. Wordsworth *Tour Scotland* in *Jrnls.* (1941) I. 297 The author-tourists have quarrelled with the architecture of it. **1835** *Court Mag.* VI. 51/1 His peculiarity as an author-actor. **1898** *Daily News* 21 May 2/2 My friend the author-statesman. **1903** *Book Lover* Sept. 4/1 The author-artist has been as successful with his pen as with his brush. **1905** *Daily Chron.* 16 Dec. 8/5 The brilliant young author-composer. **1909** *Westm. Gaz.* 15 Dec. 1/2 There was only one author-producer in his experience to whom he had given a free hand. **1922** Joyce *Ulysses* 449 *Bloom*. Well, I follow a literary occupation. Author-journalist.

¶**7.** The editor of a journal. *Obs.*

1697 *Flying Post* May 18–20 Printed by T. Snowden..for the Author. **1724** *Brit. Jrnl.* (*imprint*) London: Printed for T. Warner, at the Black Boy in Pater-Noster-Row, where Advertisements and Letters to the Author are taken in. **1753** *Jackson's Oxf. Jrnl.* 5 May 4 Printed by W. Jackson in the High-Street, Oxford: By whom Letters to the Author, Articles of News, and Advertisements are taken.

author, *v.* Revived, chiefly in the U.S.: to be the author or originator of (a book, play, remark, etc.).

1940 *Time* 15 Apr. 35/2 Her father..authored several successful plays and movies. **1957** W. C. Handy *Father of Blues* xxi. 288 He once authored the famous Ziegfeld Midnight Roof productions. [illegible] The saying was authored by some husband. **1967** *Boston Sunday Herald* 30 Apr. 2/6 She has authored a reference book on the Genus Ilex in China.

authoritarianism (ǫþǫritē³·riăniz'm). [f. Authoritarian + -ism.] Authoritarian principles.

1909 in Webster. **1914** Streeter *Restatement & Reunion* ii. 47 Logical arguments in defence of authoritarianism. **1927** *British Weekly* 30 June 285/3 A decided drift towards authoritarianism, whether Catholic or fundamentalist. **1940** *Mind* XLIX. 100 The camouflage of spirituality in which the champions of authoritarianism disguise the essential materialism of their doctrines.

authority. **3.** (Later examples of concrete use in *sing.*)

1870 *Statutes* V (Tramways Act) 491 Orders authorising the construction of tramways..may be obtained by (1) The local authority of such district. **1959** *Times* 16 Sept. 2/3 The Port of London Authority is a thoroughly practical body of men. **1951** *Good Housek. Home Encycl.* 189/2 It is usually possible to obtain the free services of one through the local Health Authority.

authorized, *ppl. a.* Add: **3.** *authorized capital* (see quots.).

1911 W. Thomson *Dict. of Banking* 39/2 *Authorised capital*, the capital of a company as authorised by its memorandum of association. **1917** W. H. Walker *Corporation Finance* ii. 25 The state..limits the number of shares or the par value of the shares which the company may issue, and this limit is known as 'authorized capital'. **1930** *Economist* 5 Apr. 784/1 Resolutions are to be considered to increase the authorised share capital to £8,000,000.

autism (ǭ·tiz'm). *Psychiatry.* [ad. mod.L. *autismus* (also used), f. Gr. αὐτός self + -ism.] A condition in which a person is morbidly self-absorbed and out of contact with reality. So **auti·stic** *a.*, of, pertaining to, or characterized by this; *sb.*, a person thus affected; also **auti·stically** *adv.*

1912 Bleuler in *Amer. Jrnl. Insanity* LXIX. 874 When we look more closely we find amongst all normal people many and important instances where thought is divorced both from logic and from reality. I have called these forms of thinking *autistic*, corresponding to the idea of schizophrenic autismus. *Ibid.* 884 The unconscious can think logically or autistically. **1912** A. Hoch *Ibid.* 888 The chief traits which had existed before the mental breakdown were those which I at that time called the shut-in tendencies—tendencies to which Professor Bleuler has recently applied the term autism. **1916** C. E. Long tr. *Jung's Coll. Papers Analyt. Psychol.* vi. 203 Autism (Bleuler) = Auto-eroticism (Freud). For some time I have employed the concept of *introversion* for this condition. **1922** Woodworth *Psychol.* xix. 508 Daydreaming..is an example of what is called 'autistic thinking', which means thinking that is sufficient unto itself, and not subjected to any criticism. Autistic thinking gratifies some desire and that is enough for it. **1962** *Guardian* 15 Nov. 6/6 London County Council has opened an experimental unit at the health centre in Guildford Street for the treatment and diagnosis of autistic children. **1963** *New Scientist* 25 Apr. 184/1 Childhood autism—sometimes inaccurately called childhood schizophrenia. **1968** *Woman* 27 Jan. 24/2 There are over 4,000 diagnosed autistics in this country. *Ibid.* 56/4 Our hospital..is by no means the only one with an autistic unit.

auto (ǭ·to), *sb.*[2] **1.** Colloq. abbrev. of *automobile *sb.* after F. *auto*; also *attrib.* and *Comb.*; **auto court** *U.S.*, a *motel. See also *auto-[2].

1899 *Boston Herald* 9 July 6/3 The accident to Mr. W. K. Vanderbilt's 'auto'. **1901** *Daily Colonist* (Victoria, B.C.) 30 Oct. 6/1 These auto enthusiasts are trying hard. **1902** *Aeronautical World* (U.S.) 1 Oct. 60/1 Cycles, autos and all other contrivances for rapid transit. **1910** *Daily Chron.* 12 Mar. 6/4 The auto-manufacturing business. **1922** *Hotel World* 11 Mar. 6/2 The chamber of commerce of Santa Barbara, Cal., fathers an 'auto camp'. **1925** *Beaver* Sept. 201/2 All our 'tourists' report..the auto camps doing a roaring business. **1927** *Blackw. Mag.* Nov. 646/2 The auto-trip to Ixtlan del Rio. **1928** *Sunday Express* 24 June 8/2, I was raised in Canada sixty odd years ago, when..the country was not safe for slick auto-tourists. **1930** R. Macaulay *Staying with Relations* xix. 274 Sign-boards.. telling travellers about hotels and auto-camps. **1934** J. M. Cain *Postman always rings Twice* i. 9 A half dozen shacks that they called an auto court. **1962** *Economist* 20 Oct. 277/2 This demand..has surprised many automen.

2. Abbrev. of Autotype 2. *rare.*

1882 Swinburne *Lett.* (1960) IV. 309 The autos gave great satisfaction.

3. Abbrev. of Autobiography. *rare.*

1940 V. Woolf *Writer's Diary* 6 Jan. (1953) 323, I suppose the origin of many of the new middle aged autos.

auto (ǭ·to), *v. U.S.* Shortened form of *automobile *v.* So **autoing** (ǭ·to,iŋ), *vbl. sb.*

1909 *Cent. Dict.* Suppl., *Autoing*, the use of automobiles for business or pleasure traveling. **1916** *Sphere* 12 Feb. 180/3, I find the following commentary on new verbs in the New York *Outlook*: '..''Thursday we autoed to the country club.'' ' **1919** Mencken *Amer. Lang.* 110 *Auto* is almost unknown [in England], and with it the verb *to auto*. **1940** Bryant & Aiken *Psychol. of English* 84 We plan to *auto* through the mountains.

auto-. Read auto-[1] and add: In free composition as a prefix element, its chief meanings are: (*a*) of oneself, one's own; self-; (*b*) self-produced or -induced (pathologically) within the body or organism; (*c*) spontaneous, self-acting, automatic (cf. esp. b below).

auto-a·bstract, a speaker's own abstract of an address or speech prepared for publication; **au:to-agglutina·tion** [G. (A. Klein 1902, in *Wiener Klinische Wochenschrift* XV. 415/1)], spontaneous agglutination, esp. produced by auto-antibodies; **auto-a·ntibody**, an antibody which is produced by an organism and which reacts against some constituent of that organism; **autocata·lysis** *Chem.*, catalysis of a reaction by one of its own products; hence **autoca·talyst**, a substance or organism capable of autocatalysis; **autocataly·tic** *a.*, **-ally** *adv.*; **autoclastic** (-klæ·stik), *a. Geol.* [Clastic *a.*], of a rock, composed of its own fragments produced by crushing or granulation; **au:to-collima·tion**, collimation of an instrument by means of its own parts; so **autoco·llimating** *ppl. a.*, that collimates by means of its own parts; **autoco·llimator**, an instrument with an autocollimating eyepiece (Webster, 1934); **autocytotoxin** (-səitotǫ·ksin) [*cytotoxin], a cytotoxin formed in the body; **au:todiagno·sis**, diagnosis of one's own disease; hence **au:to-diagno·stic** *a.*, of or pertaining to autodiagnosis; **au:to-dige·stion** = *autolysis; also, self-destruction of plant tissue; **au:to-fo·llow, -fo·llowing** *Radar*, automatic following; a system in which a radar beam automatically follows the object from which echoes are received; also *attrib.*; cf. *automatic following* (*automatic *a.* 2); **au·tofrettage** [a. F. *autofrettage*, self-hooping; cf. Frettage], the process of strengthening a tube, esp. the barrel of a gun, by applying internal pressure in order to raise the limit of strain; hence **au·tofrettaged** *a.*; **au·tograft** *Surg.*, a graft of skin or other tissue taken from a person's own body (Dorland, 1913); **au:tohypno·sis** [Hypnosis 2], **autohy·pnotism**, a self-induced hypnotic condition; so **au:tohypno·tic** *a.* and *sb.*; **au:to-hypnotiza·tion**, the inducing of hypnosis by auto-suggestion; **au:to-immu·nity**, the state produced by the presence either of auto-antibodies or of lymphoid cells sensitized against some constituent of the subject's own tissues; so **auto-immune** *a.*, characterized by auto-immunity; **au:to-immuniza·tion**, (*a*) immunization from within the body; self-immunization (Dorland, 1900); (*b*) the production of auto-immunity; **auto-infe·ctant** *sb.*, an agent of auto-infection; **auto-infe·ctive** *a.*, or of pertaining to auto-infection; **auto-into·xicant**, a toxic substance generated in the system; also *fig.*; **auto-intoxica·tion**, poisoning by or resulting from toxin produced within the body; also *fig.*; hence **auto-into·xicate** *v. trans.*; **autokinesis** (ǭ:tokəinī·sis), (*a*) = Autokinesy (Mayne, Suppl. (1860) *autocinesis*); (*b*) the apparent movement of a stationary object; hence **autokinetic** (-kəine·tik) *a.*; **auto·logous** *a.*, derived from the same organism; **au:to-lumine·scence**, the spontaneous emission of light from certain radioactive substances; **autopho·tograph** = *autoradiograph; **auto-poi·soning**, poisoning caused by a virus formed within the body; **autopoi·sonous** *a.*, that is poisonous to the organism within which it is formed; **autopsy·chic** *a. Psychol.* [ad. G. *autopsychisch* (C. Wernicke 1892, in *Pathologie des Nervensystems* (1893) 166)], of or pertaining to self-consciousness or awareness of oneself; **au:topsycho·graphy**, psychography of oneself; so **au:topsycho·graphize** *v. intr.*, to give the psychography of oneself; **auto-rota·tion** *Aeronaut.*, unpowered rotation, esp. that of the rotor of a rotorcraft; so **auto-tota·te** *v. intr.*; **au·toscript**, a communication received by a medium in the form of automatic writing; **au·to-sex** *a.*, **au·tosexed** *a.*, **au:tose·xing** *ppl. a.*, applied to any breed of poultry in which the sexes are distinguishable at hatching; so *autosex-linkage*, etc.; **auto-soteric** (-sote·rik), *a.* [Gr. σωτηρία salvation], relating to salvation by oneself; so **auto-so·terism**; **autotelic** (-te·lik) *a.*, having or being an end or purpose in itself (see quot. 1901); **au:tothe·rapy**, treatment of one's own infirmity; hence **autothe·rapist** *sb.*; **au:totoxæ·mia, -toxemia**, the presence of autotoxin in the blood; **au:totransplanta·tion**, transplantation of tissue from one site to another in the same individual; **autotrophic** (-trǫ·fik), *a.* [Gr. τροφός + -ic. Cf. Gr. τροφικός nursing, tending], self-nourishing; of a plant, as distinguished from parasitic and saprophytic; of a lake (see quot. 1927); hence **au·totroph(e**, an autotrophic organism; **autotropism** (ǭtǫ·trŏpiz'm), *Bot.* [Tropism], (see quots.).

1903 *Nature* 15 Jan. 253/2 It is pleasing to note that a considerable number of these are auto-abstracts, for this method of summarising is the only one which ensures that the really essential points in the various investigations are brought forward. **1910** *Practitioner* Feb. 234 The older the infection the more noticeable is the tendency to clump (auto-agglutination). **1960** L. Picken *Organis. Cells* iii. 66 In another phase it may aggregate spontaneously, undergoing auto-agglutination. **1910** *Lippincott's Med. Dict.* 101/1 *Auto-antibody*, an antibody against products of the individual in which it is formed. **1943** *Brit. Jrnl. Exper. Pathol.* XXIV. 122 It seems appropriate to look upon the cold agglutinin in atypical pneumonia as an auto-antibody. **1964** Humphrey & White *Immunology* (ed. 2) 448 *Auto-antibody*, antibody reacting with constituents of the subject's own tissues. **1965** *New Scientist* 11 Mar. 628/3 Since rheumatoid factor activity is directed against the individual's own IgG, it too may be regarded as 'auto-antibody'. **1891** *Jrnl. Chem. Soc.* LX. 1151 Autocatalysis...The presence of a salt of the acid, for instance the sodium salt, in the solution, retards the formation of the lactone very considerably, and the amount of free acid in the solution, as determined alkalimetrically, remains constant for days together. **1913** Dorland *Med. Dict.* (ed. 7), *Autocatalysis*..., catalysis, or alteration of the velocity of a reaction, produced by products formed during the course of the reaction. **1940** *Chambers's Techn. Dict.* 61/1 *Autocatalysis*, reaction or disintegration of a cell or tissue, due to the influence of one of its own products. **1940** C. S. Sherrington *Man on his Nature* v. 163 And the gene? A protein system containing auto-catalysts? **1952** G. H. Bourne *Cytol. & Cell Physiol.* (ed. 2) ix. 392 The suggestion has been advanced that viruses are complex auto-catalysts. **1913** Dorland *Med. Dict.* (ed. 7) Autocatalytic. **1939** *Ann. Reg. 1938* 376 Certain plant viruses..may be a new type of autocatalytic protein. **1965** Phillips & Williams *Inorg. Chem.* I. x. 370 A typical autocatalytic curve for the decomposition of a solid. **1949** *New Biol.* VII. 60 The actual flower-promoting hormone..accumulates autocatalytically at the growing points once it has been formed. **1891** H. L. Smyth in *Amer. Jrnl. Sci.* 3rd Ser. XLII. 331 The two orotechnic actions have produced great developments of autoclastic schists. That is, schists formed in place from massive rocks by crushing and squeezing, without intervening processes of disintegration or erosion, removal and deposition. **1903** Lamplugh *Geol. Isle of Man* 70 The autoclastic structure occurs where strata of different characters are in juxtaposition. **1961** J. Challinor *Dict. Geol.* 15/2 *Autoclastic rock*, a rock formed by the breaking up of a part of a rock-mass within itself. **1915** R. A. Houstoun *Treatise on Light* vii. 106 The Abbe or Auto-collimating Spectrometer..performs the functions of both telescope and collimator. **1946** *Nature* 24 Aug. 275/1 The auto-collimating spectrograph consists of solid glass prisms of 6-in. aperture. **1932** McCaw & Cazalet tr. *O. von Gruber's Photogrammetry* vii. 139 By autocollimation on one of the surfaces of the first pair the line of collimation is brought into parallelism with the picture axis. **1938** *Geogr. Jrnl.* XCI. 388 To effect auto-collimation with a scale in the focus of the telescope. **1951** *Engineering* 6 July 9/3 The probe arm..is observed by sighting an autocollimator on to a mirror fixed to [it]. **1902** *Science* 2 May 697 In a few instances autocytotoxins for blood-cells have been produced. **1894** Gould *Dict. Med.*, *Autodiagnosis*..., self-diagnosis; the morbid impression sometimes possessed by a patient that he is affected with some particular disease. **1903** *Med. Record* LXIII. 169/1 The same hand must not be used for autodiagnostic purposes all the time. **1890** Billings *Med. Dict.*, *Autodigestion*, self-digestion of stomach by gastric juice. **1896** *Jrnl. Chem. Soc.* LXX. 616 In autodigestion, xanthine-like substances are formed. **1913** S. M. Baker in *Ann. Bot.* XXVII. 172 The well-known phenomenon of 'auto-digestion' shown by the fruit bodies of most species of *Coprinus*. **1962** *Lancet* 22 Dec. 1313/1 It allows for the progress of autodigestion over a wide range of pH. **1948** Taylor & Westcott *Princ. Radar* iv. 62 Auto-follow systems..have found many applications..Any equipment which needs to track one selected target only usually benefits considerably by being fitted with auto-following. **1961** *Aeroplane* CI. 75/1 Auto-follow is manually initiated by means of 'rolling-ball' type controls, which are used to set the initial position of the auto-follow track symbols on the airways displays. **1946** *Jrnl. Inst. Electr. Engin.* XCIII. iii. 17/1 (*heading*) Auto-following Radar Sets. **1920** *Proc. U.S. Naval Inst.* Dec. 1969 Investigation..has been made at the Central Laboratory of the French Navy, under the name of 'Autofrettage', literally self-hooping or auto-hooping...The pressure of autofrettage is limited to such a pressure that..there shall not be developed any new permanent deformations in the interior layer. **1950** *Engineering* 28 Apr. 479/3 Autofrettage is the art of inducing elasticity in a tube at pressures which otherwise cause overstrain. **1933** *Jrnl. Iron & Steel Inst.* CXXVIII. 620 Some details are given of the tests applied to the autofrettaged tubes used in the construction of the pressure pipes. **1919** J. S. Davis *Plastic Surgery* ii. 16 An autograft is a graft obtained from the same individual. **1955** *Sci. News* XXXV. 104 Skin grafts are only permanently successful when the skin is taken from another area of the patient's own body—an autograft. **1903** S. S. Cohen *Syst. Physiol. Therap.* VIII. 275 Auto-hypnosis..relieves the cortex of the corrective restraint imposed in the waking condition by the contact of the senses with the outer world. **1941** 'Rebecca West' *Black Lamb & Grey Falcon* I. 209 'I will sit here and look at the maps,' said my husband, who is much given to that masculine form of auto-hypnosis. **1913** Dorland *Med. Dict.* (ed. 7), *Autohypnotic*..1. Pertaining to self-induced hypnotism. 2. One who can put himself into a hypnotic state. **1955** R. Blesh *Shining Trumpets* (ed. 3) v. 101 It is a possessive spell, seemingly autohypnotic upon the singer, which is projected. **1894** Gould *Dict. Med.*, *Auto-hypnotism*..., mental stupor induced by dwelling intensely upon some all-absorbing delusion. **1924** W. B. Selbie *Psychol. Relig.* ii. 49 He [*sc.* the Shaman] is regularly consecrated to the office through initiation and other methods, which produce a kind of auto-hypnotism, trance, or alternate personality. **1902** *Encycl. Brit.* XXXII. 53/2 They [*sc.* our witch-burning ancestors] could scarcely have reasoned otherwise..in certain cases of hysteria and autohypnotization. **1952** *Proc. Third Congr. Internat. Soc. Hematology* 120 The development of an autoimmune, antired cell mechanism. **1962** *New Scientist* 27 Sept. 668/1 A typical case..of severe autoimmune disease, systemic lupus erythematosus. **1904** *Dunglison's Dict. Med. Sci.* (ed. 23) 114/2 *Autoimmunity*, immunity..effected by the unaided powers of the organism. **1961** *New Scientist* 17 Aug. 383/2 This concept of immunization against the body's own constituents has been termed auto-immunity. **1967** J. R. Anderson et al. *Autoimmunity* ii. 33 The term autoimmunity is used..to indicate the occurrence of an immune response resulting in the production of antibody and/or sensitized lymphoid cells capable of reacting with normal endogenous antigenic body constituents. **1907** *Practitioner* Nov. 653 These cases [of pulmonary tuberculosis]..will be found..to show 'high phases' and 'low phases', corresponding with waves of auto-immunisation. **1952** *Proc. Third Congr. Internat. Soc. Hematology* 132 By some obscure mechanism the phenomenon of autoimmunization develops and becomes injurious to the person's own red cells. **1887** A. M. Brown *Anim. Alkaloids* p. iii, Their action as auto-infectants. *Ibid.* 136 The body escapes disturbance and disintegration by processes purely auto-infective. **1900** Dorland *Med. Dict.*, *Auto-intoxicant*, a poison generated within the system. **1909** *Nation* 29 Apr. 427/2 They have ceased to get very much excited since they have learned that selfishness and greed are auto-intoxicants. **1902** W. James *Var. Relig. Exper.* i. 13 If we adopt the assumption, ..Carlyle was undoubtedly auto-intoxicated by some organ or other, no matter which. **1887** A. M. Brown *Anim. Alkaloids* p. v, Lessons on Auto-intoxication in Disease. **1893** *Times* 3 Oct. 9 The practice of introspection not rarely results in autointoxication or the generation of doubts and perplexities that work like poison in the blood. **1901** H. H. Foster in *Amer. Jrnl. Psychol.* Jan. 160 The common starting point of auto-intoxication theories is the influence of certain products of decomposition of living substance upon the continuance of cell activity. **1928** Galsworthy *Swan Song* i. vii. 55 You suffer from auto-intoxication in that House [i.e. Parliament]. **1900** Dorland *Med. Dict.*, *Autokinesis*, voluntary motion. **1949** *Jrnl. R. Aeronaut. Soc.* LIII. 943/2 When only the tail light of an aircraft is visible auto-kinesis may occur. **1959** *New Scientist* 19 Feb. 389/2 An optical illusion known as autokinesis may confuse the pilots...If you stare at a stationary object for a few seconds it appears to move. *a***1884** Knight *Dict. Mech.* Suppl. 56/2 *Autokinetic telegraph*, an English name for a form of municipal telegraph for fire-alarms, police, etc. **1934** H. C. Warren *Dict. Psychol.* 26/2 Autokinetic illusion. **1966** *Daily Tel.* 16 Aug. 18/4 Autokinetic reactions..occur when a person fixes his eye on an illuminated object in an otherwise empty field of vision. **1921** Stedman *Med. Dict.* (ed. 6) 99/2 Autologous. **1958** *Immunology* I. 206 Incubated..in heparinized autologous plasma. **1962** *Lancet* 27 Jan. 194/1 The dose of autologous marrow had been so very small. **1904** *Sci. Amer.* 7 May 366/1 Prof. Barker showed a number of photographs which had been developed by the autoluminescence of the minerals which he exhibited. **1938** R. W. Lawson tr. *Hevesy & Paneth's Man. Radioactivity* (ed. 2) xxiv. 242 The rays from radium, themselves invisible, are able to excite substances to emit visible light...This is most strikingly manifested in the phenomenon of autoluminescence. **1904** *Nature* 25 Feb. 403 Thorium with less than a trace of actinium produces an auto-photograph. **1924** *Chambers's Jrnl.* 773/2 Constipation is..responsible for more ultimate disease or auto-poisoning than anything else. *a***1909** *Buck's Handbk. Med. Sci.* IV. 184 (Cent. D. Suppl.), Autopoisonous. *Ibid.* V. 27 (Cent. D. Suppl.), Consciousness is a function of the associative mechanism and may be considered in its threefold relationship to the outer world, the body and self—allopsychic, somatopsychic, and autopsychic. **1941** *Brit. Jrnl. Psychol.* Jan. 232 The amnesia was already less marked, and autopsychic orientation largely restored. *c***1833** W. H. Brookfield *Let.* in H. Tennyson *Memoir* (1897) I. v. 126 At autopsychography I am not good, if I had any idiopsychology to autopsychographize. **1940** E. Gill *Autobiogr.* 7 The only kind of autobiography I can possibly write must be an autopsychography, a record of mental experience. **1920** *Flight* XII. 1194/2 Below 15° the aerofoil remains at rest, but at high angles it auto-rotates, slowly at first, and then more quickly. **1938** *Jrnl. R. Aeronaut. Soc.* XLII. 578 The lifting screw in this case autorotates and is not engine driven. **1918** *Rep. & Mem. Advis. Committee for Aeronaut.* No. 549 p. 4 The speed of auto-rotation is very nearly proportional to the wind speed. **1935** *Times* 4 Mar. 11/3 Speed lift is obtained by autorotation and the whole of the engine power is available for speed. **1909** *Review of Reviews* Feb. 121/1 Autoscript from 'F. W. H. Myers'. **1909** *Daily Chron.* 19 Feb. 6/5 A friend of mine who has a remarkable faculty of automatic writing sends me the following autoscript which she received this morning. **1936** tr. A. L. Hagedoorn in *Scient. Rep. VIth World's Poultry Congress* III. 54 (*title*) The autosexing Barnevelder, and the autosexing Leghorn, two new breeds. *Ibid.*, The author started..to produce an autosexing Barnevelder by adding the barring factor to the ordinary laced brown Barnevelder. **1936** M. Pease *Ibid.* 59 On Brown stripe downs the barred factor in the homozygous state (male) produces a far greater effect than it does in the heterozygous state (female).. This is the principle underlying auto-sexlinkage. *Ibid.*, Directions are given for the making of new auto-sexlinking breeds. **1941** *Poultry Sci.* XX. 317/1 Auto-sex linkage is a name given to the phenomenon of simple and positive sex differences apparent among baby chicks of a fixed breed, not crossbreds. *Ibid.* 317/2 The third..method for distinguishing sex at hatching is the production of auto-sexing (self-sexing) varieties. **1951** *Catal. of Exhibits, South Bank Exhib., Festival of Britain* 112/1 Live autosexed chicks. **1894** Autosoteric [see *heterosoteric* s.v. Hetero-]. **1909** B. B. Warfield *Calvin as Theologian* iii. 31 The logic of Socinianism gave us..an auto-soteric religion. *Ibid.* ii. 18 There is nothing against which Calvinism set its face with more firmness than..auto-soterism. **1901** Baldwin *Dict. Philos. & Psychol.* I. 96/1 Autotelic is suggested as serving, in the phrases autotelic function, process, &c., the meaning indicated by the German Selbstzweck. **1932** T. S. Eliot *Sel. Essays* i. ii. 24 No exponent of criticism..has..ever made the preposterous assumption that criticism is an autotelic activity. **1958** *Times Lit. Suppl.* 12 Sept. 507/1 The autotherapists, conscious or unconscious, who compulsively at the end of each day make their confessions to their journal. **1933** H. G. Wells *Shape of Things to Come* ii. § 8. 197 This great German mind..was incapable of autotherapy, and let its sickness have its way with it. **1890** Billings *Med. Dict.*, Autotoxæmia. **1924** *Psyche* July 67 A form of blood poisoning due to the development of autotoxaemia from over indulgence in flesh foods. **1909** *Jrnl. Exper. Med.* XI. 177 Besides my own the only successful autotransplantations of these glandules in dogs are, perhaps, the two in puppies reported by Pfeiffer, Hermann, and Mayer. **1920** *Ibid.* XXXII. 113 (*title*) Homeotransplantation and autotransplantation of the spleen in rabbits. **1909** Webster, Autotrophic. **1938** S. Morgulis tr. *A. I. Oparin's Origin of Life* viii. 206 These autotrophic organisms can be compared..to the green autotrophes, even to the simplest algae. *Ibid.* 207 This fairly isolated group of living things must have arisen..at the same time when the first autotrophes capable of photosynthesis appeared. **1954** *New Biol.* XVII. 61 Organisms able to subsist entirely on inorganic nutrients are known as autotrophs ('self-feeders'); they are relatively common among the sulphur bacteria. **1959** *Chambers's Encycl.* II. 317/2 The autotrophes, which include the green plants, possess the ability to trap and harness for their own purposes energy from the inorganic world outside. **1901** I. B. Balfour in *Rep. Brit. Assoc. Advancem. Sci.* 820 The root-difference between plants and animals is one of nutrition. Plants are autotrophic, animals heterotrophic. **1927** Birge & Juday in *Amer. Philos. Soc. Proc.* LXVI. 371 Many lakes..are dependent wholly on internal sources and may be called *autotrophic*—they receive water directly from rain and from a very limited drainage which has passed through a sand filter. **1947** *Endeavour* VI. 173 Nitrogen-fixing blue-green algae are the most completely autotrophic organisms known. **1898** *Nat. Science* June 387 By autotropism is implied the inherent tendency of vegetable organs to grow in a straight line. **1908** H. Driesch *Sci. & Philos. of Organism* I. i. 158 'Autotropism', that is, the fact that branches of plants always try to reassume their proper angle with regard to their orientation on the main axis, if this orientation has been disturbed.

b. Used frequently in the names of self-acting mechanisms, machines, instruments, etc.; **auto-ala·rm**, a radio receiving device which in time of distress gives audible automatic warning of the need for help; **auto-a·nalyser**, an automatic apparatus for performing (chemical) analyses; **auto-change, -changer**, a device which automatically places a record on the turntable of a record-player when the previous record has finished playing; **au·tocode** *Computers* (see quots.); **auto-conve:rter** *Electr.* = **auto-transformer*; **au·toflare** *Aeronaut.*, a mechanism for automatically preparing an aircraft to land (see quots.); **auto-focus**, a device by means of which an enlarger, camera, or the like is automatically focused; also *attrib.* or quasi-*adj.*; hence *auto-focusing*; **auto-transfo·rmer**, a transformer or compensator in which a part of the primary coil is used as a secondary, or a part of the secondary as a primary coil.

1885 *Jrnl. Chem. Soc.* XLVIII. 854 New Pile, or Auto-accumulator. **1895** S. P. Thompson *Polyphase Electric Currents* x. 186 The auto-transformer (or 'one-coil' transformer) merely consists of a coil of wire wound on an iron core, and connected across the mains. **1902** W. J. Dibdin *Public Lighting* 176 The auto-valve, which is not affected by condensation or grit. **1904** *Electr. Rev.* 17 Sept. 459 (Cent. D. Suppl.), An 'oil-break auto-starter' switch. **1927** *Glasgow Herald* 18 July 10 The object of the auto-alarm is to ensure that the call shall be received by the smaller ships and that no wireless distress calls shall be missed by any ship owing to the operator being off duty. **1928** *B.B.C. Handbk.* 1929 427/2 *Auto-transformer*, a transformer either in radio or audio frequency in which the primary and secondary windings are formed by one and the same coil having three connections to it. **1943** *Gloss. Terms Electr. Engin.* (B.S.I.) 38 *Auto-transformer*, a transformer in which the primary and secondary windings have a common part or parts. **1944** *Gramophone* Apr. 174/3 (Advt.), H.M.V. Autochange Radiogram. *Ibid.*, Specially constructed Connoisseur Radiogram with twin auto changers. **1949** *Q. Jrnl. Forestry* XLIII. 26 A new model..auto-scythe, with independent wheel drive was demonstrated cutting bracken. **1957** *Times Survey Brit. Aviation* Sept. 2/5 Auto-stabilization is fitted to simplify the pilot's tasks. **1958** V. Drumm in M. L. Hall *Newnes Compl. Amat. Photogr.* 312 When the two systems are coupled together, auto-focus to get the approximate position, and manual focusing for the final fine touch, it is very satisfactory. *Ibid.*, A test for auto-focus enlargers. *Ibid.* 313 If the enlarger is auto-focus, try altering the focus slightly, and see if the image can be improved. **1959** *New Scientist* 25 June 1375/2 Some of the [computer] machine-makers have gone as far as building into their later machines an 'autocode' by which the advanced types can automatically adapt the programme codes of previous models. **1959** *Listener* 29 Oct. 731/1 (Advt.), *Four speed autochanger*. **1959** 'Ellis Peters' *Death Mask* i. 17 We hadn't got an auto-change on the old radiogram. **1960** *Ann. N.Y. Acad. Sci.* LXXXVII. 6/6 The development of the Auto-Analyzer has made available an instrument for rapid, precise colorimetric analysis of various biochemical components. **1960** *Times Rev. Industry* Aug. 53/2 Emission of air from the nozzles is controlled by the aircraft's electro hydraulic autostabilizer. **1962** *Engineering* 26 Jan. 137 Auto-ignition is now virtually standard on gas cookers. **1962** *Lancet* 2 June 1162/1 Urea was measured by autoanalyser ('Technicon'). **1962** *Listener* 7 June 986/2 We can just say let the letter 'A' stand for the set of instructions, and then we can, in our programme, simply write 'repeat ''A'' fifty times'. This instruction is what is called an auto-code, which the computer will itself translate into the fully detailed instructions. **1962** *New Scientist* 12 July 84/3 The autoflare system (the system which puts the aeroplane into the landing attitude at the right moment). **1963** *Times* 19 Apr. 11/6 On the Trident and VC 10 it is proposed to use the equipment in stages, the first being autoflare, in which the aircraft is brought down to the height at which it levels off for landing, leaving the pilot to put the aircraft down. **1967** Karch & Buber *Offset Processes* iv. 123 It is linked with an 'autofocussing' system.

c. *Biol.* In comp. with *-ploid*, as *autopolyploid, *autotetraploid: having sets of chromosomes derived from a single species (opp. **allo-*).

auto-[2] (ǭ·to). abbreviation of *automobile used as comb. form, chiefly in the names of vehicles, as *autobus*, *autocar*, *automotor*, etc. See also *auto *sb.*[2], *autocycle.

1895 *Daily News* 30 Nov. 5/1 To apply the new principle of the 'auto-motor' to road-waggons, heavy drags, hunting-traps, and stage-coaches. **1895** *Westm. Gaz.* 17 Dec. 3/1 We congratulate the police authorities..on having convicted the owner of an autocar for proceeding along a road at a pace exceeding three miles an hour. **1896** (*title*) The Automotor Journal. **1897** *N.Y. Herald* 19 Sept. 2/1 The introduction of an efficient autocab service in the streets of Paris. **1899** *N.Y. Jrnl.* 17 June 5/2 The New York Auto-Truck Company. **1899** *Westm. Gaz.* 4 July 6/3 The auto-conveyances of members of Parliament. **1899** *Boston Herald* 12 July 6/5 We should have the new words..auto-bus, [etc.]. **1900** *Engineering Mag.* Aug. 733 The auto-waggon, which provides just that rapid and cheap form of independent direct transport[etc.]. **1904** *Westm. Gaz.* 23 Sept. 7/3 Mr. W. K. Vanderbilt, junior's auto-boat 'Mercédès the Sixth'. **1908** A. Bennett *Buried Alive* iv. 88 Two commissionaires were helping him into an auto-cab. **1918** W. Stevens *Let.* 30 Apr. (1967) 208, I must wait..for an auto-bus back to Johnson City. **1927** *Chambers's Jrnl.* 375/2 You can..explore the French side..from end to end by safe, strong, comfortable autocar. **1927** *South America* May 137/2 The auto-coach is much needed to replace the horse-coach. **1941** Koestler *Scum of Earth* 206 The regular autobus line Bergerac–Bordeaux still functioned.

auto-analysis (ǭ:to,ănæ·lisis). [f. Auto-[1] + Analysis.] Self-analysis; *esp.* exploration by a person of his own conscious or unconscious mind. Hence **auto-a·nalyst**, one who engages in auto-analysis.

1894 W. Wallace *Proleg. to Hegel's Philos.* (ed. 2) xvii. 209 The method of mathematics or mechanics..is analytic or synthetic—but not auto-analysis or auto-synthesis. **1917** C. R. Payne tr. *Pfister's Psychoanalytic Method* xii. 346 Finally he submitted to autoanalysis and found it was the melody of Beethoven to Goethe's verse: 'Mit Männern sich geschlagen'. *Ibid.*, This also occurred to the auto-analyst in a moment. **1928** *Daily Tel.* 3 Jan. 8/5 The tradition..that the keepers of journals are more..intent on auto-analysis.

autobahn (ǭ·tobān). Pl. **autobahnen, autobahns.** [G., f. *auto* (*auto-[2]) + *bahn* path, road, f. MHG. *ban(e)*, MLG., MDu. *bāne*, usually related to OHG. *bano*, MHG. *ban(e)* death, OE. *bana* Bane *sb.*[1]] In Germany, a fast motor road; = *motorway.

1937 *Sunday Times* 18 Apr. 33/2 In the last six months or so there has sprung up from several quarters a demand for motorways, or autobahnen or autostrade—meaning.. new roads for the exclusive use of motor traffic. **1940** D. E. Stevenson *English Air* i. iv. 35 The Continental roads were the ones for speed—the long straight roads of France, and those new German *Autobahn*. **1941** *Hutchinson's Pict. Hist. War* 14 May–8 July 156 (*caption*) Autobahn bridge near Mannheim extensively damaged after a raid by the R.A.F. **1955** *Times* 14 June 7/5 It was easy for a Hitler or a Mussolini to drive great autobahns or autostradas through their respective countries.

autobasidiomycete (ǭ:tobăsi·diomaisi·t). *Biol.* [G. (O. Brefeld *Untersuchungen aus d. Gesammtgeb. d. Mykologie* (1889) viii. 185), f. *auto-[1]: see below.] A fungus belonging to the Autobasidiomycetes, a division of the Basidiomycetes, including mushrooms, toadstools, and related fungi having an autobasidium.

1895 M. C. Cooke *Introd. Study of Fungi* ii. x. 115 The Basidiomycetes are arranged in two groups: (1) The Protobasidiomycetes..(2) the Autobasidiomycetes, in which the basidia are not septate, and bear a definite number of basidiospores. **1933** *Ann. Reg. 1932* 57 Wakayama's demonstration that the chromosome numbers in thirty-four Autobasidiomycetes are either two, four, or six. **1937** Gwynne-Vaughan & Barnes *Struct. & Dev. Fungi* (ed. 2) 326 The Autobasidiomycetes include between 12,000 and 13,000 species; they are cosmopolitan in distribution.

autobasidium (ǭtobăsi·diŏm). *Biol.* [f. *auto-[1] + Basidium.] A basidium that has no septum.

1902 Webster Suppl., *Autobasidium*, an undivided basidium, characteristic of the *Autobasidiomycetes*, one of the two groups into which the Basidiomycetes are sometimes divided. **1928** C. W. Dodge tr. *Gäumann's Compar. Morphol. Fungi* xxv. 412 The basidia in which nuclear division is not followed by formation of septa are called auto- or holobasidia. **1940** *Chambers's Techn. Dict.* 60/2 *Autobasidium*, a basidium which does not become septate.

autobiographer. (Earlier example.)

1821 *Q. Rev.* XXIV. 361 He merely describes his own [thoughts]..just as a real autobiographer might do.

autobiographic, *a.* **1.** (Earlier example.)

1827 T. J. Dibdin *Reminisc.* II. viii. 326 My life, ladies and gentlemen! (that is, my autobiographic life) is now drawing towards its close.

autobiographist (ǭ:to,bəi,ǫ·grăfist). [f. *auto-[1] + Biographist.] = Autobiographer.

1840 *Fraser's Mag.* XXII. 579 We must first describe this volunteer autobiographist. **1850** *Tait's Mag.* XVII. 525/2 Our autobiographist betakes himself to London. **1920** *Westm. Gaz.* 26 Nov. 8/2 Too indiscreet a candour in autobiography..may..involve unfortunate consequences upon the autobiographist.

autobiography. (Earlier example.)

1797 *Monthly Review* 2nd Ser. XXIV. 375 It is not very usual in English to employ hybrid words partly Saxon and partly Greek: yet *autobiography* would have seemed pedantic.

autobus, autocar: see *auto-[2].

autocentric (ǭto,se·ntrik), *a.* [f. *auto-[1], after *egocentric *a.*] Centred in the self; making oneself the centre (see quot. 1889). Hence **autoce·ntrism**, the possession of an autocentric attitude or outlook.

*a***1866** J. Grote *Moral Ideals* (1876) 189 Becoming

cosmocentric instead of autocentric in our knowledge. **1889** J. VENN *Empirical Logic* xxv. 572, I should be inclined to mark the above distinction by the words *autocentric* and *coinocentric* . . the distinction between two attitudes which we may adopt in our practice. In the one case the agent is supposed to make himself the centre of his own speculation and consequent action. **1927** T. BURROW *Social Basis of Consciousness* II. vii. 188 The type of individual who upon the initial stimulus to defence has recourse to a tactic of unconditional retreat. . is the autocentric individual. **1931** K. E. KIRK *Vision of God* 489 As long ago as 1906, he had fixed upon 'autocentrism' as the clue to the complexities of Newman's character.

autochrome (ǭ·tokrōum), *a.* and *sb.* *Photogr.* (Disused.) [f. Gr. αὐτο- (see *AUTO-^{1}) + χρῶμα colour.]

A. *adj.* Defining a process and a plate used in colour-photography, invented by Messrs. Lumière of Lyons; also, a slide produced by this process.

1907 *Daily Chron.* 12 July 6/4 The new plates are called 'The Autochrome Plates'. **1907** *Brit. Jrnl. Photogr.* 2 Aug. 573/1 The Lumière 'Autochrome' Plates. **1920** *Glasgow Herald* 1 Mar. 13 A series of autochrome slides.

B. *sb.* A photograph produced by this process.

1907 *Brit. Jrnl. Photogr.* 2 Aug. 573/1 The reproduction of Lumière 'Autochromes' on 'Uto' paper. **1910** *Westm. Gaz.* 16 Apr. 14/2 An excellent collection of autochromes.

autochthon. Add: **4.** *Geol.* An autochthonous rock formation; opp. *ALLOCHTHON.

1942 M. P. BILLINGS *Struct. Geol.* x. 181 Rocks of the foreland. . are said to be *autochthonous*—that is developed where found; these rocks are sometimes called the autochthon. **1957** [see *ALLOCHTHON]. **1961** J. CHALLINOR *Dict. Geol.* 15/1 The 'autochthon' is essentially a rock-succession that, as a whole, has not been translated by tectonic movement (it forms part of the foreland); but it may show. . autochthonous folding, within itself.

5. An indigenous plant.

1893 in *Funk's Stand. Dict.* **1916** B. D. JACKSON *Gloss. Bot. Terms* (ed. 3) 40/2 *Autochthon*, . . a native plant, not an introduction.

autochthonous, *a.* Add: **c.** *Geol.* [f. G. *autochthon* (K. W. von Gümbel 1888), see *ALLOCHTHONOUS *a.*] Consisting of or formed from indigenous material (opp. *ALLOCHTHONOUS); applied to organic deposits and rock formations originating *in situ*.

1900 in J. F. KEMP *Handbk. Rocks* (ed. 2) 126. **1916** C. C. FORSAITH in *Bot. Gaz.* LXII. 33 Autochthonous peat (that type of peat which represents the amassing of successive generations of plants in. . constant, but stagnant. . water). **1928** K. E. CARPENTER *Life in Inland Waters* viii. 207 In place of the 'autochthonous' (home-produced) sediments. . we have 'allochthonous' material of drifted fragments from shore-living plants. **1935** E. B. BAILEY *Tectonic Essays* iii. 34 In tectonics, an autochthonous fold is one that is made of untravelled indigenous rocks. **1946** *Q. Jrnl. Geol. Soc.* CI. 207 It is considered that for the most part the Downtonian and Dittonian vertebrate faunas were not autochthonous, but were of freshwater origin.

autochthonously (ǭtǫ·kþǫ̆nəsli), *adv.* [-LY^2.] In an autochthonous manner.

1885 *Encycl. Brit.* XVIII. 361/2 The larger number of maladies do not arise autochthonously or 'under a whole skin', they are generated by certain morbific causes. **1953** HINSIE & SHATZKY *Psychiatric Dict.* (ed. 2) 60/2 The heart. . continues to beat for a time after it has been removed from the body. It is then said to function autochthonously.

autoclave. Add: **2.** A vessel for carrying out chemical reactions at high temperatures under pressure; an apparatus for sterilizing by steam at high pressure.

1876 *Jrnl. Chem. Soc.* XXX. 451 Saponification of Neutral Fats in Autoclaves. **1886** *Ibid.* L. 112 High Pressure Digesters (Autoclaves) for Chemical Laboratories. . . The apparatus consists of a cylindrical copper vessel, provided with a cover, which is firmly fastened down by a screw. **1919** *Modern Hospital* XII. 281 (*title*) High-pressure dressing sterilizers or autoclaves. **1931** *Forestry* V. 141 The nutrient medium. . is made up and sterilized in an autoclave. **1946** *Electronic Engin.* XVIII. 317 The phosphors. . are prepared in autoclaves at high temperature and under pressure of some thousand atmospheres.

Hence **au·toclave** *v. trans.* and *intr.* (Webster, 1934); so **au·toclaving** *vbl. sb.* and *ppl. a.*, **au·toclaved** *ppl. a.*

1929 *Biochem. Jrnl.* XXIII. 1052 The preparation of 'bios' used was an autoclaved solution of marmite. **1943** *Electronic Engin.* XV. 394 A number of steatite bodies. . were investigated by autoclaving. **1955** J. G. DAVIS *Dict. Dairying* (ed. 2) 844 Autoclave at a pressure of 15 lb. per sq. in. for 20 min[utes]. **1960** *Times* 19 Aug. 2/5 Spray drying and autoclaving techniques. **1965** *Economist* 5 June 1176/2 Siporex is an autoclaved light weight aerated concrete product.

autocopyist (ǭtokǫ·pi,ist). [f. *AUTO-^{1} + COPYIST.] An apparatus for producing facsimile copies of written matter. (Disused.)

1888 *Encycl. Brit.* XXIV. 697/2 In principle the autocopyist is like the hektograph. **1907** *Daily Chron.* 21 Jan. 3/5 Sketch maps. . have been reproduced by means of an autocopyist.

autocorrelation (ǭ:to,kǫrĭlēi·ʃən). [f. *AUTO-^{1} + CORRELATION.] A correlation between all the elements of a series and those separated from them by a given interval (see quot. 1951). Also *attrib.* Hence **au:toco·rrelator**, an apparatus for computing autocorrelations.

1950 *Jrnl. Acoust. Soc.* XXII. 677 (*title*) Autocorrelation analysis of speech sounds. **1951** *Amer. Jrnl. Psychol.* LXIV. 258 An autocorrelation is simply a correlation computed between all items of a series, and the items that follow any given item by one, two, three, or more steps, or by corresponding time-intervals. *Ibid.* 262 The autocorrelator is directly connected to the output. **1956** J. WHATMOUGH *Lang.* v. 76 The quantitative criterion. . might be established by some electrical apparatus, such as an autocorrelator. **1965** *Math. in Biol. & Med.* (*Med. Res. Council*) I. 37 The computer is being used to apply the mathematical techniques of autocorrelation and cross-correlation to the interpretation of the EEG in the treatment of temporal lobe epilepsy.

autocracy. Add: **2. c.** Autocrats collectively; the realm of autocrats.

1905 *Smart Set* Sept. 125 Of all the fish that swim or swish In ocean's deep autocracy There's none possess such haughtiness As the codfish aristocracy. **1928** *Manch. Guardian Weekly* 19 Oct. 301/2 She is credited with wiser views of Russian policy than were commonly found in the heads of that fated autocracy.

autocratism (ǭtǫ·krătiz'm). [f. AUTOCRAT + -ISM.] The principles or practices of autocrats.

1848 *Fraser's Mag.* XXXVII. 96 To liberal sentiments be united the instincts of autocratism. **1916** V. HORSLEY in S. Paget *Life* (1919) III. iii. 315 It is interesting to appreciate the sensation of an autocratism. **1952** V. GOLLANCZ *My Dear Timothy* 316 How difficult it is to overcome an autocratism native to one's temperament.

autocritical (ǭtokri·tikăl), *a.* [f. *AUTO-^{1} + CRITICAL *a.* Cf. *auto-criticism* s.v. AUTO-.] Critical of oneself or one's own work.

1642 C. HERLE *Fuller Answ. to Ferne* 14 Thats the peculiar Priviledge of Gods word to be autocritical, its own last judge. **1821** *Blackw. Mag.* X. 116 We differ. . from the autocritical junto who are willing to dictate to *us*. **1850** L. HUNT *Autobiogr.* I. p. vi, Coleridge's *Literary Life* is professedly autocritical. **1957** R. CAMPBELL *Coll. Poems* II. 132 No autocritical confession Could be extracted.

autocross (ǭ·tokrǫs). [f. *AUTO-^{2} + CROSS (-COUNTRY.] A form of cross-country racing for motor vehicles.

1965 *Observer* 5 Sept. 13/4 Rallies are a doomed sport. . . The future lies with autocross, a form of cross-country motor racing with private cars on private estates. **1966** *Pubn. Amer. Dial. Soc.* 1964 XLII. 2 *Autocross*, an automotive competition like a gymkhana, except that there are fewer obstacles and greater speeds involved.

autocue (ǭ·tokiū). [f. *AUTO-^{1} + CUE *sb.*^{2}] The name of a script device, placed out of range of the camera, which prints the words to be spoken by a speaker on television as an aid to his memory.

1958 I. BROWN *Words in our Time* 22 The autocue is. . one of the gadgets introduced by television. The speaker or recorder of news has in front of him a screen over which his text passes. Thus he can read without looking down or fidgeting with papers. **1967** P. PURSER *Twentymen* xi. 70 The revised pieces were then typed up for the Autocue prompting machine.

autocycle (ǭ·tosəik'l). [f. *AUTO-^{2} + CYCLE *sb.*] A cycle propelled by a motor; a motor-cycle. So **au·tocyclist**, one who rides an autocycle.

1905 *Daily Chron.* 21 July 5/5, 86 miles an hour. Wonderful Performance on an Autocycle. **1920** *Daily Mail* 29 Nov. 8 The Autocraft Board, which has an Auto-cycle section. **1958** *Oxford Mail* 7 Feb. 7/5 Auto-cyclist. . killed yesterday when the auto-cycle he was riding was involved in a collision with a Land Rover.

auto-da-fé. Add: **2.** Also *transf.*

1917 KIPLING *Diversity of Creatures* 148 Evidently this was their established *auto-da-fé*. *a* **1930** D. H. LAWRENCE *Last Poems* (1932) 175 Help! Help! they want to burn my pictures, They want to make an *auto da fe*!

autodidactic (ǭ:to,didæ·ktik), *a.* [f. AUTODIDACT + -IC.] Self-taught; acquired by teaching oneself; pertaining to self-teaching.

1847 S. AUSTIN tr. *Ranke's Hist. Ref.* III. 586 Autodidactic artisans. **1878** *Tinsley's Mag.* XXIII. 293 Autodidactic studies. **1886** *Contemp. Rev.* Feb. 293 He [*sc.* Menzel] was from the beginning an auto-didactic realist; he drew and painted as he saw. **1966** *English Studies* XLVII. 201 Whitman was largely a self-educated journalist whose Self, at the top of its autodidactic voice, brought into poetry many of the ideas of the time.

autodrome (ǭ·todrōum). [f. *AUTO-^{2} + Gr. δρόμος race-course.] A motor-racing track or circuit.

1935 EYSTON & LYNDON *Motor Racing* i. 7 An autodrome was opened at Linas-Montlhèry, near Paris. **1957** S. MOSS *In Track of Speed* xi. 139 The Argentine autodrome was a short twisty one of just under 2½ miles. **1963** *Times* 25 Apr. 3/7 The new Ferrari formula one car amazed experts by its speed at the Monza autodrome today.

autodyne (ǭ·todəin). *Radio.* [f. *AUTO-^{1} + -DYNE.] An electric oscillating circuit in which the same valve is used for rectification and generation of oscillations. Also *attrib.*

1918 W. H. ECCLES *Wireless Telegr.* (ed. 2) 497 Autodyne Reception, Self-Heterodyne Reception, are names given to those modes of beat reception in which the auxiliary oscillations are generated within windings and condensers that are essential parts of ordinary receiving circuits. **1943** *Gloss. Terms Telecommunic.* (*B.S.I.*) 72 *Autodyne oscillator*, a receiving device which generates the local oscillations required for beat reception in addition to performing its other functions, such as amplification or detection. **1961** *Times Rev. Industry* Apr. 18/2 The autodyne is a kind of rotary converter where 'feed back' is used to control the operation of the device.

autœcious (ǭtī·ʃəs), *a.* *Bot.* Also **autecious autoicous.** [f. Gr. αὐτός same + οἰκία house: see -IOUS.] **a.** Of parasitic fungi: inhabiting the same host throughout their life. **b.** Having both sexual organs on the same plant; monœcious.

1882 VINES *Sachs's Bot.* 332 Such forms [of parasitic fungi] as these are said to be heterœcious (metœcious), to distinguish them from those. . which inhabit the same host throughout their whole life (autœcious). **1900** B. D. JACKSON *Gloss. Bot. Terms*, *Autoicous*, in Bryophytes, the male and female inflorescences on the same plant.

auto-erotic (ǭ:to,erǫ·tik), *a.* [f. *AUTO-^{1} + EROTIC *a.*] Pertaining to auto-erotism. Hence **auto-ero·tically** *adv.*

1898 HAVELOCK ELLIS in *Alienist & Neurol.* Apr. 260 Among auto-erotic phenomena. . we must further include those religious sexual manifestations for an ideal object, of which we may find evidence in the lives of saints. **1914** *Lancet* 9 May 1303/2 When the auto-erotic impulse of the child first turns to an external object that impulse is bisexual. **1927** HENDERSON & GILLESPIE *Text-bk. Psychiatry* vi. 115 But in some individuals the development of the libido's object-direction may be arrested at an intermediate stage (the autoerotic or the homosexual). **1933** J. C. FLUGEL *Hundred Yrs. Psychol.* IV. viii. 284 A certain amount of libido always finds its satisfaction auto-erotically. **1954** KOESTLER *Invisible Writing* 362 Auto-erotic indulgences may quickly become an addiction.

auto-erotism (ǭ:to,e·rŏtiz'm). *Psychol.* Also **-ero·ticism.** [f. *AUTO-^{1} + EROTISM, EROTICISM; cf. G. *autoerotismus* (Freud 1899, *Let.* to W. Fliess 9 Dec. in *Aus d. Anfängen d. Psychoanalyse* (1950) 324).] Spontaneous erotism, self-erotism; sexual gratification aroused or obtained by oneself, i.e. not with another person; masturbation. (Cf. *ALLO-EROTISM.)

1898 HAVELOCK ELLIS in *Alienist & Neurol.* Apr. 260 (title of paper), Auto-Erotism, a Psychological Study. *Ibid.*, By 'auto-erotism' I mean the phenomena of spontaneous sexual emotion generated in the absence of an external stimulus proceeding, directly or indirectly, from another person. **1900** —— *Studies Psychol. Sex* II. 125 In a wide sense. . auto-erotism may be said to include those transformations of repressed sexual activity which are a factor of some morbid conditions as well as of the normal manifestations of art and poetry. **1916** Auto-eroticism [see *AUTISM]. **1952** V. GOLLANCZ *My Dear Timothy* 198 A common or garden tendency to masochistic autoerotism. **1954** [see *ALLO-EROTISM]. **1955** C. S. LEWIS *Surprised by Joy* xi. 160 The first and deadly error. . turning religion into a self-caressing luxury and love into auto-eroticism. **1963** in A. Heron *Towards Quaker View of Sex* ii. 16 It can firmly be said that autoeroticism (the name given by psychologists to masturbation) is a normal phase of human development.

autogamous (ǭtǫ·găməs), *a.* *Bot.* [f. Gr. αὐτο- (see *AUTO-^{1}) + γάμος marriage: see -OUS.] Characterized by self-fertilization.

1889 in *Cent. Dict.* **1925** *Glasgow Herald* 20 June 4 Distomum hepaticum is an autogamous hermaphrodite.

autogamy. Add: [First formed in G. (A. Kerner 1876, *Die Schutzmittel der Blüthen*).] (Earlier examples.)

1877 DARWIN *More Letters* (1903) II. 413, I wish that I had used some such terms as autogamy, xenogamy, etc. **1878** OGLE tr. *Kerner's Flowers* 9, I understand by autogamy the fecundation of a flower by the pollen from the andrœcium of the same flower.

2. *Biol.* Fusion of sister-cells or of pairs of nuclei within a cell. Hence **autoga·mic** *a.*

1900 K. PEARSON *Gram. Sci.* (ed. 2) 437 Is there any form of sexual selection such as autogamy, endogamy, apogamy, or homogamy, using these terms in their broadest senses? *Ibid.* 501 Variations do not occur accidentally or in isolated instances; autogamic and assortative mating are realities. **1940** *Chambers's Techn. Dict.* 61/1 *Autogamy*, the fusion of sister-cells, or of two sister-nuclei.

autogenesis (ǭtodʒe·nĭsis). [f. *AUTO-^{1} + -GENESIS.] Origination within the organism.

1890 BILLINGS *Med. Dict.*, *Autogenesis*. . , spontaneous generation.

au:togene·tic, *a.* [f. prec.; see -GENETIC.]

1. *Med.* Developed by or due to autogenesis; produced within the organism. Hence **au:toge·ne·tically** *adv.*

1886 *Brit. Med. Jrnl.* 10 Apr. 694/1 Autogenetic puerperal fevers. Mental worry from illicit pregnancies and clandestine marriages was a common source of autogenetic

empoisonment. *Ibid.*, Some septic poison, either from without or autogenetically, might cause the same series of symptoms.

2. *Physical Geogr.* (See quot.)

1902 WEBSTER Suppl., *Autogenetic* (*Phys. Geol.*), pertaining to, controlled by, or designating, a system of self-determined drainage. *Autogenetic drainage*, a system of natural drainage developed by the constituent streams through headwater erosion. *Autogenetic topography*, a system of land forms produced by the free action of rain and streams on rocks of uniform texture.

autogenic, *a.* Add: = AUTOGENOUS *a.* in various senses (in quot. 1890 = self-induced). Also *spec.* in *Ecology* (see quot. 1931).

1890 MERCIER *Sanity & Insanity* xiii. 343 The melancholy . . is a spontaneous and autogenic melancholy. **1931** [see *ALLOGENIC *a.* 2]. **1946** A. G. TANSLEY *Introd. Plant Ecology* iv. 46 The first species to occupy the area will. . in most cases give way to others. . until a relatively stable equilibrium is reached between the vegetation and its habitat. This kind of succession is called the development of vegetation (autogenic succession). *Ibid.* xiii. 157 The accumulation of humus. . is a direct reaction of the plant community itself upon its habitat (autogenic factor).

autogenous, *a.* Add: **c.** Also applied to a process of welding in which metals are united by melting their edges together, any added welding-metal being of the same composition.

1930 *Engineering* 7 Feb. 163/2 The oxygen is chiefly required. . for autogenous welding. **1961** J. N. ANDERSON *Appl. Dental Mat.* (ed. 2) xiii. 116 Sweating. . is sometimes referred to as autogenous welding.

autogiro (ǭ:to,dʒəiə·ro). Also **autogyro.** [f. *AUTO-^{1} + It., Sp. *giro* GYRE *sb.*] The proprietary name (in spelling *-giro*) of a type of aircraft, deriving its lift mainly from a system of freely rotating horizontal vanes, and capable of landing in a very small space.

1923 *Flight* 24 May 275 Some tests have recently been carried out at Getafe, near Madrid, with an extremely interesting and original type of machine, the invention of a young Spanish engineer, Don Juan de la Cierva. This machine is known as the 'Autogiro', and is what might be described as being midway between the aeroplane and the helicopter. **1927** *Observer* 17 Apr. 10/2 The wider range of control promised by such types as the 'autogyro'. **1930** *Flight* 4 Apr. 391/1 The Autogiro stands quite alone in the aircraft market, as it is the most unorthodox. **1960** STUBELIUS *Balloon* 277 Autogiros were in use up to about 1943. Today, when the design is obsolete, the word *autogiro* occurs only in historical contexts. **1962** *Listener* 21 June 1075/2 A single-seater autogyro.

autograph, *sb.* (and *a.*). Add: **A. 1. a.** (Later example.)

1892 W. W. SKEAT *12 Facs. O.E. MSS.* 10 The MS. of Piers Plowman is corrected with minute care, and seems to be an autograph of the author.

2. (Additional examples.) An *autograph-book* (or *-album*) freq. contains occasional verses, etc., as well as a person's signature.

1808 *Monthly Pantheon* I. 665/1 Another learned collector purchases a work. . because some learned man's name or *autograph*, according to the modern fashionable literary nomenclature, is written on the title page. **1841** F. A. KEMBLE *Let.* 26 Dec. in *Records of Later Life* (1882) II. 148, I am not an autograph collector. **1858** O. W. HOLMES *Autocrat* i. 6 One of the lady-boarders. . sent me her autograph-book. **1858** QUEEN VICTORIA *Let.* 15 Feb. in *Dearest Child* (1964) 45 Never mind about my sheet of the autograph album. **1870** 'F. FERN' *Ginger-Snaps* 203 If there is an intolerable nuisance, it is your persistent autograph-hunter. **1879** B. F. TAYLOR *Summer-Savory* xxvi. 209 The writer hopes the reader's name is not found in many autograph albums. **1959** I. & P. OPIE *Lore & Lang. of Schoolchildren* vii. 117 American children's film-star rhymes. . seem to be slicker, more of the autograph-album variety. **1966** *Guardian* 12 Aug. 6/6 Will it only be television that will bring the autograph-hunters to the stage door?

B. *adj.* Also used of a painting done by the painter himself, not by an imitator.

1958 *Times Lit. Suppl.* 5 Sept. 490/1 Some of the Madonnas. . must be wholly or in large part autograph.

au·tographed (-grɑft), *ppl. a.* [f. AUTOGRAPH *v.*] In senses of the verb.

1818 [see AUTOGRAPH *v.* 1]. **1901** *Daily Chron.* 10 July 3/5 Autographed photographs of celebrities. **1908** LD. ROSEBERY in Begbie *Life W. Booth* (1920) II. xxviii. 398 Send me your autographed photograph.

autographic, *a.* Add: (Earlier examples.)

1810 *Ann. Reg. 1808* 118 The manuscript. . is an autographic note, written by the late Lord Bute. **1824** *New Monthly Mag.* XI. 221 Judging from the autographic obituary before me.

2. *Path.* (See quot.)

1894 GOULD *Dict. Med.*, *Autographic. .Skin*, a condition of vaso-motor paralysis, usually in hysterical patients, in which markings made upon the skin form quite persistent and intensely red traces. *A. woman*, one with an *Autographic Skin*.

autographism (ǭtǫ·grăfiz'm). *Path.* [f. AUTOGRAPH *sb.* + -ISM.] A condition of the skin in which tracings leave an elevated mark. Hence **auto·graphist.** Cf. *AUTOGRAPHIC *a.* 2.

1890 *Globe* 13 June 6 Autographism, as this phenomenon is called, is believed to be due to a nervous susceptibility which may exist for years in a patient. **1894** GOULD *Dict. Med.*, *Autographist*. . , one who has an autographic skin.

autogravure (ǭ:to,grăviūə·ɹ). [f. *AUTO-^{1} + F. *gravure*, after *photogravure*.] A photomechanical modification of the etching process. Used *attrib.* in *autogravure process*. Also, a picture produced by this process.

1885 *Athenæum* 31 Jan. 156/3 Plates, produced by the autogravure process of the Autotype Company. **1888** *Chambers's Encycl.* I. 606 *Autogravure*, a peculiar process of photo-engraving patented by J. R. Sawyer, London, on November 12, 1884. **1889** *Guardian* 23 Jan. 137/1 An autogravure from Correggio.

autogyro, var. *AUTOGIRO.

autoharp (ǭ·tohāɹp). [f. *AUTO-^{1} + HARP *sb.*^{1}] A musical instrument of the zither type provided with dampers to facilitate the production of arpeggio effects.

1882 C. F. ZIMMERMANN *U.S. Pat.* 257,808 1/1 A harp so provided. . I term an 'autoharp'. *Ibid.*, The autoharp. . may be used for accompaniment to the voice or another musical instrument. **1893** *U.S. Pat.* 490,407, I, Christian Henry Eisenbrandt. . have invented certain new and useful Improvements in Hammers for Autoharps. . . My invention relates to means for striking chords on such stringed instruments as the autoharp. **1894** *Work* 23 June 365/2 An autoharp can be bought from almost any dealer in musical instruments. **1897** MARY KINGSLEY *W. Africa* 82 Captain Heldt. . produced an autoharp, an instrument upon which he was himself proficient. **1958** *Times* 13 Oct. 3/4 The autoharp, a kind of sweet-toned zither.

autoist (ǭ·to,ist). *U.S.* [f. *AUTO *sb.*^{2} I + -IST.] One who uses or drives an automobile, a motorist.

1903 *Sci. Amer.* 21 Feb. 134/1 Bills giving equal rights to autoists and the drivers of horses. **1904** *N.Y. Globe* 29 Mar. 4 A protest against the young lawbreakers who stone autoists.

autolithography (ǭ:toliþǫ·grăfi). [f. *AUTO-^{1} + LITHOGRAPHY.] A form of lithographic printing in which the drawing, etc., is first made on transfer paper, then put on stone for printing (see also quot. 1967). Hence **autoli·thograph** *v. trans.*, to produce or reproduce by this process; **autoli·thograph** *sb.*, a picture or print produced by autolithography; **au:tolithogra·phic** *a.*, of or belonging to this process.

1874 M. A. LENOIR (*title*) The Lenoir Collection of original French portraits at Stafford House. Auto-lithographed by Lord Ronald Gower. *Ibid.* Pref., This collection of portraits which I have attempted to reproduce in autolithography. *Ibid.*, Autolithographic copies aiding me in the production of these Autolithographs. **1895** *Daily News* 22 Nov. 6/5 An autolithograph by Mr. Whistler. **1967** E. CHAMBERS *Photolitho-offset* xvii. 264 Any image which is produced directly on to a stone, metal or plastic surface and uses lithographic principles to provide the printing medium, comes within the general classification of autolithography.

autological (ǭtolǫ·dʒikăl), *a.* [ad. G. *autologisch* (K. Grelling & L. Nelson 1907, in *Abhandl. Fries'schen* II. *Schule* 307), f. *AUTO-^{1} + LOGICAL *a.*] (See quots. 1926, 1947). Opp. *HETEROLOGICAL *a.*

1926 F. P. RAMSEY in *Proc. Lond. Math. Soc.* XXV. 358 Let us call adjectives whose meanings are predicates of them, like 'short', autological; others heterological. **1947** H. REICHENBACH *Elements Symb. Logic* (1948) vi. 220 Let us comprise as autological all properties whose names have the property they denote. **1952** R. L. WILDER *Introd. Found. Math.* iii. 75 Is the adjective 'heterological' either autological or heterological?

autolysis (ǭtǫ·lisis). *Biol.* [ad. G. *autolyse* (M. Jacoby 1900, in *Zeitschr. physiol. Chem.* XXX. 160), f. Gr. αὐτο- (see *AUTO-^{1}) + λύσις a loosening.] **a.** Self-acting disintegration of tissue. **b.** The destruction of cells of the body by the action of their own enzymes. Hence **auto·lysate**, a substance produced by autolysis; **autolyse** (ǭ·toləiz) *v. intr.* and *trans.*, to undergo or cause to undergo autolysis; **auto·lysin** [-IN^1], a substance that produces autolysis (Webster, 1934); **autolytic** (ǭtoli·tik) *a.*, belonging to or of the nature of autolysis.

1902 *Jrnl. Chem. Soc.* LXXXII. 35 After autolysis, the same organs yield solutions which hinder blood-clotting. **1902** *Science* 28 Nov. 858 Autolytic ferments. **1903** FLEXNER *Ibid.* 3 July 15 An acute lobar pneumonia in which the inflammatory exudate. . failing to autolyze perfectly cannot absorb, and hence undergoes organisation. **1911** *Encycl. Brit.* XX. 922/1 Autolysis is a disintegration of dead tissues brought about by the action of their own ferments, while degeneration takes place in the still living cell. The study of autolytic phenomena. . has thrown much light on these degenerative processes. **1927** ZINSSER & GRINNELL in *Jrnl. Bacteriol.* XIV. 307 Reactions in guinea pigs to the pneumococcus autolysate are examples of true allergic sensitization. **1932** J. S. HUXLEY *Probl. Rel. Growth* vi. 173 The autolysing agencies which operate at metamorphosis. **1933** *Times Lit. Suppl.* 12 Oct. 695/3 Injecting into animals varying quantities of a saline extract of autolysed homologous muscle. **1949** FLOREY *Antibiotics* I. i. 46 Secondary lysis is very possibly associated with the release of autolytic enzymes. **1964** M. HYNES *Med. Bacteriol.* (ed. 8) xi. 166 This process destroys an autolysin which can otherwise completely lyse a saline suspension of the organisms in a few hours.

automat (ǭ·tǒmæt). [G., f. F. *automate* AUTOMATE *sb.*] **1.** = AUTOMATON.

1671 R. BOHUN *Wind* 44 Air is a body so fluid, and tractable. . that if it once be set a going, it as a kind of perpetuall Automat, continues the motion. **1887** *Blackw. Mag.* Oct. 518 Such is not the action of flesh and blood. . but. . of some fabulous species of automat. *a* **1930** D. H. LAWRENCE *Last Poems* (1932) 186 The automat has no soul to lose.

2. A cafeteria in which food is obtained from compartments by the insertion of a coin or token. *U.S.*

1903 *Sci. Amer.* 18 July 49/2 [*in photograph of restaurant*] Automat. **1909** WEBSTER, *Automat*, a café or restaurant in which orders are automatically delivered to customers, who place coins or tokens in slots. **1919** F. HURST *Humoresque* 48 Waldorf! You've got a fine chance. You mean the Automat, and two spoons for the ice-cream. **1921** WODEHOUSE *Jill the Reckless* xiv. 208 The Automat?. . The food's quite good. You go and help yourself out of slot-machines, you know. **1951** *Amer. Speech* Oct. 166 Probably the 'Automat', a self-serving institution supplying food, gave special impetus to the extension of the suffix *-mat*.

3. An automatic slot-machine.

1908 *Daily Chron.* 11 Jan. 4/6 A very commendable feature has just been introduced into several German schools. Automats were placed in the courtyards, which, for a small coin, deliver hot or cold milk. **1958** *Times* 14 July 11/5 Next month. . an 'automat' at the station [*sc.* New York subway] will receive the passengers' soiled clothes and return them, spotlessly dry-cleaned, the same day.

automate (ǭ·tǒmēit), *v.* orig. *U.S.* [Back-formation f. next or f. *AUTOMATION; cf. AUTOMATE *sb.* and *a.*] **1.** *trans.* To apply automation to; to convert to largely automatic operation; to introduce automatic control to (the manufacture of a product, etc.).

1954 *N.Y. Times* 4 May (*heading*) Huge Sums to Be Spent to Automate Plants. **1959** *Listener* 5 Nov. 762/2 In theory. . management in steel has the right to automate the mills without interference. **1961** *Times* 3 Oct. (Computer Suppl.) ii/5 The first stage in 'automating' a production plant is to increase mechanization. **1962** *Listener* 17 May 855/1 It is natural that we should try to programme, or automate, part of the teacher's [by the work use of teaching machines].

2. *intr.* or *absol.*

1955 *Controller* Dec. 602/2 PanAm Automates. **1962** *Economist* 19 May 693/1 Those days saw the country 'automating' considerably faster than it is doing now. **1967** *Listener* 23 Feb. 248/1 The more we automate, the greater the resources needed.

au·tomated, *ppl. a.* orig. *U.S.* [Back-formation f. *AUTOMATION.] In senses of the verb (see prec.).

1952 *Cleveland* (Ohio) *Plain Dealer* 13 Apr. 12 Another 'automated' line, less spectacular than the block line, machines the cylinder head. **1954** *Economist* 6 Nov. 466 (*heading*) Automated Giants. **1962** *Observer* 15 July 3/5 An automated laundry which washes, rinses and irons in a continuous process. **1962** *Cath. Gaz.* Nov. 320/1 A society is automated when its production is dominated by machines to the extent that machines are given a priority over men in the performance of human tasks. **1967** KARCH & BUBER *Offset Processes* xii. 506 A Lawson automated-spacing power paper cutter.

automatic, *a.* Add: **2.** *automatic direction finder*, one in which the bearing is determined automatically; *automatic following* (see *AUTO-*following*); *automatic landing* (*Aeronaut.*), a landing in which the pilot is guided by instruments and not by visual observation; *automatic machine*, spec. a slot machine; *automatic observer* (*Aeronaut.*) (see quot. 1950); *automatic parachute* (see quot. 1951); *automatic pilot*, a device in an aircraft for maintaining a set course and height (cf. *AUTO-PILOT); *automatic stabilizer*, a device in an aircraft for maintaining a set attitude; cf. *autostabilizer* (*AUTO-^{1} b); *automatic train control*, a system which in appropriate circumstances provides an audible warning in the driver's cab of a locomotive; *automatic transmission* (cf. TRANSMISSION d), an automatic gear-changing system in a motor vehicle; *automatic volume control* (abbrev. *A.V.C.*), also *automatic frequency control* (abbrev. *A.F.C.*), *gain control* (*A.G.C.*): devices used in radio for regulating a frequency or signal; *automatic writing* (see *6).

1940 *Chambers's Techn. Dict.* 61/2 Automatic direction-finder. **1951** *Gloss. Aeronaut. Terms* (*B.S.I.*) III. 27 *Automatic direction finder*, an airborne equipment designed to indicate automatically the bearing of a continuous-wave ground beacon relative to the bearing of the aircraft. **1946** *Jrnl. Inst. Electr. Engin.* XCIII. III. 17/1 Automatic-following radar. . was of the essence of the A.A. gunnery successes against flying bombs. **1950** *Gloss. Terms Radar* (*B.S.I.*) 7 *Automatic following*, automatic aiming with the addition of automatic range measurement. **1935** *Proc. Inst. Radio Engineers* XXIII. 1125 (*title*) Automatic Frequency Control. *Ibid.* 1133 (*diagram*) From AFC bias. **1930** *Proc. Inst. Radio Engineers* XVIII. 633 Those components. .

which fade in and out. . can be maintained at what approaches a constant level by means of the automatic gain control. **1942** *Electronic Engin.* XV. 216 A fault of most communications receivers is the inability to use A.G.C. when the B.F.O. is switched on. **1938** *Jrnl. R. Aeronaut. Soc.* XLII. 505 No mention had been made of the automatic landing, which seemed to represent the development of this science which was now engaging the attention of American experts. **1958** *Times* 17 Oct. 3/4 More than 2,000 completely automatic landings, some in thick fog and others in strong cross-winds, have been made. **1903** SHAW *Man & Superman* I. 14 A box of matches will come out of an automatic machine when I put a penny in the slot. **1936** *Aircraft Engin.* Dec. 330/2 An 'automatic observer' was not employed from considerations of weight. **1950** *Gloss. Aeronaut. Terms* (*B.S.I.*) I. 41 *Automatic observer*, an apparatus for recording automatically the readings of a specified set of instruments in flight. [**1919** E. R. Calthrop's *Aerial Patents Book* 24 The 'Guardian Angel' Parachute, in all its different types, is instantly automatic.] **1951** *Gloss. Aeronaut. Terms* (*B.S.I.*) III. 13 *Automatic parachute*, a parachute which is withdrawn from its pack by a static line. **1916** *Aeronautics* 13 Sept. 175 The Sperry automatic pilot. **1921** *Ibid.* 3 Feb. 76/2 The automatic pilot. . enables the pilot of an aeroplane to leave the machine entirely to its own devices. **1944** 'N. SHUTE' *Pastoral* iv. 91 Marshall sat motionless at the controls, flying upon the automatic pilot. **1909** *Flight* 17 July 434/2 The automatic stabiliser must. . show a strong and immediate tendency to return to its proper normal working position under all conditions. **1950** *Gloss. Aeronaut. Terms* (*B.S.I.*) I. 41 *Automatic stabilizer*, an automatic pilot adjusted to provide increased aerodynamic stability to the aircraft. **1912** *Railway Gaz.* 12 July 40/2 Automatic Train Control Demonstration. . . The ramps are so electrically connected that either of two signals are given on the engine —a clear signal or a danger—. . as the train proceeds. **1936** *Economist* 25 Jan. 178/2 Security against collisions of this kind can only be provided by a combination of track-circuiting. . and automatic train control. **1946** W. H. CROUSE *Automotive Mechanics* xvii. 403 The Hydra-Matic drive, supplied on Cadillac and Oldsmobile cars as special equipment, combines the fluid drive with an automatic transmission that has four forward and one reverse speed. **1961** *Autocar* 29 Sept. 471/1 A conventional clutch and three-speed gearbox costs. . less than an automatic transmission. **1930** *Proc. Inst. Radio Engineers* XVIII. 321 The severe fluctuation of the signal. . indicates the desirability of some form of automatic volume control. **1933** *Pract. Wireless* 25 Nov. 545 The way in which A.V.C. operates does not seem to be widely understood.

b. Of a firearm: furnished with mechanism for successively and continuously loading, firing, and ejecting a cartridge as long as ammunition is supplied.

1877 *Independent* (U.S.) 5 July 20/2 Smith & Wesson's automatic revolvers. **1902** *Encycl. Brit.* XXX. 401/2 In the modern 'automatic' machine gun the loading, firing, extracting, and ejecting are all performed automatically by the gun itself. *Ibid.* XXXII. 649/2 No nation has yet armed her forces with an automatic rifle. *Ibid.* 658/2 The Colt Automatic Pistol, calibre ·38.

c. Of a telephone exchange or system: operated by automatic switches (opp. *manual*). Also, designating a telephone instrument fitted with a dial.

1879 M. D. CONNOLLY et al. *U.S. Pat.* 222,458, We. . have jointly invented a certain new and useful Automatic Telephone-Exchange. . so constructed and arranged that any member of the exchange may. . place himself in direct communication with any disengaged member of the exchange. **1914** W. ATKINS *Princ. Automatic Telephony* I In an automatic system of telephony it is required that the subscriber shall be able to obtain connection with any other subscriber without the intervention of an operator at the exchange. **1934** *Discovery* Mar. 58/2 The automatic telephone. . has not been an unqualified blessing. **1955** *Oxf. Jun. Encycl.* VIII. 433/2 In an automatic exchange, the connexion to the required line is made by mechanical selectors.

6. *Spiritualism.* Of or pertaining to automatism (sense *4); performed by unconscious, subconscious, or occult action.

1883 W. S. MOSES *Spirit-Teachings* Introd. 1 Automatic Writing is a well-known method of communication with the invisible world of what we loosely call Spirit. **1884** *Proc. Soc. Psychical Research* II. 226, I wished to know if I were myself an automatic writer, or so-called writing medium. **1889** BARKWORTH in *Proc. Soc. Psychical Research* Dec. 85 It is only the execution and not the initiation of the movements which is automatic, the suggestion for them being external to the subject's own personality. **1890** W. JAMES *Princ. Psychol.* I. viii. 209 Certain trance-subjects who were also automatic writers. **1934** *Archit. Rev.* LXXV. 215/1 Mr. Cooper's picture, on the other hand, might almost be a piece of automatic writing.

7. *Art.* Applied to a form of painting performed by the technique of 'automatism' (see *AUTOMATISM 5).

1951 A. HILL *Painting out Illness* xi. 74, I deny that true automatic pictures can be produced while both the eye and hand in subconscious conjunction are said to be 'employed'. **1960** E. H. GOMBRICH *Art & Illusion* x. 358 The modern painter may use what he calls 'automatic painting', the creation of Rorschach blots, in order to stimulate the mind . . towards fresh inventions.

automatic (ǭtǒmæ·tik), *sb.* [f. prec.] **1.** Abbreviation of *automatic pistol, gun*, etc.: see *AUTOMATIC *a.* 2 b.

1902 *Sears Catal.* (ed. 112) 305/2 Forehand Perfection Automatic, small frame, rebounding lock. **1914** G. ATHERTON *Perch of Devil* II. vii. 269 I've even bought an automatic. I suppose. . I should call it a gun. **1920** *Blackw. Mag.* Aug. 154/1 A German automatic hung at his side. **1945** *Diamond Track* (*Army Board, N.Z.*) 35/1 Everything was thrown into it—grenades, automatics, bayonets, and rifle butts.

2. A machine, tool, etc., that is operated automatically.

1909 in WEBSTER. **1914** *Machinery* (Engin. Ed.) XX. 468 (*title*) Making shrapnel cases on the Cleveland automatic. **1917** *Amer. Machinist* XLVII. 17 Automatics used advantageously in the making of starter parts. **1921** *Conquest* II. 125 The full advantage of automatics will only be appreciated when a large number of automatic exchanges have been erected. **1930** *Engineering* 7 Mar. 309/2 The machine tool display. . covers automatics of various types, bending machines, sheet-metal working machines, drills [etc.]. **1949** *Jrnl. R. Aeronaut. Soc.* LIII. 428/2 Failures of the automatics may be more dangerous than the human failures they are designed to prevent.

automation (ǭtǒmēi·ʃən). [irreg. f. AUTOMATIC *a.* + -ATION.] Automatic control of the manufacture of a product through a number of successive stages; the application of automatic control to any branch of industry or science; by extension, the use of electronic or other mechanical devices to replace human labour.

The example of *automation* found in some copies of the 1669 edition of S. Patrick's *Brief Account of the New Sect of Latitude-Men* is a misprint for *automaton* (see *Amer. Speech* (1959) XXXIV. 236). The coinage of the modern word is usually attributed to Delmar S. Harder (U.S.).

1948 *Amer. Machinist* 21 Oct. in *McGraw-Hill Encycl. Sci. & Technol.* I. 676/2 *Automation*, the art of applying mechanical devices to manipulate work pieces into and out of equipment, turn parts between operations, remove scrap, and to perform these tasks in timed sequence with the production equipment so that the line can be put wholly or partially under pushbutton control at strategic stations. **1953** *Manch. Guardian Weekly* 3 Dec. 15/2 Many factories are spending large sums on 'automation', that is, the adoption of automatic machines working together with little labour. **1954** *Economist* 29 May 712 Mechanisation—promoted to the exalted station of 'automation'—now consists essentially in the use of bigger and faster machine tools. **1955** *Times* 3 Aug. 8/7 The group of resolutions on automation. . says the technological advances will present the trade union movement with new opportunities, but these opportunities will be attended by new and complex human, social and economic problems. **1957** *Technology* July 182/1 Automation is now well known to be the automatic control of mechanized systems, although the term is used somewhat vaguely to cover many different aspects of control and communication, especially in the industrial situation. **1964** *Ann. Reg. 1963* 181 The demand for skilled labour and the substitution of unskilled labour by automation was increasing faster than the training and education of the Negro.

automatism. Add: **4.** Any psychic phenomenon that appears spontaneously in consciousness; any action performed subconsciously or unconsciously, undirected by the mind or will of the normal personality; also, the mental state in which these phenomena occur.

1884 MYERS in *Proc. Soc. Psychical Research* II. 223 In the graphic automatism of mental abstraction and the graphic automatism of cerebral disease, the passages written are usually very *short*. **1886** E. GURNEY et al. *Phantasms of Living* I. ii. 76 The planchette-writing obtained through the automatism of a young child. **1889** BARKWORTH in *Proc. Soc. Psychical Research* Dec. 85 We have. . instances of complete automatism in the case of the sleep-walker who goes through a variety of complicated actions entirely self-suggested. *a* **1901** MYERS *Human Personality* (1903) I. Gloss. s.v., Sensory automatism will thus include visual and auditory hallucinations; motor automatism will include messages written without intention.

5. *spec.* A technique in surrealist painting (see quot. 1958).

[**1935** D. GASCOYNE tr. A. Breton in *Short Survey Surrealism* iv. 61 *Surrealism*, pure psychic automatism, by which it is intended to express, verbally, in writing, or by other means, the real process of thought.] **1948** H. READ in *Philos. Mod. Art* (1952) I. ii. 53 Applying Freudian methods to the problems of artistic creation, Breton evolved a theory and. . practice of aesthetic automatism which is the essential feature of surréalisme. **1958** M. L. WOLF *Dict. Painting* 25 *Automatism*, in *art*, the principle of creation without the interference of thought. . . Practically, it is the unfettered stroke of *brush* or *pencil*, with no direction, will, or control exercised by the conscious mind.

automatist. Add: **2.** One who is subject to automatism; a medium; in *Art*, one whose technique is based upon automatism (sense *5). Also *attrib.*

1885 MYERS in *Proc. Soc. Psychical Research* III. 41, I have seen an automatist writing page after page in ordinary handwriting, and then a page in mirror-writing. *a* **1901** —— *Human Personality* (1903) I. 28 The ordinary consciousness of the automatist appears to be suspended; he passes into a state of trance. **1920** W. B. YEATS in *If I were Four & Twenty* (1940) 43 He gathered the opinions, as he believed, of spirits speaking through a great number of automatist and trance speakers. **1959** *Listener* 2 July 26/2 Exhibitions by their own artists, in the international idioms of pure Abstract, Abstract-Concrete, Tâchiste, Automatist [etc.].

automatization (ǭtǫ:mătəizēi·ʃən). [f. AUTOMATIZE *v.* + -ATION.] The action of making or condition of being made automatic or automatically controlled.

1924 *Psyche* July 87 The pointing method in localization . . does not tend so readily toward automatization. **1932** *Discovery* Oct. 336/2 The initial effect [of noise] tends to wear off very rapidly, mainly in consequence of automatization of the task. **1937** C. G. PHILP *Conquest of Stratosphere* x. 82 The instruments were designed to work automatically as far as possible. . and the degree of automatization reached a high standard. **1964** E. A. NIDA *Toward Sci. Transl.* vi. 128 Languages. . develop in the direction of automatization.

automatize, *v.* Restrict *rare* to sense in Dict. and add: **2.** To make or render automatic or automatically controlled; to introduce automation. So **auto·matized** *ppl. a.*, provided with or using automatic devices. orig. *U.S.*

1952 *Sci. Amer.* Sept. 152/3 The estimated cost of complete instrumentation of a new modern plant to automatize it as fully as possible. **1955** *Atlantic* June 14/2 Only a war, he believes, could automatize industry overnight. **1956** *Sci. News Let.* 7 July 9/3 Petroleum production is more highly automatized today than any other industry. **1959** *Times* 9 Oct. 20/4 It [*sc.* the plant] is so highly automatised that it employs only 350 persons. **1960** *Commentary* June 470/2 A rich, heavily automatized society.

automatograph (ǭtomæ·tŏgrɑf). [f. Gr. αὐτόματος acting of itself + -γραφος writing, writer, -GRAPH.] A scientific form of the planchette, used for recording the involuntary movements of the hand and arm. Hence **automa·togram.**

1892 J. JASTROW in *Amer. Jrnl. Psychol.* Apr. 400 This apparatus enables us to record all movement in the horizontal plane, and, inasmuch as its chief purpose is to write slight involuntary movements, we have given it the name of the *automatograph* and may speak of such a record as an *automatogram*. **1901** TITCHENER *Exper. Psychol.* I. I. 95 Automatograph. This consists of a light but strong board, . . carrying at one end a small block of wood, . . hollowed out to take the elbow, and pierced at the other end by a circular hole, through which the stylus passes.

automaton. **5.** (Earlier example.)

1785 BETSY SHERIDAN *Jrnl.* (1960) 72 Mrs Dexter. . says the Goths in her neighbourhood had the impudence to think of your playing second to that Automaton Mrs Kennan.

automatonism (ǭtǫ·mătǫniz'm). [f. AUTOMATON + -ISM.] = AUTOMATISM I.

1925 YEATS *Vision* I. iv. 78 [Shelley] was subject to an *automatonism* which he mistook for poetic invention, especially in his longer poems. **1947** *Nature* 18 Jan. 80/2 Anyone who has dealings with a tame raven will find it difficult to dispense with the word 'intelligence' in any analysis of its conduct. Is it automatonism when it tweaks the cat's tail?

automnesia (ǭtǫmnī·ziă). *Psychol.* [f. Gr. αὐτο- (see *AUTO-^{1}) + μνῆσις memory: see -IA^1.] (See quots. *a* 1901.)

1897 tr. *Ribot's Psychol. Emotions* 153 By some phenomenon of affective automnesia, this same event reproduced itself. *a* **1901** MYERS *Human Personality* (1903) I. Gloss., *Automnesia*, spontaneous revival of memories of an earlier condition of life. *Ibid.* II. 139 Littré. . describes what he calls the 'affective automnesia'—or spontaneously arising flow of emotion.

automobile (ǭ·tǒmŏbīl), *a.* and *sb.* Also (formerly) **auto-mobile.** [a. F. *automobile* (1876 in Littré Suppl.), f. Gr. αὐτός self + F. *mobile*: see *AUTO-^{1} and MOBILE *a.*] **A.** *adj.* That moves by means of mechanism and power within itself; esp. of a vehicle, self-propelling as distinguished from horse-drawn.

1883 H. GREER *Dict. Electricity* 48 There are half a dozen systems of electric traction. . in use. . . An auto-mobile car, with isolated rails. **1886** *Harper's Mag.* June 25/2 These vessels. . besides their equipment of auto-mobile torpedoes, are provided with powerful batteries. **1895** *Daily News* 15 Oct. 5/3 Signor Cleto Brena has arrived at Naples with his automobile carriage from Milan. **1902** *Encycl. Brit.* XXXI. 12/1 France has undoubtedly led in the development of the light automobile vehicle.

B. *sb.* A self-propelled vehicle; a motor vehicle.

1895 *Pall Mall Gaz.* 15 Oct. 2/3 Three miles an hour gives the automobile little chance of displaying the powers it doubtless enjoys. **1902** *Encycl. Brit.* XXXI. 11/1 On the Continent of Europe and in the United States the usual expression for these vehicles [*sc.* motor-cars] is 'automobile'. **1934** H. G. WELLS *Exper. Autobiogr.* II. viii. 543 The bicycle was the swiftest thing upon the roads in those days, there were as yet no automobiles.

attrib. **1896** *Manch. Guardian* 24 Sept. 10/1 The Automobile Club of France. **1901** *Daily News* 16 Feb. 6/1 The French have shown themselves keenly interested in the subject of auto-mobile races. **1903** [see TRIAL *sb.*^{1} 2 a].

au·tomobile, *v.* [f. prec.] *intr.* To travel or ride in an automobile or motor car; to drive. Hence **automo·biling** *vbl. sb.*, the action or practice of using an automobile; also *attrib.*

1898 *Cosmopolitan* Sept. 485/1 Mr. Tiffany assures the writer that he finds 'automobiling' more interesting than coaching. **1901** W. R. H. TROWBRIDGE *Lett. her Mother to Eliz.* xii. 54 Blanche spent the morning. . automobiling with the Vicomte and the Marquise. **1903** *N.Y. Times* 19 Dec. 9 Here is a hat for the skating girl and the automobiling woman. **1924** *Public Opinion* 24 Oct. 403/2, I have lately automobiled in Cornwall, Devonshire, Somerset and adjoining counties. **1932** *New Yorker* 9 Apr. 59/2 Yachting on the Trollfjord, . . automobiling in the Baltic Capitals.

automobilism (ǭtomōu·biliz'm). [f. *AUTOMOBILE *sb.* + -ISM after F. *automobilisme*.] The use of automobiles or motor vehicles.

1896 *Harper's Weekly* 31 Oct. 1075 We lack France's superb roads; and American automobilism is likely to be hampered in various rural sections of the country. **1898** *Cosmopolitan* Sept. 483/2 As a sport, automobilism now occupies the foremost rank. **1899** *Motor-Car World* Oct. 8/1 Automobilism will be the method of locomotion of the future. **1953** *Archit. Rev.* CXIV. 246 The shopping pedestrian, the morning coffee-taker, the cash-and-handshake dealer, certainly need protection from automobilism.

So **automobilist** (ǭtomōu·bilist), one who uses an automobile or motor vehicle, a motorist.

1897 *Daily News* 26 July 5/5 The hour's rest at St. Germain was improved by the automobilists to see that their machines were in order. **1902** C. N. & A. M. WILLIAMSON *Lightning Conductor* 114 The old peasant. . exclaimed that if all automobilists were like us there would never be complaints. **1907** 'F. MILTOUN' (*title*) The Automobilist Abroad.

automobility (ǭ:tomobi·līti). [f. *AUTOMOBILE *sb.* + -ITY.] **1.** The use of automobiles or motor vehicles as a mode of locomotion or travel.

1903 *Times Lit. Suppl.* 16 Oct. 295/3 To come. . upon a book. . packed with suggestions for the well-being of the walker is, in this year of grace and automobility, no small joy. **1926** *Daily News* 29 Oct. 6/7 The automobility of society. . has helped many Englishmen to discover England.

2. Mobility by means of an automobile or motor vehicle.

1909 *Westm. Gaz.* 18 Mar. 2/3 The interesting experiment of conveying troops by motor vehicles to Hastings. . proves what may be called the automobility of a defensive force. **1926** *Blackw. Mag.* Mar. 309/2 The small expedition, being totally inadequate to its purpose, achieved nothing but automobility.

automorphic, *a.* Add: **2.** Math. *automorphic function.* [ad. G. *automorphe funktion* (F. Klein 1890, in *Nachrichten v. d. königlichen Ges. d. Wissenschaften, Göttingen 1890* Mar. 94).] (See quot. 1892.)

1892 W. BURNSIDE in *Proc. London Math. Soc. 1891–2* XXIII. 49 On a class of automorphic functions. *Ibid.* 52, I have used the phrase 'automorphic function', as introduced by Professor Klein, to denote generally any function which is unchanged by the substitutions of a discontinuous group, whatever be the nature of the group. **1898** E. T. WHITTAKER in *Phil. Trans.* (1899) CXCII. 1 The only automorphic functions known hitherto which have been applied to uniformise forms whose genus is greater than unity, are those given by certain sub-groups of the modular group. **1902** —— *Mod. Analysis* 339 Two classes of automorphic functions are known by which this uniformisation may be effected.

3. *Cryst.* [ad. G. *automorph* (C. E. M. Rohrbach 1885, in *Mineral. u. Petrograph. Mittheilungen* VII. 18).] = IDIOMORPHIC *a.*

1888 F. H. HATCH in J. J. H. Teall *Brit. Petrogr.* 435 The mineral constituents of rocks are said to be idiomorphic in so far as they are bounded by faces peculiar to the species. The word is synonymous with Rohrbach's 'automorphic'.

automorphism. Add: **2.** *Math.* (See quots.)

1903 *Science* 5 June 904 Class of a group and degree of transitivity, automorphism, representation, index notation. **1955** L. MIRSKY *Introd. Linear Algebra* iv. 125 An automorphism of a linear manifold $\mathfrak{R}$ is an isomorphism of $\mathfrak{R}$ with itself. *Ibid.*, A linear transformation of a finite-dimensional linear manifold onto itself is an automorphism. **1959** M. HALL *Theory of Groups* vi. 86 The automorphism group *A* (*C*) is cyclic. **1959** G. & R. C. JAMES *Math. Dict.* 221/1 An isomorphism of a set with itself is an automorphism. An automorphism of a group is an inner automorphism if there is an element t such that x corresponds to x^* if and only if $x^* = t^{-1}xt$; it is an outer automorphism if it is not an inner automorphism. **1965** J. J. ROTMAN *Theory of Groups* iv. 76 *Definition*. A homomorphism $f: G \to G$ is called an endomorphism of G; an isomorphism $f: G \to G$ is called an automorphism of G.

automotive (ǭtomōu·tiv), *sb.* and *a.* [f. *AUTO-^{1} + MOTIVE *a.*] †**A.** *sb.* [a. F. *automotive*.] A self-propelled vehicle (in quots., an early kind of helicopter).

1865 C. H. TURNOR *Astra Castra* 256 This journal—this indispensable *Moniteur* to the aerial automotive. *Ibid.* 344 MM. Ponton d'Amécourt and de la Landelle were constructing their small automotives.

B. *adj.* **a.** Of or pertaining to an automobile or automobiles. **b.** Self-propelled.

1898 tr. *L. Lockert's Petroleum Motor-cars* p. ix, His work . . is read with satisfaction by all amateurs of automotive touring. **1901** (*title*) The Automotor and Horseless Vehicle Pocket-Book of Automotive Formulæ. **1922** *Public Opinion* 7 Sept. 222/1 He chugged up the street in an automotive contraption. **1938** W. WILKINSON *Puppets through America* xiv. 156 In spite of all its cars, its crowded streets and its cheap stores for automotive workers, Detroit has a splendid oasis in the white stone Merrick Public Library. **1956** A. H. COMPTON *Atomic Quest* 355 Airplane and automotive manufacture.

automotor: see *AUTO-^{2}.

autonomic, *a.* Add: **b.** *Physiol.* Functioning independently of the will; *autonomic nervous system* (see quots. 1898).

1898 J. N. LANGLEY in *Jrnl. Physiol.* XXIII. 241 The 'autonomic' nervous system means the nervous system of the glands and of the involuntary muscle. *Ibid.* 270, I propose the term 'autonomic nervous system' for the sympathetic system and the allied nervous system of the cranial and sacral nerves, and for the local nervous system of the gut. **1940** R. S. WOODWORTH *Psychol.* (ed. 12) xii. 419 *The autonomic nerves*. These are the nerves that run to the heart, blood vessels, lungs, stomach, intestines, and other viscera. . . These nerves are composed of extra-slender nerve fibres, which grow out from cells in the brain stem and cord. **1966** *Lancet* 24 Dec. 1415/1 Individuals can gain cortical control over various autonomic nervous symptoms.

B. *sb.* A part of the autonomic nervous system.

1929 F. H. GARRISON *Hist. Med.* (ed. 4) xiii. 696 Adrenalin, the hormone governing the sympathetic autonomic. *Ibid.*, The two opposing autonomics of the sympathetic system control the ductless glands and the visceral organs made up of smooth (involuntary) muscle.

autonomism (ǭtǫ·nŏmiz'm). [f. AUTONOMY + -ISM.] **1.** The principle or system of autonomy or self-government.

1874 FISKE *Cosmic Philos.* II. 205 The two feelings known to the Greeks as Pan-Hellenism and Autonomism, represented respectively by Athens and by the Doric communities.

2. *Biol.* 'The theory that the forms into which animals and plants develop are determined by an inward agency' (*Cent. Dict.* Suppl., 1909).

autonomistic (ǭtǫnŏmi·stik), *a.* [-IC.] **1.** *Politics.* Of or pertaining to autonomism (sense *1).

a **1870** GEN. R. E. LEE in *Westm. Gaz.* (1898) 25 May 3/1 The Autonomistic Cabinet.

2. *Biol.* Of or pertaining to autonomism (sense *2) or the theory of self-determined development.

1904 *Biol. Bull.* (Mass.) Sept. 201 They [*sc.* changes] resemble in character certain of the phenomena which have led Driesch to assume the existence of an autonomistic principle or entelechy governing form.

autonomously (ǭtǫ·nŏməsli), *adv.* [-LY^2.] In an autonomous position, as an autonomous state, etc.

1881 G. S. HALL *German Culture* 183 We must know and autonomously will to follow non-egoistic absolute ends as essentially our ends. **1897** GLADSTONE *Eastern Crisis* 14 Why should not Crete be autonomously united with Greece, and yet not detached in theory from the body of the Ottoman Empire?

autonym. Add: Also, one's own name as distinguished from a pseudonym, esp. the real name of an author; also *attrib.*

1895 *Daily News* 23 Dec. 5/1 Autonym works, anonymous works, and special works.

autophone (ǭ·tofōun). Also **autophon.** [f. *AUTO-^{1} + *-PHONE.] **1.** A type of barrel-organ (see quot. *a* 1875). (Disused.)

1852 *Reports Juries 1851 Exhib.* 332/2 A barrel-organ, called an Autophon. *a* **1875** KNIGHT *Dict. Mech.* I. 192/1 *Autophon*, a barrel-organ, the tunes of which are produced by means of perforated sheets of mill-board. **1878** in *Design & Work* 2 Mar. 275/3 Autophone.

2. [ad. F. *autophone* (Mahillon, 1880).] A musical instrument that produces sound through the vibration of its constituent material (see quots.).

1937 F. W. GALPIN *Textbk. Europ. Mus. Instr.* 27 The various sound-producers were classed as *Autophones* or self-vibrators. *Ibid.* 52 The representative instrument in the plucked type of *Autophones* is the *Crembalum* or Jew's-harp. **1954** *Grove's Dict. Mus.* IV. 489/2 Under Auto- or Idiophones. . we find not only the xylophone. . but also the marimba. . and the marimba gongs.

Hence **autopho·nic** *a.*^{2}

1937 F. W. GALPIN *Textbk. Europ. Mus. Instr.* 29 Autophonic Instruments or self-vibrators, i.e. instruments of solid substance which, owing to their elastic nature, have a sonority of their own, which is emitted in waves when they are struck, plucked, or stimulated by friction or air.

autopiano (ǭ:to,pi,æ·no). [f. *AUTO-^{1} + PIANO *sb.*^{2}] A piano with a playing apparatus.

1905 *Connoisseur* XIII. 41 (Advt.), The —— 'Autopiano' is a Pianoforte of excellent tone. . played with music roll. **1928** *Sunday News* 17 June 4/4 Their sale bargains including several auto-pianos.

auto-pilot, autopilot. *Aeronaut.* [f. *AUTO-^{1} + PILOT *sb.*] = *AUTOMATIC *pilot*.

1935 *Flight* 10 Jan. 42/2 No information is available concerning the degree of accuracy which the 'Autopilot' will give in bumpy weather. **1943** *Times* 21 Sept. 4/5 Because its electronic mechanism is hyper-sensitive, the auto-pilot corrects immediately the slightest deviation in the course or balance of a bomber in the midst of cross-currents. **1956** *Spaceflight* I. 24/1 Under the control of an autopilot, the whole thrust unit can be deflected up to 5 degrees from the rocket's centre-line.

‖ **autopista** (ǭ·tǫpistă). [Sp., f. *auto* (*AUTO-^{2}) + *pista* track (see PISTE^2).]

In Spain and Spanish-speaking countries, a fast motor road; = *MOTORWAY.

1955 'S. MARLOWE' *Second Longest Night* vii. 61 A four-lane superhighway [in Venezuela]. . . 'Some *autopista*,' he said. **1963** *Economist* 2 Nov. 491/1, 204 kilometres of *Autopistas* [in Spain]. **1967** *Daily Tel.* 15 Mar. 14/4 After several years' delay Spain's first *autopista*, or through motorway, linking Barcelona to the French frontier is to be built.

autoplast (ǭ·toplæst). [f. *AUTO-^{1} + -PLAST.] **a.** *Embryol.* An autogenous cell as distinguished from a cleavage cell. **b.** *Plant Physiol.* A chlorophyll granule.

1883 E. RAY LANKESTER in *Encycl. Brit.* XVI. 682/2 In addition to the layer of cleavage cells. . additional cells are formed. . , each cell having a separate origin, whence they are termed 'autoplasts'. **1885** [see *trophoplast* s.v. TROPHO-].

autoplate (ǭ·toplēit). *U.S.* [f. *AUTO-^{1} + PLATE *sb.*] A curved stereotype for newspaper printing, made by an autoplate machine; also, the machine itself; *autoplate machine*, a machine for automatically casting, shaving, and bevelling stereotypes.

1901 *Sci. Amer.* Suppl. LII. 21591/1 The Autoplate—an automatic stereotyping machine. **1902** *Westm. Gaz.* 22 May 9/1 The Autoplate, a wonderful revolution in stereotyping. **1902** *Census Bulletin* 28 June 51 (Cent. D. Suppl.), A device known as the autoplate was invented in 1900 by means of which the time required for casting plates was. . reduced. **1963** *Times* 17 Apr. 3/5 Junior Autoplate Casting Equipment is available.

autopolyploid (ǭ:topǫ·liploid). *Biol.* [f. *AUTO-^{1} + *POLYPLOID.] A polyploid having its sets of chromosomes derived from a single parent species. Hence **autopo·lyploidy**, the condition or occurrence of such a polyploid.

1928 [see *ALLOPOLYPLOIDY]. **1930** E. MÜNTZING in *Hereditas* XIII. 293 Auto-polyploids, however, are generally less stable than the diploid parents. **1930** [see *ALLOPOLYPLOIDY]. **1945** *Lancet* 14 July 47/2 Colchicine possesses the property of arresting cells in the process of mitotic division at a stage when the chromosomes have divided and the cytoplasm is undivided, so that a cell with a double complement of chromosomes results. . . This process of doubling is called autopolyploidy, and the products autopolyploids. **1952** *New Biol.* XIII. 31 These shoots have 48 chromosomes instead of the usual 24, four of each chromosome being present in place of the original pair. . . Polyploids of any sort which have thus originated from one parental species are called 'autopolyploids'.

autopsic (ǭtǫ·psik), *a.* [f. mod.L. *autopsia* AUTOPSY + -IC.] Of or pertaining to autopsy; based on personal observation; *spec.* in *Med.*, obtained by means of the post-mortem examination of a body. Also **auto·psical** *a.*

1881 L. PUTZEL in *von Ziemssen's Cycl. Med.* Suppl. 561 Basing his opinion upon the autopsical examination of five patients. **1886** *Alienist & Neurol.* July 533 The autopsic and the combined ante-mortem testimony to his insanity was not more confirmatory.

autopsy (ǭ·tǫpsi, ǭtǫ·psi), *v.* [f. AUTOPSY *sb.*] *trans.* To perform an autopsy on or make a post-mortem examination of (a body).

1900 *Jrnl. Exper. Med.* V. 257 One of the pigs was killed and autopsied, with the result that its organs. . were found to be entirely free of lesions. **1919** SACHS in S. Paget *Sir V. Horsley* II. iv. 190 It was the first walrus that had been autopsied in London for many years. **1969** *Nature* 19 Apr. 287/1 The rats. . were autopsied on the thirteenth day of pregnancy.

autoradiograph (ǭtorēi·diogrɑf), *sb.* [f. *AUTO-^{1} + RADIOGRAPH.] An image of an object produced on a sensitive film by the radioactivity of the object itself. Hence as *v. trans.*, to produce an autoradiograph of. Also **a:utoradio·graphy** [F. *autoradiographie*], the production of an autoradiograph; **au:toradiogra·phic, -ical** *adjs.*, pertaining to or involving autoradiography; **au:toradiogra·phically** *adv.* Cf. *radio-autograph*.

1903 *Dublin Rev.* July 170 Auto-radiographs of thorium. **1941** *Physical Rev.* LX. 688/1 Recently autoradiography has been of value in tracer studies of biological substances. **1943** *Amer. Mineralogist* XXVIII. 459 The elements are arranged in the order of decreasing suitability for autoradiographic study. **1947** *Nature* 9 Aug. 193/2 Suppose now that a layer of tissue one cell thick is to be autoradiographed. **1955** G. A. BOYD *Autoradiogr. in Biol. & Med.* i. 7 In 1940, Hamilton [et al.]. . used radioiodine to study the thyroid autoradiographically. **1956** *Nature* 25 Feb. 379/1 To prepare radioactive streptomycin for autoradiographical studies of its distribution in the body. **1958** *New Scientist* 2 Jan. 24/1 The tool material on the metal is sufficiently radioactive to darken the film in its neighbourhood and thus takes its own photograph. This is called an autoradiograph. **1965** G. J. WILLIAMS *Econ. Geol. N.Z.* xiii. 207/2 Autoradiographs show the radioactivity to be confined to the carbonaceous material which is clearly epigenetic in origin.

‖ **autoroute** (ǭ·torūt). [F., f. *auto* (*AUTO-^{2}) + *route* road (see ROUTE *sb.*).] In France, a fast motor road; = *MOTORWAY.

1963 H. SLESAR *Bridge of Lions* (1964) vii. 105 The lights of Paris were behind them. Ahead, on the six-lane *autoroute*, lay Versailles. **1966** *Guardian* 30 Dec. 7/6 Those. . monster housing estates that sprout along the *autoroutes* outside Paris.

autoscope. Add: **2.** (See quot. *a* 1901.)

1900 *Daily News* 10 Apr. 6/2 The so-called Divining Rod

. . is only one of many 'autoscopes', which perhaps bring into observation what is passing in the mysterious entity styled 'the subconscious self'. *a* **1901** MYERS *Human Personality* (1903) I. Gloss., *Autoscope*, any instrument which reveals a subliminal motor impulse or sensory impression; *e.g.* a divining rod, a tilting table, or a planchette.

autosemantic (ǭ:to,simæ·ntik), *a.* (*sb.*). *Philol.* [ad. G. *autosemantisch* (A. Marty 1908, *Untersuchungen zur Grundlegung d. allgemeinen Grammatik und Sprachphilosophie* II. i. 206), f. *AUTO-¹ + SEMANTIC *a.*] Of a word or phrase: having meaning outside a context; meaningful in isolation; categorematic. Opposed to *SYNSEMANTIC *a.* Also *absol.* (A distinction first proposed by the Austrian philosopher A. Marty as a correlate in linguistics of the distinction in logic between *categorematic* and *syncategorematic.*)

1929 *Year's Wk. Eng. Stud.* 1927 47 Funke. . classifies the units of speech under two heads, termed Autosemantic and Synsemantic, the former constituting the semasiological units of speech, i.e. those which give complete expression of psychic phenomena which are communicable in themselves. . . Autosemantics correspond to the three basic types of sentence and are not necessarily themselves single words. **1931** G. STERN *Meaning & Change of Meaning* iv. 86 We have to distinguish not only between autosemantic and synsemantic expressions, but also between autosemantic and synsemantic meanings. **1931** W. WORSTER tr. *Jørgensen's Treat. Formal Logic* III. xiii. 239 The word 'and' is syncategorematic or synsemantic. It *may* however, also occur as categorematic or autosemantic, as. . in the assertion: '"And" is a conjunction.' **1962** S. ULLMANN *Semantics* ii. 44 Full words are 'autosemantic', meaningful in themselves, whereas articles, prepositions, . . and the like are 'synsemantic', meaningful only when they occur in the company of other words.

autosite (ǭ·tosəit). [f. Gr. αὐτόσιτος bringing one's own provisions, f. αὐτός self + σῖτος food.] The larger twin of a double fœtal monster, which supplies nourishment to the smaller (called the parasite); also, a single monster capable of independent life. Hence **autosi·tic** *a.*, that is, or is of the nature of, an autosite.

1848 DUNGLISON *Dict. Med. Sci.* (ed. 7), *Autosite*. . , a single monster, capable of deriving nourishment from its own proper organs, in contradistinction to *Omphalosite*. **1894** GOULD *Dict. Med.*, Autosite. *Ibid.*, Autositic.

autosome (ǭ·tosōᵘm). *Biol.* [f. *AUTO-¹ + Gr. σῶμα body.] A chromosome other than a sex-chromosome. Hence **autoso·mal** *a.*, of, belonging to, or designating an autosome.

1906 T. H. MONTGOMERY in *Trans. Amer. Philos. Soc.* XXI. 97, I here employ the following nomenclature. . *Autosome* (*autosoma*), the non-aberrant chromosomes that I have previously called *ordinary chromosomes*. **1930** R. A. FISHER *Genetical Theory Nat. Selection* iii. 51 The autosomal and the sex-linked mutations. **1950** E. B. FORD *Stud. Heredity* iii. 32 The sex chromosomes. . do not differ from the autosomes except that they carry, among others, the genes which decide whether development shall proceed along male or female lines. **1967** *New Scientist* 19 Oct. 178/1 Normally, the autosomal chromosomes occur in pairs.

autostrada (ǭ·tostrādă). Pl. **-strade** (-strāde), **-stradas.** [It., = motor road, f. *auto* (*AUTO-²) + *strada* road, f. L. *stratum* (see STRATUM).] In Italy, a fast motor road; = *MOTORWAY.

1927 *Daily Tel.* 15 Nov. 7/1 It will be a 'speedway' pure and simple, somewhat on the lines of the Italian Autostrada. **1937** *Sunday Times* 10 Jan. 27/4 Ten years ago, when the first of the Italian autostrade were built. . the powers-that-be in this country might have displayed similar foresight in the planning of the new bypasses and main roads in this country. **1937** *Daily Herald* 16 Apr. 18/2 Those wonderful motor toll-roads, the autostradas, dead straight and with no cross-overs, are of non-skid concrete. **1955** [see *AUTOBAHN].

auto-suggest (ǭ:to,sŏdʒe·st), *v.* [Back-formation from next.] *trans.* To produce, remove, or influence, by auto-suggestion. Also *intr.*

1921 *Discovery* Nov. 296/2 We are suffering from toothache and wish to autosuggest it away. **1926** W. R. INGE *Lay Thoughts* III. viii. 230 The name Coué, which till lately would have auto-suggested a tale of adventure in the Australian bush, has now become famous in connection with the newest school of psychotherapy. **1936** *Discovery* Aug. 254/1 Intellectual preconceptions. . which. . autosuggest to the subject [*sc.* a dowser or diviner] some coherent. . scheme of interpretation with which his subsequent muscular reflexes. . concur. **1945** *Salt* 12 Feb. 37/2, I guessed Bill was auto-suggestin' for all he was worth.

au:to-sugge·stion. [*AUTO-¹.] Suggestion originating from oneself; *spec.* in *Psychol.*, the subconscious realization of an idea suggested to oneself for adoption. (Cf. SELF-SUGGESTION 2.)

1890 W. JAMES *Princ. Psychol.* II. xxvii. 601 When the [hypnotic] subject obeys it is by reason of the 'operator's suggestion'; when he proves refractory it is in consequence of an 'auto-suggestion' which he has made to himself. **1890** *Standard* 12 Apr. 5/5 The greatest criminals. . may be the responsible victims of a 'suggestion', or even of an 'auto-suggestion'. **1894** *Trans. Obstet. Soc.* XXXV. 246 Women may come to feel cravings by a process of auto-suggestion, especially in their first pregnancy. **1907** J. COATES (*title*) Self-Reliance: Practical Studies in Personal Magnetism, Will-power and Success, through Self-help or Auto-Suggestion. **1921** *Punch* CLX. 238/2 Though we cannot all of us be born at Newmarket, we can by auto-suggestion and psycho-analysis. . persuade ourselves that we have been born there.

So **au:tosugge·stible** *a.*, capable of being influenced by auto-suggestion; **au:tosuggestibi·lity,** the capacity of being thus influenced; **auto-sugge·stionist; auto-sugge·stive** *a.*, arising from or due to auto-suggestion.

1900 DORLAND *Med. Dict.*, *Autosuggestibility*, a peculiar mental state with loss of will, in which suggestions become easy. **1907** *Mind* XVI. 608 Hysterical persons are abnormally auto-suggestible. *Ibid.* 609 The auto-suggestibility preponderates in the severer cases. **1908** E. WORCESTER et al. *Relig. & Med.* 102 The auto-suggestionist must have the intellectual acquisitions in connection with the idea which he seeks to realize. **1910** *Athenæum* 5 Mar. 273/2 An hysterical girl under autosuggestive influences. **1926** ROGER FRY *Transformations* 1 Honesty is. . very difficult to come at, since in this matter we are all excessively auto-suggestible.

autotetraploid (ǭ:tote·trăploid). *Biol.* [f. *AUTO-¹ + *TETRAPLOID.] A tetraploid having four sets of chromosomes produced by doubling the chromosome number of a single diploid species. Hence **au:totetra·ploidy,** the condition or occurrence of such a tetraploid.

1930 J. B. S. HALDANE in *Jrnl. Genetics* XXII. 363 The self-sterile *Dahlia variabilis* behaves in some respects at least as an autotetraploid. **1942** J. S. HUXLEY *Evolution* iii. 87 The normal diploid complement of chromosomes of a species may become doubled (autotetraploidy). **1949** DARLINGTON & MATHER *Elem. Genetics* vi. 135 The allotetraploid stands at one extreme; the *autotetraploid*, the product of doubling a homozygous diploid such as a tomato, stands at the other.

autotheistic (ǭtoþi,i·stik), *a.* [f. AUTO-THEIST + -IC.] Of or pertaining to the autotheists or to autotheism.

1854 *Fraser's Mag.* XLIX. 708 There is this boundless difference between the healthy and godly 'subjective' style, and the unhealthy autotheistic subjective style. **1868** *Contemp. Rev.* IX. 75 Emerson's doctrine, too, is. . essentially autotheistic. **1872** A. M. FAIRBAIRN *Ibid.* XX. 49 In principle they [*sc.* the Hindu philosophies] might be Theistic, Auto-Theistic, Pantheistic or Atheistic.

autotomy (ǭtǫ·tŏmi). [f. *AUTO-¹ + -TOMY.] The casting off or ejection of some part or parts of the body as a reflex action peculiar to some animals when disturbed or in order to escape. Hence **auto·tomize** *v. intr.*, to practise autotomy; *trans.* to cast off; **auto·tomizing** *vbl. sb.*; **auto·tomous** *a.*, pertaining to or of the nature of autotomy.

1898 *Proc. Zool. Soc.* 1897 911 The autotomous break occurs between femur and coxa. *Ibid.*, The autotomy of the jumping-legs takes place at the femoro-trochanteric suture. **1899** *Q. Rev.* July 281 The self-mutilation (autotomy) met with in lizards. **1901** T. H. MORGAN *Regeneration* 153 In this way the arm may be autotomized, piece by piece, to its very base. **1930** *Discovery* May 166/1 An allied action is seen when the spider has cast a leg. It has lately been discovered that this familiar a[u]totomy is not, in arachnids, the reflex action it was formerly supposed to be. **1941** J. STEINBECK *Log in Sea of Cortez* (1958) xxviii. 267 Two large, hairy grapsoid crabs. . were. . difficult to catch, and when caught, battled fiercely and ended up by autotomizing.

autotoxin (ǭtotǫ·ksin). [f. *AUTO-¹ + TOXIN.] A poisonous substance formed within the body. So **autoto·xic** *a.*, of, pertaining to, or caused by an autotoxin.

1894 GOULD *Dict. Med.*, *Autotoxin*. . , any product of tissue-metamorphosis within the organism that has a toxic effect upon that organism. **1903** *Med. Record* 30 May 857 (Cent. D. Suppl.), We believe the excitant is a toxic or autotoxic agent. **1907** *Times Lit. Suppl.* 5 Dec. 371/3 Thus regarded, senility is an autotoxic phenomenon. **1938** H. BELLOC *Great Heresies* 235 We say that an organism has become 'auto-toxic' when it is beginning to poison itself.

autotypic (ǭtoti·pik), *a.* [f. AUTOTYPE + -IC.] **1.** Of, pertaining to, or reproduced by the autotype process (see AUTOTYPE 2).

1885 *Athenæum* 14 Mar. 351/3 The 'Little Devil's Bridge' . . could hardly be surpassed in autotypic reproduction.

2. Of the nature of an autotype or reproduction of an original.

1904 G. S. HALL *Adolescence* I. 124 Man. . is more autotypic than he knows, reproducing in his representation of the human form the type of his own race.

autoxidize (ǭtǫ·ksidəiz), *v. Biochem.* [f. *AUTO-¹ + OXIDIZE *v.*] *intr.* To oxidize by direct combination with oxygen at ordinary temperatures (Webster, 1909). Hence **autoxi·di·zable** *a.*, capable of oxidizing by contact with the air; **autoxid(iz)a·tion,** oxidation by direct combination with oxygen; **auto·xidator,** a substance in an active cell, oxidizable by water.

1883 *Science* 30 Mar. 229/2 Autoxidation in living vegetable cells. *Ibid.*, Autoxidizable substances, . . those bodies which, at a low temperature, and by the action of free, passive oxygen, can be oxidized. *Ibid.* 230/1 In every active cell, autoxidators are formed; that is, substances which, at a low temperature, and by the action of molecular oxygen, can be oxidized in the presence of water. **1948** *Jrnl. Brit. Interplan. Soc.* VII. 142 The so-called autoxidation process which is based on the alternate oxidation and reduction of certain organic compounds. **1956** *Nature* 21 Jan. 129/1 It is characteristic of N-dialkylhydroxylamines to autoxidize very rapidly in aqueous alkali. *Ibid.* 17 Mar. 527/2 All of these ferric chelates. . are very rapidly autoxidizable.

autozooid (ǭtozō^u·oid). *Zool.* [f. *AUTO-¹ + ZOOID *sb.*] The normal zooid in Alcyonarians, as distinguished from the siphonozooid.

1881 [see *siphonozooid* s.v. SIPHONO-]. **1888** *Athenæum* 3 Mar. 279/3 Dr. G. H. Fowler [read a paper] on a new Pennatulid from the Bahamas, the most interesting feature of which was the presence of immature autozooids at the dorsal end of the leaves. **1940** *Chambers's Techn. Dict.* 63/2 *Autozooid*, in *Anthozoa*, an ordinary typical zooid.

autumn. Add: **3.** *autumn-brown, -flowering, -sown* (earlier and later examples).

1866 WHITTIER *Snow-Bound* 21 Hill-sides autumn-brown. **1866** T. W. SANDERS *Encycl. Gardening* 86 [*Crocus*]. Plant. . autumn-flowering species in Aug. & Sept. **1765** *Museum Rusticum* IV. I. 227 The winter frosts will have broke down the clods on the roots of the autumn-sown rye. **1960** *Farmer & Stockbreeder* 2 Feb. 115/1 Grazing autumn-sown wheat the following spring reduces the subsequent yield of grain.

autumny (ǭ·tə̆mi), *a.* [f. AUTUMN *sb.* + -Y¹.] Redolent of autumn, characteristic of the autumn season.

1916 C. S. LEWIS *Let.* 27 Sept. (1966) 30 The country at home was beginning to look nice and autumny with dead leaves in the woods. **1920** *Glasgow Herald* 16 July 7 The air was distinctly 'autumny'. **1928** D. H. LAWRENCE *Let.* 12 Sept. (1962) II. 1085 Mountains are beginning to be misty and a bit damp and silent and autumny. **1941** E. BOWEN *Look at all those Roses* 112 The crushed, autumny grass.

autunite. (Earlier and later examples.)

1852 BROOKE & MILLER *Phillips' Elem. Introd. Min.* 519 *Autunite*. Found in crystals, massive, and investing other minerals, in granite at St. Symphorien near Autun. **1957** *Financial Times Ann. Rev. Brit. Industry* 39/3 Others [*sc.* secondary ore minerals] from which smaller amounts of uranium have been produced, are. . the phosphates autunite and torbernite.

auxanography (ǭksănǫ·grăfi). *Microbiol.* [ad. F. *auxanographie* (M. W. Beijerinck 1889, in *Archives Néerl. des Sci. Exactes* XXIII. 367), f. Gr. αὐξάν-ειν to increase + -OGRAPHY.] A method of studying the effects and interactions of various substances in promoting or inhibiting the growth of micro-organisms, by inducing diffusion gradients of the substances in agar plates on which the micro-organisms are grown. Hence **auxa·nogram,** a plate culture used in auxanography; **auxanogra·phic** *a.*

1905 GOULD *Dict. New Med. Terms* 108/2 *Auxanogram*, a pure plate culture of microbes which has been prepared by Beyerinck's auxanographic method. *Ibid.*, In which the colonies indicate which one of several nutrient media is best suited to their growth. *Ibid.* 109/1 *Auxanography*, a method devised by Beyerinck for ascertaining which nutrient media are suitable for a growing microbe. Plate cultures of bad media. . are stippled with drops of solutions, the nutrient properties of which are to be tested. The species of microbe which grows in the neighborhood of the drops is present. **1949** H. W. FLOREY *Antibiotics* I. i. 35 The auxanogram method consists in growing an organism by sowing the culture on the surface of a solid medium and then placing substances to be tested on it. **1946** G. PONTECORVO in *Jrnl. Gen. Microbiol.* III. 122 Sixty years ago. . Beyerinck [Beijerinck] developed a technique. . for the study of the nutritional requirements of micro-organisms. The 'auxanographic' technique, as he called it, has gone almost unnoticed. *Ibid.*, The first kind of test, based on the promotion of growth, could perhaps be termed 'positive' auxanography.

auxanometer (ǭksănǫ·mĭtəɹ). [f. Gr. αὐξάν-ειν to increase + -OMETER.] An instrument for measuring growth in plants.

1878 MASTERS *Henfrey's Elem. Bot.* (ed. 3) 612 Where great accuracy and the measurement of minute spaces [in growth] are demanded, recourse must be had to special instruments called Auxanometers. **1887** BATESON & DARWIN in *Jrnl. Linn. Soc., Bot.* XXIV. 2 The increase in length was measured by means of an auxanometer-lever.

auxesis. Add: **2.** *Biol., Plant Physiol.* (See quots.)

1848 DUNGLISON *Dict. Med. Sci.* (ed. 7) 96/1 *Auxesis*, augmentation, increase. **1900** B. D. JACKSON *Gloss. Bot. Terms* 28/1 *Auxesis*, (1) dilatation or increase in the valves of Diatoms, etc.; (2) new formation of organs (Czapek); (3) predominance of leaves, hairs, etc., on a particular side (Pfeffer). **1903** A. J. EWART tr. *Pfeffer's Physiol. of Plants* II. 73 The term 'auxesis' (*photoauxesis*) may be used to denote the predominant formation of leaves, roots, or hairs upon a particular side. **1940** *Nature* 9 Nov. 618/1 Botanists do still distinguish between *auxesis* or growth by expansion, and *merisis* or growth by cell-multiplication. **1953** *Faber Med. Dict.* 53/1 *Auxesis*, increase in size by cell expansion without cell division.

auxetic, *a.* Add: **2.** *Biol., Plant Physiol.* Of or pertaining to a substance which stimulates cell growth. Also as *sb.*, such a substance.

1913 DORLAND *Med. Dict.* (ed. 7) 123/1 *Auxetic*. 1. Stimulating cell proliferation. 2. A substance which stimulates cell proliferation. **1928** [see *AUGMENTOR 3].

auxetophone (ǭksetǒfōᵘn). [f. Gr. αὐξητός that may be increased + φωνή sound.] A pneumatic recorder for a phonograph; also, a phonograph fitted with this recorder; an amplifying instrument. (Disused.)

1904 *Daily Chron.* 9 June 4/6 The Hon. C. A. Parsons. . has recently amused himself by devising a little valve called the auxetophone which promises. . to advance the phonograph. **1907** *Ibid.* 6 Aug. 4/7 An instrument. . known as the Auxetophone, gives. . vocal and instrumental items. **1934** A. Q. CARNEGIE in G. L. Parsons *Scientific Papers of C. A. Parsons* III. 244 In its perfected form, the auxetophone gave results, thirty years ago, which in volume and quality. . have only just been approached by the electric gramophone and wireless loud-speakers of the last few years.

auxiliary, *a.* Add: **1. b.** In the names of various special ancillary services as: *Auxiliary Air Force* (formed 1917), *Auxiliary Fire Service* (formed 1937, absorbed in National Fire Service 1941), *auxiliary Territorial Service* (formed 1938, replaced by the Women's Royal Army Corps in 1949) (see also *A.T.S., separate entry).

1917 *Act 7 & 8 Geo. V*, c. 51 § 6 It shall be lawful for His Majesty to raise and maintain an Air Force Reserve and an Auxiliary Air Force. **1925** *Times* 19 Feb. 17/3 The City of London Auxiliary Air Force Squadron. . one of the first to be formed under the scheme for home defence against air attack. **1940** *Ann. Reg.* 1939 60 The fighter and general reconnaissance squadrons of the Auxiliary Air Force. **1940** *Hutchinson's Pict. Hist. War* 2 Oct.–26 Nov. 168 Men of the Auxiliary Fire Service have been fully tested by raid conditions. **1938** *Army Orders* Sept. 9 Our will and pleasure is that there shall be formed an organization to be designated the Auxiliary Territorial Service; our further will and pleasure is that women may be enrolled in this Service. **1939** *War Illustr.* 23 Sept. 9 (*title*) W.A.T.S. (Women's Auxiliary Territorial Service). **1942** *Lett. Oct.* 151 (*caption*) Dame Helen Gwynne Vaughan, Director of the Auxiliary Territorial Service.

2. a. *auxiliary language,* a language, esp. one invented for the purpose (e.g. Esperanto), used as a means of communication by speakers from two different language groups; an artificial language (see *ARTIFICIAL *a.* 5).

1905 *Westm. Gaz.* 15 Aug., Esperanto aims at becoming a generally accepted 'auxiliary language'. **1939** *Nature* 29 Apr. 717/1 The value of an auxiliary language as an agency for world peace.

auxin (ǭ·ksin). *Plant Physiol.* [G. (F. Kögl & A. J. Haagen Smit 1931, in *Proc. Sect. Sci. Kon. Akad. v. Wetenschappen* XXXIV. 1416), f. Gr. αὔξ-ειν to increase + -IN¹.] An organic substance which promotes and directs the growth of plant cells; a growth hormone.

1934 *Jrnl. Chem. Abstr.* 1332/1 Plant growth substances. *X*. Constitution of auxin-a and -*b*. **1936** *Ann. Reg.* 1935 56 Auxin *a* is an organic hydroxy-acid, Auxin *b* is a keto-acid. **1937** *Discovery* June 174/1 Putting a block of auxin-charged agar on one side of a coleoptile stump. **1937** WENT & THIMANN *Phytohormones* 4 The term auxins. . will be arbitrarily restricted to those substances which bring about the specific growth reaction which is conveniently measurable by the curvature of *Avena* coleoptiles. **1955** *Sci. News Let.* 16 July 44/1 A plant growth hormone called auxin controls the shedding of leaves in the fall and the dropping of fruit when it is ripe.

auxochrome (ǭ·ksokrōᵘm). [G. (O. N. Witt 1888, in *Ber. d. Deut. Chem. Ges.* XXI. 325), f. Gr. αὔξ-ειν to increase + -O + χρῶμα colour.] Any salt-forming group that, when combined with a chromogen, produces a dyestuff. Hence **auxochromic** (-krōᵘ·mik), **auxochromous** (-krōᵘ·məs) *adjs.*, defining such groups.

1893 *Athenæum* 15 July 100/1 They [*sc.* chromogens] become true dyestuffs by the introduction of. . what Witt terms auxochromes. **1892** COLLIN & RICHARDSON tr. *Nietzki's Chem. Org. Dyestuffs* i. 13 The chromophoric quinone group confers strong acid properties on the auxochromic hydroxyl group. **1902** *Brit. Jrnl. Photogr.* 11 Apr. 299 Auxochromic groups. **1902** *Encycl. Brit.* XXVII. 564/2 Such compounds containing chromophorous groups are termed chromogens, because, although not dyestuffs themselves, they are capable of generating such by the further introduction of salt-forming atomic groups,—*e.g.*, OH, NH₂. These Witt terms auxochromous groups. **1953** J. KARRINGTON *Mushroom & Toadstools* ix. 100 Other groups (—OH, —OCH₃, —CO₂H, NH₂), known as auxochromes, modify this basal colour in various ways.

auxospore (ǭ·ksospōᵊɹ). *Bot.* [f. Gr. αὔξ-ειν to increase + -O + σπόρος, σπορά SPORE.] A spore formed in diatoms by the union of two cells or by the excessive growth of an individual cell. Also, the resting-spore of the diatoms.

1884 MASTERS & BENNETT *Henfrey's Elem. Bot.* (ed. 4) 425 The individuals must. . constantly diminish in size, until the original size is restored by the production of auxospores. **1888** *Encycl. Brit.* XXIV. 126/2 *Diatomaceæ*. . . Reproduction, vegetative by division or by means of asexually-produced spores (auxospores). **1904** G. S. WEST *Brit. Freshwater Algæ* 269 A normal auxospore can be regarded as one produced by the conjugation of two cells (or gametes), those produced without conjugation being parthenogenetic.

avail, *v.* **5. a.** (Later examples.)

1899 *Westm. Gaz.* 25 Aug. 4/1 It is now definitely settled that the Admiralty. . will avail of the opportunity. . for the renewal of the subsidy. **1927** *Daily Tel.* 29 Aug. 8/6 The wonderful system of drainage is being availed of.

availingly (ăvēᶦ·liŋli), *adv.* [f. AVAILING *ppl. a.* + -LY².] In an availing manner; so as to avail or profit.

1853 FABER *Ess. Lives of Saints* 116 Its intrinsic beauty pleads availingly with the man of letters. **1871** *Contemp. Rev.* XIX. 136 Neither the royal Placet. . nor the right to convene synods, could be availingly employed.

avalanche, *sb.* Add: **1.** (Earlier examples.)

1771 PENNANT *Tour Scotland* III. I have seen these *spates*. . in some places the roads, as the *avalanches*, or snow-falls, do those of the *Alps*. **1787** *Monthly Rev.* LXXVII. 533 They were also apprehensive of exposing themselves to the *Avalanches*, which are frequently tumbling from the summit of the mountains.

3. *avalanche lily,* any one of several large erythroniums found near the snow-line in N. America.

1912 A. O. WHEELER *Selkirk Mts.* 74 The avalanche lilies. . which follow the edges of the glaciers. **1952** in *Jrnl. Canad. Linguist. Assoc.* (1956) II. 28 The glens were starred with white anemones and yellow avalanche lilies. **1963** W. S. AVIS et al. *Dict. Canad. Eng.* (*Intermediate*) 58/2 *Avalanche lily*, the dogwood violet of the Rockies.

avalanche (æ·vălanʃ), *v.* [f. AVALANCHE *sb.*] *intr.* To descend in or like an avalanche. Also *trans.*, to carry by or as by an avalanche.

1872 'MARK TWAIN' *Roughing It* iv. 26 We avalanched from one end of the stage[-coach] to the other. **1897** *Daily News* 29 Mar. 6/5 He was gently avalanched downwards into the street. **1899** SOMERVILLE & 'ROSS' *Irish R.M.* 244, I avalanched down the companion. **1923** *Daily Mail* 23 June 7 The boulders on the edge are continually avalanching down. **1957** CLARK & PYATT *Mountaineering in Britain* x. 178 They had nearly completed the ascent and reached a cornice when a snow step avalanched.

avalanchy (æ·vălanʃi), *a.* [f. AVALANCHE *sb.* + -Y¹.] Liable to descend in or be swept by an avalanche.

1869 *Jrnl. Tours In Alpine Valley* I. vi. 117 Rather an avalanchy place, this. **1894** *Geog. Intern. Aug.* 218 Snow-slopes, of which one was rotten and avalanchy. **1925** T. H. SOMERVELL in E. F. NORTON *Fight for Everest: 1924* III. vi. 356 The steep ice cliffs of the North Col are safer than the avalanchy slopes.

avalite (æ·vălit). *Min.* [ad. G. *avalit* (1884), f. *Avala*, name of a mountain near Belgrade, its locality + -ITE¹.] A green earthy mineral containing chromium oxide.

1889 in *Cent. Dict.* **1896** CHESTER *Dict. Min.*

avalone, var. *ABALONE.

avanc, var. *AFANC.

avant-garde. Add: **2.** The pioneers or innovators in any art in a particular period. Also *attrib.* or as *adj.* Hence **avant-ga·rdism,** the characteristic quality of such pioneering; **avant-ga·rdist(e)** (-ist), such a person; also *attrib.*

1910 *Daily Tel.* 1 July 14/6 The new men of mark in the *avant-garde*. **1925** *League of Composers' Rev.* Jan. 26 He is at once 'avant-garde'. **1929** BLACKWELL *Music* in the *avant-garde* magazines. **1940** GRAVES & HODGE *Long Week-End* xii. 197 All Paris. . British and American literary *avant-garde* dates fraternized or came to blows. **1947** *University Observer* 1 There is a terrible striving always to be *avant-garde*: to 'discover' Henry James, T. S. Eliot, Rilke and the rest, to discover modern English poets. **1947** *Horizon* Dec. 290 A literature without an *avant-garde* soon becomes a literature without a main body. **1960** G. KOESTLER in *God that Failed* 1. 31 Their policy. . in cultural matters [was] progressive to the point of *avant-gardism*. **1953** *Archit. Rev.* CXIII. 149/2 For avant-gardist architecture produces. . character-less buildings. **1967** *Spectator* 3 Nov. 547/3 What baffles me about our various well-meaning avant-gardes is their prodigious appetite for punishment. **1967** *Times* 23 Nov. 15/2 They resembled a group of avant-gardists who seemed not quite to know where they were going.

Avar (ā·vaɹ, āᵊ·vaɹ). **1.** A member of a Turkic people, prominent in south-eastern Europe from the 6th to the 9th c. A.D.; also, their language. Also *attrib.* **2.** (æ·vaɹ). Also **Awar.** A member of a people of the North Caucasus; also, their language.

1788 GIBBON *Decl. & F.* IV. xlii. 229 He imparted to the senate his resolution. . to purchase the friendship of the Avars. *Ibid.*, But the virtue or treachery of an Avar betrayed the secret enmity and ambitious designs of their countrymen. **1881** *Jrnl. R. Asiatic Soc.* XIII. 293 My ride took me through the Avār-speaking country. *Ibid.*, It is written by the one in Avār, by the other in Russian. **1882** FREEMAN *Lect. Amer. Audiences* 139 The Empire which had beaten back the Persian and the Avar lost its provinces to the Saracen and the Bulgarian. **1930** H. G. WELLS *Outline of Hist.* vi. xxxi. § 1 The Avars and Slavs struck down from the Danube country towards the Adriatic. **1934** A. TOYNBEE *Study Hist.* III. 25 The transitory Avar ascendency was the making of the Slavs. **1951** W. K. MATTHEWS *Languages of U.S.S.R.* v. 89 Awar has. . fifty phonemes, only five of which are vowels.

Hence **Avarian** (ăvēᵊ·riăn) *a.*, of or pertaining to the Avars (both senses); also *absol.*, the language of the Avars.

1875 C. HERRAGE tr. *M. von Thielmann's Journey in Caucasus* II. 286/1 Avarian guides. **1889** *Jrnl. R. Asiatic Soc.* XXI. 729 Til meant 'black' in Avarian. *Ibid.* 731 The importance of the Avarian domination. . has hardly been sufficiently appreciated. **1902** *Encycl. Brit.* XXVII. 341/2 The Avarian is a sort of inter-tribal tongue. **1909** WEBSTER s.v., *Avarian rings*, vestiges of Avarian fortifications formed by stakes surrounding a settlement.

avascular (ăvæ·skiŭlăɹ, ēᶦ-), *a. Physiol.* [f. A- 14 + VASCULAR *a.*] Lacking or deficient in vascularity or blood vessels.

1900 DORLAND *Med. Dict.* 80/1 *Avascular*, not vascular. **1908** *Practitioner* June 771 An avascular area on the bladder is exposed. **1963** *Lancet* 19 Jan. 133/1 The wound flaps were under some tension, and avascular necrosis of the wound margins developed.

avenue, *sb.* **2.** *fig.* (Further examples.)

1921 WODEHOUSE *Indiscr. Archie* ii. 12 You did not irremediably close all avenues to a peaceful settlement. **1926** *Sat. Rev.* 16 Oct. 428/2 He. . explores every avenue which may lead him to a point of vantage whence to view his life in its new meaning. **1927** *Rev. Eng. Stud.* Oct. 432 Our politicians are said to deal with *dominant issues* and to *explore avenues*.

avenued (æ·viniūd), *ppl. a.* [f. AVENUE *sb.* or *v.*] Furnished with or having an avenue or avenues.

1870 RUSKIN *Verona* § 31 One paradise of lovely pasture and avenued forest of chestnut and blossomed trees. **1899** *Daily News* 7 Nov. 3/3 An open, avenued, highly-cultivated plateau.

average, *sb.*² Add: **6. b.** spec. in *Cricket.* The mean number of runs per innings scored by a batsman, or the mean cost in runs per wicket achieved by a bowler, during a season, tour, etc.

1845 W. DENISON *Cricketer's Companion* (ed. 2) 113 Many of our finest bats have had their average diminished. *Ibid.* (*heading*) Averages for 1844. **1854** F. LILLYWHITE *Guide to Cricketers* 74 Mr. H. Lampson. Bowling average, per innings, 2 and 13 over. **1870** *Daily's Monthly Mag.* Oct. 85 In 1868 his batting average was 53 and 3 over. **1967** *Whitaker's Almanack* 982 (*heading*) Batting and bowling averages.

average, *a.* Add: **2. b.** Used with *sensual* [tr. F. (*l'homme*) *sensuel moyen*.]

1882 [see SENSUAL *a.* 4a]. **1894** G. B. SHAW in *Fortnightly Rev.* Feb. 263 The average sensual man. **1897** — *Our Theatres in Nineties* (1932) III. 28 Nothing can be more real than the average sensual man. **1937** A. HUXLEY *Ends & Means* xiv. 297 Only the disinterested mind can transcend common sense and pass beyond the boundaries of animal or average-sensual human life. **1950** — *Themes & Variations* 71 Samuel Pepys's day-by-day record of how the average sensual man comports himself. **1968** R. AMBERLEY *Incitement to Murder* i. 10 If an average sensual man, he was also an average kindly man.

3. *Comb.*, *average-sized* adj.

1851 H. MELVILLE *Moby Dick* III. ii. 340 An average-sized male. **1960** *Farmer & Stockbreeder* Suppl. 29 Mar. 9/1 They all fit an average-sized head.

average, *v.* Add: **2.** (Earlier U.S. examples.)

1769 WASHINGTON *Diaries* (1925) I. 314 At that weather. . it being imagined. . would average the sheave weight. **1806** FESSENDEN *Orig. Poems* 113 Each paper. . Will average at a hundred lies.

4. *intr.* with *out*: To work out so as to produce an average. Also *trans.* in corresponding sense.

1914 G. B. SHAW *Misalliance* 41 Averages out the human race. Makes the nigger half an Englishman. **1921** *Times Lit. Suppl.* 28 Sept. 610/4 The particular obstacles will vary from time to time and from species to species, but on the whole will average out. **1928** *Britain's Industr. Future (Liberal Ind. Inquiry)* v. xxxi. § 5. 4.6 Any attempt to average out burdens. **1934** G. B. SHAW *Too True to be Good* III. p. 97 In the army these things average themselves out.

averageness (æ·vĕrĕdʒnĕs). [f. AVERAGE *a.* + -NESS.] The fact or condition of being average; ordinariness, mediocrity.

1925 *Glasgow Herald* 18 June 9 The minister occupies a position in which the averageness of his ability is necessarily exposed to more general remark. **1961** *Encounter* Oct. 76/2 Gordon Comstock. . finally sinks in a morass of wage-earning, aspidistra-garnished, lower-middle-class averageness.

aversant, *a.* Restrict †*Obs. rare*⁻¹ to sense in Dict. and add: **2.** *Her.* Turned to show the back (said of a right hand).

1830 ROBSON *Brit. Herald* III. Gloss., *Aversant*, or *Dorsed*; as, a right hand *dorsed* or *aversant*, when turned to shew the back part. **1889** ELVIN *Dict. Her.*

aversion. Add: **7.** *attrib.*, as **aversion therapy, treatment,** therapy or treatment designed to render a particular habit repugnant to someone addicted to it.

1950 LEMERE & VOEGTLIN in *Q. Jrnl. Stud. Alcohol* XI. 199 (*title*) An evaluation of the aversion treatment of alcoholism. **1956** *Brit. Med. Jrnl.* 13 Oct. 854/2, I have been unable to find any previous record of a fetishist who responded favourably to aversion therapy. **1965** *Listener* 28 Jan. 143/1 One method of so-called aversion therapy is to associate the stimulus that one wants to get rid of with something unpleasant, such as forced vomiting or electric shocks.

aversionist (ăvɔ̄·ɹʃənist). [f. AVERSION 4 + -IST.] One who has a strong aversion or repugnance towards something.

1938 C. MORGAN *Flashing Stream* p. xix, The experience of sex is, in her case, what Nature intended it to be and what in women not aversionists it is—a delight, and not an act of self-sacrifice. **1961** P. MASON *Common Sense about Race* iv. ii. 149 A determined aversionist. . has recently argued that the dislike which he himself clearly feels must be a biological function which only emerges at puberty.

Avertin² (ăvɔ̄·ɹtin). *Med.* Also **avertin.** [G. (trade-name, 1927), f. AVERT *v.* + -IN¹; cf. G. *avertieren* to avert, ward off.] The trade-name of an anæsthetic drug.

1927 *Chem. Abstr.* 2154 (*title*) Clinical experiences with avertin. **1929** F. J. MORRIN in *Irish Jrnl. Med. Sci.* Apr. 183 Within the past year, I have used avertin anæsthesia in 54 cases. **1930** *Lancet* 7 June 1264/1 Avertin by itself procures in most cases the desired deep sleep.

avertive (ăvɔ̄·ɹtiv), *a.* [f. AVERT *v.* + -IVE.] Designed to avert or ward off.

1889 *Cornhill Mag.* May 491 A series of avertive and violent sweeps of a table-napkin. **1895** *Q. Rev.* July 211 We are not convinced that the heraldic animals were originally 'avertive' amulets.

Avesta (ăve·stă). = ZEND-AVESTA. Hence **Ave·stan, Ave·stic** *adjs.*, of or belonging to the Avesta; *sbs.*, the language of the Avesta.

1807 W. JONES *Sixth Discourse on Persians* in *Wks.* (1807) III. 113 A learned follower of Zeratusht. . assured me, that the *letters* of his prophet's book were properly called *Zend*, and the *language*, *Avestà*. **1856** W. D. WHITNEY in *Amer. Oriental Jrnl.* V. 352 The Avesta is written in a language to which. . the name of *Zend* has been given. . . If it should be regarded as still practicable to change the common usage, and give the language a more appropriate designation, none. . could be found so simple, and open to so few objections, as *Avestan*. . . The Avestan, then, is an ancient Persian language. *Ibid.*, General considerations. . refer us to a time as early as the first half of the thousand years before Christ as being that of the Avestan language. **1888** *Encycl. Brit.* XXIV. 775/2 Although the Avesta is a work of but moderate compass. . there. . exists no single MS. which gives it in its entirety. **1888** JOSEPH WRIGHT (*title*) Elements of the comparative grammar of the Indo-Germanic languages. A concise exposition of the history of Sanskrit, Old Iranian (Avestic and Old Persian), Old Armenian, . . by Karl Brugmann. **1906** L. C. CASARTELLI *Sketches in Hist.* viii. 216 Paulinus di San Bartolommeo . . not only defended the genuineness of the Avestic language, but even indicated. . its affinity with the Sanskrit. **1914** E. J. RAPSON *Ancient India* ii. 31 Avestan forms are, from the etymological point of view, manifestly later than Vedic forms. **1932** W. L. GRAFF *Lang. & Languages* x. 370 Two dialects, Old Persian and Avestan or Zend.

aviate (ēᶦ·viēᶦt), *v.* [Back-formation from *AVIATION.] *intr.* To navigate the air in an aeroplane; to fly. Also *trans.*

1887 tr. *J. Verne's Clipper of Clouds* iv. 36 Mr. Aviator. . , you who talk so much of the benefits of aviation, have you ever aviated? **1900** *Manch. Guardian* 4 Apr. 9/2 In the event of the sudden failure of the aviating mechanism. **1908** *Punch* 21 Oct. 305/2 'To aviate' is nauseous enough, and 'planing', in the atmospheric slang of country-house parties, is pretty bad. **1943** C. H. WARD-JACKSON *Piece of Cake* 11 *To aviate*, to fly, to fly flashily, to show-off when flying. **1959** W. E. JOHNS *Biggles at World's End* i. 22 The obsolete crates we're expected to aviate.

aviation (ēᶦviēᶦ·ʃən). [ad. F. *aviation*, irreg. f. L. *avis* bird + -ATION.] Aerial navigation by means of an aeroplane; the science of powered flight. Also *attrib.*

1866 J. S. HARRY tr. *Nadar's Right to Fly* xvi. 82 The cause of Aviation. **1870** *Ann. Rep. Aeronaut. Soc.* 11 Men do not consider the subject of 'aërostation' or 'aviation' to be a real science. **1891** *Daily News* 27 Aug. 5/1 The frequency of the term 'aviation' as a synonym for aerial navigation. **1908** *Daily Report* 12 Sept. 4/4 Santos-Dumont . . won the first aviation prize. **1910** R. FERRIS *How it Flies* 455 *Aviation*—Flying by means of power-propelled machines which are not buoyed up in the air, as with gas bags. **1911** *Daily Express* 13 June 1/3 The Home Secretary has issued an order prohibiting aviation in specified areas. **1916** *Sci. Amer.* Suppl. 3 June 357 (*title*) Aviation sickness; its symptoms and cure. **1920** *Flight* 26 Feb. 252/1 The Anglo-American Oil Co., Ltd., announce the present retail price of. . Pratt's aviation spirit, 4*s.* 1½*d.* per gallon. **1920** W. H. WILMER (*title*) Aviation Medicine in the A.E.F. **1928** *Lancet* 7 Apr. 714/1 Aviation or aviators' sickness. . a train of complex symptoms occurring in those who take the air. **1957** *Technology* Mar. 8/1 Some of the men from whose wisdom he will benefit have been in aviation almost from the beginning. **1960** *Times* 18 Oct. 13/7 What used to be the School of Aviation Medicine at San Antonio has been the Aerospace Medical Centre for a year now.

aviator (ēᶦ·viēᶦtəɹ). [ad. F. *aviateur*, f. L. *avis* bird + *-ateur* *-ATOR.] †**1.** A heavier-than-air aircraft. Also *attrib. Obs.*

1891 *Brooklyn Morning Jrnl.* 22 July 1/6 (Funk), Mr. Maxim's invention is called an Aviator. It is in form like a huge kite of silk, to which hangs a platform carrying the engines and the screw propellers. **1895** *Knowledge* 2 Dec. 276/1 Mr. Maxim represents gunnery and the aviator flying machine. **1908** V. SILBERER in *Aeronaut. Jrnl.* July 51/1 A flying machine or aviator, however well constructed and furnished with such a motor.

2. The pilot of an aeroplane.

In early use, as distinguished from an aeronaut, *i.e.*, a balloonist.

1887 [see *AVIATE *v.*]. **1896** *Westm. Gaz.* 15 Sept. 1/3 Intending aviators and aeronauts. **1909** *Ibid.* 26 Oct. 1/3 Other 'aviators'—the word has forced itself into the vocabulary, and it seems futile to resist it any longer—had other machines. **1911** *Yorks. Post* 3 Aug. 9/6 At height of 1,000 metres an aviator can find a submarine.

Hence **a·viatress, -trice, -trix,** a female aviator.

1910 *Daily Chron.* 5 Jan. 1/7 The aviatrice made a bad turn. **1911** *Aero* June 74/2 Various articles on the subject of 'Aviatresses' which have appeared from time to time. **1927** *Glasgow Herald* 29 Sept. 11 The English aviatrix, Miss Evelyn Spooner.

b. aviator's (or **aviators'**) **ear** = *AERO-*otitis media*; **aviator's** (or **aviators'**) **sickness,** see *AVIATION (quot. 1928).

1937 Aviator's ear [see *AERO-*otitis media*]. **1916** *Sci. Amer.* Suppl. 3 June 357/2 What distinguishes aviators' sickness from mountain sickness is that the symptoms persist during descent and are aggravated after landing.

Avicennia (ævise·niă). Also **avicennia.** [Named after *Avicenna*, Arabian physician (980–1037): see -IA¹.] A plant of the genus so named, esp. *A. tomentosa*, the white mangrove.

1829 LOUDON *Encycl. Plants* 522. **1871** KINGSLEY *At Last* II. xiii. 211 The stateliest Avicennias, or white mangroves. **1883** *Encycl. Brit.* XV. 481/2 The 'white mangrove', *Avicennia*, a verbenaceous plant.

avicolous (ăvi·kŏləs), *a.* [f. L. *avis* bird + *-colus* inhabiting + -OUS.] Living, as parasites, on birds.

1895 D. SHARP in *Cambr. Nat. Hist.* V. xv. 349 The greater portion of the avicolous species have two claws.

aviculturist (ēᶦvikɒ·ltiŭrist). [f. AVICULTURE + -IST.] One who practises aviculture; a bird-fancier.

1904 *Nature* 31 Mar. 507/2 His work forms an excellent history of these beautiful birds, and is alike interesting to the field naturalist and the aviculturist. **1923** *Glasgow Herald* 23 Oct. 6 Probably Mr Farrar thinks that only aviculturists will concern themselves with his writing.

avidin (æ·vidin). *Biochem.* [f. AVID *a.* + -IN¹, from its 'avidity' for biotin.] A protein in raw white of egg which combines with and inactivates biotin.

1941 R. E. EAKIN et al. in *Jrnl. Biol. Chem.* CXL. 536 The name 'avidin' seems appropriate for this protein; this term is suggested by its peculiar biotin-binding capacity. **1945** *Lancet* 5 May 568/1 Avidin is a protein constituent of egg-white which inactivates biotin by forming a complex with it.

aviette (ēᶦvie·t). [Fr., f. *avi*(*on* *AVION + -ETTE.] An engineless aeroplane or glider. (Disused.)

1912 *Daily News* 1 June 1/4 It is called an aviette competition, an aviette being a flying machine propelled by muscular force alone. **1923** *Westm. Gaz.* 7 Apr., The 'Aviette', as the new glider with an auxiliary motor threatens to call itself.

avifaunal (ēᶦ·vi,fǭ:năl), *a.* [f. AVIFAUNA + -AL.] Of or pertaining to avifauna.

1889 in *Cent. Dict.* **1933** *Geogr. Jrnl.* LXXXI. 315 Some influence on avifaunal distribution.

‖**avion** (avyoṅ). [F. *avion*, f. *aviateur* *AVIATOR, app. after *ballon* BALLOON *sb.*¹] A flying-machine, esp. a French war aeroplane.

1898 *Sci. Amer.* 27 Aug. 137 The 'Avion' [*sc.* flying-machine of C. Ader (1841–1925)] in its position of flight. **1915** T. F. FARMAN in *Blackw. Mag.* Apr. 435/1 How many avions the French and British military authorities possessed when war broke out. **1927** *Chambers's Jrnl.* 258/2 Less easy prey they may be for the avions. **1935** J. JOYCE *Let.* 16 July (1966) III. 369 An avion letter from Bailly.

avirulent (ævi·riŭlĕnt, ēᶦ-), *a.* [f. A- 14 + VIRULENT *a.*] Not virulent.

1900 in DORLAND *Med. Dict.* **1949** FLOREY et al. *Antibiotics* I. xxxiii. 1094 Iland (1946) found that an avirulent strain was more effective in this respect than the virulent strains. **1960** *New Biol.* XXXI. 72 In 1928. . Griffith. . discovered that an avirulent and normally harmless strain of pneumococcus was changed into a virulent strain when injected into mice together with some virulent pneumococci that had been thoroughly killed by heating.

avitaminosis (ăvəi:tăminōᵘ·sis, ēᶦ-, -vi:t-). Pl. **-oses** (-ōᵘ·sīz). [mod.L., f. A- 14 + *VITAMIN + -OSIS.] A condition resulting from a deficiency of one or more vitamins. Also used *spec.* (*avitaminosis A, B*, etc.): deficiency of the vitamin specified. Cf. *DEFICIENCY *disease*.

1914 *Lancet* 9 May 1328/2 The author [*sc.* C. Funk] states that this publication [*sc. Die Vitamine*] is the first attempt at a representation of our knowledge about vitamines and avitaminoses. . . By the word avitaminoses, coined by the author, is understood what is called in English 'deficiency diseases'. **1923** *Nature* 26 May 728/1 Contribution to the study of B avitaminosis in the pigeon. **1931** *Lancet* 14 Mar. 624/1 (*title*) A Multiple Avitaminosis. A disease which is claimed to be a multiple food deficiency. *Ibid.* 624/2 Multiple partial avitaminoses. . are probably all too common among native races the world over and are the abundant fruit of poverty.

‖**avodiré** (ævodiᵊ·re). Also **avodire.** [Fr.] The smooth-textured hardwood of light colour from a large West African tree (*Turræanthus africanus* or *T. vignei*) of the mahogany family; also, the tree itself.

1934 A. L. HOWARD *Man. Timbers World* 44 Light-weight, white wood. . offered for sale as African white mahogany. . . Among the best of these timbers it is now possible to name Avodire, which possesses a golden-yellow colour, with a firm, clean grain, capable of giving a very smooth surface. **1935** *Archit. Rev.* Oct. 160/3 A corner fitting, veneered with avodire, consisting of a radio-gramophone and copious record storage. **1936** *Nature* 9 May 790/1 The following woods amongst others have been used [for the liner *Queen Mary*]. . Avodiré (*Turræanthus africana*).

Avogadro (ævogā·dro). *Chem.* The name of the Italian scientist Count Amedeo *Avogadro* (1776–1856), used esp. *attrib.* or in the possessive case of his hypothesis that equal volumes of all gases at the same temperature and pressure contain equal numbers of molecules. Also *Avogadro('s) constant, number*: the number of molecules in a mole (*MOLE *sb.*⁷) (see quot. 1958).

1871 *Jrnl. Chem. Soc.* XXIV. 1247/1 Avogadro's law. *Ibid.* 300 *On the Law of Avogadro.*—The relation of this law —viz., that 'equal volumes of different gases contain, under like conditions, equal numbers of molecules', to the mechanical theory of gases, has lately been the subject of . . discussion. **1874** J. P. COOKE *New Chem.* ii. 37 Avogadro's law declares that all gases contain, under like conditions of temperature and pressure, the same number of molecules in the same volume. **1895** C. S. PALMER tr. *Nernst's Theoret. Chem.* v. 148 We know the gas pressure of all well-defined chemical substances, which are subject to Avogadro's rule. **1902** *Encycl. Brit.* XXVI. 739/1 Avogadro's theorem is directly applicable to dilute solutions. **1927** WEBSTER Addenda, Avogadro number. **1958** BALLENTYNE & WALKER *Dict. Named Effects* 9 *Avogadro's Number*, the number of individual atoms in a gramme-atom, of ions in a gramme-ion or of molecules in a gramme-molecule. . . The accepted value is $6{\cdot}023 \times 10^{23}$. **1959** *Chambers's Encycl.* I. 746/1 Since molecules have now lost their hypothetical character, it seems preferable to refer to Avogadro's theorem rather than Avogadro's hypothesis or law.

avoidance. Add: **7. b.** *Anthropology.* The custom prevalent among many primitive tribes by which one member of a family is forbidden to meet and address another member.

[**1865** TYLOR *Early Hist. Man.* x. 287 Their object seems to be in general the avoidance of intercourse or connexion between parents-in-law and children-in-law. . . But the reasons for this avoidance are not clear.] **1889** —— in *Jrnl. Anthrop. Inst.* XVIII. 247 If the customs of residence and the customs of avoidance were independent, or nearly so, we should expect to find their coincidence following the . . law of chance. *a* **1899** J. J. ATKINSON *Primal Law* (1903) 269 Avoidance would arise at the same time between mother-in-law and son-in-law. **1903** *Lancet* 22 Aug. 532/2 The conditions in which he lived with the female members of his own family gave origin to the curious etiquette of 'avoidances' which is still to be found amongst some savage races. **1958** A. R. RADCLIFFE-BROWN *Method Soc. Anthrop.* I. v. 122 In many primitive societies the relation established between two groups of kin by a marriage between a man of one group and a woman of the other is one which is expressed by customs of avoidance.

avondbloem(etje): see *AANDBLOM.

avuncular, *a.* Add: **b.** (Earlier example.) Also *absol.*

1832 *Fraser's Mag.* V. 85 My only good suit is at present under the avuncular protection. **1922** JOYCE *Ulysses* 417 Avuncular's got my timepiece.

Hence **avu·ncularism** (*joc.*), recourse to a pawnbroker; **avuncula·rity,** the state of being an uncle; **avu·ncularly** *adv.*, in the manner of an uncle.

1859 D. G. ROSSETTI *Let.* 15 Feb. (1965) I. 348, I have only been saved from further 'avuncularism' by a visit of old Plint, who has bought two. . drawings. **1937** A. L. ROWSE *R. Grenville* ii. 28 The pleasures of avuncularity. **1957** *Economist* 7 Sept. 824/1 The classical picture here is of Lord Woolton avuncularly presiding over the rapidly growing Young Conservatives.

avunculate (ăvɒ·ŋkiŭlĕt). *Anthropology* [f. L. *avuncul-us* maternal uncle + -ATE¹ 1 a.] The special relationship existing in some societies between a maternal uncle and his sister's son; maternal uncles regarded as a collective body.

1920 R. H. LOWIE *Prim. Soc.* v. 81 Ethnologists describe under the heading of *avunculate* the customs regulating in an altogether special way the relations of a nephew to his maternal uncle. *Ibid.* vii. 171 The Omaha are patrilineal now, but their having the avunculate proves that they once traced descent through the mother, for on no other hypothesis can such a usage be explained. **1927** *Observer* 17 July 6/4 This particular group of Melanesians is almost entirely lacking in the typical institutions of the region, the dual organisation, the avunculate. **1932** R. F. FORTUNE *Sorcerers of Dobu* i. 62 Dobu practices the avunculate, inheritance from mother's brother to sister's son. **1937** *Times Lit. Suppl.* 27 Feb. 145/1 It is a little surprising to find the mother's brother—in his collective capacity sometimes known as the 'avunculate'—so much under an obligation to do the honours.

aw (ǫ), *int.* An exclamation usually expressing mild remonstrance, entreaty, commiseration, disgust, or disapproval.

1852 STOWE *Uncle Tom* viii. 39/2 'Ugh? aw! like enough!' grunted his complaisant acquaintance. **1902** J. J. BELL *Wee Macgreegor* viii. 76 'He sudna ha'e boastit.' 'Aw, Lizzie!' **1911** J. C. LINCOLN *Cap'n Warren's Wards* xix. 312 Aw, let me alone. **1928** W. SMYTH *Jean of Tussock Country* v. 46 'Aw—yes, you want a job!' the manager muttered. *Ibid.* v. 47 'Can you ride?'. . 'Aw—you can.' **1932** 'SPINDRIFT' *Yankee Slang* 9 Aw! Shucks, forget it—to tender thanks for courtesy or kindness is to provoke this phrase.

awabi (āwā·bi). [Jap.] The Japanese abalone or sea-ear (*Haliotis gigantea*).

1889 in *Cent. Dict.* **1924** *Glasgow Herald* 12 May 11 Soochon awaibi, and a whole gamut of marvels taking days to concoct. **1934** E. BLUNDEN *Choice or Chance* 10 And ice may marry well with beer, And—what, they serve *Awabi* here? **1962** *Times Lit. Suppl.* 15 June 440/3 Sensual figures of the almost naked awabi fishergirls in another famous triptych of Utamaro's.

awakedness (ăwēᶦ·kĕdnĕs, ăwēᶦ·ktn's). [f. AWAKED *ppl. a.* + -NESS.] An awakening from sleep; the state of being awake.

1922 D. H. LAWRENCE *Aaron's Rod* xviii. 273 That click of awakedness which is the end. **1925** A. S. M. HUTCHINSON *One Incr. Purpose* III. xi. 280 Floods through the pulses surge of thoughts on which the eyes last night were closed, causing the leap to full awakedness.

awakenedness (ăwēᶦ·k'ndnĕs). [f. AWAKENED *ppl. a.* + -NESS.] The condition of being awakened.

1851 J. M. CAMPBELL *Christ the Bread of Life* 44 Some measure of awakenedness on the subject of religion. **1883** MEREDITH *Poems & Lyrics Joy of Earth* 126 With awakenedness of glee To feel stern joy her origin.

awakeningly (ăwēᶦ·k'niŋli), *adv.* [f. AWAKENING *ppl. a.* + -LY².] In an awakening manner; so as to awaken.

1855 in *Imp. Dict.* Suppl. **1865** MEREDITH *R. Fleming* III. xi. 176 'Hoy, Gammon!' he sang out, awakeningly to ordinary ears. **1866** —— *Vittoria* (1867) II. xx. 20 The intentness sent that last falling. . note travelling awakeningly through their minds.

award, *sb.* Add: **2. b.** Something conferred as a reward for merit; a prize, reward, honour.

1854 *Poultry Chron.* I. 622/2 Game (cock and two hens), £1, three entries. No award for want of merit. **1930** *Jrnl. Educ.* Sept. 720 Scholarship awards. . . The following awards have been made to candidates who have been approved for admission to a college. **1941** [see *ACADEMY 7]. **1948** *Ann. Reg.* 1947 471 For the third time since the Nobel Prizes were instituted the awards for both Physics and Chemistry have been given to British men of science.

awareness. (Earlier and later examples.)

1828 D. M. MOIR *Mansie Wauch* in *Blackw. Mag.* XXIV. 909/1 My awareness of the danger of riding in such vehicles. **1880** [see Dict.]. **1885** *Sat. Rev.* 31 Oct. 590 The most pig-headed Englishman and the most stolid German have. . an awareness (if we may be granted the word) of the existence and manners of foreign barbarians. **1904** *Daily Chron.* 16 Sept. 3/4 He has the peculiar 'awareness' of a soundly cultured Catholic. **1915** C. C. MARTINDALE *In God's Army* 125 An ultimate awareness that right is right. **1935** W. S. MAUGHAM *Don Fernando* xi. 229 The mystical experience is an awareness of a greater significance in the universe.

away, *adv.* Add: **11.** *spec.* In reference to games or matches played away from the home ground. Hence as *adj.*; also as quasi-*sb.*, a win away from home.

1893 *Abingdon Sch. Football Club Fixture Card*, Oct. 18. St. John's School. Away. **1907** *Daily Chron.* 14 Jan. 9/1 Thirteen drawn games and three away wins. **1923** *Daily Mail* 11 Jan. 9 West Ham have played so well in recent way games. **1939** M. DICKENS *One Pair of Hands* iii. 40. He was a pools maniac. . . 'When I won shixteen poundsh by a lucky shot with me four awaysh' was an anecdote I never got tired of hearing. **1962** P. VAN GREENAWAY *Crucified City* x. 107 Men dreamed opium dreams spun out of eight draws or four aways. **1968** *Sunday Express* 3 Mar. 31/4 No claims are wanted on the Eight Aways Treble Chance.

b. Used with intensive force, chiefly with advs., as *away back, down, up*, etc., = far. orig. and chiefly *U.S.* Cf. WAY *adv.* 2 b.

1818 J. PALMER *Jrnl. Trav. U.S.* 130 Perhaps away up in Canada. **1825** NEAL *Bro. Jonathan* III. 145 A. . he-yankee, from 'away down east'. *Ibid.* II. 78 The family were in a bustle, with some news about a rising 'away up, in the back settlements'. **1882** SWEET & KNOX *Texas Siftings* 45 Lawler. . shot a deer, away back in 1840, on the spot where the capitol now stands. **1895** G. H. HASWELL *Maister* i. 35 The 'Kelers' of Tynemouth were a recognised class away back in the days of the early charters. **1903** *Sun* (N.Y.) 26 Nov. 5 Turkeys are away up in price. **1906** *N.Y. Even. Post* 28 Apr., Manufacturers of all good cars are away behind in their deliveries. **1910** W. M. RAINE *B. O'Connor* 217 That . . pony in front belongs to sheriff Forbes, or I'm away wrong. **1935** *Punch* 25 Sept. 342/3, I can remember your father away back in eighty-five, long before you were born. **1958** *Times Rev. Industry* Aug. 85/1 Large towns like Liverpool. . are away down in the list [of bankruptcies].

awe, *sb.*¹ Add: **III. a.** *awe-compelling, -inspiring* (earlier example).

1880 'MARK TWAIN' *Tramp Abroad* II. xliii. 151 This awe-compelling miracle. **1889** J. H. SKRINE *Mem. E. Thring* xi. 275 The stern 'Thou shalt'. . of the law, in those awe-compelling tones. **1923** J. W. HARVEY tr. *Otto's Idea of Holy* xi. 85 'The mysterious' itself in its dual character as awe-compelling yet all-attracting. **1814** WORDSWORTH *Excursion* IV. 887 The simple Shepherd's awe-inspiring God.

aweto (ăwēᶦ·to, āwī·to). [Maori.] The vegetable caterpillar of New Zealand, consisting of a fungus which fastens upon caterpillars and mummifies them; dried and burnt it produces a pigment.

1889 [see *vegetable caterpillar* s.v. VEGETABLE *a.* 7]. **1896** A. B. JONES in *Pearson's Mag.* Sept. 290 The dye. . was a solution of burnt or powdered resin, or wood, or the aweto.

awful, *a.* Add: **4.** (Earlier examples.)

1809 FESSENDEN *Pills Polit., Poet. & Philos.* 2, I fear our . . nation Is in an awful situation. **1816** PICKERING *Vocab.* 42 *Awful*, Disagreeable, ugly. *New England.* In New England many people would call a disagreeable medicine, *awful*; an ugly woman, an awful looking woman. . . This word, however, is never used except in conversation, and is far from being so common in the *sense* as it was some years ago. **1818** KEATS *Let.* 27 Apr. (1958) I. 273 It is an awful while since you have heard from me.

b. As *adv.* = AWFULLY *adv.* 3.

1818 J. PALMER *Jrnl. Trav. U.S.* 131 [It is] awful hot. **1832** J. ROMILLY *Cambr. Diary* 24 June (1967) 17 An awful bad sermon from Hudleston. **1848** BARTLETT *Dict. Amer.* s.v., We not unfrequently hear such expressions as 'an awful cold day'. **1865** *Punch* 11 Mar. 103/1 We sailed all about the horizon Seeing sights as was awful surprisin'. **1866** A. TROLLOPE *Belton Estate* III. x. 261 It is awful lonely here, too. **1876** 'MARK TWAIN' *Tom Sawyer* ii. 17 You see, Aunt Polly's awful particular about this fence. **1923** R. D. PAINE *Comr. Rolling Ocean* iii. 39 A prairie town called Follansbee that looks awful good to me.

awheel (ā(h)wī·l), *adv.* [f. A *prep.*¹ + WHEEL *sb.*] On wheels; on a bicycle, bicycling. Cf. A-WHEELS *advb. phr.*

1887 G. B. THAYER (*title*) Pedal and Path. Across the Continent Awheel and Afoot. **1888** *Pall Mall Gaz.* 25 Oct. 5, I have looked forward to my annual trips awheel. . with . . delight. **1944** *Return to Attack* (Army Board, N.Z.) 12/2 Square mile after square mile of clanking armoured vehicles, rumbling lorries, bouncing guns and limbers, and roaring motor cycles. . . This huge force was completely awheel.

awkward, *a.* Add: **4. b.** Phr. *awkward squad*: see SQUAD *sb.*¹ 1 b.

7. c. Phr. *the awkward age*: the time of life when one is no longer a child and yet not properly grown up. (Cf. F. *l'âge ingrat.*)

1895 *Windsor Mag.* I. 705/1 She was. . at what ladies call 'the awkward age'. **1928** *Daily Mirror* 7 Dec. 11/2 'How old are you, Bobbie?' 'I'm just at the awkward age.' 'What do you call the awkward age?' 'I'm too old to cry and too young to swear.'

awn (ǭn), *v.*³ [Back-formation from AWNING.] **a.** *intr.* To hang as or like an awning. **b.** *trans.* To cover or shelter with an awning. Said also of the awning itself. Cf. AWNED *ppl. a.*² So **aw·ning** *ppl. a.*

a **1839** GALT *Demon Destiny* (1840) VII. 48 The awning clouds were as a cavern's ceil. **1844** THACKERAY *Wks.* (1900) XIII. 213 Trafalgar Square is to be awned in. **1890** *Longm. Mag.* Apr. 630 Trust Mrs. Robson for seeing that her guests are well awned on a night like this. **1904** *Daily Chron.* 8 July 7/3 The ball rested. . on the canvas that goes to awn the first floor.

awner (ǭ·nəɹ). [f. AWN *v.*¹ + -ER¹.] One who or that which awns; *spec.* a machine for removing the awns from grain.

1881 in *Imp. Dict.* **1884** J. SCOTT *Barn Implements & Machines* x. 101 Awners or Hummelling Machines. . are for the purpose of removing the awns from the barley, and are indispensable on a barley-growing farm. **1930** *Engineering* 18 July 83/1 The separated grain in the first dresser is delivered into an elevator, which feeds either into a barley awner or a distributing trough, as preferred.

axe, *sb.*¹ (The spelling *ax* is now disused in Britain.) Add: **1. c.** *the axe* (fig.): the cutting down of expenditure in the public services. Also in other extended uses, esp. the dismissal of an employee. Hence **axe** *v.*¹ *trans.*, to remove (officials, etc.) to save expenditure; to cut down (expenditure, etc.) by means of 'the axe'.

1922 *Glasgow Herald* 5 Oct. 7 Another class of military officers for some of whom assistance. . may be needed are those who are the unhappy victims of the Geddes economy 'axe'. *Ibid.*, The 'axed' officer. **1923** *Hansard Commons* CLXI. 1832 Fifteen hundred officers have been axed under the Geddes recommendation. *Ibid.* 1852 It is partly due to axing. **1923** *Times* 16 Mar. 12/1 Army and the 'axe'. Limit of safety reached. . . No fewer than 1,500 officers had fallen before the Geddes axe. **1923** *Daily Mail* 28 June 10 Lord Inchcape, the chairman of the Committee which recently 'axed' the expenditure of the Services in India. **1958** *Economist* 1 Nov. 390/1 Capital formation should never again become the first candidate for the axe when times for restraint recur.

5. *to have axes to grind*: now more commonly *to have an axe to grind.* (No longer confined to the U.S. or to politics.)

1815, 1881 [in Dict.]. **1922** JOYCE *Ulysses* 624 Skin-the-Goat. . evidently with an axe to grind, was airing his grievances. **1939** G. B. SHAW *Geneva* II. p. 30 Distinguished statesmen of different nations. . each with a national axe to grind.

6. *Comb.* **a.** *axe-edge* (example). **c. axe-adze,** a double-bladed tool, consisting of an axe and adze combined; **axe-breaker** *Austral.*, name given to a tree *Notelæa longifolia*; **axe-grinding** *vbl. sb.*, having private ends to serve (see sense *5); also *axe-grinder*; **axe-hammer,** a tool consisting of an axe and hammer combined; cf. *hammer-axe.*

1925 V. G. CHILDE *Dawn Europ. Civilization* ii. 34 From Early Minoan II the Cretans knew a curious implement with one blade parallel and the other at right angles to the shaft, called an 'axe-adze'. **1928** C. DAWSON *Age of Gods* xii. 268 The type known as an 'axe-adze', with its blades set transversely to one another, which is probably the model of the Nordic battle-axe. **1884** A. NILSON *Timber Trees N.S.W.* 133 Axebreaker. **1889** J. H. MAIDEN *Useful Native Plants* 579 (Morris), Axe-breaker. Wood hard, close-grained and firm. **1865** SWINBURNE *Chastelard* v. ii. 202 Suppose my mouth The axe-edge to bite so sweet a throat in twain. **1884** G. DOLBY *Dickens* viii. 227 Willard's Hotel. . with its *clientèle* of bar-loafers, swaggerers, drunkards, and 'axe-grinders' (a class of politician peculiar to Washington hotels). **1909** WARE *Passing English* 12/2 *Axe-grinders* (American). Men who grumble, especially political. **1916** E. V. LUCAS *Vermilion Box* lii. 57, I have known many journalists, but hardly one who was not either a cynic or an axe-grinder, or both. **1865** SALA *My Diary* I. 421 'Axe-grinding' is a term borrowed from one of the most charming stories told by the great apologist of shrewd common-sense, Benjamin Franklin. **1942** *Sphere* 27 June 409/2 Criticisms in the House of Commons and the newspapers are based largely on ignorance or axe-grinding. **1681** *New Castle Court Rec.* 476 (D.A.E.), Hee would beat him out with the ax hammer. **1928** PEAKE & FLEURE *Steppe & Sown* ii. 20 They used as their distinctive weapon a perforated axe-hammer of stone. **1950** H. L. LORIMER *Homer & Monum.* iv. 122 The bronze axe-hammer from Delphi may also come thence.

axel (æ·ksĕl). *Skating.* Also **Axel, Axel Paulsen.** [f. the name of a Norwegian skater, *Axel* Rudolf Paulsen (1855–1938).] A jumping movement in skating (see quot. 1930).

1930 T. D. RICHARDSON *Mod. Figure Skating* xv. 126 An excellent figure skater, who could do an 'Axel Paulsen' jump before he could do threes to a centre. *Ibid.* 136 The ever popular half Axel or Salchow jump. It consists of a leap from an inside back edge to an outside back, and is really a loop jump on two feet. **1959** *Times* 12 Mar. 4/4 Miss J. R. H. . . gave a brilliant display, with high axels and salchows off each foot.

axeless (æ·kslĕs), *a.* [f. AXE *sb.*¹ + -LESS.] Without an axe; having no axe.

1895 *Athenæum* 19 Jan. 77/3 The rocks were 'glazed in many places with a thin coating of ice', which he had to hammer off with stones. For the man was all the time axeless. **1938** *Antiquity* XII. 382 The axeless microlithic culture.

ax(e)-man. 1. (Earlier U.S. examples.)

1671 in *Essex Inst. Hist. Coll.* (1883) XX. 145 The time of meeting for ax men is to be by the sun half an hour high. **1777** *Maryland Jrnl.* 18 Mar. (Th.), Five hundred good carpenters, with ax-men and sawyers in proportion. **1785** A. ELLICOTT in C. V. Mathews *Life & Lett.* (1908) 45 From a Hill where our Axe-men are now Encamped we have a most beautiful Prospect.

axemanship (æ·ksmănʃip). [f. AX(E)-MAN + -SHIP 2 b.] The ability of an axeman; skill in handling an axe. Also *fig.*; cf. *AXE *sb.*¹ 1 c.

1893 *Field* 28 Jan. 117/3 Axemanship is so universal in Tasmania. . that it is said even the bank managers. . have usually one or two toes missing. **1961** *Times* 9 Feb. 5/6 When the Minister. . was appointed to a great spending department it was his axemanship that was needed.

axenic (ēᶦze·nik), *a. Biol.* [f. A- 14 (Gr. ἀ- priv.) + ξενικ-ός alien, strange.] (See quot. 1942.)

1942 BAKER & FERGUSON in *Proc. Soc. Exper. Biol. & Med.* LI. 116/1 The adjective *axenic* is introduced to denote a living organism that is free from all other demonstrable organisms. . . An axenic organism, as here defined, is a species free from any life apart from that produced by its own protoplasm. **1955** *Proc. Helminthological Soc. Washington* XXII. 49 A number of soil-dwelling nematodes, particularly of the genus *Rhabditis*, have been cultured under axenic conditions.

axial, *a.* Add: **4.** *Comb.* **axial flow**: usu. *attrib.* (with hyphen) (see quots.); **axial gradient** *Zool.*, the gradual change in the intensity of metabolism along any axis of a living organism.

1889 G. R. Bodmer *Hydraulic Motors: Turbines* ii. 33 The mixed-flow has been developed out of the inward-flow turbine by continuing the vanes into that part of the wheel where the water enters in the direction of the axis, so that the radial flow is changed into an axial flow. *Ibid.* 43 Axial-Flow Girard Turbine.. has ventilated buckets.. widened at the lower end, where the outflow takes place. **1910** *Hawkins' Electr. Dict.* 25/2 *Axial Flow*, a term applied to that class of turbine in which the fluids passes through the motor in a direction parallel to its axis.. Also termed *parallel flow*. **1942** *Flight* 9 Oct. 342/2 The combustion gases, with an excess of air, are then expanded through an axial-flow gas turbine which drives the main air compressor. **1949** *Gloss. Aeronaut. Terms (B.S.I.)* ii. 14 *Axial-flow compressor*, a compressor which functions by the action of alternate rows of fixed and rotating blades, radially mounted; the direction of flow through the compressor being axial. *Ibid.* 17 *Axial-flow turbine*, a turbine through which the general direction of flow is axial. **1911** C. M. Child in *Jrnl. Exper. Zool.* XI. 214 These facts..show very clearly that an axial gradient exists in a large number of organisms and that in many cases at least the apical or anterior region is dominant in regulation. **1926** Axial gradient [see *axiate a.].

axiate (æ·ksiēt), *a.* *Zool.* [f. L. *axi-s* Axis¹ + -ate².] = Axial *a.*

1926 L. T. Hogben *Compar. Physiol.* xii. 203 The morphological differentiation of parts (axiate pattern) with reference to a given axis is preceded by the appearance of a gradient of physiological activity (axial gradient) along this axis. **1949** Koestler *Insight & Outlook* xi. 162 In higher organisms we find axiate gradients with the growing tip..as the dominant integrative centre.

axillant (æksi·lănt), *a.* *Bot.* [f. Axilla + -ant¹.] Of or growing from the axil.

1883 *Encycl. Brit.* XVI. 641/2 A bud with its axillant leaf. **1899** *Nature* 15 June 149/1 Buds are protected by means of developments of the axillant leaf.

axiology (æksi,ǫ·lŏdʒi). *Philos.* [ad. F. *axiologie*, f. Gr. ἀξία worth, value + -ology.] The theory of value. Hence **axiological** *a.*, of or pertaining to or of the nature of axiology; **axiologically** *adv.*, according to axiological theory; **axiologist**, one who considers or treats of axiology.

[**1902** P. Lapie *Logique de la volonté* v. 385 La science de la valeur et la science du bonheur: l'Axiologie et l'Eudémonologie.] **1907** W. M. Urban in *Psychol. Rev.* XIV. 5 The term axiological (constructed on the analogy of the term *epistemological*), is here used to distinguish the problem of validity or evaluation of worth predicates from the psychological problem of their description and genesis. **1908** J. M. Baldwin *Thought & Things* II. xiv. 381 In the further reading of the context preferentially and so to speak 'axiologically'. *Ibid.*, 'Axiology' is a term suggested, I think, by Professor W. M. Urban for the science of *worth-meanings* as predicates, as contrasted with 'epistemology' which treats of predicates of fact. **1926** J. Laird *Study Moral Theory* 33 Pure axiology, the science of values of every species. **1932** A. Wolf in W. Rose *Outl. Mod. Knowl.* xiii. 547 The first axiological problem is: How many and what ultimate values are there? **1935** *Mind* XLIV. 295 There are..in the writings of all Objectivist axiologists chapters ostensibly directed to 'proving' some selected list of 'intrinsic' values. **1954** *Sc. Jrnl. Theol.* VII. 168 In a Biblical theology there is bound to be an axiological element; the judgement of value cannot be excluded.

axiomata media (æksi,ǫ·măta mī·diă), *phr.* *Philos.* [mod.L. (Bacon *Nov. Org.* (1620) I. xix. p. 51), f. pl. of Gr. ἀξίωμα Axiom + L. *medium* Medium; cf. Axiom 1 b.] (See quot. 1934.)

1843 J. S. Mill in *Monthly Repos.* VII. 669. **1843** —— *Syst. Logic* II. vi. v. 524 The principles of Ethology are properly the middle principles, the *axiomata media* (as Bacon would have said) of the science of mind...Bacon has judiciously observed that the *axiomata media* of every science principally constitute its value. **1934** Webster, *Axiomata media*, middle principles; those general principles which are above simple empirical laws, yet inferior to the highest generalizations, or to such as are taken to be fundamental. **1937** A. H. Murray *Philos. J. Ward* vii. 154 Philosophy with *axiomata media*.. It is only by such middle principles or mediating principles that it can hope to arrive at a view of the world in which all the knowledges of the various sciences will form a unified whole.

axiomatic, *a.* Add: **B.** *sb. pl.* A body of axioms; the study or use of axioms.

1927 E. T. Whittaker in *Rep. Brit. Assoc. Advancem. Sci.* 21 What we want to do, then, is to set forth the axiomatics of general relativity in the same form as we have been accustomed to give to the axiomatics of any other kind of geometry. **1959** K. R. Popper *Logic Sci. Disc.* 330 From the formal point of view of 'axiomatics', probability can be described as a two-termed functor.

axiomatize (æksi,ǫ·mătaiz), *v.* [f. Axiomat-(ic *a.* + -ize.] **1.** *intr.* To make an axiom or axioms.

1716 M. Davies *Athen. Brit.* II. i. 12 Here, the same excellent Historian Axiomatize's very opportunely.

2. *trans.* To render axiomatic, to give an axiomatic form to; to reduce to a system of axioms.

1934 in Webster. **1943** *Mind* LII. 221 Woodger's theory is axiomatised and therefore it presents us with the answer. **1956** *Trans. Philol. Soc. 1955* 97 The attempts that are made to axiomatize linguistics. **1959** K. R. Popper *Logic Sci. Disc.* iii. 72 A so-called 'axiomatized system'.

Hence **a:xiomatiza·tion.**

1931 *Jrnl. Philos.* XXVIII. 183 The Russell-Whitehead axiomatization of logic is a practical convenience. **1956** *Trans. Philol. Soc. 1955* 94 Even in mathematics the possibilities of complete axiomatization have been over-estimated.

axis¹. Add: **4. b.** *fig.* The relation between countries regarded as a common pivot on which they revolve; esp. the political association of 1936 (becoming in 1939 a military alliance) formed between Italy and Germany; later extended to that between Germany, Italy, and Japan; still later to that between other allied countries. Often used *attrib.*, as *Axis forces, powers*, and *ellipt.* for such phrases, with consequent pl. concord. Also *transf.*, of any comparable association, or connecting common interest.

1936 *Times* 3 Nov. 15/1 The 'Rome-Berlin axis' is a concept which has its momentary attractions. **1938** *New Statesman* 19 Feb. 274/2 He [sc. King Carol of Rumania] will as before try to keep one foot in the Franco-Russian camp and the other in that of the Axis Powers. **1939** *Times* 6 Feb. 11/1 (*heading*) Attacks on the Axis. **1941** *Time* 30 Apr. 34/3 The London-Washington Axis. **1942** *Times Weekly* 3 Jan. 5 The 26 anti-Axis nations are united..in the broad conduct of the war. **1952** *Economist* 19 July 145/1 The Moscow-Peking axis. *Ibid.* 22 Nov. 531/1 One (proposal) is for an Old Vic–Stratford axis, with an exchange of productions. **1959** *New Statesman* 7 Feb. 177/1 The term 'axis' is looked on with disfavour here [sc. Bonn] as a reminder of the Berlin–Rome–Tokyo axis of the Nazis.

axite (æ·ksait). [f. *-ax* in *bonax, primax*, names of sporting cartridges manufactured by Messrs. Kynoch (I.C.I. Metals Ltd.) + *-ite* of Cordite.] A smokeless powder for sporting rifles, composed of strip cordite in which a little of the gun-cotton is replaced by potassium nitrate.

1904 in *Trade Mark Jrnl.* 19 Oct. **1905** *Kynoch Jrnl.* Jan.–Mar. 86 'Axite' Service cartridge. **1929** E. de B. Barnett *Explosives* 78 Axite..is practically Cordite M.D. to which 2 per cent. of potassium nitrate has been added. **1941** *Hickie Mag.* Dec. 802/1 A high-velocity axite charge.

axle². Add: **axle-counter** (see quot.); hence *axle-counting* vbl. sb.; **axle-grease**, (*a*) grease suitable for lubricating axles; also *fig.*; (*b*) *slang*, butter.

1936 *Gloss. Terms Railw. Signalling (B.S.I.)* 7 *Axle counter*, a device for counting the number of axles of the vehicles comprising a train passing a given point. **1936** *Economist* 25 Jan. 178/2 Security against collisions of this kind can only be provided by a combination of track-circuiting (or the newer device of axle-counting) and automatic train control. **1878** *Rep. Indian Affairs* 366 (D.A.) Class 10. Miscellaneous articles.. Axle-grease—dozen boxes. **1900** A. Dillon *Greek Kalends* 31 Oh, Comedy! Laughter gives axle-grease to the wheel o' the sun. **1919** Downing *Digger Dial.* 9 *Axle-grease*, butter.

Axminster (æ·ksminstəɹ). [The name of a town in Devonshire.] Used *attrib.* in *Axminster carpet* or *rug*, a seamless carpet of a type originally manufactured at Axminster, noted for its thick and soft pile resembling that of a Turkey carpet. Also used *absol.*

1818 *Public Ledger & Daily Advertiser* 1 Apr. 1/4 Capital Axminster, Turkey, Brussels, and Kidderminster carpets. **1845** Disraeli *Sybil* I. ii. v. 97 The Axminster carpets..imparted even to this palatian chamber a lively and habitable air. **1886** J. D. Champlin *Chronicle of Coach* xii. 179 Hand-made Axminsters are still manufactured at Wilton, England, and at Tourl[coing], France. **1963** *Which?* Mar. 69/1 Axminsters..are made on looms which also put the pile tufts into position at the same time as the backing is woven... Axminster carpets always have a cut pile. **1968** D. Halliday *Dolly & Singing Bird* ix. 100 Plodding over the Axminsters to arrive at a staircase.

axo- (æ·kso). Short comb. form of Gr. ἄξω-ν axis, used in words in several scientific fields, as *axoneme, Axophyte, etc.

axon (æ·ksǫn). *Anat.* Also **axone**, pl. **axones**. [ad. Gr. ἄξων axis.] † **a.** The body axis. *Obs.* **b.** A filamentous process of a nerve cell carrying outgoing nerve impulses, usually single and often very long, in contrast to the multiple, branched, and short incoming fibres or dendrites: the essential constituent of most nerves. Hence **axo·nic** *a.*

1842 Dunglison *Dict. Med. Sci.* (ed. 3), *Axon*, axis. **1884** Wilder & Gage *Anat. Techn.* 22 *Axon*, the mesal, longitudinal skeletal axis, represented in Branchiostoma and embryos by a membrano-gelatinous notochord. **1899** L. Hill *Human Physiol.* xxix. 337 Each [nerve-]fibre consists of a soft central strand of protoplasmic substance called the axon or axis cylinder. **1902** *Encycl. Brit.* XXV. 394/1 A neurone consists of three parts..(2) A single filament or axon, starting from the perikaryon. **1939** *Nature* 23 Sept. 557/2 Such an axonic fusion is a most unusual feature in animals. **1940** R. S. Woodworth *Psychol.* (ed. 12) viii. 250 Most nerve cells have two kinds of branches, a single axon and many dendrites.

axoneme (æ·ksonīm). [f. *axo- + Gr. νῆμα thread.] One of the threads or strands in the central portion of the contractile stalk of Vorticellidæ.

1901 G. N. Calkins *Protozoa* 179 The strand has three threads which Entz calls *spasmoneme, spironeme*, and *axoneme*.

axonometric (æ:ksŏno,me·trik), *a.* [f. Axonometry + -ic.] Of or pertaining to axonometry (esp. sense *b).

1908 *R. Soc. Catal. Scient. Papers Subj. Index* I. Pure Math. 473/2 (*title*) Orthogonal axonometric foreshortened circle. **1938** *Times* 24 Jan. 19/3 Important riverside buildings, such as the Palace of Westminster, and bridges..shown in what architects call 'axonometric' projection. **1956** W. F. Grimes in R. L. S. Bruce-Mitford *Recent Archæol. Excav. in Brit.* 121 (*caption*) Axonometric reconstruction of a section at Gutter Lane, Cheapside, showing the relationship of Roman deposit to medieval pits and other later features.

axonometry. Add: **b.** *spec.* in *Geom.* 'The art of making a perspective representation of figures when the coördinates of points in them are given' (*Cent. Dict.* 1889).

axonost (æ·ksonǫst). *Ichthyol.* [f. Gr. ἄξων + ὀστ-έον bone.] In fishes, one of the interspinal bones; the basal portion of a fin-ray.

1887 E. D. Cope in *Amer. Naturalist* XXI. 1017, I first used the terms axial and basilar for these segments... Those who prefer a substantive appellation for these elements may call them *axonosts* and *basosts*... The axonost is sometimes represented by more than one segment,—*e.g.*, Dipnoi. **1898** A. S. Woodward *Outl. Vertebrate Palæont.* 21 Median fins..rods of cartilage, each segmented into three pieces... The outer (distal) element is termed a *baseost*, the inner (proximal) element an *axonost*.

axoplasm (æ·ksoplæz'm), *Anat.* [f. *axo- + Gr. πλάσμα (cf. Plasm 4).] The substance that surrounds the fibrils of an axon.

1900 in Dorland *Med. Dict.* **1937** *Nature* 13 Feb. 302/2 If a giant nerve fibre (of a cephalopod) is cut, the axoplasm flows out from the sheath, showing it to be a viscous fluid. **1963** *New Scientist* 25 Apr. 188 The deeper white matter [of the brain] consists of innumerable electrical conducting fibres—semifluid threads of 'axoplasm'.

axopodium (æksopō·diəm). *Zool.* Pl. **-dia**. [mod.L., f. *axo- + *-podium* Podium.] A pseudopodium, stiffened by an axial filament, found in some Heliozoa and Radiolaria.

1893 in *Funk's Standard Dict.* **1912** E. A. Minchin *Introd. Study Protozoa* v. 46 Pseudopodia of the radiate type are generally supported by an axial rod, a secreted structure of firm, elastic nature, and are hence known as axopodia. **1959** *Chambers's Encycl.* XI. 285/2 It has been suggested that axopodia are in origin flagella that have lost the power of rhythmic contraction.

axostyle (æ·ksostail). *Zool.* [f. *axo- + Gr. στῦλ-ος column.] 'A slender flexible rod of organic substance forming a supporting axis for the body of many Flagellates' (I. F. & W. D. Henderson, *Dict. Sci. Terms*, 1920).

1964 M. Hynes *Med. Bacteriol.* (ed. 8) xxviii. 433 The body of the parasite [*Trichomonas vaginalis*] contains an axostyle (stiffening rod) which projects like a tail spine. **1965** *New Scientist* 15 July 154/2 Some single-celled animals contain a large ribbon-shaped 'axostyle', running from one end of the cell to the other, which can bend vigorously.

Aylesbury (ē¹·lzbəri), the name of a town and vale in Buckinghamshire, used *attrib.* or *absol.* to designate a breed of white domestic ducks.

1854 *Poultry Chron.* I. 553/2 The Aylesbury ducks were good. **1855** *Ibid.* III. 394/1 If Ducks are kept, the Rouen or Aylesbury will answer best. **1916** A. Bennett *These Twain* ii. xiv. 264 A few white Aylesburys with gamboge beaks that intermittently quacked. **1960** *Farmer & Stockbreeder* 15 Mar. 133/1 The Aylesbury ducklings will require brooder heat for 10–14 days at 95 deg. F.

Aymara (ai·mārā). [Bolivian Sp.; cf. Sp. *aimará*.] (A member of) an Indian people mainly inhabiting the plateau lands of Bolivia and Peru near Lake Titicaca. Also, the language of this people. Hence **Aymara·n** *a.*

1860 W. Bollaert *Antiquarian Researches* 166 The Quichua is spoken from the equator to 18° S. (interrupted at 15° S. by the Aymara people)... The Aymara race...from 15° to 20° S., round Lake Titicaca. *Ibid.* 217 Aymará, spoken in Bolivia...and in Southern Peru bears a close resemblance to Quichua. **1902** *Encycl. Brit.* XXV. 374/1 The [linguistic] families of..South America are here given... Aymaran, Peru [etc.]. **1921** *Glasgow Herald* 17 May 3/8 The descendants of the original peoples of the Inca Empire —the Quechuas and the Aymaras. **1954** Pei & Gaynor *Dict. Linguistics* 24 *Aymará*, a family of 11 South American Indian languages, spoken by about 500,000 persons in Ecuador and Bolivia.

ayr(e, var. *air *sb.*¹

Ayurvedic (ā:yūrvē¹·dik), *a.* Also **ayurvedic.** [f. Skr. *āyur-vēda* the science of life or medicine (traditionally regarded as ancillary to the *Atharva-veda*: see Veda) + -ic.] Of or pertaining to the traditional Hindu science of medicine.

1917 *Antiseptic* (Madras) XIV. 389 (*heading*) How to improve Ayurvedic medicine. **1922** *Blackw. Mag.* July 96/2 *Hakims*, who had a smattering of the Yunani or the Ayurvedic system of medicine. **1937** *Discovery* May 142/2 Coconut roots are also employed in ayurvedic medicine in Ceylon and India. **1960** Koestler *Lotus & Robot* I. ii. 86 Every posture and exercise has both a symbolic meaning and a physiological purpose related to the tenets of ayurvedic medicine.

azaleine (ăzē¹·li,in). *Chem.* Also **-in.** [f. Azalea + -ine⁵.] = Rosaniline.

1889 in *Cent. Dict.* **1894** Gould *Dict. Med.*, *Azalein*.., same as *Rosanilin*. **1900** G. Iles *Flame, Electricity & Camera* 283 The dyes at present used in the preparation of orthochromatic plates are chiefly eosin,..azaleine, and croculeine.

Azan (ăzā·n). [Arab. *adān* invitation.] The Moslem call to public prayers, made by the crier from the minaret of a mosque.

1855 R. F. Burton *El-Medinah* II. xvii. 142 Here he prayed, hearkening to the Azan, or devotion-call, from the roof. **1922** *Blackw. Mag.* Dec. 821/2, I was wakened by the sound of the Azan from the village mosque close by.

Azande, pl. of *Zande.

azelaic (æzilē¹·ik), *a.* *Chem.* [f. Az(o- + Gr. ἔλαιον oil + -ic.] In *azelaic acid* = Anchoic *acid*.

1838 *Brit. Annual 1839* 350 Azelaic acid. **1885** Remsen *Org. Chem.* 142 Azelaic acid, $C_7H_{14}(CO_2H)_2$. **1929** *Jrnl. Chem. Soc.* 2709 The nine-carbon member of the series, azelaic acid, occurs in two crystalline modifications.

azeotrope (ăzī·ŏtrōᵘp). *Chem.* [f. Gr. ἀ- priv. (A- 14) + *zeo-* combining form f. Gr. ζεῖν to boil + *-trope* (cf. Gr. τρόπος turning).] A mixture of liquids the boiling-point of which does not change on distillation. So **azeotro·pic** *a.*; **azeo·tropism, azeo·tropy**, the stage of being an azeotrope.

1911 *Jrnl. Chem. Soc.* XCIX. 998 (*title*) Formation of azeotropic mixtures of ethyl alcohol and water. **1934** Webster, Azeotropism..azeotropy. **1936** *Nature* 22 Aug. 307/2 Mixtures which are 'azeotropic', because they form mixtures of minimum (or more rarely, maximum) boiling point. **1938** S. T. Bowden *Phase Rule & Phase Reactions* vi. 118 Ordinary ethyl alcohol and water..give rise to an azeotrope containing 96 per cent of alcohol. **1947** *Endeavour* VI. 14/2 It was in Brussels too that Lecat started his chemical researches..in the field of azeotropism.

Azerbaijani (æzəɹbaidʒā·ni). [f. *Azerbaijan*, name of a region falling partly in Iran and partly in the U.S.S.R.] One of a Turkic-speaking people in Azerbaijan; also, their language. Also **Azerbaija·n** *a.* and *sb.*, (of or pertaining to) a native or inhabitant of Azerbaijan.

1888 *Encycl. Brit.* XXIII. 661/2 Adarbaijāni Turkish. *Ibid.*, Turki dialects (Derbendi, Adarbaijāni, Krimmi, Anadoli, and Rumili). **1911** *Ibid.* XXVII. 470/1 The Turkish nomads scattered over Persian territory are often known by the name of *Azerbaijanis* or *Adharbaijanis*, though this name is strictly applicable only to the inhabitants of the province of Azerbaijan. **1954** Pei & Gaynor *Dict. Linguistics* 24 *Azerbaidjani*, a dialect of the Southern Turkic group of the Altaic sub-family of the Ural-Altaic family of languages. **1959** *Chambers's Encycl.* II. 16/2 The Azerbaijans, who are of Indo-Iranian origin, speak a Turkish language and are Moslem in religion.

azide (æ·zaid, ē¹-). *Chem.* [f. Az(o- + -ide.] A salt or ester of hydrazoic acid.

1907 *Chem. Abstr.* 431 (*heading*) Communication on syntheses by means of azides. **1910** *Encycl. Brit.* III. 83/1 It is soluble in water, and the solution dissolves many metals..with liberation of hydrogen and formation of salts (azoimides, azides or hydrazoates). **1946** *Nature* 3 Aug. 156/2 The azide retards the rate of oxidation of divalent manganese in soil. **1949** Miall & Miall *Dict. Chem.* (ed. 2) 61/2 The azides of the heavy metals (e.g. lead azide) are dangerously explosive.

Azilian (ăzi·liăn), *a.* *Archæol.* [f. *Azil* in Mas d'Azil (dept. of Ariège, France), where discoveries were made by Ed. Piette of a primitive civilization; cf. F. *époque asylienne*, etc. (f. med.L. *Asylum*: see *L'Anthropologie* (1895) VI. 151).] Of or belonging to the transition period between the palæolithic and neolithic ages. Also *sb.*

1899 A. H. Keane *Man: Past & Present* 30 M. Piette.. states..that '13 out of 23 Phoenician characters were equally Azilian graphic signs'. **1919** H. G. Wells *Outl. Hist.* I. x. § 4. 60/1 For a time there were in Southern Europe drifting communities of some little known people who are called the Azilians. *Ibid.*, These Azilian people have left behind them a multitude of pebbles, roughly daubed with markings. **1947** J. & C. Hawkes *Prehist. Brit.* i. 22 The Azilians, who brought from the south of France a microlithic culture of Magdalenian descent.

azine (æ·zin, -əin, -in, ē¹-). *Chem.* Also **azin.** [f. Az(o- + -ine⁵.] **a.** An organic compound containing the group =N—N=. **b.** One of a group of heterocyclic organic compounds with two, three, four, etc., nitrogen atoms in a six-atom ring (thus *diazine, triazine, tetrazine*, etc.), used esp. as dyes.

1887 *Jrnl. Chem. Soc.* LII. 383 Whilst the whole group of substances containing the typical ring [N ring N] should be termed quinoxaline bases, it will be convenient to divide them into two sub-groups. 1. The quinoxalines proper.. 2. The azines, in which the nitrogen-ring is in union with two other rings. **1893** *Ibid.* LXIV. i. 15 The synthesis of azine dyes. **1910** *Encycl. Brit.* VIII. 746/2 Basic colours derived from coal-tar:..azine green [etc.]. **1959** *Chambers's Encycl.* IV. 693/1 Azine, Oxazine, Thiazine, Acridine and Pyronine dyestuffs contain a central pyrazine or similar ring. *Ibid.* 693/2 The indulines and nigrosines are important azine type dyes.

azlactone (æzlæ·ktōᵘn). *Chem.* [f. Az(o- + Lactone.] A lactone of an unsaturated nitrogenous hydroxy-acid. Hence **azla:ctoniza·tion.**

1917 *Chem. Abstr.* 3267 Synthesis of aryl-substituted fatty acids..α-$C_{10}H_7CHO$ is converted into the *azlactone* $C_{20}H_{13}O_2N$, yellow needles. **1940** *Ibid.* 3716 The azlactonization and racemization of benzol-*l*-*p*-methoxyphenylalanine..were studied...NaOAc increases the rates of azlactonization and racemization in each case. **1949** Florey et al. *Antibiotics* I. iii. 189 Certain substances, such as ether-soluble keto acids, lactones, azlactones, and esters interfered with this method. *Ibid.* II. xxii. 863 It was ascertained that azlactonization of the..amino acid took place.

azo-. Add: **3. b. azo-colours, -dyes,** a group of coal-tar colours or dyes.

1879 Witt in *Jrnl. Chem. Soc.* XXXV. 181 The variety of azo-colours in the market, and the number of patents.. are constantly increasing. **1884** *Ibid.* XLVI. 237 The preparation of azo-dyes from the trisulphonic acids of β-naphthol. **1894** Gould *Dict. Med.*, *Azo-dyes*, a well defined group of the coal-tar colors, all containing the diatomic group —N = N—, bound on either side to a benzene radical. **1902** *Encycl. Brit.* XXVII. 559/1 Azo colours which are developed..upon the fibre itself (usually cotton) by the successive application of their constituent elements. **1963** R. R. A. Higham *Handbk. Papermaking* xii. 256 Azo colour pigments. Yellow, orange and red used in high-quality writings.

azoimide (æzo,i·məid). *Chem.* [f. Azo- + Imide.] Hydrazoic acid.

1891 *Athenæum* 24 Jan. 126/1 Prof. Curtius..has obtained the compound N_3H, to which he has given the name azoimide or hydrogen nitride. **1894** [see Hydrazoic *a.*].

azolla (ăzǫ·lă). [mod.L.; said to be f. Gr. ἄζειν to dry + ὀλλύναι to slay.] A plant of a genus of small floating ferns of the family Azollaceæ.

1872 Domett *Ranolf* xiv. 221 Minute azolla-stains of ruddiest hue.

azonal (ăzōᵘ·năl, ē¹-), *a.* [f. A- 14 + Zonal *a.*; cf. Gr. ἄζωνος zoneless.] Not confined to a zone, not arranged in zones; spec. in *Soil Sci.*, *azonal soil* (see quot. 1938).

1896 C. MacMillan in *Bull. Torrey Bot. Club* XXIII. 507 There are two principal types of plant arrangement in their habitats. These are (1) Zonal and (2) Azonal... When the topographic feature is devoid of..well-marked radial symmetry the plants dispose themselves according to the azonal type. **1938** *Soils & Men: Yearbook Agric. 1938* (U.S. Dept. Agric.) 1163 *Azonal soils*, any group of soils without well-developed profile characteristics. **1952** P. W. Richards *Trop. Rain Forest* ix. 210 In these cases the influence of the material from which the soil is formed overrides that of the climate giving rise to so-called azonal soils (in contrast to the climatic or zonal types).

azoospermia (ăzōᵘ:ospɜ̄·ımiă). *Path.* [mod.L., f. Gr. ἄζωος lifeless + σπέρμα seed, Sperm *sb.* + -ia¹; cf. *zoosperm*.] Absence of spermatozoa in the semen.

1889 in *Cent. Dict.* **1962** *Lancet* 27 Jan. 218/1 In three patients with oligospermia or azoospermia, we found a larger anomaly in length of the Y chromosome than normally established in our observations.

azoprotein (æzoprōᵘ·tīn). [G. (K. Landsteiner 1917, in *Z. Immun. Forschung* XXVI. 122), f. Azo- + Protein.] 'In immunology, a protein coupled with a diazonium compound so as to form an azo derivative' (MacNalty, *Brit. Med. Dict.*, 1961).

1918 *Chem. Abstr.* 2373 Altogether 23 kinds of immune sera were prepd. [by K. Landsteiner and H. Lampl] and their action on 33 azo-proteins was investigated. **1930** *Syst. Bacteriol.* (*Med. Res. Council*) VI. 237 Landsteiner and Lampl (1917) found..a general method for preparing a long series of 'azo-proteins' particularly suitable for the study of the relationship between chemical structure and specificity. *Ibid.* 238 Precipitating antisera were prepared by the injection of these azo-proteins into rabbits. **1934** *Biol. Abstr.* 10591 Certain diazo compounds or azoproteins greatly increase tissue permeability.

azotæmia (æzotī·miă). *Path.*, *Vet.* Also **azotemia.** [mod.L., f. *azot(o- + Gr. αἷμ-α blood + -ia¹.] The presence of excessive nitrogenous waste products in the blood; a disease of solipeds resulting from this; cf. *azoturia. Hence **azotæ·mic** *a.*

1900 in Dorland *Med. Dict.* **1933** *Chem. Abstr.* 2995 (*title*) Coefficient of Maillard and the azotemic ratio of the guinea pig during acute..phosphorus poisoning. **1940** *Chambers's Techn. Dict.* 62/2 *Azotæmic (azotemic) nephritis*, nephritis in which there is retention of nitrogenous products in the blood. **1961** *Lancet* 16 Sept. 656/1 Other tests..did not separate clearly hyperparathyroid subjects from euparathyroid patients with renal stones, especially when azotæmia was present.

‖**azotea** (aþotē·ā). [Sp.] The flat roof of a house, used as a place of resort in hot climates.

1824 J. R. Poinsett *Notes on Mexico* (1825) ii. 23 We next ascended to the *azotea*, (terrace roof), from whence we had a view of the harbour. **1844** J. Gregg *Commerce on Prairies* II. 102, I perceived the *azotea* of the parochial church occupied by armed men. **1921** *Chambers's Jrnl.* 257/2 Roses and carnations..trailing from every *azotea*.

azoted (æ·zōᵘtėd), *a.* ? *Obs.* [f. Azote + -ed².] Nitrogenized.

1848 Dunglison *Dict. Med. Sci.* (ed. 7), *Azoted*, nitrogenized. **1872** Aitken *Sci. & Pract. Med.* (ed. 6) II. 1061 In proportion as animals are fed on animal diet or on azoted substances, their urine becomes more and more loaded with lithic acid.

azoto-. A combining form of Azote, nitrogen; cf. Azo- in Dict.

azoturia (æzotyū·riă). *Path.*, *Vet.* [mod.L., f. *azot(o- + -uria.] A condition in which the urine contains an excess of nitrogenous matter; cf. *azotæmia.

1838 R. Willis *Urinary Diseases* I. i. 18 Of the discharge of urine which is characterized by a superabundance of Urea,—Azoturia. **1846** G. E. Day tr. *Simon's Animal Chem.* II. vii. 307 By the term azoturia Willis understands that form of disease which is usually known as diabetes insipidus. **1953** G. Brooke *Introd. Riding* xii. 133 Azoturia or haemoglobinuria is a disease peculiar to the horse tribe.

azoxy (æ·zǫksi), *a.* [f. Az(ote + Oxy(gen.] Defining a compound consisting of an azo-group into which an oxygen atom has been introduced.

1873 *Jrnl. Chem. Soc.* XXVI. 1028 Azoxybenzene is obtained when a solution of azobenzene in acetic acid is heated with chromic anhydride. **1894** *Rep. Brit. Assoc. Advancem. Sci.* 245 Very interesting in point of fastness to light are the azoxy colours. **1964** N. G. Clark *Mod. Org. Chem.* xx. 404 In the presence of bases, nitrosobenzene reacts with N-phenylhydroxylamine to give a 'bimolecular' compound, azoxybenzene.

Aztec (æ·ztek), *sb.* and *a.* One of a native American people first known (*c.* A.D. 1100) as inhabitants of the valleys of Mexico. Also *attrib.* or as *adj.* = **A·ztecan** *a.*

1787 Cullen tr. *Clavigero's Hist. Mexico* I. II. 112 The Aztecas or Mexicans, who were the last people who settled in Anahuac. **1814** H. M. Williams tr. *Humboldt's Researches* I. 81 The Toltecks, the Cicimecks..and the Aztecks.. notwithstanding their political divisions, spoke the same language. *Ibid.* 83 Tezcatlipoca, the first of the Azteck divinities after Teotl. **1824** W. Bullock *Six Months Residence in Mexico* xxvi. 375, I immediately prepared for making casts of the Aztec idols and sculptures. **1824** J. R. Poinsett *Notes on Mexico* (1825) vii. 108 They all talk the Azteck, or Ottomic, and the Spanish. **1886** *Science* 5 Nov. 403/2 Its circular ornamentation is not Aztecan. **1920** *Edin. Rev.* Oct. 350 In the Aztec mind, stone was symbolic of the atonement of sin. **1922** W. R. Inge *Outspoken Essays* Ser. II. 74 The half-civilised Aztecs practised human sacrifice and cannibalism. **1965** *Canad. Jrnl. Linguistics* Spring 78 Part of the Aztecan–Tanoan phylum. **1966** *Harper's Bazaar* Sept. 58 A stunning wool weave of Aztec checks.

azulejo. (Earlier and later examples.)

1845 R. Ford *Handbk. Spain* I. 225/1 The window..has been bricked up; it may be known by its border of *azulejos*. **1901** *Scribner's Mag.* XXIX. 449/1 The entire wall is encrusted with a mass of still brilliant tiles or 'azulejos'.

azulene (æ·ziŭlīn). *Chem.* [f. Sp. *azul* blue + -ene.] A liquid hydrocarbon, blue in colour, found in some volatile oils; = Cerulein.

The term is used to cover not only the parent hydrocarbon but the class of derivatives with the same basic structure, viz. a five-ring joined to a seven-ring, isomeric with naphthalene, which has two six-rings joined together.

1874 *Jrnl. Chem. Soc.* XXVII. 1 Pure oil of wormwood (obtained from Dr. Septimus Piesse) being submitted to distillation, the greater part passed over..close upon 200°, a portion of blue oily product being obtained at a higher temperature (the *azulene* of Piesse and cœrulein of Gladstone). **1880** [see Cerulein]. **1956** *Nature* 21 Jan. 145/2 In the fractionation of hydrocarbons found in Copaiba balsam, we have isolated..a compound with the typical reactions of azulenes, and which we propose to call 'carioazulene'.

azure, *sb.* and *a.* Add: **B.** *adj.* **4.** *Bookbinding.* Composed of horizontal parallel lines, as a tooled or stamped design; also applied to the tool used for making such a design (cf. sense A. 3).

1894 *Amer. Dict. Printing & Bookmaking* 31 Azure tools. **1922** E. J. Kimble *Vocational Vocabs.* (ed. 2) 24 *Bookbinding*..azure tools. **1953** Lemaitre & Thompson *Vocab. Bibliothec.* 198 Azure tooling (decoration with close parallel lines).

C. *azure-bright, -flaming, -lidded*, adjs.

1930 O. Gogarty *Wild Apples* 15 And gazed out at azure-bright pinnacled forms. **1929** E. Blunden *Near & Far* 19 And azure-flaming waves around rock-caves. **1820** Keats *Eve of St. Agnes* xxx, And still she slept an azure-lidded sleep.

azured, *ppl. a.* Restrict '*arch.* or *Obs.*' to senses in Dict. and add: *spec.* in *Bookbinding* = *azure *a.* 4. Also in Fr. form *azuré*.

1879 J. W. Zaehnsdorf *Art of Bookbinding* xxii. 102 The use of lined or azuré tools are a distinctive mark of the period. **1893** S. T. Prideaux *Hist. Sketch Bookbinding* i. 32 The azured tools were first used by Grolier for the bindings done in France, between 1530 and 1540; no azured tools are found on French bindings before that date. **1895** J. W. Zaehnsdorf *Short Hist. Bookbinding* 17 *Azured Tooling*, ornamentation outlined in gold and crossed by horizontal lines in the manner of indicating azure in heraldry. **1937** Rye & Quinn *Hist. & Armorial Bookbindings Exhib. in Univ. Libr.* 36 At the base is an azured border.

azurine, *sb.* Add: **2.** *Dyeing.* A base obtained from aniline black, giving a bluish black shade in printing; also the colour itself.

1878 *Jrnl. Chem. Soc.* XXXIV. 572 Azurine is characterised by the blue fluorescence of its solutions. **1894** Gould *Dict. Med.*, *Azurin*..Same as *Azulin*. **1902** *Encycl. Brit.* XXVII. 563/2 There are several oxidation products of aniline. The first..is..emeraldine..which by treatment with alkali yields a dark blue base called azurine.

azygospore (ăzəi·gospōᵊɹ, ăzi·g-, ē¹-). *Bot.* [f. A- 14 + Zygospore.] A parthenogenetic zygospore.

1889 in *Cent. Dict.* **1898** H. C. Porter tr. *Strasburger's Bot.* 348 Although the conjugating hyphæ meet in pairs, no fusion takes place, and their terminal cells become converted directly into spores, which are termed azygospores. **1952** *New Biol.* XIII. 106 Under certain circumstances the process of zygospore formation may be interrupted, whereupon a single gametangium may give rise to a thick-walled spore which resembles closely a typical zygospore with respect to size and appearance, and is termed an azygospore.

azyme. (Later *fig.* example.)

1930 T. S. Eliot tr. *St.-J. Perse's Anabasis* 69 Under the azyme of fine weather,..the whole feather of harvest!

B

B. Add: **II. 2.** In various specific applications, as (i) a blood-group (see quots.); (ii) a second-class road; (iii) a supporting film to the main feature in a cinema programme.

[**1910** W. L. Moss tr. Landsteiner in *Johns Hopkins Hosp. Bull.* XXI. 65/2 In a number of cases (Group A) the serum agglutinates the corpuscles of another Group (B), not, however, those of Group A.] **1921** *Brit. Jrnl. Exper. Path.* II. 33 The Landsteiner theory that two substances, 'A' and 'B', with their corresponding agglutinins, 'a' and 'b', are concerned in the isoagglutination of human bloods has been confirmed by absorption tests. **1927, 1928** [see A II. 7]. **1948** *New Biol.* V. 69 Human beings are classified into the four categories A, B, AB or O according as to whether certain A and B substances can be detected in their red blood cells (and incidentally in other tissues).

1921 *Autocar* 29 Oct. 829/2 It is not..intended to deal with the numbering of the B roads until that of the A roads has been completed. **1962** *Guardian* 14 July 5/2 The B road ..climbs up the Cotswolds from Stroud to Birdlip. **1967** M. Culpan *In Deadly Vein* iii. 41 A country lane of the twenties had now been widened into a B-road between A.5 and A.6.

1949 *Here & Now* (N.Z.) Oct. 29/2 There is a current reaction against the 'big' picture; and it is a sign of the times that the Academy, one of London's repertory cinemas, has been showing two American 'B' features. **1962** *Observer* 20 May 27/6 Just that something which distinguishes the good documentary from the drab British B picture. **1967** *Listener* 7 Sept. 316/2 Things are looking up on ITV, except on farmer-and-B-film-haunted Southern.

III. Add: b. (formerly also B), *Cricket*, bowled by (examples); B, breathalyser; so *B-test*; b, B, bugger (or bastard) (as a euphemism, sometimes printed b——); B, BB, BBB, black, double-, treble-black (of pencil lead); B.A.B.S. *Aeronaut.*, beam (or blind) approach beacon system, a system for approaching a landing field by means of instruments; B.A.L., British Anti-Lewisite, a drug (dimercaprol) developed as an antidote to Lewisite and used also to neutralize metallic poisons, e.g. arsenic; B. and S., brandy and soda (examples); B.A.O.R., British Army of the Rhine; B.B.C. (see as *separate entry); B.C.E., Before the Common Era; B.C.-G., B.C.G., Bacillus Calmette-Guérin, used as an anti-tuberculosis vaccine; also *attrib.*; B.D.S.T., British Double Summer Time; B.E.F., British Expeditionary Force; B.E.M., British Empire Medal; B.E.M., bug-eyed monster; BeV., B.E.V., bev, *U.S.* (billion electron volts): a thousand million (10^9) electron volts; B.F., bloody fool; B.H.C., benzene hexachloride; b.h.p., brake horse power; B.Lit(t)., Bachelor of Literature, Bachelor of Letters; B.M., British Museum; B.M.A., British Medical Association; B.O., body odour; B.O.P., Boy's Own Paper; B.P., (*a*) British Public, esp. in **G.B.P.*; (*b*) British Pharmacopœia: the title of a list of medicines and other preparations published under the direction of the General Medical Council; (*c*) before present, *i.e.*, counting backwards from A.D. 1800; B.Phil., (less commonly) B.Ph., Bachelor of Philosophy; B.S.(S.), British Standard (Specification); B.S.I., British Standards Institution; B.S.T., British Summer Time; (from 1968) British Standard Time; B.T.M., b.t.m. *colloq.*, bottom, posterior; B.T.U., Board of Trade unit; B.T.U., B.Th.U., British thermal unit; B.U., bread unit: a ration token exchangeable for bread, cakes, etc. See also *B-girl.

1744 in Nyren *Young Cricketer's Tutor* (1833) 111 Harris 0, B by Hadswell. **1822** in Pycroft *Cricket Field* (ed. 6, 1873) xiii. 292 Holloway, b. Beauclerk 0. **1968** *Times* 15 Mar. 15/1 S. Comacho c. Knott, b. Brown..87. **1967** *Daily Tel.* 21 Oct. 20/8 (*heading*) B-test driver fined. **1851** Mayhew *Lond. Lab.* I. 313/1 The poor b— is in 'stir' (prison). **1925** D. H. Lawrence *Let.* 17 Nov. (1962) 865 I'd have sent those Irish b's seven times to hell, before I'd have moved a single iota at their pencil stroke. **1952** N. Streatfeild *Aunt Clara* 14 Can't 'elp bein' sorry for the poor old B. *Ibid.* 161 'Poor little B's,' he thought. **1853** C. H. Weigall *Art of Drawing* Advts., p. 14 B, Black (for Shading, or free Sketching). BB, Softer ditto (for deep Shading). BBB, Intensely Black (for extra deep Shading). **1886** C. M. Yonge *Chantry House* I. vii. 63 We had filled whole drawing-books with..foliage in BBB marking pencil. **1931** *Boys' Mag.* XLV. 190/1 A 'B' pencil. **1945** *Amer. Speech* XX. 309 BABS. Blind Approach Beacon System. System for approaching landing field by radar means. **1951** *Gloss. Aeronaut. Terms (B.S.I.)* III. 28 Beam-approach beacon system, *abbr.* BABS. **1942** *PB 5545: Office of Sci. Res. & Devel., National Defense Res. Comm.* (E. I. Du Pont de Nemours & Co.) (*title of mimeographed report*) Report on the experimental manufacture and process study of BAL. **1943** *Hospital Corps Q.* (Washington, D.C.) XVI. 140 Persons required to enter areas specifically known to be contaminated with lewisite should apply BAL ointment sparingly to unprotected parts of the body. **1944** *Brit. Med. Jrnl.* 22 July 111/1 Recently B.A.L. ointment, which is a specific for arsenical vesicants and of great value in the treatment of the eyes, has been developed. **1964** S. Duke-Elder *Parsons' Dis. Eye* (ed. 14) xxvi. 379 Arsenical vesicants are neutralized by the local application of BAL ointment. *a* **1878** Whyte-Melville *Black but Comely* (1879) II. xxvii. 112 Now for a B-and-S, one quiet cigar, and then bed. **1882** *Punch* 11 Feb. 69/1 He'll nothing drink but 'B. & S.' and big magnums of 'the Boy'. **1945** *War Illustr.* 9 Nov. 421 (*caption*) A B.A.O.R. major examined the papers of Wehrmacht youths about to be demobilized at Staaken, near Berlin. **1881** K. Magnus *About Jews* i. 1 A glance at the Babylonian captivity (587–536 B.C.E.). **1915** M. Radin *Jews among Greeks & Romans* 15 The period..from the end of the Babylonian Captivity to the establishment of Christianity—roughly from about 450 B.C.E. to 350 C.E. **1926** *Lancet* 29 May 1030/1 It is this attenuated bacillus B.C.-G., or *Bacillus Calmette-Guérin* that be uses for his vaccinations of both calves and newborn infants. **1958** *Sunday Times* 23 Feb. 19/4 B.C.G. inoculation against tuberculosis. **1943** *Daily Tel.* 24 June 1/7 At 1.20 this morning (2.20 B.D.S.T.) the state of the parties was [etc.]. **1917** W. Owen *Let.* 9 Jan. (1967) 425 Please send the compass: 2 Manchester Regt. B.E.F. **1939** *War Illustr.* 7 Oct. 110/1 The men of the new B.E.F. have gone to the front with smiling faces. **1941** *Ibid.* 9 May 476/2 Mr. T. H. Newton, Home Guard, B.E.M., for securing a mine and rendering it safe. **1948** *Electronic Engin.* XX. 149 The bevatron (from B.E.V.—billion electron volts). **1964** *Sci. Amer.* July 44/3 The six-billion-electron-volt (6-bev) electron synchrotron in Cambridge, Mass. **1925** Fraser & Gibbons *Soldier & Sailor Words* 12 B.F...Ordinary contemptuous slang: *e.g.*, 'He's an out and out B.F.!' **1939** C. Day Lewis *Child of Misfortune* II. i. 137 You really are a B.F., Arthur. **1947** *Ann. Appl. Biol.* XXXIV. 347 The material used..was crude benzene hexachloride (B.H.C.). **1895** *Oxf. Univ. Calendar 1896* 72 B.Litt. or B.Sc. **1909** *Cent. Dict.* Suppl., B.Lit. **1937** *Discovery* Jan. ii/1 Geraldine Coster, B.Litt. (Oxon.). **1870** Geo. Eliot *Let.* 14 Sept. (1956) V. 116, I dare not expect to be in time for any address but the B[ritish] M[useum]. **1954** *Grove's Dict. Mus.* III. 168/2 An Indian bas-relief..dominated the grand staircase at the B.M. **1886** *Dict. Abbrevs.* 14 *B.M.A.*, British Medical Association. **1955** *Times* 2 June 3/6 The annual meeting of the British Medical Association, which opened yesterday at B.M.A. House. **1933** *Sat. Even. Post* 14 Jan. 91/3 Those 'B.O.' ads. I laughed at—is the joke on *me*? **1936** Auden & Isherwood *Ascent of F6* II. iii. 96 And some I know have got B.O.: But these are not for me. **1909** R. Brooke *Let.* 4 Sept. (1968) 176 Reading the B.O.P. **1932** Dylan Thomas *Let.* Dec. (1966) 8, I..contribute..funny verses to the B.O.P. **1939** 'G. Orwell' *Coming up for Air* II. ii. 58 Myself under the table with the B.O.P., making believe that the table-cloth is a tent. **1896** Ellen Terry *Let.* 26 Nov. in *E. Terry & Bernard Shaw* (1931) 129, I should say it would be a tremendous go with the B.P. **1963** 'A. Gilbert' *Ring for Noose* vi. 76 The dear B.P. doesn't worry its head much about proof. **1898** *Lancet* 6 Aug. 337/2 We hold it to be incumbent upon every medical man..to write..upon his recipes either the words B.P. 1898 or B.P. 1885 after each preparation. **1966** D. Francis *Flying Finish* vi. 69 A printed chemist's label on the front said 'Two hundred aspirin tablets B.P.' **1946** F. E. Zeuner *Dating Past* v. 142 Going backwards into the past, the radiation curve for the last 600,000 years..shows a series of three summer minima between 25,000 and 115,000 years B.P. **1923** F. W. Havcraft *Degrees & Hoods World's Universities & Colleges* 1 *B.Ph.*, Bachelor of Philosophy. **1942** Partridge *Dict. Abbrevs.* 19/2 *B.Phil.*, Bachelor of Philosophy. **1966** *Rep. Comm. Inquiry Univ. Oxf.* II. 22 The B.Phil. can be taken in a number of arts and social studies subjects. **1932** *Brit. Standards Inst. Handbk.* 45 List of B.S. Specifications. *Ibid.* 61 Subject Index to B.S. Specifications..B.S.S. No. *Ibid.* 1 Aims and Objects of the B.S.I. **1930** *Meteorol. Gloss.* (ed. 2) 36 *British Summer Time* (*B.S.T.*)..9 h. G.M.T. is the same as 10 h. B.S.T. **1960** *Farmer & Stockbreeder* 22 Mar. 72/1 (*headline*) Scots F.U. Oppose B.S.T. Change. **1968** *Guardian* 17 Feb. 16/6 Sunrise times under BST. **1919** R. Firbank *Valmouth* xi. 188 She made a sudden dash for my b-t-m. **1934** S. Beckett *More Pricks than Kicks* 52 The weary proletarians at rest on B.T.M. and elbow. **1946** *Lancet* 20 July 97/2 Each bread unit is equivalent to 7 oz. of bread, 5¼ oz. of flour, or 8 oz. of flour confectionery, so the scales of B.U. per week can be read as scales of bread in ounces per day.

baa, *sb.* Add: **b.** **baa-lamb** (examples). Also, a toy lamb.

1599 T. M[oufet] *Silkewormes* 3 Then hairy cloathes, and wooll from Baa-lambs tore. **1871** Mrs. Craik *Little Sunshine's Holiday* ii. 31 Little Sunshine was greatly charmed with the 'baa-lambs'. **1888** Mrs. H. Ward *R. Elsmere* III. vi. xxxix. 174 The baby..was..caressing..a woolly baa-lamb. **1906** (*title*) The baa lamb's ABC.

baardman (bā·ɹdmăn, ‖ bā·rtman). *S. Afr.* Also (occas.) **baa·rdmannetjie** (-mænətʃi, ‖-manəkyi). [Afrikaans, f. *baard* beard + *man* man.] Name given to various fishes with barbels on the lips, and/or chin, as the sea-fish *Sciæna capensis*, related to and very like the Mediterranean umbra, or the fresh-water fish *Barbus capensis*.

1853 L. Pappe *Edible Fishes Cape of G. H.* 16 *Umbrina capensis*..Baardmannetje. Snout obtuse.., lower jaw shortest with a barbel; dorsal fins distinct..Measures from 2 to 2½ feet, and is reputed for its delicious flesh. **1902** J. D. F. Gilchrist in *Trans. S. Afr. Philos. Soc.* XI. 226 Baardman, Baardmannetje (Pappe), Bellman (Riversdale). *Umbrina capensis.* **1907** *East London Dispatch* 5 Dec. (Pettman), *Baartman* (white-fish, barbel, catfish), a well-known ugly species of the family *Siluridae*. **1913** W. W. Thompson *Sea Fisheries Cape Col.* 158 Barbus capensis.. Moggel (Gouritz and Berg Rivers); Barbeel or Barm (Van Riebeeck's *Journals*); Witte-visch or White-fish; Baardman or Bartman. **1947** K. H. Barnard *Pict. Guide S. Afr. Fishes* 122 *Cape Baardman*..False Bay to Algoa Bay. **1955** *Cape Times* 19 Sept. 9/1 One of the..anglers..landed five kob and one baardman.

baas (bās). [Du.: see Boss *sb.*⁶] In S. Africa: a master, employer of labour. Often as a form of address.

In quot. 1625 used for a Dutch ship's captain.

1625 Purchas *Pilgrimes* I. II. 117 Our Baase (for so a Dutch Captaine is called). **1785** G. Forster tr. *Sparrman's Voy.* I. 55 The steward (or as they call him there, the *baas*) presented me with a glass of strong-bodied wine. **1850** R. G. Cumming *Hunter's Life S. Afr.* I. i. 5 At the door he is met by the baas, or master. *Ibid.* ix. 176 The Hottentot replying, 'Like so, baas.' **1920** *Chambers's Jrnl.* 25 Dec. 59/2 Jeri stated that he believed the *baas* to be bewitched. **1953** P. Lanham *Blanket Boy's Moon* I. iii. 27 'Gently, my baases,' he whispered through quivering lips.

baasskap (bā·skap). *S. Afr.* Also erron. **baaskap.** [Afrikaans, = domination, lit. mastership, f. prec.] Domination, esp. of white over non-white South Africans. Cf. *boss-ship.

1935 *Cape Times* 8 Nov. 8/4 Let us focus our attention on the fact that the primary consideration is whether Afrikanerdom will reach its ultimate destiny of bossism (baasskap) in South Africa. **1949** *Cape Argus* 11 June 8/3 The aim of the Broederbond is the baasskap of Afrikanerdom. **1955** *Times* 20 May 11/3 Stellenbosch is the intellectual stronghold of republicanism and the Nationalist Party. There, if anywhere, can be found disinterested advocacy of the doctrines of *apartheid* and *baasskap*. **1959** *Cape Times* 25 Apr. 7/6 The Prime Minister, Dr. Verwoerd, had previously said that the Natives should have baasskap within their own areas just as the Whites should have baasskap in their areas.

baba². Add: (Earlier and later examples.) Now esp. *rum baba, baba au rhum*, a rich cake soaked in a rum syrup.

1827 L. E. Ude *French Cook* 461 The oven must be moderately hot, as the *babas* must remain a long time in. **1846** Soyer *Gastron. Regenerator* 566 Take off the band of paper, turn the baba over upon a hair sieve, and serve either hot or cold. **1933** A. Christie *Lord Edgware Dies* xiv. 126 We had a delicious omelette, a sole, a chicken and a Baba au Rhum. **1939** M. Allingham *Mr. Campion & Others* 135 Accepting a rhum-baba. **1952** B. Nilson *Penguin Cookery Book* xxiv. 417 Rum Babas. **1958** W. Bickel tr. *Hering's Dict. Class. & Mod. Cookery* 668 *Baba au rhum*, savarin dough mixed with raisins and currants, baked in baba mould, soaked while still hot with hot syrup flavored with rum; served with rum-flavored apricot sauce. **1958** R. Godden *Greengage Summer* ix. 95 He explained the different kinds of cake to us: éclairs, rum babas, meringues. **1959** R. Postgate *Good Food Guide 1959–60* 212 A chocolate gâteau 'consisting of a baba au rhum with a milk chocolate couverture, far too rich, but delicious'.

babassu (bā:băsū·). [Pg. *babaçú*.] Either of two species of palm, *Orbignya martiana* or *O. oleifera*, found in north-eastern Brazil, producing a valuable oil. Also *attrib.*

1923 W. Howarth *Mod. Brazil* vi. 88 Recent years have ..added considerably to the number of oil-producing seeds and kernels, among the latter being the babassú. **1925** H. G. James *Brazil after Cent. of Independence* x. 349 The high commercial value of the oil of the babassú palm. **1933** P. Fleming *Brazil. Adv.* III. xii. 399 Only the *babassú* palms stood up proudly. **1960** A. E. Bender *Dict. Nutrition* 13/1 *Babassu oil*, edible oil from the Brazilian palm nut; similar to coconut oil and used in food, soap and cosmetics.

Babbit-metal. Add: Also **babbitt** (or **Babbitt**) **metal.** Sometimes used *ellipt.* Hence *babbitt-lined* adj.

Named after Isaac Babbitt (1799–1862).

1900 *Sci. Amer.* Suppl. XLIX. 20184/1 Almost any solid metal for lining bearings is called by the name of 'babbitt metal'...Ordinary soft lining, so-called babbitt metal, frequently is made up of four parts lead and one part antimony. *Ibid.*, Genuine babbitt will probably change its form more by reheating than the alloys of antimony and lead. **1909** *Westm. Gaz.* 30 Nov. 5/2 The nuts are formed by pouring molten Babbitt metal in the hollow bosses of the shoes. **1932** E. Wilson *Devil take Hindmost* viii. 46 The bronze and babbitt metal are scraped out of the connecting rods. *Ibid.* 49 A finished connecting rod..must be..bored with holes..the holes lined with white babbitt and bronze. **1937** *Times* 16 Sept. 6/5 The big end and camshaft bearings are babbitt-lined.

Babbitt (bæ·bit). Also *erron.* **Babbit.** [f. the name of the (hero of the) novel by Sinclair Lewis, 1922.] A type of materialistic, self-complacent business man conforming to the standards of his set. Also *attrib.* Hence **Ba·bbittism, Ba·bbit(t)ry,** the 'Philistine' behaviour associated with this type of person. Also **Ba·bbitty** *a.*

1923 *Nation* 18 Apr. 465/2 What is it I do find? A group of American business men!..A swarm of forward-lookers! A circle of Babbitts! **1925** *Spectator* 19 Dec. 1146/2 In his controversy with M. Paul Bourget [he] reveals his almost incredible 'Babbittism'. **1926** A. Huxley *Jesting Pilate* iv. 279 At all times the vast majority of human beings has consisted of Babbitts and peasants. **1928** *Daily Express* 27 Apr. 9/3 Vancouver has been inundated with the 'Babbitry' from the South. **1931** *Times Lit. Suppl.* 5 Mar. 179/3 Engrossed in intellectual matters and contemptuous of the Babbittry around them. **1931** H. G. Wells *Work, Wealth & Happiness of Mankind* (1932) x. 453 It was an Individualist's heaven, Babbitt land. **1932** *Scrutiny* I. 3 In America there is the *Hound and Horn, The Symposium,* and the *New Republic,* all of which remind us that America is not inhabited solely by Babbitts. **1942** O. Nash *Good Intentions* 130 Every party Whether Babbitty or arty. **1957** *Listener* 14 Nov. 771/1 We Americans were known as the Babbits of the nineteen-twenties...The Russians are now the Babbits of the mid-century. **1966** *Times* 20 Sept. 11/2 This was not just a bit of Babbitry—'the biggest little place in the state'—but official grading.

babble, *sb.* Add: **4.** *Telephony.* (See quots.)

1930 *Bell Syst. Techn. Jrnl.* IX. 489 Babble is the name given to the effect produced by a number of different [telephone] circuits crosstalking into a particular circuit at a given time and producing an unintelligible murmur. **1960** *Gloss. Terms Telecommunic.* (*B.S.I.*) 152 *Babble,* the aggregate crosstalk from a number of disturbing sources.

babbler. Add: **5.** [Rhyming slang f. *babbling brook* (also used).] A (camp) cook; esp. one who cooks for shepherds, musterers, or shearers in isolated districts. *Austral.* and *N.Z. slang.*

Baker *Austral. Lang.* (1945) 79 says that it 'was in Australian currency in 1906'.

1919 *Bulletin* (Sydney) 24 July 20/2 Two, and sometimes three, of these dishes (depends on the babbler's liver) are served each morning. **1919** Downing *Digger Dial.* 9 *Babbling brook, babbler,* an Army cook. Also *babblins.* **1928** *Bulletin* (Sydney) 12 Jan. 24/1, I am the man they call the babbler Sometimes known as the babbling brook. **1942** in Baker *Austral. Lang.* (1945) xvi. 284 It's also a fair cow when the *babbling brook* (cook) makes a *crook* (bad) stew. **1944** A. F. St. Bruno *Desert Daze* 33 Arch, the cook—oh, how that bold 'babbler' could curse. **1949** P. Newton *High Country Days* vi. 63 The 'babbler'..was in his element. **1963** *Weekly News* (Auckland) 5 June 37/1 That got us started on the 'babbling brooks'. I've got a few memories of station cooks. *Ibid.* 37/2 We worked it out that old babbler made 112,000 rock cakes during those four months.

babe. Add: **1.** Phr. *babe in arms.*

1912 J. N. McIlwraith *Diana of Quebec* vii. 94 She had brought three little children with her, one a babe in arms. **1967** *Woman* 23 Dec. 3/1 She can recall being held up to see a sparkly tree as a babe in arms.

3. Also, an inexperienced or guileless person; so *babes in the wood* (with overt or implied reference to the old ballad *The Children in the Wood*).

1795 W. B. Stevens *Jrnl.* 26 Mar. (1965) 245 Wishes himself at home, restrained only by Principle and a Sense of Duty—A tale to tell the Babes in the Wood! **1841** Lytton *Night & Morning* III. v. i. 98 The uncle of the babes in the wood could hardly have been more startled by the demand! **1866** Mrs. Gaskell *Wives & Daughters* I. xi. 131 Molly and her future stepmother wandered about in the gardens..hand in hand, like two babes in the wood. **1908** 'O. Henry' *Man Higher Up* in *Wks.* (1928) 249 You're both babes-in-the-wood. **1926** H. W. Fowler *Mod. Eng. Usage* 40/1 In figurative use, *babe* implies guilelessness, innocence, or ignorance. **1962** J. B. Priestley *Margin Released* III. v. 191 A big ambitious novel, in which there was to be far more social criticism than the babes-in-the-woods theme might suggest.

b. A girl or woman (often as a form of address). *slang* (chiefly *U.S.*).

1915 *Dialect Notes* IV. 231 *Babe,* a pretty girl. 'She's some babe.' **1930** C. Wittke *Tambo & Bones* iv. 143 Kiss your minstrel boy good-bye, babe, 'bye, babe, 'bye, babe. **1932** *Amer. Speech* VII. 329 Babe, a girl (usually used in direct address). **1952** S. Kauffmann *Philanderer* (1953) iv. 57 This Mrs. Adair..has such hotsy-totsy cottages... Yesterday this Adair babe has an ad in the paper.

babelish, *a.* Delete † *Obs.* and add later examples.

1825 *New Monthly Mag.* XIII. 406 He may know a Babelish confusion of languages. **1898** *Daily News* 24 Aug. 3/2 A perfectly Babelish commingling of tongues fell upon the ear.

Babi (bā·bi). Also **Babee.** [Pers., f. *Bab*-ed-Din 'gate of the faith'. The name *Bab* was assumed by Mirza Ali Mohammed ibn Radhik (d. 1850), the founder of Babi.] The name of a sect originating in Persia, whose doctrine and practice include Mohammedan, Christian, Jewish, and Zoroastrian elements; = *Baha'i. Hence **Ba·bism,** the doctrine or practice of this sect; **Ba·bist** (also *attrib.* or as *adj.*), **Ba·bite,** an adherent of Babism, a Bahai.

1850 in M. L. Sheil *Glimpses Life & Manners in Persia* (1856) xi. 178 A moolla of eminence..had been converted to Bâbeeism, but had recanted. *Ibid.* 180 This year, seven Bâbees were executed at Tehran for an alleged conspiracy... Bâbeeism had spread in Tehran. **1866** *Nation* 22 June 793/2 In Teheran in 1852..the visible remnant of the 'Babist' sect was slaughtered. *Ibid.* 794/1 The new prophet ..adopted the name of the *Bab*; that is, the gate through which alone truth is to be reached. *Ibid.* 794/2 Ali Mohammed, the founder and spiritual head of Babism. *Ibid.*, The Babists were now forbidden from making any more attempts at insurrection until the Bab should decide that the hour had come. *Ibid.* 795/1 Babism is much more in harmony with the subtle and imaginative genius of the Persian people than the Shiite Mohammedanism. **1869** *Contemp. Rev.* XII. 245 (*heading*) The Bab and Babeeism. *Ibid.* 246 He.. learned the principles and substance of the Babee doctrines. *Ibid.* 266 Faithful Babees. **1877** M. Dods *Mohammed, Buddha, & Christ* 195 The martyrs of Babism. **1896** *Daily News* 4 May 5/2 No doubt is entertained that the assassination was due to Babist fanaticism. The discontent of the Babists is due to the fact that the Shah would not allow them to succeed in establishing their religion. **1896** *Ibid.* 30 June 6/1 The Babis of Persia. **1904** *Daily Chron.* 6 Aug. 3/6 He..spoke with hope especially of the spread of the Babists. **1911** Frazer *Golden Bough* (ed. 3) I. vii. 402 The head of the great Persian sect of the Babites..is held by Frenchmen, Russians, and Americans, especially by rich American ladies, to be an incarnation of God himself.

babiche (bäbī·ʃ). *N. Amer.* [Canadian Fr., orig. Algonquian.] Thongs or thread made of raw hide, sinew, etc.

1806 S. Fraser *Let.* 30 Sept. (1960) 243 Will you be able to send me over some green skins for windows and Babich. **1836** in A. Simpson *Life Thos. Simpson* (1845) x. 189 Babiche for snowshoe lacing. **1899** O. T. Mason in *Rep. Smithsonian Inst. 1897* 732 Fine babiche or rawhide string for the webbing of the snowshoe. **1948** *Sat. Even. Post* 21 Aug. 72/4 No screws or nails are used at the joints, but *babiche* instead—rawhide thongs.

babify (bē^i^·bifəi), *v.* [f. Baby sb. + -ify.] *trans.* To render babylike or babyish. So **ba·bified** *ppl. a.*, invested with the character or attributes of a baby.

1862 M. E. Braddon *Lady Audley's Secret* xviii. 64 She looked a childish, helpless, babyfied little creature. **1891** S. J. Duncan *Amer. Girl in London* ii. 15 Good old-fashioned names, like Nancy and Betsy, that couldn't be babified. **1912** D. H. Lawrence *Let.* in F. Lawrence *'Not I, but the Wind'* (1934) 27, I think you're rather horrid to H...You make him more babified—baby-fied.

babily (bē^i^·bili), *adv.* [f. Baby *sb.* + -ly².] = Babyishly *adv.*

1891 Meredith *One of our Conquerors* I. xiv. 269 A young woman more than babily aware of the existence of one particular form of naughtiness on earth. *a* **1935** T. E. Lawrence *Mint* (1955) 98 Crying babily (the first time for years).

Babinski (bäbi·nski). The surname of J. F. F. *Babinski* (1857–1932), a French neurologist, applied *attrib.* to a reflex action in which the great toe remains extended or extends itself when the sole of the foot is stimulated.

1900 Walton & Paul in *Jrnl. Nerv. & Ment. Dis.* XXVII. 323 The Babinski reflex—deliberate and constant extension of the great toe, with or without extension and separation of other toes—is never present in health. **1901** *Jrnl. Amer. Med. Assoc.* XXXVI. 181/1 The Babinski phenomenon..consists in rather slow and deliberate extension of the great toe with or without extension..of the other toes on stroking the plantar surface. **1905** *Clinical Studies* III. 311 The Babinski sign is of importance in diagnosis, since it is 'an indication of a structural lesion of, or involving, the upper motor neuron'. **1932** Fulton & Keller *Sign of Babinski* v. 87 The Babinski response was associated with much more marked fanning and separation of hallux and second toe.

babool, var. *babul.

babotie, var. *bobotie.

babu. Var. Baboo (spelling *babu* now preferred). Similarly *babudom,* etc. Add: **babu English,** the ornate and somewhat unidiomatic English of an Indian who has learnt the language principally from books. So by extension, *babu,* attrib., excessively ornate.

1878 Geo. Eliot *Let.* 27 June (1956) VII. 33 Something more amusing—a bit of Baboo English from an Indian journal. **1889** Barrère & Leland *Dict. Slang* I. 58/2 *Baboo-English*..applied to the peculiar English which is rather written than spoken by the natives in India. **1890** Farmer *Slang* I. 86/1 *Baboo-English*...Its main peculiarity is its grandiloquence, a feature born of an attempt to adapt Western speech to Eastern imagery and hyperbole. **1907** *Westm. Gaz.* 18 Dec. 1/3 The partition of Bengal supplied the simmering discontent of Babudom with a definite grievance. **1925** *Weekly Westminster* 4 July 258 They irritate by their father Babu familiarity with West End place-names. **1926** A. Mayhew *Education of India* xii. 153 The [Indian] matriculate's mastery of English, despite all the ridicule unjustly bestowed on Babu English, is far more complete and practical than that shown by the normally intelligent and industrious English boy at the same stage. **1931** *Times Lit. Suppl.* 5 Mar. 174/3 Robin's babu clerk. **1934** R. C. Goffin *S.P.E. Tract* XLI. 23 The perpetrations universally recognized as 'babu English' with its preposterously learned pretensions. **1934** H. G. Wells *Experiment in Autobiog.* I. vi. 309 The prose was over-elaborate and with that same flavour of the Babu, to which I have called attention. **1936** C. S. Lewis *Allegory of Love* ii. 81 The fantastical 'babu' ornaments of the style [of the *De Nuptiis*] were admired.

babul, babool (bābū·l, bā·bul). *Anglo-Ind.* [Hind. *babūl, babūr,* Pers. *babūl.*] A thorny mimosa, *Acacia arabica,* common in parts of India.

1824 Heber *Narrative* 12 Dec. (1828) I. 529 A formidable underwood of cactus and bâbool. **1849** J. H. Balfour *Man. Bot.* § 852 A variety of Indian gum procured from *A[cacia] arabica,* is called Babul, or Babool-Gum; Babul-wood is used for tanning in Scinde. **1861** Bentley *Man. Bot.* 531 The barks of *A[cacia] arabica* and *A. Catechu*.. are used extensively in India under the name of Babool. **1884** Kipling *Departm. Ditties* (ed. 2, 1886) 51 In place of Putney's golden gorse The sickly *babul* blooms. **1921** *Times Lit. Suppl.* 8 Sept. 574/3 Babul seeds.

babushka (bābu·ʃkā). Chiefly *N. Amer.* [Russ., grandmother, f. *baba* (peasant) woman.] A head covering folded diagonally and tied under the chin; a head-scarf.

1938 *Chatelaine* Feb. 33/2 The babushka is a peasant-sort of hood you wear over your pretty curls. **1948** F. Brown *Murder can be Fun* (1951) vii. 106 She wore a greenish mottled *babushka* and..stringy hair..pushed out in front of it. **1959** *Encounter* Oct. 32/2 A voile scarf tied babushka-style.

baby, *sb.* Add: **I. b.** *fig.* Applied to a person's invention, achievement, concern or responsibility; so *to carry* or *hold the baby,* to be saddled with an unwelcome responsibility.

1890 *Jrnl. Soc. Arts* XLVIII. 65/1 Count Chardonnet.. was then shewing his new-born baby, which he called *soie artificielle.* **1912** *Q. Rev.* July 103 The 'bull'..becomes a 'stale' bull, and drifts into the position frequently described as 'holding the baby'—that is to say, nursing a stock or share, perhaps for months, in the vague hope of getting rid of it some day at a profit. **1927** A. E. W. Mason *No other Tiger* xxiii. 261 He certainly has had the baby to hold all his life. **1927** *Daily Express* 8 July 6 Disclaiming responsibility for all the financial misfortunes of the country, he found himself confronted by Mr. Jack Jones, who told him that he was 'carrying the baby' anyhow. **1928** *Ibid.* 14 Jan. 11/7 If other bidders enter into competition, they find themselves run up to a high figure, and are then left to 'hold the baby'. **1935** D. L. Sayers *Gaudy Night* vii. 153 Poor blighter. He *always* gets left with the baby. **1948** 'N. Shute' *No Highway* xii. 297 The Assegai, it seemed, was going to be my baby.

c. A girl; a girl-friend; a young woman: often as a form of address. *slang* (chiefly *U.S.*).

The degree of slanginess of the nineteenth-century examples is not easily determinable.

1839 F. A. Kemble *Jrnl. Residence Georgian Plantation* (1863) 30 'Oh massa!..where you get this lilly alablaster baby!'..I looked round to see if she was speaking of my baby; but no..this superlative apostrophe was elicited by the fairness of *my skin.* **1870** D. J. Kirwan *Palace & Hovel* (1963) xxii. 196 'Baby Hamilton' is another celebrity of the Half-World. Many stories are told about the recklessness of this girl. *Ibid.*, The 'Baby' had to leave Paris. **1901** A. M. Binstead *More Gal's Gossip* v. 74 One overhears a callow youth of twenty address a still fascinating belle of forty, to whom he is giving supper, as 'Baby'. **1911** H. S. Harrison *Queed* xiii. 167 Bad-eyed young men who congregate..to smirk at the working girls... 'Where you goin', baby?' **1918** C. Sandburg *Cornhuskers* 60 My baby's going to have a new dress. **1927** H. A. Vachell *Dew of Sea* 269 I'm beginning to think that baby is half vamp and half floosie. **1940** *Illustr. London News* 6 Jan. 24/1 He has a 'baby', an empty-headed little chorus-girl, whom he worships.

d. A person. *slang.*

In quots. 1919, 1953 'this baby' = the speaker himself.

1899 E. Whiteing *No. 5 John St.* xxix. 306 The bounders ..think they're such awfully warm babies. **1919** R. Lardner *Real Dope* 83 Well old pal it looks like they [*sic*] wouldn't be no front line trenchs for this baby. **1928** *Sat. Even. Post* 7 Jan. 21/1 [*wife to her husband*] 'Sure, baby,' she replied sympathetically. **1950** A. Lomax *Mr. Jelly Roll* (1952) 57 Some terrible environments..inhabited by some very tough babies. **1953** 'R. Gordon' *Dr. at Sea* vi. 70 Some skippers cook the log, but not this baby. **1968** *Listener* 31 Oct. 565/3 The dialogue is over, baby.

e. *to empty, pour,* or *throw the baby out* (or *away*) *with the bath*(*-water*) (cf. G. *das Kind mit dem Bade ausschütten*), to reject the essential with the inessential, to discard what is valuable along with what is waste or useless.

1909 G. B. Shaw *Pen Portraits & Reviews* (1932) 87 Like all reactionists, he usually empties the baby out with the bath. **1911** —— *Getting Married* Pref. 186 We shall in a very literal sense empty the baby out with the bath by abolishing an institution which needs nothing more than a little..rationalizing to make it..useful. **1922** Mrs. Patrick Campbell *Let.* 7 Jan. in *Shaw & Mrs. Campbell* (1952) 239 As to 'Relativity', I read somewhere that it is a philosophy that 'empties the baby out with the bath water'. **1939** S. Spender tr. *Toller's Pastor Hall* ii. 84 You pour the baby out with the bath. **1957** E. Gowers *H. W. Fowler* 14 We can rid ourselves of those grammarians' fetishes which make it more difficult to be intelligible without throwing the baby away with the bath-water.

7. b. Applied to a small-sized bottle, jar, etc. *slang.*

1853 *Househ. Words* VII. 123/2 A stone bottle of ardent spirits called baby for shortness and secresy. *Ibid.* 124/2 He has nabbed a 'baby'! **1909** Ware *Passing Eng.* 13 The half-bottle (2d.), which from its small size was dubbed

baby' by all men. **1958** *Spectator* 10 Jan. 40/3 Pubs refusing to serve standard-size bottles of mineral water because they only stock 'babies'.

c. *spec.* A small car.

1920 *Punch* 27 Oct. 329/3 Triumph 1920 4 h.p. Model H, also Baby, both brand new. **1931** *Star* 8 May 13/1 One of the 'babies' was setting up such a pace that the big cars were out of the hunt.

8. *transf.* The youngest or most junior of a family, group of persons, team, etc.

1897 'P. Warung' *Tales Old Régime* 243 Blake was twenty-five, Clyde was twenty-four, and Entworth forty-one...Short was the 'baby'. **1914** *Daily Express* 19 Sept. 5/1 The 'babies' of the Southern League, Croydon Common, will face Crystal Palace. *Ibid.* 20 Nov. 5/4 [He] was only eighteen years of age, and was known as 'the baby' of his company. **1925** W. Deeping *Sorrell & Son* xxiv. § 2 There was one young Pentreath,—the baby.

B. 1. b. *baby-minder, -worship* (example); **d.** *baby-language, -name, -play, -talk* (earlier and later examples); **e.** *baby-class, -clothes* (examples), *-harness, -linen* (example); **f.** *baby-faced* (earlier and later examples).

1873 C. M. Yonge *Pillars of House* I. ii. 24 That she might not..[go] down to the bottom of the baby-class. **1934** Ngaio Marsh *Man lay Dead* xii. 212 Baby-class stuff at the Yard. **1704** N. Blundell *Diary* 7 July (1952) ii. 23, I opened a box of Babby Cloths which was sent to my Wife by my Lady Webb. **1798** Jane Austen *Let.* 27 Oct. (1932) I. 25 Dame Tilbury's daughter has lain in. Shall I give her any of your baby clothes? **1866** Rossetti *Let.* 5 Jan. (1965) II. 586 Some women preparing the grave-clothes and baby-clothes at the same time. **1839** Dickens *Nickleby* liv. 536 A baby-faced chit of a girl. **1959** Ngaio Marsh *Singing in Shrouds* iv. 64 Baby-faced petulance. **1930** M. Mead *Growing Up in New Guinea* 24 The child sits quietly [in the canoe]... There are no straps, no baby harnesses to detain him in his place. **1962** *Guardian* 5 Nov. 4/3 A lightweight baby harness..used to hold a baby in a high-chair for feeding. **1892** C. M. Yonge *That Stick* xxvii. 181 That pretty precision of utterance that children sometimes acquire when baby language has not been foolishly fostered. **1923** H. Walpole *Introd.* to H. Lofting *Dr. Dolittle* p. vii, Grown-ups imagine that they can do the trick by adopting baby language and talking down to their very critical audience. **1932** *Oxf. Compan. Eng. Lit.* 756/1 It [*sc.* Journal to Stella] is a series of intimate letters..for the most part written in baby-language. **1814** Jane Austen *Mansfield Park* I. i. 6 Lady Bertram dispatched money and baby-linen. **1863** F. A. Kemble *Jrnl. Residence Georgian Plantation* App. 401 The only supervision exercised over either babies or 'baby minders' was that of the old woman left in charge of the infirmary. **1959** *Times* 8 June 13/1 Resident nannies and baby-minders. **1856** C. M. Yonge *Daisy Chain* II. v. 382 Baby-names never ought to go beyond home. **1896** G. B. Shaw in *Sat. Rev.* 6 June 575/2 Mrs Campbell's attempt at Magda is the merest baby-play in comparison with that performance. **1836** F. A. Kemble *Let.* 1 Mar. in *Rec. of Later Life* (1882) I. 43 A baby..[is] such an interruption..to any conversation but baby-talk. **1861** M. B. Chesnut *Diary* 25 Nov. (1949) 162 To me this calm, monotonous baby talk is maddening. **1933** E. Arnot Robertson *Ordinary Families* x. 215 'Pore ole pussycat!' Dru would lapse into baby-talk. **1894** M. Dyan *All in Man's Keeping* II. x. 183 His youngest sister..had brought her boy with her, for 'Uncle Dick' to do a little final baby-worship.

g. Passing into *adj.* = young; small or diminutive of its kind.

1750 H. Walpole *Let.* 25 Feb. (1903) II. 432 All the geniuses of the age are employed in designing new plans for desserts. The Duke of Newcastle's last was a baby Vauxhall, illuminated with a million of little lamps. **1819** Keats *Let.* 24 Jan. (1958) II. 36 My modest feathered Pen frizzles like baby roast beef. **1877** *Design & Work* 1 Dec. 602/2 Gas bags ..floating high over their heads, by means of baby screws.. and the like. **1878** *Ibid.* 16 Feb. 199/3 He was surprised at the strength displayed by the baby iron rope. **1899** *Daily News* 17 Oct. 6/6 [He] decided that her baby-jib must be responsible for the Shamrock's poor pointing. **1908** Findlater *Crossriggs* iv. 28 She led him out of the nursery... 'There's a babier baby than Mike,' she said. **1909** *Daily Chron.* 12 Oct. 4/4 Baby beef, which is grown so little in this country. **1917** W. Pett Ridge *Amazing Years* vii. 93 One attractive baby grand [piano] that Millwood picked up. **1918** F. A. Forbes *Let.* in G. L. Sheil *Mother F.A.F.* (1946) vi. 69, I have in my room a baby rhododendron in full bloom. **1926** *Daily Express* 30 July 9/5 Imported films for use in 'baby' cine-cameras. **1928** *Daily Express* 11 Oct. 1/1 Wonderful improvement in the 'baby car' of seven horse-power. **1929** *Melody Maker* Jan. 15/3 You can't get a harp into a Baby Austin, whereas you can garage an Austin inside a piano. **1961** C. McCullers *Clock without Hands* xiii. 249 Pounding on middle C of the baby grand piano. **1965** *Economist* 2 Oct. 70/1 Intensive beef production in the shape of barley beef or 'baby beef' started here in 1961.

2. baby act *U.S.*, (*a*) the act of a baby; (*b*) an act or statute for the protection of minors; hence *to plead the baby act,* to enter a plea that one is not legally responsible by reason of youth or inexperience; similarly, *to read the baby act*; **baby-blue** (orig. *U.S.*), a light shade of blue; **baby-blue-eyes** (orig. *U.S.*), the popular name of a plant of any of several species of the *Nemophila* family; **baby-bouncer** = *baby-jumper*; **baby buggy** (*N. Amer.*), **carriage, coach** (*U.S.*), a perambulator; **baby-doll,** (*a*) = Doll *sb.*¹ 2; (*b*) (orig. *U.S.*) a girl or woman who has the youthful and regular good looks characteristic of a doll and an ingenuous disposition; **baby-doll pyjamas,** women's pyjamas consisting of a loose-fitting top worn over short trousers; also *baby-doll nightdress*; **baby face,** (a person with) a babyish face; **baby food,** a milk-substitute or a light diet suitable for a baby; **baby-house,** a doll's house (earlier and later examples); **baby-jumper** (examples); **baby lace** (see quot.); **baby pig disease,** hypoglycæmia of newly-born pigs; **baby powder,** a skin powder for babies; **baby-ribbon,** narrow ribbon such as that used for babies' clothes; **baby's breath, babies' breath,** the popular name of any of several delicate or sweet-scented plants, esp. *Gypsophila paniculata*; **baby's head** *slang,* a steak (and kidney) pudding; **baby show** (orig. *U.S.*), a baby exhibition with an award for the 'best'; **baby-sitter,** a person engaged to be at hand to look after a young child or children in the absence of the parents; hence **baby-sit** *v. intr.*; **baby-sitting** *vbl. sb.* and *ppl. a.* (orig. *U.S.*); **baby-snatcher** *joc.*, a person who enters into an amorous relationship with a much younger member of the opposite sex; hence **baby-snatch** *v. intr.*, **baby-snatching** *vbl. sb.*; **baby walker** orig. *U.S.*, a device for assisting babies to learn to walk.

1837 *New Yorker* 9 Dec. 593/2 We do not defend the conduct of any man who pleads usury in bar of an honest man, neither would we if he should plead the baby act, or minority. **1888** *Congress. Rec.* Aug., App. 440/1 [Mr. S. S. Cox] admits the authorship..but pleads the baby act, and says he was a boy when he wrote it. **1899** 'Mark Twain' *Man corrupted Hadleyburg* (1900) 214 When politicians come out without a blush and read the baby act in this frank way, [etc.]. **1901** *Forum* Jan. 592 One minute reading the riot act of manly independence, and the next pleading the baby-act of thoughtless irresponsibility. **1904** W. H. Smith *Promoters* xv. 229 That's business honor, and anything else is the baby act. **1889** *Century Mag.* Mar. 748/2 The small, square manuscript sewed at the back with worsted of the pale tint known as 'baby-blue'. **1922** S. Lewis *Babbitt* i. 4 Faded baby-blue pyjamas. **1949** A. Wilson *Wrong Set* 29 Her corn-coloured hair and baby blue eyes. **1887** *Overland Monthly* Aug. 152/1 Then, if we could have been there, we should have seen the..'baby blue eyes' (nemophila). **1952** A. G. L. Hellyer *Sanders' Encycl. Gardening* (ed. 22) 325 *Nemophila..Menziesii.* 'Baby Blue-Eyes', white or blue, summer, spreading, California. **1968** *Times* 11 Mar. 11/2, I have him in the baby bouncer while I'm practising. **1890** *Dial. Notes* I. 411 Baby buggy. **1911** *Daily Colonist* (Victoria, B.C.) 30 Apr. 4/4 Your baby buggy..is on our fourth floor. **1934** J. T. Farrell *Young Manhood* in *Studs Lonigan* (1936) 172 He.. walked past the baby-buggy where young Horn had left his baby brother. **1866** *Leisure Hour* 2 June 348/1 The specification was not such as to prevent various modifications of three-wheeled baby-carriages. **1870** L. M. Alcott *Old-Fashioned Girl* (1874) xiii. 215 A young girl pushing a baby-carriage. **1903** *Sun* (N.Y.) 3 Nov. 7 It gave a chance for a lot of jokes which were so distinctively British that a baby carriage was called a 'perambulator'. **1909** Eaton & Underhill *Runaway Place* 155 To their astonished eyes, he seemed to slip directly through a baby carriage. **1903** *N.Y. Times* 1 Oct. 3 English baby coaches...The carriages are a distinctly English idea—they dub them 'Perambulators'. *a* **1862** Mrs. Nesbit *Let.* in D. Langley Moore *E. Nesbit* (1933) ii. 26, I shall bring you a Baby doll...It has blue Eyes and flaxen hair and is dressed like a little baby in long clothes. **1893** 'Sarah Grand' *Heavenly Twins* II. III. x. 96 It was a huge wax baby-doll, considerably battered. **1946** Tennessee Williams *27 Wagons* 11, I fell in love with this baby-doll. **1957** *Punch* 27 Mar. 418/3 The Keystone 'Baby Doll' nightdress..is barely knee-length, and has a scooped neckline trimmed with angel lace and baby ribbon. There are also 'Baby Doll' pyjamas: gossamer jacket just covering, but no more, the brief puff panties. **1959** *Listener* 2 July 35/1 The girl..appeared curled up in Baby Doll pyjamas. **1962** *Guardian* 8 July 16/6 The serious baby-doll face suddenly broke into laughter. **1967** J. Wainwright *Talent for Murder* 44 She was wearing a grimy, 'Baby Doll' nightdress. *a* **1700** Evelyn *Diary* 4 Nov. an. 1670 (1955) III. 564 That famed beauty (but in my opinion of a childish simple & baby face) Madamoiselle Quirreval. **1713** Baby face [see Baby *sb.* B. 1 d]. **1888** A. C. Gunter *Mr. Potter of Texas* xxiii. 268 'That was before her baby face came between us!' cries her ladyship, hoarsely. **1926** D. H. Lawrence *Plumed Serpent* ii. 48 Mrs. Burlap had hitched herself on to Kate, and from her silly, social baby-face was emitting searching questions. *Ibid.* 50 The couple from the Middle-West, that withered baby-face and that limping Judge. **1897** *Sears, Roebuck Catal.* 28/2 Baby Foods. **1939** *Jrnl. Amer. Med. Assoc.* 4 Mar. 843/2 Libby's Homogenized Baby Foods are..suitable for use particularly during the early months of infancy. **1946** *Nutrition & Child Welfare* Oct. 6/2 There are three types of baby food. **1726** Swift *Gulliver* I. ii. iii. 55 The Furniture of a Baby-house. **1861** H. B. Stowe *Pearl of Orr's Isl.* I. xii. 106 Mara and Sally..were revelling in apronsful of shells and seaweed, which they bustled into the other room to bestow in their spacious baby-house. **1848** *Ann. Rep. Comm. Patents 1847* (U.S.) 47 Exercising machines, under the appellation of swings, baby-jumpers, &c. **1849** J. R. Planché in *Extravaganzas* (1879) III. 290 Up and down leaping, My heart is keeping, like a baby jumper—that invention new. **1872** *N.Y. Times* 24 Apr. 8 Advt. (Hoppe), Cradle, Baby-Jumper and Nursery Chair Combined. **1882** Caulfeild & Saward *Dict. Needlework, Baby Lace,* an English pillow lace, formerly made in Bedfordshire and Buckinghamshire, and called English Lille...The name Baby Lace was given, as, on account of the narrow width of the lace, it was chiefly used for trimming babies' caps. **1942** L. L. Madsen in *Yearbk. Agric* (*U.S. Dept. Agric.*) 825 (*heading*) Acute Hypoglycemia in Newborn Pigs, or So-Called Baby Pig Disease. **1960** *Farmer & Stockbreeder* 9 Feb. 102/1 Treatment of 'Baby Pig Disease'...All the piglings in the affected litter should without delay be given glucose by mouth. **1897** *Sears, Roebuck Catal. 6th page after* p. 32, Borated Talcum Baby Powder. **1914** G. Atherton *Perch of Devil* I. 63 It was ordinary 'baby powder' for the bath. **1893** *Cassell's Fam. Mag.* Feb. 175/1 Various ways in which skilful fingers can utilise the baby ribbons. **1906** *Westm. Gaz.* 28 Apr. 14/1 Baby-ribbon velvet. **1909** *Daily Chron.* 1 June 7/5 A sandwich which was tied up with pink baby-ribbon. **1897** Mrs. C. A. Creevey *Flowers of Field, Hill & Swamp* viii. 272 Grape hyacinth. Baby's breath. Muscari botryoides. **1938** *Word Study* Dec. 4 New flowers are weeds whose names the florists have changed; 'baby's breath' is the old 'flowering spurge'. **1905** *Daily Chron.* 5 Sept. 4/7 'A baby's head and two veg.' ..The grubby waitress..brings steak pudding with potatoes and greens as trimmings. **1967** K. Giles *Death in Diamonds* v. 88 He went to the counter and ordered kidney soup and a baby's head and chips. **1854** *Chambers's Jrnl.* 30 Dec. 419/2 Baby-shows have taken place in one or two western localities. **1855** *Family Economist* Nov. 160 The stupidest exhibitions lately heard of, are the Baby-show, now being got up in this country, in foolish imitation of the same foolish thing in America. **1930** 'R. Crompton' *Wm's Happy Days* viii. 166 Oughter be in a baby show. **1967** *Guardian* 5 Aug. 12/8 The judging of the Baby Show had been completed. **1947** in *Amer. Speech* (1949) XXIV. 72 Offer to 'baby sit' with her little boy. **1950** *Punch* 15 Mar. 292/2 Well, this is the last Monday I baby-sit for the McClouds. **1937** C. R. Walker *American City* 230 There are two high-school girls in the neighborhood who hire out for twenty-five cents an evening as 'baby sitters' when the family wants to go to the movies. **1951** *Times* 18 Dec. 5/4 Everybody knows what a baby-sitter is supposed to do (though, curiously enough, hardly anybody except the baby ever sees her do it). **1947** *Richmond* (Virginia) *Times-Dispatch* 4 Mar. 4/2 The Cavalier Baby-Sitting Association, recently organized at the University of Virgina. **1949** *New Yorker* 5 Mar. 24/3 Baby-sitting having become firmly established, it was only a question of time before somebody offered to provide a similar service for dogs. **1951** *Punch* 4 Apr. 421/3 Undoubtedly one of the worst features of the baby-sitting age is the coy handing over of the fee. **1958** *Spectator* 22 Aug. 248/3 Car-owning miners and baby-sitting scholars. **1911** 'I. Hay' *Safety Match* xiv. 225 She comes trapesin' about here with a collection of middle-aged baby-snatchers. **1930** V. Sackville-West *Edwardians* v. 199 You don't imagine that he really cared about that baby-snatcher? Good gracious me, he was a year old when her daughter was born. **1933** M. Allingham *Sweet Danger* vii. 85, I meant the elder sister. You aren't baby-snatching, I hope, Wright? **1939** 'N. Blake' *Smiler with Knife* iv. 71, I don't approve of baby-snatching. **1856** *Michigan Agric. Soc. Trans.* VII. 81 Wm. Phelps..[exhibited] 1 baby walker. **1967–8** *Catal. Galt Toys* 7 Galt baby walker...Designed so as not to tip up when the baby, first learning to walk, pulls himself up on the handle.

baby, *v.* Add: **2.** *intr.* To act as if dealing with a baby.

1913 G. S. Porter *Laddie* xi. 340 I'll wager a strong young girl like the Princess will laugh at you for babying over her.

ba·by-farm. [Farm *sb.*² 7.] A derogatory term for a place where the lodging and care of babies is undertaken for profit. Hence **ba·by-fa:rming** *vbl. sb.*, the keeping of such a place; also *ppl. a.*; **ba·by-farmed** *ppl. a.*; **ba·by-fa:rmer** (see Baby *sb.* B. 2).

1868 *Brit. Med. Jrnl.* 25 Jan. 75/2 Poor-law medical officers should have a power of entering, inspecting, and regulating baby-farms. *Ibid.* 22 Feb. 174/2 Children but a few months old were fed by the less indifferent of these baby-farmers with some of their own coarse food. *Ibid.* 21 Mar. 276/1 The inquiry has for its object the determination of the extent to which Baby-Farming (as distinguished from the practice of abortion) is carried on in London. **1870** *Times* 16 June 12/4 Two children..alleged to have been brought from a baby-farming establishment. **1878** W. S. Gilbert *H.M.S. Pinafore* II. 31, I practised baby-farming. **1880** *Encycl. Brit.* XIII. 4/2 The infamous relations between the lying-in houses and the baby-farming houses of London. **1881** Ogilvie (Annandale), *Baby-farmer,* a woman who receives infants, generally illegitimate, from their parents, on the pretext of bringing them up, the object being to have the child removed from sight; one who lives by baby-farming. **1884** [see Baby *sb.* B. 2]. **1896** *Daily News* 26 Apr. 2/3 The great ledger in which are kept their patiently accumulated records of the baby farmers. **1896** *Westm. Gaz.* 30 May 3/1 The baby-farmed child. **1954** T. S. Eliot *Confid. Clerk* III. 94 You're suggesting that she ran a baby farm.

babyless (bē^i^·biles), *a.* [f. Baby *sb.* + -less.] Having no babies, without a baby.

1871 Christina G. Rossetti *Sing-Song* 125 Motherless baby and babyless mother. **1958** *Times* 2 Oct. 12/7 Being awake and babyless we lose no time crossing the gangway into the ship.

Babylonian. Add: **B.** *sb.* **2.** The language of the inhabitants of Babylon.

1870 G. C. Swayne *Herodotus* v. 78 It is in three languages—old Persian, Babylonian, and Scythian. **1888** A. H. Sayce *Hittites* ii. 23 At that time Babylonian was the international language.

bac³ (bæk): abbrev. of Baccarat.

1906 B. von Hutten *What became of Pam* III. x. 293, I know nothing so engrossing..as a game of bac. **1937** Koestler *Span. Testament* i. 23 At the Estoril Casino we came across..some playing bac, others trente-et-quarante.

bacalao. Add to def.: *spec.* dried or salted cod-fish. Also in Portuguese form *bacalhau.*

1749 Jackson *Wealth of Gt. Britain in Ocean* 54 Cod finely wet cured, are much better eating than baccalao. **1885** *Blackw. Mag.* July 104/2 Consumers of the salted cod or baccalao. **1961** M. McCarthy *On the Contrary* (1962) 116

A golden *bacalhau* (dried cod, done with egg yolks, black olives, sauté potatoes, and onions) Portugal. **1962** *Times* 13 Apr. 19/1 *Bacalhau* is as traditional [in Portugal] as the roast beef of England.

bacca (bæ·kā). Also **baccah, baccer.** Colloq. clipped forms of Tobacco. (Cf. Bacco, Baccy, Backer.)

1824 J. Wight *Mornings at Bow Street* 36 She told him her new lover 'couldn't abide the smell of *baccah*'. **1835** M. H. Barker *Tough Yarns* 10 He whips out his old quid, flings it into the fire, and we sported a fresh bit o' bacca. **1837** Dickens in *Bentley's Miscellany* Jan. 62 Are you going to put down pipes...Or trace the progress of crime to 'baccer? **1847** E. Brontë *Wuthering Heights* II. vii. 128 Joseph's 'bacca pipe is poison. **1920** D. H. Lawrence *Touch & Go* I. i. 16 They'll give you plenty to eat..and a bit of bacca.

baccalaureate. Add: **1.** (Later examples.)

1849 *Catal. Wesleyan Univ.* 22 (B. H. Hall, *College Words*), The Seniors will be examined for the Baccalaureate, four weeks before Commencement. **1895** Rashdall *Univ. Middle Ages* I. 210 The institution of the Baccalaureate or pupil-teachership. **1907** *Elem. Sch. Teacher* Mar. 372 All of these four baccalaureates extend practically the same privileges to those who have obtained them.

3. *baccalaureate degree.*

1891 D. C. Gilman *Johns Hopkins Univ.* 66 The manifold forms in which the baccalaurate [*sic*] degree is conferred. **1968** *Globe & Mail* (Toronto) 3 Feb. B7/1 (Advt.), Registered Nurses or Registered Psychiatric Nurses with baccalaureate degree or higher, and considerable experience in nursing education.

bacchiac (bækəi·æk), *a.* [ad. L. *bacchiacus,* Gr. βακχειακός.] Of the nature of a Bacchius; consisting of or characterized by bacchii.

1766 R. Lowth *Larger Confut. Bp. Hare* 36 Bacchiac and Cretic Feet. **1855** F. A. Paley *Æschylus* 204/2 The bacchiac metre also suggests τί ταῦτα for τί τάδε. **1857** E. S. Parry *Terentii Comœd.* p. lix, Cretic and bacchiac measures.

bacco, baccy. (Earlier and later examples.)

1792 W. B. Stevens *Jrnl.* 9 July (1965) 36 Thou hast got a Crescent like my 'Bacco Box'. **1793** C. Dibdin in *Britannic Mag.* I. 249/1 The tender motto written fair Upon his 'bacco-box he views. **1818** R. Wilbraham *Gloss. Cheshire* 9 *Bacco,* tobacco. **1853** Mrs. Gaskell *Cranford* xiii. 249 But the 'bacco, and the other things —. **1859** T. Hughes *Scouring White Horse* i. 8 Your bacchy's nearly out, Dick. **1906** Kipling *Puck of Pook's Hill* 251 Five and twenty ponies, Trotting through the dark—Brandy for the Parson, 'Baccy for the Clerk. **1933** 'G. Orwell' *Down & Out* xxvi. 188 [Irish tramp speaking] Dere's sixpennorth o' good baccy here!

bach (bætʃ), *sb.* [Shortened f. Bachelor 4 a.] **1.** *U.S. slang.* A bachelor; *old bach,* a confirmed bachelor. Also phr. *to keep bach* for earlier *to keep bachelor's hall* (cf. Hall *sb.* 11) = *bach *v.*

1855 *Knickerbocker* XLV. 158 The President was an 'old bach.' of some sixty-five summers. **1857** W. Chandless *Visit Salt Lake* II. vi. 236 Mormons look upon a 'bach' with great suspicion..because they consider all men should marry. **1878** I. L. Bird *Lady's Life Rocky Mts.* (1879) ix. 157 A cabin..where two brothers and 'a hired man' were 'keeping bach'. **1883** E. Eggleston *Hoosier School-Boy* xvi. 106 Don't you know..any place where we could keep 'bach' together? **1904** W. N. Harben *Georgians* 188, I.. thought now was the time fer me, old bach' that I am, to.. show them ladies I'd been about.

2. *N.Z.* Also **batch. a.** A makeshift hut, usu. one in which a man living alone fends for himself. **b.** Now esp. a small house at the seaside or at a holiday resort.

1927 J. Devanny *Old Savage* 100 Bob knocked up a bunk in Fletcher's bach. **1938** F. S. Anthony in D. M. Davin *N.Z. Short Stories* (1953) 218, I don't wonder he comes over to my batch so often—he does get a little decent cleanliness there. **1940** F. Sargeson *Man & Wife* (1944) 29 There were a few holiday baches but they were empty now that it was well on into the autumn. **1957** J. Frame *Owls do Cry* vi. 26 The Withers haven't a week-end bach. **1957** I. A. Gordon in *N.Z. Listener* 22 Nov. 4/3 From being a mere shed, the bach has become a place at the seaside, and today many a bach has a refrigerator and wall-to-wall carpets.

bach (bætʃ), *v.* *N. Amer., Austral.,* and *N.Z. colloq.* Also **batch.** [f. prec.] *intr.* Usu. of a man: to live as a bachelor; to live alone and do one's own cooking and housekeeping. Also with *it.*

a. *N. Amer.*

1870 *Repub. Daily Jrnl.* (Lawrence, Kans.) 29 Jan. (D.A.), They 'bach'. **1878** I. L. Bird *Lady's Life Rocky Mts.* (1879) ix. 156 The men don't like 'baching', as it is called in the wilds—*i.e.* 'doing for themselves'. **1888** *Century Mag.* Jan. 412/2 He had always 'bached it' (lived as a bachelor). **1898** *Lett. fr. Canada* vii. in *Times* 22 Nov. 6/1 'To batch' upon the prairies represents perhaps the *minimum* of pleasure in existence. **1927** P. H. Pearson *Prairie Vikings* 13 For two years they 'bached' as huntsmen along the creek banks.

b. *Australasian.*

1898 R. Radclyffe *Wealth & Wild Cats* viii. 71, I had a happy time at Yalgoo, 'batching' with the son of a well-known brewer and M.P. **1900** H. Lawson *Over Sliprails* 45, I hurried home to the tent—I was batching with a carpenter. **1905** C. Clyde *Pagan's Love* ii. 26 You would live in a cheap lodging-house or batch with another girl. **1930** *Bulletin* (Sydney) 19 Feb. 23/1, I allus reckoned that I'd be a-batchin' all me days. **1938** F. S. Anthony *Me & Gus* ix. 53 He batched in a tin shanty for years. **1952** J. Cleary *Sundowners* iv. 272 Where's the rest of the household? Are you batching? **1952** R. Finlayson *Schooner came to Atia* x. 56 If you're prepared to bach on your own.

Bach (bāχ). The surname of the German composer Johann Sebastian *Bach* (1685–1750), applied *attrib.* to a type of trumpet suitable for the performance of his trumpet parts.

1898 G. B. Shaw *Our Theatres in Nineties* (1932) III. 282 A whole battery of Bach trumpets. **1938** *Oxf. Compan. Music* 960/1 The Bach trumpet in A has generally given place to a shorter instrument in D.

bachelor. Add: **4. c.** *transf.* One of the young male fur-seals which are kept away from the breeding-grounds by the adult bulls. (These are the seals which may legally be killed for their fur.)

1874 *Harper's Monthly Mag.* May 801/2 To the right and left of the breeding grounds stretch sand-beaches..upon which the 'hollus-chickie', or the bachelor seals, lie by tens of thousands. **1884** H. W. Elliott *Seal Isl. Alaska* 43 The 'hollus-chickie' or 'bachelor' seals. *Ibid.* 44 Sports and pastimes of the young 'bachelors'. **1897** D'Arcy Thompson in *Parl. Papers* CII. 422 We saw a body of about 200 bachelors, mostly young or old, those of intermediate 'killable' size being very few. **1898** D. S. Jordan et al. *Fur Seals* I. 50 The bachelor seals begin to arrive at about the same time as the bulls...The older bachelors come first.

d. A size of roofing slate (see quot. 1929).

1898 in E.D.D. **1914** M. S. Gretton *Corner of Cotswolds* ix. 175 The tiles for our roofs are called, according to their sizes, long wivets, long bachelors, short bachelors, longbecks..muffities, long days, and short days. **1929** N. Lloyd *Building Craftsmanship* x. 93/1 Sizes of..slates, measuring from centre of peg hole to tail, in inches, are Long bachelors 11 Short bachelors 10½.

6. *bachelor-apartment, -cottage, -flat, -quarters*: occupied by a bachelor (cf. *bachelor-room* in Dict.); **bachelor girl, woman,** an unmarried woman who has her own income and lives independently; also (*rare*) *bachelor-lady, -maid*; **bachelor('s)** (or **bachelors') hall**: see Hall *sb.* 11; **bachelor party,** one for men only, esp. one marking the end of a bride-groom-to-be's bachelorhood; also *bachelor-dinner.*

1857 Dickens *Dorrit* II. xxx. 591 Ah, but he lived in a sweet bachelor-apartment. **1864** —— *Mut. Fr.* (1865) I. xii. 109 Mr. Mortimer Lightwood and Mr. Eugene Wrayburn..had taken a bachelor cottage near Hampton. **1894** C. C. Harrison (*title*) A Bachelor Maid. **1895** *Dialect Notes* I. 396 *Bachelor-girl,* a maiden lady. **1898** G. E. Mitton (*title*) A Bachelor Girl in London. **1898** Wright *Eng. Dial. Dict.* I. 108/2 Bachelor woman is common, spinster being unknown. **1902** *Daily Chron.* 15 Sept. 3/6 The bachelor woman who earns from two guineas a week. **1902** *Town & Country* 4 Oct. 21/2 In Baltimore, among the first events of the autumn season was the bachelor dinner given at the Baltimore Club last Saturday in honor of Mr. John T. Love, whose marriage to Miss Ellen Jenkins George will shortly take place. **1906** *Queen* 10 Nov. 808/2 The term 'old maid' is now seldom or never heard; the expression 'bachelor girl' has taken its place. **1912** Dreiser *Financier* lxxiii. 762 He occupied a bachelor apartment in North Fifteenth Street. **1921** S. McKenna *Educ. E. Lane* i. 42 If one of your brothers saw fit to invite *my* sister to a bachelor flat. **1922** Bachelor-party [see *binge *sb.*]. **1924** J. M. Murry *Voyage* xii. 229 I've been in here as a bachelor lady. **1934** H. Brighouse *Exhibit C* in *Best One-Act Plays of 1933* 35 (*stage direction*) A minimum of furnishing indicates..that this is a bachelor flat. **1954** T. S. Eliot *Confid. Clerk* II. 62 I've come to inspect the new bachelor quarters. **1955** *Amer. Speech* XXX 298. A way of living no doubt explains *bachelor girl* and not any consideration of gender. **1968** *Globe & Mail* (Toronto) 17 Feb. B3/1 (Advt.), Accommodation..from bachelor apartment to 6-bedroom home..in some of Ottawa's finest locations.

bachelorism. Add: **2.** The condition of being a bachelor; the behaviour, conduct, or nature characteristic of this. Also *old bachelorism.*

1834 *Blackw. Mag.* XXXV. 769 Bound in the triple steel of resolute bachelorism. **1838** *New Monthly Mag.* LIV. 442 This omission may be looked upon as by no means characteristic of old bachelorism. **1857** J. Aiton *Domestic Econ.* 83 A prim correct sort of bachelorism. **1884** *American* VIII. 236 The oddest theory of voluntary old bachelorism.

bachelorize, *v.* Add: **2.** = *bach *v.* *N.Z.*

The 1854 example is a nonce U.K. use.

1854 Mrs. Gaskell *Let.* ?8–14 May (1966) 282 Mr. Gaskell.. will form no plans, but bachelorize off comfortably guided by the wind of his own daily will. **1883** 'A Lady' *Facts: Experiences in N.Z.* ii. 8 Bachelorising, to use a Colonial term, is not a disagreeable life. **1905** P. Oliphant *Bill the Namer* i. 6 We had then been bachelorising, and it was decided to break up the establishment. **1906** *N.Z. Truth* 10 Mar. 5/2 Single men had either to bachelorise in their whares or board with one or another of the married men. **1924** H. T. Gibson *That Gibbie Galoot* v. 16 He promised..to show me a few more points in bachelorising.

bacillar (bāsi·lāɹ), *a.* [ad. mod.L. *bacillāris,* f. L. *bacillus*: see -ar¹.] Of, pertaining to, or of the nature of bacilli.

1883 *Lancet* 25 Aug. 317/2 The air of these regions is not absolutely fatal to bacillar life. **1884** E. Klein *Micro-organisms* 72 The bacillar growth..is thus removed from the surface. **1884** Blackmore *Sir Thos. Upmore* (ed. 2) II. xiv. 214 The great bacillar experiment. **1887** A. M. Brown *Anim. Alkaloids* 122 Bacterial and bacillar biologies.

bacillary, *a.* Add: Of, pertaining to, or caused by bacilli; spec. *bacillary white diarrhœa,* a disease attacking domestic fowl.

1884 *Weekly Med. Rev.* (Chicago) IX. 247 (*title*) On the bacillary doctrine of phthisis. **1894** *Lancet* 3 Nov. 1022/1 A further attempt on the part of the organism to protect itself against bacillary attack. *Ibid.* 1022/2 A drying process incompatible with bacillary life. **1908** R. T. Hewlett *Bacteriology* (ed. 3) 351 In one type of dysentery, the so-called epidemic or bacillary form. **1909** L. F. Rettger in *Jrnl. Med. Research* XXI. 117 The malady which for a long time I have termed fatal septicemia, and for which I would suggest the name simple or uncomplicated white diarrhea, or better still, bacillary white diarrhea, has a definite etiology. **1928** *Jrnl. Ministry Agric.* Feb. 1062 The Ministry has made arrangements for..agglutination tests for Bacillary White Diarrhœa..to be carried out at its Veterinary Laboratory. **1930** *Times* 17 Mar. 18/3 Coccidiosis, bacillary white diarrhœa, and other chicken diseases.

bacillicide (bāsi·lisəid). ?*Obs.* [f. Bacillus + -cide 1.] An agent or substance that destroys bacilli. Also *attrib.* or *adj.* = **bacillici·dal** *a.*

The usual word now is *bactericide.

1885 *Disinfectants & their Use* 19 A combination of lime with chlorine, perhaps the best of all the bacillicides. **1894** *Lancet* 3 Nov. 1021/1 Antiseptic or bacillicide measures. *Ibid.* 1021/2 Bacillicide treatment. *Ibid.* 1022/2 Sunshine and pure air are the best bacillicides. **1894** Gould *Dict. Med., Bacillicidal*.., destructive to bacilli.

bacilluria (bæsil^i^ū^ə^·riā). *Path.* [f. Bacill(us + -uria.] The presence of bacilli in the urine.

1881 *Trans. Internat. Med. Congr.* II. 157 (*title*) On Bacilluria: a Form of Urinary Disorder associated with the Discharge of Rod-shaped Bacteria with the Urine. **1906** *Practitioner* Dec. 798 A factor of extreme importance in the seasonal incidence of typhoid fever..is the large proportion..of convalescents from typhoid fever, who suffer from typhoidal bacilluria. **1962** *Lancet* 6 Jan. 46/1 (*title*) Bacilluria in pregnancy.

bacillus. Add: Freq. in *fig.* use.

1897 'Mark Twain' *Foll. Equat.* li. 494 Twenty million priests, fakeers, holy mendicants, and other sacred bacilli. **1905** W. Lee-Warner in *Anti-Slavery Reporter* June-July 63 We must not allow this bacillus of slavery to take a more virulent form. **1907** *Jewish Chron.* 15 Mar. 19/1, I refer to the existence of anti-Semitic bacilli which poison the whole air of Russia. **1918** A. Gray tr. *Grelling's The Crime* II. ii. 145 They found a fostering soil..on which the bacillus of war could develop unhindered.

bacitracin (bæsitrē^i^·sin). *Biochem.* [f. Baci(llus + Trac(y, the surname of Margaret Tracy, an American child in whom the substance was found in a wound: see -in¹.] An antibiotic obtained from organisms of the group *Bacillus subtilis.*

1945 B. A. Johnson et al. in *Science* 12 Oct. 376/1 We named this growth-antagonistic strain for the patient, 'Tracy I'. When cell-free filtrates of broth cultures of this bacillus proved to possess strong antibiotic activity and to be non-toxic, further study seemed warranted. We have called this active principle 'Bacitracin'. **1949** H. W. Florey et al. *Antibiotics* I. x. 465 There seems to be little doubt that bacitracin is a valuable agent..especially as it is active against some pencillin-resistant organisms. **1960** A. E. Bender *Dict. Nutrition* 13/2 *Bacitracin,*..of use in brewing to control infection. **1961** *Brit. Med. Dict.* 168/1 Bacitracin ..is employed in ointment form for superficial skin diseases, and in lozenges for septic throat infections.

back, *sb.*¹ Add: **3.** *back-to-back*: advb. phr. used *attrib.*, spec. (i) of houses; also *ellipt.* as *sb.*; (ii) of an aerial system or display used in radar (see quot. 1948); (iii) of a type of combination fireplace (see quots.).

1845 L. Playfair *Rep. on State of Large Towns in Lancs.* 35 Back-to-back houses cannot be considered dwellings of proper construction. **1883** [in Dict.]. **1901** B. Seebohm Rowntree *Poverty* vi. 153 Back-to-back houses in which through ventilation is impossible. **1940** 'M. Innes' *There came both Mist & Snow* xix. 202 Back-to-backs are monotonous.

1948 Taylor & Westcott *Princ. Radar* v. 79 In..airborne radars operating on the metre wave-lengths, the back-to-back display..is used. The 'split' signals are then displayed on either side of the trace and equality results in the trace bisecting the signal line. *Ibid.* 80 (*caption*) Back-to-back display..used with metric radars. **1961** *Flight* LXXX. 926/1 One of the latest Marconi long-range surveillance radars is the Type S.247, comprising two high-power (2½–3 MW) transmitters..feeding a combined back-to-back aerial system.

1951 *Good Housek. Home Encycl.* 263/2 *Back-to-back Grates,* in this variant of the combination range..the open fire is placed in the sitting-room and the cooking range at the back, in the kitchen. **1955** D. Chapman *Home & Social Status* 54 The other living-room usually has a 'back-to-back' combination fireplace 'shared' with the kitchen.

b. Used *ellipt.*, e.g. in *one-, two-, three-pair back,* a room at the back of a house on the first, second, third floor. (Cf. Front *sb.* 11.)

1836 [see *two-pair* s.v. Two IV. 2]. **1838, 1883** [see Three-pair *a.*]. **1873** J. H. Beadle *Undevel. West* xiii. 233 City pastorals, written in a third story back, by men reared in the city. **1902** Henry James *Wings of Dove* v. xii. 196 The commodious, 'handsome' room, far back in the fine old house..the rich dusk of a London 'back'.

c. *to talk out of* or *through the back of one's neck*: see *neck *sb.*¹

4. a. *Colloq.* phrases: *the back of my hand to* (something or somebody), a phrase implying

contempt and rejection (*Sc.* and *Irish dial.*); *to know* (something) *like the back of one's hand*: to be thoroughly familiar or conversant with.

1768 A. Ross *Fort. Shepherdesses* 131 The back o' my hand to the spinning o 't. **1831** S. Lover *Leg. & Stories of Ireland* 147 The back o' my hand and the sowl o' my fut to you. **1914** Joyce *Dubliners* xiv. 199 'And have you nothing for me, duckie?' 'O, you! The back of my hand to you!' said Mrs Kernan tartly. **1943** M. Millar *Wall of Eyes* 154, I know him as well as I know the back of my hand. **1944** 'M. Innes' *Weight of Evidence* x. 107, I know that book like the back of my hand. **1956** M. Stewart *Wildfire at Midnight* i. 17, I know the district like the back of my hand. **1968** 'C. Aird' *Henrietta Who?* x. 97, I know that photograph like the back of my hand.

5. b. Short for *back blocks, *back country.

1897 D. McK. Wright *Station Ballads* 57, I went poisoning out at the back. *a* **1922** C. G. Turner *Happy Wanderer* 146 These trips from 'the Back' to the town in August are stern affairs. **1932** R. A. K. Mason in *Phoenix* (N.Z.) I. II. 9 [He] came out on the 'back', as they called the main body of the sheep-run. **1949** P. Newton *High Country Days* v. 48 An occasional trip out to the 'back' to pack firewood to those of the camp sites which had no bush within easy reach.

c. *pl.* (Also with capital initial.) The gardens behind colleges bordering on the river Cam at Cambridge.

1871 *London Society* XIX. 40/2 Cambridge University Life...You wander through those lovely 'backs' of colleges, which might almost be carpeted with poetry. **1882** A. G. Hill *Tourist's Guide Cambr.* 31 Some of the most charming 'bits' in England are to be seen in spring-time at the 'backs' of the colleges, where the narrow part of the Cam winds behind S. John's, Trinity, King's, Catherine's, and Queen's colleges. **1925** W. Deeping *Sorrell & Son* xviii. 170 Sorrell and Kit made their way through Nevil's Court and across the river to the 'backs'.

21. Also in other games.

1910 *Encycl. Brit.* XIII. 555/1 (*Hockey*) RB, Right Back. LB, Left Back. **1920** E. H. Green *Hockey* i. 1 The goal-keeper and backs require a heavier stick usually than a half back or a forward. **1968** *Sunday Times* 17 Mar. 24/6 Their backs hit a succession of long passes for Hennessey and his wings to chase.

23. a. *at the back of one's mind* (rarely *head*): in the underlying or remote part of one's mind.

1895 G. du Maurier in *Idler* Dec. 420/2 *Trilby,* as a name, must have been lying *perdu* somewhere, as they say, 'at the back of my head', as important things so often do. **1903** E. Childers *Riddle of Sands* xxi. 226 At the back of such mind as was left me lodged the insistent thought: 'we must hurry on'. **1910** S. Reynolds *Alongshore* i. 5 The deep ground-rumble of London..makes one feel continually, at the back of one's mind, the presence of the great city all around. **1930** 'J. J. Connington' *2 Tickets Puzzle* v. 61 With this at the back of his mind..he had volunteered to carry the news.

b. (See also Behind *prep.* 9.)

d. Also *colloq.*, harassing, annoying. (Later examples.) So *get off my back*: stop harassing or annoying me.

1880 W. H. Patterson *Gloss. Antrim. & Down* 4 'I'm never off his back,' *i.e.* I'm always watching and correcting him. **1889** 'Mark Twain' *Yankee at Court of Arthur* x. 100, I should have had the Established Roman Catholic Church on my back in a minute. **1945** Baker *Austral. Lang.* viii. 157 *They're on your back,* they (usually officers) are over-working you, demanding too much. **1956** A. Miller *Memory of 2 Mondays* in *Coll. Plays* (1958) 349 So I buy a car, and they're all on my back—how'd I dare buy a car! **1959** J. Cary *Captive & Free* xxxiv. 146 Well, I didn't want to get into trouble and I didn't want to have him on my back either. **1961** J. Heller *Catch-22* (1962) xl. 424 Then stop picking on me, will you? Get off my back, will you? **1968** P. Marlowe *Hire me a Hearse* ix. 120 If you had tried, Wilma would have phoned Peregrine Porter and told him to get you off her back.

g. *in back of* = *back of* (see Back *adv.* 14). *U.S.*

1914 'Mark Twain' *What is Man?* (1917) 165 The picture represents a burning martyr..in back of the smoke. **1925** G. P. Krapp *Eng. Lang. in Amer.* I. 77 *Back of*..has a variant form *in back of,* which completes the analogy to *in front of.* **1952** F. Bowers in *Papers Bibliogr. Soc. Amer.* XLVI. 194 This pure form of analytical bibliography lies in back of and leads directly into two other divisions. **1958** J. Kerouac *On Road* I. xiii. 102, I was spreading mustard on my lap in back of a parking-lot john.

24. a. *to break the back of* (fig.): later examples.

1873 *Baily's Mag.* XXIV. 45 Between them they broke the back of the Eton bowling. **1883** E. G. Holtham *Eight Years in Japan* vii. 151 That I had better stick to my onward route, at any rate till I had 'broken the back' of the journey overland to Kiyōto. **1890** Kipling *Life's Handicap* (1891) I. 48 Your flanks are unprotected for two miles. I think we've broken the back of this division. **1901** *Westm. Gaz.* 9 Mar. 4/1 The back of the fire was broken at noon. **1911** H. S. Harrison *Queed* iv. 40 The very next day, the back of the morning's mail being broken, [etc.]. **1965** R. M. Erskine *Passion Flowers in Business* viii. 97 'Your labours ..must be nearing completion.' 'I've broken the back of it, oh yes!'

h. *to put one's back into,* to employ the whole strength of one's back in (rowing, lifting, hauling, etc.). Also *fig.*

1882 Stevenson *New Arab. Nts.* (1884) 301 They put their back into their work, they sang loud and louder. **1885** Rider Haggard *K. Solomon's Mines* xviii. 291 Tackle on, and put your back into it; you are as strong as two. **1889** A. Conan Doyle *Micah Clarke* iv. 28 Pull, Micah! Put your back into it! **1952** M. Laski *Village* i. 21 You can get it done in half the time if you only put your back into it.

25. a. *to be on one's back*: also, *fig.* to be prostrate, helpless.

1840 Dana *Bef. Mast* (1854) xxviii. 177 He confessed the whole matter; acknowledged that he was on his back. **1904** *McClure's Mag.* Feb. 366/1 The employers of San Francisco are flat on their backs..; when a labor leader makes a demand we give in without a word. **1938** *New Statesman* 21 May 863/2 'Speciality selling'..is the last refuge of the man who is 'on his back'.

back, *a.* Add: **1.** *back fence,* also *attrib.*; *back row*: of a chorus, line of dancers, etc.; in *Rugby Football,* the last line of forwards in a scrummage.

1887 *Lippincott's Mag.* Sept. 421 We..sot an' spit at each other, like two tom-cats on a back fence. **1912** Mrs. Woodrow *Sally Salt* 24 She could really have believed that she could transform me from an eager back-fence prowler. **1949** R. K. Merton in Lazarsfeld & Stanton *Communications Research, 1948–9* II. 206 Walter Winchell, who reports the Broadway version of intimate gossip across the backfence.

1894 W. S. Gilbert *Foggerty's Fairy* III. p. 71, I should be at once relegated to the back rows [in a ballet], among the stout ones. **1952** Granville *Dict. Theatr. Terms* 21 *She came from the back row of the chorus,* said of an actress who has risen the hard way, from the smallest beginnings.

1897 *Encycl. Sport* I. 408/1 The wheel is made on the most favourable side by the back row bringing the ball round to the front. **1906** Gallaher & Stead *Complete Rugby Footballer* vii. 100 Advantages of the New Zealand system... Two fast men in the back row [of the scrum]. *Ibid.* 105 In the back row of the scrum we [*sc.* New Zealanders] put two of the fastest forwards and two of the best collarers that we can find. *Ibid.* 106 The side-row men press on the hookers, and the back-row men push against the lock. **1959** *Times* 30 Nov. 3/7 The Swansea back-row gave Croker and Fitch room to move.

a. Also *back area*: a region behind a field of operations (esp. *Mil.*).

1923 Kipling *Irish Guards* I. p. vii, The farthest back-areas where the enemy aeroplanes harried their camps. **1937** *Times* 25 Oct. 13/4 It [*sc.* a speech] called shame on the back-area workers who shirked extra hours and thereby helped the enemy. **1940** E. C. Shepherd *Britain's Air Power* 6 The bomber..might have to travel only 50 miles from our own lines in France to the back areas of the German Army to bomb dumps. **1956** *Planning* XXII. 59 Land that has been left derelict for many years with top soil destroyed exists as a consolidated mass of weed-covered uneven land...Such areas, often referred to as 'back-areas', have to be bulldozed and scraped level.

b. *back alley, lane, road, street* (earlier, later, and *attrib.* examples).

c **1450** in M.E.D., Backe strete. **1542** Backe lane [see Meet *v.* 11 b]. **1613** L. Bayly *Pract. Piety* (ed. 3) 652 Seeke out these in the backe Lanes, and relieve them. **1638** in *Records Early Hist. Boston* (1878) III. 6 One acre and haulfe..butting south west upon the back streete. **1708** *Boston News-Let.* 22–29 Mar. 2/2 There is a..dwelling house to be lett in the back-street. **1764** J. Kirby *Suffolk Traveller* (ed. 2) 280 The Back-Road from Woodbridge to Blithborough by Snape Bridge. **1842** *Times* 12 Sept. 6/2 The city police..desert the back lanes. **1847** F. A. Kemble *Let.* 16 Dec. in *Records Later Life* (1882) III. 317 Walking up a small back street..I saw a little child..standing at a poor mean kind of pastry-cook's window. **1865** H. B. Stowe *House & Home Papers* ii. 22 A little dingy den, with a window looking out on a back-alley. **1887** *Century Mag.* July 331 The road is what is called a 'back road', and leads through woods most of the way. **1894** 'Mark Twain' *P. Wilson* xvii. 177 He said they were back-alley barbers disguised as nobilities. **1898** *Westm. Gaz.* 6 Apr. 9/3 It was indeed remarkable what a knowledge these back-street children had of flowers. **1920** W. Stevens *Let.* 2 Dec. (1967) 220 The bouquet in this month's Poetry will drive me to back alleys. **1934** E. Blunden *Mind's Eye* III. 167 A back-alley cinema. **1957** J. L. Hodson in 'C. H. Rolph' *Human Sum* x. 192 The woman..who goes to the back-street abortionist.

c. *Phonetics.* Of a sound: formed by restriction of the oral passage at the back of the mouth; formed by the back or root of the tongue. Also *Comb.*, as *back-lateral* adj.

1867 A. M. Bell *Visible Sp.* 61 The 'shut' consonants are sufficiently distinguished..by the four radical varieties 'Back', 'Front', 'Point', 'Lip'. *Ibid.* 72 The vowels..are divided into three classes of palato-lingual formations, according as the oral cavity is moulded mainly by the 'Back', the 'Front', or the 'Mixed' (Back and Front) attitudes of the tongue. The 'Back' vowels have the largest oral cavities. **1877** Sweet *Handbk. Phonetics* 11 'Back' (guttural) vowels, in which the tongue is retracted as much as possible. *Ibid.* 31 By place there are five main classes [of consonants]. (1) Back (guttural) formed by the root of the tongue and the soft palate..(2) Front (palatal)..(3) Point ..(4) Teeth..(5) Lip. **1910** *Mod. Lang. Rev.* V. 91 A back-modified glide or murmur vowel develops between a long vowel and a back-lateral. **1962** A. C. Gimson *Introd. Pronunc. Eng.* iv. 41 Naming those vowels..in which the back of the tongue is raised towards the soft palate *back* vowels.

d. Phonology. *back mutation* [Mutation 4 b], in Old English, a change in the sound of the vowels *æ, e, i,* which were diphthongized to *ea* (= *æa*), *eo, io,* when a back vowel stood in the following syllable.

1914 H. C. Wyld *Short Hist. Eng.* v. 74 *Back-, or u-Mutation.* All the O.E. dialects are to some extent subject to this change, which consists in diphthonging *i, e,* and in Mercian *æ,* when *u,* or *o* (from earlier *-an*) followed in the next syllable, e.g. **hebun* becomes *heofon.* **1927** E. V. Gordon *Introd. O. Norse* 254 The OE. equivalent of ON. fracture is the so-called back mutation. **1953** L. F. Brosnahan *Some O.E. Sound Changes* 89 The phenomenon of back-mutation is based on the same process of regressive influence of one vowel on another, but the nature and effect of the influence in this case is different from that in *i*-mutation.

3. Of rent, taxes, etc. (Earlier U.S. examples.) *back pay, payment,* payment to cover a past period of time; also *back salary, wages,* etc.

1779 W. McKendry *Jrnl.* 21 July in *Proc. Mass. Hist. Soc.* ser. II. II. 460 [He] received 500 lashes, it being back allowance due to him. **1804** in G. L. Wardle *Rep. Charges agst. Dk. York* (1809) 542 In which case, if approved, Mr Dundas would not object to allow them back-pay. **1810** G. R. Minot *Insurrections in Mass.* (ed. 2) 59 They completed an act providing for the payment of the back taxes in specifick articles. **1811** *Records Early Hist. Boston* (1908) XXXVIII. 25 Messrs. Crane & Sohier having given up their office in the Town house.—to be notified to settle their back rent with the Treasurer. **1814** J. Mayne *Jrnl.* (1909) 128 And [the postilion] when we refused told us we had passed a bridge for which he was to take back payment, one franc. **1843** Dickens *Chr. Carol* v. 159 Not a farthing less. A great many back-payments are included in it. **1888** A. C. Gunter *Mr. Potter of Texas* xviii. 225 He was paying him back wages. **1905** *N.Y. Even. Post* 29 Nov. 10 Nearly two hundred claims for back pay.

b. Belonging to past time; see also *back number.

1808 *Sporting Mag.* XXXIII. 99/2 Your neat, elegant, and conveniently sized back volumes. **1869** *Contemp. Rev.* XI. 342 It is the duty of the resident governess..to see that the knowledge of back work is carefully kept up. **1910** 'Dehan' *Dop Doctor* (1911) xxxix. 363 To its back-files I must refer those who seek a fuller account.

back, *v.* Add: **2.** (Earlier and addit. examples.)

1728 Chambers *Cycl.* I. 116/2 The French Binders..are enjoin'd by Ordonnance to back their Books with Parchment. **1880** L. Higgin *Handbk. Embroidery* iv. 35 Silk, satin, or velvet..must be backed with a fine cotton or linen lining.

b. *Photography.* To coat the back of (a plate) with some substance which will absorb light and so prevent halation. Hence **backed** *ppl. a.*; **backing** *vbl. sb.*, often *concr.* = the coating applied to the plate.

1878 W. de W. Abney *Emulsion Proc. in Photogr.* iv. 49 This defect is..cured by applying some non-actinic varnish to the back of the plate. This backing may be made as follows: [etc.]. **1882** W. K. Burton *A. B. C. Mod. Photogr.* 59 It is well to 'back' the plate; that is, to paint or otherwise cover it at the back with some substance which will absorb light. **1906** R. C. Bayley *Compl. Photogr.* 95 A thick coat of backing is quite unnecessary. *Ibid.* 320 If there be one purpose for which backed plates are more than ever necessary, it is in interiors.

c. *Bookbinding.* To force the backs of the sections of (an unbound book) outwards on each side so as to form grooves or ledges into which the edges of the binding boards will fit. The sheets are clamped in a press, after rounding, between two bevelled boards, and the backs struck with a hammer. Hence **backing** *vbl. sb.*; also *attrib.* in *backing-board, -hammer, -machine.*

1741 Chambers *Cycl.* s.v. *Book-binding,* Then the back is turned with a hammer, the book being fixed in a press between boards, called backing-boards; in order to make a groove for fixing the paste-boards. **1818** H. Parry *Art Bookbinding* 12 Place the backing-boards a little below the back on each side. *Ibid.* 13 The centre must only be well rubbed over with the face of the backing-hammer to make it smooth. **1846** Dodd *Brit. Manuf.* ser. VI. iv. 89 In the process of 'backing',..the book is laid on a bench,..and hammered near the back edge, with such a peculiar movement of the left hand as causes the back to become rounded while the hammering proceeds. *Ibid.*, When the book is 'backed'. It is placed between two pieces of plank called 'backing-boards'. **1879** Zaehnsdorf *Bookbinding* 44 The boards required for backing, called backing boards, should always be the same length as the book. *Ibid.* 167 *Backing Machine.*—A small machine introduced for backing cheap work. **1901** D. Cockerell *Bookbinding* 118 Rounding and backing are best done after the glue has ceased to be tacky.

4. c. *Mus.* To accompany (a singer). Cf. *backing *vbl. sb.* 7 b.

1961 *Jazz Jrnl.* July 4 I've heard a record or two of Lang backing a singer.

7. (Later examples.) Also *transf.* and *fig.*

1872, 1888 [see Field *sb.* 10 a]. **1878** *Chambers's Encycl.* V. 428 There are men..ready to lay against any horse and 'back' the 'field'. **1913** *Punch* 26 Mar. 250/2 The sanguine mood induced by backing two winners. **1922** C. L. Graves *Mr. Punch's Hist.* III. 15 Lord Salisbury made his remarkable speech about our having 'backed the wrong horse, i.e. Turkey, in the Crimean War'. **1940** D. Wheatley *Scarlet Imp.* ii. 20 They [*sc.* German industrialists] backed the wrong horse, for Hitler has been their taskmaster ever since. **1955** *Times* 9 June 6/3 Dr. Clark said he never backed greyhounds but confined his modest betting to horses.

8. Earlier and additional examples (cf. *backing *vbl. sb.* 6 c). Also without *up* (*Obs.*).

1767 R. Cotton in F. S. Ashley-Cooper *Hambledon Cricket Chron.* (1923) 184 Ye Fieldsmen look sharp...When the ball is return'd back it sure. **1840** A. Bunn *Stage* II. 239 Taglioni was engaged on unheard-of terms to 'back up' Malibran's '*off-nights*'. **1854** J. Pycroft in F. Lillywhite *Guide to Cricketers* 14 Point should back behind short slip. **1898** G. Giffen *With Bat & Ball* 238 Always back up another fieldsman if you possibly can.

10. (Later examples.) Now *rare.*

1881 Swinburne *Mary Stuart* I. iii. 54 If I should never more back steed alive. **1925** E. Selous in C. Waterton *Wanderings* p. ix, The future caymanestrian..backed and was shortly unbacked by a cow.

b. *intr.* (See quot. 1942.)

1934 J. Lilico *Sheep-Dog Memoirs* 26 Any keen dog.. can be trained to back when there are other dogs doing so. **1942** R. B. Kelley *Animal Breeding* xv. 142 In sheep yards the dog that will 'back' probably is the most specialized... Their function is to go forward.. over the tightly packed sheep's backs and, by barking vigorously, cause the leading sheep to keep moving.

12. b. (U.S. examples.) Also *U.S.* and *Sc.*, to address (a letter).

1825 Jamieson Suppl., *To back* (a letter), to write the direction; more generally applied merely to the manual performance. An '*ill-backit* letter'; one with the direction ill written. **1829** R. C. Sands in *Writings* (1834) II. 136 You may as well back the paper and send what loose cash you have, besides. **1859** Bartlett *Dict. Amer.* (ed. 2) s.v., To *back* a letter, is Western for to 'direct' it. **1888** Farmer *Americanisms* s.v., The frequently-heard commercial phrase of *to back*, in the sense of 'to endorse', literally, to write on the back of a letter, bill, or cheque. **1889** J. M. Barrie *Window in Thrums* ix. 87 He had written a letter to David Alexander, and wanted me to 'back' it. **1903** Clapin *Dict. Amer.* 30 *To back*, often heard, in parts of the West and the South, in sense of to address a letter.

15. a. (Earlier example.)

1781 H. Hamilton in *Hist. MSS. Comm.* 9th Rep. App. iii (1910) II. 227 The shallowness of the water obliged us to make a dam across both rivers to back the water into the swamp.

b. *to back water* (earlier U.S. example).

1808 Ashe *Trav. Amer.* xxxvii. 303, I again took the helm, and ordered the men to back water with all their might.

16. b. *to back and fill* (see Fill *v.* 4 c, d), to go backward and forward. Also *transf.* and *fig.* *U.S.* (Earlier s.v. *Backing *vbl. sb.* 4.)

1848 Durivage & Burnham *Stray Subj.* 174 The steam was well up on both boats, which lay rolling, and backing and filling, from the action of the paddles, at the dock. **1854** *Congress. Globe* 11 Dec. 57 Men will be sent to Congress who will not 'back and fill', and be on one principle for one week.. and upon another principle another week. **1869** 'Mark Twain' *Innoc. Abr.* xxiii. 229 How in the world he [*sc.* the gondolier] can back and fill, shoot straight ahead, .. is a problem to me. **1890** *Harper's Mag.* Sept. 598/2 Nimble stewards back and fill from galley to pantry. **1903** *N.Y. Even. Post* 24 Oct., The engine was backing and filling on a sidetrack.

17. Also said of a railway train, etc.

1894 *Idler* July 609 Sometimes he would pull out when he could hear the express coming, and make her back down to the next station. **1901** Guy Boothby *Myst. Clasped Hands* iv. 75 At last the train backed into the station.

18. (Earlier U.S. example.)

1807 *Deb. & Proc. U.S. Congress* 651 Our committee recommended us to recede—to back out.

19. *to back down.* Orig. *U.S.* (Earlier and later examples.)

1849 C. Lanman *Lett. Alleghany Mts.* xi. 90 When we got up about half way.. they all three of 'em backed down and said I must not keep on. **1859** Bartlett *Dict. Amer.* (ed. 2) 17 *To back down*, to withdraw a charge, eat one's own words; as 'I asked Jenkins, before witnesses, if he had called me a cheat; and he backed right down.' **1879** F. R. Stockton *Rudder Grange* x. 113 We're not going to back down. **1934** Wodehouse *Right Ho, Jeeves* vii. 73 But don't tell me that when he saw how shirty she was about it, the chump didn't back down?

20. Of a building, etc.: to be so situated that the back abuts *on* a particular piece of land or property.

1891 G. M. Fenn *Mahme Nousie* II. xv. 257 This opening backed on to the forest, and the escaping party passed in at once among the trees. **1921** *Edin. Rev.* Jan. 190 In St. James' Square, on which the club backs, the attack was more serious.

21. *to back up.* **a.** Of running water: to meet a barrier and become deeper. Of a barrier, etc.: to cause running water to accumulate and become deeper. Chiefly *U.S.* Hence **back-up** *sb.*, an accumulation of such water (Webster, 1934).

1837 *Knickerbocker* X. 409 They.. descended this river to the mouth of White river; and as this was backed up by the spring freshets, the voyagers turned their course up the stream. **1842** P. Pusey in *Jrnl. R. Agric. Soc.* III. xiii. 176 The field is subject to floods, which had backed up in the drains. **1844** W. Palin *Ibid.* V. v. 78 An obstruction to their drainage occasioned by corn mills backing up the water to a considerable extent. **1884** [see Backing *vbl. sb.* 4]. **1960** *Times* 7 Mar. 8/3 The enormous artificial lake.. will reach 300 miles southwards, 75 miles farther than the back-up behind the present dam. **1963** R. L. Carson *Silent Spring* vi. 55 Through the labour of the beavers, a lake backed up.

b. To move backwards; also *trans.*, to drive (a vehicle) backwards. orig. *U.S.*

1834 *Visit to Texas* viii. 116 A small log building.. in the rear of which a cart was backed up on the Prairie. **1872** 'Mark Twain' *Sketches* 269 He then backed up against Pompey's statue, and squared himself to receive his assailants. **1883** *Harper's Mag.* Aug. 400 The wagons were backed up against the walls. **1948** A. Cooke in *Manch. Guardian Weekly* 21 Oct. 14/2 He gave the proper signals and slowly backed up.

c. *Climbing.* (See quot. 1923.)

1909 C. E. Benson *Brit. Mountaineering* v. 139 Backing up on smooth walls is extremely tiring. **1923** G. D. Abraham *First Steps to Climbing* v. 58 It may be easy to 'back up', with feet and knees on one side and back on the other.

22. *trans.* To carry on the back. *U.S.*

1840 Dana *Bef. Mast* xx, We started off every morning.. and cut wood.. and after dinner.. carted and 'backed' it down until sunset. **1895** *Outing* (U.S.) XXVII. 47/2 These hardy woodsmen backing packs of from eighty to one hundred pounds each.

back, *adv.* Add: **13.** *back and forth.* (Earlier and additional examples, esp. of U.S. usage.) Also as *attrib. phr.*

1613 [see Forth *adv.* 1 a]. **1816** Pickering *Vocab.* s.v., *Back and Forth*, backwards and forwards. Ex. He was walking back and forth. This is a very common expression in New England; but it is used only in conversation. **1837** Dunlap *Mem. Water Drinker* (ed. 2) I. 57, I am not one of your brook trout to be played back and forth with a hair line as her husband catches um. **1857** S. H. Hammond *Wild Northern Scenes* 138 They would run here and there, back and forth, at full speed along the sands. **1899** *Congress. Rec.* Feb. 1743/1 Some of you.. remember when at Vicksburg our boys got so close to the Confederates that they talked back and forth. **1907** U. Sinclair *Industrial Republic* p. xii, I would find myself comparing.. the two eras, and transposing its leading figures back and forth. **1909** E. B. Titchener *Text-Bk. Psychol.* I. x. 47 This back-and-forth movement will continue. **1962** *Listener* 1 Mar. 377/1 Both dangers can be avoided if there is enough back-and-forth traffic of ideas, information, and influence.

14. For (In U.S.) read (Esp. in U.S.). (Earlier and later examples.) Cf. *in back of* (*Back *sb.*[1] 23 g).

1694 in *Cal. Virginia St. Papers* (1875) I. 44 We Ranged on Ackoquane and so back of the Inhabitants and ye So[u]th. **1755** L. Evans *Geogr. Ess.* 15 If the French settle back of us, the English must either submit to them, or have their throats cut. *Ibid.* 16 If we secure the Country back of Carolina in time. **1823** G. W. Ogden *Lett. fr. West* 76 The uplands back of the Oak Hills of the Ohio river. **1840** Dana *Bef. Mast* ix, The mission stands a little back of the town. **1899** *Westm. Gaz.* 3 May 3/2 He has ideas back of his words. **1949** E. Pound *Pisan Cantos* lxxx. 96 In the shade back of the jo-house. **1953** M. Laski *Vict. Chaise-Longue* 21 No one could live there, back of the railways, down by the canal.

back-. Add: **A.** **2.** *back-aching* (example), *-breaking.*

1870 'Mark Twain' *$30,000 Bequest* (1906) 214 Did you ever notice.. how back-breaking and tiresome it was? **1913** *Maclean's Mag.* May 93/2 Every Monday I have a back-breaking session with that washing machine. **1916** 'Taffrail' *Pincher Martin* iii. 34 The work of shovelling the coal into bags was back-breaking. **1940** F. Kitchen *Brother to Ox* ii. 25 What a back-aching job it was! **1958** *Times* 27 Oct. 11/3 [Potato-picking] looks so easy, and to the novice pressed into service as I am it is a back-breaking penance.

3. *back-ache* (later examples).

1907 Masefield *Tarp. Muster* xii. 132 It was 'Shift topmasts', or 'Down top-gallant yards', or some gummy backache or another all the whole day long. **1922** Joyce *Ulysses* 40 We have nothing in the house but backache pills.

5. *back-court* (see *B), *-drawing-room* (earlier example), *-garden* (examples; also *transf.* and *fig.*), *-kitchen* (later examples), *-parlour* (earlier example), *porch, shed.*

1738 *Purefoy Lett.* 8 Oct. (1931) I. viii. 191 To mend two Pannells in the back Parlour 0-03-06. **1789** Washington *Diaries* (1925) IV. 30 A back shed, which seems to be added as the family encreases. **1811** Jane Austen *Let.* 25 Apr. (1932) II. 274 We were 66.. quite enough to fill the Back Drawᵍ room. **1818** Keats *Let.* 24 Mar. (1958) I. 254 To have a sort of Philosophical Back Garden. **1832** *Chambers's Edin. Jrnl.* I. 50/1 It is true.. that nothing can match your back kitchen as a convenience to the servants. **1840** *Southern Lit. Messenger* VI. 734/1 He was led by the hand into the back porch. **1875** H. B. Stowe *We & Neighbors* liv. 480 He had leaped out of a window upon a back shed. **1889** G. B. Shaw in *Hawk* 13 Aug. 172/2 Wagner.. is 'buried in the back garden, sir, like a Newfoundland dog'. **1960** R. Williams *Border Country* i. ii. § 6.51 It had a built-on back-kitchen. **1963** *Times* 16 Jan. 13/1 After cultivating their own back gardens contentedly.. for 15 post-war years up to the beginning of 1961, steel-makers the world over are now preoccupied with export problems.

7. *back-going* (example).

1924 Buchan *Three Hostages* xxi. 299 The body was heavy, and he was clearly a back-going beast.

11. *back-blast, -eddy* (usu. *fig.*), *-flip, -flow* (fig.), *-kick* (also *fig.*), *-somersault, -swing* (also *fig.*), *-swirl.*

1923 Kipling *Irish Guards in Great War* I. 88 Our shrapnel, which had no back-blast. **1928** T. S. Eliot *Dial. Poetic Drama* p. xxii, The age of Shakespeare moved in a steady current, with back-eddies certainly, towards anarchy and chaos. **1936** J. C. Powys *Maiden Castle* (1937) i. 40 The least back-eddy of remorse or shame. **1935** W. G. Hardy *Father Abraham* i. ii. 29 His cane cut Simil-i-una.. across the calves. Its back-flip caught Abraham just as expertly across the belly. **1967** *Boston Sunday Herald* 30 Apr. i. 25/1 The Yanks would do back-flips to get Yaz, Tony C., or Petrocelli. **1940** E. Wilson *To Finland Station* i. iii. 16 The back-flow of old instincts and interests among the purposes and hopes of the new. **1924** A. J. Small *Frozen Gold* xiii. 272 Their back-kicks went out as far as their bushy tails. **1939** G. Greene *Lawless Roads* x. 261 Mexico gave me a back kick. **1940** *Chambers's Techn. Dict.* 68/1 *Back-kick*, term applied to the violent reversal of an internal-combustion engine during starting, due to a back-fire. **1591** Backe somersaut [see Somersault *sb.* a]. **1865** 'Lewis Carroll' *Alice in Wonderland* v. 64 You turned a back-somersault in at the door. **1899** H. G. Hutchinson et al. *Book of Golf & Golfers* v. 125 If the back swing is a little tied up.. the down swing and its finish are as free as we could have them. **1934** Webster, *Backswing*, the movement of the [golf] club backward to the position from which it is brought down to strike the ball. **1948** J. Harvey *Plantagenets* vii. 87 The break of dynasty and the back-swing of the pendulum. **1954** J. H. Fingleton *Ashes crown Year* iv. 45 Hole's big back-swing.. leaves him vulnerable to an in-swinger. **1966** Mills & Butler *Mod. Badminton* iv. 34 For the high singles serve.. your back-swing should go back further. **1913** D. H. Lawrence *Sons & Lovers* xiii. 365 The Trent carries bodily its back-swirls and intertwinings.

12. b. *back-fare* (example).

1832 in Mrs. A. Mathews *Mem. C. Mathews* (1839) IV. 106 No, your honour, not unless you promise me the back-fare.

c. *back-answer* (later examples); hence as *v. trans.*

1915 A. Conan Doyle *Valley of Fear* ii. ii. 187 'You have your back answer quick enough.' 'Yes, I was always quick of speech.' **1921** H. Williamson *Beautiful Years* 221 No one has ever back-answered Sol Isaacs and not paid for it. **1939** G. Greene *Confid. Agent* ii. i. 184 'No more back answers,' the policeman said. **1959** P. McCutchan *Storm South* iv. 57, I didn't want to be hauled over the coals for back-answering a passenger.

B. **back-boiler,** a boiler behind a domestic fire or cooking range; **back-brand** *dial.* = Back-log; **back-breaker,** (*b*) a back-breaking task, etc. (cf. Back *sb.*[1] 24 a and *Back- A. 2); also in extended use; **back-burn** [*Burn *sb.*[3] 1 c]: see quot. 1944; also *back-burning*; **back-comb** *sb.*, an ornamental comb worn at the back of the head; also *v. trans.* and *intr.*, to comb the underlying hairs of a strand towards the scalp; **back-court,** (*a*) (see Back- 5: examples); (*b*) in *Lawn Tennis* (see quot. 1961); also *attrib.*; **back-crawl,** in *Swimming*, a form of the crawl in which the swimmer lies on his back; **back-cross** *v. trans.* (*Biol.*), to cross (a hybrid) with one of its parents; *back-crossing* vbl. sb.; hence **back-cross** *sb.*, an instance or product of back-crossing; **back cut,** in *Cricket*, a late cut; hence *back-cut* v. trans.; **back-cutting,** in *Civil Engin.* (see quots.); **back-draught,** (*b*) a drawing in of the breath; an act of drinking or gulping down *Sc.*; (*c*) a reverse current of water, under-tow; **back-drop** *Theatr.* (orig. *U.S.*), = *Back-cloth 2; also *transf.* and *fig.*; **back-electro-motive force, back e.m.f.,** in *Electr. Engin.*, an electro-motive force which opposes that producing the current; **back-filling** (examples); so *back-fill* v. (1651 in D.O.S.T.); *back-filled* ppl. a.; **back-flash,** the act or process of flashing back (Webster, 1934); spec. (*a*) *Forestry* (see quot. 1957); (*b*) = *Flashback *sb.* 2; **back focus,** in *Photogr.* (see quots.); **back-heel** *v.*, (*b*) to kick (a football) backwards with the heel; **back-lift,** in *Cricket*, a backward lift given to the bat immediately before a stroke is played; in *Rugby* and *Assoc. Football*, a backward lift given to the leg when kicking a ball; **back-lighting,** in *Photogr.*, lighting coming from behind the subject; **back play,** in *Cricket*, a method of play in which the batsman steps back towards the wicket and plays the ball from behind the popping crease; hence *back-player*; **back-pressure** (example); also, any resistance to the flow of a liquid or gas; also *attrib.*; **back projection** *Cinematogr.* (see quot. 1933); so *back-projected* ppl. adj.; **back-saw** (see quot.); **back-shift** (example); **backshore** (see quot. 1919); **back spacer,** a typewriter key that moves the carriage one space backward; also *back-space key* and *transf.*; so **back-space** *v. intr.*, to use such a key; *back-spacing* vbl. sb.; **back-spin** = Underspin; **back-stool,** a stool with a back; **back-straight** (see Straight B. 3), the stretch along the side of a racecourse or stadium opposite to that in which the races end; **back-winter** (later example); **back-word** (examples).

1939 L. J. Overton *Dom. Hot Water Suppl.* iii. 12 The *Back-Boiler*, for setting in brickwork behind a kitchen range or at the back of an ordinary fireplace. **1844** W. Barnes *Poems Rural Life in Dorset Dial.* 201 We got a back bran', dree girt logs. **1874** Hardy *Far from Madding Crowd* I. xxii. 267 The log which was to form the back-brand of the evening fire was the uncleft trunk of a tree. **1909** Webster, *Back breaker*, a task requiring excessive exertion. **1929** F. Bowen *Sea Slang* 5 *Backbreakers*, old-fashioned ship's pumps. **1962** *Spectator* 13 Apr. 480, I don't fancy the back-breaker or the pile-driver [in wrestling]. **1944** *Soc. Amer. Foresters: Forestry Terminol.* 10/1 *Back-burn*, denotes a controlled fire burning against the wind. **1878** E. S. Elwell *Boy Colonists* 90 The 'back-burning' of the fire, which though very slow, is always the most steady and most effective. **1865** Dickens *Mut. Fr.* I. ii. xii. 268 Ladies.. twisting their back-hair.. and many of them.. carrying their back-combs in their mouths. **1955** 'C. Brown' *Lost Girls* xii. 130 She had back-combed her hair so that it stood out. **1960** *Sunday Express* 14 Aug. 12/2 Backcombing madly.. my favourite hairdresser.. built up some splendid, puffed out effects. **1960** *News Chron.* 13 Sept. 6/2 The trend in hair styles has been, basically, inflated.. set and back-combed into.. magnificent proportions. **1774** T. Pennant *Tour Scotl.* I. 1 The back courts of all these houses are level with the ground. **1784** Ld. Fife *Let.* 29 Apr. in *Ld. Fife & Factor* (1925) vii. 165, I found the Duchess of Devonshire in

the Back Court, solliciting my paistry Cook to vote for Mr Fox. **1890** L. Dod in C. G. Heathcote *Lawn Tennis* xvi. 312 For volleying, more especially than for back-court play, it is essential that the dress should be loose. **1908** *Westm. Gaz.* 11 May 8/3 Blended judiciously with her back-court game was volleying of a vigorous order. **1961** F. C. Avis *Sportsman's Gloss.* 253/2 *Backcourt*, that portion of the court extending from the service line in a direction away from the net. **1929** Handley & Howcroft *Crawl-Stroke Swimming* 77 Cinema camera pictures of a number of back-crawl exponents have shown that the majority make an eight beat kick. **1951** *Swimming* (E.S.S.A.) iii. 43 The back-crawl is the fastest of the back strokes and is the second fastest swimming stroke known. **1904** *Jrnl. R. Microsc. Soc.* Feb. 52 The correctness of Mendel's hypothesis of the purity of the germ-cells and of their production in equal numbers, is shown by back-crossing of a hybrid with one of the parental forms.. Any one of them back-crossed with the recessive parent will produce 50 p.c. pure recessives and 50 p.c. hybrids. **1915** T. H. Morgan et al. *Mechanism Mendelian Heredity* iii. 52 If the F_1 males are backcrossed to black vestigial females only two classes result. **1919** Bridges & Morgan *Contr. Genetics of Drosophila Melanogaster* 171 Two autosomal back-crosses had been completed. *Ibid.* 172 No back-cross which involved autosomal linkage had been possible. *Ibid.* 173 (*heading*) Back-Cross Test of Females, Purple Vestigial 'Coupling'. **1929** *Genetica* XI. 227 Back-crossing in peas is rather troublesome. **1931** E. B. Ford *Mendelism & Evolution* 1. i. 6 Segregation will also occur if heterozygotes are mated with either of the homozygous types. Such a mating is known as a 'back cross', for it is produced when the F_1 (heterozygous) generation is crossed back to one of the parents. **1845** 'N. Felix' *Felix on the Bat* I. ii. 12 He could not make the back cut equal to the other parts of his batting. **1898** G. Giffen *With Bat & Ball* 221 The back cut, the sweetest of strokes. **1954** J. H. Fingleton *Ashes crown Year* xi. 112 Back-cutting a no-ball from Lindwall most beautifully for 4. **1842** Back-cutting [see *side-cutting* s.v. Side *sb.*[1] 27]. **1940** *Chambers's Techn. Dict.* 68/1 *Back cutting*, earth obtained for a railway or canal bank, when the excavated earth does not suffice for a regular cut and fill. **1825** Jamieson, We was whaslin like a blastit stirk i' the back-draucht. **1874** G. M. Hopkins *Notebks.* 13 Aug. (1937) 202 It [*sc.* the wave] commonly has a pitch or lurch to one side besides its backdraught. **1887** Morris tr. *Homer's Odyssey* v. 97 The back-draft mightily Fell on him, and.. drave him out to sea. *Ibid.* xii. 218 Nor happen thou upon her [*sc.* Charybdis] when the back-draught she doth win. **1922** J. B. Salmond *Bawbee Bowden* xii. 104 Wi' the backdraucht [he] sent a moofu' o' tea up throo his nose. **1913** *Amer. Mag.* July 103/1 When the film is run off you see the back-drop right through him [*sc.* the Ghost] while he approaches Hamlet. **1947** D. M. Davin *For Rest of Lives* xliii. 215 'Antimacassars, potstands complete with ferns, occasional tables, bric-a-brac.' 'Sounds like the backdrop of a Victorian wedding photo.' **1962** *Listener* 13 Sept. 390/1 A particular contribution to any science can only be assessed against the backdrop of history. **1895** Rutherford in *Trans. N.Z. Inst.* XXVIII. 190 Since dN/dt may be called the back E.M.F. in the circuit at any instant. **1898** T. O'C. Sloane *Stand. Electr. Dict.* 156 Counter-electro-motive Force.. *Synonym*—Back Electro-motive Force. *Ibid.*, Back Electro-motive Force of Polarization. **1901** Geipel & Kilgour *Electr. Engin. Formulæ* (ed. 2) 668 When the anode and cathode are of the same metal.. there is no back E.M.F., for the back E.M.F. at the one electrode is of opposite sign to that at the other, and they cancel one another. **1936** *Discovery* July 202/2 In the very small fraction of a second that the current is flowing in one direction, the back electro-motive force which opposes it has not time to form. **1901** R. Sturgis *Dict. Archit. & Building* I. 181/1 *Back filling*,.. masonry or earth, and the like, used as a filling over the back (or extrados) of arched constructions, as tunnels and sewers. **1930** *Engineering* 29 Aug. 259/2 The trenches have to be backfilled with sand, gravel or other good clean earth. **1955** *Archit. Rev.* CXVIII. 393/3 Mineral operators, when they have finished extraction, are generally free to backfill without planning consent. **1960** *Farmer & Stockbreeder* 12 Jan. 63/2 Levelling off back-filled tile drain trenches. **1957** Cook & Welch in *Jrnl. Forestry* LV. 265/1 'Backflash'—the sickening or sudden death, for no apparent reason, of untreated trees in a stand where chemi-peeling has been done. **1958** *New Statesman* 1 Feb. 144/3 Spiro's ruinous past is displayed in a series of back-flashes. **1960** *Ecology* XLI. 56/2 Backflash is the movement of poison, through root grafts, from poisoned trees to unpoisoned trees. **1963** in Brown & Foote *Early English & Norse Studies* 133 These disorderly and almost randomly presented backflashes [in *Beowulf*] fed to the audience are.. truly confusing to a person not saturated with the material. **1897** E. J. Wall *Dict. Photogr.* (ed. 7) 295 'Back focus'.. is the distance between the posterior lens and ground-lens. **1953** Amos & Birkinshaw *Telev. Engin.* I. ix. 202 The distance between the lens and the image, known as the *back focus*. **1922** *Weekly Dispatch* 12 Nov. 10 Roberts.. back-heeled cleverly to Quinn while going at top-speed. **1912** C. B. Fry *Cricket* (*Batsmanship*) 8 Top of the back-lift for the cut. **1955** A. Ross *Australia 55* 135 He is not happy against real pace. His back lift, circular rather than straight, is high. **1960** E. S. & W. J. Higham *High Speed Rugby* ii. 28 Often enough during a game, you.. must kick immediately. Once more, be content, at first, with a short back-lift and a short follow through. **1961** *Times* 19 Jan. 3/7 This highly mobile dangerous centre forward who was prepared to shoot on sight with scarcely any backlift. **1950** R. H. Bomback *Cine Data Book* 78 'Baby' Solarspot.. small size unit for modelling, back-lighting, front and cross-lighting. *Ibid.*, General-purpose lamp, particularly suited for back-lighting. **1959** Halas & Manvell *Technique Film Anim.* 336 Back lighting is used on the rostrum camera beneath either a drawing or a celluloid to give a silhouette or a transparent effect. **1844** *Lillywhite's Illustr. Hand-Bk. Cricket* 19 Forward play.. is more pleasing and graceful.. than back play. **1897** K. S. Ranjitsinhji *Jubilee Bk. Cricket* iv. 158 All the really strong back-players draw back in making back-strokes. **1860** *Encycl. Brit.* XX. 600/2 The mean back-pressure .. exceeds the pressure of condensation. *a* **1877** Knight *Dict. Mech.* I. 206/2 *Back-pressure valve*, a ball or clack-valve in a pipe. **1930** *Engineering* 5 Dec. 699/2 To this station will be supplied all the surplus power generated by the back-pressure turbine. **1940** *Chambers's Techn. Dict.* 69/1 *Back pressure*, air pressure in pipes when it exceeds atmospheric pressure. **1962** *Times* 28 Sept. 17/3 The backgrounds, whether painted or back-projected, are handsome and spare. **1933** A. Brunel *Filmcraft* 153 *Back projection*, projection on to a transparent surface, with the projector *behind* the screen, hidden from the view of the audience in a cinema and from the view of the camera in a studio. **1939** J. Dell *Nobody Ordered Wolves* vii. 90 Back-projection.. is a process by means of which an actor in Hertfordshire can be shown in Hyde Park, or the Bois, or the plains of Tibet. By the simple expedient of placing the actor in front of a screen on which is projected a film of the required setting, and by synchronising the two cameras. *a* **1877** Knight *Dict. Mech.* I. 206/2 *Back-saw*, a saw whose web is stiffened by a metallic back of greater substance; as, a tenon saw. **1860** *Eng. & Foreign Mining Gloss.* (ed. 2) 48 *Newcastle Mining Terms*.. *Back-shift*, the second set of hewers in each day. **1919** D. W. Johnson *Shore Processes & Devel.* iv. 161 This zone is already well known as the foreshore. Back of it is the portion of the shore covered by water during exceptional storms only, which I propose to call the backshore. **1937** Wooldridge & Morgan *Physical Basis Geogr.* xxi. 322 We may thus distinguish the 'foreshore'.. from the 'backshore' lying immediately at the cliff-foot. **1915** A. J. Sylvester *Underwood Typewriter Manual* 1. 13 The back space key is at the upper left-hand side of the keyboard. **1919** B. de Bear *Typewriting from A to Z* xix. 39 Back-space once for every character and space in the title to be centred. **1907** *Jrnl. Soc. Arts* 1 Mar. 432/1 The back-spacer key.. causes the carriage to move backward one space. **1962** *Which?* Dec. 355/1 The action of the back spacer.. was a little doubtful, sometimes jumping two spaces. **1919** B. de Bear *Typewriting from A to Z* xxii. 46 It is a question of back-spacing after the first character has been written, so as to type the second on the same point. **1957** *Economist* 19 Oct. 205/1 A Stenorette [dictating machine] .. has .. automatic back-spacing and erase facilities. **1916** E. F. Benson *David Blaize* x. 200 He chipped at it [*sc.* the ball] with a lot of back-spin. **1920** E. R. Wilson in P. F. Warner *Cricket* ii. 84 Back spin is undercut applied to the back half of the ball, and is more easily put on with a low action. **1926** *Amer. Speech* I. 632/1 *Backspin*, when put on the ball [in golf] brings a special kind of stop shot. **1762** Ince & Mayhew *Univ. Syst. Houshold Furniture* 8/1 Four Designs of Back Stool Chairs.. Four more Designs of Back Stools. **1945** *Burlington Mag.* July 164/2 The single chair was called at first a 'back-stool'; for it was not a variation of the armchair by the removal of the arms, but a development of the stool, to which, in order to make it more comfortable, a back was added—hence the name back-stool. **1952** J. Gloag *Dict. Furn.* 121 Although referred to in Elizabethan inventories, back stools were not in common use until the middle years of the 17th century, and the name survived during the 18th century. **1905** *Cycling* 24 May 441/3 The teeming thousands on the terraces above the back-straight. **1952** M. Duggan in D. M. Davin *N.Z. Short Stories* (1953) 246 She saw them go into the bend and show again.. across the back-straight. **1881** W. D. Howells *Let.* 27 Feb. in *Amer. N. & Q.* (1963) 133/1 What the Canadians call the back-winter: the two months of mud and snow that precede the spring. **1841** R. W. Hamilton *Nugæ Literariæ* 357 'In consequence of her death, I was obliged to give a party who were to have dined with me backword;' that is, put them off. **1937** J. B. Priestley *I have been here Before* 1. 6 We ought to.. charge 'em a deposit when they book rooms in advance, and then if they do give backword we're not clean out o' pocket.

back-action. orig. *U.S.* [Back- A. 11.] Backward or reverse action, as in a machine. Also *attrib.* and *fig.*

1845 *Knickerbocker* XXV. 406 There lurks beneath it [*sc.* the Yankee countenance] the knowledge of.. some 'self-acting back-action sausage-stuffer'. **1862** Lowell *Biglow P.* ii. iii. i. 99 The self-cockin', back-action style o' J. D. **1873** J. H. Beadle *Undevel. West* 800 That sort of detraction has an awkward back-action about it. *a* **1877** Knight *Dict. Mech.* 205/2 Back-action Steam-engine. **1934** *Brit. Jrnl. Psychol.* Jan. 267 By 'back action' is meant the alleged fact that the conditions which follow an act affect the bond or connection responsible for the act. **1963** J. Osborne *Dental Mechanics* (ed. 5) x. 204 Back action clasps will be used on premolars.

So **back-actioned** *a.*

1875 [see Bar *sb.*[1] 29 a].

backage (bæ·kėdʒ). [f. Back *sb.*[1], after Frontage.] The back part of a building or row of buildings; the line or outlook of buildings or plots of land on the rear side.

1887 Fenn *This Man's Wife* II. ii. xviii. 92 A high wall right and left to complete the blankness of the frontage. It ought to have been called the backage; for Sir Gordon Bourne's house was very pleasant on the other side. **1892** *Pall Mall Gaz.* 22 Mar. 2/3 Converting slummy 'backages' into handsome frontages. **1894** G. Du Maurier *Trilby* II. vi. 191 Tall trees, whose lightly-falling leaves yellowed the pavement for at least a hundred yards of frontage—or backage, rather; for this was but the rear of that stately palace.

back-along (bæ·k,ǎlǫ·ŋ), *adv.phr.* *dial.* [f. Back *adv.*] Back, in direction or time. Cf. *Along *adv.* 1 c.

[**1877** *Rep. & Trans. Devonshire Assoc. Adv. Sci.* IX. 127 *Backlong* (= Backwards. Formerly). This expression is of common occurrence about Teignmouth.] **1897** T. Hardy *Well-Beloved* ii. viii. 162 And when, on my way backalong, I saw you waiting hereabout again, I slipped over the wall. **1905** E. Phillpotts *Secret Woman* i. ix. 83 Then I mind the time when Barbara Westaway wouldn't have him—backalong ten year or so. **1923** R. Frost *Let.* 12 Feb. (1964) 160 A good old-timer dating back along.

backband. Add: **2.** The outside moulding on a door or window casing.

1940 in *Chambers's Techn. Dict.* **1944** C. F. Dingman *Estimating Building Costs* (ed. 3) ix. 202 Window trim should have dimensions and descriptions of inside casings.. backbands, scotia, etc.

back bench. [Back- A. 4.] Any one of the benches in the House of Commons or similar assembly occupied by members who are not entitled to a seat on the front benches on either side. Usu. *attrib.* (with hyphen). Hence **ba:ck-be·ncher**, a member who occupies a seat on the back benches on either side of the house.

1874 *Gentl. Mag.* XII. 334 During the debate on the Education Act Amendment Bill Mr. Osborne found himself sitting on a back bench below the gangway, and rose thence to address the House. **1885** *Eng. Illustr. Mag.* Dec. 195/1 He affects the Liberal side, sometimes presenting himself from a back bench in the rear of the bishops. **1905** *Daily Chron.* 13 July 6/1 Back-bench Members on the Ministerial side of the House. **1910** *Busy Man's Mag.* Dec. 60 Some of the best men in Parliament are back-benchers. **1923** *Daily Mail* 23 May 6 Mr. Baldwin filled the modest rôle of a back-bencher in the House for many years. **1928** *Daily Tel.* 13 Mar. 9/3 The situation was obviously developing into general back-bench communal bickering. **1940** *Ann. Reg. 1939* 9 A man comparatively unknown to Parliament—.. a 'back-bencher'. **1959** *Economist* 3 July 19/2 The foolish notion that a back-bencher's life need be a useless one.

back blocks, backblocks, *sb. pl.* *Austral.* and *N.Z.* [f. Back *a.* 1 + *Block *sb.* 14 d.] Land in the remote and sparsely inhabited interior. Also, land distant or cut off from a river-front. Also (esp. in sing.) *attrib.* or *adj.*; **back-blocker**, a resident in the back blocks.

Also used in Canada: there are 1910–1956 examples in the *Dict. Canadianisms*.

1872 *Glimpses Life in Victoria* iii. 31 We were doomed to see the whole of our river-frontage selected and purchased by two gentlemen newly arrived in the colony. The back blocks which were left to us were insufficient for the support of our flocks, and deficient in permanent water-supply. **1890** E. W. Hornung *Bride fr. Bush* xix. 298 'Down in Vic' you can carry as many sheep to the acre as acres to the sheep up here in the 'back-blocks'. **1891** 'R. Boldrewood' *Sydneyside Saxon* xii. 215 One of the back-block youngsters. **1892** E. W. Hornung *Under Two Skies* 21 Sitting on his heels over the fire in an attitude peculiar to back-blockers. **1901** *N.Z. Illustr. Mag.* III. 254 The clean-limbed hardy backblocks-men mustering in the river arm. **1911** H. Foston *In Bell Bird's Lair* xi. 52 The men of the city, who had not the faintest idea of what a 'Backblock' bush road was like in the winter. **1913** B. E. Baughan *Brown Bread from Colonial Oven* vii. 140 Art comes at all times scantly to the back-blocks. **1926** A. F. Webb *Miss Peters' Special* xi. 94, I think on most back-block holdings mail day is kept as a kind of Sunday or holiday. **1930** J. Devanny *Bushman Burke* i. vii. 32 It is one of the back-blocker's pleasures, taking tea in the restaurants in town.

backbone. **4.** (Earlier example.)

1843 C. Brontë *Let.* 13 Oct. in Mrs. Gaskell *Life* I. 299 A pretty-looking.. young man, apparently constructed without a backbone.. I don't allude to his corporal spine.. but to his character.

backboned, *ppl. a.* Add: Also *fig.*

1940 H. G. Wells *All Aboard* i. 27 We'll have clear-headed, back-boned, clean-minded people.

backboneless, *a.* (Examples.)

1882 J. Payn *For Cash Only* III. xlviii. 260 Mildred's pride was of the most backboneless description, a mere social conceit. **1886** *Punch* 17 July 28/2 Think of a backboneless man, unable 'to swim against the stream'. **1903** W. T. Arnold *German Ambitions* 89 Backboneless submissiveness to Britain. **1925** *Glasgow Herald* 1 Aug. 4/2 It [*sc.* hæmoglobin] has analogues like hæmocyanin (in many backboneless animals).

back-cap, *sb.* *U.S. slang.* *to give one a back-cap*, to disclose or state something to one's detriment; to run down. Also as *v. trans.* (*Cent. Dict.*, 1889).

1883 'Mark Twain' *Life on Miss.* lii. 462 Now I didn't fear no one giving me a back-cap (exposing his past life) and running me off the job. **1903** Clapin *Dict. Amer.* 31 *To back cap*, to speak evil of some one, so as to spoil his game.

back-chat, *sb.* *colloq.* [? orig. soldiers' slang; f. Back- A. 12 c + Chat *sb.*[1]] Impertinent or impudent replies, esp. to a superior; abuse, insulting speech; altercation, heated talk; repartee. Cf. *back-answer* (Back- 12 c) and *Back-talk.

1901 *Subaltern's Lett. Wife* 108 'That'll do, Sergeant Jones', I heard one of our colonial officers remark; 'I don't want any more of your back-chat.' **1922** *Daily Mail* 31 Oct. 9 It was very amusing to hear public men described as 'cabin boys' and 'kitchen staff'; it was what children described as 'calling names' and soldiers 'back-chat'. **1922** W. J. Locke *Tale of Triona* viii. 82 'Here you are, you blackmailing thief.' 'None of your back-chat..' said the taximan. **1939** T. S. Eliot *Old Possum* 36 I'd extemporize back-chat, I knew how to gag.

Hence as *v. intr.*, to use back-chat (to).

1927 H. V. Morton in *Daily Express* 27 Apr. 11 No one back-chats to a Pacific [*sc.* railway engine] except a very shabby tank. **1930** A. Bennett *Imperial Palace* vi. 27 The bright, jostling back-chatting world of men.

back-cloth, backcloth [Back- A. 4.]

1. *Calico-printing.* A cloth placed between the fabric that is being printed and the 'blanket', in order to keep the latter clean. Also called *Back-grey.

1874 W. Crookes *Pract. Handbk. Dyeing & Calico-Printing* viii. 552 The revolution of the blanket, back-cloth, and pieces is regulated by the following contrivance. **1897** C. F. S. Rothwell *Print. Textile Fabrics* 34 The back tenter's work is to guide the piece and back cloth into the machine. **1901** A. Sansone *Print. Cott. Fabrics* (ed. 2) 420 For keeping the blanket clean, the back cloth is allowed to go between the blanket and the printing cloth.

2. *Theatr.* The painted cloth hung across the back of the stage as the principal part of the scenery. Also *transf.* and *fig.*

1886 *Cornh. Mag.* Oct. 435 They gazed awestruck at the backcloth and the flies. **1926** *Spectator* 10 July 44/2 Thirty acres or so for a stage and the whole firmament of heaven for a back-cloth. **1927** *Observer* 6 Nov. 9 The background .. is not a carefully constructed scene, but only a flimsy and dim back-cloth. **1955** *Times* 31 Aug. 9/2 hanging indeed like a backcloth to Mr. Lennox-Boyd's tour, is the policy and power of China. **1955** G. S. Fraser in J. Wain *Interpretations* 214 A river, for a Restoration poet, would be primarily.. a backcloth for pastoral.

3. *Naut.* 'A triangular piece of canvas fastened in the middle of a topsail-yard to facilitate the stowing of the bunt of the topsail' (*Cent. Dict.* 1889).

back country. Chiefly *N. Amer.*, *Austral.*, and *N.Z.* [Back *a.* 1.] The country lying towards or in the rear of a settled district.

1746 in C. R. Woodward *Ploughs* (1941) 332 In the back Country abt Paoqualin The Cattle feed is near over. **1789** Morse *Amer. Geogr.* 415 The mistletoe is common in the back country. **1831** J. M. Peck *Guide for Emigrants* iii. 292 It.. has an abundance of excellent water, and a back country as range for stock. **1840** *N.Z. Jrnl.* 29 Aug. 209/2 Hobson is founding a town at the Bay of Islands, but we have an assurance that there is no back country there. **1844** *Bytown Gaz.* (Ottawa) 8 Aug. 2/6 A road into the interior.. would not only benefit Kingston, (so much in want of a productive back country) but would open new tracts for settlement. **1860** S. Butler *First Year in Canterbury Settlement* 24 Mar. (1863) iv. 47 At last I have been really in the extreme back country. **1868** *Putnam's Mag.* Nov. 562 The hotel was a roomy log-house,.. and commanded a view of the back country—a prairie stretching off into the western horizon. **1882** [see Back *a.* 1 a]. **1902** *Westm. Gaz.* 6 Aug. 2/1 The unappropriated back-country of Tunis and Tripoli. **1936** F. Gerald *Millionaire in Memories* iii. 71 Travelling through the back country with Bill Langford was an education, a 'bushman's' education. **1964** *Weekend Mag.* (Montreal) 11 July 20/2 Devil's club [is] the thorned scourge of hikers in much of British Columbia's back country.

b. *attrib.*

1775 C. Drayton *Let.* in *South Carolina Hist. & Geneal. Mag.* (1926) XXVII. 137 His trepanned Lordship advises the back Country people not to take up arms. **1867** Lady Barker *Station Life* Aug. (1870) xx. 173 The distant 'back-country' ranges. **1871** E. Eggleston *Hoosier Schoolm.* (1872) Pref. 5 Describing life in the back-country districts. **1887** *Harper's Mag.* Jan. 328/1 In a back-country town.. there chanced to die one of the members of the community. **1901** 'Linesman' *Words by Eyewitness* (1902) 161 Imaginations so crude as those of the back-country Boers. **1948** *Coast to Coast 1947* 296, I was in a back-country farm in Central Otago. **1965** *Star Weekly* (Toronto) 2 Jan. 37/2 The nearest doctor was 70 miles away over a poor back-country road.

Hence **back-countryman.**

1796 *Gaz. U.S.* 19 Nov., Advt. (Th.), A new Ballet Dance, called the Back Countryman, or the New Settlers. **1845** W. G. Simms *Wigwam & Cabin* ser. i. iv. 22 The boatman, who.. knew by his dialect and dress that he was a back-countryman, came to his relief. **1942** Marsh & Burdon *New Zealand* 9 The shepherds, musterers, shearers and drovers, the back-countrymen of New Zealand.

back county. *U.S.* [Back *a.* 1.] A county lying in the inland part of a state. Also *attrib.*

1775 in *Maryland Hist. Mag.* X. 317 Finding it difficult to go thro' the back County.. he agreed to accept a Commission. **1788** *Massachusetts Spy* 11 Nov. (Th.), A back-county correspondent informs us that [etc.]. **1803** T. M. Harris *Jrnl. Tour* 6 June (1805) 59 In the back counties of Virginia. **1904** W. H. Smith *Promoters* xx. 290 The little book agent.. was working one of the brethren from some back county in great style.

ba·ck-date, backdate, *v.* [Back *adv.* 4.] *trans.* To affix or assign a date earlier than the actual one to (a document, book, event, etc.); to render an enactment, agreement, etc., valid retroactively from a given date.

1946 *Sun* (Baltimore) 26 Sept. 2/2 A stenographer.. testified that.. a War Assets official dictated the memorandum last April and ordered her to back-date it to December 18, 1945. **1952** *Daily Progress* (Charlottesville, Va.) 6 Feb. 1/5 Backdating of tax returns is the major irregularity.. found in the San Francisco internal revenue office. **1957** *Ann. Reg. 1956* 4 A deadlock ensued on whether the award, if accepted, should be back-dated to the beginning of the lock-out. **1958** *Listener* 19 June 1028/1 The poems cover the years 1918–58. One might back-date them in style perhaps by ten years.

back-door. Add: (Now usu. without hyphen.)

1. b. *back-door trot* (fig.); also *spec.*, diarrhœa, *dial.*

1789 D. Sillar *Poems* 57, I fear they've [*sc.* letters have] tane the back-door trott, An' miss'd the road. **1801** *Marvellous Pleasant Love-Story* I. xiii. 176 Cooky had for certain put some jalup i' the pudding, for Master George had gotten the back-door-trot with a witness! **1886** F. T. Elworthy *West Somerset Word-Bk.* 37, I be saafe, nif I was vor ate very many o' they there, twid zoon gie me the back-door trot.

2. (Earlier and later examples.)

1581 G. Pettie tr. *Guazzo's Civ. Conv.* iii. 32 b The children of the right side.. proue doltes.. contrariwise, those which come in at the backe doore.. auaunce themselves. **1907** *Daily Chron.* 12 Sept. 5/3 Several makers began to do a back-door sort of business. **1957** *Economist* 28 Sept. 1012/2 France did a backdoor deal for Egyptian cotton. **1965** *New Statesman* 30 Apr. 671/3 Having safeguarded their back door with China.. the Pakistanis may be tougher about their frontiers with India.

ba·ck-down. *colloq.* Chiefly *U.S.* [Back *v.* 19.] A complete surrender of claims; a retreat from a position taken up or from a stand made.

1862 Gray & Ropes *War Lett.* (1927) 35 The President's message.. seems to me clearly a case of back down. **1886** *Sat. Rev.* 25 Dec. 853 The present crisis, end as it may—in a general 'backdown' or a general war—is essentially a small business as compared with much that Europe has seen in its day. **1894** 'Mark Twain' *Pudd'nhead Wilson* xxi. 280 It's a clean back-down! he gives up without hitting a lick! **1961** *Guardian* 20 Sept. 16/6 Krushchev's primary aim.. was to force.. [an] American back-down at Berlin.

ba·ck-draw. *U.S.* = Drawback *sb.* 4.

1883 *Century Mag.* Oct. 815/2 There are great back-draws to the bee business, the irregularities of the flowers being chief.

backen, *v.* **2.** Delete †*Obs.* and add to def.: *spec.* = Back *v.* 17. (Later examples.)

1800 *Ann. Reg. 1798* Chron. 88/2 The wind, nearly south, .. backened towards the eastern board to about south-east. **1805** H. F. Cary tr. *Dante's Inferno* xvii. 283 As a small vessel, back'ning out from land, Her station quits. **1909** *Westm. Gaz.* 21 Aug. 2/2 It [*sc.* a squall] blew for a few minutes violently; then backened back to the south-east.

back entry. [Back *a.* 1, Back- A. 5, Entry 7.] An entry at the back of a house or houses. Also *fig.*

1677 in *Trans. Shropshire Archæol. Soc.* (1905) V. 216 For yᵉ back entry door 24 foot. **1694** W. Burnaby tr. *Petronius's Satyricon* ii. 22 We made off thro' a dirty Back-Entry. **1854** O. S. Fowler *Home for All* 127 From the center of the room toward the back-entry door to a door into the entry. **1855** [see Entry 7 *fig.*]. **1912** A. Bennett *Matador of 5 Towns* 105 He emerged at the back of the chapel and got by 'back-entries' into Aboukir Street. **1958** *Observer* 10 Aug. 2/7 Our recreation was spent in back entries.

backer, *sb.*[1] Add: **1.** *spec.* One who finances the production of a play, film, etc.

1888 G. O. Seilhamer *Hist. Amer. Theatre* I. 19 The first 'backer' of an American theatrical enterprise, to use a modern phrase, was William Hallam. **1898** J. Hollingshead *Gaiety Chron.* 363 The risk rested entirely on my individual shoulders. I.. never had a 'backer'. **1930** *Economist* 16 Aug. 312/1 In the United States.. well-known finance houses.. have become the backers of the big American film corporations.

b. *backer-up*: a person who backs or backs up (something or somebody) (see Back *v.*, esp. sense 8).

1921 *Dict. Occup. Terms* (1927) § 499 *Backer-up*.. assists sawyer to push wood up to power-driven circular saw. *Ibid.* § 524 *Backer-up*, prepares back of copper shell received from electrotyper.. by brushing it with acid. **1922** Joyce *Ulysses* 648 Given a backerup, if one were forthcoming. **1952** Sherbrooke-Walker *Khaki & Blue* vi. 56 In referring to the part-time Air Force defenders, the expressive, if homely, title of 'Backers-up' was to yield place to 'Station Personnel'. **1953** S. J. Baker *Australia Speaks* v. 124 The accomplice of a woman who *works a ginger* on a client—i.e. robs him—is a *backer-up*, and the practice is called *backing up*.

backer, backey. Add: (Earlier examples.) Also **backa, backy, bakky.** Cf. *Bacca, Bacco, Baccy.

1823 E. Moor *Suffolk Words* 13 *Backa*, tobacco. **1824** W. Oliver *Coll. of Local Songs* 9 If he's drinking gills o' yell, or axing pennies ti buy bakky. **1825** J. T. Brockett *Gloss. N. Country Words* 10 Leet thee pipe, And tyek a blast o' backy! **1828** *Night Watch* II. 159, I got this letter, which I have in my backey fob, from her. **1848** Dickens *Dombey* xxxviii. 383 'You'll smoke your pipe.. won't you?'.. 'Yes; I'll take my bit of backer.' **1885** C. M. Yonge *Nuttie's Father* I. vi. 61 'It is a smoking carriage,' said Miss Nugent. .. 'Beautiful backy—a perfect nosegay,' said Gerard.

back-fire, *sb.* [Cf. *Back-firing *vbl. sb.*] **1.** A fire purposely lighted ahead of an advancing prairie-fire or bushfire in order to deprive it of fuel and so extinguish it. *N. Amer.* and *Austral.*

1839 'Mrs. Mary Clavers' *New Home* xxix. 191 There was nothing now but to make a 'back-fire'! **1905** *Terms Forestry & Logging* s.v., The back fire is intended to burn only against the wind. **1944** *Living off Land* vii. 154 To start a backfire rapidly.. flame-throwers are used. **1959** *Globe & Mail* (Toronto) 23 July 2/1 Setting a backfire against a prairie fire.

2. A premature ignition or explosion in an internal-combustion engine, causing the piston to be driven in the opposite direction to that in which it should travel; also, an explosion in the exhaust-pipe of such an engine.

1897 G. D. Hiscox *Gas Engines* 86 Misfire or back-fire explosives. **1902** A. C. Harmsworth et al. *Motors* 169 These 'back fires' are the result of what is called 'premature ignition'. **1903** Mecredy *Dict. Motoring* s.v., An explosion in the silencer is also called, but incorrectly, a back fire. **1904** Filson Young *Compl. Motorist* ix. 215 This is necessary in order to prevent a back-fire of the engine in starting it. **1968** *Globe & Mail* (Toronto) 3 Feb. 11/2 She heard a small explosion 'like the backfire of a car'.

b. *fig.* (Also *attrib.*)

1925 A. S. M. Hutchinson *One Incr. Purpose* i. xx. 126 The outrageous sale of this upstart work.. justified it in being the subject of back-fire notices in the critical journals. **1927** *Daily Express* 22 Sept. 2/5 If Tunney, smarting under the Billingsgate backfire,.. tries to go in and knock Dempsey out.

back-fire, *v.* [f. prec. or back-formation from *Back-firing *vbl. sb.*] **1.** *intr.* To light a fire ahead of an advancing prairie-fire in order to deprive it of fuel. *N. Amer.*

1886 P. G. Ebbutt *Emigrant Life in Kansas* 54 We all.. set to work to 'back fire' from the stables, and were only just in time to save the whole place from destruction, by burning a sufficiently wide piece of grass off, and thus stopping the rush of fire. **1912** C. Dawson *Pioneer Tales* 291 Man learned to back-fire, and plow fire-guards, so but very few settlers lost their lives from prairie-fires. **1929** C. R. Cooper *Challenge of Bush* 237 At other points they back-fired, taking advantage of every change of wind.

2. Of an internal-combustion engine or its fuel: to ignite or explode prematurely. Also *transf.*, e.g. of a firearm.

1902 Kipling *Traffics & Discoveries* (1904) 184 That car.. back-fired superbly. **1906** *Westm. Gaz.* 23 Oct. 4/2 Some engines are obstinate starters. Others have a nasty tendency to back-fire. **1909** *Ibid.* 13 Oct. 7/4, I think the gas had backfired. **1938** M. K. Rawlings *Yearling* iv. 34 Penny pulled the trigger. The explosion that followed had a sizzling sound, and Penny fell backward. The gun had back-fired. **1961** *Listener* 9 Nov. 773/3 A seventh valve, called the by-pass, that takes over momentarily if any of the other valves should back-fire.

fig. **1912** C. Mathewson *Pitching* xiii. 300 One of McGraw's schemes back-fired on him. **1953** *Encounter* Nov. 70/1 But the reign of terror backfired.

3. *trans.* use of prec.

1924 *Glasgow Herald* 24 June 8 The great gun backfired its charge of 1800 lb. of powder. **1929** H. V. Morton *In Search of Scotl.* iii. 76 The horrid gelatinous mass.. back-fired a dark inky fluid into the water. **1962** *Which?* Mar. 87/2 The Wastemaster [*sc.* waste-disposer], without a splashguard, tended to backfire scraps.

back-firing, *vbl. sb.* [Back *adv.*, Back- A. 10.] **1.** The action of *Back-fire *v.* 1. *N. Amer.*

1889 *Bellevue* (Ida.) *Press* 14 Sept. 3/2 The process known as 'back-firing' was resorted to, which resulted in saving the threatened buildings. **1946** *Los Angeles Times* 9 Sept. i. 2/4 The blaze was stopped short of the Ridge Route by backfiring.

2. Premature ignition or explosion in an internal-combustion engine.

1897 G. D. Hiscox *Gas Engines* 86 Back-firings in the muffler and exhaust pipe. **1907** *Daily Chron.* 11 Nov. 7/4 The air enters the working cylinder in advance of the combustible mixture, thus preventing backfiring. **1929** *Daily Express* 12 Jan. 2 The sound of the report would.. have been mistaken for the backfiring of a motor-car.

‖ **backfisch** (bæ·kfiʃ). Also (erron.) **bachfisch.** [G., lit. 'fish for baking'.] A girl in late adolescence, a teenager.

1888 C. M. Yonge *Beechcroft at Rockstone* vii. 136 The reckless folly of the 'Bachfisch' about health. **1891** *Pall Mall Gaz.* 29 Aug. 3/2 Let us introduce the word 'Backfisch', for we have the Backfisch always with us. She ranges from fifteen to eighteen years of age, keeps a diary, climbs trees secretly, blushes on the smallest provocation, and has no conversation. **1920** Sinclair Lewis *Main Street* xiv. 161 A Gopher Prairie housewife, married a year, and yearning for a 'Prince Charming' like a bachfisch of sixteen! **1921** R. Macaulay *Dangerous Ages* xiii. 239 These were a mother and a *backfisch*, and they.. certainly did not look so prosperous and buxom as a pre-war German mother and *backfisch* would have looked. **1966** *Observer* 6 Nov. 24/4 Marlene Dietrich.. conjures up the little Berlin *backfisch*.. who auditioned for 'The Blue Angel'.

back-formation. *Philol.* [Back- A. III. (Hence G. *rückbildung*.)] The formation of what looks like a root-word from an already existing word which might be (but is not) a derivative of the former.

1889 *N.E.D.*, *Burgle*.. A back-formation from *Burglar*. **1907** *Athenæum* 5 Jan. 7/3 'Narration' is fifteenth century, 'narrative'.. and 'narrator' as early as Bacon, so that, like many verbs of a same termination, it [*sc.* 'narrate'] may have been a back-formation. **1926** Fowler *Mod. Eng. Usage* 516/1 *Scavenger*, n., is the origin, in English, from which *to scavenge* is a back-formation, the normal verb being *to scavenger*.

back-front. [Back *a.* 1, Back- A. 5.] The rear boundary line or elevation of a building.

c **1702** Celia Fiennes *Diary* (1888) 293 The back Front goes out into a garden or Court. **1726** Leoni *Alberti's Archit.* I. 39/2 From the middle of the Fore-front of the Work I draw a Line quite thro' to the Back-front. **1738** Defoe & Richardson *Tour thro' Gt. Brit.* (ed. 2) I. 124 On the Back-front of the House was to be a noble Range of Stoves for tender Exotic Plants. **1800** *Hull Advertiser* 28 June 3/4 The workmen employed in modernising the backfront of

the house. **1805** T. Thornton *Sporting Tour through France* (1806) II. 104 A Sporting palace... The back-front is decorated with a profusion of horns of stags, deer. **1907** *Westm. Gaz.* 29 Nov. 12/1 Downing built 'four plain square brick mansions' on it with 'back-fronts' to St. James's Park.

transf. **1790** *Loiterer* 16 Jan. No. 51, p. 8 The Lady seemed, when I saw only the back-front, a fine young woman. **1900** Kipling in *Times* 15 Mar. 8/1 The home Government.. maintain intimate relations with all sides, with the front, and the far more important 'back-front' which begins at Pretoria.

ba·ck-furrow, *v.* *U.S.* [Back *adv.*] *trans.* and *intr.* To plough (land) so that a second furrow-slice is laid against the face of the first by ploughing in the reverse direction. Also *back-furrow* sb., *back-furrowing* vbl. sb.

1855 *Trans. Ill. State Agric. Soc. 1853–4* I. 425 Plow first out from the row on both sides, then finish by *back-furrowing*, so as to leave the row a trifle higher than the surrounding surface. **1858** C. L. Flint *Milch Cows* 191 When arrived at the end of the piece, a back furrow is turned up to the potatoes. **1861** *Trans. Ill. State Agric. Soc. 1859–60* IV. 113 Mr. K. K. Jones.. had his land plowed deep, [and] back furrowed it in beds twenty-four feet wide. *Ibid.* 392 This mode of plowing (called 'back-furrowing') is always to be observed when the track of the road is plowed. **1873–4** *Rep. Vermont Board Agric.* II. 239, I plow and harrow again, and then in a few days back-furrow in ridges with a side hill plow.

backgame. **2.** (Earlier example.)

1750 'A. D. Philidor' *Chess Analysed* 8 As his King may retire at his Bishop's Square, it is necessary to send you to a second Back-game, which will shew you how to proceed in this Case.

back-gate. [Back *a.* 1, Back- A. 5.] A gate at the back of, or leading to the rear part of, a house or other premises.

1442 *Extr. Aberd. Reg.* (1844) I. 9 Al man that has back yettes close thaim. **1632** Massinger & Field *Fatal Dowry* iv. ii. sig. H4ᵛ Let the Coach be brought To the backe gate. **1709** [see Back- A. 5]. **1818** [see Little *a.* 1 a]. **1873** E. Eggleston *Myst. Metrop.* xxxvi. 307 When Mrs. Ferret came home from prayer-meeting she entered by the back gate. **1905** Jack London *War of the Classes* 274, I battered on the drag and slammed back gates with them, or shivered with them in box cars and city parks.

back-grey. *Calico-printing.* [Back- A. 4, Grey *sb.* 1 b.] = *Back-cloth 1, Grey-back 6.

1896 G. Duerr *Bleaching & Calico-print.* 22 Very often unbleached pieces of cloth (called back greys) are used for the purpose of keeping the blanket from getting dirty too soon;.. the 'back greys' are run between the blanket and the pieces which are being printed. **1897** C. F. S. Rothwell *Print. Textile Fabrics* 34 These pieces, or *back-greys* as they are called, are afterwards bleached and printed with their own patterns. **1963** A. J. Hall *Textile Sci.* iv. 204 Passing around this pressure cylinder.. is a thick blanket.. which is endless, a so-called 'back-grey' (unbleached calico).

background, *sb.* Add: **1. b.** (Examples in *Photogr.*) Also *attrib.*

1858 T. Sutton *Dict. Photogr.* 29 In taking portraits, it is generally necessary to place a background behind the sitter. This is made by stretching a sheet of canvas.. and painting it of an appropriate colour in distemper. **1892** *Photogr. Ann.* II. 756 Makers of exposure meters.. Background manufacturers. **1961** A. L. M. Sowerby *Dict. Photogr.* (ed. 19) 47 The following method of preparing a background and painting it will be found reliable.

c. (Further examples.) Also *attrib.*

1868 Rossetti *Let.* 7 Oct. (1965) II. 668 Immense variety of background-material for any conceivable outdoor subject. **1928** *Pref.* to *Coll. Papers H. Bradley* p. vii, These papers.. form a background without which his labours on the Dictionary cannot be truly judged. **1930** J. B. Priestley *Angel Pav.* iii. 138 That international English,.. a language without roots and background. **1940** *Camb. Bibliogr. Eng. Lit.* I. p. xv, The Middle English Period.. (ii) The Political Background.. (iii) The Social Background. **1951** G. Greene *Lost Childhood* iii. 121 A young historian.. gathering.. background material. **1956** 'M. Innes' *Old Hall, New Hall* i. v. 52 Do some background reading in published sources. **1961** B. R. Wilson *Sects & Soc.* p. ix, Some of the background literature of my subject.

d. A person's cultural knowledge, education, experience, environment, etc.; social surroundings.

1913 J. Webster *Daddy-Long-Legs* 209, I'm glad I don't belong to such a family! I should truly rather have the John Grier Home for a background. **1923** *Ladies' Home Jrnl.* Oct. 236/3 A charming girl lacking only the 'background' that wealth makes possible. **1934** A. Haskell *Balletomania* xi. 219 Had she possessed a background of orthodox technical training [etc.]. **1954** *Manch. Guardian Weekly* 9 Dec. 2/2 A military background is not a full and complete preparation for a Chief Executive. **1959** *Manchester Guardian* 1 Aug. 8/3 Pupils are selected by interview. ..Family background is a major consideration.

e. Music, sound-effects, etc., subordinated to or accompanying some other activity, esp. music used as an accompaniment to a film or broadcast programme. Chiefly *attrib.* orig. *U.S.*

1928 *B.B.C. Hand Bk. 1929* 68 A.. new design of control room has been decided upon, wherein effects, echoes, background music, etc. can be mixed together. **1934** E. B. Marks *They all Sang* xvii. 215 As the hero was wheeled to the operating room, the strains of 'You're the Cheese in My Mousetrap' would trickle into the auditorium as background music. **1937** *Printers' Ink Monthly* Apr. 49/1 *Background*, any musical or sound effect used in backing up dialog or sound. **1937** *Times* 25 Sept. 8/2 *Le Coq d'Or* was no longer a musical work 'before all else', but only a musical work heard while something else very attractive to the eye was going on. It was in fact the beginning of that process now known to the B.B.C. as 'background listening'. **1946** Scott Goddard in A. L. Bacharach *Brit. Mus.* i. 28 A rachitic progeny of background music to documentary films. **1949** L. Feather *Inside Be-Bop* ii. 17 Vocals.. with fine backgrounds and solos by Bird. **1958** *Listener* 21 Aug. 260/1 As with our own Light Programme the 'background' music or entertainment is a comfort and solace to many.

f. *background heater* (see quot. 1961), *heating.*

1939 Martin & Speight *Flat Book* ii. 25 There are excellent gas convector or background heaters. **1951** *Good Housek. Home Encycl.* 76/1 A ¼ watt per cubic foot for background heating. **1961** *Gloss. Terms Gas Industry* (*B.S.I.*) 62 *Background heater*, a heater normally in continuous use which supplies heat to a room at such a rate as will provide a general temperature slightly below comfort temperature.

2. b. *attrib.* or as *adj.* Keeping in the background; retiring. *colloq.*

1896 *Westm. Gaz.* 9 Dec. 2/1 A reticent, background kind of lover. **1904** *Daily Chron.* 11 Feb. 8/6 'Cultivate a background manner,' is the advice of a lady.. to governesses seeking situations.

3. *Electr.* Adventitious signals or effects in the reception or recording of sound. Also *attrib.*

1927 *Radio Assoc. Official Handbk.* 70 The purity of amplification.. and the silence of the 'background', owing to successful elimination of atmospherics and 'mush', is remarkable. **1942** R. C. Norris *Radio Engin.* ii. 35/1 This tends to introduce background noise, which would be entirely avoided if a really good aerial were in use. **1962** A. Nisbett *Technique Sound Studio* vii. 132 One of the problems which may arise when tape is re-used is a background 'chatter' from the previous recording.

4. *Physics.* The level of radiation arising from cosmic rays and other natural sources. Also *attrib.*, as **background count**, the result (in counts or pulses per unit time) of measuring such radiation.

1930 *Phil. Mag.* IX. 642 The aim.. was.. to deduce the relative number of electrons falling in the central spot, in one of the rings or in the continuous background. **1933** Rutherford *Coll. Papers* (1965) III. 343 This radiation gives rise to an inconvenient background 'wobble' in the output from the counting chamber. **1947** *Sci. News* II. 140 If a G-M (Geiger-Müller) counter has a background of 30 pulses per minute. **1950** F. Gaynor *Encycl. Atomic Energy* 26 Background count. **1959** C. Hodder-Williams *Chain Reaction* viii. 98 No trace of radioactivity was detected over and above the normal background count.

background, *v.* (Earlier and later examples.)

1768 S. Bentley *River Dove* 8 Far distant as Vision can go, High Weever back-grounds the gay scene. **1891** C. James *Rom. Rigmarole* 75 The 'antique spires' of the College Chapel, backgrounded with crimson sunset. **1904** *Westm. Gaz.* 17 Nov. 4/2 Planted on to the material like a raised embroidery, and backgrounded with manipulations of tulle or chiffon. **1905** *Ibid.* 4 May 4/2 Hair.. well dressed can background these [features] with such effect that the face.. may become almost beauty.

backgrounder. *U.S.* [f. Background *sb.* + -er[1].] A handout of background publicity material.

1957 J. Blish *Fallen Star* ii. 32 Stuffing into my jacket the release and the 'backgrounder'. **1959** *Manch. Guardian* 11 Aug. 6/2 The information office of the United States Embassy in London has sent to newspapers what it calls a 'backgrounder' on the events in Laos. **1967** *Observer* 30 Apr. 11/6 In a Saigon 'backgrounder', you are told about public health measures.

back-hand, *sb.* Add: **1.** Examples from *Lawn Tennis* and *Badminton*. Phr. *on the backhand*: backhanded.

1890 H. W. W. Wilberforce in C. G. Heathcote *Tennis* 261 A low ball on the back-hand is very difficult to deal with. **1948** N. H. Patterson *Compl. Lawn-Tennis Player* iv. 12 This wide service.. catches your adversary on the backhand. *Ibid.* 14 If he has a weak backhand then try to 'paste' his forehand corner until [etc.]. *Ibid.* xiv. 117 We have all seen girls trying to use both hands when attempting to take a backhand. **1966** Mills & Butler *Mod. Badminton* v. 43 Reaching to take it [*sc.* the shuttle] early on the backhand will not only save time but [etc.].

B. (Examples from *Lawn Tennis* and *Badminton.*) Also applied to the court, etc.

1886 J. Dwight *Lawn Tennis* v. 24 The backhand volley is made in much the same way. **1889** H. W. W. Wilberforce *Lawn Tennis* vi. 21 For the backhand stroke the right foot is in front. **1890** —— in C. G. Heathcote *Tennis* 268 The better volleyer of a pair should play in the back-hand court. **1908** *Westm. Gaz.* 1 May 8/3 Mr. McGregor employed his backhand volley effectively. **1921** A. W. Myers *Twenty Years Lawn Tennis* 9 He attacked the Australian's backhand corner. **1929** *Morn. Post* 13 July 16/3 (*heading*) The Backhand Grip. **1966** Mills & Butler *Mod. Badminton* v. 43 The backhand smash can be used to put away loose returns.

back-hand, *v.* In def., for 'Backhander 3' read 'Back-hander 2'.

Add: **2.** *trans.* To hit or stroke with the back of one's hand.

a **1935** T. E. Lawrence *Mint* (1955) ii. xviii. 147 Stiffy appeared, back-handing his moustache. **1962** K. Orvis *Damned & Destroyed* ix. 62 I'm lucky the goon-squad haven't back-handed me into a lane.

back-handed, *a.* Add: **2.** (Example.) Also as *adv.*

1889 H. W. W. Wilberforce *Lawn Tennis* 29 It is much more difficult to volley a high lob backhanded than fore-handed. **1890** L. Dod in C. G. Heathcote *Tennis* 310, I believe myself in changing the grip of one's racket, both for fore- and back-handed strokes. **1898** *The 'House' on Sport* I. 222 The first great step in this direction [of improving polo] was the introduction of the back-handed stroke. **1922** A. E. Crawley *Lawn Tennis Do's & Don'ts* 27 The receiver is in doubt whether to play it forehanded or back-handed. **1934** A. Christie *Murder on Orient Express* i. vii. 63 One would have to strike backhanded.

3. b. (Earlier U.S. example.)

1800 *Deb. Congress* (1851) 832/2 In a back-handed way and not in the fair regular manner.

Hence **back-ha·ndedly** *adv.*, in an indirect or back-handed manner.

1889 *Cent. Dict.*, *Backhandedly*, with the hand directed backward: as, to strike back-handedly. **1957** K. A. Wittfogel *Oriental Despotism* ix. 386 He backhandedly admitted that his statement did not include the 'barbarian' world of Oriental despotism.

back-hander. Add: **1.** (Earlier example.)

1803 *Censor* 1 Apr. 44 He will.. 'give me such a back hander, as will send me reeling a dozen yards'.

3. A back-handed stroke or blow. Cf. Back-handed *a.* 2 in Dict. and Suppl.

1890 H. W. W. Wilberforce in C. G. Heathcote *Tennis* 266 Few, if any, ladies can volley with effect, and the efforts of most to take a back-hander result in nothing better than a graceful scoop. **1958** J. Hislop *From Start to Finish* 166 *Backhander*, hitting a horse with a whip in the carrying position. **1959** *Times* 19 May 4/5 [Polo] A partial clearing gave Gracida the chance to whack in a stinging backhander.

4. A tip or bribe made surreptitiously, a secret payment. *slang.*

1960 D. Storey *Sporting Life* ii. iii. 207 That's how he hopes to get the car—from the backhander he'll get. **1962** J. Tunstall *Fishermen* iv. 109 A boy who helps with a few odd jobs.. can collect £5 or more in backhanders. *Ibid.* v. 125 Since the man's ability does not seem to qualify him, he is.. suspected of having won the job by giving a back-hander. **1968** *Listener* 7 Mar. 295/1 A bit of a backhander and, boy, you're in.

back-house. [Back *a.* 1, Back- 5.] †**1.** See Back- 5 and add earlier and later examples. *Obs.* exc. *dial.*

1557 *Edin. B. Rec.* III. 17 To abstene.. fra all melting.. of talloun within thair bak hous. **1888** B. L. Burnett *From Stable Boy* xxvii. 148, I was layving.. without going to the zider cask in the back-houze.

2. *U.S.* A privy. Also *attrib.*

1847 in Webster. **1939** C. Morley *Kitty Foyle* ii. 23 The chlorides Mother was always throwing into the little outdoor backhouse. *Ibid.* xvii. 161 None of that backhouse talk when Molly Scarf gets here.

backing, *vbl. sb.* Add: **4.** *backing and filling* (see *Back *v.* 16 b). *U.S.*

1777 in *Essex* (Salem, Mass.) *Inst. Hist. Coll.* (1906) XLII. 315 He was then order'd.. to Stillwarter, then ordered from Stillwarter to Benington... This in the Salers Frase is Backing & filling, makes but poor way a head. **1854** *N.Y. Herald* 15 June (Bartlett), There has been so much backing and filling, not only upon the Cuban question, but upon every other. **1869** Semmes *Mem. Service Afloat* 255 The reader need no longer wonder at the backing and filling of the Iroquois around the little Sumter.

attrib. **1903** Clapin *Dict. Amer.* 31 A backing and filling policy is one which is shilly-shally, trifling, irresolute. **1949** *Amer. Speech* XXIV. 172 Tickers slowed down and a series of backing and filling movements made the pattern for the day.

6. b. (Earlier example.)

1839 Ure *Dict. Arts* 369 It is necessary to turn the rim beforehand a little in the opposite direction.. an operation called in technical language, the *backing off*.

c. (Examples.) Also in other games.

1816 W. Lambert *Cricketer's Guide* (ed. 6) iii. 40 Getting behind the Wicket-keeper when the Ball is thrown in, which is called *backing up*. **1836** *New Sporting Mag.* Oct. 360 By backing up too far, he ran himself out. **1960** *Times* 14 Nov. 4/1 The age-old principle of backing-up.. was once more the basic strength of the whole [Rugby football] team.

d. *backing-out* (see Back *v.* 18). Also *attrib.*

1819 Cartwright in *Cobbett's Wkly. Pol. Reg.* 22 May col. 1112 The Duke saw his reasons for what Mr. Cobbett calls 'backing out'. **1841** *Knickerbocker* XVII. 374 Nor would her offended dignity be appeased by the self-imposed immolations of *backing-out*. **1846** S. Smith *Theatr. Apprent.* 149, I don't come from a backing out country—I must have a showing for the money that's down. **1880** Tourgée *Invis. Empire* v. 413 In explanation of the backing-out process, he says it consisted simply in not going to any more meetings.

7. b. Musical or vocal accompaniment to a singer, esp. on a recording.

1940 *Swing* Jan. 24/3 Everyone.. seems happy in the rowdy backing, which gives plenty of punch to a good old barroom song. *Ibid.* June 17 The backing is based on a riff that's been used for several other numbers lately. **1959** *Punch* 19 Aug. 60/2 Mr. B. finds the backing tedious and professes indifference to the singer's charms.

8. (Further examples.)

1880 L. Higgin *Handbk. Embroidery* iv. 35 Silk, satin, or velvet.. must be backed with a fine cotton or linen lining. The 'backing'.. is first framed. **1962** A. Nisbett *Technique Sound Studio* 241 *Backing*, the base on which the magnetic oxide coating of tape is carried. **1963** *Which?* Mar. 691/

ba·ck-line, backline. [f. Back *a.* + Line *sb.*²] A line situated behind another or others; esp. in various games, a line marking the limit of play at the end of a court, field, etc.; also [Back *sb.*¹ 21], a line of backs in football, etc.

Cf. quot. 1806 s.v. Back *a.* 1.
1890 H. W. W. Wilberforce in C. G. Heathcote *Tennis* 260 He must be content to play a defensive game, and retire about a yard behind the back-line. **1960** T. McLean *Kings of Rugby* 122 Tries which set the whole of the Lions' backline clicking on all cylinders. **1963** *Times* 27 May 5/3 Ramsden, from his own backline, galloped the length of the [Polo] ground and scored a lovely goal in four confident hits. **1966** Mills & Butler *Mod. Badminton* v. 45 You must turn your right shoulder towards the backline.

ba·cklist, back list. [*Back *a.* 3 b.] (A catalogue of) books still available for sale by a publisher or bookseller, but no longer classified by him as 'current' or 'new'. Cf. *Back number.

1946 S. Unwin *Truth about Publishing* (ed. 4) 14 The most stable firms are usually those which have a strong back list of publications with a continuous and profitable sale. **1947** *Sat. Rev. Lit.* 22 Feb. 5/1 'Ferdinand the Bull'..now in its twentieth edition, is still very much on the active backlists. **1957** *Economist* 21 Sept. 939/1 Its sales usually consist of filling small orders for a large number of individual titles, many of them from the backlist. **1959** *Manchester Guardian* 5 Aug. 10/2 Those [publishers]..with well-stocked back lists have been able to go on selling..Bibles, dictionaries, classics.

back-load, back load. [Back- A. 3.] An amount that can be carried on the back.

1725 [see Back- A. 3]. **1823** *Massachusetts Spy* 3 Dec. (Th.), A black fellow was taken up on suspicion, with a back load of live turkeys. **1854** B. Young et al. *Jrnl. Discourses* I. 255/1 You might go round exhibiting a back load of gold. **1894** *Outing* (U.S.) XXIV. 88/1 She had brought us a large back-load of wood.

back-log. Add: Canadian, Austral., and N.Z. examples.

1829 J. MacTaggart *Three Years in Canada* I. 52 A..blazing fire with a maple back log. **1895** H. L. Twisleton *Poems* 28 The back-log's yellow light. **1900** H. Lawson *On Track* 7 A fresh back-log thrown behind the fire. **1901** *N.Z. Illustr. Mag.* IV. 596 He cut up a lot of maire backlogs.

b. *fig.* Something in reserve, reserves, an accumulation.

1883 *Wheelman* I. 294 The back-log of raw eggs and milk beginning to take effect, the pace was improved. **1952** J. Steinbeck *East of Eden* 63 If one is accused of a lie and it turns out to be the truth, there is a backlog that will last a long time and protect a number of untruths. **1953** *Encounter* Oct. 55/2 This backlog of goodwill produced by a century of free immigration.

c. Arrears of unfulfilled orders, etc.

1932 *Sun* (Baltimore) 9 Feb. 16/5 From April through December..there was a steady decrease [in the number of unfulfilled orders] and the backlog at the year-end was the smallest in 21 years. **1947** *Times* 30 Apr. 2/1 It is argued that the prospects of making up this back-log of 500,000 cars cannot be considered good. **1954** *Economist* 25 Dec. 1079/1 The backlog of unfilled orders on manufacturers' books is growing at last. **1958** *Listener* 2 Jan. 19/2 Court cases of which there was a considerable back-log.

back-mark, *v.* *Sporting slang.* [Cf. next.] *trans.* To put (a competitor) back at the start of a race. Hence, to leave far behind in a contest.

1890 Farmer *Slang* s.v., *To be backmarked*..(pedestrian), in handicapping to receive less start from 'scratch' than previously given—even to being put back to 'scratch'. **1895** *Field* 24 Aug. 315/3 Jersey men eagerly awaited the issue of the race between the amateur champion of the world and their own island champions, not that they ever expected them to back mark Tyers. **1928** *Sunday Express* 17 June 20/3, I venture the opinion that he would readily back-mark any man in Scotland over a mile course.

back-marker. [f. *back mark* (Back *a.* 1, Mark *sb.*¹) + -er¹.] One who starts from 'scratch' or has the least favourable handicap in a game, match, or race.

1895 *Daily News* 5 Sept. 7/3 The back-markers were well up in the last lap. **1899** *Westm. Gaz.* 14 Mar. 10/1 One day there was an exhibition game of billiards..Captain Johnson took 150 in 300 from Cook, and had been passed by the 'back-marker'. **1957** S. Moss *In Track of Speed* i. 13, I felt it quite an honour to be the backmarker, running into third place after a lap.

back number. [*Back *a.* 3 b, Number *sb.* 6.] A number of a magazine, periodical, etc., earlier than the current one; hence *colloq.* (orig. *U.S.*), one who or a thing which is behind the times, out of date, or useless.

1812 *Niles' Reg.* I. 392/2 To reprint certain back numbers of the register. **1851** Mayhew *London Lab.* I. 308/2 The sum expended annually in the streets for back numbers of periodicals amounts to upwards of £700. **1882** G. W. Peck *Peck's Sunshine* 153 Some old back number of a girl who has no fellow who wants to go. **1890** *Harper's Mag.* Feb. 439/2 Whereas if Galen should appear among us to-day, ..he would be told he was a back number. **1907** *Westm. Gaz.* 4 Dec. 2/3 There are now so many competing forms of transport..that the steamboat seems to be doomed to be what is in current terminology called a 'back number'. **1924** Galsworthy *White Monkey* I. iv. 27 Lady Alison..finding a certain poignancy in contact with the New Age, on Fleur's copper floor. On that floor she almost felt a back number. **1945** 'G. Orwell' *Animal Farm* 36 Snowball had made a close study of some back numbers of the *Farmer and Stockbreeder*. **1961** *Times* 18 May 5/3 A veteran who..has already proved that he is no back-number.

attrib. **1902** Kipling *Traffics & Discoveries* (1904) 12 These old hand-power, back-number, flint-and-steel reaping machines.

back of (or **o'**) **beyond:** see Beyond C. b.

back of (or **o'**) **Bourke,** *phr. Austral. slang.* [f. *Bourke*, a town in the extreme west of New South Wales.] Remote inland country; the 'back of beyond'.

1918 Campbell & Nelson *Dinky-Di Soldier* 26 They smile to see me ploddin' on along the Bathurst road, The long, long road that runs to back o' Bourke. **1930** *Bulletin* (Sydney) 3 Dec. 12/2 He [*sc.* the cadger] had come to town from Back o' Bourke to see the Minister about a wheat scheme. **1933** *Ibid.* 20 Dec. 20/1 Choom had come straight from the ship to the back o' Bourke on book-keeping duty.

ba·ck-out. *U.S.* [Back *v.* 18.] An act of backing out or withdrawing from a position; inclination to back out.

1829 *Western Souvenir 1830* 314 (D.A.), There's no back out in none of my breed. **1836** Crockett *Exploits & Adv. Texas* (1837) 4 Now that idea..was a sort of cornering in which there was no back out. **1888** *Boston Weekly Globe* 28 Mar. (Farmer), Mr. Barker's back-out has not much surprised me.

back-pack, *sb.* Chiefly *U.S.* [Back- A. 3, Pack *sb.*¹ 1.] A pack carried on the back; *spec.* one consisting of a folded parachute. Also *attrib.*

1914 *Outing* (U.S.) June 312/1 By folding a blanket..it is convenient as a back-pack. **1921** *Ibid.* Mar. 254/1 How about that little back-pack tent you are going to have for your trip? **1930** C. Dixon *Parachuting* 160 (*caption*) The Back Pack is usually used. **1946** W. F. Burbidge *From Balloon to Bomber* iii. 45 One type [of parachute] is carried on the airman's lap..a 'back pack' fastens below the shoulders. **1966** *Economist* 23 July 353/1 The back-pack manoeuvering unit that was to have been tried out by Gemini 9. **1969** *Times* 22 July (Moon Rep.) p. i/3 The mass of the back-pack does have some effect on inertia.

Hence as *v. intr.*, to carry a pack on the back: used esp. of hiking, camping, etc. Also *trans.* So **back-packer, -packing.**

1916 H. C. Kephart *Camping & Woodcraft* I. 143 Back-packing is the cheapest possible way to spend one's vacation in the wilderness. **1940** W. S. Gilkison *Peaks, Packs* xiii. 102 Swagging—or, if you prefer it, back-packing—is more or less an essential part of every climbing trip. **1946** *Trail & Timberline* June 88/1 (D.A.), The deplorable housing situation could never really affect a certain group of enthusiasts known as back-packers. **1956** J. D. Leechman *Native Tribes Canada* 37 Whatever was not carried on the toboggans had to be backpacked. **1961** *Times* 18 July 11/6 Back-packing is the only means of transport.

back-pedalling, *vbl. sb.* [Back *adv.* 5.] **1.** *Bicycling.* The action of pressing down upon the pedal as it rises, in order to check the movement of the wheel. Also *attrib.* Hence (as a back-formation) **back-pedal** *v. intr.* and *trans.*; also as *sb. attrib.*

1887 Albemarle & Hillier *Cycling* 377 This form of clutch..does not admit of back-pedalling. **1890** *Ibid.* (ed. 3) 373 A break should always be used—it is absurd to do hard work back pedalling down-hill. **1897** *Encycl. Sport* I. 268/1 Do not back-pedal too suddenly, or you may break your chain. **1898** J. Pennell in *Fortn. Rev.* Jan. 60 The machine can be back-pedalled. *Ibid.* 63 An endless number of back-pedalling brakes have been brought out during the last year. **1940** Auden *Another Time* 73 She'd a bicycle with..a harsh back-pedal brake.

2. *fig.* The checking of a forward movement, the reversing of an action.

1901 'H. McHugh' *John Henry* 11 You back-pedal so hard that you grab for your hat. **1927** W. E. Collinson *Contemp. Eng.* 32 Note here the slang use of back pedal!, vulgò also come off it, hold hard. **1950** J. Dempsey *Champ. Fighting* xviii. 116 They sprout bad punching habits while concentrating on blocking, parrying, back-pedalling and the like. **1955** M. Gilbert *Sky High* viii. 109 It's not quite fair to ask me if I'm happy about his handling of the case..I've got a feeling..that he's going to back-pedal on it. **1958** *Oxf. Mag.* 29 May 466/1 Is there not a trace of cold feet in this bit of back-pedalling about Keble?

ba·ck-rest. [Back- A. 3, 4.] **1.** A contrivance to support or ease the back of a person when seated or engaged in manual work (e.g. at a lathe). Also, provision for such support.

1859 Holtzapffel *Turning* IV. 20 The turner using the pole lathe..requires the back rest to steady and support his body. **1909** *Daily Chron.* 13 Nov. 9/6 Neither does it mean that the passengers have their comfort curtailed in respect of insufficient back-rest.

2. A guide attached to the slide-rest of a turning-lathe and placed in contact with the work to steady it.

1881 in *Imp. Dict.* **1907** Perrigo *Mod. Amer. Lathe Practice* 164 There are two classes of these rests which may in a general way be called 'center rests' and 'back rests'.

The center rests usually have jaws bearing upon the work at three points spaced equally around the circle, while a back rest bears upon the work generally at the back and on top only.

3. *Weaving.* A bar over which the warp passes from the warp-beam.

1894 T. W. Fox *Mech. Weaving* 369 Each warp beam..must be parallel with back rest, breast beam, and taking-up roller.

back room. Also **back-room, backroom.** [Back *a.* 2.] A room at the back of a house or other building. Also *attrib.*

1592, 1713 [see Back *a.* 2]. **1679** [see Front *sb.* 13]. *a* **1777** Foote *Cozeners* (1778) III. i. 66 And, as you see he is violent, let him have the back room, with the barr'd windows, up two pair of stairs. **1841** *Punch* July 13/2 Paddy Green intends shortly to remove to a three-pair back-room in Little Wild-street. **1904** *Daily Chron.* 17 Dec. 6/4 Little Miss Lazarus or Solomon can generally be relied upon to look after..the little back-room home.

b. *spec.* A room where (esp. secret) research is carried out. Also in extended uses. Applied *attrib.* to one who works or wields influence 'behind the scenes'. So *backroom boy* (colloq.), a person engaged in (secret) research.

1941 Ld. Beaverbrook in *Listener* 27 Mar. 443/1 Now who is responsible for this work of development on which so much depends? To whom must the praise be given? To the boys in the back rooms. They do not sit in the limelight. But they are the men who do the work. Many of them are Civil Servants. **1943** N. Balchin (*title*) The Small Back Room. **1943** *Punch* 11 Nov. 422 (*caption*) He's one of those obscure back-room boys who have lately been so much in the limelight. **1944** *Times* 9 Aug. 2/3 The man most responsible for the development of the rocket projectile..is Group Captain John D'Arcy Bakercarr,..whose 'backroom boys' at the Ministry of Aircraft Production have worked unremittingly with him. **1945** *News Chron.* 17 May 1/1 Considerable numbers of backroom Nazis, such as bankers and industrialists, have been arrested. **1951** D. N. Chester *Lessons Brit. War Econ.* ii. 15 No group of back-room boys writing memoranda embodying ideal policies is likely to have much influence. **1958** *Times Lit. Suppl.* 20 June 337/3 It occurred to boys working in back rooms that if only neutrons could be released in the process a chain-reaction might be set up. **1959** *Oxf. Univ. Gaz.* 1 May 995/1 The staff..could not take on additional services to readers without increasing the arrears of essential cataloguing and other back-room work.

back scattering. *Physics.* [Back *adv.*] The scattering of radiation in a reverse direction from an irradiated substance. Hence (as a back-formation) **back-scatter** *v. trans.*

1940 *Physical Rev.* LVII. 29/2 (*heading*) Disregard of back scattering. It is frequently assumed that all electrons are scattered only in the forward direction. **1946** *Ibid.* LXX. 602 The backscattering of d-d neutrons was investigated for several materials. **1958** *New Scientist* 2 Jan. 23/1 Heavyweight atoms like those of lead give a bigger 'backscattering' than lightweight atoms such as those of aluminium. *Ibid.*, The beta ray source is shielded so that only radiation backscattered by the sample can enter the counter. **1960** *Gloss. Atomic Terms* (*A.E.A.*) 5 *Back scattering*, the emergence of radiation from that surface of a material through which it entered, due to its collision with and reflection from atoms in the material.

So **back scatter,** (*a*) back scattering; (*b*) the radiation scattered in this way. Hence in *Telecommunications* (see quot. 1960). Also *attrib.*

1957 *Gloss. Terms Nucl. Sci.* (Nat. Res. Council, U.S.) 16/1 Back scatter. **1959** *Manchester Guardian* 11 Aug. 7/4 The United States Navy's successful experiments with ionospheric 'back-scatter radar', capable of detecting a nuclear explosion from thousands of miles away. **1960** *Gloss. Terms Telecomm.* (*B.S.I.*) 168 A signal received by back scattering is often referred to as back scatter. **1962** *Gloss. Terms Nuclear Sci.* (*B.S.I.*) 12 *Back scatter*, the emergence of radiation from that surface of a material through which it entered..The term is also used for the actual back-scattered radiation.

ba·ck-scra:tching, *vbl. sb. colloq.* [f. Back- A. 1 + Scratching *vbl. sb.*, in allusion to the saying, 'You scratch my back, and I'll scratch yours'.] The performance of mutual services; always with the suggestion of doubt as to the legitimacy of the transactions. So **ba·ck-scra:tcher.**

(See also *back-scratcher, scratching* in Dict., s.v. Back- A. 1.)
1897 *Daily News* 9 Jan. 4/7 Does it not rather partake of the ethics of the back-scratcher and the log-roller? **1924** P. C. Macfarlane *Tongues of Flame* xxv. 223 'That is the modern method of bribery, you know—back-scratching,' she frowned savagely. '"We've scratched your back; now you scratch ours." That's the system.' **1933** *Punch* 24 May 567/1 Mr. Hugh Walpole made the startling suggestion that contemporary criticism occasionally showed symptoms of degenerating into mutual back-scratching.

back seat. [Back- A. 4.] A seat at the back of a vehicle. Also, esp. a seat at the back of a hall, etc., an inferior seat; hence *colloq.*, a position of inferiority or comparative obscurity (orig. *U.S.* in phr. *to take a back seat* (*fig.*), to occupy a subordinate place; see Seat *sb.* 27 c).

a **1832** R. C. Sands *Writings* (1834) II. 148 He had..ample room wherein to adjust himself and his properties, on the back-seat [of the coach]. **1858** Thackeray *Virginians* xx. 157 The Lady Maria had the back seat to herself. **1883** Ld. Ronald Gower *My Reminiscences* II. 234 *Le gros papa* took up all the front seat of the carriage..In the middle seats were the young Germans..and in the back seat..on which two people could have been uncomfortably seated, sat the middle-aged Englishman, Walter, and I. **1928** 'Ian Hay' *Poor Gentleman* iv. 71 I'll put you in the back seat together going home.

1859 *Harper's Mag.* June 54/2 This menagerie of..red skins, whom a score of indefatigable Coopers and Longfellows could never raise to merit a back seat in the heaven of romance. **1881** J. W. Jones in *Southern Hist. Soc. Papers* IX. 133, I tell you, those able-bodied men who are sleeping in feather beds tonight,..must be content to take back seats when we get home. **1899** H. James *Awkward Age* 180 'Don't you think that's rather a back seat, as they say, for one's best?' 'A back seat?'—she wondered with a purity! 'If you don't understand,' said her companion, 'it serves me right, as your aunt didn't leave me with you to teach you the slang of the day.' **1902** M. Beerbohm in *Around Theatres* (1924) I. 395 To enliven the milieu, on he brought a circus procession..And Oxford had to take a back seat. **1959** *Times* 22 Nov. 5/6 Those who think that the trade union movement should take a back seat in the Labour movement should think again.

b. back-seat driver, a passenger in the rear seat of a car who gives unsolicited directions to the driver. Hence, *fig.*, one who criticizes or attempts to direct without responsibility (cf. **armchair critic*); one who controls affairs from a subordinate position. Also **back-seat driving** *vbl. sb.*

1926 *Nation* (U.S.) 8 Dec. 587/1 What? No taxis..back-seat driving, back-seat kisses? **1927** *Collier's* 3 Dec. 10/1 You're just another one of them back-seat drivers. **1930** Wodehouse *Very Good, Jeeves!* ix. 245 Quite suddenly..the car..stopped in its tracks..The back-seat drivers gave tongue. 'What's the matter? What has happened?' **1948** *Economist* 8 May 748 What was often no more than vociferous, if irritating, back-seat driving has been..interpreted as attempts..to seize control of the wheel. **1955** *Times* 19 Aug. 6/4 [He] replied that it was contrary to democracy for elected members to consult 'pressure groups' and 'back-seat drivers'.

back-shop. (Earlier and later examples.)

1549 in *Abstracts of Protocols of Town Clerks of Glasgow* 14 Oct. (1894) I. 4 A fore booth, with booth adjoining, cellars, and a small chamber, commonly called 'ane bak chop'. **1852** *Blackw. Mag.* XXXII. 458 Like a show table in Rundle and Bridge's back-shop.

backsight. (Earlier and later examples.)

a. 1847 E. Cresy *Encycl. Civil Engineering* I. 802 Compound Levelling is..performed by taking back and forward sights. **1953** R. J. C. Atkinson *Field Archaeol.* (ed. 2) iii. 114 The reading to a turning-point from the *previous* level station is known as a *foresight*, and that from the *next* level station as a *backsight*.
b. 1847 [see Sight *sb.*¹ 14 b]. **1851** R. Glisan in *Jrnl. Army Life* (1874) 59 A..very heavy and costly gun, with an elevating back sight for shooting at great distances. **1859** *Musketry Instruct.* 10 Firing without using the back-sight.

ba·ck-sla:pping, *ppl. a.* and *vbl. sb.* [Back- A. 1.] Slapping the back. So **back-slapper,** a vigorously hearty person. Hence (as a back-formation) **back-slap** *v.*

1777 [see Back- A. 1 a]. **1884** E. H. Yates *Recollections* I. v. 186 A..good-natured vulgarian, of a dreadful back-slapping, Christian-name calling familiarity. **1920** Sinclair Lewis *Main Street* xxviii. § 6, Isn't the thing he lacks the back-slapping jocosity that passes for humor here? **1924** *Chicago Tribune* 1 Oct. 23/1 (*heading*) Here's a Comic Piece About a Back-Slapper. **1929** *Evening News* 4 Jan. 9/4 The back-slappers, the rib-diggers, the over hearty and the dwellers-in-the-limelight are to be avoided. **1935** Sinclair Lewis *It can't happen Here* ix. 88 He grinned and knee-patted and back-slapped. **1946** *Scrutiny* XIV. 155 The blustering, hearty, back-slapping manner..is the reverse of the discreet, self-effacing and seemly behaviour which one has a right to expect of a good undertaker. **1960** J. Fingleton *Four Chukkas to Australia* vii. 77 There were, to be true, wild jumps of elation and much back-slapping.

ba·ck-stabber. [Back *sb.*¹] One who attacks unfairly, one who attacks (a person) behind his back. Hence (as a back-formation) **back-stab** *v.*

1906 *Westm. Gaz.* 6 Jan. 15/3, I will tell you my idea of a false friend and back-stabber—to sweat the workman for personal profit and fawn on him for political profit; to promise old-age pensions for votes and, having got the votes, to refuse them. **1925** *Chambers's Jrnl.* Dec. 863/2, I, after befriending her brother..had seemed to back-stab. **1960** *Spectator* 30 Sept. 470 That party of back-stabbers and double-crossers.

back stage, backstage, *sb.* and *adv.* [f. Back *a.* 1 + Stage *sb.* 5.] The hinder or 'upper' part of a stage; the part of a theatre behind the stage containing dressing-rooms, etc.; *fig.* 'behind the scenes' (see Scene 7). Also *attrib.*

1898 Sachs & Woodrow *Mod. Opera Houses & Theatres* III. Suppl. 14 In order to obtain as much room as possible, and to increase the depth of space of the stage floor available for setting scenery, many theatres now have what is termed a 'back stage'..This floor space is most useful for distant 'scenes'. **1902** *Encycl. Brit.* XXXII. 823/1 The back stage has no openings or mechanism beyond certain trap-doors..and the necessary mechanism for raising and lowering scenery. **1916** *Stage Year Book* 37 A scene dock is provided back stage prompt. *Ibid.* 38 The speed of the back stage motor varies according to the position of the controller. **1923** A. Stevens in G. Oppenheimer *Passionate Playgoer* (1958) 154, I went backstage at the Studebaker to see Bert Williams. **1929** W. Faulkner *Sartoris* III. 177 While the ladies are backstage washing the dinner dishes. **1950** N. Cardus *Second Innings* 184 There are all sorts of..'back-stage' activities at Lord's. **1950** T. S. Eliot *Cocktail Party* III. 140 Alex is seen to enter with her and pass back-stage towards the kitchen. **1958** *Times Lit. Suppl.* 31 Jan. 67/2 Mr. Barker has done justice not only to the entertainments offered by the Coliseum but also to the troubles backstage and in the board room.

backstairs. Add: **2. b.** (Later example.)

1855 C. Kingsley *Westward Ho!* (ed. 2) II. x. 269 A priest of the Church of England, (whose business is not merely to smuggle sinful souls up the backstairs into heaven).

c. (Earlier and additional examples.) Also (in *attrib.* use) clandestine, underhand.

1663 .. Cooke in *11th Rep. Hist. MSS. Comm.* App. v. Dartmouth MSS. (1887) 11 Sends humble service to Legge's lady..Dick Lane, and all bedchamber backstair friends. **1768** Goldsmith *Good-Nat. Man* II. 21 Is not he a backstairs favourite, one that can do what he pleases with those that do what they please? **1930** G. B. Shaw *Apple Cart* I. 31 They get amusing articles, spiced with exclusive backstairs information. **1933** N. Coward *Design for Living* I. 19 We could carry on a backstairs affair for weeks without saying a word about it. **1935** T. S. Eliot *Murder Cathed.* ii. 58 This is the man who was the tradesman's son: the backstairs brat who was born in Cheapside.

backstay. Add: **2.** (Earlier and later examples.)

1846 Holtzapffel *Turning* II. 634 In cutting very long screws..the object becomes so slender, that the contrivance called a backstay, is always required for supporting the work. **1911** *Act 1 & 2 Geo. V* c. 50 § 46 A back-stay or other suitable contrivance for preventing the tub running back.

3. *local.* = Backster.

1830 *United Service Jrnl.* Mar. 359 How would he like walking over three miles of heavy shingle [at Dungeness] in ..a pair of back stays..? **1889** H. W. Lucy in *Time* Mar. 257 What the 'backstay' is to the inhabitant of the district around Lydd, the stilts are to the lonely dwellers in the Landes. **1901** *Pall Mall Gaz.* 27 May 6/1 The bird lovers..follow the local fishermen's example, and wear 'back-stays'—boards, about eight inches long and five inches wide, fastened on to the boots.

back-stone. *Lead Manuf.* [Back- A. 4.] The piece of cast iron at the back of an ore-hearth. (Cf. Pipe-stone 2, Work-*stone*, and *Fore-*stone*.)

1839 Ure *Dict. Arts* 756.

ba·ck-stop. [Back *a.*, Back- A. 4.] **a.** Something placed at the back to serve as a barrier; *spec.* a mound of earth or embankment set up behind a target on a rifle range. Also *fig.*

1851 F. Starr *20 Yrs. of Trav.'s Life* xiii. 143 The remaining shaft..broke off short, and that which when we started was a gig, was now a back stop for horses' heels. **1904** Goodchild & Tweney *Technol. & Sci. Dict.* 37/1 *Back stops* (Cotton Spinning), buffers, used to prevent the mule carriage from going beyond a certain point during its inward run. **1946** *Sports Afield* Jan. 55/1 (D.A.), Put up two targets on the backstop for the second barrel practice. **1947** *Landfall* I. 43 This being the first issue of *Landfall*, there is no previous survey to provide a convenient back-stop to discussion. **1954** W. Faulkner *Fable* (1955) 343 The railroad embankment..would serve as a backstop for what bullets neither flesh nor wood absorbed.

b. *spec.* in *Cricket.* (See Back- B.)

1819 *Suffolk Chron.* 31 July 3/2 They were deprived of two of their best bowlers, and a back-stop. **1867** *Lillywhite's Cricketer's Companion* 120 [He is] an admirable back-stop. *Ibid.* 127 T. Moore, back-stop of the eleven.

c. *Baseball.* A fence behind the catcher to stop the ball; also, the catcher himself or his position. *U.S.*

1889 *Reach's Official Base Ball Guide* 142 Backstop must be ninety feet from home base. **1890** H. C. Palmer *Stories of Base Ball Field* 46 Tom Daly was the greatest catcher..and a careful study of that back-stop's methods behind the bat to-day will help any young catcher.

back-stroke. Add: Now usu. **backstroke** (without hyphen). **c.** *Swimming.* A stroke in which the swimmer lies upon his back in the water. Also *attrib.*, and *ellipt.* for *backstroke event, race,* etc.

1876 W. Wilson *Swimming* xii. 49 The slightest motion of the legs..will sustain and at the same time propel the body. This is known as the plain back stroke. **1887** *Encycl. Brit.* XXII. 769/1 Another racing back stroke is performed by lifting hands and arms out of the water at the finish of the pull downward. **1908** *Manual of Seamanship* (*H.M.S.O.*) I. 319 Then lie on your back, hold him in front of you, and swim with the back stroke, taking care to keep his face above the water. *Ibid.* 320 This method will be useful to over-arm swimmers as well as to back-stroke swimmers. **1933** *Boys' Mag.* XLII. 83/1 The order in which you learn the strokes should be—crawl, breast-stroke, life-saving back-stroke, side, overarm. **1958** *Times* 3 Nov. 2/7 Miss C. Hussey..broke the championship record for the 100 yards junior girls' backstroke.

d. *Cricket.* A stroke made from behind the popping crease when playing back. (Usu. written as two words.) Cf. *back play* (*Back- B).

ba:cterio·phagy, the action of a bacteriophage; **bacte:riopha·gal, -pha·gic** *adjs.*, of, pertaining to, or having the action of a bacteriophage. Cf. *PHAGE.

1921 *Chem. Abstr.* 3319 The bacteriophage of Herelle. **1922** *Lancet* 2 Sept. 515/1 Acting in an attenuated state in young, actively growing, surface cultures, the bacteriophagic principle produces bare areas amidst the growth, which are analogous to colonies and are to be looked upon as colonies of the bacteriophage. **1924** *Chem. Abstr.* 4173 Bacteriophagy. **1926** G. H. SMITH tr. *d'Herelle's Bacteriophage* 5 This..gave a new proof that the dissolving principle actually regenerated in the course of the action. Further, it demonstrated that the principle was condensed in the form of active particles. It is to this principle that I have given the name Bacteriophage; the phenomenon of bacterial solution caused by it being termed Bacteriophagy. *Ibid.* 6 To attempt to reconcile 'bacteriolysis' and 'bacteriophagy', making them identical and co-extensive is a scientific absurdity. **1927** *Lancet* 10 Sept. 536/1 According to Bail, bacteriophagal action is due to the formation of minute (filter-passing) fragments from the bacterial cells. **1943** *Times* 29 Apr. 2/2 Minute organisms called bacteriophages, which destroy bacteria, have been rendered visible for the first time. **1944** *Electronic Engin.* XVI. 420 Another intestinal bacillus..is seen being attacked by a crowd of small bacteriophages, the anti-bodies present in the human system that are able to defeat infection. **1957** *New Scientist* 9 May 12/1 Once inside the bacterium the nucleic acid of the bacteriophage initiates a process which culminates in the production of 100–200 new bacteriophage particles.

bacterioscopic (bæktɪ³:rɪ,ɒskǫ·pɪk), *a.* [f. BACTERIOSCOPY + -IC.] Pertaining to bacterioscopy. Hence **bac:terioscoʼpically** *adv.*

1886 P. F. FRANKLAND in *Phil. Trans.* CLXXVIII. B. 119 The bacterioscopic examination of air. **1896** E. KLEIN *Micro-organisms* v. 67 (*heading*) Bacterioscopic Examination of Water. **1903** *Nature* 19 Feb. 370/2 The City Corporation has therefore caused a number of samples [of shell-fish] to be bacterioscopically examined by Dr. Klein.

bacteriosis (bæktɪ³rɪǭᵘ·sɪs). [mod.L.: see -OSIS.] Any disease of plants ascribed to the action of bacteria.

1899 G. MASSEE *Plant Diseases* 338 Bacteriosis of tomatoes. *Ibid.* 339 Pink bacteriosis of wheat. **1901** *Proc. R. Soc.* LXVII. 456 Arthur ascribes the action of bacteria in the bacteriosis of carnations to an enzyme.

bacteriostasis (bæktɪ³:rɪ,ɒstē¹·sɪs). [f. *bacterio-*, combining form of BACTERIUM + Gr. στάσις stopping.] Inhibition of the growth of bacteria without destroying them. So **bacteriostatic** (-stæ·tɪk) *a.*, also as *sb.*; **bacte·riostat**, an agent that causes bacteriostasis.

1912 *Jrnl. Exper. Med.* XVI. 229 Its bacteriostatic action on the violet positive organisms is evident enough. **1918** C. H. BROWNING *Appl. Bacteriol.* v. 83 This property of inhibiting proliferation, or 'bacteriostatic action' as it has been termed by Gildersleeve..is a remarkable phenomenon. **1920** *Jrnl. Amer. Med. Assoc.* 17 Jan. 145/1 The results of attempts to apply the selective bacteriostatic power of gentian violet to the treatment of infected wounds. The activity of the dye is described as bacteriostasis, the dye is called a bacteriostat, and its property is referred to as bacteriostatic. **1936** *Nature* 15 Aug. 295/1 Selective bacteriostasis (the inhibitory action of such substances as dyestuffs and the products of their own metabolism on the growth of bacteria and fungi). **1942** *Brit. Jrnl. Exper. Path.* XXIII. 123 (*title*) Proactinomycin: A 'Bacteriostatic' Produced by a Species of Proactinomyces. **1944** *Lancet* 17 June 792/2 Cultures whose growth is delayed or prevented by..one of the substances called bacteriostats.

bacteriotherapeutic (bæktɪ³:rɪ,ǫþerăpiū·tɪk), *a.* [See next and THERAPEUTIC *a.*] Of or pertaining to bacteriotherapy.

1886 *Medical News* 10 July 41 Dr. Ballagi has carefully followed the bacteriotherapeutic details advised by Cantani.

bacteriotherapy (bæktɪ³:rɪ,ǫþe·răpɪ). [Gr. θεραπεία medical treatment.] Treatment of disease by introducing bacteria into the system.

1886 *Medical News* 10 July (*heading*) Bacteriotherapy of Phthisis Pulmonalis. **1949** H. W. FLOREY *Antibiotics* I. i. 5 Cantani [1885] believed that this method of bacteriotherapy might be applied to accessible infected surfaces.

bacteriotropin (bæktɪ³:rɪ,ǫtrǫ·pɪn). [f. *bacterio-*, combining form of BACTERIUM + Gr. -τροπος turning + -IN¹.] (See quot. 1910.)

1909 *Chem. Abstr.* 2180 (*title*) Tubercular immunoopsonins (bacteriotropins). **1910** *Lancet* 25 June 1738/2 Neufeld and Rimpau drew attention to the presence in immune sera of a similar humoral substance, conditioning phagocytosis, but differing from Wright's opsonin in that it was thermostable. This they termed '*bacteriotropin*'. [In *Deut. Med. Wochenschr.* (1905) XXXI. 1610 they used the expression 'Bakteriotrope Substanzen'.—Ed.] **1961** *Brit. Med. Dict.* 169/2 *Bacteriotropin*, a heat-resisting opsonin which acts on bacteria, and which assists the phagocytes in their work. Also known as immune opsonin.

bacterium. Delete ‖ and for def., read: Any of several types of microscopic or ultramicroscopic single-celled organisms very widely distributed in nature, not only in soil, water, and air, but also on or in many parts of the tissues of plants and animals, and forming one of the main biologically interdependent groups of organisms in virtue of the chemical changes which many of them bring about, e.g. all forms of decay and the building up of nitrogen compounds in the soil. Add later examples.

1908 *Daily Chron.* 6 Aug. 6/5 Dr. Stonehouse said it was a bacteria infection. **1911** J. A. THOMSON *Biology of Seasons* II. 161 Analogous, though not 'inter-regnal', is the intimate and most profitable partnership between Bacteria-like microbes and Leguminous plants, like Clover. **1956** *Nature* 11 Feb. 279/2 This trypanosomid, first isolated in bacteria-free culture by Noguchi and Tilden. **1964** M. HYNES *Med. Bacteriol.* (ed. 8) iii. 26 The foundation of modern surgery is asepsis—the creation and preservation of a bacterium-free environment.

2. bacteria bed, a contact bed (see *CONTACT *sb.* 6).

1913 E. H. BLAKE *Drainage & Sanitation* xi. 369 This was the origin of the contact method of working bacteria beds—a method adopted with the septic tank. **1936** *Ibid.* (ed. 5) xi. 428 Such treatment is carried out in what are often called Bacteria Beds. These may be on the intermittent principle, in which case they are called Contact Beds, or on the continuous flow principle, in which case they are called Percolating Filters or Trickling Filters; in either case the cause of purification is aerobic bacteria.

bacteriuria (bæktɪ³·riyū³·riă). *Path.* Also **bacteruria.** [f. BACTERIUM + -URIA.] The presence of bacteria in the urine.

1889–90 H. E. KENDALL in *Maritime Med. News* (Halifax) II. 37 (*title*) Notes on two cases of bacteriuria. **1900** DORLAND *Med. Dict.*, Bacteriuria. **1907** *Practitioner* Aug. 328 Septic bacteruria.

bacterization (bæktĕrəizē¹·ʃən). [f. BACTERIUM + -IZATION.] The process or method of treating with bacteria. So **ba·cterize** *v. trans.*, **ba·cterized** *ppl. a.*

1902 R. L. DEVONSHIRE tr. *Vallery-Radot's Life of Pasteur* II. i. 8 Typhoid fever, bacterization! Hospital miasma, bacterization! **1915** *Chambers's Jrnl.* Dec. 830/2 He has dealt more particularly with his latest invention—bacterised peat. **1949** FLOREY *Antibiotics* I. i. 44 In the next year Novogrudsky and his collaborators (1937)..proposed to 'bacterize' flax seeds in an endeavour to reduce fungus infection. *Ibid.*, The greatest changes could be brought about in wheat seeds by the joint bacterization with this special strain and *Azotobacter*.

bacteroid, *a.* Add: **B.** *sb.* A micro-organism of bacterial character; *spec.* one found (*a*) in the root-nodules of leguminous plants, (*b*) in the body of certain insects.

1878 *Parkes' Pract. Hygiene* (ed. 5) 63 Bacteria, vibriones, or microzymes...Frequently spoken of as Bacteroids. **1887** *Phil. Trans.* CLXXVIII. B. 552 The gemmules or 'bacteroids' [in the Leguminosæ]. **1898** R. T. HEWLETT *Bacteriology* 26 If the roots of a pea, bean, or vetch be examined, numerous little nodules will be found upon them; on examining these microscopically small irregular bodies are found to be present, which have been termed bacteroids. **1940** *Chambers's Techn. Dict.* 70/1 *Bacteroids*, in *Oligochaeta*, rod-like bodies of unknown function occurring in the connective tissue. **1949** in H. W. Florey et al. *Antibiotics* II. VIII. xxxi. 1024 Certain body cells of cockroaches contain gram positive micro-organisms known as bacteroids.

Bactrian (bæ·ktrĭăn), *a.* (*sb.*) [ad. Gr. Βακτριανός, L. *Bactriānus*.] Of or belonging to Bactria, an ancient country of central Asia, lying between the Hindu-Kush and the Oxus. Also *sb.*, a native of this country.

Bactrian camel, the two-humped camel of central Asia; also *Bactrian*, ellipt.

1601 HOLLAND tr. *Pliny's Hist. World* VI. xxiii. 131 This country lieth overagainst the Bactrians. **1607** TOPSELL *Foure-Footed Beastes* 92 All those [camels] which are in India, are saide by Didimus to be bred in the Mountaines of the Bactrians,..and these are worthily called Bactrians, because they were first of all conceiued among them. *Ibid.* 94 They want hornes (I meane both the Arabian and Bactrian Camell). **1757** J. DYER *Fleece* IV. 345 O'er Bukor, Cabul, and the Bactrian vales. **1797** *Encycl. Brit.* II. 728/1 The Bactrians differed little in their manners from the Nomades. **1832** *Blackw. Mag.* XXXII. 207/2 The priests of Mithras..offered him..their Bactrian dromedaries, if he chose to depart. **1877** [see IRANIC *a.*]. **1908** *Animal Management* 276 The double humped camel, also called the 'Bactrian', is found in Turkestan, and throughout central Asia. **1912** H. G. RAWLINSON *Bactria* i. 15 The Bactrians were famous for their pithy proverbial sayings. **1927** W. J. TURNER *Aesthetes* 14 On a Bactrian Horse, T'ang period.

baculum (bæ·kiŭlə̆m). *Zool.* [mod.L., f. L., stick, staff.] The penis-bone (see PENIS b).

1939 I. F. & W. D. HENDERSON *Dict. Sci. Terms* (ed. 3) 32/1 *Baculum*, the penis bone; os priapi. **1949** A. S. ROMER *Vertebrate Body* vii. 175 A *baculum*..is a heterotopic bone, which is the skeleton of the penis, found in all insectivores, bats, rodents, and carnivores, and in all primates except man.

bad, *a.* Add: **1. b.** *bad air* (earlier example); *bad coin, penny*, etc. (additional examples). Also (chiefly *dial.*), in arrears (cf. BAD *sb.* 1 b. *b*). Also *bad egg, bad form, bad hat, bad lot*, etc.: see the sbs.

1601 SHAKS. *Jul. C.* I. ii. 252, I durst not laugh, for feare of opening my Lippes, and receyuing the bad Ayre. **1773** JOHN WESLEY *Jrnl.* 21 Aug. (1916) V. 522 Our income does not yet answer our expense. We were again near two hundred pounds bad. **1779** in *Proc. Wesley Hist. Soc.* (1930) XVII. 158 Destroyed a bad shilling. **1782** COWPER *Expostulation* in *Poet. Wks.* (1905) 52 Perjuries are common as bad pence. **1858** [see LAW *sb.*¹ 17 c (*d*)]. **1881** A. B. EVANS *Leicestershire Words* 96 'I'n got a quarter *bad* in my rent.' 'His illness threw us *bad* with the clothing-club.' **1937** A. CHRISTIE *Dumb Witness* x. 111 Always hard up—always in debt—always returning like a bad penny from all over the world. **1962** S. E. MARTIN in Householder & Saporta *Problems in Lexicography* 155 Some bad pennies of everyday conversation are shunned by dictionaries.

c. Colloq. phrases *not bad, not so bad, not half bad*: used (usu. pred.) of a state of things, the result of an effort, etc.: less bad than it might be (or have been); hence, by meiosis, fairly good, deserving some praise or congratulation.

1771 C. BURNEY *Pres. State Mus.* 65 The *intermezzo* was not bad; the music pretty, but old. **1810** C. STEWART tr. *Mirza Abu Taleb Khan's Travels* II. xxiv. 133 The gentlemen put up with bad food, and worse wine; and whenever I complained, they took great pains to persuade me the things were not so bad, or that the master of the house was not in fault. **1835** *Naut. Mag.* IV. 689 The idea of a sailor's *chemise* is not bad. **1838** in E. Eden '*Up the Country*' (1866) I. 129 These [letters] are five months old, but that is not so bad. **1839** *Mag. Dom. Econ.* May 332 Leaving out the cheese, the thing itself is not so very bad. **1860** *Englishwoman's Dom. Mag.* Oct. 26 'Not bad!' Bloomfield replied with a loud laugh. **1867** *Field* 6 Apr. 246/3 This is not half bad,..a fish a-piece, although we were clean till four o'clock. **1871, 1886** [see HALF *adv.* 3]. **1899** KIPLING *Stalky* 220 'Not half bad years, either,' said M'Turk. **1900** W. R. KENNEDY *Hurrah Life Sailor* xii. 180 We had bagged three bulls before breakfast, which was not so bad. **1925** W. DEEPING *Sorrell & Son* xvii. 163 'Did they make it rather beastly—for you?' 'O, not so bad, pater.' **1954** WODEHOUSE & BOLTON *Bring on Girls* i. 19 'What did you think of our little entertainment?'..'Not bad,' said Plum.

B. *sb.* **1. c.** *in bad*: out of favour (*with*, etc.), in bad odour. *colloq.* (orig. *U.S.*).

1911 *San Francisco Examiner* 12 June 10 You'll be on the stand today—now we're in bad if you let out anything. **1915** WODEHOUSE *Psmith, Journalist* iv. 26 Any time you're in bad. Glad to be of service. **1920** —— *Coming of Bill* II. xiv. 236, I guess this has put me in pretty bad with Mamie. **1923** E. WALLACE *Missing Million* xiii. 109 You're never satisfied till you get a man in bad. **1944** L. A. G. STRONG *Director* 63 I'm in bad with your father for the minute. **1953** K. AMIS *Lucky Jim* v. 57 This ought to put me nicely in bad with the Neddies.

C. *adv.* Delete † *Obs.* and add later examples (chiefly U.S.).

1806 in *Charges agst. the Duke of York* (1809) 413 The Regt he is in did their exercise so bad that the Duke swore at them. **1816** U. BROWN *Jrnl.* in *Maryland Hist. Mag.* (1915) X. 273 Land of not much account, farm'd bad. **1846** J. J. HOOPER *Adv. Simon Suggs* vii. 94 'Pshaw!' said Suggs, 'you aint bad hurt.' **1848** THACKERAY *Van. Fair* liv. 483 I didn't do my duty with the regiment so bad. **1854** S. HALE *Lett.* (1919) 7 The children..during that time act as bad as they can! **1870** *Trans. Ill. State Agric. Soc.* VIII. 238 This speaks bad for our application of the art. **1888** *Daily Inter-Ocean* 9 Mar. (Farmer), As the case now stands, the defense want Myers, and want him bad. **1895** 'ROSEMARY' *Under Chilterns* iii. 92 Las' week there was a job doin' up at the squire's, an' I wanted to go bad. **1901** M. E. RYAN *That Girl Montana* 3 There is one thing I want in this world, and want bad. **1938** M. K. RAWLINGS *Yearling* iii. 26 The meat's bad tore up. *Ibid.* iv. 39 She's bad hurt.

b. *bad off*: badly off; in a bad or poor condition or circumstances; esp. = POOR *a.* 1 a. *U.S. dial.*

1815 HUMPHREYS *Yankee in Eng.* 77 Bad as I am off, I wouldn't swop conditions. **1817** U. BROWN *Jrnl.* in *Maryland Hist. Mag.* (1916) XI. 371 Land full of Lime Stone..bad off for Timber and Water. **1879** TOURGÉE *Fool's Err.* xxix. 179, I told him how bad off I was. **1934** L. HELLMAN *Children's Hour* (1937) III. 121, I only came cause she's so bad off.

D. *bad-blooded, -looker, -looking, -tempered, -weather.*

1928 A. HUXLEY *Point Counter Point* vii. 115 Insolent, bad-blooded young cub! **1930** E. WAUGH *Vile Bodies* ii. 21 Not a bad-looker herself, if it comes to that. **1914** G. B. SHAW *Pygmalion* (1916) IV, Youre not bad-looking: it's quite a pleasure to look at you sometimes. **1879** ROGET *Thesaurus* (new ed.) § 901 Irascible; bad-, ill-tempered. **1922** JOYCE *Ulysses* 218 Virtuous: but occasionally they were also bad-tempered. **1883** *Man. Seamanship for Boys' Training Ships* 48 Q. What are storm trysails? A. Bad weather sails, fitted similar to other gaff sails. **1897** *Daily News* 15 Oct. 5/1 There are..some redeeming features even in cyclonic or bad weather systems. **1936** J. DESCHIN *New Ways in Photogr.* ii. 29 Bad-weather photography requires more zeal on the part of the hobbyist.

Badarian (bădē³·riăn, -ā·riăn), *a. Archæol.* [f. *Badari*, name of district in Egypt + -AN.] Of or designating the ancient pre-Dynastic culture, evidence of which was first discovered in the Badari region of Egypt in 1924.

1924 *Brit. Sch. Archæol. in Egypt, Catal. Antiq. at Qua* 3 Beyond are some of the bowls of this oldest culture known, which we may call Badarian. **1925** *Glasgow Herald* 29 Aug. 10/1 The Badarian civilisation was widespread and covered a long period. **1931** *Discovery* Sept. 293/2 Objects from the earlier Badarian and Tasian cultures included well-preserved examples of the delicate black 'ripple' ware.

baddeleyite (bæ·dəli,əit). *Min.* [f. the name *Baddeley* (see quot. 1894) + -ITE¹.] A mineral, consisting chiefly of zirconia, found in Ceylon and Brazil, and used as a refractory material.

1894 L. FLETCHER in *Mineral. Mag.* X. 158 For this new mineral I beg to propose the name *Baddeleyite*, in honour of Mr. Joseph Baddeley, by whom the interesting dense minerals of Rakwana have been brought to the notice of the mineralogical world. **1896** A. H. CHESTER *Dict. Min.* 24 *Baddeleyite*..Zirconia, in black crystals, resembling columbite. **1955** BROWN & DEY *India's Min. Wealth* (ed. 3) v. 253 Baddeleyite (or brazilite) the natural oxide of zirconium (ZrO_2).

baddy (bæ·di). *colloq.* (orig. *U.S.*). [f. BAD *a.* + -Y⁶. Prob. shortened from *BAD MAN.] A criminal or desperado (cf. *BAD MAN), *esp.* a villain in a play or film, esp. a western; hence *gen.* (as a jocular designation), a person of bad character.

1937 in *Amer. Speech* (1938) Apr. 107/1 One of the screen's consistent baddies, Bruce Cabot, feels the effect of the regime of vigorous villains. **1951** J. FRAME *Lagoon* 79 We got shouted to the pictures..where we cheered the goodies and booed the baddies. **1958** *Times Lit. Suppl.* 25 Apr. 226/1 The Communists are goodies and John L. Lewis is a baddy.

badge, *sb.* Add: **4. badge-man** (earlier example); also, *spec.* an official porter.

1804 W. COOKE *Mem. Charles Macklin* 11 After being some time in Dublin, he got settled as a badge-man in Trinity College. **1904** *Daily Chron.* 30 July 7/7 They deprive the licensed porters, or 'badge' men..of work.

badger, *sb.*¹ Add: **2.** (Earlier example.)

1833 C. F. HOFFMAN *Winter in West* (1835) I. 207 A keen eyed, leather-belted 'badger' from the mines of Ouisconsin.

4. b. *to draw the badger*: to entice (a badger, an opponent) to come into the open. Cf. DRAW *v.* 36, *badger-drawing* (BADGER *sb.*² 5).

1844 [see DRAW *v.* 36]. **1857** TROLLOPE *Three Clerks* III. iii. 50 There is a sport prevalent among the downs in Hampshire...Men and boys..congregate together on a hill side, at the mouth of a narrow hole, and proceed, with the aid of a well-trained bull-dog, to draw a badger. *Ibid.* 51 So also at Westminster—with a difference...The badger when drawn has to take his place outside the hole and fight again ..while the victorious bull-dog assumes a state of badgerdom..and in his turn submits to be baited. **1870** *Chambers's Jrnl.* 2 July 420/1 Proceed to the sick man's room, with the avowed intention of 'drawing the badger'. **1890** *Daily Chron.* 19 Sept., The Parnellite taunts regarding Balfour's indifference have at last drawn the badger.

5. badger-baiting (earlier example); **badger game**, an extortion scheme in which a man is lured usu. by a woman (the *badger* or *badger-worker*) into a compromising situation and is then surprised and blackmailed by her accomplice; **ba·dger-pied** *a.*, (of a foxhound) parti-coloured like a badger; hence **badger-pie**, such a hound; **badger plane** (see quot.); **badger softener** [SOFTENER 2], a badger-hair brush used in wood-graining and decorating; **Badger State** *U.S.*, the State of Wisconsin; **badger tongs**, tongs used to grasp the badger as it emerges from its hole, or to pull it out.

1790 *Loiterer* 6 Mar. 6 The object of this grand cavalcade had been a *Badger-baiting* on *Bullingdon-Green*. **1858** *Spirit of Times* 27 Feb. 412/2 He was the 'badger' at Moll Hodge's famous 'panel' establishment, in West Broadway and was sent up for 4 years and 8 months. **1909** in *Cent. Dict. Suppl.*, Badger game [*defined*]. **1924** G. C. HENDERSON *Keys to Crookdom* 228, I know of one case where a man alone worked a variation of the badger game on women. **1936** E. S. GARDNER *Case of Stuttering Bishop* (1937) xiv. 217 It looks too damned much like a badger game. **1962** H. KANE *Killer's Kiss* xx. 156 She stood naked...'You're in one hell of a lot of trouble,' he said. The old badger game. **1867** 'STONEHENGE' *Dogs of Brit. Islands* 101 When the colours blend or amalgamate, the hounds are said to be 'pied'. Hare, badger, red, tan, and yellow pies are the best. **1890** *Daily News* 31 Mar. 3/2 The Wiltshire squire, whose spurs were earned with the Badminton 'badger-pies'. **1845** J. MAYER *Sportsman's Directory* (ed. 7) ix. 147 Hounds are grizzled, brindled, badger-pied, &c., which colours are indicative of strength. **1922** R. LEIGHTON *Compl. Bk. Dog* IV. xvi. 245 The Sealyham..is..frequently whole white, but also white with brown, lemon, or badger-pied markings. *a* **1877** KNIGHT *Dict. Mech.* I. 207/2 *Badger Plane*, a panel plane whose mouth is cut on the skew, and from side to side, so as to work up close to a corner in making a rabbet or sinking. **1878** A. R. & P. VAN DER BURG *Imitation of Woods & Marbles* 4 The speckles must be immediately touched up by the badger softener in the form of the grain by which the pores desired are obtained. **1949** *Archit. Rev.* CVI. 244 The traditional tools of the grainer and marbler are numerous...Sash tool, jamb duster, flogger, badger softener,..all have their place in the craftsman's bag. **1850** E. S. SEYMOUR *Sk. Minnesota* 86 We have abundant reason to anticipate that Minnesota..will not lag far behind the Badger State. **1872** SCHELE DE VERE *Americanisms* 662 Wisconsin, abounding during early days in badgers, has ever since retained the name of Badger State. **1904** *N.Y. Times* 26 July 3 The Speaker will make several speeches in the Badger State. **1859** F. FRANCIS *Newton Dogvane* I. i. 11 Dog-chains, badger-tongs, rabbit-hutches. **1928** *Sunday Dispatch* 15 July 2 Mr. Tinker..got his badger-tongs..'Get the tongs round his neck—quick!' said the Master..as he pushed forward the long iron instrument. **1910** *New England Mag.* July 587/2 A woman who decoys men and then her accomplice (alleged husband) blackmails them is called a 'badger-worker'.

badger, *sb.*³ [f. BADGE *sb.* + -ER¹.] One who wears a badge (of a special kind).

1890 FARMER *Slang* I. 95/1 *Badger*..6. (Wellington School). A fellow who has got his 'badge' for play in the 2nd xv. at football. **1920** *John Bull* 28 Aug. 16/1 There are idle Badgers at Elmswell, where the job of Rate Collector was going the other day. **1925** *Glasgow Herald* 23 July 7 A Doggett badger is a man to be reckoned with in the rowing world.

Badian (bē¹·diăn), *sb.*² and *a.* Also *colloq.* **Bajan, Bajun.** [shortened from *BARBADIAN.] = *BARBADIAN *a.* and *sb.*

1910 in W. B. Whall *Ships, Sea Songs & Shanties* 83 Sally she a 'Badian bright mulatto; Seven long year I courted Sally. **1921** *Public Opinion* 30 Sept. 325/2 On the day of Gen. Baden-Powell's arrival all the Badian boatmen struck work. *Ibid.*, The Badians have an inordinate opinion of themselves and of their island. **1958** B. HAMILTON *Too much of Water* ii. 37 'I'm going to show you what a real good Bajan rum cocktail is like.' 'What's Bajan?'.. 'B'badian.' *Ibid.* 38 Did you ever hear of the Bajan who told an Englishman Bados was famous for three things?

badinage, *v.* (Earlier example.)

1804 W. COOKE *Mem. Charles Macklin* 74 And the men who chose to go and *badinage* with them, did it at the peril of their character.

bad lands. orig. and chiefly *U.S.* [tr. F. *mauvaises terres*: BAD *a.* 4.] Arid barren lands, esp. in certain parts of western U.S.A., characterized by erosion of the soft surface strata in varied and fantastic forms. Also *bad land, badland*, esp. attrib.

1852 D. D. OWEN *Geol. Surv. Wisconsin* 195 The country of the 'Bad Lands'..lying high up on White River. **1872** SCHELE DE VERE *Americanisms* 180 It is different with the *Bad Lands*, which border the Missouri for about twenty miles, and were called by the first French settlers, *Mauvaises Terres*. **1876** R. I. DODGE *Black Hills* 17 All the country south is 'bad land' or tertiary formation, much cut by deep and abrupt ravines. **1931** H. F. PRINGLE *T. Roosevelt* I. viii. 94 Until recently, buffalo had roamed the Bad Lands. **1942** C. A. COTTON *Geomorphol.* (ed. 3) 31 Bare ground is carved by rain-wash into innumerable closely-spaced, steep-sided ridges and valleys..producing an almost impassable land surface generally referred to as 'badlands' or *badland erosion*.

transf. and *fig.*

1892 *Scribner's Mag.* July 8/1 'The Bad Lands' [in Chicago] is a quarter more repellent because more pretentious than 'The Dive',..being the abode of vice and crime rather than of poverty. **1952** *Manchester Guardian Weekly* 18 Dec. 3/1 Nothing in the White House career of Harry S. Truman, pride of Boss Pendergast's political badlands of Kansas City, so well became it as his behaviour [etc.].

ba·dling², var. BADELYNG.

1924 *Chambers's Jrnl.* Nov. 714/1 A badling of duck flighting in from the sea, whirred down and settled in a reed-bound pool. **1926** *Daily Express* 4 Aug. 6/6 A badling of duck—there are seven of them..—come skimming over me.

badly, *adv.* **7.** (Further examples.)

1915 D. H. LAWRENCE *Rainbow* ii. 65 'I want my mother.'..'Ay, but she's badly.' **1966** A. E. LINDOP *I start Counting* i. 18 Your Aunt Rene Tindall says she's been badly again.

bad man. Chiefly *U.S.* [BAD *a.* 5, after Sp. *malo hombre*.] A desperado, robber, gunman; a villain (see quots.). Cf. *BADDY and quots. 1599 and 1849 *s.v.* BAD *a.* 5.

1855 *Santa Barbara* (Calif.) *Gaz.* 28 June 1/4 The 'bad man' was floored by the weight of a walking stick that the quaker had been known to carry. **1859** *Brit. Colonist* (Victoria, Brit. Columbia) 22 Jan. 1/3 It was they who tacitly aided the bad men to rescue the prisoner. **1888** ROOSEVELT *Ranch Life Far West* in *Century Mag.* Feb. 504/1 The 'bad men', or professional fighters, and man-killers, are of a different stamp, quite a number of them being, according to their light, perfectly honest. **1891** E. S. ELLIS *Check 2134* ix. 59 It is much more pleasant to relate how such a young gentleman outwitted a 'bad man'. **1906** *N.Y. Even. Post* 9 Nov. 6 He even appointed a typical 'bad man'—that is, manslayer—to office as a proof of his fondness for Arizona. **1910** MULFORD *Hopalong Cassidy* i. 9 They were good bad-men and bad bad-men, the killer by necessity and the wanton murderer. **1922** P. A. ROLLINS *Cowboy* iii. 54 The actual 'bad man' was 'short on conversation'... All actual bad men were wholly untrustworthy, were natural killers, moral and mental degenerates, inhuman brutes who would slay for personal gain or merely to gratify a whim. **1958** V. KELSEY *Brit. Columbia rides Star* 173 Graves of badmen who met justice-in-the-rough.

Badminton. 1. (Earlier example.)

1845 DISRAELI *Sybil* I. i, Waiter, bring me a tumbler of Badminton.

2. (Later examples.) Also *attrib.*

1910 *Encycl. Brit.* III. 189/2 The Badminton hall should be not less than 18 ft. high. **1966** MILLS & BUTLER *Mod. Badminton* i. 20 All badminton players are strictly amateur. *Ibid.* 22 Unfortunately, except in certain countries, badminton cannot be played outdoors because of the extreme lightness of the shuttle. *Ibid.* ii. 25 The ideal hall for a badminton court.

Baedeker (bē¹·dĕkəɹ). Any of the series of guide-books issued by Karl *Baedeker* (1801–59) at Coblenz, or by his successors; also applied loosely to any guide-book. Also *fig.*

1863 *Miss Jemima's Swiss Jrnl.* 29 June (1963) ii. 23 We followed the advice of our faithful Baedeker. **1889** G. B. SHAW *London Music 1888–89* (1937) 180 Glancing through Baedeker as I bowl along Bayreuthwards I perceive [etc.]. **1920** T. S. ELIOT *Ara Vos Prec.* 14 Burbank with a Baedeker. **1935** *Chambers's Encycl.* I. 659/2 In the 20th century 'Baedeker' had become a synonym for guidebook, and there are Baedekers not merely for the principal European countries, but for Palestine and Syria [etc.]. **1948** *Landfall* II. 58 There was no place in their ample kits for a Baedeker or a book of poetry. **1959** *Listener* 30 Apr. 767/1 *Balzac et son Monde* is an indispensable Baedeker to the *Comédie Humaine*.

b. Hence *Baedeker raid*, one of a series of raids by the German Air Force in April and May of 1942 upon places of cultural and historical importance in Britain.

1942 *Daily Mail* 1 May 4/7 The 'Baedeker' raids have put the Luftwaffe in a grave dilemma. **1942** *War Illustr.* 29 May 705 York's Guildhall as it appeared on the night of April 28 after it had been struck by incendiaries in one of Hitler's 'Baedeker raids' on Britain's cultural heritage. After the legitimate R.A.F. raids on Rostock, where the Heinkel factory and other munition works were destroyed, German officials frankly stated that the Luftwaffe would go out for every building in Britain which is marked with three stars in Baedeker's guide-books.

bael, bel (bel). Also 7 **beele, bille,** 9 **bhel, bēl, bél.** [Hindi *bel*, Marathi *bail* :– Skr. *bilva, vilva*.] An Indian tree, *Ægle marmelos*, or its aromatic fruit; the Bengal quince. Also, the medicinal extract of the fruit.

1618 in Foster *Eng. Factories India* (1906) 10 Greene ginger, mirabilons, and beeles. **1619** *Ibid.* 76 Mirabilons and billes in conserves. **1854** HOOKER *Himal. Jrnl.* I. 50 The Bhel fruit, lately introduced into English medical practice, as an astringent of great effect, in cases of diarrhœa and dysentery. **1874** GARROD & BAXTER *Mat. Med.* (ed. 4) 219 Indian bael has obtained much reputation in India in the treatment of diarrhœa and dysentery. **1877** SHUNKAR SINGH & GUNANAND tr. *Hist. Nepāl* 33 Every Newār girl, while a child, is married to a bēl-fruit, which after the ceremony is thrown into some sacred river. **1910** *Encycl. Brit.* III. 191/1 Bael fruit (*Aegle marmelos*). **1924** R. E. ENTHOVEN *Folklore of Bombay* III. 124 The *Palas* (*Butea frondosa*) and the *Bel* (*Aegle Marmelos*), a tree sacred to the god Shiva, are considered to be holy by the Hindus.

bætyl. Delete *rare* and add examples. Also in Gr.-L. form **baitylos, bætylos, -us**; also **bæty·lion** (pl. **-ia**) (Gr. βαιτύλιον). Hence **bæty·lic** *a.*, of the nature of a bætyl.

1854 *Encycl. Brit.* IV. 361/2 These *bætylia* were the objects of much veneration among the ancient heathens. **1884** *Proc. Soc. Psychical Research* II. 117 Aerolites were scouted as a kind of fetish *in excelsis*—a transcendental bætyl—'the image which fell down from Jupiter'. **1889** W. R. SMITH *Relig. Semites* 193 The use of baetylia, or small portable stones to which magical life was ascribed. **1901** A. J. EVANS in *Jrnl. Hellen. Stud.* XXI. 106 The rough pyramidal pillars of the Bhuta Spirit,..and many other rude 'baetyls' of the same kind..are commonly set up beneath holy trees. *Ibid.* 113 An artificial pillar image of the divinity, it may be even the actual 'baetylos' of remote tradition. *Ibid.* 118 The sanctity of baetylic stones and pillars is due to a variety of causes. **1903** *Amer. Jrnl. Archæol.* VII. 200 He found the stone cooled off, and recognizing that it was a baetyl, took it home with him. **1923** *Trans. Sc. Ecclesiol. Soc.* 97 The Clack or stone [at Clackmannan]..is a true Fetich or Baitylos. **1941** *Jrnl. Theol. Stud.* XLII. 60 Most of the few fragments [of coral] discovered at Gezer seem to have been unworked pieces, probably therefore prophylactic in purpose, and not *objets de luxe*. One white piece is of baetylic form.

baff (bæf), *v.*² *Sc.* [Cf. BAFF *sb.*] To beat, strike (examples in *Sc. Nat. Dict.*); *spec.* in Golf, to strike the ground with the sole of the club-head in making a stroke. So **ba·ffing-spoon** = *BAFFY.

1858 *Chambers's Jrnl.* X. 157/1 The names of the wooden-headed clubs principally used at St. Andrews..are as follows: the play-club, long-spoon, mid-spoon,..putter, and baffing-spoon. **1881** R. FORGAN *Golfer's Handbk.* 10 The 'Baffing Spoon' is so called from the sound produced by it as it smites the ground in making the stroke. *Ibid.* 33 Baff. **1890** H. G. HUTCHINSON et al. *Golf* 445 Baff.

baffle, *v.* Add: **8. d.** In technical uses.

1883 GRESLEY *Gloss. Coal Mining*, *Baffle*, to brush out or mix fire-damp with air in order to render it non-explosive; a dangerous practice, and not now allowed. **1885** *Marine Engineer* 1 Apr. 3/2 The corrugations serve to break up the volume of steam, and also baffle the water passing through the condenser. **1909** WEBSTER s.v., Plates are used for baffling the steam.

baffle, *sb.*¹ Add: **5.** Also, any shielding device or structure, in many technical uses (see quots.); *spec.* an acoustic screen.

1913 V. B. LEWIS *Oil Fuel* iv. 97 The top of the inlet tube for the sample of vapour is protected by a series of baffles. **1928** *Wireless World* 6 June 604/2 (*caption*) The baffle which greatly improves reproduction [of sound]..is incorporated in the cabinet. **1931** *B.B.C. Year-Bk.* 436/2 *Baffle*, a screen of non-resonant material, generally wood, largely used in conjunction with cone-type loud speakers instead of a horn, to ensure the radiation of the very low audible frequencies. **1939** in Henney & Dudley *Handbk. Photogr.* viii. 254 The use of baffles over the photoelectric cell [of an exposure meter]. **1951** *Gloss. Terms Plastics* (*B.S.I.*) 34 *Baffle*, a device used for the purpose of restricting the flow of hydraulic fluid in a high pressure line. It consists of a disc with a small central perforation. **1952** GRANVILLE *Dict. Theatr. Terms* 22 *Baffle*, any suitable sheet of material used to prevent a spill of light where not necessary. **1958** *Gloss. Terms High Vacuum Technol.* (*B.S.I.*) 12 *Baffle*, an obstruction placed near the mouth of a vapour pump to impede the entry of back-streaming vapour into the system. **1959**

Times 31 July 14/4 The first earth baffle..has been built by Air India at London airport to suppress engine sounds during ground running.

attrib. and *Comb.*

1909 *Cent. Dict. Suppl.*, Baffle-tube. **1926** *Chambers's Jrnl.* Apr. 270/1 The oil rises through a series of perforated baffle discs to the top of the tank. **1931** *Answers* 10 Oct. 36/2 Natural reproduction will be impoverished if a moving coil loud-speaker is used within a baffle board. **1933** *Archit. Rev.* LXXIII. 233 The six baffle-boards covering ventilation exhaust outlets. **1939** SAMUELY & HAMANN *Civil Protection* 167 Baffle wall. **1940** *Chambers's Techn. Dict.* 70/2 *Baffle tube*, a pipe of sufficient length to lower the temperature of hot gases before they enter a furnace. **1941** *New Statesman* 8 Mar. 235/1 Heavy baffle-walls have been built [in an air-raid shelter]. **1953** *Times* 10 Jan. (*headline*) Reducing Noise at Airport. Baffle Wall Nearing Completion.

baffle-plate. [BAFFLE *sb.*¹ 5.] A plate hindering or regulating the passage of fluid through an outlet or inlet, or the direction of sound. Hence **baffle-plated** *a.*, having a baffle-plate.

1882 [see BAFFLE *sb.*¹ 5]. **1906** *Daily Chron.* 3 Mar. 3/6 A system of tubes or baffle-plates which break up the hot gas. **1908** *Westm. Gaz.* 31 Dec. 5/1 In this radiator the baffle plated hood replaces the frieze and trivet. **1920** *Chambers's Jrnl.* Nov. 767/1 [Gas] is passed upwards through water dripping from a series of baffle-plates in chambers known as scrubbers.

baffy (bæ·fi). *Golf.* [f. BAFF *sb.* or *v.²] A short wooden club used to hit the ball into the air. Also *baffy spoon*.

1888 *Daily News* 22 Sept. 5/1 The old 'Baffy', a funny little toy club, with a sloped face. **1890** H. G. HUTCHINSON et al. *Golf* 59 The 'baffy', very short and stiff and with face very much laid back. *Ibid.* 335 To Allan was due..the introduction of irons and cleeks for the approach to the hole, these shots having been previously played with baffy spoons.

bag, *sb.* Add: **1. c.** A base in baseball (see quot. 1857). *U.S.*

[**1857** *Spirit of Times* 28 Feb. 420/3 The first, second, and third bases shall be canvas bags, painted white, and filled with sand or saw-dust.] **1873** *Forest & Stream* 20 Nov. 231/1 In this inning, through error, the Princetons succeeded in getting the bags full, with no men out. **1917** C. MATHEWSON *Sec. Base Sloane* xiii. 177 Hunt was two yards from the bag when the ball reached third base.

d. *fig.* A preoccupation, mode of behaviour or experience; a distinctive style or category; esp. a characteristic manner of playing jazz or similar music. Cf. *bag of tricks* (sense *17 a below). *slang* (orig. *U.S.*).

1960 J. HENDRICKS in D. Cerulli et al. *Jazz Word* (1962) 140 Lack of acceptance is a drag...Man, that's really in another bag. **1962** *Jazz Jrnl.* Mar. 30 'Bag' is a current piece of trade jargon for hip musicians, and means something between a personal style and a body of work. **1966** *Sunday Times* (Colour Suppl.) 13 Feb. 35/4 'His bag is paper sculpture', 'she's in a folk-song bag right now'. **1966** *Crescendo* Dec. 16/1 Singing blues with Basie is another kind of bag.

7. b. A diplomatic bag.

1816 H. BROUGHAM *Let.* in H. Maxwell *Creevey Papers* (1903) I. xi. 252, I think it better to trust this to the post than to any of their d—d bags. **1852** THACKERAY *Let.* 15 Mar. (1946) III. 25 When Wm. Grey goes to Paris you'll have the use of the bag again. **1964** *Times* 6 July 11/3 It [*sc.* reciprocity] should mean..that embassies and bags are inviolate.

9. *fig.* (Further examples.) Hence in *pl.* (*slang*), much, many, 'heaps'.

1858 W. H. RUSSELL *My Diary in India* 16 Mar. (1860) I. xxi. 348 The philanthropists who were cheering each other with the thought that there was sure 'to be a good bag at Lucknow', will be disappointed. **1900** *Daily News* 9 June 5/5 Our bag was 4 engines and 84 trucks, with a quantity of coal. **1917** P. H. GIBBS *Battles of Somme* 105 'We took bags of 'em [*sc.* Germans],' said an officer. **1919** W. H. DOWNING *Digger Dialects* 9 *Bags*, plenty; a large number. **1919** J. B. MORTON *Barber of Putney* xiii. 209 It's not gay, this life, but it might be bags worse. *Ibid.* xvii. 285 There's bags of good names, and yet blokes go an' call their kids Ermyntrude. **1930** BROPHY & PARTRIDGE *Songs & Slang Brit. Soldier 1914–18* 96 *Bags*, plenty, lots. E.g. 'Got any bully?' —'Yes, bags of it.' And especially *bags of room*. **1930** C. V. GRIMMETT *Getting Wickets* i. 32 It was with Prahran that I recorded my big successes in club cricket, my 'bags' in four seasons being 67, 39, 68 and 56 wickets respectively. **1940** I. HALSTEAD *Wings of Victory* I. ii. 56 What his personal 'bag' was I don't know—certainly it was over twenty. **1945** R. L. SEDDON *Whims of W.A.A.F.* 14 With 'bags' of ambition. **1955** *Times* 19 Aug. 2/5 A retrospective exhibition..an exhibition of drawings..and now a film.. this is Picasso's 'bag' for the summer of 1955. **1962** A. WESKER *Chips with Everything* I. i. 12 We 'ad bags o' fun, bags of it.

12. b. *spec.* in *Leather Industry* (see quot.).

1909 *Cent. Dict. Suppl.* 105 *Bag*, in *leather-manuf.*, fullness in the middle of a skin which prevents it from lying out flat and smooth.

c. A fold of loose skin beneath the (human) eyes. Usu. in *pl.*

1867 W. ALLINGHAM *Diary* 10 June in H. Gernsheim *J. M. Cameron* (1948) 25, I want to do a large photograph of Tennyson and he objects! Says I make bags under his eyes. **1894** SOMERVILLE & 'ROSS' *Real Charlotte* I. x. 147 The dark bags of skin under Julia Duffy's eyes became slowly red. **1910** J. BUCHAN *Prester John* xix. 316, I caught a glimpse of my face in it,..lined with blue bags below the eyes. **1938** H. G. WELLS *Apropos of Dolores* iv. 174 His large grey eyes had if anything got larger and the lower lids ower. He has bags under them.

16. *pl.* For '*vulgarly*' read '*colloq.*' Add earlier and later examples.

1853 'C. BEDE' *Verdant Green* vi. 51 Just jump into a pair of bags and Wellingtons. **1861** A. TROLLOPE in *Tales of all Countries* (ser. 2) 136 A pair of the loosest pantaloons —I might, perhaps, better describe them as bags. **1923** D. L. SAYERS *Whose Body?* iv. 82 I'll run round and change at the club. Can't feed with Freddie Arbuthnot in these bags. **1927** —— *Unnatural Death* vii. 84 Just brush my bags down, will you, old man?

16*. A disparaging term for a woman, esp. one who is unattractive or elderly; = BAGGAGE 6. *slang* (orig. *U.S.*).

1924 P. MARKS *Plastic Age* xviii. 202, I don't..chase around with filthy bags or flunk my courses. **1928** *Amer. Speech* Feb. 218 Say, Cress, who was that bag I saw you with yesterday? **1949** T. RATTIGAN *Harlequinade* (1953) 61 That's enough from you, you old bag! **1950** *Penguin New Writing* XL. 45 'It's just like you, you dreary old bag,' he would say to a blowsy old pro. **1961** M. DICKENS *Heart of London* I. 77 I've never really known a pretty girl like you. At the training college they were all bags.

17. *the whole bag of tricks* (examples). Also *bag of tricks*, stock of resources; sometimes with play on other senses of 'bag' (old woman, etc.).

1841 ELIZUR WRIGHT tr. *La Fontaine's Fables* (ed. 2) XII. xviii. 314 But fox, in arts of siege well versed, Ransacked his bag of tricks accursed. **1874** HOTTEN *Slang Dict.* 76 *Bag of tricks*, refers to the whole of a means towards a result. 'That's the whole bag of tricks.' **1889** G. B. SHAW *Lond. Mus.* (1937) 129 She relied largely for her acting on the exploitation of what is nothing but a bag of tricks. **1898** A. BENNETT *Man from North* xvi. 152 I've had three 3 A.M. midwifery cases this week—forceps, chloroform, and the whole bag of tricks. **1909** H. G. WELLS *Tono-Bungay* III. ii. § 3 301 Learn the whole bag of tricks in six months. **1922** JOYCE *Ulysses* 139 She was a nice old bag of tricks. **1924** GALSWORTHY *White Monkey* III. viii. 185 A being who completely robbed the world of its importance, 'snooped', as it were, the whole 'bag of tricks'. **1936** L. C. DOUGLAS *White Banners* xiii. 280 Men were all alike. A woman didn't have to carry a very big bag of tricks to achieve her purpose. **1942** 'P. WENTWORTH' *Danger Point* xl. 233 Fingerprints...A nice bag of tricks for our modern scientific police. You put 'em in a hat and shake 'em up, and then you put in your hand and pick your murderer. **1957** KEHOE *Technique Film & T.V. Make-Up* iii. 37 The make-up kit is the artist's tool box and bag of tricks.

b. *bag of mystery* (usu. in pl. *bags of m.*): a sausage or saveloy. *slang.*

1864 in HOTTEN *Slang Dict.* 69. **1879** W. J. BARRY *Up & Down* xvi. 163 A slice of bread was given with the 'bag of mystery', as some rowdies called the luscious saveloy. **1909** WARE *Passing Eng.* 15/2 If they're going to keep running-in polony fencers for putting rotten gee-gee into the bags of mystery, I hope they won't leave fried-fish-pushers alone. **1921** H. FOSTON *At the Front* xvi. 115 [Have you] any bags of mysteries, otherwise sausages? **1962** *John O' London's* 14 June 571/1 The bags of mystery or links of love are sausages.

c. Colloq. phr. *in the bag* (i.e. game-bag; see sense 9): (*a*) *Austral.* and *N.Z.* (see quot. 1945); (*b*) (*to be put*) *in the bag*, (to be) taken prisoner; (*c*) (orig. *U.S.*) virtually assured or secured, as good as in one's possession.

1900 J. SCOTT *Tales of Colonial Turf* 33 The neddy was in the bag in the Cup; he was no trier. **1945** BAKER *Austral. Lang.* ix. 174 A horse set to lose a race is said to be *in the bag*. **1919** J. BUCHAN *Mr. Standfast* ii. 52 Unless I went out to the Front again and got put in the bag and sent to the same Boche prison. **1956** D. M. DAVIN *Sullen Bell* II. vii. 152 When you went in the bag the chaps probably said 'Too bad about old Gus'.

1922 *San Francisco Call & Post* 19 July 17 Yes, yes, yes, but listen to me—I get this from the jock himself—this is in the bag. **1926** *Emporia* (Kan.) *Gazette* 24 Sept. 1/2 After Tunney landed with that terrific right, the fight was in the bag. **1929** *Liverpool Daily Courier* 4 Sept. 9/1 If half the members of a Talkie audience shudder every time a character on the screen says..'It's in the bag', the other half make a mental note of the expression for future use. **1932** WODEHOUSE *Hot Water* i. 32 We're sitting pretty. The thing's in the bag. **1943** B. J. HURREN *Eastern Med.* v. 51 Crete was 'in the bag' for Jerry if he wished to take it. **1957** *Economist* 30 Nov. 765/1 The message..contains a frank warning that independence is not 'in the bag'.

18. *to give* (a person) *the bag* (later examples). Also *to get the bag*: to be dismissed; *to let the cat out of the bag* (later examples).

1788 P. M. FRENEAU *Misc. Wks.* 414 He must give us the bag, Adhere to Old England, and sail with her flag. **1806** T. G. FESSENDEN *Orig. Poems* 39 'To give the bag' is an expression common with the lower classes in New England, and indicates that Miss Delia will not honour Mr. Damon with her company in a tete-a-tete conversation. *Ibid.* 73 Jonathan..tumbled, sadly, all the way, Lest he should get the bag, sir. **1825** J. NEAL *Bro. Jonathan* II. 277 Sent away, with a flea in your ear; some girl has given you the bag. **1849** C. BRONTË *Shirley* III. xiv. 300 This last epithet I choose to suppress, because it would let the cat out of the bag. **1870**, etc. [see E.D.D.]. **1871** W. S. GILBERT *Palace of Truth* I. 13 While publishing the truth He's no idea that he is doing so; And..he lets innumerable cats Out of unnumbered bags. **1913** 'IAN HAY' *Happy-go-lucky* I. 4 'Your fag, isn't he?' 'I gave him the bag two terms ago... Tiny has him now.'

20. *bag and spoon*: used *attrib.* to designate a type of dredging apparatus (see quot. 1940).

1840 *Civ. Engin. & Archit. Jrnl.* III. 30/1 Dredging with the common bag and spoon apparatus. **1940** *Chambers's Techn. Dict.* 70/2 *Bag and spoon dredger*, an implement consisting of a leather bag laced to a steel hoop;..used to dredge soft material.

V. *Comb.* and *attrib.* **1. b.** *bag-carrier, -punching* (cf. *PUNCH *sb.*²), *-snatcher.*

1890 J. WATSON *Conf. Poacher* x. 137, I had arranged with a confederate to act as bag-carrier. **1957** N. FRYE *Anat. Criticism* iii. 197 A dwarf who carries a bag of 'needments'. He is not a traitor, like the other bag-carrier Judas Iscariot. **1927** *Daily Express* 21 Sept. 1/2 Dempsey jogged some miles along the road yesterday, did bag-punching, etc. **1950** J. DEMPSEY *Championship Fighting* xxiv. 180 Bag-punching is another exercise that conditions and sharpens. **1908** *Westm. Gaz.* 20 Aug. 8/2 Sentencing a bag-snatcher.. to three months with hard labour...Prisoner snatched away the satchel of..a nurse.

2. bag-filter, a filter made of a cloth bag.

a **1877** KNIGHT *Dict. Mech.* I. 209/2 *Bag-filter*, (sugar refining) a device sometimes used in clearing saccharine solutions of feculencies and impurities mechanically suspended therein. **1910** *Encycl. Brit.* X. 346/2 A crude method [of filtration] consists of straining the liquid through cotton or other cloth..formed into long narrow bags ('bag-filters').

bag, *v.*¹ Add: **1. b.** Said esp. of trousers which become out of shape at the knees.

1893 *Scribner's Mag.* Sept. 293/1 A trouser-leg is more obstinate in its ugliness. If tight it bags at the knees. **1913** A. R. HOPE *Half & Half Tragedy* 32 The Captain of the school has a pair of new breeches..; but they bag at the knees.

5. Also *absol.*

1890 LD. LUGARD *Diaries* (1959) I. 135 In the evening I again hit several animals by the unanimous verdict of my men, but did not bag.

6. (Later examples.) To add to one's 'bag' (BAG *sb.* 9). *fig.*

1879 *Bell's Life in London* 28 June 4/2 Whom Mr. Hornby very smartly 'bagged' at mid-on. **1936** J. DOS PASSOS *Big Money* 72 He was almost bagged by a taxicab crossing the street. **1940** I. HALSTEAD *Wings of Victory* I. ii. 59 Pilot Officer Elliott..has now bagged two. **1943** WARD-JACKSON *Piece of Cake* 11 *To bag, bagged*, to hit by aerial gunfire; shot down. **1945** *Finito! Po Valley Campaign* 17 They bagged 9,000 PWs and a battery of 15-inch guns.

b. To claim; reserve. Used esp. by children (see quot. 1914 and *BAGS I). *colloq.*

1914 *Concise Oxf. Dict.* Add. 1045/2 *Bag*, (also, in school slang) claim on the ground of being the first to claim (*I b.*, but usu. *bags I* or *bags, first innings!*). **1923** J. MANCHON *Le Slang* 56, I'm going to bag the best chair. **1948** R. A. KNOX *Mass in Slow Motion* vii. 68 The other girl bagging the hot-water pipes first. **1968** *Listener* 29 Feb. 269, I bags be Anthony Wedgwood Benn.

7. To dismiss, discharge (a person). Cf. SACK *v.*¹ 5 *a.* Cf. *BAG *sb.* 18.

1848 *Chaplain's Rep., Preston House of Correction* 61 The master told him if he did not mind his work he would 'bag' him. **1895** W. WESTALL *Sons of Belial* II. xxii. 83 'Not have me at th' shop!..You surely wouldn't bag me?..Bagged, beggared, and disinherited!' he moaned. *Ibid.* 85 I'll work for nowt. Only don't bag me just like a common hand.

8. *to bag school*, to play truant. Also *to bag it*. *U.S.*

1934 J. O'HARA *Appointment in Samarra* (1935) vii. 203 She did not report him on Sunday afternoons when he 'bagged it' to go to a ball game. **1948** *Philadelphia Bulletin* 15 Jan. 14 Threatening him with castor oil, when he seemed set to bag school, never did any good.

Baganda (bāgæ·ndā), *sb.* and *a.* [Bantu name; cf. Swahili *Waganda*.] **A.** *sb.* A negroid Bantu-speaking people inhabiting Buganda, a province of Uganda on the N.W. shore of Lake Victoria. **B.** *adj.* Of or pertaining to this people.

[**1882** WILSON & FELKIN *Uganda & Egyptian Soudan* I. vii. 148 Several tribes are to be found in Uganda, living.. scattered through the country. The most important tribe, in every respect, is that of the Waganda.] **1889** R. P. ASHE *Two Kings of Uganda* vi. 47 Bloody raids for which the Baganda are infamously famous. *Ibid.* xviii. 215 (*heading*) Baganda converts and martyrs. **1902** *Encycl. Brit.* XXXIII. 541/2 The Bantu negroes..include the remarkable Baganda people. *Ibid.*, The Baganda are now mainly Christian. **1911** FRAZER *Golden Bough* (ed. 3) I. iii. 142 The Baganda believe that a barren wife infects her husband's garden with her own sterility. **1960** *Economist* 8 Oct. 136/2 The Uganda government must therefore press on with the February elections, despite the gyrations of the Baganda.

‖**bagarre** (bagār). [F.] A scuffle, tumult. Also *fig.*

1897 CONRAD *Let.* 9 Aug. in G. Jean-Aubry *J. C.* (1927) I. 209 Posterity will be busy thieving, lying..and the half-dozen men lost in that *bagarre* are more likely to weep than to smile over those [*sc.* Kipling's] masterpieces. **1955** D. BARTON *Glorious Life* vi. 73 '*Fracas* with the twins?' '*Bagarre*. You saw?' **1963** *Punch* 6 Feb. 204/3 The subtle *bagarre* triggered off by a true conflict.

bagasse. Add: (Earlier and later examples.) Also *attrib.* Cf. BEGASSE.

1833 B. SILLIMAN *Man. Sugar Cane* 15 Two boys at bagasse (ground cane) carts. **1835** J. H. INGRAHAM *South-West* I. 239 The *bagasse* or cane-trash (called in the West Indies *migass*) is..conveyed to a distance..to be burnt. **1960** *Times* 8 Jan. 7 Mauritius..solved the problems of a one-crop economy by burning 'bagasse', the fibre that remains after sugar cane is crushed. **1967** E. CHAMBERS *Photolitho-offset* xvi. 246 The fibres of straws, bamboos, sugar cane waste (bagasse) and palms can be used for paper-making.

b. = *BAGASSOSIS.

1943 *New Orleans Med. & Surg. Jrnl.* XCV. 558 (*title*) Bagasse disease of the lung. **1944** *Newsweek* 26 June 80 A new industrial disease called bagasse was reported..among sugarcane workers who handled..the fibrous material.

bagassosis (bægăsǭᵘ·sɪs). *Path.* [mod.L., f. BAGASSE + -OSIS.] A disease of the lungs, due to inhalation of the dust of bagasse.

1941 JAMISON & HOPKINS in *New Orleans Med. & Surg. Jrnl.* XCIII. 580 (*title*) Bagassosis: A fungus disease of the lung. **1944** *Dorland's Med. Dict.* (ed. 20) 199/1 *Bagassosis*, a respiratory disorder due to the inhalation of the dust of bagasse, the waste of sugar cane after the sugar has been extracted. **1947** *Thorax* II. 99 The radiographic changes in the lungs in bagassosis are illustrated.

ba·gaty, ba·ggety. *Sc.* The lump-fish.

1710 R. SIBBALD *Fife & Kinross* II. ii. 52 *Lumpus alter*.. which our Fishers call the Hush-Padle or Bagaty. **1794** in *Statist. Acc. Scotl.* XII. 521 The fish caught here are.. mackerel, baggety, sand-eel [etc.]. **1810** [see PADDLE *sb.*²]. **1828** J. FLEMING *Hist. Brit. Anim.* 190 Lump-fish..Hush, Bagaty. **1867** SMYTH *Sailor's Word-bk.*, *Lump*..the trivial name of the baggety..*Cyclopterus lumpus*.

bagel (bē¹·g'l). [ad. Yiddish *beygel*, app. (Webster) f. MHG. *böugel, whence G. dial. *beugel, bäugl*, dim. of MHG. *boug-, bouc-* ring, bracelet :—OHG. *boug* = OE. *bēag* BEE².] A hard ring-shaped salty roll of bread. Also *attrib.* Cf. *BEIGEL.

1932 L. GOLDING *Magnolia St.* x. 165 Bagels are like large wooden curtain-rings to look at...She cut them and buttered them. **1957** LUCILLE STERN *Midas Touch* III. xxiii. 174, I got lox and bagels. You hungry, Baruch? **1959** *Guardian* 3 Dec. 9/1 The 'Bronx Bagel-babies'—the middle class girls who leap feet first into the vie Bohème. *Ibid.*, Pursued by Bagel-babies..most of the genuine artists have moved East. **1961** *Encounter* XVI. No. 4. 22, I eat salty bagels in the sun. **1961** *Guardian* 25 Mar. 4/1 His haunts, like the bagel shop.

baggage. Add: **1.** (Earlier U.S. examples.)

1739 W. STEPHENS in *Colonial Rec. Georgia* (1906) IV. 410 The Captain..was carrying divers parcels of baggage to the water-side. **1842** J. S. BUCKINGHAM *Slave States* II. 173 The cart he was driving contained their baggage. **1852** C. A. BRISTED *Upper Ten Thousand* 81 An American never uses the conversational term luggage, but always speaks of his impediments as baggage.

b. *ellipt.* A baggage-car on a train. *U.S.*

1926 J. BLACK *You can't Win* vii. 82 It was a warm night, and riding the front end of the baggage was pleasant enough.

C. 1. and **2.** *baggage-cart, -check, -man, -master, -room, -smasher, -train, -wagon*: earlier and later examples. Also, *baggage-agent, -car, -hatch, -tag, -truck.*

1858 W. P. SMITH *Railw. Celebr.* 2 (Advt.), On almost all the trains will be found Baggage or Express Agents. **1833** *Amer. Railroad Jrnl.* II. 725/3 An agent is always stationed at the brake of the baggage car. **1878** B. F. TAYLOR *Between Gates* 26 A Babel of trunks is surging towards the baggage-cars. **1842** J. S. BUCKINGHAM *Slave States* II. 173 A barouche..and a baggage-cart following with five trunks. **1848** *Hunt's Merch. Mag.* (U.S.) XVIII. 334 (*caption*) Railroad Baggage checks. **1948** *Sat. Even. Post* 25 Sept. 125 *One* ticket and *one* set of baggage checks takes you straight through to your overseas destination. **1891** KIPLING *Light that Failed* xiv. 301 Cabin as close to the baggage-hatch as possible. **1858** *N.Y. Tribune* 14 Jan. 2/3 In the baggage car.. the baggage man..asked me what I wanted. **1904** *Westm. Gaz.* 4 May 5/1 A railway baggageman. **1845** *Hunt's Merch. Mag.* (U.S.) XIII. 581 The conductor..is also called baggage-master. *a* **1884** C. READE *Let.* in J. Hollingshead *Gaiety Chron.* (1898) iii. 144 He also does all utility business, and is baggage-master on tours, etc. An invaluable servant in a theatre with small pretensions. **1926** *Amer. Mercury* Mar. 340/2 The baggage-master at the college transfer remembered re-checking his trunks. **1819** P. WAKEFIELD *Excursions N. Amer.* (ed. 3) 25 On deck there are numerous conveniences, such as baggage-rooms, smoking-rooms, &c. **1963** *P.M.L.A.* Dec. p. vii/2 (*list*) U.K. left luggage: U.S. baggage room. **1851** *San Francisco True Standard* 6 Mar. 2/2 Baggage Smashers. **1879** F. R. STOCKTON *Rudder Grange* vii. 75, I went up-stairs and got a baggage tag. **1841** R. PARK *Pantology* 506 The baggage train should be placed either in the column, or so near it that it may be speedily defended. *a* **1861** T. WINTHROP *Life in Open Air* (1863) 228 The soldiers..shot overboard a heavy baggage-truck. **1936** L. C. DOUGLAS *White Banners* x. 223 Such chuff-chuffing of engines and clanging of bells and rattling of baggage-trucks. **1791** WASHINGTON *Diary* 21 Mar. (1925) IV. 149 My equipage and attendance consisted of a Charriot..a light baggage Waggon and two horses. **1849** D. NASON *Jrnl.* 78 [I] got a ride in a baggage wagon, for which I paid 37½ cents.

ba·ggageless, *a.* Having no baggage; having lost one's baggage.

1891 *Daily News* 28 Nov. 3/6 On Sunday he wandered about alone and baggageless. **1905** *Westm. Gaz.* 6 Dec. 9/1 Drenched, weary, and baggageless travellers.

baggager. Add: Also, a beast that carries baggage; a baggage horse, camel, etc.

1859 *Blackw. Mag.* Apr. 459/2 Whack goes the long whip, aimed at the rearmost baggager. The aggrieved horse flies out with a bounce. **1908** *Animal Management* 279 The ordinary pace [is] two and a half miles an hour for 'baggagers', and up to twelve miles for riding camels.

bagger. Add: (Further examples of specific uses: see quots.)

1763 *Museum Rusticum* I. 4 [The man who treads the hops into the bags] is called the bagger. **1844** H. STEPHENS *Bk. Farm* III. 1074 The best baggers use a wooden hook in the left hand to collect and bring together the cut wheat in a bundle-like shape to the ground.

b. A machine that encloses (something) in bags.

1896 *Daily News* 26 Sept. 3/5 A combined system of boring ships and baggers was invented [for canal building]. **1950** *N.Z. Jrnl. Agric.* Apr. 361/1 Mechanisation [for potato harvesting] in the form of mechanical planters, diggers, and in some cases baggers. **1960** *Farmer & Stockbreeder* 15 Mar. 104 The bagger model..has established its position as the best buy for the farmer.

baggety, var. *BAGATY.

bagging, *vbl. sb.*¹ Add: **c.** (Later examples.) Also *concr.*, that which is bagged; and with *off*.

1805 R. W. DICKSON *Pract. Agric.* II. 755 It is this..that induces the planter to make a distinction in the bagging of the article. **1900** *Daily News* 17 Sept. 2/7 *Hops*...At market to-day there was a fair supply of early baggings on offer. **1907** *Westm. Gaz.* 11 Sept. 5/1 Bagging and cartage are by no means the only charges that the merchant has to meet. **1958** *Farmer & Stockbreeder* 30 Sept. 74 Operators pick the potatoes off..the elevator-cum-picking table and throw them on the cross conveyor for discharge at the bagging-off point.

baggit. Also **baggot.** For def. read: An unbroken female salmon, one that has not shed its eggs when the spawning season is over (as distinct from a KELT or spent fish). Add later examples.

1887 F. DAY *Brit. & Irish Salmonidæ* 98 The first decision respecting which is 'clean' or 'unclean' salmon was given in December, 1885, by Mr. Fowler, who considered that a 'baggit', or gravid but unspawned fish, comes under the term 'unclean'. **1931** W. J. M. MENZIES *Salmon* (ed. 2) ii. 39 A few female fish may be found very late in the season..full of unshed ova. Such fish are usually described as 'baggots' or 'rawners'. **1963** *Times* 26 Jan. 11/4 On Tweed they [*sc.* unspawned salmon] are called baggots and many hold that they are legal quarry for the spring angler...It is extraordinarily difficult to tell whether a baggot is in fact unbroken.

baggy, *a.* Add: **1. b.** *Comb.*, as *baggy-eyed, -trousered* adjs.

1955 *Astounding Sci. Fiction* Dec. 37 He was baggy-eyed from lack of slumber. **1958** U. BLOOM *Abiding City* viii. 144 Circuit is as dull as ditch-water, with that baggy-eyed old judge. **1928** BLUNDEN *Undertones of War* xv. 163 The substitute, a tall, baggy-trousered and agreeably boastful man.

baglet (bæ·glĕt). [See -LET.] A small bag.

1885 *Graphic* 3 Jan. 18/2 He has brought back in his baglet..a dozen new-laid eggs, some milk, and a loaf of bread. **1926** *Chambers's Jrnl.* 101/2 She opened a black baglet she carried—a silken affair with an amber hoop.

bagman. Add: **1. b.** A tramp (who carries his personal effects in a bag); a swagman (SWAG *sb.* 12). *Austral.*

1896 H. LAWSON *While Billy Boils* 51 'So you're a native of Australia?' said the bagman to the grey-beard. **1954** F. J. HARDY in *Coast to Coast* 76 I'd never been a real bagman... I'd taken work when it was offering. **1955** D. NILAND *Shiralee* 26 Macauley watched him approaching and recognized him at once for what he was, a flat country bagman. **1959** S. J. BAKER *Drum* 86 *Bagman's gazette*, a non-existent publication quoted as a source of rumour, esp. in the country.

4. (See quot. 1921.)

1921 *Dict. Occup. Terms* (1927) § 759 *Bagman* (Post Office); sorts and checks returned mail bags, and issues them as required. **1927** *Daily Express* 12 Aug. 9/5 Various classes of Post Office workers..Liftmen..London porters, bagmen, [etc.].

bags I. [*BAG *v.*¹ 6 b; cf. the vulgar 'says I'.] A formula used (orig. by children) to assert a claim to an article, or the right to act in a certain way, on the ground that one is the first to speak up. Also simply *bags*, or with other extensions, as *bags not*, etc.

1866 A. DOBSON *Bob Trevor & I* II. in *Beeton's Ann.* 215 Deep in a hole..gleamed the clear water of a shallow well. 'Bags I first drink,' says Bob, according to the polite practice of schoolboys. **1886** R. HOLLAND *Gloss. County Chester*, *Bags I*,..an expression used by boys in claiming the first place in a game; or in laying claim to any treasure trove. **1899** KIPLING *Stalky* 35 There's a *Monte Cristo* in that lower shelf. I saw it. Bags I, next time we go to Aves! **1921** A. A. MILNE *Second Plays* 7 Bags I all the presents. **1926** 'R. CROMPTON' *William—the Conqueror* ix. 167 'Whose turn is it to get something next?' said Ginger. 'Bags me,' said William. **1937** J. B. PRIESTLEY *Time & Conways* I. 3 Bags I this one. **1940** M. MARPLES *Pub. School Slang* 9 Bags not..establishes exemption from anything unpleasant. **1946** B. MARSHALL *George Brown's Schooldays* xxi. 89 'Bags I do not ask about the room,' Abinger said. 'What about you doing the gassing instead of me?' 'But I bagsed-I I didn't,' Abinger protested. **1950** B. SUTTON-SMITH *Our Street* i. 25 [They] were all sit..'bagzing'. I bagz we go to the Zoo. *Ibid.* 26 He always made this bagz...I don't bagz the Zoo.

baguette. Add: **2.** *Mus.* (See quot. 1938.)

1876 STAINER & BARRETT *Dict. Mus. Terms* 43/2 Baguettes. **1938** *Oxf. Compan. Mus.* 63/2 *Baguette*, drum-stick (also sometimes used for 'baton', and as the name for the stick of a fiddle bow).

3. A gem, usu. a diamond, cut in a long rectangular shape. Also *attrib.*

1926 *National Jeweler* Nov. 33 Rings in which..baguettes, pentagons, hexagons or gems of other fancy shapes permit variety of design. **1935** *Times* 2 Dec. 19/4 In one paré and baguette diamonds form a design like two harps. The baguettes are used like a scroll. **1938** *Encycl. Brit. Bk. of Year* 276/1 Jewellers have made great use of the baguette or baton-cut diamond to produce pattern and variety in design.

bagwash (bæ·gwɒʃ). [BAG *sb.*] The rough unfinished washing of clothes; a laundry that undertakes such rough washing; the 'bag' of clothes prepared for, or given, this type of washing. Also *attrib.*

1937 EVE GARNETT *Family from One End Street* i. 4 A blue board on which was painted..'The Ideal Laundry. Careful Hand Work', and underneath..'Bag-wash'. **1944** *Our Towns* (ed. 2) ii. 61 There is as yet no good answer to the difficulties of the townswoman in congested areas except municipal washhouses or the bag-wash. **1949** *Here & Now* (N.Z.) Oct. 17/3 The bag-wash service is certainly not cheap. **1950** *Ibid.* Dec. 10/1 Some send their clothes to the bagwash and do only occasional washing at home. **1952** *Economist* 20 Sept. 673 Self-service launderettes are shops where a woman 'posts' her washing into a machine which delivers it half an hour later, washed and semi-dried... It is basically the old 'bagwash' idea. **1966** *Times* 1 Dec. 15/3 If you are interested in finding a laundry offering bagwash, you should check with the Institute of British Launderers and Cleaners, who will tell you whether there is one within easy reach.

ba·g-worm. *U.S.* [BAG *sb.* 1.] The larva of American lepidoptera of the family Psychidæ (esp. *Thyridopteryx* and *Œceticus*), injurious to trees, which builds as a portable habitation a silken case or sac covered with little twigs and leaves. Called also *basket-worm, drop-worm*.

1862 *Congress. Globe* Jan. 232 On the avenue and in the parks you will find the evergreen trees..being destroyed by the bag-worm. **1871** H. B. STOWE *Oldtown Fireside Stories* 158 The young Hokums was jest like bag-worms, the more they growed the more they eat. **1884** H. C. MCCOOK *Tenants Old Farm* 384 Basket or bag-worms. **1897** J. H. COMSTOCK *Insect Life* 204 The Bag-worms, family *Psychidæ*. **1899** *Camb. Nat. Hist.* VI. II. vi. 394 One of the North American basket- or bag-worms *Thyridopteryx ephemeraeformis*.

bahadur (bāhǭ·duɹ). *Anglo-Ind.* Also 8, 9 **behauder, baha(u)door, bahawder.** [Hindi *bahādur* hero, champion.] A great man, distinguished personage. Often affixed as a title to an officer's name.

'*Bahādur* and *Sirdār Bahādur* are also the official titles of members of the 2nd and 1st classes respectively of the Order of British India, established for native officers of the army in 1837' (Yule).

1776 *Trial of J. Fowke, F. Fowke, etc.* 1/1 Joseph Fowke of Calcutta, Gentleman,..Maha Rajah Nundocomar, Bahader, late of the same place inhabitant. **1781** J. LINDSAY in A. W. C. Lindsay *Lives* (1849) III. 296 Sheikh Hussein..tells me that our army has beat the Behauder [*sc.* Hyder Ali]. **1841** THACKERAY *Major Gahagan* iv. in *Comic Tales & Sk.* II. 80 The lips of the Bahawder are closed. *Ibid.*, Bobbachy Bahawder has seen the dreadful Feringhee. **1848** 'J. KIRKLAND' *Eerie Laird* x. 110 While he, a man of genius, must be content with the empty title of Bahadur (or knight). **1879** in T. H. S. Escott *Pillars Emp.* 275 There is nothing of the great *bahawder* about him; he is easy of access, civil, and obliging to all who approach him. **1922** *Blackw. Mag.* Oct. 519/2 He was a Bahadur, which is Indian for 'hell of a fellow'. **1957** P. KEMP *Mine were of Trouble* iii. 48, I fancied myself one of..Tamerlaine's bahadurs.

Hence **bahadur** *v. intr.*, to play the bahadur.

1860 W. H. RUSSELL *Diary India* I. 272 They had been curvetting, prancing, and bahadooring with their swords in the air.

Baha'i (bahā·ɪ). Also **Bahai, Beha'i.** [Pers.] A follower of Baha-ullah (1817–92) and his son Abdul Baha (1844–1921), propagators of a religion based on *BABISM; the doctrine and practice of this religion. Also *attrib.* Hence **Baha·ism, Baha·ist, Baha·ite.**

1889 E. G. BROWNE in *Jrnl. R. Asiatic Soc.* XXI. 518 The followers of Behá have been increasing in number and influence...Thus at the present day nearly all the Bábis are Behá'is. **1892** —— in *Relig. Syst. World* 351 So for these Behá'is, as they are now called, the writings of the Báb became an old testament. **1903** —— in M. H. Phelps *Abbas Effendi* p. xxix, The religion of Babism, or Beha'ism. *Ibid.* p. xxxiii, The Babis, who have since that time generally styled themselves..'Beha'is'. **1909** E. HAMMOND *Splendour of God* 10 Bahais..claim that Bahaism has..the Light of Love. **1909** H. DREYFUS *Univ. Religion* 26 The Bahai theology. *Ibid.* 72 When finally they left the neighbourhood of Baghdād,..the Babis had become Bahais. **1914** T. K. CHEYNE *Reconcil. Races & Relig.* 132 The confinement of the Bahaites at Acre. **1924** *Expositor* Feb. 152 The Bahaists ..have been turning their attention to the Western world and its Christianity. **1959** *Times* 23 Mar. 9/4 Baha'i is a Persian sect founded early in the last century.

Bahiric, var. *BOHAIRIC.

‖**baht** (bāt). Also **baat, bat.** [Thai *bāt*.] = TICAL; the basic monetary unit of Thailand.

1828 J. CRAWFURD *Jrnl. Embassy to Courts of Siam & Cochin China* xiii. 331 The bat, or tical, was assayed at the mint of Calcutta...The value..is about 2*s.* 6*d.* **1872** *Coin Book* 81 Baat, Siamese, Silver, value 2*s.* 6*d.*, nut shaped. **1911** *Encycl. Brit.* XXV. 6/1 The tical (*baht*) is the unit of currency and also the unit of weight [in Siam]. *Ibid.*

XXVIII. 491/1 Bat, or Tical..234 grains. **1962** R. A. G. CARSON *Coins* 551 Yet another category of silver currency in Siam is the bat or tical, a bullet-shaped piece with turned-in ends. **1963** *Whitaker's Almanack* 934/1 The exchange rate for the *Baht* is not officially fixed, but has for some time remained in the neighbourhood of *Baht* 59 = £1 sterling, with little fluctuation.

‖**bahut.** Restrict *Obs.* to sense in Dict. and add: **2.** [a. F. *bahut.*] An ornamented chest, esp. one having a rounded top; an ornamented cabinet.

1840 THACKERAY *Pict. Rhapsody* in *Fraser's Mag.* July 115/1 A chest of drawers, secrétaire, cabinet, or *bahut.* *a* **1916** H. JAMES *Ivory Tower* (1917) III. v. 207 A tall inlaid and brass-bound French *bahut.*

bahuvrihi (bahuvrī·hi), *a.* (*sb.*) *Gram.* [Skr., lit. having much rice, f. *bahú* much + *vrīhí* rice.] Of a word: composed of an adjective and a substantive so as to form, principally, a possessive adjective, like the word *bahuvrihi* itself; also *gen.*, forming a compound that is a part of speech different from its head member; *absol.*, such a compound.

1846 MONIER-WILLIAMS *Elem. Gram. Sanscr.* ix. 157 Native grammarians class compound nouns under five heads:..The 5th, *Bahuvrihi*, or those formed of any number of words associated to form an epithet to a noun. **1872** —— *Skr.-Eng. Dict.* p. xii, When a student is in doubt whether to translate compounds like *Indra-śatru* as Bahuvrihis or Tatpurushas. **1914** JESPERSEN *Mod. Eng. Gram.* II. 148 In (bahuvrihi) compounds we have formal plurals denoting one single being and used as singulars, as in *a sly-boots,..a lazybones.* **1939** L. H. GRAY *Found. Lang.* vi. 163 Possessive compounds (also called by the native Sanskrit grammatical term *bahuvrihi*..) result, in the main, from the transformation of a compound noun into an adjective with the meaning 'possessing (or possessed of) so-and-so', e.g...Greek ῥοδοδάκτυλος 'rose-fingered', ἄ-παις 'child-less'; Latin *magnanimus* 'great-minded',..and such English words as those just given as translations. **1944** *Mod. Lang. Notes* LIX. 525 Today we are witnessing a rebirth of the Bahuvrīhi type in ..*high-potency vitamins.*

Baianism (bē^i·yăniz'm). [Cf. F. *baïanisme* (1738), *baïaniste* (1720).] The heretical teaching of the Flemish theologian Michel de Bay (latinized as Baius), 1513–89, a forerunner of the Jansenists. So **Bai·anist,** a supporter of this teaching, a follower of Bay. Also *attrib.* or as *adj.*

1733 J. GORDON *Mem.* (title-p.), Wherein..the History of Baianism, Jansenism, and the Constitution *Unigenitus*, [is] impartially related. *Ibid.* 33 The Bull..did not meet with the same Reception from the Secular Clergy, who for a considerable time openly oppos'd it, and who for that Reason were called Baianists. **1928** J. BRODRICK *Card. Bellarmine* I. iv. 74 Heresies allied to Baianism. **1936** N. J. ABERCROMBIE *Origins of Jansenism* I. 90 Nothing more deplorable than the moral condition of fallen man, in the Baianist system, can be imagined. *Ibid.* 125 It may have been the influence of Jansson..that determined the young Jansen to enter the Baianist camp. *Ibid.* II. 214 The Jansenist approach to the problems of grace was..familiar inasmuch as it was Baianist.

bail, *sb.*^{3} Add: **5.** (Earlier Austral. examples; also N.Z. examples.)

1847 'A. HARRIS' *Settlers & Convicts* xiv. 287 Next get up a moderate sized stockyard...Let it contain milking bales. *Ibid.* xvi. 345, I..had seen my own cow in his bale, and the milk carried into his dairy. **1861** *Newcastle Advertiser* (N.S.W.) 24 Apr. 1 Next day I was dragged by the neck to a bail, And milked by a hairy-faced man with a pail.

1859 F. FULLER *5 Yrs. Residence in N.Z.* viii. 170 [The heifer] is secured by fastening the bails. **1874** W. M. BAINES *Narr. Edw. Crewe* x. 225 The former bovine female..would have been impossible to milk without a 'bail'. **1952** G. WILSON *Julien Ware* i. 5 In the yard outside the bail a second cow..stumbled uncertainly.

6. Of a typewriter: a hinged bar which holds the paper against the platen. Also *bail bar, roller.*

1931 M. CROOKS *Bk. Underwood Typewriter* xii. 85 The paper is normally held by the Carriage Bail. **1950** —— & DAWSON *Dict. Typewriting* (ed. 5) 23 Bail. The complete term, namely 'Paper Bail', is the name given to the swivelled arm which holds the paper down on to the paper cylinder. **1959** *Observer* 4 Oct. 15/4 Every typewriter has a 'bail-bar' to hold the paper down, and usually this has to be lifted up and down when inserting paper. **1962** *Which?* Dec. 353 (*caption*) Paper bail bar. *Ibid.* 354 The bail bar, which should hold the paper against the platen, was thin and bent easily... Bail rollers on the bail bar are an additional way of holding the paper on to the platen.

bail, *sb.*^{4} Add: **2.** (Earlier example.) Also *attrib.*, as *bail ball* = BAILER^{2}.

c **1742** J. LOVE *Cricket* III. 20 To such impetuous Might compell'd to yield The Bail. **1833** J. MITFORD in *Gentl. Mag.* Sept. 235/2 Tom Walker laid down a bail ball.

bail, *v.*^{3} **2.** Read: To secure the head of a cow in a 'bail' while she is milked. (One leg is also usually secured.) Usu. with *up. Austral.* and *N.Z.*

1846 C. J. PHARAZYN *Jrnl.* 25 Aug. (MS.) 54 Helped George in bailing up Durham [a cow]. **1853** J. ROCHFORT *Adv. Surveyor N.Z.* iii. 24 We were just in time to see his men bailing up some cattle [at the station]; *i.e.* the cow is made to put her head between two posts, when a bar slides across the space and catches her by the neck, rendering it impossible to get her head back. **1860** R. DONALDSON *Bush Lays & Rhymes* 14 A young cow must be bailed. **1888** 'R. BOLDREWOOD' *Robbery under Arms* I. vii. 89 She could frighten a wildish cow, and bail up anything that would stay in a yard with her. **1906** E. W. ELKINGTON *Adrift in N.Z.* i. 16 The cows..refused to be bailed up. **1950** *N.Z. Jrnl. Agric.* Sept. 268 After the heifer has been bailed in the normal way the board is swung into position.

3. (Orig. said of Australian bushrangers.) To 'stick up' travellers in order to rob them; to 'corner' a wild boar (or other hunted animal); (in weakened sense) to detain (a person); also *transf.* Also *intr.*, to surrender (by throwing up the arms). Usu. with *up. Austral.* and *N.Z.*

1840 *Sydney Herald* 31 Jan. 2/4 Only think of one man [*sc.* a bushranger] bailing up the master and twelve men. **1853** MRS. C. CLACY *Lady's Visit to Gold Diggings* xvi. 235 But can picture their horror when ordered to 'bail up' by a party of Australian Turpins. **1855** W. HOWITT *Land, Labour, & Gold* II. xxxix. 309 So long as that is wrong, the whole community will be wrong, — in colonial phrase, 'bailed up' at the mercy of its own tenants. **1872** LADY BARKER in D. M. Davin *N.Z. Short Stories* (1953) 25 Our head shepherd recognized an old enemy in the dead boar, and declared that he and his dogs had bailed him up unsuccessfully 'many a time and oft'. **1879** W. J. BARRY *Up & Down* xii. 112 She bailed me up, and asked me if I was going to keep my promise and marry her? **1880** [in Dict.]. **1885** H. FINCH-HATTON *Advance Australia!* vii. 105 A little further on the boar 'bailed up' on the top of a ridge. **1888** A. H. DUNCAN *Wakatipians* x. 105 We are bailed up by the snow, and look like frozen Esquimaux. **1894** J. K. ARTHUR *Kangaroo & Kauri* II. iii. 98 The pigs will oftentimes 'bale up', or stop, and with their back to a rock, tree, or other obstacle, keep two or more dogs at bay for a long time. **1894** H. NISBET *Bush Girl's Romance* 144 Reginald..acted like a wise man and 'bailed up', that is, he dropped his knife and threw up his hands as a sign of his submission. **1895** G. CHAMIER *South-Sea Siren* xiv. 205 He was immediately 'bailed up' [by the surveyors], and made to dismount against his will to partake of some refreshment. **1900** D. McK. WRIGHT *Wisps of Tussock* 50 He bailed me up straight for a shilling. **1904** M. CRADOCK *Sport in N.Z.* I. iii. 67 Unless he [*sc.* the pig] is a peculiarly obstinate beast the dogs very soon 'bail him up' again. **1911** *Chambers's Jrnl.* 4 Mar. 222/2 The dogs have 'bailed' a fighting old boar in a rock-cleft. **1930** E. WALLACE *White Face* xiii. 213 'Bail up!' It was an expression of the old Australian bushrangers. It's still used by the hold-up men in Australia. **1943** F. DAVISON in *Coast to Coast 1942* 194, I didn't like the colour of those who had him bailed up. **1946** —— *Dusty* (1947) ix. 94 The border collie's shrill barking when he had a goanna bailed up always brought him to the scene. **1960** B. CRUMP *Good Keen Man* 84, I couldn't see eye-to-eye with Jim on the matter of going in on bailed boars with a slasher.

bail, *v.*^{4} *to bail out.* Aeronaut. See *BALE *v.*^{2} 2.

bailer^{1}. Add: **2.** *attrib.* **bailer shell,** the shell of a gastropod mollusc of the genus *Cymbium* found in the south-western Pacific, also called MELON-*shell*; also, the mollusc itself.

1908 E. J. BANFIELD *Confessions of Beachcomber* I. iv. 148 The bailer shell alive is like an egg, in the fact that it is full of meat. **1926** *Austral. Encycl.* II. 135/2 The most useful of all molluscs was the Melon or Bailer-shell (*Cymbium flammeum*), which served for a canoe-bailer. **1929** *Times* 2 Aug. 14/1 Several melon or bailer shells were as large as footballs. **1947** I. L. IDRIESS *Isles of Despair* viii. 56 Many a bailer shell filled with rich turtle soup. **1955** A. ROSS *Australia 55* 121 Beautiful bailer shells, so-called because the natives use them to bail from their canoes.

bailer^{2}. (Earlier examples.)

1833 J. MITFORD in *Gentl. Mag.* Sept. 236/1 [The practice of going out to meet the ball] saves alike the fingers and the wickets from a first-rate top-bailer. **1865** *Field* 9 Sept. 189/2 Mr. Sale..at length bowled Capt. Parnell with an irresistible bailer.

Bailey bridge (bē^i·li). [f. name of Donald (later Sir D.) Coleman *Bailey* (1901–), English engineer, its designer.] A bridge of lattice steel designed for rapid assembly from prefabricated standard parts, used esp. in military operations. Also *B. bridging* and *ellipt.*

1944 *Hutchinson's Pict. Hist. War* Apr.–Sept. 269 The story of the development of the Bailey bridging equipment ..has been revealed. **1944** *Times* 14 Dec. 3/3 A Bailey bridge more than 1,000 ft. long has been built over the Chindwin near Kalewa. **1945** *Finito! Po Valley Campaign* 7 The stuff for Bailey Bridges. *Ibid.* 23 Engineers toiled ceaselessly to span the river with Baileys. **1956** W. SLIM *Defeat into Victory* xvi. 366 On the 10th December [1944] our engineers completed a floating Bailey bridge over the Chindwin. Its length was 1,154 feet, then the longest Bailey bridge in the world.

bailiwick. Add: **1. b.** *transf.* 'One's natural or proper place or sphere' (D.A.). Chiefly *U.S.*

1843 *Knickerbocker* XXI. 589 A friend..inside the southern division of Mason and Dixon's 'bailiwick'. **1892** *Outing* (U.S.) Apr. 16/1 The baggage-man stared a little when we piled our 'truck' into his bailiwick. **1911** R. D. SAUNDERS *Col. Todhunter* ix. 119 I'm skeered to the marrow,..because I'm out o' my bailiwick. **1940** *S.P.E. Tract* LVI. 216 *Bailiwick*..has given rise to the common phrase 'out of one's bailiwick', i.e. outside of one's natural sphere or function.

Baily's beads: see BEAD *sb.* 7.

bainite (bē^i·nəit). [f. Edgar C. *Bain*, U.S. metallurgist (*b.* 1891) + -ITE^{1}.] A constituent produced at a certain stage in the heat treatment of steel (see quot. 1939^{1}).

1935 A. SAUVEUR *Metallogr. & Heat Treatm. Iron & Steel* (ed. 4) xv. 235 This..constituent microscopically resembles somewhat tempered martensite although its properties are distinctly different... The associates of Bain at the Laboratory at which he was formerly located designate it as 'Bainite'. **1939** E. C. BAIN *Alloying Elem. in Steel* ii. 41 These structures known as 'bainite' formed at temperatures roughly between 1000 and 380 degrees Fahr. (500 and 150 degrees Cent.) in carbon steel have interesting mechanical properties, being tougher for the same hardness than tempered martensite. **1939** *Jrnl. Iron & Steel Inst.* CXL. 107 The extreme types differ fairly considerably from each other, for which reason certain writers have separated them into 'upper bainite' and 'lower bainite'. **1954** *Gloss. Terms Iron & Steel* (*B.S.I.*) I. 6 *Bainite*, the constituent produced when austenite transforms at a temperature below that at which pearlite is produced and above that at which martensite is formed. **1956** *Nature* 3 Mar. 420/2 Bainite reactions.. and the stabilization of austenite.

bairnhood. [BAIRN; cf. BAIRNHEID.] Childhood. *Sc.* or *affected.*

a **1835** JOSEPH GRANT in *Whistle-Binkie* (1890) II. 161 The broomy hill, Where we used to stray in bairnhood's day. **1894** R. FORD (*title*) Ballads of Bairnhood.

baister, baisting: var. ff. BASTER^{1}, BASTING *vbl. sb.*^{1}

1881 *Instr. Census Clerks* (1885) 74 Braider,..Stitcher, Baister, Button-hole Maker. **1890** *Daily News* 2 Jan. 5/3 Baisting cotton. **1902** *Daily Chron.* 5 Mar. 10/5 Coat Machinists..wanted, baisters and finishers. **1906** *Ibid.* 28 May 11/7 Tailoresses wanted; must be good coat baisters.

bait, *sb.*^{1} Add: **3.** *bait-fish, -fisher, -fishing, -tackle.*

1820 *Western Rev.* (Kentucky) II. 241 *Rutilus compressus*, a small fish..called Fall-fish, Bait-fish, Minny, etc. **1832** *Chambers's Edin. Jrnl.* 14 Apr. 44/1 The same observations apply..to bait fishing. **1835** *Ibid.* 3 Jan. 390/3 Some bait-fishers..use the smaller sorts [of hooks];..The bait-tackle ought to be loaded..with a pellet or two of lead. **1952** E. HEMINGWAY *Old Man & Sea* 35 No flying fish broke the surface and there was no scattering of bait fish.

bait, *sb.*^{2}, **bate** (bē^it), *sb.*^{6} *slang.* [f. BAIT *v.*^{1}] A fit of bad temper; a rage. Hence **ba·ity** *a.*

1857 [A. L. MAYHEW in *Eng. Dial. Dict.* s.v. *Bate* sb.^{3} 'He was in an awful bait' was common in the Clapham Grammar School, 1857]. **1882** 'F. ANSTEY' *Vice Versâ* (ed. 4) iii. 48 It would put him in no end of a bait. **1899** E. PHILLPOTTS *Human Boy* 95 I've just left Milly, and she's in a frightful bate. **1899** KIPLING *Stalky & Co.* 152 'What a bait you're in!' said Stalky. *Ibid.* 205, I got in no end of a bait. **1921** S. THOMPSON *Rough Crossing* ii. 86 Jolly lucky the C.O. didn't notice it yesterday—he gets 'baity' on these occasions. **1925** *Chambers's Jrnl.* 838/1 Now I must be going, or else dad'll be baity with me. **1953** E. TAYLOR *Sleeping Beauty* vii. 134 Flying into a bate, as we used to say at school.

baitable (bē^i·tăb'l), *a.* [f. BAIT *v.*^{1} + -ABLE.] Serviceable as cattle fodder.

1890 *Jrnl. R. Agric. Soc.* 756 At Meldon not only was there clover, but also a considerable quantity of baitable grasses. *Ibid.* 759 Grasses which are rejected when the stock has a sufficient supply of baitable food to permit of a choice.

baitless, *a.* Add: **2.** Not furnished with bait (for fish); unbaited.

1854 *Tait's Mag.* XXI. 275/1 Holding in his hand the listless line, whose baitless hook, three fathoms down, has become a jest and a laughing-stock to..the finny tribe.

baitylos: see *BÆTYL.

‖**bai-u** (bai·yū). [Jap., f. *bai* plum + *u* rain.] (A season of) rainfall in Japan in early or mid-summer (see quots.).

1910 *Bull. Central Meteor. Observ. Japan* I. v. 1 (*title*) On the Bai-u or Rainy Season in Japan. **1922** W. G. KENDREW *Climates of Cont.* III. xxii. 142 [In Japan] this rainy season is commonly called the 'Bai-u', meaning the plum-rains, as it comes when plums are getting ripe. **1945** G. T. TREWARTHA *Japan* (1947) I. ii. 42 Much cloudiness, abundant rain, high humidity, and high sensible temperatures make the so-called *bai-u* or plum rains a very..gloomy season.

baize, *sb.* **1. b.** (Further examples.)

1834 C. MATHEWS *Let.* 20 Jan. in Mrs. Mathews *Memoirs C. M.* (1839) IV. xiii. 256 He sat down contentedly before the green baize table. **1853** DICKENS *Bleak H.* xli. 405 There is an inner baize door too.

bajada (bāhā·dă). Also **bahada.** [a. Sp. *bajada* descent, slope.] In the south-west of N. America, a descent or slope; *spec.* that section of a piedmont slope formed by aggradation and composed of rock debris (detritus).

1866 *New Mexican* 17 Nov. 2/3 The road from here to Algodones by way of the Bajada is being worked. **1909** C. F. TOLMAN in *Jrnl. Geol.* XVII. 141 The *bajada*: Extending down from the rock surfaces are flanking detrital slopes, built up by terrestrial deposition. **1933** *Amer. Speech* Oct. 8/2 Special types of plant life..which can best be described as bahada inhabiting. **1937** WOOLDRIDGE & MORGAN *Physical Basis Geogr.* xx. 311 Gently sloping surfaces.. The first type is the 'bajada', plainly due to aggradation by floods debouching from the mountain valleys. Bajada plains are underlain by thick accumulations of coarse alluvia. **1960** B. W. SPARKS *Geomorphology* xi. 256 The whole slope from the range to the infilled playa lake is usually termed a piedmont. The upper part is often, but not always, a rock-cut surface, normally termed a pediment, while the lower part is an aggradation feature formed of detritus from the ranges and termed a bahada.

Bajan, Bajun, var. *BADIAN *sb*^{2} and *a.*

‖ **baju** (bā·dʒū). Also **badju.** [Malay.] A short loose jacket worn in Malaya.

1820 J. CRAWFURD *Hist. Ind. Archipelago* I. II. ii. 210 The second great portion of dress is the *coat*, called by the Malays *baju*, and by the Javanese..*rasukan*. It..may generally be described as a frock with sleeves, longer or shorter, according to the sex..of the wearer. **1834** [see SARONG 1]. **1902** *Bulletin* (Sydney) 13 Dec. 16 A painted Mongolian person..with his baju buttoned awry. **1939** *Atlantic Monthly* 858 A layette of diminutive bright calico *bajus* and sarongs. **1939** A. KEITH *Land below Wind* ix. 150 A figured baju which fell naively from his slender shoulders. **1961** CONYN & MARTEN *Bali Ballet Murder* vii. 70 The gaily printed *badjus* of the women.

bake, *v.* Add: **5. a.** Of land. (U.S. examples.)

1850 N. KINGSLEY *Diary* 12 Feb. (1914) 109 The soil looks as if it would bake hard. **1869** *Rep. U.S. Commissioner Agric.* 357 When the proportions of clay and sand are such that the soil will not bake,..it may properly be called loam. **1873** J. H. BEADLE *Undevel. West* xxxiii. 710 The land is never water-soaked, never 'bakes', and I never saw a clod as big as my fist.

b. To be made uncomfortably hot (by the sun, a fire, etc.). *colloq.*

1937 M. SHARP *Nutmeg Tree* x. 123 'I'm going to bake,' thought Julia..and indeed the plain..shimmered under a heat mist.

7. bake office, (*a*) = BAKEHOUSE; (*b*) a baker's shop (*Eng. Dial. Dict.*). *dial.*

1874 in Thirsk & Imray *Suff. Farming 19th Cent.* (1958) 121 Bake office common to the 3 tenements.

bake, *sb.* Add: **2.** (Examples.)

1565 T. COOPER *Thes.*, *Acapna thysia*, sacrifices without smoke: spoken of a simple feaste wherin is neyther bake, sodde, nor roste. **1851** *Knickerbocker* XXXVIII. 187 Saint Peter [in stained glass] is a little cracked..but I've got a first-rate bake on Paul. **1882** LEES & CLUTTERBUCK *Three in Norway* xvi. 126 After this Esau finished the oven, and accomplished a bake of bread therein. **1961** *Guardian* 27 Sept. 8/3 Many older housewives..find great satisfaction in a monster weekly 'bake'. **1963** *Listener* 21 Mar. 535/3 For *apple and pork bake* you will need..4 pork chops.

b. A social gathering at which a meal, esp. of baked food, is served; a clambake. *U.S.*

1846 *Spirit of Times* (N.Y.) 6 June 174/3 The grand 'bake' at the village hotel. **1935** STEINBECK *Tortilla Flat* viii. 148 He saw that it was a Girl Scout wienie bake.

bake-apple. *Canadian.* Also **baked-apple.** [BAKE *v.* 7.] The (dried) fruit of the cloudberry. Also *attrib.*

1775 G. CARTWRIGHT *Jrnl. Labrador* 3 Aug. (1792) II. 96, I saw the first baked apples. **1792** *Ibid.* I. p. ix, Baked Apple. The fruit of a plant so called, from the similarity of taste to that of the pulp of a roasted apple. **1829** T. C. HALIBURTON *Hist. & Stat. Acct. Nova Scotia* II. 216 The berry being nearly of the size and appearance of the yellow Antwerp raspberry,..is termed by the residents, 'bake-apple'. **1839** E. W. TUCKER *Five Months Labrador* 104 Several small shrubs are found in the country which bear fruit, the principal one of which is called the baked apple berry. **1895** *Outing* (U.S.) XXVII. 18/1 The outlying islands furnish the curlew-berry and bake-apple in profusion. **1938** *Geogr. Jrnl.* XCII. 155 On the marshes were bakeapple, pitcher plant, Labrador tea, and occasional groups of blueberry. **1950** A. P. HERBERT *Independent Member* 283 Bake-apples (little ground fruit the size of raspberries).

baked, *ppl. a.* Add: **1.** (Further example.)

1836 DICKENS in *Bell's Life in London* 17 Jan. 1/1 The baked-'tatur man has departed.

b. *baked beans*, haricot beans so cooked (now a popular tinned food, prepared in tomato sauce).

1832 L. M. CHILD *Frugal Housewife* 51 Baked beans are a very simple dish, yet few cook them well. **1873** 'SUSAN COOLIDGE' *What Katy Did* iv. 42 A breakfast of baked beans. **1920** F. E. GREEN *Hist. Eng. Agric. Labourer* ii. 47 Farmers did not have recourse to..letting their fields..as advertising sites for Somebody's Pills or Baked Beans. **1933** *Punch* 15 Feb. 170/2 All you have to do to give your family a really new tasty meal is to take a tin-opener and one tin of baked beans.

3. b. *Typogr.* (See quot. 1688.)

1683 MOXON *Mech. Exerc., Printing* II. xxii. 257 Compositers in this Case say, The Letter is Bak'd. **1688** R. HOLME *Armory* III. iii. 119/1 *Bake*, is when Letters stick together in distributing...This is called the *Letter is Baked.* **1963** KENNEISON & SPILMAN *Dict. Printing* 13 *Baked* (or *caked*), said of type which sticks together, thus making it difficult to distribute.

bakelite (bē^i·kĕləit). Also **Bakelite.** [ad. G. *bakelit* (*Chem. Zentralblatt*, 1909, 1478), f. the name of L. H. *Baekeland* (1863–1944) its inventor + -ITE^{1}.] A proprietary term for a condensation product of phenol or other phenolic bodies and formaldehyde used as a plastic and for insulating purposes. Also *attrib.*

1909 L. H. BAEKELAND in *Jrnl. Ind. & Engin. Chem.* I. 149/2 (*title*) The Synthesis, Constitution, and Uses of Bakelite. **1913** *Jrnl. Soc. Chem. Ind.* 16 June 563/2 Bakelite varnish has been suggested for use as an antiseptic in hospitals. **1921** THORPE *Dict. Applied Chem.* I. 501/1 There are..bakelite plants in Germany, France and England, and several factories where bakelite goods, such as buttons, are manufactured under licence. **1962** *Which?* Aug. 249/2 Bakelite Ltd now produce many different plastics, more than one of which is called Bakelite. **1964** N. G. CLARK *Mod. Org. Chem.* xvii. 363 A further example of thermo-setting polymers is provided by the Bakelite-type resins, from formaldehyde and phenol.

baker. Add: **2.** *U.S.* (Examples.)

1841 *Lowell* (Mass.) *Offering* I. 227 (Th.), A peep into the baker told that the potatoes were cooked. *a* **1862** THOREAU *Maine W.* (1864) 249 Somebody had left..on a deserted log ..a loaf of bread baked in a Yankee-baker. **1897** *Outing* (U.S.) XXIX. 489/1 The cooking utensils consisting of three dripping pans, one patented baker and one large coffee-pot.

5. baker's bread, bread baked by a baker (opp. home-baked).

1813 JANE AUSTEN *Let.* 3 Nov. (1932) II. 367, I suppose you will be going to Streatham, to see quiet Mr. Hill & eat very bad Baker's bread. **1860** F. NIGHTINGALE *Notes on Nursing* vii. 42, I have known patients..[who] could not eat baker's bread. **1964** M. LOCHHEAD *Vict. Househ.* vii. 104 The women..were..good bakers...Baker's bread was..an extravagance.

bakery. Add: **2.** (Earlier examples.) Also, a shop where baked products are sold. *U.S.*

c **1820** *14th Rep. Hist. MSS. Comm.* App. III (1894). 50 We marched a long forced march to the relief of Beaton, who had burnt the French backery at Marburgh. **1832** F. TROLLOPE *Dom. Manners Amer.* I. 85 There are no.. shops for eatables except *bakeries*, as they are called. **1842** DICKENS *Amer. Notes* iv. 153 Every 'Bakery', 'Grocery', and 'Bookbindery', and other kind of store. **1923** K. D. WIGGIN *Garden of Memory* 112, I bought my luncheon at a different bakery every day.

bakestone. (Further examples with spelling *bake(-)stone.*)

1741 ELLIS *Mod. Husb.* II. II. 31 In Shropshire they.. make thin Cakes on a Back or Bake-stone. **1962** *Guardian* 9 Nov. 8/4 A Welsh bakestone or a Scottish girdle.

Bakewell (bē^i·kwel). The name of a town in Derbyshire used *attrib.* to designate a baked sweet, consisting of a pastry case lined with a layer of jam and filled with a rich almond paste.

1845 E. ACTON *Mod. Cookery* xviii. 467 Bakewell Pudding ..is famous not only in Derbyshire. **1861** MRS. BEETON *Bk. Househ. Managem.* xxvii. 630 *Bakewell Pudding*..puff-paste..eggs..sugar..butter..almonds, jam. **1883** [see JAM *sb.*^{2} c]. **1906** MRS. BEETON *Househ. Managem.* xxxii. 927 Bakewell Pudding (..Bakewell Tart). **1959** R. POSTGATE *Good Food Guide 1959–60* 32 Among the dishes tested and praised by members are..braised ham and Bakewell tart.

bakhara, var. *BUCKAROO.

baking, *vbl. sb.* Add: **1.** Also *spec.* in *Typogr.* (see quot.). Cf. *BAKED *ppl. a.* 3 b.

1683 MOXON *Mech. Exerc., Printing* II. xxii. 257 This sticking together of the *Letter* is call'd *Baking* of the Letter.

3. *baking dish* (example), *-soda* (cf. SODA^{1} 1 b), *-tin*; **baking-powder** (examples), orig. *U.S.*

1856 DICKENS *Dorrit* I. v. 39 Preparing a baking-dish of beef and pudding. **1850** *Family Friend* III. App. 2/1 Chemical analyses of two of the baking powders in most general use. **1861** MRS. BEETON *Bk. Househ. Managem.* xxxv. 856, 2 teaspoonfuls of baking-powder. **1909** *Westm. Gaz.* 1 June 7/4 Baking-powder beer. **1845** E. ACTON *Mod. Cookery* xvi. 425 When the mixture has simmered..pour it out upon a delicately clean baking-tin. **1959** A. WESKER *Roots* II. i. 49 Mother, where's the bakin' tin?

baking, *ppl. a.* (Earlier and later examples.)

1786 WASHINGTON *Diaries* 9 May (1925) III. 58 The ground, by the heavy rains..and baking Winds since, had got immensely hard. **1934** 'G. ORWELL' *Burmese Days* xi. 170 It's getting beastly hot, isn't it?..Isn't it simply *baking*!

baklava (ba·klāva). Also **baclava.** [Turkish.] A dessert made from thin pieces of flaky pastry, honey, and nuts, usually cut into lozenge-shaped pieces.

1824 J. J. MORIER *Adv. Hajji Baba* I. xxiv. 258 Some rare sweetmeats, with *baklava* (sweet cake) made in the royal seraglio. **1936** M. H. BRADLEY *Five-Minute Girl* 37 People who like baklava and lamb go to Papadopoulos. **1960** *Times* 29 Oct. 9/6 Or have pilaff, kebab and baklava in an oil-lit tavern in the Turkish quarter. **1967** I. MARDER *Paris Bit* i. 20 Natasha, who had an impressive appetite, finished with a slab of *baclava.* **1968** *Woman* 20 Apr. 30/3 *Baklava*, a famous Turkish sweetmeat made from layers of wafer-thin pastry, honey and nuts.

baksheesh, *v.* (Earlier example.)

1867 *Good Words* 1 Mar. 184/1 We backshished our caffagi, and left Aiasaluck at sunrise.

baku (bē^i·ku). Also **-ou.** [Native name in Philippines.] A fine kind of straw grown in the Philippines and woven in China.

1927 *Vogue* Mar. 40 Baku, a linen-like straw. **1928** *Ibid.* 4 Apr. 26 (Advt.), Natural Bakou Straw, trimmed with a feather cockade. **1929** *Ibid.* 20 Feb. 52 Toque of beige baku straw. **1929** *Chicago Tribune* 9 May, Bakus and bali buntals.

Balaam. **2.** **Balaam-basket** (fig. example), **-box** (earlier example).

1866 *Harper's Mag.* May 816/1 There have been a number of anecdotes lying loose in the Balaam basket of memory. **1827** *Blackw. Mag.* Mar. 340 Several dozen letters on the same subject now in our Balaam-box.

Balaclava (bælăklā·vă). *Balaclava helmet* (also *cap*): a woollen covering for the head and neck worn esp. by soldiers on active service; named after the Crimean village of Balaclava near Sebastopol, the site of a battle fought in the Crimean war, 25 October 1854. Also *ellipt.*

1881 *Cliftonian* Feb. 376 Balaclava caps. **1892** C. T. DENT et al. *Mountaineering* 48 A knitted woollen Balaclava (Templar) cap is useful for sleeping out. **1900** *Queen* 10 Feb. 222/1 Balaclava Helmet in crochet. **1914** *Scotsman* 17 Sept. 8/5 The articles which at the moment will be most useful to officers and men in ships afloat are..Balaclava helmets, mufflers. **1926** *Blackw. Mag.* Mar. 398/1 Baa's balaclava helmet and woolly muffler were flaked with white.

balakhana (bælăkā·nă). Also **-khaneh, -hané.** [Pers. *bālā-khāna* upper room.] An upper room in a Persian house, in which travellers are lodged.

1840 J. B. FRASER *Trav. Koordistan* I. ii. 31 Desiring that this Khan should have such a suite of rooms..and another such a *dellân*, or *balakhaneh*. **1882** O'DONOVAN *Merv Oasis* I. xxii. 369 Above the arch was a square-topped room known as the *bala hané*, which served as quarters for the better class of travellers, as well as a kind of watch-tower. **1921** *Blackw. Mag.* Dec. 825/2 Very soon our horses were tethered, and champing good dry lucerne under the bala-khaneh of a two-storied guest-house.

balalaika (bælăləi·kă). [Russ.] A musical instrument of the guitar kind, with a triangular body, popular in Russia and other Slav countries. Also *attrib.*

1788 tr. *Staehlin's Anecd. Peter Gt.* 319 In his youth he had never heard any [music] but that of drums, fifes, balalaikas, and bagecs. **1864** ENGEL *Mus. Anc. Nat.* 55 The Russian *balalaika*, an instrument said to be of high antiquity. **1885** A. J. C. HARE *Stud. Russia* i. 27 Formerly the peasants used to dance the Barina..accompanied by the Balalaika. **1913** 'SAKI' *When William Came* iii. 43 The thrumming music of a balalaika orchestra coming up from the restaurant below. **1954** *Grove's Dict. Mus.* (ed. 5) I. 368/2 Balalaika bands have frequently visited Western Europe.

balance, *sb.* Add: **9.** (Later example.)

1881 [see HANG *v.* 17 a].

13. c. *balance of nature*, a state of equilibrium in nature produced by the interaction of living organisms; ecological balance.

1909 G. ABBEY (*title*) The balance of nature and modern conditions of cultivation. **1923** H. G. WELLS *Men like Gods* I. vi. 84 But presently he made it clear that there had been something very ancient and beautiful called the 'Balance of Nature' which the scientific methods of Utopia had destroyed. **1933** *Discovery* July 224/2 The interference of man with the balance of nature had almost always brought evil in its train. **1962** H. HANSON *Dict. Ecol.* 41 *Balance of nature*, (Ecological Balance), the state in an Ecosystem when the interrelationships of organisms..are harmonious or integrated to a considerable degree, e.g., a climax forest.

d. Phr. *balance of terror*, balance of power based on the possession of weapons of 'terror', esp. nuclear weapons.

1960 B. M. GOLDWATER *Conscience of Conservative* 91 If war is unthinkable to us but not to them, the famous 'balance of terror' is not a balance at all, but an instrument of blackmail. **1962** *Listener* 29 Mar. 548/1 There are other technical trends which seem to be contributing stability to the balance of terror between the Great Powers. **1962** *Observer* 13 May 15/4 The 'balance of terror' is always in danger of being upset by technical advance (e.g., in anti-missile missiles). **1965** H. KAHN *On Escalation* xiii. 246 A confusing thing about tactics..in a balance-of-terror situation is the great reliance on messages, symbols..and even 'spectacles'.

14. b. *spec.* of the arrangement and adjustment of sources of sound; the sound thus produced.

1929 *B.B.C. Year-Bk. 1930* 312 For balance and control.. a volume control handle varies the input to the control amplifier, and consequently the strength passed to the transmitter. **1929** *Melody Maker* Jan. 61/2 Then again balance is bad all over the place; rhythm gets lost, saxophones are often overpowering. **1933** L. E. C. HUGHES *Elem. Engin. Acoustics* vii. 141 As the sensitivities of the two ears are generally different, the balance is again taken with the receivers interchanged. **1941** *B.B.C. Gloss. Broadc. Terms* 4 *Balance*, placing of artists, speakers, or other sources of sound in relation to a microphone or microphones, or vice versa. *Ibid.*, *Balance Test*, test to establish the best balance for a particular broadcast. **1946** *Penguin Music Mag.* Dec. 48 In a *tutti* passage..I am bound to hear the brass to the almost total exclusion of the 'cellos...Therefore..we can seldom hear what you hear in the matter of the all-important balance. **1962** A. NISBETT *Technique Sound Studio* i. 13 The man responsible for balance, mixing, and control may be a 'balance engineer' or 'programme engineer'.

15. c. *off* or *out of balance.*

1881 G. W. CABLE *Mme Delphine* x. 52 As a banker, at least, he was certainly out of balance. **1954** F. C. AVIS *Boxing Ref. Dict.* 77 *Off balance*, with the weight not properly distributed over the legs, and thus in a dangerous position in regard to an opponent. **1960** E. P. C. COTTER *Tackle Croquet this Way* ix. 61 My opponent's tail was up so I decided to get him off balance if I could. **1962** *Which?* (Car Suppl.) Oct. 137/2 Two had wheels out of balance on delivery.

17. c. Phr. *on balance* (or *upon the balance*): taking everything into consideration.

1843 *Ainsworth's Mag.* IV. 308 Upon the 'balance', as the betting men say, women are quite as mercenary as men. **1861** Upon the balance [in Dict.]. **1928** *Britain's Industr. Future* II. vii. 89 We believe that these provisions would be, on balance, overwhelmingly in the interest of the investor. **1966** *Listener* 6 Jan. 14/2 The impression in India that Britain has tended, on balance, to take the Pakistan side in the Kashmir dispute.

e. *balance of payments*: the estimation of the difference of value between payments into and out of a country. (*Balance of trade* (BALANCE *sb.* 17 d), i.e. of merchandise, covers the principal items on both sides, but *balance of payments* also includes the 'invisible' items, interest on loans, tourist expenditure, etc.) Also in *attrib.* use.

1844 MILL *Ess. Pol. Econ.* i. 43 The tribute..restores the balance of payments between the two countries. **1863** *New Englander* Jan. 63 The natural effect of all this was to create for the moment a heavy balance of payments against the North. **1931** *Economist* 12 Dec. 1110/2 Study our national balance of payments, in order to see whether..restriction of imports..might..save the pound from..depreciation. **1955** *Times* 10 Sept. 6/6 Mr. Butler,..said that he thought the balance of payments problem was on the whole being kept in hand. **1958** *Listener* 31 July 151/1 So far, pressures on the balance of payments have been met successfully by foreign loans.

20. a. *balance in hand* (examples). Also *fig.*, something to spare.

[**1847** in Thirsk & Imray *Suff. Farming 19th Cent.* (1958) 163, I have written to Mr. Woodley to ask him what amount he will require in his hands to save us from the interest. We have generally had a balance in his hands of £400 or £500.] **1865** [see FORK *v.* 5]. **1876** *Coursing Calendar* 5 Wheatear..working with great smartness, won with a nice balance in hand. **1957** *Hampshire Chron.* 3 Aug., At the annual meeting of the Winchester and District Football League..a balance in hand of £1 17*s.* 9*d.* was reported.

21. (Earlier examples.) orig. *U.S.*

1788 in *Pennsylv. Mag. Hist. & Biog.* (1894) XVIII. 62 Arose early and sent off the balance of our things. **1817** S. R. BROWN *Western Gaz.* 167 The inhabitants are more than half French; the balance consists of emigrants..from various parts of Europe and America. **1845** S. JUDD *Margaret* I. xiii. 96 Deacon Hadlock himself, hearing Obed's entreaties, consented to remit the balance of the penalty.

22. balance-bob (earlier example); **balance-crane** (see quot. 1904); **balance-reef** (examples); **balance-sheet** (earlier U.S. example); **balance weight,** a counterpoise weight. In *Watchmaking* (read: see 6), *balance-cock* (examples), *-wheel* (later and additional fig. examples); also, a similar device on a sewing-machine.

1838 *Civ. Engin. & Archit. Jrnl.* I. 409/1 The heavy pump rods, balance bobs &c., attached to a mining engine. *a* **1877** KNIGHT *Dict. Mech.* I. 212/1 The balance-cock of a watch affords a bearing for the upper pivot of a watch-balance. **1962** E. BRUTON *Dict. Clocks & Watches* 19 *Balance cock*, the cock that holds the bearing, normally a shock absorber, for one end of the balance. **1824** R. STEVENSON *Bell Rock Lighthouse* 520 The Balance-Crane, constructed for building the upper part of the Bell Rock Light-house. **1904** GOODCHILD & TWENEY *Technol. & Sci. Dict.* 136/2 Some heavy weight is fixed..opposite to the point at which the jib is fixed...These arrangements constitute what is called a balance crane. **1782** P. FRENEAU *Misc. Works* (1788) 387 What he calls single, double, and balance-reef eyelet holes. **1794** D. STEEL *Rigging & Seamanship* I. 86 *Balance-reef*, a reef-band that crosses a sail diagonally, and is used to contract it in a storm. **1840** J. F. COOPER *Pathfinder* II. v. 184 By half-past two he had put a balance-reef in the sail. **1838** *U.S. Mag. & Democr. Rev.* I. 42 He becomes familiar..with trial balances, balance sheets, [etc.]. **1824** R. STEVENSON *Bell Rock Lighthouse* v. 296 The upright shaft of the new crane was to be kept in an erect position by a balance-weight acting upon the opposite end of the loaded working-beam. **1862** *Catal. Internat. Exhib.* II. x. 21 The cage returning to the upper floor for a fresh load by means of a balance weight. **1960** J. G. HORNER *Dict. Mech. Engin.* (ed. 8) 19 The weights placed in the driving-wheels of locomotives are termed balance weights. **1747** *Encycl. Brit.* XVIII. 806/1 The balance-wheel obliges the balance to vibrate backwards and forwards like a pendulum. **1958** M. E. BURTON *Lett. M. Wordsworth* p. xxviii, Mary often chooses to remain behind. She is the balance-wheel. **1961** *Which?* Nov. 277 (*diagram*) Balance wheel [of a sewing-machine].

balance, *v.* Add: **4.** *spec.* of sources of sound (cf. *BALANCE *sb.* 14 b).

1928 *B.B.C. Handbk. 1929* 68 A..new design of control room has been decided upon, wherein effects, echoes, background music, etc. can be mixed together and so balanced by a specially-trained operator. **1933** L. E. C. HUGHES *Elem. Engin. Acoustics* vii. 141 The..method is to accept the calibration curve of a moving-coil receiver, and to balance their output when supplied by a constant note. **1962** A. NISBETT *Technique Sound Studio* ii. 48 Completely dead sound is difficult to balance. *Ibid.* iii. 52 When balancing music there is..no clear-cut set of rules.

‖**balancé** (balanse). [Abbrev. of F. *pas balancé*, lit. 'balanced step'.] A dance step, now spec. in *Ballet*, comprising a swaying movement from one foot to the other; hence (see quot. 1847) the quality implicit in the step, sinuosity.

[**1728** J. ESSEX tr. *Rameau's Dancing Master* xxiv. 57, I have seen many Persons make Balances in dancing a Menuet.] *c* **1770** G.-A. GALLINI *Crit. Obs. Art Dancing* 5 Le Balancé. This is done by Sinking, then Rising as you Step forward or sideways with one foot. **1786** S. J. GARDINER *Def. Minuet-Dancing* 54 The steps generally made use of in Cotillions are Balance, Rigodon. **1847** R. BARTON tr. *Blasis's Notes upon Dancing* I. 49 The steps are to be remarked for their *balancé*, lightness, grace, and elasticity. **1952** KERSLEY & SINCLAIR *Dict. Ballet Terms* 11 *Balancé*, a rocking step in waltz-rhythm..can also be done swaying backwards and forwards instead of from side to side.

balanced, *ppl. a.* Add: **1. b.** *spec.* Of a rudder or of the control surface of an aeroplane: see quots.

1869 [see sense 1 in Dict.]. **1874** S. J. P. THEARLE *Naval Archit.* (Adv. Sci. ser.) IV. xx. 340 The balanced rudder revolves about an axis so situated that about two-thirds the area of the rudder is on the aft, and the remaining one-third on the fore side of the axis. **1920** *Flight* 5 Aug. 867/2 The aeronautical engineer needs some simple rule for the design of balanced control surfaces. **1922** *Flight* XIV. 344/1 Balanced ailerons are fitted to the top plane only. *Ibid.* The tail plains [*sic*]..consist of..two balanced elevators hinged to the stabiliser, and a balanced rudder. **1950** *Gloss. Aeronaut. Terms* (*B.S.I.*) I. 34 *Balanced surface*, a control surface in which the aerodynamic hinge moment opposing rotation has been reduced by so shaping it that part of the surface is forward of the hinge, or by fitting it to a balance tab, or by other means.

3. b. spec. *Dietetics.* Containing a suitable proportion of each type of nutriment required for adequate nutrition.

1908 *Sci. Amer.* Suppl. 24 Oct. 267/2 The early summer dinner..of lamb and peas with a salad, bread and butter, and a little fruit perhaps, makes a good example of a well-balanced meal. **1912** *Delineator* LXXX. 45 Problem of supplying a properly balanced menu. **1917** *Ladies' Home Jrnl.* Nov. 56/2 She can't understand what is meant by a 'balanced meal'. **1936** *Discovery* Apr. 98/2 A balanced diet, not merely a copious one is indicated.

5. *Naut.* Reefed with a balance-reef (see BALANCE *v.* 16).

1750 in *Essex* (Salem, Mass.) *Inst. Hist. Coll.* (1910) XLVI. 96 We were obliged to hand the jibb and lay too under balланced mainsail. **1762–9** [see BALANCE *v.* 16]. **1802** A. ELLICOTT *Jrnl.* 26 Nov. (1803) 256 We had to heave too under a ballanced main-sail.

6. Other *techn.* uses.

1902 G. S. NEWTH *Text-Bk. Inorg. Chem.* (ed. 9) I. x. 88 Reactions of this order are known as reversible or balanced actions. **1904** GOODCHILD & TWENEY *Technol. & Sci. Dict.* 38/1 *Balanced Steps*, the method of arranging the steps of geometrical stairs to increase the going (width) at the small ends of the winders, and to give a better falling line to the handrail. **1918** W. H. ECCLES *Wireless Teleg.* (ed. 2) 489 *Balanced Detectors*, a term applied to an arrangement of opposed rectifying detectors for reducing the effects of strong Xs in receiving apparatus. **1924** ROGET *Dict. Electr. Terms* 17/1 *Balanced load*, a load in a three-wire system which is equally shared by the two sides, or in a polyphase system, where it is equally shared by the various phases. *Ibid.*, Balanced modulator. **1929** WILSON & WEBB *Mod. Gramophones & Electr. Reproducers* ix. 180 There is no permanent magnetic pull on the armature when no signal is passing through the coil. This type is therefore known as a 'balanced armature'. **1930** PHILLIPS & TUTT *Mod. Gas Fitting* xi. 341 The 'Balance' radiator is so called in that the air inlet and combustion products outlets are made continuous and 'balanced'. **1933** EDGCUMBE & OCKENDEN *Indust. Electr. Measuring Instr.* xiii. 279 A strictly 'balanced' circuit is one in which the currents, voltages, and phase displacements are the same for each phase. **1938** FRITSCH & SALISBURY *Plant Form & Function* v. 49 A mixture of ions in solution in such proportions that they exert no toxic effect is called a *balanced solution.* **1940** L. HARTSHORN *Radio Freq. Meas.* iii. 48 The potential of one terminal is always as much above that of the screen (earth) as that of the other is below it. This is known as the balanced condition. **1944** *Hackh's Chem. Dict.* (ed. 3) 96/1 *Balanced reaction*, a reaction which can be made to proceed in either direction by variation of either temperature, pressure or concentration of the reactants. **1955** *Archit. Rev.* CXVII. 93/2 Balanced-flue gas heaters are provided in the kitchens and serve the sink, lavatory basin and bath.

balancer. Add: **4.** In technical uses (see quots.).

1904 S. P. THOMPSON *Dynamo-Electric Mach.* (ed. 7) I. 775 Uses of Motor-Generator Combinations...Equalization of voltages in a 3-wire or 5-wire system of distribution; in which case, the apparatus is termed a Balancer. **1923** *Daily Mail* 17 Feb. 4 The Lanchester balancer incorporated with the five-bearing crankshaft...The balancer consists of two weighted drums driven by a helical gear wheel fixed to a web of the crankshaft. They revolve at twice the crankshaft's speed and ensure its perfect balance and harmony.

5. balancer mash, meal, a supplementary food for poultry (see quot. 1951).

1950 A. CHRISTIE *Murder is Announced* i. 15 [She] was conscientiously stirring in handfuls of balancer meal to a.. basin full of cooked potato peelings. **1951** W. P. BLOUNT *Hen Batteries* viii. 122 *Balancer Meal* is not a complete feed in itself. It is intended to supply those main nutrients (proteins) deficient in ordinary household scraps. **1955** *Pract. Poultry Keeping* (ed. 5) viii. 61 A grain balancer mash should be used for this form of feeding and may be the same as for dry mash.

balanitis (bælănəi·tis). *Path.* and *Vet. Sci.* [mod.L., a. Gr. βάλανος acorn + -ITIS.] Inflammation of the glans penis in mammals.

1853 in MAYNE *Expos. Lex.* **1871** *Med. Times* (Philad.) *1870–71* I. 397 (*title*) Balanitis. **1879** *Syd. Soc. Lex.* I, *Balanitis*, inflammation of the surface of the glans penis, with purulent discharge. It may be simple or gonorrhœal. **1881** *Trans. Internat. Med. Congr. 1881* III. 142 Professor Kaposi, Vienna, mentioned that..diabetic patients are disposed to balanitis. **1922** R. LEIGHTON *Compl. Bk. Dog* xxii. 344 There are few troubles of the genital organs that need attention in either dog or bitch. What is called *Balanitis* is a slight running of pus from the organ of the male. **1955** GAIGER & DAVIES *Vet. Path. & Bacteriol.* (ed. 4) xxv. 684 Balanitis, or inflammation of the mucous membrane covering the glans penis and lining the prepuce, is not infrequently met with in animals.

balao (bălau·, ‖bala,o). Also **ballaho(o,** etc. [Sp.] The half-beak.

1854–5 *Trans. Jamaica Soc. Arts* I. 144 Hæmirhamphus Braziliensis, Ballahoo. **1858** DE VERTEUIL *Trinidad* 448 *Hemiramphus.* 1 species (s)..(s) Balaou. *Ibid.* 451 The balaou and gar-fish are commonly caught at night by torch-light, or with the seine. **1867** T. F. DE VOE *Market Assistant* 199 There are likewise to be caught in the winter season, fish, by towing over this bank, if a person has suitable bait, such as the *ballaho*, which they have generally in the West Indies. **1893** *Funk's Stand. Dict.*, *Balahoo* (W. Ind.), a halfbeak. **1896** JORDAN & EVERMANN *Fishes N. Amer.* I. 723 *Hemirhamphus balao*...If a valid species, this must be the original *balaó*, which is said to have the caudal bluish, and the common species will stand as *H. brasiliensis*. **1898** MORRIS *Austral Eng.* 16/1 *Ballahoo*, a name applied to the *Garfish*..by Sydney fishermen. The word is West Indian, and is applied there to a fast-sailing schooner; also spelled *Bullahoo* and *Ballahou*. **1922** *Outing* (U.S.) Mar. 253/1 For sail and other large fish they use a sardine-sized, sword-nosed minnow called a ballyhoo. **1949** *Esquire* Mar. 81/6 Bait—mullet, squid, balao, [etc.].

balata (bæ·lătă). Also 9 **ballata.** [S. Amer. Sp., prob. ad. a native word. So Fr. (1777).] **1.** Any of several trees of Guiana and the W. Indies belonging to the family Sapotaceæ, esp. *Mimusops balata.* Also *balata-tree.*

1858 DE VERTEUIL *Trinidad* 101 Balata (*Mimusops*). The Balata, or Bullet-wood, is one of our best and most useful timbers. **1860** *Jrnl. Soc. Arts* 24 Aug. 713/2 It appears also that the *Sapota Mulleri*, the tree producing the gutta percha of Surinam, is called Balata in French Guiana, Les Antilles, and elsewhere. **1864** GRISEBACH *Flora Brit. W. Ind. Isl.* 781 Ballata tree (Dominica): *Bumelia retusa.* **1871** KINGSLEY *At Last* I. vi. 223 A huge dark-headed Balata [*note* Mimusops Balata]. *Ibid.* vii. 257 (*page-heading*) The Balata Tree. **1922** W. SCHLICH *Man. Forestry* (ed. 4) I. II. 322 *Mimusops globosa*, the Balata, go to 120 feet high.

2. The dried juice or gum of *Mimusops balata*, used as a substitute for gutta-percha. Also *attrib.* and *Comb.*

1860 *Jrnl. Soc. Arts* 24 Aug. 713/2 A few months ago it was announced in the French journals that M. Serres, Pharmacien, at Paris, had been so fortunate as to prepare gutta percha from the Balatas of the sapotaceous plants in Guiana and Central America. **1862** *Catal. Contrib. Brit. Guiana to Internat. Exhib.* in R. Duff *Br. Guiana* (1866) 99 The leaves, branches, and trunk [of the bully tree] produce a whitish milk, forming the gum, now known as ballata. **1864** W. HOLMES in *Jrnl. Soc. Arts* 4 Mar. 245/2, I believe..that *Balata*..will supply the great want of the day, as a good insulating medium for telegraphic purposes. **1867** W. T. VENESS *El Dorado* ix. 94 The ballata-gatherer.. may thus bleed a great many trees in a single day. **1939** *Nature* 12 Aug. 300/2 Thus a polymerhomologous series of celluloses exists and the same holds also for starch, glycogen, caoutchouc and balata. **1958** J. CAREW *Wild Coast* iv. 51 He had left the balata fields. *Ibid.* 52 He was carrying a heavy balata-coated canvas bag. **1959** *B.S.I. News* June 10/2 Rubber-textile and balata-textile belting in many varieties ..is extensively used for power transmission of all kinds.

‖**balayeuse** (bal**ɛ**yöz). *Dressmaking.* [Fr., fem. of *balayeur* sweeper.] (See quot. 1882.)

1882 CAULFEILD & SAWARD *Dict. Needlework*, *Balayeuse, or Sweeper.*—A French term to signify the frilling of material or lace which lines the extreme edge of a dress skirt to keep the train clean as it sweeps along the floor. The balayeuse is allowed to project beyond the edge of the dress, so as to form a decorative as well as a useful trimming. **1894** *Daily News* 20 Jan. 5/7 Three flounces of..silk forming a richly-rustling balayeuse beneath the hem.

Balbriggan (bælbri·găn). The name of a town in Ireland, applied *attrib.* to a knitted cotton fabric manufactured there, used in hose, underwear, etc.; also to other products, and *ellipt.*

1859 G. A. SALA *Tw. round Clock* (1861) 128 A white hat with a black band, surmounting a rough coat, cord trousers, and Balbriggan boots. **1885** *Queen* 24 Oct. (Advt.), Ladies' Stockings...In Balbriggan (fourteen qualities). **1887** R. DENNIS *Industr. Irel.* ix. 104 Hosiery..is well described as having 'slipped away' from Ireland—save perhaps in the solitary instance of the Balbriggan hose, which is still in great demand. **1910** *Encycl. Brit.* III. 242/1 The town has considerable manufactures of cottons and hosiery, 'Balbriggan hose' being well known. **1922** JOYCE *Ulysses* 291 High Balbriggan buskins. **1928** A. CHRISTIE *Myst. Blue Train* xxx. 251 With a pair of good Balbriggan stockings on and sensible shoes. **1941** H. ASBURY *Underworld of Chicago* 272 Two suits of heavy underwear, one of wool and one of balbriggan.

‖ **bal costumé** (bal kostüme). [Fr., lit. 'costumed ball'.] A fancy-dress ball.

1826 M. WILMOT *Let.* 29 Feb. (1935) 234 Our Ambassadrice had plann'd to give a ball costume early in the Carnival. **1848** G. E. JEWSBURY *Let.* 4 Oct. (1892) 256 Dandified 'Church-of-Englandism', which bears the same relation to 'the Church' that a polite assemblage of ladies and gentlemen at a 'bal costumé' bears to the countries after which they have dressed themselves. **1861** G. DU MAURIER *Let.* [Jan.] in D. du Maurier *Young G. du M.* (1951) 30 On the 14th there is going to be a grand bal costumé..–150 or 200 people. **1896** BEERBOHM *Dandies* in *Wks.* 20 The clinquant corslet of the Swiss girl just survives at *bals costumés.*

bald, *a.* Add: **4. a.** (Later examples.)

1840 C. F. HOFFMAN *Greyslaer* I. x. 116 There's a ledge of bald rock to the left yonder. **1872** E. EGGLESTON *End of World* xli. 257 A large bald hill overlooking the Ohio was to be the mount of ascension. **1946** R. S. THOMAS *Stones of Field* 14 The bald Welsh hills.

c. (Examples.)

1804 J. ROBERTS *Pennsylv. Farmer* 114 What kind is the most productive,..bearded or bald? **1856** *Trans. Mich. Agric. Soc.* VII. 805 He..raises..the old-fashioned bald or bearded wheat. **1919** *Chambers's Jrnl.* Oct. 654/1 Egyptian cotton-seed..is devoid of these short fibres, and hence is commonly termed 'bald' seed.

bald, *sb.* [f. BALD *a.*] **1.** A mountain summit or region naturally bare of forest, esp. in the southern Appalachians. *U.S.*

1838 *Southern Lit. Messenger* IV. 231/2 We came to the top of the near Bald. **1885** 'C. E. CRADDOCK' *Prophet Gt. Smoky Mts.* i. 2 She paused often, and looked idly..at the great 'bald' of the mountain. **1943** R. PEATTIE et al. *Great Smokies* 154 Aboriginally, the Appalachian forests were vast in extent, clothing the mountains, except for the 'balds', from top to bottom.

2. A species of domestic pigeon; = BALD-HEAD (quot. 1867).

1854 *Poultry Chron.* I. 21/2 For the best pair of Balds or Beards. **1876** R. FULTON *Illustr. Bk. Pigeons* xi. 172 There are plenty of the pleasant-faced Balds which are all that could be desired in colour and marking. *Ibid.* 173 The Bald has so many white flights in each wing.

ba·ldachined, *a.* [f. BALDACHIN *sb.* 2 + -ED^{2}.] Canopied, covered with a baldachin.

1906 B'NESS VON HUTTEN *What became of Pam* vi. 150 In her baldachined bed with the bestarred silk curtains. **1914** T. HARDY *Satires of Circumstance* 157 The God with the baldachined altar overhead. **1966** AUDEN *About House* 42 An Emperor's baldachined..couch.

bald-coot: see COOT *sb.*^{1} 2 in Dict. and Suppl.

bald eagle, bald-eagle. *U.S.* [BALD *a.* 2.] The American eagle. (Cf. EAGLE *sb.* 1 b.)

1688 in *Phil. Trans. 1693* (1694) XVII. 989 The Second is the Bald Eagle, for the Body and part of the Neck being of a dark brown, the upper part of the Neck and Head is covered with a white sort of Down, whereby it looks very bald, whence it is so named. **1705** BEVERLEY *Virginia* II. 35 The Bald-Eagle no sooner perceives a Hawk that has taken his Prey, but he immediately pursues. **1803** [see EAGLE *sb.* 1 b]. **1868** *Amer. Naturalist* II. 194 So cautious is he lest the Bald-eagle (*Haliaetus leucocephalus*)..may approach. **1878** J. H. BEADLE *Western Wilds* xxx. 483 Small is the pleasure one can take..in the sweep of the bald eagle.

bald-face. *U.S.* [BALD *a.* 2.] **1.** The widgeon (also called *bald-head, -pate*).

1709 J. LAWSON *New Voy. Carolina* 151 The bald or white Faces are a good Fowl. *Marg.* Bald-Faces. **1768** WASHINGTON *Diaries* (1925) I. 253 Went a ducking between breakfast and dinner and kill'd 2 Mallards and 5 bald faces.

2. A variety of whisky. *slang.*

1840 *Daily Pennant* (St. Louis) 28 Apr. (Th.), He called lustily for a horn of bald-face and molasses. **1848** *Knickerbocker* XXXII. 402 What is classically denominated 'bald-face', or old brown whiskey. **1872** SCHELE DE VERE *Americanisms* 581 *Bald-face*, one of the many slang terms under which bad whiskey passes in the West.

ba·ld-faced, *a.* [BALD *a.*] Having a bald face.

1648 in *Archives of Maryland* (1887) IV. 425 A Bawld-facd heighfer. **1677** [see BALD *a.* III]. **1861** *Trans. Ill. Agric. Soc.* IV. 341 The nest of our bald-faced hornet is occasionally suspended in a house to kill off the house-flies. **1885** *Century Mag.* Nov. 60/2 He jogged along on his bald-faced bay in the bleak untempered light.

bald-head. (Earlier example as the name of a pigeon.)

1855 *Poultry Chron.* III. 491/2 Mr. Woodhouse will show his Crested Bald-head.

bald-headed, *a.* Add to def.: *bald-headed eagle*, the bald eagle.

1829 J. MACTAGGART *Three Years in Canada* I. 22, I had seen a couple of bald-headed eagles the day before. **1836** M. A. HOLLEY *Texas* v. 100 The bald-headed eagle and the Mexican eagle..are very common. **1863** [in Dict.]. **1935** *Times Lit. Suppl.* 28 Mar. 214/2 The subject..is the bald-headed eagle..the original of 'Uncle Sam's bird'.

b. colloq. phr. (orig. *U.S.*) *to go bald-headed* (*into, for, at*), to stake everything, to disregard consequences, to attack without care or thought; also *to go it bald-headed.* Hence **bald-hea·dedly** *adv.* (in similar sense).

1848 LOWELL *Biglow P.* Ser. I. vi. 79 'Pious editor' x, I scent wich pays the best, an' then Go into it baldheaded. **1867** *Ibid.* Ser. II. Introd., 'To go it bald-headed': in great haste, as where one rushes out without his hat. **1888** *Pall Mall Gaz.* 22 June 4/2 The Chicago Republicans, to use an Americanism, have gone 'baldheaded' for Protection. **1915** W. J. GORDON *Flags of World* 77 Warburg, after the colonel of the Blues, the Marquis of Granby, after a high trot of five miles led them hatless in the charge, 'going bald-headed for the enemy', and thus originated the well-known phrase. **1920** W. J. LOCKE *House of Baltazar* v. 61 Quong Ho..tried ..zealously, then desperately, then bald-headedly, but never a wild blow could pass the easy guard of his smiling master. **1942** R. G. COLLINGWOOD *New Leviathan* 116 A sensible man does not go bald-headed into a brain-twister. **1960** M. STEWART *My Brother Michael* xiv. 180 You went bald-headed for the poor chap.

balding, *a.* [f. BALD *a.* + -ING^{2}.] Going bald.

1938 *Time* 5 Dec. 12 He crams a golf cap on his balding grey head. **1941** *Time* 25 Aug. 43/1 Moonfaced, balding, bespectacled, Mort broods a great deal about his health. **1951** M. DICKENS *My Turn to make Tea* v. 91 A balding man in an evening suit. **1958** *Times* 8 Oct. 6/5 The quintessentially balding lodger in raincoat and bowler.

bald-pate. Add: *transf.* An American widgeon (*Anas americana* or *Mareca americana*). *U.S.*

1813 A. WILSON *Amer. Ornith.* VIII. 86 The..widgeons, or as they are called round the bay, bald pates. **1838** AUDUBON *Ornith. Biog.* IV. 337 In the Western Country, and in most parts of the Eastern and Middle States, it is called the *Bald Pate.* **1907** *Westm. Gaz.* 25 Nov. 8/3 The 'bald-pate', as it is familiarly called, is a gregarious bird, frequenting the North American estuaries.

Baldwin (bǭ·ldwin). *U.S.* [The personal name.] A common variety of eating apple; a tree bearing this variety of apple.

1826 *Catal. Fruits in Gard. Hortic. Soc. Lond.* 108 (Apples) Baldwin's. **1842** *Ibid.* (ed. 3) 5 Apples...Baldwin. **1848** LOWELL *Biglow P.* Ser. I. ix. 142 Looking out through my study window, I see Mr. Biglow..busy in gathering his Baldwins. **1860** *Trans. Mich. State Agric. Soc. 1858* X. 254 Baldwin is a very popular market fruit in New England. **1861** *Trans. Ill. Agric. Soc.* IV. 468 Our farmers set in the early orchards of Wisconsin a large proportion of Roxbury Russets, Baldwins,..and Spitzenbergs. **1887** M. E. WILKINS *Humble Romance* 238 On the right of the garden were two old apple-trees, a Baldwin and a Porter.

baldy (bǭ·ldi). *colloq.* [f. BALD *a.* + -Y^{6}.] A bald-headed person. Also *transf.* and as a nickname. Also as *adj.* and in *Comb.*, *baldy-headed* adj., *Sc.* and *dial.*

1863 *Rio Abajo Press* 28 Apr. 2/2 Within the perlieus [*sic*] of the Rio Chiquito and 'Old Baldy'. **1909** WEBSTER *Baldy*, somewhat bald. **1916** JOYCE *Portr. Artist* (1917) 59 He thought of the baldy head of the prefect of studies. **1925** *Scots Mag.* Dec. 196 A big, gauky stirk o' a bauldy-heided dominy! **1927** A. CLARKE *Son of Learning* III. 54 There is a baldy spot Lighting your crown. **1931** 'DEAN STIFF' *Milk & Honey Route* 199 *Baldy*, generally an old man 'with a high forehead'. **1936** F. CLUNE *Roaming the Darling* xvi. 143 A baldy and paunchy old cove, hosing down the footpath. **1936** I. L. IDRIESS *Cattle King* xi. 99 He's got a couple of good sorts of mares and a good-looking baldy chestnut. **1944** A. CLARKE *Viscount of Blarney* (1945) 35 I'll leave the lantern..for old baldy pate. **1952** B. MALAMUD *Natural* 89 Pop scratched his baldy. **1953** in WENTWORTH & FLEXNER *Dict. Amer. Slang* (1960) 15/2 Baldies are more romantic. **1959** I. & P. OPIE *Lore & Lang. Schoolch.* xvi. 359 The only thing I dinna like Is the baldy-headed master.

bale, *sb.*^{3} Add: **3.** *bale-sack*; **bale-band** (see quot.); **bale-cloth** *U.S.*, cloth used for covering bales; **bale-goods** (later examples); **bale-sling** (see quot. 1891).

1891 H. PATTERSON *Naut. Dict.*, *Bale-band*, a big shackle-shaped iron at the mast-head, supported by the cap-band, and to which the standing part of the flying jib-stay is bent on. **1797** B. HAWKINS *Let.* in *Georgia Hist. Soc. Coll.* (1916) IX. 346, 8 yds. bale cloth to Harry Dergin, at 12½c., $1.00. **1865** *Trans. Ill. Agric. Soc.* V. 159 We have pressed the sirup from the sugar through fine bale-cloth. **1800** *Asiatic Ann. Reg.* III. 41/2 One of the above ships had on board a very valuable cargo of bale goods. **1894** *Idler* Sept. 135 The original freight of the ship had been bale goods. **1883** *Century Mag.* Oct. 817/2 This man flung them into an enormous bale-sack, swinging wide-mouthed from a derrick. **1883** *Man. Seamanship for Boys* 93 There are several methods of slinging a cask, either with a pair of butt slings, bale slings, or a bowline knot. **1891** H. PATTERSON *Naut. Dict.*, *Bale-sling*, a simple strap passed round a bale or bag, the two ends meeting on top, one dipping under the other.

bale, *v.*^{2} Add: **2.** *to bale out.* [Usually so spelt, as if the action were that of letting a bundle through a trapdoor; but also (esp. *U.S.*) as *bail*, as if a use of BAIL *v.*^{4}, to lade out.] *intr.* (Of an airman) to make an emergency descent by parachute from his machine. Hence also (rare) sb. *bail-out*. orig. *U.S.*

1930 C. J. V. MURPHY *Parachute* 272 Some say the pilot 'bailed out' the moment he went into the spin. **1932** *N.Y. Times* 11 Apr. 3/2 He successfully bailed out of an airplane at an elevation of 1,500 feet. **1939** F. D. TREDREY *Pilot's Summer* 28 If you bale out and land in water..a smart rap will release the whole lot and you can swim free. **1940** *Times* 15 Aug. 4/2 He baled out before his machine crashed. **1955** *Sci. News Let.* 8 Jan. 23 The purpose..was to explore human tolerances during a high speed bailout from jet planes.

baled, *ppl. a.* (Earlier and later examples.)

1828 *Free Press* (Tarboro, N. Carolina) 7 Nov., Cotton in the seed and baled Cotton. **1872** E. EGGLESTON *End of World* xxx. 197 [The steamer] took on a new cargo of baled hay and corn and flour. **1895** *Daily News* 19 Oct. 2/4 The market for baled jute is quieter. **1898** *Westm. Gaz.* 28 Oct. 5/3 Baled hay should be stored before the winter sets in. **1918** *Nation* (N.Y.) 7 Feb. 168/1 The baled straw for the mattresses.

baleen. Add: **3. baleen whale,** a whalebone-whale; any member of the suborder Mysticeti; also *ellipt.*

1874 C. M. SCAMMON *Marine Mammals* vi. 66 This great northern baleen whale, in its principal proportions, resembles the *Balæna mysticetus.* **1878** *Cassell's Nat. Hist.* II. 257 Scammon says that three or four [killer-whales] do not hesitate to grapple with the largest Baleen Whales. **1958** *Times* 12 Nov. 11/6 There are two kinds of whale caught, the baleen and the sperm.

baler^{2}. [f. BALE *v.*^{2}] A machine or apparatus for baling hay, straw, metal, etc.

1888 *Voice* (N.Y.), Why are not balers as common as threshers? I believe it is owing to..manufacturers..not pushing the balers. **1926** *Glasgow Herald* 26 June 5/7 The baler seems a practical machine and might be tried in large nurseries. **1936** *Discovery* June 193/1 The installation of two large grass-driers and automatic balers on his Warwickshire farm. **1963** *Times* 10 Dec. 5/1 These men..with one movement of their giant balers can reduce [an old motor-car]..to a cube of metal the size of a biscuit tin.

b. One who makes up bales.

1889 in *Cent. Dict.* **1910** *Daily Chron.* 2 Feb. 5/4 One application was for three newspaper balers.

balibuntal (bælibʊ·nt'l). Also **balli-, bally-, -buntl(e** (as one word or as two); also shortened to **bal(l)i.** [Short for *Baliuag buntal*, a weave of *BUNTAL originating from Baliuag in Bulacan, Philippine Islands.] A fine straw of a very close weave, used for hats. Also *attrib.*, and short for *balibuntal hat.*

[**1911** *Philippine Jrnl. Science* Sect. C. VI. 115 The Baliuag buntal hat is..more closely woven than that of Lucban, and is consequently stronger.] **1913** T. MULLER *Industr. Fiber Plants of Philippines*, (*Philippine Islands*) *Bur. Educ. Bull.* 49, 84 Only the best weavers can make the buntal-sawalis or bali-buntals as these Baliuag hats are called..usually known there [*sc.* in U.S.] as 'Manilas'. **1918** *N.Y. Times* 25 Apr. 5 (Advt.), The Bally Buntle is one of the strictly new, extremely modish Hats. **1925** *Ibid.* 1 Mar. (Advt.), Hats of Horse Hair, Balli Buntle, Bankok. **1927** *Observer* 12 June 11/4 A bali-buntal in mauve. **1928** *N.Y. Times* 1 Apr. 7 (Advt.), Baku and Balibuntal are favorite straws for Easter Wear. *Ibid.* 11 Apr. (Advt.), This Untrimmed Ballibuntl Needs Little Adornment. **1929** [see *BAKU]. **1961** *Harper's Bazaar* Feb. 53 Hat of white balibuntal.

Balinese (bālinī·z), *a.* and *sb.* [f. *Bali* + -ESE after Du. *Balinees.*] **A.** *adj.* Of or pertaining to the island of Bali in Indonesia, or to its inhabitants. **B.** *sb.* **1.** A native of Bali; as collect. sing., the people of Bali. **2.** The language spoken by this people.

1820 J. CRAWFURD *Hist. Ind. Archipelago* I. I. iv. 69 The diversion of cock-fighting is most especially in vogue among the Malays, the people of Celebes, and the Balinese. *Ibid.* II. v. i. 27 They..have been translated into the Balinese and Malay languages. *Ibid.* II. v. iv. 69 The Balinese is the sole language of the island of Bali. **1872** *Appleton's Jrnl.* Aug. 120/2 Many of the Balinese seem to possess *sarongs* of this description. **1935** *Times Lit. Suppl.* 15 Aug. 508/4 Another important class of Balinese antiquities is richly carved water-spouts. **1954** PEI & GAYNOR *Dict. Linguistics* 25 *Balinese*, a language spoken by about 1,000,000 persons on the island of Bali; a member of the Indonesian subfamily of the Malayo-Polynesian family of languages.

baling, *vbl. sb.*^{1} Add: Also *attrib.*

1810 F. CUMING *Sk. Tour Western Country* xxvi. 165 There is [in Lexington, Ky.] one manufacturer of baling cloth. **1815** *Niles' Reg.* IX. 187/1 They are considered much superior to all other baling presses. **1922** *Outing* (U.S.) June 110/2 The time was now come..to get out..rope and baling wire. **1959** *Engineering* 9 Jan. 58/1 A scrap yard adjoins the melting shop, the handling equipment including a..baling press.

balisier (balizye). [Fr.] The West Indian musaceous plant *Heliconia bihai*, with very large leaves and brilliant orange flowers.

1858 DE VERTEUIL *Trinidad* 95 Balizier (*Heliconia*). **1871** KINGSLEY *At Last* I. vi. 222 Under the shade of great Balisiers or wild plantains. [*note*] Heliconia. *Ibid.* vii. 266 Above these, again, the Balisiers bend their long leaves, eight or ten feet long apiece. **1953** P. L. FERMOR *Violins of St-Jacques* 18 Punctilious flower-paintings of hibiscus and balisier.

balistic, -ics, var. ff. BALLISTIC, -ICS.

balk, *sb.*^{1} Add: **5. b.** *U.S.* In baseball: (see quot. 1867). Also *attrib.*

1845 in *Appleton's Ann. Cycl.* (1886) X. 77/2 A runner can not be put out..when a balk is made by the pitcher. **1867** H. CHADWICK *Beadle's Dime Base-Ball Player* 53 A balked Ball.—Should the pitcher move his foot in delivery—thereby making a 'balk'—and the Umpire call a 'balk' until the ball is returned to the pitcher, [etc.]. **1913** *Amer. Mag.* Sept. 24/1 Kilroy caught seven by his balk motion.

8. c. Of a horse: an instance of balking (cf. BALK *v.*^{1} 3).

BALK

ba·lkiness. [f. BALKY *a.*] The quality of being balky.

1894 *Outing* (U.S.) XXIV. 349/1 The mules were the very embodiment of balkiness. **1909** J. C. LINCOLN *Keziah Coffin* x. 143, I, bein' a Hammond, with some of the Hammond balkiness in me, I set my foot down as hard as his.

balky, *a.* Also **baulky.** Add to def.: reluctant to proceed; contrary, perverse. (Earlier and later examples.)

1847 J. T. HUGHES *Doniphan's Exped.* (1907) 14/1 The mules and other animals..often became refractory and balky. **1857** *Quinland* II. xvi. II. 117 One of his horses is baulky, and he can't go any farther. **1897** KIPLING *Capt. Cour.* vii. 147 Young Olley's gittin' kinder baulky an' excited. **1957** *Economist* 21 Dec. 1052/2 When federal officials hoped to proceed to more substantial matters such as public housing and public assistance, the states turned balky.

ball, *sb.*[1] Add: **1. b.** *ball and chain*, a heavy metal ball secured by a chain to the leg of a prisoner or convict, to prevent escape. Also *chain and ball* (s.v. *CHAIN *sb.* 2 a). *U.S.*

1835 J. H. INGRAHAM *South-West* II. 189 The threat of the Calaboose, or the 'ball and chain'. **1873** J. H. BEADLE *Undeveloped West* vii. 132 Those who had fined and imprisoned culprits, or sent them to work with ball and chain. **1902** W. N. HARBEN *Abner Daniel* 94 They put a ball an' chain to one of his ankles an' sent him out with the nigger gang.

4. b. *spec.* in *U.S.*, baseball.

1868 H. CHADWICK *Base Ball* 162 The National Game of ball of Americans. **1896** KNOWLES & MORTON *Baseball* 71 He saw ball played by the American Students.

c. *the balls are over*: cf. OVER *adv.* 5 c.

1870 *New Sporting Mag.* LX. 271 'The Balls are over.' Some of the umpires of the present day corrupt the four words..into '*Ver*'. **1894** E. B. Y. CHRISTIAN *At Sign of Wkt.* 75 For him who falls, His hundred made..There need no tears,..'the balls Are over' now.

6. b. Phr. *ball of fire*: (*a*) *slang* a glass of brandy; (*b*) = FIRE-BALL; (*c*) *fig.* a person of great liveliness or spirit (cf. FIRE *sb.* 13).

1821 J. BURROWES *Life in St. George's Fields* 25 *Ball of fire*, glass of brandy. **1889** BARRÈRE & LELAND *Dict. Slang* I. 70/2 A *ball* of fire in popular slang is a glass of brandy, in allusion to the fieriness and pungency of the wretchedly bad spirit sold as brandy to the lower classes. **1953** G. LAMMING *In Castle of Skin* ii. 21 The white gentl'man ..run [*sic*] like a ball of fire all the way home. **1958** J. D. MACDONALD *Executioners* (1959) x. 159 You haven't been a ball of fire around here lately. **1960** N. MITFORD *Don't tell Alfred* ii. 26 Yes, I know her. Not a ball of fire, is she? **1964** M. M. GOWING *Brit. & Atomic Energy 1939–1945* ix. 265 Sir Geoffrey [Taylor] also stimulated serious theoretical investigation into the 'ball of fire' phenomena in the explosion.

10. d. *Metallurgy.* A mass of puddled iron formed by the workman into a pasty lump, to be hammered and rolled when taken from the furnace.

1825 J. NICHOLSON *Oper. Mech.* 334 When the iron is deprived of the carbon..the furnaceman rolls it up into balls of one half or three quarters of a cwt. each. **1855** W. TRURAN *Iron Manuf.* 134/1 After 8 or 9 minutes raking of the iron, now in the condition of pasty lumps,..the puddler commences the formation of the puddle balls. **1875** [see BLOOM *sb.*[2]]. **1892** F. JOVNSON *Iron & Steel Maker* 89 The work of the puddler is..confined to..the production of the lumps or masses of metal technically called 'balls', and sometimes, though rarely in this country, 'blooms'.

e. = CLOD *sb.* 3 c.

1881 *Encycl. Brit.* XII. 239/1 The tree will then be ready to lift if carefully prized up from beneath the ball. **1885** G. NICHOLSON *Dict. Gardening* I. 153/2 *Ball.* This term is used in reference to the roots and mass of earth as they are moulded into form...The masses of roots and earth which.. must be taken intact when removing the plants, are also termed Balls. **1924** H. H. THOMAS *Compl. Amat. Gardener* iii. 18 The larger the 'ball' the more quickly will a tree or plant recover after having been transplanted.

15. b. Pl. *vulg.* The testicles; *fig.* nonsense, freq. as *int.*; hence *vbl. phr.*, *to make a balls of*, to muddle, to do badly, to make a mess of. Cf. *to ball up* (*BALL *v.*[1] 6 b).

a **1325**, *a* **1456** in M.E.D. **1889** BARRÈRE & LELAND *Dict. Slang*, s.v. *Balls*: 'To make *balls* of it', to make a mistake, to get into trouble. *Balls, all* (popular), all rubbish. **1890** FARMER *Slang* I. 109/2 *All balls*, all rubbish; nonsense. **1903** O. WISTER *Philosophy 4: A Story of Harvard Univ.* i. 10 'If I were to stop thinking about you, you'd evaporate.' 'Which is balls,' observed the second boy judicially, again in the slang of his period, 'and can be proved so. For you're not always thinking about me, and I've never evaporated once.' **1922** JOYCE *Ulysses* 134 All balls! Bulldosing the public! **1928** D. H. LAWRENCE *Lady Chatterley* xv. 263 She..gathered his balls in her hand. **1934** 'G. ORWELL' *Burmese Days* ii. 34 'We all think this idea of electing a native to the Club is absolute —' Ellis was going to have said 'absolute balls'. **1945** *Penguin New Writing* XXVI. 59 They are all of the opinion you can't write and be a soldier same time. Myself I think that's all balls. **1946** B. MARSHALL *George Brown's Schooldays* 79 What do you mean by talking all that unpatriotic balls to the Old Man yesterday? **1956** A. WILSON *Anglo-Saxon Attitudes* II. ii. 346 'Look here! this is awful balls,' said John. **1958** S. BECKETT *Endgame* 21 I've made a balls of the fly. **1960** LEONARD COOPER *Accomplices* IV. v. 243 Fanciful? Balls! It's what happens.

c. *balls-up*, a confusion, a muddle, a mess. Hence *to balls up* = *to ball up* (*BALL *v.*[1] 6 b). *vulg.*

1939 'G. ORWELL' *Coming up for Air* II. viii. 150 You couldn't go on regarding society as something eternal and unquestionable, like a pyramid. You knew it was just a balls-up. **1945** *Penguin New Writing* XXVI. 51 'What d'you make of this case, corporal?' 'Bleeding balls-up, between you and me.' **1956** R. FULLER *Image of Society* viii. 197 Stuart Blackledge made a ballsup of the valuation.

1947 E. TAYLOR *View of Harbour* v. 76 You will balls everything up with your indifference. **1961** S. PRICE *Just for Record* viii. 72 The public would laugh fit to bust if someone really ballsed-up the Civil Service.

18. *to have the ball at one's foot, feet* (earlier and later examples); *to have* or *keep one's eye on the ball*: to be, or to remain, alert; *to keep the ball up* (earlier example); *to keep the ball rolling* (examples); also *to set* (or *start*) *the ball rolling*: to begin a conversation, undertaking, etc.; *to play ball* (*with*): to act fairly (with), to co-operate (with); *to take up the ball* (earlier example).

a **1641** in E. Beveridge *Fergusson's Scot. Proverbs* (1924) MS. No. 588 He has the ball at his foot. **1693** J. HOWE *Carnality Relig. Contention* ii. 75 A mighty pleasure is taken to see the *Saw drawn*, and the *Ball kept up*. **1840** *Log Cabin & Hard Cider Melodies* 58 Virginia will keep her ball rolling. **1850** W. COLTON *Deck & Port* xiv. 390 That courageous organisation which set the ball of Anglo-Saxon supremacy rolling in California. **1857** TROLLOPE *Three Clerks* I. ix. 183 The ball is at your foot now, but it won't remain there. **1873** LYTTON *Parisians* II. v. iii. 142 The Duchesse..took up the ball of the conversation. **1884** MRS. H. WARD *Miss Bretherton* iv. 84 She has a dry original way of judging a novel which is stimulating and keeps the ball rolling. **1903** 'H. MCHUGH' *Back to Woods* vii. 100 Well, if Bunch should refuse to play ball I could send the check back to Uncle Peter. **1906** A. T. QUILLER-COUCH *Cornish Window* 127 Relief..came with his election as Fellow of Oriel..and the brilliant young scholar had..the ball at his feet. **1907** in *Screen Bk.* (1937) 102 We were forever being told 'Keep your eye on the ball'. **1911** P. V. COHN tr. *Nietzsche's Human, All too Human* II. 323 Such persons are in a position to start the ball of slander rolling. **1913** A. R. HOPE *Half & Half Trag.* 250 These amateurs failed to keep the ball rolling. **1930** C. TERRETT *Only Saps Work* 149 The police of Buffalo are too dumb—it would be redundant, I suppose, to say 'and honest'—to play ball with the hold-up mobs. **1932** *Times Lit. Suppl.* 28 Jan. 54/2 Since Malthus set the ball rolling, public opinion has undergone profound metamorphoses. **1944** L. A. G. STRONG *Director* 31 You play ball with me, and I'll see you don't regret it. **1958** *Spectator* 15 Aug. 230/2 This is an admirably professional book..; its authors keep their eyes on the ball.

b. *to have* (something), or *to be*, *on the ball*: to have special merit, to be alert. *U.S. colloq.*

to be on the ball is now also U.K.

1912 *Collier's* 13 Apr. 19/1 He's got nothing on the ball—nothing at all. **1935** *Mademoiselle* Sept. 61/3 The lass has much on the ball. **1947** A. MILLER *All my Sons* (1958) I. 65 Now you're talkin', Bert. Now you're on the ball. **1961** *Listener* 28 Dec. 1136/2 The B.B.C. are 'on the ball' as the Americans would say. **1968** *Times* 19 Apr. 13/4 Kenneth Loach's direction [of a television play] was nothing if not on the ball.

19. b. **ball-and-claw** *attrib.* = claw-and-ball (see *CLAW *sb.* 8).

1904 E. SINGLETON *French & Eng. Furniture* 237 In not one of Chippendale's drawings of chairs does the simple ball-and-claw foot occur. **1960** H. HAYWARD *Antique Coll.* 23/1 *Ball and claw foot* (or *claw and ball*), terminal to a cabriole leg...In use on English furniture from the early until the late years of the 18th cent.

20. Also formerly *three blue balls.*

1748 SMOLLETT *Rod. Rand.* xvi. 123 He..unbuckled his hanger, and shewing me the sign of three blue balls, desired me to..pawn it. **1788** V. KNOX *Winter Evenings* II. v. iv. 129 Pawn at some distant house, known by the sign of the three blue [ed. 2, 1790 golden] balls.

21. *ball-control, -field, game, ground, park, -play, -player, -playing, -stick, -team*: U.S. examples (chiefly baseball) and additional U.K. examples (football, etc.).

1928 *Sunday Dispatch* 2 Sept. 1/1 Clever ball control is returning to our football enclosures. **1951** *Swimming* (E.S.S.A.) vi. 113 You are mastering ball-control, one of the most important phases of the game [*sc.* water polo]. **1867** *Ball Players' Chron.* 13 June 4/2 Let us train up assemblages to good behavior on ball-fields. **1848** *Knickerbocker* XVIII. 216 The boys suspend their ball game while he drives over the green. **1898** S. HALE *Lett.* (1919) 335 These men were just like..Harvard men, after the ball game has gone right for us. **1944** E. BLUNDEN *Cricket Country* xii. 126 His ball-game tendencies alarmed some of the authorities. **1961** *New Left Review* July/Aug. 57/2 Another ball-games court for older boys. **1772** D. TAITT in N. D. Mereness *Trav. Amer. Colonies* (1916) 546 [I] then went to a Ball ground.. where the Eutchie and Geehaw people were playing Ball. **1856** *Spirit of Times* 13 Dec. 245/1 The Club presented their President with an elegant silver Pitcher, with a view of the ball ground carved out upon it. **1899** *Chicago Daily News* 4 Aug. 6/1 Billy Phyle..went out to the ball park. **1957** *Economist* 23 Nov. 687/1 Shortly after the war..attendance at the ball parks began to slump as a result of television. **1765** H. TIMBERLAKE *Mem.* 79, I was not a little pleased likewise with their ball-plays. **1837** J. D. WHITNEY in E. T. Brewster *Life & Lett. J. D. W.* (1909) 20 For my part, I could never make a ball player. **1827** T. L. MCKENNEY *Tour to Lakes* 181 The little naked Indian boys..were.. playing ball...This ball-playing is not unlike our game of bandy. **1775** J. ADAIR *Hist. Amer. Indians* 400 The ball-sticks are about two feet long, the lower end somewhat resembling the palm of the hand, and..worked with deer-skin thongs. Between these, they catch the ball, and throw it a great distance. **1846** J. J. HOOPER *Adv. Simon Suggs* ix. 113 They..knock down their antagonists with their ball-sticks. **1888** *Outing* (U.S.) July 356/1 The *personnel* of the average professional ball team..has improved. **1936** J. DOS PASSOS *Big Money* 17 At Exeter he was head of his class and captain of the ballteam.

22. ball boy, one who fields balls for the players at a lawn tennis tournament; **ball clay,** esp. a fine-textured clay, found mainly in south-western England, which is used in the manufacture of earthenware; **ball-cock** (earlier example); **ball-court,** an area (such as a paved yard) for the playing of ball games; *spec.* in *Archæol.*, a feature of the remains of the Maya civilization in Central America; **ball-fringe,** a decorative fringe (for a mantelpiece, etc.) consisting of ball-shaped materials hung at intervals; **ball-headed** *a.*, with the head shaped like a ball; **ball joint,** a ball-and-socket joint; **ball-lightning,** lightning in a globular form; = *globe-lightning*; **ball mill** (see quot. 1911); **ball-peen, -pome** (see PEEN *sb.*, POME *sb.* 3); **ball-planting,** a method of transplanting trees (see quot.; cf. *10 e); **ball(-point) pen,** one of which the writing point is a minute ball which is inked from an inner reservoir; also *ellipt.* **ball-point**; **ball-pointed** *a.*, having a (minute) ball-shaped point; **ball-race** (see RACE *sb.*[1] 8 g); **ball smut** = BUNT *sb.*[2] 2 (see SMUT-BALL); **ball-stone** (earlier example); **ball-trap** (see TRAP *sb.*[1] 8 and cf. *ball-valve*); **ball turret** *Aeronaut.* (see quot. 1956).

1903 *Westm. Gaz.* 29 Aug. 6/1 A black moving surface, over which red-coated ball boys dart. **1968** *Observer* 28 Apr. 22/2 The tall, dark Gonzales..telling an industrious ball-boy to calm down. **1865** E. METEYARD *Life J. Wedgwood* iv. 140 The imported clay was used as a wash, previously to firing. This was called 'Ball clay'..from being made up in heaps weighing sixty or seventy pounds each. **1903** *Daily Mail* 7 Sept. 5/5 They are the only mines in the world that produce the 'ball clay', without which the manufacture of earthenware is impossible. **1957** Ball clay [see *BLUE *a.* 12 c]. **1968** *Radio Times* 2 May 17/2 Digging for valuable ball clay makes ugly scars across the Devon countryside. **1790** J. DRING *Brit. Pat.* No. 1725 A certain improvement on all cocks, which from the nature of such improvement are termed ball cocks. **1677, 1721** Ball-court [in Dict., sense 21]. **1867** G. M. HOPKINS in *Lett. & Jrnls.* (1959) I. 159 The boys flooded the ball-court and slid and skated on it. **1912** J. W. FEWKES in *Ann. Rep. Bureau Amer. Ethnol.* XXVIII. 93 A long court extends across the whole south end of the compound. Its form suggests a ball court or course for foot races. **1959** *Listener* 12 Mar. 447/2 Huge pyramids, temples, and sacred ball courts are scattered over an area half a mile square [at Chichen Itza, Yucatan, Mexico]. **1909** H. G. WELLS *Tono-Bungay* I. ii. 64 Stuff with ball fringe along the mantel. **1902** *How to make useful Things* 48/2 With a ball-headed hammer strike the petals of the discs. **1957** R. LISTER *Decorative Wrought Ironwork* ii. 48 *Ball-headed set pins* (that is ball-headed bolts) can give a decorative appearance to a bolted joint. *a* **1884** KNIGHT *Dict. Mech.* Suppl. 70/2 *Ball-joint hinge*, one having a flexible knuckle. **1930** *Engineering* 7 Feb. 177/2 The connections with the headers made with ball joints. **1962** *Which?* (Car Suppl.) Oct. 139/2 Grease from lower front suspension ball joints. **1857** J. P. NICHOL *Cycl. Physical Sci.* 431/2 *Ball lightnings* or *globes of fire*..move slowly from the clouds to the Earth. **1930** *Discovery* Dec. 391/2 Ball lightning is probably the most interesting form of lightning discharge. **1903** R. H. RICHARDS *Ore Dressing* I. vi. 260 The Bruckner Ball Mill was the parent form..and consisted of a cylinder revolving on a horizontal axis with die plates around the circumference. The ore ground by balls, passed out through the spaces between the die plates. **1911** *Encycl. Brit.* XX. 239/1 The ball mill is a horizontal revolving cylinder with iron balls in it which do the grinding. **1930** *Engineering* 16 May 633/3 The contents of the weighing machine are discharged directly into one of four ball mills. **1963** J. OSBORNE *Dental Mechanics* (ed. 5) xi. 242 The pigments are usually impregnated to the surface of the polymer particles by means of a ball mill. *a* **1877** KNIGHT *Dict. Mech.* I. 224/2 *Ball-peen Hammer*, a metal-worker's hammer with a spherical peen. *a* **1884** *Ibid.* Suppl. 71/1 *Ball-peen Hammer*, one whose peen is round, or ball-shaped. **1946** *Esquire* (Chicago) Nov. 155 Biro who introduced the first ball-pen presents..a sensational new invention. **1958** *Times* 2 June p. vi/3 Gas is used.. in the manufacture of familiar articles such as ball pens, aircraft engines, [etc.]. **1905** *Terms Forestry & Logging* (*U.S. Dept. Agric.*) 6 *Ball planting*, a method of transplanting young trees with balls or lumps of earth around the roots. **1948** *Specifications of Inventions, Pat. Spec.* 617/176 A ball point pen of new and improved construction. **1958** *Times Rev. Industry* June 22/2 The ball-point pen has a universally inimical effect upon..handwriting. **1959** R. GANT *World in Jug* 124, I..looked at him, sitting there..holding a ball-point. **1935** *Jrnl. R. Aeronaut. Soc.* XXXIX. 1054 A few specimens of brass were measured..using a ball-pointed micrometer. **1952** 'C. BRAND' *London Particular* xvi. 217 Sergeant Bedd licked the end of his ball-pointed pen. **1922** *Weekly Dispatch* 17 Dec. 15, 1-lb. Ball-pome Hammer. **1907** *Westm. Gaz.* 18 Nov. 7/2 The ball-races fitted between the springs and the axle on which the long semi-elliptical springs are carried. **1908** *Ibid.* 30 June 4/2 Of the three Ariels [one] had the misfortune to break a ball-race in the hub. **1934** WEBSTER, Ball smut. **1950** *N.Z. Jrnl. Agric.* Dec. 508/1 The need for some type of seed treatment to control ball smut was soon recognised. **1803** J. PLYMLEY *Agric. Shropsh.* i. 56 Ballstone and earth. **1873** PARKES *Pract. Hygiene* (ed. 4) 344 The ball-trap is used in some special cases only; a ball is lifted up as the water rises, until it impinges on and closes an orifice. **1945** *Aeronautics* Feb. 43/3 Nose turrets are supplied..and the retractable Sperry-designed ball turrets. **1956** W. A. HEFLIN *U.S. Air Force Dict.* 71/1 *Ball turret*, a turret in the shape of a ball, designed to project or to be let down from the belly of an airplane, and to house the gunner.

ball, *sb.*[2] Add: **2.** For def. read: A social assembly for dancing, often of people belonging

Even silk Balloons are almost out—I have not seen a Cap since I came. **1912** C. B. Hayward *Pract. Aeronaut.* I. iii. 14 Three layers of this rubberized fabric are cemented together to form what is known as 'balloon cloth', which is about as impermeable a material as can be made without involving undue weight. **1917** *Illustr. London News* 17 Nov. 601/2, I walked inside the envelope [of a Zeppelin]...On the way I passed a tube of balloon cloth. **1865** *Mechanics' Mag.* 4 Aug. 64/2 A balloon corps should..assist in the operations of our own forces. **1919** *U.S. Navy Dept.: Type 'M' Kite Balloon Handbk.* iv. 37 All cloth used for balloon fabric..is inspected for flaws...The raw cloth is then coated with rubber. **1901** G. Nicholson *Dict. Gardening Cent. Suppl.* 607/2 *Platycodon.* Chinese Balloon Flower. **1962** *Amat. Gardening* 31 Mar. 5/1 The platycodons are called balloon flowers because the buds before they open resemble a balloon. **1853** J. W. Bond *Minnesota* 122 A little clump of shanties and balloon-frames. **1873** E. Eggleston *Myst. Metrop.* xxxv. 302 When at last he saw the familiar balloon-frame houses. **1945** *Archit. Rev.* XCVIII. 40/2 The great cities could never have arisen as quickly as they did if it were not for the invention of the balloon frame, which substituted a simple construction of nails and plates for the old craft of mortised and tenoned joints in wooden house construction. **1855** *Trans. Amer. Inst. N.Y.* 394 The balloon framing used in the Western States and California. **1940** *Chambers's Techn. Dict.* 73/2 *Balloon framing*, a cheap and rapid method of construction in which all timbers are of light scantling, and are held together entirely by nails and spikes, only the corner posts being tenoned: used in place of braced framing. **1940** A. Hocking *Wicked Flee* x. 233 Austen slowly turned his balloon glass of cognac between the palms of his hands. **1951** 'J. Wyndham' *Day of Triffids* v. 101 The plutocratic-looking balloon with the puddle of unpriceable brandy was mine. **1931** R. Aldington *Colonel's Daughter* III. 166 A large balloon goblet of very thin glass with a shallow gold deposit of brandy still in it. **1803** *Lett. Miss Riversdale* III. 202 She kept..running her balloon hat into every eye. **1870** L. Russell *Let.* 7 Nov. in *Amberley Papers* (1937) II. xvi. 454, I have had three balloon letters quite lately [from her parents in Paris]. **1816** G. Cayley in *Phil. Mag.* XLVII. 328 Balloon navigation does hold out the capabilities I have so daringly ventured to investigate. **1870** *Food Jrnl.* 1 Nov. 539 By balloon post, Oct. 19. **1885** *Encycl. Brit.* XIX. 581/2 The heroism displayed by French balloon postmen. **1899** *Cent. Dict.*, Balloon sail. **1948** R. de Kerchove *Internat. Maritime Dict.* 34/2 *Balloon sail*, a general term used for light racing and cruising sails found on yachts, such as balloon jib, balloon topsail and foresail. **1961** F. H. Burgess *Dict. Sailing* 19 *Balloon sails*, extra large sails of light material used as large jibs or spinnakers in yacht racing. **1960** *Aeroplane* XCIX. 270/3 The minute pressure of sunlight is forcing the 80-lb., 100-ft.-dia. balloon-satellite Echo steadily towards the Earth's atmosphere and ultimate destruction. **1962** *Ibid.* CII. 58/2 After being launched successfully from the Thor booster on Jan. 15, the first of the rigidized balloon-satellites on sub-orbital test ..began to inflate according to programme but disrupted before reaching its full diameter of 135 ft. **1964** *Yearbook Astr. 1965* 141 Transmissions from the United States were being sent by conventional means to Jodrell Bank for reflection to Gorky via the polar-orbiting balloon-satellite. *c* **1900** in M. Johnson *Amer. Advertising* (*a* 1960), Long sleeves, balloon shaped bottom with wide lace trimming. **1936** T. Rohan *Conf. of Dealer* (ed. 4) 8 A large 18th century English balloon-shaped bracket clock. **1907** *Westm. Gaz.* 11 Sept. 8/3 The great doors of the balloon-shed were slowly opened. **1940** G. W. Martin *Modern Camping Guide* v. 69 Terms used in connection with tent materials... *Balloon Silk* ..a fine cloth made of Sea Island or Egyptian cotton. The term is misleading, as no silk is used in its manufacture. **1940** Hemingway *For whom Bell Tolls* viii. 74 The worn, spotted green balloon silk outer covering of the five-year-old down robe. **1837** *Southern Lit. Messenger* III. 3 Women come to the spring for water in great balloon sleeves and prunella shoes. **1857** Geo. Eliot *Amos Barton* iii. in *Blackw. Mag.* Jan. 17/1 Very stiff balloon sleeves..without which a woman's dress was nought in those days. **1940** *Ann. Reg. 1939* 23 The balloon squadrons in London were now practically up to establishment. **1924** *Motor* 27 May 715/1 The low-pressure or balloon tyre manufacturers in the United States have adopted the straight-sided type of rim exclusively. **1933** *Jrnl. R. Aeronaut. Soc.* XXXVII. 795 Two types of aeroplane balloon tyres are shown. **1895** *Daily News* 1 June 3/1 The Princess Maud..here mounted on a balloon-tired 'safety'. **1836** Mrs. Lincoln *Botany* (1837) App. 84 Balloon vine, East Indies. **1901** C. T. Mohr *Plant Life Alabama*, Balloon Vine..Louisianian area, South Carolina, [etc.].

balloon, *v.* Add: **3. b.** *trans.* To puff out or cause to be inflated like a balloon; *spec.* to distend with air, gas, or water, as the abdomen in tympanites, or the rectum or vagina with specially constructed apparatus.

1889 [see *ballooning *vbl. sb.* 2]. **1906** *Macm. Mag.* Dec. 119 The wind..ballooned his cassock and carried his hat into the ditch. **1909** *Practitioner* Dec. 807 Several equal lengths of rubber tubing..were obtained, and at about the centre of some of them aneurysms..were ballooned.

4. To hit (a cricket-ball) or kick (a football) high in the air. *colloq.*

1904 *Daily Chron.* 20 July 7/2 As he did not quite get to the ball, he ballooned it to Garnett in the out-field, who brought off a well-judged catch. **1927** *Daily Tel.* 8 Feb. 16/3 As for the half-backs, they lacked method; it was not a light or flighty ball, but they were for ever ballooning it.

5. *intr.* Of an aeroplane: to rise up in the air, esp. as the result of a hard bounce on landing.

1931 P. W. F. Mills *Angles on Pract. Flying* vi. 54 Certain types of aeroplane..retain [in landing] a degree of buoyancy sufficient to cause an uncomfortable tendency to 'balloon'. **1949** J. R. Cole *It was so Late* 87 The aircraft ballooned when the wheels hit; it shot up thirty feet and seemed to hang suspended.

ballooned (bălū·nd), *ppl. a.* [f. Balloon *v.* + -ed¹.] Swollen or puffed out like a balloon.

1889 in *Cent. Dict.* **1925** W. Deeping *Sorrell & Son* ii. 20 Sorrell saw the labouring of the ballooned waistcoat. **1930** R. Campbell *Adamastor* 80 Their breasts ballooned with lust and song, The fat sopranos kick.

ballooner. 1. (Earlier example.)

1783 *Morning Herald* 15 Dec. 2/3 The sagacious King of Spain..will out-balloon all former ballooners.

balloonet (bălū·net). Also **-ette.** = *ballonet.

1902 *Westm. Gaz.* 11 Apr. 8/1 Within the main balloon there is to be a smaller balloon. This can be filled with air. If, therefore, the balloon becomes slack, it can immediately be made taut by pumping air into the balloonet. **1908** H. G. Wells *War in Air* iii. 93 A long internal balloonette of oiled and toughened silk canvas.

ballooning, *vbl. sb.* Add: **1.** (Earlier example.) Also *attrib.*

1784 H. Smeathman *Let.* 16 July in Pettigrew *Mem. of Lettsom* (1817) II. 275 This I thought might have been done by ballooning. **1961** *Daily Tel.* 1 May 13/1 President of the ballooning club in Holland.

b. *Aeronaut.* (See *balloon *v.* 5.)

1922 *Flight* 1 June 317/2 Some serious porpoising or 'ballooning'..may occur. **1935** P. W. F. Mills *Elem. Pract. Flying* vii. 103 The underlying cause of ballooning is usually too fast a gliding speed.

2. Dilatation of the walls of a cavity of the body as a symptom or for therapeutic purposes.

1889 T. Bryant in *Lancet* 5 Jan. 8/1 On the diagnostic value of 'ballooning of the rectum' in cases of stricture of the bowel...The surgeon will often find..that he has entered a cavity, the walls of which are expanded or 'ballooned'... The extent of ballooning will be found to vary in every case. **1890** Billings *Med. Dict.*, *Ballooning, vaginal*, distension of vagina as by tampons, water- or air-bags, etc. **1893** A. S. Eccles *Sciatica* 3 In the remaining nine cases there was more or less ballooning of the rectum.

b. *Path.* Distension of cells, etc. (see quot. 1913).

1913 E. M. Brockbank in *Med. Chron.* Sept. 292 When the acidity of the gastric juice reaches a certain degree..the red corpuscles are seen to distend in a most peculiar manner which for descriptive convenience I describe as 'ballooning'. **1962** *Lancet* 28 Apr. 885/2 Severe fatty change of the fine droplet type, without 'ballooning' of the cells.

3. *Spinning.* (See quots.)

1904 Goodchild & Tweney *Technol. & Sci. Dict.* 39/1 *Ballooning* (Cotton Spinning), a defect in ring spinning caused by the high velocity of the ring traveller. This has the effect of causing the spun thread to fly outwards as it winds round the bobbin. **1924** T. Lawson *Woollen Yarn Production* p. ix, *Ballooning*, extension of the arc of the axis of the spinning thread. **1963** A. J. Hall *Textile Sci.* v. 261 (*caption*) Showing how the accumulation of static electricity on a thread..can cause the individual fibres of which the thread is composed to repel each other and so cause 'ballooning' sufficient to make the manipulation of the thread difficult.

balloonist. Add: (Earlier example.) Also *fig.*

1784 *Morning Herald* 10 Aug. 3/1 Mr. Smeathman, an active and distinguished Balloonist. **1906** *Daily Chron.* 28 Aug. 4/4 If you are a balloonist in the social world, you may find a difficulty with the soup.

balloony (bălū·ni), *a.* [f. Balloon *sb.*¹ + -y¹.] Resembling a balloon or balloons.

1908 *Daily Chron.* 9 Oct. 7/1 Great balloony skirts called crinolines. **1936** 'A. Bridge' *Song in House* 76 The long balloony line of the lime avenue.

ballot, *sb.*¹ Add: **4. ballot-rigging** [Rig *v.*⁶ 2], the fraudulent manipulation of a ballot.

1959 *Daily Tel.* 1 Dec. 1/8 Ballot rigging in the union. **1961** *Times* 2 June 10/4 Newspaper charges of ballot-rigging.

ballot, *v.*¹ Add: **5.** *trans.* To procure the vote by ballot of (a body of voters) on a specific motion.

1898 *Daily News* 25 June 6/6 The Llwynpia colliers to-day rejected a motion to ballot the men for or against a sliding scale. **1899** *Ibid.* 9 Jan. 7/3 If the textile unions are to be balloted on the question of raising the age.

‖**ballotté** (balote). *Ballet.* [F., f. pa. pple. of *ballotter* to toss or shake about.] (See quot. 1957.)

1894 E. Scott *Dancing* vi. 60 *Ballotté*, 'throwing here and there step'. A movement in which the feet are crossed alternately one before or behind the other. **1913** C. d'Albert *Encycl. Dancing* 5 *Ballotté* or *Ballottés*, to toss about. When the feet are crossed alternately one before or behind the other. **1957** G. B. L. Wilson *Dict. Ballet* 46 *Ballotté*, lit. a tossing movement. A movement in which the dancer raises the left foot behind the knee of the supporting leg, and, with a slight plié, springs upwards, crossing one foot in front of the other at the cou-de-pied, alighting on the left foot, whilst extending the right in a developpé to the front.. then bringing it in to the knee.

ball-room (bǭ·lrūm). [Ball *sb.*² 4.] A room designed or suitable for dancing. Also *attrib.*, esp. in *ball-room dancing*, social dancing in a ball-room as a recreation.

1736 Fog *Jrnl.* 30 Nov. in *London Mag.* (1737) Apr. 190/2 The Conspirators..were first to blow up the Ball-Room. **1752, 1875** [see Ball *sb.*² 4]. **1894** E. Scott *Dancing* iv. 24 Not..half the people who professed to teach ball-room dancing really knew the difference. **1911** Kipling *Big Steamers* in Fletcher & Kipling *Hist. England* xii. 236 Oh, the Channel's as bright as a ball-room already. **1923** Wodehouse *Adv. Sally* ii. 54 The only thing I could do.. was ball-room dancing, so I ball-room danced. **1929** *Melody Maker* Jan. 10/1, I was given the opportunity of seeing the New Playhouse ballroom opened by the Green Brothers. *Ibid.* 10/3 Ballroom dancing has been slumping steadily.

bally (bæ·li), *a.* and *adv. slang.* A euphemism for *bloody* (see Bloody *a.* 10), used as a vague intensive of general application; 'jolly', 'confounded'. Cf. *absoballylutely* (s.v. *absolutely *adv.*).

[**1847** Thackeray in *Fraser's Mag.* Jan. 126/2 Ha! What have we here? *M. A. Titmarsh's Christmas-Book—Mrs. Perkins's Ball.* Dedicated to the Mulligan of Ballymulligan. Ballymulligan! Ballyfiddlestick!] **1885** *Sporting Times* 11 Apr. 1/4 Too bad, too bad! after getting fourteen days or forty bob, the bally rag don't even mention it. *Ibid.* 3/4 Has that bally Ptolemy won, d'ye know? **1887** S. Butler in H. F. Jones *Memoir* (1919) II. xxvi. 54 No one in those days gave him or herself any bally airs about it. **1898** Steevens *With Kitchener to Khartum* 112 I've been in this bally country five years. **1919** C. Orr *Glorious Thing* v. 56, I..talked gaily about the bally old war. **1922** H. Walpole *Cathedral* I. vi. 103 All the time behind you and them some force was insisting on places being taken, connections being formed. One was simply a bally pawn..a bally pawn. **1939** G. B. Shaw *Geneva* iv. 54 The staggering, paralyzing, jolly bally breath-bereaving point..is that the dictators have been summoned.

ballyhoo (bælihū·), *sb.* orig. *U.S.* [Etym. unknown.] A barker's touting speech; hence, blarney, bombastic nonsense; extravagant advertisement of any kind.

1901 *World's Work* Aug. 1100/2 First there is the ballyhoo —any sort of a performance outside the show, from the coon songs of the pickaninnies in front of the Old Plantation, to the tinkling tamborines of the dancers on the stage of 'Around the World'. **1914** Jackson & Hellyer *Vocab. Criminal Slang* 16 *Bally hoo*, noun. Current amongst exhibition and 'flat-joint' grafters. A free entertainment used for a decoy to attract customers. **1914** *Philad. Even. Post* 9 May, A live, little park full of side show tents..with..barkers spieling before the entrances and all the ballyhoos going at full blast. **1925** H. L. Foster *Trop. Tramp Tourists* 36 Above all, don't let them use a megaphone. It's too much like a ballyhoo. **1927** *Daily Express* 21 Sept. 1/2 Mr. Wiener, chairman of the Pennsylvania State Athletic Commission.. calls Dempsey's letter 'mere ballyhoo'. **1928** *Ibid.* 3 Mar. 9/3 Mr. McAndrew characterised Mr. Thompson's charges as 'lies and ballyhoo'. **1932** D. B. Wyndham Lewis *Emperor of West* i. 9 Mr. Christopher Hollis has dealt fittingly with some of this popular pro-Elizabethan rhetoric or ballyhoo.

Hence **ballyhoo·** *v. trans.*, to cajole by extravagant advertisement or praise (after the manner of a barker); to advertise or praise extravagantly. Also **ballyhoo·er, ballyhoo·ist,** one indulging in ballyhooing.

1901 *World's Work* Aug. 1100/2 Last of the professions on the Midway are those of the 'barker', 'ballyhooer' and 'spieler'. **1922** *Collier's* 4 Mar. 7/2, I don't like to ballyhoo myself..but here's a picture which will make you..bite your nails. **1927** *Scots Observer* 28 May, Our people will not be bullied and bally-hoed into churchgoing or anything else. **1928** *Weekly Dispatch* 6 May 15/2 How the late P. T. Barnum would have enjoyed ballyhoo-ing this new Drury Lane spectacle! **1941** *Manch. Guardian* 18 Apr. 8/5 War Weapons Week not Ballyhoo...'That, I think,' he said, 'is the complete answer to what I term the ballyhooists.' **1948** *Atlantic Monthly* Mar. 24/1 They are ballyhooed, pushed, yelled, screamed, and in every way propagandized into the consciousness of the voters. **1950** 'S. Ransome' *Deadly Miss Ashley* ii. 17 The fortunes paid to ballyhooers of phony antiseptics. **1966** *Economist* 20 Aug. 748/1 British Rail is ballyhooing the pleasures of its electrified services a bit loudly.

ballyhoo of blazes. *Nautical slang.* Also **ballyhoo.** [Etym. of first element uncertain, but perh. same word as *ballahou (see *Amer. Speech* (1945) XX. 184 ff.).] Sailors' term of contempt for a vessel which they dislike for any reason.

1836 *Knickerbocker* Aug. 203 Jack Marlinspike..couldn't get a situation afore the mast of a Ballyhoo coasting-brig. **1847** H. Melville *Omoo* lxxvi. 295 Steer clear of the likes of this ballyhoo of blazes as long as ye live. **1897** Kipling *Capt. Cour.* iii. 69 Tom Platt, this bally-hoo's not the *Ohio*. *Ibid.* ix. 205 Oh, ef it had bin even the Fish C'mmission boat instid o' this bally-hoo o' blazes. **1929** F. C. Bowen *Sea Slang* 6 *Ballyhoo of blazes*..the last word of contempt for a slovenly ship.

ballyrag, ballyragging, etc., varr. Bullyrag, -ragging (in Dict. and Suppl.).

balm, *sb.* **9. b.** *balm-wine* (earlier example).

a **1811** Marjory Fleming *Jrnl.* (1934) II. 54 Ravelston is a fine pla[ce] because I got balm win[e] and many other dain[ties].

‖**bal masqué** (bal maske). [F.] A masked ball (see Masked *ppl. a.*² 1 b).

1768 Earl of Carlisle *Let.* 31 May in J. H. Jesse *G. Selwyn & Contemp.* (1843) II. 303, I am going to a great dinner to-day..and after that to a *bal masqué* at court. **1775** H. Walpole *Let.* 23 Aug. (1904) IX. 239 On Friday he gives a *bal masqué* to the universe. **1817** H. C. Campbell *Journey to Florence* 4 Oct. (1951) 101 Afterwards there was a Bal masqué at the Pergola. **1902** G. K. Chesterton *12 Types* 4 A disguise as tawdry and deceptive as the costume of a 'bal masqué'.

balmily, *adv.* (Earlier example.)

1819 J. H. Wiffen *Aspley Wood* II. lxxii in *Aonian Hours* 107 How balmily the breeze Breathes from the sky-aspiring larch!

Balmoral. b, c. (Earlier examples.)

1857 J. E. Ritchie *Night Side London* 160 Smith with his Balmoral boots, Brown with his all-round collar. **1859** H. J. Byron *Nymph of Lurleyburg* iv, Whose morals are not tight laced—oh dear, no—Though their *bal-morals* are extremely so. **1859** *Habits of Good Society* iv. 177 Victoria has assumed the Balmoral petticoat. *Ibid.* 178 She has courageously accompanied it with the Balmoral boot.

‖**bal musette** (bal müzet). [F.] In France, a popular dance-hall (with an accordion band). Also *attrib.*

1926 Hemingway *Fiesta* (1927) I. iii. 25 The dancing-club was a *bal musette*. **1934** H. Miller *Tropic of Cancer* (1961) 295 They were urging me to accompany them to a *bal musette*. **1934** C. Lambert *Music Ho!* II. 94 The bal-musette sentimentality of the valses of Auric. **1959** 'F. Newton' *Jazz Scene* xiii. 235 Such native European forms of light entertainment as..accordion and *bal musette* music.

balmy (bā·mi), *a.* Add: **7.** 'Soft', weak-minded, idiotic. Also as *sb.* (see quot. 1903). *slang.* See also *barmy *a.*

1851 Mayhew *Lond. Labour* I. 217/2 (Street-patterers' slang) *Balmy*, insane. **1859** Hotten *Slang Dict.*, *Balmy*, insane. **1891** Farmer *Slang* II. 224/1 Balmy in one's crumpet. **1892** *Daily News* 17 Nov. 6/6 Regarding the old 'balmy' criminals, they are poor creatures, far more to be pitied than condemned. **1903** Ld. W. Nevill *Penal Servitude* 150 These are officially classed as 'W. M.'—that is, weak-minded—but are invariably known colloquially as 'balmies'. *Ibid.* 151 A man who appears to be playing 'balmy'. **1912** Masefield *Dauber* II. in *English Rev.* Oct. 350 Painting's a balmy's job [*ed.* 1913, p. 21 a balmy job] not worth a nail. **1922** 'R. Crompton' *Just—William* xi. § 1. 206 'I s'pose you're balmy on her,' he said resignedly. **1929** J. B. Priestley *Good Companions* III. i. 460 People here must have gone balmy.

8. *absol.* Sleep. *slang.* (Cf. sense 4 in Dict.)

1840 Dickens *Old C. Shop* viii, As it's rather late, I'll try and get a wink or two of the balmy.

balneary. Add: **B.** *adj.* Of or pertaining to the bath or bathing.

1883 H. James *Portr. Places* vii. 142 The French do not treat their beaches as we do ours—as..places animated simply during the balneary hours. **1924** *Weekly Westm.* 29 Aug. 530/3 In fantasy I view and loathe each balneary station—I have been down at Pebbleton-on-Sea.

balneologist (bælni,ǫ·lŏdʒist). [f. L. *balneum* bath + -ologist.] An expert in balneology.

1872 *Lancet* 20 Apr. 532/1 At Aix is to be found the most learned of living balneologists, Dr. Lerschet. **1902** *Times* 22 Jan. 2/4 Applications are invited for the appointment of balneologist under the New Zealand Government in connexion with the Thermal Springs of the Colony. **1921** *Times Lit. Suppl.* 6 Oct. 646/3 Dr. Herbert..in 1902 was appointed as Government balneologist to advise on the development of the health resorts of New Zealand.

ba:lneothe·rapy. Add examples. Hence **ba:lneotherapeu·tic** *a.*

1881 J. N. Hyde in *von Ziemssen's Cycl. Med.* Suppl. 184 Balneotherapy, when there was no mercury used, has proved of no greater worth. **1906** *Practitioner* Dec. 764 The initiation of balneo-therapeutic measures. **1907** H. & F. P. Weber (*title*) Climatotherapy and Balneotherapy.

baloney, boloney (bălōᵘ·ni, bŏ-), *sb.* and *int. slang* (orig. *U.S.*). [Commonly regarded as f. Bologna (*sausage*) but the connection remains conjectural.] Humbug; nonsense.

1928 *Sat. Even. Post* 28 Nov. 21 Gee, that's a long shot. Boloney! That's not the ball—it's the divot. **1935** *Discovery* Dec. 378/2 He even suggests that much of modern psychiatry is 'hooey' and 'baloney'. **1935** E. Weekley *Something about Words* 64 *Boloney* must surely be for *Bologna* sausage (whence also the English *polony*, dating from the 18th century), influenced perhaps by the contemptuous sense associated with the German *wurst*. **1958** *Times* 15 Nov. 5/3 That nick-name was absolute baloney. **1959** J. Braine *Vodi* vi. 93 All that baloney about going upstairs to play a harp or downstairs to roast. **1959** *Daily Mail* 15 Jan. 1/5 This is the official baloney that her family life would suffer if she were to live on a Commonwealth scale.

balsa. Also with pronunc. (bǭ·lsă). Add to def.: Also, *balsa raft.* (As date of first example, for 1778 read 1777.)

1915 'Bartimeus' *Tall Ship* i. 15 The Commander..was standing on the balsa raft.

2. A bombaceous tree of tropical America, *Ochroma lagopus*; also, the wood of this tree, used for its extreme lightness. Also *balsa tree, wood.* Cf. Corkwood 2.

1866 Lindley & Moore *Treas. Bot.* s.v. *Ochroma*, The very buoyant rafts or balsas, the unsinkable properties of which caused such surprise among the discoverers of America, are likewise made of it, whence the tree is called Balsa in some parts of America. **1917** *Sci. Amer.* Nov. 345/3 A new wood, apparently little known and called balsa wood, is exceedingly light. **1920** *Flight* XII. 1301/2 Balsa wood..is so extremely light that it can be recommended for making aerofoils. **1924** *Contemp. Rev.* Jan. 95 The balsa tree, a wood which, from its extreme lightness, is specially suitable in the construction of aeroplanes. **1933** *Discovery* June 194/1 Advance scouts had drifted on a balsa-wood raft. **1951** *Oxf. Jun. Encycl.* VII. 223/2 Balsa is only half the weight of cork, and is the lightest and softest of commercial woods.

Balt (bǭlt), *sb.* and *a.* [ad. late L. (Jordanes) *Balthæ.*] **A.** *sb.* A native or inhabitant of one of the Baltic states of Lithuania, Latvia and Estonia; *spec.* a German inhabitant of any of these states (see quot. 1937). **B.** *adj.* Of or pertaining to (a person from one of) these states.

1878 R. G. Latham in W. Smith *Dict. Gr. & Rom. Geogr.* 375/2 For a *Balt*, or an *Amal*, as real personages, we look in vain. Populations, however, to which they were *Eponymi*, we find in the two localities Baltia [Baltic] and Abalus. **1909** Webster, *Balts*, Lithuanians, who, together with the Letts, were probably the Æstii of Tacitus. **1922** E. J. Harrison *Lithuania* iii. 37 The Lithuanians together with the Letts and Old Prussians form a family of Aestians or Balts. *Ibid.* iii. 39 To-day the ancient Balts are represented only by the Lithuanians and the Letts or Latvians. **1937** *Discovery* Aug. 234/2 In 1918 the German baronial families were deprived of their estates...As late as 1918 the Baltic barons or 'balts' as they are always called, were planning to colonise Estonia with two and a half million small farmers from Germany. **1939** *Times* 1 Nov. 7/1 The exchange of documents ratifying the repatriation of Balts from Latvia is expected to take place in Berlin to-morrow. **1954** J. B. Priestley *Magicians* iii. 60 Wayland, it appeared, was part English, part Balt.

balter, *v.* Add: **1.** (Isolated later example of *baltering.*)

1952 Auden *Nones* 39 The baltering torrent Shrunk to a soodling thread.

Balthazar (bælþæ·zăɹ). Also **Balthasar, Belshazzar.** [So called in allusion to *Balthazar* (Belshazzar), 'king of Babylon', who 'made a great feast..and drank wine before the thousand' (Daniel v. 1).] A very large wine-bottle (see quots.).

1935 A. L. Simon *Dict. Wine* 22 *Balthasar*, a fancy name for a fancy bottle to hold some 16 ordinary bottles, or 12·80 litres, equal to about 2·75 gallons. **1959** *Gloss. Packaging Terms* (*B.S.I.*) 28 *Belshazzar*, a wine bottle—capacity 16 reputed quarts. **1962** *Times* 25 May 13/4 Methuselah.., Salmanasar, Balthasar, Nebuchadnezzar—'usually applied respectively to eight, twelve, sixteen, and twenty bottle containers'.

Baltic (bǭ·ltik), *a.* and *sb.* [f. med.L. *Balticus*; cf. *Balt.] **A.** *adj.* **1.** Of, pertaining to, designating or bordering upon an almost landlocked sea in N. Europe (Russ. *Baltiiskoe Móre*), called by the neighbouring Germanic countries 'East Sea' (G. *Ostsee*, etc.); *spec.* of or belonging to the states of Lithuania, Latvia, and Estonia and their inhabitants.

c **1590** A. Ashley (*title*) The Second Part of the Mariners Mirrour..with all the sounds of Denmark, & the Baltick sea. **1608** Topsell *Serpents* 236 There be also in the Sueuian-Ocean or Balthicke-sea, Serpents of thirty or forty foote in length. **1845** *Ainsworth's Mag.* VIII. 161 The failure of a long established Baltic house at Kingston upon Hull. **1920** *19th Cent.* Mar. 536 A policy aimed at the Balkanisation of the Baltic provinces. **1937** [see *Balt *sb.* and *a.*].

2. Applied to a branch of the Indo-European languages comprising Lithuanian, Lettish, and Old Prussian, usu. classified with the Slavic group (see *Balto-).

1887 Skeat *Princ. Eng. Etym.* vii. 102 Of the *Lettic* or *Baltic* group, the most interesting is the Lithuanian, spoken in parts of Eastern Prussia. **1888** [see Lettish]. **1891** A. L. Mayhew *O. E. Phonol.* p. xii, Baltic-Slavonic, including Old Prussian, Lithuanian, Lettish, and Old Bulgarian.

3. In specific combinations with sbs., as *Baltic pine* (see Pine *sb.*² 2); *Baltic shield*, the Archæan platform of Finland and E. Scandinavia.

1866 Baltic pine [see Pine *sb.*² 2]. **1891** W. Schlich *Man. Forestry* II. iv. 290 The timber of the Spruce..is known in Britain as white Baltic pine. **1906** B. C. Sollas tr. *Suess's Face of Earth* II. III. ii. 76 East of the glint lies the Archaean table-land of the gulf of Bothnia; that is, the Baltic shield.

B. *sb.* The Baltic Sea; the lands bordering upon it.

1720 *Phil. Trans. Abr.* VI. 498 Observations on the variations of the needle in the Baltic. **1776** Gibbon *Decl. & F.* (1782) I. ix. 272 Some tribes..on the coast of the Baltic, acknowledged the authority of kings, though without relinquishing the rights of men. **1935** Huxley & Haddon *We Europeans* vii. 237 The 'Windmill Hill'..type of pottery.. was brought from the Baltic by long-headed people who buried their dead in long barrows.

Baltimore. (Earlier examples.)

1669 N. Shrigley *True Rel. Virginia & Mary-land* 4 Fowle naturally to the Land are..Redbirds, the Baltenore [*sic*] bird, being black and yellow, [etc.]. **1676** J. Speed *Theatre of Empire* II. 44 That which is black and yellow is called the *Baltemore*-bird, from the colours of his Lordships Coat of Arms.

Balto- (bǭ·lto), comb. form of *Baltic*, as in *Balto-Lithuanian*; **Ba:lto-Sla·vic, -Slavo·nic** *a.* and *sb.*, the designation of the group of Indo-European languages which comprises the Baltic branch and the Slavonic branch (Russian, Polish, Czech, Serbian, Bulgarian, etc.).

1896 O. F. Emerson *Brief Hist. Eng. Lang.* i. 8 The Balto-Slavic branch consists of two divisions. **1909** Webster, Balto-Slavonic, adj. **1910** *Encycl. Brit.* XIV. 495/1 North of the Black Sea..comes the great Balto-Slavonic group. *Ibid.* XVI. 246/1 The words *mentiri*, *rōs*, *ignis* have close equivalents in Balto-Slavonic. **1922** E. J. Harrison *Lithuania* ii. 32 In 1654 the Old Prussians..renounced their Balto-Lithuanian idiom in favour of German. **1948** D. Diringer *Alphabet* II. v. 347 The new language may hold an intermediate position between..Balto-Slavonic and Greek.

Baluch (bălū·tʃ), **Baluchi** (bălū·tʃi), *sb.* and *a.* Also 7 **Bolloch, Balloch, Buluch, Boloch,** 8 **Ballowch,** 9 **Belooch(e, -ee, Bloach, Bilochee, Baloochi,** etc. [Pers. *Balūchi.*] **A.** *sb.* **1.** An inhabitant or native of Baluchistan, a region lying between the lower Indus and south-east Persia. **2.** The Iranian language of Baluchistan. **B.** *adj.* Of or pertaining to this people or their language.

1616 T. Roe *Let.* 27 Nov. (1899) II. 353 Subiect to the robberyes of the Balooches. **1617** Purchas *Pilgrimage* (ed. 3) 550 The Persian Gulfe hath beene awed, and the Arabike or Red Sea tamed, in requitall of the Turkes and Buluches Trecheries. **1619** in W. Foster *Eng. Factories India* (1906) 84 [They blame the] 'Bolloches' [for the disaster]. **1727** A. Hamilton *New Acc. E. Indies* I. 107 They were lodged in a Caravanseray, where the Ballowches came with about 300 to attack them. **1816** H. Pottinger *Trav. Beloochistan* 53 The Belooches, who form..the whole of the population throughout Beloochistan, are a people whose origin is so obscure [etc.]. **1848** L. R. Stacy *Narr. Beloochistan & Affghanistan* 72 The Brahooe and Belooche tribes. **1875** R. Caldwell *Languages of India* 4 The languages spoken on the further side of the Indian frontier, such as Belūchi, on the north-west. **1875** *Encycl. Brit.* III. 300/1 (*Baluchistan*) The country derives its name from the Baluches, but the Brahoes are considered the dominant race. **1882** Floyer *Unexpl. Baluchistan* 158 One tribe of Balūchis will 'chapao', or plunder, another tribe directly they get a chance. **1885** *Encycl. Brit.* XVIII. 655/2 Baluch, the language of Baluchistan..very closely akin to New Persian. **1886** *Ibid.* XX. 109/2 Baluchi is spoken on the same frontier, farther south, adjacent to Baluchistan. **1920** *Blackw. Mag.* Aug. 185/2 He spoke only Baluchi and indifferent Persian. **1925** *Ibid.* Apr. 490/2 He knew exactly the line of conduct which would most effectively appeal to the Baluch nature.

baluchitherium (bălūtʃiþĭᵊ·riŏm). *Palæont.* [mod.L., f. prec. + Gr. θηρίον beast.] A gigantic extinct land mammal of which remains have been found in the upper Oligocene deposits of the Bugti hills of Baluchistan.

1913 C. Forster-Cooper in *Ann. & Mag. Nat. Hist.* XII. 504, I learn that the generic name *Thaumastotherium* proposed..is preoccupied. The generic name *Baluchitherium* is therefore proposed as a substitute. **1925** *Chambers's Jrnl.* 217/2 It was a baluchitherium, the greatest of all varieties of rhinoceros. It stood thirteen feet at the shoulder, and its neck, relatively as long as a horse's, gave it an additional reach of four or five feet. **1966** C. A. W. Guggisberg *S.O.S. Rhino* ii. 26 With a length of twenty-eight feet and standing seventeen to eighteen feet at the shoulder, Baluchitherium equalled or exceeded the imperial mammoth.

baluster. Add: **6.** *attrib.* and *Comb.*: *baluster column* or *shaft* (see quot. 1853); *baluster-rail.* Esp. in sense 'having the shape of a baluster', as *baluster handle, jug, pitcher, stem, vase*; so *baluster-shaped* adj.

1844 Baluster shaft [see Baluster 1 b]. **1853** *Archit. Publ. Soc. Dict.*, *Baluster column*, the name given to a pillar used in the so-called Saxon architecture of England for a divisional support in windows. **1871** Franks *Catal. Slade Coll. Glass* 120 Goblet, with a..baluster stem. **1878** W. J. Cripps *O. Eng. Plate* x. 292 The foot is much like those of earlier cups, but the stem is different, being formed as acanthus or other leaves, the upper part of it baluster-shaped. **1906** *Daily Chron.* 28 Sept. 7/5 Going downstairs, his brother wrenched a baluster-rail out of the staircase. **1912** *Eng. Hist. Rev.* Oct. 832 A rather early baluster-stem cup. **1933** *Burlington Mag.* July 36/2 A baluster vase.. illustrates in its lotus design the type with ornament painted in outline. **1938** *Oxoniensia* III. 11 The so-called 'baluster' pitchers from London, Oxford or York. **1939** *Ibid.* IV. 122 The baluster jug from Well 9..supplies a clue to the upper limit of the group. **1956** G. Taylor *Silver* vii. 149 Saucepans..with everted lips, and a turned wooden baluster handle.

Balzacian (bælzæ·kiăn), *a.* and *sb.* [f. the name of the French novelist H. de *Balzac* (1799–1850) + -ian.] **A.** *adj.* Of or pertaining to or characteristic of Balzac or his style.

1874 Geo. Eliot *Let.* 16 June (1956) VI. 57 The Balzacian view of Paris. **1892** W. James *Let.* 11 Apr. (1920) I. 318 He is a real Balzackian figure—a regular porker, coarse, vulgar, vain, cunning, mendacious. **1927** *Observer* 1 May 6 There might seem a Balzacian pathos about it. **1958** *Spectator* 4 July 33/1 Balzacian characters set against a Boudin coast-scape.

B. *sb.* An authority on or student of the writings of Balzac.

1905 *Daily Chron.* 10 May 3/3 Mr. Helm is a discriminating Balzacian. **1923** A. Huxley *On Margin* xviii. 131 Balzacians will remember the advertisements composed by Finot and the Illustrious Gaudissard.

balzarine. (Earlier and later examples.)

1849 *The Trelawny* (Jamaica) 24 Apr. 1/2 Muslin De' Laines and Balzareens. **1854** *Chambers's Jrnl.* II. 50 Four matrons in balzarine. **1927** E. Sitwell *Rustic Elegies* 78 The Amazons wear balzarine of jonquille.

bam, *sb.* (Earlier example.)

1728 *Life & Char. Harvey the Conjuror* 21 He called the Profession of a Doctorship, in Physic, a *Bamm*, upon the World, which is a *Bite*, in modern Language.

bam (bæm), *int.* [Echoic.] An interjection imitating the sound of a hard blow.

1922 Joyce *Ulysses* 109 Bam! expires. Gone at last. **1962** *Spectator* 25 May 674/2 Think of the United States as a 3,000-mile-broad comic strip where significant occasions go bam, pop and zowie.

bambocciade. Add: Also **bambochade.**

1816 M. Bryan *Biog. & Crit. Dict. Painters* I. 317 He quitted historical subjects to paint bambochades.

bamboo, *sb.* Add: **1. b.** Cane-coloured porcelain biscuit, invented by Wedgwood. Also *bamboo ware.*

1787 Jos. Wedgwood in L. Jewitt *Wedgwoods* (1865) 310 *Bamboo*, or cane-coloured bisque porcelain, of the same nature with the porcelain No. 3 [*i.e.* a fine white biscuit ware]. **1865** L. Jewitt *Wedgwoods* 410 An open basket of bamboo. **1904** W. Burton *Hist. Eng. Earthenware* 162 The simple patterns of lines and foliage in blue and green enamel which he [*sc.* Elijah Mayer] used on his bamboo wares.

2. bamboo curtain [after *iron curtain; cf. Chick *sb.*²], a political and economic 'curtain' or barrier between territories under the control of the Communist régime in China and non-Communist countries; also *transf.*; **bamboo fern,** a fern, *Coniogramme japonica*, native to Japan; **bamboo-fish** *S. Afr.*, the sea-fish *Sarpa salpa* (family Sparidæ); also called (*striped*) *karanteen; **bamboo-grass,** bamboo, or a grass that resembles it; **bamboo joint,** an internode of bamboo (with its nodes) cut and fashioned into a vessel for carrying water, cooking rice, etc.; **bamboo palm** (see quot. 1891); **bamboo rat,** a rodent belonging to the genus *Rhizomys*, found in Malacca; **bamboo shoot,** a young, edible shoot of the bamboo; **bamboo ware** (see 1 b above).

1949 *Time* 14 Mar. 55 The Communist bosses of Peiping dropped a bamboo curtain, cutting off Peiping from the world. **1955** *Sc. Jrnl. Theol.* VIII. 422 'One Body in Christ' is written by a man who stands on the Reformed side of the theological bamboo curtain. **1957** *Listener* 6 June 903/2 We can't keep 400,000,000 people behind an economic bamboo curtain for ever. **1930** L. H. & E. Z. Bailey *Hortus* 170/2 Coniogramme..japonica (*Gymnogramma japonica*). Bamboo-Fern..Japan, Formosa. **1913** W. W. Thompson *Sea Fisheries Cape Col.* ii. 61 The pretty little bamboo-fish of the Cape is also known as stink-fish, and is the mooi nooitje of Hermanus and Struys Bay, the streepje of the Knysna and the silver karanteen of Natal. **1930** Bamboo-fish [see *karanteen]. **1909** *Cent. Dict.* Suppl., Bamboo-grass. **1920** *Blackw. Mag.* June 822/1 Coarse bamboo-grass began to fringe the banks. **1924** M. L. Milne *Home of Eastern Clan* iii. 47 Water is carried in large bamboo joints. These are pierced near the top, a string is passed through the holes, and the joints are hung on one end of a pole, which rests on the shoulders. **1866** Bamboo palm [see Palm *sb.*¹ 1 c]. **1891** *Kew Bulletin* Jan. 3 The 'Bamboo' palm, or *Raphia vinifera*, is perhaps the commonest tree in the swamps [of Lagos]. **1881** R. Hunter et al. *Encycl. Dict.*, *Bamboo-rat*, a rodent mammal belonging to Gray's genus Rhizomys. **1914** *Brit. Mus. Return* 135 An Ashy Bamboo Rat (*Rhizomys canescens*) from the Shan States. **1889** Kipling *From Sea to Sea* (1899) I. xi. 303 After the bamboo-shoots came..white beans in sweet sauce. **1940** A. Simon *Conc. Encycl. Gastron.* II. 55/2, 1 small tin bamboo shoots.

bamboo, *v.* Add: **2.** *trans.* To furnish with bamboo or bamboo laths.

1925 *Blackw. Mag.* Apr. 537/2 The roof was bambooed and ready to be thatched.

bamboos (bæmbū·s). *S. Afr.* Also **-ous.** [ad. Du. *bamboes* Bamboo.] A wooden vessel for milk, water, etc. Cf. *bamboo joint* s.v. *bamboo *sb.* 2.

1822 J. Campbell *Trav. S. Africa* I. iii. 46 They brought us three bamboosses...A bamboos is a deep wooden vessel ..cut out of a block of wood. **1842** R. Moffat *Missionary Labours S. Africa* x. 144 A kind housewife would hang a bamboos, or wooden vessel filled with milk..near my head. **1844** J. Backhouse *Narr. Visit Mauritius & S. Afr.* 565 Bambouses..are a sort of jars made of willow-wood.

bamboula (bæmbū·lă). [Creole Fr.] (See quot. 1938.)

1883 *Cent. Mag.* Nov. 45/2 In New Orleans..a minute's walk..will bring you to Congo square..where the negro slaves once held their bamboulas. *Ibid.*, Every Sunday afternoon the bamboula dancers were summoned to a wood-yard. **1938** *Oxf. Compan. Mus.* 73/1 *Bamboula*, (1) a primitive negro tambourine, formerly in use in Louisiana and still in use in the West Indies. (2) A dance to which this instrument is the accompaniment. **1956** M. Stearns *Story of Jazz* (1957) iii. 26 The Congo, as such, is no longer danced in New Orleans, but it is still danced in Haiti, along with the Bamboula.

ban, *v.* Add: **7.** *ban the bomb*: the slogan of those advocating nuclear disarmament, used (with hyphens) as attrib. phr.

1960 *News Chron.* 22 Apr. 6/3 The ban-the-bomb campaigners are well advanced with their arrangements to make a new challenge.

‖**ban** (bañ), *sb.*³ [Fr.] (See quot. 1900.)

1900 *Daily News* 14 Aug. 8/2 On the falling of the curtain, however, when a 'ban' was called for, the ice was broken. The 'ban' is the characteristic ovation of the French student. **1906** *Daily Chron.* 20 Aug. 6/5 At the end it was 'Vive' everybody, and a perfect salvo of 'triple bans' was given.

Banagher (bæ·năgəɹ). [Name of a town in Ireland, which is said to have become proverbial as a 'rotten borough'.] Phr. *to beat* (or *bang*) *Banagher*: to surpass everything. (Cf. Bang *v.* 6.)

1830 W. Carleton *Traits & Stories* I. 50 'O, by this and by that,' says he, 'but that bates Bannagher!' *a* **1845** Barham *Ingol. Leg.* (1847) Ser. III. 300 *House-warming*, Whose name..in the records of fame..beats Banagher. **1855** T. C. Haliburton *Nature & Human Nature* II. vii. 226 That would bang Banaghar. **1890** Farmer *Slang* s.v., *He beats Banaghan*, an Irish saying of one who tells wonderful stories. **1928** *Weekly Disp.* 24 June 2 'You beat Banagher Pat,' said Willie, admiringly, 'and Banagher beat the Divil.'

banak (bæ·năk). [Native name in Honduras.] A hardwood tree of the genus *Virola* grown esp. in central America; also its timber.

1921 C. Hummel *Rep. Forests Brit. Honduras* 59 Name of tree. Banak..Fairly large quantity in the south of the Colony..stems of large size, good shape and free of knots. **1948** A. L. Howard *Timbers of World* (ed. 3) 54 *Banak*, *Virola merendonia* Pittier..British Honduras and South America. A very light mahogany-coloured wood, with a straight, even, regular grain.

banal, *a.* **2.** (Earlier example.)

1840 *New Monthly Mag.* LIX. 458 All that her late companions can draw from her is the *banal* declaration, that she 'never knew what happiness was before'.

banally, *adv.* [f. Banal *a.* 2 + -ly².] In a banal manner.

1934 in Webster. **1944** *Mind* LIII. 151 If we say, somewhat banally, that it leads from arbitrary taboo and tribalism to the reasoned pursuit of what is genuinely best for everyone, [etc.].

banana. Add: **2. b.** The yellow colour of a ripe banana. Also *banan* (= F. *banane*).

1923 *Daily Mail* 3 Apr. 10 In shades of..Banana and Cream. *Ibid.* 7 May 1 In Pale Jade, Banan, Tuscan.

3. *banana-leaf* (earlier example), *-skin*, also fig.; **banana bird,** also, a name applied to certain South American and West Indian species now included in the genus *Icterus*; also = *banana quit*; **banana boat,** a boat carrying bananas; also *Mil. slang* (see quot. 1945); **banana flour,** a flour made from dried bananas; **banana fly** (see quot.); **banana liquid** (see quot.); **banana-oil,** (*a*) = *banana liquid*; (*b*) *slang*, nonsense; insincere or insane talk or behaviour; cf. **apple-sauce*; **banana quit:** see Quit *sb.*¹; **banana republic,** *colloq.* applied to a small state, esp. in central America, whose economy is almost entirely dependent on its fruit-exporting trade; **banana solution,** a solution, having the odour of bananas, used as a vehicle in applying bronze pigments; **banana split** (orig. *U.S.*), a popular concoction of ice-cream and a split banana.

1713 Sloane in Ray *Synopsis Avium* 187 *Passer coeruleofuscus.* The Bonana Bird..Arbores Bonanas dictas frequentat, unde nomen. **1734** Albin *Nat. Hist. Birds* II. 37 The Banana Bird from Jamaica..of the Bigness of our English Starling. **1756** P. Browne *Jamaica* (1789) 477 *Icterus major*..The large Banana Bird. **1847** Gosse & Hill *Birds of Jamaica* 226 Banana-bird. *Icterus leucopteryx*...Fruit is his principal diet; a ripe banana, or orange. **1916** W. A. Du Puy *Uncle Sam* 119 The skipper of the banana boat. **1943** Hunt & Pringle *Service Slang* 12 *Banana boat*, an invasion barge. **1945** Partridge *Dict. R.A.F. Slang* 14 *Banana boat*, an aircraft-carrier, according to the R.A.F. To soldiers it means an invasion barge. **1962** *Guardian* 12 Oct. 15/4 They travel by any convenient method, from..luxury liners to banana boats. **1890** Ld. Lugard *Diaries* 6 Dec. (1959) I. 409 The flour of the country is almost solely banana flour. **1951** *Gd. Housek. Home Encycl.* 345/2 Banana Flour..is ideal for flavouring cakes, buns and biscuits. **1921** *Conquest* Sept. 493/2 The common Banana Fly (*Drosophila ampelophila*). **1809** Shaw *Gen. Zool.* VII. 431 The under side of a Banana-leaf. **1916** L. A. Flemming *Practical Tanning* (ed. 3) 474 Because of its characteristic odor the amyl acetate solution is frequently called banana liquid. **1927** Wodehouse in *Sunday Express* 16 Oct. 9/5 This is pure banana oil. It is not like you to..gibber. **1934** H. Hiler *Notes Technique Painting* i. 44 The watercolour fixative..is usually known as 'banana oil' or 'banana solution'. **1960** Wodehouse *Jeeves in Offing* ix. 95 The sort of banana oil that passes between statesmen at conferences..before they tear their whiskers off and get down to cases. **1935** *Esquire* July 70/1 We strung along with Major Brown on the inhuman aspects of war in the banana republics. **1949** Koestler *Promise & Fulfilment* xiii. 143 The somewhat jerky behaviour displayed by the Central American banana republics. **1907** *Westm. Gaz.* 19 Sept. 2/3 The banana-skin trouble in this direction had seemed to be sufficient of a fresh nuisance. **1934** Wodehouse *Right Ho, Jeeves!* i. 17 Treading upon Life's banana skins. **1961** H. R. Trevor-Roper in *Encounter* July 96 The love of..laying banana-skins to disconcert the gravity and upset the balance of the orthodox. **1920** Ade *Hand-made Fables* 151, I recall many useful and interesting citizens who would walk around a banana split to get to a rickey. **1931** E. Linklater *Juan in Amer.* II. ix. 117 Brighter lights and larger jazz-bands and stronger gin and sweeter banana-splits. **1938** E. Bowen *Death of Heart* II. v. 257 They ate poached eggs on haddock and banana splits.

Bana·naland. [See quot. 1898.] A colloquial Australian name for Queensland. Hence **Bana·nalander.**

1898 Morris *Austral English*, *Banana-land*, slang name for Queensland, where bananas grow in abundance. *Bananalander*, slang for a Queenslander. **1901** *Westm. Gaz.* 2 Apr. 1/3 The Bananaland capital. **1901** *Daily Chron.* 4 Apr. 5/1 [He] was born in 'Bananaland'. *Ibid.* 18 Nov. 5/1 The Bananaland statesman. **1933** *Bulletin* (Sydney) 11 Jan. 10 Bananaland has had a fruit marketing organisation for years. **1934** T. Wood *Cobbers* 144 So let this mob of Cornstalks, Croweaters, Sandgropers, and Bananalanders go on yapping, say Victorians.

banausian (bănǭ·siăn), *a.* [f. as Banausic *a.* + -ian.] = Banausic *a.*

1901 Laurie *Training of Teachers* 235 The occupations which to the Greek sages were banausian, if not degrading.

banausic, *a.* Delete *rare* and add earlier and later examples.

1845 G. Smythe in *Oxf. & Camb. Rev.* Aug. 206 When the Banausic principle (we must coin a word from the most expressive of languages to express all its intense vulgarity) began to obtain. **1901** J. C. Collins *Ephemera Critica* 68 The one instinct in them which is not quite banausic being the conscientious thoroughness with which they impart what they have been taught. **1957** *London Mag.* July 57 A sensitive, self conscious creature..in sad revolt against uncongenially banausic employment.

‖**banco** (bæ·ŋko), *int.* [Fr., a. It. *banco* (see Bank *sb.*³).] In games of hazard such as baccarat and chemin-de-fer, a player's proposal to the banker to stake his capital on a single coup.

1789 J. Moore *Zeluco* I. viii. 38 As he shook the box, being about to throw, the Hussar officer cried, Banco; and the others took up what they had staked. **1910** *Encycl. Brit.* III. 121/1 The banker places before him the sum he wishes to stake and the punters do likewise, unless a punter desires to *go bank*, signifying his intention by saying, *Banco!* In this case he plays against the entire stake of the banker. **1964** A. Wykes *Gambling* vii. 178 Any player may bet against the whole of the banker's stake by calling 'banco'.

banco, var. *bunco *sb.*

band, *sb.*² Add: **2.** Various *spec.* uses: an identifying strip placed round the leg of a bird (cf. **bird-band*); an advertising strip round a book; a strip of paper round a cigar.

1914 *Country Life* July 36/2 These up-to-date bands are made in eight different sizes, some one of which is sure to fit the bird you wish to tag. **1937** *Brit. Birds* XXXI. 239 Any observer chancing to meet a straggler on this side of the Atlantic can learn the origin of the bird by reporting the exact order of the various bands. **1932** Q. D. Leavis *Fiction & Reading Public* I. ii. 22 An enterprising publisher will reissue the novel with a band or new dust-jacket exhibiting the caption. **1958** *Bookseller* 12 Apr. 1369/1 'Royal Ballet' Bands..—have prepared bands for the jackets of *The Sadler's Wells Ballet.* **1923** W. J. Locke *Moordius & Co.* vii. 96 He selected a cigar,..removed the band and clipped the point. **1935** A. E. W. Mason *They wouldn't be Chessmen* ix. 115 What's the use of me paying fourpence for a 'Avana cigar if I've got to take the band off before I smoke it?

9. b. *Recording.* (See quot. 1962.)

1953 *His Master's Voice Record No. ALP 1052, Elgar Symphony No. 1* (*cover*), Side 1. Band 1—First movement... Band 2—Second movement. **1957** *Records & Recording* Oct. 35/3 It looks like a normal LP, but..it plays for only half the time—the first track is on the outside, separated by a blank band from the second inner track. **1962** A. Nisbett *Technique Sound Studio* 242 *Band*, separately recorded section of a disc, of which there may be several on a side. By extension, the term may also mean an individual section of a tape recording which is bounded by spacers.

10. c. *fig.* = Range *sb.*¹ 10 b. (Cf. Band *sb.*³ 5.)

1929 A. Lloyd James in *S.P.E. Tract* XXXII. 6 We now have a certain type, or rather a carefully chosen band of types of English. *Ibid.* 9 Those who speak any one variety of the narrow band are recognised as educated speakers. **1959** *Listener* 19 Feb. 331/1 The standard of play..is at a fairly level band of skill and teamwork throughout, at least in Division One.

11. b. (Example.)

1845 *Florist's Jrnl.* Aug. 175 Bands, or Vittæ, the flattened or hollow spaces between the elevated ribs of the fruit of umbelliferous plants.

13. *Physics.* A group of closely-spaced lines, esp. in a molecular spectrum; **band spectrum,** a spectrum characterized by such bands.

1831 [see sense 10 a in Dict.]. **1869** H. E. Roscoe *Spectrum Analysis* iv. 146 In the case of bodies whose spectra change from bands to lines on increase of temperature, a recombination of the elements occurs on cooling, and the band spectrum of the compound reappears. **1885** —— *Ibid.* (ed. 4) iii. 130 Nearly all bodies..have been found to exhibit both a band and a line spectrum, the band spectrum always belonging to the lower temperature. **1903** A. M. Clerke *Problems in Astrophysics* iii. 39 Band-spectra..display no sensitiveness to pressure. **1923** Glazebrook *Dict. Appl. Physics* IV. 786/1 Bands, which are usually associated with the spectra of compounds of molecules, consist of groups of lines which converge to definite heads. **1957** *Encycl. Brit.* III. 25/2 A spectrum consists of lines showing certain definite regularities of arrangement. The so-called 'line spectra' are attributable to atoms, and band spectra are due to molecules...Such a series [of lines] appears to terminate abruptly at the point where the separation of the lines is least; this is called the head of the band, and is a prominent feature of most band spectra.

14. *Electr.* A range of frequencies or wavelengths that falls between two given limits; = **wave-band.* Also in *Comb.*, as **band-pass filter,** an electrical filter with a very low attenuation for currents within given limits of frequency; **band width, bandwidth,** the interval separating the limits of a band.

1922 A. F. Collins *Radio Amat. Handbk.* 321 When continuous waves are being sent out and..modulated by a microphone transmitter the different audio frequencies set up corresponding radio frequencies and the energy of these are emitted by the aerial; this results in waves of different lengths, or a band of waves as it is called. **1922** *Encycl. Brit.* XXXII. 712/2 The appurtenances specially developed for accomplishing this selection [of frequencies] in carrier current telephony are known as 'band-pass electrical filters'. **1929** T. E. Shea *Transmission Networks & Wave Filters* II. vii. 235 This type of filter excludes, or attenuates, all frequencies lying between its two cut-off frequencies, but..transmits..frequencies above and below this band. It is therefore commonly called a 'band elimination' filter. **1930** *Discovery* Dec. 398/2 The band-pass filter, which follows the low frequency modulator, allows the lower side-band to pass with an attenuation of six decibels. **1930** *Proc. Inst. Radio Engin.* XVIII. 168 The greater band width being required as the standard of quality [*sc.* in broadcast speech or music] becomes higher. **1931** *Daily Express* 21 Sept. 7/4 An advanced form of band-pass tuning. **1933** K. Henney *Radio Engin. Handbk.* XVII. 444 Television..requires a very wide band; high-grade program broadcasting, a fairly wide band; satisfactory speech, a somewhat narrower band. **1935** *Ibid.* (ed. 2) 171 Filters are divided into four classes, according to the frequency bands which they are intended to transmit, namely, low pass, high pass, band pass, band elimination. **1940** *Amat. Radio Handbk.* (ed. 2) 59/2 The main inductance L_2 is tuned by C_2 and C_3.. the latter of smaller capacity for band-spread purposes. **1940** *Wireless Engineer* XVII. 394/1 Band-spreading may be defined as the deliberate limitation of the frequency range covered by a tuning unit, in order to facilitate the process of tuning. **1944** *Electronic Engin.* XVI. 322 Both colour reproduction and definition would require a very much greater band-width. **1951** *Good Housek. Home Encycl.* 230/1 If the set is required for short-wave reception as well as for the long and medium bands. **1959** *Times* 16 Jan. 10/1 The most important decision the C.C.I.R. will have to make..refers to the bandwidth of television transmissions in Bands IV and V. **1969** *Times* 27 Mar. 12/8 Measurements giving the energy emitted in selected wavelength bands have been made on 300 stars.

III. band-brake, a brake consisting of a band operating on a spindle; **band ceramic** (see *Bandkeramik); **band clutch,** a clutch consisting of a band operating on a spindle or drum; **band-knife,** an 'endless' knife; **band-saw** (examples); **band-sawyer,** an operative who uses a band-saw.

1889 *Cent. Dict.*, Band-brake. **1908** *Daily Chron.* 21 Nov 9/4 A band-brake operates on the balance gear. **1910** *Cycl. Automobile Engin.* I. 210 The most usual place in which the band clutch is found is in connection with a planetary transmission. **1957** *Encycl. Brit.* V. 864/1 Band clutches are usually installed when it is necessary to transmit heavy loads accompanied by shocks. **1926** *Glasgow Herald* 12 July 8 Following the sewing machine come[s] the band-knife. **1864** Webster s.v. *Saw*, Band-saw. **1890** W. J. Gordon *Foundry* i. 30 A band-saw..which cuts through iron like cheese. **1916** *Daily Colonist* (Victoria, B.C.) 9 July 13/4 The machinery is already partly installed in the bandsaw mill. **1909** *Daily Chron.* 25 Sept. 7/6 Band Sawyer wanted.

band, *sb.*³ Add: **3. b.** A herd or flock. *N. Amer.*

1824 S. Black *Jrnl. Voy. fr. Rocky Mt.* (1955) 73 The band of Carribou is gone farther. **1824** W. H. Keating *Narr. Exped. St. Peter's River* I. viii. 379 The term *band*, as applied to a herd of buffalo, has almost become technical, being the one in use in the west. *a* **1861** T. Winthrop *John Brent* (1883) ii. 11, I had come upon a band of horses feeding on the prairie. **1872** Schele de Vere *Americanisms* 210 California for instance, forms its vast flocks of sheep into *bands*, of about a thousand each. **1920** J. M. Hunter *Trail Drivers of Texas* 319 How many of the 'band' (meaning the herd) are gone? **1953** *Canad. Geogr. Jrnl.* XLVI. 246/2 Herder and 'band' (not flock) have lost their lives in blizzards year after year.

4. b. Colloq. phrases: *when the band begins to play*, when matters become serious; *to beat the band*, lit. so as to drown the noise made by the band; hence, to exceed, surpass, or beat everything.

1890 Kipling *Barrack-r. Ballads* (1892) 6 It's 'Thank you, Mister Atkins', when the band begins to play. **1897** C. M. Flandrau *Harvard Episodes* 223, I was on the box-seat driving, you know,—lickety-split, to beat the band. **1900** G. Bonner *Hard-Pan* iii. 81 Doesn't that beat the band? **1910** W. M. Raine *B. O'Connor* ii. 24 It's send for Bucky quick when the band begins to play. *Ibid.* 236 Eating together like brothers and laughing to beat the band. **1920** Wodehouse *Coming of Bill* II. vi. 167 You certainly are working to beat the band just now. **1923** A. Christie *Murder on Links* viii. 101 Well, if that doesn't beat the band!

7. *band-leader, -leading, -playing, -room*; **band parts,** written or printed pieces of music (see Part *sb.* 10) for each member of a band of musicians.

1894 *Munsey's Mag.* XII. 411 (*heading*) Famous American band leaders. **1927** *Melody Maker* Aug. 813/1 All band leaders still find it very difficult to obtain good instrumentalists. **1961** *Times* 20 May 11/4 Another bandleader-composer in search of his youth is Count Basie. **1955** L. Feather *Encycl. Jazz* (1956) i. 18 The bandleading career that was to take many great Negro jazzmen..North. **1895** A. Roberts *Adv. by Rail* iii. 45, I had a box stolen. Amongst other things, it contained my band-parts. **1909** *Daily Chron.* 5 Oct. 1/3 An evening of excellent American band-playing. **1909** *Cent. Dict.* Suppl., *Band-room*, a store-room on a flag-ship in which the bandsmen keep their instruments and music. **1929** *Melody Maker* Jan. 59/1 A suite comprising waiting-room, band-room, engineers' room, announcer's room. **1933** P. Godfrey *Back-Stage* i. 15 The stage-manager looks at his watch: it is time to warn the 'band-room'.

band, *v.*¹ Add: **2. c.** To furnish (a bird) with an identifying band. So *banded* ppl. adj., *banding* vbl. sb. and ppl. adj. (See also **bird-banding.*) orig. *U.S.*

1914 *Lit. Digest* 17 Jan. 102/2 Last year over 150 young American and snowy egrets were banded. **1930** E. W. Hendy *Wild Exmoor* xviii. 275 The promiscuous habits of house wrens banded (i.e. ringed) in Ohio. **1930** J. S. Huxley *Bird-Watching* iii. 47 The practice of banding birds—attaching a light numbered and dated ring of metal to their legs, either when still in the nest, or after being caught in a special and harmless trap and subsequently set free again. *Ibid.*, This banding method has..been used to shed light upon other sides of bird-life. **1934** *Discovery* Apr. 111/1 It was natural that those species to be banded..should be the common frequenters of our gardens. *Ibid.* 112/2 The banding records indicate that..those that migrate through and winter in the east would be very slow to re-populate the devastated areas in the west.

4. a. Also, to form (cattle or sheep) into a herd or flock. *U.S.*

1878 B. F. Taylor *Between Gates* 266 Leave him to 'band' his sheep and herd his bees as he pleases.

‖**banda** (bæ·ndă). [Native name.] A thatched house in parts of Africa.

1920 *Blackw. Mag.* CCVII. 651/2 Opposite some tumble-down bandas, a motor-lorry is waiting. **1951** R. Campbell *Light on Dark Horse* x. 137 This banda or hut was made entirely of coconut palms.

bandabast, bandabust, varr. *bundobust.

bandar (bʊ·ndăɹ). Also **bander, bunder.** [Hind.] The Rhesus monkey, *Macaca mulatta.* Also in *Bandar-log* [Hind. *log* people], Kipling's nation of monkeys (see quot. 1894); hence *fig.*, any body of irresponsible chatterers.

1885 Kipling *Dep. Ditties* (1899) 54 It was an artless *Bandar*, and he danced upon a pine. **1894** —— *2nd Jungle Bk.* (1895) 63 Men are blood-brothers of the *Bandar-log*. **1917** 'Ian Hay' *Carrying On* xiii. 313 That is just what the *Bandar Log* overlook, when they jabber about the dreadful industrial upheaval that is coming with peace. **1922** *Chambers's Jrnl.* June 353/1 He was too used to the wordy clatter of the *bunderlogue* to notice it. *Ibid.*, Their attentions the unhappy *bunder* would gladly dispense with. **1961** *Times* 23 Nov. 17/3 He castigates the *banderlog* at all levels.

bandbox. Add: **a.** Also *fig.*, a fragile or flimsy structure or one in which the accommodation is restricted.

1875 H. Lee *Octopus* viii. 81 The vessel was lightly built —a mere bandbox of a craft.

b. (Simple *attrib.* and *Comb.* examples.) Also in senses: (*a*) resembling a bandbox; flimsy, fragile; (*b*) conspicuously neat and clean; dressed-up (cf. quots. in Dict.).

1727 Swift *Let.* 9 July (1912) III. 405 Letters..which.. I can sell at good advantage to the band-box and trunk-maker. **1838** *S. Lit. Messenger* IV. 638/2 To render fur caps and bandbox hats for winter..articles of vital necessity. **1844** Thackeray *May Gambols* in *Fraser's Mag.* June 707/2 Spick and span *bandbox* churches of the pointed Norman style. **1894** *Sketch* 13 June 355/1 The best stage-management in the world could not have made the thing lifelike.. on that bandbox stage. **1916** J. Buchan *Greenmantle* xi. 143 Better a bloody end in a street scrap than the tender mercies of that bandbox bravo. **1939** Auden & Isherwood *Journey to War* v. 130 Little band-box officers, slim and smart. **1967** V. H. Gielgud *Conduct of Member* xx. 160 She was wearing a dressing-gown..but preserved her usual band-box quality of neat fastidiousness.

c. *to look as if one came out of a bandbox* (and similar phrases): to look extremely smart and neat. orig. *U.S.*

1825 S. Woodworth *Forest Rose* I. i, He is a genteel, delightful looking fellow, neat as a starched tucker fresh from a banbox [*sic*]. **1833** *Knickerbocker* I. 198 The old gentleman..popped into the room, looking as if he had stepped out of a bandbox. **1869** 'Mark Twain' *Innoc. Abroad* xxxviii. 410 They are all..exceedingly neat and cleanly..as if they were just out of a band-box. **1888** 'R. Boldrewood' *Robb. under Arms* III. vii. 94 Starlight.. looked as if he'd just come out of a band-box. **1941** M. Marlett *Death has Thousand Doors* (1950) iv. 32 She invariably looked as though she had stepped out of the proverbial bandbox.

bandboxy (bæ·ndbǫksi), *a.* [f. Bandbox + -y¹.] Resembling a bandbox in shape, the fragility of its structure, or its restricted accommodation. (Cf. Bandboxical *a.*)

1855 B. Taylor *Pictures of Palestine* xxxvi. 428 Those bandboxy sombreros which I at first thought so ungainly. **1891** *Cornh. Mag.* Feb. 160 Houses,..light and small and bandboxy. **1896** C. Allen *Papier Mâché* 80 Sheds, and bandboxy houses all tilted up on stumps.

b. Extremely neat. *U.S.*

1870 L. M. Alcott *Old-Fash. Girl* xii. 195 The dance and the race had taken the 'band-boxy' air out of Tom's elegant array.

bandeau. Add: **a.** (Earlier and later examples.)

1706 T. Betterton *Amorous Widow* I. 4 The fairest Hair, the beautiful'st Curls do not become your Forehead, so well as a *Bando* did. **1908** [see *barrette 2]. **1959** *Sunday Times* 5 Apr. 22/5 As small as it is possible to be and still be called a hat, a bandeau and bow are caught in a cage of veiling.

c. A strip of velvet or other material generally made up in a circular form to be stitched inside the lower part of the crown of a hat that is too large for the head.

1908 *Daily Chron.* 29 Jan. 4/7 With the right sort of 'bandeau'..you need not wear a hatpin at all.

banded, *ppl. a.* Add: **3.** Esp. in specific zoological and botanical names.

1814 Mitchill *Fishes N.Y.* 427 Banded Mackerel. *Scomber zonatus.* **1823** E. James *Exped. Rocky Mts.* I. 267 *Crotalus horridus*, Banded rattlesnake. **1836** [see Kangaroo *sb.* 2]. **1848** Bartlett *Dict. Amer.* 165 *Grunter*, one of the popular names of the fish called by naturalists the *Banded Drum.* **1870** *Amer. Naturalist* IV. 102 The banded Sunfish (*Bryttus Chaetodon*). **1921** H. Guthrie-Smith *Tutira* xxiii. 207 The Banded Dottrel (*Charadrius bicinctus*). **1949** C. Longfield *Dragonflies* (ed. 2) 43 *Agrion splendens*, the Banded Agrion.

b. *Geol.* (See quots.)

1859 *Q. Jrnl. Geol. Soc.* XV. 193 Peculiar banded flints in the chalk..exhibit a central longitudinal axis or narrow stem, crossed on its middle third by numerous short parallel stripes of alternately light and dark flint. **1876** [see Dict.]. **1889** *Cent. Dict.*, *Banded structure*, the structure of a rock which is..divided into layers...The structure of a mineral made up of a series of layers.., as onyx. **1904** Goodchild & Tweney *Technol. & Sci. Dict.* 39/2 *Banded Structure.* This term is..coming into use in a more restricted sense, as descriptive of the parallel structures observable in gneisses.

banderilla. Add: Also **banderillo.** (Further examples.)

1797 [see next]. **1882** H. de Windt *On Equator* ix. 138 Each was armed with the *banderillo*, small barbed darts, about a foot long. **1927** Hemingway *Men without Women* 44 Fuentes..sank the banderillos straight down. **1939** Spender & Gili tr. *Lorca's Poems* 123 How tremendous with the final Banderillas of darkness. **1948** L. MacNeice *Holes in Sky* 61 Gaudy banderillas May quiver in our flanks.

Hence as *v. trans.* and *intr.* (also in Sp. form **banderillear**), to stick a banderilla into (a bull). So **banderi·lling** *vbl. sb.*

1932 Hemingway *Death in Aft.* xvii. 197 Bulls which will not charge..are banderilla-ed by..the media-vuelta or half-turn. *Ibid.* 198 Very rarely..a man is able to banderillear properly from both sides. **1932** R. Campbell *Taurine Provence* iii. 70 The art of banderilling is not only a matter of speed.

banderillero. Add: (Earlier and later examples.) Also *fig.*

1797 *Encycl. Brit.* III. 772/1 The..*banderilleros*..plunge into his neck..a kind of darts called *banderillas.* **1845** R. Ford *Trav. Spain* I. II. 182 The *banderilleros* go right up to him, holding the arrows at the shaft, and pointing the barbs at the bull. **1932** Hemingway *Death in Aft.* iii. 26 Men who ..place the banderillas..are called peones or banderilleros. **1934** R. Campbell *Broken Record* ii. 48, I am only a banderillero in the cuadrilla of that great matador [Wyndham Lewis].

bandersnatch (bæ·ndəɹsnætʃ). [Invented by 'L. Carroll' (C. L. Dodgson); presumably a portmanteau word like its stock epithet *frumious.*] A fleet, furious, fuming, fabulous creature, of dangerous propensities, immune to bribery and too fast to flee from; later, used vaguely to suggest any creature with such qualities.

1871 'L. Carroll' *Through Looking-Glass* i. 22 Beware the Jubjub bird, and shun The frumious Bandersnatch! **1876** —— *Hunting of Snark* vii, But the Bandersnatch merely extended its neck And grabbed at the Banker again. **1908** *Daily Chron.* 19 Aug. 7/4 'There may be a bandersnatch,' he said doubtfully, dusting the ground with his palmetto fan. **1936** C. S. Lewis *Alleg. Love* vii. 301 Always, at the critical moment, a strange knight, a swift ship, a bandersnatch or a boojum, breaks in. **1959** —— *Let.* 15 May (1966) 287 No one ever influenced Tolkien — you might as well try to influence a bandersnatch.

bandicoot. 2. (Earlier and later examples.) For (*Parameles*) read (*Perameles*).

1827 P. Cunningham *Two Years N. S. Wales* I. xvii. 316 The *bandicoot* is about four times the size of a rat, without a tail, and burrows in the ground or in hollow trees. **1865** J. G. Wood *Homes without Hands* 307 The second Building Mammal on our list is..the..Rabbit-eared Bandicoot. **1888** 'R. Boldrewood' *Robb. under Arms* I. i. 7 I'd be as miserable as a bandicoot. **1936** I. L. Idriess *Cattle King* xxiii. 203 Country that was tiptop a few years ago now wouldn't feed a bandicoot.

bandie (bæ·ndi). *Sc.* and *north.* Also **bandy.** [perh. derived from BANSTICKLE.] The stickleback, *Gasterosteus aculeatus.*
1825 in JAMIESON. **1854** in A. E. BAKER *Gloss. Northampt. Words.* **1876** SMILES *Sc. Natur.* i. 14 'Mother,' said he, 'where are my crabs and bandies?' **1921** *Glasgow Herald* 25 Mar. 7/2 Little boys poking in their depths for bandies.

banding, *vbl. sb.*¹ Add: **2.** (Earlier example.)
1739 LABELYE *Piers Westm. Br.* 41 Each of the Piers of the intended Bridge will be built..with the same Bandings, the same Cementing and Crampings, as if built upon dry Ground.
4. = BAND *sb.*² 3.
1892 *Daily News* 2 July 6/7 White sailor hat, with blue ribbon banding.

bandit. Add: **a.** Also, in modern use, = *GANGSTER 1. See also *one-armed bandit.*
1935 E. WEEKLEY *Something about Words* ii. 44 *Bandits* were formerly Italians, picturesque in costume and impressive in armament; now that they are revived, they ride in motor-cars. **1935** *Amer. Speech* Apr. 120/1 The world of crime has its significant distinctions: *hijacker, car bandit,* [etc.]. **1944** AUDEN *For Time Being* (1945) 116 The bandit who is good to his mother.
d. *transf.* A hostile aircraft (see quot. 1943).
1942 I. GLEED *Arise to Conquer* xii. 111 One bandit shot down in sea about ten miles out. **1943** C. H. WARD-JACKSON *Piece of Cake* 12 *Bandit*, enemy aircraft. It is used throughout Fighter Command, and elsewhere...The term was originally a code word.

banditism (bæ·nditiz'm). [f. BANDIT + -ISM. Cf. F. *banditisme* (Flaubert, 1853).] The practices of bandits.
1885 *Manch. Guardian* 25 May 8 In England they re-establish the fusillade, in Russia torture, in Germany banditism. **1921** *19th Cent.* Sept. 407 The resistance of the peasantry to the Communist system of food supply is one of the causes of the development of banditism. **1941** KOESTLER *Scum of Earth* 116 Cyrano's régime had been old-fashioned, nineteenth-century banditism.

banditry (bæ·nditri). [f. BANDIT + -RY.] The practices of bandits.
1922 *Q. Rev.* July 157 Already they have made short work of banditry. **1924** *Glasgow Herald* 29 Oct. 8 The ponderous banditry which the Chinese dignify by the name of war.

‖**Bandkeramik** (ba·ntkera:mik). *Archæol.* Also **band ceramic.** [G. (F. Klopfleisch 1883, in *Vorgeschichtliche Alterthümer der Provinz Sachsen* 1. 92).] A type of neolithic pottery with banded decoration. Also *attrib.*
1921 J. M. TYLER *New Stone Age North. Europe* vii. 155 The pottery of northern Europe can be distributed into a few groups or general types.. 1. Banded pottery..*Bandkeramik.* **1936** *Discovery* Sept. 289/1 The pottery is similar to the 'band ceramic' of central Europe. **1940** C. F. C. HAWKES *Prehist. Found. Europe* iv. 117 The great settlement of Lindenthal near Cologne, in which the whole history of the *Bandkeramik* culture in this far north-western region has been exposed in detail. **1953** J. PROCHÁZKA tr. *Hrozný's Anc. Hist. W. Asia, India & Crete* xvii. 220 The Illyrian and Thracian peoples are responsible for the band ceramics.

bandobast, bandobust, var. *BUNDOBUST.

bandolero (bændolē³·ro). [Sp.] A highwayman or robber. Also *attrib.*
[**1645** J. HOWELL *Ep. Ho-Elianæ* I. xxiii. 43 These parts of the Pyreneys that border upon the Mediterranean, are never without Theeves by Land (call'd *Bandoleros*).] **1832** W. IRVING *Alhambra* (1896) 9 The solitary *bandolero*, armed to the teeth..hovers about them. **1927** *Chambers's Jrnl.* 733/2, I was soon rigged out in the costume of a Spanish villager; that is, rope-soled shoes, white woollen stockings, bright yellow knickerbockers, a frilled shirt, and a bandolero hat.

bandoliered (bændǒlī³·ɹd), *a.* [f. BANDOLEER, -IER + -ED².] Wearing a bandolier.
1900 *Daily News* 29 Jan. 7/5 The dashing bandoliered invader. **1921** *Blackw. Mag.* Oct. 458/1 A bandoliered and sheep-skin cloaked policeman.

bandoline, *sb.* (Earlier and later examples.)
1846 'A LADY' *Jewish Manual* 206 As a bandoline to make the hair set close .. linseed .. boiling water .. rose-water. **1957** V. J. KEHOE *Technique Film & T.V. Make-Up* iii. 31 *Bandoline*, a viscous liquid used for setting hair stiffly in place.

bandoline (bæ·ndǒlin), *v.* [f. the sb.] To dress (the hair) with bandoline. Hence **ba·ndolined** *ppl. a.*, **ba·ndolining** *vbl. sb.*
1856 [in Dict.]. **1866** DICKENS *Rugby Junction* iii, in *Christmas Stories* (1956) 516 You should see our Bandolining Room..where Our Missis and our young ladies Bandolines their hair... You should see 'em at it,..Bandolining away, as if they was anointing themselves for the combat. **1876** *Tinsley's Mag.* XIX. 401 Moustaches..carefully bandolined into a ring at each end. **1933** MRS. C. S. PEEL *Life's Enchanted Cup* i. 6, I remember, too, that the smooth, parted front hair was 'bandolined' with a stick of some white greasy composition.

ba·nd-wagon. orig. *U.S.* Also **-waggon.** [BAND *sb.*²] A large wagon, capable of carrying the band in a procession. Freq. *fig.*, as of one conveying a 'band' of usu. successful (political) leaders. Hence *to climb, hop, jump,* etc., *on the band-wagon*, to join in what seems likely to be a successful enterprise, to strive to join the winning side.
1855 BARNUM *Life* 205 At Vicksburg we sold all our land conveyances excepting four horses and the 'band wagon'. **1893** *Congress. Rec.* 25 Aug. 897/1 It is a lamentable fact that..our commercial enemy..should come along with a band wagon loaded with hobgoblins. **1899** T. ROOSEVELT *Let.* 28 Apr. (1951) II. 999 When I once became sure of one majority they tumbled over each other to get aboard the band wagon. **1905** *N.Y. Even. Post* 21 Oct. 1 Jerome's band wagon began to move over the town to-day. It bears on its sides announcements of these mass meetings on Monday night. **1906** *N.Y. Even. Post* 5 Sept. 4 Many of those Democrats..who rushed into the Bryan band-wagon.. will now be seen crawling out over the tailboard. **1927** J. NEEDHAM *Man a Machine* iii. 67 Neo-Lamarckism, whose 'band-wagon' has been so kindly thrown open to all by an eminent English friend of Sig. Rignano's. **1933** *Amer. Speech* VIII. 1. 22/2 The socialists climbed on the alliteration band-wagon with *Sail safely and surely with Socialism.* **1940** H. G. WELLS *New World Order* § 7. 104, I find most of these United States of Europe movements are now jumping on to the Federation band-waggon. **1941** A. L. ROWSE *Tudor Cornwall* iv. 86 A few who were forward Protestants..may be said to have jumped on the band-waggon. **1950** *Hansard Commons* CDLXXII. 970 The Tory party are now trying to climb on to the band wagon. **1959** *Observer* 1 Feb. 10/4 He is surrounded by yes-men who have jumped on to the bandwagon.
attrib. **1908** *Nation* 16 Apr. 343/1 We shall now hear the 'band-wagon' argument for Taft more confidently than ever. **1958** *Times Lit. Suppl.* 15 Aug. p. xxxii/1, The whipping-up of public emotions has been made the excuse for the cult of band-waggon personalities.
Hence **band-wagoner, band-wagoning.**
1949 *Here & Now* (N.Z.) Oct. 13/1 Not having had the wit to see that Labour never amounted to anything in terms of real power, he is paying the penalty now for his band-waggoning. **1958** C. S. LEWIS *Reflections on Psalms* vii. 70 Here is the perfect band-wagoner. Immediately on the decision 'This is a revolting tyranny', follows the question 'How can I as quickly as possible cease to be one of the victims and become one of the tyrants?' *Ibid.* 71 There are subtler, more social or intellectual forms of band-wagoning which might deceive us. **1958** *Times* 1 Dec. 10/2 The bandwagoners—latter-day 'Gaullists'..who have found it convenient for electoral and other purposes to fight under U.N.R. colours.

bandwidth: see *BAND *sb.*³ 14.

bandy, *sb.*¹ Add: **6.** *bandy-wicket* (see quots.). *dial.*
1749 ELLIS *Syst. Improvem. Sheep* II. iv. 199 The bad Example of others, who..play at Bandy-wicket..on the Sabbath-Day. **1823** E. MOOR *Suffolk Words* 14 *Bandy-wicket*, a game with bats, or sticks, and ball, like cricket—but with bricks..or..hats, instead of bales and stumps, for wickets. *a* **1825** FORBY *Vocab. E. Anglia* (1830) I. 14 *Bandy-wicket*, the game of cricket. Of the several games at ball played with a bandy, that in which a ball is aimed by one player at a wicket, defended by the adversary with his bandy, must be allowed to be very appropriately called *bandy-wicket.*

bandy, var. *BANDIE.

bandy-bandy (bæ·ndi bæ:ndi). The native name in Australia of a nocturnal snake, *Furina annulata*, marked with black and white bands; also called *bandy-snake.*
1926 *Austral. Encycl.* II. 473/2 *Furina annulata* (Bandy-bandy, Black-and-white-ringed Snake), another common venomous kind, sometimes grows to 30 inches...This species is better known as *F. occipitalis*, but the specific name *annulata* has priority. **1958** *Ibid.* VIII. 167/1 The black-and-white ringed snake, or bandy-bandy,..is very widely known and is often referred to as being deadly. **1959** J. WRIGHT *Generations of Men* 19 The curious pink bandy-snakes that scuttled in and out of the bushes in the garden.

bandyman (bæ·ndimæn). [f. BANDY *sb.*⁴ + MAN *sb.*¹] The driver of a bandy (see BANDY *sb.*⁴).
a **1881** CALDWELL (Ogilvie), When also, as all over India, our white kinsmen speak of bandymen and bandies, the word thus anglicized is simply the old Tamilian one. **1922** *Other Lands* Apr. 76/2 The shout of 'Ho, Bandyman'.

banesman (bēⁱ·nzmæn). *pseudo-arch.* Also **baneman.** [Rendering of ON. *banamaðr*, f. *bana*, gen. of *bani* BANE *sb.*¹ + *maðr* MAN *sb.*¹] A murderer.
1870 MAGNÚSSON & MORRIS tr. *Völsunga Saga* ii. 3 He had slain all his father's banesmen. **1926** *Trans. Scott. Ecclesiol. Soc.* 74 He prayed, not only for himself and his friends, but still more for 'his foes and banemen'.

bang, *v.*¹ Add: **2.** (Further example.)
1894 G. DU MAURIER *Trilby* I. 31 He strummed: 'Messieurs les étudiants'..striking wrong notes, and banging out a bass in a different key.
5. a. Hence *banged up* ppl. adj., knocked about. *U.S. colloq.*
1886 E. L. DORSEY *Midshipman Bob* II. vii. 172 Then Young dragged himself on those banged up legs ever so far.. to the Life-Saving Station. **1886** *Harper's Mag.* June 107 Even the trig, irreproachable commercial drummer actually looks banged up and nothing of a man. **1934** J. M. CAIN *Postman always rings Twice* viii. 76, I began to fool with her blouse, to bust the buttons, so that she would look banged up.
c. *Stock Exchange.* To depress (prices, the market). Cf. HAMMER *v.* 2 d (*b*).
1884 [see Dict.]. **1907** *Daily Chron.* 10 Dec. 5/4 What prompted the selling is unknown. It appears like an attempt to bang the price. **1927** *Sunday Times* 13 Feb. 2 Attempts to bang prices failed to induce much selling. **1938** *New Statesman* 30 Apr. 750/1 Oil shares were banged in the 'Street' on Tuesday night, the leaders falling by about 5s.
8. a. b. c. (Further examples.) So *bang off*, immediately, without delay. Cf. *BANG *adv.*
1832 *Blackw. Mag.* XXXII. 31 A 32 lb. shot struck us bang on the quarter. **1868** *Punch* 5 Dec. 235/1 Mun, a had na' been the-erre abune Twa Hoours when—Bang—went Saxpence! **1886** W. JAMES *Let.* 29 Aug. in R. B. Perry *Thought & Char. W. J.* (1935) I. 602 The moment I get interested in anything, bang goes my sleep. **1886** BAUMANN *Londinismen* 7/1 *Bang-off*..he wrote it ~ *er schrieb's in einem Zuge.* **1895** H. JAMES *Notebks.* 14 Feb. (1947) 188 This thing has for my bang-off purpose the immense merit of having no prescribed or imposed length. **1895** G. B. SHAW *Let.* 28 Nov. in *E. Terry & Shaw* (1931) 20 Somebody will give a surreptitious performance of it: and then bang goes my copyright. **1896** H. JAMES *Spoils Poynton* (1897) viii, She may..think I may want to make her reply bang off? **1909** T. E. LAWRENCE *Lett.* (1938) 79, I am afraid I have to drive from here to Urfa (Edessa) which is going to cost me about £7: so bang go my proposed purchases in Damascus. **1912** D. H. LAWRENCE *Let.* 2 Sept. (1932) 55 Then bang-slap went my heart.
d. *humorous* (with allusion to 'bang goes saxpence' as in 8 c): to spend ('saxpence') all at once in a fit of extravagance. Hence **ba·nging** *vbl. sb.*
1897 *Westm. Gaz.* 17 May 10/1 The desirability of avoiding any unnecessary banging of saxpences. **1901** *Daily Chron.* 11 Nov. 3/7 Our Northern friends look twice before they 'bang' their 'saxpences'.
9. b. bang-about *a.* (cf. KNOCK-ABOUT *a.*), rough, boisterous.
1933 E. A. ROBERTSON *Ordinary Families* ix. 199 Dru, that devilish sailor and bang-about good sort! **1960** V. GIELGUD *To Bed at Noon* III. ix. 222 She keeps the wild-bull bang-about side of Rupert in hand.
10. Slang phr. *bang to rights*, of a criminal: (caught) red-handed. Also *banged to rights.*
1904 'No. 1500' *Life in Sing Sing* 255/1 *Bang to Rights*, caught in the act. **1932** A. R. L. GARDNER *Tinker's Kitchen* (Gloss.) 281 *Banged to rights*, found in possession of stolen property. **1958** F. NORMAN *Bang to Rights* I. 33 One night a screw looked through his spie hole and captured him bang to rights. **1962** *New Statesman* 21 Dec. 897/2 If I was making a book on the chances of my being banged to rights, you or any other punter could have 100 to eight to any amount.
11. *trans.* and *intr.* To copulate (with), to make love (with). *slang.*
1937 in PARTRIDGE *Dict. Slang.* **1957** J. KEROUAC *On Road* (1958) vii. 42 He rushes from Marylou to Camille ..and bangs her once. *Ibid.* 43 Marylou's all for it [*sc.* divorce], but she insists on banging in the interim. **1962** K. ORVIS *Damned & Destroyed* xiv. 90 We banged twice more after you left.

bang, *sb.*¹ Add: **2.** With allusion to T. S. Eliot's line (see quot. 1925).
1925 T. S. ELIOT *Hollow Men* v. 31 in *Poems* 99 This is the way the world ends Not with a bang but a whimper. **1931** R. ALDINGTON *Colonel's Daughter* I. 56, I wish you'd all shoot yourselves with a bang, instead of continuing to whimper. **1953** 'M. INNES' *Christmas at Candleshoe* i. 16 Benison is going to end not with a bang but a whimper. **1959** *Times* 16 Dec. 3/2 Here the world ends neither with a bang nor a whimper, but with a slow, resigned sigh at its own criminal imbecility.
b. *spec.* A nuclear explosion.
[**1951** *Time* 16 Apr. 17/1 (*title*) Bang! The Day when A-Bomb hit Hiroshima...The old woman neither heard bang nor felt shock, but both ceiling and roof fell down.] **1955** *Times* 17 May 11/3 Even if these bangs are let off with disgust, not gusto, they can rock international friendship and confidence. **1957** J. OSBORNE *Look back in Anger* III. i, If the big bang does come, and we all get killed off, it won't be in aid of the old-fashioned grand design.
c. Short for *sonic bang* (see *SONIC *a.*). Also *attrib.*
1955 *Britannica Bk. of Year* 489/2 A new piece of R.A.F. slang emerged in *Bang-Book*, a register that pilots were required to sign in the event of their having broken through the sound-barrier. **1955** *Times* 12 July 8/6 If the pilot produces a 'bang' accidentally..he must report this immediately to flying control, who notify the command headquarters. **1963** *Aeroplane* CV. 5/3 He described the strip of country subjected to sonic bangs from aircraft as the 'bang swath', whose boundary of area was the envelope of the bang rays.
4. *colloq.* A 'thumping' lie, a banger; *bang-words*: explosive epithets, 'swear' words.
1879 MEREDITH *Egoist* xxix, Every crack and bang in a boy's vocabulary. **1906** *Westm. Gaz.* 20 Jan. 2/1 When the recipient of a letter has to..go in for a comparative analysis of the different letters..he is justified in using bang words.
5. [Cf. *BANG *sb.*²] Excitement, pleasure; a 'kick'. *U.S. slang.*
1931 D. RUNYON *Guys & Dolls* (1932) vi. 129 He seems to be getting a great bang out of the doings. **1951** J. D. SALINGER *Catcher in Rye* iv. 37, I hate the movies like poison, but I get a bang imitating them.
6. [Cf. *BANG *v.*¹ 11.] An act of sexual intercourse. *slang.*
1937 in PARTRIDGE *Dict. Slang.* **1965** A. PRIOR *Interrogators* xii. 239 Isn't it amazing..what a quick bang does for old Lance. **1968** J. UPDIKE *Couples* i. 35, I bet she even gives him a bang now and then.

bang, *sb.*³ (Earlier and later examples.)
1878 F. M. A. ROE *Army Lett.* (1909) 186 It had a heavy bang of fiery red hair. **1936** M. H. BRADLEY *Five-Minute Girl* ii. 23 The straight dark hair, with its heavy bang across her childish forehead.

bang, *sb.*⁴ Revived (*U.S. slang*) and often treated as if a slang sense of BANG *sb.*¹ Also, a 'shot' (of cocaine, etc.).
[**1922** E. F. MURPHY *Black Candle* (1926) I. vi. 61 He resorts to 'a shot' of morphine, or 'a bhang' of cocaine.] **1929** C. G. GIVENS in *Sat. Even. Post* 13 Apr. 54/4 An addict is..a bangster, and a bang is a load, a charge or a hyp of the drug he uses. **1933** *Amer. Speech* Apr. 27/2 The injection of dope is referred to as *a bang in the arm* or *a shot in the arm.* **1962** K. ORVIS *Damned & Destroyed* xii. 82 He .. talked me into sampling a bang.

bang, *adv.* [See BANG *v.*¹ 8, and cf. SLAM-BANG.] Thoroughly, completely; exactly. orig. *dial.* and *U.S.*, now *colloq.*
1828 *Night Watch* II. 27, I fetched way bang overboard into the trawl. **1885** TENNYSON *Tiresias* 109 Steevie be right good manners bang thruf to the tip o' the taail. **1907** G. B. SHAW *How he Lied* in *Works* (1930) XI. 192 Do..you propose that we should walk right bang up to Teddy and tell him we're going away together? **1924** A. J. SMALL *Frozen Gold* I. 28 Here they were right bang on hand—and.. they might as well be a thousand miles away. **1931** L. A. G. STRONG *Garden* xix. 170 Bang opposite him..hung a..blue cylinder. **1938** G. GREENE *Brighton Rock* II. i. 74 He..led the way bang straight down Frank's stairs.
b. *bang on*, exactly on. (Cf. quot. 1832 s.v. *BANG *v.*¹ 8.) Used as *adj.*, exactly right, extremely apposite, excellent.
1936 *Punch* 22 Apr. 461/3 'Quiet garden square near Hyde Park. Real hot water. Bang on Tubes.' *Newspaper Advt.* **1943** HUNT & PRINGLE *Service Slang* 12 *Bang on*, bomber slang for 'O.K.' or 'Everything's all right'. **1945** PARTRIDGE *Dict. R.A.F. Slang* 14 *Bang on!*, All right! Correct! In Bomber Command: from a bomb dropped *bang on* (exactly on) the target. **1945** C. H. WARD-JACKSON *Piece of Cake* (ed. 2) 12 *Bang-on*, perfect, excellent. **1949** E. POUND *Pisan Cantos* lxxx. 94 And he dumped all his old stock of calicos plumb bang on the germans. **1958** *Times* 4 Jan. 6/1 'Steering by the sun from earlier fixes we came bang on the base,' Sir Edmund Hillary reported. **1958** *Spectator* 14 Feb. 210/3 As a realistic tale of low life in London, it is bang on. **1958** *Oxf. Mag.* 27 Feb. 324/2 It [*sc.* a play] has enough quality and sense of the theatre to suggest that before long he will land one bang on the target. **1958** J. WAIN *Contenders* 6 I'd been to Brighton for a holiday, and I thought it was bang-on.

bangalay (bæŋæ·li). *Austral.* Also (*slang*) **bang alley.** [Native name.] Bastard mahogany.
1884 A. NILSON *Timber Trees N.S.W.* 133 Bangalay. Eucalyptus botryoides. Myrtaceæ. **1898** MORRIS *Austral English* 18/1 *Bangalay*, a Sydney workmen's name for the timber of *Eucalyptus botrioides*, Smith...The name is aboriginal, and by workmen is always pronounced Bang Alley. **1935** *Bulletin* (Sydney) 2 Jan. 21/4 Here in the Waikato (M.L.) district, *E. botryoides* (bangalay) has grown from a seedling to 15 ft. 2 in. in 18 months.

Bangalore torpedo (bæ·ŋgălǭ:ɹ tǫɹpī·do). *Mil.* [f. *Bangalore*, south Indian city.] A tube containing explosive used by troops for blowing up wire entanglements.
1913 R. L. MCCLINTOCK in *R.E. Jrnl.* Mar. 135 (*caption*) The Bangalore 'Torpedo' for destroying wire entanglements. *Ibid.* 137 The 'Bangalore Torpedo' system is both quick to load and easy to carry up and place. **1925** FRASER & GIBBONS *Soldier & Sailor Words*, *Bangalore torpedo*, a device for clearing a pathway through a barbed wire entanglement, ..Introduced on the Western Front early in 1915. **1941** *Illustr. London News* CXCVIII. 80/2 The Bangalore Torpedo, first used by an inventive R.E. officer at Bangalore, consists of sectional tubes, each packed with explosive, having a pointed nose and a hollow cone at the base, into which fits the nose of the next section.

bangalow (bæ·ŋgălōᵘ). Also **-alo.** [Native name.] Either of two Australian palms of the genus *Archontophœnix*, esp. *A. cunninghamiana*, having feathery leaves. Also *attrib.*
[**1826** J. ATKINSON *Agric. & Grazing N.S.W.* 4 The bangally, much resembling the cabbage tree in appearance, but having some long and wide leaves of a thick and tenacious texture; these the natives tie up at each end in the form of a boat, and use them for carrying water.] **1851** J. HENDERSON *Excurs. N.S.W.* II. ix. 229 The Bangalo ..is a palm, and a native of the brushes...Its bunch of large leaves, surmounting a fine, straight stem, has a very beautiful effect. **1878** W. R. GUILFOYLE *Austral. Bot.* 18 The Aborigines of New South Wales and Queensland..eat the young leaves of the cabbage and bangalow palms. *a* **1882** H. C. KENDALL *Poems* (1886) 193 You see, he was bred in a bangalow wood, And bangalow pith was the principal food. **1905** *Westm. Gaz.* 22 Mar. 2/1 The forest of lofty gums and iron-barks, and clumps of graceful bangalow palms.

banger¹. Add: (Earlier and later examples.)
1657 A. C[ROWTHER] & T. V[INCENT SADLER] *Daily Exercise, of the Devout Rosarists* 374 To omit any necessary Ceremonie, or commit any notorious banger, in saying Mass, or administering any Sacrament. **1899** *Daily News* 23 Sept. 6/3 In London the unfortunate lady banger would have been run in as a nuisance. **1959** I. & P. OPIE *Lore & Lang. Schoolchildren* xii. 283 They classify their fireworks as 'bangers', 'pretty ones', 'hoppers', and rockets.
2. A bludgeon. *U.S. slang* (at Yale).
1849 (*title*) The Yale Banger (Th.). **1856** *Yale Lit. Mag.* XXI. 282 (Th.), Brandishing a banger above my head, I came on to the stage with a yell. **1906** *Springfield Weekly Republ.* 10 May 1 He..has rescued from some museum of Yale antiquities..his old 'banger' of student days.
3. A kiss, esp. a violent one. *slang.*
1898 *Daily News* 25 Jan. 7/3 There were hieroglyphics in the form of crosses for kisses; in one case the cross occupying a very large space..having written against it 'Oh, what a banger'. **1959** H. HOBSON *Mission House Murder* ii. 19 'Here—give us a banger first.' Honeypuss .. obediently offered him her lips.
4. A sausage. *slang.*
1919 W. H. DOWNING *Digger Dialects* 10 *Banger*, a sausage. **1928** *Weekly Dispatch* 27 May 14 Away they [*sc.* the boats] go fully laden with men and tea—not forgetting bangers (sausages) to cook after bathing. **1949** M. DICKENS *Flowers on Grass* vii. 182 The chap had bought him tea and bangers and mash.. **1959** *Times* 5 Nov. 13/6 There is.. nothing exclusive about the childish use of 'banger' for sausage.

†**banger,** *sb.*² *U.S.* var. BANJO. *Obs.*
1775 J. ADAIR *Hist. Amer. Indians* 175 One of their old sacred musical instruments..resembled the Negroe-Banger in shape. **1803** J. DAVIS *Travels in U.S.A.* 379 My young master..made me learn to play the Banger. **1849** in *Amer. Speech* (1951) XXVI. 182 About a dozen [*sic*] negroes.. enjoying the innocent amusement of playing the banger.

banghy, var. *BANGY.

banging (bæ·ŋiŋ), *vbl. sb.*³ [f. BANG *v.*² + -ING¹.] The action of BANG *v.*²
1877 F. H. BURNETT *Pretty Polly Pemberton* (1878) i. 12 Her hair is cut, on her round white forehead, Sir Peter Lely fashion, (they call it banging, I believe).

Bangkok (bæŋkǫk). (Stress variable.) [Name of the capital of Thailand.] A kind of woven straw for hats.
1920 *Tatler* 16 June p. xxv (Advt.), Becoming Hat for Country Wear. In Bangkok, crown and edge of printed union. **1924** *Sketch* 9 Apr. p. xii, Charming Hat in Chinese bangkok. **1927** *Observer* 28 Aug. 17/2 It matters little whether it [*sc.* a hat-shape] be carried out in felt or bangkok or manilla.

bango (bæ·ŋgo). An East African reed.
1899 WERNER *Captain of Locusts* 15 His dug-out canoe.. was found..among the *bango* in a sheltered backwater. *Ibid.* 66 Sleepy natives..were turning over on their *bango* mats. **1907** *Macm. Mag.* May 525 All the bango reed was dead and burnt.

Bangorian (bæŋgōᵊ·riăn), *a.* [f. *Bangor* + -IAN.] Of or pertaining to Bangor, N. Wales, in *Bangorian controversy*, a religious controversy raised by a sermon preached before the king in 1717 by Benjamin Hoadly, Bishop of Bangor, directed against the non-jurors.
1718 C. NORRIS (*title*) The Reconciler, or the Bangorian Controversy, abridg'd. **1759** GOLDSMITH *Bee* VII. 201 They are informed of the excellence of the Bangorian controversy. **1875** *Encycl. Brit.* III. 280/2 It was the year [1717] in which Bishop Hoadley preached that famous sermon on *The Kingdom of Christ*, which gave rise to the..theological war known as the 'Bangorian controversy'. **1881** *Ibid.* XII. 29/1 A war of pamphlets known as the Bangorian controversy.
Hence as *sb.*, a supporter of Benjamin Hoadly.
1721 [see CONSTITUTIONER 2].

bang-tail. Add: Also, a horse or other animal (in Australia, esp. cattle) whose tail has been cut in this way; in *U.S. slang*, any horse, esp. a race-horse; *bang-tail muster*: in Australia, a round-up of cattle during which the tuft at the end of the tail is cut straight across as the cattle are counted.
1887 W. S. S. TYRWHITT *New Chum in Queensland Bush* iii. 61 Every third or fourth year on a cattle station, they have what is called a 'bang tail muster'; that is to say, all the cattle are brought into the yards, and have the long hairs at the end of the tail cut off square, with knives or sheep-shears...The object of it is, to..find out the actual number of cattle on the run. **1921** *Collier's* 27 Aug. 20/2 If by some miracle the bangtail wins—beat it! **1927** M. M. BENNETT *Christison* xxv. 223 In a bang-tail muster each animal was put through the yards and counted, the tip of hair on its tail being cut off square.

bang-up, *adj. phr.*, (*advb. phr.*), and *sb.* Add: **A.** *adj.* (Earlier and later examples.) See also quot. 1811.
1810 C. I. M. DIBDIN *Bang Up!* 6 'Bang-up' seems the watchword to be, From one tip-top driver to t'other. **1811** *Lex. Balatronicum*, *Bang Up*, (*Whip*) quite the thing, hellish fine...A bang-up cove; a dashing fellow who spends his money freely. To bang up prime: to bring your horses up in a dashing or fine style: as the swell's rattler and prads are bang up prime; the gentleman sports an elegant carriage and fine horses. **1865** DICKENS *Mut. Fr.* I. II. viii. 239 A slap-up gal in a bang-up chariot. **1923** R. D. PAINE *Comr. Rolling Ocean* vii. 118 The salaries are bang-up nowadays. **1938** I. KUHN *Assigned to Adv.* iv. 39 There'd be a bang-up dinner, then a drive to some place outside the city where they could dance.
B. Illustration of the advb. phr. *bang up* = quite close up (to), right up (to). In modern use, freq. in *attrib.* phrases.
1819 J. H. VAUX *Mem.* II. 154 A person, whose dress..is in the first style of perfection, is declared to be *bang up to the mark.* A man who has behaved with extraordinary spirit.. is also said to have come *bang up to the mark.* **1858** LYTTON *What will he do with it?* I. i, Smart and sharp, bang up to the day. **1907** [see *BANG *adv.*]. **1914** JOYCE *Dubliners* 260 Drive bang up against Trinity College gates. **1963** *Times Lit. Suppl.* 8 Feb. 92/5 Bang-up-to-date neutron know-how. **1963** *Listener* 21 Feb. 353/2 They had a bang-up-to-the-minute subject in Mr. Harold Wilson's election to the Labour Party leadership.
C. *sb.* **a.** A man of fashion; a dandy. *Obs.*
1811 *Lex. Balatronicum* Pref., We trust..that the whole tribe of second-rate Bang ups will feel grateful [etc.]. **1824** *Examiner* 613/1 Our Corinthians, *Roués* and Bang-ups. **1882** *Punch* 22 Apr. 185/1 The Trio turned into the Arcade and saw a number of gay sparks and fair ones promenading. ..These then are the dandies, the fops, the goes, and the bang-ups, these the Corinthians of to-day.
b. A heavy overcoat (see quot. 1903). *U.S.*
[**1810** *Sporting Mag.* Dec. 127/1 One article was..a *bang-up* great coat.] **1835** *Fraser's Mag.* XI. 298 Dames in bang-ups, Shawls swath'd round men. **1842** *Spirit of Times* (Philad.) 13 Jan. (Th.), A gentleman dressed in a dark coloured fashionable bang-up. **1843** LEVER *J. Hinton* xxi, A green coat..over which he wore..a white 'bang-up', as it was called. **1846** J. KEEGAN *Leg. & Poems* (1907) 396 His old grey frieze bang-up. **1853** *Public Ledger* (Philad.) 11 June (De Vere), He was attired in an old bang-up, black vest, grey pants, and straw hat. **1903** CLAPIN *Dict. Amer.* 36 *Bang-up*, an old word for a heavy overcoat, still surviving in some parts of the Union.

bangy, banghy (ba·ŋgi). *India.* [ad. Hind. *bahangi*, Marathi *bangi* (Skr. *vihamgika*).] A yoke for carrying loads; such a yoke with its pair of baskets or boxes; hence, parcel post.
1789 R. BROOME *Lett. Simpkin the Second* 21 I'll give them Two Thousand, with Bangies and Coolies. **1809** VISCT. VALENTIA *Voy. & Trav. India* I. 67 We take..., in six bangys, sufficient changes of linen. **1810** T. WILLIAMSON *E. India Vade-Mecum* I. 325 The bangy-wollah, that is, the bearer who carries the bangy. **1837** T. BACON *First Impr. Hindostan* II. vii. 218 A weighty banghi-load of provisions. **1840** *Narr. Three Months' March India* vii. 200 Hindoos carrying small baskets..in bangy-fashion. **1842** G. ARTHUR *Let.* 27 Oct. in Ld. Ellenborough *Ind. Administr.* (1874) 221, I will forward with this, by bhangy dâk, a copy of [a book].

bani(y)a, var. *BUNNIA.

banjo. Add: (Earlier U.S. examples of mod. form.)
1774 P. V. FITHIAN *Jrnl.* (1900) 103 In the School-Room, ..several Negroes & Ben & Harry are playing on a Banjo & dancing. **1801** *Port Folio* (Philad.) I. 270 (Th.), The sound of Banneker's banjo would be as tuneable as Gallatin's spoken French. **1836** *Southern Lit. Messenger* II. 162, I found a crowd of negroes..dancing..to the music of a banjo.
Also with distinguishing term.
1934 S. R. NELSON *All about Jazz* ii. 53 The banjolin was.. soon discarded in favour of the Tenor Banjo. **1961** A. BIRCH in A. Baines *Mus. Instruments* vii. 182 For the violinist who wished to play some more suitable instrument in the dance-band of the twenties there was the 'tenor-banjo' tuned in fifths.
2. *transf.* Applied to contrivances of the shape of a banjo: see quots.
1867 SMYTH *Sailor's Word-bk.*, *Banjo*, the brass frame in which the screw-propeller of a steamer works, and is hung for hoisting the screw on deck...The banjo is essential to lifting the screw. **1902** *Encycl. Brit.* XXXII. 147/2 (*Railways*) The enclosed disc signal, commonly called a 'banjo', is a circular box..with a glass-covered opening, behind which a red disc is shown to indicate stop. **1964** *Gloss. Letterpress Rotary Print. Terms* (*B.S.I.*) 9 *Banjo*, the end adjusting duct screw.
b. *spec.* A shovel; an entrenching tool. Chiefly *Austral.* and *N.Z.*
1918 *N.Z. Exped. Force Chron.* 30 Aug. 57/1 We are still wielding the old 'Banjo' in good style. **1925** in FRASER & GIBBONS *Soldier & Sailor Words.* **1931** 'DEAN STIFF' *Milk & Honey Route* 199 Banjo, a short-handled shovel. **1933** *Bulletin* (Sydney) 22 Nov. 21/1 With banjo, pick and barrow I toil. **1954** *Coast to Coast* 67 All swung their banjos so hard that production costs dropped. **1960** *N.Z. Listener* 21 Oct. 7/3 'It's interesting work—anything's better than the banjo.' 'The banjo?' 'A shovel. Some people call it a Mexican side-loader.' **1966** T. H. SHEPPARD *Dict. Railway Slang* (ed. 2) 3 *Banjo*, fireman's shovel.
c. *attrib.*, with the meaning 'banjo-shaped', as *banjo axle, union*; **banjo-clock** *U.S.*, a clock in a case shaped like a banjo; **banjo-frame** (see quot. 1888).
1888 *Lockwood's Dict. Mech. Engin.*, *Bow Connecting-Rod*..or *Banjo Frame*, a form of connecting-rod employed in steam pumps. **1903** F. C. MORSE *Furniture of Olden Time* xi. 328 Simon Willard patented in 1802 an 'improved timepiece' which Mr. Howard says is the clock now known as the 'banjo' clock from its shape. **1922** *Autocar* 10 Nov. 962 A vertical banjo type of back axle. *Ibid.* 982 A neat banjo axle. **1958** *Times* 10 Oct. 8/3 He had found a leak in the banjo union of the cylinder [*sc.* of a car]. *Ibid.*, He had tightened the banjo union nut one-sixth of a turn. **1959** *Engineering* 6 Feb. 187/3 The air from the compressor.. then passes through a banjo type junction.

banjolin (bæ·ndʒǒlin). Also **-ine.** [f. BANJO + *-lin* of MANDOLIN.] A musical instrument combining the characteristics of the banjo and mandolin.
1889 *Pall Mall Gaz.* 4 July 5/1 An instrument that appeared to be a cross between a banjo and a mandolin—dubbed on the spot a 'banjolin'. **1923** *Ashore & Afloat* Aug. (Advt.), Banjolins or Jazz Banjos. **1950** J. VEDEY *Band Leaders* 39 One playing drums and the other the banjoline, an instrument which was enjoying popularity in dance bands but which is now obsolete. **1961** A. BIRCH in

A. Baines *Mus. Instruments* vii. 182 The mandolinist could play the double-strung 'banjo-mandoline' or the single strings of the 'banjoline'.

banjulele (bændʒūlē¹·li). Also **banjo-.** [f. BANJ(O + *-ulele* of *UKULELE.] A stringed musical instrument of a type between a banjo and a ukulele.
1925 *Glasgow Herald* 7 Apr. 8 A native band playing on banjoleles, a sort of zither, from which the tones are provided by gourds filled with varying amounts of water. **1926** *Bulletin* 27 Feb. 10 Alvin D. Keech, from Hawaii, has secured a British patent for his invention, the banjulele. **1926** *Westm. Gaz.* 1 July, A youth..strumming on a banjulele. **1937** *Sunday Times* 21 Feb. 4/4 Driven to try to pawn his precious banjolele.

bank, *sb.*¹ Add: **2. b.** *spec.* on a railway track.
1893 in *Funk's Stand. Dict.* **1908** *Railway Mag.* Feb. 111/1 The fastest work I ever recorded with the same loads up the Hemerdon and Burlescombe banks.
8. c. Phr. *bank and bank* (see quot. 1933). *N.Z.*
1863 S. BUTLER *First Year in Canterbury Settlement* vi. 79 Half a dozen times in a year, the river is what is called bank and bank, that is to say, one mass of water from one side to the other. **1933** L. G. D. ACLAND in *Press* (Christchurch) 9 Sept. 15/7 *Bank and bank.* A river runs b. and b. when all the streams join into one; i.e., when in high flood.
10. (Earlier and later examples in *Mining.*)
1708 J. C. *Compl. Collier* (1845) 32 Horses to draw your Coals to Bank (or Day) out of the Pit. **1892** *Daily News* 11 Mar. 5/7 The preparations which are made at many pits to bring horses and ponies 'to bank'.
12. *Aeronaut.* The lateral inclination of an aeroplane when turning or rounding a curve.
1913 C. MELLOR *Airman* vi. Illustration 29 A left-handed turn with plenty of bank. **1928** C. F. S. GAMBLE *North Sea Air Station* xiii. 225 When turning with a heavy bank. **1955** *Times* 13 July 4/6 The maximum of the bank was about 40 deg. When the bank was jerked off there was a crash and he saw the port wing was gone.
III. bank-bait, the may-fly; **bank-barn** *N. Amer.* (see quot. 1909); **bank beaver** (see *BEAVER¹ 1 b); **bank-engine,** (*b*) a locomotive used to assist in taking a heavy load up a steep incline (= *BANKER³ 5); **bank-fence** (example); *bank-fisherman, -fishery, -fishing* (examples).
1879 E. P. WRIGHT *Animal Life* 485 A great many [may-flies] fall into the water a prey to fishes...Hence the name bank-bait. **1894** *Congress. Rec.* Jan. 1036/1 On my father's farm, when I was a boy, there stood a big bank-barn. **1906** 'R. CONNOR' *Doctor* 26 The foundation of the bank-barn. **1909** *Cent. Dict.* Suppl., *Bank-barn*, a barn built on a hill-side or sloping ground, so that three sides of the lower story are surrounded by earth, the fourth being unbanked. **1968** *Globe & Mail* (Toronto) 13 Feb. 31/6 (Advt.), New bank barn 36 × 80, equipped for 38 cows. **1893** *Funk's Stand. Dict.*, Bank-engine. **1707** MORTIMER *Husb.* 231 The Bank-fence is likewise a good shelter for the Land and the Cattle. **1782** ST. J. DE CRÈVECŒUR *Lett. Amer. Farmer* iv. 146 Nantucket is a great nursery of seamen, pilots, coasters, and bank-fishermen. **1777** in *9th Rep. Hist. MSS. Comm.* App. III. II. (1910) 69 Without a large force is sent out to me the bank fishery is at a stand. **1861** *New Amer. Cycl.* XII. 302/1 About..80 [vessels are employed] in the coast and bank fisheries. **1797** *Encycl. Brit.* XIII. 27/1 The bank fishing season, begins the 10th of May. **1955** *Bulletin* (Bridgewater, N.S.) 9 Mar. 9/2 A number of fishermen..have gone sealing out of Halifax, or engaged in bank fishing.

bank, *sb.*² Add: **5. b.** A row of keys on a typewriter.
1875 [see TYPEWRITER 1]. **1959** *Chambers's Encycl.* XIV. 66/2 The keyboard had four straight rows, or 'banks', of eleven keys each.
7. b. *optical bank*: an optical bench; a graduated bench, usually of steel, on which the holders of lenses, prisms, etc., may be set up in alignment.
1888 *Electrician* 21 Sept., To order expensive and highly polished optical banks and other apparatus from an instrument maker.
8. b. (Further examples.) Cf. *pot-bank* (POT *sb.*¹ 14).
1902 A. BENNETT *Anna of Five Towns* iii. 61 What's amiss with this bank is that it wants pullin' down. **1903** —— *Leonora* iii. 91 He's somewhere on the bank, sir—speaking to the mouldmaker.
10. A set of similar pieces of apparatus or units of equipment grouped together. In various spec. uses: **a.** *Electr.* Lights arranged in rows or tiers.
1902 WEBSTER Add., *Bank*, a group or series of objects arranged near together; as, a *bank* of electric lamps, etc. **1911** D. S. HULFISH *Cycl. Motion-Pict. Work* II. 102 A bank of mercury-vapor lamps is suspended from the ceiling. **1935** *Discovery* Apr. 111/1 'Drifting' news signs, in which a series of letters is caused to drift across the face of a bank of lamps. **1958** *Times* 24 Sept. 13/4 A huge bank of lights dominates this rebuilt sound stage.
b. In *Automatic Telephony*: a series of fixed contacts in a selector or switch.
1904 M. M. KIRKMAN *Telegraph & Telephone* II. iii. 50 The contact banks consist of ten layers of contacts...The top bank is known as the 'private' bank, while the two lower banks are called 'line' banks. **1922** *Encycl. Brit.* XXXII. 708/2 Banks of contacts for 200 and 500 [telephone] lines respectively are employed. **1926** *Gloss. Electr. Terms* (Brit. Engin. Stand.) 174 *Wiper*, that portion of the moving member of a selector or other similar device which engages with the contacts of a bank.

bank, *sb.*³ Add: **7. d.** Delete †*Obs.* and add further examples.
a **1747** S. CIBBER *Let.* in D. Garrick *Private Corresp.* (1831) I. 50 She is the greatest coquet in England, and has half-a-dozen husbands in bank, in case of your death. **1844** *Knickerbocker* XXIII. 14 [She] claimed..among her chattels sundry shares in bank. **1902** O. WISTER *Virginian* xxiii. 273 Take my land away to-morrow, and I'd still have my savings in bank.
e. Phr. *in the Bank* (see quot. 1930).
1930 M. CLARK *Home Trade* xxix. 234 In the language of the Money Market, money becomes tight and, maybe, the loans available from the banks are insufficient to go round. Then the assistance of the Bank of England is sought and the Money Market is said to be 'in the Bank'. **1955** *Times* 30 June 16/7 Credit was tighter in Lombard Street yesterday, and one or two houses were 'in the Bank' for a very small amount.
f. *transf.* A store of things for future use, a reserve supply: *spec.* of blood for transfusion, tissue for grafting, or the like. Cf. *blood-bank* (s.v. *BLOOD *sb.* 19), *eye bank* (s.v. *EYE *sb.*¹ 28). orig. *U.S.*
1938 *Life* 28 Feb. 33/2 The bank maintains a positive balance of blood. **1944** *Reader's Digest* XLV. 25 (*title*) Banks for Human 'Spare Parts'. **1945** *Ann. Reg. 1944* 379 Nerve 'banks' for the supply of quick frozen and dried nerve fragments for nerve grafts were developed. **1947** *Lancet* 12 Apr. 493/2 The value of a bone bank. **1959** *Times* 13 Mar. 15/2 A foetal tissue bank is being established at the Royal Marsden Hospital, comparable to the eye banks. **1963** *Times* 16 Feb. 9/2 We must be prepared for the day when 'banks' of different organs..will be integral parts of all major hospitals.
8. a. *bank-account, -deposit, -depositor, -monger, -president, -robber, -robbery, -snatcher.*
1799 C. B. BROWN *A. Mervyn* iv. 39 Have I not seen his bank account. His deposits..amount to not less than half a million. **1929** D. H. LAWRENCE *Pansies* 33 With bank accounts and insurance policies Don't sympathise. **1832** *Chambers's Edinb. Jrnl.* I. 186/1 After exhausting his bank deposits, he still felt himself in difficulties. **1940** G. CROWTHER *Outl. Money* ii. 40 The ordinary bank depositor keeps all his money in the bank and makes his daily payments out of it. **1814** JEFFERSON *Let.* 24 Jan. in *Writings* (1854) VI. 305, I was derided as a maniac by the tribe of bank-mongers. **1853** *Harper's Mag.* Jan. 193 As if some invalid clergyman or bank-president, in white cravat, wished sedately to have his carriage called. **1799** *Aurora* (Philad.) 15 Mar. (Th.), Groups of pickpockets, bank-robbers, and hen-pecked dotards. **1894** NISBET *Bush Girl's Rom.* 277 Wildrake was the real criminal and bank-robber. **1854** DICKENS *Hard T.* III. viii. 339, I have suspected young Mr. Tom of this bank-robbery from the first. **1890** *Harper's Mag.* Feb. 472 One of the most daring bank snatchers in the city effected two robberies in the course of a single day.
b. bank-clerk, a clerk (see CLERK *sb.* 6 b) in a bank; hence *bank-clerkly* adj.; **bank-paper,** (*b*) (see quot. 1888); **bank-parlour** (earlier example); **bank-post** (earlier example); **bank-roll** *U.S.*, a roll of bank-notes; hence as *v. trans. colloq.*, to support financially.
1829 *Harlequin* 13 June 35 This burlesque was written by Mr. Rhodes, who was 'only a bank-clerk'. **1859** SALA *Tw. round Clock* 42 From sober Hackney, and Dalston, and Kingsland, bank-clerk beloved. **1920** E. POUND *H. S. Mauberley* 19 But in Ealing With the most bank-clerkly of Englishmen? **1959** *Listener* 23 July 153/3 The same sort of boyish fun—or bank-clerkly humour—induced inordinate pleasure [etc.]. **1888** C. T. JACOBI *Printers' Vocab.* 6 *Bank paper*, a thin paper mostly used for foreign letter or note paper to save cost of postage. **1859** SALA *Tw. round Clock* 160 Tremendous bank-partners..pay a farewell visit to the bank parlour. **1854** C. TOMLINSON *Cycl. Useful Arts* II. 369/1 *Names, dimensions*, and *weight per ream* of Writing and Drawing Papers... Bank post 19 by 15½ [inches] 7 [lb.]. **1887** *Courier-Jrnl.* (U.S.) 23 Jan. 5/3 One night a well-dressed stranger went over and won the bank roll. **1928** J. P. MCEVOY *Showgirl* vii. 106 That's the one Milton wants to bankroll, isn't it? **1938** *Amer. Speech* XIII. 196 Bank-rolling. **1944** D. RUNYON *R. à la Carte* (1946) vi. 92, I am not able to bank-roll you to a very large start. **1959** *Times Lit. Suppl.* 30 Oct. 628/4 He was the big banker of New York crime; he kept the bankroll. **1959** J. R. MACDONALD *Galton Case* (1960) xxi. 165 Who's bankrolling you? Dr. Howell? **1967** *Boston Sunday Herald* 26 Mar. IV. 5/4 The network has ..bankrolled the Broadway production of 'My Fair Lady'. **1968** *Globe & Mail* (Toronto) Mag. 17 Feb. 9/2 An angry Texan is said to have bet his $45,000 bank roll that Jowett couldn't do it.

bank, *v.*¹ Add: **8.** (Earlier example.) Also *fig.*
1712 J. JAMES tr. *Dézallier d'Argenville's Gardening* II. ii. 111 You bank it up by causing Earth to be laid about the Foot of it. **1920** *Round Table* Dec. 88 The steady march of Russia eastwards, and her threat to..the independence and future of Japan was rapidly banking up the clouds of war.
b. *spec.* To pile up (logs) at a landing, etc., for transport by water or rail. *U.S.* and *Canada.*
1856 *Trans. Mich. Agric. Soc.* VII. 828 There will be logs enough cut and 'banked' for 100,000,000 feet of lumber. We are informed that the amount now banked daily, will amount to 2,500,000 feet. **1888** *B.C. Moon* 21 Apr., Wright & Davis..have purchased the logs banked at West Superior. **1904** S. E. WHITE *Blazed Trail Stories* iii. 40 The firm agreed to pay..for all saw-logs banked at a rollway.
9. (Example of *refl.* use.) Also, to pile *up*, accumulate. Also *absol.*
1889 *Good Words* Mar. 154/1 The driven snow was now banking itself up in wreaths. **1889** E. RANDOLPH *New Eve* I. i. 40 The fleecy clouds..are already banking to prepare the sun his couch. **1922** *Glasgow Herald* 24 Nov. 9 Imagine ..a gravelled roadway flanked by redcoats, and a crowd banking and shelving on both sides. **1955** *Times* 24 May 19/4 The six months' gap..had allowed the traffic to bank up.
10. Also *to bank down* (example is *fig.*).
1923 D. H. LAWRENCE *Birds, Beasts & Flowers* 82 His fires of wrath are banked down.
12. *trans.* **a.** To ascend (an inclined surface). **b.** To cause to travel an ascending track; also in *vbl. sb.* (attrib. in *banking engine* = *bank-engine*, *BANK *sb.*¹ III).
1892 *Livestock Jrnl. Alm.* 34 They ascended a steep hill, banking field after field to a flag-post at the top. **1908** *Model Engineer & Electr.* 11 June 570 Where banking engines are employed for assisting trains on inclines. *Ibid.*, Wherever banking assistance is taken. *Ibid.*, The practice of banking trains out of Euston, up the Camden incline.
13. *trans.* In *Aeronautics*, to tilt (an aeroplane) sideways in turning. Also *intr.*, to incline sideways in turning. Also with *up.*
1911 GRAHAME-WHITE & HARPER *Aeroplane* 133 He 'banked' his biplane over too sharply. **1913** C. MELLOR *Airman* vi. 29 We swung round left-handed and the machine 'banked' up to the right. **1920** *Blackw. Mag.* July 72/2 Adam Smythe..then banked left-handed towards Delhi Fort.

bank, *v.*² Add: **1.** (Later example.)
1912 E. V. LUCAS *Wand. in Florence* v. 60 Giovanni [de' Medici] had been a banker before everything, Cosimo an administrator... Lorenzo continued to bank but mismanaged the work and lost heavily.
3. (Earlier example.)
1838 *Actors by Daylight* I. 55 After having 'banked' their cash.
4. b. To 'put one's money' *upon*; to count or rely *on* with confidence or assurance. Also const. *that.* *colloq.* (orig. *U.S.*).
1884 NYE *Baled Hay* 127 The man who ranks as a dignified snoozer, and banks on winning wealth and a deathless name. **1892** *Congress. Rec.* Apr., App. 249/2, I am not banking heavily on [him]..as an honest man. **1898** *Sun* (N.Y.) 14 Sept., The Democrats are banking upon this movement to help them out this fall. **1903** A. ADAMS *Log Cowboy* vi. 79, I was banking plenty strong, that next year.. I'd take her home with me. **1910** W. M. RAINE *B. O'Connor* 58 The one friend you would have banked on to a finish. **1949** DODIE SMITH *I capture Castle* I. ix. 152 'Don't bank on things too much,' I begged. 'Simon may not have the faintest idea of proposing.'
5. *trans.* To store (blood, tissue, or the like) for future use. orig. *U.S.* Cf. *BANK *sb.*³ 7 f.
1938 *Life* 28 Feb. 33/2 Here you see how blood is taken in, banked and given out. **1963** *Guardian* 17 July 1/1 Aortic heart valves from young victims of road crashes are being 'banked' by a team of doctors to replace those of older people.

bankable, *a.* (U.S. examples.)
1818 M. & S. C. AUSTIN *Austin Papers* (1924) 329 The money must be *Bankable*, as none other will do. **1832** *Register Deb. Congress* 7 Mar. 204 A currency..perhaps not bankable at all places. *Ibid.*, Any broker..would make it bankable any where in the Union. **1848** BARTLETT *Dict. Amer.* 22 In New York, at auction sales, the auctioneer.. invariably states, that the money must be bankable. **1921** *Collier's* 19 Feb. 17/1 (*heading*) Good will a bankable asset.

banker¹. Delete †*Obs.* and add: Now *arch.* or *dial.*
1870 BOTTRELL *Trad. W. Cornwall* 257 The cosy, old, panelled settle, but now without the bankers and dorsars, or the cushions, for the seats and back. **1890** W. MORRIS in *Eng. Illustr. Mag.* July 755 Some went to the chests and brought out the rich hangings, the goodly bankers and dorsars.

banker². Add: **4.** A card game in which the banker divides the pack into a number of piles placed face downward, and each punter bets on the chance that the bottom card of the pile chosen by him is higher than the bottom card of the pile left to the banker.
1891 *Daily News* 3 Dec. 7/6 A boy..was charged with gambling with cards at 'banker'. **1903** *Daily Chron.* 14 Mar. 7/5 He got £150 during the voyage home on the troopship by playing 'banker' and 'the crown and anchor'.
5. In *Football Pools*: a result which one forecasts consistently in a series of entries. Cf. *BANK *v.*² 4 b.
1947 *Answers* 30 Aug. 9/3 This is the method of entry for a 14-match pool, allowing for six permutated matches with eight bankers. In 12- and 13-match pools, the bankers required will be six and seven respectively. **1958** *Punch* 27 Aug. 265/3 No pools investor of quality would seek advice from hacks who write: Wolves have banker look.

banker³. Add: **1.** (Later examples.)
1815 in J. Q. ADAMS *Duplicate Lett.* (1822) 219 Those descriptions of vessels are not so valuable as the bankers, more particularly those that go from the District of Maine, Connecticut, and Rhode Island. **1880** *Harper's Mag.* Aug. 350/1 The long voyages of the bankers. **1960** *Atlantic Advocate* Nov. 30 New vessels—craft of fifty tons or less, and less than half the size of latter-day bankers.
b. A fisherman on the Bank of Newfoundland.
1861 *Harper's Mag.* Mar. 461/2 On the banks of Newfoundland..some of the old bankers predicted a gale. **1907** J. G. MILLAIS *Newfoundland* vii. 154 The fishermen of all lands have to encounter the perils of the deep, but none have to face the risks that the 'bankers' do.

4. (Examples.)
1890 *Cassell's Picturesque Australasia* III. 175 The Murrumbidgee was running a 'banker'—water right up to the banks. **1936** M. FRANKLIN *All that Swagger* x. 88 Every stream Delacy..met was a banker. **1936** I. L. IDRIESS *Cattle King* xvi. 162 Occasionally a river or creek in a raging banker would defeat the little time-tables of man.
5. *banker* (*engine*), a locomotive used to assist in taking a heavy load up a steep slope. (Cf. *bank-engine*, *BANK *sb.*¹ III.)
1907 *Westm. Gaz.* 3 Dec. 7/2 A 'banker' engine had been fixed to the rear of the mineral train to help to take the load up a steep incline. After going some 300 yards the 'banker' was cast off.

banker⁴. Add: **b.** *attrib.* **banker-mark** (see quots. 1888 and 1928).
1888 T. W. WHITLEY in *Leamington Spa Courier* 11 Aug. 7/6 Each man as he finishes his work at the banker, places his mark upon the stone before it leaves the shed. The banker is the stone bed or bench upon which a mason works. ..These marks have hence been called banker marks, and perhaps the name is more appropriate than that of masons marks. **1910** *Encycl. Brit.* III. 319/2 Banker-marks are..to be found on all old buildings of consequence, ecclesiastical or otherwise. **1928** G. G. COULTON *Art & Reformation* viii. 143 'Banker-marks', that is, the mason's sign-manual which he set on his finished stone before it left the banker, or working-bench.

banket (bæŋke·t). [Du., a confection resembling almond hardbake (see quot. 1887).] The name given by the early gold-diggers of the Transvaal to the gold-bearing conglomerates of the Witwatersrand, later extended to similar conglomerates elsewhere.
1887 *Chambers's Jrnl.* Apr. 284 The conglomerate..is a peculiar formation of almond-shaped pebbles, pressed into a solid mass in a bed of rock of an igneous nature, and is called 'Banket' on account of its resemblance to a favourite Dutch sweetmeat known in England as almond rock. The 'Banket' is also rich in gold. **1897** BRYCE *Impressions S. Africa* 217 In 1885 the conglomerate or *banket* beds of the Witwatersrand were discovered. **1900** *Times* 13 Feb. 13/3 The banket formation of the Witwatersrand. **1902** *Encycl. Brit.* XXIX. 17/2 Deposits similar to the Witwatersrand banket occur in Zululand, and also on the Gold Coast of Africa. **1932** WATERMEYER & HOFFENBERG *Witwatersrand Mining Practice* ii. 28 The world-wide fame of the gold deposits of the Witwatersrand has resulted in the term 'banket' or 'banket reef' being adopted for gold-bearing conglomerates in several other parts of the world. **1954** D. DIVINE *Golden Fool* vi. 57 Another smaller outcrop of the same stone—banket, he called it—a sort of fruit cake of pebbles and small pieces in a conglomerate.

bank holiday. Add: Also *attrib.*; and as *adj.*, (as if) enjoying a bank holiday; festive. Hence **bank-ho·lidayish** *a.*
1871 *Act* 34 *Vict.* xvii. 7 This act may be cited for all purposes as the Bank Holidays Act, 1871. **1897** *Westm. Gaz.* 7 Aug. (Advt.), Bank holiday attractions. *Ibid.*, Bank holiday programme. **1899** CONAN DOYLE *Duet* (1909) 9/2 If he had to travel all the way from Edinburgh with a Bank-holiday crowd. **1926** S. T. WARNER *Lolly Willowes* III. 244 Grass that has been laid upon has always a rather bank-holidayish look. **1930** R. LEHMANN *Note in Music* vii. 294 'We look very bank-holiday,' she said. **1938** L. MACNEICE *Zoo* xiv. 227 We felt pleasantly Bank Holidayish.

banking, *vbl. sb.* Add: **2.** (Earlier and later examples.)
1610 FOLKINGHAM *Art of Survey* II. ii. 50 Sometimes this Compound Boundage implies a mutuall propertie or duty participable to the Conterminants, as bancking, balking, [etc.]. **1712** J. JAMES tr. *Dézallier d'Argenville's Gardening* II. ii. 106 It is call'd Earthing or Banking up a Stake, when you cause Earth to be brought, and make a Bank about the foot of it. **1845** C. M. KIRKLAND *Western Clearings* 103 'Banking up'..consists in piling earth round the foundations. **1865** *Nation* (U.S.) I. 683 He has so much a task for banking.
3. (Earlier examples.)
1776 G. SEMPLE *Building in Water* xv. 101 The banking.. is to extend from Pier to Pier. *Ibid.* 107 The banking or artificial Bed of the River.
7. On a road or track: see quot. 1904.
1904 GOODCHILD & TWENEY *Technol. & Sci. Dict.* 40/1 *Banking of cycle tracks*, the raising of the outer edge of a bend in the track to counteract the centrifugal force. **1905** *Cycling* 26 Apr. 362/1 Herne Hill track is now in very good condition. The banking is not so steep as the Palace track. **1937** *Times* 13 Apr. p. viii/2 The danger of vehicles leaving the road at bends..has during recent years been counter-balanced..by the process known as 'banking' or super-elevating.

ba·nking, *ppl. a.*² [f. BANK *v.*¹ 9.] Forming into banks.
c **1867** 'MARK TWAIN' *Sketches* in *Wks.* XIX. 372 A weird picture, that small company of frantic men fighting the banking snows.

Banksia. Add: **2.** *Banksia rose*: the Banksian rose (see next).
1890 [see TIME *sb.* 15 b]. **1905** 'Q' *Shining Ferry* i. 2 A climbing Banksia rose overgrew the sill and ran up the mullions. **1934** *Bulletin* (Sydney) 21 Feb. 21/3 A Banksia-rose bush planted as a cutting 50 years ago last August.

Banksian (bæ·ŋksiăn), *a.* [f. the name *Banks*: see below and -IAN.] Epithet of: (*a*) a Chinese species of climbing rose, bearing small white or yellow flowers in clusters, named after Lady Banks; (*b*) the Labrador, Grey, or Jack Pine, *Pinus banksiana*, named after Sir Joseph Banks.
1837 [see ROSE *sb.* 3 a]. **1841** MRS. LOUDON *Ladies' Comp. Fl. Gard.* 252/1 The Banksian Roses..which are of two kinds. **1864** S. HIBBERD *Rose Bk.* 31 The Double White Banksian was introduced in 1807, and was so named in honour of Lady Banks. **1872** W. F. BUTLER *Great Lone Land* (ed. 2) xxii. 346 The juniper, the banksian pine, and the black spruce. **1886** *Encycl. Brit.* XX. 851 The Banksian Rose is a Chinese climbing species, with small white or fawn-coloured flowers of great beauty. **1920** *19th Cent.* July 175 The little yellow Banksian is still incomparable.

banky, *a.* Delete *Obs.* or *dial.* and add later examples.
1868 *Jrnl. R. Agric. Soc.* IV. 266, 120 acres of banky grass land. **1931** *Observer* 4 Oct. 26/5 The banky marge of a woodland..will be a good site for a large colony.

banner, *sb.*¹ Add: **1. e.** A flag awarded as a distinction. (See quot. 1840 and cf. *6 b.) *U.S.*
1840 *Log Cabin* (N.Y.) 5 Dec. 2/3 It is known that the Ladies of New Orleans early in the late contest offered a splendid Banner to the State which should give the largest relative majority for Harrison and Tyler in its popular vote for Presidential Electors. **1900** *Century Mag.* LIX. 636/1 Local authorities..united in the belief that..Ashtabula County might be accorded the banner.
2. c. A headline in large type, esp. one running across a whole page in a newspaper. Also *attrib.*, esp. *banner headline.* orig. *U.S.*
1913 W. G. BLEYER *Newsp. Writing & Editing* xi. 305 Important news may be given as a head of one, two, or three parts extending across the whole front page. Such a head is often called a 'banner'. **1915** G. M. HYDE *Newspaper Editing* I. v. 167 Banner headlines have undergone a strange evolution. They were invented to assist in street sales by advertising the news. **1952** V. GOLLANCZ *My Dear Timothy* xx. 275 The *Daily Herald* came out with a huge banner headline, in letters half an inch high, on its opening page. **1957** *Listener* 1 Aug. 165/1 A five-column streamer in looming Gothic, followed by a banner line in some sort of spidery italic. **1965** *Ibid.* 9 Sept. 374/2 There is no longer a morning newspaper bravely bearing the name of Manchester on its banner-head.
3. b. (*a*) Each of the eight divisions, with distinguishing flags or banners, into which the Manchu army was divided. (*b*) A military subdivision of Mongolian tribes. Cf. *BANNERMAN 2.
1842 K. S. MACKENZIE *Narr. Second Campaign in China* ix. 140 The army is divided into eight divisions, distinguished by the colour of their respective flag; the yellow or Imperial..being the highest, next the white banner, red and blue banners. **1848** S. W. WILLIAMS *Middle Kingd.* I. vii. 333 The Manchu army..was assisted by Mongols and Chinese, the three nations were divided into eight corps or 'banners'. **1880** J. ROSS *Manchus* xvi. 610 Up till 1613 the Manchus were divided into four banners—yellow, red, blue and white; but they had become so large an army, that for efficiency in manoeuvring they were sub-divided into as many more—bordered yellow, bordered red, bordered blue, and bordered white. *Ibid.* 611 In 1635, the Mongols were separated from the Manchu banners, under eight banners of their own. **1894** *New Review* Nov. 528 The Banner troops received donations from the Emperor.
6. *banner-flying, -towing* (in aerial advertising); **banner cloud,** a cloud that streams outwards from the lee side of a mountain peak; **banner headline** (see sense *2 c).
1909 *Cent. Dict.* Suppl., Banner-cloud. **1957** *Meteorol. Gloss.* (*Air Ministry*) (ed. 3) 32 The Helm Cloud over Crossfell Range, the Table Cloth over Table Mountain..are.. well-known examples of banner cloud. **1930** *Flight* 7 Feb. 195/1 Another profitable use to which N.A.L. have been putting their machines is banner-flying. **1960** *Guardian* 29 Dec. 3/1 The rates for banner-towing are between £25 and £30 an hour.
b. *attrib.* or quasi-*adj.* Entitled to a banner as a distinction (orig. in *banner state, county*); hence, pre-eminent, supreme. *U.S.*
1840 *Niles' Reg.* 5 Dec. 210/1 Which is the Banner State? —The Whigs..proposed to designate whichever state should give the Harrison ticket the largest majority, as the banner state. *Ibid.*, The banner county. Designation is claimed by Worcester, Massachusetts. **1843** *Knickerbocker* XXII. 431 He who was for many years the banner-veteran of our worthies. **1886** *Harper's Mag.* June 78/2 The Magnolia Plantation..which claims to be one of the banner plantations of the State. **1887** *Ibid.* July 237/2 She had the banner crop of tobacco in that county last year. **1903** *N.Y. Even. Post* 21 Sept., The earnings of all Vanderbilt lines had a banner month in August. **1911** S. E. WHITE *Bobby Orde* (1916) x. 128 On his banner day he brought down two fox-squirrels. **1931** G. T. CLARK *Leland Stanford* x. 361 That was the farm's banner year, for at the close of the season it held world trotting records for yearlings..and stallions. **1967** *Boston Sunday Herald* 26 Mar. III. 3/2 Next season will be a banner year for the browns, reports the American Wool Council.

banner, *v.* Add: **3.** *trans.* To announce in a banner headline (see *BANNER *sb.*¹ 2 c).
1951 *Manch. Guardian Weekly* 15 Mar. 10/4 It prominently reported what the Miami 'Daily News' had bannered the night before. **1959** D. BEATY *Cone of Silence* xix. 209 The name of the aircraft that for years they strove to perfect has been bannered across the papers of the world with the most tragic connotations.

bannered, *ppl. a.* Add: **1.** (Later examples.) Also *fig.*
1869 'MARK TWAIN' *Innoc. Abroad* xxx. 317 Clothes-lines ..waving their bannered raggedness over the swarms of people below. **1897** *Westm. Gaz.* 9 Dec. 2/1 Through arched and bannered woodlands. **1908** *Daily Chron.* 2 June 3/2 The English Prince daily contemplated the bannered lines which he had been born to lead. **1952** R. CAMPBELL tr. *Poems of Baudelaire* 166 Our torch-like hearts their bannered flames unroll.
2. (Earlier example.)
1796 SOUTHEY *Joan of Arc* v. 130 The banner'd Lion waved on Gergeau's wall.

banneret², var. BANNERETTE. (Further examples.)
1914 YEATS *Responsibilities* 38 Packed his marriage day With banneret and pennon. **1916** E. POUND *Lustra* 101 The crackling of small fires, the bannerets, The lazy leopards on the largest banner.

bannerman. Restrict *Sc. arch.* to sense in Dict. and add: **2.** A soldier belonging to any of the eight banners of the Manchu army.
1899 MRS. H. FRASER *Dipl. Wife in Japan* II. 133 Shinno-Shiko nodded to his green bannermen to slay them. **1908** *Westm. Gaz.* 17 Dec. 7/3 The Manchu 'Bannermen' (soldiery) of Pekin.

banquette. Add: **2.** (Earlier example.)
1772 C. HUTTON *Bridges* v. 83 The banquet or raised foot way on each side, leaving a sufficient breadth in the middle for horses and carriages.
4. An upholstered bench-like seat.
1851 [see SALA¹]. **1933** *Archit. Rev.* LXXIV. 242 (*caption*) A view across the counter from the serving side shows the wide recess round which runs a long low banquette covered in a brilliant red tartan. **1953** *Ibid.* CXIII. 168/1 (*caption*) Banquette seating in the staff cafeteria. **1958** E. DUNDY *Dud Avocado* I. vi. 92 Along the walls [of the café] ran a banquette upholstered in very old red plush.

bantam. Add: **1. c.** *bantam weight* (Boxing): see quot. 1954.
1884 *Times* 10 Apr. 6/4 The competitions were this year five in number, or one more than usual, a cup having been instituted for 'bantam weights', or men not exceeding 8 st. 4 lb. **1894** *Outing* (U.S.) XXIV. 71/2 An attempt to match the coxswains for a bantam-weight contest. **1897** *Encycl. Sport* May 139/2 Boxers are divided by the Amateur Boxing Association into five classes, according to their weights, as follows:—Bantam Weight, not exceeding 8 st. 4 lb. **1903, 1910** [see WELTER WEIGHT 2]. **1954** F. C. AVIS *Boxing Ref. Dict.* 8 *Bantamweight*, a standard weight division for professional boxers weighing more than 8 st. but not more than 8 st. 6 lb.; amateurs 8 st. and 8 st. 7 lb. respectively.
d. Applied to battalions, etc., of small-sized soldiers.
1914 *Daily Express* 20 Nov. 5/5 'Bigland's Bantams' will probably be the pet name of a battalion which is being raised of men who are just too short to enlist under the ordinary conditions... The Bantams Battalion has been recognised by the War Office. **1914** *Scotsman* 11 Dec. 7/4 The Edinburgh Rotary Club..has now completed arrangements for the raising of a 'Bantam Battalion'. **1927** *Observer* 30 Oct. 7 The 35th was a bantam division which went out to France very early in 1916.

banteng, banting (bæ·nteŋ, bæ·ntiŋ). [Malay.] A species of wild ox; = TSINE.
1817 T. S. RAFFLES *Hist. Java* I. i. 49 Wild quadrupeds, the rhinoceros, and *bánteng*, or wild Javan ox. **1860** *Chambers's Encycl.* I. 674/2 *Banteng*..a native of Java and Borneo..black, with white legs. **1880** *Encycl. Brit.* XII. 742/2 The *tsine* or *banting* (*Bibos sondaicus*), found in Burmah. **1894** *Field* 9 June 815/3 Blood-curdling tales are told in Batavia of the ferocity of the rhinoceros and banteng. **1900, 1903** [see TSINE]. **1961** *Times* 20 May 6/3 Whipsnade Zoo today announced the birth of a female banteng calf.

banter, *sb.* **4.** (Examples.)
1835 LONGSTREET *Georgia Scenes* 26 No, said Peter, you made the banter, now make your pass. *a* **1861** WINTHROP *John Brent* (1883) ii. 16 I'm goan to make yer a fair banter. **1872** SCHELE DE VERE *Americanisms* 439 We had a fine banter, but the match was postponed till spring.

banter, *v.* **6.** (Examples.)
1810 F. CUMING *Sk. Tour Western Country* 135 Two hunters..bantered each other to go out and kill a deer. **1834** CARRUTHERS *Kentuckian in N.Y.* I. 183, I was thinking of walking out into the country and bantering somebody for a footrace. **1836** D. CROCKETT *Exploits in Texas* 83 The black-leg set to work with his thimble again, and bantered me to bet. **1860** *Knickerbocker* Aug. LVI. 221 The farmer again bantered him to buy his berries. **1872** E. EGGLESTON *End of World* xxvi. 177 The cards were put face down, and the company was bantered to bet the wine. **1902** HARBEN *Abner Daniel* 163 Colonel Barclay has..bantered me for a trade time an' again.

banting, var. *BANTENG.

Bantu (bæ·ntu, bæntū·), *a.* and *sb.* [In certain Bantu dialects, pl. (also *abantu*, etc.) of *-ntu* man.] **A.** *adj.* Of or pertaining to an extensive group of negroid peoples inhabiting the equatorial and southern regions of Africa, and of the languages spoken by them; originally given to the family of languages by W. H. I. Bleek. **B.** *sb.* **1.** *ellipt.* as *pl.* Bantu people.

2. Any of the languages spoken by them.
1862 BLEEK *Compar. Gram. S. Afr. Lang.* i. 4 The South African division of the Bâ-ntu family of languages consists of one large middle body, occupying almost the whole known territory between the tropic of Capricorn and the equator. *Ibid.* 11 The abandonment in the Bâ-ntu languages of such syllabic elements as are more difficult of pronunciation. 1886 *Encycl. Brit.* XIII. 820/2. 1884 *Ibid.* XVII. 319/1. 1928 *Contemp. Rev.* Mar. 298 The Bantu, or Native races, numbering 4,698,000. 1948 M. GUTHRIE *Classif. Bantu Lang.* 19 Although languages of this kind cannot be called Bantu owing to their not having the complete prefix system we have described as a criterion, their relationship to the Bantu languages is sufficiently close for them to be taken into account. We shall therefore call them 'Sub-Bantu'. 1951 R. FIRTH *Elem. Social Organization* i. 8 In parts of Bantu Africa, to give a person an ancestor's name means that he is believed to reincarnate that ancestor's spirit.

banya, var. *BUNNIA.

banzai (bænzai·), *int.* [Jap., literally, ten thousand years.] **A.** A shout or cheer used by the Japanese in greeting the emperor or in battle. Also as *sb.*
1893 E. ARNOLD *Adzuma* II. i, At the departure of the Imperial train, the citizens raise loyal cries of 'banzai! banzai!' 1904 *Daily Chron.* 12 Feb. 5/3 The crowd sang national songs, and shouted 'Banzai' continuously. 1905 *Times* 7 Oct., Enthusiastic banzais were given for King Edward and the Emperor of Japan.
B. *attrib.* or as *adj.* = (as if) shouting 'banzai', uproarious, jollificatory. *slang.*
1929 F. BOWEN *Sea Slang* 7 *Banzai Party*, a party of naval men going ashore for a spree. 1932 KIPLING *Limits & Renewals* 199 That's how it was till the Squadron returned. The banzai-parties came ashore, all hats and bosannas like a tax-payers' treat.
b. Applied to a reckless attack by Japanese servicemen.
1945 *Coast to Coast 1944* 106 Out in the glaring sky a Zero started its Banzai run. 1945 *San Francisco News* 13 June 3/1 Smashed desperate 'banzai' charges by doomed enemy survivors.

baptism. Add: **2. b.** *baptism of fire*: after eccl. Gr. βάπτισμα πυρός (e.g. Macarius Ægyptius *Hom.* xxvii. 17; cf. Matt. iii. 11), (*a*) the grace of the Holy Spirit imparted through baptism, as distinguished from the sacrament or rite; (*b*) martyrdom, esp. by fire; (*c*) the undergoing of any severe ordeal or painful experience; (*d*) a soldier's first experience 'under fire' in battle (so F. *baptême de feu*). Cf. *fire-baptism* s.v. FIRE *sb.* B. 3 a.
[1822 B. E. O'MEARA *Napoleon in Exile* I. 107, I love a brave soldier who has undergone, *le baptême du feu*, whatever nation he may belong to.] 1857 G. LAWRENCE *Guy Livingstone* xii, It's only in their baptism of fire that the young ones shrink and start. 1881 R. HUNTER et al. *Encycl. Dict.* s.v., When during the Franco-German war of 1870, Prince Louis Napoleon..was first exposed, by direction of his father, Napoleon III, and with his own consent, to the fire of the enemy at Saarbrück, the event was called a 'baptism of fire'.

bar, *sb.*[1] Add: **3.** Also of chocolate.
1906 *Daily Chron.* 25 July 6/4 A shop-worn chocolate-cream bar. 1959 *Elizabethan* Apr. 10/1, I gave you a bar of chocolate on the train from London.
b. Used as a standard of weight or a denomination of currency. Cf. BAHAR, BARR(E.
1732 in F. MOORE *Trav. Inland Afr.* (1738) App. II. 9 Barr, or Sixteenth Part of an Ounce of Gold. 1738 *Ibid.* 45 A Barr is a Denomination given to a certain Quantity of Goods of any kind; which Quantity was of equal Value among the Natives to a Barr of Iron, when this River was first traded to. Thus..an Ounce of Silver is but a Barr. 1732 *Hist. of Voy. to New Calabar River, 1699, taken from Jrnl. of James Barbot* in *Coll. Voy. & Trav.* V. 460/1 We were obliged to make the reduction of our merchandize into bars of iron, as the standard coin, viz. One bunch of beads, one bar.. One piece broad Hamborough, one ditto.. One piece Nicanees, three ditto..And so pro rata, for every other sort of goods. 1737 J. ATKINS *Voy. to Guinea* (ed. 2) 40 They all keep Gromettas (Negro Servants) which they hire from Sherbro-River, at two Accys or Bars a Month. 1755 [see FLET].
c. A pound; esp. in *half a bar*, ten shillings. *slang.*
1911 J. W. HORSLEY *I Remember* xi. 254 Others [slang words] were new to me, such as..'bar' for a sovereign. 1938 F. D. SHARPE *S. of Flying Squad* 331 *Half a bar*, ten shillings. 1939 J. B. PRIESTLEY *Let People Sing* x. 256 Knocker brought out some money and examined it... 'A nicker, half a bar, a caser an' a hole.' 1958 M. PUCKLE *Wilderness of Monkeys* 77 Half a bar, or what you call ten bob!
5. *stars and bars*: see STAR *sb.*[1] 6 b.
b. *bar of Michael Angelo*, the superciliary ridge or prominence of the frontal bone at the base of the forehead, characteristic of the heads of Michael Angelo's statues.
1850 TENNYSON *In Mem.* lxxxv. 127 And over those ethereal eyes The bar of Michael Angelo.
9. b. *spec.* in *pl.* A set of wooden rails which may be withdrawn to afford an opening through a fence or wall. (Cf. DRAW-BAR.) *U.S.*
1639 in *Conn. Hist. Soc. Coll.* (1897) VI. 5 All the fences & gates..to the bares shall be sufitiently mad up. 1660 in *Rec. Providence, R.I.* III. 139 Provided that they Keepe a Sufficient inlett of Barres at Each End of the highway for A cart to passe through. 1703 *Ibid.* V. 209 [He] shall set up a Gate, or inlett of Barrs in said fence. 1887 M. E. WILKINS *Humble Romance* 315 The younger of the two old women let down the bars which separated the blooming field from the road, and they passed through.
13. d. (Also in Fr. form *barre*.) A horizontal bar fixed to the wall at waist level, serving as a support for dancers in certain of their exercises.
1883 DUTTON COOK *On Stage* II. i. 4 Then the pupil is taught to stand on one leg while extending the other until the foot rests upon a horizontal bar raised some four or five feet above the floor. 1922 BEAUMONT & IDZIKOWSKI *Manual Class. Theatr. Dancing* II. i. 13 Generally the bar is of wood, and is fixed to the walls of the practice room in a horizontal position. 1936 A. L. HASKELL *Prelude to Ballet* ii. 7 She performs her complicated routine, first hanging on to the barre, wooden counterpart to the hands of a partner. 1958 J. WINEARLS *Mod. Dance* (ed. 2) ii. 17 Exercises at the barre should be done daily in the same order.
16. *attrib.* in *bar-line*.
1927 *Grove Dict. Mus.* (ed. 3) I. 219/1 A bar..is, literally, the straight line drawn across the stave to mark the metrical accent... In ordinary parlance that is now called the 'bar-line'. 1959 WIMSATT & BEARDSLEY in *PMLA* Dec. 589 Music—or at least music with bar-lines—is precisely a time-measuring notation.
25. b. The counsel retained in a particular case.
1891 E. KINGLAKE *Australian at Home* 36 He had as strong a Bar as could be retained on his side. 1892 *Daily News* 13 Oct. 5/2 There has rarely been such a Bar in any modern case, either for quantity or for quality. Sir Charles Russell, the Attorney-General, the Solicitor-General, Sir Henry James, Mr. Inderwick, and Mr. Tindal Atkinson were but a few of them.
28. b. With defining word: a shop counter at which a particular item or group of items is sold. orig. *U.S.*
1954 *Word Study* Feb. 4/1 In a New Jersey suburban town someone has opened what she calls 'Mi-Lady's Corset Bar'. From its wares mi-lady probably gets food for thought. 1966 *Harper's Bazaar* May 57 (Advt.), The Fashion Bar.
29. a. *bar-magnet* (earlier example); **c.** *bar-counter, -girl, -loafer* (hence *-loafing* adj.), *-stool*.
1842 DICKENS *Amer. Notes* I. vi. 218 He finishes by leaping gloriously upon the bar-counter, and calling for something to drink. 1945 KOESTLER *Twilight Bar* i. (stage direction) At front, left, semicircular bar counter with high bar-stools. 1857 C. KINGSLEY *Two Y. Ago* III. vi. 165 The bar-girl, who knew his humour, came forward. 1870 D. J. KIRWAN *Palace & Hovel* (1963) viii. 82 A little girl, with a bold face, acted as bar-girl. 1962 *Times Lit. Suppl.* 19 Apr. 262/4 Changes in postwar Japan: the popularity of 'bar-girls', the modern substitute for geisha. 1968 *Listener* 23 May 657/1 Few Americans now are seen in central Saigon; respectable girls will now venture down To Do Street, which the bar girls have abandoned. 1889 G. B. SHAW in *Star* 30 Aug. 4/3 The mere bar loafers at these concerts. 1868 F. WATKINS *Electro-Magnetism* 18 A fundamental principle of magnetism may be shown by freely suspending on its centre of gravity an artificial bar magnet. 1922 JOYCE *Ulysses* 446 Bob Doran, toppling from a high barstool, sways over the munching spaniel.
30. *bar-armature Electr. Engin.*, a bar-wound armature; **bar-bell**, a steel bar weighted with a ball of iron at each end, used as a dumb-bell; **bar-boat**, (*b*) a boat adapted for carrying goods across the bar of a river; **bar-cutter**, a shearing-machine for cutting metallic bars into lengths; a workman who passes the metal through the machine; **bar-fly**, a frequenter of bars; **bar-frame** (or **-framed**) *a.*, of a bee-hive: fitted with bars instead of sections; **bar-headed goose**, a goose found in India and Central Asia, *Anser indicus*; **bar-hive**, a bar-framed beehive; **bar-keel**, one composed of rectangular bars of iron or steel; **bar-keep** *U.S.*, a bar-keeper (for refreshments); **bar-movement**, a type of watch movement in which the upper pivots are carried in bars; **bar-point**, the point or division nearest the bar in the outer 'table' of a backgammon board; **bar-room** (earlier *U.S.* example); also *attrib.*; **bar-soap**, soap made up into bars as distinguished from soap in cakes or tablets; also *attrib.*; **bar-super** (see quot.); **bar tacker** (see quot.); so **bar tack**, *-tacking*; **bar winding** *Electr. Engin.*, an armature-winding consisting of metallic bars instead of wires; **bar-wound** *a.*, of an armature: fitted with bars instead of wires.
1888 S. P. THOMPSON *Dynamo-Electric Machinery* (ed. 3) 646 (Index) Bar Armatures. 1904 GOODCHILD & TWENEY *Technol. & Sci. Dict.* 40/1 *Bar armature*, a large armature in which the windings are built up of copper bars instead of continuous wire. 1887 *Howe Lisl* I. 17 A complete set of dumb-bell, bar-bell, marching and running exercises. 1895 *Cal. Univ. Nebraska 1895–6* 252 The gymnasium..is well equipped with clubs, wands, bar bells, and dumb bells. 1883 MOLONEY *W. African Fisheries* 17 (Fish. Exhib. Publ.) Bar-boats of seven to eight tons have been used at Lagos. 1897 M. KINGSLEY *W. Africa* 635 [It is] too bad a bar for boats to cross; but a steamer on the Lagos bar boat plan might manage it. *a* 1877 KNIGHT *Dict. Mech.* II. 229/2 *Bar-cutter* (Metal-working), a shearing-machine which cuts metallic bars into lengths. 1904 *Westm. Gaz.* 13 June 7/2 He gives bar-cutters an advance of a halfpenny per ton. 1910 *Sat. Even. Post* 16 Sept. 5/2 Then, after having confessed to so much money, he hastened out, for he wud not be stung by bar-flies. 1932 C. ISHERWOOD *Memorial* iii. ii. 201 A small market-town, inhabited by commercial travellers,..and other bar-flies. 1944 AUDEN *For Time Being* (1945) 80 The pay-off lines of limericks in which The weak resentful bar-fly shows his sting. 1881 *Gardening Illustr.* 7 May 123/1 There would be no difficulty whatever in putting swarms of bees into a bar-frame hive, provided it has a movable top and floor-board. 1892 Bar frame [see SUPER *sb.* 4]. 1905 *Daily Chron.* 18 June 6/6 Bar-frame beehives. *Ibid.* 1 Sept. 6/4 Bar-framed hives. 1924 *Glasgow Herald* 29 July 8 The barhead goose and the ruddy sheldrake collect in flocks on the Tibetan swamps. 1879 *Encycl. Brit.* X. 777/2 The Bar-headed Goose (*Anser indicus*). 1884 PHIN *Dict. Apiculture* 70 *Bars*, strips of wood to which combs are attached, and from which they hang in bar-hives. 1874 THEARLE *Naval Archit.* (Advt. Sc. ser.) IV. xvii. 268 The Bar Keel..is generally of hammered iron, made in pieces as long as can be conveniently forged. 1846 *Spirit of Times* (N.Y.) 4 July 218/2 We embarked..in company with..a barkeep to mix the l-q-rs. 1900 KIPLING *Captive* in *Traffics & Discoveries* (1904) 8 Take away his hair and his gun and he'd make a first-class Schenectady bar-keep. 1918 H. A. VACHELL *Some Happenings* I. 2 Hobo listened attentively to the bar-keep. 1905 F. J. GARRARD *Watch Repairing* i. 1 To describe, in general terms, the mechanism of a watch..a 'Geneva bar' movement will be used as an illustration..the 'bar' construction enables all the wheelwork to be seen. 1962 E. BRUTON *Dict. Clocks & Watches* 21 *Bar movement*, early form of partly machine-made watch movement in which bars, or bridges and cocks, are used to hold bearings for one pivot of each wheel, for easy dismantling. 1743 HOYLE *Back-gammon* 10 The next best Point (after you have gained your Cinq Point) is to make your Bar Point. 1840 Bar point [see RUN]. 1793 J. HOLTZHEIMER *Diary* 28 July (1803) 245 Seidel's conversation in the barroom started from a spring in his garden, through pipes into his bar-room. 1839 'MRS. MARY CLAVERS' *New Home* i. 9 When my husband..drew with a piece of chalk on the bar-room table at Danforth's the plan of a village. 1946 AUDEN *Introd.* in H. James *Amer. Scene* p. v, One can easily imagine Stendhal or Tolstoi or Dostoevsky becoming involved in a bar-room fight, but James, never. 1826 *Catawba Jrnl.* (N.C.) 26 Oct. 10 [dozen] Bar Soap. 1879 MARK TWAIN *Tramp Abroad* iv. 42 A piece of yellow bar-soap. 1893 EARL DUNMORE *Pamirs* I. 64 Some common yellow bar-soap. 1906 *Westm. Gaz.* 27 Oct. 7/2 Bar-soap soiled. 1884 PHIN *Dict. Apiculture* 70 A bar super is simply a case or crate in which the honeycomb is hung from bars. 1921 *Dict. Occup. Terms* (1927) § 419 *Bar tacker*, bastes or tacks round buttonholes of tailored garments, to keep parts together before holes are cut. 1955 I. E. LIBERTY *Pract. Tailoring* (ed. 2) xi. 204 The Bar Tack is usually done with buttonhole twist. A bar of two or three stitches is formed..It is then worked with a small over stitching from end to end. 1904 GOODCHILD & TWENEY *Technol. & Sci. Dict.* 40/1 *Bar winding*, conductors in the form of bars are slipped into the slots from the end, and then connected up into coils by means of separate V-end connectors all the same size and shape. 1902 *Encycl. Brit.* XXVII. 284/2 If, however, the current is reversed..the former-wound conductors which are sufficiently thin to be inserted, after shaping in a suitable jig. 1940 *Chambers's Techn. Dict.* 76/1 *Bar-wound armature*, an armature in which the conductors consist of copper bars.

bar, *sb.*[5] *U.S.* Short for *mosquito bar*. (Cf. MOSQUITO 6.)
1847 C. LANMAN *Summer in Wilderness* xxiv. 143 Had I not taken with me..bar netting..the creatures would have eaten me. 1866 C. H. GREGG *Life in Army* xv. 140 Nothing can exceed the luxury of lying down inside your 'bars' of a midsummer night, and feeling secure from these voracious bills. 1894 'MARK TWAIN' *Those Twins* 415 Get their bed ready, and see that you drive all the mosquitoes out of their bar.

bar (bɑː), *sb.*[6] [f. Gr. βάρος weight; cf. ISOBAR.] **1.** A unit of pressure equivalent to one dyne per square centimetre.
1903 RICHARDS & STULL *New Method determining Compressibility* 43 Might not the pressure of a dyne per square centimeter be suitably called a 'bar'?
2. A unit of barometric pressure equivalent to a pressure of 29·53 inches or 750·1 mm. of mercury at 0° C. in latitude 45°. (See also quot. 1918.)
1910 V. BJERKNES et al. *Dynamic Meteorol. & Hydrography* I. i. 7 It will be necessary for us to have names for the employed units of pressure. ... as names for the units of pressure...I employed these expressions for the first time in a paper published in 1906 [in *Beiträge zur Physik der freien Atmosphäre*, Strassburg]. 1917 A. MCADIE in *Ann. Astron. Observ. Harvard* LXXXIII. 47 The term millibar was unfortunately used by Bjerknes in his 'Dynamic Meteorology and Hydrography'. He defined the *bar* as a megadyne atmosphere, seemingly unaware of a

prior use of the word by Richards and others in its proper sense. 1918 *Meteorol. Gloss.* 43 Bar..was introduced into practical meteorology by V. Bjerknes, and objection has been raised by McAdie..on the ground that the name had been previously appropriated by chemists to the C.G.S. unit of pressure, the dyne per square centimetre. The meteorological bar is thus one million chemical bars, and what chemists call a *bar* we should call a *microbar*.

bar, *v.* **9.** (Later examples.)
1903 WODEHOUSE *Prefect's Uncle* i. 11, I bar the man. He's slimy. 1935 C. MACKENZIE *Sinister St.* I. ii. xvii. 433 Why, my dear girl, he's absolutely barred. He's as unpopular as anything.

bar, *prep.* Add: *bar none*, with no exceptions.
1866 M. E. BRADDON *Lady's Mile* (ed. 4) II. vi. 192 Your 'Aspasia' is the greatest picture that ever was painted—'bar none', as Mr. Lobyer would say.
b. Esp. in *Betting*, indicating the number of horses excluded from odds being offered.
1860 HOTTEN *Slang Dict.* (ed. 2), *Bar*,..in common use in the betting-ring; 'I bet against the field bar two'. 1864 *Ibid.* (ed. 3) s.v., 'Two to one bar one', *i.e.*, two to one against all the horses bar one.

baræsthesiometer (bæresθisi,ɒ·mitəɹ). Also **bares-**. [f. Gr. βάρος weight + αἴσθησις perception + -OMETER.] An instrument for measuring the sense of pressure. Hence **baræsthesiometric** *a.*
1876 H. POWER tr. *von Ziemssen's Cycl. Med.* XI. 213 Eulenburg has materially facilitated the testing of the sense of pressure by constructing a baræsthesiometer,..on which, by varying pressure upon a spring, different degrees of pressure may be read off on a dial. 1885 STERLING tr. *Landois' Human Physiol.* 1092 In order to avoid the necessity of changing the weights, A. Eulenburg invented his baræsthesiometer, which is constructed on the same principle as a spiral spring paper-clip or balance. 1900 DORLAND *Med. Dict.* 102/2 Baresthesiometer.

Bara Sahib: see *BURRA SAHIB.

barasingha (barasi·ŋgă). Also **barasinga**, **barasingh**. [Hind. *bārah singā* lit. twelve-horn.] The east Indian deer *Cervus wallichii*; also, *C. duvaucelii*, the Swamp Deer.
1860 HUTTON *Sporting Sk.* ... 1873 ... 1876 *Encycl. Brit.* VII. 23 ... 1892 *Kipling & Balestier Naulahka* 189 The barasingh ... 1901 *Field* 20 Apr. 554/1 ... 1907 *Westm. Gaz.* 5 Apr. ... 1968 *Times* (Pakistan Suppl.) 6 Apr. p. viii/5 The tiger's principal prey, the gaur (jungle bison) and barasingha (marsh deer) were soon wiped out.

barathea (bærăþī·a). [Origin unknown.] A cloth of a fine texture composed of a silk warp and woollen weft, also of cotton and wool and entirely of wool.
1862 *Internat. Exhib., Catal. Industr. Dept.* II. No. 3958, Coburgs, paramattas, baratheas, reps, cords, cloths. 1897 *Daily News* 29 Oct. 6/5 Venetian crape, has taken the place of the old baratheas, Balmorals, bombazines, &c. 1923 K. ANSELL *Lady Jim* 12. 56 His lavender barathea trousers swayed gracefully with his walk. 1963 *Guardian* 15 Mar. 1/4 All ranks will receive a second suit of the No. 2 khaki service dress, of 22 oz. barathea.

‖**baraza** (bārā·zā). [Swahili.] A place of public audience or reception; hence also a meeting, reception, palaver.
1894 LD. LUGARD *Diary* 17 Mar. (1959) III. 98 Big Baraza, most of the big chiefs present. 1926 *Blackw. Mag.* Apr. 549/1 In the big baraza or native council-house we noticed a native weaving cloth. 1958 *Daily Express* 1 Oct. 11/4 Hundreds of native tribal chiefs..have come here from all parts of East African fastnesses to participate in the great Baraza (pow-wow) for the Prince yesterday. 1959 *Times* 23 Feb. 12/7 The three native rulers of Uganda's western province will be among those who will greet her to-morrow at a baraza.

barb, *sb.*[3] Add: **3.** A black kelpie (see *KELPIE[2]). *Austral.*
1926 K. S. PRICHARD *Working Bullocks* 209 The barb had never fought in a ring before. 1946 F. D. DAVISON *Dusty* (1947) iii. 33 Fine dogs,..black kelpies and red, barbs and border collies.

barb, *v.* Add: **6.** To bend into hook form the points of wire teeth used in carding textile fibres.
1890 NASMITH *Mod. Cotton Spinning Mach.* 94 There are two evils to be guarded against—the barbing or hooking of the wire points and the striation of the sides of the teeth. 1946 *Ibid.* 95 Striated sides and barbed points are common in this series.

Barbadian (baɹbē·diăn), *a.* and *sb.* [f. BARBAD(OES + -IAN.] **A.** *adj.* Of or pertaining to Barbados or its inhabitants. **B.** *sb.* An inhabitant of Barbados. Cf. *BADIAN *sb.*[2] and *a.*

1732 *Cal. State Papers, Col. Ser., Amer. & W. Indies* (1939) 6 'Tis not more than a leather in the cap, but 'tis what the Barbadians like. 1741 *Caribbeana* I. 56 To our Lovely Country-Women, who are single, the Barbadian Batchelors, and Widowers, send lovingly Greeting. 1875 *Encycl. Brit.* III. 361/1 In the meantime Barbadian affairs had attracted notice in Parliament. *Ibid.*, Nor were the Barbadians themselves backward in stating their grievances. 1960 *Times* 11 Jan. 16/6 Not for nothing is Worrell known as the prince of Barbadian batsmen.

Barbadoes. Add: Now usu. **Barbados**. **Barbadoes gooseberry** (see quot. 1876).
1756 P. BROWNE *Civil & Nat. Hist. Jamaica* 237 Cactus 1 ..The Gooseberry, or Barbadoes Gooseberry Bush. 1876 *Encycl. Brit.* IV. 626/2 *Pereskia aculeata*, or Barbados Gooseberry, the Cactus *Pereskia* of Linnaeus. 1902 *Daily Chron.* 15 May 5/1 Attention is again called..to the want of uniformity in the spelling of Barbados, many papers spelling it Barbadoes... All official documents emanating from the Colony, and all the stamps for upwards of half a century, have the word Barbados.

barbarianess (baɹbē·riănĕs). [f. BARBARIAN *sb.* + -ESS[1].] A female barbarian.
1849 C. M. ARNOLD *Let.* 5 Feb. (1895) I. 387 As a very charming Barbarianess, Lady Portarlington, expresses a great desire to make my acquaintance, I daresay the race will bear no malice. 1885 *Salem* (N.Y.) 4 June 261/3 This genuine feat [*sc.* that of eating peas with a knife] he has, in person, contemplated as performed by a charming Viennese barbarianess.

‖**barbasco** (baɹbæ·sko). [Amer. Sp., app. alt. of Sp. *verbasco*, ad. L. *verbascum* mullein.] The popular name of a variety of S. American plants, chiefly of the family *Lonchocarpus*, the roots of which yield a poison; hence, the poison obtained from such a plant.
1860 MAYNE REID *Odd People* 33 They (*sc.* the Amazonian Indians)..fish with a harpoon spear..and sometimes by poisoning the water with the juice of a vine called barbasco. *Ibid.* 61 The deadly poisons of barbasco and *mararucu*. 1934 *Discovery* Sept. 251/1 Wherein such insect controls as Derris root..and Barbasco (*Lonchocarpus nicou*) are definitely exotic, pyrethrum is at home almost anywhere in the temperate zone. 1957 DORLAND *Med. Dict.* (ed. 23) 166/1 *Barbasco*, tropical plants, *Jacquinia paramentica* and *Paullinia pinnata*, used as fish poisons.

barbecue, *sb.* **4.** Delete '(in *U.S.*)' and add earlier and later examples. Also *attrib.*
1733 B. LYNDE *Diary* (1880) 138 Fair and hot; Browne, barbecue; Sabbath. 1813 *Salem* (Mass.) *Gaz.* 30 June 3/2 An elegant Barbecue Dinner. 1881 H. PIERSON *In Brush* 36 On any occasion where the barbecue feast was to be the agreeable conclusion. 1933 *Wodeh. Mar.* 6/2 Today the American countryside is heavily sprinkled with barbecue stands. 1938 D. KENYON *Take it Easy* 301 They are down in Florida running a barbecue stand. 1957 *Daily Mail* 5 Sept. 11/5 Anywhere they (*sc.* Americans) can find a clearing with a barbecue-pit set up, they bring out masses of steak, and the gas-fired barbecue to make the fire. 1968 *Globe & Mail* (Toronto) 3 Feb. 41/3 (Advt.), Lovely covered patio with built-in barbecue. 1968 *Peace News* 21 June 2/4 (Advt.), London WC 1. 7.30 p.m. 28 Great James Street. Summer Peace Party and Barbecue.

barbecue, *v.* **2.** (Later U.S. example.)
1920 J. M. HUNTER *Trail Drivers of Texas* 82 We killed and barbecued a beef.

barbecuing, *vbl. sb.* (Earlier and later examples.)
1705 R. BEVERLEY *Virginia* III. iv. 13 They [the Indians] have two ways of Broyling, viz. one by laying the Meat..upon Sticks rais'd upon Forks at some distance above the live Coals..this they, and we also from them, call Barbacueing. 1968 *Globe & Mail* (Toronto) 17 Feb. 45 (Advt.), Tree shaded patio for summer barbecueing.

barbed, *ppl. a.*[1] Add: **4. b.** *barbed wire*: see WIRE *sb.* 1 e.

barber, *sb.* Add: **1. c.** *transf.* Applied *colloq.* to a bitterly cold wind which seems to 'cut' the face. Chiefly *Canada* and *N.Z.* Also, *frost-smoke* (see FROST *sb.* 7 c).
1830 W. S. MOORSOM *Let. fr. Nova Scotia* 131 Frozen particles of the atmosphere, aptly termed by the natives 'the barber', sweep the surface of the water. 1832 J. MCGREGOR *British Amer.* I. 135 The keen north-west wind, during winter, is often called the 'Barber' in America. 1867 SMYTH *Sailor's Word-bk.* 78 In meteorology, *barber* is a singular vapour rising in streams from the sea surface—owing probably to exhalations being condensed into a visible form, on entering a cold atmosphere. It is well known on the shores of Nova Scotia. 1899 *Westm. Gaz.* 1 Mar. 10/1 Our searching east winds are nothing in comparison with the Canadian 'barber'. 1924 J. S. ANGUS *Glass, Shetland Dial.*, *Barber*, a haar which rises from the surface of water with a very keen freezing. 1933 L. ACLAND in *Press* (Christchurch) 16 Sept. 15/7 *The barber*, a very bitter cold wind which blows down the gorge of the Mawhera and affects..Greymouth. The miners and drovers used the word in the 'sixties, and it is still in use. 1937 J. ELLIOTT *Firth of Wellington* ... 1 There [Grey Gorge] in winter raged an icy blast known as 'the barber', for it was as keen as the sharpest razor.
3. *barber's itch* or *rash*: a disease affecting the face and neck, caused by a fungoid organism resulting from the use of insanitary shaving apparatus.
1890 BILLINGS *Med. Dict.*, *Barber's itch*, tinea barbæ. 1906 *Daily Chron.* 2 Jan. 6/2 A skin complaint known as 'barber's rash'.

barber, *v.* Delete *rare* and add later examples.
1877 E. PEACOCK *Gloss. Manley & Corringham, Linc.* 15/1, I alus barber my son o' Setterda' neet. 1904 *Windsor Mag.* Jan. 288/1 These latter [*sc.* landlords] will 'barber' a dog for a couple of francs. 1948 *Penguin New Writing* XXXIV. 29 He advised Father Eudes to barber his armpits. 1963 *Times* 12 Mar. p. vi/4 Light's noble belt of parklands around the capital has been so barbered and coiffured that rose gardens and ornamental ponds begin to replace gum-trees and native grasses.
Hence **ba·rbered** *ppl. a.*, trimmed or groomed by or as by a barber; also in extended use, of grass, etc.: cut closely.
1910 *Daily Chron.* 29 Jan. 6/1 Arnold..dismissed him [Emerson]..as a barbered and exquisite Carlyle. 1922 *Glasgow Herald* 22 Apr. 4 Bijou villas in barbered demesnes. 1947 W. DE LA MARE *Coll. Stories Childr.* 166 The trim barbered lawns.

barber-shop. [BARBER *sb.* 2.] **1.** A shop where a barber's services may be had. Now chiefly *N. Amer.*
1579, 1872 [see BARBER *sb.* 2]. 1916 W. A. DU PUY *Uncle Sam* 55 His dark locks brushed forth odors of the lotions of cheap barber shops. 1968 *Globe & Mail* (Toronto) 3 Feb. 26/9 (Advt.), Barber shop wanted to purchase.
2. *attrib.*
1905 *Daily Chron.* 25 Jan. 3/3 The portrait [*sc.* of Hawthorne] with which we are all familiar—a curled barber-shop affair—gives no idea of the..charm of his face. 1931 W. FAULKNER *Sanctuary* iv. 36 Don't think I spent last night with a couple of your barber-shop jellies for nothing.
b. Designating music of simple or 'close' harmony, esp. for a male vocal quartet, or a musical ensemble playing or singing such music. Cf. *barber's music* (see BARBER *sb.* 3). *colloq.* (orig. *U.S.*).
1910 WILLIAM TRACEY (title) Play that barber shop chord. 1912 W. STEVENS *Let.* 26 Aug. (1967) 177 It was a chorus of barber-shop harmonies. 1926 H. C. WITWER *Roughly Speaking* 302 The barber shop quartette harmonizes and moonlight necking parties on the boat deck. 1934 C. LAMBERT *Music Ho!* iii. 205 The phrase 'barber-shop chord'..denotes a chord of unusual succulence. 1947 *Penguin Music Mag.* May 28 In his own work he [Tippett] has..taken away the barber-shop harmonies and substituted pure harmonies. 1956 W. MARLOWE *Hallo Adonis* 46 Alphabet 99 That brass band, those lutes and barber-shop ensembles. 1966 *Crescendo* Nov. 13/2 Those guys play what they call a barbershop harmony.

Barberton (bā·ɹbəɹtən). [The name of a town in the Transvaal, S. Africa.] *Barberton daisy*: the Transvaal daisy (see TRANSVAAL). Also *ellipt.*
1906 R. STONEMAN *Plants S. Afr.* 115 When several bracts surround a head of flowers, as in *Protea*, the Barberton daisy, and others of their tribe, they form an involucre. 1932 R. MARLOTH *Flora S. Afr.* III. ii. 28 The best known species is *Gerbera Jamesonii* from the Eastern Transvaal, commonly called Barberton Daisy. 1956 *Amat. Gardening* ... various varieties of colour and foliage. 1956 *H. Gdn.* ... 1960 C. LIGHTON *Cape Floral Kingdom* ... The Barberton daisy is..the daisy from the Barberton district of the Transvaal, but outside the Union our Barbertons are commonly called gerberas.

barbierite (bā·ɹbiəɹait). *Min.* [f. the name of P. Barbier, French chemist: see -ITE. Cf. Fr. *barbiérite*.] (See quots.)
1910 W. T. SCHALLER in *Amer. Jrnl. Sci.* XXX. 358, I propose to call this particular monoclinic soda feldspar *barbierite*, in honor of Prof. Ph. Barbier, of the University of Lyons, France. 1913 *Min. Mag.* XVI. 354 Barbierite..Monoclinic soda-felspar ($NaAlSi_3O_8$) isomorphous with orthoclase and dimorphous with albite. 1957 *Encycl. Brit.* IX. 183/2 Soda-orthoclase, which is more likely to be a pseudomorph..has been described as barbierite..was earlier proposed..and given the name barbierite, but its reality as a mineral has been disputed.

barbiers (bā·ɹbiəɹz). [f. alteration of BERIBERI.] A paralytic disease common in India.
1698 FRYER *Acc. E. India & P.* 68 Whence follows Fluxes, Dropsy, Scurvy, Barbiers. 1768 LIND *Ess. Dis. Hot Climates* 260 The barbiers, a species of the palsy... This disease is generally sudden, and entirely deprives the limbs of their motion. 1822 GOOD *Study Med.* III. 451 [heading] Beriberi, Barbiers. 1833 *Cycl. Pract. Med.* I. 243 Barbiers is generally a chronic disease.

barbital (bā·ɹbităl). The equivalent in the U.S. Pharmacopoeia of *BARBITONE.
1919 *Jrnl. Amer. Med. Assoc.* 1918 XL. 54 During the year 1918, the laboratory examined several specimens of barbital... Barbital was introduced into medicine under the proprietary name 'Veronal'. *Ibid.* 54 Barbital is described..as diethylbarbituric acid. 1922 E. PHILLPOTTS *Grey Room* viii. 183 A woman, who had taken morphine and barbital, was found apparently dead. 1948 T. SOLLMANN *Man. Pharmacol.* (ed. 7) 757 Barbital (veronal; diethyl-barbituric acid). This one, being the earliest, has been most widely used, and has..a more prolonged action. Fischer and Mering, 1903 and 1905... Barbital is used in nervous insomnia, and delirium...

barbitone (bā·ɹbitōn). [f. as next + -ONE.] A hypnotic drug, diethyl barbituric acid, often known by the trade name VERONAL.
1914 *Brit. Pharmacopœia* 62 Barbitonum. Barbitone. Synonyms—Diethyl-barbituric Acid: Malonurea: Diethylmalonyl-urea..$C_8H_{12}N_2O_3$. 1918 *Nomencl. Dis.* (ed. 5), Chloral and other Hypnotic Drugs. Sulphonal, Trional, Tetronal, Barbitone, Proponal. 1923 *Brit. Pharmac. Codex*

175 Barbitone was introduced and first tested, pharmacologically and in clinical practice, under the trade-name Veronal. 1927 *Daily Express* 30 Dec. 7/7 The chemist..supplied a bottle containing twenty-five tablets of barbitone, which was the equivalent of veronal. 1957 J. BRAINE *Room at Top* xxii. 187 Its silence as different from real silence as a barbitone trance from natural sleep.

barbiturate (baɹbi·tiŭrĕt, -ēit, baɹbitiū·rĕt). *Chem.* [f. *BARBITUR(IC *a.* + -ATE[1] 1 c.] A salt of barbituric acid; one of the group of hypnotic and sedative substances derived from barbituric acid. Also *attrib.* or as *adj.*
1928 *Biol. Abstr.* 17323 (*title*) Anesthesia in the dog of allyl-isopropyl-diethylamine barbiturate. 1930 *Lancet* 1 Mar. 473/2 G. Link, who has used isoamylethyl barbiturate in goitre surgery. *Ibid.* 15 Nov. 1086/1 (*title*) The Barbiturates in Labour Cases. 1931 *Ibid.* 14 Mar. 586/1 A symposium on the *Barbiturates in Anæsthesia* was opened by Dr. M. Nierenstein, who described in detail the chemistry of barbituric acid and its derivatives. 1955 *Sci. News Let.* 9 July 19/2 The barbiturate family of sleeping medicines. 1957 *N.Z. Listener* 26 July 15/3 Some of you must already be addicts to barbiturates... Ultimately not only physical but moral health deteriorates if you become a barbiturate habitue. 1963 *Times* 21 Dec. 6/2 The pathologist's view was that the overdose of barbiturate she had taken would have proved fatal. 1967 *Spectator* 7 July 10/1 As regards addiction, it is now definitely established that people can become dependent on barbiturates.

barbituric (baɹbitiū·rik), *a.* *Chem.* [ad. F. *barbiturique* (Ann. de Chimie et de Physique, 1865), f. G. *barbitur* in *barbitursäure* (Baeyer 1863, in *Ann. d. Chemie und Pharm.*), f. *Barbara*, a woman's name.] *barbituric acid*, an acid ($C_4H_4O_3N_2$) from which various hypnotic and sedative drugs are derived; malonyl urea.
1866 ODLING *Anim. Chem.* 128 Baeyer has increased the list of compounds by his discovery of..the violuric and barbituric acids. 1885 REMSEN *Org. Chem.* xii. 204 Barbituric acid..is a product obtained from uric acid. 1927 *Proc. R. Soc. Med.* 20 (*title*) The Clinical and Pathological Effects of Hypnotic Drugs of the Barbituric Acid and Sulphonal Groups. 1941 T. S. ELIOT *Dry Salvages* v. 14 To..fiddle with pentagrams Or barbituric acids.

Barbizon (bā·ɹbizɒ̃n), *a.* [The name of a village near the forest of Fontainebleau, France.] Epithet of a French school of painting of the middle of the 19th century associated with the village of Barbizon.
1890 D. C. THOMSON (*title*) The Barbizon School of Painters. 1896 *McClure's Mag.* VI. 471/2 Corot, Daubigny, Dupré, Troyon, Diaz, Jacque, and others who, with our mania for classification, we call the 'Barbizon school'. 1967 *Listener* 12 Jan. 51/2 At the time he painted this picture the Impressionists were still..followers of the Barbizon painters, Courbet or Boudin.

barbola (baɹbōᵘ·lă). [Proprietary term, arbitrarily f. BARBOTINE.] In full *barbola work*, decorative work composed chiefly of flowers and fruit modelled in a plastic paste and coloured, used to embellish small articles of wood, glass, pulp, etc.
1927 *Daily Express* 26 Sept. 5/5 Tiny porcelain-looking flowers..for the newest buttonholes..which are really a form of barbola work, are arranged in small round posies. 1928 *Ibid.* 22 Feb. 5/3 The latest designs in barbola work mirrors... The barbola or other wood-carving decoration is sparingly used—just a cluster of the painted raised flowers.

barbone (bāɹbōᵘ·ni). *Veterinary Path.* [It., lit. 'thick or long beard', augmentative form of *barba* (see BEARD *sb.*).] An infectious disease of buffaloes and cattle; hæmorrhagic septicæmia.
1907 F. S. H. BALDREY in *Jrnl. Trop. Vet. Sci.* II. 287, I obtained a culture of 'Barbone' or 'Hemorrhagic Septicæmia'... Nomenclature. Hemorrhagic Septicæmia, Pasturella des Bouffles (France), Barbone (Italy) .. Ghotu, Golghotu or Ghotina (India). *Ibid.* 304 A peculiarity of Barbone, as it occurs in India, is that it seldom occurs in a violently epidemic form. 1929 *Veterinary Record* 24 Aug. 709/1 Hæmorrhagic septicæmia. .. The more common alternative terms applied to the condition are Barbone, Buffalo Disease, Buffalo Cholera, Pasteurellosis and, in Malaya..Hawa Kerbau.

barbotine. Add to def.: Also, pottery ornamented with barbotine. Also *attrib.*
1888 *Atalanta* Sept. 692/2 'Barbotine' work is like very rough painting with thick *impasto*. 1891 *Mag. Art* 353/1 There is 'barbotine' for those who like it—rich in colour. 1914 *Oxf. Univ. Gaz.* 573/1 (Rep. Ashmolean Mus.) Fragments of barbotine ware, showing that this type of pottery was popular in Memphis as well as in the Sudan.

barb wire: see WIRE *sb.* 1 e.

Barcelona. Add: (Earlier example.)
1761 in E. Singleton *Social New York* (1902) 236 Black Barcelona handkerchiefs.
2. Used *attrib.* in *Barcelona nut* or as short for this.
1851 MAYHEW *Lond. Labour* I. 89/1 The 'Barcelonas' are from 4½*d.* to 6*d.* a quart to the street-sellers. 1880 *Encycl. Brit.* XI. 549/1 Hazel-nuts, under the name of Barcelona or Spanish nuts, are largely exported from France and Portugal, and especially Tarragona. 1899 [see PETROUS *a.* 1]. 1951 *Good Housek. Home Encycl.* 346/2 *Barcelona Nut*..usually kiln-dried to make it keep well. *Ibid.* 505/1 Cultivated varieties include the Filbert, Cob and Barcelona.

barchan (baɹkā·n). Also **barchane, barkhan.** [Native word.] A crescent-shaped dune of shifting sand such as occur in the deserts of Turkestan. Also *attrib.*
1888 *Encycl. Brit.* XXIII. 511/2 Shifting sands blown into *barkhans*, or elongated hills, sometimes 50 and 60 feet in height. 1897 [see MEDANO]. 1900 *Geogr. Jrnl.* XV. 24 On the higher portions of the foreland, to leeward and further from the river barchans (or medanos) occur. 1953 *Sci. News* XXVII. 14 The *barchan types* [of dune]..found only where the wind is predominantly from a single direction.

Barcoo (bāɹkū·). The name of a river in Queensland, Australia, used *attrib.* and *ellipt.* to designate certain local conditions, phenomena, etc. (see quots.).
1889 J. H. MAIDEN *Useful Native Plants Australia* 76 *Anthistiria membranacea*, Lindl...'Barcoo Grass' of Queensland; also called 'Landsborough Grass'. One of the best pasture grasses in Queensland. 1898 MORRIS *Austral Engl.* 19 *Barcoo Rot*,..persistent ulceration of the skin, chiefly on the back of the hands, and often originating in abrasions. *Ibid.* 20 *Barcoo Vomit*, a sickness occurring in inhabitants of various parts of the high land of the interior of Australia. 1903 'T. COLLINS' *Such is Life* 30 Grand remedy for scurvy, and Barcoo rot, and the hundreds of natural diseases that flesh is subject to. 1934 *Bulletin* (Sydney) 21 Nov. 21/1 The electrical storms there [*sc.* Longreach, Queensland]—called Barcoos—are terrific and take toll of both human and stock life. 1945 BAKER *Austral. Lang.* iii. 62 *Barcoo spew*, *Barcoo vomit*..a sickness characterized by vomiting after food is taken. 1958 D. NILAND *Call me when Cross turns over* 52 Don't try the Barcoo spews. A cow of a thing. 1967 *Sunday Mail Mag.* (Brisbane) 10 Sept. 3/2 Young Arthur, fortunately dodged the typhoid, but did collect a man-sized dose of what was popularly called Barcoo rot.

bard, *sb.*[3] (Earlier and later examples.)
1706 PHILLIPS *New World Words* (ed. 6), *Menehout*..a peculiar manner of baking Meat, cover'd with Bards, or thin Slices of Bacon, in an Oven between two Fires. 1960 *Times* 1 Aug. 9/3 Barding. Covering with thin slices of fat bacon or pork (bards), poultry, game or meat that is deficient in natural fat and is to be either oven or pot roasted. Bards are also used to line..a *terrine*.

bardolatry (baɹdǫ·lătri). [f. BARD *sb.*[1] (sense 4) + -OLATRY.] Worship of the 'Bard of Avon', i.e. Shakespeare. (Occas. used of other writers.) So **bardolater** (-ǫ·lătəɹ) [-OLATER], a worshipper of the Bard, a Shakespearolater; **bardo·latrous** *a.*, tending to or characterized by bardolatry.
1901 G. B. SHAW *Plays for Puritans* Pref. p. xxxi, So much for Bardolatry! 1903 —— *Man & Superman* Ep. Ded. 30 Foolish Bardolaters make a virtue of this after their fashion. 1905 —— in *Sat. Rev.* 11 Feb. 170/2 The word 'pity' does not reach even the third row of the stalls, much less the gaping bardolatrous pit. 1911 *Times Lit. Suppl.* 9 Nov. 440/3 Playing for the sympathy of the 'bardolaters'. 1914 G. B. SHAW *Dark Lady* Pref. 112 The familiar plea of the Bardolatrous ignoramus, that Shakespear's coarseness was part of the manners of his time. 1952 *Scrutiny* XIX. 13 The writer who found it 'extremely salutary' that Milton's eminence should be questioned appears now..a resolute bardolater. 1958 *Listener* 19 June 1031/1 Could this bud of Bardolatrous love be proved a flower? 1963 *Punch* 26 June 934/3 Those whose Bardolatry..falls this side the '*Dream*'.

Bardolphian (baɹdǫ·lfiăn), *a.* [f. *Bardolph* (see below) + -IAN.] Resembling or characteristic of Bardolph, a character in Shakespeare's *Henry IV*, *Henry V*, and *Merry Wives of Windsor*, noted for his red nose.
1756 TOLDERVY *Hist. 2 Orphans* I. x. 81 The tears very seasonably rolled over his Bardolphian cheeks. *Ibid.* IV. xxiv. 134 His Bardolphian nose. 1827 J. WIGHT *More Mornings at Bow St.* 19 His face did bear certain *Bardolphian* tokens of late hours, and 'healths five fathoms deep'. 1899 *Manch. Guardian* 22 June 8/2 We do not agree with those who think that M. Coquelin has made too much of his nose [in 'Cyrano de Bergerac']. It is not Bardolphian or acquired. 1950 G. WILSON *Brave Comp.* iii. 30 That red face and large, pitted Bardolphian nose.

bardy (bā·ɹdi). *Austral.* Also **bardee, bardi.** [Aboriginal.] An edible Australian wood-boring grub (*Bardistus cibarius*) or its larvæ. Used locally in *starve the bardies!*, an exclamatory phr. of surprise or disgust.
1926 R. J. TILLYARD *Insects Austral. & N.Z.* xx. 233 Coleoptera..The 'bardee' of Western Australia, *Bardistus cibarius* Newm., ranges right across to New South Wales; its larvae are found in the stems of grass-trees and 'black-boys' (*Xanthorrhoea*) and are eaten both by aborigines and white people. 1941 BAKER *Dict. Austral. Slang* 8 *Starve the bardies!*, a popular W.A. ejaculation, synonymous with 'Stone the crows!' *a* 1951 X. HERBERT in Murdoch & Drake-Brockman *Austral. Short Stories* 299 Kaijek paused to look among the broken roots for bardies. 1959 A. UPFIELD *Bony & Black Virgin* xvii. 150 The bardee grubs came up from tree-trunks and..tree-roots to split their skins and emerge as great winged moths the size of a man's hand. *a* 1963 J. FOUNTAIN in B. James *Austral. Short Stories* (1963) 270 Fisher..sneaking round the corner..with a jar of bardi grubs under his arm.

bare, *a.* Add: **6. c.** *bare pole(s)* (earlier examples). Also *bare-poled* adj., having bare poles; also *transf.*, of trees lacking leaves or branches.
1762 *Let.* 11 Nov. in *Ann. Reg. 1762* (ed. 5, 1787) 117/2 The ship sprung a leak, and we were obliged to lie to under bare poles. 1774 N. CRESSWELL *Jrnl.* 5 Aug. (1925) iii. 31 To furl the F.S. [*i.e.* fore-sail] and scud under bare pole. 1851 MELVILLE *Moby Dick* III. xxxiii. 197 The *Pequod*..bare-poled was left to fight a Typhoon. 1864 J. A. GRANT *Walk across Africa* 33 Open forests of bare-poled trees.
IV. a. **bare-belly** *Austral.* and *N.Z.*, a sheep with no wool on the belly; so *bare-bellied* adj.; **bare-fallow**, land left fallow for a whole year; = *naked fallow* (see NAKED *a.* 9 d); hence as *v.*, to leave land fallow for a whole year; **bare-fallowing** *vbl. sb.*; **c.** *bare-arsed* (later examples), *-bosom(ed)*, *-knuckled*.
1928 D. H. LAWRENCE *Lady Chatt.* xv. 276 Nowt but a bare-arsed lass. 1950 HEMINGWAY *Across River* v. 29 Beyond the Sile there was nothing but bare-assed plain. *c* 1875 G. L. MEREDITH in A. J. Harrop *Adventuring in Maoriland* (1935) xiii. 143 Naturally, the easiest-shorn sheep—'bare-bellies' and 'bare-points'—are selected first. 1878 E. S. ELWELL *Boy Colonists* 109 The ewes have many of them at shearing-time no wool on the legs or under the belly, and hence are called 'bare-bellies'. Of course, these 'bare-bellied' ewes..are very quickly shorn. 1956 G. BOWEN *Wool Away!* (ed. 2) 155 *Bare-belly*, a sheep with all the wool scraped or dropped off its belly. 1965 J. S. GUNN *Terminol. Shearing Industry* 1. 6 *Barebelly*, a sheep with defective wool growth caused by a break in the fibre structure. This causes the wool to fall off the belly and legs, and the rest of the wool can be removed with a few blows or even with the hands. 1927 D. H. LAWRENCE *Morn. in Mex.* 43 A bare-bosom, black-browed girl. 1809 M. BERRY *Jrnl.* 31 May in 'Lewis Melville' *Berry Papers* (1914) vi. 291 Such an over-dressed, bare-bosomed, painted eye-browed figure one never saw! 1855 WHITMAN *Leaves of Grass* 27 Press close barebosomed night! 1831 J. C. LOUDON *Encycl. Agric.* (ed. 2) III. vi. 801 The expediency or inexpediency of pulverising and cleaning the soil by a bare fallow, is a question that can be determined only by experience, and not by argument. 1831 *Q. Jrnl. Agric.* II. 101 Some writers maintain, that bare fallowing is not necessary on any kind of soil, as judicious management will prevent an influx of weeds. 1891 R. WALLACE *Rural Econ. Austral. & N.Z.* ix. 161 Bare-fallowing of land in place of manuring was too much practised as a means of restoration. 1961 *Agric. Hist. Rev.* IX. 3 A larger proportion of land was bare-fallowed. 1924 MASEFIELD *Sard Harker* I. 5 The truth man learns Fighting bare-knuckled Nature in the ring.
C. *sb.* Delete †*Obs.*, and add to 1:
1906 *Westm. Gaz.* 29 Dec. 16/1 The downtrodden slippers tied on with string, toes out, and hardly any sole: the child is walking 'on the bare', as the saying is.

bareback, *a.* and *adv.* (Later examples.)
1768 in *Essex Inst. Hist. Coll.* (Salem, Mass.) XIV. 262 Went to Boston in sley..; came home bare back. 1870 O. LOGAN *Before Footlights* 369 When the 'bare-back rider' returns from the ring. 1880 'MARK TWAIN' *Tramp Abroad* xxvi. 269 He could ride bareback and know and feel that he was safe. 1923 J. H. COOK *50 Years on Old Frontier* 48 Bareback riding was the order of the day.

barefoot, *a.* Add: *U.S.* in fig. use (see quots. and cf. *BARE-FOOTED *a.* c).
1866 LOWELL *Biglow P.* Ser. II. Introd., 'I take my tea barfoot,' said a backwoodsman when asked if he would have cream and sugar. 1888 *Chicago Herald* (Farmer), Never touch coffee unless you like it barefoot, that is, without sugar or milk.

bare-footed, *a.* Add: Of a horse: having a shoeless foot. Hence **barefoo·tedness.**
1756 TOLDERVY *Hist. 2 Orphans* I. 74 Many worthy gentlemen are become egregious sufferers, both by the barefootedness of their horses and the loss of their hares. 1891 *Athenæum* 28 Nov. 714/1 The barefootedness of the women and children. 1906 SOMERVILLE & 'ROSS' *Some Irish Yesterdays* 88 'Ye're barefooted,' he said. I found that I [*i.e.* my hunter] had indeed lost a foreshoe.
c. *U.S.* (See quots. and cf. *BAREFOOT *a.*)
1847 PAULDING *Amer. Comedies* 194, I thought even a Yankee knew that 'stone fence barefooted' is the polite English for whisky uncontaminated,—pure, sir! 1878 J. H. BEADLE *Western Wilds* xii. 183 It was sod corn [*sc.* whisky] barefooted.

barège. Add: **1.** Also **barege, barège.** (Earlier and later examples.)
1828 DISRAELI *Voyage Capt. Popanilla* xiv. 176 He had no doubt his mistress would look most charmingly in a barege. 1845 C. RIDLEY *Lett.* (1958) 192, I have got..a sort of *barège* for evening, which I shall make very warm by lining with silk. 1848 MRS. GASKELL *M. Barton* I. x. 191 The gay-coloured barège shawl. 1901 C. MORRIS *Life on Stage* (1902) 3, I had a tormenting barège veil over my face. 1927 E. SITWELL *Rustic Elegies* 79 On their deep barouche pillows In cashmere Alvandar, barège Isabelle.

ba·rehanded, *a.* and *adv.* [see BARE *a.* 6.] Having nothing in or covering the hands, esp. carrying no weapon.
c 1440 *Lyfe of Ipom.* 2057 They hewyd the gloves of his hand; All bare-handyd faught þis knyght. 1874 CLEASBY & VIGFUSSON *Icelandic-Eng. Dict.* 60/2 *Ber-hendr*, adj. barehanded. 1895 *Educ. Rev.* Sept. 125 The civilized man has no superiority over the savage in bare-handed strength. 1919 F. HURST *Humoresque* 142 I'll go barehanded to a snowball feast rather than wear your duds. 1930 G. SURREY *Shack in Coulee* xi. 93 His revolver was still in its holster. He was barehanded. 1949 N. LLOYD *Young Liberators* ii. 32 There are lots of tricks for fighting barehanded. 1958 *Times*

3 Dec. 5/1 Lilani Daka, a villager, went to the hut with a gun, but finding that it was not loaded he attacked the lion barehanded.

baren (bæ·ren). [Jap.] A pad used in wood-block printing.
1895 W. ANDERSON *Jap. Wood Engr.* 63 Impressions are taken upon specially prepared paper by rubbing with a flat disc (*baren*) worked by hand pressure. 1916 F. M. FLETCHER *Wood-block Printing* ii. 12 Pressure is then applied to the back of the paper as it lies on the wet block. This is done by a round pad called the *baren* by the Japanese. It is made of a coil of cord covered by bamboo sheath. 1924 H. FURST *Mod. Woodcut* 94 The Japanese manner of designing in map-like spaces and printing from the plank by rubbing with the 'baren'..became general.

Barff (bāɹf). *Metallurgy.* [The name of F. S. *Barff*, an American engineer.] A process for protecting iron and steel from rust by means of a coating of iron oxide produced by the application of steam to the heated surface. Hence **Barff** *v. trans.*, to treat by this process (Webster, 1934). See *BOWER-BARFF.
1880 *Sci. Amer.* Suppl. 31 Jan. 3393/2 (*title*) The Barff Process for Preserving Iron. 1884 [see *BOWER-BARFF]. 1951 *Good Housek. Home Encycl.* 41/1 In districts where the water is acid in reaction.., the metal is usually specially processed or 'Barffed'.

bargain, *sb.*[1] Add: **3. b.** In certain coalfields in England, a piece of work let to the workmen making the lowest offer. Also Comb., as *bargain-letting, -man, -taker, -work* (see quot. 1851 in sense 8 in Dict.).
1825 E. MACKENZIE *Hist. Northumb.* (ed. 2) I. 100 These bargains are taken in partnerships, consisting of from two to eight men. 1858 SIMMONDS *Dict. Trade*, Bargain-men. 1881 *Instr. Census Clerks* (1885) 84 Copper Miner... Bargain Man. 1897 *Daily News* 18 Mar. 3/4 They cannot take special bits of work, known as bargains, but must go to the bargain-takers and ask them for rock to cut up. 1897 *Westm. Gaz.* 1 Sept. 7/1 Bargain-letting recommenced to-day at Lord Penrhyn's Carnarvonshire slate quarries.
c. *esp.* An article of which the price is professedly reduced for the purpose of a special sale in a shop or stores; also *attrib.* and *Comb.* designating persons and things associated with the practice of offering goods for sale in this way, e.g. *bargain counter, -day, -hunt* (also as *v. intr.*), *-hunter, -price, -sale, -seeker*; **bargain basement**, a basement floor where bargains are displayed; also *transf.* and *fig.*
1899 *Chicago Daily News* 25 May 19/6 There is not room in our Bargain Basement for all the bargains we have created for to-morrow. 1927 *Star* 1 June 6/1 Selfridge's Bargain Basement. 1932 *Daily Express* 27 June 11/3 Every railway company, every travel bureau, every steamship line has its own 'bargain basement'. 1935 *Brit. Jrnl. Psychol.* Oct. 190 Ghostly raiment which to them is none the less resplendent through being acquired in the bargain basement of compromise and expediency. 1959 *Economist* 10 Jan. 100/1 Any such bargain-basement nuclear deterrent as we could hope to produce and mount in this country will be minimally effective. 1888 *Scribner's Mag.* Jan. 65/2 Ladies..in all the finery that the 'bargain counters' of Fourteenth Street could furnish. 1904 *Post Express* (Rochester, N.Y.) 22 July 4 The bargain counter rates for steerage traffic. 1908 G. BURGESS *Maxims of Methus.* xvi. 14 From the bargain counter she selecteth her gloves. 1887 *Puck* 23 Nov. 210/2 Football's too tame...You've never seen the women at Macy's on bargain-day. 1902 A. BENNETT *Anna of 5 Towns* xiii. 350 The bargain-hunt was up.., always second-hand, but always good. 1937 M. HILLIS *Orchids on Budget* (1938) iii. 39 If you *must* bargain-hunt, do it late in the season at a good shop. 1791 J. LACKINGTON *Memoirs* xix. 143 These very bargain hunters have given me double the price that I now charge. 1838 *Mag. Domestic Econ.* III. 201 If you leave it to their integrity..you will generally be supplied with better provisions..than by becoming a 'bargain hunter'. 1886 *Longm. Mag.* VII. 447 Last month was a lucky one for bargain-hunters. 1938 *Encycl. Brit. Bk. of Yr. 1938* 423/1 Endeavours to lure the impoverished, bargain hunting public of the United States to the box office. 1904 *Westm. Gaz.* 7 Jan. 8/1 Jewellery and chiffons and laces are there, going at bargain prices. 1898 C. A. BATES *Clothing Bk.* No. 5211 Garments for which you pay the additional price at widely advertised 'bargain' sales. 1907 'O. HENRY' *Trimmed Lamp* (1916) 114 'Have I ever chucked any bargain sale stuff at you, Moll?' asked the Kid, with calm dignity. 1834 *Chambers's Edin. Jrnl.* III. 145/3 The unrespective hands of brokers and bargain-seekers.
7. *to make the best of a bad bargain* (earlier examples).
1670 RAY *Prov.* 61 Make the *best* of a bad bargain. 1712 ARBUTHNOT *John Bull* ii. 11 Matters have not been carried on with due secrecy; however, we must make the best of a bad Bargain.
8. *bargain-driving.*
1902 *Spectator* 27 Dec. 1023/2 Some men are, it might almost be said, victims of the habit of bargain-driving. 1930 T. S. ELIOT tr. *St.-J. Perse's Anabasis* 69 Manœuvres over field to ravish a woman, bargain-driving and plots.

bargain, *v.* **3.** (Earlier example.)
1801 JANE AUSTEN *Let.* 3 Jan. (1952) 103 My Mother bargains for having no trouble at all in furnishing our house in Bath.

barge, *sb.*[1] Add: **6. b.** A large carriage. *U.S.*
1882 HOWELLS *Modern Instance* xxvii. 328 Marcia watched him drive off toward the station in the hotel barge. 1903 *Boston Herald* 19 Aug., The visitors were conveyed in barges to the crest of High Pole hill. 1907 *Springfield Weekly Republ.* 21 Feb. 16 [A sleigh-ride] which required every four-horse barge in the north half of the county.
7. *barge-load.*
1609 *Act 7 James I* xviii, The..landing of every Barge-load..of the said Sand. 1960 *Times* 30 Apr. 9/1 A small party of Dutch soldiers was smuggled through the water-gate, concealed in a bargeload of peat.

barge, *sb.*[2] (See quot. 1908.)
1908 *Animal Management* 206 A piece of leather..running from the fore-wale *beneath* the after-wale [of a horse's collar] is known as the 'barge'. 1946 N. WYMER *Eng. Country Crafts* v. 48 For the body he will make a leather throat-piece and stitch it..together with one edge of a piece of woollen cloth, to his barge.

barge, *sb.*[3] *slang.* [f. *BARGE *v.*[2]] An argument, dispute.
1934 *Bulletin* (Sydney) 5 Dec. 36/2 The player-writer business has been causing a lot of barge in Brisbane. 1948 *Punch* 24 Nov. 491/3 Mr. Attlee and Mr. Churchill had a 'barge' on the subject of European Federation.

barge, *v.*[1] Add: **1. b.** *intr.* To travel by barge.
1909 in WEBSTER. 1962 N. MAXWELL *Witch-Doctor's Appr.* ix. 111 Stretched on soft blankets in the shade of the pamacari..I was as comfortable as Cleopatra barging down the Nile.
3. *intr.* To bump heavily *into* (a person), to knock roughly *against*; to go roughly and heavily *through, into, along, about*, or *against* (a place, etc.); also with advs. *about, around*. Also *to barge one's way*.
1888 *Boy's Own Paper* Christmas No. 56/2 Dig your heels in, old chap,..and barge into the bank! 1890 FARMER *Slang* I. 124/1 *Barge*..(Uppingham School.)—To knock against a person; to come into collision with. 1899 *Daily News* 10 July 9/2 Defendant denied that the cocks were fighting. They were merely 'barging' as it was called in Lancashire. 1904 KIPLING *Traffics & Discov.* 318 You ought to have summoned me for trespass when I barged through your woods. 1904 —— in *Windsor Mag.* Jan. 234/2, I remember..the dropped jaw of the midshipman in her whaler when we barged fairly into him. 1911 C. E. W. BEAN *'Dreadnought' of Darling* xxxviii. 339 A crowd of men came barging into the hut. 1915 W. J. LOCKE *Jaffery* xvii. 232 He ..barged mightily down Fleet Street. 1919 C. ORR *Glorious Thing* vi. 63 His first mishap was to barge into Isabel's fiancé, his second to be barged into by Isabel. 1924 *Blackw. Mag.* June 751/2 The camel..is an awkward and troublesome creature on a narrow road, especially if he takes fright, barging about to the danger of everybody. 1924 *Chambers's Jrnl.* 294/2 Pattering donkeys barge their way through the crowded alleys. 1961 *Times* 13 Feb. 4/2 Place kick..given for barging in the line-out.
b. *transf.* and *fig.*
1923 *Chambers's Jrnl.* 718/2, I hadn't barged about the world then. 1928 *Observer* 11 Mar. 14/5 There is a sort of masterful way in which a theme is made to barge its way through its surroundings. 1930 W. S. MAUGHAM *Cakes & Ale* iv. 50 He'll hate having a lot of strangers barging in on him.
c. *trans.* To cause to move forcibly or heavily.
1903 WODEHOUSE *Tales St. Austin's* 4 There was something wonderfully entertaining in the process of 'bargeing' the end man off the edge of the form into space. 1923 *Public Opinion* 19 Jan. 61/2 Heaven knows I'm always trying to barge it at you. 1924 W. J. LOCKE *Coming of Amos* viii. 93 By degrees he edged (or barged) his huge frame to the front rank. 1927 *Observer* 3 Apr. 27/6 Scotland bore down in a body, and Morton barged the ball past Brown.

barge (bāɹdʒ), *v.*[2] *slang.* ? *dial.* [? Back-formation f. BARGEE, as if 'to use the language of a bargee'.] *trans.* To abuse or 'slang' (a person).
a 1860 ALB. SMITH *Med. Student* (1861) 102 Whereupon they all began to barge the master at once; one saying 'his coffee was all snuff and chickweed.' 1881 J. F. T. KEANE *Six Months in Meccah* iv. 98 My informer..blaming those 'Shaitan' English, and barging them in choice Arabic.

bargemaster. (Further examples.)
1664–5 in C. Welch *Hist. Company of Pewterers* (1902) II. 132 The salary of Mr. Pike, barge master, was fixed on 16th March at 3*l.* yearly. 1892 W. C. HAZLITT *Livery Companies* 475 The duties of the Bargemaster, which, after the discontinuance of water-pageants, have been confined to the supervision of the swans on the Thames. 1898 *Daily News* 21 Apr. 2/2 The Fishmongers' Company's late Bargemaster. —The office of bargemaster is one of the most ancient in the annals of the City Companies.

barge-pole. A long pole with which a barge is propelled (see also quot. 1890); frequent in colloq. phr. (and variants) *I wouldn't touch him* (*it*) *with* (*the end of*) *a barge-pole*, I refuse to have anything to do with him (it).
1890 FARMER *Slang* I. 124 *Barge-Pole.* (Winchester College.) A large stick or thick bough, of which there was one in each fagot. Also generally used for any large piece of wood. 1893 LADY MONKSWELL *Diary* Dec. (1944) 239 It will be a long while before any political party touches Home Rule again with the end of a barge pole. 1914 H. A. VACHELL *Quinneys'* II. xvi. 229 If he was a fool I wouldn't touch him with a barge-pole. 1915 A. H. GIBBS *Persistent Lovers* li. Hideous little beast! I wouldn't touch him with the end of a barge pole. 1918 MRS. H. WARD *War & Elizabeth* v. 86 If he tries to leave me this funny old place..there are two can play at that game. I wouldn't touch it with a barge-pole.

barging (bā·ɹdʒiŋ), *vbl. sb.* [f. BARGE *sb.*[1] + -ING[1]. Cf. BARGE *v.*[1]] Transport by barge.
1901 *Fielden's Mag.* IV. 433/2 The cartage or barging of the material to sites. 1904 *Daily Chron.* 29 Jan. 3/6 The barging is done by contract. 1905 *Westm. Gaz.* 9 Mar. 9/2 The cost of 'barging' and transport.

barhal, var. *BHARAL.

Barisal (bærisā·l). [The name of a town in East Bengal.] *Barisal guns*, booming sounds heard in Barisal and certain other regions, esp. on or near water.
1896 *Nature* 2 Jan. 197, I first heard the Barisal Guns in December 1871, on my way to Assam from Calcutta through the Sunderbans. 1902 *Encycl. Brit.* XXVI. 147/2 Barisal has given its name to a curious physical phenomenon, known as the 'Barisal guns', the cause of which has not yet been satisfactorily explained. These are noises, like the report of cannon, frequently heard in the channels of the delta of the Brahmaputra. 1936 *Discovery* Aug. 252/2 Mysterious booming sounds, like distant gunfire, which may be heard in the Australian desert and suitable places all over the world. In Australia they are known as Barisal guns, probably because so many white men from India retire over here.

baritone. Now the usual form of BARYTONE. Add: Of an instrument: occurring in a 'family' of instruments between the tenor and the bass. Also *ellipt.*, a baritone saxophone. So **baritone horn** (*a*) = BARYTONE A. 3 b; (*b*) *U.S.*, a baritone saxophone.
1876 STAINER & BARRETT *Dict. Mus. Terms* 385/2 s.v. *Saxophones*, brass wind instruments, the invention of M. Sax. ..They are six in number, the high, soprano, alto, tenor, baritone and bass. 1883 GROVE *Dict. Mus.* III. 233/2 It [*sc.* the saxophone] is made in a number of sizes..the contralto in F and E♭; the barytone in C and E♭; the bass in F and E♭, [etc.]. *Ibid.*, Those most used are the contralto and barytone varieties [of saxophone]. 1926 WHITEMAN & MCBRIDE *Jazz* ix. 193 One baritone saxophone is equal in sonorousness to a section of nine or ten cellos. 1938 *Oxf. Compan. Mus.* 626/2 *Heckelphone*, baritone oboe, basset oboe. 1949 L. FEATHER *Inside Be-Bop* ii. 11 He played baritone horn in the school band. 1952 B. ULANOV *Hist. Jazz* (1958) vi. 52 The scoring of two altos, two tenors, and a baritone in the present-day dance band.
Hence **ba·ritonist**, a baritone-saxophone player.
1958 K. GOODWIN in P. Gammond *Decca Bk. of Jazz* xiii. 157 He has..transferred his..technique..to the baritone-saxophone, and..rates as the most important baritonist on the coast.

barium. Add: **barium meal**, a mixture containing barium sulphate, a white compound that is opaque to X-rays, used in radiological examination of the alimentary tract. So also *barium enema, swallow*.
1913 *Lancet* 31 May 1529/2 Dr. J. Delpratt Harris gave a demonstration of a Barium Meal in a case of Gastro-enterostomy. 1930 D. A. RHINEHART *Roentgenographic Technique* xviii. 346 Postero-anterior views of the colon filled with a barium enema. 1940 *Med. Ann.* 507 A patient..was submitted..for a barium enema examination of the colon. 1946 *Brit. Jrnl. Radiology* Mar. 114/2 On examination with a barium swallow and meal it was seen that there was..dilatation..of the œsophagus. 1962 *Lancet* 13 Jan. 67/1 Barium-swallow examinations may fail to demonstrate œsophageal varices. 1966 *Listener* 24 Feb. 273/1 They performed all sorts of tests on me with barium meals.

bark, *sb.*[1] Add: **9. a.** *bark-cloth* (earlier and later examples).
1864 J. A. GRANT *Walk across Africa* 138 Or an Arab cloak or shawl of bark-cloth hung from his shoulder, reaching below the knee. 1951 R. FIRTH *Elements of Social Organization* ii. 51 The people still wear their traditional bark-cloth, made from the fibrous inner bark of the paper-mulberry tree.
10. bark-beetle, any beetle of the family *Scolytidæ*, the members of which burrow beneath the bark of woody plants; **bark-borer** *U.S.*, a species of bark-beetle; **bark-bound** *a.* (later example); **bark canoe** *U.S.*, a canoe made of birch-bark; **bark-house** (later examples); **bark-hut**, a hut built with the bark of trees; **bark-louse** *U.S.*, any one of a number of aphids infesting the bark of trees; **bark-mill** *U.S.*, a mill in which tanning bark is ground; **bark-peeler** (*a*) a person who peels bark from a tree; (*b*) an implement for peeling bark; **bark stone** = CASTOR[1] 2.
1862 T. W. HARRIS *Insects Injur. Veget.* 85 Though these cylindrical bark-beetles are of small size, they multiply very fast. 1953 H. L. EDLIN *Forester's Handbk.* xvi. 264 A very large group known as bark beetles, because they feed and breed in and beneath the bark of living or dead trees. 1859 *Trans. Ill. Agric. Soc. 1857–8* III. 345 Another species is that sometimes called the bark borer, from its feeding exclusively upon the cambium immediately beneath the bark. 1831 *On Planting* II. vi. 63 in *Libr. Useful Knowl.*, To secure against any chill or sudden effect of cold, so as to bring about..the injury of being bark-bound, the most effectual impediment to growth either in height or thickness. 1725 in *Lancaster* (Mass.) *Rec.* (1884) 232, 27 day we traveld down the river and found a bark cannow. 1888 *Harper's Mag.* Mar. 537/1 But only a bark canoe now and then comes along from one of the thirteen rivers. 1660

Boston Rec.. 155 Henry Bridgam..sett part of his barke house upon part of the townes land. **1721** *Boston Selectmen* 83 Liberty..to erect a bark house near Snow Hill. **1824** in Thornton *Amer. Gloss.* Suppl., A bark-house, and a good iron bark-mill. **1744** F. Moore *Voy. Georgia* 123 Some bark-huts, which our friendly Indians had some time since built for their lodging. **1843** 'R. Carlton' *New Purchase* i. 2 We..talked of bark huts and bows and arrows. **1890** 'R. Boldrewood' *Miner's Right* vi. 61 Bark-huts, of which both sides, and sometimes doors are composed of sheets of the flattened eucalyptus bark. **1841** in Johnson *Farm Encycl.* (1868) 137/2 (D.A.E.), The bark lice are found apparently torpid..sticking..closely to the bark. **1959** Southwood & Leston *Land & Water Bugs* vi. 153 They [sc. Emesinæ] may also feed on insects already trapped in spiders' webs and they can attack bark-lice (Psocids) through their web. **1749** J. Eliot *Field-Husb.* (1934) ii. 37 Take your Clover Hay to a Tanners Bark-mill, where they use a Stone Wheel, grind it. **1885** Bark-mill [see sense 9 a]. **1862** *Rep. Comm. Patents: Agric.* (U.S.) 414 Tanneries sprang into existence..and the bark-peelers and teamsters ..made the whole region one of active and prosperous industry. **1902** *Encycl. Brit.* XXV. 376/1 All textile work [of American Indians] was done by hand; the only devices known were the bark-peeler, the shredder [etc.]. **1925** Hemingway *In our Time* (1926) i. 18 The shanties where the Indian bark-peelers lived. **1961** M. W. Barley *Eng. Farmhouse & Cottage* i. ii. 36 In Cumberland some sixty years ago barkpeelers built themselves huts which consisted of four poles lashed in pairs to support a ridge piece. **1805** M. Lewis in *Jrnls. Lewis & Clark Exped.* (1904) III. 319 The male beaver has six stones, two of which..are called the bark stones or castors. **1817** *Ann. Reg. 1816* 551/2 To prepare beaver-bait, the castor or bark stone is first gently pressed from the bladder-like bag which contains it. *Ibid.* 552/1 The bark stones are two inches in length.

bark, *sb.*[3] **2. b.** Phr. *his bark is worse than his bite* (earlier example).
1816 Scott *Antiq.* II. vii. 186 'Monkbarns's bark,' said Miss Griselda Oldbuck..'is muckle waur than his bite.'

bark, *v.*[1] Add: **2.** *fig.* (Later example.) *to bark up the wrong tree* (earlier and later examples). For 'in U.S.' read 'orig. U.S.'
1832 J. Hall *Leg. West* 46 You are barking up the wrong tree, Johnson. **1887** *N. & Q.* 17 Sept. 221 Mr. Rye is barking up the wrong tree. **1899** [see Lumber *v.*[1] 1]. **1961** *Technology* Feb. 31/1 Her researches show the Government to be barking up the wrong tree.

b. To call out or 'spiel' at the entrance of a cheap shop or show to attract customers. (Cf. *barker[1] 2.) *U.S.*
1904 G. V. Hobart *Jim Hickey* v. 84 We could make sandwich money in front of a bootchy-kooch palace, barking at the Rubes. **1908** K. McGaffey *Sorrows of Show-Girl* 16 By gum, I'd take a job barking for a snake race. **1917** D. G. Phillips *Susan Lenox* I. xiii. 224 Pat, ready to take tickets, was 'barking' vigorously...addressing crowd. **1948** *Time* 19 July 90/2 [It] was another triumph for Liberty's brand of mass production plus carnival barking.

3. b. To emit an explosive sound, esp. of a fire-arm.
1853 F. W. Thomas *J. Randolph* 132 These boats bark so you can hardly hear yourself talk. **1907** S. E. White *Arizona Nights* III. xiii. 342 The Colt's forty-five barked once, and then again.

5. Delete †*Obs.* and add examples.
1821 Clare *Vill. Minstr.* II. 33 The dog bark'd a welcome. **1864** Tennyson *Enoch Arden &c.* 170 Bark an answer, Britain's raven! **1916** 'Boyd Cable' *Action Front* 131 Both the muzzles tilted a little and barked off another flight of shells. **1922** *N. & Q.* 12th Ser. XI. 206/2 A stable secret which has leaked out and is common property is referred to as 'the dogs are barking it'.

6. To drive *away* or *back* by barking.
1829 M. Mitford in *The Gem* 195 Frisk's own doggish exploit in barking away a set of pilferers. **1891** Meredith *One of our Conq.* II. iii. 42 The Dog..would have barked the breathing intruder an hundredfold back to earth.

ba·rken, *a.* Chiefly *poet.* [f. Bark *sb.*[1] + -en[4].] Made or consisting of bark.
1755 T. Forbes in C. Gist *Jrnls.* (1893) 148 Easter Tuesday we embarked..in about 300 Batteaus or Canoes (not barken). **1835** R. M. Bird *Hawks of Hawk-h.* I. v. 61 Perhaps some tall and tawny hunter..may yet..urge his barken canoe over some cypress-fringed pool. **1864** G. M. Hopkins *Pilate* in *Note-Books* (1937) 15 Then knot a barken band To hold me quite fix'd in the selfsame plight. **1890** *Harper's Mag.* Aug. 365 A sword-lunge of assailant thunder Slashed down thy barken mail.

barker[1]. **2.** Also, one who 'barks' at a cheap shop or show: see *bark *v.*[1] 2 b. Chiefly *U.S.*
1699 B. E. *Dict. Cant. Crew*, *Barker*, a Salesman's Servant that walks before the Shop, and cries, Cloaks, Coats, or Gowns, what d'ye lack, Sir. **1822** [in Dict.]. **1859** Hotten *Slang Dict.*, *Barker*, a man employed to cry at the doors of gaffs, shows, and puffing shops, to entice people inside. **1897** Howells *Landlord at Lion's Head* 247 The barker began to fill the night with hoarse cries of 'Miss Lynde's carriage; carriage for Miss Lynde!' **1903** *N.Y. Times* 28 Aug., Conditions were so dull that barkers had to be enlisted to call the public's attention to the boats. **1910** H. A. Franck *Vagabond Journ.* 276 The secretary was a man..with the voice of a side-show barker.

barkey. Add: Also **barkie, barky.** (Earlier and later examples.)
1703 J. Campbell *Jrnl.* 27 Apr. in *Proc. Mass. Hist. Soc.* (1867) IX. 487 The Lark..In her Passage home meett 3 french Barkies. **1805** W. Hunter in *Naval Chron.* XIII. 36 Had we been acquainted with the real state of our Barkie's Bows. **1832** *Fraser's Mag.* IV. 464 Fears for the safety of his barky. **1930** *Sea Breezes* 73 This proved the old 'barkie's' undoing, as..she was..sunk off Picton by the Germans.

Barkhausen (bā·ɹkauz'n, -hauz'n). [The name of the German scientist H. *Barkhausen* (1881–1956).] **Barkhausen effect**: the name given to the series of sudden changes in the magnetization of a substance when the magnetizing force is gradually altered. **Barkhausen oscillation, oscillator** (see quots.).
1924 E. P. T. Tyndall in *Physical Rev.* XXIV. 439 Barkhausen effect for silicon steel...Barkhausen and Van der Pol consider the effect as due to the sudden re-orientation of groups or chains of molecular magnets. **1925** *Chem. Abstr.* XIX. 1659 Occurrence of Barkhausen oscillations in triodes. **1936** R. S. Glasgow *Radio Engin.* x. 310 (*caption*) Circuit for generating Barkhausen oscillations. **1940** *Chambers's Techn. Dict.* 76/2 *Barkhausen oscillator*, a generator of very high frequency oscillation (above 10[8] cps.), whose action depends on the transit time of the electrons from cathode to anode of a three-electrode valve. *Barkhausen–Kurz oscillations*, oscillations of very high frequency generated in a triode valve with a positive grid, the anode potential being approximately the same as the cathode. The oscillations are those of electrons passing to and fro through the grid. **1944** W. L. Emery *Radio Engin.* 284/1 (*Index*) Barkhausen–Kurz oscillator. **1957** *Encycl. Brit.* XIV. 646/1 *Barkhausen Effect.* In 1919 H. Barkhausen discovered the effect known by his name, and interpreted it as demonstrating that the magnetization of iron proceeds by steps and not in a continuous manner.

barking, *vbl. sb.*[2] **2.** *barking-irons* (example).
1845 S. Judd *Margaret* I. iii. 12 Here were a draw-shave, a cross-cut saw..barking irons, a scythe.

barking, *ppl. a.*[1] Add: **2. b. barking deer,** the Indian muntjac, *Cervulus muntjac*, found in India, Burma, and Tibet; so named from its call; **barking iron** (earlier examples); **barking spider** (see quot. 1952); **barking wolf** *U.S.* = Coyote.
1880 *Encycl. Brit.* XII. 742/2 The barking deer or muntjac (*Cervulus vaginalis*). **1898** *Geogr. Jrnl.* XI. 502 The small barking deer, called 'pause' by the natives. **1939** A. Keith *Land below Wind* xiv. 259 There is no meat as sweet as barking deer. **1785** Grose *Dict. Vulgar T.*, *Barking Irons*, pistols, from their explosion resembling the bow-wow or barking of a dog. (*Irish*). **1825** Paulding *John Bull in Amer.* 56 Seeing the barking iron [he] shrunk back. **1934** *Bulletin* (Sydney) 24 Jan. 21/4 The barking spider of Central Australia and other stridulating species can probably hear quite well. **1952** A. G. Mitchell in *Chambers's Shorter Eng. Dict.* Suppl. 790/1 *Barking spider*, a large reddish-brown spider of Central Australia, said to produce a barking or whistling sound. **1826** J. D. Godman *Amer. Nat. Hist.* I. 260 The Prairie or Barking Wolf..frequents the prairies..of the west. **1867** *Amer. Naturalist* I. 289 The Prairie or Barking Wolf (*Canis latrans*, Say), is by far the most abundant carnivorous animal in Arizona, as it is also in almost every part of the West.

barkle (bā·ɹk'l), *v. dial.* [f. Bark *v.*[2] (see sense 4) + -le 3.] To cake, encrust (with dirt, etc.). So **ba·rkled** *ppl. a.*
1819, etc. in *Eng. Dial. Dict.* **1913** D. H. Lawrence *Sons & Lovers* ii. 34 'Aven't you got a drink, Missis, for a man when he comes home barkled up from the pit? *Ibid* xii. 311 Their barkled shoes hung heavy on their steps.

barkless (bā·ɹklės), *a.*[2] [f. Bark *sb.*[3] + -less.] Having no bark: applied to a dog which rarely or never barks.
1938 *Tail-Wagger* May 172/1 About four years ago the Basenji or Congo Dog was introduced into England...The breed is barkless, being mute. **1950** A. C. Smith *About our Dogs* xxiii. 353 The interest aroused by the 'Barkless Dogs' revealed the fact that one was exhibited at Crufts in 1895. **1957** B. Vesey-Fitzgerald *Domestic Dog* ix. 181 *The Basenji.* First introduced..in 1934, and widely known as the 'barkless dog'.

barkometer (baɹkǫ·mitəɹ). [f. Bark *sb.*[1] + -ometer.] A tanner's hydrometer for testing the strength of bark infusions.
1821 J. Swan (*title*) Explanation of improved mode of Tanning;..intended to accompany the New Invented Barktrometer [*sic*]. *Ibid.* 40 Though you show us, by the assistance of your new Barktrometer that there is a something derived from the bark of which it shows the quantity. **1852** Morfit *Tanning & Currying* (1853) 329 A barkometer ..is specially adapted to testing the strength of bark lyes. **1882** [see *tan-liquor* s.v. Tan *sb.*[1] C. 1].

barley. Add: **B. 1. c.** *barley-harvest, -meal* (later examples), *-mill, -seed* (example), *-straw, -stubble.* **d.** *barley-beer, -broth, -soup.*
1901 Kipling *Kim* xiv. 367 A drink of *chang*—the barley-beer that comes from Ladakh-way. **1906** *Daily Chron.* 16 Mar. 7/7 Not less than 85 per cent. of the total saccharine yielding materials shall be barley malt. This beer is to be known as 'barley beer'. **1723** J. Nott *Cook's & Confectioner's Dict.* BA 26, *Barley Broth.* Boil a Pound of French Barley in three Quarts of Water, with some whole Spice..put in Raisins..Butter, Rose-water, and Sugar, and so eat it. *c* **1806** D. Wordsworth *Tour Scotland* in *Jrnls.* (1941) I. 281 A fowl stewing in barley-broth. **1930** T. S. Eliot tr. *St.-J. Perse's Anabasis* 27 Man goes out at barley harvest. **1832** *Scoreby Farm Rep.* in *Brit. Husb.* (1840) III. 19 They have for the first fortnight boiled potatoes only, then a little barley-meal is added. **1797** Johnston tr. *Beckmann's Inventions* I. 266 Mills by which grain is only freed from the husk and rounded, are called barley-mills...The first kind of barley-mills is a German invention. **1840** J. C. Loudon *Cottager's Manual Husb.* i. 9 The husking can only be well done at a barley-mill. **1831** *Sutherland Farm Rep.* in *Brit. Husb.* (1840) III. 72 The gates are again shut, until the completion of the barley-seed. **1747** Mrs. H. Glasse *Art of Cookery* ix. 78 (*heading*) A Barley Soup. **1935** Browne & Williams *World's Best Recipes* 22 (*heading*) Barley Soup. **1678** Ray *Prov.* 51 Barly-straw 's good fodder when the cow gives water. **1794** *Trans. Soc. Promotion Agric.* (U.S.) I. 145 Barley-straw is hearty fodder for horned cattle in the winter. **1837** *Flemish Husb.* in *Brit. Husb.* (1840) 59 This is their food during the whole winter with a little wheat or barley-straw. **1922** Joyce *Ulysses* 263 We'll put a barleystraw in that Judas Iscariot's ear this time. **1733** Tull *Horse-Hoing Husb.* xi. 50 It has brought as good a Crop of Wheat on Barley Stubble. **1913** Masefield *Daffodil Fields* 67 Westward was barley-stubble not yet cleared.

2. barley-grass, meadow barley; **barley-itch** (see quots.); **barley-milk** (later example); **barley-mow** (substitute quot. 1714 for *c* 1732 in Dict.); **barley-sugar,** also *attrib.*, as *barley-sugar drop, stick*; *spec.* used to designate features of architecture, furniture, etc., which in shape resemble a twisted barley-sugar stick; **barley-wine** (earlier example); also, a strong English ale.
1795 W. Winterbotham *Hist. View U.S.* III. 401 Those which are found most common are..Barley grass, *Hordeum pratense.* **1891** R. Wallace *Rural Econ. Austral. & N.Z.* xxii. 294 Barley grass..Throughout Colonies, except Tasmania. **1928** *Daily Express* 28 Nov. 4 It [*sc.* baker's itch] is similar to the affection from which millers suffer under the name of 'barley itch', due to a small parasite which preys on the larvae of small moths which infest the grain. **1961** *Brit. Med. Dict.* 772 *Barley itch*, 1. Grain itch..2. A sensitivity to the plant *Mucuna pruriens*, which grows among the barley, and sheds irritant hairs. **1846** 'A Lady' *Jewish Manual* 178 *Barley Milk.* Boil..pearl barley..in new milk..sweeten. **1714** Gay *Sheph. Week* Pastoral v. 75 Whenever by yon Barley Mow I pass. **1819** Keats *Let.* 12 Apr. (1958) II. 52 As fine as barley sugar drops are to a schoolboy's tongue. **1859** L. Wraxall tr. *Robert-Houdin's Mem.* II. xii. 257 The confectioner had exchanged the barley-sugar stick for the magician's wand. **1936** W. de la Mare *Wind Blows Over* 239 The barley-sugar-legged walnut prie-dieu at her bedside. **1937** *Archit. Rev.* LXXXII. 63/2 The altars and chapels are there in both cases, but in Lecce the 'barley sugar' columns are of stone. **1939** G. Greene *Lawless Roads* x. 233 Barley-sugar pillars up the façade. **1728** E. Smith *Compleat Housewife* (ed. 2) 208 To make Barley-wine. Take..Barley..boil it..mix it with..White Wine..Borage water..Cluny-water..Lemons..Sugar.. bottle it up. **1820** F. Accum *Art of Brewing* 1 The ancient Greek writers gave the name of barley wine to malt liquors. **1940** *Chambers's Techn. Dict.* 21/1 Burton and barley-wine are strong ales. **1953** *Word for Word* (Whitbread & Co.) 11/2 *Barley broth*, a form of strong ale (also barley wine..).

barley-corn. Add: **5.** A gun-sight shaped like an inverted V.
1896 W. W. Greener *Gun & its Devel.* (ed. 6) xxxi. 736 Military Match Rifles..The fore-sight must be the simple reversed V, or barley-corn military pattern. **1901** *Kynoch Jrnl.* Apr.–May 76/2 In our opinion the best sights for target purposes are the barleycorn foresight and tangent backsight. **1904** *Daily Chron.* 10 Dec. 7/6 The Small Arms Committee..were of opinion that the barley-corn should be adhered to pending further trials with the other patterns of foresight.

barlow knife. *U.S.* [From the name of the maker.] A large single-bladed pocket-knife.
1779 in *Documents Revol. Hist. New Jersey* (1906) III. 676 Barlow penknives. **1819** *Massachusetts Spy* 29 Dec. 3/2 A barlow knife, bloody, and another knife, rusty, lay along side of him. **1876** 'Mark Twain' *Tom Sawyer* iv. 30 Mary gave him a brand-new 'Barlow' knife. **1896** J. C. Harris *Sister Jane* 229 There on the side of the pew were the letters W. W. which many years ago I had carved with my barlow knife.
ellipt. **1890** *Congress. Rec.* Aug. 8818/2 He did not want to carry a cheap and nasty knife, but the little fellow has to carry a ten cent Barlow.

barm, *sb.*[2] Add: **3.** *barm-man, -pot* (also *fig.*); **barm-stick** *fig.*, a feeble-minded person (E.D.D.); also *attrib.*
1913 D. H. Lawrence *Sons & Lovers* iii. 49 She was looking down the alley..for the barm-man. **1924** —— & Skinner *Boy in Bush* 278 We've heard that barm-stick yarn before. **1925** *British Wkly.* 5 Mar. 557/1 The man who sold 'barm' (brewers' yeast) drove round the district..one saw women hastening over the heath, jug in hand, to catch the 'barm-man'. **1951** J. B. Priestley *Fest. Farbridge* I. i. 34 Lady Barth, a rich old barmpot. **1957** Mankowitz & Haggar *Encycl. Eng. Pott. & Porc.* 16/2 *Barm pot*, a pot for storing barm or yeast. **1963** T. & P. Morris *Pentonville* ix. 196 Thus a harmless schizophrenic will be classified by the staff as a 'barmpot' and by the prisoners as a 'nutter'.

barmaid. Add: (Earlier and later examples.) Also *attrib.* and *fig.*
a **1658** [see Stroke *v.*[1] 4]. **1880** Lady Jebb *Let.* 3 June (1960) 159 She is lovelier than ever, in an exquisite highbred way, which throws such a barmaid beauty as Mrs. Fred. Myers completely in the shade. **1911** Masefield *Everl. Mercy* 24 And then men ask, Are Barmaids chaste?
barmaidenly, *a.* (Further examples.)
1905 G. B. Shaw in *Sat. Rev.* 11 Feb. 170/2 The merry lady with her barmaidenly repartees. **1932** *Times Lit. Suppl.* 27 Oct. 776/4 She was what some one once styled a 'barmaidenly maiden'.

barminess (bā·ɹminės). [f. Barmy *a.* 2 b + -ness.] Weakness of intellect.
1896 [see *barmy *a.* 2 b]. **1908** E. V. Lucas *Over Bemerton's* vi. 60, I heard what sounded like a sarcastic sniff deprecative of her uncle's barminess.

Bar-mitzvah (bāɹ,mi·tsvā). *Judaism.* Also **-mitsvah, -mitzva.** [Heb., lit. 'son of commandment, man of duty'.] **a.** A Jewish boy who has reached the age of thirteen, regarded as the age of religious responsibility. **b.** The 'confirmation' ceremony in a synagogue on this occasion. Also *attrib.*
1861 *Children's Jewish Advocate* VII. 84 You look like my Samuel, whom I have not seen since he was *Barmitsva*. **1877** *Encycl. Brit.* VII. 134/1 This dull routine..continued till Emanuel was thirteen years old, when he returned to Neisse, to solemnize his religious majority (Bar-mitzva). **1891** M. Friedländer *Jewish Rel.* 481 The boy when thirteen years old is *bar-mitsvah*..bound to obey the Law, and responsible for his deeds. **1892** I. Zangwill *Childr. Ghetto* (1893) I. x. 97 My Ebenezer is *Bar-mitzvah* next Shabbos week. **1941** B. Schulberg *What makes Sammy Run?* ix. 181 *Bar-mitzvah* is the Hebrew ceremony celebrating a boy's reaching the state of manhood at the age of thirteen. **1960** *Commentary* June 493/2 Its willingness to accept boys over eleven for Bar Mitzvah preparation.

barmy, *a.* Add: **2. b.** = *balmy *a.* 7 (of which it is an altered form, after Barm *sb.*[2]). *slang.*
1892 *Answers* 27 Feb. 242/1 One plan is..to give foolish answers when asked questions. By this means the shammer gets known as being 'barmy' (weak-minded) among his shipmates. **1896** *Westm. Gaz.* 30 May 8/1 Should not 'balmy' be 'barmy'? I have known a person of weak intellect called 'Barmy Billy'...The prisoner..meant to simulate semi-idiocy, or 'barminess', not 'balminess'. **1902** *Ibid.* 8 Nov. 2/1 All the boys think him barmy.

barn, *sb.* Add: **1. c.** A stable or cattle-house. *U.S.*
1770 J. R. Forster tr. *P. Kalm's Trav. N. America* I. 223 The *barns* had a peculiar kind of construction...In the middle was the threshing floor..on one side were stables for the horses, and on the other for the cows. **1828** Mrs. Royall *Black Bk.* II. 71 Every farmer has his small wooden barn, under which name they include stables. **1901** M. D. Babcock *Thoughts* 17 Locking a barn seems no longer commonplace when the horse is stolen. **1904** *N.Y. Even. Post* 28 Jan. 1 Cattle were found frozen stiff in the barns by farmers this morning.

d. [Said to have originated in the phrase 'as big as a barn'.] In nuclear physics, 10^{-24} sq. cm., a unit of area used in the measurement of the cross-section of a nucleus.
1947 R. D. Evans in C. Goodman *Sci. & Engin. Nucl. Power* i. 15 This area has been dubbed the 'barn', 1 barn = 10^{-24} cm²/nucleus. **1947** C. Goodman *Ibid.* ix. 290 The nuclear cross section, σ, in barns. **1950** S. Glasstone *Sourcebk. Atomic Energy* x. 264/2 A unit, called a *barn*, equal to 10^{-24} sq. cm. per nucleus, has been adopted. *note.* The term 'barn' was proposed in 1942 by the American physicists M. G. Holloway and C. P. Baker, as a result of a broadly humorous association of ideas. It served the purpose of a code word..and seemed appropriate because 'a cross section of 10^{-24} sq. cm. for nuclear processes was really as big as a barn'. **1957** *Sci. News* XLV. 106 The cross sections of gadolinium, samarium, and europium are 30,000, 4,250, and 2,500 barns respectively.

2. barn-ball, a children's game of the United States (see quot. 1879); **barn-boss** *U.S.*, a horse-keeper; **barn chamber** *U.S.*, a loft above a barn; **barn-gallon** (examples); **barn-lot** *U.S.*, a piece of ground for or about a barn (see Lot *sb.* 6 a); **barn-raising** *U.S.*, 'the erection of the frame of a barn with the help of neighbours; a social gathering on this occasion' (*D.A.*); **barnstormer** (earlier and later examples); (*b*) in *Aeronaut.* (see quot. 1928 and *barnstorm *v.* 3) *U.S.*
1841 *New Orl. Picayune* 25 May 2/2 Who has not played 'barn ball' in his boyhood? **1879** B. F. Taylor *Summer-Savoury* 122 The writer knew a boy..who never got farther than 'barn-ball', which means throwing a ball at the gable and catching it when it returns. **1901** W. Churchill *Crisis* ix. 196 A tall man in his shirt sleeves was playing barn-ball with some boys. **1902** S. E. White *Blazed Trail* xxix. 201 So Shearer had picked out a barn-boss of his own. **1838** H. Colman *Mass. Agric. Surv. Rep.* 16 The best method of curing it [*sc.* herds grass]..is to..tie it in bundles; and set it upright in a barn chamber. **1858** Simmonds *Dict. Trade* 27/2 *Barn-gallon*, a double gallon of milk. **1865** W. White *East. England* II. xv. 217 The gallon being a 'barn-gallon' of seventeen pints. **1724** *New Hampsh. Prob. Rec.* II. 250, I give to my Daughters..the other half part of my afores[aid] barn Lott in Salsbury. **1932** W. Faulkner *Light in Aug.* (1933) i. 12 When the wagon passes the house, goes on toward the barnlot. **1856** T. D. Price *MS. Diary* 28 Apr. (D.A.), Went to D. D. Keller's barn raising. **1952** *Economist* 9 Aug. 340/1 The old custom of 'barn-raising', at which neighbours and friends volunteered their services, is coming back into fashion [in the U.S.A.]. **1859** Hotten *Dict. Mod. Slang* 3 *Barn stormers*, theatrical performers who travel the country and act in barns, selecting short and frantic pieces to suit the rustic taste. **1928** *Daily Mail* 7 May 6/4 *Barnstormers*, itinerant flyers, appearing at fairs and race tracks, like Lindbergh in his earlier years. **1930** *Punch* 19 Mar. 330/1 Those barn-stormers who tore the play's passions and the spectators' heart-strings to shreds.

barnacled, *ppl. a.* Add: **a.** Also *fig.* **b.** (Earlier examples.)
1890 *Jrnl. Educ.* 1 Mar. 149/2 With their brains barnacled over with shellfish facts. **1900** J. L. Allen *Incr. Purpose* xvi. 222 There might have been..fewer creeds barnacled in the World's Ship of Souls.
1725 *New Canting Dict.* s.v. *Barnacles, The Cuffin Quire with his Nose Barnacled*..the Justice of Peace with his Spectacles on. **1829** *Olio* III. 55/2 The general reader and spokesman was ready barnacled for his office.

barnbrack (bā·ɹnbræk). *Ir.* [f. Ir. *bairghean* cake of bread + *breac* speckled.] A cake or loaf of bread containing currants. Cf. var. Barmbrack.
[**1772** Vallancey *Ess. Antiq. Irish Lang.* 22 On St. Briget's Eve every Farmer's Wife in Ireland makes a Cake called bairin-breac.] **1867** P. Kennedy *Banks of Boro* xlii. 349 Piles of hot griddle-baked wheaten cakes.., barnbracks, and other varieties of the staff of life. **1901** G. B. Shaw *Devil's Disciple* I. 12 Two green ware plates, on one of which she puts a barnbrack. **1928** *Universe* 3 Feb., A loaf of curious, very sweet currant bread is made and sold for All Souls Day. Even the poorest household manages to secure one of these Barn-bracks.]

ba·rn dance. orig. *U.S.* [Barn *sb.*] A dance danced in a barn; applied *spec.* to a kind of schottische in which the partners advance side by side for a few steps and then dance a waltz or schottische step.
1892 *Daily News* 28 Sept. 5/5 The now inevitable barn dance figured once on the programme. **1895** L. Grove et al. *Dancing* 424 'Barn dance' is an American designation; but as many other dances take place in barns out West, it is difficult to see why the title is specially applied to this Scotch lilt and schottische hops. **1900** C. C. Munn *Uncle Terry* 20 If it was a barn dance she was always there and never lacked partners.
allusively. **1898** Kipling *Fleet in Being* 7 We of the light horse did barn-dances about the windy floors.

barn-door. Add: In *attrib.* use: as large as, or resembling, a barn-door; often in humorous application (see note in Dict.). Also in *Comb.*
1829 *Blackw. Mag.* Dec. 856 Such is the birthright of Britons—to open their mouths barn-door wide... The aforesaid barn-door-wide mouths. **1865** Mrs. Whitney *Gayworthys* xxvi. 256 Skirts were trodden on..and there was more than one 'barn-door' rent. **1868** C. Box *Theory & Pract. Cricket* 91 The Gentlemen's wickets were 27 in. by 8; the Players' 36 in. by 12 in. This was called the Barn-Door Match. **1898** G. Giffen *With Bat & Ball* v. 64 It was almost painful to watch a giant of six feet and a half playing the barndoor game. **1932** Blunden *Face of Eng.* 76 He met each ball with a barn-door bat.

2. *transf.* A door-like shutter in front of a lamp used for photographic purposes.
1942 Berrey & Van den Bark *Amer. Thes. Slang* 600 *Barn door*, a doorlike attachment on a lens to control the light. **1959** W. S. Sharps *Dict. Cinematogr.* 79/1 *Barn-doors*, adjustable flaps fitted to the front of a lighting unit, in order to control the spread of the light beam. **1960** O. Skilbeck *ABC Film & TV Terms* 16 *Barn door*, a variable mask designed to be used in front of studio lamps..and consisting of 'doors' opening longitudinally and/or vertically.

barney. Substitute for def.: **1.** In various *dial.* and *slang* uses. **a.** A lark or spree.
1859 Hotten *Slang Dict.* 3 *Barney*, a lark, spree, rough enjoyment; 'get up a *barney*', to have a lark. *Barney*, a mob, a crowd.

b. Humbug, cheating; *spec.* an unfair sporting contest.
1865, 1882 [see Dict.]. **1885** *Bell's Life* 3 Jan. 3/4 Few genuine matches have taken place this season,..though exhibitions and barney contests have been plentiful.

c. A noisy dispute or altercation. Also *Austral.* and *N.Z.*
1864 C. R. Thatcher *Colonial Songster* 86 A barney first commenced, in one of their *Celestial* revels. **1885** in Elworthy *W. Somerset Word-bk.* (1886) 45 I'll warn't there'll be a barney over thick job. **1888** 'R. Boldrewood' *Robbery under Arms* II. xiii. 216 We had long talks and barneys over the whole thing. **1895** *Daily News* 4 Jan. 3/7 (E.D.D.), Selby runs out, and goes to get another knife, but I stops him, and the barney was all over. **1916** *Bulletin* (Sydney) 17 Feb. 48/4 We did 'ave a bit of a barney. She tole me I'd better be goin'. **1929** J. B. Priestley *Good Companions* I. iv. 112 He has a bit of a barney with the other two partners, decides to have a split. **1958** *Encounter* May 12/2 There was a right barney at the other end of the shop. **1958** D. Niland *Call me when Cross turns over* 200 There should never have been a barney between them in the first place. **1959** *Spectator* 29 May 752/3 That a team will go forward without Maori players..is the only positive conclusion that can be drawn from the barney. **1964** J. Burke *Hard Day's Night* i. 20 'If you're going to have a barney,' said John, 'I'll hold your coats.'

2. *Mining.* (See Dict.)

3. Phr. *Barney's bull*, a worthless person or thing. Also *Austral.* (see quot. 1945).
1908 Masefield *Capt. Margaret* iv. 98 Your trading lay..is all Barney's bull. It's got more bugs than brains, as you might say. **1929** —— *Hawbucks* 76 Those girls, without any mother, only that old Barney's bull of a da. **1945** Baker *Austral. Lang.* iv. 88 *All behind like Barney's bull*, delayed, backward (a play on the word *behind*, although the expression is probably as alliterative as it is biological).

barney (bā·ɹni), *v.* [f. prec.] **1.** *intr.* To argue, dispute. *dial.*, *Austral.* and *N.Z.*
1880 *Evening Post* (N.Z.) 7 Jan. 13/2 He saw the prisoner and prosecutor 'barneying' in the middle of some strangers. **1890** 'R. Boldrewood' *Col. Reformer* II. xvi. 42 If you go barneying about calves..he'll best ye. **1913** A. Bathgate *Sodger Sandy's Bairn* x. 77 But I can't barney with you all day.

2. To act or play unfairly; to cheat.
1905 *Daily Chron.* 17 Aug. 4/4 The thirst for money prizes and the gambling craze has brought in 'barneying' and the buying and selling of falls in professional [wrestling] rings. **1908** *Westmorld. Gaz.* 22 Aug. 8/3 Heavyweight wrestling..Third round..T. T. and G. M. were both blown out for barneying.

barngun (bā·ɹngɒn). *dial.* [f. *barn*, dial. var. Burn *v.*[1] + *gun*, dial. var. Gound (OE. *gund* matter, pus).] An eruption of the skin; shingles.
1746 *Exmoor Courtship* in *Gentl. Mag.* XVI. 300/1 Vorewey struck out and came to a barngun. **1746** 'Devoniensis' *Vocab. Exmoor*, *Ibid.* 405/2 *Barngun*, a breaking out in small pimples, or pustles in the skin. **1869** R. D. Blackmore *Lorna Doone* I. xviii. 206 'Thou art not come to me,' she said..'to be struck for bone-shave, nor to be blessed for barn-gun.' **1892** H. C. O'Neill *Devon. Idyls* 87 When I were bad with the barngun.

barnstorm (bā·ɹnstǫɹm), *v.* [f. Barn *sb.* + Storm *v.* 7.] *intr.* **1.** *Theatr.* To tour (rural districts), giving theatrical performances of a popular kind (formerly often in barns).
1883 *Mercury* (N.Y.) in Ware *Passing Eng.* (1909) 20/1 Miss Helen Bancroft, who recently played in this city, was announced as with a barn-storming company. **1883** *Sporting Life* 29 Apr. 5/3 Footlight Flickerings. Owen Fawcett will go barn-storming through Michigan this summer. **1884** [see Barn *sb.* 2]. **1936** Auden & Isherwood *Ascent of F6* II. 79, I have dreamed of a threadbare barnstorming actor, and he was a national symbol. **1949** *Here & Now* (N.Z.) Oct. 30/1 The National Orchestra..has this year been barnstorming with opera, and latterly with concert programmes, taking in provincial centres.

2. To make a rapid tour holding meetings for propaganda or election purposes. *U.S.*
1896 *Congress. Rec.* 7 Apr. 3661/1 [He] was barnstorming down in Georgia in favor of a gold monometallism. **1944** *Chicago Daily News* 27 Oct. 16/2 President Roosevelt indignantly denies that his New York and Chicago barnstorming trips violate his pledge not to campaign in the usual sense. **1947** *Time* 10 Mar. 21/1 Barnstorming Presidential Candidate Harold Stassen whirled through Belgium in one day.

3. *Aeronaut.* To give informal exhibition and sightseeing flights; to perform stunts at local functions. Chiefly *N. Amer.*
1928 L. Gravatt *Pioneers of Air* 210 He gave exhibitions at county fairs, or barnstormed as the pilots say. **1933** F. B. Willoughby *Alaskans All* 90, I had an old army plane, and as a gypsy flyer I barnstormed thirty-seven states. **1936** *Christian Sci. Monitor* 20 Aug. 3/4 *Barnstorm*, to fly about from town to town taking passengers up for a small sum. **1936** F. Clune *Roaming round Darling* xiii. 116 They were flying from town to town, barnstorming as they called it. **1938** *Times* 19 July 14/5 Mr. Steve Reich, who had barnstormed with him in 1931, was as surprised as everybody else to hear he had flown to Ireland.

Barnum (bā·ɹnǝm). [The name of Phineas Taylor *Barnum* (1810–91): see Barnumize *v.*]
1. Humbug, nonsense; showmanship.
1856 G. D. Brewerton *War in Kansas* 17 He believed the whole affair to be a 'Barnum'—*alias* humbug, of the most unmitigated kind. *a* **1914** Joyce *Stephen Hero* (1944) xxi. 117 It's absurd: it's Barnum. He comes into the world God knows how, walks on the water. **1937** *Daily Express* 7 Jan. 10/5 There is a touch of Barnum about scientists which sets them whooping when any one of them has found something new.
2. *Comb.*
1874 W. James *Let.* 8 Aug. (1920) I. vii. 182 The dreadful feeling of wounded pride and Barnum-born resentment may with time fade away. **1899** *Westm. Gaz.* 28 June 2/2 Was there ever a more Barnum-like sort of exhibition?
Hence **Barnumese** [see -ese]; **Barnumesque** *a.* [see -esque]. Cf. Barnumize *v.*, Barnumism.
1888 Farmer *Americanisms* 40/1 We get words like *barnumese* and *telegraphese*, to signify exaggeration of style. **1890** W. J. Gordon *Foundry* 189 A florid prospectus in the best Barnumese. **1890** *Cornh. Mag.* Dec. 629 A perfectly Barnumesque profusion of quaint monstrosities.

barocco (bārǫ·ko), *a.* and *sb.* [It.] = *baroque.
1877 Symonds *Renaiss. in Italy* III. ii. 99 Palladio was followed by the violent reactionaries of the *barocco* mannerism. *Ibid.* x. 503 Stucco, fresco, and gilding in a style only just removed from the *barocco*. **1881** S. Butler *Alps & Sanct. Piedmont* 332 The architecture is late, and *barocco*, not to say *rococo*, reigns everywhere. *a* **1902** —— *Note-Bks.* (1912) 260 The band played the barocco music on the barocco little piazza and we were all barocco together. **1923** L. Pullan *Religion Since Reform.* vii. 221 Peter the Great..pronounced the name of his new capital as if it were Dutch, and his architects built it after the manner of the Dutch and German barocco cities. **1955** E. Pound *Section: Rock-Drill* lxxxvii. 33 Then false fronts, barocco.

baroceptor (bærose·ptǫɹ). *Physiol.* [f. Gr. βάρο-ς weight + *re)ceptor.] A receptor sensitive to changes in pressure. Also *attrib.* Cf. *baroreceptor.
1949 C. L. Evans *Princ. Human Physiol.* (ed. 10) V. xxxii. 595 The nerves originating in the carotid sinus are stimulated by stretch, i.e. they originate from special stretch receptors (baroceptors) in its walls. **1961** *Brit. Med. Dict.* 1218/1 *Pressure receptor*, often called pressoreceptor or baroceptor. **1961** *Lancet* 19 Aug. 439/2 A local modulator of baroceptor response.

barocyclonometer (bæ:rosəiklonǫ·mitəɹ). [f. Gr. βάρο-ς weight + Cyclone + -meter.] An aneroid barometer with diagrams and directions for detecting the existence of a storm at a distance of several hundred miles.
1906 *Month* Dec. 562 The ships that sail the seas of the Far East are equipped with the Barocyclonometer.

Barolo (bărō[u]·lo). [Named from *Barolo*, Piedmont, the region of its production.] A full-bodied red wine of Piedmont.
1875 H. Vizetelly *Wines of World* IX. i. 136 The popular Barolo, a red wine produced from the Nebiolo grape. **1877** *Encycl. Brit.* VI. 272/2 In the vicinity [of Cuneo] a good wine is made called Barolo. **1902** A. Bennett *Grand Babylon Hotel* xxii. 251 He proceeded to the Italian cellar, and descanted upon the excellence of Barolo from Piedmont. **1959** W. James *Word-bk. Wines* 17 *Barolo*,..is high in alcohol (some 14 degrees or 24·5 proof).

barometric, *a.* Add: *barometric altimeter* = *altimeter 2; *barometric tendency* (see quot.).
1930 *Proc. Inst. Radio Engin.* XVIII. 832 The barometric altimeter is adjusted to read altitude above the [landing] field. **1962** *Gloss. Aeronaut. Terms (B.S.I.)* v. 11 *Barometric altimeter*, an aneroid barometer graduated to indicate altitude according to a standard atmosphere. **1916** *Weather Map (Meteorol. Office)* 36 Barometric tendency ..is the change in the pressure..within the three hours immediately preceding the fixed hours of observation.

baron. Add: **2. b.** A magnate in commerce, finance, or the like; a great merchant in a certain commodity, usu. defined by a qualifying word, as *beef baron, coal baron.* (Cf. King *sb.* 6 a.) orig. *U.S.*
[**1776** J. Adams *Familiar Lett.* (1876) 154 But the spirit of these Barons [*sc.* North Carolina gentry] is coming down, and it must submit.] **1818** *Niles' Reg.* XIV. 226/1 The name of a *Jew* and '*rag-baron*' is synonimous. **1885** *Century Mag.* Sept. 804 Who is responsible for [the depression]?..Is it the 'silver barons' or the 'gold bugs'? **1887** *Harper's Mag.* Apr. 822/1 When the great 'coal barons'..deliberately combine to put up the price of coal. **1894** J. L. Ford *Lit. Shop* (1896) iv. 46 Magazine barons. **1902** *Westm. Gaz.* 17 May 5/2 Nearly 150,000 coalminers are on strike in the anthracite region of Pennsylvania...The coal barons have built barricades and barbed wire fences around the shafts. **1904** 'O. Henry' *Cabbages & Kings* 4 A rubber prince, a sarsaparilla, indigo, and mahogany baron. **1907** *Westm. Gaz.* 20 Dec. 9/4 The American beef barons, Armour's, Swift's, and Morris, are..the greatest captains of commerce in any market. **1932** *Ann. Reg. 1931* 15 The 'money barons' were using the whole of their influence to restrict the raising of money for national development. **1932** Wodehouse *Louder & Funnier* 24 A couple of great film barons. **1947** J. Hayward *Prose Lit. since 1939* 47 Newspaper barons and film magnates have shown that it is not difficult to bend authors to their own desires. **1958** *Listener* 18 Dec. 1030/2 He became one of those barons who advance their fortunes under the slogan of National Trade.

c. Applied *gen.* to a person having power or influence in any sphere; in *Prison slang*: see quot. 1950.
1876 R. L. Stevenson *Walking Tours* in *Cornh. Mag.* June 687 The great barons of the mind will not rally to the standard. **1919** *Athenæum* 15 Aug. 759/1 Slang in War-time. ..Baron, army commander. **1950** P. Tempest *Lag's Lexicon* 10 A baron is one who always has plenty of money and/or tobacco. **1958** F. Norman *Bang to Rights* I. 23 If people can't be barons with out going around punching little geezers up in the air. **1959** *Listener* 10 Dec. 1050/1 A round dozen barons of the microscope, from Cambridge, London, Edinburgh, Paris, Pennsylvania, Wisconsin, California, Moscow.

baronetize (bæ·rǫnėtəiz), *v.* [f. Baronet + -ize.] *trans.* To confer a baronetcy on, make a baronet. Also *transf.*
1860 Meredith *Evan Harrington* xiv. 164 *Third* son! Don't commit yourself there. We dare not baronetize him. **1892** T. G. Hake *Mem. 80 Years* lxv. 276 We have no 'Sir Barry' nowadays; premiers do not build houses, so that they do not baronetize architecture. **1967** K. Giles *Death & Mr. Prettyman* i. 32 The late Sir Ebenezer had been a jerry builder..whom Lloyd George had baronetized for war services.

baronian (bărō[u]·niăn), *a.* [f. Baron(y + -ian.] = Baronial *a.*
1656 Prynne *Fund. Rights of Eng. Freemen* 31 The particulars whereof would amount to many Baronian Tomes, if at large recorded. **1905** *Daily Chron.* 14 June 4/7 The castle is built of granite, in the Scotch Baronian style.

baroque, *a.* (*sb.*) Add: Applied to a florid style of architectural decoration which arose in Italy in the late Renaissance and became prevalent in Europe during the 18th century. Also *transf.* in reference to other arts.
This term and *rococo* are not infrequently used without distinction for styles of ornament characterized by profusion, oddity of combinations, or abnormal features generally.
1765 H. Fuseli tr. *Winkelmann's Refl. Painting & Sculpt. of Greeks* 122 This style in decorations got the epithet of *Barroque* taste, derived from a word signifying pearls and teeth of unequal size. **1846** *Athenæum* 17 Jan. 58/2 Sometimes *baroque*, Mr. Browning is never ignoble: pushing versification to the extremity of all rational allowances, and sometimes beyond it, with a hardihood of rhythm and cadence little short of Hudibrastic. **1877** *Baedeker's Central Italy & Rome* (ed. 5) p. lix, The authors of the degenerated Renaissance known as Baroque were really Vignola (1507–73) and Fontana's nephew Carlo Maderna (1556–1639)...An undoubted vigour in the disposition of detail, a feeling for vastness and pomp, together with an internal decoration which spared neither colour nor costly material to secure an effect of dazzling splendour: such are the distinguishing attributes of the Baroque style. **1921** B. F. Fletcher *Hist. Archit.* (ed. 6) I. 546 In the fullness of time the Renaissance ..passed into the Baroque, which at the beginning of the seventeenth century gave expression once again to the human side in architecture, for it was a spontaneous breaking away from orthodoxy in plan, design, and treatment. **1928** *Times Lit. Suppl.* 15 Mar. 188/2 French-Canadian art.. is being recognized..as a baroque style which is other than the European baroques. **1938** W. S. Maugham *Summing Up* 28 The sonorous periods and the baroque massiveness of Jacobean language. **1938** *Mod. Lang. Notes* Oct. 547 The period of literature..described as 'baroque' ends about 1690, when German baroque architecture..is beginning to develop. **1949** *Times Lit. Suppl.* 10 June 376/4 The word 'baroque'..has come to be accepted as a convenient portmanteau term which covers the music composed between 1580 and 1750 and the plastic arts of an era which begins and ends slightly earlier. **1953** O. de Mourgues (*title*) Metaphysical, Baroque and Précieux Poetry. **1953** J. N. Summerson *Archit. Brit. 1530 to 1830* III. 125 (*heading*) Wren and the Baroque (1660–1710). *Ibid.* xvii. 172 At Blenheim the English Baroque culminates. *Ibid.* 178 Its spirit is the emotional spirit of English Baroque, and it was that which touched Burlington's antipathies. **1954** L. D. Ettlinger in *Listener* 2 Dec. 954/1 The robustness of the Baroque gives way [in the 18th cent.] to the gentler graces of Rococo. **1957** T. S. Eliot *On Poetry & Poets* 167 The conjunction of Christian and classical imagery [in *Lycidas*] is in accord with a baroque taste which did not please the eighteenth century.

baroreceptor (bæ:ro,rise·ptǫɹ). *Physiol.* [f. Gr. βαρό-ς weight + *receptor.] = *baroceptor.
1951 *Jrnl. Physiol.* CXIII. 450 The carotid sinus nerve is a mixed afferent nerve containing afferents from the baroreceptors of the carotid sinus. **1955** *Sci. News* XXXV. 81 Preliminary investigations have not proved that this plexus can function as a baroreceptor, i.e. that it is sensitive to changes in pressure and capable of transmitting such information.

barostat (bæ·rostæt). [f. Gr. βάρο-ς weight + -stat.] An automatic apparatus for regulating pressure; *spec.* a device that controls the pressure of fuel.
1929 F. Daniels et al. *Exper. Phys. Chem.* xviii. 301 The Bureau of Mines has developed a barostat for use in vacuum distillations. **1946** *Jrnl. R. Aeronaut. Soc.* L. 363/2 Failures of fuel pumps, barostats and throttle valves were still far too common. **1951** *Engineering* 30 Mar. 368/1 A barostat-controlled timing device automatically opens the pilot's parachute. **1956** *Flight* 8 June 696/2 The pilot is automatically released from his seat after three seconds, whereupon the barostat in his parachute pack will ensure free fall down to 15,000 feet before the rip-cord is pulled by a spring.

barotaxis (bærotæ·ksis). *Physiol.* [a. G. (M. Verworn, *Allgemeine Physiologie* (1894)), f. Gr. βάρο-ς weight + τάξις arrangement.] 'Stimulation of living matter by change of the pressure relations under which it exists' (Dorland).
1899 F. S. Lee tr. *Verworn's Gen. Physiol.* v. 440. **1901** Calkins *Protozoa* 300 A general result of mechanical stimulation is a motor response followed by the tendency to turn away from the source, and the general reaction, whether positive or negative, since it deals with the question of pressure in some form or other, is called *barotaxis* (Verworn). **1912** E. A. Minchin *Introd. Study of Protozoa* x. 202 We can distinguish..Chemotaxis, or reactions to chemical stimuli..Barotaxis, or reactions to mechanical stimuli.

barotaxy (bæ·rotæksi). *Physiol.* [f. as prec. + -y[3].] = prec.
1906 M. Hartog in *Cambr. Nat. Hist.* I. 20 Most living beings are able to maintain their level in water by floating or crawling against gravity, and they react in virtue of the same power against centrifugal force. This mode of irritability is termed (negative) 'geotaxy' or 'barotaxy'.

barothermograph (bæ:ro,þɔ̄·ɹmograf). *Meteorol.* [f. Baro(graph + Thermograph.] (See quots.)
1896 *Monthly Weather Rev.* Sept. 323/2 Early in August [1895] a baro-thermograph..was constructed and observations begun with the new instrument. **1911** R. M. Pierce *Dict. Aviation* 42 *Barothermograph*, an apparatus for recording simultaneously the pressure and the temperature of the atmosfere, often used on kites and sounding-balloons; a combination of a barograf and thermograf. **1923** Glazebrook *Dict. Appl. Physics* III. 192/2 *Barothermograph*, a self-recording instrument which records the temperature as a function of the pressure.

Barotse (bārǫ·tsi). The name of a negroid race inhabiting the region of the Upper Zambezi basin in southern Africa, and of the language spoken by them.
Barotse represents the pronunciation of the name in former times by the inhabitants of Basutoland and is now replaced in local official usage by *Balozi*, the language being called *Silozi*. *Barotse* and *Balozi* are based upon the same root pronounced differently by different people. (M. Guthrie, private letter, 1959.)
1851 Livingstone *Let.* 1 Oct. (1940) 151 We proceeded.. to see the Sesheke or river of the Barotse. **1857** —— *Trav.* iv. 87 The black men referred to were the Barotse, or, as they term themselves, Baloiána. **1888** *Encycl. Brit.* XXIV. 828/1 Language..Zambesi Group..Ba-Rotse. **1910** *Ibid.* III. 424/2 Barotse..is also applied to contiguous subject tribes, Barotseland being the country over which the Barotse paramount chief exercises authority. **1936** P. M. Clark *Autobiogr. Old Drifter* ix. 127 At this time I knew nothing of the Barotse language, but got along with some of the natives who could talk what was called Kitchen Kaffir.

barouche. In etym., for *birōtus* read *birotus.* Add: (Earlier examples.)
1801 E. Fremantle *Diary* 27 Sept. in *Wynne Diaries* (1940) III. 63 Drove out with Ly. Westmoreland in her Baroutsch. **1805** *Naval Chron.* XIV. 284 The Duke of Sussex arrived in his barouche. **1811** Jane Austen *Sense & Sens.* I. iii. 32 It would have quieted her ambition to see him driving a barouche.
attrib. and *Comb.*
1805 C. Wilmot *Let.* 31 Aug. in Londonderry & Hyde *Russ. Jrnls.* (1934) II. 178 My 2 protectors..ride on the Barooch seat. **1807** C. Sedley (*title*) The barouche-driver and his wife: a tale for haut ton. **1815** [in Dict.]. **1820** G. Colman (Jun.) *X.Y.Z.* I. i, I went in a Brutus wig, and a barouche great-coat. **1960** Cunnington & Beard *Dict. Eng. Costume* 11/1 *Barouche coat.* 1809. A ⅞-length close-fitting outdoor coat, with round bosom and full sleeves; fastened down the front with barrel-snaps and a buckled girdle round the waist.

barouchet. Add: Also **barouchette.** (Earlier examples.)
1807 C. L. Birch *Brit. Pat.* 3063, An Improvement in the Construction of the Roofs and Upper Quarters of Barouches, Barouchets, Curricles, and other Carriages. **1816** Peacock *Headlong Hall* xi. 144 Every chariot, coach, barouche and barouchette..was in motion. **1852** J. W. Carlyle *Let.* 5 Aug. (1883) II. 170 A gentleman who was preparing to start in a barouchette with two horses.

barrack, *sb.*[1] Add: **2. c.** Also *attrib.*: **barrack school,** a disparaging term formerly applied to a large district school for poor-law children.
1894 E. Hart in *Brit. Med. Jrnl.* 21 Apr. 879/2 The system of pauper district schools organised on the 'barrack' principle should be mended or ended as soon as possible. *Ibid.* 28 Apr. 928 Poor Law Barrack Schools. **1902** *Encycl. Brit.* XXXI. 835/1 Adverse criticism..in 1874..has been directed against these large, or, as they are invidiously called, barrack schools.

d. *transf.* (*sing.* or *pl.*). A large plain building or range of buildings, tenements, or flats in which a number of people are housed; also, any strikingly plain-looking building suggestive of a military barracks. Also *attrib.* as *barrack-flat.*
1862 Mrs. Gaskell *Let.* 23 July (1966) 926 We went to the Hotel de Sévigné, her old town house,..an immense barrack of an old half-fortified house. **1880** Trollope *Duke's Children* III. xix. 215 'What a nice little room..' said Isabel. 'It's a beastly great barrack,' said Silverbridge. **1886** *Contemp. Rev.* Sept. 329 The railway has come close under the walls of the *château*, while an ugly barrack has sprung up on the other side. **1909** *Westm. Gaz.* 20 Oct. 1/3 A feature of German housing which reformers desire to abolish: that is, of the many-storied barrack-flat system. **1922** Joyce *Ulysses* 750, I dont like being alone in this big barracks of a place. **1956** M. Duggan *Immanuel's Land* 108 The secret drink, I've heard it said, of old Ignatz, back at the barracks.

e. *spec.* The regular quarters of the Salvation Army.
1887 W. Booth in A. R. Wiggins *Hist. Salvation Army* (1964) IV. v. 189 No barracks are..to be let or used for Political Meetings of any kind. **1907** G. B. Shaw *Maj. Barbara* II. 224 She's gone to Canning Town, to our barracks there.

3. *barrack-field, -like* (or *barracks-like*), *-room* (earlier and *attrib.* examples), *-shed, -wing, -yard* (earlier example); **barrack-master** (earlier example); **barrack-rat** (see quot.); **barrack-square,** the square near military barracks, where drill, parades, etc., take place; also *fig.*, = strictness, rigorous training; also *attrib.*
1769 J. Wesley *Jrnl.* 13 July (1916) V. 328, I was driven to the barrack-field, where were twice as many as the hall could have contained. **1915** R. Lankester *Divers. of Naturalist* 164 The latter barrack-like edifices. **1951** S. Spender *World within World* 221 Albacete was a barracks-like town in a dull plain. **1715** Addison *Let.* 9 Mar. (1941) 313 The Barrack-Master of Waterford..may not be thought proper to be continued in that Station. **1936** F. Richards *Old Soldier Sahib* viii. 160 Children born in Barracks were referred to as 'barrack-rats'. **1777** in *New Hampshire Hist. Soc. Coll.* VII. 68 That..he leave the paying for Barrack room..to a special Committee. **1787** W. Dyott *Diary* 9 July (1907) I. 30 We were not able to get into our barrack-rooms. **1892** Kipling (*title*) Barrack-room ballads and other verses. **1945** A. J. P. Taylor *Course German Hist.* 100 Roon had the typical barrack-room mentality. **1749** in J. S. McLennan *Louisbourg* (1918) 410 Two Barrack Sheds of hundred feet long each. **1932** *Statesman* (Calcutta) 21 July, Our cricket..needs a bit of the barrack square. **1958** S. Race in P. Gammond *Decca Bk. of Jazz* x. 124 The Goodman band was the first to combine barrack-square precision with solo freedom. **1901** Kipling *Kim* vi. 143 Spreading his cloth in the shade of a deserted barrack-wing. **1760** J. Wesley *Jrnl.* 28 June (1913) IV. 395 The colonel..gave me the liberty of preaching in the barrack-yard.

barrack, *sb.*[2] *Austral.* and *N.Z.* [f. *barrack *v.*[2]] An act, or the action, of barracking.
c **1890** D. McK. Wright in A. E. Woodhouse *N.Z. Farm & Station Verse* (1950) 33 There's the 'barrack' at the table and the clever things are said. **1931** V. Palmer *Separate Lives* 13 They received him with shouts and good-natured barrack, as if he were one of the crowd. **1948** —— *Golconda* viii. 60 His flood of good-humoured barrack made the newcomers feel at home. **1949** P. Newton *High Country Days* 46 The other four, full of noisy barrack, were playing pitch and toss with a set of old horse shoes.

barrack (bæ·răk), *v.*[2] [app. orig. Australian (? alteration of *borak), but E.D.D. cites *barrack* 'to brag, to be boastful of one's fighting powers', *barracker* 'a braggart', and *barracking* 'bragging, boastfulness' from northern

Ireland.] *intr.* To shout jocular or derisive remarks or words of advice as partisans against a person, esp. a person, or side collectively, engaged in a contest. Also, with *for*, to support (a player, speaker, etc.) (esp. by shouting). (Said of a section of the crowd of spectators, orig. Australian.) Also *transf.* **b.** *trans.* To shout in this way at (a player, speaker, etc.). Hence **ba·rracking** *vbl. sb.* and *ppl. a.*; also **ba·rracker** *sb.*, one who barracks.
1885 in Baker *Austral. Lang.* (1945) xvii. 309 Barracking. **1889** *Cricket* 99 Junior clubs [in Sydney] have their armies of what are known as 'barrackers', who follow and howl for their side. **1890** Farmer *Slang*, *Barracking* (Australian), banter, chaff. **1890** *Melbourne Punch* 14 Aug. 106/3 To use a football phrase, they all to a man 'barrack' for the British Lion. **1893** *The Age* 27 June 6/6 (Morris), People were afraid to go to them [*sc.* football matches] on account of the conduct of the crowd of 'barrackers'. *Ibid.*, The 'barracking' that was carried on at football matches. **1895** *Westm. Gaz.* 1 Mar. 5/1 A spontaneous burst of cheering and 'barracking', with loud cries of 'Bravo, Stoddart!' were heard. **1900** H. Lawson *On Track* 89, I was too shy to go in where there was a boy wanted and barrack for myself properly. **1901** *Westm. Gaz.* 19 Aug. 6/2 The crowd had absolutely no right ..to 'barrack' the players by yelling in concert now and again, at a critical moment. **1904** Warner *How we recovered Ashes* 73 Hayward and myself had to undergo some 'barracking' for playing slowly. *Ibid.* 167 They will grow up into the type of man who 'barracked' Crockett so disgracefully at Sydney. **1911** E. M. Clowes *On the Wallaby* iii. 47 It seems as if I was 'barracking' for Australia as against England. **1926** *Chambers's Jrnl.* 543/2 Only once.. was a querulous barracking voice raised. **1934** A. E. Mulgan *Spur of Morning* 87 'Varsity barrackers gave up hope. **1943** *Coast to Coast 1942* 165 Dingo kept his pipe in his mouth all evening, not saying a word till even Ward barracked him. **1955** *Glasgow Herald* 25 July, It has helped to correct the poor impression he has of the supporters of Scottish Rugby, who never barrack. **1963** *Times* 18 Feb. 20/4 It was a hot day but the heat did not deter the barrackers as England struggled for runs. *Ibid.* 11 May 5/1 When Miss Truman led 4–1 in the first set, the crowd began to barrack every point she scored and to encourage the Italian girl with prolonged cheering.

barracoat, var. *barrow-coat*: see Barrow *sb.*[4]

barracuda, barracouta. [Amer. Sp. *barracuda.*] Add: Australian and New Zealand examples (*Thyrsites atun*). In S. Afr. the same fish is called the Snoek.
1839 J. Heberley *Autobiogr.* (MS.) 81 This is a piece of red wood with a nail driven through it and bent round in the shape of a hook commonly called a Barracoota hook. **1845** J. Deans *Lett.* (1939) 94 Baricouta are also to be got in great abundance and are very easily caught. **1874** A. Bathgate *Colonial Experiences* iii. 27 The barracouta (*thyrsites atun*), a rather coarse fish,..a great favourite with boarding-house keepers. **1880** W. Senior *Trav. & Trout in Antipodes* 295 The barracuda is..well known throughout the colonies. **1897** W. Saville-Kent *Naturalist in Australia* 170 The well-known Barracouta of the Adelaide, Melbourne, Sydney and Hobart markets. **1960** Doogue & Moreland *N.Z. Sea Anglers' Guide* 262 There are four kinds of snake-mackerel in New Zealand waters of which the barracouta is best known.

barragan. (Earlier example.) Also *attrib.*
1677 W. Cunningham *Diary* 12 Feb. (1887) 89, 11 ells Barragon to be a cloak. **1735** *London Even. Post* 25–27 Nov. 3/1 In his Walk [the missing man] stoops in his Shoulders, about 24 or 25 Years of Age, had on a light Barragon Coat and short brown Wigg. **1854** *Poultry Chron.* I. 114 On that barragon sleeve a human dew-drop glistened.

barrage (bæ·rāʒ, -ėdʒ, ‖barā·ʒ), *sb.* Add: In modern military use: a barrier of continuous artillery or machine-gun fire concentrated in a given area, used to prevent the advance or retreat of enemy troops, to protect troops advancing against the enemy, to repulse attacks by aircraft, and for destructive purposes; *creeping* or *moving barrage*, a curtain of fire moving before and directed from behind advancing troops. More explicitly *barrage fire.*
1916 H. W. Yoxall *Let.* 22 Sept. in *Fashion of Life* (1966) iv. 32 The Bosche barrage was almost as good as ours. **1917** *Times* 11 July 5/5 The German barrage fire on the trenches, though it lasted so short a time, was of extreme severity. *Ibid.* 26 Sept. 7/6 The barrage was effective in keeping the raiders at a great height. **1918** 'Boyd Cable' *Air Men o' War* ix. 118 Next instant he plunged at, into, and through the barrage, his machine rocking and pitching and rolling in the turmoil of shell-torn air. **1922** *Encycl. Brit.* XXX. 250/1 This breakdown of communications..led to the general introduction of the 'creeping barrage' (French *barrage roulant*, German *Feuerwalze*). Briefly, it is a screen of shells bursting on and close to the ground, which is moved forward across the country by short leaps according to a pre-determined time-table. **1938** *Times* 7 Sept. 9/1 Aeroplanes, balloons, guns, and searchlights, and the barrage itself had their part to play in the air defences of London.

b. *transf.* and *fig.*
1917 W. Owen *Let.* 25 May (1967) 464 You too have walked slowly through many barrages..from resisting Life. **1918** *Daily Chron.* 20 Dec. (Weekley), Field-Marshal Sir Douglas Haig passed into London through a creeping barrage of cheers. **1920** W. J. Locke *House of Baltazar* xxiv. 300 If the barrage of silence is maintained. **1927** M. Diver *But Yesterday* xxvii. 301 All that barrage about his health. **1942** E. Colby *Army Talk* 20 The word 'Barrage' ..also gets into slang, in the service..to describe a flood of conversation purposely put up to deceive and divert. **1958** F. C. Avis *Boxing Ref. Dict.* 8 *Barrage*, a succession of fast and hard punches by one boxer against another. **1964** *Observer* 5 Jan. 2/5 The barrage of official statements.

c. In *Sport*: a heat for the selection of contestants. Also, the deciding jump-off, run-off, etc., in the event of a draw.
1955 *Times* 23 July 3/2 Twelve competitors out of a representative international field qualified for the first barrage for the puissance jumping. **1956** *Times* 1 Dec. 10/1 Miss G. Sheen, successful in the final barrage of the women's individual foils at Melbourne yesterday. **1963** Bloodgood & Santini *Horseman's Dict.* 16 *Barrage*, jump-off in a horse-show.

2. barrage balloon, a captive balloon forming part of a balloon barrage. Cf. *balloon barrage* s.v. *balloon *sb.*[1] 10.
1923 *Aviation* 25 June 690/2 The employment of barrage balloons..permits the realization of economy in the distribution of 'active means' of anti-aircraft defense. **1925** *Sci. Amer.* July 61/1 The Barrage Balloon..is designed to ascend to a great height, trailing fine steel wires extending to the ground and wrecking any hostile airplane. **1943** *Roof over Britain* 5 The 'plane had struck a barrage balloon cable.
Hence **barrage** *v. trans.*, to put a barrage upon; *intr.* to put down a barrage. Also *fig.*
a **1917** E. A. Mackintosh (1918) *War, the Liberator* 124 The artillery will barrage at x. 20. **1918** 'Boyd Cable' *Air Men o' War* xvii. 230 The only bit of the whole line they managed to barrage properly. **1960** *New Left Rev.* May–June 64/1 She..barraged *me* with questions.

barramundi (bæ:rămɒ·ndi). Also **burramundi, -munda.** [Aboriginal.] The native name in Australia for any of various freshwater fish (see quots. 1880, 1898).
1873 Trollope *Australia* I. xi. 189 There is a fish too at Rockhampton called the Burra Mundi,—I hope I spell the name rightly,—which is very commendable. **1880** Günther *Stud. Fishes* 357 Two species, C[eratodus] *forsteri* and *C. miolepis*, are known from fresh waters of Queensland...The aborigines [call it] 'Barramunda', a name which they appear to apply also to other large-scaled freshwater fishes, as the Osteoglossum leichardti. **1898** Morris *Austral Eng.* 67/2 *Burramundi* or *Barramunda*..is also incorrectly applied by the colonists to the large tidal perch of the Fitzroy River, Queensland, *Lates calcarifer*, Gunth., a widely distributed fish in the East Indies. **1909** Webster, Barramundi. **1930** E. R. B. Gribble *Forty Yrs. with Aborigines* xvi. 165 John left at once,..returning in the evening with a fine barramundi. **1938** *Bulletin* (Sydney) 6 Dec. 21/2 The waters teem with barramundi (king of Australian fishes).

barranca. Add: Also **barranco.** (Earlier and later examples.)
a **1691** Boyle *Hist. Air* (1692) xix. 178 The Canary-Birds ..breed in the Barrancos of Gills, which the Water has fretted away in the Mountains. **1816** C. Smith *Jrnl.* 9 Apr. in J. K. Tuckey *Narr. Exped. R. Zaire* (1818) 245 Another barranco was winding its course down to *Publico grande.* **1829** W. Irving *Granada* I. xii. 112 A barranco, or deep rocky valley. *Ibid.* II. lxxvii. 226 Threads of water.. tinkling down the bottoms of the barrancas or ravines. **1939** G. Greene *Lawless Roads* viii. 213 The nearly dry barranca where the women did their washing.

barre (bāɹ). [F., lit. 'bar'.] **1.** *Mus.* = *capo tasto; also, the placing of the forefinger of the left hand across the finger-board to act as a capo tasto. So ‖**barré** (bāre) *a.* [*ppl. a.* of F. *barrer* to bar], (of a chord) played with all strings stopped in this way.
1876 Stainer & Barrett *Dict. Mus. Terms* 51/1 *Barré.* **1900** Grove *Dict. Mus.* I. 135/1 'Barre' designates the false nut made by placing the first finger of the left hand across the whole of the strings at certain lengths from the bridge to effect transposition. **1927** *Melody Maker* Aug. 756/3 Remember the most-used chord (the common chord) root position on the 'G' [banjo] is merely a barre, whilst it is a big stretch on the tenor [banjo]. **1934** S. R. Nelson *All about Jazz* ii. 55 Placing the index finger across the entire fingerboard in what is known as a 'barré' position. **1965** *Melody Maker* 10 July 12/4 Girls with small hands can play barre chords on a normal Spanish guitar.
2. *Dancing.* (See *bar *sb.*[1] 13 d.)

barred, *ppl. a.* Add: **3.** *spec.* in *Zool.* names (see quots.). Also *Comb.*, as *barred-breasted* adj.
1809 Shaw *Gen. Zool.* VII. I. 113 Barred-Breasted Buzzard. Falco lineatus. **1811** Wilson *Amer. Ornith.* IV. 61 Barred Owl, *Strix nebulosa*,..is frequently observed flying during day. **1885** Swainson *Prov. Names Birds* 98 Lesser spotted woodpecker (*Dendrocopus minor*). Also called Little wood pie (Hants). Barred woodpecker. **1961** R. South *Moths of Brit. Isles* (ed. 4) II. 147 The Barred Carpet (*Perizoma taeniata* Steph.). *Ibid.* 150 The Barred Rivulet (*Perizoma bifaciata* Haw.). *Ibid.* 168 The Barred Straw (*Lygris pyraliata* Schiff.).

barrel, *sb.* Add: **1. c.** In slang phr. *over a barrel* [app. in allusion to the state of a person placed over a barrel to clear his lungs of water after being rescued from drowning], helpless, in someone's power. Chiefly *U.S.*
1939 R. Chandler *Big Sleep* xxx. 270 We keep a file on unidentified bullets nowadays. Some day you might use that gun again. Then you'd be over a barrel. **1945** L. McClung *Stream runs Fast* xv. 129 You sure have me over a barrel. You caught me red-handed. **1963** N. Freeling *Because of Cats* xi. 175 If he's been fool enough to get himself over a barrel, that's his business.

5. b. (Earlier example.)
1591 in *Mod. Lang. Rev.* (1951) XLVI. 323 For peisinge ye spindle of ye howre wheele & ii newe bushes for ye barrell of ye same wheele.

d. A cylindrical button used in conjunction with a loop of braid as an ornamental fastening of a coat.
1898 *Daily News* 12 Nov. 6/3 Frogs and barrels are adapted to the otherwise plain red coats. **1900** *Ibid.* 29 Dec. 6/6 The fronts are fitted out with braided frogs and barrels.

6. c. *slang.* (A nickname for) a corpulent person.
1909 Ld. Harcourt in *Daily Chron.* 16 July 1/5 When that measure reached the House of Lords it was met with a barricade of barrels insufficiently disguised in robes and coronets. **1959** I. & P. Opie *Lore & Lang. Schoolch.* ix. 168 The unfortunate fat boy..is known as:..barrel, barrel-belly, [etc.].

10. c. *barrel-bodied, -churn* (examples).
1894 M. Dyan *All in Man's Keeping* I. xiv. 219 Nipper, fat, puffing, and barrel-bodied. **1743** Ellis *Mod. Husb.* May 101 When the Cream is ready, carefully strain it..into the Barrel Churn. **1805** R. W. Dickson *Pract. Agric.* I. pl. XXII, The improved barrel-churn with wheel. **1853** Barrel-churn [see Dasher 2]. **1853** 'A Lady' *Facts: Experiences Rec. Colonist N.Z.* iv. 34 The second churn for the larger quantity had better be a barrel churn fixed on a stand.

11. barrel bolt, a bolt made to slide into a barrel-shaped socket; **barrel-bulk** (examples); **barrel cactus,** the genus *Echinocactus*; **barrel-chair** *U.S.*, an upholstered chair having a back shaped like a barrel; **barrel-chest** (see quot. 1907); so *barrel-chested* adj.; **barrel distortion** *Opt.* (see quot. 1953); **barrel-fish,** the black rudder-fish, *Lirus perciformis*, found off the U.S. coast; **barrel-plating** (see quot.); **barrel roll,** an aeronautical feat in which an aeroplane makes a complete revolution about the longitudinal axis; so as *v. intr.*, to make a barrel-roll; **barrel-scraping** *vbl. sb.*, the action or result of 'scraping the barrel' (see *scrape *v.*); so *barrel-scraper*; *barrel-scraping* ppl. a.; **barrel shop** *U.S.*, a low-class drinking-saloon; *barrel-vaulted* (earlier example); **barrel winding** *Electr.*, an armature winding in which the end connections lie flat upon a cylindrical surface.
1909 Webster, Barrel bolt. **1911** *Encycl. Brit.* XV. 482/2 Among locks and fastenings the ordinary barrel or tower bolt needs no description. **1930** *Engineering* 2 May 564/2 Turned barrel bolts were used for connecting the brackets to the post webs. **1720** J. Steuart *Letter-Book* 22 Aug. (1915) 122 If William Binie Desires two barell bulk home, let him have it. **1853** *Harper's Mag.* VI. 581/2 Freights from Detroit to Lake Superior ports are usually $1 the barrel bulk. **1881** *Amer. Naturalist* XV. 984 The 'nigger-head' or 'barrel' cactus. **1850** A. J. Downing *Arch. Country Ho.* (1853) xii. 414 Another of the cheapest and simplest seats for a cottage is the *barrel-chair*. **1947** *Downtown Shop. News* (Chicago) 10 Feb. 15 (D.A.), Curved back barrel chairs foster that new look in fine furniture. **1907** Gould *Practitioner's Med. Dict.* 267/2 *Barrel-chest*, a peculiar formation of the chest observed in cases of long-standing emphysema of the lungs. **1961** Webster, Barrel-chested. **1961** J. Heller *Catch-22* (1962) xxi. 212 The wing commander was a blunt, chunky, barrel-chested man in his early fifties. **1889** Barrel distortion [see View *sb.* 19 b]. **1903** Beck & Andrews *Photogr. Lenses* iii. 73 These two types of distortion found in single lenses are known as 'barrel' and 'pin-cushion' distortion. **1929** G. H. Gliddon *Opt. Replica Human Eye* 21 At 20 degrees a trace of barrel distortion developed. **1937** A. T. Witts *Television Cycl.* 16 *Barrel Distortion*, the type of distortion produced by defective optical systems that causes a square image to have rounded (convex) sides, similar to the form of a barrel. **1953** Amos & Birkinshaw *Television Engin.* I. ix. 187 When distortion is present, the image of a rectangular object has sides which curve inwards (pincushion distortion) or outwards (barrel distortion), depending whether the magnification increases or decreases with distance from the centre of the image field. **1884** Goode et al. *Nat. Hist. Useful Aquatic Anim.* I. 334 This fish is also called by the fishermen 'Log-fish' and 'Barrel-fish'. **1885** J. S. Kingsley *Stand. Nat. Hist.* (1888) III. 191 The fishermen call them barrel-fish, though the most usual name is rudder-fish. **1936** J. T. Jenkins *Fishes Brit. Is.* (ed. 2) 71 The barrel-fish..secretes itself in floating barrels or boxes. **1911** *Handbk. Electro-Plating, Polishing* (ed. 4) 58 The Mechanical Plating of small articles in a barrel, revolving in a solution of the metal to be deposited, has reached a stage where developments in Barrel Plating can be looked forward to. **1927** in *Amer. Speech* (1930) V. 210 She looped, barrel-rolled, tail-spinned..and then swooped some. **1932** Auden *Orators* II. 69 Went into a barrel roll at 8000 ft. and never came out. **1959** *Observer* 11 Jan. 18/6 A makeweight in a volume of barrel-scrapings. *Ibid.* 15 Feb. 21/6 These over-edited, barrel-scraping volumes. *Ibid.* 15 Mar. 20/8 As for the Sunday-night offerings [on television], they were both barrel-scrapers. **1961** *Guardian* 22 June 8/7 The significance which critical barrel-scraping has bestowed on him. **1904** *N.Y. Tribune* 12 Oct. 1 A poisonous substitute for whiskey sold in the low 'barrel shops' along Tenth Avenue. **1825** *Ann. Reg. 1824* 276*/2 Three large connected chambers, all barrel-vaulted. **1902** *Encycl. Brit.* XXVII. 583/2 For multipolar armatures with two or more layers of inductors, 'surface' or '*barrel*' winding is now extensively used.

barrelage (bæ·rėlėdʒ). [f. Barrel *sb.* + -age.] The total amount of any commodity,

especially beer, as measured by barrels in a specified period; output estimated in barrels.
1890 *Certificate of valuation of Breweries*, The barrelage for the same period was, for the year 1888/9..312,292 bls. 1893 *Star* 10 Jan. 3/5 The increase in barrelage since the company was formed in 1887 amounted to 42,400. 1917 *Times* 26 Feb. 9/6 It is..impossible for us to guarantee the food of this country without making a very much deeper cut into the barrelage of the country. 1960 *Economist* 9 July 123/1 Not enough barrelage in prospect is the reason given for the lack of development in the hotel industry.

barrel-house. orig. *U.S.* [f. BARREL *sb.* 1.] **1.** A low-class drinking saloon, often incorporating a lodging-house or brothel.
1883 G. W. PECK *Peck's Bad Boy* 120 After I had put a few things in his brandy he concluded it was cheaper to buy it, and he is now patronizing a barrel-house. 1888 *Missouri Republican* 11 Feb. (Farmer), The West-side police are still arresting barrel-house loafers. 1934 J. A. & A. LOMAX *Amer. Ballads* (1960) p. xxxi, Jazz reigns happily and completely in the barrel-house regions. 1950 A. LOMAX *Mr. Jelly Roll* (1952) 79 Any musician who wasn't too proud to earn a dollar in a barrelhouse.
2. *attrib.*, passing into *adj.* Designating an unrestrained type of jazz such as is played in barrel-houses, characterized by a forceful rhythm. Also *absol.*
1926 H. O. OSGOOD *So this is Jazz* 99 Trumpets and trombones..in that semi-muffled voice aptly described by the term 'barrel-house tone'. 1937 *Amer. Speech* XII. 182/1 The barrelhouse style is very African. 1938 *Manch. Guardian Weekly* 2 Sept. 188/3 It may start 'barrelhouse' (free improvisation), blending into 'screwball' (fast free improvisation). 1958 P. GAMMOND *Decca Bk. Jazz* xv. 181 Barrelhouse retains a close allegiance with ragtime.

barrel-organ. Add to def.: Also extended to similar instruments not of the organ type but producing the notes by means of metal tongues which are struck by pins fixed in the barrel. The tone resembles that of a piano; hence they are distinguished as 'piano organs'. Hence **barrel-organ** *v. intr.*, to play a barrel-organ.
1842 DICKENS *Amer. Notes* I. vi. 209, I remember [seeing in New York]..one barrel-organ and a dancing-monkey. 1871 LE FANU *Checkmate* II. xiii. 128 The guitarring, singing, barrel-organing,..all make a curious and merry Babel. 1928 *Times* 27 Dec. 15/7 Those who have barrel-organed in a hard December. 1938 *Oxf. Compan. Mus.* 554/1 The application of the colloquial term 'Barrel Organ' to this instrument [the street piano] is evidently due to mere association of ideas at the time it became popular.

barremian (bărē·miăn), *a. Geol.* Also **-ien.** [f. F. *Barrême*, canton in department of Basses-Alpes, France + -IAN.] = URGONIAN *a.*
1903 GEIKIE *Text-bk. Geol.* (ed. 4) 1197 The abandonment of the term Urgonian and the adoption in its place of 'Barremien'. 1914 *Brit. Mus. Return* 210 The Barremian Beds of Sandown Bay, I. of Wight. 1963 D. W. & E. E. HUMPHRIES tr. *Termier's Erosion & Sedimentation* i. 30 In the Cretaceous, from the Barremian-Aptian to the Lower Senonian, the widespread transgression which led to a uniform marine fauna also favored the development of forests.

barren, *a.* Add: **4. b.** *Barren Grounds*, the district lying between Hudson Bay and Mackenzie River in Canada, used attrib. in *Barren-Ground bear* (see quots.); *Barren-Ground caribou, reindeer*, any of the several varieties of reindeer found in the Barren Grounds and Greenland, esp. *Rangifer arcticus* and *R. grœnlandicus*.
1691 KELSEY in Doughty & Martin *Papers* 23 Aug. (1929) 13 Now ye manner of their hunting these Beast on ye Barren ground is..they surround them. 1781 *Cumberland House Jrnls.* (1952) 226 The Indians lying Dead about the Barren Ground like rotten sheep. 1825 J. RICHARDSON in *Parry's Jrnl. 2nd Voy.* App. 329 The *cetrariæ, cornicularæ*, and *cenomyces*, which clothe the barren-grounds like a carpet. *Footnote*, The Barren Grounds ('Hi lichene obsiti campi quos Terram Damnatam diceret peregrinus', Flor. Lapp., p. 374). 1829 —— *Fauna Bor.-Amer.* I. 23, I have given this one the *ad interim* name of Barren-ground Bear, until its difference from, or identity with, the *Ursus arctos* of Linnæus be fully established. *Ibid.* 241 Cervus tarandus var. α, arctica. Barren Ground Caribou. 1836 R. KING *Narr. Journey Arctic Ocean* I. vi. 149 The barren-ground rein-deer, or *caribou*..is of small size compared to the other deer. 1877 *Encycl. Brit.* VII. 25/1 The Barren Ground Caribou, and the Woodland Caribou. 1897 LYDEKKER in *Proc. Zool. Soc.* 425 *Ursus arctus Richardsoni*—Barren-Ground Bear... This Bear..differs from the Grizzly in the shorter skull.
B. *sb.* **3. a.** *attrib.* as **barrens oak** (see quot.).
1832 D. J. BROWNE *Sylva Amer.* 269 In New Jersey and Pennsylvania it [*sc.* the Black Jack Oak] is called Barrens Oak.

‖ **barrera** (bare·rā). [Sp., barricade, barrier: see BARRIER.] A barrier, *spec.* that encircling a bull-ring; also the row of seats nearest to it.
1924 HEMINGWAY *In our Time* 22 He [*sc.* a horse] cantered jerkily along the barrera. 1926 —— *Sun also Rises* xiv. 155 Carpenters replaced weakened or cracked planks in the barrera. *Ibid.* xv. 167, I had taken six seats..three of them were barreras, the first row at the ringside. 1932 —— *Death in Afternoon* iii. 35 A red wooden fence a little over four feet high..called a barrera.

barrette (bare·t). Also **barette.** [a. F. *barrette*, dim. of *barre* BAR *sb.*[1]] **1.** The crossbar of a fencing foil or the hilt of a rapier.
1909 *Cent. Dict.* Suppl.
2. A bar for supporting a woman's back hair; also, a hair-ornament.
1901 *Daily Colonist* (Victoria, B.C.) 10 Oct. 5/6 Shell Hair Barrettes, each 10 c. 1908 *Westm. Gaz.* 29 Feb. 13/2 The front part of the barette forms a bandeau. 1908 *Ladies' Field* 4 Apr. 197/1 The favourite [hair-ornament] seems to be Greek barrettes over hair combed low. 1952 M. McCARTHY *Groves of Academe* (1953) ii. 21 [Girls with] jewelled barrettes in their new-washed hair.

barretter (bāre·təɹ). *Electr.* [Said to be alt. of OF. *barateor* BARRATOR, exchanger.] **a.** An early device for detecting radio waves by means of the change in resistance in a metal filament. **b.** A modern adaptation used to stabilize an electrical current.
1903 R. A. FESSENDEN *U.S. Pat.* 727,331, The hot-wire receiver described..which has been termed a 'barretter'. 1905 *Trans. Internat. Electr. Congress* III. 417 The device selected [as a detector for wireless reception]..was the 'solid barretter'... This consists of a minute loop of very fine platinum wire or filament. 1923 [see *hot wire* s.v. *HOT *a.*]. 1935 *Wireless World* 15 Nov. 522/1 A barretter is a device in which the resistance varies according to the current. 1940 *Chambers's Techn. Dict.* 77/2 *Barretter*, an iron wire resistance mounted in a glass bulb containing hydrogen, and having a temperature coefficient so arranged that the variation of resistance produced ensures that the current in the circuit to which it is connected remains constant over a wide range of voltage. 1942 *Electronic Engin.* XV. 36 The filament heating of the master oscillator is stabilised by a barretter which also contributes to frequency stability. 1943 *Ibid.* 384 The barretter tube, which although fundamentally a current stabiliser does not actually regulate the voltage.

barricading, *vbl. sb.* Add: **b.** *concr.* A barricade or the materials of a barricade.
1890 J. WATSON *Conf. Poacher* 159 Quietly as we could, we undid the barricading. 1910 *Westm. Gaz.* 24 Jan. 8/1 Some of the barricading was thrown among the crowd.

barrier, *sb.* Add: **3. b.** The mass of ice which fringes the Antarctic coast, occas. spec. *the Barrier*.
1847 J. C. ROSS *Voy. Discovery & Research* I. viii. 222 Some of the numerous fragments of the barrier that were about us. 1905 R. F. SCOTT *Voy. 'Discovery'* I. v. 168 The barrier edge, in shadow, looked like a narrowing black ribbon. 1930 *Discovery* Mar. 91/1 Flat fields of snow-ice, strictly speaking neither land-ice nor sea-ice, which, under the name of 'barriers' fringe almost the whole circumference of the continent.
c. *Physics.* The region of high potential energy through which a charged particle must pass on leaving or entering a nucleus; also *attrib.*
1929 RUTHERFORD in *Proc. R. Soc.* A. CXXIII. 378 The nucleus, supposed spherical in shape, is thus surrounded by a very high potential barrier. 1938 R. W. LAWSON tr. *Hevesy & Paneth's Man. Radioactivity* (ed. 2) ix. 98 The nucleus is thus protected by a barrier at an exceedingly high potential. 1955 J. A. WHEELER in W. Pauli *Niels Bohr* 166 Odd nuclei have a higher barrier against fission than corresponding even nuclei. *Ibid.* 167 The barrier penetration exponent is roughly proportional to the barrier height. 1962 SIMPSON & RICHARDS *Junction Transistors* ii. 23 Because electrons flowing from *n*- to *p*-type material or holes flowing from *p*- to *n*-type material require excess energy to mount the potential step, the latter is sometimes called the 'barrier'.
4. b. Freq. with defining word; *sound barrier* (see *SOUND *sb.*[3]).
1923 *Internat. Jrnl. Psycho-analysis* Oct. 495 (*sub-heading*) The Incest Barrier. 1957 *Ann. Reg. 1956* 346 Problems of producing aircraft for still higher speeds, with special attention for the next main obstacle—the heat barrier. 1958 F. KENYON *Our Bible & Anc. MSS.* (ed. 5) i. 36 They [*sc.* the Dead Sea Scrolls]..enable us..to 'penetrate the Massoretic barrier' by centuries.
5. barrier cream, a protective cream for the skin; **barrier layer** in *Physics*, an electrical layer lying between two different metals or between a metal and a semiconductor; also *attrib.*
1950 *N.Y. Times* 16 Apr. E9/5 A British chemical firm protects workers who handle dyestuffs, chemicals, explosives, oils, acids, alkalis and grease by means of 'barrier creams'. The cream is spread over the hands until an invisible 'glove' is formed. 1958 *Sunday Times* 20 July 16/5 A reputable barrier cream against sunburn. 1934 *Physical Rev.* XLVI. 1051 A barrier layer photo-cell..consists of a closed conducting circuit which comprises a semi-conductor forming a link in a metallic circuit. 1944 C. E. K. MEES *Theory of Photographic Process* xvii. 663 Numerous densitometers consist..merely of a light source, a barrier layer cell, and a microammeter. 1948 R. C. WALKER *Photoelectr. Cells in Indust.* i. 35 The rectifier type of light sensitive cell, often referred to by alternative names such as barrier layer, blocking layer, or self-generative cells.

barrikin (bæ·rikin). = BARAGOUIN.
1851 MAYHEW *London Lab.* I. 15/1 The high words in a tragedy we calls jaw-breakers, and say we can't tumble to that barrikin.

barring, *vbl. sb.* Add: **c.** (Later examples.) Also *attrib.*
1922 R. C. PUNNETT *Mendelism* (ed. 6) ix. 88 The cock in a pure breeding strain of Plymouth Rocks is homozygous for the barring factor. 1961 BANNERMAN *Birds Brit. Isles* IX. 2 The breast is also russet in colour, merging into black barring on the abdomen.
f. The division of music into bars: see BAR *sb.*[1] 16.
1874 [in Dict.]. 1920 *Times Lit. Suppl.* 6 Aug. 493/3 Latter-day music..shuns symmetrical barring and full closes. 1959 D. COOKE *Lang. Mus.* iv. 187 The same key, tempo, dynamics, phrasing, and barring.

barristress (bæ·ristrès). Also **-eress.** [f. BARRISTER + -ESS.] A woman barrister.
1898 *Fortn. Rev.* LXIV. 127 The modern barristress. 1899 *Westm. Gaz.* 8 June 1/2 Barristeress-at-law.

barrow, *sb.*[1] Add: **3.** *attrib.* **barrow-wight** (see quot. 1891); so *barrow-wightish* adj. (*nonce*).
1891 A. LANG *Ess. in Little* 146 In the graves where treasures were hoarded the Barrowwights dwelt, ghosts that were sentinels over the gold. 1954 TOLKIEN *Fellowship of Ring* I. viii. 151 The dreadful spells of the Barrow-wights about which whispered tales spoke. *Ibid.* xi. 197 It has a —well, rather a barrow-wightish look.

barrow, *sb.*[3] Add: **4. barrow boy,** a costermonger.
1939 J. WORBY *Spiv's Progress* viii. 72, I then gave him a brief I had got from one of the barrow boys' to take him back to Manchester. 1949 *Hansard Commons* CDLXVII. 2970 A lesson might be learnt from that humble member of the trading community, the barrow boy.

barrow (bæ·rou), *v.*[2] *Austral.* and *N.Z.* [Etym. unkn.; ? cf. Gaelic *bearradh*, shearing, clipping.] (See quot. 1933.) Hence **barrowing** *vbl. sb.*; **barrowman**[2], one who barrows; also *barrower*.
1933 L. ACLAND in *Press* (N.Z.) 16 Sept. 15/7 *Barrow*, to shear or partly shear a sheep for a shearer. 'No barrowing allowed on the board' was at one time a rule which the Shearers' Union got into the award. 1940 E. C. STUDHOLME *Te Waimate* (1954) I. xv. 129 There were always a certain number of 'barrowmen' and 'learners' on the board. [*footnote*] Friends of the shearers who came on to the board for a chat and..shore a few sheep for their pals. 1949 P. NEWTON *High Country Days* v. 54 The wool rollers and fleece pickers..tried their hand at shearing, a practice known as 'barrowing'. 1959 H. P. TRITTON *Time means Tucker* 26/1, I spent most of the afternoon watching and 'barrowing'; that is, finishing-off. 1965 J. S. GUNN *Terminol. Shearing Ind.* I. 6 *Barrowing*, a term to describe the action of a shed-hand or other learner who finishes off a sheep for a shearer after the bell has gone for the end of the run. *Ibid.*, Barrowers could cause considerable union trouble because they delay the other shedhands who have to clean up after the last sheep shorn in a run. Thus an accepted rule in most sheds to-day is 'no barrowing on the board'.

Barsac, barsac (bā·ɹsæk). Also †**barasack, barsack.** [a. F. (see below).] A sweet white wine from the district of Barsac, department of Gironde, France.
1728 J. STEUART *Letter-Bk.* 21 Dec. (1915) 302 A Tun of Barasack white wine in hogsheads. 1820 *Blackw. Mag.* VIII. 44 To lawyers I would give the sharp *Bar-sac*. 1827 *Mirror of Lit.* II. 164/1 Elephants don't drink barsack. 1845 *Tait's Mag.* XII. 698 Barsac, champagne, hock, and claret. 1951 *Good Housek. Home Encycl.* 360/2 Such white wines as Sauternes, Graves and Barsac.

Bart. (Earlier and later examples.)
a 1771 T. MARTIN *Hist. Thetford* (1779) xxii. 285 The rev. Sir John Cullum, bart. of Hardwicke. 1793 W. B. STEVENS *Jrnl.* 14 Mar. (1965) 72 Dined at Foremark—the Bart delighted with the carnage of the French. 1887 W. S. GILBERT *Ruddigore* (1963) I. 104 When I'm a bad Bart, I will tell taradiddles! 1906 'Q' *Mayor of Troy* xiv. 188 But if you ask me what was peculiar about the man, he was called Bart.—Sir Samuel Brooks Bart. 1958 B. HAMILTON *Too much of Water* iii. 46 We've really got a frightfully distinguished company on board. There's the Bart of course.

bartender (bā·ɹte:ndəɹ). orig. *U.S.* Also **bar-tender.** [f. BAR *sb.*[1] + TENDER *sb.*[1]] **a.** A keeper or manager of a refreshment bar. **b.** A bar-attendant or barman.
1836 *Franklin Repository* (Chambersburg, Pa.) 5 Apr. 3/3 My bar-tender..has become a drunkard according to the course of trade. 1864 G. A. SALA in *Daily Tel.* 21 Nov., The bar-tender is a person of great gravity of countenance. 1876 J. HARTLEY *Seets i' Lundun* iv. 53 Aw axed th' bartender if he'd onny. 1883 [see TENDER *sb.*[1] 2]. 1884 [see BAR *sb.*[1] 29 c]. 1961 *Spectator* 11 Aug. 199 A dish-washer and part-time bartender. 1968 *Globe & Mail* (Toronto) 17 Feb. 51 (Advt.), Bartender, for dining lounge.

barterable (bā·ɹtərăb'l), *a.* [f. BARTER *v.* + -ABLE.] Capable of being bartered; suitable for trading by exchange.
1852 MUNDY *Antipodes* (ed. 2) II. 234 They found it a barterable commodity. 1890 H. M. STANLEY *Darkest Africa* I. vii. 155 Trifles..were easily barterable for sugar-cane.

Barthian (bā·ɹtiăn), *a.* and *sb.* [f. name of Karl *Barth*, Swiss theologian (1886–1968) + -IAN.] **A.** *adj.* Of, pertaining to, or characteristic of Barth, or his writings. **B.** *sb.*

A follower of Barth. Hence **Ba·rthianism,** the substance of Barthian theology.
1929 E. BRUNNER *Theology of Crisis* iv. 87 The 'Barthian theology' had as its origin the teaching of men like Blumhardt, Ragaz, and Kutter. 1931 J. McCONNACHIE *Significance of K. B.* ii. 60 He [*sc.* Barth] says ruefully that he is punished for the success of his books by the existence of regular 'Barthians'. 1934 *Times Lit. Suppl.* 25 Oct. 724/2 The main currents in German Protestant theology..with particular attention to Barthianism. 1939 V. A. DEMANT *Religious Prospect* viii. 200 In their [*sc.* Hegel's and Luther's] following we have liberal modernism on the one hand and the Barthian school on the other. 1945 *Mind* LIV. 183 What..are moral philosophers to make of the Barthian distinction between moral sin and religious sin?

Barton[2] (bā·ɹtən). The name of a district near Lymington, Hampshire, applied *attrib.* to deposits of a division of the Eocene period in Britain. Hence **Barto·nian** *a.*, pertaining to or resembling the deposits at Barton.
1847 J. PRESTWICH in *Q. Jrnl. Geol. Soc.* III. 357 Having thus associated the London clay with the Barton beds..the organic remains..were taken as belonging to one and the same deposit. 1893 GEIKIE *Textbk. Geol.* (ed. 3) VI. iv. 980 In..the northern Apennines Professor Sacco regards as Eocene a mass of strata..which he subdivides as follows:—Bartonian. 100 metres [etc.]. 1902 A. J. JUKES-BROWNE *Stratigr. Geol.* xvi. 487 The Calcaire Grossier of the Paris Basin is succeeded by a group of sands... The upper part corresponds to our Barton Sands..containing Bartonian fossils. 1957 *Encycl. Brit.* VIII. 632/2 Overlying these in the Bartonian stage are the Barton clay and sands.

barukhzy (baru·kzi). [ad. *Bārakzi*, name of an Afghan people.] The Afghan hound.
1895 *Daily News* 11 Dec. 2/1 An Afghan Barukhzy hound. 1927 *Daily Tel.* 17 May 15/6 Every sort of dog—from the barukhzy of Cabul..to my lady's Pekingese.

Barum = Barnstaple (Devon). Used *attrib.*
1899 *Westm. Gaz.* 6 Mar. 9/2 They presented to 'Barumites in London' one of the choicest production of the Royal Barum ware pottery.

baryon (bæ·riǫn). *Nucl. Physics.* [f. Gr. βαρύ-ς heavy + *-on* of *ELECTRON[2].] A collective name for the heavier elementary particles. Also *attrib.*
1953 *Prog. Theor. Physics* X. 457 It seems practical to have a collective name for these particles and other which possibly may still be discovered... It is proposed to use the fitting name 'baryon' for this purpose. 1958 *Physical Rev.* CXII. 640/1 It is at this stage that we first explicitly meet S and N, the baryon number. 1959 *Sci. News* LII. 99 Nucleons and hyperons, or baryons as they are called collectively. 1967 *New Scientist* 22 June 724/1 Mesons could be composed of a quark plus an antiquark, baryons of three quarks.

barysphere (bæ·risfīɹ). [f. Gr. βαρύ-ς heavy + σφαῖρα SPHERE *sb.*] The internal substance of the earth enclosed by the lithosphere.
1901 *Science* 15 Nov. 747 It was argued by Posepny that the ores came from the barysphere. 1926 *Encycl. Brit.* New Suppl. II. 172 The bulk of the earth consists of a nickel-iron mass, the barysphere, which is enclosed by a rocky crust, the lithosphere.

baryta. Add: **baryta paper,** paper coated with an emulsion of barium sulphate; **baryta white** = permanent white (see PERMANENT *a.* 1 d).
1900 *Photogr. Ann.* 235 *Baryta Paper*, (Faced Paper)..Fine paper, faced with an emulsion of barium sulphate, rendered partly insoluble by chrome alum. 1954 *Spalding & Hodge's Paper Terminol.* 13 *Baryta Paper*, a form of metallic paper on which marks can be made with a metal point or stylus... It..is used in certain types of automatic recording apparatus. 1885 *Encycl. Brit.* XIX. 86/2 *Baryta White*..is prepared by grinding to a fine powder the pure white native sulphate of baryta (heavy spar). 1886 H. C. STANDAGE *Artists' Man. Pigments* i. 1 *Baryta White* (also known as Constant or Permanent White).

barytone: see *BARITONE.

basal, *a.* Add: **2.** (Later examples.)
1918 F. WOOD-JONES *Probl. Man's Ancestry* 34 A being, whose body is replete with features of basal mammalian simplicity. 1963 *Publishers' Weekly* 2 Sept. 44/2 To what extent are available paperback books adequate as compared with basal readers and early reading textbooks.
3. basal anæsthesia, administration of a light but long-acting anæsthetic which may be supplemented when necessary during an operation by a deeper short-acting anæsthetic; so *basal anæsthetic*; **basal area,** (i) the area situated at or forming the base; spec. in *Ecol.* (see quot. 1962); (ii) *Forestry*, the area of the cross-section of a tree at a height of 4·5 feet above the ground, or the sum of such areas; **basal complex** *Geol.* = *basement complex*; **basal conglomerate** *Geol.*, a conglomerate forming the lowest of a particular series of strata; **basal cover** *Ecol.* = *basal area* (i); **basal ganglion,** a ganglion situated at the base of the cerebrum (*Cent. Dict.* 1889); esp. in *pl.* **basal ganglia,** the thalamus, lentiform, and caudate nuclei considered as a group; **basal granule,** a granule found at the base of a flagellum or cilium in certain protozoa; **basal metabolism,** the metabolism of an organism in a fasting and resting state; so *basal metabolic rate* (see quots.); **basal narcosis** (see quots.); cf. *basal anæsthesia*; **basal rot** *Hort.* (see quot. 1954).
1934 H. W. FEATHERSTONE in *Brit. Med. Jrnl.* 24 Feb. 322/2 The pioneer work in basal anaesthesia. *Ibid.* 326/1 Basal anaesthetics are a boon to patients. 1610 W. FOLKINGHAM *Art of Survey* II. viii. 62 Multiply the Basall Area by 6. 1895 W. SCHLICH *Man. Forestry* III. i. 46 The sample tree should show a basal area which corresponds exactly to the mean section of the class. 1938 J. R. CARPENTER *Ecol. Gloss.* 39 *Basal area*, (1)..The sum of the basal areas of trees in a forest stand is the basal area of the stand. (2)..Ground cover, basal cover. 1962 H. HANSON *Dict. Ecol.* 43 *Basal Area*..the surface of the soil actually covered or occupied by a plant, especially the basal part, as compared to the full spread of the herbage. 1897 Basal complex [see *BASEMENT 1]. 1900 *Geol. Mag.* VII. 564 On the western shore of Ullswater..a good section has recently been exposed of the basal conglomerates variously ascribed to the Old Red or lowermost Carboniferous age. 1923 J. T. SARVIS *U.S. Dept. Agric. Bull.* MCLXX. 20 Basal cover, as used here, means the extent of ground surface actually covered by plants after the foliage has been removed by grazing or clipping. 1913 *Cunningham's Text-bk. Anat.* (ed. 4) 637 (*heading*) Basal Ganglia of the Cerebral Hemisphere. 1945 *New Biol.* I. 82 Deep seated primitive components of the brain which include..the basal ganglia. 1920 I. F. & W. D. HENDERSON *Dict. Scientific Terms* 28/2 Basal granule. 1951 *New Biol.* XI. 114 Beneath the pellicle [of paramecium] the cilia connect with small round bodies known as basal granules or kinetosomes. 1922 *Jrnl. Amer. Med. Assoc.* LXXVIII. 1887/2 (*title*) Correlation of Basal Metabolic Rate with Pulse Rate and Pulse Pressure. Extreme deviations from the normal basal metabolic rate are accompanied by..alterations in other functions. 1951 M. ABERCROMBIE et al. *Dict. Biol.* 138 In man, basal metabolic rate (B.M.R.) is expressed as the output of Calories per square metre of body surface per hour. 1952 *New Biol.* XIII. 64 The standard index of the latter [i.e. the metabolism of the body] being the 'basal metabolic rate' (B.M.R.), which is essentially the rate at which oxygen is consumed by the body when it is in a resting condition. 1914 *Jrnl. Biol. Chem.* XIX. 239 (*title*) Basal metabolism and creatinin elimination. 1920 *Lancet* 7 Aug. 289/2 Basal metabolism—i.e., the metabolism of the individual during complete muscular repose, where a period of twelve hours had been allowed to elapse since taking the last meal. 1963 M. McCARTHY *Group* xiii. 303 The result of her basal metabolism [test] seemed to have restored her natural optimism. 1935 *Brit. Jrnl. Anaesthesia* XIII. 15 This preliminary narcosis, or basal narcosis as it has been called. 1938 *Encycl. Brit. Bk. of Year* 39/1 'Basal narcosis'..means the giving, at various times before the anaesthetic itself, of a sedative or hypnotic drug, which will reinforce the action of the gas. 1896 *Daily News* 15 Apr. 715 Dr. Crawford [spoke] on basal rot. 1950 *N.Z. Jrnl. Agric.* Feb. 176/2 The lily is subject to certain diseases; the most common are leaf spot..and basal rot (*fusarium*). 1954 A. G. L. HELLYER *Encycl. Garden Work* 18/1 *Basal rot*, a name rather loosely applied to several quite distinct diseases which all attack bulbs, causing a decay of the base of the bulb... The decay is caused by fungal attack.
B. *sb.* (Earlier example.)
1620 W. FOLKINGHAM *Brachigraphy* sig. A 6v, The third are Basals, falling from the Head of the Line, and landing below the Foot.

bas bleu. For Dict. '1801' read '? 1786'. Add additional and later examples. Hence **bas-bleu·ism.**
1788 H. WALPOLE *Let.* 12 July (1961) XXXI. 271 You have gratified my self-love so amply in your '*Bas Bleu*'. 1812 *Dramatic Censor for 1811* 360 Lady Bab Blue is a literary woman of fashion, and a vestige of the Bas bleu club. 1842 *Times* 2 Sept. 2/6 The aristocratic *bas-bleus* are at present very much devoted to the abolition of slavery. 1871 J. C. YOUNG *Mem. C. M. Young* I. v. 200 They are utterly devoid of pedantry, or Bas-Bleu-ism, or any other ism.

Bascology (baskǫ·lŏdʒi). Also **Bask-.** [f. *Basc-*, used as comb. f. BASQUE + -OLOGY.] The study of the Basques and their language. **Basco·logist,** a student of, or one versed in Bascology.
1896 *Academy* 18 July 52/2 This Bascologist is incorrect. 1901 E. S. DODGSON in *N. & Q.* 9th Ser. VIII. 377/2 The following complementary details concerning the book..will be of interest to Baskologists. 1903 *Ibid.* XI. 355/2 Mr. Hutchinson has done good service to Bascology.

base, *sb.*[1] Add: **11. b.** *Cosmetics.* A substance used as a foundation.
1932 *Woman's Pictorial* 23 Apr. 8/2 Use a pure vanishing cream..for the powder base. 1950 J. EMERALD *Photographic Make-Up* IV. 115 It should be applied to the surface of the features..over an invisible make-up base.
14. (Earlier examples.)
1845 *Proc. Philol. Soc.* I. 28 The *p* is indicatory, signifying that, for the radical vowel of the base, a Guna letter or diphthong is to be substituted. 1848 *Ibid.* III. 58 This being the characteristic of the case itself, has of course nothing to do with the base of the word.
15. c. esp. in *Baseball*, each of the four stations at the angles of the 'diamond', all of which the batsman has to touch in succession in order to score a run. Phr. *base on balls*, an advance to first base allowed to the batsman when the pitcher has delivered four balls outside the strike zone.
1845 in *Appleton's Ann. Cycl.* (1886) X. 77/2 No ace or base can be made on a foul strike. 1868 CHADWICK *Game Base Ball* 34 He has also to look sharp after a base runner, when he is trying to run to second base. 1874 —— *Base Ball Manual* 92 The bases must be four in number, and they must be placed and securely fastened upon each corner of a square whose sides are respectively thirty yards... The first, second, and third bases shall be canvas bags, painted white, and filled with some soft material; the home base shall consist of white marble or stone. 1886 [see HOME *sb.*[1] B. 4]. 1891 N. CRANE *Baseball* x. 79 *Base on balls*. When a batsman is awarded first base by the umpire on 'four balls' called on the pitcher, the batsman is said to 'take his base on balls'.
d. *fig.*, in various expressions in *U.S. slang*, as *off one's base*, wildly mistaken, crazy, mad; *to get to first base*: to achieve the first step towards one's objective.
1882 G. W. PECK *Peck's Sunshine* 42 The Boston lady held up her hands in holy horror, and was going to explain..how she was off her base. 1888 'MARK TWAIN' in *Century Mag.* Jan. 463 It's about the gaudiest thing in the book, if you boom it right along and don't get left on a base. 1907 M. C. HARRIS *Tents of Wickedness* III. iii. 251 Mrs. Butterbeans was so off her base about it, it was ludicrous. 1938 F. SCOTT FITZGERALD *Let.* May (1964) 31, I thought I'd read Italian to read Dante and didn't get to first base. 1962 WODEHOUSE *Service with Smile* x. 157 She gives you the feeling that you'll never get to first base with her.
16. b. Similarly, an air or naval station.
1896 *19th Cent.* Mar. 461 Tactical considerations demand a strong naval base, which we already possess in Gibraltar and Malta. 1909 R. P. HEARN *Aerial Warfare* x. 134 (*caption*) Airship base. 1914 *War Illustr.* 5 Dec. 384 Three Englishmen..on November 23rd..made a bomb attack on the Zeppelin workshops... Two..adventurers succeeded in flying back to their base. 1940 *Economist* 7 Sept. 299/1 The leasing to the United States of air and naval bases in British possessions in America. 1947 AUDEN *Age of Anxiety* (1948) I. 19 While we hurried on to our home bases.
19. In sense 16, *base camp, censor, port, ship, wallah*.
1898 *Daily News* 27 May 7/5 The boats will be used as base camps. 1900 *Blackw. Mag.* Sept. 442/1 If..the senior officer commanding the Channel Squadron..brings his fleet into one of its natural base-ports. 1915 *Daily Express* 12 Nov. 5/3 It had been resealed in the customary way with the printed label showing that it had been 'Examined by Base Censor'. 1919 W. DEEPING *Second Youth* xxix. 251 This Base-wallah of a doctor. 1928 *Daily Express* 13 June 1 It was sent by wireless to the expedition's baseship. 1937 *Discovery* Dec. 376/1 He..established a base-camp where supplies could be concentrated before he advanced. 1962 P. PURSER *Peregrination 22* xv. 69 Some of the chaps are going to cross an ice-cap... Not me..Strictly a base-wallah.
20. baseboard, a board situated at or forming the base; *spec.* (chiefly U.S.) a skirting-board; **baseboard heating,** heating of a room by means of heat supplied to the skirting-boards; **base box** = *basis box* (*BASIS III); **base hospital** *Mil.*, a hospital at some place distant from the area of active operations; **base level,** *spec.* in *Phys. Geogr.* (see quots.); hence **base-levelled** *ppl. a.*, brought to base level; **base-line** (earlier examples); also *fig.*; cf. also quot. 1802; **base-load** (see *LOAD *sb.*); **base rate** (see quot.); **base-ring,** (*b*) a projecting circular base; applied *attrib.* to a type of late Bronze Age pottery from Cyprus.
1854 O. S. FOWLER *Home for All* 159 After mop or base-boards are nailed on..fill in between these boards..with stone or mortar. 1899 *Nature* 15 June 149/2 The author conceived the idea of using on a base-board a rotary disc to represent a crank-shaft. 1935 A. SQUIRE *Sing Sing Doctor* xiv. 210 A broad baseboard curves inward on either side [of the electric chair] to form a single wide center leg. 1943 *Electronic Engin.* XV. 390 The chassis is fixed vertically to a wooden baseboard so that the components are easily accessible. 1958 M. L. HALL *Newnes Complete Amat. Photogr.* iii. 42 The scale is engraved on the base-board of the camera. 1954 *Archit. Rev.* CXVI. 343 One last variety is skirting panel heating, which is used extensively in America where it is known as 'baseboard' heating. 1925 A. H. MUNDEY *Tin & Tin Industry* 95 There was hot-rolled a total of 213,940 base boxes. 1956 Base box [see *basis box* s.v. *BASIS III]. 1895 *Daily News* 29 Apr. 5/3 It was intended to divide the 'Base Hospital' among several of the larger stations within easy reach of the frontier. 1895 *Westm. Gaz.* 30 Dec. 5/1 There are no dangerous cases of illness at the base hospital. 1875 J. W. POWELL *Explor. Colorado River* II. xii. 203 We may consider the level of the sea to be a grand base level, below which the dry lands cannot be eroded. 1939 BAILEY & WEIR *Introd. Geol.* xxxii. 188 Sea-level is often called the base level of stream erosion, although rivers do cut a little below sea-level. 1949 W. G. MOORE *Dict. Geogr.*, The permanent base-level is the level of the sea; a lake provides a temporary base-level, but the sediment deposited in it by the stream destroys its effect. 1925 J. JOLY *Surface-Hist. Earth* v. 81 The base-levelled stumps of Archaean mountains cover two millions of square miles in Canada. *Ibid.* vii. 114 The base-levelled remains of pre-Cambrian mountains or of the ancestral Rockies. 1750 M. MACKENZIE *Orcades* 3/1 The Direction of this Base-line having been exactly taken with a Magnetic Needle, from each of its Extremities, with a good Theodelite, the Angles were observed, contained between the Base and visual rays connecting the Beacons. 1802 C. JAMES *Mil. Dict.*, *Base-line*, the line on which troops in column move, the first division that marches into the alignement forms the base line, which each successive division prolongs. 1810 *Ibid.*, *Base-line* also signifies the line on which all the magazines and means of supply of an army are established, and from whence the lines of operation proceed. 1902 *Daily Chron.* 25 July 3/3 A definite base-line from which the future progress of Irish industry can be measured. 1961 *Lancet* 29 July 227/2 The distribution of sensitivities found in various parts of the country served as a baseline. 1923 J. D. HACKETT *Labor Terms* in *Management Engineering* May, *Base Rate*,

the ordinary day rate of wages guaranteed, in scientific management, whether the standard task is accomplished or not. 1899 MYRES & OHNEFALSCH-RICHTER *Catal. Cyprus Museum* 16 With the exception of a few late and distinct fabrics, the vessels have no foot or base-ring to enable them to stand upright. *Ibid.* 37 *Base-Ring Ware*..is confined to the later Bronze Age, and does not appear much before the Mykenaean vases. 1905 H. B. WALTERS *Hist. Anc. Pottery* I. vi. 242 Base-ring ware..is marked off from other Bronze-Age types by its flat-ringed base in all cases. 1949 W. F. ALBRIGHT *Archæol. Palestine* v. 99 Base-ring ware from Cyprus..died out rapidly after the beginning of the thirteenth century.
b. In *Baseball*, as *base-bag, -player, -playing, -stealer, -stealing* (cf. STEAL *v.*[1] 5 g); **base hit,** a hit from which the ball cannot be fielded in time to prevent the batsman from reaching his base; **base-line** (*a*) the line, three feet wide, marked on the turf from base to base of a baseball field; (*b*) the line at each end of a lawn-tennis court, also *attrib.*, as *base-line game, driver*; hence **base-liner,** one who drives from the base-line, also a base-line drive; **base-runner,** in *Baseball*, the player who, having made a fair hit or in other contingencies specified in the rules, is running the bases; so *base running* vbl. sb.
1864 *Wilkes' Spirit of Times* 10 Dec. 229/1 The player shall not be declared out if he maintains his position with any part of his person on the place where the base-bag belongs. 1867 CHADWICK *Beadle's Dime Base Ball Player* 10 The rule makes the base-bag the base, not the post to which it is fastened. 1874 —— *Base Ball Manual* 83 All ordinary errors, such as dropped flyballs, bad muffs, wild throws, and failures on the part of base players to hold balls thrown to them—all count in preventing base hits being made. 1867 *Ball Players' Chron.* 4 July 5/1 Foul balls are those striking the ground or a player back of the base lines. 1875 *Encycl. Brit.* III. 406/2 [*Base ball*] The position of the bases and base lines may be likened to a 90 feet square shaped diamond. 1875 'STONEHENGE' *Brit. Rural Sports* 690/1 (*Lawn tennis*) The player who wins choice of courts serves; *i.e.* delivers the ball, standing with one foot outside the base line. 1878 *Laws of Lawn-tennis* 7 At each end of the Court, parallel with the net, and at a distance of 39 ft. from it, are drawn Base-Lines. 1905 *Westm. Gaz.* 29 June 5/2 Unsurpassed as a base-line driver. 1929 W. E. COLLINSON *Spoken Eng.* 90 You'd better stand clear of the base-line or you may foot-fault. 1903 *Westm. Gaz.* 27 July 10/2 Ritchie displayed remarkable activity and resource in getting back most of his opponent's lightning 'base-liners'. 1867 CHADWICK *Base Ball Player's Bk.* 23 A base player taking a ball from a fielder. 1868 —— *Game of Base Ball* 34 We now come to base playing, and we propose to show that each position has its peculiar points of play. 1867 —— *Beadle's Dime Base Ball Player* 129 The base-runner ceased to be forced to leave the base. 1875 *Encycl. Brit.* III. 407/1 Directly a striker has fairly struck a fair ball he becomes a base-runner. 1867 *Ball Players' Chron.* 6 June 4/2 Prohibiting base running on called balls. 1886 CHADWICK (*title*) The Art of Batting and Base Running. 1896 —— *Spalding's Base Ball Guide* 88 We have made up a record of the most successful base-stealers of the twelve League clubs who have a record of total stolen bases for at least three seasons. 1912 MATHEWSON *Pitching in a Pinch* 272 Merkle..is a great base stealer because he has acquired the knack of 'getting away'. 1886 *Chicago Tribune* 14 May 3/3 The fifth inning was characterized by a cheeky bit of base-stealing by Dalrymple. 1917 MATHEWSON *Sec. Base Sloan* xi. 145 He got to first and gave a very pretty exhibition of base-stealing a moment later.
c. *Chem.* **base exchange,** a reaction in which atoms of one chemical base in a compound substance are replaced by those of another; also *attrib.* (Also called *CATION *exchange*.) So **base-exchanger,** a substance capable of such a reaction; *base-exchanging* adj.; **base-rich** *a.*, rich in basic ions.
1922 *Chem. Abstr.* 3042 (*title*) Base exchange in silicates. I. Exchange of alkalies and ammonia in..permutite. 1930 *Discovery* Oct. 340/1 Work on the base-exchange (zeolite) method of water-softening was..described. 1949 P. C. CARMAN *Chem. Constit. & Prop. Engin. Materials* xiii. 385 A potassium-kaolinite can be prepared by base-exchange and shows no reluctance to exchange K+ for other cations. 1943 *Thorpe's Dict. Appl. Chem.* VI. 218/1 Processes involving base-exchangers include water-softening, total demineralising of water, purification of sugar juices, [etc.]. 1917 *Chem. Abstr.* 80 Base-exchanging substances, such as Na zeolites. 1958 *New Biol.* XXVI. 91 Fen peat formed under conditions which are base-rich (i.e. containing relatively substantial concentrations of cations such as those of calcium, magnesium, potassium and sodium) and alkaline. 1959 A. R. CLAPHAM et al. *Excursion Flora Brit. Isles* 328 Base-rich water.
21. *Electr.* One of the three electrodes of a transistor. Also *attrib.*
1948 *Physical Rev.* LXXIV. 230/1 The transistor..consists of three electrodes... The third is a large area low resistance contact on the base. 1957 *Encycl. Brit.* XXII. 404/1 In a typical transistor,..the charge carriers are..controlled by a signal between the emitter and a control electrode (base). 1959 *Electronic Engin.* XXXI. 331 The base current of the transistor under test is changed in steps.

baseball. Add: **b.** *attrib.*, as *baseball club, field, game, player*.
1855 (*title*) Atlantic Base Ball Club, Jamaica, N.Y. (D.A.E.). 1857 in Chadwick *Base Ball Manual* (1874) 7 The National Association of Base Ball Players. 1860 CHADWICK (*title*) Beadle's Dime Base Ball Player comprising [etc.]. 1874 —— *Base Ball Manual* 9 A base ball field. 1886 *Baltimore Amer.* in *Boston Jrnl.* 21 July 2/3 (D.A.E.), Any respectable base ball game can attract from 2000 to 7000. 1911 H. HARRISON *Queed* xviii. 225 On the following Saturday,..he took Miss Weyland to another base-ball game.
Hence **ba·seballer, -ballist.**
1868 (*title*) New England Base Ballist. A weekly journal. 1886 *Congress. Rec.* 2 Apr. 3043/2 [He is well known] as a baseballist among constitutional lawyers, and a constitutional lawyer among baseballists. 1888 *Battle Creek Jrnl.* 12 Dec., Western Base Ballers. 1896 KNOWLES & MORTON *Baseball* 79 He..at once became a London baseballer.

base-burner. *U.S.* (See BASE *sb.*[1] 20 and quots.)
a 1877 KNIGHT *Dict. Mech.* I. 242/1. 1877 BARTLETT *Dict. Amer.* (ed. 4) 32 *Base-burner*, a sheet-iron stove for burning anthracite coal, which is only fed at the top, while the fire is confined to the base, or lower part of the stove. 1893 'O. THANET' *Stories Western Town* 34 We got to keep a fire in the base-burner good, all night, or the plants will freeze. 1908 S. E. WHITE *Riverman* xxi. 188 When the very cold weather came and they had to light the base-burner stove. *Ibid.* 189 He shook down the base-burner vigorously. 1922 TITUS '*Timber*' ii. 28 A gaunt man..was putting wood in the base burner.
So **base-burning** *a.*, applied to a furnace or stove.
a 1877 KNIGHT *Dict. Mech.* I. 242/1.

based (bēɪst), *pa. pple.* [f. BASE *sb.*[1] 16 + -ED[2].] Established at a base.
1925 E. F. NORTON *Fight for Everest 1924* 57 No. 1 party was to..remain based there for the purpose of getting the next camp on to the North Col.
Hence in Comb. with preceding sbs.
1927 *Daily Tel.* 22 Mar. 10/7 Limiting the number of shore-based aircraft..in commission. 1933 *Bulletin* (Glasgow) 21 Sept. 18/4 The flying boats..have the assistance of land-based aircraft. 1935 Carrier-based [see *CARRIER 7]. 1943 in *Amer. Speech* (1945) XX. 142 Mediterranean-based. 1944 *Ibid.*, Russia-based. 1965 *Listener* 10 June 857/2, I..doubt whether ten years from now this 'science-based' or technological university will be very different from any other.

baseman (bēɪ·smæn). [f. BASE *sb.*[1] 15 c + MAN *sb.*[1]] Each of the fielders who stand near the first, second, and third bases in baseball.
1857 *Spirit of Times* 28 Nov. 196/1 Logan, as third base, is..probably the best baseman on the club. 1874 CHADWICK *Base Ball Manual* 13 The positions in the field are as follows:—Catcher, pitcher, first baseman, second baseman, third baseman, short-stop, left fielder, centre fielder, and right fielder. 1917 MATHEWSON *Sec. Base Sloan* xi. 145 A dexterous hook-slide that kept him far out of reach of the baseman's sweep. 1968 *Washington Post* 4 July C1/7 The Yankee second baseman charged Mike Epstein's slow hopper up the middle.

basement. Add: **1. basement complex** *Geol.* (see quot. 1961).
1897 W. B. SCOTT *Introd. Geol.* IV. xxi. 358 The Archaean includes the most ancient rocks, often spoken of as the 'basement, or basal complex'. 1940 R. A. DALY *Strength & Struct. of Earth* i. 17 The average density of the Basement (Archean) Complex of each continent. 1961 CHALLINOR *Dict. Geol.* 19/1 *Basement complex*, a regional mass of Archaean metamorphic crystalline rocks (gneisses and schists) and igneous intrusions on which lie veneers of obviously sedimentary rocks (of ages from the later Pre-Cambrian onwards)... 'Fundamental complex' may be said to be synonymous, and 'Archaean' is an equivalent name.
3. b. (Later examples.)
1917 T. S. ELIOT *Prufrock & other Observations* 31 They are rattling breakfast plates in basement kitchens. 1939 —— *Old Possum's Pract. Cats* 29 A small basement flat. 1962 J. G. BENNETT *Witness* xv. 178 My wife and I had found a small basement flat in Bayswater. 1966 *Listener* 3 Feb. 176/2 A despondent servant-girl from the country in each basement kitchen.

basenji (bäse·ndʒi). [Bantu.] An African breed of smallish hunting dog, native to the inner Congo regions, which rarely barks.
1933 O. BURN in *Field* 4 Nov. 1157/1 Three years ago I imported to England five Baseiyis—the smooth-coated chestnut dog of terrier size used by the native chiefs for hunting antelope etc. on the Congo plateaux. 1935 *Ibid.* 18 May 1251/1 Basentis don't bark! 1937 *Daily Herald* 11 Feb. 3/6 The Basenjis—the most puzzled dogs in the show [*sc.* Crufts']. They cannot bark, and could not understand all the other noisy dogs... They come from the Congo, are used for hunting by Central African tribes, and are believed to descend from dogs depicted on old Egyptian tombs. 1938, 1957 [see *BARKLESS *a.*[2]].

bash (bæʃ), *v.*[3] *local.* 'To fill with rubbish the spaces from which the coal has been worked away' (Gresley *Gloss. Coal-m.* 1883). Hence **ba·shing** *vbl. sb.*
1905 *Daily Chron.* 26 June 6/5 A 'bashing'—a barricade of coal and rubbish.

bash, *sb.* Add: Now in *gen.* use, a heavy blow. (Later examples.)
1949 C. FRY *Lady's not for Burning* 86 If he wants to fight me, let him come out in the garden. Whatever happens I shall have one bash at him. 1959 *Listener* 8 Jan. 77/1 A weak, wan lad..escaped with no worse than a bash and a hang-over.
2. In various slang uses: (i) an attempt, esp. in *to have a bash* (*at*); (ii) a good time; a spree; a party (see quot. 1948, *Amer. Speech*) *U.S.*; phr. *on the bash*, on a drinking bout (app. *Sc.* and *N.Z.*); also (examples are U.K.), soliciting as a prostitute (quots. 1936 and 1959[1]); (iii) in *Jazz*, a 'jam session' (only *U.S.*?).
1901 G. DOUGLAS *House with Green Shutters* xxi. 222 Let us go out and do a bash! 1919 J. BUCHAN *Mr. Standfast* viii. 167 Ye ken what a man's like when he's been on the bash. 1924 *Kelso Chron.* 12 Sept. 2/8 The village tailor..had an unfortunate weakness for getting terribly 'on the bash' perhaps twice a year. 1936 J. CURTIS *Gilt Kid* ii. 23 Most of the time she's on the bash round the flash bars. 1948 *Amer. Speech* XXIII. 219 One could store or *stash* food for a big *bash*. This involved eating two or three days' rations at one time. 1948 D. BALLANTYNE in *Landfall* II. 111 He figured what he spent on beer weeknights would total no more than what most jokers spent on their Saturday bashes. 1948 PARTRIDGE *Dict. Forces' Slang 1939–45* 11 *Have a bash at*, to make an attempt. 1949 L. FEATHER *Inside Be-Bop* vi. 42 One jazz concert promoter, who in previous years had presented nothing but Dixieland bashes. 1950 *Home* (U.S.) 2 Apr. 8 (Wentworth & Flexner), Some of these bashes were impromptu at 4 in the morning by trumpet players. 1950 C. MACINNES *To Victors the Spoils* I. 135 He's decided to have a bash at tightening up the discipline. 1957 J. BRAINE *Room at Top* xxiii. 192 I'll have a bash just the same. 1957 I. MURDOCH *Sandcastle* iii. 38 Come on..have a bash. You can translate the first word anyway. 1959 *Streetwalker* iii. 58 From the hours you keep..I'd say you were on the bash. 1959 M. SHADBOLT *New Zealanders* 156 Jack and I went on the bash every Saturday... Drink all day and pay a visit to the local house at night. 1959 *Times* 26 May 12/6 Tried some anti-rust oil? Worth a bash. 1961 A. BERKMAN *Singers' Gloss. Show Business* 7 Bash, a ball; party.

basha (bæ·ʃā). [Assamese.] A hut made of bamboo with a thatched roof.
1921 *Blackw. Mag.* Feb. 257/1, I was awakened by my servant entering the basha with the tea-tray. 1947 'N. SHUTE' *Chequer Board* v. 107 They drove on around the house to a small bashah made of bamboo, palm leaf matting, and palm thatch. 1956 W. SLIM *Defeat into Victory* vii. 143 The troops lived in tents or *bashas*, the bamboo huts, thatched with leaves. *Ibid.* xiii. 287 There were now orderly *basha* camps, hutted hospitals. 1969 *Sunday Mail Mag.* (Brisbane) 18 May 1/3 At his 'basha'..he has a radio with a 700-mile range by his bed.

bashaw. Add: **3.** Local name for a very large catfish of the species *Pylodictis olivaris*; the mud cat. *U.S.*
1888 G. B. GOODE *Amer. Fishes* 378 *Leptops olivaris*, the 'Mud Cat', 'Yellow Cat', 'Goujon', or 'Bashaw' is found in all the large rivers of the West and South. 1923 *Public Opinion* 12 Oct. 357/3 A good-sized fish, itself Carnivorous, called a basha.

bashed (bæʃt), *ppl. a.*[2] [f. BASH *v.*[2] + -ED[1].] Having the surface beaten or smashed in.
1830 J. WILSON *Noct. Ambr.* xxv. in *Wks.* (1856) III. 16 Like a heap o' bashed and birzed paddocks. 1898 *Westm. Gaz.* 31 Mar. 7/2 A bashed mask.

basher[1] (bæ·ʃəɹ). [f. BASH *v.*[2] + -ER[1].] **1.** One who gives, or strikes with, a smashing blow; *spec.* a pugilist.
1882 *Daily Tel.* 16 Dec. 2/6 The professed 'basher'. 1886 *World* 11 Aug. 8 This bruiser of the police court, this basher of a little foreign Jew. 1888 J. GREENWOOD *Policeman's Lantern* 35 (*title*) The Man-Basher. *Ibid.* 44 'Does that mean for simply knocking him down?'..'It means whatever you mean,' returned the professional 'basher'..; 'it wont make no odds to me, when I'm once set about him.' 1927 *Daily Tel.* 11 Oct. 12/6 He fights as if he enjoys fighting; and there is much of a basher in him.
2. *Services' slang* (see quot. 1943). Also, more generally, a man, often with defining word. Cf. *BIBLE-*basher*.
1942 F. RHODES *Let.* in Partridge *Dict. Slang* (1961) II. 991/1 *Buster* or *basher* is very common for mechanics, as in *compass basher, instrument basher*. 1943 HUNT & PRINGLE *Service Slang* 13 *Bashers*, instructors in physical training. 1945 *Gen* 30 June 51/2 One of the cookhouse bashers that came off at five. 1946 E. H. PARTRIDGE in Raymond & Langton *Slipstream* 62 The term basher, as in clock basher, stores basher..now means no more than fellow, chap, man.

basher[2] (bæ·ʃəɹ). *slang.* [Perh. same as prec.] A straw hat, a boater.
1901 *Westm. Gaz.* 9 Aug. 8/1 They parade in the grounds of the Exhibition with well-cut clothes and straw 'bashers'. 1933 *Bulletin* (Glasgow) 11 July 5/5, I am inclined to think that there are certain kinds of clothes that do not blend with a 'basher'.

bashing (bæ·ʃiŋ), *vbl. sb.* [f. BASH *v.*[2] + -ING[1].] **1.** The action of striking so as to dint, bruise, or crush; an instance of this.
1733 W. ELLIS *Chiltern & Vale Farm.* 45 Its hard and crusty Surface, so made by the Weights and Bashings of the heavy Rains. 1921 *Public Opinion* 21 Jan. 62/2 One would have expected to find a studious cultivation of more scientific methods of bashing. 1956 E. GRIERSON *Second Man* ix. 201 Street brawling—what the Aussies call 'Bashing'. 1958 J. CANNAN *And be a Villain* iii. 49 The first murder, other than slum-area bashings, that had occurred.
2. A flogging (with the cat-o'-nine-tails).
1877 W. H. THOMSON *Five Years' Penal Servitude* iii. 157 There were the evidences of former floggings, or 'bashings', as the prisoners call them. 1898 *Daily News* 23 Feb. 3/5, I got a bashing as well.—What do you mean by that?—Oh, I had the 'cat'.
transf. and *fig.*
1948 PARTRIDGE *Dict. Forces' Slang* 11 *Take, get, a bashing*, to suffer heavy losses. 1959 *Listener* 29 Jan. 219/1 In their 1911 bashing of Mr. Lloyd George the profession far excelled their efforts of 1948.

3. Used in *Services' slang* to denote any arduous task.
1940 *Daily Mail* 7 Sept. 3/8 Here are some current military phrases, interpreted:..'Spud-bashing'—Potato peeling. 1942 *Horizon* VI. 114 C.B. for a cert. Scrubbing, or mebbe a spot of spud bashing. 1943 WARD-JACKSON *Piece of Cake* 56 *Square bashing*, marching, drilling or walking. 1946 G. KERSH *Clean, Bright & Slightly Oiled* i. 3 Poor old Gerald done fourteen drills that week, plus a nice basinful of spud-bashing.

bashlik (bæ·ʃlik). Also **bashluik, beshlik,** etc. [ad. Russ. *bashlýk*.] A kind of hood with long side-pieces worn by Russians in inclement weather as a protective covering for the head. Also *transf.*, a light covering for the head, worn by women in the U.S. (*Cent. Dict.*).
1881 A. LESLIE tr. *Nordenskiöld's Voy. of Vega* I. x. 477, I had bought the baschliks in St. Petersburg. 1882 O'DONOVAN *Merv Oasis* I. iv. 74 Hanging between the shoulders, and knotted around the neck, is the *bashlik*, or hood, worn during bad weather. 1885 A. FORBES *Souvenirs of some Continents* 330 The flames danced..on the sombre hoods of beshliks. 1904 *Daily Chron.* 28 Mar. 3/1 The bashluik, or hood, worn to protect the ears.

basic, *a.* Add: **1.** Applied *spec.* to an industry which plays a major role in the national economy.
1928 *Rep. Liberal Industr. Inquiry* i. 12 The great basic exporting industries of Great Britain—coal, metallurgy and textiles—have been in a bad way. 1929 *Times* 25 May 9/4 The industry [*sc.* shipbuilding]..still had the highest percentage of unemployment of any of the basic industries in the country. 1940 *R. Comm. on Distrib. Industr. Pop.* (Cmd. 6153) iii. 28 The industries which, for the purposes of exchange, send their products to places outside the area in which they are situated, may be termed 'basic' industries.
b. That is or constitutes a standard minimum amount in a scale of remuneration or the like. Also *ellipt.* as *sb.*
1922 *Encycl. Brit.* XXXI. 392/1 When trade unions fix the length of the working day, they mean the basic working-day, with a higher rate of pay for overtime. 1923 J. D. HACKETT in *Management Engineering* May, *Basic Eight-Hour Day*, that period of time for which a specific wage rate is paid, beyond which a higher rate, generally 'time and a half', is paid. 1939 *Times* 25 Oct. 5/2 The next issue of basic petrol ration books will be made for a period beginning November 23. 1940 *Ibid.* 27 Feb. 5/6 At any time the unit of the basic petrol ration may be made smaller. 1948 *Observer* 11 Apr., The right way to check abuses of E and S petrol would be to confine it strictly to essential needs, leaving basic to fulfil its supposed purpose of allowing all motorists some pleasure trips. 1949 *Ann. Reg. 1948* 447 The 'basic allowance' of foreign currency for tourists outside the sterling area was restored. 1958 A. HACKNEY *Private Life* viii. 78 There's a job in Stores and Packing. Hundred and eighty-nine shillings basic.
c. *Gram.* Of or belonging to the base or theme of a word. See BASE *sb.*[1] 14.
1885 A. S. COOK tr. *Sievers' O.E. Gram.* § 86. 38 The development of the basic vowel into a diphthong.
d. Applied to a limited, 'essential' vocabulary in any language; spec. **Basic English,** a variety of the English language, comprising a select vocabulary of 850 words, invented by C. K. Ogden, of Cambridge, and intended for use as a medium of international communication; also ellipt. *Basic*.
1929 C. K. OGDEN in *Psyche* IX. III. 4 It is the continuous approximation of East and West, as a result of the analytic character of Chinese and English..which makes this particular [Panoptic] form of English *basic* for the whole world. Many special captions or trade-marks for the system have been suggested, but *Basic* = British American Scientific International Commercial (English)—is for the time being as good as any. *Ibid.* 97 (*title*) Translation into Basic English. 1930 C. K. OGDEN (*title*) Basic English. 1931 *Routledge Autumn Books*, Basic English is a system in which 850 English words do all the work of over 20,000, and so give to everyone a second or international language which will take as little of the learner's time as possible. 1933 *Discovery* Sept. 280/1 Science itself..might go forward with greatly increased efficiency if the language barrier were removed by the adoption of Basic for Abstracts and Congresses. 1933 H. G. WELLS *Shape of Things to Come* v. § 7. 419 It was more difficult to train English speakers to restrict themselves to the forms and words selected than to teach outsiders the whole of Basic. 1935 *N. & Q.* CLXIX. 145/2 The Swedish Anglic and our own Basic English. 1944 H. G. WELLS *'42 to '44* 141 'Basic' English, Russian or Italian is the minimum vocabulary necessary to talk understandably in any of these tongues. 1965 *New Statesman* 2 July 20/2 The tale is told in a Basic American style to represent the rudimentary nature of his mind.
e. *spec.* in *Philos.* Applied to a statement, proposition, etc. (see quots.).
1933 *Proc. Arist. Soc.* XXXIII. 80 The facts upon which all facts which are the immediate reference of a true proposition are based..may..be called basic facts. 1936 *Mind* XLV. 273 A basic proposition is one which asserts that an object has a particular property or that a particular relation holds between two objects, e.g. 'this is red', 'this is earlier than that'. 1937 A. J. AYER in *Proc. Arist. Soc.* XXXVII. 138 Propositions which need not wait upon other propositions for the determination of their truth or falsehood, but are such that they can be directly confronted with the given facts..I propose to call basic propositions. 1939 *Mind* XLVIII. 485 (*title*) On the class of 'basic' sentences. 1959 K. R. POPPER *Logic Sci. Discovery* § 7 p. 43 What I call a 'basic statement' or a 'basic proposition' is a statement which can serve as a premise in an empirical falsification; in brief, a statement of a singular fact. 1961 *Proc. Arist. Soc.* LXI. 180 We must..treat some concepts as not requiring reduction to others; these I shall call *basic concepts*.
f. *basic box*, see *basis box* (*BASIS III).
1914 J. H. JONES *Tinplate Industry* 141 Orders are often given for the equivalent of a specific number of basic boxes, for example, 10,000 boxes of 20/14.
2. c. *basic refractory*, a refractory material with a high content of basic oxides.
1917 R. MOLDENKE *Princ. Iron Founding* viii. 277 The iron foundry has so far had but little to do with basic refractories. 1944 GREGORY & SIMONS *Steel Manuf.* (ed. 3) xxi. 164 For this reason silica bricks cannot be used, and therefore magnesite, dolomite or other basic refractories have to be employed.
d. *basic slag*, slag from the basic or Bessemer process of steel manufacture, used as a fertilizer when finely ground.
1888 *Chambers's Jrnl.* 28 July 478/2 The value of basic slag as a manure. 1920 *Conquest* Aug. 487/2 Owing to the cattle grazing..phosphates and lime are withdrawn from the soil, but a dressing of basic slag..replaces this loss.
e. *basic dye*, a dye consisting of salts of bases containing aromatic amino- and substituted amino-groups.
1891 [see MORDANT *sb.* 3 c]. 1905 CAIN & THORPE *Synth. Dyestuffs & Intermed. Prod.* vii. 35 Wool takes up the basic dyes in a very uniform manner without the aid of any addition to the dye bath. 1952 J. R. BAKER in G. H. Bourne *Cytol. & Cell Physiol.* (ed. 2) i. 2 When no further structure can be discovered in unstained cells, the effect of dyes of small toxicity should be tried. Many basic dyes are suitable.

basically (bēɪ·sikăli), *adv.* [f. BASIC *a.*: see -ICALLY.] As a basic or fundamental principle, condition, matter, etc.; essentially; fundamentally.
1903 *Academy* 11 Apr. 364/1 Accent..though usual, is not invariable, not basically indispensable. 1905 CHESTERTON *Heretics* 270 The basically democratic quality which belongs to a hereditary despotism. 1927 *Observer* 20 May 17/7 The author's thesis that airship navigation is basically impracticable.

basichromatin (bēɪ:sikrōu·mătin). *Biol.* [G. (M. Heidenhain 1894, in *Arch. f. Mikrosk. Anatomie* XLIII. 543), f. Gr. βάσι-ς base + CHROMATIN.] Chromatin that stains readily with a basic dye; basophilic chromatin.
1902 E. B. WILSON *Cell in Devel. & Inheritance* (ed. 2) i. 38 Heidenhain..distinguishes..*lanthanin*- or *oxychromatin*-granules from the *basichromatin*-granules of the chromatic network. *Ibid.* 438 *Basichromatin*,..that portion of the nuclear network stained by basic tar-colours. 1903 J. B. FARMER in E. R. Lankester *Treatise Zool.* I. II. 13 In addition to the true nuclein chromatin there have been described other inclusions within the linin known as..oxy- or basi-chromatin, which appear to be related to the nuclein series. 1952 R. J. LUDFORD in G. H. Bourne *Cytol. & Cell Physiol.* (ed. 2) ix. 398 Lucas has sought to correlate the degree of association between basichromatin and virus bodies with the cellular specificity of the viruses.

basicity. Add: **2.** (See quot.)
1888 J. G. HORNER *Lockwood's Dict. Terms Mech. Engin.*, *Basicity*. Refers to the proportion of metallic oxide present in, and the absence of silicon from a furnace lining, a cinder, flux, or slag. The fettling of a puddling furnace depends for its efficacy on its basic character.

basilar, *a.* Add: *basilar membrane*, spec. the membrane in the cochlea that bears the organ of Corti.
1867 QUAIN *Elem. Anat.* (ed. 7) II. ii. 760 Cochlea..At its inner angle is a structure named limbus laminæ spiralis, and in its interior, resting on the basilar membrane, is the organ of Corti. 1934 *Brit. Jrnl. Psychol.* Oct. 137 The analysis of sound stimuli at the basilar membrane cannot possibly provide a basis for the psychological experience [of hearing].
b. Of or belonging to a low moral nature or condition.
1884 H. W. BEECHER in *Chr. World Pulpit* XXV. 235/2 Rousing men and lifting them out of the basilar into this higher spiritual condition. 1899 *Westm. Gaz.* 9 June 2/1 He makes evident the disgust of his own higher nature for the past excesses of his basilar passions.

Basilian (bāzi·liăn), *a.* [f. L. *Basilius*, St. Basil the Great + -AN.] Of or pertaining to St. Basil or to the order of monks and nuns following his monastic rule.
1780 A. BUTLER *Lives Saints* (ed. 2) VII. 344 A..convent of Basilian monks. 1848 DE LIANCOURT & MANNING *Pius IX* II. ii. 59 The Basilian nuns of Minsk. 1890 *Athenæum* 8 Feb. 182/2 The Basilian monk and savant Father Cozza. 1923 L. PULLAN *Relig. since Reform.* App. 278 In it there were seven Greek Basilian monasteries in the fifteenth century.

basin, *sb.* Add: **9.** *spec.* A submarine hollow or cavity.
1881 J. F. WILLIAMS *Geogr. Oceans* III. i. 142 The eastern and north-western basins have average depths of 2,500 and 3,000 fathoms respectively. 1904 A. KNOX *Gloss. Geogr. & Topogr. Terms* 37 *Basin*,..used, in sub-oceanic relief, for a depression of approximately round form.
12. (Earlier example.)
1804 C. B. BROWN tr. *Volney's View of Soil & Climate U.S.A.* 67 Their basins, the vallies which supply them, are of a greater elevation.
15. *Hort.* The depression at the apex of a pomaceous fruit, in which is situated the calyx or eye.
1909 in *Cent. Dict.* Suppl.
III. basinful, also *transf.* an excessive amount, (more than) enough *slang*; **basin irrigation** (see quot. 1961).
1935 G. INGRAM *Cockney Cavalcade* 136 My mother's had a 'basinful', if anyone has, I can tell you. 1957 'J. WYNDHAM' *Midwich Cuckoos* iii. 24 That there Miss Ogle ain't 'alf goin' to cop 'erself a basinful of 'Er Majesty's displeasure over this little lot. 1960 *News Chron.* 27 June 4/8 I've had a basinful of bowler-hat and furled-umbrella parts. 1903 *Westm. Gaz.* 29 Sept. 1/3 The old basin irrigation will be transformed into a perennial system. 1961 L. D. STAMP *Gloss. Geogr. Terms* 54/2 *Basin irrigation*: the type of irrigation associated especially with the Nile in Egypt and the Sudan whereby flood waters are led off into specially prepared 'basins' which vary in size from a few dozen acres to many square miles. The basins are separated from one another by earth banks.

basioccipital (bēɪ:siǫksi·pităl), *a.* and *sb.* [f. L. *basi-s* base + OCCIPITAL *a.*] Pertaining or belonging to the base of the occiput or the basilar part of the occipital bone. Also as *sb.*
1853 in MAYNE *Expos. Lex.* 1878 T. DUNMAN *Gloss. Biol., Anat. Terms* 15 *Basioccipital*, a bone of the skull which in Human Anatomy is represented by the basilar process of the occipital bone. 1933 *Jrnl. R. Anthrop. Inst.* LX. 20 The Calotte..is made up by the bones of the calvaria deprived of the basioccipital, the ethmoid and the basal parts of the sphenoid and temporal bones.

basion (bēɪ·siǫn). *Anat.* [mod.L., f. Gr. βάσι-ς BASE *sb.*[1]] The middle of the anterior border of the occipital foramen.
1878 BARTLEY tr. *Topinard's Anthrop.* 234.

basipetal, *a.* Add: (Earlier and later examples.) Hence **basi·petally** *adv.*
1869 *Student* II. 13 The well ascertained basipetal formation of many foliary parts. 1939 *Nature* 18 Mar. 486/2 There is no suggestion that this procambial activity contributes basipetally to an enlargement of the vascular system. 1946 *Ibid.* 23 Nov. 737/2 The supposed basipetal development of the leaf traces.

basis. Add: **III. basis box,** also *basic box*, the unit of area in the tinplate industry (see quot. 1956); **basis wine,** a fermented liquor obtained chiefly from raisins or concentrated must and used as the basis or main constituent in the manufacture of various wines.
1926 *Jrnl. Iron & Steel Inst.* CXIII. 602 There are two gas-fired furnaces each with a capacity of about 30 tons, or 600 basis boxes per charge. 1956 W. E. HOARE *Tinplate Handbk.* (ed. 3) iv. 13 The unit of area used by the tinplate industries is the basis box or base box, originally defined as 112 sheets each 20 in. × 14 in. 1958 A. D. MERRIMAN *Dict. Metallurgy* 16/1 The weight may vary from 55 to 275 lb. per basis box, but is usually between 100 and 150 lb. 1905 A. CHAMBERLAIN in *Hansard* 10 Apr. CXLIV. 1049 Wine manufactured..partly from imported grape must, more largely from currants and raisins..known to the trade as 'basis' wine. 1905 *Daily Chron.* 21 Dec. 6/3 Fraudulent wines, containing what was called 'basis' wine.

bask, *sb.* Add: (Earlier examples.)
1790 T. WILKINSON *Memoirs* I. 239 However, that bask of sunshine did not last. 1847 STODDART *Angler's Compan.* 308 Pike..when on the bask, or in sunning humour, distribute themselves along the margin..of floating weeds.
b. *fig.* The 'sunshine' or 'warmth' (of favour, popularity, etc.).
1762 H. WALPOLE in Vertue *Anecdotes of Painting* I. p. ix, Milton and Fontaine did not write in the bask of court-favour. 1779 —— *Let.* 16 Sept. (1904) 26 Lord Temple..had grown up in the bask of Lord Chatham's glory.

Baskerville (bæ·skəɹvil). The name of John *Baskerville* (1706–75), type-founder and printer, applied to types of his founding re-cut.
1802 C. WILMOT *Let.* 28 Nov. in T.U. Sadleir *Irish Peer* (1920) 121 Lady Mount Cashell bought a good many books from Bodini (*sic*)... His type surpasses Baskerville. 1828 E. FRY in T. B. Reed *Old Eng. Letter Foundries* (1887) xv. 310 The Baskerville and Caslon imitations..were laid by for ever. 1887 T. B. REED *Ibid.* xv. 299 The complete series of Romans, cut after the Baskerville models. 1922 D. B. UPDIKE *Printing Types* II. xvii. 107 The slight touch of over-delicacy which the Baskerville letter possessed. 1928 R. B. McKERROW *Introd. Bibliogr.* 302 The rather weak and grey descendants of Caslon plus Baskerville.

basket, *sb.* Add: **1.** *fig.* A group, category; a range.
1916 H. WALPOLE *Dark Forest* II. i. 211 Semyonov at this time flung Nikitin, Andrew Vassilievitch, Trenchard and myself into one basket. We were all 'crazy romantics'. 1959 *Times* 9 Mar. (Suppl. Britain's Food) p. i/4 The types of goods on sale are continually changing: so also is the way in which housewives choose to spend their money on this changing basket of available commodities. 1962 *Economist* 26 May 799/1 Might various categories of exports be grouped in comparable 'baskets'.
2. (Earlier and later examples.)
1631 in *Rec. Mass. Bay* (1853) I. 92 Plastowe shall (for stealing 4 basketts of corne from the Indians) returne them 8 basketts againe. 1908 *Westm. Gaz.* 26 Sept. 8/2 Scotland trout-fishing remains open..and some nice baskets are being made. 1965 A. MACLEOD *Blessed above Women* vii. 80

'The fishing's pretty good here, I believe?'..'I've certainly been lucky in getting good baskets.'

b. Phr. *a basket of chips*, used allusively in comparisons. Chiefly *U.S.*

1788 Grose *Dict. Vulgar T.* (ed. 2), *Basket*..He grins like a basket of chips; a saying of one who is on the broad grin. 1819 Moore *Tom Crib's Mem. Congress* 25 On which the whole Populace flash'd the white grin Like a basket of chips. 1827 *Massachusetts Spy* 28 Nov. (Th.), The Yankee will say of a young lady, 'She is a real pretty girl, but she is as homely as a basket of chips.' 1892 *Congress. Rec.* Mar. 2367/1 My ticket was handed to me at once and the seller looked as pleasant as a basket of chips. 1938 *Amer. Speech* XIII. 74/1 As illustrations of the type of expression familiar to me in Ohio, I may list..'polite as a basket of chips'.

†5. b. A *colloq.* name for a part of the auditorium of a theatre. (See quots.) *Obs.*

1812 *Dramatic Censor 1811* 99 There shall not be the nuisance called a *basket*, which is the constant rendezvous of tumult and profligacy. 1812 G. Wyatt *Design for Theatre* ii. 13 The deep recess of Boxes, commonly called 'the Basket' (which has generally been permitted to extend to a great depth at the back of the circle), as well as the public Saloon..have been entirely excluded. *Ibid.* 14 This..does not include the deep chasm of 'the basket', which reached still farther backwards, to an extent of nearly 15 feet. 1815 L. Simond *Jrnl. Tour Gt. Brit.* I. 104 The pit of Covent-Garden is nearly square..certain back boxes, called the basket, may hold one hundred and eighty people at 7s. 1831 *Examiner* 390/1 A sort of black hole (we believe called the basket at the back of the front boxes, in which all the raff of the house have place).

10. *Aeronaut.* A structure suspended from the envelope of a balloon to carry passengers, ballast, etc.

1783 Franklin *Let.* 8 Oct. (1907) IX. 106 The Basket [of the Montgolfier balloon] contain'd a Sheep, a Duck & a Cock. 1784 *London Chron.* LVI. 219/2 The balloon being filled..he seated himself in the basket. 1909 *Flight* 29 May 310/2 Thanks to Mr. Brewer's insisting on the envelope being immediately over the basket in the course of its swayings during the trying of the lift of the balloon, we were able to carry easily two more bags of ballast. 1910 R. Ferris *How it Flies* 1. 16 In its usual form [the balloon] is spherical, with a car or basket suspended below it.

11. Euphem. alteration of *bastard* (esp. in sense *A. 1 c). *slang.*

1936 N. Coward *Still Life* iii. 73 Come on, Johnnie—don't argue with the poor little basket. 1958 'J. Gillespie' *Disobedience* i. 10 He's a nice old basket really.

B. 1. b. *basket-car, -fire, -grate* (examples); **basket-bodied** *a.*, having a wicker body.

1903 *Daily Chron.* 19 Oct. 3/2 The wax-headed basket-bodied lady in a draper's window. 1908 *Ibid.* 25 Aug. 7/3 The slender-wheeled basket-bodied chairs. 1922 Joyce *Ulysses* 699 Roadster cycle with side basketcar attached. 1962 *Times* 26 Oct. Suppl. p. xiv/4 The basket-car weighbridge. 1932 G. Bottomley *Bower of Wandel* in *Lyric Plays* 64 (*stage direction*) An arras-hung interior is seen, lit by a basket-fire. 1889 *Cent. Dict.* I. 468/2 Basket grate. 1942 Cary *To be Pilgrim* xxiv. 51 Lucy's room..still has the iron basket grate.

2. basket cell, a nerve cell having a basket-like network of fibrils; **basket chair,** one made of wickerwork, a wicker chair; **basket clause** *N. Amer.*, a clause of a general or comprehensive nature; **basket coil, winding** *Electr.* (see quots.); **basket dinner, lunch, picnic** *U.S.*, one for which the provisions are brought in a basket; **basket fern** *dial.* = male fern (Male *a.* 2 b); **Basket Maker** *Archæol.*, a member of an ancient culture in south-western U.S. preceding the Pueblo culture and characterized by basket-work; also applied *attrib.* to this culture or period; **basket mat** (see *mat *sb.*[1]); **basket-meeting** *U.S.* (see quots.); **basket-plant,** an orchid of the genus *Stanhopea*, often grown in baskets through which the flowers protrude; **basket-shell,** a bivalve mollusc, esp. of the genus *Corbula*, having valves of unequal size; **basket sugar** (see quot.); **basket weave,** a style of weave in which the pattern resembles basket-work (Webster, 1911); a fabric woven in this pattern.

1901 *Gray's Anat.* (ed. 15) viii. 671 [Cells] of the inner layer run for some distance horizontally .. giving off at intervals collaterals, which pass..towards the cell-bodies of Purkinje's corpuscles, around which they become enlarged, and ramify like a basket. Hence these cells of the inner layer are named *basket-cells*. 1910 *Encycl. Brit.* IV. 395/1 The..*basket-cells*..have a rounded body giving off many branching dendrons. a 1631 Donne *Elegies* in *Poems* (1912) I. 79 When he swolne, and pamper'd with great fare, Sits downe, and snorts, cag'd in his basket chaire. a 1782 E. Parkman *Diary* 298, 1 great chair 3/- Six lath basket chairs 24/-. 1896 Mrs. Caffyn *Quaker Grandmother* 20 He sank into a well-cushioned basket-chair. 1883 *Congress. Rec.* Feb. 2580/1 This basket-clause seems to be a sort of Prophetic fine-comb with us. 1897 *Ibid.* Mar. 367/2 If we strike..[an item] from the dutiable list, we transfer it to the 'basket clause' at 25 per cent. 1968 *Globe & Mail* (Toronto) 13 Feb. B 4/6 The basket clause permits trust companies to invest up to 15 per cent of capital. 1923 *Daily Mail* 28 Apr. 5 Those compact multilayer coils called 'basket', 'pancake', or 'honeycomb' coils. 1960 H. Carter *Dict. Electronics* 25 *Basket coil*, an inductor consisting of a self-supporting spiral winding, adjacent turns being disposed in criss-cross fashion to minimize self-capacitance. 1892 *Illinois Kentuckian* (Lexington) Dec. 88/3 (D.A.), This is a noted place for picnics where pies and cake and basket dinners prevail. 1883 W. H. Cope *Gloss. Hampsh. Words & Phr.* 5 Basket Fern. 1945 E. Step *Wayside & Woodl. Ferns* (ed. 2) 76 There are several other names [for the male fern] in use locally, such as Basket-fern in Cornwall and Hampshire. 1905 *Springfield Weekly Republ.* 11 Aug. 14 At noon a bountiful basket lunch was served under the trees in the park. 1897 *Harper's Mag.* June 61/1 Over the head a small flat basket, and a great finely woven basket over all—such was their burial fashion...The Wetherills [i.e. R. Wetherill and brothers] soon recognized the ethnological importance of their discovery, and have provisionally named the people who buried in these older graves the 'Basket-Makers'. 1904 A. J. Burdick *Mystic Mid-Region, Deserts of Southwest* 92 The basket-makers of that time had all the skill that is known to their descendants to-day. 1938 *Southwestern Lore* June 5 (D.A.), We now place the beginning of the Basket Maker II period as about the opening of the Christian Era and its end about 400 A.D., with the Basket Maker III period lasting up to 700 A.D. 1948 A. L. Kroeber *Anthropology* (ed. 2) xviii. 806 The Basket Maker and Pueblo cultures are continuous in character as well as in time...The ..sequence begins with Basket Maker II, stage I having been left for an as yet undiscovered culture which was inferred as earlier. 1859 Bartlett *Dict. Amer.* (ed. 2), *Basket-meeting*, in the West, a sort of pic-nic, generally with some religious 'exercises'. 1872 Schele de Vere *Americanisms* 191 A corn-husking is announced..and the neighbors from far and near assemble, each bringing his provisions in a basket. From the latter feature these pic-nics derive their names of Basket-meetings. 1882 in G. Foreman *Last Trek of Indians* (1946) 198 This year witnessed a good, old-fashioned basket picnic. 1904 *Boston Herald* 22 Aug. 6 A long political speech in the open air..at a basket picnic meeting in Ohio. 1865 *Trans. Ill. Agric. Soc.* V. The *stanhopea sigrina* (or basket plant) is exceedingly rare and fragrant. 1713 J. Petiver *Aquat. Anim. Amboinæ* 1/2 *Cassis rubra*..Red Basket-shell. 1945 E. Step *Shell Life* ix. 150 The Basket-shells (*Corbula*), though a large genus, are represented in this country by a single species, the Common Basket-shell (*C. gibba*). *Ibid.* 151 There are..other Basket-shells beside those of the genus *Corbula*..the Pointed Basket-shell (*Neæra cuspidata*). *Ibid.* 152 The Short Basket-shell (*N. abbreviata*)...The Fine-ribbed Basket-shell (*N. costellata*). 1902 *Encycl. Brit.* XXXIII. 45/1 In certain districts, notably in the Straits Settlements..it [*sc.* syrup] is slowly boiled up in open double-bottom pans... The sugar thus produced..forms a species of small-grained concrete. It is called 'Basket Sugar'...In the Straits Settlements the 'Basket Sugar' factories are of considerable importance. 1925 *New Yorker* 4 Apr. 22/2 His attached silk collar, which he wore plain with basket-weave tie. 1940 *Chambers's Techn. Dict.* 80/1 *Basket weave*, woollen fabrics of hopsack or mat-weave, with a basket effect. 1960 *Times* 21 Jan. 14/4 Fabrics were interesting in texture, including basket-weave woollens. 1910 H. M. Hobart *Dict. Electrical Engin.*, *Winding, Basket*, more often known as the chain type of winding.., in which the coils belonging to the three separate phases are laid out in two ranges, the centre of one coil or set of coils being occupied by the side or sides of the adjacent coil or coils on opposite sides.

ba·sket-ba:ll. Also **basketball.** A game played indoors or out of doors with a large inflated ball, which is thrown from player to player, the object being to score by casting it into one's opponents' goal, a basket fixed ten feet above the ground at each end of the field of play.

1892 J. Naismith in *Triangle* (U.S.) Jan. 144 (*heading*) Basket Ball. We present to our readers a new game of ball. 1893 *Birkenhead News* 9 Dec. 7/5 Interesting Basket-Ball Match. 1898 *Daily News* 8 June 5/2 Vassar, Syracuse, Cornell, Wellesley, and Rosemary Hall have each their teams of girl basket-ball players. 1901 *Westm. Gaz.* 1 May 7/1 A game of 'basket ball', played by ten over-heated and dishevelled ladies in—bloomers! 1926 *Encycl. Brit.* New Suppl. I. 337/2 Basketball has become the national indoor game of the United States.

basketry. Add: **b.** The art or craft of basket-making.

1893 K. Sanborn *Truthful Woman S. Calif.* viii. 104 The art of basketry is rapidly deteriorating [among Indians]. 1905 *Daily Chron.* 24 May 3/5 The Red Indians of America had their own particular craft of basketry. 1957 *Encycl. Brit.* III. 180/2 No satisfactory automatic basket-weaving machinery was perfected. True basketry, therefore, remains a handicraft.

Baskish, var. *Basquish.

Ba·skonize, *v. trans.* To turn into Basque.

1904 *N. & Q.* 10th Ser. I. 18/2 Castilian *voz* = voice would be baskonized by *boz*.

basophil (bēˈ·sǒfil), *a.* and *sb.* *Biol.* Also **-phile** (-fəil). [f. Gr. βάσι-ς Base *sb.*[1] + φίλος loving: see -phil, -phile.] **A.** *adj.* Applied to a cell or other structure having an affinity for basic substances; that stains readily with a basic dye. **B.** *sb.* A cell, etc., of this nature. Hence **basophi·lic, baso·philous** *adjs.*

1890 Billings *Med. Dict.*, *Basophile*, staining well with basic aniline dyes. 1894 Gould *Dict. Med.*, *Basophilic*.., combining readily with bases; stainable by means of basic dyes. *Basophilous*.., stained by basic rather than by acid dyes (applied to certain cells and tissue-elements). 1898 *Allbutt's Syst. Med.* V. 402 The further division of the latter [*sc.* the granular form of white blood-corpuscles] into basophil and oxyphil. 1900 Dorland *Med. Dict.*, *Basophil*, a basophilic element. 1907 *Practitioner* Sept. 455 The nuclei of all these varieties of colourless blood corpuscles are basophil, which means that they have a strong affinity for basic aniline dyes such as methylene blue. 1908 *Ibid.* Dec. 839 The red corpuscles of those affected by lead poisoning are basophilic. 1932 Fuller & Conard tr. *Braun-Blanquet's Plant Sociol.* xiii. 311 Those basophiles which cannot endure high concentration of salts. *Ibid.* 314 In cool, humid climates the development of vegetation becomes essentially a struggle of the acidophilous against the neutrophilous and basophilous species. 1962 *Lancet* 27 Jan. 206/2 In the early stages of the production pathway, the lymphocytes are large and possess an intensely basophilic cytoplasm. 1966 *Lancet* 31 Dec. 1457/1 Mast cells in the rabbit exist mainly in its blood as the mast leucocytes, or basophils.

basophilia (bēˈsǒfi·liă). Also **baso·phily.** [f. as prec.: see -ia[1].] **a.** A tendency to stain readily with a basic dye. **b.** A condition of the blood marked by the formation and accumulation of an excess of basophil cells.

1905 J. H. Drysdale in *Allbutt's Syst. Med.* I. 1. 669 (*title*) Granular Basophilia (Degeneration) of the Red Cells. 1946 *Nature* 27 July 120/2 Normal grain showed a gradual increase in the size of the nucleolus..accompanied by a rapid decrease in the basophily of the cytoplasm. 1960 L. Picken *Organiz. of Cells* v. 182 A cytochemical consequence of this presence of nucleic acids is the long familiar basophilia (affinity for basic dyes) of the cytoplasm of embryonic and other actively growing cells. 1962 *Lancet* 27 Jan. 206/2 A cell with a large faintly staining nucleus and pale cytoplasm, with hardly any basophilia.

Basque. Add: **B.** *Basque beret, cap*, a soft, close-fitting cap resembling that worn by Basque peasants.

1926 Hemingway *Fiesta* (1927) xiii. 154 Brett was wearing a Basque beret. 1928 Galsworthy *Swan Song* ii. v. 146 Everything now depended on the Basque cap. If women took to them, shingling would stay. 1941 Koestler *Scum of Earth* 132 He wore a strange sort of Basque beret.

basqueless (ba·sklės), *a.* [f. Basque *sb.* + -less.] Without a basque (Basque *sb.* 4).

1895 *Daily News* 3 Sept. 6/3 A tight-fitting coat.., basqueless in front and at the sides, but with a full one at the back. 1909 *Daily Chron.* 8 Dec. 9/2 A basqueless blouse.

Basquish, *a.* and *sb.* Delete †*Obs.* and add later examples. Also **Baskish.**

1706 J. Stevens *Sp. & Eng. Dict.* Pref., This Basquish is so different from all the other European Languages. *Ibid.*, *Arriugurridca*, it signifies red Stones in the Basquish Language. 1904 E. S. Dodgson (*title*) A Synopsis..of..Forms of the Verb..as found in the Baskish New Testament ..printed in 1571. 1918 *Trans. Scott. Eccles. Soc.* 229 They ..spoke a language probably akin to the Baskish.

bass, *sb.*[2] Add: **1. c.** A fibre obtained from the leaf-bases or leaf-stalks of certain palms, used in the manufacture of brushes, ropes, etc.; also the palm from which this fibre is obtained. Also Comb.: **bass-broom** (cf. *bast *sb.*[1] 1 b).

1881 *Instr. Census Clerks* (1885) 79 Bass Broom Maker. 1883 *Cassell's Fam. Mag.* 222/1 The coarse familiar bass-brooms. 1891 *Kew Bulletin* Jan. 3 (*heading*) Extract of Minute by the Governor of Lagos on the Bass fibre of the Bamboo palm (*Raphia vinifera*). *Ibid.*, The 'African Bass' is..a stiff and wiry fibre, varying in colour from dark brown to light red. 1895 *Army & Navy Co-op. Soc. Price List* p. xx, Bass Brooms. *Ibid.* 182 Bass Heads... Handles for Bass, each extra, 0/2. 1902 Hannan *Textile Fibres Commerce* 145 Bass, Monkey, or Grass, Leopoldinia Piassava or Para Piassava.

bass, *a.* and *sb.*[5] Add: **A.** *adj.* **3. b.** *bass drum* (earlier and later examples). For *bass clarinet, sax, tuba*, see the sbs.

1804 W. Bentley *Diary* (1911) III. 68 The instrument Music..consisted of the Bass Drum, Bassoon, Clarinet & flute. 1884 G. W. Cable *Dr. Sevier* II. liv. 171 The soft boom of a bass-drum.

B. *sb.* **4. b.** A double-bass.

1927 *Melody Maker* Sept. 926/3 The bass, being a rhythm instrument, must conform to the rhythm set by the rhythm section. 1962 *Oxf. Mail* 22 June 5/5 The rules allow competitors to use bass and drum accompaniment.

Bass (bæs), *sb.*[6] [Proper name: see below.] Bass's ale or beer, the 'India Pale Ale' or 'Bitter Beer' manufactured by Messrs. Bass & Co. of Burton-on-Trent. Also with *a* and *pl.*, a bottle of Bass.

1849 *Illustr. Lond. News* 26 May 336 The consumption of Bass..is beyond belief. 1850 Thackeray *Pendennis* II. v. 42 Many a day I've drunk a dozen of Bass at Calcutta. 1909 *Daily Chron.* 3 Feb. 4/6 The first signs of returning vitality given by the Prince was to ask feebly for a bottle of 'Bass'. 1953 Dylan Thomas *Let.* 24 Aug. (1966) 412 Oh, to bask unasked in a Bass cask.

Bassalia (bāsēˈ·liă). *Zoogeogr.* [mod.L., f. late L. *bassus* deep, Base *a.* + Gr. ἁλία assembly, with allusion to ἅλς sea.] The region of the deep sea. Hence **Bassalian** (bāsēˈ·liăn), *a.*, of, pertaining to, or found in this region.

1884 *Science* 23 May 621/1 A special realm, which has been called 'Bassalia' or the 'Bassalian realm'. 1885 T. Gill in *Rep. Nat. Acad. Sci. Washington 1884* 6 On the Ichthyological Peculiarities of the Bassalian Realm.

bassarid (bæ·sărid). [ad. L. *Bassarid-*, *Bassaris*, a. Gr. βασσαρίς a Thracian bacchanal, lit. a fox, prob. from their dress, made of fox-skins (βασσάρα a fox).] A Thracian bacchanal; a bacchante.

1865 Swinburne *Atalanta* 108 And Pan by noon and Bacchus by night,..fills with delight The Mænad and the Bassarid. 1871 —— *Songs before Sunrise* 5 Hills that hid The blood-feasts of the Bassarid. 1948 R. Graves *Coll. Poems* 239 Her Bassarids now bed With the ignoble usurer. 1949 E. Pound *Pisan Cantos* lxxix. 77 Shake the castagnettes of the bassarids.

bass-court, var. Base-court.

‖**basse danse** (bas dāns). Also **basse dance.** [Fr., lit. 'low dance'.] A court dance in duple or triple time which originated in France in the 15th century; = Base-dance. Also *attrib.*

1789 Burney *Hist. Mus.* III. 263 John D'Etrée..published four books of *Danseries*. .. The editor of these books tells us, that they contained .. *ballets, voltes, basses dances,..allemandes*. Printed at Paris, 1564. 1899 Stainer in *Musical Times* 1 July 461/2 The 'basse danse', as it was termed in French, was so named by way of contrast to the 'danse par haut' or 'danse sautée'. 1927 *Daily Tel.* 30 Aug. 10/6 There was the Basse Dance, queen of measures in the fifteenth century. 1954 Grove *Dict. Mus.* (ed. 5) III. 821/1 Ortiz's first example is a well-known *basse-danse* tenor.

basset, *sb.*[1] Add: *attrib.* in *basset-hound*.

1883 *Illustr. Sporting & Dram. News* 23 June 371/2 The last new club is 'The Basset-hound Club', its objects being to promote the breeding of pure Basset-hounds d'Artois, rough and smooth. 1928 *Morning Post* 20 Oct. 6/1 Basset Hounds—those wholly delightful miniatures of the classic Bloodhound.

‖**basse-taille** (bas,tāyˈ). [Fr. *basse*, fem. of *bas* low (see Base *a.*) + *taille* cut.] (See quots.)

1899 H. Cunynghame *Art-Enamelling upon Metals* i. 12 Gold was also used for champlevé work...The later style of this is called 'basse-taille', and consists in carving out a subject in a thick plate of gold in low relief. 1956 G. Taylor *Silver* i. 17 *Bassetaille* is a rather more sophisticated form of enamelling in which the ground is decorated with engraving. 1962 *Internat. Art Treasures Exhib.* (*V. & A. Museum*) 45/2 A Louis XV gold eye-glass case..the exterior decorated in coloured 'basse taille enamels' with diamond push-piece.

bassine (bæsī·n). [f. Bass *sb.*[2] + -ine[4].] A fibre obtained from the leaf-bases of the palmyra *Borassus flabellifer*, used in the manufacture of brooms, ropes, etc. Also *attrib.*, as *bassine broom*.

1902 Hannan *Textile Fibres Commerce* 149 Bassine or Palmyra Fibre. 1923 *Daily Mail* 16 June 10 Best Hair broom, bassine broom [etc.]. 1969 *Sunday Mail* (Brisbane) 26 Jan. 25 Bassine yard broom.

bassist. Add: **2.** One who plays a bass instrument, spec. a double-bass (see *bass *sb.*[5] 4 b).

1909 in *Cent. Dict.* Suppl. 1955 *Jazzbook 1955* 15 Within the ranks of this band he met up with tenor saxist Chu Berry, trombonist Tyree Glenn, bassist Milton Hinton. 1956 S. Traill *Play that Music* 13 He..was offered the part of principal bassist to the Los Angeles Symphony Orchestra.

bassman (bēˈ·s,mæn). orig. *U.S.* [f. Bass *sb.*[5] + Man *sb.*[1]] One who plays the double-bass.

1952 B. Ulanov *Hist. Jazz in Amer.* (1958) xix. 241 The best bassman jazz has ever known. 1958 K. Goodwin in P. Gammond *Decca Bk. of Jazz* xiii. 154 One of the most important bassmen..is Monty Budwig.

‖**basso.** Add: **basso-buffo** (-bu·fo) [It., = comic bass], a bass singer who plays the comic part in opera; **basso cantante** (-kanta·nte) [It., lit. 'singing bass'], a male voice of fine quality and power of expression in the upper register of the bass range; also, a singer having such a voice; **basso ostinato** (-ǫstinā·to) [It., lit. 'persistent bass'], a musical structure in which a figure is repeated successively throughout a work, with or without variation, usually in the bass part; ground-bass (cf. Ground *sb.* 6 c, 18 a); also *transf.*

1909 in *Cent. Dict.* Suppl., Basso-buffo. 1960 *Times* 15 July 16/3 Mr. Forbes Robinson..may seem an unlikely basso buffo. 1876 Stainer & Barrett *Dict. Mus. Terms* 52/1 Basso cantante. 1889 G. B. Shaw in *Star* 11 Oct. 2/3 He was by no means the deep, powerful *basso cantante* his brother Edouard now is. 1916 —— in *Nation* 6 May 158/1 A good rough *basso cantante*. 1963 *Times* 28 Feb. 16/2 Voice that seemed most like a *basso cantante*. 1876 Stainer & Barrett *Dict. Mus. Terms* 211/1 The connecting link between one form of the falso-bordone and the basso ostinato. 1935 A. Galsworthy in J. Galsworthy *End of Chapter* Foreword p. v, This is not by any means the whole of the plan, or story, but is a sort of *basso ostinato* to it.

‖**basso-profondo,** usually **-profundo** (bæ·so,profǫ·ndo). *Mus.* [It., = deep bass.] A deep bass voice, having a compass of about two octaves above the D below the bass stave; also, a singer having a voice of this compass. Also *attrib.*

1860 Thackeray *Round. Papers* (1863) 34 Why not a singing artist? Why not a basso-profondo? 1909 *Daily Chron.* 19 Jan. 4/4 The basso-profondo thunderer of the church choir. 1924 *N. & Q.* 22 Mar. 221/1, I heard a song sung by a *basso profundo*, once a professional. 1965 G. McInnes *Road to Gundagai* xi. 195, I..had to be..prompted ..in the basso profundo of an over zealous Latin teacher.

basso-relievo. (Earlier and later examples.)

a 1666 Evelyn *Diary* 19 Nov. an. 1644 (1955) II. 259 A deepe basso-relievo, a l'antique. 1930 R. Campbell *Poems* 15 We know the veld..This basso-relievo of a land. 1960 R. Lister *Decorative Cast Ironwork Gt. Brit.* iv. 83 Giving some firebacks the appearance of borders from title pages translated into basso-relievo.

bass-wood. (Earlier Amer. examples.)

1670 *Rowley Rec.* (1894) 210 The Northwest Angle is a basswood tree. 1728 *Rec. Early Hist. Boston* 25 June (1883) VIII. 222 We are of Opinion That no..Black ash, Basswood, or Ceder Shall be Corded up.

bast, *sb.*[1] Add: **1. b.** *bast-broom* (cf. *bass *sb.*[2] 1 c).

1867 *Ure's Dict. Arts* (ed. 6) I. 282 *Bass* or *Bast*...The name is also used for the bark or tough fibres of the flax and hemp plants of which Bast brooms are made. 1877 *Design & Work* 11 Aug. 235/2, I have been making some bast brooms.

bast, *sb.*[4] [Persian.] Sanctuary, refuge, asylum. So **ba·sti,** a refugee.

1856 M. L. Sheil *Glimpses Life & Manners in Persia* x. 165 An extraordinary device adopted by the moollas..for restoring the right of bast, or sanctuary, to its ancient vigour. 1894 Safar Nameh *Persian Pict.* 32 Across the gateways a chain is drawn, denoting that the garden is Bast—sanctuary—and into these the European may not go. 1923 *Blackw. Mag.* Jan. 49/1 In Persia the system of 'bast', or asylum, was the only means whereby the oppressed could do something to right..his grievances. *Ibid.* 49/2 The refusal of bast to any one with a legitimate grievance would be an affront to public opinion. *Ibid.* 56/1 The Bastis had not come to the Consulate by my invitation.

bastard, *sb.* and *a.* Add: **A.** *sb.* **1. b** (or **3. b**). Also **bastaard.** [Afrikaans *bastaard* (now *baster*).] A person of mixed breed; a Griqua. *S. Afr.*

1790 E. Helme tr. *Le Vaillant's Travels into Africa* II. viii. 163, I mean to speak of the natural children which have sprung from an intercourse of the Whites with the female Hottentots, or between these same women and the negroes. They are commonly known at the Cape under the appellation of Bastards. 1806 J. Barrow *Voy. to Cochinchina* 377 They came..to a second horde of Bastaards and Bosjesmans. 1814 W. Brown *Hist. Propag. Christianity* II. ix. 425 The term *Bastard* applied to a Hottentot, does not mean that he is illegitimate, but merely that he is of mixed breed. 1866 J. Leyland *Adv. Far Interior S. Afr.* ii. 32 The Griquas or Bastards. 1900 A. H. Keane *Boer States* vi. 85 Many are in fact 'Bastaards', that is to say, Hottentot-Dutch half breeds.

c. Used vulgarly as a term of abuse for a man or boy, and, with weakened force, as the equivalent of 'fellow', 'chap'; also trivially for 'thing', esp. something bad or annoying.

1830 Scatcherd *Hist. Morley* 339 *Bastard*, a term of reproach for a mischievous or worthless boy. 1833 C. Lamb *Let.* 27 Apr. (1935) III. 367 We have had a sick child, who sleeping, or not sleeping, next me with a pasteboard partition between, killed my sleep. The little bastard is gone. 1917 J. Masefield *Old Front Line* iv. 58 For all their bloody talk the bastards couldn't bring it down. 1927 T. Wolfe *Lett.* (1956) 118 Joe, you old bastard, how the hell are you! 1919 H. Crane *Let.* 22 Nov. (1965) 24 —— is a poor ignorant bastard of some kind. 1931 R. Aldington *Colonel's Daughter* i. 56 The smug bastards who declare that all is for the best in this best of all possible worlds. 1936 J. Curtis *Gilt Kid* ii. 18 What's Wandsworth like? Proper bastard, I suppose. 1937 J. A. Lee *Civilian into Soldier* i. 29 'He's a bastard.' Guy used the term not for its dictionary meaning, but because among New Zealanders no term expressed greater contempt. 1938 J. Masefield *Dead Ned* 233 It's a bastard being in quad with no blunt. 1940 H. G. Wells *Babes in Darkling Wood* i. iv. 104 Serve the cocky little bastard right. 1942 T. Rattigan *Flare Path* iii. 164 Johnny, you old bastard! Are you all right? 1960 *Observer* 18 Dec. 10/3 Australians pride themselves on their imperviousness to excitement. The phrase 'she's a bastard' is usually regarded as adequate for most dramas from four-year droughts to bush fires. 1961 J. MacLaren-Ross *Doomsday Book* ii. i. 108 This bastard of a bump on the back of my head. 1968 K. Weatherly *Roo Shooter* 23 'You're not a bad bastard, Hunter,' he said, 'in spite of your lousy cooking.'

12. = *bastarda.

1928 S. Morison *German Incunabula in Brit. Mus.* 10 The Bastards are scripts answering to the need for a speedy letter appropriate for the copying of books or documents of minor value or importance. *Ibid.* The Bastard of the 36-line Indulgence.

B. *adj.* **2. b.** In South Africa: of or pertaining to a person of mixed breed. Cf. sense A. 1 b above.

1792 E. Riou tr. *Van Reenen's Jrnl. Journey from Cape of Good Hope* 28 A village of bastaard Christians, who were descended from people shipwrecked on that coast...The Cape people..call all those Hottentots, bastaard Hottentots, whose race has been intermixed with the slaves brought from the East Indies, [etc.].

6. c. Of the Gothic script known as *bastard* (sense *A. 12) or *bastarda*.

1888 J. H. Hessels in *Encycl. Brit.* XXIII. 694/1 Bastard Italian or bastard Roman was introduced in 1454 at Mainz in the 31-line and 30-line indulgence.

7. b. bastard sugar (earlier example).

1833 B. Silliman *Man. Sugar Cane* 93 The language of the sugar refiners appears to be tolerably uniform, in applying the term bastard sugar to that which is formed from the first dripping of the lump.

8. *bastard trenching* in *Hort.* (see quots.). Hence (as a back-formation) *bastard-trench* v. trans.

1842 Loudon *Suburban Hort.* ii. iii. 230 'Double digging' is in horticulture what subsoil ploughing is in agriculture; the surface soil is kept on the surface, but the bottom of the trench is dug over as the work proceeds...By many this is called 'bastard trenching'. 1909 *Daily Chron.* 23 Oct. 3/5 The best method is to dig out the surface soil..and to carry it to the far end of the bed, to work the manure into the lower soil now exposed..and cover it with the upper soil of the next yard; and so on down the bed, filling the last hole with the soil taken from the first. This is known as 'bastard-trenching'. 1923 W. Deeping *Secret Sanctuary* x. 104 Stretton set himself to bastard-trench a quarter of an acre. 1933 *Jrnl. R. Hort. Soc.* LVIII. 98, I invariably endeavour to bastard trench all the ground to be occupied by chrysanthemums.

‖**bastarda** (bastā·ɪdā). [It., f. *bastardo* (see Bastard *sb.*).] A Gothic script used in France and Germany during the 14th and 15th centuries.

1934 A. F. Johnson *Type Design* i. 6 For..theological texts a less formal and rounded letter was used.., for works in the vernacular, a still less formal and cursive hand known as Bastarda. *Ibid.* 31 The French 'lettre bâtarde' then passed out of use..but in Germany the Bastarda has remained the national type. 1963 *Times* 7 Feb. 16/4 Illustrated manuscripts..with Gothic bastarda script.

bastardy. Add: **5.** *attrib.*, as *bastardy law*; **bastardy order,** an order made by a magistrate for the support of an illegitimate child by the putative father.

1867 *Law Rep. Queen's Bench* II. 468 In *Slater's Case* it was held that an original bastardy order could not be annulled by a subsequent order. 1872 *Act* 35 & 36 *Vict.* c. 65 An Act to amend the Bastardy Laws. 1880 *Act* 43 & 44 *Vict.* c. 32 § 2 This Act may be cited as the Bastardy Orders Act, 1880. 1912 G. Stone *Quest. National Insur.* 100 Will the fact that the mother..is receiving maternity benefit make any difference to the amount of the bastardy order?

bastnäsite (bæ·stnɛ̄ˈsəit). *Min.* [a. F. *bastnaesite* (1841), f. the place-name *Bastnäs*, Sweden, its locality: see -ite[1] 2 b.] A fluo-carbonate of cerium, occurring in small embedded masses, with a greasy lustre and wax-yellow colour.

1872 G. J. Brush in *Dana's Syst. Min.* (ed. 5) App. 1914 *Brit. Mus. Return* 229 Crystals of betafite, bastnasite, columbite [etc.]. 1967 *New Scientist* 3 Aug. 253/1 All the europium obtained in Britain is extracted from an ore known as bastnaesite, imported from deposits at Mountain Pass, California.

bastose (bæ·stōᵘs). *Chem.* [f. Bast *sb.*[1] + -ose[1].] = *Lignocellulose* (see Ligno-). Also *attrib.*

1882 Cross & Bevan in *Jrnl. Chem. Soc.* XLI. 102 We propose to give the name of *Bastose* to this transition modification of cellulose, at the same time observing that as there are many celluloses, so there would be necessarily corresponding forms of bastose. 1902 Hannan *Textile Fibres* 17 The blending of the bastose and plumose fibres to make one complete yarn is not feasible to begin with.

Basuto (bāsū·to). A member of a S. African people of the Bantu stock. Also *attrib.* or *adj.*

1835 A. Smith *Diary* 7 Dec. (1939) I. 163 A Basutu said the hamerkop gives rain. 1837 F. Owen *Diary* (1926) 64 Sintuala chief of the Mantitees or Basutoos. *Ibid.* 70 The *Boers* in conjunction with the Busutoos had made another attack on him. 1842 R. Moffat *Missionary Labours S. Afr.* xv. 242 The Basuto country, once the theatre of plunder and bloodshed, is now studded with missionary stations. 1876 *Encycl. Brit.* V. 47/1 The Basutos, sometimes called Mountain Bechwanas, the fragments of several broken tribes of the Bechwana Kaffres. 1878 P. Gillmore *Great Thirst Land* 268 My poor little Basuto pony, the last of my Natal stud, has died of horsesickness. 1892 Widdicombe *14 Yrs. Basutoland* ii. 20 The Basuto Christians in communion with the English Church are called *Machurche*. 1926 *Blackw. Mag.* June 826/1 He is a Basuto. 1958 *Yr. Bk. & Guide S. Afr.* 205 The once-famous Basuto pony has deteriorated, but steps are being taken to rehabilitate the breed.

bat, *sb.*[1] Add: **1. b.** Colloq. phr. (*to have*) *bats in the belfry*: (to be) crazy or eccentric. Similarly (rare) *to take the bats*. Hence *bats* = *batty *a.*, used esp. as adj. complement.

c1901 G. W. Peck *Peck's Red-Headed Boy* 82 They all thought a crazy man with bats in his belfry had got loose. 1907 A. Bierce in *Cosmopolitan Mag.* July 335/2 He was especially charmed with the phrase 'bats in the belfry', and would indubitably substitute it for 'possessed of a devil', the Scriptural diagnosis of insanity. 1919 F. Hurst *Humoresque* viii. 314 'Are you bats?' she said. 1927 A. E. W. Mason *No other Tiger* xix. 197 On this sort of expedition!' Phyllis Harmer exclaimed, looking at Strickland as if he were a natural. 'Dear man, you've got bats in the belfry.' 1927 *Chambers's Jrnl.* 740/2 Have you taken the 'bats' or what? 1928 *Blackw. Mag.* Jan. 17/2 The sahib had bats in his belfry, and must be humoured. 1938 E. Bowen *Death of Heart* ii. vi. 285 You're completely bats. 1948 *Daily Express* 8 Oct. 2/5 The Secret Life of Walter Mitty..was written by James Thurber, whose bats viewpoint on life can be summed up by a story about him.

c. Slang phr. (*to go*) *like a bat out of hell*, (to go) very quickly.

1921 J. Dos Passos *Three Soldiers* (1922) ii. ii. 67 We went like a bat out of hell along a good state road. 1925 Fraser & Gibbons *Soldier & Sailor Words* 19 *To go like a bat out of hell*, to go at extreme high speed (Air Force). 1939

I. Baird *Waste Heritage*, iv. 52 When it started to move I hared off an' picked out my car an' beat it like a bat out of hell. 1961 I. Fleming *Thunderball* viii. 87 The motor cyclist..had gone like a bat out of hell towards Baker Street.

2. a. *bat-flight*; **bat-light** *poet.*, darkness or gloom. **b.** *bat-eared*; **bat-eyed** (later example); **bat-wing, bats-wing** (also **bat's-wing**) (earlier and later examples); also applied to that part of the human face which surrounds the eyes and nose, and to a long sleeve having a deep armhole and fitting closely at the cuff (Webster, 1934); **bat-winged** (additional and fig. examples).

1903 *Daily Chron.* 25 May 5/2 Several of the bat-eared French bull-dogs. 1927 *Glasgow Herald* 24 Oct. 10 A recumbent area of, say, six feet in diameter would be sufficient for the most bat-eyed foozler. 1927 E. Wallace *Feathered Serpent* xviii. 226 I'd had a couple of drinks that night, and naturally I was a bit bat-eyed. 1934 T. S. Eliot *Rock* ii. 84 The twilight over stagnant pools at batflight. 1871 G. M. Hopkins *Let.* 2 Aug. (1935) 27, I live in bat-light and shoot at a venture. 1946 C. Fry *Firstborn* 10, I was out before daybreak. It's a good marksman who hunts by batlight. 1838 *Penny Cycl.* XI. 88/2 The burners are of many different forms... The batswing is a thin sheet of gas produced by its passing through a fine saw-cut in a hollow globe. 1846 Holtzapffel *Turning* II. 753 The gas-burners designated as *bat's-wing burners* have a narrow slit through which the gas issues: these are cut..by thin circular saws. 1872 *Young Englishwoman* Oct. 547/1 The batswing skirt is made in all colours...The best is seamless; the second..in seams..[is] cheaper than the seamless batswing. 1904 *Daily Chron.* 23 Aug. 8/1 In the red straw hat there are batswing bows. 1908 *Practitioner* Jan. 22 The bat's-wing area of the face. 1823 *Local & Pers. Acts* I. 128 Any Light or Lights, or Argand, Cockspur, Batwing or any other Kind of Burner. 1959 *Guardian* 28 Aug. 3/5 The new Balenciaga coat has very wide batwing armholes. 1961 *Harper's Bazaar* June 22/2 Loosely-fitting top with batwing sleeves. 1911 Fletcher & Kipling *Hist. England* i. 9, I remember the bat-winged lizard birds. 1923 D. H. Lawrence *Birds, Beasts & Flowers* 76 Bat-winged heart of man.

bat, *sb.*[2] Add: **3. b.** (Earlier example.)

1756 *Connoisseur* 5 Aug. 796 His greatest excellence is cricket-playing, in which he is reckoned as good a bat as either of the Bennets.

c. (Earlier and later examples.) Also †*bat's end*, a local term for 'point' (see Point *sb.*[1] B. 11 a) (*Obs.*); *with the bat*, in batting; as a batsman; *from* or *off the bat*: of runs scored from actual hits (opp. 'extras'); *to carry one's bat*: see *carry *v.* 53 c.

1742 in H. T. Waghorn *Cricket Scores* (1899) 29 The bets on the Slendon man's head that he got 40 notches off his own bat were lost. 1786 *County Mag.* Nov. 171 These two things then you next must do, Place one at middle wick't, at batt's end two. 1832 *P. Egan's Bk. Sports* 345/2 Pilch.. showed great capabilities, both in the field and with the bat. 1862 *Baily's Mag.* Aug. 83 Out of the 204 runs scored from the bat by Oxford, 90..were contributed by Mr. Mitchell. 1863 *Lillywhite's Cricket Scores* III. 97 Hodgson got more runs in his one innings than Rochdale did in their twenty-two innings *off the bat*. 1887 F. Gale *Game Cricket* 45 Seventy years ago .. He played as substitute for an absent mate and was placed at 'bat's-end', as point was always called. 1939 T. S. Eliot *Old Possum's Pract. Cats* 30 All his Inventions are off his own bat.

d. In baseball, the implement used to strike the ball or the act of using it; esp. in phrases *at bat, hot* (or *right*) *off the bat, to* (*the*) *bat*; also *fig.* *U.S.*

1856 *Spirit of Times* (N.Y.) 6 Dec. 229/1 The bat or club [used in baseball] is of hickory or ash, about 3 feet long, tapering..and round. 1868 *Iowa State Reporter* (Des Moines) 21 Oct. 2/4 The penny was flipped to see who should go first to the bat. 1875 *Chicago Tribune* 18 Aug. 5/6 The fine play of the home nine..both in the field and at the bat. 1881 *Sun-beam* (Terre-Haute, Ind.) June 5/1 Picking up a base-ball bat. 1884 E. W. Nye *Baled Hay* 52 Common decency ought to govern conversation without its being necessary to hire an umpire to announce who is at bat. 1888 *Outing* (U.S.) May 118/2 Ferguson..sent the Cincinnatis to the bat. 1888 'M. Twain' *Meisterschaft* 459 Whoever may ask us a Meisterschaft question shall get a Meisterschaft answer—and hot from the bat! 1889 —— *Conn. Yankee* xi. 516 Step to the bat, it's your innings. 1914 *Maclean's Mag.* Feb. 135/2 Get one that chums-up with your spirit right off the bat, natural like. 1955 *New Yorker* 21 May 76/3 You can tell right off the bat that they're wicked, because they keep eating grapes indolently.

e. In the game of two-up (see quot. 1945).

1917 *N.Z.E.F. Chrons.* 16 May 137/2 The big brown paw that held the 'bat' Was trembling like a leaf. 1945 Baker *Austral. Lang.* ix. 176 The small piece of board upon which the two pennies are rested for spinning is called the *kip, stick, bat* or *kiley*.

f. Usu. in pl., the objects resembling table-tennis bats used to guide aircraft landing (e.g. on a ship's deck). Hence used *colloq.* as a name for one who signals with these bats; = *batsman 2.

1943 *Fleet Air Arm* (Min. of Information) v. 32 (*caption*) The Deck-Landing Control Officer guides the Seafire pilot in with his 'bats'. 1943 T. Horsley *Find, Fix & Strike* v. 45 The control officer 'bats', which are now fitted with small electric bulbs, are clearly seen against the background of 'glim' lights. *Ibid.* x. 80 (*caption*) The 'Bats' Officer, in charge of the landing, is about to give the pilot the signal to cut his engine. 1948 Partridge *Dict. Forces' Slang 1939–45* 12 *Bats*, the Ward-room name for the Deck Landing Officer on an aircraft carrier.

8. b. *Pottery.* (*a*) = Stilt *sb.* 4 f; (*b*) a piece of unfired clay (see quot. 1825[2]).

1825 J. Nicholson *Operat. Mechanic* 273 Pieces of clay, called stilts, pins, bats [etc.] are put to keep them apart. *Ibid.* 466 The piece is then laid on a flat surface of board, or plaster, and the workman with a heavy lump of clay, with a level under-surface, adapted for holding in the hand, beats the clay to the thinness the vessel is intended to form. These pieces of clay are technically called *bats*. 1961 M. Jones *Potbank* viii. 30 A tool..comes down to press the lump out into a..pancake. The maker puts the clay—now called a bat—in the mould.

14. b. A movement of the eyelids (see Bat *v.*[2] 2).

1932 E. Caldwell *Tobacco Road* iv. 41 Almost as quickly as the bat of an eye. 1941 'M. Home' *Place of Little Birds* ii. 21 He didn't show by the bat of an eyelid that you were a friend. 1948 C. Fry *Thor, with Angels* 7 We were at the boy in the bat of an eye.

15. (Later examples.)

1888 'R. Boldrewood' *Robb. under Arms* I. xxi. 293 We could hear a horse coming along at a pretty good bat. *Ibid.* II. xvi. 247 A cove comes tearing up full batt. 1949 'J. Tey' *Brat Farrar* xv. 138 [The horse] took Felix under an oak, going an awful bat. 1961 J. Welcome *Beware of Midnight* ii. 20 We turned on to the main..road and started going a hell of a bat across the Cotswolds.

V. bat-willow, a species of willow from which cricket bats are made. See also *cricket-bat willow.

1907 *Kew Bulletin* No. 8. 311 The supplies of the best 'Bat Willow' have become seriously limited. 1910 *Westm. Gaz.* 6 Apr. 4/2 The fast growing bat-willow..a first-cross between two common varieties of willow..appeared in Norfolk about 1700. It is still chiefly obtained from East Anglia.

bat, *sb.*[4] Also **batt.** *slang* (orig. *U.S.*). [Of obscure origin: cf. *batter *sb.*[4]] A spree or binge.

1848 Durivage & Burnham *Stray Subj.* 102 (Th.), Zenas had been on 'a bat' during the night previous. 1869 W. T. Washburn *Fair Harvard* 69 (Th.), I went to a 'bat' in S.'s room, and we smoked and drank till three. 1891 *Harper's Mag.* Oct. 778/1 He had been on a bat, and all on earth that ailed him was that spree. 1901 *House Party* 188 We defied the Head and went off on the meekest and stupidest little bat you ever saw. 1942 E. Waugh *Put out more Flags* iii. § 4. 187 Why don't you switch to rum? It's much better for you...When did you start on this bat?

bat (bat), *sb.*[5] [Hindi, = speech, language, word.] *the bat*: the colloquial speech of a foreign country; chiefly in phr. *to sling the bat*.

1887 Kipling *3 Musketeers* in *Plain Tales* (1888) 62 T' *Sahib* doesn't speak t' *bat*. 1889 —— *Barrack-r. Ballads* (1892) 67 An' ow they would admire for to hear us sling the *bat*. 1919 *War Terms* in *Athenæum* 8 Aug. 729/1 A variant for 'sling the bat' (speak the lingo) is 'spin the bat'. 1924 *Glasgow Herald* 14 Apr. 10 He continued eagerly..'that in the bat of the Arab "Shmallock" and "Amenak" mean "left" and "right".'

bat, *v.*[1] Add: **2.** (Earlier example.) Also *fig.*

1745 in H. T. Waghorn *Cricket Scores* (1899) 36 The girls bowled, batted, ran, and catched..as well as most men could do in that game. 1959 *Observer* 18 Jan. 19/2 The healer, who went in to bat last, was lured into the last ditch of philosophical idealism. 1961 *Listener* 2 Nov. 737/3 Two contributors, finally, bat for Christianity.

b. *to bat on a sticky wicket*: see *sticky *a.*[2]

4. To go or move; to wander, to potter. Usu. with advb. extension, *along, around, away*, etc. Chiefly *dial.* and *U.S.*

a 1898 *Old Radicals & Young Reformers* 13 (E.D.D.), Heaw they stared when they seed Billy battin away across a fielt. 1907 W. D. Howells *Let.* 3 Oct. in *Mark Twain–Howells Lett.* (1960) II. 826 [She] was in England..batting round with two other girls, and having a great time. 1926 *S.P.E. Tract* XXIV. 119 *Bat round*, have a good time, go from place to place (in quest of pleasure). 'We've been batting round all evening.' 1929 E. Rice *Street Scene* (1930) 1, I want 'em [*sc.* the kids] home, instead o' battin' around the streets. 1938 *Reader's Digest* Mar. 13/2 A Department Sanitation truck was batting along as fast as it could go. 1959 *Encounter* Aug. 30/2 So I batted along, and I tried to make conversation with the kiddo. 1959 I. & P. Opie *Lore & Lang. Schoolchildren* x. 192 Expressions inviting a person's departure..bat off, beat it, [etc.].

bat, *v.*[2] Add: **2.** (Earlier example.) Also *intr.*

1838 Holloway *Prov. Dict.* 9/1 *Bat*, to wink..*Derby*. 1846 J. J. Hooper *Adv. Simon Suggs* xii. 143, I didn't say nuthin, but jist batted my eye at old Chamblin.

Also freq. in colloq. phr. (normally in negative form), *not to bat an eye, eyelid*, etc. (i) not to sleep a wink; (ii) to betray no emotion. orig. *U.S.*

In quot. 1950 the phr. means contextually 'I didn't open my eyes (*i.e.* I slept heavily)'.

1889 'Craddock' *Broomsedge Cove* xii. 208 If my patient can't sleep, not a soul in the house shall bat an eye all night. 1904 *Sun* (N.Y.) 7 Aug. 1 The Judge would say: 'That's interesting..I hadn't heard of it.' But, as they say out West, 'he wouldn't bat an eye'. 1910 'O. Henry' *Whirligigs* viii. 113 I've stood by you without batting an eye in earthquakes, fire and flood. 1930 *English Jrnl.* XIX. 607 We do want the facts, and we are willing to look them straight in the eye without batting a lash. 1950 J. Cannan *Murder Incl.* vi. 109, I was tired..and I never batted an eyelid until Beatrice brought in my breakfast. 1959 *News Chron.* 14 July 4/6 [Japan] slipped from..'past to..present without, you might say, batting an eyelid.

Batak (bæ·tāk), *a.* and *sb.*[1] Also **Battak.** = *Batta[2].

1811 W. Marsden *Hist. Sumatra* (ed. 3) 365 By many regarded as having the strongest claims to originality, is the nation of the *Battas* (properly *Batak*), whose remarkable dissimilitude to the other inhabitants..renders it necessary that..attention should be paid to their description. 1875 *Encycl. Brit.* III. 442/2 (*heading*) Batta or Batak language. *Ibid.* 443/1 The Batak literature consists chiefly in books on witchcraft, [etc.]. 1933 Bloomfield *Language* xviii. 310 The eight normal types of correspondence will appear sufficiently if we consider three languages: Tagalog.., Javanese, and Batak (on the island of Sumatra). 1957 *Encycl. Brit.* XXI. 550/2 The Batak inhabit the mountainous region about Lake Toba and a part of the east coast [of Sumatra]. 1958 A. Toynbee *East to West* xviii. 54, I was told..that the Batak Church has more theological students in training than are to be found in all the Anglican theological colleges in Britain.

Batak (bātā·k), *sb.*[2] [Native name.] (A member of) a people on the island of Palawan, Philippines.

1904 W. A. Reed in *Ethnol. Survey Pubns. Manila* II. 1. 22 The Tagbanua and Batak of Paragua...The Tagbanua.. can not be classed with the Negritos...The Batak who inhabit the territory from the Bay of Ulugan north to Caruray and Barbacan may be so classed, although they are by no means of pure blood. 1905 E. Y. Miller *Ibid.* iii. 183 The peculiar primitive people known as Bataks are to be found in the mountains in the interior of Palawan. 1957 *Encycl. Brit.* III. 198/2 *Batak*..Short statured and brown to dark skinned they appear to be a composite of Negrito and Malayan type elements.

Batavia (bătēˈ·viă). [Former name of Djakarta, capital of Indonesia.] A kind of shot silk material. Also *attrib.*

1907 *Daily Chron.* 10 June 8/1 A new shot silk, known as 'Batavia'...Smart little coats carried out in this Batavia silk look particularly well. 1909 *Cent. Dict.* Suppl. s.v. *Weave*, *Batavia weave*, a kind of twilled armure weave made on four harnesses.

batch, *sb.*[1] Add: **6. c.** The quantity of stricks or bundles of jute laid out at one time for treatment. (Cf. *batch *v.* 1.)

1880 *Encycl. Brit.* XIII. 803/1 These batches [of jute], which generally contained from 4 to 5 tons each, were allowed to lie from twenty-four to forty-eight hours. 1893 W. Leggatt *Jute Spinning* 21 The batch put down for ordinary hessian warps should be composed of six bales.

d. *Calico-printing* and *Dyeing.* The mass of material collected in 'batching' (cf. *batch *v.* 2). Also *attrib.*

1911 Trotman & Thorp *Princ. Bleaching and Finishing Cotton* xxviii. 304 The chief use of the machine is to transfer cloth from the lap to the batch form. *Ibid.*, The bearings of the batch roller work in vertical slides under the pressure of springs or weights, and can be raised and held with a pawl to unroll a part of the batch for detailed inspection.

e. *Glass-making* = Frit *sb.*[2] 1.

a 1877 Knight *Dict. Mech.* I. 246/1 *Batch*,..the frit of a glass-maker compounded and sifted for use, ready for the glass-pot or crucible. 1933 *Antiquity* VII. 420 Glass workers use the term 'batch' to denote the raw ingredients of the glass before fusing.

7. In various combinations denoting a machine or process that treats a batch of material (esp. as opposed to a continuous process).

1940 *Chambers's Techn. Dict.* 80/2 *Batch furnace*, a furnace in which the charge is placed and heated to the requisite temperature, subsequently being withdrawn... Distinguished from continuous furnace. 1954 *Economist* 20 Feb. 537/2 Strenuous attempts to convert their batch processes to the continuous processes which have so much lower labour costs. 1955 J. G. Davis *Dict. Dairying* (ed. 2) 136 Batch pasteurisation is a satisfactory and economical method for small factories. 1957 *Economist* 16 Nov. 621/1 Glass for spectacles..has to be of very precise physical qualities... Until recently this has meant making it in a form of batch process known as the pot roast method. 1957 *Ibid.* 23 Nov. 645 Continuous batch testing with the latest scientific equipment. 1958 A. Hackney *Private Life* vii. 65 Nummums and Chokers are still in batch production, but the Bumper Bars themselves are in continuous-flow production.

batch (bætʃ), *v.* [f. Batch *sb.*[1]] **1.** *trans.* To treat (bundles of raw fibre of jute, wool, etc.) in batches for various purposes. Hence **ba·tching** *vbl. sb.*, the action or process by which this is done; also *attrib.* Also **ba·tcher,** an operative who does this.

1880 *Chemical News* XLII. 77/2 The fibres are matted together by a resinous constituent, and in order to fit them for the operations of combing they are subjected to the preliminary treatment of 'batching'. This consists in moistening the fibre with a mixture of oil, [etc.]. 1880 *Encycl. Brit.* XIII. 802/2 (Jute) Batching or Softening. 1881 *Instr. Census Clerks* (1885) 71 Jute preparing: Preparer. Batcher. 1882 P. Sharp *Flax, Tow, & Jute Spinning* 158 In this system the jute is batched in the same way as tow. 1893 W. Leggatt *Jute Spinning* 19 The batchers..break up the large streaks or heads into streaks of about two pounds each. *Ibid.*, The jute warehouse, which will be seen from a reference to the ground plan to adjoin the batching house. 1921 Mackinnon *Social & Industr. Hist. Scot.* ii. 115 This is done in the 'batching room', where the raw material is sprinkled with oil and water. 1921 *Dict. Occup. Terms* (1927) § 398 Batcher (flax and hemp); (i) arranges different qualities of tow in layers to facilitate mixing of tow for

carder; (ii) applies water and oil to opened out hemp, to soften fibre and to prevent it from 'licking up' during.. carding. 1943 J. S. Huxley *TVA* 23 From the quarry the rock was carried to the crusher, thence to the screening plant, the batching plant and concrete mixers. 1950 *Engineering* 20 Jan. 81/3 In the batching process, the quantities of cement and..aggregate are best measured by weight. 1952 *Electronic Engin.* XXIV. 205 Electronics can accurately control a mechanical batching device.

2. *Calico-printing* and *Dyeing.* To collect into a 'batch' or mass. Hence **ba·tching** *vbl. sb.* (also *attrib.*).

1876 *Encycl. Brit.* IV. 685/2 In printing, the white calico is batched at C, and the cloth D passes inwards over tension rails. 1911 H. R. Carter *Bleaching, Dyeing of Flax*, etc. 90 The cloth..is..'batched' on to a larger roller, upon which it is removed for subsequent processes—washing, drying, &c. 1911 Trotman & Thorp *Princ. Bleaching & Finishing Cotton* xxviii. 304 When wound up, the bearings of the batching roller are racked up and the roll unwound or removed. *Ibid.*, The winding-on frame, or canroy, is a frame with tensioning staves, rollers, and scrimp rails, and batching tackle.

batch, var. *bach *sb.* 2, *bach *v.*

batchy (bæ·tʃi), *a. slang.* = *batty *a.*; dotty.

1898 F. T. Bullen *Cruise of Cachalot* v. 39 They [*sc.* two of the crew] had gone quite 'batchy' with fright. 1931 *Punch* 14 Oct. 402 Anyone who continues to keep a car in London these days in preference to a horse is batchy. 1960 J. Stroud *Shorn Lamb* xii. 140 Perhaps Egbert's all right and we're all a bit batchy?

bate, *sb.*[6] See *bait *sb.*[2]

batea (bǣti·ă). *U.S. mining.* [a. Sp. *batea* tray, trough.] A shallow wooden vessel used in the washing of ores in California and Mexico.

1864 S. Mowry *Arizona & Sonora* 44 In the rubbish which was thrown out of the old mine a comfortable subsistence is gained by washing in bateas [ed. 1859, battas]. 1874 Raymond *Sixth Rep. Mines* 315 In these they wash the gravel and earth, by means of wooden bowls or Bateas. 1897 *Engineering Mag.* XVI. 51 Wooden bateas about eighteen inches in diameter (probably used to carry the ore out of the mine).

bateau. Add: (Earlier and later examples.) Also *U.S.* Cf. *battoe.

1711 in *N.J. Archives* (1882) IV. 137 If you have not engaged Sloops for Albany they may go from hence in Bateaux. *Ibid.* 318, I pressed all the carpenters in the place ..for the dispatch of these batteaux. 1905 G. E. Cole *Early Oregon* i. 12 The batteau which was to bring our luggage would not be there for several hours. 1947 *Amer. Dial. Soc. Pubn.* VIII. 34 *Bateau*: In southern Virginia this is not a small boat but a flat-bottomed freight boat.

2. *attrib.* and *Comb.*, as **bateau-mouche,** a boat which takes sightseers on the Seine in Paris; **bateau (neck-) line,** in *Dressmaking*, a décolletage having a bow-shaped curve from shoulder to shoulder; hence *bateau-necked* adj.; cf. *boat neck(-line).

1903 H. James *Ambassadors* xii. xxxiv. 434 An hour with his present friend on a *bateau-mouche*. 1923 *Daily Mail* 20 Feb. 14 The bateau neck-line...This décolletage still retains its popularity. *Ibid.* 1 Mar. 15 The bateau line is still seen on smart gowns. 1938 W. S. Maugham *Summing Up* 19, I used to go down with him every Sunday by the Seine on a *bateau-mouche*. 1959 *Times* 31 July 8/4 A simple bodice often bateau-necked. 1961 *Guardian* 30 Mar. 9/3 The very latest knitted jumpers have..bateau necklines.

bateless, *a.* Add: **2.** Unabating.

1886 *Harper's Mag.* May 884 From heaven of heavens above God speaketh with bateless breath.

bateleur (ba·t'lǒr). [Fr., = mountebank, juggler.] In full *bateleur eagle*: a short-tailed African and Arabian eagle, *Terathopius* (*Helotarsus*) *ecaudatus*. Cf. *berghaan, *dassie-vanger 1.

1864 J. A. Grant *Walk across Africa* 306 The largest birds were the Batteleur eagle and the Buceros. a 1867 K. J. Andersson *Notes Birds Damara Land* (1872) 10 Helotarsus ecaudatus..Rufous-backed Bateleur eagle. 1895 Lydekker *R. Nat. Hist.* IV. 211 The handsomely coloured bird known as the bateleur eagle..differs from the sea-eagles..by the extreme shortness of its tail. 1940 V. Pohl *Bushveld Adventures* xi. 233 A couple of bateleur eagles that were circling round a spot. 1955 K. A. Thompson *Great House* iii. 106 The trousered bateleurs.

batement. Add: **b.** *batement light* [Light *sb.* 10], an upper window or opening with a sloping or curved sill, i.e. abated or cut off at the low end to accommodate the curve of an arch, etc.

1445 in J. A. Nichols *Descr. Beauchamp Chapel* (1838) 29 In the est windowe be vij lights..also sixe batements lights, every light conteininge ij foote. 1844 R. Willis *Archit. Nomencl. Mid. Ages* iv. 51 They are lights with the lower corner cut off, or, in the language of workmen, they are lights with a *batement* .. hence the term 'batement lights'. 1901 R. Sturgis *Dict. Archit. & Building* 238/2 *Batement light*, in English Gothic traceried windows, a peculiar form of light, or subdivision of a window; one comprised between two mullions, and having a curved or inclined bottom caused by the arched heads of other lights below.

Batesian (bēˈ·tsiăn), *a.* [f. the name of H. W. Bates (1825–92), English naturalist: see -ian.] **Batesian mimicry** *Zool.*, a form of mimicry in which an edible species is protected by its resemblance to one which is avoided by predators. So *Batesian mimic*, the species which is so protected.

1896 E. B. Poulton *Darwin* xxv. 214 Many cases which have been up to the present explained under the theory of true (Batesian) mimicry are now believed to come under that which we owe to F. Müller. 1899 D. Sharp in *Cambr. Nat. Hist.* VI. vi. 339 The comparatively simple, hypothetical explanation, originally promulgated by Bates, is sometimes called Batesian mimicry; while the 'inedible association' hypothesis is termed Mullerian mimicry. 1933 *Discovery* Nov. 357/2 The cuckoo's egg, a clear instance of Batesian mimicry deceiving the selective-agents, i.e., the foster-parents. 1951 *New Biol.* X. 73 Mullerian and Batesian mimics may be involved together in nature in a complex mimetic association.

batey (bēˈ·ti), *a.* var. *baity adj. (s.v. *bait *sb.*[2]).

1946 B. Marshall *G. Brown's Schooldays* 66 You'll make me batey if you say rotten caddish things like that. 1954 P. H. Johnson *Impossible Marriage* 11 I'd better roll the damned thing in or Mater will be batey.

bath, *sb.*[1] Add: **9. b.** In the hydropathic treatment of disease, any yielding medium, as water (natural or medicated), mud, sand, etc., in which the body is bathed or immersed, or with which it is sprayed or showered: for examples see Douche-bath, Mud-bath, Needle-bath, Sand-bath 2, Shower-bath, Turkish bath.

12. Now esp. a room where one may bathe, a bathroom.

In Britain chiefly used as an advertising term for *bathroom*; in somewhat more general use in N. Amer.

1922 S. Lewis *Babbitt* x. 141, I think I can let you have a room with bath. 1939 O. Lancaster *Homes Sweet Homes* 66 Three bed, two bath, a kitchen and all the usual offices. 1953 G. V. Carey *Amer. into Eng.* 34 *Bath*, sometimes denotes in American (scarcely ever in English) 'bathroom': 'He went into the bath for a shower.' 'In the bath there was no shower but only a bathtub.' 1967 R. Raine *Wreath for America* vii. 55 You've got a room booked for me, a single room with bath. 1968 *Globe & Mail* (Toronto) 3 Feb. 43/8 (Advt.), Don Mills townhouse, 3 bedrooms, 1½ baths.

16. b. *spec.* in *dung-bath*: see Dung *sb.* 5 c; *long bath*, a dilute bath in which chemical action is comparatively slow; *short bath*, a concentrated bath; *single bath*, one in which the whole operation is completed; *standing bath*, one that is used continuously; *white bath*: see White *a.* 11 e.

VI. *bath-bed, -brush, -gown, -mat, -night, oil, -powder, -sponge, -time, -tub* (earlier example), *-waste*; **bath cubes, essence, salts,** toilet preparations for softening or perfuming bath-water; **bath house,** a building equipped with facilities for bathing, occas. public baths; *U.S.*, a place where one may change into beach clothes at the seaside, etc.; **bath-robe** orig. *U.S.*, a dressing-gown, esp. one made of towelling; **bath-sheet,** a large bath-towel; **bath-towel,** a large towel; hence *bath-towelling*; **bath-tub gin,** a concoction of spirits simulating gin (orig. used to designate illicitly manufactured liquor); **bath vat** *poet.* = Bath *sb.*[1] 11.

1894 *Daily News* 8 Oct. 7/1 The appliances for treatment of special diseases, such as bath beds for typhoid. 1895 *Montgomery Ward Catal.* 103/2 Combined Bath and Flesh Brushes..for wet or dry use. 1900 H. Lawson *On Track* 72 In the other hand she carried her tooth-brush and bath-brush, and soap. 1959 A. Wesker *Roots* ii. ii. 53 These bath cubes smell beautiful. 1911 Beerbohm *Zuleika D.* xxi. 302 Eau de Violettes was the bath-essence that Zuleika always had. 1954 E. Jenkins *Tortoise & Hare* viii. 75 The bland scent of expensive soap and bath essence. 1909 *Westm. Gaz.* 19 Aug. 5/3 The dressing- and bath-gowns. 1705, 1882 Bath-house [in Dict.]. 1800 W. Bentley *Diary* (1907) II. 399, I bathed in the river this evening, and the Bath House was opened for the first time. 1851 C. Cist *Cincinnati* 167 There are several public bath houses. 1895 *Army & Navy Co-op. Soc. Price List* 190 Felt Bath Mats. 1927 M. de la Roche *Jalna* vii. 82 She stepped dripping on to the thick bath mat. 1921 D. H. Lawrence *Phoenix* (1936) 15 It was Saturday night—bath-night. 1925 W. de la Mare *Broomsticks* 367 Even though it was 'bath-night' on Saturday. 1962 *Guardian* 5 Dec. 6/4 'Arpège' Bath Oil, 43s. 6d., will do 30 baths. 1968 K. Bird *Smash Glass Image* iii. 43 My array of bath oils and talcum powder. 1907 *Yesterday's Shopping* (1969) 536/2 Parma violet bath powder and water softener. 1923 W. A. Poucher *Perfumes & Cosmetics* 327 Bath Powders are generally..highly perfumed. 1940 N. Marsh *Surfeit of Lampreys* (1941) viii. 115 Her round face shone and she smelt of bath powder. 1902 *Sears Catal.* (ed. 112) 846/2 Fine Terry Cloth Bath Robe with Hood. 1924 A. D. Sedgwick *Little French Girl* ii. v. 139 One undressed in one's room and ran out over the cliff-top in *espadrilles* and bath-robe. 1907 *Yesterday's Shopping* (1969) 536/1 Bath Salts (Fragrant)..will instantly soften the hardest water. 1920 A. Huxley *Limbo* 236 A very hot bath with lots of verbena bath-salts. 1930 J. B. Priestley *Angel Pavement* viii. 411 My worthy employeess's terribly expensive bath salts. 1899 *Westm. Gaz.* 3 Jan. 3/2 The bath-sheets..seem to become masses of vibrating silver. 1912 E. M. Dell *Way of Eagle* (1927) lvi. 221 He stepped from the tent, clad loosely in a bath-sheet. 1967 *Guardian* 5 Sept. 4/6 Enormous great bath towels..as big as the bath sheets in Swiss hotels. 1889 *Cent. Dict.* I. 474/3 Bath sponge. 1927 Haldane & Huxley *Animal Biol.* xii. 266 Our bath-sponges..are colonial, composed of a large number of sponge-bodies aggregated together. 1907 A. Bennett *Grim Smile Five T.* 40 He isn't used to you at bath-time. 1863 Dickens *Uncommercial Traveller* in *All the Year Round* 24 Oct. 206/1 A piece of sculptured drapery resembling the effigy of Titbull's bath-towel. 1903 *Town & Country* 24 Jan. 2/1 There's..a bath room with hot and cold water, soap, and *real* bath towels. 1880 Bath towelling [see Towelling 1]. 1926 'R. Crompton' *Wm.—the Conqueror* iv. 65 Henry's was made of bath towelling and was rather conspicuous in design. 1869 'Mark Twain' *Innoc. Abroad* xix. 187 They were going to put all three of us in one bath-tub. 1930 C. Beaton *Diary* Dec. in *Wandering Years* (1961) 196 Five-star scotch or bathtub gin. 1967 *Boston Sunday Herald* 26 Mar. iv. 1/1 Marijuana has become the Bathtub Gin of the 1960s. 1874 Swinburne *Bothwell* ii. xviii. 216 A cover for his bath-vat. 1887 Morris *Odyssey* viii. 456 He gat him up out of the bath-vat. 1936 *Discovery* Aug. 244/1 The trunk sprang from near the cement bath-waste channel... A large, shining drop of water splashed from the bath waste pipe. 1958 Betjeman *Coll. Poems* 142 All the bells of all the churches Sounded in the bath-waste running out into the frosty air.

Bath, *sb.*[2] Add: **1. b.** As a place of consignment for a person one does not wish to see again, in the phrase *to go to Bath*, chiefly used imperatively.

1837 Barham in *Ingol. Leg.* (1840) Ser. i. 83 'Go to Bath!' said the Baron. 1858 Thackeray *Virgin.* I. xvi. 127 She may go to Bath, or she may go to Jericho, for me. 1908 *Daily Chron.* 27 Oct. 6/6 A later use of the expression was 'Go to Bath and get your head shaved'.

2. a. *Bath water*(*s*) (earlier examples). **b.** *Bath-bun, -ring* (earlier examples). **c. Bath asparagus,** a variety of Star of Bethlehem (*Ornithogalum pyrenaicum*) with edible shoots; **Bath-chair** (earlier examples); **Bath coup,** a method of play in whist or bridge in which a player, who holds the ace, knave and another card of the same suit, refrains from taking his adversary's king; **Bath-metal** (earlier and later examples); **Bath Oliver,** an unsweetened biscuit said to have been invented by William Oliver (1695–1764), a physician of Bath; formerly simply *Oliver*; **Bath-post** (earlier examples); **Bath White (butterfly),** a rare European pierid butterfly, *Pontia daplidice*, having the under side of the hind wing of a greenish colour spotted with white.

1791 Collinson *Hist. Somerset* I. p. xx, Wood Star of Bethlehem or Bath Asparagus...The young shoots of it are eaten by the common people as asparagus. 1959 A. R. Clapham et al. *Excursion Flora Brit. Isles* 445 Very locally abundant in woods and scrub. Bath Asparagus. 1801 Jane Austen *Let.* 3 Jan. (1952) 101 Disordering my Stomach with Bath bunns. 1823 M. Wilmot *Let.* 30 Apr. (1935) 188 My dearest Mother, do you not make use of a Bath Chair? 1847 M. Edgeworth *Tour Connemara* (1950) 109 She cannot walk now beside Honora's Bath chair. 1897 R. F. Foster *Complete Hoyle* 613 Bath Coup, holding up Ace Jack on a King led by an adversary. 1899 W. M. Butler *Whist Reference Book* 51/1 *The Bath Coup.* A strategic play at whist which originated at Bath, England, in the time of Hoyle. The fourth hand, holding ace, jack, and others, refuses to take a king when it is led. 1936 E. Culbertson *Contr. Bridge Complete* xl. 456 The Bath Coup gains time by causing the opponents to waste an entry before they can establish their suit. 1714 *Boston News-Let.* 7 June 2/2 A blew Coat..with wrought Bath metal Buttons. 1738 J. Wesley *Let.* 1 Dec. (1931) I. 276 Could not you purchase for me half a dozen Bath-metal tea-spoons? 1958 A. D. Merriman *Dict. Metallurgy* 16/1 *Bath metal*, a silvery-white copper-zinc alloy which at one time was favoured for cheap table-ware. Approximate composition: Zn 45%, Cu 55%. 1878 *Official Guide & Album Cunard Steamship Co.* 156/1 Bath Oliver. 1900 in Ware *Passing Eng.* (1909) 21/1 'Bobs' fights on 'Bath Olivers'. 1915 A. D. Gillespie *Let.* 14 June in *Lett. from Flanders* (1916) 320 The sausages [bombs] are rather like a Bath Oliver biscuit tin—only not quite so big—full of old nails and rusty scrap-iron. 1928 Compton Mackenzie *Extremes Meet* 123 We had a periscope scare about five hours out from Malta... This time it was a Bath Oliver biscuit tin. 1837 Dickens in *Bentley's Misc.* I. 53 Four sides of closely-written, gilt-edged, hot-pressed Bath post letter paper. 1845 *Ainsworth's Mag.* VII. 26 They beheld their father..repiling the quires of Bath post. 1771 Smollett *Humph. Cl.* I. 118, I send you two dozen of Bath rings...I don't know how you will approve of the mottoes. 1673 J. Locke *Let.* 14 Feb. in Fox Bourne *Life* (1876) I. 317 You may possibly bring home with you a new use of our Bath waters. 1693 E. Clarke *Let.* 2 Aug. in Rand *Locke & Clarke* (1927) 379 Whether advisable to drink the Bath water, if she be desirous of it, whilst she is with child. *Ibid.*, Frequently asks if the Bath waters may not be drank here at home with good success. 1795 W. Lewin *Insects Gt. Britain* 62 Bath White. Daplidice. *Linnæus.* This is a rare butterfly in England..named the Bath white, from a piece of needle work, executed at Bath, by a young lady, from a specimen of this insect, said to be taken near that place. 1832 J. Rennie *Consp. Butterfl. & M.* Index, Bath White B. 1885 Kane *European Butterflies* 9 *Pieris Daplidice* L. The Bath White.

bathe, *v.* **5. b.** (Later example.)

1855 F. W. Faber *Growth in Holiness* xi. 169 Youth.. bathing in devotional sweetness.

bather. Add: **3.** *pl.* Bathing trunks; a bathing suit. esp. *Austral.*

1945 BAKER *Austral. Lang.* ix. 183 Swimming costumes are known variously as *togs, bathers.* 1957 'N. SHUTE' *On the Beach* 3 You join us there for a swim. I could take the trailer and your bathers. 1965 J. R. STOW *Merry-go-round in Sea* i. 23 His mother sat by the water, on a red towel, in blue bathers. 1966 *Southerly* XXVI. 175 The girl wore a blouse over her bathers. 1968 N. MARSH *Clutch of Constables* viii. 204 It was nothing, really...I'm a Sydneysider, don't forget, and I *was* in my bathers.

bathinette (bɑːþine·t). *U.S.* Also **-et.** [Proprietary name.] A small bath used in bathing infants.

1936 *Sears, Roebuck Catal.* (ed. 173) 173 Genuine bathinette...Dupont rubber coated fabric tub. 1938 *Capital* (Topeka, Kan.) 25 Nov. 3 (Advt.), Genuine baby bathinette. 1954 F. G. PATTON *Good Morning, Miss Dove* (1955) 5 Stop at the Burnhams' and pick up the bathinet they're lending us. 1963 M. MCCARTHY *Group* xiv. 326 She did not possess a baby scales or a bathinet. He was bathed in the wash-basin.

bathing, *vbl. sb.* Add: **1. b.** The conditions under which bathing can be carried on at a watering-place, etc. (including the quality of the water, the character of the beach, accessibility, and the like).

c 1830 F. COGHLAN *Coast Compan.* 37 The prospects are fine, and the walks and rides excellent; and the bathing is remarkably good. 1877 *Seaside Watering Places* xxiv. 46 The bathing is excellent and especially safe. 1881 *Dict. Watering Places* 108 The sands and bathing [at Westward Ho!] are good, machines and tents abundant.

2. *bathing beach, -cabin, -cap, -costume, -drawers, -dress* (earlier example), *-house, -hut, -machine* (later examples), *-robe, -room* (examples), *-season, -shed, -suit, trunks, -tub* (later examples), *-woman* (earlier example).

1926 *Daily Colonist* (Victoria, B.C.) 2 July 5/3 The lure of the bathing beach and wooded lane. 1943 J. S. HUXLEY *TVA* ix. 63 There are bathing beaches, a wading pool for children, and a diving tower. 1905 W. J. LOCKE *Morals M. Ordeyne* xii. 145 Below the terrace are the bathing-cabins. 1968 L. DURRELL *Tunc* ii. 51 The cubicles of the girls, somewhat like a row of bathing cabins. 1867 *Belgravia* III. 355 A bathing-cap is *de rigueur.* 1937 J. LAVER *Taste & Fashion* xvi. 220 In 1919..the bathing cap has made its appearance..not yet made of rubber. 1830 F. TROLLOPE *Domestic Manners Amer.* (1960) 273 Her petticoat, the most important part of her bathing costume, dropped off. 1839 DICKENS *Tugg's at Ramsgate* in *Sk. Boz* 369 One of the young ladies..in her bathing costume. 1928 H. WILLIAMSON *Pathway* xii. 275 'We came without bathing costumes.' 'Bathing *suits,*' Ronnie corrected him, with a self-assured air. 'Only board-school boys say *costumes.*' 1940 GRAVES & HODGE *Long Week-end* viii. 122 The conjunction of a negro band with white girls in bathing costumes. 1893 Bathing-drawers [see DRAWERS]. 1931 W. DE LA MARE *Seven Short Stories* 109 In one of the huge 'stores' in search of bathing-drawers. 1774 J. SCHAW *Jrnl. Lady of Quality* (1921) i. 69 When we are to leave the cabin in our bathing dress, all the people quit the deck. 1598 FLORIO *Worlde of Wordes* 36/2 *Bagno*, a bathe, a baine, or bathing house. 1695 in LADY G. BAILLIE *Household Bk.* (1911) 3 For baithing in Rees bathing hows. 1754 in *N. & Q.* (1865) VIII. 178 Notice is hereby Given, That the Bathing-House in this Place [*sc.* Portsmouth] will be finished and fit for Use, by about the 10th Day of May. 1835 J. H. INGRAHAM *South-West* I. xvi. 174 The long white bathing-houses, which stretched along the south side of the pier. 1838 F. A. KEMBLE *Let.* 10 Aug. in *Records of Later Life* (1882) I. 158 There are two little stationary bathing-huts. 1922 A. HUXLEY *Mortal Coils* 180 Running up towards his bathing hut. 1860 C. A. COLLINS *Eye-Witness* ii. 15 Four bathing-machine boys. 1906 *Westm. Gaz.* 26 July 10/1 A writer in *T.P.'s Weekly* states, that the first bathing-machine was seen at Margate..in 1750. 1924 A. D. SEDGWICK *Little French Girl* II. ix. 168 When they came to the rock where, with safety, the bathing-robes might be deposited. a 1700 EVELYN *Diary* 10 Sept. an. 1677 (1955) IV. 118 There are bathing roomes, Elaboratorie, Dispensatorie, what not. c 1702 C. FIENNES *Journeys* (1947) 100 A batheing roome, the walls all with blew and white marble..the bath is one entire marble. a 1828 D. WORDSWORTH *Tour Continent* in *Jrnls.* (1941) II. 270 The commencement of the bathing-season. 1932 N. PALMER *Talking it Over* 139 A few large bathing-sheds tucked against the cliffs. 1873 *Young Englishwoman* June 280/2 Very pretty cases for holding the complete bathing suit are easily made of American cloth. 1881 MARSHALL *Through Amer.* (1882) 398 There had appeared in the *Salt Lake Daily Tribune,* in the morning, the following announcement: 'Bathing Suits to order in six hours.' 1906 *Daily Chron.* 18 Aug. 4/6 The 'Bathing Suit Dance' is the latest idea..at American summer resorts. 1922 JOYCE *Ulysses* 723 If all women were her sort down on bathingsuits and lownecks. 1895 *Montgomery Ward Catal.* 488/2 Knee Tights..also make good bathing trunks. 1956 F. CASTLE *Violent Hours* (1966) x. 98 He wore bathing trunks, wet and clinging. 1633 MASSINGER *Guardian* II. iv, The Silver Bathing Tub. 1841 CATLIN *N. Amer. Indians* I. 97 A crib or basket, much in the shape of a bathing-tub. 1789 F. BURNEY *Diary* 30 June (1842) V. 33 Even the bathing-women had it [*sc.* 'God save the King'] in large coarse girdles round their waists.

bathing, *ppl. a.* Add: *bathing beauty, belle,* an attractive woman in a bathing suit, used esp. of one taking part in a beauty contest; also *attrib.* and *transf.*

1920 B. LEVY in *Stage Year Bk.* 69 It would be hard to throw a stone..upon the American Vaudeville stage without hitting what the eight-sheets describe as 'California Bathing Beauties'. 1926 A. HUXLEY *Jesting Pilate* IV. 266 Mack Sennett Bathing Beauties by the hundred. 1946 J. IRVING *Royal Navalese* 28 *Bathing Beauty,* strawberry- or raspberry-flavoured pink blancmange. 1967 MCLUHAN *Medium is Massage* 18 Today's television child is attuned to 'adult' news—inflation, rioting, war,..bathing beauties. 1924 P. MACDONALD *Rasp* v. 57 Minister murdered by Bathing Belle. 1928 *Oxford Poetry* 39 The others all Were giggling blond bathingbelles. 1948 'J. TEY' *Franchise Affair* xv. 162 Coloured publications with bathing-belle covers.

bathless (bɑ·þlės), *a.* [f. BATH *sb.*[1] + -LESS.] Without a bath or baths; not having had a bath. Hence **ba·thlessness,** bathless condition.

1889 *Century Mag.* Aug. 503/2 A bathless, breakfastless Mexican smokes his cigarette. 1907 *Daily Chron.* 14 Oct. 4/4 A crowd who have worked for a hot week, bathless. 1909 *Cornhill Mag.* Aug. 230 About the fact—of the bathlessness of the Pickwickians—we may infer that there is no doubt whatever. 1963 *Guardian* 13 Feb. 8/4 Birmingham is..preparing to mount a broad-fronted attack on bathlessness. ..30,000 to 50,000 homes lack bathrooms.

bathmic (bæ·þmik), *a.* [f. Gr. βαθμός: see *BATHMISM + -IC.] Of or pertaining to bathmism; exhibiting or caused by bathmism as a form of evolution.

1872 in COPE *Origin of Fittest* (1887) i. 26 It is a nice point of phylogeny (or the science of genealogy) to ascertain whether adaptive or strictly 'bathmic' (or embryonic grade) characters came first in time in a given group. 1879 *Ibid.* (1887) vi. 229, I compared the transmission of bathmic force to that of the phenomenon of combustion. 1905 G. A. REID *Princ. Heredity* i. 12 The Bathmic [doctrine of racial change], which attributes it to an 'inherent adaptive growth-force'.

bathmism (bæ·þmiz'm). [f. Gr. βαθμός step, threshold (f. root of βαίνειν to walk, step) + -ISM.] A term invented by E. D. Cope to denote a supposed form of chemical force which is active in the processes of growth.

1871 in COPE *Origin of Fittest* (1887) v. 205 The Vital forces are (nerve-force) Neurism, (growth-force) Bathmism, and (thought-force) Phrenism. 1904 G. S. HALL *Adolescence* II. x. 90 It is the age of bathmism, or most rapid variation.

batholith (bæ·þolip). *Geol.* Also **batholite, bathylith.** [a. G. *batholith* (Suess 1892, *Das Antlitz der Erde* I. 219), f. Gr. βάθο-ς depth + -LITH.] A large dome-shaped mass of intrusive igneous rock without a visible foundation. Hence **ba:tholi·thic** *a.*

1903 R. A. DALY in *Amer. Jrnl. Sci.* XV. 270 'Batholiths'..characteristically occur in regions of great structural complexity. 1904 SOLLAS tr. *Suess's Face of Earth* I. iv. 168 The magma simply filled the space..forming a cake of rock or true *batholite.* 1905 CHAMBERLIN & SALISBURY *Geol.* I. 477 Great masses of irregular or undetermined forms..are called *bathyliths.* 1906 *Ibid.* II. 131 The Archean batholiths. 1912 R. A. DALY *Geol. N. Amer. Cordillera* II. 726 The batholithic axes may have indefinite relations to axes of earlier and later crustal deformation. 1937 WOOLDRIDGE & MORGAN *Physical Basis Geogr.* viii. 106 The larger bodies of intrusive rock. These are the so-called bathyliths. 1960 L. D. STAMP *Britain's Struct.* (ed. 5) ix. 80 The great igneous bosses or batholiths of plutonic rock..stand up as hill masses.

Bathonian, *a.* Add: **2.** *Geol.* [ad. F. *bathonien* (J. J. D'O. D'Halloy *Précis El. Géol.* (1843) 471.] Denoting a subdivision of the Jurassic, of which the formations at Bath are typical.

1858 *Q. Jrnl. Geol. Soc.* XIV. 100 The Oolitic rocks..may be thus tabulated..Portland Oolite [is] Portlandian..Cornbrash, Forest Marble, Bradford Clay, Great Oolite, Stonesfield Slate, Fuller's Earth [are] Bathonian. 1903 GEIKIE *Text-bk. Geol.* (ed. 4) 1140 The Great Oolite (Bathonian), between Dorset and Somerset on the west and Oxfordshire on the east. 1914 *Brit. Mus. Return* 201 Ophiuroidea from..the Bathonian of Ardèche. 1960 L. D. STAMP *Britain's Struct.* (ed. 5) xii. 133 The Inferior Oolite or Bajocian beds are represented..by limestones, the succeeding Great Oolite or Bathonian by deltaic deposits of coarse sand.

bathotic, *a.* Delete *nonce-wd.* and add further examples. Also *absol.*

1874 E. S. NADAL in *Scribner's Monthly* VII. 486/1 The cadencies of this bathotic expression rolled among the arches of the cathedral. 1903 A. MCNEILL *Egreg. Eng.* 35 In England the bathotic has always had the majority in its grip. *Ibid.* 36 On that stop—the bathotic stop—the English journalist makes a point of playing. 1952 R. CAMPBELL *Lorca* 49 The gipsy..comes right down with a bathotic lapse at the end.

bathroom. [BATH *sb.*[1]] A room containing a bath and often other toilet facilities. Hence also euphem. for a lavatory.

1780 [see BATH *sb.*[1] VI]. 1888 BARRIE *When a Man's Single* xv. 242 What are politics when the pipes in the bathroom burst? 1934 J. O'HARA *Appointment in Samarra* (1935) iv. 108 Julian wanted to go to the bathroom after the dinner party stood up, and on his way to the men's locker room he had to pass Mrs. Gorman's table. 1946 'J. TEY' *Miss Pym Disposes* ii. 9 It would show on the bathroom scales at the end of the week, but who cared? 1956 A. HUXLEY *Adonis & Alphabet* 260 The Prince of Venosa could never go to the bathroom (*cacare non poterat*) unless he had first been flogged. 1960 *Times* 14 Sept. 12/7 It is necessary [in U.S.A.]..to realize just what a rest room or bathroom is. I did find it odd..when told that a small day school..had a bathroom on every floor. 1962 J. BRAINE *Life at Top* xiii. 144 The bathroom cabinet crammed with eau-de-Cologne and talcum and bath salts. 1967 *House & Garden* June 88/4 Bathroom scales..with magnified reading panel, 56s. 9d. 1968 *Listener* 22 Feb. 228/2 An unfortunate amount of the humour in other light British programmes tends to centre around the bathroom.

Bathurst burr (bæ·þəɹst,bɔ̄ɹ). [f. *Bathurst,* a town in N.S. Wales, Australia + BUR *sb.*] A plant of the genus *Xanthium,* found esp. in Australia.

1855 W. HOWITT *Land, Labour & Gold* I. 261 The Bathurst bur (*Xanthium spinosum*), a plant with long triple spines like the barberry, and burs which are ruinous to the wool of sheep. 1904 'S. RUDD' *Sandy's Selection* 4 The remaining hundred and fifty-six were under scrub, prickly-pear, wallaby-bush and Bathurst burr. 1921 H. GUTHRIE-SMITH *Tutira* xxx. 291 A dense crop of..Bathurst burr. 1926 J. DOONE *Timely Tips* Gloss., *Bathurst Burr,* a spiny, bean-shaped pod which grows on a weed and is notorious for attaching itself to the wool of passing sheep. 1934 *Bulletin* (Sydney) 12 Dec. 25/1 He 'ad whiskers all over him and they was as stiff as the spines on a Bathurst burr. 1967 *Southerly* XXVII. 199 Weeds flourished, wait-a-whiles, bull-burrs, wild blackberries, Bathurst burrs.

bathy- (bæ·þi), comb. form of Gr. βαθύς deep, as in BATHYMETRIC *a.*: **ba:thyæsthe·sia,** muscle sensation produced by muscular movement (Dorland, *Med. Dict.* 1901); **ba·thyal** *a.* [-AL[1]], pertaining to the zone between the continental shelf and the abyssal zone (see quots.); **bathyco·lpian** *a.,* a more normal form of BATHUKOLPIAN *a.*; **ba:thylimne·tic** [Gr. λίμνη pool], pertaining to or inhabiting the depths of freshwater lakes (Webster, Suppl. 1902); **ba·thylite** [-LITE], = *BATHOLITH; **bathy·meter** = BATHOMETER (Cent. Dict. 1889); **ba:thyorogra·phic, -gra·phical** *adjs.,* applied to representations of the contours of the surface of the earth and of the sea-bed; so **ba:thyorogra·phically** *adv.*; **ba:thypela·gic** *a.,* pertaining to or inhabiting the intermediate depths of the sea (opp. *abyssal, pelagic*); **ba:thythe·rmograph,** an instrument which records automatically the temperature of water at various depths; so **ba:thythermogra·phic** *a.,* pertaining or relating to a bathythermograph or its recordings.

1926 W. H. TWENHOFEL *Treat. Sedimentation* vii. 612 The bathyal environment of the sea bottom is that portion between 100 and 1000 fathoms. *Ibid.* 613 The same processes noted in connection with the neritic environment are operative in the waters and sediments of the bathyal, but probably with less intensity. 1944 A. HOLMES *Princ. Physical. Geol.* II. xv. 314 The muds, etc., of the continental slope, and of similar depths around oceanic islands, belong to the *bathyal* zone; while the oozes of the deep ocean floor belong to the *abyssal* zone. 1961 CHALLINOR *Dict. Geol.* 19/2 *Bathyal,* applied to the moderately deep seas beyond the continental shelf and down to an indefinite depth, various authorities taking 6,000 ft., 10,000 ft. &c., but approximately to the limit of the accumulation of the terrigenous deposits, beyond which are the abyssal depths. 1858 O. W. HOLMES *Aut. Breakf.-t.* (1859) iv. 66 The bathycolpian Here..sent down Iris. 1902 *Encycl. Brit.* XXVI. 528/2 The Rocky Mountains..comprise, however, in their central parts..great masses of granite that have welled up as 'bathylites' along the axis of elevation. 1938 *Geogr. Jrnl.* XCI. 475 A new bathy-orographic map of the sea. 1911 *Bartholomew's Physical Atlas* V. pl. 1 Note to bathy-orographical map. *Ibid.* pl. 1 Bathy-orographical configuration. 1932 *Nelson's World Gaz.* 543/1 A *bathyorographical* map shows depths of the oceans and heights of the land. 1921 *Times Lit. Suppl.* 6 Oct. 646/3 This map of the Pacific, on Mollweide's homolographic projection, is bathyorographically coloured in shades of blue and brown. 1909 WEBSTER Bathypelagic. 1912 MURRAY & HJORT *Depths of Ocean* ix. 563 We shall..simply use the term 'bathypelagic' to denote those animals that live deep in the intermediate layers. 1936 J. T. JENKINS *Fishes Brit. Is.* (ed. 2) 249 They [*sc. Argentina sphyræna*] are also bathypelagic, that is, found floating at intermediate depths and not at or near the surface. 1958 *New Biol.* XXV. 125 Other deep sea anglers [*i.e.* fish] are..bathypelagic, living well above the bottom. 1938 A. F. SPILHAUS in *Jrnl. Marine Res.* I. 97 Measurements with the bathythermograph were carried out hourly..and remarkably sudden discontinuities in the thermocline and inversions of the thermocline were revealed. *Ibid.*, The bathythermograph records were traced directly on to this diagram from photographic enlargements of the original traces. 1947 *Sci. News* IV. 84 The bathythermograph..scratches a temperature depth record on a smoked glass slide as it sinks almost vertically on a thin wire running very freely from a small winch. 1952 *Jane's Fighting Ships 1951–52* 109 *New Liskeard*..is now used for Bathythermographic duties a Halifax.

bathybial (bāþi·biāl), *a.* [f. BATHYBIUS + -AL.] Of or pertaining to bathybius or the depths at which it is found; belonging to or living in the deepest parts of the sea. So **bathy·bian, bathy·bic** *adjs.*

1876 *Encycl. Brit.* XXI. 774/2 [Sharks] known to belong to the bathybial fauna. 1881 *Arc. Cruise of Corwin* 14 (Cent. Dict.), The use of the dredge resulted in finding the usual bathybian forms. 1891 G. W. FIELD tr. *Haeckel's Planktonic Studies* in *Rep. U.S. Fish. Comm.* XVII. 582 The deepest part of this zonary fauna forms the bathybic plankton. 1898 *Athenæum* 4 June 729/3 Bathybial and pelagic life.

bathyscaph(e (bæ·þiskæf). [a. Fr. *bathyscaphe* (A. Piccard), f. Gr. βαθύ-ς (*BATHY-) + σκάφος ship.] The name given by the Swiss

scientist Professor Auguste Piccard to his deep-sea diving vessel; also applied gen. to other similar vessels. (See quots.)

1947 *Time* 18 Aug. 38/1 Last week 63-year-old Scientist Piccard told the North American Newspaper Alliance about the 'bathyscaphe'..his submarine balloon which will descend into the sea suspended from a..'gas bag' full of lightweight gasoline. 1953 J. Y. COUSTEAU *Silent World* ix. 87 The elderly scientific extremist [Auguste Piccard] had designed the *Bathyscaphe* ('Depth-craft') a decade before and, after the delay of a world war, it had been built by a brilliant Belgian physicist, Dr. Max Cosyns. *Ibid.* 89 The *Bathyscaphe* was to navigate twenty-five times as deep as conventional submarines. 1958 *New Scientist* 17 Apr. 31/3 The bathyscaph is an untethered and self-contained craft..capable of diving to the deep sea bed and rising again by its own devices. 1959 *Sunday Times* 5 Apr. 7/4 (*caption*) The bathyscaphe Trieste in which..United States scientists are to explore the ocean depths of the Pacific.

bathysphere (bæ·þisfiə̆ɹ). [f. Gr. βαθύ-ς (*BATHY-) + SPHERE.] A spherical diving apparatus for deep-sea observation.

1930 W. BEEBE in *Bull. N.Y. Zool. Soc.* XXXIII. 203/1 When I was writing the name of a deep-sea fish—*Bathytroctes*—the appropriateness of the Greek prefix occurred to me: I coined the word *Bathysphere,* and the name has stuck. 1931 *New Statesman* 21 Mar. 142/1 They crossed the yellow beam of the search-light from the bathysphere. 1947 *Sci. News* V. 67 'Bathyspheres'—closed diving chambers maintained at normal atmospheric pressure—..are provided with a tripod-type base to give stability, and the job of the occupant is to chart the variation in the strength of gravity, when he is moved from one position to another over the sea bed.

batik (bæ·tik, bāti·k). Also **battik.** [Javanese, lit. 'painted'.] **1.** The Javanese art and method (introduced into England by way of Holland) of executing designs on textiles by covering the material with wax in a pattern, dyeing the parts left exposed, and then removing the wax, the process being repeated when more than one dye is used. Also, (a garment made of) a fabric dyed in this way; the kind of pattern, consisting of a medley of colours, characteristic of this art.

1880 *Encycl. Brit.* XIII. 604/2 Another mode is to cover with melted wax or damar the part of the cloth not intended to receive the dye...The 'battiks', as the cloths thus treated are called, are in request by the wealthier classes. 1906 *Daily Chron.* 26 Jan. 4/6 Java is the home of Batik, which is a kind of colour-printing on fabrics. 1922 *Daily Mail* 8 Aug. 3/3 Exact in detail, faultless in design, brilliant in colour, good batik is a joy. 1924 *Times Lit. Suppl.* 20 Nov. 777/1 'Batik' is a Javanese word, and the art was originally practised in Java, whence the Dutch brought it to Europe over three hundred years ago. 1936 J. DOS PASSOS *Big Money* 428 Ladies in flowing batiks. 1958 *Times* 13 Aug. 12/5 Real wax batiks are now being sold to meet the demand for gay summer frocks.

2. *attrib.* passing into *adj.* Executed by the art or method of batik; of, or ornamented with, batik work; hence, loosely, characterized by a fantastic colour-pattern.

1914 *Studio* 14 Mar. 154/2 Examples of embroidery and batik work. *Ibid.* 155/1 Batik Shawl. 1920 *Brit. Mus. Return* 64 A piece of batik cloth, dyed with Chinese designs, from Java. 1920 S. LEWIS *Main St.* xvii. 215 An arty arrangement of batik scarfs and heavy tables. 1923 *Advt.* (Nonesuch Press) in *Times Lit. Suppl.* 27 Sept. 627 [Volumes] bound in quarter-vellum with batik sides. 1927 *Punch* 20 Apr. 424/3 He removed his horn-rims and began polishing them vigorously, producing for the purpose a large silk handkerchief of chaste design. There is nothing batik about Edward. 1967 *Spectator* 14 July 53/2 'Digs' from Hampstead to Earls Court are littered with batik-patterned fabrics.

Hence **ba·t(t)iking** *vbl. sb.,* the production of batik work.

1880 *Encycl. Brit.* XIII. 604/2 A more rapid process of battiking by means of hand stamps has begun to be employed both by native and Chinese workers.

bating, *prep.* (Later example.)

1859 MACAULAY *Jrnl.* 16 Dec. in G. O. Trevelyan *Life & Lett.* (1959) xv. 684 Bating the irregularity of the pulse, I suffered all that I suffered..in 1852.

batman[2]. Add: Now generally, an officer's servant. (Earlier and later examples.)

1755 in S. M. HAMILTON *Lett. to Washington* I. 96 They have taken..another man who was batman to Doct. Craik. 1941 *Aeronautics* Oct. 60/3 R.A.F. officers in the future are to have the services of members of the W.A.A.F. for duties which have been carried out hitherto by batmen. 1955 *Times* 18 Aug. 5/1 Men employed as outside batmen in the married quarters were expected to clean and polish the houses, clean windows, cut lawns, fetch coal, and run errands. 1966 *Times* 9 July 9/7 Command Orders say a batman must now be dignified as an 'orderly'.

batologist (bătǫ·lŏdʒist). [f. Gr. βάτο-ς bramble + *-ologist*: see -OLOGY.] One who makes a botanical study of the genus *Rubus,* esp. the blackberry bramble. So **bato·logy; batolo·gical** *a.*

1889 in *Cent. Dict.* 1897 G. C. DRUCE *Flora Berks.* 187 Boar's Hill, that very interesting locality to the batologist. 1899 HANBURY & MARSHALL *Flora of Kent* 109 Our leading English batologist.

baton, *sb.* Add: **1. c.** *transf.* Of bread: a long loaf; also, a thin short stick.

1857 GEO. ELIOT *Amos Barton* ii. in *Blackw. Mag.* LXXXI. 11/1 Chubby, who is making a round O of her mouth to receive a bit of papa's 'baton'. 1901 *Westm. Gaz.* 22 Apr. 7/2 The crowd consumed..17,000 batons. 1959 M. STEEN *Tower* I. v. 76 A *bâton* of French bread and some cheese.

2. *attrib.* esp. in *baton charge,* a charge made by police constables with drawn truncheons.

1890 *Rev. of Reviews* Nov. 489/1 As police officer, having headed a bâton charge upon them. 1900 *Westm. Gaz.* 12 Nov. 5/1 There a large force of police was gathered, and, a baton charge taking place, several people were badly injured. 1906 *Daily Chron.* 2 Nov. 7/6 Bleeding..from stick and baton wounds.

b. *Athletics.* The short stick or rod passed from one runner to another in a relay race.

1920 *Isis* 13 Oct. 2/2 Ten yards is allotted each side of the starting line in which to pass the baton to the next competitor. 1927 W. DEEPING *Kitty* xxiv, You snatched the baton from the failing hand of the past. 1958 *Times* 25 Aug. 4/1 The British sprint relay teams won their silver medals through immaculate baton changing.

4. (Earlier examples.) Also in the phr. *under the baton of,* under the conductorship of, conducted by.

1785 C. BURNEY in *Mus. Performances in Westm. Abbey 1784* 14 So numerous a band moving in such exact measure, without..a *Coryphæus* to beat the time, either with a roll of paper, or a noisy *baton.* 1829 *Morning Post* 27 May 3/2 Mr. Mendlessohn conducted his Sinfonia with a *baton,* as is customary in Germany, France, etc. 1877 G. B. SHAW in *Hornet* 27 June 330/2 Its [*sc.* the opera's] new aspect under the *bâton* of Signor Vianesi. *Ibid.* 1 Aug. 378/2 He is the only chief under whose baton orchestras display good training. 1962 *Observer* 15 July 22/3 The work was played by another British orchestra under the composer's baton.

batonless (bæ·tənlės), *a.* [f. BATON *sb.* + -LESS.] Without a baton.

1885 [in Dict. s.v. BATON *sb.* 6]. 1906 *Daily Chron.* 7 May 4/5 Batonless conductor. 1909 *Westm. Gaz.* 22 Oct. 5/2 M. Safonoff conducted, and the chorus-singers seemed a little bewildered by his batonless movements.

‖**bâtonné** (batone). *Philately.* [Fr., pa. pple. of *bâtonner* to beat with a stick, f. *bâton* stick.] (See quot. 1897.)

1892 R. B. EARÉE *Album Weeds* (ed. 2) p. ix. 1897 O. FIRTH *Postage Stamps* 15 The thin laid paper commonly called 'foreign note' is known in the stamp world as *batonné;* if it be plain between the lines for writing on, it is known as 'wove *bâtonné*'; if filled with the usual lines, 'laid *bâtonné*'. *Ibid.* 16 The local Mexican stamps of Guadalajara..are to be found printed upon these *bâtonné* and *quadrillé* papers of various colours. 1928 *Stanley Gibbons' Catal. Stamps, Foreign Countries* 12.

‖**bâtonnier** (batonye). [Fr., = 'staff-bearer', f. *bâton* staff.] The title of the doyen of the Bar of Paris and of Quebec.

1907 *Westm. Gaz.* 26 Nov. 10/1 It is from circumstances connected with St. Catherine's festivals that the 'bâtonnier' of the Paris Bar derives his curious designation. 1918 E. CLARKE *Story of my Life* xxviii. 368, I was entertained at dinner by the Quebec Bar; Mr. Donald MacMaster, the batonnier, presided. 1921 *Q. Rev.* Jan. 23 The riding in which he [*sc.* Sir W. Laurier] had practised as a lawyer; in which he achieved the only prize in his profession that ever fell to him—election as batonnier by the Bar of the county.

batrachiate (bătrēⁱ·kiĕⁱt), *a.* and *sb.* [f. BATRACHIA + -ATE[2].] = BATRACHIAN.

1902 *Encycl. Brit.* XXVIII. 139/2 In the batrachiate Amphibia, Héron Royer succeeded in 1883 in rearing..a few hybrids.

bats, batts *sb. pl.,* Sc. form of *bots, botts* (see BOT in Dict. and Suppl.).

batsman. Add: *batsman's wicket,* a cricket pitch more favourable to the batsman than to the bowler.

1876 *Baily's Mag.* June 415 Much..will depend on..whether it is a batsman's or a bowler's wicket on the day of the match. 1927 G. A. TERRILL *Out in Glare* v. 84 It would be a dumbfounding disgrace..if wickets went on falling in this way—on this 'batsman's wicket'.

b. *U.S.* One who wields the bat in baseball.

1856 *Spirit of Times* (N.Y.) 6 Dec. 229/1 He who strikes it [*sc.* a fast ball by Stevens] fairly must be a fine batsman.

2. A man who signals an aircraft with a pair of bats. Cf. *BAT *sb.*[2] 3 f.

1943 *Fleet Air Arm* (Ministry of Information) iv. 30/1 The Deck-Landing Control Officer is known as 'the Batsman'—from the implements with which he signals to the pilots. 1949 *Aeronautics* July 35 When a pilot is high the batsman stands with the bats above his shoulders.

batsmanship (bæ·tsmənʃip). [f. BATSMAN + -SHIP.] The batsman's art; the art of batting at cricket; batting performance.

1907 *Westm. Gaz.* 30 May 10/1 The Rev. F. H. Gillingham ..has done splendid service for Essex by his vigorous batsmanship. 1912 C. B. FRY (*title*) Cricket: batsmanship.

Batta[3] (bæ·tā). *a.* and *sb.* [Native name.] **A.** *adj.* Of or pertaining to a people of the northern part of Sumatra. **B.** *sb.* A member of this people or its language. Cf. *BATAK *a.* and *sb.*[1]

1779 C. MILLER in *Phil. Trans. R. Soc. 1778* LXVIII. 162 In the Batta country, immediately under the line, I have seen it [*i.e.* thermometer] frequently at six A.M. as low as 61°. *Ibid.* 165 The country..is well inhabited by a people called Battas, who differ from all the other inhabitants of Sumatra in language, manners, and customs. 1783 W. MARSDEN *Hist. Sumatra* 165 The principal internal languages of Sumatra, are the Rejang and the Batta. *Ibid.* 296 The *Battas* are in their persons rather below the stature of the *Malay,* and their complexions are fairer. *Ibid.* 300 Human flesh is eaten..on the island of Sumatra, by the *Batta* people. 1860 MAYNE REID *Odd People* 408 Ptolemy..may have referred to Sumatra and its Battas—who *are* cannibals beyond a doubt. 1875 [see *BATAK *a.* and *sb.*[1]]. 1890 FRAZER *Golden Bough* I. i. 28 During a tempest the inhabitants of a Batta village in Sumatra have been seen to rush from their houses armed with sword and lance.

battal·ion. Add. **1. a.** In, or with allusion to, the saying: *God* (or *Providence*) *is on the side of the big battalions.*

[1702 E. BOURSAULT *Lettres Nouvelles* 356 La Reine-Mere dit un jour au Maréchal de la Ferté..les Ennemis sont plus forts que nous..mais nous avons le bon droit pour nous... *Madame,* lui répondit-il, *ne vous y fiez pas: j'ay toûjours vû Dieu du coté des gros Bataillons.*] 1842 [see PROVIDENCE *sb.* 4 a]. 1862 [in Dict.]. 1914 G. B. SHAW in *New Statesman* 14 Nov. Suppl., A Militarist is a person who believes that..Providence is on the side of the big battalions. 1944 BLUNDEN *Shells by Stream* 37 Four hundred thousand men there lay, The Big Battalions blessed the day.

batteau, var. BATEAU. Cf. *BATTOE.

‖**battement** (batəmaṅ). *Dancing.* [Fr., lit. 'beating'.] A rhythmic movement of one leg or foot (see quot. 1830). Hence *grand battement, b. tendu,* etc. Cf. *BEAT *sb.*[1] 1 b.

1830 R. BARTON tr. *Blasis's Code Terpsichore* II. 100 *Battements* consist of the motions of one leg in the air, whilst the other supports the body. They are of three kinds, viz., *grands battements, petits battements,* and *battements* on the instep. 1869 W. S. GILBERT *'Bab' Ballads* 93 The Bishop buckled to his task With *battements,* cuts, and *pas de basque.* 1917 FIRBANK *Caprice* xi. 95 A dance all fearless somersaults and quivering *battements.* 1952 KERSLEY & SINCLAIR *Dict. Ballet Terms* 27 *Battements tendus*..in which one leg slides out along the floor until the foot is fully pointed..and then steadily closes again. 1953 *Ballet Ann.* VII. 66/1 The soloists in the centre and the *corps* on the three sides doing simple *battements tendus.*

batten, *sb.*[1] Add: **1. b.** *spec.* A strip of wood carrying gas or electric lamps; esp. *Theatr.,* one carrying a series of lamps for lighting a stage; also, such a bar used for supporting scenery, curtains, etc. Also *attrib.,* as *batten (lamp-) holder,* a lamp-holder fitted with a support which enables it to be screwed on to a flat surface.

1881 [see *ground-row,* GROUND *sb.* 18 a]. 1899 *Army & Navy Auxiliary C.S.L. Catal.* 384 Batten lampholders. For use on low ceilings, such as lavatories, &c. 1902 *Daily Chron.* 19 July 6/1 That electric light 'battens' should be substituted for gas. 1920 F. HAMILTON *Days bef. Yest.* vi. 160 Thin screens of coloured silk over the gas-battens in the flies. 1921 G. B. SHAW in *Times Lit. Suppl.* 17 Mar. 178/2 Take your ambers out of your number one batten. 1926 J. A. FLEMING *Electr. Educator* I. 138/1 The Batten Holder ..is just a simpler..form of backplate holder. 1933 P. GODFREY *Back-Stage* i. 16 The sky-cloth..leaps to dazzling life as the 'floods' and 'battens' throw their massed beams upon its surface. 1967 *Times Rev. Industry* Feb. 41/3 In smaller factories management will often 'put in batten fittings without reflectors'.

batter, *sb.*[1] **1. c.** *batter-cake* (earlier examples).

1830 F. TROLLOPE in *Dom. Manners Amer.* (1960) 427 Waffles. Batter cakes. 1833 in *Maryland Hist. Mag.* (1918) XIII. 319 Hot muffins and corn batter cakes.

batter, *sb.*[3] (Earlier example.)

1773 J. DUNCOMBE *Surry Triumphant* st. xxxiii, in R. Freeman *Kentish Poets* (1821) II. 368 At last, Sir Horace took the field, A batter of great might.

ba·tter, *sb.*[4] *slang.* [Of obscure origin: cf. *BAT *sb.*[4]] A spree, debauch, esp. in phr. *on the batter.* (See also quot. 1890.)

1839 A. RODGER in *Whistle-Binkie* Ser. II. 17 My hat was smash'd..Ae night when on the batter. 1856 *Knickerbocker* XLVIII. 502 Ellis had..just returned from a prolonged batter in Paris. 1865 *N. & Q.* Ser. III. VIII. 369/2 It was among working-men that I first heard 'on the batter' employed as an equivalent for going 'on the spree'. 1890 FARMER *Slang* I. 143/2 *To go on the batter, i.e.,* to walk the streets for purposes of prostitution. 1899 R. WHITEING *No. 5 John St.* xxi. 214 D'ye call that goin' on the batter?..I call it goin' out with the governiss. 1957 J. OSBORNE *Entertainer* v. 35 Have you been on the batter, you old gubbins! 1966 A. PRIOR *Operators* iii. 25 The mothers and sisters on the batter—any man on the streets.

battered, *ppl. a.* Add: **b. battered baby,** an infant exhibiting symptoms (the *battered baby syndrome*) resulting from repeated injuries inflicted upon it over a period.

[1962 C. H. KEMPE et al. in *Jrnl. Amer. Med. Assoc.* 7 July 17/1 The battered-child syndrome is a term used by us to characterize a clinical condition in young children who have received serious physical abuse, generally from a parent or foster parent.] 1963 *Brit. Med. Jrnl.* 21 Dec.

1558 (*heading*) Multiple epiphysial injuries in babies ('battered baby' syndrome). *Ibid.* 1560/1 The *x*-ray changes in the 'battered baby' are..like those often described in infantile scurvy.

‖**batterie** (batəri). [Fr., see BATTERY.] **1.** *Dancing.* A movement in which the dancer's feet or calves are beaten together during a leap.

1712 J. WEAVER *Hist. Dancing* vii. 163 Notwithstanding there are some Steps peculiarly adapted to this Sort of Dancing, viz. Capers, and Cross-Capers of all kinds; Pirouttes [*sic*], Batteries, and indeed almost all Steps from the Ground. 1949 *Ballet Ann.* III. 54 *Temps de batterie,* or beaten steps. *Ibid.* 116/2 Articulate and well-arched feet that helped to give a brilliant display of *petite batterie.* 1950 *Ibid.* IV. 131/1 One of the best classical dancers, with a *batterie* unequalled for brilliance. 1952 KERSLEY & SINCLAIR *Dict. Ballet Terms* 17 *Batterie,*..this word added to the name of a step does not imply any alteration in the basic movement, but that the dancer beats his calves together..and possibly also changes the position of his legs while doing so.

2. *batterie de cuisine* (see BATTERY 12).

1773 H. WALPOLE *Let.* 26 Oct. (1857) VI. 1 Unless he carries his *batterie de cuisine,* cook and camp equipage, I doubt he must eat the game raw. 1819 [see BATTERY 12]. 1906 I. BEETON *Househ. Managem.* lxii. 1654 *Batterie de cuisine,* complete set of cooking utensils and apparatus. 1934 'A. BRIDGE' *Ginger Griffin* vii. 84 Some different and highly burnished article of her *batterie de cuisine.*

3. = *BATTERY 16.

1934 S. R. NELSON *All about Jazz* ii. 50 The drummer was banging away at his *batterie!*

Battersea (bæ·təɹsi). The name of a district of London, used to designate articles ornamented with a decorative enamel produced at York House, Battersea, in the 18th century.

1869 LADY C. SCHREIBER *Jrnl.* (1911) I. 2 One or two pieces of Battersea enamel. *Ibid.* 16 A pair of Battersea enamel candlesticks. *Ibid.* 27 Battersea needlecase with birds in pink. 1924 A. E. W. MASON *House of Arrow* xiii. 153 Pen-tray, candlestick, sand-castor and all were of the pink Battersea enamel. 1949 A. CHRISTIE *Crooked House* x. 74 A small Battersea enamel box.

battery. Add: **4. c.** In baseball, applied to the pitcher and catcher (orig. used of the pitcher alone). Also *attrib. U.S.*

1867 *Ball Players' Chron.* 6 June 2/2 He soon resumed his position, once more facing the battery of Lovett. 1886 CHADWICK *Art of Pitching* 8 The 'battery' of a club's team, that is the pitcher and catcher. 1897 *Daily News* 29 July 9/2 So good was the fielding and battery work..that no scoring took place. 1967 *Boston Sunday Herald* 14 May II. 5/1 Pitcher Ed McGrath went the distance and battery mate, Tony Carderelli, drove in four runs in the Boston State victory.

5. c. An oblong box or boat submerged to the brim, used in wild-fowl shooting; = SINK *sb.*[1] 12 a. Also *attrib.,* as *battery-gunner, -shooting. U.S.*

a 1841 W. HAWES *Sporting Scenes* (1842) I. 198 A machine, or battery, is a wooden box of the necessary dimensions to let a man lie down upon his back, just tightly fitting enough to let him rise again. 1859 [see SINK *sb.*[1] 12 a]. 1866 *Game Laws Va.* in *Fur, Fin & Feather* (1872) 144 Any person shooting or using a skiff, box or battery while hunting wild fowl. 1874 J. W. LONG *Wild-Fowl Shooting* 71, I shall describe that in reference to battery-shooting. 1875 *Fur, Fin & Feather* 120 But this is nothing to the numbers slain by the battery; and we have known one battery to kill over three hundred fowl in a tide. *Ibid.* 122 The battery gunner ..has a great advantage over the fowler who shoots from the shore.

7. (Earlier U.S. examples.)

1853 *Harper's Mag.* VI. 578/2 Openings in this opposite each cover or battery of stamps. 1872 'MARK TWAIN' *Roughing It* in *Writings* (1900) VII. xxxvi. 278 These [six rods] rose and fell, one after the other,..in an iron box called a 'battery'...One of us stood by the battery all day long.

10. b. *attrib.* and *Comb.,* as *battery-receiver, -set, -unit; battery-operated, -powered* adjs.

1930 *Wireless World* 11 June 601/1 Few battery-operated portables consume more than the economical limit of 10 mA. 1957 *BBC Handbk.* 52 Small self-contained battery-powered tape recorders. 1964 T. L. KINSEY *Audio-Typing & Electric Typewriters* iii. 16 Battery-operated machines are also available. 1928 *Wireless World* 5 Dec. 754/3 A battery-receiver unit...The H.T. battery can therefore be adapted to serve as the actual receiver, by incorporating in it the necessary coupling unit. 1930 *Ibid.* 16 July 71/3 (*heading*) An Ambitious Battery Set. 1933 *Boys' Mag.* XLVII. 135/2 The Class 'B' valve..is of great importance to users of battery sets. 1939 *War Illustr.* 9 Dec. p. ii/3 This trade, which consists in selling torches that are loaded with worn-out battery-units from wireless sets. 1958 M. L. HALL et al. *Newnes Compl. Amat. Photogr.* x. 112 A flashgun consists of a socket to hold the flashbulb, a case to contain the battery unit and a reflector.

12. Also *batterie de cuisine* (see *BATTERIE 2).

12*. Used *gen.* for a collection of similar pieces of apparatus grouped together as a set (see quots.).

1885 [see TACHE *sb.*[2] 1]. 1911 D. S. HULFISH *Cycl. Motion-Pict. Work* II. 137 The remaining proportion of light may be supplied by lighting a partial battery of lamps. 1920 *Sci. Amer.* 2 Oct. 346 (*caption*) A battery of projectors and a stereopticon machine in the operator's booth of a leading motion-picture theatre. 1931 *Economist* 25 Apr. 888/2 There are batteries of forty looms each. 1958 *Engineering* 18 Apr. 509/1 When the Coal Board took over there were 45 ovens ..in two batteries of 15 and 30.

b. A series of psychological or clinical tests.

1921 *Mem. Nat. Acad. Sci.* XV. II. ii. 316 If a test correlated almost perfectly with some other test in the battery, it could be omitted without loss. 1928 K. J. HOLZINGER *Statist. Methods in Educ.* xv. 310 The writers..eliminate certain tests..and thus obtain a shorter and possibly as good a test with unweighted items as with the whole battery. 1930 —— *Statist. Resumé Spearman Two-Factor Theory* 42 Two arithmetic tests or two opposite tests should not be used in a given battery. 1940 *Brit. Jrnl. Psychol.* Apr. 358 This is the estimated reliability of the battery formed by the straight sum of its constituent tests. 1961 *Lancet* 26 Aug. 487/1, I also heartily endorse his call for a refinement in the constituent subtests of psychological batteries.

c. A series of hutches, cages, or nesting-boxes in which laying hens are confined for intensive laying or poultry reared and fattened. Later extended to denote accommodation for fattening cattle; freq. *attrib.,* as *battery hen, system.*

1931 G. W. WRENTMORE *Battery System of Poultry Keeping* 27 The system is best carried out with birds that have been in Batteries from the start. *Ibid.*, The battery is commenced at a temperature of 90–94 degrees and gradually lowered to a temperature of about 65 degrees as the size of the chicks increases. 1938 *Reader's Digest* Mar. 93/1 The smallest complete battery..fills a 14-by-20-foot space. 1940 N. MITFORD *Pigeon Pie* ix. 138 Those raising hens on the battery system. 1953 A. WATKYN *For Better, For Worse* II. i, It ain't right to ask 'uman beings to live like Battery Hens. 1958 *Observer* 19 Oct. 17/4 They [*sc.* geese] defy horrible human plans to mass-produce in battery and deep scratch. 1960 *Farmer & Stockbreeder* 16 Feb. 78/1 We are going to hear more of battery beef. 1960 *News Chron.* 8 June 4/4 The broiler battery..may be good business, but is..revolting. 1961 *Guardian* 17 May 3/5 It was doubtful if..battery eggs were less nutritious than other eggs.

16. [F. *batterie*; cf. *BATTERIE 3.] The percussion section of an orchestra.

1926 WHITEMAN & MCBRIDE *Jazz* ix. 197 The battery of an orchestra includes so many instruments.... Perhaps the most important instruments of the battery are the timpani or kettle drums. 1955 R. BLESH *Shining Trumpets* (ed. 3) ii. 30 A part of the rhythm is kept separate in drum battery. 1960 *Times* 23 June 17/3 A percussion player..proceeded to exploit a battery including cowbells and finger-cymbals.

batting, *vbl. sb.* Add: **1. b.** Also in *Baseball.* Also *attrib.,* as *batting average, glove.*

1856 *Househ. Words* 2 Feb. 59/2 Some tubular batting-gloves. 1867 *Ball Players' Chron.* 12 Dec. 5/3 The best players are those making the best batting and fielding average. 1870 [see *AVERAGE *sb.*[2] 6 b]. 1910 *Westm. Gaz.* 14 Apr. 7/4 Cricketers will have to pay a trifle more for batting-gloves and bat handles.

battle, *sb.* Add: **12.** *line-of-battle ship* (earlier and later examples); also irreg. *line of battle-ship.*

1705 *Admiralty Sec. In-Lett.* 5249 (P.R.O.), The capital ships and line-of-battle ships are often laid up in the winter. 1800 NELSON in A. Duncan *Nelson* (1806) 121, I saw the Alexander in chase of a line of battle ship. 1894 *Times* (weekly ed.) 19 Jan. 50/1 A heavily armoured line-of-battleship. 1899 R. ROUTLEDGE *Discov. & Invent. 19th Cent.* (ed. 13) 167 Before the close of 1894, the British navy possessed no fewer than eight of the largest armoured line of battle-ships.

13. a. *battle-grimed, -hardened, -tried, -weary* adjs.; **b.** *battle-fleet, -front, -hymn* (examples), *-practice* (also *attrib.*), *-smoke* (example), *-training, -zone.*

1862 [J. W. HOWE] in *Atlantic Monthly* Feb. 145 *Battle Hymn of the Republic.* Mine eyes have seen the glory of the coming of the Lord: [etc.]. 1877 TENNYSON *Harold* III. i. 87 A ghostly horn Blowing continually, and faint battle-hymns. 1897 *Trans. Inst. Naval Archit.* XXXVIII. 50 These two ships form still part of the German battle fleet. 1898 KIPLING *Fleet in Being* ii. 17 That a cruiser at 7.30 that morning had reported to the Battle Fleet..'Enemy to the Westward'. 1899 R. MEINERTZHAGEN *Army Diary* 1 July (1960) 15 The hopelessly out-of-date battle-training of my battalion. 1900 *Blackw. Mag.* Dec. 931/2 No amount of battle-smoke can hide the red stain of pure unadulterated murder. 1902 *Westm. Gaz.* 6 Jan. 4/3 So long as our battle-fleet is able to keep the seas. 1905 L. BINYON *Penthesilea* 38 So now the battle-weary Greeks prepared their meal. 1905 *Westm. Gaz.* 20 June 2/2 Battle-practice has for 1905 been ordered on more regular and practical lines than hitherto. 1907 *Daily Chron.* 14 Oct. 4/4 Winchelsea sees battle-grimed French and Spaniards scale her walls. 1909 *Westm. Gaz.* 18 Jan. 3/1 These men fought in the ranks of battle-tried battalions. *Ibid.* 9 Feb. 1/2 Mr. Solano, the inventor of the new system of battle-practice targets, which has received the approval of the War Office. 1914 *Scotsman* 6 Oct. 4/1 The latest news from the two great battle-fronts affords no ground for dissatisfaction. 1931 *Times Lit. Suppl.* 21 May 399/3 Soult..decided to make the main defence in the rear line, turning this into the 'battle zone', as it was called in France in 1918. 1937 KOESTLER *Span. Testament* iv. 82 Every town along the enemy's line of retreat and all the areas behind the enemy lines are to be considered as battle zones. 1940 W. TEMPLE *Thoughts in War-time* vii. 43 If we pray as Christ taught us to pray, we pray in perfect unity on both sides of the battle-front. 1944 *Ann. Reg. 1943* 19 In training the most notable innovation had been the introduction of realistic battle training. 1945 *Finito! Po Valley Campaign* 31 Exhausted, battle-weary crews. 1949 KOESTLER *Promise & Fulf.* III. i. 299 The battle-hardened workers of the Soviet Union. 1949 E. POUND *Pisan Cantos* lxxx. 88 Following the Battle Hymn of the Republic.

14. battle bowler *slang* = *TIN HAT 1; **battle-lantern,** a lantern used on a ship; formerly one placed at each gun to light up the deck during a night engagement, a fighting-lantern; **battle-school,** a military establishment providing training under conditions resembling those of battle; **battle-twig** *dial.* [corruption of *beetle-wig;* cf. BEETLE *sb.*[2], EARWIG *sb.*], an earwig; see also quot. 1942; **battle-wag(g)on** *slang* (orig. *U.S.*), (*a*) a battleship; (*b*) (see quot. 1926); (*c*) an armed or armoured vehicle.

1925 FRASER & GIBBONS *Soldier & Sailor Words* 19 Battle-bowler. 1940 N. MITFORD *Pigeon Pie* iv. 75 She lunched alone at the Ritz yesterday in a black wig, a battle bowler and her silver foxes. 1830 J. F. COOPER *Water Witch* III. vii. 206 Lifting a lighted battle-lantern to his face, he saw that he slept. 1938 MASEFIELD *Dead Ned* 235 He had a battle-lantern with him, a ship's lamp with a strong reflector. 1942 *Hutchinson's Pict. Hist. War* 18 Mar.–9 June 219 Attack is the spirit of the battle school. 1787 GROSE *Prov. Gloss., Battle-twig,* an earwig. Derb. 1885 TENNYSON *Tiresias,* etc. 111 'Twur as bad as a battle-twig 'ere i' my oän blue chaumber to me. 1929 D. H. LAWRENCE *Pansies* 144 There isn't a damn thing in 'em..they haven't the spunk of a battle-twig. 1942 *Archit. Rev.* XCII. 154/2 He has needles of several sizes and of two different types: the small 'battletwig' (i.e. earwig) needle for tiny work, the regular netting needle for string and rope nets. 1926 *Amer. Speech* I. 650/2 *Battle wagon,* an iron coal car. 1927 *Ibid.* III. 452 *Battle wagon,* warship. 1938 *Newsweek* 14 Nov. 11/2 The Navy had sent out bids..for three new 35,000-ton battlewagons. 1945 *Penguin New Writing* XXIII. 10 But the battle-wagon—her great guns swing up in a silent arc. 1949 F. MACLEAN *Eastern Appr.* II. ii. 200 The 'battle-wagon'..was a new, cut-down Ford station waggon...It was fitted with mountings for two machine guns in front and two behind.

battle-axe. Add: (Spelling with *-ax* now chiefly U.S.)

3. *Archæol.* A type of prehistoric stone weapon; hence applied *attrib.* to a neolithic culture characterized by this weapon.

1859 *Proc. Soc. Antiq. Scotl. 1856–7* II. 306 Stone Hammer or Battle Axe, formed of fine-grained mica schist. 1880 DAWKINS *Early Man* x. 390 A bronze battle-axe fifteen inches in length and seven pounds in weight. 1925 CHILDE *Dawn Europ. Civiliz.* xii. 198 The most conspicuous weapons, the battle-axes and hammer-axes..go back to the copper age and prove the dominance of the battle-axe element in the population. 1928 C. DAWSON *Age of Gods* xii. 268 The Battle-axe Culture. *Ibid.* 270 The Battle-axe People. 1931 *Jrnl. R. Anthrop. Inst.* LXI. 345 The process of assimilation was accelerated in Sweden and replaced in the East Baltic regions by intrusions of the battle-axe folk...Battle-axes and even battle-axe graves in the dwelling-places are the proofs of their advent. 1950 H. L. LORIMER *Homer & Monum.* i. 6 The Battle-axe culture commonly regarded as Indo-European.

4. *fig.* A formidable or domineering woman. orig. *U.S. slang,* now *colloq.*

1896 ADE *Artie* ix. 81 Say, there was a battle-ax if ever you see one. She had a face on her that'd fade flowers. 1938 'E. QUEEN' *Four of Hearts* I. ii. 25 These old female battle-axes don't feaze me. 1957 C. BROOKE-ROSE *Languages of Love* 8 Do I look like a female novelist? I thought they were all battle-axes. 1959 *Punch* 21 Jan. 135/3 Though slim as an arrow A girl can wax In the course of time To a battle-axe.

ba·ttle-crui·ser. A heavily armoured cruiser or cruiser-battleship (see *BATTLESHIP b).

1911 *Times* 22 Nov. 6/3 In order to distinguish the armoured cruisers of earlier dates from those of the 'Invincible' and later types the latter vessels are to be..classified as battle cruisers. 1914 *Daily Express* 9 Sept. 1/2 The Nuremberg, a German cruiser, is said to be flying from a British battle cruiser. 1946 *Jane's Fighting Ships 1944–45* 432 The repair and refit of this battle-cruiser.

battle-dress. [f. BATTLE *sb.* + DRESS *sb.* 2.] A soldier's normal 'undress' khaki uniform, consisting essentially of a tunic and trousers. Also *attrib.* and *transf.*

1938 *Times* 1 June 7/3 The experimental 'battle-dress' is being tried on a large scale this year. 1940 *Hutchinson's Pict. Hist. War* 14 Feb.–9 Apr. 93 The men have discarded their kilts for the more suitable battle dress. 1963 *Guardian* 29 Jan. 7/7 A plain, natural shantung dress has a matching battledress top.

ba·ttleplane. [f. BATTLE *sb.* + *PLANE *sb.*[3], after *battleship.*] An aeroplane designed for use in warfare, carrying a gun or guns, bombs, etc.

[1915 *Daily Express* 15 Oct. 4/3 The French Government has now no objection to the world at large knowing that battle-aeroplanes of large size are being built in France.] *Ibid.* 16 Dec. 4/5 Genuine battleplanes were first heard of officially in the German communiqués. 1917 'CONTACT' *Airman's Outings* 180 The Fokkers were satisfactorily dealt with by the de Haviland and the F. E. 'battleplane', as the newspapers of the period delighted to call it. 1917 *Observer* 11 Mar. 10/3 The effort of each side is bent on producing not giant battleplanes but faster and better climbing small craft.

battler[1]. Add: **2.** *spec.* A swagman (cf. SWAG *sb.* 12 (b)). *Austral.*

1900 H. LAWSON *Over Sliprails* 6 We're only 'battlers', and me and my mate, pickin' up crumbs by the wayside. 1928 *Bulletin* (Sydney) 15 Feb. 21/1 Last time the poor old battler jus' struggled to our place. 1961 'J. DANVERS' *Living came First* vii. 116 'Nice bloke, Marty,' added the swagman.., 'always got a crust for a battler, an' a kind word.'

b. Used in Australia and New Zealand in various other senses and shades of meaning (see quots.), esp. a person struggling against odds. Cf. quot. 1862 under sense 1.

1898 *Bulletin* (Sydney) 17 Dec. (*Red Page*), A *bludger* is about the lowest grade of human thing, and is a brothel bully... A *battler* is the feminine. 1941 K. TENNANT *Battlers* xvi. 165 They were a new sort of people, the travellers; and he belonged to them...He was a 'battler'...They would still have a job on their hands clearing out the battlers; men and women who could face a desert and live off the country, travelling in small mobs. 1943 J. A. W. BENNETT in *Amer. Speech* XVIII. 88 [In N.Z.] a *toiler* or *battler* is a hard, conscientious worker—both are used with a shade of condescension. 1943 BAKER *Dict. Austral. Slang* (ed. 3) 9 *Battler,* a small-time hawker; a hard-up horse trainer struggling along in the game; a broken-down punter who still continues betting;..anyone who struggles for existence. 1962 *Listener* 18 Jan. 128/1 Ditchburn is another Australian, a 'battler' as we say, a man who knows what he wants. 1964 D. HORNE *Lucky Country: Australia* 25 Australians love a 'battler', an underdog who is fighting the top dog, although their veneration for him is likely to pass if he comes out from under.

battleship (bæ·t'lʃip). orig. *U.S.* [Shortening of *line-of-battle ship*: see BATTLE *sb.* 12.] A line-of-battle ship; a warship of the largest and most heavily armoured class.

1794 D. HUMPHREYS *Poem on Industry* 20 The dock equips, With batt'ries black and strong, the battleships. 1834 GLASCOCK *Naval Sk. Bk.* I. 185 A bluff weather-beaten captain of a battle ship. *Ibid.* 235 A battle-ship's bowsprit. 1845 R. FORD *Handbk. Trav. Spain* I. III. 363 Like the spars of a storm-wrecked battle-ship. 1884 *Marine Engineer* 1 Apr. 4/2 The very heavily-armed battle-ship. 1959 *Chambers's Encycl.* II. 162/1 The battleship has been the basic unit in all navies.

b. *cruiser-battleship* or *battleship cruiser*: a battleship of the type designed for speed, less heavily armoured than a ship of the line.

1909 *Whitaker's Alm.* 681/2 Modern vessels of this class are no longer, save in official phraseology, 'armoured cruisers', but 'cruiser-battleships', or 'cruiser-Dreadnoughts'. 1909 *Westm. Gaz.* 18 Mar. 7/2 Armoured cruisers, or 'battleship cruisers', as they have been popularly termed.

c. *attrib.* and *Comb.* **battleship grey** (or **gray**), a slightly bluish grey colour often used in painting battleships.

1834 GLASCOCK *Naval Sk. Bk.* I. 154 To ascertain 'the difference 'twixt the rigging and palaver of a methody parson, and the togs and talk of a reg'lar-built battle-ship preacher'. 1901 *Westm. Gaz.* 8 Aug. 6/1 To choose his own time and place for the battleship action. 1904 *Ibid.* 11 Feb. 8/2 The battleship strength of the Russians at Port Arthur. 1908 *Ibid.* 11 Feb. 3/3 When the Dreadnought appeared battleship building in foreign yards paused. 1908 *Daily Chron.* 21 Aug. 1/7 The American battleship fleet. 1916 *Daily Colonist* (Victoria, B.C.) 1 July 12/3 The colors include battleship grey. 1938 L. MACNEICE *Earth Compels* 54 Nights with stars or closely interleaved with battleship grey or plum.

battlesome, *a.*[1] (Later U.S. example.)

1888 *Harper's Mag.* Apr. 690 Of battlesome wit..the author has so much of his own that he has given some to all of his characters.

battle-worthy (bæ·t'lwɔ̄ːɹði), *a.* [f. BATTLE *sb.* + -WORTHY; after *seaworthy.*] Fit for use in battle. Hence **ba·ttle-wo:rthiness.**

1889 MUNDELLA in *Hansard* CCCXXXIII. 119 Really good,..battle-worthy swords. 1892 *Pall Mall G.* 13 Sept. 2/2 If the ship is not battle-worthy, the best thing is to sell her or break her up at once. 1904 *Times* (weekly ed.) 1 Jan. 7/1 Battle-worthiness—a..compendious term, including all the essential qualities required of a vessel intended to fight in line. 1940 *Illustr. London News* CXCVI. 840/1 What the Germans did was to concentrate all their tanks which were still battle-worthy on one or two roads. 1943 *Jane's All World's Aircraft* p. iii/2 Experience played no small part in making the..aeroplanes and aero-engines battle-worthy.

battling, *vbl. sb.*[4] Add: *battling-bench, -board, -stick.*

1846 J. J. HOOPER *Taking Census* in *Adv. Simon Suggs* etc. (1851) II. 183 John Green's sister..goes to her battlin bench. 1848 —— *Ride with Old Kit Kuncker* in *Widow Rugby's Husb.* etc. (1851) 96 What a devil of a paddlin' the old woman gin him with the battlin'-stick. 1887 *Harper's Mag.* July 272/1 The splay legged battling-boards fastened themselves into the earth under the blows of the bats.

ba·ttoe, freq. U.S. var. of BATEAU, *BATTEAU. *Obs.*

1711 *Boston News-Let.* 23–30 July 2/1 All our Battoes are finished. 1770 WASHINGTON *Diaries* (1925) I. 381 Got the Battoe, and the two Boats round to the Mill with Stone. 1801 *Austin Papers* (1924) I. 72 Passed a Battoe in from New Orleans.

Hence **battoe man, battoeing** *vbl. sb.*

1756 in *Doc. Hist. N.Y. State* (1849) I. 477 Upwards of 500 Battoe Men were sent different Ways into the Woods. 1760 in *Essex Inst. Hist. Coll.* XX. 199 This Day ye Battoue men marched off. *Ibid.*, To Day there was a Draught out of our Company for battowing from fort Miller to fort Edward.

batty (bæ·ti), *a. colloq.* or *slang.* [f. BAT *sb.*[1] + -Y[1].] 'Balmy', 'dotty'. (Cf. *BAT *sb.*[1] 1 b.)

1903 A. L. KLEBERG *Slang Fables from Afar* iii. 23 She..acted so queer..that he decided she was Batty. 1917 J. FARNOL *Definite Object* iii. 32, I mean, is 'e batty? 1926 *British Weekly* 25 Nov. 242/3 He's a bit batty every now and anon. 1934 R. MACAULAY *Going Abroad* vii. 57 He'd go batty with rage.

ba·twoman [after *BATMAN[2].] A member of one of the women's auxiliary services performing the duties of a batman.

1941 *Aeronautics* Oct. 60/3 R.A.F. officers in future are to have the services of members of the W.A.A.F. for duties which have been carried out hitherto by batmen. These 'batwomen' are being trained in Service methods. 1944 'N. SHUTE' *Pastoral* i. 4 The W.A.A.F. batwoman found him sitting so when she brought in his tea.

batyl alcohol (bæ·til æ·lkŏhǫl). [ad. G. *batylalkohol* (M. Tsujimoto and Y. Toyama 1922, in *Chem. Umschau* XXIX. 36/1), f. mod.L. *Batis* genus of fishes (Gr. βατ-ίς flat fish) + -YL.] A colourless crystalline alcohol, $C_{21}H_{44}O_3$, found in many shark and ray liver oils.

1922 *Chem. Abstr.* 1513 The formula $C_{20}H_{42}O_3$ was selected to represent this satd. alc. and the name *batyl alcohol* given to it. 1955 *Sci. News Let.* 19 Feb. 121/3 Batyl alcohol promises to become a prized material for possible protection against atomic radiation.

baud (bōᵘd). *Telegr.* [From the name of J. M. E. *Baudot* (1845–1903), a French engineer, who invented a telegraph printing system (*Baudot code, system*).] The unit of speed of telegraphic code transmission, the number of bauds being the reciprocal of the duration in seconds of the unit interval.

[1929 *Documents de la deuxième réunion du comité consultatif international des communications télégraphiques à Berlin 1929* 115 L'avis que la vitesse de transmission soit exprimée par l'inverse de la valeur de l'intervalle élémentaire mesurée en secondes; que la vitesse de transmission d'un intervalle par seconde soit appelée baud, pour honorer la mémoire du grand télégraphiste Emile Baudot.] 1932 *P.O. Electr. Engin. Jrnl.* XXV. 262/2 It has been proposed to use the duration of the 'unit' element to define the speed of transmission of successive signals. According to this suggestion the speed is defined as N 'Bauds' where the duration of the unit element is equal to 1/N seconds. 1934 A. T. STARR *Electr. Circuits & Wave Filters* xii. 345 The most usual telegraph system of to-day is the Baudot code... The speed of the system is said to be (1/τ) bauds: thus if the duration of an element is 1/60 sec., the speed is 60 bauds. 1954 *Electronic Engin.* XXVI. 228/1 For speeds up to 120 bauds twelve channels..are normally grouped together. 1961 *Flight* LXXX. 428/1 Of particular interest..is the use of..land lines to achieve high information rates of up to 3,600 bauds.

Baudelairean, -ian (bōᵘdəlēᵊ·riăn), *a.* Also occas. **-ien.** [f. the name of Charles *Baudelaire* (1821–67), French poet and critic + -IAN.] Of, or pertaining to, or characteristic of Charles Baudelaire or his poetry. Hence as *sb.,* an admirer or imitator of Baudelaire's work; also **Baudelai·rianism.**

1899 'G. F. MONKSHOOD' *R. Kipling* 181 'Love-o'-Woman', a Baudelairean study of physical passion, disease and Death. 1909 *Daily Chron.* 30 Aug. 3/3 The little Baudelairiens who have an itch to seem Satanic. 1921 *Glasgow Herald* 9 Apr. 6 The moral aspect of Baudelairianism. 1962 A. HUXLEY *Island* iii. 20 In a Baudelairian sort of way it's rather beautiful. 1967 *Listener* 27 July 109/2 André Gide, who was a great Baudelairean.

bauera (bɑu·ərā). [mod.L., f. the name of Franz and Ferdinand *Bauer,* botanical draughtsmen.] The Tasmanian name for a shrub of the species *Bauera rubioides,* one of the three Australasian species of the family Saxifragaceæ. Also *attrib.*

1835 *Ross's Hobart Town Alm.* 70 *Bauera rubiæfolia.* Madder leaved Bauera. 1888 R. M. JOHNSTON *Geol. Tasmania* Introd. 6 The Bauera scrub..is a tiny, beautiful shrub. 1891 [see ROSE *sb.* 2 b]. 1927 *Chambers's Jrnl.* May 345/1 An impenetrable thicket of bauera.

Bauhaus (bɑu·hɑus). [Ger., lit. 'architecture house', f. *bau* building (*bauen* to build) + *haus* house.] The name of a school of design founded in Weimar, Germany, in 1919 by Walter Gropius (1883–1969); used for the principles or traditions characteristic of the Bauhaus.

1923 H. G. SCHEFFAUER in *Freeman* 5 Dec. 304 The Staatliche Bauhaus of Weimar represents one of the most interesting and significant enterprises in the vivifying of modern arts and crafts. 1932 D. M. HOFFMANN tr. *Moholy-Nagy's New Vision* i. 18 The Bauhaus became in Germany the focussing point of the new creative forces accepting the challenge of the time and technical progress. *Ibid.*, Systematic work toward 'standardized production' did not, however, form the first step in Bauhaus instruction. 1935 P. M. SHAND tr. *Gropius' New Architecture & the Bauhaus* 37 The *Bauhaus* represented a school of thought which believes that the difference between industry and handicraft is due..to subdivision of labour in the one and undivided control by a single workman in the other. 1958 *Times* 11 Oct. 7/6 The contemporary Russian style [of architecture], which is a blend of Bauhaus and post-Waterloo station grandiose.

bauhinia (bōᵘhi·niă). [mod.L. (Linnæus 1737), named after Jean (1541–1613) and Gaspard (1560–1624) *Bauhin.*] A plant of the genus *Bauhinia* (family Leguminosæ) of which there are many tropical species. Also *attrib.*

1790 J. BRUCE *Trav. Source of Nile* V. App. 59 Mr. Jussieu says this bauhinia is by Mr. Bruce taken for an acacia. 1833 *Penny Cycl.* I. 447/1 (*America*) Bauhinias..cling round the trees like enormous cables. 1849 C. STURT *Narr. Exped. C. Australia* I. 359 The Bauhinia here grew to the height of 16 to 20 feet. 1887 MOLONEY *Forestry W. Africa* 187 Plaintain and Bauhinia fibres. 1922 *Chambers's Jrnl.* Dec. 859/2 Trees..with a good deal of Bauhinia creeper all over. 1934 *Bulletin* (Sydney) 27 June 24/1 Scattered cabbage-gum, kurrajong, bauhinia and other fodder trees. 1967 'A. CORDELL' *Bright Cantonese* xiii. 150 The bauhinia trees were etched in darkness against a round China moon.

Baumé (bo·me). Also erron. **Beaumé.** The name of a French chemist, Antoine *Baumé* (1728–1804), the inventor, in 1768, of a hydrometer, of which the scale is uniformly graduated. Used in the possessive, also *attrib.* and *ellipt.,* to denote the scale introduced by him.

1844 G. FOWNES *Man. Chem.* 545 Hydrometer Tables. Comparison of the degrees of Baumé's Hydrometer, with the real specific gravities. a 1877 KNIGHT *Dict. Mech.* II 1153/2 The following table shows the specific gravity corresponding to different degrees of Beaumé. 1877 *Proc. Amer. Chem. Soc.* I. II. 11 The degree Beaumé..is equal to the fractional part of the number expressing the corresponding specific gravity. *Ibid.* 13, I have adopted an even scale of which 1°.805 is equal to 1°, Beaumé. *Ibid.*, The coarse Beaumé areometer..is commonly used in the arts. 1884 *Mem. Nat. Acad. Sci.* III. 63 The degrees on the Baumé scale are entirely arbitrary, and bear no obvious relation to the specific gravity of the liquid. 1886 *Harper's Mag.* Jan. 248/1 All that is below 60° B. (Baumé, standard of density) ..is turned into a tank for kerosene distillates. 1951 *Good Housek. Home Encycl.* 486/1 If a hydrometer is available, test the strength of the syrup, which should register 25° Beaumé or 37° Balling.

baum marten (bɑu·m mā:ɹtĕn). [Partial tr. G. *baummarder,* f. *baum* tree (see BEAM *sb.*[1]) + *marder,* MARTER[1], marten.] The pine marten or its fur.

1879 *Encycl. Brit.* IX. 838/2 *Baum Marten,*..Found in Europe and Asia, of fine overhair, but woolly fur, of a brownish colour, approaching that of the American marten. 1906 *Westm. Gaz.* 3 Nov. 31/1 There are many kinds of sable, the Russian, Canadian, the baum-marten, and the stone-marten. 1909 *Daily Chron.* 23 Feb. 7/3 The Hudson Bay sables and the dyed baum marten. 1954 *Economist* 20 Feb. 567/1 Sable and ermine..silver and blue fox, baum marten and squirrel.

bauxite [F. *bauxite,* 1821], var. BEAUXITE.

1861 H. SAINTE-CLAIRE DEVILLE in *Chem. News* 9 Nov. 241/1 This substance, obtained from Baux, near Arles, consisted of small round grains buried in a perfectly crystallised pure limestone...I then recognised it as the mineral to which M. Berthier gave the name of Bauxite. It is a hydrate of alumina which M. Dufrénoy has ranked with gibsite, or rather diaspore. 1872 *Jrnl. Chem. Soc.* XXV. 467 Bauxite from the Wochein (Austria). 1883 *Ibid.* XLIV. 397 The practical value of bauxite depends on the high amount of alumina in proportion to the silica. 1922 *Blackw. Mag.* July 18/2 A pink vein of bauxite ore. 1940 *Geogr. Jrnl.* XCVI. 45 The supply of aluminium is also dependent upon external sources for the bauxite and cryolite from which it is extracted.

Bavarian (bāvēᵊ·riăn), *a.* [f. *Bavaria* + -AN.] Of or belonging to Bavaria; also in special collocations (see quots.).

1638 *Invasions of Germanie* §10 On the Bavarian, and Imperialists side, slaine some 250. 1700 RYCAUT *Turkish Hist.* III. 201/1 The Bavarian Troops approaching towards Pesth, enter'd it without any opposition. 1815 *Wynne Diaries* 14 Sept. (1940) III. 382 We found the Inn occupied by Bavarian Officers. 1880 WEBSTER Suppl. s.v. *Cream, Bavarian cream,* a preparation of gelatine, milk, cream, and eggs, flavored, and eaten cold. 1892 T. F. GARRETT *Encycl. Cookery* I. 9/2 Almond Bavaroise, or Bavarian Cream. *Ibid.* 87/2 *Bavarian beer*...The average beer brewed is not of a very high alcoholic character, and is said to obtain some of its characteristic flavour from the pitch used to line the casks. *Ibid., Bavarian creams,* a great variety of these are to be found under the name of Bavaroises. 1893 *Ibid.* II. 385/2 Bavarian Sauce. 1893 E. KNECHT et al. *Man. Dyeing* 479 Diphenylamine-blue. $C_{37}H_{30}N_3$.Cl. Bavarian Blue..gives a finer blue shade than..any other blue. 1934 PRIEBSCH & COLLINSON *German Lang.* vii. 327 The Bavarian-Austrian group includes the dialects of Old Bavaria with a southern set of dialects. 1968 *Globe & Mail* (Toronto) 17 Feb. 43/1 *Happy Valley* (Walkerton)—Excellent with three inches new on packed base of 6 inches. T-bar, Bavarian lift.

bavaroise (bavarwā·z). Also **bavarois.** [Fr.] A cream dessert containing gelatine and whipped cream, and served cold.

1846 A. SOYER *Gastronomic Regenerator* 529 Bavaroise aux Fraises. 1868 E. ACTON *Mod. Cookery* (rev. ed.) xxiii. 478 By mingling the cream..the preparation becomes what is called *un Fromage Bavarois,* or Bavarian cream, sometimes simply, *une Bavaroise.* 1877 E. S. DALLAS *Kettner's Book of Table* 56 Bavarois—Bavarian cheese. Boil as much of the best milk as will half fill the mould...Sugar it, and add whatever flavour may be chosen...Mix in..yolks of eggs...Being cool, it is to be mixed well with..whipt cream, poured..into a mould..and given over to the ice-box. 1892 [see *BAVARIAN *a.*]. 1964 *Observer* 11 Dec. 34/5 *Bavaroise* is just a Crème Bavaroise in a mould. 1968 *Woman* 7 Dec. 26/1 This wonderful sweet [*sc.* charlotte russe] filled with a rich bavaroise mixture.

bawl, *v.* Add: **3. c.** With *out*: To reprove or reprimand loudly or severely. orig. *U.S.* Also *ball out* (see *BALL *v.*[5]).

1908 R. BEACH *Barrier* xvii. 270 If you'll go back on your word like this you'll 'bawl me out' before the priest. 1917 MATHEWSON *Sec. Base Sloan* xv. 203 You'll get bawled out when you pull a boner. *Ibid.* xviii. 239 Wayne thought that the manager's 'bawling out' that forenoon had done good. 1922 H. L. FOSTER *Adv. Trop. Tramp* ix. 119 In private Griffis bawled me out for my rashness. 1933 *New Statesman* 18 Mar. 331/2 All the *plats du jour* were 'off' and we bawled out the head waiter. 1942 L. A. G. STRONG *Unpractised Heart* xii. 82 He bawled him out. Gave him such a tongue lashing as the louse will remember to his dying day.

Hence **bawl-out** *sb.* a reprimand. *U.S.*

1926 J. BLACK *You can't Win* vi. 70, I..don't want to.. give myself a bawl-out in front of the woman.

bawley (bǭ·li). *local.* Also **bauley, baully.** [Of obscure origin.] A fishing-smack peculiar to the coasts of Essex and Kent. Also *attrib.*, as *bawley-boat.*

1887 PARISH & SHAW *Dict. Kentish Dial.* 9 *Bawley*, a small fishing smack used on the coasts of Kent and Essex, about the mouth of the Thames and Medway. 1888 *Sat. Rev.* 24 Mar. 349 A little creek where barges and bawley-boats can ride. 1895 RYE *Gloss. E. Anglia, Bauley Boats*, Harwich fishing-smacks. 1921 *Spectator* 7 May 589/1 Leigh-on-Sea is the producer of the bawley, a type of craft which has a loose-footed mainsail, but would otherwise be called a cutter. 1948 H. BENHAM *Last Strongholds of Sail* xxiii. 194 They are comely little bawleys, well kept, power-driven, and obviously prosperous.

bawneen, bauneen (bǭ·nīn). [ad. Irish *báinín* undyed flannel, f. *bán* (*báin-*) white.] In Ireland: a sleeved waistcoat made from undyed flannel worn by farm-labourers. Also *attrib.*; **bawneen yarn,** a coarse light fawn unwashed knitting-wool.

1910 *Month* CXV. 66 The last rays of the setting sun fell slanting, lighting up the white bauneen of the turf cutters. 1919 YEATS *Two Plays for Dancers* 2 He seems an Aran fisher, for he wears The flannel bawneen and the cow-hide shoe. 1958 *Observer* 19 Jan. 11/2 Tweed car-coats will be lined in firm handknits, some in the 'bawneen' yarn which washes so magnificently. 1967 N. FITZGERALD *Affairs of Death* x. 164, I bought cigars, a bawneen sweater and..a cuddly toy.

baw-ways, bawways (bǭ·wēɪz), *adv.* and quasi-*adj. Anglo-Irish.* [Origin of first element uncertain; cf. Sc. *baw(w)aw* side-glance of contempt or scorn. For the formation cf. *edgeways, endways*, etc.] Crookedly; leaning awkwardly, leaning to one side; sideways.

1907 J. JOYCE *Let.* 1 Mar. (1966) II. 218 Scholz's five crown cloak hung bawways on me. 1922 —— *Ulysses* 297 Little Alf was knocked bawways. 1944 *Béaloideas* XIV. 164 [South-West Dublin Glossary] *Baw-ways* (aw as in *bawl*). Crooked or leaning to one side. 'That hay-reek is badly built; it's all baw-ways.' 1947 *Ibid.* XVII. 264 [North-County Dublin Glossary] *Baw-ways*. Crooked. Awkwardly. He did it *baw-ways* = wrong.

bay, *sb.*[1] **4.** Add: **bay-gall** *U.S.* (*a*) = BAY *sb.*[2] 4 b (and see Suppl.); (*b*) *Bot.*, the red bay (see RED *a.* 17 d); **bay-rum** (earlier U.S. examples); **bay-swamp** *U.S.* = BAY *sb.*[2] 4 b.

1775 ROMANS *Florida* 15 Swamps, marshes, and bay, or cypress galls. 1861 A. WOOD *Class-bk. Botany* 620 Red Bay ..Bay Galls..Wood of a fine rose-color, once used in cabinet-work. 1872 SCHELE DE VERE *Americanisms* 440 Bay-Galls are large, gloomy, almost impenetrable swamps in Florida, full of deer, bear, and catamount. 1840 *Knickerbocker* XVI. 34 Perfumed 'as to our locks' with the bay-rum or fragrant cologne. 1859 BARTLETT *Dict. Amer.* (ed. 2) 25 *Bay Rum*..is chiefly used for the purposes of the toilet. 1741 in *Colonial Rec. Georgia* (1908) IV. Suppl. 237 The Land in these parts, setting aside the Pine-Barren, and some Bay-swamps,..with proper cultivation, will yield a reasonable Increase. 1832 D. J. BROWNE *Sylva Amer.* 164 These spots are entirely covered with the loblolly bay, and are called Bay Swamps.

bay, *sb.*[2] Add: **4. a.** (Examples.)

1850 W. COLTON *Three Years Calif.* 370 Still, in some of its bays, the evidences of fertility exist. 1874 B. F. TAYLOR *World on Wheels* 17 In the bottom of a bay of land bounded ..by wooded hills.

b. (Add earlier examples and transfer this sense to BAY *sb.*[1] 2 c.)

1795 F. ASBURY *Journal* (1821) II. 285 This country [*sc.* S. Carolina] abounds with bays, swamps, and drains. 1845 W. G. SIMMS *Wigwam & Cabin* 17 He wandered along the edges of a dense bay or swamp-bottom.

5. bay-bird *U.S.*, a shore-bird that frequents the bays and estuaries of the Atlantic coast; **bay-craft** *U.S.*, a vessel or vessels used in the navigation of bays; **bay-ice** (earlier example); **bay-like** *a.*, resembling a bay; **bay-snipe** *U.S.* = *bay-bird*; **Bay State** *U.S.* (earlier examples); hence *Bay Stater*; **bay-vessel** *U.S.* = *bay-craft*; **bay whaling,** a method of whaling in the shore waters of Australia and New Zealand using land-based stations; formerly called **shore whaling*; hence **bay whaler,** a boat used in bay whaling; **bay whale,** a whale that frequents bays, spec. the Southern Right Whale (*Balæna australis*).

1889 *Cent. Dict.* Bay-birds. 1725 in *New Eng. Quarterly* (1929) II. 660 We met a Ship which they took and burnt then sending away what Prisoners they thought fit in a Bay Craft. 1835 C. J. LATROBE *Rambler in N. Amer.* II. vi. 102 Many a settler loads his small bay-craft with planks and shingles in the spring of the year. 1818 SCORESBY in *Ann. Reg. 1817* 534/2 This is termed bay-ice. 1874 DISRAELI in Buckle *Life* (1920) V. ix. 354 The waters glittering in the bay-like coast. 1897 M. KINGSLEY *W. Africa* 130 The great forest sweeping away in a bay-like curve. 1856 *Spirit of Times* 6 Sept. 9/1 Bay Snipe shooting is at its acme, and can be enjoyed everywhere on Long Island. 1875 *Fur, Fin & Feather* 121 It is also a capital place for bay-snipe shooting in summer. 1789 S. DAVIS *Jrnl.* 1 Sept. in *Proc. Mass. Hist. Soc. 1869* (1871) XI. 14 The style of building varies somewhat from that of the Bay State, as they term Massachusetts. 1837 R. M. BIRD *Nick of Woods* I. 132 He was from the Down-East country; a representative of the Bay State. 1845 *St. Louis Reveille* 14 May 2/4 The inhabitants of..Massachusets [are called] Bay Staters. 1789 *Maryland Jrnl.* 24 Feb. (Th.), I will exchange a small Bay Vessel for a large one, and give the difference. 1850 H. T. CHEEVER *Whaleman's Adv.* vi. 79 Bay whaling, which destroys the cows about the time of calving. 1853 G. B. EARP *New Zealand* vii. 95 [The] whale fishery [was] carried on chiefly on the coasts of the northern and middle island by means of boats—a method technically called 'bay-whaling', these animals frequenting the coasts in great numbers during the breeding season. 1905 W. B. *Where White Man Treads* 36 The lawless pakeha bay-whaler. *Ibid.* 75 An old-time bay whaling station consisted..of at least two boats. 1913 R. MCNAB *Old Whaling Days* i. 6 During the following month—November—the remaining bay whalers returned to Sydney. 1933 F. D. OMMANEY *Whaling in N.Z.* in *Discovery Rep.* VII. 243 Much of the Right whale industry was carried on by the method known as 'bay whaling'. This branch of the fishery derived its name from the Right whales' habit of entering shallow bays and inlets of the coast for the purpose of..calving. 1947 A. H. CLARK in H. Belshaw *New Zealand* 32 Fur seals and bay whales (i.e., the right whales) which once frequented the island are now virtually extinct.

bay, *sb.*[3] Add: **3.** *bomb bay*: see *BOMB *sb.* 6; *sick bay* (see SICK *a.* 10 a in Dict. and Suppl.).

5. b. *Mil.* A section of a trench in which the line is modified in order to allow more space for passing.

1916 'BOYD CABLE' *Action Front* 73 The trenches..with bays and niches cut deep in the side to permit the passing of anyone meeting a line of pack-burdened men in the shoulder-wide alley-way. *a* 1917 E. A. MACKINTOSH *War, the Liberator* (1918) 154 'Don't put it in so high up, boys,' he said. 'They'll see it and knock this bay to hell.'

6. A side or subordinate line of railway at a station; also *attrib.* in **bay-line,** a line at the side of, and terminating in, a railway station.

1906 *Westm. Gaz.* 20 Sept. 7/1 A bay line out of Grantham Station crosses some of the roads. 1907 *Ibid.* 10 June 9/1 The passenger station has not been interfered with, except in No. 1 bay-line. 1939 A. CHRISTIE *Murder is Easy* i. 13 A train..came slowly puffing in and deposited itself in a modest bay.

7. A series of racks in a telephone exchange on which equipment is mounted.

1906 J. POOLE *Pract. Telephone Handbk.* (ed. 3) xxi. 291 The incoming and outgoing junction and other..lines are also accommodated on the arrester frame, special bays being reserved for them. 1920 *Post Office Electr. Engin. Jrnl.* Jan. 205 The apparatus is placed on the various bays of the racks.

bay, *a.*[1] (and *sb.*). Add: **2. b.** *The Bays*: see quots.

1837 R. CANNON *Hist. Records Brit. Army, Second Dragoon Guards* 64 About this period [*sc.* 1767] the regiment was mounted on Bay Horses; and as the other regiments of heavy cavalry were mounted on black horses (except the Scots Greys) the Queen's Dragoon Guards were commonly styled the Queen's Bays. 1878 R. TRIMEN *Regiments Brit. Army* 14 Second Dragoon Guards...Being mounted on bay horses about 1767 caused it to be called the 'Queen's Bays'. ..It is now commonly called 'the Bays'.

bay, *v.*[6] [f. BAY *sb.*[2]] *intr.* To spread *out* in a bay-like form.

1906 'A. HOPE' *Sophy of Kravonia* viii, The town was no more than one long street, which bayed out at the farther end into a market-place.

bayadère. Add: (Later example.)

1859 SALA *Tw. round Clock* (1861) 240 A gay audience shouting applause to mimes and jesters and painted bayadères.

2. Any textile fabric having stripes running across the material. Also *attrib.* or as *adj.*, defining a material striped in this way.

1856 *Illustr. London News* 27 Dec. 653/1 Bayadere skirts. 1879 *Cassell's Fam. Mag.* Apr. 312/1 Pompadour silks with Bayadère stripes beneath the bunches of flowers are somewhat new. 1960 *Guardian* 22 Apr. 8/4 Exquisite evening shoes of glazed Bayadere material.

bayberry. 2. (Earlier examples.)

1687 in *Manchester* (Mass.) *Rec.* 32 The sd. tree being near Vincsons baiberry medow. 1769 *Massachusetts Gaz.* 21 Dec., Advt. (Th.), Bayberry-wax candles. 1792 J. BELKNAP *New Hampsh.* III. 123 The bay berry (*myrica cerifera*), the leaves of which yield an agreeable perfume, and the fruit a delicate green wax, which is made into candles.

Bayer (bəi·əɹ). [From the name of the inventor, K. J. *Bayer*, a German metallurgist.] **Bayer process:** a process for the production of aluminium from bauxite.

1910 *Metall. & Chem. Engin.* Aug. 498/1 Two processes are used in France to produce alumina from bauxite. The Bayer process is used at Gardanne, where the ore is attacked under pressure with caustic soda solution. 1911 *Ibid.* Mar. 147/1 The alumina is purified from the bauxite by the Bayer process—solution in caustic soda and precipitation therefrom. 1949 P. C. CARMAN *Chem. Const. & Prop. Engin. Materials* vii. 229 To get pure Al_2O_3, the Bayer process is used, in which bauxite is treated with 45% NaOH heated to 100 p.s.i. 1956 *Nature* 10 Mar. 462/2 The purification of the mineral by the Bayer process.

Bayer 205 (bəi·əɹ). [G. *205 Bayer* (L. Haendel and K. W. Joetten 1920, in *Berliner Klin. Wochenschr.* 30 Aug. 821/1), f. the name of F. Bayer & Co., a German chemical firm.] A synthetic trypanocidal drug, $C_{51}H_{34}N_6Na_6O_{23}S_6$.

1922 *Chem. Abstr.* XVI. 3128 Chemotherapeutic experiments with 'Bayer 205', a new trypanocidal agent of special activity. 1933 *Discovery* Apr. 116/1 In sleeping sickness drugs like arsenic and 'Bayer 205' are relatively effective.

bay-man, bayman[1]. [BAY *sb.*[2] 5.] **1.** A resident beside a bay (usu. some specific bay); one accustomed to navigating a bay. *U.S. Hist.*, an inhabitant or native of Massachusetts Bay.

1641 in *New Plymouth Colony Records* II. 23 That clause ..which concerned the boundes from Narragansetts Bay to..Pockanockett, in regard the Bay men would haue had Sicquncke from us. 18.. *Shore Birds* 43 (Cent. D.), When the birds are traveling with the wind, or as baymen call it, a 'free wind'. 1904 *N.Y. Even. Post* 11 June, Somers P'int, as the baymen call it, is one of several very attractive summer resorts that have grown up about the bay in recent years.

†**2.** A mahogany-cutter of the Bay of Honduras. *Obs.*

1715 *Boston News-Let.* 10 Oct. 2/2 Huntingdon and.. Holder..report that 250 of the Bay-Men..were designed to Campeche, to burn the Shipping there. 1781 *Ann. Reg. 1780* *211/2 The baymen on the Musquito and bay of Honduras shores, (as the logwood cutters are called). *a* 1821 C. BIDDLE *Autobiogr.* (1884) i. 17 The baymen at this time would frequently sell their wood to two or three different captains.

bayman[2]. *U.S.* [BAY *sb.*[3] 3.] A sick-bay nurse.

1888 CHURCHWARD *'Blackbirding'* 25, I stole a beautiful knife from the sick bayman's locker. 1891 H. PATTERSON *Naut. Dict.* 346 *Bayman*, a hospital nurse.

bayonet, *sb.* Add: **6. bayonet cap,** a cap on an electric light bulb for insertion in a bayonet socket; **bayonet-capped** *a.*, fitted with a cap for fastening in a socket as a bayonet joint; **bayonet catch,** the spring catch by which a bayonet is secured to a rifle; **bayonet grass,** a popular name for a New Zealand umbelliferous plant of the genus *Aciphylla*; **bayonet-socket,** a socket with which a bayonet-capped fitting engages.

1914 S. C. BATSTONE *Electric Light Fitting* vii. 136 In Fig. 142 the finished lamps are shown, No. 1 with what is known as the 'bayonet cap', B.C. 1943 *Electronic Engin.* XVI. 247 The activated electrodes at either end are connected to a two-pin bayonet cap. 1904 *Daily Chron.* 18 June 2/6 Bayonet-capped [electric] lamps. 1901 'LINESMAN' *Words Eyewitness* (1902) 317 They have struck as true and hard as any, right up to the bayonet-catch. 1868 W. L. LINDSAY *Contrib. N.Z. Bot.* xii. 49 *Aciphylla*..The larger species are familiar to the settler as 'Spear-grass', or 'Bayonette-grass' ..in allusion to their very rigid, strong, poniard-like, sharp-pointed leaves. 1899 T. KIRK *Students' Flora N.Z.* 207 Bayonet-grass. 1946 *Jrnl. Polynesian Soc.* LV. 158 *Taramea* ..spear-grass, bayonet-grass, spaniard: a hill and mountain plant from whose spiny blades the Maori by heat and torsion extracted a valued scent. 1892 F. C. ALLSOP *Pract. Electric-Light Fitting* vi. 75 The..lamp..is used with the bayonet socket holders, which is certainly the most convenient and efficient method of making connection between lamp and conducting wires. 1955 *Times* 13 July 4/1 The rod..was then removed by a turn of a bayonet socket.

bayou. Add: Also **byo, bayoue, bayeau,** *pl.* **bayoux.** For etym. read: [Amer. Fr., f. Choctaw *bayuk*.] (Earlier and later examples.)

1766 H. GORDON in N. D. Mereness *Travels Amer. Col.* (1916) 484 We left New Orleans..and lay that night in the Bayoue. 1808 ASHE *Trav. Amer.* xl. 323 Below the Red River, five miles, is one of the most dangerous bayeaus on the Mississippi. 1814 BRACKENRIDGE *Views Louisiana* 162 On some of these bayoux the land is sufficiently high to admit of settlements. 1834 CROCKETT *Narr. Life* vi. 53 A small byo, across which there was a log. 1872 *Amer. Naturalist* VI. 725 A peculiar feature of the bottom lands of the western and southern rivers, locally termed bayous. 1901 S. E. WHITE *Westerners* xiii. 93 In a word the broad sea of the wilderness has shrunken to bayous and bays.

attrib. 1850 H. C. LEWIS *Louisiana Swamp Doctor* 161, I saw the dust up the bayou road shaken up by a half-naked negro. 1886 *Harper's Mag.* Aug. 483/1 The following bayou version of one of the negro folk-lore stories.

Bayreuth (bəi·roit). The name of a Bavarian town in which festivals of the music of Richard Wagner have been held since 1876 in a theatre specially built for the production of his operas. Also *attrib.* and *ellipt.* Hence **Bayreu·thian,** pertaining to or characteristic

of Bayreuth. **Bayreu·ther,** an inhabitant of Bayreuth.

1876 *All Year Round* 23 Sept. 39/2 The Bayreuth Performances. There has probably never been an event in.. the history of music which has caused so much excitement as the recent first performances, at Bayreuth,..of Richard Wagner's..'Der Ring des Nibelungen'. 1876 *Nation* 7 Sept. 148/1 Dazzling spectacles, in an opera-house.., were not strange to the eyes of the Bayreuthers. 1885 G. B. SHAW *How to become Musical Critic* (1960) 101 Singers of Bayreuth music-dramas. 1889 *Ibid.* 151 As to the peculiar merits of the Bayreuth mode of performance, they are simply the direct results of scrupulous reverence for Wagner. 1896 *Ibid.* 243 And this, if you please, is Bayreuthian fidelity to 'The Meister'. 1896 —— *Our Theatres in Nineties* (1932) II. 67 We could build it [*sc.* the Stage Festival Playhouse] for ourselves better and cheaper than the Bayreuthers built it for Wagner. 1959 *Listener* 29 Jan. 216/1 The error, common in the past, against which Bayreuth now ensures the spectator is that of believing that Wagner used Teutonic myth and medieval legend simply as a vehicle for crude theories of German nationalism.

bay-windowed. (Earlier example.)

1836 T. HOOK *G. Gurney* III. vi. 331 In the bay windowed drawing-room.

bazaar. Add: **2.** Also used of a shop, or arcade of shops, displaying an assortment of fancy goods (see quot. 1889).

1807 SOUTHEY *Lett. from England* I. vii. 82 My way..took me through a place called Exeter Change, which is precisely a *Bazar*, a sort of street under cover, or large long room, with a row of shops on either hand, and a thoroughfare between them; the shops being furnished with such articles as might..remind a passenger of his wants. 1816 [in Dict.]. 1835 *Penny Cycl.* IV. 76/1 Paternoster-row with its books, Monmouth-street with its shoes..are more properly bazaars than the miscellaneous shops assembled under cover, which are in London designated by the name. 1888 E. BELLAMY *Looking Backward* (1889) v. 42 The great city bazaar crushed its country rivals with branch stores. 1889 *Cent. Dict.* s.v., Marts bearing the name of bazaars, for the sale of miscellaneous articles, chiefly fancy goods, are now to be found in most European and American cities. 1930 *Economist* 3 May 1008/2 A preliminary statement issued by this progressive company of bazaar proprietors reports excellent results for the year ended March 31, 1930.

bazoo (băzū·). *U.S. slang.* [Origin unknown; cf. Du. *bazuin* trumpet.] = KAZOO; also *transf.*, mouth.

1877 BARTLETT *Dict. Amer.* (ed. 4) 49 *Blowin' his bazoo*, gasconade; braggadocio. Tennessee. 1884 E. W. NYE *Baled Hay* 237 People..listen to the silvery tinkle of his bazoo. 1888 W. WHITMAN *November Boughs* 407 Among the far-west newspapers, have been, or are,..*The Bazoo*, of Missouri. 1909 WEBSTER, *Bazoo*, a kind of wind instrument; also, the mouth. 1928 *Collier's* 18 Aug. 34/3 After the orators have been braying away on their bazoos, futilely naming obscure statesmen for Vice President. 1932 J. T. FARRELL *Young Lonigan* (1936) iii. 53 His mother was always blowing off her bazoo about him being her blue-eyed baby. 1948 *Sat. Even. Post* 24 Apr. 173/2 Shut yer big bazoo!

bazooka (băzū·kă). orig. *U.S.* [app. f. prec.]

1. (See quot. 1935.)

1935 *Newsweek* 14 Dec. 29/2 Burns peps up his lengthy yarns with periodic outbursts on his own invention, the bazooka, a trombone-like instrument confected of two gas-pipes and a whisky funnel. 1945 *N. & Q.* 19 May 215/2 Each broadcast Mr. Burns renders a bazooka solo. 1956 R. NETTEL *Seven Centuries Pop. Song* xiv. 229 During the troubles of 1926 a group of miners..formed themselves into a band. The instruments they used were called in the musical trade 'kazoos' and by the miners themselves 'bazoukas'.

2. A tubular anti-tank rocket-launcher.

1943 *War Illustr.* 15 Oct. 297 (*caption*) American anti-tank rocket thrower.., known as the 'bazooka'. 1944 *Britannica Bk. of Year* 769/2 *Bazooka*, a firearm, consisting of a metal tube slightly over 50 inches long and under 3 inches in diameter. 1945 *N. & Q.* 10 Mar. 106/2 The world's fastest fighting plane..carries six bazooka tubes.

bazouki, see *BOUZOUKI.

B.B.C. (bībīsī·). Initial letters of *British Broadcasting Corporation*, a public corporation orig. having the monopoly of broadcasting in Gt. Britain, financed by a grant-in-aid from Parliament; established 1927 by royal charter to carry on work previously performed by the British Broadcasting Company; hence **B.B.C. English,** standard English as maintained by B.B.C. announcers; so *B.B.C. pronunciation*, etc.

1923 *Radio Times* 28 Sept. 12/1 It seems to me that the B.B.C. are mainly catering for the 'listeners' who own expensive sets. 1925 *Punch* 22 Apr. 440/1 The daily wireless programme of the B.B.C. 1926 *Encycl. Brit.* Suppl. I. 454/2 The 'B.B.C.' is constituted as a limited company, the shareholders being wireless manufacturers and traders. 1928 *Times* 13 Jan. 8/5 B.B.C. English. Mr. Lawrence omits from his list of solecisms in pronunciation perpetrated by the B.B.C.'s 'Advisory Committee on Spoken English' the crowning horror. 1932 *Listener* 13 Jan. 45/1 Critics who enjoy making fun of what they are pleased to call 'B.B.C. English' might with profit pay occasional visits to the other side of the Atlantic, in order to hear examples of our language as broadcast where there are no official 'recommendations to announcers'. 1936 W. HOLTBY *South Riding* i. 18 She talked B.B.C. English to her employer..and Yorkshire dialect to old milkmen. 1938 P. THORESBY JONES *Welsh Border Country* viii. 95 The educated and older local (as opposed to the 'board-school' and B.B.C.) pronunciation of the town's name is *Shrozebury*, not Shroozbury. 1944 *Penguin New Writing* XXII. 47 Her accent was impeccably B.B.C. 1956 A. WILSON *Anglo-Saxon Attitudes* II. ii. 338 B.B.C. officials—programme planners, features-producers, poetry readers.

be, *v.* Add: **6. c.** *been and* (*gone and*) — : vulgar or facetious expletive amplification of the pa. pple. of a verb, used to express surprise or annoyance at the act specified.

1836 DICKENS *Pickw.* xxvi, Lauk, Mrs. Bardell,..see what you've been and done! 1847 THACKERAY *Van. Fair* xv, Sir Pitt has been and proposed for to marry Miss Sharp. 1869 W. S. GILBERT *Bab Ballads* 218 The padre said, 'Whatever have you been and gone and done?' 1891 [see GO *v.* 32 c]. 1920 R. MACAULAY *Potterism* II. i. 61 She's been and gone and done it. She's got engaged. 1926 D. L. SAYERS *Clouds of Witness* ii. 48, I say, Helen, old Gerald's been an' gone an' done it this time, what?

9. d. *colloq.* With idiomatic repetition of the verb in the following clause. (Further examples in Visser *Hist. Syntax* I. 55.) Cf. IT 4 c.

1828 M. MITFORD *Our Village*, III. 202 He's a sad pickle is Sam! 1928 R. MACAULAY *Keeping up Appearances* iv. § 1. 35 She's very sympathetic, Daphne is. 1930 BELLOC *Wolsey* iv. 58 It is a rare function, is industry upon this level. 1932 R. KNOX *Broadcast Minds* vii. 156 Yes, he is true to type, is Mr. Heard. 1958 P. GALLICO *Steadfast Man* ii. 43 He was an honest man, was Patrick.

be-. Add: **2. bechill** (? nonce-wd.).

1952 AUDEN *Nones* 54 The spreading ache bechills the rampant glow Of fortune-hunting blood.

4. bejuggle (later example.)

1851 H. MELVILLE *Moby Dick* III. xlvii. 268 No matter how many .. thou may'st have bejuggled and destroyed before.

7. *be-aureoled, bedragoned, befezzed, befrilled* (earlier example), *begoggled, bejacketed, bejeaned, bemotored, beribboned* (later examples), *beskirted, betabbed, betrousered, bevillaed.*

1923 D. H. LAWRENCE *Birds, Beasts & Flowers* 29 Your.. idealism, Like a be-aureoled bleached skeleton. 1945 DYLAN THOMAS *Let.* 30 July (1966) 282 Like a be-aureoled bleached skeleton hovering its cage-ribs in the social heaven. 1925 E. BOWEN in *Lond. Mercury* July 249 Eleanor.. followed the be-dragoned back. 1928 BLUNDEN *Japanese Garland* 22 A wind like fifty winds at once Through the bedragoned kingdom runs. 1960 T. HUGHES *Lupercal* 29 Prehistoric bedragonned times Crawl that darkness with Latin names. 1905 *Daily Chron.* 2 June 6/7 The Arab, a befezzed, well-knit young man. 1932 *Times Lit. Suppl.* 24 Nov. 867/2 The befezzed junior officer. 1843 *Fraser's Mag.* XXVIII. 569 The close bepuckered and befrilled gowns and collars. 1903 *Daily Chron.* 30 June 3/7 These begoggled road demons. 1914 G. B. SHAW *Misalliance* 44 The Passenger, also begoggled, comes in. 1837 C. RIDLEY *Lett.* (1958) 23 We are also *bejacketted* in a very tight fitting style. 1958 *Manch. Guardian* 7 Mar. 8/5 A sulky adolescent and his be-jeaned girl-friend. 1960 *Bookseller* 28 May 2070/1 The bejeaned and pony-tailed young ladies of my village. 1908 *Westm. Gaz.* 5 June 4/2 The be-motored and dusty road. 1931 *Times Lit. Suppl.* 12 Feb. 117/4 The boys, hip-flasked and be-motored. 1951 S. SPENDER *World within World* 5 My father, removing with a flourish his beribboned spectacles. 1954 DYLAN THOMAS *Quite Early one Morning* 16 Autograph albums with a lock of limp and colourless beribboned hair lolling out between the thick black boards. 1936 O. NASH *Primrose Path* 165 And all the trim and not so trim ladies who have been be-trousered begin thank God once more to be be-skirted. 1959 *Times* 16 Nov. 4/1 Even the deployment of the animals,..high-stepping horses and beskirted chimpanzees, is done with tact and artistic propriety. 1920 *Chambers's Jrnl.* Sept. 569/1 The whole party indulging in explosive merriment over the quips of the betabbed gentleman in the centre. 1930 BLUNDEN *De Bello Germanico* i. 5 We joined the avalanche with which the betabbed encyclopædia was dealing. 1934 WEBSTER, Betrousered. 1937 *Daily Express* 3 Mar. 14/2 Betrousered, she hacks in the park at Arundel and once hid when Queen Mary passed by in case her boyish clothes were frowned on. 1866 G. M. HOPKINS *Let.* 30 June (1938) 10 A great fashionable town, and even the country round much bevillaed.

beach, *sb.* Add: **3.** The shore of a lake or large river. (Examples.)

1756 in *Doc. Hist. State N.Y.* (1849) I. 478 Upwards of 1000 French and Indians appeared upon the Beech [of Lake Ontario]. 1771 PENNANT *Tour in Scotl.* 201 A little isle, in a small loch in Badenoch, was totally reversed and flung on the beach. 1843 N. BOONE *Jrnl.* in L. Pelzer *Marches of Dragoons* (1917) 189 Captain Boone..commenced his march from the beach of Grand River opposite Fort Gibson.

b. *Naut.* The shore, any part of the coast-line off which a ship is at anchor; hence *on the beach*, ashore; retired (*the beach* = land, civilian life); *to take the beach*, to go on shore leave. By extension *on the beach* is used to mean 'beachcombing, unemployed'; also (occas.) penniless, 'broke'.

1903 J. LONDON *People of Abyss* xi. 127 England is always crowded with sailormen on the beach. 1915 'BARTIMEUS' *Tall Ship* iv. 71 The head of the Officer of the Watch appeared...'Wake up, you Weary Willies. There's a boat to the beach at seven-bells.' 1916 'WARD-ROOM' *H.M.S.* ii. 29 The captain and the commander had 'taken the beach' for the afternoon. 1923 *Daily Mail* 15 May 8 Hundreds of the trawlermen to-day find themselves 'on the beach', owing to the distressed condition of the fishing industry. 1925 'BARTIMEUS' *Great Security* iii. § 2. 181 Hitherto he had been accustomed to view 'the Beach' as an incident in his normal life, an environment that asked nothing of him and gave nothing in return. 1925 WODEHOUSE *Let.* 11 Jan. in *Perf. Flea* (1953) 30 The world is full of poor devils on the brink of being chucked out of jobs and put on the beach. 1935 'G. ORWELL' *Clergyman's Daughter* ii. § 1. 99 You on the beach, kid?...On the bum?...What I mean to say, kid —have you got any money? 1938 W. S. CHURCHILL *Great Contemporaries* 336 The somewhat pathetic appeal of a retired officer mouldering on the beach.

4. *beach-bag, -pyjamas, -robe, sandals, shirt, -wear, -wrap*; **beach-ball,** a large inflated ball for use at a beach; **beach boy,** a male beach attendant; a play-boy on a beach; **beach-comber** (earlier and additional examples); **beach crab,** any crab of a species living on sea-beaches, esp. *Ocepoda arenaria*; **beach cusp,** a cusp of gravel or sand found at intervals of about 20 to 30 feet on a beach; **beach gown,** a bathing-wrap; **beach-grass** (earlier U.S. example); **beachhead,** also **beach-head** *Mil.* [illogically formed after *bridge-head*], a fortified position of troops landed on a beach; also *transf.* and *fig.*; **beach-man,** (*b*) (see quot. *a* 1865); **beach-plum** *U.S.* (see quots.); **beach-rock,** a conglomeration of calcareous beach sand cemented by chalk into rock formation, found on coral reefs; **beach-wagon** (U.S. examples); a station-wagon.

1934 R. MACAULAY *Going Abroad* iv. 40 They gathered up their beach-bags and went up to the bar. 1940 'N. BLAKE' *Malice in Wonderland* I. vi. 73 The combatants were assaulting each other with beach-balls. 1939 *Time* 12 June 87/1 Kilt-style skirt worn over shorts (already fashionable among Florida's rich beach boys). 1965 'W. HAGGARD' *Hard Sell* ii. 20 Penniless Sicilian barons were six a penny among the beach-boys. 1965 *N.Z. Listener* 17 Dec. 5/2 The Beach Boys are at the front of this wave of frantic surf music. 1840 R. H. DANA *Bef. Mast* (1841) xix. 46/1 In the twinkling of an eye I was transformed from a sailor into a 'beach-comber' and a hide-curer. 1845 E. J. WAKEFIELD *Adv. in N.Z.* I. xi. 339 Idle, drunken, vagabond..he wanders about without any fixed object, cannot get employed by the whaler or any one else, as it is out of his power to do a day's work; and he is universally known as the 'beach-comber'. 1859 A. S. THOMSON *Story of N.Z.* I. II. iii. 297 The Pakeha Maori must not be confounded with the idlers and beach-combers who loitered about Kororareka. 1919 W. S. MAUGHAM *Moon & Sixpence* xlvii. 206 The corpse of a nameless beach-comber would be fished out of the dirty water of the harbour. 1909 WEBSTER, Beach crab. 1900 J. C. BRANNER in *Jrnl. Geol.* VIII. 481 (*title*) The origin of beach cusps. 1900 *Geogr. Jrnl.* Dec. 704 The author comes to the conclusion that beach cusps are produced by the interference of two sets of waves of translation on the beach. 1952 F. P. SHEPARD in *Bull. Amer. Assoc. Petrol. Geol.* XXXVI. 1909 Following common practice, these relatively small features are being referred to as 'beach cusps'. 1928 *Sunday Dispatch* 8 July 16 When getting a bathing suit, don't forget a beach-gown, which is now worn with a monk's hood, on the shore. 1681 in *Rec. of Town* (East Hampton, N.Y.) II. 102 Thomas Bee doth..maintaine a sofisient three raile fence one the beach..down so low as any Beach grass groues. 1940 in *Amer. Speech* (1942) XVII. 122/1 The second theory [of the Germans], to harass communications and airports and beachheads so effectively that landings could be undertaken. 1941 *Time* 25 Aug. 22/2 Marines.. trained in the terrible job of establishing a beach-head. 1944 *Times* 24 Jan. 4/3 A substantial beachhead was seized. This beachhead has been widened and deepened. 1949 KOESTLER *Promise & Fulf.* xvi. 176 This commonwealth of immigrants would have become a beachhead of European tradition and democracy in the Levant. 1965 C. WALSH in J. Gibb *Light on C. S. Lewis* 110 *The Screwtape Letters* established Lewis's beachhead in America. *a* 1865 SMYTH *Sailor's Word-Bk.* (1867) 88 *Beach-man*, a person on the coast of Africa who acts as interpreter to shipmasters, and assists them in conducting the trade. 1929 F. BOWEN *Sea Slang* 9 *Beach Men*, West African surf men and interpreters. 1784 in *Mem. Amer. Acad. Arts & Sci.* (1785) I. 449 The Beach, or Sea-Side Plumb. 1877 BARTLETT *Dict. Amer.* (ed. 4) 550 *Sand-Plum*,..a *beach-plum*. A plum growing on plum-trees whose habitat is sandy beaches. 1928 *Vanity Fair* Aug. 74 These beach pyjamas of figured foulard are slipped on over the bathing suit. 1936 M. LASKI in *Cherwell* 7 Mar., They paint their toenails, stroll down the High in beach pyjamas. 1959 *New Statesman* 13 June 838/2 A lady visitor to an East Coast resort..is discovered dead on the shore..wearing beach-pyjamas at early dawn. 1952 *Manch. Guardian Weekly* 21 Feb. 5/1 The Admiral Lord Rodney, got up much like Nero in a cool beachrobe or toga. 1919 R. A. DALY in *Carnegie Inst. Year Bk.* 192 Cementation of beach sand by calcium carbonate is very common in .. tropical seas generally. The product may be called 'beach-rock'. 1940 *Geogr. Jrnl.* XCV. 30 On tropical beaches the sands, or even coarse shingle, are commonly cemented into a hard rock, called beach-rock or conglomerate, by carbonate of lime precipitated from sea-water. 1934 W. PLOMER *Invaders* xiv. 262 He was wearing only..trousers and beach sandals. 1966 'A. YORK' *Eliminator* iii. 49 She..thrust her feet into gold beach sandals. 1966 G. LYALL *Shooting Script* viii. 47 There was just one man alone, wearing a vivid beach shirt 1868 L. M. ALCOTT *Little Women* (1869) II. 35, I shall hire a beach-wagon. 1935 H. NICOLSON *Let.* 17 July (1966) 209 There was the beach-wagon going down to the village. 1948 *Chicago Tribune* 9 May II. 10/4 Use of the parking facilities is restricted to automobiles, beach wagons, and motor cycles. 1928 *Men's Wear* (U.S.) Oct. 10 (*caption*) Fantastic Beach Wear at Juan les Pins. 1952 *Vogue* June 6 (Advt.), Our gay, new beachwear. 1927 *Star* 30 May 8/1 Women's Beach Wraps, made of..coloured towelling.

bea·ch-co:mbing, *vbl. sb.* **1.** Living as, or following the occupation of, a beach-comber

(see BEACH *sb.* 4); also, the material found by a beach-comber.

a 1865 SMYTH *Sailor's Word-Bk.* (1867) 88 *Beach-combing*, loafing about a port to filch small things. 1910 S. REYNOLDS *Alongshore* IV. xxiii. 257 Our own beachcombings are odds and ends, flotsam and jetsam, to eke out a living. 1918 W. J. LOCKE *Rough Road* iv, An old hand who knew the whole thing backwards, from company promoting to beach combing.

2. *Mining.* Working the sands on a beach for gold, tin, or platinum.

1900 *Coal & Metal Miners' Pocketbk.* (ed. 6) Gloss.

beached, *ppl. a.* Add: **1.** *beached-up.*

1889 *Sci.-Gossip* XXV. 162 The fossils..were not in good preservation, owing to the beached-up condition of the formation.

2. (Example.)

1871 H. B. FORMAN *Living Poets* 392 Lying in the dark shadow of his beached Argo.

3. *fig.* Laid aside, discarded; unemployed (cf. *BEACH *sb.* 3 b).

[1925 FRASER & GIBBONS *Soldier & Sailor Words* 20 To be beached. To be turned adrift. Put out of employment. (Navy).] 1955 M. MCCARTHY *Charmed Life* (1956) i. 22 All the beached failures and second-raters of the twenties. 1962 J. TUNSTALL *Fishermen* xii. 259 Retired or 'beached' fishermen, as they forlornly call themselves.

beacher (bī·tʃəɹ). [f. BEACH *sb.* + -ER[1].] Used occas. in various (chiefly *slang*) senses (see quots.).

1923 U. L. SILBERRAD *Lett. J. Armiter* ii. 37 She does nothing with a passivity wonderful in a girl of her age, neither looking at her fellow beachers..or playing with the children. 1930 J. R. ELDER *Goldseekers in N.Z.* iv. 155 On this..beach he found..a number of miners who worked.. amid the black-iron sands to be found on the beaches...Many of these 'beachers' or 'surfacers' had prospected the whole southern part. 1946 J. IRVING *Royal Navalese* 28 *Beacher*, a trip ashore, on liberty. 1956 S. HOPE *Diggers' Paradise* xviii. 166 This body-surfing is exclusively an Australian development...It needs..practice if you want to show off.. shooting a 'beacher' on your back. The beacher..takes you from deep water right to the shallows.

beaching, *vbl. sb.*[1] Add: Also *attrib.*, as *beaching gear*, an appliance for hauling a seaplane to or from the beach.

1936 M. B. GARBER *Mod. Mil. Dict.* 42 *Beaching gear*, an arrangement of wheels to be attached to the hull of a seaplane to permit landing ashore. 1939 *Meccano Mag.* Aug. 470/3 When the hull was ready it was placed on a specially built 'beaching gear', a 15-ton 8-wheeled cradle of structural steel with a series of air tanks that give it buoyancy in the water.

Beach-la-mar (bītʃ la māɹ). Also **biche-.** [Alteration of Pg. *bicho do mar* *BÊCHE-DE-MER.] The jargon English used in the Western Pacific.

1911 W. CHURCHILL (*title*) Beach-la-mar. 1912 R. J. FLETCHER *Let.* 10 Sept. in *Isles of Illusion* (1923) 61, I will omit the Biche-la-mar and give you the gist of his story.

bea·chward, -wards, *advs.* [see -WARD, -WARDS.] In the direction of or towards the beach.

1831 *Blackw. Mag.* XXIX. 883 A neap-tide comes flowing beachward. 1895 KIPLING *Seven Seas* (1896) 83 The grandam of my grandam was the Lyre..That the Stealer stooping beachward filled with fire. 1909 *Daily Chron.* 21 Sept. 7/4 He put the dinghey about and ran beachwards.

beacon, *sb.* Add: **3. c.** = *BELISHA BEACON.

6. c. *Aeronaut.* A light placed at or in the vicinity of an aerodrome for the guidance of pilots; also *attrib.*

1918 [see *aerial lighthouse* s.v. *AERIAL 5]. 1927 V. W. PAGE *Mod. Aircraft* (1928) xvii. 700 These beacons are usually mounted on towers of sufficient height so that they will be above obstacles which might obstruct the beam. 1930 *Aircraft Engin.* II. 211 Night Beacons for Night Flying. 1957 *Encycl. Brit.* I. 228/2 Beacon lights, which mark major airways and airports..consist of a searchlight which throws a beam of white light elevated several degrees above the horizontal.

d. A radio transmitter enabling pilots to fix their position or the aerodrome staff to locate, identify, or guide aircraft; also *marker, radio beacon*, and *attrib.*

1919 *Pop. Sci. Monthly* Oct. 49/2 What is a radio beacon? 1922 Radio-beacon [see *RADIOPHARE]. 1929 *Techn. News Bull., Bureau of Standards* Nov. 108 The Bureau has found it necessary to give..attention to marker beacons, for two purposes. One is the adaptation of the marker beacon principle for use .. on the airways in connection with the visual beacon system. Another is the provision of marker beacons as auxiliaries..with fog-landing equipment. 1932 F. E. TERMAN *Radio Engineering* xvi. 593 The instant when the received beacon signal goes through zero. 1937 *Jrnl. R. Aeronaut. Soc.* July 591 An analysis of the flight of an aeroplane, when directed by means of a radio beacon. 1962 *Times* 21 Feb. 10/2 Direct communication with the capsule was briefly lost as it reentered the atmosphere because of an ionization process, but beacon signals were received.

beaconed, *a.* (Later examples.)

1890 *Harper's Mag.* Sept. 596/2 The beaconed ledges and buoyed reefs..slip helplessly into the Powder Fog. 1931 *Flight* 27 Mar. 275/2 Blind flying..has..its limit in the event of no beaconed route..being able to give the pilot a chance to unbroken control of his navigation.

Beaconite (bī·kənəit). [f. *beacon* in the title of I. Crewdson's *A Beacon to the Society of Friends*, 1835 + -ITE[1].] In the history of the Quakers, one of a party who, following the lead of Isaac Crewdson, held that the current Quaker doctrines were contrary to Scripture and seceded in 1836.

1835 M. HOWITT *Let.* 29 Dec. in *Autobiogr.* (1889) I. vii. 246, I should like to know whether any persons in whom I am interested in Liverpool join the Beaconites. 1836 J. SOUTHALL in *Jrnl. Friends Hist. Soc.* (1920) XVII. 82 The Beaconites preach up charity so exclusively. *Ibid.* 83 A..man..labouring under disadvantage from Beaconite associations. 1921 R. M. JONES *Later Periods Quakerism* I. ix. 310 The position of the 'Beaconites'—*i.e.* the position of literalism and ultra-orthodoxy.

bead, *sb.* Add: **5. a.** *spec.* of sweat, esp. on the face.

1596 [in Dict.]. 1888 A. C. GUNTER *Mr. Potter of Texas* v, [He] wipes the great beads of exhausted toil from his forehead. *Ibid.*, The beads of perspiration.

d. (Earlier and later examples.) Also *fig.*

1831 AUDUBON *Ornith. Biogr.* I. 294 He raised his piece until the *bead* (that being the name given by the Kentuckians to the *sight*) of the barrel was brought to a line with the spot he intended to hit. 1919 *Chambers's Jrnl.* June 399/1 I'd got a lovely bead on her with one of my own torpedoes. 1929 G. MITCHELL *Myst. Butcher's Shop* xii. 132 You've got a bead on your man all right.

6. a. In the names of various ornamental designs, as *bead and butt, flush, reel*, etc.

1799, 1842 Bead and flush [see *FLUSH *a.*[1] 5 b]. 1904 P. MACQUOID *Hist. Eng. Furniture* vii. 191 The sides are inlaid with the bead and spindle, or husk design so popular at this time. 1909 WEBSTER, Bead and butt, bead and reel. 1937 W. ROSE *Village Carpenter* iv. 42 The outside doors to the ordinary house were made to the still well-known orders of 'bead and butt' or 'bead and flush'. 1955 R. FASTNEDGE *Eng. Furniture Styles* 280 *Bead and reel*, a decorative border found in the form of inlay in the sixteenth and seventeenth centuries.

c. Used of any thickened, rounded edge (cf. **bead rim*).

1962 *Gloss. Terms Glass Ind.* (*B.S.I.*) 27 *Bead*, an enlarged, rounded edge of a glass article, or any raised section extending around the article.

8. *bead-berry, -note*; *bead-bonny, -brown, -dark, -eyed* adjs.; **bead lightning** (see quot. 1901); cf. **beaded lightning*; **bead-plane,** a carpenter's plane for running a bead on moulding; **bead rim,** a thickened, rounded rim (cf. *6 c); also *attrib.*; so *bead-rimmed* adj.; **bead screen,** beaded screen (see *BEADED *ppl. a.* 1 d); **bead-stone** *Zool.*, (*b*) (see quot. 1896); **bead-work,** (*a*) ornamental work with beads; also *attrib.*; so *bead-worked* adj.; (*b*) bead moulding (BEAD *sb.* 6).

1923 D. H. LAWRENCE *Birds, Beasts & Flowers* 107 He squatted and looked at me. With sticking-out, bead-berry eyes. 1881 G. M. HOPKINS *Poems* (1918) 53 The beadbonny ash that sits over the burn. 1912 W. DE LA MARE *Listeners* 14 Her small bead-brown eyes. 1937 —— *This Year: Next Year*, Low head outstretched, and bead-dark eyes. 1835 F. A. BUTLER *Jrnl.* II. 179 A fat, good tempered, rosy, bead-eyed, wet-haired, shining-faced looking man accosted me. 1901 *Rep. Smithsonian Inst. 1899* 129 Bead lightning..is a very beautiful luminous appearance, like a string of beads hung in a cloud. 1938 W. DE LA MARE *Memory* 60 Bead-note of bird where earth and elfland meet. 1858 SIMMONDS *Dict. Trade, Bead-plane*, a moulding plane of semicylindrical contour. 1936 *Antiquaries Jrnl.* XVI. 270 The bead rim and the countersunk handle. 1943 R. E. M. WHEELER *Maiden Cas.* III. 204 The rolled or 'bead' rim, which is predominantly characteristic of the 'Wessex hill-fort B' pottery, is in origin a device, not of the potter, but of the metal-worker, who thus gave rigidity to the lip of a vessel of thin metal. 1940 V. G. CHILDE *Prehist. Communities Brit. Isles* xiii. 251 The ceramic industry was industrialized, specialist potters turning out *en masse* bead-rimmed vases of Continental pattern. 1934 *Amat. Cine World* May 10/1 The most satisfactory type of screen so far evolved..is that known as the glass, bead or crystal. 1938 G. H. SEWELL *Amat. Film-Making* vi. 60 The silver and bead screens are an attempt to secure the utmost reflection of light from the projector. 1896 *Trans. Victoria Inst.* XXVIII. 206 Bead-stone is also called St. Cuthbert's beads, Fairy beads..and St. Boniface's money. They are the ring-like transverse sections of the so-called Vertical Column of stalked Echinoderms. 1751 DORRINGTON *Hermit* III. 260 He..makes 24 of those Plaits, which he weaves together, making a flat Piece of Bead-work. 1840, 1881 Bead-work [in Dict.]. 1919 R. FRY in *Athenæum* 27 June 529/2 As we look at Leech's drawings, or sit in a bead-work chair. 1909 *Daily Chron.* 18 Nov. 4/5 Painted and bead-worked lampshades. 1920 GALSWORTHY *In Chancery* I. xii, A gilt chair with a bead-worked seat.

bead, *v.* Add: **4.** Also *intr.*, const. *out.*

1938 G. GREENE *Brighton Rock* VII. vii. 326 The Brighton lamps beaded out towards Worthing.

5. To aim at (cf. BEAD *sb.* 5 d).

1888 *Eng. Illustr. Mag.* Dec. 214/1 Never fire until you have beaded your man.

beaded, *ppl. a.* Add: **1. d.** Applied to a projection screen (see quot. 1959).

1936 *Amat. Cine World* Mar. 552/2 Beaded screens are.. brighter than white and some makes are brighter than silver. 1938 G. H. SEWELL *Amat. Film-Making* vi. 60 There are three types of screen available, .. white .. silver .. beaded. 1959 W. S. SHARPS *Dict. Cinematogr.* 127/2 *Beaded screen*, a beaded projection screen is surfaced with small glass beads and a brilliant image is obtained that is strongly directional towards the projector.

2. b. Strung out like beads.

1937 WOOLDRIDGE & MORGAN *Physical Basis Geogr.* xxii. 388 Thus a series of fans or deltas may be strung along the 'feeding esker', like beads on a string. To such a form the term 'beaded esker' is applied. 1956 J. C. SWAYNE *Gloss. Geogr. Terms* 22 *Beaded lakes*, strings of long, narrow lakes between sand dunes.

4. b. Applied to a thickened, rounded rim, = *bead rim* (see *BEAD *sb.* 8).

1917 BULLEID & GRAY *Glastonbury Lake Village* II. xvi. 519 Bowls with beaded rims, a type of pottery belonging to the latter part of the La Tène III period.

5. *beaded-edge*, defining a kind of pneumatic tyre with bead or thickened edge that fits into the turned-over rim of the wheel; *beaded lightning*, bead lightning (see *BEAD *sb.* 8).

1902 *Windsor Mag.* May (Advt.), Wired or beaded edges optional. 1904 *Westm. Gaz.* 19 Sept. 9/1 The Bartlett beaded-edge—or, as it is more generally known, the Clincher—tyre. 1889 *Morning Post* 29 June 5/7 Photographs of lightning..arranged..according to their form— stream,..beaded..lightning.

beader. Add: **2. a.** One who sews beads on fabrics. **b.** One who puts a beading on an article.

1881 *Instr. Census Clerks* (1885) 105 Beader (Coach), Beader (Lace), Beader on Net. 1886 BESANT *Childr. Gibeon* II. iv, She was a beader: she was that clever with her fingers she could do all kinds of things. 1921 *Dict. Occup. Terms* (1927) § 254 *Beader*..a press hand who places machine folded body of tin..on block of beading machine and operates lever to form a beaded edge. *Ibid.* 407 *Beader, bead trimmer, bead worker*,..sews beads, by hand, upon a fabric foundation. *Ibid.* 414 *Beader*..applies waxy mixture..round edge of sole [etc.].

beading, *vbl. sb.* Add: **2. b.** A lace-like edging made of loops; also, an edging with openings through which ribbon, etc., may be run.

1900 *Amer. Mail Order Fashions* in *Americana Review* (1961) 16 Corset Cover..armholes followed with beading and narrow ribbon insertion. 1932 D. C. MINTER *Mod. Needlecraft* (*caption*) 137 Beading Stitch. 1964 *McCall's Sewing* xiii. 236/1 Beading is a very effective trim but tedious to do.

3. *spec.* in *Arch.* and *Joinery*, a bead; also collectively, the beads of a moulding. Also *attrib.*, as *beading plane* = *bead-plane* (see *BEAD *sb.* 8).

1858 SIMMONDS *Dict. Trade.* 1875 *Carpentry and Join.* 27 The beading planes are made of various widths.

beady, *a.* Add: **1. b.** *Comb.*, as **beady-eyed** *a.*

1873 M. F. MAHONY *Chron. Fermors* I. xiii. 249 That beady-eyed,..lanthorn-cheeked little lady. 1897 KIPLING in *Pearson's Mag.* Dec. 615/2 The 600 quick-footed, beady-eyed rank and file. 1904 *Daily Chron.* 15 Sept. 4/4 A beady-eyed inscrutable stare.

2. b. Of a garment, material, etc.

1892 *Longman's Mag.* May 76 Beady and buttony girls. 1897 *Daily News* 25 June 3/1 Cheap black, sheeny cloth, very beady, much trimmed.

beagling (bī·gliŋ), *vbl. sb.* [f. BEAGLE *sb.* + -ING[1].] The sport of hunting on foot with beagles. Hence **bea·gle** *v. intr.*, to hunt with beagles; **bea·gler.**

1824 *Sporting Mag.* XIV. 312/2 There is no prettier sport for youth than rabbit beagling. 1841 C. RIDLEY *Lett.* (1958) 65 Besides the *beaglers*, there was a horseman in white trousers. 1889 *Pall Mall G.* 24 Oct. 7/1, I am in the habit of accompanying my brothers out shooting and beagling. 1896 *Daily News* 20 July 8/2 Coursing—which, with 'beagling' and 'harriers', takes up a large part of the book. 1910 *Humanitarian* Mar. 21/1 Beaglers' manners. 1915 *Oxford Mag.* 5 Mar., He beagled and played cricket. 1923 J. O. PAGET *Beagles & Beagling* i. 15 The sport of beagling is every day becoming more popular.

beak, *sb.*[1] Add: **1. c.** *transf.* The long snout of a fish.

1889 in *Cent. Dict.* 1927 *Observer* 24 July 22 A 38-lb. cock-salmon...I was obliged to measure it, and I made it over 50 inches, from tail to beak.

III. *beak-mouth*; *beak-leaved, -sharp* adjs.

188. G. M. HOPKINS *Poems* (1918) 52 Only the beak-leaved boughs. 1921 D. H. LAWRENCE *Tortoises* 9 To open your tiny beak-mouth. 1933 W. DE LA MARE *Fleeting* 152 Beak-sharp nose.

beak, *sb.*[3] Add: (Further examples.)

1799 in *Spirit Pub. Jrnls.* III. 352 Took a gentle walk to the [police] office..paid my respects to Sir William, and the rest of the beaks. 1879 E. WALFORD *Londoniana* I. 233 We hope and trust [they] were brought before the 'beak' and duly punished.

b. *transf.* (*Schoolboys' slang.*) A schoolmaster.

1888 *Pall Mall G.* 9 Feb. 5/1 One of the Eton masters, or 'beaks', if we may be allowed to use a schoolboy phrase. 1916 E. F. BENSON *David Blaize* ix, You can have your crib spread out..and he won't see it. He didn't ought to be a beak at all. 1926 *Spectator* 11 Sept. 383/1 This scholarly Irishman and Eton 'beak'. 1960 BETJEMAN *Summoned by Bells* vii. 69 Comparing bruises, other boys could show Far worse ones that the beaks and prefects made.

beak, *v.* Add: **4.** *trans.* To ram (a vessel). Also *absol.* Hence **bea·king** *vbl. sb.*

1898 KIPLING *Fleet in Being* 55 We could run in and beak 'em while it's thick. I believe in beaking...Oh, he'd beak like a shot, if he saw his chance.

beaked, *ppl. a.* Add: **2. c. beaked whale** (see quot.).

1877 *Encycl. Brit.* VII. 280/2 The *Hyperoodontidæ*, or beaked-whales, are widely distributed in northern, southern, and tropical seas.

beaker. Add: **1. c.** *spec.* in *Archæol.* A type of tall wide-mouthed vessel found in the graves of a people who came to Britain from Central Europe in the early Bronze Age; hence *attrib.*, as *beaker-folk, -maker, people.*

1902 J. ABERCROMBY in *Jrnl. Anthrop. Inst.* XXXII. 374, I propose to substitute for the double-barrelled name 'drinking cup', the compacter term 'beaker'. 1906 *Archaeol. Æliana* II. 147 The Dilstor Park find..also proves the great difficulty of attempting to fix any relevant dates of Bronze Age beakers by a comparison either of their shape or ornamentation. 1916 *Jrnl. Anthrop. Inst.* XLVI. 117 The Borreby, or Beaker-Maker Type...Probably tall and often fair, light eyed, broad headed, short faced. 1922 *Ibid.* LII. 45 The culture is that known in Britain as that of the Beaker-folk. 1932 *Discovery* Aug. 270/1 The Bronze Age in England began roughly about 2,000 B.C. with the arrival of the 'Beaker' people on the east coast—people who used bronze and copper knives and pots of a special type. 1935 *Proc. Prehist. Soc.* I. 83 A solitary beaker from a cist at Barroose, Lonan, and three polished flint discoidal knives from the neighbourhood of Peel, are the only indications of the beaker people on the island. 1963 H. N. SAVORY in Foster & Alcock *Culture & Environment* iii. 26 The barbed and tanged arrowheads which were undoubtedly introduced into south Wales by the 'Beaker Folk'.

beak-iron. Add to def.: Also, an anvil with two projecting taper ends. (Add further examples of *bick-iron.*)

1828 SPEARMAN *Brit. Gunner* (ed. 2) 109 Anvil and Bick Iron. 1941 *Archit. Rev.* XC. 97/2 The bick iron, a narrow anvil, mounted about 3 feet 6 inches high, is for rounding, boring and cutting the iron hoops.

beaky, *a.* Add: **b.** *Comb.*, **beaky-nosed** *a.*, having a nose shaped like a beak.

1923 R. GRAVES *Feather Bed* 20 A red-haired beaky-nosed burly nun. 1937 C. DAY LEWIS *Starting Point* I. ii. 24 There was another girl, too—a beaky-nosed creature.

beam, *sb.*[1] Add: **16. b.** *transf.* The (width of the) hips or buttocks; esp. in colloq. phr. *broad in the beam.*

1929 H. WALPOLE *Hans Frost* II. vi. 177 He stood watching disgustedly Bigges' broad beam. 1944 MRS. HICKS-BEACH *Annabel & M. Verena* xxiv. 92 A cast-off of Jim's. He's grown too broad in the beam for it. 1960 I. CROSS *Backward Sex* i. 24 'I'm too broad around the beam.' 'What do you mean?..' 'My hips, silly..I've got wide hips.'

19. d. A directed flow of radiation or particles; freq. *attrib.*

1906 RUTHERFORD in *Phil. Mag.* XI. 168 Radium served as a source of α rays. A narrow beam of rays, after passing through a parallel slit, fell on a photographic plate. 1908 H. GEIGER in *Proc. R. Soc.* A. LXXXI. 174 Scattering is well known in the case of β-particles. A narrow pencil of β-rays emerges after passing through a metal plate as an ill-defined beam. 1933 *Discovery* Aug. 236/2 The new research will be termed 'beam' therapy and will be carried on at the London Radium Institute in Portland Place. 1937 *Ibid.* Nov. 329/2 The mechanical difficulties are obviated by using a beam of electrons for scanning. 1938 R. W. LAWSON tr. *Hevesy & Paneth's Man. Radioactivity* (ed. 2) xxvi. 288 Cockcroft and Walton succeeded in 1932 in the transmutation of light elements, by bombarding them with a beam of fast protons. *Ibid.* 291 The focusing action results in the generation of beams of ions. 1959 *Times* 19 Jan. 8/2 A brain operation in which for the first time a beam of protons was used instead of surgical instruments has been performed at the Werner Institute for Nuclear Chemistry at Uppsala.

24. *Radio Communication.* In full *radio, wireless beam*: radio waves transmitted as a beam, i.e. undispersed, from a special aerial system, part of which acts as a reflector; usu. *attrib.*

[1899 MARCONI in *Jrnl. Inst. Electrical Engin.* XXVIII. 274 Should it be necessary to direct a beam of rays in one given direction I prefer to use an arrangement similar to a Righi oscillator placed in the focal line of a suitable cylindrical parabolic reflector.] 1924 (July 2) MARCONI in *Jrnl. R. Soc. Arts* 25 July 607/2 The transmission and reception of intelligible signals over a distance of 1¼ miles of a beam system employing short waves and reflectors. 1924 *Westm. Gaz.* 24 July, Arrangements had been made for the erection of a beam station in Canada. 1924 *Daily News* 14 Nov. 7/4 Beam wireless. 1925 *Telegr. & Teleph. Jrnl.* Mar. 107/1 Australian papers give the following..information regarding the 'Beam' Radio Stations. 1927 *Daily Mail* 2 Dec. 11/1 Cable Companies and The Beam. 1928 *Morning Post* 23 Mar. 14 Beam-cable fusion [of the Eastern Exchange Cable Company and Marconi Company]...The beam system of telegraphy. 1934 *Nature* 24 Feb. 297/2 One advantage of short-wave transmission in radio communication is that it is possible to concentrate the radiation..in one direction, thus forming a beam of waves...A series of vertical aerials.. (called a beam array) will act like Hertz's reflector. 1946 *Proc. Inst. Radio Engineers* XXXIV. 335 (*title*) A Current Distribution for Broadside Arrays which Optimizes the Relationship between Beam Width and Side-Lobe Level. 1958 *Engineering* 31 Jan. 157/3 The aerial has been designed to give a sharp beamwidth at high signal strength. At 20 db the horizontal beamwidth extends about 0·65 deg.

b. *Aeronaut.* A directional radio transmission used to guide aircraft or missiles; freq. *attrib.*, as *beam approach.*

1927 *Sci. Amer.* Jan. 32 Radio along the Airways. Invisible beams guide birdmen in flights between European cities. 1929 *Aviation* 28 Dec. 1277/1 A landing beam transmitter buried flush with the ground in the center of the field. *Ibid.*, After finding the beam the pilot glides down it toward the field. 1933 K. HENNEY *Radio Engin. Handbk.* 537 The diminution of intensity as the airplane drops below the inclined axis is compensated by the increase of intensity due to approaching the beam transmitter. 1941 *Tee Emm* Oct. 8/1 The policy is that Beam Approach training shall be introduced..as equipment becomes available. 1949 A. R. WEYL *Guided Missiles* i. 26 The 'beam-rider' system in which the missile flies along a radar beam. 1958 *Times* 9 Oct. 10/2 It carries a warhead with proximity fuse and a beam guidance system. 1963 *Oxf. Mag.* 9 May 280/2 The Germans with their beam flying provided us with targets that would have been peculiarly favourable to aerial mine defence.

c. *to be on the beam*, to be on the course indicated by a radio beam; hence *fig.* (*colloq.*) to be on the right track, right, sane. So *to be off the beam.*

1941 *Amer. Speech* XVI. 238/2 Expressions disparaging a person's mental state..*off his beam* (airplane). 1941 *Daring Detective* Nov. 7/2 'Now we know we're on the beam,' said Brubach. 'Sex maniacs and drinking companions are definitely eliminated.' 1943 HUNT & PRINGLE *Service Slang* 49 *On the beam*, I follow what you are saying. (R.A.F.). 1948 *Observer* 18 Jan. 2/3 Hugh Burden, as Barnaby, was right on the beam from the start. 1949 *Jrnl. Brit. Interplanet. Soc.* VIII. 143 Thus rocket aeroplanes might seem 'off the beam' of true astronautics. 1954 'N. BLAKE' *Whisper in Gloom* I. iv. 58 Never heard of him. You're off the beam.

IV. beam-action, the action of a beam-engine; **beam aerial, antenna,** a directional aerial for transmitting a radio beam; **beam-blind** *a.*, uncritical of oneself (cf. sense 3 c in Dict.); so *beam-blindness*; **beam-engine,** a steam engine having a vibrating beam through which the piston effort is transmitted to the crank; **beam-splitter** *Photogr.*, a device consisting of a prismatic arrangement of mirrors (see quots.); so *beam-splitting* adj.

1896 *Daily News* 7 Jan. 6/5 The..beam-action locomotives of that extremely primitive line. 1926 *Gloss. Terms Electr. Engin.* (*B.S.I.*) *Beam Aerial System*, a combination of aerials with their earthing, tuning and reflecting arrangements so disposed as to concentrate the available radiated energy into a beam. 1945 *Electronic Engin.* XVII. 719 Indicating the target by means of a rotating beam aerial. 1935 K. HENNEY *Radio Engin. Handbk.* (ed. 2) 774 The elements of the Walmsley beam antenna. 1879 G. M. HOPKINS *Poems* (1918) 47 What hinders? Are you beam-blind, yet to a fault In a neighbour deft-handed. 1932 S. C. CARPENTER *Supernat. Relig.* iv. 94 The difference between his [*sc.* our Lord's] moteless eye and the beam-blindness of the Pharisees. 1844 H. STEPHENS *Bk. Farm* II. 312 The beam-engine of 6-horse power. *a* 1877 KNIGHT *Dict. Mech.* 257/1 *Beam-engine*.., an engine with an oscillating beam, to whose respective ends the connecting-rod from the piston and the pitman from the crank are attached. 1935 *Discovery* July 189/1 Iceland spar..beam splitter..entirely free from parallax. This system may be employed..to obtain three-colour negatives. 1940 *Chambers's Techn. Dict.* 83/1 *Beam-splitter* or *beam-splitting camera*, a camera for colour photography, in which the beam of light from the object is separated into three components by means of a complicated prism. 1956 A. L. M. SOWERBY *Dict. Photogr.* (ed. 18) 52 *Beamsplitter*, a combination of four mirrors.. used either for taking a stereoscopic pair of pictures with an ordinary single-lens camera, or for projecting or viewing a stereo pair so made. 1958 M. L. HALL et al. *Newnes Complete Amat. Photogr.* xxvi. 235 Stereos taken on the same frame by a beam-splitting device.

beam, *v.* Add: **3. b.** Of a person: to smile radiantly, broadly, or good-naturedly. Freq. const. *adv.*

1893 *Illustr. Sporting & Dram. News* 11 Mar. 28/2, I.. tried to beam on a friend—albeit, a poor beamer at the best. 1900 E. GLYN *Visits Eliz.* (1906) 14 He..jumped off his horse and beamed—just as if we had parted the best of friends. 1922 H. CRANE *Let.* 7 Aug. (1965) 95, I realize that he insulted me while he was here..but I shall beam on until all hope of his getting Bill an audience has vanished. 1936 R. A. FREEMAN *Penrose Myst.* I. i. 15 Mr. Polton took the coffee-pot in his..hands, beamed on it approvingly. 1937 A. J. CRONIN *Citadel* II. ii. 123 He beamed at the watch, for he could contemplate even inanimate objects..with that bland cordiality which was especially his own.

III. *trans.* **7.** To direct (a radio transmission) to a specific area. (Cf. *BEAM *sb.*[1] 24.)

1927 *Daily Express* 5 July 3/3 The King's message..was almost instantaneously 'beamed' to the receiving station at Milnerton, seven miles from Capetown. 1955 *Times* 29 July 3/1 A message announcing the inauguration of 28 new high-power radio transmitters..was 'beamed' to New York..and other regions. 1964 *Ibid.* 2 Apr. 20/7 Microwave equipment which will beam large numbers of telephone calls between Bloemfontein..and East London.

8. *fig.* To direct at, towards.

1956 L. ASHTON in A. Pryce-Jones *New Outline Mod. Knowl.* III. 300 Modern invention is often only a re-alignment, but for the great artists it is a realignment 'beamed', to use a modern word, on the future. 1959 *Listener* 24 Sept. 469/2 Mr. Khrushchev's tour is beamed— so to speak—at Radio Peking.

beamage (bī·mėdʒ). [f. BEAM *sb.*[1] + -AGE.] A deduction for loss of weight by evaporation in cooling, made in weighing the dressed carcass of a beast.

1902 in WEBSTER Suppl.

beamer. Add: Also, one who beams; in *Cricket* (see quot. 1961).

1857 DICKENS *Dorrit* II. xxxii. 603 The form of words which that benevolent beamer generally employed..'everything had been satisfactory to all parties'. 1893 [see *BEAM *v.* 3 b]. 1961 F. C. AVIS *Sportsman's Gloss.* 111/2 *Beamer*, a ball aimed high by the bowler, often to the height of the batsman's head. 1962 *Times* 21 Mar. 4/3 Let us not confuse orthodox fast bowling with bouncers and beamers.

beaming, *vbl. sb.* Add: **2.** *Comb.*, as *beaming-tool*; **beaming machine** (*a*) a machine for beaming and working hides; (*b*) a machine for filling the beams of looms with yarn; also called *beaming mill.*

a 1877 KNIGHT *Dict. Mech.* 257/2 Beaming-machine. (Leather.) *Ibid.*, The beaming-tool *f* is brought in contact with the leather or raised therefrom by means of the eccentric-rods. 1878 *Sci. Amer.* Suppl. 1938 Self-stopping Beaming Machine. *a* 1884 KNIGHT *Dict. Mech.* Suppl. 85/1 *Beaming Machine*, one for filling yarn beams of looms.

bean, *sb.* Add: **4. Queensland** or **Leichhardt's bean,** Australian names for a tall climbing leguminous plant *Entada scandens*, bearing long scimitar-shaped pods, which are used to make match-boxes, snuff-boxes, etc.: the seed is also called *match-box bean* (see MATCH *sb.*[2] 5) and *scimitar-pod* (see SCIMITAR 3).

1889 J. H. MAIDEN *Useful Native Plants Australia* 175 *Entada scandens*..'Queensland Bean'. 'Leichhardt Bean'.

5. b. The head. *slang* (*orig. U.S.*). **bean ball** *Baseball*, a ball pitched at the batter's head.

c 1905 C. DRYDEN *Champion Athletics* 16 While pitching Mr. Bender places much reliance on the bean ball. 1908 H. GREEN *Maison de Shine* 130 Pop swung on a guy an' come near knockin' his bean offa him. 1910 EVERS & FULLERTON *Touching Second* vi. 92 One of the greatest and most effective balls pitched is the 'bean ball'. 'Bean' is baseball for 'head'. 1912 A. H. LEWIS *Apaches N.Y.* 20 Beat it, before I bump me black-jack off your bean! 1923 R. D. PAINE *Comr. Rolling Ocean* x. 168 If these Dutchmen get nasty, bang their blighted beans together. 1924 WODEHOUSE *Bill the Conqueror* ii. 63 Have I got to clump you one on the side of the bean?

6. d. (Further examples of phr. (*to know*) *how many beans make five.*)

1886 ELWORTHY *W. Somerset Word-Bk.* 51 He knows how many beans make five, is a very common description of a cute, clever fellow. 1898 in M. Beerbohm *Around Theatres* (1924) I. 25 You say you've never heard How many beans make five? It's time you knew a thing or two— You don't know you're alive! 1958 'A. GILBERT' *Death against Clock* vi. 86 Mr. Crook knew how many beans make five.

e. Slang phrases: *not to know beans* (U.S.): not to know something, to be not well informed; *not to care beans* (U.S.), not to care at all; *a hill of beans* (orig. U.S.): a thing of little value (cf. sense 6 a in Dict.); *to spill the beans* (see *SPILL *v.*[1]); *to be full of beans*: to be full of energy, and in high spirits (cf. BEANY *a.*) (see also quot. 1874); *to give* (a person) *beans* (orig. U.S.): to deal severely with, to punish heavily; so *to get beans*; *old bean* (cf. OLD *a.* 8 a), a familiar form of address.

1833 A. GREENE *Life & Adv. D. Duckworth* II. 66 He don't know beans. 1855 *Yale Lit. Mag.* XX. 192 (Th.), Whatever he knows of Euclid and Greek, In Latin he don't know beans. 1857 *Knickerbocker* Feb. XLIX. 138 (Th.), I don't care beans for the railroad. 1863 'E. KIRKE' *My Southern Friends* v. 80, I..karn't take Preston's note—'taint wuth a hill o' beans. 1901 HARBEN *Westerfelt* 5 He didn't care a hill o' beans fer no gal. 1926 D. H. LAWRENCE *Let.* 4 Jan. (1962) II. 876 Saying my say and seeing other people sup it up doesn't amount to a hill o' beans, as far as I go.

1854 SURTEES *Handley Cross* xxxii. 254 'Ounds, 'osses, and men, are in a glorious state of excitement! Full o' beans and benevolence! 1874 HOTTEN *Slang Dict.* 171 *Full of beans*, arrogant, purseproud. A person whom sudden prosperity has made offensive and conceited, is said to be too 'full of beans'. Originally stable slang. 1875 DISRAELI *Let.* 20 Aug. in *Lett. to Lady Bradford* (1929) I. xvi. 275 The Sultan..was full of beans. 1911 GALSWORTHY *Patrician* I. x, Versatile, 'full of beans'. 1927 J. ELDER *Thomasina Toddy* xxiii. 226 We start off—oh, full of beans—and then we stop.

1835 in *Amer. Speech* (1965) XL. 127, I pose you heard ob de battle New Orleans, Whare Ole Gineral Jackson gib de British Beans. 1892 *Punch* 24 Sept. 133 Bad enough if you 'ave to wolf one, but it fair gives yer beans when 'tis two. 1893 *Pick-me-up* 5 Aug. 302 He would get beans at Bedford. 1900 *Daily News* 5 June 3/4 We started shelling them in the open, and gave them beans. 1914 *Evening News* 29 Sept. 2/2 We can't get them in the open, only on very rare occasions, and when we do, by gum, they don't half get beans! 1946 WODEHOUSE *Joy in Morning* xvii. 145 He wanted to give me beans, but Florence wouldn't let him. She said 'Father you are not to touch him. It was a pure misunderstanding.'

1917 'CONTACT' *Airman's Outings* 231 Chorus—'Goodnight, old bean.' 1918 *Blighty Christmas No.* 27 'What made you join the air service?' 'No earthly reason, old bean!' 1920 *Punch* 1 Sept. 168/1 The anxiety of the 'Bewildered Parent' who complains of the child of two and a-half years who addressed her learned parent as 'Old bean'. 1955 J. THOMAS *No Banners* xxix. 286, I say, old bean, let's stick together.

f. Formerly, a sovereign or a guinea; now only in phraseological use, a coin, a bit of money (*not a bean*, no money whatever, not a cent). *slang*.

1811 *Lex. Balatr.*, Bean, a guinea. 1834 AINSWORTH *Rookwood* III. vi, As yellow as a bean. 1837 *Ibid.* (rev. ed.) xxviii. 245 Offering a *bean* to *half a quid* (in other words, a guinea to a half-guinea). 1885 D. C. MURRAY *Rainbow Gold* v. vi, 'Here's some of the beans,' he continued figuratively, as he drew five sovereigns from the same pocket. 1928 GALSWORTHY *Swan Song* II. iv, They..never saved a bean, having no beans to save. 1928 D. L. SAYERS *Unpleasantness at Bellona Club* iii, None of the Fentimans ever had a bean, as I believe one says nowadays.

7. d. *bean-land* [OE. *bēan-land*], *-pod*, *-porridge*, *-row*, *-seed* [OE. *bēan-sǣd*], *-soup*, *-stubble* (examples), *-weevil* (examples).

1744 W. ELLIS *Mod. Husbandm.* Jan. iv. 46 Keeping the Bean Land clear of the Thistle. 1960 *Times* 5 July Agric. Suppl. iii/4 The traditional wheat and beanlands—now usually minus the beans—no longer produce the highest yields. 1733 W. ELLIS *Chiltern & Vale Farm.* xli. 348 Here is free access even to the minutest part of the Stalk-blossom or Bean-pod. 1913 D. H. LAWRENCE *Love Poems* 37 You ..who fall to earth At last like a bean-pod. 1821 R. B. THOMAS *Farmer's Almanack 1822* 2–6 June, Uncle Jeremy.. never turned up his nose at a bowl of bean-porridge. 1878 B. F. TAYLOR *Between Gates* 286 Two days more would.. ripen bean-porridge to the fine perfection of 'nine days old'. 1890 YEATS *Lake Isle of Innisfree* i, Nine bean-rows will I have there, a hive for the honey-bee. 1744 W. ELLIS *Mod. Husbandm.* Jan. iv. 44 They begin .. to sow their Bean Seed in two Forms or Methods. 1934 HALDANE & HUXLEY *Animal Biol.* x. 218 The weights of bean-seeds. 1837 J. C. NEAL *Charcoal Sk.* (1838) 98 Hollering oysters and bean soup has guv' me a splendid voice. 1743 W. ELLIS *Mod. Husbandm.* June iii. 20 To prepare a Bean Stubble for Turneps. 1833 *Ridgemont Farm Rep.* in *Brit. Husbandry* (1840) III. I. 138. The process of cultivation as thus described on fallow lands, is pursued in the same manner when the wheat has been sown on .. a bean-stubble. 1870 in *Mass. Agric. Rep.* I. 370, I sent specimens of the bean weevil..to Dr. G. A. Horn..who pronounces it to be..a native species (*B. varicornis* of Leconte). 1959 E. F. LINSSEN *Beetles Brit. Isles* II. 176 (*title*) Phytophaga, Bruchidae—Pea 'Weevils' and Bean 'Weevils'.

e. in the names of various machines for harvesting field beans and preparing them for use, as *bean-harvester*, *-mill*, *-sheller*, *-thresher*.

1858 SIMMONDS *Dict. Trade*, Bean-mill. *a* 1877 KNIGHT *Dict. Mech.* 257/2 Bean-harvester..Bean-sheller.

8. bean-bag, a small bag filled with beans, used esp. in children's games; **bean cake,** a material consisting of compressed beans, or some substitute, deprived of oil, used in China as a food and fertilizer; **bean curd, paste,** a thick jelly or paste made from beans, eaten in north China and adjacent countries; **bean-fed** *a. fig.*, living on the best of everything; **bean metal** *Salt-mining*, marl in the form of granules (cf. *beany metal*, s.v. *BEANY *a.* 2); **bean oil,** oil expressed from beans in the manufacture of bean cake, used as an illuminant; **bean-pole** (earlier and later examples); **bean-shooter** *U.S.*, 'a toy for shooting beans, shot, or other small missiles; a pea-shooter' (*Cent. Dict.* 1889).

1871 L. M. ALCOTT *Little Men* xxi. 321 A spirited exhibition of gymnastics..there was some danger of his..sending his bean-bags whizzing among the audience. 1895 KIPLING *Day's Work* (1898) xiii. 361 Here are your bean-bags for the Ladies' Competition. 1929 E. BOWEN *Last September* x. 122 At school sports; in a team race, one of those things with bean bags. 1968 *Guardian* 19 Apr. 1/1 The children, all receiving free meals, hurled beanbags at the women serving them. 1887 *Encycl. Brit.* XXII. 733/2 Bean-cake ..is also imported in large quantities from New-chwang, Chefoo [etc.]. 1909 WEBSTER, Bean curd. 1967 O. WYND *Walk Softly, Men Praying* vii. 110 The cry of the bean-curd seller. 1889 KIPLING *From Sea to Sea* vi, in *Wks.* (1900) I. 289 The bean-fed, well-groomed subaltern with the light coat and fox-terrier. 1892 *Cornh. Mag.* Sept. 263 A shaft is sunk till the 'flag' or 'bean metal' has been pierced. 1908 *Westm. Gaz.* 15 Oct. 13/2 According to reports from Chefoo, ..Bean-oil is giving place to kerosene. 1904 R. J. FARRER *Garden of Asia* 146 Buns stuffed with the cloying mustiness of bean-paste. 1798 T. B. HAZARD *Nailer Tom's Diary* 10 May (1930) 210/2, I dug[d] up Dung for to Put on the Garden and gott Bean and Pee Poles. 1900 E. BRUNCKEN *N. Amer. Forests* 61 Hop poles, bean poles, Christmas trees. 1890 *Congress. Rec.* Mar. 1920/1 Some boy, with a bean-shooter..struck Mr. Benjamin.

bean (bīn), *v. slang* (chiefly *U.S.*). [f. prec., sense *5 b.] *trans.* To hit on the head.

1910 *Amer. Mag.* 398/2 He is in extreme danger of being 'beaned', which, in baseball, means hit in the head. 1924 WODEHOUSE *Bill the Conqueror* v. 93 Why did you not bean him with a shoe before he could make his getaway? 1939 C. MORLEY *Kitty Foyle* xii. 124 She was beaned by a copy of *A Girl of the Limberlost* that fell from the third floor.

beanery (bī·nəri). *U.S. slang.* [f. BEAN *sb.* + -ERY.] A cheap restaurant (orig. one where beans were served).

1887 *Grip* (Toronto) 26 Feb. 8/2 Go to, illustrious reader; get thee to a beanery. 1888 *Texas Siftings* 7 Jan. 4/2 In a beanery you get biscuit fresh from the nest, Canada grouse from New Jersey and immigrant waiters. 1894 *San Francisco Midwinter Appeal* 10 Feb. 4/1 Papa Peakers gave a baked bean blowout to..the Press Men at his beanery. 1933 HEMINGWAY *Torrents of Spring* I. v. 50 Inside the door of the beanery Scripps O'Neil looked around him. 1955 E. POUND *Section: Rock-Drill* xci. 74 The american beaneries. 1967 WODEHOUSE *Company for Henry* iv. 66 He used to wander around town trying to find some beanery where they'd have their own specials.

bean-feast. Add earlier, later, and *attrib.* examples. Also, (*colloq.*) any festive occasion.

The earlier examples give conclusive evidence of the origin of the word from BEAN.

1805 *Sporting Mag.* XXVI. 304/2 At a late bean feast, a Gentleman Taylor, celebrated for his liberality, gave a rich treat to his men, at his occasional country residence. It was called a Bean Feast; but, exclusive of the beans, the table literally groaned with bacon [etc.]. 1875 W. T. VINCENT *Warlike Woolwich* 49 The holiday on the second Saturday in July, which is a special and extra holiday, known as 'Bean-feast day', and is usually spent in excursions to some country place and a dinner, at which beans form an indispensable dish. 1897 *Daily News* 23 Sept. 8/5 A beanfeast party from Woolwich. 1897 [see WAYZGOOSE b].

b. *fig.*

1899 R. WHITEING *No. 5 John St.* xxv. 250 The day was for the Empire-makers...To-night is the bean-feast of the 'hands', of the myriads in collar. 1906 W. J. LOCKE *Beloved Vagabond* iii, He..alluded to 'the regular beanfeast' they would have when they were married. 1922 MRS. A. SIDGWICK *Victorian* xxix, 'Won't life be a beanfeast?' she said. 'We shall enjoy ourselves.'

beanie (bī·ni). orig. *U.S.* [cf. *BEAN *sb.* 5 b.] A small close-fitting hat worn off the face.

1943 *Mademoiselle* Nov. 196 Matching felt beanie pins. 1944 *Sears Roebuck Midsummer Book* 22 *Beanie*, a clever little dink to perch atop your curls. 1958 *Vogue* Mid-Sept. 58 Back-of-the-head beanies..are a piece of evening prettiness. 1962 M. MCCARTHY *On Contrary* 188 Courageous Mrs. C...wears the same beanies every year regardless of the milliners. 1966 *Punch* 16 Mar. p. xv, Model girls snapping up..tiny beanie hats.

beano (bī·no). *slang.* orig. Printers' abbreviation of BEANFEAST. Later, in gen. use, a festive entertainment frequently ending in rowdyism.

1888 JACOBI *Printers' Vocab.*, *Beano*, a slang abbreviation for 'beanfeast', which is, however, usually termed 'goose' or wayzgoose by compositors. 1897 *N. & Q.* 8th Ser. XII. 175/1, I can remember hearing a street song, some years ago, in which one of the lines ran something to the effect that We don't have a beano every day. 1914 *Evening News* 15 Oct. 4/6, I wish you could hear the men on their first morning in the hospitals...'Fighting's a beano with this sort of thing for dessert.' 1920 W. J. LOCKE *House of Baltazar* ii, I'll get hold of Dorothy, and you and I and she'll have a little beano at the Carlton. 1924 *Blackw. Mag.* Oct. 494/1 There had been one of those awful last-night-of-leave beanos in town. *a* 1930 D. H. LAWRENCE *Last Poems* (1932) 269 O Pino What a bean-o! When we printed Lady C.! 1967 *Listener* 17 Aug. 205/3 Dear-heart, I fear we will have to make a token appearance at the beano those thrusting young String-Along's are giving tonight.

beany, *a.* Add: **2.** In salt-mining: *beany marl* or *metal* = *bean metal* (*BEAN *sb.* 8).

1886 R. HOLLAND *Gloss. Words Chester*, Beany marl. 1892 *Cornh. Mag.* Sept. 258 After a while what is locally termed 'beany metal' is reached.

bear, *sb.*[1] Add: **1. a.** (Examples of *black bear*.)

1781 PENNANT *Hist. Quadrup.* II. II. § II. xx. 286 The black bears of America form a very distinct variety. *c* 1880 *Cassell's Nat. Hist.* II. 167 The American Black Bear, *Ursus americanus*.

b. *like a bear with a sore head* and similar phrases: used with reference to bad-tempered behaviour.

1785 GROSE *Dict. Vulgar T.* s.v. *Grumble*, He grumbled like a bear with a sore ear. 1830 [in Dict.]. 1922 S. J. WEYMAN *Ovington's Bank* v. 49 He's like a bear with a sore head.

c. Also *spec.*, Russia.

[1794 W. B. STEVENS *Jrnl.* 15 Dec. (1965) II. 214 Those Russian Bears after having devoured the Unhappy Poles are..to direct their fell tusks against France.] 1804 M. WILMOT *Let.* 24 July in Londonderry & Hyde *Russ. Jrnls.* (1934) I. 147 Take the two Nations..and trust me the Bears would triumph. 1831 T. CAMPBELL *Wks.* (1907) 220 France turns from her abandoned friends afresh, And soothes the Bear that prowls for patriot flesh. 1853 *Punch* XXIV. 222 We recommend the Bear to hug himself as comfortably as he likes, in his own security, but we would advise him to keep his paws off from Turkey. 1939 W. S. CHURCHILL *Into Battle* (1941) 145 The left paw of the Bear bars Germany from the Black Sea. 1967 *Observer* 15 Jan. 32/8 When he allowed himself to be flown back to Moscow..he was consciously putting his head in the Bear's mouth.

d. A child's toy, esp. TEDDY *bear*.

1907 [see TEDDY]. 1928 A. A. MILNE *House at Pooh Corner* x. 178 In that enchanted place on the top of the Forest a little boy and his Bear will always be playing.

2. *to play the bear with* (colloq.): to play the deuce with, inflict great damage upon. ? *obs.*

1854 A. E. BAKER *Gloss. Northampt. Words* I. 38 A market-gardener says, 'A wet Saturday plays the bear with us'; *i.e.* keeps our customers away, and injures our goods. 1888 'R. BOLDREWOOD' *Robb. under Arms* II. ii. 26 Chaps that have got something on their minds can't stand idleness, it plays the bear with them. 1891 J. M. DIXON *Dict. Idiom. Phr.* s.v., The last storm has played the bear with my crops.

4. (Earlier example.)

1827 [see KOOLAH].

9. *bear-fat*, *-fur*, *-hunt*; *bear-furred* adj.

1780 W. FLEMING in N. D. Mereness *Trav. Amer. Col.* (1916) 640 Bear fat is preserved sweet and pure. 1920 D. H. LAWRENCE *Lost Girl* i. 11 Winter coats..flourished their bear-fur cuffs. 1926 E. SITWELL *Elegy on Dead Fashion* 3 Nor walk within vast bear-furred woods. 1803 *Lit. Mag.* (Philadelphia) Oct. 64 A grand bear hunt is proposed on the third Wednesday in October. 1845 MRS. KIRKLAND *Western Clearings* 125 They were going to have a bear-hunt out there.

10. bear animalcule, a microscopic animal of the group *Tardigrada*, a water-bear; **bear-covering** *vbl. sb.* (see quot. 1930); **bear dance,** a kind of dance practised by American Indians; **bear-fight** *fig.*, a riotous scuffle; a humorous name for a social gathering; hence as *v. intr.*, to behave rowdily; **bear-grass** *U.S.*, one of various species of Yucca (esp. *Yucca filamentosa*), or of some similar plants; **bear-lead** *v. trans.*, to act as bear-leader or travelling tutor to (a youth); to conduct or lead; hence, to supervise the conduct of, arrange the affairs of; hence *bear-leading* vbl. sb. and ppl. adj.; **bear-leader** (later example); also, a captor, custodian; **bear oak** *U.S.*, the black scrub-oak, *Quercus ilicifolia*; **bear-pit,** a sunken enclosure in which bears are kept for exhibition; **Bear State** *U.S.*, the state of Arkansas (occas. also California); **bear-wallow** *U.S.*, a hollow in the ground attributed to the wallowing of bears.

1889 *Cent. Dict.* s.v. *Arctisca*, Bear-animalcules. 1929 *Observer* 17 Nov. 4/2 'Bear-covering'..and the lowering of the New York discount rate..were..partly responsible for this..partial recovery. 1930 M. CLARK *Home Trade* 271 On the other hand, the bear who does not see prices fall in accordance with his hopes may also have to cut his loss and buy the shares he has already sold when not in possession of them at the best price possible. He enters the market as a buyer and by his buying sends up the price of the securities in which he is dealing. This buying is known as 'bear covering'. 1969 *Daily Tel.* 5 Mar. 4/1 The share-price, helped also by a measure of bear-covering, rose 1s 6d. *a* 1820 in *Western Rev.* II. 161 There are a number of other [Indian] dances, such as the bear dance, the Turkey dance. 1855 W. G. SIMMS *Forayers* xxxii. 373 It'll be a bear fight first, I tell you—tooth and nail! 1861 C. M. YONGE *Stokesley Secret* xiii. 209 'Oh, Papa, may we drink tea with you?'..'Yes, to be sure, if you won't make a bear fight..for your uncle.' 1873 —— *Pillars of House* IV. xlvi. 298, I shall..come back at dark when the bear fight is over. 1892 *Daily News* 23 Feb. 4/8 Did she bear-fight and play practical jokes on the Lord Admiral? 1900 *Ibid.* 24 Oct. 10/2 A don called Bedmynster used to bearfight in hall. 1750 T. WALKER *Jrnl.* 12 Apr. in J. S. Johnstone *First Explor. Kentucky* (1898), On the Banks is some Bear-Grass. 1909 *Cent. Dict.* Suppl., *Bear-grass* 2. a bunch-grass, *Stipa setigera*, ranging from the mountains of California, where it is considered valuable, to Oregon and Texas. 1891 C. DUNSTAN *Quita* I. i, I do call it hard .. to condemn me to bear-lead a savage. 1898 *Daily News* 31 Aug. 5/5 It was he who bear-led Admiral Avellane and his officers. *a* 1935 T. E. LAWRENCE *Mint* (1955) III. xi. 188 The Adjutant bear-led poor enduring Dolly down the three interminable lanes of us dressed hambones. 1968 *Listener* 21 Mar. 381/1 The Establishment, bear-led by the Hearst Press, had decided that this turbulent man..must..go. 1794 W. GODWIN *Caleb Williams* III. vi. 114 My bear-leaders were considerably surprised with my firmness. 1846 THACKERAY *Snobs of England* in *Punch* 18 Apr. 166/1 They pounced upon the stray nobility, and seized young lords travelling with their bear leaders. 1955 J. MASTERS *Coromandel!* ii. 87 She's picked up dancers and bearleaders and fiddlers and bullies. 1901 'LINESMAN' *Words Eyewitness* (1902) 289 A guard of honour, and a bear-leading general officer to see the creature safely and comfortably down to the sea. 1928 *Daily Tel.* 7 Aug. 6/5 There are.. more applicants for what used to be called 'bear-leading' than there are bears to be led. 1810 in *D.A.E.*, Bear oak. 1832 D. J. BROWNE *Sylva Amer.* 263 This diminutive species is known in the Northern and Middle States by the name of Bear Oak. 1849 MRS. F. L. MORTIMER *Near Home* 297 In the city [*sc.* Berne] there is a bear pit with three fat lazy living bears. 1862 F. LOCKER *London Lyrics* 66 (*title*) The Bear Pit at the Zoological Gardens. 1848 BARTLETT *Dict. Amer.* 392, I once asked a Western man if Arkansas abounded in bears, that it should be designated as the 'Bear State'. 1872 SCHELE DE VERE *Americanisms* 658 Arkansas is called the *Bear State*, though..the name is pronounced *Bar* State... California enjoys the same title. 1766 in *Amer. Speech* (1940) XV. 155/1 Two White Oaks Saplings by the Bear Wallow Drains. 1787 *Ibid.*, Standing by two Ironwood trees nigh a bear wallow. 1891 M. E. RYAN *Pagan of Alleghanies* v. 62 He rode..on through the columns of white-oak, whose feet are caressed by feathers and fern in the long, desolate 'bear-wallow'.

bear, *sb.*[6] *U.S.* Also **baire.** [Perh. the same as BEAR *sb.*[4], but cf. *BAR *sb.*[5]] (See quot. 1775.)

1775 B. ROMANS *Florida* 228 Baires are a kind of tent made of a light coarse cloth, like canvas gauze, called by the French *villemontiers*. 1797 F. BAILY *Jrnl. Tour* (1856) 309 Over the whole [bed] there is a large gauze net (called a *bear*), which is intended as a defence against the mosquitos. 1798 A. ELLICOTT in C. V. Mathews *Life & Lett.* (1908) 158 Our beds are all surrounded with a kind of thin curtains called bears to keep them off when we go to rest. 1895 G. KING *New Orleans* 65 The oarsmen [*c* 1726] made their mosquito *baires* for them.

bear, *v.*[1] Add: **11. b.** Of ice: to support the weight of a person, etc.

1768 WASHINGTON *Diaries* (1925) I. 246 Attempted to go into the Neck on the Ice, but it wd. not bear. 1833 *Chambers's Edin. Jrnl.* XI. 381/2 [He] succeeded in forming a curling rink weeks before the ice would bear in any other quarter. 1917 O. VINEY *Let.* 27 Dec. in E. V. Lucas *Post-Bag Diversions* (1934) 58 We cycled to Hartwell lake..as we heard that it was bearing.

16. For † *Obs.* read *Obs.* exc. *arch.*

1874 SWINBURNE *Bothwell* (1882) II. i. 97 It may be you do well to bear me hard.

23. a. Phr. *to bear date*: to be dated (as specified).

1440, *c* 1448, 1463 [in *M.E.D.* s.v. *date* n. (2) 2 (a)]. 1446 in *Trevelyan Papers* (1857) 27 Your letters..beryng date at Westm the xx day of Juyl the xxv yere of your.. regne. 1503–4 [in Dict.]. 1712, 1837 [see DATE *sb.*[2] 1]. 1875 *Encycl. Brit.* II. 593/2 Bearing date the 16th April 1871.

bear, *v.*[2] Add: *transf.* and *fig.* Also, to affect (the market) in this way. Hence **bea·ring** *vbl. sb.*, in quot. *attrib.*

1848 W. ARMSTRONG *Stocks* 19 This is perhaps the grand theatre for bulling and bearing stocks. 1861 *N.Y. Tribune* 29 Nov. (Bartlett), His Lordship is wholly guiltless of the charge which the 'Herald' in its anxiety to bear the market has brought against him. 1869 *Trans. Ill. Agric. Soc.* VII. 431 The strong influences which were used to 'bear' the hog market. 1887 *Pall Mall Gaz.* 26 Nov. 12 Bulling and Bearing Men's Lives. 1887 *Century Mag.* XXIII. 500 Even the best regulated thermometer will have its vagaries, and there is no protection against it when it does 'bear' the weather. 1897 *Daily News* 26 Apr. 5/4 'Bearing' influences are encouraged, and those whose interest is in forcing down prices are very active in that direction.

bearability (bɛ̄ᵊ:răbi·lĭti). *rare.* [f. BEARABLE *a.* + -ITY.] **a.** Ability to support the weight of a person, etc. **b.** The quality or state of being bearable.

1926 J. C. LINCOLN *Big Mogul* ii. 21 A boardwalk, some of its boards in the last stages of bearability. 1958 W. T. O'DEA *Soc. Hist. Lighting* 19 Tallow came in various grades, but even the best was near to the minimum standard of bearability.

bearable, *a.* Add: **2.** Capable of bearing or supporting a weight; esp. of ice that is safe for skating, etc. *rare.*

1895 *Daily News* 11 Feb. 5/4 The Serpentine being covered with bearable ice.

bearably, *adv.* (Examples.)

1892 *Black & White* 17 Dec. 710/1 The water was bearably cold. 1909 *Daily Chron.* 23 Jan. 4/4 Five out of every ten Americans..make you shiver when they open their mouths, three speak more or less bearably.

bearance. 1. (Later examples.)

1877 R. D. BLACKMORE *Erema* I. xvi. 190 'You are not what I thought of you,' I cried, being vexed beyond bearance by such words. 1923 D. H. LAWRENCE *Kangaroo* vi. 133 He was glum, silent, tortured by them all a bit beyond bearance.

bea·r-cat. [BEAR *sb.*[1]] **1.** The panda or red bear-cat (see PANDA).

1889 in *Cent. Dict.* 1901 [see PANDA].

2. The binturong.

1895 H. N. RIDLEY *Nat. Sci.* VI. 93 The Bear-cat (*Arctictis binturong*), the 'Bintūrong' or 'Menūrong' of the Malays, is generally obtained in Malacca, and is sometimes kept as a pet. 1910 *Encycl. Brit.* III. 952/1 This animal, also called the bear-cat, is allied to the palm-civets. 1934 *Times Educ. Suppl.* 14 Apr. p. iv/3 The binturong, or 'bear-cat', is a native of South-East Asia.

3. *transf.* An aggressive or forceful person; one of great energy or ability. *U.S. slang.*

1916 *Amer. Mag.* Apr. 77/1 The director said he had 'discovered' a bearcat, and engaged Alec at once. 1920 WODEHOUSE *Coming of Bill* I. i. 17 'Were they healthy?' 'Fit as fiddles.' 'And your grandparents?' 'Perfect bear-cats.' 1932 J. T. FARRELL *Young Lonigan* (1936) vii. 147 George was a bearcat at forging handwriting.

beard, *v.* Add: **3.** Esp. in fig. phr. *to beard the lion in his den* or *lair*.

1808 [see LION *sb.* 2 a]. 1882 W. S. GILBERT *Iolanthe* II, Beard the lion in his lair—None but the brave deserve the fair. 1894 R. D. BLACKMORE *Perlycross* II. iv. 68 Nothing less would satisfy her than to beard—if the metaphor applies to ladies—the lion in the den, the arch-accuser, in the very court of judgment.

bearded, *ppl. a.* Add: **1.** *bearded collie*.

1880 H. DALZIEL *Brit. Dogs* II. i. 208 In the west of Scotland there is a rough-faced and very shaggy-coated dog called the bearded colley, differing mainly from the true colley in being rough-faced, rather heavier built, altogether less elegant, and with a shaggier and harsher coat. 1928 F. T. BARTON *Kennel Encycl.* 91 *The Bearded Collie.*—In certain parts of Scotland there is a type of sheep-dog or collie which has a beard, or profusion of hair on the muzzle.

beardedness (bĭᵊ·ɹdėdnės). [f. BEARDED *ppl. a.* + -NESS.] Bearded condition.

1888 *Times* 18 Aug. 9/1 The beardedness of the chief. 1917 W. J. LOCKE *Red Planet* xii, Their composite paunchiness, beardedness, scragginess, .. impressed me unfavourably.

beardie. Add: Also **beardy. 1. b.** The bearded collie. Cf. *BEARDED *ppl. a.* 1.

1907 R. LEIGHTON et al. *New Bk. Dog* 102/2 Peeblesshire is regarded as the true home of the Beardie. 1931 A. CROXTON SMITH *About our Dogs* xviii. 269 The hero of Alfred Ollivant's 'Owd Bob' is said to have been a 'beardie'. 1945 BAKER *Austral. Lang.* iii. 73 The *beardy* or *bearded collie* is a type of sheep dog with long hair that resembles the hair of a Skye terrier.

2. (A nickname for) a bearded man. *colloq.* Also as *adj.*: bearded.

[1808 SCOTT in Lockhart *Life* (1837) I. i. 3 My father's grandfather was Walter Scott, well known in Tiviotdale by the surname of *Beardie*...*Beardie*..derived his cognomen from a venerable beard, which he wore unblemished by razor or scissors.] 1941 BAKER *Dict. Austral. Slang* 9 *Beardy*, a person with a beard or with long hair. 1959 I. & P. OPIE *Lore & Lang. Schoolchildren* iii. 54 Someone with a beard has 'Beardie!' or 'Fungus face!' shouted after him. 1960 *Spectator* 22 Apr. 569 There were more than forty thousand of us—weirdies and beardies, colonels and conchies, Communists and Liberals. 1961 *Observer* 28 May 1/5 The beardy weirdies with their querulous bleatings.

b. A local Australian nickname applied to a body of Southcotians, followers of John Wroe, who called themselves Christian Israelites.

1875 *Melbourne Spectator* 21 Aug. 190/1 The Beardies or Christian Israelites of Ballarat. 1905 *Daily Chron.* 8 Mar. 4/7 There is only one founder of a religion buried in Australia —John Wroe, who started the 'Christian Israelites', nicknamed the 'Beardies', since they never cut their hair. 1961 W. H. G. ARMYTAGE *Heavens Below* III. viii. 275 [John] Wroe..let his beard grow (his followers were known as the beardies).

beardlet. Add: Also, a small beard on a man's face.

1909 in *Cent. Dict.* Suppl. 1928 *Daily Express* 21 July 15/2 The beardlet adorning the under-lip of Lord Bertie of Thame. 1930 H. BELLOC *Richelieu* v. 73 The military moustache and beardlet of that pointed face.

bearer. Add: **1. f.** *bearer company*, *corps*, a medical division for military field service.

1878 (*title*) Stretcher-Bearers and Bearer-Companies. Manual of Exercises for, Horse Guards. 1899 *Daily News* 27 Nov. 5/3 More assistance will be necessary, both in skilled attention to wounded, and in carrying them from the field. This is provided for by 'bearer companies'. 1901 *Empire Rev.* I. 432 The work of the bearer company is intermittent, having to collect the sick and wounded after a fight, and to bring them to the field hospital. 1902 *Encycl. Brit.* XXV. 352/1 The bearer company..is composed of three officers, thirteen non-commissioned officers, and forty-eight privates of the Royal Army Medical Corps, with a detachment of the Army Service Corps for transport duties. 1909 WEBSTER, Bearer corps.

3. b. *bearer security*, an unregistered security the title to which is vested in its possessor and is transferred by simple delivery. So *bearer bond*, *stock*, *warrant*.

1897 *Westm. Gaz.* 5 Aug. 6/3 These bearer-warrants are, we venture to imagine, part of the £140,000 in fully-paid shares. 1906 *Ibid.* 12 July 9/2 All kinds of bearer securities. 1911 *Encycl. Brit.* XXV. 933/2 It has been found necessary to convert a part of the stocks into bearer bonds or shares. 1964 *Financial Times* 10 Feb. 10/7 Holders of Bearer Stock, to obtain this dividend, must deposit Coupon No. 251.

IV. 12. *Geom.* A space of any sort which is the locus of a collection of geometrical objects of any sort; e.g. a straight line is the *bearer* of the set of all points on it.

1909 in *Cent. Dict.* Suppl.

bearing, *vbl. sb.* Add: **13. a.** Also, the direction of an arriving radio wave or radar echo determined by a direction-finding system.

1907 *Electr. Engin.* II. 775/2 The bearings of quite a number of wireless telegraph stations in the Channel have been determined. 1920 *Discovery* May 131/2 By means of the direction-finding apparatus a bearing to the source from which the wireless waves are coming may be obtained. 1942 *Electronic Engin.* XV. 9 Long wave pulse transmitter.. to enable bearings on the aircraft to be obtained on the ground. 1950 *Gloss. Terms Radar (B.S.I.)* 7 *Automatic range, bearing*.., the automatic determination of range, bearing..by a mechanism actuated by the echo. 1952 F. J. WYLIE *Use of Radar at Sea* viii. 121 Taking a radar bearing of the side of a prominent land feature.

14. c. *line of bearing*: the direction in which a thing lies or moves.

1839 URE *Dict. Arts* 967 When the line of dip, and consequently the line of bearing which is at right angles to it, are unknown, they are sought for by making three bores. 1920 *Discovery* Nov. 330/1 Collingwood signalled to some of his ships to spread out.., the result being a partial resemblance to what is called a line of bearing.

17. bearing metal, metal having anti-frictional properties, used for bearings (sense 12); **bearing pile** (see PILE *sb.*[1] 3 b).

1893 T. E. THORPE *Dict. Appl. Chem.* III. 1052/1 Part of the tin in gun metal and in bearing metal is frequently replaced by zinc. 1905 *Engineering Mag.* XXIX. 592/1 A paper was presented by Professor Melvin Price, upon the microstructure and frictional characteristics in bearing metals. 1923 GLAZEBROOK *Dict. Appl. Physics* V. II. 233/1 The antimony-tin alloys are rarely used by themselves, but form the basis of bearing metals. 1906 *Westm. Gaz.* 23 Aug. 10/1 The foundations of each pier are formed by 145 bearing-piles.

bearing, *ppl. a.*[1] Add: **5.** Also of years, etc.

1742 ELLIS *Timber-Tree* (ed. 3) II. xxi. 130 The second [kind of cherry]..may be enjoyed at so cheap a Rate, in a bearing Year, as a Penny a Pound. 1860 *Trans. Mich. Agric. Soc.* X. 614 Of fruits, although it was not a 'bearing year', the exhibition was very fine.

Béarnaise (bēᴵăɹnē̆ᴵ·z, ‖bearnē̄z). [Fr., fem. of *béarnais* of Béarn, a region of south-western France.] *Béarnaise sauce*, also in Fr. form *sauce béarnaise*, a rich white sauce flavoured with tarragon.

1877 E. S. DALLAS *Kettner's Book of Table* 58 Béarnaise Sauce is made with yolks of egg and ounces of butter in equal numbers. 1885 MAJOR L— *Pytchley Bk. Refined Cookery* ii. 112 *Béarnaise*...This sauce should be the substance of and resemble Mayonnaise Sauce. 1906 I. BEETON *Househ. Managem.* x. 220 Béarnaise Sauce. 1926 RADCLYFFE HALL *Adam's Breed* II. vi. § 6. 202 Get that Sauce Béarnaise remade, and be quick! 1936 WODEHOUSE *Young Men in Spats* ix. 238 You wouldn't know a martyred proletariat if they brought it to you on a skewer with Béarnaise sauce.

bearskinned (bɛ̄ᵊ·ɹskind), *a.* [f. BEARSKIN + -ED[2].] Wearing a bearskin.

1694 T. ROGERS *Loyal & Impartial Satyrist* 14 When Bearskin'd Men in Floating Castles land. 1909 *Daily Chron.* 1 Oct. 4/6 The bearskinned heroes. 1966 *Listener* 17 Nov. 715/3 The bearskinned guards who stand outside the doors of the palace.

beast, *sb.* Add: **1. c.** (Later examples.)

1850 TENNYSON *In Mem.* cxviii. 27 Move upward, working out the beast, And let the ape and tiger die. 1895 W. D. HOWELLS *My Literary Passions* xvi. 110 The base of the mind is bestial, and so far the beast in us has insisted upon having its full say. 1948 J. THURBER (*title*) The Beast in Me and Other Animals.

4. b. *the beast with two backs*, a man and woman in the act of copulation. (Cf. Rabelais's *faire la bête à deux dos*.)

1604 SHAKES. *Oth.* I. i. 117 Your Daughter and the Moore are making the Beast with two backs. *a* 1693 *Urquhart's Rabelais* I. iii. 18 These two did often times do the two backed beast together. 1785 GROSE *Dict. Vulgar T.*, *Beast with two backs*, a man and woman in the act of copulation. 1921 A. HUXLEY *Crome Yellow* x. 94 There they were, Anne and Gombauld, moving together as though they were a single supple creature. The beast with two backs. 1922 JOYCE *Ulysses* 546 The beast that has two backs at midnight. 1950 AUDEN *Enchafèd Flood* (1951) iii. 115 It is possible to attach too much importance..to the sexual symbolism of the Whale as being at once the *vagina dentata* and the Beast with two backs or the parents-in-bed.

5. Now freq. in weaker sense.

1899 KIPLING *Stalky & Co.* 49 He's an awfully sensitive beast. 1923 D. L. SAYERS *Whose Body?* i. 13 I'll..try and console the poor little beast.

b. *fig.* Applied to things; also in colloq. phr. *a beast of a* . . .: an abominable or disgusting . . ., a beastly . . . .

1862 S. HALE *Lett.* (1919) 13 One thing shall I rejoice at,—my own bed,—for this husk thing we sleep on is a beast. 1878 W. S. GILBERT *Pinafore* I, It's a beast of a name. 1891 H. C. BUNNER *Zadoc Pine* 196 I've got to stay and finish my grind. It's a beast. 1898 *Westm. Gaz.* 4 May 2/2 I've had a beast of a time.

7. Phr. *the mark of the beast*: see MARK *sb.*[1] 11 c.

1382 [see MARK *sb.*[1] 11 c]. 1833 M. EDGEWORTH *Thoughts on Bores* in *Tales & Novels* XVII. 327 Lord Chesterfield set the mark of the beast, as he called it, on certain vulgarisms in pronunciation. 1847 J. BATES *Seventh Day Sabbath* (ed. 2) 59 Is it not clear that the first day of the week for the Sabbath or holy day is a mark of the beast. 1847 E. G. WHITE in J. White *Word to 'Little Flock'* 19 All we were required to do, was to give up God's Sabbath, and keep the Pope's, and then we should have the mark of the Beast. 1874 [see MARK *sb.*[1] 11 c]. 1888 E. G. WHITE *Great Controversy* xxv. 449 They will thereby accept the sign of allegiance to Rome—'the mark of the beast'. 1891 KIPLING *Life's Handicap* 208 (*title*) The mark of the beast. 1954 A. S. C. ROSS in *Neuphilologische Mitteilungen* LV. 22 At the beginning of the century many slang phrases were used to designate non-gentlemen e.g...*showing* (or similar introduction) *the mark of the beast*.

IV. c. *beast-dance*, *-epic* (cf. G. *tier-epos*), *-figure*, *-tale*; **beastman** *local*, a cattleman.

1900 J. C. LAWSON in *Ann. Brit. Sch. Athens* VI. 125 A beast-dance in Scyros. 1889 J. JACOBS *Fables of Aesop* I. 159 The so-called Beast-Epic of Reynard the Fox. 1908 *Publ. Mod. Lang. Assoc. Amer.* XXIII. 497 There was the distinctively medieval set of [animal] stories, told because of their own intrinsic power of affording amusement, to which is generally given the name 'beast epic'. Of this.. set of beast tales..which is so well represented in the branches of the French *Roman de Renard*, English offers few specimens. 1924 *Ibid.* XXXIX. 764 Chaucer's readers were educated to expect satirical hits and some more than didactic pieces of allegory in their beast epic and beast fable. 1932 E. WEEKLEY *Words & Names* ix. 121 The beast-epic obviously belongs to very primitive races. 1933 E. K. CHAMBERS *Eng. Folk-Play* 215, I am inclined to think that there must have been an early variant of the *ludus*, in which a single beast-figure was alone represented. 1311 in *N. & Q.* (1963) July 250/2 Bestemon. 1899 *Yorkshire Post* 26 Dec., Wanted, beastman,..on farm, near Hull. 1921 *Dict. Occupational Terms* (1927) § 22, *Cattleman*, beastman, byreman. 1888 J. JACOBS *Fables of Bidpai* p. xxxix, Benfey.. claims a Western (Greek) origin for beast-tales in which animals act 'as sich'. 1951 DICKINS & WILSON *Early M.E. Texts* 62 The comparative lack of beast tales in ME is particularly surprising when contrasted with their popularity abroad.

beastie. Add: (Further examples of non-Scottish use.) Also applied jocularly to insects.

1852 C. M. YONGE *Two Guardians* vii. 118 Having the little beastie down to show the company. 1952 *Arena* (N.Z.) XXXI. 14 But the leaving open of all doors and windows.. often lets in innumerable small beasties. 1955 E. POUND *Classic Anthol.* III. 181 Beasties on wing Time's dart shall touch presently.

beastish, *a.* (Later example.)

1923 A. HUXLEY *On Margin* 206 Human beings..he takes as he finds, noble and beastish, but..decent.

beastly, *a.* Add: **5.** Also with weakened emphasis, as a mere expletive.

1888 A. C. GUNTER *Mr. Potter* xxiii, I..came to tell her that story of her father's flying to escape arrest is all beastly rot. 1961 B. E. FERGUSSON *Watery Maze* xi. 278 The second major item was of equally beastly complexity.

beat, *v.*[1] Add: **3. b.** *to beat one's way*: to travel, or make one's way, *spec.* by illicit means. *U.S.*

1883 G. W. PECK *Peck's Sun* (Milwaukee) 16 June 1/2 He started home, beating his way on the trains. 1891 C. ROBERTS *Adrift Amer.* 53 To beat one's way, or to beat the conductor or the railroad, are equivalent phrases for travelling in the cars without paying any fare. *Ibid.* 195 There was nothing for it but to start out and beat my way there. 1904 *N.Y. Tribune* 8 May 10 [They should] stop trying to 'beat their way' by stealing a right of way that belongs to other people. 1926 J. BLACK *You can't Win* vi. 75 'Traveling?' he asked...'Beating it.'

c. *to beat it*: to go away, to 'clear out'. orig. *U.S.*

1906 H. GREEN *Actors' Boarding House* 108, I told 'em to beat it. 1908 A. RUHL *Other Americans* ii. 10 He'll be beatin' it for Paris pretty soon where the rest of 'em all went. 1917 MATHEWSON *Sec. Base Sloan* xiv. 193 You get your boss to let you off for that long, beat it over to Harrisville tomorrow night. *Ibid.* xxi. 283 Beat it! Get out of here. 1926 LEACOCK *Winnowed Wisd.* 79 'To your posts, all of you!' she cried, 'Beat it,' she honked. 1928 C. F. S. GAMBLE *North Sea Air Station* xii. 170 We were all awakened at 1.30 a.m., and told to beat it to the air station. 1930 P. WYNDHAM LEWIS *Apes of God* XII. vi. 469 That's enough! Don't waste my time but beat it... Get to hell out of this! 1951 'J. WYNDHAM' *Day of Triffids* ii. 38 Fedor had not waited once the plane was down. He had switched off the lights, and beat it.

10. (Further examples.) Phr. *to beat all*, *anything*, *everything*, etc., has been common in the U.S. since the second quarter of the 19th cent. (examples from 1831 onward in D.A. and D.A.E.).

1839 C. BRONTË *Let.* in Mrs. Gaskell *Life of C.B.* (1857) I. 199 Well! thought I, I have heard of love at first sight, but this beats all! 1863 DICKENS in *All Year Round* (Christmas No.) 7/2 'Well!' I says, 'if this don't beat everything!'

spec. in *Cricket*. (See quots.)

1867 G. H. SELKIRK *Guide Cricket Ground* ii. 22 The striker is said to be beat when he receives a ball so good that he is unable to play it properly and without a mistake. 1891 GRACE *Cricket* ix. 246 Try to have sufficient command of the ball so that if it beat the batsman it will hit the wicket. 1925 *Times* 27 Aug. 6/1 Douglas..beat the bat once or twice with balls that broke back and kept low.

b. (Later example.) Also, to baffle, perplex. Phrases *to beat the band*, *rap*: see the sbs.

1930 W. DE LA MARE *On the Edge* 135 Why you should have taken so much trouble about it simply beats me.

c. (Later example.)

1876 HARDY *Ethelberta* Sequel II. 309 She threatened to run away from him..and being the woman, of course she was sure to beat in the long run.

d. To get the better of (one) by trickery; to cheat or defraud. *U.S.*

1873 *Newton* (Kansas) *Kansan* 1 May 2/2 Johnson..left.. for the east, after having beat several creditors. 1886 *Century Mag.* Feb. 513/2 How do I know you ain't tryin to beat me? 1888 *Daily Inter-Ocean* 23 Mar. (Farmer), Two boys..were each fined twenty-five dollars...They have been beating boarding-houses all over the West Side. 1891 [see 3 b above]. 1904 *Columbus Even. Dispatch* 29 June 4 The..people who try to beat the street car conductors out of their fare.

e. To get ahead of; so, *to beat* (one) *to it*: to anticipate in doing something. orig. *U.S.*

1847 E. BRONTË *Wuthering Heights* II. xvii. 327 She would gladly have gathered it [*sc.* a letter] up..but Hareton beat her; he seized, and put it in his waistcoat. 1898 H. S. CANFIELD *Maid of Frontier* i. 21 He's watching the rangers, ..and will probably try to beat them here. 1904 *McClure's Mag.* Mar. 556/2 'They simply beat us to it,' complained Barrett, as we rode south. 1937 M. ALLINGHAM *Dancers in Mourning* xvi. 203 Poor old Chloe! I never thought she'd beat me to it.

f. Slang phr. *to have* (a person) *beat*: to be sure of his defeat; hence *gen.* to have got the better of; to baffle.

1916 'BOYD CABLE' *Action Front* 30 Why..you can't make your hands do what your tongue says 'as me beat. 1945 *Coast to Coast 1944* 103 Well, he's got me beat.

g. Slang phr. *can you beat it?*: an expression of surprise or amazement.

1917 WODEHOUSE *Uneasy Money* vii. 79 They pay me money for that!..Can you beat it? 1926 *S.P.E. Tract* XXIV. 121 *Can you beat it?* can you imagine anything worse than that? 1951 H. HASTINGS *Seagulls over Sorrento* III. iii, Oh, boy, can you beat it! Fourteen days leave... And we thought it was gonna be thirty days' cells.

14. b. (Later example.)

1962 A. NISBETT *Technique Sound Studio* 242 If two tones which are within about fifteen cycles per second of each other are played together the combined signal is heard to pulsate or beat at the difference frequency.

30. d. Phr. *to beat it out*: in Jazz (see quots.). Cf. *BEAT *sb.*[1] 4.

1945 L. SHELLY *Hepcats Jive Talk Dict.* 21 *Beat it out*, play it hot. 1947 *The Beat* July–Aug. 10/2 *Beat it out*, play 'hot' music with plenty of rhythm in the background. 1948 *Penguin Music Mag.* V. Feb. 64 In the style of a couple of rhythm boys beating it out.

40. f. To knock about savagely, to thrash. orig. *U.S.* Cf. *BEATING *vbl. sb.* 1 b.

1907 'O. HENRY' *Trimmed Lamp* (1916) 157, I wouldn't have a man..that didn't beat me up at least once a week. 1912 MULFORD & CLAY *Buck Peters* i. 24, I found that I'd beat up a couple of policemen when I was drunk. 1928 E. WALLACE *Flying Squad* i. 14, I don't say they intended killing him, but they certainly beat him up. 1938 E. AMBLER *Cause for Alarm* ix. 155 'Is he drunk?'..'No—beaten up.' 1939 *War Illustr.* 21 Oct. 190 We heard the police in the room next door beating up another prisoner. 1958 *Times Lit. Suppl.* 8 Aug. 450/2 Mr. Szabo was captured by the AVO and beaten up.

h. *to beat it up*: = to 'whoop it up' (see WHOOP *v.* 1 e). *slang.*

1933 *Times Lit. Suppl.* 19 Oct. 713/2 James, the son, grows up, 'beats it up' a little in Paris, and finally gets a job in Malaya. 1958 *Daily Tel.* 1 July 11/1 What sort of noise did the neighbours complain about? Did the Purdoms and their friends beat it up a little in the evenings?

i. *Aeronaut. slang.* (See quots.)

1940 *Bulletins from Britain* 11 Dec. 3 in *Amer. Speech* (1941) XVI. 76/1 *To beat up*, to dive on to a friendly flying field as practice, a gesture of triumph or sheer joie-de-vivre. 1942 T. RATTIGAN *Flare Path* 1, I put the old Wimpey into a dive and beat him up—you know, pulled out only a few feet above his head and stooged round him.

beat, *sb.*[1] Add: **1. b.** *Ballet.* = *BATTEMENT.

1913 C. D'ALBERT *Dancing* 6 Ailes de Pigeon...These two beats are performed with both feet off the floor. 1931 C. W. BEAUMONT *Dict. Techn. Terms Classical Ballet* 14 The noun *entrechat* is qualified..according to the number of crossings required; in this calculation the beats by each foot are included. 1950 *Ballet Ann.* IV. 69 An admirable facility for the execution of beats. 1952 KERSLEY & SINCLAIR *Dict. Ballet Terms* 17 *Beats*, the dancer executes a beat in the course of a jumping step when he strikes both calves sharply together so that they rebound. The legs are then ready to beat again, to change places before beating again, or to continue the movement.

4. (Later examples.) Also *spec.*, the strongly-marked rhythm of jazz and popular music.

1911 *Encycl. Brit.* XXIII. 279/1 Simple time is that in which the normal subdivision of its beats is by two, whether the number of the beats themselves is duple or triple. Compound time is that in which the beats are regularly divided by three. 1933 S. MOUGIN in *Hot News* (1935) June 16/1 Swing..is the balance found between the strong beat and the weak beat or beats in any bar. 1939 W. HOBSON *Amer. Jazz Music* iii. 49 To make this matter of beat and rhythm, so far as jazz is concerned, somewhat clearer for the layman, it may be pointed out that often in a jazz performance the only instruments playing regularly *on the beat* are, say, the bass drum and string bass; the rest are playing rhythms variously suspended around the beat. 1954 L. ARMSTRONG in *Grove's Dict. Mus.* IV. 600/2 Anything played with beat and soul is jazz. 1958 B. ULANOV *Hist. Jazz in Amer.* xxv. 349 *Beat*, jazz time; more meaningful to jazz musicians as an honorific description of rhythmic skill ('he gets a fine beat') than as a description of an underlying 2/4 or 4/4..or any other time. 1959 *Punch* 19 Aug. 60/2 Miss A likes it [*sc.* a pop record]. Oh, yes, it's got that beat and will sell. 1964 *Daily Tel.* 20 Feb. 22/6 Who dares to say that the cult of the beat groups by the young for the young is not vastly superior to the flood of pulp literature and horror comics pumped out for them by their commercially minded elders? 1967 *Crescendo* Dec. 33/3 The strange sounds emanating from an upstairs room revealed just who the Jazz Messengers were—yes, a beat group!

5. *beat-rhythm* (see quot.)

c 1873–4 G. M. HOPKINS *Note-books* (1937) 235 We have said that rhythm may be accentual or quantitative, that is go by beat or by time...The Saturnian..must have been chanted, as the beats often disagree with the word-accents. This beat-rhythm allows of development as much as time-rhythm.

7. *in* or *out of beat*, *off the beat*: making a regular or irregular succession of strokes.

1860 B. DENISON *Clocks, Watches, & Bells* (ed. 4) 101 The proper way to try whether a clock is in beat is to let the pendulum swing only just far enough for the escape, and then you will easily hear if the beats are unequal. 1874 BECKETT *Ibid.* (ed. 6) 73 When a clock with any kind of anchor escapement..sounds 'out of beat', it wants either one side lifting or the crutch bending. 1889 HASLUCK *Clock Jobber's Handybk.* v. 94 Put on the pendulum, and set the clock 'in beat'. The meaning of 'in beat' is, that the escape takes place at equal distances each side of the pendulum's centre of gravity...When 'in beat' the tick sounds regular, and nearly equal, differences of the drop making it slightly uneven.

8. b. *Radio.* The periodic variation of amplitude produced by the combination of oscillations of different frequencies. Also *attrib.*

1918 W. H. ECCLES *Wireless Telegr.* Gloss., *Beats* occur when two oscillations of differing frequencies occur simultaneously in the same system. The gradual change of phase difference causes the amplitudes to be opposed at one instant, and to concur at a later instant, with all the intermediate stages in the interval; the time between two successive oppositions, *i.e.* between two instants of minimum resultant amplitude, is called the time of a beat. The beat frequency is therefore equal to the difference between the frequencies of the two oscillations. *Ibid.*, *Beat Reception* (or Interference Reception) is the process of making high-frequency oscillations received by an antenna audibly evident by combining with them other oscillations of suitably different frequency. 1921 L. B. TURNER *Wireless Telegr.* 74 During a signal, the two oscillations are combined, with the interference or beat effect familiar in acoustics when two musical tones of slightly different pitch are mingled. 1942 *Electronic Engin.* XV. 120 It is often necessary to retune the oscillator after a short while to obtain the correct beat frequency.

10. (Earlier U.S. example.)

1721 *New-Eng. Courant* 2–9 Oct. 2/2 The several Clerks of the Train-Bands made a strict Enquiry at all the Houses within their respective Beats.

b. (Earlier and additional examples.)

1786 W. COWPER *Let.* 1 May (1904) III. 27 The Chesters, the Throckmortons, the Wrightes, are all of them good-natured agreeable people, and I rejoice, for your sake, that they lie all within your beat. 1854 MRS. GASKELL *Let.* 27 Oct. (1966) 318 She [*sc.* Florence Nightingale] said, 'The prostitutes come in perpetually—poor creatures staggering off their beat!' 1872 'MARK TWAIN' *Roughing It* vi. 28 His [*sc.* superintendent of a stage company's] beat or jurisdiction..was called a 'division'. 1937 N. MARSH *Vintage Murder* xxii. 245, I am very busy—consulting-room hours in town, and a wide country beat. 1965 *New Statesman* 7 May 715/2 The world is James Cameron's beat; he has visited every country but three covering the great events of our time, from the Allied victory in Germany to Vietnam, for press and television.

c. *U.S.* (See quot. 1857.)

[1736 in *Smithtown Rec.* (N.Y., 1898) 229 The place called the Horse beat.] 1834 AUDUBON *Ornith. Biog.* II. 433 When we went to look for the other [moose]..we found that he had..gone to the 'beat'. 1857 *Harper's Mag.* Nov. 819/1 The bear goes to and from his den..by certain paths called 'beats'...A bear will use the same 'beat' for years.

d. 'In Alabama and Mississippi, the principal subdivision of a county; a voting-precinct' (*Cent. Dict.* 1889).

1860 J. F. H. CLAIBORNE *Sam. Dale* x. 166 Governor Holmes appointed me..commissioner to take the census and organize beats or precincts. 1893 *Congress. Rec.* Feb. 2298/1 The evidence shows that his tickets were brought to the polls by friends of Turpin, and peddled there by them. This is shown to have been the case at Steep Creek beat,.. at Hopewell beat, in Loundes County. 1896 *Ibid.* Mar. 2788/1 Testimony was taken to show that fraud was committed in certain beats,—the River beat, Union, and one or two others.

e. The stretch of country assigned to a musterer (of sheep or cattle). *Austral.* and *N.Z.*

1873 J. E. TINNE *Wonderland of Antipodes* 38 As they complete each flock, it is turned over to a shepherd, who would drive it off with the aid of his dogs to a beat; possibly ten or twenty miles distant. 1941 BAKER *Dict. Austral. Slang* 9 *Beat*, the area patrolled by a sheep or cattle musterer. 1953 B. STRONACH *Musterer on Molesworth* ii. 13 Boy.. hunted them all [*sc.* the sheep] on to the next man's beat. 1958 J. PASCOE *N.Z. Sheep-Station* 22 Getting the sheep off the mountain is more difficult. Usually one man and his dogs will climb well above the sheep to what is known as the 'top beat'...The man half way down the slopes has what is called the 'middle beat'.

14. The action or an act of beating in order to rouse game.

1876 A. A. A. KINLOCH *Large Game Shooting* II. i. 2 The howdah elephants on which the sportsmen are mounted are distributed at intervals along the line, and as the beat progresses, some commotion may be observed as various species of game are roused. 1897 *Encycl. Sport* I. 84/2 The Sloth Bear..except when driven out in the course of a beat ..will not be observed during the day.

15. *U.S.* (chiefly *dial.*). **a.** That which surpasses, excels, or outdoes (something). Only in phr. *to see*, or *hear*, *the beat* (*of*).

c 1827 R. M. BIRD *News of Night* in E. H. O'Neill *Cowled Lover* (1941) 147 Did you ever see the beat o' that? 1833 S. SMITH *Major Downing* 129, I never see the beat of it. 1847 *Great Kalamazoo Hunt* (Philad.) 100 (Th.), You don't tell me so! Did I ever hear the beat o' that! 1878 H. B. STOWE *Poganuc P.* x. 86 That Bill is saassy enough to physic a hornbug. I never see the beat of him. 1888 'C. E. CRADDOCK' *Broomsedge Cove* v. 80 Waal, sir, eatin' supper by a tallow dip—who ever hearn the beat! 1907 'O. HENRY' *Trimmed Lamp* (1916) 209 Count Fernando Mazzini was his name. I never saw the beat of him for elegance.

b. *to get a beat on*: (see quot.).

1888 FARMER *Americanisms* s.v., *To get a beat on* is to get the advantage of... As used by thieves and their associates, *to get a beat on one*..also implies that the point has been scored by underhand, secret, or unlawful means.

c. A success scored against rivals by a reporter or newspaper; an item of news secured and published in advance of competitors.

1873 *Harper's Mag.* July 231/1 One of these 'enterprising' individuals secured his first 'beat' by riding in .. on a horse not his own, and taking news of the disaster to Philadelphia by rail, before an injunction was laid on the transmission of the truth. 1887 *Detroit Tribune* 27 June 3/2 They finally succeeded, and cheered lustily as the Ocean King steamed for New York with a big 'beat' for the Times. The office was safely reached, and the 'beat' appeared that morning. 1905 E. WALLACE *Four Just Men* i, The obedient reporter went forth. He returned in an hour in that state of mysterious agitation peculiar to the reporter who has got a 'beat'. 1940 GRAVES & HODGE *Long Week-end* xvii. 283 The newspapers paid well for 'beats', as 'scoops' were now called. 1969 S. GREENLEE *Spook who sat by Door* xvii. 152, I have a beat for you...That is the right word, beat? They stopped using scoop in the movies in the thirties.

16. [f. BEAT *ppl. a.*] An idle, worthless, or shiftless fellow. (Cf. DEAD-BEAT *sb.*[2]) *U.S.*

1865 *Canteen Songster* (1868) 26 Before 'this cruel war' broke out, he was what's termed 'a beat'. 1887 J. D. BILLINGS *Hard Tack & Coffee* 95 (Th.), The original idea of a beat was that of a lazy man or a shirk who would by hook or by crook get rid of all military or fatigue duty that he could. 1887 *Harper's Mag.* Dec. 107/1 The inevitable squad of 'beats' with bleary eyes and wolfish faces infesting the doorways of the saloons. 1903 *Boston Herald* 19 Aug., He would not loan money to policemen or firemen, stating that they were the biggest beats in the country.

beat, *sb.*[4]: see *BEAT GENERATION.

beat, *ppl. a.* Add: **1. b.** (Earlier and additional examples.) *beat out*, *up*, worn out, exhausted; *dead beat*: see DEAD BEAT *ppl. a.* in Dict. and Suppl.; *beat generation*: see as separate entry.

1758 in *Essex Inst. Hist. Coll.* (Salem, Mass.) XVIII. 92 Some was very much beat out by their march from Northampton. 1833 S. SMITH *Major Downing* 127 At last he got so beat out he couldn't only wrinkle his forehead and wink. 1914 *Daily Express* 2 Sept. 3/1 We were all beat up after four days of the hardest soldiering you ever dreamt of. 1945 L. SHELLY *Hepcats Jive Talk Dict.* 7 *Beat*, worn out. 1954 P. FRANKAU *Wreath for Enemy* III. iv. 191, I was too beat and hazy to take anything in. 1956 J. HEARNE *Stranger at Gate* xii. 92 'You look beat up.'..'I couldn't look as beat up as I feel.'

2. *beat elbow*, *hand*, *knee*: injuries incident to miners caused by the jarring and friction of the pick. Cf. *miner's elbow* s.v. MINER 6 b.

1905 *Daily Chron.* 17 Mar. 5/6 Judge Greenwell decided that 'beat hand' could not be classed as an accident...He found similarly in a claim with respect to 'beat knee'. 1907 *Ibid.* 17 May 5/5 'Beat hand', 'beat knee', and 'beat elbow'. 1935 A. J. CRONIN *Stars look Down* III. xii. 588 He worked with this committee on nystagmus, beat knee and the incidence of silicosis in non-metalliferous mines.

beatable, *a.* (Later example.)

1892 *Atlantic Monthly* Mar. 331 A more beatable child than Samuel Taylor it would be hard to find.

bea·temest, *a. U.S. dial.* Also **-omest.** [app. f. *beat 'em* + -EST *superl. suffix*.] Most excellent or splendid.

1831 *Boston Evening Transcript* 9 Aug. 2/3 Beatemest feller with a gun ever you seed. 1838 B. DRAKE *Tales* 30 Your the beatomest shakes I ever seed. 1851 S. JUDD *Margaret* (ed. 2) III. 245 Take it by and large fifty head a season, and she is the beatomest.

beaten, *ppl. a.* Add: **12.** Special Combs.: **beaten biscuit** *U.S.*, a small hard biscuit, the dough for which is thoroughly beaten and frequently folded; **beaten zone,** the area of ground struck by gun-fire; also *transf.*

1876 M. F. HENDERSON *Pract. Cooking* 69 Little machines ..for the purpose of making beaten biscuit. 1902 G. C. EGGLESTON *Dorothy South* 312 Slipping a surreptitious beaten biscuit into his pocket. 1918 E. S. FARROW *Dict. Mil. Terms* 67 Beaten Zone. 1962 *Times* 10 Oct. 13/7 Sticks are infinitely more efficient than stones for knocking down conkers: their 'beaten zone', as it were, is greater.

bea·tenest, *a. U.S. dial.* Also **beatin'est, -enes'.** [f. BEATEN *ppl. a.* or *beating* pres. pple. of BEAT *v.*[1] + -EST *superl. suffix*.] Most excellent or splendid; most unusual.

1860 *Harper's Mag.* 135 A countryman..attracted by the white slab..exclaimed, 'Well, if this ain't the beatenest town I ever saw!' 1874 E. EGGLESTON *Circuit Rider* 147, I reckon I am the beatin'est man to ax questions in this neck of timber. 1884 'MARK TWAIN' *Huck. Finn* xiii. 114 It's the beatenest thing I ever struck. 1889 *Harper's Mag.* Dec. 120/1 Thet thar rock house o' his'n,..I 'low it's the beatenes' house in creation. 1909 'E. C. HALL' *Aunt Jane of Kentucky* ii. 33 Of the preachers that ever I heard, he certainly is the beatenest.

beater. Add: **2. b.** *Paper-manuf.* A pulping machine. Also *attrib.* and *Comb.*, as *beater plate*, *roll*; **beater-man,** the operative who has charge of a pulping machine or beating-engine.

1825 J. NICHOLSON *Operat. Mechanic* 368 (Paper manufacture) The only difference between the washing-engine and the beater is that the teeth of the latter are finer. 1880 J. DUNBAR *Pract. Papermaker* 44 Two chests should be used, the beaterman emptying into the one, and the machine-man working from the other. *Ibid.* 71 The beater roll and plate should..be in good order.

3. A person or thing that beats or excels others. *U.S. colloq.*

1845 S. JUDD *Margaret* II. v. 283 Take it by and large.. and she is the beater of all. 1886 *Harper's Mag.* Sept. 580/2 I've heerd news that beats the beater! *Ibid.* Nov. 835/1 Well, for getting sunthin outer northing, she's a beater!

beat generation. [Etym. disputed; the evidence suggests that it is f. *BEAT *ppl. a.* 1 b, perh. infl. by *BEAT *sb.*[1] 4, but the coiner of the expression, J. Kerouac (1922–69), connected *beat* with BEAT(ITUDE (see quot. 1958[1]).] An expression applied at first to a group of young people, predominantly writers, artists, and their adherents, in San Francisco, later to similar groups elsewhere, adopting unconventional dress, manners, habits, etc., as a means of self-expression and social protest.

1952 J. C. HOLMES in *N.Y. Times* 16 Nov., Mag. 10/2 It was the face of a Beat Generation...It was John Kerouac.. who..several years ago..said 'You know, this is really a *beat* generation'. The origins of the word beat are obscure, but the meaning is only too clear to most Americans. More than the feeling of weariness, it implies the feeling of having been used, of being raw. It involves a sort of nakedness of mind. 1955 'JEAN-LOUIS' [= J. Kerouac] in *New World Writing* 10 Jazz of the Beat Generation...Here we were dealing with the pit and prune juice of poor beat life itself and the pathos of people in the Godawful streets. 1958 J. C. HOLMES in *Esquire* Feb. 35/2 'The Beat Generation,' he [*sc.* J. Kerouac] said, 'is basically a religious generation.' And later, in another interview, Kerouac amplified '..Beat means beatitude, not beat up.' 1958 *Sunday Times* 25 May 6/4 'Where go? What do? What for?..' That is the cry and philosophy of the beat generation, if a philosophy can be distilled from bewilderment, aimlessness and apathy. 1958 *Observer* 14 Sept. 4/5 The 'beat generation' is beginning to acquire the same kind of dubious place in American culture as the Young Angries in Britain. 1963 *New Statesman* 8 Feb. 202/3 The Beat Generation have come and gone, making a lot of noise but little real impact.

Hence **beat** *sb.*[4], **beatnik** (bī·tnik) [*-nik* arbitrarily after *SPUT)NIK, infl. by Yiddish *-nik*], **bea·tster**[2], one of the beat generation, one who leads a 'beat' life (cf. *BEAT *sb.*[1] 16, from which *beat* in the present sense cannot be entirely dissociated); **beat** *attrib.*, or as *adj.*, **bea·tniky** *a.*, of, or characteristic of, the beat generation or of a beatnik; **bea·tness,** such a state.

1955 [see above]. 1958 *Daily Express* 23 July 4/2 This [*sc.* San Francisco] is the home and the haunt of America's Beat generation and these are the Beatniks—or new barbarians. 1958 J. KEROUAC *On the Road* ix. 87 The beat countermen and dishgirls who made no bones about their beatness. 1958 *New Statesman* 6 Sept. 294/3 The 'beats' reached literary respectability in Jack Kerouac's *On the Road*. 1959 *Encounter* June 42 Portrait of the Beatnik...The Beatnik in his bold rebellion against American Bourgeois Values. *Ibid.* July 56/2 'Beat Zen' followers take dope or alcohol to reach a giggly state of ecstasy. 1959 *New Statesman* 6 June 795/1 The post-Presley teenage beatsters. 1959 *Guardian* 14 Oct. 7/3 He calls a flat a 'pad' (Beatnik language). 1964 *Listener* 5 Nov. 709/1 Suddenly everybody was slightly beatniky. 1965 *Spectator* 22 Jan. 98/2 One of the first changes he noticed was that the beats, instead of writing poems, were making films. 1966 *English Studies* XLVII. 154 In the mid-twentieth century the typical Bohemian has become the beatnik poet or pseudo-philosopher.

beating, *vbl. sb.* Add: **1. b.** *beating-up* (cf. *BEAT *v.*[1] 40 f.).

1915 A. CONAN DOYLE *Valley of Fear* II. vii. 629 I've a mind to send a couple of the boys round before evening to give him a beating up and see what they can get from him. 1939 G. GREENE *Confid. Agent* I. ii. 70 If they began with a beating-up, their next attempt was likely to be drastic. 1947 *Penguin New Writing* XXX. 127 The director seems determined to equal in savagery the beatings-up in *The Glass Key*.

6. beating-engine, a machine (*a*) for preparing the materials used in the manufacture of paper, strawboard, millboard, etc.; (*b*) for opening, beating, and cleaning cotton in cotton-manufacture.

1825 J. NICHOLSON *Operat. Mechanic* 368 (Paper-manufacture) The beating-engines..are seldom provided with these waste-pipes. 1846 DODD *Brit. Manuf.* VI. 21 The rags are..conveyed to the washing-engine,..then let off into the beating-engine. 1880 J. DUNBAR *Pract. Papermaker* 61 Alum intended for the beating-engine should be perfectly pure.

beatitude. Add: **1. b.** An honorific title (rendering Gr. μακαριότης) applied esp. to ecclesiastics of the Eastern Church of patriarchal rank.

1658 J. BURBURY tr. *Gualdo Priorato's Hist. Christina Q. Swedland* 127 When his Beatitude had declared four extraordinary Nuntii, to receive her on the confines of the Ecclesiastical State. 1751 CHAMBERS *Cycl.* s.v. *Holyness*, S. Gregory compliments some of his cotemporary bishops with, your beatitude, and your holyness. 1925 *Glasgow Herald* 7 Sept. 11 His Beatitude the Lord Photios, Patriarch and Pope of Alexandria.

beatitudinous (bi,æ:titiū·dinəs), *a. rare.* [f. L. *beātitūdin-* stem of *beātitūdo* blessedness + -OUS.] Characteristic of blessedness or happiness.

1926 D. H. LAWRENCE *Plumed Serpent* xiv. 229 Her childish..slightly imbecile face would take on a black, arch, beatitudinous look.

Beatle (bī·t'l). [f. 'The Beatles', the name of a group of pop singers from Liverpool.] Applied *attrib.* to the hair-style or other characteristics of 'The Beatles' or of their imitators. Also, **Beatlema·nia** [-MANIA], addiction to the Beatles and their characteristics; the frenzied behaviour of their admirers. Hence other nonce-formations (see quots.).

1963 *The Beatles* 3 Their unique hair-styles have forced hair-dressers to follow the Beatle cut—for girls as well as boys. 1963 *Daily Tel.* 1 Nov. 15/6 Beatle fans hold up premier's car. 1963 *Times* 27 Dec. 4/6 The social phenomenon of Beatlemania, which finds expression in handbags, balloons and other articles bearing the likeness of the loved ones, or in the hysterical screaming of young girls whenever the Beatle Quartet performs in public. 1964 *Daily Tel.* 10 Feb. 1/3 Outside, hundreds of squealing Beatlemaniacs carried such signs as 'We love you—never leave us'. *Ibid.* 20 Feb. 22/7 The first export consignment of Beatle wall-paper will be flown from Manchester to Canada to-day. 1964 *Daily Mail* 11 Feb. 6/8, I hope someone explains to the Americans that we wear our Beatle wigs and bowlers on separate occasions. 1968 *Courier-Mail* (Brisbane) 22 June 3/3 Of the Beatles he [*sc.* Maharishi Yogi] said: '..They weren't prepared to end their beatledom for meditation.'

beatnik: see *BEAT GENERATION.

Beatrician (bĭ,ătri·ʃən), *a.* [f. name of *Beatrice*, character in Dante's *Vita Nuova* and *Divina Commedia*, + -IAN.] Of, pertaining to, or resembling (Dante's vision of) Beatrice; of or concerning a revelatory or transcendental vision, experience, etc.

1943 C. WILLIAMS *Figure of Beatrice* 123 The Beatrician moment is a moment of revelation and communicated conversion by means of a girl. 1948 C. S. LEWIS in C. Williams *Arthurian Torso* II. iii. 116 The Beatrician experience may be defined as the recovery (in respect to one human being) of that vision of reality which would have been common to all men in respect to all things if Man had never fallen. 1958 *Times* 8 Nov. 7/3 Moments of Beatrician vision are experienced by a great many people when they first fall in love.

beatster[1] (bī·tstəɹ). *local.* Also **beetster.** [f. BEET *v.* 1 + -STER.] A mender or mounter of fishing-nets. (Sense of 1575 quot. uncertain.)

1575 *Aldeburgh Rec.* in *N. & Q.* (1920) VII. 226/2 To iiii[or] poore foulks for carienge y[e] beetster to hir buryall..viii[d]. 1641 S. SMITH *Herring-Busse Trade* 4 For the Tanning and Beetsters work. 1808 *Norfolk Tour* (ed. 6) 43 This fishery gives bread..to about 2,000 fishermen, and 4,000 braiders, beetsters, towers, rivers, ferry-men [etc.]. 1858 *N. & Q.* V. 116/1 Modern terms, such as one..hears among the fishermen of this coast [e.g. *Beatsters*, i.e. net menders (females)]. 1866 J. G. NALL *Great Yarmouth & Lowestoft* 292 In a long loft adjoining that of the ransackers, is the work-room of the *beetsters*, women and girls engaged in *betting* or mending up the nets. 1920 *N. & Q.* VII. 267/2 A notice-board at Lowestoft..'2 Beatsters wanted'. 1921 *Dict. Occup. Terms* (1927) § 398 *Beatster*, *beetster*; fishing net mender, net mender, net repairer, .. sometimes also a net mounter. 1932 *Daily Mail* 22 Sept. 4/6 Beatsters on the drift nets—stabbing their needles through a myriad meshes with rhythm.

beatster[2]: see *BEAT GENERATION.

beat-up, *sb.* and *a. slang.* Chiefly *U.S.* [cf. *BEAT *v.*[1] 40 f, i.] **a.** *sb.* (See quot. 1940.) **b.** *adj.* Worn out, shabby, showing signs of over-use.

1940 N. MONKS *Squadrons Up!* i. 14 Comes back over the 'drome, above the heads..twenty feet off the ground...The boys call this a 'beat-up'. 1946 *Amer. Speech* XXI. 251 *Beat up*, an adjectival phrase of all work meaning damaged, worn-out or of unimpressive appearance. 1951 J. D. SALINGER *Catcher in Rye* ii. 11 He showed us this old beat-up Navajo blanket. 1953 W. R. BURNETT *Vanity Row* xviii. 135 The girl was sitting once more in the beat-up leather chair.

Beatus (be,ā·tʊs). [L.] Esp. in *Beatus page*: the page of a psalter containing the initial B of Psalm 1 (Beatus vir...), which was frequently decorated in medieval manuscripts, often with elaborate designs.

[1892 J. H. MIDDLETON *Illum. MSS. Classical & Mediaeval Times* p. viii, The first letter B of the beginning of the Psalms (*Beatus vir* etc.) is in this and some other illuminated *Psalters*..of such size and elaboration that it occupies most of the first page.] 1911 J. A. HERBERT *Illum. MSS.* x. 180 The *Beatus vir* page is of an unusual and amusing type. The 'B', made of narrow entwined ribbons on a gold field, forms a small and rather insignificant foundation; but round about it..are eight gold medallions containing delightful figures of animal musicians. 1926 E. G. MILLAR *Eng. Illum. MSS. X to XIII Cent.* 91 David harping, and Beatus page; the medallions at bottom and top of the stem of the B contain respectively David slaying Goliath, and bringing Goliath's head to Saul. 1954 M. RICKERT *Painting in Brit.: Middle Ages* iv. 80 The identification of the Edmund artist with the illuminator of the Albani *Beatus* initial seems to be confirmed by a comparison..with the very lively single combat scene above the B on the same *Beatus* page of the Psalter. *Ibid.* 98 A full-page *Beatus* and eight large initials to the psalms.

beaty (bī·ti), *a.* [f. *BEAT *sb.*[1] 4 + -Y[1].] (Of jazz, etc.) full of 'beat', swinging.

1956 S. TRAILL *Play that Music* vi. 60 Drumming.. reached an altogether higher standard of crisp beaty playing. 1960 *Melody Maker* 31 Dec. 6/3 Dean sings with the requisite fervour to a beaty accompaniment. 1963 *Today* 30 Nov. 18/4 Ringo ad-libs a beaty blues.

beau, *sb.* Add: **3.** *attrib.*, † **beau-catcher,** a kiss-curl. *Obs.*

1818 *Publ. Ledger* 18 Apr. 3/2 A girl..twisting her hair into rings, which they term 'beau-catchers'. 1857 M. J. HOLMES *Meadow-Brook* ii, Arranging just in front of her ears two spit curls, sometimes called 'beau catchers'. 1909 WARE *Passing Eng.* 23/1 *Beau-catcher*, a flat hook-shaped curl, after the Spanish manner, gummed on each temple, and made of the short temple hair, spelt sometimes *bow-catcher*...Now obsolete on this side of the Pyrenees.

Beaufort scale (bōᵘ·fǫ̈ɹt). *Meteorol.* [Devised by Admiral Sir Francis *Beaufort* (1774–1857).] A series of numbers from 0 to 12 assigned by Captain (later Admiral Sir Francis) Beaufort to indicate the strength of the wind from a calm, force 0, to a hurricane, force 12.

The Beaufort numbers and their limits of velocity have varied in different countries since the scale was devised *c* 1805. The specification of the Beaufort scale adopted now by the Meteorological Office allows for 17 Beaufort numbers ranging in velocity from 0–0·2 metres per second (0) to 56·1–61·2 metres per second (12).

BEAU GESTE

1858 [see STORM sb. 1 b]. 1906 *Beaufort Scale of Wind-Force* (Rep. Meteorol. Office) 5 Assume first that the velocity of motion of the air is the ultimate measurement to which the numbers of the Beaufort scale shall be referred. 1933 *Jane's Fighting Ships* 22 *Trials*..151,000 S.H.P. = 32·07 kts. (run in bad weather, wind force 6 Beaufort scale). 1961 *Times* 5 Aug. 3/1 South-westerly winds between force 5 and force 6 on the Beaufort scale meant rigorous yachting conditions.

‖**beau geste** (bo ʒɛst). [Fr., = fine gesture.] A display of magnanimity.

1920 *John O'London's Weekly* 21 Feb. 562/1 The *beau geste* is less a physical than a mental and literary revelation. 1922 *Westm. Gaz.* 9 Dec., I think Germany can pay,..and Italy is too poor to make the *beau geste* of abandoning her claims. 1925 W. DEEPING *Sorrell & Son* xv. § 3 He gave in... He was not going to quarrel with a *beau geste*.

Beau-gregory (bōᵘgre·gōri). Also **Beau Gregory, Beau Gregoire** (bo grigwa·r). [Etym. unknown.] A brightly coloured fish (*Eupomacentrus leucostictus*) of the family *Pomacentridæ*, found in Florida and the West Indies; the cockeye pilot.

1847 R. H. SCHOMBURGK *Hist. Barbados* III. 674 Pomacentrus leucostictus, *Müll. et Tr.* nov. spec...Black Pilot. Beau Gregory...In the younger specimens the white dots are much more distinct, and this may have induced the fishermen to give them the name of Beau Gregory; the full-grown specimen is called Black Pilot. 1929 C. M. BREDER *Field Bk. Marine Fishes* 199 Beau-gregory occurs in the West Indies north to Florida. 1930 D. S. JORDAN et al. *Check List of Fishes of N. & Middle Amer.* 413 Cockeye pilot; Beau Gregoire; Black pilot. West Indies to Snapper Banks, West Florida. 1933 BEEBE & TEE-VAN *Shore Fishes of Bermuda* 191 Yellow Belly; Beau Gregory.

Beaujolais (bɔʒolɛ). [Name of a former district of France.] The name given to several light Burgundies, both red and white, produced in the Beaujolais district.

1863 T. G. SHAW *Wine* viii. 258 All common cheap French red wines seem now to have got the name of Beaujolais, as white have that of Chablis. 1879 STEVENSON *Trav. Cevennes* 33, I uncorked my bottle of Beaujolais, and asked the host to join me. 1960 *Farmer & Stockbreeder* 22 Mar. Suppl. 6/2 The next stew is a party dish and is best made with topside of beef and a respectable wine like Beaujolais.

Beaumé: see *BAUMÉ.

beaumontage (bōᵘmɒntɑ̄·ʒ). Also **-montague.** [?f. name of Élie de *Beaumont* (1798–1874), French geologist.] A composition (of various mixtures) used by cabinet-makers, pattern-makers, and iron-founders for the concealment of cracks and holes in wood or metal work. Also *fig.* (see quot. 1895).

1886 ELWORTHY *W. Somerset Word-bk.*, Boman teg. 1888 *Chambers's Encycl.* II. 3 Beaumontague is a composition..which is used to fill up cracks in an iron casting. 1895 BREWER *Dict. Phrase & Fable*, *Beaumontage* (pronounced *bo-mon-taig*), bad work, especially ill-fitting carpenter's work; literary padding; paste and scissors literature. 1899 HASLUCK *Wood Finishing* ix. 67 Beaumontage..is generally called 'stopping-out wax'.

Beaune (bōᵘn). [The name of a town in the department of Côte d'Or, France.] A red wine of Burgundy, produced in the district around Beaune.

1818 MOORE *Fudge Fam. Paris* iii. 51 Some glasses of *Beaune*. 1841 THACKERAY *Mem. Gormandising* Misc. Essays, &c. (1885) 386 Always drink red wine with beefsteaks..good Beaune, say. 1863 T. G. SHAW *Wine* viii. 255 Beaune, 1858, first growth, high flavour, much body, fine. 1875 *Ure's Dict. Arts* III. 1143 Haute-Burgogne..produces the most famous wines in Burgundy...Here grow the renowned *Volnay, Pomard, Beaune, Nuits*. 1958 A. L. SIMON *Dict. Wines* 27/2 *Beaune*..Among the many fine red wines of the vineyards of Beaune, some of the better known are: Grèves, Fèves, Clos des Mouches [etc.].

‖**beau rôle** (bo rōᵘl). [Fr., = fine rôle.] A fine acting part; the leading part; also *transf.*

1887 *Athenæum* 29 Oct. 561/3 Each assumed the moral government of the world without appealing to any revelation. This assumption, of course, gives the *beau rôle* to a prophet. 1927 in D. MCCARTHY *Drama* (1940) 102 In Ibsen the woman nearly always has the *beau rôle*. 1938 *Times Lit. Suppl.* 10 Sept. 581/4 The *beau rôle* at Fontenoy was that of the Irish Brigade. 1939 A. TOYNBEE *Study of Hist.* IV. 221 The beau rôle of 'the prisoner of the Vatican'.

‖**beau sabreur** (bo sabrō·r). [Fr., lit. 'handsome (or fine) swordsman'; cf. SABREUR.] A sobriquet for Joachim Murat (1767–1815), a famous French cavalry officer, brother-in-law of Napoleon I; hence used *transf.*: a fine soldier, a handsome or dashing adventurer.

1834 *Baboo & Other Tales Descr. Soc. India* I. vii. 113 Handsome, gallant, and young, he held the place that Murat did in the armies of Italy, and might have been called our 'beau sabreur'. Although, like Murat, without many pretensions to genius, his soldierly qualities invested him with a kind of romance. 1865 'OUIDA' *Strathmore* I. i. 9 The Beau Sabreur, as he had been nicknamed, à la Murat, was as soft as silk in the hands of a beauty. 1888 *Athenæum* 5 May 573/1 [His] long fair hair, bound in braids about his head, after the fashion of his people (a fashion revived by the *beaux sabreurs* of Napoleon's time). 1919 G. B. SHAW *Playlets of War* 234 We women admire..the heroic warrior, the *beau sabreur*. 1947 WODEHOUSE *Full Moon* i. 14 The latter's sister..considered that *beau sabreur* and man about town a blot on the escutcheon of a proud family.

beaut (biūt). Chiefly *U.S., Austral.* and *N.Z. slang*. Also (now rare) **bute.** [Abbrev. of BEAUTY *sb.* 5.] A beautiful or outstanding person or thing. Also *attrib.*, passing into *adj.*

1866 'F. KIRKLAND' *Bk. Anecdotes* 178 Hopeful is not a beauty..and though some of the rustic wits call him 'Beaut', he is well aware that they intend it for irony. 1896 ADE *Artie* i. 5 They was beauts too. 1903 CLAPIN *Dict. Amer.*, Bute. 1909 T. H. THOMPSON *Ballads about Business* 79 Well, I guess she ain't a bute. 1910 O. JOHNSON *Varmint* i. 16 The tin one was easier, but it's a beaut. *Ibid.* v. 67 Some of the fellows have perfect beauts. 1929 W. SMYTH *Girl from Mason Creek* xv. 161 We didn't quite git th'hang of things first go off, but now I reckon it will be a 'beaut'. 1934 T. WOOD *Cobbers* v. 68 The last of the big sticks in this sector, and a real beaut. 1936 WODEHOUSE *Laughing Gas* xiii. 139, I seemed to be getting a lot of steam behind the punch...I got those two bozoes a couple of beauts! 1944 *Coast to Coast* 98 'We're a pair of beauts,' he said with the semblance of a grin. 1952 *Arena* (N.Z.) XXXII. 5, I scratched her a beaut. 1952 J. CLEARY *Sundowners* i. 6 Ain't that a beaut farm? 1957 'N. SHUTE' *On Beach* ii. 71 It's been a beaut evening. 1968 K. WEATHERLY *Roo Shooter* 22 The bushie grabbed a plate and headed for the camp oven. 'You beaut,' he said. 'Coffee 'll do.'

‖**beauté du diable** (botɛ,dü,dyabl). [Fr., lit. 'devil's beauty'.] Superficial attractiveness; captivating charm.

[1825 H. WILSON *Memoirs* III. 8 She possessed, what the French term, la beauté du diable, namely, youth, and a particularly youthful appearance.] 1863 *Fraser's Mag.* Mar. 309/1 The increasing number of barristers who find themselves unable to resist the *beauté du diable* of the fair daughters and sisters of the proscribed race [sc. attorneys]. 1870 R. BROUGHTON *Red as Rose* I. viii. 157 Hers is essentially *beauté du diable*..one of those little faces that have been at the bottom of half the mischiefs the world has seen. 1936 E. BOWEN in Verschoyle *Eng. Novelists* 104 Henry Crawford is more energetic, dashing and unscrupulous. He has a certain *beauté du diable*. 1967 H. MCCLOY *Further Side of Fear* iii. 41 He studied both photographs. 'They can't mean ugly as sin! They must mean *beauté du diable*.'

beautician (biūti·ʃən). orig. *U.S.* [f. BEAUTY + -ICIAN.] One who runs a beauty parlour; a beauty specialist.

1924 *Cleveland Teleph. Directory* (Mencken, Suppl. I, 572), Very efficient beauticians. 1926 *Glasgow Herald* 12 June 8 The immense growth of 'beauty parlors' in the United States has added to the American language the word 'beautician'. 1934 *Southport Visiter* 8 Dec. (Advt.), J. H...beautician..modern hairdressing..rejuvenating facials. 1948 M. SHARP *Foolish Gentlewoman* xviii. 190 'I'm going to be a Beautician.'..'They get good money..and it's nice work.' 1959 *Observer* 15 Mar. 14/4 Most of the beautician's clients form the habit early. Others drop in late in life as a result of some personal crisis.

beautiful, *a.* Add: **1. c.** Used for emphasis or ironically, after the noun it qualifies.

The expression *the House Beautiful* is taken from Bunyan's *Pilgrim's Progress*, where Beautiful is to be regarded as a proper name.

1857 C. M. YONGE *Dynevor Terrace* II. xx. 306 'I must look in at the House Beautiful,' said Louis. 1880 W. ROBINSON (*title*) God's Acre Beautiful or the Cemeteries of the Future. 1883 O. WILDE *Lett.* (1962) 151 (*lecture-title*) The House Beautiful. 1899 W. ROBINSON *Eng. Flower Garden* (ed. 7) 181 The Summer Garden Beautiful. *Ibid.* 378 The Orchard Beautiful. 1904 KIPLING *Traffics & Discoveries* 357 The front of the House Beautiful. 1917 *Ladies' Home Jrnl.* Feb. 78/4 Send two cent stamp for 'The Body Beautiful' and trial plan today. 1929 *Melody Maker* Feb. 128 (Advt.), The new song-sensation by the writer of 'Charmaine'. 'Anita'. 'The waltz beautiful.' 1941 L. MACNEICE *Poetry of W. B. Yeats* iv. 82 Pater's doctrine of the Body Beautiful.

2, 3. (Further examples of *colloq.*, passing into *slang*, uses.) Also, *beautiful people* (occas. written with capital initials), 'flower' people, hippies.

1824 *Sporting Mag.* XIV. 165/2 Spring made some beautiful stops, left and right [in boxing]. 1899 KIPLING *Stalky & Co.* 60 Everybody paid in full—beautiful feelin'. 1929 M. DE LA ROCHE *Whiteoaks* xvi. 226 I'll never forget how beautiful you were to me. 1967 *N.Y. Times* (Internat. Ed.) 11–12 Feb. 9/2 The beautiful part of it..is that the trade is across the board, not limited to certain big sectors. 1967 *Boston Sunday Herald Mag.* 30 Apr. 32/3 'We had one guy,' he said, 'he was so beautiful. A jazz musician who also wrote children's books.' 1967 *Spectator* 4 Aug. 131/1 Far from being one of the Beautiful People, I was in an ugly frame of mind. 1968 *Crescendo* June 12/2 Maynard was a great leader...He was beautiful for the whole spirit of the band.

beautify, *v.* Add: **2.** *intr.* To make oneself beautiful.

1902 H. JAMES *Wings of Dove* v. xiv. 224 How tremendously Susie must be beautifying!

beauty, *sb.* Add: **5.** (Further examples.) Applied colloquially to an exceptionally good specimen of something (as a ball in cricket, a blow, etc.); cf. *BEAUT.

1882 *Australians in England 1882* 46 Spofforth was bowled by a 'beauty' from Mycroft. 1897 I. SCOTT *How I stole 10,000 Sheep in Austral.* iii. 11 Our own dogs..turned out to be 'beauties'. 1899 J. BELL *Shadow of Bush* viii. 46, I saw a beauty of a two-bladed knife at Buncombe's store. 1923 J. MANDER *Strange Attraction* vi. 71, I had a beauty of a little boat. 1924 WODEHOUSE *Bill the Conqueror* viii. 147 She..swung her right and plugged Slingsby a perfect beauty in the eye.

c. In various collectors' names of butterflies and moths.

1766 M. HARRIS *Aurelian: Nat. Hist. Moths & Butterflies* 19 The Brindled Beauty...prettily diamonded on the Back with black, and spotted with yellow. 1832 J. RENNIE *Consp. Butterfl. & M.* 104 The Oak Beauty (*Biston prodromarius*, Leach). The Brindled Beauty (*Biston hirtarius*, Leach). 1847 [see CAMBERWELL BEAUTY]. 1921 *Conquest* Sept. 495/3 The Pale Brindled Beauty (*Phigalia pilosaria*)..has a particularly handsome black form.

6. b. *pl.* In the titles of collections of the beautiful or choice passages of a writer or speaker, or examples of art.

1737 (*title*) The Beauties of the English Stage, consisting of all the celebrated passages, soliloquies, similies, descriptions and other poetical beauties in the English plays, *etc.* 1752 W. DODD (*title*) The Beauties of Shakespear, regularly selected from each play. 1767 (*title*) The Beauties of English Poesy. Selected by Oliver Goldsmith. 1786 (*title*) The beauties of the British Senate, taken from the debates of the Lords and Commons. 1860 *Athenæum* 31 Mar. 442/1 It might have been fancied that the days of 'Beauties', 'Gems', 'Anthologies' were over. 1865 (*title*) Beauties of Poetry and Art.

III. a. *beauty-bow, -crest, -drunken, -in-the-ghost.* **b. beauty contest,** a competition of women for a prize or distinction awarded to the most beautiful; **beauty culture** chiefly *U.S.*, use of cosmetics, etc., to improve a person's appearance; hence *beauty culturist*; also *beauty doctor, specialist*; †**beauty-man** *Obs.*, a handsome fellow, a dandy, a lady's man; **beauty parlour** orig. *U.S.*, an establishment in which the trade of a beauty specialist is carried on; **beauty queen** orig. *U.S.*, name given to the winner of a beauty contest; **beauty shop** *U.S.* = *beauty parlour*; **beauty show** = *beauty contest*; **beauty treatment,** the use of cosmetics, etc., in order to improve personal beauty.

a1889 G. M. HOPKINS *Poems* (1918) 83 Rough-Robin or five-lipped campion clear For a beauty-bow to his hat. 1809 A. M. BINSTEAD *Gal's Gossip* v. 77 Just the sort of woman who could apparently hold her own either in a beauty contest, a political debate, or a scrape. 1933 J. B. PRIESTLEY *Wonder Hero* iv. 129 She had won a beauty contest, and was probably easily the prettiest girl staying in the hotel. 1818 KEATS *Endym.* i. 363 To nightly call Vesper, the beauty-crest of summer weather. 1909 *Harper's Bazaar* Feb. 172 (*title*) Modern beauty culture. 1911 W. A. WOODBURY (*title*) Beauty Culture: A Practical Handbook on the Care of the Person. *Ibid.* 14 The successful beauty culturist must, above all, be modest, tactful, and discreet. 1928 *Punch* 5 Sept. 280/1 Valuable information which is afforded about domestic economy, feminine attire, cookery, beauty-culture. 1933 *Times Lit. Suppl.* 29 June 448/4 A small American town, whose main interests are bridge, poker, 'beauty-culture' and gossip. 1919 *Honey Pot* I. iii. 40 Dr. Caissarate, that wonderful beauty culturist. 1905 E. WHARTON *House of Mirth* II. ix. 43 A strange throng of hangers-on—manicures, beauty-doctors, hair-dressers. 1921 *Dict. Occup. Terms* (1927) § 920 Beauty specialist, beauty doctor. 1928 YEATS *Tower* 62 The Great Mother, mourning for her daughter And beauty-drunken by the water. a1889 G. M. HOPKINS *Poems* (1918) 56 Beauty-in-the-ghost, deliver it, early now, long before death Give beauty back. 1837 LYTTON *Ernest Maltrav.* I. II. ii. 181 The Beauty-man is, nine times out of ten, little more than the oracle of his aunts, and the 'sitch a love' of the housemaids. 1860 *Temple Bar Mag.* I. 68 A beauty-man, who rides and dances well. 1888 F. HUME *Mystery of Hansom Cab* xix. 130 A clergyman ..preached a sermon to prove that good looks and crime were closely connected, and that both Judas Iscariot and Nero were beauty-men. 1908 *Harper's Weekly* 24 Oct. 22/1 The 'beauty parlors' of a large department store. There are a number of booths divided off by wooden partitions. 1932 *Daily Express* 20 Sept. 5/5, I have decided to go into a beauty parlour when I grow up. 1938 E. BOWEN *Death of Heart* II. iv. 244 She was the receptionist in Southstone's biggest beauty parlour. 1922 *N.Y. Times* 5 Sept. 19/6 The winning beauty will be heralded as America's 'Beauty Queen'. 1933 J. B. PRIESTLEY *Wonder Hero* iii. 86 The girl who's just won the *Morning Pictorial*'s beauty competition ..the Beauty Queen. 1960 *Guardian* 11 June 7/4 Seventeen national beauty queens compete for the title 'Miss Europe'. 1901 *Current Lit.* Apr. 446/1 The Oldest Beauty-Shop. 1939 A. HUXLEY *After many a Summer* I. i. 6 Next door to the beauty shoppe was a Western Union office. 1948 MENCKEN *Amer. Lang.* Suppl. I. 573 *Beauty-parlor* began to appear before World War I, and soon afterward it was displaced by *beauty-shop*. Sometimes the latter is spelled *beauty-shoppe*, or even *beauté-shoppe*. 1969 B. KNOX *Tallyman* vii. 132 Janey Milton..on her way to have her hair set at a local beauty shop. 1896 C. S. *Leaves from Diary in Lower Bengal* v. 74 The idea occurred to him to have a Beauty Show of our servants..the prize to be given to the ugliest. 1907 G. B. SHAW *Let.* 7 Sept. (1941) 37 You would make me a curtain-raiser for a beauty show. 1907 M. E. BRADDON *Dead Love has Chains* vii. 148 She must have beauty-specialists, massage, electricity. 1938 N. MARSH *Death in White Tie* xxix. 303 Mrs. Halcut-Hackett..looking like a beauty-specialist's mistake. 1928 B. BUSHBY (*title*) Postal Tuition Course of Beauty Treatment. 1934 R. MACAULAY *Going Abroad* xxxiii. 282 Beauty treatment is never cheap.

beautyhood (biū·tihud). Also **beautihood.** [f. BEAUTY *sb.* + -HOOD.] A woman's 'reign' as a beauty; society of beauties, also beauties collectively.

1889 M. E. MARTIN *Common Clay* III. xv. 240 The short season of her beautihood in London. 1889 H. F. WOOD *Englishman Rue Cain* xvi, Initiation into the Fashionable Beauty-hood.

Beauvais (bove). The name of a town in northern France, used *attrib.* to designate things manufactured there, as *Beauvais tapestry*.

1885 L. J. DAVIS tr. *Müntz's Short Hist. Tapestry* xiv. 335 (*caption*) Louis XV. Fauteuil, covered in Beauvais tapestry. 1899 R. GLAZIER *Man. Hist. Ornament* 118 There are some fine Gobelin and Beauvais tapestries in Windsor Castle. 1961 *Connoisseur* Dec. p. xxxviii, A fine Louis XV Beauvais tapestry.

‖**beaux arts** (bozar). Also **beaux-arts.** [Fr. *beaux-arts*.] The fine arts. Also used *ellipt.* for the *École des Beaux-Arts* in Paris; *freq. attrib.*, of the standards of architecture and art maintained by the École in France in the nineteenth century and early part of the twentieth century and imitated elsewhere, characterized by the influence of older styles and a reliance on decorative and period design.

1821 M. WILMOT *Jrnl.* 26 Oct. (1935) 134 Went to the top of the tower to see the view of all Venice... Next the Academy of beaux arts. 1833 J. S. MILL *Let.* 5 Sept. in *Wks.* (1963) XII. 177 The beaux-arts (what beaux-arts!) which had been the glory of the *siècle de Louis 14.* 1854 THOREAU *Walden* 52 So are made the *belles-lettres* and the *beaux-arts* and their professors. 1924 *Arts & Decoration* July 10/1 The French or Beaux Arts method of developing the finished drawing, or 'project' on a competition is followed by the Institute [sc. Beaux-Arts Institute of Design, N.Y.]. 1931 tr. *Willy & Colette's Claudine in Paris* xi. 17 He's at the Beaux Arts and he's going to be a great artist. 1945 *Encycl. Brit.* II. 273/1 The largest school of architecture in England is that of the Architectural Association and the next largest that of Liverpool university. Both these schools show definite tendencies away from the *Beaux Arts* influence and towards contemporary architecture, period design having been dropped almost entirely. 1956 K. CLARK *Nude* ix. 351 The classic nude with its exhausted *beaux-arts* associations. 1961 *Listener* 10 Aug. 201/2 He [sc. the architect] has failed to make proper use of technology, and is perpetuating *beaux arts* attitudes to design. 1967 E. WYMARK *As Good as Gold* xv. 219, I left the Beaux Arts and went back to New York.

‖**beaux yeux** (bozyø̄). [Fr., = fine eyes.] Beautiful eyes; attractive beauty; admiring glances, favour.

[1825 H. WILSON *Memoirs* I. 110, I tried this method of making a little whig of myself, pour les beaux yeux de mi Lord Ponsonby.] 1828 LYTTON *Pelham* I. xxii. 182 He will scratch out the lady's *beaux yeux*. 1850 THACKERAY *Pendennis* I. xxvii. 262 The poor fellow is mad for your *beaux yeux*, I believe. 1908 MRS. H. WARD *Diana Mallory* II. xii. 252 Oliver condemn himself to the simple life..for the sake of the *beaux yeux* of Diana Mallory! 1931 *Times Lit. Suppl.* 5 Mar. 165/1 The *beaux yeux* of Victor Emmanuel and Cavour were being expensively purchased at the price of thousands of French lives.

beaver¹. Add: **1. b.** *bank beaver*, a beaver living in a burrow apart from the colony. *N. Amer.*

1903 *Outing* (U.S.) Mar. 669/1 You find the bank beaver mostly on lakes, or large rivers, which they are unable to dam. 1953 *Canadian Geogr. Jrnl.* Sept. 88/2 All along the river we came upon bank beaver.

c. *coll.* Beavers. Chiefly *U.S.*

a1649 J. WINTHROP *Jrnl.* (1908) I. 61 A sagamore...offered to give them yearly eighty skins of beaver. 1770 WASHINGTON *Diaries* I. 441 Then Bever catch it in there way up. 1778 J. KING *Jrnl.* Sept. in *Cook Jrnls.* (1967) III. ii. 1439 Their jackets were principally of Deer Skins, edg'd with the Skin of other Animals, as Wolves, foxes, beaver &c. 1789 MORSE *Amer. Geogr.* 198 In this country are..beaver, otters, sables. 1840 C. F. HOFFMAN *Greyslaer* I. v. 60, I had gone clean up to Racket Lake..hoping to get a few beaver. 1890 L. C. D'OYLE *Notches* 66, I knew that beaver were plentiful.

d. *Phr. to work like a beaver*: to work hard. orig. *U.S.*

1741 in H. M. Brooks *Days of Spinning-Wheel* (1886) II. 31 To be sold..., the very best negro woman in this town, who..will work like a beaver. 1877 RAYMOND *Mines* 225 Mr. Baldwin..has worked like a beaver since he assumed the management of the mine. 1915 *Lit. Digest* 21 Aug. 347/2 Every one knows what 'working like a beaver' means.

2. b. *beaver hat* (later examples).

1862 G. BORROW *Wild Wales* I. i. 7 He made his appearance very respectably dressed, in a beaver hat, blue surtout [etc.]. 1928 D. BYRNE *Destiny Bay* iii. 236 There comes an old fellow in a beaver hat.

c. A shade of brown resembling that of the fur of a beaver; more explicitly *beaver-brown*; also *beaver-coloured, -hued* adjs.

13.. *Gaw. & Gr. Knt.* 845 Brode, bryȝt, watȝ his berde, and al beuer-hwed. 1888 *Cassell's Family Mag.* Apr. 313/1 Many new colourings..Beaver is such a becoming tone. 1895 *Ibid.* Feb. 234/1 A picture hat..of beaver brown velvet. 1895 *Windsor Mag.* I. 340/2 The cloak is in beaver-coloured velvet. 1914 *Scotsman* 26 Oct. 12/2 The shades are Steel, Oxford and Parson Grey, Mole, Beaver, Fawn.

6. beaver-board, a trade-mark (U.S.) for a kind of wood-fibre building board; **beaver cloth** (cf. sense 4); **beaver-dam,** a dam made by beavers; **beaver-eater** (see quot. 1771); **beaver finish,** a finish giving a resemblance to beaver fur; hence, a finish in which the fibres are all laid in one direction; so *beaver-finished* adj.; **beaver lamb,** lambskin cut and dyed to resemble beaver fur; also *attrib.*; **beaver-poison** *U.S.*, the water-hemlock, *Cicuta sp.*; **beaver-root** *N. Amer.*, a pond-lily, *Nymphæa odorata*; **beaver-tail,** the tail of a beaver; also *transf.*; **beaver-tree,** *Magnolia virginiana*, the sweet or white bay of the U.S.; **beaver-wood,** (*a*) the hackberry tree of the U.S., *Celtis occidentalis*; (*b*) the beaver-tree; the wood of this tree.

1909 *Sat. Even. Post* 20 Feb. 35/1 [Advt. Beaver Manuf. Co., Buffalo, N.Y.], Beaver Board..Takes Place of Both Lath and Plaster. 1933 D. L. SAYERS *Murder must Advertise* viii. 145 The thinness of the beaverboard partition between Mr. Hankin's room and Mr. Copley's. 1948 'N. SHUTE' *No Highway* i. 7 A shabby little room of glass and beaverboard. 1858 SIMMONDS *Dict. Trade*, *Beaver-cloths*, a species of felted woollens made in America. 1904 GOODCHILD & TWENEY *Technol. & Sci. Dict.* 46/2 *Beaver Cloth*, a thick woollen fabric covered with fibre or nap. The best qualities are made in the West of England; medium and lower qualities in the heavy woollen districts of Yorkshire. 1968 IRONSIDE *Fashion Alphabet* 213 *Beaver cloth*, a heavily-napped overcoating resembling the beaver fur. 1638 in *Amer. Speech* (1940) XV. 155/1 Upon the branches of a swamp running North west up into the woods from the head of the said Vlyes Creeke out of a Bever dam. 1703 in *Cal. Virginia St. Papers* I. 83 To three white oakes, by the East side of the Tuckahoe Bever Dam. 1903 S. E. WHITE *Forest* xvi. 231 He knows the beaver-dams, how many animals each harbors. 1931 W. CATHER *Shadows on Rock* (1932) II. i. 46 A group of little boys played just below, building 'beaver-dams' in the gutter to catch the overflow. 1968 *Globe & Mail* (Toronto) 3 Feb. 43/2 (Advt.), Large trout stream 15 acres flooded by beaver dam. 1771 PENNANT *Syn. Quadrup.* 197 Wolverene...in America is called the Beaver-Eater. 1791 J. LONG *Voy. & Trav. Indian Interpreter* 41 The country everywhere abounds with wild animals, particularly bears, beaver eaters, [etc.]. 1804–5 LEWIS & CLARK *Jrnls.* (1905) VI. ii. 107 Carkajous, wolverine or Beaver Eaters..or Links. 1904 GOODCHILD & TWENEY *Technol. & Sci. Dict.* 46/2 Beaver Finish. 1909 R. BEAUMONT *Finishing Textile Fabrics* 11 The lateral surfaces of the fibres mainly resist the friction in the beaver-finished cloth. 1939 *Fur Times* 12 May 2/3 Beaver lamb is shown in swaggers and short jackets. *Ibid.* 25 Aug. 8/1 Beaver lamb coats..are now offered at eleven-and-a-half guineas. 1953 *Economist* 9 May 346/2 Thanks to the postwar prices of woollen (cloth) coats, 'beaver lamb' coats have provided the English fur manufacturers with a mass market. 1960 A. BURGESS *Right to Answer* ix, Alice in her beaver lamb. 1857 A. GRAY *First Less. Bot.* 157 Spotted cowbane. Musquash-root. Beaver-poison...The root is a deadly poison. 1832 W. D. WILLIAMSON *Hist. Maine* I. 126 Of the Lily tribe, we have several species..such as the yellow water-lily, or dog-lily, or beaver-root. 1856 W. E. CORMACK *Narr. Journey Newfoundland* (1874) 27 They also subsist on the large roots of the white waterlily..called by the Indians beaver-root. a1811 J. J. HENRY *Campaign against Quebec* (1812) 23 They returned two fresh beaver tails, which when boiled, renewed ideas. 1909 WARE *Passing Eng.* 23/2 *Beaver-tail* (Mid.-class, 1860). A feminine mode of wearing the back-hair, turned up loose in a fine thread net..which fell well on to the shoulders...Obviously from the shape of the netted hair [compared] to a beaver's flat and comparatively shapeless tail. 1937 *Discovery* Sept. 259/2 The last coach [of the L.N.E.R. 'Coronation' streamlined train] now has a streamlined 'beaver' tail, and is an observation car. 1960 *Aeroplane* XCVIII. 521/2 Major assembly has begun on the first AW 660 for the R.A.F. This version has a 100,000-lb. gross weight, 'beaver-tail' rear doors, and no nose doors. 1756 P. KALM *Resa till N. Amer.* II. 324 *Magnolia*,..Beaver-Tree. 1866 LINDLEY & MOORE *Treas. Bot.* s.v. *Magnolia*, *Magnolia glauca*..is also known by the name of Beaver-tree, because the root is eaten by beavers. 1901 C. T. MOHR *Plant Life Ala.* 505 White Bay. Sweet Bay. Beaver Tree...The bark is used medicinally. 1810 F. MICHAUX *Hist. des Arbres* I. 33 Small magnolia..ou Beaver wood. 1813 H. MUHLENBERG *Cat. Plantarum* 95 Beaver wood [or] hoop ash. 1880 *Encycl. Brit.* XI. 360/1 [The hackberry tree] is also known under the name of 'beaver-wood', and 'nettle tree'.

beaver⁴ (bī·vəɹ). *slang*. [Etym. uncertain.] (*a*) A beard; (*b*) a bearded person; (*c*) a game, in which points are scored in various ways by 'spotting' beards.

In the U.S., *beaver* (prob. BEAVER¹) is recorded in the sense 'a person' (not necessarily one with a beard): 1850–66 examples in D.A.

1910 F. RICHARDSON *Whiskers & Soda* 211 He provided a list of celebrated clean-shaven men and also of celebrated beavers, as bearded men are technically termed. 1922 J. KETTELWELL *Beaver* 58 The outlines of the game itself are so simple..that the question of rules scarcely arises. A bearded man is a Beaver, claim him, crying aloud, as musically as possible, 'Beaver, fifteen love'—or appropriately to the score. If both players cry aloud simultaneously it is a 'no ball'. 1922 *New Statesman* 12 Aug. 511/1 This amazing game of Beaver..is played..by two persons, and the points are scored as in tennis. Whichever of the two first cries 'Beaver!' as a beard heaves into sight, scores. 1922 *Westm. Gaz.* 12 Oct. 6/5 He was a beaver of a pronounced type, wore horn-rimmed spectacles, and two huge opal rings. 1927 W. E. COLLINSON *Contemp. Eng.* 63 About three years ago a fashion was started among men of wearing beards. These were greeted with the cry: Beaver!, a term now often applied to the beard itself [o]n 'a beaver'. 1936 WODEHOUSE *Laughing Gas* xxiv. 255, I had fallen among a band of criminals who were not wilful beavers, but had merely assumed the fungus for purposes of disguise. 1959 I. & P. OPIE *Lore & Lang. Schoolchildren* iii. 54 The cry 'Beaver!' is a thing of the past.

bea·ver, *v.* **1.** *local*. [cf. E.D.D. *beaver* sb.³ (Lincs.), a term applied to fine woad.] (See quots.)

1799 A. YOUNG *Gen. View Agric. County Lincoln* vii. 155 Without these attentions the woad will not beaver well, a term descriptive of the fineness of the capillary filaments into which it draws out when broken between the finger and thumb. 1805 R. W. DICKSON *Pract. Agric.* II. xi. 771 When this attention is neglected, the woad will not, on being broken between the finger and thumb, draw out into fine hair-like filaments, or, in the language of the manufacturer, *beaver* well. 1942 J. M. DOWSETT *Romance of England's Forests* viii. 210 This powder was spread on the floor, moistened with water and allowed to 'couch' (ferment), a process which required very considerable care if the material was to 'beaver' well.

2. [f. BEAVER¹.] Const. *away*: to work like a beaver.

1946 *Time* 22 Apr. 49 He found time to dash off five other books while beavering away at his vast *History*. 1966 M. R. D. FOOT *SOE in France* p. xxi, The gaullists beavered away at their own plans, irrespective of the prospects of putting them into action. 1967 *Spectator* 15 Dec. 740/2 The Germans beaver away at their scheme for 'entry by stages'.

beavered, *ppl. a.* Add: **c.** [f. *BEAVER⁴ + -ED².] Bearded.

1928 *Sunday Express* 6 May 2/3 My beavered friend became more lugubrious than ever.

Beaverette (bīvəre·t). [f. BEAVER¹ + -ETTE.] An imitation beaver fur; also, cloth with a beaver finish (see *BEAVER¹ 6).

1922 *Daily Mail* 1 Dec. 1 Pelt in superb quality pulled Beaverette. 1923 *Ibid.* 16 Jan. 1 Beaverette stoles. *Ibid.* 29 Jan. 1 Rubber on strong Beaverette Cloth.

bebop (bī·bɒp). orig. *U.S.* Also **rebop,** and abbrev. ***bop** *sb.*² (q.v.). [App. imitative (see quot. 1955).] A development of jazz, begun in the U.S. at the end of the 1939–45 war, characterized by complex harmony, dissonant chords and highly syncopated rhythm. Also *attrib.* Hence **be·bopper,** a performer or supporter of bebop.

P. Tamony in *Jazz* (1959) II. 114–19 cites instances of *bebop, rebop, bop,* etc., as onomatopœic or nonsense syllables, from jazz recordings of 1928 on: e.g. McKinney's Cotton Pickers, *Four or Five Times* (1928), Bop-do-de-de-do-do.. Bebop one, bebop two, bebop three.

1945 *Down Beat* 1 Aug. 8/2 (*title of gramophone record*) Dizzy Gillespie. Salted Peanuts/Be-Bop. *Ibid.* 1 Sept. 3/3 Much controversy has arisen of late over the claims of altoist Charlie Parker and trumpeter Dizzy Gillespie to the origination of their fantastic and exciting 're-bop' style. 1946 *Melody Maker* 31 Aug. 5/4 Whenever Be-Boppers and the others got together, trouble was..the outcome. *Ibid.* 19 Oct. 5/2 The music isn't rebop stuff. *Ibid.* 26 Oct. 3/1 The re-bop (or, if you prefer it, be-bop: they're using both names for it in America) is clearly evident. 1948 *Life* 11 Oct. 139/2 Boppers go gaga over such bebop classics as *OO Bop Sha Bam*. 1950 E. HYAMS *From Waste Land* 203 A kind of bebop or boogy-woogy. 1955 L. FEATHER *Encycl. Jazz* (1956) 27 As musicians gathered outside the clubs along Fifty-Second Street to discuss the music of Charlie Parker.. or of Dizzy Gillespie,.. they would use an onomatopoetic expression to describe a typical phrase played by these musicians: 'rebop' or 'bebop' they would say. Eventually the word became shortened and 'bop' was accepted as the name for the new branch of jazz. 1959 J. CARY *Captive & Free* xiv. 71 He was all for the teenager, except the Teddy boys and the bebop parties. 1966 *Crescendo* Jan. 3/1 The young aspiring beboppers..were trying to play the new thing.

‖**bebung** (bē·buŋ). *Mus.* [G., lit. 'trembling'.] A pulsating or trembling effect given to a sustained note, a tremolo; *spec.* such an effect produced on the clavichord.

1879 GROVE *Dict. Mus.* I. 160/1 *Bebung*, a certain pulsation or trembling effect given to a sustained note in either vocal or instrumental music, for the sake of expression. 1885 *Encycl. Brit.* XIX. 67/1 The clavichord was obedient to a peculiarity of touch possible on no other keyboard instrument... It is the 'Bebung'..gained by a repeated movement of the fleshy end of the finger while the key was still held down.

because, *adv.* and *conj.* Add: **A. 1.** *because why* used interrogatively, = 'why?' (cf. *cause why?* s.v. CAUSE *sb.* 3 c); chiefly *dial.*

1887 PARISH & SHAW *Dict. Kentish Dial.* 10 *Because why*, why? wherefore? A very common controversy amongst boys:—'No it ain't'—'Cos why?'—'Cos it ain't.' 1921 D. H. LAWRENCE *Sea & Sardinia* i. 12 The painters try to paint her [sc. Etna]..in vain. Because why? 1937 D. L. SAYERS *Busman's Honeymoon* xiv. 253 It's a cur'ous thing you askin' about that there sun-dial, because why? The very man wot sold vicar the chimbley-pot, 'e wos found dead in his own 'ouse only yesterday. 1961 S. CHAPLIN *Day of Sardine* viii. 164, I know a lot of people that rant on about their religion and it doesn't do any good. Because why? Because they're trying to convince themselves, maybe?

C. Used elliptically in answer to a question, implying that a fuller reply is being withheld for some reason.

1871 'L. CARROLL' *Through Looking-Glass* vi. in *Wks.* (1939) 200 The little fishes of the sea, They sent an answer back to me. The little fishes' answer was 'We cannot do it, Sir, because—'. 1924 W. JAMES' *Letters* II. 'Cos vi. 6 'Mustn't we? Why not?' 'Oh, Warren! Because.' She might have been the fifteen-year-old child again. 1967 C. FREMLIN *Prisoner's Base* vii. 48 'Why do you go out with him, Helen?' 'Because.' Helen didn't mean to be cheeky. 1968 M. CARROL *Dead Trouble* x. 175 'Why didn't you leave the bottle?' 'Because!' I said shortly. 'I wasn't going to explain my feelings on the matter.'

bêche-de-mer. (Later examples.) For etym. read: Fr., altered from *bicho de mar*, a Pg. *bicho do mar*, lit. sea-worm. Add: Also *BEACH-LA-MAR.

1849 W. T. POWER *Sketches in N.Z.* xxiii. 216 The *bêche-de-mer* and shark-fins, both of which..are valuable articles of trade with China. 1861 [see SEA-SLUG 1]. 1921 J. LONDON *Cruise of Snark* xvi, Bêche de mer English was the product of conditions and circumstances. 1947 L. L. JOHNES *Isles of Despair* xxxviii. 251 The venturesome crews had found a rich patch of bêche-de-mer. 1961 *Amer. Speech* XXVI. 25 The pidgin of Hawaii differs markedly from the pidgin English..of the China coast, the pidgin of Melanesia, and the bêche-de-mer of the western Pacific.

Bechuana (betʃuā·nă). Also bekiuā·nă). Pl. **Bechuana, -anas** (-z). Also **Bechwana,** and other variants. A member of a negroid people inhabiting the country between the Orange and Zambezi rivers in Southern Africa, speaking Tswana (formerly called *Sechuana*), a Bantu language. Also *attrib.*

1804 J. BARROW *Trav. S. Afr.* II. ii. 114 The city *Leetakoo*, the capital of a tribe of Kaffers called *Booshuanas*, situate at the distance of sixteen days' journey beyond the Orange River, in the direction of north-east from the Cape. 1822 W. J. BURCHELL *Trav.* I. 581 *Bichuana* [or] the country of the Bichuanas (Bitjuana) nations. 1826 R. MOFFAT (title) & Bechuana Catechism. 1850 J. W. APPLEYARD *Kafir Language* 31 The terms Bechuana (a variation of Bachuana) and Bechuana, are different forms of the same verbal root, the former referring to people, and the latter to language... By the Hottentot tribes, the Bechuanas are called *Briqua*, the goat-people. 1875 *Encycl. Brit.* III. 478/1 The Bechwana are divided into numerous tribes. 1936 *Discovery* June 172/1 The Southern Bantu comprise the Shona peoples, the Xulu-Xosa...the Suto-Chawana (Bechwana), Basuto, etc.), and the Herero-Ovambo. 1954 D. HUXLEY *Back Street New Worlds* xii. 125 A group of Bechuana girls photographed by air...into a big London hospital.

beckite: see *BEEKITE.

Becquerel (be·krɛl). The name of a French physicist, Antoine Henri Becquerel (1852–1908), used *attrib.* in *Becquerel('s) rays*, formerly a general term for the radiation from radioactive substances.

1897 S. P. THOMPSON in *Phil. Mag.* July 105 While agreeing with the Röntgen rays in the property of penetrating aluminium [etc.], the Becquerel rays differ in the circumstance that they can be refracted and polarized. 1897 ...*Light Visible & Invis.* 279 Becquerel's rays possess the property of dislectrifying charged bodies. 1898 *Physical Rev.* Apr. 239 Becquerel rays, or uranium rays, as Becquerel himself called them. 1931 *Discovery* July 212/1 When X-rays and the Becquerel rays from uranium were discovered it was demonstrated that these rays could make air into a conductor.

becuiba (bikwī·bă). Also **bicuiba.** [Pg. *bicuiba*, f. Tupi *bicuiba, bicuhyba*.] A Brazilian timber tree, *Virola bicuhyba*; used *attrib.* in **becuiba nut,** the fruit of an aromatic Brazilian tree (*Myristica bicuhyba*) of the nutmeg family; **becuiba tallow** or **fat,** a balsamic product of the becuiba nut.

1842 DUNGLISON *Dict. Med. Sci.* (ed. 3) 87. Ibicuiba 37 1/2 *Bicuiba*, or *Bicuiba nux*, a species of nut from Brazil, the emulsive kernel of which is ranked amongst balsamic remedies. 1884 *Encycl. Brit.* XVII. 744/2 Becuiba tallow. [Source] *Myristica bicuhyba*. [Principal use] medicine; candles. 1889 *Cent. Dict.*, Becuiba-nut. 1934 WEBSTER, *Becuiba*, a Brazilian timber tree (*Virola bicuhyba*), family Myristicaceae. 1955 *Nomencl. Commerc. Timbers (B.S.I.)* 18 Bicuiba.

bed, *sb.* Add: **1. c.** Chiefly used as an advertising term: bedroom.

1926 R. MACAULAY *Crewe Train* II. ix. 172 How many bed and recep.? 1939 [see *BATH sb.¹ 12]. 1961 WODEHOUSE *Ice in Bedroom* xxii. 177 A joyous suburban villa equipped with main drainage,..four bed, two sit and the usual domestic offices.

f. *bed and breakfast*: the provision of a bed for a night and breakfast the following morning; an arrangement offered by hotels, boarding houses, etc. Also *attrib.*

1910 *Bradshaw's Railway Guide* Apr. 1125/1 Residential Hotel...Bed and breakfast from 4/-. 1930 *Morning Post* 12 June 18/5 (Advt.), Married couple for bed and breakfast house; Kitchen Man and House-Parlourmaid. 1936 J. L. HODSON *Our Two Englands* x. 174 It is true that I have seen the signs 'Bed, breakfast and garage'—a new form which the historian should make a note of. 1967 *Listener* 10 Aug. 178/1, I had previously booked bed and breakfast somewhere in Bloomsbury.

6. c. In phrases implying sexual intimacy. (Recent examples.)

1945 A. HUXLEY *Time must have Stop* iv. 46 How much less awful the man would be..if only he sometimes lost his temper...or went to bed with his secretary. 1962 J. WAIN *Strike Father Dead* vi. 264 If you go to bed with a man, he always thinks he owns you. 1966 'It's like your knows that.' 1968 A. HERON *Towards Quaker View of Sex* v. 44 A young doctor..may think it all right to propose 'going to bed' to a nurse he has only just met.

10. *to put to bed* (of a newspaper, journal, etc.), to go to press (cf. sense 11), start printing. Also, *to see, put (a paper) to bed.*

1891 *Cent. Dict.* s.v., *Round Clock* 35 For the 'Times'—the mighty 'Times'—has 'gone to bed'. 1899 *Daily News* 30 Sept. 6/1 Night by night the paper was 'put to bed' in the last..seeing the paper to bed (to use the old-fashioned phrase), and examining the first copies printed. 1933 M. LUTYENS *Forthcoming Marriages* 197 He went straight from the office to..see the other till after midnight when he saw the paper 'went to bed'. 1955 A. CHRISTIE *Hickory Dickory Dock* ix, We went to press, or, as we liked to say in our..of the newspaper office. First..we put the paper to bed.

12. f. The body of a cart or wagon. *dial.* and *U.S.*

c1700 KENNETT *B.M. MS. Lansdowne 1033*, Bedd of a cart, the body of it. 1851 MAYHEW *London Lab.* I. 26/2 Other commodities are 'hawked' with a body..of the cart. 1854 A. E. BAKER *Gloss. Northampt. Words, Bed*...3. The body of a cart or wagon. 1873 J. H. BEADLE *Undevel. West* xxv. 491 In this [journey, we encountered dangerous whirls and jump-offs, the wagon often plunging so as to throw the bed. 1904 W. H. SMITH *Promoters* xvii. 279 Some of these beds will hold more than a hundred bushels. 1953 S. CLOETE *Curve & Tusk* (1953) i. 21 The ten stood near the truck because when they cleaned its bed, bits of meat often fell on the ground.

18. *bed-candle, -cap, -carriage, -clothing, -knob, -mat, -pal, -quilt* (earlier examples), *-rug* (examples), *-sheet* (later example), *-steps* (example), *-thane; bed-bound.*

1822 *Daily Mail* 30 Nov. 14 It is often, however, a problem to know what to choose that will most amuse the girl or boy who is bed-bound. 1961 *Guardian* 30 June 8/3 Another bed-bound reader edits the religious page. 1850 THACKERAY *Pendennis* II. xv. 146 Martha brought the bed-candle and the breakfast. 1832 ... *Engl.* May 14. Mrs. Beecher ... 1864 MRS. GASKELL *French Life* i, in *Fraser's Mag.* Apr. 438/1 When it's grease from our party ... 1839 ... 1960 ... In a high trilled bed-cap, swaying balloon-like skirts. 1869 TROLLOPE *He knew he was Right* II. xcv. 354 We got a bed-carriage [on a train] for him at Dover. 1869 F. E. GRETTON *Memory's Harkback* iv. 69 He never walked again, but was drawn about lying at full length in a bed-carriage. 1843 H. B. STOWE *Uncle Tom's Cabin* xxxii, A tattered blanket, formed his only bed-clothing. 1883 W. DE LA MARE *Sleep & Nonsense* 74 A visage, with eyes like brass bed-knobs. 1931 ... *Women* ... four women, women's baskets on their bed-knobs. 1888 ... near the bedpal. 1922 JOYCE *Ulysses* 40 Papa's little bedpal. Lump of love. 1765 in E. Singleton *Social N.Y.* (1902) 334 Knotting for Bed Quilts or Toilets. 1803 MRS. E. BOWNE *Let.* (1887) 184 Our poor bed quilt is all I have towards housekeeping. 1847 ... *Prelude Rec.* (*Essex Co., Mass.*) (1916) I. 75 An old Straw bed and old Quilts over the door and the bed-rug. 1850 *Knickerbocker* XXXVI. 75 Open the door and the gentle breeze from without will lap the blue woollen 'bed-rug'. 1864 H. LAWRENCE *Last Poems* (1932) 217 Who is't amongst the bed-sheets? 1853 LOUDON *Encycl. Archit.* § 2125 A set of bed-steps, with two of the steps arranged as cupboards. 1922 JOYCE *Ulysses* 379 Trout bedthanes they twain are, for Home holding warriest ward.

19. bed-book, a book suitable for reading in bed; **bed-bottle,** a bottle for urination for the use of male patients in bed; also, a bottle for heating a bed; **bed-cord** (later examples); **bed-cover,** (*a*) a covering or case for a feather- or flock-bed, etc.; (*b*) a bed-quilt (Webster 1828); **bed-jacket,** a short jacket worn by women sitting up in bed; **bed-joint** (*b*) *Building*, a horizontal joint (cf. sense 10); **bedlock** *nonce-wd.* = WEDLOCK; **bed-plane** *Geol.*, the junction between two layers or strata; **bed-plate** (examples); **bed-rest,** (*a*) a support for a person in bed; (*b*) confinement of a sick person to bed; **bedroll,** † (*a*) a list of women to sleep with *Obs.*; (*b*) *U.S.* and *N.Z.* bedding rolled into a bundle for carrying; **bed-sack,** 'a sack to be filled with straw or hay, used in army camps for convenience of carrying them' (D.A.E.); **bed-settee,** a settee that can be converted into a bed; **bed-sock,** a sock worn in bed; **bed table,** a small table or tray usually with a ledge at the back and sides, adapted for placing on or over the bed of an invalid person; **bed-vow** (modern example); **bed-warmer,** a device for warming a bed; **bed-wetting** *vbl. sb.*, incontinence of urine while in bed; hence *bed-wet v.*, *bed-wetter*; **bed-winch** (example); **bed-worthy** *a. colloq.*, sexually attractive; hence *bed-worthiness*.

1906 A. BENNETT *Whom God hath Joined* i. 4 'Dip into it anywhere. It's a bed-book'. Book 'A *Lawrence* and Mark Pattison. Bed Books of the Month'. 1928 *Publishers' Weekly* 9 June 2373 'Cranford' was one of the favorite bed-books of her day. 1907 N. MUNRO *Daft Days* i. 1 The burgh town turned on its pillows, drew up its feet from the bed-bottles, last night hot, now turned to chilly stone. 1950 'G. ORWELL' *Shooting an Elephant* 21 The bedbottles and the grim bedpan. 1874 J. W. LONG *Wild-Fowl Shooting* 107 All you need to carry besides your ordinary bed-clothes is a common bed-cord. 1886 *Harper's Mag.* June 58/2 Traces are made of hickory or papaw, as also are bed-cords. 1828 WEBSTER, *Rug*,..in America, I believe,..is applied only to a bed cover for ordinary beds, and to a covering before a fire-place. 1837 *United Service Jrnl.* June 107 Those bed-covers which our grandmamas were delighted to cobble together. 1967 E. SHORT *Embroidery & Fabric Collage* iii. 67 Three of the most popular forms of embroidery for bed-covers, namely quilting, patchwork and candlewick. 1914 MILLICENT, DUCHESS OF SUTHERLAND *Six Weeks at War* ii. 21 Our nurses cut out red flannel bed-jackets and tried to take photographs. 1919 H. WALPOLE *Secret City* III. xii. 405 An old quilted bed-jacket of a purple green colour. 1876 *Encycl. Brit.* IV. 305/2 The *joints*, or bed-joints, are the surfaces separating the voussoirs. 1922 JOYCE *Ulysses* 483 Born out of bedlock hereditary epilepsy is present, the consequence of unbridled lust. 1895 DANA *Man. Geol.* (ed. 4) 111 Bed-plane faults are still another kind in which the plane of displacement is that between two layers or strata. 1850 *Ann. Rep. Commissioner Patents 1849* (U.S.) 211 This is combined with the bed plate for guiding and keeping the edge of the tire true. 1850 N. KINGSLEY *Diary* (1914) 107 Putting in timbers to set the bed-plates to the Engines. 1959 *Times* 19 Nov. 14/6 That there which'd lift straight out of its bedplate. 1964 *McCall's Sewing* v. 69/2 Close the bed plate. 1968 *Gloss. Terms Offset Lithogr. Printing* (*B.S.I.*) 36 *Bed plate*, the component upon which is mounted the main structure of a printing machine. 1872 GEO. ELIOT *Middlem.* II. xiv. 233 His uncle..propped up comfortably on a bed-rest. 1899 R. WHITEING *No. 5 John St.* xvii. 172 This spring lowers the bed-rest, so as he can sit up an' read. 1950 *New Engl. Jrnl. Med.* CCXLIII. 486 Bed rest has long been the *sine qua non* in the treatment of myocardial infarction. 1958 *Times* 15 Oct. 19/2 So effective is the modern treatment of pulmonary tuberculosis that the traditional sanatorium treatment..with its prolonged bed-rest..is seldom necessary. a1652 R. BROME *Madd Couple Well Matcht* III. i, But why me up in your bedroll George? 1910 S. E. WHITE *Rules of Game* vi. 203 Thus instead of his 'turkey'—or duffle-bag—he speaks of his 'bed-roll', and by that term means not only his sleeping equipment but often all his worldly goods. 1916 H. KEPHART *Camping & Woodcr.* II. 136 A bed roll made with flaps and sides and ends is best for this purpose. 1950 G. WILSON *Brave Company* 20, I can distinguish the long vermouth stain above Hope's bedroll. 1958 *Landfall* XII. 115 When he had spread out his bedroll, the little man came out of his cell. 1861 in *Probate Rec.* (*Essex Co., Mass.*) (1916) I. 323 A bead sacke,..a cheste. 1861 *Revised Regulations U.S. Army* 169 Bed-sacks are provided for troops in garrison. 1933 *Discovery* July 219/1 There is a demand for less and smaller furniture, and for the dual-purpose piece—the table-bookcase, table-stool, even bed-settee, which converts the living room into a bedroom for the unexpected guest. 1958 *House & Garden* Feb. 21 (Advt.), Here at last is a bed-settee...To convert the settee into a bed the back is simply swung down. 1961 *B.S.I. News* Nov. 20/2 Bed-settee mattresses. 1870 L. M. ALCOTT *Old-Fashioned Girl* iii. 43 She began to knit a pretty pair of white bed-socks..for her mother. 1914 W. OWEN *Let.* 1 Jan. (1967) 225 And only one bed-sock. 1934 L. A. G. STRONG *Corporal Tune* 266 Continuing to read even while she pulled great clumsy bedsocks up his shrunken legs. 1811 *London Cabinet-Makers Book of Prices* 314 A Bed-Table. All solid. —Two feet five inches long. 1853 MRS. GASKELL *Cranford* xi. 208 Lady Glenmire..rummaged up all Mrs. Jamieson's medicine glasses, and spoons, and bed-tables. 1894 *Daily News* 28 June 6/3 In white enamelled wood are some bed tables...They are for resting on the knees while sitting up in bed. 1948 L. A. G. STRONG *Trevannion* 167 Mrs. Bracegirdle drew up a rosewood bed-table, which fitted right across the bed, and slid her tray upon it. 1922 JOYCE *Ulysses* 200 But she, the giglot wanton, did not break a bedvow. 1931 *Kansas City Star* 2 Sept., One electric bed warmer. 1933 W. DE LA MARE *Lord Fish* 101 A ship's compass, a brass cannon, a bed-warmer,...—such curios as that. 1952 S. SPENDER *Learning Laughter* viii. 112 The children don't only suck their thumbs. They also bed-wet. 1938 *Time* 16 May 72/3 Bed-Wetters Belled. 1940 MRS. ST. L. STRACHEY *Borrowed Children* v. 69 As the younger evacuee was a bed-wetter..the billet-mother said she could not keep the younger. 1969 *New Scientist* 16 Jan. 148/3 Mousepie, still made in country districts for bed-wetters. 1890 J. MC-GREGOR-ROBERTSON *Househ. Physic.* xv. 463 *Bed-wetting*...is due very often to the presence of some irritant in the bowels. 1940 HARRISSON & MADGE *War begins at Home* xii. 310 Bed-wetting was one of the main faults found with evacuee children. 1848 DICKENS *Dombey* lix. 592 The men with the carpet-caps gather up their screw-drivers and bed-winches into bags. 1959 T. GIRTIN *Unnatural Break* xxi. 72 Certificate of Bedworthiness, eh? 1936 A. HUXLEY *Eyeless in Gaza* liv. 608 In fact, thoroughly bed-worthy. Or at least he looks it. Because one never really knows till one's tried, does one? 1954 J. B. PRIESTLEY *Magicians* ii. 38 Bed-worthy luscious blonde.

bed, *v.* Add: **4. a. b.** Also with *down*.

1906 E. DYSON *Fact'ry 'Ands* ix. 117 Ther Firm and its missus was preparin' t' bed down. 1914 B. M. BOWER *Flying U Ranch* 7 Throw out your war-bag and make yourself to home, Mig-u-ell; some of the boys'll show you where to bed down. 1938 N. MARSH *Artists in Crime* ix. 120 She meant to come back and bed down with Garcia..you know—to spend the night with him. 1944 G. NETHERWOOD *Desert Squadron* 115 Being forced to spend the night in the desert there was nothing else to do but 'bed-down'.

c. Of cattle: to be bedded *down* for the night. *U.S.*

1903 A. ADAMS *Log Cowboy* viii. 110 Not a hoof would bed down. 1920 MULFORD *J. Nelson* xvi. 267 After..the great herd had bedded down.

10. b. Also with *down*.

1892 STEVENSON & OSBOURNE *Wrecker* xiii. 213 When she [sc. a boat] first struck and before she bedded down, seven or eight hours' work would have got this hooker off.

17. *refl.* and *intr.* Of eels (see quots.).

1653 [see Dict. sense 14]. 1746 R. GRIFFITHS *Ess. Jurisdict. Thames* 194 [Eels] get into the soft Earth or Mud ..and bed themselves. 1883 G. C. DAVIES *Norfolk Broads* xxxi. 213 Big and little [eels], start on this singular voyage, ..and 'bed' themselves. 1902 CORNISH *Naturalist Thames* 216 In winter the eels 'bed', *i.e.* bury themselves in the mud.

bedazement (bidē·zmënt). [f. BEDAZE *v.* + -MENT.] Bedazed condition.

1887 M. ARNOLD *Ess. Crit.* (1888) Ser. II. ix. 311 This bedazement with the infinite. 1903 *Westm. Gaz.* 30 June 7/2 The Unionist Party is being caught in the Protectionist toils before it can recover from its bedazement. 1918 A. T. QUILLER-COUCH *Foe-Farrell* x. 175 He followed me..in mere bedazement, speechless.

bed-bug. [BED *sb.*] = BUG *sb.*² 2 a.

Freq. in U.S., on account of the use of *bug* in the general sense of 'insect'.

1809 *Farmer's Almanack* (Boston) July 1 Ladies, for mercy's sake, see about the bed bugs. 1813 [see BED *sb.* 18]. 1861 MAYHEW *Lond. Lab.* III. 35/1 The bed-bug is not the only one of its congeners which preys upon man. 1909 *Cent. Dict. Suppl.*, *Bedbug-hunter*, a reduvid bug,..which inhabits houses, where it preys upon bedbugs. 1964 M. HYNES *Med. Bacteriol.* (ed. 8) xxx. 462 Bed-bugs live in hiding places such as the crevices of furniture and behind pictures, from which they emerge at night to suck blood.

beddable (be·dăb'l), *a.* [f. BED + -ABLE.] That may be taken to bed; sexually attractive.

1941 *Sat. Rev. Lit.* 29 Nov. 13 Byron only tolerated brains in women who were too old to be beddable. 1952 E. GRIERSON *Reputation for Song* xxi. 172 Favours granted him..by ladies just as succulently beddable. 1957 W. CAMP *Prospects of Love* II. iii, There must be something about her ..which screams that she's beddable. A girl doesn't have to be pretty to be sexy.

bedder² (be·dəɹ). [See -ER¹.] One who goes to bed. In collocations *early, late (go-to-)bedder.*

1908 *Daily Chron.* 27 Oct. 6/7 The late-go-to-bedder and the early riser. 1921 *Glasgow Herald* 19 May 6 Our forefathers were 'earlier bedders' and risers than the present artificial age. 1961 Y. OLSSON *Syntax Eng. Verb* vi. 125 He's a late bedder and a later riser.

bedder³ (be·dəɹ). *slang.* [See *-ER⁶.] A bed-room.

1897 *Westm. Gaz.* 2 Feb. 1/3 She'd want to come up to my bedder and give me Somebody's beastly food for infants. 1908 D. COKE *House Prefect* xvii. 219 He's been nabbed, and shut up in his bedder.

bedding, *vbl. sb.* Add: **III. bedding fault** *Geol.*, a fault parallel to a bedding-plane; **bedding-ground** *U.S.* = *BED-GROUND; **bedding-plane** = *bed-plane* (*BED *sb.* 19); **bedding-plate** = *bed-plate* (BED *sb.* 19); **bedding-roll,** a roll of bedding (sense 1).

1909 *Cent. Dict. Suppl.*, Bedding fault. 1961 J. CHALLINOR *Dict. Geol.* 21/1 *Bedding-fault*, the result of bedding-plane slip. 1884 W. SHEPHERD *Prairie Exper.* 199 For the bedding-ground a bare open spot..away from damp. 1920 J. M. HUNTER *Trail Drivers of Texas* 215 It looked like a 'round up' when turning them off of the bedding ground. 1897 *Q. Jrnl. Geol. Soc. Index I–L* 33/2 Bedding-planes comp. w. cleavage-planes in N. Wales. 1908 *Daily Chron.* 15 Dec. 1/5 Originally..horizontal, the bedding plane now dips gently to the south. 1920 L. V. PIRSSON *Physical Geol.* (ed. 2) 271 The two layers will be separated by a distinct juncture plane; this is stratification, and the juncture plane is called a bedding plane. 1879 *Man. Artill. Exerc.* 255 The racer is secured to the bedding-plate by steel bolts with wrought-iron nuts. 1901 KIPLING *Kim* vii. 179 His new bullock-trunk..and bedding-roll lay in the empty sleeping-room. 1963 *Times* 5 Mar. 12/7 His bedding-roll in one hand.

beddy (be·di), *a.* [f. BED *sb.* + -Y¹.] Of stone: having natural cleavages, with liability to split.

1709 T. ROBINSON *Nat. Hist. Westmld. & Cumbld.* xiv. 79 Under this white Metal, comes in a beddy Free-stone, which is always of a grey Colour. 1829 S. GLOVER *Hist. Derby* I. 88 The freestones,..often called building stone.. (in distinction from beddy stone, flags or paviers). 1888 BARING-GOULD *R. Cable* xlii, He may discover, when he's half cut it [sc. granite], that it's beddy (liable to split). 1892 HESLOP *Northumb. Words*, 'Beddy freestone' is thus distinguished from a compact, granular deposit.

beddy-byes (be·di,bɔi:z). Also **-bye.** [f. BED *sb.* + -Y⁶ + BYE(-BYE¹).] A nursery and facetious term for 'bed' or 'sleep' (BYE-BYE¹).

1906 E. DYSON *Fact'ry 'Ands* xvii. 234 Booser M'Gunn..went ter beddy-bye. 1926 P. GIBBS *Young Anarchy* x. 59 Come on, little one. Bedibyes. 1930 WODEHOUSE *Very Good, Jeeves!* vii. 181 And then off to beddy-bye. 1946 'S. RUSSELL' *To Bed with Grand Music* iv. 62 Mrs. Chalmers rolled up her knitting and said she supposed it was time for beddy-byes.

Bedford (be·dfəɹd). *Bedford cord*, a woven fabric with prominent cords running in the direction of the warp. Hence in *pl.*, trousers made of Bedford cord. Also *Bedfords* ellipt.

1862 *Catal. Internat. Exhib., Brit.* II. No. 4152 Woollen cords, Bedford cords, and velveteens. 1912 *Encycl. Brit.* III. 278/1 *Bedford cord*, a cheaper variety of piqué in which the stripes run the length of the piece. 1914 W. S. MURPHY *Mod. Drapery* IV. 20 *Bedford cord*, a kind of ribbed cotton fabric, woven with heavy warp, which composes the main face of the cloth, the weft floating at the back of the structure. 1954 W. FAULKNER *Fable* (1955) 33 In the Bedford cords of a British officer. *Ibid.* 83 Official slacks in place of pink Bedfords and long boots. 1966 'E. PETERS' *Piper on Mountain* viii. 136 He had on Bedford cords and a tweed jacket.

bedgown. (Earlier examples.)

1. 1756 TOLDERVY *Hist. 2 Orphans* I. xiii. 120 No night-cap,..fetid bed-gown, or greasy sack, were appendages to the bodies of these agreeable women.

2. c1806 D. WORDSWORTH *Tour Scotl.* in *Jrnls.* (1941) I. 257 The maid-servant..was..dressed in a white bed-gown. a1823 —— *Second Tour Scotl.* in *Ibid.* II. 341 Pity the dress of the young at harvest work is so slovenly. Loose Bed-gowns mostly of white calico.

bed-ground. *U.S.* [BED *sb.* or *v.*] Ground on which cattle are bedded for the night.

1880 *Scribner's Monthly* Mar. 770/2 There the cattle are huddled together or 'rounded up' in as small a compass as possible, called 'the bed-ground', and the herders stand guard over them..until morning comes. 1920 MULFORD *J. Nelson* xxvi. 267 The bed ground was well chosen and the night promised to be a good one. 1921 —— *Bar-20 Three* xvii. 220 The best bed-ground on the ranch.

bedlamer² (be·dlāməɹ). [? ad. F. *bête de la mer* sea-beast.] A seal-hunters' name for a hooded seal of one year old and a harp-seal of two years old.

1773 G. CARTWRIGHT *Jrnl.* 15 Nov. (1792) I. 284 He saw a shoal of bedlamers before his door one day last week. 1895 *Outing* (U.S.) XXVII. 22/1 The harp seals..post a sentinel, usually a bedlamer or two-year-old harp, on one of the highest pinnacles of the floe. *Ibid.* 23/2 The 'Sculp'..of the harp and bedlamer will average one hundred and fifty pounds.

bedless, *a.* (Earlier example.)

1707 J. STEVENS tr. *Quevedo's Com. Wks.* 430, I am careless and bedless.

Bedlington (be·dliŋtən). [Named after *Bedlington* in Northumberland.] In full *Bedlington terrier*: a short-haired terrier characterized by a narrow head, short body, and longish legs.

[1867 *Field* 3 Aug. 102/1 In the dog show a local breed of terriers was distinguished by a separate class. They are named after the town of Bedlington.] *Ibid.* 102/2 (*heading*) Prize List..Bedlington Terriers. 1869 *Ibid.* 27 Mar. 265/3 In 1825 she [sc. 'Coates's Phoebe'] was mated with Anderson's Piper, and the fruit of this union was the Bedlington terrier ..The model Bedlington should be rather long and small in the jaw, but withal muscular. 1875 in V. Shaw *Illustr. Bk. Dog* (1881) xix. 144 The Bedlington Terrier should be broad in the nostril, with a flesh-coloured nose. 1881 *Ibid.* 143 The support of the public at large has not hitherto extended much beyond affording prizes for Bedlingtons.

bed-pan. 2. (Earlier examples.)

1678 in *Rec. New Castle Del. Court* (Penn.) (1904) 361 Two Earthen bed Pans. 1711 in *Essex Inst. Hist. Coll.* IV. 187/1 A bed pan and stool Pan. 1865 *Med. Jrnl.* XIV. 254 The urine has continued to flow involuntarily, but a few spoons full have been occasionally voided into the bed pan.

bed-post. Add: Phr. *between you and me and the bed-post*: in all confidence or secrecy.

1830 MRS. ROYALL *Lett. fr. Alabama* 136 Between you and I and the bed post, I begin to think it all a plot of the priests. 1832 [see POST *sb.*¹ 1 c]. 1882 BLACKMORE *Christowell* xiii, Between you and me and the bed-post, Short—as the old ladies say—I don't want Jack to have her.

bed-rock. orig. *U.S.* (See BED *sb.* 19.)

1850 N. KINGSLEY *Diary* (1914) 154 We are in for seeing the bed rock all along the bottom. 1872 SCHELE DE VERE *Americanisms* 171 The miner..hopes to reach bed-rock, where gold is found in quantities. 1873 G. A. LAWRENCE *Silverland* 181 The material invariably waxes greatly richer as the 'bed-rock' is neared. 1879 [see BED *sb.* 19].

fig. 1869 S. BOWLES *Our New West* v. 99 We came down to 'bed-rock' as the miners say, i.e. an extra flannel shirt and a pocket-comb. 1881, 1883 [see BED *sb.* 19]. 1884 NYE *Baled Hay* 65 Whenever the dead-beat poet strikes bedrock. 1900 'O. HENRY' *Roads of Destiny* iii. 52 From general topics the conversation concentrated to the bed-rock of grim personalities. 1912 J. S. HUXLEY *Individ. in Animal Kingdom* vi. 151 We have come down to the bed-rock questions of biology.

bedroom. Add: **3.** *attrib.* and *Comb.*

1813 SOUTHEY *Life Nelson* I. i. 7 He was lowered down at night from the bed-room window by some sheets. 1816 SCOTT *Antiq.* I. ix. 208 The Antiquary took up a bedroom candlestick of massive silver and antique form. 1834 DICKENS *Sk. Boz* (1836) ser. I. I. 211 Plated bedroom candlesticks. 1905 F. H. BURNETT *Little Princess* viii. 120 She shuffled across the attic in her bedroom slippers. 1928 J. CARY *Castle Corner* 455 He had a full inventory of the marriage chamber, including the bedroom china, decorated with primroses. 1967 E. SHORT *Embroidery & Fabric Collage* iii. 80 For bedroom cushions finer fabrics and technique could be used.

b. In extended use, usu. implying sexual relationship; sexy.

[1915 E. POUND in *Lett. J. Joyce* (1966) II. 366 Bed rooms scenes where the audience can be tittivated, eroticised.] 1924 M. ARLEN *Green Hat* iv. 112, I thought the editor might take objection to certain passages, as there is some strong bedroom stuff in it. 1947 AUDEN *Age of Anxiety* (1948) ii. 41 Here a dean sits Making bedroom eyes at a beef steak. 1953 *Time* 6 July 72/1 It has also been described as a bedroom farce set in a living room. 1959 M. CHAMBERLAIN *Dear Friends & Darling Romans* (1960) iv. 41 Italians are bedroom-eyed gigolos. 1967 J. POTTER *Foul Play* i. 11 George's wife had blue bedroom eyes.

bedside. Add: **b.** *attrib.*, as *bedside book, literature*; **bedside manner,** the deportment of a medical man towards his patient.

1837 DICKENS *Pickw.* xxxviii. 416 A female servant came out..to shake some bed-side carpets. 1860 F. NIGHTINGALE *Notes on Nursing* viii. 46 If a patient can turn on his side, he will eat more comfortably from a bed-side table. 1869 PRENTISS *Stepping Heavenward* (1870) xxv. 237 He was her 'pet-doctor', he had such 'sweet, bed-side manners'. 1879 C. M. YONGE *Magnum Bonum* II. xxvii. 570 Poor Janet found the thing in the back of the bedside table-drawer. 1884 *Punch* 15 Mar. 121 *Lady Visitor*. 'Oh that's your Doctor, is it? What sort of a doctor is he?' *Lady Resident.* 'Oh well, I don't know much about his ability; but he's got a very good bedside manner!' 1907 *Brit. Med. Jrnl.* 28 Dec. 1845/1 The ordinary notion is that a good bedside manner consists of suavity carried to the verge of civility. 1920 *Cornh. Mag.* July 63 Bedside Books. *Ibid.* 64 Bedside literature. 1949 D. SMITH *I capture Castle* I. ii. 12, I keep my bedside candlestick on a battered tin trunk.

bed-sitting-room. [f. the sbs.] A room serving both as a bedroom and as a sitting-room. Also (orig. *University slang*), **bed-sitter,** whence abbrev. **bed-sit, bedsit.**

1892 ZANGWILL *Childr. Ghetto* I. xvi. 29 The bed-sitting-room which they rented was turned into a salon of reception. 1905 G. B. SHAW *Passion, Poison & Petrifaction* in *H. Furniss's Annual* 11/1 In a bed-sitting room in a fashionable quarter of London. 1927 W. E. COLLINSON *Contemp. Eng.* 89 The problem of house-hunting brought with it various technical terms or rather revived interest in them viz...bed-sitter (bedroom and living-room combined). 1933 C. DAY LEWIS *Magnetic Mountain* 12 Then book your bed-sitter at the station hotel. 1933 *Archit. Rev.* LXXIV. 23/1 No one who has seen this will be content with the unhandiness and drab squalor of the usual bed-sitting-room method of living. 1938 N. MARSH *Artists in Crime* xv. 225 I'm going to spend the..morning in a chorus lady's bed-sit. in Chelsea. 1957 *Oxford Mag.* 293 But pressure for admissions does mean that some freshmen sit disconsolate in a 'bedsit' far down the Iffley Road. 1959 *House & Garden* July 44/1 When the children are older, their play space and bedrooms will be planned as two large bed-sitting-rooms. 1967 *Spectator* 14 July 53/2 Permanent homes are no longer necessities and a bed-sit mentality has taken over.

bedspread (be·dspred). orig. *U.S.* [f. BED *sb.* + SPREAD *sb.* 8 a; cf. Du. *bed(de)sprei*.] A light coverlet for a bed, usually removed when the bed is occupied.

1845 W. T. PORTER *Big Bear Arkansas* 30, I made a bed-spread of his skin. 1848 BARTLETT *Dict. Amer.*, *Bed-spread*, in the interior parts of the country, the common name for a bed-quilt, or coverlet. 1887 *Queen* 29 Oct. 558/1 For bed-spread and chair-seats in your pink room. 1914 *Evening News* 15 Oct. 7/6 You will have a pretty, light, and warm bedspread at a cost of 1s. 7½d. 1926 *British Weekly* 24 June 250/5 Silk bedspreads in rich colourings, with shot effects.

bedtime. Add: (Later example.)

1837 DICKENS *Pickwick* xxxvii. 406 Master opens it, and reads the label, 'Draught to be taken at bedtime'.

attrib. **bedtime story,** a story told (to a child) at bedtime; also *fig.*

1894 S. FISKE *Jack's Partner* iv. § iii. 80 Rose had suggested this partnership to John during a bed-time conference. 1899 (*title*, Blackie & Son) Bedtime Stories, a picture-book for little folk. 1926 A. HUXLEY *Jesting Pilate* III. 254 Racing results and bed-time stories. 1940 A. HOCKING *Wicked Flee* viii. 173 If we come to the conclusion that it's all just a bedtime story, you're left to assume that Lavinia is unhinged.

Bedu (be·dū), *sb.* and *a.* [Arab. *badw* desert, Bedouins, *badawī* Bedouin: see BEDOUIN.] (Of or pertaining to) a Bedouin.

1912 T. E. LAWRENCE *Home Lett.* (1954) 188 Egypt is not to be ranked with Syria for a moment: you have only to think of fellahin—and ours are Bedu. 1917 —— *Lett.* (1938) 234 The Bedu had their own food with them in their saddle bags. 1940 F. STARK *Winter in Arabia* iii. 15 We.. took with us a bedu wounded in the foot by stones. 1959 *Listener* 19 Nov. 893/2 The vanishing life of his Bedu friends.

bee¹. Add: **3. c.** 'A lump of a yeast (*Saccharomyces pyriformis*) intermittently rising and releasing bubbles in brew;—usually in *pl.*' (Webster 1934). So **bee wine** (see quot. 1960).

1923 *Harmsworth's Househ. Encycl.* I. 312/2 Bee wine is a modern name for the fermented drink produced by what was known as the ginger beer plant. *Ibid.*, In its dry, inert condition the bee is a shapeless mass of gelatinous material. 1938 R. GRAVES *Count Belisarius* x. 206 The Massagetic Huns carried with them what is called a 'bee', a sort of yeast that they put into mare's milk to make it ferment. 1960 A. E. BENDER *Dict. Nutrition* 16/1 *Bee wine*, wine produced by the usual alcoholic fermentation of sugar, but using yeast in the form of a clump of yeast and lactic bacteria. The clump rises and falls with bubbles of carbon dioxide produced, hence the 'bee'.

4. (Earlier example.) *lynching bee*: see *LYNCHING *vbl. sb.*

1769 *Boston Gaz.* 16 Oct. (Th.), Last Thursday about twenty young Ladies met at the house of Mr. L. on purpose for a Spinning Match; (or what is called in the Country a Bee).

5. b. *bee's knee*: (*a*) a type of something small or insignificant; (*b*) *pl.* (*slang*, orig. *U.S.*), the acme of excellence; 'the cat's whiskers'; *to put the bee on* (slang, chiefly *U.S.*): (*a*) to quash, put an end to; to beat; (*b*) to ask for a loan from, to borrow money from (cf. *STING *v.*¹ 2 e).

1797 Mrs. Townley Ward Let. 27 June in N. & Q. (1896) X. 260 It cannot be as big as a bee's knee. 1870 G. M. Hopkins Jrnl. (1937) 133 Br. Yates gave me the following Irish expressions...As weak as a bee's knee. 1894 G. F. Northall Folk-phrases 7 As big as a bee's knee. 1923 H. C. Witwer Fighting Blood iii. 101 You're the bee's knees, for a fact! 1936 H. L. Mencken Amer. Lang. (ed. 4) 561 The flea's eyebrows, the bee's knees and the canary's tusks will be recalled. 1958 Times 15 Aug. 9/4 Lord Montgomery..holds that to label anything the 'cat's whiskers' is to confer on it the highest honour, and the 'bee's knees' is not far behind it as a compliment.

1918 H. C. Witwer Baseball to Boches 131 It's always open season for Americans over here. They sure know how to put the bee on you too. 1923 L. J. Vance Baroque xxvii. 264 I've heard a heap of fairy tales in my time..but this puts the bee on the lot. 1927 Wodehouse in Sunday Express 23 Oct. 9 The old boy..got the idea that I was off my rocker, and put the bee on the proceedings. 1929 Amer. Speech IV. 338 'To put the bee on', means to beg. 1931 G. Irwin Amer. Tramp & Underworld Slang 25 To say 'I put the bee on him' usually means that the donor has been 'stung', when he gives up the loan, since seldom is it repaid. 1936 J. Curtis Gilt Kid v. 47 If a bloke had come up and put the bee on him all the handout would have been..a lousy tanner.

c. bees and honey: rhyming slang for 'money'.
1892 Answers 10 Sept. 276/1 'Bees and honey'..for 'money'. 1935 'L. Luard' Conq. Seas iii. 47 A skipper's life ain't all bees and honey. 1944 Amer. Speech XIX. 191/1. 1960 J. Ashford Counsel for Defence v. 65 D'you reckon we'd waste good bees and honey on a slump like you for nothing.

6. bee-farming, -hunt, -hunter (later example), -hunting (example), -keeper (examples); bee-infested, -studded, -thronged, -winged adjs.
1908 Wodehouse & Westbrook Globe by Way Book 124 Lord Sangazure has tired already of his latest hobby, bee-farming. 1835 W. Irving Tour on Prairies ix. 61 (heading) A bee hunt. 1837 —— Capt. Bonneville I. ii. 52 These frontier settlers..prepare for a bee hunt. 1954 J. R. R. Tolkien Two Towers 67 Bear bee-hunter, boar the fighter. 1824 W. N. Blane Excursion U.S. 239 It is a favourite amusement..to go bee-hunting. a1882 Emerson Fragm. Nature in Poems (1904) 343 Bee-infested quince or plum. 1950 D. Gascoyne Vagrant 61 A loud Bee-Infested Lion-skin. 1817 Kirby & Spence Entomol. II. xx. 211 It is a saying of bee-keepers in Holland, that [etc.]. 1937 Discovery June 191/2 The most enthusiastic bee-keepers. 1881 Wilde Burden of Itys in Poems 68 Brown bee-studded orchids. 1910 Kipling Rewards & Fairies p. x, Winter's bee-thronged ivy-bloom. 1923 E. Sitwell Bucolic Comedies 40 The bee-wing'd warm afternoon.

7. bee-fold (later example); **bee-gum,** add to def.: orig. a hollow gum tree or log housing a swarm of bees; (earlier examples); **bee-hawk (-moth)** (earlier example); **bee-line** (earlier U.S. example); **bee-loud** a., resonant with the hum of bees; **bee-louse,** an insect of the family Braulidæ, parasitic on bees, esp. Braula cæca; **bee-man,** a bee-keeper; **bee-moth** U.S., Galleria mellonella: = wax-moth (see Wax sb.¹ 12); **bee-range** U.S., a row of beehives; **bee-smoker,** a bee-keeper's apparatus for driving smoke into a hive to stupefy the bees while the honeycomb is being removed; **bee-tree** (earlier U.S. examples).
1940 C. Day Lewis tr. Virgil's Georgics iv. 82 Shoo the drones—that work-shy gang—away from the bee-folds. 1817 M. L. Weems Lett. (1929) III. 215 To be run..round & round the circumference of a Bee-Gum. 1859 Bartlett Dict. Amer. (ed. 2), Bee-gum, in the South and West, a term originally applied to a species of the gum-tree from which bee-hives were made; and now to beehives made of any kind of boards. 1815 Kirby & Spence Entomol. I. vi. 207 The bee-hawk-moth (Sesia apiformis, F.)..feeds upon the poplar. 1830 Massachusetts Spy 24 Nov. (Th.), The squirrel took a bee line, and reached the ground six feet ahead. 1890 Yeats Lake Isle of Innisfree i, [I will] live alone in the bee-loud glade. 1840 J. & M. Loudon tr. Kollar's Treat. Insects i. 74 A bee infested with a bee-louse, endeavours..to get rid of such an unwished-for guest. 1875 J. Hunter Man. Bee-keeping xxx. 198 On the Continent of Europe a small insect known as the Bee-louse, Braula Cæca, often infects the Bees. 1861 Trans. Ill. Agric. Soc. IV. 82 Our best bee men. 1928 Daily Tel. 11 May 19/5 Uncontrolled swarming ..is not permitted by the experienced modern beeman. 1829 Massachusetts Spy 27 May (Th.), Instinct teaches the bee-moth to secrete herself, during the day, in the corners of the hive. 1838 H. Colman Rep. Agric. Mass. 71 The bee moth is to be guarded against by making the crevices of the hive tight with putty or glue. 1862 T. W. Harris Insects Injur. Veget. 489 The group called Crambidæ, or Crambians, among which the bee-moth or wax-moth is to be placed. 1845 S. Judd Margaret iii. 402 In the garden is a large Bee-range. 1897 Westm. Gaz. 7 Aug. 8/1 A bee-smoker filled with tobacco and brown paper. 1782 St. J. de Crèvecœur Lett. 37 If we find..what is called a bee-tree, we must mark it. 1834 Brackenridge Recoll. xii. 129 A harmless fellow, who followed hunting bee trees on the mountains for a living.

bee³. The name of the letter B, used for 'bloody' (see Bloody A. 10 and B. 2); so bee aitch, bloody hell; bee eff, bloody fool. slang.
1926 Galsworthy Silver Spoon I. iii. 21 This is a bee map. ..Quite the bee-est map I ever saw. Ibid. III. ii. 230 It's a bee nuisance. Ibid. 231 We have the best goods..and we must bee well deliver them. 1928 —— Swan Song I. vii, Mr. Blythe's continual remark: 'What the bee aitch are they all about?' 1960 M. Cecil Something in Common i. 22 'Your mother's relations,' he muttered, 'bee effs, every one of 'em.'

beech. Add: **3.** beech-bole, -nut (earlier example), -wood (earlier example); -green adj. (example; also as sb.).
1681 Grew Musæum III. § I. ii. 269 A Stone..which looks like a piece of Beech-wood. 1739 E. Smith Compl. Housew. (ed. 9) 316 A quarter of a spoonful of oil of beech-nuts. Ibid. 317 The beech-nut-oil. 1908 R. South Moths II. 184 The Beech-Green Carpet (Amoebe olivata). 1909 L. Huxley in Westm. Gaz. 22 Jan. 2/3 The silver beechboles burn to gold. 1916 D. H. Lawrence Amores 109 The woods where the beech-green spurts Like a storm of emerald snow.
4. beech disease, beech-wood sugar (see quots.); **beech-drops** (earlier U.S. example); **beech-finch** (example).
1905 Daily Chron. 3 July 9/1 Beech disease (cryptococcus fagi) is widely distributed throughout England. 1815 Drake Cincinnati ii. 86 Beech drops. 1816 Stephens in Shaw's Gen. Zool. IX. II. 444 It [the Chaffinch] is called by various names in this country, such as..Flax-finch, Beech-finch, [etc.]. 1913 Dorland Med. Dict. (ed. 7) 919/1 Beechwood sugar, xylose.

bee-eater. (Earlier and later examples.)
1668 Charleton Onomast. 87 Merops..Apiaster, the Bee-eater. 1896 H. A. Bryden Tales S. Afr. vi. 129 Gem-like bee-eaters flashed among the reeds. 1955 Bannerman Birds Brit. Isles IV. 43 The European bee-eater and the blue-cheeked bee-eater, which has once wandered to Britain, ..have overlapping migration routes in Egypt, Syria and Persia.

beef, sb. Add: **2. b.** Also, strength, muscular power; effort. Cf. *beef v. 2.
1851 Melville Moby Dick II. xxxix. 261 Oh, do pile on the beef...Oh! my lads, do spring. 1863 [in Dict.]. 1968 R. M. Patterson Finlay's River 25 We on the other hand, were paddling our canoe and putting all the beef we could behind it.
c. Slang phr. beef to the heel(s): of a person: massive, bulky, brawny.
1867 R. Broughton Cometh Up I. xviii. 294 Dolly was not a fine person, as they say, at all; not beef to the heels, by any means. 1922 Joyce Ulysses 65 We did great biz yesterday. Fair day and all the beef to the heels were in. Ibid. 366 Transparent stockings, stretched to breaking point. Not like..the one in Grafton street. White. Wow! Beef to the heel. 1958 A. Wall Queen's English xxxi. 110 'Beef' and 'beefy' are not complimentary terms, and 'beef to the heels' is even less so.
3. b. Common in U.S.
1758 in Essex Inst. Hist. Coll. (1881) XVIII. 93 Six men.. put under Guard on suspicion of killing a young Beef and 2 Calves. 1878 Southern Hist. Soc. Papers VI. 212 Behind these came a beef, driven by soldiers...The beef was immediately shot at and butchered. 1904 J. H. Claiborne Seventy-five Yrs. Old Virginia 289 A beef..was driven up and shot.
c. collect. Cattle. U.S.
c1706 Voy. Maryland in Amer. Hist. Rev. (1907) XII. 329 They have fatt Beefe and fatt Porke Comes home to their Doores without giving 'em any Corne. 1840 J. Buel Farmer's Compan. (ed. 2) 318/2 The fattening of beef. 1907 S. E. White Arizona Nights ii. 30 We was just gettin' back from drivin' some beef up to the troops.
4. beef animal, -cattle (U.S. examples), -house (example).
1838 H. Colman Rep. Agric. Mass. 73 They agree to pay 32 cents. for the offal of every beef animal there slaughtered. 1758 T. Walker Let. 14 Aug. in Amer. Antiq. Soc. Proc. (1932) XLI. 130 Five Hundred Beef Cattle are to set off tomorrow. 1899 Scribner's Mag. XXV. 116/2 We passed a herd of fine beef cattle on their way to Santiago. 1807 J. Beresford Miseries II. xix. 211 Taking your dinner from an a-la-mode beef-house.
5. beefburger (bī·fbɜːgəɹ) [*burger], a hamburger (orig. U.S.); **beefcake** (bī·fkeᶦk) (slang, orig. U.S.) [humorous, after *cheesecake], (a display of) sturdy masculine physique; **beef chit** Naut. slang, a menu; **beef-dairy** a., applied to a cross between beef and dairy cattle; **beef essence** (see quot. 1890); **beef-extract,** an extract of the soluble fibrin of beef; **beef-fed** a., fed on beef; esp. as an epithet for a typical Englishman; **beef-headed** a., = beef-brained; **beef stroganof(f)** (see *stroganoff); **beef trip** Naval slang, (a) a routine trip in the ship's boat assigned to bring the meat ration from the shore; (b) an escort of food ships in convoy in the 1914–18 war.
1940 in Amer. Speech XV. 452/1 Hamburgers are out, beefburgers are in! 1959 Observer 8 Nov. 3/5 We can expect the kind of hamburger proliferation that exists in America, with beefburgers, eggburgers, [etc.]. 1949 in Amer. Speech (1954) XXIX. 282 Alan Ladd has a beef—about 'beefcake', the new Hollywood trend toward exposing the male chest. 1954 Sunday Pictorial 18 Apr. 15/1, I learned that beefcake was bunk because one of my jobs in the circus was lacing the strongman's shoes. 1963 Guardian 29 June 5/4 The other poster..shows Albert Finney in a beefcake pose with his shirt slit to the navel. 1911 'Guns' et al. Middle Watch Musings 54 The Commander (having failed to arouse the Wardroom Waiters from their lethargy; sitting for five minutes without being offered the beef chit). 1962 John o'London's 14 June 571/1 The beef chit, or [wardroom] menu. 1960 Farmer & Stockbreeder 16 Feb. 15/1 Beef-dairy crosses include crosses between beef breeds and Friesians or Dairy Shorthorns. 1853 Dunglison Dict. Med. Sci. (ed. 9) 130/1 Beef essence. 1890 Billings Med. Dict., Beef essence, cooked juice of beef, prepared by introducing lean beef in small pieces into a bottle, and subjecting it to heat of a boiling water-bath for an hour. 1890 W. James Princ. Psychol. II. xxiii. 380 The stimulating effects of tobacco, smoke, alcohol, beef-extract (which is innutritious), etc., etc., may be partly due to a dynamogenic action of this sort. 1903 Kipling Five Nations 78 And beef-fed zealots threaten in To Buddha and Kamakura. 1903 A. McNeill Egreg. English viii. 79 A beef-fed army and a beef-fed navy are what Englishmen believe they get for their money. 1934 E. H. W. Meyerstein Let. 3 Apr. (1959) 159 Of course you (i.e. beef-fed materialistic Englishman (woman)) can argue. 1828 Mrs. Royall Black Book II. 114 Such a great beef-headed fellow as you editor of a paper. 1900 E. Terry Let. 28 Jan. (1931) 372 The war—and the disgrace of it (beef-headed Buller's doings). 1919 'Taffrail' H.M.S. Anonymous vii. § 2. 139 We called it the 'Beef trip' because in some remote fashion it reminded us of our midshipman days, when, in charge of the 'blood boat' we left our ships.. to bring off from the shore the fresh meat which would..be served out to the ship's company. Ibid. 142 In spite of all precautions on our part the German flotillas from Zeebrugge ..might altogether have put a stop to the 'beef trip' carried on under their very noses. Ibid. § 3. 152 The 'beef trip' was over. Yet another consignment of margarine and Dutch cheeses had been brought into the United Kingdom. 1925 Fraser & Gibbons Soldier & Sailor Words 21 On the beef trip, a Navy term in the War, used of the vessels carrying meat to the Grand Fleet at Scapa.
6. A protest, (ground for) complaint, grievance. slang (orig. U.S.). Cf. *beef v. 4.
1899 Ade Fables in Slang (1900) 80 He made a Horrible Beef because he couldn't get Loaf Sugar for his Coffee. 1931 D. Runyon Guys & Dolls (1932) iv. 79 To round himself up with his ever-loving wife in case of a beef from her over keeping the baby out in the night air. 1945 Daily Express 22 May 2/6 The beef is, Why should every battle we fight have to be a 'Battle of Britain'? 1946 Wodehouse Money in Bank xii. 95 What's your beef about taking ninety per? 1949 [see *beefcake above].

beef, v. Add: **2.** To put more muscle into, to drive harder. U.S. (College slang.)
1860 Yale Lit. Mag. XXVI. 83 (Th.), The first boat in is the winner of the race, so round they turn, and 'beef her' for the home stretch.
b. slang. To strengthen; to add vigour, power, or importance to. Const. up (occas. out). Hence beefed-up (or -out), ppl. a. Cf. *beef sb. 2 b.
1941 A. O. Pollard Bombers over Reich xi. 155 When the Fortresses reach Britain from the United States certain alterations are made; the larger guns are..'beefed up' so as to give them a rate of fire of 900 rounds a minute. 1944 Time 24 Jan. 22/3 The Eighth [Air Force] proved that it had beefed up its reserves. 1958 Observer 2 Feb. 1/2 This main stage—the 'beefed-up' Redstone—uses a secret formula 'exotic' fuel. 1960 Times 21 Jan. 13/6 Economic planners.. may be tempted to beef-up a project to make it ambitious enough to qualify for large-scale aid. 1963 Language XXXIX. 242 A new generation was fed the nonsense of Marr, only partially beefed out with surreptitious teachings. 1966 New Scientist 10 Mar. 618/1 The Defense Department has spent $50 billion building and beefing up the non-nuclear elements of the armed forces.
3. To slaughter (an ox, etc.) for beef. Also transf., to knock down. U.S. slang.
1869 Harper's Mag. Jan. 159/1 They [sc. the buffalo] were to be beefed and sent East or put into cattle-cars, and killed after they had arrived in the Eastern cities. 1916 B. M. Bower Phantom Herd xvi. 268, I calc'late I'd better beef another critter. 1926 J. Black You can't Win xiii. 185 When one of them got peeved..some hard-fisted miner beefed him like an ox. 1934 A. Hyder Black-Girl, White-Lady xxi. 309 'Yo' kills niggers?' 'Like flies,' Charley assured her. 'You want me to beef a few for you?'
4. intr. To complain, grumble, protest. slang (orig. U.S.). Hence bee·fing vbl. sb.
1888 N.Y. World 13 May (Farmer), He'll beef an' kick like a steer an' let on he won't never wear 'em. 1889 The Road (Denver) 28 Dec. 4/3 He will be coming down town again soon on crutches, 'beefing' about cancer of the stomach. 1922 S. Lewis Babbitt v. 60 Course I wouldn't beef about it to the fellows at the Roughnecks' Table there. 1930 Wodehouse Very Good, Jeeves! iii. 74 You have beefed about Miss Wickham. 1931 E. Linklater Juan in America II. xvii. 182 'So I'm a slave, am I? Like hell I am!' 'Stop beefing,' said Spider. 1939 J. B. Priestley Let People Sing viii. 210 Too much arguin' an' beefin', majority an' minority nonsense, talkin' shop stuff. 1957 H. Croome Forgotten Place xi. 138 Stop beefing, Frank. You'll be seeing her again soon enough.

beefeater. Add: **2. b.** A style of hat worn by women, resembling that worn by a Yeoman of the Guard. Also attrib.
1785 Betsy Sheridan Jrnl. 5 July (1960) ii. 59 Hats tied under the chin, beefeaters etc are not worn by fashionable people. 1895 Daily News 3 Sept. 6/3 The black straw sailor hat, with Beefeater crown. 1909 Westm. Gaz. 4 Jan. 5/2 A style of hat which was a favourite with us a year or two ago—viz., a velvet beefeater—has been brought in again.

beefer (bī·fəɹ). [f. Beef sb. + -er¹.] An animal bred for beef.
1660 May Accompl. Cook 113 Take the blood of the beefer when it is warm, put in some salt, and then strain it. 1809 T. Batchelor Anal. Eng. Lang. v. 127 A beefer; a familiar name for a calf. 1938 J. Moses Nine Miles from Gundigai 17 Milkers, too, very fine, And beefers off the Considine.

beefish (bī·fiʃ), a. [f. Beef sb. + -ish¹.] **a.** Beefy (with favourable implication). **b.** = beef-fed (*beef sb. 5).
1882 G. M. Hopkins Let. 3 Apr. (1935) 143, I hope..you are now..well; strong, vigorous, lusty, beefish. 1887 Andover Rev. VII. 32 This degeneracy has turned him into that 'beefish, porterish', bellowing sort of a John Bull.

beef-steak. b. Add: **beef-steak fungus,** a fungus, Fistulina hepatica, somewhat resembling a beef-steak in appearance. Also beef-steak mushroom.
1886 J. Stevenson Brit. Fungi II. 183 Fistula..F. hepatica..Edible and nourishing, but rather coarse, and not a very pleasant flavour. Known as the beef-steak fungus. Slices exactly like beef-steak. 1895 W. H. Gibson Our Edible Toadstools & Mushrooms 213 The beefsteak mushroom—Fistulina hepatica...Its upper surface was dark meaty red or liver colored. 1953 J. Ramsbottom Mushrooms & Toadstools xix. 231 Fistulina hepatica, the Beef-Steak Fungus, looking somewhat like an ox-tongue with a glutinous upper surface, yellow tubes and streaky flesh exuding a red juice.

beefy, a. Add: (Earlier and later examples.) Also, brawny.
1743 H. Walpole Let. 20 Aug. (1903) I. 373, I here every day see men, who are mountains of roast beef...I have an Aunt..who..is as beefy as her neighbours. 1865 Slang Dict., Beefy, unduly thick or fat, commonly said of women's ancles. 1889 Cent. Dict., Beefy..2. Brawny; muscular; hardy. 1890 Kipling Barrack-r. Ballads (1892) 52 Beefy face an' grubby 'and—Law! wot do they understand?
b. Schoolboys' slang. Good, clever, fair, attractive.
1903 Wodehouse Prefect's Uncle xv. 213, I doubled across to bring off a beefy c-and-b. 1905 H. A. Vachell Hill i. 15 Comparing these [sc. other rooms] with his own apartment, John said shyly—'It's not very beefy.' 'Beefy? You smell of a private school.'

beega(h, varr. *bigha.

beehive. Add: **1.** Further fig. examples, esp. in the sense of 'a place swarming with busy people'. Also attrib. (Cf. Hive sb. 3.)
1725 New-Eng. Courant 8 Mar. 1 If we could..make that Building a Bee-Hive of Business. 1940 Manch. Guardian Weekly 15 Mar. 216 He works in a great hospital now, a beehive of a place—swarming with people. 1949 Koestler Promise & Fulf. III. iv. 324 The bustling bee-hive activity of the whole country.
b. A hat shaped like a beehive. Cf. beehive-hat below (sense 3).
1909 Westm. Gaz. 2 Mar. 5/2 A useful hat..is one of the new shape, which some milliners are calling the 'Beehive'. 1937 N. Coward Present Indic. I. xix. 66 Gertie Millar in Our Miss Gibbs, wearing a beehive.
c. A high beehive-shaped hair-style.
1960 Guardian 22 July 7/2 Three East Berlin peroxide girls whose beehives tower over provocative curves. Ibid. 26 Aug. 6/4 Exaggerated beehive hair-styles. 1962 New Statesman 18 May 708/2 Rube, neck stiff so as not to shake her beehive, stares sultry round the packed pub.
3. beehive-hat; **beehive coke,** coke produced in a beehive oven; **beehive tomb,** a dome-shaped tomb of the Mycenæan age in Greece, cut in a hillside.
1909 Webster, Beehive coke. 1914 J. S. S. Brame Fuel vi. 97 It is found that the lower portions of beehive coke are more dense than the upper. 1909 Daily Chron. 22 Feb. 7/5 The latest beehive hat. 1961 Guardian 19 Jan. 9/7 Worn with a..beehive hat. 1887 Mahaffy Rambles Greece xv. (ed. 3) 417 A race..who constructed great beehive tombs. 1957 Childe Dawn Europ. Civ. (ed. 6) v. 80 These [sc. chieftains] celebrated their elevation by erecting stately beehive tombs or tholoi.

beekite (bī·kəit). Min. Also **beckite** (be·kəit). [f. the name of Henry Beeke, dean of Bristol (d. 1837): see -ite¹.] (See quots.)
[1847 A. Dufrénoy Traité de Minéralogie III. 750 Beckite, petites masses concrétionnées, à structure botrioïde, d'un gris rougeâtre, d'un gris sale, qui rayent le verre.] 1856 W. Pengelly in Rep. Brit. Assoc. Advancem. Sci. (Trans.) 74 The most interesting things found in the red Triassic conglomerates of Torbay are the Beekites, so named from the late Dr. Beeke, Dean of Bristol, by whom, it is believed, they were first noticed. Ibid., The interior of the Beekite is calcareous. 1896 A. H. Chester Dict. Min. 29 Beckite,..a chalcedonic pseudomorph after corals or shells, often called beckite. The specimens are locally known as beekites, the name probably being used prior to the reference given [i.e. 1847, cited above]. 1950 New Biol. VIII. 85 Replacement by silica may also start from various centres on the shells of molluscs, producing an imperfect crystallization in rings (known as beekite structure).

been-to (bī·ntū). Also **bintu.** [f. Been pa. pple. + To prep.] A term used in Africa and Asia for a person who has been to England, usu. for education. Also attrib.
1960 M. Laurence This Side Jordan iv. 82 Who did he think he was, a 'been-to' man, educated abroad, like Victor? Ibid. xiv. 251 She was obviously a 'been-to', probably trained in England. 1961 Time 3 Nov. 65/1 An educated modern African..sometimes calls himself a Bintu, meaning 'been to' Oxford or Cambridge.

beep (bīp), sb. [Imitative.] The sound made by a horn on a motor car or other vehicle; a short high-pitched sound such as is emitted by an echo-sounder, a radar device, etc. Also attrib. So **beep-beep,** such a (reduplicated) sound; also concr., a horn.
1929 E. Wilson I thought of Daisy v. 299 Did you ever hear them talk about auto-horns?..There's a toot-toot, and a beep-beep. 1937 N. Marsh Vintage Murder xxii. 250 This horn is called a 'beep-beep'. 1951 A. C. Clarke Sands of Mars iv. 45 The carrier wave..now modulated into an endless string of 'beep-beep-beeps'. 1952 [see *beeper]. 1956 J. Potts Diehard vii. 109 The ritual beep-beep of his horn...Two brief, muted beeps. 1957 Economist 2 Nov. 399/1 A four-stage research rocket..sent back beeps. 1962 A. Nisbett Technique Sound Studio vii. 131 Time signal pips or 'beep' tones: the very words 'pip' and 'beep' suggest the effect I mean.

beep (bīp), v. [Imitative; cf. prec.] **a.** trans. To sound (a horn); to make (something) emit a short high-pitched sound; to indicate by sounding beeps. **b.** intr. To emit beeps. Hence **beeped, bee·ping,** ppl. adjs.
1936 O. Nash Primrose Path 101 Beep the horn and howl the klaxon For Hebrew, Latin and Anglo-Saxon; Howling klaxon, beeping horn. 1953 Pohl & Kornbluth Space Merch. (1955) xiii. 140 He went pale, but beeped his alarm, and went down in a tangle of fists and boots. 1958 New Scientist 16 Oct. 1060/1 The first Russian moon rose beeping into the sky a year ago. 1962 Flight Internat. LXXXI. 299/1 An Aero Commander was fitted with Lear L-5B autopilot (modified for close pitch control, decrabbing and 'beeped' pitch trim).

beeper (bī·pəɹ). [f. *beep v. + -er¹.] A device that emits beeps (see also quot. 1946); also attrib.
1946 Sun (Baltimore) 7 Aug. 1/3 Planes called 'Beepers'. Ibid., Mother planes were piloted by Capt. Ragner Carlson... Radio operators of the drone-controlled planes, or 'beepers', were Capt. John Evans, [etc.]. 1952 A. C. Clarke Islands in Sky iv. 72 A Beeper..is a tiny radar set..used to locate objects that have drifted away from the Station...When its beam hits anything you hear a series of 'Beeps'. 1957 Time 2 Sept. 32/2 Drowning out racy epithets with an electronic beeper signal.

beer, sb.¹ Add: **1. b.** Also, on the beer: on a bout of drinking (cf. on the booze s.v. *booze sb. 2 b); beer and skittles: see Skittle sb. 1 b.
1887 F. T. Havergal Herefordshire Words 6/2 When a man is in a fuddled state it is said that 'he is on the beer'. 1959 Listener 4 June 980/2 A riotous evening on the beer.
3. a. beer-swilling ppl. a. and vbl. sb. **d.** beer-bottle, -mat.
1839 C. Sinclair Holiday House i. 13 If I want to draw a cork out of a beer bottle. 1968 B. K. Martin Editor xi. 237 Many of the men..wore chains round their necks made of the tops of beer-bottles. 1939 C. Day Lewis Child of Misf. III. v. 323 The beer-mats on the mantlepiece (what odd things for her to have brought out of Germany). 1849 T. Martin Ballads 165 In beer-swilling Copenhagen I have drunk your Danesman blind. 1908 Westm. Gaz. 28 Sept. 3/2 Beer-swilling in Hyde Park on a Sunday afternoon was not ..one of the regular sights of London.
4. beer-boy = Pot-boy; **beer-cellar,** (a) an underground room for storing beer; (b) a beer-shop in a cellar or basement; **beer drink** S. Afr., a native gathering for the purpose of drinking Kaffir beer; **beer-garden** (earlier example); **beer-hall** S. Afr., a public hall where Kaffir beer is sold to non-whites; **beer money** (earlier and later examples); also gen., a small amount of money earned or allowed; **beer-off** slang, an off-licence; **beer-parlor, -parlour** [cf. Parlour 3] Canadian, a room in a hotel or tavern where beer is served; **beer-up,** a drinking-bout or -party.
1840 Dickens Old C. Shop xxxiv. 285 A beer-boy happened to pass. 1908 Westm. Gaz. 5 Feb. 12/1 His silk hat..had been ruined by the beer-boy spilling a pint of ale into it. 1732 S. Carolina Gaz. 28/2 At the Beer Cellar, over against Mr Elliot's Bridge. 1817 J. K. Paulding Lett. from South II. 57 Diving into stews and beer-cellars. 1820 Sporting Mag. VI. 225/1 The coffee-houses and beer-cellars of the Palais-Royal [in Paris] teem indiscriminately with both sexes. 1865 [in Dict., sense 3 d]. 1952 A. Bullock Hitler i. 43 He [sc. Hitler] spent much time in cafés and beer-cellars, devouring the newspapers and arguing about politics. 1895 Scully Kafir Stories 182 Lukwazi rode..from beer-drink to beer-drink, and he was drunk most of his days. 1899 W. H. Brown On S. Afr. Frontier xv. 213 When the bride reaches her new home the event is celebrated with a big dance and a 'beer drink'. 1959 Rand Daily Mail 21 Jan. 9/6 Police investigating housebreaking on a farm stumbled on an illicit beer-drink and arrested 14 Africans. 1863 Trollope in Tales of all Countries (ser. 2) 302 They all passed their evenings together in the beer-garden. 1900 Statute Law of S. Rhodesia 1899 II. 90 'Beer Hall Licenses' may be issued to brewers of beer manufactured in Southern Rhodesia for the sale of such beer by retail. 1942 Cape Times 24 Dec. 7 The shebeeners of large townships naturally press for the abolition of municipal beer-halls. 1958 N. Gordimer World of Strangers xiii. 193 The people..read of strikes, of beer-hall riots. 1827 J. Wight More Mornings at Bow St. 24 Recommending him to spend his beer-money at home instead of abroad. 1961 A. Miller Misfits xi. 125 There's hardly beer money in it for six, Gay. 1965 G. Melly Owning-Up iv. 38 We, poor fools, made nothing but a little beer money. 1939 Nottingham Jrnl. 15 Mar. 4/4 Children and beer-offs. 1958 A. Sillitoe Sat. Night & Sun. Morning vii. 98 Bill..had called at the beer-off by the street-end. 1925 in E. Tuck Brief Hist. Pouce Coupe Village, B.C. 14 Beer parlour was opened in Pouce Coupe. 1935 Calgary Typogr. News 26 Apr. 3/1 There is a growing agitation, both in Calgary and Edmonton, against the segregation of women customers of the beer parlors of these two cities in separate rooms. 1937 I. Baird John xix. 226 He took Albert and Red..to the beer parlour and stood them a bottle apiece. 1968 Globe & Mail (Toronto) 11 July 5/2 The prostitutes..who hang around the beer parlors are having poor pickings. 1919 W. H. Downing Digger Dialects 10 Beer-up, a drunken orgy. 1941 K. Tennant Battlers xxviii. 314 'If he's on a real proper beer-up,' the Stray whispered, 'he may go on for days.' 1945 E. Taylor At Mrs. Lippincote's x. 89 Does you good to have a bit of a beer-up now and then.

beerage (bīə·rėdʒ). slang. [blend of Beer sb.¹ and Peerage.] Brewers collectively, esp. those who are made into peers; the beer industry; also, the British peerage (viewed as containing a large number of brewers).
1891 Pall Mall Gaz. 14 Dec. 1/3 The peerage is becoming the beerage. 1901 Westm. Gaz. 17 May 6/2 The following members of the Upper House are interested in Public House property... Another link..between the peerage and the beerage. 1958 Observer 14 Dec. 15/2 Mrs. Dot, the elegant, brewing millionairess—a figure of some social significance in 1908 when the capture of society by the beerage was not yet consolidated.

beeriness. (Examples.)
1877 F. Hall Eng. Adj. in -able 158 Our distant forefathers who..divided their energy pretty impartially between bloodshed and beeriness. 1921 W. De Morgan Old Man's Youth xxiii. 202 A negligible quantity of beeriness.

beerless (bīᵊ·ɹlės), a. [f. Beer sb.¹ + -less.] Without or unprovided with beer.
1846 R. Ford Gather. Spain xv. 169 This..tealess, beerless, beefless land. 1909 Daily Chron. 24 June 4/4 A beerless public-house, where sugary non-alcoholic beverages are sold. 1970 Daily Tel. 19 Oct. 15/8 Some villages..are left beerless.

Beer stone (bīᵊɹ stōᵘn). (Also with lower-case initial letter.) A kind of limestone used in building, obtained from Beer Head in Devonshire, England.
1871 Q. Jrnl. Geol. Soc. XXVII. 98 Cliff west of Beer Head...An old quarry in the high cliff above the western end of the great landslip is in what I take to be the 'Beer stone'. 1894 Daily News 16 Mar. 5/2 A handsome building of white brick and beer-stone. 1910 Encycl. Brit. V. 806/2 The more important building stones are 'Beer stone', from Beer Head in Devonshire [etc.].

beery, a. (Earlier examples.)
1848 Thackeray Van. Fair lxvi. 600 Idlers, playing cards or dominoes on the sloppy, beery tables. 1854 J. R. Planché Once Upon Time II. ii. 37, I'm not jolly—even when I'm beery.

beestie: see Bheesty.
1922 Blackw. Mag. Sept. 408/2 A beestie..splashed a little water from his mussack on his upturned face.

beestings. 1. (Later examples.)
1844 H. Stephens Bk. Farm II. 458 The first milk that comes from the cow after calving is of a thick consistence and yellow colour, and is called biestings. 1847 W. C. L. Martin Ox 173/2 Some farmers refuse the first milk or beastings to the calf, ignorant that it is a purgative expressly intended by nature for this purpose. 1950 N.Z. Jrnl. Agric. July 4/2 The [cow's] first milk known as colostrum or 'beastings' contains substances which increase a calf's resistance to scours.

beet. 1. Add to def.: Now usu. in sing. form, but the pl. form is still current in the U.S. (Further examples.)
1844 H. Stephens Bk. Farm III. 1038 The beets should always be small, evenly sized, straight and even. 1883 Rep. Indian Affairs 17 (D.A.E.), They will raise about 100 bushels of beets. 1917 U.S. Dept. Agric. Yearbk. 405 The development of desirable strains of beets.

Beethovenian (bēᶦthōvī·niăn), a. and sb. [-ian.] **A.** adj. Of or pertaining to the German composer Ludwig van Beethoven (1770–1827), his music and theories of musical composition. **B.** sb. An admirer or adherent of Beethoven; an interpreter of his works.
1894 G. B. Shaw in World 2 May 26/1 The Beethovenian music. 1947 A. Einstein Music in Romantic Era ix. 90 Schubert..enlarged this ensemble along Beethovenian lines. 1947 C. Gray Contingencies vi. 130 Many devout Beethovenians regard the last movement of the Ninth as the summit of the later master's achievement. 1958 Times 10 Nov. 14/4 Miss Eleanor Fine..is an unstylish Bach pianist, and a capable but dull Beethovenian.
Also **Beethovene·sque** a., resembling the style of Beethoven; **Bee·thovenish** a., = *Beethovenian a.; **Bee·thovenized** ppl. a., made to resemble the style of Beethoven.
1879 Grove Dict. Music I. 201/2 Some will find it hard to place the Quartet in F minor, which Mendelssohn thought the most Beethovenish of all Beethoven's works, in anything but the third style. 1927 G. B. Shaw Pen Portraits & Reviews (1932) 35 Jazz, by the way, is the old dance band Beethovenized. 1935 D. F. Tovey Ess. Mus. Anal. 86 The accompaniment of a very Beethovenish rhythm. 1955 R. Blesh Shining Trumpets (ed. 3) v. 108 The harmonic changes, simple, rugged, and Beethovenesque.

beetle, sb.² Add: **2. b.** A dice game having as its object the drawing or assembly of a beetle-shaped figure. So beetle drive (after whist drive).
1936 B. Stanley Games for Party 124 (Advt.), Have you played the Beetle Game?..Simple rules..allow for the game to be played either in pairs..or arranged as a Beetle Drive. 1959 Guardian 23 Dec. 4/5 Parents play whist and 'beetle' madly throughout the winter. 1960 Ibid. 10 Nov. 6/7 A more civilised Britain..will not be built on a foundation of beetle drives and bingo.

5. beetle-droning, -eyed (later example), -like adjs.; **beetle-back,** a back shaped like the wings of a beetle; so beetle-backed adj.; **beetle-crusher, -squasher** slang, a boot or foot, esp. a big one; an infantry soldier (Farmer Slang); so **beetle-crushing** a., wearing big boots; belonging to the infantry; **beetle-stone** = Septarium 2.
1933 J. E. Liberty Pract. Tailoring vii. 113 Beetle back. This is an extended back, with rounded corners. The bottom edge of the back being longer than the lines of the edge on the foreparts. 1958 Vogue Sept. 103 Falling loose at the back and following the line of the spine in a just discernible beetle-back curve. 1959 S. Gibbons Pink Front Door vii. 87 What a thrilling two-piece..that's the first beetle-backed jacket I've seen actually on anyone. 1962 Times 10 Apr. 6/5 A little beetle-backed 1,200 c.c. car. 1860 Hotten Slang Dict. (ed. 2) 94 Beetle-crushers, or squashers, large flat feet. 1869 'Wat Bradwood' The O.V.H. xxi, The infliction which the beetle-crusher of a recent arrival had just inflicted on his pet corn. 1870 R. Broughton Red as Rose xxxv, What howible boots! Whoever could have had the atwocity to fwame such beetle-cwushers? 1897 Punch 30 Oct. 195/1 If you need a meal, you can boil your beetle-crushers. 1958 'A. Gilbert' Death agst. Clock 155 He looked down..at his own enormous beetle-crushers in bright tan Oxfords. 1871 G. A. Lawrence Anteros xiv, The possibility..of exchange into a sedate, beetle-crushing corps. 1917 Masefield Lollingdon Downs 31 This beetle-droning downland. 1843 Ainsworth's Mag. III. 563 On stealthily raising his eyes.., he descried a toothless beetle-eyed antique. 1908 Westm. Gaz. 22 Aug. 16/3 A great beetle-like shape. 1859 Page Handbk. Geol. Terms s.v. Septarium, Such..nodules..when split up.. exhibit very curiously marked sections; hence the names beetle-stones, turtle-stones.

beetle (bī·t'l), v.³ colloq. [f. Beetle sb.²] intr. To fly off; to go, make one's way, move (like a beetle); freq. with off, away, etc.
1919 W. H. Downing Digger Dialects 11 Beetle about, fly aimlessly (of an aeroplane). 1920 W. Noble With Bristol Fighter i. 20 We were on our side of the line, and the Huns had beetled away eastwards. 1923 Wodehouse Good Morning, Bill! I. 19 'What are you doing about two weeks from now?'..'Nothing in particular. Just beetling around.' Ibid. II. 104 Did he specify that you were to come beetling in at midnight? 1925 Fraser & Gibbons Soldier & Sailor Words 21 To beetle-off: Air Force slang. To fly straight. To go off direct, as a beetle flies; e.g., 'I just beetled off home.' 1943 P. Brennan et al. Spitfires over Malta i. 31, I beetled about the mess and the bastions. 1948 C. Day Lewis Otterbury Incident vi. 74 E. Sidebotham beetled up the ladder to examine the window-sill. 1952 N. Coward Rel. Values I. ii. 29 There was..a terrible scene..and Freda beetled off to America.

beeve (bīv). [sing. form derived from beeves, pl. of Beef sb.] An ox; = Beef sb. 3.
1847 Whittier Drovers 61 Each stately beeve bespeaks the hand That fed him unrepining. 1899 Daily News 4 Dec. 8/3 Herefords are a noble type in all the points that are expected in the modern beeve.

beezer (bī·zəɹ). [Origin obscure.] **1.** [App. orig. Sc.] A smart fellow (quot. 1914); a person, a 'chap'. See Sc. Nat. Dict.
1914 J. L. Waugh Cracks wi' Robbie Doo ix. 110 Weel dune, Robin Hood; dash it, man, but you're a beeser. 1935 'N. Blake' Question of Proof viii. 156 He's not a bad sort of beezer at all, if he wasn't such an absolute ass. 1963 'G. Carr' Lewker in Norway ii. 30 Get this beezer outa my sight, someone!
2. (Perhaps a different word.) Nose. slang.
1915 C. E. Van Loan in C. Grayson New Stories for Men (1941) 550 Perhaps it should be explained that, in the patois of the [boxing] profession..the nose becomes a beezer. 1920 Ade Hand-made Fables i. 12 One brilliant Man about Town, with a Beezer that never could have been coloured by the use of Malted Milk. 1932 Wodehouse Hot Water vi. 112 Back home..a cop..don't hardly even notice it if you sock him on the beezer. 1960 —— Jeeves in Offing v. 52 It is virtually impossible to write a novel of suspense without getting a certain amount of ink on the beezer.

before. Add: **A. 5. c.** Used in contrast with after in various locutions to designate a set of two contrasting pictures, cartoons, etc., esp. illustrating the efficacy of a remedy, product, etc., alleged to produce a remarkable change for the better. Hence allusively.
1768 W. Hogarth in Trusler's Hogarth Moralized (Index), A List of Prints published by Mr. Hogarth...Before and After. 1846 Punch XI. 243/1 (captions) Before. After. Ibid., Here are two portraits, both of myself: the one before, the other after the cold Brandy-and-Water Cure. 1853 Ibid. XXV. 45 (title of cartoon representing a difference of opinion between cabman and fare) Before and after. 1889 Puck 3 July 307/1 I'm working a 'before and after' racket for a hair-renewer advertisement. 1902 Little Folks II. 432/1 Those restaurants which advertise by means of looking-glasses labelled 'before' and 'after'. As you go in you behold yourself very thin..as you go out..fat and well-satisfied. 1938 N. Marsh Artists in Crime iii. 24 You're not doing a 'before and after', like a strip advertisement.
B. 11. (Later examples.)
1897 C. Garnett tr. Turgenev's Torrents of Spring xliv. 240 Then Gemma..wished him before everything peace and a tranquil spirit. 1911 D. H. Lawrence White Peacock II. ii. 231, I was a good animal before everything, and I've got some children.

D. 2. (Later example.)
1897 Daily News 6 Mar. 6/1 One who has witnessed the before and after of the abolition of pain.
E. 1. b. The prep. in comb. with a sb., used attrib.
1865 C. M. Yonge Clever Woman of Family II. xiii. 248, I have just lighted on poor little Rosie's before-breakfast composition. 1898 Daily News 28 Sept. 5/1 The 'before luncheon' rehearsal. 1902 M. Barnes-Grundy Thames Camp 83 These before-breakfast expeditions. 1919 Wodehouse Damsel in Distress iv, A fellow with the appearance of a before using advertisement of an anti-fat medicine. 1926 D. H. Lawrence Glad Ghosts 64 The tender before-dawn freshness of a new understanding. 1966 'W. Cooper' Memoirs of New Man III. ii. 214 We were going upstairs to have our before-dinner drink in the library. 1968 D. Torr Treason Line 24 He had had his first before-breakfast swim of the year.
c. The prep. in comb. with a sb., as before-life.
1927 D. H. Lawrence Morn. Mex. 154 They were the lords of shadow, the intermediate twilight, the place of after-life and before-life.

beforeness. (Further example.)
1908 'Mrs. Willett' Let. 9 Oct. in Proc. Soc. Psychical Res. (1935) XLIII. 49 The words seemed to form in my brain before the pen set them down, just before, as if tripping on the written word—a sort of hair's-breadth beforeness.

befrogged (bĭfrǫ·gd), a. [f. Be- 7 + Frogged ppl. a.] Decorated with frogging. So **befro·gging** vbl. sb.
1843 Blackw. Mag. LIV. 65 The betasseling, the befrogging, the flaunting attempts at 'costuming'. 1895 A. M. Stoddart J. S. Blackie (ed. 3) I. ix. 205 A lustrous gown, befrogged and ample. 1958 E. Hyams Taking it Easy 292 A crimson flannel dressing-gown heavily befrogged.

befuddle (bĭfʌ·d'l), v. [f. Be- 4.] To make stupid with tippling; also, to confuse, to stupefy. Hence **befu·ddlement,** intoxication; confusion, stupefaction.
1887 in Dict. s.v. Be- 4. 1905 Academy 11 Feb. 132/2, I am not aiming at a general befuddlement..when I call it [sc. matter]..an expression of a relation. 1922 Public Opinion 11 Aug. 135/2 Freedom from the befuddlement of drink. 1926 Contemp. Rev. July 40 Their propaganda has befuddled public opinion. 1945 A. J. P. Taylor Course German Hist. 7 The only ones to escape from complete befuddlement are the Marxists.

befur, v. **2.** (Earlier examples.)
1470–85 Malory Arthur I. xvii, And Merlyn was so disguysed that kynge Arthur knewe hym not for he was al be furred in black shepe skynnes and a grete payre of bootes. 1842 Fraser's Mag. XXVI. 362 A..middle-aged man, bespectacled and befurred.

beg, v. Add: **2. a.** Phr. to beg, borrow or steal.
[a1340 (see Borrow v.¹ 1). c1386 (see Indigence 2 a).] 1760 Sterne Tr. Shandy III. xxxviii. 177 He has taken in, Sir, the whole subject..begging, borrowing, and stealing, as he went along. 1794 F. Burney Let. 22 Mar. (1905) V. 240 He has been drawing a plan for it, which I intend to beg, borrow, or steal (all one), to give you some idea how seriously he studies. 1859 Macmillan's Mag. Dec. 118/1 Not because they want to beg, borrow, or steal thoughts which are not theirs.
h. Card-playing. In All Fours (U.S. Seven-up), to ask for a point, or three additional cards and a new trump (said of the elder hand).
1793 Sporting Mag. II. 160/1 The Duke of York and Lord Barrymore were playing the game of All Fours...The Duke ..overlooked his cards, and..begged one, which was granted, though he held the ace, deuce, and jack of trumps. 1800 Hoyle & Jones Hoyle's Games Improved 308 The Game of All-Fours...If the eldest [hand] don't like his cards, he may, for once in a hand, say, I beg, when the dealer must either give a point or three more cards to each..player. 1897 R. F. Foster Complete Hoyle 289 If the eldest hand is not satisfied, he says: I beg; and the dealer, after examining his own hand, has the option of giving him a point or running the cards. Ibid. 290 Begging is resorted to by a player who holds no trumps.
i. Said of a dog trained to sit up and hold up its fore paws when told to beg.
1816 Sporting Mag. XLVIII. 23/2 He began to teach me [sc. a puppy] to beg, and to fetch and carry. 1837 Jas. Prior Life Goldsmith II. 33 Teaching a favourite dog to sit upright upon its haunches, or as is commonly said, to beg. 1854 J. G. Wood Anim. Life 98 Four cats..had taught themselves the art of begging like a dog...They waited until they saw the dog sit up in the begging position, and immediately assumed the same attitude. 1884 Century Mag. Dec. 198/2 To squat back and raise his front legs from the ground, much in the position of a 'begging' poodle. 1927 E. V. Lucas More I see of Men iv. 32 He begs even when there is no meal in progress.
3. b. ellipt. for beg leave.
1767 Wilkes Corr. (1805) III. 197, I shall very soon beg to call the public attention to some points of national importance. 1898 Westm. Gaz. 29 Oct. 7/3 You say, 'I beg to take exception', which, of course, is not English at all. You mean, 'I beg (leave) to take exception'.
Also ellipt. in epistolary formulas of good-will: to desire to send, to offer.
1755 H. Walpole Let. 7 Jan. (1903) III. 277 He would.. beg his compliments to Miss Montagu. 1816 A. Constable Let. 9 July in J. Constable's Corr. (1962) 137, I am glad to hear so good an account of Miss B's health, I beg best regards. 1836 Dickens Let. 27 Feb. (1965) I. 134 Begging my best remembrances to Mrs. Thomson. 1839 —— Let. ? Apr. (1965) I. 546 Begging my best compliments at home.

5. b. (Later examples.)
1854 E. Ruskin Let. 28 Feb. in M. Lutyens Millais & Ruskins (1967) 145 What does John do but..say that he wishes to beg off to dine with his Father and Mother. 1966 Amer. Speech XLI. 174 He begged off because he had just moved from his librarianship at Kassel.
7. To make (one's way) begging.
1840 Dickens Old C. Shop xliv, To-morrow we will beg our way to some quiet part of the country.

beg, sb.² [f. Beg v.] **a.** An act of begging.
1814 J. Mayne Jrnl. 13 Sept. (1909) iii. 60 This was an invocation to some saint for a prosperous beg. 1912 Gambier-Parry Alleg. of Land iv. 134 Letters simply mean begs, bothers, and bills. 1920 Punch 30 June 508/1 Agenda and minutes and constituents' grievances, and charitable appeals and ordinary begs.
b. spec. in All Fours (U.S. Seven-up) (see *beg v. 2 h).
1897 R. F. Foster Complete Hoyle 289 (Seven-up) No second beg is allowed, but when only two play, if either player is dissatisfied with the new trump he may propose to bunch the cards.

begasse. Add: Also **begass.**
1839 Ure Dict. Arts 1195 Liquor which is speedily absorbed by the spongy begass. Ibid. 1197 The begass.. is tied up in bundles, and..is finally stored in the begass-house. 1887 Harper's Mag. Feb. 336 An old begass chimney against the sky.

bège, var. *beige.

begem, v. Add: (Earlier and later examples.) Hence **bege·mmed** ppl. a. (also absol.).
1749 L. Pilkington Mem. I. 151 With glittering Glow-worms begemming each Tree. 1916 Blunden Harbingers 60 The lopped tree..begems its leafits in a year. 1927 W. de la Mare Told Again 101 The three twigs of the trees—the silver and the gold and the begemmed.

beggar, sb. Add: **6. d.** A person who is remarkably adept at or keen on a particular pursuit, subject, etc. Const. for or with inf. colloq.
1859 Chambers's Jrnl. 29 Jan. 77/1 He was..what Mr. Leech's miner denominated the conciliating curate, 'a beggar to argue', a stiffish one to tackle upon any mortal subject. 1949 Partridge Dict. Slang (ed. 3) 987/2 A beggar for work, a constant hard worker: coll.: late C. 19–20.
7. a. beggar-beard, -child, -clan, -king.
1955 J. R. R. Tolkien Return of King 354 Here's your fee, beggar-beard. 1938 L. MacNeice Earth Compels 54 The shapes of mist like hooded beggar-children. a1821 Keats The Poet, Poorest of the beggar-clan. 1938 Yeats New Poems 32 Some boast of beggar-kings.

beggar, v. Add: **3. beggar-my-neighbour** (earlier example). Also fig.
1734 Poor Robin's Almanack C 6 The Lawyers play at beggar my Neighbour. 1930 Times Educ. Suppl. 20 Dec. 511/2 The..argument..that the scale for Middlesex should be nearer to that for London..was merely playing 'beggar-my-neighbour'. 1958 Spectator 1 Aug. 174/3 The continuous concentration on national selfishnesses and beggar-my-neighbour is distasteful.

begging, vbl. sb. **4.** begging-bowl; begging-box, -letter, -letter-writer (earlier examples).
1894 Kipling 2nd Jungle Bk. (1895) 35 Let the begging-bowl be placed outside the shrine. 1668 Dryden Evening's Love (1671) III. i. 32, I must be fain to take up their Questions in a cleft-Cane, or a Begging-box, as they do Charity in Prisons. 1846 Times 12 Feb. 6/6 (heading) Begging-letter writers. 1849 C. Brontë Shirley III. xiv. 299 He sent out begging-letters far and wide. 1852 Dickens Bleak Ho. xii. 117 He was a begging-letter writer.

begin, v.¹ Add: **1. e.** Usu. with preceding negative: To make any (or the least) approach to, to come anywhere near. colloq. (orig. U.S.).
1833 Niles' Reg. XLIV. 348/1 The one in Bleecker street ..cost ten thousand dollars, and that does not begin to be as expensive as this. 1865 Congress. Globe Feb. 664/1 New York does not begin to have sixty-nine thousand square miles. 1888 Harper's Mag. Sept. 545/2 He got Bret to take her picture,..and he said it didn't begin to do her justice. 1907 Howells Through Eye of Needle 43 Often there's a.. dinner that you couldn't begin to get for the same price anywhere. 1915 W. Raleigh Let. 30 Mar. (1926) II. 420, I can't begin to tell about America. 1957 R. W. Zandvoort Handbook Eng. Gram. I. ii. 27, I felt I did not begin to understand her. (footnote) American, but spreading in England. 1963 Listener 24 Jan. 168/2 The Aeneid is not an Augusteid, because Augustus could not begin to embody Virgil's feelings. 1968 Observer 22 Dec. 8/5 Dollar for dollar, man in space does not begin to be cost-effective.
f. To compare in any degree with. U.S.
1862 O. W. Norton Army Lett. (1903) 47 There is no other man whom I would be so much pleased to have taken as..Floyd. Jeff Davis wouldn't begin. 1877 'Mark Twain' in Atlantic Nov. 590 There ain't a book that begins with it. 1897 —— Following Equator xxxviii. 347 Indeed, our working-women cannot begin with her as a road-decoration.

beginner. Add: **2. b.** Phr. beginner's luck: the good luck supposed to attend a novice at betting, games, etc.
1897 Kipling Capt. Courageous iii. 55 'Beginner's luck,' said Dan...'He's all of a hundred [sc. pounds, in weight].' 1902 A. Bennett Anna of Five Towns x. 227 'You have covered yourself with glory.'..'How?' 'By not being ill.' 'That's always the beginner's luck.' 1966 J. Potts Footsteps on Stairs ix. 115 Beginner's luck. It went smooth as silk all the way.

4. Arch. The lower part of a mullion worked on the stone forming the sill.
1886 Willis & Clark Cambridge II. 514 The mullions of the four-light window..do not correspond with the 'beginners' on the sill.

beginning, ppl. a. Add: **1.** spec. Of a course of study, book, student, etc.: preceding others in a series; elementary. N. Amer.
1923 E. M. Roberts (title) The beginning telegrapher. 1928 Almack & Lang (title) The beginning teacher. 1962 W. S. Avis et al. (title) Dictionary of Canadian English. The Beginning Dictionary. 1962 S. E. Martin in Householder & Saporta Problems in Lexicography 153 In a sense, the beginning student needs something very similar. 1964 Amer. Speech XXXIX. 51 It is intended as a textbook for a beginning course.

beglamour (bĭglæ·məɹ), v. [f. Be- 6 + Glamour sb.] trans. To invest with glamour; = *glamorize v.; also, to deceive or impress with glamour. So **begla·mour(iz)ed, begla·mouring** ppl. adjs.
1840 Fraser's Mag. XXI. 8 A new Pactolus..seemed to the beglamoured eyes of the prosperity-men to have overflowed the land. 1926 Chambers's Jrnl. July 437/2 He's beglamoured her—she worships courage—and he's a brave man. 1932 F. R. Leavis New Bearings i. 16 The appropriate metaphor would suggest something not only beglamoured, but also ritualistic and religiose. 1948 A. Waugh Unclouded Summer xii. 231 It had been a bondage, yes, but a beglamoured bondage. Ibid. xiv. 246 It was the very fact that they had not been equal that had so beglamoured him. 1953 H. P. Collins in Ess. in Crit. Jan. 61 The passion, the Protestant fury, the beglamouring naturalism of Charlotte Brontë. 1958 Times Lit. Suppl. 21 Feb. 101/2 The city is beglamourized.

begob (bĭgǫ·b), int. Also **begobs.** = *begorra.
1889 St. James's Gaz. 1 Aug. 7/1 It's Irish, begobs! 1892 Barlow Irish Idylls ii. 34 No begob; I'll just be keepin' the feel of it in me hand for this night. 1907 G. B. Shaw John Bull's Other Island iv, Begob, it just tore the town in two. 1958 Betjeman Coll. Poems 70 The feast is spread out, and begob! what a sight.

begorra (bĭgǫ·ră), int. Also **begarra, begorrah.** Anglo-Irish alteration of the expletive by God (see God sb. 13); cf. Begar, and dial. begor(z.
Rarely heard in current speech.
1839 Carleton Fardorougha xvi, Begarra, Captain dear, it seems that good people is scarce. 1843 Lever J. Hinton ii, 'Begorra, you're in it', was the answer. 1856 —— Martins of Cro' M. x, Be gorra! when a man would give four hundred for a bull, there's no saying what he'd stop at. 1895 J. Barlow Strangers at Lisconnel i, Fine company they'd be for anybody begorrah.

beg-pa·rdon, sb. Austral. and N.Z. colloq. [f. phr. to beg pardon: see Beg v. 3 and Pardon sb.¹ 6.] An expression of apology.
1906 E. Dyson Fact'ry 'Ands xv. 198 'Twas quick business down below here, 'n' no beg-pardons with Bunyip. 1916 J. B. Cooper Coo-oo-ee i. 11 Then without a 'beg pardon', off she goes again. 1965 F. Sargeson Memoirs of Peon vi. 187 Tony, after a maternally-directed beg pardon ..rapped out a command. 1967 Sunday Mail Mag. (Brisbane) 8 Jan. 6/2 There were no beg-pardons about Mrs. Hodges (or Debbie, as she insisted I call her). 1969 Sun-Herald (Sydney) 13 July 45/7 Rucking was heavy and there were no 'beg pardons' as each pack used its weight.

begrudgingly (bĭ,grʌ·dʒiŋli), adv. [f. begrudging ppl. adj. + -ly².] In a grudging manner or spirit.
1853 Fraser's Mag. XLVIII. 159, I looked begrudgingly on them as they occupied the whole pool. 1878 Hardy Ret. Native vi. iv, The original owners..cackled begrudgingly at sight of such a quantity of their old clothes. 1890 Illustr. Lond. News 29 Nov. 686/1 It was a narrow little way begrudgingly left between these sullen hedges. 1968 Listener 11 July 44/1 Begrudgingly, the situation was accepted.

‖ **béguin** (begæ̃n). [colloq. Fr.] An infatuation; a fancy.
1919 W. S. Maugham Moon & Sixpence li. 223 It appears that she has a béguin for you...She's willing if you are. 1930 Times Lit. Suppl. 30 Oct. 886/2 Lady Bettine herself conceives a béguin for Arden. 1942 E. Waugh Put out More Flags i. 34 Cedric Lyne..used to tell himself..that a béguin like that could not possibly last.

beguine² (bėgī·n). [Amer. Fr., f. Fr. béguin (see prec.).] A kind of popular dance, orig. associated with Martinique; also applied to a kind of syncopated dance rhythm.
1935 Cole Porter (song title) Begin the Beguine. 1939 W. Hobson Amer. Jazz Music 53 Jazz rhythms have been partly affected by West Indian rhythms, such as the rhumba, beguine, etc. 1950 J. Vedey Band Leaders xviii. 131 The 'Beguine' from Mexico..has an essentially Afro-Spanish flavour. 1961 A. Berkman Singers' Gloss. Show Bus. 8 Beguine, a rhythm used to give a certain amount of 'lift' to a sustained melody by doubling up the rhythmic beats...In the Beguine the rhythm instruments repeat the same rhythmic pattern over and over, without variation.

Behari, var. *Bihari a.

behaviour. Add: **1. b.** (Later examples.)
a1763 'Geo. Psalmanazar' Mem. (1764) 186, I could see..thro' all his artifices and different behaviours. 1959 Camb. Rev. 7 Mar. 405/1 We must surely accept that the pattern of associated behaviours first noticed by Weber was one of the most brilliantly successful suggestions in the whole history of intellectual endeavour.
6. attrib. and Comb., esp. in Psychol., as behaviour-cycle, data, -study, -system, -trend; **behaviour pattern,** a set or series of acts regarded as a unified whole; **behaviour segment,** a part of a behaviour pattern; **behaviour therapy,** a method of treating neurotic disorders (see quots.).
1921 B. Russell Anal. Mind iii. 65 A 'behaviour-cycle' is a series of voluntary or reflex movements of an animal, tending to cause a certain result, and continuing until that result is caused, unless they are interrupted by death, accident, or some new behaviour-cycle. 1913 J. B. Watson in Psychol. Rev. XX. 158 On this assumption, behavior data (including under this term everything which goes under the name of comparative psychology) have no value per se. 1926 Psychol. Rev. XXXIII. 51 Is this modification of activity the result of environmentally conditioned learning or of the maturing of certain innate behavior patterns or 'instincts'? 1929 B. Russell Marriage & Morals ii. 19 Where human beings are concerned we do not have the precise behaviour-patterns which are to be found among other animals. 1956 Evolution X. 421 (title) A gene mutation which changes a behavior pattern. 1960 20th Cent. Apr. 372 As far as behaviour patterns are concerned, I feel..a greater affinity with the working-class Briton than with the middle-class man. 1934 H. C. Warren Dict. Psychol. 31/1 Behavior segment. 1936 J. Kantor Objective Psychol. Gram. vi. 74 B..now becomes speaker. His speaking behaviour constitutes his second linguistic behaviour segment, his first being his audient response. 1953 N. Tinbergen Herring Gull's World vii. 64 A man who does not have the patience simply to sit and watch for hours, days,.. is not the type of man to undertake a behaviour-study. 1927 G. A. de Laguna Speech vi. 132 The behavior-system of one species differs from that of another. 1938 A. N. Whitehead Modes of Thought i. 20 There is no one behaviour-system belonging to the essential character of the universe, as the universal moral ideal. 1958 A. R. Radcliffe-Brown Method in Social Anthrop. I. iv. 103 Psychology is here taken to mean the study of the mental or psychic systems—if you will, the behaviour systems—of organisms. 1959 H. J. Eysenck in Jrnl. Mental Sci. CV. 66, I have called these methods [of treatment] 'behaviour therapy' to contrast them with methods of psychotherapy. ..Psychoanalysts show a preoccupation with psychological methods involving mainly speech, while behaviour therapy concentrates on actual behaviour as most likely to lead to the extinction of the unadaptive conditioned responses. 1961 Guardian 12 May 6/6 This new approach, which owes much to J. B. Watson,..and to J. Wolpe, the well-known South African psychologist, has been christened Behaviour Therapy...Behaviour therapy..tries to understand neurotic symptoms..in terms of..experimentally established facts of human and animal behaviour. 1949 G. Ryle Concept of Mind iv. 110 To explain an action as done from a certain motive is..to subsume it under a..behaviour-trend.

behavioural (bĭhēᶦ·vyərăl), a. [f. Behaviour + -al.] Concerned with, or forming part of, behaviour. Hence **beha·viourally** adv.
a1927 E. B. Titchener Systematic Psychol. (1929) iii. 263 All biological facts, we propose to say, are 'behavioural'. 1936 J. Kantor Objective Psychol. Gram. xv. 213 Vocabulary phenomena..., though remote from things,..operate behaviourally in a definite adjustmental manner. 1946 C. W. Morris Signs, Lang. & Behavior i. 4 A behavioral theory of signs. Ibid. i. 21 Vagueness shows itself behaviorally in an uncertain and hesitant response to an object to which the organism has been directed by a sign. 1956 Camb. Rev. LXXVII. 301 Some contrasting of C and I behavioural and linguistic patterns is possible. 1958 New Statesman 6 Sept. 300/2 The so-called 'behavioural sciences' ..—sociology, social psychology, social anthropology—have been much pushed by the foundations.

behaviourism (bĭhēᶦ·vyəriz'm). Psychol. [f. Behaviour + -ism.] A theory and method of psychological investigation based on the study and analysis of behaviour. Hence **beha·viourist,** one who practises this method; also attrib.; **behaviouri·stic** a., of or belonging to the behaviourists; characterized by behaviourism; also gen., pertaining or relating to behaviour; **behaviouri·stically** adv.; **behaviouri·stics** sb. pl., the study of the responses of organisms to their environment.
1913 J. B. Watson in Psychol. Rev. XX. 158 Psychology as the behaviorist views it is a purely objective experimental branch of natural science. Its theoretical goal is the prediction and control of behavior. Ibid. 166, I feel that behaviorism is the only consistent and logical functionalism. 1914 E. B. Titchener in Proc. Amer. Philos. Soc. LIII. 3 Most of the essential problems with which psychology as an introspective science now concerns itself are open to behaviorist treatment. Ibid. 13 The facts of psychology.. are also to be carried, by way of behavioristic substitution, to the bodily periphery. 1916 Boston Even. Transcript 26 July 116 A behavioristic psychology. 1920 A. N. Whitehead Concept of Nature ix. 185 Our attitude towards nature is purely 'behaviouristic'. 1921 Edin. Rev. Apr. 351 Psychologists are divided into several camps, one of which, the American 'Behaviourists', cares very little for the social aspects of the subject. 1922 Times Lit. Suppl. 20 July 478/4 The determinist is logically driven to 'behaviourism'. 1924 J. B. Watson in Psyche July 11 Behavioristic psychology. 1933 Mind XLII. 381 The Communist studies religion, as a social phenomenon, behaviouristically. 1936 E. E. Evans-Pritchard Ess. Soc. Anthrop. (1962) viii. 196 We treat them [sc. the ideas] behaviouristically as ritual responses and do not attempt to create for them an ideology that will explain them by seeming to cause them. 1940 Bryant & Aiken Psychol. of English i. 5 The English language and grammar are the products of the group thinking of billions of people whose minds have worked psychologically rather than logically; and the fruit..is a system which reflects behavioristic patterns rather than formal regularity. 1941 O. Neurath in Proc. Arist. Soc. XLI. 128 There is a trend to build up a Lingua Franca..which would enable us to pass from the theory of behaviour ('behaviouristics') to geology, biology and mechanics without any alteration of the type of our expressions. 1945 Mind LIV. 193 Scientific psychology is either behaviouristic or physiological. 1953 J. B. Carroll Stud. Lang. iii. 107 The kind of analysis suggested by Miller and Frick (1949) in their paper on what they call 'statistical behavioristics'. Statistical behavioristics is the theory of stochastic processes applied to the study of sequences of responses. 1960 Times 5 Feb. 3/6 The third difficulty was that some mosquitoes were developing what is technically known as behaviouristic resistance to insecticides. What this means is that, after entering a house and feeding on the occupants, they escaped to outdoor resting places, thus avoiding the lethal effects of the insecticide sprayed on the walls of the house.

behemothian (bĭhĭmōᵘ·þiăn), a. Chiefly poet. [f. Behemoth + -ian.] Monstrously large; of or belonging to a large animal.
1910 W. de la Mare Three Mulla-Mulgars iii. 47 A behemothian bull-Elephanto. 1911 H. S. Harrison Queed i. 3 Down the street came a girl and a dog, rather a small girl, and quite a behemothian dog. 1946 H. Read Coll. Poems 145 Faced by the behemothian jaws.

behind. Add: **A.** adv. **7. d.** ellipt. for behind the scenes (see Behind B. 6 c).
1824 J. Decastro Memoirs 8 To visit the theatre whenever he was so disposed, either in front or behind. 1856 Dickens Dorrit I. xx. 283 But the idea, Amy, of you coming behind! I never did! 1885 G. B. Shaw in Works (1932) VI. 202, I am going to take a peep behind: that is, if non-performers may be admitted. 1926 R. Macaulay Crewe Train II. ii. 64 Leonard's fate will be settled by the time the curtain goes up. He's gone behind, poor Leonard.
B. prep. **3. c.** To be behind the times: see Time sb. 5 a. Also attrib.
1905 Daily Chron. 14 Feb. 6/3 A slow-going, old-fashioned, behind-the-times country.
6. c. behind-the-scenes attrib. phr.; also behind-scene. See also Scene 7.
1841 E. FitzGerald Let. 16 Jan. (1889) I. 64 And go right through it [sc. a picture] into some behind-scene world on the other side. 1933 Essays & Studies XVIII. 156 They [sc. 'stream of consciousness' novels] have..a strong behind-the-scenes interest. 1961 John o'London's 5 Oct. 374/2 His political novel tells the story of the behind-the-scenes struggle for power. 1968 J. W. Wainwright Web of Silence 100 It hit the headlines..but it didn't help the behind-scene manœuvring.
C. as sb. (Later examples.)
1926 D. H. Lawrence Let. 19 Jan. (1932) 647 Lucky I'm not a professional behind-kicker. 1928 G. B. Shaw Intell. Woman's Guide lxxiv. 362 You can say 'If I catch you doing that again I will..smack your behind'.
2. Australian National Football. A scoring kick that earns one point (see quot. 1968). Also attrib.
1888 Pall Mall G. 23 July 6/2 The visitors won by five goals and ten behinds to four goals and eight behinds. 1890 Melbourne Punch 14 Aug. 107/2 South Melbourne 3 goals 10 behinds. 1968 Eagleson & McKie Terminology Austral. Nat. Football I. 14 A behind is scored when the ball, after being kicked, is touched by or touches any player before passing through the goal posts; or when it touches a goal post; or when it passes immediately above a goal post or between a goal post and a behind post; or when it is kicked or knocked through the goal posts by one of the defending players. Ibid. 15 Behind line, the line between the goal and behind posts. Ibid., Behind post, a post seven yards to the side of a goal post, and not as tall as the goal post.
b. (See quots.)
1898 Encycl. Sport II. 143 (Eton football) Each side consists of the 'bully', outsides, and behinds, but all except the behinds are commonly spoken of as 'the bully'. Ibid., The 'behinds' are 'short' and 'long behind' and 'goals'.

behinder (bĭhəi·ndəɹ). [f. Behind adv. + -er¹.] An operative in certain trades, as a tinplate worker whose work lies behind the rolling-mill, and the man who works at the back of a welding-furnace in a tube mill.
1881 Instr. Census Clerks (1885) 105 Behinder. Tin Plate Worker. 1906 Westm. Gaz. 10 June 2/1 Behinders [tinplate millmen].

Behmenism, -ist, see Boehmenism.

behove, behoove, v. The spelling with -oo- is now restricted to the United States. (Later U.S. examples.)
1952 M. McCarthy Groves of Academe (1953) iii. 40 It behooved him to tread warily with Domna. 1955 Sci. Amer. Aug. 71/1 It behooves us to know as much as possible about this problem.

behovely, a. (Later poet. example.)
1942 T. S. Eliot Little Gidding iii. 13 Sin is Behovely, but All shall be well.

beidellite (bəide·ləit). Min. [f. Beidell (see def.) + -ite¹.] A clay mineral from Beidell, Colorado.
1925 Larsen & Wherry in Jrnl. Washington Acad. Sci. XV. 465 Beidellite, a new mineral name..we now propose ..from the locality of the first occurrence described in

detail, Beidell, Colorado..$Al_2O_3.3SiO_2.XH_2O$. **1932** E. S. DANA *Textbk. Mineral.* v. vii. 682 Beidellite, $Al_2O_3.3SiO_2.4H_2O$. Probably orthorhombic..*Iron-beidellite* is a variety with considerable amount of Fe_2O_3. **1955** BROWN & DEY *India's Min. Wealth* (ed. 3) xiii. 532 Beidellite, anauxite and others, in which the proportion of water varies and the ratio of the silica to the alumina changes. **1963** D. W. & E. E. HUMPHRIES tr. *Termier's Erosion & Sedimentation* vi. 135 More often..the clay formed is beidellite (Al, Mg)$_4$ (Si Al)$_8$ $O_{22}(OH)_{20}$ which is closely related to montmorillonite.

beige (bēɪʒ), *sb.* and *a.* Also formerly **bège.** [a. F. *beige* adj.] **A.** *sb.* **1.** A fine woollen fabric used as a dress-material, originally left in its natural colour but later dyed in various colours. Also *beige cloth.*

1858 SIMMONDS *Dict. Trade, Beige,* a French coarse cloth. **1879** *Cassell's Fam. Mag.* Sept. 634/2 The young lady..is in bège and silk. *Ibid.* Nov. 755/1 Her skirt is of silk and beige cloth. **1882** CAULFEILD & SAWARD *Dict. Needlework* s.v. *Beige* or *Bège,* Beige is made of undyed wool, is an extremely soft textile, graceful in draping, and employed for morning and out-door wear...There is a description of this textile, called snowflake beige, of a neutral ground.

2. A shade of colour like that of undyed and unbleached wool; yellowish-grey. Also *beige colour,* whence *beige-coloured* adj.

1879 *Cassell's Fam. Mag.* Mar. 249/1 Beige shades go with moss-green. *Ibid.* 250/2 The hat..is of bège-coloured plush. **1896** *Daily News* 9 May 8/6 The colour of grass lawn is technically known as beige. **1899** *Ibid.* 19 Aug. 7/4 Beige is the coolest possible colour.

B. *adj.* Of wool or woollen and other fabrics, etc.: of a natural yellowish-grey colour.

1879 [implied in 2 above]. **1899** *Daily News* 20 Mar. 8/7 The creamy lace..will be deep enough in tint to be beige. **1926** *British Weekly* 24 June 250/5 The dress of beige lace is very much liked just now.

beigel, var. *BAGEL.

[**1892** ZANGWILL *Childr. Ghetto* I. iii. 96 Moses..treating his children to some *Beuglich,* or circular twisted rolls.] **1919** *Century Mag.* July 381/2 The bread-rings called beigel. **1959** *Times* 8 Dec. 15/4 Six taxi drivers on night duty went to an East End bakery to buy bread rolls known as beigels. **1967** L. DEIGHTON *London Dossier* 135 An old woman selling *beigels.*

beignet (bẹnye). *Cookery.* [Fr.] A fritter.

1835 IRVING *Tour Prairies* xxxiii. 306 We..supped heartily upon stewed buffalo meat,..beignets, or fritters of flour fried in bear's lard. **1892** T. F. GARRETT *Encycl. Cookery* I. 132/1, II. 34/2. **1901** *Daily Chron.* 7 Sept. 8/4 Cheese beignets.

Beilby (bēɪ·lbi). [Surname of Sir George Thomas *Beilby* (1850–1924), Scottish industrial chemist.] **Beilby layer** *Metallurgy* (see quot. 1958).

1930 N. K. ADAM *Phys. & Chem. Surfaces* vi. 172 The mechanical processes of grinding always result in the formation of a certain amount of the amorphous 'Beilby' layer which is obtained by polishing. **1937** *Ann. Reg. 1936* 63 Electron diffraction examination of engine cylinders showed that a substantial Beilby layer is formed by the 'running in' process. **1958** A. D. MERRIMAN *Dict. Metallurgy* 16/2 *Beilby Layer.* Beilby's experiments led him to conclude that the action of polishing a metal surface caused the surface layer to flow like a liquid and then to solidify without recrystallisation, forming an amorphous layer.

bein, *a.* **3. b.** For '1859 *Parish Mag.* Oct. 149' read '1847 MRS. GASKELL *Sexton's Hero* in *Howitt's Jrnl.* II. 151/1'.

be-in (bi·ˌin), *sb.* [f. BE *v.* + IN *adv.* after *teach-in,* etc.] A public gathering of hippies.

1967 *Daily Tel.* 23 Mar. 18/8 Thousands of people with painted faces and chests and love on their minds pranced through New York's Central Park yesterday to celebrate Easter Sunday with a 'be-in'. **1967** *Nova* Oct. 115/1 Activities at be-ins have included chanting Hindu prayers, carrying crosses, ringing bells, striking gongs, uttering the word 'banana', staring into space, examining other people's beaded necklaces.

being, *vbl. sb.* Add: **4. d.** Phrases in *Philos.,* formed mainly to translate the corresponding Ger. and Fr. expressions, as **being-for-(it)self,** conscious being; being as actuality; **being-in-(it)self,** being that lacks conscious awareness; being as mere potentiality; **being-itself,** pure being, regarded as infinite and uncharacterizable; **being-with,** human existence, regarded as membership of the community of persons.

1854 FERRIER *Inst. Metaph.* 525 Our alleged ignorance of 'Being in itself'. **1865** J. H. STIRLING *Secret of Hegel* II. III. 8 *Being-for-self* is the literal rendering of *Fürsichseyn*; which, indeed, cannot be translated otherwise. **1874** G. S. MORRIS tr. *F. Ueberweg's Hist. Philos.* II. III. 241 The Idea runs through a series of stages, from its abstract being-out-of-self in space and time to the being-in-self of individuality in the animal organism, their succession depending on the progressive realization of the tendency to being-for-self, or to subjectivity. **1892** E. S. HALDANE tr. *Hegel's Lect. Hist. Philos.* I. 20 Two different states must be distinguished. The first is what is known as capacity, power, what I call being-in-itself..; the second principle is that of being-for-itself, actuality. *Ibid.* 24 Being-in-self and being-for-self are the moments present in action. **1892** W. WALLACE tr. *Hegel's Logic* vii. 179 The readiest instance of Being-for-self is found in the 'I'. We know ourselves as existents. **1945** *Mind* LIV. 177 Since the subject realises itself as a subject, it has being-for-itself and therefore also possesses being-in-itself. **1956** F. COPLESTON *Contemp. Philos.* xi. 180 Being-in-the-world is being-with (*Mitsein*). **1957** *Sc. Jrnl. Theol.* X. 236 Being-itself, for Tillich, is the only non-symbolic or literal definition of God. **1962** R. G. OLSON *Existentialism* ii. 38 In Sartre's system the noumenal world ..is named 'being-in-itself' or sometimes simply 'the in-itself'. **1963** *Times Lit. Suppl.* 24 May 376/5 A certain..complacency..seems..to pervade this world of mutual 'being-with'.

being, *ppl. a.* **2.** For 'See BE *v.* B. II.' read 'See BE *v.* B. I. 3.'

beingness. Delete † *Obs.* and add later examples.

1897 J. H. STIRLING *Secret of Hegel* (ed. 2) 374 One gets a vivid glance of the direct beingness which immediacy amounts to. **1933** *Mind* XLII. 319 It may be possible to isolate certain aspects of the Aristotelian doctrine of 'beingness' or essence which have an obvious affinity with the ideas connoted by the word 'substance'. **1957** J. F. HORNER *Summary of Scientology* 57 The term, 'Thetan', refers to the single unit of beingness which each person is.

beisa (bəi·sä). [Native name.] An African antelope, *Oryx gazella beisa.*

1850 *Proc. Zool. Soc.* XVIII. 134 Oryx Beisa. The Beisa. Horns straight; throat without any bunch of hairs; black face-streaks separate...Inhabits Abyssinia. **1902** *Encycl. Brit.* XXV. 454/2 Widely different..is the African group of *Hippotraginæ*...Among these are..the straight-horned gemsbok and beisa (*Oryx*). **1921** *Edin. Rev.* July 105 Others ..are driving ostriches, gemsbok, and beisa antelopes. **1969** *Times* 30 Jan. p. iii/2 The proposed park will contain no unique mammals, but harbours large herds of Beisa oryx.

Beja (be·dʒä). Also **Bedja.** A nomadic people of Hamitic extraction living between the Nile and the Red Sea; the Cushitic language of this people. Also *attrib.*

1819 J. L. BURCKHARDT *Trav. Nubia* 503 In battle the Bedja pursue each other with their camels. *Ibid.* 526 People were found among the Djidda inhabitants who spoke Bedja. **1884** *Encycl. Brit.* XVII. 611/2 The Hamitic Beja. **1955** *Times* 8 July 7/2 The Beja is the tribe immortalized by Kipling under the name of Fuzzy Wuzzies after they had broken or nearly broken a British square at the battle of Tamai in 1884. **1961** *Listener* 16 Nov. 797/1 The Beja people of the Red Sea coast.

bejab(b)ers (bɪdʒēɪ·bəɪz, -æ·-), *int.* Also **be (by) jappers,** etc. [Corruption of *by Jesus.*] A *dial.* (esp. Irish) expletive.

In quot. 1962 used as a nonce-verb.

1821 D. HAGGART *Life* 118 By jappers, we were tould he was the boy. **1866** MAYNE REID *Headless Horseman* v. 30 'Be japers!' he exclaimed. **1890** HUME NISBET *Bail Up!* xxxviii. 265 A head wind, be jabbers! **1892** —— *Bushranger's Sweetheart* xx. 152 Arrah, be jabbers! but that's the foinest song I have listened to since I left Ould Oirland. **1895** J. BARLOW *Strangers at Lisconnel* iv. 58 Bejabers, you've got it now. **1939** *Times Lit. Suppl.* 25 Nov. 688/1 A Londoner who speaks nothing but cockney feels it [*sc.* the sentence] ought to be rounded off with 'bejabers'. **1962** *Listener* 9 Aug. 214/2 To live as an Irishman in England is to be forced to play a part. To begorra, to bejaper, and to be always after having a drink.

bejel (bēɪ·dʒēl, be·-). [Arab.] (See quot. 1961.)

1928 E. H. HUDSON in *U.S. Nav. Med. Bull.* XXVI. 818 The Bedouins have a disease which they call bejel. The practitioners of medicine..refer to this disease without hesitation as syphilis...Bejel is usually contracted in childhood, and apparent recovery follows quickly. **1945** *Lancet* 21 Apr. 505/2 Why..is syphilis spread venereally, while yaws and bejel are spread either by contact or by various insects feeding on the infected discharging sores? **1961** *Brit. Med. Dict.* 184/1 *Bejel,* a non-venereal form of syphilis which is endemic among the Arabs (particularly the children) of the valleys along the Middle Euphrates. The infecting organism is morphologically indistinguishable from *Treponema pallidum.*

bejesus (bɪdʒɪ·zəs), *int.* Also (esp. Anglo-Irish) **bejasus** (-dʒēɪ·zəs). An alteration of the oath *by Jesus.* Also as *sb.* in phr. *to beat the bejesus out of,* to give a good hiding to.

1908 *Dialect Notes* III. 290 *Bejasus,* interj., a mild expletive: often used in the phrase 'Faith and bejasus!' **1934** A. WOOLLCOTT *While Rome Burns* 258 My partner.. will..beat the be-Jesus out of you. **1949** 'J. TEY' *Brat Farrar* xv. 136, I know men who'd beat the bejasus out of you for that. **1956** J. DICKSON CARR *P. Butler for Defence* iv. 38 'Oh, bejasus!' roared the Irishman. **1960** *Guardian* 28 Feb. 8/3 She would put her fingers in her ears and scream: 'Bejasus, will ye stop it?'

bejewelled, *ppl. a.* (Earlier and later examples.)

1848 *Fraser's Mag.* XXXVII. 404 Bearing in his hand a bejewelled club. **1922** C. E. MONTAGUE *Disenchantment* xii. 168 Now the men would be rising..with smoking breath and bejewelled eyebrows.

bejuco (behū·ko). [Sp.] A liana, esp. the vine *Hippocratea scandens* of tropical America.

1848 WHITTIER *Slaves of Martinique* 19 As the serpent-like bejuco winds his spiral fold on fold Round the tall and stately ceiba, till it withers in its hold.

bekko (be·ko). [Jap.] Tortoise-shell.

1889 REIN *Industries Japan* 421 Bekkô, tortoise-shell, comes principally from *Chelonia imbricata,* L., the genuine loggerhead turtle.

bel (bel). *Electr. Communic.* [f. name of A. G. *Bell* (1847–1922), inventor of the telephone.] A unit, equivalent to ten decibels (see *DECIBEL), used in the comparison of two levels of power in an electrical communication circuit. (Non-technical sense in quot. 1958.)

The *bel* as a unit of power level is defined logarithmically as $N = \log_{10}(P_1/P_2)$, where *N* designates the number of bels and P_1 and P_2 the two amounts of power.

1929 W. H. MARTIN in *Bell System Techn. Jrnl.* VIII. 2 It was further suggested that the naperian unit be called the 'neper' and that the fundamental decimal unit be called the 'bel', these names being derived from..Napier..and Alexander Graham Bell. **1930** *Gloss. Terms Electr. Engin.* (*B.S.I.*) 13 The bel is a unit used in the comparison of the magnitudes of power, voltages or currents at two different points in a network of lines or apparatus. **1937** *Nature* 11 Sept. 447/1 The unit adopted..is the *bel,* which is a ratio signifying a 10-fold increase in intensity, power or energy. Two bels signify a 100-fold increase, three bels a 1000-fold increase, and so on. **1958** K. AMIS *I like it Here* vi. 70 Relaying a girls' choir at a volume of a couple of bels.

bel: see *BAEL.

belah (bɪ·lā). Also **belar,** † **beela, beal.** [Aboriginal name.] The Australian name for various trees, chiefly of the genus *Casuarina*; also the wood of these trees.

1862 H. C. KENDALL *Poems, Kooroora* 14 A voice in the beela grows wild in its wail. **1868** J. A. B. *Meta* 19 Blazing fire of beal. **1873** RANKEN *Dom. Australia* vi. 110 These scrubs..sometimes crown the watersheds as 'belar'. **1911** C. E. W. BEAN *'Dreadnought' of Darling* xix. 188 Mulga trees, and belah. **1933** *Bulletin* (Sydney) 20 Sept. 28/2 The casuarinas—she-oak, silky oak, belar, forest oak and creek or river oak—are all valuable for foliage, timber and bark. **1936** F. CLUNE *Roaming round Darling* xiv. 118 Plenty of timber: wilga, box, cypress, pine, and belah. **1944** F. D. DAVISON in *Coast to Coast 1943* 228 The line of wallaby snares in the belah scrub at the back of his selection.

belatedly (bɪlēɪ·tėdli), *adv.* [f. BELATED *ppl. a.* + -LY².] In a belated manner.

1896 *Westm. Gaz.* 21 Sept. 3/3 A fact..which her allies.. appear now somewhat belatedly to recognise. **1910** H. G. WELLS *Hist. Mr. Polly* ix. 240 He came belatedly in. **1917** CHESTERTON *Short Hist. Eng.* 219 Gladstone..rather belatedly realized that the freedom he loved in Greece and Italy had its rights nearer home. **1950** *Engineering* 2 June 625/1 A reversal of the trend..is now belatedly in evidence.

belatedness. Delete † *Obs.* and add:

1922 *Glasgow Herald* 12 Oct. 9 Considerable comment is being aroused by the long delay.., but this belatedness is, I understand, unavoidable.

belay, *v.* **5.** Delete '(The only current sense.)' in def. and add recent examples in *Mountaineering.* Hence **belay·ed** *ppl. a.*

1910 J. M. ARCHER-THOMSON *Climbing in Ogwen District* viii. 79 After belaying the rope to a bollard on the right, the second man can assist the leader to start. **1957** CLARK & PYATT *Mountaineering in Brit.* ix. 160 Belayed by the third man, the second steadies the leader's foot. **1957** R. G. COLLOMB *Dict. Mountaineering* 28 *Belay,* to tie oneself, as a stationary member of a roped party, to a firm rock projection..or to a piton, etc...in order to secure oneself and to afford a safeguard to the moving climber.

belay (bɪlēɪ·), *sb. Mountaineering.* [f. BELAY *v.*] A turn or fastening of a rope by belaying (see *BELAY *v.* 5). Also *attrib.* and *Comb.*

1908 *Westm. Gaz.* 12 June 5/1 A special knowledge of knots and roping method and belays might be their [*sc.* mountaineers'] only salvation. **1920** G. WINTHROP YOUNG *Mountain Craft* v. 226 A very common position upon steep rock..is to turn face inward, and pass the rope round some belay-point from one hand to the other. **1957** CLARK & PYATT *Mountaineering in Brit.* xiii. 212 The use of the shoulder belay, and the technique that went with it, became standardised.

belaying, *vbl. sb.* Add: **2.** Also *attrib.* in *Mountaineering,* as *belaying-pin, -point.*

1903 *Climbers' Club Jrnl.* VI. 5 So excellent was the anchorage afforded by this colossal belaying-pin that [etc.]. **1920** G. WINTHROP YOUNG *Mountain Craft* v. 220 A direct belaying-point which only leaves a short run-out.

bel canto (bẹl ka·nto). [It., = fine song.] Singing characterized by full, rich, and broad tone.

1894 G. DU MAURIER *Trilby* I. i. 46 It was lost, the *bel canto*—but I found it. **1908** *Daily Chron.* 9 May 4/4 In New York musical critics complain that audiences do not want Wagner,..and that the public flocks to the Italian bel canto. **1920** *Glasgow Herald* 14 May 8 For pure bel canto the English blackbird is hard to beat. **1938** *Oxf. Compan. Music* 85 *Bel canto*..This comprehensive term covers the vocal qualities of the great singers of the seventeenth and eighteenth centuries—the palmy days of Italian singing.

Belcher². (Earlier examples.)

1805 *Sporting Mag.* XXVII. 126/1 Their opponents were decked in the yellow stripe [handkerchief], which has acquired the appellation of the Belcher. **1809** *Monthly Pantheon* XIV. 546/1 If there be any of them who will spar with each other, let them wear the appropriate *Belcher handkerchief.*

bel-esprit. 2. (Earlier example.)

1806 M. EDGEWORTH *Leonora* II. lxv. 107 In these times a woman has no choice at a certain period but politics, or bel esprit.

belfry. Add: **4. e.** The head. See also *BAT *sb.*¹ 1 b. *slang.*

1907 N. MUNRO *Daft Days* xxxii. 267 When they've got cobwebs in their little brilliantined belfries, I'm full of the songs of spring. **1907, 1911,** etc. [see *BAT *sb.*¹ 1 b]. **1911** H. S. HARRISON *Queed* vii. 84 Something loose in his belfry.

Belgic, *a.* Add: **b.** Of or pertaining to the Belgæ (see BELGIUM). Hence **Be·lgicized** *ppl. a.* (esp. of pottery), made Belgic in form, appearance, etc. So **Be:lgiciza·tion.**

1589 A. FLEMING tr. *Virgil's Georg.* III. 43 And he shall better beare and draw *Belgic* coches with His gentle soft or tender necke. **1740** STUKELEY *Stonehenge* xi. 47 It seems not improbable, that the Wansdike was made, when this Belgic kingdom was at its height. **1743** —— *Abury* vii. 28 The great belgic rampart, the Wansdike, licks all the southern horizon. **1835** *Penny Cycl.* IV. 177/2 The whole southern coast from Suffolk to Devonshire was occupied by Belgic tribes. **1941** *Oxoniensia* VI. 87 Belgicized pottery ..dating just before or just after the Roman conquest. **1942** *Ibid.* VII. 59 Sherds from a quarry on Akeman Street. ..The cordoned forms indicate fairly recent Belgicisation, but may prove to be survivals into the post-conquest period. **1947** J. & C. HAWKES *Prehist. Britain* vi. 121 The Belgæ.. had overflowed into south-eastern Britain, and had made the lands they had settled provinces of Belgic culture.

Belgravia (belgrēɪ·viă). [f. *Belgrave* Square, named after Belgrave, a town in Leicestershire + -IA.] A fashionable residential district in London, south of Knightsbridge. Hence **Belgra·vian** *a.,* pertaining to, or characteristic of, Belgravia; as *sb.,* a resident of Belgravia.

1848 THACKERAY *Van. F.* li, Ask the Reverend Mr. Thurifer if Belgravia is not a sounding brass, and Tyburnia a tinkling cymbal. *Ibid.,* Her [*sc.* Semele's] myth ought to be taken to heart amongst..the Belgravians. **1849** —— *Pendennis* I. xxxvii. 358 The most elderly Belgravian Venus, or inveterate Mayfair Jezebel. **1850** C. KINGSLEY *Alton Locke* I. ii. 34 Shriek not in your Belgravian saloons. **1851** KNIGHT *Cycl. Lond.* 758 Architecture..in the Belgravian style. *a* **1852** MAYHEW *Lond. Labour* (1861) II. 395/2 The patrician squares of what has been called Belgravia and Tyburnia. **1864** M. ARNOLD *Let.* 10 May (1895) I. 232, I just get here, within reach of the Belgravian paradise. **1882** *Encycl. Brit.* XIV. 851/1 The fashionable Belgravia was built about 1825. **1891** *Athenæum* 27 June 824/2 That ineffable Belgravian, Lady Galbraith. *Ibid.,* The De Moleyns are excellent conventional Belgravians.

believe, *v.* Add: **1. b.** (Later example.)

1948 G. VIDAL *City & Pillar* (1949) I. v. § 2. 117 Sullivan believed in exercise.

5. Phrases. *I believe you,* an expression of emphatic agreement. Also in colloq. phrases strengthening an assertion, as *believe it or not, believe (you) me, would you believe it?* (see WILL *v.*¹ 43), *you'd better believe* (see *BETTER *a.* 4 b).

1646 CRASHAW *Delights of Muses* 130 The modest front of this small floore, Beleeve mee, Reader can say more Then many a braver Marble can. **1743** FIELDING *J. Wild* III. iii. 208 Believe me, Lad, the Tongue of a Viper is less hurtful than that of a Slanderer. **1776** H. MORE *Let.* (1925) 33 Would you believe it? In the midst of all the pomps and vanities of this wicked town, I have taken it into my head to study like a dragon. **1790** WALPOLE *Let.* 11 Dec. (1944) XI. 158 Believe me it is not for my own sake that I desire this. **1820** MOORE *Irish Melodies* 51 Believe me, if all those endearing young charms..Were to change by to-morrow [etc.]. **1832** DICKENS *Let.* 4 Feb. (1965) I. 3 Believe me Yours Truly Charles Dickens. **1834** —— *Sk. Boz* (1836) ser. I. I. 175 'Were you not a little surprised?' 'I b'lieve you!' **1860** TROLLOPE *Framley P.* II. ii. 35 Now, would you believe it? I have used up three lifts of notepaper already. **1910** KIPLING *Rewards & Fairies* 233 'The tides run something furious here.' 'I believe you,' said the Archbishop. **1918** C. SANDBURG *Cornhuskers* 30 Pike's Peak is a big old stone, believe me. **1926** *S.P.E. Tract* XXIV. 119 Believe *me* (sometimes expanded to 'believe you me')—take my word for it. **1929** R. L. RIPLEY (*title*) Believe it or not! **1931** L. STEFFENS *Autobiogr.* III. xxxvi. 617 But the only individual he ever exposed was Martin Lomasy, who, believe it or not, was one of the best men I met in Boston. **1943** 'E. M. DELAFIELD' *Late & Soon* iv. 63 Believe you me, that's no hardship. *Ibid.* ix. 123 Believe you me, in all the years, and all the adventures I've deliberately sought out..it's never been like this. **1951** L. MACNEICE tr. *Goethe's Faust* II. IV. p. 256 No, you shall win it, believe you me. It's you to-day are C. in C. **1967** 'O. MILLS' *Death enters Lists* v. 47 *Someone's* making a good thing out of the contracts, believe you me. **1968** *Sunday Express* 8 Dec. 8/1 Having died for a minute and a half I suppose I am one of those believe-it-or-not Ripley characters.

Belisha beacon (bɪlɪ·ʃă bɪ·kən). [f. surname of Leslie Hore-*Belisha,* Minister of Transport 1931–7 + BEACON *sb.*] A post about seven feet high surmounted by a flashing amber-coloured globe and erected on the pavement at officially recognized pedestrian crossings of the highway. Also **Belisha.** Hence *Belisha crossing.*

1934 *Punch* 21 Nov. 583/1 One of the clever people who have been going about stealing and even shooting the Belisha Beacon globes—(1) 'as a protest against their futility'; (2) 'because they slow down the traffic'. *Ibid.* 5 Dec. 617 (*caption*) Why not be in the movement, Sir, and 'ave a Belisha? **1936** *N. & Q.* CLXXI. 355/1 With a view to learning what people in general called these crossings, I asked two intelligent young working-women. One said 'Belisha crossing', the other 'pedestrian crossing'. **1942** *Motor Driving Made Easy* (Autocar) (ed. 7) ix. 121 Once he [*sc.* the pedestrian] has left the pavement at a Belisha beacon, motor and other traffic must yield to him. **1958** L. BLIGHT *Love & Idleness* iii. 29 The yellow belishas going on and off at the zebra crossings.

belittle, *v.* **1.** (Earlier U.S. example.)

1782 JEFFERSON *Notes Virginia* (1787) 107 So far the Count de Buffon has carried this new theory of the tendency of nature to belittle her productions on this side the Atlantic.

3. (Earlier U.S. examples.)

1797 *Independent Chron.* (U.S.) 30 Mar., [He] is..an honorable man,..let the writers..endeavour to belittle him as much as they please. **1837** HALIBURTON *Clockm.* Ser. I. xxii. 226 When..they began to raise my dander, by belittleing the Yankees. **1843** —— *Attaché* II. xviii. 39, I won't stay here and see you belittle Uncle Sam, for nothin'.

belittler (bɪlɪ·t'ləɪ). [f. BELITTLE *v.* + -ER¹.] One who belittles or depreciates.

1887 *Daily News* 27 Jan. 6/3 The belittlers more than half confirm the story they would be delighted to contradict. **1898** *Pop. Sci. Monthly* LIII. 396 His belittlers emphasize Gerarde's ignorance of the classic writers on botany. **1920** *Contemp. Rev.* Aug. 171, I protest against all the patriotic belittlers of their own nations.

bell, *sb.*¹ Add: **6. b.** (Earlier and later examples.)

1806 BUSBY *Dict. Mus.* (ed. 2), *Bell of a Horn,* the large, open part of the instrument, from which the sound immediately issues. **1926** WHITEMAN & MCBRIDE *Jazz* ix. 201 The players [of cornets] got that effect by inverting glass tumblers over the bells of the instruments. **1966** *Crescendo* Oct. 22/3 Sitting only three feet from the..bell of Jimmy Heath's tenor [saxophone].

d. (Further examples of techn. uses.)

1893 *Funk's Stand. Dict., Bell,* the movable cap at the top of a modern blast-furnace, which is lifted to put in the charge of ore, etc. **1930** *Engineering* 2 May 589/1, 10 per cent of the total gas made was lost owing to the use of single bells on the blast-furnaces. **1944** *Gloss. Terms Gas Industry* (*B.S.I.*) 26 *Bell,* the hollow cylinder closed at its upper end which forms the gas container.

7. c. *to ring the bell* (colloq.): to carry off the prize; to be the best of a lot: in allusion to the ringing of the bell attached to a strength-testing machine.

1900 J. M. BARRIE *Tommy & Grizel* viii. 102 It was a shot that rang the bell. **1928** *Publishers' Weekly* 26 May 2094 This [book] liberally illustrated, with a great jacket, rings the bell.

d. *to ring a bell* (colloq.): to awaken the memory, to set one remembering.

[**1934** A. HUXLEY *Beyond Mexique Bay* 51 Why should the Local Pavlov have chosen to ring just those particular bells which happen to be rung?] **1939** N. MONSARRAT *This is Schoolroom* II. xi. 239 The things we talked about meant nothing to them: they rang no bell. **1945** M. AGATE *Madame Sarah* xii. 171 They would be quick to recognize a great performance, but would it 'ring a bell' with them, I wonder? **1945** M. ALLINGHAM *Coroner's Pidgin* xxiii. 203 That's where I saw the name, then...It rang only a very faint bell. **1957** P. FRANKAU *Bridge* 64 'Do you remember..the stage-hands who built the set?' 'That rings a bell.'

11. c. *bell-barrow* [BARROW *sb.*¹ 3], *-beaker* [*BEAKER 1 c], *bell-button* (earlier example), *-mouth* (later examples); *bell-crowned* (earlier and later examples), *-mouthed* (earlier and later examples).

[**1743** STUKELY *Abury* ix. 41 Barrows with ditches round them. These are..generally of an elegantly turn'd bell-form.] **1812** R. COLT HOARE *Anc. Hist. S. Wilts.* 21 *Bell Barrow.* This, from the elegance of its form, seems to have been a refinement on the Bowl Barrow: they abound in the neighbourhood of Stonehenge. **1928** CRAWFORD & KEILLER *Wessex from Air* iv. 13 Some *Bell-barrows* certainly belong to the very beginning of the period...The name is, of course, derived from the form. The mound is high, and is usually separated from the surrounding ditch by a narrow shelf or berm of natural soil. **1902** *Jrnl. Anthrop. Inst.* XXXII. 390 The different class of beaker..is often called the 'Bell-beaker', from its caliciform or bell-shaped form. **1925** V. G. CHILDE *Dawn Europ. Civilization* xix. 293 In Bavaria and Thuringia as elsewhere, the bell-beaker folk had used metal. **1967** *Antiquaries Jrnl.* XLVII. 182 Crichel Down in Dorset..where one sherd of rusticated ware was found associated with small sherds of the Bell-Beaker class. **1775** ADAIR *Amer. Ind.* 7 The beaus..choose bell-buttons, to give a greater sound. **1823** J. DODDRIDGE *Dial. Backwoodsman & Dandy* in *Logan* (1868), Your bell crowned hat. **1910** J. FARNOL *Broad Highway* I. xvi. 86 It's precious lucky for you as you are a-wearin' that there bell-crowned 'at! **1896** *Daily News* 29 May 6/7 The huge brass bellmouth apparatus fixed on the prompt side of the opera proscenium is an electrophone. **1930** *Engineering* 4 Apr. 436/1 The lining at both extremities of the cutout ends, including the bell-mouth. **1958** J. S. SCOTT *Dict. Civ. Engin.* 22 *Bellmouth overflow,* an overflow from a reservoir through a tower built up from the bed to the overflow level. **1967** *Jane's Surface Skimmer Systems 1967–68* 106/1 There is a dam for solid water separation and four right angle turns before the air reaches the engine bellmouths. **1797** J. CURR *Coal Viewer* 47 For the common engine..[cylinders] must be bell-mouthed ½ of an inch on each side. **1946** *Nature* 2 Nov. 635/2 A bell-mouthed estuary.

Examples of recent *poet.* Combs.

1879 G. M. HOPKINS *Poems* (1918) 41 Cuckoo-echoing, bell-swarmèd, lark-charmèd. **1888** —— *Poems* (1918) 89 Bellbright bodies huddling out. **1889** YEATS *Wand. Oisin* III. 38, I gazed on the bell-branch, sleep's forebear. **1919** W. DE LA MARE *Flora* 33 On the mirroring sands Bell-shrill the oyster-catchers. **1935** W. EMPSON *Poems* 24 Drowned under flounces and bell-calm of trees. **1945** W. DE LA MARE *Burning-glass* 75 Its secret bell-clear song.

12. bell-animalcule (examples); **bell-bottomed** *a.,* of trousers, having a considerable increase in width from below the knee to the bottom of the leg, giving a bell-shaped appearance; also, of an individual, esp. of a class accustomed to wear such trousers; **bell-bottoms** *sb. pl.,* bell-bottomed trousers; **bell-buoy** (examples); **bell captain** *U.S.,* one who supervises a group of bell-boys; **bell cord,** a cord to be pulled in order to ring a bell, spec. in a passenger vehicle; **bell-crater** *Gr. Antiq.,* a bell-shaped bowl (see CRATER 1); **bell-hanger** (earlier examples); **bell-hop, -hopper** *U.S.* and *Canada,* a hotel page-boy (cf. *BELL-BOY 2); **bell-horse** (examples in sense 'leader of a pack-train'); **bell-mouth** *v. trans.,* to furnish with a bell-mouth; **bell-punch** (see quot. *a* 1884); **bell-push,** the button that is pushed to ring an electric bell; also, a table bell to be rung in this way; **bell-shade,** a bell-shaped light-shade; **bell-sheep** *Austral.* (see quot.); **bell-signal,** a signal transmitted by the bell-telegraph; so *bell-signalling* (both *disused*); **bell-skirt,** a bell-shaped skirt of a garment; **bell-sleeve,** a long sleeve flared at the lower edge; **bell strap** = **bell cord* (Funk, 1893); **bell-telegraph,** an instrument in which two bells are used to transmit a message, one indicating (by its tone) the movement of the needle to the right, the other, to the left (*disused*); **bell-topper,** a 'topper', or top-hat, esp. one of old-fashioned type with a bell-shaped crown.

1875 *Med. Trainer & Gaz.* 8 May 495/1 A species of *Amphileptus* has been observed to swallow—or rather envelope—a stalked bell-animalcule (*Vorticella*). **1959** J. CLEGG *Freshwater Life* (ed. 2) v. 100 The Bell Animalcules, as the various species of *Vorticella* are popularly called, are found in dense masses at times. **1891** *Argus* 5 Dec. 13/2 (Morris), The bell-bottomed ballottee. **1904** *Daily Chron.* 6 June 4/5 The crews were not less smart in.. blue jackets and white bell-bottomed ducks. **1955** 'NEVIL SHUTE' *Requiem for Wren* 83 A naval cutter manned by two Wrens in jerseys and bell-bottomed trousers. **1898** J. D. BRAYSHAW *Slum Silhouettes* 220 They 'ad a crease right dahn 'em, an' bell bottoms. **1929** *Daily News* 10 Apr. 7/3 An official [of the Admiralty] said: '..Bell-bottoms are the Navy. Abolish them and you abolish the Navy.' **1838** E. FLAGG *Far West* in R. G. Thwaites *Early Western Trav.* (1908) XXVI. 52 The hated clang of the bell-boy [*sic*] was soon after heard..throughout the cabins. **1884** Bell-buoy [see BUOY *sb.* 1]. *a* **1910** 'O. HENRY' *Trimmed Lamp* (1916) 60 A pale-faced, fat man huskily enveloped him with a raised, red fist, and the voice of a bell buoy. **1926** *Sat. Rev. Lit.* 20 Nov. 315 His experienced demeanor among hotel porters and ticket agents and..Bell Captains. **1944** *Reader's Digest* Mar. 108 The coruscating bell captain commands an army of bell-boys. **1843** *Knickerbocker* XXI. 332 He found no bell-cord to pull. **1875** *Chicago Tribune* 2 July 8/1 When a half-drunken man has hold of the bell-cord instead of the strap to steady himself. **1921** *Brit. Mus. Return* 61 Bell-crater, late red-figure style. **1939** PENDLEBURY *Archaeol. Crete* vi. 309 The bell-kraters are deeper and less swelling. **1540** in W. H. Turner *Select. Records of Oxford* (1880) 160 John Payne, bellehanger. **1789** BETSY SHERIDAN *Jrnl.* (1960) 180, I have scribbled thus far in the midst of the noise of Bell-hangers who are rectifying all negligence of our predecessors. **1910** ADE *I knew him When* 14 He is not a bell hop—the boys used to dress like that. **1919** D. L. CADY *Rhymes Vermont Rural Life* (1923) 99 The traveler saw no bellhops hop. **1925** A. LOOS *Gentlemen prefer Blondes* (1926) 83 The boy friend of ours who is the bell hop, waked me up at ten o'clock. **1900** ADE *More Fables* 5 When he got back to his Room the Bell-Hopper came round and asked him if he cared to Sit in a Quiet Game. **1775** ADAIR *Amer. Ind.* 337 But they [*sc.* the Choctaw Indians]..stole one of the bell horses. **1891** *Harper's Mag.* Nov. 890/1 The pack trains consisted of a 'bell-horse' and boy, and six horses following. **1880** *Encycl. Brit.* XII. 463/2 It is often desirable to bell-mouth the ends of pipes. **1877** BARTLETT *Dict. Amer.* 253 *Gong-Punch,* an instrument used by conductors [etc.]..a bell-punch. *a* **1884** KNIGHT *Dict. Mech.* Suppl., *Bell Punch,* a hand-punch, for perforating a ticket or trip slip. It secures the piece punched out, and rings a bell; in some instruments it also registers the fares collected. **1894** *Westm. Gaz.* 26 Sept. 2/2 Thanks to the bell-punch, the number of passengers carried can now be estimated with tolerable accuracy. **1884** D. P. HEAP *Report Internat. Exhib. Electricity, Paris 1881* 26 This little battery..is contained in the ordinary bell-push. **1887** *Cassell's Fam. Mag.* 703/2 An indicating disc, which by its vibrations tells the bell push.. of the person who has rung up. **1921** *Blackw. Mag.* June 739/2 Pressing the bell-push to call the steward. **1890** W. J. GORDON *Foundry* 140 At one chair here some large bell-shades for lamps are being made. **1900** LAWSON *On Track* (1945) xvi. 99 He times himself to get so many sheep out of the pen *before* the bell goes [for breakfast or dinner], and *one more*—the 'bell-sheep'—as it is ringing. **1905** *Westm. Gaz.* 19 Apr. 8/1 He announced the fact..by sending to him the block telegraph bell-signal 'Vehicles running away on right line'. **1905** *Daily Chron.* 16 Nov. 7/3 Bell-signalling between torpedo craft and submarines in night-time. *a* **1910** 'O. HENRY' *Strictly Business* xx, The latest thing in suitings with side vents and bell skirt. **1940** W. EMPSON *Gathering Storm* 35 One swing of the bell skirt. **1892** *Daily News* 26 Oct. 2/1 The 'bell' sleeves are turned back with white silk. **1960** C. W. CUNNINGTON et al. *Dict. Eng. Costume* 15/1 *Bell sleeve,* second half 19th c. Close-fitting to mid-forearm and there expanding into a bell-shaped opening. **1922** JOYCE *Ulysses* 219 The conductor pulled the bellstrap. *a* **1877** KNIGHT *Dict. Mech., Bell-telegraph,* .. invented by Sir

Charles Bright. **1858** W. KELLY *Life in Victoria* I. xvi. 268 Bell-topper was the derisive name given by diggers to [an] old style hat, supposed to indicate the dandy swell. *Ibid.,* Merchants ventured to the Chamber of Commerce in the regular British 'bell-topper', some of the nattier going the length of sporting kid gloves. **1871** *Fav. Reciter* (Simpkin) 6 A bell-topper hat. **1888** B. L. FARJEON *Miser Farebrother* II. i. 3 On his head the shiniest of belltoppers. **1936** I. L. IDRIESS *Cattle King* xxxi. 274 Sitting next to him was a 'swell' whose belltopper Kidman was quietly admiring.

bell, *v.*⁵ Add: **2. b.** *intr.* with *out.* To spread *out* like the mouth of a bell. So *belled-out* ppl. a.

1922 *Blackw. Mag.* June 731/2 The skirt belled out like an inverted campanula bloom. **1959** *New Scientist* 11 June 1291/1 Shafts can be dug, 'belled out' at the base to get a larger load bearing area. *Ibid.* 1291/2 A concrete cylinder with a 'belled out' foot.

belladonna. Add: **3.** *attrib.*

1856 *Med. Times & Gaz.* XIII. 513 Case of poisoning from the application of belladonna plaster to the skin. **1869** G. LAWSON *Dis. Eye* iv. 126 A fold of lint..kept moist with..the belladonna lotion. **1885** *Buck's Handbk. Med. Sci.* I. 486/2 The clinical history of a case of belladonna poisoning. *Ibid.,* The patient..had eaten..about thirty belladonna berries. **1890** BILLINGS *Med. Dict.,* Belladonna-leaves...B. plaster. **1896** *Daily News* 10 Sept. 2/6 Belladonna poisoning. *Ibid.,* The belladonna liniment. **1968** *Times* 3 Dec. 10/8 Drugs of the belladonna group.

bell-bird. Add: (Earlier and additional examples.) Also used as the name of various birds with a clear ringing call.

1802 BARRINGTON *Hist. New S. Wales* viii. 284 The cry of the bell-bird seems to be unknown here. **1825** WATERTON *Wand. S. Amer.* 117 The celebrated Campanero of the Spaniards, called..bell-bird by the English. **1828** WORDSWORTH *On Power of Sound* ii, Toll from thy loftiest perch, lone bell-bird, toll. **1845** E. J. WAKEFIELD *Adv. N. Zealand* I. 23 The melodious chimes of the bell-bird were especially distinct. **1865** *Ibis* I. 90 The Costa-Rican Bell-bird (Chasmorhynchus tricarunculatus). **1882** W. L. BULLER *Man. Birds N.Z.* 11 Anthornis melanura..Bell-bird. Mocker. Korimako. Makomako. **1887** *Ibid.* (ed. 2) I. 92 Anthornis Melanocephala. (Chatham-Island Bell-Bird.) **1903** *Westm. Gaz.* 28 Oct. 12/2 The Banded Bell-Bird..(*Cotinga cincta*). **1966** G. M. DURRELL *Two in Bush* i. 45 A Bellbird..entertained us with a concert of wonderful, flute-like notes, wild, liquid and beautiful.

bell-boy. [BELL *sb.*¹ 1.] **1.** A boy who rings a bell.

1851 MELVILLE *Moby Dick* I. xxxix. 274 Eight bells there! d'ye hear, bell-boy? *Ibid.* III. xxxix. 227 'Who art thou, boy?' 'Bell-boy, sir; ship's-crier. Ding, dong, ding!'

2. A hotel page-boy. *U.S.*

1861 G. F. BERKELEY *Eng. Sportsman* 366 'What are you, then, young fellow?' 'I'm bell-boy.' **1897** KIPLING *Capt. Courageous* ix. 196 Hotel piazzas where the ingenuous young of the wealthy play with or revile the bell-boys. **1932** E. WILSON *Devil take Hindmost* xxiii. 245 Glimpses as a bellboy of the luxurious life of the hotel.

belle, *a.* Add: **2.** *belle laide,* an attractively ugly woman.

1908 W. S. MAUGHAM *Magician* ii. 19 She was one of those plain women whose plainness does not matter. A gallant Frenchman had..called her a *belle laide.* **1946** 'J. TEY' *Miss Pym Disposes* xiv. 152 A woman with the makings of a *belle laide.* **1956** L. E. JONES *Edwardian Youth* i. 6 Conscious of the physical failings of that fascinating *belle-laide.*

Belleek (belɪ·k). The name of a town in Fermanagh, Ireland, used *attrib.* or *absol.* to designate a kind of pottery produced there (see quot. 1960).

1869 *Arts Jrnl.* May 149 (*title*) The Belleek Pottery. *Ibid.* 151/1 Neither of these glazes..can compare with the beauty of Belleek ware... The most welcome of the patrons of Belleek is the Prince of Wales. **1935** *Discovery* July 205/2 Thin section of Belleek china showing layer of crystals found between body and glaze. **1960** H. HAYWARD *Antique Coll.* 29/1 *Belleek,* a light, fragile feldspathic porcelain cast in moulds, with lustrous pearly glaze.

belleter (be·lėtəɪ). *Hist.* [= *bell-yetter* (BELL *sb.*¹ 12).] A bell-founder.

1891 *Athenæum* 12 Sept. 360/2 On the tenor at Great Bradley we recognize the time-honoured name of a belleter whom Mr. Stahlschmidt..restored to renown. **1898** *Ibid.* 16 July 103 The Van den Ghens, of Louvain and Malines, were belleters of renown.

belletrist. Add: (Later example.) *attrib.* or as *adj.* = BELLETRISTIC *a.* So **belletri·stical** *a.,* **belletri·stically** *adv.*

1840 *Fraser's Mag.* XXI. 199 No historical, biographical, geographical, or *belle lettristical* book..would now have much chance. **1889** J. M. ROBERTSON *Ess. Crit. Method* 40 The belletrist essays of Mr. Lowell have had a wide public. **1894** J. DAVIDSON *Ballads & Songs* 38 Heed not bellethrist [*sic*] jargon. **1894** *Westm. Gaz.* 25 June 3/1 He trifles bellettristically around ghosts,..and the relation of ghosts to religion. **1923** J. M. MURRY *Pencillings* 108 No amount of sedulous apery or word-mosaic will make a writer of the dilettante bellelettrist.

Hence **belle·trism,** the study or composition of belles-lettres.

1938 *Scrutiny* VII. 208 Her university..could only be a breeding-ground for boudoir scholarship..and belletrism. **1961** *Essays in Crit.* XI. 196 Some of the assumptions he has innocently taken over from stock Victorian-romantic theory ..serve only for effusions of belletrism such as this.

bellowed, *ppl. a.* (Earlier example.)

1806 J. BERESFORD *Miseries Hum. Life* I. vi. 130 Your bellowed intreaties that he would stop.

bellows, *sb.* Add: **3.** *fig. bellows to mend* (examples).

1853 R. S. SURTEES *Sponge's Sp. Tour* lvi. 322 Is old bellows-to-mend gone to bed? **1854** 'C. BEDE' *Further Adv. Verdant Green* iv. 31 To one gentleman he would pleasantly observe, as he tapped him on the chest, 'Bellows to mend for you, my buck!' **1888** F. W. J. HENNING *Recoll. Prize Ring* 156 As the two were sent up it was a case of 'bellows to mend', especially with Grant. **1923** J. M. MURRY *Pencillings* 248 Johnny Keats always did have bellows to mend.

6. bellows pocket (see quot. 1960).

1922 JOYCE *Ulysses* 435 Mrs. Breen in man's frieze overcoat with loose bellows pockets. **1960** C. W. CUNNINGTON et al. *Dict. Eng. Costume* 14/2 *Bellows pocket,* late 19th c. A patch-pocket with side folds capable of expanding or lying flat, like a bellows. Common in Norfolk jackets from 1890 on.

bellum (be·lǝ̆m), and variants. [a. Pers. *balam*; cf. *balaum* s.v. *Baloon* in Yule *Hobson-Jobson.*] A small boat or canoe used in ports along the shores of the Persian Gulf.

1901 *Wide World Mag.* VI. 464/1 Queenie was landed easily enough in a *bellum.* **1916** V. HORSLEY *Let.* 18 May in S. Paget *Life* (1919) 319 The whole real traffic of the place is done..by 'bellums': these are exactly like dugouts. **1916** T. E. LAWRENCE *Home Lett.* (1954) 319 The native boats give a character to Basra...You..shop or pay your calls in a 'Bellam'. A bellam is a sort of gondola, thirty or forty feet long, about four feet wide, and shallow. Two men work them, either by sculls, or by poling along with a light bamboo...The bellam is the passenger boat. **1919** *Chambers's Jrnl.* Jan. 23/2 The *bellem*..is a down-river boat peculiar to the city of pale-pink Persian roses—Basra.

bellwort. (Earlier examples.)

1785 *Memoirs Amer. Acad. Arts & Sci.* I. 434 *Uvularia,*.. Bellwort, sweet-smelling Solomon's Seal. **1845** A. WOOD *Class-bk. Bot.* 396 Uvularia..1. *U. Sessilifolia*..Bell-wort. Wild Oats. 2. *U. perfoliata*..Perfoliate Bell-wort.

belly, *sb.* Add: **4. c.** In full *belly-wool.* The wool from a sheep's belly, of inferior quality to the main fleece. *Austral.* and *N.Z.*

1851 F. A. WELD *Hints to Sheep-farmers in N.Z.* 10 Their mothers do not lose the belly wool, as they would do by lambing in spring. **1900** LAWSON *Over Sliprails* 32, I had just slipped a light ragged fleece into the belly-wool and 'bits' basket. **1911** E. M. CLOWES *On the Wallaby* iv. 101 Clipping the belly-wool. **1933** L. G. D. ACLAND in *Press* (N.Z.) 9 Sept. 15/7 *Belly*..(2) the wool off a sheep's belly, pressed and sold as 'bellies'. **1956** G. BOWEN *Wool Away!* (ed. 2) vii. 94 A bale of bulk ewe bellies can be depreciated a grade because of the inclusion of a few wethers' bellies from which urine stains have not been removed.

d. Applied to the under part of the fuselage of an aeroplane. Hence *belly-landing,* a crash-landing of an aeroplane on its belly without use of the undercarriage; also (as a back-formation) *belly-land* v.; *belly tank,* a reserve fuel tank carried in the 'belly' of an aeroplane.

1917 'CONTACT' *Airman's Outings* 50 They saw an enemy plane turn over, show a white, gleaming belly, and drop in zigzags. **1939** *Flight* 19 Oct. 314/1 A bellylanding can be made without personal damage in almost any reasonable terrain. **1940** R. HARTLEY *Aeronaut. Dict.* 16 Belly Tank. **1942** *Aircraft of Fighting Powers* III. 66/1 The S-OO fighters could pursue them over long distances due to their long range of 1,600 miles when fitted with the external streamlined belly tank. **1943** T. D. GORDON *Coastal Command at War* x. 94 The Hudsons..went down to sea-level, to guard themselves from a belly attack. **1944** *Sun* (Baltimore) 10 June 2/3 Johnson bellylanded his plane safely near the spot where he had seen Allied troops from the air. **1958** *Times* 15 Oct. 16/6 A Jordanian Air Force pilot..climbed without a scratch..from a Vampire jet aircraft yesterday after a 'belly-landing' at nearly 200 m.p.h.

14. The sound-board of a piano. Also *attrib.,* as *belly-bar, -bridge*; **bellyman,** the workman who makes and fits the 'belly'.

1845 G. DODD *Brit. Manuf.* IV. 155 The 'bellyman' or 'sounding-board maker'. **1905** HASLUCK *Pianos* 21 Prick through the belly about every 2 in. with a small bradawl; this will help in putting on the belly bridges. **1905** *Sci. Amer.* Suppl. 6 May 24536 The sound-board..barred beneath with batons..technically 'belly-bars', which strengthen the belly. **1910** *Daily Chron.* 19 Jan. 12/7 Pianos. Bellyman and marker-off contractor wanted.

17. belly-button *colloq.,* the navel (Bartlett, 1877); **belly-dance,** an erotic oriental dance performed by women, involving abdominal contortions; hence *belly-dancing* vbl. sb.; *belly-dancer*; **belly-flop** *colloq.,* (of troops) a sudden drop to the ground to avoid enemy fire; (of a swimmer) a dive that brings one's body flat on the water (also *belly-flopper*); hence as *v.,* and *transf.*; **belly-laugh** *colloq.,* a deep, unrestrained laugh.

1934 KIPLING in *Strand Mag.* Apr. 350/1 Why waste time fighting atomies who do not come up to your belly-button? **1946** J. B. PRIESTLEY *Bright Day* iii. 66 If you'd ever gone to school with your belly-button knockin' against your backbone. **1899** MORROW *Bohem. Paris* iv. 95 The *danse du ventre* (literally, belly-dance) is of Turkish origin. **1943** KOESTLER *Arrival & Departure* 54 The loud-speakers blared a hot belly-dance with drums and castanets. **1931** C. BEATON in *Wandering Years* (1961) 217 The wow of the evening was Carmen, the belly-dancer. **1957** R. CAMPBELL *Portugal* ix. 192 The *lundum*..was a highly sensual song accompanied by much belly-dancing. **1925** FRASER & GIBBONS *Soldier & Sailor Words* 21 *Belly flopping,* a term for the sectional rushes of troops in an attack in which the men advance in a crouching posture. **1931** BROPHY & PARTRIDGE *Songs & Slang 1914–18* (ed. 3) 282 In France, whenever one heard a shell coming uncomfortably close, one *belly-flopped* or *did a belly-flop.* **1937** 'R. HYDE' *Wednesday's Children* ii. 49 'It hurt,' she added..'So I didn't do any more worshipful belly-flops.' **1945** PARTRIDGE *Dict. R.A.F. Slang* 15 *Belly-flop,* a landing effected with the wheels unlowered. **1953** R. LEHMANN *Echoing Grove* 238 The nasty expectation of a belly-flop:..the water has become so huge. **1916** E. F. BENSON *David Blaize* iv. 81 A loud, flat smack was heard as he fell into the water. And Ferrers said to Mullins: 'I say, can your pater take belly-floppers?' **1941** *Lancet* 12 July 31/2 Perhaps the nearest experience that many of us may have had to blast-concussion is the flat dive—the 'belly-flopper' of our schooldays. **1960** BETJEMAN *Summoned by Bells* v. 50 The belly-floppers from the punt. **1921** S. H. ADAMS *Success* III. iv. 397 'I'm after the laugh that starts down here.' He laid hand upon his rotund waistcoat. 'The belly-laugh.' **1931** E. LINKLATER *Juan in Amer.* v. ii. 376 He laughed, deep belly-laughs. **1957** R. HOGGART *Uses of Literacy* vii. 191 Sensationalism.. is as much without a belly-laugh as it is without bowels of compassion.

belly, *v.* Add: **4.** *trans.* To remove the wool on the belly of (a sheep) before shearing. *Australia.*

1909 in WEBSTER.

belly-ache, *sb.* (Later example.)

1967 A. WILSON *No Laughing Matter* II. 76 Don't blame me if you all get the bellyache.

be·lly-ache, *v. slang.* orig. *U.S.* [f. BELLY-ACHE *sb.*] *intr.* To complain querulously or unreasonably; to whine, grizzle. Hence as *sb.,* a querulous complaint. So **belly-aching** *vbl. sb.* and *ppl. a.*; **belly-acher,** one who so complains.

1888 FARMER *Americanisms* 50/2 Employés *bellyache* at being overworked, or when they fancy themselves underfed. A vulgarism. **1923** U. L. SILBERRAD *Lett. J. Armiter* vi. 132 Not a bad sort of woman, except that she's always belly-aching about money. **1930** W. R. BURNETT *Iron Man* I. i. 9 'Now start your belly-ache,' said Regan. 'To hear you talk you'd think you really did some training.' **1930** S. HENRY *Conquering Amer. Plains* xvii. 221 These voluble doubters were commonly called old croakers, backbiters, 'belly-achers'. **1931** E. LINKLATER *Juan in Amer.* IV. i. 272 Another of these belly-aching German war books. Who started the War, anyway? **1933** E. CALDWELL *God's Little Acre* i. 8, I reckon there's enough to complain about these days if a fellow wants to belly-ache some. **1934** 'G. ORWELL' *Burmese Days* xv. 230 Other people..will listen to their belly-achings with sympathy. *Ibid.* xv. 232 It was an impertinence to go belly-aching on and on about myself. **1958** *Listener* 13 Nov. 791/2 The subordinate who argued about orders was always 'a bellyacher'. **1959** *Times* 22 July 5/4 The 'belly-aching' was over the pay cut, not the principle of negotiation.

belly-band. Add: **4.** A wide belt, such as a cholera-belt or (*slang*) corset.

1888 KIPLING *Soldiers Three* (1889) 54 He ran forward.. an' swung a Paythan clear off his feet by the belly-band av the brute. **1922** JOYCE *Ulysses* 231 She was well primed with a good load of Delahunt's port under her bellyband. **1925** FRASER & GIBBONS *Soldier & Sailor Words* 21 *Belly band,* the flannel cholera belt issued to the troops.

beloid (bɪ·loid), *a.* [f. Gr. βέλος arrow, dart + -OID.] Arrow-shaped; spec. in *Craniometry,* having a broad occiput and narrow frontal region.

1901 G. SERGI *Mediterranean Race* v. 110 These six photographs present one [skull] of beloid shape.

belong, *v.* Add: **4. a.** Also const. *to,* †*unto.* **b.** Substitute: With an adv. or advb. phr. (esp. *here, where* = to this or these, to which), also with various preps. or without const.: to be related or connected; to have a certain connection indicated or implied in the context; to fit a certain environment, group, etc. orig. *U.S.*

1822 COOPER *Spy* xxvi, I have never known whether he belonged above or below. **1861** [in Dict.]. **1867** A. WILSON *St. Elmo* x, To replace it in the glass box where it belongs. **1889** WALT WHITMAN in *Century Mag.* (1911) 11 Jan. 256/2 He was not a closet man, belonged out-of-doors. **1897** *N.E.D.* s.v. *Fit v.*¹ 2 The first examples given under.. 3 may belong here. **1922** D. H. LAWRENCE *England* (1924) 232 He was still in the choir of Morley Chapel—not very regular. He belonged just because he had a tenor voice, and enjoyed singing. **1924** A. D. SEDGWICK *Little French Girl* I. x, I saw you took to each other. I saw you belonged with each other. *Ibid.* II. xiv, From the first moment I saw her I felt that she belonged. **1936** WODEHOUSE *Laughing Gas* iii. 31, I looked as if I belonged in Whipsnade. **1942** M. MCCARTHY *Company she Keeps* (1943) v. 164 It was the Moscow trials that made him know, for the first time, that he did not really 'belong'. **1949** *Scrutiny* XVI. 9 This remark of Eliot's..suggests that Byron doesn't quite 'belong'. **1960** *Guardian* 4 Mar. 8/7 People also feel they want to belong and matter.

c. With inf.: to be accustomed, ought; to seem, intend. *U.S. dial.*

1901–7 in H. WENTWORTH *Amer. Dial. Dict.* (1944) 53/1 John Henry belongs to folla aftch Sayrah. **1935** A. C. BAUGH *Hist. Eng. Lang.* xi. 453 The expression reported from South Dakota, 'I got up at six o'clock this morning although I don't *belong* to get up until seven.' **1938** M. K. RAWLINGS *Yearling* iv. 29 You belong to figger..a man.. cain't out-run a bear, but he's a sorry hunter if he cain't out-study him. *Ibid.* 35 When it back-fired, that belongs to mean the mainspring's got weak.

belonging, *vbl. sb.* **II.** Add after def.: *Esp.* a person's membership in, and acceptance by, a group or society (cf. *BELONG *v.* 4 b).

1934 W. PLOMER *Invaders* ii. § 4. 43 He had little sense of belonging, of being necessary to the world he lived in. **1958** H. REILLY *Ding Dong Bell* (1959) i. 16 What the child needs is a settled home, a feeling of permanence, security, of *belonging.*

Comb., belonging-together(*ness*) (cf. *BELONGINGNESS 2).

1890 W. JAMES *Princ. Psychol.* I. x. 337 It seems as if our description of the belonging-together of the various selves, as a belonging-together which is merely *represented,* in a later pulse of thought, had knocked the bottom out of the matter. *Ibid.* II. xxviii. 671 *Any* really inward belonging-together of the sequent terms, if discovered, would be accepted as what the word cause was meant to stand for. **1938** *Mind* XLVII. 380 From the outset our perceptual world is a continuum organised into 'belonging-togethernesses'. **1939** *Ibid.* XLVIII. 247 This *belonging together* is the basis of 'Gestalt' psychology.

belo·ngingness. [f. BELONGING *vbl. sb.* + -NESS.] †**1.** The state of having the properties appropriate to something. *Obs.*

1656 BLOUNT *Glossogr., Radicality,* the belongingness of a thing to a root. *Ibid., Seminality*..a belongingness to seed.

2. [Cf. G. *zugehörigkeit.*] The state or condition of belonging.

1931 E. L. THORNDIKE *Human Learning* ii. 23 With only a fourth as many repetitions the greater belongingness results in..nearly twice as many correct responses. **1938** E. LEEN *Why the Cross?* ix. 355 By the holocaust man signified to God his utter 'belongingness' to Him. **1951** D. RIESMAN in A. W. Loos *Relig. Faith & World Culture* 69 We must skeptically question the demands for greater social participation and belongingness among the group-minded. **1957** W. H. WHYTE *Organization Man* 7 A belief in 'belongingness' as the ultimate need of the individual.

Belorussian (belorʊ·ʃăn), *a.* and *sb.* Also **Byelorussian** (byelo-). [f. Russ. *Byelorussiya* Belorussia, f. *byelo-* white + RUSSIA + -AN.] **A.** *adj.* Of or pertaining to Belorussia, one of the constituent republics of the Soviet Union, its people or its language. **B.** *sb.* = *White Russian* (see WHITE *a.* 11 e).

[**1911** *Encycl. Brit.* XXIII. 884/2 The White Russians (the Byelorusses).] **1944** G. VERNADSKY *Hist. Russia* i. 3 The Polish influence to which the Ukrainians and Byelo-Russians had been subjected for several centuries. *Ibid.* xvi. 319 Other Socialist Soviet republics were founded—the White Russian (Byelo-Russian) [etc.]. **1948** J. TOWSTER *Polit. Power in U.S.S.R.* iv. 90 The Ukrainian and Belorussian Republics. **1949** *Amer. Slavic & East Europ. Rev.* VIII. 205 The Belorussians 'borrowed' many important characteristics..from the Slavs. **1950** *Ibid.* IX. 140 The absence of special contributions to..Byelorussian literature is an unfortunate omission. **1956** *Archivum Linguisticum* VIII. II. 174 The same problem exists in..Belorussian. **1958** *Economist* 1 Nov. 424/2 General Uborevich, commander of the Byelorussian district.

belote (bělǫ·t). Also **belotte.** [a. F. *belote.* f. the name of F. *Belot,* a Frenchman who perfected the game.] A game of cards resembling pinocle, played with a 32-card pack.

1941 KOESTLER *Scum of Earth* 7 Soldiers—grumbling, drinking red wine, playing belotte, and bored. **1944** W. S. MAUGHAM *Razor's Edge* iii. 93 There were men with sweaty faces round tables playing belote with loud shouts. **1959** *Sunday Times* 1 Mar. 4/2 Belote, or 'Klabrias', as it is named in Germany..is certainly the best and wittiest card game...In France it has become the most popular national card game, as in Switzerland.

below, *adv.* Add: **1.** Phr. *Below there!* a warning addressed to persons to beware of a descending object.

1896 *Bow Bells* 6 Mar. 264/2 One man was tying a paint-pot to the rung of the ladder, the cord slipped, down came the pot, the man singing out 'Below there!'

3. (Earlier and later U.S. examples.)

1645 in *Springfield Rec.* (1898) I. 164 The Brooke in the longe meddow..and the Brooke a little below on the other side. **1817** in *Essex Inst. Hist. Coll.* VIII. 238 Mr. Lefavour of Salem, who was living below, had come up here on business. **1845** SIMMS *Wigwam & Cabin* Ser. I. 16 There's ne'er a house either above or below for a matter of fifteen miles.

5. = below zero (see B. 6 a).

1795 E. P. SIMCOE *Diary* 12 Jan. (1911) 265 Thermometer 10 degrees below. **1896** W. D. HOWELLS *Impressions* 9 The frosts of ten and twenty below. **1904** M. E. WALLER *Wood-Carver* ii. 82 [It was] twenty-seven below this morning. **1968** *Globe & Mail* (Toronto) 17 Feb. 1/6 The predicted high [temperature] for Elliot Lake today is 5 below; the low tonight 20 below.

B. *prep.* **8.** *Comb.* **below-ground** (*a*) *adj.,* that is below the ground; (*b*) *sb.,* a place or places below the ground; **below-the-belt** *a.,* unfair, underhand (see BELT *sb.*¹ 1 d).

1941 L. A. G. STRONG *Bay* viii. 183 The soft pressure, the below-the-belt, if-you-love-me stuff that women employ so readily. **1960** *Farmer & Stockbreeder* (Suppl.) 16 Feb. 35/1 A below-ground, outside concrete tank. **1960** E. BOWEN *Time in Rome* iv. 118 From Rome's below-ground I must not omit the Catacombs.

‖ **bel paese** (bel pa,e·ze). [Italian proprietary term, f. It. *bel* beautiful + *paese* country, region.] A rich, white, creamy cheese of mild flavour originally made in Italy.

1935 O. BURDETT *Bk. Cheese* iv. 54 Bel Paese, another round, flat, spongy cheese, is also a shade less mild than Port Salut. **1948** W. S. MAUGHAM *Here & There* ii. 25 The maid brought us *bel paese* cheese and a plate of figs. **1951** 'M. INNES' *Operation Pax* III. iii. 91, I have been able to buy some *salami*—and a *bel-paese* too.

Belsen (be·lsən). The name of a concentration camp in Germany in the war of 1939–45, applied hyperbolically to any very unpleasant place. Also *attrib.* and *Comb.*

1948 PARTRIDGE *Dict. Forces' Slang* 14 *Belsen.* Most army camps sooner or later acquired, with or without justification, the name of this most notorious of Nazi concentration camps. **1953** L. A. G. STRONG *Hill of Howth* ix. 194 It was a Belsen urge, a blast from the pit. **1956** 'M. INNES' *Appleby plays Chicken* I. i. 16 My public school was a regular old Belsen. **1959** *News Chron.* 12 Dec. 7/1 Allegations about 'hoodlum warders' with 'Belsen-type mentalities' at Exeter prison.

Belshazzar, see *BALTHAZAR.

belt, *sb.*¹ Add: **1. b.** (Examples in *Boxing* and *Wrestling.*) For examples in *Judo* see *BLACK BELT 3.

1812 'ONE OF THE FANCY' *Boxiana* 422 We understand that an emblematical Belt has been some time preparing for the Champion, but not yet presented to him. **1822** *Sporting Mag.* X. 106/2 Cribb was decorated with the *belt,* in the front of which are a couple of silver fists, and on each side are two large circles of silver plate, with inscriptions engraved on them. **1850** [in Dict.]. **1872** R. D. BLACKMORE *Maid of Sker* II. xxvii. 10 He had held the belt seven years.. for wrestling, as well as for bruising. **1889** E. B. MICHELL *Boxing & Sparring* (Badm. Libr.) i. 125 The belt—the emblem of modern championship in the P.R.

d. *to hit* (or *strike*) *below the belt* (literal and *fig.* examples).

1889 E. B. MICHELL *Boxing & Sparring* (Badm. Libr.) i. 125 The rule against hitting below the belt. **1890** FARMER *Slang* I. 175/1 To strike a man *below the belt*..is akin with 'To stab a man in the back'. **1891** *Chambers's Encycl.* VIII. 485/2 No man might be struck below the belt—the belt in practice being a handkerchief tied tightly round the waist. **1891** [see HITTING *vbl. sb.*]. **1903** *Punch* 1 July 453 Call this *Fair* Trade, hitting me below the Belt?

e. *to tighten* (etc.) one's *belt*: to stave off hunger, to bear the pangs of hunger philosophically. Also *fig.*

[**1841** LEVER *C. O'Malley* xx, 'Perhaps not' lisped Melville, tightening his belt; 'but it's devilish convivial.'] **1887** KIPLING *Life's Handicap* (1891) 290, I also was once starved, and tightened my belt on the sharp belly-pinch. **1907** MULFORD *Bar-20* v. 46 They's three things that's good for famine... Yu can pull in your belt, yu can drink, an' yu can eat. **1927** *Observer* 24 Apr. 15/3 A travelling troupe who quoted Corneille while tightening their belts.

f. Colloq. phr. *under one's belt,* in one's stomach. Also *fig.*

1839 *Spirit of Times* 21 Dec. 498/3 Away we went, each bearing, under his belt, his full share of the antifogmatical ..compound. **1938** CRAIGIE & HULBERT *Dict. Amer. Eng.* I. 193/2 *Belt,* v...To put under one's belt; to swallow. **1954** *Manch. Guardian Weekly* 12 Aug. 12 His wife had 135,000 miles driving in the States under her belt..but was still failed. **1954** WODEHOUSE in *Encounter* Oct. 19/1 Just as you have got Hamlet and Macbeth under your belt. **1962** J. WAIN *Strike Father Dead* v. 216 He wanted me to get plenty of Latin and Greek under the belt so that I could be like him.

g. One used to support the figure; a suspender-belt; a corset.

1880 *Drapers' Jrnl.* 24 June p. iii (Advt.), Important to ladies. The new model figure belts. *Ibid.,* Our special ladies supporting belts are strongly recommended..for giving great support both before and after accouchement. **1932** R. LEHMANN *Invit. Waltz* I. ix. 98 Etty wears..just her belt and knickers. **1952** C. W. CUNNINGTON *Eng. Women's Clothes* vii. 238 Slim-fitting knickers to replace panties and belt. **1961** *Housewife* Apr. 114/2 Dainty bra.. matching belt with adjustable suspenders.

3. b. In a machine gun, a length of woven fabric or of metal plates pinned together, fitted with cartridges and revolving on the feed-block.

1902 *Encycl. Brit.* XXX. 403/1 Figs. 7 and 8 show the feed-block and method of packing the cartridge belts. The greatest number usually carried in a belt is 250. *Ibid.* 406/1 A belt of cartridges..has been placed on the feed-wheel. **1914** *Scotsman* 26 Sept. 5/6 The belt of the gun [*sc.* a Maxim] was still charged.

c. An assembly-belt or conveyor. Also *transf.* orig. *U.S.*

1908, 1909 [see **belt conveyor*]. **1936** B. & S. SPEWACK *Boy meets Girl* I. 13 We are dealing here with a factory that manufactures entertainment in approved sizes; that puts the seven arts right on the belt. **1937** U. SINCLAIR *Flivver King* (1938) lxxiv. 194 I'd rather take my chance on the belt. **1938** *Reader's Digest* Jan. 123/2 Automobiles leaving the belt as finished products.

5. a. Also, a zone or district, usu. with defining term denoting the principal product or characteristic. Cf. **Bible belt,* *BLACK BELT 1, *GREEN BELT. orig. *U.S.*

1869 *Overland Monthly* III. 12 Between the short-staple [cotton] belt and the rice and long-staple belt of the coast. **1871** R. SOMERS *Southern States* xxxvii. 263 The 'Cotton Belt' of the Southern States. **1875,** etc. [see *BLACK BELT 1]. **1891** *Harper's Mag.* Aug. 446/2 A fierce storm swept over the whole gold belt. **1903** A. B. HART *Actual Govt.* 116 Illinois is divided into a wheat belt, a corn belt, and the city of Chicago. **1960** *Spectator* 29 July 173 The Copper Belt is necessary to Southern Rhodesia. **1960** *Observer* 25 Dec. 7/6 A stately great drag headed for the stockbrokers' belt. **1968** D. E. ALLEN *Brit. Tastes* ii. 43 Affluent couples..on the coast who leapfrog the dormitory belt altogether. **1968** C. FORSYTE *Murder with Minarets* iii. 11 One could live abroad and still be only on the tattered fringes of the servant belt.

6. belt conveyor (see quot. 1909) = **conveyor belt*; **belt-driven** *a.* (*Mech.*), driven by means of a flexible endless belt; hence **belt drive,** driving mechanism of this type; **belt-driving** *vbl. sb.*; **belt-knife,** (*a*) a knife carried in a belt for use as a weapon, hunting-knife, etc.; (*b*) *U.S.,* a revolving knife on the band-saw principle, used in splitting hides or skins; **belt line** (*U.S.*), a railway, tram-line, or road that makes a complete circuit of a city; also *attrib.*; so *belt tram.*

1908 *Engineering Mag.* Dec. 440 The original belt conveyor consisted of a very wide belt running on straight idlers and carrying a small amount of material distributed along the middle. **1909** *Cent. Dict.* Suppl. s.v. *conveyer,* A *belt-conveyer* employs a broad endless belt of leather, rubber, or similar fabric, running over stationary rollers placed at equal distances apart. When in operation a stream of light material, such as grain, can be fed to the belt and be transported at high speed in large quantities. **1930** *Engineering* 13 June 783/1 (*heading*) Portable Belt Conveyor. **1906** *Motor Cycles & how to manage Them* (ed. 10) 89 Types of Belt Drive. **1907** *Daily Chron.* 17 Oct. 8/2 A motor cycle.. with belt drive. **1934** *Discovery* Nov. 324/2 Two-stroke petrol engine, connected by belt-drive to a 200 watt alternator. **1893** *Jrnl. Soc. Arts* 5 May 622/1 Seventeen dynamos, all belt-driven. **1906** *Westm. Gaz.* 26 Mar. 10/2 The old belt-driven Benz [motor car] of a prehistoric past. **1908** *Ibid.* 30 Jan. 4/1 A belt-driven fan. **1892** NASMITH *Students' Cotton Spinning* Index 428 Belt Driving. **1902** H. STURMEY in A. C. Harmsworth *Motors* x. 186 Belt driving is quite silent in running. *Ibid.* 195 Belt-driving cars have usually two belts running on pulleys of different sizes. **1840** C. F. HOFFMAN *Greyslaer* I. 234 The hand of the Mohawk clutched the belt-knife..half drawn from its sheath. **1895** *Daily News* 21 Mar. 5/7 That they had not..so much as a belt knife. **1909** H. G. BENNETT *Manuf. Leather* 279 There are three types of splitting machine, the 'union', the vibrating knife, and the band-knife (or belt-knife) machine. **1894** *Cent. Mag.* Dec. 290 De Belt Line stables ain't no Hoffman House. **1900** *Engineering Mag.* XIX. 698 The belt-line railway, running largely in cuttings and underground. **1903** *N.Y. Times* 24 Oct. 2 George B. McClellan and Edward M. Grout were scheduled for a belt line tour of speechmaking. **1922** *Daily Ardmoreite* (Ardmore, Okla.) 6 Jan. 6/5 A belt line of gravel highway is now under construction around the city of Ardmore. **1966** *Economist* 26 Mar. 1234/1 Circumferential 'belt lines' around cities could be built more quickly than radial routes leading into their older centres. **1894** J. DALE *Round the World* 333 The belt tram took us round the city, 8 miles.

belt, *sb.*⁴ *colloq.* [f. BELT *v.* (sense 4).] A heavy blow or stroke.

1899 SOMERVILLE & 'ROSS' *Exper. Irish R.M.* 217 Will I give him [*sc.* a horse] a couple o' belts, your Honour? **1911** MASEFIELD *Everl. Mercy* 33 I'd like to hit the world a belt. **1927** *Glasgow Herald* 26 Aug. 11 [He] took three mighty belts at the ball. **1934** in *Sc. Nat. Dict.,* He gave the man a belt on the jaw. **1953** L. A. G. STRONG *Hill of Howth* 68 He'd give Moo a belt in the puss.

belt, *v.* Add: **2.** *spec.* to girdle (a tree) by stripping off the bark. *U.S.*

1812 H. MARSHALL *Hist. Kentucky* 14 These improvements..consisted principally in..belting the larger trees. **1853** P. P. KENNEDY *Blackwater Chron.* xiv. 216 One man,.. in a hundred days, would belt or deaden one hundred acres.

4. b. Various slang uses: (*a*) To hit; to attack; (*b*) *to belt* (*the bottle*), to drink heavily; (*c*) to sing, play, or speak with great vigour (const. *out*).

1838 J. C. NEAL *Charcoal Sketches* 46 He intends to belt me, does he? Take a stick—. **1884** 'MARK TWAIN' *Huck. Finn* 21 They [*sc.* genies] don't think nothing of..belting a Sunday-School superintendent over the head. **1909** WARE *Passing Eng.* 25/2 There comes that old maid; belt her. **1912** GALSWORTHY *Pigeon* II. 55 Megan'll get his mates to belt him. **1957** I. CROSS *God Boy* (1958) xviii. 152 It [*sc.* a dog] cut right across the road..and was belted by a wool truck.

1931 RUNYON *Guys & Dolls* (1932) ii. 43 Jack takes to belting the old grape right freely to get his zing back. **1960** *Observer* 18 Sept. 19/2 He is given to belting the bottle.

1953 *Sat. Rev. Lit.* 12 Dec. 55/1 Standing there..belting out the sophisticated sweetness of Porter's 'Get Out of Town'. **1959** J. STEINBECK *Once there was War* xix, One of the finest jazz combos I ever heard was belting out pure ecstasy.

6. *trans.* To connect with a machine-driven belt. Also *absol.*

1902 J. S. THOMPSON *Mech. Linotype* (1908) xxiv. 169 The size of the pulley on the motor to which it is belted will decrease the speed. *Ibid.* 172 To drive the machine by belting directly to the intermediate shaft.

7. *intr.* To hurry, to rush. *slang* (orig. *dial.* and *U.S.*).

1890 J. D. ROBERTSON *Gloss. Dial. Gloucester, Belt v.* to racket or bustle about. **1894** *Outing* (U.S.) XXIV. 57/2, I belted along as fast as the waders and treacherous footing would allow. **1942** *We speak from Air* vii. 24 A picture of two Me. 109's belting down on your tail from out of the sun. **1949** D. M. Davin *Roads from Home* III. iii. 227 Looked like the one that raced us on the way up this morning...He's belting it out by the look of him. **1958** K. Amis *I like it Here* 179 Getting up as she often did to..switch off the immersion heater, belting downstairs to let that sod of a dog in. **1962** *New Statesman* 18 May 710/1 Cor, we used to belt along that road.

8. Slang phr. *to belt up*: to be quiet, 'shut up'. Usu. as *imp.*

1949 Partridge *Dict. Slang* (ed. 3) Add., *Belt up*: Shut up!: R.A.F.: since ca. 1937. **1958** *News Chron.* 22 May, Belt up is just another way of saying be quiet. **1959** M. Pugh *Chancer* v. 57 Why don't you belt up?..Go and boil your can. **1969** *Listener* 30 Jan. 147/3 May we hope that Hamilton will do a service to art by belting up and going back to school?

be·ltful. [f. Belt *sb.*[1] + -ful.] As many cartridges, etc., as are contained in a belt.

1916 'Boyd Cable' *Action Front* 131 The first [anti-aircraft] gun opened with a trial beltful.

belting, *vbl. sb.* Add: **1. b.** Beating, thrashing.

1854 A. E. Baker *Gloss. Northampt. Words* s.v. *Belt*, 'He got a good belting.' **1896** A. Morrison *Child of Jago* 33 The belting was bad..very bad. **1907** *Westm. Gaz.* 16 Aug. 5/2 He is a very bad boy...After a 'belting' he seems worse.

2. (Later examples.)

1884 A. Daniell *Princ. Phys.* 162 Belting.—There is a very interesting and familiar case in which friction serves as a means for the transmission of energy—that is, transmission by machine-belting. **1898** *Westm. Gaz.* 14 Jan. 9/3 Men mounted the belting of the ship. **1910** *Ibid.* 11 Jan. 5/2 The snapping of some of the steel belting. **1964** *McCall's Sewing* xii. 225/1 Commercial belting gives the most professional looking belt and never loses its shape.

bematist (bī·mătist). [ad. Gr. βηματιστής, f. βηματίζειν to measure by paces, f. βῆμα pace, step.] An official road-measurer or surveyor in the time of Alexander the Great and the Ptolemies.

1875 *Encycl. Brit.* II. 748/2 The bematists or surveyors of Alexander and the Ptolemies. **1886** Sheldon tr. *Flaubert's Salammbô* x. 242 The bematists of Euergates, who measured the heaven by calculating the number of their paces.

Bembo (be·mbo). *Typogr.* The name given to a type face cut in 1929 by the Monotype Corporation after that used by Aldus Manutius (see *Aldine) in his edition of *De Ætna*, a tract by the Italian cardinal and scholar Pietro Bembo (1470–1547).

1930 *Fleuron* VII. 178 The consummate distribution of the shading over the finely proportioned skeleton gives the new Bembo a singular 'presence' on the page. **1935** Berry & Johnson *Catal. Specimens Printing Types* I, The extension of the ascenders makes the 12-point size [of Perpetua] look 'small' in comparison with the 12 point of Fournier, Baskerville, and the Aldine Bembo. **1945** O. Simon *Introd. Typogr.* iii. 12 The roman lower-case letters of Scotch and Baskerville, for instance, are wide and generous, whilst, at the other extreme, Fournier and Bembo occupy considerably less width.

bemusedly (bĭmyū·zėdli), *adv.* [f. Bemused *ppl. a.* + -ly[2].] In a bemused or bewildered manner or condition.

1896 A. Morrison *Child of Jago* 141 'Lor,' she said bemusedly. **1921** A. S. M. Hutchinson *If Winter Comes* III. ii. § 1 He spoke bemusedly. No need for caution that he could see.

bemusement (bĭmyū·zmėnt). [f. Bemuse *v.* + -ment.] Bemused condition.

1907 Galsworthy *Country House* I. viii, The devilry, mockery, admiration, bemusement, had gone out of his face. **1921** A. S. M. Hutchinson *If Winter Comes* II. iv. § 9 She threw away the bemusement in which she had sat.

benab (be·næb). Also **benaboo.** [Arawak (*u*)*bannabuhu*.] In Guyana, a shelter made of a framework of poles, covered with branches and leaves.

1867 W. T. Veness *El Dorado* 141 'Buck Houses', or models of the huts of the Indians of British Guiana, containing in miniature all the articles usually found in Indian 'benaboos'. **1875** —— *Mission Life Brit. Guiana* ii. 21 They [sc. the Indians] have no burying ground, the grave is always made in the benab occupied by him during his lifetime. **1899** Rodway *Guiana Wilds* 251 The benab was the property of the Mission, and had been put up for the accommodation of visitors. **1922** *Blackw. Mag.* July 10/1 There are Indian benabs, close by in the forest. **1959** P. Capon *Amongst those Missing* 159 Women sat outside the benabs, suckling babies and grinding cassava.

benatura (benatū·ra). [app. alteration of *bénitier on some wrong analogy.] A holy-water stoup, bénitier.

1873 J. Maclean *Hist. Trigg Minor* I. 508 On the eastern side of the south door is a fine Benatura,..well carved. **1891** *Athenæum* 28 Mar. 412/3 Dr. Fryer..reported the discovery of the site of an ancient Benatura in the south porch of St. Mary Redcliffe.

Bence-Jones (be·ns dʒōᵘ·nz). The name of Henry *Bence-Jones* (1813–1873), English physician, applied *attrib.* to a protein found esp. in the urine in the disease myelomatosis, and characterized by precipitation on moderate heating followed by re-solution at higher temperatures.

1902 Anders & Boston in *Trans. Coll. Physicians* (Philad.) XXIV. 175 Bence-Jones Albumosuria...In 1847 Henry Bence-Jones presented the first recorded instance before the Royal Society of London. *Ibid.* 192 The urine contained Bence-Jones albumose and albumin. **1961** *Lancet* 5 Aug. 291/2 Bence Jones protein which behaves classically in the heat test.

bench, *sb.* Add: **1. c.** *Baseball, Football*, etc. A seat provided for the members of a team who are waiting to bat, play, etc. *N. Amer.*

1912 C. Mathewson *Pitching in Pinch* 93 The *bench*! To many fans..this is a long, hooded structure from which the next batter emerges and where the players sit while their club is at bat. **1916** *Spalding's Baseball Guide* 316 All players and substitutes of the side at bat must be seated on their team's bench, except the batsman, base-runners and such as are legally assigned to coach base-runners. **1967** *Boston Globe* 5 Apr. 51/1 For the first play-off game..Auerbach sat next to the Boston bench. **1968** *Globe & Mail* (Toronto) 15 Jan. 19/2 The Boston team then became involved in a fight with some fans behind its bench. *Ibid.* 13 Feb. 28/4 Davis seldom went to his bench and his players appeared very tired in the final quarter.

7. (Examples of U.S. use.) Hence *bench-land*.

1811 *Deb. Congress* (1853) 2116 Towards the left flank this bench of high land widened considerably. **1857** W. Chandless *Visit Salt Lake* II. x. 326 Bench-land fifty or hundred feet above the water-level. **1873** J. H Beadle *Undevel. West* xxiv. 481 We turn south-west, rising by successive 'benches' to a vast barren table land. **1920** Mulford *J. Nelson* xii. 129 Right on them benches on th' east end of th' mountain.

b. *Geol.* A natural terrace marking the outcrop of a harder seam or stratum.

1884 *Science* 13 June 729/1 On this rest argillaceous, splendent, siliceous talc schists..; and on these, three benches of conglomerates, tuffs, and argillaceous schists and lime-stones. **1898** *Westm. Gaz.* 20 May 9/1 The same feature will doubtless exist as each bench is worked.

c. The ledge or floor upon which the retorts stand in a retort-house; also, a set of retorts; also, the complete furnace or oven containing a set of retorts.

1841 *Civil Eng. & Archit. Jrnl.* IV. 100/2 The works are laid out in eight distinct sections of ten 'benches', or thirty retorts each. **1920** *Conquest* May 320/3 In a large retort-house the settings are built in benches containing as many as 150 'through' retorts.

9. bench-end, the end of a seat in a church, freq. ornamented (see Poppy-head 2); **bench-hook,** (*b*) (see quot. 1888); **bench-man,** an operative who works at a bench, in various trades, as a joiner; also, a cabinet-maker who assists at bench-work, a labourer who works at a coke bench; **bench-root** *Agric.*, 'an abnormal root development due to the presence of tough seed coats; the roots are often twisted together or badly formed' (Webster *Addenda* 1918); **bench-show** (U.S. examples); **bench test,** a test of a motor engine carried out in a workshop before fitting it to a motor body; **bench-warrant** (later examples).

1898 A. Jessopp in *19th Cent.* Jan. 56 We get a payment entered for the carving of the bench-ends in a little church, 500 years ago. **1960** *Times* 12 Aug. 12/7 This church..is full of interest, with exceptionally attractive poppyhead benchends. **1888** J. G. Horner *Dict. Mech. Engin.* 28 *Bench hook*, a stop for sawing light work on the bench without damaging the bench itself. It is a block of wood about 12 in. long, furnished with a projecting stop at each end placed on opposite faces of a central web. One face of the web being laid on the bench, the lower stop is pressed against the bench edge, while the upper one takes the thrust of the wood which is being cut by the saw. **1901** *Daily Chron.* 7 May 10/6 Boot trade.—Good benchman wanted for repairs. **1961** *Oxford Mail* 5 Oct. 2/2 Benchman required..footwear. **1874** *Forest & Stream* 29 Oct. 182/1 The first regular bench show of dogs. **1887** *Harper's Mag.* May 934/1 The American bench shows furnish an opportunity to most readers to see the best mastiffs in the country. **1909** *Westm. Gaz.* 6 Apr. 4/2 Their..38 h.p. motors which recently underwent a 132 hours' continuous running bench test at Coventry. **1878** J. H. Beadle *Western Wilds* xxxii. 514 Another called for the immediate arrest of Brigham on a bench warrant before he could fly the country. **1883** *Wharton's Law-Lexicon* (ed. 7) 92/1 *Bench warrant*, an attachment issued by order of a criminal court against an individual..a warrant signed by a judge, or two justices of the peace, to apprehend a prisoner charged with an offence. **1959** Jowitt *Dict. Eng. Law* I. 225 *Bench warrant*, a warrant issued by a court of record, *sedente curia*, for the arrest of a person against whom an indictment for treason, felony or misdemeanour has been found..; if such person does not appear to answer the indictment or articles.

bench, *v.* Add: **3. c.** *trans.* To put (a dog) on a show-bench for exhibition; to exhibit at a dog-show.

1891 *Times* 28 Oct. 11/5 Almost every breed of spaniel is benched. **1898** *Standard* 1 Dec. 2/6 Possibly the soundest coloured Chow ever benched. **1924** *Westm. Gaz.* 22 Oct., Among those benched at the Toy Dog Show..will be black-and-tan miniature dogs.

5. *trans. N. Amer.* In Baseball, Football, etc., to remove a player from a game or prevent him from taking part in it. Cf. *Bench *sb.* 1 c.

1917 Mathewson *Sec. Base Sloan* xvi. 224 Some of you stuffed sausages will be benched mighty quick if you don't wake up. **1947** *Harper's Mag.* June 560/2, I should have benched him long ago, he got only nineteen hits all season. **1967** *Boston Herald* 8 May 16/5 Second baseman Woody Woodward, benched because he was in a hitting slump, returned to Atlanta's starting lineup Sunday.

bench-legged, *a. U.S.* [Bench *sb.* 1.] Having the fore-legs wide apart.

1866 C. H. Smith *Bill Arp* 159 Dodds says,..he'd have his soul transmigrated to a bench-leg'd fice [sc. dog]. **1889** *Harper's Mag.* Aug. 485/2 Selling his little bench-legged mule to a travelling showman. **1903** A. Adams *Log Cowboy* xii. 79 The Indians' little bench-legged ponies were no match for them. *Ibid.* xv. 100 A long bench-legged black dog with a Dutch name.

bench-mark. Add: (Earlier example.)

1842 Francis *Dict. Arts*, *Bench marks*, in surveying, fixed points left on a line of survey for reference at a future time, consisting of cuts in trees, pegs driven into the ground, etc.

b. *transf.* and *fig.* A point of reference; a criterion, touchstone.

1884 *Science* IV. 202/1 These star-places..are the reference-points and bench-marks of the universe. **1957** R. K. Merton *Student-Physician* III. 195 Standards represent 'benchmarks' with which students compare their ability and performance. **1963** *Economist* 18 May 663/2 Foreign firms have failed to get..orders unless they have offered a price advantage of at least 50 per cent. This is the 'benchmark'.

bench-table. *Hist.* [f. Bench *sb.* + Table *sb.* 8 c.] An official body of benchers of the Inner Temple: see quot. 1896.

1673 *Cal. Inner Temple Rec.* (1901) III. 92 Ordered at the bench table that Kenricke Eyton..and Ralph Sumner..do by the beginning of Michaelmas term next write a court hand. **1692** W. Sherlock (*title*) A Sermon Preached at the Temple-Church, May 29. 1692. And Printed at the Desire of the Bench-Table of the Honourable Society of the Inner-Temple. **1896** *Cal. Inner Temple Rec.* I. Introd. p. xxxiii, The officers of the house..met together frequently..at what was and is known as the Bench Table, when orders were made for the governance of the Inn.

bend, *sb.*[4] Add: **2. b.** *the bends*: the acute attacks of pain in muscles and joints suffered on over-rapid reduction of the surrounding air pressure, chiefly by workers in compressed air who are decompressed too quickly, with consequent liberation of dissolved nitrogen from the body tissues. Also, more loosely, the whole disease (also called *caisson-disease*) produced by decompression.

1894 *Westm. Gaz.* 16 Oct. 3/2 The pressure..is quite enough to give the men a dose of the 'bend' [*sic*] as it is called. **1902** *Idler* July 485 That..terrible air-pressure disease known as the 'bends'. **1913** Pembrey & Ritchie *Gen. Path.* 494 These pains [in Caisson disease] pass off in a few hours, and are known to the workmen as 'bends', apparently because of the flexed positions which they induce. **1962** *Listener* 29 Mar. 562/1 Nitrogen narcosis must not be confused with decompression sickness, commonly known as the bends.

5. b. The curve of a gun-stock, shaped to fit the arm of the person for whose use it is made.

1859 'Stonehenge' *Shot-gun & Sporting Rifle* IV. i. 229 In addition to the adaptation in length and bend of the stock, it is also..bent sideways. **1892** W. W. Greener *Breech-Loader* 73 The distance from A to heel, and from B to comb. This is the bend.

c. A curved drain-pipe.

a **1884** Knight *Dict. Mech.* Suppl., *Bend*, a flexed pipe, changing the direction. **1908** *Animal Managem.* 53 Any change of direction being made by curved pipes or 'bends'.

d. In a carding machine, the semicircular frame which carries the brackets in which the rollers are borne; also, in a carding machine of the revolving flat type, the curved surface which sustains the chain of flats.

1882 *Spon's Encycl. Industr. Arts* V. 2073 As the periphery passes round to the cylinder, the teeth are then in the act of ascending (the bend being thus in the opposite direction), and presenting facilities for being stripped of the wool they have acquired. **1890** Nasmith *Mod. Cotton Spinn. Mach.* 64 The phrase 'bend' should only be applied to that portion of the mechanism upon which the flats actually travel. **1892** —— *Students' Cotton Spinning* 101 The whole of the rollers are borne in brackets fixed to a semicircular frame bolted on the lower frame P, and known as the 'bend', the brackets having open bearings formed at their heads.

10. Phrases. *on the bend*: by means that are not straightforward, 'crookedly'; *to go on the* (or *a*) *bend*: to go 'on the spree'; also *to have a bend*; cf. *Bender 5 b. *slang.*

1863 Jeaffreson *Live it Down* xxviii, I'll order my executor to buy my coffin off the square. He shall get it on the bend, somehow or other. **1879** Kidston in *Proc. Gen. Assembly Free Church Scotl.* 62 'Going on the spree' or 'having a bend'. **1887** F. Francis Jr. *Saddle & Mocassin* 84 They do say as he was 'customed to go on a scoop—on a bend, occasionally, as it were. **1891** Kipling *City Dreadf. Nt.* 71 The gallant apprentice may be a wild youth with an earnest desire to go occasionally 'upon the bend'. **1891** —— *Life's Handicap* 60, I went on the bend with a intimate friend. **1936** L. A. G. Strong *Last Enemy* I. x. 152 Been on the bend, 'aven't you?

b. *above one's bend*: beyond one's powers. *U.S.*

1835 Crockett *Tour down East* 44, I shall not attempt to describe the curiosities here [sc. at Peale's Museum]; it is above my bend. **1848** J. F. Cooper *Oak Open.* (De Vere), It would be above my bend to attempt telling you all we saw among the Redskins. **1872** Schele de Vere *Americanisms* 577 *Above one's bend* means, above one's power of bending all his strength to a certain purpose.

c. *round the bend*: crazy, insane. *colloq.*

1929 F. C. Bowen *Sea Slang* 114 *Round the bend*, an old naval term for anybody who is mad. **1951** 'N. Shute' *Round the Bend* xi. 361 People are saying that I've been out in the East too long, and I've gone round the bend. **1955** J. I. M. Stewart *Guardians* vii. 78 Right round the bend..I mean..as mad as a hatter.

bend, *v.* Add: **9. a.** *to bend over*: also, to put oneself into position to receive a beating; also as *v. trans.*; *to bend over backwards*: see *backwards *adv.* A.

1889 in Barrère & Leland *Dict. Slang* I. 107/2. **1946** B. Marshall *George Brown's Schooldays* ii. 6 They make you bend over again and the second time they often draw blood. *Ibid.* vii. 36 'Bend him over,' the Bruiser order[ed]. He took a great run and smote the tight little bottom mightily. **1948** C. Day Lewis *Otterb. Incident* iii. 24 He doesn't offer to bend over when one of *us* is going to be beaten. **1960** Betjeman *Summoned by Bells* v. 49 Bravely I answered, 'Please, sir, it was me.' 'All right. Bend over.'

d. *to catch* (a person) *bending*: to catch (someone) at a disadvantage. *colloq.*

1910 Wodehouse *Psmith in City* xviii. 163 If any tactless person were to publish those..speeches..our revered chief would be more or less caught bending..as regards his chances of getting in as Unionist candidate at Kenningford. **1938** —— *Code of Woosters* iv. 97 You'll get the poor bird unfrocked...It's something they do to parsons when they catch them bending. **1967** A. Wilson *No Laughing Matter* II. 139 He then goes off singing, 'My word, if I catch you bending, my word, if I catch you bending.'

15. Revived in mod. *slang*: *spec.* to use for 'crooked' or wrongful purposes; to steal; to 'throw' (a contest, etc.).

1864 O. W. Norton *Army Lett.* (1903) 242 Perhaps you think it bending the Sabbath to build while I should be at church. **1930** *Amer. Mercury* XXI. 454/2 We bend a boat to hist the hooch. **1958** *Observer* 30 Nov. 13/8 There are honest landladies in districts like Victoria who let a flat to someone they think is an ordinary girl, who then proceeds to 'bend' it: uses it for prostitution. **1960** *Sunday Express* 16 Oct. 1/7 Watford players shared £110 given to them by the Brighton players to 'bend' a home and away game with them.

Ben Day (ben dēⁱ). *U.S.* Applied *attrib.* to a process in photo-engraving developed by Benjamin Day (1838–1916), a New York printer. (See quot. 1949.)

1912 F. Weitenkampf *Amer. Graphic Art* 217 The 'Ben Day' process of quick mechanical production of tints by 'rapid shading mediums' has also been a time-saver. **1949** *Manual of Style* (Chicago Univ. Press) 245 *Ben Day Process*, an engraving process for producing a variety of shaded tints by the use of gelatin films, particularly in connection with line (zinc) etchings.

bender. Add: **4. b.** A leg or knee. *slang* (orig. *U.S.*).

1849 Longf. *Kavanagh* xii, Young ladies are not allowed to cross their benders in school. **1925** A. S. M. Hutchinson *One Increasing Purpose* III. xi, They say family prayers there with the servants every night, all down on their benders.

5. b. A bout of drinking; a riotous party. *slang* (orig. *U.S.*).

1846 D. Corcoran *Pickings from Portfolio* 62, I was on an almighty big bender last night..and the way we *did* walk into the highly concentrated hard cider. **1887** J. Hatton *Old Ho. Sandwich* I. II. iv. 82 The boss of Drummond's Gulch may be said to have begun his 'bender', as a bout of drunken dissipation was called in these regions. **1929** K. S. Prichard *Coonardoo* 7 And I've warned Paddy Hanson to look after Hughie if Sam does get on a bender. **1933** *Bulletin* (Sydney) 14 June 11/4 Being on a strenuous bender, he had forgotten to sign a cheque. **1951** Wodehouse *Old Reliable* iv. 64 Where's the harm in an occasional bender? Boys will be boys.

7. A big or good specimen of its kind; a 'whopper', 'corker' (*s.w. dial.*).

1842 Daniel *Bride of Scio* 190 (E.D.D.), Ma vice [= fist] es wat I kal a bendur. **1891** Chope *Dial. Hartland* s.v., 'A proper bender, an' no mistake!' **1895** Kipling *Day's Work* (1898) 180 By Jove, it's a bender of a night.

bending, *vbl. sb.*[1] Add: **2. b.** *spec.* The curvature of a beam. So *bending moment*, the moment tending to produce curvature in a beam; *bending stress*, the stress that causes curvature in a bar, beam, etc.

1858 Bending moment [see Moment *sb.* 8 b]. **1876** *Encycl. Brit.* IV. 290/1 The moment of this couple must be equal to the moment of the couple tending to bend the beam at this section, or to what is called the bending moment. **1887** *Ibid.* XXII. 604/1 The strain produced by bending stress in a bar or beam. **1888** *Lockwood's Dict. Terms Mech. Engin.*, *Bending*, or *Flexure*, the curvature of a beam about its axis or central plane. **1961** C. C. T. Baker *Dict. Math.* 29 The bending moment at any section of a beam is the couple which one part of a beam exerts on the other part at that section.

7. The action or process of shaping wood, iron, or other material by pressure instead of by cutting or casting. Chiefly *attrib.* and *Comb.*: *bending machine, rolls*; **bending cradle** (see quot. 1874); **bending form,** a kind of bench anvil used in bending steam, gas, and water pipes; **bending slab,** an iron floor upon which ships' frames are bent.

1874 Thearle *Naval Archit.* 132 The 'bending cradle' is composed of a pair of stout iron vertical frames, between the bars of which transverse beams of iron, bent to the necessary curvature, are secured. *a* **1884** Knight *Dict. Mech.* Suppl. 97/2 A French bending machine, for cart and wagon tires. **1888** *Lockwood's Dict. Terms Mech. Engin.*, *Bending Rolls*, heavy rollers of cast iron or steel set in strong standards, and used either for the straightening of crooked plates or for bending them into arcs of circles or into complete cylinders. *Ibid.*, *Straightening machine*,..a machine in which channel, angle, and bar iron are straightened or bent, in boiler and smiths' shops, by squeezing...Since it will bend as well as straighten it is also called a bending machine. **1890** W. J. Gordon *Foundry* 60 The 'bending-slab'—a pavement of square masses of iron,..large enough to take any frame required in the ship to be built.

8. *Horsemanship.* (See quot. 1891.)

1891 J. M. Brown *Polo* 323 The 'bending' course is a capital institution, of which I believe the Earl of Harrington was the originator about eight years ago. Two lines of sticks should be set up 20 yards apart and parallel to each other. The sticks should be about 7 feet high and 8 yards apart...Then begin by cantering your pony up one side and down the other zigzagging between the posts. *Ibid.*, I may here remark that this 'bending' competition is the most invaluable practical test of a really good polo pony. **1900** *Daily News* 16 Mar. 3/4 The bending competition, in which the considerations of pace, precision, riding, and 'make' [of a pony] are joined in equal importance. **1922** *Times* 20 June 7/2 Both [polo ponies] being particularly handy at turning and bending.

bendy (be·ndi). Also **bendee, bendi, bindy, banda.** [Hind. (Hindi) *bhiṇḍī*.] The plant *Hibiscus* (or *Abelmoschus*) *esculentus*, also the fruit of this plant. Also *attrib.* = Okro.

1812 M. Graham *Jrnl. Resid. India* 24 The bendy, called in the West Indies *okree*, is a pretty plant, resembling a dwarf holyhock. **1813** Jas. Forbes *Oriental Mem.* I. 32 The banda (hibiscus esculentus, Linn.) is a nutritious oriental vegetable. **1866** Lindley & Moore *Treas. Bot.* 1135/1 *Tetrameles*...This tree is the Jungle-bendy of India, and the Weenong of Java. **1927** *Blackw. Mag.* Nov. 678/2 Flanking the western slopes..were the dense bendi scrubs.

bendy-tree. (See quot.)

1886 Yule & Burnell *Hobson-Jobson*, *Bendy-tree*, this, according to Sir G. Birdwood, is the *Thespesia populnea*.

bene. Quot. 1881 is merely an allusion to:

1807 Wordsw. *Force of Prayer* i, What is good for a bootless bene?

beneaped, *ppl. a.* Add: **b.** *fig.*

1913 T. Hardy *Places in Coll. Poems* (1930) 332 One there is..To whom to-day is beneaped and stale.

Benedictiness (benĭdi·ktinės). [f. Benedictine + -ess.] A nun of the Benedictine order.

1872 J. Morris *Condit. Cath. under Jas. I* (ed. 2) p. ccvii, Father Gerard..took one [image] to Ghent, which he gave to the English Benedictinesses there. **1909** *Dublin Rev.* Jan. 61 We have Benedictinesses at East Bergholt [etc.].

benedictional (benĭdi·kʃənăl), *a.* [f. Benediction + -al.] Of or pertaining to the pronouncing of a benediction.

1902 *Encycl. Brit.* XXVII. 238/2 Small benedictional crosses belong to each altar, and processional crosses are common [in the Coptic Church].

benedictionally (benĭdi·kʃənăli), *adv.* [f. prec. + -ly[2].] In or by way of benediction.

1911 W. De Morgan *Likely Story* 208 The old lady..kissed her benedictionally.

bene esse: see *esse 2 b.

benefact (be·nĭfækt), *v.* [Back-formation f. Benefactor.] *trans.* To help or endow as a benefactor.

[**1594** O. B. *Questions* 3 Whose benefacting..extended chieflie to their supposed children and Paramoures.] **1898** E. W. B. Nicholson in *Westm. Gaz.* 10 June 2/3 Mr. Gladstone..offered to try to get one of the richest men in the world to benefact us. **1923** *Chambers's Jrnl.* Aug. 507/2 He did not want to benefact Hamborne in any kind of way.

benefactive (benĭfæ·ktiv), *a.* [ad. L. *benefact-us* capable of giving + -ive.] Used of an affix or verbal aspect, esp. in various American Indian languages, to indicate that a benefit is conferred on someone. Also as *sb.*, a benefactive form or set of forms.

1943 W. L. Wonderley *Notes on Zoque Gram.* 92 The suffix *-hay-*..at times does not parallel indirect or benefactive uses in other languages. **1946** E. A. Nida *Morphology* viii. 160 (*caption*) Benefactive indirective. **1947** *Internat. Jrnl. Amer. Linguistics* XIII. 28/2 The benefactive suffix -y is suffixed to themes containing stems ending in a vowel or in w. **1964** E. A. Nida *Toward Sci. Transl.* ix. 200 Voice specifies the relationship between the participants and the event indicated in the verb; e.g...benefactive (*worked for him*). **1965** *Amer. Speech* XL. 116 In combining Zoque *-hay-*, a benefactive morpheme, with *-pa*, a present-tense morpheme, one should get only *-hapya*.

benefactrix. Delete †*Obs. rare*[−1] and add earlier and later examples.

1615 J. Day *Festivals* xi. 301 The Widowes that wept so much for the Death of Dorcas their Benefactrix. **1775** Challoner in E. H. Burton *Life* (1909) II. xxviii. 116 God reward the good Lady, our benefactrix. **1907** W. De Morgan *Alice-for-Short* xxxii, The expedient of merging her personality in that of an imagined benefactrix.

‖**beneficiaire** (benefisyę̄r). [a. F. *bénéficiaire*, f. *bénéfice* benefit.] A player who is taking a benefit (Benefit *sb.* 4 a in Dict. and Suppl.).

1840 *Fistiana* 49 Some dozens of noblemen and persons of high rank, whose liberal contributions..added greatly to the receipts of the *beneficiare*. **1843** *Bell's Life* 27 Aug. 3/6 The *beneficiaire* was, no doubt, rather unnerved by the applause. **1862** *Sporting Mag.* XL. 11 A *bénéficiaire* very often cannot prevent that. **1905** *Daily Chron.* 13 July 3/5 Fifty baskets of flowers..were brought on the stage and surrounded the beneficiaire. **1927** *Daily Tel.* 23 Aug. 5/3 Sandham's Benefit Match...The beneficiaire was not destined to do well.

benefit, *sb.* Add: **3. a.** *benefit of the doubt*: see *doubt *sb.*[1] 1 b.

4. a. Hence, any entertainment or display the receipts from which are given to a particular player or company; also, the proceeds from such an entertainment.

1802 *Sporting Mag.* XXI. 17/1 The grand display of pugilistic dexterity, advertised by Belcher, for his benefit, at Sadler's Wells. *Ibid.* 17/2 Spectators at Mr. Belcher's benefit. [**1846** *Spirit of Times* (N.Y.) 18 Apr. 91/2 A disposition prevails to give him a sort of 'benefit' as an offset to the late loss he sustained by fire.] **1854** *Lillywhite's Guide to Cricketers* 85 Those noblemen, gentlemen, and clubs..who were kind enough to honour him with a 'donation' towards his benefit. **1890** Conan Doyle *Sign of Four* v. 74 Don't you remember the amateur who fought three rounds with you..on the night of your benefit four years back?

d. That which a person is entitled to in the way of pecuniary assistance, medical or other attendance, pension, and the like, under the National Insurance Act of 1911 and similar subsequent Acts, or as a member of a benefit (or friendly) society; more explicitly *maternity, medical, sick*(*ness*) *benefit*.

1875 *Act* 38 & 39 *Vict.* c. 60 § 24 (8), Provided as follows as respects friendly societies: (*a*) No special resolution..is valid..without the written consent of every person for the time being receiving or entitled to any relief..or other benefit from the funds of the society. **1891** [see Sickness 6]. **1895** *Daily News* 7 June 2/3 Contributions..providing for confinement benefit. **1901** B. S. Rowntree *Poverty* 362 In addition to the sick benefit there is a funeral benefit of £10. **1911** *Act* 1 & 2 *Geo. V*, c. 55 § 8 Payment in the case of the confinement of the wife..of an insured person..of a sum of thirty shillings (in this Act called 'maternity benefit'). *Ibid.*, In the case of insured persons who have attained the age of seventy, the right to sickness benefit and disablement benefit shall cease. *Ibid.* § 10 His right to medical benefit, sanatorium benefit, and maternity benefit shall be suspended. **1927** Carr-Saunders & Jones *Soc. Struct. Eng. & Wales* 150 The applicant may at the discretion of the Ministry of Labour receive 'extended' benefit for the remainder of the benefit year. **1958** *Listener* 23 Oct. 634/1 Unemployment and sickness benefit were put on a new basis [in 1946].

5. *benefit-bill, match, -social*; **benefit-society** (earlier examples); **benefit-ticket,** (*b*) a ticket for a benefit (sense 4 a).

1755 Mrs. C. Charke *Life* 56 To inform all my Acquaintance, that I was the Person so set down in Mrs. Thurmond's Benefit-Bills. **1844** J. Cowell *Thirty Yrs. Among Players* (1845) I. ix. 23 For so he used to designate himself in his benefit-bills. **1871** *Baily's Mag.* June 169 The great compliment of a benefit match at Lord's has been granted to Willsher by the Committee of the M.C.C. **1929** *Collier's* 12 Jan. 10/2 He met her at a church benefit-social the very same day he arrived at Bellwood. **1801** H. More *Let.* in J. Aitken *Eng. Lett. XIX cent.* (1946) 52, I have instituted..friendly benefit societies for poor women, which have proved a great relief to the sick and lying-in. **1837** Dickens in *Sk. Boz.* 2nd Ser. 322 There is not a parlour, or club-room, or benefit society..without its red-faced man. **1746** Garrick *Corresp.* (1831) I. 41 A terrible 'Row' ensued, between the few who paid ready money, and those who brought in his benefit-tickets. **1859** Sala *Tw. round Clock* (1861) 257 Solicitations for engagements, cards, bills, and applications for benefit tickets.

Benelux (be·nilɒks). [f. *Be*lgium, *Ne*therlands, *Lux*embourg.] The customs union of Belgium, the Netherlands, and Luxembourg formed in October 1947. Also *attrib.*

1947 *Foreign Affairs* July 692 The Secretariat prepared a common tariff for the 'Benelux Union'. **1947** *Spectator* 10 Oct. 454/1 Success would make 'Benelux'..the third trading power in the world. **1960** *New Statesman* 26 Mar. 434/3 This means that British exports will be discriminated against, especially in the German and Benelux countries.

Beneventan (benĭve·ntăn), *a.* Also (formerly) **Beneventine.** [It. *beneventano*, f. med.L. *Beneventanus*, f. *Beneventum* Benevento, province of Italy + -an.] Of or pertaining to a medieval script principally of southern Italy.

1882 *Catal. Add. MSS. Brit. Mus. 1876–81* 70 [Exultet Roll]..written in Italy in Lombardic or Beneventine characters of the twelfth century. **1912** E. M. Thompson *Introd. Gr. & Lat. Palaeogr.* xvi. 348 Although the title of Lombardic is applied as a general term to the writing of Italy in the early middle ages, that title might be more properly restricted to its particular developement in the south, to which the titles [*sic*] of Beneventan is also given. **1914** E. A. Loew *Beneventan Script* ii. 22 The peculiar script which grew up and flourished within the ancient duchy of Benevento..we shall consistently call..Beneventan. **1959** *Chambers's Encycl.* X. 358/1 It is known by its mediaeval name of Beneventan script (*littera Beneventana*; the term 'Lombardic' by which it was known until quite recently should be avoided).

benevolence. Add: **2. b.** *love of benevolence*: see Love *sb.* 2.

Bengal. Add: **2. Bengal fire, flash** = *Bengal light*; **Bengal isinglass** = *agar-agar; **Bengal light** (earlier examples); also *fig.*; (see also quot. 1899).

1941 J. Cary *House Childr.* xliv. 191 If you have a sea battle, Harry, we must get you a Bengal fire—it's the finest thing in the world for a ship blowing up after a battle. **1946** Koestler *Thieves in Night* 68 In the reddish bengal flashes which accompanied the detonations the silhouette of the barbed wire emerged. **1863** Bengal isinglass [see *agar-agar]. **1791** *Aris's Gazette* (Birmingham) 5 Sept. 3/5 A Bengola Light. **1818** in *Pall Mall Gaz.* (1885) 5 Nov. 4/2 Superior Fireworks...A Bengal light. ***1852** Geo. Eliot *Lett.* (1954) II. 54 Froude is good—writes very judiciously and pleasantly, except that at the end he brings on Bengal lights and goes off in a Carlylian flourish. **1899** *Connorton's Tobacco Brand Directory U.S.* 550 Bengal Lights (cigarettes and cheroots).

Bengalese (beŋgǭlī·z), *a.* and *sb.* [f. Bengal + -ese.] = Bengali *a.* and *sb.*

1778 Halhed *Gram. Bengal Lang.* p. xxi, The native Bengalese. **1872** Calverley *Fly Leaves* (1903) 91 A patient of Skey's, Who is prone to catch chills, like all old Bengalese.

benignant, *a.* Add: **2. b.** Of a disease: not malignant or recurrent; = Benign *a.* 5 b.

1897 [see Siphoned *a.* 2]. **1932** *Discovery* Dec. 376/2 Similar rays..are given out..by cancerous growths, but not by so-called 'benignant' growths.

Benin (bĕnī·n). Also **Beni·, Bini** (bī·ni·). (A member of) a Negro people of southern Nigeria, noted for their production of fine bronzes and carved ivories. Also *attrib.* or as *adj.* So **Benine·se** *sb. pl.*

1875 *Encycl. Brit.* III. 573/1 The Beninese weave their cotton into a fine kind of muslin. **1893** *Geogr. Jrnl.* I. 127 The Benin people, the regular inhabitants of the vast Benin kingdom..are somewhat superior to the neighbouring coast tribes. **1897** R. H. S. Bacon *Benin City of Blood* i. 14 One of the Jijus of the Beni is never to cross water. **1902** *Encycl. Brit.* XXXI. 115/1 *Niger-Benue Groups*—Benin; Ibo; [etc.]. **1903** H. L. Roth *Great Benin* i. 3 The Jekri had a most profound fear of the Binis' knowledge and use of poisons. **1925** A. Huxley *Along Road* III. 191 A statuette that was sixth-century Greek, subtly mingled with Benin. **1963** *Times* 12 Feb. 12/4 The same purchaser gave £750 for a cast bronze head of a Benin chieftain, about A.D. 1800.

benish (bĭnī·ʃ). Also 8 **beniche,** 9 **beneesh.** [Turkish *biniş* (properly = riding-habit), f. *binmek* to mount a horse.] An outer garment of cloth with very full sleeves.

1797 *Encycl. Brit.* VI. 403/2 [The Mamlouks] have an outer covering called the *beniche*, which is the cloak or robe of ceremony...Thus when the beniche and other accoutrements are on, the whole body appears like a long sack. **1836** Lane *Mod. Egypt.* I. 34 A *beneesh*, or *benish*; which is a robe of cloth, with long sleeves. **1840** J. B. Fraser *Koordistan* II. 404 The furred *kuirks* and flowing *benishes* of former days.

‖**bénitier** (benitye). [Fr., f. *bénit*(*e* blessed + *-ier* -er[2] 2.] A vessel to contain holy water.

1853 C. Brontë *Villette* I. xiv. 255 In the presence of *bénitier*, candle, and crucifix. **1858** Simmonds *Dict. Trade*, *Benitier*, a holy-water pot or vessel, sometimes a large shell, used in Catholic countries. **1907** *Connoisseur* 1 Oct. 4/6 For benitiers—especially of the domestic type—Flanders will safely bear the palm. **1908** B. Harraden *Interplay* II. i, A photographic series of bénitiers and drinking horns. **1923** *Blackw. Mag.* Aug. 156/2 Marie took it to the church and dipped it into the bénitier.

benitoite (benī·toəit). *Min.* [f. San *Benito* County, California, where found: see -ite[1].] A sapphire-blue crystallized barium titanosilicate.

1907 G. D. Louderback in *Bull. Dept. Geol. Univ. Calif.* V. 149 It is a new mineral species, it has been called benitoite, as it occurs near the head waters of the San Benito River in San Benito County. **1912** *Brit. Museum Return* 196 Benitoite crystals. **1965** Phillips & Williams *Inorg. Chem.* I. xiv. 541 This ion occurs in benitoite $BaTiSi_3O_9$.

benj. (Examples.)

1839 E. W. Lane tr. *Arab. Nts.* I. ii. 107 She contriveth to defraud him by means of the cup of wine..putting benj into it. **1850** C. Kingsley *Alton Locke* I. xxi. 291 Mesmerism and magic-lanterns, benj and opium, winna explain all facts. **1934** F. Stark *Valley of Assassins* v. 284 An aromatic sage-like plant they call generically *Benj*.

benjamin[2]. (Earlier examples.)

1810 *Sporting Mag.* Dec. 127/1 One article was an *upper benjamin*, eight guineas. **1812** [see Jarvey *sb.* 1].

Be·njamin[3]. The name of the patriarch Jacob's youngest son. Hence *allusively*, the youngest (and, consequently, favourite) son of a family; also *transf.*; *Benjamin's mess* or *portion*: the largest share (with allusion to Gen. xliii. 34).

1840 G. A. Lundie *Jrnl.* May in M. Duncan *Missionary Life in Samoa* (1846) xvii. 109 Our share was ten live pigs, (a truly Benjamin's portion). **1852** H. W. Greville *Leaves fr. Diary* (1883) 417 Another person was chaffing Lady Colchester..upon the Government being designated 'Benjamin's Mess' [with allusion to Benjamin Disraeli]. **1913** A. R. Hope *Half and Half Tragedy* 219 The new uncle being my granny's Benjamin. **1925** W. Deeping *Sorrell & Son* xviii. § 1 Kit was his Benjamin of pupils.

Benjamite (be·ndʒămɔit). [see prec. and -ite[1].] A descendant of Benjamin. Also as *adj.*, used allusively (see prec.).

1611 Bible *Judges* iii. 15 Ehud son of Gera, a Benjamite. **1625** T. Godwin *Moses & Aaron* I. x. 46 S. Paul was a Beniamite. **1857** Dufferin *Lett. High Lat.* 61 The dinner was excellent, and..we were helped in Benjamite proportions. **1867** C. M. Yonge *Pupils St. John* ii. 12 That Cilician Benjamite, at home called by his Jewish name of Saul.

benne (be·ni). Also **bene, beni, benni, benny.** [ad. Mende (Sierra Leone) *bene*.] Sesame, the plant *Sesamum indicum*. Chiefly *attrib.*, as *benne-oil, -seed*.

1769 in *Early Proc. Amer. Philos. Soc.* (1884) 44 On cutting & gathering the Bene seed. **1775** Romans *Florida* 130 The negroes use it as food either raw, toasted, or boiled in their soups and are very fond of it; they call it *Benni*. **1853** Mayne *Expos. Lex.*, Benne oil. **1874** Flückiger & Hanbury *Pharmacogr.* 425 Oleum Sesami. Sesamé Oil, Gingeli,..Til or Teel Oil, Benné Oil. **1885** *Buck's Handbk. Med. Sci.* I. 487/1 The Benné plant is a native of Africa, and probably also of Asia. *Ibid.* 487/2 The leaves of Benné are very mild and mucilaginous. **1887** Moloney *Forestry W. Afr.* 50 Beni-seed (*Sesamum indicum*). **1941** *N. & Q.* CLXXXI. 204/2 Some ten years or so ago, Sierra Leone began to include benniseed among its exports. **1960** *Times* 29 Sept. (Nigeria No.) x/6 Benniseed and soya beans are grown.

Bennettitales (be:nĕtitēⁱ·līz). *Palæobotany.* [mod.L. (A. Engler 1897, in *Engler & Prantl's Natürl. Pflanzenfam.* Nachtrag to Parts II–IV. 5), f. next + *-ales* pl. of *-alis* Lat. suffix (see -al).] A class of gymnospermous fossil plants found chiefly in Mesozoic rocks. Hence **Be:nnettita·lean** *a.*, of or belonging to the Bennettitales.

1907 *Jrnl. Linnean Soc.* XXXVIII. 52 Engler..adopts the derivative Bennettitales. **1910** A. C. Seward *Fossil Plants* II. xxii. 396 The microsporophylls of the Mesozoic Bennettitales produced their spores in sporangial compartments. **1917** *Ibid.* III. xxxvi. 366 The term Bennettitales is used by Engler, Nathorst, and several other authors as a class-designation for a large number of Mesozoic Cycads agreeing in their..morphological characters with the Lower Cretaceous stems on which Carruthers founded the genus Bennettites. *Ibid.* 386 The memoir by Carruthers..contains the first account of the morphological features of Bennettitalean flowers based on petrified material.

Bennettites (be:nĕtəi·tīz). *Palæobotany.* [mod.L., f. surname of John Joseph *Bennett* (1801–1876), English botanist, + *-ites* (see -ite[1]).] A genus of gymnospermous fossil plants having seeds borne on long stalks.

1871 W. Carruthers in *Trans. Linnean Soc.* XXVI. 695 To this genus I have given the name *Bennettites*, after my distinguished colleague. **1911** *Encycl. Brit.* XX. 546/1 The best preserved specimens of the true *Bennettites* type..are from the Lower Greensand and Wealden of England, and from Upper Mesozoic strata in North America, Italy and France. *Ibid.* 547/1 It is clear that *Bennettites* differed in many essential respects from the few modern survivors of the Cycadophyta.

benniseed: see *benne.

benny[1] (be·ni). *U.S. slang.* [app. shortening of Benjamin[2]; cf. -y[6].] A sack coat; an overcoat.

1903 R. L. McCardell *Conversations of Chorus Girl* 29 He had on one of them dust-proof Bennys that delegates to Granger conventions wear. **1914** Jackson & Hellyer *Vocab. Criminal Slang* 17 Benny. General usage. A sack coat; derived from Benjamin, some say the biblical character, while others say the New York manufacturer of men's garments. **1931** 'Dean Stiff' *Milk & Honey Route* xiii. 145 The benny, or overcoat, he should have for at least four months in winter. **1945** L. Shelly *Hepcats Jive Talk Dict.* 7/2 *Benny*, an overcoat. **1955** D. W. Maurer in *Publ. Amer. Dial. Soc.* XXIV. 126 An overcoat of full size and weight is called a *benny*.

benny[2] (be·ni). Slang abbrev. of *benzedrine. orig. *U.S.*

1955 *Amer. Speech* XXX. 89. **1956** S. Longstreet *Real Jazz* xviii. 146 Of course, you can take bennies (Benzedrine)..but they make me too nervous. **1957** J. Kerouac *On Road* (1958) I. i. 6 You've got to stick to it with the energy of a benny addict. **1967** A. Diment *Dolly Dolly Spy* i. 11 The benny was starting to wear out and I was hot, thirsty and exhausted.

beno (bī·no). [Pseudo-phonetic representation of Sp. *vino* wine, with bilabial *v*.] In the Philippine Islands: = *aguardiente. Also *Comb.*

1903 *Med. Record* 4 Apr. 547 The vile whiskey and 'beno' selling dens.

Bensonian (bensōᵘ·niăn), *a.* and *sb.* [f. the name of Sir F. R. *Benson* (1858–1939), actor-manager: see -ian.] **A.** *adj.* Of or belonging to Sir F. R. Benson or his Shakespearian company. **B.** *sb.* A member of this company.

1901 M. Beerbohm *Around Theatres* in *Wks.* (1924) VIII. 237 Last year, when the Bensonians were figuring in the Lyceum Theatre. *Ibid.*, The removal of the Bensonian venue from the Lyceum to the Comedy is a welcome change. **1917** W. Owen *Let.* 30 July (1967) 479 She is an old Bensonian, like her husband, who played Shylock. *Ibid.* 15 Aug. 484 My first make up was done with the real Bensonian touch. **1927** *Daily Express* 2 Sept. 3/5 Sir Frank Benson, who has produced Shakespeare for more than forty years, and has trained famous actors and actresses by the dozen, all of whom still proudly call them selves 'Old Bensonians' will be joined by Gerald Lawrence in the management of a company known as 'The Bensonians'. **1928** *Daily Tel.* 19 July 21/1 The cast will include..Harcourt Williams, the Bensonian actor. **1958** *Times* 31 Oct. 3/6 By degrees the Bensonians became the nursery of the English stage.

bent, *sb.*[2] Add: **10.** *Building, Carpentry*, etc.: a section of a framework or framed building. orig. *U.S.*

1815 *Niles' Reg.* IX. 200/2 On each of them [sc. the floats] were raised two bents or frames. *Ibid.*, This made sixteen bents, on which the grand and enormous structure was raised. *a* **1877** Knight *Dict. Mech.*, *Bent*, one section of the frame of a building, which is put together on the ground..and then raised. **1898** *Engineering Mag.* XVI. 91 The cradle is composed of forty-three inverted bents, twelve feet apart. **1952** *Archit. Rev.* CXI. 179 At the top of the boom may be seen the steel cables, attached to the [timber] bent.

bent, *ppl. a.* Add: **1. d.** In the names of articles, work, etc., in which the materials are bent to shape, as **bent iron work,** the making of ornamental ironwork as a home occupation, by bending strips of iron to form the various parts of the design; also, the ornamental ironwork thus made; **bent-panel,** one that is bent to shape instead of framed; in quot. *attrib.*

1858 Simmonds *Dict. Trade*, *Bent-timber Manufacturer*, a shaper of timber by steam and pressure. **1902** P. N. Hasluck (*title*) Bent Iron Work. **1909** *Stratford-on-Avon Herald* 7 May 4/3 For sale, excellent Bent-panel dog cart.

5. *fig.* (cf. Crooked *a.* 3). In various slang uses: **a.** Dishonest, 'crooked', criminal. Also as *sb.* orig. *U.S.* **b.** Illegal; stolen. *orig. U.S.* **c.** Of things: out of order, spoiled. Of persons: eccentric, perverted; *spec.* homosexual (also as *sb.*). (In quot. 1958 'faithless'.)

a. 1914 Jackson & Hellyer *Vocab. Criminal Slang* 17 *Bent*, crooked; larcenous. Example: His kisser shows that he's bent. **1948** *Sunday Pictorial* 29 Aug. 6/5 A 'bent screw'..a crooked warder who is prepared to traffic with a prisoner. **1958** *Times* 14 Feb. 3/5 What made the witness think the two officers were offering a bribe? Mitchell replied, 'I had known for years that certain members of the Brighton police force were what we call bent.' *Ibid.*, There were plenty of ways in which bents could help. **1963** *Ibid.* 2 Feb. 9/6 Successful crime preventing does not make criminals give up; they simply change their methods, or as Mr. Brown said: 'They stay bent but alter their tactics.'

b. 1930 E. H. Lavine *Third Degree* (1931) iv. 39 For having sold a stolen or *bent* car to a complainant. **1955** P. Wildeblood *Against Law* 151 He had got a short sentence for receiving stolen goods, which he swore he had not known to be 'bent'.

c. 1930 Brophy & Partridge *Songs & Slang 1914–18* (ed. 2) 210 *Bent*, spoiled, ruined, e.g. 'a good man bent' or even 'good tea bent'. **1942** Berrey & Van den Bark *Amer. Thes. Slang* § 143/4 *Eccentric.* Balmy, bats, bent, [etc.]. *Ibid.* § 152/5 *Insane; crazy*...bent. **1956** I. Asimov 9 *Tomorrows* (1963) iii. 87 He's gone crazy...He was always a little bent. Now he's broken. **1957** Rawnsley & Wright *Night Fighter* v. 75 Whenever a set became unserviceable in the air the code word used to notify ground control was to say that the weapon was 'bent'. **1957** A. Wilson *Bit off Map* 29 'I shouldn't think you did know any Teddy boys, but if you did, I know what they'd call you—a f—bent, see.'..Mr. Fleet..reddened with fury; his reputation as a womaniser was known to everyone. **1958** F. Norman *Bang to Rights* III. 72 My bird's gone bent...She went case with some geezer now she's liveing [*sic*] with him. **1959** C. MacInnes *Absolute Beginners* 64 No one..cares..if you're boy, or girl, or bent, or versatile, or what you are. **1960** F. Raphael *Limits of Love* I. v. 70 'Great thing about gay people...' 'Gay?' Tessa said. 'Bent, queer, you know. Homosexual.'

Benthamism. (Earlier examples.)

1826 *Times* 8 Nov. 2/6 The Ghost of Miltiades came at night, And he stood by the bed of the Benthamite. **1829** J. Sterling *Let.* in *Ess. & Tales* (1848) I. p. xxix, It was the first help I had in getting out of the slough of Benthamism. **1829** Mill in *Autobiogr.* (1924) App. 301 Those who are ignorant enough to fancy that there is a Benthamite sect. *Ibid.*, What are vulgarly considered to be the Benthamite doctrines. **1832** —— *Let.* 22 Oct. in *Wks.* XII. (1963) 128 You know our Benthamic Utilitarians.

benthos (be·nþɒs). *Biol.* [a. Gr. βένθος depth of the sea.] Haeckel's name for the flora and fauna at or near the bottom of the sea. So *abyssal benthos*, plants and animals of the deep sea; *littoral benthos*, those of the sea near the coast. Hence **be·nthic, bentho·al, bentho·nic** *adjs.*

1891 G. W. Field tr. *Haeckel's Planktonic Studies* in *Rep. U.S. Fish. Comm.* XVII. 582 The abyssal benthos. **1895** *Nat. Sci.* July 29 The greatest part of the..discoveries..concerns the Benthos. **1897** T. J. Parker & Haswell *Zool.* II. 600 Others [sc. marine animals]..are either permanently fixed, like Zoophytes and Stalked Crinoids, or move by creeping over the sea-bottom, like Starfishes, Holothurians, Chætopods, etc.; such forms constitute the *Benthos*, or 'bottom-fauna'. **1902** *Encycl. Brit.* XXXIII. 933 Each of the three benthic groups is well characterized by a special fauna. **1905** *Q. Jrnl. Geol. Soc.* LXI. Proc. p. lxxiv, The benthoal organisms existing in tracts where the physical conditions..vary rapidly, are limited as to their horizontal range by the distribution of those conditions which determine their station. **1909** Webster, Benthonic. **1913** J. Murray *Ocean* viii. 160 Pelagic larvæ of benthonic animals are abundant near shore in shallow water, but become less numerous farther out to sea. **1921** *Discovery* Oct. 265/1 Marine organisms can be roughly divided into..the *plankton* or drifters, the *nekton* or swimmers, and the *benthos* or fixed organisms. **1923** W. A. Herdman *Founders Oceanogr.* 327 The demersion upon which hordes of benthonic animals can browse. **1942** H. U. Sverdrup *Oceans* viii. 276 The benthic division may be subdivided into..the *littoral* and the *deep-sea systems*...[This] is divided into an upper (*archibenthic*) and a lower (*abyssalbenthic*) zone. **1956** *Nature* 25 Feb. 375/1 The animals were only about 5 mm. long, and as lancelets are benthic forms it is most unlikely that the scattering layer consisted of this species. **1959** A. Hardy *Fish & Fisheries* v. 94 All these other forms of life, referred to collectively as the *benthos*..are vitally important to the fish either as their prey or as voracious competitors for limited supplies of food.

bentonite (be·ntənəit). *Min.* [f. *Benton* (see below) + -ite[1].] A clay found in the Fort Benton strata of the Cretaceous of Wyoming. Also, any of several clayey deposits containing montmorillonite which have various practical applications (see quots.).

1898 W. C. Knight in *Engineering & Mining Jrnl.* 22 Oct. 491/1 In a recent article..the writer described briefly a new variety of clay found in Wyoming and suggested the name Taylorite...It has since been learned that the name Taylorite is preoccupied; consequently the clay will hereafter be known as Bentonite. **1939** *Jrnl. R. Aeronaut. Soc.* XLIII. 647 Highly diluted water dispersions of natural bentonite have proved very suitable for such work [sc. detection of turbulence of a liquid by means of polarization]. **1943** *Electronic Engin.* XV. 322 X-ray analysis has..made possible the gaining of valuable knowledge of the crystals forming the extremely plastic and interesting substance 'bentonite'. **1956** *New Biol.* XXI. 20 Another method of attacking this difficult problem [sc. haze formation in beer] is by the adsorption of the undesirable material on to an inert material such as bentonite. **1959** *New Scientist* 2 July 21/2 The manufacture of British bentonites used in foundry moulding sands, paints, oil-well drilling muds and other things. **1964** R. F. Ficchi *Electr. Interference* viii. 159 The anodes were placed in a 8-in. diameter well..then back-filled with bentonite.

‖**ben trovato** (ben trovā·to), *adj. phr.* [It., = well found.] Of a story, etc.: appropriate; happily invented if untrue.

Se non è vero, è molto ben trovato 'if it is not true, it is a happy invention' was app. a common saying in the 16th cent. It is found, for example, in Giordano Bruno (1585).

[**1771** Smollett *Humph. Cl.* II. 1 Your fable of the monkey and the pig, is what the Italians call *ben trovata*.] **1883** W. Pater *Let.* 22 July (1970) 78 The title of your proposed volume is I think ben trovato. **1884** *N. & Q.* 29 Mar. 244/2 It must be admitted that all this has a *ben trovato* character about it. **1886** Havelock Ellis in Landor *Imag. Conv.* Introd. p. xv, A story told of him by the Italians round Fiesole, where he lived for many years, is at least *ben trovato*. **1952** R. M. Hare *Lang. Morals* 53 We might say that to tell a story about someone, which every one knows is *ben trovato*, is not *lying*. **1966** *Punch* 16 Nov. 726/1 Who..are *we* and *they* can be illustrated by a story, true or *ben trovato*, told of Rose Macaulay.

benturong, var. *binturong.

bent-wood. Add: Also **bentwood. 1.** (Earlier and later examples.)

1862 *Illustr. Times* 15 Nov. 473/1 Messrs. Thouet Brothers 'Bent-Wood' Furniture at the Great Exhibition. The manufactory of these bent-wood chairs and couches is in the Carpathian Mountains, between Hungary and Moravia, in the district of Koritchan. **1933** *Archit. Rev.* LXXIV. 77 A typical example of bent-wood chairs for children. **1959** *House & Garden* Sept. 71/2 Chair in foreground, bentwood upholstered in black leather.

benzaldehyde (benzæ·ldĭhəid). *Chem.* Also formerly **benz-aldehyd.** [ad. G. *benzaldehyd*, f. Benzo- + Aldehyde.] A colourless liquid aldehyde, C_6H_5CHO, having the odour of bitter almonds and used in the manufacture of perfumes, dyes, etc.

1866 W. Odling *Animal Chem.* 117 Some tolerably simple well-characterised substance—such..as benzaldehyd, or essential oil of bitter almonds. **1882** [see Benzo-]. **1920** *Lancet* 18 Sept. 615/2 He also found that the anæsthetic action which he had shown to be marked for benzyl alcohol is exhibited in a high degree by benzaldehyde. **1962** *Ibid.* 1 Dec. 1168/1 Some substances such as benzaldehyde and camphor excite trigeminal afferents as well as olfactory afferents in the nasal passages.

benzedrine (be·nzĕdrīn). [Proprietary term, f. Benzo- + *-edrine* as in *ephedrine.] A preparation of amphetamine. Also *ellipt.*, a dose of benzedrine in tablet form.

1933 *Lancet* 16 Dec. 1383/1 A new drug has recently been introduced into rhinology under the name of benzedrine. It is a synthetically prepared compound, the carbonate of benzyl-methyl-carbinamine, which is described as a racemic mixture of bases having the formula $C_6H_5CH_2.CH.NH_2.CH_3$. **1935** *Trade Marks Jrnl.* 22 May 634/1 *Benzedrine*, a medicated preparation consisting of benzyl-methyl-carbinamine, oil of lavender and menthol..Smith, Kline & French..Laboratories..Pennsylvania. **1938** *Times* 22 June 17/4 The victims [of hay-fever]..fumble for the benzedrine. **1938, 1939** [see *amphetamine]. **1956** J. Hearne *Stranger at Gate* v. 38 He went in and bought a benzedrine inhaler: the sort one uses for colds. **1958** *Listener* 3 July 15/2 They..swallow a benzedrine.

benzene, benzine. Add: **3. benzene hexachloride,** a compound, the gamma isomer of which is used as an insecticide, = *Gammexane; abbrev. *B.H.C.*

1884 *Jrnl. Chem. Soc.* XLVI. 887 Determination of the vapour-density of the new isomeride of benzene hexachloride..shows that its formula is $C_6H_6Cl_6$. **1945** *Times* 9 Mar. 2/5 Gammexane..is the gamma isomer of benzene hexachloride. **1952** *Oxf. Jun. Encycl.* VI. 248 BHC (*benzene hexa-chloride*) can penetrate the insect's skin.

benzenoid (be·nzĭnoid), *a.* [f. Benzene + -oid.] Derived from, related to, or pertaining to benzene.

1887 *Standard* 16 Sept. 3/3 Transition from tars of the paraffinoid to those of the benzenoid or ordinary gas tar varieties. **1900** *Rep. Brit. Assoc. Advancem. Sci.* 166 Like hexamethylene, tetrahydrobenzene shows no selective absorption. The examination of these two substances thus confirms the conclusion previously reached, that the banded spectrum is shown only by substances which possess the true benzenoid structure. **1964** N. G. Clark *Mod. Org. Chem.* ii. 20 The first group is generally known by its original name 'aromatic', although the alternative 'benzenoid' is more accurate, for it comprises derivatives of the hydrocarbon benzene. **1967** *Times Rev. Industry* July 66/3 The very big benzenoid chemical market of America.

benzine (be·nzīn). *Chem.* Also **-in. a.** The name originally given to Benzene. **b.** An inflammable liquid (petroleum ether) prepared by purifying, deodorizing, and distilling natural petroleum, and used as a solvent; it is a mixture of hydrocarbons of the paraffin series.

1853 Mayne *Expos. Lex.*, Benzin. **1865** *Med. & Surg. Reporter* (Philad.) XIII. 188/1 In the vicinity of refineries, where this gas exists..Benzine gas is known to be very inflammable, highly volatile, and to permeate almost all substances. **1885** *Buck's Handbk. Med. Sci.* I. 487/2 Benzin is not used in medicine, but is useful to the pharmacist for its solvent powers over fats, resins, volatile oils, and other bodies. **1887** J. W. Richards *Aluminium* 236 Mourey..recommends the employment of benzine in the melting of all the noble metals. **1895** *Bloxam's Chem.* (ed. 8) 515 Benzine (sp. gr. 0·74), a solvent which must not be confounded with benzene, the coal-tar product. **1908** *Chemist & Druggist* 25 July 144/2 Benzine..in U.S. means always petroleum benzin, while benzole is the name for the coal-tar product C_6H_6. **1912** G. Martin *Ind. Chem., Org.* 8 Coal-Tar Naphtha, Benzene or Benzol, C_6H_6, is quite a different product from benzine or petroleum naphtha, being obtained by distilling coal-tar. **1928** *Observer* 26 Aug. 6/4 Another fuel..called Steiger-Brennstoff..contains no benzin or benzol at all.

benzo-. Add: **benzidine** (examples); **benzocaine** (be·nzokēⁱn) [*-caine*, after Cocaine], a white crystalline powder, ethyl para-aminobenzoate, used mainly as a local anæsthetic; **benzpyrene** (be:nzpəiᵊ·rīn) [Pyrene[2]], a carcinogenic hydrocarbon found in coal-tar, etc.; **benzyl** (later examples).

1878 *Jrnl. Chem. Soc.* XXXIV. 668 This glyoxal derivative of benzidine differs from the preceding compounds. **1916** Cross & Bevan *Paper-Making* (ed. 4) ix. 261 Benzidine is para-diamino-diphenyl, composed of two benzene rings joined together, each containing the amino radical NH_2. **1946** *Nature* 30 Nov. 791/2 Such soil extracts gave a strong blue coloration with benzidine. **1966** G. P. Ellis *Mod. Textbk. Org. Chem.* xiv. 366 Benzidine (4, 4'-diaminodiphenyl)..is obtained by an interesting isomerization (the benzidine rearrangement) of hydrazobenzene..when it is heated with an acid. **1967** Karch & Buber *Offset Processes* vii. 268 Benzidine yellows also are very strong, but not as fast to light as the Hansa yellows, but are more transparent. **1922** *Chem. Abstr.* 1487 Benzocaine..was formerly made in Germany and sold under the name 'anesthesine'. The Council on Pharmacy and Chemistry of the Am. Med. Assoc. adopted the newer name. **1959** *Sunday Graphic* 25 Jan. 4 (Advt.), De Witt's Antibiotic Throat Lozenges also employ benzocaine, the swift pain-killer. **1927** A. W. Stewart *Rec. Adv. Org. Chem.* (ed. 5) I. vi. 123 The polyketide derivatives yield members of the pyrone, benzopyrone, benzpyrene, pyridine, and isoquinolone series. **1951** M. J. D. White in G. H. Bourne *Cytology* (ed. 2) v. 221, 1, 2, 5, 6-dibenzanthracene and benzpyrene..have been shown to produce chromosome rearrangements in *Drosophila*. **1920** *Lancet* 4 Sept. 512/1 A man..complained of persistent hiccough...He was given a 20 per cent. solution of benzyl benzoate. **1932** *Times* 1 Nov. (Brit. Industries no.) p. xxiv/4 A recent development in the plastics industry involves the use of esters and derivatives of cellulose or wood fibre, such as benzyl cellulose. **1947** Arnstein & Cook in *Brit. Jrnl. Exper. Path.* XXVIII. 94 The present paper records..experiments leading to the final identification of the major constituent of the material as penicillin-II (penicillin-G or benzylpenicillin). **1951** Yarsley & Kitchen in H. M. Langton *Synth. Resins* (ed. 3) ii. 107 Benzyl cellulose..is thermoplastic and can be heated up to 180° C without fear of decomposition.

bepuzzled (bĭpɒ·z'ld), *ppl. a.* [f. Bepuzzle *v.* + -ed[1].] Utterly puzzled.

1826 *Examiner* 647/2 The bepuzzled scribblers. **1888** *Westm. Rev.* July 2 The bepuzzled claimant for a vote. **1909** J. London *Let.* 2 Aug. (1966) 282, I..gave eminent satisfaction to the be-puzzled inquirers.

ber (bēᵊɹ). Also **bher, bir.** [Hindi.] The Chinese date or jujube (genus *Zizyphus*). Also *attrib.*, as *ber-fruit, -tree*.

1860 in H. F. C. Cleghorn *Forests & Gardens S. India* (1861) 60 The bér tree (*Zizyphus jujuba*) is approved for saddletrees. **1861** —— *Ibid.* 244 The wild bèr tree, common almost everywhere. *Ibid.* 281 The Ber-fruit tree..is used for native sandals. **1874** Stewart & Brandis *Forest Flora India* 87 All Ber trees of North and Central India. **1886** Yule & Burnell *Hobson-Jobson*, Bear-tree, Bair, &c. **1887** Moloney *Forestry W. Afr.* 299 Jujube or Ber Tree. **1895** Mrs. Croker *Village Tales* (1896) 22 The sahibs shall sit above in the old bher tree. **1908** *New Reformer* I. 414 The *Zezyphus Jujuba*, the Bir universally known in India. **1924** *Blackw. Mag.* Oct. 478/1 Thickets of *ber* and acacias. **1925** *Ibid.* Jan. 66/2 These [bears] had fallen out to-night over their supper of *ber*-fruit. **1969** *Hindu* 28 July 6/5 The most striking thing about cuscuta is that it is notoriously partial to ber.

berate, *v.* (Later U.K. examples.)

1871 Meredith *H. Richmond* liii, What! You think he was not punished enough when he was berated and torn to shreds in your presence? **1893** *Times* 1 Feb. 9/5 The famous allocution in which he [sc. Sir James Mathew] berated Lord Clanricarde before a single witness had been heard. **1952** M. Laski *Village* xvii. 238 She perceived that Miss Evadne was not antagonistic, had not sent for her to berate her. **1965** *Times Lit. Suppl.* 23 Sept. 834/1 'She is..an assiduous toady, and a petty thief.' Further to berate her or the book would be supererogatory.

berberine[1]. Add: **berberine tree,** an African tree, *Xylopia polycarpa*, which yields a yellow dye containing berberine.

1861 Bentley *Man. Bot.* 440 The Berberine or Yellow-dye tree of Soudan.

Berberine[2] (bɜ̄ɹbərī·n). [prop. pl. used as sing., f. Berber + Arab. pl. suffix *-īn* (cf. *fellaheen*, pl. of Fellah).] A Berber. Also *attrib.* Also **Be·rberin** *pl.*, Berbers; **Berberee·, Berberi,** a Berber; **be·rberize** *v. trans.*, to impart a Berber character to; **be·rberized** *ppl. a.*

1852 B. St. John *Village Life in Egypt* II. i. 9 Berberi race, black and well-featured. *Ibid.* iii. 30 In the neighbourhood of Essouan..is a curious race of people..distinct from the Berberis, although confounded with them by many travellers. **1875** *Encycl. Brit.* I. 260/2 The Barâbra or Berberines are a people well known in Egypt. **1900** Conan Doyle *Green Flag* 270 In front rode the three Berberee body-servants upon donkeys. **1906** *Daily Chron.* 22 Sept. 2/7 The municipality of Alexandria are now endeavouring to induce the surplus Berberin, &c., to return to their own country, on the upper reaches of the Nile. **1914** *Eng. Hist. Rev.* Oct. 786 Many Arabs..had settled down [in Tunisia, Algeria, and Morocco] and become in part Berberized. **1928** *Blackw. Mag.* Mar. 406/1 This..was given to me..by a grateful Berberine. **1930** C. G. Seligman *Races of Afr.* vi. 137 It is always difficult to decide whether any particular people..are to be regarded as arabized Berbers or as berberized Arabs. **1938** *Times Lit. Suppl.* 8 Oct. 645/2 A lady's maid who managed Berberine servants with grim efficiency.

Berbice (bɜɹbī·s). The name of a river and a county in Guyana, applied to a type of long chair (see quots.).

1951 E. Mittelholzer *Shadows Move* II. ii. 172 An easy chair provided with long projecting arms for resting one's outstretched legs (a Berbice chair, he had heard it called). **1959** 'A. Glyn' *I can take it All* xiii. 241 Berbice chairs—the ones with long wooden extensions in the arms for you to hitch your legs over.

‖**berceau** (bę̄rso). [Fr., 'arbour, bower'; lit. 'cradle'.] An arbour, bower; a shaded or foliage-covered walk. Also *attrib.*

1699 M. Lister *Journ. to Paris* 209 The small leaved Horne-Beam; which serves for Arcades, Berceaus. **1771** Pennant *Tour Scotl.* 1769 77 The Berceau walk [at Taymouth] is very magnificent, composed of great trees, forming a fine gothic arch. *a* **1794** E. Gibbon *Memoirs* in *Misc. Wks.* (1796) I. 182, I took several turns in a *berceau*, or covered walk of acacias. *a* **1828** D. Wordsworth *Tour Cont.* in *Jrnls.* (1941) II. 61 The country richer than ever—Berceaus of vines—yards and courts roofed with vines. **1828** —— *Tour Isle of Man* (1941) 415 He had contrived to bury his house among trees and..to make the approach to it (a long berceau) as dark as a dungeon alley. **1853** C. Brontë *Villette* I. viii. 142 Under the vast and vine-draped berceau madame would take her seat on summer afternoons. **1960** M. Sharp *Something Light* viii. 72 No head-high *berceaux* of Gloire de Dijon roses.

‖**berceuse** (bę̄rsöz). *Mus.* [Fr., f. *bercer* to rock + fem. agent-suffix *-euse*.] A cradle-song, lullaby; an instrumental piece with a lulling rhythm.

1876 Stainer & Barrett *Dict. Mus. Terms* 58/1 *Berceuse*, a cradle song. **1879** Grove *Dict. Mus.* I. 229/2 His [sc. Schumann's] 'Schlummerlied' is a berceuse in all but name. **1889** G. B. Shaw in *Star* 13 Dec. 2/4 Composes fantasias, berceuses, serenades, etc., with great facility. **1931** E. Dannreuther *Oxf. Hist. Mus.* (ed. 2) VI. xi. 257 In the Barcarolle, the Berceuse, and the Ballades..Chopin discovered a form of expression peculiar to himself.

berdache (bɜɹdæ·ʃ). Also **berdash.** [ad. F. *bardache*: see Bardash.] Among N. American Indians: a transvestite (see esp. quot. 1955).

1806 A. Henry *Jrnl.* 21 July in E. Coues *New Light Hist. Greater Northwest* (1897) I. 348 The Mandanes..often prefer a young man to a woman. They have many berdashes amongst them, who make it their business to satisfy such beastly passions. **1843** H. S. Lloyd tr. *Maximilian's Trav. N. Amer.* xxv. 351 Among all the North American Indian nations there are men dressed and treated like women, called, by the Canadians, Bardaches. **1906** R. G. Thwaites *Early Western Trav.* XXIII. 284 The berdash was noted by

BÉRET

most early travellers among Western Indians. **1912** *Anthrop. Pap. Amer. Mus. Nat. Hist.* IX. 226 Berdaches naturally associate with girls and pretend to have sweethearts among men. **1949** M. Mead *Male & Female* vi. 129 Among many American Indian tribes the *berdache*, the man who dressed and lived as a woman, was a recognized social institution. **1955** Angelino & Shedd in *Amer. Anthropologist* LVIII. 125 In view of the data we propose that berdache be characterized as an individual of a definite physiological sex (male or female) who assumes the role and status of the opposite sex, and who is viewed by the community..as having assumed the role and status of the opposite sex.

béret. Add: Now usu. **beret** (be·rĕi). **b.** A cap resembling the Basque béret, worn by men and women, esp. for casual or holiday wear; also, such a cap forming part of many British service and other uniforms.

1827 *Lady's Mag.* Feb. 117/2 Berets of black velvet, decorated with gold lace,..are much in request at evening parties. *Ibid.* June 343/2 Beret-hats are more in request than the beret-turbans, at the opera. **1832** F. Kemble *Jrnl.* (1835) I. 68 Saw a woman riding to-day; but she has gotten a black velvet beret upon her head.—Only think of a beret on horse-back! **1894** G. du Maurier *Trilby* I. i. 18 He..wore a red *béret* and a large velveteen cloak. **1901** *Daily Chron.* 3 Aug. 10/3 The beret so specially becoming to a young face. *Ibid.* 7 Sept. 8/3 The beret shape is always modish on the moors. **1909** *Ibid.* 18 Jan. 7/5 A beret of sable. **1948** A. Baron *From City* 38 The corporals rose and pulled their berets from under their shoulder-straps. **1960** O. Lancaster in *Daily Express* 21 Jan. 1/3 Isn't it strange the way a beret always seems to *do* something to generals?

berg[2] (bɜːɹg). *S. Afr.* [Afrikaans, f. Du., = OE. *beorg*, etc., Barrow *sb.*[1]] A mountain.

1840 B. Shaw *Mem. S. Afr.* i. 27 To Cape Town school—o'er *bergs* and *knowes*, They sent the tawney-coloured boy. **1865** T. Leask *S. Afr. Diary* 12 June (1954) 2 The wind was blowing down the berg, almost cutting us thro'. **1902** De Wet *Three Years' War* 25 As there was no water to be obtained nearer than a mile from the berg, we suffered greatly from thirst. **1929** D. Reitz *Commando* xiii. 121 Having left..for Waterval-onder below the berg.

b. *attrib.*, as *berg-top*; **berg adder,** a South African adder, *Bitis atropos*, found chiefly on high ground and the hillsides; **berg cypress,** a mountain shrub, *Widdringtonia cupressoides*, found growing from Cape Town to Natal; **Berg Damara,** see *Damara; **berg wind,** a hot, arid wind coming from the mountains, prevalent in several coastal districts of the Cape Province and Natal at various times of the year.

1818 Latrobe *Jrnl.* (1905) v. 89 A wood-keeper..had lately lost his life by the bite of a Berg-adder. **1912** Fitzsimons *Snakes S. Afr.* 243 The Berg adder is as venomous as the Puff Adder. **1905** *Westm. Gaz.* 9 Oct. 10/1 Patches of berg cypress..afford splendid cover for that magnificent antelope the eland. **1953** *Cape Times* 4 July 3/1 Berg-top rescue ends in romance. **1905** *Nature* 2 Nov. 19/2 Remarkable winds, locally called 'Berg winds', blew from the plateau. **1959** *Cape Times* 27 June 9/3 Berg winds blow in different months in different areas. In winter they are most frequent along the west and south coasts extending to beyond Knysna.

Be·rgamasque, *a.* [Fr., f. It. *Bergamasco*: see Bergamask.] Of or pertaining to the province or city of Bergamo in northern Italy; *spec.* pertaining to or characteristic of a school of painting originating in Bergamo.

1879 *Encycl. Brit.* VIII. 214/1 The lower Alps feed large herds of cows, the upper are let to Bergamasque shepherds, who travel thither every summer with their flocks. **1938** *Burlington Mag.* LXXII. 33/2 Other paintings of the Bergamasque school. **1939** *Ibid.* LXXV. 141/2 His particular Bergamasque variant of Giorgionesque sentiment. **1959** J. Chapin tr. *Giovannetti's We have Pope* i. 8 The Bergamasque shrines of our Lady.

bergamot[1]. Add: **4.** Applied in Britain most commonly to *Monarda didyma*. Also, U.S. example of *Wild Bergamot*.

1843 J. Torrey *N.Y. Nat. Hist. Surv.: Flora* II. 59 *Monarda Fistulosa*..Horse Mint. Wild Bergamot. **1958** *Popular Gardening* 15 Feb. 34 All the charm of an old-world garden is fast locked in the heart of the Bergamot, botanically Monarda didyma.

Bergenia (bəɹgī·niă). *Bot.* [mod.L. (N. J. Necker 1790, *Elem. Bot.* II. 108), f. the name of K. A. von *Bergen* (1704–60), German physician and botanist, + -ia[1].] A genus of perennial herbs of the family Saxifragaceæ, having large, thick leaves and usually pink, red, or purple flowers; also (with lower-case initial), a plant of this genus.

1838 C. S. Rafinesque *Sylva Telluriana* 102 Bergenia was Necker's name for G. Cuphea. **1886** *Encycl. Brit.* XXI. 350/2 The Bergenias or Megaseas with their large fleshy leaves and copious panicles of rosy or pink flowers. **1962** *Amateur Gardening* 17 Mar. 4/2 Bergenias..are useful..the fat stems packed with purplish pink flowers often appearing in late February or early March.

‖ **bergère** (bɛ̃rʒɛ̃r). Also 8 **bergier, burgair, burjair,** etc. [Fr., lit. 'shepherdess'.] **1.** A large easy chair of a style fashionable in the eighteenth century (see quot. 1952); also, a kind of couch.

1762 Ince & Mayhew *Univ. Syst. Houshold Furn.* Pl. lx, Burjairs. *Ibid.* 8/1 Two Designs of Birjairs, or half Couches. **1773** in E. Singleton *Social N.Y.* (1902) 83 All sorts of..settees, couches, burgairs. **1784–87** in J. Gloag *Short Dict. Furnit.* (1952) 141 A large and handsome mahogany bergier, stuffed back in green morocco. **1803** T. Sheraton *Cabinet Dict.* 19 Arm-chair, No. 5, is a bergere, having a caned back and arms. Sometimes the seats are caned, having loose cushions. **1814** M. Edgeworth *Patronage* I. v. 145 Miss Hauton seated herself..upon a *bergère*. **1934** H. Nicolson *Curzon: Last Phase* 299 There were palms in the room upon high white stands, and bergère chairs, and little round tables with glass ashtrays. **1942** N. Marsh *Death & Dancing Footman* ii. 27 The armchairs and the bergére [*sic*] sofas. **1942** J. Cary *To be a Pilgrim* cxlvi. 324 The armchair, a tattered bergère in white and gilt. **1952** J. Gloag *Short Dict. Furnit.* 141 *Bergère*, an armchair with canework sides, back, and seat, with a loose cushion on the seat, or an upholstered seat.

2. *bergère hat*, a large straw hat.

1873 *Young Englishwoman* July 339/1 Bergère hat with gauze veil. **1905** *Daily Chron.* 13 Feb. 8/1 A pretty bergère hat made of fine straw.

‖ **bergerette** (bēᵊɹʒəre·t, *French* bɛ̃rʒərɛt). *Mus.* [Fr., see Bargeret.] = Bargeret.

[**1876** Stainer & Barrett *Dict. Mus. Terms* 50/2 *Barginet, berginet, bargaret*, or *bergeret*. Shepherd's songs, to accompany dances. Songs relating to pastoral matters.] **1924** E. Sitwell *Sleeping Beauty* xxi. 79 A florid bergerette. **1959** *Times* 15 Sept. 14/3 There were occasional glimpses of the old finesse: in the bogus *bergerette*, for example.

Berger rhythm (bɜ̄·ɹgəɹ). [f. name of Hans *Berger* (d. 1944), German neurologist.] = *alpha rhythm* (*alpha 3 f).

1934 Adrian & Matthews in *Brain* Dec. 355 (*heading*) The Berger rhythm: potential changes from the occipital lobes in man. *Ibid.* 356 Since the effect is so characteristic we shall refer to it in future as the Berger rhythm. **1936** [see *alpha 3 f]. **1938** *Jrnl. Neurol. & Psychiatry* I. 366 Berger's first papers described two electrical rhythms. The first, which he called 'alpha waves' and which was referred to five years later by Adrian and his collaborators as the 'Berger rhythm', consists of an almost sinusoidal discharge with a frequency of about 10 per second and with a potential varying irregularly from zero to about 100 microvolts. **1941** [see *alpha 3 f].

berghaan (bɛ·rχhān). *S. Afr.* [Afrikaans, f. *berg[2] + *haan* cock.] A South African eagle, esp. the bateleur eagle, *Terathopius ecaudatus*.

1867 E. L. Layard *Birds S. Afr.* 11 *Aquila Verreauxii*..*Dassie Vanger* and *Berghaan* of Colonists...It is called '*Dassie Vanger*' (coney-eater) and '*Berghaan*' (mountain-cock) by the colonists, from feeding principally on the coney, or rock-rabbit (*Hyrax capensis*). **1889** H. A. Bryden *Kloof & Karroo* 273 Suddenly..comes..a great black mountain eagle. We know him at once for a berghaan. **1893** Newton *Dict. Birds*, *Berghaan* (Mountain-cock), the name given to some of the larger Eagles, and especially to the beautiful *Helotarsus ecaudatus*..by the Dutch colonists in South Africa. **1910** J. Buchan *Prester John* viii, A brace of white *berghaan* circled far up in the blue.

bergschrund (bɛ·ɹkʃrunt). *Phys. Geogr.* [G., f. *berg* (see Barrow *sb.*[1]) + *schrund* cleft, crevice.] A crevasse or series of crevasses often found near the head of a mountain glacier.

1843 J. D. Forbes *Trav. through Alps* 298, I perceived an enormous *Berg-schrund*, or well defined crevass, which separated the higher summits from the glacier steep. **1860** J. Tyndall *Glaciers of Alps* i. xiv. 98 This slope was intersected by a so-called Bergschrund, the lower portion of the slope being torn away from its upper portion to form a crevasse. **1871** [see Fringe *sb.* 2 a]. **1957** J. Masters *Mountain Peak* 99 There was the usual bergschrund between the snow and the warmer rock of the gendarme.

Bergsonian (bəɹgsōᵘ·niăn), *a.* and *sb.* [f. the name *Bergson* (see below) + -ian.] **A.** *adj.* Of, pertaining to, or characteristic of the French philosopher, Henri Bergson (1859–1941). **B.** *sb.* A follower or adherent of Bergson. So **Bergsonism** (bɜ̄·ɹgsŏniz'm), the philosophical doctrine of Bergson.

1909 W. James *Plural. Univ.* v. 215, I must..give some preliminary account of the bergsonian philosophy. *Ibid.* vi. 266 They are now Bergsonians..and possess the principal thoughts of the master all at once. *Ibid.* vii. 277 Philosophy, you will say, doesn't lie flat on its belly in the middle of experience, in the very thick of its sand and gravel, as this Bergsonism does. **1920** H. Begbie *Wm. Booth* I. 146 Bergsonism has here a most admirable example of its thesis. **1944** G. B. Shaw *Everybody's Political What's What* viii. 62 When will the royalist lie down with the republican..the Bergsonian with the Darwinian? **1955** D. Davie *Artic. Energy* i. 6 Much modern criticism is Bergsonian, perhaps without knowing it.

bergy, *a.* Add: **b.** spec. *bergy bit*: a large piece of ice that has broken away from an iceberg (see also quot. 1958).

1935 *Geogr. Jrnl.* LXXXVI. 301 The channel carrying ice-floes and bergy bits. **1958** *New Scientist* 10 July 358/3 As it weathers away, a berg becomes known as a 'bergy bit' (when about the size of a small house).

beriberi. Add: Also (all *obs.*) **beriberii, beribery, berri berri.** (Earlier examples.)

1703 tr. *Nieuhoff's Voy.* in A. & J. Churchill *Voy.* (1704) II. 340/2 They [*sc.* the shrubs] have a peculiar Virtue..to cure the Indian Gout or Barrenness, called Beribery. **1769** tr. *Bontius's Acc. Diseases of East Indies* i. 1, The inhabitants of the East Indies are much afflicted with a troublesome disorder which they call the Beriberii (a word signifying a sheep). The disease has, probably, received this denomination on account that those who are seized with it..exhibit to the fancy a representation of the gait of that animal. **1832** H. S. Fleming in *Fort St. George Gaz.* 12 May 1 On the disease called 'Beriberi'.

berk (bɜːɹk). *slang.* Also **birk, burk(e.** [Abbrev. of *Berkeley* (or *Berkshire*) *Hunt*, rhyming slang for *cunt*.] A fool.

1936 J. Curtis *Gilt Kid* vi. 66 'The berk.' Jealousy and savage contempt blended in the Gilt Kid's tone. **1938** W. Greenwood *Only Mugs Work* vii. 49 'Stick the burke in a taxi,' he said. **1954** 'N. Blake' *Whisper in Gloom* ii. xiv. 197 'Don't be a little berk,' he said, as Foxy showed signs of recalcitrance. **1959** J. Osborne *Paul Slickey* i. iv, The Tories were burglars, berks and bloodlusters. **1960** H. Pinter *Dumb Waiter* in *Birthday Party & other Plays* 141 You mutt...You birk! **1963** *Sunday Express* 10 Mar. 22/5 All my mates thought I was a burk to try to break away: now they know they were the burks.

Berkefeld (bɜ̄·ɹkfeld, -elt). Also (erron.) **Berkefield.** The name of W. *Berkefeld* (1836–1897), German mine-owner, used to designate a bacterial filter containing diatomaceous earth.

1894 *Brit. Med. Jrnl.* 29 Dec. 1489/2 We must conclude that the Berkefeld filters may afford an efficient safeguard against the passage of disease germs. *Ibid.* 1488/2 The Berkefeld Filter Company Ltd. **1902** *Encycl. Brit.* XXXIII. 790/1 The Berkefield filter, constructed of baked infusorial earth. **1946** *Nature* 17 Aug. 217/2 Preparing from the eggs Berkefeld filtrates. **1951** R. J. Ludford in G. H. Bourne *Cytology* (ed. 2) ix. 402 A filterable tumour..yields by Berkefeld-filtration a cell-free filtrate.

berkelium (bəɹkī·liəm, bɜ̄·ɹkliəm). *Chem.* [mod.L., f. *Berkeley*, California, where the element was first made + -ium.] A metallic radioactive transuranic element not occurring in nature but made artificially; symbol Bk; atomic number 97.

1950 S. G. Thompson et al. in *Physical Rev.* LXXVII. 838/2 It is suggested that element 97 be given the name berkelium (symbol Bk), after the city of Berkeley. **1951** J. R. Partington *Gen. & Inorg. Chem.* (ed. 2) xxvi. 760 Berkelium..and californium..are formed by bombarding americium and curium, respectively, with high-energy helium ions.

Berkshire (bā·ɹkʃəɹ). Name of an English county, applied to a famous breed of pig.

1811 R. Henderson *Treat. Breeding Swine* i. 13 The Berkshire pig, is generally allowed to be a good kind. **1814** —— *Ibid.* (ed. 2) iii. 29, I would give the preference to the Cheshire, or rather the Berkshire hog. **1831** Loudon *Encycl. Agric.* (ed. 2) iii. 1069 The old Irish breed are a long-legged ..unprofitable sort of swine; but when they have been crossed with the Berkshire, they are considerably improved. **1842** D. Low *Dom. Anim. Brit. Isl.* I. 17 The true Berkshires are of the larger races of Swine, though they fall short in size of some of the older breeds. *Ibid.*, The Berkshire breed has..been crossed and recrossed with the Chinese. **1855** Morton *Cycl. Agric.* II. 941/2 The Berkshire breed of pigs has probably been the best known, and had in the highest estimation of any of our British breeds. **1953** A. Jobson *Household Crafts* vi. 65 Almost every county in England has produced its own breed of pigs, and we have.. Berkshire, Essex Black, [etc.].

berley (bɜ̄·ɹli). *Austral.* Also **burley.** [Of unknown origin.] Ground-bait.

1874 E. S. Hill in J. E. Tenison-Woods *Fish & Fisheries of N.S.W.* (1882) iii. 75 The bait should be crabs. It is usual to wrench legs and shell off the back, and cast them out for berley. **1896** *Badminton Mag.* Aug. 201 Sometimes adding bait chopped small to serve for what Australian fishermen call Berley.

b. (See quots.)

1941 Baker *Dict. Austral. Slang* 15 *Burley*, humbug, nonsense. **1943** *Ibid.* (ed. 3) 9 *Berley*, nonsense, humbug; e.g. 'a bit of berley'.

Berlin. Add: **1.** (Earlier examples.)

1694 Earl of Perth *Let.* 17 June (1845) 30 A woman with a maid following her came to the Berline side (this is a kind of traveling coach used here). **1717** Lady Montagu *Let.* 29 May (1763) 110 The meadows being full of all sorts of garden flowers, and sweet herbs, my berlin perfumed the air as it pressed them.

4. (Examples.) Also **Berlin pattern,** a pattern in Berlin work; **Berlin spirit** (see quot. 1878).

1795 R. Kirwan *Elem. Min.* (ed. 2) I. 491 The Berlin blue, so I call the Prussian blue of the shops, is not pure Prussiated iron, but a mixture of this with embryon alum. **1829** R. C. Sands *Writings* (1834) II. 163 Her girdle was fastened in front with a massive shining clasp of Berlin ware. **1841** Lady Wilton *Art of Needlework* xxv. 397 The style of modern embroidery, now so fashionable, from the Berlin patterns, dates from the commencement of the present century. *Ibid.* 398 The 'Berlin wools'...These yarns, however, are only dyed in Berlin, being manufactured at Gotha. **1845** G. Dodd *Brit. Manuf.* IV. iv. 110 The 'Berlin' patterns now so well known. *c* **1845** C. Brontë *Professor* (1857) I. xvii. 285 You can work with Berlin wools. **1853** E. M. Sewell *Exper. of Life* xiii. 131 The..footstool, worked in

the homely period between mediæval tapestry and modern Berlin patterns. **1854** *Encycl. Brit.* IV. 667/2 Its [*sc.* Berlin's] principal branches of industry..are porcelain, silks ..Berlin iron, &c. **1854** C. M. Yonge *Castle Builders* vi. 78 Their purse netting and Berlin work. **1862** —— *Countess Kate* iii. 52 She had a bunch of flowers in Berlin wool which she was supposed to be grounding. **1863** G. M. Hopkins *Note-Bks.* (1937) 8 She abominated the Berlin wool shop. **1878** *Chambers's Encycl.* II. 52/2 *Berlin spirit*, a coarse whisky made chiefly from beetroot, potatoes, &c.

Berliner (bəɹli·nəɹ). [a. G. *Berliner*, f. Berlin + -er[1].] A native or inhabitant of Berlin, Germany.

1859 L. Wraxall tr. *Robert-Houdin's Mem.* II. vi. 172 The reception I obtained from the Berliner will ever remain one of my pleasantest reminiscences. **1959** *Times* 18 Feb. 14/4 It would be a mistake to assume that only east Berliners trade in the west. **1963** V. Nabokov *Gift* i. 35 The perpetual fetters that chain a Berliner to the door lock.

Berliozian (bēᵊɹliōᵘ·ziăn), *a.* and *sb.* [f. the name of *Berlioz* (see below) + -ian.] **A.** *adj.* Of, pertaining to, resembling, or characteristic of Hector Berlioz (1803–1869), French composer, or his music. **B.** *sb.* An admirer of Berlioz; an interpreter of his work.

1910 *Westm. Gaz.* 9 Mar. 4/2 Robert Houdin—whose Berliozian portrait adorns this volume. **1936** *Scrutiny* Dec. 270 That extraordinarily subtle and flexible organism the Berliozian melody. **1951** J. Barzun *Berlioz* II. 303 The Berliozians could do little. **1961** *Times* 7 June 17/1 Mr. Davis, a perceptive Berliozian. **1962** *Times* 18 Apr. 7/5 This is scored, with Berliozian grandeur, for 90 percussion instruments.

berm. Add: **1. b.** *spec.* in *Geol.* (See quots.)

1931 F. Bascom in *Science* LXXIV. 172/1 The word *berm* ..should be used to distinguish those terraces which originate from the interruption of an erosion cycle with rejuvenation of a stream in the mature stage of its development. **1942** C. A. Cotton *Geomorphology* (ed. 3) xviii. 242 The term 'berm' has been introduced for any remnant of a surface developed to full maturity in a cycle that has since been interrupted. Though it has been said to be a kind of 'terrace', a berm may include more than the valley floor which becomes a true river terrace. **1942** O. D. von Engeln *Geomorphology* xii. 221 Such a remnantal flat, which has a surface slope downstream, may be called a strath or sometimes, together with the valley shoulder, a berm.

3. A ledge or flat of land bordering either bank of the Nile and inundated when the river overflows.

1891 *Daily News* 31 Oct. 6/4 To raise the Nile at the apex of the Delta to a level sufficient to flood the islands and berms of the two branches in the Delta. **1900** *Westm. Gaz.* 10 July 2/2 The water level in the winter was some fifty centimetres below the general level of the berm.

Bermuda. Add: **Bermuda(s cedar,** a species of juniper, *Juniperus bermudiana*; **Bermuda grass** (earlier and later U.S. examples); **Bermuda lily,** a lily of the variety *Lilium longiflorum eximium*, also known as *Lilium harrisii*, originally obtained from Bermuda; **Bermuda rig** = *Bermudian *rig*; **Bermuda shorts,** knee-length shorts; also *ellipt.* as *Bermudas*.

1756 P. Browne *Civil & Nat. Hist. Jamaica* (1789) ii. 362 Juniperus 1...The Bermudas Cedar. **1794** [see Cedar 3]. **1829** Loudon *Encycl. Plants* 848 *Juniperus bermudiana*, Bermudas Cedar. **1876** *Encycl. Brit.* V. 286 The Bermuda cedar..used in joinery and in the manufacture of pencils. **1808** H. Muhlenberg *Let.* 5 July in Rowland *Life W. Dunbar* (1930) 199 The *Bermuda grass* is..the same with Cumberland grass. **1906** F. Lynde *Quickening* 11 The dooryard with its thick turf of uncut Bermuda grass. **1899** G. Jekyll *Wood & Garden* ix. 106 The Bermuda Lilies (*Harrisi*) are intergrouped with *L. speciosum*. **1911** *Sutton's Amateur's Guide in Hort.* 195 *Lilium Harrisii* (Bermuda Lily). A large and elegant pure white Lily, adapted for forcing or growing in pots. **1853** R. Kipping *Mast-making & Rigging* ii. 5 Brig Forward, Common, and Bermuda Rig. **1928** *Daily Mail* 9 Aug. 19/7 Most sailing men agree that the Bermuda rig is preferable for smaller boats. **1953** R. Chandler *Long Good-Bye* iii. 17 Loafing around one of the swimming pools in Bermuda shorts. **1961** *Times* 11 July 12/6 No right-thinking suburbanite would be seen on his lawn without wearing Bermudas.

Bermudan (bəɹmiū·dăn), *a.* [f. Bermuda + -an.] = *Bermudian *a.*

1895 *Outing* (U.S.) XXVII. 240/2 The oval top of the Bermudan ocean peak. **1923** *Public Opinion* 24 Aug. 180/2 Her lofty Bermudan main-sail. **1928** *Daily Express* 20 July 10/2 The Astra and Cambria..carry the Bermudan-rig—a high tapering sail like the wing of a gigantic bird.

Bermudian (bəɹmiū·diăn), *a.* and *sb.* [f. Bermuda + -ian.] **A.** *adj.* Of or pertaining to the Bermudas or their inhabitants. **Bermudian rig,** a rig for a yacht, carrying a high tapering sail, called a *Bermudian mainsail*. So *Bermudian-rigged* adj., fitted with a rig and sail of this kind. **B.** *sb.* An inhabitant of the Bermudas; a Bermudian ship; a Bermudian-rigged ship.

1777 J. Adams *Let.* in *Mass. Hist. Soc. Coll.* (1917) LXXII. 313 Many french Vessells have arrived there, some Bermudians, and some of their own. **1803** A. Ellicott *Jrnl.* x. 287 Bermudian mulberry, (callicarpa americana). **1821** G. Gleig *Campaigns Brit. Army* vi. 73 You may perhaps consider me as too severe upon the Bermudians. **1895** *Boy's Own Paper* XVII. 429/3 One of the Manchester boats being a schooner and another a Bermudian. **1915** C. P. Lucas *Brit. Emp.* 163 The little Bermudian Assembly is..the oldest Parliamentary institution in the British Empire outside the United Kingdom. **1926** *Glasgow Herald* 20 Aug. 9 She is Bermudian rigged. **1926** *Blackw. Mag.* Nov. 698/1 By implicitly obeying the Bermudian, we escaped this danger. **1928** *Observer* 15 Apr. 29/5 The 12-metre yachts..with a Bermudian rig of moderate area, can be sailed efficiently with four paid hands. **1928** *Daily Tel.* 1 May 16/6 The Cambria will be Bermudian rigged, carrying a Bermudian mainsail. **1933** *Amer. Speech* VIII. 3/1 Bermudians would be justified in taking pride in their speech. **1953** 'N. Shute' *In Wet* 92 She's a Bermudian cutter, five and a half ton.

Bernese (bɜːɹnī·z), *a.* and *sb.* [f. *Bern(e)* (see below) + -ese.] **A.** *adj.* Of or pertaining to Bern (or Berne), a city and canton of Switzerland, or its inhabitants. **B.** *sb.* A native or inhabitant of Bern(e); also *collect.* as *pl.*; also, one of a Swiss breed of large, long-coated, black dogs (in full *Bernese mountain dog*).

1806 W. Guthrie *New Geogr. Gram.* (ed. 20) 533 The defeat of the Bernese was followed by the submission of nearly the whole of Switzerland. **1816** Byron *Let.* 30 Sept. (1830) II. 13, I have lately been over all the Bernese Alps and their lakes. **1822** L. Simond *Switzerland* I. 218, I never saw such a proud looking set of men as the Bernese peasantry. **1863** *Miss Jemima's Swiss Jrnl.* 3 July (1963) ii. 49 Stuffed bears, suspended by some patriotic Bernese under the eaves of his chalet. **1904** J. M. Stone *Reform. & Renaiss.* viii. 315 If the Bernese gained a footing in Geneva. *Ibid.*, The Bernese army smashed all the statues of saints. **1935** *Hutchinson's Dog Encycl.* III. 1800/1 The Bernese Mountain Dog..is now generally considered the best-looking of all Mountain Dogs. **1936** *Dog World Ann.* 62 Swiss Mountain Dogs...The Bernese, from the Canton of Berne..is the only one with a long coat. **1939** *S.P.E. Tract* LII. 83 We speak of the Bernese Oberland.

Bernician (bəɹni·ʃiăn), *sb.* and *a.* [f. med.L. *Bernicia* (cf. OE. *Beornice* inhabitants of Bernicia) + -an.] **A.** *sb.* A native or inhabitant of Bernicia, an Anglian kingdom founded in the 6th cent. A.D., extending from the Tyne to the Forth and eventually united with Deira to form Northumbria. **B.** *adj.* Of or pertaining to Bernicia or its inhabitants; spec. *Geol.*, designating the carboniferous limestone rocks of Northumberland and its borders.

1819 J. Lingard *Hist. Engl.* I. ii. 89 The Bernicians submitted cheerfully to the good fortune of the son of Ælla. **1856** S. P. Woodward *Man. Mollusca* III. iii. 409 (*heading*) Geological table..Bernician. **1878** G. A. Lebour *Geol. Northumberland* i. 2 Lower carboniferous...Bernician series. *Ibid.* viii. 32 The Bernician Rocks...The series consists essentially of numerous beds of limestone. **1907** H. M. Chadwick *Orig. Eng. Nation* vii. 182 These persons again were nearly related to the ancestors of the Bernician royal family. **1932** G. Sheldon *Transit. Roman Britain* viii. 134 A member of the Bernician royal house and a pretender to the throne. *Ibid.* 144 The ever-smouldering feud between Bernician and Deiran.

Bernkasteler (bɜ̄·ɹnkɑstələɹ). [Ger.] Any of a group of Moselle wines produced in the villages of Bernkastel and Cues, of which the best known is *Bernkasteler Doctor*.

[**1875** H. Vizetelly *Wines of World* 51 Graach, Zeltinger, and the Berncastle wine, known locally as 'the doctor', all three of such excellent quality as to secure a couple of medals to their exhibitors.] **1891** in C. Ray *Compleat Imbiber* (1967) IX. 122 Moselle...Berncastler Doctor. **1920** G. Saintsbury *Notes on Cellar-Bk.* vi. 85 Nor did I ever much affect the loudly-trumpeted Berncastler Doktor. **1967** A. Lichine *Encycl. Wines & Spirits* 365/1 Some Bernkastelers, and especially Bernkasteler Doctor, have a slight smoky under-taste.

† **Bernois(e.** *Obs.* [Fr.] = *Bernese *sb.*

1687 I. Spon *Hist. Geneva* ii. 86 The Bernoises..banished the Roman Catholick Religion from their City. **1761** G. Keate *Short Acc. Geneva* 53 The Bernois coming to the Assistance of Geneva, drove away the Troops of the Duke and Bishop. **1832** W. Liddiard *Three Months' Tour* 88, I have already had more than one application from different Bernois, looking out..to get fares on their return to Berne.

Bernoulli (bəɹnū·li). The name of a Swiss family which in the 17th and 18th centuries contained several eminent mathematicians and scientists, applied to various principles, theorems, etc., formulated by them (see quots.). Hence **Bernou·llian** *a.*

B.'s formula and *B.'s theorem* in hydrodynamics were proposed by Daniel Bernoulli (1700–1782); *B.'s numbers* and *B.'s theorem* in statistics were proposed by Jacob (also known as James) Bernoulli (1654–1705).

1749 J. Stirling *Differential Method* 94 The first series is not extended to those cases in which the first ordinate touches the curve, nor does Bernoulli's series extend to those cases wherein the last ordinate touches the curve. **1842** A. De Morgan *Diff. & Integr. Calculus* xiii. 247 The values of U, U', &c. are called the numbers of Bernoulli; and though they do not follow a visibly regular law, yet the connexion between them is simple. *Ibid.* 248 The development of tan x by Bernoulli's numbers. **1865** I. Todhunter *Math. Theory Probability* vii. 71 In the fourth part of the *Ars Conjectandi* is the enunciation and investigation of what we now call *Bernoulli's theorem*. *Ibid.* xi. 226 Let x denote the age expressed in years; let ξ denote the number who survive at that age out of a given number who were born; let s denote the number of these survivors who have not had the small-pox...Daniel Bernoulli's formula then gives the value of s. **1875** *Encycl. Brit.* I. 114/1 The basis of Bernoulli's [*sic*] Theory of Pipes. **1876** *Messenger Math.* VI. 49 Bernoullian and Eulerian numbers. **1888** *Encycl. Brit.* XXIII. 14/1 Bernoullian numbers. **1920** L. Bairstow *Appl. Aerodynamics* vi. 281 The simple form of Bernoulli's equation developed in the chapter on fluid motion may be applied separately to the two parts of streamlines which are separated by the actuator disc. **1922** Glazebrook *Dict. Appl. Physics* I. 26/2 *Bernoulli's theorem*. Along any stream line in a liquid subject only to gravity $p + g\rho z + \frac{1}{2}\rho v^2$ = constant, p being the pressure at a point at a depth z below the plane of reference, ρ the density, and v the velocity. **1937** *Mind* XLVI. 488 The least rigid of these suggested conditions is that the series must be 'Bernoullian'.

berrugate (berugē[i]·t). [f. Sp. *verruga* wart. Cf. Verruga.] A fish, *Verrugato pacificus*, found on the Pacific coast of Central America, used as a food.

1898 Jordan & Evermann *Fishes N. & Mid. Amer.* iii. 2858 Abundant at Panama, where it is known as *Berrugate*.

berry, *sb.*[1] Add: **1. c.** *slang* (*U.S.*). A dollar; also (in *U.K.*), a pound. Usu. in *pl.* Hence *the berries*: an excellent person or thing; 'the cat's whiskers'.

1918 H. C. Witwer *From Baseball to Boches* iv. ii. 147 When..I go back to baseball, I can drag down six thousand berries a year. **1920** 'B. L. Standish' *Man on First* 127 It don't take the shine off your little performance. You were there with the berries. **1922** S. Lewis *Babbitt* vii. 103 A fellow that..pulls down fifteen thousand berries a year! **1925** H. Foster *Trop. Tramp Tourists* 300 You think you're the berries, don't you? Well, you might have been once, but you're a flat-tire these days! **1926** *S.P.E. Tract* xxiv. 120 *That's the berries*, that's just right. **1934** *Humorist* 26 May 482/1 An attachment worth ten thousand berries in the open market. **1936** J. Dos Passos *Big Money* 43 He had what was left of the three hundred berries Hedwig coughed up. **1943** Wyndham Lewis *Let.* 9 Nov. (1963) 369 No intelligent book *could* get accepted by a N.Y. publisher, except perhaps a little publisher, who would give you a maximum of a thousand berries.

4. *berry-bearing* (earlier and later examples); **berry-bearing alder** (example); **berry wax,** wax obtained from the wax-berry (*Myrica* spp.), used for making candles and polishing floors (cf. *bayberry-wax* in quot. 1769 s.v. *Bayberry 2).

1742 W. Ellis *Timber-tree* II. xxiv. 140 A bacciferous, or berry-bearing, Tree or Shrub. **1796** W. H. Marshall *Planting* II. 313 Frangula, or Berry-bearing Alder. **1933** *Jrnl. R. Hort. Soc.* LVIII. 400 *Wilsonii* with leaves as large as *Marnockii*, but dull green and spiny, also berry-bearing. **1897** Edmonds & Marloth *Elem. Bot. S. Afr.* xvii. 169 The genus *Myrica*, of which *M. cordifolia* and others supply the berry-wax. **1913** R. Marloth *Flora S. Afr.* I. 133 The layer of wax on the berries of some species [of *Myrica*] is so considerable that it is technically exploited. The farmers boil the berries with water, strain the hot mixture and allow the melted wax to solidify. The *berry wax* (myrica wax) is of a pale greenish colour and considerably harder than beeswax.

berryless (be·rilės), *a.* [f. Berry *sb.*[1] + -less.] Without producing berries; not berried or furnished with berries.

1887 *Sat. Rev.* 30 Apr. 624 The female plant..berryless, may be said to have suffered a grass-widowhood of some eighty years. **1924** *Glasgow Herald* 3 Jan. 6 Berryless holly. **1942** H. J. Massingham *Field Fellowship* xii. 110 A bleached and berryless powder with the courtesy title of flour.

‖ **bersagliere** (bēᵊɹsālyēᵊ·re). Usu. in pl. **bersaglieri** (-*i*). [It., f. *bersaglio* target, mark.] A rifleman or sharpshooter in the Italian army.

1862 Crown Princess of Prussia *Let.* 8 Nov. in R. Fulford *Dearest Mama* (1968) 129 As one is not safe from Banditti—General La Marmora gave us an escort of Bersaglieri and Gendarmes. **1875** *Encycl. Brit.* II. 612/2 The Italian army consists of 80 regiments of the line, 10 of *bersaglieri* (riflemen), [etc.]. **1883** *Daily News* 7 Sept. 3/1 The same war cry would resound from a battalion of dark-plumed Bersaglieri as they dashed up a bank at their peculiar pace. **1929** Hemingway *Farewell to Arms* viii. 47 The drivers..wearing red fezzes..were bersaglieri.

b. *attrib.* and *Comb.* **bersaglieri hat,** a hat with a dark plume of cock's feathers, as worn by the bersaglieri.

1875 *Encycl. Brit.* II. 613/1, 4 *bersaglieri* battalions. **1946** Koestler *Thieves in Night* 10 About half of them wore ..Bersaglieri hats which made their faces look even more adolescent.

berserk. Add: (Also pronounced bəɹsɜ̄·ɹk, bəɹzɜ̄·ɹk.) Now usu. as *adj.*, frenzied, furiously or madly violent; esp. in phr. *to go berserk*.

1867 H. Kingsley *Silcote* I. xii. 136 With her kindly, uncontrollable vivacity, in the brisk winter air she became more 'berserk' as she went on. **1887** E. C. Dawson *Bp. Hannington* v. 57 He..was filled with a Berserk rage and thirst for retribution. **1908** Kipling *Diversity of Creatures* (1917) 264 You went Berserk. I've read all about it in *Hypatia*..you'll probably be liable to fits of it all your life. **1940** *Chicago Daily Tribune* 20 Nov. 10/3 America goes berserk. *Ibid.*, The recent addition of the word 'berserk', as a synonym for crackpot behaviour, to the slang of the young and untutored...American stenographers..are telling one another not to be 'berserk'. **1944** 'P. Quentin' *Puzzle for Puppets* xvii. 121 Edwina [*sc.* an elephant], had gone berserk. **1961** G. Smith *Business of Loving* iii. 124 Hammond

converted and Shallerton came back as if berserk. *Ibid.* 132, I think Ken Heppel will go berserk. **1962** P. Brickhill *Deadline* xviii. 213, I went berserk, kicking his head again and again. *Ibid.* 214 In that berserk mood I think I could have bent an iron bar. **1964** J. Symons *End of S. Grundy* i. i. 27 If you have chaps like old Sol going berserk, it's enough to break up any party.

berserkly (bɜ̄ɹsɜ̄·ɹkli), *adv.* [f. Berserk + -ly[2].] In a berserk manner; madly.

1963 *Economist* 28 Dec. 1318/2 It is berserkly dangerous. **1967** 'C. Franklin' *Death in East* ii. 27 The headlamps illuminated a tree which seemed to be leaping berserkly towards her.

berth, *sb.* **6. b.** (Earlier and later examples.)

1806 Z. M. Pike *Acc. Exped. Mississippi* (1810) 81 We returned to the chief's lodge, and found a birth provided for each of us. **1838** *Amer. Railroad Jrnl.* VII. 328 If you travel in the night you go to rest in a pleasant berth. **1967** *Gloss. Caravan Terms* (*B.S.I.*) 3 A caravan with two double beds, or one double and two singles, is a four-berth caravan. **1968** *Globe & Mail* (Toronto) 17 Feb. 33 (Advt.), First Class all-inclusive fare, including lower berth and all meals.

berth, *v.*[1] Add: **2. b.** *intr.* To occupy a berth or berths.

1886 Stevenson *Kidnapped* vii. 61 The round-house, where he berthed and served. **1902** *Westm. Gaz.* 13 Sept. 6/2 The accommodation is very simple, consisting of berthing in two tiers in the women's ward, and feeding and living in a separate saloon.

bertha, berthe. (Earlier examples.)

1842 C. Ridley *Let.* in Ridley *Cecilia* (1958) ix. 109, I shall be very glad..of the bugle flowers and bertha and rosettes. **1842** *Illustr. Lond. News* 24 Dec. 525/1 The berthe had a double row of point d'Argentan.

Bertha[2] (bɜ̄·ɹþă). [Named after Frau *Bertha* Krupp von Bohlen und Halbach, owner of the Krupp steel works in Germany from 1903 to 1943.] Soldiers' name for a German gun or mortar of large bore, used in the war of 1914–18; freq. *Big Bertha*.

1914 *Scotsman* 30 Oct. 9/6 This mortar of 42 centimeters was made at the Krupp works, and for this reason the Germans have baptised it '*Bertha—die fleissige*' (Bertha, the Zealous), Bertha being the name of Madame Krupp von Bohlen. *Ibid.*, 'Bertha' is not the delicate plaything that it has sometimes been represented to be, and the maximum of 150 shells that they say can be fired from the gun is below the truth. **1918** *Sphere* 20 July 48/2 Big Bertha spoke for the first time on March 23, and at the sound of her voice Paris was intensely surprised. **1958** Hayward & Harari tr. *Pasternak's Dr. Zhivago* i. iv. 111 That's a Bertha, a German sixteen-inch.

berthage. Add: **b.** The dues payable for mooring a vessel.

1893 in *Funk's Stand. Dict.*

berthing, *vbl. sb.*[1] Add: **1. b.** The occupation of a berth or mooring position; also, mooring position.

1891 *Daily News* 15 July 5/8 There being a high wind from the north north-east..the berthing was very uncomfortable for the..launches. **1906** *Westm. Gaz.* 9 July 4/2 Berthing accommodation will be provided for about 300 boats. **1908** *Ibid.* 26 May 9/1 The..Railway Company's boats..have changed their berthing from North Wall to Carlisle Pier. **1909** *Daily Chron.* 6 Dec. 6/7 They..came up practically to the berthing which the Ellan Vannin used to occupy.

2. The arrangement of berths or the provision of sleeping accommodation; accommodation in berths.

1863 Luce *Seamanship* (ed. 2) xvi. 297 Berthing requires the earliest attention, and the operation may be facilitated by having a plan of the decks. *a* **1871** C. F. Hall *Polar Exp.* (1876) 123 The special object of these [changes] was the economy of fuel, and the berthing of the whole crew below deck.

Berthon (bɜ̄·ɹþɒn). The name of the Revd. Edward Lyon *Berthon* (1813–1899), used *attrib.* to designate a small collapsible boat invented by him. (Freq. with lower-case initial.)

1878 G. S. Nares *Voy. to Polar Sea* I. i. 20 The light canvas Berthon boats, available for one or two persons, proved of great service. **1898** W. A. Beauclerk in W. A. Morgan '*House*' *on Sport* 444 A berthon canoe..is very useful for cripples [*sc.* birds] when they get into shallow water. **1901** *Boy's Own Paper* 19 Oct. 48/2 A small Berthon dinghy that you can carry on board is much better.

‖ **Bertillon** (bɛ̃rtiyoṅ). [See next.] *Bertillon system* = *Bertillonage. So *Bertillon measurement*.

1896 W. A. McCorn in *Amer. Jrnl. Insanity* LIII. 47 (*title*) Degeneration in criminals as shown by the Bertillon system of measurement and photographs. *Ibid.* 50 The Bertillon system does not take account of the facial angle. **1902** *Encycl. Brit.* XXV. 468/1 Previously to the introduction of the Bertillon system the means relied upon were (1) the memory of warders [etc.]. **1928** G. B. Shaw *Crude Criminology* in *Doctors' Delusions* (1932) 246 We were assured that the Bertillon measurements were infallible until the fingerprint method was substituted.

‖ **bertillonage** (bɛ̃rtiyonā·ʒ). [Fr., f. the name of the inventor (see below).] The system of identification of criminals by anthropometric measurements, finger-prints, etc., invented by Alphonse Bertillon (1853–1914), French criminologist.

1892 F. Galton *Finger Prints* 156, 562 prisoners who gave false names in the year 1890 were recognised by *Bertillonage*. **1892** *Athenæum* 24 Dec. 893/3 An interesting part of the work is the criticism of Bertillonage. **1901** E. R. Henry *Finger Prints* (ed. 2) 63 The system suggested by Mr. Galton..failed to deal as effectively as Bertillonage with primary classification. **1920** *Discovery* Apr. 124/1 Extended examples are given of the possibilities of variation curves, variation 'steps', and 'gradation curves'; the combination of such data, constituting a quantitative diagnosis of a species, is defined as 'Bertillonage'.

bertrandite (bɜ̄·ɹtrăndəit). *Min.* [a. F. *Bertrandite* (A. Damour 1883, in *Bull. Soc. Min.* VI. 254), after E. *Bertrand*, who first described it: see -ite[1].] Hydrous silicate of beryllium.

1887 Dana *Man. Min. & Petrog.* (ed. 4) 275 Bertrandite is related to phenacite in composition. **1911** *Encycl. Brit.* VII. 578/2 Examples of other substances belonging to this [*sc.* pyramidal] class are..Bertrandite ($H_2Be_4Si_2O_9$). **1914** *Parl. Papers* LXXI. 414 Twinned crystals of felspar from Portugal and of bertrandite from Cornwall have been measured and described.

berycoid (be·rikoid), *a.* and *sb.* *Ichthyol.* [f. mod.L. *Berycoidei*, f. *Beryc-*, *Beryx*, generic name: see -oid.] **A.** *adj.* Belonging to the group Berycoidei. **B.** *sb.* A fish of this group.

1880 Günther *Fishes* 420 Fossil Berycoids show a still greater diversity of form than living. *Ibid.*, Berycoid fishes have a wide horizontal range. **1887** *Athenæum* 9 July 58/3 The genus had hitherto been erroneously associated with the percoids and berycoids. **1955** E. A. Lechner in *Proc. U.S. Nat. Mus.* CV. (*title*) Populations of the Berycoid Fish Family Polymixiidae.

beryl. Add: **B. 1. b.** *Comb.*, as *beryl-blue*, *-covered* adjs.

1881 W. D. Hay *300 Years Hence* vii. 141 The great basin of beryl-blue water. **1871** G. M. Hopkins *Poems* (1930) 144 Long reefs of violets In beryl-covered fens so dim.

beryllium. (Further examples.)

1955 *Sci. News Let.* 16 Apr. 249/1 There is one other sort of beryllium, isotope 10, with a half life of 2,500,000 years, that it is expected will be found to be cosmic-ray produced. **1958** *Chambers's Techn. Dict.* 960/1 *Beryllium*, used in nuclear reactors, as it reflects neutrons. **1961** *Times* 11 July 7/2 It is expected that beryllium-clad fuel elements, more resistant to corrosion, will be developed.

beryllonite (bĕri·lŏnəit). *Min.* [f. Beryl-l(ium + -on, mod.L. form of Gr. -ον + -ite[1].] A phosphate of beryllium and sodium, $NaBePO_4$.

1888 E. S. Dana in *Amer. Jrnl. Sci. & Arts* XXXVI. 291, I would suggest the name *Beryllonite*, in allusion to the fact that it contains the rare element beryllium. **1896** A. H. Chester *Dict. Min.* 30 *Beryllonite*, so named because it contains the rare element beryllium. An anhydrous phosphate of glucinum (beryllium) and sodium, occurring in transparent, colorless, orthorhombic crystals. **1957** *Encycl. Brit.* III. 469/2 *Beryllonite*, a mineral phosphate of beryllium and sodium..found..at Stoneham and Newry [Maine].

besom, *sb.*[1] Add: **7.** A contemptuous or jocular designation for a woman. *Sc.* and *dial.*

1808–1929 in *Sc. Nat. Dict.* **1816** [see Besom *sb.*[2]]. **1930** J. Buchan *Castle Gay* xvi. 255 She's a determined besom. **1936** W. Holtby *South Riding* v. i. 276 Gosh! The little besom!

besom, *sb.*[2] Delete entry in Dict. The distinction *sb.*[1], *sb.*[2] is invalid (see prec.).

bespeak, *sb.* Add: **1.** (Earlier example.)

1807 A. C. Holbrook *Memoirs of Actress* 24 The first families in the neighbourhood, who had come to attend a bespeak that evening.

2. An application made to a lending library for the loan of a book when it shall become available.

1922 *Glasgow Herald* 28 Dec. 9 In the past year the number of 'bespeaks' was 4153, an increase of 650 on the previous year.

bespell (bĭ,spe·l), *v.* [f. Be- 6 + Spell *sb.*[1] 3.] *trans.* To cast a spell on; to bewitch literally or figuratively. Hence **bespe·lled** *ppl. a.*

1894 Meredith *Ld. Ormont* iv, If his glory bespells her. **1897** A. Nutt in K. Meyer *Voy. Bran* II. 6 A bard..had threatened to bespell his land. **1902** —— *Leg. Holy Grail* 58 The bespelled kinsman appears in a hideous guise. **1926** *Contemp. Rev.* Oct. 503 We were gazing, bespelled at the fair scene.

bespoke (bĭ,spōᵘ·k), *ppl. a.* [See Bespeak *v.*] = Bespoken *ppl. a.* 2; *spec.* of goods: ordered to be made, as distinguished from Ready-made; also said of a tradesman who makes goods to order. Also *sb.*, a bespoke article.

1755 Mrs. C. Charke *Life* 203 At length the bespoke Play was to be enacted. **1866** *Chambers's Encycl.* VIII. 691/1 The shoemaking trade..is divided into two departments—the bespoke and the ready-made or sale business. **1884** [see Bespoken *ppl. a.* 2]. **1907** W. De Morgan *Alice-for-Short* xlvii, His boots may have been 'bespokes' for anybody, except himself. **1908** *Daily Chron.* 13 June 4/7 A 'bespoke tailor'. **1928** *Punch* 30 May p. xv (Advt.), Lotus Bespoke Model Shoes. **1965** *Ibid.* 29 Sept. 478/1 A 'Special Collection' which is, in harsh reality, a collection of bargain bespokes. **1966** *Economist* 16 July 239/2 Although there is a lot of bespoke work in this [steel] plant, management would prefer some element of payment by results.

besport (bĭ,spōᵊ·ɹt), *v.* [Alteration of Disport *v.*, by exchange of prefix (Be- 1).] *refl.* To disport oneself. Also *intr.* (for *refl.*).

1855 *Chambers's Jrnl.* IV. 65 Living gems of light, besporting themselves on the bosom of the lake. **1906** *Westm. Gaz.* 7 June 9/2 And surely Shylock never besported himself more ostentatiously. **1925** A. S. Alexander *Tramps across Watersheds* viii. 256 The famous golf course, where ladies and gentlemen besported in the glowing sun.

Bessarabian (besărē[i]·biăn), *a.* and *sb.* [f. *Bessarabia*, a province of south-west Russia + -an.] **A.** *adj.* Of or pertaining to Bessarabia. **B.** *sb.* An inhabitant of Bessarabia.

1835 *Penny Cycl.* IV. 340/1 Its [*sc.* the Dniester's] chief tributaries on the Bessarabian side are, the Reut..and the Botna. *Ibid.* 341/1 The majority of the Bessarabians are Moldowans or Moldavians. **1852** H. W. Dulcken tr. *Pfeiffer's Visit to Holy Land* ii. 37 Tehussa, a Bessarabian village of most miserable appearance. **1908** Kipling *Lett. Travel* (1920) 160 You are always talking of the second generation of your Smyrniotes and Bessarabians. **1925** *Chambers's Jrnl.* Apr. 214/2 The traditional Bessarabian hospitality. **1959** E. H. Carr *Socialism in One Country* II. xxiii. 394 When a Bessarabian division was enrolled in the Ukraine, desertions amounted to 50 per cent. **1963** *House & Garden* Mar. 73/2 This room, with its handsome Bessarabian rug.

Bessel (be·sĕl). The surname of F. W. *Bessel* (1784–1846), a German astronomer, used *attrib.* or in the possessive in *Bessel('s) function*: a solution of Bessel's differential equation $x^2\frac{d^2y}{dx^2} + x\frac{dy}{dx} + \left(x^2 - n^2\right)y = 0$. So **Besse·lian** *a.* (Cent. Dict., 1889).

1872 *Phil. Mag.* XLIV. 328 The value of Bessel's functions is becoming generally recognized. **1894** *Ann. Math.* IX. 27 (*heading*) Roots of the Second Bessel Function. **1939** *Nature* 22 July 162/2 Results calculated..by the Besselian method. **1963** G. Troup *Masers & Lasers* (ed. 2) vii. 117 The cut-off parameters u_{om} for these modes are the mth roots of the zero order Bessel function J_0.

bessemerize (be·sĭmərəiz), *v.* [f. Bessemer + -ize.] *trans.* To treat by the Bessemer process.

1888 in *Amer. Speech* (1960) XXXV. 267. **1901** *Prospectus Mond Nickel Co.*, Plant of the most modern type for Roasting, Smelting and Bessemerizing this ore.

best, *a.* and *adv.* Add: **A.** *adj.* **1. b.** Applied to a room that is furnished especially well, often one reserved for special use. In U.S., *best room*, spec. a parlour.

1702 C. Fiennes *Journeys* (1947) i. ix. 56 These doores open through to the end one way the best bed chamber.. the other side a visto. **1719** S. Sewall *Diary* 30 Oct. in *Mass. Hist. Soc. Coll.* (1882) 5th Ser. VII. 233 Had a very good Dinner..in the best room. **1800** Jane Austen *Let.* 8 Nov. (1952) 83 The little Table..has most conveniently taken itself off into the best bedroom. **1839** F. A. Kemble *Jrnl. Residence on Georgian Plantation* Jan. 80 In all establishments..some disparity exists between the comforts of the drawing-room and best bed-rooms, and the servant's hall and attics. **1872** *Harper's Mag.* Oct. 684/2 A parlor, or 'best room'. **1911** Beerbohm *Zuleika D.* ii. 7 The sun streamed through the bay-window of a 'best' bedroom in the Warden's house. **1967** J. Speight *Till Death us do Part* i. 22 I've always wanted a best room, somewhere you could put all your decent bits of furniture.

c. *spec.* denoting a certain grade of wrought iron; also *best best* adj., of a higher quality than best; *best best best* adj., of the highest quality.

1888 *Lockwood's Dict. Terms Mech. Engin.*, *Best, or B.*—A brand of wrought-iron plate or bar equivalent to No. 3 quality, or that grade which is only just superior to the commonest...The 'best' plates of the first-class houses are, however, equal to the 'best, best' and 'treble best' of other firms. *Ibid.*, *Best, Best, or B.B.*..obtained by piling, reheating, and rerolling 'best' or No. 3 bars. *Ibid.*, *Best, Best, Best, or Treble Best, B.B.B.* **1940** *Chambers's Techn. Dict.* 67/1 *B.* and *B.B.*, brand-marks signifying *Best* and *Best Best*, placed on wrought-iron to indicate the maker's opinion of its quality.

d. *spec.* in *best boat*, a racing boat on the Thames, as distinguished from the heavier types; esp. used attrib. in *best-boat race, racing, regatta, rowing*. Also *best-and-best boat, punt*: a boat of the lightest and narrowest construction, used for racing.

1890 *Lock to Lock Times* 23 Aug. 162/2 This year the amateur oarsmen have ungrudgingly given up their boats to the watermen, who have been rowing in the past regatta; wager, or 'best' boats, being in special demand by the fleet of scullers, who were attracted by the liberal programme. **1894** *Daily News* 5 May 6/5 The race will..be decided in best-and-best punts. **1897** *Ibid.* 30 Aug. 5/1 No best-boat racing will be attempted. **1927** *Observer* 24 July 26/2 The best-boat rowing season..is all too short.

2. b. *best friend*, one's favourite friend (of persons and *transf.* of things); also in colloq. phr. *to be 'best friends' with* (a person).

c **1374** Chaucer *Troylus* ii. 412 When he, that for my beste freend I wende. **1609** Beaumont & Fletcher *Kn.*

Burn. Pestle iii, *Luce.* No my best friend. **1671** [see Friend *sb.* 5 d]. **1796** M. Edgeworth *Parent's Assistant* (ed. 2) I. 23 Ay, go along with its best friend, and take care it does not get into a scrape. **1885** C. M. Yonge *Two Sides of Shield* II. vii. 130 'I want to be friends with both. One can have two friends.' 'No! no! no! not two *best* friends. And you are my best friend, Nysie.' **1905** F. H. Burnett *Little Princess* iii. 36 Lavinia and Jessie are 'best friends'...I wish we could be 'best friends'. **1967** E. McGirr *Here lies my Wife* v. 149 Our mothers are on best-friend terms.

c. *best girl* (Girl *sb.* 2 c), a man's favourite female companion. orig. *U.S.* Also *ellipt.* (U.S. slang) for *best girl*.

1887 *Texas Siftings* 7 May 11/2 You can't convince a young man whose best girl has just said 'Yes' that this country is going to wreck and ruin. **1904** *Democrat* (Celina, Ohio) in *Daily Chron.* 23 Feb. 5/2 Wm. Londen and best were at church Sunday evening. **1904** *Indicator* (Merriweather, Georgia) in *Daily Chron.* 21 May 4/5 Messrs. Bub Peary and Pierce Biggers called on their best girls last Sunday afternoon. **1918** *Punch* 16 Jan. 38/3 When he [*sc.* a Tommy] gave me an unprovoked smile..I hoped that perhaps I reminded him of his best girl. **1944** *Sat. Rev.* 8 July 16/2 To pluck a bouquet for his best girl.

5. *best end*, the end of a neck of lamb, mutton, or veal consisting of the ribs; opp. *scrag-end*.

1728, 1874 Best end [see Scrag *sb.*[1] 2]. **1936** Mencken *Amer. Lang.* (ed. 4) vi. 236 What we call the *rib* chops are [in England] the *best end of the neck* or *best end*. **1962** *House & Garden* Nov. 93/1 *Crown roast of lamb*. Ingredients: 2 best ends of neck (there are usually 6 cutlets in each).

b. *to give* (a person or thing) *best*: to admit the superiority of, give way to. Also *absol.*

1888 'R. Boldrewood' *Robbery under Arms* I. vii. 94, I could hardly stand for laughing till the calf gave him best and walked. **1895** A. B. Paterson *Man from Snowy River* (1906) 71 My country joker, are you going to give it best? Are you frightened of the fences? **1911** J. Masefield *Everlasting Mercy* 31 In all the show from birth to rest I give the poor dumb cattle best. **1927** T. E. Lawrence *Let.* 1 Dec. (1938) 549 Those people who have compared the versions generally give me best with the new one. **1959** *Times* 20 Nov. 18/5 It was now the turn of the young professionals to give best.

7. b. *spec.* = Record *sb.* 5 d.

1881 [in Dict.]. **1930** F. A. M. Webster *Athletics of to-day for Women* vi. 86 Miss M. A. Gunn..ripped through her heat in 14⅘ secs., thus equalling the world's record and beating Miss Hatt's British best. **1963** G. F. D. Pearson et al. *Athletics* i. ii. 40 Snell was obviously not deterred by having to set personal bests *en route* to..the Gold Medal in Rome. *Ibid.* iv. 85 Lord Burghley, who produced a best of 14·5 sec. in 1930. *Ibid.* ii. vii. 154 A stupendous new world's best of 63 ft. 1½ in.

c. *all the best*: an expression of goodwill, used as a toast or a valediction.

1937 M. Sharp *Nutmeg Tree* xv. 192 All the best, Fred. **1951** J. B. Priestley *Fest. Farbridge* i. ii. 78 Let's drink to it. All the best!

8. e. *Ellipt.* for 'best wishes'; usu. *one's best*. *U.S. colloq.*

1922 F. Scott Fitzgerald *Let.* Jan. (1964) 332 Zelda sends best. **1953** W. Stevens *Let.* 28 Jan. (1967) 770 My best to Marianne Moore. **1969** *Rolling Stone* 28 June 13 Billy Jean came here tonight, too, and she sends her best.

10. a. †*at the best* (later slang example: see quot.); †*at the best hand* (see Hand *sb.* 10 b); *at best*: also in *Finance*, at the best possible pitch or price; *at the best of times*: even in the most advantageous conditions or circumstances.

1812 Vaux *Flash Lang.* 155 *To get your money at the best*, signifies to live by dishonest or fraudulent practices, without labour or industry. **1938** *Times* 29 Sept. 19/1 Numerous orders to sell 'at best' have been given recently in respect of small amounts of rubber.

1936 Bentley & Allen *Trent's Own Case* iv. 30 A shifty-looking character at the best of times, thought the Inspector; and now looking sick and frightened. **1966** M. Kelly *Dead Corse* x. 162 Allie could not take drink at the best of times.

d. (Additional examples.) In colloq. phrases: (i) (a specified number, esp. *six*) *of the best*, a thrashing; (ii) referring to a sum of money: pounds or dollars; (iii) *one of the best*: a good fellow.

1338 R. Brunne *Chron.* (1725) I. 183 Ten sergeanz of þe best his targe gan him bere. *c* **1510** Skelton *E. Rummyng* (1568) line 265 And dame Elinoure entreaet To byrle them of the best. **1534** in Edw. Peacock *Eng. Ch. Furniture* (1866) 185 Itm ix cusshions of the best wherof.. be backyd wt red lether.

1912 'Saki' *Unbearable Bassington* ii. 29 You'll get six of the very best, over the back of a chair. **1923** Kipling *Land & Sea Tales* 110, I got nine cuts of the best from the Senior Sub for occupying the bathroom ten seconds too long. **1929** Wodehouse *Mr. Mulliner Speaking* ix. 297 He was.. an officious little devil who needed six of the best with a fives-bat.

1911 E. Wallace *Sanders of River* v. 103, I wouldn't take a half-share of the trouble he's going to find for five hundred of the best. **1920** Wodehouse *Let.* 28 Feb. in *Perf. Flea* (1953) 15 Heaven knows what a women's magazine wants with my sort of stuff, but they are giving me fifteen thousand of the best for it. **1928** Galsworthy *Swan Song* ii. xii. 204 The walrus put down five thousand of the best.

1917 'Contact' *Airman's Outings* 238 The brigades are directed by the General-Officer-Commanding.., one of the best, who treats us like brothers. **1959** *Listener* 22 Jan. 155/1 Harry Truman was one of the best.

f. *with the best* (*of them*): as well as anyone.

1748 M. W. Montagu *Let.* 5 Jan. (1966) II. 393, I can ass[ure you] she jumps and Gallops with the best of them. **1864** [see Dict., sense 6]. **1935** 'L. Luard' *Conquering Seas* 32, I can make tea with the best of 'em, Peter. **1957** I. Cross *God Boy* (1958) xxii. 187, I can drag my share out of the old girls with the best of them.

11. a. *Comb.* with sbs. used attrib., as *best-quality*.

1906 *Westm. Gaz.* 12 Nov. 2/3 It is on the best-quality produce, not the second quality..that the British agriculturalist can make the most money. **1960** *Farmer & Stockbr.* 29 Mar. 116/3 The production of the best-quality wood.

b. *Comb.* with advbs. used attrib., as *best-ever*.

1959 *News Chron.* 23 Nov. 3/4 Twelve of Duke Ellington's best-ever pop up on..a set of re-issues from the early 'forties. **1960** *Times* 6 Feb. 9/1 Many a game book received its best-ever entry.

B. *adv.* **2. a.** *best-dressed* (examples); *best-kept*.

1903 A. Bennett *Leonora* viii. 206 We are relying on you to be the best-dressed woman in the place. **1964** *McCall's Sewing* i. 1/1 There are several ways to earn a place in the list of 'best-dressed' women in your town. **1956** *Nature* 4 Feb. 214/1 In a night they [*sc.* worms] may cover a best-kept green with little piles of earth. **1968** *Guardian* 10 Aug. 6/2 You would certainly never win a best-kept village competition that way.

b. *best-abused, -described, -discussed, -hated.*

1848 *Punch* XV. 155/1 We have heard certain persons whose claims to distinction arose from their being the 'best abused men in the world'. **1869** *Church Times* 26 Feb., [Pusey] for years and years the best abused man in England. **1900** *Daily News* 16 Feb. 7/2 Englishmen are the best-hated people in the world. **1903** 'A. McNeill' *Egregious Eng.* 100 The best-discussed and best-described man in England.

3. b. best seller (orig. *U.S.*), one of the books having the largest sale of the year or the season; also, a writer of such books; also *transf.* and *attrib.*; hence (as a back-formation) **best-sell** *v. intr.*, to be or become a best seller; **best-sellerdom, -sellership,** the state or achievement of being a best seller; **best-sellerism,** concentration on best sellers; **best-selling** *a.*, that is a best seller.

1889 *Kansas Times & Star* 25 Apr., Kansas City's literary tone is improving. The six best sellers here last week were 'Fools of Nature' [etc.]. **1895** *Bookman* July 429/2 The best selling new book is Mr. Stockton's *Adventures of Captain Horn*. **1905** *Out West* (U.S.) Sept. 363 To be able to discuss the Six Best Sellers has become as much an article of faith as any in the Longer Catechism. **1913** R. W. Chambers *Gay Rebellion* ii. 28 Go on with the dolly dialogue..you third-rate best seller. **1920** *Chambers's Jrnl.* 8 May 363/2 [He] promised to reach the doubtful altitude of best-sellership without difficulty. **1922** Joyce *Ulysses* 450 He has cribbed some of my bestselling books. **1925** *Punch* 20 May 560/1 Fiction of best-seller quality and little more. **1926** W. R. Inge *Lay Thoughts* 190 Best-selling novels. **1928** *Daily Express* 1 June 5/3 'A house can be full of sweet scents..' writes one of our best-sellers, and he is perfectly right. *Ibid.* 16 July 8 A 'best-selling' biography. **1928** *Publishers' Weekly* 24 Nov. 2184/2 Two books that almost made the Best Seller List. **1928** Weseen *Dict. Eng. Gram.* 120 *Coined Words*...Bestsellerdom. **1931** O. H. Cheney *Economic Surv. Book Industry* i. iii. 126 Best-sellerism is an intolerable curse on the industry. **1937** *Observer* 8 Aug. 6/5 And, in the words of the elderly, this is a thoroughly 'nice' book. I think it may best-sell. **1938** H. Granville-Barker *Quality* 11 Authors..who set out in a businesslike way to supply what is supposed to be most in demand..taking example by the latest best seller. **1946** *Sat. Rev. Lit.* 19 Jan. 7/1 It is one of those rare books which, fated for bestsellerdom, will at the same time interest..the serious, adult reader. **1956** A. Wilson *Anglo-Saxon Attitudes* ii. i. 200 All the bad writers and painters who aren't even best-sellers.

bester. (Earlier and later examples.)

a **1852** Mayhew *Lond. Labour* Extra vol. (1862) 32/1 The '*Bouncers*' and '*Besters*' obtain what they want by betting, intimidating, or talking people out of their property. **1906** *Daily Chron.* 15 Nov. 6/5 An Action for slander had been commenced against him by the Hoopers for calling them 'two besters'. **1941** Baker *Dict. Austral. Slang* 9 *Bester*, a fraudulent bookmaker.

bestest (be·stĕst), *a.* and *adv.* *dial.* and *joc.* [f. Best *a.* + -est.] Used emphatically as superl. of Good *a.* and of Well *adv.*: very best.

1868 S. Lysons *Vulgar Tongue* 49 Our word 'best' is a syncope of 'beltistos'; and our Gloucestershire people stick to the 'bestest', which is nearer to the Greek than the modern pronunciation. **1905** E. Phillpotts *Secret Woman* i. iv. 36 It all comes back to Him, though you may try your bestest to make a case. **1922** Joyce *Ulysses* 419 The bestest puttiest longbreak yet.

besweat, *pa. pple.* Revived by William Morris.

1876 Morris *Sigurd* iv. 316 Lo, lord, our spurs are bloody, and our brows besweat with haste.

bet, *sb.* Add: **1.** *a good bet, best bet* (fig.) (orig. *U.S.*): a satisfactory choice; the person, thing, or course most likely to succeed; *my bet is* (colloq.): = 'I bet', my opinion is.

1906 H. Green *Actors' Boarding House* 224 He hailed his juicy jay as the one best bet of a good afternoon. **1923** Wodehouse *Inimit. Jeeves* vi. 71 'The fact is, Sir Roderick is being rather troublesome.' 'Thinks I'm not a good bet?' Wants to scratch the fixture? Well, perhaps he's right.' **1941** L. A. G. Strong *Bay* viii. 190 Keep that up...It's our best bet. **1947** 'N. Blake' *Minute for Murder* viii. 168 On the face of it, he's the best bet...He provided the poison. **1958** *Times Rev. Industry* May 52/3 Speculating on possible markets...Packaging film and pipes are considered the 'best bet'. **1954** J. B. Priestley *Magicians* vi. 134 My bet is there's been a fair amount of swift dirty work round here while poor little hubby's been kept at his lab. **1958** *Listener* 25 Sept. 468/2 My bet is that the listening audience would decrease rapidly.

b. An amount staked on the result of a card-game; *spec.* in faro (see quot. 1909); *heeled bet*: see *Heeled *ppl. a.*

1796 C. Jones *Hoyle's Games Improved* (new ed.) 276 *The Game of Faro*...He may masque his Bets, or change his Cards whenever he pleases. **1810** *Sporting Mag.* XXXVI. 32/1 The dealer [at Lansquenet]..proceeds to play, having first *covered*, that is, placed an equal sum near the sums staked by the players, to demonstrate to each that he has accepted his bet. **1844** [see *Ante *sb.*[1]]. **1880** J. Blackbridge *Compl. Poker-Player* xix. 129 When a player makes a bet, the next player must either *see him*..or *go better, i.e.*, make the previous bet good..or he must pass out. **1891** 'L. Hoffmann' *Cycl. Card Games* 202 If a player bets, or raises a bet, and no other player goes better, or calls him, he wins the pool. **1909** *Cent. Dict. Suppl.*, *Bet*, in faro, a card which is a case, that is, the only one of that denomination remaining in the box: so called because the player cannot be split.

bet, *v.* Add: **b.** *you bet* (earlier and later examples); also *you bet you*.

1857 *Phoenix* (Sacramento) 22 Nov. 2/2, I saw all the 'boys', and distributed to them the papers and 'you bet', they were in great demand. **1868** [see sense *c]. **1872** 'Mark Twain' *Roughing It* xx. 152 'I'll get you there on time' and you bet you he did, too. **1910** S. E. White *Rules of Game* v. xxxiv, 'He's a quick thinker, then,' said Bob.. 'You bet you!' **1928** D. L. Sayers *Lord Peter views Body* iv. 68 'Can you handle this outfit?' 'You bet,' said the scout.

c. In various (orig. *U.S.*) slang asseverative phrases meaning: to stake everything or all one's resources (upon the truth of an assertion).

1852 *San Francisco Sun. Dispatch* 18 Jan. 1/5 He's around when there's money in the pipe—bet your life on t-h-a-t. **1856** *Spirit of Times* (N.Y.) 6 Sept. 3/3 You may bet your old boots on that. **1865** Bret Harte *Poet. Wks.* (1872) 81 Smart! You bet your life 'twas that! **1866,** etc. [see *bottom dollar* s.v. *Bottom *sb.* 19]. **1868** *All Year Round* 31 Oct. 489/2 'You bet' or 'You bet yer life', or 'You bet yer bones', while to 'bet your boots' is confirmation strong as holy writ—in the mines, at least. **1913** Wodehouse *Little Nugget* i. i. 14 'You will order yourself something substantial, marvel-child?' 'Bet your life,' said the son and heir tersely. **1933** M. Lowry *Ultramarine* 39 'You bet your boots,' he replied. **1957** P. Frank *Seven Days to Never* i. 22 He would bet his bottom dollar..that his target would be one of those bases.

Also in corrupt forms (*I*, *you*, etc.) **betcha, betcher,** representing colloq. pronunciation of *bet you* or *your* (*life*).

1922 'R. Crompton' *Just—William* ix. 174 You betcher life! **1936** Wodehouse *Laughing Gas* v. 61 'You're homesick, what?' 'You betcher.' **1936** F. Clune *Roaming round Darling* xxii. 219 Andrew Hume..had said one day to the governor of the prison, 'Betcher I know where to find the sole survivor of the Leichhardt expedition.' 'Betcher!' mocked the governor. **1940** G. Butler *Kiss Blood off Hands* v. 82, I collared a kid..and asked him if he wanted to earn a shilling. 'You betcha, mister,' he said. **1962** J. Ludwig in R. Weaver *First Five Years* 29 Your tea's cold, I betcha.

beta. Add: **2. e.** *Math.* *beta distribution*, a distribution of a variate x of the form

$$dF = x^{p-1}(1-x)^{q-1}dx/B(p, q),$$

where $0 \leqslant x \leqslant 1$, $p, q > 0$; *beta function*, the function $B(p,q) = \int_0^1 x^{p-1}(1-x)^{q-1}dx$, where $p, q > 0$; also called the *complete beta function* and the *first Eulerian integral*; also, the function $B_x(p, q) = \int_0^x x^{p-1}(1-x)^{q-1}dx$ (the *incomplete beta function*).

1888 W. W. R. Ball *Hist. Math.* xviii. 368 The Beta and Gamma functions were invented by Euler. **1924** C. H. Forsyth *Introd. Math. Anal. Statistics* iv. 61 The value of any Beta function for positive values of m and n can be expressed directly in terms of Gamma functions. **1941** *Biometrika* XXXII. 153 The special character of the Beta-distribution..makes it necessary to vary the method of computation in different parts of the range of variables covered. **1967** A. Battersby *Network Analysis* (ed. 2) 319 The method of calculating the expected value and variance of the duration of a job depends on the assumption that the duration follows a Beta distribution. The Beta distribution is derived from the Beta function.

f. *Physics.* *beta* (or β) *radiation*, the second of three types of radiation emitted by radioactive substances, having greater penetrating power than alpha radiation. So *beta* (or β) *particle, ray*.

1899 Rutherford in *Phil. Mag.* XLVII. 116 These experiments show that uranium radiation is complex, and that there are present at least two distinct types of radiation —one that is very readily absorbed, which will be termed for convenience the α radiation, and the other of a more penetrative character, which will be termed the β radiation. **1902** Rutherford & Grier in *Phil. Mag.* IV. 325 For brevity and convenience we will call the non-deviable rays of *all* radioactive substances α *rays* and the deviable rays β *rays*. **1904** Rutherford in *Nature* 10 Mar. 437/1 It is to be expected that Röntgen rays would be set up at the sudden *starting* as well as the sudden stopping of the electron or β particle. **1909** Soddy *Interpr. Radium* 55 In ordinary circumstances radium appears to be expelling both α- and β-particles together. **1922** A. S. Eddington *Theory of Relativity* 19 The β particles shot off from radioactive substances are negative electrons which sometimes attain speeds of 100,000 miles a second. **1933** *Discovery* Feb. 45/1 Beta radiations, which are streams of electrons. **1957** *Technology* Mar. 14/2 Both beta and gamma rays can also be used for gauging thickness or height in the examination, for instance, of red-hot sheets of steel.

Hence denoting a process in which, or a substance from which, beta particles are emitted, as *beta decay, disintegration, emission, emitter*.

1931 Rutherford in *Proc. R. Soc.* A. CXXXII. 674 There will be a broadening of the line by the Doppler effect due to recoil of the nucleus from the immediately preceding β-disintegration. **1933** C. D. Ellis & Mott *Ibid.* CXLI. 502 The..upper limit of the β-ray spectrum is a significant parameter with which to classify a β-disintegration. **1934** G. Gamow *Ibid.* CXLVI. 217 Every theory treating the β-decay as the transformation of a nuclear neutron into a proton. **1935** *Internat. Conference Physics 1934* I. 41 The final state of the nucleus after a β-emission is an excited one. **1947** C. D. Coryell in C. Goodman *Sci. & Engin. Nuclear Power* I. 248 (*heading*) Long-lived β emitters without γ radiation. **1955** *Gloss. Terms Radiology* (*B.S.I.*) 6 *Beta decay*, the radioactive decay process characterized by the emission of a beta particle. The parent atom involved is called a beta emitter.

g. *Metallurgy.* (*a*) Applied to the second of a series of allotropic forms of iron (see quot. 1949); (*b*) applied to the second of a range of alloys, as *beta brass*, the second of a series of alloys of copper and zinc, containing less copper than in alpha brass. Cf. *alpha 3 d.

1885, etc. [see *alpha 3 d]. **1895** *Jrnl. Iron & Steel Inst.* XLVIII. 258 A special strong, hard, brittle, allotropic state called β iron. **1914** *Jrnl. Inst. Metals* XI. 105 Plane sections through a bees' honeycomb..a benzene water-foam, and an over-annealed β brass are all very similar in appearance. **1949** R. T. Rolfe *Dict. Metallogr.* (ed. 2) 24 *Beta iron*, formerly, a supposed allotropic form of i[ron] stable between..768° and about 900° C., of body-centred cubic form, and only very feebly magnetic. In reality, the change in magnetic susceptibility of α-iron occurring at 768° C. is not associated with any structural transformation and β-iron is no longer recognized as a distinct variety.

h. *beta rhythm, waves*, fast activity of the brain recorded in an electroencephalogram, having a frequency of fourteen to twenty-four cycles per second. Cf. *alpha 3 f.

[**1935** *Brain* LVIII. 349 Berger describes a definite series of small oscillations—the *b* waves—superimposed on the primary *a* waves of the 10 a second rhythm. The frequency of the *b* waves is given as 35 a second.] **1936** *Archives Neurol. & Psychiatry* XXXVI. 1215 Less prominent, because lower in voltage, is a faster rhythm, with a frequency of about 25 waves a second, called by Berger the 'beta waves'. *Ibid.* 1220 A prominent fast (beta) rhythm. **1943** *Electronic Engin.* XVI. 237 Faster rhythms, sometimes called 'beta' rhythms, have been described in the EEG.

3. An examiner's second-class mark. Also *transf.*

1902 [see *alpha 4]. **1936** 'N. Blake' *Thou Shell of Death* vi. 104 He's quite a decent old trot, but definitely in the Beta class. **1958** *Oxf. Mag.* 13 Mar. 362/1, I have mentioned twenty recent pictures..four have some *alpha* about them; the rest are gamma minus to beta plus (I have not mentioned deltas).

betacism (bi·tăsiz'm). [ad. mod.L. *betacismus*, f. *beta* (see Beta), after L. *iotacismus*, etc.] (See quot. 1926.)

[**1885** *Amer. Jrnl. Philol.* VI. 501 Even these forms were threatened with destruction by the spread of Betacismus, whereby *amavit* was pronounced like *amabit* and vice versa.] *Ibid.*, In Africa, especially where betacism flourished, the need of new substitutes for the future made itself early felt. **1926** Plater & White *Gram. Vulgate* 43 What is called betacism (the interchange of *b* and *v*) causes great confusion in verbs of the first conjugation.

betafite (bi·tăfəit). *Min.* [a. F. *betafite* (A. Lacroix 1912, in *Compt. Rend. Acad. Sci.* CLIV. 1042), f. *Betafo* in Madagascar, one of its localities + -ite[1].] A niobate and titanate of uranium, etc.

1912 *Jrnl. Chem. Soc.* CII. ii. 567 Radioactive Uraniferous Columbotantalotitanites from Pegmatites of Madagascar...Betafite, D 4·17, is dark green, and shows rudimentary cubic faces. **1921** *Brit. Mus. Return* 155 Minerals from.. Madagascar (chrysoberyl, betafite, columbite, etc.).

betatron (bi·tătrǫn). [f. Beta + *-tron.] A machine designed for the acceleration of β particles.

Kerst first spoke of the machine (1940) as an 'Induction Accelerator'.

1941 *Time* 29 Dec. 44/3 A new electron-hurling machine.. its brilliant young inventor, Donald William Kerst, 30, who calls the machine a 'betatron'. *Ibid.*, 'The heart of the betatron,' explains Inventor Kerst, 'is a doughnut-shaped glass vacuum tube between the poles of a large electromagnet.' **1942** D. W. Kerst in *Rev. Sci. Instr.* XIII. 387 (*title*) A 20-Million Electron-Volt Betatron or Induction Accelerator. **1960** *Gloss. Atomic Terms* (*H.M.S.O.*) 6 *Betatron*, a machine for accelerating a continuous beam of electrons to high speeds in a circular path by means of the electric field produced by a changing magnetic flux.

betcha, betcher: see *bet *v.* c.

betel. Add: **2. betel(nut) palm**, the Areca palm, *Areca catechu*, from which the 'betel nut' or areca nut is obtained.

1861 Bentley *Man. Bot.* 684 The Catechu or Betel Nut Palm. **1875** *Encycl. Brit.* III. 616/2 The betel nut is the fruit of the Areca or betel palm, *Areca Catechu*.

Betelgeuse (be·tĕldʒūz, ‖-elʒöz, bi-). *Astr.* Also **Betelgeux** (-ʒö). [Fr. *Bételgeuse*, f. Arab.] A yellowish-red variable star of the first magnitude, the brightest in the constellation of Orion.

1796 C. Hutton *Math. & Philos. Dict.* I. 205/1 *Betelgeuse*, a fixed star of the first magnitude in the right shoulder of Orion. **1820** H. Brooke *Guide to Stars* 12 A line from the Pleiades..will give Bellatrix in the western shoulder of Orion, and the bright star immediately to the eastward is Betelgeux in the right or eastern shoulder. **1866** J. N. Lockyer tr. *Guillemin's Heavens* ii. viii. 355 Betelgeuse, the brightest star in Orion, and Aldebaran show a decided red tint. *a* **1930** D. H. Lawrence *Last Poems* (1932) 49 The heavens are wandering The moon and the earth, the sun, Saturn and Betelgeuse. **1938** R. Graves *Coll. Poems* 121 We have no truck either with the forebeings Of Betelgeux or with the atom's git. **1955** *Oxf. Jun. Encycl.* III. 109/2 On Orion's shoulder the giant orange star Betelgeuse shines. **1962** *Listener* 31 May 950/1 Betelgeux in Orion..is a typical Red Giant.

‖ **bête noire** (bę̄t nwār, bę̄ᶦt nwāɹ). [Fr., lit. 'black beast'; *fig.* insufferable person.] A person or thing that is the bane of a person or his life; an insufferable person or thing; an object of aversion.

1844 Thackeray *Barry Lyndon* xvii. in *Fraser's Mag.* XXX. 235/1 Calling me her bête noire, her dark spirit, her murderous adorer, and a thousand other names indicative of her extreme disquietude and terror. **1850** *Household Wds.* 6 July 359/1 You or any one else's *bête noire* is apt to get polished off with a few extra touches of blacking. **1860** W. H. Russell *Diary India* I. 209 Jung Bahadoor, who is evidently the present *bête noir* of our General's life. **1866** Mrs. H. Wood *Elster's Folly* xiv, It was the bête noire of Clerk Gum's life. **1905** *Spectator* (Lit. Suppl.) 28 Jan. 118/2 [His] *bête-noire* is the submission of military affairs to the control of 'political exigencies'. **1961** L. Mumford *City in History* x. 290 As for the disposal of ordure, this has always been the bête noire of close urban settlements.

‖ **bête rouge** (bę̄t rūʒ, bę̄ᶦt rūʒ). [Fr., lit. 'red beast'.] A species of *Trombidium* (see quots.).

1909 *Cent. Dict.* Suppl., *Bête-rouge*,..probably a larval *Trombidium*, which in the West Indies corresponds to the so-called 'red bug' or 'jigger' of the southern United States. **1932** Dorland *Med. Dict.* (ed. 16) 191/1 *Bête rouge*, the red mite of Martinique and Honduras which burrows into the skin. It is probably a species of *Trombidium*. **1934** E. Waugh *Handful of Dust* v. 267 He had picked up bêtes rouges in the bush and they were crawling and burrowing under his skin.

Beth Din (bēᶦt dī·n). Also **Beth-Din, Bethdin.** [Heb. *bēth* house, *dīn* judgement.] A Jewish court composed of the Chief Rabbi and two or more assistants, responsible for matters of Jewish ecclesiastical law and the settlement of disputes between Jews.

1795 William Scott in J. Haggard *Rep. Cases Consist. Crt.* (1822) I. 244 The Answers of the Bethdin to the Questions proposed by the Court. **1897** *Daily News* 30 Jan. 3/3 Are you in competition with the Chief Rabbi in his Court the Beth Din? **1928** *Sunday Dispatch* 30 Dec. 7/2 Should the proceedings of London's Jewish House of Judgment—the Beth-Din—be kept secret?

Bethell (be·pĕl). The name of John *Bethell*, a nineteenth-century American inventor, used *attrib.* or in the possessive to designate a process of wood preservation (see quot. 1940).

1838 *Brit. Pat.* 7731 (*title*) Rendering wood, cork, and other articles more durable, &c. Bethell's specification. **1901** *Trans. Amer. Soc. Civil Engin.* XLV. 503 All the French roads [*i.e.* railways] but one have creosoting by the 'Bethell' process. **1940** *Chambers's Techn. Dict.* 88/2 *Bethell's process*, a process for preserving timber, which is first dried, then subjected to a partial vacuum within a special cylinder, and finally impregnated with creosote under pressure. **1962** W. P. K. Findlay *Preservation of Timber* iii. 38 Bethell's process is often called the Full Cell Process because it results in filling the cells of the treated zone with a liquid.

Beth Hamidrash (bēᶦt hamidrā·ʃ). Also **Bet Hamidrash, Beth Hamedrash, Beth Hammidrash.** [Heb. *bēth hammidhrāsh*, lit. 'house of study'.] (See quot. 1962.)

1874 *Jewish Directory* 17 Beth Hamedrash. Smith's Buildings, 75, Leadenhall Street, E.C. For the study and exposition of the Scriptures and Commentaries and Hebrew Literature generally. **1887** *Encycl. Brit.* XXII. 811/2 The school (*bēth hammidrāsh*) trained scholars, but the synagogue..was the means of popular instruction. **1892** Zangwill *Childr. Ghetto* (1893) i. v. 57 The learned Rabbonim, who sat shaking themselves all day in the Beth Hamidrash. **1962** Bridger & Wolk *New Jewish Encycl.* 51/1 *Bet Hamidrash*, Hebrew term meaning 'house of study', applied to a place where Jews congregate for prayer and study... The term is also applied to a rabbinical school for higher studies.

‖ **bêtise** (bę̄tiz, bēᶦti·z). [Fr. = stupidity, f. *bête* foolish, OF. *beste* Beast.] A foolish, ill-timed remark or action; a piece of folly.

1827 Lady Granville *Lett.* (1894) I. 406, I think it is an *enfantillage*, a *bêtise*, on the part of the Austrian Government. **1831** Disraeli *Young Duke* II. iii. xix. 256 This *bêtise* of a war has made us all serious. **1849** Geo. Eliot *Let.* 9 Feb. (1954) I. 278, I will not dilate on the book or on your criticism for I am so sleepy that I should write nothing but bêtises. **1893** Henry James *Let.* 2 May in C. Mackenzie *Life* (1963) II. 309 If your public..can't see any of that charm, and wants such a *bêtise* instead, we are engaged in a blind-alley. **1948** 'J. Tey' *Franchise Affair* i. 9 In spite of her *bêtise* about striped suits he saw her point of view.

Betjeman (be·tʃəmăn). The surname of John *Betjeman* (1906–), English author and poet, applied *attrib.* to a style of Victorian architecture that he is known to admire. Hence **Be·tjemanic, Be:tjemane·sque, Be·tjemanish,** *adjs.*, of or pertaining to Betjeman; resembling the style admired by Betjeman; **Be·tjemanite**, a supporter of Betjeman or his views.

1956 L. E. Jones *Edwardian Youth* viii. 182 We emerged from those grim Betjemanic halls. **1958** *Observer* 14 Dec. 9/2 Betjeman belongs to that eccentric group of people.. whose names have become adjectives. Betjeman country, Betjeman houses, and Betjeman streets seem to have changed from gloomy Victoriana into charming antiquities since he started writing about them. **1959** M. Bradbury *Eating People is Wrong* i. 22 He found himself getting more and more a Betjemanesque *frisson* from Victoriana. **1959** *Spectator* 20 Feb. 258/3 The Beauregard Hotel..is a frowsty, tragi-comic, Betjemanish place with permanent 'residents', stained-glass lozenge windows, and the aspidistras and gentility and ghastliness no more exaggerated than you find them in such places. **1959** *Listener* 11 June 1031/2 The neo-Victorian Age is upon us. The Law Courts have become a shrine for the Betjemanites. **1963** *Punch* 9 Oct. 527/2 A Betjemanesque society dedicated to the preservation of fairground lettering.

betray, *v.* Add: **4. b.** *spec.* To induce (a woman) to surrender her chastity by false promises; to seduce. Also *absol.*

1766 Goldsmith *Vicar* xxix, When lovely Woman stoops to folly, And finds too late that men betray. **1891** Hardy *Tess* xvi, A conviction not so entirely unknown to the 'betrayed' as some amiable theorists would have us believe. **1926** J. Black *You can't Win* v. 52 Betrayed and deserted, she stole enough of her father's money to take her to the city and into a hospital where her baby was born.

Betsy (be·tsi). *slang* or *dial.* (orig. *U.S.*). Also **Bessy, Betsey,** and with lower-case initial. [Var. Betty *sb.*] A gun or pistol; *spec.* one's favourite gun; freq. *old Betsy*.

1856 *Spirit of Age* (Sacramento) 4 Nov. 3/1 Jest let them raise that check agin me, and if I don't shoot why old Betsy won't blizzard. **1869** *New No. West* (Deer Lodge, Mont.) 20 Aug. 2/7 Mr Fredericks proceeded immediately on the horse, loaded 'Betsey', (his shot gun). **1895** P. H. Emerson *Marsh Leaves* liv. 137 Potter had a long-barrelled gun, his 'old Bessy', and I took my eyes along with me. **1965** J. P. Carstairs *Concrete Kimono* v. 48 'You've noticed I'm toting a Betsy.' 'Betsy?' 'Equalizer, rod, gat, iron.' **1969** *Win* 15 May 12/2 Emotionally, the satisfying thing to fantasize is grabbing ole betsy and going to join my black brothers in Wayne County.

better, *a.* Add: **A.** *adj.* **1. c.** *Comb.* with sbs. used attrib., as *better-class, -quality, -type*.

1890 W. F. Butler *Sir Chas. Napier* 7 Merchants or local better-class farmers. **1965** K. Giles *Some Beasts No More* ii. 30 The better-class houses of 1910 vintage with area steps now unwhitened. **1908** *Westm. Gaz.* 12 Aug. 11/3 Better-quality teas. **1958** *Spectator* 8 Aug. 196/1 The better-type delivery library.

d. In phrases used *attrib.*, as *better-than-average, better-than-chance*.

1922 A. Bennett *Let.* 8 Nov. (1966) I. 317 The respectable better-than-average English plays. **1938** J. Hilton *To You, Mr. Chips* i. 21 The school was perhaps a better-than-average example, both educationally and structurally, of its type. **1964** *Language* XL. 262 A better-than-chance probability that a language will possess certain features.

4. b. *you'd better believe*: you may be assured. *colloq.* (orig. *U.S.*).

1856 *Yale Lit. Mag.* XXI. 171 (Th.), You'd better believe, I'll live in the clover. **1872** O. W. Holmes *Poet. Breakf.-t.* x, My old gentleman means to be Mayor or Governor or President..you'd better b'lieve. **1968** *Globe & Mail* (Toronto) 17 Feb. 42 (Advt.), You'd better believe it... We've got 'em.

(*b*) With *had* omitted (occas. with pronoun also omitted). *colloq.* (orig. *U.S.*).

1831 S. Smith *Major Downing* (1834) 65 My clothes had got so shabby, I thought I better hire out a few days and get slicked up a little. **1846** J. J. Hooper *Some Adv. Simon Suggs* (1851) i. 154 You better mind the holes in them ere rocks. **1865** *Major Jack Downing of Downingville Militia* (1867) ix. 70 You better believe we've been in an awful excitement here. **1904** A. Bennett *Great Man* xii. 120 Miss Foster she says her name is. Better show her in here, hadn't I? **1910** C. E. Mulford *Hopalong Cassidy* xi. 73, I reckon you better pull out—you ain't needed around here. **1946** K. Tennant *Lost Haven* (1947) i. 15 Well, we better get back to the house. *Ibid.* iii. 53 Brace better meet that train. **1968** *Listener* 9 May 596/2 You better get those two guys inside.

5. Phr. *no better than* (one) *should* (or *ought to*) *be*, of doubtful moral character, (usu.) of easy virtue.

1604 *Pasquils Jests* sig. C, A Man, whose wife was no better then she should be. **1764, 1780** [see Shall *v.* 18]. **1815** C. Lamb *Let.* 28 Apr. (1935) II. 158 To term her a poor outcast seems as much as to say that poor Susan is no

better than she should be. **1871** Trollope *Eust. Diamonds* (1873) I. xxiii. 305 He..almost believed that she was not now, and hadn't been before her marriage, any better than she should be. **1882** [in Dict. sense 1 a]. **1904** A. Bennett *Great Man* xiii. 139 Her..suspicion that the self-styled Miss Foster was no better than she ought to be. **1937** A. L. Rowse *Sir R. Grenville* xix. 346 The one was a brute, and the other a vixen, and no better than she should be into the bargain.

6. *for better, for worse*: see Worse B. 3 a; *to go (one) better*: see Go *v.* 36.

B. *adv.* **7.** *better off*: see Off *adv.* 11. Also as *adj. phr.* and *absol.* as *sb.*

1865 [see Off *adv.* 11]. **1865** Mill *Auguste Comte* 167 The better-off classes. **1940** F. Kitchen *Brother to Ox* i. 15 The 'better offs' sat in the chairs. **1959** B. Wootton et al. *Soc. Sci. & Soc. Path.* 366 The better-off, or those who still have better-off standards, can deal with the unmarried mother and the illegitimate child more easily.

betterment. Add: **2.** (Earlier example.)

1785 in *Vermont State P.* (1823) 501 The value of the improvements and betterments made on such lands by such possessor or possessors.

b. Enhanced value (of real property) by reason of local improvements. So *betterment tax*, an assessment made on this.

1890 *Spectator* 28 June 893/2 It was hoped that the principle of 'betterment' might become the means for raising considerable sums of money. **1890** *Contemp. Rev.* May 644 The betterment tax. **1896** J. H. B. Browne & Allan *Compensation* 683 'Betterment' as a principle is only a proposal to tax the increment when it clearly and directly arises from an improvement carried out by a public authority and at the public expense. *Ibid.* 684 A betterment charge in respect of improvements effected by local authorities. **1902** *Encycl. Brit.* XXVII. 182/2 The principle of 'betterment', according to which promoters would be allowed to set off against an owner's claim for compensation any enhancement of the value of his land by the use of the land taken by the promoters. **1960** *Spectator* 15 July 88 The 1947 Silkin Town and Country Planning Act..introduced a system of development charges—or betterment tax, based on the value of a particular planning permission.

better-to-do, *adj. phr.* [Comparative of Well-to-do.] Above the well-to-do in social condition or worldly circumstances; more prosperous. Also *absol.*

1890 W. Booth *In Darkest England* ii. vi. 217 His better-to-do neighbour will often assist him by providing the capital. **1898** *Daily News* 6 Dec. 5/6 To the new L.C.C. buildings there had come many better-to-do workmen. **1905** *Daily Chron.* 1 Feb. 6/2 Children..of the better-to-do. **1940** *Manch. Guardian Weekly* 9 Feb. 116 Even among the better-to-do there is greater simplicity of living and independence of service than among their compeers in England. **1965** in P. Jennings *Living Village* (1968) 133 Many of the better-to-do people send their children away to school.

betting, *vbl. sb.*[1] Add: **2. betting-book** (earlier examples); **betting-house** (example); **betting-man** (earlier and later examples); **betting office**, an establishment licensed to handle bets (on horse-races, dog-races, etc.); **betting ring** = Ring *sb.*[1] 13 b; **betting room**, a room where betting is carried on; **betting shop** *colloq.* = **betting office*; **betting-slip**, a slip of paper on which a bet is entered.

1813 *Sporting Mag.* XLII. 141/1 By a recent exposition of his betting-book at Tattersall's, it appeared that he lost 1200 guineas. **1836** R. S. Surtees *Let.* in E. K. Mathews *Mem. C. Mathews* (1839) IV. 194 As soon as ever he got to town he would advertise for a man that could keep a betting-book. **1853** *Act* 16 & 17 *Vict.* c. 119 (*title*) An Act for the Suppression of Betting Houses. **1819** *Sporting Mag.* V. 3/2 The betting men have all agreed to pay and receive accordingly. **1825** P. Egan *Life of Actor* iv. 140 Sir Henry was..not..acknowledged as a betting man in the sporting circles. **1939** A. Christie *Ten Little Niggers* xi. 162 I'm not a betting man. And anyway if you were dead I wouldn't get paid. **1852** *Chambers's Edin. Jrnl.* XVIII. 57/1 'Commission Office', 'Racing Bank',..'Betting-Office', are the styles of announcement adopted by speculators who open what low people call Betting-shops. **1951** *R. Commission on Betting, Lotteries & Gaming 1949–51 Rep.* (*Cmd.* 8190) 65 We were told by one police witness that the licensing of betting offices would merely be regularising what is a fact in his city. **1961** *Economist* 6 May 565/1 Nobody has any real idea of how many bookmakers in Britain have taken advantage of the new Betting and Gaming Act..to open betting shops—or..Licensed Betting Offices. **1822** *Sporting Mag.* X. 4/1 He was well known in the betting ring. **1954** W. Faulkner *Fable* 57 A hard-faced jockey-sized man who seemed to have brought on his warped legs..something of hard, light, razor-edge horses and betting-rings. **1793** *Sporting Mag.* II. 124/1 In the betting-room, at Tattersall's the horse was so little thought of, that he had never been mentioned. **1857** Dickens & W. Collins in *Househ. Words* XVI. 410/1 A complete choke and stoppage of the thoroughfare outside the Betting Rooms. **1852** Betting-shop [see **betting-office* above]. **1961** *Daily Tel.* 2 May 13/5 The Betting and Gaming Act came into force yesterday, but..betting shops are not easy to find. **1927** W. E. Collinson *Contemp. Eng.* 37 Betting agents or touts are often had up before the courts for passing betting-slips in the streets to would-be backers. **1959** J. Braine *Vodi* xxv. 264 The action was surreptitiously executed, as if he were passing a betting-slip in the street.

Betty, *sb.* Add: **5.** *Betty lamp* (see quots.).

1893 A. Earle *Customs Old New England* iv. 125 Betty lamps..were a shallow receptacle..with a projecting nose an inch or two long..they were filled with tallow or grease, and a wick..was placed so that the lighted end could hang on the nose. **1952** J. Gloag *Dict. Furn.* iii. 141 *Betty lamp*, a type of lamp used in America during the 18th century... It resembles the Scottish type of hanging lamp known as a cruisie. **1957** *Encycl. Brit.* XIV. 101/2 The Betty lamp of the Pilgrims (1620)..was equipped for hanging from mantelpieces or shelves.

Betty Martin. In slang phr. (*all*) *my eye (and) Betty Martin*: 'all my eye' (see Eye *sb.*[1] 2 h).

1781 S. Crispe *Let.* 16 Oct. in W. H. Hutton *Burford Papers* (1905) iv. 69 Physic, to old, crazy Frames, like ours, *is all my eye and Betty Martin*—(a sea phrase that Admiral Jemm frequently makes use of). **1785** [see Eye *sb.*[1] 2 h]. **1812** J. & H. Smith *Rej. Addr.* 104 The knife that I thought that I saw, Was nought but my Eye Betty Martin. **1859** Hotten *Dict. Slang* 1 All my eye, answer of astonishment to an improbable story; All my eye and Betty Martin, a vulgar phrase with similar meaning. **1930** W. de la Mare *On Edge* 108 You might be suggesting that both shape and scarecrow too were all my eye and Betty Martin.

between, *prep.* and *adv.* Add: **B.** *adv.* **4. between-lens shutter** *Photogr.*, a type of shutter that is fitted between the components of a double lens; **between-time** (delete † and add later examples); **between-times** *adv.*, in the intervals of time: = Between-whiles; **between-war(s)** *adj. phr.*, of or belonging to the period between two wars, spec. the world wars of 1914–18 and 1939–45.

1909 *Westm. Gaz.* 20 Feb. 14/2 The popular between-lens shutters. **1950** G. L. Wakefield *Your Camera Lens & Shutter* 86 A between-lens shutter exposes the whole of the film at once; the focal-plane shutter..exposes it piecemeal in strips. **1909** *Westm. Gaz.* 10 Aug. 6/3 There is no question of any between-time. Immediately the day ends, the night's work begins. **1952** E. Bowen *Demon Lover* 222 These are between-time stories. **1907** *Westm. Gaz.* 13 Sept. 2/1 The Fairy Queen sat between-times, when she was aweary, beside the Fairy King. **1941** 'G. Orwell' *Lion & Unicorn* 44 The stagnation of the Empire in the between-war years. **1945** J. B. Priestley *Three Men in New Suits* viii. 142 Those bright successful novels of the between-wars period.

C. 2. (Earlier and later examples.)

1849 Longfellow *Kavanagh* v. 22 If I do not like the sizes, he offers to exchange them for others, either sharps or betweens. **1955** J. E. Liberty *Pract. Tailoring* (ed. 2) i. 2 There are several types of needles, chief of which (for tailoring) are ground downs, betweens and sharps...Betweens, as the name implies, are medium pointed.

betwee·n-maid. A maidservant who assists both the cook and the housemaid; = Tweeny. Also *between girl, servant.*

1890 *Daily News* 23 June 7/5 Between servant..to assist cook and housemaid. *Ibid.* 9 Dec. 7/6 Between-maid wanted. **1891** *Ibid.* 11 Nov. 8/4 Wanted..good Plain Cook, also Between Girl. **1914** J. M. Barrie *Admirable Crichton* i, A tweeny; that is to say, my lady, she is not at present, strictly speaking, anything; she is a between maid; she helps the vegetable maid. **1923** *Daily Mail* 25 Jan. 5 All of the general servant class masquerading as parlour-maids, cooks, between-maids, or others of experience.

betweenness (bitwi·n,nės). [f. Between + -ness.] The condition or fact of being between; *spec.* in *Math.*, the ordinal quality possessed by one of every three points on a straight line, in virtue of which it may be said to lie between the other two points.

1892 *Monist* II. 243 In reality there are not two things and, in addition to them a betweenness of the two things. **1904** *Science* 11 Mar. 410/2 Of Hilbert's betweenness assumptions, axioms of order, one of the five is redundant. **1966** *Mathematical Rev.* Jan. 18/1 The author lists fifteen possibilities for a betweenness relation on a lattice.

beudantite. (Examples.)

1826 A. Levy in *Ann. Philos.* II. 194 The two [specimens] ..belong, I believe, to two new species. For one of them I propose the name of Beudantite, in honour of Mr. Beudant. **1850** *Phil. Mag.* XXXVII. 167 Beudantite is *chiefly* composed of sesquioxide of iron and oxide of lead. **1960** *Mineral. Mag.* June 423 Carminite and beudantite are both rare minerals for which very few localities have hitherto been recorded.

‖ **beurre** (bör). [Fr., butter.] In phrases: *beurre manié* (manie) [F. *manié* handled], a mixture of flour and butter used for thickening sauces or soups; *beurre noir* (nwār) [F. *noir* black], a sauce made by heating butter until it is brown, usually mixing it with vinegar; = **black butter* (b).

1856 E. Acton *Mod. Cookery* ii. 53 Skate..is..sauced with *beurre-noir* (burned or browned butter). **1939** A. L. Simon *Concise Encycl. Gastronomy* 8/1 A 'beurre manié'..is a piece of butter the size of a large walnut, rolled and indeed 'squashed' in flour. **1960** *Times* 1 Aug. 9/3 Beurre Manié. For thickening a sauce quickly just before serving. *Ibid.*, Beurre Noir. Heat 1½ oz. butter till it is brown, but not black.

BeV, B.E.V., bev: see * B III.

bevatron (be·vătrǫn). [f. *BeV* (see *B III) + *-tron.] A synchrotron designed to accelerate protons to energy levels in the BeV range.

1947 E. O. Lawrence in *Centennial of Sheffield Sci. School* (1950) 24 The Bevatron..a high-energy multi-billion volt proton accelerator. **1947** *Sci. News Let.* 25 Oct. 259/1 A new monster atom-smashing 'bevatron' to attain 10 billion electron volts was also shown.

bevel, *a.* and *sb.*[1] Add: **A.** *adj.* **2.** (Later examples.)

1943 L. MacNeice in *Penguin New Writing* XVI. 41 A bevel hill with..a cairn of stones. **1944** Auden *For Time Being* (1945) 55 With bisson eye and bevel course.

B. *sb.* **4.** *Typogr.* (A piece of metal used by stereotypers to form) the bevelled edge of a plate.

a **1877** Knight *Dict. Mech.* I. 278/2 *Bevel*.., a slug cast nearly type-high and with chamfered edges. Used by stereotypers. **1900** H. Hart *Cent. Typogr.* 8 The above and also the plate used as a Frontispiece to the 1695..and 1706 *Specimens*, have no flanges or bevels, but are almost straight cut at the edges.

beveller (be·vĕləɹ). [f. Bevel *v.* + -er[1].] An operative in various trades, esp. one who bevels the edges of plate glass or steel plates; one who cuts stereotype, electrotype, or process plates and bevels them; one who cuts, bevels, and (when necessary) gilds cards.

1881 *Instr. Census Clerks* (1885) 55 Plate and Looking Glass Silverer: Beveller, Dipper. *Ibid.* 83 Card, Pattern-card..Perforator, Beveller. **1890** W. J. Gordon *Foundry* 131 We may as well follow the plate to the bevellers, where the edges are again ground and smoothened. **1921** *Dict. Occup. Terms* (1927) § 558 Card beveller..arranges a pile of cards so that edge of pile slopes at required bevel angle.. and scrapes sloped edge with sharp knife. **1961** *Evening Standard* 19 July 19/4 (Advt.), Wanted Beveller..Gt. Yarmouth.

beverage. Add: **6.** *attrib.*, as **beverage room**, in Canada, a bar-room in which beer is served; = **beer-parlor*.

1936 L. A. McKay in F. R. Scott & A. J. M. Smith *Blasted Pine* (1957) 23 'Taprooms' and 'taverns' and 'pubs' are absurd; Give us..A respectable name like 'Beverage Room'. **1968** *Globe & Mail* (Toronto) 10 July 6/4 All the hotel beverage rooms within the City of Sudbury are operating with ale and lager supplied by the local brewery.

Beveren (be·vĕrĕn). [Name of a town in Belgium.] A breed of rabbit.

1919 *Bazaar, Exchange & Mart* 18 Jan. Suppl. 14 Beverens for sale, bucks and does. **1920** *Ibid.* 19 Mar. Suppl. 24/2 Unique snow white Beverens, with sky blue eyes. **1946** *Nature* 13 July 68/1 A small feeding trial was started with twenty young Beveren rabbits.

Bevin (be·vin). [f. the name of Ernest *Bevin* (1881–1951), Minister of Labour and National Service 1940–5.] Used *attrib.* in **Bevin boy**, a young man of age for military service, but selected by lot to work in a coal-mine.

1944 *Manpower* 53 The 'Bevin boys' selected by ballot undergo preliminary training. **1947** L. MacNeice *Dark Tower*, etc. 158 *Hare.* Excuse me. Are you a Bevin Boy? *Miner.* I am not. I am a professional. **1969** *Peace News* 5 Dec. 9/2 The Bevin boys who were conscripted to go down the mines.

bevvy (be·vi). *slang.* Also **bevali, bevie, bevy.** [f. Bev(erage + -y[6].] A drink, esp. beer.

1889 Barrère & Leland *Dict. Slang* I. 110/1 *Bevy* or *bevali* (common), beer; abbreviation of beverage. **1925** Fraser & Gibbons *Soldier & Sailor Words* 22 *Bevvy*, beer. Any drink. **1934** P. Allingham *Cheapjack* vi. 59 'I think this calls for a bevvy,' I said, and we walked off to the nearest pub together. **1960** K. Amis *Take Girl like You* vi. 87 Din-din. The old scoff. Bevvy too. **1963** *New Statesman* 18 Oct. 537/1 The police..accuse the public of a lack of civic responsibility—a phrase that's good for a laugh over a bevy.

Hence as *v. intr.*, to drink. So **be·vvied** *a.*, drunk, intoxicated.

1934 P. Allingham *Cheapjack* x. 111 They just sit in the 'ouse and talk and bevvy till all the money's gone. **1960** J. MacLaren-Ross *Until Day she Dies* ii. 28 Lon and Carl coming backstage, both of them bevvied. **1966** 'L. Lane' *ABZ of Scouse* 8 The Scouser's favourite excuse for an act of hooliganism is *I wuz bevvied*. **1966** F. Shaw et al. *Lern yerself Scouse* 77 Ard cases who could bevvy by the jug.

bewer, var. *buer.

bewrite, *v.* Restrict †*Obs.* to sense in Dict., and add: **2.** *trans.* To write about. In *pa. pple.*

1875 Swinburne in *Times Lit. Suppl.* (1909) 3 June 204/2, I have probably been more be-written and belied than any man since Byron. **1895** *Critic* 1 June 396/2 Tammany has been much be-written of late. **1931** *Times Lit. Suppl.* 5 Nov. 864/3 The track covered by Mr. Waugh has been..thoroughly bewritten in recent years.

beyond, *adv.* and *prep.* Add: **B. 5. b.** (Further examples.)

1928 E. O'Neill *Strange Interlude* vii. 250 Why Gordon should take such a fancy to that old sissy is beyond me. **1966** *Listener* 12 May 699/1 How someone like Anthony de Lotbinière, its producer, could make Japan boring is beyond me, but he succeeded.

D. beyond-man, an early synonym of Superman.

1896 A. Tille tr. *Nietzsche's Thus spake Zarathustra* in *Wks.* VIII. 5 Behold, I teach you beyond-man! Beyond-man is the significance of earth. Your will shall say: beyond-man shall be the significance of earth. *Ibid.* 129 Never yet beyond-man existed. I have seen them both naked,

the greatest and the smallest man. **1896** T. Common tr. *Nietzsche's Twilight of Idols* in *Wks.* XI. 198 To be set up ..as a 'higher man', as a kind of beyond-man. **1908** *Athenæum* 13 June 729/1 The 'Super-tramp' is..the opposite of the 'oversoul' or 'beyond-man'. *a* **1917** G. B. Shaw *Let.* in *Trans. Philol. Soc. 1916–20* (1932) 8 Some of our most felicitous writers..had been using such desperate and unspeakable forms as Beyondman, when the glib Superman was staring them in the face all the time.

beyondness (bĭ,yǫ·ndnės). [f. Beyond *adv.* and *prep.* + -ness.] The condition or quality of being beyond.

[**1848** A. Steinmetz *Hist. Jesuits* I. iii. 283 Oh! 'twas a glorious prospect—a spirit-stirring something-beyondness!] **1923** W. Deeping *Secret Sanct.* xvi. 164 There was the beyondness of growing cabbages, the secret life of a cabbage patch. **1963** *Times* 26 Apr. 8/1 Will it not still be a relation to a Beyond? Call it deep down, but it always means a Beyondness.

B-girl (bī·gɜ̄ɹl). *U.S. slang.* [Abbrev. of *bar-girl* (*bar *sb.*[1] 29 c).] A woman employed to encourage customers to buy drinks at a bar.

1936 *San Francisco News* 31 July 3/6 No B Girls buzzing around you here to sip tea you think is a highball. **1938** *True Story* Dec. 66/2 'B-girls?' I repeated, 'What are they?' ..'That's just short for bar girls.' **1964** 'F. Archer' *Out of Blue* (1966) v. 76 If I stand here, I'm a waitress, see? If I sit down, I'm a B-girl, and this joint doesn't pay for that kind of protection.

‖ **bhajan** (bɒ·dʒən). *Hinduism.* Also **bhajana, bhajjan.** [Skr.] A devotional song.

1914 A. H. Fox-Strangways *Music of Hindostan* xi. 286 The oldest forms of Hindu song are the *Pada* and *Bhajana*. **1925** A. B. Fyzee-Rahamin *Music of India* iii. 32 Bharat Math..sang religious songs called *Bhajans*. **1968** *Observer* 25 Feb. 3/5 The singing of devotional songs called *bhajans*.

‖ **bhakta** (bɒ·ktā). *Hinduism.* [Skr.] A religious devotee; a worshipper, believer.

1828 H. H. Wilson in *Asiatic Researches* XVI. 12 The *Bhâktas* worshipped Vishnu as Vásudeva, and wore no characteristic marks. **1886** *Encycl. Brit.* XXI. 290/2 The *Bhaktas*, a Vaishṇava sect founded, towards the end of the 15th century, by Chaitanya. **1910** *Ibid.* XIII. 511/1 His doctrine of Bhakti distinguishes five grades of devotional feeling in the *Bhaktas*, or faithful adherents. **1930** V. Elwin *R. Rolle* i. 7 Rolle himself is a typical *bhakta*..his wandering sannyāsī life; his glowing, burning heart—all mark him as a *bhakta* after the mind of Jesus.

‖ **bhakti** (bɒ·kti). *Hinduism.* [Skr.] Religious devotion, piety, or devoted faith, as a means of salvation.

1832 H. H. Wilson in *Jrnl. Asiatic Soc.* (Bengal) I. 220 Expatiating upon the different degrees of *Bhakti*, or faith, and the various kinds of *Mukti*, or salvation. **1877** Monier-Williams *Hinduism* ix. 115 The doctrine of *bhakti*, or 'salvation by faith'. **1911** E. B. Havell *Ideals Ind. Art* 117 *Bhakti* is the moving spirit in all great religious art, in the West as in the East. **1915** N. Macnicol *Ind. Theism* 30 Those in India whose hearts are filled with *bhakti* or 'loving faith'. **1921** T. R. Glover *Jesus Exper. Men* iii. 41 Tuka and other mystics of India believed that from this a man might be saved by *Bhakti*, by self-annihilating devotion to a friendly god. *Karma* and *Bhakti* are the two poles of Indian religious thought. **1953** E. M. Forster *Hill of Devi* 117 Sivaji..believed in bhakti, in our union with the Divine through love.

b. *attrib.* and *Comb.*, as **bhakti-marga** [Skr. *mārga* path], the way to salvation through religious devotion or faith.

1936 *Times Lit. Suppl.* 18 Jan. 57/4 The *bhakti* movement, the movement of emotional religion. **1937** A. Huxley *Ends & Means* xiii. 234 In India it is known as *bhakti-marga*, the path of devotional faith, as opposed to *karma-marga*, the path of duty or works, and *jñana-marga*, the path of knowledge. **1959** *Chambers's Encycl.* XIV. 798/2 *Bhakti yoga* or devotion to God is also advocated as an additional means. **1963** *Listener* 28 Mar. 550/2 In the last centuries B.C. a second way, *Bhakti-marga* or the way of devotion came into prominence, seeking to obtain union with the chosen god through love and service.

bharal (bɒ·răl). Also **b(h)arhal, burhel, burrell, burrhal, burrhel.** [Hindi *bharal*.] The wild or blue sheep of the Himalayas and Tibet, *Pseudois nayaur*.

1838 *Proc. Zool. Soc.* 79 A skin of the Burrhal Sheep from the Himalaya Mountains. **1840** *Ibid.* 68 The Burrhel would seem to inhabit a much loftier region of the Himalaya than the Nahoor. **1881** *Encycl. Brit.* XII. 742/2 The *barhal*, or blue wild sheep. **1886** *Ibid.* XXI. 785/1 A species..with smaller and more spreading horns, the burrhel, *O[vis] nahoor*. **1887** *Field* 19 Feb. 243/2, I had a shot at a small burrell and missed. **1924** H. Whistler *In High Himalayas* xiii. 165 We found the tracks of burhel on some bare soil. **1937** H. W. Tilman *Ascent of Nanda Devi* xi. 129 A herd of bharal, or wild sheep, was seen traversing..the..cliffs. **1965** S. H. Prater *Bk. Indian Animals* 252 In structure and habits the Bharal holds a place intermediate between sheep and goats.

‖ **bhikku** (bi·kʊ). [Pali *bhikkhu*, f. Skr. *bhikṣú* (see next).] A Buddhist mendicant or religious devotee. Cf. next.

1846 *Jrnl. R. Asiatic Soc.* (*Ceylon Branch*) 84 The original formulary of introduction into the priesthood, after the shaving of the head and beard and putting on the yellow garment, being simply 'Approach, O Bikkhu', the Bikkhu being necessarily from his appellation a religious mendicant. **1903** *Nature* 17 Dec. 163/2 Bhikkus of occidental nationalities..would be able to return to their own countries, there to spread the knowledge of Buddhism. **1927** J. P. Arendzen *Whom do You Say—?* v. ii. 284 Buddhist Bhikkus when in England would..assume this title [Reverend]. **1959** *Times* 3 Feb. 11/6 There is now a separate order of English *bhikkus* near Swiss Cottage.

‖ **bhikshu** (bi·kʃu). Pl. **-shook, -shu, -shus.** [Skr. *bhikṣú* a beggar, f. *bhikṣ* to beg, ask alms.] A Brahmanical or Buddhist mendicant or religious devotee.

1811 W. Ward *Acc. Hindoos* III. vi. 400 He becomes a Bhikshookŭ, literally a beggar, but his person is supposed to be distinguished by his having become..insensible to all human things, and to be absorbed in thinking on God. **1827** A. Steele *Law & Custom of Hindoo Castes* ii. 86 Bhikshook or professional beggars may be Gruhust, and may marry and have children, which the Sunyasees may not. **1876** *Encycl. Brit.* IV. 204/2 *Sannyāsin* (or *bhikshu*), or religious mendicant. **1924** W. B. Selbie *Psychol. Relig.* vii. 138 This transforming experience..made men Sannyasi or Renouncers, and Bhikshu or beggars. **1962** *Times Lit. Suppl.* 10 Aug. 609/3 The self-complacency which he had previously remarked in Hinayana bhikshus.

Bhil (bīl). Also **Bheel.** [Hindi *Bhīl*, f. Skr. *Bhilla*.] (A member of) a central Indian people. Also *attrib.*

1823 J. Malcolm *Mem. Central India* I. xi. 517 The Bheels are quite a distinct race from any other Indian tribe. *Ibid.* xii. 551 The Bheel chiefs have a power over the lives and property of their own subjects. **1848** J. D. Hooker *Let.* in L. Huxley *Life* (1918) I. xii. 246 The monkey, man, Bhil, and all the other common tropical animals. **1891** Kipling *From Sea to Sea* (1899) I. vii. 54 The little Bhil is an aborigine. **1937** L. Bromfield *Rains Came* i. v. 35 The wild Bhils who inhabited the rocky hills. **1964** R. Perry *World of Tiger* xii. 177 The Gonds and Bhils and other aboriginal hunting tribes. *Ibid.* xiv. 210 The elephant was mounted, and with a Bheel walking by his side to track, proceeded into cover.

‖ **Bhoodan** (būdā·n). Also **bhoo-dan, bhoodan.** [Hindi, f. *bhū* earth, land + *dān* gift.] A movement in India initiated by Vinoba Bhave in 1951 for landowners to make free gifts of land to the poor.

1953 *Times* 15 Apr. 7/3 The *bhoodan* (land-gift) movement..is attempting to help the thirty million to forty million landless cultivators of India. **1958** *Times of India* 3 June 6 Many of them had not even heard of bhoo-dan or gram-dan before Vinoba and his men came along and asked them to give a part of their land as a gift or pool their land in the interest of the whole village. **1960** Koestler *Lotus & Robot* i. i. 19 His..solemn pledge to devote the rest of his life to the service of the Bhoodan movement.

‖ **bhoosa** (bū·sā). *India.* Also **bhoos, bhusa.** [Hindi *bhus, bhusā*.] Husks and broken straw, used as food for cattle. Also *attrib.*

1829 Tod *Annals* I. 700 Bhoos stacks are erected to provide provender for the cattle in scanty rainy seasons. **1888** Kipling *Plain Tales* 148 A pinch of *bhusa* or cattle-food. **1901** —— *Kim* iii. 80 Grain and cotton and timber, bhoosa, lime and hides. **1920** *Blackw. Mag.* Oct. 449/2 Having to wait weary hours sitting on a bhoosa bale. **1954** J. Masters *Bhowani Junction* xxvi. 222 The explosives that had been found..under bhoosa stacked in cattle sheds.

bhoot, var. *bhut.

B-horizon, see *horizon *sb.*

Bhotanese, var. *Bhutanese *sb.* and *a.*

Bhotia (bōᵘ·tiă), *sb.* (and *a.*). Also **Bhooteah, Bhooteea, Bhotea, Bhotiya, Bhutia.** [ad. Skr. *Bhoṭiya* Tibetan, f. *Bhoṭa* Tibet.] A native or inhabitant of Bhutan (see *Bhutanese *sb.* and *a.*) or southern Tibet; the language(s) of this region; also, (one of) a breed of pony native to the Himalayas. Also *attrib.* or quasi-*adj.* Hence **Bho·tian** *a.*

[**1826** F. C. G. Schrœter *Dict. Bhotanta Lang.* (*heading*) A Grammar of the Bhotan Language. *Ibid.*, A Dictionary of the Bhota Language.] *a* **1826** R. Heber *Narr. Journey Upper Provinces of India* (1828) I. xviii. 493 The Bhooteahs, a Mongolian tribe, worshippers of the Delai Lama. **1835** R. B. Pemberton *Rep. Eastern Frontier Brit. India* ii. 81 The Bhooteeas imported woollen cloths, gold-dust, salt, musk, horses, the celebrated Thibet chowries, and Chinese silk...In 1833, two Bhooteea merchants only came down from the hills..of Thibet and Bhootan. **1848** J. D. Hooker *Let.* 10 June in L. Huxley *Life* (1918) I. xiii. 253 The people (Bhoteas) are a disagreeable and morose race. *Ibid.* 25 Apr. xiv. 287 My lower extremities cased in Bhotea boots. **1875** *Encycl. Brit.* III. 631/2 Physically the Bhutiás are a fine race, although dirty in their habits and persons. *Ibid.* 632/2 Several raids in British territory headed by Bhutiá officials. **1884** *Ibid.* XVII. 341 The Bhotiyas..are..Buddhists. **1899** L. J. Trotter *Hist. India* vii. ii. 383 The Bhotia highlanders had made frequent inroads into British ground. **1937** H. W. Tilman *Ascent of Nanda Devi* i. 7 The Bhotia.. has little respect for things sacred. **1939** R. Godden *Black Narcissus* i. 8 They rode on Bhotiya ponies that were small and thick-set like barbs. *Ibid.* 11 They were only ponies like the Bhotiyas. **1948** D. Diringer *Alphabet* ii. vi. 352 The Indian term Bhotia has been accepted..to designate the group of languages, of which Tibetan is a member; other Bhotian dialects are spoken in Bhutan, [etc.]. **1963** *Times* 13 June 13/6 The Bhutias, people from Tibet and Bhutan who have been Sikkimese for centuries.

b'hoy. *colloq.* (orig. *U.S.*). Also **bhoy, bo-hoy.** [Supposed Irish pronunc. of Boy *sb.*[1]] = *boy *sb.*[1] 6 b; a gay or spirited fellow.

1846 *Knickerbocker* XXVII. 467 A smile on his lip peculiar to 'one of the bo-boys'. *Ibid.* XXVIII. 557 A friend of ours ..was much struck..by '*The B'hoys*', returning from their Sunday drives. **1850** C. A. Bristed *Upper Ten Thousand* (1852) 29 Of not-to-be-mistaken Bowery cut—veritable 'bhoys'. **1866** *Atlantic Monthly* Dec. 727 'I want', said the stranger, 'to see a b'hoy,—a real b'hoy.' **1929** R. Aldington *Death of Hero* i. i. 39 George Augustus..thought himself rather a hell of a bhoy because he occasionally sneaked off to a play or a whore.

‖ **bhut** (būt). Also **bhoot, bhuta.** [Hindi *bhūt*, f. Skr. *bhūtá* lit. 'become, been, existing'.] In India, a spirit; a demon or goblin.

1785 C. Wilkins tr. *Bhăgvăt-Gēētā* ix, The servants of the *Bhōōts*, or spirits, go unto the *Bhōōts*; and they who worship me go unto me. **1838** H. H. Wilson in *Jrnl. R. Asiatic Soc.* V. 290 These superterrene realms, the Lokas or spheres, inhabited by various orders of beings, as the Bhútas. **1883** D. C. J. Ibbetson *Rep. Rev. Settlt. Karnal* ix. 154 People who have died violent deaths..are specially likely to become *bhúts*. **1909** *Westm. Gaz.* 11 Aug. 8/3 These 'bhutas' are the devils or goblins of our mediæval fairy tales. **1939** R. Godden *Black Narcissus* xvi. 147 'Probably He's a dragon, a bhût in the shape of a child...' They were all afraid of bhûts, Hindus and Buddhists alike.

Bhutanese (būtănī·z), *sb.* and *a.* Also **Bhotanese.** Pl. **-ese.** [f. *Bhutan* (see below) + -ese.] **A.** *sb.* A native or inhabitant of Bhutan, a country in the south-east of the Himalayas; the language of Bhutan. **B.** *adj.* Of or pertaining to Bhutan, its inhabitants or its language.

1813 C. Stewart *Hist. Bengal* 47 Kurmputtun..was inhabited by Brahmans and Bootaneers. **1848** J. D. Hooker *Let.* 10 June in L. Huxley *Life* (1918) I. xiii. 254 Trisected..by the Goorkhas, Bhotanese and Thibetans. **1854** —— *Himalayan Jrnls.* I. v. 136 The Bhotanese, natives of Bhotan, or of the Dhurma country, are called Dhurma people. **1880** J. T. Wheeler *Short Hist. India* iii. xxvi. 674 In reality the Bhutanese authorities did not want to receive a mission at all. **1960** 'S. Harvester' *Chinese Hammer* xxv. 203 They were married according to a mixture of Tibetan and Bhutanese custom.

Bhutia, var. *Bhotia.

Biafran (bi,æ·frăn, bəi-), *a.* and *sb.* [f. *Biafra* + -an.] **A.** *adj.* Of or pertaining to Biafra, a region in West Africa that seceded from federal Nigeria in 1967 and was reunited with it in 1970. **B.** *sb.* A native or inhabitant of Biafra.

1967 *Guardian* 19 June 1/4 The Biafrans naively thought them dead, ambled up, and were themselves killed. **1967** *Times* 27 July 4/2 Enugu radio claimed today that the Biafran air force..had killed about 1,000 federal troops. **1969** *Listener* 8 May 634/2 They have British armoured cars which are almost invulnerable to Biafran weapons. They have so many big guns and mortars that they can lay continuous barrages on the Biafrans.

‖ **bianco sopra bianco** (bia·nko sǫ·prā bia·nko). Also **sopra bianco.** [It., lit. 'white upon white'.] A form of white decoration on white porcelain (see quots.). Also *attrib.*

1856 Owen Jones *Grammar of Ornament* xvii, *Sopra-bianco* was a painting in white upon a white-lead ground, with green or blue borders round the margin of the plate. **1870** W. Chaffers *Marks & Monogr. Pott.* (ed. 3) 55 *Sopra bianco*, white upon white, palmette ornaments of opaque white enamel upon milky white ground. **1877** A. Beckwith *Majolica & Fayence* 51 The work is notable for exquisite finish..*Bianco sopra Bianco* ornament was much used. **1896** C. D. E. Fortnum *Maiolica* 273 A large plate..painted in blue relieved with white on a white ground; on the upright sides a diaper pattern in *bianco sopra bianco*. **1932** W. Chaffers *Marks & Monogr. Pott.* (ed. 14) 873 A peculiarity to be noticed in the Bristol ware is a charming effect produced by painting part of the design in white opaque enamel on the white ground of greenish tint. This process has been called *bianco sopra bianco*.

bias, *a.*, *sb.*, and *adv.* Add: **A. 1. b.** (Earlier example.)

1815 *La Belle Assemblée* in Jane Austen *Persuasion* (Chapman, 1933) 274 This pelisse is trimmed..with byas white satin laid on in folds.

B. 2. c. *Cricket.* The turning of a ball in its course from the leg side towards the off after pitching. Also *attrib.*

1833 J. Mitford in *Gent. Mag.* Sept. 238 The plan adopted by good batters against slow bias bowling was successful. *Ibid.* 239/1 His balls..have a very perplexing bias. **1921** Ld. Harris *Few Short Runs* v. 135 The bowler was not slow to take advantage of the opening by developing anew the old 'bias' or break from leg.

3. b. *Statistics.* A systematic distortion of an expected statistical result due to a factor not allowed for in its derivation; also, a tendency to produce such distortion.

1900 *Phil. Mag.* L. 167 The results show a bias from the theoretical results, 5 and 6 points occurring more frequently than they should do. **1911** G. U. Yule *Introd. Theory Statistics* xiv. 277 Such an examination may be of service.. as indicating one possible source of bias, viz. great heterogeneity in the original material. **1943** M. G. Kendall *Adv.*

Theory Statistics I. viii. 189 If the observer was unbiased the digits should appear in approximately equal numbers; but there is a bias in favour of all the even numbers and against the odd numbers 1, 3 and 9.

6. *Telegr.* (See quot. 1940.)

1885 W. Williams *Man. Telegr.* c. 47 The force of restitution is no longer effected simply by the bias of the tongue but by an opposite current drawing it back. **1902** *Encycl. Brit.* XXXIII. 221/2 With the tongue set neutral, having no bias either to the spacing or marking side—the relay will give good signals with 1½ milliamperes of current. **1940** *Chambers's Techn. Dict.* 89/1 *Bias*, the adjustment of a telegraph relay so that it operates for currents greater than a given current (against which it is biased), or for a current of one polarity.

7. *Electr.* A steady voltage or current applied to an electronic device (see quot. 1960); also *attrib.*

1922 *Electrical Review* 30 June 928/1 Security from undesired operation is obtained by the introduction of a controlling bias, and distinction is made between earth faults.. and phase faults. **1926** [see *grid 5 b]. **1932** F. E. Terman *Radio Engin.* xi. 392 The regeneration increases as the total amplification A_1A_2 and the bias impedance *Zc* are increased. **1940** *Chambers's Techn. Dict.* 89/1 *Bias voltage*, generally, the mean potential of any electrode in a thermionic tube, measured with respect to the cathode. Specially applied to that of the control grid. **1942** *Electronic Engin.* XV. 9 The sensitivity of the receiver is adjusted by varying the bias of the..amplifiers. **1960** H. Carter *Dict. Electronics* 28 *Bias*, steady direct voltage applied between the cathode and control electrode of a thermionic tube in order to determine its working point. *Ibid.* 119 *Grid bias*, steady negative potential applied to the control grid of a thermionic valve or other tube in order to pre-set the no-signal value of the cathode current.

D. bias binding, a narrow strip of cloth cut on the bias and used for binding (cf. sense A 1 b); **bias-cut** *a.*, cut on the bias.

1927 *New Butterick Dressmaker* xix. 174 Bias bindings make attractive finishes either in the same or in a contrasting material or color. **1931** *Times Lit. Suppl.* 29 Oct. 844/3 The teacher..may perhaps think that in these days of 'bias bindings' it is not worth while to put children to the pains of making curved hems. **1960** *Woman's Realm* 2 Apr. 36/4 Bind napkin all round with red bias binding. **1960** *Lebende Sprachen* V. 35/1 Bias-cut. **1969** *Guardian* 30 July 7/4 Bias cut skirts flare out from the hips.

bias, *v.* Add: **4.** *trans.* To cut bias. *U.S.*

1883 *Century Mag.* XXVI. 960/1 You may baste, you may bias the Gore if you will.

5. *Electr.* To apply bias (*bias *sb.* 7) to. Hence **bi·as(s)ing** *ppl. a.* and *vbl. sb.*

1922 *Electrical Review* 30 June 928/1 The equipment for each end of the line of a 3-phase system comprises three protective transformers, a biasing transformer, an earth fault relay, [etc.]. **1923** Meare & Neale *Electr. Engin. Practice* (ed. 4) xv. 511 Illustrating the principle of the biassed transformer. *Ibid.*, The 'restraining' or 'biassing' winding *BB* produces a flux as shown on the dotted line. **1930** *Engineering* 31 Oct. 543/2 The multiplier circuits are heavily biassed. **1944** *Electronic Engin.* XVI. 336 Bias the valve so that the anode current is normally zero or small. **1953** Amos & Birkinshaw *Telev. Engin.* (1957) I. vi. 117 The signal plate is biased approximately 30 volts positive with respect to the cathode. **1962** Simpson & Richards *Junction Transistors* ix. 213 (*heading*) Other biasing methods.

biased, *ppl. a.* Add: Also **biassed. 2. b.** spec. in *Statistics.* Containing a bias or error which will not balance itself out on average.

1911 G. U. Yule *Introd. Theory of Statistics* xiv. 276 Any sample, taken in the way supposed, is likely to be definitely *biassed*, in the sense that it will not tend to include, even in the long run, equal proportions of the A's and α's in the original material. **1931** *Economist* 18 July 127/2 What statisticians describe as 'biassed errors' in company accounting (the term is used without any derogatory sense) are made cumulative.

3. Of fabric: (see quot.). Cf. Bias *sb.* 1.

1919 W. B. Faraday *Dict. Aeronaut. Terms* 55 *Biassed fabric*, fabric laid with its threads inclined to the length of the piece of cloth.

bias-wise, *adv.* Delete †*Obs.* and add later example.

1888 M. W. Hungerford *Mrs. Vereker* II. xi. 129 They have pulled it bias-wise, and cross-wise, and straight-wise.

‖ **biathlon** (bəi,æ·plǫn). [f. Bi-[2] II + Gr. ἆθλον contest, after Pentathlon.] An athletic event consisting of ski-ing and shooting performed by the same contestants.

1958 *Times* 6 Nov. 14/4 Events new to the Olympic Games are the biathlon, a 20 kilometres combined skiing and shooting competition, and a speed skating contest for women. **1966** *Daily Mail* 22 Dec. 11/5 The biathlon is a 20-kilometre race with a pause every five kilometres, when the competitor has to fire five rounds at a target 150 yards away.

biaxial, *a.* Add: **2.** *Geom.* (See quot.)

1889 Cayley *Math. Papers* (1897) XIII. 13 Each conic meets each axis in two points; and at each of these points the axis and the tangent to the conic form with the lines to I, J a harmonic pencil. The binodal quartic is in this case said to be biaxial.

bib, *sb.*[1] Add: **b.** *best bib and tucker* (earlier examples); *bib-and-brace*, applied *attrib.* to a type of overall having the upper part supported by braces.

1747 tr. *Marquis d'Argens' New Memoirs establishing a True Knowledge of Mankind* II. 264 The Country-woman ..minds nothing on Sundays so much as her best Bib and Tucker. **1793** [see Tucker *sb.* 3]. **1855** *Sharpe's Lond. Mag.* V. 92/2 All-London..puts on its best bib and tucker. **1959** *Manchester Guardian* 13 July 5/6 The engine-driver, a dapper figure in bib-and-brace overalls.

2. A patch of feathers resembling a bib under the bill of a bird.

1854 *Poultry Chron.* I. 391/1 A dark bib under the bill. **1948** *Brit. Birds* XLI. 156 A distinct bib..contrasted with white under-parts.

bi·bble-ba:bble, *v.* [f. the sb.] *trans.* and *intr.* To indulge in bibble-babble or idle talk; to babble.

1888 Doughty *Trav. Arabia Deserta* I. 256 They all love to bibble-babble their infirmities in the wholesome ears of the hakim. *Ibid.* 596 They sat out long hours bibble-babbling.

bib-cock. [? f. Bib *sb.*[1]] A cock or tap with a turned-down nozzle, as distinguished from a stop-cock. So *bib-nozzle, -valve*.

1797 J. Curr *Coal Viewer* 63, 1 Piston cock (bib)..1 Air cock (bib) to fix in the sink pipe. **1853** *Dict. Archit.*, *Bib-cock*, a term used..to distinguish a cock which delivers fluids out of a pipe, from a stop-cock which is placed in the length of a pipe. *a* **1877** Knight *Dict. Mech.* s.v., *c* is a bib-valve, the closure being by a reciprocating slide. **1909** Webster, Bib nozzle. **1951** *Good Housek. Home Encycl.* 281/2 *Screw-down bib cock*: this is the best-known type of tap.

‖ **bibelot** (biblo). [Fr.] A small curio or article of *virtù*.

1873 Lady C. Schreiber *Jrnl.* (1911) I. 185 After 3, walked up to Lady Hopetoun's to amuse her with some of our little bibelots. *Ibid.* 231 Van Minden had a lovely little bibelot of Cupid with drums. **1882** 'Ouida' *Resurgo* vi. in *Belgravia* XLVI. 453 His soul never rises above brocades and *bibelôts*! **1886** F. M. Crawford *Tale Lonely Parish* xviii, The spectacle of her pictures, her furniture and her bibelots. **1900** H. Harland *Cardinal's Snuff-Box* xix, The table, with its books and bibelots. **1936** *Burlington Mag.* Apr. 186/1 A Chinese *bibelot* on the table shows the influence of the fashion for lacquer.

‖ **biberon**[2] (bibroṅ). [Fr.] A drinking-vessel with elongated spout, formerly used by travellers, invalids, and children.

a **1853** Mayne *Expos. Lex.* (1860) 120/2 *Biberon*, name proposed for a feeding-bottle for infants. **1862** *Catal. Internat. Exhib., Brit.* II. ii. xvii. 127 The Biberon, a new patent invention, adapted for a lady's travelling companion. **1905** *Westm. Gaz.* 27 May 15/3 The biberon is wrought in rock-crystal..and is in the form of a grotesque animal, of which the head is utilised as the spout.

‖ **bibi** (bi·bī), var. Beebee.

1842 in Yule & Burnell *Hobson-Jobson* (1886) s.v. *Beebee*. **1921** *Spectator* 14 May 627/2 They get..one or two tomans when the carpet is finished..according to whether the Bibi (the chief's wife) is pleased with the result or not. **1935** *Discovery* June 164/2 *Bibi* (mistress of the household).

Bible. Add: **1. a.** *the open Bible*: the Bible accessible to all in the vernacular.

1837 *Ch. Eng. Mag.* 9 Dec., Bear witness, martyrs of the olden days, How your true hearts the open Bible priz'd. **1908** Tilby *Eng. People Overseas* I. 48 The persecution of Mary and the open Bible worked a change.

4. *Nautical slang.* (See quots.) Cf. Prayer-book 2.

1867 Smyth *Sailor's Word-Bk.* 98 *Bible*,..a squared piece of freestone to grind the deck with sand in cleaning it; a small holystone, so called from seamen using them kneeling. **1906** *Voy. of 'Scotia'* iii. 25 Holy-stones for polishing decks ..are commonly known to sailors as 'Bibles'.

III. *Bible-black* adj., *-reader* (earlier example), *-reading* vbl. sb., *-story*; **Bible-banger, -basher** *Austral.* and *N.Z. slang*, = **Bible-pounder*; hence *Bible-banging, -bashing* ppl. adjs.; **Bible belt**, a designation of those parts of the United States reputed to be fanatically puritan or fundamentalist; also *attrib.*; **Bible-bigot, -moth**, a person who is obstinately devoted to the Bible; esp. applied to the Methodists, sometimes contemptuously (cf. quot. 1820 in Dict.); **Bible-box**, a box with a flat lid, esp. one of the 17th century, large enough to hold a family Bible; **Bible class** (examples); **Bible-leaf**, (*a*) a thin slice of whale-blubber for heating in a try-pot; (*b*) *U.S.* (see quot. 1931); **Bible-moth**: see **Bible-bigot*; **Bible paper**, a very thin but strong opaque printing-paper; cf. India paper 2; **Bible-pounder, -puncher** *slang*, one who expounds or follows the Bible in a vigorous or aggressive manner, esp. a clergyman; hence **Bible-pounding, -punching** *ppl. adjs.* and *vbl. sbs.*; **Bible-thumper** *slang*, = **Bible-pounder*; **Bible-woman** (examples).

1942 M. Harcourt *Parson in Prison* iii. 35 We don't want any damned Bible bangers around here! **1945** Baker *Austral. Lang.* vi. 131 Here are some self-explanatory Australianisms: *bible banger, bible basher*, [etc.]. **1948** D. Ballantyne *Cunninghams* (1963) i. xxiv. 106 Gil reckoned Kent was better than most of the Bible-bangers who visited the hospital. **1964** O. E. Middleton *Walk on Beach* 40 That Bible-banging, psalm-singing old crawler. **1958** R. Stow *To Islands* iv. 74 They were Bible-bashers and humourless clods. **1944** L. Glassop *We were Rats* xxi. 124, I doan want any bible-bashing bastard..mumblin' any bull—over me. **1926** *Amer. Mercury* Feb. 141/2 The *Baptist Record*, of Jackson, Miss., [is] in the heart of the Bible and Lynching Belt. **1926** S. Lewis in G. Frankau *My Unsent. Journey* xi. 148 I'm collecting parsons, Gilbert...That's why I've been living in Kansas City. It's the centre of the Bible belt. **1960** *20th Cent.* Dec. 558 Nashville is..the Bible Belt capital. **1766** J. Wesley *Jrnl.* (1938) 5 June, I am a Bible-bigot. I follow it in all things, both great and small. **1789** —— *Sermon* no. xlviii, in *Arminian Mag.* Jan. 7 They were continually reproached for this very thing: some terming them in derision, *Bible-bigots*; others *Bible-moths*: feeding, they said, upon the Bible, as moths do upon cloth. **1953** Dylan Thomas *Under Milk Wood* (1954) 1 It is spring, moonless night in the small town, starless and bible-black. **1904** P. Macquoid *Hist. Eng. Furniture* I. vii. 190 Boxes..with a flat lid..were termed Bible or lace boxes, and used for these purposes. **1966** A. W. Lewis *Gloss. Woodworking Terms* 48 Jacobean period (1603–1649)...Gate-legged tables and Bible boxes first appeared. **1824** *Amer. Baptist Mag.* IV. 371, I intimated my intention to establish..Bible classes. **1888** C. M. Yonge *Our New Mistress* xvii. 161 She was so fond of her church and her Bible class. **1851** Melville *Moby Dick* III. ix. 69 Arrayed in decent black; occupying a conspicuous pulpit; intent on 'Bible leaves'..what a lad for a Pope were this mincer! **1931** W. N. Clute *Plants* 60 More appealing names are..bible-leaf (*Chrysanthemum balsamita*) for that fragrant-leaved plant whose leaves were often carried to church. **1926** *Paper Terminol.* (*Spalding & Hodge*) 2 *Bible paper*, see *India paper*. **1889** Barrère & Leland *Dict. Slang* I. 111/1 *Bible-pounder* (popular), a parson. **1890** Farmer *Slang* I. 186/2 *Bible-pounder*, a clergyman..from the practice indulged in by some excitable exponents, of pounding or beating their hands upon the book or desk while preaching. **1951** C. Simak *Time & Again* (1956) xxxvi. 173 The old-time Bible-pounding preachers. **1917** *Tiki Talk* (N.Z.) 10 Synonyms: clergyman, 'parson.. preacher, bible-puncher. **1938** X. Herbert *Capricornia* (1939) xxxii. 486 You horse-faced bible-puncher. **1933** P. Fleming *Braz. Adv.* ii. iii. 205 It seemed..that bible-punching was a bit of a racket. **1961** *John o'London's* 13 Apr. 415/2 An earnest evangelist..denouncing the devil with bible-punching relish. **1538** Bale *Thre Lawes* 1204 Then I holde it best that we alwayes condempne The Byble readers. **1827** Cunningham *N.S. Wales* II. xxx. 252 None remained but the old *fence*, who continued Bible-reading to the end of the voyage. **1863** Bible-reading [in Dict.]. **1888** C. M. Yonge *Our New Mistress* xvi. 152 The tinies..answered about their Bible stories as prettily as could be wished. **1967** 'M. Underwood' *Man who died on Friday* xv. 139 He could remember a picture, in a children's book of Bible stories, of..'Elijah being taken up to heaven'. **1923** J. Manchon *Le Slang* 61 *Bible-pounder* (or *-thumper*), un prêtre. **1942** A. L. Rowse *Cornish Childhood* 31 It's always the Bible-thumpers who are the greatest hypocrites. **1859** Mrs. Gaskell *Let.* 7 Nov. (1966) 587 Just before I left Manchester I heard of search being made for a 'Bible-woman' to work there; doubtless suggested by the success in London. **1862** Mayhew *London Labour* Extra vol. p. xix, This void has been admirably supplied by the 'Bible women' of the nineteenth century. **1908** *Daily Chron.* 27 Mar. 27/2 £200 to Mrs. Wilson, her Biblewoman in London.

bi·ble-back. *U.S.* and *dial.* [Bible III.] A hump-backed or round-shouldered person.

1873 G. W. Perrie *Buckskin Mose* x. 145 Pitching our camp in the Lassen Meadows, at La Due Very's, generally known as Old Bible-back. **1896** Northall *Warwickshire Word-Bk.*, *Bible-back*, a person with broad, rounded shoulders. 'Here comes old bibleback!' **1903** *Dial. Notes* II. 349 *Bible-back*, Rockland Co. N.Y. Bergen Co. N.J.

So **bible-backed** *a.*

1857 *Olympia* (W.T.) *Pioneer & Democrat* 11 Dec. (Th.), We might, in consequence, become somewhat round-shouldered and 'bible-backed'. **1873** *N. & Q.* XII. 227 What kind of shoulders? Rather high. Anything else? He was humpy or bible-backed.

Bible-clerk. b. (Later examples.)

1845 J. Pycroft *Collegian's Guide* 42 In All Souls there are none but fellows who have already graduated, except four Bible Clerks. **1889** [see Clerk *sb.* 2 b]. **1965** *Oxf. Univ. Calendar* 213 Hertford College...Bible Clerks.

biblical, *a.* Add: *biblical theology*: see *theology.

biblicality. Delete *nonce-wd.* and add further examples.

1920 Galsworthy *In Chancery* i. x. 91 'I must live again—live and move and have my being.' And in answer to that queer biblicality church-bells chimed the call to evening prayer. **1946** *Scrutiny* XIV. 103 The wastes of biblicality and fervid enthusiasm..devoted to Mordecai.

biblioclast. Add: (Earlier example.) Also **biblioclastic** *a.*

1880 Blades *Enemies Bks.* 96 Such a wicked old biblioclast as John Bagford. **1887** *Longman's Mag.* Dec. 239 The biblioclastic dead.

bi·bliograph, *v.* [Back-formation from Bibliography.] *trans.* To compile a bibliography of (an author or subject).

1961 B. Malamud *New Life* (1962) 120 Goddam, imagine bibliographing Civil War fiction.

bibliographer. 2. (Earlier example.)

1810 E. Brydges (*title*) The British Bibliographer. Vol. 1.

bibliographic, *a.* (Further examples.)

1890 A. W. Verrall in *Class. Rev.* XII. 109/1 All of them originated in times which we should now call pre-historic, or at least pre-bibliographic. **1948** H. E. Bliss in *Rev. Docum.* XV. 93 Books and articles on the organization of knowledge and bibliographic classification (a select bibliography).

bibliographical, *a.* (Earlier and additional examples.)

1679 (*title*) Baconiana, Or Certain Genuine Remains of Sr. Francis Bacon..in Arguments Civil and Moral, Natural, ..and Bibliographical. **1846** Geo. Eliot tr. *Strauss's Life of Jesus* I. p. xii, The author has not..aimed to give a complete bibliographical view of this department of theological literature, but..has adhered to the chief works in each separate class of opinions. **1961** T. Landau *Encycl. Librarianship* (ed. 2) 38/2 *Bibliographical note.* 1. A note, often a footnote, containing a reference to one or more works used as sources for the work. 2. A note in a catalogue or a bibliography, relating to the bibliographical history or a description of a book.

bibliography. 4. (Earlier and later examples.)

1814 T. H. Horne *Introd. Study Bibliography* I. II. iii. § 4. 365 *Professional*, or..*special bibliography*, has reference only to one class of books, and comprehends every work published on the subject of which it treats...Special bibliographies may be disposed either alphabetically, or systematically. **1930** K. Malone in *English Jrnl.* XIX. 646 The term *bibliography* is also used, unhappily, I think, to mean 'list of writings germane to a given topic'.

bibliomane. (Earlier and later examples.)

1809 Ferriar *Bibliomania* 5 The Bibliomane exclaims.. 'No Margin!' turns in haste, and scorns to buy. **1960** C. Day Lewis *Buried Day* ii. 32 He has been writing..to other bibliomanes.

bibliomania. (Earlier example.)

1734 T. Hearne *Diary* 9 Nov. (1921) XI. 389, I should have been tempted to have laid out a pretty deal of money without thinking my self at all touched with Bibliomania.

bibliope·gia. *rare.* = Bibliopegy.

1835 J. A. Arnett (*title*) Bibliopegia; or, the art of bookbinding, in all its branches.

bibliope·gically, *adv.* [f. Bibliopegic *a.*: see -ically.] As regards bookbinding. So **biblio·pegism** = Bibliopegy.

1817 Dibdin *Decam.* II. 513 The art and craft of bibliopegism. **1896** Gosse in *Contemp. Rev.* Jan. 98 His own library..was not conspicuous bibliopegically.

bibliopegy (and derivs.). (Earlier and later examples.)

1835 J. A. Arnett *Bibliopegia* p. iv, A record of the present state of the Bibliopegistic Art. **1876** *Encycl. Brit.* IV. 42/1 Contemporary masterpieces of French, Italian, and German bibliopegy. **1958** 'M. Innes' *Long Farewell* 72 Appleby, although hazy about bibliopegy, was quite certain he wasn't a distinguished student of it. **1968** *Oxf. Mag.* 22 Nov. 88 (*caption*) A sad case of bibliopegistical elephantiasis.

bibliophily. (Earlier example.)

1877 Swinburne *Let.* 9 Oct. (1960) IV. 24, I have lately had two noble windfalls in the way of dramatic bibliophily (if there is such a word).

bibliothecarial (bi:bliǫþīkẽə·riăl), *a.* [-al.] = Bibliothecary *a.*

1889 *Athenæum* 27 Apr. 539/2 Of the two sets of terms that are so constantly confounded, one may be said to be bibliographical or scientific, the other bibliothecarial or practical. **1893** F. Madan *Bks. in MS.* vii. 75 The one great bibliothecarial fact of antiquity is the Library of Alexandria. **1918** *Bodleian Q. Rec.* II. 156 So little has happened in Bodley in this Long Vacation that it is lawful to extravagate a little in Oxonian (and not purely bibliothecarial) pastures.

bibliothecary, *sb.* **2.** (Later U.S. example.)

1887 O. W. Holmes *Hundred Days* vi. 218 These two experts in books, the bibliopole and the bibliothecary.

bibliotherapy (bi:bliǫþe·răpi). [f. Biblio- + Therapy.] The use of reading matter for therapeutic purposes in the treatment of nervous disorders.

1920 C. Morley *Haunted Bookshop* 10 The young man had heard of none of these books prescribed by the practitioner of bibliotherapy. **1929** *U.S. Veterans' Bur. Med. Bull.* V. 440 (*title*) Bibliotherapy; Use of Books as Form of Treatment in Neuropsychiatric Hospital. **1931** *Library Assoc. Rec.* June 198 The articles have been roughly arranged.. according to the particular aspect of hospital library work with which they deal:—1. General Works. 2. Bibliotherapy. **1961** *New Scientist* 26 Oct. 253/3 Books on science..and ..science fiction, appear to be the most effective types of reading matter in bibliotherapy.

bibliothetic (bi:bliǫþe·tik), *a.* [f. Biblio- + Gr. θετικός concerned with placing or setting.] Relating to the placing and arrangement of books on the shelves of a library.

1901 E. C. Richardson *Classification* 93 The systems for classifying books on the shelves of a library, which, if you need a technical name, may be called 'bibliothetic'. **1959** in L. M. Harrod *Librarians' Gloss.* (ed. 2) 37.

bibovine (bəibō̆ᵘ·vəin), *a.* *Zool.* [f. Bi-^{2} + Bovine *a.*] Designation of the group of bovine ruminants which includes the Oriental species, gaur, gayal, and banteng (forming the subgenus *Bibos*).

1889 Nicholson & Lydekker *Palæontol.* (ed. 3) II. lviii. 1351 The genus *Bos*..may be divided into the Bibovine and Taurine groups. In the former are included the wild Oxen of India and Burma. **1891** Flower & Lydekker *Mammals* ix. 365 The Bibovine group.

bibulo·sity. [See Bibulous *a.* and -osity.] Addiction to tippling. Also **bi·bulousness** (cf. Bibulous *a.* 1, 2).

1844 *Jrnl. R. Agric. Soc.* V. I. 125 Their colour;..their bibulousness; all these properties conspire to the determination of the temperature of a given soil. **1901** Rashdall & Rait *New College* 207 The common-room which Spence occasionally adorned could not have been wholly given up to uncultivated bibulosity. **1928** *Daily Express* 2 Oct. 6 He caught the man under the arm and affected a slight bibulousness himself.

bicarb (bəikā·ɹb). Colloq. abbrev. of Bicarbonate (of soda).

1922 S. Lewis *Babbitt* xxiii. 361 Bad indigestion? Shall I get you some bicarb? **1947** Wodehouse *Full Moon* ii. 35 The sort of hangover which makes a man lose interest in anything but bicarb of soda. **1952** M. Steen *Phoenix Rising* iii. 61 Tell me where I can find the bi-carb., and I'll mix you a charge that will jolt your ridiculous stomach into order.

bicentennial, *a.* Add: (Examples.) Also, of or pertaining to a 200th anniversary; = Bicentenary *a.*

1883 'Mark Twain' *Life on Miss.* 297 The bicentennial anniversary of this illustrious event. **1901** Goldwin Smith *Let.* 18 Oct. in *Oxf. Univ. Gaz.* 12 Nov. 140/2 On my way to your [*sc.* Yale University's] bicentennial celebration I am unfortunately arrested by sickness. **1959** *Times* 13 May 5/1 She paid her homage to bi-centennial Handel.

bichromate. Add: **b.** *attrib.*, as **bichromate battery** or **cell,** a zinc-carbon cell containing an electrolyte of an acid bichromate solution.

1873 *Eclectic Engin. Mag.* VIII. 220/1 Next, I try the bichromate battery, without porous cells; small resistance, high potential, but soon losing its power..destroying itself by a single night's action. **1875** J. T. Sprague *Electr.* iv. 108 The bichromate cell..furnishes a most powerful current for a very short time. *Ibid.* 111 The cost per unit is..2·16 with the simple bichromate solution. **1886** L. M. Fishback tr. *Niaudet's Electric Batteries* (ed. 4) viii. 213 (*heading*) Chemical action in the bichromate battery. **1892** *Daily News* 25 Oct. 3/1 Casting zinc rods for the bichromate cells.

bichrome (bəi·krō̆ᵘm), *sb.*^{1} *Chem.* [Shortened form of Bichromate.] Bichromate of potassium or sodium. Also *attrib.* (see quot. 1904).

1896 G. Duerr *Bleaching* iv. 56 The following colours.. require to be passed through a solution of bichrome in order to develop or raise the colours. **1904** *Westm. Gaz.* 31 Dec. 14/2 Another improvement..is the 'Bichrome' lamp, in which the light is filtered through a tank containing a solution of bichromate with certain selected aniline dyes.

bichrome (bəi·krō̆ᵘm), *a.* and *sb.*^{2} [f. Bi-^{2} + Gr. χρῶμα colour.] **A.** *adj.* Having two colours. **B.** *sb.* A two-coloured design, etc.

1921 M. C. Burkitt *Prehistory* xix. 265 Bichromes produced by restoring an animal painted in one colour in another. **1929** *Gryphon* (Univ. Leeds) Feb. 155 To hurl the completed bichrome palimpsest in my face with a hideous oath. **1948** Kroeber *Anthrop.* xviii. 830 Mochica ceramics are only bichrome. **1959** J. D. Clark *Prehist. S. Afr.* x. 278 This monochrome group is followed by shaded polychromes and bichromes. **1960** K. M. Kenyon *Archæol. in Holy Land* viii. 200 The decoration is bichrome, nearly always red and black.

bici·rcular, *a.* *Math.* [Bi-^{2} 1 and 2.] Applied to a class of quartic curves each of which passes twice through each of the circular points at infinity, and thus resembles (analytically, and sometimes in form) a pair of circles.

1867 J. Casey *Bicircular Quartics* in *Trans. R. Ir. Acad.* XXIV. I. 458 A bicircular quartic; that is,..a quartic having the two circular points at infinity as double points. **1872** B. Williamson *Diff. Calc.* xi. 195 A curve of the fourth degree of the class called 'bicircular quartics'. **1879** G. Salmon *Higher Plane Curves* (ed. 3) 241 Quartics with two nodes, in the case where these are the circular points at infinity, have been extensively studied under the name of bicircular quartics. **1880** [see Bi-^{2} 1].

bick. Short for *bick-iron* (see Beak-iron) or Bickern.

1896 *Farriers' Price List* Suppl., Farriers' Anvils, with Steeled Bick. **1953** A. Jobson *Househ. Crafts* x. 112 The real hoops are made on a bick or beck-iron, which is an anvil mounted about 3 feet 6 inches high. **1957** R. Lister *Dec. Wrought Ironwork* ii. 12 The pointed cone at the opposite end is variously called a *beak-iron, bickern, pike, bick, beak* or *horn*.

bickiron. Delete 'obs.' Cf. *beak-iron.

bicky, bikky (bi·ki). *colloq.* [dim. of Biscuit; see -y^6.] A diminutive or affectedly childish form of Biscuit.

1930 R. Blaker *Medal without Bar* iv. 303 Sammy..said there was some breakfast. 'Only "bikkies" again, I'm afraid,' he apologised. **1947** *Mod. Lang. Rev.* July 356 Eng[lish] stands almost alone with *bicky* for 'biscuit'. **1954** C. P. Snow *New Men* viii. 56 The Minister pulled out a bag...It contained grey oatmeal cakes. 'Bikkies,' explained the Minister. **1955** B. Moore *Judith Hearne* i. 11 Have a bikky, Bernie? **1963** M. Allingham *China Governess* xiv. 171 He's going to be asked to share Eustace's boiled milk and bickies.

bicollateral (bəi:kǫlæ·tĕrăl), *a.* *Bot.* [Bi-^{2}.] Of a vascular bundle: having two masses of phloem on opposite sides of the xylem. (See Collateral A. 1 e.) Hence **bi:colla:tera·lity.**

1881 *Encycl. Brit.* XII. 18/1 In *Cucurbita, Solanum*, and others the bundles are 'bicollateral', there being an additional phloem portion inside the xylem. **1882** Vines tr. *Sachs's Bot.* 949 *Bicollateral bundles*, in which..there is a layer of phloëm on the inner as well as on the outer side of the xylem. **1884** Bower & Scott tr. *De Bary's Phaner.* 319 A special subordinate form, to be called the double collateral or bicollateral, is distinguished from the usual one by the fact that two phloem groups lie on opposite sides of one xylem group. **1900** W. Wallace in *Ann. Bot.* Dec. 640 Most Cucurbitaceæ have bicollateral bundles. *Ibid.* 641 Weiss adopts this conception of bicollaterality. **1953** K. Esau *Plant Anat.* xv. 355 The presence of phloem on both sides of the xylem makes the bundle bicollateral.

bicolour, -color (bəi·kʌləɹ), *a.* and *sb.* [ad. L. *bicolor* or F. *bicolore* adj.: see Bi-^{2}.] **A.** *adj.* Having two colours. **B.** *sb.* A two-coloured blossom or animal.

1889 *Cent. Dict.*, *Bicolor* a., same as *bicolored*. **1898** C'tess von Arnim *Elizabeth & Germ. Gard.* 56 The Persian Yellows and Bicolors have been, as I predicted, a mistake among the teas. **1911** R. C. Punnett *Mendelism* (ed. 3) vii. 74 The proportion in which the three classes of purples appeared was 9 bicolors, 3 deep purples, 4 picotees. **1931** *Antiquity* V. 430 The second variety of new faience beads are the bicolour inverted drop pendants. **1967** *Guardian* 18 Dec. 5/1 The Bi Colour Longhair, long observed about Britain, has only been recognised as a breed since 1966.

biconditional (bəi:kǫndi·ʃənăl). *Logic.* [f. Bi-^{2} + Conditional *sb.*] The relation between two propositions when one is true only if the other is true, or false if the other is false; the symbolic or verbal representation of such a relation. Also *attrib.* or as *adj.*

1940 W. V. Quine *Math. Logic* i. § 3. 20 This binary mode of composition may be called the biconditional. **1941** —— in P. A. Schilpp *Philos. A. N. Whitehead* II. iv. 141 '≡' is the *biconditional* sign. **1950** —— *Methods of Logic* § 3. p. 16 The idiom '*p* if and only if *q*', called the biconditional, amounts obviously to the conjunction of two conditionals, 'if *p* then *q*' and 'if *q* then *p*'. **1961** R. R. Stoll *Sets, Logic, &c.* ii. 58 The words 'if and only if' are used to obtain from two sentences a biconditional sentence. **1962** *Gloss. Automatic Data Processing* (*B.S.I.*) 30 Equivalence operation ..bi-conditional.

bicone (bəi·kō̆ᵘn). [f. Bi-^{2} + Cone *sb.*^{1}] An object having the form of two cones placed base to base; esp. of beads. Also *attrib.*

1928 H. C. Beck in *Archæologia* LXXVII. 7 Bicone. Beads in which the profile consists of two straight lines, at an angle to one another, which meet the perforation. *Ibid.* 25 (*caption*) Bicone fluted bead. Agate, Ur. **1931** *Antiquity* V. 428 The faience beads include all the simple forms such as oblate, spherical..bicone..and cylindrical. **1937** Mond & Myers *Cemeteries of Armant* I. v. 70 It is often difficult to tell whether a bead is..a long barrel [or]..a long truncated convex bicone.

bico·nical, *a.* [f. Bi-^{2} + Conical *a.*] Similar in shape to two cones placed base to base.

1870 [see Bi- *pref.*^{2} 2]. **1875** *Encycl. Brit.* III. 775/1 Eggs of the Grebes..which also have both ends nearly alike but pointed, are so wide in the middle as to present a biconical appearance. **1928** V. G. Childe *Most Anc. East* viii. 191 Biconical faceted beads. **1963** H. N. Savory in Foster & Alcock *Culture & Environment* iii. 42 The biconical Cinerary Urn of Hilversum affinity which accompanied the primary burial in the round cairn.

bicultural (bəikʌ·ltiŭrăl), *a.* [f. Bi-^{2} + Cultural *a.*] Having or combining two cultures. So **bicu·lturalism, bicu:ltura·tion.**

1940 C. F. Ware *Cultural Approach Hist.* II. 67 The assumption of a single culture with a common language, rather than the bicultural tradition of Quebec. **1953** *Internat. Jrnl. Amer. Linguistics Memoir* VIII. 40 Weinreich considered Lounsbury's point important for bilingualism and biculturalism. **1955** W. E. Collinson in *Archivum Linguisticum* VII. 143 In..the socio-cultural setting of linguistic contact all kinds of different factors emerge, e.g... biculturation, religious differences, [etc.]. **1958** *Jrnl. Abnormal & Soc. Psychol.* LVI. 242/1 A comparison was made between those who acquired their two languages in geographically distinct cultures (bicultural experience) and those who acquired both languages in separated contexts within one geographical region (unicultural experience). **1963** *Times* 25 Feb. (Canada Suppl.) p. v/7 Officially Canada has been a bilingual and a bicultural country since 1867, but in fact this bilingualism and this biculturalism do not exist.

bicursal (bəikʊ̄·ɹsăl), *a.* *Geom.* [f. Bi-^{2} + L. *cursus* course + -al.] Of a curve: having two paths.

1873 Cayley *Math. Papers* (1895) VIII. 181 A curve of deficiency 1 may be termed bicursal. **1876** *Ibid.* (1896) IX. 582 The bicursal sextic has in general 9 nodes.

bicuspidal (bəikʌ·spidăl), *a.* *Geom.* [f. Bi-^{2} + Cuspidal *a.*] Of a curve: having two cusps.

1867 Cayley *Math. Papers* (1893) VI. 528 A bicuspidal or Cartesian curve. **1879** G. Salmon *Higher Plane Curves* (ed. 3) 241 Quartics having the two circular points as cusps have also been much studied under the name of Cartesians, the properties of which may similarly be generalized and stated as properties of bicuspidal quartics.

bicycle, *sb.* Add: **b.** *attrib.* and *Comb.*, as *bicycle-basket, -gymkhana, -lamp, -polo, -pump, -race, -ride, -riding, -shed, -trade, -tyre*; **bicycle-chain,** the chain which transmits the driving power from the pedals of a bicycle to its rear wheel; also, one used as a weapon in gang-warfare, etc.; **bicycle-clip,** a clip used to hold a cyclist's trouser-leg firmly around the ankle; **bicycle-rickshaw,** a rickshaw drawn by a person on a bicycle.

1870 *Belgravia* X. 445 A novel feature in connection with bicycle races is taken from a donkey race...A prize is frequently given for the rider who can go, say, 100 yards in the *longest time*. *Ibid.* 446 Physicians of all nations have had time to judge of the effects of bicycle-riding on the human frame. **1882** Bicycle trade [in Dict.]. **1896** Bicycle gymkhana [see Gymkhana]. **1897** *Illustr. London News* 27 Mar. 433 (Advt.), The 20th century bicycle lamp. **1898** G. B. Shaw *Let.* 7 May in J. Dunbar *Mrs. G.B.S.* (1963) x. 169, I went for a bicycle ride. **1899** M. Beerbohm *Around Theatres* (1924) I. 59 We watched four..men riding round the stage on bicycles and trying to project with their front-wheels a small ball into one or other of two small goals. 'Bicycle Polo' was the name given to this cheerful pastime. **1899** W. James *Talks* 222 Just as a bicycle-chain may be too tight, so may one's carefulness and conscientiousness be so tense as to hinder the running of one's mind. **1905** E. M. Forster *Where Angels fear to Tread* iv. 104 There were bicycle gymkanas, and on the 30th Mrs Herriton would be holding the annual bazaar..for the C.M.S. **1907** *Yesterday's Shopping* (1969) 124/1 Bicycle Baskets, for fixing to handle-bar of Bicycle. **1908** *Sears, Roebuck Catal.* 170/2 Bicycle Pumps..including hose connection which will fit all valves. **1922** H. Williamson *Dandelion Days* 157 If you had a pin stuck..in a bicycle tyre. **1923** R. Macaulay *Told by Idiot* x. 97 Her bicycle basket was full of primroses. *a* **1930** D. H. Lawrence *Phoenix II* (1968) 177 Do you think he'd flown over, with bicycle-clips round his trouser-ankles and spots of mud on his nose? **1936** *Punch* 12 Aug. 190/2 He [*sc.* the Headmaster] was venturing to ask friends of the School for subscriptions towards the cost of a new bicycle-shed to replace the old one. **1939** C. Day Lewis *Child of Misfortune* II. i. 144, I could hear you playing bicycle-polo in the yard below. **1960** Koestler *Lotus & Robot* I. i. 78 The bicycle-rickshaw—whose half-starved operator must pedal single-legged up to four passengers. **1960** C. Day Lewis *Buried Day* vi. 124 Thomas Hardy..was living only a bicycle-ride away. **1962** *John o'London's* 1 Nov. 404/2 The bicycle-chain Teds. **1963** P. M. Hubbard *Flush as May* i. 11 The constable..took out a pair of bicycle clips, which he adjusted round his ankles.

bicyclette (bəisikle·t). [Fr., dim. of *bicycle.*] A low-wheeled or safety bicycle (see Safety 9).

1895 *Columbus* (Ohio) *Dispatch* 17 Jan. 4/3 There were two-seated and four-seated vehicles and a petroleum bicyclette. **1895** *Blackw. Mag.* Nov. 646/2 If Nature..be more to you than bicyclettes and German bands.

bicyclist. (Earlier example.)

1869 *Porcupine* XI. 72/3 Jeered and hooted by the mob, the 'bicyclists' pushed along.

bid, *v.* Add: **3. c.** *Card-playing.* (*a*) *intr.* To make a bid (see *bid *sb.* 2). (*b*) *trans.* To make a bid of or in (a number of tricks, a specified suit, etc.). Cf. *declare *v.* 11 c.

1880 'Trumps' *American Hoyle* 229 When the eldest hand makes a bid of five or more tricks, and another player bids the *same number* of tricks, the eldest hand may bid over him, or abandon his bid. **1897** R. F. Foster *Complete Hoyle* 270 If a player proposes to win all five tricks he bids *nap*, which is the highest bid possible. **1908** L. Hoffmann *Five Hundred* 14 A player who has once 'passed' cannot again bid. **1910** *Encycl. Brit.* IX. 878/1 *Bid Euchre*...Each player 'bids', *i.e.* declares and makes a certain number of tricks. **1929** [see *biddable *a.* 2]. **1933** C. Vandyck *Contract Contracted* i. 15 If there have been two no-bids before your turn to bid, you should [etc.]. *Ibid.* ii. 17 Always bid a suit in preference to No Trumps. **1958** [see *bidding *vbl. sb.* 1 b].

bid, *sb.* Add: **b.** Phr. *to make a bid for*: to make an attempt to secure; to 'have a try' at getting. Hence the simple sb. is freq. used, esp. in journalese, for: an attempt to win or secure something.

1885 *Century Mag.* Dec. 179/2 He was a little ashamed of making such a bare-faced bid for her sympathy. **1893** *Cassell's Fam. Mag.* Apr. 357/2 The Colonel makes a good bid for the hole with the odd, and lies by the rim. **1895** *Geogr. Jrnl.* May 415 To make a bid himself for the throne. **1935** *Punch* 21 Aug. 204/2 'Britain's Bid for War-Plane Supremacy.'—*Daily Mirror*. ..'Campbell's Bid for Record.' *All the Papers.* **1942** *Sunday Express* 14 June 1/6 Here they [*sc.* the German army] are now making a determined bid to move on Acroma.

2. *Card-playing.* The statement of an undertaking which a player makes; *spec.* in *Bridge*, an announcement of the number of tricks in a specified suit or 'no-trumps' by which a player proposes to beat his opponents. Cf. *contract *sb.*^{1} 1 g, *declaration 8 b, and see *approach *sb.* 12.

1880, 1897 [see *bid *v.* 3 c]. **1908** L. Hoffmann *Five Hundred* 14 A player who has been over-bidden is entitled to make a further bid. **1913** W. Dalton *Royal Auction Bridge* ii. 48 You should never make a bid unless you are prepared to play the hand with that suit as trumps. **1928** [see *score *sb.* 21].

bidarka (bəidā·ɹkă). Also **baidarka, baidarke, bidarkee.** [ad. Russ. *baidárka*, pl. *-ki*, dim. of *baidára* an oomiak.] In Alaska and adjacent regions, a portable canoe for one or more persons; a kayak.

1834 W. F. Tolmie *Jrnl.* 20 June (1963) 283 The youth took his departure—soon after a bidarka approached—paddled by two men—the officer sitting in the middle seat. **1868** F. Whymper *Trav. Alaska* 137 Their 'baidarkes' are similar to the Greenland 'kyack'. **1894** *Outing* (U.S.) XXIII. 389/1 A one-hatch *bidarka*, or hunting boat, of the Aleutian Islands. **1898** *Century Mag.* LV. 672 Their kayaks and bidarkees. **1967** J. C. Beaglehole in *Cook's Jrnls.* (1967) III. I. 412 The small *baidarka* [was] the same as the Eskimo *kayak*.

bidax(e, dial. var. *beat-axe* (see Beat *sb.*^{3}).

1778 Pryce *Min. Cornub.* IV. i. 226 Then it is digged and broken to pieces with a bidax, or hedging tool. **1806** R. Polwhele *Hist. Cornwall: Population* 31 Where a man of York or Kent would employ a shovel or a spade, a Cornishman uses a bidaxe.

bi:ddabi·lity. [f. Biddable *a.* (see -ility).] The state or quality of being biddable; obedience.

1947 W. F. Brown *Field Trials* xxii. 148 There is not the pronounced emphasis on biddability, kindly handling response. **1948** Lt. A. G. Strong *Trevannion* iii. 49 There were limits to Mrs. Wishart's biddability. **1958** *Times* 13 Dec. 7/6 The modern Labrador..was selectively bred.. with speed, style, and biddability in mind.

biddable, *a.* Add: **2.** *Card-playing.* Of a hand, suit, etc.: capable of being bid; strong enough to warrant a bid.

1926 M. C. Work *Auction Bridge Complete* I. vi. 56 Either suit is biddable because both have length. **1929** —— *Compl. Contract Bridge* v. 79 If he has a biddable suit,..he unhesitatingly bids two of his best suit. **1959** *Listener* 22 Jan. 189/2 The slam is hardly biddable.

bidder. Add: **5.** *Card-playing.* One who makes a bid (see *bid *sb.* 2) (see also quot. 1908). Cf. *declarer 3 b.

1880 'Trumps' *American Hoyle* 230 The bidder may at his option call for a lead of any suit from the player whose proper lead it is. **1897** R. F. Foster *Complete Hoyle* 271 The hands are usually abandoned when the bidder succeeds in his undertaking. **1908** L. Hoffmann *Five Hundred* 14 The player who has bid highest is thenceforth known as 'the bidder' for that hand. **1929** M. C. Work *Compl. Contract Bridge* ii. 9 The exact strength of the bidder's hand is of vital importance.

bidding, *vbl. sb.* Add: **1. b.** *Card-playing.* The act or process of making a bid or bids (see *bid *sb.* 2). Also *attrib.*

1880 'Trumps' *American Hoyle* 229 A player having the highest bid, must declare the suit he plays in as soon as the bidding ceases. **1908** W. Dalton *Auction Bridge* vii. 103 The bidding is quite the most interesting, as well as the most exciting, feature of Auction Bridge. **1908** L. Hoffmann *Five Hundred* 14 The bidding is continued till no player will go any further. **1928** [see *score *sb.* 21]. **1929** M. C. Work *Compl. Contract Bridge* p. xii, To claim that any sound bidding system can be reduced to the simplicity of a multiplication table. **1958** *Listener* 6 Nov. 753/2 Over Two Diamonds I bid Two Hearts rather than Three Diamonds to preserve bidding space.

biddle, dial. var. Beetle *sb.*^{1}

1876 G. L. Gower *Surrey Prov.* 82 A 'stake-*biddle*' is that which is used for driving stakes, a long or dumb-*biddle* for cleaving wood. **1937** *John o'London's* 30 Apr. 197 The 'biddle' itself is a huge wooden mallet. *Ibid.*, Peculiar to Devon, sometimes used as a term of endearment, but not always, is 'biddle head'. **1941** H. J. Massingham *Remembrance* xiii. 133 He was a thatcher..with biddle, leggatt, [etc.].

biddy^1. Add: (Examples.)

[**1708** Swift (*title of poem*) To Mrs. Biddy Floyd. **1858** O. W. Holmes *Aut. Breakf.-t.* xii. 363 Poor Bridget, or Biddy, our red-armed maid of all work!] *a* **1861** T. Winthrop *Life in Open Air* (1863) 261 Our thousand [*sc.* soldiers] did the Capitol little harm that a corporal's guard of Biddies with mops and tubs could not repair in a forenoon's campaign. **1902** Bret Harte in *Harper's Mag.* Apr. 733 He's..puttin' up a high-toned house on the hill..with a Chinese cook, and a Biddy. *a* **1953** E. O'Neill *More Stately Mansions* (1965) II. iii. 120 A mother of children, our Irish biddy nurse girl and house servant.

2. A woman, usu. with derogatory implication (see quots.). *slang.*

1785 Grose *Vulg. Tongue*, *Biddy*, or *Chick-a-biddy*, a chicken, and figuratively a young wench. **1947** *Horizon* XVI. 205 Few of our chaps have a biddy [*sc.* Italian girl] up the alley here. **1958** *Observer* 10 Aug. 11/3 He is already installed as the local delinquent, using his considerable charm on the local biddys. **1960** C. P. Snow *Affair* xl. 368, I believe she's the bloodiest awful specimen of a party biddy.

biddy^2 (bi·di). = **red biddy*.

1940 Dylan Thomas *Portrait of Artist* 228 He used to drink a pint of biddy before his breakfast.

biddy-biddy. Also **bid-a-bid, biddy-bid, bidi-bid, bidy-bidy.** [Corruption of Maori *piripiri*.] The New Zealand name for the troublesome burr of the piripiri. Also *attrib.*

1866 G. Mueller *My Dear Bannie* (1958) iv. 110 We shall have..after that 'bidy-bidy' tea..without sugar. **1880** T. H. Potts in *N.Z. Country Jrnl.* XII. 195 (Morris), Piri-piri (*acaena sanguisorbe*) by the settlers has been converted or corrupted into *biddy-biddy*. **1896** *Otago Witness* 23 Jan. II. 36 (Morris), Biddybids detract very materially from the value of the wool, and the plant should not be allowed to seed where sheep are depastured. **1944** D. Stewart in *Coast to Coast 1943* 96 Scrape them off like sand or bid-a-bid seeds. **1951** J. Frame *Lagoon* 19 Father shook the bidi-bids off the big red and grey rug.

bidet. Add: **2.** (Examples.) Now usu. a shallow oval basin fitted in a bathroom, used for washing the perineum.

1766 Smollett *Trav.* I. v. 64 Will custom exempt from the imputation of gross indecency a French lady, who shifts her frousy smock in presence of a male visitant, and talks to him of her *lavement*, her *medecine*, and her *bidet*! **1785** Grose *Dict. Vulg. Tongue*, *Bidet*, commonly pronounced biddy, a kind of tub, contrived for ladies to wash themselves, for which purpose they bestride it like a little French poney, or post horse, called in France bidets. **1801** *Ann. Reg. 1799* 401/1 A machine answering the purposes of a portable water-closet, or bidet, and easy chair. **1959** R. Gant *World in Jug* 60 We checked into our hotel near the Opéra and the boys joked about the bidets.

bi:dime·nsional, *a.* [Bi-^{2}.] Having or perceived in two dimensions. Also *transf.*

1927 N. T. Burrow *Social Basis of Consciousness* I. v. 92 Depending on whether the child 'succeeds' or 'fails' as judged by the bidimensional standard of good and bad. **1937** H. F. Brandt in *Amer. Jrnl. Psychol.* XLIX. 666 A bidimensional eye-movement camera. *Ibid.* 669 This new instrument provides laboratory equipment which is essential for all types of ocular photography in the bidimensional plane.

bi:dire·ctional, *a.* [Bi-^{2}.] Functioning in two directions; *spec.* of a microphone (see quot. 1962^1).

1941 A. E. Craig *Speech Arts* (ed. 2) xxxix. 509 A bidirectional microphone, with the speech directed into either, or both, the openings or faces, is the kind deemed best for the production of radio plays. **1960** *Aerodrome Lighting* (*B.S.I.*) 29 Light fittings..may be bi-directional or uni-directional according to whether they are to cover both directions of landing and take off or not. **1962** A. Nisbett *Technique Sound Studio* 243 *Bi-directional microphone*, one which is live on the front face and back, but which is dead at the sides and above and below. **1962** Simpson & Richards *Junction Transistors* xvi. 384 A bi-directional switch using four silicon junction diodes.

‖ **bidon** (bidoṅ). [Fr.] A container for liquids: *spec.* (*a*) a wooden cup; (*b*) a bottle or canteen for water, wine, etc.; (*c*) an oil drum or petrol tin. Hence **bidonville** [F. *ville* town], a shanty-town built of oil drums or petrol tins.

1867 'Ouida' *Under Two Flags* II. i. 13 'Take a draught of my burgundy...' He..took the *bidon*. **1922** E. E. Cummings *Enormous Room* iii. 48 Tilting their *bidons* on high and absorbing the thin streams which spurted therefrom. *Ibid.*, He remarked: 'Bread without wine doesn't taste good,' and proffered his *bidon*. **1933** 'G. Orwell' *Down & Out* xv. 115 She had picked up an empty oil *bidon*. **1955** *Times* 21 July 6/6 Huge fires destroyed two Moroccan bidonvilles (shanty towns) in the suburbs of Fedala. **1961** *Times* 6 May 9/7 Elsewhere the *bidonvilles*, or petrol can shanty towns, spread like a rash. **1964** E. Ambler *Kind of Anger* vi. 164 Get the car filled up. Adèle left me two *bidons* for emergencies.

bieberite (bī·bĕrəit). *Min.* [ad. G. *bieberit* (Haidinger 1845), f. *Bieber*, near Hanau, Germany: see -ite^1.] Sulphate of cobalt, found as a red crust on other minerals.

1854 Dana *Syst. Min.* 385 *Bieberite*...Lustre vitreous. Color flesh and rose-red. **1882** —— *Mineral. & Lithol.* 168 Bieberite or Cobalt Vitriol.

Biedermeier (bī·dəɹməiəɹ). Also **-maier, -meyer.** [The name of a fictitious poet, Gottlieb *Biedermaier* (see below).] Applied *attrib.* to the period between 1815 and 1848 in Germany and to styles, furnishings, etc., characteristic of that period, esp. to a type of furniture derived from the French Empire style. Also *transf.* with derogatory implication: conventional, bourgeois.

[**1854** [L. Eichrodt] in *Fliegende Blätter* XXI. 102/1 (*title*) Auserlesene Gedichte von Weiland Gottlieb Biedermaier, Schulmeister in Schwaben. *Ibid.* 103/1 Die ästhetischen Begriffe des Biederschönen und Bidermaiern.] **1905** A. S. Levetus *Imperial Vienna* xvi. 245 The Biedermaier period also produced landscape-painters. **1914** Eberlein & McClure *Pract. Bk. Period Furniture* xv. 323 Following the prevalence of the Empire style we see the advent of the Biedermeyer type of painted decoration. **1924** G. B. Stern *Tents of Israel* i. 8 A true son of the Biedemeyer period, in his long, tight pantaloons, his cambric frills. **1956** *English Studies* XXXVII. 183 The examples..from the early Victorian age may be regarded as contradictions of the 'Biedermeier' dominance in literature. **1957** *Encycl. Brit.* IX. 948E/2 The Biedermeier style..is characterized by chairs with curved legs and sofas with rolled arms and generous upholstery. **1963** *Listener* 24 Jan. 161/1 A hill close to the centre of the city [*sc.* Darmstadt] where there was a piece of Biedermeier park and a Russian church.

Bielid (bī·lid). *Astr.* [f. the name *Biela* (see below) + *-id^2, as in *Leonid*, etc.] An Andromede. So called because supposed to come from the remains of Biela's comet (named after

the Austrian astronomer, W. von Biela). Also *attrib.*

1885 *Science* VI. II. 496/2 The Bielid meteors were observed here in considerable numbers. **1899** *Edin. Rev.* Oct. 318 Displacements of the Bielid orbit are no abnormal events. **1899** *Sci. Amer.* 9 Dec. 379/3 A Shower of Bielids. A well-marked shower of Bielid meteors was observed at Princeton on the evening of November 24.

‖ **bien** (bįeṅ), *adv.* The French word for 'well'; used in certain French phrases:

bien entendu (bįeṅ aṅtaṅdü). [F. *entendu*, pa. pple. of *entendre* to hear, understand.] Of course; that goes without saying.

1844 Thackeray *Barry Lyndon* vii. in *Fraser's Mag.* Apr. 396/1 Burgundy and velvet are the best, *bien entendu*. **1863** C. Reade *Hard Cash* I. vi. 190 And I was not penniless, bien entendu. **1904** H. O. Sturgis *Belchamber* x. 136, I could never call her that to her face, *bien entendu*. **1927** B. Malinowski *Sex & Repression in Savage Society* I. ix. 77 First of all—and that has, *bien entendu*, nothing to do with matriliny—there is no condemnation of sex.

bien-être (biaṅętr). [Fr. (16th c.); *être* to be.] A state of well-being; comfort.

1849 Geo. Eliot *Let.* 26 Oct. in J. W. Cross *Life* (1884) I. iv. 189, I never enjoyed a more complete *bien être* in my life than during the last fortnight. **1873** W. James *Let.* 11 May in R. B. Perry *Thought & Char. W.J.* (1935) I. 344 Your letter..seemed to reveal a great physical *bien être*. **1883** C. M. Yonge *Stray Pearls* I. x. 112 There was an indefinable bliss and *bien-être* in their very presence. **1946** Wodehouse *Joy in Morning* i. 3 No inkling of the soup into which I was to be plunged came to mar my perfect *bien-être*.

bien pensant (bįeṅ paṅsaṅ), *a.* [Fr. *pensant*, pr. pple. of *penser* to think.] Right-thinking; orthodox, conservative. Also as *sb.*

1923 A. Huxley *On Margin* 113 If you are rich, of good family and *bien pensant*. **1938** *Times Lit. Suppl.* 1 Oct. 622/3 A Nationalist rising in Spain, which not only M. Maurras but all the French *bien-pensants* supported. **1958** *Listener* 7 Aug. 209/2 The pitiful, hasty funeral from which the local *bien-pensants* remain away. **1962** N. Mitford *Water Beetle* 136 In her world, Catholic, royalist, *bien pensant*.

‖ **bierhaus** (bīə·ɹhaus). [G.] A German public house or ale-house.

1930 G. Greene *Name of Action* viii. 219 A bierhaus with an open door. **1963** —— *Sense of Reality* 83, I imagined myself drinking my way through the festival itself in some *bierhaus* decorated with holly. **1967** J. Leasor *Passport in Suspense* 2 His father ran a *bierhaus* behind Hamburg railway station.

‖ **bierstube** (bīə·ɹstūbə, -ʃt-). Also **Bierstube.** Pl. **-ben** or **-bes.** [G.] A German tavern, taproom or bar.

1909 in Webster. **1922** *Glasgow Herald* 6 May 9 Here [*sc.* in Berlin], in the streets, hotels, and bierstubens, many languages..can be heard. **1952** Koestler *Arrow in Blue* x. 84 Our fraternity marched..to our favourite Bierstube. **1956** E. Berckmann *Strange Bedfellow* xii. 111 She would wait for him in the *Bierstube*.

bif, biff, *int.* orig. *U.S.* An exclamation uttered when something strikes an object, or a sound imitative of such a blow. Cf. *biff *v.* 3.

1847 J. S. Robb *Squatter Life* 137, I hit him, *biff*, alongside his smeller. **1905** H. G. Wells *Kipps* I. iv. § 1 When I go to turn, if I don't remember, Bif!—and I'm *in* to something. **1917** H. A. Vachell *Fishpingle* xi, Hamlin bowled straight and true for the middle stump. The youth smote and missed. 'Bif!' yelled Lionel. *a* **1930** D. H. Lawrence *Phoenix II* (1968) 109 He..took the poker with satisfaction. Biff! A well-aimed blow.

biface (bəi·fēⁱs). *Archæol.* [f. Bi-^{2} + Face *sb.*] A type of prehistoric stone implement flaked on both faces. Also *attrib.*

1934 A. C. Haddon *Hist. Anthropol.* vii. 89 Throughout all Africa a core-tool, or 'biface', is the oldest known technique, and the same applies to India; in Africa it developed from a chipped-pebble industry. **1937** *Proc. Prehist. Soc.* III. 15 The lower Palaeolithic industries may be roughly divided into two main groups:— biface cultures and flake cultures. **1946** F. E. Zeuner *Dating the Past* vi. 164 The bifaces are mostly made on flakes. **1953** J. Procházka tr. *B. Hrozný's Anc. Hist. W. Asia, India & Crete* iv. 12 The so-called flint flakes and more massive bifacial tools ('bifaces') have been uncovered in the early palaeolithic strata.

bifa·cial, *a.* [Bi-^{2}.] Having two faces; spec. *Bot.*, having distinct dorsal and ventral surfaces; *Archæol.*, of a flint: worked on both sides. Hence **bifa·cially** *adv.*

1884 [see Bi-^{2} 1]. **1952** V. G. Childe *New Light on Most Anc. East* ii. 19 Bifacial trimming in which the primary flake is thinned and shaped by the removal of fine secondary flakes from both sides or faces and along both edges. **1953** [see prec.]. **1957** V. G. Childe *Dawn Europ. Civ.* (ed. 6) vii. 111 The Bükkians..made from the volcanic glass knives and scrapers, but no bifacially worked arrow-heads.

biff (bif), *v.* *slang.* [Imitative.] **1.** *trans.* To hit, strike. Also *to biff* (a person) *one*.

1888 *Judy* 18 Jan. 27 He playfully biffed him with a brick. **1894** Kipling *Day's Work* (1898) 71 Ef we all biffed you now, these same men..'u'd call us off. **1903** Clapin *Dict. Amer.* 50 *Bif, biff, bift*, current in several parts of the States in sense of to strike, and especially to give a quick blow. 'He biffed him on the ear.' **1908** G. H. Lorimer *J. Spurlock* i. 19 If she had biffed me one it would have been all right. **1950** A. Baron *There's no Home* 181 Where'd you get that bruise on your forehead? Girl friend been biffing you with the old rolling pin?

b. *fig.* To deal a blow to, to refute, to 'stump' (see also quot. 1895).

1895 W. C. Gore in *Inlander* Nov. 60 *Biff*, to refuse; to repulse; to slight. **1914** T. E. Lawrence *Let.* June (1938) 174 To biff that, Woolley & I walked over next day to Mesopotamia. **1915** C. Mackenzie *Guy & Pauline* v. 263 'Oh, it's in old French, is it?' said Brydone in a disappointed voice. 'That would biff me.' A silence fell upon the room, a silence that seemed to symbolize the 'biffing' of the doctor's son by old French.

c. To throw. Also *intr.* *Austral.* and *N.Z.*

1941 Baker *Dict. Austral. Slang* 9 *Biff*, to throw. **1964** *N.Z. Listener* 1 May 4/2 'All I can do is biff.' 'Then just biff—as hard as you can. You're a natural [at putting the shot].'

2. *intr.* To go, proceed. Esp. with *off*, to leave, depart.

1923 Wodehouse *Inimit. Jeeves* i. 9 Bingo biffs about London on a pretty comfortable allowance. *Ibid.* xv. 189, I can't go chucking all my engagements..in order to biff down to Twing. **1929** —— *Mr. Mulliner Speaking* viii. 250 Biffed off a trifle abruptly.

3. The verb used adverbially with *go*, in the sense of 'with a violent blow'. Cf. *bif, biff *int.*

1904 *Daily Chron.* 15 Dec. 4/7 There might go biff through the glass the occasional plank.

Hence **bi·ffing** *vbl. sb.*, the action of the verb *biff*.

[**1894** Kipling in *Century Mag.* Dec. XLIX. 295/2 S'pose we want the Back Pasture turned into a biffin'-ground on our only day er rest?] **1915** [see sense 1 b above]. **1959** I. & P. Opie *Lore & Lang. Schoolch.* xvii. 374 The miscreant receives..a..biffing.

biff, *sb.* *slang.* Also **bif.** [f. the vb.] A blow, whack. Also *fig.* Cf. Baff *sb.*

1889 Barrère & Leland *Dict. Slang*, *Biff* (Americanism), to give [one] a 'biff in the jaw'. **1890** *Dialect Notes* I. 72 *Bif*,..oftenest used in such phrases as 'to give one a biff in the ear'. **1904** W. H. Smith *Promoters* x. 165 What an idiot a man can be when he gets a biff that takes his wind. **1926** *Blackw. Mag.* Feb. 224/1 Contusions on top of head.. biff on the — bruise on inside of knee. **1935** L. A. G. Strong *Tuesday Afternoon* 45 A biff that made him bite his tongue.

biff, *int.*: see *bif *int.*

bifilar, *a.* Add: (Earlier examples.) Also as *sb.*

[**1836** W. S. Harris in *Trans. R. Soc.* CXXVI. 417 A new species of balance; it may be termed, from the peculiar mechanical principle on which it depends, a bifile balance.] **1839** Whewell *Let. to Quételet* 17 Jan. (1876) II. 275 Airy has just put up his Gaussian apparatus..at Greenwich, including a *Bifilar*. **1846** *Penny Cycl.* Suppl. II. 256/1 The horizontal force magnetometer at the Greenwich Observatory is bifilar. **1848** W. S. Harris *Rudimentary Electricity* iii. 83 *The Bifilar Balance.* In this electrometer..a re-active force is obtained by means of a lever at the extremity of two parallel and vertical threads of unspun silk.

biflecnode (bəifle·knō̆ᵘd). *Math.* [f. Bi-^{2} + Flecnode.] A point on a curve that is both a node and a point of inflexion for each branch passing through it. Hence **biflecno·dal** *a.*

1874 G. Salmon *Analyt. Geom.* (ed. 3) p. xvii, Number of points at which two tangents are biflecnodal. *Ibid.* 521 Biflecnodal points. **1879** —— *Higher Plane Curves* (ed. 3) 217 Such a node may be considered as the union of an ordinary node with (in the first case) a point of inflexion, and with (in the second case) two points of inflexion; and the node may be termed a flecnode or a biflecnode in the two cases respectively.

bifo·cal, *a.* and *sb.* orig. *U.S.* [Bi-^{2}.] **A.** *adj.* Having two foci; *spec.* designating spectacles with two segments of different focal lengths.

1888 G. M. Gould in *Med. & Surg. Reporter* LIX. 545 (*title*) A new style of bi-focal lenses. **1895** *Pop. Sci. Monthly* Aug. 470 Ordinary bifocal glasses. **1938** E. Ambler *Cause for Alarm* ii. 29 A..man of about fifty with rimless, bi-focal spectacles. **1968** *New Scientist* 21 Mar. 619/2 Bifocal and trifocal spectacles have been a blessing.

B. *sb. pl.* Bifocal spectacles.

1899 *Ophthalmic Rev.* XVIII. 270 Dr. G. C. Harlan..had shown..a pair of bi-focals made by Borsch in which the reading part is formed of a circular lens 15 mm. in diameter.. sunk into the distant lens. **1936** G. B. Shaw *Simpleton* 31 Wears glasses. Bi-focals. **1960** J. Wain *Nuncle* 6 Harley Street types looking at me through their bifocals.

bifolium (bəifō̆ᵘ·liə̆m). *Palæogr.* Pl. **-ia.** [f. Bi-^{2} + L. *folium* leaf.] A pair of conjoint leaves.

1938 E. A. Lowe *Codices Latini Antiquiores* III. 283 Foll. 3 ..composed of a bifolium and a single leaf. *Ibid.* 284 Ruling on the hair-side central opening, four bifolia at a time before folding. **1957** N. R. Ker *Catal. MSS. containing Anglo-Saxon* 407 A bifolium followed by three singletons. **1960** E. A. Lowe *English Uncial* 11 The following Insular, at least non-Italian, features were enumerated: pricking and ruling of the bifolia after folding, [etc.].

‖ **bifteck** (biftęk). Also **biftek.** [Fr., adaptation of Beef-steak.] = Beef-steak.

[**1825** *French Domestic Cookery* 50 Bifteck au Beurre d'Anchois.—*Beef-Steaks with Anchovy Butter.*] **1861** Mrs. Beeton *Bk. Househ. Managem.* 264 There is in every *carte* at every French restaurant..*biftek d'Angleterre*. **1941** Koestler *Scum of Earth* 16 To become a Deputy or Minister was only a form of earning one's *bifteck* and rather a fat one at that. **1969** *Guardian* 30 Aug. 3/4 With meat the price it is, it is unlikely that it will be a very good biftek in a 27s 6d meal.

big, *a.* Add: **3. e.** Fig. phr. *to get, grow*, etc., *too big for one's boots* (*breeches*, etc.), to become conceited, put on airs.

1835 D. Crockett *Tour to North* 152 When a man gets too big for his breeches, I say Good-bye. **1879** [see *boot *sb.*^{3} 1 c]. **1893** H. Maxwell *Life of W. H. Smith* I. ii. 57 Sometimes a young man, 'too big for his boots', would..sniff at being put in charge of a railway bookstall. **1905** H. G. Wells *Kipps* III. ii. § 1 He's getting too big for 'is britches. **1929** W. Faulkner *Sound & Fury* 270 You're getting a little too big for your pants. **1952** M. Laski *Village* xv, A young man who was getting too big for his boots.

f. In the collocations *big brother, sister*, elder (cf. 3 b). Also *transf.* (see *Big Brother below). Hence *big-brotherly* adj.

1851 Mayhew *London Lab.* I. 151/1 I've a big sister, and a brother and a sister younger than I am. **1863** *Harper's Mag.* Apr. 693/1 In a modern story, a big brother would have kicked the noble lord out of the front door. **1873** J. H. H. St. John *Pakeha Rambles through Maori Lands* viii. 149 Little mud volcanoes..aped the customs of their big brothers, and blobbed out their stenches with as much complacency as Tongariro. **1902** A. Bennett *Anna of Five Towns* i. 4 Your big sister isn't out of school yet? **1922** Joyce *Ulysses* 497 Big Brother up there, Mr. President, you hear what I done just been saying to you. **1937** H. G. Wells *Star Begotten* v. 84 If there is such a thing as a Martian..he's humanity's big brother. *Ibid.* viii. 157 Out of these cravings come all these impulses towards slavish subjection to Gods, Kings, leaders, heroes, bosses, mystical personifications like the People, My Country Right or Wrong, the Church, the Party, the Masses, the Proletariat. Our imaginations hang on to some such Big Brother idea almost to the end. **1947** *Landfall* I. 293 Lachlan was in splendid fettle..radiating big-brotherly goodwill. **1961** *Listener* 24 Aug. 293/3 For the Czechs and Slovaks.. Russia..was the big sister to whom they looked..for deliverance from the Austrian and Hungarian oppressors.

g. *Big Brother Movement* (see quots.).

1925 *Times* 24 Oct. 9/5 Mr. Richard Linton, the originator of the 'Big Brother' Movement. *Ibid.*, Mr. Linton had conceived the idea of providing 'Big Brothers' for boys who were emigrating to Australia. **1958** *Oxf. Mail* 14 Aug. 3/1 The Big Brother movement—a voluntary organisation founded in 1925 to assist the settlement of British boys in Australia.

h. Of a letter: capital. Cf. *capital *a.* 5 b.

1874 Trollope *Way we Live* I. i. 1 She spoke of herself.. as a woman devoted to Literature, always spelling the word with a big L. **1894** Mrs. H. Ward *Marcella* III. IV. vi. 374 'You had spoken of "*marriage*"!' she said. 'Marriage in the abstract, with a big *M*.' **1964** *Times* 11 Jan. 5/5 An attempt to impose Culture, with a big 'C', on the..people.

7. b. *spec.* In phrases with numerals, as *big three, four, five*, designating a combination of three, etc., important things, persons, companies, nations, etc. (see quots.); also *attrib.* orig. *U.S.*

1886 *Outing* (U.S.) Nov. 156/1 The trial races..proved beyond a doubt that the *Mayflower* was the queen of the 'big four'. **1890** Wheeler & Cardwill *W. A. W.* p. vi, Big 4 Route..Best Modern Day Coaches on all Trains. **1922** *Encycl. Brit.* XXX. 397/2 The resources of the 'big five' [*sc.* banks] were..very substantial. **1924** *Golfers' Guide to Happy Holidays* 25 Golf's 'Big Four' are: Skill, Temperament, Experience, Luck. **1931** H. G. Wells *Work, Wealth & Happiness of Mankind* (1932) xii. 635 Both China and Japan are members of the League [of Nations] and Japan is one of the Big Five. The Council..has five permanent members (Britain, France, Germany, Italy, and Japan). **1932** Wodehouse *Louder & Funnier* 283 The Big Four at Scotland Yard..hold a round-up of the novelists. **1934** *Ann. Reg. 1933* II. 68 The 'big five' banks—Barclay's, Lloyds, the Midland, the National, and the Westminster. **1934** Webster, *Big Three, Gr. Brit.*, the industrial alliance (formed 1919) of the miners' federation, the national transport workers' federation, and the national union of railwaymen. **1945** *Ann. Reg. 1944* 314 A further 'Big Three' meeting which could concert policy on common problems. **1946** *News Chron.* 8 July 1 The decisions laboriously reached..by the Big Four. **1968** *Globe & Mail* (Toronto) 17 Feb. 26 Toronto is already in the first line of orchestras after the big five in the United States.

c. In designations of operatives, denoting the chief or senior men.

1910 *Westm. Gaz.* 9 Mar. 4/1 There are over 40,000 male cotton-piecers who earn from 8s. 6d. to 15s. 6d. as little-piecers (youths), and from 13s. 6d. to 22s. 6d. as big-piecers (young men). **1921** *Dict. Occup. Terms* (1927) § 363 Big tenter. *Ibid.* § 365 Big piecer..big spinner.

8. b. (Later examples.)

1841 [see Talk *v.* 3 c]. **1865** Dickens *Mut. Fr.* II. xii. 274 You talk big, you do, but things look pretty black against yourself.

9. *colloq.* (orig. *U.S.*). Generous, magnanimous; freq. ironical, esp. in phr. *that's big of you*.

1934 Webster s.v. *big*, He is big enough to overlook the slight; he can be depended upon to do the big thing. **1942** N. Coward *Blithe Spirit* I. ii. 39 *Ruth.* You can come in and say good night to me if you feel like it. *Elvira.* That's big of her, I must say. **1951** N. Marsh *Opening Night* iv. 92 You will be really generous won't you? Really big? You won't bring me into it, will you? **1959** H. Hamilton *Answer in Negative* vi. 82 That's big of you...I ought to lead a.. sober life, and if I do you'll consent to be seen about with me. Thank you.

10. *slang* (orig. *U.S.*). In various quasi-*advb.* uses. **a.** With pronounced success, esp. in phr. *to go* (*over*) *big* (see *go *v.* 19 b and 87 h). **b.** Feelingly; emotionally. Cf. sense 8 b.

1912 J. London *Let.* 7 Sept. (1966) 363 That the book should sell big, I have all the confidence in the world. **1932** Wodehouse *Hot Water* xvii. 300, I see now why you took it so big when I mentioned that Soup Slattery was in the neighbourhood. **1936** M. Allingham *Flowers for Judge* i. 12 The ridiculous task of putting himself over big. **1958** I. Cross *God Boy* xiv. 110 My actually seeing God would put me in big with Father Gilligan.

B. 1. a. *big-brained, -eye(d), -hearted* adjs.; *-heartedness.*

1915 E. R. Lankester *Divers. Nat.* xxix. 270 We compare the actual mental accomplishments of the highest civilized races of man with those of big-brained savages. **1818** *Amer. Jrnl. Sci.* I. 79 The big-eye herring (*Clupea megalops*) begin to be seen at the fish-market. **1885** J. S. Kingsley *Standard Nat. Hist.* III. 187 The big-eyed scad,..the *Trachurops crumenophthalmus* of naturalists. **1957** T. Hughes *Hawk in Rain* 23 Now he rides the morning mist With a big-eyed hawk on his fist. **1868** Trollope *He knew he was Right* I. xiii. 106 Had he not been so manly and big-hearted, he would have taken such pressure as a sign that she wished him to ask her again. **1914** D. H. Lawrence *Let.* 18 Dec. (1962) I. 300 You are so bighearted, we think of you with great affection. **1872** W. F. Butler *Great Lone Land* (ed. 2) xvii. 282 After such a present no man can possibly entertain ..a doubt upon the subject of the big-heartedness of the donor. **1953** *Scrutiny* XIX. 143 Implying that you are a rather inferior creature if you do not share his manly big-heartedness.

2. *big bird, board, deal, drink, hand, idea, noise, pot, shot, talk, way* (see the sbs.); **Big Apple,** a ballroom dance for a group of people, popular in the 1930's; **big band,** a large band of musicians playing jazz, dance-music or the like (as distinct from a small group or 'combo'); freq. *attrib.*; **big bang,** a great or loud explosion; *spec.* the explosion of a single compact mass, in which (according to one cosmological theory) the universe originated; freq. *attrib.*; **big-bellied** (further examples); **big bore,** a rifle of large calibre *U.S.*; also *attrib.* or quasi-*adj.*, having a large calibre, large-bore; **big boy** *colloq.* (orig. *U.S.*) = *big bug* (see Bug *sb.*^{1} 1 b); freq. used as an ironical form of address; also *transf.*; **big bud,** a disease of plants caused by the gall-mite; **big bug**: see Bug *sb.*^{1} 1 b in Dict. and Suppl.; **big business** (orig. *U.S.*), (those in control of) large mercantile organizations or transactions collectively; also *attrib.*, esp. *big-business man*; **big city** (orig. *U.S.*), a large city; esp. *attrib.* or quasi-*adj.*, of, pertaining to, or characteristic of a large city; **Big Daddy,** a paternal, dominating, or influential person; also *attrib.*; **Big Dipper,** (*a*) *U.S.* = Dipper 5 b; (*b*) (also with lower-case initials) a switchback at a place of amusement; also *fig.*, *transf.*, and *attrib.*; **big dog** (earlier U.S. examples); **big end,** the end of the connecting-rod that encircles the crank-pin, esp. in a motor-vehicle engine; also *attrib.*; cf. **small end*; **big-endian** (also **Big-endian**), substitute for def.: in Swift's 'Gulliver's Travels', a person who believed that eggs should be broken at the larger end before they are eaten (cf. *little-endian* s.v. Little *a.* 14); hence allusively and as *adj.*; (add further examples); **big figure** *U.S. slang* (see quot. 1848); **big game,** large animals hunted as game; freq. *attrib.*; **big gun**: see Gun *sb.* 7 b in Dict. and Suppl.; **big-horn, bighorn** (earlier and later examples); **big inch,** also *attrib.*, (of) an oil or natural gas pipe-line 20 inches in diameter or larger; **big laurel** *U.S.*, (*a*) a species of large magnolia, *M. grandiflora*; (*b*) the rose-bay *Rhododendron maximum*; **big league** *U.S.*, a major American baseball league (hence *big leaguer*); also *transf.* and *attrib.*; **(the) big lie** [cf. G. *grosse lüge*], (an instance of) falsehood on a large scale, *spec.* used in (Nazi) propaganda; **big man** (see sense 7); **big money** (orig. *U.S.*), a large amount of money; high salary, large profit; also *attrib.*; **big mouth** (orig. and chiefly *U.S.*), a very talkative or boastful person; also, loquacity, boastful talk; **big-mouthed** *a.*, (*a*) having a big mouth; (*b*) loquacious or boastful (cf. prec.); **big name** (orig. *U.S.*), a famous or celebrated person, esp. in the field of entertainment; also *attrib.*; **big picture** *colloq.*, the main film in a cinema programme; **big science,** term used of scientific and technological investigation that requires large resources; **big smoke,** (*a*) *Austral.*, an Aboriginal name for a town or city; (*b*) London (see Smoke *sb.* 1 d); **big stick** (orig. *U.S.*), (a display of) force or power; hence *big-sticker, -stickism*; **big stuff,** in various *slang* uses (see quots.); **big thing** *colloq.*, a promising affair, a good prospect; something magnanimous; **big time** (orig. and chiefly *U.S.*), an excellent time; hence (often *attrib.*), (of) the best kind or the highest rank; so **big-timer,** a top-ranker; **big top** (orig. *U.S.*), the main tent of a circus; the circus in general; also *transf.*; **big tree** *U.S.* (earlier examples); **big wheel,** (*a*) a Ferris wheel; (*b*) *slang* (orig. and chiefly *U.S.*) = *big shot* (see *shot *sb.*^{1} 22 c).

1937 *Dancing Times* Nov. 170/1 The rage of the winter is the Big Apple and its related steps...Such steps as the Shag, the Flea Hop, the Strut, and the Walk, are combined with the new Big Apple notes. **1958** *Publ. Amer. Dial. Soc.* XXX. 48 This term [Apple = New York] gave its name to that defunct dance, the Big Apple. **1926** *Melody Maker* Feb. 35 [It] gives the lie to those who say that a 'big band' is unwieldy. **1947** R. de Toledano *Frontiers of Jazz* xiii. 137 It remains the best big band jazz. **1955** *Jazzbook 1955* 6 He [*sc.* Duke Ellington] is the only man who has consistently created big band jazz of more than ephemeral value. *Ibid.* 7 In James Crawford the band possessed perhaps the finest of big band drummers. **1950** F. Hoyle *Nature of Universe* v. 102 One [idea] was that the Universe started its life a finite time ago in a single huge explosion...This big bang idea seemed to me to be unsatisfactory. **1957** Big bang [see *bang *sb.*^{1} 2 b]. **1964** *Listener* 3 Sept. 340/2 The man who comes to astrophysics with a belief in after-life might be expected to have some thoughts about the 'big-bang theory'. **1969** *Observer* (Colour Suppl.) 2 Feb. 16/1 A unique and inscrutable primeval atom, out of which all matter was born in some mysterious First Explosion—the famous 'Big Bang'. *Ibid.* 16/4 A mysterious radio hiss, which may be the dying echoes of the original big bang. *c* **1660** *Sea Crabb* in *Bp. Percy's Folio MS. Loose & Humorous Songs* (1963) 99 This goodwiffe was bigbellyed, & with a lad. **1812** J. Sinclair *Syst. Husb. Scot.* I. 357 When they [*sc.* calves] are allowed to drink much water at an early age, they will become big-bellied. **1916** R. Graves *Goliath & David* 14 Big-bellied, spectacled, crop-haired. **1843** 'R. Carlton' *New Purchase* II. xxxvi. 31, I had a powerful big bore to fix for a feller going out West. **1960** *Farmer & Stockbreeder* 23 Feb. 45/2 Big-bore, short-stroke engine—high power at low piston speeds. **1918** A. P. McKishnie *Willow* xx. 279 'Hold on, Big Boy,' he called. **1925** Fraser & Gibbons *Soldier & Sailor Words* 22 *Big boys*, large guns. 'Heavies'. **1929** W. R. Burnett *Little Caesar* I. i. 12 The Big Boy can't fix murder. He can fix anything but murder. **1939** J. B. Priestley *Let People Sing* iv. 94 'Am I right, sirs?' 'You sure are, big boy.' **1940** *Times Weekly* 27 Nov. 10 We had one of the really 'big boys' on board—the brother of the one I dropped at Munich. **1900** G. Nicholson *Dict. Gardening* Suppl. 274/1 Currant-Bud Mite (*Phytoptus ribis*), or Black Currant Gall Mite, is responsible for the condition of the bushes known as 'Big Bud'. **1958** *Spectator* 10 Jan. 58/2 Mites cause the swelling of blackcurrant buds at this time of year and give rise to what is commonly called Big Bud, a destructive thing resulting in a poor crop. **1905** *McClure's Mag.* 49 The stench of the vice graft did not repel, it attracted big business. **1912** J. H. Moore *Ethics & Educ.* vi. 32 These are big-business times. **1913** T. Roosevelt *Autobiogr.* (Appendix A) 615 We demand that big business give the people a square deal; in return we must insist that when any one engaged in big business honestly endeavors to do right he shall himself be given a square deal. **1922** J. M. Murry *Things we Are* 23 Mr. Thomson, whom nature had modelled after the physical pattern of the dour American big-business man. **1930** G. B. Shaw *Apple Cart* I. 22 The political encroachments of big business. **1909** 'O. Henry' *Options* 110 The big city is like a mother's knee to many who have strayed far and found the roads rough beneath their uncertain feet. **1929** D. Hammett *Red Harvest* xxiii. 235 A big-city paper or two are sending in special correspondents. **1960** *New Left Rev.* Mar.–Apr. 45/1 A big-city airport. [**1955** T. Williams *Cat on Hot Tin Roof* (1966) p. xi, Characters of the play...Big Daddy, [etc.].] **1958** *Spectator* 29 Aug. 278/2 Mr. Francis Williams, journalism's Big Daddy. **1959** N. Mailer *Advt. for Myself* (1961) 348 The Fuehrer's tone will be heard in the Twenty-First Century as the Big Daddy voice of all virile and velvety broadcasters. **1869** 'Mark Twain' *Innoc. Abr.* 19 Constellations that never associate with the 'Big Dipper'. **1936** J. L. Hodson *Our Two Englands* vi. 111 [They] spend a pound a day on the Big Dipper when on holiday at Blackpool. **1957** R. Hoggart *Uses of Literacy* v. 127 Thus one arrives at the 'big-dipper' style of singing, the style used by working-class entertainers...Here the voice takes enormous lifts and dips to fill out the lines of a lush emotional journey. **1960** *Spectator* 21 Oct. 601 It has been a week on the big dipper—plunging below the surface of farce into tragedy and then jerking back up again. **1833** J. S. Jones *Green Mt. Boy* I. iii, For the rale genuine grammar larnin' I am a six-horse team and a big dog under the wagon. **1846** J. J. Hooper *Adv. Simon Suggs* x. 126 Pointing to the reverend gentleman who..was the 'big dog of the tanyard'. **1906** W. P. Adams *Motor-Car Mechanism* 160 *Big end*, the lower end of an engine connecting-rod enlarged to carry the big end bearing or brasses which surround the crank pin. **1909** *Westm. Gaz.* 4 Feb. 4/2 The big-end bearings can be examined and adjusted. **1726** Swift *Gulliver* I. I. vii. 120 All the People of that Empire, who would not immediately forsake the *Big-Endian* Heresy. **1726** Mrs. Howard *Let.* Nov. in *Swift's Wks.* (1824) XVII. 81 Many disputes have arisen here, whether the big-endians, and lesser-endians, ever differed in opinion about the breaking of eggs, when they were to be either buttered or poached? **1888** Big-endian [see *little-endian* s.v. Little *a.* 14]. **1905** *Westm. Gaz.* 29 Aug. 10/2 The quarrel between the Big-Endians and the Little-Endians promises to be a merry one. **1961** Y. Olsson *Syntax Eng. Verb* ii. 18 Though what Jonathan Swift might have called the Small-Endian view seems to be in a certain vogue, the procedure here followed is Big-Endian. **1831** *Boston Transcript* 28 Oct. 2/2 The opponents of the existing militia system..are 'going it' at New York 'on the big figure'. **1848** Bartlett *Dict. Amer.*, *Big figure.* To do things on the big figure, means to do them on a large scale. **1864** Big game [in Dict.]. **1890** Ld. Lugard *Diary* 10 Nov. (1959) I. 362 Endless big-game tracks led westwards now. **1905** *Daily Chron.* 25 May 5/2 President Roosevelt's love of big-game hunting. **1966** J. Bingham *Double Agent* ii. 32 In the old days, a jilted man might go big game shooting. **1805** W. Clark *Jrnl.* 4 Apr. in *Lewis & Clark Exped.* (1905) I. vii. 240 The horns of the mountain ram, or big horn. **1926** *Chambers's Jrnl.* July 449/1 It had been the regular route traversed by large droves of bighorn. **1943** *Times* 19 May 3/5 An oil line known as the 'Big Inch', which was laid from Texas oilfields to Illinois..was broken to-day. **1960** *Times Rev. Industry* Dec. 79/2 The construction of two big-inch pipelines from the North Sea ports of Wilhelmshaven and Rotterdam... The first pipeline..is 28 in. dia. ... The second..is 24 in. dia. **1810** F. A. Michaux *Hist. des Arbres* I. 32 The large magnolia..[or] Big laurel. **1853** D. H. Strother *Blackwater Chron.* vii. 89 This dale is girt round..by a broad belt of the Rhododendron—commonly called the *big laurel* out here. **1899** *Chicago Daily News* 10 May 6/1 One of Chicago's crack players new in the big league. **1946** *Ibid.* 31 Aug. 6/6 Let's train our men to be big leaguers if they are to compete, as our representatives, in the big leagues! **1947** *Time* 14 Apr. 66/3 They announced a prize book contest baited with enough cash to make big-league authors sit up and take notice. **1910** *Amer. Mag.* May 6/1 College players stop eight out of nine grounders and big leaguers stop 15 out of 16. **1946** 'G. Orwell' in *Polemic* Jan. 7 The friends of totalitarianism in this country usually tend to argue that since absolute truth is not attainable, a big lie is no worse than a little lie. **1948** *News* (San Francisco) 30 July 2/5 This is a continuation of the Nazi theory of the 'big lie', expounded by Hitler and Goebbels, that the bigger the lie and the more frequently it is told the more people who would accept it. **1951** in *Amer. Speech* XXVI. 293/2 Gloomy Washington prophets are forecasting a period of 'the big lie', of the furtive informer...They lump the whole under the term McCarthyism. **1874** Trollope *Way we Live* II. liii. 14 A moment's private conversation with the big man. **1886** *Lantern* 3 Nov. 3/1 Who is supposed to be a big man in the boot and shoe trade. **1936** *Amer. Speech* XI. 118/2 *Big man*, the brains behind a dope ring; the one who seldom takes the rap...The *big man* wholesales dope to peddlers. **1880** in *Amer. Speech* (1942) XVII. 65/2 'Big money' to be made in cattle and sheep. **1907** *Westm. Gaz.* 16 Aug. 2/1 The skilled artisan has always been..well paid in Belfast. He makes what he calls 'big money'. **1950** J. Dempsey *Championship Fighting* 10 My five big-money bouts. **1889** Barrère & Leland *Dict. Slang* I. 113/2 *Big mouth* (American), a very common expression applied to any man who talks too much. **1890** Farmer *Slang* I. 190/2 *Big mouth* (American), excessive talkativeness; loquacity. **1938** M. K. Rawlings *Yearling* xxiii. 288 Now mister impudent big-mouth. **1940** R. Chandler *Farewell, My Lovely* xxxiii. 154, I didn't have any idea of getting tough in the first place, except just the routine big mouth. **1951** E. Coxhead *One Green Bottle* i. 9 He was a big mouth. He picked up strangers..and told them the story of his life. **1642** Big-mouth'd [see Big *a.* B. 1 a]. **1914** C. Mackenzie *Sinister Street* II. IV. iv. 930 Fancy going off..with that big-mouthed five-to-two. **1919** Masefield *Reynard the Fox* 27 Bill, that big-mouthed smiler. **1926** *Amer. Cinematographer* Dec. 5 A 'big name' actor. **1932** Q. D. Leavis *Fiction & Reading Public* I. ii. 32 *Nash's* in search of 'big names' ran the last Forsyte epic as a serial. **1942** Berrey & Van den Bark *Amer. Thes. Slang* § 144/7 Famous, *big-name, blown to the skies*, [etc.]. **1949** L. Feather *Inside Be-Bop* vi. 43 A big-name policy at the Roost. **1927** *Melody Maker* Sept. 933/3 A score specially composed for each 'big picture'. **1961** A. M. Weinberg in *Science* 21 July 161/1 She [*sc.* history] will find in the monuments of Big Science—the huge rockets, the high-energy accelerators, the high-flux research reactors—symbols of our time. **1848** H. W. Haygarth *Bush Life Australia* i. 6 As he gradually leaves behind him the 'big smoke' (as the aborigines picturesquely call the town). **1917** E. Miller *Diary* 24 Apr. in *Camps, Tramps & Trenches* (1939) x. 61 My first day in 'the Big Smoke'. **1968** *Tel.* (Brisbane) 14 Aug. 54/1 He falls for a beautiful blonde who wants him to stay in the Big Smoke—but city life has no appeal. [**1900** T. Roosevelt *Let.* 26 Jan. (1951) II. 1141, I have always been fond of the West African proverb: 'Speak softly and carry a big stick; you will go far.'] **1904** *Springfield Weekly Republ.* 26 Aug. 1 Happily the revolution in Paraguay is too far down in South America to arouse the 'big stick' in Washington. **1912** Kipling *Diversity of Creatures* (1917) 171 The secret of power..is not the big stick. It's the liftable stick. **1956** A. Wilson *Anglo-Saxon Attitudes* I. iii. 65 Elvira's on the side of the big stick in these things. **1905** *Springfield Weekly Republ.* 9 June 1 A big-sticker after Mr. Roosevelt's own heart. **1926** *Spectator* 2 Jan. 19/1 There is nothing in the British record to compare with Roosevelt's robust big-stickism in the Alaska boundary case. **1925** Fraser & Gibbons *Soldier & Sailor Words* 22 *Big stuff*, heavy shells. **1927** R. W. Hinds in *Flynn's Weekly* 19 Feb. 9/2 'Bagler's big stuff.' I got his slang. Big stuff meant that Bagler was a crook who conducted extensive deals. **1942** Berrey & Van den Bark *Amer. Thes. Slang* § 20/4 Something important..*big stuff*. *Ibid.* § 388/1 *Big stuff*..persons of importance. **1948** Partridge *Dict. Forces' Slang* 15 *Big stuff*. In the Navy, a battleship or an aircraft carrier; or collectively.—In the Army, heavy guns, and the exploding shells from them. Heavy vehicles such as tanks. **1953** R. Lehmann *Echoing Grove* 36, I played the lead, and it was big stuff; supporting roles are less rewarding. **1859** G. W. Matsell *Vocabulum*, *Big thing*, a rich booty. **1862** *Campfire Songster* 48 (D.A.), There's a big thing coming, boys. **1935** *Times* 13 Feb. 7/7 He appealed to the Minister of Labour, who had no responsibility for the muddle and who had done the big thing, to go on doing the big thing in his constructive proposals. **1863** O. W. Norton *Army Lett.* (1903) 183 The brigade was flying round, getting into line, drums beating and a big time generally. *Ibid.* 190 We had big times that night for fires. **1910** R. Grau *Business Man in Amusement World* 36 The 'big time', as such theatres as Percy Williams' and William Morris' are termed. **1914** in *Amer. Speech* (1957) XXXII. 209 Here are huge ledgers that tell the past movements and future bookings of every good act and every artist deemed worthy of 'big time'. *Ibid.*, They buy and sell for all 'big time' acts and all 'big time' theaters. **1921** *Collier's* 25 June 3/3 Like as not I will have to go back pitchin' baseball in some bush league on the account I am too old for the Big Time. **1935**

M. M. Atwater *Murder in Midsummer* xxviii. 261 Of the big-time news-hawks who had gathered in Keedora, only Matter remained. **1936** *Amer. Speech* XI. 117 Big-time gangsters, racketeers, and the criminal aristocrats do not use narcotics. **1951** J. B. Priestley *Fest. Farbridge* 52 From now on it's Big Time stuff. **1966** *Crescendo* Feb. 9 (Advt.), Scores of drummers who hit the big time play Premier. **1932** E. Wallace *When Gangs came to London* xxiii. 233 Only the big timers—I'll interpret that, gentlemen: it means the more important armies—would be employed. **1959** F. Usher *Death in Error* vii. 99 Do you know what sort of criminals he associates with?..Big-timers? **1895** *McClure's Mag.* V. 49/2 Having settled where the 'big top' will stand, the location of the other eleven tents is determined with mathematical precision. **1946** *Univ. of Chicago Mag.* Dec. 9/2 The fact is that he was not really a good showman under the academic big-top. **1962** *Radio Times* 2 Aug. 37/1 Whatever type of act you prefer under the Big Top, whether it's clowns, jugglers or acrobats. **1853** *Placer Times* (S.F.) 27 June 2/2 The Big Tree at the World's Fair. **1865** J. M. Hutchings *Scenes Calif.* i. 13 The starting-point for the Big-Tree Grove. *Ibid.* iv. 77 A sketching tour to the Big Trees. **1908** *Chambers's Jrnl.* Mar. 271/1 Sensational amusements invariably associated with exhibitions, such as the water-chute, big-wheel, [etc.]. **1942** Berrey & Van den Bark *Amer. Thes. Slang* § 388/2 Person of importance or self-importance..*big cog* or *wheel*. **1952** Auden *Nones* 57 They met some big wheels, and do not Let you forget it. **1958** C. Fremlin *Hours before Dawn* xx. 168 The Big Wheel, which rose with strange dignity into the quiet sky. **1958** M. Dickens *Man Overboard* i. 17 He was evidently quite a big wheel at the studio.

b. In collocations used *attrib.*

1909 *Westm. Gaz.* 29 Jan. 2/2 Whether we be 'big-Navy' men or 'little-Navy' men. *Ibid.* 27 Apr. 11/2 Big-print headlines in newspapers. **1909** *Daily Chron.* 7 May 1/4 The big-fleet party in Vienna. **1930** *Economist* 13 Dec. 1110/1 To ascertain the view of manufacturers with regard to big-scale amalgamations. **1947** 'G. Orwell' *English People* 30 The big-circulation newspapers. **1961** *Sunday Express* 2 Apr. 18/6 The film..is not a big-budget one. **1962** 'J. Le Carré' *Murder of Quality* v. 62 Smiley was fascinated by Fielding..by his whole big-screen style.

‖**bigarade** (bigarād). [Fr.] The Spanish bitter orange. Freq. *attrib.*, as *bigarade sauce*, a sauce flavoured with this orange; also applied to a dish (usu. roast duck) having this sauce as a dressing.

1703 tr. *H. van Oosten's Dutch Gardener* v. xiii. 289 Among the sour ones [*sc.* oranges], the Bigarade are the best, the handsomest, and the most esteemed. **1753** N. Torriano tr. *Chomel's Hist. Diss. Sore Throat* 78 The Bigarades, (*i.e.* large Oranges),..are very wholesome in such Cases. **1757** A. Cooper *Compl. Distiller* III. xxxix. 205 Take of the outer or yellow Part of the Rinds of fourteen Bigarades. **1877** E. S. Dallas *Kettner's Bk. of Table* 68 *Bigarade sauce.*—Bigarade is the French name for a Seville orange, and the sauce is described under its English name of Orange Gravy Sauce. **1911** A. Filippini *Internat. Cook Bk.* 651 *Bigarade duckling*..roast forty minutes...Pour a bigarade sauce over. **1930** *Times Educ. Suppl.* 25 Jan. 31/2 The bitter or bigarade oranges are used for making marmalade. **1957** R. Postgate *Good Food Guide 1957–1958* 297 Duck Bigarade.

bigarreau. (Earlier and later examples.)

1629 Parkinson *Parad.* III. xii. 572 The Biguarre Cherrie is..sometimes discoloured halfe white and halfe reddish. **1925** Blunden *Eng. Poems* 117 From this fine orchard, branches sprawl And bigarreaus and blackhearts fall.

Big Brother. [Cf. *Big *a.* 3 f.] The head of state in George Orwell's novel *1984*; hence, an apparently benevolent, but ruthlessly omnipotent, state authority. Also *attrib.* Hence **Big Brotherdom**, the rule or position of Big Brother; **Big Brotherism**, behaviour characteristic of Big Brother; **Big Brotherly** *a.*, of or pertaining to Big Brother.

1949 'G. Orwell' *1984* i. 5 On each landing..the poster with the enormous face gazed from the wall..Big Brother is watching you, the caption beneath it ran. *Ibid.* ii. 209 One could infer..the general structure of Oceanic society. At the apex of the pyramid comes Big Brother. Big Brother is infallible and all-powerful...Nobody has ever seen Big Brother. He is a face on the hoardings, a voice on the telescreen. **1953** *Economist* 12 Sept. 683/1 The distrust of the concierge who is also a police spy, of the admirable focusing device which the big block provides for the watchful eye of Big Brother. **1957** *Ibid.* 19 Oct. 208/2 The reporting..of any evidence discovered..of 'misconduct in the administration of security organisations' would usefully discourage the Big Brother mentality. **1958** *Listener* 9 Oct. 556/3 Mr. Harold Wilson's ingenious but ridiculous variant of 'big-brotherism'. **1959** *Glasgow Herald* 17 Apr. 8/5 One sight of that terrible big-brotherly finger and we will redouble our efforts to find a litter basket. **1964** *Punch* 7 Oct. 547/2 The growth of Big Brotherdom. **1967** *Boston Sunday Globe* 9 Apr. 28 We must take these measures to protect every citizen's privacy before 'Big Brotherism' and the police state take over.

bigeneric (baidʒĕne·rik), *a.* *Bot.* [f. Bi-² + Generic *a.*] Of, pertaining to, or produced from two distinct genera.

1883 *Gardeners' Chron.* XXIII. 631/3 How will these bigeneric crosses affect the stability of the genera as at present circumscribed? **1952** A. G. L. Hellyer *Sanders's Encycl. Gardening* (ed. 22) 205 *Gaulnettya*..Bigeneric hybrid between Pernettya and Gaultheria. **1964** *Guardian* 6 June 5/5 Bigeneric hybrids are plants whose parents belong to entirely separate genera..this type of cross seldom occurs in nature.

biggie (bi·gi). *slang.* [f. Big *a.* + -ie.] **1.** An important person; a 'big shot'. orig. *U.S.*

1931 H. Mutschmann *Gloss. Americanisms* 8/1 *Biggie*, important person, celebrity. **1941** Auden *New Year Let.* III. 69 Each biggie in the Canning Ring An unrobust lone Fisher-King. **1969** *Melody Maker* 13 Sept. 12/7 It's time for me to be a biggie..My aim now is to get..on to the front page.

2. *pl.* A children's word or euphemism for 'excrement'.

1953 E. Simon *Past Masters* iv. iii. 233 You're used to having biggies all over your floor, aren't you? **1967** A. Wilson *No Laughing Matter* ii. 115 He's a bit erratic where he does his biggies, now he's a grown up parrot.

biggity (bi·giti), *a.* orig. and chiefly *U.S. dial.* Also **biggety**. [f. Big *a.* + -y¹, -ety, -ity, unexplained, but cf. Bigoted *a.* and *uppity *a.*] Vain, conceited, boastful; assertive, impudent. Also as *adv.*

1880 J. C. Harris *Uncle Remus* xviii, Like po'in' spring water on one er deze yer biggity fices. **1898** *Westm. Gaz.* 11 Mar. 3/1 John Bull naturally talks 'biggity' about his Navy. **1938** M. K. Rawlings *Yearling* xxxi. 382 You mighty biggety now you're a yearlin'. **1960** C. McCullers *Heart is Lonely Hunter* (1961) I. iii. 42 From the biggity way you been acting lately I'd seem to me like you already got one toe in the pit. **1966** Mezzrow & Wolfe *Really Blues* i. 5 We had a yen..to strut and act biggity.

bigha (bi·gā). Also **beega(h, begah, biggah**. [Hindi.] A measure of land-area in India varying locally from ¼ acre to 1 acre.

1763 in Gleig *Mem. Hastings* (1841) I. 129, I never seized a boga [*sic*] of a beewa of the land belonging to Calcutta. **1823** Malcolm *Central India* II. 15 A Beegah has been computed at one third of an acre, but its size differs in almost every province. The smallest Begah may perhaps be computed at one third, and the largest at two thirds of an acre. *a* **1876** Meadows Taylor *My Life* (1878) xii. 151 It would be a noble sheet of water, and very profitable, as it would irrigate upwards of 10,000 beegahs of rice. **1895** Mrs. Croker *Village Tales* (1896) 167 Two biggahs of land, planted in rape and linseed. **1905** *Statesman* 22 Aug. 2/4 All that piece or parcel of vacant land containing by measurement one bigha fifteen cottahs more or less.

big-head. Also **big head, bighead.** [Big *a.* 3.] **1.** Any of certain diseases of cattle, horses, sheep, etc., characterized by swelling of the head. orig. *U.S.*

1805 *Lancaster* (Pa.) *Intelligencer* 3 Dec. Advt. (Th.), A Brown Steer, having 'what they call the big Head'. **1868** *Rep. Iowa Agric. Soc. 1867* 130, I have not lost but two of the bovine species..; one in 1861 had big head or jaw. **1888** in Farmer *Americanisms*. **1955** Gaiger & Davies *Vet. Path. & Bacteriol.* (ed. 4) xxxvii. 711 The condition [*sc.* osteofibrosis] is also known..in England, as 'big head'.

2. *U.S.* and *Austral.* A popular name for various fish having large heads (see quots.).

1820 C. S. Rafinesque *Ichthyologia Ohiensis* 49 Vulgar names, Chub, Big-mouth, and Big-head. **1889** *Cent. Dict.*, *Bighead*, a local name of a Californian species of sculpin, *Scorpænichthys marmoratus*, a fish of the family *Cottidæ*. **1909** *Ibid. Suppl.*, *Bighead*, a fish, *Electris nudiceps*, of the family *Gobiidæ*. [Australia.]

3. *fig.* 'Swelled head'; an inflated opinion of oneself; conceit, arrogance. *U.S. colloq.*

1850 H. C. Lewis *Louisiana 'Swamp Doctor'* 157 Pride..lets human nature die of the big-head before common sense can bleed freely. **1857** B. Young in *Jrnl. Discourses* IV. 69/1 They need to be careful, or they will have the 'big head', and become as dead..as old pumpkins. **1869** *Congress. Rec.* Mar. 3030/2 [Such] men holding subordinate places in the government of the U.S. to-day, have got the 'big head'. **1901** G. H. Lorimer *Lett. Self-Made Merchant* 218 A boss with a case of big head will fill an office full of sore heads.

b. A conceited or arrogant person. *colloq.*

[**1846** *Warsaw* (Ill.) *Signal* 6 Feb. 3/1 A certain Jackmormon of Hancock county, we won't call him big-head, (but the Saints used to) is in the habit of shaving the hair off his forehead, in order to give it an intellectual appearance. **1863** W. Stokes tr. *Creation of World* 107 in *Trans. Philol. Soc. 1864* The Father's anger hath gone with me For slaying Abel (the) big-head.] **1931** J. Cary *Aissa Saved* 106 He is simply a fool, a conceited bighead who thinks he knows better than the oldest and most experienced men in the whole country. **1955** E. Blishen *Roaring Boys* iv. 138 Saying..'This man was a bighead,' in baffled parody of Shakespeare's funeral speeches.

4. (A person wearing) a large and grotesque mask covering the head.

1890 Roberts & Morton *Adv.* vi. 73 Everybody gagged, even the supers and the big-heads. **1933** P. Godfrey *Back-Stage* iv. 53 'You 'ad real pantos.—'Arlequinades, and a chorus of Big 'Eads.' From the litter of the bygone he produces a specimen 'Big Head'. It is a fantastic object nearly three feet high.

big-headed, *a.* [Big *a.* 3; cf. prec.]

1. Having a large head.

1869 A. R. Wallace *Malay Archipelago* I. xvi. 376, I killed..ravens [*sic*] of a very abundant species, big-headed and of a bright green colour. **1921** E. Step *Brit. Insect Life* vii. 251 A common type of these devouring Insects is the Big-headed Brocon..., ... **1966** ... *Wild Animals* 155 Their form is more graceful and not 'big headed'.

2. Conceited, arrogant. *colloq.*

1942 in Berrey & Van den Bark *Amer. Thes. Slang* §301/5. **1959** J. Braine *Vodi* xii. 166 The selfish big-headed, hard-hearted young lover. **1960** P. Hastings *Sandals for my Feet* I. vii. 71 It was the duty of a sister not to let her sister get big-headed.

big house. [Big *a.* 3.] **1.** A large house, a mansion; *spec.* (freq. with capital initials) the principal house of an estate. Also *attrib.* orig. *U.S.*

1823 J. F. Cooper *Pioneers* xii, The big house has rung with merriment this month past! **1848** A. Somerville *Autobiogr. Working Man* i. 13 The stack-yard..and cattle sheds..were at Thriepland Hill, while the 'big house', the stables, and some other offices, were at Branxton. **1867** Trollope *Claverings* II. xxiii. 265 We might as well stay here as that the Big House should be to let. **1890** J. Devanny *Butcher Shop* 15 A passage leading to the 'big house' as the homestead is generally called. **1942** L. MacNeice *Poetry of Yeats* v. 102 Yeats's cult of the Big House is to be correlated with his distaste for democracy. **1959** L. Lee *Cider with Rosie* vii. 138 There were other stories of Big House life.

2. The workhouse. (Cf. House *sb.*¹ 2 a.) *slang.*

1851 Mayhew *London Lab.* I. 48/2 As long as they kept out of 'the big house' (the workhouse) she would not complain. **1868** J. Hartley *Yorks. Ditties* (1873) 13 But him 'ats as poor as a mease..He mun point his noas up to th' big hause.

b. A prison. *slang* (orig. *U.S.*).

1916 *Ld. Digest* 19 Aug. 424/3 The malefactor is sent away to the big house. **1940** F. Scott Fitzgerald *Let.* 11 Apr. (1964) 68, I feel like a criminal who has been in a hideout, been caught, and has to go back to the Big House. **1942** 'D. Hume' *Destiny is my Name* ii. 21 You'll land yourself in the big house for fourteen years.

bight. Add: **3. a.** Also, an indentation or bay in a mass of ice.

1818 W. Scoresby in *Memoirs Wernerian Nat. Hist. Soc.* II. 266 A bight signifies a bay or sinuosity, on the border of any large mass or body of ice. **1905** Armitage & Roberts *Illustr. Ice Gloss.* 5 *Bight*, an extensive crescent-shaped indentation in the ice-edge, formed either by wind or current.

bignonia. (Earlier examples.)

1786 H. Marshall *Amer. Grove* 22 Ever-green Bignonia, or Yellow Jasmine. **1800** M. Cutler in W. P. & J. P. Cutler *Life & Corresp.* (1888) II. 104 Magnolias, bignonias, liriodendrons. **1826** T. Flint *Recoll.* i. 255 Beautiful stone cottages, clustered with the bignonia in full flower.

bigonial (bigōᵘ·niăl), *a.* [f. Bi-² 5 + mod.L. *goni-on* the outer point of the jaw-bone (Gr. γωνία angle) + -al.] Of or relating to the distance between the points at the angle of the lower jaw on each side.

1895 tr. *Lombroso & Ferrero's Female Offender* i. 25 The bigonial diameter of female criminals is much greater than in women and men of moral lives. **1937** *Nature* 16 Jan. 120/2 The bigonial breadth, 108·6 mm. is below the Eskimo measurement of 130 mm.

big-side. (See quot. 1900.)

1845 *Rules Foot. Rugby School* 21 Two big-side balls must always be in the Close during a match or big-side. **1848** C. H. Newmarch *Recoll. Rugby* 133 The 'Laws of Football', which were sanctioned by a Levee of Big Side', on the 7th of September, 1846. **1896** Goulburn *Bk. Rugby School* 154 Allowing the blood to flow out freely in that most glorious channel—a Big Side Match. *Ibid.* 157 The tuft of trees bordering Big-side. **1900** J. S. Farmer *Public Sch. Word-Bk.* 21 *Big-side* (Rugby and elsewhere). The combination of all the bigger fellows in the school in one and the same game or fun. Also the ground specially used for the game so denominated.

biguanide (baigwā·naid). *Chem.* [f. Bi-² + Guan(idine + -ide.] = *diguanide.

1881 *Jrnl. Chem. Soc.* XL. 890 By the action of an ammoniacal solution of cupric oxide on dicyandiamide, biguanide is produced. **1960** *Diabetes* IX. 227/1 Phenformin is a biguanide which..has a blood-sugar-lowering effect in certain diabetic patients.

bigu·ttate, *a.* [Bi-².] Having two guttate spots.

1887 Phillips *Brit. Discomycetes* 161 Sporidia..shortly fusiform, biguttate.

bigu·ttulate, *a.* [Bi-².] Having two drops or spots.

1887 Phillips *Brit. Discomycetes* 25 Sporidia..biguttulate. **1969** *Gardener's Chron. Dict.* 90/1 *Biguttulate*, containing two vacuoles or two oil drops.

bigwig. (Earlier and later examples.)

a **1731** E. Ward in Partridge *Dict. Slang* (context not given). **1781** G. Selwyn *Let.* in *15th Rep. Hist. MSS. Commission* (1897) App. vi. 526 A new point of discussion for the bigwigs; for they are big wigs, for their Lordships. **1945** Auden *Coll. Poetry* 121 O beggar, bigwig, mugwump.

Bihari (bihā·ri). [Hindi *bihārī*, f. *Bihār* the state of Bihar or Behar in north-eastern India.] **a.** (Also **Behari**.) A native or inhabitant of Bihar. **b.** An Indo-Aryan language of north-eastern India.

1883 G. A. Grierson *Let.* 24 Jan. in *Seven Grammars of the Bihārī Lang.* (1883) Pref., The first or introductory part of the series of Bihari grammars which I was deputed to compile in the early part of 1880. **1897** *Jrnl. R. Asiatic Soc.* XXIX. 471 The Behari women's ceremony for producing rain. **1899** Kipling *From Sea to Sea* I. v. 238 The street..ought to have been full of Beharis. **1910** *Encycl. Brit.* III. 924/1 *Bihari*..the name of the most western of the four forms of speech which comprise the Eastern Group of modern Indo-Aryan Languages. *Ibid.* 924/2 The Bihari vocabulary calls for few remarks. **1933** Bloomfield *Lang.* iv. 63 Indic..include[s] such great languages as..*Bihari* (36 millions), *Bengali* (50 millions).

bijou. Add: (Earlier and additional examples.) Loosely as *adj.*: small and elegant, luxurious (applied esp. to houses).

1668 Lady Chaworth in *12th Rep. Hist. MSS. Commission* (1889) App. v. 10 Perfumed gloves, fans, and all sorts of delicate bijoux for each lady to take att her pleasure. **1747** H. Walpole *Let.* 5 June (1857) II. 86, I..may retire to a little new farm..This little rural bijou was Mrs. Chevenix's. **1860** *Players* I. i. 7 The new theatre will be a perfect bijou temple of Thalia. **1865** 'Ouida' *Strathmore* I. xxii. 311 Her bijou theatre. ... **1894** *Society in New Reign* iii. 23 The London pied-à-terre consisted..of a bijou residence in Mayfair. **1921** A. Bennett *Cardi* iii. An apartment so tiny that an auctioneer would have been justified in terming it 'bijou'. **1952** Wodehouse *Pigs have Wings* i. 14 The Empress lived in a bijou residence not far from the kitchen garden.

bijouterie. (Earlier examples.)

1815 J. Scott *Visit to Paris* (ed. 2) 315 They have improved every article of *bijouterie* to the highest pitch of excellence. **1826** Disraeli *Vivian Grey* II. ii. 97 Her Ladyship was not remarkable for anything, save a correct taste for poodles, parrots, and bijouterie.

bike (baik), *sb.*¹ **1.** Colloq. abbrev. of Bicycle. Also used for *motor bike* (see *motor *sb.* 6).

1882 *Wheelman* I. 189 Much I should like To know why you..have such a hearty spite 'Gainst the little 'bike'. **1888** W. A. Raleigh *Let.* 8 Oct. (1926) I. 96 The people..dress in European fashion and go about in bikes and parasols. **1913** G. B. Shaw *Let.* 20 Mar. in *Corresp. B. Shaw & Mrs. P. Campbell* (1952) 96 In the morning I brave the bike [i.e. motor-bicycle] at last. **1924** T. E. Lawrence *Let.* 1 Mar. in D. Garnett *Lett.* (1938) 456 ... **1968** *Guardian* 1 Apr. 5/3 The ceremonial riding of the last Automobile Association motor-cycle patrol combination... The bikes were fine and ideally suited to the times in which they served.

b. *to get off one's bike* (usu. in negative contexts), to get rattled, to get annoyed. *Austral.* and *N.Z. colloq.*

1940 ... **1943** J. A. W. Bennett in *Amer. Speech* XVIII. 90 'Don't get off your bike' (from 'don't get off your horse' via a music-hall song) means 'don't get excited'. **1944** J. H. Fullarton *Troop Target* ix. 75 Don't get off your bike, Joe.

2. *attrib.*

1899 A. C. Gunter *M. S. Bradford Special* ii. xiv. 185 The ting-ting-ting of the bike bells. **1921** E. Cornaby *One Green Bottle* vi. 173 If I climbed to please you, you'd have to..do a few bike-rides to please me. **1958** Betjeman *Coll. Poems* 133 As slow the weary clergyman subsides Tired with his bike-ride from the parish church.

Hence **bike** *v. intr.*, to ride a bike; also *to bike it*. So **bi·ker, bi·king**.

1883 *Wheelman* I. 336 Nature had rallied all her forces for one grand attack on us three poor, miserable 'bikers'. *Ibid.* We very modestly declined, informing them that 'biking' and drinking are inconsistent. **1895** *Daily News* 10 Sept. 6/2 Have you come? I biked it from the pier. **1895** *Westm. Gaz.* 8 Jan. 2/3 Those young girls..who are learning to 'bike'. **1896** *Ibid.* 5 Oct. 2/2 The 'biking' craze seems to grip..Lady Warwick..was one of the 'bikers'. **1897** R. S. S. Baden-Powell *Matabele Campaign* xix, The other..can use his revolver—which cannot be done by single bikers. **1901** W. H. H. Trowbridge *Lett. Her Mother to Eliz.* vi. 23 Daisy biked over to Exeter this morning with Mr. Fraser. **1931** M. Franklin *Old Blastus of Bandicoot* xv. 188 How did you come? I biked it. **1940** Auden *Another Time* 74 She was biking through a field of corn. **1960** Betjeman *Summoned by Bells* v. 48 All that was crumbling, picturesque and quaint..sent me biking off..for churches.

bikini (bikī·ni). [f. *Bikini*, name of an atoll in the Marshall Islands where an atomic bomb test was carried out in July, 1946.] **a.** A large explosion. (? *Obs.*)

1947 *Daily Tel.* 19 Apr. 6/6 Those in [the troopship] must have had a splendid view of what..was a miniature Bikini.

b. [a. Fr., app. f. prec. sense.] A scanty two-piece beach garment worn by women. Also *attrib.* and *Comb.* Hence **biki·nied** *a.*, clad in a bikini.

[**1947** *Le Monde Illustré* Août 9/2 Bikini, ce mot singulier ...] **1948** *Newsweek* 14 June 33 This..French beauty..shows the 1948 counterbrand against the skimpy 'Bikini' style..which swept French beaches and beauty contests last year. **1950** *News of World* 25 June 5/6 ——made an unsuccessful attempt yesterday to swim in a Hampstead Heath pond in her home-made 'Bikini' costume. **1955** *Times* 8 July 12/7 The Watch Committee there..has never yet lost a night's sleep over a bikini. **1957** *Times* 2 July 12/6 'What is a bikini?' 'A small pair of pants and a brassiere.' **1959** *N.Z. Listener* 9 Jan. 4/2 The bikinied beauties of the Redfield Pool. **1959** *Times* (Suppl. Britain's Food) 9 Mar. p. xii/5 The wrapping is not a bikini-sized tissue. **1965** *Economist* 3 Aug. 422/1 Our bikini-clad assaults on Spanish beaches. **1967** C. Drummond *Death at Furlong Post* xi. 131 He accidentally lodged a ball-bearing in the navel of a bikinied damsel.

bikky: see *bicky.

bilabial, *a.* Add: **2.** *Phonetics.* Of certain consonants (e.g. *p, b, m, v, w*): produced by the juncture or apposition of the lips. Also *sb.*, a bilabial consonant.

1894 W. M. Lindsay *Lat. Lang.* 47 At some time before the fifth cent. A.D..initial v..seems to have passed from the bilabial spirant (Spanish b) to the labio-dental spirant (our v). **1899** W. Kipfmann *Elem. Phonetics* 82 The labio-dentals and the bilabials. *Ibid.* 90 In general the lip stops are bilabial, i.e. formed by a closure of both lips. **1911** *Encycl. Brit.* XXVII. 893/1 The Latin V..was not a labio-dental spirant like the English v, but a bilabial equivalent like the English w.

bilateral, *a.* Add: **c.** Pertaining to or concerning two countries (only), esp. of the trade and financial agreements made between them. Cf. *multilateral *a.*

1935 *Economist* 2 Feb. 279/2 Those countries with which bilateral trade agreements have been concluded. **1937** *Sunday Times* 24 Jan. 19/3 The Duce now pins his faith to bilateral agreements.

d. *Education.* Of a secondary school or its educational system: providing two of the three possible types of course ('grammar', 'modern', or 'technical').

1938 *Min. Educ. Circul.* CXLIV. 1/2 Combinations of two or more types of secondary education are often referred to as bilateral, multilateral or comprehensive...A bilateral school means one which is organized to provide for any two of the three main elements of secondary education, i.e. modern, technical or grammar. **1957** *Economist* 11 Oct. 103/1 Surrey is developing many grammar school courses in 'bilateral schools'.

Also **bilateralism** (examples of sense *c); **bilaterality** (examples).

1884 W. James in *Mind* Apr. 282 Neither rightness nor leftness, except through bi-laterality. **1901** G. N. Calkins *Protozoa* vi. 172 In Didinium..where bilaterality and asymmetry are well established. **1949** *Times* 10 Sept. 4 Bilateralism ..has acquired a special meaning from..recent trade and financial agreements between this country and other individual countries..marked by two main principles: (a) that an equal value of trade on each side is aimed at; (b) that the proceeds of each country's exports can be spent only in the country concerned. **1959** *Times Rev. Industry* Sept. 100/2 Bilateralism..aims at securing a market for each of the partners' products.

bilby (bi·lbi). [Prob. Austral. Aboriginal name.] An Australian nocturnal burrowing marsupial with blue-grey fur; rabbit-bandicoot. Also *attrib.*, as *bilby-hole*.

1903 'T. Collins' *Such is Life* 240 There happened to be a bilby-hole close in front, and she fell in the sort of trough. **1945** Baker *Austral. Lang.* iii. 58 Holes made by the Australian marsupial 'rabbit' or bilby.

‖**Bildungsroman** (bi·ldunzrōmān). [G., f. *bildung* education + *roman* novel.] A novel that has as its main theme the formative years or spiritual education of one person (a type of novel traditional in German literature). Hence **Bildungs(roman)-hero**, the main character in such a novel.

1910 *Encycl. Brit.* XII. 185/1 Its flashes of intuitive criticism and its weighty aphorisms add to its value as a *Bildungsroman* in the best sense of that word. **1938** *Times Lit. Suppl.* 10 Sept. 10/1 The story is then a kind of *Bildungsroman*. **1956** C. Wilson *Outsider* iii. 51 The *Bildungsroman* is a sort of laboratory in which the hero conducts an experiment in living. **1964** *Listener* 7 May 778/1 The *Bildungsroman* heroes lead in a fairly straight line from Simplicissimus to the ks of Kafka. *Ibid.* 778/2 Musil's heroes..are all *Bildungsroman* heroes.

bilge, *sb.* Add: **1. c.** Nonsense, 'rubbish', 'rot'. *slang.*

1908 G. Coke *House Prefect* viii. 99 Let's go..This [*sc.* an excursion] is awful bilge. **1921** A. S. M. Hutchinson *If Winter Comes* ii. vii, I'd say they didn't talk any of this bilge about fighting us in England. **1954** Wodehouse *Jeeves & Feudal Spirit* i. 1 The stuff..was well received by the intelligentsia, who notoriously enjoy the most frightful bilge.

3. *bilge-block* (examples).

1862 *Catal. Internat. Exhib. Industrial Dept.* II. xii. 7 A new method of adjusting bilge-blocks on the arms of the cradle. **1961** F. H. Burgess *Dict. Sailing* 27 *Bilge blocks*, supports placed at the bilges when a ship docks.

bilge-water. Add: **b.** = *bilge *sb.* 1 c. *slang.*

1878 G. M. Hopkins *Let.* 13 May (1935) 50 Write no bilgewater about it.

bilgy, *a.* Add: (Examples.) Also *fig.*

1883 G. M. Hopkins *Let.* 11 May (1935) 51 You drew off your criticisms all stinking..and bilgy. **1890** *Harper's Mag.* Sept. 594/2 Living in quarters as crowded and bilgy as a slaver's hold.

Bilharzia, bilharzia (bilhā·ziă). [mod.L., f. the name of Theodor *Bilharz* (1825–62), a German physician who discovered the parasite in 1851.] **a.** A genus of trematode worms parasitic in the veins of the pelvic region and urinary organs of human beings, esp. in Egypt and other parts of Africa. Also *attrib.*

1859 T. S. Cobbold in *Trans. Linnean Soc.* XXII. iv. 364 *Bilharzia magna* (mihi)...Up to the time of Bilharz's discovery of *Distoma hæmatobium*, all the flukes were considered hermaphrodite...Trematodes possessing so marked a structural peculiarity as *D. hæmatobium*..deserve, I think, to be generically separated from Distomata properly so called, and I have therefore employed the indefatigable discoverer's name for that purpose. **1864** —— *Entozoa* 203 The blood..forms the proper habitat of the *Bilharzia*. **1877** Huxley *Anat. Inv. Anim.* 202 Dioecious Trematodes are very rare, the most important being the formidable *Bilharzia*. **1916** *Brit. Med. Jrnl.* 18 Mar. 411/1 Three species..were infected with certain cercariæ, which developed in mice under experimental conditions into bilharzia worms. **1963** *Times* 16 May 14/7 Perennial streams which are clear and cool to the eye but full of bilharzia.

b. By extension = *bilharziasis.

1889 *Brit. Med. Jrnl.* 20 Apr. 891/2 Cases of Bilharzia. Mr. Anthony Bowlby showed..organs..from two cases of this disease. **1937** C. L. Leipoldt *Bushveld Doc.* ii. 44 To-day bilharzia is as easily curable as malaria. **1961** *New Scientist* 21 Sept. 734/1 Bilharzia..is a widespread disease in many tropical countries.

Hence **bilha·rzial, bilha·rzic** *adjs.*; **bilharzia·sis, -o·sis**, the disease produced by the presence of this worm in the bladder; = *schistosomiasis.

1893 *Lancet* 9 Sept. 622/1 The rules I laid down many years ago for the prevention of bilharzial infection..are still applicable. **1900** Dorland *Med. Dict.* 100/1 *Bilharziosis*, bilharziasis. **1903** *Philad. Med. Jrnl.* 31 Jan. 208/1 Bilharziosis of the kidney may show primary bilharzic lesions and secondary bilharzic lesions. **1906** *Lancet* 27 Oct. 1178/2 The liver proved to be a typical specimen of advanced bilharzial cirrhosis. *Ibid.*, A remarkable case of bilharziasis. **1913** *Glasgow Med. Jrnl.* ... Egypt has well been termed the 'home of bilharziasis'. **1930** J. Clegg *Freshwater Life* (ed. 2) xix. 312 Bilharziasis..in some parts of the world has been said to affect every third person.

bilian (bi·liăn). Also **bahan, billian, bulean.** [Malay.] The ironwood of Borneo.

1812 J. Hunt in Moor *Ind. Archipelago* (1837) ... The woods of which..for carpentry and joinery are, kayu bahan, ... *Ibid.* ... The Chinese export considerable quantities of the ironwood from Sambas and Pontiana, ... **1882** De Windt *Equator* 65 Loading 'bilian' or iron wood. **1887** *Daily News* 19 Dec. 3/2 Forests of bilian..in British North Borneo. **1902** *Encycl. Brit.* XXXII. 417/1 With the exception of iron wood (bilian), from the river Rejang, very little is exported. **1906** *Macm. Mag.* Nov. 42 That wonderful wood bilian, the Bornean ironwood.

bilinear, *a.* Delete *rare* and add further examples.

1851 *Phil. Mag.* I. 132 There are three points of contact..where the two right lines meet. This point is a case of triple contact. I distinguish it by the name of bilinear contact. **1866** *Math. Questions from 'Educational Times'* V. 76 Any equation expressed in Cartesian language may immediately be transposed into another expressed in bilinear (perpendicular) coordinates. **1885** S. Newcomb *Elem. Analytic Geom.* ii. 13 (*heading*) Cartesian or bilinear co-ordinates.

2. Linear in two ways; chiefly in *Math.*: linear and homogeneous in each of two sets of independent variables. Also *absol.*, a bilinear form.

1886 G. S. Carr *Synopsis Pure & Appl. Math.* I. ii. 851/1 Bilinear forms. *Ibid.*, Bilinear functions. **1933** ... *Theory Determinants* IV. xx. 428 The corresponding theorem for the general bilinear is easily anticipated. **1938** A. D. Campbell *Adv. Analytic Geom.* viii. 121 This choice of new fundamental circles amounts to a so-called bilinear transformation of the parameter. **1967** Gruenberg & Weir *Linear Geom.* v. 91 A bilinear form is orthosymmetric if, and only if, it is either symmetric or skew-symmetric.

bilingual, *a.* Add: **1. b.** In extended use: speaking two forms of the same language.

1951 T. H. Pear *Eng. Social Diff.* iii. 97 In London, many waitresses and shop assistants are 'bilingual'. **1961** Partridge *Dict. Slang Suppl.* 997/2 He's bilingual—speaks both English and American.

c. As *sb.*, one who can speak two languages.

1949 *Archivum Linguisticum* I. 155 'Foreign accent' in the speech of bilinguals. **1955** J. C. Catford in Quirk & Smith *Teaching of English* vi. 164 The teaching of English or of any second language as a foreign language may be described as a process of creating bilinguals.

2. (Earlier example.)

1848 *Proc. Philol. Soc.* I. 193 The bilingual inscriptions furnish the meanings of ... Lycian words.

bili·ngualism. [f. Bilingual *a.* + -ism.] Ability to speak two languages; the habitual use of two languages colloquially. Hence **bili·ngualist** = Bilinguist.

1873 J. Earle *Philol. Eng. Tongue* (ed. 2) 85 The Bilingualism of King's English...Words run much in couples, the one being English and the other French. **1887** J. A. Butler tr. *Ratzel's Hist. Mankind* II. 20 The bilingualism of the Carib tribes, whose women speak Arawak and the men Carib. **1926** *Encycl. Brit.* III. 114 The effects of the bilingualism deserve attention, especially in those areas where permanent contact is established between two different families of speech, as in India. **1927** *British Weekly* 12 May 125/4 Welsh bilingualists [decreased] by more than two per cent. **1960** *Guardian* 7 Mar. 5/4 The progress in bilingualism shown by the Welsh children.

b. The use of two forms of the same language. (See also quot. 1940.)

1940 J. H. Jagger *Eng. in Future* ii. 38 It [*sc.* Elizabethan English] suffered from two literary fashions...The first..is known to literary historians as Bilingualism, and consisted in the incessant employment of a pair of synonyms to convey a single idea. **1955** T. H. Pear *Eng. Social Diff.* iii. 105 For the great majority of pupils, bilingualism is to be preferred to the replacement of their native speech by a colourless standard dialect.

bilirubin (bi:lirū·bin). *Biochem.* [a. G. *bilirubin* (G. Städeler 1864, in *Ann. d. Chem. und Pharm.* CXXXII. 326), f. Bili- + L. *ruber* red + -in¹.] A reddish pigment, $C_{33}H_{36}O_6N_4$, occurring in bile. Also *attrib.*

1871 *Jrnl. Chem. Soc.* IX. 419 An alcoholic solution of biliverdin or of bilirubin and ammonia, was decomposed. **1878** [see Bili-]. **1961** *Lancet* 12 Aug. 376/2 The table lists all the bilirubin levels available.

biliteral, *a.* Add: **2.** Of an inscription: written in two different scripts. Also *sb.*, a biliteral inscription.

1877 Skene *Celtic Scot.* II. 450 Several of those [*sc.* ogham inscriptions] in Wales being biliteral, and having a corresponding inscription in debased Roman characters. **1878** Petrie & Stokes *Christian Inscript.* II. 165 The biliteral inscriptions on our stone monuments. *a* **1886** S. Ferguson *Ogham Inscript.* (1887) 114 Other biliterals of.. mere British origin.

-bility [F. *-bilité*, L. *-bilitātem*], a termination forming abstract substantives from adjectives in -ble.

biliverdin (bi:livǝ·ɹdin). *Biochem.* Also **-dine.** [a. G. *biliverdin* (Berzelius 1840, in *Ann. d. Chem. und Pharm.* XXXIII. 140), f. Bili- + *verd-* (as in Verdure) + -in¹.] A green pigment, $C_{33}H_{34}O_6N_4$, occurring in bile.

1845 Wood & Bache *Dispensatory of U.S.A.* (ed. 6) 1285 Biliverdin, a green colouring matter resulting from the absorption of oxygen. **1856** *Fownes' Man. Chem.* (ed. 6) 672 *Biliverdin*..is dark green, amorphous, without taste or smell. **1863** W. O. Markham tr. *Neubauer & Vogel's Anal. Urine* xxiv. 86 *Biliverdine* (green-pigment) is the form of pigment into which cholepyrrhine often passes, and into which, indeed, it may be converted. **1878** [see Bili-]. **1963** *New Scientist* 25 July 182/1 Biliverdin is formed by biological degradation of the red blood pigment hemin.

bilker (bi·lkǝɹ). *slang.* [f. Bilk *v.* + -er¹.] One who practises cheating; *esp.* one who evades payment of a cabman's fare.

1717 *Daily Courant* 27 Dec., A Dramatick Entertainment of Dancing, call'd, The Cheats, or, The Tavern Bilkers. **1887** *Tit-Bits* 25 June 172 Cab Bilkers. **1892** *Daily News* 6 May 6/5 Fares who tried to evade payment, known as 'bilkers'. **1898** *Ibid.* 8 Jan. 3/5 The London Cab Act, 59 and 60 Victoria, chapter 27, commonly known as the Bilkers' Act. **1909** H. G. Wells *Ann Veronica* xvi. 336 Ann Veronica, you're a bilker!

bill, *sb.*² Add: **5.** *bill-clappering, -fencing, -snapping.*

1937 *Brit. Birds* XXXI. 239 The weird bill-clappering and strident cries of old and young Herons. **1957** Bannerman *Birds Brit. Isles* VI. 13 One or two days later..courtship [of white storks] begins. This takes the form of bill-clappering and the assumption of curious postures. **1936** *Brit. Birds* XXIX. 327, I have had the opportunity of seeing 'bill-fencing'. **1949** *Ibid.* XLII. 286 Bill-snapping: loud snapping of the mandibles is used as a threat when attempting to drive other birds—of any species—from food, or as a prelude to attack.

bill, *sb.*³ Add: **5. e.** At Harrow School, properly, the list of boys; hence, the calling-over of the list. Also *attrib.*

1814 W. C. Trevelyan *Diary* 25 Jan. in G. T. Warner *Harrow in Prose & Verse* (1913) 107 Locked up at 8 o'clock, placed at the 2 bill. **1815** *Ibid.* 2 Mar. 113 Dr. B showed us a bill of Harrow School for 1770. **1818** N. Carlisle *Descr. Endowed Gram. Sch. Eng. & Wales* II. 148 The Absences for Bills..are regularly called every Two hours by The Head Master. **1894** *Harrovian* 15 Nov. 103/2 From the Bill-yard to the Old Station. **1894** R. C. Welch *Harrow Sch. Reg.* 17 Our present Bill Books. **1898** Howson & Warner *Harrow School* 275 The 4 bill in summer is double, part on the cricket-ground (cricket-bill)...At yard-bill the boys pass one by one in bill order before the bill-master of the week, answer their names, touch their hats, and walk on.

f. At Eton College, the punishment-list.

a **1877** Swinburne *Lesbia Brandon* (1952) iii. 53, I and another fellow..were so used to our flogging every after twelve, the praepostor of our division wouldn't have dared show up a bill without our two names in it. **1883** J. Brinsley-Richards *Seven Yrs. at Eton* v. 46 A perfume which told its tale to their tutors, and caused them to be put in the Bill.

8. c. A list of the items on a (theatre) programme; hence, the entertainment itself; a group of entertaining items. orig. *U.S.* Cf. quot. 1666 for sense 8 a in Dict., and Programme *sb.* 2 a (esp. quot. 1823). So *double bill*, a programme consisting of two plays, films, etc.; so *triple bill*; also *transf.*

1851 H. Melville *Moby Dick* I. i. 8 This part of the bill.. must have run something like this: 'Grand Contested Election for the Presidency of the United States. Whaling Voyage by one Ishmael. Bloody Battle in Affghanistan.' **1855** W. B. Wood *Recoll. Stage* 155 Warren had prepared..a strong bill..and announced the 'Foundling of the Forest', with the 'Budget of Blunders'. **1875** *Chicago Tribune* 13 Sept. 7/3 Hooley's Theatre..California Minstrels, in a Bill of Rare Ethiopian Sketches. **1891** G. B. Shaw *Quint. of Ibsenism* 147 Mr Thorne..was the first leading manager who ventured to put a play of Ibsen's into his evening bill. **1895** —— in *Sat. Rev.* 11 May, I hope I have not conveyed an impression that the triple bill makes a bad evening's entertainment. **1938** *Encycl. Brit. Bk. of Yr.* 423/1 Theatres widely adopted the 'double bill policy', which meant offering two picture dramas on a programme for a single admission. **1956** C. B. Purdom in *Shaw's Lett. to Granville Barker* 184 The dramatist was proposing that the piece should go with Galsworthy's *Eldest Son* into the bill at the Kingsway. **1959** T. S. Eliot *Elder Statesman* II. 73 It's a very long time since the name of Maisie Mountjoy Topped the bill in revue. **1962** *Listener* 27 Sept. 495/2 Apple jam by itself can be dull, but there are plenty of variations... For a triple bill there is apple, pear, and quince.

d. *to fill the bill*: to fulfil the necessary requirements; to come up to the requisite standard. orig. *U.S.* (Cf. Fill *v.* 7 c.)

1861 *Trans. Ill. Agric. Soc. 1860* IV. 471 Austin..Seedling, Dr. W. hopes well from because of its great vigor, but doubts if it fills the bill. **1880** A. A. Hayes *New Colorado* (1881) ii. 23 With this requirement in view does Colorado 'fill the Bill'? **1890** *Harper's Mag.* Feb. 441/1 They filled the bill according to their lights. **1904** W. H. Smith *Promoters* i. 20, I don't think I ever saw a word used that..filled the bill quite so completely as this word 'experimentally' will do for us. **1954** W. S. Maugham *Ten Novels* iv. 75 He felt that he owed it to himself to have a mistress he could love, and whose position would add to his prestige. He decided that Alexandrine Daru, Pierre's wife, would fill the bill.

e. *slang.* (See quots.)

1930 'A. Armstrong' *Taxi!* xii. 164 The 'bill' is the driver's licence, generally with reference to the original held by the man's proprietor, and to 'draw your bill' is to ask your employer for your licence, i.e., leave his employment. **1939** H. Hodge *Cab, Sir?* xv. 217 'The Bill' is the Metropolitan Police Cab-driver's licence—as distinct from the ordinary County Council driving licence.

9. c. = Note *sb.*² 18 a. Cf. *dollar-bill. orig. and chiefly *U.S.*

1682 in A. McF. Davis *Tracts Mass. Bay* (1902) 7 Credit pass'd in Fund, by Book, & Bills, (as afore) will fully supply the defect of Money. **1790** *Deb. Congress* II. 2055 The bills and notes of the bank..shall be receivable in all payments to the United States. **1852** Mrs. Stowe *Uncle Tom's C.* 438 She unlocked the desk, took from it a roll of bills, which she counted over rapidly. **1854** A. E. Baker *Gloss. Northants Words* I. 49 *Bills*, bank notes; all kinds of paper money. **1924** E. O'Neill *Welded* II. 145 He takes a bill from his pocket and holds it out to her—contemptuously. **1947** —— *Iceman Cometh* I. 46, I t'rows down a fifty-dollar bill like it was trash paper. **1953** *Manch. Guardian Weekly* 29 Oct. 7 This bold proclamation can be found on every dollar bill.

10. *bill of costs* (see quot. 1911); *bill of goods* (orig. and chiefly U.S.), a consignment of merchandise; so in colloq. phr. *to sell* (someone) *a bill of goods*: to persuade (someone) to accept something undesirable; to swindle.

1815 (*title*) Bills of costs and allowances, in the Court of King's Bench, according to the present scale of allowance; with a precedent for affidavits of increase. **1826** *New Monthly Mag.* II. 109 [He] made the invitation an item in his bill of costs. **1869** *Yng. Men Gt. Brit.* 14 May 260/2 An attorney, about to furnish a bill of costs. **1911** *Encycl. Brit.* III. 932/2 A *Bill of Costs* is an account setting forth the charges and disbursements incurred by a solicitor in the conduct of his client's business. The delivery of a bill of costs is by statute a condition necessary before the solicitor can sue upon it. **1927** E. O'Neill *Marco Millions* I. 41 Selling a big bill of goods hereabouts, I'll wager, you old rascals? **1945** E. Ford *Larry Scott* xii. 120 Don't let Jim sell you a bill of goods. **1968** *Globe & Mail* (Toronto) 17 Feb. 3/3 There was no production bonus...We were sold a bill of goods.

11. (Sense 3) *bill-drafter, -drafting*; (sense 8) *bill-posting, -sticking* (earlier example); **bill-book**, *spec.* in U.S., a pocket-book; **bill chamber** (examples); **bill-clerk** *U.S.*, a clerk who deals with the bills at a hotel; **billfold(er** *U.S.*, a note-case or wallet; **bill-head** (examples); **bill-holder**, (*a*) a person who holds a bill or acceptance; (*b*) *U.S.* a container for bank-notes, etc.

1895 *Montgomery Ward Catal.* 101/1 Seal grain leather bill book, size 3½ × 8 inches. **1699** Lord Reay *Let.* 10 Oct. in *Pepys Corr.* (1926) I. 190 Direct it to the care of Mr Robert Menzies, to be found att the Bill Chamber in Edinburgh. **1838** W. Bell *Dict. Law Scotl.* 99 By 53 Geo. III. c. 64, § 2, it is enacted that the junior judge in the Court of Session shall officiate permanently in the Bill Chamber during the sitting of the court. **1901** *Daily Chron.* 10 Oct. 3/5 Paying such surplusage to the bill-clerk, or the head-porter. **1886** P. Robinson *Teetotum Trees* 65 The most experienced of bill-draughters. **1894** *Westm. Gaz.* 6 Feb. 2/2 The kind of mistake for which even a Bill-drafter can hardly be expected to provide. **1895** *Ibid.* 8 Jan. 2/2 Ministers have led..a quiet, if busy and Bill-drafting, existence. **1895** *Montgomery Ward Catal.* 100/3 Pocket books..4 pockets and bill fold. **1914** C. E. Walk *Green Seal* xxii. 283 Here, perhaps, was a surer passport to my goal than the tickets reposing in my bill-fold. **1919** H. Leverage *White Cipher* 43 Sir Richard drew from his inner vest-pocket a thin bill-fold. **1951** *Catal. of Exhibits, Festival of Britain* 57/2 Woman's billfold. **1961** J. Steinbeck *Winter of our Discontent* I. 40 From his side pocket he brought a billfold, a rich and beautiful affair of pin seal. **1909** 'O. Henry' *Options* (1916) 35 He drew out his bill-folder to pay the cabman. **1845** J. W. Norris *Chicago Directory* 136 (D.A.E.), Circulars, Bill Heads, Hand Bills. **1910** A. Bennett *Clayhanger* I. iii. 21 Nobody heard romance in the puffing of the hidden steam-engine multiplying catalogues and billheads. **1830** G. Colman *Random Rec.* I. viii. 271 The bill-holders..were to be paid. **1890** Webster, *Bill holder*,..a device by means of which bills, etc., are held. **1929** E. Rice *Street Scene* II. 176 He..carries a large black-covered bill-holder. **1869** J. R. Browne *Adv. Apache Country* 352 Bill-posting is one of the fine arts. **1895** in *N. & Q.* (1941) CLXXXI. 159/1 One of the largest Bill-posting Firms in the World. **1851** Dickens *Bill-sticking* in *Househ. Words* 22 Mar. 604/1 Several bill-sticking companies have started.

bill, *v.*³ **2.** Delete †*Obs.* and insert: *Obs.* exc. *arch.*

1910 Masefield *Pompey* 22 Send out your press. Bill every able-bodied man. Bill the women if the men won't come.

4. Also *fig.*

1934 E. Bowen *Cat Jumps* 159 She was billed, it appeared, for yet another confession. **1963** *Listener* 21 Mar. 484/2 He [*sc.* Kenyatta] is so much more the Kikuyu, and the older Kikuyu, than the national leader he is billed to become.

5. (Earlier example.)

1821 P. Egan *Real Life in London* ix. 158 The practice of advertising and billing the town has become so common.

6. To send a bill or account to; to charge.

1867 Trollope *Last Chron.* xxxii, As for billing my first cousin, which your wife is, I should as soon think of sending in an account to my own. **1953** 'S. Ransome' *Hear no Evil* (1954) vi. 65 If the money doesn't come through, please bill me. **1956** 'N. Shute' *Beyond Black Stump* 40 If there's a scratch on it, I'll bill you for it. **1960** *Times* 12 Aug. 14/4 It [*sc.* photocopying] is used by doctors to bill their patients.

billabong (bi·lăbɒŋ). *Austral.* Also 9 **billibong, billy-bonn.** [ad. *Billibang*, Aboriginal name of Bell River, f. *billa* water + *bang* of uncertain meaning.] A branch or effluent of a river, forming a blind channel, backwater, or stagnant pool.

[**1836** T. L. Mitchell *Jrnl.* 11 Apr. in *Three Exped. Interior E. Australia* (1838) II. ii. 30 A small watercourse, then dry, and named Billibang, skirts the eastern side of the hill.] **1865** W. Howitt *Discov. Australia* I. 298 Their way on was intercepted..by..what the Major calls..anabranches of the river, but which the natives call billibongs, channels coming out of a stream and returning into it again. **1866** W. Landsborough *Explor. Australia* vi. 31 In the south, such a creek as the Macadam is termed a *billy-bonn*, from the circumstance of the water carrier returning from it with his pitcher (*billy*) empty (*bong*, literally *dead*). **1896** H. Lawson *While Billy Boils* (1897) 91 While out boating one Sunday afternoon on a billabong across the river. **1911** C. E. W. Bean *'Dreadnought' of Darling* xiv. 133 Australian rivers.. often have a long arm or two wandering off into the plain, and either coming back to the river again in the end or else straying off for perhaps a hundred miles and ending nowhere in particular. Those arms are what they call billabongs. *a* **1917** A. B. Paterson *Waltzing Matilda* 1 Once a jolly swagman camped by a billabong. **1950** 'N. Shute' *Town like Alice* iii. 81 You get a sandy billabong, and you'll get water there by digging.

bill-board, billboard (bi·lbōᵊɹd). orig. *U.S.* [f. Bill *sb.* 8 + Board *sb.* 2 b.] **1.** A board to which notices, posters, advertisements, etc., are affixed; a notice-board or hoarding.

1851 W. K. Northall *Bef. & Behind Curtain* 15 The bill-boards of the Park..still continued to style the Park 'The Theatre'. *a* **1877** in Bartlett *Dict. Amer.* (ed. 4) 44 People who fail to edit a bill-board are firmly convinced that they could edit a newspaper. **1950** *Manch. Guardian Weekly* 19 Jan. 10/3 Billboards bearing the strange advice: 'Go East, Young Man, Go East!' **1950** O. Nash *Fam. Reunion* 131, I think that I shall never see A billboard lovely as a tree. Indeed, unless the billboards fall I'll never see a tree at all. **1954** N. Coward *Future Indef.* iv. 247 Outside a canteen was a billboard announcing 'To-Nite Bing Crosby and Dorothy Lamour'.

b. *attrib.*; **bill-board pass**, a free theatre ticket issued to tradespeople who allow advertisements to be displayed on their premises.

1890 'Biff' Hall *Turnover Club* 100 The holder of a bill-board pass addressed the Count Bozenta as 'Mr. Modjeska'. **1902** W. N. Harben *Abner Daniel* 58 Jack Ass in bill-board letters would come nearer to it than anything that occurs to me now. **1963** *New Yorker* 1 June 24/1 Exquisitely faithful duplications of bill-board lettering. **1967** *Listener* 3 Aug. 156/2 Garish and absurd as a billboard likeness.

2. An aerial array resembling a bill-board in appearance, consisting of a number of dipoles mounted in front of a large vertical metal reflector. Freq. *attrib.*

1950 J. D. Kraus *Antennas* iv. 118 A 'billboard' type of array..having many dipole antennas carrying equal currents. *Ibid.* xii. 327 (*caption*) Array of ½-wavelength elements with flat sheet reflector (billboard or mattress antenna). **1963** *Life* 1 Mar. 20/1 Dye Main's 60-foot-high radio antennas—the technicians call them 'billboards'—make sure these warning signals get through. **1966** *Electronics* 17 Oct. 138/3 The tropo routes use 60-foot billboard antennas. At some of the stations in Turkey, the tropo paths require 120-foot billboards.

billed (bild), *ppl. a.*² [f. Bill *v.*³]

1. Entered in a bill or list; *spec.* (see quot.).

1873 W. W. Knollys *Dict. Mil. Terms*, *Billed*, a term exclusively confined to the Foot Guards. It means that a man's name is placed in the list or bill of those who are to undergo drill and confinement to barracks. Hence a 'billed man'.

2. Announced or advertised by a bill. (See Bill *v.*³ 4.)

1895 *Westm. Gaz.* 18 July 8/1 The..train reached Aberdeen at 6.46..or fourteen minutes before the billed time of arrival. **1897** *Daily News* 22 May 5/7 The establishment of a boldly billed enclosure for lost children.

billet, *sb.*¹ Add: **4.** (Later examples.)

1837 [see Bullet *sb.*¹ 7]. **1922** Joyce *Ulysses* 366 The ball rolled down to her as if it understood. Every bullet has its billet.

b. A place in which a soldier is billeted; a soldier's lodging or quarters. Also *gen.*, spec. in the war of 1939–45, quarters assigned to civilian evacuees.

1830 E. S. N. Campbell *Dict. Mil. Sci.* 26 Any Justice may, at the request of the Officer or Non-commissioned Officer commanding any Soldiers requiring Billets, extend the Routes or enlarge the District within which Billets shall be required. **1858** *Billets & Bivouacs* 246 Thou long-suffering reader,..tarry with me yet a moment in the last billet we shall ever share together—my dark abode on the dreary shores of Old Father Thames. **1875** *Encycl. Brit.* II. 571/2 The army lived in barracks, camps, or billets. **1919** Wodehouse *Damsel in Distress* vii, So George took up his abode..in the plainly-furnished but not uncomfortable cottage...He might have found a worse billet. **1942** E. Waugh *Put out more Flags* iii. 161 Getting harder to find billets, particularly since the anti-aircraft battery had come to South Grappling. **1946** [see *billetee, billettee]. **1968** A. Sillitoe in B. S. Johnson *Evacuees* 238 We were allotted to four different houses, and taken by car to our separate billets.

c. *transf.* An appointment, post, or 'berth'.

1870 *Cassell's Mag.* II. 58/1 The private..doesn't see why ..he should give up his billet as pioneer or canteen waiter. **1887** 'Hopeful' '*Taken-in*' 135 If you are out of work, or out of a situation, as we say, here [*sc.* in N. Zealand] it is called 'out of billet'. **1890** Hornung *Bride fr. Bush* xvii, If ever she went back to Australia, she'd remember my young man, and get him a good billet. **1890** *Lippincott's Monthly Mag.* XLV. 384 A billet as clerk in the recruiting rendezvous. **1941** Baker *Dict. Austral. Slang* 10 *Billet*, a position or job.

billet, *sb.*² Add: **4.** (Earlier and later examples of *attrib.* use.)

1845 R. Ford *Handbk. Trav. Spain* II. xiii. 980/2 Observe the singular billet patterns on the arches. **1926** R. Glazier *Man. Hist. Ornament* (ed. 4) 45 Later Norman work is very rich, the mouldings being well carved with..the Chevron,.. Billet,..or Key Patterns.

9. *billet mill* (see quot. 1910).

1910 H. P. Tiemann *Iron & Steel* 282 Billets, blooms, small slabs, and sheet bars are produced on a mill called a billet mill, blooming mill, or cogging mill. **1958** *Times* 7 Jan. 15/1 The billet mill should be completed before the end of the current financial year.

billet, *v.* **2. b.** Add to def.: *spec.*, in the war of 1939–45, to assign quarters to (civilian evacuees). Freq. const. *on*.

1939 *Punch* 6 Dec. 631/3 It was when you first heard that little Sidney and the others were to be billeted on you. **1942** E. Waugh *Put out more Flags* iii. 165 Wait till the gentleman billets you. **1955** *Times* 2 Aug. 8/6, I dimly recognized one of the children who had been billeted on us during the wartime evacuation, and who had gone back to London 10 years ago.

billetee, billettee (bileti·). [f. Billet *v.* + -ee.] A person who is billeted.

1939 *Times* 12 Oct. 9/4 Mrs. M[iniver]..will cope in a wonderful manner with refractory billetees. **1946** A. Phelps '*I couldn't care Less*' xv. 120, I sought out the billet that had been arranged for me. I am not a good billettee. **1958** D. Wallace *Forty Years On* xv. 185 A lot of our dear billettees weren't even house-trained.

billeter. Add: **d.** A person who receives a billetee into his or her household.

1939 *Picture Post* 18 Nov. 44 The Government pays billeters 10s. a week for one child or 8s. 6d. each for more than one. **1966** *New Statesman* 11 Feb. 202/2 First, there's evacuation. 'Are they coloured?' asks a billeter. 'I can't take eight people,' says another.

billeting, *vbl. sb.* Add: **1. b.** The provision of quarters for civilian evacuees, esp. in the war of 1939–45. Freq. *attrib.*

1936 in R. M. Titmuss *Probl. Social Policy* (1950) iii. 25 The evacuation of London needs to be thought out in terms, not of transport only but of reception, housing (by compulsory billeting if necessary) and feeding. **1938** *Hansard Commons* 12 May 1703, I assume that my hon. Friend has in mind what are sometimes called billeting surveys. **1939** *Punch* 11 Oct. 400/2 The children were deposited on her doorstep by an apologetic but helpless billeting-officer. **1940** *Ann. Reg. 1939* 112 Evacuation necessarily caused a good deal of inconvenience...There were many 'misfits' in the billeting. **1969** A. Calder *People's War* ii. 45 In spite of all the fuss, there had been many cases of successful billeting. *Ibid.* vi. 409 There had been many allegations that billeting officers, usually 'respectable' people themselves, had let other 'respectable' people off lightly.

billian, var. *bilian.

bi·lliardist. [f. Billiards + -ist.] A billiard-player.

1865 *Philadelphia Sun. Mercury* 15 Oct. 6/3 Nelms, the great billiardist, covered himself with glory. **1888** *St. Louis Globe Democrat* 24 Jan. (Farmer), Each competitor has put up fifty dollars entrance money in Billiardist Daly's hands. **1901** *Daily Chron.* 3 Apr. 6/6 A player, like a billiardist, would require practice to be successful. **1921** *Glasgow Herald* 18 Jan. 6 Smith's rise to fame as a billiardist.

billiards. Add: **2.** *billiard-hall, -player* (examples), *-playing, -room* (earlier examples), *-stick* (earlier examples); **billiard-table** (earlier example); also used *attrib.* to describe a perfectly smooth green, road, etc.

1873 *Winfield* (Kansas) *Courier* 18 Jan. 3/2 The lower room will soon be occupied..as a saloon and billiard hall. **1939** Joyce *Finnegans Wake* (1964) 125 And uses noclass billiardhalls with an upandown ladder? **1790** T. Wilkinson *Mem.* IV. 228 This was of infinite service to Mr. Fleetwood as a billiard player. **1872** Geo. Eliot *Middlem.* I. II. xviii. 321 The Vicar was a first-rate billiard-player. **1807** J. R. Shaw *Life* (1930) 76 Men of diabolical principles..employed in..horse-racing, billiard-playing, [etc.]. **1890** W. James *Princ. Psychol.* I. xiii. 509 Billiard-playing, rifle-shooting, tight-rope-dancing, demand the most delicate appreciation of minute disparities of sensation. *c* **1702** C. Fiennes *Journeys* (1947) III. iv. 172 Out of the billiard roome the first was with gravell walks and a large fountaine. **1814** Jane Austen *Mansf. Park.* I. xiii. 259 Maria, Julia, Henry Crawford, and Mr. Yates, were in the billiard room. **1588** in *N. & Q.* (1915) XI. 227 Three billyard stickes and one porte and ij balles of yvery. *a* **1700** Evelyn *Diary* 4 Dec. (1955) IV. 190 They for the most part use the sharp & small end of the billiard-stick, which is shod with brasse or silver. **1641** in *N. & Q.* (1915) XI. 227 A billiard table and three bearers. [**1851** J. Pycroft *Cricket Field* vii. 107 We do not play cricket..on billiard tables.] **1887** F. Gale *Game Cricket* 183 Pougher..got seven wickets..for 116 runs, on a billiard-table ground. **1905** *Westm. Gaz.* 6 June 4/2 A land of billiard-table roads.

billiken (bi·likin). Also **-kin.** [Prob. f. Billy² + -kin.] A small, squat, smiling figure used as a mascot.

1914 Wodehouse *Man Upstairs* 257 When you send a girl three bouquets, a bracelet, and a gold Billiken with ruby eyes, you do not expect an entire absence of recognition. **1927** A. Huxley *Proper Studies* 225 We now worship, under the name of mascots, lucky pigs, billikens, swastikas, and the like, a whole pantheon of fetishes which stand for nothing beyond themselves. **1927** W. E. Collinson *Contemp. Eng.* 100 Before the war..there was a craze for Billikins or little oriental gods of luck. **1932** *Times Lit. Suppl.* 10 Nov. 829/3 Are there not modern airmen who will not go up without the moral support..from some grinning Billiken or 'Fumsup'?

billing, *vbl. sb.*³ Add: **b.** Announcing or advertising by bill or poster; publicity. Also *attrib.* So *top billing*, the most prominent position on a poster, advertisement, or the like; stardom.

1875 *Chambers's Jrnl.* 132/1 Advertisements, which most likely took the form of 'billing'. **1900** *Daily News* 4 Dec. 3/2 No billing is attempted in the City, and hence the urgent necessity for a thorough distribution of election literature. **1902** *Daily Chron.* 20 Mar. 7/4 If plaintiff did not send..particulars of her performance for billing purposes. **1905** *Ibid.* 28 June 6/2 The defence was that Mr. Elen did not send on 'billing matter' two weeks in advance of his engagement. **1957** L. F. R. Williams *State of Israel* 145 English plays of recent date, in Hebrew translations, figure more frequently in the billings. **1968** *Globe & Mail* (Toronto) 13 Jan. 24/1 He made his Broadway debut as Lancelot in Camelot, with billing below the title; now, he is returning to Broadway, with top billing. **1969** *Times* 14 Feb. 10/7 He and other dance band leaders..vied with Gracie Fields for the top billing on the music halls and radio.

c. *Comm.* The act of making out or sending a bill, invoice, etc. Also *attrib.*, as *billing machine*.

1908 *Westm. Gaz.* 26 Feb. 9/2 The billing machine..will produce half a dozen business documents at one operation, and will total up automatically the various items entered. *Ibid.* 29 Feb. 7/1 The special Oliver for Billing and Invoicing. ..Combined Billing and Adding Machines. **1959** *Economist* 4 Apr. 68/2 The two industries might profitably co-operate in such things as meter reading, billing and collection of accounts.

d. The total amount of business or investment, esp. of an advertising agency, during a given period. *U.S.*

1958 M. Mayer *Madison Ave.* iii. 53 We've grown from $2,900,000 in billings in 1940 to about $100,000,000 in 1957. **1959** I. Ross *Image Merch.* (1960) viii. 134 Their gross billing was $150,000 in 1951.

billion. 2. (For examples 1834– see *D.A.E.*) Since 1951 the U.S. value, a thousand millions, has been used by some writers in Britain, but the older sense 'a million millions' still prevails.

billitonite (bi·litǫnait). *Min.* [G. (Suess 1901, in *Jahrb. d. K.K. Geol. Reichsanstalt 1900* L. 194), f. *Billiton* the name of an Indonesian island + -ite¹ 2 b.] A form of tektite found in the East Indies.

1909 Sollas tr. *Suess's Face of Earth* IV. v. xv, The acid felspar rocks of the earth are represented among the meteorites..by the once completely molten tektites (Moldavite, Australite, Billitonite). **1944** Palache et al. *Dana's Syst. Min.* (ed. 7) 121 *Tektites*..include..the billitonites of the East Indies.

billow, *sb.* Add: **4.** *billow-like* adj.

1851 H. Melville *Moby Dick* I. ix. 65 How billow-like and boisterously grand! **1904** W. de la Mare *Henry Brocken* 95 The hosts of our pursuers paused, billow-like, reared, and scattered.

bi·llowlet. [See -let.] A small billow.

1867 J. MacGregor *Rob Roy Baltic* 99 The dancing billowlets of the lake. **1892** *Temple Bar* Oct. 292 The bay.. with its sheltered succession of crescent-shaped billowlets.

billy². **1. b.** Add to def.: A bludgeon; also (*U.S.*), a policeman's truncheon; see also quot. 1848. Also *billy club*.

1848 'Ned Buntline' *Myst. N.Y.* iv. 49 The foremost villain..broke down her guard with a short iron crowbar, or 'billy' as the burglars term it. **1856** *Santa Barbara* (Calif.) *Gaz.* 14 Feb. 2/5 He was knocked down by a blow from a 'billy'. **1889** *Weekly Scotsman* 21 Sept. 3/5 He was felled by Marshal Glade's billy, which stunned him. **1903** *N.Y. Times* 11 Sept., Eight men set upon a policeman this morning,.. taking his revolver and billy away from him. **1931** L. Steffens *Autobiogr.* I. II. v. 214, I felt the ache they [*sc.* policemen] conveyed to rap some head with the handy little billy. **1949** *Amer. Speech* XXIV. 262 By 1859 the billy club was demanding order in American cities. **1967** M. Procter *Exercise Hoodwink* ix. 67 Lying in the other bunk was a shark club or 'billy'. *Ibid.*, He had not taken his shark billy. **1969** *Listener* 3 July 11/2 He thought it necessary, this pig, this racist, to take his billy club out and crush her skull.

d. = Billy-goat.

1849 C. J. Pharazyn *Jrnl.* 17 Dec. (MS.) 137 Hunting goats Robin shot a Billy. **1928** *Daily Tel.* 9 Oct. 4/6 The Ministry of Agriculture, through the agency of its stud goat scheme, has now placed 103 'billies' at the disposal of smallholders.

billy³ (bi·li). *Austral.* and *N.Z.* [Origin uncertain: perh. f. Austral. Aboriginal *billa* river, water (cf. *billabong).] A cylindrical container, usu. of tin or enamel ware, with a close-fitting lid and a wire handle, used for making tea and for cooking over fires in the open, and for carrying food or liquid.

1839 J. Heberley *Autobiogr.* (MS.) 87 [We] boiled the Billy and made some Tea out of tawa bark. **1853** J. Rochfort *Adv. Surveyor in N.Z.* viii. 63 We must needs purchase a 'billy' (a tin pot for boiling tea, coffee, meat or anything you may have the luck to get). **1858** *Jrnl. R. Geogr. Soc.* XXVIII. 310 We are..boiling our flesh or fowl in our tea-can (called a *billy*). **1872, 1881** [see Billy² 1]. **1940** F. Sargeson *Man & Wife* (1944) 10 She was letting them have milk at half the town price...And my last job each day was to take a billy up to the back fence. **1943** D. Stewart in *Coast to Coast 1942* 212 The girls began to pluck handfuls of the berries and carry them to the billies the boys were filling. **1950** G. Wilson *Brave Company* xi. 179 The billy is boiled and the tea made.

attrib. Also **bi·llyful.**

1865 G. Mueller *My Dear Bannie* (1958) ii. 55 Price had a billy lid for a plate. **1866** A. H. Williamson in R. P. Whitworth *Martin's Bay Settlement* 23/1 Gathered a billyful of mussels..which we had for supper. **1897** D. McK. Wright *Station Ballads* 17 The spuds and meat were nicely done, the billy tea was made. **1934** *Bulletin* (Sydney) 12 Sept. 9/2 'Billy tea,' says a paragraph issued by the Melbourne Centenary Committee,' will be served to H.R.H. the Duke of Gloucester in the East Gippsland forest.' **1950** G. Wilson *Brave Company* xi. 179 First we boil a billyful of water.

Billy Bunter (bi:li bʊ·ntǝɹ). The name of a schoolboy in stories written between 1908 and 1961 by 'Frank Richards' (Charles Hamilton, 1875–1961), used allusively to indicate fatness, gluttony, clumsiness, etc.

1939 L. MacNeice *Autumn Jrnl.* x. 41 Our little jokes of Billy Bunter dirt. **1959** *Times* 7 Mar. 7/4 'It is better for Britain,' declared an M.P. recently, 'to have a few more Billy Bunters than the kind of hunger which was caused to many children before milk in schools.' **1961** A. Wilson *Old Men at Zoo* ii. 72 Strawson, whose fat, self-satisfied Billy Bunter form waddled towards me at that moment. **1966** B. Cooper *Drown him Deep* xviii. 147 The full moon.. stared down with a Billy Bunter's curious face.

billy-can (bi·likæ:n). *Austral.* and *N.Z.* [f. *billy³ + Can *sb.*¹] = *billy³.

1885 C. Praed *Head Stat.* xxiii, I lost the billy-can lid. **1899** 'S. Rudd' *On our Selection* 96 He carried no swag or billycan. **1933** *Bulletin* (Sydney) 26 Apr. 21/4 It contains half a dozen fire-blackened billy-cans. **1951** *Here & Now* (N.Z.) May 27/1 In many suburbs billy cans are still used for deliveries, the milk being ladled out of a container that is open to germs and dust.

Billy Fairplay or **Playfair.** *Coal-mining.* (See quot. 1883.)

1876 *Brit. Pat.* 4387, Machines for weighing small coal known by the name of 'Billy fair play'. **1883** Gresley *Gloss. Coal Mining*, *Billy Playfair* or *Fair-play*,..a mechanical contrivance for weighing coal, consisting of an iron trough with a sort of hopper bottom, into which all the small coal passing through the screen is conducted and weighed off and emptied from time to time. **1889** *Law Rep.* (*Appeal Cases*) XIV. 231 The coal was shot down on a screen called Billy Fairplay.

billy-o. *colloq.* Also **-oh.** Used in the intensive phr. *like billy-o* (see Like *adv.* 1 b) = like the devil (see Devil *sb.* 16); also in other intensive phrases.

1885 *Referee* 9 Aug. 7/4 Shure it'll rain like billy-oh! **1914** W. J. Locke *Fortunate Youth* iii, 'And they fight?' 'Like billy-o,' said Paul. **1927** *Punch* 25 May 584/3 He would take a team anywhere... He would take a team to Billy-O. **1927** *Observer* 11 Dec. 26 The Holy Rollers were going it like billy-oh to a tune which sounded like 'My old man's a fireman'. **1934** *Bulletin* (Sydney) 25 July 21/4 With the price of copra and rubber all to billyoh, Papua has been in the doldrums lately. **1969** *Sunday Mail Mag.* (Brisbane) 1 June 6/3 There was Amundsen..with his dogs going like billyo for the Pole.

biloculine (bailǫ·kiūlǝin), *a.* *Zool.* [ad. mod.L. *Biloculina*, f. *bi-* Bi-² + L. *loculus*, dim. of *locus* place: see -ine¹.] Having two chambers or compartments; characteristic of the foraminiferous genus *Biloculina*.

1898 Sedgwick *Zool.* I. 9 In the genus *Biloculina* (*Miliolidæ*), while the mode of growth of the megalospheric form is..on the biloculine plan from the first, that of the microspheric form is at first on the quinqueloculine plan, and it is not until many chambers have been formed that the biloculine plan, characteristic of the genus, is assumed.

bilophodont (bailǫ·fǭdǫnt), *a. Zool.* [f. Bi-[2] + *lophodont* (see Lopho-).] Of the molar teeth in certain ungulates: having two transverse crests or ridges on the grinding surface.
1868 Owen *Anat. Vertebr.* III. 343 The 'bilophodont' sub-type becomes more marked in *Dinotherium*. *Ibid.* 358 Certain huge fossil bilophodont grinders, which seemed to indicate a gigantic Tapir. 1891 Flower & Lydekker *Mammals* 373 Molars brachydont and bilophodont.

Bim[1] (bim). Also **Bimm.** Colloq. name for an inhabitant of Barbados. So **Bi·mshire,** Barbados.
1852 C. W. Day *Five Yrs. W. Indies* I. 15 The Barbadians are familiarly known as 'Bimms'. 1859 Trollope *W. Indies* xiii. 207 One soon learns to know a — Bim. That is the name in which they themselves delight, and therefore, though there is a sound of slang about it, I give it here. *Ibid.* 208 The Bims..are generally stout fellows. 1887 W. A. Paton *Down the Islands* (1888) xii. 135 Barbadoes is known to the initiated as Bimshire—a Barbadian as a Bim. 1902 *Daily Chron.* 1 Feb. 5/1 Mr. Bosanquet's team of English cricketers, which has lately been playing in Barbados, and making but an indifferent show before the doughty 'Bims'.

bim[2]. *U.S. slang.* [Abbrev. of *bimbo[2].] A girl, woman; a whore.
1925 *Eng. Jrnl.* Nov. 700 John took his bim to a dance. 1935 J. T. Farrell *Studs Lonigan* (1936) iii. i. 467 Studs Lonigan copped off a bim whose old man is lousy with dough. 1953 W. R. Burnett *Vanity Row* ix. 71 I'd hate to turn you loose on a desert island with that big bim.

bim[3]. *slang.* = Bum *sb.*[1] 1.
1935 I. Miller *School Tie* vi. 59 Fligg never hits you properly on the bim, always below it. 1948 C. Day Lewis *Otterbury Inc.* ix. 119 He slid gracefully down it on his bim.

bimanual (baimæ·niuăl), *a.* [f. Bi-[2] + Manual *a.*] Performed with both hands; in which both hands are employed. Hence **bima·nually** *adv.*, by means of both hands.
1898 G. E. Herman *Dis. Women* 59 Bimanual examination. *Ibid.* 63 There are cases in which..you cannot bimanually feel the distinction. 1902 D. J. Cunningham *Anat.* 1233 By the bimanual examination the pelvic organs are steadied and pushed downwards towards the pelvic outlet by the pressure of the left hand applied in the hypogastric region.

bimbashi (bimbɑ·ʃi). Also **bimbashee, binbashi.** [Turk., lit. 'one who is head of a thousand' (*bin* thousand, *baş* head; cf. Bashaw).] A Turkish major, naval commander, or squadron-leader. Also formerly in Egypt, an English officer in the service of the Khedive.
1819 T. Hope *Anastasius* II. xv. 329 A Bimbashee [*note* Turkish colonel], with about eighteen hundred men. 1876 Gladstone *Bulgarian Horrors* 31 Their Bimbashis and their Yuzbachis, their Kaimakams and their Pashas. 1896 *Daily News* 28 Mar. 5/3 'Bimbashi' is the general name for the English officers attached to the Egyptian army. 1899 Kipling *Stalky* 197 He did not know that Wake ..would be a bimbashi of the Egyptian Army ere his thirtieth year. 1922 *Blackw. Mag.* Apr. 519/1 Let me be the Bimbashi of the Hakim's river watchers.

bi·mbo[2]. *slang* (orig. *U.S.*). Pl. **-os, -oes.** [It., = little child, baby; cf. Bambino.] **a.** A fellow, chap; usu. contemptuous.
1919 *Amer. Mag.* Nov. 69/1 Nothing but the most heroic measures will save the poor bimbo. 1924 Wodehouse *Bill the Conqueror* xx. 285 The bimbo Pyke arrived. 1936 R. Chandler *Killer in Rain* (1964) 53 There's a thousand berries on that bimbo. A bank stick-up, ain't he? 1947 Wodehouse *Full Moon* v. 90 Bimbos who went about the place making passes at innocent girls after discarding their wives.
b. A woman; esp. a whore.
1929 *Amer. Speech* IV. 338 *Bimbo*, a woman. 1937 *Detective Fiction Weekly* 3 Apr. 20/2 We found Durken and Frenchy LaSeur, seated at a table..with a pair of blonde bimboes beside them. 1952 S. Kauffmann *Philanderer* (1953) xii. 194 Not that you were just a bimbo to me...I've discovered that I'm a little in love with you, too.

bimeby (bai·mbai), *adv. dial.* Also **bimby(e, bimebye, bymeby,** etc. Reduced form of By and by A. 4.
[1549 [see By and by A. 4]. 1708 *Deplorable State of New Eng.* 35 Altho'..one Day it was voted, That the Fort should be attack'd, it was by 'nd by, unvoted again.] 1722 in O. E. Winslow *Broadside Verse* (1930) 115/2 Indian bimeby take Captain Westbrook's fort. 1824 *Nantucket Inquirer* 5 Jan. (Th.), Well, bimeby he took notion to hab my daughter... Well, bimeby I found em out. 1825 J. Jennings *Obs. Dial. West of Eng.* 26 *Bimeby*, by-and-by; some time hence. 1825 etc. [see By and by A. 4]. 1850 'N. Hogg' *Lett. in Devonshire Dial.* (ed. 2) 6 Bim-bye, in com'd a wacking hoss. 1857 J. G. Holland *Bay-Path* xiv. 156 It's a thing that'll come round byme-by. 1873 'Mark Twain' & Warner *Gilded Age* ii. 32 But bymeby she roused up like, and looked around wild. 1899 R. Whiteing *No. 5 John St.* vii. 64, I s'pose everybody all over the world 'll know our patter bimeby. 1907 E. Wharton *Fruit of Tree* I. ii. 27 A small boy who said breathlessly: 'Mr. Truscomb wants you to come down bimeby.' 1952 *Amer. Speech* XXVI. 26 Common expressions in the pidgin of Hawaii are..*bimeby* (by-and-by), *wasamatta* (what's the matter?), [etc.].

bi-metal (bai·metăl). [f. Bi-[2] + Metal *sb.*] A bimetallic material or object. Freq. *attrib.* or as *adj.*, = next.
1935 *Discovery* Dec. 369/1 The word 'bi-metal' is used for a composition of two metals with different expansion coefficients. *Ibid.*, The 'Rototherm' bi-metal thermometer. *Ibid.* 369/2 Making the bi-metal in the form of a single spiral. 1938 *Jrnl. Marine Res.* I. 95 A Bathythermograph... Mounted on the movable end of the pressure element is a straight bi-metal strip, and thus motion with pressure is at right angles to the deflection of the strip with temperature. 1946 D. de Carle *Pract. Watch Repairing* xii. 119 A bi-metal balance of steel with brass fused on the outer edge is used. 1953 *Electronic Engin.* XXV. 78 The loop is provided by a spiral bi-metal strip which tends to coil and uncoil with change of temperature.

bimetallic, *a.* Add: **2.** Composed of two different metals; *bimetallic strip*: a sensitive element in some thermostats, made of two bands of different metals, one of which expands more than the other when the temperature rises, and so causes the strip to bend.
1907 *U.S. Pat.* 851,684, The process..of making bi-metallic products such as ingots, sheets, tubes, wires and the like, of different metals welded together. 1922 *Jrnl. Amer. Chem. Soc.* XLIV. 2504 Bimetallic electrode systems in electrometric analysis. I. Systems comprising two dissimilar metals. 1930 *Engineering* 31 Oct. 543/1 The smaller ovens are controlled by a bimetallic-strip thermostat.

bimetalli·stic, *a.* [f. Bimetallist + -ic.] Relating or inclining to bimetallism.
1889 in *Cent. Dict.* (citing *Contemp. Rev.*). 1897 *Daily News* 5 Feb. 10/5 The Agriculturists, with their..bimetallistic hankerings.

bimodal (baimō^u·dăl), *a.* [f. Bi-[2] + *modal *a.*] Having two modes, esp. in *Statistics* (*mode *sb.* 8 d).
1903 *Amer. Naturalist* 302 The tendency to bilateral symmetry here in the number of teeth is very strong and so gives a bimodal curve with apices at 10 and at 12. 1903 *Bot. Gaz.* Dec. 477 Field studies showed that the stamens vary from 0 to 13, forming a bimodal curve with principal maximum on 3 and secondary minimum on 5. 1925 F. C. Mills *Statistical Methods* iv. 129 In Table 9, representing the distribution of wages, there are two definite modal points. A distribution of this type is called bi-modal; when plotted, a frequency curve having two humps is obtained. 1935 R. S. Woodworth *Psychology* (ed. 10) iii. 37 When only a small number of individuals is tested, the distribution often looks bimodal. 1968 *Brit. Med. Bull.* XXIV. 212/1 The corresponding histogram of conjugated bilirubin is markedly bimodal.

bimodality (baimodæ·liti). [f. Bi-[2] + *modality.] The quality or fact of being bimodal, esp. in *Statistics*.
1903 *Amer. Naturalist* 308 As in the case of the curve for the prosternal teeth, which is strongly bimodal, the bimodality is due to the tendency towards bilateral symmetry. 1928 S. A. Rice *Quantitative Methods in Politics* viii. 113 Bimodality is invariably regarded as evidence of absence of homogeneity in the data which is distributed. 1932 J. S. Huxley *Problems of Rel. Growth* ii. 70 A similar but less-marked bimodality occurs in this species.

bimolecular (baimole·kiŭlăɪ), *a.* [Bi-[2].] Involving two molecules.
1899 J. Walker *Introd. Physical Chem.* 256 Saponification of ethereal salts by alkalies affords us an example of a bimolecular reaction. *Ibid.* 257 A bimolecular reaction, which is, strictly speaking, a balanced action, but proceeds very nearly to an end in aqueous solution, is the formation of urea from ammonium cyanate. 1904 J. W. Mellor *Chem. Stat. & Dynam.* 35 The hydrolysis of ethyl acetate by sodium hydroxide is a bimolecular reaction. 1946 *Nature* 20 July 94/1 Those bimolecular and unimolecular substitutions in which a nucleophilic reagent displaces halogen as halide ion from an alkyl halide. 1957 *Encycl. Brit.* XIX. 3D/1 If a kinetic study shows that two molecules (of the same or of different compounds) react with one another directly, then the reaction is said to be bimolecular.

bi-monthly, *a., sb.,* and *adv.* [Bi-[2] 4.] (Occurring or produced) every two months; *sb.*, a bi-monthly periodical. Also used = occurring (etc.) twice a month (cf. *semi-monthly* s.v. Semi- 5 b).
1846 *Punch* 31 Jan. 61/1 Bi-monthly breakneck expresses between the Punjaub and Cheapside. 1864, etc. [see Bi-[2] 4]. 1926 (*title*) Vogue Fashion Bi-Monthly. 1959 *Times* 10 Jan. 7/7 A bi-monthly called *Recorded Folk Music*.

bimorph (bai·mǭɹf). *Electr.* [f. Bi-[2] + Gr. μορφή form.] A device consisting of two piezo-electric crystals, used in sound reproduction.
1932 *Electronics* May 166/2 The bimorph or opposed action types [of crystal units] have been adapted and used for loudspeakers, phonograph pickups and microphones. 1962 A. Nisbett *Technique Sound Studio* 247 A crystal bimorph—two plates cut from different planes of a crystal of Rochelle Salt and held together in the form of a sandwich.

bimorphemic (baimǫɹfi·mik), *a.* [f. Bi-[2] + *morphemic *a.*] Consisting of or pertaining to two morphemes.
1942 *Amer. Speech* XVII. 168 In 168 oscillograph records of such homophonic pairs [as *tide, tied*; *rise, rye's*]..the average duration of the vowel in the bimorphemic words is about ·030 sec. greater than that of the vowel in the monomorphemic words. 1947 *Language* XXIII. 339 Some may..prefer to reinterpret portmanteaus as bimorphic as well as bimorphemic, even though to do so one must extend the definition of 'morph' to cover elements of other than overt phonemic content. 1964 R. H. Robins *Gen. Linguistics* v. 205 The bimorphemic structure of all English noun plurals.

bin, *sb.* Add: **6.** One of a number of receptacles in a wool-shed where wool is stowed by classes after sorting. *Austral.* and *N.Z.*
1865 Lady Barker *Let.* 1 Dec. in *Station Life N.Z.* (1870) v. 33 Armfulls of rolled-up fleeces [were] laid on the tables before the wool-sorters who..pronounced..to which bin they belonged. 1891 R. Wallace *Rural Econ. Austral. & N.Z.* xxix. 383 Bins of fleeces awaiting pressing. 1933 L. G. D. Acland in *Press* (Christchurch) 23 Sept. 13/7 A man or boy..carries the fleeces from the wool table to their bins after the wool-classer has classed them.
7. Short for *loony bin* (see *loony *a.* and *sb.*). *slang.*
1938 E. Waugh *Scoop* I. i. 10 To my certain knowledge she's driven three men into the bin. 1942 L. A. G. Strong *Unpractised Heart* 77 The chaps who certified you and popped you in the bin.

binant (bai·nănt). [f. Bin- + *-ant* as in Quadrant.] A half of a circle or circular body; *binant electrometer*, an electrometer in which the index moves through half a circle.
1908 *Chem. Abstr.* 2492 (*title*) A Binant Electrometer for Pointer and Meter Reading... Description of a new form of electrometer which has many advantages over the quadrant form. 1936 *Nature* 21 Nov. 866/1 After the Great War, still more sensitive electrometers were designed; for example.. a binant electrometer of high sensitivity by Hofmann. 1938 R. W. Lawson tr. *Hevesy & Paneth's Man. Radioactivity* (ed. 2) 290 Semicircular hollow plates or dees, resembling the duants or binants of an electrometer. 1952 F. K. Harris *Electr. Measurements* x. 447 Electrometers (Hoffman) have been constructed using binants instead of quadrants and having an unsymmetrical needle position... Instability increases with increasing voltage between the binants.

binary, *a.* Add: **f.** Examples of uses in *Math.* Cf. *j below.
1796 C. Hutton *Math. & Philos. Dict.* I. 206/1 *Binary number*, that which is composed of two numbers. *Binary arithmetic*, that in which two figures or characters, viz, 1 and 0, only are used; the cipher multiplying every thing by 2, as in the common arithmetic by 10: thus, 1 is one, 10 is 2, 11 is 3, 100 is 4, 101 is 5. 1823 J. Mitchell *Dict. Math. & Phys. Sci.* 327/2 Therefore 1810 = 11100010010 in the binary scale. 1858 Cayley *Math. Papers* (1889) II. 527 Certain binary quantics, viz. the quadric, the cubic, and the quartic. 1860 [see Ternary *a.* 1 e]. 1861 Binary scale [see Scale *sb.*[3] 6 b]. 1871 Cayley in *Messenger of Math.* V. 53 A covariant of the binary form U. 1898 —— *Math. Papers* (Suppl.) 78 Tables of Binary Quadratic Forms. 1929 L. E. Dickson *Introd. Theory Numbers* v. 63 The function $q = ax^2 + bxy + cy^2$ is called a *binary quadratic form*. 1941 D. H. Lehmer *Guide to Tables in Theory of Numbers* 37 These tables give the primes also in the binary scale, or rather in a condensed form of binary scale. 1957 *Encycl. Brit.* I. 607/2 Forms in two variables are called binary... Consider a binary quadric:—$f(x) = x_1^2 - x_2^2$.
g. *binary colour* U.S. (see quots.).
1876 S. R. Koehler tr. *von Bezold's Theory of Color* p. xxiii, The other colors of the spectrum are called secondary or binary, since..each is composed of two primary colors. 1919 H. D. Eberlein et al. *Interior Decorating* II. ii. 192 The primary colours are yellow, red and blue, and the binary colours (those composed of two) are orange (yellow and red), violet (red and blue) and green (yellow and blue). 1937 *Discovery* July 217/1 Binary colours, that is to say, colours obtained by combining two of the above [*sc.* the primary colours].
h. In *Biol.*, *binary fission*: the division of a cell or organism into two (see Fission 2).
1897 Parker & Haswell *Textbk. Zool.* I. i. 14 It is mainly by this simple process of division into two, or *binary fission* as it is called, that reproduction or multiplication takes place in the Amœba. 1910 *Encycl. Brit.* IX. 317/1 The division of the nucleus is effected by the process called binary fission; that is to say, it divides into two, then each of these divides simultaneously again into two, giving four nuclei; each of these after a pause again simultaneously divides into two.
i. *Logic.* Involving two items or terms.
1933 *Mind* XLII. 203, *a × b*, or simply *ab*, is the result of a binary operation on *a* and *b*. 1951 Quine *Math. Logic* i. 13 Conjunction and alternation are *binary*, in that they combine statements two at a time. 1956 A. Church *Introd. Math. Logic* 33 A connective is called singulary, binary, ternary, etc., according to the number of its operands.
j. *binary digit*, one of the two digits (conventionally 0 or 1) in a binary system of notation; also = *bit *sb.*[4] (Also examples of *binary number, system*, with reference to electronic computers.) Cf. *f above.
1946 A. W. Burks et al. *Electr. Computing Instr.* in *von Neumann's Works* (1963) V. 42 An accuracy of ten decimal digits requires at least 40 binary digits. 1946 *Electronic Engin.* XVIII. 373/1 The internal working of the machine will be entirely in the binary system, in which a number is represented by a series of 1's and 0's, the 1's being pulses and the 0's the spaces between them. 1948 *Ibid.* XX. 209 For mathematical reasons..it was decided to make provision for 34 binary digits in each number. 1948 [see *bit *sb.*[4]]. 1951 *Electronic Engin.* XXIII. 140 Binary numbers can be conveniently represented in electrical circuits. 1953 *Sci. News* XXX. 83 Elemental areas of the

track magnetized with positive and negative polarity represent the binary digits. 1957 *Encycl. Brit.* XII. 350A/1 The simplest type of choice is a choice from two possibilities, each with probability ½... It is convenient to use the amount of information produced by such a choice as the basic unit and it has been called a binary digit or, more briefly, a 'bit'.
k. *Linguistics.* Having two elements or aspects.
1948 *Language* XXIV. 32 In any of certain dialects, there are two segments X and Y occurring in the dyad XY, and a segment A that is homotypical with either X or Y and congruent with XY...Such a dyad XY constitutes a single binary segment. 1953 C. E. Bazell *Linguistic Form* 6 The modern fashionability of the binary opposition. 1953 J. B. Carroll *Stud. Lang.* ii. 34 These distinctive features [of the phoneme] consist of binary, two-valued oppositions or contrasts, such as *voiced/unvoiced, high/low, rounded/unrounded*.
l. Designating a system of higher education (see quot. 1965).
1965 *Sun* 2 Nov. 11/5 The present Government have decided..to establish what they call a binary system of education. On the one hand would be the universities and the colleges of education. On the other hand would be the local authority regional and area colleges of technology, cut off from the universities and getting their degrees and other diplomas and certificates from a quite independent source. 1969 *Listener* 27 Mar. 414/1 People complain of what they call the binary system of higher education. It's quite wrong, we're told, to set the universities apart from other institutions of advanced education.

bina·tional, *a.* [Bi-[2].] Concerning or consisting of two nations. Hence **bina·tionalist** *a.*
1888 *Blackw. Mag.* Sept. 391 Another piece of binational work. 1922 *Glasgow Herald* 15 July 4 Alsace and its historic capital..bi-lingual and bi-national..as they are. 1949 Koestler *Promise & Fulf.* iii. i. 298 An Arab-Jewish binational state in which the 'toiling masses' of the two races were to unite against the Arab effendis and the Jewish capitalists...The idealists of the Zionist left.. dropped all binationalist pretences.

binaural, *a.* Add: **1.** (Earlier example.)
1878 *Engineering* XXIV. 151/3 Mr. S. P. Thompson.. read his paper 'On Binaural Audition'.
2. Applied to a system of sound reproduction that uses two separated microphones and two transmission channels to achieve a stereophonic effect; *esp.* one in which the sound is delivered to each ear separately by earphones.
1933 *Electronics* July 196/2 Ready-made equipment for an experiment with binaural transmission. 1952 *N.Y. Times* 26 Oct. II. 11/6 In binaural transmission, the sound is broadcast on both the standard radio wave length and on FM. In the home two receivers are necessary to achieve the effect of 'three dimensional' sound. 1958 *N.Z. Listener* 10 Oct. 21/3 In a binaural recording of an orchestral performance, two microphones would be placed some distance apart among the players and the output from each microphone recorded separately. Both sound tracks would then go onto a single tape or onto a single record.

Binche (bæ̃nʃ). The name of a town in Belgium, applied *attrib.* or *absol.* to a type of lace originally made there.
1888 A. S. Cole tr. *Lefébure's Embroidery & Lace* 325 List of laces according to their names and local origins... Binche. 1902 B. Palliser *Hist. Lace* vii. 135 The first Binche lace has the character of Flanders lace. *Ibid.*, The characteristic peculiarities of Binche are, that there is either no cordonnet at all outlining the pattern, or that the cordonnet is scarcely a thicker thread than that which makes the *toilé*. 1908 N. H. Moore *Lace Book* iii. 108 Binche Lace of the old make resembles the old Valenciennes very closely. ..Modern Binche lace is machine-made net with bobbin sprigs applied. 1966 'E. Kyle' *Love is for Living* iii. 27, I just bought a few handkerchiefs bordered with Binche. *Ibid.* vi. 51 Binche..is made at a small village on the other side of Brussels...Binche makes the best bordering.

bind, *v.* Add: **10. a.** spec. in *Cookery*.
1861 Mrs. Beeton *Bk. Househ. Managem.* x. 199 Pound well, and bind with 1 or 2 eggs. 1962 *Listener* 29 Nov. 943/2 You will need..beaten egg (to bind). 1963 *Ibid.* 14 Mar. 479/1 Mix together the stuffing ingredients, binding with beaten egg.
17. b. Freq. *to bind over* (used without following *to* and dependent phrase): to oblige (a person) to undertake to do, or abstain from, a particular act; *spec.* to make (someone) give a recognizance not to commit a breach of the peace, usu. for a specified period.
1610 W. Yonge *Diary* 28 May (1848) 21 Being asked by Mr. Drake wherefore he was bound over, he said for a private grudge borne him by the bishop. 1764 T. Cunningham *Law-Dict.* s.v. *Good behaviour*. This statute..seems to impower them, not only to bind over those, who seem to be notoriously troublesome,..but also those who are publickly scandalous. 1820 Tomlins *Law-Dict.* (ed. 3) s.v. *Surety of the peace*. A Peer or Peeress cannot be bound over in any other place than the Courts of King's Bench or Chancery. *Ibid.*, A justice may bind over all night-walkers. 1917 *Oxford Jrnl. Illustr.* 17 Jan. 8/3 The prisoner was bound over for twelve months. 1970 *Oxf. Mail* 6 Jan. 5 Judge Mais told him that in binding him over there was no possible slur cast upon him.
V. *slang.* **22.** *trans.* To bore, weary.
1929 T. E. Lawrence *Lett.* (1938) 640 Voyages are binding things, & I'm lucky to have had this job to keep me busy. 1943 C. H. Ward-Jackson *Piece of Cake* 13 *Bind*, a depressing job or situation, one who is a bore. Used as a verb it suggests a petty or depressing order or regulation...Thus, 'It's a perfect bind, old boy', or 'Smith's got his tapes: I suppose he'll be binding everyone now', or 'He's the worst binder I ever served under'...*Binds you rigid, binds you stiff*, bores you completely. 1944 E. Partridge in *19th Cent.* CXXXV. 184 To *bind* a person is to bore him stiff; probably from the ill-temper that so often results from being *bound* or constipated; but perhaps from the fact that such a duty as is a *tie* may easily become boring.
23. *intr.* To complain. (In quot. 1942 used transitively, to complain to, to admonish.)
1942 I. Gleed *Arise to Conquer* i. 11 We stayed sitting in our cockpits listening to Pat binding the ground station on the radio telephone. 1943 *Penguin New Writing* XVII. 18 'Stop binding,' said Ginger mechanically...'When I get back to civvy street I'll never moan about my job again.' 1948 *Jrnl. R. Aeronaut. Soc.* LII. 719 It is useless to expect information on maintenance difficulties..because usually the operator 'binds' to himself or his mates on encountering a snag. 1955 W. Fink in P. Chambers & A. Landreth *Called Up* vi. 115, I dare say we would have binded more about all this bull if it hadn't been that there was a system by which the best hut got a forty-eight hour pass. 1959 'D. Buckingham' *Wind Tunnel* viii. 73 Eddy's been binding to Vic about you.

bind, *sb.* Add: **7.** A bruise on a horse's foot caused by the pressure of a nail on the sensitive parts.
1908 *Animal Managem.* 239 'Pricks', wounds from nails driven into the fleshy part, and 'presses', or 'binds', from their being so close that they bruise them.
8. A bore, nuisance (cf. *bind *v.* 22). *slang.*
1930 T. E. Lawrence *Lett.* (1938) 676 Letter writing is what the R.A.F. call a 'bind'. 1942 *Penguin New Writing* XIII. 24 Occasionally..it publishes a serious article...But this is regarded as a 'bind', i.e. boring and tiresome. 1943 [see *bind *v.* 22]. 1953 'N. Shute' *In Wet* viii. 272 But it's an awful bind for you, at such a time as this.

bindei, var. *bindi-eye.

binder. Add: **4. a.** (Later example.)
1919 Masefield *Reynard the Fox* 55 The binders crashed as hounds went over.
5. b. (Examples.) Also, a separate machine used for binding up the grain (Knight, *Dict. Mech.* Suppl.).
1857 *Trans. Illinois Agric. Soc.* II. 120 A self raker, and even a binder, may be just as simple..as some hand raker. a 1877 Knight *Dict. Mech.* II. 1891/2 1850. Heath's binder, with a reciprocating rake beneath the platform. 1851. Watson's automatic binder. 1891 R. Wallace *Rural Econ. Austral. & N.Z.* v. 89 The light steel binder..is produced wholesale in..[Canada] for £28 each. 1945 'G. Orwell' *Animal Farm* v. 39 Electricity..could operate threshing machines..and reapers and binders.
e. (Examples.)
1837 *U.S. Patent* 26 Dec., Binder for newspapers. 1964 *Which?* July 224/1 All new members get an index when they join CA (which can be kept in the binder pocket).
7. Delete † *Obs.* and add later examples.
1940 [see *binding agent* s.v. *binding *ppl. a.* 3]. 1957 *Brit. Commonw. Forest Terminol.* 23 *Binder*,..the component of a glue primarily responsible for the adhesive forces.
b. In road-making (see quot. 1911).
1901 G. W. Tillson *Street Pavements* viii. 230 It was deemed best to change..the method of construction and for the cushion-coat to substitute a so-called binder, made up of coarse stones held together by asphaltic cement. This binder has been laid of different thicknesses, sometimes 1½ or even 2 inches. Its object..is simply to serve as a medium between the wearing surface and the concrete. 1908 *Chambers's Jrnl.* 29 Aug. 624/1 A light binder of clay and gravel is added to the second coating of stone. 1911 *Encycl. Brit.* XXIII. 392/1 Upon the concrete foundation is first spread a layer of fine bituminous concrete called 'binder', 1½ in. thick, to unite the wearing surface to the concrete foundation. Upon the binder the asphalte is laid to a thickness of 2 in. 1937 *Times* (Brit. Motor No.) 13 Apr. viii/3 The principle is the same as that of the ordinary tar-macadam road, but cement mortar is used as the 'binder' in the place of tar.
c. A binding medium used in painting to cause pigment to cohere and set.
1922 M. Toch *How to paint Permanent Pictures* 33 The decorations made by the Egyptians were made without any binder other than the lime naturally found in the soil, and in a few cases the glue was used. *Ibid.* 34 There are some really wonderful decorative [Egyptian] paintings in which binders were used. The portraits..in the second century were done with wax and resins. 1934 H. Hiler *Notes on Technique of Painting* iii. 144 These media, or binders, each require a different manner and method in their use. 1951 R. Mayer *Artist's Handbk. Materials & Techniques* i. 19 Tempera paint films are adequately strong and durable, but when dry the volume of binder in relation to the volume of pigment is less than that of oil paints.
8. binder twine, twine used in a binder or for binding.
1890 *Moose Jaw* (Sask.) *Times* 18 July 4/6 The local situation in binder twine is somewhat interesting. 1899 *Westm. Gaz.* 12 June 9/2 The fibre was chiefly used in the United States for binder twine for harvesting. 1950 *N.Z. Jrnl. Agric.* Mar. 227/3 Tying them [*sc.* the leaders of a vine] with binder twine is preferable to twisting the leaders around the wires [of the trellis].
9. A large quantity, esp. of food; a satisfying meal. *dial.* and *N.Z. colloq.*
1881 H. & C. R. Smith *Isle of Wight Words* 46 *Binder*, a quantity. 'A pretty good binder of it.' 1917 *Chrons. N.Z.E.F.* 19 Sept. 55/1, I was hungry so I turned my eyes away from the promising binder. 1943 F. Sargeson in *Penguin New Writing* XVII. 59, I shouted him a bob dinner and I could tell by the way he ate he was in need of a binder. 1967 —— *Hangover* vi. 42, I am so devilish hungry I must first spend half an hour in the kitchen putting away a binder.
10. A last drink (see also quot. 1953). *slang.*
1899 A. Binstead *Gal's Gossip* 15 He joyfully fell in with her suggestion to step inside and take a 'binder'. 1953 *Word for Word* (Whitbread & Co.) 13/1 *Binder*, colloquial expression for the last drink...Also used to describe the person who orders a drink after closing time.
11. *slang.* **a.** A boring person or thing (see *bind *v.* 22). **b.** One who 'binds' or complains (see *bind *v.* 23).
1930 T. E. Lawrence *Lett.* (1938) 679 More books go to you almost at once. You'll find some of the packets have quite decent things amongst them—though generally they are what the troops call my 'binders'. 1937 Partridge *Dict. Slang* 54/1 *Binder*, a bore (person). 1943 [see *bind *v.* 22]. 1944 Partridge in *19th Cent.* CXXXV. 184 A person..who is a grouser or a fault-finder is termed a *binder*.
12. *pl.* Brakes. *slang.*
1942 'H. W.' *What's the Gen?* 19 *To jump on the binders*, to apply the brakes. 1962 *Amer. Speech* XXXVII. 267 *Binders*, brakes. Most often used in referring to emergency stops. 'Hit the binders!'

bindery. (Earlier U.S. example.)
1810 I. Thomas *Hist. Printing in Amer.* I. 402 At Worcester, he..set up a bindery.

bindi-eye (bi·ndi,ai:). Also **bindei, bindyi.** [Origin unknown.] The popular name in Australia of a small perennial herb, *Calotis cuneifolia*, with burr-like fruit.
1911 C. E. W. Bean *'Dreadnought' of Darling* v. 50 They swung their legs over the saddle and rode off over the grass and the twigs and the bindyi. 1923 *Census Plants of Victoria & Vernacular Names* 64 Calotis cuneifolia..Bindi-eye. N. W. 1925 *Austral. Encycl.* I. 231/2 Calotis... *C. cuneifolia* (with blue flowers) and *C. lappulacea* (with yellow flowers) are known in the west as bindieyes (or bindeis—the spelling is doubtful). Their burr-like flower-heads become entangled by the barbed bristles in the wool of sheep. 1934 *Bulletin* (Sydney) 13 June 19/3 Bindei burr, which was one of their favourite foods. 1941 K. Tennant *Battlers* xiii. 147 Some comfortable little parking-place that was neither..full of bindi-eyes and other assorted prickles, nor the home of ants.

binding, *vbl. sb.* Add: **5. d.** An arrangement of straps or springs holding the ski to the boot.
1911 *Encycl. Brit.* XXV. 186/2 The feet..are fastened to the middle of the skis by an arrangement of straps, called the binding. 1924 *Tourist* (Winter Sports) 12/1 There are many varieties of 'rigid' or 'spring' bindings, which hold the ski to the foot.
6. binding copy (see quot. 1951).
1936 *Rogers' Catalogue* (*Newcastle upon Tyne*), [Item] 249 ..is really a binding copy. 1951 J. Carter *ABC for Book-Collectors* 33 *Binding copy*, this means..a copy whose covers are in a very poor state, but whose interior is clean and which is worth re-binding.

binding, *ppl. a.* Add: **3. binding agent,** a substance that assists cohesion (cf. Bind *v.* 10 in Dict. and Suppl.); **binding energy** *Physics*, the energy required to break up a nucleus into its constituent particles; **binding-screw** (earlier examples); **binding-twine,** twine used for binding sheaves of corn; **binding wire,** wire used for binding.
1933 *Burlington Mag.* Mar. 123/2 Egg, milk, or size, was used as a binding-agent. 1940 *Chambers's Techn. Dict.* 91/1 *Binding agent* or *binder*, the basic material of disc records, chiefly shellac, which causes the various materials to adhere together and form, after heating, a solid mass. 1932 J. Chadwick in *Proc. R. Soc.* A. CXXXVI. 708 This suggests that the neutron consists of a proton and an electron in close combination, the binding energy being about 1 to 2×10^6 electron volts. 1933 E. Wigner in *Physical Rev.* XLIII. 253/2 One of the remarkable facts about the mass defects in the very first elements is the very great binding energy of the He nucleus. *Ibid.* 256/1 We now come to the calculation of the binding energy of the He nucleus, 1 and 2 are neutrons, 3 and 4 protons. 1957 *Encycl. Brit.* XVI. 593D/1 The binding energy is a negative energy in the sense that one has to add energy to decompose a helium nucleus into neutrons and protons. 1958 W. K. Mansfield *Elem. Nucl. Physics* ii. 15 It should require, roughly, 28/4 = 7 MeV to detach a single nucleon. This amount of energy is known as the average *binding energy* per nucleon. 1828 F. Watkins *Electro-Magnetism or Electro-Dynamics* 74 Two wooden..troughs are secured by binding-screws. 1839 *Annals Electr., Magn. & Chem.* IV. 65 The result is a method whereby solder..and binding screws are dispensed with. 1890 *U.S. Congress. Record* Aug. 9260/1 The observations of..[Senator Davis] in respect to binding-twine are very important to the people of La. 1767 H. M. Brooks *Olden Time Series* (1886) IV. 31 Binding wire; Brass and Iron ditto. 1843 C. & J. J. Holtzapffel *Turning* I. xx. 424 The soft iron binding-wire used in soldering. 1920 *Whittaker's Electr. Engin. Pocket-Bk.* (ed. 4) 363 The binding wire should be wrapped round the conductor where it rests against the insulator.

bindle[2] (bi·nd'l). *U.S.* and *Canadian slang.* [Prob. a corruption of Bundle *sb.*, but cf. Bindle in Dict.]
a. A bundle containing clothes and possessions, *esp.* a bedding-roll carried by a tramp. Hence **bindle-man, -stiff,** a tramp who carries such a bundle.
1900 'Flynt' *Itinerant Policeman* 167 Among the 'Bindle Men', 'Mush Fakers', and 'Turnpikers' of the middle West, the East, and Canada, there exists a crude system of marking 'good' houses. 1901 J. London *Let.* 6 Dec. (1966) 126 Wyckoff only knows the workingman, the stake-man, and the bindle-stiff. 1925 *Forum* Aug. 232 Carrying his 'bindle roll' or roll of blankets on his back, he is prepared to make his home wherever night finds him. *Ibid.* 235 *Bindle stiff*, a western hobo, who carries his blankets in a roll or bindle. 1927 *Glasgow Herald* 24 July 8 In his stride he took almost all the experiences that can befall bums, bindle stiffs..and all other variously designated knights of the moonlight. 1937 J. Steinbeck *Of Mice & Men* 4 George unslung his bindle and dropped it gently on the bank. 1952 —— *East of Eden* vii. 46 Before he knew it he was a bindlestiff himself.
b. Any package or bundle, spec. one containing narcotics (see quot. 1923).
1916 *Lit. Digest* 19 Aug. 425/1 A package is a 'bindle'. 1922 E. Murphy *Black Candle* (1926) xi. 214 The pedlars would buy it [*sc.* morphine] in big amounts; would..dish it out in small packages. We would call these 'decks', but some people call them 'bindles'. 1923 *Dialect Notes* VI. vi. 246 *Bindle*, a package containing either morphine or cocaine. 'Give me a bindle of snow.'

Binet–Simon (bine,simon). The names of the French psychologists A. *Binet* (1857–1911) and T. *Simon* (1873–1961), applied *attrib.* to a form of intelligence test which they devised. *Binet–Simon scale*, the measurement of intelligence by such a test. Also *Binet('s) scale, test*. Also *ellipt.*
1910 G. M. Whipple *Man. Mental & Physical Tests* 278 There are marked differences in the *amount of work* (number of additions) done..*e.g.*, in Binet's tests, from 40 to 96 numbers. *Ibid.* 491 (*heading*) Application of the 1905 Binet–Simon Tests to Defectives. 1912 *Jrnl. Educ. Psychol.* III. 70 The following chart..throws into relief the periods of rapid, normal and slow mental growth as measured by the Binet scale. *Ibid.* 63 The Binet tests of 1908 are the only set hitherto devised covering any considerable variety of functions. 1914 M. Drummond in W. B. Drummond tr. *Binet & Simon's Mentally Defective Children* 147 Part of the interest of this work on defective children consists in the fact that in it we find the origin of those ideas and investigations which culminated in the formation of the Binet–Simon Scale of Intelligence, now so widely known throughout Europe and America. 1921 C. L. Burt *Mental & Scholastic Tests* 3 The Binet–Simon scale consists of about sixty graded tests for measuring the intelligence of school children. 1963 M. McCarthy *Group* iv. 87 She wished he would let her give him the Binet and some of the personality tests she had tried on the group at Vassar.

bing (biŋ), *sb.*[3] and *int. dial.* and *colloq.* [Echoic, representing a sudden banging noise or blow.] **a.** *sb.* A thump or blow. *dial.* (See *E.D.D.*) **b.** *int.* All of a sudden; in a flash; with a bang.
1922 Joyce *Ulysses* 506 Now I do this kind of thing On the wing, on the wing! Bing! 1924 Wodehouse *Ukridge* iii. 57 Always getting ideas—*bing*—like a flash. 1925 C. Morley *Thunder on Left* xi. 140 And then the train.. smashes into a lot of people, bing! 1946 Wodehouse *Joy in Morning* xxix. 264 She looked round and—*bing*—a pillar of salt. 1959 'A. Gilbert' *Death takes Wife* xv. 187 Money saved..and then bing! everything vanished into thin air.
Also **bing-bang** *colloq.*, an onomatopœic reduplication expressing a repeated heavy thump or a continued banging noise.
1914 W. J. Locke *Fortunate Youth* xiii, Lets 'em have it bing-bang in the eye. Don't he, Jane? 1926 *Glasgow Herald* 1 Mar. 10 Bing-bang, cling-clang clatter.

binge (bindʒ), *sb. slang* (orig. *dial.*: see *E.D.D.*). [Special use of dial. *binge* to soak (a wooden vessel).] A heavy drinking-bout; hence, a spree.
1854 A. E. Baker *Gloss. Northampt. Words* s.v., A man goes to the alehouse to get a good binge, or to binge himself. 1889 Barrère & Leland *Dict. Slang*, *Binge* (Oxford), a big drinking bout. 1922 *Chambers's Jrnl.* Sept. 569/1 This is only a binge—just a jolly old bachelor-party. 1928 Wodehouse in *Strand Mag.* July 4 Eh? What about our Monte Carlo binge?
So **binge** *v. refl.* and *pass.*, to drink heavily, 'soak'; *intr.* to have a 'binge'; *trans.* to enliven, const. *up* (cf. 'pep up').
1854 [see *sb.*]. 1910 Belloc *On Something* xxii. 195 It is plainly evident that they know how to binge. 1914 Earl Beatty *Let.* 17 Nov. in W. S. Chalmers *Life & Lett.* (1951) 163 And one has to be cheerful and encouraging to the others and binge them up to live in hope every time that this is *the* time. 1925 *Sunday at Home* Dec. 146/1 One man was so binged in drink and so enchained by the craving for it. 1935 E. Bagnold *National Velvet* xiv. 224 The information having been looked over and binged up here and toned down there..Reuter sent round the world the following message.

binghi, Binghi (bi·ŋgai). *Austral. slang.* Also **bingy,** etc. [Aboriginal.] An Aboriginal.
Recorded earlier in the general sense 'brother': *e.g.* 1831 R. Dawson *Present State of Australia* 221, I was received by them with the greatest cordiality, and greeted by the term bingeye, or brother.
1933 *Bulletin* (Sydney) 15 Feb. 20 One good idea that far inlanders..learnt from Binghi. *Ibid.* 2 Aug. 21/4 Binghi's legend is some evidence that it existed there. *Ibid.* 13 Dec. 37/1 A binghi wife the Chinaman had. 1934 T. Wood *Cobbers* 34 The home for a floating population of Japanese, Chinese, Malays..binghis, half-castes and whites.

bingle (bi·ŋg'l). [f. *b(ob *sb.*[1] 5 b + *sh)ingle *sb.*[1] 1 e.] A short hairstyle for women, between a bob and a shingle. Hence as *v. trans.*, to cut (hair) in this style. So **bi·ngled** *ppl. a.*, **bi·ngling** *vbl. sb.*
1925 *Punch* 11 Mar. 256 This lady complains that you have—ah—completely bungled her bingle. 1925 *Westm. Gaz.* 13 Apr. 5/3 The shingled or bingled head is firmly established in popularity. 1927 *Daily Express* 2 Mar. 3/7 Of course, shingled hair goes with any of this kit, or it may be bobbed, Eton cropped, or bingled. 1927 W. E. Collinson *Contemp. Eng.* 63 Bobbing..shingling..and a combination of the two, bingling. 1959 W. Golding *Free Fall* xii. 228 Bobbing, bingling, Eton crops. 1960 *News Chron.* 11 Oct. 8/3 Bingle it [*sc.* your hair] yourself.

bingo[1]. (Earlier examples.)
1699 B.E. *Dict. Cant. Crew*, *Bingo*, c. Brandy. *Bingo-boy*, c. a great Drinker or Lover thereof. *Bingo-club*, c. a set of Rakes, Lovers of that Liquor. 1750 B.-M. Carew *Apol. for Life* 337 *Bingo-Mort*, a female drunkard, a she brandy drinker. 1839 G. W. M. Reynolds *Pickwick Abroad* xxvi. 224 From morn to night we'll booze a ken, And we'll pass the bingo round.

bingo[2] (bi·ŋgo). [Of obscure origin, but cf. next.] A modern development of Lotto (sense 1), often played in public halls, etc., for prizes. Also *attrib.*
1936 *Time* 21 Dec. 26/2 In many a U.S. Catholic diocese during the past few years the simple gambling game of bingo..has served as a prime money-raiser. 1949 N. Streatfeild *Painted Garden* vi. 57 Such heavenly things were happening on deck...There was a game called Bingo. 1953 *Oxford Mail* 21 Nov. 4/9 For some time now Bingo drives have been gaining in popularity at Oxford social gatherings. Bingo..is very closely akin to what used to be known as 'Housey Housey'. 1958 *Economist* 29 Nov. 798/1 While the essential elements of the game have provided amusement and diversion for centuries, bingo was named and introduced as a modern parlour game in 1929 by a toy manufacturer, Mr. Edwin Lowe. Each player has a card with 25 numbers between 1 and 75 arranged in a square, and the first to get five in a row as the numbers are called out cries 'bingo'. 1964 A. Wykes *Gambling* x. 250 British entrepreneurs have converted most of the nation's failing cinemas into thriving bingo halls.

bingo (bi·ŋgo), *int.* [Echoic, cf. *bing.] = *bing b; *spec.* an exclamation made by someone winning a game of bingo (see prec.). Also in the colloq. intensive phr. *like bingo* (see Like *adv.* 1 b), = 'like billy-o'.
1927 E. Wallace *Ringer* (1952) xxiii. 93, I just laid my hands on him when—bingo! I was on the ground with four inches of good knife in me. 1931 Wodehouse *If I were You* xi. 125 What ensued? Bingo! Eaten by bears! 1933 M. Allingham *Sweet Danger* iv. 50, I just 'ad one look at 'im and came back 'ere like bingo. 1951 J. B. Priestley *Fest. Farbridge* ii. ii. 222 If he can't do it, I will. Do it on my head—bingo! 1958 [see prec.]. 1959 N. Marsh *False Scent* (1960) i. 33 I've been drinking with The Management. Only two small ones, but on an empty tum: Bingo!

bingy (bi·ndʒi). *Austral. slang.* Also **bingee, bingie, bingey, binjy.** [Aboriginal.] The stomach, belly. Also *attrib.*
[1851 Rev. D. Mackenzie *Ten Years in Australia* xiv. 140 They lay rolling themselves on the ground, heavily groaning in pain,..exclaiming, 'Cabonn buggel along bingee' (that is, I am very sick in the stomach).] 1859 H. Kingsley *Recoll. G. Hamlyn* II. vi. 94 Don't you fret your bingy, boss; he'll be as good a man as his father yet. 1908 E. J. Banfield *Confessions of Beachcomber* I. iv. 143 Bingie (belly) belonga you, sore fella. *Ibid.* II. ii. 301 His 'debil-debil all the same like dead man', had 'sat down' in 'Little Jinny's bingey'. 1924 D. H. Lawrence & Skinner *Boy in Bush* 85 Success is t'grow a big bingy like a bloke from town. 1929 K. S. Prichard *Coonardoo* xv. 150 Give them a bit of pain killer or a dose of castor oil when they've got a bingee ache. 1931 I. L. Idriess *Lasseter's Last Ride* ii. 18 Micky landed, whoof! on his bingy. 1963 *Australasian Post* 20 June 44 Plenty tucker here. Just look at those binjies!

Bini, var. *Benin.

biniou (bi·niū). [Breton, pl. of *benvek* tool, (musical) instrument.] The form of bagpipe used in Brittany.
1902 *Westm. Gaz.* 20 Mar. 10/1 A toast was followed by such music as the native biniou can produce. 1925 *Blackw. Mag.* Sept. 359/2 When Lommic took the biniou under his arm he made the pipes sing.

binit (bi·nit). [Abbrev. of *bin(ary dig)it.] A binary digit (occas. distinguished from *bit *sb.*[4]); also, a probability expressed as the logarithm to the base two of the odds (also called *binit weight*).
1953 S. Goldman *Information Theory* i. 4 We shall use the contractions *decit* for decimal digit and *binit* for binary digit. *Ibid.* 5 The author does not know who originated the term *binit*, but he first heard it used by Dr. E. W. Samson. *Ibid.* 7 The total amount of information necessary to locate the cow is thus 6 binits. 1959 *Science* CXXX. 956/3 It is conventional practice in information theory to use logarithms to the base 2 of the probabilities expressed in the form of the 'odds', for or against. The units are known as 'binits'. 1963 N. Abramson *Information Theory & Coding* i. 7 It is important to make a distinction between the binit (binary digit) and the bit. *Ibid.* ii. 18 If the 0s and 1s are not equally probable, then the amount of information provided by a given binit will be either less than or greater than 1 bit, depending upon its probability. 1965 H. B. Newcombe in *Math. in Biol. & Med.* (*Med. Res. Council*) II. 46 We have found logarithms to the base 2 especially convenient and have called these 'binit weights'.

binitarian (bainitē^ə·riăn), *a.* and *sb. Theol.* [f. L. *bini* (pl.) twofold, double, after Trinitarian. Cf. G. *binitarisch* (Loofs 1898).] **A.** *adj.* Of or belonging to a belief in a Godhead of two persons only. **B.** *sb.* A believer in this doctrine. Hence **binita·rianism.**
1908 W. R. Nicoll in T. H. Darlow *Life* (1925) 360 There are Trinitarians, Binitarians, Arians, and Unitarians. 1910 Sanday *Christologies Anc. & Mod.* i. 12 The same alternation of Trinitarian and Binitarian language (the conjunction of Father, Son, and Spirit by the side of Father and Son). 1928 K. E. Kirk in A. E. J. Rawlinson *Ess. Trin. & Incarn.* 207 The two strains of thought—the binitarian and the trinitarian. *Ibid.* 215 The binitarianism of Tertullian's earlier Catholic thought.

bink. 5. (Later examples in form *benk*.)
1797 J. Curr *Coal Viewer* 15 The long way of working collieries, where the roads along the benk faces are narrow. 1940 *Chambers's Techn. Dict.* 87/1 *Benk*, the place underground where coal is being broken from the face of the coal seam.

binocular, *a.* and *sb.* Add: **2.** In *Photogr.* = stereoscopic.
1889 E. J. Wall *Dict. Photogr.* 18 *Binocular Camera*, another name for Stereoscopic Camera. 1901 *Amateur Photographer* 26 July 70/1 The binocular portraits of M. Fred. Boissonnas. *Ibid.*, Binocular photography. *Ibid.* 70/2 Binocular photographs.
B. *sb.* Now usu. *pl.*
1928 E. O'Neill *Strange Interlude* viii. 272 Looking up the river through a pair of binoculars. *Ibid.* 295 Raising the binoculars as he goes to the rail, he looks up the river. 1932 *Bird Lore* XXXIV. 124 Tests to determine quality of binoculars. 1935 *Discovery* Nov. 330/2 We were busy watching the parties of cormorant through the binoculars.

bino·culared, *ppl. a.* [f. Binocular *sb.* + -ed[2].] Furnished or provided with binoculars.
1959 *Landfall* XIII. 130 Becket, grey-suited and binoculared, seemed to stand beside him. 1962 *Punch* 26 Sept. 458/2 Spied on by binoculared office workers. 1962 *Times* 20 Dec. 5/4 A binoculared naval man.

binode (bai·nō^u d). *Geom.* [Bi-[2].] A point on a surface at which there are two tangent planes.
1869 Cayley in *Phil. Trans. R. Soc.* CLIX. 201 Conical and biplanar nodes, or, as I call them, cnicnodes, and binodes. 1874 G. Salmon *Analyt. Geom.* (ed. 3) 457 The quadric cone may degenerate into a pair of planes. Such a node may be called a *binode*. 1934 D. M. Y. Sommerville *Anal. Geom. Three Dimensions* xvii. 375 If the double-point at *D* is a binode, $ax^2 + \ldots = 0$ breaks up into two planes and the discriminant Δ of this quadratic vanishes.

binomial, *a.* and *sb.* Add: **A.** *adj.* **1.** Also, relating to or derived from the binomial theorem or the binomial distribution; *binomial coefficient*: a coefficient of a term in a binomial expansion; *binomial distribution*: a frequency distribution of the possible number of successful outcomes in a given number of trials in each of which there is the same probability of success; *binomial equation*: an equation reducible to the form $x^n - A = 0$; *binomial expansion*: an expansion of a power of a binomial; *binomial series*: an infinite series obtained by expanding $(x+y)^n$, where *n* is not a positive integer or zero; also, a binomial expansion; *binomial theorem* (earlier example).
1755 J. Landen *Math. Lucubrations* ix. 132 By the Binomial Theorem

$$\frac{p}{1+x} \text{ is } = 1 + px + \frac{p \cdot p-1}{2}x^2 + \frac{p \cdot p-1 \cdot p-2}{2 \cdot 3}x^3 \text{ \&c.}$$

1796 C. Hutton *Math. & Philos. Dict.* I. 208/2 He [*sc.* Newton] happily discovered that, by considering powers and roots in a continued series,..the same binomial series would serve for them all, whether the index should be fractional or integral. 1814 P. Barlow *New Math. & Philos. Dict.*, *Binomial equation*, is any equation of two terms, but more commonly applied to the higher order of equations of the form $x^n = 1$. 1848 A. De Morgan in *Camb. & Dublin Math. Jrnl.* III. 239 Use the binomial expansion up to the term in $(x-b)^{-(n-1)}$. 1889 *Cent. Dict.*, Binomial coefficient. 1911 G. U. Yule *Introd. Theory Statistics* xv. 305 The binomial distribution,..only becomes approximately normal when *n* is large, and this limitation must be remembered in applying the table..to cases in which the distribution is strictly binomial. 1914 *Biometrika* X. 36 Binomial frequencies belong to the teetotum class of chances. 1925 R. A. Fisher *Statistical Methods* iii. 65 The binomial distribution is well known as the first example of a theoretical distribution to be established. 1948 J. V. Uspensky *Theory Equations* i. 26 The particular binomial equation $x^n = 1$, defining the so-called roots of unity of degree *n*, is of special interest. 1949 W. L. Ferrar *Higher Algebra Schools* v. 73 The binomial expansion can be used to find the value of expressions such as $(1{\cdot}002)^{12}$, $(1{\cdot}01)^7$ to any desired number of significant figures. 1959 G. & R. C. James *Math. Dict.* 31/1 The $(r+1)$th binomial coefficient of order *n* (*n* a positive integer) is $n!/[r!(n-r)!]$, the number of combinations of *n* things *r* at a time. 1961 P. G. Guest *Numer. Meth. Curve Fitting* iv. 72 The binomial distribution for rare events.. approximates to the Poisson form. 1966 *McGraw-Hill Encycl. Sci. & Technol.* XII. 191/2 One of the most important power series is the binomial series.
B. *sb.* (Later example.) Also *ellipt.* for *binomial distribution, theorem*, etc.

1835 *Penny Cycl.* IV. 412/2 The general theorems of which the binomial is a particular case. 1914 *Biometrika* X. 64 The binomial is built up on the assumption of the repetition *n* times of a number of independent events, of which the chance of occurrence is identical and equal to *q*. 1966 Meyer & Hanlon *Fun with New Math* viii. 105 Perhaps you have forgotten how to multiply a binomial (an algebraic expression of two terms) by a binomial. 1968 P. A. P. Moran *Introd. Probability Theory* ii. 77 If we require a distribution with a variance smaller than that of a binomial we can use a hypergeometric distribution.
2. *Biol.* The two-part Latin name of a plant or an animal.
1945 *Rhodora* XLVII. 372 Binomials had appeared sporadically in the publications of Bauhin, Cornut and others. 1957 W. T. Stearn *Linnæus's Species Plantarum* i. 2 In Linnaeus's work the binomial became more than a two-word label; it functioned as a point of reference within a vast logically devised and integrated system.
3. *Philol.* An expression consisting of two words of the same form-class.
1959 *Lingua* VIII. 113 In the typical newspaper headline *Cold and snow grip the nation* it is proper to set off the segment *cold and snow* as a binomial. 1964 *Linguistics* May 69 In A-Tokharian texts we find numerous combinations of two synonymous words or words expressing nearly related—or, in some cases, opposite—conceptions. The meaning of such a binomial obviously approaches that of a dvandva.

binovular (bin-, bainō^u·viulăɪ), *a. Biol.* [f. Bin- (or L. *bini* two by two) + Ovular *a.*] Pertaining to or derived from two ova.
1900 in Dorland. 1917 C. Berkeley *Midwifery* xxxii. 376 *Binovular twins* are developed from two ova. 1961 *Lancet* 30 Sept. 743/2 Two cases of goitre are described in the binovular twins of a mother who for years had taken 'Felsol' powders.

bint (bint), *sb.*[2] *colloq.* [Arab. *bint* daughter.] A girl or woman (usu. derog.); girl-friend.
The term was in common use by British servicemen in Egypt and neighbouring countries in the wars of 1914–18 and 1939–45.
1855 R. F. Burton *Pers. Narr. Pilgrimage to Meccah* I. v. 121 'Allah! upon Allah! O daughter!' cry the by-standers, when the obstinate 'bint' of sixty years seizes their hands. 1888 C. M. Doughty *Trav. Arabia Deserta* I. viii. 231 Hirfa sighed for motherhood: she had been these two years with an husband and was yet *bint*, as the nomads say, 'in her girlhood'. *Ibid.* xiii. 374 The homesick Beduin bint. 1919 *Athenæum* 25 July 664/2 *Bint*, girl. 1930 E. Raymond *Jesting Army* i. ii. 24 Damned jolly little bint, that one, too! 1938 'R. Hyde' *Godwits Fly* xi. 169 Fancy turning in a smoke for a bint. 1941 *New Statesman* 30 Aug. (*list of war slang*) *Bint*—Girl friend. 1942 N. Streatfeild *Table for Six* 151 I'd like her to grow up a lush bint. 1946 *Penguin New Writing* XXVIII. 175 What are the bints like round here, Tom? 1958 K. Amis *I like it Here* xiii. 162 As the R.A.F. friend would have put it, you could never tell with these foreign bints.

binturong (bi·ntiurǫŋ). Also **benturong.** [Malay.] A prehensile-tailed civet, *Arctictis binturong*, found in southern Asia.
1822 T. S. Raffles in *Trans. Linnean Soc.* XIII. 253 Intermediate between *Viverra* and *Ursus* is an animal called Binturong, found at Malacca by Major Farquhar. 1838 *Penny Cycl.* XII. 434/1 The newly discovered forms, the Benturong and the Panda. 1883 *Encycl. Brit.* XV. 436/1 The Binturong, an inhabitant of southern Asia from Nepal through the Malay peninsula to the islands of Sumatra and Java. 1930 *Times Educ. Suppl.* 16 Aug. iv/2 A female binturong with two young. 1964 G. Durrell *Menagerie Manor* i. 16 The binturong..suggests a badly-made hearth-rug, to one end of which has been attached a curiously oriental-like head with long ear-tufts and circular, protuberant and somewhat vacant eyes.

biny (bai·ni), *a.* [f. Bine + -y[1].] Of hops: abounding in bine, running much to bine.
1881 C. Whitehead *Hops* 36 It will be found in 'biny' years that many of the hops are light and 'white livered' from sheer lack of sunshine.

bio (bai·o), colloq. or slang abbrev. of Biography (occas. of Biographical *a.*).
1961 in Webster. 1965 J. M. Cain *Magician's Wife* (1966) v. 37 Fisher's credit department did me your bio sketch. 1967 *Sat. Rev.* 25 Nov. 50/3 Paul Gallico's bios are, as one would expect of such a gifted journalist, clear and revealing.

bio-. Add: **bio-assa·y** = *biological assay* (see *assay *sb.* 6 b); **bio-astronau·tic** *a.*, of or pertaining to bio-astronautics; **bio-astronau·tics** *sb. pl.* (see quots.); **bio-biblio·graphy,** a bibliography containing biographical information about the author or authors; **bi·ochrome,** any natural colouring matter of plants or animals; hence **biochro·mic** *a.*, **biochro·my** (see quot. 1944[1]); **bioco·lloid** *Chem.* [contracted form of *biological colloid*: cf. G. *biokolloid*], a colloid derived from an organic substance; **biodegra·dable** *a.*, susceptible to the decomposing action of living organisms, esp. of bacteria; occas., broken down by biochemical processes in the body; so **bi:odegradabi·lity, biodegrada·tion; bi:odeteriora·tion,** deterioration of a substance or material caused by the action of living organisms; so **biode·teriorative** *a.*; **bio-eco·logy** (see quot. 1957); **bioelectric(al** (bai̯o,īle·ktrik, -ăl) *a.*, of or pertaining to electrical phenomena produced in living organisms; hence **bioelectri·city,** electrical phenomena produced in living organisms; **bio-energe·tics,** the study of the transformation of energy in living organisms; **bio-enginee·ring,** the application of engineering techniques to biological processes; so **bio-engineer; biofla·vonoid** = *flavonoid; **biogeoche·mistry** [f. *geochemistry, after Russ. *biogeokhimiya*], the branch of biochemistry that deals with the relation of chemicals found in the soil to living organisms; the biological application of geochemistry; hence **biogeoche·mical** *a.*; **bioherm** (bai·o,hɜɹm), *Geol.* [Gr. ἕρμα sunken rock, reef], a reef (as a coral reef) formed from organic material; **bi·olith** *Geol.* [ad. G. *biolith* (C. G. Ehrenberg, 1854): -lith], a rock formed from organic material; **bi·omass,** the total weight of the organic substance (as plankton) or organisms in a given area; **biomathema·tics** (see quots.); **biome·dical** *a.*, pertaining or relating to both biology and medicine; *spec.* pertaining to the biological effects of space-travel; **biomo·lecule** (see quot.); hence **biomole·cular** *a.* (1909 in *Cent. Dict.* Suppl.); **bi·osatellite** (orig. *U.S.*), an artificial satellite containing living organisms for experimental purposes; **bi·oscience,** a collective term for the biological sciences; hence **bioscientist; bioso·cial** *a.*, of or pertaining to the interaction of biological and social factors; relating to both biology and sociology; **biosocio·logy** (see quot. 1901); so **biosociolo·gical** *a.*, involving both biological and sociological factors; **biospelæo·logy, -eology** [ad. F. *biospéléologie* (A. Vire 1904, in *Comptes rendus Acad. Sci. Paris* CXXXIX. 992), f. Spelæology], the study of the fauna of caves; hence **biospelæolo·gical** *a.*; **biostrome** (bai·ǒstrō^u m), *Geol.* [Gr. στρῶμα mattress, bed]: see quots.; **biosy·nthesis,** the production of a chemical substance by a living organism; hence **biosynthe·tic** *a.*, pertaining to biosynthesis; (of a substance) produced by a living organism; **biote·chnics,** the practical application of discoveries in the biological sciences; so **biote·chnic** *a.*; **bi·ozone** *Geol.* (see quot.).
1912 H. C. Wood in *Jrnl. Amer. Med. Assoc.* LIX. 1433/2 The underlying principle of physiologic standardization, or as those engaged in this work prefer to call it, the bio-assay, is to determine the quantity of a given sample of drug required to produce some easily recognizable effect on a lower animal. *Ibid.* 1434/1 The bio-assay must be acknowledged as a great step forward in the direction of accuracy in our materia medica. 1939 *Nature* 8 July 76/2 A bioassay on the isolated benzoate..gave an 80 per cent response in a group of ten mice, each receiving a single dose of 0·5 γ in nut oil. 1955 *New Biol.* XIX. 52 The suitability of an inbred strain of mice for the bio-assay of oestrogens. 1961 *Flight* LXXIX. 344/2 American bio-astronautic activities. 1957 *N.Y. Times Mag.* 20 Oct. 12/1 The third phase has a wholly new name, 'bioastronautics'—the study of man's capabilities and needs, and the means of meeting those needs, for travel in outer space. 1959 H. Strughold in *Space Weapons* 136 Space medicine..can ..be considered a branch of the more general field which we call 'bioastronautics', the investigation of every aspect of life in the universe, as it may be encountered by men from this planet in the future. 1959 L. M. Harrod *Librarians' Gloss.* (ed. 2) 39 *Biobibliography*, a bibliography which contains brief biographical details about the authors. 1944 D. L. Fox in *Science* C. 470/2 From the standpoint of the biochemist..concerned with the metabolic significance of natural coloring matters, the designation of these by a discriminating scientific term has long been desirable. In response to an inquiry, Dr. George M. Calhoun..suggested to the writer [in a personal communication, 1936] the descriptive and self-explanatory term *biochrome*, with an adjective *biochromatic* (*biochromic* is perhaps preferable), and a collective noun *biochromy*. *Ibid.* 471/1 Biochromes.. possess two chief characteristics in common, *i.e.*, their origin and occurrence in living organisms and their reflection of a fundamental chemical property, the selective absorption of light waves in the visible spectrum. *Ibid.*, Our present understanding of the parts played by..biochromic compounds in the metabolic economy of organisms leaves much to be desired. 1954 *Sci. News* XXXIV. 91 [Porphyrins] probably serve in this role, as well as a biochrome, in insects. 1917 *Chem. Abstr.* 3283 (*title*) Effect of acids and salts on 'bio-colloids'. *Ibid.*, An extension of the tests..was made to include mixts. of the 'bio-colloid' agar. 1932 *Ibid.* 19 (*title*) Studies of the combination of biocolloids. 1956 *Nature* 24 Mar. 562/2 General behaviour of biocolloids and polyelectrolytes in aqueous solution. 1960 *Biochim. & Biophys. Acta* XLIII. 348 (*heading*) Biodegradation of dehydro-L-ascorbic acid; 2, 3-diketo-aldonic acid decarboxylase from rat liver. 1961 Nelson et al. in *Devel. Industrial Microbiology* II. 93 The present study was initiated to provide information on the effect of molecular structure on the biodegradability of synthetic detergents. *Ibid.* 101 Compounds with strictly linear side chains and those containing one or two methyl branches on the carbon atom attached to the benzene ring are readily biodegradable. 1962 *Chem. & Engin. News* 10 Sept. 38/3 Feeds tock.. suitable for the production of a completely biodegradable detergent. *Ibid.* 24 Dec. 36/1 Biodegradability has so far been tested mainly in flask tests. 1969 *Nature* 19 July 230/2 Biodegradable detergents are now a reality. 1961 W. M. Bejuki in *Devel. Industrial Microbiology* II. 265 The bacteria..are microorganisms being newly introduced into biodeterioration evaluations. 1964 *Nature* 5 Sept. 1084/1 These thermophilic fungi may play a significant part in the biodeterioration of oil palm produce during processing and storage. 1966 D. G. Coursey in *Microbiological Deterioration in Tropics* (Soc. Chem. Industry Monograph xxiii) 44 (*title*) Biodeteriorative processes in palm oil stored in West Africa. *Ibid.* 55 Purely chemical changes.. are slow compared with the biodeteriorative process. 1923 F. E. Clements et al. in *Carnegie Inst. Washington Year Book 1922* 355 The concept of the biome was advanced in 1915 to emphasize the importance of treating plants and animals together as mutually interacting members of a community...The entire field of bio-ecology has been sketched in outline to serve as a guide for further work. 1927 W. P. Taylor in *Ecology* VIII. 280 If the botanists persist in appropriating the term ecology as synonymous with plant ecology we shall be forced to domesticate the new term bio-ecology to take its place as referring to the whole field. 1939 F. E. Clements & Shelford (*title*) Bio-Ecology. 1957 *Encycl. Brit.* XVIII. 29B/1 Other aspects of general ecology are animal ecology, human ecology and bio-ecology, which is defined provisionally as the ecology of plant and animal interrelations. 1918 *Science* 22 Nov. 518/2 Bioelectric phenomena constitute a group of facts for which adequate and satisfactory explanations have hitherto been lacking. 1927 Lund & Kenyon in *Jrnl. Exper. Zoöl.* XLVIII. 333 (*title*) Relation between continuous bio-electric currents and cell respiration. 1943 *Electronic Engin.* XV. 519 The above facts suggest that the alpha rhythm..may possibly be due to the same cause as that producing slow bio-electric rhythms in other organs. 1962 *IRE Trans. Bio-Medical Electronics* IX. 85/1 Bioelectrical brain reactions arising in response to certain stimuli are often very poorly expressed. 1949 *Blakiston's New Gould Med. Dict.* 136/1 *Bio-electricity*, electric phenomena occurring in living tissues; effects of electric currents upon living tissues. 1969 *Listener* 15 May 688/1 A bioelectricity, which originates in the abdominal cavity. 1945 S. Brody *Bioenergetics & Growth* ii. 12 Bioenergetics, concerned with energy transformations in living things, is a branch of general energetics generalized briefly by the first and second laws of thermodynamics. 1947 *Ann. Reg. 1946* 388 The Timiriazeff Institute [U.S.S.R.], which is divided into sections dealing with photosynthesis,..comparative physiology, bioenergetics. 1960 *Appl. Microbiol.* VIII. 122/1 The real problem may exist near the cell surfaces where flow..may depend upon capillary action. Herein may lie one of the real challenges for the bioengineer. *Ibid.* 124/1 This bioengineering approach has been most fruitfully applied to microbiological processes. 1964 *New Scientist* 30 July 262/1 Bio-engineering to help in converting new crops into human food. 1952 *Chem. Abstr.* 10215/1 (*title*) Effect of bioflavonoids on radiosensitivity of transplanted tumors. 1955 *Sci. News Let.* 26 Feb. 141/1 Nature's yellow dyestuff and related chemicals..promise to help fight certain artery troubles and other ailments of man. The chemicals are now called bioflavonoids. 1938 tr. *Vernadsky's On some Fundamental Probl. Biogeochem.* 5 Life in the biogeochemical aspect is the living matter of the biosphere, that is, the total of all the living organisms present in the biosphere at a given moment. 1956 *Nature* 25 Feb. 353/1 To further the study of biogeochemical cycles. 1938 tr. *Vernadsky's On some Fundamental Probl. Biogeochem.* 5 Biogeochemistry, which is a part of geochemistry and has peculiar methods and peculiar problems of its own, may be finally reduced to a precise quantitative mathematical expression of the living nature in its indissoluble connection with the external medium, in which the living nature exists. 1940 *Chem. Abstr.* 3769 (*title*), Problems of biogeochemistry. 1928 E. R. Cumings & R. R. Shrock in *Bull. Amer. Geol. Soc.* XXXIX. 599 *Coral reef* encourages the misconception that reefs are largely made of coral, whereas many of them were formed by other organisms...The authors have for some time used the term 'bioherm'. 1939 W. H. Twenhofel *Princ. Sedimentation* ix. 329 Colonial corals have been among the chief contributions to the building of the reefs or bioherms. 1954 W. D. Thornbury *Princ. Geomorphology* xviii. 480 The term 'bioherm' (organic mould) is a more appropriate name for such structures, but they have been called reefs so long that the name persists despite its inadequacy. 1906 H. Potonié in *Rep. Brit. Assoc. Advancem. Sci.* 748, I set aside those combustible biolithes which are pyromonimites, such as amber. 1920 A. W. Grabau *Textbk. Geol.* (1921) I. xii. 269 The organic or biogenic rocks...The true organic rocks are conveniently termed *bioliths*. 1934 B. G. Bogorov in *Jrnl. Marine Biol. Assoc.* XIX. 585 Weighing is the main method of obtaining the biomass. The conception 'biomass' may be applicable also to the quantity of substance, characteristic for a given species at its different stages of development. 1937 W. Pickles in *Jrnl. Animal Ecol.* VI. 54 In order to determine the biomass of each species, it was necessary to weigh the ants, larvae, pupae, etc., of each nest. *Ibid.* 60 The success of species..can also be expressed as the biomass, i.e. the total weight per unit area. 1957 *Encycl. Brit.* VII. 919/2 The biomass could be determined for each link of a food chain, or for each stratum of a community or for the whole community. *Ibid.* XVIII. 237/1 This method uses weights of population samples (biomass) instead of counts of individuals. 1923 W. M. Feldman *Biomath.* i. 1 Biomathematics is the science and art of rapid and accurate computation applied to the study and investigation of biological problems. 1963 *New Scientist* 10 Jan. 72/2 There are now strong indications that biomathematics and medicomathematics are emerging as distinctive scientific disciplines. 1955 *Bull. Atomic Sci.* May 200/2 The only biomedical data which remains classified is in piecemeal or incomplete form and therefore inadequate for use by the medical profession. 1962 S. Carpenter in *Into Orbit* 160 Next, you run a check on the intercom and the bio-medical leads to make sure they are working. 1963 C. D. Green in J. H. U. Brown *Physiol. of Man in Space* 257 (*title*) Biomedical capsules. 1946 *Nature* 28 Sept. 440/2 A later technique is beautifully illustrated by the exhibit from the Department of Biomolecular Structure of the University of Leeds. 1901 J. G. McKendrick in *Rep. Brit. Assoc. Advancem. Sci.* 810 The conception of a

biomolecule, or living molecule, that is to say the smallest quantity of living matter that can exhibit some of the chemical phenomena of life. 1958 *Newsweek* 17 Feb. 95/1 The NACA sphere could be ready within a year as a bio-satellite to test space stresses on small laboratory animals. 1966 *Daily Tel.* 19 Dec. 1/4 They were millions of insects and bacteria in a 'biosatellite' that went astray. 1964 (*title of periodical*) BioScience. 1966 I. Asimov *Fantastic Voyage* i. 12 Tell him I want some crumbs for the bio-sciences. 1959 *Listener* 14 May 834/2 Most physical scientists can be rather precise in their measurements and in the predictions based on them. Not so the bio-scientists. [1894 J. Izoulet *La Cité Moderne* I. i. 1 (*title*) Livre Premier. Exposé de notre hypothèse bio-sociale.] 1927 L. L. Bernard *Introd. Social Psychol.* I. v. 79 The physico-social and bio-social environments are intimately connected with human behavior. 1935 L. Bloomfield in *Language* XI. 98 The biophysical description, in terms of speakers' movements, or of sound-waves,..must be left to the laboratory. The biosocial description consists in a statement of the combinations in which these phonemes occur. 1937 C. W. Morris *Log. Positivism* ii. 22 Pragmatism..is not falsely described as a biosocial positivism. 1942 C. F. Hockett in *Language* XVIII. 6 Phonological equivalence involves, generally, the two factors of a-phonetic similarity and biosocial equivalence. 1936 E. van Loo tr. *Bouger's Introd. Criminol.* iv. 77 'Every crime is the resultant of individual, physical and social conditions.'..This view was shared by several later authors...Its adherents form what is called the 'biosociological school'. 1949 Koestler *Insight & Outlook* xvi. 221 The biosociological significance of these upheavals is rarely appreciated. [1894 J. Izoulet *La Cité Moderne* iv. i. 544 Je crois que la psycho-morale et la bio-sociologie ne peuvent coexister parallèlement, sans agir et réagir l'une sur l'autre.] 1901 G. Gissing *Our Friend Charlatan* ii. 22 'It's uncommonly suggestive,' said Dyce..'The best social theory I know. He [*sc.* J. Izoulet] calls his system Bio-Sociology, a theory of society founded on the facts of biology.' 1953 Hazelton & Glennie in C. H. D. Cullingford *Brit. Caving* ix. 248 British biospeleological literature is necessarily scanty. 1947 *New Biol.* III. 103 Small bands of enthusiastic workers, like those who laid the foundations of biospelæology, can hardly be expected to produce rapid results. 1965 B. E. Freeman tr. A. Vandel (*title*) Biospeleology: the biology of cavernicolous animals. 1932 E. R. Cumings in *Bull. Geol. Soc. Amer.* XLIII. 334 For purely bedded structures, such as shell beds, crinoid beds, coral beds, etcetera, consisting of and built..by sedentary organisms.. I propose the name *biostrome*...*Biostrome* means literally an organic layer. 1961 J. Challinor *Dict. Geol.* 23/1 Where the biolith is a more or less extensive bed it is sometimes called a 'biostrome'. 1930 *Brit. Chem. Abstr.* A. 479/2 The closely related structure of natural anthocyanins, [etc.]..is discussed on the basis of biosynthesis from the simple components formaldehyde and its condensation products. 1949 E. Chain in H. W. Florey *Antibiotics* II. xvii. 695 The strain 1984. A is more exacting in its requirements for the biosynthesis of penicillin. 1959 *Sci. News* LII. 17 The presence of iron is essential to the biosynthesis of chlorophyll. 1948 *Biol. Abstr.* 304 The dominant opinion was that biosynthetic processes were to be considered as cascades. 1938 L. Mumford *Culture of Cities* 495 Biotechnic, refers to an emergent economy..pointing to a civilization in which the biological sciences will be freely applied to technology... The application of bacteriology to medicine and sanitation, [etc.]..are further marks of this order. 1925 Geddes & Thomson *Biol.* 245 (*heading*) The Applications of Biology (Biotechnics). 1902 J. J. Buckman in *Geol. Mag.* IX. 556 Why cannot we call this a *biozone*, using the term to signify the range of organisms in time as indicated by their entombment in the strata? Thus we might have the biozone of a species, of a genus, of a family, or of a larger group.

bioce·ntral, *a.* = *biocentric *a.*
1905 *Athenæum* 8 Apr. 436/2 The study of synthetical chemistry from the biocentral point of view.

biocentric (bəiŏse·ntrik), *a.* [f. Bio- + Centric *a.*] Centring in life; regarding or treating life as a central fact. Hence **bioce·ntrically** *adv.*, from a biocentric point of view.
1889 in *Cent. Dict.* 1899 R. Meldola in *Nature* 5 Jan. 217/1 In brief, there has arisen a set of ideas which are even broader than 'anthropocentric', and which might fairly be designated *biocentric*. 1904 —— *Vital Products* I. 6 Organic Chemistry from the Biocentric Standpoint. 1913 L. J. Henderson *Fitness Environm.* 110 Water is the one fit substance for its place in the process of universal evolution, when we regard that process biocentrically. *Ibid.* 312 The biologist may now rightly regard the universe in its very essence as biocentric. 1952 V. A. Demant *Relig. & Decl. Capitalism* vii. 169 Bio-centric living supplants logo-centric living when the latter claims too much.

biochemical (bəiŏke·mikăl), *a.* [f. Bio- + Chemical, after G. *biochemisch*.] Of or pertaining to biochemistry. Hence **bioche·mically** *adv.*
1867 R. E. Wallis tr. *Delitzsch's Bibl. Psychol.* 273 Without wishing to substitute for Scripture a knowledge of that bio-chemical process. 1887 A. M. Brown *Anim. Alkaloids* 147 That which might seem *a priori* evident, is now biochemically proved to demonstration. 1897 A. B. Griffiths *Respir. Proteids* 107 The foundation of all life, the biochemical processes. 1904 R. Meldola *Vital Products* I. 5 It is doubtful whether this compound [*sc.* furfural] is really a biochemical product. 1969 N. W. Pirie *Food Resources* iv. 122 Processes of this type are sometimes referred to as 'chemurgy'; 'biochemical engineering' seems a preferable term.

biochemist (bəiŏke·mist). [f. next.] One who is versed in biochemistry.
1897 *Amer. Naturalist* XXXI. 272 In bacteriology the biochemist has a wide field of work. 1913 [see *biophysics]. 1935 *Discovery* Apr. 98/2 A problem for a physical chemist, assisted by a bio-chemist. 1957 *Encycl. Brit.* III. 591/1 The biochemist..must know the chemical structure of the myriads of substances present in living material. *Ibid.* 591/2 Biochemists have long been interested in the chemical composition of the food of animals.

biochemistry (bəiŏke·mistri). [f. Bio- + Chemistry, after *biochemical *a.* Cf. G. *biochemie*.] The science dealing with the substances present in living organisms and with their relation to each other and to the life of the organism; biological or physiological chemistry.
1881 M. D. Walker tr. *Schüssler's Cure of Diptheria by Biochemic Treatment* 16 My system, or method of Procedure, is direct Biochemistry, because I use *only* tissue-cell-salts. 1902 *Nature* 14 Aug. 381/2 The chair of biochemistry recently founded in University College [Liverpool]. 1904 R. Meldola *Vital Products* I. p. v, I am not without hope that it may be found of service as a step towards the foundation of a more exact science of Biochemistry. 1926 *Encycl. Brit.* I. 373/2 Biochemistry may be defined as the study of the chemical or physico-chemical processes which play a part in the life phenomena of plants and animals. 1943 *Endeavour* Apr. 41/1 Dramatic progress..continues to be made in the applications of biochemistry to the eradication of disease.

bi·ochore (-kǭɹ). *Ecol.* [G. *biochore* (W. Köppen 1900, in *Geogr. Zeitschr.* VI. 675).] **a.** (In the use of Köppen and Raunkiaer) the climatic boundary of a region (see quot. 1934). **b.** A group of similar biotopes; the largest division or region of animal and plant environment, as forest, desert, grassland, etc.
1913 *Jrnl. Ecology* I. 20 These climate-zones..are limited by *biochores* or plant-climate boundaries as suggested by Köppen. 1934 A. G. Tansley in tr. *Raunkiaer's Life Forms of Plants* p. xiii, The 'biochore'.., originally Köppen's term and analogous to the climatological terms isotherm and isohyet, is the line passing through all the areas whose floras show the same percentage number of a characteristic life-form: biochores can only be determined by the analysis of a large number of adjacent local floras. 1937 Allee & Schmidt tr. *Hesse's Ecol. Animal Geogr.* ix. 137 Regions with a general similarity of habitus are recognized as biochores, which may be united as superbiochores or subdivided into subbiochores; areas within the biochores which have uniform external habitat conditions are biotopes. 1957 P. Dansereau *Biogeography* iii. 125 The biochore is the geographical environment where certain dominant life-forms appear to be adapted to a particular conjunction of meteorological factors...Each biochore is characterized by a major type of vegetation...Within each biochore there develops one or more formations which extend over an entire province.

biocide (bəi·ŏsəid). [f. Bio- + -cide.] (See quots.); *spec.* a pesticide.
1947 E. M. Greenberg in *Sat. Rev. Lit.* 30 Aug. 21/3 'Biocide', or destruction of the tissues of the human body, is caused by all the unphysiologic habits that civilization has imposed upon the human race. 1963 W. G. Holbrook in *Sat. Rev.* 19 Jan. 23, I have encountered the term 'biocide' used in a context similar to genocide..namely, the destruction of living species by indirect chemical, or, for that matter, nuclear, poisons. 1963 R. L. Carson *Silent Spring* ii. 7 Can anyone believe it is possible to lay down such a barrage of poisons on the surface of the earth without making it unfit for all life? They should not be called 'insecticides', but 'biocides'. 1968 *Proc. 1st Int. Biodeterioration Symposium* ix. 511 The toxicity and hazards of the biocide to higher forms of life must be considered.

bioclimatic (bəio,kləimæ·tik), *a.* [f. Bio- + Climatic *a.*] Pertaining to the study of climate in relation to the seasonal activities and geographical distribution of living organisms. Hence **bioclima·tics** (= next).
1918 A. D. Hopkins in *Monthly Weather Rev. Suppl.* IX. 9/1 We have in this proposed bioclimatic law of latitude, longitude and altitude, a fundamental guide to lines of research and practice as related to any periodical phenomena in life activity including that of man. 1921 —— in *Jrnl. Washington Acad. Sci.* XI. 223 (*title*) International problems in bioclimatics with special reference to natural and artificial distribution of plants and animals. *Ibid.* 223 *The bioclimatic law* is a law of life and climate as related to the geographical coordinates *latitude, longitude* and *altitude*. 1929 W. C. Cook in *Ecology* X. 283 In bioclimatic studies it is impossible to demonstrate the effect of climate upon insect distribution outside the economic range of the species. 1938 A. D. Hopkins (*title*) Bioclimatics. A Science of Life and Climatic Regions. 1938 *Nature* 25 June 1146/1 The bioclimatic zone and zonal types of a place, area, or local region are the most reliable indexes to the species of animals and plants, and to the types of agriculture, that are best adapted to the local conditions. 1959 *Times* 9 Nov. (Ghana Suppl.) p. iv/4 Ghana's agricultural production naturally is related directly to her major bioclimatic regions.

bioclimato·logy. [f. Bio- + Climatology.] The study of climate in relation to living organisms and esp. to human health. Hence **bioclimatolo·gical** *a.*, of or pertaining to bioclimatology.
1922 *Readers' Guide Periodical Lit.* 1919–21 V. 146/1 (*heading*) Bioclimatology...Bioclimatics. 1949 W. G. Moore *Dict. Geogr.* 22 *Bioclimatology*, the study of climate in relation to life and health, one of its objects being to determine the climatic conditions most favourable to human habitation, especially for invalids, and to define the areas where such climates exist. 1961 *Lancet* 23 Sept. 706/2 The factor which primarily actuates this seasonal variation [of poliomyelitis] is unknown; but it is evidently bioclimatological.

biocœnosis (bəio,sīnōᵘ·sis). *Ecol.* Also **bio-cenose, biocœnose.** [mod.L., ad. Ger. *biocönose* (K. Möbius 1877, *Die Auster und die Austernwirthschaft*), f. Bio- + Gr. κοίνωσις sharing (κοινός common).] An association of organisms forming a biotic community; the relationship that exists between such organisms.
1883 H. J. Rice tr. *Möbius' Oyster & Oyster Culture* in *Rep. U.S. Comm. Fish & Fisheries 1880* VIII. 723 Science possesses, as yet, no word..for a community where the sum of species and individuals..have, by means of transmission, continued in possession of a certain definite territory. I propose the word *Biocœnosis* for such a community. Any change in any of the relative factors of a biocönose produces changes in other factors of the same. 1925 I. W. Bailey in *Bot. Gaz.* LXXX. 93 (*title*) The 'Spruce Budworm' Biocoenose. 1957 *Encycl. Brit.* III. 607/2 The idea of the animal community and of the fully integrated biological community or biocoenosis is largely the result of the study of shallow water populations in the sea. *Ibid.* VII. 917/2 Such assemblages..have been called biocoenoses, if the inter-relations are especially closely knit, and they are known generally as communities. 1957 *Sci. News* XLIII. 71 A fossil 'community'..is seldom identical with the original biocoenose (or life community).
Hence **biocœno·logy,** the study of biocœnosis; **biocœno·tic** *a.*, of or pertaining to biocœnosis.
1927 *Biol. Abstr.* 593/1 Small associations should be known as biocoenotic complexes. 1931 R. N. Chapman *Animal Ecol.* ix. 203 From the viewpoint of biocenology the population, or biocenose, may be classified on the basis of the bond which is of primary importance in holding the population together as an ecological unit. 1939 Clements & Shelford *Bio-Ecology* i. 10 Gams (1918). To him, 'vegetation research' is synonymous with his new term 'biocenology'. 1957 *Encycl. Brit.* III. 607/1 The first contributions to what is now called *synecology*, or perhaps better *biocoenology*, or the study of biological communities..were devoted solely to plants. 1967 *Oceanogr. & Marine Biol.* V. 489 The whole population of the *Posidonia* meadows.. cannot be considered as a single biocoenotic unit or entity.

biog (bəi·ǫg), colloq. or slang abbrev. of Biography.
1942 Berrey & Van den Bark *Amer. Thes. Slang* § 522/11 Biography, *biog, close-up*. *Ibid.*, *Puff biog*, a laudatory biographical sketch. 1967 P. McGerr *Murder is Absurd* iii. 44 If he should drop out—well, 'replaced Mark Kendall' will sound good in your biog.

biogen (bəi·ŏdʒen). *Biol.* Also **-gene.** [See Bio- and -gen.]
1. (See in Dict. under Bio-.)
2. [G. *biogen* (M. Verworn 1895, *Allgemeine Physiologie* vi. 468).] A hypothetical protoplasmic unit. (Cf. *biophor.)
1899 F. S. Lee tr. *Verworn's Gen. Physiol.* vi. 481 In order to distinguish this body..from dead proteid and to indicate its high significance in the occurrence of vital phenomena, it appears fitting to replace the term 'living proteid' with that of *biogen*. 1909 *Westm. Gaz.* 23 Oct. 6/3 First, the biogene,..secondly, the cell, composed of a great number of biogenes. 1911 *Encycl. Brit.* XXI. 557/2 These combinations have been termed 'biogens'. 1915 W. M. Bayliss *Physiol.* i. 18 A certain theory, that of 'biogen molecules', has attracted many investigators. 1916 A. P. Mathews *Physiol. Chem.* 261 A molecule which may be called a biogen. *Ibid.* 844 The biogenes or large molecules of the cell.

biogenesis. Add: **2.** = *biosynthesis* (see *bio-).
1933 *Chem. Rev.* XIII. 479 Attempts to explain the biogenesis of the terpenes on the basis of occurrence side by side in one or more species of plants have not been wanting. 1959 *New Scientist* 27 Aug. 302/2 The mode of biogenesis of cellulose still remains one of the major unsolved problems of carbohydrate chemistry.
3. The biochemical evolution of living matter, esp. its hypothetical development from complex but inanimate substances.
1960 *Science* 19 Feb. 479 The possibility of recurring biogenesis and the abiotic origin of optical activity are considered. 1965 J. B. S. Haldane in S. W. Fox *Orig. Prebiological Syst.* 11 Some particular group may have been the limiting factor in biogenesis. Perhaps amino acids..were present in adequate amounts, but high-energy phosphates were only rarely so. 1967 J. D. Bernal *Orig. Life* 192 The molecular biology of biogenesis consists of successive stages of topochemical reactions.

biogenetic, *a.* Add: Also, of or pertaining to biogenesis; **biogenetic law** (G. *biogenetisches grundgesetz*): the theory, formulated by Haeckel, that evolutionary stages are repeated in the growth of a young animal; also called *recapitulation theory* (see Recapitulation¹ 1 b).
1882 R. Meldola tr. *Weismann's Studies Theory Descent* III. 611 A corollary to the 'fundamental biogenetic law' first enunciated by..Haeckel. 1933 *Chem. Rev.* XIII. 481 The coëxistence of compounds in a few instances out of many..may indicate nothing as to their biogenetic relationship. 1934 *Nature* 10 Feb. 199/1 In our judgment the formulation of this biogenetic law was the greatest service which Haeckel did to the science of zoology. 1962 E. M. Kosower *Molec. Biochem.* i. 47 The correlation of the structures of many classes of natural products by 'biogenetic' schemes has long been a successful intellectual pastime in organic chemistry.

biogenic (bəiŏdʒe·nik), *a.* [f. Biogeny + -ic.]
1. = Biogenetic *a.*
1904 G. S. Hall *Adolescence* I. ii. 55 The great biogenic law that the individual recapitulates the growth stages of his race.

2. Produced by living organisms.
1913 A. W. Grabau *Princ. Stratigraphy* x. 384 Biogenic rocks, or Bioliths as Chr. G. Ehrenberg has termed them, are deposits of organic material, or material formed through the physiological activities of the organisms. 1957 *Encycl. Brit.* XX. 274/2 *Foraminifera* and *Radiolaria* are examples of biogenic sediments. 1958 F. E. Zeuner *Dating Past* (ed. 4) 430 He..upholds the biogenic character of *Corycium*, possibly an alga, from the pre-Cambrian of Finland.

bi:ogeo·graphy. [Bio-.] The science of the geographical distribution of living things, animal and vegetable. Hence **biogeo·grapher**, one versed in biogeography; **biogeogra·phic, -ical** *adjs.*
1895 C. H. T. Townsend in *Geogr. Jrnl.* (1896) VII. 333 (*title*) On the bio-geography of Mexico, Texas..and Arizona, with special reference to the limits of the life areas, and a provisional synopsis of the bio-geographic divisions of America. 1899 H. R. Mill *Internat. Geogr.* 4 Living things possess the world, and the purpose of Biogeography is to trace out the reasons why particular species occupy the regions where they are now found. 1909 Webster, Biogeographic. 1910 H. R. Mill *Guide Geogr. Bks.* 110 This is also the aspect of biogeography usually dealt with in ordinary text-books. 1922 L. R. Dice in *Science* LV. 335/2 Biogeographers have long made use of floral and faunal areas for the classification of distribution. 1928 *Funk's Stand. Dict.*, Biogeographical. 1958 *New Scientist* 10 Apr. 12/2 The southern Pacific area presents several fascinating biogeographical problems. One..is the affinity between the flora and fauna of South America and that of Australasia. 1961 *Ibid.* 15 June 636/3 The biogeographer's task is to explain how this contact [between plants and animals] was achieved, and how it gave rise to the present pattern of distribution.

biograph, *sb.* Add: **2.** An earlier form of cinematograph, introduced from the U.S. (Cf. *animatograph.) Also *attrib.*
1897 *Westm. Gaz.* 19 Mar. 3/3 The exhibition of the American biograph at the Palace Theatre..is another step towards the perfection of the animatograph. *Ibid.*, It cannot be claimed that the biograph is free from vibration, nor do the pictures rest as steadily on the screen as they might. 1898 *Brit. Jrnl. Photogr. Alm.* 655 The exhibition of animated photographs on a larger scale than usual, by the biograph, the invention of an American, Mr. Casler. 1901 W. R. H. Trowbridge *Lett. her Mother to Eliz.* xviii. 89 The Palace [music-hall] in place of the ballet [at the Empire] has a Biograph, which wiggles and makes you feel cross-eyed. 1912 C. N. & A. M. Williamson *Heather Moon* II. iv, They're going to take photographs of a Gretna Green wedding..for a biograph show.

biograph, *v.* Add: **2.** *trans.* To make a biograph (sense *2) of.
1898 *Westm. Gaz.* 27 May 4/2 The finish of the Derby on Wednesday was 'biographed' for the Palace Theatre. 1901 *Ibid.* 14 Jan. 8/2 The experiment of biographing a battle is not absolutely new.

biographize, *v.* Add: Also *intr.*
1897 W. Raleigh *Let.* 13 Mar. (1926) I. 192 If I biographize it should be on a literary gent.

bio·graphy, *v.* [f. the sb.] *trans.* To write the life of; to make the subject of a biography. So **bio·graphied** *ppl. a.* (also *absol.*).
1844 *Fraser's Mag.* XXX. 518/1 If..the biographied be of low birth. 1858 [see Biographying *vbl. sb.*]. 1887 *L'pool Merc.* 10 Mar. 4/5 Captain Burton is to be biographied. 1908 *Fabian News* XIX. 22/2, I cannot help a sneaking regret that G. B. S. has mounted above us into the regions of the biographied. 1909 *Westm. Gaz.* 29 May 5/3 Diane de Poitiers, Lola Montez [etc.] are nearly all much-biographied ladies.

biological, *a.* Add: In various specific combs.: *biological assay* (see *assay *sb.* 6 b); *biological chemistry* = *biochemistry; *biological clock*, an innate mechanism that regulates various cyclic and rhythmic activities of an organism; *biological control* (see quot. 1930); *biological hole*, a cavity in a nuclear reactor designed to permit the placing of animals or plants near the core in order to test the biological effects of radiation; *biological race, species* (see quot. 1916); *biological shield*, a shield round a nuclear reactor or cyclotron as a protection against radiation; *biological spectrum* (see *spectrum); *biological value* [G. *biologische wertigkeit*, K. Thomas 1909, in *Archiv f. Physiologie* 219]: of a protein (see quots.); *biological warfare*, warfare involving the use of toxins, germs, etc., harmful to plants, animals, or human beings; a more comprehensive term than **bacteriological warfare*.
1897 A. B. Griffiths (*title*) Respiratory proteids. Researches in biological chemistry. 1957 *Encycl. Brit.* III. 589/2 The terms 'biochemistry', 'biological chemistry' and 'physiological chemistry' are often used interchangeably. 1955 *Proc. Nat. Acad. Sci.* XLI. 99 Such a biological clock would seem theoretically to be a necessary component in the mechanism of the normally precise rhythms of solar and lunar frequency. 1967 *Guardian* 1 Sept. 16/4 We do not have one biological clock, but many working together, and after a long flight to east or west or after a change of working shift, they may take from one to five days to synchronize again. 1923 *Jrnl. Econ. Entomology* XVI. 506 (*title*) What may we expect from Biological Control? 1926 A. D. Imms in *Ann. Appl. Biol.* XIII. 402 (*title*) The biological control of insects and injurious plants in the Hawaiian Islands. 1930 *Times Lit. Suppl.* 18 Dec. 1078/3 The saving of the copra industry is as complete an example of what is now called 'biological control', the subduing of a living pest by the introduction of a living enemy of the pest, as the saving of the citrus industry of California by the introduction of a lady-beetle from Australia. 1957 *Gloss. Terms Nucl. Sci.* (*A.S.A.*) 19/2 Biological hole. 1916 B. D. Jackson *Gloss. Bot. Terms* (ed. 3) 50/1 *Biological* (*races*, or) *species*, those species which differ only by their physiological behaviour, being morphologically identical. 1941 J. S. Huxley *Uniqueness of Man* v. 157 What are called 'biological races' of parasites adapted to different hosts. 1942 —— *Evolution* v. 166 Groups..remain separate in spite of the complete or almost complete absence of morphological differences. In many such cases (e.g. in 'biological' or 'physiological races'), the allocation of specific rank must be a mere matter of opinion and convenience. 1957 *Financ. Times Ann. Rev. Brit. Ind.* 1 July 66/1 Reinforced concrete also provides the vital barrier against the insidious dangers of radiation. This 'biological shield' must totally enclose the reactor. 1902 H. M. Ward in *Ann. Bot.* XVI. 236 Various names have been proposed for these physiologically, but as yet not morphologically, different varieties...Rostrup names them 'Biological species' (biologische Arten). 1924 H. H. Mitchell in *Jrnl. Biol. Chem.* LVIII. 878 The 'biological value' of a protein, as the term was applied originally by Karl Thomas, referred to the utilization by the body of the products of protein digestion. *Ibid.* 901 The biological value of the protein is taken as the percentage of the absorbed nitrogen that is not eliminated. 1960 A. E. Bender *Dict. Nutrition* 17/2 The biological value of a protein is the amount, when fed under standard conditions, that is retained in the body for synthesis of body protein. It is expressed on the percentage scale; 100% is the perfect protein. 1946 *Life* 18 Nov. 118/2 Biological warfare, using scourges of disease and famine as weapons, is as dreadful as the atomic bomb and far more difficult to control. 1951 'J. Wyndham' *Day of Triffids* ii. 31 The United States Government took the suggestion seriously enough to deny emphatically that it controlled any satellites designed to conduct biological warfare directly upon human beings. 1959 Biological warfare [see **bacteriological warfare*].
B. as *sb.* (After *chemical*, chiefly in *pl.*) A biological product used for therapeutic purposes.
1921 *Lancet* 5 Mar. 497/2 A report..to consider the effective control of therapeutic substances which cannot be tested by direct chemical means. On the analogy of 'chemicals' we propose, for want of a better term, to call these substances 'biologicals'. 1937 *Times* 28 July 16/5 The manufacture of fine chemicals and biologicals. 1957 *New Scientist* 21 Nov. 12/3 In recent years there has been a shortage of these biologicals.

biologism (bəiǫ·lŏdʒiz'm). [f. Biology + -ism.] The interpretation of human life from a strictly biological point of view.
Motley's use of the word (= Electro-biology 2) is *Obs. rare*.
1852 Motley *Corr.* 18 May (1889) I. 143 Whenever a charlatan can't find any to believe in his tricks of mesmerism or biologism, or whatever may be the latest neologism. 1924 *Public Opinion* 25 Jan. 81/3 When we try to force all the facts of human society into frameworks of zoology we are guilty of a biologism. 1926 J. A. Thomson *Man in Light Evol.* 32 A biologism is an attempt to ignore the uniqueness of man by forcing his activities in their entirety into the framework of mammalian physiology...It is a biologism to picture an average man as the slave of his hormones. 1967 *Guardian* 20 Oct. 6/3 Much of the biologism of this and other of his recent books is mere accretion concealing his own experiential progress.
Hence **biologi·stic** *a.*, of or relating to biologism.
1948 K. Davis *Human Society* (1959) xx. 556 So ingrained is the old biologistic approach to human fertility. 1959 *New Scientist* 17 Dec. 1266/1 A Neo-Freudian who has tried to correct Freud's biologistic approach to behaviour by stressing some of its cultural determinants.

bi:olumine·scence. *Biol.* [Bio-.] The emission of light by living organisms; also, the light so produced. Hence **bi:olumine·scent** *a.*
1916 E. N. Harvey in *Amer. Jrnl. Physiol.* XLI. 449 (*title*) Studies on Bioluminescence. 1929 *Biol. Abstr.* (Index) 2366/3 Bioluminescent bacteria. 1936 *Mind* XLV. 69 Bio-luminescent fluids. 1953 J. Ramsbottom *Mushrooms & Toadstools* xiv. 157 The phenomenon of bioluminescence is widespread in the animal kingdom, examples occurring in at least thirty-nine orders. 1954 *Lancet* 25 Dec. 1317/1 Several substances, under the appropriate stimulus, can emit light, and so far the compounds responsible for bioluminescence have not been conclusively identified.

biolysis (bəiǫ·lisis). [f. Bio- + Gr. λύσις dissolution.] Chemical decomposition of organic matter brought about by bacteria, etc.
1897 Scott-Moncrieff in *Prof. Papers Corps R. Engineers* XXIII. 213 (*heading*) Purification of Sewage by Biolysis. 1922 H. E. Babbitt *Sewerage & S. Treatment* xii. 366 The biolysis of sewage is the term applied to the changes through which its organic constituents pass due to the metabolism of bacterial life. *Ibid.* 367 In the biolysis of sewage there is no destruction of matter.

biome (bəi·ōᵘm). *Ecol.* [f. Gr. βίος life + *-ome.] A biotic community of plants and animals; *spec.* such a community in a prehistoric period.
1916 F. E. Clements *Plant Succession* 319 Human evidence of past climates and biotic communities, or biomes, must come to be of very great value. 1918 —— in *Bull. Geol. Soc. Amer.* XXIX. 372 In paleo-ecology the concept of the biome, or biotic community, seems to have peculiar value, as it directs special attention to the causal relations and reactions of the three elements—habitat, plant, and animal. *Ibid.* 374 Climatic changes and biotic movements..are involved in..the appearance of new biomes and the disappearance of preceding ones. 1931 *Ecology* XII. 456 Because the larger and more influent animals tend to range throughout units of largest (formational) size including their seral stages, the biome or biotic formation is the natural ecological unit with some properties which are well illustrated by comparison with an organism. 1935 *Ibid.* XVI. 399 Migratory species..are frequently seasonal influents classed in seasonal communities in two biomes. 1957 G. Clark *Archæol. & Society* (ed. 3) vi. 175 All the forms of life, including man himself, that together constitute the biome. 1957 *Encycl. Brit.* VII. 434/2 A biome is based on the concept that the plants and animals associated together in the same area have a certain unity. The biome is thus a plant-plus-animal formation that is composed of a plant matrix together with all associated animals.

bi:omecha·nics. [Bio-.] The study of the mechanical laws relating to the movement or structure of living organisms. So **bi:omecha·nical** *a.*, pertaining or relating to biomechanics.
1933 Stedman *Med. Dict.* (ed. 12) 134/1 *Biomechanics*, the science of the action of forces, internal or external, on the living body. 1941 J. S. Huxley *Uniqueness of Man* i. 9 Specialization is one-sided progress, and after a..time, reaches a biomechanical limit. The horse stock cannot reduce its digits below one;..feathered flight cannot become aerodynamically more efficient than in existing birds. 1951 *Engineering* 23 Feb. 226/2 Their subject covers what have been described as biotechnology, biomechanics, human engineering,..engineering psychology. 1970 *Physics Bull.* June 264/1 A biomechanical investigation of the human hearing mechanism.
b. *Russ. Theatre.* [cf. Russ. *biomekhanika*.] Also **biome·chanism.** (See quots.)
1924 H. Carter *New Theatre & Cinema in Soviet Russia* vi. 67 He [*sc.* Meierhold] established a studio of bio-mechanics [in 1921]. *Ibid.* 70 The principles of bio-mechanics..became systematically applied by him in the R.S.F.S.R. theatre in Petrograd from 1918 to 1922. *Ibid.*, The laws of bio-mechanics are founded on the study of the physiological construction of man. The system aims to produce men who understand the mechanism and laws of their structure and can, therefore, use it perfectly. It has established a principle of analysis by which each movement of the body can be differentiated and made fully expressive. 1930 P. England tr. *Fulöp-Miller & Gregor's Russ. Theatre* 68 The classical example of bio-mechanism is the comedy *The Magnificent Cuckold*. In this the movements of the performers were so standardized that they seemed to obey some geometric law... Large revolving wheels were also employed, in order to register the various emotions that prevailed from time to time in the breasts of the actors.

biometric (bəiŏme·trik), *a.* and *sb.* [f. Bio- + Metric *a.*¹]
A. *adj.* Of or pertaining to biometry. So **biome·trical** *a.*
1901 *Biometrika* I. 2 The collection of biometric data. *Ibid.* 6 Biometric workers. 1902 *Encycl. Brit.* XXVIII. 344/2 Biometric investigation. *Ibid.*, Biometrical inquiry. 1936 *Discovery* Jan. 32/2 The application of biometric methods to the study of the composition of ethnic groups. 1937 R. H. Lowie *Hist. Ethnol. Theory* ix. 130 He has done a full-fledged professional's work in physical anthropology ..investigating growth by biometrical techniques.
B. *sb. pl.* = *biometry 2.
1902 in *Encycl. Brit.* XXVIII. 344/1. 1923 W. M. Feldman *Biomath.* xxi. 327 Biometrics is the application of modern statistical methods to the measurements of biological (variable) objects. 1935 Huxley & Haddon *We Europeans* ii. 61 The work of Karl Pearson and his school in the study of statistical methods as applied to biological and anthropological problems ('Biometrics').

biometrician (bəi:ŏmetri·ʃən). [f. *biometric *a.* + -ian: cf. **metrician*.] One who is versed in biometry; one who applies statistics to the problems of biology, esp. that of variation. So **biome·tricist.**
1901 F. Galton in *Biometrika* I. 10 Biology could soon be raised to the status of a more exact science..., if each of many biometricians would thoroughly work out his own particular plot. 1902 *Encycl. Brit.* XXVIII. 344/2 The biometricist demands statistics to show the range of the variations and the mode of their distribution. 1914 K. Pearson (*title*) Tables for Statisticians and Biometricians. 1922 W. R. Inge *Outspoken Ess.* 263 It seems to be established by the biometricians that children who are born after their fathers are fifty seldom attain distinction. 1948 *Mind* LVII. 294 They [*sc.* two cats] certainly differ slightly in various measurements, and it is the business of a biometrician to describe such differences.

biometry (bəiǫ·metri). [See Bio- and -metry.] †**1.** (See in Dict. under Bio-.) *Obs.*
1831 Whewell *Let.* 12 Nov. (1876) II. 135 By the way there is a problem in *Biometry* (if you choose to call your calculations on lives by a Greek name) which may perhaps be included. 1875 *Med. Rec.* X. 481 Biometry: its relations to the practice of medicine. 1881 *Index-Catal. Libr. Surg.-Gen. U.S.* II. 66/2 Biometry, *see* anthropometry; longevity.
2. The application of mathematics to biology, esp. the study of resemblances between living things by statistical methods.
1901 F. Galton in *Biometrika* I. 9 The primary object of Biometry is to afford material that shall be exact enough for the discovery of incipient changes in evolution which are too small to be otherwise apparent. 1927 Haldane & Huxley *Anim. Biol.* ii. 72 When we take the averages of large numbers these irregularities become smoothed out, and we find a strong average resemblance, due to heredity, between parent and offspring, or between brothers and sisters. The science of biometry deals with studies of this sort. 1942 *Nature* 18 Apr. 428/2 Thus with Pearson defending biometry and Bateson advocating Mendelism a feud arose.

biomorph (bəi·ŏmǭɹf). [f. Bio- + Gr. μορφή form.] A decorative form representing a living object. Hence **biomo·rphic** *a.*
1895 Haddon *Evol. Art* 126 The biomorph is the representation of anything living in contradistinction to the skeuomorph. *Ibid.* 188 Biomorphic Pottery. 1954 *Archit. Rev.* CXV. 272/2 Leaden echoes of lively biomorphs with a sinister absence of meaning. 1959 P. & L. Murray *Dict. Art* 27 *Biomorphic art* is a form of abstract art which purports to take its abstract forms from living organisms rather than from the geometrical basis of such abstract movements as constructivism. 1960 H. Read *Forms of Things Unknown* III. x. 165 Some are biomorphic, reminiscent of cellular structures as revealed by the microscope.

bionics (bəi,ǫ·niks). [f. Bio- + *elec-tr)onics.] (See quots.)
1960 J. E. Keto in *Proc. Nat. Aeronaut. Electronics Conference* 218 The title of this session is Bionics. This is a new term referring to a relatively new but rapidly expanding area of activity—the study of systems which function after the manner of or in a manner characteristic of or resembling living systems. 1962 *New Scientist* 29 Mar. 737/3 Much of the emphasis in 'bionics'—study of animals by engineers in an attempt to learn 'dodges' from nature—has been on sense organs and nervous systems. 1966 *Ibid.* 10 Mar. 626/1 Engineering bionics is a relatively new science aimed at making machines which behave like living systems.

bionomic (bəiŏnǫ·mik), *a.* and *sb.* [f. Bio-, after Economic.] **A.** *adj.* Of or pertaining to the conditions under which an organism lives in its natural habitat; of or pertaining to bionomics (see B).
1899 J. A. Thomson in H. R. Mill *Internat. Geogr.* 85 Bionomic Relations. 1920 *Rep. Brit. Assoc. Advancem. Sci.* 1919 207 The formation of a great collection of butterflies—a most valuable asset for bionomic research. 1956 *Nature* 25 Feb. 359/1 A number of his systematic papers contain bionomic and ecological observations.
B. *sb. pl.* The branch of biology which deals with the mode of life of organisms in their natural habitat, their adaptation to their surroundings, etc.; ecology.
1888 E. R. Lankester in *Encycl. Brit.* XXIV. 803/2 Bionomics..including thremmatology, or the science of breeding. 1895 *Naturalist* 337 The old out-door ornithologists are to-day's sportsmen, who study the bionomics of their quarry that they may outwit them. 1902 *Westm. Gaz.* 6 Dec. 8/1 The bionomics of the North Sea; in other words, the relationships of the various forms of life to one another and to their surroundings. 1947 *New Biol.* III. 57 This study of the mode of life of any animal and of its relationships to its surroundings and to other animals is called the study of the bionomics of the animal.

biono·mical, *a.* Of or pertaining to bionomics (see prec. B); ecological.
1902 *Encycl. Brit.* XXVIII. 344/1 Bionomical investigators. 1909 *Athenæum* 17 July 72/1 The bionomical stores contained in its pages. 1930 *Jrnl. Ecol.* XVIII. 81 (*title*) Bionomical investigation of English mosquito larvae with special reference to their algal food.

biophor (bəi·ŏfǭɹ). *Biol.* Also **-phore.** [G. *biophor*, f. Gr. βίος life + -φορος -bearing.] In Weismann's theory of heredity: a supposed ultimate unit of living protoplasm.
1893 Parker & Rönnfeldt tr. *Weismann's Germ-Plasm* I. i. 40 We have to imagine that..protoplasm..consists of groups of molecules, each of which is composed of different kinds of chemical molecules. I shall call these units the 'bearers of vitality' ('Lebensträger') or 'biophors', because they are the smallest units which exhibit the primary vital forces, viz. assimilation and metabolism, growth, and multiplication by fission. 1894 *Nat. Sci.* Sept. 185 His [*sc.* Weismann's] germ-plasm is made up of primary units, the biophores, which are built up into determinants, a determinant existing for each independently varying cell or group of cells in the adult organism. 1959 A. Hughes *Hist. Cytol.* iv. 83 His [*sc.* Weismann's] ideas on development..are as outdated as the peculiar terminology in which the postulated structure of the germ-plasm was expressed. The 'Biophor' has long been extinct, and the 'Id' has acquired a new meaning in another science.

biophy·sics. The science dealing with the mechanical and electrical properties of the parts of living organisms. Hence **biophy·sical** *a.*, of or pertaining to biophysics; **biophy·sicist,** one who is versed in biophysics.
1892 K. Pearson *Gram. Sci.* 470 This branch of science which endeavours to show that the facts of Biology..constitute particular cases of general physical laws has been termed *Ætiology*. It would perhaps be better to call it *Bio-physics*. 1913 L. J. Henderson *Fitness Environm.* 310 While biophysicists like Professor Schäfer follow Spencer in assuming a gradual evolution of the organic from the inorganic, biochemists are more than ever unable to perceive how such a process is possible. 1926 *Encycl. Brit. Suppl.* II. 675/1 His main work is on the fundamental problem of biophysics—the stimulation of living matter. 1926 J. Pryde *Rec. Advances Biochem.* v. 137 (*heading*) Biophysical Aspects of the Lipides and their Significance in Biology. 1958 A. R. Radcliffe-Brown *Meth. Soc. Anthrop.* I. iv. 103 The nature of multicellular organisms is determined by the nature of the living cell which it is the business of the cytologist, the biochemist and the biophysicist to study. 1959 *New Biol.* XXX. 112 The study of the forces which mould a developing egg..is a comparatively new branch of biophysics. 1961 *Lancet* 22 July 190/2 Our American colleagues are fully aware of the importance of integrating basic biochemical, biophysical, and physiological effort.

biopoesis (bəiopōᵘ,i·sis). [f. Bio- + Gr. ποίησις making.] The (hypothetical) origination or evolution of living or lifelike structures from lifeless matter; abiogenesis. Hence **biopoe·ic** *a.*, pertaining to, causing, or resulting from biopoesis.
1953 N. W. Pirie in *Discovery* Aug. 239/1 Biopoesis (the cognate adjective is biopoeic) seems a more suitable word than Huxley's abiogenesis for the making of life de novo. Words ending in genesis carry a sense of generation as in animals; the making of like by like. *Ibid.*, Perhaps..there is no need to postulate any special biopoeic agency at all. 1967 J. D. Bernal *Orig. Life* viii. 137 The modern view of biopoesis includes elements both of deterministic character and of the older teleological character.

biopsy (bəi·ǫpsi, bəi,ǫ·psi). [ad. F. *biopsie* (Besnier), f. Gr. βίος life + ὄψις sight. After Necropsy.] Examination of tissue, etc., removed from the living body; also, the removal of such tissue.
1895 E. Besnier in J. J. Pringle *Pict. Atlas Skin Dis.* 88 Clinical teaching will now find in a 'biopsy' a valuable means of confirming or invalidating a dubious diagnosis. 1902 *Brit. Jrnl. Dermatol.* XIV. 56 A certain diagnosis was probably impossible without a biopsy. 1910 *Practitioner* Apr. 453 More definite evidence of the existence of gastrostaxis, apart from gastric ulcer, is obtained from biopsies and necropsies. 1955 *Brit. Med. Jrnl.* 31 Dec. 1590/1 (*title*) Serial biopsy in ulcerative colitis. 1966 *New Statesman* 11 Mar. 328/1 This is called a biopsy, wherein a cone of the infected cervical cell-tissue is surgically removed.

bios (bəi·ǒs). *Biochem.* [Gr. βίος life.] The name given by E. Wildiers to the factor or factors stimulating the growth of yeast.
[1901 E. Wildiers in *Cellule* XVIII. 322 Qu'il nous soit permis d'adopter un nom commode pour désigner cette substance inconnue..nommons-la 'bios'.] 1907 *Chem. Abstr.* 1311 The active principle of the 'Bios' is contained in lecithin. 1919 *Jrnl. Biol. Chem.* XXXVIII. 466 The possibility presented itself that the substance which Wildiers called 'bios' might be the same as the vitamine known to be contained in yeast. 1936, 1937 [see *biotin]. 1953 J. Ramsbottom *Mushrooms & Toadstools* viii. 81 It was found that bios is a complex substance and components bios I, bios II, [etc.]..were recognised; later they were identified with known growth substances.

bioscope (bəi·ŏskōᵘp). [See Bio- and -scope.] †**1.** A view or survey of life. *Obs.*
1812 G. Penn (*title*) The Bioscope, or Dial of Life, explained. 1824 [see under Bio-].
2. An earlier form of cinematograph (cf. *biograph *sb.* 2); retained in South Africa as the usual term for a cinema or a moving film.
1897 *Wall's Dict. Photogr.* (ed. 7) 75 Bioscope. See Zoetrope. 1901 *Daily Chron.* 27 Dec. 6/1 The Bioscope views are not so much incidents of the day as exciting adventures. 1904 J. Joyce *Let.* 28 Dec. (1966) II. 75 The other evening we went to a bioscope. There were a series of pictures about betrayed Gretchen. 1925 *Public Opinion* 4 Sept. 234/3 No use to try and blame your poor digestion Nor yet the bioscope's unnerving scene. 1968 D. Braithwaite *Fairground Archit.* 21 In the early 1900's there were 'Bioscope Shows'.
1908 *Cape Times* 14 Jan. 6 The Electro-Chrono ('King of Bioscopes'). A Machine for exhibiting Living Pictures of exceptional brilliance. *Ibid.* 22 Jan 6 The Bioscope will project 'The Boxer Rebellion in China'. 1915 *The Cape* I. no. 10. p. 11 They extended to her facetious invitations to the theatre and the bioscope. 1919 S. G. Millin *Dark River* xxii. § 2. 148 The Novelty was one of the Cape Town bioscopes that boasted a small orchestra and occasional variety turns. 1935 *Quandary* Sept. 20 When a bioscope company comes to the village. 1950 L. G. Green *Land of Afternoon* i. 12 He denounced newspapers and bioscopes. 1958 *Observer* 12 Jan. 15/4, I saw the film..on the quivering screen of a Central African bioscope.
Hence **biosco·pic** *a.*
1903 *Athenæum* 12 Dec. 799/1 The utility of the bioscopic camera in zoological work. 1928 *Daily Tel.* 18 Sept. 6 It is excellent to have bioscopic presentations of the eminent, but Herr Ludwig's film is too long.

biose (bəi·ōᵘs). *Chem.* [f. Bi-² + -ose².] (See quots.)
1892 E. F. Smith tr. *Richter's Org. Chem.* (ed. 2) 507 Only the disaccharides of the hexoses, $C_6H_{12}O_6$, are known. They consist of two molecules of the glucoses or monoses..and therefore are called bioses. 1903 *Jrnl. Chem. Soc.* LXXXIV. I. 713 By the action of alcoholic hydrogen chloride on bioses, hydrolysis occurred and glucosides of the monoses were formed. 1916 A. P. Mathews *Physiol. Chem.* 18 The monosaccharides are in their turn classified by the number of carbon atoms, or more properly by the aldehyde, ketone and alcohol groups they contain into bioses, trioses, tetroses, pentoses, hexoses, heptoses, octoses, nonoses, etc.

biosphere (bəi·ŏsfīəɹ). [ad. G. *biosphäre* (E. Suess 1875, *Entstehung d. Alpen* 159), f. Bio- + Sphere.] The regions of the earth's crust and atmosphere that are occupied by living organisms; occas., the living organisms themselves; also *transf.*
1899 H. R. Mill *Internat. Geogr.* 4 Some geographers even bring in the layer of living matter to complete four parts of the physical globe—the lithosphere, hydrosphere, atmosphere and biosphere. 1909 H. B. C. Sollas tr. *Suess's Face of Earth* IV. 637 It brings with it the idea of a biosphere, which assigns to life a place above the lithosphere. 1931 J. A. Thomson in W. Rose *Outl. Mod. Knowl.* vi. 204 The Biosphere of living organisms, both plants and animals. 1949 W. C. Allee et al. *Princ. Animal Ecol.* xxxv. 729 Life and habitat are integrated into an evolving ecosystem, ultimately incorporating the entire biosphere of the earth. 1957 P. Dansereau *Biogeography* iii. 125 The biosphere is that part of the earth's crust and atmosphere which is favorable to at least some form of life. 1962 F. I. Ordway et al. *Basic Astronautics* vi. 297 A mature civilization..would possibly create an artificial biosphere at a comfortable distance from its star to sustain a greatly expanded population. 1968 *Guardian* 16 Sept. 1/8 The biosphere, i.e. living things and the milieu in which they exist.

biota (bəi,ōᵘ·tă). *Ecol.* [mod.L., cf. Gr. βιοτή life.] A collective term for the animal and plant life of a region.
1901 L. Stejneger in *Amer. Naturalist* XXXV. 89 The author, like many other writers..has felt the need of a comprehensive term to include both fauna and flora which will..designate the total of animal and plant life of a given region or period..also any treatise upon the animals and plants of any geographical area... As such a term I would suggest Biota. 1939 *Nature* 25 Mar. 504/2 In Natal, the population [of animals and algæ] consists partly of 'Cape' species, but a high percentage of the biota belongs to the great tropical Indo-West-Pacific group, and includes reef corals and many species which characteristically inhabit coral reefs. 1957 *Ibid.* 4 May 892/2 The need for planned observational work on the airborne biota still remains.

bi:otechno·logy. [Bio-.] The branch of technology concerned with the development and exploitation of machines in relation to the various needs of human beings. Hence **bi:otechnolo·gical** *a.*
1947 *Science* CV. 217/2 Hours of work, on-the-job feeding, rest periods, etc. are also phases of the physiology of work which form an important part of a comprehensive biotechnology. *Ibid.* 218/2 Biotechnological functions either are being performed by inadequately or narrowly trained individuals or are neglected entirely. 1969 *Sci. Jrnl.* June 50/1 Biotechnology is just as concerned with the provision of tools for medical research as with the development of equipment for medical service.

biotic, *a.* Restrict *rare* to sense 1 and add: **2.** (Further examples.) Also, pertaining to, produced, or influenced by living organisms, esp. in their ecological relations.
[Reverse the order of quots. 1874, 1847 in Dict. and for date 1874 substitute 1868.] 1907 F. E. Clements *Plant Physiol. & Ecol.* i. 5 Cases of abnormal response are due to biotic factors, particularly parasitic fungi and insects. 1923 A. G. Tansley *Pract. Plant Ecol.* iv. 48 Man is constantly stopping or modifying the development... Where he has introduced a more or less permanent modifying factor or set of factors, we have biotic (anthropogenic) climaxes. 1949 *New Biol.* VI. 54 British grasslands represent neither a climatic nor an edaphic climax, but are a *biotic climax*. They are maintained as grassland *per se* under the influence exerted upon them by the biota, i.e. by living organisms, in particular by man and his domestic animals.

biotin (bəi·ŏtin). *Biochem.* [a. G. *biotin* (F. Kögl & B. Tönnis 1936, in *Zeitschr. physiol. Chem.* CCXLII. 50), f. Gr. βίοτος (βίος) life (see *bios) + -in¹.] A crystalline growth vitamin, $C_{10}H_{16}O_3N_2S$, occurring esp. in yeast and egg-yolk, that is inactivated by combination with avidin; vitamin H.
1936 *Chem. Abstr.* XXX. 7624 (*heading*) The bios problem. Isolation of crystalline biotin from egg yolk. 1937 *Nature* 24 July 161/1 'Biotin', a sulphur-containing substance of great potency which may be the chief component of 'bios', originally described by Wildiers as a necessary growth factor for yeast. Biotin has also been shown to have an effect on the growth of higher plants. 1944 *Lancet* 24 June 826/2 The egg-white injury demonstrated in rats is not a direct toxic effect but the result of biotin deficiency.

biotope (bəi·ŏtōᵘp). *Ecol.* Also **biotop.** [ad. G. *biotop*, f. Gr. τόπος place.] The smallest subdivision of a habitat, characterized by a high degree of uniformity in its environmental conditions and in its plant and animal life.
1927 *Biol. Abstr.* 593/1 A true association is the population of a biotop. Distinction is made between biotop and habitat. 1936 *Proc. Prehist. Soc.* II. 64 Taken as a whole this assemblage suggests a tundra biotope. 1937 Allee & Schmidt tr. *Hesse's Ecol. Animal Geogr.* ix. 135 The primary topographic unit is the 'niche' or 'biotope'. Such a unit is an area of which the principal habitat conditions and the living forms which are adapted to them are uniform. 1937 [see *biochore]. 1962 H. Hanson *Dict. Ecol.* 55 *Biotope*, the smallest geographic unit of a habitat:..e.g., a decaying stump, a sandy beach.

biotype (bəi·ŏtəip). *Biol.* [f. Bio- + Type *sb.*¹] A group of organisms having a common genotype. Hence **bi:otypo·logy** [-ology], the study of biotypes.
1906 W. Johannsen in *Rep. Third Internat. Conference Genetics* 98 It remains quite uncertain whether the numbers

..contain a multitude, or a few, or only one single 'sort' of organism—'biotypes' as I have called them. *Ibid.* 111 To elucidate the origin and conduct of the veritable types of organism, the 'biotypes'. 1937 *Nature* 11 Sept. 471/2 One section of the International Congress was devoted to matters coming under such heads as biometry, biotypology and heredity. 1955 *New Biol.* XVIII. 114 Evidently the East Anglian colony [of *Nymphæa alba*] contains a far greater range of biotypes than this particular Irish one. 1957 *Encycl. Brit.* XXI. 177/2 *Biotype*..is used for a population of individuals with identical genotypical constitution under similar conditions. .. Parallel mutation may possibly account for some biotypes.

bi·-pack. [Bi-².] In colour photography, a pack of two sensitive plates or films used to obtain colour separation.
1924 *Brit. Jrnl. Photogr.* Suppl., Sept. 35/2 The staining of the front member of a bi-pack with aniline orange. 1935 *Discovery* July 192/1 Using two films in the form known as 'bi-pack'.

biparental, *a.* [Bi-².] Of, pertaining to, or derived from, two parents.
1900 K. Pearson *Gram. Sci.* (ed. 2) 469 These are the questions of bi-parental heredity. *Ibid.* 473 The variability of the individual makes itself felt not only in bi-parental reproduction but in autogamic and parthenogenetic reproduction. 1903 *Biometrika* Nov. 396 The existence of this assortative mating most substantially modifies the form of biparental inheritance. 1929 B. Russell *Marriage & Morals* xvi. 187 Easy divorce..must be regarded as a transitional stage on the way from the bi-parental to the purely maternal family. 1955 *New Biol.* XIX. 7 Biparental reproduction is the main cause of variation in most species.

bipa:rtisa·n, *a.* [Bi-².] Of, representing, or composed of members of, two (political or other) parties.
1909 in Webster. 1920 *Glasgow Herald* 13 Feb. 10 The 14 Republican reservations, as modified by the Bipartisan Conference. 1947 *Times* 13 Dec. 5/2 That rare thing in American public life a bi-partisan policy. 1959 *Ann. Reg. 1958* 32 Mr. Gaitskell..showed a determination to cling as long as possible to a bipartisan foreign policy.
Hence **bipartisanship.**
1950 *Congressional Dig.* Feb. 36 Bipartisanship in the balance. 1952 *Oxf. Mag.* 27 Nov. 108/1 This House sees no sufficient cause to abandon bipartisanship in British foreign policy.

bipartite, *a.* Add: **1. d.** *Math.* Of a curve (see quot. 1879).
1858 Cayley *Math. Papers* (1889) II. 527 Bipartite binary quantics. 1869 *Ibid.* (1893) VI. 464 The quantic is unipartite, bipartite, tripartite, &c., according as the number of sets is one, two, three, &c. 1879 G. Salmon *Higher Plane Curves* (ed. 3) 168 We shall then call the curve we have been considering a bipartite curve, as consisting of two distinct continuous series of points.

bi·-pa:rty. Used *attrib.* = consisting of, or representing, two (political or other) parties.
1898 *Westm. Gaz.* 17 Feb. 3/1 British institutions presuppose the British character, with..its faithful adherence to the bi-party system. 1900 *Ibid.* 2 July 3/1 Philosophers cling with the same pathetic insistence as members of Parliament to their traditional bi-party system. 1926 *Contemp. Rev.* Apr. 454 The bi-party system [in Canada].

bipaschal (bəipɑ·skăl), *a.* *Theol.* [f. Bi-² + Paschal *a.*] Including two consecutive passover feasts: applied to the view that limits Christ's public ministry to a little over one year.
1883 Schaff *Hist. Ch., Apostol. Chr.* I. 131 The bipaschal scheme confines the public ministry to one year and a few weeks or months. 1893 Broadus *Harm. Gospels* 243 The Bi-paschal theory makes the time of the public life of Jesus one year, allowing only two Passovers to the Gospel of John. 1908 J. Hastings *Dict. Christ* II. 185/1 Between these two Passovers lay the whole ministry, hence this theory is called the *bipaschal* view.

bipedal, *a.* **2.** Add to def.: *spec.* designating a reptile that uses its two hind feet for walking or running; also denoting this method of movement. Hence **bipe·dalism**; **bipe·dally** *adv.* (Add further examples, substituting the 1870 quot. for Dict.'s 1830 example.)
1854 *Fraser's Mag.* Sept. 273/2 Clergymen may still be found..in an easy association with their flocks, quadrupedal and bipedal. 1870 T. H. Huxley in *Q. Jrnl. Geol. Soc.* XXVI. p. xlviii, These Bird-reptiles or Reptile-birds were more or less completely bipedal. 1896 *Nature* 27 Feb. 397/1 The absurdly grotesque appearances these lizards presented when progressing in this bipedal fashion. 1898 *Ibid.* 10 Feb. 341/1 The Mexican Iguanoid Lizard..also runs bipedally. *Ibid.* 341/2 Several varieties of these bipedal lizards are now on view in the Reptile House. 1907 *Proc. Zool. Soc.* Feb. 231 The most primitive Dinosaurs are bipedal in their habits. *Ibid.* 234 A bipedal animal never could or did develop a patagium without giving up bipedalism. 1908 *Westm. Gaz.* 17 Oct. 5/1 Having once lost the prize in the overture With his bipedal rival. 1969 *Times* 6 Feb. 10/8 Palaeo-anthropologists still do not know what factors influenced the adoption of bipedalism.

bipe·rsonal, *a.* Of the godhead: existing in two persons; also, relating to this system of belief, binitarian.
1928 K. E. Kirk in A. E. J. Rawlinson *Ess. Trin. & Incarn.* 162 'Bipersonal', 'ditheistic', or 'dyarchian' are occasionally employed also as synonyms [of binitarian]. *Ibid.* 196 The cosmogony of *Poimandres* is fantastically pluralist..; its religion is strictly bipersonal.

bipha·sic, *a.* [Bi-².] Having two phases.
1909 in *Cent. Dict.* Suppl. 1956 *Nature* 3 Mar. 431/2 Fluoride changed only the slope, but not the linearity, except for sodium, which developed a biphasic curve. 1961 *Lancet* 23 Sept. 684/2 Case 5 showed a biphasic neurological illness. 1964 M. Hynes *Med. Bacteriol.* (ed. 8) xxv. 395 When disease does appear, it is biphasic, a viræmia with influenza-like symptoms subsiding to be followed by signs of CNS involvement.

biplanar (bəiplē¹·năɹ), *a.* (and *sb.*). *Math.* [f. next + -ar¹: cf. *planar*.] Lying or situated in two tangent planes. Also *sb.*, a biplanar node.
1849 G. Salmon in *Cambr. & Dublin Math. Jrnl.* IV. 257 The two tangent planes at the biplanar point. *Ibid.*, One [plane] through both conical points (1 × 4), one through the biplanar alone (3). 1927 H. P. Hudson *Cremona Transformations* xii. 238 With a d[ouble] p[oint], the parts are both planes, and the singularity is called a binode or biplanar point.

biplane (bəi·plē¹n). [f. Bi-² + Plane *sb.*³] **1.** *Math.* Each of the pair of planes tangential to a surface at a binode.
1870 Cayley in *Phil. Trans. R. Soc.* CLIX. 206 The binode ..is a biplanar node, where instead of the proper quadricone we have two planes; these may be called the biplanes, and their line of intersection, the edge of the binode. The biplanes form a plane-pair. 1905 W. H. Blythe *Models Cubic Surfaces* p. xii, When the biplanes coincide the node is said to be uniplanar.
2. An aeroplane having two 'planes' or main supporting surfaces, one above the other.
1874 D. S. Brown in *Ann. Rep. Aeronaut. Soc.* 14 The bi-plane when elevated will proceed for a considerable distance. 1908 *Westm. Gaz.* 14 Aug. 10/2 Mr. Wilbur Wright is busy experimenting with his greatly successful bi-plane. 1910 R. Ferris *How it Flies* 12 In March [1908], Delagrange flew in a Voisin biplane 453 feet. 1911 Grahame-White & Harper *Aeroplane* 17 He [*sc.* Lilienthal] abandoned the monoplane type of glider in favour of a biplane. *Ibid.* 18 One of Chanute's biplane gliders was fitted with a stern rudder. 1957 P. Kemp *Mine were of Trouble* iv. 70 A formation of Fiat CR 42 bi-planes.

biplate (bəi·plē¹t). *Optics.* [Bi-².] A combination of two glass plates used in polariscopic observations.
1890 T. Preston *Th. Light* 121 Bi-plates.—A beam of light may be subdivided by refraction through two plates, of the same nature and equal thickness, placed at an angle. 1904 A. Schuster *Th. Optics* iv. 80 The 'bi-plate'..serves either to separate or to bring together two parallel beams of light. It consists of two plane parallel plates of glass cemented together at an angle.

bipod (bəi·pǫd). [f. Bi-² + -*pod*, after Tripod.] A two-legged support or structure, esp. one for a light machine-gun. Also *attrib.*
1938 *Jrnl. R. Aeronaut. Soc.* XLII. 868 Later I introduced a gaff mainsail..and still later (1930) I rigged a bipod mast. 1940 *War Illustr.* 2 Feb. 41 It [*sc.* the Bren gun] may be mounted either on a tripod or a bipod, or fired direct from the shoulder. 1950 *Archit. Rev.* CVII. 262/2 The first floor consists of a series of boxes suspended from steel bipods, and connected by corridors in the form of enclosed bridges.

bipolar, *a.* Add: **b.** Of, pertaining to, or occurring in both polar regions.
1896 A. E. Ortmann in *Proc. Acad. Nat. Sci. Philad.* 1895 191 The examples of bipolar distribution of crustacea. 1951 *Jrnl. Ecol.* XXXIX. 209 The term 'bipolar' is not precise. What is intended is taxa with widely discontinuous ranges to the north and to the south of the tropics.

bipolarity. Add: Also *transf.* and *fig.*
1917 C. R. Payne tr. *Pfister's Psychoanal. Method* I. xii. 361 As a result of the 'bipolarity of all psychic phenomena', each of the two possible interpretations..may be correct. 1955 M. Beloff *Foreign Policy* 9 We have been taught to view things in the light of the rivalry between two powers and two powers only..'bipolarity', as it is called.
2. The occurrence of the same species in each of the polar regions but not in the intervening zones (see also quot. 1959).
1896 A. E. Ortmann in *Proc. Acad. Nat. Sci. Philad.* 1895 191 Neither in this species is..a bipolarity of the genus probable. 1959 *New Biol.* XXIX. 117 Bipolarity..has undergone some change of meaning. .. In a very short period of time, the sharpness of its meaning was lost and bipolarity came to include those cases where organisms occurred in the north and south temperate zones but were absent from the tropics. The term underwent still further changes and was used to cover not only species, but genera and even higher classification categories provided they had a discontinuous distribution with northern and southern forms.

bipp (bip). *Med.* Also **bip, B.I.P.P.** [f. the initials of *B*ismuth *I*odoform *P*araffin *P*aste.] A paste containing bismuth subnitrate, iodoform, and (sometimes) paraffin, used as a dressing for wounds. Also *attrib.* Hence as *v. trans.*, to treat or cover with bipp.
1916 R. Morison in *Lancet* 12 Aug. 268 The paste: Bismuth subnitrate, 1 oz. by weight; iodoform, 2 oz. by weight; paraffin liq. q.s. to make thick paste. I suggest for this the name 'Bipp'. 1917 *Lancet* 10 Mar. 378/1 Excellent results with 'Bipp' in compound comminuted fractures. *Ibid.*, I 'Bipped' the sloughing surface and renewed the 'Bipp' every ten days. 1933 *Ibid.* 4 Feb. 250/1 The wound was swabbed with ether and smeared with B.I.P.P., the surplus being removed on a dry gauze swab. 1939 *Times* 11 Jan. 12/3 The 'Bip' treatment of wounds.

bi·prism. [Bi-².] A glass prism with a refracting angle of nearly 180 degrees, used in observing the interference of light.
1884 A. Daniell *Princ. Physics* 486 Such a beam may be divided in two parts..by refraction through a biprism. 1890 T. Preston *Th. Light* 119 Fresnel's Bi-prism... Let CDE represent a glass prism with a very obtuse angle E... The whole prism is as if made up of two prisms CE and DE of very small angle (at C and D) placed base to base at E, and hence the name bi-prism. 1904 A. Schuster *Th. Optics* iv. 60 (*heading*) Fresnel's Biprism.

bipunctual, *a.* Add: **2.** *Math.* Having certain specified properties in relation to two points.
1878 *Amer. Jrnl. Math.* I. 152 As long as the direction of reference remains fixed, the only change that can be made in a system of bipunctual coordinates is an alteration in the position of the initials. 1922 H. F. Baker *Princ. Geom.* II. iii. 123 To speak of the latter conic as being bipunctual.

biquartz (bəi·kwǭɹts). *Optics.* [Bi-².] A double quartz plate used for detecting polarization.
1878 Lockyer *Stargazing* 452 There is also another instrument for detecting polarization which is perhaps more commonly used than the biquartz. 1884 A. Daniell *Princ. Physics* 510 A double-quartz plate, or Biquartz. 1890 T. Preston *Th. Light* 373 The biquartz..consists of two semicircular plates of quartz placed in juxtaposition. 1911 *Encycl. Brit.* XXI. 938/1 The chief features of Soleil's saccharimeter are the biquartz and the compensator.

biquaternion (bəikwǫtə̄·ɹniǫn). *Math.* [Bi-².] **1.** A quaternion with complex coefficients.
1852 *Rep. Brit. Assoc.* (1853) Notices & Abstracts 2 On Biquaternions. By Sir William R. Hamilton. 1853 W. R. Hamilton *Lect. Quatern.* p. lxviii, A biquaternion may be considered generally as the sum of a biscalar and a bivector. 1905 C. J. Joly *Man. Quaternions* ii. 21 The tensor of a biquaternion may vanish.
2. (See quots.)
1873 W. K. Clifford in *Proc. London Math. Soc.* IV. 386 In this way the ratio $\frac{\gamma+\omega\delta}{\alpha+\omega\beta}$ is expressed in the form $q+\omega r$, which expression may conveniently be called a biquaternion. 1876 —— *Math. Papers* (1882) 394 The ratio of two motors is a quantity of the form..$s+\omega t$ (if $2s = p+q$, $t = 2p-q$), where p, q, s, t are quaternions. This combination of two quaternions I have called a Biquaternion. 1910 *Encycl. Brit.* I. 614/2 Clifford's biquaternions are quantities $\xi q+\eta r$, where q, r are quaternions, and ξ, η are symbols (commutative with quaternions) obeying the laws $\xi^2 = \xi$, $\eta^2 = \eta$, $\xi\eta = \eta\xi = 0$.

biqui·nary, *a.* [Bi-².] Designating a type of code for use in computers or the like (see quot. 1962).
1946 O. Cesareo in *Bell Lab. Rec.* XXIV. 457/2 In bi-quinary notation, a digit is always represented by two numbers: a quinary part consisting of one of the digits 0 to 4 inclusive, and a binary part consisting of either 00 or 5. 1949 D. R. Hartree *Calculating Instrum.* vii. 80 The machines work in a scale of ten, with number [*sic*] expressed in the so-called 'bi-quinary' system of coding decimal digits. 1962 *Gloss. Terms Autom. Data Processing* (*B.S.I.*) 21 *Biquinary code*, a code in which a decimal numeral n is represented by the pair of numbers a and b where: $n = 5a+b$ in which $a = 0$ or 1 and $b = 0, 1, 2, 3$ or 4.

bi-ra·cial, *a.* [Bi-².] Concerning or containing (members of) two races.
1922 *Contemp. Rev.* Feb. 243 Belgium is bi-lingual and bi-racial. 1960 *Guardian* 11 Apr. 1/7 Decent business men form bi-racial committees.

biramose (bəirē¹·mōᵘs), *a.* *Zool.* [f. Bi-² + Ramose *a.*] = Biramous *a.*
1877 H. Woodward in *Encycl. Brit.* VI. 652/1 Behind these are six pairs of powerful biramose natatory feet. 1888 Rolleston & Jackson *Anim. Life* 539 With a bivalve shell and large natatory biramose second antennæ.

birational (bəiræ·ʃənăl), *a.* *Math.* [f. Bi-² + Rational *a.*] Applied to a transformation in which each of two sets of variables is expressed rationally in terms of the other, or to a correspondence between two sets of variables that can be so related. Hence **bira·tionally** *adv.*
1894 C. A. Scott *Introd. Plane Analyt. Geom.* xi. 240 The transformations hitherto considered are birational transformations of the whole plane. 1915 R. A. P. Rogers *Salmon's Analyt. Geom.* (ed. 5) II. 268 Such a correspondence is called birational. *Ibid.*, Any surface which can be birationally transformed into a unicursal surface is itself unicursal. 1954 Hodge & Pedoe *Methods Algebraic Geom.* III. xviii. 224 An irreducible correspondence between two varieties U and U' in which to a generic point of each there corresponds a unique point of the other is called a birational correspondence.

birch, *sb.* Add: **4.** *birch-tree* (later examples); **birch bark,** the bark of a birch-tree (also *attrib.*); (*U.S.*) a birch-bark canoe; **birch beer** *U.S.*, a beverage of slight alcoholic content prepared with an extract from the birch-tree; also a carbonated soft drink flavoured to resemble this; **birch partridge,** a North American name for the ruffed grouse, *Bonasa umbellus*.

1643 Williams *Key Lang. Amer.* 67 Others make slighter doores of Burch or Chestnut barke. **1771** Pennant *Tour Scotl.* 97 The materials [of the nest were] moss, worsted, and birch bark. **1829** J. MacTaggart *Three Yrs. Canada* II. 54 Thus we can run a rapid of the Rideau River with a birch-bark canoe heavily laden. **1843** 'R. Carlton' *New Purchase* 175 To float in birch bark canoes on..free waters. **1868** F. Whymper *Trav. & Adv. in Alaska* 212 Birch-barks are..easily navigated. **1927** E. V. Gordon *Introd. Old Norse* 225 They used birch-bark for leggings. **1883** *Wheelman* (U.S.) I. 392 We reached Bushkill at 12:30.P.M., stopping—for birch beer—at odd places. **1933** E. C. Guillet *Early Life in Upper Canada* iv. 100 There was..a considerable manufacture of birch beer, a very popular drink among those who did not aspire to social heights. **1823** *Gen. Descrip. Nova Scotia* iii. 31 A list of most of the known birds of the Province with their popular names...Birch Partridge. **1834** *Chambers's Edinb. Jrnl.* 21 June 168/1 A bird, called the partridge,..is found all over the American continent; they are of two sorts, the spruce and the birch, so called from the different buds which they select for their food. **1880** *Encycl. Brit.* XI. 223/2 *B. umbellus*, the Ruffed Grouse or Birch-Partridge. **1578** Lyte *Dodoens* vi. lxxvii. 758 The Birche tree hath taglettes or Chattons for his blossom, lyke as the Hazell. **1924** C. Oman *Road Royal* xiv. § 2 She was made ready for bed and was all white as a moonlit birch tree.

bircher (bɜ̄·ɹtʃəɹ). [f. Birch *v.* + -er^1.] One who administers a birch-rod; a flogger.

1888 *Ch. Times* 13 July 622 The renowned bircher [*sc.* Dr. Keate]. **1908** *Daily Chron.* 18 June 4/7 The vicarious cane is considered by birchers of experience to be an unsatisfactory substitute for the birch rod itself. **1953** L. Paul *Sir T. More* ii. 30 Erasmus relates that often after a common meal a schoolmaster would pick out a boy and hand him over for punishment to a bircher.

Bircher (bɜ̄·ɹtʃəɹ). [f. the name of John *Birch*, a U.S.A.F. officer killed in 1945, sometimes referred to as 'the first casualty of the Cold War'.] A member or supporter of the John Birch Society, an extreme right-wing American organization founded in 1958 by Robert Welch. Also **Bi·rchite** *sb.* and *a.*; **John Bircher.**

1961 in *Amer. Speech* (1962) XXXVII. 145 'Birchers' active in S.F. *Ibid.*, In Birchite circles. *Ibid.* 146 We have a lot of near John Birchers right here in Nebraska. **1961** *Guardian* 25 Nov. 7/2 Robert Welch, the Birchites' leader. **1962** *Economist* 17 Nov. 666/1 Two veterans of Congress whose Birchite connections had been unknown in previous elections. **1963** *Ibid.* 5 Oct. 43/2 Shrugging off the Birchers has not prevented the Senator [*sc.* Mr. Goldwater] from continuing to denounce some of the President's leading advisors. **1968** J. Updike *Couples* iv. 299 A lot of melancholy fill-in about Dallas the Birchers' paradise.

bird, *sb.* Add: **1. d.** In mod. (revived) use: a girl, woman (often used familiarly or disparagingly). *slang.*

1915 P. MacGill *Amat. Army* v. 62 There's another bird there—and cawfee! **1927** Collinson *Contemp. Eng.* 96 Bird (used like Ger. *Biene* especially for a more flirtatious or less reputable type of girl). **1935** 'G. Orwell' *Clerg. Daughter* ii. 161 He kept a sharp eye open for the 'birds'. **1958** *Observer* 14 Dec. 7/6 The birds..get 'emselves into trouble, let 'em get 'emselves out. **1960** *News Chron.* 16 Feb. 6 Hundreds more geezers were taking their birds to 'The Hostage' and 'Make me an Offer'. **1961** *New Statesman* 26 May 830/2 Victor is an ex-seaman in his twenties, who deserted in South Africa and got in law trouble out there for shacking up with a coloured bird.

e. *jocularly.* A man, a 'cove'; esp. in *old bird* (see Old *a.* 5 b). Cf. quots. 1799 and 1875 under sense 4 in Dict., and sense 4 b below.

Often in the punning collocation *downy bird*: see Downy *a.*^{2} 5.

1852 Bristed *Upper Ten Th.* vi. 128 The same reason..kept Mr. Simpson, and other 'birds' of his set, out of the exclusive society. **1853** 'C. Bede' *Verdant Green* vi, I suppose the old bird was your governor. **1873** [see Downy *a.*^{2} 5]. **1877, 1890** [see Old *a.* 5 b]. *a* **1885** 'H. Conway' *Living or Dead* (1886) vii, After all, Philip,..your father must be a queer bird—excuse slang, mother. **1899** Kipling *Stalky* 131 The Head 's a downy bird. **1928** [see *coffin-nail*, *coffin *sb.* 13]. **1930** E. Pound *XXX Cantos* xi. 49 And that gay bird Piero della Bella. **1939** J. B. Priestley *Let People Sing* 358 He's one of them queer birds that aren't human until they're properly pickled.

4. Also, referring to a (pretended) private or secret source of information; esp. in phr. *a little bird.*

1546 J. Heywood *Dial. cont. Prouerbes* folio Hiii. *recto*, I dyd lately here..By one byrd, that in myne eare was late chauntyng. **1597** [see Dict.]. **1711** Swift *Jrnl. to Stella* 23 May (1948) I. 277 You quarrelled this morning..: I heard the little bird say so. **1833** [see Dict.]. **1853** C. Brontë *Villette* III. xxxi. 47 'Who told you I was called Carl David?' 'A little bird, monsieur.' **1872** Geo. Eliot *Middlemarch* III. vi. lix. 314, I know all about it. I have a confidential little bird.

b. An exceptionally smart or accomplished person (freq. ironical); a first-rate animal or thing. *U.S. slang.*

1839 *Spirit of Times* 21 Dec. 498/2 If you jist could see one man what the Gineral Government sent out with an office to these parts,—he is a bird! **1840** *Ibid.* 27 June 199 An Ivanhoe has been winner and a second. Kendall has made a good beginning, and Sufferer may yet prove a 'bird'. **1852** *Knickerbocker* Oct. 320 Talking of fast men, that Williams is a bird. **1856** *Ibid.* Apr. 429 A sleigh, drawn by a 'perfect bird' of a three-mile bay mare. **1907** S. E. White *Arizona Nights* I. vii. 129 A little place..in the Colorado mountains. Fellows, she was a bird. **1911** H. Quick *Yellowstone N.* ix. 230 He's got a disguise that's a bird.

c. A prison sentence; prison. Cf. *bird-lime 2.

1924 E. Wallace *Room 13* ix. 90 He's just out of 'bird'—that's jail. **1931** *Police Jrnl.* Oct. 501 This, with Jack's previous convictions (bird), caused him to be sentenced..to five years' penal servitude..at Parkhurst. **1938** J. Curtis *They drive by Night* ii. 22 Hell of a long time the next bit of bird was going to be unless he got done for suspect. **1953** *Listener* 3 Sept. 366 Having done his bird, as imprisonment is called in the best circles. **1961** [see *bird-lime 2].

d. An aeroplane. Also, a guided missile, rocket, or space-craft. *slang.*

1933 H. G. Wells *Shape of Things to Come* III § 11. 332 We had no gas masks on our bird, so I didn't take part in the landing party which seized the new 'planes. **1951** *Time* 21 May 34/1 The military phrase of the day is 'guided missiles'... These 'birds' (so the missilemen call them) are the heirs presumptive of war. **1962** A. Shepard in *Into Orbit* 100, I really enjoy looking at a bird that is getting ready to go.

5. b. Phr. *the big bird*, cf. Goose *sb.* 1 g; esp. in *to get the (big) bird*: of an actor, to be hissed by the audience; hence *gen.* to be dismissed, get the sack. Similarly, *to give* (a person) *the bird*, *to get the bird*. orig. *Theatr. slang.*

1825 P. Egan *Life of Actor* p. xii, And the end of their folly marked by the attacks of the big birds (*geese*) driving them off the stage. **1846** Planché *Aristophanes' Birds* 6 So hear him patiently before you frown Nor let his first shot bring the 'Big Bird' down. **1865** Hotten *Slang Dict.* s.v., 'To get the big-bird', *i.e.* to be hissed, as actors occasionally are by the 'gods'. **1884** in Ware *Passing Eng.* (1909) s.v., Professor Grant, Q.C., had both 'the bird' and 'the needle' at the Royal on Monday. **1886** *Graphic* 10 Apr. 399/2 To be 'goosed', or, as it is sometimes phrased, to 'get the big bird', is occasionally a compliment to the actor's power of representing villainy. **1895** *People* 13 Jan. 7/2 Three or four of the most prominent artistes..have been..threatened with..'the bird'—that is, hissing. **1924** Galsworthy *White Monkey* 56 Mr. Danby had 'given him the bird'. *Ibid.* 255 When you were ill, I stole for you. I got the bird for it. **1927** *Daily Express* 4 Feb. 6/4 Britons in Hollywood will get what is locally known as the 'razzberry', which may be translated as 'the bird'. **1928** Wodehouse *Money for Nothing* vii. 137 Would a Rudge audience have given me the bird a few years ago? **1957** P. Kemp *Mine were of Trouble* iii. 35 She gave him the bird—finally and for good. So he came to Spain to forget his broken heart.

c. *like a bird*: with swift and easy motion onwards; easily; without resistance, difficulty, or hesitation.

1825 J. Neal *Bro. Jonathan* I. xiii. 421 Away she went, like a bird. **1873** W. S. Gilbert *More Bab Ballads* 118 'Miss Emily, I love you—Will you marry? Say the word!' And Emily said 'Certainly, Alphonso, like a bird!' **1879** *Cassell's Fam. Mag.* Mar. 197/2 Over he went like a bird, and, with his fair burthen yet in the saddle. **1914** G. B. Shaw *Fanny's First Play* Induction, I told him Trotter would feel lonely without him; so he promised like a bird.

d. Phr. (*strictly*) *for the birds*: trivial, worthless; appealing only to gullible people. orig. and chiefly *U.S. colloq.*

1951 J. D. Salinger *Catcher in Rye* i. 6 'Since 1888 we have been moulding boys into splendid, clear-thinking young men.' Strictly for the birds. **1953** *Time* 7 Sept. 2/1 Kinsey's book is strictly for the birds. **1955** J. Potts *Death of Stray Cat* vii. 79 He's the jerk that brought you home tonight, isn't he? Strictly for the birds. **1957** *Amer. Speech* XXXII. 240 In 1942, when I entered the U.S. Army..the disparaging term *that's for the birds* was in common use among officers and enlisted men...The metaphor alludes to birds eating droppings from horses and cattle. **1958** Osborne & Creighton *Epit. G. Dillon* II. 49 'You aren't very impressed with Geoffrey..?' 'Right. What the Americans call "strictly for the birds".' **1963** *Listener* 14 Feb. 301/2 Our answer, at that age, would have been that Stanley Matthews was for the birds. Football was just not mobile enough.

7. Bird of Freedom *U.S.*, the emblematic bird of the U.S. (see Eagle *sb.* 1); **bird of paradise** (earlier examples), also *fig.*; (*b*) *Astron.* (see quot. 1659); (*c*) **bird-of-paradise flower,** a perennial musaceous plant, *Strelitzia reginæ*, having scapes of purple and orange flowers; **bird of passage** (see also Passage *sb.* 1 e); **bird of prey** (see Prey *sb.* 4 b).

1848 Lowell *Biglow P.* Ser. I. ii. 28 Yourn, Birdofredom Sawin. **1854** B. F. Taylor *Jan. & June* 120 The Bird of Freedom inclined his body forward. **1906** *Harper's Mag.* Mar. 638 The short story is peculiarly an American institution, and we are as proud of it as we are of the 'Bird of Freedom'. **1606** J. Carpenter *Solomon's Solace* xxi. 86 The bird of Paradise, which being taken in a snare is neuer quiet. *a* **1649** Crashaw *Carmen Deo Nostro* (1652) f. a. ij *verso*, With heauenly riches: which had wholy call'd His thoughtes from earth, to liue aboue in'th aire A very bird of paradise. **1659** Moxon *Tutor Astron.* I. iii. § 10. 19 There are in Heaven yet twelve Constellations more, posited about the South Pole, which were added by Frederico Houtmanno, ..who..named them as follows..4 The *Peacock*, 5 The *Bird of Paradice*, 6 The *Fly*. **1663** T. Killigrew *Parsons Wedding* III. ii, *Wild.* A Bird of Paradise, what's that? *Capt.* A Girl of Fifteen, smooth as Satten, White as her Sunday Apron, Plump, and of the first down. **1884** W. Miller *Dict. Eng. Names of Plants* 253/2 *Strelitzia.* Bird's-tongue-Flower, Bird-of-Paradise-Flower. **1926** D. H. Lawrence *Glad Ghosts* 5 She showed off a bit, it is true, playing bird of paradise among the pigeons.

8. a. *bird-catcher* (later example), *-echoing* (example), *-keeper*, *-lover*, *-marking*, *-snaring*, *-stuffer* (example). **b.** *bird-haunted*. **c.** *bird-black*, *-blithe*, *-clear*, *-eyed* (earlier and later examples), *-headed*, *-high*, *-voiced*, *-winged*. **d.** *bird-claw*, *-migration*, *-note*, *-skin*, *-voice*; *bird-box*, *-rock*.

1918 E. Sitwell *Clowns' Houses* 10 Amid thick leaves I saw the wink Of bird-black eyes. **1917** D. H. Lawrence *Look! We have come Through!* 84 You bird-blithe, lovely Angel in disguise. **1869** A. R. Wallace *Malay Archipelago* I. iii. 46 We took with us..insect and bird boxes,..guns and ammunition. **1914** W. M. Webb in *Bird-Lover* I. 3 The requests for bird-boxes..led to the keeper being employed in autumn and winter in making them. **1909** *Westm. Gaz.* 8 July 8/1 The cruel trade carried on by the bird-catcher. **1889** W. B. Yeats *Wanderings of Oisin* III. 38 Golden the nails of his bird-claws. **1922** W. de la Mare *Down-adown-Derry* 8 Never so much as a bird-claw print Of footfall to be seen. **1838** —— *Memory* 16 Her glistening bird-clear eyes. **1928** C. Day Lewis *Country Comets* 24 We let The brave, bird-echoing sunlight in. **1564** Bulleïn *Dial. agst Pest* f. 43 *recto*, He is a burde iyed iade, I warrant you. **1933** W. de la Mare *Fleeting* 80 In your bird-eyed wonder. **1852** Arnold *Empedocles*, etc. 88 Some wet bird-haunted English lawn. **1897** *Daily News* 29 June 6/3 The bird-haunted estuary of a river falling into Lake Tanganyika. **1916** *Jrnl. R. Anthrop. Inst.* XLVI. 184 The human figure with the two supporting bird-headed *manaias*, a design recurring..through the whole field of Maori carving. **1958** Bird-headed [see *bird-mask* under sense *9]. **1920** E. Sitwell *Wooden Pegasus* 81 Bird-high voices shrill and chatter. **1880** Jefferies *Gt. Estate* i. 4 Should anyone in authority ask where that gun went off, the labourer 'thenks it wur th' birdkippur up in th' Dree Vurlong'. **1938** *British Birds* XXXI. 375 A Common Partridge..was taken to St. James's Park and handed over to the bird-keeper. **1866** C. M. Yonge *Dove in Eagle's Nest* I. iv. 83 The gentle Minnesinger and bird lover, Walther von Vogelweide. **1938** *British Birds* XXXI. 383 Richmond Park has much to interest the bird-lover. **1909** *Ibid.* 1 Apr. 364 To show the value of bird-marking I conclude by giving short summaries of the results obtained at Rossitten. **1908** *Westm. Gaz.* 1 Feb. 16/1 The bird-migration routes. **1908** *Ibid.* 26 June 4/1 A different bird-note seems to come from every bush and corner. **1898** R. Kearton *Wild Life at Home* iii. 99 Ailsa Craig..is a capital bird-rock. **1778** J. Cook *Jrnl.* 26 Oct. (1967) III. ii. 1142 Their dress consists of a bird skin Frock. **1904** *Daily Chron.* 7 May 5/6 Birdskin garments from the islands of the South Pacific. **1933** *Brit. Birds* XXVII. 174 For over forty years I have been collecting bird-skins for various scientific purposes. **1899** *Daily News* 10 June 8/5 Go bird-nesting or bird-snaring in one of the parks. **1838** Audubon *Ornith. Biogr.* IV. 271 A bird-stuffer whom I knew at Camden had many fine specimens. **1904** W. H. Hudson *Green Mansions* iii. 41 Hearing nothing except the usual bird voices. **1923** R. Graves *Whipperginny* 68 Now a bird-voice sings, And a loud throat bellows. **1925** E. Sitwell *Troy Park* 48/1 Bird-voiced fire screams. **1869** A. R. Wallace *Malay Archipelago* II. xxx. 199 One of the most magnificent insects the world contains, the great bird-winged butterfly, Ornithoptera poseidon. **1903** Kipling *Five Nations* 41 The blazoned, bird-winged butterflies. **1936** *Discovery* Dec. 367/1 The *Troides* or *Ornithoptera*, the magnificent bird-winged butterflies, are the glory of the eastern tropics.

9. bird-band, an identifying band fixed on to a bird's leg; so *bird-bander*, *-banding*; **bird bath,** a vessel, usu. in a garden, for birds to bathe in; **bird brain,** (a person with) a small brain; so **bird-brained** *a.*, having a small brain; *fig.* inattentive, flighty (cf. *bird-witted*); **bird-catching, -eating spider** = **bird-spider*; **bird dog** *U.S.*, a gun dog trained to retrieve birds; **bird-glasses,** field-glasses for observing birds; **bird-land,** the land or realm of birds; so **bird-folk, bird-world**; **bird-life** = Avifauna; also, the life and habits of birds; **bird-lore,** the facts and beliefs concerning birds and their life and habits (see also Lore *sb.*^{1} 5 a); **bird-louse,** any of the small wingless insects of the family *Mallophaga*, parasitic on birds and mammals; **bird-mask,** a mask resembling a bird's face; hence *bird-masked* adj.; **bird-organ** (example); **bird-ringing** = **bird-banding*; so *bird-ringer*; **bird sanctuary,** a piece of land where birds are protected, and encouraged to build and breed; **bird-seed** (examples); **bird-snake,** a S. African tree-snake, *Thelotornis capensis*, that preys on birds; **bird-song** (see Song 3); **bird-spider,** a large hairy spider (*Mygale* or *Avicularia*) of tropical America, which kills small birds; also *bird-catching spider*; **bird-spider y** (see quot. 1844); **bird-stone** *U.S.* (see quot. 1907); **bird-strike** (see quots.); **bird table,** a raised platform on which food for birds is placed; **bird-tick,** a fly (*Olfersia*) parasitic upon birds (*Riverside Nat. Hist.* 1888 II. 433); **bird-watching,** the study of birds in their homes and haunts by a naturalist or bird-lover; also as *ppl. a.*; also **bird-watcher,** one who thus observes the ways of birds; **bird-wittedness,**

the condition of being bird-witted (see Dict.); **bird-world** (see **bird-land* above).

1939 *Nat. Geogr. Mag.* Mar. 368 My companion placed bird bands around their legs and released them one by one. **1914** *Country Life* July 36/2 One of the first questions asked of the bird-bander, is 'How do you get hold of the bird in order to band it?' **1912** *Technical World Mag.* 214/1 The Bird Banding Association of America has just been formed in New York. **1954** Bannerman *Birds Brit. Isles* III. 209 As we have learned from the results of bird-banding, the ring-ouzel..has many miles to travel before it reaches our shores. **1895** *Montgomery Ward Catal.* 545 Bird baths, white or opal for large sized birds. **1912** Jekyll & Weaver *Gardens for Small Country Houses* xiv. 165 A shallow bird bath made of lead. **1933** *Boys' Mag.* xlvii. 106/1 Some gardens boast of superb and ornamental bird baths. **1943** *Gen* 2 Jan. 28/1 There are more birdbrains and dim-wits outside the boxing ring..as ever stepped around in it. **1948** H. Lawrence *Death of Doll* iv. 87 Jewel has a bird brain. You know what a bird brain is? Teeny-weeny. **1922** M. A. von Arnim *Enchanted April* xii. 176 Bird-brained...Not an idea in her head. **1805** Bingley *Anim. Biog.* III. 489 The Bird-catching Spider. **1871** Kingsley *At Last* xvii, A live 'Tarantula', or bird-catching spider. **1888** in *Amer. Speech* (1960) XXXV. 267 *Bird-dog*, a dog which accompanies a bird-hunter for the purpose of scenting the game and picking it up after it falls. **1923** J. Miner *J. M. & Birds* xv. 61 He went for his bird dog, thus giving the drake nearly an hour to make his escape. **1792** M. Riddell *Voy. Madeira* 73 The *aranea avicularia*, and the *aranea venatoria*, the *great hairy bird-eating spider* of Brasil. **1934** Haldane & Huxley *Animal Biol.* xii. 277 South American bird-eating spider. **1901** F. T. Bullen *Sack of Shakings* 10 The hovering bird-folk gathered in myriads. **1924** A. D. Sedgwick *Little French Girl* I. vi, Yes, I can see him... And with his bird-glasses. He would have been watching birds. **1924** R. Macaulay *Orphan Island* iii, William..produced his bird-glasses and said no, it would, on the other hand, be a frigate-bird. **1900** O. G. Pike (*title*) In Bird-Land with Field Glass and Camera. **1940** C. Day Lewis tr. *Virgil's Georgics* II. p. 48 And blood-red berries like rubies adorn the untilled bird-land. **1874** E. Coues *Field Ornith.* iii. 30 Bird-life is too beautiful a thing to destroy to no purpose. **1910** W. H. Hudson *Shepherd's Life* p. x, Bird-life on the Downs. **1934** *Discovery* Oct. 293/1 It is not true..that bird-life in this country is decreasing and that the future of British bird-life is one of depleted numbers. **1830** 'Dilettante' [= W. Barnes] in *Gent. Mag.* June 503/2 Ornithology, should be birdlore. **1826** Kirby & Spence *Entomol.* III. xxix. 97 The egg..of some bird-louse attached to the golden pheasant..resembles the purest wax. **1887** *Buck's Handbk. Med. Sci.* V. 751/1 *Mallophaga* (Bird-lice). Distinguished from the lice by possessing three- to five-jointed antennæ. **1930** E. Sitwell *Coll. Poems* 88 The Sun, that blackamoor, Comes in a bird-mask With a bird-flute. **1958** E. A. Armstrong *Folklore of Birds* i. 11 Some of the other human figures are indeed bird-headed and presumably wearing bird masks. *Ibid.* i. 16 A group of people..including apparently bird-masked officiants. **1851** Bird organ [see *minuet-tune*, Minuet *sb.* 3]. **1949** *British Birds* XLII. 102 Bird ringers generally state the size of the brood when ringing nestlings. **1927** *Ibid.* XXI. 134 (*title*) Traps for Bird Ringing. **1936** *Ibid.* XXIX. 226 The bird-ringing trap in the garden..was missing a great deal of the visible small bird migration. **1887** *Leisure Hour* 69/1 Mr. Maynard..has suggested the establishment of 'bird sanctuaries' along Cape Cod and other coast regions. **1922** (*title*) Bird sanctuaries in Royal Parks. **1840** Dickens *Old C. Shop* I. xiii. 163, I can buy some birdseed. **1909** *Daily Chron.* 25 Sept. 5/3 A bird-seed company. **1910** F. W. Fitzsimons *Snakes S. Afr.* iii. 57/1 *Thelotornis kirtlandii.* Bird Snake...Distribution Tropical Africa. All over South Africa. **1954** J. A. Pringle *Common Snakes* 4 The Bird-snake has a long thin body which is admirably suited for its arboreal life...It is common in the Lowveld of the Transvaal. **1896** R. L. Stevenson *Songs of Travel* 15 Bird-song at morning. **1927** E. M. Nicholson *How Birds Live* iv. 39 It has become impossible now to treat the subject of territory without taking into consideration bird-song. **1826** Kirby & Spence *Entomol.* III. xxviii. 40 In the class *Arachnida*, the bird-spiders (*Mygale*) are amongst the principal giants. **1840** J. & M. Loudon tr. *Köllar's Treat. Insects* I. 67 The Bird Spider-Fly..is scarcely half as large as the forest fly. **1844** H. Stephens *Bk. Farm* III. 856 The smallest insect on the same cut is the Bird-spider fly, *Ornithomyia avicularia*. **1881** C. C. Abbott *Primitive Industry* 365 The curious carved 'bird stones', common to our Atlantic coast states. **1907** F. W. Hodge et al. *Amer. Indians* I. 148/1 *Bird-stones*, a name given to a class of prehistoric stone objects of undetermined purpose, usually resembling or remotely suggesting the form of a bird. **1963** *Daily Tel.* 19 June 19/2 The..search..to find the answer to the growing menace of what they call 'birdstrike', the paralysing, sometimes fatal, effect of a bird being drawn into a jet engine. **1967** *Idle Moments* Oct. 15 The (RN) Air Station at Lossiemouth..had the worst record of any airfield in the United Kingdom for birdstrikes—collisions between birds and aircraft. [**1894** *Nature Notes* V. 5 A pair of hedge-sparrows frequented the shrubs surrounding the table..content to pick up the crumbs underneath, dropped by the other birds.] **1905** *Country-side* I. 82/3 A cock chaffinch will often play the bully at your bird-table. **1931** *Boys' Mag.* XLV. 140/2 The large platform a little way beneath the bird table is a safeguard against cats. **1905** E. Selous (*title*) The Bird Watcher in the Shetlands. **1930** J. S. Huxley *Bird-Watching* i. 13 From the bird-watcher pure and simple it is but a step to the bird-watcher naturalist. **1901** E. Selous (*title*) Bird Watching. **1920** *Edin. Rev.* Jan. 63 Bird-Watching as a Hobby. **1930** J. S. Huxley *Bird-Watching* iii. 52 Accompanying Mr. Eliot Howard..on his bird-watching rounds. *Ibid.* iv. 64 A party of bird-watching friends. **1904** Raymont *Princ. Educ.* x. 231 The deliberate cultivation of mere irrelevance and bird-wittedness. **1910** 'Mark Rutherford' *More Pages fr. Jrnl.* 90 The birdwittedness, the absence of resistance and of difficulty, were intolerable. **1881** A. Leslie tr. *Nordenskiöld's Voyage of Vega* II. xi. 42 The acquaintance I had made..with the bird-world of the high north. **1904** *Westm. Gaz.* 20 Aug. 2/3 In the bird-world the lark has an exclusive and singular reputation for early rising. **1907** *Daily Chron.* 16 Feb. 6/7 The albinos—those freaks of nature in the bird-world.

bird, *v.* Add: **3.** *trans.* To give the bird to (an actor, entertainer, etc.): see *bird *sb.* 5 b. *colloq.*

1936 N. Coward *Red Peppers* 93 Lot of hooligans birding the poor old man.

bird-cage. Add: **1.** *bird-cage clock* (see quot.).

1952 Gloag *Dict. Furniture* 144 *Birdcage clock*, a modern term sometimes used to describe the iron or brass weight-driven wall clocks which were made in large numbers during the 17th century.

2. Also, a similar paddock at other race-courses. (N.Z. examples.)

1934 *Evening Post* (Wellington) 13 June 6/3 The stewards' stand, however, has not yet been touched, and only the birdcage will be used at this portion of the course. **1940** F. Sargeson *Man & Wife* (1944) 64 And didn't Fred and me get a kick out of taking the horses into the birdcages and leading them round. **1952** M. Duggan in D. M. Davin *N.Z. Short Stories* (1953) 246 In the birdcage the horses for the first race circled and danced.

birdeen (bɜ̄ɹdī·n, bə·ɹdīn). [f. Bird *sb.* + *-een*, Ir. *-in* dim. suffix: cf. *colleen*.] A young bird; also, a young girl (used as a playful form of address).

1895 'Fiona Macleod' *Mount. Lovers* 47 She was like a birdeen lured by the dancing sunrays. *Ibid.* 61, I am sure, birdeen. **1916** C. A. Renshaw *England's Boys* 58, I can hear your sleepy birdeens in their swaying moonlit nest. **1922** I. J. Postgate (*title*) Robin in khaki: a book of birdeens.

birdie. Add: **1. b.** = Bird *sb.* 1 d (see Dict. and Suppl.). Cf. *birdeen.

1889 *Cent. Dict.*, *Birdie*..., a term of endearment for a child or a young woman. **1915** P. MacGill *Amateur Army* v. 62 'Cup of coffee, birdie!' he cried,..trying to grip her [*sc.* the waitress's] hand.

2. In golf: the fact of doing a hole in one under the par score. (Cf. *bird *sb.* 4 b.)

[**1911** *Maclean's Mag.* Sept. 205/1 Lanesborough followed with a 'bird' straight down the course, about two hundred and fifteen yards.] **1921** *Glasgow Herald* 14 Sept. 10 Brown squared with a 'birdie' 3 at the second. **1923** *Daily Mail* 13 June 10 Then he went all out to 'shoot birdies'..the American colloquialism for aiming at doing holes in a stroke under the par scores. **1960** *Times* 23 May 5/1 She..at last got down a birdie putt.

birdless. (Later examples.)

1906 *Westm. Gaz.* 14 Apr. 18/1 The Yellow Cat's paw is out, after a birdless winter. **1955** P. Larkin *Less Deceived* 20 A huge and birdless silence.

bird-lime. Add: **2.** *Rhyming slang.* Time; often *spec.* a term of imprisonment (cf. Time *sb.* 7 b (*d*)).

1857 'Ducange Anglicus' *Vulgar Tongue* 2 *Bird-lime*, time. **1961** *John o' London's* 16 Nov. 551/2 Time is *bird lime*, but this is shortened to *bird* and..imprisonment is often called *bird*. **1962** *Radio Times* 20 Sept. 35 In the past Charley's done his 'birdlime' but he was given time off for good behaviour.

bird-man, birdman. Also **birdsman. 1.** A man concerned with birds; an ornithologist.

1697–1796 [see Bird *sb.* 8 d]. **1844** C. Waterton *Ess. Nat. Hist.* (ed. 3) p. lxviii, These birdmen outwardly had the appearance of Italian banditti, but it was all outside and nothing more. **1969** *Observer* 16 Feb. 7/5 The Smithsonian staff itself is..envious of the birdmen for getting so much money. Those involved in the ornithological expedition have had to receive military clearance.

2. An aviator, airman. *colloq.*

1909 *Daily Chron.* 27 Oct. 4/4, I shall say: 'I saw the first bird-men in England, my dears.' **1917** 'Contact' *Airman's Outings* 244 Even intrepid birdmen (war correspondentese for flying officers) tire of trying to be offensive on a patrol. **1957** J. Braine *Room at Top* iv. 44 You were both intrepid birdmen, weren't you?

bird's-eye, *sb.* and *a.* **2.** (Earlier and later examples.)

1857 Trollope *Three Clerks* I. ii. 29 Mary, my dear, a screw of bird's-eye! *Ibid.* 33 An ample allowance of gin-and-water and bird's-eye tobacco. **1922** Joyce *Ulysses* 444 Gaudy dollwomen loll in the lighted doorways..smoking birdseye cigarettes.

4. *bird's-eye maple* (earlier and later examples).

1793 E. P. Simcoe *Diary* 26 Apr. (1911) xi. 161 Capt. Shaw has given me a tea-chest in bird's-eye maple. **1848** Mrs. Gaskell *Mary Barton* I. iii. 37 Her business was duly announced in gold letters..enclosed in a bird's-eye maple frame. **1916** A. Bennett *Lion's Share* xxx. 221 A table of bird's-eye maple.

bird's-eyed, *a.* [f. Bird's-eye.] Of maple: having small eye-like markings.

1855 *Trans. Mich. Agric. Soc.* VI. 528 Much of this maple timber is curled and some bird's-eyed.

bird's-foot. Add: **1. d.** Applied to the delta of a river (see quot. 1944).

1939 A. K. Lobeck *Geomorphology* vii. 230 (*caption*) Three types of delta; arctuate, estuarine, and bird's foot. *Ibid.* 231 Bird's foot delta. Still a third type of delta is built by streams carrying large amounts of extremely fine material. **1944** A. Holmes *Princ. Physical Geol.* x. 171 The Mississippi..extends its mouths seaward by way of deep channels..which are outstretched like fingers. This part of its delta is the chief example of the *bird's foot* type.

bird's-nest, *sb.* Add: **1. b.** *transf.* Also *attrib.*

1931 R. Campbell *Georgiad* ii. 32 A bird's-nest wig, a melancholy face. **1950** A. Christie *Murder is Announced* i. 16 Her curly bird's nest of grey hair was in a good deal of disorder. **1961** *Sunday Express* 19 Feb. 14/3 Worn with dark legs and a bird's nest hair style.

5. *Path.* (See quots.)

1879 *St. George's Hosp. Rep.* IX. 500 A tubular epithelioma..consisting of numerous tracks of epithelial cells (without any birds' nests). **1879** *Syd. Soc. Lex.*, *Bird's nest bodies*, a name given to a condition of arrangement of the scaly cells of epithelial cancer, in which they are placed in nest fashion around a circular central space, which contains amorphous colloid matter or degenerated cells. **1885** *Buck's Handbk. Med. Sci.* I. 771/1 Concentric formations known as 'birds' nests', which characterize the typical squamous epithelioma. **1894** Gould *Dict. Med.*, *Bird's-nest Bodies*, or *Cells*, the cells of certain forms of carcinoma, distinguished by the concentric arrangement of their cell-walls.

6. bird's-nest spider = *bird-spider* (*bird *sb.* 9).

1910 *Encycl. Brit.* II. 307/1 The larger species of Bird's Nest Spiders (*Avicularia*)..undoubtedly attack young birds.

bi·rd's-ne:ster. One who goes bird's-nesting.

c **1825** (*title*) Ben the Birdsnester. **1896** *Brit. Birds, Their Nests & Eggs* I. 35 A trick which..renders the birds-nester more satisfied that he is on the right scent.

bird's-nesting, bird-nesting, *vbl. sb.* Add: **2.** Furring of a passage.

1901 *Daily Mail* 26 Aug. 3/5 The cap ferrules of the Minerva's tubes were found to be partially choked from 'bird-nesting'. **1930** *Engineering* 9 May 597/3 To overcome the trouble of 'birdnesting' originally experienced in the lower rows of boiler tubes.

birefringence (bəirīfri·ndʒĕns). *Optics.* [Cf. F. *biréfringence*, *-ent.*] = *double refraction* (see Refraction 2). Also **birefri·ngent** *a.*

1889 *Cent. Dict.*, Birefringent. **1898** E. S. Dana *Text-bk Min.* (ed. 2) II. 172 The wide separation of the two refracted rays by calcite..is a consequence of the large difference in the values of its indices of refraction, in other words, as technically expressed, it is due to the *strength* of its double refraction, or its *birefringence*. **1949** *New Biol.* VII. 73 The spindle is birefringent (i.e. it has different refractive indices in different transecting planes). **1957** *Ibid.* XXII. 67 The polarizing microscope..is a device for detecting and measuring optical anisotropy, commonly referred to as birefringence (or double refraction), because the value of the refractive index varies systematically in different directions.

birk, var. *berk.

Birman, var. *Burman.

Biro (bəi^ə·ro). [f. the name of László *Biró*, the Hungarian inventor.] The proprietary name of a particular make of ball-point pen; also (with lower-case initial) applied loosely to any ball-point pen. Hence **bi·ro** *v. trans.*, to write (something) with a biro; **bi·roed** *ppl. a.*, written with a biro.

1947 *Trade Marks Jrnl.* 29 Oct. 660/2 (*heading*) Biro. Writing instruments and parts thereof, not included in other classes. **1948** 'G. Orwell' *Let.* 2 Jan. in *Coll. Ess.* (1968) IV. 393 Thanks ever so for sending the pen...It'll do just as well as a Biro. **1952** J. Cannan *Body in Beck* ii. 39 Price jumped and dropped his Biro pen. **1958** *Oxf. Univ. Gaz.* 1 Dec. 299/1 A table 7 ft. 6 in. long, and biro pens were installed. **1962** L. Deighton *Ipcress File* i. 12 The biro'd message, 'Inquiries third floor..'. **1962** *Observer* 4 Mar. 13/2 Most White Papers, editorials and poems are now not penned, but biro'ed. **1967** M. Drabble *Jerusalem the Golden* vii. 169 Even the sight of a broken biro on his windowsill was of interest to her.

birotation (bəirōtē^i·ʃən). *Physical Chem.* [Bi-^{2}.] The power possessed by certain sugars of changing their rate of optical rotation according to the length of time they have been in solution. Now usually called *mutarotation.

1882 Robb & Veley tr. *Landolt's Handbk. Polariscope* iv. 62 A peculiar case of variation of specific rotation occurs in crystallized milk-sugar...As the initial rotation is nearly twice as great as the constant rotation, the name of *birotation* has been applied to this peculiarity. **1893** *Jrnl. Chem. Soc.* LXIV. I. 125 Birotation and Hydrazone Formation of some Sugars. **1896** *Ibid.* LXX. II. 139 The birotation phenomena of glucose. **1907** J. B. Cohen *Org. Chem.* 100 Freshly prepared solutions of certain sugars change in rotation on standing. The specific rotation of glucose sinks to one-half before becoming constant. The phenomenon was therefore termed *birotation*. **1955** J. G. Davis *Dict. Dairying* (ed. 2) 78 *Birotation ratio*..certain sugars when dissolved in water gradually change in their optical rotation.

Birrell (bi·rĕl). The name of Augustine *Birrell* (1850–1933), English writer and politician, used in various formations (see quots.) alluding to his easy and discursive style, his beliefs, etc.

1894 *Westm. Gaz.* 4 Apr. 3/1 'Birrelling.' Mr. Birrell followed with a speech of quips and cranks, pleasantly diverting in a Scotch debate. **1896** *Ibid.* 6 Feb. 2/3 How would he like the future chronicler to have to say that Birrellism and Trilbyism were the marked features of 1896? *Ibid.* 18 June 2/3 He was in the habit..of 'birrelling' his examiners. **1898** G. W. E. Russell *Coll. & Recoll.* 212 To 'birrell' is now a verb as firmly established as to 'boycott',

and it signifies a style, light, easy, playful, pretty, rather discursive, perhaps a little superficial. Its characteristic note is grace. **1906** *Daily Chron.* 16 Feb. 4/7 It's a two to one chance..Religion, Irreligion, and Birreligion. **1906** *Westm. Gaz.* 23 Feb. 3/1 The House rejoiced to hear a new Birrellism. **1906** *Daily Chron.* 29 Dec. 3/5 A Birrelline Christmas card. **1937** *Times Lit. Suppl.* 10 Apr. 271 Birrelled out of his pose of sympathy. *Ibid.*, The Birrellian *obiter dictum*, fatal to a political reputation. **1960** *Times* 9 May 16/2 Barrie-isms became even more popular than once Birrellisms had been.

birth, *sb.*^{1} Add: **12.** *birth-carol* (later example), *-notice*, *-trauma*. **b.** *birth-mark* (*fig.* example); hence *birth-marked* adj.

1935 C. Day Lewis *Time to Dance* 11 Birth-carol of spring. **1869** Swinburne in *Fortnightly Rev.* July 74 Here too the birth-mark of the great race is visible. **1934** 'G. Orwell' *Burmese Days* ii. 21 His birthmarked cheek. **1925** L. P. Smith *Words & Idioms* v, The birth-notices in *The Times*. **1929** O. Rank *Trauma of Birth* p. xiii, The to all appearances purely physical birth trauma with its prodigious psychical consequences for the whole development of mankind.

13. birth certificate, a certificate of birth (Certificate *sb.* 3); **birth control,** the prevention of conception; also *attrib.*; hence *birth controller*; **birth-rate,** the ratio of the number of births to the population (usually calculated per thousand); **birth-root** (U.S. examples).

1900 *Whitaker's Almanack* 1901 4 (*Index*) Birth certificates. **1928** W. C. & J. C. Hall *Law of Adoption* I. 47 The production of a birth certificate is necessary for candidates for nearly all important examinations and for many kinds of employment. **1914** *The Woman Rebel* June 39/2 (*heading*) The Birth Control League. **1922** Ld. Dawson (*title*) Love—Marriage—Birth Control. **1936** D. V. Glass *Struggle for Population* iii. 35 Condoms are listed as preventatives of disease and not as birth-control appliances, and are thus easily available. **1937** V. Gollancz in 'G. Orwell' *Road to Wigan Pier* p. xvii, Birth control clinics up and down the country. **1921** A. Huxley *Crome Yellow* v. 45 The indignation of a convinced birth-controller. **1923** M. Edge *Artificial Birth Control* 15 Another reason the Birth Controller gives in support of his theory is, that we are over-populated. **1859** *Ann. Rep. Registr. Gen.* p. ii, The marriage rate and the birth rate were above, the death rate was below the average. **1919** W. R. Inge *Outspoken Ess.* 69 In the Rhondda Valley the birth-rate is still about forty. **1822** *Amer. Jrnl. Science* IV. 62 Plants collected..around the great Lakes..[include] Birth root. **1867** A. Gray *Man. Bot.* (ed. 5) 523 *Trillium erectum*, Purple T. or Birth-root.

birth, *v.* Add: **2.** *trans.* To give birth to; to give rise to. Chiefly *dial.* and *U.S. dial.*

1906 V. McNabb in F. Valentine *Fr. V. McNabb* III. iii. § 7. 300 Salute the sail that birthed you free. **1928,** etc. in H. Wentworth *Amer. Dial. Dict.* **1945** in *Amer. Speech* (1946) XXI. 303 The plan for UNO was birthed at Dumbarton Oaks. **1958** J. Carew *Wild Coast* ii. 22 The night his mother birthed him.

birthday. Add: **3.** *birthday boy*, *card* (examples), *dress*, *feast*, *finery*, *girl*, *ode*, *party*, *present* (examples), *treat*; **birthday book:** add examples, and add to def.: usu. having quotations for each day from the works of a particular author; also, a book suitable for a birthday present; a book compiled to honour someone's birthday; **birthday honours,** the titles of honour conferred by the sovereign on each anniversary of his or her official birthday; so *birthday gazette*; **birthday suit,** also (humorous): the bare skin; so *birthday attire*, *clothes*, *gear*.

1860 'G. & P. Wharton' *Wits & Beaux of Society* I. 127 In 'birthday attire'. **1859** (*title*) The boy's birth-day book; a collection of tales, essays, and narratives of adventure. **1881** C. Dixon (*title*) The Longfellow birthday book. **1884** *Spectator* 12 Jan. 58/1 Mr. Arnold does not describe classes of men sufficiently numerous and various to supply apt quotations for a birthday book. **1967** *Listener* 25 May 690/1 Garlands, festschrifts and birthday books tend to be tedious. **1959** H. Pinter *Birthday Party* II, Shine it on the birthday boy. **1902** *Little Folks* Oct. 287/2 Miss Shaw seemed to appreciate Moya's birthday card very much. **1961** H. S. Turner *Something Extraordinary* ii. 39 My birthday was on Sunday...I have got six birthday cards. **1732** Ld. Egmont *Diary* 2 June (1920) 279 Mr. Spencer was a man of extraordinary breeding to acknowledge the favour of a common visit in his birthday clothes. **1801** M. Edgeworth *Belinda* I. v. 148 (*title*) Birthday Dresses. **1856** C. M. Yonge *Daisy Chain* I. xxiii. 244 Mrs. Charles Wilmot's little girl was to have a birth-day feast, at which Mary, Blanche, and Aubrey were to appear. **1749** J. Cleland *Mem. Woman of Pleasure* II. 84, I now stood..in all the truth of nature...The figure I made..outshone all other birth-day finery whatever. **1910** Beatrix Gatacre *General Gatacre* xvii. 262 On the 24th in the *Birthday Gazette*, his name appeared as a recipient of the Gold Medal of a New Order, the Kaiser-i-Hind. **1734** Swift *Strephon & Chloe* in *Poems* (1958) II. 591 To see some radiant nymph appear In all her glitt'ring birthday gear. **1909** E. Nesbit *Daphne in Fitzroy St.* i. 3 (*title*) Birthday Girl. **1886** Kipling *Departm. Ditties* (ed. 2) 8 Then the Birthday Honours came. Sad to state and sad to see, Stood against the Rajah's name nothing more than *C.I.E.*! **1897** *Sat. Rev.* 5 June 626/2 In any list of Birthday honours or New Year honours one expects to see the name of at least one distinguished man. **1803** C. Wilmot *Let.* 6 Mar. in *Irish Peer* (1920) 146 The composition of a birthday Ode. **1942** E. Blom *Music in England* vi. 104 Purcell in his birth-day-ode and welcome-song mood. **1852** Mrs. Gaskell *Let.* 4 Sept. (1966) 198 Yesterday..we had a birthday party for Julia. **1796** M. Edgeworth *Parent's Assistant* (ed. 2) II. ii. (*title*) The Birth-Day Present. **1854** C. M. Yonge *Castle Builders* xxii. 342 Her immense wax-doll, a birthday present from her papa. **1753** Smollett *Ferd. Count Fathom* II. xli. 43 He made an apology for receiving the count in his birth-day suit. **1809** [see Suit *sb.* 19 e]. **1922** Blunden *Bonadventure* xxvii. 167 A dancing saloon, where birthday suits were the fashion. **1851** *Sharpe's London Jrnl.* XIII. 200/1 Father alwaies gives us a birthdaye treat. **1902** A. Bennett *Grand Babylon Hotel* i. 12 I'll be content this year with the cheapest birthday treat you ever gave me.

biscuit. Add: **1.** (Examples of U.S. sense.)

1818 J. Palmer *Jrnl. Trav. N. Amer.* ix. 125 Hot short cakes, called biscuits. **1828** Webster, *Biscuit*,..a composition of flour and butter, made and baked in private families. **1843** 'R. Carlton' *New Purchase* v. 27 Hot rolls.., a novelty then, but much like biscuits in parts of the Far West. **1903** *N.Y. Sun* 1 Dec. 8 Did he never spread cream ham gravy on his hot biscuits?

b. *transf.* (*a*) (See quot.)

1881 *Encycl. Brit.* XII. 836 The flat rounded cakes of [South American] rubber made in this manner are known in the London market as 'biscuits'.

(*b*) *Military slang.* A square brown palliasse or mattress.

1915 'Ian Hay' *First Hundred Thous.* v. 33 'Got the biscuits here, Sergeant-Major?'..The Sergeant-Major dives into a pile of brown blankets and presently extracts three small brown mattresses, each two feet square. **1917** *Times* 21 Nov. 11/4 Sleeping on the floor on army 'biscuits'—as they had already learned to call the military mattress. **1919** *Athenæum* 22 Aug. 791/2 'Biscuits.' These were the palliasses—square in form and brown as to colour—of which three went to each bed. *a* **1935** T. E. Lawrence *Mint* (1936) III. v. 173 The mattresses are three little square brown canvas cushions, rammed solid with coir. Biscuits they call them.

c. The characteristic light-brown colour of biscuits; biscuit colour; often *attrib.* = of this colour.

1884 Biscuit colour, biscuit satin [see Dict. s.v. sense 3]. **1892** *Daily News* 16 Sept. 3/3 A biscuit straw hat. **1896** *Ibid.* 18 July 6/3 White, cream, and biscuit-fawn. **1897** *Ibid.* 15 Sept. 6/6 Biscuit-tinted lace insertion. **1923** *Daily Mail* 15 Jan. 6 Newest shades, including: Pale Pink..Scarlet Fuchsia, Biscuit, Mauve.

d. Colloq. phr. *to take the biscuit*: to 'take the cake' (see Cake *sb.* 7 in Dict. and Suppl.).

1907 G. B. Shaw *John Bull's other Island* III. 76 All you know is ah to ahl [howl] abaht it. You take the biscuit at that, you do. **1930** Wodehouse *Very Good, Jeeves!* vi. 167 Of all the absolutely foul sights I have ever seen, this took the biscuit with ridiculous ease. **1961** *Listener* 16 Nov. 825/2 For the sheerest idiocy, it's the comparative 'as contemporary as...' that takes the biscuit.

3. a. (in sense 1) *biscuit-tin*; (in sense 2) *biscuit oven* (examples), *warehouse* (hence *b. warehouseman*). **b.** *biscuit-toss* (examples). **biscuit barrel,** a barrel used to contain biscuits; *spec.* a small domestic barrel-shaped container for biscuits; **biscuit root** *N. Amer.*, the quamash (*Camassia esculenta*) of North America, or other esculent roots similar to this.

1787 T. Jefferson *Let.* 18 Sept. in *Writings* (1853) II. 269 He has contrived a varnish..for lining biscuit barrels. **1840** N. Hawthorne in *Harper's Mag.* XLV. 690/2 Sometimes I..warmed myself by a red-hot stove, among biscuit-barrels..and kettles. **1886** F. H. Burnett *Little Lord Fauntleroy* (1887) ii. 21 You said..that you wouldn't have them sitting 'round on your biscuit barrels. **1935** D. Sayers *Gaudy Night* xi. 174 People *still* give plated biscuit-barrels! **1768** J. Wedgwood *Let.* 13 June (1965) 65 We..cannot keep Pace with the Biscuit oven. **1902** A. Bennett *Anna of Five Towns* viii. 172 There's the biscuit oven, but we can't inspect it because it's just being drawn. **1904** Goodchild & Tweney *Technol. & Sci. Dict.* 52/2 *Biscuit oven*, is that in which the clay articles are placed, and the heat of which renders them more or less vitrified. **1837** W. Irving *Capt. Bonneville* III. i. 9 The cowish, also, or biscuit root, about the size of a walnut, which they reduce to a very palatable flour. **1847** De Smet *Oregon Missions* 116 The bitter root..grows in light dry, sandy soil as also the caious or biscuit root. **1901** Kipling *Five Nations* (1903) 114 We stumble on refuse of rations, The beef and the biscuit-tins. **1891** *Pall Mall Gaz.* 26 Jan. 3/1 This was but a biscuit-toss from Crown Office-row. **1896** Kipling *Seven Seas* 29 And north, amid the hummocks, A biscuit-toss below. **1960** *Listener* 18 Aug. 250/2 Daughters of 'rails' and raised within biscuit-toss of the 'big rust'—the main line. **1902** A. Bennett *Anna of Five Towns* viii. 173 Mynors took the plate..to the biscuit-warehouse...A solitary biscuit-warehouseman was examining the ware.

biscuity (bi·sketi), *a.* [f. Biscuit + -y^1.] Biscuit-like in texture, flavour, colour, etc.

1892 *Eng. Illustr. Mag.* IX. 859 The pleasant 'biscuity' flavour of the atmosphere. **1925** *British Weekly* 22 Oct. 127/1 In Germany and Austria..*Zwieback* comprises a vast variety of small crunchy biscuity cakes. **1961** *Guardian* 27 Sept. 3/6 Warm, biscuity, hand-shaped stone.

bisect (bəi·sekt), *sb. Ecol.* [f. Bisect *v.*] (See quots.)

1916 F. E. Clements *Plant Succession* xv. 432 It is proposed to indicate the vertical and lateral relations of individuals by means of a cross-section showing both shoots and roots in their normal position. Such a cross-section may be termed a *bisect*. **1919** J. E. Weaver *Ecol. Relations Roots* 32 Schematic bisect showing the root and stem relations of..prairie plants. **1926** Tansley & Chipp *Stud. Vegetation* iv. 66 When a profile chart includes the root systems of the plants it is called a *bisect*. **1962** H. Hanson *Dict. Ecol.* 55 *Bisect*, a line transect which shows the vertical and lateral distribution of roots along the side of a trench in the soil and the above-ground parts of the plants along the line.

bisensory (bəise·nsōri), *a.* [Bi-^{2}.] Of an hallucination: affecting two senses.

1894 *Proc. Soc. Psych. Research* X. 194 In these 'bisensory' or 'trisensory' cases, the constructive imagination seems to have reached a higher point than in simple visual or auditory hallucinations. *a* **1901** F. W. H. Myers *Hum. Pers.* (1903) I. 254 Visual, auditory, bisensory, and trisensory hallucinations.

bish (biʃ), *sb. slang.* [Etym. unknown.] A mistake, blunder.

1937 Partridge *Dict. Slang* 55/2 *Bish*,..a mistake: Seaford Preparatory School: from ca. 1925. **1955** F. Swinnerton *Sumner Intrigue* xx. 198 He's always making bishes like this! **1956** B. Goolden *At Foot of Hil's* x. 236 She..suddenly realised she'd made an [*sic*] complete bish.

bish (biʃ), *v. Austral.* and *N.Z. slang.* [Imitative.] *trans.* To throw.

1940 N. Marsh *Surfeit of Lampreys* (1941) x. § 1. 142 They'd just sort of bished them into the cupboard. **1945** Baker *Austral. Lang.* xi. 205 To throw is *to bish*, *biff* or *peg*.

bish (biʃ). Jocular abbrev. of Bishop *sb.* and *v.*

1875 A. Porson *Quaint Words* 20 To be *bish'd*..confirmed. **1927** Wodehouse *Meet Mr. Mulliner* iii. 88 He turned appealingly from one to the other. 'Vicar! Bish!' **1930** *Outlook & Independent* 29 Oct. 328/3 Some one..asked in amazement, 'When does the Bishop bish?' **1958** I. Murdoch *Bell* 247, I suppose I'd better leave the way clear for the bish's Rolls Royce.

bishop, *sb.* Add: **6. b.** = **bishop-bird*.

1934 in Webster. **1953** D. A. Bannerman *Birds W. & Equat. Afr.* II. 1415 When not breeding the Fire-crowned Bishops go about in small flocks.

10. bishop-bird, any of various African weaver-birds the males of which have scarlet, orange or black plumage during the breeding-season; **bishop('s) sleeve,** a sleeve in a woman's dress made full and gathered in at the wrist like the lawn sleeves of an Anglican bishop.

1884 Layard & Sharpe *Birds S. Afr.* 462 Pyromelana Oryx (*Linn.*). Red Bishop Bird. *Ibid.* 463 We have not thought it necessary to separate these two Bishop Birds specifically. *Ibid.*, Pyromelana Capensis, *Linn.* Black and Yellow Bishop Bird. **1931** R. C. Bolster *Land & Sea Birds S.W. Cape* 133 The Bishop Birds are also known as 'Kaffir Finks'. **1829** *Mirror of Lit.* 26 Sept. 205/2 The mistress..conceived the idea of bishops' sleeves, an article of dress which precludes all hope..of imitation in the kitchen. **1846** D. Corcoran *Pickings from 'Picayune'* 56 Instead of making the sleeves tight they are the old fashioned *bishop* sleeves. **1861** *Englishwoman's Dom. Mag.* III. 69/1. 1. A bishop sleeve, with a narrow wristband. 2. A bishop sleeve, the fulness gathered in to a band the length of the arm, this band being shown on the upper part of the sleeve. **1958** *Vogue* Sept. 109 The bishop sleeves are a widely-seen revival.

biskop (bi·skop). *S. Afr.* [Afrikaans, lit. Bishop.] Either of two species of S. African fishes, namely the white biskop, *Sparodon durbanensis*, and the black, or blue, biskop, *Cymatoceps nasutus*.

1902 J. D. F. Gilchrist in *Trans. S. Afr. Philos. Soc.* XI. 227 Biscop, Poeskop..*Chrysophrys*, *sp.* **1930** C. L. Biden *Sea-Angling Fishes of Cape* 257 *The Biskops or Musselcrushers*..White Biskop..Black Biskop. **1957** S. Schoeman *Strike!* 60 In False Bay and Hermanus it is called white musselcracker or white biskop, or simply biskop.

Bismarck (bi·zmāɹk). [The name of the German statesman Prince Otto von *Bismarck* (1815–98).]

1. In full *Bismarck brown*: = Vesuvin.

1885 J. J. Hummel *Dyeing Textile Fabrics* xix. 413 Phenylene Brown..also bears the commercial names: Bismarck Brown, Vesuvine, Cannelle, Manchester Brown. **1886, 1897** [see Vesuvin].

2. A drink consisting of a mixture of champagne and stout.

1910 *Daily Chron.* 7 Apr. 6 The formidable drink compounded of champagne and stout, and known in the Colonies as 'Bismarck' has..an intimate connection with the German Chancellor. **1958** A. L. Simon *Dict. Wines* 31/1 *Bismarck*, another name for *Black Velvet*—half Champagne and half stout; a pick-me-up which Bismarck is said to have favoured.

3. *Bismarck herring*: a marinaded herring served cold. Hence *Bismarcked* ppl. adj.

1931 D. Runyon *Guys & Dolls* (1932) vii. 155, I set him down and get him a cup of coffee and a Bismarck herring to revive him. **1945** *Times* 2 Oct. 5/6 Herrings, which appear in various forms including fresh, kippered, bismarcked, klondyked, and home smoked.

Bismarckian (bizmā·ɹkiăn), *a.* [f. prec.: see -ian.] Of, designating, or typical of Bismarck or his policy.

1878 H. J. Coleridge in *Month* XIII. 263 The principles of State persecution—principles which have been repudiated by the civilization of modern Europe, even though the new Bismarckian phase of that civilization seems to tend to their re-adoption. **1901** G. B. Shaw *Three Plays for Puritans* p. xx, The strong-minded Bismarckian man of action. **1914** *Chambers's Jrnl.* 3 Jan. 78/1 Germany's policy is the Bismarckian policy of 'blood and iron'.

Also **Bisma·rckianism**; **Bi·smarckism.**

1870 *Fortn. Rev.* 1 Dec. 640 To substitute Bismarckism for Napoleonism would be a very small gain to civilisation. **1889** *Voice* (N.Y.) Feb. 7 The general expression of the public loss..may serve to check a too rampant Bismarckianism. **1931** *Times Lit. Suppl.* 19 Feb. 122/1 Commercial and territorial expansion—Bismarckism in Germany, Imperialism in England. **1954** A. J. P. Taylor *Struggle for Mastery in Europe 1848–1918* xx. 467 It was Bismarckianism run mad for Kiderlen to suppose that Caillaux would find it easier to compromise if he was first threatened.

bismuth. b. bismuth-glance (example); **bismuth-ochre** (example).

1839 Ure *Dict. Arts* 119 Bismuth..is a rare substance, occurring native..as a sulphuret, called bismuth glance. **1796** R. Kirwan *Elem. Min.* (ed. 2) II. 265 *Bismuth.* Mineralized by Oxygen, with or without Sulphur. Bismuth Ochre.

bisontine (bəi·sǫntəin), *a.* [ad. L. **bisontinus*, f. *bison*, *bisont-* Bison, after *elephantinus* Elephantine *a.*] Pertaining to or resembling the bison.

1885 E. Balfour *Cycl. India* (ed. 3) III. 1104/1 The general aspect of the yak is distinctly bisontine. **1887** *Blackw. Mag.* CXLI. 795 They belong to the same sub-family, Cattle (Bovinæ), but are members of the Bisontine group.

bisque^2. Add: **2.** Also *attrib.*, as *bisque oven*.

1853 Ure *Dict. Arts* (ed. 4) II. 454 The quantity of coals necessary for a 'bisque' oven is from 16 to 20 tons.

3. A light brown colour or tint. (Cf. *biscuit 1 c.)

1922 *Glasgow Herald* 17 Aug. 6 A dress of pale bisque. **1923** *Daily Mail* 15 Jan. 1 Navy, Bisque, Rust, Champagne, Orchid, Flesh, Silver, Nattier Blue, Black, Jade & Ivory.

bistable (bəistē^i·b'l), *a.* [Bi-^{2}.] Having two stable states.

1949 D. Sayre in B. Chance et al. *Waveforms* v. 164 The basic plate-to-grid-coupled bistable multivibrator is illustrated. *Ibid.*, Bistable circuits have also been called 'scaling' circuits, 'scale-of-two' circuits, and in England 'lockover' circuits. **1951** *Wireless Engineer* XXVIII. 101/2 Two-position (bistable) trigger circuits. **1952** *Proc. Inst. Radio Engineers* XL. 1531/2 The operation of this circuit may be monostable, bistable, or astable (oscillatory), depending on the nature of the emitter load. **1962** *Gloss. Automatic Data Processing* (*B.S.I.*) 58 *Toggle*, bistable trigger circuit, a trigger circuit which has two quasi-stable or stable states and which requires an appropriate excitation in each state to cause a transition to the other. **1970** *Nature* 24 Oct. 319/2 It is, however, possible to obtain bistable switching action in which the device remains indefinitely in the desired 'on' or 'off' state until specifically switched out of it.

‖**bistro** (bistro). Also **bistrot.** [Fr.] A small wine-shop, bar, or restaurant.

1922 C. Bell *Since Cézanne* 213 Perhaps the best painter in France, one of the best musicians, and an obscure journalist were sitting in a small *bistrot* on the Boulevard St. Germain. **1924** *Blackw. Mag.* Nov. 649/1 The cook of our little bistro restaurant. **1941** 'R. West' *Black Lamb* I. 144 A restaurant which though small was not a mere bistrot. **1954** Koestler *Invis. Writing* 249 He bought me a real hot dinner in a *bistro*. **1959** *Spectator* 31 July 134/3 Candle-lit Chelsea *bistros*.

bit, *sb.*^{1} Add: **8. d.** *on the bit*: (of a horse) pulling at the bit or ridden on a tight rein; also *transf.*; *off the bit*: ridden on a loose rein; *up to the bit*: up to the full speed allowed by the degree of restraint in which a horse is held by the bit; also *transf.* (Cf. Bridle *sb.* 1.)

1859 J. S. Rarey *Modern Art of Taming Wild Horses* (ed. 2) viii. 126 Your legs are to be used to force your horse forward up to the bit...Unless a horse rides up to the bit you have no control over him. **1878** Whyte Melville *Riding Recoll.* v. 92 Keeping the rebel up to his bit with legs and spurs if necessary. **1889** Lady C. Campbell *D. Blake* v. 95 'It's the only way to treat women,' he thought; 'they will always come up to the bit if you show who is the master.' **1890** *Field* 6 Sept. 393 At the top of the hill we cast off our leader, the remaining four go in their collars and up to their bits. **1928** *Daily Tel.* 16 Oct. 19/1 He was going so strongly and 'on the bit' that it took Dines nearly two furlongs to pull him up in his own time. **1958** J. Hislop *Start to Finish* vi. 35 All work except trials or semi-trials..is done 'on the bit'. This means that the horse is being ridden on a tight rein and is not galloping flat out. *Ibid.* viii. 66 There is a vast difference between sitting against a horse when he is on the bit and riding him out when he is off the bit. **1962** *Listener* 1 Nov. 739/3 North remained on the bit for so long [in Bridge bidding] that his partner's interest in a slam could not be awakened.

e. *bit-chain.*

1902 *Daily Chron.* 26 Sept. 3/6 Something wrong with the bit-chain. **1908** *Westm. Gaz.* 5 June 10/1 Bombardier Mueller deposed that Thamme had struck him several times with his fist and bit-chains.

9. The copper head of a soldering iron (exampl es).

a **1877** Knight *Dict. Mech.* I. 289/2 The copper piece of a soldering-tool riveted to an iron shank. A copper-bit. **1893** [see Soldering *vbl. sb.* 5]. **1944** *Electronic Engin.* XVI. 343 The handle does not tend to become unduly hot in spite of the user's hand being so near the copper bit.

bit, *sb.*^{2} Add: **3. b.** *bits and bats* (or *bobs*, *pieces*): fragments, oddments, odds and ends; small articles, personal belongings, bric-à-brac; (*all*) *to bits*: (reduced) to the condition of fragments; *to go to bits*: to go to pieces (see Piece *sb.* 1 c).

a **1720** [see To *prep.* 11 a]. **1859** E. Fitzgerald *Omar Khayyam* lxxiii. 16 This sorry Scheme of Things entire, Would not we shatter it to bits. **1896** *Yorks. Weekly Post* 7 June, Ah've sammed up a toathry oddments—bits an' bats mi mother ud call em. **1896** Northall *Warwicks. Word-Bk.* 29 Gather up your bits-and-bobs, and let me lay the tea. **1905** G. B. Shaw *Let.* 31 July (1946) 21 We have withdrawn the play there because it went to bits. **1931** S. Jameson *Richer Dust* xv. 427 His right leg and his stomach had been full of what he called bits and pieces. **1933** W. S. Maugham *Sheppey* III. p. 91 I'm not going to stand by and see you sneak his bits and pieces. **1936** G. Pollett *Song for Sixpence* III. iv. 239 A shop of artistic 'bits and pieces' standing alongside the close. **1940** *New Statesman* 9 Nov. 466/1 There was still only the rubble, the bits and bats of broken furniture. **1955** 'E. C. R. Lorac' *Ask Policeman* ii. 25 She'd taken all her bits and pieces to uncle long ago. **1958** M. Allingham *Hide my Eyes* xv. 151 Give those chaps half an hour in here with their bits and bobs and there's no telling what they might be able to find. **1959** *Listener* 16 July 107/1 Is it all bits and pieces, a mosaic of images?

4. c. Add to def.: *a bit of* (*a*) is also used in understatements, in the sense of 'a considerable; quite a (lot of)'. Cf. *to take a bit of doing* (sense 4 h, below). (Further examples.)

c **1771** S. Foote *Maid of Bath* I. 20 Your beauty is a little bit of a jilt. **1836** T. Hook *G. Gurney* I. 7 My young companion was a bit of a poet, a bit of an artist, a bit of a musician, and..a bit of an actor. **1866** J. E. Brogden *Prov. Words in Lincolnshire* 229 We'd six little mouths ta fill...'twer a-bit-on-a-pill [pull]. **1913** 'G. A. Birmingham' *Gen. John Regan* vi. 99 He may have been a bit of a lad in his early days. **1930** 'J. J. Connington' *Two Tickets* viii. 109 'That was a bit of a job,' he said ruefully. **1966** G. W. Turner *Eng. Lang. Austral. & N.Z.* iii. 48 Faced with tragedy, what can one say except 'It's a bit of a cow'?

d. = Tit-bit b; esp. in *pl.*, a number of short items in a popular periodical.

1884 (*title*) Illustrated bits. **1892** *Idler* May 483 Once on board ship, I tried a course of 'bits', pictorial, philosophical, scientific, conundrumistic, and otherwise. **1896** *Daily News* 4 Nov. 2/7 This is a weekly journal called 'Gems'. As its title suggests, the new paper will be of the 'bits' order. **1928** *Granta* 30 Nov. 172/1 If the editor of the *Review* were to ask me to write a little bit about Christmas I should laugh in his face.

e. *a bit*: a sum of money; money. (Cf. 8 a.) *slang.*

1894 A. Morrison *Mean Streets* 214 Bill had been 'left a bit', and 'a bit' means money, all the world over. **1909** Ware *Passing Eng.* 31/1 *To have a bit on* (Sporting), to have a bet on—a 'bit' of money on—a race. **1928** Galsworthy *Swan Song* II. iii. 124 He knew that everybody had 'a bit on' something now-a-days. For one person who ever went racing there were twenty..who didn't, and yet knew at least enough to lose their money.

f. A woman or girl (perh. ellipt. for *a bit of muslin*, etc.: see *4 h); cf. Piece *sb.* 9 b. *slang.*

1923 in J. Manchon *Le Slang* 62. **1931** Galsworthy *Maid in Waiting* v. 34 He had seen the girl pass twice or three times—had noticed her specially as a 'nice bit'. **1953** B. Goolden *China Pig* xiii. 195 If I want a common little bit for a best girl that's my look-out, too.

g. A small part in a play or film; freq. *attrib.* and *Comb.*, as *bit part*, *player*; also *transf.* orig. *U.S.*

1926 *Amer. Speech* I. 437/1 *Bit*, any small part. It may be a thinking part or the character may have a few lines. Occasionally a bit in the hands of a capable actor..is the outstanding bit of the show. **1936** F. Scully *Bedside Manna* 155 So I got a job as a bit-player in Hollywood. **1937** *Kansas City Star* 17 Oct. 4 The bass horn player..getting ready to do a bit part. **1940** *Manch. Guardian Weekly* 14 June 469 He spotted Cagney when Cagney was a 'bit' player, and was early in recognising Bette Davis's..talents. **1946** J. B. Priestley *Bright Day* v. 138 They gave her a test and a couple of bit parts. **1954** M. Allingham *No Love Lost* 108 Francia had been spotted playing 'bits' in Sweden.

h. In various colloq. and slang uses containing *a bit of* = a small amount or a small specimen of (see quots.). So *a bit of blood*: a mettlesome horse; *a bit of cavalry*: a horse; *a (little) bit of all right*: something or somebody regarded as highly satisfactory; esp. applied to a pretty or obliging woman; *a bit of muslin*: see Muslin 2; similarly *a bit of fluff*, *a bit of goods*, *a bit of mutton* (see *fluff *sb.*^{1} 1 d, Mutton 4); *to take a bit of doing*: see *doing *vbl. sb.* 1 a; *not a bit of it* = 'not a bit' (see sense 5 in Dict.). Also *to do one's bit*: to play one's part; to fulfil one's responsibilities or obligations; to make one's contribution to a cause or the like, esp. by serving in the armed forces.

1787 [in Dict., sense 4 a]. *c* **1810** W. Hickey *Mem.* (1960) xx. 329 If you be a lord, it must be a bit of bastard business. **1819** Moore *Tom Crib* (ed. 3) 11 A showy, but hot and unsound, *bit of blood*. **1821** Moncrieff *Tom & Jerry* I. vi, I shall here buy a bit of cavalry—that is a prad. **1828** *Subaltern's Log Bk.* II. 164, I entered the house in great spirits, fancying myself, to make use of a slang phrase, a very good bit of stuff. **1844** Dickens *Mart. Chuz.* xxxvi, We rather put the bits of blood upon their mettle. **1847** A. & H. Mayhew *Greatest Plague* vii. 86 When the conceited bit of goods came after the situation, she looked so clean, tidy, and respectable. **1853** C. R. Read *Austral. Gold Fields* 123 A good paddock, in which are a few *bits of blood*. **1854** Lever *Dodd Fam. Abr.* xxii, 'I'm sorry,' said he, 'that the "bit of stiff",' meaning the bill, 'wasn't for five thousand francs.' **1855** *Harper's Mag.* May 791/1 'She is too good for me, or for the world!' 'Not a bit of it, Philip.' **1874** Hotten *Slang Dict.*, *Bit-of-Stuff*, overdressed man; a man with full confidence in his appearance and abilities; a young woman, who is also called a *bit of muslin*. **1882** *Papers for People* 7 June 102 The convicts almost to a man set such a high value on a 'bit of leaf'. **1889** Barrère & Leland *Dict. Slang*, *Bit of blood*, a spirited horse that has some blood. *Bit of cavalry*, a saddle horse. *Bit of leaf* (prison), a small quantity of tobacco. *Bit of mutton* (common), a nice woman, generally in a questionable sense...*Bit of sticks* (sporting), a copse...*Bits of stiff* (popular), bank notes. **1889** G. B. Shaw in *Star* 13 Apr. 4/3 The generosity with which 'the industrial classes' audience applaud you if they think you have 'done your bit' heartily. **1898** J. D. Brayshaw *Slum Silhouettes* 109 She was a nice little bit o' goods. There's her portrait...A nice little bit of orlright, ain't she? **1902** Kipling in *Windsor Mag.* Dec. 4/2 'This,' said Pyecroft..'is a bit of all right.' **1903** Yeats *Let.* 1 Aug. (1954) III. 408 Archbishop Walsh has 'a little bit of all right' for the Chester Cup. **1904** A. Bennett *Great Man* xii. 120 It's a girl...She's a bit of all right. **1905** H. G. Wells *Kipps* I. iv. § 1 That was a Fair Bit of All Right. **1910** —— *Mr. Polly* ix, This beef is a Bit of All Right, Ma'm. **1911** Kipling in *Windsor Mag.* Aug. 261/2 'He's a raging blight!' 'Not a bit of it,' said Stalky cheerfully. **1915** *Punch* 12 May 362/1 For these our Army does its bit. **1917** *Ladies' Home Jrnl.* June 78/2 Every man beyond the military age can and should do his 'bit'. **1919** Galsworthy *Saint's Progress* III. ii. § 2 These young women are 'doing their bit', as you call it; bringing refreshments to all those who are serving their country. **1919** G. B. Shaw (*title*) Augustus does his bit. **1932** T. S. Eliot *Sweeney Agonistes* 18 Yes we did our bit, as you folks say, I'll tell the world we got the Hun on the run. **1957** *Listener* 24 Oct. 671/1 The B.B.C. has done its bit to restore the balance.

i. (See quot. 1960.) *slang* (orig. *U.S.*).

1958 G. Lea *Somewhere there's Music* 69 What's the Mister Musician bit? **1959** F. Astaire *Steps in Time* (1960) vi. 45 We were in Detroit—stranded—and that is where Mother did the pawning-of-the-jewels bit. **1960** Wentworth & Flexner *Dict. Amer. Slang* 39/2 *Bit*, any expected or well-defined action, plan, series of events, or attitudes, usu., but not necessarily, of short duration; one's attitude, personality, or way of life; fig., the role which one assumes in a specif. situation or in life. Orig. bop and cool use. **1961** 'I. T. Ross' *Requiem for Schoolgirl* xi. 179 'The Gestapo bit,' she told her friends bitterly. *Ibid.* 182 'I don't dig the bit,' Monty muttered. **1968** *Scottish Daily Mail* 3 Jan. 6 They're crazy about this psychological bit. **1969** *Gandalf's Garden* IV. 9/1, I was originally on the jazz scene and in a terrible state. You know, doing the whole bit, being on the phoney junkie trip which nearly every jazz musician was on.

5. *a bit much*: a little too much (to have to endure); something excessive or very annoying.

1939 'N. Blake' *Smiler with Knife* iv. 70 This is a bit much, isn't it, darling?..I don't approve of baby-snatching. **1954** I. Murdoch *Under Net* xviii. 248 This is a bit much! ..I was asleep.

6. c. A prison sentence. *slang.*

1871 *Session Paper: Central Criminal Court* 10 July 156 Bill, how do they know of your bit in Dover? **1884** A. Griffiths *Chron. Newgate* II. ix. 407 All three..passed on..to Leicester, where they did their 'bit'. **1917** *New Republic* 13 Jan. 294/1 Ferrati, whose 'bit' was three to seven years. **1951** J. H. Smyth *I, Mobster* xii. 133 The only question was how much of a bit Lucky would get.

8. b. Add to def.: Now usu. applied in the U.S. to a unit of value equivalent to an eighth of a dollar; used only in even multiples, as *four bits*, *six bits*.

1873 Barry & Patten *Men & Mem. San Francisco* xiii. 132 Two bits for a cup of coffee; two bits for a piece of pie; or if hunger and economy were to be considered, two doughnuts for a quarter of a dollar. **1909** Webster, *Bit*..now, usually, the sum of 12½ cents (generally in the phrases *two bits*, *four bits*, *six bits*). **1938** Runyon *Furthermore* xiv. 281 She has..a smile like six bits. **1939** Steinbeck *Grapes of Wrath* xvi. 224 If you wanta pull in here an' camp it'll cost you four bits.

bit (bit), *sb.*^{4} [Abbrev. of **binary digit*.] A unit of information derived from a choice between two equally probable alternatives or 'events'; such a unit stored electronically in a computer.

1948 C. E. Shannon in *Bell Syst. Techn. Jrnl.* July 380 The choice of a logarithmic base corresponds to the choice of a unit for measuring information. If the base 2 is used the resulting units may be called binary digits, or more briefly *bits*, a word suggested by J. W. Tukey. **1952** *Sci. Amer.* Sept. 135/1 It is almost certain that 'bit' will become common parlance in the field of information, as 'horsepower' is in the motor field. **1957** *New Scientist* 9 May 14/1 One 'bit' is the smallest amount of data which can exist, and corresponds to the answer to a yes-or-no question. On this basis, a decimal numeral can be described with four bits and an alphabetic letter with five...Existing electronic computers can store, in their normal memories, up to about one million bits.

bitangent (baitæ·ndʒĕnt). *Geom.* [Bi-^{2}. Cf. F. *bitangente.*] A line that touches a curve or surface at two different points; a double tangent. Hence **bitange·ntial** *a.*, designating a curve that passes through the points of contact of the bitangents of a given curve; also *ellipt.*

1873 G. Salmon *Treat. Higher Plane Curves* (ed. 2) vi. 218 Through the four points of contact of any two bitangents we can describe five conics. *Ibid.* ix. 336 The equation of the bitangential curve of a quartic. *Ibid.* 340 If the equation of this tangential curve were once formed, then, by forming the condition that the given tangent should touch this curve, we should immediately have the equation of the bitangential. **1948** J. W. Archbold *Introd. Algebr. Geom. Plane* vii. 271 The only 'multiple lines' of the tangent envelope C̄ are β bitangents of C̄ and the tangents at ι simple inflexions.

bitch, *sb.*^{1} Add: **2. a.** (Further examples.) In mod. use, *esp.* a malicious or treacherous woman; of things: something outstandingly difficult or unpleasant. (See also *son of a bitch.)

1814 *Byron Let.* 15 Oct. (1830) I. 586 It is well that *one* of us is of such fame, since there is a sad deficit in the *morale* of that article upon my part,—all owing to my 'bitch of a star', as Captain Tranchemont says of his planet. **1904** Kipling *Traffics & Discoveries* 165 After eight years, my father, cheated by your bitch of a country, he found out who was the upper dog in South Africa. **1913** D. H. Lawrence *Sons & Lovers* I. iv. 60 'Look at the children, you nasty little bitch!' he sneered. **1931** T. E. Lawrence *Let.* 10 June (1938) 722 'She' says the incarnate sailor, stroking the gangway of the *Iron Duke*, 'can be a perfect bitch in a cross-sea.' **1931** R. Aldington *Colonel's Daughter* I. 50 What a preposterous old bitch that woman is. **1944** Wyndham Lewis *Let.* 20 Aug. (1963) 378 For it may be a bitch of a Peace. **1956** S. Beckett *Godot* I. p. 37 That's how it is on this bitch of an earth.

b. (Later examples.)

1893 Stevenson *Catriona* xi. 123 Ay, Davie, ye're a queer character..a queer bitch after a', and I have no mind of meeting with the like of ye. **1916** Joyce *Portrait of Artist* v. 203 Is your lazy bitch of a brother gone yet?

c. A primitive form of lamp used in Alaska and Canada.

1904 E. Robins *Magnetic North* I. 233 'I'll light a piece of fat pine,' shouted the Boy...'Where's your bitch?' said Dillon...'Haven't you got a condensed milk can with some bacon grease in it, and a rag wick?' **1927** C. M. Russell *Trails plowed Under* 159 In the long winter nights their light was coal oil lamps or candles—sometimes they were forced to use a 'bitch', which was a tin cup filled with bacon grease and a twisted rag wick. **1961** *Canadian Geogr. Jrnl.* Jan. 14/2 'Office' work was done by candlelight and sometimes by nothing better than a 'bitch'—a wick in a shallow tin of tallow.

3. bitch-goddess, in William James's phr. (see quot. 1906). Cf. Success *sb.* 3.

1906 W. James *Let.* 11 Sept. (1920) II. 260 A symptom of the moral flabbiness born of the exclusive worship of the bitch-goddess *success*. **1928** D. H. Lawrence *Lady Chatterley* ix. 125 He realized now that the bitch-goddess of success had two main appetites: one for flattery, adulation, stroking and tickling such as writers and artists gave her; but the other a grimmer appetite for meat and bones. **1960** *Cambr. Mag.* 21 May 554/2 The pursuit, in our time, in the University and outside, of the Bitch Goddess Success, of a 'better living' without quality.

bitch, *v.*^{1} Restrict †*Obs.* to senses in Dict. and add: **2.** *trans.* and *intr.* To behave bitchily towards (a person); to be spiteful, malicious, or unfair (to); to deceive (in sexual matters).

1934 E. Waugh *Handful of Dust* ii. 58, I quite enjoy coping—in fact I'm bitching him rather. **1936** D. Powell *Turn, Magic Wheel* II. 125 She'd been bitched by everyone else..so that she wouldn't be surprised to have me leave her there in the rain. **1948** G. Greene *Heart of Matter* II. III. i. 199 She said, 'I thought you were never coming. I bitched you so.' **1955** P. Wildeblood *Against Law* III. 105 They're absolute hell, all having affairs with the Officers and bitching everybody like mad. **1958** B. Hamilton *Too Much of Water* xi. 248 She'd started bitching him from [illegible]. **1963** *Listener* 7 Feb. 263/1 A very Mayfair cast of a new play bitched and intrigued at a rehearsal [illegible].

bitch, *v.*^{2} Add to def.: Cf. Botch *v.*^{1} Also const. *up*. Hence **bi·tched(-up)** *ppl. a.*, spoilt, bungled. *colloq.*

1823 'Jon Bee' *Slang* 10 To bitch a business, to spoil it, by awkwardness, fear, or want of spirit. **1856** C. Reade *Never too Late to Mend* III. iii. 18 You will bitch my schemes and lose your fifty pounds. **1928** Hecht & MacArthur *Front Page* III. 150 You've just bitched up my whole life! **1929** J. B. Priestley *Good Companions* II. vi. 409 She [*sc.* a bus-engine] was bitching 'erself up all along. **1932** Kipling *Limits & Renewals* 197 'Reg'lar lower-deck palaver. Jemmy damned 'em all for bitchin' the evolution.' **1932** H. Nicolson *Public Faces* viii. 201 Yes, that charming M. Cocquebert has, for you will excuse the word, bitched the whole business. **1937** E. St. V. Millay *Conv. at Midnight* iii. 58 Can't you let the poor benighted bitched-up country go to hell in peace? **1958** J. Wain *Contenders* 188, I was in a hurry to..begin the hunt for Ned, before Robert had a chance to do anything to bitch it up. **1960** R. Daniel *Death by Drowning* xv. 225 [illegible] bitching his escape route..he would be in France.

3. *intr.* To grumble, to complain. *colloq.* (orig. *U.S.*). Hence **bi·tching** *vbl. sb.*

1930 *Amer. Speech* V. 238 He bitched about the course. **1941** B. Schulberg *What makes Sammy Run?* vi. 96 What the hell have you got to bitch about when I'm putting the money in your pocket? **1953** C. M. Kornbluth *Syndic* (1964) xiv. 145 If you let me set off the explosion, I'll quit my bitching. **1954** J. Christopher *22nd Cent.* 68 Tele did a damned fine job. It'll be a long time before I bitch at Saguki again. **1961** B. Crump *Hang on a Minute* i. 12 Nothing to do but go home and listen to the bitching. *Ibid.* xiii. 147 Couples bitching at each other is human nature.

bitchery. Delete †*Obs.* and add later examples.

1929 A. Huxley *Do what you Will* 85 Here is a happy mean between bitchery and virginity. **1932** W. Faulkner *Light in Aug.* (1933) vi. 119 Womansinning and bitchery.

2. Malice, vindictiveness, bitchiness.

1936 C. Day Lewis *Friendly Tree* i. 22 How I hate the Womanly woman and all her bitchery! **1962** *Times* 27 Sept. 15/2 Fausto's fiancée is an admirable study in feminine bitchery.

bitchily (bi·tʃili), *adv.* [f. next.] In the manner of a bitch; sensually; also, maliciously.

1938 G. Greene *Brighton Rock* II. ii. 83 A little Jewess sniffed at him bitchily and then talked him over with another little Jewess on a settee. **1961** J. B. Priestley *Saturn over Water* viii. 119 All no good. That's why I've behaved so bitchily. **1963** I. Fleming *On H.M. Secret Service* vi. 66 He still, rather bitchily, flirted with her.

bitchy (bi·tʃi), *a.* [f. Bitch *sb.*^{1}] **1.** Belonging to or resembling a bitch.

1948 J. Creasey in B. Vesey-Fitzgerald *Bk. of Dog* II. 694 The full-grown dog is termed 'bitchy' in head. **1961** C. H. D. Todd *Pop. Whippet* iv. 55 Avoid a bitchy 'pretty pretty dog' [for stud] no matter how brilliant his show record may be.

2. *transf.* **a.** Sensual, sexually provocative. **b.** Malicious, catty.

1928 A. Huxley *Point Counter Point* xii. 215 It was different in the past...The liveliness wasn't..so exclusively bitchy, to put it bluntly. **1933** E. A. Robertson *Ordinary Families* iv. 65 Ronald said proudly that the bits not dull were bawdy...Lester pronounced them 'bitchie', having..acquired the term from an American art critic. **1941** *Time* 13 Oct. 100/1 Two bitchy strip queens are murdered with their own G-strings. **1947** J. Steinbeck *Wayward Bus* vii. 94 There wasn't anything mean or bitchy about her. **1958** P. Mortimer *Daddy's gone A-Hunting* xxxiii. 186 There's no need to be bitchy.

Hence **bi·tchiness,** the quality of being 'bitchy'; sensuality; maliciousness.

1934 Dylan Thomas *Let.* 9 May (1966) 120 Let me raise one nasty growl about your unparalleled bitchiness in pinching my letter. **1951** E. Ambler *Judgment on Deltchev* xvii. 198 Do I detect a note of bitchiness and distrust? **1954** Koestler *Invis. Writing* xv. 174 A great lady..who constantly fought a losing battle against her own innate bitchiness. **1957** W. Camp *Prospects of Love* III. i. 150 A girl..with none of Lucinda's exhausting 'come-and-get-me' bitchiness.

bite, *v.* Add: **1. b.** *to bite back*: to restrain (speech) by biting the lips.

1881 Mrs. Riddell *Senior Partner* II. xi. 221 Hot and strong was the reply which rose to Robert's lips, but he bit it back. **1923** 'Joan Sutherland' *Garland of Olive* xxv, Hunt bit back his sharply released breath.

2. b. *to bite on* (fig.): to 'get one's teeth into', to take or get hold of (something substantial).

1904 W. H. Smith *Promoters* i. 20 They'll bite on anything that promises water west of either of those places. **1920** *Times Lit. Suppl.* 25 Nov. 770/2 These two writers are not..'Academics'..and there is plenty to 'bite on' in their criticisms.

6. b. (Later examples.) Also *absol.*

1917 Wodehouse *Uneasy Money* ix. 99 'Nutty, he's bitten.'..'Good gracious! What by?' 'You don't understand. What I meant was that I invited your Mr. Chalmers to help me open a hive, and he said "Rather!"' **1948** D. Ballantyne *Cunninghams* 48 He'd just touched Sydney to start a scrap, but Sydney wouldn't bite.

12. g. Of a cricket ball: to get a grip of the surface of the ground on pitching.

1867 J. Lillywhite's *Cricketers' Companion* 7 If the [illegible] best, the ball hangs or [illegible] *Ashes* iii. 56 When [illegible] the ball 'bite' a bit. [illegible] *Revisited* 231 The [illegible]

h. *Palæogr.* Of strokes of part of two letters: to converge (cf. *biting *vbl. sb.* 1 b).

1957 N. R. Ker *Catal. MSS. containing Anglo-Saxon* p. xix, *d* and *o* occasionally 'bite'..but not apparently *d* [illegible]

13. To exercise, excite; to worry, perturb; esp. in phr. *what's biting you?* *colloq.* (orig. *U.S.*).

[illegible] what's biting you? [illegible] xii. 173 'What's biting [illegible] grievances were so [illegible] *Queue* v. 59 'Where the [illegible] behind you,' she said, [illegible] C. Williams *Greater* [illegible] what's biting him now. [illegible] xvii. 220 'What's biting [illegible] if he hears you.'

15. b. To cadge or borrow (money, etc.) from. *Austral. slang.*

1919 W. H. Downing *Digger Dialects* 11 *Bite*, to borrow. [illegible] (Sydney) 7 Nov. 10/1 [illegible] **1935** *Ibid.* 6 Mar. [illegible] swagmen sallied forth to [illegible] K. Tennant *Battlers* [illegible] her lack of success in 'biting' housewives. It was only dole-day, and those wise women knew that travellers could not be really destitute.

16. *to bite one's head* (or *nose*) *off*: to snap one's head off (see Snap *v.* 7 b); *to bite the dust, ground* (earlier and later examples); also, to fall to the ground, to fall wounded; to be abased; also *fig.*; *to bite the hand that feeds one*: to injure a benefactor; to act ungratefully; *to bite off more than one can chew* (orig. *U.S.*): to undertake too much, to be too ambitious; *to bite one's ear*: also (*slang*), to borrow money from (someone); cf. *15 b and *bite *sb.* 1 i; *to bite (on) the bullet*: to behave courageously; to avoid showing fear or distress.

1599 [see Nose *sb.* 9 c]. **1697** Dryden *Virg. Æneid* xi. 528 So many Valiant Heros bite the Ground. **1750** Smollett *Gil Blas* I. iii. ii. 223 We made two of them bite the dust, and the others betake themselves to flight. **1770** Burke *Pres. Discont.* 3 This..proposition..that we set ourselves to bite the hand that feeds us; that with..insanity we oppose the measures..whose sole object is our own peace and prosperity. **1855** *Golden Era* 18 Mar. 1/5 We..made one bite the dust every crack. **1856** [see Dust *sb.*^{1} 4]. **1857** Trollope *Barchester T.* III. xiii. 229 That ecclesiastical knight before whose lance Mr. Slope was to fall and bite the dust. **1857** Lytton *What will he do with It?* III. iv. in *Blackw. Mag.* Sept. 301/2 He will never bite the hand that feeds him now. **1857** Dickens *Dorrit* II. xv. 456 You make one bite your head off, when one wants to be soothing beyond everything. **1878** J. H. Beadle *Western Wilds* ii. 42 You've bit off more'n you can chaw. **1879** *Macm. Mag.* Oct. 502/1 He used to want to bite my ear (borrow) too often. **1887** *Lantern* (New Orleans) 5 Mar. 3/1 If he ain't careful, he'll bite the dust [get broke and go begging]. **1891** Kipling *Light that Failed* xi. 219 Bite on the bullet, old man, and don't let them think you're afraid. **1916** G. B. Shaw *Pygmalion* 195 The mistake we describe metaphorically as 'biting off more than they can chew'. **1923** Wodehouse *Inimit. Jeeves* ii. 27 Brace up and bite the bullet. I'm afraid I've bad news for you. **1925** —— *Carry on, Jeeves!* ii. 36 His principal source of income..was derived from biting the ear of a rich uncle. **1927** H. T. Lowe-Porter tr. *Mann's Magic Mountain* I. v. 282 Thus ungrateful is immature youth! It takes all that is offered, and bites the hand that feeds it. **1940** Wodehouse *Eggs, Beans & Crumpets* 229, I was not dreaming of biting your ear...What I require is something far beyond your power to supply. Five pounds at least. **1943** S. V. Benét *Western Star* 70 The treacherous redskins always bite the dust. **1946** R. Lehmann *Gipsy's Baby* 126 He'd bite my head off if I asked him to come on the committee. **1955** *Times* 11 May 14/6 'Better men than Mr. Strachey have bit the dust in Dundee,' was how the Tory hopes were summed up by one Conservative. *a* **1960** J. L. Austin *Sense & Sensibilia* (1962) i. 1 They [*sc.* doctrines] all bite off more than they can chew.

bite, *sb.* Add: **1. f.** = *occlusion 1 b. Also, the imprint of the occlusion in a plastic material.

1848 J. Tomes *Lect. Dental Physiol. & Surg.* xvi. 367 The *bite*, or closure of the upper and under teeth, must be adjusted. **1878** C. Hunter *Mech. Dentistry* vi. 74 (*heading*) Taking the 'Bite' in wax and plaster. *Ibid.* 79 Bites..may be cast by first cutting a perpendicular groove on the back of the model, then filling plaster into the impression of the teeth in the wax block. **1880** N. W. Kingsley *Treatise on Oral Deformities* v. 84 The object was, not to protrude the lower teeth, but to change or jump the bite in the case of an excessively retreating lower jaw. **1902** [see *occlusion]. **1904** V. H. Jackson *Orthodontia* xi. 201 An apparatus..was utilized for opening the bite and moving the upper incisors outward. **1968** J. Woodforde *False Teeth* 126 Having decided which is the proper relationship, or bite, the dentist attaches the upper and lower casts of the jaw to a machine called an articulator, which reproduces chewing motions.

g. *Cricket.* The quality in a cricket-pitch that helps a ball to 'bite' (see *bite *v.* 12 g).

1905 *Daily Chron.* 5 May 8/3 Aided by the trifle of 'bite' in the pitch, the Surrey bowler always appeared likely to get wickets.

h. *fig.* Incisiveness, pungency; point or cogency of style, language, etc.

1899 R. Whiteing *No. 5 John St.* xxiii. 228 There seems no 'bite' in their pretty ways, their soft voices, their allusive turns of phrase. **1921** J. Agate in *Sat. Rev.* 24 Dec. 708/2, I want to hear..common Spanish speech interpreted and made real. I want tang and bite which I can translate into actuality. **1939** *Punch* 11 Oct. 398/1 There is an unexpected and genuine satirical bite in the whole treatment of the story. **1946** H. Foss in A.L. Bacharach *Brit. Music* iv. 70 The quality of 'bite', of urgency, or, as I would call it, of single-minded artistic sincerity. **1957** *Listener* 19 Sept. 416/1 The party's election propaganda lacked bite and purpose.

i. Slang phr. *to put the bite on*: to borrow money from (someone); to ask (someone) for a loan; also, to threaten, to blackmail, to extort money from. orig. and chiefly *U.S.*

1933 D. Runyon *Furthermore* (1938) v. 81 He once tries to put the bite on Sorrowful for a sawbuck. **1934** Wodehouse *Thank You, Jeeves* v. 61 For years and years I have been trying to lend him of my plenty, but he has always steadfastly refused to put the bite on me. **1939** R. Chandler *Big Sleep* xxvi. 232 You can put the bite on the peeper and be on your way. **1950** 'S. Ransome' *Deadly Miss Ashley* iii. 33 Everybody keeps putting the bite on me for money I haven't got.

2. a. Add to def.: Also, a small meal; a snack. (Later examples.)

1899 Ade *Fables in Slang* (1900) 188 He would be reading the Menu Card to her, and telling her how different it is when you have Some One to join you in a Bite. **1929** 'G. Daviot' *Man in Queue* iv. 46 Have a bite before you go to bed. **1952** M. Laski *Village* xi. 166 Come back for a bite, just a scratch meal. **1959** 'C. Carnac' *Death of Lady Killer* xi. 124, I..had a bite with my friend at the fish and chips stall.

biter. 2. (Further examples.)

1693 D'Urfey *Richmond Heiress* Epil. p. 64 Is't not fit Once in an age the Biter should be bit. **1710** E. Ward *Nuptial Dial.* II. ix. 179, I think she merits equal Praise That has the Wit to bite the Biter. **1826** M. Kelly *Reminiscences* (ed. 2) I. 336 Delighted at seeing the biter bit.

biting, *vbl. sb.* **1. b.** Add to def.: *spec.* in *Palæogr.* The convergence of part of two letters, esp. *d* with *e* or *o*, in some medieval manuscripts. Cf. *bite *v.* 12 h.

1957 N. R. Ker *Catal. MSS. containing Anglo-Saxon* p. xix, The earliest dated English example of biting of *d* and *e* known to me is in the form of profession of Bishop Sefrid of Chichester in 1180 (Muniments of the Dean and Chapter of Canterbury, C.115, no. 43). **1960** —— *English MSS. Cent. after Norman Conquest* 39 This *de* ligature is perhaps the earliest of the numerous bitings of converging strokes which characterize the formal book-hand throughout the rest of its long existence.

bi·tingness. [f. Biting *ppl. a.* + -ness.] Biting quality.

1894 *Contemp. Rev.* July 43 His article, by its very sharpness and bitingness, will help to wake our people up.

bitonality (bəitonæ·liti). *Mus.* [f. Bi-^{2} + Tonality.] The simultaneous use of two keys in a musical composition. So **bito·nal** *a.*, characterized by bitonality.

1927 A. E. Hull *Mus. Class. Romantic & Mod.* xvii. 317 *Atonality*, *Bi-tonality* and *Polytonality* are the technical catch-words of the day. .. Bi-tonality is a richer kind of counterpoint (or harmony), produced by two melodies (or harmonic streams) which proceed simultaneously, each in its own key. *Ibid.* 318 Bi-tonality in two harmonic streams arose from the isolated bi-tonal chords, so frequently found in the works of Strauss and the French impressionists of the period 1900–10. **1945** *Times* 28 Sept. 7/5 His [*sc.* Bartok's] Fourteen Bagatelles (1908), trifles maybe, but exploiting modern devices like bitonality and irregular rhythm, are fairly well known. **1956** W. Mellers in A. Pryce Jones *New Outl. Mod. Knowl.* 360 The result is a bitonal ambiguity between major and minor.

bitsy (bi·tsi), *a.* *colloq.* Also **bittsie.** [f. pl. of Bit *sb.*^{2} or f. *bitty *a.*: see -sy.] = *itsy-bitsy; tiny, (charmingly) small. Freq. preceded by 'little'.

1905 *Dialect Notes* III. 87 They was little-bitsy fellers. **1906** 'O. Henry' *Four Million* 110 Oh, oo's um oodlum, doodlum, woodlum, toodlum, bitsy-witsy skoodlums? **1919** F. Hurst *Humoresque* 147 Wouldn't I just love to..hang some great big headlight earrings in them little bittsie ears. **1959** W. Golding *Free Fall* xiv. 251 A bitsy village with reed thatch and wrought-iron work.

bitten, *ppl. a.* Add: **5.** In engraving, *bitten-in*: see Bite *v.* 11 b.

1878 Abney *Treat. Photogr.* 183 The plate has to be..again heated to slightly melt the bitumen, so as to allow it to flow down the sides of the bitten-in lines.

6. *bitten-off*: abruptly terminated (as if by the action of biting).

1829 T. Castle *Introd. Bot.* 20 The eroded or bitten-off appearance. **1937** V. Woolf *Years* 288 Baxter gave a queer little bitten-off smile.

bitter, *a.* and *sb.*^{1} Add: **A. 1. c. bitter lake** = *salt lake* (see *salt *sb.*^{1}); *spec.* as the name of certain lakes in Egypt; (see also quot. 1882).

1843 E. Clarkson *Suez Navigable Canal* 7 The Bitter Lakes would fill up at any time from the Red Sea. **1882** A. Geikie *Text-bk. Geol.* III. II. 395 Saline lakes, considered chemically, may be grouped as *salt lakes*, where the chief constituents are sodium and magnesium chlorides with magnesium and calcium sulphates: and *bitter lakes*, which usually are distinguished by their large percentage of sodium carbonate as well as chloride and sulphate. **1957** *Encycl. Brit.* XXI. 517/2 Finally in the summer of 1869 the waters of the Mediterranean and the Red Sea were united [by the Suez Canal] in the Bitter lakes.

2. b. (Examples.) Cf. Bible *Prov.* v. 4.

1849 *Congress. Globe* 12 Dec. 23, I am unfortunately among those who voted for the gentleman from Indiana, even 'to the bitter end'. **1850** *Ibid.* 9 Apr., App. 434 Our defence is a just one, and will be maintained by us to the 'bitter end'. **1921** L. Strachey *Q. Victoria* vi. 210 He would go on, working to the utmost and striving for the highest, to the bitter end. **1955** G. Greene *Loser takes All* I. ix. 62 A wife ought to believe in her husband to the bitter end.

B. 1. b. A bitter part.

1860 O. W. Holmes *Prof. Breakf.-t.* iv. 104 When you can get the bitter out of the partridge's thigh, you can make an enlightened commonwealth of Indians.

3. (A glass of) bitter beer. *colloq.*

1857 'C. Bede' *Verdant Green Married* x. 78 Mr. Verdant Green and Mr. Bouncer..turned into the coffee-room of 'The Mitre' to 'do bitters', as Mr. Bouncer phrased the act of drinking bitter beer. **1862** Trollope *Rachel Ray* I. iii. 41 Going into Parliament..just as they pleased, like the modern heroes of the bitter cask. **1865** Hotten *Slang Dict.*, *Bitters*, 'to do bitters', to drink beer.—*Oxford*. **1874** *Ibid.* s.v., 'To do a bitter', to drink beer.—Originally Oxford, but now general. **1894** G. Moore *Esther Waters* xxx. 236 A dozen pots of beer..and a few glasses of bitter. **1896** H. G. Wells *Wheels of Chance* ix, Every public-house..meant a lemonade and a dash of bitter. *Ibid.* xviii, A lemonade and bitter, please. **1901** *Westm. Gaz.* 8 June 6/3 A bitter having been brought, he quaffed it to his second's health. **1942** *Penguin New Writing* XIV. 133 The barmaid..replied there was nothing left but bitter and old-and-mild.

bitter-, *a.* and *adv.* in *combination*. Add: **2. bitter bark,** a popular name for any of various shrubs and small trees (see quots.); **bitter beer** (earlier example); also *fig.*; **bitter-ender** *colloq.*, one who fights or holds out 'to the bitter end' (see Bitter *a.* 2 b); one who refuses to yield, give way, or compromise; hence *bitter-enderism*; **bitter-nut** (example); also *bitter-nut hickory*; **bitter pit,** a disease of apples, characterized by brown spots; **bitter root,** a popular name for a plant of the species *Apocynum androsæmifolium*; also a plant of the N. American species *Lewisia rediviva*; (see also quot. 1909).

1884 W. Miller *Dict. Eng. Names Plants* 231/1 *Pinckneya pubens*, bitter-bark-tree, Fever-tree of Georgia. **1889** J. H. Maiden *Native Plants Australia* 198 Petalostigma quadriloculare, *F.v.M.*..Bitter Bark...The bark contains a very powerful bitter, said to have the same properties as cinchona. **1898** Morris *Austral Eng.* 31/2 *Bitter-bark*, an Australian tree, *Petalostigma quadriloculare*,..N.O. *Euphorbiaceæ*...The name is also applied to *Tabernæmontana orientalis*..N.O. *Apocyneæ*, and to *Alstonia constricta*..N.O. *Apocynaceæ*. **1955** *Times* 26 May 9/2 *Alstonia constricta*, or 'bitter bark', a new source of supply of the drug known as 'reserpine'...Supplies have also been found in the roots of 'bitter barks' grown in India. **1850** Thackeray *Pendennis* II. xx. 200 The bitter beer hot and undrinkable. **1890** Kipling *Barrack-r. Ballads* (1892) 4 'I've drunk 'is beer a score o' times,' said Files-on-Parade. ''E's drinkin' bitter beer alone,' the Colour-Sergeant said. **1850** *Congress. Globe* 12 Mar., App. 303 The disunionist looks forward to a southern confederacy; the bitter-ender to the triumph of his party. **1906** H. Spender *General Botha* vii. 127 In these discussions their leaders still proved the most obstinate 'Bitter-enders'—just as now..they still resist most obstinately the mingling of their racial influence with that of the British stock. **1926** *Contemp. Rev.* June 687 The trade union world lies shattered and in ruins [after the General Strike]. There is fierce controversy between volunteers, 'scabs', 'hands-uppers', and 'bitter-enders'. **1918** *N.Y. Times* 15 Sept. III. 2/1 An unreasoning 'bitter enderism'. **1810** Michaux *Arbres* I. 19 Bitter nut hickory.., seul nom en usage dans N.Y. **1832** D. J. Browne *Sylva Amer.* 170 This species is generally known in New Jersey by the name of Bitternut Hickory. *Ibid.*, The inhabitants of New Jersey give it the name of Bitternut, which..indicates one of the peculiar properties of the fruit. **1898** *Agric. Gaz. N.S. Wales* I. 683 *Bitter Pit.* This disease appears in the form of small sunken brown pits having a bitter taste. **1960** *New Scientist* 24 Nov. 1392/3 The incidence of bitter pit [in Australia]..has been reduced in a spectacular manner by spraying with calcium salts. **1838** S. Parker *Expl. Tour* 204 The racine amère, or bitter root, which grows on dry ground, fusiform, and though not pleasant to the taste, yet it is very conducive to health. **1909** *Cent. Dict.* Suppl. 138/3 *Natal bitter root*, a climbing vine of the gourd family, *Gerrardanthus macrorhiza*, having tuberous roots..which are intensely bitter and are used by the natives in medicine. *a* **1918** G. Stuart *40 Years on Frontier* (1925) I. 178 The wild flax and bitter-root are in full bloom. **1945** B. MacDonald *Egg & I* (1946) 18 The bitter-root daisies, the Montana State flower.

bitterling (bi·təɹliŋ). [G., f. *bitter* Bitter *a.* (transl. L. *amārus*) + *-ling*, -ling^1.] A small carp-like freshwater fish, *Rhodeus amarus*, of Central Europe.

1880 Günther *Fishes* 601 The European species is known in Germany by the name of 'Bitterling'. **1911** *Encycl. Brit.* XX. 794/2 *Rhodeus amarus*, the bitterling,..injects its eggs into the mantle-cavity of pond-mussels, where the fry develop. **1959** J. Clegg *Freshwater Life Brit. Isles* (ed. 2) xvi. 266 The Bitterling..uses one species [of mussel] in particular, the Painter's Mussel..as a nursery for its young.

bittern, *sb.*^{1} Add: With qualifying adj., as **American bittern,** *Botaurus lentiginosus* of N. America; **least bittern,** *Ixobrychus exilis* of N. America; **little bittern,** any of several small bitterns of the genus *Ixobrychus*; **sun bittern** (see Sun *sb.* 13 b).

1776 T. Pennant *Brit. Zool.* II. 537 The Little Bittern..was shot as it perched on one of the trees in the Quarry or public walks in Shrewsbury. **1813** Wilson *Amer. Ornith.* VIII. 35 American Bittern, *Ardea Minor*..is another nocturnal species, common to all our sea and river marshes. *Ibid.* 37 Least Bittern: *Ardea exilis*..is commonly found in fresh water meadows. **1831** Least bittern [see Least *a.* 1 c]. **1957** Bannerman *Birds Brit. Isles* VI. 115 It seems to be generally conceded that the whole European population of little bitterns winters in Africa. *Ibid.* 134 *American bittern*..An Accidental Visitor which has occurred about fifty times in Britain.

bitters. Add: Also *fig.*, esp. in *U.S.* colloq. phr. *to get one's bitters*: to get one's deserts. (Cf. quot. 1836 in Dict.) *Obs.*

1812 in *Maryland Hist. Mag.* (1914) IX. 70 You might get your bitters in Baltimore Town. *a* **1846** in W. T. Porter *Quarter Race Ky.* (1846) 194 The seal soon got his bitters, and the captin cut a big hunk off the tail eend.

bitty (bi·ti), *a.* [f. Bit *sb.*^{2} + -y^1.]

1. Made up of little bits (used disparagingly); consisting of (too many) unrelated parts; scrappy.

1892 *Idler* May 484 It is an awful thing to be shut up for a week with all the 'bitty' papers of the day. *c* **1909** in C. W. Cunnington *Eng. Women's Clothes* (1952) iii. 91 The bitty dress is entirely relegated to the mercantile mind. **1924** N. Coward *Rat Trap* II. p. 30 She doesn't dress smartly enough—too bitty. **1928** *B.B.C. Handbk.* 71/2 At the risk of the programme being called 'bitty' by sections of the audience with specially marked and developed tastes—it is necessary to compound the programmes..of diverse elements. **1952** E. R. Janes *Flower Garden* 31 Unless there is enough width, a double path..gives a bitty appearance. **1963** *Times* 18 May 5/7 The costumes are gaudier..and the whole thing is rather bitty.

2. Covered with or containing bits or scraps (of a material).

1902 *Little Folks* Jan. 58/2 'Oh, Ted; you are *bitty*!' she remarked..while she..picked off the shavings that clung to his clothes. **1953** *Gloss. Paint Terms* (*B.S.I.*) 5 *Bitty*, a description applied to paints or varnishes containing bits of skin or flocculated material which project above the surface when the paint or varnish is applied.

3. = *bitsy *a.* *U.S. colloq.*

1905 *Dialect Notes* III. 87 They was little-bitty fellers. **1908** *Ibid.* 291. **1940** R. Chandler *Farewell, My Lovely* x. 78 That toy...It's just a little bitty gun, a butterfly gun.

So **bi·ttiness,** the state of being bitty, scrappiness.

1950 M. Peake *Gormenghast* 182 Unknown to her the 'bittiness' that resulted gave to the salon a certain informality far from her intentions. **1957** R. Hoggart *Uses of Literacy* vii. 179 Little tablets of curious information are scattered throughout...This is bittiness run riot.

Bitumastic (bi:tiŭmæ·stik). Also **bitumastic.** [Trade-name.] A protective coating (see quot. 1889). Also *attrib.* or as *adj.*

1889 *Trade Marks Jrnl.* 23 Jan. 77 *Bitumastic.* Asphaltic and Composite Paints for Protecting from Corrosion the interior parts of Iron or Steel Ships..Bridges, and other structures of Iron and Steel. **1933** *Archit. Rev.* LXXIII. p. liv (Advt.), The flat roofs are constructed from..bitumastic felt sheeting. **1940** *Manch. Guardian Weekly* 12 Apr. 292 The enameller did not take sufficient care to see that the test cock hole was kept clear of bitumastic. **1960** *Farmer & Stockbreeder* 19 Jan. Suppl. 19/2 Painting houses with bitumastic paint greatly helps to keep them clean and free from pests.

bitumen. Add: **2. b.** *the bitumen*: a tarred road; *spec.* the road from Darwin to Alice Springs. *Austral. colloq.*

1953 Baker *Australia Speaks* v. 137 *One for the bitumen*, a final round of drinks, i.e. 'one for the road'. **1954** B. Miles *Stars my Blanket* xxv. 219 We had not seen another vehicle since we turned off the bitumen at Elliott. **1958** A. J. Toynbee *East to West* xiv. 44 'The bitumen', running dead straight for hundreds of miles without a swerve, is an impressive symbol of the modern world. **1963** V. B. Cranley *27,000 Miles through Australia* ii. 19 The minute you left the 'bitumen', as tarred roads are called here, you were back in the bush.

5. bitumen process, a photographic process using a metal plate coated with bitumen which is rendered insoluble by the action of light.

1858 T. Sutton *Dict. Photogr.* 329 Mr. Macpherson is not the *inventor* of the bitumen process; M. Nicéphore Niépce first used bitumen in photography. **1960** A. L. M. Sowerby *Dict. Photogr.* (ed. 19) 40 Half-tone or Process Blocks by the Bitumen Process. A copper or zinc plate is coated with bitumen, and it is exposed...After development, or the removal of the superfluous bitumen, the plate is etched.

biune (bəi·yūn), *a.* *rare.* [f. Bi-^{2} + L. *ūnus* one, after *triune*.] Two in one.

1897 F. Thompson *Orient Ode* in *New Poems* 34 Yea, biune in imploring dumb, Essential Heavens and corporal Earth await. **1900** G. W. Allen *Mission of Evil* 75 This Edenic Man was..male and female in one; that is, 'bi-une'.

biurate (bəiyūə·rĕt). *Chem.* [f. Bi-^{2} + Urate.] An acid salt of uric acid.

1891 F. Taylor *Pract. Med.* (ed. 2) 702 They [*sc.* urates] are decomposed in the presence of water into free uric acid and biurates. **1908** *Practitioner* Apr. 517 The writer is doubtful that the cause of the pain, in gouty arthritis, is the result of the deposit of crystals of sodium biurate.

biuret (bəi·yuret). *Chem.* [a. G. *biuret* (G. Wiedemann 1847, in *Jrnl. f. prakt. Chem.* XLII. 256), f. Bi-^{2} + Urea: see -uret.] A compound formed by heating urea. Also *attrib.*, as *biuret base, test*; **biuret reaction,** the reaction of biuret, proteins, and some other substances containing the —CO · NH— group with an alkaline solution of cupric sulphate, which gives a red or violet coloration.

1869 Roscoe *Elem. Chem.* xxxv. 354 When heated to 120°, urea fuses and begins to decompose, forming substances termed ammeline and biuret. **1877** Roscoe & Schorlemmer *Chem.* I. 652 Biuret, $C_2O_2H_5N_3$. This compound, discovered by Wiedemann, is formed when urea is heated for some time to 150–160°. **1883** *Jrnl. Chem. Soc.* XLIV. 1019 The author..described a substance..obtained from peptones, which gives conspicuously the biuret reaction of those bodies. **1895** D. C. Black *Urine in Health & Dis.* iv. 153 The albumen is separated by filtration and the biuret test applied to the filtrate. **1907** J. B. Cohen *Org. Chem.* 410 Triglycylglycine (biuret base). **1916** A. P. Mathews *Physiol. Chem.* 145 Biuret, a substance..formed by the condensation of two molecules of urea (hence biurea or biuret) with the elimination of ammonia. **1956** *Nature* 4 Feb. 234/2 A number of amino-acids..which appear to constitute a polypeptide moiety yielding a weak biuret reaction.

bivalence (bi·vălĕns, bəi:vē^1·lĕns). [f. Bivalent *a.*; see -ence.] In various subjects: the quality of being bivalent; bivalency; *spec.* in *Logic, principle of bivalence* (see quots.).

1889 *Cent. Dict.*, *Bivalence* (Chem.), a valence or saturating power which is double that of the hydrogen atom. **1909**

Ibid. Suppl., *Bivalence of the chromosomes*, the double as distinguished from the single (univalent) condition of the chromosomes, for example, in the oöcytes and spermatocytes of one of the forms of *Ascaris megalocephala*. **1951** J. Łukasiewicz *Aristotle's Syllogistic* iv. 82 The so-called principle of bivalence, which states that every proposition is either true or false, i.e. that it has one and only one of two possible truth-values: truth and falsity. **1962** W. & M. Kneale *Devel. Logic* ii. 47 The principle that every statement is true or false is called the Principle of Bivalence.

bivalent, *a.* Add: **2.** *Biol.* Applied to homologous chromosomes united in pairs at the first division of meiosis. Hence as *sb.* (Webster, 1934).

1899 J. H. McGregor in *Jrnl. Morphology* XV. (Suppl.) 81 The only possibility of a true reduction..is by conceiving the chromosomes to be 'bivalent'..and to assume that the halves of these bivalent chromosomes form new connections before the final mitosis. **1937** *Nature* 30 Oct. 760/2 When the bivalent..attempts to divide, the 'dicentric' chromatid forms a bridge between the daughter nuclei. **1949** Darlington & Mather *Elements of Genetics* i. 30 The compact paired, or *bivalent*, chromosomes..lie evenly in the nucleus. **1952** *New Biol.* XIII. 35 Sometimes during the nuclear divisions, four chromosomes similar to each other all 'pair' together, forming a quadrivalent, instead of the usual two pairs of bivalents.

bivallate (bəivæ·lē^{i}t, -ĕt), *a.* [f. Bi-^{2} + Vallate *a.*] Having two encircling ramparts.

1930 *Antiquity* IV. 198 This 'bivallate ditch' is a good example of a type of earthwork found commonly in Wessex. **1936** *Proc. Prehist. Soc.* II. 216 The enclosure appeared to have been built on to a ditch of the bivallate meandering variety. **1963** G. Daniel in Foster & Alcock *Culture & Environment* ii. 20 The building of the bivallate and multivallate hillforts of Wales.

bivalve, *a.* **1.** Delete †*Obs.* and add later examples.

a **1877** Knight *Dict. Mech.* s.v. *Speculum*, Webber's magnifying bivalve ear-speculum. **1908** *Practitioner* Aug. 284 There was no appreciable difference obtained in the results by using a bivalve speculum in order to avoid contact with the external meatus.

bivariant (bəivē^ə·riănt), *a.* *Physical Chem.* [Bi-^{2}.] Having two degrees of freedom (see *freedom 10 b).

1902 [see Trivariant *a.*]. **1904** A. Findlay *Phase Rule* ii. 16 We may also speak of the variability or variance of a system, and describe a system as being invariant, univariant, bivariant, multivariant, according as the number of degrees of freedom is nought, one, two, or more than two.

bivariate (bəivē^ə·riĕt), *a.* *Statistics.* [f. Bi-^{2} + Variate *a.*] Involving or depending upon two variables.

1920 *Biometrika* XIII. 37 Thus in 1885 Galton had completed the theory of bi-variate normal correlation. **1923** *Ibid.* XV. 77 (*title*) On the general forms of bivariate frequency distributions. **1951** D. R. Whitney *Ann. Math. Statistics* XXI. 274 (*title*) A bivariate extension of the *U* statistic.

bivium. (Earlier example.)

1870 [see Trivium 2].

bivoltine (bəivǫ·ltin), *a.* Also **-in.** [a. F. *bivoltin*, f. *bi-*, Bi-^{2} + It. *volta* time.] Of certain silkworms: producing two broods per annum. Also as *sb.*

1874 [see *multivoltine* s.v. Multi- 1 a]. **1888** E. A. Butler *Silkworms* 69 Most of the other species produce two, three, four, six, or even eight broods per annum, and in the commercial world are..distinguished as 'bivoltins', 'trivoltins', 'quadrivoltins', etc. **1932** *Proc. 6th Internat. Congress Genetics* I. 377 The bivoltine breeds also show less growth than the univoltine ones.

bivvy, bivy (bi·vi). *Army slang.* [Shortened f. Bivouac *sb.*] A temporary shelter for troops; a small tent.

1916 *Anzac Book* 142 We lays down in the open W'en our 'bivvies' isn't dug. **1918** *N.Z.E.F. Chrons.* 13 Feb. 9/1 We arrived at our allotted spot, somewhere in Palestine, and erected our bivvies. **1920** *Blackw. Mag.* May 596/2 The Egyptian Camel Corps and Gurkhas arrived, bringing 'Bivies' and other luxuries. *Ibid.* 606/1 Tying his horse to my bivy-pole. **1925** *Glasgow Herald* 30 Mar. 10 That word was 'tambu', meaning a rough and ready shelter made of branches, planks, corrugated iron, a 'bivvy', in fact. **1947** D. M. Davin *Gorse blooms Pale* 199 Snow and me were sitting outside the bivvy.

biz (biz). Colloq. abbrev., orig. *U.S.*, of Business. So *show biz* (see *show *sb.*^{1}).

1862 C. F. Browne *Artemus Ward His Bk.* 222, I must forth to my Biz. **1873** Leland *Egypt. Sketch-Bk.* 35 No. IV. made it his 'biz' to fall upon the old hands. **1878** J. H. Beadle *Western Wilds* iii. 46 He had what he called a 'big biz' at each successive terminus town. **1890** *Harper's Mag.* Apr. 813/1 I'll go Down to the sale, if I can leave my biz. **1909** Wodehouse *Mike* xx. 115 As the poet has it, 'Pleasure is pleasure, and biz is biz, and kep' in a sepyrit jug.' **1941** F. Scott Fitzgerald *Last Tycoon* (1947) iii. 36 The cleverest plagiarist in the biz.

biza·rrely, *adv.* [f. Bizarre *a.* + -ly^2.] In a bizarre manner. So **biza·rreness** = Bizarrerie.

1884 Rider Haggard *Dawn* v, This woman so bizarrely beautiful. **1920** *Edin. Rev.* Oct. 255 Of his faults he was perfectly conscious—the over-intensities, the dualities, sometimes the bizarreness, which we have shadowed. **1953** Quine *From Logical Point of View* iii. 54 A reaction suggesting bizarreness of idiom. **1960** *20th Cent.* Oct. 366 Rilke waffles bizarrely on about Arabian poetry.

‖**blaasop** (blā·sǫp). *S. Afr.* [Afrikaans, f. *blaas* to blow + *op* up.] A puff-fish: see Puff *sb.* 9 b.

[**1853** L. Pappe *Edible Fishes Cape Good Hope* 8 This fish (*Blaasopvisch*; *Balloonfish*; *Toadfish*) is never found in Table Bay, but is very common in the bays to the east of it.] **1902** J. D. F. Gilchrist in *Trans. S. Afr. Philos. Soc.* XI. 227 Blassop, Toad-fish (E. London). *Tetrodon honkenyi*. **1913** W. W. Thompson *Sea-Fisheries Cape Colony* 159 Tetrodon honkenyi..Blaasop; Toad-fish (East London). **1947** K. H. Barnard *Pict. Guide S. Afr. Fishes* 209 *Globe-Fishes or Puffers*. Known at the Cape as Blaasops and in England and America as Puffers from their habit of 'blowing up' or puffing out the body; called Tobies in Natal. **1947** L. G. Green *Tavern of Seas* iv. 37 The blaasop's liver contains a deadly poison.

blab, *v.*^{1} Add: **5.** *blab-mouth* (further example); so *blab-mouthed* adj.; **blab-off** *attrib.* (see quot. 1953).

1958 A. Sillitoe *Sat. Night & Sun. Morn.* vi. 82 She was known as a blab-mouth in all the pubs. **1913** D. H. Lawrence *Sons & Lovers* I. i. 16 He was blab-mouthed, a tongue-wagger. **1953** *Daily Mail* 20 Nov. 7/2 A member suggested that the Chamber of the House [of Commons] should be fitted with 'blab-off switches'...These 'popular devices enabled American viewers [of television] to switch off advertisers' announcements'. **1962** *Punch* 5 Dec. 834/2 Others..ignore the commercial interference..without benefit of blab-off devices.

blabber-mouth, blabbermouth (blæ·bəɹmauþ). *slang* (orig. *U.S.*). = Blabber *sb.*

1936 Steinbeck *In Dubious Battle* xiii. 249 One minute he's a blabber-mouth kid. **1956** D. Karp *All Honorable Men* 174 No, Burney isn't a blabbermouth. He tells you a lot less than he knows. **1959** I. & P. Opie *Lore & Lang. Schoolchildren* x. 189 One who blabs to a teacher or to a senior is a 'blabber-mouth'.

black, *a.* Add: **3. c.** Of coffee: see Coffee 1.

1796, 1867 [see Coffee 1 γ]. **1913** *Lancet* 29 Nov. 1563/2 Coffee is often made with a generous proportion of the powdered bean as in the case of after dinner 'black' coffee, the view being that the secret of good coffee is to make it strong. **1940** Auden *Another Time* 77 Kept awake with black coffee.

6. b. In the names of artificial flies used in fly-fishing.

1496 [see Louper]. **1655** Walton *Angler* (ed. 2) v. 145 There are twelve Kinds of Artificial made Flies to Angle with upon the top of the water...The sixth is, the black-fly.., the body made of black-wool and lapt about with the herle of a Peacocks tail. **1799** [see Midge 2]. **1837** Kirkbride *Northern Angler* 35 The Black Midge, or Gnat,..cannot be made too fine and small. *Ibid.* 51 The Black Palmer..is made with a body of black ostrich harle, ribbed with silver thread. *Ibid.* 57 The Black Spinner..has acquired a high reputation, both as a lake and a river-fly. **1923** *Daily Mail* 11 Aug. 7 The Lea should yield some good specimens of these fish to the black gnat.

8. b. *spec.* Of comedy: macabre. Cf. Fr. *pièce noire*.

1963 *Listener* 14 Feb. 310/1, I prefer my black comedy a little blacker. **1964** *Ibid.* 13 Feb. 287/3 His recent group of Swiftean black comedies. **1964** *Guardian* 8 July 7/6 [It is] 'black comedy'. Death's kind of fashionable nowadays. **1965** *Listener* 11 Mar. 379/3 The whole form of 'Berck-Plage', a poem about a seaside funeral, is a most remarkable piece of black cinema.

11. b. Short for *black-leg 3 c: of persons or of work performed by 'blackleg' labour. Hence in extended use, of work boycotted by trade unions during a dispute, also of products, supplies, etc., which they refuse to handle.

1927 *Daily Tel.* 22 Nov. 12/2 The Waterside Workers' Federation to-day declared as 'black' the steamer Kakakiri. **1935** *Economist* 8 June 1302/2 [Work books] will facilitate the control of the labour market and the prevention of 'double-earning' and 'black labour'. **1956** *Times* 24 Aug. 8/4 Two more freighters..were declared 'black' to-day by the boilermakers' union because of work being done by members of the crew. **1959** *New Statesman* 11 July 34/1 Meanwhile, as magazines and provincial papers progressively close down, tension is increasing in those offices where arrangements have been made to produce 'black' editions. **1963** *Times* 28 May 5/1 When supervisory staff took over maintenance and transport driving duties, all the firm's production men quickly classified their work as 'black' and walked out.

c. Existing in contravention of economic regulations, as the 'black market'; hence, bought or sold by illicit trading.

1937 S. & B. Webb *Soviet Communism* (ed. 2) II. 1199 There was always in Moscow and Leningrad, Kharkov and Kiev, the so-called 'Black Exchange', where native speculators illegally offered to the tourist, for American dollars or British pound notes, five, ten or twenty times as many roubles as the State Bank would give. **1938** *Baltimore Sun* 18 Mar. 3/1 We're expected to buy 'black' rubles or bootleg rubles which are available..at a fraction of the standard rate. We used to get them in what we call the Black Bourse...It's..a name applied to the 'bootleg' circles, to channels through which we bought black rubles. **1942** *Ann. Reg. 1941* 175 Contravention of war economic decrees, *e.g.* black slaughtering, were frequently punished with death. **1946** *Times* 18 Apr. 5/6 There are some things—food bought 'black' and services—which are not to be had for Reichsmarks.

14. black and tan. (Earlier and additional examples.)

1850 C. M. Yonge *Langley School* vi. 41 Oscar, the blood-hound, that monster of a black-and-tan dog. **1870** D. J. Kirwan *Palace & Hovel* (1963) xiii. 118 The dog-fancier may be noticed with..a black and tan under one arm and a spaniel under the other. **1948** C. L. B. Hubbard *Dogs in Britain* III. xx. 271 Since 1925 Black-and-Tan Miniature Terriers may exceed the previous weight limit of 7 pounds.

b. *black and tan*: a drink composed of porter (or stout) and ale. *slang.*

1889 in Barrère & Leland *Dict. Slang.* **1955** G. Freeman *Liberty Man* I. iv. 60 He asked for a pint of black and tan, and had to explain to the waiter what it was. **1958** A. Sillitoe *Sat. Night & Sun. Morn.* vi. 93 Gin-and-orange? I'll have a black-and-tan.

c. *Black and Tans*: popular name for an armed force specially recruited to combat the Sinn-Feiners in 1921, so named from the mixture (black and khaki) of constabulary and military uniforms worn by them. Also *attrib.* Hence *black-and-tannery*: the principles or activities of the Black and Tans.

1921 *Times* 28 Jan. 7/5 'God Save the King' and 'God Bless the Black and Tans'. **1921** Ld. Braye in *Hansard Lords* XLIV. 792, I rise to ask His Majesty's Government..whether they will..recall the Black and Tans. **1922** W. B. Yeats *Lett.* (1954) 680 The Black and Tans flogged young men and then tied them to their lorries by the heels. **1923** *Weekly Disp.* 4 Mar. 9 'How can I,' concluded Sir John Simon, 'defend black-and-tannery?' **1958** *Spectator* 8 Aug. 183/2 The only way would be a reversion to Black-and-tannery, to forcible coercion and repression.

15. a. (Earlier and later examples.) Also applied *spec.* to a type of house painted white and having black timbers.

1612 W. Strachey *Hist. Trav. Virginia* (1849) I. x. 123 Squirrells they have..some blackish, or black and white. **1811** A. Wilson *Amer. Ornith.* III. 23 Black and White Creeper. *Certhia Maculata.* **1819** Keats *Let.* 3 Jan. (1958) II. 29 Mrs. Dilke has two cats...The Mother is a tabby and the daughter a black and white. **1959** *New Biol.* XXX. 50 The Black and White Hawk-Eagle, *Spizaetus alboniger*.

c **1830** E. Grosvenor in G. Huxley *Lady Eliz. & Grosvenors* (1965) ii. 45 The house black and white outside and good carved oak within. **1848** Mrs. Gaskell *Mary Barton* I. i. 2 One of those old-world, gabled, black and white houses. **1936** M. Allis *Eng. Prelude* xv. 116 The same holds true of old houses, in which Shrewsbury is very rich, especially the type called 'black and white'. **1968** 'R. Simons' *Death on Display* xi. 171 High Tor was a black and white two-car-garage type, with a weedy drive, and..a large wooden cross above the front porch.

c. Also *attrib.*, as in *black-and-white art, artist, drawing, sketch*.

1889 *Pall Mall Gaz.* 2 Jan., The modern work in which black-and-white art is seen in its most perfect stage of development...Many of the best black-and-white men of the day are represented. **1892** C. G. Harper *Eng. Pen Artists* 60 Such excellent black-and-white renderings of dog life. *Ibid.* 178 The arts of black-and-white drawing. *Ibid.*, Colourists or black-and-white artists. **1895** Pennell *Mod. Illustr.* Introd. 3 The amount of black-and-white work which he [*sc.* Dürer] produced. **1896** *Daily News* 29 Sept. 6/6 The well-known black-and-white artist, Mr. Fred Barnard.

d. *Photogr.* and *Television.* Applied to monochrome photographs, photographic reproductions, films, etc., opp. 'colour'.

1890 *Portfolio* XXI. Art Chron. xviii/2 In so far as small photographic black-and-white versions can be satisfactory they are very good. **1940** Graves & Hodge *Long Week-End* xxv. 434 American advertising carried far more photographs than British, both coloured and black and white. **1958** *Amateur Photographer* 31 Dec. 914/2 A 35-mm camera, with which it is undesirable to use the very fastest black-and-white films unless forced. **1961** *Guardian* 7 Feb. 6/4 Black-and-white television in cinemas.

16. *black-green.*

1849 D. Campbell *Inorg. Chem.* 257 This oxide separates after some time as a black-green hydride. **1923** D. H. Lawrence *Birds, Beasts & Flowers* 180 The black-green skirts of a yellow-green old Mexican woman.

17. (Further examples.)

1910 J. Buchan *Prester John* v. 97, I'll admit the truth to you, Davie. I'm black afraid. **1915** —— *Salute to Adventurers* iii. 47, I had been sore at my imprisonment, I was black angry at this manner of release. **1938** E. Bowen *Death of Heart* I. i. 24 His father was black depressed.

18. *black-backed* (examples), *-bearded* (examples), *-bordered* (example), *-breasted*, *-clothed*, *-draped*, *-hearted* (earlier and later examples), *-maned*, *-necked*, *-orbed*, *-rimmed*, *-robed* (examples), *-scarved*, *-shawled*, *-souled* (also *absol.*), *-spotted*, *-striped*, *-tailed*, *-throated* (earlier example), *-veiled*, *-veined*, *-winged* adjs.

1776 Pennant *Brit. Zool.* II. 445 Black backed Gull..Larus marinus. **1858** *U.S. War Dept. Rep. Explor. Railroad Route* IX. 98 Black-backed Three-toed Woodpecker. **1874** Coues *Birds N.W.* 368 Black-backed Eagle. **1833** Tennyson *Dream Fair Women* in *Poems* 130 The stern blackbearded kings with wolfish eyes. **1881** Wilde *Poems* 107 Grim watchmen on their lofty seats..strained black-bearded throats across the dusky parapet. **1838** Dickens *Nickleby* i. 2 A black-bordered letter to inform him how his uncle..was dead. **1815** Stephens in Shaw *Gen. Zool.* IX. I. 205 Black-breasted woodpecker. **1854** *Poultry Chron.* I. 544/2 Game (Black-breasted and other Reds). **1873** W. Cory *Lett. & Jrnls.* (1897) 322 Our parson bends his black-clothed back in the sun. **1898** *Westm. Gaz.* 28 May 10/1 The bowed, black-draped figure passing sadly from the shadows in the Abbey. **1904** *Daily Chron.* 8 Dec. 3/2 The black-draped scaffold at Whitehall. **1849** Eastwick *Dry Leaves* 175 The black-hearted gibes of a portion of our English press. **1872** *Rep. Vermont Board Agric.* I. 94 A dry, dead knot is left when they [*sc.* the branches] are cut off, which sometimes kills the center of the tree, making it 'black-hearted'. **1932** A. J. Worrall *Eng. Idioms* iii. 20 He is a black-hearted villain. **1850** Cumming *5 Yrs. in S. Afr.* (1902) 115/2 The most magnificent old black-maned lion. **1839** W. B. O. Peabody in *Mass. Zool. & Bot. Surv.*, *Rep. Ornith.* 358 The Black Necked Stilt, *Himantopus nigricollis*. **1910** F. W. FitzSimons *Snakes of S. Afr.* 73/2 Naia nigricollis. Black-necked Cobra...Average length 5 to 6 ft. Distribution: Natal, Zululand, Transvaal, [etc.]. **1934** *Discovery* Oct. 294/2 There has been a marked increase in the nesting of the black-necked grebes in Britain and Ireland. **1938** W. de la Mare *Memory* 73 Prowling, black-orbed, disconsolate, Questing antennae, quivering wing. **1729** Black-rimm'd [see Rimmed *a.* 1]. **1922** Joyce *Ulysses* 128 Staring through his blackrimmed spectacles. **1858** M. Arnold *Merope* 9 This black-rob'd train. **1863** Black-robed [see Black *sb.* 5 c]. **1917** D. H. Lawrence *Look! We have come Through!* 81 Black-scarved faces of womenfolk. **1929** R. Graves *Poems* 27 The black-shawled peasant woman. **1840** Whitman *Uncoll. Poetry & Prose* (1921) I. 15 All lie earth's spreading arms within, The pure, the black-souled, proud and low. **1944** E. Blunden *Shells by Stream* 43 And forth from black-souled hurricanes Conjures glad day. **1591** *Troub. Raigne K. John* I. iv, Black-spotted Periure as he is. **1662** R. Smith (*title*) A Wonder of Wonders: ..an Invective against Black-spotted Faces. **1783** Latham *Gen. Syn. Birds* II. II. 633 Black-spotted P[igeon]. **1901** *Westm. Gaz.* 27 June 3/2 Black-spotted white foulard. **1932** D. Gascoyne *Roman Balcony* 9 A tattered projection of black-spotted leaves On a branch. **1908** *Westm. Gaz.* 21 Nov. 16/2 The finest perch of the week (2½ lb.) also hails from this part, though some nice specimens of the black-striped fish have been secured from the Thames. **1923** D. H. Lawrence *Birds, Beasts & Flowers* 167 She [*sc.* the goat]..reaches her black-striped face up like a snake. **1806** Lewis in *L. & Clark Exped.* (1905) IV. 87 The black tailed fallow deer. **1863** Black tailed [see Godwit]. **1916** E. Blunden *Harbingers* 32 And black-tailed chub still shoal below. **1785** Pennant *Arctic Zool.* II. 363 Black-throated Bunting. **1631** Blacke veyled [in Dict.]. **1906** *Westm. Gaz.* 18 Aug. 11/2 These black-veiled children of the East. **1906** B'ness von Hutten *What became of Pam* III. i, A black-veiled nursing-sister. **1775** M. Harris *Eng. Lepidoptera* 7 *Papilio*. English Names... White, black veined. **1903** *Westm. Gaz.* 1 Dec. 8/2 Black-veined marble. **1817** Shelley *Laon & Cythna* I. xxx. 16 Black winged demon forms. **1953** E. Sitwell *Gardeners & Astronomers* 41 Only her sisters, harpy winds, black-winged like flies.

19. black-about *a.* (*nonce-wd.*), black all around; **black arm,** a bacterial disease of cotton plants characterized by angular discolorations; **black bean,** (*a*) a bean of the genus *Phaseolus*, having black seeds; (*b*) an Australian hardwood tree (*Castanospermum australe*); also its timber; **black body** *Physics*, a body or surface that absorbs all radiation falling upon it; also *attrib.*, esp. in *black-body radiation*, the radiation emitted by a black body; **black bomber** *slang*, an amphetamine tablet (see quot. 1964); **black box,** orig. Royal Air Force slang for a navigational instrument in an aeroplane; later extended to denote any automatic apparatus performing intricate functions (cf. quot. 1674 under Black *a.* 1 a); also *fig.* and *attrib.*; **black bread:** see Bread *sb.* 2 e; **black-buck,** a name used by sportsmen for the antelope proper (*Antilope cervicapra*); also, the South African *Hippotragus niger*; **black butter,** (*a*) † = *apple-butter*; (*b*) butter browned in a pan and mixed with vinegar and parsley to make a sauce; **black-clock** [Clock *sb.*^{3}], any black beetle, esp. the cockroach (*north. dial.*); **black-coat,** (*b*) any black-coated worker (see **black-coated* adj.); also *attrib.*; **black-coated** *a.*, spec. defining clerical or professional as distinguished from industrial or commercial occupations; hence, of or composed of persons engaged in such occupations; **black code** *U.S.*, a code or body of laws relating to Negroes in some Southern States, esp. before the abolition of slavery; **black cotton ground** or **soil,** a dark, rich soil found esp. in the Deccan region of India, produced by the disintegration of a black lava; = *regur; **black curlew** (see quot.); **black disease,** a fatal disease of sheep characterized by necrosis of the liver (see quot. 1918); **black doctor** (see quot.); **black dress** *attrib.* (see quot.); **black earth,** (*a*) = *chernozem; also *attrib.*; (*b*) *attrib.* in *black-earth country* = *black belt 2; **black fast** *Eccl.*, a fast which includes abstinence from milk and eggs (cf. *black fasting* under sense 17 in Dict.); **black fellow,** also *gen.* = *black man 1; **blackfellow's bread** = *native bread* (see Bread *sb.*^{1} 2 e); **black-figure** *Archæol.*, applied *attrib.* to a type of Greek pottery ornamented with figures in black silhouette; also *absol.*; so *black-figured* adj.; **black fox,** the red fox, *Vulpes fulva*, of northern America, during a colour phase in which its fur is black; **black frost**: see Frost *sb.* 1; **black gang,** a gang employed on such work as coaling, stoking, etc.; **black gold** *colloq.* (chiefly *U.S.*), oil; also *transf.*; **black gown** *U.S.*, an Indian name for a Roman Catholic priest; **black growth** *U.S.* (see quots.); **Black Hand,** (*a*) defining a Spanish revolutionary society of anarchists; (*b*) defining a secret society of Italian immigrants in U.S., concerned chiefly in levying blackmail; hence **Black Hander,** a member of such a society; **black heart** (earlier example); (*b*) *Metallurgy*, used *attrib.* or *ellipt.* of a type of malleable cast iron having a core of graphite or the process for making this; **black heat** (see quots.); **black helmet,** a mollusc shell used in the making of shell-cameos; **black ice,** a thin hard transparent ice; **black Jew** (see *Jew *sb.*); **black job** *slang* (*Obs.*), a funeral; **black joke**: see Joke *sb.* 1 in Dict. and Suppl.; **black knot,** (*a*) any of certain kinds of fungus in the U.S.; (*b*) a fast knot as distinguished from a running knot (Ogilvie, 1881); **black larch** *U.S.*, an American variety of larch, the hackmatack; **black-lark** (see quots.); **black level** *Television*, the level of the picture signal that corresponds to black in the transmitted picture; also *attrib.*; **black light,** light-rays beyond the two ends of the visible spectrum; invisible ultra-violet or infra-red light; **black liquor,** acetate of iron used instead of green copperas as a mordant in dyeing (*Cent. Dict.* 1889); **black magic** (see Magic *sb.* 1 b); **Black Museum,** the name given to a collection preserved at Scotland Yard of exhibits connected with crimes of the past; **Black Muslim,** a member of the Nation of Islam, an American Negro sect, established in 1931 by 'Wallace Farad' and developed from 1934 by Elijah Muhammad, which preaches a form of Islam and proposes principally the separation of Negroes and Whites; **black oil,** any of various dark-coloured oils, *spec.* heavy crude oil used for lubrication; **Black Panther,** a member of an American Negro organization which adopts a militant attitude to the promotion of the Negro cause; **black-plate** (see quot. 1858); **black pod,** a disease of the cocoa tree caused by the fungus *Phytophthora faberi*; **black pope:** see Pope *sb.*^{1} 1; **black power,** power for black people; used as a slogan of varying implication by, or in support of, Negro civil rights workers and organizations; **black print** *Photogr.* [Print *sb.* 13], a print giving black lines on a white ground; **black quarter** (earlier example); **black root,** any of various American plants with dark-coloured roots; **black root rot,** any of various diseases of plants characterized by black lesions of the root; spec. a disease of tobacco and other plants caused by the fungus *Thielaviopsis basicola*; **black rot,** any of various diseases of plants characterized by dark discoloration and decay; **black rubber vine,** an African plant producing a black juice which is used as rubber; **black rust** [Rust *sb.*^{1} 6 b], a plant rust producing black discoloration; **Black Sash,** applied *attrib.* to a women's anti-apartheid organization in South Africa; **black scab** (see quot. 1915); **black scale,** a destructive scale (*Lecanium oleæ*) infesting olive, citrus, and other cultivated plants; **black scour(s)** [Scour *sb.*^{2} 5], a hæmorrhagic diarrhœa of sheep, pigs, and cattle; **black sheet,** (a piece of) ungalvanized sheet iron; **black silk,** used *attrib.* to define a period of mourning during which black silk is worn instead of crape; **black soil** = *chernozem; **black south-easter** *S. Afr.* (see quots.); **black spring** *Austral.* (see quot.); **black step** *Printing* (see quot.); **black stripe** (example); (*b*) a disease of the tomato-plant, produced by the fungus *Alternaria solani*; **black tie,** *spec.* a man's black bow-tie worn with a dinner-jacket; also *ellipt.*, evening dress including a black tie; so **black-tied** *a.*, wearing a black tie; **black tongue** chiefly *U.S.*, any of various diseases in human beings or animals characterized by a dark discoloration of the tongue; **black tooth** *U.S.*, a condition of pigs in which the teeth become black; **black tracker** *Austral.*, an Aboriginal employed to help the police in tracking down fugitives or persons lost in the bush; **black tripe,** unbleached tripe; **black turf (soil),** a dark-coloured soil found in the Transvaal (see quots.); **black velvet,** (*a*) a drink made by mixing stout and champagne; (*b*) *Austral.* and *N.Z. slang*, a black-skinned or coloured woman; such women collectively; **black wart** = *black scab*; **black widow (spider)** (see quots.); **black willow,** any of various willows (see quots.), esp. the *Salix nigra* of N. America; **black work,** (*a*) iron-work; also spec. (see quot. 1888); (*b*) blacked leather; (*c*) a kind of embroidery (see quot. 1910); (*d*) undertakers' work (cf. *black man 3 and *blackmaster).

1876 G. M. Hopkins *Wr. Deutschland* (1918) st. xxiv, She to the black-about air..Was calling. **1907** W. A. Orton *U.S. Farmers' Bull.* cccii. 41 There is a bacterial disease of cotton (*Bacterium malvacearum*)..producing various symptoms and receiving various names, such as angular leaf-spot, black-arm, boll-spot, etc. **1939** *Nature* 14 Oct. 676/2 (*title*) Genetics of Blackarm Resistance in Cotton. **1792** E. Riou tr. *J. van Reenen's Jrnl. Journey from Cape Good Hope* 29 Extensive handsome gardens, planted with kaffer corn, maize,..black beans. **1863** A. Gray *Man. Botany* (ed. 4) p. xliv, Egyptian or Black Bean, cultivated for ornament, rarely for its beans..seeds black or tawny with a white scar. **1895** J. H. Maiden in *Agric. Gaz. N.S.W.* V. 1 Because of the dark colour of the wood, and partly by way of distinction from the red bean, it is usually known by timber merchants as black bean. **1912** C. H. B. Quennell in L. Weaver *Home & Equipm.* 58 The bookcase illustrated has been made in a new Australian wood called 'black bean'. **1937** E. Hemingway *To have & have Not* I. ii. 35, I had black bean soup and a beef stew. **1710** Black body [see Black *a.* 1 a]. **1895** H. F. Reid in *Astrophysical Jrnl.* II. 161 It is curious that a really black body, the most important in the theory of radiation, has not so far been used by experimenters. **1919** T. Preston *Theory of Heat* (ed. 3) vi. 559 The *intensity* of black-body radiation. **1923** Glazebrook *Dict. Appl. Physics* IV. 568/1 Suppose that we have a substance with the property that all radiation falling on its surface is completely absorbed. This is called a *black body*. **1928** F. K. Richtmyer *Mod. Physics* vii. 183 The 'black body' may radiate, as well as absorb, energy. Such radiation is called 'black body' radiation. **1957** *Encycl. Brit.* III. 679/1 The black-body temperature of a source like the sun is the temperature at which a black body would emit radiation of the same intensity. **1963** *Daily Tel.* 16 Oct. 23/1 Police who raided the house found..100 methydrine tablets and 99 'Black Bomber' pills. **1964** *Lancet* 29 Aug. 452/1 The preparations in circulation apparently included 'black bombers' ('Durophet', a mixture of the two amphetamines). **1945** Partridge *Dict. R.A.F. Slang* 16 *Black box* or *gen box*, instrument that enables bomb-aimer to see through clouds or in the dark. **1947** *Jrnl. R. Aeronaut. Soc.* LI. 432/1 These British night fighters were crammed with 'black boxes' all of which had to be operated by the pilot or his navigator. **1948** A. P. Rowe *One Story of Radar* i. 6 For many years the 'death ray' had been a hardy annual among optimistic inventors. The usual claim was that by means of a ray emanating from a secret device (known to us in the Air Ministry as a Black Box) the inventor had killed rabbits at short distances. **1953** *Jrnl. Brit. Interplan. Soc.* XII. 223 As far as the layman is concerned, a phantastron is a 'black box' which will divide the frequency of its output pulses by any integral number between 2 and 20. **1956** W. R. Ashby *Introd. Cybernetics* vi. 86 Black Box theory is..even wider in application than these professional studies. *Ibid.*, In our daily lives we are confronted at every turn with systems whose internal mechanisms are not fully open to inspection, and which must be treated by the methods appropriate to the Black Box. **1962** *Daily Tel.* 14 Nov. 1/3 Russia advocated the use of unmanned seismic detection stations, known as 'black boxes' to Western scientists, as a means of avoiding inspection. **1964** *Ibid.* 3 July 25/5 The flight recorder is an indestructible 'black box' which automatically records the key functions in the aircraft...The 'black box' can..tell what went wrong in a crash. **1964** *Economist* 25 July 356/1 Silent 'black boxes'—electronic computers which measure the varying demand for power of the *Savannah*. **1888** W. T. Blanford *Mammalia* 521 *Antilope cervicapra.* The Indian Antilope or black Buck. **1902** *Encycl. Brit.* XXV. 453/2 Antelope..properly denotes the Indian black buck, which alone constitutes the genus *Antilope*, with the title of *A. cervicapra*. **1808** Jane Austen *Let.* 27 Dec. (1932) I. 241 Our black butter..was neither solid nor entirely sweet... Miss Austen had said she did not think it had been boiled enough. **1877** E. S. Dallas *Kettner's Bk. of Table* 96 *Black Butter*—the French Beurre noir, much used for skate, for calf's brains, and the like. **1895** G. A. Sala *Thorough Good Cook* 470/1 Black butter sauce. **1906** Mrs. Beeton *Bk. Househ. Managem.* x. 273 Black Butter Sauce. Ingredients—1½ ozs. of butter, 1 teaspoonful of chopped parsley, ½ a teaspoonful of vinegar. **1962** *Listener* 23 Aug. 299/1 Serve them [*sc.* mackerel] hot with black-butter sauce. To make the sauce..melt 2 ounces of butter until it is dark brown but not burnt, and then throw chopped parsley into it and let it fry for a few seconds only. **1620** Black Clock [see Clock *sb.*^{3}]. **1886** in F. Podmore *Apparitions & Thought Transf.* (1894) vii. 169 He found a 'blackclock' (*i.e.* cockroach) floating in his coffee. **1944** S. Chaplin in *Penguin New Writing* XXII. 106 Living in that rat, mice, bug and blackclock infested area. **1920** *Times Weekly* 19 Nov. 427 (*heading*) Rights of 'Black Coats'. **1940** *Archit. Rev.* LXXXVII. 110/1 Both these sorts of public house are essentially working class; the blackcoat's pub is another sort of place. **1945** H. J. Massingham in F. Thompson *Lark Rise to Candleford* p. ix, It was the vanguard of the city black-coats and proletariat, governed by the mass-mind. **1893** *Jrnl. Soc. Arts* 14 Apr. 506/1 The young fellow who will devote himself to agriculture..in New Zealand..may do even better than the youth who wins his way to the black-coated servitude of a bank. **1902** *Westm. Gaz.* 11 Aug. 7/3 A serried rank of the black-coated. **1928** *Britain's Industr. Future* III. xiv. § 6. 158 The workers are apt to be suspicious of 'black-coated' unions. **1932** D. L. Sayers *Have his Carcase* i. 12 He can't be a fisherman or anything of that kind; they don't waste time snoozing. Only the black-coated brigade does that. 1937 *Daily Herald* 5 Jan. 3 London's blackcoated workers are to hold a mass meeting in the Albert Hall. 1840 *Daily Picayune* (New Orleans) 30 July 2/1 A black man..[will] be tried before Judge Preval, under the Black Code. 1876 *Congress. Rec.* 9 Aug. 5347/2, I hold in my hand the laws of the Legislature of South Carolina passed in the session of 1865–'66...Among the very first acts that they passed was the act which is known all over this country and all over the world as the 'black code' of South Carolina, a code that should disgrace every one of its authors. [*a* 1826 R. HEBER *Narr. Journey India* (1828) II. xxv. 168 A black soil, with many deep cracks, chiefly cultivated in cotton.] 1838 *Penny Cycl.* XII. 205/1 On the whole surface of the table-land [of the Deccan] a black soil prevails, which, from being favourable to the growth of cotton, has been called the black cotton ground, or *regur*. 1844 W. H. SLEEMAN *Rambles & Recoll.* I. xiv. 123 The soil of the valley of the Nerbudda..is formed for the most part of the detritus of trap-rocks...This basaltic detritus forms what is called the black *cotton* soil by the English, for what reason I know not. 1882 GEIKIE *Text-bk. Geol.* III. II. 442 The 'regur', or rich black cotton soil of India. 1935 THOMAS & SCOTT *Uganda* vii. 116 The top soil is very sticky when wet and cracks badly on drying...In its extreme form it is almost uncultivable and is known (quite erroneously since it will not grow cotton) as 'black cotton soil'. 1960 M. PERHAM *Lugard 1898–1945* iii. 51 The black cotton-soil plains of Bornu [Nigeria], flat dry land patched with thorn-scrub. 1889 H. SAUNDERS *Man. Brit. Birds* 379 The Glossy Ibis. *Plegadis falcinellus* (Linnæus)..was known to gunners and fishermen as the 'Black Curlew'. 1911 M. HENRY in *Agric. Gaz. N. S. W.* XXII. 111 The post-mortem appearances in general are all indicative of 'black disease'. *Ibid.*, In 'black disease', so far as is known, sheep about 20 months and 2 years old are principally affected. 1918 S. DODD in *Jrnl. Comp. Path. & Therap.* XXXI. 2 Two explanations of the origin of the term black disease are given, viz., (1) on account of the dark appearance of the liver of animals dead of the disease; (2) because of the dark colour assumed by the under surface of the skin. 1932 *Discovery* Dec. 401/2 Research on the black disease of sheep..showed that the malady is caused through infection following injury to the liver by the young fluke. 1909 *Westm. Gaz.* 22 Oct. 4/2 When..there is no possibility of catching a salmon except by that engine of death, the 'Black Doctor'—the three big hooks tied back to back and dragged along the floor of a pool. 1899 *Daily News* 3 July 5/6 The convict Billinge is what is known as a 'Black dress' man, being thus distinguished because of his bad conduct. 1842 *Proc. Geol. Soc.* III. 712 (*title*) On the Tchornoi Zem, or Black Earth of Central Russia. 1905 *Athenæum* 5 Aug. 175/1 The black-earth country of the south. 1935 HUXLEY & HADDON *We Europeans* vii. 198 The black-earth belt of Russia. 1955 G. V. JACKS *Multiling. Vocab. Soil Sci.* 310 *Black earth*, general term including chernozem and dark plastic clays of tropics. 1577 R. BARNES *Charge* 16 in *Newcastle Tracts* (1847) VI. 16 That no..superfluous faste be vsed as those called the Lady fast saint trinyons fast, the black faste. 1738 F. MOORE *Trav. Africa* 191 Natives, who had been got up together at the Persuasion of a Black Fellow. 1828 *New Monthly Mag.* XXIII. 220 The fish are so abundant that a black fellow with a seine, can load a bullock-cart at one or two hauls. 1925 *Austral. Encycl.* I. 496/1 Near decaying stumps or roots of eucalypts rounded lumps varying in size from a pin's head to a human head, and known to most people as 'blackfellows' bread' are often found. 1935 *Bulletin* (Sydney) 8 May 21/1 Quite good pipes can be carved out of a fungus known as 'blackfellow's bread', which can be dug up almost anywhere on the far South Coast of N.S.W. 1891 *Jrnl. Hellenic Stud.* XII. 164 Small fragment of a black-figure cylix. 1930 J. D. BEAZLEY in *Proc. Brit. Acad.* XIV. 217 (*title*) Attic Black-Figure. 1948 A. LANE *Gk. Pottery* iii. 19 This *orientalising* phase of the seventh century saw also the introduction of polychrome painting and the 'black figure' technique, wherein black-painted silhouette figures were enriched with detail incised in the yet unfired clay. 1858 S. BIRCH *Hist. Anc. Pottery* I. iii. 224 The subjects represented on the black figured vases. 1890 VERRALL & HARRISON *Anc. Athens* 432 It [*sc.* the vase] is of the finest early black-figured style, not later than the time of the sixth and fifth centuries B.C. 1602 J. BRERETON *Discov. Virginia* 13 Beares. Luzernes. Blacke Foxes. 1826 J. D. GODMAN *Amer. Nat. Hist.* I. 276 The black fox is found throughout the northern parts of America..where it is considered among the richest and most valuable of furs. 1922 C. T. BARNES *Mammals of Utah* 120 Black Fox, Silver Fox, *Vulpes fulva argentata*..[is] uniform lustrous black with a distinct white tip to the tail. 1918 L. E. RUGGLES *Navy Explained* 24 The engineer's division is always known as the 'black gang'. 1923 R. D. PAINE *Comr. Rolling Ocean* iii. 44 There wasn't a smarter man in the black gang. 1910 *Sunset* XXV. 173 (*title*) California's black gold: the romance of the oil gushers. 1926 H. C. WITWER *Roughly Speaking* 27 Gold—black gold—this farm is a hotbed of *oil!* 1948 *South Bend* (Ind.) *Tribune* 3 Apr. 5/1 The 'Black Gold' of the farm belt—grunting hogs—is beginning to tarnish. 1969 *Punch* 19 Mar. 401/3 Tankers which leave the Gulf with 200,000 tons of black gold come to Milford Haven with nothing but a cargo of pullulating bacteria. 1804 C. B. BROWN tr. *Volney's Climate & Soil U.S.A.* 409 This is as difficult to the *black gowns* as to ourselves. 1872 *Amer. Naturalist* VI. 94 Everywhere among the western Indians the Jesuits were known by the name of Blackgowns. 1814 in *Mass. Hist. Soc. Coll.* (1815) III. 121 The wood is chiefly black growth, viz. hemlock and spruce. 1851 S. JUDD *Margaret* (ed. 2) I. xvii. 215 Yet there are, what by a kind of provincial misnomer is called the black growth, pines and firs. 1898 *Harper's Weekly* 12 Nov. 1100/2 The secret society called—from its picturesque method of warning its victims—the 'Black Hand' is a political organization known to have existed in the island before the beginning of the American war with Spain. 1904 *N.Y. Tribune* 31 July 2 For months the black hand society has been forcing Italians to contribute to its treasury with threats of death. 1905 *Westm. Gaz.* 5 Jan. 6/3 The notorious Italian blackmailing gang which has been given the name of the 'Black Hand Society'. 1906 *Ibid.* 2 June 9/3 *La Mano Negra*, the famous revolutionary Black Hand Society. 1923 L. J. VANCE *Baroque* viii. 49 The Wop detective that used to play horse with the Black Handers. 1707 MORTIMER *Husb.* 546 In June are ripe the White, Red, Black and Bleeding Hearts. 1910 *Jrnl. Iron & Steel Inst.* LXXXII. 477 In the American black-heart process low sulphur-irons are used. 1917 *Ibid.* XCVI. 330 For the most part, Americans do not machine their malleable castings, and hence they use 'black-heart', which is ordinary cast iron when the decarburised skin is taken off. ..'Black-heart' castings were not suitable for certain purposes. 1962 *B.S.I. News* June 12/2 Methods of test for blackheart, whiteheart and pearlitic malleable iron. 1910 H. P. TIEMANN *Iron & Steel* 22 *Black heat*; *black hot*, a temperature just below a visible red. 1958 *Good Housekeeping* Oct. 133/2 Most electric convectors give what is known as a black heat, which means that the element is made from a special, low resistance, thick wire which does not glow. 1962 *Newnes Encycl. Electr. Engin.* 365/2 Infra-red rays may be produced electrically by utilizing a 'dull emitter' resistance element operating at black heat. 1861 *Chambers's Encycl.* s.v. *Cameo*, The Black Helmet..has a dark onyx ground. 1876 *Encycl. Brit.* IV. 740 The black helmet (*Cassis tuberosa*) of the West Indian seas. 1829 J. MACTAGGART *Three Years in Canada* I. 66 The most compact black ice anywhere to be found. 1922 WRIGHT & PRIESTLEY *Glaciology* x. 325 Large sheets of fresh 'black' ice a very few inches thick..had evidently formed between the floes. 1941 T. A. H. PEACOCKE *Mountaineering* v. 64 There is white ice which is really a form of very hard snow, and there is black ice, the real thing. *Ibid.*, Black ice is seldom met with..except..when the snow has been stripped from the peaks by the fierce heat of the sun. [Black] ice slopes can generally be detected by a glint or gleam, for they reflect the sun's rays much like a piece of glass. 1961 *Guardian* 28 Dec. 1/3 Roads in many places were covered with black ice which was hard to detect at night. 1849 THACKERAY *Lett.* (1887) 91 An expatriated parson..who gets his living by black jobs entirely and attends all the funerals of our country-men. 1876 LORD LENNOX *Celebrities* I. xii. 313 The 'black job' business. 1851 P. BARRY *Fruit Garden* 364 The Plum Wart or Black Knot..originates..from an imperfect circulation of the sap. 1884 W. MILLER *Dict. Names Plants* 252 *Sphæria mortosa*, 'Black knot' fungus. 1915 *Board Agric. & Fisheries Leaflet* No. 213 Gooseberry Black-knot (*Plowrightia ribesia*, Sacc.). The fungus..causing this disease is closely related to *Plowrightia morbosa*, Sacc., the widely distributed 'black-knot' of plum and cherry trees in the United States and Canada. [1785 H. MARSHALL *Amer. Grove* 104 Black American Larch-Tree.] 1803 LAMBERT *Descr. Genus Pinus* I. 56 Black Larch, *P*(*inus*) *pendula*,..shews itself only in the cold mountainous parts of North America. 1907 *Westm. Gaz.* 11 Mar. 13/1 At the last meeting of the British Ornithologists' Club, three examples of the black-lark (*Melanocorypha yeltoniensis*) were exhibited. 1953 BANNERMAN *Birds Brit. Isles* II. 4 *Black lark*..is another very large lark, more closely resembling a blackbird in size but with a short stumpy bill and a short tail. 1936 E. J. G. LEWIS *Television* 80 In high definition systems signals below about 37·5 per cent (Baird) or 30 per cent (Marconi–E.M.I.) of the peak carrier represent synchronizing signals and occur below the 'artificial black' level. 1938 *Jrnl. Inst. Electr. Engineers* LXXXIII. 744/2 The intermediate amplifier received the output from the master oscillator and raised it to a level of 1·2 kW at black level. 1942 *Electronic Engin.* XV. 127 Among the novel features of the design are..keyed diodes for black-level setting. 1953 AMOS & BIRKINSHAW *Television Engin.* I. i. 17 The picture signal voltage at any point in a television system varies between two limits, one representing the maximum brightness it is intended to transmit and known as *white level* and the other representing zero tonal value or black and known as *black level*. 1927 *Daily Tel.* 21 Feb. 7/4 Mr. John L. Baird..described what he calls 'black light'...This..makes it possible to see without any visible light. 1933 *Trans. Illumin. Engineering Soc.* XXVIII. 618 The 'black bulb' ultraviolet lamp..can be easily concealed..projecting invisible radiations of 'black light' on to luminous designs painted on the walls. 1937 *Discovery* Nov. 348/2 'Black light' contains a large proportion of ultra-violet rays, which, when they impinge upon specially prepared surfaces, cause these to luminesce. 1957 *Encycl. Brit.* IX. 426D (Plate 1: *caption*) Photograph of white girl made in the dark by black light. 1877 *Littell's Living Age* 24 Nov. 501/1 In one of the houses in Scotland Yard..is the 'Black Museum', in which, during the last three years, *pièces de conviction*, which until then had been kept indiscriminately with the other property of criminals, have been arranged and labelled, forming a ghastly, squalid, and suggestive show. 1902 A. GRIFFITHS *Myst. Police & Crime* III. xxxiii. 124 Some very beautiful implements are now exhibited in the Black Museum of New Scotland Yard..Amongst these..are some of the tools used by the notorious Charles Peace. 1926 G. DILNOT *Story Scotl. Yard* xxxvii. 303 Round about 1875 the Black Museum came into being, ostensibly for the instruction of young officers, though for long afterwards it was, in fact, little more than a show place, wherein privileged visitors might see the grim relics of great crimes. 1960 *San Francisco News-Call Bulletin* 11 Oct. 4/2 Black Muslim leaders concede there are many ex-convicts, former prostitutes, drug addicts, and alcoholics in the movement, but they put a different interpretation on this than the police. 1961 C. E. LINCOLN *Black Muslims in Amer.* p. iv, The racial emphases peculiar to this rapidly growing, Chicago-centred movement suggested the descriptive phrase 'Black Muslims', which I coined in 1956...Theretofore they had been variously known as the 'Temple People', 'the Muhammadans', 'the Muslims', 'the Voodoo Cult' and 'the Nation of Islam'. 1964 *Punch* 1 Apr. 505/1 Black Muslim claims to an African heritage. 1968 E. CLEAVER *Soul on Ice* ii. 96 A racist Black Muslim heavyweight champion is a bitter pill for racist white America to swallow. 1968 M. A. MALIK *From Michael de Freitas to Michael X* xv. 131, I read as much as I could about the Black Muslims in the Press, but they were presented more or less as a bunch of lunatics. 1874 W. GREGOR *Echo of Olden Time* iii. 22 The oil used was made from the livers of the haddock, cod, ling, and other fish caught on the coast, and was distinguished by the name of *black oil*. 1896 B. REDWOOD *Petroleum* II. IX. 532 Lubricating Oils..'pale' oils..'black' oils..'compound' oils. *Ibid.* 532 Black oils should be free from solid matter in suspension. 1904 GOODCHILD & TWENEY *Technol. & Sci. Dict.* 54/1 *Black oils*, mineral oils which have not been chemically purified. 1962 *Listener* 10 May 796/1 The consumption of oil, to be exact of the so-called black oils, rose 249 per cent. 1965 *San Francisco Examiner* 20 Dec. 53/8 SNCC has chosen a black panther to adorn its party emblem—a requirement of parties in Alabama..SNCC's 'black pather' movement is viewed by some as a form of black nationalism at the local level. 1966 *Economist* 19 Nov. 807/3 Good sense (or fear of white retaliation) led Negroes in Lowndes County, Alabama, to reject the 'Black Panthers', who stand for 'black power'. 1969 E. CLEAVER *Post-Prison Writings* 23, I fell in love with the Black Panther Party immediately upon my first encounter with it. 1970 *Idiot International* Jan. 11/1 The Black Panthers put before the June Convention a proposal for community control of police. 1858 SIMMONDS *Dict. Trade*, *Black plates*,..thin sheets of iron not coated with tin. 1962 *Engineering* 16 Mar. 372/1 Continuous annealing of blackplate has been made more than ten times faster. 1927 *Bull. Dept. Agric. Gold Coast* No. 11, p. 15 In the early morning..pods suffering from black pod disease are coated by mildew. 1945 *Rep. Cocoa Research Conf.* 76 'Black pod' accounted for a loss in Nigeria now of not much less than 10,000 tons...Places which were most susceptible to 'black pod' were the wetter areas in Ife and Oyo Provinces. 1953 *Economist* 19 Dec. 917/1 The outlook for the Nigerian [cocoa] crop is uncertain. ..The poor weather..increases the risk of black pod disease. 1966 *Times* 27 June 1/4 Young Negroes.., supporters of the 'black power' group led by Mr Stokely Carmichael. 1966 *Guardian* 28 June 8/1 The emergence of the slogan 'black power', proclaimed by the supporters of the Student Non-violent Coordinating Committee. 1967 CARMICHAEL & HAMILTON *Black Power* ii. 44 The concept of Black Power..is a call for black people in this country to unite, to recognize their heritage, to build a sense of community. It is a call for black people to begin to define their own goals, to lead their own organizations and to support those organizations. It is a call to reject the racist institutions and values of this society. 1967 *Listener* 5 Oct. 423/1 The word 'black power'..for most Negroes..means the exercise of legitimate power of numbers at the polls, in the market-place and in the community, whereas for many whites it means an open threat to use physical force to gain objectives. 1968 W. SAFIRE *New Lang. Politics* 40/1 *Black power*, a deliberately ambiguous Negro slogan, meaning antiwhite rebellion to some, the use of political and economic 'muscle' to others. 1888 *Lockwood's Dict. Terms Mech. Engin.*, Black Print. 1834 W. YOUATT *Cattle* xi. 356 Inflammatory fever..is termed black quarter, quarter evil, [etc.]. 1709 W. BYRD *Secret Diary* 13 Feb. (1941) 4, I sent him some blackroot..for the gripes. 1833 A. EATON *Man. Bot.* (ed. 6) 288 *Pterocaulon pycnostachya*, black root. 1843 T. TALBOT *Jrnl.* 8 Sept. (1931) 45 We traded some Kooyah or Black root,..a black, sticky, suspicious looking compound, of very disagreeable odor. 1851 C. CIST *Sk. & Stat. Cincinnati* 211 Concentrated extracts of vegetable medical articles..such as..leptandrin or black-root extracts. 1910 G. MASSEE *Dis. Cultiv. Plants* 160 *Black root rot* (*Thielavia basicola*,..) was first met with in England on the roots of peas. 1954 A. G. L. HELLYER *Encycl. Gard. Work & Terms* 208/2 Black root rot..is found most commonly on peas, violas and violets. 1850 *Rep. Comm. Patents: Agric. 1849* (U.S.) 438 In the southern part of the State winter apples are very liable to the black-rot, spots, [etc.]. 1957 *Encycl. Brit.* X. 640/1 Europeans imported the American varieties [of grape-vine], and thereby unwittingly imported phylloxera..and black rot (*Guignardia bidwellii*). 1887 MOLONEY *Forestry W. Afr.* 90 The black-rubber vine, known to the natives as 'Duah Kurrie'. 1790 S. DEANE *New-Eng. Farmer* 20/1 The pods are liable to be hurt by a black rust, if they are exposed much to the sun. 1931 *Discovery* Aug. 260/1 The problem in Kenya with regard to wheat is that two different strains of black rust occur in the main wheat-growing areas. 1957 *Ann. Reg.* 1956 91 The 'Black Sash' movement, consisting of women pledged to 'defence of the Constitution', organized 'vigils' of protest in the larger towns. 1908 [see SCAB *sb.* 2 b]. 1915 *Board Agric. & Fisheries Leaflet* (1916) No. 105. 1 Wart Disease (Black Scab) of Potatoes. (*Synchytrium endobioticum*.) ..In recent years a variety of other names such as Black Wart and Potato Wart have been given to it. 1881 J. H. COMSTOCK in *Rep. U.S. Dept. Agric. 1880* 336 Lecanium Oleae Bernard. The Black Scale of California. 1930 *Discovery* Apr. 135/2 The little ladybird has been enlisted to fight the black-scale pest in California. 1942 C. G. GREY & DALE *U.S. Dept. Agric. Farmers' Bull.* MCMXIV. 15 Swine Dysentery is also known as..bloody diarrhea, bloody scours, bloody dysentery, bloody flux, black scours, and colitis. 1947 *New Biol.* III. 59 Species of Trichostrongylus often cause a blackish-green diarrhoea, which is called 'black scour' by some farmers. 1895 *Daily News* 18 Feb. 2/5 Black sheets for galvanisers. 1930 *Engineering* 21 Feb. 254/3 In the black-sheet trade. 1894 *Daily News* 22 Nov. 8/1 Deep crape and distinctive headgear have been dropped at the end of six months, the period known technically as 'black silk' then setting in. 1845 R. I. MURCHISON et al. *Geol. Russia in Europe* I. xxii. 559 The black soil does not, however, occupy all the vast country alluded to. 1908 N. M. TULAIKOFF in *Jrnl. Agric. Sci.* III. 82 *Black soils* (*Tchernozem*).. cover generally the grassy steppe or prairies of the temperate zone. 1914 H. I. JENSEN *Soils of N.S.W.* x. 47 In Australia we can find..black-soil plains (pasture and agriculture, maize and gourds). 1936 W. LAWSON *When Cobb & Co. was King* xiv. 263 The black-soil plains, which can be so bare in drought-time..were green with short grass. 1959 M. HASTINGS *Hour-Glass to Eternity* I. ii. 40 What we call black soil areas. They're covered in kunai grass..no trees will grow in that soil. [1861 LADY DUFF-GORDON *Let.* 18 Nov. in *Last Lett. fr. Egypt* (1875) 227 Alas! next day came the south-easter—blacker, colder, more cutting than ever.] 1870 *Cape Monthly Mag.* Dec. 357 Then we are sometimes blessed with a *black* south-easter...The mountain is then *quite buried* in cloud, the air is laden with moisture, the winds run howling hither and thither,..the rain descends. 1950 L. G. GREEN *Daybreak for Isles* vi. 74 In the middle of January 1844 a 'black south-easter', the wind most feared in summer on this coast, forced most of the fleet to run for open sea. 1848 H. W. HAYGARTH *Bush Life Australia* ix. 98 A narrow rill, rising out of some rich dark soil, known in Australia as a 'black spring'. 1946 A. MONKMAN in H. Whetton *Pract. Printing* iv. 56/2 Another style of signature which has become popular in bookwork offices is known as the black-step method...A piece of rule, about 6 points thick and 24 points long, is placed between the first and last pages (or spine) of each section...For the first section the rule would be positioned opposite the top line of text. The next section would have the rule stepped down, say 24 points, and so on through the sections. 1880 Black stripe [see STRIPE *sb.* 9]. 1927 *Smallholder* 26 Mar. 104/2 If

any [*sc.* tomato plants] have very finely cut, lace-like leaves, take them out...These plants have contracted Black Stripe disease. 1856 C. M. YONGE *Daisy Chain* II. xxiii. 603 Tom.. had sent on his black tie and agate studs. 1933 J. B. PRIESTLEY *Wonder Hero* iv. 102 Not tails, I think...Not in character. Essentially a black tie part. 1936 R. C. K. ENSOR *England, 1870–1914* xv. 555 Till some time after 1914 there was no rigorous division between a white-tie and a black-tie *ensemble*, such as now compels gentlemen to keep two sets of evening dress. 1951 I. SHAW *Troubled Air* x. 165 Mary Lowell called to ask us..for dinner next Wednesday. Black tie. 1848 A. H. CLOUGH *Bothie of Tober-na-Fuosich* 5 Hope was the first, black-tied, white-waistcoated. 1834 *Amer. Railroad Jrnl.* III. 120/3 A disease in horses and cattle [is] called the *Black Tongue* or *Burnt Tongue*. 1845 W. BAGLEY *Let.* 20 Mar. in N. E. Eliason *Tarheel Talk* (1956) 260 A very fatal disease has been ravaging the counties of Edgecomb & Northampton, it is called the 'black tongue', the patient dies in six hours after he is taken & very few cures are effected. 1858 C. FLINT *Milch Cows* (1860) 281 Typhoid fever..is sometimes followed by diseases known as black tongue, black leg, or quarter evil. 1919 J. P. DUNN *Indiana* II. 804 In 1842–3 epidemic erysipelas prevailed in a number of counties in southern Indiana, and was known by a number of popular names, as 'black tongue', 'sore throat', etc. 1960 RUNNELLS & MONLUX *Princ. Vet. Path.* (ed. 6) 65/2 Dogs fed exclusively on a cereal diet..develop a niacin deficiency disease called blacktongue. 1877 *Rep. Vermont Dairym. Assoc.* VIII. 107 Black tooth is a popular disease of swine. 1867 J. MORISON *Australia as it Is* 88 The native police, or 'black trackers',..are a body of aborigines trained to act as policemen. 1952 R. E. ROBINSON in *Coast to Coast 1951–52* 213 The black-tracker, in his police boy's shirt, trousers, and hat, was with them. 1937 'G. ORWELL' *Road to Wigan Pier* I. i. 7 The grey flocculent stuff known as 'black tripe'. 1961 *Guardian* 29 Sept. 13/6 Yorkshiremen know how succulent it [*sc.* tripe] is—and how varied:..there are..'black' tripe (unbleached), [etc.]. 1939 A. L. DU TOIT *Geol. S. Afr.* (ed. 2) xix. 442 *Black 'Cotton' Soils*..This type is known to the Dutch farmer under the name of '*Black Turf*' from its peaty aspect. *Ibid.* 443 (*heading*) Black turf soils. 1955 J. H. WELLINGTON *S. Africa* I. II. xi. 329 The norite soils are heavy black clays with a much lower magnesium content (12 mg.). These 'black turf' soils occur along the outcrop of the Great Dyke. 1929 K. S. PRICHARD *Coonardoo* vii. 79 'No "black velvet" for you, I suppose?' 'I'm goin' to marry white and stick white,' Hugh said. 1930 H. CRADDOCK *Savoy Cocktail Bk.* I. 29 *Black velvet*. Use long tumbler. ½ Guinness Stout. ½ Champagne. 1942 E. WAUGH *Put out more Flags* 45 Young men..gulping Black Velvet. 1948 D. BALLANTYNE *Cunninghams* 263 I'd like a nice piece of black velvet. One of those quarter-castes, boy. 1861 *Trans. Ill. Agric. Soc.* IV. 454 It is sometimes affected by the black wart. 1915 *Jrnl. Parasitology* I. 107 The not-infrequent occurrence of the notorious 'black widow' spider, *Latrodectes mactans*, in the vicinity of Stanford University. 1927 *Daily Express* 21 July 2/7 A small black spider known to entomologists as '*latrodectus mactans*', and commonly called the 'black widow' or 'shoe-button', which has been introduced from Oriental ports into North America. 1803 A. ELLICOTT *Jrnl.* x. 284 Black-willow..becomes more scarce as you descend the river. 1841 W. A. LEIGHTON *Flora of Shropshire* 485 *Salix Pentandra*, Linn. *Sweet Bay-leaved Willow*... Much sought after by the Irish harvestmen, who call it 'the black willow', and cut it for their *shillelahs*. 1884 C. S. SARGENT *Rep. Forests N. Amer.* 170 *Salix flavescens*.. Black Willow. 1569–1595 Blak wark (work), see D.O.S.T. 1859 SALA *Gaslight & Daylight* xxvi, A florid man who.. sometimes takes a spell in the black work, or undertaking line of business. *a* 1877 KNIGHT *Dict. Mech.*, *Black-work*, the work of the blacksmith in contradistinction to bright-work or the work of the silversmith. 1888 *Lockwood's Dict. Terms Mech. Engin.*, *Black work*, work which has not been machined or polished. In some instances the term would apply to metal work which had been machined on a bearing section, but not elsewhere. And in other cases where no portion, working or otherwise, had been machined. 1910 *Encycl. Brit.* IX. 311/2 A kind of embroidery known as 'black work', done in black silk on linen, was popular during the same reign [*i.e.* of Elizabeth I].

b. Esp. in *Ceramics*.

1766 WEDGWOOD in L. Jewitt *Wedgwoods* (1865) 187 Basaltes or black ware; a black porcelain biscuit. 1787 *Ibid.* 332 The black basaltes having the appearance of antique bronze..is excellently adapted for busts, sphynxes, small statues, etc. 1832 [see BASALT 2]. 1865 [see *BLUE PRINTING, *vbl. sb.*]. 1875 METEYARD *Wedgwood Handbk.* 391 *Black Marble*. A crystalline terra-cotta body. The colour is black shaded. *Black painted*. Single stems and flowers painted on black glazed ware...*Black printed*. Cream-ware printed over or under the glaze with patterns in black. *a* 1884 KNIGHT *Dict. Mech.* Suppl., *Black-basalt Ware*. 1885 *Encycl. Brit.* XIX. 615/1 Etruscan Black Ware. *Ibid.* 618/2 Black Pottery is usually made from a very silicious or sandy clay. *Ibid.* 619/1 Roman black ware decorated with groups of dots in relief. 1934 V. G. CHILDE *New Light on most Anc. East* iii. 79 The rare Black Incised ware, a fabric principally found in Nubia. 1936 *Discovery* Sept. 289/1 This 'black burnished' pottery is identical with the black ware found in Mohenjodaro and not dissimilar to the black ware..of the earlier Chalcolithic periods of Baluchistan and Persia.

black, *sb.* Add: **2. d.** *to be in the black*: to show a profit; to have a credit balance. (From the practice of recording credit items and balances in black.) Cf. *in the red* (*RED *sb.*[1]). orig. *U.S.*

1928 *N.Y. Times* 11 Mar. VIII. 6/2 *In the black*, showing a profit. 1940 *Jrnl. R. Aeronaut. Soc.* XLIV. 734 With but few exceptions American air carriers are..operating 'in the black'. 1955 *Times* 30 Aug. 9/5 On an overdraft, understandably, the banks charge their customers a percentage; but when we are 'in the black' nothing is ever added. 1958 *Times* 24 July 5/1 All the independent television companies are well in the black. A.T.V. made a profit of over £3,500,000 for the past year.

3. c. A deposit of dirt on the body, esp. under the finger-nails. *dial.*

1753 WASHINGTON *Diaries* (1925) I. 49 You say this land belongs to you, but there is not the Black of my Nail yours. 1870 J. P. ROBSON *Evangeline* Introd. (E.D.D.), Aw ha'e wesht baith maw feet frae the black. 1889 *Brighouse News* 14 Sept. (E.D.D.), He weant pairt wi' t'black afore his finger-nails.

7. d. A pigeon of a black variety. **e.** A black postage stamp.

1855 *Poultry Chron.* II. 515/2 A pen of short-faced bald head Tumblers, Blues, Blacks,..Almond, of rare quality. 1890 S. C. SKIPTON *Auction Epitome 1889* 28, 1d. black sheet of 240, used. 1907 *Daily Chron.* 12 Dec. 6/6 A 12d. black of Canada, 1851. 1936 R. GRAVES *Antigua, Penny, Puce* x. 149, I specialize in the archetype and grandmother of all stamps—the Penny Black of 1840. 1970 *Times* (Sat. Rev.) 31 Jan. p. v/2 (Advt.), Over 250 lots of One Penny Blacks, incl. all Plates, many with matched 1d. Reds.

8. In Italian history. **a.** A member or supporter of the political faction of the Neri, the opponents of the Bianchi (Whites), in a feud which began in Pistoia in 1300 and later spread to Florence. **b.** In Rome, a supporter of the Vatican as opponents of the Italian monarchy; one of the clericals.

1680, etc. [see WHITE *sb.* 19]. 1802 C. WILMOT *Let.* 17 Dec. in *Irish Peer* (1920) 134 The blacks and the whites form'd opposite parties which totally disorganized the Republick. 1877 *Encycl. Brit.* VI. 811/2 A quarrel had arisen in Pistoia between the two branches of the Cancellieri,—the Bianchi and Neri, the Whites and the Blacks. The quarrel spread to Florence, the Donati took the side of the Blacks, and Cerchi of the Whites. 1902 *Academy* 11 Oct. 390/1 Were you a White and for the people, or a Black and for the nobles? 1903 [see WHITE *a.* 6 b]. 1909 *Daily Chron.* 29 Jan. 4/6 Most of the skaters are of the Vatican party...'Black' is the local name.

9. A black horse.

1846 J. J. HOOPER *Taking Census* in *Adv. Simon Suggs* I. 153 Mounting our old black, we determined to give the old soul a parting fire. *a* 1861 T. WINTHROP *John Brent* (1883) iii. 26 The black was within the corral, pawing the ground, neighing, and whinnying. 1874 *Rep. Vermont Board Agric.* II. 211 They are able to ride in grand carriages with their three minute blacks hitched thereto.

10. Short for BLACK-LEG 3. (Cf. *BLACK *a.* 11 b.)

1866 *Sat. Rev.* 20 July 59 The lists of 'blacks', and the victims of the picket system.

11. Short for BLACKMAIL 2. *to put the black* (*on a person*), to blackmail. *slang.*

1923 E. WALLACE *Missing Million* xii. 95 I've been paying 'black' for years. *Ibid.* xiv. 115, I don't know the 'black' people; they're not the kind of folk I like to meet. 'Black's' dirty and always will be. 1924 —— *Room 13* iii. 38 Are you trying to put the black on me? 1928 —— *Gunner* xxiii. 190 I've never known you put the black before. 1951 J. B. PRIESTLEY *Fest. Farbridge* III. iii. 561 Got a lovely pub..and yet wants to start putting the black on people!

12. A serious mistake or blunder, esp. *to put up a black*: to make a blunder. *colloq.* or *slang* (orig. *Services'*). (Cf. BLACK MARK.)

1939 *Flight* 26 Oct. 335/2 Last week's special black. The ex-civil pilot..who thought that the ripcord of his parachute was a carrying handle specially provided by the kind manufacturers. 1941 *English Digest* Feb. 38/1 A glaring error is a 'black'. 'I have put up a black,' they will say. 1943 *Word Study* Apr. 6/1 Far from committing the black they expected, she showed great heroism. 1946 G. GIBSON *Enemy Coast Ahead* xi. 142 One day she put up a black... She had fried our salmon in batter. 1948 'N. SHUTE' *No Highway* v. 142 Probably I should have to..leave Government service altogether, having put up such a black as that.

black, *v.* Add: **2. a.** Also with *up*; spec. *intr.*, to colour one's face black in order to play the role of a Negro. orig. *U.S.*

1877 W. R. ALGER *Life of E. Forrest* I. 109 He blacked himself up and rigged his costume quite to his content. 1890 'BIFF' HALL *Turnover Club* 197 They barely had time to get back to the theater to black up for the evening performance. 1934 WODEHOUSE *Thank You, Jeeves* xv. 212 Old Glossop isn't blacking up?

d. *to black* (*a person's*) *eye*: to bruise or discolour the eye by a blow (cf. BLACK EYE 2).

1902 E. NESBIT *Five Children & It* viii. 203 The baker's boy blacked his other eye. 1950 G. GREENE *Third Man* ii. 26 I'd rather make you look the fool you are than black your bloody eye.

e. To blackmail (cf. *BLACK *sb.* 11). *slang.*

1928 E. WALLACE *Gunner* xxx. 244 If I 'blacked' you after this I should be cutting my own throat. 1964 G. SIMS *Dreadful Door* xxiii. 124 He..took naughty photos of them and then blacked them.

3. b. (Later example.)

1905 *Westm. Gaz.* 2 Mar. 10/2 A memorial..urging that betting news should be 'blacked out' from the newspapers in the libraries...Ultimately the Committee decided to 'black-out' horse-racing news.

c. *to black out* (*trans.* and *intr.*), to extinguish or obscure (lights), esp. during a stage performance, or as a precaution against air-raids; also *intr.* of lights, etc.: to be so extinguished or obscured. Also in extended uses.

1921 GALSWORTHY *Six Short Plays* 127 Mr. Foreson!.. Black out! *The lights go out.* 1928 *Amer. Speech* IV. 70 To *black out* is to cut off all light, footlights, borders, and spot. 1934 *Baltimore Sun* 15 Aug. 4/6 There will be a burst of music, and the lights will 'black out'. This will form the prelude to the pageant. 1935 N. MARSH *Enter a Murderer* v. 62 We black out for a little before the curtain goes up. *Ibid.* xxii. 264, I am not going to black-out the lights. 1939 *Daily Mail* 12 Sept. 5/3 It took about three visits from courteous wardens before my house was properly blacked out. 1939 E. AUGUST *Black-out Book* 18/2 (*heading*) You can't black-out the Stars. 1940 *Ann. Reg. 1939* 377 In many countries the lamps of science were dimmed, and in Austria, Czechoslovakia, Poland, and Finland they were blacked out. 1944 in Zandvoort *Wartime Eng.* (1957) 36 Look how the windows are hastily blacking out. 1965 in M. McLuhan *Medium is Massage* (1967) 149 The largest power failure in history blacked out nearly all of New York City.

4. b. To declare to be 'black' (see *BLACK *a.* 11 b). Hence *blacked* ppl. adj.

1958 *Times* 20 Jan. 5/4 The firm's 1,500 employees are 'blacking' work in the fettling shop. 1960 *Guardian* 21 Dec. 2/4 Four men who refused to repair 'blacked' machinery. 1961 *Daily Tel.* 11 Dec. 11/6 (*heading*) Equity 'blacks' TV programme.

6. *to black out*, to suffer a 'black-out' (see *BLACK-OUT 4); esp. in flying, to be temporarily blinded through the effect of a sudden sharp turn or acceleration; to lose consciousness. Cf. *BLACKING *vbl. sb.* 1 c.

1940 *Illustr. London News* CXCVI. 449/1 The blood in his head seeks to fly outwards, and, as his body and feet are away from the centre of the turn, runs towards his legs and drains from behind his eyes, so that he becomes temporarily blind, or 'blacks out'. 1940 *Times* 30 Mar. 9/6 The pilot's body weight would soar, to six, eight, or more times normal. At 8*g* he would black-out, and at 10*g* faint. 1940 *Flight* 7 Nov. 387/1 So I went into a steep left-hand turn and blacked out. 1957 'C. E. MAINE' *High Vacuum* x. 84 'How did you react to the take-off?' 'I blacked out. Don't remember much about it.' 1958 P. MORTIMER *Daddy's gone A-Hunting* vii. 36 The child, dizzy with speed, was blacking out.

black ash. *U.S.* [BLACK *a.* 6.] A North American species of ash (*Fraxinus nigra*), also called *basket-ash* and *hoop-ash*.

1673 in *Essex Inst. Hist. Coll.* (1864) VI. 178/2 A forked Black ash which is alsoe osmund trask his bound. 1728 *Boston Town Rec.* (1883) III. 222 We are of Opinion That no Popler,..Black ash, Basswood, or Ceder Shall be Corded up. 1872 *Rep. Vermont Board Agric.* I. 154 An experiment had been tried by a Cornwall farmer, packing butter in spruce, oak and black ash tubs.

black-ball, *sb.* Add: **3.** A hard sweetmeat. *dial.* Also, in N.Z., *spec.* a humbug.

1851 MAYHEW *Lond. Labour* I. 203/2 'Hard-bake',.. 'black balls',..and 'squibs' are all made of treacle. 1877 *N. & Q.* VIII. 481 'Black-ball' is a delicacy compounded of black treacle and sugar boiled together in a pan. 1943 *Amer. Speech* XVIII. 87 The peppermints described in England as *humbugs* become *blackballs* in New Zealand. 1957 J. FRAME *Owls do Cry* xi. 47 Also a sixpenny shout from her pay for blackballs or acid drops or aniseed balls.

black belt. [BLACK *a.*] **1.** That portion of the southern United States (see quot. 1905) in which the coloured population is most numerous. Also *transf.*

1875 *Congress. Rec.* Jan. 342/1 During this campaign I made a number of speeches in Georgia. I spoke in what is known as the 'Black Belt'. 1898 *Ibid.* Feb. 1594/1 The Fourth Alabama Congressional district..was composed entirely of Black Belt counties. 1905 *N.Y. Even. Post* 21 Nov. 3 The Black Belt has a curiously irregular shape. Extending from Virginia across North and South Carolina, Georgia, Alabama, to Mississippi and Southern Louisiana, it stretches a narrow arm across the river and up into southern and central Arkansas. 1960 *Guardian* 8 Dec. 8/3 Would it have been possible for you to have lived with your white wife in the North reasonably unmolested? If we had kept within the Black Belt.

2. A belt of fertile black soil.

1870 *Q. Jrnl. Geol. Soc.* XXVI. 120 The black soil lies in an extensive bed..the black belt running from the head of the Gulf in a north-westerly direction. 1883 E. A. SMITH *Rep. Alabama Geol. Surv. 1881–82* 268 The Black Belt or Canebrake Region. This division of the prairie region is underlaid by the Rotten Limestone before described. 1901 B. T. WASHINGTON *Up from Slavery* 108 The term 'Black Belt'..was first used to designate a part of the country which was distinguished by the colour of the soil.

3. The belt worn by one who has attained a certain degree of proficiency in judo; also, a person qualified to wear such a belt.

1913 E. J. HARRISON *Fighting Spirit of Japan* iv. 59 Costume in the practice of *judo*...The only outward distinction of rank in the Kano school lies in the colour of the belt...From *shodan* upwards a black belt is worn. 1954 E. DOMINY *Teach yourself Judo* vii. 75 If unable to obtain the usual grip, you will find it almost as effective—many Black Belts consider it better—to place one or both your arms outside your opponent's arms. 1960 *Oxf. Mail* 10 Mar. 8/2 The Cambridge captain, D. J. Mason..a black belt of 1st dan (teacher grade). 1967 'J. MUNRO' *Money that Money can't Buy* vii. 81 His black belt Loomis knew all about, but what dan he held was a secret.

black birch. *U.S.* [BLACK *a.* 6.] One or other of several North American species of birch, esp. *Betula lenta* and *B. occidentalis*.

1674 JOSSELYN *Two Voyages* 69 The Birch-tree is of two kinds, ordinary Birch, and black Birch. 1711 *Boston Town Rec.* (1883) III. 86 A black burch on the Side of a Swamp. 1832 D. J. BROWNE *Sylva Amer.* 118 Wherever it grows in the United States, it is known by the name of Black Birch. 1957 M. HADFIELD *Brit. Trees* 174 The wood of some birches..produces wood alcohol. For this purpose the sweet or black birch..of the north-eastern U.S.A. is much used.

blackbirder (blæ·kbɜːdəɹ). [f. BLACKBIRDING + -ER[1].] A man or a vessel engaged in BLACKBIRDING or slave-traffic.

1883 *All Year Round* 22 Sept. 355 Blackbirders, the

kidnappers for labour purposes on the islands of the Pacific. 1888 W. B. CHURCHWARD *Blackbirding* 126 You see the harm those cussed blackbirders do in the islands. *Ibid.* 163 That chap whose throat I cut on board the blackbirder. 1928 *New Statesman* 28 July 507/2 Polynesians in their wild state..were..shipped to Australia by enterprising gentlemen called blackbirders. 1969 *Sunday Mail Mag.* (Brisbane) 9 Feb. 4/1 After 20 years as a 'blackbirder', Buttray had returned to the more respectable coastal service.

blackbirding, *vbl. sb.* (Earlier and later examples.) Hence (as a back-formation) **blackbird** *v.*

1873 HOTTEN *Slang Dict.* 84 *Blackbirding*, slave-catching. Term most applied nowadays to the Polynesian coolie traffic. 1888 W. B. CHURCHWARD (*title*) 'Blackbirding' in the South Pacific. *a* 1889 in BARRÈRE & LELAND *Dict. Slang* I. 125/1 But sometimes—we are glad to say in the past—iniquitously blackbirded or kidnapped, and practically sold into slavery. 1897 W. C. MORROW *The Ape, the Idiot* (1898) 314 'Blackbirding' (which is kidnapping Gilbert Islanders and selling them to the coffee-planters of Central America). 1908 *Daily Chron.* 6 Nov. 4/6 'Blackbirding'..is not yet an entirely extinct industry [in the Pacific Islands]. 1946 K. TENNANT *Lost Haven* (1947) 4 He had really made it [*sc.* money] blackbirding natives from the Pacific Islands to the Queensland sugar plantations.

blackboard. Add to def.: Now also used of a similar board in any colour. **b.** *attrib.*, as **blackboard jungle**, an undisciplined or unpleasant school.

1870 MEDBERY *Men & Myst. Wall St.* 21 The 'marker' or black-board clerk writes off the prices upon the tablet. 1880 *Plain Hints Needlework* 33 These three strips can be sewed..together, and thus form a dish-cloth or black-board rubber. 1890 W. J. GORDON *Foundry* 89 To see the service complete we should have been in the great hall, and heard the blackboard lesson. 1895 *Daily News* 26 Apr. 3/2 'Black-board Drawing'..seemed to him to be useful..for..the student, the teacher, and the child. 1905 H. H. STEPHENS (*title*) Black-board and free-arm drawing. 1954 E. HUNTER (*title*) Blackboard jungle. 1958 *Economist* 29 Nov. 784/1 This is a variation, *andante moderato*, on the theme of the blackboard jungle. Mr Townsend describes the teacher's end of the stick—now used more and more sparingly—in the new system of secondary modern education.

black bottom. 1. [BLACK *a.* 1 c, BOTTOM *sb.* 5.] A low-lying area inhabited by a coloured population. *U.S.*

1915 *Lit. Digest* 4 Sept. 500/2 Uncle Mose aspired to the elective office of justice of the peace in the 'black bottom' part of town.

2. The name of a dance, esp. popular in and for a time after 1926. Also as *v.* orig. *U.S.*

1926 *N.Y. Times* 19 Dec. VII. 4/6 It occurred to the producer that if you could dance before the beat you would have a new rhythm...The result is the Black Bottom. 1927 *Observer* 6 Feb. 15/7 The accounts of the new dances are discouraging. There is the Black Bottom, the very name of which spoils a spring morning. 1927 *Daily Express* 25 May, Miss Bradhurst had black bottomed nineteen miles ..before she collapsed. 1928 'SAPPER' *Female of Species* v. 76 'What matter that his Black Bottom is the best in London.' 'My Gawd! sir,' gasped the other. 'His 'ow much?' 1968 D. BRAITHWAITE *Fairground Archit.* viii. 137 The mock elegance and good taste of the Victorian *soirée* gave place to the rumbustious 'Charleston' and the 'Black Bottom'.

black boy. Also **black-boy, blackboy. 1.** A boy having a black or very dark skin; spec. (*a*) = *BLACK MAN 1; (*b*) a Negro manservant (cf. BOY *sb.*[1] 3).

1635 *Relation of Maryland* v. 28 The Children live with their Parents; the Boyes untill they come to the full growth of men..then they are put into the number of Bow-men, and are called Black-boyes. 1666–7, etc. [see BLACK *a.* 1 c]. 1681 [see BOY *sb.*[1] 3 c]. 1738 F. MOORE *Trav. Inland Afr.* 64 *Creague* saved himself in his Boat, with the Help of his black Boy. *Ibid.* 175 As he was chastising his BlackBoy. 1774 J. HARROWER *Diary* 24 May in *Amer. Hist. Rev.* (1900) VI. 78 There was a gray Mare started with the Bay a black boy ridder but was far distant the last heat. 1829 B. HALL *Trav. N. Amer.* I. 122 His first assistant was a sharp-faced, well-tanned, old woman in spectacles; next came a black boy. 1923 D. H. LAWRENCE *Birds, Beasts & Flowers* 178 The fully fathomless, shining eyes of an Australian black-boy. 1948 V. PALMER *Golconda* xviii. 148, I had a trip over there with a blackboy and a couple of camels.

2. An Australian grass-tree of the genus *Xanthorrhœa*, having a thick dark trunk and a head of grass-like leaves, esp. *Xanthorrhœa preissii*.

1840 *Jrnl. Bot.* II. 344 One of the most striking plants to a stranger is our common *Blackboy*, a fine arborescent species of *Xanthorrhœa*. 1840 PEREIRA *Elem. Mat. Med.* II. 658 A red resin, probably from *X*[*anthorrœa*] *arborea*.. .has been recently imported under the name of *black-boy gum*. 1846 J. L. STOKES *Discov. in Australia* II. 132 Black boy.. gum on the spear—resin on the trunk. *Ibid.* 380 *note*, These trees, called Blackboys by the colonists, from the resemblance they bear, in the distance, to natives. 1870 W. H. KNIGHT *W. Australia* 46 A resin got from the 'black-boy', or grass tree, (*xanthorrhea drummondii*). 1956 S. HOPE *Diggers' Paradise* 59 The 'blackboys' several times taller than a man with twisted coal-black stems crowned by massive raffia-like tufts in green and white.

3. A kind of black tobacco.

1898 *Daily News* 8 Oct. 3/3 She had been addicted to the use of the weed, in the specific shape of 'black boy', for over forty years. 1908 *Daily Chron.* 10 Apr. 4/7 A black boy tobacco much in favour with mariners.

black-browed, *a.* Add: **b.** *black-browed albatross*, an albatross, *Diomedea melanophrys*, found in the south Atlantic Ocean.

[1865 GOULD *Handbk. Birds Australia* II. 438 *Diomedea Melanophrys*..Black-eyebrowed Albatros.] 1883 *Encycl. Brit.* XV. 334/2 All the rest of the plumage is white, except a dusky superciliary streak, whence its name of Black-browed Albatros. 1968 T. STOKES *Birds of Atlantic Ocean* 26/2 The black-browed albatross is a beautiful bird of regular, clear-cut marking.

Blackbu·rnian, *a. U.S.* [f. the name of Mrs. Hugh *Blackburn*, fl. 18th c.] *Blackburnian warbler*, a North American warbler (*Dendroica blackburniæ*). Also *ellipt.* as *sb.*

1783 LATHAM *Gen. Syn. Birds* II. II. 461 Blackburnian Warbler. 1868 *Amer. Naturalist* II. 179 The Blackburnian Warbler (*D. Blackburniae*) is one of the most beautiful. 1893 B. TORREY *Footpath-Way* 6 We..soon were in the old forest listening to bay-breasted warblers, Blackburnians, black-polls, and so on. *Ibid.* 16 A Blackburnian warbler perched as usual, at the very top of a tall spruce. 1936 T. S. ELIOT *Coll. Poems 1909–35* 152 Leave to chance The Blackburnian warbler, the shy one. 1947 *Harper's Mag.* May 480/1 The orange flittings of a Blackburnian in the treetops.

blackbutt (blæ·kbɒt). *Austral.* Also **black butt, black-butt.** [BLACK *a.*, BUTT *sb.*[3] 2.] An Australian timber tree, *Eucalyptus pilularis*. Earlier, *black-butted gum*.

1801 C. GRIMES in *Hist. Rec. Austral.* (1915) 1st Ser. III. 414 The finest stringy-bark and black-butted blue-gum trees I ever saw. 1820 J. OXLEY *Jrnls. Two Exped. N.S.W.* 331 The timber was chiefly black butted gum. 1847 LEICHHARDT *Jrnl.* ii. 49 The range..having, with the exception of the Blackbutt, all the trees..of Moreton Bay. 1901 *Daily Chron.* 3 Sept. 7/4 Fencing post of blackbutt, forty-five years in the ground. 1934 *Archit. Rev.* LXXVI. 70/2 In addition jarrah, blackbutt, mountain ash and blackwood from Australia..are all good flooring timbers.

black-cap. Add: **4. b.** A halved apple baked with the flat side downwards and a 'cap' of sugar.

1723 J. NOTT *Cook's & Confectioner's Dict.* No. 103 B, *Black Cap.* Take..large Pippins..cut them in halves.. squeeze a Lemon..over them..grate over them..Sugar ..put them into a quick Oven. 1958 *Listener* 27 Nov. 903/1 Sprinkle a 'cap' of caster sugar on each [*sc.* apple]. Bake for twenty or twenty-five minutes above the chicken, and when the bird goes to the top of the oven, put the 'blackcaps' in its place.

5. A variety of raspberry; the Black Raspberry, *Rubus occidentalis*. *U.S.*

1847 W. DARLINGTON *Agric. Bot.* 127 *Rubus occidentalis*.. Wild or Black Raspberry. Thimble-berry. Black Caps. 1886 *Harper's Mag.* July 281/2 There is another American species of raspberry (*Rubus occidentalis*) that is almost as dear to memory as the wild strawberry—the thimble-berry, or black-cap. 1946 B. MACDONALD *Egg & I* (1947) vi. 85 Alder, salal, wild raspberry and blackcaps.

black-capped, *a.* **1.** [Cf. BLACK CAP 3.] Of birds: having the top of the head black.

1781 J. LATHAM *Gen. Synop. Birds* I. 727 Black-Capped Cr[eeper]..head is black. *Ibid.* 748 Black-Capped H. B. [Humming-Bird]. 1808 A. WILSON *Amer. Ornith.* I. 134 [The] Black-Capt Titmouse, *Parus Atricapillus*,..is one of our resident birds. 1868 *Amer. Naturalist* II. 175 The Green Black-capped Warbler (*Myiodioctes pusillus*). 1870 *Ibid.* III. 74 The cries and habits of all these black-capped species. *Ibid.* IV. 543 In the depths of winter they and..the Black-capped Titmouse (*Parus Atricapillus*), enliven the woods.

2. Wearing a black cap.

1908 *Westm. Gaz.* 15 Jan. 3/1 Moors in turbans..and black-capped Jews. 1930 BLUNDEN *Summer's Fancy* 22 And black-capped and gowned The sour-tongued master stared.

black currant. Add: (Earlier example.)

1629 PARKINSON *Parad.* III. ii. 558 The blacke Curran bush riseth higher than the white.

b. *black currant* (*gall*) *mite*, an insect pest that attacks currant and gooseberry bushes, but especially the black currant bush, living chiefly inside the buds and producing the disease known as 'big bud' (*BIG *a.* B. 2).

1890 E. A. ORMEROD *Man. Injurious Insects* (ed. 2) III. 303 The details of those of the Black Currant Gall Mite have not been described. 1918 *Board Agric. & Fisheries Leaflet* No. 1. The disease known as 'big bud' in Black Currants, caused by the Black Currant Mite, was first recorded in these islands in the year 1869.

blacked, *ppl. a.* Add: **2.** *blacked out*, (*a*) obliterated with black; also *fig.*; (*b*) with lights extinguished or obscured; (*c*) of an aircraft pilot, temporarily blinded (see *BLACK *v.* 6).

1919 *Illustr. London News* 27 Sept. 482 (*caption*) No longer 'blacked out': London herself again. 1930 E. POUND *XXX Cantos* vii. 26 Time blacked out with the rubber. 1933 *Flight* 2 Feb. 100/1 Tight turns had been made for much longer periods than 10 sec., while the pilot was 'blacked out', without any tendency whatsoever to lose consciousness. 1938 *San Francisco Examiner* 27 Sept. 3/7 (*headline*) City blacked out in raid test. 1939 *Punch* 8 Nov. 504/2 We must still endure the totally blacked-out nights. 1940 C. DAY LEWIS tr. *Virgil's Georgics* I. 23 At the South pole.. it's dead of night,..the shadows shrouded in night, blacked out for ever. 1940 *New Statesman* 19 Oct. 371 Most unfavourable conditions—crowded shelters, stuffy, blacked-out factories and offices. 1967 V. NABOKOV *Speak, Memory* (ed. 2) xiv. 292 Opaque curtains separated me from blacked-out Paris.

blacketeer: see *BLACK MARKET.

black eye. Add: **2.** (Later examples.)

1886 C. COBORN (*title of song*) Two lovely black eyes. 1926 GALSWORTHY *Silver Spoon* II. iii. 132, I thought they'd have a row...Hadn't Michael a black eye? 1937 D. L. SAYERS *Busman's Honeymoon* i. 54 At this moment I could have been tramping at your heels with five babies and a black eye.

b. *fig.* A severe blow or rebuff.

1744 C. CIBBER *Another Occas. Let. to Mr. P*[*ope*] 8 If you had not been a blinder Booby, than myself, you would have sate down quietly with the last black Eye I gave you. 1813 J. ORROK *Let.* 1 Oct. (1927) 167 The young folks gave the Jelly and Jam a black eye. 1900 *Congress. Rec.* Jan. 1004/2, I hope the Pension Committee will give a black eye to every bill of that kind. 1962 *Listener* 29 Nov. 895/1 A black eye for the Indian Army.

black-eyed, *a.* [BLACK *a.* 18.]

1. Having black eyes. Cf. BLACK EYE 1.

1598 CHAPMAN *Iliad* I. p. 4 Nor will containe..his heauie hand before: The blacke eyde virgin be releast. *a* 1667, 1775 [see BLACK EYE 1].

b. *spec.* Of a variety of pea: having a black speck.

1728 W. BYRD *Hist. Dividing Line* in *Writings* (1901) 74 Each Cell [of N.C. pine cone] contains a Seed of the Size and Figure of a black-ey'd Pea. 1786 WASHINGTON *Diaries* (1925) III. 56 They proceeded to sow the small black eyed pea. 1857 *Texas Almanac* 13 Plant Black-Eyed..Peas. 1862 *Chambers's Encycl.* s.v. *Dolichos*, *D. sphaerospermus* (Calavana or Black-eyed Pea), a native of the West Indies.

2. *black-eyed Susan*: **a.** A name applied to various plants having pale flowers with dark centres, esp. *Thunbergia alata* and *Rudbeckia hirta*.

1891 *Cent. Dict.* s.v. *Thunbergia*, The hardy annual *T. alata*, known locally by the name *black-eyed Susan* from its buff, orange, or white flowers with a purplish black center. 1900 C. BENNETT in W. D. Drury *Bk. Gardening* xv. 696 *T. alata*, with a buff-yellow corolla, and a very dark eye nearly approaching black (hence the common name Black-eyed Susan). 1906 H. D. PITTMAN *Belle of Blue Grass C.* xvi. 241, I found all of the waste places now covered with black-eyed susans. 1956 A. M. COATS *Flowers & their Histories* 226 'Black-eyed Susan' (the annual *R. hirta*) immigrated in 1714.

b. *U.S. slang.* (See quots.)

1869 in M. Mathews *Beginnings Amer. Eng.* (1931) 153 Among names of revolvers I remember the following: Meat in the Pot,..Black-eyed Susan. 1888 FARMER *Americanisms*, *Black-eyed Susan*.—Texan for a revolver.

black-face. Add: (Also without hyphen.) **2.** †(*a*) A Negro. *Obs. U.S.* (*b*) The exaggerated make-up worn by an actor playing a Negro role. Also *attrib.* orig. *U.S.*

(*a*) 1704 S. KNIGHT *Jrnl.* (1825) 40 Order the master to pay 40s to black face. 1899 H. B. CUSHMAN *Hist. Choctaw Indians* 312 Do they not even now kick and strike us as they do their black-faces?

(*b*) 1869 F. DUMONT *Benedict's Congress Songster* 9 Lew made his first bow before the public at the Metropolitan Theatre in a black face. 1895 *N.Y. Dramatic News* 16 Nov. 7/4 An old-time black-face actor died in Bellevue hospital. 1969 C. HIMES *Blind Man with Pistol* xii. 129 Two big rugged black men clad in belted leather coats, looking for all the world like Nazi SS troopers in blackface.

black-faced, *a.* Add: **1.** Of sheep (further examples).

1801 *Farmer's Mag.* Nov. 394 The Cheviot breed of sheep will succeed fully as well in the North, if not better, than the common Black-faced. 1886 C. SCOTT *Sheep-Farming* 113 Blackfaced stocks [are letting] at about 2s 6d per sheep.

black fish. Add: **1.** (U.S. examples.)

1765 R. ROGERS *Acct. N. Amer.* 68 In the sea adjacent to this island [*sc.* Long Island] are sea-bass and black-fish in great plenty. 1888 GOODE *Amer. Fishes* 39 The Sea Bass is also known south of Cape Hatteras as the 'Blackfish'. 1897 *Outing* (U.S.) XXX. 160/2 These fishermen, who think that God made blackfish for them exclusively.

2. (Earlier example.)

1688 S. SEWALL *Diary* (1878) I. 239 [We saw] a number of Fishes called Bottle-noses. Some say they are Cow-fish or Black-fish.

4. The freshwater fish *Gadopsis marmoratus*, found in Australia.

1850 CLUTTERBUCK *Port Phillip* iii. 44 The Schnapper, blackfish and eel. 1879 McCOY *Zool. Victoria* I. III. 39 *Gadopsis gracilis*...The Yarra Blackfish. 1890 *Melbourne Argus* 9 Aug. 4/5 You could catch a few blackfish in the pools. 1962 L. WEDLICK *Fishing in Australia* III. 76 The blackfish is a poor fighter.

black flag. (Further examples.)

1720 DEFOE *Capt. Singleton* xiii, We let them soon see who we were, for we hoisted a black flag, with two cross daggers on it. 1821 SCOTT *Pirate* xl, Up goes the Jolly Hodge, the old black flag, with the death's head and hour-glass. 1827 P. CUNNINGHAM *N.S.W.* II. xiii. 302 Every effort ought to be made to impress..upon..the multitude the terrible nature of our punishments, without permitting them to be..spectators...The having..a black flag with emblems of death hung out [etc.]. 1870 BREWER *Dict. Phrase & Fable* 301/1 *To unfurl the black flag*, to declare war. 1887 *Longm. Mag.* Nov. 105 Mr. Brander Matthews alleges that certain English publishers can also run up the 'Jolly Roger' on occasion, and sail under the Black Flag.

black fly. [BLACK *a.* 6.] **1.** Any of various dark-coloured flying insects.

1608 [see FLY *sb.*[1] 1 a]. 1716 J. LOWTHORP *Phil. Trans. Abridged* II. 338 In the Cross-Bath the Guides have observed a certain Black Fly with Sealed Wings, in the form of a Lody-Cow, but somewhat bigger. 1789 B. LINCOLN in *Mass. Hist. Soc. Coll.* (1795) IV. 148 Objections have been made to these counties, on account of the black flies, and other insects. 1876 *Fur, Fin & Feather* Sept. 139 At Calais, Maine, last fall rugged grouse were as thick as black flies in August. 1889 [see FLY *sb.*[1] 2]. 1959 *Times* 30 Oct. 15/1 Blackfly or buffalo gnat, but coming to be known in West Africa by the ominous name of 'eye flies'.

2. A name given to various insects infesting plants, esp. to an insect of the species *Frankliniella robusta* (formerly *Thrips pisivora*) and *Aphis rumicis*: used *collect.* in *sing.* for thripses or aphides when infesting certain plants and giving a blackish appearance to the part affected.

1743 ELLIS *Mod. Husb.* July iv. 31 But when the Black Bug, or Fly..takes the Ear..they do a great deal of harm. 1850 *Working Man's Friend* 13 July 55/1 To Remove the Aphides, or black fly. 1884 *Sutton's Cult. Veget. & Flowers* 269 The Blue and the Black Fly are common plagues of the Peach-house and the orchard. 1919 *Board Agric. & Fisheries Leaflet* No. 104. 2 The Bean Aphis, known variously as Black Fly, Collier, and Black Dolphin. 1951 *New Biol.* XI. 50 Many a garden crop of beans..has been saved from the devastating 'black-fly' by early pinching-out of the growing tips.

black foot. Add: **1.** (Earlier and later examples.) Also, an Indian of this tribe. (*b*) The language of this tribe. Also *attrib.*

1794 *Mass. Hist. Soc. Coll.* (1810) 1st Ser. III. 24 The Blackfeet tribe. 1834 J. TOWNSEND *Jrnl.* 17 July in A. B. Hulbert *Overland to Pacific* (1934) IV. 192 The Blackfoot is a sworn and determined enemy to all white men. 1836 IRVING *Astoria* I. xv. 246 Colter..[had] some knowledge of the Blackfoot language. 1933 BLOOMFIELD *Lang.* iv. 72 A few detached languages in the west: Blackfoot, Cheyenne, and Arapaho. 1962 *Listener* 26 Apr. 721/2 He was captured by Blackfeet Indians. 1969 *Observer* (Colour Suppl.) 18 May 22/4 Among the Blackfoot, stealing an enemy's weapons was the highest exploit. *Ibid.* 25/2 Yet calling a Blackfoot, for example, 'warlike' reveals nothing. The entire Blackfoot tribe did not habitually engage in war because individual members possessed 'warlike' personalities.

black-grass. [BLACK *a.* 1.]

1. A species of rush (*Juncus Gerardi*) growing in salt-marshes. *U.S.*

1782 H. ST. JOHN DE CRÈVECŒUR *Lett. of American Farmer* vi. 159 The best mowing grounds in the island, yielding four tons of black grass per acre. 1872 SCHELE DE VERE *Americanisms* 408 *Salt-hay*, a very important product of salt-marshes, is of two principal sorts, called *salt-grass* and *black-grass*.

2. A species of foxtail grass, *Alopecurus agrestis*.

1798 W. CURTIS *Flora Lond.* II. Plate 16 *note*, The Farmer also distinguishes the *Alopecurus agrestis*..by the name of Black Grass. 1840 *Farming at Scoreby* in *Husbandry* (*Libr. Useful Knowl.*) III. 6 'The black grass'..is a very noxious weed.

blackguardism. 1. (Earlier example.)

1785 G. A. BELLAMY *Apology* I. lxii. 311 He had the meanness, servility, and blackguardism of a Buckhorse.

blackguardry. Add: Also **blackguardery** *U.S.*

2. = BLACKGUARDISM 1.

1881 G. M. HOPKINS *Let.* 14 May (1935) 129, I have in me a great vein of blackguardry and have long known I am no gentleman. 1883 —— *Let.* 14 Jan. (1956) 251 Blackguardry stamps my whole behaviour. 1936 M. MITCHELL *Gone with Wind* x. 202 The blackguardery of the Irish who were being enticed into the Yankee army by bounty money.

black gum. *U.S.* (Also hyphened and as one word.) [BLACK *a.* 6.] A North American tree of the genus *Nyssa*. Also *attrib.*

1709 LAWSON *Carolina* 95 Of the Black Gum there grows, with us, two sorts. 1785 WASHINGTON *Diary* 28 Feb. (1925) II. 346 Planted all the..Blackgums in my Serpentine Walks. 1868 [see GUM *sb.*[2] 5]. 1868 *Amer. Naturalist* II. 122 When mast is not plenty, they [*sc.* bears] lap black-gum berries. 1901 C. MOHR *Plant Life Alabama* 32 Chestnut..and black gum (*Nyssa sylvatica*) are common.

black-head. Add: **4.** = COMEDO.

1837 J. F. PALMER *Gloss.* in 'A Lady' *Dialogue Devonshire Dial.* 30 *Blackhead*, a boil or pinswell, from the black spot which appears at the apex. 1885 [see COMEDO]. 1910 *Daily Chron.* 15 Jan. 9/1 Sometimes blackheads gather in the forehead lines,..they must be pressed out with the blackhead instrument, if steaming and massage fail to remove them. 1949 E. DE MAUNY *Huntsman in Career* III. iii. 160 His skin was tanned deep brown, which somehow accentuated the blackheads.

5. *Vet.* An infectious disease of turkeys and other birds caused by the protozoon parasite *Histomonas meleagridis*, producing a bluish-black discoloration of the head.

1902 *U.S. Dept. Agric. Experiment Station Rec.* XIII. 287. 1906 *U.S. Dept. Agric. Bureau Animal Ind. Circular* No. 5 (*title*) Blackhead, or infectious entero-hepatitis in turkeys. 1955 GAIGER & DAVIES *Vet. Path. & Bacteriol.* (ed. 4) xxiii. 468 In dealing with blackhead emphasis should be placed on prevention.

6. *attrib.* = *BLACK-HEADED *a.*

1827 J. L. WILLIAMS *View W. Florida* 30 Black-head fly catcher. 1835 J. MARTIN *Gaz. Virginia* 483 A great variety of ducks as..the red head shoveler, the black head shoveler.

bla·ck-hea:ded, *a.* Having a black head: used in the names of animals, esp. birds.

1774 [see BLACK *a.* 18]. 1783 LATHAM *Gen. Syn. Birds* II. I. 198 Black-headed B[unting]. *Emberiza melanocephala*. 1785 PENNANT *Arct. Zool.* II. 398 Black-headed Warbler. 1870 *Amer. Naturalist* III. 75 The habits of the Black-headed grosbeak are quite different. *Ibid.* 234 The Black-headed Gull, a southern and somewhat rare species. 1902 *Encycl. Brit.* XXXII. 109/2 The Black-headed Rock Snake (*Aspidiotes*), one of the Pythons. 1931 *Discovery* Mar. 87/1 Black-headed gulls, the little red-legged, red-beaked sea birds so familiar to Londoners as winter visitors to the metropolitan parks and riverside. 1964 C. WILLOCK *Enormous Zoo* ix. 165 A black-headed heron picks for grubs in the garden.

black house. †**1.** A prison; also, a place of business where working-hours are long and wages are very small. *Obs. slang.*

1848 *Flash Dict.* in *Sinks of London laid Open* 99/1 *Black houses*, prisons. 1861 MAYHEW *Lond. Labour* III. 224/2 The black houses, or linendrapers at the west end of London, were principally supplied from the east end.

2. *Sc.* Also *black-house*. (*a*) A turf house; (*b*) a house built of unmortared stone, found esp. in north-western Scotland and the Hebrides.

1824 J. MACCULLOCH *Highlands & Western Isles* I. 112 The genuine, pure black house is built entirely of turf; walls and roof: it is a 'good black house' when the roof is of thatch. 1870 *Proc. Soc. Antiq. Scotl.* VII. 154 The native structure is a Black-house or Tigh-dubh. 1911 W. C. MACKENZIE in N. Munro *Home Life of Highlanders* 38 In some of the outlying districts..there are phases of life that have apparently remained unaltered since the Middle Ages. They are typified by the 'black houses', many of which are still to be found in the Long Island. 1931 W. C. MACKENZIE *Sc. Place-Names* viii, 'Black-houses' of the Outer Hebrides, *i.e.* the houses built of turf [as contrasted with] the 'white' or stone houses, by which they are rapidly being replaced. 1951 *Antiquity* XXV. 200 The roof had doubtless been the normal blackhouse roof with no smoke-hole.

blackie, var. BLACKY *sb.*

blacking, *vbl. sb.* Add: **1. b.** *blacking-out*: see BLACK *v.* 3 b and *3 c.

1905 *Academy* 27 May 562/1 Maternal censorship is rigid, the Russian blacking-out system not more so. 1919 *Illustr. London News* 27 Sept. 482 After the war-years.., the great 'blacking-out' to deceive enemy aircraft, London is herself again.

c. *blacking-out*: see *BLACK *v.* 6.

1932 *Flight* 1 Apr. 272/1 With the speed of modern fighters when on the dive, it might happen that the fighter pilots would experience 'blacking out' at the critical moment of the turn and zoom. 1933 *Jrnl. R. Aeronaut. Soc.* XXXVII. 397 The effect most frequently noticed is the well-known phenomenon of 'blacking-out', wherein the subject without losing consciousness becomes momentarily blind.

3. c. (Further examples.) **blacking leather,** leather which is to be blacked, as distinct from patent leather.

1837 DICKENS *O. Twist* (1838) I. v. 78 His 'prentice..sent 'em some medicine in a blacking-bottle. 1850 —— *Dav. Copp.* xxvii. 283 His blacking-brushes and blacking were among his books. 1893 YEATS *Celtic Twilight* 184 A fiddle made apparently of an old blacking-box. 1896 *Daily News* 2 July 8/7 High glacé boots have patent leather goloshes, and others have plain 'blacking' leather.

black jack, black-jack. 1. Delete ? *Obs.* and add later examples.

1854 C. M. YONGE *Heartsease* I. ii. 31 She was enchanted with St. Cross..in raptures at the black jacks, dole of bread and beer. 1921 O. BAKER *Black Jacks & Leather Bottells* iii. 67 The black jack was..a kind of leathern pitcher or jug, always lined with pitch or metal. 1965 *Harper's Bazaar* 49 Black Jack jugs in two sizes, £4 7s. 6d. and 6 gns. (waterproof leather).

3. (Earlier examples.)

1765 J. BARTRAM *Diary* 31 July in *Trans. Amer. Philos. Soc.* (1942) XXXIII. 17/1 Yᵉ oaks black which is reconed yᵉ best fire wood thay have thay call them black Jacks seldom grow above A foot diameter. 1782 JEFFERSON *Notes Virginia* (1787) 62 Black jack oak. Quercus aquatica.

6. The black caterpillar of the turnip saw-fly, *Athalia centifolia* or *A. spinarum*.

1840 [see NIGGER *sb.* 2].

7. The South African plant *Bidens pilosa*; also its hooked seed.

1877 LADY BARKER *Year's Housekeeping S. Africa* vii. 130 An innocent-looking plant..bearing a most aggravating tuft of little black spires, which lose no opportunity of sticking to one's petticoats in myriads. They are familiarly known as 'black jacks'. 1932 *Discovery* Jan. 24/2 The seeds of the horrible weed 'Black Jack' are eaten by a serin-finch.

8. = BLACK FLAG as the ensign of a pirate. Cf. JACK *sb.*[3]

1867 in SMYTH *Sailor's Word-bk.*

9. A weapon consisting of a weighted head and short pliable shaft, used as a bludgeon. Hence as *vb.*, to strike with a blackjack. *U.S.*

1889 in *Cent. Dict.* 1895 *Denver Times* 5 Mar. 8/5 During the scuffle Miss Alderfer, Knapp's niece, saw the 'black jack' up his sleeve,..and as a result, swore out the concealed weapons charge. 1904 *N.Y. Even. Post* 10 Mar. 1 This position..was not such as the body would have taken had Newman been struck with a blackjack or other weapon. 1905 *Ibid.* 2 Sept., 'I got a partner there [*sc.* in the penitentiary],' Red said,..'blackjacked a man.' 1934 J. M. CAIN *Postman always rings Twice* iv. 31 She was to..clip him from behind with a blackjack I had made for her out of a sugar bag with ball bearings wadded down in the end. 1946 'P. QUENTIN' *Puzzle for Fiends* (1947) xv. 106 Perhaps you gave a ride to a hitchhiker who blackjacked you.

10. = VINGT-ET-UN.

1910 R. W. SERVICE *Trail of '98* (1911) IV. iv. 316 A tall fair-moustached man whom I recognized as a black-jack booster. 1931 *Kansas City Star* 28 Dec. 16 The governor knows his politics and is too poor a black jack player to mingle with gobs, anyway. 1937 J. STEINBECK *Of Mice & Men* 117, I seen guys nearly crazy with loneliness for land, but ever' time a whore-house or a blackjack game took what it takes. 1954 *Encounter* Oct. 8/1 Roulette and dice and blackjack were available. 1967 *Guardian* 25 Feb. 8/1 The proprietors must not be allowed..unfair advantage of their customers by keeping the bank to themselves in..roulette, chemin de fer and blackjack.

blacklead, *v.* Add: **b.** In *Electrotyping*, to brush (the mould) over with blacklead.

1888 *Encycl. Brit.* XXIII. 703/2 The mould, having been finished, has to be blackleaded...To facilitate this operation, a blackleading machine is used in large establishments... The process of depositing a copper solution upon the black-leaded surface of the mould is continued until a solid plate is formed.

black-leg, *sb.* Also **blackleg.** Add: **1. b.** Any of various diseases that attack vegetables (see quots.).

1880 *Encycl. Brit.* XII. 281/1 To prevent the cauliflowers from getting the disease of 'black legs'. 1916 B. D. JACKSON *Gloss. Bot. Terms* (ed. 3) 52/1 *Black Leg*, a bacterial disease of potatoes due to *Bacillus phytophthorus*. 1918 *Board Agric. & Fisheries Leaflet* No. 23. 17 Diseases of the Potato ..Black-Leg or Black-Stalk Rot. 1956 *Nature* 10 Mar. 465/2 *Phoma betae*..the cause of black leg in sugar beet.

3. Now *gen.* (Earlier and later examples.)

1875 R. J. HINTON *Eng. Rad. Leaders* IV. xix. 333 The police were used to watch the strikers or to protect the black-legs, as those are called who work outside the Union movement. 1890 W. BOOTH *In Darkest England* I. iv. 34 Men hungering to death..are the materials from which 'black-legs' are made.

b. A person who fails or refuses to join his fellows in combination for a given purpose, or breaks the rules of a particular trade or group.

1889 *Pall Mall Gaz.* 21 Nov. 5/1 The question of the preparation of a list of master-baker 'blacklegs' was also touched upon. These men are selling bread at 4½d. the quartern, and at even a lower rate. 1913 'A. R. HOPE' *Half & Half* 275 We abused him as a 'blackleg' because his industry set too high a standard to the rest. 1955 *Times* 3 Aug. 9/4 Any milkman delivering outside his boundaries would be regarded as 'a blackleg' (their own expression). 1971 *Times* 28 Jan. 7/8 The clause was known as the black-legs' charter.

c. *attrib.* or as *adj.*

1890 *Daily News* 8 Sept. 6/1 There were hundreds of men ..being subjected to blackleg competition. 1894 *Ibid.* 31 May 7/5 There were two 'blackleg' cabs discovered on the rank. 1955 *Times* 3 June 6/6 Members of the Amalgamated Engineering Union in the Swindon Railway Works have been recommended by their district committee not to do any 'blackleg' work such as repairing and servicing engines and rolling stock.

blackleg (blæ·kleg), *v.* [f. BLACK-LEG *sb.*]

a. *trans.* To take the place of (a worker who is on strike), thereby helping the employer to carry on his business and defeat the ends of the strike.

1897 *Daily News* 1 Sept. 2/4 They would be able to get many German engineers to 'blackleg' their English brethren. 1904 *Daily Chron.* 23 Apr. 5/7 The employers are permitted to persuade other workmen to 'blackleg' the men on strike. *fig.* 1893 *Daily News* 10 July 5/2 Is it fair to 'blackleg' these industrious men, as it were,..and adapt their researches to the needs and purposes of romance? 1906 G. B. SHAW in *Fabian News* XVII. 2/2 One result is that the dead dramatist blacklegs the live one: Shakespeare can charge no royalty.

b. *intr.* To act or work as a blackleg. So *to blackleg it*: to return to work before a strike is settled (said of a trade-union workman).

1888 *Baltimore Herald* 6 May (Farmer), Knights of Labor who had determined to blackleg it, regardless of the jeers and threats of their companions. 1892 *Chambers's Jrnl.* 17 Feb. 98/1 Volunteers were also obliged..to satisfy the leaders that they never 'black-legged' in any Australian strike. 1907 *Daily Chron.* 9 Mar. 6/6 If the alternative is to 'blackleg' or to starve. 1952 *Granta* 15 Nov. 12/2 In the General Strike of 1926, for instance, most Cambridge and Oxford students either blacklegged or were prepared to blackleg.

So **blacklegging** *vbl. sb.* [partly from BLACK-LEG *sb.*]

1839 THACKERAY in *Misc.* (1856) II. 454 Blacklegging is as bad a trade as can be. 1894 G. B. SHAW *Let.* 20 Mar. (1965) 421 As there is no Trade Union of dramatic authors, and I am going into the business, I can only avoid black-legging by asking you what are the list prices, so to speak, for the work. 1909 H. G. WELLS *Ann Veronica* x, It's a sort of blacklegging to want to have a life of one's own. 1955 *Times* 19 July 9/6 Should all good trade unionists boycott any blacklegging?

black list, *sb.* Add: **1.** Also *transf.* (Cf. sense 2 in Dict.)

c 1619 MASSINGER *Unnat. Combat* (1639) II. i, The blacke list of those That have nor fire nor spirit of their owne. 1788 GIBBON *Decl. & F.* V. xlviii. 82 His memory was stored with a black list of enemies and rivals. 1920 *Nature* 27 May 392/2 A chapter is devoted to beasts which the author would place in a black list as having many undesirable proclivities

b. *Naut.* A list of delinquents to whom extra duty is assigned as a punishment. Also, the punishment of being put on the black list.
1834 'Old Sailor' *Tough Yarns* 34 Almost every ship had a black list as long as the main-top bowline. 1837 *United Service Jrnl.* ii. 10 The cleaning, polishing, and black-list methods of wasting time. 1902 W. Kennedy *Sport in Navy* 76 Ten days black list for the boat's crew for not giving way. 1914 L. C. Beresford *Mem.* I. 110 Such a process [sc. the spit-and-polish system] involves perpetual extra bother and worry and black-list.

c. (a) An employers' list of workmen whom it is considered undesirable to employ. (b) A trade-union list of employers for whom their members are instructed not to work.
1888 *Atlantic Monthly* Nov. 611/2 He had got his name taken off from the black-list. 1888 *Encycl. Brit.* XXIII. 786/2 The..dreaded weapon known as the 'black list', by which combinations of employers..drove employees inclined to 'agitation' out of employment. 1923 *Management Engin.* May 343/1 Blacklist, a list of union workmen circulated by employers to prevent such workers from being hired.

d. A list of persons convicted as habitual drunkards under the Licensing Act of 1902. Hence **black-lister**, one who is put on the black list.
[Cf. 1902 *Act 2 Edw. VII* c. 28 § 6 [Habitual drunkards] Whether an order of detention is made or not, the court shall order that the conviction..] 1903 *Daily Chron.* 7 Jan. 5/2 ..suitable word is wanted by magistrates..to denote a drunkard on the 'black list' under the new Act..The word 'blacklister' is one that readily occurs. 1903 *Ibid.* 19 Jan. 2/7 The first number of the 'Black List'..issued under the new Licensing Act, was sent out from Scotland-yard on Saturday. 1904 *Ibid.* 16 Feb. 6/7 It was suggested that on the approach of a known 'black-lister' the police should give warning to the publican.

black-list, *v.* Add: (Further examples.) So **black-listed** *ppl. a.*; **black-listing** *vbl. sb.* (also *attrib.*)
1884 *Milnor* (Dakota) *Teller* 30 July, All the clerks making application [for work at other stores] were spotted and blacklisted and many have been dismissed. 1888 *Atlantic Monthly* Nov. 608/1 The manufacturers..had retaliated for some 'labor troubles'..by 'black-listing' about thirty men. *Ibid.* 'Mr. H. informed us that he was a 'black-listed' man. *Ibid.* 608/2 The increase of evil in the world thus resulting from the black-listing scheme. 1892 *Daily News* 13 Feb. 6/1 Calling on members of the Society to 'black list' men for life just because they offended their employers. 1892 *Pall Mall Gaz.* 29 Nov. 2/2 There are heavy penalties, too, for black-listing, or in any other way trying to induce persons to leave their employment (in Russia). 1899 *Daily News* 18 Jan. 4/5 The Plasterers' Union..prohibiting their members from working for certain black-listed firms. 1903 *Westm. Gaz.* 8 Jan. 2/2 A stoker in the Royal Navy, who has been black-listed for several weeks as a deserter. 1955 *Times* 5 May 11/3 The recent blacklisting of two British and one Italian cargo steamers by the Alexandria customs administration.

blackly, *adv.* (Further examples.)
1876 Ruskin *Fors Clav.* VI. lxix. 283 A perfect example of the strictly Black Country Englishman now developing generally in England; more or less blackly pulpous and swollen in all the features. 1894 H. Nisbet *Bush Girl's Rom.* 195 Mysterious ledges high up, which yawned blackly at the spectator below. 1898 H. S. Watt in *Dict. Nat. Biogr.* LVI. 256/2 His writings are blackly pessimistic in tone. 1940 H. Lawson *Over Slipperails* 27 He went and stood before the fire..and looked blackly at a print against the wall before his face.

black mail, *sb.* Now usu. **blackmail.** Add: **2.** (Earlier and later examples.) Now usu. a payment extorted by threats or pressure, esp. by threatening to reveal a discreditable secret; the practice of such a payment. Also *fig.*
a 1826 R. Heber *Narr. Journey Upper Prov. India* (1828) I. xvi. 421 The country is burdened with a crowd of lazy, profligate, self-called sawars, who..obtain for the most part a precarious livelihood by sponging on the industrious tradesmen and farmers, on whom they levy a sort of 'black-mail'. 1927 F. W. Crofts *Inspector French & Starvel Tragedy* viii. 121 Was it credible that a man would really pay blackmail for fear of having an obviously forged confession produced? *Ibid.* 122 Philpot's admission that he had submitted to blackmail was actually in his favour. 1940 A. Christie *One, Two, Buckle my Shoe* 70 He's believed to have done a spot or two of blackmail. 1969 *Woman* 19 Apr. 20/1 That old moan: 'You'll be sorry when I've gone' is a very common one: It's an emotional blackmail to keep loved ones on their toes.

black-mail, *v.* Now usu. **blackmail.** Add to def.: *spec.* to extort money from (a person) by threatening to reveal a discreditable secret. Hence **blackmailed** *ppl. a.*, that is subjected to blackmail; also *absol.* (with *the*), the person on whom blackmail is levied.
1895 *Westm. Gaz.* 18 Feb. 2/1 There must be a distinct threat, a direct menace of the blackmailed by the blackmailer. 1898 *Ibid.* 10 June 6/1 The blackmailer naturally keeps his own counsel as well as the money, and the blackmailed submits in silence. 1950 A. Christie *Murder is Announced* xxiii. 238 He hadn't the faintest idea he knew anything to blackmail her about.

black man.
1. A man having a black or very dark skin. (Cf. quot. 1815 for Black *a.* 3 a.)
[1592 (see Black *a.* 1 c].] 1738 F. Moore *Trav. Africa* 102 After which came on Shore the Captain, four Writers, one Apprentice,..and one Black Man. 1820 *Hist. N. Amer.* II. 10 Two slaves, the one a black man, a native of the place. 1894 F. A. Steel *Potter's Thumb* iv, He wondered..what the tyrannical black man meant..by the condited distrust of the black man.

2. An evil spirit; also, the evil one, the devil; also, a spirit or bogey invoked in order to terrify children. *colloq.* or *dial.*
[1591 in Pitcairn *Crim. Trials Scotl.* (1833) I. 246 Þe Dewill start vp in ye pulpett, lyke ane meikill blak man, with ane blak baird stikand out lyke ane gettis baird.] 1698 tr. *Bergerac's Satyr. Char.* xii. 48, I send the Goblins, the nightbats, the black men. 1851 *Fraser's Mag.* Feb. 140/2 'The foregoer', who is to the farmers what the black man was to the sweep of nurse-maids are to children. 1862 G. Meredith *Evan Harrington* iii, Rich as Croesus, and as wicked as the black man below! 1873 J. Ogg *Willie Waly* 123 Nor will the black man get ye. 1878 *Folk-Lore Rec.* I. 19 What nights of misery does that name, the black man, bring back to my memory! 1886 Elworthy *W. Somerset Word-bk.*, Black-man, a terrible object; a bogy, a nursery terror. (Very com.) Now you be good children, else the black-man 'll come down the chimley arter ye. 1969 J. C. Faris in Halpert & Story *Christmas Mumming in Newfoundland* 138 One would seldom, even in jest, call someone..a Devil..circumlocutions are used, such as 'Black Man', 'etc.

3. A local equivalent of *blackmaster.
1921 *Dict. Occup. Terms* (1927) § 925 Undertaker; blackman (Lancashire).

Black Maria. *colloq.*
1. A van for the conveyance of prisoners. *orig. U.S.*
1847 *Boston Even. Traveller* 25 Sept. 2/3 A new Black Maria,..a new wagon for the conveyance of prisoners to and from the courts of justice. 1869 J. Greenwood *Seven Curses of London* vi. 87 The van that conveys prisoners to gaol—Black Maria. 1924 *Punch* 17 Sept. 309 A Black Maria containing twelve prisoners.

2. A name used by soldiers in the war of 1914–18 for a German shell that on bursting emitted volumes of dense smoke, and for a German gun.
1914 *Scotsman* 12 Oct. 10/5 The 16-inch 'Black-Maria' shells of the heaviest German artillery. 1916 'Ian Hay' *First 100,000* ..The enemy were all this time firing..shells..'Black Marias'.

black market. [Black *a.* 11.] Unauthorized dealing in commodities that are rationed or of which the supply is otherwise restricted. Freq. *attrib.* Hence **black marketeer**, (less commonly) **marketer**, one engaged in such dealing; also abbrev. **blacketeer** (cf. *racketeer). Hence also **black-marke·t**, **black-marketee·ring**, **-marketing** *vbl. sbs.*, to deal, dealing, in the black market.
1931 *Economist* 10 Jan. 60/2 The growth of an unofficial or 'black' market in sterling exchange. 1936 *Ibid.* 5 Oct. 66/2 'Black market' quotations of Lit. 65–75 are reported in London. 1942 *New Statesman* 25 Apr. 272/2 There is evidence of a 'black market' where food can be bought without restriction of the price is high enough. 1942 *Ann. Reg.* 1941 96 The Government..took active steps to suppress the so-called 'black market' transactions in foodstuffs (etc.), which were also a matter of grave scandal in the country, particularly to those who could not afford 'black market' prices. 1942 *Times* 29 Jul. The British people..favour outright imprisonment of guilty 'blacketeers'. 1942 *New Republic* 13 Apr. 490/1 The most delicious black marketeer..is the sort of man who used to be a confidence man or other City hanger-on. 1942 *New Statesman* 11 July 25/1 If Jewish black-marketeers flourish in England, it is an indication of the inefficiency of the Government in checking this pursuit. 1942 *Times Lit. Suppl.* 15 Aug. 401/3 The profits of black marketeering. 1943 *Ann. Reg.* 1942 22 The penalties for 'black marketing' were also made more severe. 1943 E. M. Almedingen *Frossia* ii. 71 People black-marketed in currency, in timber, in leather and in bread. 1944 *Daily Express* 6 Dec. 4/3 Alleged blacketeering by Paris 'Gestapo'. 1955 *Economist* 5 Sept. 688/1 By some means or other this increased pressure must be offset..to make it more attractive to sell through legal channels than through the black market. Vigorous use of powers of deportation against the black marketeers would help. 1957 *New Statesman* 2 Nov. 554/1 Scruffy blackmarketeers..offered me wads of escudos for my ticket.

black mass. [Mass *sb.*[1]] A mass for the dead, at which the vestments and drapings are black. Also, a travesty of the mass, used in the cult of Satanism.
1893 *Nation* 2 Feb. 84/3 He recorded to a sort of benevolent parody or the Christian sacrifice—a sort of White Mass, as one might call it, in contradistinction to the Black Mass of the Kabalists. 1896 A. E. Waite *Devil-worship in France* 4 To say..that Black Masses are celebrated..will not enhance the credibility or establish the intelligence of the speaker. 1904 Rolfe *Hadrian the Seventh* (1929) 61 My first Mass must be a black mass, Excellency. 1961 A. Christie *Pale Horse* v. 64 She's very occult...Goes in for spiritualism and trances, and magic. Not quite black masses, but that sort of thing.

blackmaster (blæ·kmāːstəɹ). ?*Obs.* [Black *sb.* 5.] A funeral furnisher, an undertaker.

1901 *Daily Chron.* 14 May 12/2 To Blackmasters and Undertakers. 1904 *Ibid.* 12 Nov. 6/7 'What is a black-master?'..'That's a well-known term in London,' said the applicant, 'and means an undertaker.' 1905 *Ibid.* 11 Feb. 9/3 Goodwill of a Black Master's Business.

black-mouthed, *a.* Add: **b.** Applied to a fish having a black mouth: *black-mouthed dog-fish* (*Pristiurus melanostomus*).
1880 J. Couch *Hist. Fishes Brit. Isl.* I. 18 The Black-mouthed or Eyed Dogfish is better known in the Mediterranean than with us. 1903 G. Sim *Vertebrate Fauna of 'Dee'* 272 Pristiurus melanostomus. Black-mouthed Dog-fish...One was caught by trawl four miles off Aberdeen.

black-out (blæ·kaut). [f. phr. *to black out* (Black *v.* 3 b).] **1.** *Theatr.* The darkening of a stage during a performance; a darkened stage.
1913 G. B. Shaw *Let.* 3 Apr. (1956) 188 The more I think of that revolving business the less I see how it can be done... There will have to be a black-out. 1928 E. T. Dent in R. Brooke *Coll. Poems* p. xxxii, The elder generation were scandalized almost before the play began: no scenery..no music, no footlights, frequent 'black-outs'. 1932 A. J. Worrall *Eng. Idioms* iii. 20 Immediately after the murder there is a black-out; when the lights come on again the stage is empty. 1952 Granville *Dict. Theatr. Terms* 28 In revue a *black out* serves as a quick curtain to a sketch.

2. *transf.* and *fig.* A condition of (temporary) obscuration; *spec.* (a) a temporary loss of memory; (b) temporary suppression of information or news; (c) loss of a radio signal (because of an electrical storm, etc.).
1934 Galsworthy *Forest* iii. i, What do you think death really is?..Change of trains, or a black-out, eh? 1934 *Atlantic Monthly* Mar. 350/1 'Black-out'..is now used of a temporary loss of memory, or failure of the electric light. 1935 C. Day Lewis *Time to Dance* 55 The arctic winter and frost of your creation..dark. 1937 C. Wells *Baker in Darkling Wood* iv. iii. 356 There's not a trace now of the original concussion...But there is still a black-out, is there? 1941 Wyndham Lewis *Let.* 17 Oct. (1963) 301 If my eyes go I go too. Loathsome as the world is, I do like to see it. That sort of blackout I could not live in. 1942 *Mind* LI. 257 It is only when we reach the rarefied air of high theology that this particular intellectual black-out occurs. 1945 *Daily Express* 19 Apr. 1/1 There is still a news blackout in Moscow. 1958 *Listener* 25 Dec. 1072/1 The normal reflections from the F-layer [of the ionosphere] cease and there is prolonged disturbance of short-wave reception. This is what has been called the radio 'black-out'.

3. The action of extinguishing, covering, or obscuring lights as a precaution against air-raids, etc.; the resulting darkness; the time or period of compulsory covering of lights; the material used to obscure the lights. Also *attrib.*
1935 *Lancet* 3 Aug. 281/1 Mr. Harcourt Johnstone asked the Prime Minister whether instructions for compulsory 'black-outs' in districts where experiments were being carried out against air attacks were issued by authority of any Government department. 1939 L. MacNeice *Autumn Jrnl.* xi. 48 Black-out practice and A.R.P. 1939 *Archit. Rev.* LXXXVI. 105/1 Blackout curtains were..in use. 1939 *Sun. Times* Aug. 1939 127 Of the inconveniences the most serious continued to be the 'black-out'. 1940 *Flight* 8 Aug. 102/2 Also fitted are special 'black-out blinds' controlled from the pilot's compartment, which completely cover all the windows within the passenger cabin. 1940 'N. Shute' *Landfall* i. 7 No parking allowed on these roads in the black-out. 1940 *New Statesman* 17 Aug. 154/2 Six people cannot be nursed in wards with broken windows or with damaged black-out. 1942 E. Waugh *Put out More Flags* 252 It's like walking in the black-out with a shaded torch. You can see just as far as the beam they're taking. 1960 J. Lehmann *I am my Brother* i. 3 That night—the first night of black-out—I struggled out in the slithery rainy darkness to dine up in Highgate.

4. Temporary complete loss of consciousness; also, in flying, temporary blindness resulting from physiological derangement produced by a sudden sharp turn or acceleration.
1940 *Hutchinson's Pict. Hist. War* 14 Feb.–9 Apr. 217 The problems of a fighter pilot..The physical strain of 'black-out' causes..1946 ..to such an extent the pilot may have a black-out..1957 *Sun* 1950 *Daily Mail* 2 Apr. 10/5 The man-carrying centrifuge at the Farnborough Institute of Aviation Medicine tests pilots for 'black-outs'. 1961 *Lancet* 5 Aug. 322/1 He had had several 'black-outs' resembling epilepsy.

black-poll. *U.S.* [Black *a.* 1.] A North American warbler (*Dendroica striata*), the male of which has a black head when in full plumage. In full, *black-poll warbler*.
1783 J. Latham *Gen. Synop. Birds* II. 460 Black-poll Warbler...Crown black...Found at New York and Newfoundland. 1868 *Amer. Naturalist* II. 180 It is very quick, scarcely less so than the black-poll. 1872 Coues *Key N. Amer. Birds* 100 Blackpoll Warbler..*Dendrœca striata.* 1893 B. Torrey *Footpath-Way* 6 We..were in the old forest listening to bay-breasted warblers, Blackburnians, black-polls and so on.

black-pot. Delete †*Obs.* and add: **2.** A black pudding. (Cf. Pot *sb.*[1] 8.) *s.w. dial.*
1825 Jennings *Observ. Dial. W. Eng.*, *Black-pot*, black-pudding. 1880 Hardy *Trumpet-Major* xvi, Seventy rings of

black-pot. 1891 —— *Tess* i. i, I should like for supper,—well, lamb's fry if they can get it; and if they can't, black-pot. 1895 —— *Jude* vi. viii, I shall have to make black-pot and sausages.

3. Coarse Danish crockery-ware exposed to dense smoke in baking as a substitute for glazing.
1889 in *Cent. Dict.* 1904 in R. Hunter et al. *Encycl. Dict.* Suppl.

black rubric. An inaccurate term for the declaration explanatory of the rubric concerning kneeling at the reception of the Holy Communion, which was first inserted at the end of the Communion Office in the Book of Common Prayer of 1552 (omitted in 1559, and restored in 1662) (see quot. 1957).
1866 J. H. Blunt *Annot. Bk. Com. Prayer* 199 They retained the protest against Transubstantiation, whilst they removed all risk of the Declaration, or 'Black rubric' as it was sometimes called. 1877 E. Daniel *Prayer-Book* 348 The Black Rubric explains the intention of the rubric, which prescribes that communicants should receive the Holy Communion kneeling. 1957 *Oxf. Dict. Chr. Ch.* 176/1 The expression 'Black Rubric' dates only from the 19th cent. when the practice of printing the BCP with the rubrics in red was introduced and the fact that the 'Declaration' was really not a rubric at all was marked by printing it in black. In modern two-colour reprints of the BCP it will be found, however, printed in red.

4. A place or area of trouble, anxiety, or danger; esp. a dangerous section of a road.
1935 *Daily Herald* 30 June 2/3 Against general and protracted depression..the black spots should..set..improvement in the trades. 1936 *Discovery* Nov. 355/2 The development of newer industries is vital to the recovery of our distressed areas, which remain the one black spot in the otherwise remarkable position of Great Britain. 1937 *Daily Herald* 5 Jan. 7/4 Road-crossing improvements at 'black spots' throughout London. 1960 R. Morrish *Police & Crime-Detection* i. 20 'Black spots' are easily recognized by recording the districts and streets where, for example, housebreaking takes place. 1959 *Listener* 6 Aug. 208/2 You begin to see what are the black spots, that is to say, where accidents do in fact happen.

black sand. [Black *a.* 1.] Dark-coloured sand, esp. that on beaches in Australia and New Zealand; also *attrib.* Hence **black-sander**, a beach-comber who washes the black sand of the beach for gold; **black-sanding** *vbl. sb.*, this process or occupation.
1778 D. Samwell *Jrnl.* 27 June in Cook *Jrnls.* (1967) III. ii. 1120 We came to an Anchor in about 10 fathm Water with bottom of black sand and shells. 1849 *Rep. Comm. Patents* (U.S.) (1850) 205 Much of the black sand now in use is colored, and is common siliceous sand, but the best article is a ferruginous sand. 1874 A. Bathgate *Colonial Experiences* xi. 147 It promised well from the quantity of black sand visible in the dish. 1882 *Jrnl. Iron & Steel Inst.* I. 251 It was found very difficult to work this black sand, on account of the large percentage of titanic acid it contains. 1906 in P. Galvin *N.Z. Mining Handbk.* 122 (*heading*) Black-sand beaches on the West Coast. *Ibid.*, At present there are small parties who earn a good average living by what is known as 'blacksanding'..between Jackson's Bay and Karamea. *Ibid.* 124 The 'black-sanders' reaped their harvest. *Ibid.*, What old leads there were were then washed out in detail and most of the gold secured by the busy 'blacksanders' in the process. 1965 L. Williams *Econ. Geol. N.Z.* ix. 130/2 A 'blacksand' is material containing opaque iron-bearing grains in sufficient quantity to impart at least a grey colour to the whole.

black shirt, black shirt. [tr. It. *camicia nera*.] A black shirt as the distinctive mark of the uniform of the Fascist party of Italy; hence *transf.* and *gen.* = *Fascist. So **black-shirted** *a.*
1922 *Youth's Companion* 30 Nov. (ed. L. Thompson, 1954) 84 The 'black shirts' have beaten down Communism by force of arms. 1923 P. Phillips (*title*) The 'Red' Dragon and the Black Shirts. 1923 *Weekly Dispatch* 28 Jan. 9 The 'scotching' of Bolshevism in Italy by the black-shirted Fascisti. a 1929 D. H. Lawrence *Phoenix II* (1968) 534 She cares terribly because far-off, invisible, hypothetical Italians wear black shirts. 1934 H. G. Wells *Exper. Autobiogr.* ii. ix. 782 It was a gathering of Mosley's black-shirts. 1939 'G. Orwell' *Coming up for Air* 1. 1 The wearing of a distinctive uniform—in this case a black shirt. *Ibid.* 51 Next day the whole of England was ringing with the cry of 'blackshirt brutality'.

blacksmith. Add: **b.** *blacksmith's* (U.S. *blacksmith*) *shop*, a smithy.
1752 in *New Hampshire Probate Rec.* (1915) III. 454 The Blacksmiths Shop which stand[s] on the home lot. 1795 P. M. Freneau *Poems* 421 Unless the [stage-]driver..Has made some business for the black-smith-shop. 1845 F. Douglass *Life* (1846) 94 Fred, go to the blacksmith's shop and get a new pair of tongs. 1896 *Harper's Mag.* Aug. 390/1 Perhaps he had better ride over to the blacksmith shop.

blacksmithery (blæ·ksmɪþəri). [f. Blacksmith + -ery; cf. Smithery.] A smithy. *U.S.*
a 1854 J. F. Kelly *Humors of Falconbridge* (1856) 303 The town—of some four houses, six 'groceries', a *store* and blacksmithery—was aroused. 1872 *U.S. Office Indian Affairs Rep.* 1871 337 The fixed property of this reservation consists of agents quarters, 3 dwellings, 2 blacksmitheries.

2. Blacksmith's work. *orig. U.S.*
1869 *Overland Monthly* III. 10 In North Carolina, as in the North, blacksmithery, wagon-making, coopery, and other sorts of hard-handed industry, were in noisy blast. 1936 J. Gerrard *High Failure* vi. 134 Altering its raise—a tricky piece of blacksmithery. 1958 *Antiquity* XII. 8 The blacksmithery of their simple ironwork.

black-snake. Add: **2.** (Earlier example.) Also *attrib.*
1864 J. Fisk *Exped. Rocky Mts.* 5 A 'black snake'..brought him on his legs. 1901 S. E. White *Westerners* xxxiv. 310 He snapped the lash of his black-snake whip.
Hence **bla·cksnake** *v. trans.*, to flog with a black-snake whip.
1870 'Mark Twain' in *Galaxy* Dec. 878/2, I lay I'll black-snake you within an inch of your life!

black spot. **1.** Any of various diseases of plants or animals, producing black spots upon the diseased portions; esp. a disease of roses (see quot. 1889).
1889 *Cent. Dict.* I. 571/1 *Black-spot*, a disease of rose-bushes, characterized by diffuse, dark-colored spots on the upper surface of the leaves. It is caused by a parasitic fungus, *Asteroma Rosae*. 1906 M. C. Cooke *Fungoid Pests* 50 Chrysanthemum Leaf Spot. The black spot [*Cylindrosporium Chrysanthemi*] is very destructive in Canada. *Ibid.* 134 Peach Freckle..is known in the United States as 'scab' or 'black spot'. The fungus is a kind of black mould which attacks ripe Peaches. 1910 F. C. Stewart in *N.Y. Agric. Exper. Station Bull.* CCCXXVIII. 364 Maple, Silver..Black Spot, *Rhytisma acerinum*..The characteristic tar-coloured spots of which may be detected at a considerable distance. 1933 *Jrnl. R. Hort. Soc.* LVIII. 253 Black spot, the worst enemy of the rose. 1966 *Punch* 30 Mar. 462/1 Huge stocks of the mutations of bacterial warfare, of mildews and rust, black spot, botrytis and scab, are being made ready for immediate use at call.

black-tail. Add: **3.** The black-tailed deer. Also *attrib.* with *deer, buck.* *U.S.*
1828 H. G. Rogers *Jrnl.* 20 June in H. C. Dale *Ashley-Smith Explor.* (1918) 259, I killed a fine black tail buck. 1848 J. W. Abert *Exam. New Mexico* 23 Of the latter there are two varieties, the common deer, and the black tail. 1872 Schele de Vere *Americanisms* 372 The variety found on the Pacific coast (*Cervus columbianus*) is more commonly designated as the Black-tail Deer, from the black tip to its tail. 1890 *Outing* (U.S.) XXVII. 45/2 We..had two large black-tail bucks in camp.
4. = *dassie *sb.*[1] *S. Afr.*
1905 *East London Dispatch* 19 July 7/3 A few friends, fishing in the Buffalo River, had some excellent sport, taking black-tail of about 2½ lbs. 1906 *Ibid.* 6 Mar. 7/5 The biggest I caught on these rocks was a fine *dasje* (black-tail) weighing 7¼ lbs.

black-thorn. Add: **1. c.** *blackthorn winter* (later examples). Also *Comb.*, as *blackthorn-fleeced* adj.
1856 C. M. Yonge *John Keble's Parishes* xvi. 214 Black-thorn winter is supposed to bring fresh cold in spring, when the blackthorn looks as if covered by hoar-frost. 1905 *Daily Chron.* 11 Apr. 4/7 'Blackthorn winter' is the Gloucestershire name for the fit of January weather that almost always follows the flowering of the beautiful wild sloe blossom. 1946 L. B. Lowe *Rough Walk Home* 28 Pray blackthorn-winter's down nay-blackthorn week. 1948 C. Day Lewis *Poems 1943–47* 84 Walking by blackthorn-fleeced Hedges to church.

2. (Earlier examples.)
1737 J. Brickell *Nat. Hist. N. Carolina* 79 The Black Thorn..grows plentifully in several parts of this Province. 1798 *Trans. N.Y. State Soc. Agric.* I. III. 18 The black thorn has been destroyed by a worm that gnaws its twigs. 1848 A. Gray *Man. Bot.* 128 Black or Pear Thorn.

black-throat. *U.S.* [Black *a.* 1.] A black-throated warbler. In full, *black-throat warbler*.
1786 Pennant *Arct. Zool.* II. 399 Black-throat Warbler..Inhabits, during summer, Canada and other parts of America, to the south. 1903 S. E. White *Blazed Trail* xlii. 296 Wilson's warblers..pine creepers, black-throats..passed silently or noisily.

black-top. *U.S.* [Black *a.* 1.] A blackish material used for surfacing roads, etc.; a road, etc., surfaced with this material. Freq. *attrib.* Hence as *v.*, to surface (a road) with black-top; **black-topping** *vbl. sb.*
1931 *Amer. City* Oct. 112/1 We are trying to adopt a black border for the letters and other characters, which for black top pavements are yellow. 1940 *Topeka Jrnl.* 22 Apr. 1 Strewed some of the victims on the blacktop. 1947 J. Steinbeck *Wayward Bus* iii. The turned right into the black-top road. 1956 'F. Richards' *Practise to Deceive* ii. 18 Next spring would be time enough to think of black-topping. *Ibid.* 26 She took it easy, backing up the narrow blacktop.

black-wash, *sb.* Add: **3.** The opposite of Whitewash *sb.* 3. Also **bla·ck-wa:shing** *vbl. sb.*
1859 Kingsley *Misc.* II. 46 To remove..the modern layers of 'black-wash', and to let the man himself, fair or foul, be seen. 1877 *Chambers's Jrnl.* 28 Apr. 258/1 A skilful counsel..using as much whitewash as he can for the accused, and applying plentiful blackwash to the witnesses for the prosecution. 1890 *Literary World* XLI. 388/1 By giving Cromwell another plentiful coating of black-wash. 1905 *Contemp. Rev.* Oct. 552 (*title*) The Blackwashing of Dante.

black water.
1. A stream stained brown by the peat of the mosses from which it flows.
1676 Cotton *Walton's Angler* ii. ii. 14 This River [Trent] from its head for a Mile or two is a black water. *Ibid.* 16 The River Wye..a black water too at the Fountain...Derwent, a black water too.

2. A disease incident to sheep and cows.
1800 Tuke *Agric. N. Riding Yorks.* 272 There is another disorder to which lambs are liable in autumn; it is called the black-water. 1825 Loudon *Encycl. Agric.* § 6263 The black water is only the aggravated and latter stages of [red water]. 1879 J. Lucas in *Zoologist* 3rd Ser. III. 356 Many [sheep] die in cold nights, when they contract a disease known as 'blackwater'.

3. In full *blackwater fever*. A tropical fever to which chiefly white people are subject, characterized by a brown or blue-black colour of the urine.
1884 J. F. Easmon (*title*) The nature and treatment of blackwater fever. 1897 Mary Kingsley *W. Africa* 645 Two extremely deadly forms of fever have come into notice here, malarial typhoid and blackwater. 1926 *Blackw. Mag.* Nov. 663/1 The doctor had died three weeks before of blackwater. 1965 B. Sweet-Escott *Baker St. Irreg.* viii. 249 Musgrave had known Burma well before the war, and had twice suffered from blackwater fever.

black-wood. Add: **b.** = *black growth* (see *Black *a.* 19). Also *attrib.* *U.S.*
1812 F. A. Michaux *Hist. Arbres de l'Amérique* II. 220 On désigne ces terreins sous le nom de *Black wood lands*, terreins à essence noire. 1848 Bartlett *Dict. Amer.* 33 *Black wood*, hemlock, pine, spruce, and fir. Maine. 1872 Schele de Vere *Americanisms* 420 Black Wood is, in the Northern States, used as a generic term for the evergreens, hemlock, pine, spruce, and fir.

Bla·ckwood. *Bridge.* The name of an American, E. F. *Blackwood*, used *attrib.* or *absol.* of a bidding system that he devised, whereby a bid of four no-trumps is used as an asking bid to which the partner's reply in five of a suit shows the number of aces that he holds.
1938 E. Culbertson *Contract Bridge Complete* (1939) xxvi. 321 A four and five no-trump convention known in most parts of the country as the Blackwood Convention has a number of sturdy adherents. 1958 *Listener* 16 Oct. 611/2 Assuming the use of Blackwood, my auction would be [etc.].

blacky, *sb.* **2.** For '*Sc.* and *north.*' read '*dial.* and *colloq.*' (Add examples.)
1876 Mrs. Francis in W. W. Skeat *S. Warwickshire Words* 124 *Blackie*, a blackbird. 1890 A. J. Armstrong *Ingleside Musings* 84, I listen to the blackie's note. 1940 F. Kitchen *Brother to Ox* i. 3 Every bush along the carriage drive owned a blackie or a thrush.

blad, *sb.*[2] Add: **2.** A portfolio (Jamieson, 1808).
1813 E. Picken *Poems* II. 132 He staps in his warks in his pouch in a blink, Flang by a' his warklooms, his blaud an' his ink.
b. A blotting-pad or writing-pad.
1837 *Tait's Mag.* IV. 103/2 As if I were merely amusing myself with my pen on my blad. 1923 G. Watson *Roxb. Word-bk.* 58 *Blad*, a blotting-pad.
c. [cf. Sw., Du. *blad* leaf.] In non-dial. use: see quots.
1933 Partridge *Slang Today & Yesterday* iii. iii. 181 Blad..is applied to a sheaf of specimen pages or to other 'illustrative matter' liked by the bookseller, especially the bookseller resident abroad. 1960 G. A. Glaister *Gloss. of Bk.* 28 *Blad*, a sample of a book, made up for the publisher's traveller to show to the trade. It usually consists of the first thirty-two pages, including prelims, bound up in the same cloth as the finished book.

bladder, *sb.* Add: **6. c.** *bladder of lard*: a bald-headed or fat person. *slang.*
1864 in Hotten *Slang Dict.* 1886 *Athenæum* 31 July 142/1 An elderly Jew money-lender, whom she afterwards describes to her admiring friends as a 'bladder of lard', a graceful reference to his baldness and tendency to stoutness. 1943 W. de la Mare *Magic Jacket* 24 Here's that bladder-of-lard, schoolmaster Smiles, saying exactly the same thing.
10. bladder-fern (examples); **bladder lard,** lard put up for sale in bladders, used spec. as the trade name for lard of the best quality; **bladder worm,** the larva of a tapeworm in its encysted state; a hydatid.
1828 J. E. Smith *Eng. Flora* IV. 297 *Cystea*. Bladder-fern. *Ibid.* 298 *C. fragilis*. Brittle Bladder-fern. 1961 R. W. Butcher *Brit. Flora* I. 154 The Common Bladder Fern grows in walls and rocky woods on basic soils commonly in highland Britain and rarely in the eastern and southern portions. 1872 *Eng. Mech.* 11 Oct. 82/3 The fourth [sample] (a bladder lard) contained 10 per cent. of water. 1858 Bladder worm [see Cysticercal *a.*]. 1949 *New Biol.* VII. 116 If pigs swallowed the eggs of *Taenia solium*, bladder-worms (*Cysticercus cellulosae*) developed in their flesh.

blade, *sb.* Add: **5. c.** A vane upon the circumference of a revolving cylinder or disc of a turbine.
1887 *Encycl. Brit.* XXII. 517/1 Attempts have been made..to devise steam-engines of the turbine class, where rotation of a wheel is produced..by impact of a jet [of steam] upon revolving blades. 1900 [see Turbine 1 b]. 1904 Goodchild & Tweney *Technol. & Sci. Dict.* 795/2 [In an impulse turbine]..the water rushes with high velocity through a series of jets or nozzles and impinges on suitably shaped blades.

d. *Aeronaut.* A part of the propeller of an aeroplane or rotary-wing aircraft which acts upon the air. Also *attrib.*
1907 W. M. Varley tr. *Moedebeck's Aeronautics* xxv. 421 A screw-propeller is built up usually of two or more blades. 1916 H. Barber *Aeroplane Speaks* iv. 115 The propeller screws through the air, and its blades..secure a reaction. 1920 *Flight* XII. 1309/1 Sketch of a sectioned blade root. 1958 *Chambers's Techn. Dict.* 960/2 *Blade loading*, the thrust of a helicopter rotor divided by the total area of the blades.
e. = *switch-blade* (see Switch *sb.* 8).
1920 *Whittaker's Electr. Engin. Pocket-Bk.* (ed. 4) 323 The blades of isolating switches should be locked in position.
6. b. *Archæol.* A long, narrow flint-flake, used esp. as a tool in prehistoric times (see quot. 1959). Freq. *attrib.*, as *blade-axe, -culture, -tool.*
1921 M. C. Burkitt *Prehistory* iv. 65 While long flakes or blades are being struck off a core, vibrations transverse to the blow itself are set up in the flake or blade. 1926 R. A. Smith *Flints* 29 (*caption*) A blade-scraper. *Ibid.* 30 Graving-tools..are generally made from flakes, preferably blades (flakes with the side-edges parallel and longer than the ends). 1935 Huxley & Haddon *We Europeans* ii. 53 Various types of Homo sapiens from Africa and Asia whose implements are typically..blade-tools, a form of flake technique. 1937 *Discovery* Sept. 287/2 Blade-axes, chisels, spear-heads. 1943 J. & C. Hawkes *Prehist. Britain* i. 21 The late Palæolithic hunters had a much more delicate and specialized equipment than their predecessors, that is distinguished from the core and flake forms by the general name of 'blade' culture. 1959 J. D. Clark *Prehist. S. Afr.* v. 111 The lump is called a *core*..and the pieces removed are known as *flakes* or, if they are long and narrow, they are called *blades*.
7. c. Usu. *pl.* Hand shears for shearing sheep. Also *attrib.* *Austral.* and *N.Z.*
1905 A. B. Paterson *Old Bush Songs* 26 All among the wool, boys, Keep your wide blades full, boys. 1917 *N.Z. Jrnl. Agric.* 20 Sept. 134 The majority of the larger sheep-owners..have come to recognize the advantages of 'machines' over 'blades'. 1923 W. Perry et al. *Sheep Farming in N.Z.* vi. 83 In some parts there is a reaction in favour of blade shearing. 1945 Baker *Austral. Lang.* iii. 64 Shearers have also been called..*blades*men, although the last term is going out of fashion because most shearing is now done with machines.
12. blade-consonant *Phonetics*, a consonant formed with the blade of the tongue; also **blade-point** (see quot. 1890); **blade-spring,** a form of spring used to hold piston rings in place; **blade-work,** work done with the blade of an oar.
1877 Blade consonant [see Dict., sense 5 b]. 1890 Sweet *Primer of Spoken Eng.* 8 The *blade*-consonant *s* is formed with the 'blade' or flattened point of the tongue; if the tongue is retracted from this position, and the point raised, we get the *blade-point* consonant ʃ in 'fish'. 1863 Campin *Mech. Engin.* 130 Four arms, which serve a double purpose, connecting the boss with the top and bottom of the piston, and carrying at their extremities the blade-springs. 1905 *Daily Chron.* 30 Mar. 6/7 Their blade work is not pretty, the finish not being clean, and the feather frequently under water. 1959 *Times* 13 Mar. 18/1 They [sc. the Cambridge crew] went up quickly, in spite of some dirty bladework.

bladeless (blēɪ·dlės), *a.* [f. Blade *sb.* + -less.] Without or lacking a blade.
1857 *Blackw. Mag.* Oct. 486/1 Only the bladeless [*i.e.* grassless] rocks rang hollow mocking replies to his despairing 'Allah!' 1896 J. A. Mitchell *Amos Judd* viii. 161 A bladeless jack-knife. 1927 *Glasgow Herald* 16 Apr. 9 A bladeless safety razor.

blading, *vbl. sb.* Add: **3.** *collect.* The blades of a turbine.
1905 W. J. A. London *Brit. Pat.* 13,883, The blading of the high pressure section is mounted on a drum. 1955 *Times* 12 July 4/4 By extensive use of ceramics and liquid-cooled blading..the engine will operate at much higher temperatures than any other marine gas turbine being developed.

blady, *a.* Add: (Later example.) Esp. in *blady grass*; spec. the coarse grass *Imperata arundinacea* of Australia, Indonesia, etc.
1827 P. M. Cunningham *N.S. Wales* II. xxvii. 170 The roof being..thatched with blady grass. 1847 L. Leichhardt *Jrnl. Exped. Australia* iii. 59 The blady-grass had..begun to show its young shoots. 1864 G. M. Hopkins *Poems* (1948) 120 I'll take in hand the blady stone And to my palm the point apply. 1918 *Chambers's Jrnl.* Feb. 140/1 Utilising the lalang or 'blady' grass, for the manufacture of paper-pulp.

blag (blæg), *sb.* *slang.* [Etym. unknown.] Robbery (with violence); theft.
1885 *Session Paper* 30 July 471 There has been another *blag* down round here. 1936 J. Curtis *Gilt Kid* 22, I don't want to say 'O.K.' and then find out that I've let myself in for..doing a blag on the crown jewels. 1960 *Observer* 24 Jan. 5/1 The top screwing teams, the ones who went in for the really big blags, violent robberies.
Hence **blag** *v.*, to rob (with violence); to steal. So **bla·gger**, one who blags.
1933 C. E. Leach *On Top of Underworld* x. 137 *Blag*, snatch a watch chain right off. 1938 F. D. Sharpe *S. of Flying Squad* i. 15 Blaggers, screwsmen, [etc.]..abound in the Underworld of London. *Ibid.* 329 'Johnny blagged the till'—Johnny took the till. 1945 J. Henry *What Price Crime?* 93 Another is known as the 'blagger' or 'snatcher'. These are usually young louts who specialize in snatching ladies' handbags and bolting off.

‖**blagueur** (blagöʳ). [Fr. (Robert, 1808), f. *blague* Blague *sb.* + *-eur* -er[1].] One who talks pretentiously; a joker or teller of tall stories.
1883 *Sat. Rev.* 14 Apr. 467/2 [It is] not the laughter of the true humourist, but that of the professional *blagueur*. c 1903 Wyndham Lewis *Lett.* (1963) 8, I went to a cabman in the rank, and he asked 5/–; so I left him as a blageur [*sic*]; but it appears that it is without the 4 mile radius, and it is 1/- a mile. 1950 A. L. Rowse *Eng. of Elizabeth* xii. 515 That marvellous adventurer and *blagueur* Anthony Shirley of All Souls.

blah (blā), *sb.* *colloq.* (orig. *U.S.*). Also **bla, blaa.** [Imitative.] Meaningless, insincere, or pretentious talk or writing; nonsense, bunkum. Also used as a derisive interjection. Freq. reduplicated.
1918 *Wine, Women & War* (1926) 136 [He] pulled old blah about 'service', 'doing one's bit', etc. 1921 *Collier's* 15 Jan. 10/3 Then a special announcer begin a long debate with himself which was mostly blah blah. 1922 S. Lewis *Babbitt* xxx. 359 Why the dickens they want to put in their time listening to all that blaa. 1924 M. Arlen *Green Hat* vi. 164 So you heard about it from that *femme fatale*, did you? Damn that man! Bla, bla, bla! 1927 *Observer* 10 July 28/1 England isn't fooling anyone with so much 'blah' about the world's greatest tournament. 1943 'G. Orwell' in *New Road* 151 Exactly similar blah is being written about the Red Army at this moment. 1958 E. H. Clements *Uncommon Cold* 229 A good deal of blah about waste of public money.
Hence **blah, blah-blah** *v. intr.*, to talk or write 'blah'.
1924 'W. Fabian' *Sailors' Wives* iv. 56, I was just blahing to hear myself blah. 1931 St. John Ervine in *Time & Tide* 12 Sept. 1057 That schoolmasters should permit him to blah-blah about his ignorance is a crime against the mind. 1942 'G. Orwell' *Diary* 18 Apr. in *Coll. Ess.* (1968) II. 419 The tactless utterances of Americans who for years have been blahing about 'Indian freedom' and British imperialism. 1945 —— in *Tribune* 14 Dec. 10/3 Instead of blah-blahing about the clean, healthy rivalry of the football field..it is more useful to inquire how and why this modern cult of sport arose.

blah, *a.* *slang.* [Cf. *blah *sb.*]
1. Mad. (? orig. *U.S.*)
1924 *Telegr. & Teleph. Jrnl.* X. 68/2 The third class is hopeless...It consists of the people who, in New York slang, have gone 'blah'. 1928 A. E. W. Mason *Prisoner in Opal* xxiii. 276 More of your questions and I am blah.
2. Dull, unexciting; pretentious.
1937 N. Marsh *Vintage Murder* xi. 122 That fascinating blah stuff of hers goes down with the nitwits. 1955 H. Roth *Sleeper* ix. 68 You must..have come to realize how blank and blah he made himself. 1959 J. Verney *Friday's Tunnel* iv. 49 One of those blah sneery voices like a butler in a film.

Blake (blēɪk). The name of Lyman Blake, the inventor of a sewing-machine for boots and shoes, as in *Blake-sewn*, whence *Blake-sewer*, etc.
1895 *Daily News* 26 Mar. 6/7 A Levant goat quarter, whole satin golosh, blake sewn. 1921 *Dict. Occup. Terms* (1927) § 414 *Machine operator, Blake; Blake sewer, Blake sole sewer*, a sole attacher who stitches sole of boot or shoe on to insole..with a Blake sewing machine, i.e. heavy sewing machine worked by power.

Blakeian (blēɪ·kiăn), *a.* Also **Blakean, Blakian.** [-ian.] Of, pertaining to, or characteristic of William Blake (1757–1827), poet and painter, or his work.
1906 E. J. Ellis *Poet. Wks. W. Blake* 337 In more Blakean terms, they are Albion as Ijim, bearing Tiriel on his shoulders between East and West. 1907 *Daily Chron.* 27 June 3/2 There is something Blakean in his artistic conception of the human form as typifying abstract energy. 1930 T. E. Lawrence *Lett.* (1938) 687 The idea of 'genitals being beauty' in the Blakian sense. 1952 S. Spender *Shelley* v. 40 The beautiful poem *To William Shelley*..has a Blakian simplicity and innocence. 1954 *Essays & Studies* VII. 110 Blake's 'Jerusalem'..is betrayed by Parry's tune, the inimitable Blakeian imagery being obliterated by the prevalent vociferation.

blame, *v.* Add: **5.** Also, *to blame* (something) *on* (or *on to*) (someone): (cf. quot. 1835 in Dict.).
1903 Kipling *5 Nations* 22 We will blame it on the deep. 1910 —— *Rewards & Fairies* 175 If you can keep your head when all about you Are losing theirs and blaming it on you. 1955 A. L. Rowse *Expans. Eliz. Eng.* viii. 283 Naval historians have all blamed the failure on to the Queen.
7. *dial.* and *U.S.* Used as an imprecation.
a. In the imperative mood.
1830 R. Forby *Vocab. E. Anglia* I. 27 *Blame*, *v.* a very decent and commendable evasion of the horrible word *damn*. Ex. '*Blame* me', or 'I will be *blamed*, if', &c...Nobody ever heard so mild an imprecation as '*blame* you for a rascal', or 'John Smith be *blamed* for a fool'. a 1832 R. C. Sands *Writings* (1834) II. 149 Goy blame it all. 1876 'Mark Twain' *Tom Sawyer* vii. 67 Blame it, I ain't going to stir him much. 1886 *Harper's Mag.* June 52/2 Blame me if them ain't the darnedest beans *I* ever seen! 1897 Kipling *Capt. Cour.* ix. 210 Blame that boy! He never told. I'd ha' listened to *that*. 1905 G. H. Gibson in A. B. Paterson *Old Bush Songs* 70 But blame my cats if I know what else They'll find for Bill to do.

b. In passive, in phr. (*I'm*) *blamed if* (etc.).
1830 [see 7 a above]. 1844 'J. Slick' *High Life N.Y.* I. iii. 31 I'll be blamed if you've the least bit of Yankee in you. 1876 'Max Adeler' *Elbow-Room* xv, Blamed if I haven't forgotten that word. 1905 H. R. Martin *Tillie* 322 I'm blamed if I dare adwise you.
c. as *adv.* = *blamed 2.
1843 'R. Carlton' *New Purchase* xvii. 134 'Blame close, stranger,' said the old hero. 1876 'Mark Twain' *Tom Sawyer* vii. 67 He's my tick and I'll do what I blame please with him or die! 1905 G. H. Lorimer *Old Gorgon Graham* 160 You know blame well that I don't understand any French.
d. as *adj.* = *blamed 1.
1876 'Mark Twain' *Tom Sawyer* vi. 65 Talk about trying to cure warts with spunk-water such a blame-fool way as that! 1897 Kipling *Capt. Cour.* vi. 129 You..go hoggin' the road on the high seas with no blame consideration fer your neighbours. 1917 —— *Diversity of Creatures* 263 The blame thing jarred off—spiteful as a rattler! 1934 *Amer. Ballads & Folk Songs* 155 Them blame little children is all crying yet.

blamed (blēɪmd), *ppl. a.* and *adv.* *dial.* and *U.S.* [Cf. *blame *v.* 7.]
1. *ppl. a.* = Blasted *ppl. a.* 3.
1840 Haliburton *Clockm.* 3rd Ser. vi. 84 Yes, John Bull is a blamed blockhead. 1876 'Mark Twain' *Tom Sawyer* ix, Drunk, same as usual, likely—blamed old rip! 1905 G. H. Lorimer *Old Gorgon Graham* 200 They've an ache or a pain in every blamed joint. 1946 K. Tennant *Lost Haven* (1947) i. 14 I've stood the ugly, blamed thing long enough.
2. *adv.* Confoundedly, excessively.
1833 *Knickerbocker* I. 303 Which I now look upon as a blamed foolish notion. 1876 'Mark Twain' *Tom Sawyer* viii. 79 Well, it's blamed mean—that's all. 1884 W. Cudworth *Yorks. Dial. Sk.* 27, I knaw they wor blamed nice. 1904 G. Stratton-Porter *Freckles* 51, I am..so blamed ignorant I don't know which ones go in pairs.

blanc. Add: **1.** *blanc fixe*: sulphate of barium used esp. as an extender in paint.
1866 Roscoe *Elem. Chem.* xx. 179 *Barium Sulphate*..is used as a paint, and the precipitated salt is termed *blanc fixé*. 1904 Goodchild & Tweney *Technol. & Sci. Dict.* 54/1 *Blanc fixe*, an artificial sulphate of barium, very white and fine and somewhat heavy. Principally used in papermaking and in the manufacture of wallpaper colours. 1963 R. Higham *Handbk. Papermaking* iv. 90 Blanc fixe is still used in the preparation of photographic base papers because of its soft powder structure and slightly higher opacity than the barytes form.
2. (Examples.)
1845 Bregion & Miller *Pract. Cook* 40 *Blanc*, a rich broth or gravy, in which the French cook palates lamb's head, and many other things. 1869 M. Jewry *Warne's Model Cookery* 20/1 First-rate cooks preserve the whiteness of their boiled meats..by..using..a sort of broth..called *blanc*. 1952 F. White *Good Eng. Food* II. iii. 130 To Make a Blanc..do not take off the fat as the *blanc* cooks.

‖**blanc-bec** (blaṅ bek). [Fr., lit. 'white beak'.] A raw youngster, greenhorn.
c 1845 C. Brontë *Professor* (1857) I. xii. 219 It was nonsense for her to think of taking such a 'blanc-bec' as a husband, since she must be at least ten years older than I. 1853 —— *Villette* I. ix. 172 You should have seen what a *blanc-bec* he looked..how he hesitated and blushed. 1923 Conrad *Rover* viii. 132, I may be *disparu* but I am too solid yet for any blancbec that loses his temper.

‖**blanc de Chine, chine** (blaṅ də ʃin). [Fr., lit. 'white of China'.] White glazed porcelain made at Te-hua in south-eastern China, esp. during the Ming period.
1888 F. Hirth *Anc. Porcelain* xi. 44 Tê-hua has since the Ming dynasty furnished porcelains of a fascinating creamy white, a distinct class, which I believe constitutes together with the Ting-chou white pottery the article known as 'blanc de Chine'. 1910 S. W. Bushell *Chinese Art* (ed. 2) II. 26 The velvety white porcelain sometimes known as *blanc de Chine*. 1955 *Times* 19 July 10/7 An eighteenth-century Chinese blanc-de-chine vase.

‖**blanc de perle** (blaṅ də pɛ̄rl). [Fr.] Pearl-white (see Pearl *sb.*[1] 18).
1881 *Queen* 12 Mar. (Advt.), Eyebrow pencils, 12 stamps;..Blanc de Perle, 30. 1897 M. Corelli *Ziska* viii. 164 She managed to produce a delicate shudder of her white shoulders without cracking the *blanc de perle* enamel. 1919 Firbank *Valmouth* xi. 189, I gave her my little precious volume of *blanc de perle* in order to rub her nose.

blanch, *v.*[1] Add: **1. c.** *spec.* in coining money.
1868 [see Blanching *vbl. sb.*[1]]. 1883 *Encycl. Brit.* XVI. 489/2 The removal of a small portion of the alloying metal in this way constitutes 'blanching' or 'pickling' the coin.

blanching, *vbl. sb.*[1] Add to def.: Also in *Cookery*: see Blanch *v.*[1] 2 a.
1951 *Good. Housek. Home Encycl.* 357/2 *Blanching*..in which food is treated with boiling water. 1960 *Times Rev. Industry* Aug. 22/3 Another operation applied to vegetables is blanching—immersion in near-boiling water for a few minutes.

blanco (blæ·ŋko). [Trade name, f. F. *blanc* white.] A white preparation for whitening accoutrements; also, a similar preparation of khaki colouring. Hence **bla·nco** *v. trans.*, to treat with blanco; **bla·ncoed** *ppl. a.*
1895 in *Army & Navy Co-op. Soc. Price-list.* 1906 *Daily Chron.* 30 Mar. 3/7 The sleeves get covered with 'blanco'

off the belt. 1912 E. Wallace *Pte. Selby* viii. 75 One unhappy mortal, 'warned' for guard..was lugubriously 'blancoing' his straps. a 1918 W. Owen *Coll. Poems* (1963) 79 Boots dubbined; rifles clean and oiled; Belts blancoed. 1948 A. Baron *From City* 32 For half a crown he would blanco the whole of a man's equipment. a 1953 Dylan Thomas *Quite Early One Morning* (1954) 30 Gym-shoes white as the blanco'd snow. 1955 *Times* 25 June 6/5 Since the war they have worn British battle-dress and blancoed web equipment. 1966 D. Holbrook *Flesh Wounds* 37 They covered their webbing with the watery yellow blanco.

blandander (blændæ·ndəɹ), *v.* *colloq.* [cf. Ir. *blanndar* dissimulation, flattery.] *trans.* To tempt by blandishment (*into*); to cajole. Hence **blanda·ndering** *ppl. a.*
1888 Kipling *Soldiers Three* (1895) 70 I've blandanthered thim through the night somehow. 1896 G. B. Shaw in *Sat. Rev.* 1 Feb. 123/1 Boucicault was a coaxing, blandandhering sort of liar. 1898 *Link* 3/1 When you bullied and blandandered me into learning how to ride. 1914 *Times Lit. Suppl.* 4 June 267/2 [European diplomacy] refused to be blandandered by King Nicholas.

blanditude. (Later example.)
1922 *Observer* 16 Apr. 10/5 Then Mr. Lloyd George rose ..in wise fun and paternal blanditude.

blank, *a.* Add: **4. c.** Also const. *of*.
1924 A. E. W. Mason *House of Arrow* viii. 97 Hanaud's eyes..were blank of all expression.
7. c. *Cards.* Unsupported by other cards of the same suit (see quots.).
1895 Manson *Sporting Dict.*, *Blank*, a card in hand is said to be blank when there is no other card of the same suit in hand with it. The term is also applied to a king and queen of the same suit, in which case the twenty of that suit is blank. 1934 [see *blank *v.* 8]. 1958 *Listener* 16 Oct. 611/3 The blank honour combination in diamonds takes the bloom off West's hand.
9. *blank-faced* adj.
1887 G. Meredith *Ballads & Poems* 61 He raced..across a ground Flint of breast, blank-faced. 1921 V. Woolf *Writer's Diary* 8 Apr. (1953) 31 This diary being a kindly blankfaced old confidante.
10. blank cheque (see also Cheque, check 3); **blank flange** (see quot. 1940); **blank wall,** (*a*) a wall without an opening in it; (*b*) an apparently impenetrable obstacle, esp. in phr. *to come up against a blank wall* (cf. *brick wall *sb.*[1] 2).
1940 *Chambers's Techn. Dict.* 94/2 *Blank flange*, a disc, or solid flange, used to blank off the end of a pipe. 1904 Goodchild & Tweney *Technol. & Sci. Dict.* 54/1 *Blank wall*, a wall without an opening in it. 1930 *Economist* 9 Aug. 290/1 Restriction proposals in the rubber-growing industry have apparently come up against a blank wall. 1958 *Times* 17 Nov. 11/3 The Advisory Council on Standards for Consumer Goods..has continued to come up against a blank wall of opposition on matters which were clearly worrying it a year ago.

blank, *sb.* Add: **4.** *to draw a blank*: see Draw *v.* 52 b in Dict. and Suppl.
5. c. *in blank* [after F. *en blanc*]: with blank spaces for the filling in of details.
c1814 A. Rees *Cycl.* (1819) XXVIII. sig. Hh[1]/2 Inconvenience having been experienced from having marine insurances in blank. 1842 Stephen *Comm. Laws Eng.* II. ii. v. 164 A bill, payable to order, is indorsed in *blank* by the payee. 1858 J. W. Smith *Law of Bills* 27 Another way in which the holder of a bill or note indorsed to him *in blank* may transfer it without incurring personal liability, is by writing over the indorser's signature the words, 'Pay A. B. or order'. 1882 Bithell *Counting-ho. Dict.* 45 Bills and Notes of this description are said to be drawn 'in Blank'.
6. Delete † and add: Now chiefly *U.S.*, freq. with defining term.
1805 D. McClure *Diary* (1899) 83 Dr. Wheelock..had given us blanks, for bills of Exchange, on the School's funds, in Scotland. 1860 Holland *Miss Gilbert's Career* x. 181 You've had to write the whole of this. How long have we been out of blanks? 1904 P. H. Hanus *Mod. School* 128 By means of question blanks sent to the parents, much information..is secured by the teachers. 1904 *N.Y. Times* 24 Nov. 1 The messages were always on regular telegraph blanks. 1963 *P.M.L.A.* LXXVIII. iv. ii. 15/1 These committees required applicants to send in a completed application blank together with academic transcripts.
c. An unprinted leaf of a book.
1952 J. Carter *ABC for Book-Collectors* 36 Blanks sometimes occur at the beginning of the book, sometimes at the end of a clearly marked division, more often at the end of the last gathering.
9. b. (See quot. 1962.)
1899 *Daily News* 14 June 9/1 In the blowing of the opal glass surface of the 'blank' a blister or bubble may be caused. 1962 *Gloss. Terms Glass Ind.* (*B.S.I.*) 23 *Blank*, any article of glass on which subsequent processing is required.
c. *Electric recording.* (See quot.)
1940 *Chambers's Techn. Dict.* 94/2 *Blank*, the shaved wax ready for placing on a recording machine for making wax records with a stylus.
12. (Examples.)
1773 Goldsmith *Stoops to Conq.* v. 106 Anthony Lumpkin, Esquire, of blank place. 1818 Scott *Rob Roy* xi, A dispensation..to Diana Vernon to marry *Blank* Osbaldistone, Esq. 1836 Dickens *Pickw.* xxiv. 251 Here's my authority. Blank Tupman, blank Pickwick—against the peace of our sufferin Lord the King. 1861 *Two Cosmos* II. iv. ix. 83 She was married, though—blank—years younger than Miss Lambert. 1888 *Co-operative News* 24 Mar. 266 Our adjoining neighbours at blank—the place shall be nameless. 1968 *Listener* 4 Apr. 445/2 It was not practicable to print anecdotes week after week about Lord Blank's personality and foolish views.
b. Used euphemistically as a verbal representation of a dash put instead of an oath or profane word. Cf. Blank *v.* 5 c. So (as *adjs.* or *advs.*) **blankety** (blæ·ŋkėti), which represents an adj. derivative, such as *bloody*; less freq. **blanked** (blæŋkt), **blanky** (blæ·ŋki).
1854 'C. Bede' *Further Adv. Verdant Green* iv. 28, I wouldn't give a blank for such a blank blank. I'm blank, if he doesn't look as if he'd swallowed a blank codfish. 1874 M. Clarke *His Natural Life* II. iii. xiii. 236 'My blank!' cried Burgess. 'You blank blank, is that your blank game? I'll blank soon cure you of that!' 1876 Bret Harte *Gabriel Conroy* vi. vi, Blank me, if she was inclined to show some attention to Colonel Starbottle. *Ibid.* vii, But what in blank are you waiting for? 1886 Baumann *Londinismen* 11/1 Blanked. 1888 *Troy Daily Times* 3 Feb. (Farmer), He had known vessels to be hindered thirty days; yes, even three months, by that blankety blankety bar! 1892 *Photogr. Ann.* II. 42 The blankety blankness that ensues when the tyro..finds that he has forgotten..his tripod top, &c.! 1896 *Daily News* 15 Feb. 5/1 He..is called 'a blanky capitalist'. 1900 *Westm. Gaz.* 25 May 2/2 Then he lay back, swearing at the blankety blank young blanker. 1900 H. Lawson *On Track* 31 And do you think I'd tell you a blanky lie? *Ibid.* 40 'What does he want to do that for?' 'To get it blanky well analysed! You ass!' 1902 —— *Children of Bush* 12 Let 'em go to ——! I'm blanked if I give a sprat. 1908 M. Diver *Great Amulet* xviii, Colonel Stanham Buckley..inquired picturesquely of a passing official when the blank this blankety blank train was supposed to start. 1914 W. J. Locke *Fortunate Youth* i, What cared he for the blankety little blanks who gibed at him? 1952 A. Grimble *Pattern of Islands* 8 Never springing a blanky leak anywhere. 1959 in N. Mailer *Advts. for Myself* (1961) 345 Mailer? He's an incorrigible blank.
13. A zero score in a game. *U.S.*
1867 *Ball Players' Chron.* 6 June 2/2 Sharp fielding should have disposed of the Harvards for a blank. 1888 *Outing* (U.S.) May 119/2 The tenth innings had seen both sides retire for blanks.
14. Short for *blank cartridge* (see Blank *a.* 10).
1896 Kipling *Seven Seas* 202 For it's 'Three rounds blank' an' follow me. 1935 N. Marsh *Enter Murderer* iv. 52 The report comes from the wings. A blank was never used on the stage, as it would have scorched Surbonadier's clothes.

blank, *v.* Add: **3. b.** To dismiss (a sports team) without a score; to prevent from scoring. *N.Amer.*
1870 *N.Y. Herald* 2 July 5/4 St. John..again blanked the Mutuals and then scored two. 1887 *Courier-Jrnl.* 26 May 2/6 In the eighth and ninth innings both [baseball] teams were blanked. 1951 *Amer. Speech* XXVI. 230 Michigan Normal blanks Ball State. 1968 *Globe & Mail* (Toronto) 5 Feb. 18/7 Oakville Oaks came up with their best effort of the season last night to blank Toronto Marlboros 3–0.
5. a. Also const. *out* (see quot. 1881 in Dict.).
1937 *Times* 16 Apr. 8/6 A smoke screen by howitzers to blank out hostile observation posts and machine-guns.
d. To seal or render inoperative. Const. *off, up.*
1928 *Daily Tel.* 6 Mar. 5/6 The steam inlet and exhaust passages are blanked up and the cylinder subjected to a hydraulic pressure. 1932 *Amer. Speech* VII. 264 *Blank off*, to case off (a portion of an oil sand). 1940 [see *blank flange* s.v. *blank *a.* 10]. 1963 *Guardian* 7 Mar. 3/2 Three-wheeled vehicles with the reverse gear 'blanked off'..remain Group 'A' vehicles..attracting a higher rate of duty.
e. *intr.* To become blank or empty.
1955 J. Christopher *Year of Comet* i. 11 The callscreen blanked.
8. *Cards.* To leave (a card) unsupported by another card of the same suit. Cf. *blank *a.* 7 c.
1884 'Cavendish' *Whist* (ed. 14) 100 It is dangerous to unguard an honour, or to blank an ace. 1934 *Amer. Speech* IX. 11/1 A *blank* King is a King without a guard and to *blank* a King is to discard its guard.
9. To cut or prepare a blank (see Blank *sb.* 9); often const. *out*. Hence *blanked* ppl. a., *blanking* vbl. sb.
1914 *Amer. Machinist* 19 Feb. 342/1 The finished articles..are pierced, embossed, bent at a right-angle, and blanked in one operation. *Ibid.* (European ed.) 28 Mar. 66E/2 Describing their guard [for power presses] the firm classify press operations under blanking, clipping, and raising. 1943 F. D. Jones *Engineering Encycl.* (ed. 2) 136 A blanking die consists essentially of: a die-block..which has an opening that conforms to the shape of the part to be cut or blanked out. 1958 C. D. Hanbury *Industr. Efficiency Rural Labour* viii. 89 *Small dish blanking*...Strip metal is..fed into the machine, which blanks one disk every time a foot pedal is operated.

blank book, blank-book. orig. *U.S.* [Blank *a.* 2.] A book of clean writing-paper in which to make entries, keep accounts, etc.
1713 C. Mather *Diary* 12 Apr. in *Mass. Hist. Soc. Coll.* (1912) VIII. 199 To transcribe..into a blank Book, some instructive Passage. 1802 in C. Cist *Cincinnati: Early Annals*, etc. 193 For a blank-book to record ordinances. 1831 *Boston Directory* 18 (Advt.), Blank book manufacturers. 1886 S. W. Mitchell *R. Blake* ii. 14 He spent a few minutes more over the details of daily duty set out in a little blank-book. 1952 J. B. Oldham *Eng. Blind-Stamped Bindings* i. 2 Sometimes what appears at first to be a blank book is not really one at all. *Ibid.* iii. 34 Blank-book binderies.

blanked: see *blank *sb.* 12 b.

blanket, *sb.* Add: **2. c.** A large number of bombs dropped to cover a wide area.
1944 *Flight* 22 June 661/2 A blanket of bombs is dropped to smother the whole of the area.
d. *Nuclear Engin.* (See quot. 1960.)
1955 S. Visner in *Reactor Handbk.: Engin.* iv. 533 The conversion ratio increases with decreasing core diameter owing to increasing neutron leakage from the core to the blanket. 1960 *Gloss. Atomic Terms* (*H.M.S.O.*) 7 *Blanket*, fertile material put round a reactor core to breed new fuel, e.g. thorium becomes uranium by absorption of spare neutrons.
3. Also (*on*) *the right side of the blanket*.
1842 Marryat *P. Keene* II. i. 100 The Captain marrying and having children on the right side of the blanket as they call it. 1919 D. Ashford *Young Visiters* (1951) v. 34 An old friend of mine not quite the right side of the blanket as they say.
7. *blanket-hidden, -tossed, -wrapped* adjs.; *blanket-tossing*; **blanket bath,** the washing of an invalid in bed; **blanket bog** (see quot.); **blanket coat** *N. Amer.* (see quot. 1872); **blanket finish,** a finish of a race in which the contestants are so close together that they could be covered with a blanket; **blanket fish** *U.S.* (see quot.); **blanket flower,** popular name of the Gaillardia; **blanket overcoat** *U.S.* = **blanket coat*; **blanket pack,** a pedestrian traveller's kit with his blanket rolled about it; **blanket-piece** = Blanket *sb.* 6; **blanket-roll** *Mil.* (*U.S.*), a soldier's equipment of blanket and kit made into a roll for use on active service; **blanket shawl,** a thick woollen shawl; **blanket sheet** *U.S.*, a newspaper in folio form; also *attrib.*; **blanket stitch,** a buttonhole stitch worked on the edge of a blanket or other material too thick to be hemmed; hence **blanket-stitched** *a.*, sewn with a blanket stitch.
1917 A. M. Ashdown *Compl. Syst. Nursing* ii. 11 Points to bear in mind when giving a blanket bath. 1962 C. Watson *Hopjoy was Here* xvi. 180 An ambulent case, no bed pans or blanket baths. 1939 A. G. Tansley *Brit. Islands* xxxiv. 676 Where the rainfall is high and the air so constantly moist that bog is the *climatic formation*, not necessarily arising in fen basins but covering the land continuously except on steep slopes and outcrops of rocks. This is the third type of bog met with in the British Isles and may be called *blanket bog*, because it covers the whole land surface like a blanket. 1761 A. Henry *Trav. & Adv. Canada* (1901) iii. 35 A molton, or blanket coat. 1775 *Pennsylvania Even. Post* 31 Oct. 497/1 Our people had taken from the regulars some blanket coats. 1872 Schele de Vere *Americanisms* 194 Mackinaw blankets..being very thick and well made,..served not only for beds but also for overcoats, which were called Blanket-Coats. [1793 *Sporting Mag.* II. 52/1 Of the nineteen [race-horses] that started, the judge could only place the first four, for not only those, but four or five others, might have been nearly covered with a blanket.] 1934 Webster s.v., A blanket finish. 1960 *Times* 1 Sept. 4/3 Black was beaten by a hairsbreadth in a blanket finish for the silver and bronze medals. 1870 *Amer. Naturalist* IV. 597 Large numbers of 'blanket fish' (a species of *Thymallus*) were to be seen ascending the small rivers. 1879 T. Meehan *Native Flowers of U.S.* II. 182 In the settled parts of Texas..the *Gaillardia* is known as the 'Blanket-flower'. 1884 W. Miller *Dict. Eng. Plants* 14/2 *Blanket flower*, blunt-toothed. *Gaillardia amblyodon* [etc.]. 1963 *Oxf. Bk. Garden Flowers* 146/2 *Gaillardia*,..they are sometimes called 'blanket flower' because the margins of the petals are like the edging of a blanket. 1903 Kipling *Five Nations* 131 The funerals through the market (Blanket-hidden bodies). 1822 J. A. Quitman *Let.* 16 Jan. in J. F. H. Claiborne *Life J. A. Q.* (1860) I. iv. 72 In winter coarse shoes and blanket overcoats. 1920 *Chambers's Jrnl.* Apr. 220/1 Only one small tin of corned beef remained in his blanket-pack. 1851 H. Melville *Moby Dick* II. xxv. 181 The longer upper strip, called a blanket-piece. 1854 *Chambers's Jrnl.* I. 54 The strip of blubber thus in course of separation is about four feet in length, and is called a blanket-piece. 1891 *Harper's Mag.* June 8/1 His bridle hand is raised by the blanket roll. 1899 *Scribner's Mag.* XXV. 27/1 These men..were..making ready to disembark, carrying their blanket-rolls and rifles with them. 1837 *Southern Lit. Messenger* III. 660 The blanket shawls with their varied coloring looked pretty and comfortable. 1839 *Boston Wkly. Mag.* 2 Feb. 175 [The] Baltimore Athenæum..is a *blanket sheet*, very well executed. 1870 A. Maverick *Raymond & N.Y. Press* 36 The heavy, old-fashioned, 'blanket sheet' newspaper. 1880 L. Higgin *Handbk. Embroidery* iii. 23 Blanket stitch is used for working the edges of table-covers, mantel valances, blankets, etc...It is simply a button-hole stitch. 1882 Caulfeild & Saward *Dict. Needlework*, *Blanket Stitch*, used in crewel work and other embroideries for edging woollen, linen, and silk materials, and for forming ornamental lines. It is a variety of button-hole worked wide apart in long loops. 1960 *House & Garden* May 113/1 The quilt..is in a soft lime green..with a blanket-stitched edge. 1881 Duffield *Don Quixote* I. 374 The blanket-tossed Squire. *Ibid.* 369 Never a word did he say of the blanket tossing. 1923 D. H. Lawrence *Birds, Beasts & Flowers* 197 Mountains blanket-wrapped Round a white hearth of desert.
b. Used *attrib.*, (*a*) to designate American Indians who use the blanket as a garment, remaining in a primitive state of civilization and keeping tenaciously to their old tribal customs.
1859 Bartlett *Dict. Amer.* (ed. 2), *Blanket-Indian*, a wild Indian, whose principal article of dress is the blanket. 1866 *Rep. Indian Affairs* (U.S.) 173 There is..great ignorance concerning the location of the prairie or blanket tribes.

BLANKET

1891 M. E. Ryan *Told in Hills* iii. iii. 166 You should hear her talking Chinook to a blanket brave. 1906 *Atlantic Monthly* Mar. 328/2 Only 26,000 blanket Indians are left in the United States.

(*b*) Similarly in South Africa to designate Africans who wear a blanket; hence, by extension, *blanket vote*, the vote of such Africans.

1892 Mitford *'Tween Snow & Fire* xxxvi, There were a few muttered jeers about..getting into the Assembly on the strength of 'blanket votes'. 1904 *Daily Chron.* 13 May 3/3 The 'compound' system is essentially degrading even for 'blanket' Kaffirs.

c. Used adjectivally in the sense: covering or including all, or a number of, cases, contingencies, requirements, things, etc.; all-embracing; indiscriminate. orig. *U.S.*

1886 *Rep. Sec. U.S. Treasury* I. p. xli, Suitable annual appropriations..require no blanket-clause to justify or cover them. 1896 *Congress. Rec.* May 4783 Messrs. Morgan & Co. had given a blanket bid to cover the whole amount... Under the terms of the blanket bid, which covered all bids [etc.]. 1908 W. James *Let.* 2 Jan. (1920) II. 300 The general blanket-word pragmatism covers so many different opinions. 1926 *Mind* XXXV. 251 The German habit of ascribing everything to '*Vorstellungen*', and using this vague and futile blanket-term to cover all the manifestations of mind. 1930 M. Mead *Growing up in New Guinea* xvi. 277 Once we lose faith in the blanket formula of education.. we can turn our attention to the vital matter of developing individuals. 1933 *Brit. Jrnl. Psychol.* Oct. 161 Observers.. generalize and agglomerate under a few 'blanket' epithets the immense complexities and inconsistencies of behaviour. 1935 *Economist* 13 Apr. 848/1 The..Appropriations Bill..is virtually a blanket authorisation for the President to expend $5,000 millions for any purposes which he considers desirable. 1949 'G. Orwell' *1984* 305 Countless other words such as honour, justice, morality..had simply ceased to exist. A few blanket words covered them, and, in covering them, abolished them. 1951 *Ann. Reg. 1950* 68 With..the 'blanket-powers' under regulation 55, the country would approach the Reichstag method of government by order.

blanket, *v.* Add: **1.** (Further examples of fig. use.)

1875 Tennyson *Queen Mary* iii. ii. 122 Blanketed In ever-closing fog. 1895 Kipling *2nd Jungle Bk.* 198 The face of the water was blanketed with wild bees buzzing sullenly and stinging all they found. 1910 J. Simon in *Times* 11 Oct. 10/2 It was a free country,..and he had no intention to blanket his opinion. 1962 *Economist* 3 Nov. 456/1 The energetic campaign..with which Mr Kennedy was blanketing the country.

2. Also *fig.*

1900 Ade *More Fables* (1902) 44 She had her Upper Rigging set, and was trying to Blanket everything on the Street. 1923 *Weekly Dispatch* 1 Apr. 2 Lord Curzon's chief ambition has been to become Prime Minister, and he has been known to complain to his intimates that he has always been blanketed by Arthur Balfour, who was just a little ahead of him.

b. To exclude (a radio signal) from reception by the use of a stronger signal. Const. *out.*

1938 *Nation* 12 Mar. 301/1 The blanketing out of American broadcasts to South America by Berlin and Rome. 1952 *Economist* 26 July 235/1 The Soviet Union had just extended its jamming operations to blanket out not only BBC programmes in Russian but also those in Polish and Finnish.

4. To supply with blankets; to furnish with blankets.

1874 *Contemp. Rev.* XXIII. 466 Schemes of clothing and blanketing whole districts. 1899 *Daily News* 21 July 8/6 The beds are amply blanketed hammocks.

blanketing, *sb.* Add: **1.** Also, as a dress material.

1903 *Daily Chron.* 20 June 8/4 Wraps..made of fine cream blanketing with big sleeves brought into high cavalier cuffs. 1908 *Ibid.* 21 Sept. 7/2 The warm Witney blanketing ..makes exceedingly cosy coats for girls.

5. The action of covering with, or as with, a blanket.

1896 *Pop. Sci. Monthly* L. 245 There's a blanketing of the earth's heat.

blanketing (blæ·ŋkétiŋ), *ppl. a.* [f. Blanket *v.* + -ing².] That covers as with a blanket. Also *transf.*

1904 Farrer *Garden Asia* 244 The smoke descends densely upon the volcano in blanketing clouds. 1925 Joly *Surface-Hist. Earth* vi. 103 The blanketing effects of continental radioactivity.

blankety, blanky: see *blank *sb.* 12 b.

‖**blanquette** (blañket). Also 8 **blanquet.** [Fr. (see Blanket *sb.*).] A dish of light meat (esp. veal) in a white sauce.

1747 H. Glasse *Art of Cookery* ii. 29 Veal Blanquets. Roast a Piece of Veal..cut it into little thin Bits..put in some good Broth... Keep it stirring. 1846 'A Lady' *Jewish Man. Cookery* p. xiii, *Blanquette*, a kind of fricassee with a white sauce. 1906 Mrs. Beeton *Bk. Househ. Managem.* xv. 435 Blanquette of veal. *Ibid.* xviii. 560 Blanquette of lamb. *Ibid.* 719 Blanquette of turkey. 1923 *Harmsworth's Househ. Encycl.* III. 2404/3 *Lamb Blanquette.* This is an entrée composed of scallops of cold roast lamb beated in a rich white sauce, with the addition of mushrooms. 1960 *Times* 1 Aug. 9/3 It is hard to distinguish clearly between a *blanquette* and a *fricassee*. Both should be made from fresh meat (usually chicken, veal or lamb) cooked in a flavoured white stock... For a *blanquette* the meat is more often blanched with herbs and seasoning.

Blanquism (bla·ñkiz'm). [From the name of Louis Auguste *Blanqui* (1805–81), French revolutionary communist + -ism.] The doctrine or practice of Blanqui and his followers (see quots.). So **Bla·nquist,** one who advocates Blanquism; also as *adj.*

1879 *Nation* 15 May 333/2 There was, last Sunday, another trial of electoral strength, out of which the Blanquists emerged the victors. 1922 R. W. Postgate *Out of Past* 7 Very attentive students of modern revolutionary literature may have noticed recently a few references— generally ill-informed—to 'Blanquism'. *Ibid.* 67 The independent history of the Blanquist Party begins with the fall of the Parish Commune... Almost immediately after the Commune's fall and the scattering of the Blanquists, questions of policy arose. 1949 I. Deutscher *Stalin* iii. 55 To the Menshevists this sounded like an ominous repetition of Blanquism, the doctrine of the leader of the Paris Commune, who believed that the only method of achieving revolution was direct action by a small conspiratorial minority ignoring the will of the majority. 1950 E. H. Carr *Bolshevik Rev.* I. i. 18 'Blanquism' in nineteenth-century revolutionary parlance meant addiction to the isolated revolutionary conspiracy or *putsch* and neglect of methodical organization.

blare, *sb.*¹ **2.** (Earlier and later examples.)

1880 Tennyson *Ballads & other Poems* 149 Lured by the light from afar,..Lured by the glare and the blare. 1957 C. Day Lewis *Pegasus* 18 A pure pale blare of distance.

blarney, *sb.* Add: (Earlier and later examples.) Also, nonsense.

[1766 Goldsmith *Vicar* I. xi. 95 Our two great acquaintances from town, Lady Blarny and Miss Carolina..Skeggs!] 1796 Scott *Let.* 26 Sept. (1932) I. 55, I hold it (so to speak) to be all *Blarney*. 1925 D. H. Lawrence *Refl. Death Porcupine* 178 Perfect love, I suppose, means that a married man and woman never contradict one another... What blarney! 1955 *Times* 18 June 6/6 You do not want to come here every day to listen to a lot of blarney.

blast, *sb.*¹ Add: **4. c.** *in blast* (earlier example); *in full blast*: also *transf.*; also *full blast*: at full pitch; *esp.* very loudly.

1780 in *Virginia State Papers* (1875) I. 370 If Mr. Ross can get in Blast time enough..he shall be paid for Shot Twenty five pounds pr: ton. 1839 Marryat *Diary* II. 229 *In full blast*—something in the extreme. 'When she came to meeting, with her yellow hat and feathers, was'n't she in *full blast*?' 1853 A. Bunn *Old & New England* I. v. 86 Oyster-saloons..to use an American phrase 'in full blast' (*Anglicè*, having a great run of business). 1936 H. Miller *Black Spring* (1938) 84 A penny arcade is going full blast. 1938 E. Bowen *Death of Heart* ii. ii. 191 Even when the wireless was not on full blast, Daphne often shouted as though it were. 1957 I. Cross *God Boy* (1958) xxi. 183 If it had come through a radio going full blast [etc.].

d. *fig.* A severe or violent reprimand, outburst, or the like. *colloq.* (orig. *U.S.*).

1874 'Mark Twain' *Let.* 4 Sept. (1917) I. 226, I gave the P.O. Department a blast in the papers. 1930 E. Raymond *Jesting Army* ii. iv. 203 If he but heard them, he charged up to revile them... Fred Roberts came under his blast. 1935 *Time* 11 Mar. 23/3 Despite blast and counterblast between President Roosevelt and Soviet Foreign Minister Litvinoff. 1936 *Variety* 1 July 35/5 Would You Like a Nice 15-Minute Blast at President Roosevelt? 1954 *New Yorker* 31 July 48/2 A typical blast comes from the West Renfrewshire group, which concludes its resolution by saying tartly that the 'blind folly' of the official Party shows that the leadership is completely divorced from the feelings of the majority of the Labour movement.

8. c. A destructive wave of highly compressed air spreading outwards from an explosion. Also *attrib.* and *Comb.*, as *blast wall* (see quot. 1852), *wave*; *blast-proof* adj.

1852 *Harper's Mag.* Apr. 644/2 A structure of black timber..set up in the shape of an acute angle. This is a 'blast-wall', intended to offer some resistance to a rush of air in case of an explosion [at the powder-mill]. 1923 [see *back-blast* s.v. *back- A.11]. 1939 *Jrnl. R. Aeronaut. Soc.* XLIII. 225 Blast is a non-translational shock wave that is transmitted through the air to considerable distances from an exploding bomb... In the blast wave a phase of positive pressure is followed by a phase of negative pressure. 1940 Graves & Hodge *Long Week-end* xxiv. 420 The Government was planning..to provide blast-proof steel shelters for every house in the country. 1941 *Flight* 10 Apr. 272/2 Even some of the machines..are protected against bombing by blast walls.

d. A party, esp. one that is very noisy or wild. *slang.*

1959 *Times* 9 Mar. 13/4 *A blast*, a great party. 1967 W. Murray *Sweet Ride* vi. 89 Man, they're throwing a monster blast over on East Latego later... Everybody's going.

10. blast-lamp, (*a*) see quot. *a.* 1884; (*b*) a lamp in which the flame is driven on to a surface by a current of air; a blow-lamp; **blast-pot** (see quot.).

a 1884 Knight *Dict. Mech.* Suppl., *Blast Lamp*, one with an artificially produced draft of air to aid combustion. 1902 Mrs. Barnes-Grundy *Thames Camp* 57 A benzoline blast lamp which would fetch off any varnish in the world. 1887 *Harper's Mag.* Apr. 670/2 Before the war only seven small furnaces—'blast-pots' they were called—having a total capacity of 20,000 tons, were in operation in all Tennessee.

blast (blæst), *sb.*² [f. Gr. βλαστ-ός: see -blast.] A primitive undifferentiated blood-cell, esp. one found in acute leukæmia. In full *blast cell.*

1947 H. A. Christian *Osler's Princ. & Practice Med.* (ed. 16) 965 Monocytic Leucemia... The cell is a monocyte or monocyte blast. 1952 *Science* CXV. 357/2 The inhibitor activity of the primitive blast cells from acute leukemia was close to zero. 1961 *Lancet* 9 Sept. 603/1 Leukæmic cells, particularly the blast cells, contain no inhibitor whatever.

blast, *v.* Add: **5. b.** *intr.* Of a rocket or spacecraft: to take off, be launched into space; usu. const. *off.* Also used of any powered phase of flight.

1951 R. Bradbury *Silver Locusts* 190 You could still smell the hard, scorched smell where the last rocket blasted off when it went back to Earth. 1953 H. Haber *Man in Space* 262 The moment the big ship blasts off there is no allowance for the slightest failure. 1956 R. Heinlein *Double Star* (1958) ii. 38, I was spacesick..as soon as the rocket ship quit blasting and went into free fall. 1969 *Times* 17 May 8/1 It only remains for three veteran space travellers ..to blast off on Sunday.

9. (Later U.S. examples.)

1748 J. Eliot *Field-Husb. New-England* (1760) i. 14, I have been told that Summer Wheat sowed with Barley is not apt to blast. 1838 E. Flagg *Far West* II. 217 All of the smaller grains..being liable to blast before the harvesting.

10. Freq. in imprecations in the imperative or optative form (for *God blast*...); also as an exclamation of annoyance.

a 1634 Chapman *Revenge for Hon.* v, And thus I kiss'd my last breath. Blast you all. 1752 Fielding *Amelia* IV. x. v, But, blast my reputation, if I had received such a letter, if I would not have searched the world to have found the writer. 1762 Goldsmith *Cit. W.* cv, 'Blast me!' cries Tibbs, 'if that be all, there is no need of paying for that.' 1793 T. Hastings *Regal Rambler* 74 Leaving all the ladies below to blast or bless their eyes, no matter which. 1824 Scott *St. Ronan's* viii, 'As I think, he laid hands on your body...' 'Hands,..no, blast him—not so bad as that neither.' 1916 E. F. Benson *David Blaize* ix. 158 'I say, Blazes, there's extra confirmation class this evening.'..'Oh, blast!' said David. 1936 Auden & Isherwood *Ascent of F6* i. iii, Give it here, blast your eyes! 1955 N. Marsh *Scales of Justice* ix. 209 'Damnation, blast and bloody hell!' Alleyn said.

blasted, *ppl. a.* Add: **3.** Also used adverbially.

1854 M. J. Holmes *Tempest & Sunshine* (1858) xv. 204 Lord's sake be spry, for I'm blasted hungry! 1886 *Leslie's Pop. Monthly* Jan. 67/2 He's too blasted smart for an Indian.

blaster. Add: **9.** *Science Fiction.* A weapon that emits a destructive blast.

1950 I. Asimov *Pebble in Sky* xvii. 179 It was a full-size blaster that could shred a man to atoms. 1958 *Listener* 13 Nov. 775/2 Elijah Baley, the human detective, with a blaster-pistol.

blast-furnace. Add: Also *attrib.*

1877 *Practical Mag.* VII. 248 (*title*) The utilization of blast furnace slag. 1902 *Encycl. Brit.* XXVIII. 601/2 (*heading*) Blast-furnace gases. 1906 *Westm. Gaz.* 19 July 8/2 The blast-furnacemen at Workington. 1930 *Engineering* 17 Jan. 68/1 Mortar, with the exception of blast-furnace cement, ceased to change in strength.

blasting, *vbl. sb.* Add: **2. b.** *Radio.* (See quot. 1926.)

1926 S. O. Pearson *Dict. Wireless Techn. Terms*, *Blasting*, term used to denote the distortion which takes place in loud-speaker or telephone signals on extra loud notes, due to working beyond the straight portion of valve characteristic. 1928 *Observer* 29 Jan. 22/5 When the definite minimum level of sensitivity is given, so that the softer passages are not lost, heavy passages cause most distressing blasting.

5. Also, the operation of breaking up ice.

1856 E. K. Kane *Arctic Explor.* I. xxvi. 340 The blasting had succeeded; one canister cracked and uplifted two hundred square yards of ice with but five pounds of powder.

6. blasting cartridge, a cartridge containing a blasting charge, usually exploded by electricity.

1889 in *Cent. Dict.* 1899 *Westm. Gaz.* 1 June 4/1 Several cavalry horses have been injured by blasting cartridges exploded under their feet.

blasto-. Add: **blastomyco·sis** [Mycosis], a disease caused by infection with pathogenic fungi, affecting either the skin or the organs generally; **blasto·phagine** *a.*, of or belonging to the *Blastophaga*, a genus of fig-insects; **blasto·poral** *a.*, of or pertaining to the blastopore.

1900 Dorland *Med. Dict.* 113/1 Blastomycosis. 1901 H. T. Ricketts in *Jrnl. Boston Soc. Med. Sci.* V. 453 (*title*) A new mould-fungus as the cause of so-called blastomycosis or oidiomycosis of the skin. 1903 *Brit. Jrnl. Dermatology* XV. 121 A case of blastomycosis. 1921 *Brit. Mus. Return* 133 Notes on Fig Insects, including..a new Blastophagine Genus. 1881 *Encycl. Brit.* XII. 557 The ciliated 'planula'.. fixes itself, probably by the blastoporal pole. 1933 *Discovery* Feb. 55/2 A small piece of the blastoporal lip [was] ..cut out of the gastrula of a species of newt.

bla·st-off. [f. *blast *v.* 5 b.] The initial thrust required to launch a rocket or the like into space; the launching of the rocket itself. Also *attrib.*

1951 M. Greenberg *Travelers of Space* 20 *Blast-off*, the initial expenditure of energy by a space ship leaving a planet, or in emergency takeoffs. 1952 A. C. Clarke *Islands in Sky* viii. 125 We were supposed to keep out of the pilot's way at blast-off. 1958 *Observer* 2 Feb. 1/3 This stage developed a blast-off thrust of 78,000 lb.

blastogenesis (blæsto,dʒe·nėsis). *Biol.* [f. Blasto- + Genesis.] **1.** Reproduction by gemmation or budding.

1889 in *Cent. Dict.* 1966 *Immunology* X. 281 Chapman and Dutton..demonstrated a similar blastogenesis when cells from spleens or lymph nodes of two non-related rabbits were cultured together.

2. The theory of the transmission of inherited characters by germ-plasm, as distinguished from 'pangenesis'.

1893 in *Funk's Stand. Dict.*

blastogenic (blæsto,dʒe·nik), *a. Biol.* [ad. G. *blastogene* (Weismann 1888, in *Biologisches Centralblatt* VIII. 106); f. Blasto- + *-genic.] Of or pertaining to blastogenesis; pertaining to origin from, or that originates in, the germ-cell or germ-plasm.

1889 E. B. Poulton et al. tr. *Weismann's Ess. Heredity* vii. 412 'Acquired characters'..we might also call..'*somatogenic*'..; while all other characters might be contrasted as '*blastogenic*', because they include all those characters in the body which have arisen from changes in the germ. *Ibid.* 413 Among the *blastogenic* characters, we include not only all the changes produced by natural selection operating upon variations in the germ, but all other characters which result from this latter cause. 1912 A. Dendy *Outl. Evol. Biol.* xi. 157 Blastogenic modifications are from their very nature as attributes of the germ cells handed on by heredity. 1966 *Immunology* X. 283 An intrinsic difference may exist in the blastogenic potential of adult rabbit and rat thymus glands.

blastoid (blæ·stoid), *a.* and *sb.* [f. mod.L. *Blastoidea*, f. Gr. βλαστός sprout, germ + εἶδος form: see Blasto- and -oid.] **A.** *adj.* Of or belonging to the *Blastoidea*, a group of fossil echinoderms. **B.** *sb.* An echinoderm of this group.

1876 *Q. Jrnl. Geol. Soc.* XXXII. 112 The Blastoid affinities of *Astrocrinites* (= *Zygocrinus*) have been pointed out. 1882 Geikie *Text-bk. Geol.* iv. ii. 722 The blastoids or pentremites, which now took the place in the Carboniferous waters that in Silurian times had been filled by the Cystideans. 1914 *Brit. Mus. Return* 202 Newly-described Blastoids from Somerset. 1962 D. Nichols *Echinoderms* viii. 96 Tubercles have not been found in cystoids, blastoids and heterosteles. *Ibid.* xii. 149 The *steganoblastidae* are superficially blastoid-like.

blastophthoria (blæsto,fþǭ·riă). *Path.* [mod.L., f. Blasto- + Gr. φθορά destruction (φθείρειν to destroy): see -ia.] The hypothetical degeneration of germ cells caused by chronic poisoning as from alcoholism or other diseases. Hence **blastophtho·ric** *a.*, of or relating to blastophthoria.

1908 *Index Medicus* VI. Index 112/2 Blastophthoria. 1913 Stedman *Med. Dict.* (ed. 2) 114/2 Blastophthoria ... Blastophthoric. 1914 *Proc. Soc. Exper. Biol. & Med.* XIV. 14 The continued administration of small doses of lead produces a definite blastophthoric effect in male guinea pigs; and..the lead blastophthoria thus induced manifests itself [etc.]. 1938 *Nature* 16 July 107/2 Forel's contributions to the study of the alcohol problem were then considered under the headings of blastophthoria, alcoholism and the sexual question.

blastula (blæ·stiŭlă). *Embryol.* [mod.L., f. Gr. βλαστός sprout + dim. suffix (see -ule). Cf. Blastule.] An embryo, typically composed of cells arranged in a sphere enclosing the blastocœle. Hence **blastula·tion,** the formation of the blastula.

1887 A. C. Haddon *Introd. Study Embryol.* ii. 21 The result of segmentation is the formation of a multicellular body, usually enclosing a central cavity—'Segmentation cavity' or 'Blastocœl'. The body itself is variously called 'Blastula' or 'Blastosphere'. *Ibid.* ii. 50 Segmentation..of Sponges..results in the formation of a hollow blastula. 1889 *Cent. Dict.*, Blastulation. 1893 Tuckey *Hatschek's Amphioxus* 43 An equal segmentation leading to a blastula without any well-defined main axis. 1924 E. W. MacBride *Stud. Heredity* iii. 70 When the [sea-urchin's] egg has divided into about 1000 cells these form a little hollow balloon or vesicle known as the blastula. 1931 J. Needham *Chem. Embryol.* II. iii. iv. 642 Vles concluded that blastulation involves a change of some kind in the metabolism of the embryo. 1962 D. Nichols *Echinoderms* x. 120 The resulting blastula is oval, hollow and ciliated all over.

blat, *v.* orig. and chiefly *U.S.* Also **blatt.** [Imitative.] **1.** *intr.* To bleat, or make similar sounds. Also *fig.*, to talk noisily or impulsively. Hence **bla·tting** *ppl. a.*

1846 in W. K. Northall *Recoll. Yankee Hill* (1850) 102 Your fellow-countrymen..are not allowed to emigrate north of the Columbia River, on account of a raging he-calf who is bla-ting on the other side. 1884 'Mark Twain' *Huck. Finn* xxv. 248 He blatted along, and managed to inquire about pretty much everybody. 1888 *San Francisco News Let.* 4 Feb. (Farmer), One of these insects of an hour rears up and blatts. 1890 L. C. D'Oyle *Notches* 34 The poor 'blatting' creatures were dragged over to the fire. 1902 Kipling *Traffics & Discoveries* (1904) 22 He'd wipe his long thin moustache..and blat off into a long 'a-aah'. 1932 W. Faulkner *Light in Aug.* (1933) xiii. 271 Others came out from town in racing and blatting cars. 1951 L. Hobson *Celebrity* (1953) iv. 46 She has more self-control... I'd have blatted to the first customer. 1959 I. Jefferies 13 *Days* x. 151, I blatted up to the little shed [on a motor-bicycle].

2. *trans.* To blurt *out*; to emit (a shrill noise).

1879 Howells *Lady of Aroostook* I. v. 50 If I have anything on my mind, I have to blat it right out. 1931 F. D. Davison *Man-Shy* (1932) xiii. 136 Calves blatted their shrill fear. 1942 R. Chandler *High Window* (1943) viii. 66 The radio..was still blatting the baseball game.

blat, *sb.* orig. *U.S.* [Imitative.] A bleating or shrill sound.

1904 M. E. Waller *Wood-carver* 71 Not a sound outside except..the thin blat of a sheep beneath the barn. 1925 *Glasgow Herald* 8 Sept. 6 A bold spirit fired off his gun, and the 'blat' of the shot betrayed the cheat.

blatant, *a.* Add: **2. c.** In recent usage: obtrusive to the eye (rather than to the ear as in orig. senses); glaringly or defiantly conspicuous; palpably prominent or obvious.

1889 W. S. Gilbert *Gondoliers* ii, I write letters blatant On medicines patent. 1903 G. Gissing *Private Papers H. Ryecroft* 274 The blatant upstart who builds a church, lays out his money in that way not merely to win social consideration. 1912 G. B. Shaw *Let.* 19 Aug. in *Shaw & Mrs. P. Campbell* (1952) 38 You don't loathe the scenery for being prosy and mediocre in spite of its blatant picturesqueness as you do in Switzerland. 1930 Sayers & Eustace *Documents in Case* li. 246 The blatant way in which he had marked his trail..[etc.] were actions entirely inconsistent with the carelessness of an innocent man. 1937 H. Nicolson *Helen's Tower* ix. 191 If they were kept in the Museum.. their blatant lack of human interest had caused me to pass them by. 1942 *New Statesman* 11 July 26/1 Mankind, he said, is led by half-truths or blatant lies. 1957 A. E. Coppard *It's Me, O Lord!* v. 55 The colonel..clad in a suit of blatant check, spats, and a monocle. 1957 *Times* 19 Dec. 4/3 A blatant piece of late tackling.

blatantly, *adv.* Add: **2.** Obtrusively, unashamedly, defiantly; as an obvious untruth. (Cf. prec.)

1878 [in Dict.]. 1911 E. Wallace *Sanders of River* viii. 160 Sanders was blatantly unrepentant. 1928 A. Waugh *Nor Many Waters* ii. 70 His features were delicate, but not blatantly aristocratic. 1941 H. G. Wells *You can't be too Careful* v. i. 237 The professional Jewish 'champions' set themselves..to ignore as blatantly as possible the common need for a world settlement. 1959 *Sunday Times* 22 Mar. 9 The Turkish-Cypriot leader..has announced that his T.M.T. organisation has no arms to surrender—rather blatantly, considering the number of Greek Cypriots who were fired upon by Turkish Cypriots. 1959 *Observer* 14 June 17/6 'No waiting' signs are often habitually and blatantly ignored by the motorist.

blather (blæ·ðəɪ), *v.* orig. *dial.* [Variant of Blether *v.*] *intr.* To talk foolishly, talk nonsense. Often in *ppl. a.* Hence **bla·therer,** a foolish talker (1866 in *E.D.D.*).

1825 Brockett *Gloss. N. Country Words* 18 He blathers and talks, is a common phrase where much is said to little purpose. A person of this kind is..styled a *blathering hash*. 1891 Kipling *Light that Failed* xv. 59 If you were only a mass of blathering vanity,..I wouldn't mind. 1892 *Cassell's Fam. Mag.* Dec. 11/1 Hold your tongues, you blathering idiots. 1900 *Century Mag.* Feb. 504/1 Morland had a blathering contempt for nobility and society. 1920 D. H. Lawrence *Touch & Go* iii. i. 76 They've got a set of loud-mouthed blatherers and agitators among them. 1951 *Essays in Criticism* I. 17, I sympathize with O'Casey's character Who called him 'a tired-out old blatherer'. 1951 'J. Wyndham' *Day of Triffids* i. 15 Gentlemanly tones which blathered about this 'magnificent spectacle' and 'unique phenomenon'.

blather, *sb.*: see Blether *sb.*

bla·tting, *vbl. sb.* The action of the verb *blat.

1935 S. Lewis *It can't happen Here* xxix. 315 The nervous blatting of tremendous traffic. 1946 F. Davison *Dusty* vi. 61 There was much bellowing and blatting [of cows]. 1955 M. Millar *Beast in View* iii. 46 Anyway, you didn't come here to listen to my blatting.

Blattnerphone (blæ·tnəɪfōᵘn). Also **b-.** [Named after the inventor L. *Blattner* (1881–1935) + *-phone.] An instrument for recording sound on magnetic tape. Also *attrib.*

1931 *Wireless World* 17 June 664/1 A few days ago I was privileged to see and hear the Blattnerphone in operation. 1932 *B.B.C. Techn. Tables & Glossary* 45/2 *Blattnerphone*, a system of recording speech or music on a steel wire or tape by means of variable magnetisation along the length of the wire or tape. 1932 *BBC Year-Bk.* 101 In some ways the most important event of the year has been the adoption by the BBC of the Blattnerphone recording apparatus. 1937 *Brit. Jrnl. Psychol.* Jan. 282 If there are any blattnerphone records..of psychoanalysis they are unlikely to be long or numerous because of their great expense.

blay (blē¹), *a.* Variant of Blae *a.*, esp. in sense 5 (= unbleached), frequent in Irish use.

1785 *Ann. Reg. 1783* (Useful Projects) 85/1 When I mention white flax, I do it in opposition to that, which..has the appellation of blay. 1861 O'Curry *MS. Materials* 38 Some with..green or blay or white cloaks. 1869 [see Blae *a.* 5]. 1913 'G. A. Birmingham' *Gen. John Regan* xvii. They draped it..in a large sheet of blay calico of a light yellowish colour.

blaze, *sb.*¹ Add: **2. b.** (Earlier and additional examples.) Also *blue blazes.*

1818 A. Burton *Johnny Newcome* 41 They thought he must be mad as blazes. 1818 M. L. Weems *Drunkard's Looking Glass* (ed. 6) 49 Ye steep down gulphs of liquid fire! Ye blue blazes of damnation! 1837 Dickens *Pickw.* liv. 587 How the blazes you can stand the head-work you do, is a mystery to me. 1853 —— in *Househ. Words* 5 Feb. 483/2 Letting the teeth go (to Blazes, he observed indefinitely). 1858 S. A. Hammett *Piney Woods Tavern* 37 And the two Jacobs swore like blue blazes agin him. 1861 Dickens *Gt. Expect.* I. x. 160 What the Blue Blazes is he? 1924 G. L. Mallory *Let.* 7 May in E. F. Norton *Fight for Everest* (1925) 231 The *moral* of A party had gone to blazes. 1925 W. Deeping *Sorrell & Son* xiii. § 1. 119 When you have found out what you want to do—then go at it like blazes. 1948 C. Day Lewis *Otterbury Incident* ix. 121 What the blue blazes is all this?

7. *Physiol.* An electric current passing along living tissue in response to mechanical stimulus. Also *attrib.* in *blaze current.* Hence *blaze reaction, response*, reaction or response so obtained.

1902 *Nature* 18 Sept. 491/2 The blaze reaction..requires short strong currents for its manifestation. 1903 *Ibid.* 9 July 238 This 'blaze' response is the algebraic sum of post-anodic and post-kathodic currents. 1903 *Jrnl. R. Microsc. Soc.* Oct. 599 A crystalline lens is a good object upon which to study the nature of blaze-currents.

8. In poker: (see quot.). *U.S.*

1880 'Trumps' *Amer. Hoyle* (ed. 13) 197 *Blaze.* This hand consists of five court cards, and, when it is played, beats two pairs.

blaze, *sb.*² Add: **1.** Also of other animals.

1850 Cumming *S. Afr.* ix. 53/1 The blesbok..is one of the true antelopes... A broad white band, or 'blaze', adorns the entire length of its face. 1952 C. L. B. Hubbard *Pembrokeshire Corgi Handbk.* x. 108 *Blaze*, a white (usually bulbous) marking running up the centre of the head.

2. (Earlier example.) Also *attrib.* in *blaze-mark.*

1662 in *Groton Rec.* (1880) 7 The meetinge house shall be set..by a small whit oak marked at the souwest side with two notches and a blaze. 1885 Mrs. C. Praed *Head Station* xlvi, Here were new blaze-marks; and here, upon a bottle-tree,—the bark unhealed—that old trace of Durnford's tomahawk.

blaze, *v.*¹ Add: **8.** (Earlier and later examples.) Also *to blaze (out) at.*

1776 *Battle of Brooklyn* (1873) ii. p. i, We bid them stand and blazed away like brave boys. 1883 'Mark Twain' *Life on Mississippi* lix. 531 The elements..banged and blazed away in the most blind and frantic manner. 1909 T. E. Lawrence *Lett.* (1938) 76 He had just bought a Mauser, and blazed at everything with it. 1914 *Ibid.* 173 They all grabbed rifles & revolvers, & through the windows blazed out at everyone they saw.

9. *trans.* *to blaze* (*off*): to cause (the grease) to flash in the operation of tempering steel; also, of the grease, to flash; to temper (steel) by this process. Hence **bla·zing** (**off**) *vbl. sb.*

1823 *New Monthly Mag.* IX. 121 The cast steel articles ..may be quenched in this composition, in order to harden them; and then be blazed off. 1860 Tomlinson *Useful Arts & Manuf.* Ser. ii. 36 Large saws..are moved backwards and forwards over the fire till the unctuous matter adhering to the surface of the saw begins to ignite or 'blaze off'. *Ibid.* 37 During this 'blazing off', the saw is removed from the furnace and allowed to cool. 1885 *Spons' Mech. Own Bk.* 66 They are then heated..till the grease inflames. This is called being 'blazed'.

blaze, *v.*³ Add: (Earlier examples.) Also *transf.* and *fig.*, esp. in phr. *to blaze the way* (*trail*, etc.).

1750 T. Walker *Jrnl. Exploration* 30 Apr. (1888) 50, I Blazed a way from our House to the River. *Ibid.* 23 May 56, I Blazed several Trees in the fork. 1841 in Thornton *Amer. Gloss.* (1912) I. 70, I desire to new blaze landmarks which.. have divided Federal and Democratic parties. 1850 *Southern Quarterly Rev.* XVIII. 418 Champollion..having done little more than 'blaze out' the road to be travelled by others. 1902 L. Mead *World-Coinage* vi, Professor Bréal has blazed the way for future explorers in the wilderness of philology. 1904 *Daily Chron.* 29 Nov. 4/4 So intricate a maze that an old warder of long standing used to 'blaze' his way through the corridors with the help of a piece of chalk. 1937 *Discovery* Sept. p. lxxiii (Advt.), Dufaycolor blazes a new trail!

blazed, *ppl. a.*³ **1.** (Later U.S. examples.)

1787 Washington *Diaries* (1925) III. 155 A sorrel mare, blazed face, off hind foot white. 1869 *Overland Monthly* III. 126, I had seen..an old gray mare, considerably flea-bitten, with a blazed face and a docked tail.

blazer¹. Add: **1. b.** Now usu. an unlined jacket of lightweight material (often flannel), freq. with coloured stripes, decorated edges, or a badge on the breast-pocket, worn esp. with sports clothes or as part of a school uniform. (Add further examples.)

1889 *Daily News* 22 Aug. 6/6 In your article of to-day.. you speak of 'a striped red and black blazer', 'the blazer', also of 'the pale toned' ones... A blazer is the red flannel boating jacket worn by the Lady Margaret, St. John's College, Cambridge, Boat Club. When I was at Cambridge it meant that and nothing else. It seems from your article that a blazer now means a coloured flannel jacket, whether for cricket, tennis, boating, or seaside wear. 1892 Kipling *Lett. Travel* (1920) 93 More does he wonder still at the city clerk in a blazer. 1963 *Times* 23 Jan. 12/4 A simple short blazer-type jacket. 1964 *Which?* Mar. 94/2 More navy blue blazers are sold than any other colour. *Ibid.* 96/1 Blazer cloth was not originally intended for hard daily wear, but for more occasional use—sitting in a cricket pavilion, for instance.

2. Someone or something that attracts attention. Chiefly *U.S.*

1845 Mrs. Kirkland *Western Clearings* (1846) 127 T'other gal is likely enough, but the mother's a blazer! 1892 'H. Lawson' in *Penguin Bk. Austral. Ballads* (1964) 154 You must prove that you're a blazer—you must prove that you have grit. 1903 A. Adams *Log Cowboy* vi. 81 Are you sure you wasn't running a blazer yourself? 1906 *Springfield Weekly Republ.* 19 Apr. 1 The Kaiser's telegram..recalls some of his blazers in the past.

3. A small cooking apparatus. *N. Amer.*

1889 in *Cent. Dict.* 1895 *Harper's Mag.* May 885/1 Delicacies which Tom prided himself on being able to prepare on a blazer. 1967 *Canadian Antiques Collector* Apr. 17/1 Cooking can be done in the blazer pan over a direct flame.

4. A sheet of metal placed against the bars of a grate to create a draught and cause the fire to blaze; a blower. *north.*

1892 Heslop *Northumb. Words*, *Bleezer*, a hood to blow up a fire... Put the bleezer up, and let's hev a lowe. 1924 T. Burt *Autobiogr.* (1924) 177 Women and children..armed with 'blazers' and tin-cans..used as cymbals.

bla·zered, *a.* [f. Blazer¹ 1 b + -ed².] Wearing a blazer.

1928 *Times Lit. Suppl.* 14 May 386/4 When his blazered team play the village. 1959 *Times* 12 June 13/7 The blazered cheer-leader writes a vast dial on the wall chart.

blazing, *ppl. a.*¹ Add: **4.** Used as a substitute for a profane epithet.

1888 Kipling *Plain Tales fr. Hills* (1890) 19 Once I said, 'What's the blazing hurry, Major?' 1916 'Boyd Cable' *Action Front* 66 You have the blazing cheek to keep me lying here in the filthy muck.

blazing star. 4. For *Alteris farinosa* read *Aletris farinosa.* Also used of other plants with star-shaped flowers. (Examples.)

1789 *Trans. Amer. Philos. Soc.* III. 72, The root of *Alteris farinosa*..is called star-root, blazing star, devil's bit.. 1822 A. Eaton *Man. Bot.* (ed. 3) 303 Blazing star; false unicorn root. 1836 A. H. Lincoln *Fam. Lect. Bot.* (ed. 5) App. 110 *Liatris scariosa*, blue blazing star. 1947 *Desert Mag.* May 28/3 Visitors will find..blazing star..in the Valley of Fire and other rocky areas.

bleach, *sb.*¹ Add: **3.** A bleaching process; also, a bleached condition.

1887 *Sci. Amer.* 16 Apr. 249/1 What is known as 'the three-quarter bleach' with flax. 1920 *Discovery* Mar. 86/2 A perfect bleach is almost impossible to secure.

4. *Bleach-field* (later examples).

1866 *Gaz. Scotl.* 339/2 The excellence of its water for bleaching processes has induced many to establish extensive printfields and bleachfields on its banks. 1957 R. Watson-Watt *Three Steps Victory* v. 33 A polychromatic stream which served..as a carrier of my bleachfield days.

5. A bleaching liquor or powder.

1898 *Daily News* 15 Dec. 6/5 A quantity of bleach escaped from a tank at one of the paper mills. 1929 *Daily Chron.* 23 Apr. 7/3 There are several good bleaches that are safe to use. 1970 *Which?* May 149/1 All the scouring powders contained some bleach.

bleach, *v.*¹ Add: **2. c.** *Photogr.* To remove the silver image from (a negative or print) after development; *bleach-out process*, a system of colour printing, now disused, whereby dyes are decolourized by being exposed through transparencies.

1889 R. Meldola *Chem. of Photogr.* vi. 209 A solution of potassium iodide also bleaches the darkened product of photo-decomposition under the influence of light. 1896 C. F. Townsend *Chem. for Photographers* (ed. 2) vi. 54 The image is first bleached with mercuric chloride, which converts the black silver image into a white double silver-mercurous chloride. 1911 E. J. Wall *Dict. Photogr.* (ed. 9) [illegible] The negative is bleached in an acidified solution of bichromate salt, and then re-developed. 1914 G. L. Johnson *Photogr. in Colours* (ed. 2) xi. 169 The difficulty of reproducing colour transparencies..has at length been more or less overcome by the Bleach-out Process of colour printing. 1925 F. J. Mortimer *Wall's Dict. Photogr.* (ed. 11) 101 The print may be bleached after the final wash. 1936 J. Deschin *New Ways in Photogr.* xvii. 244 A photograph may be converted into a drawing by..the bleach-out process.

bleacher. Add: **3.** One of a roofless set of benches for spectators at outdoor events such as baseball and football games; also an occupant of these benches. Usu. in *pl.* Also *attrib.* orig. and chiefly *U.S.* Hence **blea·cherite** *U.S.*, a frequenter of bleachers; **blea·chery** *U.S.*, an open-air stand for spectators.

1889 *Chicago Tribune* 18 May 6/1 The grand stand and bleachers were well filled with something over 2,000 spectators. 1896 *Ibid.* 3 July 9/3 The money for it is being subscribed by the bleacherites. 1900 *Cent. Dict.* Suppl., Bleacherite, Bleachery. 1917 Mathewson *Sec. Base Sloan* xvii. 237 The two boys settled themselves in their places on the bleachers. *Ibid.* 238 More than half of the bleacher seats were empty. *Ibid.* 239 'Sadie' was a grand favourite with the bleacherites. 1964 *Sun-Herald* (Brisbane) 21 June 136/5 Those empty bleachers at the Beatles concerts on Thursday and Friday. 1968 *Listener* 29 Feb. 265/2 The many innocent spectators in the south-east bleachers.

4. *Photogr.* An agent for bleaching a negative or print (see *bleach *v.*¹ 2 c).

1911 B. E. Jones *Cassell's Cycl. Photogr.* 196/1 Any reducer or bleacher may be used. 1912 F. J. Mortimer *Wall's Dict. Photogr.* (ed. 9) 113 If the print is a big one it is generally as well to wet it..before placing it in the bleacher.

bleachy (blī·tʃi), *a.*¹ *dial.* [f. Bleach *sb.*¹ or *v.*¹ + -y¹.] Bleached, pale.

1821 Clare *Vill. Minstrel* II. 194 Nodding lands of wheat in bleachy brown. 1835 —— *Rural Muse* 106 The mowers swept the bleachy corn.

bleachy (blī·tʃi), *a.*² *dial.* Brackish.

1825 Jennings *Observ. Dial. W. Eng.* 1879 Hardy *Wessex T.* (1888) II. 148 It makes the stuff [sc. rum] taste bleachy.

bleakly, *adv.* Add: **4.** Cheerlessly.

1928 E. Waugh *Scoop* ii. iv. 203 The Minister..smiled bleakly, the wry smile of one heroically resisting an emotion of almost overwhelming repulsion. 1941 *New Review* [illegible] He spoke bleakly and with the sense of futility. 1959 *Times* 1 May 17/1 The treasury bleakly announced that no extra allowance or increases would be granted to Army Officers.

blear-eyed, *a.* Add: **1.** (Later examples.)

1798 [see *coiffe-woman* s.v. Coiffe 1 1]. 1935 W. S. Maugham *Don Fernando* xi. 232 You have painted me ugly and blear-eyed.

2. (Later examples.)

1687 Dryden *Hind & Panther* ii. 61 That ev'n the blear-ey'd sects may find her out. 1937 Yeats *October Blast* 15 Nor blear-eyed wisdom out of midnight oil.

blearily (blī·rili), *adv.* [f. Bleary *a.*] In a bleary manner.

1936 W. Owen *Coll. Poems* (1963) 72 He..Just blinked at my revolver, blearily. 1958 *Blackw. Mag.* Mar. 274/1 Thomas Owen looked up blearily from his task.

bleary, *a.* Add: **b.** *Comb.*, as **bleary-eyed** *a.*, = Blear-eyed *a.* 1.

1927 *New Republic* 9 Mar. 71. 1930 Saffir *Finger of Fate* xvi. An unshaven, bleary-eyed man. 1957 R. Campbell *Portugal* 33 The weirdest lot of bleary-eyed thugs.

bleat, *sb.* Add: **d.** A (feeble) complaint. *colloq.*

1936 'Taffrail' in *Royal Mag.* Dec. 99/2 'Got a bleat, 'ave yer?' growled an unsympathetic Petty Officer. 1948 'N. Shute' *No Highway* 107 He had heard that nothing..in reply to his signal stating Mr. Honey's bleat.

bleaty (blī·ti), *a.* [f. Bleat *sb.* + -y¹.] Having a sound like the bleat of a beast.

1928 W. G. Whittaker *Class-Singing* 77 'A' (as in 'ant') is apt to be 'bleaty'.

bled (bled), *ppl. a.* [f. Bleed *v.*] **1.** Drained of blood or sap.

1894 *Pop. Sci. Monthly* June 184 A series of tests and examinations of bled and unbled trees has been carried on.

2. Of a page, illustration, etc.: see *bleed *v.* 13.

1940 Graves & Hodge *Long Week-end* xxv. 434 'Bled' photographs..printed to cover the whole of the page, without a margin. 1942 H. A. Maddox *Dict. Stationery* 15 Bled-off pictures cause a larger size of paper to be used to accommodate the protruding blocks. 1959 *Camb. Rev.* 21 Nov. 187/2 Its front cover is..a 'bled' on photograph of Great St. Mary's.

bleed, *v.* Add: **2. b.** *transf.* Of a dye: to 'run' or become diluted when wetted.

1862 C. O'Neill *Dict. Calico Printing & Dyeing* 34/1 Woollen articles [are] worked in it until saturated with colour, then washed in cold water to begin to 'bleed', that is until the washing water begins to remove the blue and become tinged with it. 1893 E. Knecht et al. *Man. Dyeing* 714 Fastness to washing and to bleeding or running should be determined with water alone and with soap. *Ibid.* 725 Most of the direct cotton colours bleed very much when dyed on cotton.

c. 'To leak; especially, to leak an iron-stained liquid, as the seams of a boiler' (*Funk's Standard Dict.* 1893). Hence **bleed·ing** *vbl. sb.* (see quot.).

1888 *Lockwood's Dict. Terms Mech. Engin.*, *Bleeding*, the red streaks of rust which weep through the scale adherent to the insides of boilers, and which reveal the presence of corrosion in the plates underneath.

3. (Later examples.)

1874 *Rep. Vermont Board Agric.* II. 289 If pruned later the trees will often 'bleed', though it is stated that a perfectly healthy tree will not bleed if pruned at any season. 1966 Bell & Coombe tr. *Strasburger's Textbk. Bot.* 237 Many plants bleed only in the spring and at a certain stage of development, in others bleeding can occur at almost any time.

10. *to bleed white*: see White *a.* 5 b.

13. *Bookbinding* and *Printing.* To cut into the print of (a book) in trimming the margin; to print (an illustration) so that it reaches beyond the normal margin to the edge of the page. Also *intr.* (see quots.). So *to bleed off* or *on.*

1835 'J. A. Arnett' *Bibliopegia* 203 A work is said to bleed, if cut into the print. 1876 *Daily Tel.* 9 June 2/1 (Farmer), The pages bleed in many places—i.e. the binder's knife when cutting the edges has also cut away portions of the printed matter. 1887 Knight *Dict. Mech.* Suppl. 107/1 *Bleeding*, cutting into the printed matter of a book when cutting the edges. 1917 *Graphic Typography* [illegible] A margin so that This plate can then be printed in color on gummed paper and the paper trimmed so as to 'bleed' off the edges of the printed background. 1942 H. A. Maddox *Dict. Stationery* 14 If a guillotine operator slightly cuts into the type of a printed forme when trimming the edges, he is said to have bled the edges. The term Bleed-off is a modern adaptation applied to the style of..magazine pages in which illustrations are actually run right off the edge of the paper. The term employed by some printers is..'cut to bleed'. 1948 *Words into Type* (*Appleton-Century-Crofts Inc., N.Y.*) 533 Printing is said to bleed when the margins are overcut in trimming and the printing mutilated. 1962 *Listener* 26 July 142/3 It looks brilliant to place a detail from one Cézanne beneath the whole of another and to make the detail 'bleed' off.

14. To allow (liquid) to drain away or (gas) to escape through a cock, valve, or the like. Also with *off.* Cf. *bleeding *vbl. sb.* 3.

1886 *Cent. Dict.* s.v., *To bleed the brakes*, in a locomotive, to relieve the pressure on the air-brakes by opening the bleed-ing valve or cock. 1948 *Chambers's Techn. Dict.* 89/1 [illegible] bleed..off. 1958 *Motor Manual* (ed. 36) iii. 61 A very rich mixture of fuel, and a little air bled in by the air bleed jet at the well. 1962 A. Sheppard in *Into Orbit* 104 The technicians found they were able to bleed off the excess pressure by turning some of the valves by remote control. 1963 *Which? Car Suppl.* Oct. 138/1 The sponginess of the brake system could not be entirely overcome by bleeding the hydraulic system.

bleed, *sb.* [f. the *v.*] **1.** = Bleeding *vbl. sb.* 1 a.

a 1585 A. Montgomerie *Flyting betwene Montgomerie & Polwart* 309 The bleid[s] and bellithrow. 1863–6 [see *bleed* [illegible]]. 1922 T. Hardy *Late Lyrics* 26 The silent bleed of a world decaying.

2. The action of *bleed *v.* 13; a page or illustration that is printed or trimmed so as to leave no margin. Also *attrib.*

1930 in Webster *Add.* 1948 *Words into Type* (*Appleton-Century-Crofts Inc., N.Y.*) 533 Illustrations which extend to the edges of the page when printed are called bleed cuts. 1967 Karch & Buber *Offset Processes* 531 *Bleed*, a printing image or illustration, extending beyond any one or more edges of a sheet.

3. = *bleeding *vbl. sb.* 3; also, the cock, valve, or the like through which bleeding occurs. Freq. *attrib.* So *bleed-off.*

1948 *Gloss. Aeronaut. Terms* (*B.S.I.*) II. 21 *Bleed*, to eliminate air from a hydraulic system, e.g., by undoing bleed screws. 1958 *Times Rev. Industry* July 26/1 The range comprises turbo-jet, air-bleed, free turbine, and shaft-drive units. 1959 [see *bleeding *vbl. sb.* 3]. 1969 *New Scientist* 23 Jan. 174/2 An opacity-measuring instrument..controls a continuous bleed-off.

bleeder. Add: **1. b.** One who extorts money. *U.S.*

1858 *Snell's Night Guide* 114/1 *Bleeder*, a sponger. 1894 *Columbus* (*Ohio*) *Dispatch* 5 Oct. The police of New York were not the only bleeders.

3. *Slang.* A very stupid, unpleasant, or contemptible person; also *transf.*; also used inoffensively, preceded by *little, poor*, etc., = Devil *sb.* 4 c.

1887 *Sessions' Papers* 26 Apr. 16 Look out, Carry; don't let the bleeder take me. 1889 Barrère & Leland *Dict. Slang* 134/2 (University), a 'regular bleeder' was a superlative duffer. 1902 J. H. M. Abbott *T. Cornstalk* xiii. 168 'Bleeder'..seemed to be reserved for a typical low-class Londoner—a slum-dragger. 1923 D. H. Lawrence *Sons & Lovers* xiii. 343 'What'll you have?'..'Nowt' wi' a bleeder like you!' 1936 J. Curtis *Gilt Kid* xxiv. 241 Give me the damn gun, you bleeder. 1958 Ismay *Rep.* III. MacNeice *Earth Compels* 19 His brother caught three bleeders of crows. 'There the bleeders back in the sea. 1952 A. Baron *With Hope, Farewell* 44 She'll kill the poor little bleeder.

4. A pipe, valve, or the like used for 'bleeding' (see *bleeding *vbl. sb.* 3). Also *attrib.*

1893 *Funk's Stand. Dict.*, *Bleeder*..a pipe fitted on engines by which steam may be passed directly from the main steam-pipe to the condenser. Called also *bleeder-pipe*. 1930 *Hawkins's Electr. Dict.* 38/2 *Bleeder*, a small cock or valve to draw off water of condensation from a steam pipe.

b. *Electr.* A resistor, permanently connected across a voltage source to improve voltage regulation. In full, *bleeder resistance* or *resistor*. Also *attrib.*

1935 Nilson & Hornung *Pract. Radio Communication* xxiii. 609 The value of the bleeder resistance is frequently chosen to draw appreciable current in the circuit. A value of bleeder current equal to 25 or 30 per cent of the total drain on the rectifier unit is used in many installations. 1953 *Electronic Engin.* XXV. 202/1 The voltage across a 'bleed' part of the..bleeder resistor.

bleeding, *vbl. sb.* Add: **2. bleeding-bowl** (see quot. 1916).

1911 C. J. Jackson *Hist. Eng. Plate* I. i. ix. 264 Surgeon's bleeding-bowl, with pierced handle. 1916 *Connoisseur* Dec. 207/2 The so-called Bleeding Bowl, or Cupping Dish. Such bowls were used by surgeons when bleeding their patients.

3. The draining of liquid, gas, etc., through a cock, valve, or the like (see *bleed *v.* 14). Also *attrib.*

1889 [see *bleed *v.* 14]. 1928 *Daily Tel.* 17 Jan. 5 Where this 'bleeding' is resorted to for feed heating. 1959 *Engineering* 2 Jan. 26/3 The bleeding of air from the hydraulic system. 1970 *Which?* Oct. 226/2 Bleeding the system is..provided with bleeding screws to allow the escape of air [illegible].

4. Of a dye: see *bleed *v.* 2 b.

1893 [see *bleed *v.* 2 b]. 1959 *Which?* Oct. 129/2 The washed samples [of carpet] were then examined for any changes in colour and in particular for any bleeding of colours into the white ground.

b. The penetration of a coat of paint or the like by colour or other substance from an underlying surface.

1926 *Paint, Oil & Chem. Rev.* 25 Nov. 14/1 *Bleeding*, a defect of pigments by which they slowly soak into or penetrate an overlying coat, or the dissolving of a pigment in the vehicle. 1951 R. Mayer *Artist's Hand-Bk.* ii. 91 An obstacle in the way of the adoption of coal tar colours of the property of bleeding or striking through when used with oil or oily mediums.

c. (See quot.)

1914 H. P. Boulnois *Gloss. Road Terms* 14 *Bleeding*, the exudation of pitch or bitumen from the road surface or from the prepared material. 1954 *Highway Engin. Terms* (*B.S.I.*) 30 *Bleeding*, a road condition in which free binder rises to the surface of a bituminous..road in hot weather. *Ibid.* 36 *Bleeding*, the formation of a thin layer of water on the surface of concrete during the finishing processes.

bleeding, *ppl. a.* Add: **5. bleeding heart** (examples); (*b*) *fig.* an excessively soft-hearted or sympathetic person (*colloq.*); also *attrib.*; **bleeding tooth,** the shell of a gastropod mollusc, *Nerita peloronta*, which has a red mark on the columella tooth.

1691 Evelyn *Kal. Hort.* (ed. 8) 171 Cherries. Bleeding Heart. 1803 [see Heart *sb.* 30]. 1825 Britton *Beauties of Wilts.* III. 371 *Bleeding-heart*, the wall-flower. 1863 *Cent. Mag.* July 351 The white-hearts (related to the bleeding-hearts of the gardens, and absurdly called 'Dutchman's breeches'). 1946 D. H. Lawrence *Amores* 134 The newly-filled Flowers of the bleeding-heart. 1938 J. Bingham *Murder Plan Six* iii. 74 You want to think straight, Victor. You want to control this bleeding-heart trouble of yours. 1962 'T. Hinde' *Hunter on Ground* (1963) vi. 56 A lot of bleeding-hearts got the idea they knew about everything. 1881 *Proc. Acad. Nat. Sci. Philad.* 168, I did not find the 'Bleeding Tooth' very common at Key West. 1954 R. T. Abbott *Amer. Seashells* 128 *Nerita peloronta* Linné. Bleeding Tooth. Southeast Florida, Bermuda and the West Indies.

6. A substitute for Bloody *a.* 10 (and *adv.* 2). *low slang.*

1858 Furnivall in *Athenæum* 24 July 118 Costermongers have lately substituted the participle 'bleeding' for the adjective '[bloody]'. 'My bleeding barrow' is the latest phrase in vogue. 1876 Bryant & Rice *Son of Vulcan* II. xxiii, When he isn't up to one dodge he is up to another. You make no bleeding error. 1884 *Sessions' Paper* 8 Jan. 315 'I said you don't [sic] bleeding well fetch me.' 1890 A. Morrison *Child of Jago* i. 4 This is a bleed'n' unsocial sort o' evenin' party. 1922 [see Blimey *int.*]. 1967 *Times* 17 Nov. 8/4 Why don't you bleeding do something about it? *Ibid.*, He wants to take bleeding care that that light duty work is continuous.

bleep (blīp), *sb.* [Echoic.] A thin, high-pitched blipping sound, *spec.* that of a radio signal. Hence as *v. intr.*, to make such a sound or signal; so **blee·ping** *ppl. a.* and *vbl. sb.*

1953 *N.Y. Herald-Tribune* 2 Aug. ii. 5 The bleeps of Geiger counters make 'penny stocks' on the country's exchanges tabletop-alert to investors' bonanzas. 1957 *Daily Mail* 7 Oct. 1/1 From now on we shall be living in a different world, for science fiction has become fact... Those eerie 'bleeps' from outer space tell us that. *Ibid.* 1/2 Instead of the quick bleep-bleep sound from it. 1957 *Daily Tel.* 7 Oct. 2/2 The bleeps were heard on one of..the first stations to pick up the signals. 1957 *Times* 11 Oct. 11/6 There is a Soviet satellite bleeping round the earth. 1957 *Observer* 13 Oct. 12/3 The suggestion has been that..the bleeping of the most..signals come from batteries to power the bleeping radio transmitters. 1958 *Times* 29 July 10/7 The driver brushed his horn. 'Bleep, bo-o-oop, parp.' 1958 *Daily Mail* 13 Aug. 4/3 Every dinghy was fitted with a bleeper. 1965 *New Scientist* 25 Feb. 497/1 The automatic bleeper gives a strong response, with a sharp cut-off at the bleeping. The bleeper..sends out a bleep of sound. 1966 *New Scientist* 24 Feb. 497/1 The acoustic gives a strong response..with a sharp cut-off at the bleeping..and other deliberate bleeping..signal. *Ibid.* 497/2 The bleeping frequency..the loudspeaker..suddenly changes. 1968 R. Nobbs *Hazard of Spain* xvii. 187 The bleeping grew fainter or stronger depending on the bearing.

ble:pharoconjunctivi·tis. *Path.* [mod.L., f. Blepharo- + Conjunctivitis.] Inflammation of the eyelids and conjunctiva.

1899 in Billings *Med. Dict.* I. 171/1. 1962 *Lancet* 28 Apr. 877/1 He presented..a typical picture of primary ocular vaccinial blepharoconjunctivitis.

blepharoplast (ble·fărəplast). *Biol.* [f. Blepharo- + -plast.] **a.** A centrosome-like protoplasmic body found in the sperm-cells of certain plants. **b.** In protozoans, a minute granule at the base of each flagellum.

1897 I. Webber in *Bot. Gaz.* Oct. 233, I would..suggest the name *blepharoplast* to distinguish them from other organs of the same..character. 1908 *Fortnightly Rev.* 227 Flagellated forms [possess] a nucleus and blepharoplast. 1964 M. Hyman *Med. Bacteriol.* 21 The blepharoplast..The anterior half of the cell (Trichomonas vaginalis) are the nucleus and the blepharoplast from which arises a flagellum..the border of the undulating membrane.

ble:pharopto·sis. *Path.* [mod.L., f. Blepharo- + Ptosis.] = Ptosis *a.*

1807 Morris & Kendrick *Edinb. Med. & Phys. Dict.* I, *Blepharoptosis*, a prolapsus, or falling down of the upper eyelid, so as to cover the cornea. 1868 Brittan tr. *Malgaigne's Man. Opt. Surg.* 379 Blepharoptosis may arise from two principal causes. 1961 *Brit. Med. Dict.* 204/2 Blepharoptosis.

blemya (ble·miă). Usu. in pl. **blemyae.** [L. *Blem(m)yæ*, f. Gr. Βλέμυες, an Ethiopian people, 'acc. to the fable (Mel. 8), without head and eyes, and with the mouth in the breast' (Lewis and Short).] A type of medieval carved image (see quot. 1915).

1915 C. J. Bruce in *Archæological Jrnl.* LXXII. 138 A pair of Blemyae are carved as wing-subjects upon one of the misereres in Norwich cathedral church... Each of them is represented as a headless man with his face upon his breast. 1955 M. D. Anderson *Imagery Brit. Churches* iii. 181 The Blemyae, whose faces are in their stomachs, are carved upon misericords at Ripon and in Norwich Cathedral.

blend, *sb.* Add: **1. b.** *spec.* A mixture of different kinds of woollen or other fibres (see also quot. 1959).

1884 W. S. B. McLaren *Spinning* ix. 184 The quantity used varies very much, but for blends half wool and half shoddy to 1/6 of oil per 100 lb. of wool is a common allowance. *Ibid.* After this operation the blend is again spread on the floor. 1888 R. Beaumont *Woollen & Worsted Cloth Manuf.* iii. 49 A layer of teased cotton is, in such blends, first spread for a foundation, then layers of wool and cotton alternately. 1958 *Eng. Gaz.* 25 Aug. 8/1, A blend varies in size and weight from 1 pack upwards. 1911 *Encycl. Brit.* XXVIII. 810/2 A blending of various materials..to obtain a cheap blend which may be spun into a satisfactory warp or weft yarn. 1959 *Chambers's Encycl.* XIV. 661/1 If a mixture yarn is required, the necessary proportions of dyed and undyed wool are built up in layers in a stack (usually known as the 'blend') and passed into a machine which..mixes the materials.

2. *Philol.* A word or phrase formed by blending (see *blending *vbl. sb.* 2); so *blend-word.*

1909 *Cent. Dict. Suppl.*, Blend-word. 1911 *Mod. Philol.* IX. 197 All the so-called 'stretchformen' may not be blends. 1914 L. Pound *Blends: Their Relation to Eng. Word Formation* 1. 1 Blend-words, amalgams, or fusions may be defined as two or more words, often of cognate sense, telescoped as it were into a single word. 1925 A. C. Baugh *Hist. Eng. Lang.* x. 373 Words of the type of *electrocute*..are often called *portmanteau* words, or better, blends.

blending, *vbl. sb.* Add: **1. b.** *spec.* The action or process of mixing materials used in woollen manufacture.

1884 W. S. B. McLaren *Spinning* ix. 184 Blending..is one of the most important operations in the whole manufacture. *Ibid.* Blending may mean many things. It may be different colours of dyed wool, or wool and shoddy, mungo or flocks, or wool and cotton, or cotton and silk, or cotton, or grain wool, or all of them.

2. *Philol.* = *contamination 1 d.

1900 H. Sweet *New Engl. Gram.* § 1 Grammatical and logical anomalies often arise through the blending of two different constructions... The plural *themselves* may be regarded as a blending of *himself* and *ourselves*. 1924 Jespersen *Philos. Gram.* i. 149 Contaminations or blendings of two synonymous (or nearly synonymous) expressions are frequent in all languages. *Ibid.* 196 The blending may also be due to the fact that what was grammatically the subject of one verb was logically the subject of another verb. 1963 G. A. Miller *Lang. & Communication* x. 285 Blending..phrase of Spoonerisms or of the Synonymous Cognate Expressions in English. 1969 H. Marchand *Categories Process-day Eng. Word-Formation* 451 Blending can be considered relevant to word-formation only insofar as it is an intentional process of forming words. We shall use the term here to designate the method of merging parts of words into one new word, as when *smoke* and *fog* derive *smog*.

Blenheim. Add: **a.** (Earlier and later examples.)

1830 C. Sinclair *Holiday House* xv. 332 She..had taken into the berth beside her a little Blenheim spaniel. *c* 1845 *Mrs. Rundell's Dom. Cookery* [illegible]. 1865 A. Hartley *Merrie Profiteer* (1857) ii. xiv. 358 To lead in a ribbon ten Blenheim spaniel of Italian greyhound. 1957 *Encycl. Brit.* VII. 497/1 *English Toy Spaniel*..There are several varieties of this breed, including the Prince Charles..Ruby and Blenheim, the main variations being in colour.

b. *Blenheim pippin.* Also *ellipt.*

1862 R. Hogg *Fruit Man.* (ed. 2) [?] Blenheim Pippin (Woodstock Pippin; Northwick Pippin). 1877 E. S. Dallas *Kettner's Bk. of Table* 24 Dessert Apples..Blenheim Pippin. 1925 Blunden *Eng. Poems* 17 And the sweet smell of Blenheims lapped in straw.

blessed, blest, *ppl. a.* Add: **4. c.** *blessed word*: (applied to) a long and high-sounding word, erroneously or ironically taken to be of great significance. Cf. Mesopotamia 2.

1910 *N. & Q.* 9th Ser. I. 458/2 That blessed word Mesopotamia. 1919 G. B. Shaw in *Irish Statesman* 25 Oct. 427/2 There is at first sight something to be said for that blessed word Devolution. 1928 D. L. Sayers *Unpleasantness at Bellona Club* ii. 125 Nothing..except Blessed Word..that blessed word hippopotamus.

blessing, *vbl. sb.* Add: **4. c.** Phr. *a blessing in disguise*: an apparent misfortune that works to the eventual good of the recipient.

1746 Hervey *Refl. Flower-Garden* 76 Ev'n Crosses from his sov'reign Hand Are Blessings in Disguise. 1873 *Cassell's Mag.* VII. 295/2 Like many another misfortune, this great calamity was in truth only a blessing in disguise. 1900 *Jrnl. Soc. Arts* 25 June 595 We find that the Pacific cable scheme has really been a blessing in disguise because it has delayed it until now. 1967 *Westm. Gaz.* 4 May 2/1 Religion would gain greatly if the clergy would make a more sparing use of the blessing-in-disguise argument.

blethering, *ppl. a.* Add: **2.** = *blinking *ppl. a.* 4. Cf. *blithering *ppl. a.* *colloq.*

1915 A. Kinross in *Times Red Cross Story Bk.* 148 If my boy ever gets married on the quiet and plays the fool, I'll break his blethering neck for him.

bletherskate and variants. Add: **b.** Foolish talk; nonsense.

1825 [in Dict.]. 1861 *N.Y. Tribune* 25 Dec., To wit, our proving, not by verbal blatherskite, but by facts, that the C.S.A. is dependent on us. 1892 J. Barlow *Irish Idylls* 82 Wid your little black book full o' blatherumskite. 1900 *Daily News* 29 May 5/1 It is still..six hundred pages of sheer blood and blatherumskite. 1907 J. H. Hyslop *John Bull's Other Island* iv, There's too much blatherumskite in Irish politics. 1966 C. Wilson *Outsider* xiv. 275 For Nietzsche..there is no such thing as blatherskite: there is only useful knowledge and unprofitable blatherskite.

blethery (ble·ðəri), *a. rare*⁻¹. [f. Blether *sb.* + -y¹.] Characterized by blether.

1889 G. M. Hopkins *Let.* 29 Apr. (1935) 304 Rot about babies, a blethery bathos.

bleu (blö). [Fr., see Blue *a.*] Used in the names of various French cheeses with veins of blue mould; **bleu cheese** [part-tr. Fr. *fromage bleu*] = *blue cheese* (see *blue *a.* 13).

1914 *U.S. Dept. Agric. Bull.* 608 6 The names that Bleu and Fromage Bleu are applied to several kinds of hard, rennet cheese made from cows' milk in imitation of Roquefort cheese. *Ibid.*, Several more or less distinct kinds..are designated Bleu d'Auvergne, Gautal..and St. Flour. 1952 W. Plomer *Museum Pieces* xix. 117 His favourite cheese was Bleu d'Auvergne. 1957 O. Nash *You can't get there from Here* 86 Every time the menu lists bleu cheese I want to order *fromage bleu*.

‖**bleu-de-roi** (blö də rwa). *Ceramics.* Also **-du-.** [Fr. = 'king's blue'.] The ultramarine blue of Sèvres porcelain; also called *bleu de Sèvres*. Also *attrib.*

1848 H. R. Forster *Stowe Catal.* 38 A coffee-cup and saucer—bleu du Roi. 1868 *Rolls House & Ss. Paris Exhib.* xiii. 153 A Sèvres vase, *bleu-de-roi*. 1902 *Connoisseur* Jan. 30/1 The famous *bleu de roi* has been traced..a rich dark blue enamel. 1959 G. Savage *Antique Coll. Handbk.* 63 A rich blue enamel ground, the *bleu de Roi*, was introduced in 1749.

blew, var. *blue *v.*²

blight, *sb.* Add: **4. b.** *spec.* An unsightly urban area (cf. *blighted *ppl. a.* 1 b).

1935 R. Mumford *Culture of Cities* 270 ..face the accumulated physical and social results of that disruption: ravaged landscapes, disorderly urban districts, patches of blight, mile upon mile of standardized slums. 1952 M. Lock et al. *Bedford by River* i. 13 Blight elsewhere will affect another 4,100 people who will be displaced from the main clearance areas. *Ibid.* 13/2 Treatment of the blight areas..

5. blight-bird *Austral.* and *N.Z.*, an early settlers' name for a bird belonging to the Australian genus *Zosterops.*

1870 R. Taylor *Ika a Maui* (ed. 2) 17/2 Kanohimowhiti, or Tauhau, white eye or blight bird (*Zosterops lateralis*) was first observed at the Bay of Islands in 1856. 1888 T. H. Potts *Out in Open* 130 The white-eye or blight-bird appears to..migratory..of small insect pests. 1888 Newton in *Encycl. Brit.* XXIV. 824/1 In 1856 it..was noticed..as occurring in the South Island of New Zealand, where it became known..to the English settlers as the 'Blight-bird'. 1964 *Austral. Encycl.* VIII. 129/2 Silvereyes.. do much good by destroying scale insects and other pests, and have thereby earned the name of blight-birds.

blighted, *ppl. a.* Add: **1. b.** Applied to an urban area (see quot. 1951).

1951 R. A. Cordingley in *Town Planning Rev.* XXI. 283 The unbearable municipal burden of blighted areas. 1951 R. J. Collins *Development Plans Explained* 37 Blighted areas, which are characterized by conditions of bad layout or obsolete development. 1952 M. Lock et al. *Bedford by River* 9/1 The primary object of the housing survey was to find out to what extent the town is blighted..and the housing or other reason requiring to be demolished.

2. Used as a mild substitute for Blasted *ppl. a.* 2.

1915 Locke *Jaffery* xii. 171, I think he's a blighted malingerer. 1916 Wodehouse *Joy in Morning* iv. 15, I never heard the blighted name before in my life. 1932 D. L. Sayers *Have his Carcase* xxi. 288 What's the blighted Syndrift.

blighter. Add: **2.** *slang.* A contemptible or unpleasant person; often merely as an extravagant substitute for 'fellow'. Also *transf.*

1896 *Idler* Mar. 282/1 'Larry,' says they, 'you ain't going to let that blighter chew you.' 1900 *Westm. Gaz.* 28 Mar. 9/3 Down with the dirty blighters who will not remove their hats. 1904 Kipling *Traffics & Disc.* 349 'There's another of 'em, I 'ope.' 'There's an accommodation' blighter for you!' said Pyecroft. 1920 Locke *House of Baltazar* xxiii. 218 He could buy up this old blighter of a lord twice over. 1922 *Daily Mail* 3 Nov. 15, I never dreamt the little blighter would get in with such a hurry. 1952 R. Firbank *Valmouth* Lord's Caprice 1 Also 101 Mina birds. 'Noisy blighters!' he said. 1957 I. M. Stewart *Use of Riches* 16 'What we have to contrive,' he said, 'is fair blighters—or rowdies.' 1964 *Daily Mirror* 1 Oct. 1/1 Mr. Chapman suddenly became aware of the loud clatter of a cat 'blighter' 1972 in Shetland—Blooming, the blighter.

blighting, *ppl. a.* Add: **2.** = *blighted *ppl. a.* 2. *slang.*

1916 'Boyd Cable' *Action Front* 187 There's that blighting ..blighting shell. 1934 T. S. Eliot *Rock* i. 32 'E showed up the 'ole blightin' swindle.

blighty (blǝi·ti), *a.* [f. BLIGHT *sb.* + -Y[1].] † **a.** = BLIGHTING *ppl. a. Obs.*

1731 SWITZER *Pract. Fruit-Gard.* (ed. 2) 287 Those blighty Airs rise, and by Elasticity are driven in Columns against Fruit-Trees.

b. Affected with blight; blighted.

1900 *Standard* 9 Nov., The acorn crop is an abundant one, the fruit being rather blighty and undersized.

Blighty, blighty (blǝi·ti), *sb. Army slang.* [Contracted form, originating in the Indian army, of Hind. *bilāyati* = *wilāyati* foreign, and esp. European, f. *wilāyat* prop. Arabic, inhabited country, dominion, district, VILAYET, in Hind. esp. foreign country (cf. Arab. *wali* governor of a province, VALI, WALI).

Cf. *Bilayutee pawnee*, *Bilātee panee*. The adject. *bilāyatī*..is applied specifically to a variety of exotic articles,.. and most especially *bilāyati pānī*, 'European water', the usual name of soda-water in Anglo-India (Yule & Burnell, *Hobson-Jobson*).]

England, home. (Used by soldiers on foreign service.)

[**1886** KIPLING *From Sea to Sea* (1899) II. 358 Let the town hear of the wonders which I have seen in Belait.] **1915** *Times* (weekly ed.) 8 Oct. 852 The only thing they looked forward to was getting back to 'Blighty' again. **1916** *N. & Q.* 19 Feb. 151/1 One poem I have recently seen begins:—Oh, send me back to Blighty. **1917** P. MACGILL *Gt. Push* xix. 238 I'll send out the money and fags when I go back to blighty. **1968** J. R. ACKERLEY *My Father & Myself* vii. 60, I was not happy in Blighty.

b. *attrib.* or *adj.* 'Home', as distinguished from 'foreign'.

1918 *Aussie* Aug. 9/2 The C.O. endeavours to persuade Private Hardcase to accept Blighty Leave. **1926** *Morn. Post* 8 Dec., An Exhibition and Sale of Blighty Industries.

c. In the war of 1914–18 applied to a wound that secured return to England. Also *attrib.*

1916 *N. & Q.* 4 Mar. 194/2, I believe that 'B.B.' is the regular, though unofficial description of any non-fatal wound serious enough to send its victim back to a base hospital—Blighty Boy. **1916** *Daily Mail* 1 Nov. 4/4 So-and-so stopped some shrapnel and is back at the base in hospital,.. he wasn't lucky enough to get a blighty. **1917** 'CONTACT' *Airman's Outings* 29 A Blighty bullet sent him back to England and gave him a mention in the casualty list. **1918** LOCKE *Rough Road* xix, Mo says he's blistering glad you're out of it and safe in your perishing bed with a Blighty one. **1927** *Daily Express* 18 Oct. 1/1 Soldiers are visiting the battlefields..in the hope of finding trenches, dug-outs, or the exact spot where they received their 'blighties'. **1934** V. M. YEATES *Winged Victory* I. xii. 104 Marsden..had had his left arm damaged by a bullet and had gone to hospital very pleased with himself for having picked out of the dip the ideal Blighty.

blimey (blǝi·mi), *int.* Also **bli' me, blime, blymy.** Vulgar corruption of the imprecation *blind me!* or *blame me!* (*BLAME *v.* 7 a). (Cf. *GORBLIMEY.)

1889 in BARRÈRE & LELAND *Dict. Slang.* **1894** *Punch* 27 Oct. 193/1 Blymy, you're a knockout! **1897** W. S. MAUGHAM *Liza of Lambeth* ix. 153 Bli' me if I know wot yer all talkin' abaht. **1922** JOYCE *Ulysses* 305 God blimey if she aint a clinker... Blimey it makes me kind of bleeding cry. **1932** *Punch Almanack 1933* 7 Nov. [18/1] 'Your mentality is erroneous and—er—soggy. Blime! *what* a cod!' he concluded. **1954** 'R. CROMPTON' *William & Moon Rocket* viii. 235 'Blimey!' said Charlemagne. 'Pardon him, dear,' said Miss Milton in a shaking voice. 'He doesn't often use bad language.'

blimp (blimp). [Of uncertain origin. Said to have been coined by the aviator Horace Shortt (see quot. 1918) or by Lieut. A. D. Cunningham (1951 *Aeroplane* 5 Oct.), and to have been based on the adj. LIMP.]

1. A small non-rigid airship orig. consisting of a gas-bag with the fuselage of an aeroplane slung underneath; in the war of 1939–45 the name was sometimes applied to a barrage balloon.

1916 ROSHER *In R.N.A.S.* 11 Feb. 146 Visited the Blimps..this afternoon at Capel. **1918** *Illustr. Lond. News* 27 July 96 Nobody in the R.N.A.S. ever called them anything but 'Blimps', an onomatopœic name invented by that genius for apposite nomenclature, the late Horace Shortt. **1926** J. R. R. TOLKIEN in *Year's Wk. Eng. Stud. 1924* 52 It is perhaps more in accordance with their looks, history, and the way in which words are built out of the suggestions of others in the mind, if we guess that *blimp* was the progeny of *blister* + *lump*, and that the vowel ı not *u* was chosen because of its diminutive significance—typical of war-humour. **1928** GAMBLE *North Sea Air Station* x. 149 The Submarine Scout non-rigid type. The name was abbreviated to S.S. airships, but they were generally known as 'Blimps'. **1934** *Discovery* Jan. 14/2 Excellent photographs..could probably be secured next summer from a small 'blimp' carrying a pilot and a photographer and directed by wireless telephony. **1939** *War Illustr.* 29 Dec. 538/1 The term 'blimp' originated in the last war, when British lighter-than-air aircraft were divided into A-rigid, and B-limp (i.e. without rigid internal framework). The modern barrage balloon may therefore be classed as a blimp. **1940** HARRISSON & MADGE *War begins at Home* v. 125 The [barrage] balloons, so suitably called blimps, became a major symbol in the first three months of the war.

b. *Cinematogr.* A sound-proof covering for a ciné camera. *colloq.*

1936 C. B. DE MILLE in *Words* Oct. 6/1 A 'blimp' in studio jargon is..a soundproof covering for the camera. **1959** *John o' London's* 3 Dec. 287/2 If I asked you to fetch me a 'blimp'..you might toss over..a light cover to deaden the sound of a..cine camera, so that its whirring didn't get recorded on the sound track.

2. (Colonel) **Blimp,** a character invented by David Low (1891–1963), cartoonist and caricaturist, pictured as a rotund pompous ex-officer voicing a rooted hatred of new ideas. Hence **blimp,** a person of this type; also *attrib.* Also **bli·mpery, bli·mpishness, bli·mpism,** behaviour or speech characteristic of a blimp; **bli·mpian, bli·mpish** *adjs.*, typical of a blimp.

1934 *Evening Standard* 28 May 10 Prime Minister Blimp: 'Gad, sir, the Air League is right. We must oppose all proposals for the abolition of military aviation.' **1935** *Ibid.* 2 Sept. 15 Blimpian 'statesmanship'—'Gad, madam, you can't lock up the explosives! That's a warlike act!' **1937** 'G. ORWELL' *Road to Wigan Pier* x. 197 Easy to laugh at.. the Old School Tie and Colonel Blimp. **1937** F. P. CROZIER *Men I Killed* 13 Blimp still reigns, unfortunately, in places of greater responsibility where he can make a fool of himself more easily. *Ibid.* vii. 137 Our new system of rearmament is at least serving the purpose of encouraging our Colonel Blimps to hide their heads..in the sand. **1937** *New Statesman* 18 Dec. 1055/1 All to the good if people would be careful to send the full context and not send other people's ironical remarks or conscious jokes as if they were Blimpisms. **1938** *Ibid.* 5 Nov. 715/2 The modern clothes *Hamlet* at the Old Vic has excited a lot of Blimpish indignation. **1941** 'G. ORWELL' *Lion and Unicorn* 44 Thirty years ago the Blimp class was already losing its vitality. **1941** N. MARSH *Surfeit of Lampreys* viii. 115 [He's] very nice... Sort of old-world without any Blimpishness. **1942** *Times* 26 Feb. 8/3 The essence of blimpery is a refusal to entertain new ideas, and a determination to keep the bottom dog permanently in his place. **1943** C. BEATON *Near East* ii. 29 Blimpism, plus the Cairene climate, are two of Hitler's strongest weapons. **1944** H. G. WELLS *'42 to '44* 147 The more Blimp-like officers began to exclude this 'dangerous' topic. **1958** *Times* 15 Oct. 8/7 His innate Blimpishness suddenly arrests itself and he sings a jolly nostalgic song about the cricket fields of Harrow. **1968** *Daily Tel.* 13 Dec. 16/4 His usual comic character of pub pundit or cockney blimp.

blin (blin). Pl. **blini, bliny** (bli·ni), **bli·nis.** [Russ., pancake.] (See quots.)

1889 *Harper's Mag.* LXXVIII. 854/1 The terrible Russian General absorbs before his soup a dozen *blinies*—which are heavy pancakes stuffed with caviare and seasoned with hot melted butter. **1892** T. F. GARRETT *Encycl. Pract. Cookery* II. 157/2 *Blinis*, small meal cakes which are eaten in Russia during Lent. **1920** M. W. DAVIS *Open Gates to Russia* xvi. 247 *Bliny*, the crisp Russian pancakes. **1943** E. M. ALMEDINGEN *Frossia* x. 376 They remembered *bliny*, those famous Carnival pancakes, thin and golden, eaten sometimes with sour cream, and oftener with caviare. **1945** E. WAUGH *Brideshead Revisited* I. vi. 152 The maitre d'hotel was turning the blinis over in the pan.

blind, *a.* (and *adv.*). Add: **1. a.** In comparisons, as *blind as a bat* or †*brickbat*; also *blind as a beetle, mole, stone* (see the sbs.); *to turn a blind eye*: see *EYE *sb.*[1] 5 e.

1588 J. HARVEY *Disc. Probl.* 40 As blinde, as moules, or bats. **1639** J. CLARKE *Parœm.* 52 As blind as a bat at noone. **1840** in *Amer. Speech* (1965) XL. 127 For blind as a bat he was. **1850** DICKENS *Dav. Copp.* xlii, The old Scholar..is as blind as a brickbat. **1925** W. DEEPING *Sorrell & Son* xviii. § 2, I was blind as a bat. Pushed into a job by my people. **1943** E. CALDWELL *Georgia Boy* iii. 50 I've done gone and got as blind as a bat. I can't see nothing at all.

f. (See quot.)

1879 *Encycl. Brit.* IX. 292/2 The side [of a flat-fish] which is turned towards the bottom..is generally colourless, and called 'blind', from the absence of an eye on this side.

g. Short for *blind drunk* (see 15).

1630 J. TAYLOR *Water-Cormorant* in *Wks.* III. 5/1 For though he be as drunke as any Rat, He hath but catcht a Foxe, or Whipt the Cat. Or some say hee's bewitcht, or scratcht, or blinde, Which are the fittest tearmes that I can finde. **1845** A. M. HALL *Whiteboy* v. 49 They'll be all *blind* by the time they get home from G. F.'s wedding. **1903** 'A. MCNEILL' *Egreg. Eng.* 145 The artisan..improves the shining hours, by 'getting blind', to use his own elegant phrase. **1930** W. S. MAUGHAM *Gent. in Parlour* xliii. 263 On the night he arrived in London he would get blind, he hadn't been drunk for twenty years.

h. In Poker, *to go blind*: to put up a blind (see *BLIND *sb.* 9); hence *to go* (a specified stake) *blind*. Hence applied to forms of Poker in which this is done; so *blind hand*.

1872 [see i below]. **1882** *Poker* 88 For some reasons players never give the blind hand credit for a good or even an average hand. **1885** *Encycl. Brit.* XIX. 283/2 The age is sometimes allowed to *go blind*, i.e., to raise the ante before he sees his cards. **1885** [see STRADDLE *v.* 7].

i. fig. *to go it blind*: to act without previous investigation of the circumstances; to plunge without regard to the risks involved. Also *to go blind (on ——)*. orig. *U.S.*

1840 *Spirit of Times* (N.Y.) 14 Mar. 18 Don't think of 'going it blind', but according to Walker! **1846** *Congress. Globe* App. 120 All I ask of him is that he will not 'go it blind' upon Oregon. **1848** LOWELL *Biglow P.* Ser. I. viii. 155 It gives a Party Platform, tu, jest level with the mind.. Of..honest folks thet mean to go it blind. **1872** SCHELE DE VERE *Americanisms* 328 Poker, when played by betting before looking at one's hand, is called *Blind Poker*, and this has given rise to the very common phrase, *to go it blind*, used whenever an enterprise is undertaken without previous enquiry. **1875** GEN. SHERMAN *Mem.* I. 342, I know that in Washington I am incomprehensible, because at the outset of the war I would not go it blind. **1892** KIPLING *Barrack-r. Ballads* 63 And faith he went the pace and went it blind. **1909** *Daily Chron.* 8 Feb. 4/4 If that be true, shall we be quite wise to 'go blind' on Dreadnoughts alone? **1924** GALSWORTHY *White Monkey* III. viii, Yes; they go it blind; it's the only logical way now.

j. *Aeronaut.* Applied to flying and aerial bombing executed by means of instruments without direct observation. Also as *adv.* Hence *blind-bombing, -flying, -landing* attrib.; **blind approach** (see *16).

1919 *Aviation* 1 Feb. 22 Flying in fog, clouds or darkness —which may be called 'blind flying'—involves difficulties not encountered in clear air. **1928** A. KLEMIN *If you want to Fly* 54 We have all sorts of instruments to tell us whether the plane is flying over an even keel, but it seems almost impossible to fly 'blind'. **1930** *Techn. News Bull., Bureau of Standards* June 61/1 Various combinations of the three elements making up the blind-landing system have been tried. **1936** *Nature* 23 May 863/2 His development of blind-flying instruments employing gyroscopic principles. **1937** *Times* 16 Oct. 11/5 A machine which..can be landed 'blind' from any height. **1940** W. S. CHURCHILL *Into Battle* (1941) 284 The diminution of the damage done by blind bombing. **1944** *Times* 31 Jan. 4/6 Frankfurt was covered by cloud and 'blind bombing' methods were used. **1958** *Times* 17 Oct. 3/4 The blind landing experimental unit have now shown that it is possible to complete the landing automatically without the pilot touching the controls or even seeing the ground.

k. *a blind*: used *colloq.* with a following sb. to mean 'a single ——', 'the least or slightest ——'.

1938 J. CURTIS *They drive by Night* xvii. 193, I don't want a blind word out of either of you. **1941** *Penguin New Writing* II. 87 There's not a blind thing you can do about it. **1966** 'L. LANE' *ABZ of Scouse* 9 'E wooden take a blind bit er notice of me. *Ibid.*, Nobody could get a blind bit er sense outer 'im.

2. c. (Earlier example.) Also, the side on which the view is obstructed from sight. In *Rugby Football*, the side of a scrum opposite to that on which the main line of the opponents' backs is ranged.

1606 CHAPMAN *Gentleman Usher* I. i, For that, we'll follow the blind side of him. **1917** 'CONTACT' *Airman's Outings* 212 An instinct for the 'blind side' of whatever Hun machine he had in view, made him a master in the art of approaching unobserved. **1959** *Times* 30 Nov. 3/6 Cambridge were in peril now, but they came again, worked the blind side.

7. d. *Bookbinding.* Ungilt; cf. *blind-blocking, -tooling* in 16 in Dict., *BLIND *v.* 8.

1835 'J. A. ARNETT' *Bibliopegia* 125 Graining may be properly considered as a blind ornament. **1846** DODD *Brit. Manuf.* VI. 105 The block..imprints the device; whether it be gilt or 'blind'.

9. b. Applied to a corner or other feature where the road or course ahead is concealed from view.

1927 W. E. COLLINSON *Contemp. Eng.* 34 A further safety first proposal..is to eliminate the blind corners, i.e. corners where oncoming traffic cannot be seen. **1954** WODEHOUSE *Jeeves & Feudal Spirit* xvi. 161 No speeding? No passing on blind corners? **1957** S. MOSS *In Track of Speed* xii. 148 We could keep at 170 over blind hillbrows.

10. c. Of an alphabetic letter: written or printed with the loop closed or filled in: *spec.* in *Typogr.* defining the paragraph mark with a closed loop.

1820 KEATS *Lett.* (1958) II. 262 The fault is in the Quill: I have mended it and still it is very much inclin'd to make blind es. **1888** JACOBI *Printers' Vocab.*, *Blind P.* A paragraph mark ¶ so called from the loop of the P being closed. **1905** F. H. COLLINS *Author & Printer* 34 Blind ¶, paragraph mark.

d. *Cookery.* Applied to a pastry case baked before the filling is added (see quots.). Also as *adv.*

1943 A. SIMON *Concise Encycl. Gastronomy* IV. 92/2 Bake 'blind' till golden... 'Blind' means pricking paste well and filling with tissue paper and beans to stop crust rising. **1952** B. NILSON *Penguin Cookery Book* 313 If the flan is to be baked without a filling (i.e. 'blind'), prick the bottom well. **1958** *Woman* 4 Oct. 23/1 'Baking blind', simply means lining the pastry with a round of greaseproof paper, then weighting this down with rice or beans to prevent it rising during cooking.

11. c. Applied to a geographical feature, as a spur, reef, or valley, that terminates abruptly.

1848 C. J. PHARAZYN *Jrnl.* 15 Jan. (MS.) 97 Teddy and W. to lambs to drive them to pen, smother'd 10 in a blind gully. **1861** J. VON HAAST *Rep. Topogr. & Geol. Expl. Nelson Prov.* i. 8 We had selected a so-called blind spur, which fell abruptly into a deep gully. **1882** W. D. HAY *Brighter Britain* II. viii. 299 Eventually it proved that the find was but a 'blind reef', a 'pocket', a mere isolated dribble from the main continuous vein we had at first supposed we had struck. **1898** J. GEIKIE *Earth Sculpture* xiii. 217 Not less characteristic features of the karst-lands are the so-called blind-valleys and dry-valleys. Through the former a river flows to disappear into a tunnel at the closed or blind end. **1942** O. D. VON ENGELN *Geomorphology* xxii. 578 Blind valleys differ from pocket valleys in that the latter develop where underground water emerges in greater or less volume.

d. Of a baggage-car on a train: see quot. 1901. *U.S.*

1893 *Daily Ardmoreite* (Oklahoma) 8 Nov. 1/4 They didn't even have an opportunity to fire a tramp off the blind

baggage. **1901** *Scribner's Mag.* XXIX. 429/1 The train's got a blind baggage-car on... That's a car that ain't got no door in the end that's next the engine. **1926** J. BLACK *You can't Win* ix. 120 With much caution I made my way..till I got near enough to the depot to get aboard the blind end of a baggage car.

15. *blind-born* (later example), *blind-drunk* (examples), *blind-eyed, -weary.*

1935 *Mind* XLIV. 434 Blind-born persons, such as Helen Keller, talk *intelligently* about colours. *c* **1775** *Sandman's Wedding: A Cantata*, She being blind drunk Sir, Joe drove her away in his Cart. **1827** W. CLARKE *Every Night Book* 191 Whenever I've been tipsy in your company, you have always been blind drunk. **1845** DISRAELI *Sybil* III. iv. x. 132 Hang me if I wasn't blind drunk at the end of it. **1953** E. O'NEILL *Moon for Misbegotten* IV. 140 Sure, I got so blind drunk at the Inn I forgot all about our scheme. **1887** MORRIS *Odyssey* x. 493 Tiresias..The blind-eyed, the foreseer. **1902** *Daily Chron.* 18 Feb. 3/2 Are we to suppose, then, that Goethe..was a blind-eyed fool? **1923** D. H. LAWRENCE *Birds, Beasts & Flowers* 107 [The bat] flying slower, Seeming to stumble, to fall in air. Blind-weary.

16. blind advertisement, also **blind ad.** *U.S.* (see quot. 1948); **blind approach** *Aeronaut.*, an approach made without direct observation (see *1 j); applied *attrib.* to a radio navigation system controlling such an approach; **blind back,** applied *attrib.* to a type of house that has no back door; **blind booking,** the booking of films by cinema proprietors without previous selection on their merits; **blind creek** (see quot.); **blind date** [*DATE *sb.*[2] 2 c] orig. and chiefly *U.S. colloq.* (see quot. 1929); also, the person with whom such a 'date' is arranged; **blind hazard, hole** *Golf* (see quots.); **blind ink** (see quot.); **blind-pig** *U.S. colloq.*, a place where liquor is illicitly sold; hence *blind-pigger, -pigging*; **blind poker** *U.S.* (see *1 h); **blind printing** (see quots.); **blind roller** [ROLLER *sb.*[1] 15] (see quot. 1948); **blind-seed disease,** a fungal disease of rye grass in which the seed fails to germinate; so *blind-seed fungus*; **blind spot,** (*b*) *Cricket*, that spot of ground in front of a batsman where a ball pitched by the bowler leaves the batsman in doubt whether to play forward or back; (*c*) *Radio* (see quot. 1923); (*d*) *transf.* and *fig.*; **blind staggers** (see STAGGER *sb.*[1] 2); **blind-stamping** = *blind-blocking* (see Dict.); hence *blind-stamped* ppl. a.; **blind stitch,** a stitch taken on one side of the material so as to be invisible on the other; hence as *v. trans.*, to sew or fasten with blind stitch; **blind tiger** *U.S.* = *blind-pig*; **blind tooling** (earlier example).

1948 MENCKEN *Amer. Lang.* Suppl. II. xi. 731 *Blind ad.*, an unsigned newspaper or magazine advertisement. **1962** 'A. A. FAIR' *Stop at Red Light* ii. 25 The insurance company have been running a blind ad..offering one hundred dollars for any witness..to an accident..at Seventh and Main Streets. **1842** *Ainsworth's Mag.* II. 43 The Puff Preliminary is known to..'printer's devils', by the less euphonious title of 'a blind advertisement'. **1936** C. B. ALLEN *Wonder Book of Air* 300 To aid him in this, the present set-up in the radio 'blind approach' system..includes the use of a string of powerful and distinctive lights. **1947** R. A. SMITH *Radio Aids to Navig.* xi. 88 It is generally felt now that some form of glide path is preferable to an altimeter for blind approach. **1937** 'G. ORWELL' *Road to Wigan Pier* iv. 53 House in Wallgate quarter. Blind back type. One up, one down. *Ibid.* 56 Houses of what is called the 'blind back' type..in which the builder has omitted to put in a back door. **1927** *Glasgow Herald* 20 Jan. 7 'Blind-booking' was responsible for a very large percentage of the machine-made pictures which came from America to this country. **1927** *Daily Tel.* 15 Mar. 9/2 The Labour-Socialists..are in favour of the proposals for the abolition of blind booking and restrictions on advance booking. **1886** J. W. ANDERSON *Prospector's Handbk.* 115 Blind Creek, a creek, dry, except during wet weather. **1925** *Lit. Digest* 14 Mar. 65/1 No, got a blind date on to-night. **1929** *Amer. Speech* IV. 420 *A blind date*, a date with someone whom the datee does not know but which is arranged by a third person. **1947** *Chicago Tribune* 14 June 18/8 In describing your blind date, I would say she has a wonderful personality. **1900** A. E. T. WATSON *Yng. Sportsman* 315 A Blind hazard is also a hazard which is hidden from his view. *Ibid.*, A Blind hole is one of which the putting-green is not visible to the player as he plays his shot. *a* **1884** KNIGHT *Dict. Mech.* Suppl., *Blind Ink.* Invented by Edison. An ink which..swells up into relief on the paper. **1887** *Minnesota Gen. Statutes Suppl.* (1888) 248 Whoever shall attempt to evade or violate any of the laws of this state ..by means of the artifice or contrivance known as the 'Blind Pig' or 'Hole in the Wall'..shall..be punished. **1903** *N.Y. Even. Post* 23 Sept., But a 'blind pig' is at best but a sordid institution. **1961** *Spectator* 28 July 135 Blind pigs—establishments with anonymous blank façades entered by a basement front door with a peep-hole. **1894** *Voice* (N.Y.) 6 Dec. 1/5 Headed by one of the blind-piggers who was under arrest. **1918** WEBSTER Add., *Blind-pigging*, n. **1927** *Blackw. Mag.* June 833/1 Amongst the common herd two crimes ranked as serious—'blind-pigging' and 'high-grading'. **1904** E. F. STRANGE *Japanese Colour Prints* xi. 110 In addition to the blocks for various colours, an effect of blind printing (*gauffrage*) was often secured by the use of an additional printing from a clean block. **1926** H. HUBBARD *How to distinguish Prints* 21 Charpentier and others..experimented with 'blind printing', that is, the use of an uninked relief block that merely embossed the paper with its engraved design. **1888** G. O. PRESHAW *Banking under Difficulties* xxv. 155 'Blind rollers' often rising and swamping a boat. **1948** R. DE KERCHOVE *Internat. Maritime Dict.*, *Blind rollers*, relatively heavy and often dangerous ocean swell caused by water in motion meeting lesser depth as it passes over shoals or approaches land. **1939** J. C. NEILL & E. O. C. HYDE in *N.Z. Jrnl. Sci. & Technol.* Feb. 283A, Low-germination trouble of otherwise apparently sound, well-harvested crops [of rye-grass] is caused by a pathogenic fungus... So unnoticeable are the symptoms that it has not yet even received a common name... It is proposed that it be called *Blind-seed Disease*. *Ibid.* 288A, The blind-seed fungus appears to be allied to *Helotium herbarum* Fries. **1956** *Nature* 10 Mar. 466/1 Blind seed disease of ryegrass caused by *Phialea temulenta*. **1864** *Baily's Monthly Mag.* Sept. 301 Now the great difficulty of slows, besides being (as they ought all to be) 'in the blind spot', consists in the elevation..of a dropping ball. **1891** GRACE *Cricket* iii. 73 Too often would come a ball on the blind spot. **1907** G. B. SHAW *John Bull's Other Is.* Pref. p. xli, You find that there is a blind spot on their moral retina, and that this blind spot is the military spot. **1910** GALSWORTHY *Justice* I. 19 No doing anything with them... They've got a blind spot. **1923** *Daily Mail* 13 Feb. 7 Wireless blind spots, where distant broadcasting is heard more clearly than that nearer at hand. **1932** E. V. LUCAS *Reading, Writing & Remembering* viii. 150 He was too full of prejudices and had too many blind spots, to be the perfect critic. **1784** J. LEWIS *Diary* 4 Mar. in *New Jersey Hist. Soc. Proc.* (1941) LIX. 169 We discovered that my horse had a distemper called the blind staggers. **1874** *Rep. Vermont Board Agric.* II. 341 The disease is frequently called 'blind staggers'. **1931** *Library* XI. 395 Long after blind-stamped pictorial panels had gone out of use in Paris. *Ibid.* 425 Business in the Gothic blind-stamped bindings. **1910** *Encycl. Brit.* IV. 217/1 English binders excelled in this art of 'blind' stamping, that is, without the use of gold leaf. *a* **1884** KNIGHT *Dict. Mech.* Suppl., *Blind Stitch.* (Harness)... A stitch that is shown on one side only of the leather. **1909** *Daily Chron.* 14 Jan. 7/5 Facings should be blind-stitched into place. **1857** *Spirit of Times* 23 May 182/1, I sees a kinder pigeon-hole cut in the side of a house, and over the hole, in big writin', 'Blind Tiger, ten cents a sight.'.. That 'blind tiger' was an arrangement to evade the law, which won't let 'em sell licker there, except by the gallon. **1884** *Arkansas Digest Laws 1883* 493 Any person..who shall sell..any alcohol..by such device as is known as 'the blind tiger',..shall be deemed guilty of a misdemeanor. **1818** *Art Bk.-binding* 31 In addition to the gilt back, rolled plain, that is, blind-tooling, on the inside of the board.

blind, *v.* Add: **1. c.** Used in vulgar imprecations, as *blind me!* Cf. *BLIMEY *int.*

1890 FARMER *Slang* I. 230/1 *Blimey*, a corruption of 'blind me!'; an expression little enough understood by those who constantly have it in their mouths. **1923** E. O'NEILL *Hairy Ape* v. 47 There's a 'ole mob of 'em like 'er, Gawd blind 'em! *a* **1953** —— *In Zone* (1955) 523 'E ain't arf a sly one wiv 'is talk of submarines, gawd blind 'im.

2. b. *intr.* To go blindly or heedlessly; to drive very fast. *slang.* Also *fig.*

1923 *Daily Mail* 21 June 12 Motor-cyclists who blind along the road. **1928** *Daily Express* 19 May 10/6 By recreation I do not mean blinding along the Brighton road at fifty miles an hour. **1935** *Punch* 21 Aug. 198/1 It is far better to get a little work done which is perfectly planned and organised than to let people go blinding on without anyone in authority knowing what they are doing. **1937** M. ALLINGHAM *Dancers in Mourning* iv. 58, I was blinding... Didn't see her until I was over her. **1954** C. FRY *Dark is Light Enough* II. 71 A trap, they've set for us. Who's got our pistols? Gone blinding into it.

8. *trans.* In *Bookbinding*, to stamp *in* (a pattern) without gilding.

1901 COCKERELL *Bookbinding* 212 The pattern is blinded in through the leather.

9. To cover the surface of (a newly made road) with fine material. cf. *BLINDING *vbl. sb.* 4.

1812 J. SINCLAIR *Syst. Husb. Scot.* I. 66 No large stones to be employed..nor sand, earth or other matter, on pretence of *blinding* [the road]. **1880** W. H. PATTERSON *Gloss. Antrim & Down* s.v., To 'blind a road' = to spread small stones or cinders so as to cover up the large stones, with which a new road has been 'pitched', and to fill the interstices.

blind, *sb.* Add: **2.** (Earlier examples.)

1730 Window-Blind [see WINDOW *sb.* 5 a]. **1771** J. S. COPLEY in *Copley–Pelham Lett.* (1914) 142 Those Windows having new fassioned Blinds such as you see in Mr Clarke's Keeping room.

3. (Examples.)

1828 WEBSTER s.v., A blind..for a horse. **1848** *Congress. Globe* 30 June, App. 820/1 [Mr. Polk] was worked into the Presidency with Oregon and Texas on either side, as a horse is worked with blinds. **1901** *Munsey's Mag.* XXV. 739/1 A halter has a soft leather covered bit, and is without blinds.

5. Delete † *Obs.* and add: *spec.* a hiding-place in which a hunter conceals himself from the game. *U.S.*

1818 *Niles' Reg.* XV. 64/2 Col. Boon rode to a deer lick, seated himself within a blind raised to conceal him from the game. **1869** *Game Laws (Penn.)* in *Fur, Fin & Feather* (1872) 94 No person shall..build blinds for the purpose of killing..any wild turkey. **1874** J. W. LONG *Wild-Fowl Shooting* 45 Ingenuity in the providing of proper ambush, or blind, as all such hiding places are generally termed by wild-fowlers. **1894** *Outing* (U.S.) XXIV. 73/2 A glance..discloses the fact that no time should be wasted in getting started for the blinds.

6. *spec.* (See quot. 1929.)

1929 *Amer. Speech* IV. 338 *Blind*, a legitimate business used to conceal an illegitimate one. **1938** F. D. SHARPE *Sharpe of Flying Squad* viii. 107 Another man..used to trade dogs as a blind.

8. *blind-maker*; **blind-bridle** (earlier U.S. example).

1833 J. HALL *Harpe's Head* 30 Some rode with blind-bridles. **1837** N. WHITTOCK *Bk. Trades* 488 Blind (Venetian) ..maker. **1894** *Daily News* 12 Nov. 7/2 William Wilshaw.., a blindmaker, was sentenced to death.

9. In Poker, a stake put up by a player before seeing his cards (see quots.); cf. *BLIND *a.* 1 h. Also *attrib.* in *blind-money*. Also *fig.*

1857 *Hoyle's Games* (Amer. ed.) 289 Should a party see fit to call the blind, [he] must put twice the number in the pool. **1872** 'MARK TWAIN' *Innoc. at Home* (1882) ii. 268 Now you talk! You see my blind and straddle it like a man. **1882** *Poker* 49 The straddle is nothing more than a double blind. *Ibid.* 91 It is an error on the part of the Age to fill the Blind simply because he has already invested the Blind-money. **1885** *Encycl. Brit.* XIX. 283/2 The next player [to the age] may *double the blind*, i.e., raise to double what the age staked; the next may *straddle the blind*, i.e. double again,.. and so on. Only the age can start a blind. **1888** FARMER *Americanisms*, *Blind* (in poker), the ante deposited by the age previous to the deal... To make a blind good costs double the amount of the ante, and to make a straddle good costs four times the amount of the blind. **1894** *Congress. Rec.* May 4408/2 Put up your blind. It's my deal.

10. A blind baggage car on a train (see *BLIND *a.* 11 d). *U.S.*

1893 *Chicago Record* 14 July 11/3 In hobo language 'beating the blinds' means to steal a ride on the mail car next to the engine. **1895** *Dialect Notes* I. 390 *Jump the blind*, to steal a ride on platform of baggage-car. **1948** *Sat. Even. Post* 31 July 89/1 If there were any hobos on the blind, they would step off into my arms.

11. [f. *BLIND *a.* 1 g.] A drunken bout or orgy; a binge.

1917 A. WAUGH *Loom of Youth* I. iii. 37 For six weeks we'll train like Hades, and then, when we've got the cup, we'll have a blind. **1936** G. GREENE *Journey Without Maps* II. iv. 189 It became more and more like a blind in Paris..in the Montparnasse bar. **1943** J. B. PRIESTLEY *Daylight on Sat.* x. 67 I'm not off on a blind, if that's what you're worrying about. **1943** *Mind* LII. 280 How much of my reaction against a 'blind' on a Saturday night is due to my puritanical upbringing?

blind alley. † **a.** An out-of-the-way or secret alley. *Obs.* **b.** An alley closed at one end (see BLIND *a.* 11); a cul-de-sac; also *fig.*, a course of action that fails to effect its purpose or from which there is no resultant benefit.

1583 [see BLIND *a.* 8]. **1662** [see BLIND *a.* 11]. **1681** DRYDEN *Spanish Fryar* I. 7 I'll e'en go lose my self in some blind Alley; and try if any courteous Damsel will think me worth the finding. **1854** BAYLE ST. JOHN *Purple Tints Paris* II. 2 When..he..is compelled to become one of the blind-alleys of the species. **1882** P. H. FITZGERALD *Recreat. Literary Man* I. ii. 12 A familiarity with all the blind alleys.. and passages of letters. **1898** E. GREY in *Westm. Gaz.* 6 June 2/3 Many of Lord Salisbury's concessions were blind alleys which led to nowhere. **1925** W. DEEPING *Sorrell & Son* iii. § 2 No blind alleys, or office stools.

b. *attrib.* and *fig.*, applied to something that 'leads nowhere', esp. employment that offers no opportunities for promotion or advancement.

1902 *Westm. Gaz.* 10 June 2/1 The second act, with its long 'blind alley' episode of Paolo's wife, lets down the piece severely. **1909** *Ibid.* 27 Oct. 2/1 'Blind alley' employments. **1910** *Ibid.* 6 Jan. 3/3 The number of lads between the age of fifteen and twenty who are engaged in what have come to be known as 'blind-alley occupations'. **1937** 'G. ORWELL' *Road to Wigan Pier* v. 88 The youth who leaves school at fourteen and gets a blind-alley job is out of work at twenty. **1957** J. S. HUXLEY *Relig. without Revel.* (new ed.) ix. 215 Through this radiating fan of restricted improvements and blind-alley specialisations there runs a trend towards major advance.

blinder. Add: **1. b.** Something 'dazzlingly' good or difficult, esp. an excellent piece of play in Rugby Football or Cricket. *colloq.*

1950 W. HAMMOND *Cricketers' School* iii. 35 Striking out at an innocent-looking ball, I've sent a blinder—dead into the fieldsman's hands. **1960** D. STOREY *This Sporting Life* I. ii. 17 You played a blinder... It was the best game I ever saw. **1963** *Times* 16 Feb. 3/3 They dropped one easy catch and caught three blinders.

blindfold, *a.* Add: **1. c.** Of a match at chess: conducted by a player without seeing the board but not necessarily blindfolded; hence *blindfold player*.

1790 'MR. PHILIDOR' (*title*) Analysis of the Game of Chess..to which is added several Parties, Played Blindfold, against three adversaries. **1850** BROWNING *Christm.-Eve* 90 Playing pawns at blindfold chess. **1900** *Westm. Gaz.* 31 May 2/1 In almost all these blindfold games I have the move...Occasionally I overlook something..in a blindfold match... The blindfold player..cannot see so far ahead as if he were looking at the board.

blinding, *vbl. sb.* Add: **4.** The process of covering the surface of a newly made road with fine material to fill up the spaces between the stones; also, the material used for this purpose. Cf. *BLIND *v.* 9.

1843 *Civil Engin. & Arch. Jrnl.* VI. 274/1 The stoning and blinding is conducted in the same manner as in ordinary roads. *a* **1877** KNIGHT *Dict. Mech.* **1926** *Missionary Rec.*

United Free Ch. Scotl. May 224/1 The fierce wind tore the blinding from the hard clay roads.

blindness. Add: **4.** Of a plant: abortiveness.

1921 *Times Lit. Suppl.* 24 Feb. 130/3 The Frit Fly, which produces blindness in oats.

blini: see *BLIN.

blink, *v.* Add: **2. b.** *trans.* (*Coursing.*) To elude (the dogs) temporarily.

1876 *Coursing Calendar* 197 The hare blinked Grace at the fence. *Ibid.* 252 Hylactor and Blue Sea ran very evenly for some distance, but, as puss blinked them in a hollow, Hylactor was so well placed that he made a few weak points before effecting the death.

4. d. *trans.* *to blink* (tears) *away* or *back*: to send (tears) away, to avoid shedding (tears), by blinking.

1905 E. GLYN *Viciss. Evang.* 215 Tears kept rising in my eyes, and I did not even worry to blink them away. **1919** F. HURST *Humoresque* 146 She blinked back the ever-recurring tears. **1924** ROSE MACAULAY *Orphan Island* xxi. § 1 Rosamond blinked away tears, with the salt Pacific, from her eyes. **1945** 'BRAHMS' & SIMON *Six Curtains for Stroganova* xx. 162 Above her flowers Dourakova bowed, smiled, and blinked back her tears.

10. To look upon with the evil eye, to bewitch. *Sc.* and *Irish.* Cf. 7 a in Dict.

1880 W. H. PATTERSON *Gloss. Antrim & Down* 9 Cow's milk is said to be *blinked* when it does not produce butter, in consequence of some supposed charm having been worked. **1886** *Folk-Lore Jrnl.* IV. 255 Cattle can be fairy-struck or bewitched..the first is called 'sheetin' and the second 'blinked'. **1892** *Ballymena Observer* (E.D.D.), *Blink*, to bewitch cattle and cause them to have little or no milk and butter. **1926** *Blackw. Mag.* Apr. 479/1 Perhaps we are bewitched or blinked, as Shamus Byrne would say. **1927** *Scots Observer* 15 Oct. 2/5 Mrs. Hazelton..had indeed blinked William Blair's cows. *Ibid.*, He had set fire to the wisp of straw and had put it under a blinked cow's nose.

blink, *sb.*[2] Add: **1. d.** *on the blink*: on the point of becoming extinguished; in a bad state, out of order. *slang* (orig. *U.S.*).

1901 'H. MCHUGH' *John Henry* 83 A stranglehold line of business that will put Looey Harrison on the blink. **1904** 'O. HENRY' *Cabbages & Kings* iii. 51 This café looks on the blink, but I guess it can set out something wet. **1912** WODEHOUSE *Prince & Betty* xiii. 176 That punt-pole's on the blink. I tried it yesterday, and it creaked. **1934** —— *Right Ho, Jeeves* xi. 136 All those years he spent in making millions in the Far East put his digestion on the blink. **1960** J. ASHFORD *Counsel for Defence* vi. 68 No good, David. The 'frig. is on the blink again.

2. b. The action or an act of blinking.

1924 GALSWORTHY *White Monkey* I. xii, He did not miss the shift and blink in the manager's eyes.

5. blink microscope [G. *blinkmikroskop*], an instrument for viewing two photographs of the same section of the sky alternately in rapid succession. Also called **blink comparator.**

[**1910** C. PULFRICH in *Zeitschr. f. Instrumentenkunde* XXX. 1 Die Anwendung des Blinkmikroskops.] **1911** C. PULFRICH in *Encycl. Brit.* XXV. 900/2 Since 1904 binocular observation of stellar plates..has been gradually discarded for the method devised by Pulfrich, which consists in the monocular observation of the two plates.. with the assistance of the so-called 'blink' microscope. **1930** *Discovery* Aug. 252/2 It [*sc.* the planet Pluto] was identified from its motion past the numerous fixed stars as revealed on plates of the same star field while being compared under the blink comparator.

blink, *sb.*[3] *U.S.* A fisherman's name for the mackerel when about a year old.

1856 [see TINKER *sb.* 3]. **1888** ATWOOD in Goode *Amer. Fishes* 174 Fish of this size are sometimes called 'Spikes'... The next year I think they are the 'Blinks', being one year old.

blink, *a.* Add: **1. blink-eyed** *a.* (Earlier and later examples.) Also *fig.*

1575 GASCOIGNE *Hearbes* 152 Remembre Batte the foolish blink eyed boye. **1846** DICKENS *Pictures from Italy* 9 Blink-eyed little casements. **1929** J. B. PRIESTLEY *Good Compan.* III. i, A piebald blink-eyed,..little pierrot show.

blinker. Add: **4.** A sporting dog that refuses to see and mark the position of game. Cf. *BLINKING *vbl. sb.* 3.

1814 W. DOBSON *Kunopædia* 98 We shall, I fear, be compelled to class him along with the blinker of a very different nature, the brute of perverse..yet snivelling disposition. **1845** YOUATT *Dog* iii. 91 The chastisement..would make the setter disgusted..and leave him a mere *blinker*. **1848** W. N. HUTCHINSON *Dog Breaking* 94 Excess of punishment has made many a dog of good promise a confirmed blinker.

5. An intermittent flash-light. In full, *blinker light*. *U.S.*

1923 R. D. PAINE *Comr. Rolling Ocean* xvi. 285 Take this bug-light [*sc.* an electric torch] and use it as a blinker. You learned the Morse code at Camp Stuart. **1943** J. STEINBECK *Once there was a War* (1959) I. 52 The second in command takes up the blinker and signals. **1959** N. MAILER *Advts. for Myself* (1961) 432 Motels, blinker lights, salt-eroded billboards. **1964** 'M. E. CHABER' *Six who Ran* (1965) i. 17 Three passenger cars were stolen... One of them was converted into a State Police car with the aid of a blinker light.

blinkered (bli·ŋkǝɹd), *ppl. a.* [f. BLINKER *sb.* or *v.* + -ED.] Of a horse: provided with blinkers. Also *fig.*, having a limited range of outlook.

1867 HARDY *Time's Laughingstocks* (1909) 53 A century which..Will show..A scope above this blinkered time. **1897** *Daily News* 19 June 9/1 The colt..could not quite withstand the rush of the blinkered El Diablo. **1898** *Westm. Gaz.* 27 Aug. 1/3 The padded, blinkered life of her spinster sister. **1962** *Daily Tel.* 8 May 14/2 This is merely a case of legal procedure grinding along on its blinkered way.

blinking, *vbl. sb.* Add: **3.** The faulty action, in a sporting dog, of refusing to see and mark the position of game. Cf. *BLINKER 4.

1814 W. DOBSON *Kunopædia* 89 (*heading*) On Blinking. Defect in blood, a suspected cause. **1848** W. N. HUTCHINSON *Dog Breaking* 10 The unreasonableness of not always giving initiatory Lessons. Causes Blinking. **1865** *Ibid.* (ed. 4) 202 Some argue that blinking arises from a defective nose, not from punishment. **1897** H. DALZIEL *Brit. Dogs* (ed. 2) III. 336 Blinking..is caused..by undue severity or punishment administered for chasing game or poultry.

blinking, *ppl. a.* Add: **4.** Used as a substitute for a strong expletive. *slang.*

1914 *Scotsman* 12 Oct. 7/5 One..Guardsman..declared.. that His Majesty seemed to carry the 'blinking Army List in his 'ead'. **1927** *Observer* 21 Aug. 17/5 The type of golfer who..hurls the bag of clubs after it, accompanied by the remark, 'Go on, have the blinking lot'.

blintze (blints), and variants. [Yiddish *blintse*, f. Russ. *blinets*, dim. of *BLIN.] = *BLIN.

1903 *Jewish Encycl.* IV. 256/1 The *kasha* and *blintzes* of the Russian Jews..are dishes adopted by the Jews from their Gentile neighbors. **1932** L. GOLDING *Magnolia St.* I. ii. 32 Wouldn't it be nice to make a few *blintsies* for Mr. Emmanuel. **1958** W. BICKEL tr. *Hering's Dict. Class. & Mod. Cookery* 10 *Cheese blinzes*.., crêpes au fromage à la juive. **1961** *Woman* 21 Jan. 16/3 Blintzes are cheese-filled pancakes served with jam.

bliny: see *BLIN.

blip (blip), *v.* [Echoic.] **1. a.** *trans.* To strike with a brisk rap or tap. **b.** *intr.* To make a quick popping sound. Hence **bli·pping** *ppl. a.*

1924 A. A. MILNE *When we were Very Young* 93 They pulled him out and dried him, and they blipped him on the head. **1946** K. TENNANT *Lost Haven* (1968) xviii. 311 A big moth flopped into the room and blipped about. **1952** 'C. BRAND' *London Particular* vi. 64 Some horrible burglar.. blipped him on the head and killed him. **1955** W. GOLDING *Inheritors* iii. 60 Little bubbles bulged out of the scum, wandered and blipped out. **1957** 'C. E. MAINE' *High Vacuum* xiii. 108 The signal blipped hollowly from the speaker, fading and sporadic. **1963** P. MCCUTCHAN *Man from Moscow* xx. 205 The only noise was that of the blipping tyres.

2. *slang.* To switch an aeroplane engine on and off. Hence as *sb.* Also **bli·pping** *vbl. sb.*

1925 FRASER & GIBBONS *Soldier & Sailor Words* 26 *To blip*, to switch an aeroplane engine on and off. **1931** *Vanity Fair* Nov. 78/2 *Blipping* is the flippant term for nonchalantly and rapidly switching the ignition off and on while in flight. **1958** 'N. SHUTE' *Rainbow & Rose* iii. 79 They brought back memories of slow-revving engines blipping on the switch. *Ibid.* 84, I..flew her over to the hangar in little blips of engine on the switch.

blip (blip), *sb.* [Echoic.] **1.** Any sudden brisk blow or twitch; a quick popping sound.

1894 'MARK TWAIN' in *St. Nicholas* Apr. 540/1 We took him a blip in the back and knocked him off. **1927** A. A. MILNE *Now we are Six* 23 It wasn't that he did not care For blips and buffetings and such. **1932** AUDEN *Orators* II. 56 Three warnings of enemy attack—depression in the mornings—rheumatic twinges—blips on the face. **1947** CROWTHER & WHIDDINGTON *Science at War* 16 A 'blip' or 'break' which marks the moment of emission of the pulse. **1958** *Times* 29 July 10/6 A series of preliminary blips [of a motor-horn]. **1962** A. NISBETT *Technique Sound Studio* vii. 130 A burst of applause lasting, say, ten seconds is very difficult to cut down to five without a slight 'blip' at the join.

2. A small elongated mark projected on a radar screen.

1945 *Electronic Engin.* XVII. 716 Note the calibration scales, in this case formed of small and larger 'blips', not bright dots. **1957** *Times* 11 Oct. 10/2 The first 'blip' appeared on a blue trace which was crossing the screen of an ex-military radar set.

blissless, *a.* (Modern example.)

1952 *Antiquity* XXVI. 95 In our present blissless ignorance.

blister, *sb.* Add: **2. b.** A disease incident to peach-trees, caused by the fungus *Exoascus deformans*, which produces a distortion of the leaves.

1864 *Ohio Agric. Rep.* XVIII. 460 For some years, in this country, the disease which produces the 'Blister and Curl' in the peach leaf, and decay in the peach fruit, has.. produced extensive ravages. **1919** *Board Agric. & Fisheries Leaflet* No. 120 'Curl' or 'leaf blister' proves very injurious to peaches and nectarines during certain seasons.

c. *Naut. colloq.* An outer covering fitted to a vessel to provide protection against torpedoes and mines or to improve stability. Hence, a ship so protected. Also *attrib.*

1919 *Chambers's Jrnl.* 26 July 543/1 Immunity from the evil effects of torpedoes and mines is sought by the provision of a swelling, commonly called a 'bulge' or a 'blister', below the water-line on each side. **1921** *Flight* XIII. 584/1 By suitable methods of 'blister' construction of ships, however, this mining effect could be reduced to something as negligible as the direct hit. **1923** W. S. CHURCHILL *World Crisis* II. i. 23 When at last Monitors, 'Blisters' and Tanks had been devised and built. **1948** R. DE KERCHOVE *Internat. Maritime Dict.* 64/1 Blisters which communicate with the open sea, are fitted on some ships as an anti-rolling device.

d. A rounded compartment protruding from the body of an aeroplane. Also *attrib.*

1939 *Meccano Mag.* Sept. 517/2 There are five machine gun positions, one in the fuselage nose, and four others in the form of streamlined 'blisters' on the top, bottom, and sides of the fuselage. **1941** *Illustr. London News* CXCVIII. 516 An air-gunner in one of the carefully stream-lined blister gun turrets of a 'Catalina'. **1943** 'T. DUDLEY-GORDON' *Coastal Command at War* xv. 140 Two big gun blisters, with sliding, rounded panels of perspex, in which two gunners are always on watch.

4. blister blight, a disease of the tea-plant caused by the fungus *Exobasidium vexans*, which produces blisters on the leaves; **blister-fly** (earlier examples); **blister furnace,** a furnace for the conversion of copper regulus or matte into blister-copper; **blister gas,** a poison gas which causes blisters on or intense irritation of the skin; **blister pearl,** a pearly excrescence of irregular shape found on the shell of a pearl oyster.

1877 S. BAILDON *Tea in Assam* 45, I do not know whether it has been really ascertained what causes 'Blister blight'. A leaf gets a small speck upon it at first, which, as it enlarges, assumes the appearance of a blister. **1949** *Ann. Reg. 1948* 144 Tea bushes were attacked by blister blight. **1585** Blister fly [see WHELK[1] 1 (*Comb.*)]. **1842** Blister-fly [see SPANISH *a.* 8 b]. **1902** *Encycl. Brit.* XXVII. 237/1 The multiple system anodes are sometimes cast directly from the blister furnace or the converter. **1936** *Current Hist.* July 61/2 In Class D come the *vesicants* or blister gases. **1938** *Encycl. Brit. Bk. of Year 1938* 144/1 The blister gases..penetrate nearly every material except glass, porcelain, and unglazed metals. **1885** *Encycl. Brit.* XVIII. 446/2 The mollusc..depositing nacreous matter.., thus forming a hollow body of irregular shape known as a 'blister pearl'. **1910** *Daily Chron.* 4 Apr. 4/5 Overdress of gauze encrusted with blister pearls.

5. A derogatory term for a person, esp. an annoying one. *slang.*

1806 J. BERESFORD *Miseries* I. vii. 145 A perpetual blister; — alias, a sociable next-door-neighbour, who has taken a violent affection for *you*. **1880** W. H. PATTERSON *Gloss. Antrim & Down* 9 *Blister*, an annoying person. **1914** 'I HAY' *Lighter Side School Life* iii. 81 Mr. Wellings' reputation throughout the school..was that of a 'chronic blister'. **1930** WODEHOUSE *Very Good, Jeeves!* xi. 308 Women are a wash-out. I see no future for the sex, Bertie. Blisters, all of them.

6. A summons. *slang.*

1903 *Sessions Paper* 17 Nov. 33, I was served with four *blisters* yesterday. **1906** *Daily Chron.* 20 Mar. 5/6 'Have you never had a "blister"?'.. The solicitor explained to the Court that a 'blister' was a summons for 'scorching'. **1947** F. SARGESON in *Penguin New Writing* XXIX. 62 He'd been paying off a few bob every time he had a few to spare... And then he gets a blister!

blister, *v.* Add: **5.** *trans.* Used as an imprecation. *slang.*

1840 COCKTON *Val. Vox* xxvi, Blister 'em! Where can the scoundrels be got to? **1964** WODEHOUSE *Frozen Assets* iii. 46 Why didn't they send it up before, blister their insides? I've been in agonies of suspense.

6. Of a policeman: to record a person's name for an alleged offence; esp. in *pass.*, to have one's name recorded in this way; to be summoned or punished for an offence (cf. *BLISTER *sb.* 6). *slang.*

1909 WARE *Passing Eng.* 34/2 *To blister*..Used chiefly by cabmen in relation to magisterial fines, *e.g.*, 'I was blistered at Bow Street to-day for twenty hog.' **1938** F. D. SHARPE *Sharpe of Flying Squad* 329 *Blistered*, served with a summons. **1939** H. HODGE *Cab, Sir?* xvi. 225 When the policeman puts his notebook away again, we've usually been 'blistered'. During recent years, policemen have been blistering us over three thousand times in a twelvemonth.

blistered, *ppl. a.* Add: **1. b.** (Earlier U.S. examples of *blistered steel*.)

1744 in *Maryland Hist. Mag.* (1926) XXI. 243, 3 flaggotts Blistered Steel. **1770** *Carroll Papers* in *Maryland Hist. Mag.* (1918) XIII. 65 My Smiths say the Bristol or Blister'd steel sent to us is very bad.

c. Provided with a 'blister' (see *BLISTER *sb.* 2 c).

1923 W. S. CHURCHILL *World Crisis* II. i. 23 The Monitor and the 'bulged' or 'blistered ship' were the beginning of the torpedo-proof fleet. **1928** *Daily Express* 21 Nov. 1/1 The methods of defence employed by surface craft consist of anti-aircraft guns..; of reinforced or 'blistered' bottoms..; and of manœuvre.

blistering, *ppl. a.* Add: **2.** Used as a substitute for a strong expletive. *slang.*

1900 *Daily News* 30 July 6/4 One blistering young woman actually unstraps her kodak and begins operations upon the great white mountain.

blither (bli·ðǝɹ), *v.* (*dial.* or) *colloq.* [var. BLETHER *v.*] *intr.* To talk nonsense; = BLETHER *v.* 1. Hence **bli·therer,** one who blithers; **bli·thering** *vbl. sb.*

1868 VERNEY *Stone Edge* i, What did the imp come blitherin' and botherin' there for? **1902** W. RALEIGH *Let.* 6 Jan. (1926) I. 237 A 'Civic Society' of earnest burgesses and blitherers. **1903** J. K. JEROME *Tea-Table Talk* i. 20 If he was to blither, it was only fair that she should bleat back. **1916** W. OWEN *Let.* Aug. (1967) 402 One old blitherer let his bullet off by accident. **1921** *Blackw. Mag.* Oct. 455/1 The inevitable pasty-faced babu waddled up, blithering about the delay. **1925** *Public Opinion* 13 Mar. 258/2 He ignores all their blithering.

blither (bli·ðǝɹ), *sb.* (*dial.* or) *colloq.* [var. BLETHER *sb.*] Nonsense. Cf. BLETHER *sb.*

1866 *Banffshire Jrnl.* 27 Mar. 3 Some lightly [*i.e.* belittle] Scotland their auld mither An' ca' her tongue a vulgar blither. **1901** *Daily Chron.* 10 Aug. 4/7 We have heard a lot of blither (and, perhaps, a little sense). **1911** E. FERBER *Dawn O'Hara* v. 68 'What utter blither!' I scoffed.

blithering (bli·ðǝriŋ), *ppl. a. colloq.* [f. *BLITHER *v.* + -ING[2].] Senselessly discursive or talkative, babbling; esp. of a person, used chiefly as an intensive adjective, with the meaning 'consummate' (freq. in *blithering idiot*); also more widely = despicable, contemptible.

1889 *Punch* 9 Feb. 65 I'll state pretty clearly that his son is a blithering idiot. **1895** *Ibid.* 30 Mar. 153, I had thought that you..would have had a soul above blithering detail. **1903** 'A. MCNEILL' *Egreg. Eng.* 179 These songs..are of the most blithering and bathotic nature. **1923** *Blackw. Mag.* Jan. 70/2, I was cursing myself for the blitheringest ass that ever was born. **1926** G. FRANKAU *Masterson* xxix, I was a blithering idiot to get in—knowing you as well as I do. **1928** 'REBECCA WEST' *Strange Necessity* 310 The blithering incompetence of English statesmen during the War of Independence.

bli·thesomely, *adv.* [f. BLITHESOME *a.* + -LY[2].] In a blithesome manner; cheerily. So **bli·thesomeness.**

1858 WHITTIER *Pipes at Lucknow* in *Writings* (1888) I. 185 Full tenderly and blithesomely The pipes of rescue blew! **1886** *New Princeton Rev.* II. 78 A glad blithesomeness belonged to her, potent to conquer even ill health and suffering. **1888** A. S. SWAN *Doris Cheyne* xvii, 'I should scold you.., but I am so glad to see you that I have not the heart,' she said blithesomely.

blitz (blits), *sb.* [Short for *BLITZKRIEG.] An attack or offensive launched suddenly with great violence with the object of reducing the defences immediately; *spec.* an air-raid or a series of them conducted in this way, esp. the series of air-raids made on London in 1940. Also *attrib.*

1940 *Daily Express* 9 Sept. 1 Blitz bombing of London goes on all night. *Ibid.* 10 Sept. 1/1 In his three-day blitz on London Goering has now lost 140 planes. **1940** *Daily Sketch* 21 Sept. 8/3 Neighbourhood Theatre braved the blitz and yesterday presented a new play. **1941** *New Statesman* 15 Feb. 160/3 The Home Guard of young architects who spent the night of the City 'blitz' battering out Hitler's incendiaries. **1944** *Ourselves in Wartime* viii. 177 All under five, many born since the beginning of the war, 'blitz' babies knew instinctively that the ground floor was safest. **1958** *New Statesman* 22 Feb. 223/3 Depopulation..began even before the war and the blitz.

b. *transf.* and *fig.*

1940 *Topeka Jrnl.* 19 Apr. 4/4 Setting the stage for a 'blitz' comeback. **1941** WYNDHAM LEWIS *Let.* 16 Apr. (1963) 288 At the time I was going through a minor economic Blitz of my own. **1960** *Guardian* 30 Dec. 10/5 The women did only the bare essentials of housework during the week, with a 'blitz' at weekends.

blitz (blits), *v.* [f. prec.] To attack with a blitz; to hit, blast, destroy, etc., by an air-raid. Hence **blitzed** (blitst) *ppl. a.*, attacked or destroyed by a blitz; **blitzing** (bli·tsiŋ) *vbl. sb.*, attacking in this way.

1939 in *Amer. Speech* (1940) XV. 110/2 Formal committee chairmen must have known how the poor Poles felt when the German blitzkrieg suddenly started 'blitzing' around their ears yesterday noon. **1940** *Daily Sketch* 2 Sept. 7 We 'blitz' hun planes in week-end raids. **1940** *Daily Express* 6 Dec. 1/6 A south coast town felt the heaviest weight of last night's Nazi blitzing. **1941** *War Illustr.* 30 Dec. 377 For the past few months demolition squads have been working on this heavily blitzed site. **1942** *Ann. Reg. 1941* 100, 70,000 meals had to be provided by the Emergency kitchens for people 'blitzed' out of their homes. **1945** 'G. ORWELL' in *Contemp. Jewish Record* VIII. 165 The Jewish quarter of Whitechapel was one of the first areas to be heavily blitzed. **1955** *Times* 25 Aug. 5/4 Local authorities..decided to recondition 30,000 sub-standard houses in 'blitzed or blighted' areas.

‖ **Blitzkrieg, blitzkrieg** (bli·tskrig, -krīk). [G., f. *blitz* lightning + *krieg* war.] (See *BLITZ *sb.*)

1939 *War Illustr.* 7 Oct. 108/1 In the opening stage of the war all eyes were turned on Poland, where the German military machine was engaged in *Blitz-Krieg*—lightning war—with a view to ending as soon as possible. *Ibid.* 9 Dec. 386/3 Everything was ready for the opening of the 'Blitz-krieg' on the West. **1940** *Ann. Reg. 1939* 217 The complete failure of the Soviet *Blitzkrieg* [on Finland]. *Ibid.* 225 Germany's 'Blitzkrieg' methods had proved only too effective [in Poland]. **1940** *New Statesman* 19 Oct. 371/2 Since the Blitzkrieg started and people went underground. **1955** *Times* 25 June 6/2 It opened with the new type of blitzkrieg. Widespread 'atomic attacks' were made on major targets.

b. *transf.* and *fig.*

1939 *N.Y. Daily News* 19 Oct., Now that that blitzkrieg of puns is over. **1941** H. G. WELLS *You can't be too Careful* I. vi. 41 She had brought herself down to a vulnerability that gave any old germs a fair chance with her. Their blitzkrieg was swift and successful. **1963** *Times* 23 Feb. 9/6 Hours elapsed before the spate came jostling down the valley. Today it is a blitzkrieg.

blizzard. Transfer *U.S.* to sense 1.

1. Alter def. to: A sharp blow or knock; a shot. Also *fig.*

1829 *Virginia Lit. Museum* 16 Dec. 418 *Blizzard*, a violent blow. **1856** *Sacramento City* (Cal.) *Item* (Th.), When some true archer, from the upper tier, Gave him a 'blizzard' on the nearest ear. **1872** SCHELE DE VERE *Americanisms* 443 *Blizzard*..means in the West a stunning blow or an overwhelming argument.

2. (Earlier and later examples.) Also *attrib.* and *Comb.* orig. *U.S.*

1859 L. B. WOLF *Diary* 1 Dec. in *Kansas Hist. Q.* (1932) I. 205 A blizzard had come upon us about midnight... Shot 7 horses that were so chilled could not get up. **1876** *Monthly Weather Rev.* Dec. 424 The very severe storms known in local parlance as 'blizzards' were reported on the 8th as prevailing in Iowa and Wisconsin. **1888** T. WATTS in *Athenæum* 18 Aug. 224/2 By Ferrol Bay those galleys stoop To blasts more dire than breath of Orkney blizzard. **1902** R. F. SCOTT *Jrnl.* 12 Aug. in *Voyage of 'Discovery'* (1905) I. ix. 383 Another blizzard, so thick that one cannot see one's hand before one's face. **1903** *Ibid.* 12 Dec. II. xviii. 276 Our long stay in the blizzard camp. **1912** —— *Jrnl.* Mar. in *Scott's Last Expedition* (1913) I. xx. 592 It was blowing a blizzard. He [*sc.* Captain Oates] said, 'I am just going outside and may be some time.' He went out into the blizzard and we have not seen him since. **1963** D. W. & E. E. HUMPHRIES tr. *Termier's Erosion & Sedimentation* i. 8 The coldest, blizzard-swept regions of the world. **1969** *Times* 8 Feb. 1/2 Blizzards and icy winds swept across Britain yesterday.

Hence **bli·zzarded** *pa. pple.*, a more emphatic form of 'blowed' (BLOW *v.*[1] 29); **bli·zzardy** *a.*, characterized by, or resembling, a blizzard or blizzards.

1892 GUNTER *Miss Dividends* I. vi. 67 Then he suddenly ejaculates 'Well I'm blizzarded!' **1888** *San Francisco News Let.* (Farmer), I should like to have seen the Colonel's face when he got that very cold blizzardy letter. **1946** *Chicago Daily News* 5 Mar. 8/4 [It] would ruin the disposition of the throngs..especially on blizzardy nights.

bloat (blōut), *sb.* [f. BLOAT *a.*[2]]

1. Bloatedness.

1905 G. B. SHAW *Irrational Knot* xi. 226 He..had noted with aversion a certain unhealthy bloat in her face. **1911** F. H. BURNETT *Secret Garden* xxvii. 300 If he took his food natural sir, you'd think he was putting on flesh—but we're afraid it may be a sort of bloat.

b. *spec.* in *Veterinary Path.* A disease of livestock characterized by an accumulation of gases in the stomach; = HOOVE.

1878 G. H. DADD *Amer. Cattle Doctor* iii. 65 *Tympanites rumenites* signifies distension of the rumen in the bovine species—the ox and cow—and, in the phraseology of the grazier, is known as bloat or hoven. **1962** R. SEIDEN *Livestock Health Encycl.* (ed. 2) 65/2 *Bloat*, hoven, tympanites, or tympany in livestock (especially in cattle, sheep and goats)..is primarily caused by vigorous fermentation in the rumen.

2. A conceited or contemptible person (see also quot. 1888). *U.S. slang.*

1860 in *Amer. Speech* (1947) XXII. 299/1, I considered such an old bloat not worth minding. **1871** *Congress. Globe* Feb., App. 129/1 Wife whippers, penitentiary birds, street vagabonds, beastly bloats, and convicted felons. **1888** FARMER *Americanisms* 64/1 *Bloat* (cant), a drowned body; also a drunkard.

bloat, *v.*[2] Add: **1. b.** *fig.*

1896 J. A. HOBSON *Probl. Unempl.* 1 Well-meaning.. social reformers stretch the term ['unemployment'] and bloat it out to gigantic proportions.

blob, *sb.*[1] Add: **3.** Also *fig.*

1905 W. RALEIGH *Lett.* (1926) II. 286 Christianity caught on by its very rumness; all the loose floating blobs of superstition ran to it.

4. b. *Cricket colloq.* A batsman's score of no runs, so called from the zero placed against his name in the score-sheet; = DUCK'S EGG b. Hence *fig.* generally, 'nought'; also, a senseless error, a blunder.

1889 *Cricket* 18 Apr. 60/2 St. Leonard's were all out for 25... Seven 'blobs' was the total. **1903** *Punch* 27 May 366 To come home and be treated as if I'd made a brace of blobs. **1912** J. B. HOBBS *Recov. Ashes* 19 Mr. Foster.. taking the first wicket of the tour by bowling Mr. E. R. Mayne for a 'blob'. **1952** M. ALLINGHAM *Tiger in Smoke* i. 17 Crooks can be..wanting in the top storey, but I've never heard of one who'd make a blob like that. **1958** B. HAMILTON *Too Much of Water* vi. 134 A cricketer..may make a string of blobs, and then hit a couple of hundreds. **1960** L. COOPER *Accomplices* II. ii. 86 He'd been in trouble with us before and he knew that another blob would about finish him.

blob, *sb.*[2] *local.* A bait used in fishing for eels, consisting of a worm strung on a worsted thread. (Cf. BOB *sb.*[1] 7.)

1874 E. PEACOCK *J. Markenfield* vii, Along o' my runnin' away wi' her crewell fish, and makin' a blobb for eels wi' it. **1905** *Westm. Gaz.* 28 Apr. 3/1 Your plan is now, when the eel is thus grabbing the worm, to lift the 'blob' very gently.

blob, *v.* Add: **4.** *intr.* To make a 'blob' (see *BLOB *sb.*[1] 4 b); to score no runs. *Cricket colloq.*

1905 N. GALE *More Cricket Songs* 34 I've tasted Sweet and bitter supplied by Luck,.. Whether I blobbed or whether I stuck.

blobbing (blǫ·biŋ), *vbl. sb. local.* [f. *BLOB *sb.*[2]] The method of fishing with a bob for eels (cf. BOBBING *vbl. sb.* 3).

1877 E. PEACOCK *Gloss. Manley*, *Blobbing*, a method of catching eels by means of worms strung on a worsted thread. **1905** *Westm. Gaz.* 28 Apr. 3/1 That is 'blobbing'.

‖ **bloc** (blǫk). [Fr., = BLOCK *sb.*] In Continental politics, a combination of divergent political parties which supports the government in power. Also *transf.*, a combination of persons, groups, parties, or nations formed to foster a particular interest. Also *attrib.*

1903 *Ann. Reg. 1902* 272 The Government remained in the hands of the *bloc*; and the Radical-Socialist party was free to proceed with the enforcement of the law with regard to the Congregations. **1905** *Spectator* 14 Jan. 38/1 It is more probable..that when the votes are open the members of the 'Bloc' will shrink from turning out M. Combes. **1907** *Westm. Gaz.* 5 Dec. 5/1 Prince Bülow..summoned the leaders of the 'bloc' parties to a meeting. **1908** *Daily Chron.* 9 Oct. 4/6 Clémenceau, who was..the man of the Bloc, who had invented the name. **1921** *Nation* 21 Sept. 308/2 The formation of the agricultural bloc which came close to holding the balance of power in the Senate. **1923** *Daily Mail* 29 Mar. 6 France also can stand apart from Great Britain. She can join a Continental *bloc* with Italy and the Little Entente and Belgium. **1946** *Times* 22 May 5/5 The Soviet Union has been..forming a compact and well-organized political and economic *bloc*, stretching from the Baltic to the Black Sea. **1957** *Observer* 29 Dec. 1/4 Military circles in Jakarta have indicated that they believe in the superiority of American to Soviet bloc weapons.

block, *sb.* Add: **4. d.** Later examples (all *slang*), esp. in *to knock one's block off*. So *off one's block*: angry, insane. Also *to lose* or *do (in) one's block* (chiefly Austral. and N.Z.), to become angry, excited, or anxious.

1862 H. KINGSLEY *Ravenshoe* II. ix. 86, I cleaned a groom's boots on Toosday, and he punched my block because I blacked the tops. **1913** A. J. REES *Merry Marauders* ii. 23 When I seen that 'earse of yours comin' a-plungin' down straight towards me..I admit I fair lost me block. **1916** J. B. COOPER *Coo-oo-ee!* i. 19 Mrs. Muller went on no end! Did in her block, thinking of the night's ride Nipper had given her. *Ibid.* xiv. 211 Sandy Pilkins..always lost his block at the local football match. **1916** C. J. DENNIS *Songs Sentimental Bloke* 118 *To lose or do in the block*, to become flustered; excited; angry; to lose confidence. *To keep the block*, to remain calm, dispassionate. **1918** *N.Z.E.F. Chrons.* 21 June 22/1 If you can keep your block, while those about you are losing theirs. **1925** FRASER & GIBBONS *Soldier & Sailor Words* 26 *Block*, head, face. 'Off his block', angry, off his head. **1928** T. E. LAWRENCE *Lett.* (1938) 572 If I had any side, they would knock off what they call my 'block'. **1931** V. PALMER *Separate Lives* 220 There was a sheelah back in Salisbury who did her block on me. **1933** *Bulletin* (Sydney) 15 Nov. 33/1 The Cossack done his block again—slapped Plato on the cheek. **1939** H. G. WELLS *Holy Terror* I. i. 12 Many suggestions were made, from 'Knock his little block off', to 'Give him more love'. **1939** A. UPFIELD *Myst. Swordfish Reef* (1943) vi. 52 Some will lose their blocks, get excited, and then something has to go west. **1956** E. GRIERSON *Second Man* vii. 144, I did my block—panicked, I reckon you'd say. **1966** 'L. LANE' *ABZ of Scouse* 10 *I done me block*; I went off my head; I lost my temper.

5. c. *block and tackle*, = TACKLE *sb.* 3; also *fig.*

1838 *Knickerbocker* XII. 373 The diver began to don his submarine habiliments, which were swung inward from the vessel's side..by means of a block-and-tackle. **1864** O. W. NORTON *Army Lett.* (1903) 221 General Birney seems to consider the Eighth as..block and tackle by which to hoist his favorites into place and power. **1910** *Hawkins' Electr. Dict.* 39/1 *Block and Tackle*, a term including the block and the rope wove through it, for hoisting or obtaining a purchase. **1936** *Discovery* Aug. 238/1 The camera is lowered on its mount by means of special block and tackle installation fixed to the cabin roof.

6. e. A shaped piece of wood forming part of a shoeblack's equipment, on which a customer places his foot.

1872 *Cassell's Mag.* V. 84/1 The boys are provided with their uniforms, their block, blacking, and brushes by the society.

f. *Drapery.* A roll of material wound on a board: now called *piece*.

1905 H. G. WELLS *Kipps* I. vi. § 4 Being subsequently engaged in serving cretonnes, and desiring to push a number of rejected blocks up the counter. **1934** —— *Exper. in Autobiogr.* I. iv. 149 Wrappered blocks labelled incomprehensibly Hard Book or Turkey Twill.

10. b. (Earlier and later U.S. examples.)

1870 MEDBERY *Men & Myst. Wall St.* 134 Block. A number of shares, say 5,000 or 10,000, massed together, and sold or bought in a lump. **1901** MERWIN & WEBSTER *Calumet 'K'* iii. 40 A big block of treasury stock.

c. = PAD *sb.*[3] 4. Also *attrib.*, as *block calendar*.

1865 G. M. HOPKINS *Note-Bks. & Papers* (1937) 54 Shewing a sketching-block, he asked if there would be any objection to his sketching there. **1893** Sketch-block [see SKETCH *sb.* 6]. **1908** Scribbling-block [see SCRIBBLING *vbl. sb.*[1] 4]. **1910** *Brit. Empire Paper Trades' Jrnl.* Feb. 66/2 Blotting pads, books, etc., note books, note blocks, address books, etc. *Ibid.* May 186 Portrait and Figure Studies, with Daily Tear-off Block and Quotations for every day in the year. *Ibid.* 187/1 A new series of larger size shilling block calendars. **1940** H. G. Wells *Babes in Darkling Wood* II. i. 128 He pushed the paper-block back and began writing on the blotting-paper before him.

d. The carcass of a bullock; also *attrib.*, as *block test*, ascertainment of the dead weight of a beast when on the butcher's block for cutting up.

1893 *Westm. Gaz.* 8 Mar. 9/1 The 'block test'..used by Mr. McJannet, of Stirling...A set of tables by which, when the weight of the live bullock is ascertained, the weight of the 'block' as it hangs up in the flesher's shop for sale can be established within about a couple of pounds. **1902** *Encycl. Brit.* XXV. 188/1 In 1895 the Smithfield Club instituted a carcase competition...The cattle and sheep entered for this competition are shown alive on the first day, at the close of which they are slaughtered and the carcases hung up for exhibition, with details of live and dead weights. The competition thus constitutes what is termed a 'block test'.

e. = *BLOC.

1925 M. Eastman *Since Lenin Died* ii. 19 To perfect and solidify the block which they had already formed against him among the leaders of the party. **1940** *Manch. Guardian Weekly* 12 July 25 Working for the entry of France into a vast Continental block with Germany and Italy as the chief partners. **1957** *Economist* 7 Sept. 744/1 The extent to which the Soviet block is believed to have settled down again.

f. *Computers.* A set of data or instructions: (i) a group of successive locations in a memory and the data they contain; (ii) a group of words treated as a single unit in a program or by a computer.

1948 *Math. Tables & other Aids to Computation* III. 11 Characterizing a block of data on a table tape by a single block number. *Ibid.* 71 If a block has been called for, the machine..automatically transfers the first number in this block into the table register. **1958** Gotlieb & Hume *High-Speed Data Processing* ix. 179 A block is an integral number of words stored in a single register and transferred as a whole from one part of the machine to another. **1964** A. Lytel *Fund. Data Processing* iv. 202 A block is a series of declarations and statements enclosed between the words *begin* and *end*. **1964** *Ann. N.Y. Acad. Sci.* CXV. 663 Reels of 3/4 inch-wide tape which contain 512 consecutively numbered blocks, each capable of storing 256 12-bit numbers. **1968** Lehman & Bailey *Digital Computing* x. 200/2 The common area of memory may be divided into any number of separate regions, or blocks.

11. b. A piece of wood or other material placed in front of a wheel of an aeroplane to prevent it from moving forward; a chock. Also *attrib.*, as *block time* (see quot. 1950), etc.

[**1918** W. G. McMinnies *Pract. Flying* 228 *Chocks*, wooden blocks placed in front of the wheels of a machine to prevent it moving when the engine is started.] **1930** P. White *How to fly an Airplane* xiii. 189 You must remove the blocks from under the wheels. **1936** *Jrnl. R. Aeronaut. Soc.* XL. 852 On short ferry services the time occupied at each end in taking off and landing had a curiously levelling effect, so that the block-to-block speed was much the same as for high or low speed aircraft. **1948** *Ibid.* LII. 624/2 This method of arriving at stage distance—by taking a block time and by relating that time to the distance which would be covered at the block speed in the wind conditions. **1950** *Gloss. Aeronaut. Terms* (*B.S.I.*) I. 11 *Block time*, the period from the time the chocks are withdrawn, brakes released or moorings dropped, to the time of return to rest, or taking up of moorings after flight.

12. c. *Archit.* Each of the squared pieces above and sometimes below the columns of a chimney-piece.

1775 J. Wedgwood *Let.* 14 Jan. in E. Meteyard *Life* (1866) II. 321, I..stand in need of your directions relative to the blocks & ovals to the Tablets. **1875** E. Meteyard *Wedgwood Handbk.* Gloss. s.v., The more ordinary chimney-pieces had only a tablet and blocks...Occasionally the base of the columns had blocks also.

14. (Earlier and later examples.) orig. *U.S.*

1796 *Aurora* (Philad.) 13 Dec. (Th.), The whole block of buildings included between that slip, Front Street, and the Fly Market. **1817** S. R. Brown *Western Gaz.* 101 Each block of lots has the advantage of two 16 feet alleys. **1837** *Knickerbocker* IX. 72 Paved thoroughfares and manufacturing or commercial blocks. **1843** Dickens *Mart. Chuz.* xvi. 203 A neighbouring bar-room, which..was 'only in the next block'. **1963** J. T. Story *Something for Nothing* vi. 214 Outside the office there was nowhere to park...She cruised round the block twice, then finally double-parked and hurried into the building.

c. A fashionable promenade outside a particular block of buildings and shops in some Australian cities; hence *on the block*, on the promenade; *to do the block*, to lounge or saunter on the fashionable promenade.

1869 M. Clarke *Peripatetic Philosopher* 13 (Morris), If our Victorian youth showed their appreciation for domestic virtues, Victorian womanhood would 'do the Block' less frequently. **1872** 'Resident' *Glimpses of Life in Victoria* 349 A certain portion of Collins Street, lined by the best drapers' and jewellers' shops..is known as 'The Block'. **1896** *Argus* 17 July 4/7 (Morris), Just as the busy man, who generally walks quickly, has to go slowly in the crowd on the Block. **1902** *Daily Chron.* 20 Oct. 5/2 Sundowners ..who once enjoyed fat berths in Melbourne, 'doing the block' every afternoon in Collins street. **1906** E. Dyson *Fact'ry 'Ands* viii. 96 Man 'n' woman ud come jiggin' erlong in their block clobber. **1916** J. B. Cooper *Coo-oo-ee!* i. 14 When I go to town I'm crowded on the Block, I can't breathe.

d. Each of the large lots into which land is divided for settlement and development; any (fairly) large area of land; also (*Austral.*), a building section or lot. See also *BACK BLOCKS. *Austral.* and *N.Z.* (For N. Amer. examples see *D.A.*)

1840 in *Hist. Rec. Austral.* (1924) 1st Ser. XX. 744 If a Person should pay at once for 5,120 Acres or 8 Square Miles, We propose to allow him the privilege of demanding a Survey of that Quantity in one Block in any part of the Colony he may choose. **1841** W. Deans *Let.* 25 Mar. (1937) 31 We are insured our land, but are bound..to take the 110,000 acres for the principal settlement in one continuous block, round Port Nicholson...The boundaries of the block are now pretty well defined. **1858** Simmonds *Dict. Trade*, *Block*,.. in colonial parlance a piece of land. **1866–7** *Acts S. Australia* 30 *Vict.* No. 21 § 6 No..lease shall be granted at a lower yearly rent than at the rate of Ten Shillings for every block or Section, and no such block or Section shall exceed one square mile of land therein comprised. **1872** *Ibid.* 35 & 36 *Vict.* No. 18 § 24 No person shall hold..more than three separate and detached blocks of land; and such area shall.. be comprised in one block. **1909** B. R. Wise *Commw. Australia* 100 Some far-seeing men..would have set aside defined blocks for agricultural settlement. *Ibid.* 113, 27,000 acres, in forty-acre blocks. *Ibid.* 127 There has.. been some 'dummying', but the majority of the block-holders are *bonâ fide* occupiers. **1933** *Bulletin* (Sydney) 11 Jan. 25 My block of 200 acres had previously been subdivided for cropping. **1950** *N.Z. Jrnl. Agric.* Oct. 293/2 Old fences were removed..and new ones erected to subdivide the run into blocks of convenient size. **1969** *Advertiser* (Adelaide) 8 Feb. 42/1 (Advt.), Older Style Marino Home and 2 established adjoining building blocks, corner position, all blocks enjoy magnificent coastal views. **1969** *Age* (Melbourne) 24 May 4/4 Mrs. Smith said that the Normanton Aborigines had a reserve of 82 blocks on a hill overlooking the town, separated by a salt pan. **1969** *Northern Territory News* 11 July 4/3 (Advt.), Freehold Residential Holiday Sites. N. T. Real Estate Pty. Ltd. have great pleasure in introducing the sale of these excellent freehold blocks.

†**e.** A blockhouse. *U.S. Obs.*

1829 J. F. Cooper *Wish-ton-wish* vii. 108 He that is wise, however, will take but little of the latter, until his head be safely housed within some such building as yon block. **1845** W. G. Simms *Wigwam & Cabin* Ser. I. 57 As it was only a short mile and a half from the block, and we could hear of no Indians.

f. A large single building, esp. one containing a number of flats or offices; *block dweller, dwellings*: see *23.

1849 in Mrs. F. L. Adams *Pioneer Hist. Ingham County* (1923) 149 A little old 'corner grocer' building occupied the corner where Pratt & Millsapugh's block now stands. **1903** G. Gissing *Private Papers H. Ryecroft* 256 The sixth floor of a 'block' in Shoreditch. **1915** J. Buchan *39 Steps* i. 13 My flat was the first floor in a new block behind Langham Place. **1933** *Discovery* Aug. 254/1 The designers of large blocks of flats. **1969** *Times* 20 Oct. 9/5 Like the great bulk of Glasgow's nineteenth century housing Bernard Street consists of four-storey tenement blocks.

19. d. *Neurol.* Obstruction of the passage of a nervous or muscular impulse; an instance of this. Cf. *BLOCK *v.* 4 b and *heart-block* (*HEART *sb.* 56).

1882 W. H. Gaskell in *Phil. Trans.* CLXXIII. 1031 If the section is severe the block will be complete; no contractions will pass. **1883** —— in *Jrnl. Physiology* IV. 66 The contraction wave is unable to pass, then the 'block' is complete...Romanes..has..made use of the term 'block' to express any artificial hindrance to the passage of the contraction. I therefore make use of the same term in speaking of the results of experiments on the cardiac muscle which are very similar to those which he performed on the muscle of the Medusa. **1905** *Amer. Jrnl. Physiology* XIII. p. xxvii, A condition of partial 'block' in which the rhythm of the auricle is to the rhythm of the ventricle as 3 is to 1. **1957** *Encycl. Brit.* I. 864/2 When the local anaesthetic agent is injected so that it blocks the nerve supply, the method is known as 'block' or 'conduction' anaesthesia. *Ibid.*, Spinal anaesthesia is a special form of block anaesthesia in which the local anaesthetic agent is deposited in the spinal fluid. **1963** G. B. Carter et al. *Dict. Midwifery* 404/2 *P*[*udendal*] *block*, infiltration of the pudendal nerve with a local anaesthetic..preparatory to forceps delivery or to repair of perineal rupture.

e. *Psychol.*, spec. = *BLOCKING *vbl. sb.* 1 f.

1931 A. G. Bills in *Amer. Jrnl. Psychol.* XLIII. 230 The term 'block'..refers to those periods, experienced by mental workers, when they seem unable to respond and cannot, even by an effort, continue until a short time has elapsed. **1946** W. S. Knickerbocker *20th Cent. Eng.* 187 Some 'block' in his mental make-up. **1949** *Brit. Jrnl. Psychol.* Sept. 38 Occasional breakdowns of response, or 'blocks' occurred. **1969** *Times* (Sat. Rev.) 15 Nov. p. iv/3 Henry James..freeing himself from emotional and work blocks.

20. (Examples.) Also, a stroke of the bat to block a ball.

1825 *New Monthly Mag.* XIII. 498, I..admired the dexterity of the block at hand, which frustrated the perilous three-quarter ball. **1837** D. Walker *Games & Sports* 215 The popping crease..having in its middle..a hole called the block hole. **1845** Wanostrocht *Felix on Bat* I. ii. 12, I next recommended him to take the block for the middle stump, about five inches behind the popping crease. **1851** [see Work *sb.* 2 b]. **1866** 'Captain Crawley' *Cricket* 8 The block is the spot where the batsman grounds his bat when prepared to play. **1956** N. Cardus *Close of Play* 11 After he had taken guard and marked his block-hole, he would pat the crease neatly.

23. block ball *Baseball*, a ball, either hit or thrown, which is handled or stopped by a non-player; **block-board,** a plywood board having a core of thin wooden strips with the grain at right angles to the adjacent veneers; **block bond** *Bricklaying* (see quots. and Bond *sb.*¹ 13 a); **block-buster,** an aerial bomb capable of destroying a whole block of buildings; also *transf.* and *fig.*; **block-busting** *a.*, of the nature of a block-buster (*fig.*); also as *vbl. sb.* and *attrib.*, *U.S. colloq.* (see quot. 1959); **block capital,** a capital letter written or printed without serifs; **block chain,** an endless chain composed of alternate blocks and links; **block coal,** coal that splits easily into blocks; *spec.* an American bituminous furnace coal; also, coal in large lumps; **block coefficient** *Naval Archit.* (see quot.); **block diagram,** (*a*) a type of relief model (see quot. 1924); (*b*) a diagram in which squares and other conventional symbols show the order and arrangement of parts of an apparatus; **block dwellings** *pl.*, dwellings consisting of flats for working-class families in large barrack-like buildings several storeys high; hence *block dweller*; **block-faulting** *Geol.*, faulting which divides a region into blocks; **block lava** *Geol.*, a lava field composed of angular blocks; = *AA²; **block letter,** also = **block capital*; **block model** *Shipbuilding*, a model of a ship shaped from a block made up of flat pieces of wood fastened together, the lines of junction showing, on a reduced scale, the water-lines of the vessel to be built; **block mountain,** a mountain formed by faulting of the earth's crust; **block plan,** an outline plan or sketch, esp. of a building-site; **block plane** (see quots.); **block test** (see sense *10 d); **block train,** a railway train of which the component parts are kept permanently made up; **block universe** *Philos.*, the universe conceived as being like a block, as a unitary closed system of interlocking parts in which there is neither genuine plurality nor room for alternative possibilities; **block welding** (see quot. 1952); **block working,** the working of railway traffic on the block system (see Block *sb.* 19 c).

1891 N. Crane *Baseball* 79 *Block ball*, a batted or thrown ball handled by an outsider. **1932** A. Mora *Plywood* 86 (*caption*) Method of cross cutting core pieces for block boards. **1936** *Archit. Rev.* LXXIX. 76/1 The only structural difference between a sheet of three-ply and a laminated board or blockboard is that in both the latter the core is of composite vertical formation instead of being a single or continuous horizontal sheet of veneer. **1939** *Ibid.* LXXXVI. 129/4 Blockboard plywood, i.e., plywood with an interior composed of strips of solid timber, was first industrially produced by Kümmel's at Rehfelde near Berlin about 1902. **1865** Webster s.v. *Bond*, English or block bond. **1901** Sturgis *Dict. Archit.* s.v. *Bond*, Block Bond. Same as Flemish Bond. **1942** *Time* 14 Sept. 29/1 Inside a sturdy observation tower a mile from the exploding block busters which the Army is now testing. **1943** *Times* 22 Dec. 4/5 Bombs were falling..many 8,000 lb. and 4,000 lb. 'block-busters' among them. **1957** G. Smith *Friends* vi a. 114 One day I had what seemed to me like a blockbuster of an idea for a musical play. **1967** *Spectator* 6 Oct. 394/2 The 'block-buster' is a figure in American urban life who has yet to emerge in this country. He is a property dealer who by subterfuge introduces black residents into all-white neighbourhoods. **1943** *New Statesman* 20 Nov. 332/3 Those English Labour M.P.s who in their speeches show only a blockbusting contempt for Ulster. **1959** *Economist* 31 Jan. 415/2 Once a single negro moves into a block..the houses on both sides of the street from corner to corner are bound to become..Negro... Such 'block-busting' [etc.]. **1961** *Nation* 7 Oct. 223/2 The block-busting real-estate men show homes in integrated districts..only to prospective Negro buyers. **1902** *Combined Training* (War Office) 61 Names of places and persons will be written in block capitals. **1924** *Contemp. Rev.* Apr. 482, I published an article in the *Novoie Vremya* under the title 'Bread', which appeared in large block capitals. **1896** A. Sharp *Bicycles & Tricycles* xxvi. 431 The 'Roller' has the advantage over the 'Humber', or block chain, that its rubbing surface is very much larger. **1904** Goodchild & Tweney *Technol. & Sci. Dict.*, Block Chain (*Cycles*). **1871** *Amer. Naturalist* V. 177 A visit will be made to the celebrated Block-coal field (iron smelting coal). **1874** *Amer. Cycl.* IV. 726/1 We shall herein denote the prominent varieties as..non-caking or block coal, and caking or coking coals. **1883** Gresley *Gloss. Coal-m.*, Block coal. **1901** *Feilden's Mag.* IV. 421/1 In every case it is advisable also to calculate the value of the 'block coefficient' or so-called coefficient of fineness. **1902** *Encycl. Brit.* XXXII. 551/1 The block coefficient is the ratio of the volume of the immersed portion of the ship to the volume of the parallelopipedon. **1924** A. K. Lobeck *Block Diagrams* I. i. 1 A block diagram presents the relationship between the surface of the ground and the underground structure by representing an imaginary block cut out from the earth's crust. **1944** *Electronic Engin.* June 24/3 Fig. 2 shows a block diagram of the circuit technique used. **1962** *Gloss. Terms Autom. Data Proc.* (*B.S.I.*) 59 *Block diagram*, a conventional drawing of a system, instrument, computer or program in which all portions are represented by annotated boxes. **1902** C. F. G. Masterman *From Abyss* iii. 37 No..dreams of impossible millennium will haunt the block dweller of the future. **1899** *Daily News* 17 Jan. 3/2 The slums are sickening,..and the block dwellings often more like warehouses than homes. **1902** *Encycl. Brit.* XXXII. 673/2 Slum dwellings have been cleared under Cross's Acts 1875–82, and the Housing of the Working Classes Act, 1890; and..block dwellings have been erected. **1921** A. W. Grabau *Textbk. Geol.* I. xxi. 689 Block faulting, the dropping down or raising of circumscribed blocks, is not uncommon. **1940** E. S. Hills *Structural Geol.* 63 Regions which are divided by

faults into a number of differentially elevated or depressed blocks are said to exhibit *block faulting*. **1965** G. J. Williams *Econ. Geol. N.Z.* v. 42/1 The schist..must have been steeply tilted by block-faulting in some places. **1914** R. A. Daly *Igneous Rocks* xiii. 291 The vesiculation of pahoehoe or ropy lava was found to be more evenly developed than in the aa or block lava. **1920** A. Holmes *Nomencl. Petrology* 47 *Block-lava*, a term applied to lava flows which occur as a tumultuous assemblage of angular blocks having extremely rough surfaces due to the abundant development of large vesicles; = *aa*-lava or *aphrolithic* lava. **1908** *Installation News* II. 115/2 The cost of current for a large block-letter sign is frequently six or seven pounds a week. **1929** *Humorist* 5 Jan. 682 Name and address should be written here in plain block letters. **1901** *Feilden's Mag.* IV. 426/1 The angle of entrance..may either be measured from the block-model or calculated. **1896** J. W. Gregory *Great Rift Valley* xii. 221 (*caption*) Section across a 'Block Mountain'. **1901** C. R. Dryer *Lessons Physical Geogr.* xiv. 181 *Block Mountains*, probably the simplest mountains in existence are those of southern Oregon and northern California and Nevada. **1929** L. J. Wills *Physiogr. Evolution Brit.* I. ii. 11 Differential elevation of a block relative to its surroundings may produce a *plateau* or a *block-mountain*. **1909** Webster, Block plan. **1941** *City of Oxford Building Byelaws* No. 134 A person who intends to erect a building..shall send or deliver to the clerk or surveyor..a block plan of the building. *a* **1884** Knight *Dict. Mech.* Suppl., *Block Plane*, a plane, the bit of which is set at a very acute angle to the working surface, to enable it to plane across the grain of the wood. **1966** A. W. Lewis *Gloss. Woodworking Terms* 66 *Block plane*, small low-angled (20°) plane, not more than 8 in long, used for cleaning up mitres and end grain. **1902** *Encycl. Brit.* XXXII. 154/2 In block-trains, where the component coaches are permanently coupled together, one dynamo sometimes lights all the carriages. **1963** *Times* 18 Feb. 6/5 Instead of its being carried in mixed freight trains, the block trains will carry nothing but coal. **1881** T. Davidson *Let.* 24 Dec. in R. B. Perry *Tht. & Char. W. James* (1935) I. 736 That last remnant of mythology and scholasticism, *viz.*, theism and a block-universe. **1884** W. James *Will to Believe* (1897) 181 Is not the notion of eternity being given at a stroke to omniscience only just another way of whacking upon us the block-universe? *a* **1910** —— *Some Probl. Philos.* (1911) xii. 191 It is the famous 'principle of causality' which, when combined with the next two principles, is supposed to establish the block-universe, and to render the pluralistic hypothesis absurd. **1943** W. G. John in *Electr. Welding in Shipbuilding* (*H.M.S.O.*) 187 Block welding should be used for thicknesses above ⅜", the blocks starting from the centre of a seam and alternate block welds made from the centre outwards. **1952** *Gloss. Welding & Cutting of Metals* (*B.S.I.*) 41 *Block welding*, welding in which increments of the weld are made by superimposing a number of runs up to the full section before proceeding with the next increment. **1904** *Westm. Gaz.* 19 Jan. 10/2 With block working, only about 25,000 miles..are at present worked in America on the block system.

24. *attrib.* and *Comb.*, with the meaning 'in a block or mass', 'inclusive', 'solid', etc.: **block-booking,** (*a*) (see quot. 1925); hence *block-booker*; (*b*) the booking of a block of reservations; **block closure,** the legislative closure of the clauses of a measure in a block or in blocks; **block grant,** a fixed inclusive grant of money made by the Exchequer to a local authority, esp. for education; also *attrib.* and *transf.*; **block heater,** see **block-storage heater*; **block rate,** a uniform rate charged in a given area, etc.; **block release,** used esp. *attrib.* of a system whereby a person is released from his work for a stated period in order to pursue a course of study; **block (-storage) heater,** a heating unit which accumulates warmth during the night and gives it off during the day; so *block-storage heating*; **block vote,** (*a*) the vote of a considerable number of people used for a particular end; (*b*) a method of voting at a congress, conference, or the like, whereby a delegate's vote has the value of the number of members he represents; such a vote; so *block voting*.

1927 *Daily Express* 25 July 6 The Wicked 'Block-booker'. **1925** *Weekly Westm.* 29 Aug. 444/3 Block-booking is the system whereby American producers refuse to let the English exhibitors have one important film unless they take also a 'block' of others, which they may never have seen. **1926** *Manch. Guardian Weekly* Mar. 221/3 The Government are prepared to remove the block-booking grievance by legislation if necessary. **1939** J. Phillips in T. Harrisson & C. Madge *War begins at Home* (1940) ix. 237 The B.B.C. gradually took back the bands. They started off with block bookings. **1960** *Guardian* 17 Mar. 9/1 Sorting out a 'double' block booking in the dress circle. **1963** *Ibid.* 23 Jan. 7/7 Among the first block bookings [on a car ferry] was one for a party of 1,300 French Boy Scouts. **1901** *Daily Chron.* 14 Aug. 3/6 There was a block closure to which the Government could resort. **1900** *Westm. Gaz.* 24 Mar. 2/2 Schools earning the very highest grants will suffer because their block grant will be less than the old variable grant. **1901** *Daily Mail Year Bk.* 226/2 The Code for 1900 created a revolution in the method by which grants had been paid to schools...This is called the 'block grant system'. **1928** Block grant [see *PERCENTAGE]. **1966** *Rep. Comm. Inquiry Univ. Oxf.* I. 424 The total block grant for libraries should be allocated between the libraries by the General Board. **1958** *Engineering* 4 July 32/3 The houses will be equipped.. with block heaters in the living room. **1909** *Westm. Gaz.* 1 Mar. 10/3 The insurance offices are reducing the block rate of £1 per cent. that has been charged in the fire-zone of the City. **1958** *Engineering* 28 Feb. 279/3 The theoretical side of apprentice training can best be covered by block-release courses instead of by part-time day releases. **1963** *Higher Educ.* (*Cmnd. 2154*) 318 Students who are released by their firms to take a series of short periods of full-time study..are counted..as part-time day students. Their courses are known as *block releases*. **1960** *Engineering* 17 June 810/2 Block storage heaters are growing in popularity in the United Kingdom. **1962** *Ibid.* 5 Oct. 460/2 Block storage heating is being installed in many offices. The principle is simple: a large block of refractory material is heated by an element during the night and gives off heat to the surroundings during the day. **1901** *Daily Chron.* 15 July 7/3 Welsh miners, who, by the block vote, were enabled to return a Welshman to the Victorian Parliament. **1955** *Times* 12 July 9/3 Resolutions from the branches are critical of the way in which the block vote of the union is used at the Trades Union Congress and Labour Party conferences. **1928** *Daily Sketch* 10 Aug. 2/1 Glasgow Corporation, through block voting by Labour members, had to refuse a civic welcome to the officers.

b. *Mus.* Applied to a succession of chords in which all the parts change in the same rhythm.

1934 Webster, Block chords. **1942** E. Blom *Mus. in Engl.* i. 17 It may have..made use of its harmony vertically, in block chords. *Ibid.* v. 81 The choruses that often move impressively in solid block harmony. **1948** *Down Beat* 19 May 14 The final chorus is git [*sc.* guitar] and block chords. **1955** G. Abraham in H. van Thal *Fanfare for E. Newman* 15 Block changes of harmony at each half-bar are now avoided. *Ibid.* 22 Chromatic and diatonic passing-notes.. are used to break up some of the earlier block-chords. **1967** *Listener* 2 Mar. 288/2 In the glee, one voice had the tune, and the others accompanied it, mainly in block chords.

block, *v.* Add: **3. c.** *Draughts.* To force (one's opponent's men) into such a position that they cannot move.

1850 H. G. Bohn *Handbk. Games* 408 The game is won by him who can first succeed in capturing, or blocking up, all his adversary's men. **1877** *Encycl. Brit.* VII. 445/2 The game [*sc.* draughts] proceeds until one of the players has all his men and kings taken, or has all those left on the board blocked.

d. *Cards.* (See quots.)

1884 W. B. Dick *Dick's Games of Patience* 7 *Available cards.*—Those which are not 'covered' or 'blocked' by other cards; that is, not forbidden by the particular rules of each game to be used. **1885** 'Cavendish' *Whist Devel.* Pref. p. x, All good players know that it is disadvantageous to block their partner's long suit. *Ibid.* 57 If the lead was from ace, queen, knave, ten only, B would block his own suit. **1898** Melrose *Scientific Whist* 24 The retaining of a high card against a partner's command is termed 'blocking' or 'obstructing' his suit. **1901** 'Tarbart' *Games of Patience* I A Patience is said to be *blocked* when, before completion, no further card is playable.

4. (Earlier U.S. example.)

1844 G. W. Kendall *Santa Fé Exped.* II. xiii. 260 Soon after [he] ordered his own men to leave the gambling cot of the leper, and by this means 'blocked the game'.

b. *Neurol.* To obstruct the passage of a nervous or muscular impulse. Cf. *BLOCK *sb.* 19 d.

1876 *Proc. R. Soc.* XXIV. 148 At whatever point in a contractile strip that is being progressively elongated by section the contractile wave becomes blocked, the blocking is sure to take place completely and exclusively at that point. **1877** *Ibid.* XXV. 483 Anæsthetics block spasmodic waves [in *Medusæ*]. **1906** [see *heart-block*, *HEART *sb.* 56]. **1957** [see *BLOCK *sb.* 19 d].

c. *to block off*: to stop, to head off. *U.S. colloq.*

1893 W. K. Post *Harvard Stories* 86 The two opposing crowds..swept across the diamond 'blocking off' the owners of the two dogs. **1899** A. H. Quinn *Pennsylv. Stories* 190, I tried to fix up two or three things with Miss Fitzgerald and she blocked me off each time, very nicely, it is true, but still she blocked me off.

d. To restrict the use or conversion of currency or other assets. Cf. *FREEZE *v.* 5 e.

1932 *Economist* 30 Jan. 221/2 Unsecured cash advances may be converted into 'blocked' investments in Germany. **1937** E. Ambler *Uncommon Danger* i. 30 At that time all German bonds were 'blocked' and not negotiable abroad. **1959** *Chambers's Encycl.* VII. 689/1 This is the 'blocked' currency. This method exerts pressure on the foreigner to buy goods from the blocking country which he would not otherwise wish to take at the price.

5. (Earlier example, used of *wicket*.)

1772 in H. T. Waghorn *Dawn of Cricket* (1906) 35 As the lame man could only block his wicket.

9. *to block out* (earlier and later examples).

1753 Washington *Diaries* (1925) I. 59 They..told 50 [canoes]..besides many others which were blocked-out, in Readiness to make. **1829** *Massachusetts Spy* 16 Dec. (Th.), There are portions [of the message] which bear the marks of having been 'blocked out' by General Jackson. **1911** H. S. Harrison *Queed* iv. 49, I wish you would block out a series of articles..designed to prepare the public mind for a thorough-going reform.

b. *Theatr.* (See quots. 1961.) Also without *out*.

1961 A. Berkman *Singers' Gloss. Show Bus. Jargon* 9 In a stage play the director *blocks* the scene when he designates the position and action of each of the players, as well as the location on stage of all the props. **1961** Bowman & Ball *Theatre Lang.* 32 *Block out*, to work out the principal business, positions and movements of actors, including their entrances and exits, during rehearsals. **1967** P. McGerr *Murder is Absurd* vi. 73 Wednesday Warren finished blocking the first act. **1967** *Listener* 24 Aug. 240/1 At the first rehearsal..we were blocking out the moves.

10. b. *Drapery.* To make into a block (*BLOCK *sb.* 6 f).

1905 H. G. Wells *Kipps* I. iii. § 2 With hands much exercised in rolling and blocking. *Ibid.* III. iii, One whole piece most exquisitely blocked of every possible width of tape.

11. b. To pave (a street) with blocks.

1891 *Argus* (Melbourne) 25 Nov. 7/8 Only those streets in which the most traffic takes place will be blocked.

blockade, *sb.* Add: **3.** *spec.* in U.S., a stoppage or block on a railway by snow or some accident.

1856 *N.Y. Herald* 8 Jan. 1/4 The railroads are being slowly relieved from the blockade of snow. **1873** J. H. Beadle *Undevel. West* xix. 350 With the snow sheds since constructed, and other precautions, we may reasonably expect no more blockades. **1881** [see Dict.]. **1910** *N.Y. Even. Post* 13 Dec. 3 When a drawbar on the middle car of a third Avenue elevated train broke..a long blockade began.

b. A barrier on a river. *U.S.*

1871 *Game Laws* (*N.C.*) in *Fur, Fin & Feather* (1872) 153 It shall not be lawful for any person to draw a seine.. between the blockade near Hill's Point..and the falls at Wm. S. Battle's factory.

4. blockade-runner (earlier example); **blockade-running** (example).

1863 *Rep. Secr. of Navy* (*U.S.*) p. v, Not a single blockade-runner has succeeded in reaching the city [of Charleston] for months. **1864** *Ibid.* p. xvii, This species of illicit traffic and blockade running.

blockade, *v.* Add: **2.** *spec.* in U.S., to block (a road or railway, etc.). Cf. *BLOCKADE *sb.* 3.

1816 *Niles' Reg.* X. 216/2 In consequence of the vast body of ice with which it [*sc.* the harbour] is yet *blockaded*, they were unable to get in. **1872** *Rep. Vermont Board Agric.* I. 24 An exceedingly severe snow storm..completely blockaded many of the roads.

blocker. Add: **1.** Also in *Hat-making* (cf. Block *sb.* 4 a, *v.* 8).

1894 *Daily News* 26 Apr. 2/5 A straw hat blocker. **1898** *Daily Chron.* 14 Oct. 10/5 Blocker..for felt and straw hats.

4. A cricketer who habitually blocks the bowling. (Cf. Block *v.* 5.)

1851 J. Pycroft *Cricket Field* vii. 14 One of the most awkward, poking, vexatious blockers that ever produced a counterfeit of cricket defied Bayley and Cobbett at Oxford in 1836,—three hours, and made five and thirty runs.

5. *slang.* A bowler hat.

1934 J. Brophy *Waterfront* xi. 204 He'd worn one at the wedding, the first 'blocker' he'd ever had. **1966** F. Shaw et al. *Lern Yerself Scouse* 58 Foremen traditionally wore bowler-hats, or 'blockers'.

blockhouse, *sb.* Add: **1. e.** A reinforced concrete shelter used as an observation point, etc.

1953 *Monsanto Mag.* July 4 This blockhouse is as close as anyone gets to a missile at take off. **1962** A. Shepard in *Into Orbit* 104 All the training we had gone through with the blockhouse crew and booster crew was really paying off.

3. blockhouse system, the system of separating the theatre of war by chains of blockhouses, devised by Lord Kitchener in the later stages of the South African war, 1899–1902, and also used elsewhere.

1901 *Daily Chron.* 4 Sept. 5/3 The section of the line south of Pienaars River..is not yet fully protected by blockhouses, the blockhouse system having been first applied to those sections most requiring such protection. **1946** *Ann. Reg.* 1945 270 The Communists were..moving freely through the gaps in the Japanese block-house system.

blockhouse (blǫ·khaus), *v.* [f. the sb.] *trans.* To cut off from occupation or attack by a line of blockhouses under the **blockhouse system*.

1901 *Daily Chron.* 4 Sept. 5/3 Two thousand miles of railway are already 'blockhoused'. *Ibid.* 27 Dec. 4/5 General De Wet..is doing his best to prevent the 'blockhousing' of the north-eastern angle of the Orange River Colony.

blocking, *vbl. sb.* Add: **1.** *attrib.* (Examples of uses in *Cricket*: see Block *v.* 5.)

1836 *New Sporting Mag.* Oct. 360 Mr. Massey..in his anxiety to preserve his wicket, has carried the blocking system too far. **1867** G. H. Selkirk *Guide to Cricket-Ground* vii. 120 Some men adopt what they call a blocking game, never hitting at a ball unless it is decidedly wide of the wicket.

b. Signalling by the 'block system' (see Block *sb.* 19 c). Also *attrib.*, as *blocking inspector*.

1903 *Westm. Gaz.* 26 Oct. 13/1 Blocking inspector between Eastwood and Bradford. **1908** Tratman *Railway Track* (ed. 3) 266 In some cases a third position [of the semaphore arm] is introduced for permissive blocking. **1908** W. G. Raymond *Railroad Engin.* 121 The foregoing description introduces no distant signals, and in early automatic blocking there were none. **1909** H. R. Wilson *Safety Brit. Railways* 59 The signalman must give what is known as the blocking-back signal.

c. *Bookbinding.* The action of the verb (sense 8 c); also *attrib.*, as *blocking-machine, -press, -shop*.

1846 Dodd *Brit. Manuf.* VI. 103 This is..effected..by means of a small blocking-press. *Ibid.* 105 The punches or small devices..are fixed..to a metallic plate, and thus impressed on the book at one blow by a press. This is then called 'blocking'. In the 'blocking-shop' are [etc.]. **1902** *Encycl. Brit.* XXVI. 302/2 Blocking Machine...A blocking

press is now, in consequence of the size of many of the blocks, a large and cumbersome machine.

d. *Hat-making.* The shaping of a hat on the block (see Block *sb.* 4 a); also *attrib.*, as *blocking-kettle, -machine* (hence *-machinist*).

1845 Dodd *Brit. Manuf.* V. 170 The shaping of hats, from the first rough 'blocking' to the production of a flat and smooth-edged brim. *a* **1877** Knight *Dict. Mech.*, *Blocking-kettle*, a hot bath in which hats are softened in the process of manufacture, so as to be drawn over blocks. *a* **1884** *Ibid.* Suppl., *Blocking machine*, a machine in which the crude cone-shaped hat-body is brought to shape. **1921** *Dict. Occupational Terms* (1927) § 409 *Blocking machinist*: (i) (felt hats) pulls felt hood by hand on to block [etc.]; (ii) (straw hats) presses hat..in a gas-heated machine [etc.].

e. *Bootmaking.* (See quot. *a* 1877.)

a **1877** Knight *Dict. Mech.*, *Blocking*,..the process of bending leather for boot-fronts to the required shapes. **1889** G. B. Shaw *Fabian Ess. Socialism* 67 A visitor to a shoe factory to-day will see the following machines..for blocking.

f. *Psychol.* The inhibition of a train of thought or association, etc.

1890 W. James *Princ. Psychol.* II. xxvi. 527 We shall study anon the blocking and its release. Our higher thought is full of it. **1930** *Psychol. Abstr.* IV. 120/2 Concerning the question of associative blockings. **1943** *Mind* LII. 364 Köhler believes that there may be 'blocking' processes of an electro-chemical character.

g. *Theatr.* The action of the verb (sense *9 b).

1961 A. Berkman *Singers' Gloss. Show Bus. Jargon* 9 *Blocking*, the act of planning the stage action of a play. **1967** P. McGerr *Murder is Absurd* vii. 92 On Friday the blocking was completed and Mark was beginning to feel secure in his movements and the handling of his props.

5. Blocks collectively.

1901 *Daily Chron.* 3 Sept. 7/4 Karri blocking.

blockman² (blǫ·kmæn). [f. Block *sb.*] An assistant in a butcher's or fishmonger's shop employed chiefly at the block in cutting up meat, filleting fish, etc.

1903 *Daily Chron.* 16 Sept. 8/5 Fishmongers.—Good blockman wanted.

blocky (blǫ·ki), *a.* [f. Block *sb.* + -y¹.] **1.** Of the nature of or resembling a block, *esp.* **a.** Of a person or animal: of solid build, stocky. **b.** Defining a commercial grade of shellac.

1879 G. F. Jackson *Shropsh. Word-bk.*, *Blocky*,..short and stout. **1895** *Daily News* 4 Sept. 7/5 Shellac...Ordinary to good second orange cakey and blocky 115s. to 119s. **1900** *Westm. Gaz.* 11 Dec. 7/3 A most shapely beast [*sc.* a heifer], ..being grandly filled up, and of a most blocky type throughout. **1904** *N.Y. Tribune* 15 May, A blocky pair of forty-seven inch ponies. **1958** *Times* 1 July ii/5 The level blocky beef animal, which delighted the farmer's eye in the past. **1968** H. Harmar *Chihuahua Guide* 232 *Blocky*,..used to describe a head which is cube-shaped, as in the Boston Terrier.

2. *Photogr.* Having the appearance of being printed in blocks, from an unequal distribution of light and shade.

1889 in *Cent. Dict.*

blodge (blǫdʒ). [Imitative; cf. Blotch, Splodge *sb.*] = Splotch *sb.*

1930 'R. Crompton' *William's Happy Days* viii. 192 Sixpennoth of cream blodges, please. Big 'uns. **1951** S. Spender *World within World* 153 The pink blodges, small black dots, lines like brackets, characteristic of this style of decoration.

bloke, *sb.* Add: **b.** *Naval slang.* The ship's commander.

1914 'Watch-keeper' *Five Minutes to One Bell* 7 Also the sailor will come to say of you, 'Oh! once you're taken before '*Im*, you're safe to see "The Bloke" in the morning.' **1919** W. Lang *Sea Lawyer's Log* I. vi. 69 If you gets noisy and boisterous-like you sees the Bloke in the morning. **1946** J. Irving *Royal Navalese* 33 *Bloke*..the traditional Lower-Deck name for a warship's Commander and Executive Officer.

blond, blonde. Add: (The spelling *blonde* is now in British usage the commoner in all senses.) **A.** *adj.* **d. blond(e) beast** [tr. G. *blonde bestie*], a man of the Nordic type.

[**1887** F. W. Nietzsche *Zur Genealogie der Moral* I. 21 Das Raubthier, die prachtvolle nach Beute und Sieg lüstern schweifende blonde Bestie.] **1907** G. B. Shaw *Maj. Barbara* Pref. 151 Nietzsche..is the victim in England of a single much quoted..phrase 'big blonde beast'. **1911** G. K. Chesterton *Innoc. Father Brown* x. 266 The man..was a magnificent creature...In structure he was the blonde beast of Nietzsche. **1949** G. B. Shaw *16 Self Sketches* xiv. 105 My auburn hair was never really Highland red like my sister Agnes's. But I was a 'blonde beast' of Danish type unmistakably. **1967** M. Procter *Exercise Hoodwink* x. 73 If he did not throw in his lot with this big blond beast he would go to prison anyway.

Blondin, blondin (blǫ·ndin). [F. *blondin* cable-way, f. the name of '*Blondin*' (J. F. Gravelet, 1824–97), a French tight-rope walker.] A tight-rope, a cable-way. Also *fig.*, a tight-rope walker. Also *attrib.*

1863 Dickens *Uncommercial Trav.* in *All Year Round* 15 Aug. 588/2 An appalling accident happened at the People's Park near Birmingham..the enterprising Directors ..hanging the Blondin rope as high as they possibly could hang it. **1906** *Westm. Gaz.* 27 Mar. 5/2 Scores of cranes, blondin and other transporters..are continually at work. **1934** *Discovery* Oct. 299/2 A hair-raising trip in the 'Blondin' carrier down to the quarry face. **1934** *Essays & Stud.* XIX. 123 The tight-rope of serio-comic wit from which so many literary Blondins have been ignominiously precipitated. **1937** *Nature* 21 Aug. 312/2 The crossings of the Tigris and the Euphrates had to be made by 'Blondins' (overhead cable ways). **1958** *Times* 3 Dec. 20/2 Hoppers of concrete being carried by blondin cables to each section [of a dam] in turn.

blondine (blǫndi·n). *U.S.* [f. Blond *a.* + -ine⁴.] **a.** A bleach for the hair. **b.** A woman with hair made blonde by bleaching. Also *attrib.* Hence as *vb.*, to bleach the hair with blondine; also **blondining** *vbl. sb.*, **blondined** *ppl. a.*

1888 in *Amer. Speech* (1960) XXXV. 267 *Blondine*, a preparation for dying the hair a blond color. *Blondining*, dyeing the hair by the application of blondine. **1890** S. J. Duncan *Social Departure* xxxvii. 349 A fair woman, *passée*, blondined, in widow's weeds, with red eyes. **1894** in G. Langford *Alias O. Henry* (1957) 256 She [*sc.* the New Woman] wears shirts and ties, and either cuts her hair short or blondines it. **1908** S. Ford *Side-Stepping with Shorty* vi. 95 'There, now!' says the blondine. **1909** F. Calhoun *Miss Minerva* 73 Jimmy..returned with a big bottle of a powerful 'blondine' in one hand. **1920** S. Lewis *Main Street* xxv. 312 Swell dame with blondine hair? **1957** V. Kehoe *Technique Film & T.V. Make-Up* vi. 85 Hair should be darkened or blondeened if gray is present.

blondinette (blǫndine·t). [Fr., lit. 'little blonde girl', f. Blond + -ine² + -ette.] A breed of oriental frilled pigeons.

1879 L. Wright *Pract. Pigeon Keeper* 177 Blondinettes ..are..peak-crested and plain-headed. **1881** J. C. Lyell *Fancy Pigeons* 234 The blondinette has been produced in recent years, according to Mr. Caridia...The blondinettes bear the same relation to the satinettes, in their several varieties, as the *schietti* or whole-coloured Triganicas do to the *gazzi* or pied ones. *Ibid.*, The blondinettes are grouse-legged and generally peak crested.

blondish (blǫ·ndiʃ), *a.* [See -ish.] Somewhat blonde; rather light-coloured.

1961 in Webster. **1968** 'R. Macdonald' *Instant Enemy* ii. 15 'What colour is his hair?' 'Blondish.'

blondism (blǫ·ndiz'm). [f. Blond, Blonde *a.* and *sb.* + -ism.] The state of being blonde; blondness.

1939 C. S. Coon *Races of Europe* iv. 98 The blondism of Hetep-Heres II apparently belonged to the Delta. **1965** Coon & Hunt *Living Races of Man* i. 8 Blondism in Europe may be less than 10,000 years old.

blood, *sb.* Add: **1. f.** Phrase. (*You cannot get*) *blood out of* (or *from*) *a stone* or *turnip*: (you cannot achieve) the impossible, esp. pity from the hard-hearted, or money from the avaricious.

1662 G. Torriano *Second Alphabet* 148/1 To go about to fetch bloud out of stones, *viz.* to attempt what is impossible. *Ibid.* 165/1 To go about to fetch bloud out of a turnip, *viz.* to attempt impossibilities. **1836** F. Marryat *Japhet* I. iv. 41 There's no getting blood out of a turnip. **1849** Dickens *Dav. Copp.* xi. 114 Blood cannot be obtained from a stone, neither can anything on account be obtained at present..from Mr. Micawber. **1889** Stevenson & Osbourne *Wrong Box* 8 'You cannot get blood from a stone,' observed the lawyer. **1929** *Observer* 21 July 15 This Court has no machinery that I know of for extracting blood from stones. **1938** R. D. Finlayson *Brown Man's Burden* 27 You may as well try to get blood out of a stone as evidence out of a native! **1940** A. E. Hertzler *Doctors & Patients* (1941) x. 261 The age old difficulty of getting blood out of a turnip.

3. d. *blood and thunder*, bloodshed and violence; used *attrib.* in *blood-and-thunder book, tale*, etc., one describing the murderous exploits of desperadoes. (orig. *U.S.*) Also shortened to *blood* (esp. in *pl.*), as in *blood books*, (*penny*) *bloods*.

1852 *Lantern* (N.Y.) II. 67/2 Most, however, of these 'blood, thunder, and whiskey articles', are written by raw lads. **1857** *Quinland* II. 76 Mrs. Bill, left to herself, resumed reading a blood and thunder romance. **1894** *Daily News* 29 May 6/4 'Blood and thunder books';..'blood books'—brief and brutal—is the expression in general use. **1897** W. J. Locke *Derelicts* iii, A writer of 'penny bloods'. **1925** W. de la Mare *Two Tales* 32 The penny blood concealed in his 'Arithmetic'. **1935** *Discovery* Nov. 346/2 As exciting reading as any blood-and-thunder novelette.

e. *blood and iron* [G. *blut und eisen*], military force as distinguished from diplomacy, esp. in *the man of blood and iron*, Prince Bismarck, who advocated the use of this as his policy.

1869 *Porcupine* 18 Sept. 235/1, I don't wonder at that man of 'blood and iron'—Bismarck—openly expressing himself an admirer of La good-looking Lucca. **1872** *New Dominion Monthly* Oct. 195/1 You will find him indeed a man of 'blood and iron'. **1889** H. P. Hughes *Social Chr.* v. 74 'There,' they are saying, 'nothing succeeds like a blood-and-iron policy.' **1898** A. J. Butler tr. *Bismarck's Refl. & Remin.* I. 310 We should be unable to avoid a serious contest, a contest which could only be settled by blood and iron. **1922** C. E. Montague *Disenchantment* v. 69 In the whole blood-and-iron province of talk he would.. outshine any actual combatant.

f. *blood and soil* [G. *blut und boden*], a Nazi catch-phrase, used *attrib.* to denote Nazi members or ideology.

1940 Auden *I Believe* 22 The success of Fascist blood-and-soil ideology. **1957** M. K. Joseph *I'll soldier no More* (1958) xii. 223 There was a pile of stuff..put out by some kind of East-Prussian patriotic blood-and-soil gang.

5. *one's blood runs cold* (examples); *his blood is up* (earlier example).

1727 [see Run *v.* 20 c]. **1818** *Edin. Rev.* XXX. 238 Her whole appearance, gestures, voice and dress, made De Courcy's blood run cold within him. **1829** G. Griffin *Collegians* II. xviii. 55 To use a vulgar but forcible expression, the blood of Hardress was now completely up. **1852** Stowe *Uncle Tom* i. 4 It kinder makes my blood run cold to think on't.

8. *blue blood*: also, a person with blue blood; an aristocrat; *new blood* = *fresh blood*.

1853 Lytton *My Novel* II. v. ii. 9 Long may the new blood circulate through the veins of the mighty giantess [*sc.* England]. **1880** *Baily's Mag.* Oct. 149 There was a good deal of change in the judicial bench...New blood was infused through Colonel —, Mr. —, and Major —. **1894** *Daily News* 16 Apr. 3/6 Many an aristocratic blue-blood.. was glad to marry a rich burgher's daughter. **1920** Galsworthy *In Chancery* II. i. 128 Round Crum were still gathered a forlorn hope of blue-bloods with a plutocratic following.

9. Proverb. *blood is thicker than water*: the tie of relationship is strong. (Examples.)

1815 Scott *Guy Mannering* II. xxxviii. 318 Weel—blood's thicker than water—she's welcome to the cheeses. **1867** Trollope *Chron. Barset* xxxii. 271 'I am aware that there is a family tie, or I should not have ventured to trouble you.' 'Blood is thicker than water, isn't it?' **1920** A. Huxley *Leda* 35 For Blood, as all men know, than Water's thicker, But Water's wider, thank the Lord, than Blood.

12. b. (Earlier examples.) Also with a qualifying word. Cf. *bit of blood* (*bit *sb.*² 4 h).

1792 *Sporting Mag.* I. 101/1 The Wold dogs beat the blood of the Norfolks, as some of the best bred of the late Lord Orford were completely worsted. **1793** *Ibid.* II. 334/1 That famous horse *Eclipse*, whose excellence in speed, blood, pedigree, and progeny, will be, perhaps, transmitted to the end of time. **1895** C. B. Lowe *Breeding Racehorses* 180 He will always do best with a strong return to his stout Blacklock, Bird-catcher, and Glencoe blood. **1902** *Encycl. Brit.* XXV. 190/2 When Shorthorn breeders of to-day talk of 'Booth blood', or of 'Bates blood', they refer to animals descended from the respective herds of Thomas Booth and Thomas Bates.

c. (Earlier example.)

1794 *Sporting Mag.* IV. 31/1 He [*sc.* a race-horse] is now a stallion..at 3 gs a mare, and 5s. the groom, blood mares.

15. c. At public schools and universities applied to those who are regarded as setting the fashion in habits and dress; also, a youthful member of a party, etc.

1892 *Pall Mall Gaz.* 8 Mar. 7/1 The result was that the new party won by 127 to 103...A great triumph for the Bloods—as we are accustomed to call them—who mustered in great force to defeat Mr. Childers. **1893** *Granta* 9 June 374/2 A Committee, consisting of a blood, a Girtonian, and a resident married M.A., shall supervise all flirtations. **1896** *Ibid.* 16 May 310/1 Mifflin and 'is friends talked..an' said 'ow much better Cambridge'd be if there wasn't no 'bloods' to spoil things. **1955** *Times* 25 Aug. 11/5 The rugger match dinner at the Trocadero with a select club of 'bloods'.

d. A passenger on a ship. *Ship's stewards' slang.*

1929 Bowen *Sea Slang* 14 *Bloods*, the modern steward's name for the passengers—used only when they are regarded kindly. **1962** *Harper's Bazaar* Dec. 74/3 Stewards will help you...Behind your back they will call you a 'blood'— ..they themselves being 'wingers'—and wonder how much 'rent' you will pay them at the end of the voyage.

17*. *ellipt.* = **blood orange* (see *19).

1907 N. Munro *Daft Days* i. 6 Oranges! Oranges!—rale New Year oranges, three a penny; bloods, a bawbee each!

17.** A tribe of North American Indians belonging to the Blackfoot confederacy; a member of this tribe. Also *attrib.*

1794 *Mass. Hist. Soc. Coll.* (1810) 1st Ser. III. 24 The tribes of Indians which he passed through, were called.. Blood Indians, the Blackfeet tribe..and several others. **1863** *Trans. Amer. Phil. Soc.* XII. 249 The Blackfeet inhabit a portion of country farther north than the Bloods. **1957** *Encycl. Canadiana* I. 403/1 Together with the Piegan and the Blood, they [*sc.* the Blackfoot] covered an enormous area of the western prairies and lower foothills of the Rockies.

18. Selected additional examples. (Such combs. are especially common in the writings of D. H. Lawrence.) *blood-beat* (example), *-being, -bondage, -compensation, -consciousness, -corpuscle* (example), *-covenant, -curdler, -desire, disease, -drop* (earlier and later examples), *-flow, -freezer, -gout, -kin, -knowing, -knowledge, -passion, -pride, -revenge, -sacrifice* (examples), *-soul, -spoor, -supply, -system, -tie, -vengeance*; *blood-bedabbled* (later example), *-black, -conscious, -curdling* (example), *-dabbled, -dark, -freezing, -masked, -soaked, -sodden*, adjs.

1947 E. Sitwell *Shadow of Cain* 11 The blood-beat of the Bird. **1895** Yeats *Poems* 22 Along the blood-bedabbled plains. **1915** D. H. Lawrence *Let.* 8 Dec. (1962) 394 All living things, even plants, have a blood-being. If a lizard

falls on the breast of a pregnant woman, then the blood-being of the lizard passes with a shock into the blood-being of the woman, and is transferred to the foetus...We have a blood-being, a blood-consciousness, a blood-soul, complete and apart from the mental and nerve consciousness. *a* **1918** W. Owen *Poems* (1963) 69 Sunlight seems a blood-smear; night comes blood-black. **1926** D. H. Lawrence *Plumed Serpent* ix. 154 And there was a thin little thread of blood-bondage between them. **1958** Middleton & Tait *Tribes without Rulers* 27 Such features are..chiefship, blood-compensation and non-empirical, religious sanctions. **1923** D. H. Lawrence *Stud. Class. Amer. Lit.* vii. 126 They [*sc.* Americans] *admire* the blood-conscious spontaneity. **1845** Blood-corpuscle [see Corpuscle 2]. **1886** *Encycl. Brit.* XXI. 137/2 The sacramental rites of mystical sacrifice are a form of blood-covenant. **1889** Barrère & Leland *Dict. Slang* s.v., It will contain..a blood-curdler, by the murder-man. **1906** E. Dyson *Fact'ry 'Ands* xv. 197 That one yowl was er blood-curdler. **1934** *Essays & Stud.* XIX. 13 The blood-curdling nature of the cry. **1904** W. H. Hudson *Green Mansions* xxi. 306 Cla-cla's wrinkled dead face and white, blood-dabbled locks. **1885** Yeats in *Dublin Univ. Rev.* July, A star-lit rapier, half blood-dark. **1958** R. S. Thomas *Poetry for Supper* 38 You were born on a blood-dark tide. **1930** D. H. Lawrence *A Propos of Lady Chatterley's Lover* 46 The two blood-streams are brought into contact, in man and woman, just the same as in the urge of blood-passion and blood-desire. **1902** *Encycl. Brit.* XXXI. 557/1 Anaemia is often used as a generic term for all blood diseases. *c* **1390**, *a* **1400** Blood-drop [in *M.E.D.*] **1923** D. H. Lawrence *Birds, Beasts & Flowers* 91 Away with a pæan of derision You winged blood-drop. **1873** T. H. Green *Introd. Path.* (ed. 2) 329 Thrombosis from Retardation of the Blood-flow. **1886** H. Baumann *Londinismen* 12/1 *Blood-freezer*.. Schauerroman. **1902** W. James *Varieties Relig. Exp.* vi. 162 The grisly blood-freezing heart-palsying sensation of it [*sc.* evil] close upon one. **1800** Bloud-gout [see Gout *sb.*¹ 5 a]. **1952** R. Campbell tr. *Poems of Baudelaire* 45 Sabres bleak With crimson blood-gouts lit the air above. **1880** 'Mark Twain' *Tramp Abroad* 173 The seven hundred inhabitants are all blood-kin to each other. **1937** R. H. Lowie *Hist. Ethnol. Theory* vi. 65 The classification of kin survives from a period in which the closest blood-kin regularly cohabited. **1915** D. H. Lawrence *Let.* 8 Dec. (1962) I. 394 When I take a woman, then the blood-percept is supreme, my blood-knowing is overwhelming. *Ibid.*, Some tribes no doubt really *were* kangaroos; they contained the blood-knowledge of the kangaroo. **1928** Blunden *Undertones of War* 5 Others lay near him, also bloodmasked. **1923** D. H. Lawrence *Kangaroo* xvii. 367 The brave, silent blood-pride. **1932** W. Faulkner *Light in August* (1933) i. 4 The bleak heritage of his bloodpride. **1855** R. Martineau tr. *Gregorovius's Corsica* I. x. 144 Many a case is known of one bandit having ..slain another..for blood-revenge. **1877** *Gentl. Mag.* Apr. 478 The vendetta or duty of blood-revenge. **1591** Shakes. *1 Hen. VI* v. iii. 20 Cannot my body, nor blood-sacrifice, Intreate you to your wonted furtherance? **1866** Swinburne *Poems & Ballads* 70 With offering and blood-sacrifice of tears. **1905** *Daily Chron.* 16 Jan. 3/4 A staggering, blood-soaked figure. **1929** Yeats *Winding Stair* 11 The heart in his blood-sodden breast. **1863** W. C. Baldwin *Afr. Hunting* v. 129 The blood-spoor of one of the wounded koodoos. **1884** J. L. Corning *Brain Exhaustion* ii. 45 The demands of an abnormally developed muscular system upon an already insufficient blood-supply. **1962** *Lancet* 2 June 1166/2 The blood-supply to the central nervous system. **1912** J. S. Huxley *Indiv. in Animal Kingdom* ii. 63 Cyclosis..performs the same general functions for the organism as does a blood-system. **1876** J. F. McLennan *Stud. Anc. Hist.* II. ii. 373 The system of blood-ties and the system of addresses would begin to grow up together. **1965** G. McInnes *Road to Gundagai* xvi. 282 The call of the Anglo-Saxon or Anglo-Celtic bloodties. **1893** *Funk's Standard Dict.*, Blood-vengeance. **1926** A. Møller tr. *Pedersen's Israel* I. II. 269 Blood-vengeance, which in its old form is one of the most pronounced outcomes of the solidarity of the family.

19. blood agar *Bacteriol.*, a culture medium containing blood and nutrient agar; **blood-alley** [Ally *sb.*²], a boy's white marble marked with red spiral lines; **blood-bank,** a place where a supply of blood for transfusion is stored (cf. *BANK *sb.*³ 7 f); also, a reserve of blood so stored; **blood-beet,** the red beet-root; **blood boat** (see quot. 1914); also *attrib.*; **blood-brother,** (*a*) a brother by birth; (*b*) one who has been bound to another in solemn friendship by a ceremonial mingling of blood; so *blood-brotherhood*; also *fig.*; **blood cell,** any of the cells or cell-types that circulate in the blood; **blood chit** *colloq.* (see quots.); **blood count,** (the determination of) the number of blood cells contained in a given volume of blood; **blood culture** [Culture *sb.* 3 c] *Bacteriol.*, a culture of a sample of blood to detect micro-organisms in it; **blood disc,** a red blood-corpuscle; also = **blood plaque*; **blood donor,** one who gives blood for transfusion; **blood-drinker,** a primitive savage who killed and feasted on his 'kill'; *fig.* one who has a lust for blood; **blood-drinking** *a.*, that drinks blood; also *fig.*; **blood dust,** a collective name for the minute refractive bodies found floating free in the blood plasma; **blood-eagle** [ON. *blóð-ǫrn*], a Viking method of killing someone, usually the slayer of a man's father, by cutting out the ribs in the shape of an eagle; **blood film,** a smear of blood (cf. Smear *sb.* 3 b); **blood fluke** [Fluke *sb.*¹ 2], = *SCHISTOSOME; **blood-gas,** the gas or gases present in the blood; also *attrib.*; **blood-groove,** a groove cut in the head or the shaft of an arrow or spear, supposed to increase the flow of blood from the wound made by the weapon; **blood group,** one of the genetically determined types into which human blood may be divided on the basis of its compatibility with the blood of other individuals; *esp.* one of the four original red cell groups; **blood grouping,** the determination of the blood group of a person or of a sample of blood; also = **blood group*; **blood-knot,** a knot tied in a rope to draw blood when it is used as a whip; **blood line** = Line *sb.*² 24, 25; (see also quot. 1909); **blood-lust,** lust for the shedding of blood; **blood meal** [G. *blutmehl*], dried blood used for feeding animals and as a fertilizer; **bloo·d-mobile** [formed after *AUTOMOBILE *sb.*] *U.S.* (see quot. 1961); **blood orange,** a variety of orange having the pulp streaked with red; earlier *blood-red orange*; **blood pheasant,** a species of pheasant (see quot. 1864) marked with red on the throat and breast; **blood picture,** (*a*) (see quot. 1881) *disused*; (*b*) the condition of the blood as determined by chemical or microscopical analysis; **blood plaque, plate, platelet,** a minute disc-shaped body found in large numbers in mammalian blood; **blood plasma,** the fluid part of the blood in which the cells and platelets are suspended (Billings 1890); **blood plum:** see Plum *sb.* 3 b; **blood-poisoning,** a morbid condition of the blood formerly thought to be caused by the absorption of putrefying matter but now recognized as being due to infection; *spec.*, septicæmia or pyæmia; **blood pressure,** the pressure of circulating blood on the walls of the blood-vessels, esp. the systemic arteries; **blood pudding,** delete † and add later *transf.* example; **blood-pump,** (*a*) see quot. 1902; (*b*) *Pugilistic slang*, the heart; **blood purge,** one that is achieved by bloodshed; also *fig.*; **blood-red orange** = **blood orange*; **blood-sausage** (examples); **blood-sister,** a woman bound to another in the same manner as blood-brothers (see also quot. 1933); so *blood-sisterhood*; **blood sports** *sb. pl.*, sports involving the killing of animals, esp. sports of the chase; **blood-stream,** the stream of blood circulating through the human or animal system; **blood-striking,** a disease incident to cattle (see Striking *vbl. sb.* 2 b); **blood sugar,** glucose contained in the blood; **blood-tax** *fig.*, a tax paid by bloodshed; *spec.* a derogatory term for military conscription; **blood test,** a test performed on blood for some specific purpose; also *fig.*; hence *blood-tested* adj.; **blood transfusion** = Transfusion 2; also *attrib.* and *fig.*; **blood-tub** *slang*, †(*a*), a rough or rowdy (*U.S. obs.*); (*b*) a theatre presenting lurid melodrama; **blood type** = **blood group*; hence *blood-typing* vbl. sb. and *attrib.*; **blood-urea,** the concentration of urea in the blood; **blood wagon** *slang*, an ambulance (see also quot. 1969); **blood-wealth** *Anthrop.*, money or goods given as compensation for a murder; **blood-wood** (earlier example).

1899 T. Bowhill *Man. Bacteriol. Technique* II. 54 (*heading*) Blood Agar. **1927** R. A. Kelser *Man. Vet. Bacteriol.* v. 57 *Blood Agar.* Add 10 per cent of sterile defibrinated horse, sheep or rabbit blood to a definite amount of sterile nutrient agar..which has been completely melted and allowed to cool to 45 °C. **1854** A. E. Baker *Gloss. Northampt. Words* 10 Blood alleys. **1881** A. B. Evans *Leic. Words* 91 If streaked with red veins it is called a 'blood alley'. **1963** Compton Mackenzie *Life & Times* II. 49 A blood alley was a pinkish stone with some deeper pink markings. **1938** *N.Y. Post Digest* April 23/1 Bellevue and Kings County Hospitals in New York have now followed the lead of Chicago and Philadelphia hospitals in the establishment of 'blood banks'. **1942** *Times of India* 2 May 6/2 The appeal..for blood donors to come forward and contribute to the blood bank which has been opened in the city. **1829** *Free Press* (Tarboro, N.C.) 20 Feb., Blood Beets. **1831** Peck *Guide for Emigrants* II. 141 The blood beet [is] less deeply colored. **1889** Barrère & Leland *Dict. Slang*, *Blood boat* (naval). **1899** Bullen *Log Sea-Waif* 230 Half the crew..looked as if all the ways of 'Western Ocean blood-boats' were familiar to them. **1914** Ld. Chas. Beresford *Mem.* I. i. 2, I was only the 'blood-boat' (the jolly-boat bringing beef to the ship) midshipman of a man-of-war. **1890** Ld. Lugard *Diary* 7 Oct. (1959) I. 314 His camp..where Dualla had blood brothers. **1933** E. E. Evans-Pritchard in *Africa* VI. 370 If you exchange blood with a member of the Akowe clan all other members of this clan rank as your blood-brothers. **1885** H. M. Stanley *Congo Free State* II. xxvi. 23 The next day we made blood brotherhood. The fetish-man pricked each of our right arms, pressed the blood out..and the black and white arms were mutually rubbed together. **1909** *Daily Chron.* 11 Sept. 4/6 His Majesty..will doubtless..drink a toast or two with especial reference to that blood-brotherhood between the two nations. **1937** E. Snow *Red Star over China* v. 195 They swore blood brotherhood in the tribal manner. **1846** T. W. Jones in *Philos. Trans. R. Soc.* CXXXVI. 65 The red oval corpuscle of the blood of the Skate under consideration, I propose to name nucleated blood-cell, in contradistinction to the granule-cell, which..in consequence of its being filled with granules is otherwise well designated granule blood-cell. **1866** Aitken *Pract. Med.* (ed. 4) II. 57 Salts..in which the blood-cells ultimately become deficient. **1878** F. J. Bell tr. *Gegenbaur's Comp. Anat.* 172 In many Nemertina the blood-cells have a red colour. **1943** C. H. Ward-Jackson *Piece of Cake* 15 *Blood chit*, any written authorisation supplied to an individual to 'cover' him; or, more originally, a ransom note supplied to pilots over possibly hostile territory in the East. **1954** *Neuphilologische Mitteilungen* LV. 47 *Blood-chit*, 'document, signed by a non-R.A.F. passenger before a flight, which exonerates the R.A.F. in the event that they damage or destroy him'. **1900** Dorland *Med. Dict.*, Blood-count. **1907** *Practitioner* Dec. 852 A blood-count, made two years ago, showed: Red cells, 5,000,000 [etc.]. **1947** *Mind* LVI. 246 Healthy children with a normal red blood count. **1899** *Jrnl. Exper. Med.* IV. 429 Blood cultures made in cases of pneumonia ..indicate..a general blood invasion. **1963** *Lancet* 5 Jan. 55/2 Blood-cultures were negative. **1840** *Philos. Trans. R. Soc.* CXXX. 596 Whether the globules into which the blood-disc resolves itself, are the foundations of new corpuscles of the blood, I do not know. **1845** Todd & Bowman *Phys. Anat.* I. 60 Certain particles, the blood-discs, which float in it [*sc.* the blood] in great numbers. **1879** *Jrnl. Anat. & Physiol.* XIV. 295 All cells..from which red blood-discs may spring. **1921** *Lancet* 26 Nov. 1123/1 In a recent number of the *Guy's Hospital Gazette* the editor protests against the too free use of students as blood-donors. **1958** *Times* 7 July xxi/4 The unselfish help of many thousands of voluntary blood donors. **1968** *Brit. Med. Bull.* XXIV. 210/2 Histograms of readings of serum chloride and urea taken from 1,000 blood donors are shown. **1898** Meredith *Odes Fr. Hist.* 16 The blood-drinker's madness fast upon her. **1899** *Daily News* 28 June 8/4 The primitive 'food group' of hunters, who, like the beasts they killed, were 'blood-drinkers'. **1588** Shakes. *Tit. A.* II. iii. 224 In this detested, darke, blood-drinking pit. **1591** —— *1 Hen. VI* II. iv. 108 My blood-drinking hate. **1903** *Edin. Rev.* Oct. 307 Blood-drinking savages. **1900** Dorland *Med. Dict.* s.v. *Blood*, Blood-dust, or hemoconiæ. **1839** G. Stephens tr. *Tegnér's Frithiof's Saga* xvi. 158 Blood-eagle lines on Thy foe shall be flowing. **1922** *Cambr. Med. Hist.* III. xiii. 329 Cutting the blood-eagle in the back of the fallen foe is well known from the vengeance for their father taken by the sons of Ragnarr Loðbrók. **1964** G. Turville-Petre *Myth & Religion of North* xiii. 254 A peculiarly revolting form of human sacrifice was that of cutting the 'blood-eagle' (*blóðǫrn rista*). The ribs were cut from the back and the lungs drawn out. **1904** Blood film [see Smear *sb.* 3 b]. **1961** *Lancet* 5 Aug. 315/1 To stain a blood-film the solution is added drop by drop till it entirely covers it. **1872** *Brit. Med. Jrnl.* 27 July 89/1 She was suffering from the effects of an invasion by the African blood-fluke, a parasite now generally recognized by the generic and specific title of *Bilharzia hæmatobia*. **1962** *New Scientist* 6 Sept. 490/3 The disease known as bilharziasis, caused by three species of bloodflukes or schistosomes, is carried by various types of water snail. **1908** J. Barcroft in *Jrnl. Physiol.* XXXVII. 12 (*title*) Differential Method in Blood-Gas Analysis. **1922** J. S. Haldane *Respiration* i. 11 Careful blood-gas determinations showed that when apnoea had been produced by forced ventilation of the lungs the arterial blood contained.. more oxygen. **1966** *Lancet* 24 Dec. 1394/2 The hypothetical absorption of significant amounts of oxygen from the skin or upper respiratory tract has never been substantiated by arterial blood-gas measurements. **1897** *Geogr. Jrnl.* X. 156 Arrowheads and spears, many of them curiously barbed and twisted, and some showing a knowledge of the value of the 'blood-groove'. **1916** W. V. Brem in *Jrnl. Amer. Med. Assoc.* LXVII. 190/2 Isohemolysins cannot be used, therefore, in determining blood groups. **1917** *Brit. Med. Jrnl.* 24 Nov. 696 The selection of an ideal donor—that is, one belonging to the same blood group as the patient—may be made. **1935** Huxley & Haddon *We Europ.* iv. 127 We plot the distribution of the three blood-group genes on a map. **1958** *Listener* 28 Aug. 320/2 It is now thought to be possible to discover the blood groups of ancient peoples from their bones. **1916** W. V. Brem in *Jrnl. Amer. Med. Assoc.* LXVII. 190/2 We have been able to modify the technic of blood grouping so that it can be done easily and accurately within a few moments' time. **1962** *Lancet* 15 Dec. 1279/2 In 16·6% of myoadenoma patients and 11·0% of the carcinoma patients the blood-grouping was not recorded. **1901** 'L. Malet' *Hist. Sir R. Calmady* III. viii. 233 Blood-knots in the whip-lash. **1910** J. Masefield *Lost Endeavour* I. iv. 45 A 'teaser', or blood-knot of hard, tarred spunyarn. **1968** E. Franklin *Dict. Knots* 9 *Blood knot*, a multiple Overhand Knot tied in the end of a rope..used as a weapon or for inflicting punishment. **1909** *Cent. Dict.* Suppl., *Blood line*, a particular transmissible character in an animal, or, analogically, in a plant. **1927** *Daily Tel.* 6 Dec. 9/1 [The steer] reflected his excellent bloodline descent in his shapely contour. **1948** C. L. B. Hubbard *Dogs in Britain* 31 He [*sc.* the breeder] will doubtless endeavour to improve or alter his blood-lines as the necessities arise. **1848** Lytton *Harold* I. III. ii. 175 Hear me, thou with the vulture's blood-lust. **1942** Wyndham Lewis *Let.* 25 Oct. (1963) 338 He [a soldier] would be disgusted and amazed to find us all foaming at the mouth, our eyes full of bloodlust. **1887** F. H. Storer *Agric. Rel. with Chem.* I. xiv. 399 Nitrate of soda and horn-meal did well, and fermented blood-meal also. *Ibid.* 383 Fresh blood is made to separate into liquid serum..and into a solid clot, which, when dried and ground, is the substance used as a fertilizer and known as dried blood or blood-meal. **1956** Gillespie & Hathway *Textbk. Gen. Agric.* xvii. 225 Blood-meal, another residue from the slaughterhouse, is a favourite ingredient of rations for pigs and poultry. **1948** *Sci. News Let.* 10 July 26 (*caption*) Red Cross Bloodmobile. **1961** *Guardian* 28 Jan. 12/7 'An automobile equipped for collecting blood from volunteer donors'..bloodmobile. **1855** E. Acton *Mod. Cookery* xxviii. 571 Tangerine Oranges. ..There is another variety of this fruit known commonly as the *blood-orange*. **1862** Hereman *Vine & Fruit Tree Cultiv.* 43 Maltese Blood Oranges. **1892** *Granta* 14 May 321/1 The piles of blood oranges were diminishing rapidly. **1864** Jerdon *Birds India* III. 522 *Ithaginis cruentus*..the

Green Blood-Pheasant. **1884** *Encycl. Brit.* XVII. 341/1 Among the birds [of Nepal] are .. the blood pheasant (*Ithaginis cruentus*),..&c. **1964** E. P. Gee *Wild Life India* xiv. 117 All along that magnificent country in Sikkim, Bhutan and N.E.F.A...Marvellous birds, including horned, monal and blood pheasants, live up there. **1881** *Syd. Soc. Lex.* I, *Blood pictures*, the network formed by the adhesion of the red corpuscles to each other on a slide under the microscope, and supposed to be of a different pattern in human blood to that formed in the blood of other animals. **1908** *Practitioner* Feb. 234 The blood-picture of pernicious anaemia is presented. **1961** *Lancet* 16 Sept. 617/1 The blood-picture returned gradually to normal in seven days. **1889** *Buck's Handbk. Med. Sci.* VIII. 626/2 Blood-plaques, methods of studying. **1907** *Practitioner* Aug. 195 The alkalinity of the inorganic constituents of the blood-plasma is increased in cancer. **1945** *Daily Tel.* 3 July, Administering penicillin and blood plasma. **1885** *Buck's Handbk. Med. Sci.* I. 554/1 The blood-plates are not products of the degeneration of white corpuscles. **1898** W. S. L. Barlow *Gen. Pathol.* 153 The number of blood-platelets in normal blood has been variously estimated from 180,000 to 500,000 per cubic millimetre. **1962** *Lancet* 22 Dec. 1316/1 The part that blood-platelets play in stopping bleeding from an injured blood-vessel is now well established. **1863** *Illustr. Times* 17 Oct. 243 The alleged cases of 'blood-poisoning' in Bethnal Green. **1886** *Encycl. Brit.* XXI. 666/2 After a wound..blood-poisoning may occur. **1874** Garrod *Mat. Med.* (ed. 4) 123 Small doses raise the blood-pressure. **1888** *Encycl. Brit.* XXIV. 97/2 The blood-pressure gradually diminishes from the heart to the periphery. **1919** Wodehouse *Damsel in Distress* iv, His blood-pressure at a far higher figure than his doctor would have approved of. **1916** D. H. Lawrence *Lett.* (1962) I. 492 We have read the 'Cavalleria Rusticana': a veritable blood-pudding of *passion*! **1898** *Daily News* 15 Nov. 8/4 Ryan [i.e. a boxer] kept to work at his little target over the blood-pump. **1902** *Encycl. Brit.* XXX. 379/1 For the purpose of his researches on the gases in the blood, he [sc. C. F. W. Ludwig] designed the mercurial blood-pump. **1935** *Mind* XLIV. 438 This blood-purge in philosophy must rest..with the logical positivists. **1959** *Encounter* July 78/1 The murderous days of the blood-purges. **1838** Loudon *Arbor. et Fruticetum Brit.* I. 396 The kinds are, the common, Seville, and blood-red orange. **1868** W. James *Let.* 4 Mar. (1920) I. 136 The sausages (liver sausages, blood sausages, and more). **1965** *House & Garden* Jan. 60 Some blood sausages are cooked and cured and ready to eat; others must be grilled. **1933** E. E. Evans-Pritchard in *Africa* VI. 370 The term *nakurëmi*, my blood-sister, is occasionally used for the wife of a blood-brother. **1957** V. W. Turner *Schism & Continuity* x. 307 A special form of friendship similar to blood-brotherhood and blood-sisterhood. **1895** *Humanity* Oct. 58 If a poll could be taken, we believe that blood sports would be condemned by a larger number of persons than could be mustered on any other humanitarian issue. **1937** *Discovery* July 226/1 He has tended to limit his blood-sports. **1873** T. H. Green *Introd. Pathol.* (ed. 2) 109 The reproduction of the malignant growth in distant tissues is..owing to the entry of its elements into the blood-stream. **1913** *Field* 30 Aug. 493/3 Infection does not impair the health of the cow...The responsible organism gets into the blood stream both by the alimentary and genital tracts. **1928** *Daily Express* 6 June 5/3 Alkaline citrates, which are changed into alkaline carbonates in the blood stream. **1834** Youatt *Cattle* xi. 356 Inflammatory fever..is termed..blood-striking, shewt of blood, &c. **1861** Blood-striking [see Striking *vbl. sb.* 2 b]. **1927** Haldane & Huxley *Anim. Biol.* vii. 156 A dose of 100 grains will make the blood-sugar rise above 0.17 per cent. **1961** *Times* 22 Dec. 5/7 Insulin reduces the blood-sugar level. **1890** H. P. Hughes *Philanthropy of God* v. 75 France is the mother of Conscription. What has she gained by that blood-tax? **1901** Shee *Briton's 1st Duty* 250, I appeal to the working men of Great Britain..not to be misled by catch-phrases about 'the liberty of the subject' and the ridiculous cant about a 'Blood Tax'! **1910** W. James *Mem. & Stud.* (1911) xi. 291 They would have paid their blood-tax, done their own part in the immemorial human warfare against nature. **1912** Kipling *As Easy as ABC* 4 Democracy is Disease. I've proved it by the blood-test, every time. **1969** *Times* 27 Jan. 10/8 Blood tests will be required from all except fully credited herds. **1960** *Farmer & Stockbreeder* 2 Feb. 132/1 All our chicks are bred from 100% blood-tested breeding stock. **1879** E. A. Schafer in *Med. Times & Gaz.* 702 (*title*) Report..to determine..by what methods the operation of blood-transfusion may best be performed. **1916** *Lancet* 2 Sept. 429/2 (*title*) Employment of blood transfusion in war surgery. **1943** *Ann. Reg. 1942* 354 The blood transfusion service revealed interesting differences in the blood grouping of people of various localities. **1958** *Times* 1 July vii/3 British agriculture has long benefited by money earned in industry being reinvested in the form of capital equipment for farm land. It may be that forestry will benefit..by a similar blood transfusion. **1861** in F. Moore *Rebellion Rec.* (1862) I. III. 73/1 'Blood-tubs' and 'Plug-Uglies', and others galore, Are sick for a thrashing in sweet Baltimore. **1910** A. Bennett *Clayhanger* II. ii. § 1, The Bulgarian Atrocities had served to give new life to all penny gaffs and blood-tubs. **1963** *Times* 25 Apr. 16/2 In its 145 years of life..making successive come-backs as a burlesque house, a home of melodrama, a notorious blood-tub, and a centre of moral uplift. **1932** *Jrnl. Amer. Med. Assoc.* XCIX. 2129/1 There are particular features in the distribution of the blood types among athletes. **1928** *Jrnl. Laboratory & Clin. Med.* XIII. 774 (*caption*) Blood-typing plate. *Ibid.* 774 The macroscopic slide agglutination method of blood typing is considered. **1960** *Farmer & Stockbreeder* 8 Mar. 101/1 Blood typing, protein typing and the use of genetic markers may revolutionize sire selection in the seventies. **1915** *Jrnl. Exper. Med.* XXII. 213 When the blood urea remains constant the rate times the square root of the concentration in the urine remains constant. *Ibid.* 231 A comparison of the chloride results with the blood urea figures. **1966** *Lancet* 24 Dec. 1384/2 Blood-urea 55 mg. per 100 ml. **1922** *Flight* XIV. 34/2 'The old blood wagon', as the air ambulance..was generally called. **1957** Stirling Moss *In Track of Speed* vi. 82 Out came the 'blood wagon' and back to the ambulance station in the paddock I went. **1969** R. Petrie *Despatch of Dove* III. 152 A full-length stretcher sledge. A blood-wagon, as the laconic skiers dub it. **1951** E. E. Evans-Pritchard *Kinship among Nuer* iii. 98 Blood-wealth, the payment of cattle in compensation for homicide.

1724 in F. Moore *Trav. Inner Afr.* (1738) 267, I shall now describe the *Pau de Sangue*, or Bloodwood, so call'd from a Red Gum which issues from it.

blood, *v.* Add: **3. b.** To smear the face of (a novice at hunting) with the blood of a fox, etc., after the kill. Also *transf.* and *fig.* So **bloo·ding** *vbl. sb.*

1908 L. C. R. Cameron *Otters & Otter-Hunting* xvi. 198 *Blooding*,..marking a boy or girl on the brow and cheeks with a small piece of Otter's flesh to 'enter' him or her to the sport. **1922** M. Arlen *'Piracy'* I. iv. § 4. 55 He rolled and wallowed in it, he let life 'blood' him. **1926** J. Fairfax-Blakeborough *Hunting & Sporting Reminisc. H. W. S. Lowndes* ii. 8 Harry Lowndes was blooded to hounds at the early age of four and a half..by Tom Whitmore, the celebrated Oakley huntsman. **1927** W. E. Collinson *Contemp. Eng.* 37 A year or so ago..I met the word blooding in an illustrated weekly, depicting the ceremony of smearing with the fox's blood a girl taking part in her first hunt. **1929** J. Masefield *Hawbucks* 93 Carrie had had the brush and been blooded. **1955** *Times* 24 May 15/5 These largely mining areas are of little use to Conservatism save as battle courses for blooding new candidates. **1959** *20th Cent.* Dec. 484 Lee is, of course, blooded—'he criss-crossed her cheek thoroughly'. Blooding is an invariable feature of pony books.

5. To apply a coat of blood to (leather) in leather-colouring, in order to obtain a good black.

190. *Mod. Amer. Tanning* 110 (Cent. Dict. Suppl.).

6. *intr.* with *it*: to play the 'blood' (see *Blood *sb.* 15 c).

1922 Joad *Highbrows* v. 179 When I wasn't 'blooding' it with the second-year men..your scout..used to bring your lunch down into my rooms.

bloo·d-horse. Chiefly *U.S.* [Blood *sb.* 12 b.] A thoroughbred or pedigree horse.

1794 *Sporting Mag.* IV. 35/2 The various judicious crosses that have brought the breed of blood horses into such a state of unprecedented perfection. **1800** [see Blood *sb.* 12 c]. **1841** H. S. Foote *Texas & Texans* II. 383 They [sc. wild horses]..are..inferior to the American blood-horse in volume of muscle. **1856** Mrs. Stowe *Dred* II. 145 The fleet blood-horse was whirling Harry and Lizette past bush and tree. **1860** O. W. Holmes *Prof. Breakf.-t.* iii. 65 Admiral Sir Issac Coffin sent out two fine blood-horses.

bloo·dhound, *v. rare.* [f. the sb.] *trans.* To pursue ruthlessly.

1935 G. Barker *Janus* i. 15, I am by blood bloodhounded out of doors. **1961** *Guardian* 25 May 10/4 Naively imagining that MI 5 was only bloodhounding those with supposed Cliveden or Mosley ideas.

blood-relation. (Earlier example.)

1709 W. Steuart *Coll. conc. Govt. Church Scotl.* II. iv. § 8, Those of our Houshold and Blood-Relations, as Husband with Wife, and Father with Children or the like.

blood-stain, *sb.* (Earlier example.)

1820 Shelley *Prometheus Unbound*, etc. 194 Hide the blood-stains now.

bloodstock (blʌ·dstɒk). [f. Blood *sb.* 12 c + Stock *sb.*¹ 54.] Thoroughbred or pedigree horses collectively. Also *attrib.*

1830 *Sporting Mag.* 2nd Ser. I. 333/2 We cannot trace the pedigrees of a large proportion of our blood stock up to foreign mares. **1888** W. Day *Horse* 9 Those engaged in breeding blood-stock only. **1909** *Westm. Gaz.* 15 July 12/2 Messrs. Tattersall's bloodstock sales at Newmarket. **1912** (*title*) The bloodstock breeder's review.

blood-sucker. Add: **1.** *spec.* **a.** A lizard belonging to the species *Lacerta cristata*, the individuals of which change their colour, especially about the neck, from grey to dark red. **b.** A lizard of the genus *Amphibolurus* (*Grammatophora*), esp. of the species *A. muricata*.

a. 1819 J. Morton *Poet. Rem. J. Leyden* p. xc, A large lizard, termed a blood-sucker. **1882** *Encycl. Brit.* XIV. 736/2 *Calotes* is another genus of agamoids peculiar to the East Indies; it comprises numerous species well known in India by the name of 'blood-suckers'. **b. 1852** L. A. Meredith *My Home in Tasmania* II. 37 Another description of lizard is here vulgarly called the 'blood-sucker'. **1886** F. McCoy *Zool. Victoria* II. XII. 47 Grammatophora muricata...The Blood-sucker.

bloody, *a.* and *adv.* Add: **A.** *adj.* **10.** (Additional examples, esp. as a mere intensive.)

1785 *Fifth Session Old Bailey* May 772/1 The prisoner Fennell swore an oath, if he had a knife he would cut his bloody fingers off. **1950** *Landfall* IV. 23 You mind your own bloody business. **1950** G. Wilson *Brave Company* v. 83 Thrilling? Hell's bloody bells!

b. Bad, unpleasant, deplorable; perverse. Cf. *bloody-minded *a.* 2. Hence as *sb.*, an unpleasant person.

1934 *Neuphilologische Mitteilungen* XXXV. 129 Modern slang (e.g. *bloody* as in *Jones is bloody* 'Jones is objectionable', [etc.]). **1936** R. Lehmann *Weather in Streets* ii. 48 He developed my nastiness from a mere seed into a great jungle. He made me so mean and bloody...Well, I just am a bloody character, I suppose. **1939** R. W. Chambers *Man's Unconquerable Mind* xii. 391 'It's bloody,' I said. 'To call it bloody,' Ker replied, slowly and sadly, 'is fulsome flattery.' **1954** A. Heckstall-Smith *Eighteen Months* xiii. 164 Why go out of your way to be bloody about Archie when I'm trying to help him? **1960** D. Potter *Glittering Coffin* vii. 106 A few bloodys were provoked into their usual braying. **1960** *20th Cent.* Nov. 492 The new generation of college bloodies out on the cobbles.

B. *adv.* **2.** (Later examples, esp. as a mere intensive.) Also in tmesis.

1801 M. Edgeworth *Belinda* I. vii. 208 Sir Philip writes a *bloody* bad hand. **1914** Shaw *Pygmalion* III, *Liza.* Walk! Not bloody likely. (*Sensation*). I am going in a taxi. **1923** Manchon *Le Slang* 65 Half bloody dead...stoney bloody broke..like any-bloody-thing..a handi-bloody-cap ..hoorah! hoo-bloody-rah! **1935** Abso-bloody-lutely [see *absolutely *adv.* ¶]. **1937** 'J. Bell' *Murder in Hospital* viii. 156 I've always thought her a bloody awful great brute. **1951** E. Taylor *Game of Hide-and-Seek* I. iii. 61 Serve them bloody right. **1953** —— *Sleeping Beauty* ix. 156 You bloody know you didn't. **1963** L. Meynell *Virgin Luck* v. 101 Remember the *News Chronicle*?..On sale one day. Amalga-bloody-mated the next.

C. 2. bloody back *U.S. slang*, a contemptuous name for a British soldier (*Obs.* exc. *Hist.*); cf. Bloody-*backed*, Lobster¹ 3 a, Redcoat 1 a; **Bloody Mary**, a drink containing vodka and tomato juice; **bloody-nosed beetle** (see quots.).

1770 *Mass. Gazette Extraord.* 21 June 2/2 You Rascals, you bloody Backs, you Lobster Scoundrels; fire if you dare. **1962** *Times* 24 Apr. 12/7 We dislike hearing the Mall echo to the notes that 'bloodybacks' heard. **1956** *Punch* 15 Aug. 191/1 Those two..are eating raw steaks and drinking Bloody Marys. **1961** *House & Garden* Apr. 93/2 Vodka is..mixed with tomato juice ('Bloody Mary'). **1880** *Encycl. Brit.* XIII. 150/1 *Timarcha* (the Bloody-nosed Beetle). **1959** E. F. Linssen *Beetles Brit. Isles* II. 89 A species..which owing to its method of 'bleeding' is called the Bloody-nosed Beetle.

b. bloody cardinal = Cardinal-flower.

1851 D. G. Mitchell *Dream Life* (1852) 199 The bloody cardinal of the swamp-lands.

bloo:dy-mi·nded, *a.* [Bloody *a.* C. 1.] **1.** Inclined to bloodshed; bloodthirsty, cruel.

1584 R. Greene *Gwydonius* f. 57ᵛ, I will neither bee so bloudie minded as to breede thy bane. **1593** [see Bloody *a.* and *adv.* C. 1]. **1647** Beaumont & Fletcher *Lawes of Candy* v. 67/2 She is bloudy minded, And turnes the justice of the Law to rigor. **1739** Hume *Hum. Nat.* II. ii. § iii, He is bloody-minded, and takes a pleasure in death and destruction. **1845** [see Bloody *a.* C. 1]. **1920** R. Macaulay *Potterism* II. i. 58 To make out that all our combatants were full of sweet reasonableness .. and all our non-combatants bloody-minded savages.

2. Perverse, tiresome, cantankerous; stubbornly intransigent or obstructive.

1935 J. Agate in *Sunday Times* 17 Mar. 6/2 A man says to a presumed lady, 'What a bloody-minded woman you are!' **1944** 'N. Shute' *Pastoral* vii. 148 They'll lose this fishing, and they'll all get bloody-minded about that. **1959** *Spectator* 1 May 604/1 The building unions..have never been as bloody-minded about demarcation as the shipbuilders and others.

bloo:dy-mi·ndedness. [f. prec. + -ness.] **1.** Inclination to bloodshed; bloodthirstiness, cruelty.

1789 G. Vassa *Life* I. v. 218 By the 329th Act, page 125, of the Assembly of Barbadoes, it is enacted 'That..if any man shall out of wantonness, or only of bloody-mindedness..wilfully kill a negro..he shall pay into the public treasury fifteen pounds sterling'. **1822** Carlyle *Let.* July (1886) II. 108 The persecution of the Albigenses has little to distinguish it from other persecutions..except a darker tinge of bloody-mindedness. **1870** [see Bloody *a.* C. 1].

2. Perversity, contrariness, cantankerousness.

1910 *Daily Chron.* 15 Mar. 6/7 Sparrows..do not eat the flowerets of the larch, but merely peck them off out of sheer devilry and 'bloodymindedness'. **1953** R. Lehmann *Echoing Grove* 69 They all agreed, too monstrous, your bloody-mindedness, insufferable rudeness to me. **1959** *Spectator* 19 June 875/1 We can only hope that the intransigence so far shown by both sides [in the printing dispute] will not develop into bloody-mindedness, as such disputes sometimes do.

blooey (blū·i), *a.* *U.S. slang.* Also **blooie.** [Etym. unknown.] Awry, amiss; usu. in *to go blooey*. Also as *int.*

1920 G. Ade *Hand-made Fables* iii. 30 The Orders stopped coming. Collections went blooey. **1926** Whiteman & McBride *Jazz* ii. 32, I spilled the salt. It rained. At rehearsal my fiddle went blooey. **1928** [see *G III. f]. **1929** *Collier's* 5 Jan. 40/4 But in 1920 I weakened. As a result, my bank roll went blooie. **1948** J. Blish in D. Knight *Dark Side* (1966) 40 One minute I was..all set to shoot him and get my wife out of there; and then blooey! **1961** J. Updike *Rabbit, Run* 142 A clear image suddenly in the water wavering like a blooey television set.

bloom, *sb.*¹ Add: **4. c.** Also, the cloudy appearance on a varnished surface (= Blooming *vbl. sb.*¹ 2).

1886 in Ogilvie *Imperial Dict.* **1953** *Gloss. Paint Terms* (*B.S.I.*) 6 *Bloom*, a thin film which sometimes forms on glossy paint or varnish films, thereby dimming their lustre or veiling their depth of colour.

d. (See quots.)

1908 *Animal Managem.* 134 In no case should..the work [be] so hard, that the colt's 'bloom', *i.e.*, his round and glossy sleekness, disappears. **1952** C. L. B. Hubbard *Pembrokeshire Corgi Handbook* 108 *Bloom*, glossiness or good sheen of coat.

e. = *water-bloom.

1939 B. M. Griffiths in *Proc. Linn. Soc.* CLI. 13 In 1838 Drummond gave a description of the greenish bloom on Glaslough, Co. Monaghan. **1948** *Hydrobiologia* I. 1 About 65 species [of algae] are known to cause 'blooms' in lakes. **1957** *Penguin New Biol.* XXIII. 87 At least two early chroniclers have recorded that the occurrence of a red bloom..forewarned of the untimely death of William Rufus. **1969** *Daily Colonist* (Victoria, B.C.) 16 May 32/1 Dr. Parsons said that although the deep Pacific was uneconomical, similar plankton blooms had been found at the juncture of the Fraser River and the sea.

8. The perfume exhaled from wine, bouquet.

1888 *Encycl. Brit.* XXIV. 602/2 The smell common to all wines (which remains in an empty wine cask after the bloom proper has gone).

bloomed, *a.* Add: **2.** Of a photographic lens: covered with a 'bloom'; see *blooming *vbl. sb.*¹ 4.

1945 A. Cox *Optics* (ed. 4) 224 This latter [*sc.* straining] gives not a bloomed effect to the surface of the lens but a *frosted* effect. **1957** Amos & Birkinshaw *Television Engin.* I. ix. 190 One way of reducing the reflection at lens surfaces is by coating them with a very thin film of silica or evaporated fluorite...Lenses treated in this way are termed *coated* or *bloomed* and have a characteristic blue appearance.

bloomer¹. Add: **3.** [for *blooming letter*: see *blooming *ppl. a.* 6.] A floriated initial letter of the alphabet.

1899 Mackail *W. Morris* II. 256 The large floriated initials, or 'bloomers', in the slang of the press. **1901** *Sunday Times* 21 Apr., This book will be in the new type, with a fresh set of 'bloomers' specially designed for it.

4. *slang.* [See quot. 1889.] A very great mistake: chiefly in phr. *to make a bloomer.*

1889 Barrère & Leland *Dict. Slang*, *Bloomer* (Australian), prison slang for a mistake. Abbreviated from the expression 'a blooming error'. **1902** *Westm. Gaz.* 25 Nov. 8/1 The defendant replied, 'You have made a bloomer, old chap; you never made a greater mistake in your life.' **1928** *Daily Express* 19 Apr. 15 His mount, Clearmount, made a terrible bloomer at the last fence. **1959** *Economist* 6 June 920/1 'The Times'..has this week made a bloomer about a minister.

bloomer². Add: **1.** (Earlier examples.)

1851 *Boston Transcript* 27 May 2/3 (Th.), The *Bee* says the daughter of Dr. Hanson, of this city, appeared in the Bloomer suit..last week. **1851** *Ibid.* 29 May 2/4 (Th.), The first 'Bloomer' made its appearance in our city yesterday. *Ibid.*, Quite a large number of young ladies in that city have made arrangements to attend church tomorrow in the Bloomer costume. **1853** 'Old Etonian' *Alphabet Annotated* 61 Until they all the Bloomer dress assume. **1855** *Kansas Tribune* (Th.), Perhaps Lawrence [Kansas] is the only city in America where the majority of the ladies wear Bloomers.

b. Regularly in *pl.* Loose trousers reaching to the knee or knickerbockers worn by women for bicycling, gymnasium practice, etc.; called also 'rational dress'. Also, a woman's knee-length undergarment (the usual sense in later usage).

In quots. 1862, 1863, a man's attire.

1862 Dickens *Somebody's Luggage* in *All Year Round* (Christmas) 9/1 A corporal of his country's army, in..the line of his waist, the broadest line of his Bloomer trousers, and their narrowest line at the calf of his leg. **1863** *Miss Jemima's Swiss Jrnl.* (1963) i. 7 Soldiers in many uniforms, some of them most laughable, their capacious bloomer costume reaching to the centre of the calf of the leg. **1895** *Westm. Gaz.* 25 July 8/1 Female teachers who have been riding bicycles in male attire, commonly called bloomers. **1906** Galsworthy *Man of Property* III. i. 289 She could tell you..what they were doing in Paris about wearing bloomers. **1909** *Daily Chron.* 5 Aug. 7/3 [The hockey girl] in her short skirts, bloomers, flannel blouse, and cloth cap. **1967** *Observer* 30 Apr. 29/3 The tireless search for something else has paused for a moment to put matching bloomers or shorts under the skirt.

2. (Earlier examples.)

1851 *Punch* 30 Aug. 103/2 If women assume the dress of men, let them undertake men's duties; hence, every Bloomer shall be liable to be drawn for the militia, without benefit of substitute. **1853** R. S. Surtees *Mr. Sponge's Sporting Tour* xliii. 238 (*heading*) A Literary Bloomer. *Ibid.* 239 Lucy,—a young lady of a certain age—say liberal thirty—an ardent Bloomer.

Hence **bloo·mered** *a.*, wearing bloomer costume; **bloo·merism** (earlier example); **bloo·merize** *v. trans.*, to dress in bloomers; also *fig.*

1895 *Advance* (Chicago) 6 June 1287 The 'new woman',..though not necessarily of the bloomered type, is marching..to victory in Wisconsin. **1900** *Daily News* 4 Dec. 6/5 In these days of Lady Wranglers and bloomered 'sportswomen'. **1907** *Westm. Gaz.* 2 Oct. 1/3 The outer lines were marked by companies of..bloomered Tirailleurs. **1960** *Times* 27 Sept. 13/6 A bloomered girl in the foreground. **1851** (*title*) 'Bloomerism', or The New Female Costume of 1851. **1885** Mrs. Lynn Linton *Chr. Kirkland* III. i. 18 Doubtful whether they were girls Bloomerized or boys in feminine tunics. **1897** *Daily News* 3 Feb. 6/4 The New Woman..has fairly bloomerised journalism in the United States.

bloo·mful, *a.* [f. Bloom *sb.*¹] Rich in bloom.

1890 *Lippincott's Mag.* Mar. 392 Bloomful maidens. **1913** Mrs. G. S. Porter *Laddie* xv, Then she leant toward me all wavery, and shining eyed, and bloomful.

bloominess (blū·min'ɛs). [f. Bloomy *a.*¹ + -ness.] The condition of being covered with bloom or having a bloom-like surface.

1889 'Theo Gift' *Not for Night-time* 75 A canary-coloured gown, with a white bloominess on the edges of it. **1895** *Westm. Gaz.* 22 Apr. 3/1 In the lanes where the woodbine leaves are showing the brown bloominess on the backs of their unfolding leaves.

blooming, *vbl. sb.*¹ Add: **3.** *Television.* (See quots.)

1940 D. G. Fink *Princ. Telev. Engin.* iii. 72 The brightness contrast is limited by halation and saturation of the luminescent screen and by the defocusing effect ('blooming') associated with the electron gun when large signals are impressed upon it. **1945** N. M. Cooke *Electronics Dict.* 37/2 *Blooming*, a fuzzy effect in a reproduced television picture.

4. *Photogr.* Coating with a 'bloom'; the process of coating a photographic lens with a metallic fluoride in order to reduce surface reflection.

1943 A. Cox *Optics* 267 In the cases of some older lenses which had acquired a *bloom* on their surfaces with the passing of time the amount of reflection was cut down at each of these surfaces. But the blooming could not be produced regularly as a practical proposition. **1944** *Photogr. Jrnl.* LXXXIII. 225/2 It is on account of this colour that the process is often called 'blooming'. **1957** T. L. J. Bentley *Man. Miniat. Camera* (ed. 5) v. 59 The 'blooming' treatment gives to the surface of new lenses an even purple coating by deposition of magnesium fluoride, to a minute and finely adjusted thickness.

blooming, *vbl. sb.*² *attrib.* Add: **blooming machine, blooming rolls** (see quots.).

1869 *Spons' Dict. Engin.* I. 367 The blooming machine, invented by Jeremiah Brown..consists of three large eccentric rolls..placed horizontally in the strong holsters.., the centres of the rolls being arranged in a triangular position, and the bottom roll..being nearly central between the two top rolls. **1888** *Lockwood's Dict. Terms Mech. Engin.*, *Blooming Rolls*, see Puddling Rolls...*Puddling Rolls*, or *Forge Train*, the first set of rolls through which a shingled bloom is passed.

blooming, *ppl. a.* Add: **6.** *blooming (initial) letter*: a floriated initial letter of the alphabet; = *bloomer¹ 3.

1713 Jas. Watson in *Hist. Printing* Pref. 9 Curiously cut Head-Pieces, Finis's, Blooming-Letters [etc.]. **1785** W. Herbert *Typogr. Antiq.* 527 The chronicle of each reign begins with a blooming letter. **1902** Sayle *Init. Lett. Early Eng. Printed Bks.* in *Trans. Bibliogr. Soc.* VII. 18 The bloomers or blooming initials as Herbert calls them.

Bloomsburian (blūmzbiū·riăn). [f. *Bloomsbury* (see below) + -an.] A dweller in Bloomsbury, London; *spec.* = *Bloomsburyite.

1902 *Daily Chron.* 24 Nov. 4/4 The flippant playwright who made one of his characters say that 'the Bloomsburians live mainly on a dish called "smoked 'addick"'. **1927** *Weekly Dispatch* 23 Oct. 11/2 A horde of minor Bloomsburians. **1967** *Guardian* 4 Oct. 7/2 A Jacques like some last of the Bloomsburians in a faded white alpaca suit.

Bloomsbury (blū·mzbĕri). A school of writers and aesthetes living in or associated with Bloomsbury (see prec.), that flourished in the early 20th century; a member of this school. Also *attrib.*, associated with, or having similar intellectual pretensions to, the Bloomsbury school. So *Bloomsbury group, set*, the above-mentioned school of writers; *Bloomsbury gang*, a political party that appeared in July 1765, led by the 4th Duke of Bedford, and otherwise known as 'the Bedford party'. Hence **Bloo·msburyite.**

1910 *Encycl. Brit.* III. 619/1 Bloomsbury gang. **1914** J. M. Keynes *Let.* 2 July in R. F. Harrod *Life* (1951) iv. 171 She..is asking no one but a few of my so-called 'Bloomsbury set'! **1920** D. H. Lawrence *Let.* 7 May (1962) I. 628 Nothing will happen to the world: Bloomsbury will go on enjoying itself in Paris and elsewhere. **1928** A. P. Herbert *Trials of Topsy* 8 My dear *too* Bloomsbury for anything, and such *forests* of hair both of them. **1931** R. Campbell *Georgiad* i. 12 It was a voice of 1930 model And in a Bloomsbury accent it could yodel. **1933** *Times Lit. Suppl.* 2 Mar. 148/4 The descriptions..of young earnest and well-meaning Bloomsburyites..could not be bettered. **1948** Wyndham Lewis *Let.* 9 Oct. (1963) 460 It was written by Raymond Mortimer—an old Bloomsbury. **1951** R. F. Harrod *Life of J. M. Keynes* v. 184 The Bloomsbury voice was a distinct contribution. It was based on Lytton Strachey's, consisting not so much in a special pronunciation of words as in the cadences of sentences. **1954** J. K. Johnstone (*title*) The Bloomsbury Group.

bloop (blūp), *sb.* [Echoic.] **a.** A plopping sound; a howling sound. **b.** A blooping patch that prevents such a sound (see next).

1931 L. Cowan *Recording Sound for Motion Pictures* 360 *Bloop*, dull thud emitted in sound reproduction, due to a poorly made Blooping Patch. **1953** K. Reisz *Technique Film Editing* 278 *Bloop*, small opaque patch over a splice in a positive sound-track designed to smother any intrusive noise which the splice might otherwise produce. **1962** A. Nisbett *Technique Sound Studio* xii. 219 Sounding like a series of pips, beeps, and bloops.

bloop (blūp), *v.* [Echoic.] **a.** *intr.* To make a howling noise; to operate a radio set in such a way that it emits such a noise. **b.** *trans.* To cover a splice on a sound-track with a blooping patch (see quot. 1931²). So **bloo·ping** *vbl. sb.*, the process of covering such a splice.

1926 *Radio Rev.* Jan. 23 That neighbor who constantly 'bloops' and sends out his cat-calls just at the time all ears are focussed on the umpth symphony of Rubenthoven. **1927** *People's Pop. Monthly* Mar. 281 You can make any type of receiver regenerate or 'bloop' to more or less extent. **1931** L. Cowan *Recording Sound for Motion Pictures* 360 *Bloop*, to prepare a Blooping Patch. *Ibid.*, *Blooping Patch*, triangular or oval black section introduced over a splice in the positive sound track, to prevent the noise which the splice would otherwise cause. **1934** [see *blooper 1]. **1936** *Words* Oct. 6/1 Blooping is dying a film splice with ink sprayed from an airgun so that the splice will run through the projectors without making unnecessary noise. **1959** W. S. Sharps *Dict. Cinematogr.* 80/2 *Blooping ink*, the ink used to paint a sound track in blooping. **1960** O. Skilbeck *Film & TV Working Terms* 19 Dubbing Tracks printed from an Edited sound negative are often electrically blooped by the Labs. with a device which automatically brings a special fogging light up and then down at joins.

blooper (blū·pəɹ). *colloq.* [f. *bloop *v.*] **1.** (See quots.)

1926 *Dairy Farmer* (Des Moines, Iowa) 15 May 12 It is this type of set which is commonly criticized for causing interference and when improperly operated soon becomes known as a 'blooper'. **1934** Webster, *Blooper*, a radio receiving set which generates current of radio frequency, thus causing radiation from the receiving antenna. This radiation, under certain circumstances, causes near-by sets to bloop or howl.

2. A blunder or howler, esp. of a public or politically embarrassing kind; = *bloomer¹ 4.

1947 *Partisan Rev.* XIV. 550/2 Why does Farrell write so badly? Why in his latest book such bloopers as 'art..presents *imaginary images* of life'. **1951** in Wentworth & Flexner *Dict. Amer. Slang* (1960) 44/2 But on one subject—'security'—he may have felt he pulled a political blooper. **1956** *Amer. Speech* XXXI. 252 In radio and television broadcasting, where the spoken word is of utmost interest and importance, lapses are known as *bloopers, fluffs, slips* and *boners*. **1961** *Daily Tel.* 28 Jan. 11/5 The Administration had made a 'blooper' over the custom of allowing members of Congress to provide constituents with guided tours of the White House.

blossom, *sb.* Add: **1. b.** Also said of grain, grass, etc.

1769 Washington *Diaries* (1925) I. 327 The head [of the wheat] was shot out, and in many places in Blossom. **1869** *Rep. U.S. Commissioner Agric. 1868* 423 F. V. Stewart..states that his practice has been to commence cutting the grass when in the 'second blossom'.

3. a. (Further example.) Also, radiated quartz. *U.S.*

1819 H. R. Schoolcraft *View of Lead Mines of Missouri* 71 This variety of quartz has acquired the popular name of blossom of lead, or mineral blossom. *Ibid.* 91 In searching for ore, the soil, the slope of the hills, spar, blossom, trees, etc., are taken as guides. **1882** J. H. Beadle *Western Wilds* (ed. 2) xxxiv. 560 We are certainly near the outcrop from which the 'blossom' was broken.

4. blossom-rock *U.S.* (see sense 3 a).

1871 *Colorado Gaz.* 156 The 'blossom rock' (quartz stained with metallic oxides), which indicates the proximity of mineral deposits, differs but little in gold and silver lodes. **1878** J. H. Beadle *Western Wilds* xxx. 479 Men were let down from above to 'prospect', a crevice was found with blossom rock.

blossom, *v.*¹ Add: **2. c.** Const. *out.*

1894 G. du Maurier *Trilby* I. II. 162 He..blossomed out into beautiful and costly clothes..so that people would turn round and stare at him. **1929** J. B. Priestley *Good Compan.* II. vii. 447 Very shortly, you'll see, I shall be Blossoming Out—and then I shall expect a cable from S. Africa. **1947** Auden *Age of Anxiety* (1948) ii. 39 Black umbrellas blossom out.

blossomry (blǫ·səmri). [f. Blossom *sb.* + -ry.] Blossoms collectively.

1901 Abp. of Armagh in *Westm. Gaz.* 3 Jan. 8/2 Such souls predestined have not one high range, One climate's blossomry alone.

blot, *sb.*¹ Add: **1. a.** *spec.* A set of ink-blots made on a piece of paper as a basis for the composition of an imaginary landscape, according to the technique invented by Alex. Cozens (d. 1786). (Cf. *Rorschach.) Also *attrib.*

a **1786** A. Cozens *New Method in Drawing Landscape* 7 A blot is an assemblage of dark shapes or masses made with ink upon a piece of paper. All the shapes are rude and unmeaning, as they are formed with the swiftest hand. But at the same time there appears a general disposition of these masses, producing one comprehensive form. **1931** *Times* 24 Mar. 19/6 Examples of Cozens's 'blots' have long been known. But it was only the other day that five blots accompanied by the five drawings made from them, were discovered. **1962** *Listener* 19 July 95/2 The manipulation of accident in the blot landscapes of Alexander Cozens.

c. Esp. in phr. *a blot (up)on the landscape*; also *fig.*

1912 T. E. Lawrence *Let.* 20 Feb. (1938) 137 His two Kufti people..will be rather a blot on the landscape. **1960** Wodehouse *Jeeves in Offing* i. 8 'And a rousing toodle-oo to you, you young blot on the landscape,' she replied cordially. **1962** *Listener* 11 Jan. 90/2 Charabancs and monstrous hordes of hikers are blots upon the landscape.

d. *Painting.* (See quot.)

1910 *Edin. Rev.* Apr. 371 Painters are accustomed to speak of the 'Blot' of a picture, meaning its immediate appearance as colour, line, massing, or flat space.

2. a. Also in extended use: a person who is a disgrace (Webster, 1909).

1938 E. Bowen *Death of Heart* I. iii. 60 Those four chaps were a blot. **1961** M. Kelly *Spoilt Kill* ii. 116 The silly blot forgot to switch on.

blotch, *sb.* Add: **1. c.** A disease of fruit or leaves, characterized by the formation of spots; *sooty blotch*, a disease of the apple.

1909 *Cent. Dict.* Suppl.

blotter. Add: **4.** A record of arrests and charges in a police office; a charge-sheet; also *gen.* a record-book or list. *U.S.*

1887 *Harper's Mag.* Mar. 500/2 Every item of police duty, and of civil or criminal occurrence, is inscribed on the 'blotter'. **1906** *Atlantic Monthly* Feb. 264 It was necessary..to examine the day-book or blotter in the chief clerk's office [at the Patent Office]. **1910** *Washington Times* 14 Dec. 1 Three more additions were made yesterday to the hospital blotters. **1926** J. Black *You can't Win* xix. 299, I never put his name, which is my name, on a police blotter or a prison register while he was alive. **1965** 'R. L. Pike' (*title*) Police blotter.

blottesque, *a.* (Earlier and later examples.)

1856 Ruskin *Mod. Paint.* IV. v. v. Pl. 27, Modern, or, Blottesque. *Ibid.* 80, I have given, Fig. 1., a Durerseque, and Fig 3. a Blottesque, version of the intermediate wall. **1866** Swinburne *Lett.* (1959) I. 165 'Guy Deverell' I think too hasty—too blurred and 'blottesque'. **1934** H. G. Wells *Exper. in Autobiogr.* I. vi. 345 My notes and comments were sometimes more blottesque than edifying. *Ibid.* II. vii. 424, I sat down..to my blottesque red corrections again.

blotting, *vbl. sb.* Add: **3. blotting-book** (examples).

1598 Florio *Worlde of Wordes* 39/3 *Bastardolo*, a blotting booke, a daie booke, a house booke to write in euery thing. **1848** H. R. Forster *Stowe Catalogue* 28, Item 453 A blotting book. **1874** H. H. Cole *Catal. Ind. Art S. Kens. Mus.* 321 A blotting-book cover.

4. *techn.* Material for blotting-paper; also, the finished article.

1872 *Eng. Mech.* 15 Nov. 228/3 [To] give to used..or dry..blotting its original absorbent power. **1880** J. Dunbar *Pract. Papermaker* 72 For pink blottings furnish two thirds of white cottons and one third of turkey reds. *Ibid.*, In this way the author has made blotting which was considered a good article. **1920** *Printers' & Stat. Yr. Bk.* 13 Calf Papers, Blottings. Enamelled Blottings.

blotting-paper. Add: (Later examples.) Also *attrib.*

1853 Mrs. Gaskell *Cranford* ix. 163 The same confused sort of way in which written words run together on blotting-paper. **1894** H. Nisbet *Bush Girl's Rom.* vii. 68 Acid with the soup, salt and warmth with the fish to intensify its brain-feeding and blotting-paper qualities. **1904** *Daily Chron.* 6 Feb. 9/1 Blotting-paper pink and turtle-dove drab. **1937** E. J. Labarre *Dict. Paper* 26/1 *Blotting paper*..is, peculiarly enough, produced from the same materials as the most expensive papers, viz: rags...It is now also largely made from chemical wood pulp, esparto, etc. **1967** E. Short *Embroidery & Fabric Collage* iv. 108 Once the work is complete it should be stretched face upwards over damp blotting paper.

blotto (blǫ·to), *a.* *slang.* [Obscurely f. Blot.] Fuddled with liquor, intoxicated.

1917 W. Muir *Observations of Orderly* xiv. 230 The words for drunkenness are innumerable—'jingled', 'oiled', 'tanked to the wide', 'well sprung', 'up the pole', 'blotto', etc. **1919** *Winter's Pie* Pl. 17. **1921** *Blackw. Mag.* Feb. 157/1 The evening's potations left him, by bed-time, in a state somewhere between 'blotto' and 'blithero-blotto'. **1923** *Daily Mail* 13 June 12, I got properly 'blotto' and don't know what I did. **1951** Wodehouse *Old Reliable* xiii. 149 Did you ever see a blotto butler before?

blouse, *sb.* Add: **1.** (Earlier example.)

1828 *Engl. in France* II. 100 (Stanford), Neither wearers of plaid, nor devourers of porridge, but *blousses* and *soupe maigre* well supplied the want.

4. A loosely-fitting bodice worn by women and girls, usually tucked inside the skirt at the waist. Also, formerly called *blouse-bodice* (cf. *blousée, *shirt-blouse*).

1870 *Young Ladies' Jrnl.* 1 Oct. 626/1 A costume for a young lady...A blouse, or tunic, of white and pink-striped silk. **1887** *Cassell's Fam. Mag.* Sept. 631/1 Frenchwomen show just the same favour to the blouse bodices as their sisters across the Channel. **1899** *Daily News* 14 Jan. 2/4 The fashionable blouse, or 'blouse-bodice'. **1909** *Westm. Gaz.* 15 Feb. 5/2 Blouses will be more in evidence this season than ever, for the mere strap that forms the bodice of the newest gowns, and the high waists, and bretelles, all demand the finishing touch of a blouse.

b. *attrib.* and *Comb.*, as *blouse-clad* adj., *-maker*, *-making* (also *attrib.*); **blouse coat**, a blouse to be worn outside the skirt at the waist; **blouse length**, a piece of material sufficient for the making of a blouse, cut off for sale; **blouse slip** (see Slip *sb.*³ 4 c); **blouse suit**, a blouse and skirt finished complete, and ready to wear.

1874, 1883 [see Dict., 3]. **1892** *Daily News* 2 July 6/7 A comparison between the blouse-clad ladies and those who wear habit-bodices. **1898** *Westm. Gaz.* 16 June 3/2 Blouses and blouse coats of Irish lace. **1910** *Ibid.* 29 Jan. 3/3 Who..would venture to buy a blouse-length with the fate of the Budget hanging in the balance? **1905** *Daily Chron.* 18 Oct. 3/5 A wife who was employed at a blouse-maker's. **1905** *Pearson's Mag.* July 105/1 The blouse-making industry. **1908** *Westm. Gaz.* 21 Aug. 8/2 Working at blouse-making. **1907** *Ibid.* 21 Feb. 8/2 How to make a Blouse-Slip. **1909** *Daily Chron.* 22 Feb. 7/5 With these cashmere gowns, blouse-slips of various kinds can be worn. **1905** *Ibid.* 20 Oct. 8/5 Blouse-Suits in Velveteens, Silks, and Various Cloths.

blouse (blauz), *v.* [f. the sb.] **1.** *intr.* To assume a blouse-like form (Webster, 1909).

1938 J. Rice *Somers Inheritance* IV. vii. 270 Her hat fell off and her hair bloused about her ears.

2. *trans.* To make (the bodice of a garment) full, like a blouse.

1934 *Times* 12 Nov. 17/5 She blouses the bodices.

bloused, *ppl. a.* Add: **2.** Of a bodice: made full, like a blouse.

1935 *Times* 11 Feb. 15/6 The bloused bodice is shirred into shape in front and at the back. **1958** *Vogue* May 122 Lanvin-Castillo's spare little dress, the top lightly bloused and buttoned behind.

blousée (blū·ze), *a.* [f. Blouse + Fr. *-ée*.] In *blousée bodice* = *blouse *sb.* 4.

1899 *Westm. Gaz.* 24 Aug. 3/2 The simple blouse, or I might say *blousée* bodice, which will probably be the favourite for indoor wear. **1903** *Ibid.* 19 Feb. 4/2 The becoming *blousée* bodice for gowns.

‖**blouson** (blūzoṅ). [Fr.] A short jacket shaped like a blouse. Also *attrib.*

1904 *Westm. Gaz.* 27 Oct. 4/2 No doubt there will be many fur coats quite short in the bolero blouson build and in the sac species. **1958** *Woman's Own* 6 Aug. 20/1 Crisply fresh at the office in a white drip-dry cotton 'blouson'. **1959** *Ibid.* 20 June 12/2 Drip dry poplin hooded blouson with drawstring waist. **1960** *Guardian* 22 July 8/4 The line leaves the shoulders in a blouson effect to end smoothly at hip-line. **1970** *Ibid.* 23 July 9/4 Suits with raglan *blouson* tops.

blousy, var. Blowzy *a.*

1922 *Blackw. Mag.* Mar. 353 That frousy, blousy, lousy tug. **1931** S. Benson *Tobit Transplanted* iii. 33 His mother's large blousy bun of hair was always coming down.

blow, *v.*¹ Add: **1. b.** Phr. *blow high, blow low*: whatever may happen. *U.S.*

1774 P. V. Fithian *Jrnl.* (1900) 235 Ben is in a wonderful Fluster lest he shall have no company to-morrow at the Dance—But blow high, blow low, he need not be afraid; Virginians..will dance or die! [**1776** C. Dibdin *Seraglio* I. 11 Blow high, blow low, let tempests tear The mainmast by the board.] *a* **1861** T. Winthrop *John Brent* (1883) vi. 52 I've booked Brother John fur Paradise; Brother Joseph's got a white robe fur him, blow high, blow low! **1923** R. D. Paine *Comr. Rolling Ocean* x. 171 There were three musketeers..who were blithely resolved to stand by each other through thick and thin, blow high, blow low.

6. a. Further examples of *Austral.*, *N.Z.*, and *S. Afr.* usage.

1873 Lady Barker *Station Amusements in N.Z.* xiv. 237, I..mention this, not out of any desire to 'blow' about our sheep. **1885** Mrs. Praed *Austral. Life* ii. 45 He was famous for 'blowing', in Australian parlance,..of his exploits. **1896** H. A. Bryden *Tales S. Afr.* iii. 68 You know I don't 'blow', Jim, or spout tall yarns. **1917** *N.Z.E.F. Chrons.* 5 Sept. 35/2, I don't care about blowin' about it.

9. b. (Further examples.) See also Cloud *sb.* 5 b.

1848 *Sessions* Mar. 847, I could sit down and *blow* my *'bacco*. **1856** Mayhew *Great World of London* 5 Smoke a *pipe* of *baccer*..blow your yard of *tripe* of *nosey-me-knacker*.

c. To lay out or get through (money) in a lavish manner; to squander; = *blue *v.*² 1. *slang.* Also *refl.* (U.S. dial.): see quot. 1896.

1874 Hotten *Slang Dict.*, *Blew*, or *blow*,..to lose or spend money. **1892** *Daily News* 5 Sept. 6/3 Sometimes you'll blow a little money..but another week you may make a lot. **1896** *Dialect Notes* I. 412 'To blow oneself', to spend money freely. **1904** W. H. Smith *Promoters* v. 100 The Church people in England were the folks that had the money to blow. **1932** H. Simpson *Boomerang* x. 244 A thousand pounds, which she proposed..to blow in a couple of months' high living. **1957** *Economist* 21 Dec. 1030/1 He will probably feel able to blow with a clear conscience the £2,000.

d. *to blow in*: to spend, squander. Also *absol.* *slang* (chiefly *U.S.*).

1886 in *Amer. Speech* (1950) XXV. 30/2 When Davis has a dollar he's dead bent on blowing it all in. **1887** F. Francis Jr. *Saddle & Mocassin* 144 'Sam went off on a bend.' 'To blow in?' Jake laughed assent. **1903** *Outlook* 7 Nov. 586 He had blown in all his earnings in a grand frolic. **1934** J. T. Farrell *Young Manhood* (1936) i. 163 Last winter I got sixty bucks from him for tuition and books and blew it in. **1938** R. D. Finlayson *Brown Man's Burden* 24 After breakfast I'll borrow Henare's car and we'll blow in the cash. **1946** F. Sargeson *That Summer* 35 Then he'd go to town and blow his money in, usually at the races.

12. b. Also (*U.S. colloq.*), to move as if carried or impelled by the wind.

1844 *Knickerbocker* XXIII. 51, I was half awake..when Bob came in, blew about the room for a while, and cried out. **1868** S. Hale *Lett.* (1919) 42 She is a picturesque looking creature...Why she blows up and down the Nile year in and year out,..I dunno.

d. *to blow in*: to appear or turn up unexpectedly; to drop in. *colloq.* (orig. *U.S.*).

1895 F. Remington *Pony Tracks* 104 We were all very busy when William 'blew in' with a great sputtering. **1904** G. H. Lorimer *Old Gorgon Graham* 47 Yesterday our old college friend, Clarence, blew in from Monte Carlo. **1913** R. Brooke *Let.* 6 Sept. (1968) 505, I 'blew in' here yesterday, & found about nine letters from you. **1940** *War Illustr.* 16 Feb. ii/3 He just blew in out of the black-out and asked if he might use the telephone.

e. To go away, to leave hurriedly. *slang* (orig. *U.S.*).

[**1902** *B. Burgundy's Lett.* 50 Then we had another and blew the joint.] **1912** Ade *Knocking the Neighbors* 93 She..tied up the Geranium and took the unfinished Tatting and Blew. **1936** J. Curtis *Gilt Kid* 130 Sorry and all that, but I'm afraid I must blow now. **1937** E. Linklater *Juan in China* xxv. 315 'And what's happened to Rocco?'..'He's blown. He's gone up north.' **1961** J. I. M. Stewart *Man who won Pools* iii. 38 All I want is that all these people should blow.

14. e. To play jazz on (any instrument). Also *intr.* *colloq.* (orig. *U.S.*).

1949 L. Feather *Inside Be-Bop* II. 72 Nobody ever gave Diz or Bird a lesson in the art of blowing a jazz chorus. **1962** *John o' London's* 3 May 433/1 A blowin' session is a general term used to describe that form of jazz where men get together for the pleasure of making free and spontaneous music. **1962** *Radio Times* 17 May 43/3 A jazz musician never *plays* an instrument—he *blows* it, whether it be drums, piano, bass, or horn. Should he 'blow' with feeling, or great excitement ('like wild') he is either 'way out' or 'wailing'. **1966** *Crescendo* Sept. 27/1 The not-so-advanced suffered from insufficient outlet, and opportunity to blow and to improve. **1966** *Melody Maker* 15 Oct. 6 Dave Gelly is a school librarian who also blows jazz tenor with the New Jazz Orchestra. **1968** *Jazz Monthly* Apr. 23/2 His style was hard to fit into the standards of hard bop blowing sessions.

15. b. *to blow up*: to sound a whistle (as a signal).

1889 Barrère & Leland *Dict. Slang* 135/2 *To blow up* (*i.e.*, to sound the whistle), is to call the men to work; used by foremen and ga[n]gers. **1954** J. B. G. Thomas *On Tour* vi. 65 The referee blew up to see who was actually lying on the ball.

19. b. *to blow on*: (*trans.*) to solder on by means of the blow-lamp.

1893 *Spons' Mechanics' Own Bk.* (ed. 4) 641 Lead the pipe away to the main supply, and 'blow' it on by means of a union suited to the case.

c. *Electr.* (*intr.*) Of a fuse: to melt under an abnormally high electric current; to fuse. Also with *out*. Also *trans.*, to cause (a fuse) to melt. Hence *fig.* (*colloq.*).

1902 in Webster Suppl. **1908** *Installation News* II. 86/1 A fuse has blown owing to a fault on the circuit. **1925** Wodehouse *Carry On, Jeeves!* viii. 188 However firmly and confidently he started off, somewhere around the third bar a fuse would blow out. **1949** S. J. Perelman *Listen to Mocking Bird* x. 120 Relax..or you'll blow a fuse. **1962** A. Battersby *Guide to Stock Control* ix. 81 If a fuse blows at home, the effect is immediate—the lights go out. **1969** *Woman's Own* 1 Nov. 15/1 This means plugging all three appliances into one 13 amp. socket. Are you likely to blow a fuse?

22. c. *Veterinary Path.* To cause (the stomach of an animal) to swell. (Cf. *blowing *vbl. sb.*¹ 2 d.) Also *refl.*

1778 W. H. Marshall *Minutes Agric.* 16 Jan., sig. Z2 *verso*, Perhaps it was the core of a cabbage which blowed the cow. **1893** Dartnell & Goddard *Gloss. Wilts.* 14 Sheep and cattle 'blow' themselves, or get 'blowed', from over-eating when turned out into very heavy grass or clover, the fermentation of which often kills them on the spot, their bodies becoming terribly inflated with wind.

23. c. In *Photogr.*, *to blow up*: to enlarge (a photograph, etc.). *colloq.*

1930 *Popular Sci.* Jan. 27/3 The little sixteen-millimeter film of such cameras is too small for projection..and when it is 'blown up' to standard size, the images often blur. **1959** *Punch* 21 Oct. 337/3, I blew up the two faces on an epidiascope. **1961** G. Millerson *Telev. Production* iii. 29 Reducing the lens angle 'blows up' a proportionately smaller area of the centre of the shot, to fill the whole screen.

24. a. *to blow to atoms* (example); also, *to blow to bits* (in quot. *fig.*).

1802 *Monthly Mag.* XIV. 31/1 [Robert Fulton] proposes to blow men of war to atoms. **1936** *Discovery* July 229/1 The old..story that Fulton was the father of the steamboat, blown to bits in favour of the cruelly forgotten John Fitch.

c. *to blow from (the mouth of) a gun*, etc.: to execute summary justice on (a traitor, etc.) by binding him to the muzzle of a gun and firing the piece.

[**1776** *Monthly Rev.* LV. 276, I ordered..the artillery officers to prepare to blow them away.] **1857** *Blackw. Mag.* Sept. 389/1 That if there be 10,000 who refuse to use them, they are to be blown away from cannon. **1885** *Century Mag.* Jan. 411/2 That Black Idol..Was..Blown hellward

from the cannon's mouth. **1893** W. S. Gilbert *Rosencrantz & Guild.* II, The Bench of Bishops seize you..And blow you from a gun.

d. *Sporting.* To shatter (a game bird) in shooting; more explicitly *to blow to pieces.* So (U.S. slang) *to blow apart.*

1892 W. W. Greener *Breech-Loader* 275, I defy any one, after a day's shooting, to point out a single bird that has been 'blown'. My experience has proved beyond doubt that the choke-bore does not blow pheasants or any other game to pieces at 20 yards. **1920** Mulford *J. Nelson* iv. 45 'Squint,' said his captor in a hard, level voice, 'if you give me th' least excuse I'll blow you apart.'

e. To remove by the force of an explosion; with advbs. *away, back, off.*

1899 *Westm. Gaz.* 13 July 7/2 Cartridges in which the caps have been blown back when the rifle was discharged.

f. *to blow the lid off* (fig.): to expose (a state of affairs). orig. *U.S.*

1928 *Daily Tel.* 1 May 9/5 He 'blew the lid' off a notorious national condition of affairs.

g. *to blow out*: (of a cock, valve, etc.) to be driven out by the expansive force of gas or vapour.

1909 in Webster.

h. To produce by blowing or shooting.

1871 Bret Harte *East & West: Poems* 18 Walker of Murphy's blew a hole through Peters For telling him that he lied. **1891** C. Roberts *Adrift Amer.* 153 If you talk to me like that I'll blow a hole through you.

i. *to blow one's top* (less commonly *topper*): to lose control of oneself through anger, excitement, etc. (see also quot. 1938).

1928 R. J. Tasker *Grimhaven* iii. 28 Unless you say 'blew his top' or 'blow his topper'; then it means to go crazy. **1938** *New Yorker* 12 Mar. 47/2 If he smokes to excess, he blows his top; that is, he gets sick. **1941** *Amer. Speech* XVI. 163/2 *To blow your top*: vocal loss of temper. **1947** Steinbeck *Wayward Bus* 90 He blew his top and lost his job and came bellyaching to Loraine. **1958** *Economist* 1 Nov. 387/2 This was not just a newly retired officer blowing his top after years of enforced silence.

j. *to blow* (a person's) *mind*, to induce hallucinatory experiences (in a person) by means of drugs, esp. LSD; hence *transf.*, to produce (in a person) a pleasurable (or shocking) sensation.

[**1966** *San Francisco Examiner & Chron.* 12 June 33/3 The Barry Goldberg Blues Band..does an LP called 'Blowin' My Mind'.] **1967** *San Francisco Examiner* 12 Sept. 26/3 On a hip acid (LSD) trip you can blow your mind sky-high. **1967** *San Francisco Chron.* 2 Oct. 49/3 Because when the Red Sox rallied to beat the Minneapolis Twins..Boston fans blew their minds. **1968** J. D. MacDonald *Pale Grey for Guilt* (1969) xii. 152 They had some new short acid from the Coast that never gives you a down trip and blows your mind for an hour only. **1970** *Rolling Stone* 30 Jan. 1/2 Blue blazer, grey flannel pants, shirt and a beautiful scarf with a chunky Mexican turquoise/silver bracelet and ring which blew the white-shirted jury's minds.

25. b. (Earlier examples.)

1710 Duchess of Marlborough *Let.* in A. T. Thomson *Mem.* (1839) II. viii. 173 This plainly showed that the cabal had been blowing her up, but that she could not, however, contradict her own order. **1807** Geo. Colman *Let.* 10 Dec. in J. C. Young *Mem. C. M. Young* (1871) I. ii. 47 Now for this I will blow you up!

c. To go to pieces, give out, fail. Chiefly *U.S. slang.*

1934 in Webster. **1957** N. Frye *Anat. Criticism* iii. 178 An epilogue in Plautus informs us that the slave-actor who has blown up in his lines will now be flogged.

26. c. Said also of veins of metals; and *fig.*

1857 *Trans. Ill. Agric. Soc.* II. 364 Quantities of the public lands were entered to cover the supposed copper mines. It has all blown out. **1873** J. H. Beadle *Undevel. West* xviii. 333 The fluid turned aside to existing crevices, or 'blew out' through hollow chambers.

27. (Further examples.) Also *to blow the gab* or *gaff* (see Gab *sb.*² 2, Gaff *sb.*³ 2).

1925 E. Wallace *Mind of Mr. J. G. Reeder* v. 165 This officer 'blew' the raid to Tommy. **1958** G. Greene *Our Man in Havana* II. 99 We can't risk blowing him now. *Ibid.* III. i. 111 They're anxious you should take no risk of being blown. It doesn't matter so much if I'm blown. **1961** 'B. Wells' *Day Earth caught Fire* vii. 114 You can't go on blaming him for blowing the story you gave him.

b. (Earlier and later examples.)

1848 E. Judson *Mysteries N.Y.* II. 48 To 'blow' would be to tell of some of my stealing. **1928** E. Wallace *Gunner* xxxiii. 281 If..Mr. Morell..has blown—has told the story of Taffany's, every boat will be watched. **1967** C. Drummond *Death at Furlong Post* xii. 154 If Ada blows she gets life, they don't like Crown Evidence these days.

29. Also with the implication of ignoring or disregarding. (Earlier and additional examples.) Also *blow!*: used *absol.* as an exclamation of anger or vexation; *blow me tight!* (cf. sense 22).

1781 G. Parker *View of Society* I. 48 Blow me up (says he) if I have had a fellow with such *rum toggys* cross my company these many a day. **1819** T. Moore *Tom Crib's Mem. to Congress* 46 Says Bill, 'there's nothing like a Bull: And *blow me tight.*' **1821** P. Egan *Life in London* iii. 225 Blow me tight if ever I saw such a thing in my life before. **1827** J. Wight *More Mornings at Bow St.* 55 *Blow me* if I do! **1836** Dickens *Sk. Boz.* in *Morning Chron.* 11 Oct. 3/4 The said Thomas Sludberry repeated the aforesaid expression, 'You be blowed!' **1859** —— *T. Two Cities* II. i. 36 One blowed thing and another. **1865** —— *Mut. Fr.* II. iv. xv. 287 Blowed if I shouldn't have left out lakes. **1871** *Blackw. Mag.* May 551/2 Oh, blow it, governor. **1922** F. Hamilton *P. J., Secret Service Boy* ii. 70 I'm absolutely blowed if I know what to do. *Ibid.* 84 Oh, blow! And I go back to school in ten days. **1933** P. Macdonald *Myst. Dead Police* i. 6 'Blow me *tight*!' said Sergeant Guilfoil. For things were certainly happening in Farnley. **1957** I. Cross *God Boy* (1958) xv. 124 Then blow me if Dr Hutchinson..didn't come padding round the post office corner. **1963** *Listener* 28 Mar. 540/1 It is no longer proper to use as our second national motto in education 'Blow you, Jack, our top five per cent. are absolutely splendid'.

30. Also const. *on.*

1844 *Spirit of Times* 20 Jan. 557/2 Go! Get off; I'll not blow on you. **1877** J. Greenwood *Dick Temple* II. i. 10 She ain't got nobody but me to keep a secret for her, and I've been and blowed on her. **1916** E. Wallace *Clue Twisted Candle* (1918) xvii. 197, I'm not going to blow on it, if it's going to get me into trouble, but if you'll promise me that it won't, I'll tell you the whole story. **1960** W. Haggard *Closed Circuit* viii. 94 There could be only one explanation: Menderez had blown on him.

32. To treat (someone) *to. U.S. slang.* Also *refl.* (See also quot. 1889.)

1889 Barrère & Leland *Dict. Slang* I. 134/1 *To blow off*, to treat to drinks. **1896** Ade *Artie* xvii. 155 'I noticed that you'd been talking bicycle lately, but I didn't know you were going to get one.' '..I'm goin' to do the sucker act and blow myself.' **1903** C. L. Burnham *Jewel* 97 Father took me to the horse show...He told mother he was going to blow me to it. **1949** A. Miller *Death of Salesman* II, Tell Dad, we want to blow him to a good meal.

blow, *sb.*¹ Add: **1. c.** A stroke of the shears in shearing sheep. *Austral.* and *N.Z.*

1878 'Ironbark' *Southerly Busters* 180 If coves would let me 'open out', And take a bigger 'blow'. **1890** *Melbourne Argus* 20 Sept. 13/7 The shearers must make their clip clean and thorough. If..a 'second blow' is needed, the fleece is hacked. **1956** G. Bowen *Wool Away!* (ed. 2) iii. 23 The first blow starts at the top of the brisket. **1959** H. P. Tritton *Time means Tucker* 26/1 Bill would shear to the whipping side, then pass the sheep to me, and..seven or eight blows would complete the job.

d. An outcrop of mineral. *Austral.* and *N.Z.*

1879 W. J. Barry *Up & Down* xxii. 224, I then came to the conclusion that the lode was not a permanent one, but only what is called a 'blow' of quartz. **1922** *Chambers's Jrnl.* Nov. 704/1 Silver-lead (galena) outcrops standing out in great 'blows'. **1934** *Geogr. Jrnl.* LXXXIV. 504 Quartz blows and ironstone reefs were noted. **1965** G. J. Williams *Econ. Geol. N.Z.* iii. 29/2 A large blow of quartz was found on the surface of the Rainy Creek property in 1872.

5. blow-by-blow (orig. *U.S.*), used *attrib.* to designate a detailed account of the sequence of punches given in a boxing match; freq. *transf.*

1933 *Amer. Speech* Oct. 34/2 Radio announcers..describe the struggle as they see it, give the blow by blow account of its progress. **1939** *Time* 23 Jan. 30/3 Joe Louis v. John Henry Lewis...Blow-by-blow report by Clem McCarthy and Edwin C. Hill. **1948** H. Lawrence *Death of Doll* ix. 225 Maybe we'd better get together. Blow by blow description. **1962** *Times* 7 June 17/3 Blow-by-blow descriptions of Jonson-Jones masques.

blow, *sb.*² **1. a.** Add to def. = A breath of fresh air; a 'breather' (sense 3). *colloq.*

1849 *Theatrical Programme* 9 July 48/2 A 'blow' upon the river. **1856** Dickens *Out of Season* in *Househ. Words* 28 June 554/1, I really ought to go out and take a walk in the wind...I was under a moral obligation to have a blow. **1888** C. M. Yonge *Beechcroft at Rockstone* II. xv. 40 'You must be tired out!'..'Oh no, Aunt Ada! Quite freshened by that blow on the common.' **1936** R. Lehmann *Weather in Streets* IV. i. 386 Been for a blow?..It freshens you up.

c. Add to def.: Also used of other musical instruments; a musical session. (Later examples.) Cf. *blow *v.*¹ 14 e.

1962 *Sunday Times* (Suppl.) 10 June 3 He is now rarely heard having a relaxed blow in the clubs. **1965** G. Melly *Owning-Up* vii. 75 We played three one-hour sessions and relied on musicians who wanted a blow to fill in the gaps. **1966** *Crescendo* Oct. 31/3 Ernie Garside scoured the city for a set of vibes, but he was unlucky! So Gary didn't get to have a 'blow'. **1968** *Ibid.* May 31/1 He told me that he was drastically curtailing the activities of his big band. The musicians might come together now and again for a blow, but to all intents and purposes it's finished as a regular aggregation.

blow-. Add: **blow-away** *a.* (cf. Blow *v.*¹ 12 a), susceptible to the influence of a current of air, hence fragile; also *fig.*; **blow-back,** the act or process of blowing back; spec. (*a*) the back-pressure in a boiler or internal-combustion engine; (*b*) in fire-arms (cf. *blow *v.*¹ 24 e); **blow-by** (see quot. 1959); also *attrib.*; **blow-fish,** a popular name for any of several fishes which inflate their bodies when stimulated by fear, etc.; **blow-lamp,** a lamp designed to give a condensed hot flame, directed by air-pressure or gas on the spot to be heated; **blow-off,** (*a*) the action or operation of ridding a boiler of water or sediment by the force of steam; also *attrib.*; (*b*) *fig.* an outburst; **blow-over** *Glass-making*, the surplus glass forced out of the mould in blowing (Knight *Dict. Mech.*, *a* 1877); **blow-room** = *blowing-room*: see *blowing *vbl. sb.*¹ 5; **blow-through,** (*a*) the process of blowing steam through the cylinder of an engine, etc., to clear it of air; used *attrib.* in *blow-through cock, valve*; (*b*) defining a jet or lamp for the blowing of oxygen through a flame against lime or magnesium to produce a brilliant light; **blow-torch** = **blow-lamp*; **blow-well** (earlier and later examples). See also *blow-out, *blow-up.

1858 *Jrnl. R. Agric. Soc.* XIX. 257 It is represented as being at that time a blow-away sand. **1908** *Daily Chron.* 22 June 3/4 The Commissioners got little of the right sort of evidence..and were dealing..with 'a mass of blow-away stuff'. **1938** F. Scott Fitzgerald *Let.* 18 Nov. (1964) 42 You unveiled the story of the blow-away pink slip. **1958** *Vogue* July 86 The blowaway prettiness of an Empire dress. **1883** J. D. Fulton *Sam Hobart* 224 The flames originated from the '*blow back*' on the engine. **1899** *Westm. Gaz.* 13 July 7/2 Foul barrels were the cause of ineffectual shots and 'blow-back' caps. **1905** *Kynoch Jrnl.* Jan.–Mar. 10 The cracks..will at once develop into fissures when the cartridge is fired, cause 'blow-backs', and seriously reduce the velocity of the projectile. **1929** *Evening News* 18 Nov. 14/3 He received his injuries as the result of a blow-back in one of the boilers when pulverised fuel was being lit. **1940** *Chambers's Techn. Dict.* 98/2 *Blow back*, the return, at low speeds, of some of the induced mixture through the carburettor of a petrol-engine; due to late closing of the inlet valve during compression. **1933** Streeter & Lichty *Internal Combustion Engines* (ed. 4) xv. 432 The oil film is blown out by the high-pressure gases, and blow-by occurs. **1959** *N.Y. Times* 6 Dec. 67/5 The gasoline vapors, called 'blowby' by automotive engineers, are hydrocarbons forced between the cylinder wall and the piston during the compression stroke. **1961** *Times* 19 Dec. 2/6 A method of making the engine consume its own 'blowby' vapours. **1893** *Funk's Standard Dict.*, *Blowfish*, the wall-eyed pike-perch. **1933** A. Davenport *Country Holiday* viii. 240 Theer'll be hundreds o' great fish followin' every shoal—blow-fish, porpoise, [etc.]. **1934** *Bulletin* (Sydney) 1 Aug. 21/4 North Australian waters contain..the blowfish...When it is hooked it inflates itself to many times its normal size. Among its *aliases* are balloon-fish, puffball and wind-bag. **1896** J. W. Hart *External Plumbing Wk.* 259 The tinning can be done best by means of a blow-lamp of some kind. **1909** *Westm. Gaz.* 27 Apr. 12/4 He was filling a blow lamp with motor spirit when it exploded. **1842** *Civil Engin. & Archit. Jrnl.* V. 367/2 Blow-off cocks. **1859** Rankine *Steam Engine* § 305, The blow-off apparatus consists, in fresh-water boilers, simply of a large cock at the bottom of the boiler. *a* **1877** Knight *Dict. Mech.*, Blow-off Pipe. **1898** S. M. Fergusson in '*House' on Sport* I. 174 A blow-off in this wise [*i.e.* swearing at golf] does one good now and then. **1899** *Daily News* 9 Mar. 5/3 The blow-off pipes into the sea being broken, the chief engineer..had to let the water run out of the boilers into the hold. **1934** J. M. Cain *Postman always rings Twice* xii. 93 What we had the big blow-off over was the beer licence. **1908** *Daily Chron.* 5 Aug. 1/6 The card and blow room hands. *a* **1877** Knight *Dict. Mech.*, Blow-through valve. **1892** *Photogr. Ann.* II. 231 Blow-through jet. **1906** *Westm. Gaz.* 24 Nov. 14/2 A blow-through lamp. **1909** *Century Dict.* Suppl., Blow-torch. **1921** R. D. Paine *Comr. Rolling Ocean* xvi. 274 Down to the engine-room went Briscoe and returned with..a blow-torch. **1963** J. Osborne *Dental Mechanics* (ed. 5) xv. 340 The face of the blowtorch nozzle. **1799** A. Young *Agric. Linc.* i. 15 In the parishes of Tetney, Fulstow, and that vicinity, blow-wells, which are deep flowing pits of clear water, which flow in considerable streams. **1924** Kendall & Wroot *Geol. Yorks.* I. xlvii. 626 On the warp-lands of Holderness and the Humber estuary there bubble up with great vigour a series of springs known as 'blow-wells'...A 'blow-well' is a natural artesian discharge.

blow, *sb.*³ **1.** (Earlier and later examples.)

1744 in *Amer. Speech* (1940) XV. 226/2 The aple trees are full in the blow. **1866** M. Arnold *Thyrsis* in *New Poems* (1867) 76 And stocks in fragrant blow. **1895** Meredith *Amazing Marr.* I. iv. 45 Purple crocuses in bud and blow.

blower¹. Add:

3. b. *esp.* An apparatus for creating an artificial current of air by pressure, used as a ventilator, dryer, etc., and to produce a blast of air in a furnace, etc. Also *attrib.*, as *blower-engine, -fan, -pipe.*

1858 Simmonds *Dict. Trade*, *Blower*,..a fan used on board American river steam-boats, to increase the current of air. **1875** *Encycl. Brit.* III. 552/2 The rotary blower, invented by Messrs. Root of Connersville, Ind., is one which has of late years found extensive use both in America and Europe. **1900** *Everybody's Mag.* III. 528 The waves, breaking over the blower-pipes, poured through in such quantities as to dampen the belts of the blower-engine. **1903** *Daily Chron.* 9 Dec. 6/3 The air is first filtered...Afterwards it passes through a blower-fan. **1958** *Times* 19 Aug. 11/7 The car..was fitted with a heater..and had a plastic knob for the blower. **1959** *Listener* 22 Jan. 159/2 Blowers which blast the snow into banks at the roadside.

c. A blowing-machine: (*a*) for cleansing and 'opening' the fibres of cotton or wool; (*b*) for cleansing and separating rabbit fur.

1867 *Chambers's Encycl.* IX. 46/2 The cotton or wool..is..taken to the 'blower' or 'opener', and being put into a shaft, is there acted upon by a stream of air..which blows it forward, removes extraneous matters, [etc.]. *a* **1877** Knight *Dict. Mech.*, *Blower*, a machine for separating the hair from the fur fibres [in hat-making].

d. In an aeroplane engine, esp. a supercharger.

1920 *Flight* XII. 1004/2 A blower driven off the Renault engine was provided. **1922** *Advisory Comm. Aeronaut. Techn. Rep. 1918–19* 30 In connection with high altitude flying a large number of experiments have been carried out on different forms of blowers for aircraft engines. **1933** *Jrnl. R. Aeronaut. Soc.* XXXVII. 86 Further developments in scavenging and supercharging by auxiliary blowers. 1946 B. SUTTON *Jungle Pilot* xiii. 66 At this height I found I could climb appreciably faster than the Tomahawks. This was due to my two-speed 'blower'.

e. A speaking-tube or telephone. *colloq.*

1922 E. WALLACE *Flying Fifty-Five* xxx. 182 The club enjoyed a 'blower service'. The 'blower' is difficult for the outsider to understand. 1926 —— *More Educ. Evans* v. 121, I heard it on the telephone...They got that price from the blower round at the Arts Club. 1945 PARTRIDGE *Dict. R.A.F. Slang*, *Blower*, from the Naval sense, 'telephone' or, rather, the speaking-tube connecting Bridge and Engine-Room. 1951 'N. SHUTE' *Round the Bend* vii. 217 'Somebody must have got on the blower from Bahrein.'..He meant the radio telephone that connects the aerodromes all down the eastern route. 1957 'J. WYNDHAM' *Midwich Cuckoos* iii. 24 I'd of said the old girl was *always* listenin' when there was anyone on the blower.

blow-fly. Add: (Earlier example.)

1821 T. NUTTALL *Trav. Arkansa* xiv. 201 The green blow-flies, attracted by the meat brought to our camp, exceeded everything that can be conceived.

b. *blow-fly strike*: a disease affecting sheep. (Cf. STRIKE *v.* 45 d.)

1933 *Council Sci. & Ind. Res.* Pamphlet No. 37 ii. 17 Though death following blowfly strike is not uncommon, its actual causation has never been thoroughly investigated. 1955 J. H. WELLINGTON *Southern Africa* II. I. v. 74 Other points of superiority in the new strain [of sheep] include.. a low liability to blowfly strike.

blow-hard, *a.* and *sb.* *colloq.* (orig. *U.S.*). [BLOW *v.*^{1} 6.]

A. *adj.* Boastful, blustering.

1855 *Oregon Weekly Times* 21 July (Th.), The *Oregonian* ..has a blowhard article on the subject. 1921 R. D. PAINE *Comr. Rolling Ocean* vi. 105 He don't get by with his blow-hard stuff, but I'll have to say he is entertaining. 1946 K. TENNANT *Lost Haven* (1947) xvi. 248 He's a blow-hard bugger.

B. *sb.* A blustering person; a braggart.

1857 DUFFERIN *Lett. High Lat.* 161 He is a fine old grey-headed blow-hard of fifty odd. 1880 HARDY *Trumpet-Major* xxxi, Bob's temper began to rise. 'Don't you talk so large, about your pinioning, my man....' 'Now or never, young blow-hard,' interrupted his informant. 1964 *Punch* 8 Apr. 514/2 'Old *John Bull*' might be a blabbermouth and blowhard.

blow-hole. 2. Delete †*Obs.* and add:

1878 *Ure's Dict. Arts* IV. 835 The following experiments were made in order to prepare solid steel without blow-holes by the crucible process. 1906 *Westm. Gaz.* 29 Aug. 2/1 The way to prevent steel rusting..is to have no blow-holes. 1960 *Jrnl. Iron & Steel Inst.* CXCV. 223/2 (*title*) Blowholes in castings, and methods of preventing their occurrence.

3. (Additional examples: see quots.) Also = **puffing-hole*.

1858 JOHN SHAW *Gallop to Antipodes* vii. 131 That is a blowhole. Don't you see the water forced into the air, like a whale spouting?..You will see the water first and afterwards you will hear the report. 1872 [see *PUFFING *vbl. sb.* 5]. 1891 R. WALLACE *Rural Econ. Austral. & N.Z.* i. 24 Here also the cavernous condition of the subjacent rocks extended, and every here and there was a large pit or depression like a deserted quarry which had fallen in. From some of these so-called 'blow-holes' water could be pumped for the stock in dry seasons. 1938 *Times* 7 Sept. 9/1 He.. crawled in through the 'blow-hole' [of a barrage-balloon] to look for leaks and examine the pressure valve. 1958 ARMSTRONG & ROBERTS *Illustr. Ice Gloss.* II. 91 *Blow hole*, opening through a snow bridge into a crevasse or system of crevasses which are otherwise sealed by snow bridges. 1961 J. CHALLINOR *Dict. Geol.* 24/1 *Blow-hole*, a natural chimney, on a coast, reaching from the inner end of a cave to the surface of the ground above. 1962 *Listener* 11 Jan. 57/2 The whole thermal area of the North Island [of New Zealand], with its hot pools and geysers and blowholes.

blowing, *vbl. sb.*^{1} Add: **1. b.** *Cotton-manuf.* The cleansing of cotton. Cf. *BLOWER^1 3 c (a).

1844 [see SCUTCHING *vbl. sb.*^{1} 1]. 1875 *Ure's Dict. Arts* I. 955 The willowing, scutching, or blowing, an operation which removes the seeds and dirt, and prepares the material in the form of a continuous lap.

c. The shattering (of a bird) in shooting. (Cf. *BLOW *v.*^{1} 24 d.)

1892 W. W. GREENER *Breech-Loader* 271 No one could detect the 'blowing' of a single bird.

d. (See quot.)

1881 R. HUNTER et al. *Encycl. Dict.*, *Blowing of Firearms* (Gunnery), the art or operation of constructing firearms in such a way that the vent or touch-hole is run or 'gullied', and becomes wide, allowing the powder to blaze out.

e. *blowing off*: the action or process of firing (a rifle) to cleanse the barrel.

1893 *Daily News* 2 Feb. 2/6 The ammunition..was issued in packets of ten, thus allowing for blowing off and for one compulsory sighting shot.

f. *fig.* Boasting, bragging. *U.S.*, *Austral.*, etc. Cf. BLOW *v.*^{1} 6 a in Dict. and Suppl.

1840 *Congress. Globe* App., 9 Jan. 50/1 [I advise them] to treat with contempt and scorn, all the blasting, blowing, blustering, and bullying displays they may see. 1873 TROLLOPE *Australia & N.Z.* I. 387 A fine art much cultivated in the colonies, for which the colonial phrase of 'blowing' has been created. 1878 J. H. BEADLE *Western Wilds* ix. 134 Its bright and saucy editorials excelled all specimens extant of Kansas blowing.

2. c. The formation of bubbles or blisters in the texture of a manufactured article.

1872 *Spons' Dict. Engin.* v. 1547 As the pressure is kept on the metal until it is well set and solid, such a thing as blowing will never, or very seldom take place.

d. = *BLOAT *sb.* 1 b. (Cf. *BLOW *v.*^{1} 22 c.)

1891 R. WALLACE *Rural Econ. Austral. & N.Z.* xxii. 301 Cattle not accustomed to an abundant supply of green food are liable to suffer from 'hoven' or 'blowing' after eating it.

e. Of a tin of food (see quot.).

1950 J. G. DAVIS *Dict. Dairying* 43 '*Blowing*' *of condensed milk*, the bulging or bursting of tins of condensed milk usually due to the production of carbon dioxide by fermentation of the sugar by micro-organisms, usually yeasts.

4. (Earlier example of colloq. use.)

1820 G. SIMPSON *Jrnl.* 18 Aug. in *Hudson's Bay Record Soc. Publication* (1938) I. 17 Mr. Clarke gave him what is vulgarly called 'a good blowing up'.

5. blowing-cone, a volcanic cone by which vapours escape from a subterranean molten lake; **blowing-engine** (add example); **blowing-machine,** (*a*) any mechanical contrivance for producing a blast of air; (*b*) a machine for cleansing and separating fur in hat-manufacture; (*c*) an apparatus forming part of the machine for cleansing the cotton in cotton-manufacture; (*d*) an apparatus for blowing glass-ware (*Cent. Dict.*, 1889); **blowing-pot,** a vessel containing clay paste which the workman 'throws' on pottery by blowing; **blowing-room,** a room in which the cleansing or separating of materials is done, esp. in a cotton factory; also *attrib.*, as (*card and*) *blowing-room hand, machinery, operative*.

1895 DANA *Man. Geol.* (ed. 4) 279 In cases, outside of the lava-lakes, where the bubbles are bursting beneath an opening in the bottom of the crater, the vapors and lava driblets escape from the aperture with a rush and a roar. The driblet-cone, thus made, is sometimes called a blowing-cone. 1845 DODD *Brit. Manuf.* V. 159 Blowing-Engine. 1835 URE *Philos. Manuf.* II. ii. 111 The blowing machine for thoroughly opening out the cotton into clean individual fibres. 1839 —— *Dict. Arts* I. 345 Batting (*beating*), scutching, and blowing machines. 1845 DODD *Brit. Manuf.* V. 158 The 'blowing-machines'..act as follows. 1868 F. H. JOYNSON *Metals* i. 18 The blast from the blowing machine is conducted into the furnace by means of the tuyers. 1940 *Chambers's Techn. Dict.* 99/1 *Blowing machines*, machines for forming molten glass into articles by the use of air under pressure. 1875 *Ure's Dict. Arts* III. 622 Common earthenware is coloured by means of..the blowing-pot. 1845 DODD *Brit. Manuf.* V. 158 The fur..is conveyed to the blowing-room, finally to effect the separation [of the coarse from the fine fur]. 1887 J. E. HOLME *Cotton Spinning* Introd., The blowing-room machinery should be kept well oiled. 1892 *Daily News* 14 Apr. 3/4 The card and blowing-room hands. 1898 *Westm. Gaz.* 23 Dec. 7/3 The Card and Blowing-room Operatives Society.

b. With advs.: **blowing-in, -out**: the action or process of putting a blast furnace into or out of action (see BLOW *v.* 19); **blowing-off, -through** = *blow-off, -through* (see *BLOW-).

1925 *Jrnl. Iron & Steel Inst.* CXII. 409 The methods pursued..in the blowing in, damping down, and blowing out of blast furnaces. 1944 *Ibid.* CL. 378P From blowing-in to blowing-out furnace operation is handicapped, because we do not know what is going on inside the furnace. 1863 ATKINSON tr. *Ganot's Physics* VI. 328 Blowing off taps, for use when the pistons are in motion. *a* 1877 KNIGHT *Dict. Mech.*, Blowing-off. ..Blowing-through.

blown, *ppl. a.*^{1} Add:

5. Applied spec. to glassware formed by forcing air into molten glass. (Recent examples.)

1961 *Listener* 26 Oct. 674/1 Not many people are even aware of the difference between blown glass and pressed glass. 1962 *Gloss. Terms Glass Ind.* (*B.S.I.*) 36 *Blown glassware*, glassware formed by air pressure produced by mouth or by a machine.

b. Recent examples in sense 'exaggerated'.

1958 A. J. P. TAYLOR in *New Statesman* 15 Nov. 664/1 The talk about homosexuals corrupting young boys is blown-up nonsense, a perverted wish-fulfilment. 1961 *Listener* 23 Nov. 887/1 The sorrows of a blown-up 'pop singer' became moving.

c. *Veterinary Path.* (See *BLOW *v.*^{1} 22 c.)

1833 YOUATT *Compl. Grazier* (ed. 6) VI. i. 316 No distemper is of more frequent occurrence among cattle than that of being swollen, *blown*, or *hoven*. 1898 in *E.D.D.* 1950 *N.Z. Jrnl. Agric.* Feb. 167/2 The combination of them causes fermentation, with disastrous results—blown cows and digestive disorders.

d. Filled with bubbles or the like. (Cf. *BLOWING *vbl. sb.*^{1} 2 c.)

1872 *Spons' Dict. Engin.* v. 1548 The..result is what moulders term a *blown* casting; that is, its surface becomes filled with bubbles of air, rendering its texture porous and weak.

e. Of a tin of food: swollen or inflated (see *BLOWING *vbl. sb.*^{1} 2 e).

1899 *Daily News* 16 Jan. 3/3 The sanitary inspector explained that the contents of a 'blown' tin would 'not taste so bad after the gas had come out'. 1955 J. G. DAVIS *Dict. Dairying* (ed. 2) 79 Blown cans of evaporated milk and canned cream.

f. *colloq.* = *SUPERCHARGED *ppl. a.* (Cf. *BLOWER^1 3 d.)

1934 *Neuphilologische Mitteilungen* XXXV. 131 Motoring ..a blown model 'model fitted with a supercharger'. 1940 *Chambers's Techn. Dict.* 99/1 *Blown*, a colloquial term applied to a supercharged petrol-engine. 1959 *Motor* 2 Sept. 92/2 Sears'..blown 4½-litre model. *Ibid.* 11 Nov. 532/3 Won the..Handicap at 125 m.p.h. in a blown Bugatti.

g. *blown oil*: oil that has had air blown through it in order to increase its viscosity.

1887 *Analyst* xii. 33 Blown or thickened oils are prepared from various seed oils, by heating the oil in a suitable tank to 70° C., and then blowing air into them through a perforated tube. 1951 R. MAYER *Artist's Hand-bk. Mat. & Techniques* iii. 109 Heavy, viscous blown oils are prepared commercially in large quantities.

6. a. *Plumbing.* Of a joint: made with a blow-pipe. **b.** *Electr.* Of a fuse (see *BLOW *v.*^{1} 19 c).

1904 GOODCHILD & TWENEY *Technol. & Sci. Dict.*, Blown Joint. 1909 *Installation News* III. 137 We will allow 33⅓ per cent. of the nett price for all 'blown' cartridge fuses of our manufacture.

7. With advs., as *blown-back, -down, -up.* For *blown in* and *out* see BLOW *v.*^{1} 19.

1864 TROLLOPE *Small Ho. at Allington* II. xvi. 165 The clerks in this room would not infrequently be blown up.. and..the blown-up young man would refer Mr. Kissing to his enemy. 1866 Blown-up [see sense 5 b]. 1878 W. MORRIS in J. W. Mackail *Life* (1899) I. 360 The blown-down tree was a fine branchy tree. 1905 *Westm. Gaz.* 13 Sept. 6/3 The blown-up steamer. 1926 E. BOWEN *Ann Lee's* 196 Her forehead and blown-back hair.

blow-out. [BLOW- 1.] **1.** An outbreak of anger; a quarrel, disturbance, row. *dial.* and *U.S.*

1825 J. K. PAULDING *John Bull in Amer.* 137 We had a *blow out* here last Sunday, and half a dozen troublesome fellows..were done for by the brave *rowdies*. 1826 [see BLOW- 1]. 1842 *Spirit of Times* (Philad.) 15 Feb. (Th.), I've had five breezes, seven blow-outs, nine shindies, and a dozen ructions on this $1 Relief note.

2. A dinner, supper, or other entertainment for which an abundant supply of food and drink is provided or at which it is consumed; a 'feast' or 'feed'. *colloq.*

1824 SCOTT *St. Ronan's* xxxiii, 'She sent me a card for her blow-out,' said Mowbray, 'and so I am resolved to go.' 1840, 1856 [see BLOW- 1]. 1930 R. LEHMANN *Note in Music* vii. 264 Have a nice blowout and a good sleep afterwards. 1966 'J. HACKSTON' *Father clears Out* 172 Saint's menu in between such blood-building blowouts was rabbit, with pollard as a savoury.

3. a. *Mining.* A portion of a lode where the mineral appears to have been dislodged by some eruptive force. Also *fig. U.S.*

1873 J. H. BEADLE *Undevel. West* xviii. 333 All the strange terms in mining parlance: 'true lodes,..blow-outs'. 1901 S. E. WHITE *Westerners* xxiii. 214 He saw that..a third [claim]..gave indications of being nothing but a blow-out.

b. A butte, the top of which has been blown out by the wind until it resembles the crater of a volcano; a hollow in an area of shifting sand, or light soil, caused by the action of the wind. orig. *U.S.* Also *spec.* a water-hole (see quot. 1935).

1893 SMITH & POUND in *Bot. Surv. Nebraska* II. 8 If a spot on a dry hill becomes bare, the loose sand is blown away, a small hollow is made... Such blow outs were seen 100 surrounding meters in diameter and 15..meters deep. 1895 P. A. RYDBERG in *U.S. Nat. Herb. Contributions* III. 135 It sometimes happens that settlers [in the sand-hill region] a few years after breaking their land find a field transformed into a big blowout. 1897 POUND & CLEMENTS *Phytogeogr. Nebraska* (1898) 248 A small number of these grasses are especially adapted to these localities and are uniformly to be found in such blow outs... These grasses mark a second formation, which may be called the blow out formation. *Ibid.*, The blow out grasses..bind the sand together with their roots. 1911 F. O. BOWER *Plant Life* ix. 151 Close by an effete Dune with its grasses weakened in growth is being attacked by the wind, and eroded into hollows or 'Blow-outs'. 1935 *Discovery* Dec. 359/2 'Blow outs', i.e. dried up water holes, from which the wind has eroded the sands.

4. A burst in a pneumatic rubber tyre caused by air-pressure from the inside. Also *fig.*

1908 *Westm. Gaz.* 7 Jan. 4/1 Miraculum will not seal a blow-out. 1915 *Lit. Digest* (N.Y.) 21 Aug. 387/1 The Goodyear Cord Tires, without a blow-out, took the car back to Detroit. *Ibid.* 4 Sept. 482/1 (Advt.), No domestic punctures or blow-outs—just easy running over smooth roads to Health and Happiness. 1967 I. HAMILTON *Man with Brown Paper Face* vii. 94 This road..safe enough if you don't have a blow-out.

5. *Electr.* The suppression of an arc in an electric circuit; a device for producing this. Also *attrib.*

1902 *Encycl. Brit.* XXVIII. 97/2 A magnetic blow-out is simply a small magnet so arranged that the arc caused by breaking the circuit takes place in the magnetic field. 1916 *Standardization Rules of Amer. Inst. Electr. Engin.* 28 June § 731 Fuses of the magnetic blow-out type. 1930 *Engineering* 7 Feb. 173/3 Anything which tends to produce sudden speed-changes, such as..magnetic blow outs.

blowser (blau·zəɹ). *local.* Also **blouser.** A landsman who assists in working the seine nets at pilchard-fishing time. Also **blow·sing** *sb.*

1816 PARIS *Mount's Bay* 91 It [*sc.* the net] is moored, or where the shore is sandy and shelving, as in Saint Ives Bay, drawn into shallow water, by a number of men, who are called *Blowsers*. 1877 *Fraser's Mag.* XV. 220 The whole is warped nearer to the shore by blousers, who are landsmen ..employed on land during the pilchard season for general purposes. 1879 *Encycl. Brit.* IX. 254/2 Landing and

carrying the fish to the curing houses is done by men termed 'blowsers'. 1880 M. A. COURTNEY *W. Cornw. Gloss.*, *Blowsing*, working in seine boats.

blow-up. [BLOW- 1.]

1. = *BLOW-OUT 1.

1809 W. GELL *Let.* 22 Jan. in C. K. Sharpe *Lett.* (1888) I. 355 There won't be any quarrel, so you need not fear. The only chance is Keppel making a blow up when she abuses me. 1813, 1834 [see BLOW- 1]. 1846 SOL. SMITH *Theatrical Apprent.* 132 When we had got their jealousy and hatred excited to a proper pitch, it was agreed that a regular 'blow up' between the two should end the joke. 1900 H. LAWSON *Over Sliprails* 46 Some others were making a night of it..as they'd been doing pretty often lately—and went on doing till there was a blow-up about it. 1947 STEINBECK *Wayward Bus* 56 After the initial blow-up the subject had never verbally come up again, but her mother disapproved with her face.

b. An explosion.

1807 W. IRVING et al. *Salmagundi* (1811) xiii. 58 Our citizens did not refuse the invitation of the society to the *blow up*. 1852 DICKENS *Bleak Ho.* xxvi. 260 It was after the case-filling blow-up, when I first see you. 1867 M. ARNOLD *Let.* 14 Dec. (1895) I. 376 Every one is full of the Clerkenwell blow-up.

2. *Sugar-manuf.* The place where the raw sugar is dissolved; also *attrib.*, as *blow-up cistern, pan*, a vessel used in dissolving raw sugar by 'blowing' or forcing steam through it.

1833 B. SILLIMAN *Man. Sugar Cane* 77 The vats, or blow-ups, as they are called, containing the sugar,..are heated by steam tubes passing through them. 1845 DODD *Brit. Manuf.* V. 108 A steam-pipe, in communication with a boiler, is enclosed within the 'blow-up cistern' [for dissolving sugar]; and..steam is forced or 'blown' by its own pressure into the solution. *a* 1877 KNIGHT *Dict. Mech.*, Blow-up Pan. (*Sugar-Machinery.*) 1886 *Harper's Mag.* June 82/2 These 'mixers' or 'blow-ups' are really great stew-pans set in the ground. 1935 *Discovery* Dec. 363/2 The resultant mixture..becomes a syrup known as 'blow-up thick juice'.

4. A photographic enlargement (cf. *BLOW *v.*^{1} 23 c). *colloq.* (orig. *U.S.*).

1945 *Life* 9 July 100/2 (*caption*) The wall behind bar has blowups of the arm insignia of every U.S. division in Europe. 1946 *Electronics* Sept. 157 (*caption*) Big Blow-up... Electronics and Photography work together to produce records like this 47,500-diameter magnification. 1957 *New Yorker* 5 Oct. 34/2 Gigantic photographic blowups of the more important candidates.

blowzily (blau·zili), *adv.* Also **blowsily.** [f. BLOWZY *a.* + -LY^2.] In a blowzy way, coarsely. So **blow·ziness, blow·siness.**

1866 M. E. BRADDON *Lady's Mile* xxii, She was a very handsome lady..in spite of a little blowsiness. 1905 *Daily Chron.* 5 Apr. 4/7 Daintiness would certainly pay better than the present blowziness [of the flower-seller]. 1914 W. J. LOCKE *Fortunate Youth* i, Once blowsily good-looking. 1933 *Times Lit. Suppl.* 28 Sept. 653/3 The much-emphasized charm of the slatternly Kate is perhaps less convincing than her equally emphasized blowsiness. 1959 *Times* 23 Feb. 12/3 Drawings of..blowsily ample girls undressing.

blub, *v.* Add: **4.** Shortened form of BLUBBER *v.* in sense 3, to weep, cry. *colloq.*

1866 A. DOBSON in *Beeton's Annual* 211 Bob privately confessed to me that he always felt inclined to 'blub' over those whipper-tunes. 1884 BLACKMORE *T. Upmore* vi, Keep up your spirits, young fellow, and don't blub. 1899 KIPLING *Stalky* 48 Stalky .. pretended to blub. .. Then I blubbed, too. 1928 TEMPLE THURSTON *Portrait of Spy, Paris* § 1 He was blubbing in her arms.

blub (blɒb), *sb.* [f. *BLUB *v.* 4.] A fit or spell of weeping.

1894 G. DU MAURIER *Trilby* II. 255 It would do him good once more to have a good blub. 1968 *Listener* 11 July 38/3, I had a good blub down the first five miles of the dual carriageway.

blubber, *sb.*^{1} Add: **6.** *blubber-hook, -knife*; **blubber-boiler** *slang*, a whaling ship; **blubber-spade** (example).

1820 SCORESBY *Acc. Arctic Reg.* II. 299 The harpooners.. divide the fat into oblong pieces or 'slips', by means of 'blubber-spades', and 'blubber-knives'. 1851 H. MELVILLE *Moby Dick* II. xi. 75 They..repeat gamesome stuff about 'spouters' and 'blubber-boilers'. *Ibid.* II. xxx. 204 The blubber-hook was inserted into the original hole.

bludge (blɒdʒ), *v.* *slang.* [Back-formation from *BLUDGER.] *intr.* **a.** To act as a prostitute's pimp. **b.** *Austral.* and *N.Z.* To shirk responsibility or hard work; to impose *on*. Also *trans.*, to cadge or scrounge.

1919 DOWNING *Digger Dial.* 12 *Bludge on the flag*, to fail to justify one's existence as a soldier. 1931 V. PALMER *Separate Lives* 264 I've stood you too long already, loafing around here, and bludging on your mother. 1937 L. MANN *Murder in Sydney* 222 A bludger is a ruffian living on the earnings of an immoral woman. Britannia may be..an immoral woman, but there's no need for Australia to bludge on her. 1939 K. TENNANT *Foveaux* II. iv. 176 Everybody bludges and robs. 1941 —— *Battlers* iii. 26 You..bludging little mongrel! 1944 J. H. FULLARTON *Troop Target* xxvi. 186 You were one of the 95 per cent who bludged at base in Enzed or England or Yankee-land. 1944 L. GLASSOP *We were Rats* xxv. 147 Probably a Free Frenchman bludging a lift. 1945 *Southern Cross* (London) 15 Dec. 4/1 Place seemed to be full of..sisters and cousins, all staying with the blokes Herbert had aimed to bludge on. 1945 E. G. WEBBER *Johnny Enzed in Middle East* 43 Those oysters you bludged from me. 1967 I. HAMILTON *Man with Brown Paper Face* vi. 73 He bludged three cigarettes off me. 1967 *Southerly* XXVII. 199 The bludging, dirty mong to whom she had.. entrusted heart and hand.

Hence as *sb.*, an easy job or assignment; a period of loafing.

1945 BAKER *Austral. Lang.* viii. 156 *Bludge*, a soft job. 1949 J. CLEARY *Long Shadow* 195 He was happy in his job, it was a good bludge. 1969 *West Australian* 16 Jan. 13/3 Prime Minister Gorton..quoted..as saying..he was coming to..the Commonwealth Prime Ministers' conference 'on a bit of a bludge'.

bludgeon, *v.* Add: **b.** *fig.* To strike heavily, as with a bludgeon. Const. *in*: to drive in as with a bludgeon. Hence **blu·dgeoning** *vbl. sb.* and *ppl. a.*

1888 W. E. HENLEY *Bk. of Verses* 56 Under the bludgeonings of chance My head is bloody, but unbowed. 1892 STEVENSON & OSBOURNE *Wrecker* xviii. 273 Repentance bludgeoned me. 1894 *Athenæum* 14 July 55/1 It is not..the artful bludgeoning that gets the praise. 1906 R. WHITEING *Ring in New* 238 The militant knifing and bludgeoning men. 1928 E. BLOM *Limit. Music* 37 A truth that has no need of literal bludgeoning-in.

blu·dgeoned, *ppl. a.* [f. BLUDGEON *v.* + -ED^1.] Struck down or wounded with or as with a bludgeon.

1887 STEVENSON *Misadv. J. Nicholson* vi, Next bludgeoned vanity raised its head again, with twenty mortal gashes.

bludgeoner. (Earlier example.)

1842 R. OASTLER *Fleet Papers* II. 8 They have set their hired bludgeoners at me.

bludgeonist (blɒ·dʒənist). [f. BLUDGEON *sb.* + -IST.] One who strikes with or as with a bludgeon.

1811 *Ann. Reg.* 1809 680 No hired bludgeonists astound him. 1903 *Daily Chron.* 25 May 3/3 Critic is a mild word to use. Bludgeonist would suit the case better.

bludger (blɒ·dʒəɹ). *slang.* [Shortened from BLUDGEONER.] **1.** = BLUDGEONER; *spec.* a prostitute's pimp.

1856 [see *stick-slinger*, STICK *sb.*^{1} 16]. 1898 *Bulletin* (Sydney) 17 Dec. Red Page/2 A *bludger* is about the lowest grade of human thing, and is a brothel bully. 1936 'R. HYDE' *Passport to Hell* vi. 103 The male defenders of the prostitutes' quarters, known to the dictionary as *souteneurs*, and to the troops as bludgers. 1937 [see *BLUDGE *v.*]. 1960 *Observer* 28 Aug. 28/2 They are strikingly different to the white prostitutes who ply their trade for coloured bludgers.

2. A parasite or hanger-on; a loafer. Now often in weakened sense. *Austral.* and *N.Z.* *slang.*

1939 X. HERBERT *Capricornia* (1949) 81 All the men here are loafers and bludgers. 1941 BAKER *Dict. Austral. Slang* 11 *Bludger*, a loafer, idler, ne'er-do-well; one who imposes on others. 1942 *2 N.Z.E.F. Times* 21 Dec. 17/4 Some bludger's got my tin hat. 1953 R. BRADDON in I. Bevan *Sunburnt Country* 130 A *bludger* is a shirker. 1969 *Courier-Mail* (Brisbane) 12 June 2/5 Surely if one is willing to give a good day's work for a good day's pay one should be given a chance to earn. I'm no bludger.

blue, *a.* Add: **1. b.** *Magnetism.* Defining the south pole of a magnet (of a steel-blue colour) as distinguished from the north (red) pole; also, the magnetism of this pole.

1893 SLOANE *Electr. Dict.* 345 A two-fluid theory of magnetism has been evolved...It assumes a north fluid or 'red magnetism' and a south fluid or 'blue magnetism'.

f. See also GREY-BLUE, POWDER-BLUE 2, *smalt-blue* (SMALT *sb.* 4), STONE-BLUE 2.

g. Defining a quality of sheep's wool (see quot.).

1888 *Encycl. Brit.* XXIV. 656/2 In the worsted trade the classification [of wool] goes..in descending series, from fine, blue, neat, brown, breech, downright, seconds, to abb. ..The greater proportion of good English long wool will be classified as blue, neat, and brown.

h. Applied to animals with fur of a bluish-grey colour. So *blue hare* (see *12 a).

1863, 1884 [see *blue fox*, BLUE *a.* 12 a]. 1887 P. M. RULE *Cat* v. 66 *Blue or Silver Tabby.*.. The ground-colour is a silver grey, with the stripes of a darker shade. 1909 *Daily Chron.* 15 Dec. 3/5 Cats of all colours, from 'blue' (which stands for smoke-grey in the cat-world) to 'cream'. 1968 H. HARMAR *Chihuahua Guide* 232 *Blue*, a blue-gray, such color as might be seen in the whippet or Bedlington.

i. blue and white *a.*, having a surface diversified with blue and white; *spec.* of china. Hence *ellipt.* as *sb.*

1719 LADY FERMANAGH *Let.* 19 Mar. in M. M. Verney *Verney Lett. 18th Cent.* (1930) II. II. xxiii. 60, I shall want a dozen of blew and white china plates. 1753 *Aris's Birmingham Gaz.* Nov. in E. Meteyard *Life of J. Wedgwood* (1865) I. 243 This is to give notice to all painters in the blue and white potting way, and enamellers on china ware, that [etc.]. 1830 P. NEILSON *Recoll. Six Years' Residence in U.S.* 320 Trowsers and frocks made of common drugget, (or blue and white as it is called). 1848 H. R. FORSTER *Stowe Catal.* 1 Twelve fruit dishes, of old blue and white. *Ibid.* 215 A blue-and-white bowl and covers. 1856 C. M. YONGE *Daisy Chain* I. i. 2 The fire-place..was ornamented with blue and white Dutch tiles bearing marvellous representations of Scripture history. 1875 *Cornh. Mag.* XXXI. 538 The blue-and-white porcelain of Nankin. 1941 E. BOWEN *Look at all those Roses* 236 Blue-and-white plates, in metal clamps, hung in lines up the walls. 1968 K. WEATHERLY *Roo Shooter* 120 The blue-and-white roo had chosen..a day camp.

j. *blue china.* Cf. BLUE *sb.* 4.

1833 LAMB *Old China* in *Last Ess.* 218 A set of extraordinary old blue china. 1881 TROLLOPE *Ayala's Angel* i, A few little dinner parties to show off his blue china. 1937 V. WOOLF *Years* 9 A Dutch cabinet with blue china on the shelves.

2. b. Phr. *blue* (*in the face*): livid with effort, excitement, etc. Used hyperbolically.

1864 TROLLOPE *Small Ho. at Allington* II. xvii. 175 You may talk to her till you're both blue in the face, if you please. 1917 'H. H. RICHARDSON' *Fortunes R. Mahoney* II. viii. 175 He alone must argue himself blue in the face over it. 1928 N. COWARD *C. B. Cochran's 1928 Revue* I. v. 10 Beckon and coo Till you are blue, Mermaids have got no dam chance at all. 1934 F. BALDWIN *Innoc. Byst.* (1935) xi. 212 I've talked myself blue in the face to him. 1959 'P. QUENTIN' *Shadow of Guilt* x. 89 Swear till you're blue in the face that Chuck was with you all day. 1968 *Observer* 3 Nov. 3/3 I've been looking into..cases of dealers' rings..until I'm blue in the face.

3. *blue fear*, a variant of *blue funk*. Also *blue-funk school*, a jocular perversion of 'blue-water school'.

1883 STEVENSON in *Longm. Mag.* Apr. 683 The very name of Paris put her in a blue fear. 1908 *Daily Chron.* 24 Feb. 4/6 The identification by Mr. Harvey, M.P., of the 'blue-water school' with the 'blue funk school'. *Ibid.* 20 July 4/3 The Jingo..is a nobler being than the disciples of our 'blue-funk' school.

b. Intoxicated. *slang* (chiefly *U.S.*).

1818 M. L. WEEMS *Drunkard's Looking Glass* (ed. 6) 4 The patient goes by a variety of nicknames..such as boozy—groggy—blue—damp. 1860 [see sense 10]. 1945 BAKER *Austral. Lang.* ix. 166 A man who is drunk is said to be..*blue*.

c. Of affairs, circumstances, prospects: dismal, unpromising, depressing. Chiefly in *a blue look-out, to look blue.*

1833 *Mirror of Lit.* 25 May 350 'Why it's a blue look out, Master,' said he. 1857 TROLLOPE *Three Clerks* xxix, Charley replied that neither had he any money at home. 'That's blue,' said the man. 1879 HARTIGAN & WALKER *Stray Leaves* Ser. II. xv. 257 If our present officers are like them..it's a blue look-out for the Afghans! 1888 'R. BOLDREWOOD' *Robb. under Arms* II. xvii. 258 It seemed a rather blue look-out. 1966 WODEHOUSE *Plum Pie* iii. 84 You don't want Freddie's whole future to turn blue at the edges.., do you?

d. Of, pertaining to, or characteristic of the singing or playing of the *BLUES. So *blue note*: a minor interval occurring where a major would be expected; an off-pitch note; also *transf.* and *fig.*

1919 *Dancing Times* Aug. 482/2 From the clarionet began to flow the wierdest blue notes ever heard. 1928 *Melody Maker* Oct. 1093/1 Gramophone enthusiasts will be able to determine..the class of rhythm of the rendering, i.e. Blue, slow, or fast Fox-trot. 1930 E. RICE *Voy. Purilia* x. 134 In the jazz symphony of life, there are many blue notes. 1945 V. THOMSON *Musical Scene* i. 30 The greatest master of 'blue', or off-pitch, notes. 1955 H. KURNITZ *Invasion of Privacy* (1956) vi. 49 Her voice trailed off in a forlorn blue-note fashion. 1969 C. BOOKER *Neophiliacs* iii. 68 The 'blue notes' and 'flattened chords'..which provide sensations in jazz.

8. *blue murder*, used in intensive phrases: see MURDER *sb.* 3 in Dict. and Suppl.

9. (Examples.) Cf. *BLUE *sb.* 14.

1864 HOTTEN *Slang Dict.* 78 *Blue*, said of talk that is smutty or indecent. 1935 *Economist* 16 Mar. 584/2 The songs sounded not vulgar exactly, but .. 'a bit on the blue side'. 1959 *Spectator* 14 Aug. 180/1 It meant that the theatre-going public were deprived of..outstanding contemporary plays, yet allowed to visit 'blue' variety shows. 1965 *Punch* 2 June 799/1 He also wanted to see a blue movie.

b. (See quot. 1890.)

1890 FARMER *Slang* I. 256/1 *To make the air blue*, to curse; to swear; to use profane language. 1924 'R. CROMPTON' *William—the Fourth* iv. 72 A man in his shirt-sleeves whose language is turning the air blue for miles around.

11. Selected additional examples: *blue-bleak, -blooded* (earlier example), *-bloused* (examples), *-brilliant, -cheeked, -clad, -haired* (recent example), *-hearted, -nailed, -rinsed, -shirted, -suited, -ticked, -white* (example), *-winged* (earlier and additional examples).

1877 G. M. HOPKINS *Poems* (1918) 29 And blue-bleak embers..Fall, gall themselves. 1853 MRS. GASKELL *Cranford* vii. 128 The old blue-blooded inhabitants of Cranford. 1885 WARREN & CLEVERLY *Wand. of Beetle* i. 3 The women ..attended by blue-bloused admirers. 1949 BLUNDEN *After Bombing* 48 And blue-bloused workmen in the yards at meals. 1923 D. H. LAWRENCE *Birds, Beasts & Flowers* 143 The peacock..struts blue-brilliant out of the far East. 1787 LATHAM *Synopsis of Birds* Suppl. 93 Blue-cheeked C[urucui]. 1956 *Nature* 3 Mar. 404/2 The blue-cheeked bee-eater, added to the British list only in June 1951. 1871 WHITMAN *Passage to India* 91 Disperse, ye blue-clad soldiers! 1956 R. FULLER *Image of Society* vii. 178 Rose conferred with a blue-haired saleswoman. 1894 *Geol. Mag.* Oct. 463 A blue-hearted limestone. 1920 T. S. ELIOT *Ara Vus Prec* 15 A meagre, blue-nailed, phthisic hand. 1958 J. CANNAN *And be a Villain* ii. 47 He thought with satisfaction of his own blue-rinsed little-black-frocked mother-in-law. 1968 *Listener* 31 Oct. 565/3 'Is this what you call an education?' said a blue-rinsed Manhattan matron. 1907 *Westm. Gaz.* 6 May 7/1 The voiceless millions of blue-shirted fellaheen.

1934 *Times* 19 Feb. 17/3 General O'Duffy was met by 2,000 blue-shirted men and women yesterday. 1925 BLUNDEN *English Poems* 87 Blue-suited Labour's hoeing By Labour's graves. 1961 *Guardian* 12 June 7/1 Blue-suited young organisation men. 1908 *Daily Chron.* 29 Aug. 7/4 The blue-ticked dog. 1909 *Westm. Gaz.* 1 May 2/3 With half-mad eyes and blue-white quivering lips. 1637 MORTON *New Canaan* II. iv, Teales there are of two sorts, green winged and blew winged. 1732 Blue-wing'd [see SHOVELLER^2 2]. 1789 MORSE *Amer. Geogr.* 59 The Blue winged Teal. *Ibid.*, The Blue winged shoveller. 1931 *Hardy's Anglers' Guide* 60 Dry Flies..Blue Winged Olive, Male.

12. a. blue-bill *U.S.* and *dial.*, = SCAUP-DUCK; **blue cat,** also a North American species of cat-fish; **blue fly,** a blue-bottle fly; **blue-grey** (see quot. 1902); **blue hare,** the varying hare (see VARYING *ppl. a.* 3); **blue heeler,** an Australian cattle-dog with a dark blue speckled body; **blue ling,** popular name of a kind of ling, *Molva byrkelange* (earlier called *lesser ling*); **blue pointer,** the popular name of a shark of either of two species found esp. in Australasian waters (see quot. 1953); **bluetail,** (*a*) *dial.* the fieldfare; (*b*) *red-flanked bluetail*: an Asian bird, *Tarsiger cyanurus*; **blue-throat,** substitute: any of various birds esp. of the genus *Cyanecula* or *Cyanosylvia* (see quots.); **blue whale,** a bluish-grey rorqual, *Sibbaldus musculus*; **blue-wing,** *spec.* an American variety of teal.

1813 WILSON *Ornith.* VIII. 84 Scaup Duck..better known among us by the name of the Blue Bill. 1890 J. WATSON *Nature & Woodcraft* vii. 83 The fishermen hereabout call them [*sc.* scaups] 'dowkers' and 'bluebills'. 1909 R. W. CHAMBERS *Firing Line* x, The little blue-bill ducks came swimming in scores. 1965 *Jrnl. Lancs. Dial. Soc.* Jan. 13 *Scaup* (bluebill, cockleduck: Morecambe Bay, 1892). 1835 J. MARTIN *Descr. Virginia* 347 Fine fish, particularly the mud and blue cat. 1877 R. I. DODGE *Hunting Grounds Gt. West* 250 The blue cat is also common in all the plain streams, attaining sometimes a weight of fifteen to twenty-five pounds. 1759 GOLDSM. *Bee* No. 4 ¶ 30 A large blue fly fell into the snare. 1856 STOWE *Dred* 160 He just puts me in mind of one of these blue-flies. 1910 W. DE LA MARE *Three Mulla-Mulgars* xi. 151 That howl brings half the forest against me, like blue-flies to meat. 1902 *Encycl. Brit.* XXV. 191/2 The cross between the Shorthorn and the Aberdeen-Angus [breed of cattle], known as the 'Blue Grey'. 1960 *Times* 25 Jan. 19/1 The original Galloways were put to a Shorthorn bull. This gave us a Bluegrey cow. 1895 R. LYDEKKER *Handbk. Brit. Mammalia* 226 The geographical distribution of the Mountain, Alpine, Blue, Irish, or Polar, Hare, as the animal is variously called, is very extensive. 1962 M. BURTON *Dict. Mammals of World* 92 Varying Hare..also known as Blue Hare, Scottish Hare or Alpine Hare..has bluish tinge in spring and autumn, at change of coat. 1946 F. D. DAVISON *Dusty* (1947) xvii. 196 One of the dogs was a blue heeler, a cattle dog. 1916 A. MEEK *Migrations of Fish* xviii. 237 Genus Molva..*M. byrkelange.* Blue or lesser ling. *Ibid.* 238 The blue ling is occasionally landed at our ports from boats which have been fishing in deep water to the north. 1882 J. E. TENISON-WOODS *Fish of N.S.W.* 95 (Morris), On the appearance of a 'blue pointer' among boats fishing for schnapper outside, the general cry is raised, 'Look out for the blue pointer'. 1917 *Chambers's Jrnl.* Sept. 588/1 Those that the Sydney fisherman knows best,..and destroy human life, are the tiger shark, the blue pointer [etc.]. 1953 J. L. B. SMITH *Sea Fishes S. Afr.* 49 *Carcharodon carcharis*..Blue-Pointer (Durban)...This swift and voracious shark is a terror to all who venture on or in the water. *Ibid.* 50 *Isurus glaucus*..Blue Pointer, Mako Shark...A famous angling fish, abundant in Australasian waters, very swift and powerful. 1866 J. BOWRING in *Trans. Devonshire Assoc. Advancem. Sci.* I. v. 18 The *bluetail*..and many more will probably fly away. 1878 H. E. DRESSER *Hist. Birds of Europe* II. LXVII. 2 Nemura Cyanura (Red-flanked Bluetail)..this richly coloured Asiatic bird is found throughout Asia. 1885 SWAINSON *Prov. Names Birds* 5 *Fieldfare*..From the predominant bluish tinge of its upper plumage are derived:—Blue tail (Midlands; West Riding), [etc.]. 1948 S. BRUCE in *Scottish Naturalist* LX. 6 (*title*) The Red-Flanked Bluetail in Shetland. *Ibid.*, The bluetail will henceforward appear on the British list...The bluetail is related to the redstarts, bluethroats, robins, and chats. 1873 W. YARRELL *Hist. Brit. Birds* (ed. 4) I. 321 The Bluethroat *Phœnicura Suecica*. *Ibid.* 323 The majority of Bluethroats which come to the rest of Continental Europe ..were..first distinguished by Brehm as *Cyanecula leucocyana*. 1954 BANNERMAN *Birds Brit. Isles* III. 307 Bluethroat, *Cyanosylvia svecica*. .. Red-spotted Bluethroat, *Cyanosylvia svecica svecica*. .. White-spotted Bluethroat, *Cyanosylvia svecica cyanecula*. 1851 H. MELVILLE *Moby Dick* I. xxxi. 229 There are a rabble of uncertain, fugitive, half-fabulous whales..the Iceberg Whale..the Blue Whale; &c. 1888 *Encycl. Brit.* XXIV. 524/2 The 'blue whale', the largest of all known animals, attains a length of 80 or even sometimes 85 feet. 1937 *Discovery* Nov. 357/2 The largest animal known, the blue whale or Sibbald's rorqual. 1709 LAWSON *Carolina* 148 The Blue-Wings are less than a Duck, but fine Meat. 1754 [see TEAL 2]. 1768 WASHINGTON *Diaries* (1925) I. 294 Went into the Neck and up the Creek after Blew Wings. 1874 J. W. LONG *Amer. Wild-fowl Shooting* xv. 192 They are a trifle smaller than the blue-wings. 1895 *Outing* (U.S.) Oct. XXVII. 43/1 A bunch of blue-wing teal rose from the ice-pond.

b. blue ash *U.S.*, a North American variety of ash; **blue-bush,** a popular name of any of several varieties of shrub, esp. the Australian *Kochia pyramidata*; **blue star** (see quot.).

1783 W. FLEMING in N. D. Mereness *Trav. Amer. Col.* (1916) 667 Blue Ash a spieces of the White Ash and called so from the bark tinging water of that colour. 1819 D. THOMAS *Trav.* 93 The blue ash..is a fine stately tree of two or three feet diameter, generally of a straight grain, and may be easily split into rails. 1832 D. J. BROWNE *Sylva Amer.* 156 The Blue Ash is unknown to the Atlantic parts of the United States. 1876 W. HARCUS *South Australia* 124 (Morris), Thickly grassed with short fine grass, salt and blue bush, and geranium and other herbs. 1933 *Bulletin* (Sydney) 7 June 25/2 Blue bush is far-famed among cattlemen—a stout shrub of 2ft. or 3ft. high, with short thick leaves. 1934 *Ibid.* 5 Dec. 20/2 Sandalwood-tree..grows over the whole s.-w. of Queensland, though not always in commercial quantities, and is generally known as 'plum' or 'bluebush'. 1936 F. CLUNE *Roaming round Darling* xviii. 177 All about this country grows a bluebush shrub (actually more green than blue), with closely assembled leaves, very good for sheep. 1958 L. VAN DER POST *Lost World of Kalahari* i. 24 A cleft over-grown and purple with the shadow of blue-bush. 1886 *Encycl. Brit.* XX. 174/1 The blue star [grass, of Queensland], *Chloris ventricosa*.

c. blue clay, a clay of this colour, esp. (*a*) that used in pottery manufacture (see quot. 1957); (*b*) = *blue ground* (see sense *13); **blue copper:** see also COPPER *sb.* 1 b; **blue metal,** also, argillaceous shale of a bluish colour, used esp. in road-making.

1733 ELLIS *Chiltern & Vale Farming* 17 Black or Blue Clays—I am now come to touch on the very best of all Clays. 1784 R. KIRWAN *Elem. Min.* viii. 78 Blue clays. These sometimes lose their colour and become white when heated. 1886 G. A. FARINI *Through Kalahari Desert* iii. 31 Two, three, four blasts followed..and some tons of hard blue clay were loosened ready to be carried to the 'floors'. 1938 *Thorpe's Dict. Appl. Chem.* II. 309/1 Diamonds occur sparingly in river beds or embedded in 'blue clay'. 1957 *Encycl. Brit.* VII. 842/2 *Ball Clay.* This is found principally in Devon and in Dorsetshire, and is sometimes known as blue clay, owing to its greyish-blue colour, which is due to organic matter. When fired..it becomes white. 1808 Blue metal [see METAL *sb.* 10]. *a* 1835 J. PHILLIPS *Geol.* in *Encycl. Metrop.* (1845) VI. 628/2 The least fragment of jet or morsel of bituminous shale, especially if accompanied by 'blue metal' is enough to make a credulous proprietor listen to an ignorant collier. 1892 R. O. HESLOP *Northumberland Words* 69 *Blue metal*, indurated argillaceous shale, of a bluish purple colour, resembling that of blue slates. 1909 A. H. DAVIS *On our Selection* (1953) vi. 30 We had been accustomed to pelt her with potatoes and blue-metal. 1966 'J. HACKSTON' *Father clears Out* 102 They came close enough into the college to send messages to the teacher by showering blue metal on the roof of their Alma Mater.

13. blue baby, an infant suffering from congenital cyanosis; **blue bag,** a barrister's (orig. a solicitor's) brief-bag of blue stuff; hence, one carrying such a bag; so (*nonce*) *to forget the blue bag*, to ignore (the indications of) one's rank; **blue band,** a band of glacier ice of a blue colour due to the absence of air-bubbles; **blue boy** *slang*, a policeman (usu. in *pl.*): cf. *boys in blue* s.v. *BLUE *sb.* 3 b; so **Blue Force**; **blue brick** (see quot. 1889); **blue brittleness,** the brittleness of steel at blue heat; **blue butter** = **blue ointment*; **blue cheese,** a cheese marked with veins of blue mould (cf. *BLEU); **blue-collar** *attrib.* (chiefly *U.S.*), designating a manual or industrial worker, as distinct from a 'white-collar' worker; so *blue-collared* adj.; **blue comb (disease),** a disease affecting young pullets in which there is cyanosis of the comb; **blue-domer** *colloq.*, one who does not go to church, preferring to worship beneath the 'blue dome' of heaven; hence **blue-domeism,** this attitude to worship, **blue-domeist,** one who has such an attitude; **blue earth** = **blue ground*; **Blue Force** (see **blue boy*); **blue gas** (see quot.); **blue ground,** the dark soil, normally greyish-blue, in which diamonds are found; **Blue Guide** [cf. F. *Guide Bleu*], one of a series of popular guide-books with blue covers; **blue-jean** *attrib.*, made of blue jean; as *sb. pl.*, trousers made of blue jean; also *transf.*; hence *blue-jeaned* adj.; **blue laws** (earlier examples); **blue measure** (see quot.) *Obs.*; **blue Monday** (earlier examples); **blue moon** (earlier example); **blue-mouldy** *a.*, covered with blue mould; also *fig.*; **Blue Mountain (coffee),** a type of Jamaican coffee; **blue mud,** a marine deposit coloured by organic matter and iron sulphide; **Blue Nun** = CONCEPTIONIST 2; **blue oil** (see quot.); **blue ointment,** mercurial ointment (see MERCURIAL *a.* 5); **blue pencil,** a blue 'lead' pencil used chiefly in marking corrections, obliterations, and the like; **blue-pencil** *v. trans.*, to mark, score through, or obliterate with a blue pencil; hence, to make 'cuts' in; to censor; also used euphemistically (cf. *BLANK *sb.* 12 b); hence **blue-pencilling** *vbl. sb.*; **Blue Peter,** (*b*) *U.S.* (see quots.); **blue pigeon** *slang*, †(*a*) (the act of stealing) lead, esp. that used for roofing; also *attrib.*; hence *to fly the blue pigeon*: to steal lead; (*b*) *Naut.* the sounding lead; see PIGEON *sb.* 4 b; **blue pill,** (*b*) *U.S. slang*, a bullet; **blue-plate,** *U.S.* (see quot. 1961); **blue process** (see quot.); **blue (process) paper,** a sensitized paper used for copying maps and plans, made by saturating the paper with potassium ferrocyanide; **Blue Shirt, Blueshirt,** one who wears a blue shirt as a sign of allegiance to or membership of a particular group, party, etc.; **blue streak** *colloq.* (orig. *U.S.*), (*a*) something resembling a flash of lightning in speed, vividness, etc.; (*b*) a constant stream of words; esp. in phr. *to talk a blue streak*; **Blue Train** [cf. F. *Train Bleu*] (see quot. 1951); **blue-vinnied** *a. dial.* [cf. FINNY, VINNY *a.*^{2}], having a blue mould; hence **blue-vinn(e)y,** a blue-mould cheese made in Dorsetshire; **blue-washed** *a.*, (*a*) washed by the blue sea; (*b*) covered with a blue wash; **blue water-gas** = **blue gas*; **blue-water school,** a collective term applied to politicians or political thinkers who regard a strong navy and the command of the sea as essential to the security of the country, or as the chief or the only sufficient defence.

1903 R. H. BABCOCK *Dis. Heart* xxix. 692 The little patient had been a *blue baby* from birth. 1964 M. MCLUHAN *Understanding Media* II. xxviii. 282 Watching a blue-baby heart operation on TV. 1809 MALKIN tr. *Le Sage's Gil Blas* II. v. i. 343 Said I to myself, every now and then, when they forgot the blue-bag: this is the way of the world! Every one fancies himself to be..superior to his neighbour. 1817 *Black Dwarf* 31 Dec. 814 Black legs, blue-bags, learned wig-blocks. 1837 DICKENS *Pickw.* xxxiii, Mr. Pickwick.. followed Mr. Perker and the blue bag out of court. 1852 —— *Bleak Ho.* i, A battery of blue bags is loaded with heavy charges of papers and carried off by clerks. 1910 Blue bag [see *BRIEF *sb.* 11]. 1895 DANA *Man. Geol.* (ed. 4) 243 Lamellar or straticulate structure of glacier-ice modified by ..the 'blue bands', or 'veined structure'. 1937 *Discovery* Feb. 36/1 There are 'blue bands' or layers of a very much harder ice, containing small bubbles of air. 1883 JAS. GREENWOOD *Odd People in Odd Places* 68 The instrumental 'blue boys' belonging to several metropolitan divisions. 1912 A. S. M. HUTCHINSON *Happy Warrior* v. vii. § 2 The Blue Boys from the police-station..scoured the country. 1850 R. PROSSER in E. Dobson *Manuf. Bricks & Tiles* I. iv. 99 An ordinary blue brick weighs, wet from the mould, 12 lbs. 4 oz. 1889 E. J. BURRELL *Elem. Building Construction* xiv. 218 *Staffordshire blue bricks.* These are composed of clay containing a large percentage of oxide of iron, which is converted into the black oxide by intense heat, giving a characteristic dark-blue colour to the bricks. 1940 *Chambers's Techn. Dict.* 99/1 *Blue bricks*,..(made chiefly in Staffordshire and North Wales)..form the best quality engineering bricks. 1919 *Jrnl. Iron & Steel Inst.* C. 521 Fettweis.. explains the phenomenon of blue-brittleness and the aging of steel. 1959 *Ibid.* CXCI. 100/2 It was found that the best heat treatment for reducing blue brittleness in 0·55% C steel was subcritical annealing at 550° C. 1874 HOTTEN *Slang Dict.*, *Blue Butter*, mercurial ointment used for the destruction of parasites. 1925 L. J. LORD *Pract. Butter & Cheese Making* xxvi. 145 This kind [*sc.* Stilton] and that next described are full-milk blue cheeses. 1950 *Tuscaloosa News* 25 Nov. 1/5 'Blue collar' workers also include helpers, laborers, and supervisors. 1958 *Listener* 23 Oct. 631/1 The blue-collar people, the machine operators. 1959 *Observer* 1 Nov. 8/2 The split is no longer between white- and blue-collar workers, but between those with and those without the college diplomas. 1967 *Boston Herald* 1 Apr. 5/8 He thought his main appeal would be to the 'little business men', clerks, auto workers, building and trade workers and similar blue collar groups. 1959 V. PACKARD *Status Seekers* (1960) ii. 25 The recent great gain of white-collared workers over blue-collared ones. 1967 *Guardian* 16 May 2/4 The traditional 'iron curtain' between white and blue collared civil servants. 1941 *Amer. Jrnl. Vet. Res.* II. 261/2 Depression is marked and the comb may be wilted and cyanotic, hence the popular name 'blue comb'. 1950 *N.Z. Jrnl. Agric.* May 442/1 The condition of poultry known as pullet disease, blue-comb disease, new wheat disease, or X disease has been known in Great Britain and America for several years. [1820 SHELLEY *Cloud* vi, And the winds and sunbeams with their convex gleams, Build up the blue dome of air.] 1945 A. HUXLEY *Time must have a Stop* xxx. 289 Those of us who ..have found that humanism and blue-domeism are not enough. 1952 J. MASEFIELD *So Long to Learn* II. 100 There were..Celtic Fringers,..Blue Dome-ists, [etc.]. 1961 PARTRIDGE *Dict. Slang* (ed. 5) 1005/1 *Blue-domer*, an officer that absents himself from church parade. 1962 *Times Lit. Suppl.* 3 Aug. 554/4 'Blue-domer' parents in whose eyes the most damnable thing was to be 'pi'. 1897 *Westm. Gaz.* 22 May 3/2 The well-known 'blue earth' of the diamond mines. 1927 *Daily Express* 27 Dec. 1 The 'Blue Force',—that is, the uniform branch of the police. 1902 *Encycl. Brit.* XXVIII. 602/1 Water gas in its original state is called 'blue gas', because it burns with a blue, non-luminous flame. 1886 J. NOBLE *Handbk. Cape Good Hope* 192 The 'blue ground'.. far from being barren of diamonds..yielded even better returns than the upper layers of 'yellow ground'. 1911 *Daily Chron.* 25 Mar. 3/2 Of very great rarity is a piece of 'blue ground' with a diamond embedded in it...The 'blue ground' ..is..not uncommonly black or brown. 1918 (*title*) The Blue Guides: London & its Environs. 1935 *Discovery* June 156/2 The Blue Guides..bear the mark of an original and scholarly mind. 1855 H. MELVILLE *Israel Potter* xx. 216 Across the otherwise blue-jean career of Israel, Paul Jones flits and re-flits like a crimson thread. 1932 F. L. WRIGHT *Autobiogr.* I. 16 Blue-jean overalls with blue cotton suspenders. 1957 Blue jeans [see *JEAN]. 1960 *20th Cent.* Dec. 557 A simple provincial gaucheness, a corny blue-jeans innocence. 1956 J. POTTS *Diehard* viii. 138 She.. swung her blue-jeaned legs. 1781 S. PETERS *Hist. Connecticut* (1877) 44 Even the religious fanatics of Boston and the mad zealots of Hertford..christened them the 'Blue Laws'. 1842 DICKENS *Amer. Notes* I. v. 175 The local legislature of Connecticut, which sage body enacted..the renowned code of 'Blue Laws'. 1891 *Daily News* 27 Apr. 3/2 The 'Blue' Measure...A measure called the blue (which contained two-thirds of a pint, and was universally used in Wales). 1801 W. RENDER *Tour through Germany* I. 100

Germany is indebted to this wise emperor [*sc.* Joseph II], among many other abolitions, to the two following in particular; namely, *Der blaue Montag*, 'the blue Monday', and the 'infamy of certain trades'. 1840 *Boston Transcript* 2 Mar. 2/2 This was blue Monday in the House. 1844 W. HOWITT tr. *Holthaus's Wand. Journeyman Tailor* ix. 109, I did not omit on Sundays, and sometimes too on blue Mondays, to go about and observe the life and manners of this great city [*sc.* Constantinople]. 1821 P. EGAN *Real Life in London* I. xiv. 249 'How's Harry and Ben?—haven't seen you this *blue moon*.' [*Footnote*] *Blue Moon*—This is usually intended to imply a long time. 1874 HARDY *Madding Crowd* II. iii. 39 Stales [*sc.* cakes], that were all but blue-mouldy, but not quite. 1900 *Daily News* 3 Apr. 2/5, I was 'blue-mouldy for want of a batin',' as they say in certain parts of the Empire. 1922 JOYCE *Ulysses* 293, I was blue mouldy for the want of that pint. 1922 W. H. UKERS *All about Coffee* xxiv. 350 Jamaica produces two distinct types of coffee, the highland and the lowland growths. Among the first named is the celebrated Blue Mountain coffee, which has a well developed pale blue-green bean that makes..a pleasantly aromatic cup. 1963 L. DEIGHTON *Horse under Water* xxxiii. 130, I was drinking a second cup of Blue Mountain. 1876 J. MURRAY in *Proc. R. Soc.* XXIV. 499 A blue mud containing:—A great many pelagic Foraminifera and some Pteropod shells. 1882 GEIKIE *Text-Bk. Geol.* III. ii. 438 The deposits were found to consist of blue and green muds derived from the degradation of older crystalline rocks. 1885 Blue mud [see MUD *sb.*^{1} 1 c]. *a* 1700 in *Harleian Misc.* (1744) I. 425/2 A Monastery of Visitation-Nuns, otherwise *Blue Nuns*. *a* 1848 E. PETRE *Notices of Eng. Colleges* (1849) p. v, Nuns of the Conception, *or Blue Nuns*, at Paris. 1931 *N. & Q.* 22 Aug. 143/2 The most prominent of them is Elizabeth Anne, who helped to establish the Blue Nuns in Paris, and became their Abbess. 1885 *Encycl. Brit.* XVIII. 242/2 The oil from which hard and soft paraffin are separated ..exhibits a blue fluorescence, and is hence called blue oil. 1839 *R. Hooper's Lex. Med.* (ed. 7) 268/1 *Blue ointment*, the mercurial ointment. 1888 *N.Y. Herald* 29 July (Farmer), The editor of the *Century Magazine* blue pencils magazine articles by the bushel. 1893 KIPLING *Many Invent.* 167 The blue pencil plunged remorselessly through the slips. 1899 *Daily News* 17 Feb., The actor will have a better chance after the blue pencil has eliminated the unnecessary verbiage in the dialogue. 1904 S. E. WHITE *Blazed Trail Stories* iii. 48 One log had not been blue-pencilled across the end. 1914 G. CANNAN *Old Mole* 39 He blue-pencilled false quantities in Latin verse. 1925 *Mus. Assoc. Proc.* 60 He would blue-pencil an unprepared chord of the seventh in a motet in the style of Palestrina. 1940 *New Statesman* 13 Apr. 491/2 They certainly are blue-pencil unpopular with a lot of people in wartime. 1904 A. BENNETT *Great Man* xxviii. 331 The blue-pencilling of the play proceeded. 1709 LAWSON *Carolina* 151 Blue-Peters, the same as you call Water-Hens in England, are here very numerous. 1917 *Birds of Amer.* I. 214 Coot. *Fulica americana.* .. Other names.—American Coot..Blue Peter. 1732 *Select Trials Old-Bailey* (1735) II. 468/1 We could find no better Business than stealing Lead—We call it the *Blue Pigeon*, or the *Buff-Lay*. 1781 G. PARKER *View of Society* II. i. 63 *Blue Pigeon-Flyer.* These are Journeymen Plumbers and Glaziers who repair houses, and Running Dustmen. To *fly the Blue Pigeon* is cutting off lead from what they call a Prayer Book up to a Bible. 1789 —— *Life's Painter* xv. 165 *Blue pigeon flying.* Fellows who steal lead off houses, or cut pipes away. 1856 *Harper's Mag.* XIII. 589/1 The speed [of the ship] was slackened, and the 'blue pigeon' kept constantly moving. 1887 *Judy* 27 Apr. 200 A burglar whose particular 'lay' was flying the blue pigeon, i.e., stealing lead. 1897 Blue pigeon [see PIGEON *sb.* 4 b]. 1834 'J. DOWNING' *Andrew Jackson* 111 They saw no hopes from fitin, they wern't fond of blue pills. 1945 S. LEWIS *C. Timberlane* (1946) xix. 112 They were taking the Blue Plate Dinner. 1952 AUDEN *Nones* 27 Having finished the Blue-plate Special And reached the coffee stage. 1961 WEBSTER *Blue Plate.* 1. A restaurant dinner plate divided into compartments for serving several kinds of food as a single order. 2. A main course (as of meat and vegetable) served as a single menu item. *a* 1884 KNIGHT *Dict. Mech.* Suppl., *Blue Process for Copying*, a mode of copying tracings in lieu of re-tracing them. 1933 YEATS *Let.* 13 July in *New Statesman* (1965) 19 Mar. 441/2 Our chosen colour is blue and Blueshirts are marching about all over the country. 1937 E. SNOW *Red Star over China* i. 30 Chiang Kai-shek's own Blueshirt gendarmes. *Ibid.*, Over 300 Communists were imprisoned.. and the Blueshirts were hunting for more. 1938 *Ann. Reg. 1937* 274 He also demanded that the Blueshirts, a sort of private army of the Wafd, should be dissolved. 1965 C. D. EBY *Siege of Alcázar* (1966) ii. 48 Moscardó discovered the sixty Falangists hiding in the photography laboratory... Vela argued that if fighting broke out, the Blue Shirts would be useful to the Alcázar. 1830 *Kentuckian* 14 May, To pass..with such rapidity as not even to leave a 'blue streak' behind him. 1847 *Knickerbocker* XXX. 178 Interspersing his vehement comments with a 'blue streak' of oaths. 1895 S. HALE *Lett.* (1919) 289, I..drove in her sort of..carryall..talking a blue streak two miles to her house. 1937 RUNYON *More than Somewhat* iii. 64 She hears..a guy cussing a blue streak. 1949 *Landfall* III. 236 Sid was talking a blue streak to Jean. 1968 'R. RAINE' *Night of Hawk* xxii. 109, I was talking a blue streak, my expression like thunder. 1928 A. CHRISTIE *Mystery of Blue Train* ix. 65 *The* best train is what they call 'The Blue Train'. You avoid the tiresome Customs business at Calais. 1951 *Oxf. Jun. Encycl.* IV. 46/2 *Blue Train.* This train, named from the colour of its cars, was originally the Calais–Mediterranean express, which carried passengers to and from the sea-side towns of the French Riviera. More recently, however, the main Blue Train has worked between Paris and Mentone..though sleeping cars run through from Calais to Mentone without a change. 1893 DARTNELL & GODDARD *Wiltshire Gloss.* 14 *Blue-vinnied*, covered with blue mould. 1863 Blue vinny [see FINNY *a.*^{2}]. 1886 W. BARNES *Gloss. Dorset Dial.* 50 *Blue-vinny, or vinnied, cheese*, blue mouldy. 1895 'C. HARE' *Down Village Street* 231 Us do want..a pen'orth o' blue-vinny cheese. 1955 J. G. DAVIS *Dict. Dairying* (ed. 2) 190 Dorset Blue cheese (also known as Blue Vinney).. is a skimmed-milk cheese. 1905 *Westm. Gaz.* 18 Nov. 6/3 A small stone city, set round a blue-washed bay. 1906 *Ibid.* 13 Jan. 4/3 Out in the blue-washed bay. *Ibid.* 14 Nov. 2/1 From the blue-washed wall an unshaded lamp shone brilliantly. 1921 *Flight* XIII. 387/2 There is also a hydrogen plant, of the blue water gas type, having a capacity of 200,000 cu. ft. per day of hydrogen at not less than 99·2 per cent. purity. 1944 *Gloss. Terms Gas Industry* (*B.S.I.*) 20 *Blue water-gas* (*B.W.G.*), gas consisting almost exclusively of carbon monoxide and hydrogen in nearly equal proportions. 1902 *Westm. Gaz.* 13 Jan. 3/2 Mr. Roosevelt is revealed..as convinced a member of the blue-water school as Mr. Clowes in England. 1908 [see sense *3]. 1966 *New Statesman* 25 Feb. 247/1 The differences between the Powellites and the Blue Water school on China, Vietnam, Aden [etc.].

blue, *sb.* Add: **2. a.** (Earlier examples.) See also BLUE *a.* 1 f.

1724, etc. [see PRUSSIAN *a.* 2 a]. 1835 [see COBALT 2 a].

c. A cake or ball of blue powder for laundry use; also *attrib.* in **blue bag,** a bag containing one of these for such use (see also quot. 1869). See also POWDER-BLUE 1, STONE-BLUE 1. Hence *blue-bagged* adj.

1836 *Mag. Dom. Econ.* I. 6 Make also a proper flannel 'blue-bag'. 1869 MRS. BEETON *Bk. Househ. Managem.* 1098 To remove a Bee Sting, pull the sting out at once.. wipe the place, suck it, and then apply the blue-bag. 1904 H. G. WELLS *Food of Gods* I. ii. § 5 She 'athen't even got a Blue Bag, Thir. 1928 *Daily Express* 9 Aug. 3/3 The blue bag—that truly rural remedy for wasp stings. 1953 DYLAN THOMAS *Under Milk Wood* (1954) 72 She sees in the still middle of the bluebagged bay Nogood Boyo fishing from the Zanzibar. 1961 *Coast to Coast 1959–60* 128 Knobs of blue.

3. esp. of blue uniform, e.g. of policemen, wounded soldiers in hospital, (*U.S.*) the Northern or Union soldiers (contrasted with the Southern or Confederate grey).

1831 *The Olio* June 341/2 A young off'sir—a likely youngster he would have been in blue, 'stead o' red. 1867 F. M. FINCH in *Atlantic Monthly* XX. 369 (*title*) The Blue and the Grey. 1879 TOURGÉE *Fool's Err.* xxv. 153 He wore the blue. 1884 *Eng. Illustr. Mag.* Dec. 167 Answered the warder, 'He's in blue, so he's in his last year.' 1893 *Cassell's Fam. Mag.* Apr. 338/2 My little friend in blue [i.e. messenger boy]. 1918 *Reveille* Aug. 94 The discharged men..no longer cut the handsome gaudy figure of the man in blue. 1930 T. E. LAWRENCE *Let.* 8 Jan. (1938) 677, I haven't drawn my blue yet! 1952 R. SHERBROOKE-WALKER *Khaki & Blue* i. 9 Why 'Khaki' found so much to criticise in 'Blue'. *Ibid.* ii. 15 From these early 'soldiers in blue'..there was ultimately evolved..the R.A.F. Regiment.

b. *the men* (*gentlemen* or *boys*) *in blue*: (*a*) policemen; (*b*) sailors; (*c*) American Federal troops.

1851 F. STARR *Twenty Yrs. Trav. Life* xxxii. 317, I was not long managing my exit..whilst the 'gentleman in blue' was busy examining other tickets. 1857 H. LAWRENCE in W. Brock *Biogr. Sk. Sir H. Havelock* (1858) viii. 125 The gentlemen in blue—the sailors. 1866 L. P. BROCKETT *Camp, Battle Field*, etc. II. 264 Tread lightly, oh! loyal-hearted, the boys in blue are lying there. 1866 *Congress. Globe* 27 Jan. 460/1 The brave 'boys in blue' fought manfully and through their efforts the Union has been preserved. 1868 *Cassell's Mag.* 8 Aug. 240/1 The man in fustian lies under certain obligations to the man in blue. 1882 BESANT *All Sorts* xliii, You must now begin to think seriously about handcuffs and prison, and men in blue. 1943 WYNDHAM LEWIS *Let.* 26 Jan. (1963) 344 He will choose..*The Navy.* He will plump for the boys-in-blue every time.

4. (Further examples.) Also applied to police and sailors.

1844 HOOD *Whimsicalities* II. 186 Whether this here mobbing..Will grow to such a riot that the Oxford Blues [*i.e.* policemen] must quell it. 1860 HOTTEN *Dict. Slang* (ed. 2), *Blue*, a policeman. 1877 W. H. THOMSON *Five Years' Penal Servitude* iv. 257 'Bilking the blues,'—evading the police. 1898 KIPLING *Fleet in Being* iii. 40 Next time you see the 'blue' ashore you do not stare unintelligently. 1903 F. SIMPSON *Bk. Cat* xi. 126/2 As tiny kittens blues frequently exhibit tabby markings. 1968 K. WEATHERLY *Roo Shooter* 62 The last one [*sc.* kangaroo] was a tremendous blue.

b. A blueberry or bilberry. *U.S.*

1587 R. HAKLUYT tr. *Laudonnière's Notable Hist.* 2^r, There are Raspisses, and a little bearie which we call among us Blues, which are very good to eate. 1709 J. LAWSON *New Voy. Carolina* 104 The Hurts, Huckle-Berries, or Blues of this Country are four sorts...The first sort is the same Blue or Bilberry, that grows plentifully in the North of England.

c. A 'blue chip' in the game of poker. *U.S.*

1890 BIFF HALL *Turnover Club* 206 When turned to faro, You sometimes caused 'a stack of blues' to win. 1920 MULFORD *J. Nelson* xxi. 228 'Two pairs...Well I'll see it an' add a blue.' 'Any time you raise a blue, you got two pairs, all right!'

5. a. Phrases: *out of the blue*, 'out of a clear sky' (cf. *SKY *sb.*^{1}), without warning, unexpectedly; *a bolt from* (or *out of*) *the blue*, something unexpected, a complete surprise.

1837 CARLYLE *Fr. Rev.* III. VI. i. 347 Arrestment, sudden really as a bolt out of the Blue, has hit strange victims. 1910 O. W. HOLMES *Let.* 19 Dec. (1964) 58, I got an encouragement out of the blue..in the form of an honorary degree from Berlin. 1911 W. F. BUTLER *Autobiogr.* xxi. 380 Like a bolt from the blue came the news of the Jameson raid. 1919 W. T. GRENFELL *Labrador Doctor* (1920) xx. 334 Nothing in my life ever came more 'out of the blue' than my marriage. 1925 'SAPPER' *Out of the Blue* i. 10 Suddenly —out of the blue—comes one disconnected event. 1930 W. S. CHURCHILL *My Early Life* xiii. 178 Spontaneously and 'out of the blue', he formed a wish to make the acquaintance of its author. 1933 *Mind* XLII. 289, I do not state it out of the blue, with no reason at all. 1953 X. FIELDING *Stronghold* III. i. 175 Then, suddenly, out of the blue, the leading question. 1970 *Times* 30 Apr. 5/8 Out of the blue comes a request for information about the work of the Estimates Committee.

11. (Earlier example.)

1781 F. BURNEY *Diary* 12 June (1904) I. 498 A whole tribe of *blues*, with Mrs Montagu at their head.

12. (Earlier and later U.K. examples.)

1741 D. GARRICK *Let.* 11 July (1963) I. 26, I am far from being quite well, tho not troubled wth y^e Blews as I have been. 1856 WHYTE MELVILLE *K. Coventry* viii. 89 The moat alone is enough to give one the 'blues'. 1887 'J. S. WINTER' *That Imp* ii. 11, I wonder you and Betty don't die of the blues. 1960 *New Statesman* 27 Feb. 274/2 The post-election blues are beginning.

14. = BLUENESS 4. Cf. BLUE *a.* 9 in Dict. and Suppl.

1824 MACTAGGART *Gallov. Encycl.* 446 *Thread o' Blue*, any little smutty touch in song-singing, chatting, or piece of writing. 1889 BARRÈRE & LELAND *Dict. Slang* I. 145/2 'A bit of *blue*', an obscene or libidinous anecdote.

15. *U.S.* The style or mood of the *BLUES. Cf. *BLUE *a.* 3 d.

1924 G. GERSHWIN (*title of music*) Rhapsody in blue.

16. *Army slang.* **a.** (See quot.)

1925 FRASER & GIBBONS *Soldier & Sailor Words* 28 *In the blue*, failure. Something gone wrong. An attack that broke down, or troops who got out of touch, would be said to be 'in the blue'.

b. *the blue*: the desert, *spec.* in N. Africa in the 1939–45 war.

1944 *Return to Attack* (*Army Board, N.Z.*) 9/1 Although their business address was the Western Desert, the New Zealanders were not yet ready for action. They trained hard. At intervals they disappeared 'into the blue' for a few days. 1958 L. VAN DER POST *Lost World of Kalahari* v. 83 You ought to have letters from us to our people in the blue just in case of need. 1963 *Times* 26 Jan. 9/6 In the 20 years since the tide of battle finally receded, 'up the blue', as the Desert Rats called it, has been tidied of war debris.

17. *Austral.* and *N.Z. slang.* **a.** A summons (cf. *BLUEY *sb.* 4). **b.** An argument; a fight or brawl. **c.** A mistake or blunder. Cf. *BLACK *sb.* 12.

1941 in BAKER *Dict. Austral. Slang* 11. 1944 in L. GLASSOP *We were Rats* (Partridge *Slang* Suppl.). 1946 J. MORRISON in *Coast to Coast* 156 There's a bit of a blue on, and he can hear nearly every word they say. 1953 K. TENNANT *Joyful Condemned* iv. 37 Every time Rene comes round there's some kind of a blue. 1957 'N. CULOTTA' *They're Weird Mob* (1958) iv. 54 When you get into a blue do yer pull knives? 1941 BAKER *Dict. Austral. Slang* 11 *Blue*, an error or mistake; a loss. 1957 'N. SHUTE' *On Beach* i. 23, I put up a blue right away by ordering a pink gin. 1961 B. CRUMP *Hang on a Minute, Mate* 17 Trouble with you blokes is you won't admit when you've made a blue.

blue (blū), *v.*^{2} *slang.* Also **blew. 1.** *trans.* To spend or get through (money) lavishly or extravagantly; = *BLOW *v.*^{1} 9 c.

1846 *Swell's Night Guide* 76 The coves..vot we blues a bob or a tanner to see. 1859 HOTTEN *Dict. Slang*, s.v. *Blewed*, 'I blewed all my blunt last night', I spent all my money. 1867 T. W. ROBERTSON *Caste* III, 'So Papa Eccles had the money?' 'And blued it!' 1884 *Daily Tel.* 28 May 5/1 He took to horses, and blewed the blooming lot [i.e. £1,700] in eighteen months. 1888 FARJEON *Miser Farebrother* III. i. 5 You brought down two thousand pounds with you, and you blued it. 1930 W. DE LA MARE *On Edge* 228 She had taken a holiday and just blued some of her savings. 1959 *Observer* 17 May 8/5 Men in cotton shirts and corduroys met there to 'blue' their cheques on supplies and on fiery colonial rum.

2. To make a mess of, spoil, ruin.

1880 *Punch's Almanack* 2 This top coat would blue it.

blueback (blū·bæk). [f. BLUE *a.* + BACK *sb.*^{1}]

1. A species of fish, or bird, having a bluish back. Also *attrib.*

c 1532 [see BLUE *a.* 12 a]. 1812 W. BENTLEY *Diary* 22 Oct. (1914) IV. 125 Mr Osgood..had taken great numbers of the Herrings called Bluebacks..this season. 1843 J. E. DEKAY *Zool.* in *Nat. Hist. N.Y.* v. 24 There is a variety of the Lobster, termed Bluebacks, on account of their dark bluish colour. 1871 *Game Laws* in *Fur, Fin, & Feather* (1872) 158 This section shall not apply to the taking of blue back trout in Franklin and Oxford counties. 1881 *Amer. Naturalist* XV. 178 Of these species, the blue-back predominates in Frazer's river. *Ibid.* 180 Little blue-backs [*sc.* salmon] of every size down to six inches are also found in the Upper Columbia in the fall. 1883 [see BLUE *a.* 12 a]. 1965 *Jrnl. Lancs. Dial. Soc.* Jan. 9 *Fieldfare*,..blueback: 'Fylde',.. Warrington, Liverpool.

2. A legal-tender note issued by the Confederacy during the Civil War. *U.S.*

1869 in MATHEWS *Beginnings Amer. English* (1931) 156 The Rebels had their 'bluebacks' for money. 1872 SCHELE DE VERE *Americanisms* 47 During the Civil War..the original Blue Backs of the Confederacy (so-called in opposition to the Green Backs of the Union) soon became known as Shucks.

3. Chiefly *pl.* Paper money issued by the Transvaal Government in 1865. *S. Afr.*

1878 TROLLOPE *South Africa* II. 225 The blue-backs as they were called were printed. 1900 *Daily News* 17 Sept. 5/2 President Kruger has deserted them, taking all the gold, and leaving them only 'blue-backs'.

blue bag: see *BLUE *a.* 13, and *sb.* 2 c.

blue-bird. Add: **1. b.** *fig.* Happiness (in allusion to the title of a play, *L'Oiseau bleu* by Maurice Maeterlinck).

1909 A. T. DE MATTOS tr. *Maeterlinck's Blue Bird* III. ii. 131 You are looking for the Blue Bird, that is to say, the great secret of things and of happiness. **1909** M. BEERBOHM *L'Oiseau Bleu* in *Yet Again* 300 All these ladies..have tried to catch this same Blue Bird. **1968** K. O'HARA *Bird-Cage* vi. 46 A little blue dickey-bird. A bluebird of happiness, a love-bird of happiness.

blue book. Add examples of general use:
1633 S. GOFFE *Let.* 7 Nov. in *Cal. State Papers* (1863) 279 Dr. Ames's death has put them in hope that they shall not be troubled so much with blue books as heretofore. **1793** *Gent. Mag.* LXIII. 59/2 A work which, among the Roman Catholicks, is known by the name of 'The Third Blue Book'. **1848** (*title*) The Blue Book of the British Manufacturers. **1909** B. WARD *Dawn Cath. Revival* I. i. 2 It is described in the Third Blue Book as follows. *Note*, The Blue Books were the official publications of the Catholic Committee. **1912** CONAN DOYLE *Lost World* vi. 89, I was the flail of the Lord up in those parts, I may tell you, though you won't find it in any Blue-book. **1928** *Daily Tel.* 5 May 7/6 The 'Blue Book', the directory..of the electrical engineering and allied trades. **1967** *Listener* 10 Aug. 173/3 What was new was the copious Blue Book documentation that Marx gave.
b. *U.S.* (Examples.)
1836 BENTON in *Deb. Congress* June 1719 An array of names more numerous..than was to be found in the 'Blue Book'. **1964** *Brooklyn Eagle* 7 Sept. 5 The bulky Blue Book of the present day..is in two volumes, each of more than 1000 pages.
Hence **blue-booky** *a.*, **blue-bookiness.**
1894 *Speaker* 9 June 649/1 A dry, blue-booky article. **1909** *Fabian News* XX. 76/1 His volume is..readable, notwithstanding the complete blue bookiness of its contents.

bluebottle. Add: **2.** (Further examples.)
1846 G. W. M. REYNOLDS *Myst. London* I. ii. 5/2 Them chaps in..the House of Commons gets on their legs and praises the bluebottles up to the skies as the most acutest police in the world. **1946** WODEHOUSE *Joy in Morning* xi. 87 'Stilton turns out to be the village bluebottle.'..'A policeman, sir?'
4. A Portuguese man-o'-war (MAN-OF-WAR 4). *Austral.* and *S. Afr.*
1933 *Bulletin* (Sydney) 29 Nov. 21/3 The stings of the pestilent 'bluebottle', or Portuguese man-o'-war. **1947** L. G. GREEN *Tavern of Seas* xv. 143 A seashore strewn with physalia, better known as blue-bottles or Portuguese men-o'-war, is no place for bare feet. **1956** S. HOPE *Diggers' Paradise* 164 The risk to swimmers and surfers in the southern seas is mainly from..the Portuguese man-o'-war which the Aussies call a 'bluebottle'. **1964** *Cape Times* (Week-End Mag.) 11 Jan. 4/3 The Portuguese Man o' War is small...The common name is bluebottle.

blue buck. Add: Also, a small South African antelope of the genus *Cephalophus*.
1835 J. W. D. MOODIE *Ten Years S. Africa* II. 139 The woods also abound with..an elegant little antelope,..called the '*blaawe bock*' or blue buck. **1893** J. NOBLE *Handbk. Cape & S. Afr.* 62 The diminutive bluebuck, not bigger than a rabbit.

blue-cap. Add: **8.** (See *CAP *sb.*[1] 11 c.)

blue chip. orig. *U.S.* [BLUE *a.* + CHIP *sb.*[1] 2 d.] A blue counter used in Poker, usu. of high value. Also *transf.*, spec. *Stock-Exchange*, a stock issue considered to be a fairly reliable investment, though less secure than gilt-edged; hence any reliable enterprise, etc. Also *attrib.* or quasi-*adj.* Cf. *BLUE *sb.* 4 c.
1904 S. E. WHITE *Blazed Trail Stories* viii. 146, I reckon I don't stack up very high in th' blue chips. **1929** *Sun* (Baltimore) 30 Oct. 12/2 Agriculture still has too much to reclaim..to become a 'blue chip' in the near future. **1932** *Ibid.* 16 Apr. 15/8 Investors do not see as much intrinsic value in lower-priced issues as they do in the so-called 'blue chips'. **1935** *Economist* 17 Aug. 339/1 The result is not a 'blue-chip' boom. **1954** *Ibid.* 20 Feb. 568/2 Tobacco equities seem rather more vulnerable than other industrial 'blue-chips'. **1957** R. POWELL *Philadelphian* 185 One of the gilt-edged prep schools..or even the blue-chip ones. **1958** *Times* 2 Jan. 12/2 Demand was again concentrated on the 'blue chip' issues and a comparatively few store..and chemical shares. **1967** *Listener* 5 Oct. 423 (Advt.), M & G General Trust Fund is a very big Fund..with a portfolio largely of British 'blue chips', plus a number of overseas stocks.

blue coat, blue-coat. Add: **1. b.** A bluish colour of the coat in deer at a certain period.
1870 *Amer. Naturalist* IV. 190 The spike-horn was shot just as deer were attaining the 'blue coat'.
2. a. Also, a policeman.
a **1852** MAYHEW *Lond. Labour* (1861) II. 369/1, I thinks them Chartists are a weak-minded set..a hundred o' them would run away from one blue-coat. **1875** *Chicago Tribune* 29 Aug. 5/4 One of the blue coats would attempt to put back the crowd. **1932** J. FARRELL *Young Lonigan* (1936) iv. § 2. 74 No cop could think that he was going to get away with pushing his son. And he told the damn bluecoat that..he'd punch him all over the corner.
c. A soldier in the Federal army during the Civil War. *U.S.*
1865 G. W. NICHOLS *Story Gt. March* xxiii. 154 So we jogged on for awhile, and then..we descried a blue coat and a white-eared mule approaching. **1879** TOURGÉE *Fool's Err.* xxi. 122 But only wait until the States are restored and the 'Blue Coats' are out of the way.
3. *blue-coat girl.*
1695 PEPYS *Let.* 20 Sept. (1926) I. 110 Two wealthy citizens are lately dead, and left their estates, one to a *Blewcoat-Boy* and the other to a *Blew-Coat-Girl* in Christ's Hospital. **1894** *Daily News* 30 Mar. 5/1 To many..the notion of a Bluecoat 'girl' will be somewhat strange. It appears, nevertheless, that the Hertford establishment now shelters no fewer than 112 scholars of that sex.

blue devil. Add: **2. a.** (Earlier and later examples.)
1781 F. BURNEY *Diary* 2 July (1904) II. 17 Thinking..that generous wine will destroy even the blue devils. **1880** *Weekly Dispatch* 8 Feb. 12/2 He got discontented and had fits of blue devils.
Hence **blue-de·vil** *v. trans.*, to affect with the 'blue devils'; **blue-de·villed** *ppl. a.*; **blue-de·vilish, blue-de·villy** *adjs.*
1817 T. L. PEACOCK *Melincourt* I. iii. 32 When Mr. Hippy was blue-devilled, old Harry was vapourish. **1824** *New Monthly Mag.* XI. 427 Our blue-devilish and hypochondriacal countrymen. **1832** WHEWELL *Let.* 1 July in Todhunter *W. W.* (1876) II. 143 Do not wait till I become blue-devilled with hard work. **1836** in *Jrnl. Southern Hist.* (1935) I. 364 To be hemmed up in a strange place without..anything to interest you,..is enough to Blue Devil one. **1871** L. W. L. LOCKHART *Fair to See* I. viii. 208 The pine-trees loomed through stagnant mists with a dejected, blue-devilly aspect.

blue eye. Add: **d.** An Australian species of honey-eater (see quots.).
1848 J. GOULD *Birds Australia* IV. pl. 68 *Entomyza cyanotis*, Swains. Blue-faced Entomyza. Blue-eye of the Colonists. **1861** *Chambers's Encycl.* II. 171/1 Blue-eye (*Entomyza cyanotis*), a beautiful little bird, abundant..in New South Wales.

blue-eyed, *a.* (See Dict. s.v. BLUE EYE.) *blue-eyed grass* (earlier *U.S.* example).
1784 CUTLER in *Mem. Amer. Acad. Arts & Sci.* I. 487 *Sisyrinchium*. Blue-Eyed Grass. Blossoms blue.
b. *fig.* Innocent, ingenuous; favoured, esp. in phr. *blue-eyed boy*.
1924 WODEHOUSE *Bill the Conqueror* xix. 278 If ever there was a blue-eyed boy, you will be it. **1935** AUDEN & ISHERWOOD *Dog beneath Skin* I. ii, You can keep that blue-eyed stuff for the others. **1959** *Manch. Guardian* 4 Aug. 4/6 The blue-eyed enthusiasm of a writer. **1963** *Times* 5 Mar. 7/1 During this period, farmers were 'blue-eyed boys'.

blue grass, blue-grass. Chiefly *U.S.* [BLUE *a.* 12 b.]
1. a. A field-grass (*Poa compressa* or *pratensis*), characteristic especially of Kentucky and Virginia. (See also quot. 1903.)
1751 C. GIST *Jrnls.* (1893) 47 It is..full of beautiful natural Meadows covered with wild Rye, blue Grass and Clover. **1879** [see BLUE *a.* 12 b]. **1889** *Harper's Mag.* Jan. 259/1 The blue-grass..blooms toward the middle of June in a bluish, almost a peacock blue, blossom, which gives to the fields an exquisite hue. **1903** CLAPIN *Dict. Amer.* 60 In Maryland the term *blue grass* is applied to a species of grass very injurious to wheat and clover, and hard to eradicate.
b. The region of the blue grass; *spec.* the State of Kentucky.
1872 SCHELE DE VERE *Americanisms* 407 Both the region where it grows naturally, and the settlers there are known as Blue Grass simply, and hence the State of Kentucky especially is often thus designated. **1887** *Harper's Mag.* June 48/2 To that pleasant land of the Kentuckian, the 'Blue-grass'.
c. A type of American folk-music. Also *attrib.*
1958 in *Amer. Speech* (1960) XXXV. 223. **1960** *20th Cent.* Dec. 559 A new compromise simulacrum, the bluegrass band..five-string banjo, mandolin, guitar, bass, fiddle and a holler-vocalist. **1969** *Rolling Stone* 28 June 11/3 Hillman has watched the rock scene change over the years (as a bluegrass mandolin player, then one of the original Byrds).
d. *attrib.* So **Blue Grass State,** Kentucky.
1772 *Pennsylvania Gaz.* 16 Apr. 4/3 Timothy and blue grass hay to be sold. **1863** E. DICEY *Six Months in Federal States* II. 64 What is called the Blue Grass region of Kentucky. **1883** [see BLUE *a.* 12 b]. **1886** *Chicago Wkly. News* 29 Apr. 4/3 Kentucky..is also called the Blue Grass state.
e. *ellipt.* as *adj.* Belonging to or characteristic of the blue-grass country.
1889 *Harper's Mag.* Aug. 459/2 Bud rode into the yard on Mollie... 'Blue-grass all over. I wonder how he came by her.' **1913** G. STRATTON-PORTER *Laddie* x. 274 There's a strain of Arab in the father,..and the mother is bluegrass.
2. Applied to other grasses.
1811 W. AITON *Gen. View Agric. Ayr* 305 Carices, sedge-grasses..are, by the Ayrshire farmers, called blue, sour, one-pointed grasses. **1845** *New Statist. Acc. Scot.* VII. 518 We have but very little of the blue or star-grass (*carex panicea*). **1902** *Encycl. Brit.* XXXII. 108/2 On the extensive plains [of Queensland]..may be seen..'Blue grass' (*Andropogon sericeus, A. pertusus, A. refractus*, and *A. erianthoides*).

Blue Hen. *U.S.* The state of Delaware; chiefly in *Blue Hen's chickens* (see quot. 1840), *Blue Hen State*.
1830 MRS. ROYALL *Lett. fr. Alabama* 69 He told one of our party he was 'One of the blue hen's chickens'. **1840** *Niles' Reg.* 9 May 154/3 In the revolutionary war..Captain Caldwell [of Delaware] had a company called by the rest 'Caldwell's game cocks', and the regiment after a time in Carolina was nicknamed from this, 'the blue hen's chickens' and the 'blue chickens'...But after they had been distinguished in the south the name of the *Blue Hen* was applied to the state. **1864** *Congress. Globe* June 2968/2, I remember the early history of the Blue Hen's Chickens and it is a proud one. **1867** *Trübner's Amer. Lit. Rec.* 1 Aug., [Delaware] is popularly known..as The Blue Hen or Diamond State. **1897** *Congress. Rec.* Mar., App. 68/2, [I thank] the gentleman from the 'Blue Hen State' for his suggestion.

blueing, bluing *vbl. sb.* **2.** Delete 'In U.S.' and add examples.
1669 SIR THOMAS BROWNE *Let.* 28 Apr. (1946) 51 Enquire after smalt, a stone wherof they make bleweing for paynting & starch. **1889** *Cent. Dict., Bluing*..The indigo, soluble Prussian blue, or other material, used in the laundry to give a bluish tint to linen. **1895** *Cal. Univ. Nebraska 1895–6* 174 Soaps, washing powders, polishing powders, bluing, etc. **1897** *Star* 17 Sept. 4/7 Dust the glass with powdered bluing done up in a bit of cheese cloth.

blueness. Add: **4.** (Further example.)
1891 *Sat. Rev.* 8 Aug. 168/2 That tinge of 'blueness' which repels English propriety.
5. A state of depression or melancholy.
1867 W. JAMES *Let.* 7 Nov. (1920) I. 120, I am in a mood of indigestion and blueness. **1911** J. C. LINCOLN *Cap'n Warren's Wards* ix. 147 Before evening his blueness had disappeared.
b. The fact or quality of music being 'blue' (see *BLUE *a.* 3 d). orig. *U.S.*
1949 R. BLESH *Shining Trumpets* II. viii. 193 The beautiful, unveiled tone that on other occasions could transmit a blueness difficult to imagine. **1959** W. RUSSELL in M. T. Williams *Art of Jazz* (1960) 36 Those devices that gave a feeling of blueness to his harmony.

blue-nose. Add: **2. b.** A Canadian, esp. Nova Scotian, ship.
1889 in *Cent. Dict.* **1899** *Daily News* 2 Dec. 6/4 For sheer brutality commend me to the life on board a 'Bluenose', as we call the Canadian whalers. **1949** LAWSON & SWEET *Our New Brunswick Story* 229 In those old days all the ships of the Maritime Provinces were known among sailors the world over as Bluenoses.
4. A person who is excessively puritanical. Also *attrib.* *U.S.*
1927 H. C. BROWN *In Golden Nineties* v. 187 With a lot of blue noses on the Board,..this concession was not secured without great diplomacy. **1929** *Variety* 3 Apr. 11/4 That this picture may aggravate blue nose censors is not beyond the bounds of possibility. **1945** *Chicago Daily News* 2 Aug. 10/7 Our bluenoses are doing a grave injury to the men serving overseas, who have got the impression that married women are running wild.

blue-nosed, *a.* [BLUE *a.* 1.] Having a blue nose; also (chiefly *U.S.*) *transf.* and *fig.*, esp. as a term of disparagement or contempt; *spec.* priggish, puritanical (cf. *BLUE-NOSE 4).
1809 W. IRVING *Knickerb.* III. ii, A goodly, blue-nosed, skim'd milk, New-England cheese. **1866** C. H. SMITH *Bill Arp* 87 The blue-nosed Yankees were to pollute our sacred soil the next morning. **1890** *Amer. N. & Q.* V. 6 Can you tell me why Presbyterians are sometimes called 'blue-nosed'? **1959** F. BOWERS *Textual & Lit. Crit.* iii. 112 It is an ignorant view..that bibliography acts as a blue-nosed and puritanical censor to take all the joy out of the textual critic's life. **1959** P. H. JOHNSON *Humbler Creation* xxxv. 235 Let's not be blue-nosed, but..he's doing it under the rose.

blue point, bluepoint. *U.S.* [The name of a headland near Great South Bay, Long Island.] Used *attrib.* or *ellipt.* to designate a small oyster from the south shore of Long Island; orig. one from a bed off Blue Point.
1789 in *Mass. Hist. Soc. Proc.* (1869) XI. 24 Judge Hobart..treated us with Blue Point oysters from the shell. **1868** G. ROSE *Great Country* 25 [Oysters] are called by many names; 'saddle rocks', 'blue points', and 'Shrewsburys', being the most popular. **1897** *Daily News* 19 Nov. 5/2 The menu will include such national dishes as..blue points, and pumpkin pie. **1909** 'O. HENRY' *Roads of Destiny* xx. 343 He got along fine with the olives and celery and the bluepoints. **1951** *Good Housek. Home Encycl.* 578/1 The large American Blue Points (now cultivated in the British Isles)..are bought for cooking.

blue-print, blueprint, *sb.* Also **blue print.** [BLUE *a.* 1.]
1. (A process for making) a photographic print composed of white lines on a blue ground or of blue lines on a white ground, used chiefly in copying plans, machine-drawings, etc.; also, a blue-toned photograph.
1886 *Electrician* XVI. 466/2 The Western Edison Light Company of Chicago..have adopted an arrangement for taking blue print by electric light. **1887** *U.S. Postal Laws & Reg.* § 364. 152 'Blue prints' reproduced only as copies of the original. **1892** *Photogr. Ann.* II. 102 Blue Prints can be obtained on albumenised paper. **1936** *Economist* 15 Feb. 351/2 Blue-prints were drawn for 'mastodon' capital ships of 60,000 tons, and for cruisers of 20,000 tons. **1943** J. S. HUXLEY *TVA* 141 Specifications for TVA Demountable Defense Houses (pp. v+42, illus. and blueprints).
b. *attrib.*
1895 *Montgomery Ward Catal.* 211/2 Blue print paper. **1909** *Cent. Dict.* Suppl. I. 147/2 *Electric blue-print machine*, a machine in which the sensitized paper is exposed behind tracings to a series of electric lights instead of to sunlight. **1923** H. A. MADDOX *Dict. Stationery* 14 *Blueprint paper*, a type of photographic paper used for taking rapid duplicates from machine plans, etc., the lines of the drawing developing out in white relief against a solid blue background or *vice versa*.
2. *fig.* A (detailed) plan or scheme; a pattern.
1926 *Spectator* 11 Sept. 385/1 Surely he can complete his life by giving us the blue-prints of the millennium. **1939** *Times Lit. Suppl.* 11 Feb. 87/2 Blue-prints of a new society come with better grace from Mr. Wells than from any other contemporary novelist. **1942** J. S. HUXLEY in *Fortune* Dec. 152/2 To nail some elaborate blue-print of international organization to our masthead.

blue-printing, *vbl. sb.* **a.** (The process of) printing in blue beneath the glaze on china, as distinguished from 'black-printing'. **b.** The action or method of making a photographic blue-print (*Cent. Dict.*, 1889). So **blue-printer.** Also (as a back-formation or f. **blue-print* sb.) **blue-print, blueprint** *v.*, lit. and fig. (see prec.).
1825 J. NICHOLSON *Operat. Mechanic* 470 *Blue-printing* is the impressions taken from engraved copper-plates by means of a rolling-press. *Ibid.*, The blue-printer lays the plate upon a stove. *Ibid.* 482 The *blue-printed pottery* is a very popular kind. **1865** L. JEWITT *Wedgwoods* 391 Blue Printing was introduced at Etruria at an early date, and has, of course, with black, etc., been continued to the present day. **1939** J. B. PRIESTLEY *Let People Sing* ix. 215 Though these girls kept away, thousands of others were only too eager to be blue-printed, numbered and air-conditioned. **1942** H. A. WALLACE *Cent. Common Man* 28 Dec. (1944) 36 We cannot now blueprint all the details, but we can begin now to think about some of the guiding principles of this world-wide new democracy. **1948** A. O'RAHILLY *Soc. Principles* viii. 55 The tidy bureaucratic minds of planners and blue-printers. **1960** AUDEN *Homage to Clio* I. p. 24 Do-it-yourself America Prophetically blue-printed this Palace kitchen. **1966** R. JEFFRIES *Death in Coverts* ix. 97 The first death might or might not have been part of a pattern, but the second death blueprinted that pattern.

blues (blūz). orig. *U.S.* [A use of the pl. *blues* (see BLUE *sb.* 12), treated as either sing. or pl.] A melody of a mournful and haunting character, originating among the Negroes of the Southern U.S., freq. in a twelve-bar sequence using many 'blue' notes. Also *attrib.*
1912 W. C. HANDY (*tune-title*) Memphis Blues. **1917** *Lit. Digest* 25 Aug. 29/2 'Must I Hesitate?', 'The Blues', 'Franky and Johnny', and other classics of the levee underworld. **1921** *Outward Bound* May 58/2 These 'labour songs',..like the 'blues' of to-day, were rather humorous. **1923** *Daily Mail* 28 July 7 Noisy 'jazz' music..is being driven out..by the soft pulsing of muffled melody in new tunes known as 'Blues'. *Ibid.*, A special 'Blues Trot' has been devised for dancing with the tunes, which are slower than [those of] a fox-trot. **1927** *Melody Maker* Sept. 865/2 The Yale..is danced to 'blues' tempo. **1928** *Oxf. Mag.* 1 Nov. 84/2 The use of a blues for the slow movement is interesting. **1949** R. BLESH *Shining Trumpets* I. v. 107 The blues scale..enters into and colors all singing and playing by American Negroes. *Ibid.* II. xi. 247 The blues-shouting trombone. **1956** MORGAN & HORRICKS *Modern Jazz* 16 The twelve-bar blues, long a source of expression for the outpouring of emotion, underwent a startling change with the introduction of the riff in jazz.

blue sky, blue-sky. [BLUE *a.* 1.]
1. *attrib.* Having the pleasant appearance of a blue sky; by extension, with difficulties ignored, unrealistic. Spec. *blue-sky book* (orig. *U.S.*): see quot. 1956.
1895 *Daily News* 2 Nov. 6/3 Paintings of Venice..which may be described as a blue sky series, so cheerful and gay is the aspect under which this artist generally views the picturesque city. **1920** W. HARD *Raymond Robins' Own Story* iii. 136 Lenin and Trotzky never gave me any blue-sky talk. They never promised unless they had the will and the *power* to deliver. **1956** *Observer* 15 Jan., Blue Sky Books are those literary works which, without any specific technique or expert knowledge, tell us their authors' views on the nature and purpose and proper conduct of human life. **1957** H. CROOME *Forgotten Place* iv. 55 The manuscript in this case is a blue-sky book, about to regenerate family life.
2. Used *attrib.* to denote dealing in doubtful or worthless securities, or legislation relating to this. (The allusion is supposed to be to one ready to sell the 'blue sky' to a credulous buyer.)
1906 C. R. WOOLDRIDGE *Grafters of Amer.* 48 They were what I would term 'blue sky and hot air' securities. **1912** *N.Y. Even. Post* 13 Jan., Financ. Suppl. 3 The 'Blue Sky' law of Kansas prohibits the sale..of stock or bonds of any company chartered outside the State, unless [etc.]. **1921** *Springfield Weekly Republ.* 3 Mar. 1 The origin of 'blue sky law'—legislation providing for State regulation of the sale of corporate securities—is middle western. **1948** MENCKEN *Amer. Lang.* Suppl. II. vii. 149 Carruth..suggested that *blue-sky*, to indicate a bad investment, might be from the German *blauer dunst*.

bluestone, blue stone. [BLUE *a.* 12 c.]
1. a. Copper sulphate.
1651 BIGGS *New Disp.* 27 That divine blew stone, is but Roman vitriol, that is to be had at every Druggist and Apothecaries shops. **1883** [see BLUE *a.* 12 c]. **1932** R. F. FORTUNE *Sorcerers of Dobu* 46, I doctored a leg-sore with bluestone.
b. Gin or whisky of so poor a quality as to be comparable with vitriol (sense 4). *slang.* Cf. *blue ruin* s.v. BLUE *a.* 13.
1880 *Blackw. Mag.* June 786 The effects of the mixture of spirits of wine, bluestone, and tobacco-juice. **1882** *N. & Q.* V. 348 'It wasn't whiskey,' he said, 'it was nothing but bluestone.'
2. a. A building or paving stone of a bluish grey colour; spec. *U.S.*, a bluish argillaceous sandstone quarried near the Hudson river. Also *attrib.*
1709 LAWSON *Carolina* 50 This [river]..affords as good blue Stone for Mill-Stones, as that from Cologn. **1852** A. CARY *Clovernook* 106 A little puddle of water had..soiled the bluestone hearth. **1895** DANA *Man. Geol.* (ed. 4) 593 The thicker layers [of flagstones] are called bluestone, from the bluish gray color. **1906** CHAMBERLIN & SALISBURY *Geol.* II. 559 Bluestone formation...Purple shale and thin red sandstone, with calcareous beds.
b. A basalt used for building and other industrial purposes. *Austral.* and *N.Z.*
1850 *Australasian* Oct. 138 A stone not unlike the trap or bluestone around Melbourne. **1886** J. HECTOR *Handbk. N.Z.* (ed. 4) 58 Basalts, locally called 'blue-stones', occur of a quality useful for road-metal, house-blocks, and ordinary rubble masonry. **1961** *Coast to Coast 1959–60* 36 He built the bluestone house near that little chapel you can see from here. **1969** *Melbourne Truth* 12 July 2/3 The hotel..has been renovated by putting brick veneer over the original bluestone.
c. Term applied to the stones, geologically foreign to the site, forming the inner circle and inner horseshoe of Stonehenge; later identified as spotted dolerite from the Prescelly Hills in Pembrokeshire. Also *attrib.*
1812 R. C. HOARE *Anc. Hist. S. Wilts.* v. 127 One of the blue stones of Stonehenge. **1865** J. LUBBOCK *Pre-historic Times* ii. 52 Fragments, not only of Sarcen stones, but also of the blue stones which form the inner circle at Stonehenge. **1926** *Man* XXVI. 42 Stonehenge—the supposed blue stone trilithon. **1957** CHILDE *Dawn Europ. Civilization* (ed. 6) xviii. 331 From the Presely Mountains in South-West Wales huge blocks of spotted dolerite (Bluestone) were transported to Salisbury Plain.

blue tongue, blue-tongue.
1. [Afrikaans *bloutong*.] A virus disease affecting horses and sheep, in which the tongue becomes swollen and blue. orig. *S. Afr.*
1863 J. S. DOBIE *S. Afr. Jrnl.* 19 Apr. (1945) 86 Blue tongue here also (which is a swelling of the tongue and lips which prevents the poor beast from eating). **1887** RIDER HAGGARD *Jess* viii, It's a beautiful veldt..no horse sickness, no blue-tongue. **1905** *Nature* 4 Sept. 502/2 Catarrhal Fever of Sheep: Blue Tongue. **1963** *Times* 22 Apr. 2/6 Blue tongue, though it exists no nearer to the United Kingdom than Spain and Portugal, is so feared in Australia that it has led to a ban on all livestock imports from Britain. **1966** *New Scientist* 3 Nov. 250/3 The most serious potential threat is probably from the bluetongue virus to which Merinos are highly susceptible.
2. An Australian lizard of the genus *Tiliqua*, belonging to the family Scincidæ.
1883 F. MCCOY *Zool. Victoria* I. VIII. 15 These Lizards are very sluggish, so that the popular name 'Sleepy Lizard' as well as 'Blue-tongue' comes to be applied to both. **1904** *Daily Chron.* 28 Jan. 6/2 The 'Australian Blue-Tongue' is a..lizard, with a curious habit of thrusting forth a long tongue as bright as turquoise blue. **1963** E. WORRELL *Reptiles of Australia* 34 The best-known lizards are the Bluetongue group (*Tiliqua*).
b. = ROUSTABOUT 2. *Austral. slang.*
1943 in BAKER *Austral. Slang* (ed. 3) 12. **1955** G. BOWEN *Wool Away!* 157 The poor old rouseabout...Australians sometimes call this hand a 'blue tongue'.
Hence **blue-tongued** *a.*, having a blue tongue; *blue-tongued lizard* = 2 above.
1883 F. MCCOY *Zool. Victoria* I. VIII. 13 Cyclodus Gigas...The Northern Blue-tongued Lizard. **1887** *Ibid.* II. 120 'Blue-tongued Lizard' or 'Sleepy Lizard'.

bluette (blue·t). A breed of oriental frilled pigeons having a white body and blue or silver wings.
1879 L. WRIGHT *Pigeon Keeper* 180 This is almost exactly the marking of the Bluette. **1891** R. WOODS *Pigeon-Culture* 138 Brunettes, Bluettes and Silverettes are..subvarieties of the Satinette, differing only in colour and markings.

blue-veined, *a.* [BLUE *a.* 11 b.]
a. Having blue veins.
1593, 1797 [see BLUE *a.* 11b]. **1953** R. GRAVES *Poems* 10 In the warm blue-veined nook Of your elbow crook.
b. spec. *blue-veined cheese* = *blue cheese* (see *BLUE *a.* 13); so *blue vein, blue veiny* (*cheese*). (Cf. **blue-vinnied* adj.)
1892 I. BEETON *Househ. Managem.* xli. 894 Dorsetshire yields us a very good cheese called *Blue veiny*, from its blue-veined appearance. **1950** J. G. DAVIS *Dict. Dairying* 43 *Blue-veined cheese*, a term applied to Stilton, Wensleydale, Roquefort, and Gorgonzola cheese, in which the blue-green mould, *Penicillium roqueforti*, grows in cracks and is an important ripening agent. **1959** MCLINTOCK *Descr. Atlas N.Z.* 58 The main kind of cheese produced [in N.Z.] is 'cheddar', with one factory at Eltham producing 'blue vein' cheese.

bluey, *a.* and *sb.* Add: **A.** *adj.* (Further examples.) Also *Comb.*
1889 *Pall Mall Gaz.* 1 May 3/1 An exquisite effect in bluey grey. **1900** *Daily News* 11 May 5/2 A bluey green colour. **1908** *Westm. Gaz.* 6 Feb. 4/3 There are few furs so eminently satisfactory..as the pretty, soft, bluey-grey squirrel.
B. *sb.* (Examples.) Esp. (of tramps or hikers) in the phr. *to hump bluey*, to travel with a swag, to hit the trail.
1890 *Melbourne Argus* 16 Aug. 13/2 'We shall have to hump "bluey" again,' said Tom...Early November saw us again on the track. **1891** R. WALLACE *Rural Econ. Austral. & N.Z.* 73 'Humping bluey' is for a workman to walk in search of work. **1896** C. J. O'REGAN *Poems* 29 You can fancy, old mate, how I'm longing..To flee with a 'billy' and 'bluey'. **1903** H. B. KING *Bill's Philos.* 28 It ain't for bluey-humpers to travel on *his* road. **1911** E. M. CLOWES *On Wallaby* xi. 278 An expression used for what in England we call 'tramping', is 'going on the wallaby', otherwise 'humping the swag', or 'the bluey', or 'sundowning'. **1966** G. MCINNES (*title*) Humping my Bluey. **1968** K. WEATHERLY *Roo Shooter* 43 The shooter walked fifteen or twenty miles carrying his swag, had a few at the pub, then found somewhere to roll his bluey.
2. Lead. (Cf. BLUE *a.* 12 c.) See also quot. 1897. *slang.*
a **1852** MAYHEW *Lond. Labour* Extra vol. (1862) 26 'Bluey-hunters', or those who purloin lead from the tops of houses. **1859** HOTTEN *Dict. Slang* 9 *Bluey*, lead. **1897** R. H. SHERARD *White Slaves of Eng.* v. 188 When the lead is cold, it is carried by the blue-bed women to the stack-house...They are called blue-bed women because after the molten lead has cooled in the moulds it assumes a bluish hue. The blueys can sometimes earn two shillings a day.
3. A rough outer garment, or the material of which such a coat is made. *Austral.*
1891 W. TILLEY *Wild West Tasmania* 29 (Morris), Miners, with their swags, surveyors in their 'blueys'. **1898** MORRIS *Austral Eng.* 39/2 *Bluey*. In the wet wildernesses of Western Tasmania a rough shirt or blouse..worn over the coat like an English smock-frock. Sailors and fishermen in England call it a 'Baltic shirt'. **1934** T. WOOD *Cobbers* 86 A seamless coat of 'Tasmanian bluey', cut low round the neck and under the arms.
4. A summons. Also *transf.* (Cf. *BLUE *sb.* 17.) *Austral.* and *N.Z.* *colloq.*
1909 T. H. THOMPSON *Ballads about Business* 13 I'll show you walls papered with blueys. *Ibid.* 49 How they served the 'blueys'. **1941** in BAKER *Dict. Austral. Slang* 11. **1942** 2 *N.Z.E.F. Times* 16 Mar. 6/5 That speed cop, who gave me my last bluey on point duty. **1945** O. BURTON *In Prison* i. 9 A summons, vulgarly known as a 'bluey'. **1951** J. FRAME *Lagoon* 122 The headmaster sent Mum a bluey saying action will be taken unless [etc.]. **1965** G. MCINNES *Road to Gundagai* xiii. 242 A uniformed John Hop..handed me the dreaded 'bluey', the summons for riding a bike without lights.
5. A blue Australian cattle-dog. *colloq.*
1933 *Bulletin* (Sydney) 13 Sept. 21/1 The third [dog] was a common little bluey I bought..as a pup. **1945** BAKER *Austral. Lang.* iii. 73 The *bluey* is a type of cattle dog: originally a cross of the smooth-haired Scotch sheep dog and the dingo.

bluff, *a.* Add: **1. c.** Of other objects: see quots.
?**1637** *Trav. P. Mundy* (MS. Rawl. A. 315), Peeces of wood..bluffe or blunt att both ends. **1920** *Discovery* Mar. 78/1 The streamline form for a kite balloon is rather bluff..and short compared with an airship. **1949** O. G. SUTTON *Science of Flight* ii. 48 Bodies which have sharply truncated tails (such as a shell) leave behind them a large disorderly wake and have large form drag. Such shapes are called *bluff bodies*.
3. *bluff-browed, -chested.*
1851 MELVILLE *Moby Dick* I. xli. 304 When Adam walked..bluff-browed and fearless. *Ibid.* 303 Milk-white charger..small-headed, bluff-chested.

bluff, *sb.*[1] Add: (Earlier examples.)
1687 in *S. Carol. Hist. & Gen. Mag.* (July 1929) 131 We landed on a Bluffe where some shads were. **1707** *S. Carol. Warrants for Lands* (1915) 206 One hundred acres of land on the uppermost bluff.
b. *attrib.*
1666 in *S. Carol. Hist. Soc. Coll.* V. 62 The North East side is a bluffe land, rounding from the River. **1768** J. LEES *Jrnl.* (1911) 33 With rather blowing Weather went round that large Bluff point. **1880** DANA *Man. Geol.* (ed. 3) 549 The loess...In the Mississippi valley, it covers the 'Orange sand', forming with it the 'Bluff formation'—so called because standing in bluffs in Missouri. **1882** *Encycl. Brit.* XIV. 42/2 (Kentucky) A homogeneous buff-coloured silicious loam known as the 'bluff' or loess formation. **1884** *Ibid.* XVII. 309/2 (Nebraska) These so-called 'bluff lands', composed of loess materials.
2. *Canada.* A grove or clump of trees.
c **1752** W. COATS *Geogr. Hudson's Bay* (1852) 52 Near the same latitude, on the west main, is a bluff of wood, caled Point Mourning, from buriing one of Captain James men there. **1895** *Times* 13 Sept. 12/3 [Manitoba.] Level or rolling prairie with scattered 'bluffs' (small patches of wood) in some districts. **1912** N. L. MCCLUNG *Black Creek Stopping-House* 209 The Fourth came in a glorious day..with birds singing in every poplar bluff. **1959** *Country Guide* (Winnipeg) Apr. 71/1 The cabin still stood in the bluff above the pass.

bluff, *sb.*[2] Add: **3.** (Earlier example: = POKER *sb.*[4]) So *to call one's* (or *the*) *bluff* (orig. *U.S.*); *fig.* to make a person show his 'hand'; so, to accept a challenge or invite a show-down; cf. CALL *v.* 1 e (*b*); *to run a bluff on* (U.S.): to use bluff against or practise deception on (someone).
1846 S. SMITH *Theatr. Apprent.* 148 The game of 'bluff' or 'poker', as it is more generally called. **1876** BRET HARTE *Two Men of Sandy Bar* 17 But suppose that he sees that little bluff, and calls ye. **1887** F. FRANCIS Jr. *Saddle & Mocassin* 130 'You got the stock, though?' 'Oh,—yes! I run a bluff on 'em.' **1891** *Chambers's Jrnl.* 26 Sept. 618/2 A man may have a big name as a fighting man and yet not be able always to run a bluff on people so easily. **1896** *Congress. Rec.* 26 Mar. 3248 Where shall we be when the bluff is called? **1898** *Times* 4 Nov. 3/6 The policy of the Russian Foreign Office..has been a series of prodigious bluffs. Some of these Lord Salisbury has seen fit to 'call'; others he has refrained from 'calling'. **1933** *Discovery* Sept. 282/2 The childish questioner will very quickly call our bluff.
b. *attrib.* as *bluff game.*
1846 J. J. HOOPER *Adv. Simon Suggs* x. 129 'No!' said Simon.., 'the bluff game ain't played here!' **1888** *St. Louis*

Globe Democrat 29 Apr. (Farmer), They..thought he could outnumber them. It was a bluff game, and he won.
4. A bluffer. *U.S.*
1904 S. E. WHITE *Blazed Trail Stories* ii. 27 'You're a bluff!' said he, insultingly.

bluff, *v.*[1] Add: **2.** (Earlier examples.)
1846 S. SMITH *Theatr. Apprent.* 149 Inasmuch as I believe you only trying to bluff me off, I go two hundred. *a* **1859** *N.Y. Spirit of Times* (Bartlett), 'I goes you five dollars this time,' says Jim...'I..goes you ten better,' said Bill; 'you ain't agoin' to bluff dis child.'
3. (Earlier example.)
1854 *Congress. Globe* 3 May 1070 We both know how that game [of brag] is played. I thought I would bluff back on him.

bluffer. Add: **2.** (Examples.) orig. *U.S.*
1888 *Detroit Free Press* 5 May (Farmer), Is Uncle Sam a bluffer? **1891** 'L. HOFFMANN' *Cycl. Card & Table Games* 197 (*Poker*) Unless E has the reputation (a very undesirable one) of a habitual bluffer, B will probably begin to feel alarmed. **1894** *Westm. Gaz.* 24 Sept. 2/1 The 'bluffer', she who trades on the gilded European's illusion that all American girls are heiresses, and catches him under this false pretence, only to discover that she has fallen into the hands of a male 'bluffer' whose pretensions are equally baseless. **1946** *Sci. News Let.* 23 Mar. 188/1 The *puff adder*..is one of the world's biggest bluffers.

blu·ffing, *vbl. sb.* [f. BLUFF *v.*[1] 2 and 3.] The action of using bluff; also *attrib.* or as *ppl. a.*
1850 *Congress. Globe* 14 May, App. 606/1, I cannot look upon the effort of Texas in any other light than a bluffing, brow-beating game. *c* **1850** *Southern Sketches* 137 (Bartlett 1859), Jim..tried the bluffing system. **1871** [see BLUFF *v.*[1] 2]. **1948** *Redbook* (U.S.) Mar. 23/1 Andy was a blustering, bluffing, not too straitlaced fellow.

bluggy (blʌ·gi), *a.* Pretended infantile pronunciation of BLOODY (see quot. 1876), used subsequently as a euphemistic pronunciation of 'bloody', esp. to define literature of the blood-and-thunder type (see *BLOOD *sb.* 3 d.)
1876 J. HABBERTON *Helen's Babies* 68 Bliaff's head was all bluggy, an' David's sword was all bluggy—bluggy as everyfing. **1890** *Scott. Leader* 17 July 4 The shriek with which Mr. Hall Caine greeted Mr. Grant Allen's protest against bluggy novels. **1921** *19th Cent.* May 770 East End ambuscades where the air is thick with the smell of fried fish and Yiddish and bluggy rhetoric. **1966** 'K. NICHOLSON' *Hook, Line & Sinker* vi. 74 Bluggy Joan of Arc, with her messages.
Hence **blu·gginess.**
1894 *Punch* 30 June 305 It is not poesy, culture, wisdom, wit, That make the literary world go round. Much 'blugginess' has more to do with it. **1901** *Academy* 2 Nov. 404 In the close it falls sheer into rant and 'blugginess'.

blunder, *v.* Add: **7. b.** *trans.* To bring or cause to fall *into* a state by clumsy or inept behaviour; to use blunderingly.
1901 *Westm. Gaz.* 27 June 4/3 They believe that Mr. Chamberlain has blundered the country into war. **1912** A. S. M. HUTCHINSON *Happy Warrior* v. x. § 1. 347 He..hated to have blundered all his dullness on so rare and exquisite a thing.

blunt, *a.* Add: **2.** spec. *blunt instrument*, a loose term covering any large, heavy object that might be used as a murder weapon; *transf.*, a crude threat.
1923 D. L. SAYERS *Whose Body?* vi. 147 The blow..had been made with a heavy, blunt instrument. **1966** *Observer* 13 Nov. 6/7 He chided the Government for using words like 'freeze' as blunt instruments.
6. *blunt-featured, -fingered, -pointed* (examples), *-topped*; **blunt head,** a snake belonging to the family Amblycephalidæ.
1916 JOYCE *Portrait of Artist* v. 232 MacCann's flushed bluntfeatured face. **1858** A. IRVINE *Handbk. Brit. Plants* 440 *Veronica triphyllos*. Blunt-fingered Speedwell. **1869** GILLMORE tr. *Figuier's Reptiles & Birds* (1870) 51 The *Amblycephalidæ*, or Blunt Heads comprise a few species of moderate or small size. **1881** *Trans. Obstetr. Soc.* XXII. 37 A curved, blunt-pointed bistoury. **1937** *Discovery* Jan. 13/2 A straight-rooted, blunt-pointed tooth. **1908** E. STEP *Wayside Ferns* 108 Blunt-topped Horsetail. *Equisetum pratense.* **1963** *Times* 25 May 9/6 The hills are old and blunt-topped.

blurb (blɜ̄ɪb). *slang* (orig. *U.S.*). [See note below.] A brief descriptive paragraph or note of the contents or character of a book, printed as a commendatory advertisement, on the jacket or wrapper of a newly published book. Hence in extended use: a descriptive or commendatory paragraph. Also *Comb.*
Said to have been originated in 1907 by Gelett Burgess in a comic book jacket embellished with a drawing of a pulchritudinous young lady whom he facetiously dubbed Miss Blinda Blurb. (D.A.) See Mencken *Amer. Lang.* Suppl. I. 329.
1914 G. BURGESS *Burgess Unabridged* 7 *Blurb*, 1. A flamboyant advertisement; an inspired testimonial. 2. Fulsome praise; a sound like a publisher...On the 'jacket' of the 'latest' fiction, we find the blurb; abounding in agile adjectives and adverbs, attesting that this book is the 'sensation of the year'. **1918** *Wine, Women & War* (1926) 106 Americans, despite blurb in home press, [have] not yet succeeded in revolutionizing art of war. **1923** *Nation* 1 Aug. 121/2 The publishers..clapped on a jacket containing a blurb. **1924** *Spectator* 27 Sept. 426 The note of vanity is ominously accentuated by the publisher's blurb on the dust-cover, as silly and vulgar as the present writer has ever seen. **1926** *Times Lit. Suppl.* 21 Oct. 710/2 The paragraph briefly setting forth the merits of the book (known in 'the trade' as a 'blurb'). **1934** A. HUXLEY *Beyond Mexique Bay* 2 The blurb-writers promise to take you into the very heart of all these variegated delights. **1947** *Penguin Music Mag.* Sept. 60 The cast, the 'blurb' tells us, includes the pick of the younger generation of Italian operatic singers. **1948** *Ibid.* Oct. 22 Her name appeared recently [in concert advertisements]..mixed up with a blurb about 'the greatest living exponent' and so on. **1955** *Times* 4 Aug. 9/5 For why must publishers prefix to novels of this school a blurb in which much of the substance of the thriller is already revealed?

blurriness (blɜ̄·rinės). [f. BLURRY *a.* + -NESS.] The quality of being blurry; blurredness.
1937 *Burlington Mag.* Jan. p. xiv/2 The days of romantic blurriness are finished, and with them the indeterminate, fuzzy attitude towards art. **1958** *Listener* 6 Nov. 730/1 The increasing redness and blurriness of their later paintings.

blush, *sb.* Add: **2.** Phr. *at first blush* (examples).
1629 QUARLES *Argalus & P.* i. 33 And at first blush, she seemes, as if it were Some curious statue on a Sepulchre. **1886** *Bibliotheca Sacra* XLIII. 618 This sounds, at first blush, very neat, if not even very profound. **1955** *Times* 11 May 18/1 It may, at first blush, seem invidious to single out anyone in particular for special comment.
C. *blush-making* adj.; **blush-rose** (earlier examples); also *attrib.* or as *adj.*
1924 *Spectator* 12 Apr. 604/1 Otherwise we should have been spared those blush-making passages. **1944** T. RATTIGAN *While Sun Shines* III. i, Some idiotic, blushmaking, sentimental slush. **1629** PARKINSON *Parad.* cix. 420 The flowers are small single blush Roses, of little or no sent at all. **1707** MORTIMER *Husb.* xviii. 478 The *Blush Rose*, that differs in nothing from the other [*sc.* White Rose], but in the Colour of the Flowers. **1888** W. D. HAY *Blood* xi. 47 Soft dimpling blush-rose cheeks.

blusher. Add: **2.** The popular name for the mushroom *Amanita rubescens*.
1887 W. D. HAY *Elem. Text-bk. Brit. Fungi* vii. 50 The Blusher is first-rate eating, perfectly wholesome, and makes a remarkably good ketchup. **1921** *Nature* 25 Aug. 822/2 Mr. Britten has watched them [*sc.* squirrels] feeding on the Blusher (*Amanita rubescens*). **1966** F. H. BRIGHTMAN *Oxf. Bk. Flowerless Plants* 130/2 Amanita rubescens ('Blusher') is found commonly in all kinds of woodland.
3. A cosmetic used to give an artificial colour to the face.
1965 *Vogue* 1 Apr. 27 (Advt.), 'Ultima II' blushing creme...You can do no wrong with this 'transparesscent' blusher—it is so blendable, so edgeless, so *atmospheric* in texture. **1969** *Woman* 11 Jan. (Make-up Booklet) 8 Blushers and shaders are large-size block rouges in variety of shades to add colour and shape. **1969** *Guardian* 9 Sept. 7/2 The famous Roman Glow cream blusher..an unusual terracotta colour that is blended on chin and forehead as well as on cheeks.

blushing, *ppl. a.* Add: **1.** Also used, often somewhat facetiously, with *bride*.
1858 [see BRIDE *sb.*[1] 1]. **1955** D. LESSING *Proper Marriage* I. i. 25 Martha remarked irritably that Dr. Stern was something of an old woman, 'sitting all wrapped up behind his desk like a parcel in white tissue paper, being tactful to a blushing bride'.
b. *blushing bride*: the S. African proteaceous plant *Serruria florida*; also its flower.
1917 R. MARLOTH *Flora S. Afr.* 13 *Blushing bride. Serruria florida*..from one of the valleys of the upper Berg-river (Franschhoek). Flowering in winter. **1955** K. A. THOMPSON *Great House* vii. 213 The small, frail-looking pink protea called the Blushing Bride..was supposed to be extinct when he found it flowering in a tiny kloof in the Fransch Hoek Mountains.
3. Surviving in *blushing honours*, in sense 'causing blushes', in imitation of Shakespeare.
1613 [see HONOUR *sb.* 6 b]. **1806** M. EDGEWORTH *Leonora* II. xlii. 7 To be handed into her own coach with all the blushing honours of a bride. **1857** TROLLOPE *Three Clerks* I. xiii. 275 Alaric, with his blushing honours thick upon him, was left alone.

bluster, *v.* **4.** Delete † *Obs.* and add later examples.
1891 C. GRAVES *Field of Tares* 181 He blustered out another oath as he looked at her. **1913** W. OWEN *Let.* 2 Nov. (1967) 205 She blustered 'it was now too late to go out!' **1955** A. CLARKE *Later Poems* (1961) 48 A child of clay Had blustered it away.

blusterously, *adv.* (Later example.)
1905 G. R. GISSING *Will Warburton* xl. 273 'The counter; *my* counter!'..shouted Will blusterously.

blustery, *a.* Add: **1.** (Earlier Amer. example.)
1774 P. V. FITHIAN *Jrnl.* (1900) 105 The day very blustry & cold.
2. (Earlier example.)
1739 G. OGLE *Gualth. & Gris.* L'Envoy l. 2468 Why, to his blustry Oath, such Def'rence paid?

‖ **blutwurst** (blʊ·tvʊ̄ɹst). orig. *U.S.* [G.] = BLOOD-*sausage*.
1856 *Spirit of Age* (Sacramento, Calif.) 27 Mar. 3/1 'What is the German diet?' 'Sourkrout, pretzels, plutworst, and lager beer.' **1911** *Cosmopolitan* Feb. 314/1 See that I have blutwurst, brattwurst, and mettwurst every day for breakfast. **1951** M. LOWRY *Let.* Feb. (1967) 233 Or is perhaps Daseinsaussage his translation of my *boudin*—a sort of cousin to blutwurst? **1969** *Coast to Coast 1967–8* 189 She bought leber-wurst and long thin met-wursts and big hunks of blut-wurst speckled with rice, delicious when they were fried.

blymy, see *BLIMEY *int.*

bo (bōᵘ). *slang.* Chiefly *U.S.* [Cf. HOBO and *BOZO, but since it is recorded earlier *bo* is perh. more likely to have been orig. a shortening of *boy*.] A familiar form of address.
1825 W. N. GLASCOCK *Naval Sketch-Bk.* (1826) I. 148 Small helm bo—steady—ey-a. **1874** M. CLARKE *His Natural Life* I. v. 74 Half a gallon a day, bo', and no more. **1879** *Punch* 24 May 239/2 *Traveller*..'Which is the—quickest Way—for me to get to the Station?' *Street Arab.* 'Wh' run bo'!' **1893** *Chicago Record* 14 July 11/3 An' den w'en ye meets one uv yer own kind ye feels like old pals, 'cause he calls ye 'Ho' an' ye calls him 'Bo'. See? **1916** *Lincoln Even. News* 1 Jan. in *Dialect Notes* IV. 272 The swaggerest rag you can put on, bo, is one of the non-skid, full-dress shirt bosoms. **1919** *Judge* (ibid.), The man who tells the bootblack 'Keep the change, bo'.

board, *sb.* Add: **1. b.** *spec.* = *surf-board*, esp. in *attrib.* uses.
1779 J. KING *Jrnl.* Mar. in *Cook's Voy.* (1785) III. v. vii. 145 Twenty or thirty of the natives [of the Sandwich Islands], taking each a long narrow board, rounded at the ends, set out together from the shore. *Ibid.* 146 As soon as they have gained..the smooth water beyond the surf, they lay themselves at length on their board, and..place themselves on the summit of the largest surge, by which they are driven along with amazing rapidity toward the shore. **1898** [see *surf-riding*, SURF *sb.* 3]. **1962** *Austral. Women's Weekly* Suppl. 24 Oct. 3/1 Special sections of most beaches are now reserved for board-riders. **1963** *Observer* 13 Oct. 15/6, I hate to think of the next kid that gets stoked on board riding..and wins a world championship and nobody even knows him. **1966** *Weekly News* (N.Z.) 19 Jan. 10/4 Lyall Bay..has a regular and well-shaped wave suitable for both swimmer and board rider.
c. (Earlier and later examples.) Cf. TREAD *v.* 1 b.
1768 A. MURPHY *Let.* 14 Mar. in *Private Corr. Garrick* (1831) I. 291 Mrs. Yates would have died on the boards sooner than have served me in that manner. **1948** W. S. MAUGHAM *Catalina* xxxii. 220 Nor must you think that you demean yourself by treading the boards.
2. c. Also, the target in the game of darts.
1936 R. CROFT-COOKE *Darts* ii. 10 However boards may vary in size, in the arrangement of double spaces, the numbers are always placed in the same order. **1940** N. MARSH *Death at Bar* ii. 28 Cubitt hurled his last dart at the board. *Ibid.* 31 If you'll stretch your hand out flat on the board I'll outline it with darts. **1969** *Punch* 25 Nov. 808/2, I am the man who gets a double 20 with his opening dart, then never again even reaches the board.
d. *Austral.* and *N.Z.* (See quots. 1890 and 1941.)
[**1857** R. B. PAUL *Lett. fr. Canterbury* vi. 90 A tarpaulin or a few boards to shear on.] **1878** 'R. BOLDREWOOD' *Ups & Downs* vii. 74 Next year I hope we shall have fifty thousand to shear, and..I don't see why there shouldn't be a hundred thousand on the board before you sell out. **1890** *Chambers's Jrnl.* 17 May 310/2 Down each side [of the Australian wool-shed] is a clear space some ten feet in width, technically known as 'the board'. Here the shearers work. **1925** R. REES *Lake of Enchantment* viii. 113 The shed hands with brooms [swept] the 'board' clear. **1941** BAKER *Dict. Austral. Slang* 11 *Board*, the floor of a shearing shed; the whole number of shearers employed in a single shearing shed. **1956** G. BOWEN *Wool Away!* (ed. 2) vii. 96 If a 'sheepo' wants to stay popular with the board of shearers he will be fair at all times.
e. *across the board*: see *ACROSS B. 2 c.
3. b. In *pl.*, playing-cards. *slang.*
1923 S. T. FELSTEAD *Underworld of London* i. 11 The..steward [at the Cardsharpers' Club] is a well-known criminal famous for his skill with the 'boards'. **1927** E. WALLACE *Mixer* i. 7 The greatest and most amazingly clever card-sharp that ever handled the 'boards'.
5. c. Also often *transf.* and *fig.*, to carry off all the stakes or prizes; hence, to carry off all the honours.
1882 H. SMART in *Li-quor Christmas Ann.* II. 7/1 We have swept the board so far [in racing]. **1884** *Livestock Jrnl.* 25 July 83/3 Mr. Parry Thomas swept the board in Any Variety Sheep-dogs with his Sir Guy and Welsh Boy. **1905** HORNUNG *Thief in Night* 256 The bloated Guillemard usually sweeps the board with his fancy flyers [*sc.* horses].
8. c. *U.S.* (*a*) The stock exchange; also *attrib.*; *big board* (colloq.), *spec.* the New York Stock Exchange or a quotation board for securities listed there. (*b*) (See quot. 1909.)
1837 *Hennepin* (Illinois) *Jrnl.* 26 Oct. 1/4 The sales of specie to-day, at the Board, were $1,700 in American gold. **1905** *Daily Chron.* 28 Apr. 4/4 None of the 'board members'—as the Stock Exchange men are called—ever appears on the kerb...While fortunes are made and lost on the kerb, it does not seem so serious a business as 'on the board'. **1909** *Ibid.* 3 May 4/6 A 'board' appears to be a 'committee' in the United States, while their word committee is applied to what we should call a board. **1929** *Times* 30 Oct. 14/1 Just before the close of the market on the 'big board'. **1969** *Daily Tel.* 6 Feb. 3/6 The New York Stock Exchange, known as the 'big board', does about twice the amount of business, but is less speculative than the American Stock Exchange.
12. e. *board and board*: (sailing) by a succession of close tacks.
1926 R. CLEMENTS *Stately Southerner* 156 She met with a severe hammering off the Horn, but clawed her way to windward, and, after a week of board and board, managed to slip round.

14. a. Hence *board-ship* used attrib. or as adj.
1852 *Life in Bombay* 216 The board-ship habit of taking brandy and water at night. **1894** M. DYAN *Man's Keeping* II. iii. 62 The liberal allowance of 'board-ship' flirtation. **1924** *Blackw. Mag.* June 743/2 In the curiously intimate routine of a board-ship life..we became very friendly.
c. For 'U.S.' read 'orig. U.S.' Also, in or into an aircraft.
1915 *Sphere* 6 Feb. 151 The forward end of the front gondola of a Zeppelin is screened to protect the pilot...Searchlights..are carried on board to be used when necessary. **1969** *Times* 28 Nov. 1/4 A four-jet B.O.A.C. VC 10 airliner with 69 people on board.
d. Of drink: having been consumed (by a person). *slang.*
1800 R. LOWTH *Billesdon Coplow* 2 Well sous'd by their dip, on they brush'd o'er the bottom, With liquor on board enough to besot 'em. **1840** MARRYAT *Poor Jack* xxiii. 164 The captain..had his grog on board, and was as brave as brass. **1940** N. MARSH *Death at Bar* iv. 72 With a brandy like this on board, I'd face the devil himself.
16. *board-* (or *bord-*)*and-pillar* (examples).
1854 *North of England Inst. Mining Engineers* II. 252 It is the practice here..to arrange board and pillar workings so that the goaf may lay on the dip of the face of the work. **1877** *Encycl. Brit.* VI. 64/1 In the former [*sc.* pillar work], which is also known as..'bord and pillar' in the north of England,..the field is divided into strips. **1960** *Times Rev. Industry* Oct. 34/3 Bord-and-pillar working.
17. *board-room* (later example); also *transf.*, the members of a board.
1935 G. GREENE *England made Me* ii. 47 The monogram had been designed by Sweden's leading artist..E.K. in the board-room; E.K. in the restaurants. **1959** *Times* 5 Oct. 2/6 The work involves..convincing the boardroom, management and operatives.
18. board-and-bat, -batten, applied *attrib.*, esp. to a building constructed from wide boards, normally in conjunction with narrow battens; **board-bill** orig. *U.S.*, the charge made for board (sense 7); **board-fence** *U.S.*, a close fence made with boards; so *board-fencing*; **board foot,** the volume of wood in a piece of timber 1 ft. square and 1 in. thick; **board-game,** a game played on a board (sense 2 c); **Board of Trade unit,** the commercial unit of electrical energy, equivalent to one kilowatt-hour of current; abbrev. B.T.U.; **board-work, -worker** (see quots.).
1902 G. ELLIS *Mod. Pract. Joinery* xxiii. 350 *Board and batten*, a method of forming the walls of wooden houses with a thick and thin board placed alternately. **1918** H. A. VACHELL *Some Happenings* x. 157 A collection of the worst-looking board-and-batten shacks between Shasta and San Diego. **1939** STEINBECK *Grapes of Wrath* xv. 182 Board-and-bat shacks. **1833** E. T. COKE *Subaltern's Furlough* ii, He has gone away without paying..his board bill. **1952** KOESTLER *Arrow in Blue* xv. 129 My board-bill in the Pension Glaser was often overdue for several weeks. **1718** in *Records Early Hist. Boston* XIII. 48 They..shall..maintain a substanciall board fence..from the Barn to three rods distant southerly from the dwelling house. **1860** O. W. HOLMES *Prof. Breakf.-t.* xi. 331 When the boys used to make pictures of me with chalk on the board-fences. **1917** MATHEWSON *Sec. Base Sloan* xi. 143 It had a board fence around it. **1870** *Trans. Ill. Agric. Soc.* VIII. 232 By means of hedging and movable board fencing, keep up a great deal of pasturage. **1896** *Vermont Board of Agric. Rep.* XV. 83 About 24 cubic feet per acre is added..annually—this means about 150 board feet. **1934** *Discovery* Oct. 287/2 Among Vikings as well as Celts board-games of this type are widely known. **1913** *Metal Industry Handbk.* 41/1 The Board of Trade Unit is the commercial standard for purposes of public supply, and is measured by the product at the rate of doing work into the hours divided by 1,000; hence 1 B.T.U. = 1,000 Watt hours. **1887** CREER (*title*) Board work, or the Art of Wig making. *Ibid.* Introd., Board-work, in the fullest extent of its signification, means all that which is done by clever hairdressers and wig-makers in the workshop and at the work table. **1927** *Daily Express* 5 July 5/4 Students are taught the general principles of dressing hair and the technique of board-work, or preparation of hair for making transformations, wigs, and curls. **1921** *Dict. Occup. Terms* (1927) § 428 *Board worker*; hair-dresser's model maker, perruquier, postiche worker, posticheur, wig-maker; general terms for all workers engaged in..making wigs [etc.].

board, *v.* **2. b.** Substitute for def.: *transf.* To enter (a vehicle, railway train, aircraft, etc.). orig. *U.S.* (Earlier and later examples.) Also *absol.*
1848 J. BURNS *Notes of Tour in U.S. & Canada* vi. 108 We..then *boarded*, to use a Yankeeism, the stage for Cleveland. **1935** *Discovery* Feb. 58/2 London bus-conductors..are having a busy time dissuading would-be passengers from trying to board their buses. **1959** I. FLEMING *Goldfinger* xii. 164 Going to ask both to board the plane before the car. **1968** A. HAILEY *Airport* II. iv. 184 A would-be stowaway merely boarded an aeroplane..and sat quietly, waiting for departure. *Ibid.* III. viii. 399 The gate agent who had been in charge at gate forty-seven when Flight Two left..did not remember Guerrero boarding.
7. d. To treat (leather) with a graining-board.
1860 URE *Dict. Arts* (ed. 5) 691 The stiffer parts being boarded both on the grain and flesh sides. **1882** *Encycl. Brit.* XIV. 387/1 When dry enough for the purpose, the skin is boarded,..the effect of which is to bring up the grain,..and also to make it supple. **1968** J. IRONSIDE *Fashion Alphabet* 237 *Box leather*: This is the result of 'boarding' the leather (i.e. breaking up the natural grain surface by close parallel creases) and is a process used on high-grade smooth leathers.
8. b. To put up and feed (an animal). orig. *U.S.*
1875 *Cincinnati Daily Times* 1 July 2/8 Metropolitan Livery, Boarding and Sale Stables..Special attention given monthly and day boarding horses. **1880** W. D. HOWELLS *Undiscovered Country* 261 The mare..was consequently boarded out of town a good deal. **1905** *N.Y. Even. Post* 24 Feb. 1 The owner of a large stable..said that..he had recently had some seventy horses to board. **1969** *Times* 14 Nov. 20/6 (Advt.), Two adorable scottie pups...Will board 4s. per day till Christmas.
9. b. *to board round* or *around*: to board in succession in different houses. *U.S.*
1828 *Ladies' Mag.* (Boston) I. 215, I boarded *round*, as they termed it, that is, I boarded with every family in proportion to the number of scholars they sent. **1831** *Ibid.* IV. 557 [There was] a custom..that the instructor should 'board around' as it is called. That is board a short period in each family who sent children to the school—the length of time regulated by the number of scholars sent. **1833** *Niles' Reg.* XLIV. 347/1 Our schoolmasters are..'boarded round', so as to save the drawing the pay of the schoolmaster's board from the school fund. **1869** [see *AROUND *adv.* 5 a, b]. **1871** C. M. YONGE *Pioneers & Founders* vi. 165 The system was that of 'boarding round'—*i.e.* the young mistress had to live a week alternately at each house, and went from thence to her school.
11. To call before a selection board, medical board, or the like. Usu. in *pass.*
1917 W. OWEN *Let.* 25 Sept. (1967) 496, I am to be boarded today, and am waiting to be called in at any moment. **1917** G. S. GORDON *Let.* 1 Dec. (1943) 82 He has never been boarded. **1964** *New Society* 16 Apr. 13/1 Of the 715 candidates boarded, 104 were selected.

boarded, *ppl. a.* Add: **1.** *boarded up*: see BOARD *v.* 7.
1890 *Harper's Mag.* Dec. 102/2 It was not long before the dismal little, boarded-up, spidery coquina house was as clean as a whistle. **1905** *Daily Chron.* 26 Jan. 5/1 The boarded-up shop fronts and broken windows. **1961** K. REISZ *Technique Film Editing* (ed. 9) ii. 199 People standing in front of boarded-up shops.
3. Also *boarded-out* (see BOARD *v.* 10).
1897 *Daily News* 4 Apr. 3/1 Amongst boarded-out children ophthalmia is almost unknown.

boarder. Add: **1. b.** A horse that is put up and fed. (Cf. *BOARD *v.* 8 b.)
1806 SIR W. SCOTT *Let.* 11 Aug. (1932) I. 314, I wish I could promise to add to your convenience by accommodating the boarder, but our grass has been so scanty that we could not do him justice. **1903** *N.Y. Tribune* 20 Sept., (Advt.), Boarders wanted at Rochville Boarding Stable.

boarding, *vbl. sb.* Add: **7.** *boarding card, kennel, knife, party, pass, ticket*; **boarding-car** *U.S.*, a railway carriage fitted with sleeping, cooking, and dining accommodation; **boarding foreman** (see quot.); **boarding-house** (earlier example); **boarding-master,** a keeper of a boarding-house for seamen; **boarding officer** = **boarding foreman*; **boarding-place** *U.S.*, a boarding-house; **boarding-stable** *U.S.*, a livery stable.
1867 *Commerc. & Financ. Chron.* (U.S.) 8 June 726 The equipment of the road consists of..4 boarding and 8 wrecking cars. **1891** C. ROBERTS *Adrift Amer.* 87 The boarding cars had to be fitted up, the cooking appliances put in order, &c. **1958** *Spectator* 22 Aug. 251/3 It is now necessary, when travelling abroad by air, to carry passports, travel tickets, embarkation cards, boarding cards, luggage vouchers, bus tickets and airport passes. **1962** L. DEIGHTON *Ipcress File* v. 30 A steward took my..boarding card. **1921** *Dict. Occup. Terms* (1927) § 731 *Boarding foreman, boarding officer*, is responsible to harbour or dock authority for seeing that ship's papers, etc., are in order before allowing her to enter dock or harbour. **1728** A. ROGERS *Let.* 16 Aug. in *Mem. of Royal Chaplain* (1905) 110 Mr Gill has a Daughter..who..has for many Years last past kept a Boarding-House at Yarmouth. **1969** R. RENDELL *Best Man to Die* ii. 17 The dog had awakened him..with long-drawn howls...'This isn't a boarding kennels, you know,' he said. **1807** in *Maryland Hist. Mag.* (1910) V. 176 She proved to be a small Pilot Boat.., with 50 Muskets ready loaded, about 18 Boarding knives, &c. **1869** *Porcupine* 217/2 The moment a sailor is shipped he obtains his advance-note..and the cupidity of the boarding-master is called into action. **1904** *Daily Chron.* 31 Mar. 7/6 A Greek boarding-master boarded a vessel in the Roath Dock with sailors. **1881** *Instr. Census Clerks* (1885) 36 Boarding Officer. **1882** Boarding party [see BOARD *v.* 1]. **1946** C. S. FORESTER *Lord Hornblower* vi. 55 Have the boarding-party secured as they come on board. **1969** *Daily Express* 17 Nov. 13/7 She had to stop to have her boarding pass checked. **1970** *Sunday Times* 18 Jan. 13/6 He gave me the boarding pass which got me aboard one of the last relief planes. **1854** M. J. HOLMES *Tempest & Sunshine* iv. 21 Mr. Middleton set off for Frankfort to find 'as smart a boarding-place for his gals as anybody had'. **1911** H. S. HARRISON *Queed* iii. 28 She runs this boarding-place, and people of various kinds come to her. **1903** Boarding stable [see *BOARDER 1 b]. **1967** M. DRABBLE *Jerusalem the Golden* viii. 206 Clara..obtained her boarding ticket, and went and sat down on a plastic covered seat to await..the announcement of her flight.

board-walk. orig. *U.S.* [BOARD *sb.* 1.] A footway or walking-path constructed of boarding.
1872 F. M. A. ROE *Army Lett.* (1909) 53 We reached a narrow board-walk that was supposed to run along by her side fence. **1895** C. KING *Fort Frayne* vii. 98 The back gate stood open..and the board walk leading from it to the rear door was visible for half its length. **1906** *Harper's Mag.* June 61 A few days later, on the board-walk at the sea-shore, she came face to face with Hugh Wilberding. **1951** *Archit. Rev.* CX. 100 This board-walk has a row of tables along its outer edge and thus combines the functions of open-air café and river promenade.

boardy (bōᵊ·ɹdi), *a.* [f. BOARD *sb.* + -Y[1].]
1. (See quot.)
1893 JACOBI in *Trans. Bibliogr. Soc.* I. 197 Thickness has to be studied—a stout paper being clumsy, and handling very 'boardy'—an expression applied to cardboards.
2. Of the wire teeth of card-clothing when worn down: stiff, not pliable.
1909 in *Cent. Dict.* Suppl.

boast, *v.*[2] Add form **bost,** and:
3. *Wood-carving.* To model roughly the details of (the design). So with *in* or *out.* Also **boa·sted** *ppl. a.*, **boa·sting** *vbl. sb.*
1867 G. A. ROGERS *Wood Carving* 11 When the stalk and leaves have been bosted into the agreeable curves they assume in nature. **1875** T. SEATON *Fret Cutting* 36 Every piece of work of any importance must go through the three stages—blocked out, bosted, finished. *Ibid.* 97 The leaf having been bosted it must now be finished, so let my readers take pencil in hand, and..let them sketch on the bosted mass each leaflet, division, and part. **1890** C. G. LELAND *Wood Carving* 50 The three stages of blocking out, bosting, and finishing. *Ibid.* 56 In commencing or bosting out this pattern. **1907** E. ROWE *Pract. Wood-Carving* vi. 68 The next step is to *bost in* the ornament. **1970** H. BRAUN *Parish Churches* vii. 90 The mason would leave a rough lump called a 'boasting' as a basis for the intended carving.

boast (bōᵘst), *v.*[3] and *sb.*[2] *Tennis* and *Rackets.* [? f. F. *bosse* the place where the ball hits the wall. (Cf. BOAST *v.*[2])] *trans.* To hit (the ball) so that it strikes either of the side-walls before it strikes the end-wall; also to make (a stroke of this kind), and *intr.* Also *sb.* = the stroke. So **boa·sted** *ppl. a.*, **boa·sting** *vbl. sb.*
1878 J. MARSHALL *Ann. Tennis* 156 This stroke is called *a boast*, or *boasted stroke*. *Ibid.* 176 What is called *boasting* the ball. *Ibid.*, We..call them all *boasts* or *boasted* balls indifferently, whether struck from one or the other side of the net. **1898** KENNEDY & COHEN *Tennis* in '*House*' *on Sport* I. 421 The service was boasted under the winning gallery wall. **1902** E. MILES *Racquets*, etc. 194 This is called 'boasting', and it gives the ball a powerful twist. *Ibid.* 231 The Boasted Volley, the Volley hit direct onto the Side-wall. **1959** *Times* 4 Mar. 4/5 A liberal use of the boasted stroke. **1963** *Ibid.* 8 Jan. 3/5 This Birms did with his usual touch strokes, boasting with precision. *Ibid.*, This reaching and turning repeatedly for boasts is the most tiring way to be forced to run.

boat, *sb.* Add: **1. d.** *to be in the same boat* (examples); *to miss the boat* (see *MISS *v.*[1]); *to rock the boat* (fig.): to disturb the equilibrium (of a situation, etc.).
1845 DICKENS *Cricket on Hearth* i. 40 You'll come to the wedding? We're in the same boat. **1921** H. CRANE *Let.* 17 Oct. (1965) 68 He..made me feel myself, as a poet, as being 'in the same boat' with him. **1931** F. L. ALLEN *Only Yesterday* vi. 156 Unfortunate publicity had a tendency to rock the boat. **1944** AUDEN *For Time Being* (1945) iii. 33 Some stranger vision of the large loud liberty violently rocking yet never, he is persuaded, finally upsetting the jolly crowded boat. **1958** *Punch* 15 Jan. 111/1 The trouble with these people who nail their colours to the mast—they always rock the boat.
e. Short for **boat-race* b, rhyming slang for 'face'.
1958 F. NORMAN *Bang to Rights* 35 As soon as he had stripped this crank down the boat. **1962** R. COOK *Crust on its Uppers* i. 26 We've seen the new boat of the proletariat, all gleaming eyes.
3. *boat-builder* (later examples), *-building* (earlier example), *-hand* [HAND *sb.* 8], *-load* (examples); **boat-axe** *Archæol.* [Sw. *båtyx*], a boat-shaped battle-axe of the neolithic period in Scandinavia, applied *attrib.* to the culture characterized by these axes and to the people of this culture; **boat-bearer,** a man or boy who carries the incense-boat, in attendance on the thurifer; **boat-boy,** (*a*) a boy engaged to help to manage a boat; (*b*) a boy who carries an incense-boat; **boat-bridge** (examples); **boat-car,** (*a*) a wheeled car used for launching or beaching a boat; (*b*) an airship-car built like a boat; **boat-cradle** [CRADLE 9], a cradle for holding a boat; **boat-deck,** the deck of a ship from which lifeboats are launched; **boat-drill,** practice by a ship's crew and passengers in the launching and manning of lifeboats; **boat-express,** an express train timed to meet a boat; **boat-hat** = *BOATER 2; **boat-house** (earlier and later examples); **boat neck(-line)** = **bateau* (*neck-*) *line*; cf. **boat-shaped*; **boat-quarters,** the quarters occupied by members of the crew detailed to man the boats; **boat-race** (earlier example); *spec.* a race between crews of the universities of Oxford and Cambridge, rowed annually over a course from Putney to Mortlake; also *attrib.*; (*b*) *rhyming*

slang for 'face'; **boat-shaped** *a.*, shaped like a boat; esp. applied to a wide neck-line curving downwards from the shoulders; **boat-sponge,** a fine sponge of the Bahamas and Florida (see quot.); **boat-steerer** (see quot. 1845); **boat-stretcher** = Stretcher 7; **boat-yard** (orig. *U.S.*), a yard in which boats are built and stored.

1919 *Jrnl. R. Anthrop. Inst.* XLIX. 197 Almgren has shown..that Scandinavian culture of a type which he calls *batyx* or boat-axe culture, which is found in the passage-graves of Sweden, had reached the coastal areas of Finland before the close of the Neolithic Period. 1957 Childe *Dawn Europ. Civilization* (ed. 6) ix. 162 The distribution of these graves..leaves no doubt that the Boat-axe folk were intruders. 1899 P. Dearmer *Parson's Handbk.* 128 The thurifer and boat-bearer enter with the censer and boat. 1918 A. Fortescue *Cerem. Rom. Rite* 25 *note*, The boat-bearer will stand or kneel at the thurifer's left. 1889 Gretton *Memory's Harkback* 78 One day I set out with a boat-boy to sail and row to Ely. 1902 P. Dearmer *Parson's Handbk.* (ed. 4) 245 There is no English authority for a 'boat-boy' to accompany the thurifer. 1794 J. B. S. Morritt *Let.* 22 May (1914) ii. 29 The Danube..which we crossed on one of the *ponts volants* or boat bridges you have heard me mention as on the Rhine. 1959 E. Pound *Thrones* xcvi. 9 By the boat-bridge over Euphrates. 1708 Sewall *Diary* (1879) II. 236 A smith's shop..and a Boat-builder's Shed. 1878 *Harper's Mag.* Feb. 324/2 The inhabitants are fishermen, farmers, and boat-builders. 1963 *Times* 24 Apr. 17/3 The film tells the story of a firm of family boatbuilders. 1780 in *Virginia State P.* (1875) I. 391 It is absolutely necessary that the 'Boat-building business' be pushed. 1907 *Westm. Gaz.* 11 Sept. 8/3 A dozen soldiers hung on by the boat-car..to keep it down. 1909 *Daily Chron.* 1 June 1/2 The two boat-cars which carry the engines. 1948 R. de Kerchove *Internat. Maritime Dict.* 69/1 *Boat cradle*. 1960 *Guardian* 19 Sept. 2/4 For sailing families, boat-cradle, mast support, and long tow-bar can be bought as extras [to caravans]. 1927 G. Bradford *Gloss. Sea Terms* 17/1 *Boat Deck*, that upon which the lifeboats are secured. 1938 *British Birds* XXXII. 113 Two adult starlings with one young one..settled on the boat-deck. 1906 *Act 6 Edward VII* xlviii. § 9 The Master of every British ship shall enter..a statement..of every occasion on which boat drill is practised on board the ship. 1943 H. Pearson *Conan Doyle* x. 139 Unfortunately [on the *Titanic*] there had been neither boat-muster nor boat-drill. 1910 *Daily Chron.* 9 Apr. 1/4 A boat-express passenger. 1911 Fletcher & Kipling *School Hist. England* xii. 247 The boat-express is waiting your command! 1821 *Deb. Congress* I. 46 This admiralty jurisdiction had done much to ruin those who were engaged in..[steamboat] navigation, by making the boat-hands unfaithful. 1936 *Discovery* Dec. 380/1 There had been a great demand for boat-hands. 1889 Gretton *Memory's Harkback* 310 The dandy of that time in Anglesea boat hat, blue coat with brass buttons, high velvet collar, and swallow-tails. 1722 Swift *Let.* 22 Dec. (1963) II. 440 Is this a Time to build Boat-Houses, or pay for Carriage? 1801 D. Wordsworth *Grasmere Jrnl.* 8 Dec. in *Jrnls.* (1941) I. 89 Mary and William walked to the boat house at Rydale. 1925 Dreiser *Amer. Tragedy* (1926) III. xix. 230 And next to him, the boathouse-keeper who had rented him the boat. 1680 in *New Castle* (Pennsylvania) *Court Rec.* (1904) 442 Wee have sent away a boat load. 1745 J. MacSparran *Letter Book* (1899) 27 In the last Boat-load..I lost my dear Servant. 1807 R. Southey *Lett. from England* (1951) 480 Here they are selling the sublime and beautiful by the boat-load! 1956 K. Clark *Nude* iii. 76 Boat-loads of tourists. [1932 R. Lehmann *Invit. Waltz* i. iii. 50 What about the bodice, now? V, rounded, boat, or square?] 1960 *Times* 18 Jan. 15/5 There are sweaters with V necks, and wide boat necks. 1968 *Woman* 14 Dec. 8/2 *Cuddly popover*..has a wide boat neckline. 1904 *Westm. Gaz.* 29 Dec. 8/2 The crew were kept at boat-quarters in readiness for immediate launching. 1791 H. Walpole *Let.* 23 Aug. (1944) XI. 341 On Monday was the boat-race. I was in the great room..to see the boats start. 1834 *Bell's Life in London* 25 May 2/5 We are sorry to state that the proposed boat race between the two Universities will not take place. 1889 E. Dowson *Let.* 24 Mar. (1967) 54, I am sick of 'varsity men & shall leave on Friday for Brighton & so escape whiskey & whores on boat race night. 1914 G. B. Shaw *Fanny's First Play* ii, Everyone said it was a bit of English fun, and talked about last year's boat-race night when it had been a great deal worse. 1958 F. Norman *Bang to Rights* 36 A big bandage round his thumb and a big smile on his boat race. 1959 I. & P. Opie *Lore & Lang. Schoolchildren* xvi. 349 Until the last war the event which most excited children's loyalties was the Boat Race. 1785 *Daily Universal Register* 1 Jan. 3/2 Boat-shaped snuffer pans. 1855 C. Kingsley *Glaucus* 159 That wondrous bug the Notonecta,..rowing about his boat-shaped body. 1937 E. V. Gordon tr. *Shetelig & Falk's Scandinavian Archaeol.* v. 68 The boat-shaped battle-axe, which appears in the later part of the passage-grave period. 1959 *Times* 24 Aug. 6/6 The dresses, with boat-shaped necks and bell-shaped skirts. *Ibid.* 26 Sept. 8/3 A gown of white grosgrain with a boat-shaped neckline. 1883 S. Kent in A. J. Adderley *Fisheries Bahamas* 47 The so-called Velvet, Abacco-velvet, or Boat-sponge (*Spongia equina*, var. *meandriniformis*). 1845 E. J. Wakefield *Adv. N.Z.* I. xi. 317 The *boat-steerer* pulls the oar nearest the bow of the boat, fastens to the whale with the harpoon, and takes his name from having to steer the boat under the headsman's directions while the latter kills the whale. 1851 Melville *Moby Dick* I. xxvi. 188 His boat-steerer or harpooner. 1888 Churchward *Blackbirding* xii. 220 If they tried to shirk rowing, the chap in the bows or stern would fetch them a crack with the boat-stretcher. 1847 L. Collins *Hist. Sketches Kentucky* (1850) 335 *West Point* is situated at the mouth of Salt river..and recently there has been an extensive *boat yard* established. 1960 Delmar-Morgan *Cruising Yacht Equipment* vii. 83 Stores in boatyards..are only comparatively dry.

boatel, var. *botel.

boater. Add: **2.** A stiff straw hat with a flat crown and brim, originally one suitable to wear when boating. Also *boater straw*.

1896 *River* 8 Aug. 5/1 The news has gone forth from one end of Wood-street to the other that the Prince of Wales has purchased a 'boater', and will wear it..for the rest of the summer. 1905 *Daily Chron.* 2 June 4/7 In a shop on Ludgate-hill, there are placards announcing 'straw boaters'. 1910 H. G. Wells *Mr. Polly* vii. 151 Boater straws, imitation Panamas, [etc.]. 1926 *Blackw. Mag.* June 735/1 That horrible and obsolete form of head gear..known as a 'boater'. 1963 C. Mackenzie *My Life & Times* II. 120 The word 'boater' was never used, and did not exist outside the catalogue of a firm like Ponting's..'Boater' was intended to suggest to maidservants that they were as familiar with Maidenhead as Gaiety girls.

boatswain. Add: **3. boatswain's chair, cradle,** a board on which a sailor (or other workman) sits when at work aloft.

1894 [see *bosun]. 1928 *Daily Express* 19 Nov. 11/7 When you have been sitting up in a bo'sun's cradle playing hell with a pneumatic riveter.

bob, *sb.*[1] Add: **2. d.** A short sleigh-runner. *N. Amer.*

1857 *Knickerbocker* XLIX. 67 The 'stage' consisted of a rickety pair of bobs [etc.]. 1888 Farmer *Americanisms*, *Bob* or *Bob Sled* or *Bob Sleigh*, a sleigh used in the West for conveying large timber, its special characteristic being two pairs of *bobs* or short runners. 1927 *Atlantic Monthly* Mar. 335 We sat on some boards nailed on the front bob of his old bobsled. 1964 *Canad. Geogr. Jrnl.* Mar. 86/2 Many carriages could have their wheels and sled runners or 'bobs' bolted on to replace them in winter.

e. Abbrev. of Bob-sled, -sleigh; also *attrib.*, as *bob-run*. orig. *U.S.*

1856 M. Y. Jackson *Diary* 27 Jan. in *Minn. Farmers' Diaries* (1939) 145 Went to Kinnik-kinnik yesterday with the bobs. Had to stay over night & return to day with part of a load of lumber. 1887 *Harper's Mag.* Dec. 113/2 The Captain's sleigh went townward toward evening, and the butcher's 'bob' tore an ugly groove along the lower edge. 1888 *Ibid.* May 973/1 Telling the little ones how they might have been mangled by one of the swift 'bobs'. 1897 *Sears, Roebuck Catal.* 63/2 Bob Woods..Bob Runners..Bob Sleigh gearing..Bob Beams. 1906 *Daily Colonist* (Victoria, B.C.) 13 Jan. 2/3 Robertson's bob crashed with terrific momentum into a horse and cutter. 1906 *N.Y. Even. Post* 19 May 9 The same spruce and hemlock logs drawn on bobs. 1927 *Observer* 18 Dec. 9/4 Long and well-made bob-runs. 1963 *Times* 1 Feb. 4/3 The No. 2 British bob was the first to crash. *Ibid.* 7 Feb. 3/7 The fastest time ever recorded on the St. Moritz bob run.

f. In plural form = prec. *U.S.*

1911 S. E. White *Bobby Orde* (1916) xvii. 194 At last Bobby saw..a magnificent bobs that had not before appeared. *Ibid.* 195 If the bobs upset, or the horse went too fast.

5. b. A style of cutting women's hair short and even all round. (See *bob *v.*[5] 2.) Also, hair cut in this way.

1926 Galsworthy *Silver Spoon* III. xi. 312 Her hair, again in its more natural 'bob', gleamed lustrously under the light. 1940 M. Dickens *Mariana* v. 141 'Haven't you cut it rather short?'..'Oh no, it's a lovely bob.'

7. (Later N.Z. example.) Also *U.S.* (See quot. 1883.) Cf. *bobbing *vbl. sb.*[1] 3.

1874 A. Bathgate *Colonial Experiences* xvii. 243 The eels bolt the 'bob', and are readily pulled out of the water, the same bait serving again and again. 1883 *Century Mag.* July 383/1 The 'bob', which is formed by tying three hooks together, back to back, and covering their shanks with a portion of a deer's tail.

bob, *sb.*[7] Add: **dry-bob** (earlier example).

1844 Disraeli *Coningsby* I. ix. 102 'The match to-morrow shall be between Aquatics and Drybobs,' said a senior boy.

2. Short for Bob-white.

1883 *Century Mag.* Aug. 483/2 The European partridge ..weighs twice as much as Bob White, but he has not Bob's sturdy, rapid..flight. 1902 Sandys & Van Dyke *Upland Game Birds* 9 Then brave, brown Bob..enters Love's fateful lists.

3. Slang phr. *Bob's (bob's) your uncle*: everything is all right.

1937 in Partridge *Dict. Slang* (ed. 2) 981/2. 1946 S. Spender *Europ. Witness* 143 He mixes up phrases such as 'Oh boy, oh boy', with cockney such as 'Bob's-your-uncle'. 1949 'N. Blake' *Head of Traveller* iv. 60 Three curves and a twiddle, label it 'Object', and bob's your uncle.

bob, *sb.*[8] Add: (Earlier and later examples.)

1789 *Sessions' Papers* June 550/1 *Bulls* and *half bulls* are crowns and half crowns, in coiner's language, and a *bob* is a shilling. 1915 'Bartimeus' *Tall Ship* iv. 73 Have I got time to borrow five bob from the messman before the boat shoves off?

2. Phrases. *bob a job*: the slogan of the Boy Scout organization in an annual effort to raise money for funds by doing jobs at a shilling a time; also (with hyphens) as attrib. phr.; *bob a nob* [Nob *sb.*[2]] *colloq.*: a payment of a shilling a head; also *attrib.*; *bob in* (*Austral.* and *N.Z. colloq.*): a subscription of a shilling to a common fund.

1823 'Jon Bee' *Dict.* 13 *A bob a nob*, a shilling a head. 1851 Mayhew *London Lab.* I. 313/2 These he would engage at 'a bob a nob' (one shilling each). 1889 W. Davidson *Stories N.Z. Life* 5 From tricks at cards, the fun changed to 'a bob in' the winner shouting. 1933 *Bulletin* (Sydney) 28 June 36/3 What say we rig a few bob-ins for the poor cow? 1944 *Times* 19 May 2/3 Among the younger boys the effort is already known as the 'Bob-a-Job' or 'Bob-a-Nob' Day. 1945 Baker *Austral. Lang.* ix. 172 Just as the *shout* is an institution in this part of the world so are the *bob in*, *two bob in*, [etc.]..all of which concern the creation of a jack-pot, usually with the object of buying drinks. 1954 'N. Blake' *Whisper in Gloom* I. iii. 41 Clean your car, sir? Bob a job. Kensington Scouts. 1958 'R. Crompton' *William's Telev. Show* vi. 162 William took his 'Bob-a-job' book from his pocket and studied it complacently. 1959 *Daily Tel.* 28 July 12/4 A shilling-a-head subscription, popularly known as the 'bob-a-nob', for some form of testimonial was launched to-day. 1966 J. Hackston *Father clears Out* 113 The Red Range Federal Capital Site Committee (bob in and the winner shout) met on a Sunday.

bob (bɒb), *sb.*[10] Altered form of God, used in oaths and exclamatory phrases.

1823 'Jon Bee' *Dict.* 13 'So help me *bob*,' is an oath to deceive the hearer, doubly; for a *bob* is but a shilling, and not a fit thing to swear by. 1842 [see S'elp]. 1905 W. Raleigh *Let.* 17 Nov. (1926) II. 284 It's all very well Jesus Christ being imagination, but when it comes to nothing but menageries of howling giants—s'welp me Bob! I can't show it up, I don't know enough.

bob, *v.*[3] Add: **1. c.** *to bob on*: to await anxiously (the turn of events). *slang.*

1925 Fraser & Gibbons *Soldier & Sailor Words* 29 *Bobbing on*, anticipating or expecting something..with the sense of looking forward to something unpleasant. *a*1935 T. E. Lawrence *Mint* (1955) I. xxiv. 85 I'm bobbing on not getting that intelligent job from him.

2. b. To come or go *in*, *into*, *up*, etc.

Quot. 1835 may belong to sense 2 a.

1835 Dickens in *Sk. Boz* (1836) ser. I. II. 27 'Please, Sir, missis has made tea,' said a middle-aged female servant, bobbing into the room. 1890 *Texas Siftings* 8 Nov. 7/1 The straws man bobs up serenely at the regular time every four years. 1924 Galsworthy *White Monkey* I. viii, Thanks, old man, awfully good of you—will you bob in, then? 1928 *Public Opinion* 19 Oct. 371/1 Everything but the kitchen stove, as our idiom has it, is likely to bob up for notice.

4. bob-a-cherry *transf. attrib.* (cf. *bob-cherry*); **bob-apple** (later examples); **bob-up** *attrib.*, that bobs up.

1899 T. S. Moore *Vinedresser*, etc. 19 Kisses sadly blown across the sea..Bob-a-cherry kisses 'neath a tree—'O, give me one.' 1926 Masefield *Odtaa* ii. 22 They're bob-a-cherry birds. You always see them working their lower jaws as though to get the cherry in. 1940 F. Kitchen *Brother to Ox* i. 15 We children..played snapdragon and bob-apple. 1959 I. & P. Opie *Lore & Lang. Schoolchildren* xii. 271 In Liverpool Hallowe'en is known as 'Duck Apple'..in Pontypool 'Bob Apple' or 'Crab Apple Night'. *Ibid.* 272 *Bob Apple* is also known as 'Snap Apple' or 'Apple on the Line'. 1935 *Discovery* Jan. 10/1 This type of mechanism has come to be known as the 'bob-up' type.

bob (bɒb), *v.*[5] [f. Bob *sb.*[1] (sense 5). Cf. Bob *a.* 1 and Bobbed *a.*]

1. To dock, cut short (a horse's tail, etc.). Also with *off*. ? *U.S.*

1822 J. Fowler *Jrnl.* (1898) 112 Two of them [*sc.* wild horses] must have been in Hands, as their tails were Bobed short. 1889 *Cent. Dict.*, *Bob*, v. 2. To cut short; dock: often with *off*: as, to *bob* or *bob off* a horse's tail.

2. To cut (the hair of a woman or girl) short and even all round.

1918 *Punch* 25 Sept. 193 Alarming spread of bobbing. 1919 *Home Notes* 8 Feb. 130, I went to a hairdresser's..He bobbed my hair. 1920 R. Macaulay *Potterism* I. i, When the time came to bob the hair, she bobbed it. 1940 M. Dickens *Mariana* v. 140, I want to have my hair bobbed, please.

bob (bɒb), *v.*[6] [f. Bob *sb.*[6]] *trans.* To polish (metal) with a bob (see Bob *sb.*[6]).

1879 *Cassell's Techn. Educ.* IV. 414/1 Our spoons..will be first 'bobbed' with fine sand on an ordinary buff-covered polishing wheel.

bob, *v.*[7] [See *bob *sb.*[1] 2 e.] **a.** *trans.* To carry on a bob-sleigh. **b.** *intr.* To ride on a bob-sleigh.

1880 *Wisconsin Rep.* 254 Injuries suffered..by collision with persons 'bobbing' or 'coasting' on such street. 1909 *Cent. Dict.* Suppl., *Bob*, v.t. To transport (a load, as of logs) on a bob or sled.

bobac. (Earlier and later examples.)

1697 Locke *Let. to Bp. of Worcester* 195 Three *Bobaques* are all true and real *Bobaques*, supposing the Name of that Species of Animals belongs to them. 1901 *Westm. Gaz.* 31 Dec. 5/3 The bobac, the still living marmot of the Siberian steppes. 1968 F. Kertesz *Lang. Nuclear Sci.* 12 The underground nuclear tests at Los Alamos were designated first by burrowing mammals such as..Bobac.

bobachee (bɒ·bātʃi). *Anglo-Ind.* Also **babachy, bobarchee, bobba-, bobberjee,** etc. [Corruption of Hindi *bāwarchī*.] A male cook.

1810 T. Williamson *E. Ind. Vade-Mecum* I. 238 It being in the power of the *babachy* to baste any part with great precision. 1863 G. O. Trevelyan *Dawk Bungalow* 18 And every night and morning The Bobarchee he shall kill The sempiternal Moorghye, And we'll all have a grill. 1883 E. H. Aitken *Tribes on Frontier* 35 He was only a *maistry*, or more vulgarly a *bobberjee*. 1920 *Outward Bound* Nov. 67/1 Here, you, bobbajee, khansammeh, send up a sack of potatoes. 1958 *Manch. Guardian* 11 June 3/2 Sergeant Frederick McManus, who joined the army in 1937, said, '..I miss the cooks—bobajees we used to call them.'

bobbed, *a.* Add: (Later U.S. examples.)

1768 Washington *Diaries* (1925) I. 256 Hunting again, and catchd a fox with a bobd Tail and cut Ears. 1894 *Outing* (U.S.) XXIV. 193/1 There is nothing more hideous than a bobbed jib. 1897 *Ibid.* XXIX. 464/2 Small boys strutting about in tall beavers with bobbed-off coats, looking..like tailless sparrows.

b. Of a woman's or girl's hair: cut short and even all round (see *bob *v.*[5] 2). Hence **bobbed-hair, -haired** *a.*, with bobbed hair.

1918 *Home Chat* 3 Aug. 106/1 There is quite a craze for 'bobbed' hair, for big and little girls alike. 1924 Galsworthy *White Monkey* I. viii. 63 Her bobbed black hair, crinkly towards the ends. 1928 *Punch* 30 May 589/3 The new bobbed-haired lady cat-burglar. 1952 S. Kauffmann *Philanderer* (1953) ix. 141 She was a young, bright, snub-nosed, bobbed-hair cartoon character.

bobber[3] (bɒ·bəɹ). [f. *bob *v.*[6]] One who polishes articles on a bob or polishing-disc.

1881 *Instr. Census Clerks* (1885) 53 Locksmith, Bellhanger. Bobber and Grinder. 1906 *Daily Chron.* 23 Apr. 10/6 Good emery bobber: must be used to cycle work. 1921 *Dict. Occup. Terms* (1927) § 237 Emery bobber (foundry). *Ibid.* 238 Bobber and polisher.

bobber[4] (bɒ·bəɹ). [f. *bob *v.*[7]] One who rides on a bob-sleigh.

1904 *Windsor Mag.* Jan. 248/2 Thus spoiling many a race and importing many a risk to the bobber. 1906 *Tribune* 16 Jan. 3/7 Curlers and skaters have been enjoying themselves, but skiers and bobbers..are grumbling.

bobber[5] (bɒ·bəɹ). [perh. f. Bob *sb.*[8] (see quot. 1921).] In full *fish bobber*: a workman who unloads fish from trawlers and drifters to the quay. So **bo·bbing** *vbl. sb.*[3], working as a 'bobber'.

1921 *Dict. Occup. Terms* (1927) § 745 *Bobber, fish bobber* (East Coast term which originated when fish porters were paid 1s. per hour). 1934 'Taffrail' *Seventy North* ix. § i. 185 Men called 'bobbers', hired by the owners, unloaded all the fish in baskets. 1959 *Times* 20 Jan. 6/3 The fish bobbers—men who unload trawlers. *Ibid.*, The bobbers are members of the General and Municipal Workers' Union. 1962 J. Tunstall *Fishermen* iii. 81 Bobbing..can easily be done by men in their fifties.

bobbery (bɒ·bəri), *a.* Chiefly *Anglo-Ind.* Also **bobberee.** [? f. Bobbery *sb.*] Of a pack of hounds: miscellaneous in breed or quality. Also more widely: poor, of indifferent quality, 'scratch'. See also quot. 1934.

1873 'A. Cheem' *Lays of Ind.* 2nd Ser. 86 Other servants bobbery pack, Drinking up what master leave, sir. 1878 *Life in the Mofussil* I. 142 On the mornings when the 'bobberee' pack went out. 1894 M. Dyan *Man's Keeping* I, The daily drills and the hunts with the bobbery packs. 1901 'Linesman' *Words by Eyewitness* (1902) 234 A mounted officer pushing forward here, another there trying to turn his 'bobbery' horse to get back. 1934 *S.P.E. Tract* xli. 20 The word [bobbery] is scarcely used by Anglo-India, except in the phrase 'bobbery pack', where *bobbery* is an adjective, bearing the derived meaning 'mixed', 'mongrel', i.e. not composed of fox-hounds, &c.; also, occasionally (= fidgety), of a horse or infant. 1968 'J. Welcome' *Hell is where you find It* xi. 146 And what the hell do I do..besides breaking my neck in a bobbery second-class steeplechase?

bobbin, *sb.*[1] Add **5.**: *bobbin-boy*.

1871 J. R. Lowell *My Study Windows* 45 Not everybody has the genius to be a Bobbin-Boy! 1921 *Dict. Occup. Terms* (1927) § 399, *Bobbin boy* (jute)..*bobbin boy* (rope and twine).

bobbin (bɒ·bin), *sb.*[3] [Etym. obscure.] Applied *attrib.* to the payment made to a fish porter at Billingsgate market.

1936 J. L. Hodson *Our Two Englands* xii. 239 The carrying prices, or bobbin rates, are all fixed in the trade. 1953 *Times* 3 Jan. 3/2 They heard of 'bobbin charges', paid by the customer to the porter for carrying the fish from the merchant.

bobbing, *vbl. sb.*[1] Add: **3.** (Later N.Z. example.) Also *U.S.* (see *bob *sb.*[1] 7).

1815 S. L. Mitchill in *Trans. Lit. & Philos. Soc. N.Y.* I. 360 The eel..is also taken..by a bunch of tough bait, after a manner called *bobbing*. 1874 A. Bathgate *Colonial Experiences* xvii. 243 The manner most frequently adopted for catching eels, called 'bobbing', is primitive enough, and is by means of a line formed of narrow stripes of flax leaves, to the end of which is tied a bundle of earth worms. 1883 *Century Mag.* July 383/1 Bobbing has been practiced in Florida for more than a century, and is a very simple but remarkably 'killing' method of fishing. 1888 Goode *Amer. Fishes* 62 Trolling with the rod, 'skittering', and 'bobbing' are other modes of local popularity.

bo·bbing, *vbl. sb.*[2] [f. *bob *sb.*[1] 2 e.] The action or pastime of riding on bob-sleighs. Also *attrib.*

1888 *Troy Daily Times* 31 Jan. (Farmer), All the village bobbing clubs will participate in the carnival at Albany tomorrow..There are seventy-eight entries for the bobbing parade. 1888 *Harper's Mag.* May 973/1 Ruby and Ned displeased their mother by joining a 'bobbing' party on a neighbouring 'hill' street. 1904 *Windsor Mag.* Jan. 245/1 Bobbing can be practised on any high-road having a sufficient length of suitable slope. 1928 *Evening Standard* 5 Jan. 6/2 Bobbing is carried out either on bobs (five passengers) or boblets (three passengers). 1942 A. Koroleff *Efficiency in Skidding Wood* 5/2 Other methods in which log carriers of some kind are used (such as 'bobbing', 'draying', 'boganning', etc.).

bobbing, *vbl. sb.*[3]: see *bobber[5].

bobbin-net, bobbinet and variants. (Earlier examples.)

1814 Jane Austen *Let.* 22 Nov. (1952) 413 Mrs. Clement walks about in a new Black velvet Pelisse..& a white Bobbin-net-veil. 1819 H. Busk *Vestriad* iii. 85 Ye kerchief'd damsels, who encushion'd fret, ..your stiffen'd bobinet. 1824 M. Wilmot *Let.* 19 Jan. (1935) 204 The finest bobbinett dresses sprigged to imitate brussels lace.

bobbish, *a.* (Earlier and later examples.)

1780 R. Tomlinson *Slang Pastoral* 3, I was so good-natur'd, so bobbish and gay. 1862 Trollope *Rachel Ray* II. xiii. 268 Pretty bobbish, thankee, Mr. Rowan; and how's yourself? 1928 Galsworthy *Swan Song* III. vi. 258 He did not feel so 'bobbish' before this third encounter with that fellow.

bobbishly, *adv.* (Later example.)

1936 Wodehouse *Laughing Gas* xxix. 304 'Oh, so you know about that, too?' She laughed, though not too bobbishly.

bobble, *v.* Add: (Later examples.) Also *transf.* and *refl.* So **bo·bbling** *ppl. a.*

1836 T. Hook *G. Gurney* III. vi. 316 The..Ship..was comfortably bobbling herself about at Spithead. 1892 *Daily News* 20 Sept. 6/4 If they must shoot something, let them pop away at bottles..in a bobbling sea. 1896 W. Park *Game of Golf* x. 202 It is absolutely essential that a putting-green be firm and smooth, and the turf close and short, so that the ball will roll on it and not 'bobble' or jump, as it certainly will if the turf be brushy and uneven. 1942 W. Faulkner *Go Down, Moses* 18 He never even bobbled; he knocked Uncle Buck down and then caught him..without even stopping. 1960 I. Cross *Backward Sex* iii. 79 The stretch and wriggle of her shoulders bobbled her body.

bobble (bɒ·b'l), *sb.*[2] [dimin. of Bob *sb.*[1]: see -le 1.] A small woolly ball used as an ornament or trimming. Also *attrib.*

1923 *Daily Mail* 13 Feb. 1 Knitted Wool Jumpers, finished cord and bobbles. *Ibid.* 17 Apr. 1 Trimmed with bobbles of own material arranged in sets. 1927 Russell Thorndike *Slype* vi, A black gaberdine with bobbles on it. 1938 G. Greene *Brighton Rock* II. i. 76 A stand lamp in a red silk shade with a bobble fringe. 1948 M. Sharp *Flowering Thorn* II. i. 55 There's a lot of stuff I think must be rep—it's got bobbles all round the edge. 1968 *Guardian* 20 Sept. 1/6 A man in an orange bobble hat walked across the hill.

bobbled (bɒ·b'ld), *a.* [f. *bobble *sb.*[2] + -ed[2].] Ornamented with bobbles.

1955 E. Bowen *World of Love* iv. 76 The bobbled edge of the curtain. 1960 L. R. Banks *L-Shaped Room* vi. 90 There was a bookshelf, garnished with a bobbled fringe.

bobbly (bɒ·bli), *a. colloq.* [f. Bobble *v.* + -y[1].] Jumpy, uneven; knobbly.

1909 *Westm. Gaz.* 1 Feb. 12/2 The hard ground, the bobbly putting, and all the other horrors of golf in frosty weather. 1921 *Chambers's Jrnl.* Jan. 3/1 The fully-opened door revealed the white coat and bobbly trousers of a veritable bed-room steward. 1969 P. Dickinson *Pride of Heroes* 51 The first-floor windows mirrored one bobbly cloudlet.

bobby, *sb.* Add: **2.** (Earlier example.)

1844 *Sessions' Paper* June 341, I heard her say..'a bobby'..it was a signal to let them know a policeman was coming.

3. *bobby wren* (*dial.*): see quot.

1885 Swainson *Prov. Names Birds* 35 *Wren*..From its short bob-tail it has the names of Cutty or Cut..Bobby wren (Norfolk), [etc.].

4. In full *bobby calf*, an unweaned calf slaughtered soon after birth. Cf. *bob-veal.

1927 *Meat Trades Jrnl.* 16 June 1209/7 Central Meat Market..*Veal*: Moderate supply of fat calves..bobbies very short and dear. 1928 *Auckland Weekly News* 26 July 58/3, 30,000 bobby calves, mostly from the Waikato, were exported..last season as boneless veal. 1929 *Daily Express* 7 Nov. 15/5 Veal..specials 8s.–9s., bobbies 4s.–5s. 1938 *N.Z. Statutory Regulations* 334 'Bobby calf' means a calf which is intended for slaughter for human consumption as boneless veal. 1960 *Farmer & Stockbreeder* 16 Feb. 11/1 Scotch bobbies, 1s. 4d. to 1s. 6d. 1965 *Economist* 2 Oct. 70/2 In the past, the calf of a pure dairy cow, unless she was a heifer and so a potential milk-factory, was classed as a useless 'bobby' calf fit only for veal.

bobby-dazzler (bɒ·bi dæ·zləɹ). orig. and chiefly *dial.* [Dazzler.] Something striking or excellent; a strikingly-dressed person. Hence **bobby-dazzling** *sb.* and *a.*

1866 *N. & Q.* X. 290/2 *Morgan-rattlers*..In Cornwall the word is frequently applied to..things that are particularly striking or excellent of their kind. What a Lancashire man would sometimes call a 'regular bobby-dazzler', a Cornishman would call a 'regular morgan-rattler'. 1891 J. Baron *Blegburn Dickshonary* 15 *Bobby-dazzler* (a grand one). A lass is a bobby-dazzler when hoo's getten a new frock on. 1913 J. L. Waugh *Robbie Doo* (1914) ii. 22 And, at last, wi' a perfect bobby-dazzler, he landed trampie yin below the jaw that made him spin like a peerie. 1913 D. H. Lawrence *Sons & Lovers* I. vi. 122 She had got a new cotton blouse on..'Oh, my stars!' he exclaimed. 'What a bobby-dazzler!'..'It's not a bobby-dazzler at all!' she replied. 'It's very quiet.' *Ibid.* xii. 334 There isn't room in *this* house for two such bobby-dazzlers. 1915 —— *Rainbow* ix. 237 That's a bobby-dazzlin' posy you've brought! 1931 V. Palmer *Separate Lives* 177 There's a filly..She'd be good to win a race or two with—a real bobby-dazzler, she might turn out to be. 1944 *Gen* 1 Jan. 19/2 Matthews' bobby-dazzling wouldn't be wanted in a line containing authentic inside forwards. 1949 E. de Mauny *Huntsman in Career* 155 A real bobbydazzler of a morning. 1959 J. Braine *Vodi* xviii. 215 By God, you're what my old Nanny used to call a bobby-dazzler in that dress.

bo·bby pin. *U.S.* [Etym. uncertain, but cf. *bob *v.*[5] and *bobby sock.] A kind of sprung hair-pin or small clip, orig. for use with bobbed hair.

1936 J. Lawrence *If I have Four Apples* iv. vii. 201 She wondered whether she had lost all the bobby-pins from her marcelled hair. 1939 C. Morley *Kitty Foyle* viii. 81 How patient he was picking up bobby pins and the caps of coke bottles. 1952 M. McCarthy *Groves of Academe* (1953) iii. 32 Her dark, straight, glossy hair was worn short and loose, without so much as a bobby-pin. 1956 R. Heinlein *Double Star* (1958) v. 87 Penny girl, got a bobby pin?

bo·bby sock. orig. *U.S.* [Etym. uncertain, but cf. *bob *v.*[5] to cut short.] (Usu. in pl., **bobby socks, sox**), socks reaching just above the ankle, esp. those worn by girls in their teens; also *attrib.* Hence **bo·bby-socker,** (more commonly) **-soxer,** an adolescent girl, esp. one in her early teens, wearing bobby socks.

1943 *Time* 5 July 26/1 Hundreds of..girls in bobby socks sat transfixed. 1944 *Newsweek* 6 Mar. 88 In New York City last week, the 'Bobby-sox Brigade' had swelled to such alarming proportions..that police imposed an unofficial curfew. 1944 *Birmingham* (U.S.) *News-Age-Herald* 19 Nov. 11 A/4 About 6,000 bobby soxers attended the concert. 1944 *Life* 27 Nov. 76/2 When bobby-socker squeals, sound is carried to loudspeaker. 1945 Sinclair Lewis *C. Timberlane* (1946) xlvii. 333 Small white wool socks, 'bobby socks' they were called, to be worn with bare legs that were made-up to look tanned. 1945 *This Week* (U.S.) 15 Apr. 21/2 Comments on teen-age bobby-sock girls afflicted with dementia-Sinatra. 1946 *Jazz Writings* 22/1 George Webb's Dixielanders..had the satisfaction of laying a mainly bobby-sox audience flat in the aisles. 1948 *Observer* 8 Feb. 3/2 His air of smiling tolerance could deceive none but a bobby-soxer. 1958 *Times Lit. Suppl.* 10 Oct. 573/2 The recent spate of elopement scandals and bobby-sox novelists.

bobcat. *N. Amer.* [Bob *sb.*[1] 5, in allusion to the short tail.] The bay lynx, *Lynx rufus*, a small North American lynx with a spotted reddish-brown coat. Also *transf.*

1888 T. Roosevelt in *Century Mag.* Mar. 656 We also keep hens, which, in spite of the damaging inroads of hawks, bob-cats, and foxes, supply us with eggs. 1906 *Springfield Weekly Republ.* 20 June 2 The creation of this new commonwealth on virgin soil, where for centuries the wolves, bob-cats and Indians had leisurely roamed. 1911 O. Wister *Members of Family* vii. 250, I came for a box of matches, y'u bawlin' bobcat.

bob-haired, *a.* [See Bob *a.* 1 and *bob *v.*[5] 2.] Of a woman or girl: having the hair cut short all round.

1923 C. Sidgwick *None-Go-By* xvii, The bob-haired female gazed at him as if she wished to probe into his soul. 1925 *Daily News* 7 Sept. 5/4 A bob-haired brunette.

boblet (bɒ·blét). [f. *bob *sb.*[1] 2 e + -let.] A bob-sleigh for two persons. Also *attrib.*

1914 *Winter Sports Ann.* 15 (*caption*) Boblet..Race on Schatzalp, at Davos, 1913. 1923 *Daily Mail* 11 Jan. 9 Boblet Races Results. 1958 *Times* 27 Dec. 3/5 Lord Kimberley arrived here [*sc.* at St. Moritz] to-day to start training for the world two-man boblet championships.

bobolink. *U.S.* Add: (Earlier examples.) Also **bob-a-link, bob-o'-linck, bob-a-linkum.**

1774 J. Adams *Diary* 24 Oct. in *Wks.* (1865) II. 401 Young Ned Rutledge is a perfect Bob-o-Lincoln. 1783 [see Conquedle]. 1826 T. Flint *Recoll.* 243 Those merry and chattering birds, that we call bob-a-link, or French blackbird. 1840 C. F. Hoffman *Greyslaer* III. i. 104 There he goes..singing for all the world like a Bob-a-linkum on the wing. 1847 *Tom Pepper* I. 145 (Th.), I heard her voice, which was sweeter than a bob-o'-linck's.

‖ **bobotie** (bobo·ti). *S. Afr.* Also **babotie** and variants. [Afrikaans, prob. of Malay or Javanese origin.] A dish of curried minced meat with a variety of ingredients.

1870 *Cape Monthly Mag.* Oct. 224 '*Babootie*' and '*frickadel*' and '*potatoe-pie*' are great improvements upon the minced meats of England. 1883 O. Schreiner *Afr. Farm* I. viii. 133 Till there is not one bone left in your old body that is not broken as fine as bobootie-meat, you old beggar. 1919 M. Higham *Household Cookery in S. Afr.* (ed. 2) 90 Bobotee. 1 lb. beef or mutton..Mince the meat; chop onions finely. *a*1920 O. Schreiner *From Man to Man* (1926) ii. ix. 318 For dinner and supper she made sosatie and bobotie. 1950 *Cape Times* 6 June 16/2 Something typically South African like babotie. 1950 L. G. Green *Land of Afternoon* iv. 61 Mrs. Dykman made her bobotie from minced meat..bread, butter, onion, pounded almonds, lemon juice, eggs, curry powder and a small bunch of orange leaves. The mixture was baked and served with boiled rice.

Bob's-a-dying. *dial.* Also in contracted form **bobsy-die** (quots. are N.Z.). A great fuss; pandemonium.

1829 W. N. Glascock *Sailors & Saints* II. iv. 80 Nothing but dining, and dancing, and Bob's-a-dying on deck from daylight till dark. 1872 Hardy *Greenw. Tree* I. II. vi. 204 She used to kick up Bob's-a-dying at the least thing in the world. 1876 —— *Ethelberta* Sequel II. 309 She threatened to run away from him, and kicked up Bob's-a-dying, and I don't know what all. 1892 R. O. Heslop *Northumb. Words* I. 72 'What a *Bob's-a-dying* they made!' means 'What a row they kicked up'. 1940 N. Marsh *Surfeit of Lampreys* (1941) xvi. 251 If she's right..it plays Bobs-a-dying with the whole blooming case. 1952 *Here & Now* (N.Z.) II. iv. 25 Or Dido kicked up quite such bobsy-die if she had had a squint in her left eye. 1960 T. McLean *Kings of Rugby* vii. 50 By generally kicking up bobsy-die.

bob-sled, -sleigh. Add examples. Also **bob-sleighing** *vbl. sb.*, the sport of riding on a bob-sleigh.

1839 in *Amer. Speech* (1948) XXIII. 44/1 Then there would be wear and tear of bob-sled. 1888 [see *bob *sb.*[1] 2 d]. 1892 *Eng. Illustr. Mag.* Sept. 882 The logs are loaded on bob-sleds and taken to the mills. 1894 H. Gibson *Tobogganing* 163 Bobsleighing is another amusing form of tobogganing which has become very popular during the last few years. *Ibid.*, A Bobsleigh consists of two machines..connected together endways by a board, upon which the crew sit or lie. *Ibid.*, The Bobsleigh, Bobsled, or double-ripper. 1910 *Daily Chron.* 12 Jan. 7/2 An English bobsleigh team..dashed down the bobsleigh run and route above Grindelwald at a great speed. 1959 *Maclean's Mag.* 9 May 82/4 A winded horse, his sleighbells jingling, struggled to pull a real bobsled up the icy slope. 1968 *Listener* 12 Sept. 335/2 The night that we first transmitted our £60,000 production of *Hamlet at Elsinore*, they gave the front cover to the world bob-sleigh finalists.

bob-tail. Add: **B.** *attrib.* **bobtail car** *U.S.* (see quot. 1888); **bobtail discharge** *U.S.* (see quot.); **bobtail plough** (see quot. 1743).

1875 *Chicago Tribune* 8 Sept. 8/2 The bobtail cars ought to be taken off the streets right away, or conductors put on them. 1883 [in Dict.]. 1888 Farmer *Americanisms*, *Bobtail car*, the popular name for a small tram-car horsed by a single animal, and on which the only official is a driver, whose office it is to collect fares and generally perform the duties of conductor in addition to his own. 1886 *Outing* (U.S.) Dec. 227/2 Upon the expiration of his first enlistment, he was given what is called a *bobtail* discharge, a discharge without character. 1733 Ellis *Chiltern & Vale Farm.* xxxviii. 319 Either the Swing, Bobtail, or Foot Plough. 1743 —— *Mod. Husb.* May i. 6 The Creeper or Bobtail Plough, is a two-Wheel Plough, differing but very little from the Fallow Wheel Plough.

C. *sb.* **1. c.** = *bobtail discharge. *U.S.*

1915 *Recruiters' Bulletin* (U.S.) June 8/1 (*title*) Bobtails. 1926 R. A. Bennet *Boss of Diamond A* viii. 72 My discharge is in that trunk of mine. I'll show you it's no bobtail! 1937 *Amer. Speech* XII. 75/1 *Bobtail*, (yellow-colored) dishonorable discharge.

4. Also *pl.* (cf. quot. 1785 in Dict.).

1926 J. Black *You can't Win* xiii. 180 She poured liquor into the bums, beggars, ragtags and bobtails that hung around the saloons.

5. In full *bobtail flush, straight*. A four-card flush or straight in the game of poker. *U.S.*

1875 *Cincinnati Enquirer* 2 July 2/3 The gentlemen..can not 'bluff' to any success on their 'bob-tail flush'. 1887 *Courier Jrnl.* 20 Jan. 4/1 The Confederates would..bet away the Washington monument on a 'bob-tail flush'. 1901 R. F. Foster *Poker* 26 The value of the 'bobtail', as a four-card flush is called, is nothing unless it can be improved in the draw.

bob-tailed, *a.* Add: spec. (*b*) *bobtailed flush, straight* = *bob-tail C. 5; (*c*) *bobtailed car* = *bobtail car* (see above). *U.S.*

1873 *Winfield* (Kansas) *Courier* 15 Feb. 1/5 Well up in all the mysteries of a 'pair' or a 'bobtailed flush'. 1875 *Chicago Tribune* 16 Oct. 11/3 In regard to public feeling being against bobtailed cars. 1901 S. E. White *Westerners* i. 4 He owned ..a light, two-wheeled wagon of the bob-tailed type. 1944 *Amer. N. & Q.* Sept. 85/1 Bobtailed Straight or Flush.

bob-veal. *U.S.* [Bob *sb.*[1]; cf. *staggering bob* (Staggering *ppl. a.* 1 d).] The veal of a very young calf. Cf. *bobby 4.

1855 in De Voe *Market Assistant* (1867) 421 Butchers call this *bob-veal*. 1888 *San Francisco Weekly Exam.* 22 Mar. (Farmer), It is time this traffic in..bob-veal was stopped. 1911 *N.Y. Even. Post* 13 Oct. 1 A former butcher was sentenced..for..shipping..the carcasses of five bob veal calves to this city.

bob-white. (Earlier and later examples.)

[1812 Wilson *Amer. Ornith.* VI. 25 The quail..will sometimes sit, repeating, at short intervals, 'Bob White', for half an hour at a time.] 1819 J. F. Stephens in *Shaw's Gen. Zool.* XI. ii. 380 By the natives of New England they [*sc.* Northern Colins] are called *bob-white*. 1907 *Westm. Gaz.* 24 Aug. 16/1 The Virginian quail is more familiarly known by the name of 'Bob White', being the nearest rendering of the clear and loud call of the male bird. 1936 T. S. Eliot *Coll. Poems 1909–35* 152 The bob-white Dodging by bay-bush.

bob-wire. *U.S.* colloq. alteration of *barb wire* (see Wire *sb.* 1 e).

1929 J. F. Dobie *Vaquero of Brush Country* 116 Chaos demanded order, and the means of order were to be barbed wire—'bob-wire', most people used to call it. 1939 Steinbeck *Grapes of Wrath* iv. 26 Some sisters took to beatin' theirselves with a three-foot shag of bobwire. 1940 C. McCullers *Heart is Lonely Hunter* (1943) II. xi. 233 We got to climb the bob-wire fence.

bocage. Add: **2.** The representation of silvan scenery in ceramics. Also *attrib.*

1902 W. Burton *Hist. & Descr. Eng. Porcelain* v. 49 Little figures were produced with wreaths of flowers and foliage, and, finally, the fully developed boscage, or *bocage* pieces. *Ibid.* vii. 73 Little figures on stands, with *bocages*, and nozzles for candlesticks, were also produced at Bow. 1950 *Antiquity* XXIV. 111 The 19th century ushers in John Walton, with his well-known 'bocage', which has been described as 'the art of the pastry-cook'. 1961 *Connoisseur New Guide to Antique Eng. Pott., Porc. & Glass* 66 Ralph Salt of Hanley specialised in the rather more costly *bocage* pieces, sporting dogs, and sheep with hand-raised wool. 1961 *Times* 8 Apr. 11/6 The leafy arbour of *bocage* groups associated with exquisite porcelain.

boccaro (bɒ·karo). Also **bucaro, buccaro.** [Prob. ad. Pg. *búcaro* clay cup:—L. *poculum* cup.] A scented red earthenware brought originally by the Portuguese from Mexico and similar earthenware made in Portugal and Spain from the 16th to the 18th century; the name is also applied to unglazed Yi-hsing ware of China.

1872 J. F. Riano *Catal. Art Objects of Spanish Production in S. Kens. Mus.* p. xxix, This author [*sc.* Mme D'Aulnoy] also mentions the red earthen pottery, bucaros of Toledo. 1898 M. L. Solon *Pottery Worship* 13 Guadalaxara, in Mexico, sent over the most highly-scented Buccaros. *Ibid.* 17 In 1568 Buccaros were made at Talavera, in Spain, as the imitation of those..imported from Portugal. *Ibid.* 30 True Buccaros never came from China..and..must not be looked for out of the pottery sent over from Central America, or the Portuguese imitations. 1910 *Encycl. Brit.* V. 741/2 From the 16th to the 18th century a special kind of unglazed pottery vessels known as *buccaros* were extensively made both in Spain and Portugal..The vessels were delicately scented, like a ware imported from Mexico. 1945 W. B. Honey *Ceramic Art of China* ii. 136 The Yi-hsing ware..made of various brown, buff and reddish clays,..is unglazed..In Europe, from their similarity to a Spanish-American ware they were sometimes called '*boccaro*' (or '*buccaro*') ware.

bocconia (bɒkōᵘ·niă). [mod.L.; named after the Sicilian botanist, Paolo Boccone (1633–1704).] A tall herbaceous plant or shrub of the papaveraceous genus of the name, with large lobed leaves and panicles of flowers. The favourite species is *B. cordata*, with cream-coloured flowers growing on tall stems.

1867 K. Prosper in *Floral World* Apr. 107, I have had the honour to introduce the Bocconia to cultivation on the continent..This plant is of most quick growth, the great leaves are deeply notched, and with much of whitish colour. The flowers are like poppies..and the colour rose, or pink, or white. 1902 W. P. Wright *Cassell's Dict. Gard.* I. 128 The herbaceous Bocconias like a rich, heavy soil. 1961 *Times* 2 Dec. 11/4 Herbaceous plants such as..bocconias.

Boche (bɒʃ). Also *erron.* **Bosch(e.** [Fr. slang, = rascal, German, said to be shortened from *caboche* head, or from *Alboche*, modification of *Allemand* German.] The (French) soldiers' name for a German. Also *attrib.* or *adj.*, German.

[1887 A. Barrère *Argot & Slang*, *Boche*..rake..*Tete de boche*, German.] 1914 *Daily Express* 30 Sept. 4/2 Monsieur had better come under cover. The 'Boches' are still firing this way. 1915 *National Rev.* Mar. 21 Easter passed without the promised bonne-bouche for the Bosches. 1917 G. S. Gordon *Let.* 12 Mar. (1943) 72 Our Archibalds are peppering a Boche aeroplane. 1919 'Boyd Cable' *Old Contemptibles* viii. 130 A Boche..proceeded to drop bombs all over the place. 1920 *Punch* 20 Oct. 305/1 Bosch aeroplane observers. 1940 W. S. Churchill *Into Battle* (1941) 296 All Europe, if he [*sc.* Hitler] has his way, will be reduced to one uniform Boche-land. 1952 E. F. Davies *Illyrian Venture* ii. 43 If the Boche wanted a rough-house he could rely on Pickering to give it to him.

bock (bɒk). [Fr., a. G. *bock*, in full *bockbier*, shortened f. *Eimbockbier*, now *Einbecker bier*, f. *Einbeck*, *Eimbeck*, a town in Hanover.] In full *bock beer*. A strong dark-coloured variety of German beer. Also, a glass of this or any other beer.

1856 *Ill. State Reg.* 26 June 4/3 There is a Bavarian lager beer which is called 'bock'. 1867 Simmonds *Dict. Trade* Suppl., *Bock-beer*, a favourite Bavarian double strong malt beverage, of the best lager description. 1879 Sala *Paris Herself Again* I. 183 A 'bock', or glass of light and frothy beer. *Ibid.* 186 Allsopp at fifty centimes the 'bock'. 1899 Wilde *Lett.* (1962) 784, I go there on foot daily and have a *bock*. 1917 T. S. Eliot *Prufrock* 19 Let us..sit for half an hour and drink our bocks.

bod (bɒd). Slang abbrev. of Body *sb.* in various senses; esp. a person. (In Sc. quots. 1788 and 1813, perh. abbrev. of *bodach.)

1788 J. Macaulay *Poems* 143 While aulder bods were babees birling, We took our slide. 1813 E. Picken *Poems* II. 131 Sae he made a lang blaw about graces, an' gods, Like Vulcan, an' Bacchus, an' ither sic bods. 1933 R. Strachey *Many Happy Returns* i. 33 The Red, the White and the Blue is unfurled and covers the dead bod, the corpse of a grocer given for England. 1953 'N. Shute' *In Wet* vi. 187 The rest go to the camp with the R.A.A.F. bods. 1954 C. P. Snow *New Men* xxii. 151 'Bods' meant bodies, people, any kind of staff: scientists were bods, so were floor cleaners. 1955 P. Sharp in Chambers & Landreth *Called Up* iii. 73 To join a station you have to get these cards signed by odd bods all over the place. 1966 *Crescendo* Aug. 8/1 The show-tune formula is quite simple—I know dozens of bods who make a living using it.

bodach (bɒ·daχ). *Ir.* Also **-agh.** [Gaelic and Ir. *bodach*.] A peasant, churl; also (*Sc.*) a spectre.

1814 Scott *Wav.* III. xii. 157, I have seen the Bodach Glas [*i.e.*, the Grey Spectre]. 1827 —— *Highl. Widow* i, Oh! then the mystery is out. There is a bogle or a brownie, ..a bodach or a fairy, in the case? 1830 W. Carleton *Traits Irish Peasantry* I. 28 Hut! he's none iv yer proud, stingy, upsthart *bodaghs*. 1865 *Dublin Rev.* July 73 It admits of no doubt that the Gaelic is withering away..The coarse Bodach almost alone retains it as the language of common life. 1903 *Westm. Gaz.* 18 May 4/3 Father Dinneen's play 'The Enchanted Well' dealt with a rivalry in love between a rich 'bodach' and a poor poet. 1903 W. B. Yeats *Hour-Glass* 13, I met a bodach on the road yesterday.

bodacious (bŏdēⁱ·ʃəs), *a. U.S. dial.* Also **bow-.** [Perh. a variant of Eng. dial. *boldacious*, a combination of *bold* and *audacious*.] Complete, thorough, arrant. Also as *adv.*

1845 W. T. Thompson *Chron. Pineville* 178 She's so bowdacious unreasonable when she's raised. 1887 'C. E. Craddock' *Keedon Bluffs* 153 Air ye turned a bodacious idjit, Skimp? 1941 *Time* 17 Mar. 23/2 The bodacious blurt did him no good with his brother Navy men.

boda·ciously, *adv. U.S. dial.* Also **bodyaciously.** [See prec.] Completely, thoroughly.

α. 1837 A. Sherwood *Gaz. Georgia* (ed. 3) 69 *Bodaciously*, wholly. 1840 *Congress. Globe* 20 July 545 (Th.), [It has been proclaimed abroad] that the Administration is bodaciously used up. 1843 'Carlton' *New Purchase* 143 It was now snowing so bodaciously fast. 1887 'C. E. Craddock' *Keedon Bluffs* 104 He drug two needles bodaciously out an spiled fower rows. 1904 Harben *Georgians* vii. 69 You was ruinin' yoreselves bodaciously.

β. 1833 J. Hall *Leg. West* 82 It seems like it would jist use me up bodyaciously. 1833 *Sk. D. Crockett* Pref. p. v, A constant repetition of the terms bodyaciously, teetotaciously, obflisticated, &c.

Bode (bōᵘ·də). *Astr.* [The name of Johann Elert *Bode* (1747–1826), German astronomer.] *Bode's law*: see Law *sb.*[1] 17 c.

1833 [see Law *sb.*[1] 17 c]. 1959 *Listener* 30 July 171/1 During the eighteenth century a German named Titius worked out a numerical relationship in connection with the distances of the various planets; in 1772 this relationship was discussed by another German, Bode, and is generally, though rather unfairly, known as Bode's Law.

bodega. (Earlier examples.)

1846 R. Ford *Gatherings fr. Spain* xiii. 148 The best vineyards and *bodegas* or cellars are those which did belong to Don Carlos. 1872 Lady C. Schreiber *Jrnl.* (1911) I. 144 He took us..on foot, to the bodega of Messrs. Gonzalez.. where we were asked to taste a number of different wines.

‖ **bodegon** (bodego·n). Pl. **-ones** (-o·nes). [Sp., f. *bodega* Bodega.] Properly, a Spanish tavern-piece; hence, any Spanish picture representing still-life or a *genre* subject.

[1748 H. Johnson tr. *Lozano's Earthquake at Lima* (ed. 2) 317 Those who keep the Bodegones, (a Sort of Taverns, which are no better than a Chandler's-shop; for besides Wine, they sell Candles, Fish, Salt, Cheese and Bacon).] 1843 *Penny Cycl.* XXVI. 189/1 Beginning with subjects of still-life..such as meat, vegetables, and kitchen utensils: hence the generic name *Bodegones*, by which they are still known. 1848 W. Stirling-Maxwell *Artists of Spain* II. 580 These 'bodegones' of his early days are worthy of the best pencils of Flanders. 1896 W. Armstrong *Life Velazquez* 12 His earliest independent works were *bodegones*—kitchen and tavern scenes.

bodger[3] (bɒ·dʒəɹ). *dial.* In full *chair bodger*. A local name in Buckinghamshire for a chair-leg turner. Hence (*chair-*)*bodgering*, the action or process of chair-leg turning.

1911 G. Eland *Chilterns & Vale* vi. 136 The men who thus work in the woods are called 'chair-bodgers'. *Ibid.* 137 The purchaser then employs the 'bodger' to turn it [*sc.* a 'fall' of beech] into chair-legs. 1921 K. S. Woods *Rural Industries round Oxford* II. i. 102 Most village turners or 'chair bodgers' confine themselves to the making of legs which they sell to the factories, mainly at Wycombe. 1939 D. Hartley *Made in England* i. 23 The shed for bodgering jobs may be left standing the whole year.

bodger (bɒ·dʒəɹ), *a. Austral. slang.* Inferior, worthless; (of names) false, assumed.

1945 Baker *Austral. Lang.* viii. 156 *Bodger*, worthless, second-rate. (This term is apparently related to English dialect in which *bodge* means to botch or work clumsily.) 1950 F. J. Hardy *Power without Glory* 383 This entailed the addition of as many more 'bodger' votes as possible. 1954 —— in *Coast to Coast 1953–4* 76 We stuck together all through the war—we was in under bodger names.

bodgie (bɒ·dʒi). *Austral.* and *N.Z.* [Perh. f. *bodger *a.* + -ie.] A teenage boy who conforms to certain fashions in dress, etc. (see quots. 1952); the Australasian counterpart of a *teddy-boy. Hence **bo·dgieism,** the state or condition of being a 'bodgie'.

1952 *Sunday Chron.* 6 Jan. 4/8 *Bodgie*, boy with crew-cut and zoot-suit, playing juke-box in milk-bar. 1952 H. Boyle in Wentworth & Flexner *Dict. Amer. Slang* (1960) 49/1 A bodgie is a jitterbug-crazy boy 'who wears his hair curled and long and a sport coat too big for him'. 1958 A. E. Manning *The Bodgie* vii. 89/2 These 'Bodgies' and 'Widgies' are not *bad* however repellent their conduct may appear. 1958 *N.Z. Listener* 3 Oct. 10/3 Every psychologist who has talked with bodgies will know that fear of an uncertain future is one of the factors in youthful misconduct. 1958 *N.Z. Herald* 5 Nov. 14/7 Some of New Zealand's suggested cures for 'bodgieism' were not only 'wide of the mark' but surprisingly 'vindictive', says Dr. D. P. Ausubel.

bo·dging, *vbl. sb.*[2] (*dial.*) = *bodgering; see *bodger[3].

1953 A. Jobson *Househ. & Country Crafts* xx. 178 Of all the woodland crafts, that of chair-bodging seems the most rural. 1957 *Times* 2 July (Agric. Suppl.) p. viii/7 The demonstration of chair leg bodging.

‖ **Bodhisattva** (bɒdisa·tvă). *Buddhism.* Also **Bodhisat, Bodhisatta, Bodhisatwa.** [Skr., 'one whose essence is perfect knowledge', f. *bodhi* perfect knowledge (*budh* to know: see Buddha) + *sattvá* being, reality.] One destined to become a Buddha; a Buddhist saint who, having only one birth to undergo before attaining Nirvana, consents to be reborn for the sake of suffering mankind; a superhuman being of infinite wisdom and compassion. Hence **bodhisa·ttvahood, -ship, -sa·tship,** the state of a Bodhisattva.

1828 *Asiatic Res.* XVI. 422 The form of the works..is that of a lecture, or lesson, delivered by a Buddha to his Bodhisatwas, or disciples. 1850 R. Spence Hardy *Eastern Monachism* xviii. 170 The Jâtakan, containing an account of 550 births of the Bódhisat who afterwards became Gótama Budha. 1889 M. Monier-Williams *Buddhism* v. 98 He [*sc.* Buddha] gave to every being destined to become a Buddha the title Bodhi-sattva (Bodhi-satta), 'one having knowledge derived from *self-enlightening intellect* for his essence'. *Ibid.* 181 He had transferred the Bodhi-sattvaship to Maitreya. 1895 L. A. Waddell *Buddhism of Tibet* vi. 138 The Arhats being dead cannot be active, the Bodhisattvas as living beings can. 1926 *Encycl. Brit.* XVI. 96/2 The leaders of the Great Vehicle urged their followers to seek to attain, not so much to Arahatship, which would involve only their own salvation, but to Bodhisatship. 1927 A. Huxley *Proper Studies* 181 Buddhism became an entirely new religion, with a pantheon of Bodhisattvas. 1951 H. Zimmer *Philos. India* 510 His [*sc.* Fa Hian's, c. 400 A.D.] account of the cities further west reveals that at Mathurā the Bodhisattvas Mañjuśrī and Avalokiteśvara received worship as divinities—which is a Mahāyāna feature. *Ibid.* 553 Within the hearts of all creatures compassion is present as the sign of their potential Bodhisattva-hood. 1957 B. J. Gould *Jewel in Lotus* 210 The Dalai Lamas are Bodhisattvas in whom is incarnate Chenrezi, the God of Mercy.

bodice. Add: **2.** Now only in the sense 'the upper part of a woman's dress, above the waist'.

bodiliness. (Later example.)

1847 J. C. & A. W. Hare *Guesses at Truth* (series 1, ed. 3) 55 The living energy and definiteness and bodiliness of Homer's characters.

bodingly, *adv.* (Earlier example.)

1811 Shelley *St. Irvyne* i, They bodingly presag'd destruction and woe.

Bodoni (bodōᵘ·ni). A book produced by the celebrated Italian printer Giambattista Bodoni (1740–1813); a modern type based on that of Bodoni. Also *attrib.* in *Bodoni type*.

1880 [see *Aldine *a.* and *sb.*]. 1922 D. B. Updike *Printing Types* II. 235 The 'Bodoni' type of commerce is a composite picture of many of Bodoni's fonts, rather than a reproduction of any one of them. 1928 *Scholartis Press Catal.* June, Printed by the Glasgow University Press in Bodoni type. 1967 E. Chambers *Photolitho-offset* ii. 12 Some families have dozens of different type faces (Caslon, Bodoni, Baskerville, Plantin, etc.).

bod-stick, var. *bott-stick*: see *bott 2.

1883 T. D. West *Amer. Foundry Pract.* (ed. 2) 331 The melter..runs the bod-stick without any clay on it into the running iron. 1900 [see *bott 2].

body, *sb.* Add: **1. d.** (Usu. hyperbolic.) Phr. *over my* (etc.) *dead body*.

1833 Seba Smith *Life & Writings J. Downing* (1834) 137 You don't go through this door to-night, without you pass over the dead body of Jack Downing. 1936 H. Brighouse *New Leisure* in *Best One-Act Plays 1936* 81 Elsie Dixon doing confidential secretary! Over my dead body. 1963 *Times* 27 May 6/2 If the number of distinguished gentlemen who cry 'Over my dead body' really mean what they say, this will be a fairly lethal summer in Whitehall.

7. b. The foundation of a felt or silk hat. Also Comb. *body-maker*.

1845 Dodd *Brit. Manuf.* V. 159 The 'body', or 'foundation', of a good beaver hat is..made of eight parts rabbits' fur [etc.]. 1880 *Encycl. Brit.* XI. 519/2 A silk hat consists of a light stiff body covered with a plush of silk. 1881 *Instr. Census Clerks* (1885) 74 Silk Hat Making: Body Maker. Finisher. Shaper. 1906 Watson Smith *Chem. Hat Manuf.* 65 The stiffening and proofing of hat forms or 'bodies'. 1921 *Dict. Occup. Terms* (1927) § 409 Body-maker [of hats].

8. b. (*a*) Later examples. (*b*) Used for the corresponding part in a motor-car and in an aeroplane. (*c*) *attrib.* and *Comb.*

(*a*) 1761, 1794 [see Carriage 28]. 1881 J. W. Burgess *Coach-building* 42 The body is a species of box, fitted with doors and windows, and lined and wadded for the purpose of comfort. 1897 J. Philipson *Coachbuilding* 2 The body is the most essential part of the carriage.

(*b*) 1896 *Horseless Age* May 20 Width of body [of motor-car] 32 inches; length of body 8 feet 6 inches. 1906 *Motors* 52 The Tonneau body was till lately most popular. 1909 A. Berget *Conquest of Air* 166 The body..is the space designed to carry the motor, propeller, and the aviator. 1920 Jones & Frier *Aeroplane Design* 109 The main function of the body or fuselage is to provide accommodation for cargo, pilot, passengers, flying instruments, and a reliable bearing for the power unit.

(*c*) 1611 [see Body 29]. 1802 *Sporting Mag.* XIX. 205/1 In the first shop [of a coach manufactory] the body-makers are employed. 1846 Dodd *Brit. Manuf.* VI. 113 'Body-makers' [are] employed principally on delicate framework and panelling. 1884 [see Body 29]. 1891 *Daily News* 29 Dec. 6/4 The body-making and harness departments. 1908 *Westm. Gaz.* 19 Mar. 4/1 A large number of chassis..fitted with every class of bodywork. 1909 *Ibid.* 17 June 4/1 The body-painting, smithy, and upholstery shops. 1914 C. W. Terry *Motor Body-building* 58 Materials used by body-builders. 1920 Jones & Frier *Aeroplane Design* 99 The outer, inner, and body struts. 1963 *Times* 14 May 7/3 The bodywork is virtually as good as new after years of operation. 1967 *Autocar* 28 Dec. 2/1 The Hillman Super Minx, with which the previous Vogue had shared a common body-shell.

21. b. *spec.* A mass or deposit of metalliferous ore.

1909 *Westm. Gaz.* 9 Feb. 11/4 The opening of an entirely new body carrying on an average 3 per cent. copper and 15 ounces of silver to the ton. 1929 *Times* 25 Jan. 12/3 A number of areas [in Great Britain] are worth prospecting in the hope of discovering new ore bodies.

22. c. The paste or clay (of a particular kind) used in the manufacture of porcelain.

1774 J. Wedgwood *Let.* 21 July (1965) 163 At one time the body is white and fine as it should be, the next we make ..is a Cinamon color. 1839 Mortar body [see Mortar *sb.*[1] 5 b]. 1893 E. A. Barber *Pott. & Porc. U.S.* 127 The proportion of phosphate of lime..being..a very much smaller percentage than in the English bone body. 1902 A. Bennett *Anna of Five Towns* viii. 169 The four sorts of clay used in a common 'body'—ball clay, China clay, flint clay and stone clay.

29. b. (*a*) *body-blow* (earlier, later, and *fig.* examples), *-build, -odour, -swing, -weight.*

1792 *Sporting Mag.* I. 43/2 After sparring some time.. Stanyard put in a body blow. 1908 *Daily Chron.* 24 Aug. 5/5 Its latest action is a body-blow to the growers. 1950 Dempsey *Champ. Fighting* 39 Face-punches and body-blows. 1958 *Times* 22 Apr. 6/7 It is a body blow. I am carrying the can for somebody else. 1923 C. B. Davenport (*title*) Body-build and its inheritance. 1961 *Lancet* 12 Aug. 341/2 On admission her weight was 129 lb., which was proportional to her height and body-build. 1933 D. L. Sayers *Murder must Advertise* iv. 69 Do you ever ask yourself about Body-Odour? 1896 *Daily News* 13 Mar. 8/4 The form of the men at the slow stroke was admirable, body-swing and feather alike being capital. 1950 W. Hammond *Cricketers' School* v. 54 Thus gaining the sort of body-swing that won Maurice Tate his wickets. 1873 C. H. Ralfe *Physiol. Chem.* 81 Men excrete little more urea in proportion to their body weight than women. 1907 *Westm. Gaz.* 23 Aug. 4/2 He [*sc.* the batsman] throws his body-weight on the left, the forward foot. 1966 *Lancet* 24 Dec. 1380/1 No patient was preselected on the basis of serum-lipid level or body-weight.

30. body belt, a belt worn close to the body; **body-box,** a brood-box, brood-chamber; **body-building,** the feeding and strengthening of the human frame by diet and exercise; also *attrib.* or *adj.*; **body carpet, carpeting,** carpeting manufactured in strips that are joined together to form the required size; also *body* ellipt.; **body-cavity** *Zool.*, the cœlom; **body cell** *Biol.*, a somatic cell; **body-centred** *a.*, applied to a type of crystal structure in which an atom or ion occurs at each corner and in the centre of a cubic unit cell; **body-check,** a movement in lacrosse (see quot. 1892); also, a similar movement in ice-hockey; hence as *v.*; so **bo·dy-che:cking** *vbl. sb.*; **body drop,** a throw in ju-jitsu; **body-face** = *body-type*; **body-image** *Psychol.*, the subjective picture or mental image of one's own body; **body-line bowling,** fast bowling delivered persistently on the leg side so as to be likely to strike the batsman's body; also in other collocations; **body-mark, stroke** *Printing*, the stem or 'thick-stroke' of the face of a type-letter; **body-mind** *Philos.* (see quots.); **body-rope** *Naut.* (see quots.); **body-schema** = *body-image*; **body-soul,** body and soul regarded as a unified whole; also *attrib.* or *adj.*; **body stocking** (see quot. 1968); **body strike** [*strike *sb.*]: see quot. 1957; **body-type,** the type used for printing the text of a book; **body-urge,** sexual passion; **body-wall** *Zool.*, the general envelope of an animal body; the cell-wall of a lower organism; **bodywork** (see sense *8 b (*c*)).

1911 *Alfred Weeks's Sales Catal.* All Wool Body Belts..to clear 6½d. 1962 T. C. H. Jacobs *Red Net* xviii. 178 A wide body-belt which Carlo had worn next to his skin..Two rows of tiny pockets ran its entire length. 1881 T. W. Cowan *Bee-keeper's Guide Bk.* 37 A second hive, having eight frames the same size as those used in the body-box, is provided for use on the top of the other. 1895 *Modern Bee-Keeping* (ed. 8) 22 The body-box or brood-chamber. 1904 *Daily Chron.* 4 May 4/3 Proteid, or the body-building element. 1904 E. Sandow (*title*) Body Building, or Man in the Making. 1962 *Times* 14 Nov. 3/7 The swimming club has 45 minutes on body-building. 1946 *Carpet Rev.* Oct. 15/2 A large quantity of special heavy Wilton body carpet..was also made by the firm for the cabin class. 1947 *Ibid.* May 23/1 As for piece goods, there is quite a demand for ¾ stair and runners, but one often sells body goods to fill the want, as stair carpets are in short supply. A good deal of plain body is sold. 1947 J. F. C. Brinton *Carpets* v. 37 Originally, carpeting was only made in body or filling 27 in. wide, and border 22½ or 18 in. wide. 1957 *Times* 14 Oct. 13/5 For close-carpeting it is more economical to buy 'body' carpeting for alcoves, fireplaces and other odd corners. 1963 *Which?* Mar. 71/1 This..carpet..is made on a broad loom (6 ft. wide or more) instead of the usual 'body' carpet which is in rolls 18–54 in. wide. 1875 Huxley & Martin *Elem. Biol.* 100 The body-cavity [in *Hydra*]. 1888 Rolleston & Jackson *Anim. Life* Introd. p. xxix, The cavity, or series of cavities, known as body cavities or coelome. 1927 Haldane & Huxley *Anim. Biol.* i. 10 The stomach and intestine lie in a space, the general body-cavity or coelom. 1896 E. B. Wilson *Cell in Devel. & Inherit.* ix. 329 Whether these variations first arise in the idioplasm of the germ-cells..or whether they may arise in the body-cells and then be reflected back upon the idioplasm. 1926 J. S. Huxley *Essays Pop. Sci.* i. 7 The nucleus of an ordinary body-cell. 1921 *Physical Rev.* XVII. 574 The calculated spacings are those of a body centered cubic lattice, the side of the cube being 2·895 Å., and the distance between nearest atoms 2·508 Å. 1925 *Jrnl. Iron & Steel Inst.* CXII. 502 There are only two polymorphous phases..a cubic body-centred modification..and a cubic face-centred modification. 1944 *Electronic Engin.* XVII. 142 In the case of iron the atoms form a regular cubic pattern known as 'body-centred' in which, if we consider an elemental cube of the crystal or unit cell, there will be an atom at each corner of the cube and another at the centre. 1892 *Lacrosse: Laws* 6 Body-check is the placing one's body in the way of an approaching opponent, so that the latter is simply impeded. No checker shall use force in the body-check. 1901 *Encycl. Sport* I. 608/1 When a player is dodging, no notice should be taken of his crosse, the *Checker* simply taking care to place his body in the way of the dodger. This is known as the *body check*, and no force may be imparted to it, or it becomes a charge, which is forbidden. Body checking is of most use out of the field. 1909 *Westm. Gaz.* 17 Dec. 12/2 It might be a hint to Forster to body check more efficiently [in lacrosse]. 1962 *Amer. Speech* XXXVII. 126 Bodychecks and unpermitted blows occurred [in ice-hockey]. 1936 *Times Lit. Suppl.* 14 Mar. 227/2 It [ice hockey] offers to the eye the clash of bodies (for all the rules about body-checking). 1960 *Times* 29 Nov. 17/5 Oxford and Cambridge completed their lacrosse programmes for the term..Cambridge started slowly. Their defence was open, there was little body-checking. 1948 G. Koizumi *Twelve Judo Throws* 26 Body-drop (Taiotoshi)..is one of the hand throws. 1960 *Oxf. Mail* 10 Mar. 8/3 Buley scored a point with a body drop throw, but Garnett scored points with hip and body drop throws. 1898 J. Southward *Mod. Printing* I. 134 Body or text faces. 1934 P. Schilder in *Proc. Assoc. Res. Nerv. Ment. Dis.* XIII. 466 (*title*) Localization of the body image. 1935 —— *Image & Appearance of Human Body* 11 The body schema is the tridimensional image everybody has about himself. We may call it 'body-image'. 1950 *Lancet* 25 Feb. 335/1 The expression body-image..refers to the mental idea which an individual possesses as to his own body and its physical and æsthetic attributes. 1933 *Times* 19 Jan. 12/6 The Australian Cricket Board of Control has sent the following telegram to the M.C.C.: 'Body-line bowling has assumed such proportions as to menace the best interests of the game, making the protection of his body by a batsman his main consideration [etc.].' 1955 I. Peebles *Ashes 1954–55* ix. 93 Voce did so [*sc.* took ten wickets], aided by a wet wicket, and Larwood by a bodyline field. 1896 De Vinne *Moxon's Mech. Exerc., Printing* 414 Stem is the thick-stroke of a letter, sometimes called by type-founders the body-mark. 1877 G. H. Lewes *Physical Basis of Mind* III. iii. 350 We know ourselves as Body-Mind; we do not know ourselves as Body *and* Mind, if by that be meant two coexistent independent Existents. 1945 *Mind* LIV. 58 The defenders of this view have recently been making quite a point of speaking not of 'a mind' and 'a body' but of a 'body-mind', seeking to emphasize by the hyphen in this compound word the monistic identity or inseparableness of the mental and bodily components. 1883 *Man. Seamanship Boys' Training Ships* 41 The ropes [for royals] are of two sizes only—viz., head rope from earring to earring, and a body rope on the foot and leeches. *Ibid.* 46 The largest or body rope..and the head rope. 1935 Body-schema [see *body-image*]. 1942 *Brit. Jrnl. Psychol.* Apr. 280 The notion of the 'body schema' (or 'body image') has..come to enjoy something of a vogue in contemporary neuropsychiatry. 1962 Hoenig & Hamilton tr. *Jaspers' Gen. Psychopathol.* i. 89 The car I drive, if I am a good driver, becomes part of my body-schema or image and is like an extended body which I invest fully with my own senses. *a*1897 W. Wallace *Lect.* (1898) 152 As if we were to say of a human being (and it is what perhaps we dare say of the fewest), that he or she was body-soul: the body the transparent and perfect temple of the spirit. 1956 E. L. Mascall *Christ. Theol. & Nat. Sci.* vii. 271 The doctrine that the human soul..is only one part of the twofold body-soul unity of the man. 1958 D. M. Baillie *Out of Nazareth* II. i. 150 Man is a body-soul organism. 1965 *Vogue* 15 Apr. 93 New body stocking.. body-coloured, body-shaped Lycra, just about invisible. 1968 J. Ironside *Fashion Alphabet* 66 *Body-stocking*, covering for the entire body from neck to feet, sometimes with sleeves; they are the same as a dancer's leotard..They are worn as a single undergarment in place of brassière, pants and stockings. Nowadays body-stockings are often flesh coloured and only opaque over the actual body, to give an impression of nudity under a transparent dress. 1969 P. Roth *Portnoy's Complaint* 203 Walking into a restaurant with a long-legged *kurveh* on his arm! An easy lay in a body stocking! 1937 Belschner & Seddon *Stud. Sheep Blowfly Problem* in *Dept. Agric. N.S. Wales Sci. Bulletin* No. 54 (*title*) Observations on Fleece Rot and Body Strike in Sheep. *Ibid.* 24 Body strike follows the development in the fleece of the condition known as fleece rot. 1957 *New Biol.* XXII. 94 Strike starting on some other part of the body [of sheep] such as the shoulder or over the back, is referred to as body strike. 1959 S. J. Baker *Drum* 110 *Flystrike*, infestation of sheep by blowflies..Also, body-strike. 1898 J. Southward *Mod. Printing* I. 140 The thick lines..are called the body strokes. *Ibid.* 134 Body or text types, used for plain paragraph matter. **1961** T. LANDAU *Encycl. Librarianship* (ed. 2) 43/2 *Body type*, type suitable for reading matter (8–14 point) as in the text of a book, as distinguished from display type, used in headings, display lines in advertisements, etc. **1930** Body-urge [see *IT *pron.* 1 f]. **1932** S. GIBBONS *Cold Comfort Farm* xvii. 237 Teck's a good kid .. but he's got no body-urge. **1888** ROLLESTON & JACKSON *Anim. Life* 357 The body wall [in Vertebrata]. **1898** A. SEDGWICK *Zool.* I. 549 The soft part of the body-wall [in Polyzoa], which consists of ectoderm and mesoderm. **1909** *Westm. Gaz.* 28 Aug. 13/2 A special series of muscles in the body-wall. **1959** E. F. LINSSEN *Beetles* I. 14 Another characteristic feature of insects is the hard, horny body-wall consisting of plates..composed mostly of chitin.

Boehm (bōm). The name of Theobald *Böhm* (1794–1881), German musician, applied *attrib.* to the system of keys and fingering which he invented in 1832. Hence applied to a flute which he designed, and to other wood-wind instruments with similar features.

1845 R. CARTE (*title*) A Complete Course of Instructions for the Boehm flute...Preceded by an analysis of the Boehm flute and of the old eight keyed flute. **1889** G. B. SHAW *London Music 1888–89* (1937) 77 In spite of the splendors of the Boehm flute, it is often lost in passages where the old flute used to tell when violins were less numerous. **1905** *Harmsworth Encycl.* IV. 2510/2 Nearly all concert flutes now in use are constructed upon the Boehm system, or modifications of it. **1935** *Chambers's Encycl.* IV. 727/2 The modern cylinder flute..when combined with the Boehm fingering..forms a nearly perfect instrument. **1954** *Grove's Dict. Mus.* (ed. 5) II. 324/1 In 1844 H. Klosé.. and A. Buffet devised an entirely new clarinet..now generally known as the Boehm clarinet.

Bœhmenism. Add: So **Behmenish, Behmenist, Behmenistic** *adjs.*; **Boehmist (Behmist)** *sb.* and *a.*

1731 SWIFT *Let.* 10 Sept. in Pope *Wks.* (1757) IX. 142 A very profound Behmist assures me, the style is poetic. **1739** JOHN WESLEY *Jrnl.* 23 Oct. (1938) II. 297, I read over Mr. Law's book on the New Birth..Behmenish, void, and vain! **1854** *Encycl. Brit.* IV. 805/1 The sect of Boehmists. **1912** W. C. BRAITHWAITE *Beginnings of Quakerism* ii. 38 We are reminded of the similar Familist and Boehmist teaching with respect to perfectionism. **1919** P. H. OSMOND *Myst. Poets Eng. Church* viii. 258 The Behmenistic view is too much influenced by the thought of the caprice and passion associated with human anger. **1961** W. H. G. ARMYTAGE *Heavens Below* I. iv. 38 Law channelled Behmenist thought into the main current of English mysticism.

Boer. Add: Now often with pronunc. (bōᵘəɹ) or (bōəɹ), Afrikaans (būɹ).

b. *S. Afr.* In special *Comb.* signifying made, produced, used by, or typical of Boers; often also ‖ **boere-** (bū·rə): *boerbeskuit, boer biscuit, boermeal, boer(e)musiek, boer-rusk, boer(e)wors* (-vǭɹs), *boerwyn* (-vēⁱn).

1943 'B. KNIGHT' *Covenant* (1944) IV. xix. 232 There was a good supply of..*biltong* and Boerbeskuit and a couple of roasted fowls. **1882** MRS. HECKFORD *Lady Trader in Transvaal* xxviii. 309 Hendrik managed to get some Boer biscuits from this man. *a* **1920** O. SCHREINER *From Man to Man* (1926) iii. 111 One morning..Baby-Bertie was kneeling in the pantry, making Boer biscuits. **1873** F. BOYLE *To the Cape for Diamonds* x. 141 Boer meal (cheap at this moment (42s per *muid* (200 lbs)). [**1878** *Roche's 'On Trek in the Transvaal'* 110 (Pettman), Bread we could not get, only the *Boer's meal*, i.e. the flour of the country.] **1949** *Cape Times* 24 Sept. 8/7 We used to get on farms the boermeal bread made from wheat, and nothing but wheat. **1952** *Ibid.* 27 Sept. 4/4 The music blared forth real *boeremusiek* which sent the whole crowd dancing. **1937** S. CLOETE *Turning Wheels* xxv. 392 Going without food except for the Boer rusks and biltong that he carried. **1944** V. POHL *Adv. Boer Fam.* i. 12 Saddlebags stuffed to bursting with boer-rusks, bread and biltong. **1948** *Cape Times* 18 Sept. 9/6 An expert on *sosaties, boerwors*, and *braaivleis*. **1950** L. G. GREEN *Land of Afternoon* iv. 63 Boerewors is another farm product which some still make in the old way; a sausage in which the meat has been pounded with a wooden stamper rather than minced. Modern boerewors..is usually a mixture of lean beef with pork fat, seasoned with wine or vinegar. **1947** *Cape Times* 14 May 3/6 Boerwyn—the unfortified wine sold to coloured patrons.

boffin (bǫ·fin). *slang.* [Etym. unknown. Numerous conjectures have been made about the origin of the word but all lack foundation.] **1.** An 'elderly' naval officer.

1941 C. GRAVES *Life Line* 143 Their ages are as youthful as air crews. Thirty-two is considered the maximum...In H.M.S. *Wasps' Nest*, anyone aged thirty-two is officially a 'boffin'. There is even a song about them. .'He glares at us hard and he scowls, For we're the Flotilla Boffins.' **1942** 'SEA-WRACK' *Random Soundings* 71 We were 'Old Boffins', the Pay. and I. He had been in the Bank of England for many years, and in the R.N.R. almost as long...I hadn't been to sea in a professional capacity for some eighteen years.

2. A person engaged in 'back-room' scientific or technical research. Hence **bo·ffin(e)ry**, boffins collectively; also, the activity of a boffin.

The term seems to have been first applied by members of the Royal Air Force to scientists working on radar.

1945 *Times* 15 Sept. 5/4 A band of scientific men who performed their wartime wonders at Malvern and apparently called themselves 'the boffins'. **1948** 'N. SHUTE' *No Highway* iii. 61 'What's a boffin?' 'The man from Farnborough. Everybody calls them boffins. Didn't you know?' ..'Why are they called that?'..'Because they behave like boffins, I suppose.' **1948** LORD TEDDER in A. P. Rowe *One Story of Radar* p. vii, I was fortunate in having considerable dealings in 1938–40 with the 'Boffins' (as the Royal Air Force affectionately dubbed the scientists). **1952** *Picture Post* 30 Aug. 20/1 Only a backroom boffin out of touch with the classroom could hold this pious belief. **1954** *Economist* 19 June Suppl. 6/3 The graduate from research—roughly.. the boffin of industry. **1957** R. WATSON-WATT *Three Steps to Victory* xxxiii. 201 The proud title of Boffin was first conferred on a few radar scientists by Royal Air Force officers with whom they worked in close co-operation...I am not quite sure about the true origins of this name of Boffin. It certainly has something to do with an obsolete type of aircraft called the Baffin, something to do with that odd bird, the Puffin; I am sure it has nothing at all to do with that first literary Back Room Boy, the claustrophiliac Colonel Boffin. **1958** *Times Lit. Suppl.* 14 Feb. 83/3 In one of those diverting interludes..he writes an anatomy of Boffinry. **1958** *Economist* 25 Oct. 298/1 The unexpected success of the boffins' conference at Geneva..ending in agreement on the feasibility of controlling a nuclear test suspension. **1960** J. MACLAREN-ROSS *Until Day* viii. 132, I was engaged in some boffinery in a blasted back-room unit.

Bofors (bōᵘ·fǭɹz). [Site of munition works in Örebro, Sweden.] Used *attrib.* or *ellipt.* of a type of light anti-aircraft gun.

1933 *Jane's Fighting Ships* 406/2 Armament: 5–4·7 inch, 50 cal., Bofors. **1939** *War Illustr.* 25 Nov. 330 A quick-firing two-pounder Bofors anti-aircraft gun of Swedish design, built by licence in Great Britain. Such guns were introduced into the British Army in 1938, but they soon formed the standard equipment of light anti-aircraft units whose job it is to deal with low-flying..aircraft. **1941** W. S. CHURCHILL *Secret Session Speeches* (1946) 31 Already 200 Bofors or their equivalents have been ordered to be made available by A.D.G.B. and the factories.

bog, *sb.*^{1} Add: **2.** *bog-peat, -water* (earlier example); *bog-black* adj.

1953 DYLAN THOMAS *Under Milk Wood* (1954) 48 The lickerish bog-black tea. **1743** ELLIS *Mod. Husb.* III. I. 113 In a low Meadow..there is a Peat dug called *Bog-peat*. **1958** *New Biol.* XXVI. 91 Peat is of two main types (with intermediates): (1) fen peat..(2) bog peat formed under acid conditions. **1847** E. BRONTË *Wuthering Heights* II. xiv. 280 The bog-water got into her head, and she would have run home, quite flighty, but I fixed her.

3. bog berry (examples); **bog moss** (examples); **bog onion**: see ONION *sb.* 2 b; **bog rush** (examples); also *U.S.*, a plant of the genus *Juncus*.

1760 J. LEE *Introd. Bot.* 297/1 Bogberries, *Vaccinium*. **1892** J. BARLOW *Irish Idylls* viii. 217 She made a feint of looking for bogberries. **1785** MARTYN tr. *Rousseau's Bot.* xxxii. 493 Sphagnum, or Bog-moss, has the capsule covered with a lid. **1840** [see SPHAGNUM]. **1959** J. CLEGG *Freshwater Life* (ed. 2) iii. 63 Of the mosses, the most important are the species of Bog Mosses (*Sphagnum*)..which form soft carpets of vegetation in the damp areas. **1760** J. LEE *Introd. Bot.* 315/1 Rush, Round black-headed, Marsh or Bog, *Schœnus*. **1843** J. TORREY *N.Y. Nat. Hist. Surv.: Flora* II. 325 *Juncus effusus*..Bog-rush. **1855** Lesser Bog Rush [see RUSH *sb.*^{1} 4 a]. **1898** C. M. YONGE *John Keble's Parishes* xvi. 231 *Bogrush* (*L. campestris*).—Little rush.

4. bog-garden, a piece of ground laid out and irrigated to grow plants whose habitat is bog-land and a peaty soil; **bog-hole,** a natural hole with a swampy bottom; **bog-land** (later examples); **bog Latin,** a spurious form of Latin; cf. *dog-Latin*; also = SHELTA; **bog oak** (earlier and later examples).

1883 W. ROBINSON *Engl. Flower Garden* p. lxiii/1 A more perfect bog garden is made by forming a basin of brickwork and Portland cement, about one foot in depth. **1908** R. FARRER *Alpines & Bog-plants* 154 The prime..necessity of the bog-garden is the most perfect drainage. **1788** G. O'NIAL *The Minor* II. III. iii. 14 He and his horse had bounced into a bog-hole. **1839** 'MRS. M. CLAVERS' *New Home* i. 15 Down came our good horse to the very chin in a bog-hole. **1936** P. FLEMING *News from Tartary* III. vi. 138 Slippery flats, pitted with bog-holes. **1745** ELLIS *Mod. Husb.* VI. II. 47 A proper Plough to plow Bog lands. **1940** L. MACNEICE *Last Ditch* 9 The night came down upon the bogland. **1785** GROSE *Dict. Vulgar T.*, *Bog latin*, (*Irish*) barbarous latin. **1891** [see SHELTA]. **1927** J. JOYCE *Let.* 20 May (1957) 254 *Dulce et decorum est propre mare sedere*—boglatin for it is a sweet and seemly thing to sit down by the sea. **1813** M. EDGEWORTH *Let.* 24 Oct. in S. H. Romilly *Romilly–Edgeworth Lett.* (1936) 60 This necklace, and bracelets, are genuine Irish—made of *bog-oak*—that is, of oak found in our bogs. **1940** L. MACNEICE *Last Ditch* 9 Stumps of hoary bog-oak.

bog, *sb.*^{4} (Examples.)

a **1789** in J. HOWARD *Lazarettos* (1789) 181 That no dirt.. be thrown out of any window, or down the bogs. **1864** HOTTEN *Slang Dict.* 79 *Bog*, or *bog-house*, a privy as distinguished from a water-closet. **1929** H. WILLIAMSON *Beautiful Years* xxiii. 165 His headquarters are in The Bog,..where the kids go and hide, locking themselves in when they think those cads are after them. **1959** W. GOLDING *Free Fall* i. 23 Our lodger had our upstairs, use of the stove, our tap and our bog. **1960** *New Left Rev.* May–June 61/1 Toilet paper in the bogs. **1962** P. PURSER *Peregrination 22* xv. 71 Rolls of brown bog paper.

bog, *v.*^{1} Add: **1. b.** Also with *down*; chiefly *fig.*

1928 *Amer. Speech* IV. 132 To be 'bogged down' or 'mired down' is to be mired, generally in the 'wet valleys' in the spring. **1953** A. UPFIELD *Murder must Wait* xii. 108 The investigation stopped when bogged down by official impatience. **1955** *Times* 11 May 10/3 His approach to them would not be bogged down by 'minor points of protocol'.

2. (Later examples.) Also with *down* and *fig.*

1900 SMITHWICK *Evol. State* 325 The animal had bogged in crossing the little creek. **1903** A. ADAMS *Log Cowboy* xii. 77 Bob Blades attempted to ride out of the river below the crossing, when his horse bogged down. **1928** *Sat. Even. Post* 12 May 184/2 On a clean sheet of paper he wrote the words 'We know', and there he bogged down. **1937** *Times* 22 Nov. 12/5 Congress..may bog down and do nothing. **1951** E. TAYLOR *Game Hide-&-Seek* I. iii. 70 'I wish Tiny and Kitty would come,' anxious young hostesses would think when parties bogged down.

bogey (bōᵘ·gi). *Golf.* Also **bogy, bogie.** [The following story reproduces the current account of the origin of the term:—

One popular song at least has left its permanent effect on the game of golf. That song is 'The Bogey Man'. In 1890 Dr. Thos. Browne, R.N., the hon. secretary of the Great Yarmouth Club, was playing against a Major Wellman, the match being against the 'ground score', which was the name given to the scratch value of each hole. The system of playing against the 'ground score' was new to Major Wellman, and he exclaimed, thinking of the song of the moment, that his mysterious and well-nigh invincible opponent was a regular 'bogey-man'. The name 'caught on' at Great Yarmouth, and to-day 'Bogey' is one of the most feared opponents on all the courses that acknowledge him (1908 *M.A.P.* 25 July 78/1).]

The number of strokes a good player may be reckoned to need for the course or for a hole.

1892 *Field* 2 Jan. 6/1 A novelty was introduced in shape of a Bogey tournament for a prize. .. Fourteen couples started, but the Bogey defeated all. **1903** *Westm. Gaz.* 21 Feb. 6/2 Jones, with a handicap of 17, receives an allowance against Bogey of 13 strokes. **1910** *Encycl. Brit.* XII. 221/2 There is also a species of competition called 'bogey' play, in which each man plays against a 'bogey' score—a score fixed for each hole in the round before starting.

b. *transf.* and *fig.*

1922 WODEHOUSE *Clicking of Cuthbert* iii. 80 'Weren't you giving yourself rather a large family?'..'Was I?' he said, dully. 'I don't know. What's bogey?' **1958** J. A. BARLOW *Elem. Rifle Shooting* (ed. 5) iii. 43 It is a good plan to set oneself a definite score below which one must never fall. In other words, a bogey score for the practice or shoot. **1959** *Listener* 5 Nov. 802/1 Par Contract is a way of playing bridge against bogey.

Boghead, boghead (bǫ·ghed). The name of an estate near Bathgate in West Lothian applied *attrib.* to a deep brown shale found there. See TORBANITE.

1858, 1867 [see TORBANITE]. **1919** *Chambers's Jrnl.* June 390/2 The famous oil-shale of Torbanehill, torbanite or bog-head cannel, is often regarded as a variety of cannel coal. **1937** *Nature* 20 Feb. 340/1 Boghead coal is a comparatively rare and valuable material yielding gas and paraffin on distillation. **1960** *Gloss. Coal Terms* (*B.S.I.*) 6 *Boghead coal, Torbanite*, coal resembling cannel coal in physical appearance and properties, but distinguished microscopically by the presence of the remains of algae.

bog-house. (Earlier example.)

1666 R. HEAD *Eng. Rogue* x. 85 Fearing I should catch cold, they out of pitty covered me warm in a Bogg-house.

Bogomil, -mile (bōᵘ·gǫmil, bǫ-, -məil). *Hist.* [ad. med.Gr. Βογόμιλος, of disputed origin; the first syllable may represent Russ. *Bog* God.] A member of a heretical Bulgarian sect which arose in the 10th or 11th century, whose main tenet was that God the Father had two sons, Satan and Christ. Hence **Bogomi·lian** *a.* and *sb.*, **Bo·gomilism, Bo·gomilist.**

[**1574** R. BRISTOW *Plaine Wayes Truthe* To Rdr. *iiij b, In the twelfth hundred [year after Christ], the Bogomili, the Petrobusians, the Apostolikes.] **1841** *Penny Cycl.* XX. 271/1 The sect of the *Bogomiles*, which was well known in the Greek empire. **1852** J. TORREY tr. *Neander's Ch. Hist.* VIII. 356 The Bogomilian view of the Trinity. *a* **1875** FINLAY *Hist. Greece* (1877) III. 69 A Bogomilian who was put to the torture by the imperial officers. **1887** M. GASTER *Greeko-Slavonic Lit.* 17 Bogomilism..ruled Bulgaria for not less than five centuries. *Ibid.* 20 Nikita, the bishop of the Bogomils of Constantinople. *Ibid.* 35, I will here give the version which comes nearest to the original Bogomilist form. **1920** *Q. Rev.* Jan. 71 He invited the Bogomile heretic, Basil, to a private colloquy. **1939** A. J. TOYNBEE *Stud. Hist.* IV. iii. 633 The Bogomil Church is a monument of the successful execution of this project. **1941** 'R. WEST' *Black Lamb & Grey Falcon* I. 172 The Puritan heresy known as Paulicianism or Patarenism or Bogomilism or Catharism.

bogong, var. *BUGONG.

bogue (bōᵘg). [Fr. (16 c.), f. OPr. *boga*, f. med.L. *boca*, L. *bōx* (see BOCE^{1}), Gr. *βῶξ*, contraction of *βόαξ* a grunting fish.] = BOCE^{1}.

1862 J. COUCH *Fishes Brit. Isl.* I. 225 In some parts of the European side of the Mediterranean the Bogue is a common fish. The first British example we have a record of was caught..at St. Mawes. *Ibid.* 226 The Bogue grows to the length of eight or nine inches. **1880** F. DAY *Fishes Gt. Brit.* I. 28 *Sparidæ*: Box vulgaris..*Bogue*..*Habitat.* From the southern shores of the British isles as far south as Madeira and the Canaries... Common throughout the Mediterranean. **1936** J. T. JENKINS *Fishes Brit. Isl.* (ed. 2) 41 The Bogue (*Box vulgaris*) is also a Mediterranean species [of Sea Bream] which occasionally migrates into British waters.

bogus, *sb.*^{1} and *a.* Add: †**1. b.** *bogus press, machine* = sense 1 in Dict.

1844 *Spirit of Times* (Philad.) 12 Oct. (Th.), A bogus press for making counterfeit money. **1850** *Frontier Guardian*

(ed. O. Hyde) 23 Jan. (Th.), We employed that same Bill Hickman to ferret out a bogus press and a gang of counterfeiters. .. A part of the bogus machine has been found.

†**c.** Counterfeit coin. Also *Comb.*

1842 *Life in West* 297 They had attempted to pass bogus (base coin). **1844** *Nauvoo Neighbor* 12 June (Th.), To bolster up the interests of blacklegs and bogus-makers. **1854** B. YOUNG in *Jrnl. Discourses* I. 270 The Magicians of Egypt..produced a very good bogus, but it was not quite the true coin.

2. *adj.* (Earlier and U.K. examples.)

1839 MRS. KIRKLAND *New Home* xxxii. 212 The boxes ..contained..half-dollars 'principally bogus'. **1859** 'DOW JR.' *New Patent Sermons* 216 Crocodile tears are bogus. **1892** G. B. SHAW *Pen Portraits & Rev.* (1932) 243 Mr Frederic Harrison deliberately talked bogus Shelleyism to the reporters. **1940** N. MITFORD *Pigeon Pie* xvi. 244, I also hoped they would show people here that the whole thing was bogus. **1942** E. WAUGH *Put out more Flags* i. § 6. 68 You..haven't got any of those bogus regional connections like the Scots and Irish and Welsh.

bogusly (bōᵘ·gəsli), *adv.* orig. *U.S.* [f. BOGUS *a.* + -LY^{2}.] In a bogus manner; spuriously.

1862 *N.Y. Herald* 2 May, When..the oath was sent to us, we filed it bogously [*sic*], and sent it to Richmond without swearing to it. **1932** AUDEN *Orators* I. i. 19 The bogusly cheerful—the games master. **1941** *Mind* L. 111 To reply.. 'By analogy' is very apt to be bogusly soothing and thus misleading. **1964** I. FLEMING *You only live Twice* v. 61 The bogusly important-looking entrance.

bogusness (bōᵘ·gəsnės). [f. BOGUS *a.* + -NESS.] The fact or quality of being bogus.

1921 H. L. MENCKEN *Prejudices* 2nd Ser. I. ix. 79 The old bogusness hangs about them [*sc.* the intelligentsia], as about the fashionable aristocrats of the society columns. *c* **1935** R. SHERIDAN *Let.* in C. Sheridan *To Four Winds* (1957) xxvii. 298 The pettiness, dirtiness, meanness, bogusness of it all. **1953** F. SWINNERTON *Month in Gordon Square* 9 The affectation was part of his general bogusness. **1958** *Times Lit. Suppl.* 24 Jan. 47/3 Produces the same sort of bogusness as do those actors on the stage who cannot represent a member of the rural lower classes without bent knees.

bogy^{1}, bogey^{1}. Add: **4.** *Criminals' slang.* A detective; a policeman.

1924 S. SCOTT *Human Side* i. 23 Men will listen to the vilest epithets, but call them 'bogey', 'brassey', 'copper', or 'policeman', and they will be at your throat. **1931** W. F. BROWN in *Police Jrnl.* Oct. 501 She told a detective (bogey) she knew that Jack was in the brothel (case). **1936** J. CURTIS *Gilt Kid* 17 One of the bogies from Vine Street reckernizes me. **1960** *Observer* 24 Jan. 7/2 Suppose..a bogy did get it up for a villain now and again by making sure that some gear was found in his flat?

5. A piece of dried nasal mucus. *colloq.*

1937 in PARTRIDGE *Dict. Slang* 853/1 s.v. *sweep*. **1955** K. AMIS *That Uncertain Feeling* xii. 158 'You've got a bogey on your nose. Improves your looks no end.' I was near the mirror...I peeped in and saw the bogey. It was large and vermiform and clung to the wing of my right nostril. *Ibid.* xiii. 176, I felt my nostrils carefully, testing for bogeys. **1967** D. PINNER *Ritual* xv. 148 He..removed wax from ears, bogeys from nose, blackheads from chin.

6. An unidentified aircraft; an enemy aeroplane. *slang.*

1943 WARD-JACKSON *Piece of Cake* 16 *Bogey*, a friendly aircraft [corrected in ed. 1945 to:] a suspect aircraft. **1944** *Life* 17 July 20 Before supper was over this evening, several 'bogeys'—as unidentified planes are called under such circumstances—were seen approaching from different directions.

7. *attrib.* and *Comb.*, as *bogy-man, -word*, etc.

1863 KINGSLEY *Water Bab.* iv. 146 The old German bogy-painters. *c* **1890** Bogey man [see note s.v. *BOGEY]. **1912** G. B. SHAW in *Christian Globe* 22 Feb. 433/4 It was manlier than clinging to Britannia's skirts for protection against the Bogey Man with the triple tiara. **1919** J. L. GARVIN *Econ. Found. Peace* 112 To confuse or weaken the Allies by using 'Bolshevism' as a bogey-word. **1926** FOWLER *Mod. Eng. Usage* 559/1 Bogy-haunted creatures who for fear of splitting an infinitive abstain from doing something quite different. **1954** J. R. R. TOLKIEN *Fellowship of Ring* I. vi. 121 The old bogey-stories Fatty's nurses used to tell him. **1959** *Listener* 16 Apr. 657/2 Black children were brought up to believe that if they were naughty the white bogy-man would come and gobble them up.

bogy^{2}, bogey^{2} (bōᵘ·gi). *Austral. slang.* Also **bogie.** [App. Aboriginal word.] **a.** A bathe. **b.** A bathing-place, a bath. Also *attrib.* Hence as *v. intr.*, to bathe; so **bo·gying** *vbl. sb.*

1849 A. HARRIS *Emigrant Family* viii. 145 'Bogie,' I suppose must be aboriginal also... Its signification is *a bathe*. **1893** K. MACKAY *Out Back* iv. 50, I don't care to bogey in our drinking tank. **1928** 'BRENT OF BIN BIN' *Up Country* (1966) ii. 24 They..took her for bogeys in the swimming hole. **1934** *Bulletin* (Sydney) 13 June 19/4 Blacks on the tidal creeks and rivers of Queensland prefer to bogey when the tide is on the ebb. **1941** BAKER *Dict. Austral. Slang* 11 *Bogie*, a swim, a bath, or wash. (2) A swimming hole, a bath. Also, 'bogiehole', 'bogiehouse'. **1945** —— *Austral. Lang.* xiii. 223 Then there are the aboriginal words which we have borrowed and extended in meaning, e.g. *bogie* or *bogey*, to bathe, from which we have taken *bogiehole*, a swimming hole, *bogiehouse*, a bathroom, and *bogieing*, bathing. **1946** F. D. DAVISON *Dusty* viii. 82 They went down for a bogey on warm days.

boh (bōᵘ). *India.* [Burmese *bo.*] A chief or leader of dacoits.

1888 [see *DAH]. **1890** KIPLING *Departm. Ditties* (ed. 4) 82 Bohs that were brave departed. **1923** *Chambers's Jrnl.* Oct. 650/2 He would ask which was their Boh, or leader. **1926** *Ibid.* Feb. 85/2 A man, obviously the Boh, or Chief.

Bohairic (bohəiᵊ·rik), *a.* and *sb.* Also **Bahiric.** [f. *Bohairah, Bahirah* (*Boheira, Beherah*), the Arabic name of Lower Egypt (Arab. *buḥaira* lake).] The designation of the classical or standard form of Coptic spoken in Alexandria and the north-western Delta, and of the version of the Bible (the official version of the Bible of the Coptic Church) written in this language. (= MEMPHITIC.)

1830 H. TATTAM *Gram. Egypt. Lang.* 135 The Coptic, or, as it has been called, the Bahiric, but more properly the Memphitic, was the Dialect of Lower Egypt. **1874** LIGHTFOOT in Scrivener *Crit. N.T.* (ed. 2) 327 The Bahiric or Memphitic Version. **1898** G. W. HORNER (*title*) The Coptic Version of the New Testament in the Northern Dialect otherwise called Memphitic and Bohairic. **1958** F. KENYON *Our Bible* (ed. 5) viii. 234 Originally, however, the dialect in which it is written belonged only to the coast district near Alexandria, and another dialect was in use in Memphis itself; hence it is better to avoid the term Memphitic, and use the more strictly accurate name *Bohairic*.

bohemianize, *v.* Add: *trans.* To make bohemian in life and habits.

1898 WATTS-DUNTON *Aylwin* ii. 52 To run the risk of becoming bohemianised like Cyril Aylwin.

bohereen, bohireen, varr. BOREEN.

1899 SOMERVILLE & 'ROSS' *Irish R.M.* 138, I thought you were a dead man when you faced him at the bohereen. **1920** *Cornh. Mag.* Oct. 494 A campaign among bogs and bohireens. **1921** *Blackw. Mag.* Jan. 11/1 The grass-grown bohereen leading over the crest of the hill. **1953** N. FITZGERALD *Midsummer Malice* xv. 189 The quarter of a mile of bohereen that connected the stables to the main road.

Bohr (bōᵊɹ). The name of Niels Henrik David *Bohr* (1885–1962), Danish physicist, used to designate certain theories, measurements, etc., formulated by him; as *Bohr atom*: the atom as described by the Bohr theory; *Bohr('s) magneton*: see *MAGNETON; *Bohr model*: a model of the structure of the hydrogen atom; *Bohr radius*: the radius of the smallest electron orbit in the Bohr model; *Bohr('s) theory*: a theory of the structure of the hydrogen atom (and hence of other atoms).

1923 KRAMERS & HOLST *The Atom & the Bohr Theory of its Structure* v. 119 The Bohr model of the hydrogen atom in the simplified form. *Ibid.* 120 In the Bohr atom, likewise, the frequency of revolution *w* of the electron in its stationary orbit has no direct connection with the frequency of the radiation emitted when the electron passes from this orbit. **1923** H. L. BROSE tr. *Sommerfeld's Atomic Structure* iv. 212 The times of revolution in the Bohr circles are proportional to the cubes of the quantum numbers. *Ibid.* 249 We shall call the value (15) the Bohr magneton. **1927** [see *MAGNETON]. **1936** *Nature* 1 Feb. 187/2 The Bohr formula for the Rydberg constant is not correct. **1938** R. A. HOUSTOUN *Treat. Light* (ed. 7) xviii. 335 Bohr's Theory of the Hydrogen Spectrum... Bohr supposed that the electron was circling round the proton, and that the centrifugal force was balanced by the electrostatic attraction. **1956** *Nature* 4 Feb. 241/2 *R* is the Bohr radius, 0·529 A. **1958** *Chambers's Techn. Dict.* 101/1 *Bohr theory*, a combination of the Rutherford conception of the atom as a central, positively-charged nucleus surrounded by planetary electrons, with the quantum theory, which restricts the permissible orbits in which the electrons can revolve. **1968** J. J. C. SMART *Betw. Sci. & Philos.* iii. 74 The Bohr theory of the atom may appeal to us because it somehow reduces the behavior of atoms to the more familiar behavior of the solar system.

bohreen, var. BOREEN.

bohunk (bōᵘ·hʌŋk). *N. Amer. slang.* [App. f. BO(HEMIAN + *-hunk*, alt. of HUNG(ARIAN.] A Hungarian; an immigrant from central or south-eastern Europe, esp. one of inferior class; hence, a low rough fellow, a lout. Also *attrib.*

1903 *Cincinnati Enquirer* 9 May 13/1 Bohunk—A Bohemian; a foreigner. **1910** *Butte* (Mont.) *Even. News* 24 July 1 The bohunk miner is a low grade foreigner who buys his job from the foreman and pays him for keeping it. **1914** G. ATHERTON *Perch of Devil* i. 58 The 'Bohunks', or 'dark men', an inferior class of Southern Europeans. **1921** *Chambers's Jrnl.* 848/2 To condemn his girl to a yellow bohunk like Milroy. **1927** *Daily Tel.* 24 May 17/6 A contractor for railroad ties employing hundreds of the roughest of bohunks to hew and carry them for him. **1930** J. DOS PASSOS *42nd Parallel* I. 82 Bohunk and polak kids put stones in their snowballs. **1963** H. GARNER in R. Weaver *Canad. Short Stories* 2nd Ser. (1968) 52 If you and that bohunk partner of yours had checked the gang press as you're supposed to, Colby mightn't have spoiled so many pieces.

boil, *sb.*^{2} Add: **1.** *spec.* in *N.Z.* The act of making tea in a billy. Cf. *BOIL *v.* 4 c and *BOIL-UP.

1940 W. S. GILKISON *Peaks, Packs & Mountain Tracks* 82 The best part of a tramp on the home hills is the midday 'boil'. **1953** B. STRONACH *Musterer on Molesworth* iv. 26 We had a boil and then started the ascent.

2. b. *boil-off*, the evaporative loss from liquefied natural gases, esp. methane. Also *attrib.*

1956 *Trans. Inst. Gas Engin.* 1955–56 986 Storage tanks for liquid methane. .. The heat losses from the spherical tanks were found to be lower than anticipated and the measured rate of 'boil off' was equivalent to 0·23 per cent per day of the tank content. *Ibid.* 988 A vapour connexion to carry away the 'boil off' gas. **1960** *Times* 8 Mar. 17/3 It [*sc.* liquid methane] has been kept in special containers, the 'boil-off' being pumped to the board's Romford works.

4. *U.S.* (See quots.)

1805 CLARK in *Lewis & Clark Exped.* (1905) III. 151 In those narrows the water was agitated in..boils, swells, & whorlpools. **1853** *Putnam's Monthly* Aug. 188/2 These 'boils', as the boatmen call them, are immense upheavings of the moving waters [of the Mississippi]. **1883** 'MARK TWAIN' *Life Mississippi* ix. 120 Those tumbling 'boils' show a dissolving bar and a changing channel there.

5. *Angling.* A sudden bold rise of a fish at a fly. (Cf. *BOIL *v.* 11.)

1893 *Field* 17 June 894/3 A fish had made a boil at my fly. **1894** *Daily News* 22 Aug. 5/1 The only hopeful sign is when salmon now and then break the surface of the water with a 'boil', a movement like the rise of a trout.

boil, *v.* Add: **1.** Also with *up*. Also *fig.*

1612 W. STRACHEY *Travaile into Virginia* (1849) II. 184/2 To boyle up, potopotawh tawh. **1747** H. GLASSE *Art of Cookery* ix. 94 Let it boil up till it is thorough hot. **1818** W. TUCKER *Fam. Dyer & Scourer* (ed. 2) 61 When this has boiled up, cool down your copper. **1871** L. W. M. LOCKHART *Fair to See* (1872) II. xiii. 62 Just as things were boiling up very satisfactorily to a climax, in came Mrs M'Killop..and cleared them off. *Ibid.* III. xxxv. 150 All this time Mrs M'Killop..had been bridling and boiling up on her sofa, waiting for recognition.

4. c. *to boil up*: *absol.* to boil or wash clothes. *U.S.* and *N.Z. colloq.* Also (*Canad.* and *N.Z.*), to make tea. Also *to boil the billy*, to make tea (*Austral.* and *N.Z. colloq.*).

1839 [see *BILLY^{3}]. **1881** [see BILLY^{2} 1]. **1891** [see DAMPER 6]. **1902** B. BAYNTON in Murdoch & Drake-Brockman *Austral. Short Stories* (1951) 42 Blest if I evven fergot t' bile th' billy. **1911** W. H. KOEBEL *In Maoriland Bush* x. 150 When you boil the 'billy', you make tea, *voila tout*. **1929** *Amer. Speech* IV. 338 *Boil up*, to wash your clothes. **1933** E. MERRICK *True North* 30 At three we boiled up again. **1947** P. NEWTON *Wayleggo* (1949) iii. 34 Musterers.. carry billies..and 'boil-up'..when circumstances permit.

11. *Angling.* Of a fish: to rise boldly at a fly.

1898 J. A. GIBBS *Cotswold Village* viii. 164, I see one [*sc.* a trout] boil up just above that mess of weed.

boiled, *ppl. a.* Add: **b.** (Earlier and later examples.)

1804 M. EDGEWORTH *Pop. Tales* I. v. 277 Mr. Hill commenced a practice..of going..into the kitchen..to take a slice from the roast or the boiled before it went up to table. **1834, 1856** [see ROAST *sb.* 2]. **1861** [see STEWED *ppl.a.*^{1}].

c. Intoxicated. *slang.* Also phr. *as drunk as a boiled owl.*

[**1885** *Referee* 31 May 3/3 Twiss..had just the boiled-owlish appearance that is gained by working all night in a printing-office.] **1886** J. A. PORTER *Sks. Yale Life* 156 There is a balm for a headache caused by last night's debauch to have it said you were 'slightly cheered' or 'slewed' or 'boiled'. **1892** *Daily Tel.* 12 Dec. 5/4 The expression, 'Intoxicated as a boiled owl', is a gross libel upon a highly respectable teetotal bird. **1922** JOYCE *Ulysses* 300 He brought him home as drunk as a boiled owl. **1928** *Amer. Speech* IV. 102 Expressions synonymous with or circumlocutory for 'drunk'..blotto, boiled. **1940** 'H. PENTECOST' *24th Horse* (1951) v. 45 He's boiled to the ears.

2. Special Combs.: **boiled crow** (see CROW *sb.*^{1} 3); **boiled dinner** (orig. *U.S.*), a dinner of meat and vegetables boiled together; **boiled oil,** a preparation of linseed oil used as a drying-oil; **boiled shirt,** (*a*) *U.S.* a white linen shirt (see SHIRT *sb.* 1); (*b*) a man's dress shirt.

1805 *Pocumtuc Housewife* (1906) 9 Directions for a Boiled Dinner may seem unnecessary. **1897** HOWELLS *Landlord Lion's Head* iii. 13 The woman brought in a good boiled dinner of corned beef, potatoes, turnips, and carrots. **1858** SIMMONDS *Dict. Trade* 43/2 *Boiled oil*, a drying oil made by boiling a small quantity of litharge in linseed oil, till it is dissolved. **1887** F. B. GARDNER *Painters' Encycl.* 53 *Boiled oil*,..an oil which has been brought by the action of heat and of oxidising materials into a state of greater activity, in fact—into a state of incipient slow oxidation. **1921** C. WORTH *Yacht Cruising* (ed. 2) 407 Boiled oil is darker and thicker than raw linseed oil and dries more quickly. **1853** in *Amer. Speech* (1954) XXIX. 7 When I get shaved and get a 'boiled shirt' on, which I have not had on since I left home, for we don't boil our shirts here, for we think cold water quite enough in a country where there is no female society. **1854** [see SHIRT *sb.* 1]. **1872** *Dublin Univ. Mag.* Feb. 219 Every man arrays himself in 'store-clothes' and boiled shirts. **1920** 'SAPPER' *Bulldog Drummond* ii. 47 If one goes about..in boiled shirts while pretending merely to be out for the afternoon, people have doubts as to one's intellect. **1928** D. L. SAYERS *Unpleasantness at Bellona Club* xxi. 266 You've sent for a bloke in a boiled shirt to take your place, I suppose?

boiler. Add: **2. c.** *to bu(r)st one's boiler* (fig.), to come, or bring, to grief. *U.S. colloq.*

1824 in Thornton *Amer. Gloss.* (1912). **1834** CARRUTHERS *Kentuckian in N.Y.* I. 218 That'll make them think somebody's busted their biler. **1847** PAULDING *Madmen All* 189 May my boiler be eternally busted, if there isn't that are lady.

5. boiler-deck *U.S.*, the lower deck of a steamer, lying immediately above the boilers; **boiler-iron** (examples); **boiler plate** *transf.*, (*a*) *U.S.* stereotyped or formulaic writing; *spec.*

syndicated matter issued to the newspaper press; (*b*) *Mountaineering* (see quot. 1957); **boiler suit,** an outer garment combining overalls and shirt, worn to protect clothing.

1840 S. A. HOWLAND *Steamboat Disasters* 131 Boiler-deck, —being that part of the upper deck situated immediately over the boilers. **1877** HABBERTON *Jericho Road* i. 10 The new hand reached the boiler-deck, and reported to the mate. **1851** C. CIST *Cincinnati* 213 The yearly products are..five hundred tons boiler-iron, heads, etc. **1877** R. W. RAYMOND *Statistics Mines* 29 It [*sc.* the water-jacket] is constructed of the heaviest boiler-iron. **1893** *Congress. Rec.* Aug. 465/1 The country weeklies have been sent tons of 'boiler plates' accompanied by..letters asking the editors to use the matter as news. **1905** D. G. PHILLIPS *Plum Tree* 190 He attended to the subsidizing of news agencies that supplied thousands of country papers with boiler-plate matter to fill their inside pages. **1924** J. BUCHAN *Three Hostages* xi. 165 Left me to finish my ascent by way of some very loose screes and unpleasant boiler-plates. **1957** R. G. COLLOMB *Dict. Mountaineering* 32 *Boiler Plates*, Overlapping, undercut slabs of rock; convex slabs are usual. **1965** 'E. MCBAIN' *Doll* (1966) viii. 113 The rest of the will was boilerplate. Meyer scanned it quickly. **1969** *Word Study* Apr. 3/2 Other examples of standard slang are *debugging* and *boiler-plate* (used in regard to formula-type language standard for all reports). **1928** *Sunday Express* 28 Oct. 3/7 The students had thoughtfully attired themselves in boiler suits. **1949** F. MACLEAN *Eastern Approaches* III. ix. 412 The Prime Minister..was wearing a bright blue boiler-suit.

6. In the West Indies and Bermudas, a sunken coral reef into which the sea breaks with foam and spray.

1909 in *Cent. Dict.* Suppl.

boiler (boi·ləɹ), *v. trans.* To furnish (a steamship) with its boiler or boilers. So **boi·lering** *vbl. sb.*

1890 *Whitby Times* 3 Jan. 4/4 The steamers built and launched at this port have to go elsewhere to be engined and boilered. **1897** *United Service Mag.* June 226 The special conditions which..govern the boilering of warships.

boiling, *vbl. sb.* Add: **1. b.** With *down*: the process of boiling or heating something to reduce its bulk or to liberate oil or the like. Also *attrib.*

1848 H. W. HAYGARTH *Bush Life in Australia* vi. 71 The process of 'boiling down', or converting the whole carcase into tallow. **1859** F. FULLER *5 Years' Resid. in N.Z.* viii. 166 A boiling-down price for wethers would be reached in a few years. **1903** *Westm. Gaz.* 31 Jan. 8/1 A Grimsby fishing vessel..if properly equipped with boiling down works could gather the oil [from seals].

3. b. With *down*: the process of condensing or abridging literary matter; *concr.* a condensation or epitome. (See BOIL *v.* 8.)

1898 *Daily News* 27 Jan. 8/4 The book is little more than a boiling-down of the vast literature on the subject. But the boiling-down is well done.

boiling, *ppl. a.* Add: **1. b.** Hyperbolically: extremely hot. *colloq.*

1930 R. LEHMANN *Note in Music* 34 He was the sort of boy who would..declare on the coldest day that he was boiling.

boil-up (boi·lʌp). [f. BOIL *sb.*^{2} + UP *adv.*^{1}] **a.** = BOIL *sb.*^{2} 1; the act of boiling or washing clothes, (*Canad.* and *N.Z.*) making tea, etc. Also *attrib.*

1728 E. SMITH *Compleat Housewife* (ed. 2) 24 Strain out some of the liquor..give it a boil up. **1861** MRS. BEETON *Bk. Househ. Managem.* 103 Mix with it the cream and milk. Give one boil up, at the same time adding the tails. **1933** E. MERRICK *True North* 262 Some of the peas we save for.. one of the boil-ups during the day. **1934** *Detective Fiction Weekly* 21 Apr. 108/2 The boil-up can is generally an oil can with the top removed. **1938** J. ROBERTSON *With Cameliers in Palestine* xiii. 113 No opportunity was lost of having a 'boil up'. **1949** *Here & Now* (N.Z.) Oct. 17/3 Once a week or once a fortnight there's a copper boil-up. **1958** *Tararua* XII. 26 The billy brings to our mind the inevitable *boil-up*, the boiling of the billy for a cup of tea.

b. *transf.* and *fig.*

1871 L. W. M. LOCKHART *Fair to See* (1872) III. xxxiii. 85 Americans, filled with envious admiration by the costliness of the spectacle, were reminded of the superior though somewhat similar 'boil-up' of Mrs Thaddeus G. Cass of Boston, U.S. **1941** BAKER *Dict. Austral. Slang* 11 *Boil-up*, a row or argument. **1963** *Listener* 14 Mar. 477/3 The girl..swings into a symphonic boil-up of a popular number.

‖ **boina** (bo·ina). [Sp.] A flat cap worn in northern Spain.

1904 GALLICHAN *Fishing in Spain* 19 In his blue boina, a cap resembling the tam-o'-shanter, he looked like a 'braw Scot'. **1922** *Chambers's Jrnl.* 15 July 519/2 He wore a hat of soft black felt or else a *Boina* (the Biscayan bonnet). **1937** *Times* 16 Nov. 17/6 All over Spain the traditionalists donned their red *boinas* and rose in the name of Don Carlos to battle with innovation. **1957** R. CAMPBELL *Coll. Poems* II. 153 The scarlet boinas trickled to the plain.

boing (boiŋ), *int.* Also **boing boing.** [Echoic.] The noise made when a compressed spring is suddenly released; a reverberating sound.

1952 B. CLEARY *Henry & Beezus* i. 21 'Boi-i-ing!' shouted the two boys together. **1957** MANVELL & HUNTLEY *Technique Film Music* iii. 156 Gerald is the despair of his parents since he cannot talk; he only speaks sound effects of which the most formidable is the familiar 'boing' of the cartoon world, a noise made by a short length of strong wire spring attached to a sounding board. *Ibid.* 157 He didn't talk words. He went 'boing-boing' instead. **1967** I. HAMILTON *Man with Brown Paper Face* iv. 60 He had big coil springs fitted to the heels of his boots and with a merry *boing boing* he would spring out from the night shadows to confront ladies on the street. **1967** A. J. MARSHALL in L. Deighton *London Dossier* 141 The newest and nicest tower block is Centre Point—boing!

‖ **bois brûlé** (bwa brüle). *N. Amer.* [Fr., 'burnt wood'.] An American Indian half-breed, esp. one of French and Indian extraction.

In popular use corrupted into *bob ruly*.

1805 LEWIS in *Doc. Congress* 1047 The Tetons Bois brûlé killed and took about 60 of them last summer. **1840** C. F. HOFFMAN *Greyslaer* III. xiv. 260 Of Guisbert or Guise, as the 'Bois-brulé', or half-blood child was generally called, we have as yet been enabled to gather but few traditions. **1878** J. H. BEADLE *Western Wilds* xxix. 380 Most of the drivers were of the pure Bois Brules stock. **1960** *Press* (Vancouver) Dec. 14 The result of the Bay proclamations was to provoke civil disobedience on the part of the boisbrules and active opposition by the Nor' westers.

‖ **bois d'arc** (bwa dark). *N. Amer.* Also **bow-dark,** ‖ **bodok.** [Fr., 'wood of bow'.] The wood of the osage orange, used by American Indians for making bows.

1805 *Ann. 9th Congr. 2 Sess.* 1138 At this place Mr. Dunbar obtained one or two slips of the '*bois d'arc*'. **1848** BARTLETT *Dict. Amer.*, *Bow-dark tree* (Fr. *bois d'arc*), a western tree, the wood of which is used to make bows with. **1853** R. B. MARCY & MCCLELLAN *Expl. of Red River* 98 (Bartlett), The bows [of the Comanches] are made of the tough and elastic wood of the bois d'arc, or osage orange. **1877** R. I. DODGE *Hunting Grounds Gt. West* 348 The best wood [for bows] is the Osage Orange (*bois d'arc* of the old French trappers, corrupted into 'bow dark' by plains Americans).

‖ **boiserie** (bwazəri). [Fr.] Wainscoting, wooden panelling.

1832 S. AUSTIN tr. *Pückler-Muskau's Tour German Prince* III. 196 The walls of the dining room are covered with oaken boiserie. **1833** J. DALLAWAY *Discourses Archit. Eng.* v. 312 The walls of the state-chambers were painted, and sometimes lined with wainscot, of curious carved *boisserie* on the panels. **1909** *Daily Chron.* 20 Mar. 4/5 The ball-room..is panelled with dark French boiserie work, the wood embellished with gilding. **1940** *Horizon* II. Nov. 263 A shadowy, oval room with gilt boiseries.

‖ **boîte** (bwat). [Fr., lit. 'box'.] A small restaurant or night-club. Also *boîte de nuit.*

1922 W. S. MAUGHAM *Writer's Notebk.* (1949) 188 They'd gone to *boite* after *boite* till every penny was spent. **1925** A. HUXLEY *Along Road* I. 6 In a Montmartre *boîte*.. sat three young American girls. **1932** J. LAVER *Nymph Errant* v. 101 *Boîtes de nuit* where the sham *apaches* were paid so much a night to impress the gaping visitor. **1942** D. POWELL *Time to be Born* (1943) i. 26 'Jean's', the new little *boite*..with red-checked tablecloths and a sixty-five cent dinner. **1960** *Harper's Bazaar* Aug. 60 The *boîtes* of the Left Bank. **1968** *Times* 13 Nov. 10/4 The opening of this first boite de nuit in the new shopping and entertainment centre of Kaliningsk Prospekt.

bokay (bōᵘkēⁱ). Repr. a vulgar pronunciation of BOUQUET.

1848 THACKERAY *Van. Fair* xxix. 248 Bedad it's him.. and that's the very bokay he bought in the Marshy aux Flures! **1904** E. NESBIT *Phoenix & Carpet* iii. 62 'What! all them lovely bokays for me!' exclaimed the enraptured cook. **1947** N. MARSH *Final Curtain* xvi. 244 She..asked in that ghastly voice for what she called her bokay.

Bokhara (bokā·rā). Also **Bukhara.** The name of a town and district in the Uzbek Soviet Socialist Republic, applied *attrib.* to a rug or carpet made there. Also *absol.*, a rug or carpet of this kind.

1913 T. E. LAWRENCE *Home Lett.* (1954) 246 The little Bokhara you have was a very exceptional find. **1918** W. S. CATHER *My Antonia* (1919) II. xiv. 265 The deep, velvety red that is in Bokhara carpets. **1936** A. CHRISTIE *Cards on Table* xv. 149 He'd got some good rugs. Two Bokharas and three or four really good Persian ones. **1941** L. MACNEICE *Plant & Phantom* 53 Flicking the ash On the Bokhara rug. **1966** *Guardian* 16 June 8/3 We owned also two Bokharas, one white and one red.

‖ **bokmakierie** (bǫkmaki·ri). *S. Afr.* Also **bac-,** and variants. [Afrikaans, onomatopœic.] The shrike *Telephorus zeylonus* of southern Africa.

1834 *C.G.H. Lit. Gazette* IV. 52 (Pettman), The..cry of the backbacery. **1852** GODLONTON & IRVING *Kaffir War, 1851–2* 247 (Pettman), The cry of the back-my-keerie (whip-poor-will). **1867** E. L. LAYARD *Birds S. Afr.* 161 Its loud call of 'bacbakiri', its imitative powers, and bright plumage, render it one of the most conspicuous birds of the colony. **1949** *Cape Argus* 15 Oct. (Magazine) 2/6 One of the most familiar birdcalls in this country is that of the bokmakierie. The loud, ringing note of this handsome yellow and green shrike may be heard almost anywhere. **1966** E. PALMER *Plains of Camdeboo* xii. 197 Here soon would be shrikes in numbers, the yellow, green and black bokmakierie with its clear, melodious calls, and that shocking little cannibal, the butcher bird or fiscal shrike.

boko (bōᵘ·ko). *slang.* Also (*U.S.*) **boke.** [Origin unknown.] The nose.

1859 G. W. MATSELL *Vocabulum* 13/2 *Boke*, the nose. **1874** HOTTEN *Slang Dict.*, *Boko*, the nose. Originally pugilistic slang, but now general. **1879** BESANT & RICE *Seamy Side* (1880) i, 'Conk or boko,' said Nicholas the vulgar. 'It's all the same.' **1901** G. B. SHAW *Admir. Bashville* II. i. 313 Flush on the boko napped your footman's left. **1910** C. E. MONTAGUE *Hind let Loose* i. 12 Brumby got one for his nob...Bellona's other bridegroom took it on the boko.

bold, *a.* Add: **8. b.** *Typogr.* Of type = *BOLD-FACE. Also *ellipt.* or as *sb.*

c **1871** V. & J. FIGGINS *Types Specimen Bk.*, Pica Bold Italic. Long Primer Bold Italic. Brevier Bold Italic. Nonpareil Bold Italic. **1884** *Type in Use at Messrs. Parker's Printing Office, Oxford* May 25 Pica bold..brevier bold.. nonpareil bold..pearl bold. **1933** D. L. SAYERS *Murder must Advertise* viii. 132 Can you cut away the headline and re-set in Goudy Bold? **1962** *Which? Car Suppl.* Oct. 114/1 Version tested [*sc.* is printed] in **bold**.

bold-face, -faced. Add: Used to designate type with a thick or 'fat' face, such as 'Clarendon' or 'antique'.

1889 *Cent. Dict.*, Bold-face. **1890** WEBSTER, Bold-faced. **1902** DE VINNE *Pract. Typogr., Processes Type-making* (ed. 2) i. 50 A bold-face with hair-lines and serifs too weak. *Ibid.* xii. 331 A bold-faced clarendon with strong bracketed serifs. **1905** F. H. COLLINS *Authors' & Printers' Dict.*, *Bold-face type*..indicated in MS. by wavy underlining. **1931** *Boys' Mag.* XLV. 148/1 To use Roman, *italic*, and bold face capitals, and lower case as well as numerals, signs, borders, etc., one needs no fewer than 225 characters.

boldo (bǫ·ldo). Also **boldu.** [Sp., a. the native Chilean word.] An evergreen tree of Chile, *Peumus boldus*; also, a medicinal preparation of the leaves of this tree, formerly used as a tonic.

1717 tr. *Frézier's Voy. to South-Sea* 78 The woods are full of Aromatick Trees, as..*Boldu*, the Leaf whereof smells like Frankincense. **1872** *Pharm. Jrnl. & Trans.* 26 Oct. 323/2 The boldo is a tree indigenous to Chili..and belongs to the order *Monimiaceæ*. **1908** *Practitioner* Aug. 339 Slight cholagogues, such as..boldo. **1924** RECORD & MELL *Timbers Tropical Amer.* II. 172 *Peumus Boldus* Mol...is an aromatic evergreen tree 40 to 60 feet high, growing on the dry sunny hills in Chile, where it is known as the 'boldo' or 'boldu'. **1934** *Brit. Pharm. Codex* 255 Boldo is now used principally in the form of tincture, as a diuretic and supposed liver stimulant. **1934** *Tropical Woods* XXXIX. 20 The leaves of *Peumus* yield a product known under such names as 'Folia Boldo' and 'Boldo Leaf Oil', from which a stomachic is derived; this was formerly an article of commerce, but by 1900 had apparently become obsolete. **1951** G. H. M. LAWRENCE *Taxon. Vascular Plants* 512 Chilean boldo wood ..sometimes reaches our markets [in the U.S.A.] as a rarity for cabinet work.

Hence **bo·ldin** [-IN^{1}], a glucoside having hypnotic properties found in the boldo; **bo·ldine** (also †*boldeine*, †*boldina*) [a. F. *boldine* (Bourgoin & Verne 1872, in *Jrnl. de Pharm. et de Chim.* XVI. 191): see -INE^{5}], a bitter, light-sensitive alkaloid, $C_{19}H_{21}O_4N$, found in the boldo and formerly used as a hypnotic; **boldo-glu·cin** = *BOLDIN.

1872 *Pharm. Jrnl. & Trans.* 26 Oct. 323/2 The leaves.. contain an essential oil and an organic alkali, to which the authors propose to give the name of boldine. **1907** *Brit. Pharm. Codex* 167 The leaves also contain a bitter alkaloid, boldine..and a glucoside, boldin or boldoglucin. **1928** SOLIS-COHEN & GITHENS *Pharmacotherapeutics* 1189 Boldo ..contains..a glucoside (boldin or boldoglucin), an alkaloid (boldeine or boldine), gum and tannin. **1939** T. A. HENRY *Plant Alkaloids* (ed. 3) 324 Boldine, like bulbocapnine, antagonises but does not invert the action of adrenaline in raising blood pressure.

bolero. Add: (Also bǫlē·ro.) **2.** A short jacket coming barely to the waist; worn by men in Spain; applied to a similar garment worn by women elsewhere, usually over a blouse or bodice. Also *attrib.*

1892 [see TOREADOR c]. **1893** *Daily News* 1 Apr. 2/4 The Zouave is quite as popular as it was last year...Sometimes it is pure bolero. **1899** *Westm. Gaz.* 6 July 3/2 Robbing the coat of its basque has created..the bolero corsage, really an actual bodice, though appearing a bolero coat and skirt. **1909** 'O. HENRY' *Roads of Destiny* ix. 151 He wore a suit of coarse brown ducking, the coat being a sort of rakish bolero. **1924** *Countries of World* 2495/2 Farmers and peasants ..with wide-brimmed black felt hats, boleros, coloured sashes, and tight-fitting trousers. **1941** 'R. WEST' *Black Lamb* I. 407 The boleros the women wore over their white linen blouses. **1968** J. IRONSIDE *Fashion Alphabet* 35 *Bolero*, a short jacket reaching to the waist, worn open over a blouse..sleeved or sleeveless..worn by Spanish dancers and bullfighters.

Bolivian (bŏli·viăn), *a.* and *sb.* [f. *Bolivia*, a republic in South America, founded in 1825: named after Simon *Bolivar* (1783–1830), South American soldier and statesman.] **A.** *adj.* Of or pertaining to Bolivia. **B.** *sb.* A native or inhabitant of Bolivia.

1835 *Jrnl. R. Geogr. Soc.* V. 71 The great emporium of the foreign trade to the Bolivian provinces. **1854** L. H. DE BONELLI *Trav. Bolivia* I. ii. 49 On the following day the two Bolivians started. **1876** *Encycl. Brit.* IV. 16/1 It [*sc.* the constitution] guarantees to the Bolivians civil liberty..and equality of rights. **1934** *Times Lit. Suppl.* 28 June 456/2 The Bolivian poet Ricardo Jaimes Freyre. **1967** *Listener* 10 Aug. 182/3 Régis Debray's..capture the following month by the Bolivian army.

boliviano (bŏliviā·no). [Sp.] A monetary unit of Bolivia.

1872 I. S. HOMANS *Coin Book* 83 *Boliviano*, the new peso or dollar of Bolivia, equal to five francs. **1957** *Encycl. Brit.*

III. 820/2 The monetary unit, as accepted by the International Monetary Fund on April 8, 1950, is the gold boliviano, equal to..0·0148112 g. of fine gold. **1962** R. A. G. CARSON *Coins* 441 A decimal coinage, introduced in 1863, had as its unit the peso or boliviano in silver...The coinage of 1951 consists of 1, 5 and 10 boliviano pieces. **1964** *Whitaker's Almanack* 828/2 From Jan. 1, 1963, the *Boliviano* is replaced by the *Peso Boliviano* at the rate of *Bs.* 1,000 = *Peso* 1 ($b. 1).

boll, *sb.*^{1} Add: **6. boll-weevil** (in full *cotton-boll weevil*), a weevil (*Anthonomus grandis*) destructive to the cotton-plant; also *fig.*; **boll-worm** (examples).

1895 *Insect Life* Mar. 295 Report on the Mexican Cotton-Boll Weevil in Texas..by C. H. Tyler Townsend..[dated] December 20, 1894. **1903** *Westm. Gaz.* 26 Nov. 12/1 The boll weevil..has caused America a loss of £14,000,000. **1906** *Ibid.* 19 Dec. 1/3 The Mexican cotton boll-weevil. **1906** *Springfield Weekly Republ.* 19 July 16 The 'boll-weevil democrats' is the term of opprobrium which a southern paper applies to democrats who favour Hearst. **1928** *Manch. Guardian Weekly* 31 Aug. 180/1 Reports of boll weevil damage and crop deterioration. **1950** A. LOMAX *Mr. Jelly Roll* (1952) 113 The longshoremen had two parades —one for the union men and one for the boll weevils, the scabs. **1848** *Rep. Secretary U.S. Dept. Agric. 1847* 171 The destruction caused by the boll worm. **1888** *Congress. Rec.* 12 May 4070 Then comes the 'army worm', and then the 'boll-worm'.

bollard. Add: **b.** A post on a traffic island.

1948 *Archit. Rev.* CIV. 80 Demarcation of pavement..is left to discretion of designer. In this drawing a two-foot band of setts is used with bollards at intervals. **1955** *Times* 23 June 6/6 The woman was waiting between the bollards in the middle of the crossing.

bollix (bǫ·liks), *v. low slang.* Also **bollux.** [Alt. of *bollocks*, pl. of *BOLLOCK.] *trans.* To bungle, make a mess of, confuse; also with *up*. So **bo·llixed** *ppl. a.*

1937 J. WEIDMAN *I can get it for you Wholesale* 36 Watch your script...You're getting your cues all bollixed up. **1945** A. J. LIEBLING in *Best Amer. Short Stories* (1946) 275 He explained that a ratfest was 'a rat race, but all bollixed up'. **1945** A. KOBER *Parm Me* 139 So rather I should tell it and maybe bollix up the whole thing, it's better that it should come from Billie. **1952** STEINBECK *East of Eden* I. 478 He'd made a mess of things. He wondered if he'd bollixed up the breaks. **1953** C. ARMSTRONG *Catch-As-Catch-Can* xxii. 177 That skunk..bolluxed Pearl's radio.

Also as *sb.*, a mess, confusion.

1935 DYLAN THOMAS *Let.* July (1966) 156 I've been meaning..to learn about..the bollix of the old gang. **1936** 'G. ORWELL' *Let.* 3 Apr. (1968) I. 215 My novel..would have been out a month ago if it had not been for all that bollux about libel. **1957** J. BLISH *Fallen Star* v. 62 Some kind of intra-departmental bollix.

bollock (bǫ·lək). [var. BALLOCK.] **1.** *pl.* The testicles. Cf. BALLOCK 1.

1744 *School of Venus* in D. Thomas *Long Time Burning* (1969) 362 You..can now without blushing call prick, stones, bollocks, cunt, tarse and the like names. **1763** WILKES & POTTER *Essay on Woman* (1871) 19 Prick, cunt, and bollocks in convulsions hurl'd. **1874** *Lett. fr. Friend in Paris* II. 158 At the same time handling the noble bollocks. **1968** *Landfall* XXII. 17 Fine specimen of a lad, my Monty. All bollocks and beef.

2. *Naut.* Either of two blocks fastened to the topsail-yard, for the topsail-ties to reeve through.

1889 *Cent. Dict.*, Bollock-block. **1898** ANSTED *Dict. Sea Terms*, *Bollocks*, blocks secured to the middle of the topsail yards in large ships; the topsail ties pass through them, and thereby gain an increase of power in lifting the yards.

3. *pl.* (See quot. 1919); also, a mess, a muddle; nonsense (also as *int.*). As *adj.*, naked. *low slang.* Cf. *BALLOCK 2.

1919 DOWNING *Digger Dial.* 12 Bollocks (n. or adj.), absurd; an absurdity. **1950** G. WILSON *Brave Company* (1951) viii. 142 He's stark bollock. *Ibid.* ix. 159 Christ, what a bollocks. **1969** *It* 11–24 Apr. 15/2 Bollocks, nobody at all wants to know. *Ibid.* 15/3 It's really a load of bollocks.

bolo^{1} (bōᵘ·lo). [Sp.] A kind of cutlass used in the Philippine Islands for agricultural and domestic work and as a fighting weapon. *Comb.*, *bolo-maker*; **bolo-man,** one armed with a bolo.

1901 *Westm. Gaz.* 1 Oct. 10/1, 400 bolomen attacked the garrison of seventy-two Americans. **1901** *Daily Chron.* 11 Oct. 5/7 Insurgents, who were armed only with bolos. **1905** LE ROY *Philippine Life* 85 The local blacksmith and bolo-maker, as he generally is (for the bolo is the chief working implement of the Filipino in the thicket or the field or about the house, as well as being, in another form, his principal weapon). **1906** *Macmillan's Mag.* Oct. 903 We had never visited the place before, and, for all we knew to the contrary, it was full of *bolo*-men.

Bolo^{2} (bōᵘ·lo). **1.** One who pursues anti-patriotic 'underground' activities like those of Paul Bolo, a French adventurer shot for treason, 17 April 1918; one suspected of engaging in pacifist propaganda in favour of Germany during the war of 1914–18.

1917 *19th Cent.* Dec. 1265 The activity of those whom we might call the British and German Bolos, naturalised or not. *Ibid.* 1266 As far back as 1907 these Bolos were working under the guidance of German agents.

2. Misused for: a Bolshevik. Also *collect. sing.* = the Bolshevists. Also *attrib.*

1919 *Blackw. Mag.* June 742/1 The Bolo was ignorant of the fate of our raft gun. *Ibid.* Nov. 722/1 A party of Bolo prisoners. **1920** *Ibid.* Mar. 396/1 The missing Bolos from Selmenga. **1924** *Ibid.* Sept. 387/2 Bolo and anti-Bolo propaganda. **1937** C. MADGE in C. Day Lewis *Mind in Chains* 158 Attempts have also been made to prove that Broadcasting House is the headquarters of a conspiracy of 'bolos'. **1963** *Spectator* 27 Sept. 377/1 The police, of course, had passed this bolo under the illusion that he was only some simple patriotic goon.

bolo^{3} (bōᵘ·lo). *U.S.* Also *bolo punch.* An upper cut in pugilism.

1950 DEMPSEY *Championship Fighting* xvii. 114 A long-range right uppercut called the 'bolo' punch. *Ibid.*, The *bolo* is more showy than explosive. **1954** F. C. AVIS *Boxing Ref. Dict.* 13 *Bolo*, a very powerful upper cut, brought from the rear of the body and taking the full weight of the boxer using it.

bologna. Add: Short for *Bologna sausage. U.S.*

1850 *Knickerbocker* XXXV. 23 Relishing 'Bolognas', will he plead that a jelly-eyed roaster is disgusting? **1916** C. SANDBURG *Chicago Poems* 24 The dago shovelman finishes the dry bread and bologna.

Bologne·se, *a.* and *sb.* [a. It. *Bolognese*: see BOLOGNA + -ESE.]

A. *adj.* Of or belonging to, made in or obtained from, Bologna in the north of Italy, or its school of painting.

1821 M. WILMOT *Jrnl.* 1 Nov. in *More Lett.* (1935) 138 We saw..some of the chef d'œuvres of the Bolognese school. **1826** M. KELLY *Reminiscences* (ed. 2) I. 121, I had a letter to deliver to a Bolognese nobleman. **1926** *Glasgow Herald* 8 Apr. 4 Bolognese paper of about 1473, made entirely of linen. **1968** *Listener* 20 June 813/2 During the second half of the 17th century..the Bolognese composers ..played an important part in..Italian music.

B. *absol.* or as *sb.* **1.** A native or an inhabitant of Bologna. (Unchanged for *pl.*) **b.** The dialect of Italian used by the Bolognese. **2.** *The Bolognese*: the territory of Bologna. **3.** A type of toy dog.

1756 A. BUTLER *Lives Saints* II. 206 In the Romagna, the marquisate of Ancona, Tuscany, the Bolognese, and the Milanese. **1818** SHELLEY *Let.* 9 Nov. (1964) II. 51 There is another painter here called Franceschini, a Bolognese. **1839** *Dublin Rev.* VI. 3 We can sufficiently understand them..to put ourselves at ease with the peasants of any district. *Bolognese*, however, is yet too much for us. **1876** *Encycl. Brit.* IV. 19/2 In the Crusades the Bolognese took an active share. **1905** H. DE BYLANDT *Dogs all Nations* II. 729 Bolognese, a ladies' toy dog, somewhat dolent. **1935** *Hutchinson's Dog Encycl.* I. 161 In 1880 the Toy Dog known as the Bolognese was greatly admired in Central Europe.

bolograph (bōᵘ·lǒgraf). [See BOLOMETER and -GRAPH.] An automatic record of the indications of a bolometer. Hence **bologra·phic** *a.*, pertaining to this.

1903 S. P. LANGLEY in *Astrophys. Jrnl.* Mar. 93 A study of the yearly variations of the selective absorption of the Earth's atmosphere by the aid of a long series of bolographs. **1903** AGNES M. CLERKE *Probl. Astrophysics* 22 Using a 'bolographic' method, in which the camera registers what the bolometer *feels*.

Boloism (bōᵘ·lo,iz'm). [See *BOLO^{2}.] Nefarious anti-patriotic activities resembling those of Paul Bolo; *esp.* the engineering of underground pacifist propaganda in favour of an enemy country. Also *erron.*, Bolshevism.

1917 LLOYD GEORGE in *Times* 23 Oct. 6/2 See what has happened in France..and look out for Boloism in all its shapes and forms. **1919** *War Slang* in *Athenæum* 11 July 582/2 In his [*sc.* the British soldier's] linguistic haziness he used 'Boloism' as a synonym for 'Bolshevism', and found an etymology for both in 'Bolo', a variant of 'diabolo', the devil.

boloney, var. *BALONEY.

bolsh (bǫlʃ), *v. intr.* A jocular back-formation from *BOLSHEVIK and derivs.

1921 D. H. LAWRENCE *Let.* 2 Mar. (1962) II. 644, I think Italy will not revolute or bolsh any more. *Ibid.* 645 There will be no definite revolution or bolshing at all. **1924** *British Wkly.* 6 Nov. 134/4 *On Rowdyism.* If people want to 'bolsh' they will 'bolsh', rain or no rain.

Bolshevik (bǫ·lʃivik), *sb.* and *a.* Also rarely **-ic.** [a. Russ. *bol'shevik*, f. *ból'she* more, f. *bol'shói* big. The Russ. pl. *bol'sheviki* has been used by some English writers.]

A. *sb.* A member of that part of the Russian Social-Democratic Party which took Lenin's side in the split that followed the second Congress of the party in 1903, seized power in the 'October' Revolution of 1917, and was subsequently renamed the (Russian) Communist Party. Cf. *MENSHEVIK.

1917 *19th Cent.* July 141 The Mensheviki or Minimalists (Moderate Socialists)...The Bolsheviki (Extreme Socialists). **1917** *New Europe* 6 Dec. 236 It was from this Conference [of Socialists at Stockholm, 1903] that the cleavage between Bolševiks and Menševiks dates, the former being those who held a 'majority' at the Conference, the latter a 'minority'. **1918** E. J. DILLON *Eclipse of Russia* 10 The Bolsheviks at once outbid the Cadets. *Ibid.* 33 After the triumph of the Bolsheviki. **1918** C. E. RUSSELL *Unchained Russia* 253 The Bolshevik was in Russia the most natural fruitage. *a* **1968** M. RICHLER in R. Weaver *Canad. Short Stories* 2nd Ser. (1968) 196 The crazed man behind him, another old bolshevik, repeats over and over again, 'This could never happen in a socialist country.'

b. *transf.* and *fig.* A person of subversive or revolutionary views; an out-and-out opponent of the existing social order or accepted codes.

1926 W. R. INGE *Lay Thoughts* 29 The cliques of literary Bolsheviks, who seem to be inspired by a destructive hatred ..of civilisation.

B. *adj.* = *BOLSHEVIST *a.*

1907 I. ZANGWILL *Ghetto Comedies* 401 Go to the Labour Parties...Not the Bolshewiki faction..but the Menshewiki. **1917** *New Europe* 8 Nov. 112 The Bolševik mentality. **1918** C. E. RUSSELL *Unchained Russia* 261 The Bolshevic movement. **1919** J. POLLOCK *Bolshevik Adv.* p. xx, The part played by the Jews in Bolshevik Russia. **1952** *Economist* 25 Oct. 240/1 The party is no longer Bolshevik, even in name. The Politburo has been dropped.

Hence **Bo·lshevikism** = *BOLSHEVISM; **Bo·lshevikize** *v.* = *BOLSHEVIZE *v.*

1918 *Nation* (N.Y.) 7 Feb. 135/1 What Germany is resolved upon is that these lands shall not be Bolshevikized economically. **1919** H. S. KING *Russia during War* 49 The rising tide of Bolshevikism.

Bolshevism (bǫ·lʃiviz'm). [a. Russ. *bol'shevizm*: see *BOLSHEVIK.] The doctrines and practices of the Bolsheviks; the communistic form of government adopted in Russia since the Bolshevik Revolution of October (November), 1917.

1917 *New Europe* 8 Nov. 112 The good sense of Russian democracy threw off the yoke of Bolševism. **1918** THEO. ROOSEVELT in *Metropolitan Mag.* June 9/1 The most powerful indictment of the corrupt and inefficient tyranny of the Romanoffs..is that it produced Bolshevism. *Ibid.* 9/2 Parlor or pink-tea bolshevism dear to the hearts of so many..who like to think of themselves as intellectuals. **1918** *Times* 7 Dec. 9/6 There must be no Bolshevism in [British] land policy. **1926** W. R. INGE *Lay Thoughts* 159 Bolshevism is no new thing there [*sc.* in Russia]. **1926** D. H. LAWRENCE *Plumed Serpent* ii. 47 Bolshevism only smashed your house or your business or your skull, but Americanism smashes your soul.

Hence **Bolshevi·sia** [-IA^{1}], **Bo·lshevy** [after *Muscovy*], the land of Bolshevism; Russia under the rule of the Bolsheviks.

1919 J. POLLOCK *Bolshevik Adv.* 195 A producer [of stage-plays] in Bolshevisia. **1920** *Chambers's Jrnl.* 514/1 Life in Bolshevisia—such as it was in July 1918. **1921** *Times Lit. Suppl.* 18 Mar., Bolshevy from within. **1922** *Times* 18 Apr. 21/1 The Bondmaid of Bolshevy.

Bolshevist (bǫ·lʃivist). [a. Russ. *bol'shevist* (now disused) *BOLSHEVIK.] A Bolshevik; a supporter of Bolshevism. Also *transf.*, esp. as a term of reproach for an out-and-out revolutionary. Also *attrib.* or as *adj.*

1917 *19th Cent.* Dec. 1106 The reign of Bolshevists and Terrorists. **1920** *Edin. Rev.* July 33 The Government of Styria, which, having a large industrial population of its own, is particularly sensitive to the 'Bolshevist Peril'. **1920** *Chambers's Jrnl.* Aug. 513/1 Packing the meetings with Bolshevist agents. **1922** E. WALLACE *Flying Fifty-Five* xxxiv. 204, I call a horse a Bolshevist when he doesn't run twice alike. **1926** W. J. TODD *Port* v. 83 Cheese (of not too bolshevist a nature) prepares the palate, but cheese is too violent a preparation for a fine old Port. **1940** *Tablet* 4 May 417/1 Under the Bolshevist-Nazi dictatorship, two hundred million human beings are forced to live deprived of the foundations on which Western civilization was built.

Hence **Bolshevi·stic** *a.*, of, pertaining to, or characteristic of, the Bolshevists; **Bo:lshevi·stically** *adv.*

1920 *Glasgow Herald* 14 May 9 London has established the alarming precedent of applying such words as 'horrible' and 'Bolshevistic' to the well-intentioned proposals of certain pious Churchmen. **1920** *Punch* 13 Oct. 282/1 In these Bolshevistic days I should have preferred of course to have started off with 'Comrade' or 'Brother'. **1923** D. H. LAWRENCE *Birds, Beasts & Flowers* 69 Never, bolshevistically To be able to stand for all these!

Bolshevize (bǫ·lʃivəiz), *v.* [f. *BOLSHEV(IK + -IZE.] *trans.* To make Bolshevist in character; to convert (a country) to a Bolshevistic form of government.

1919 *Observer* 23 Mar. 10/5 Unless we want to bolshevise the Boche and the world. **1920** *19th Cent.* Mar. 536 Taking her chance that Germany will be Bolshevised before Russia is Germanised. **1920** *Contemp. Rev.* Aug. 268 Thereby preventing these people from being Bolshevized.

Hence **Bo·lshevized** *ppl. a.*; **Bo·lshevizing** *vbl. sb.* and *ppl. a.* Also **Bo:lsheviza·tion,** the process of making, or becoming, Bolshevist.

1920 *19th Cent.* Mar. 536 Great Britain has been accused ..of pursuing a policy aimed at the Balkanisation of the Baltic provinces. It may appear that we have only achieved their Bolshevisation. **1920** *Glasgow Herald* 15 Apr. 7/1 Syndicalists of a Bolshevik or Bolshevising hue. *Ibid.* 2 Oct. 7/2 The withdrawal of the Bolshevised troops from Asia Minor. **1927** *Daily Tel.* 16 Aug. 11/5 With a view to the Bolshevising of the 400,000,000 of Chinese. *Ibid.*, A Bolshevised China would be the world's greatest peril. **1940** *Manch. Guardian Weekly* 22 Mar. 225 Since 1918, when the German Army took some small part in freeing Finland from Bolshevisation.

Bolshy, Bolshie (bǫ·lʃi). [-Y^{6}.] A jocular or contemptuous name for a Bolshevik. Also *transf.* and *fig.*

BOLSON

1920 *John Bull* 28 Aug. 1 We can quite understand the Bolshies' repugnance to labour—except with the jaw. 1920 *Punch* 10 Nov. 373/1 Mr. Stanton could think of no better retort than the stereotyped 'Bolshie!' 1921 HICHENS *Spirit of Time* xii, Those Russians..breed more princesses than we do..but they seem to be giving them toko over there—the Bolshies, I mean. 1923 D. H. LAWRENCE *Kangaroo* i. 1 A comical-looking bloke! Perhaps a Bolshy. 1940 'G. ORWELL' in *Horizon* Mar. 187 If the Russian Revolution is anywhere referred to it will be indirectly, in the word 'Bolshy' (meaning a person of violent disagreeable habits).

Hence *attrib*. or as *adj*. **a.** = *BOLSHEVISTIC *a*. **b.** *transf*. and *fig*. Left-wing; uncooperative, recalcitrant.

1918 D. H. LAWRENCE *Lett*. (1962) I. 565 The railway people, when one travels, seem rather independent and Bolshy. 1922 T. E. LAWRENCE *Lett*. (1939) 382 My nature doesn't second the demands of discipline very well, and unless I keep working at something I get Bolshie! 1930 H. G. WELLS *Autocr. Mr. Parham* III. vi. 208 The young men of science, the clever ones, are all going Bolshy or worse. *Ibid*. IV. iv. 296 Street-corner boys and Bolshie agitators. 1934 G. B. SHAW *On the Rocks* II. 231 Lenin and Stalin and Trotsky and all that Bolshy lot. 1966 *New Statesman* 1 Apr. 473/1 The neglected dullards not unnaturally turn bolshy. 1969 D. CLARK *Nobody's Perfect* v. 156 You can stop looking so bolshie, because I think you've done what you did well enough.

bolson (bōᵘ·lsǒn, ‖bolsō·n). *U.S.* [Sp., augmentative of *bolsa* purse.] In the south-western U.S. and Mexico, a basin-shaped depression surrounded by mountains. Also *bolson-plain*.

1838 A. GANILH *Mexico v. Texas* 9 A desert known, in the maps [of Mexico], under the name of 'Bolson of Mapimi'. 1847 RUXTON *Adv. Mexico* xiii. 96 The sun was fast sinking behind the rugged crest of the 'Bolson', tinging the serrated ridge..with a golden flood of light. 1904 *Amer. Geol*. Sept. 164 The bolson plains may be considered as sections of an upraised peneplain surface in its earliest infancy, at a stage in which they are as yet untouched by stream-action. 1964 F. O'ROURKE *Mule for Marquesa* 15 It would be..the Bolsón, ten thousand square miles of the high desert, cupped within the waterless mountains.

bolster, *sb.*[1] Add: **6.** *attrib*. and *Comb*.: *bolster-shaped* adj.; **bolster collar,** a bolster-shaped collar of a woman's coat or cloak.

1900 *Westm. Gaz*. 16 May 2/1 The bolster-shaped kit bag. 1904 *Windsor Mag*. Jan. 238/2 Long bolster-shaped sweets. 1923 *Daily Mail* 19 June 1 New short coat..with smart finely pleated bolster collar.

bolster (bōᵘ·lstəɹ), *sb.*[2] [? cf. BOASTER[2].] A type of chisel used by bricklayers for cutting bricks.

1908 C. F. MITCHELL *Brickwork & Masonry* (ed. 2) ii. 96 *Brick-Cutting Tools*..The club hammer and bolster, for cutting with greater exactitude than with the trowel. 1924 H. L. BRIGGS *Pract. Bricklaying* iv. 39 For finer work the brick-cutting chisel, or 'set' or 'bolster', as it is usually called, is used. 1936 *Archit. Rev*. LXXIX. 240/4 (*caption*) The bolster, a long-edged, cold chisel used for cutting bricks and also for cleaning joints.

bolstered, *ppl. a*. Add: **2. b.** With *up* in sense of BOLSTER *v*. 4 b.

1901 *Westm. Gaz*. 17 Sept. 7/3 There is a fearful lot of over-trading and bolstered-up credit..in German industrials.

bolt, *sb.*[1] Add: **7. b.** A sliding metal rod in the breech mechanism of a rifle which opens and closes the bore and positions the cartridge.

1859 'STONEHENGE' *Shot-gun & Sporting Rifle* IV. iii. 259 The barrel-maker has to braze on with great care two lumps of iron to the lower sides of the barrels, one of which..forms about three-fifths of the socket in which the circular bolt fixed in the stock revolves. In order to understand the exact form of the bolt, a gun on this principle must be examined, and moreover as scarcely any two makers adopt the same shape, the description of one would not suffice for all. 1881 W. W. GREENER *Gun & its Devel*. 129 The piece is cocked by the thumb, as is the needle-gun; the bolt is then turned one-quarter of a circle to the left, and drawn back: the cartridge is put in and pushed home by the bolt; this bolt is turned back one-quarter of a circle to the right; the piece is then ready for firing. 1930 G. BURRARD *In Gunroom* 122 It is extremely difficult to insert an incorrectly assembled bolt into the action, and this can only be done by a combination of undue force and careful manipulation. 1969 [see *bolt action* in sense *15].

15. bolt action (see quot. 1969[1]); also (with hyphen) *attrib*.; also *ellipt*. = a bolt-action gun.

1871 W. W. GREENER *Modern Breech-Loaders* 52 In all cases where they exploded caps with 'bolt' actions, the mark of the striker was found upon the cap. 1896 —— *Gun & its Devel*. (ed. 6) xxx. 702 (*heading*) Bolt-action systems. 1958 R. ARNOLD *Automatic & Repeating Shot-guns* ii. 46 In the United States the bolt-action has given excellent results at both Down-the-line and Skeet. 1969 D. C. FORBES *Sporting Gun* 137 *Bolt action*, refers to a gun with a breech which is opened by a bolt being turned and slid back. The cartridge is laid in front of the bolt and the bolt slid forward to push the cartridge into the breech. 1969 *New Yorker* 29 Nov. 150/2 He was carrying a First World War Czech bolt-action rifle.

bolt, *sb.*[3] Delete †*Obs*. and add: **2.** A hypothetical law case propounded and argued for practice by students of the Inns of Court. (Cf. BOLTING *vbl. sb.*[1] 2 b.)

1556 *Black Books of Lincoln's Inn* (1897) I. 316 Everi daye (except Sondayes and festifall dayes, when ther is a mote or a bolte). 1570 in R. J. Fletcher *Pension-bk. Gray's Inn* (1901) I. 4 Item it is ordered..that upon the other dayes not appointed for the moting it shalbe lawfull to the utter baristers to keepe bolts. 1593–. in Douthwaite *Gray's Inn* (1886) 83 None shall be called to the barr but such as..have put Cases at Bolts in Term six times. 1880 *Encycl. Brit*. XIII. 89/1 Bolts were of an analogous character, though deemed inferior to moots. Both had fallen into desuetude until lately. 1956 A. L. ROWSE *Early Churchills* ii. 14 The readings, moots and bolts—the public exercises that tested the knowledge acquired by the students from their seniors.

bolt, *v.*[2] Add: **2. c.** *Hawking*. (See quot.)

1855 SALVIN & BRODRICK *Falconry Brit. Isles* Gloss. 149 *Bolt*, to fly straight from the fist at game, as Goshawks and Sparrow-Hawks do.

d. *Horticulture*. To 'run to seed' prematurely.

1889 in *Cent. Dict*. 1961 *Amateur Gardening* 16 Sept. 12/2 In April or early May many of the plants 'bolted'.

3. a. *spec*. Of a rabbit, fox, etc.: to escape from its burrow or earth.

1575 TURBERV. *Venerie* 179 Put in a Ferret close musseled, and she will make the Conies bolte out againe into your pursenets. *Ibid*., It will make the Conies bolte out of the earth. 1851 [see *BOLT-HOLE 1]. 1900 A. E. T. WATSON *Young Sportsman* 234 A rabbit will bolt much sooner from a ferret that is free. *Ibid*., A rabbit will sometimes decline to bolt, and will be killed in the burrow.

c. (Earlier examples.)

1821 in E. S. Brown *Missouri Compromises* (1926) 43 Parker of Virginia, & some others, bolted. 1871 *St. Louis Democrat* 3 Apr. (De Vere), Several of our contemporaries have announced..that Carl Schurz has bolted from the Republican party.

4. b. *spec*. To cause (a fox, rabbit, otter, etc.) to retreat from its hole or burrow. Also *transf*. and *fig*.

1638 GUILLIM *Heraldry* (ed. 3) III. xiv. 176 You shall say Bowlt the Conie. 1863 ATKINSON *Stanton Grange* 201 He intended to dig at his leisure until he bolted him [*sc*. an otter]. 1892 MRS. J. E. H. GORDON *Eunice Anscombe* 176 The terrier..was put into the hole to 'bolt' the otter. 1902 *Daily Chron*. 13 Mar. 8/2 A brace of foxes were next bolted from an artificial earth. 1914 R. CURLE *Life is a Dream* 229 The dogs became wildly excited, pawing at the sand around the hole, bolting the crab, and then biting it. 1922 E. PHILLPOTTS *Grey Room* vii. 172 He'll bolt it [*sc*. the evil spirit] yet,..like a ferret bolts a rat.

7. (Earlier example.)

1813 *Portsmouth* (N.H.) *Oracle* 20 Nov. 2/3 (Th.), Others, ..without sufficient courage to do their duty, bolted the question.

9. c. *absol*. or *intr*.

1847 MARRYAT *Childr. N. Forest* ii, We can bolt and bar.

d. *intr*. for *pass*.

1907 *Smart Set* Feb. 77/1 The door bolts on the inside.

bolter[2]. Add: **1. b.** A fugitive from justice. Now *Austral. hist*.

1699 B. E. *Dict. Cant. Crew*, *Bolter of White Friers*, one that Peeps out, but dares not venture abroad, as a Coney bolts out of the Hole in a Warren, and starts back again. 1737 DYCHE & PARDON *Dict*. (ed. 2), *Bolter*, a Cant Name for one who hides himself in his own House, or some priviledged Place, and dares only peep, but not go out of his Retreat. 1855 *Argus* (Melbourne) 19 Jan. 5/4 The arrest in London of a well-known bolter from Melbourne. 1897 'P. WARUNG' *Tales Old Régime* 160 One of the 'bolters', advertised by Mr. F. A. Hely, J.P., Principal Superintendent of Convicts, in the *Sydney Gazette*, in the year 1827. 1964 R. WARD *Penguin Bk. Austral. Ballads* 23 Some convicts 'took to the bush' as 'bolters' or bushrangers.

3. One who swallows (food) hastily or whole.

1825 *New Monthly Mag*. XVI. 178 Pettifogging strainers at gnats and bolters of camels. 1833 T. HAMILTON *Men & Manners in Amer*. I. ii. 44 The most expeditious bolters of dinner.

bolt-hole. [BOLT *v.*[2]] **1.** = *bolting-hole* (see BOLTING *vbl. sb.*[2] 5).

1851 STERNBERG *Dial. Northampt*., *Bolt-hole*, the hole from which the rabbit makes its escape; or, in the phraseology of the craft, '*bolts*'. 1877 E. PEACOCK *Gloss. Manley*, *Bolt-hole*, (1) the hole by which a rabbit makes its escape when the ferret pursues it. (2) Any unknown hole by which a person makes his way into or out of a house. 1887 KIPLING *Lett. Marque* v. in *From Sea to Sea* (1900) 49 A figure in saffron came out of a dark arch into the sunlight, almost falling into the arms of one in pink...The pink and saffron figures..disappeared into separate bolt-holes. 1891 *Pall Mall Gaz*. 4 June 6/2 We left them no bolt-hole,..so every man fought till he was killed. 1924 E. MARSH tr. *42 Fables of La Fontaine* 71 [The hare] heard a rustle, And took the hint to bustle Off to his bolt-hole. 1932 H. SIMPSON *Boomerang* xii. 306 A girl who had been jilted might choose any bolt-hole to hide her shame. 1967 *Listener* 30 Nov. 705/3 With all the mental bolt-holes stopped up, a clear look at England around 1930 produced traumatic results.

2. *Coal-mining*. A short connecting heading or opening.

1839 URE *Dict. Arts* 981 Two, three, or even four bolt-holes open into a side of a work, according to its extent. 1860 *Eng. & For. Mining Gloss*. (ed. 2) 69 *Bolt hole*, a passage ..from the gate-road into a side of work. 1877 *Encycl. Brit*. VI. 66/1 From the gate road a heading called a bolt-hole is opened. 1963 *Gloss. Mining Terms* (*B.S.I.*) II. 8 *Bolt hole*. 1. A short connecting passage made for ventilation purposes.

bolting, *vbl. sb.*[2] Add: **2. b.** *Horticulture*. The action of *BOLT *v.*[2] 2 d.

1933 *Discovery* Mar. 76/1 The often troublesome 'bolting' of lettuce may be prevented by short periods of light. 1961 *Amateur Gardening* 16 Sept. 12/2 'Bolting' is a gardening term meaning that the plants throw up seed heads instead of forming the tight hearts we aim at producing.

bolting (bōᵘ·ltiŋ), *ppl. a*. [f. BOLT *v.*[2] + -ING[2].] That bolts or runs to cover.

1907 *Daily Chron*. 28 Nov. 8/1 Some of the bolting escort were returning. 1908 *Westm. Gaz*. 31 Jan. 4/2 The bolting rabbit.

b. *Horticulture*. That is prematurely 'running to seed'. Cf. *BOLT *v.*[2] 2 d.

1961 *Auckland Weekly News* 5 July 41/1 Bolting cabbages and savoys can be restored to normal if the stem is pierced one inch above soil level with a sharp penknife and a small pebble inserted in the slit.

bo·ltless, *a*. [BOLT *sb.*[1]] Without a bolt or bolts (in various senses).

1832 I. TAYLOR *Saturday Even*. xv. 248 The boltless thunders of the mere man of rhetoric. *c* 1877 THEODORE WATTS in *Westm. Gaz*. (1900) 15 Feb. 10/1 'Mid boltless thunder. 1902 'MONKSHOOD' & GAMBLE *R. Kipling* 259 The boltless door. 1903 KIPLING *Five Nations* 75 We shall go back by boltless doors, To the life unaltered our childhood knew.

Bolton (bōᵘ·ltǫn). The name of a county borough in Lancashire used attrib. in: **Bolton bay,** a variety of fowl, gold and silver in colour and minutely marked; **Bolton sheeting,** a twill cotton fabric containing a condenser weft and used for curtains, linings, etc.; **Bolton twill,** a twill cotton fabric manufactured in Bolton.

1854 *Poultry Chron*. II. 295/1 Provincial terms, as Minorcas, Bolton Bays,..Pheasants, Chitteprats. 1861 MRS. BEETON *Bk. Househ. Managem*. 475 *Pencilled Hamburg*. This variety of the Hamburg fowl is..also known, in different parts of the country, as Chitteprats, Creoles, or Corals, Bolton bays and grays. 1880 L. HIGGIN *Handbk. Embroidery* ii. 12 (*heading*) Bolton, or workhouse sheeting, is a coarse twilled cotton fabric, seventy-two inches wide, of a beautiful soft creamy colour. 1938 D. GARNETT in *T. E. Lawrence Lett*. 40 He hung the walls of his study with green Bolton sheeting which made it very quiet. 1967 *J. W. Coates & Co. Catal*. Spring 11/1 Heavy Quality, hemmed Bolton Twill Sheets..39/6 pr.

Boltzmann (bǫ·ltsmăn). The surname of Ludwig *Boltzmann* (1844–1906), an Austrian physicist, used *attrib*. or in the possessive of various laws, phenomena, etc., in *Physics*. **Boltzmann's constant** (see quot. 1959).

1879 *Trans. Cambridge Phil. Soc*. XII. 547 (*title*) On Boltzmann's Theorem on the average distribution of energy in a system of material points. 1910 *Sci. Abstr*. A. XIII. 184 The value of Boltzmann's universal constant is found to be $K = 7{\cdot}5 \times 10^{-17}$. 1915 *Proc. Roy. Soc*. A. XCI. 529 Where ϵ is the charge and m the mass of an electron and k is Boltzmann's constant. 1958 *Chambers's Techn. Dict*. 961/1 *Boltzmann equation*, algebraic equation stating that the entropy of a system of particles depends on the logarithm of the probability of its macroscopic state. 1959 *Chambers's Encycl*. XI. 398/1 The constant k is known as Boltzmann's constant, sometimes called the entropy constant. It represents the gas constant divided by the number of molecules in the gas (gas constant per molecule). 1962 SIMPSON & RICHARDS *Junction Transistors* i. 11 In transistor technology these conditions apply almost invariably so that the Boltzmann distribution is generally used. 1968 P. A. P. MORAN *Introd. Probability Theory* vii. 341 The density of the probability distribution of their potential energies will be proportional to $\exp\left(-\frac{U}{kT}\right)$, where U is the potential energy, k is Boltzmann's constant, and T is the absolute temperature.

bom (bǫm). [Of imitative origin.] The sound caused by the discharge of a gun, less deep and sonorous than a 'boom'. Also, the sound of a heavy object falling.

1906 *Westm. Gaz*. 11 June 8/2 A faint distant Bom! and everybody murmurs with one accord, 'First Gun!' 1922 JOYCE *Ulysses* 97 Bom! Upset. A coffin bumped out on to the road.

boma (bōᵘ·mā). *E. Afr*. [Swahili.] **a.** An enclosure or stockade used for herding beasts and for defensive purposes. **b.** A police post. **c.** A district commissioner's or magistrates' office; an administrative centre associated with such an office.

1878 STANLEY *Dark Cont*. I. vi. 137 From the staked bomas..there rise to my hearing the bleating of young calves. 1898 *Geogr. Jrnl*. (R.G.S.) XI. 389, I went out on a sandspit into the lake and camped, cutting down the bush and placing it across the shore end of the bank so as to form a boma. 1903 STORDY in *Jrnl. Soc. Arts* 10 July 691/2 The construction of the boma employed fifty hands for the space of nine weeks. 1920 *Blackw. Mag*. Jan. 59/1 It [*sc*. the fort] was a typical Boma, built of bricks and plaster. 1961 *New Scientist* 24 Aug. 451 The innumerable trails that radiate from each cattle boma. 1964 C. WILLOCK *Enormous Zoo* iii. 49 A boma..that would have held a herd of a hundred or more zebra. 1967 L. KAYIRA *Looming Shadow* (1968) v. 60 He did not have the vaguest notion about the implications of the Boma and the law. 1969 N. CARR *White Impala* ii. 20 The hunter was awarded the 'ground' tusk.., while the other was to be handed in to the Boma. *Ibid*., Nowadays the word 'Boma'..is used to mean a Government administrative centre. But more particularly it means the District Commissioner's office and all the services that go with it—dispensary, post office, agricultural officer's office, and so on.

bomah (bōᵘ·mā). Also **boomah.** [Cf. Zulu *imboma* aloe-berry.] *bomah nut*, the fruit of a southern African shrub *Pycnocoma macrophylla*, used in tanning.

1874 LINDLEY & MOORE *Treas. Bot*. II. 943/1 Boomah Nuts. 1887 MOLONEY *Forestry W. Africa* 417 The Bomah Nut..is extensively cultivated by the natives near the Victoria Falls.

bomb, *sb*. Add: **2.** In modern use: a case filled with explosive, inflammable material, poison gas, or smoke, etc., fired from a gun, dropped from aircraft, or thrown or deposited by hand. Also freq. in Comb., as *atomic*, *flying*, *gas*, *incendiary bomb*, etc. (see under the first elements).

1914 *Times* 9 Oct. 6/5 A German aeroplane flew over the outskirts of Paris early this morning and threw several bombs. 1914 *Whitaker's Almanack 1915* 821/2 Lieut. Marix, who also made an attack on the Dusseldorf shed, and by means of a bomb destroyed it. 1940 *Times* 15 Aug. 4/3 Numerous direct hits with heavy calibre bombs were scored..and the crew of one aircraft whose bombs fell in a line across the main buildings, reported that one had hit and destroyed the main power house. *Ibid*. 17 Oct. 4/4 Almost as soon as the sirens sounded bombs fell in some districts in the London area. 1943 *Times Weekly* 18 Aug. 18/4 About 10 tons of light demolition bombs, besides incendiary bombs, were scattered over a wide area.

b. (**the**) **bomb** (also *Bomb*), a pregnant expression for the atomic or hydrogen bomb, as used or to be used by any country as a weapon of war, and regarded as unique because of its utterly destructive effects.

[1932 H. NICOLSON *Public Faces* I. 23 True it was that their acute distaste for the bomb..did credit to their humanity, to their state of civilisation.] 1945 *Times* 15 Aug. 5/5 (*headline*) Victory and the Bomb. 1959 *Sunday Times* 5 Apr. 19/5 Twenty years ago, I mean: before the war, the Bomb, the satellites, the space-travellers and the nudist paradises. 1966 *Listener* 20 Jan. 83/1 One of the most persistent fallacies of the..debate about nuclear weapons is the proposition that 'the bomb', as it has come to be called almost with affection, has put an end to war.

c. Short for *radium bomb* (*RADIUM b).

1930 *Brit. Med. Jrnl*. 8 Feb. 232/2 The apparatus to be described was designed..to meet the problem of making.. use of the 4 grams of national radium that had been lent to the hospital for use as a 'bomb'.

d. A success (esp. in entertainment); also *U.S.*, a failure. So phr. *like a bomb* and varr., with great speed; with considerable effectiveness or success. *colloq*.

1954 *Amer. Speech* XXIX. 99 *Like a bomb*,..very fast. 1961 *New Yorker* 28 Oct. 43/2 What had once been called a failure became a 'bomb'. 1962 *Listener* 11 Oct. 581/2 Leslie Crowther, introducing *The Black and White Minstrel Show* ..from the Victoria Palace, remarked, 'We're going like a bomb here.' 1963 *The Beatles* 5 Once, Paul McCartney and I played Reading as the Nurk Twins. Went down a bomb, I recall. 1967 A. DIMENT *Dolly Dolly Spy* ii. 15 His straight-backed 'visitor's' chair, which for pure discomfort would have gone down a bomb with the Gestapo.

e. A large sum of money. *slang*.

1958 F. NORMAN *Bang to Rights* II. 79 There are not many bent screws the reason being that many of them are to honest or to scared to do any trafficing but the ones that do make a bomb. 1963 M. LEVINSON *Taxi!* x. 122 Large original oil paintings..which, in cab-driver's language, looked as though they were 'worth a bomb'. 1969 A. E. LINDOP *Sight Unseen* xxiv. 202 Can I have that instead of the five pounds? I might flog it for a bomb in me old age.

4. b. An old car (see also quot. 1953). *Austral*. and *N.Z. slang*.

1953 BAKER *Australia Speaks* iv. 106 *Bomb*, an old car or motor cycle. 1961 *Coast to Coast 1959–60* 120 Get out, buy yourself a car...Do as I did, start with a bomb and keep adding a bit and trading it in till you've got what you want. 1965 M. SHADBOLT *Among Cinders* xviii. 163 The car.. wasn't much more than an old bomb. 1967 F. SARGESON *Hangover* vii. 53 We had a job shoving her into the bomb.

5. (Earlier examples; also later examples without *Volcanic*.)

1798, 1833 [see VOLCANIC *a*. 1 c]. 1956 W. EDWARDS in D. L. Linton *Sheffield* 8 Tuffs are especially well seen in the Ashaver anticline,..and are dark green, brown, and purple well-bedded basaltic ashes with 'bombs' of basalt, limestone, and chert. 1969 *Nature* 8 Nov. 557/2 Volcanic activity occurred at several places along this fracture and fresh bombs rest on the ice close to the edge of the chasm.

6. In many obvious comb., now esp. relating to aerial bombs, as *bomb-aimer*, *-aiming*, *-carrier*, *-carrying*, *-crater*, *-damage*, *-dropper*, *-dropping*, *-dump* [*DUMP *sb.*[4] 1 c], *-load*, *-maker*, *-raid*, *-release*, *-thrower*, *-throwing*; *bomb-damaged*, *-pitted*, *-shattered* adjs.; **bomb alley** *Service slang*, an area repeatedly attacked by bombing; **bomb bay,** a compartment in an aircraft for holding bombs; **bomb calorimeter** (see quot. 1928); **bomb-disposal,** the removal and detonation of unexploded and delayed-action bombs; usu. *attrib*., esp. in *bomb-disposal squad*; **bomb door,** usu. in *pl*.: the movable covering of a bomb bay; **bomb-happy** *a. colloq*. [*-HAPPY], mentally affected by exposure to a bomb or shell explosion at close quarters; shell-shocked; hence *bomb-happiness*; **bomb-lance** (examples); **bomb line** (see quot. 1944); **bomb-rack,** a rack (in an aircraft) for holding bombs; **bomb run** (also *bombing run*), the line of flight of bombing aircraft over the target; **bomb sight** (also *bombing sight*), a device for sighting the target in bombing from an aeroplane; so *bomb-sighting* vbl. sb.; **bomb-site,** ground on which buildings, etc., have been destroyed by aerial bombing.

1935 *Meccano Mag*. Oct. 577/2 The telegraphist is also acting as bomb-aimer. 1937 *Jrnl. R. Aeronaut. Soc*. XLI. 423 A stable platform for bomb-aiming. 1942 W. SIMPSON *One of our Pilots is Safe* ii. 18 We were..given an extra gun—firing downwards and backwards out of the bomb-aiming hatch. 1942 *Hutchinson's Pict. Hist. War* 10 June–1 Sept. 263 The narrow stretch of water between the island of Pantellaria and Sicily, officially known as the Sicilian Channel, but called Bomb Alley by the Navy and Merchant Service. 1945 *Ann. Reg. 1944* 76 The inhabitants of the so-called 'bomb alley' had also suffered considerably. 1918 *Aeronaut. Insp. Directorate Data Bk*. (Handley Page V/1500) 3 The bomb bay is rectangular, and built entirely of spruce. 1934 *Flight* 15 Feb. 156/1 Bombs are internally stowed in a bomb bay closed by doors controlled by the bomber. 1902 *Encycl. Brit*. XXVII. 444/1 The potential energy..is measured by the heat of combustion..in the bomb calorimeter. 1928 A. B. CALLOW *Food & Health* 29 The apparatus used for this laboratory oxidation is a small calorimeter which is known as the bomb calorimeter, because the oxidation takes place inside a thick-walled vessel which in some ways resembles a bomb. 1928 GAMBLE *Story N. Sea Air Station* vii. 109 The only standard bomb-carriers in service at this period were the single 16-lb bomb and the 20-lb gear for two Hale 20-lb bombs. *Ibid*., Equipping all the machines with bomb-carrying and release gear. 1920 *Blackw. Mag*. July 76/1 Ploughed up with bomb-craters. 1941 *Partisan Rev*. Nov.–Dec. 498, I can see no bomb damage anywhere, except for a few churches. 1942 *Ann. Reg. 1941* 328 Four galleries..being closed on account of bomb damage. 1945 W. S. CHURCHILL *Victory* (1946) 157 Bomb-damaged houses in the London area. 1954 R. MACAULAY *Last Lett*. (1962) 150 It was one of the City churches very little bomb-damaged, I think. 1954 'N. BLAKE' *Whisper in Gloom* I. vi. 77 A row of bomb-damaged houses. 1940 *War Illustr*. 4 Oct. 338 The Bomb Disposal Sections of the Royal Engineers whose job it is to dig up and destroy the time-bombs. *Ibid*. 6 Dec. 612 The bomb-disposal squads of the Royal Engineers..described..the removal of delayed-action bombs. 1939 *Meccano Mag*. Mar. 150/3 Large bomb doors cover the bottom of the fuselage. 1928 GAMBLE *Story N. Sea Air Station* vii. 104 The last three machines were classified as 'Gun Machines' and 'Bomb Droppers'. 1910 R. FERRIS *How It Flies* xvii. 372 There have been many contests by aviators in 'bomb-dropping'. 1939 R. CAMPBELL *Flowering Rifle* VI. 148 Keep safe his bomb-dump while our patience lasts. 1941 *Illustr. London News* CXCVIII. 728/2 Bombing up the 'planes is the work of the armourers, who also have charge of the vast station bomb-dump. 1944 J. H. FULLARTON *Troop Target* iii. 29 Now, when 'bomb-happiness' and the 'jitterbugs' threatened to touch the troop with palsied fingers. 1943 *San Francisco Chron*. 1 Dec. 2/2 A barrage so incessant..that many troops of the crack 65th Nazy Division were rendered 'bomb happy' and fell easy prisoners. 1944 A. JACOB *Traveller's War* iv. 68, I was, in fact, slightly 'bomb happy'. 1901 F. T. BULLEN *Sack of Shakings* 18 He took..an extra supply of bomb-lances, in the use of which he was an acknowledged expert. 1917 *Chambers's Jrnl*. Sept. 590/1 This monster..was killed by a bomb-lance from a whale-boat. 1943 G. L. CHESHIRE *Bomber Pilot* i. 9 On the large map on the wall was a red line; they called it the bomblin[e], and it was supposed to represent the area behind which we could [not] bomb. 1944 *Times* 17 Apr. 3/3 The 'bomb line'—that is, the line ahead of the troops behind which aircraft supporting the ground forces should not drop their bombs. 1961 W. VAUGHAN-THOMAS *Anzio* ii. 23 We could not..pick up any feature we could recognise...We turned tail and flew disconsolately back over the bomb line. 1921 *Aeronaut. Jrnl*. Mar. 166 The bomb load of the standard..four-engined machines amounted to 3,000 kilograms. 1905 *Westm. Gaz*. 19 June 5/1 The bomb-makers..were inextricably trapped. 1943 T. HORSLEY *Find, Fix & Strike* 15 Operating from bomb-pitted aerodromes. 1917 *Advis. Comm. Aeronaut., Rep. & Mem*. No. 378 Tests were made up at the request of the Air Board, who supplied drawings of a bomb rack to carry two 112-lb. bombs. 1918 *Times* (Engineering Suppl.) 26 Apr. 88/1 The bomb-racks in the covered-in passage..are capable of holding five 25-pounder bombs. 1944 *Times* 23 May 4/2 The Spitfires..saw the German aircraft pass them at high speed with bomb-racks full. 1917 'CONTACT' *Airman's Outings* 259 A daylight bomb raid is seldom a complete failure. 1945 W. S. CHURCHILL *Victory* (1946) 29 Bomb-raid damage repairs in London. 1928 GAMBLE *Story N. Sea Air Station* v. 87 Very little work had been done with bomb releases. *Ibid*. vii. 110 Bomb-release gears. *Ibid*. xiii. 224 Bomb release-slips. 1941 *Science Digest* Nov. 53/1 The pilot controls the 'bomb run', which is the line of flight of the plane. 1944 *Hutchinson's Pict. Hist. War* 12 Apr.–26 Sept. 30/2 The deliberate bomb run through the target flak itself. 1945 H. READ *Coat of Many Colours* lxvi. 319 The bomb-shattered ruins of human tenderness and faith. 1949 M. LASKI *Little Boy Lost* III. xvi. 209 The emptiness of the bomb-shattered square. 1917 'CONTACT' *Airman's Outings* vii. 176 Owing to the difficulty of correct aim, before the advent of modern bomb-sights, all the early raids were carried out from a low altitude. 1928 GAMBLE *Story N. Sea Air Station* v. 87 Of the available bomb sights, the most practical and successful were those invented by Lieutenant Scott, U.S.N., and Zeiss. 1931 *Air Annual Brit. Empire* 230 The platform is filled with navigation and bomb-sighting equipment for the observer's use. 1945 *Daily Express* 23 May 3/3 (*caption*) British workmen from a bomb-site about 100 yards away. 1959 *Times* 8 Dec. 5/6 Many of the bomb-site parks in central London are seldom full. 1891 *Pall Mall Gaz*. 14 Dec. 5/2 The bomb-thrower, who lost his life in attempting that of Mr. Russell Sage. 1916 'BOYD CABLE' *Action Front* 24 The bomb-thrower seized the missile quickly,..threw the bomb, and jumped back under cover. 1905 *Westm. Gaz*. 13 May 7/2 The workmen [of St. Petersburg] practised shooting and bomb-throwing. 1908 *Daily Chron*. 14 Aug. 4/4 The attack from bomb-throwing airships is very little, if any, more alarming than from a gunboat.

bomb, *v*. Add: **1. b.** To attack with an explosive bomb placed or thrown for the purpose of destruction; (of aircraft) to attack with bombs from the air; to drop a bomb or bombs upon. So *to bomb one's way*: to advance by bombing; *to bomb out*: to clear by bombing; esp. in pa. pple. *bombed out*, driven by bombs out of a building, etc. Also *transf*.

1909 *Daily Chron*. 25 Feb. 1/6 Attempts had been made.. to bomb trains known to contain Europeans. 1915 *Draconian* Apr. 1683/1 They bombed us periodically during the day and night. 1916 'BOYD CABLE' *Action Front* 174 He himself had known a line bombed out. *a* 1917 E. A. MACKINTOSH *War, the Liberator* (1918) 97 When we're bombing our way up the streets of Berlin. *Ibid*. 133 He turned to bomb the big dug-out. 1940 *New Statesman* 19 Oct. 373/1 What..is happening to the host of shopkeepers.. driven from their shops..or bombed out of them?..Our ruined or bombed-out shopkeeper will pay out of his income. 1940 *War Weekly* 25 Oct. 1313 (*heading*) Story of a bombed-out family. 1969 *Daily Tel*. 17 Dec. 3/8 Hundreds of pigeons congregated in the area, ruining washing and 'bombing' children.

3. *to bomb up*: to load (aircraft) with bombs.

1939 *Flight* 28 Sept., (*caption*) 'Bombing up' a squadron of Ju 87s at Kitzingen-on-Main aerodrome. 1940 *Illustr. London News* CXCVII. 308/1 The order has been given to 'bomb up' this hardy squadron of 'Whitley' bombers. 1943 G. L. CHESHIRE *Bomber Pilot* i. 9 When we landed, the armourers were standing by to bomb up.

4. *intr*. To fail. Also const. *out*. *slang* (orig. *U.S.*).

1963 *Amer. Speech* XXXVIII. 168 To fail to pass an examination: flunk..., flag, blow, bomb. 1966 *Listener* 9 June 838/3 When a machine goes wrong it 'bombs out' and has to be 'debugged'. 1968 *TV Times* (Austral.) 10 Apr. 11/4 Everyone had expected it to be [good], so when it bombed it was a shock.

bombable (bǫ·măb'l), *a*. [f. BOMB *v*. + -ABLE.] Open to attack by bombing.

1930 *New Statesman* 9 Aug. 570/2 It is very hard to say what part of a nation's wartime industrial organisation is not bombable. 1938 *Ibid*. 24 Sept. 452/1 They [*sc*. the working class] are more numerous, live closer together, and are therefore more bombable, than other classes.

bombard, *sb*. Add: **4. b.** [Also in Fr. form.] A foot reed-stop of an organ.

1876 STAINER & BARRETT *Dict. Mus. Terms*. 1884 *Encycl. Brit*. XVII. 833/1, 32 contra trombone, posaune, bombarde, sackbut (reed).

bombard, *v*. Add: **3.** (Earlier example.)

1747 H. GLASSE *Art of Cookery* ii. 28 Bombarded veal.

4. *Physics*. To subject to a stream of ions or sub-atomic particles.

1907 J. J. THOMSON in *Phil. Mag*. XIII. 562 All gums &c. when bombarded by the rays are liable to give off gas. 1913 *Chem. Abstr*. 1442 The most abundant supply of these gases was obtained by the process of bombarding with cathode rays metals and other substances. 1932 *Discovery* May 139/1 The tiny nuclei at the centre of certain light elements were bombarded with the swift massive particles spontaneously ejected by the element polonium. 1941 *Ann. Reg. 1940* 352 Nishing and others..examined the products obtained by bombarding uranium with fast neutrons. 1969 *Times* 11 Feb. 12/4 Elements beyond uranium are not known to occur naturally, although several have been prepared in the laboratory by bombarding heavy nuclei with atomic particles in circumstances that encourage fusion.

bombardier. Add: **2. c.** A bomb-aimer in an aircraft. *U.S.*

1932 in H. I. T. CRESWELL et al. *Dict. Mil. Terms* 55. 1942 *Time* 2 Feb. 35/2 At 30,000 feet, flying 200 m.p.h., a bombardier at best has only about 60 seconds in which to locate his target.

bombardment. Add: **2.** *Physics*. Subjection to a stream of particles (see *BOMBARD *v*. 4).

1898 J. J. THOMSON *Discharge Electricity* 191 The thermal effects are readily explained on the corpuscular theory by the heating of the substance by the bombardment with the particles. 1932 *Discovery* May 139/1 From a bombardment of nitrogen, aluminium and other elements in this way, the alchemist's dream of changing one kind of element into another has been undoubtedly effected on an exceedingly minute scale. 1945 *Electronic Engin*. XVII. 669 The use of α-particles as projectiles for nuclear bombardment was responsible for the next great step forward. 1969 *Times* 5 Feb. 13/6 The Xi particles were formed in the bombardment of a tank of liquid hydrogen with very fast particles emitted by the accelerator.

Bombay (bǫmbē̆ⁱ·). The name of a city in India, used attrib. in: **Bombay chair** (cf. *Bombay furniture*); **Bombay duck** (see DUCK *sb.*[1] 10); **Bombay furniture,** a style of furniture combining European forms with Indian ornamentation; **Bombay hemp** (see HEMP 5); **Bombay pearl** (see quot.); **Bombay shell,** the bull's-mouth shell, *Cassis rufa*, used for cutting shell cameos.

1896 *Pall Mall Mag*. Mar. 399 [They] succeeded in installing themselves in two immense Bombay Chairs. 1910 *Encycl. Brit*. IV. 185/2 The workmen who manufacture the most artistic Bombay furniture are a special class with inherited traditions. 1967 M. PEGLER *Dict. Interior Design* 59 *Bombay furniture*, furniture manufactured in India after 1740...The furniture is a conglomeration of French and Portuguese styles and forms which are overlaid with typically elaborate and minute Indian carving. 1885 *Encycl. Brit*. XVIII. 447/1 Most of the products of this fishery [*sc*. in the Persian Gulf] are known as 'Bombay pearls', from the fact that many of the best are sold there. 1858 SIMMONDS *Dict. Trade*, Bombay Shells.

bombe (boṅb). *Cookery*. [Fr.; see BOMB *sb*.] A conical or cup-shaped confection, freq. frozen. Also *attrib*.

1892 T. F. GARRETT *Encycl. Pract. Cookery* I. 40/2 Apricot Bombe with Maraschino...Set two freezing-pots and a bombe-mould in some pounded ice and bay-salt. 1902 *Daily Chron*. 24 May 5/4 Fish bombes are made with any kind of cooked white fish. *Ibid*., Turn it out on a hot dish, place four little 'bombes' round it. 1960 J. BETJEMAN *Summoned by Bells* ix. 101 Claret and *tournedos*; a *bombe surprise*.

‖**bombé** (boṅbe), *a*. [Fr., pa. pple. of *bomber* to swell out; cf. BOMBED.] Having an outward swelling curve, esp. of furniture. Also in *Comb*.

1904 R. D. BENN *Style in Furniture* 109 A chest of drawers, or 'commode', and wardrobe, in which the *bombé* form is introduced. 1941 *Burlington Mag*. Feb. 59/2 The shaping of a base of a cabinet or chest of drawers (often termed today *bombé*)..was called in the eighteenth century 'comode shaped'. 1955 *Times* 21 May 5/4 Marquetry cabinet, in the French style, with slightly *bombé* front and curved and splayed sides. 1962 *Times* 1 June 8/5 A pair of eighteenth-century painted bombé-shaped Venetian commodes.

bombed, *ppl. a*. (Further example.)

1896 A. H. KEANE *Ethnology* I. viii. 185 The small flat concave [nose] is usually correlated with high cheek-bones..; the short with wide nostrils and depressed root, with everted lips and bombed frontal bone (Negro).

bomber (bǫ·məɹ). [f. BOMB *sb*. or *v*. + -ER[1].] **1.** One who throws a bomb; *esp*. in military use, one of a bombing party.

1915 J. BUCHAN *Nelson's Hist. War* V. 25 The bombers.. seizing one of these rocket-like bombs from their belts.. hurl them high above the parapet. 1927 *Daily Tel*. 8 Mar. 12/2 A fifth attempt to dynamite the Roman Catholic Church of S. Peter and Paul, San Francisco, ended yesterday in the killing of one bomber.

2. An aircraft equipped with bombs for bombing an enemy, his positions, territory, etc.

1917 'CONTACT' *Airman's Outings* 260 The fighters guard the bombers until the eggs are dropped. 1923 *Daily Mail* 19 Mar. 9 Goods-'planes—all capable of transformation into bombers. 1940 *Times Weekly* 27 Nov. 10/3 Our medium bombers were the first to arrive. 1941 *Ibid*. 30 July 7/1 For 27 hours on end we were subjected to continuous attacks from bombers—high level, dive and torpedo. 1956 A. H. COMPTON *Atomic Quest* 224 American bombers attacked..the power station.

b. *attrib*. and *Comb*., as *bomber aircraft*, *base*, *force*; **Bomber Command,** an organization of bomber aircraft forming part of the Royal Air Force.

1935 C. G. BURGE *Complete Bk. Aviation* 193/1 Bomber aircraft. 1959 *Listener* 22 Jan. 156/1 American bomber bases. 1939 *Flight* 7 Dec. 455/1 Bomber Command..pounces when a worthy objective is in sight. 1945 W. S. CHURCHILL *Victory* (1946) 174 The glorious part..played by Bomber Command in forging the victory. 1940 —— *Into battle* (1941) 231 We have a very large bomber force also.

3. A marijuana cigarette. *U.S. slang*.

1952 *Amer. Speech* XXVII. 24 *Bomber*, giant-size marijuana cigarette. 1957 J. KEROUAC *On Road* (1958) IV. v. 283 Victor proceeded to roll the biggest bomber anybody ever saw.

b. *slang*. A barbiturate drug. Cf. **black bomber*.

1962 K. ORVIS *Damned & Destroyed* xiii. 87 Last year, eighteen tons..of barbiturates were sold in this country. Kids all over are going ga-ga over them. They call them goof-balls and bombers—and they hold bomber parties. 1966 'K. NICHOLSON' *Hook, Line & Sinker* x. 116, I was planning to go back on bombers today.

‖**bombilla** (bǫmbi·lʸă). [Sp., dim. of *bomba*.] A vessel from which maté is drunk in South America.

1866 *Commissioners of Patents Jrnl*. 24 July 1094/1 Patent 1895 [for] apparatus used for imbibing or sucking in liquids, known as 'bombillas'. 1921 *Glasgow Herald* 11 Feb. 9 A bombilla manufactured specially for the South American market. 1921 *Chambers's Jrnl*. June 382/1 *Maté* circulates in a small gourd-shaped vessel, each person taking a pull at the *bombilla*, a kind of spoon-strainer perforated to let the liquid pass.

bombing, *vbl. sb*. Add: The action or operation of throwing or dropping bombs. Also *transf*.

1915 *Daily Mail* 18 May 6/4 Three Air Raids on Calais. Futile bombing by 3 Zeppelins. 1940 'GUN BUSTER' *Return via Dunkirk* II. xi. 167 The bombing had been mercilessly thorough. 1953 N. TINBERGEN *Herring Gull's World* iv. 31, I watched Hooded Crows picking up the mussels and dropping them on the road...I could not but conclude that they had concentrated their 'bombing' on the road. 1956 A. H. COMPTON *Atomic Quest* 228 These bombings were reducing Japan's fighting machine to impotence. 1968 LUND & LUDLAM *PQ17* vi. 127 During all this time the bombing had continued, hour after hour.

Also *attrib*., as *bombing attack*, *offensive*, *party*, *plane*, *raid*, *sight*, *squad*; **bombing run** = **bomb run*. Also **bombing up,** the loading of aircraft with bombs.

1942 *Times Weekly* 2 Dec. 6/3 While Stukas were occasionally used for level bombing attacks 'they did not carry out one proper dive-bombing attack during the entire campaign'. 1917 'CONTACT' *Airman's Outings* 171 After a tribute has been paid to the bombing offensives for which the Naval Air Service has always been famous. 1915 *Morning Post* 29 May 8/1 A bombing party presents a weird sight...The men carry 5 or 6 grenades each. 1934 WEBSTER, Bombing plane. 1936 *Discovery* Feb. 33/1 The Air Ministry..use them as practice ranges for bombing planes. 1917 *Flying* 14 Feb. 89/1 In bombing raids..it is necessary for aeroplanes to fly in a fixed formation. 1941 J. HAMMERTON *ABC of RAF* 23 An aspect of training for night-flying: after a practice 'bombing run' with a camera obscura the pupils check up results. 1946 *News Chron*. 2 Mar. 3/3 Trent completed his bombing run..and dropped his bombs in the target area. 1920 *Flight* XII. 744/2 An invention of bombing sights for aeroplanes. 1944 *Times* 20 Apr. 5/7 One factory turned from making fountain pens and propelling pencils to the production of aircraft bombing sights. 1923 KIPLING *Irish Guards* I. II. 125 O'Brien started to train fresh bombing-squads with the Mills bomb. 1932 *Flight* 10 June 508/1 (*caption*) *Bombing up*: Fixing a 550-lb. bomb in its rack. 1939 *Times* 3 Nov. 8/5 In another hangar the King watched a demonstration of 'bombing up'. On low trucks bombs were run under the body of the aeroplane and there by mechanical means loaded into the machine.

bombora (bǫmbō·ră). *Austral*. (and *N.Z.*). Also **bomboora.** [Aboriginal.] A dangerous stretch of water where the waves break over a submerged reef of rocks.

1933 *Bulletin* (Sydney) 24 May 27/1 'Bombora' is an aboriginal word applied to the high-crested wave which breaks, even on windless days, over submerged rocks near the coastline and in some cases at entrances to coastal harbors and inlets. 1945 BAKER *Austral. Lang*. xiii. 224 *Bomboora* to describe a dangerous reef of rocks usually found at the foot of cliffs. This is naturally confined to coastal regions. 1966 *Weekly News* (N.Z.) 19 Jan. 6/5 Surf sizes range from small 'hotdogging' waves with excellent shapes to 'bombora' surf. 1968 W. WARWICK *Surfriding in N.Z.* 34/1 Two left and right bomboras (large reef breaks), best at high tide.

bomb-proof. Add: **A.** *adj*. (Earlier and additional examples.) Also *transf*. and *fig*.

1702 T. MARWOOD in *Cath. Rec. Soc. Publ*. VII. 124 Walls & Vaults all Bombe proof [of Fort]. 1734 A. WILLIAMSON *Diary* (1912) 85 So that it is bomb proof as I believe all the arched cellers in the Towers are. 1915 G. B. SHAW in *New Statesman* 13 Mar. 559/2 In Dunkirk, Düsseldorf, Freiburg and other towns where the women and children, being foreign, are conventionally assumed by us to be naturally bomb-proof. 1918 *Flying* 6 Feb. 90/1 A bomb penetrated the building..in what was regarded as a more or less bomb-proof shelter. 1940 N. MARSH *Surfeit of Lampreys* (1941) vii. 104 Each locked up inside his mental bomb-proof shelter.

b. Not exposed to the dangers of war. *U.S.*

1867 *Harper's Wkly*. 6 Apr. 211/1 The 'bomb-proof' editors will probably continue to repeat the heroics of the war. 1868 *Putnam's Mag*. I. 715/1 During the late war the Simminses did their share of the fighting, for..none of them had influence to get 'bomb-proof' places, and keep in the rear. 1928 S. V. BENÉT *John Brown's Body* 155 Muddy Washington..full of..'Bombproof' officers, veterans back on leave.

B. *sb*. (Earlier and additional examples.)

1780 W. HEATH *Let*. 2 Feb. in *Coll. Mass. Hist. Soc*. (1905) V. 28 The bunks and lineing of the bomb proof were taken out. 1918 E. M. ROBERTS *Flying Fighter* iii. 26, I was obliged to find shelter in a bomb-proof.

b. One who avoids exposure to the dangers of war. Also *bo·mb-proofer*. orig. *U.S.*

1869 *Overland Monthly* III. 128 In the cis-Mississippi States they were generally dubbed 'bomb-proofs'. 1872 SCHELE DE VERE *Americanisms* 281 Officials who were not expected to expose themselves to the fire of the enemy, like quartermasters, commissaries, etc. were nicknamed *bomb-proofs*. 1925 FRASER & GIBBONS *Soldier & Sailor Words* 32 *Bomb-proofer*, a man given to scheming methods of evading duty on dangerous occasions. 1936 M. MITCHELL *Gone with Wind* xviii. 314 Things must be in a desperate pass if this rabble of bombproofers, old men and little boys were being called out! 1950 R. CHANDLER *Let*. 18 May (1962) 88 Doesn't he [*sc*. Partridge] overlook some of the most commonly used words of soldier-slang? E.g. 'bomb-proofer', 'cushy job', 'bivvy'.

bomb-shell. Add: Often *fig*. (or in fig. phr.). A shattering or devastating act, event, etc. Phr. *blonde bomb-shell*, a fair-haired person, esp. a woman, of startling vitality or physique.

1926 MAY SUTHERLAND *One o' the Herd* vii, Do you think it was kind to let her think she had plenty and then drop down on her like this? It's a regular bomb-shell. 1928 *Manch. Guardian Weekly* 26 Oct. 337/1 The letters do not drop any historical bombshells. 1942 BERREY & VAN DEN BARK §184/14 Blonde Bombshell (as a nickname). 1949 A. HYND *Public Enemies* 41 (Wentworth & Flexner), Bonnie Parker was a rootin', tootin', whisky-drinking blonde bombshell. 1955 M. HASTINGS *Cork & Serpent* ii. 29 'What was she like?'..'A blonde bombshell I should call her, sir.' 1951 E. COXHEAD *One Green Bottle* iii. 86 Cathy..now had standing and a soubriquet. She became the Merseyside Menace, or, alternatively, the Birkenhead Bombshell. 1965 *Times Lit. Suppl*. 25 Nov. 1072/3 The bombshell effects..of the intellectual and social crises of late antiquity.

bon, *a*. Add: **bon enfant** (bon aṅfaṅ). lit. 'good child'; an agreeable or jolly companion.

1836 H. GREVILLE *Leaves fr. Diary* (1883) 105 He is very merry and *bon enfant*, and quickly enters into conversation. 1848 THACKERAY *Van. Fair* lxiv, Look, Madame Crawley, you were always *bon enfant*, and I have an interest in you, *parole d'honneur*. 1883 *Sat. Rev*. 6 Jan. 3/1 He was always and to every one *bon enfant*.

bon gré mal gré *adv. phr*., willingly or unwillingly. (See BONGRE.)

1818 E. BLAQUIÈRE tr. *Pananti's Narr. Residence in Algiers* xii. 233 The mother is constantly in attendance, to enforce their being devoured *bon gré malgré*. 1818 LADY MORGAN *Fl. Macarthy* I. iii. 153 And now, you may depend upon it, bon gré, malgré, we shall be fated to stop at this Lis—something. 1825 H. WILSON *Mem*. I. 172 Your sister Amy..drove me, bongré malgré, to her house, and then insisted on my walking up stairs, and supping with her. 1848 TROLLOPE *Kellys & O'Kellys* II. xi. 249 They were going to drag him into the box *bongré malgré*. 1870 C. M. YONGE *Caged Lion* x. 177 She must submit, *bon gré, mal gré*, to become the wife of the Scottish prince.

bon jour (boṅ ʒūr). lit. 'good day'; a form of salutation on meeting in the daytime; hence, a civil greeting.

a 1577 GASCOIGNE *Fable of Ieronimi* in *Wks*. (1587) 260 Who after theyr Boniure dyd all seeme to lament the sicknesse of Ferdinando. 1588 SHAKES. *Tit. A*. I. i. 494 To morrow..With horne and Hound, Weele giue your Grace *Bon iour*. 1595 GOODWINE *Blanchardine* K2, After Sadony had royally presented himselfe to the veiw of all, giuing a princely boniure to the Lordes. 1823 SCOTT *Quentin D*. iii, The landlord entered,—answered Maitre Pierre's *bon jour* with a reverence. 1853 THACKERAY *Newcomes* I. v. 50 The sea being recommended to him Mrs. Newcome.. transferred him to his maternal aunt at Brighton. Then it was *bonjour*.

bon-ton. (Earlier example.)

1747 CHESTERFIELD *Let*. 1 Dec. in *Wks*. (1777) III. 173 Leipsig is not the place to give him that *bon ton*, which I know he wants.

bon-vivant. (Earlier example.)

a 1695 HALIFAX *Works* (1912) 164 The truth is, the habit of such *Bons vivants*, which is the fashionable word, maketh a suspicion so likely.

bon viveur (boṅ vivǒr). [F. *viveur* a living person.] A pseudo-French substitute for BON VIVANT.

1865 'OUIDA' *Strathmore* I. xx. 297 A cosy bachelor-villa that had been long inhabited by an old English *bon viveur*. 1888 *Athenæum* 11 Feb. 171/3 Mr. Rogers has said and eaten as many good things as those excellent *bons viveurs*.

bon voyage (boṅ vwayāʒ): see BOON *a*. 2. 'Pleasant journey': used esp. as an expression of farewell to a traveller. (Examples of use of French phrase.)

1680 R. L'ESTRANGE tr. *Erasmus sel. Colloq*. iii. 51 After this, a swinging Glass was put about, to the *Bon Voyage*. 1825 H. WILSON *Mem*. II. 149, I coolly wished him un bon voyage, and..jumped into the carriage. 1848 [see BOON *a*. 2]. 1936 *Times Lit. Suppl*. 20 June 512/4 The reader will wish her *bon voyage*, many adventures..and a safe return. 1954 W. STEVENS *Let*. 6 Apr. (1967) 824 Good luck and bon voyage!

bona fides. Add: ¶ **2.** Erroneously treated as pl. form of *bona fide* (assumed to be *sb. sing*.): guarantees of good faith.

1944 *N. & Q*. CLXXXVI. 246/1, I notice in one of our best sellers the remark, 'If Mina's *bona fides* are once questioned'. 1962 B. E. WALLACE *Device* xxvii. 195 All strangers whose bona fides are not completely verified must be immediately reported to your superior officer.

bonanza. 1. (Earlier and additional, incl. *fig*., examples.)

[1842 'A. T. MYRTHE' *Ambrosio de Letinez* II. 127 This was what Mexicans term a *bonanza*.] 1844 J. GREGG *Commerce of Prairies* I. 170 When the Placer was in its greatest *bonanza*—yielding very large profits to those engaged in the business. 1847 RUXTON *Adv. Mexico* xi. 79 The..famous black vein of Sombrerete yielded the greatest *bonanzas* of any mine on the continent of America. 1875 *Scribner's Mag*. July 272 But a bonanza with 'millions in it' is not struck every week. 1892 STEVENSON & OSBOURNE *Wrecker* ix. 139 It's not a bonanza, but there's boodle in it; we'll try it on. 1955 *Bull. Atomic Sci*. Mar. 88/2 At the time Steen's diamond-core drill struck the bonanza, he was financially hard-pressed. 1963 *Listener* 10 Jan. 98/3 The show is still, as topical entertainment, a real bonanza.

‖**bona vacantia** (bōᵘ·nă vǎkæ·ntiă). [L., ownerless goods (*vacāre* to be ownerless).] Goods without any apparent owner.

1756 W. BLACKSTONE *Comm. Laws Engl*. I. viii. 288 Besides the particular reasons before given why the King should have the several revenues of royal fish, shipwrecks, treasure-trove, waifs, and estrays, there is also one general reason which holds for them all; and that is, because they are *bona vacantia*, or goods in which no one else can claim a property. 1927 F. A. ENEVER *Bona Vacantia* i. 13 Bona vacantia are chattels real and personal which not being previously Crown property become through lack of other ownership the property of the Crown. 1959 *Listener* 23 Apr. 709/2 What is known as *bona vacantia*, that is to say, goods without an owner.

bon-bon. Add: **1.** (Earlier example.) Also *attrib*.

1796 MADAME D'ARBLAY *Camilla* III. VI. iii. 171 Clarendel, lounging upon a chair in the middle of the shop, sat eating *bon bons*. 1819 M. WILMOT *Let*. 26 Nov. (1935) 32 The

pretty papers with which the bon bon plates are covered. 1886 A. T. RICHIE *Let*. 1 Jan. (1924) x. 192 The bonbon tongs had an immense success. 1911 *Daily Colonist* (Victoria, B.C.) 29 Apr. 4/4 (Advt.), Cut Glass..Handled Bon Bon Dishes $3.00. 1964 M. LASKI in S. Nowell-Smith *Edwardian England* iv. 195 Innumerable bon-bon dishes and table napkins.

transf. and *fig*.

1856 *Farmer's Mag*. Nov. 426 A good thing, quite a bon-bon. 1955 *Times* 30 Aug. 5/4 They opened this morning with a programme of French bon-bons.

3. In full *cracker bon-bon*: see CRACKER 6 b. Also *attrib*.

1846 DICKENS *Pictures from Italy* 170 What with this green, and the intolerable reds and crimsons, and gold borders..the whole concern looked like a stupendous Bonbon. 1894 H. NISBET *Bush Girl's Rom*. 287 Gilt paper and coloured bon-bon stuff. 1901 *Daily Chron*. 10 Aug. 10/3 Frieze suits in the loveliest bon-bon shades of blue and red.

bonbonnière. (Earlier example.)

1818 LADY MORGAN *Autobiogr*. (1859) 176, I have sent you several letters and shabby little trifles:—a scarf, books, music, shoes, workboxes, bonbonnières.

bonce. Add: **2.** The head. *slang*.

1889 in BARRÈRE & LELAND *Dict. Slang* 158/1. 1909 J. R. WARE *Passing English* 42/1 Look out, or I'll fetch you a whack across the bonse. 1959 I. & P. OPIE *Lore & Lang. Schoolchildren* ix. 155 Children go in for short sharp words, as..'bonce', 'block', and 'dome' for head. 1962 L. DEIGHTON *Ipcress File* xv. 91 This threat..[is] going to be forever hanging over your bonce like Damocles' chopper.

bond, *sb.*[1] Add: **13. a.** *English bond* (examples); also *English cross bond* (see quots.).

1825 G. A. SMEATON *Builder's Pocket Man*. I. iv. 100 The principal methods of brick-laying are known under the appellation of English bond and Flemish bond. *Ibid*., The English bond is composed of alternate courses of headers and stretchers. 1842 N. WHITTOCK et al. *Bk. Trades* 75 The disposition of bricks in a building where there are alternate courses of headers and stretchers, is called English bond. 1876 *Encycl. Brit*. IV. 461/1 English bond should have preference when the greatest degree of strength and compactness is considered. 1888 C. F. MITCHELL *Building Constr. & Drawing* ii. 37 *English Cross Bond*, a class of English bond. Every other stretching course has a header placed next the quoin stretcher, and the heading course has closers placed in the usual manner. 1909 WEBSTER 251/3 *English cross bond*, called also *cross bond*, is a modification of English bond in which the stretcher courses break joints with each other. 1936 *Archit. Rev*. LXXIX. 242/3 English Cross bond is a slight deviation from pure English bond, and has a header laid, as second brick from the angle, in each alternate stretcher course; the stretchers therefore 'break joint', and there is a little more play in the pattern of the bond. 1964 C. DENT *Quantity Surveying by Computer* v. 52 One brick wall in Flettons in English bond in cement mortar.

d. *Electr*. A metallic connection between conductors forming part of an electric circuit, as between the abutting or adjoining rails of an electric railway line.

1903 [see *BONDER[1] 3]. 1904 *Westm. Gaz*. 14 Dec. 10/2 To provide electric continuity [both] are connected together by flexible strips of copper called 'bonds'.

e. *Chem*. = LINKAGE. Also *attrib*.

1884 FRANKLAND & JAPP *Inorg. Chem*. viii. 58 Each unit of atom-fixing power will be named a *bond*,—a term which involves no hypothesis as to the nature of the connexion. 1936 *Discovery* Nov. 339/1 It is convenient in chemistry to show the linking between any two atoms by means of a line or lines, commonly called bonds. 1938 *Ann. Reg. 1937* 346 The view [was] advanced that spontaneous mutations are mono-molecular reactions produced by thermal agitation when this oversteps the energy threshold of the chemical bonds. 1962 S. GLASSTONE *Textbk. Physical Chem*. (ed. 2) viii. 588 The bond energy..is the average amount of energy required to dissociate bonds of the same type in 1 mole of a given compound.

14. (sense 10) *bond-salesman*; **bond paper,** a paper of superior manufacture used for bonds and other documents; also simply *bond* in some trade-names of writing-paper; **bond washing** (see quots.); hence *bond washer*.

a 1877 KNIGHT *Dict. Mech*., Bond-paper. 1909 *Buckeye Informer* IX. 214/2 A small quantity of Japan dryer added to heavy black inks will accelerate their drying on linen and bond papers. 1952 A. CHRISTIE *Mrs. McGinty's Dead* vii. 48 She stretched up to a top shelf for notepaper and envelopes... 'Here you are, sir, that's a nice blue Bond, and envelopes to match.' 1925 F. SCOTT FITZGERALD *Great Gatsby* (1926) iii. 68, I knew the other clerks and young bond-salesmen by their first names. 1959 *Times* 8 Apr. 17/2 The bond-washer would buy shares, notably gilt-edged bonds, cum dividend and sell them ex-dividend with the gross investor reclaiming the tax. 1937 *Economist* 24 Apr. 220/1 The Chancellor proposes to abolish 'bondwashing'—a term used to describe operations by which the owner of securities sells them at a price which covers accrued dividend, and repurchases them ex dividend. 1966 *Ibid*. 29 Jan. 439/2 Bond-washing also refers to the conversion of (taxable) dividend income into (tax-free) capital gains.

bond, *v*. Add: **6.** *Electr*. To connect with an electrical bond. (See *BOND *sb.*[1] 13 d.)

1904 *Jrnl. Franklin Inst*. Apr. 287 My experience in bonding rails is that [etc.]. 1908 *Installation News* II. 103/2 Care should be taken to maintain the continuity of the run ..by means of bonding round the block with a piece of copper wire.

bonded, *ppl. a*. Add: **bonded warehouse** (earlier example).

1846 C. DICKENS *Pict. from Italy* 56 Goods brought in from foreign countries pay no duty until they are sold and taken out, as in a bonded warehouse in England.

bonder[1]. Add: **3.** One who fixes or adjusts the metallic bonds of an electric circuit. (Cf. *BOND *sb.*[1] 13 d and *v*. 6.)

1903 *Westm. Gaz*. 20 Jan. 9/2 The bonders being told off to attend to the copper bonds which make the electrical connexion between each of the three rails.

Bonderized (bǫ·ndəraiz'd), *a*. Also **bonderized.** [Trade-mark.] Of metal: coated with a patented solution that acts as a surface protection and primer. **Bo·nderizing** (bo·nderizing) *vbl. sb*., the process of coating with such a solution. Hence (as back-formation) **bo·nderize** *v*.

1932 A. R. PAGE in *Metal Ind*. XL. 369/1 The new processes are known as Parkerizing and Bonderizing. 1932 —— *Ibid*. 369/2 The..insoluble basic phosphates ..are chemically combined with the ferrous surface. The Parkerized or Bonderized surface, therefore, cannot be removed by peeling off. 1938 *Steel* 22 Aug. 42 Bonderizes automatically..Once the assembled windows..are put on the conveyor system, no hand touches them until after they have been washed, bonderized [etc.]. 1956 *Archit. Rev*. CXX. 348/1 The cabinets are bonderized aluminium. 1959 *Engineering* 13 Feb. 210/2 The steel is either bonderised or electro-zinc-coated on the reverse side.

bondholder[2]. (Earlier example.)

1823 *Drama* IV. 403 The next claim to be reduced was that of the bond-holders, to the amount of £24,416.

‖**bondieuserie** (boṅdiöʒēri). [Fr., f. *bon* good + *Dieu* God.] A collective term for church ornaments or devotional objects, esp. those of little artistic merit; an ornament or object of this kind.

1941 'R. WEST' *Black Lamb* I. 164 It is defaced by hideous bondieuseries of the modern Roman Catholic church. 1958 *Times* 22 May 13/2 The touched-up photographs..the sugary smiles on the plaster statues, all this *bondieuserie* has obscured the strength and simplicity of Thérèse Martin's character and doctrine. 1961 *Encounter* XVII. 32 The saccharine *bondieuseries* which desecrate almost every religious building one enters.

bonding, *vbl. sb*. Add: **1.** Also, the binding or connecting together of any substances esp. by adhesion. Also *attrib*. Hence *concr*., a material or substance used for bonding.

1917 *Machinery's Encycl*. V. 318/2 Bonding or junction coatings can be freely applied to brick, metal, or porcelain connections with the metal spray. 1958 *Spectator* 30 May 691/2 Its special undercoating provides a bonding of immense strength. 1962 A. NISBETT *Technique Sound Studio* iv. 84 The iron oxide is distributed throughout the bonding material. *Ibid*. 244 The two ends [of the magnetic tape] are overlapped and fixed together by an adhesive or bonding agent.

c. *Electr*. The connecting of metal parts with an electrical bond (cf. *BOND *sb.*[1] 13 d); spec. in aircraft (see quot. 1940).

1910 A. HAY *Electr. Distributing Networks* xv. 206 A third form of bonding device..consists of a tinned copper strip soldered to the lead sheathing and screwed to the side of the sealing chamber of the joint-box. 1911 *Whittaker's Electr. Engin. Pocket-bk*. (ed. 3) 527 *Bonding* may have some minor advantages, but it is better for all systems of armoured cables to be split up into sections, the metallic sheathing of each section being isolated. 1922 *Flight* XIV. 519/2 *Electrical Bonding* (in the case of aircraft fitted with *Wireless Apparatus*).—Methods of jointing—Points where bonding is necessary and position of same. 1940 *Chambers's Techn. Dict*. 102/2 *Bonding*, electrical connexion of all available metal on an aircraft, in order to give earth-capacity for wireless purposes.

bo·ndmaid, -maiden. [BOND *a*.] A slave girl. So **bo·ndservant, -service.**

1526 TINDALE *Gal*. iv. 22 Abraham had two sonnes, the one by a bonde mayde, the other by a fre woman. 1535 COVERDALE *Lev*. xxv. 44 Yf thou wylt haue bonde seruauntes and maydens. 1552 ABP. HAMILTON *Catech*. 193 Behald the boundmaidin of our Lord. 1591 SPENSER *Virg. Gnat* 489 Th'one was rauisht of his owne bondmaide. 1596 SHAKES. *Tam. Shr*. II. i. 2 To make a bondmaide and a slaue of mee. 1611 BIBLE 1 *Kings* ix. 21 A tribute of bond-seruice. 1815 SCOTT *Lord of Isles* II. xxv, Like a..bond-maid at her master's gate.

bondon (boṅdoṅ). [Fr., = bung.] A soft Neufchâtel cheese made to resemble a bung in shape.

1894 G. DU MAURIER *Trilby* I. I. 58 A little cylindrical cheese called 'bondon de Neufchâtel'. 1902 *Encycl. Brit*. XXVII. 355/2 The Neufchâtel, Gervais, and Bondon cheeses are soft varieties intended to be eaten quite fresh, like cream cheese. 1951 *Good Housek. Home Encycl*. 359/2 *Bondon*, a small, soft, whole-milk cheese made in France, shaped in the form of a bun[g].

bone, *sb*. Add: **1. e.** *spec*. One used by Australian Aborigines when pronouncing certain spells intended to cause sickness or death; cf. *BONE *v.*[1] 5.

1904 *Daily Chron*. 16 July 3/3 From one of two causes, either the abduction of a woman or the 'giving of a bone', the members of two groups will be at enmity. 1934 A. RUSSELL *Tramp-Royal* x. 78 'Milbuka point bone at Boss Tuck's bruder,' she muttered stealthily to Tuck in her native English. 1962 *John o' London's* 22 Mar. 266/1 An aborigine stockboy..is found to have been the victim of 'bone-pointing'.

2. Phr. *to* (*live to*) *make old bones*: (*a*) with negative: not to live to an old age; (*b*) to seem or feel old.

1872 C. READE *Wandering Heir* (1905) ix. 175 She..will ne'er make old bones. 1873 [in Dict.]. 1886 M. E. BRADDON *One Thing Needful* i, Lord Lashmar would never make old bones. 1924 R. H. MOTTRAM *Spanish Farm* I. 75 Poor old father, he's making old bones; it's the boys he misses. 1953 'N. SHUTE' *In Wet* viii. 259 Edward the Seventh and George the Fifth—they neither of them made old bones.

3. b. Phrases *to feel* (etc.) *in one's bones*: to have a sure intuition of (something); *to work one's fingers* (or *oneself*) *to the bone*: to work extremely hard; *near* (occas. *close to*) *the bone*: (*a*) miserly, niggardly; (*b*) hard up, destitute; also *on one's bones*, (N.Z.) *on the bone*; (*c*) 'near the knuckle' (see *KNUCKLE *sb*.).

1841 DICKENS *Barnaby Rudge* liii. 242, I seem to hear it, Muster Gashford, in my wery bones. 1844 in T. W. Barnes *Mem. Thurlow Weed* (1883) 123 It was in my bones all summer. 1853 [see WORK *v*. 19 b]. 1853 LYTTON *My Novel* V. xviii. 80 I'll work my fingers to the bone till I pay back the other five. 1855 M. THOMSON *Doesticks* xvi. 130 For I know that I am charitable; I feel it in my bones, like rheumatism. 1901 *Aberdeen Wkly. Free Press* 30 Mar. (E.D.D.), I hate yer near the bane wyes. 1912 MULFORD & CLAY *Buck Peters* iii. 57 Son, there's a big time due in these parts; I feel it in my bones. 1922 GALSWORTHY *Loyalties* I. i. 13 Ronny Dancy's on his bones again, I'm afraid. 1923 WODEHOUSE *Inimit. Jeeves* x. 100 That poor, misguided pinhead worked himself to the bone over it. 1933 in *Sc. Nat. Dict*. s.v. *bane*, He's awfu' near the bane. 1935 'J. GUTHRIE' *Little Country* x. 178 Mr. Winks, who had no other living but politics, had 'been on the bone'.., scratching an existence from commission work. 1941 A. L. ROWSE *Tudor Cornwall* xiii. 337 Charging him..with having 'two harlots begotten with child in his own house'...This was getting pretty near the bone. 1947 N. CARDUS *Autobiogr*. 9 He declined, in a family which was always living close to the bone, to take on any job. 1950 A. L. ROWSE *Engl. of Eliz*. v. 162 One sees how much nearer the bone they were in medieval circumstances.

5. b. (Earlier example used as a name for the player.)

1846 R. FORD *Gatherings from Spain* xxiii. 325 Many a performer, dusky as a Moor, rivals Ethiopian 'Bones' himself.

f. *Golf*. (See quot.)

1900 A. E. T. WATSON *Young Sportsman* 315 *Bone*, the piece of horn, vulcanite, or other material let into the sole of wooden clubs to protect the lower edge of the face.

8. Add to def.: Now usu. with negative. (Later examples.)

1885 W. E. NORRIS *Adrian Vidal* III. xxxiv. 117, I didn't quite like to draw out my money so long as Pilkington held on; but I shall make no bones about it with this fellow. 1955 *Bull. Atomic Sci*. Sept. 256/1 On the other hand, Dr. Libby makes no bones about the catastrophe of a nuclear war.

12. b. Usu. *pl*. The 'skeleton' of a novel, play, or other literary work; basis of literary style; *the bare bones*: the mere essentials.

1888 *Sat. Rev*. 15 Dec. 714/2 There are 'the bones of' something like a novel of some merit in *The Jewel Reputation*. 1905 *Westm. Gaz*. 23 Mar. 7/3 Counsel did not allege that Mr. Tanner had copied plaintiff's dialogue, but the 'bone' was the same. 1915 J. BUCHAN *39 Steps* iv. 81 The bare bones of the tale were all that was in the book. 1928 *Publishers' Weekly* 9 June 2373 My own bed-book is Mrs. Gaskell's 'Cranford', and as I read it again that night I could find no bones in it at all. 1962 A. NISBETT *Technique Sound Studio* xiii. 233 He will need time..to present the bare bones of the argument.

c. A hardness of the ground due to frost.

1906 *Westm. Gaz*. 23 Feb. 5/1 The night's frost had left a great deal of 'bone' in the ground. 1927 *Observer* 18 Dec. 25/2 The ground had been protected by straw, but there seemed to be a good deal of bone in it.

14. c. A dollar. *N. Amer. slang*.

1896 ADE *Artie* ii. 10, I guess I saw as much as two bones change hands. 1921 *Daily Colonist* (Victoria, B.C.) 6 Apr. 4/3 Tory and Grit, Tom Uphill too, Declared that they were speaking true, Two thousand bones to each was due, Or else they would be scabbing.

16. (Selected additional examples.)

a 1825 R. FORBY *Vocab. E. Anglia* (1830) I. II. 32 *Bone-dry*, adj. perfectly dry; as dry as a bone long bleached in the weather. 1896 KIPLING *Seven Seas* 73 Bone-bleached my decks. 1919 'BOYD CABLE' *Old Contemptibles* xvi. 268 They was like a good long drink to a bone-dry man. 1920 R. GRAVES *Country Sentiment* 51 Honest men..with glaring eyes, Bone-chilled. 1924 A. J. SMALL *Frozen Gold* i. 23 The bone-hard stamp of starvation. 1926 J. S. HUXLEY *Ess. Pop. Sci*. xviii. 278 A bone-structure in the foot which is still more or less adapted to an arboreal life. 1936 L. B. LYON *Bright Feather Fading* 24 One bone-grey sun-ray between dreamer and dreamer! 1939 D. CECIL *Young Melbourne* vi. 171 Ghastly pale, bone-thin..she looked insane. 1951 S. SPENDER *World within World* 229 A full moon..exposing walls of bone-coloured palaces. 1954 J. R. R. TOLKIEN *Two Towers* IV. iii. 253 Arms and legs almost bone-white and bone-thin. 1959 *Listener* 19 Mar. 516/3 The bone-hard ground. 1969 *Jane's Freight Containers 1968–69* 530/2 At one container exhibition, York light-heartedly moored a standard container at sea for a week. It was found to be bone-dry inside when retrieved.

d. Used adverbially, 'to the bone'; used as an intensive, usu. with adjs., as *bone-idle*, *-lazy* (so *-laziness*), *-tired*, etc.

a 1825 R. FORBY *Vocab. E. Anglia* (1830) I. II. 32 Bone-

lazy, bone-sore, bone-tired, adj. so lazy, sore, or tired, that the laziness, the soreness, or the fatigue, seem to have penetrated the very bones. 1836 CARLYLE *New Lett.* (1904) I. 8 For the last three weeks I have been going what you call *bone*-idle. 1891 KIPLING *Light that Failed* vi. 98 Bone-idle, is he? Careless, and touched in the temper? 1849 G. E. JEWSBURY *Let.* 29 Mar. in *Sel. Lett. to J. W. Carlyle* (1892) 287, I am 'bone lazy', as my nurse used to phrase it. 1899 C. J. CUTCLIFFE HYNE *Further Adv. Capt. Kettle* vii. 136 This the bone-weary crew were but feebly competent to give. 1912 A. S. M. HUTCHINSON *Happy Warrior* I. i. § 3 Egbert was bone-tired. 1925 M. WILTSHIRE *Thursday's Child* iii. 72 It's nothing in the world but bone-laziness that makes you shy at it. 1939 G. GREENE *Confid. Agent* I. ii. 78 That doesn't mean a thing to me... I'm bone-ignorant. 1942 'R. WEST' *Black Lamb* II. 478 Lazy, bone-lazy, they wish to believe that life is lived simply by living. 1965 G. MCINNES *Road to Gundagai* x. 178 The Baas..strode up and down the line of bone-tired scouts.

17. bone-ache, -ague, add to def.: *spec.* venereal disease; delete † before *bone-ache* and add earlier and later examples; **bone-bed** (earlier example); **bone-boiler,** a workman who performs the operation of boiling or steaming bones used in the manufacture of glue, bone meal, etc.; **bone-brown** (example); **bone-cell, -corpuscle,** an osteoblast; **bone china,** china-ware made of clay mixed with bone-dust or phosphate of lime; **bone-earth** (examples); **bone-fat,** fatty matter extracted from fresh bones for use in the manufacture of soap, etc.; **bone-flour,** a flour obtained by sifting ground bones, used as a fertilizer; **bone-forceps,** a pair of short strong shears for cutting bone during surgical procedures; **bone-head** *slang* (orig. *U.S.*), a block-head; also *attrib.* or as *adj.*; **bone-headed** *a. slang* (orig. *U.S.*), thick-headed, stupid; hence **bone-headedness**; **bone-heap,** a heap of bones; *spec.* in *Archæol.*, a refuse pile of bones, etc., of a prehistoric village (*Funk's Standard Dict.* 1893); also *fig.*; **bone (manure) man** (see quot. 1921); **bone-marrow** = MARROW *sb.*[1] 1; **bone-meal** (orig. *U.S.*), the coarser siftings of ground bones, used as a fertilizer; **bone phosphate,** a commercial name for tricalcium phosphate, the phosphate that forms bone-tissue; **bone-picker** (example); **bone-pit** *U.S.* (see quot.); **bone porcelain** = **bone china*; **bone powder,** powdered bone, used as a fertilizer; **bone-saw,** a surgical saw for cutting bone; **bone-seed** (see quot.); **bone-seeking** *a. Med.*, tending to be deposited in the bones following entry into the animal body; hence **bone-seeker**; **bone-shark,** the basking-shark; **bone-throwing** *vbl. sb.*, the throwing of bones; *spec.* that practised by some primitive peoples as a means of divination; hence as *ppl. a.*; so **bone-thrower**; **bone turquoise** (see TURQUOISE 4).

1398 TREVISA tr. *Bartholomeus Anglicus's De Propr. Rerum* 63 a (M.E.D.), Þe boneache is I rotid. *c* 1520 SKELTON *Magnyfycence* 1907 To cry out of the bone ake. 15.. BECON *Jewel of Joy* in *Catechism*, etc. (1844) 464 Grieved with bone-ache. 1900 *Daily News* 15 Nov. 6/5 He was attacked with headache, bone-ache, lassitude, [etc.]. 1841 LYELL *Elem. Geol.* (ed. 2) II. xxii. 82 There intervenes..a dark-coloured stratum, well known by the name of the 'bone-bed'. 1843 *Civil Eng. & Arch. Jrnl.* VI. 216/2 Any trade or business such as..bone-boiler. 1906 *Daily Chron.* 26 May 2/7 Bone boilers and tallow melters. 1895 *Montgomery Ward Catal.* 252/3 Artists Tube Oil Colors ..Blue Black—Bone Brown—Brown Pink. *a* 1847 *Todd's Cycl. Anat.* III. 856/1 The bone-cells..form the outer layer of cells in the Haversian system. 1903 L. M. E. SOLON *Old Eng. Porcelain* 220 Josiah Spode..composed a new china body which..from the nature of its chief constituent.. received the vulgar name of 'Bone China'. *Ibid.*, This evergreen 'bone china' has remained unaltered ever since the first pieces of it came out of Spode's oven. 1965 FINER & SAVAGE in *J. Wedgwood's Lett.* 179 It abandoned true porcelain for a bone-china body. 1875 *Encycl. Brit.* I. 854/1 The lacunæ look like solid, black bodies, and..were erroneously called by the earlier observers bone-corpuscles. 1851 *Fraser's Mag.* Feb. 246/2 They have a cheap substitute in superphosphate of lime, a soluble form of bone-earth. 1886 A. H. CHURCH *Eng. Porcelain* iii. 29 There can be no difficulty in identifying the earth produced by the calcination of certain animal and vegetable matters with bone-earth, that is, calcined bones which consist mainly of phosphate of lime. 1873 SPON *Workshop Rec.* Ser. I. 373/2 For purifying bone fat, melt the fat and a small quantity of saltpetre together. 1887 *Jrnl. Soc. Chem. Industry* VI. 825/1 Loss of Nitrogen in the Manufacture of Bone-fat and the Analysis of Bone-fat. 1888 *Ibid.* VII. 81/2 The bones..are first broken up more or less finely, and go to produce what are known as—½ inch bones. ¼ inch bones. Crushed bones. Bonedust. Bonemeal. Bone flour. 1879 *St. George's Hosp. Rep.* IX. 259 One of these presented a bony growth..the end of which was cut off with bone-forceps. 1908 C. DRYDEN in *Chicago Daily Trib.* 24 Sept. 12/1 Then came the bone-head finish which left the bugs puzzled and wondering. 1909 R. BEACH *Silver Horde* xx. 271 What's the use?.. That bone-head wouldn't understand! 1915 J. LONDON *Let.* 5 Nov. (1966) 463 Now, why be serious with this bone-head world? 1917 CONAN DOYLE *His Last Bow* viii. 292 James was a bonehead—I give you that. 1958 J. & W. HAWKINS *Death Watch* (1959) iii. 82 The best of us have made a bonehead mistake or two. 1903 *Smart Set* IX. 96 You talk like a bone-headed fool! 1915 WODEHOUSE *Something Fresh* v. 149 You blanked bone-headed boob! 1923 —— *Adv. Sally* xiii. 212 The wilful bone-headedness of our fellows. 1940 'G. ORWELL' *Inside Whale* 142 The ancient boneheap of Europe, where every grain of soil has passed through innumerable human bodies. 1899 *Daily News* 21 July 5/2 Defendant gave instructions for the bone man to take away the bad meat. 1921 *Dict. Occup. Terms* (1927) § 149 *Bone manure man*, general term for any unskilled worker, other than a maintenance man, employed in a bone manure factory. 1908 W. E. C. DICKSON (*title*) The Bone-Marrow. 1927 HALDANE & HUXLEY *Anim. Biol.* ix. 189 An extra production of red blood-corpuscles by the bone-marrow. 1850 *New Engl. Farmer* II. 44 On Mr. Preston's farm..they began to use bone-meal. 1933 *Jrnl. R. Hort. Soc.* LVIII. 119 Prepared soil..may consist of equal parts of loam and leaf-soil to which has been added some bone-meal. 1849 CRAIG, Bone-phosphate. 1871 ROSCOE *Elem. Chem.* 219 Calcium phosphate, or bone phosphate. 1861 MAYHEW *Lond. Labour* Extra vol. (1862) 314/2 The people who usually lodge here are crossing-sweepers, bonepickers, and shoeblacks. 1872 SCHELE DE VERE *Americanisms* i. 25 In the State of New York and in Canada there are..many places..where the Indians buried their dead, and these are known as *bonepits*. *a* 1884 KNIGHT *Dict. Mech.* Suppl., *Bone Porcelain*, a ware into the composition of which enters phosphate of lime in the form of bone dust. 1888 *Jrnl. Soc. Chem. Industry* VII. 133/1 The bone-powders of commerce are not always products of manufacture solely derived from the grinding of bones. *a* 1884 KNIGHT *Dict. Mech.* Suppl., Bone Saw. 1908 HARDY *Dynasts* III. VII. v, A surgeon's horse..laden with bone-saws,..and other surgical instruments. 1866 LINDLEY & MOORE *Treas. Bot.*, *Bone-seed*, the common name for *Osteospermum*. 1948 M. HELLER in W. Bloom *Histopathol. of Irrad.* 101 In young rats treated with strontium 89 (and probably also other 'bone seekers') an initial cessation of growth was followed by a period of growth resumption. 1955 *Sci. Amer.* Aug. 37/1 Most fission products are known as 'bone-seekers': they tend to concentrate in the skeleton. 1947 *Radiology* XLIX. 347/1 In the case of rats, the effect of the bone-seeking isotopes is a bit different. 1802 in *Mass. Hist. Soc. Coll.* VIII. 199 There is a large shark in the harbour, named the bone-shark, and similar in shape to the man-eating shark, but harmless. 1917 *Chambers's Jrnl.* Sept. 590/1 The tiger shark of the Indian Ocean and the 'bone shark'. 1953 P. ABRAHAMS *Return to Goli* v. 180 In desperation she went to a 'bone-thrower'. 1927 E. V. GORDON *Introd. Old Norse* 185 The Danish practice of bone-throwing is heard of in actual history. 1931 J. MOCKFORD *Khama* xxxiii. 236 First he had found it necessary to fight with his bone-throwing father.

bone, *v.*[1] Add: **4.** *intr.* To study hard; to apply oneself diligently or determinedly. Freq. with *down*, *in*, and esp. *up* (*on*). *slang* (orig. and chiefly *U.S.*).

1841 GREELEY in W. M. Griswold *Corr. R. W. Griswold* (1898) 53 Webb..has been round boring every big-bug in the State to bone for him. *a* 1861 T. WINTHROP *Life in open Air* etc. (1863) 148 We was about sick of putty-heads and sneaks that..didn't dare to make us stand round and bone in. 1883 H. A. BEERS in *Century Mag.* June 273/2 I'm going to bone right down to it. 1887 E. B. CUSTER *Tenting on Plains* (1889) ix. 286, I have known the General to 'bone-up', as his West Point phrase expressed it, on the smallest details of some question at issue. 1959 *Punch* 14 Oct. 309/1 The Wrens..boned up on Russian to be ahead for the next war. 1968 *Daily Tel.* (Colour Suppl.) 29 Nov. 67/4 Mr Robert Powell,..who is on the set as technical adviser but who wastes no opportunity to bone up on his hobby—Romanesque architecture.

5. *trans.* To point a bone at (someone) as part of a spell or curse; cf. *BONE *sb.* 1 e.

1904 *Daily Chron.* 16 July 3/3 They will be 'boned' by men of the moiety of the tribe. 1934 A. RUSSELL *Tramp-Royal* x. 80 One of their number 'boned' him; that is, pointed the magic bone at him. Eight weeks later he was dead. 1936 M. FRANKLIN *All that Swagger* 384 The whole world is paralysed by the mumbo-jumbo of banking jargon, like a binghi 'boned' by a medicine man.

boneen (bǫnī·n). *Irish.* Also **bonyeen.** [ad. Ir. *bainbhín* sucking-pig, f. *banbh* pig + *-ín* dim. suffix.] A young pig.

1841 LEVER *C. O'Malley* lxxxv, What's that you have dragging there behind you? A boneen, sir. 1855 MUNDY *Antipodes* (ed. 3) xviii. 438 The wretched shieling of poor Paddy, with his dudeen, his caubeen, his boneen. 1892 EMILY LAWLESS *Grania* I. iv. 22 The relative number of cows, turkeys, feather-beds, boneens, black pots and the like. 1892 YEATS *C'tess Kathleen* iii. 47, I in image of a nine-monthed bonyeen. 1949 E. POUND *Pisan Cantos* lxxx. 106 A sow with nine boneen.

bone-fish. Chiefly *U.S.* [BONE *sb.*] A name applied to various fishes (see quots.).

1734, 1809 [see BONE *sb.* 17]. 1884 GOODE *Nat. Hist. Aquatic Anim.* 279 The 'Bone-fish' of Key West, according to Stearns, belongs, to this [sturgeon] family and genus. *Ibid.* 612 With us it [*sc. Albula vulpes*] is usually called the 'Lady-fish'; in the Bermudas the 'Bone-fish', or Grubber. *Ibid.* 674 In Southern New England this fish [*sc.* dogfish] is called the 'Bone-fish'. 1897 *Outing* (U.S.) XXIX. 331/1 The bone-fish somewhat resembles a whiting in shape, with the mouth of a sucker and no teeth.

bonelessness (bōᵘ·nlėsnės). [f. BONELESS *a.* + -NESS.] Boneless condition. Also *transf.* and *fig.*

1885 *Spectator* 20 June 805 They weary of the besetting defect of modern English statesmen, bonelessness. 1895 *Athenæum* 2 Mar. 289/1 The partial bonelessness..is [perhaps] due to them. 1928 *Daily Express* 9 Nov. 15/1 The softness of this chair!..And its bonelessness and bendability!

boner[2] (bōᵘ·nəɹ). *slang* (orig. *U.S.*). [f. BONE *sb.* + -ER; cf. **bone-head*.] A mistake, a blunder.

1912 *Amer. Mag.* June 200/1 Boner—a stupid play; a blunder in the science of the game. 1913 *Ibid.* Sept. 94/3 Got his signals mixed and pulled a boner. 1933 *Passing Show* 9 Dec. 9/3 Poor Carol... She made a boner to-night... Ronnie was simply livid. 1960 *Spectator* 7 Oct. 509 This Government has made about every boner possible.

bone-yard, boneyard. orig. *U.S.* [BONE *sb.*] **a.** A yard or place where the bones of dead animals are collected for subsequent use; **b.** (see quot. 1903); **c.** a cemetery (*slang*).

1854 J. R. PLANCHÉ *Once upon Time* I. i in *Extravag.* (1879) 331 Or in a bone-yard with some knacker dwell. 1866 'MARK TWAIN' *Lett. fr. Hawaii* (1967) 277 It's one of them infernal old ancient graveyards... Yes, likely. I suppose you didn't know that bone yard was there. 1902 W. J. LONG *Beasts of Field* 70, I have met men..who speak of 'bone yards' which they have discovered... They say that the caribou go there to die. 1903 *N.Y. Even. Post* 7 Oct. 7 A dilapidated horse saved from the bone-yard. 1945 R. FINLAYSON in F. Sargeson *Speaking for Ourselves* 17 He wouldn't have gone home late at night past the old bone-yard.

bong (bǫŋ). *U.S.* [Imitative.] 'A deep resonant sound as of a bell' (Webster 1934). Hence as *v.*, to emit such a sound.

1924 in *Dial. Notes* V. 263. 1936 STEINBECK *In Dubious Battle* vi. 94 The pump bonged with a deep, throaty voice. 1960 V. NABOKOV *Invit. Beheading* iv. 43 The merciless bong of the clock. 1961 STEINBECK *Winter of Discontent* 181 The clock bell of the firehouse began bonging.

bong, var. *BUNG *a.*[2]

bongo[1] (bǫ·ŋgo). [Cf. Bangi *mbangani*, Lingala *mongu*.] An antelope, *Boocercus eurycerus*, found in central Africa, from Sierra Leone to Kenya.

1861 P. B. DU CHAILLU *Equat. Afr.* xvi. facing p. 306 (*caption*) The Bongo Antelope...The chief features of the animal are the stripes on each side. 1902 O. THOMAS in *Ann. & Mag. Nat. Hist.* X. 309 No evidence as yet exists as to whether the true western Bongo has horns in the female. 1910 *Westm. Gaz.* 1 Mar. 11/1 Next come nine white rhinoceroses and a couple of bongos, a specimen of the latter animal never before having fallen to the gun of a white man. 1911 ROOSEVELT in Ld. Charnwood *T. Roosevelt* (1923) 243 He had killed a bongo, a bull. 1958 E. S. WARNER *Silk-Cotton Tree* xvii. 175 The chiefs..were announced..with great blasts blown on bongo horns. 1964 E. P. WALKER et al. *Mammals of World* II. 1418/1 Bongos live in the densest, most tangled parts of the forest.

bongo[2] (bǫ·ŋgo). orig. *U.S.* Pl. **bongo(e)s.** [Amer. Sp. *bongó*.] One of a pair of small (Cuban) drums, usu. held between the knees and played with the fingers; in full *bongo drum*.

1920 J. HERGESHEIMER *San Cristóbal de la Habana* 232 My head filled with the resonant bos and bongos of ñañiguismo. 1928 *Vanity Fair* Nov. 72 A fashionable evening event along the Havana water-front is a concert by black boys with their primitive instruments, the bongó..and claves. 1934 S. R. NELSON *All about Jazz* 167 Then we have the Bongos, which are a kind of tom-tom, used in pairs. 1952 *New Yorker* 1 Nov. 6/2 Candido, whose peculiar fancy is bongo drums. 1956 *Ibid.* 11 Feb. 92/2 She's been doing her calypso act—with bongo accompaniment—since 1941.

bonham. *Irish.* [dial. var. (in Ulster and Connaught) of Ir. *banbh* pig.] A sucking-pig. Cf. *BONEEN.

1880 W. H. PATTERSON *Gloss. Antrim & Down* 11 Bonham. 1938 J. CARY *Castle Corner* v. 279 She's as fat as a bonham.

‖ **bonheur du jour** (bonœr dü ʒūr). [Fr.] A small writing-table or desk, sometimes fitted to hold toilet accessories.

1878 B. PALLISER tr. *Jacquemart's Hist. Furnit.* I. iii. 66 In small objects for ladies, such as 'bonheurs-du-jour' étagères, work-tables, nothing more elegant can be seen. 1902 *Provisional Catal. Furnit. etc. in Wallace Coll.* 283 Small Bureau of the 'Bonheur du Jour' type, in marqueterie of various natural and stained woods. 1926 H. SCHMITZ *Encycl. Furnit.* xxii. 53 The writing-table for ladies, known as the *bonheur du jour*, receives a rectangular upper part, the panels of which imitate book backs. 1955 *Times* 21 May 5/4 A magnificent Louis XVI black lacquer *bonheur-du-jour*, on slightly square, curved and tapering legs. 1967 *Times* 7 Mar. 21/5 (Advt.), A Hepplewhite bonheur du jour.

bonhomous (bǫ·nǒməs), *a.* Also (*rare*) **bonhomm(i)ous.** [f. BONHOM(IE + -OUS.] Full of bonhomie or good-fellowship. Hence **bo·nhomously** *adv.*

1905 *Spectator* 18 Feb. 257/2 A delightful bonhommious person. 1914 C. MACKENZIE *Sinister Street* II. III. v. 593 'He's not a very bonhomous lad,' said Lonsdale. 1927 *Observer* 10 July 6 The hearty and bonhomous J. D. Marstock. 1928 E. WAUGH *Decline & Fall* x. 109 Lady C.'s hardly what you might call bonhommous. 1960 *Spectator* 5 Feb. 185 An air of bonhomous candour. 1965 *Economist* 5 June 1126/1 The American Secretary of Defence departs ..waving bonhomously.

bonism (bǫ·niz'm). [f. L. *bonus* good + -ISM.] The doctrine that the world is good, but not the best possible. So **bo·nist**, one who favours this doctrine; hence **boni·stic** *a.*

1882 J. W. BARLOW *Ultim. Pessimism* 5 So we see that Optimist and Pessimist are no longer suitable names..; and the positive forms Bonist and Malist would certainly be more appropriate. 1893 MYERS *Sci. & Future Life* 10 The view of the universe loosely styled optimism, but which some now term *bonism*, with no greater barbarism in the form of the word, and more accuracy in its meaning. 1895 TOLLEMACHE *Jowett* 91 Jowett's optimism verges on pessimism, or, let us say, his *bonism* verges on *malism*. *Ibid.* 95 After putting side by side the bonistic and malistic sayings of Jowett.

bonkers (bǫ·ŋkəɹz), *a. slang.* [Origin unknown.] **a.** (See quot. 1948). **b.** (The usual current sense.) Mad, 'crackers'.

1948 PARTRIDGE *Dict. Forces' Slang* 23 *Bonkers*, light in the head; slightly drunk. (Navy.) Perhaps from *bonk*, a blow or punch on the *bonce* or head. 1957 J. OSBORNE *Entertainer* v. 37 One Catholic poet who went bonkers. 1961 SIMPSON & GALTON *Four Hancock Scripts* 60/2 By half-past three he'll be raving bonkers. 1967 *Spectator* 4 Aug. 130/3 I think you're bonkers 'cos your attitude's all wrong. You're more than bonkers. You're sick.

Bonnaz (bǫ·næz). [Name of J. *Bonnaz*, of Lyons, inventor (1863) of a chain-stitch embroidery machine.] A kind of machine-made embroidery.

1881 *Instr. Census Clerks* (1885) 75 Bonnaz Machinist. 1903 *Daily Chron.* 2 Mar. 9/6 (Advt.), Bonnaz Machinists... Bonnaz Braiding Machines. 1921 *Dict. Occup. Terms* (1927) § 278 Bonnaz hook maker.

bonne. Add: **C. bonne femme** (bon fam). [F. *bonne* good, *femme* wife.] Also (*à la*) *bonne femme*. In the manner of a good housewife: applied esp. postpositively to food prepared in various ways.

1824 BYRON *Don Juan* XV. lxiii. 36 There was a goodly 'soupe à la *bonne femme*'. 1869 M. JEWRY *Warne's Model Cookery* 35/2 Eggs à la Bonne Femme. 1959 R. POSTGATE *Good Food Guide* 148 Sole bonne femme.

bonne fortune (bon fǫrtün). [Fr.] A lady's favours, as a thing to boast of or pride oneself on; also *transf.*

1823 BYRON *Don Juan* XIV. lxiv. 149 'Tis best to pause, and think..If that a '*bonne fortune*' be really '*bonne*'. 1825 H. WILSON *Memoirs* I. 314 Beauclerc was, in due time, tired of his bonne fortune. *Ibid.* II. 86 She flattered their vanity with the idea, that her acquaintance was an unusual bonne fortune. 1939 *Times Lit. Suppl.* 4 Nov. 637/1 London drawing-rooms, whose *bonnes fortunes* were many and gratifying even by Regency standards.

bonner (bǫ·nəɹ). *University slang.* [f. BONFIRE + *-ER[6].] A bonfire. *Obsolescent.*

1898 *Oxford Mag.* 19 May, While the Bonner burns, make haste to pass The Revellers firework-scattered on the Grass. 1898 *Westm. Gaz.* 6 Dec. 10/1 The authorities of a certain college gave permission for the turf of the quadrangle to be taken up in order to make better provision for the 'bonner'. 1927 *Daily Tel.* 15 Mar. 10/4 We should be rightly apprehensive if we were to hear of no more 'rags' or 'bonners'.

bonnet, *sb.* Add: **f.** A hinged cover over the engine (in some makes over the luggage compartment) at the front of a motor vehicle.

1902 HARMSWORTH *Motors & Motor-Driving* 181 If you have a flare-up,..take off bonnet to save the paint, and smother the flames. 1904 YOUNG *Compl. Motorist* 111 The engine, instead of being placed vertically under a bonnet in front of the car, lies horizontally immediately in front of the dash-board. 1955 *Times* 21 June 5/4 The forward view from the driving seat is better than the rather high bonnet would suggest.

10. *bonnet-box* (earlier example); **bonnet-cap,** a cap worn inside a bonnet.

1834 DICKENS *Sk. Boz* (1836) 1 Ser. I. 152 Trunks, bonnet-boxes, muff-boxes. 1835 *Ibid.* 318 The derangement of their curls and bonnet-caps. 1872 GEO. ELIOT *Middlem.* VIII. lxxiv. 212 She brushed her hair down and put on a plain bonnet-cap, which made her look suddenly like an early Methodist.

bonnet, *v.* Add: **4.** To provide (a safety lamp) with an iron bonnet or shield.

1901 *Jrnl. Inst. Elect. Engin.* XXX. 834 In appearance it is a bonneted Mueseler lamp.

bonobo (bǫ·nobo). [Native name for the animal.] The pygmy chimpanzee, *Pan paniscus*, found south of the Congo river.

1954 *Zool. Record* XVIII. 101/2 (*summary of German paper*) Generic separation of P[*an*] t[*roglodytes*] *paniscus* as *Bonobo* not justified. 1962 *Internat. Zoo Yearbk.* IV. 77/1 The bonobo comes exclusively from south of the great curve of the Congo. 1966 R. & D. MORRIS *Men & Apes* viii. 255 The strange bonobo, or pygmy chimpanzee..was originally described in 1929 as a small, black-faced race of the common chimpanzee.

‖ **bonsai** (bǫ·nsai, bōᵘ·nsai). [Jap.] A Japanese potted plant or small tree, intentionally dwarfed; the method of cultivating such a plant. Also *attrib.*

1950 N. KOBAYASHI (*title*) Bonsai—miniature potted trees. 1960 KOESTLER *Lotus & Robot* II. vii. 191 The cultivation of bon-sai, that is, trees planted in pots..and made to grow into genuine dwarfs. 1960 *Times* 4 Jan. 13/4 Sales of bonsai plants. 1963 *Times* 6 Feb. 12/3 *Bonsai* is now attracting many admirers in Britain.

bonsense (bǫ·nsēns). [f. BON *a.* + SENSE *sb.*, after *nonsense*.] Good sense.

1681 S. COLVIL *Whiggs Supplication* II. 38 A *Cerberus* with three-heads: Neither of which barks any bon-sence. 1714 CRAIG in R. Smith *Poems of Controversy* 11 As to thy Poetry its nonsense, and therein not a word of bonsense. 1900 H. HARLAND *Cardinal's Snuffbox* xxi, Nonsense or bonsense, that is the sober truth.

‖ **bontbok** (bǫ·ntbǫk). = BONTEBOK.

1959 *Cape Argus* 22 Aug. 8/8 Moving the..herd of bontbok to its new pasture. 1960 *Times* 30 Mar. 9/4 The well-known Bredasdorp Bontbok herd.

bont tick. Also 9 **bonte tick.** [Afrikaans, f. Du. *bont* variegated + TICK *sb.*[1]] A South African tick, *Amblyomma hebræum*, parasitic on cattle, ostriches, and man.

1881 DOUGLASS *Ostrich Farming* 17 The large Bonte tick that..produces terrible sores on all animals. 1900 C. P. LOUNSBURY in *Proc. Assoc. Econ. Ent.* 41 (Cent. Dict. Suppl.), The tick of greatest importance, because of its injuries to stock, is *Amblyomma hebræum* Koch, commonly known as the bont tick. 1952 *Cape Times* 20 Feb. 2/6 The dreaded 'heartwater' disease which was caused by the bont tick.

bony, *a.* Add: **2. b.** Of coal: containing a considerable amount of slate or shale. *U.S.*

1857 *Harper's Mag.* Sept. 463/1 Much of the slate and 'bony coal' that occurs in the vein is separated.

3. bony-fish *U.S.*, the menhaden or moss-bunker (*Brevoortia tyrannus*).

1815 *Trans. Lit. & Philos. Soc. N.Y.* I. 453 Bony-fish, Hard-heads, or Marsbankers. 1871 *Game Laws N.Y.* in *Fur, Fin & Feather* (1872) 21 Bony fish, or moss bunkers.. are exempted from the operation of this section. 1884 L. F. ALLEN *New Amer. Farm Bk.* 80 The moss-bonker, or bony-fish [etc.]..are caught in seines, and sold to the farmers by the wagon load.

bony (bōᵘ·ni), *sb. U.S. mining.* [f. the adj.] = BONE *sb.* 13.

1874 RAYMOND *Statist. Mines & Mining* 39 It [*sc.* the coal] is interstratified with sand-rock and shale. In some of the mines the roof consists of a mixture of the two, called by the men 'bony'. *Ibid.* 41 The Black Diamond vein has for roof and floor shale, slate, and 'bony'.

bonzer (bǫ·nzəɹ), *a. Austral.* (and *N.Z.*) *slang.* Also **bonza, bonser,** etc. [Perhaps alteration of BONANZA.] Excellent, extremely good. Also *absol.* as *sb. Obsolescent.*

1904 *Bulletin* (Sydney) 14 Apr. 29 A bonser or bonster is comparatively superior to a bons. 1906 E. DYSON *Fact'ry 'Ands* xiv. 182 What's er bonzer like you doin' spreadin' sour paste fer yer daily? *Ibid.* xviii. 246 'E had er bonzer nose. He'd fair run t' nose. 1915 C. J. DENNIS *Sentimental Bloke* 18 An' on the air a sad, sweet music breaves A bonzer song. 1922 *Glasgow Herald* 8 Nov. 8, I have..heard Australians described as 'Bonzers', probably owing to the way they have of overworking that strange compound. It is used in the various States in place of the adjectives good, great, and splendid, as, for example, 'Massenger is a bonzer player; he scored a bonzer try.' 1927 FRANCIS BRETT YOUNG *Portrait of Clare* 653 She came from Toowoomba, in Queensland, a bonza spot. 1936 F. CLUNE *Roaming round Darling* xxiv. 244 It's bonzer in the water. Aren't you coming in? 1947 D. M. DAVIN *Gorse blooms Pale* 200 An Iti bint..with bonzer black eyes. 1949 F. SARGESON *I saw in Dream* viii. 60 A bonzer afternoon. 1969 *Advertiser* (Adelaide) 30 Dec. 1/3 (*heading*) 1970? She'll be bonzer.

Bonzo (bǫ·nzo). [Arbitrarily formed by Capt. Bruce S. Ingram, editor of 'The Sketch'.] The figure of a comically-shaped puppy which came into vogue through a series of drawings by G. E. Studdy (the first of which appeared in 'The Sketch' 8 Nov. 1922), and is used in various forms, as toys, etc. Hence (*trivial*) **Bo·nzoid** *a.*

1927 *Bulletin* 11 Oct. 5/4 The craze for vanity bags in the form of Teddy Bears, Bonzos, and other zoological specimens. 1928 GALSWORTHY *Swan Song* II. vi. 152 You're 'for it', as they say in this Bonzoid age. 1934 'G. ORWELL' *Burmese Days* ii. 22 For ornament there were a number of 'Bonzo' pictures. 1949 A. WILSON *Wrong Set* 40 A side table ..embellished with china models of Bonzo and Felix the cat.

bonzoline (bǫ·nzǒlīn). A composition used as a substitute for ivory in the manufacture of billiard balls.

1896 *Daily News* 18 Apr. 6/1 During the last two or three years the balls have been made of a new composition called bonzoline. 1899 *Westm. Gaz.* 10 Nov. 5/2 Bonzoline billiard balls.

boo, *v.* Add: **2.** *trans.* To assail with cries of 'boo!' as an expression of dissatisfaction or disapproval; to condemn by booing. Hence **boo·er.**

1893 *Daily News* 27 Jan. 5/1 It was not a booing that booed the piece to immediate perdition. 1904 *Daily Chron.* 19 Nov. 4/6 The 'booer', she argues, does not make himself heard at the opening of all new hotels, banks, and public buildings. What indeed would be thought of the man who 'booed' his dinner at the Carlton because his appetite failed? 1923 *Daily Mail* 28 Sept. 6/3 Then..the booer will learn to give vent to his feelings in a more civilised way. 1955 *Times* 4 Aug. 10/4 Some of the audience booed the performance on the first night. 1967 *Listener* 12 Jan. 45/2 François Mauriac ..tried to boo the singer off the stage.

boo-ay (bū·ˌəi). *N.Z.* and *Austral. colloq.* Also **boo-ai, booeye, boohai.** [? f. place-name *Puhoi*, N. Auckland.] Remote rural districts. Phr. *up the boo-ay*, completely astray, altogether wrong.

The word is much older but has not been found in printed sources earlier than 1955.

1955 D. NILAND *Shiralee* 67 He could have been out there in the boo-eye, thrusting his body into the wind and rain. 1959 G. SLATTER *Gun in Hand* viii. 86 Slingin' off at me he was. You're up the boo-ay he told me. 1963 R. CASEY *As Short Spring* xxi. 244 You fancy them up here in the boo-eye, mortgaged up to their necks for the rest of their lives.

boob (būb), *sb. slang* (orig. *U.S.*). [Shortened f. BOOBY *sb.*]

1. = *BOOBY *sb.* 1 e.

1908 J. M. SULLIVAN *Crim. Slang* 4 *Boob*, the lockup, station house, or city prison. 1911 G. BRONSON-HOWARD *Enemy to Society* v. 159 Stooling for the coppers and swearing many a right guy into the boob. 1923 E. WALLACE *Missing Million* xiii. 108, I thought you were in 'boob'. 1941 *Coast to Coast 1941* 232 Seeing Don get chucked out of the Ballarat and carted off to the boob. 1958 M. PUGH *Wilderness of Monkeys* xiv. 177 'Got six months in jail,' May added. 'Half a stretch in the boob,' Maguire said.

2. = BOOBY *sb.* 1 a.

1909 *Sat. Even. Post* 27 Mar. 7/3, I had to tell her the boob had gone for the day. 1915 [see *BONE-*headed a.*]. 1920 *Chambers's Jrnl.* May 282/1 Of course war is wrong—any boob knows that. 1923 W. DEEPING *Secret Sanctuary* xvi. 160 And not a soul to speak to but that boob of a boy once in the day. 1930 G. B. SHAW *Apple Cart* I. 26 You gave it away, like the boobs you are, to the Pentland Forth Syndicate. 1961 [see *BOO-BOO].

3. A foolish mistake or blunder. Also *attrib.*, foolish, inane.

1934 R. STOUT *Fer-de-Lance* (1935) i. 8 A boob thing to say. 1959 'O. MILLS' *Stairway to Murder* xvi. 167 Boob Number Two... The prison service isn't quite like the Army, Colonel Clive. 1966 P. MOLONEY *Plea for Mersey* 55 Newspapers have I read in every town And many a boob and misprint have I seen. 1969 *Daily Tel.* (Colour Suppl.) 10 Jan. 7 (U.S.A.) A factory hand released by technology into ever greater leisure may be content to watch the 'boob toob' (TV) or go fishing.

4. [prob. shortening of *BOOBY *sb.*[2]] *pl.* The breasts. *slang.* (orig. *U.S.*)

1949 H. MILLER *Sexus* (1969) xiii. 305, I felt her sloshy boobs joggling me but I was too intent on pursuing the ramifications of Coleridge's amazing mind to let her vegetable appendages disturb me. 1955 T. WILLIAMS *Cat on Hot Tin Roof* (1956) I. 7 He always drops his eyes down my body when I'm talkin' to him, drops his eyes to my boobs an' licks his old chops! 1968 *Daily Mirror* 27 Aug. 7/5 If people insist on talking about her boobs, she would rather they called them boobs, which is a way-out word,..rather than breasts.

boob (būb), *v. slang.* [cf. sense 3 of prec.] *intr.* To make a foolish mistake or blunder. Also *trans.* (see quot. 1941).

1935 T. E. LAWRENCE *Lett.* (1938) 851 The camera seems wholly in place as journalism: but when it tries to re-create it boobs. 1941 *Amer. Speech* XVI. 76/1 Slang of the R.A.F. ..*He boobed it*, he made a mess of it. 1951 'N. SHUTE' *Round Bend* v. 132 If I boob on this one it'll mean the finish of the business.

booboisie (būbwazī·). *U.S. slang.* [Jocular formation f. *BOOB *sb.* 2 + *-oisie*, after *bourgeoisie*.] 'Boobs' as a class.

1922 in MENCKEN *Amer. Lang.* (ed. 4, 1946) xi. 560. 1927 ST. JOHN ERVINE in *Observer* 24 Apr. 15/1, I would not for the world rob the booboisie of their entertainments. Let them have their footling stories..on the screen. 1959 J. THURBER *Years with Ross* iv. 61 Those well-known.. objects of..scorn and ridicule, the booboisie. 1961 *20th Cent.* Jan. 72 They wanted to be as eccentric as they pleased ..and..could not do that among the *booboisie* of America.

boo-boo (bū·bū). *slang* (orig. *U.S.*). [Prob. redupl. of *BOOB *sb.*] = *BOOB *sb.* 3.

1954 *Los Angeles Times* 17 Oct. II. 1 Defense Secretary Wilson, whose recent boo-boo..threatens to become historic. 1957 *Harper's* Aug. 51/1 The original boo-boo that started all this confusion. 1961 *Guardian* 2 Nov. 7/6 Mr Melvin Lasky, though certainly no boob..has made a booboo. 1967 'O. MILLS' *Death enters Lists* iv. 33 My fault, I'm afraid. I've just made what the Yanks call a boo-boo.

boobook (bū·buk, bū·būk). Also **bu·ckbuck.** [Imitative of the bird's note.] The name given in Australia to a medium-sized brown spotted owl, *Ninox novæ-seelandiæ boobook*; the mopoke. In full, **boobook owl.**

1801 LATHAM *Gen. Synop. Birds* Suppl. II. 64 *Boobook O[wl]*, size of the Brown Owl, and at first sight somewhat resembles it, especially on the upper parts. This inhabits New Holland, where it is known by the name of *Boobook*. 1826 *Trans. Linnean Soc.* XV. 188 The native name of this bird, as Mr. Caley informs us, is Buck'buck. 1916 J. B. COOPER *Coo-oo-ee* ii. 31 A boo-book owl calling 'Mopoke!' 1934 A. RUSSELL *Tramp-Royal* 11 The insistent 'morepork' of the boobook owl. 1936 M. FRANKLIN *All that Swagger* vii. 66 At night there were the unfailing boobooks and dingoes to augment the curlews' concert. 1968 V. SERVENTY *Wild-life Austral.* 79 The best known of Australian owls is the boobook, named after its call.

booby, *sb.*[1] Add: **1. a.** (Recent example.) Also, *spec.* a cry-baby (*dial.* or *children's colloq.*).

1616 [in Dict.]. 1891 R. P. CHOPE *Dial. Hartland, Dev.* 29 *Booby*, a big child given to crying. 1945 E. WAUGH *Brideshead Revisited* I. viii. 189 'Poor simple monk,' I thought, 'poor booby.' 1959 I. & P. OPIE *Lore & Lang. Schoolchildren* x. 186 Gloucestershire children comment: you babby, big baby,..booby, [etc.].

e. Shortened f. *booby-hutch*: a lock-up or cell. *slang.*

1938 F. D. SHARPE *Sharpe of Flying Squad* 329 Booby or Booby Hutch, the cell.

3. booby-hack *U.S.* = *booby-hut*; **booby hatch,** (*b*) a lock-up or gaol (*U.S. slang*); (*c*) a home for the insane (*U.S. slang*); **booby-hutch** (earlier example); also, a lock-up or cell; a police station; **booby-prize,** a prize awarded in ridicule or fun to the player with the lowest score; **booby-trap** (earlier example); also *Mil. colloq.*, a harmless-looking object concealing an explosive charge, designed to go off if the object is disturbed; hence as *v. trans.*, to set with a booby-trap; so *booby-trapped* ppl. adj.

1888 *Boston Daily Globe* (Farmer), They collided with Crowley's booby hack, knocking the horse down and demolishing the front of the vehicle. 1859 G. W. MATSELL *Vocabulum* 13/2 *Booby-hatch*, station-house. 1897 ADE *Pink Marsh* 3 They'd have him in the booby-hatch in about two hours. 1923 *Time* 10 Mar. 15/3 Liane,..whose specialty is driving lover after lover to ruin, death, or the booby-hatch. 1931 G. IRWIN *Amer. Tramp & Underworld Slang* 33 *Booby Hatch*, a police station or village gaol. 1936 WODEHOUSE *Laughing Gas* xi. 111 What, tell people you're me and I'm you. Sure we could, if you don't mind being put in the booby-hatch. 1720 *Weekly Jrnl.* 4 June 1623/2 Not a Raw bon'd Jade, or a Booby-Hutch in City and Suburbs but will be hacked out to City Apprentices. 1889 BARRÈRE & LELAND *Dict. Slang* I. 161/1 *Booby-hutch* (thieves), the police-station. 1938 Booby hutch [see sense *1 e]. 1889 *Puck* (U.S.) 17 July 378/1 Into infinitesimal shreds he tears A beautiful Booby Prize. 1900 E. T. FOWLER *Farringdons* iii. 55 Your prize would have been no better than a booby-prize. 1929 G. STOWELL *Hist. Button Hill* I. 64 The incorrigible Mr. Denworthy presented as a booby-prize a small sample bottle of Worcester sauce. 1850 F. E. SMEDLEY *Frank Fairlegh* iii. 28 The construction of what he called a 'booby-trap'. 1917 E. F. WOOD *Note-Bk. Intelligence Officer* xix. 271 'Booby' traps were sprinkled about the country in the form of bombs. 1918 P. GIBBS *From Bapaume to Passchendaele* 4 The enemy left..'booby-traps' to blow a man to bits or blind him for life if he touched a harmless-looking stick or opened the lid of a box. 1943 *Illustr. London News* 1 May 483 Doors and windows are easily booby-trapped. 1959 'M. DERBY' *Tigress* iii. 130 The ugly foreground, mined and booby-trapped and ambushed.

booby (bū·bi), *sb.*[2] = BUBBY[1]. *slang* (orig. *U.S.*).

1934 H. MILLER *Tropic of Cancer* 315 She was lying on the divan with her boobies in her hands. 1968 *Guardian* 2 Sept. 4/6 The characters were constantly referring to her large bosom (even descending to calling them 'big boobies').

boobyalla (būbiæ·lă). *Austral.* Also **boobialla.** [Native name.] **1.** The native name in Australia for the Australian wattle. **2.** Any of various shrubs or trees of the genus *Myoporum*, having clusters of small white flowers.

1835 *Hobart Town Almanack* 63 *Acacia sophora*, Sophora podded Acacia or Booby-aloe. 1843 J. BACKHOUSE *Narr. Visit to Austral. Colonies* 59 The sandbanks at the mouth of Macquarie Harbour are covered with Boobialla, a species of *Acacia*. 1861 L. A. MEREDITH *Over Straits* ii. 62 Boobyalla bushes lay within the dash of the ceaseless spray. 1934 *Antiquity* VIII. 182 Sand dunes clothed with boobyallas and other indigenous shrubs.

boodie-rat (bū·diˌræt). *local* (*S.W.*) *Austral.* Also **boodee, boody-rat.** [Native name.] A species of rat-kangaroo. Also called **boodie.**

[1863 J. GOULD *Mammals of Australia* II. 74 Hypsiprymnus (Bettongia) Lesueuri..*Boor-dee*..The Boor-dee is exclusively a nocturnal feeder, and..one of the most destructive animals to the garden of the settler in Western Australia.] 1910 G. C. SHORTRIDGE in *Proc. Zool. Soc.* 1909 823 Bettongia Lesueuri..kangaroo-rats..Boodee Rat of Colonists. 'Boodee' of natives (S.W.). 1924 D. H. LAWRENCE & SKINNER *Boy in Bush* vii. 96 Little hunts of wallabies or bandicoots or bungarras, or boody-rats. 1963 *New Scientist* 3 Jan. 12 Boodies apparently emit a series of grunts and chuckles when chased.

boodle[1]. Add: **1.** (Further U.S. examples.) Cf. *CABOODLE.

1833 J. NEAL *Down-Easters* I. 61, I know a feller 'twould whip the whool boodle of 'em an' give 'em six. 1847 D. P. THOMPSON *Locke Amsden* 76 [He] stumped all the rest to come on, one at a time, and there wasn't a soul of the whole boodle that dared do it. 1895 [see *KIT *sb.*[1] 3].

2. Substitute for def.: **a.** Counterfeit money. *U.S. slang.*

1858 *Harper's Weekly* 3 Apr. (Th.), Boodle is a flash term used by counterfeiters... The leaders [of the gang] were the manufacturers and bankers of the boodle.

b. Money acquired or spent illegally or improperly, esp. in connection with the obtaining or holding of public offices; the material means or gains of bribery and corruption; also, money in general. *slang* (orig. *U.S.*).

1883 *Judge* (U.S.) 3 Mar. 7/1 As long as the people pay, As long as there's 'boodle' to capture. 1884 *Mag. Amer. Hist.* XII. 566/2 *Boodle*..has come to mean a large roll of bills such as political managers are supposed to divide among their retainers. 1887 *Nation* (N.Y.) 14 Apr. 307/3 New York is better known all over the..world for boodle Aldermen and municipal rings than for anything else. 1888 *St. Louis Globe Democrat* (Farmer), She presented a draft for 2000 dols. drawn upon a Detroit bank, and received the cash. Immediately after that she left for the East with the boodle. 1892 STEVENSON & OSBOURNE *Wrecker* ix. 138 It took my fancy; it was so romantic, and then I saw there was boodle in the thing. 1917 CONAN DOYLE *His Last Bow* viii. 294 'What about the dough?' he asked... 'The boodle. The reward. The £500.' 1922 JOYCE *Ulysses* 600 Ready to decamp with whatever boodle they could.

Hence **boo·dle** *v. trans.*, to bribe; *intr.*, to practise bribery; **boo·dler,** one who practises boodling; **boo·dling** *vbl. sb.*, bribery, bribing.

1872 G. P. BURNHAM *Mem. U.S. Secret Service* 347 He ventured to collar Wightman, knowing him..to be grand Sachem among these 'boodlers' or confidence men. 1888 in *Amer. Speech* (1960) XXXV. 267 *Boodling*, taking money unlawfully by way of bribe or otherwise. 1894 *Congress. Rec.* Aug., App. 1229/1 What we call the government of the people..has become a government of boodlers. 1903 *N.Y. Even. Post* 31 Aug., We fancy that the people of Missouri will conclude to diminish boodling rather by taking from than adding to the powers of the dominant machine. 1904 W. H. SMITH *Promoters* iv. 86 If you're going to boodle you've got to do it on a party basis. If I wanted to boodle an Illinois legislature, [etc.]. 1909 C. F. G. MASTERMAN *Condition of England* iv. 153 The average citizen..is not yet convinced that its [*sc.* Socialism's] adherents will make a better job of it than the 'boodlers' and 'blood suckers'.

boogie[1] (bū·gi). *U.S. slang.* [Perh. alteration of BOGY.] A derogatory term for a Negro.

1923 *Confessions of Bank Burglar* vii. 40 Three coons came into the barn..the three of them took a drink and then put the bottle in the hay... At noon the 'boogies' came in for another shot. 1925 *Flynn's* 1 Aug. 572/1 One of the cops..caught two boogies. We picked up the two hard-lookin' young negroes. *Ibid.*, The boogie jus' got up and grinned. 1937 HEMINGWAY *To have & have Not* III. xiv. 205, I seen that big boogie there mopping it up.

boogie-woogie (bū·giˌwū·gi). orig. *U.S.* [Obscure.] A style of playing blues (usu. on the piano) marked by a persistent bass rhythm. Also (abbrev.) **boogie**[2].

1928 (*title*) Pinetop's Boogie Woogie. 1935 *Swing Music* Aug. 153 This side [of the record] might be an instruction disc in Swedish Drill..only it's a lesson in Boogie Woogie. 1938 *Manchester Guardian Wkly.* 2 Sept. 188/3 Sometimes they play 'boogie-woogie' (on the heavy brass instruments). 1940 *Time* 4 Mar. 48/2 Boogie-woogie is a kind of blues piano playing in which the left hand drones a set bass phrase over and over, while the right hand goes to town with whatever variations the player can think up. 1941 *N.Y. Times* x. 7/3 The twelve-bar structure which is the basis of almost all boogie-woogie music is also one of the mainstays of hot jazz as a whole. 1941 *Brunswick Records Catal.* 6 (*title*) Scrub me, Mama with a Boogie Beat. 1950 S. MORTON *Unfamiliar Name* v. 76 They wanted the jive and boogie-woogie boys. 1960 *20th Cent.* Dec. 560 Negro boogie pianists. 1966 *Crescendo* Nov. 14/3 I'd listen to a lot of boogie woogie records and copy them off by ear. This led to me making up my own boogie woogies.

boojery, var. *BUDGEREE *a.*

boojum (bū·dʒəm). [Invented by 'Lewis Carroll' (C. L. Dodgson) in *The Hunting of the Snark* (1876).] An imaginary animal, a particularly dangerous kind of 'snark'.

1904 B'NESS VON HUTTEN *Pam* III. vi. 146 We shall see a good deal of each other. I am a boojum, and I know. 1922 *Edin. Rev.* Oct. 241 Both these beautiful abstractions are in reality boojums. 1925 *Blackw. Mag.* Mar. 345/1 A solitary Boojum-like person. 1950 AUDEN *Enchafèd Flood* (1951) i. 42 The dreadful Boojum of Nothingness.

book, *sb.* Add: **3. d.** *transf.* Of things composed of 'leaves' or 'plates'.

1840 R. H. DANA *Bef. Mast* (1841) xxix. 85/1 A large 'book' was made of some twenty-five to fifty hides, doubled at the backs, and put into one another, like the leaves of a book. 1859 C. TOMLINSON *Illustr. Useful Arts* 21 (*caption*) Book of Silk from China. 1885 J. S. KINGSLEY in *Q. Jrnl. Microsc. Sci.* Oct. 538 The primary stigma formed by the insinking of the respiratory book is not the functional one of the adult. 1892 *Photogr. Ann.* II. 327 To put the book in camera, the camera is tilted front up. *Ibid.* 328 The book of plate-holders. 1910 *Sessions' Paper* 17 Nov. 21, I..pulled out my cigarette book to make a cigarette. 1937 *Popular Sci.* Nov. 68/2 (*caption*) Match Books Get Foolproof Cover. 1962 J. BRAINE *Life at Top* v. 88 One of the books of matches I'd taken away from the Savoy.

e. An angler's pocket-book for fishing-tackle.

1824 *Sporting Mag.* XV. N.S. 147/1 The fisherman, who has got a book full of good ready-made flies. 1847 STODDART *Angler's Comp.* 61 Angler's trouting book.

f. A magazine. Now *colloq.* and *vulg.*

1800 H. MORE *Let.* 11 Sept. (1925) 177 The Anti-Jacobin Magazine, which is spreading more mischief over the land than almost any other book. 1873 *Young Englishwoman* Aug. 416/2 She has taken in *The Young Englishwoman* for four years, and thinks it is the best of books for young ladies. 1937 A. THIRKELL *Summer Half* ix. 254 Rose was reduced to reading a book, by which..she meant an illustrated weekly. 1942 A. P. JEPHCOTT *Girls Growing Up* v. 98 With..working-class girls, you can create a common interest..by offering..some 'girls' books'. These 'books' are what one would ordinarily call papers or magazines. 1959 I. & P. OPIE *Lore & Lang. Schoolchildren* ix. 161 A good 'book' (i.e. a magazine) is said to be 'smashin''.

4. d. A telephone directory.

1925 A. CHRISTIE *Secret of Chimneys* iv. 30 You might ring up a number for me now. Look it up in the book. 1960 K. AMIS *Take a Girl* vi. 87 Let's just hypothesise that I give you a ring, shall we? You're in the book, eh?

8. b. The script of a play.

1598 FLORIO *Worlde of Wordes* 51/3 Buriasso,..a prompter, or one that keepes the book for plaiers. 1879 D. K. T. RANOUS *Diary* 5 Sept. in *Diary of Daly Déb.* (1910) 6 Old Mr. Moore held the book of the play, and the actors moved slowly about the stage with manuscript copies of their *rôles* in their hands. 1895 G. B. SHAW in *Sat. Rev.* 2 Feb. 151/2 The play was..pulled to pieces in order that some bad scenery..might destroy all the illusion which the simple stage directions in the book create. 1923 T. E. LAWRENCE *Lett.* (1938) 442 Your people had no technique, no arts and graces, to put between their 'book' and us. 1933 P. GODFREY *Back-Stage* vi. 82 By the time that the principals are rehearsing regularly again the company are working without their books.

c. The repertoire or sheet music of an orchestra or musician.

1939 *New Republic* 17 May 47 If the road man wanted to pack Ziggy's book half an hour before the show broke he would say yeah, yeah, yeah. 1958 *New Statesman* 23 Aug. 221/1 There he could learn..how to handle a large 'book' (repertoire) and difficult arrangements.

9. c. *West Africa.* (See quots.)

1863 *Fraser's Mag.* LXVII. 146/1 It was resolved..to renew his 'book'. 1897 M. KINGSLEY *W. Africa* x. 203 In order to..simplify this goods traffic, a written piece of paper is employed—practically a cheque, which is called a 'bou' or 'book' and these 'bous' are cashed—i.e. gooded, at the store. *Ibid.* xii. 286, I would give the creditor a book on Hatton and Cookson for the coat.

d. The total of charges that can be made against an accused person. Phr. *to throw the book at* (a person): to accuse of all the possible crimes; to award the maximum penalty. So *to get* or *do the book* (*U.S. slang*), to suffer the maximum penalty.

[1926 G. H. MAINES *Wise-Crack Dict.* 5/2 *Book*, crook's term for a stretch of life in the penitentiary.] 1928 R. J. TASKER *Grimhaven* (1929) i. 11 I'm doing one life jolt, and two one-to-fiftys..—yes sir, doing the book. 1932 *Flynn's* 6 Feb. 125/1 The prosecuting attorney..determined to try to get the trial judge to 'throw the book' at him, (which means give him the limit). 1961 J. HELLER *Catch-22* (1962) viii. 74 He was formally charged with 'breaking ranks while in formation, felonious assault, indiscriminate behaviour, mopery, high treason, provoking, being a smart guy, listening to classical music, and so on'. In short, they threw the book at him. 1962 'B. GRAEME' *Undetective* iv. 45 They'll dig out some old act that hasn't been repealed..and then they'll throw the book at him.

10. (Earlier examples.)

1812 *Sporting Mag.* XL. 70/1 This is the exact statement of my bets, as my book left with Mr. Smith, Clerk of the Subscription-Betting-Room, at Tattersall's, for the inspection of the public, will prove. 1836 DISRAELI *Henrietta Temple* III. xv. 196 Go and take all the odds you can get upon Goshawk. Come, now, tomorrow you will tell me you have a very pretty book.

b. = BOOK-MAKER 3 (cf. *BOOKIE[2]). *Austral.* and *N.Z. colloq.*

1900 J. SCOTT *Tales Colonial Turf* 134 If the 'books' did not see the horse was doing good work, they would not bite. 1915 C. J. DENNIS *Songs Sentim. Bloke* (1916) 118 *Book*, a bookie.

13. *Book of the Dead* [tr. G. *totenbuch*], in *Egyptology* (see quot. 1906); *book of the* (or *a*) *film*: the reproduction in book form of the script or story of a cinema film; also, the book upon which a film is based; *book of the month*: a book chosen as the most outstanding during a particular month; also as *attrib. phr.*; *book of* (*the*) *words*: (*a*) the libretto of an opera, etc. (cf. BOOK *sb.* 8); (*b*) *colloq.*, any written or printed record or set of rules; also in extended uses.

1853 L. & J. B. HORNER tr. *Lepsius's Introd. Chronol. Egyptians* 392, I can as little agree with the opinion that the great Book of the Dead of the Egyptians was one of the ten books of the Stolistes. 1885 G. B. SHAW *How to become Musical Critic* (1960) 108 A gentleman who carries a bundle of white pamphlets, and cries incessantly 'Book of the words! Program! Book of the words!' 1906 J. H. BREASTED *Hist. Egypt* xiii. 249 The magical formulæ by which the dead are to triumph in the hereafter become more and more numerous, so that it is no longer possible to record them on the inside of the coffin, but they must be written on papyrus and the roll placed in the tomb. As the selection of the most important of these texts came to be more and more uniform, the 'Book of the Dead' began to take form. 1929 J. B. PRIESTLEY *Good Companions* II. ii. 288 Don't say these things. Think 'em but don't put 'em in the book of words. 1929 H. G. WELLS *King who was King* (*sub-title*) The Book of a Film. 1930 E. GEORGE *Down Our Street* i, There was nothink ag'inst me, but I was known to frequent bad comp'ny. That's 'ow they put it in the book of words. *a* 1932 in Q. D. LEAVIS *Fiction & Reading Public* I. i. 16 The filmgoer wishes also to read the book of the film, and the reader to see the picture. 1933 F. R. LEAVIS & THOMPSON *Culture & Environment* 42 The Committee select their 'books of the month'. 1940 GRAVES & HODGE *Long Week-end* iv. 52 The mezzo-brow 'Book of the Month' choice of the dailies. 1946 WODEHOUSE *Money in Bank* iii. 28 There's nothing in the book of the words to prevent Mrs. Cork having the bozo..stowed away in the cooler, if he's a thief. 1947 *Horizon* XVI. Oct. 4 Book-of-the-month clubs. 1955 *Times Lit. Suppl.* 23 Sept. 560/3 Those who would give us the great classics of Greece and Rome in the 'rude langage' shorn of all 'superflue' of some rewritten book-of-the-film.

14. *by the book* (later example); also, according to the rules; *in* (a person's) *good books* (example); *in* (one's) *book*: in the opinion of, according to the judgement of (a person); *in the book(s)*, recorded, in existence (*colloq.*); *like a book*: see *LIKE *adv.*

1839 DICKENS *Nickleby* xxxi. 304 If you want to keep in the good books in that quarter, you had better not call her the old lady. 1841 POE *Murders in Rue Morgue* in *Tales* (1845) 118 To have a retentive memory, and to proceed by 'the book', are points commonly regarded as the sum total of good playing. 1958 J. KEROUAC *On Road* vi. 148 That night Marylou took everything [i.e. every drink] in the books. 1959 *Listener* 31 Dec. 1172/3 Every human evil in the book was thrown at us—cancer, paralysis, famine, gas-chambers. 1960 J. WAIN *Nuncle* 49 They'll soak me for defamation of character and everything else in the book. 1964 W. MARKFIELD *To Early Grave* (1965) vii. 130 In my book, you're still *putz*. 1966 S. JACKMAN *Davidson Affair* ii. 13 In his book the function of television was to edify, not to entertain. 1968 A. DIMENT *Gt. Spy Race* I. iii. 32 They keep rigidly to the book down here.

15. *to bring to book* (earlier examples); *to take a leaf out of* (a person's) *book* (earlier examples); *to suit* (a person's) *book* (orig. a bookmaker's phrase: see sense 10): to fall in with his plans or answer his requirements; to be agreeable to.

1804 *Sporting Mag.* XXIII. 265/2 'Tis not my business to examine your accounts, Sir—but should I bring you to book..there is something in that sly countenance that tells me you have *sometimes* staked your credit at too great a venture. 1809 [see LEAF *sb.* 7 b]. 1851 R. I. MURCHISON *Let.* 14 Apr. in Lady Prestwich *Life Sir J. Prestwich* (1899) 83 Would it suit your book to make a run of a day or two to the other side of the Weald? 1852 F. E. SMEDLEY *Lewis Arundel* vi, By which time he expects to be so hard up that he must marry somebody, and as there will be plenty of the needful she will suit his book as well as any other. 1865 DICKENS *Mut. Fr.* II. III. xv. 131 I'll bring this young man to book. 1868 L. M. ALCOTT *Lit. Women* (1872) II. vii. 292 I'll take a leaf out of her book. 1955 *Times* 30 Aug. 6/1 Plainly the strikes suit the Communists' book.

17. a. *book-auction* (earlier and later examples), *-auctioneer* (earlier example), *-box*, *-cover* (earlier example), *-jacket*, *-knowledge* (earlier examples), *-list*, *-load*, *-market* (earlier example), *-sale* (examples), *-stall* (later example), *-stock*, *-trade* (earlier example), *-unit*, *-wrapper*.

1790 *Pennsylv. Packet* 31 Dec. 3/4 Book-auction will continue during the season. 1903 (*title*) Book-Auction Records. 1781 *Salem Gaz.* 19 June, The Book-Auctionier intends also to exhibit a Collection of Books by Auction. 1885 C. M. YONGE *Two Sides of Shield* I. i. 7 Dolores, who had in her mother's time been allowed a pretty free range of 'book-box'. 1900 —— *Mod. Broods* iv. 35 Agatha was struggling with the straps of a book box. 1961 T. LANDAU *Encycl. Librarianship* (ed. 2) 45/1 *Book boxes*, boxes, usually of standard sizes, in which books are transported between libraries, often made in plywood for lightness. 1845 POE *Purloined Let.* in *Tales* 206 We also measured the thickness of every book-*cover*. 1928 *Publishers' Weekly* 9 June 2359 Several specimens of modern trade book jackets. 1665 NEEDHAM *Med. Medicinæ* vii. 253 A Book-knowledge of Hippocrates, Galen, and the rest that are counted Classick. 1833 MILL in *Tait's Edin. Mag.* III. 348 Great natural powers..have supplied the place of a more extensive book-knowledge. 1937 *Discovery* June 192/2 (*heading*) A Scientific and Technical Book List. 1907 *Daily Chron.* 19 Dec. 3/5 The heroine, who is the best of the book-load. 1836 J. S. MILL *Ess. Politics & Culture* (1962) 72 The bulk of the purchasers in the book-market. 1797 W. B. STEVENS *Jrnl.* 11 Apr. (1965) v. 421 To Ashby for a book sale. 1952 R. CAMPBELL *Lorca* 7 Sincerity is so often subordinated to book-sales. 1913 'A. R. HOPE' *Half & Half Tragedy* 70, I had something to get at the railway bookstall. 1957 *BBC Handbook* 102 At the Monitoring Service Centre the bookstock's accent is on politics and biography. 1833 in *N. & Q.* (1942) CLXXXII. 143/2 Times are now so altered in the book trade, that we can no longer venture on such a work single handed. 1933 *Discovery* Aug. 254/2 Along one wall [are] two plywood book units. 1932 *Book-Collector's Q.* Apr.–June 10 The book-wrapper is relatively and absolutely an upstart, and a good number of people have not yet decided whether they wish to encourage it or not.

b. *book-buyer* (earlier examples), *-buying*, *-fancier* (earlier example), *-mindedness*, *-notice* [NOTICE *sb.* 8 a], *-reading*, *-review*, *-reviewer*, *-reviewing*, *rights*.

1693 LOCKE *Let.* 2 Jan. in B. Rand *Locke & Clarke* (1927) 366, I wish you would have some care of book-buyers as well as all of booksellers. 1838 DICKENS *Let.* 16 Nov. (1965) I. 455 Not book-buyers I am sorry to say. 1832 MILL in *Tait's Edin. Mag.* II. 343 The wisdom of the book-buying public. 1853 in *N. & Q.* (1962) CCVII. 84/2 Free Trade has been also a great hinderance to my bookbuying. 1810 *Irish Mag.* Mar. 127/1 The valuable collection made by this book-fancier. 1805 WORDSWORTH *Prelude* iii. 395 Antiquity and stedfast truth And strong book-mindedness. 1903 J. MORLEY *Life of Gladstone* I. II. ii. 117 His bookmindedness is unabated. 1868 W. JAMES *Let.* 24 May in R. B. Perry *Thought & Char. W. J.* (1935) I. 277, I have written a few book notices lately. 1938 *Times Lit. Suppl.* 17 Sept. 591/1 The majority of the book-notices..were unpaid until the 'fifties. 1832 *Times* 6 Jan. [3]/3 To be sure, the book-reading lovers of antiquity would cry 'horrible'. 1910 A. BENNETT *Clayhanger* III. i. 325 A new series of sixpenny reprints which had considerably excited the book-selling and book-reading worlds. 1861 A. J. GRAHAM *Phonographic Odds & Ends* 71 (*heading*) Book-Review. 1904 W. JAMES *Mem. & Stud.* (1911) iv. 66 His contributions to literature were all anonymous, book-reviews chiefly. 1965 *Times Lit. Suppl.* 25 Nov. 1035/2 The President could not have read all those..book-reviews he was said to. 1898 M. BEERBOHM in *Variety of Things* (1928) 224 Even the book-reviewers could no longer assert that he did not know how to draw. 1873 H. JAMES *Let.* 31 May in R. B. Perry *Thought & Char. W. J.* (1935) I. 348, I am struggling through long delay to get at something better than book reviewing. 1956 *N. & Q.* CCI. 190/2 Here is the true dawn of book-reviewing in England. 1904 A. BENNETT *Let.* 26 Nov. (1966) I. 58 You have also the six short stories, & the book rights on these. 1915 J. JOYCE *Let.* 1 Apr. (1966) II. 338 That [agreement] for my novel should be conditional on Mr Grant Richards' refusal of the book rights.

c. *book-fed*, *-filled*, *-lined* (examples), *-walled*.

1932 V. WOOLF *Let. to Young Poet* 24 As if they had neither ears nor eyes..but only honest enterprising book-fed brains. 1965 F. SARGESON *Memoirs of Peon* ii. 32 The back book-filled room. 1897 *Daily News* 18 June 8/4 Warm and cosy, with book-lined walls. 1947 AUDEN *Age of Anxiety* (1948) iii. 75 But in book-lined rooms at the back Committees meet. 1904 W. DE LA MARE *Henry Brocken* i. 2 Half my youthful days passed in that low, book-walled chamber.

18. book account *U.S.*, a statement of accounts recorded in a book; **book agent** *U.S.*, one who promotes the sale of books; **book-board** (earlier example); **book-boy,** a boy employed to fetch books for readers in a library; **book-burning,** the destruction of writings regarded as harmful or subversive; so *book-burner*; **book canvasser,** one who canvasses schools, public offices, etc., for the sale of books, esp. on the subscription system; **book-cloth,** a cloth manufactured for the bindings of books; **book-club,** delete (*obs.*) and add example; (*b*) a society which produces books for its members, such as the Warton Club, the Roxburghe Club, the First Edition Club; **book concern** *U.S.*, an establishment engaged in the printing and sale of books; **book-credit** (earlier example); **book-edge gilder, marbler,** a bookbinder's workman who gilds or marbles the edges of books; **book end,** one of a pair of (ornamental) book props (see below), used to keep a row of unshelved books in an upright position; **booketeria** (buˌkētiə·riă) *U.S.* [after *CAF)ETERIA], a self-service book-store or library; **book-farmer,** one who farms with knowledge acquired from books; so **book-farming**; **book-folder,** (*a*) a printer's, bookbinder's, or stationer's employee who folds the paper sheets into page-size; (*b*) the paper wrapper of a newly published book; **book-form,** in advb. phrase *in book form* (see sense 16); also *attrib.*; **book-gill** = *gill-book* (s.v. *GILL *sb.*[1] 5 b); **book hand,** the hand or writing used by the official transcribers of books before the invention of printing; **book-hunt** *v. intr.*, to follow the pursuit of a book-hunter or searcher of old and rare volumes; **book-label,** a label bearing the title and author's name, the owner's name, etc., affixed to the cover of a book; **book-length** *attrib.*, of the length of a book; **book-louse** (U.S. example); **book-lung,** the lamellate respiratory organs of a scorpion; **book-mark, -marker** (earlier examples); **book match,** one of a set of tear-off matches, sold in packets hinged at one end like a book (cf. sense 3 d above); **book mite** (see quot.); **bookmobile** (bu·kmǒbiːl) *U.S.* [after *AUTO)MOBILE *sb.*], see quot. 1941; **book-muslin** (earlier examples); **book name,** a name of a plant or animal, other than the scientific name, used only in books; also *transf.*; **book-number,** 'in library-cataloguing, a particular number (or a number and a letter) designating the book in its proper sequence in the smallest division to which it belongs' (Cent. Dict. Suppl. 1909); **book-page,** (*a*) a page of a book; (*b*) in a newspaper or journal: a page that is devoted to reviews and notices of books; **book piles,** a type of book-plate in which piles of books are used as the design; **book pocket** (see quot. 1955); **book-post** (earlier example); **book prop, support,** an angular support for the end of a row of books, esp. in a partly-filled shelf; **book-shy** *a.*, reluctant or unwilling to read books; **book-society,** also (in modern use) = **book-club* (*b*); **book-stack,** see *STACK *sb.*; **book-stamp,** a stamp for embossing the covers of books; **book-stand** (examples); **book-table,** a table intended solely or mainly for books; **book token,** a voucher exchangeable at a bookseller's for a book or books; **book-tray,** a tray for carrying books; **book-trough** (see quot. 1961); **book type** (see quot.); **book value** *Book-keeping*, the value of a commodity as shown by a firm's books, as distinguished from its market value; **book-work,** (*b*) the printing of books or similar matter, as opposed to job-work (JOB *sb.*[3] 1 c); **book-world,** the world of literature; the affairs of life as described in literature.

1672 *Mass. Col. Laws* (1887) 39 Inconveniences..through want of seasonable examination..of Book accompts. 1741 *New Hamp. Probate Rec.* III. 86 My book accompt standing against James Ried. 1883 C. F. WILDER *Sister Ridnour* 232 The treasurer..keeps a correct book-account of all moneys taken and expended. 1830 *Williams's N.-Y. Ann. Reg. 1830* 299 John Emory and Beverly Waugh, Book Agents, New-York. 1886 *Harper's Mag.* Dec. 162/1 They may both be glad to invoke the aid of the despised book agent, who carries literature from door to door. 1910 MULFORD *Hopalong Cassidy* vii. 50 Was you ever an auctioneer..or a book agent? 1847 C. M. YONGE *Scenes & Char.* ix. 110 She put her arm on the book board, while rising from kneeling. 1903 *Daily Chron.* 13 Feb. 5/1 His first situation was as book-boy in the library of the Bristol Law Society. 1899 *Book-Lover* 86/2 The virtuous Romans appear to have been greater book-burners than the Greeks. 1951 I. SHAW *Troubled Air* xxii. 389 The censors and book-burners. 1892 J. A. FARRER *Books condemned to be Burnt* 170 The custom of book-burning, never formally abolished, died out at last from a gradual decline of public belief in its efficacy. 1954 *Ann. Reg. 1953* 169 The destruction of many of these [Communist] books provoked a general outcry against 'book-burning'. 1848 *Philadelphia Almanac* 2 (Advt.), Book canvassers and agents wanted. 1921 *Dict. Occup. Terms* (1927) § 774 *Book canvasser*, canvasses schools, public offices, etc., for sale of technical or other books on subscription system. 1891 *Pall Mall Gaz.* 30 Nov. 7/1 Makers of book-cloth. 1905 *Times* 20 Sept. 5/5 The privileges of *The Times* Book Club are offered to those only who subscribe to *The Times* for a year. 1929 H. WILLIAMS (*title*) Book Clubs and Printing Societies of Great Britain and Ireland..published by the First Edition Club. *Ibid.* 7 The prototype of the book club, the Roxburghe. 1937 V. GOLLANCZ in 'G. Orwell' *Road to Wigan Pier* p. xi, The three selectors of the Left Book Club Choices. 1968 *Listener* 6 June 725/3 There are plenty of people who belong to book clubs but otherwise buy almost no books. 1786 F. ASBURY *Jrnl.* 26 Apr. (1852) I. 511 Arrived in Baltimore, and was occupied..in collecting money for the books, and inspecting the accounts of the Book-Concern. 1872 *Congress. Globe* May 3909/3 Every book published by the Methodist Book Concern..is published on sized paper. 1844 MILL in *Westm. Rev.* XLI. 592 In almost all other transactions between dealers, bank notes are already superseded by cheques, or book credits. 1858 SIMMONDS *Dict. Trade*, *Book and card-edge gilder and marbler*, a workman who ornaments and finishes off the edges of books, etc. 1898 *Daily Chron.* 24 Sept. 10/6 Book-edge gilders wanted. 1921 *Dict. Occup. Terms* (1927) § 548 Book-edge marbler. 1907 *Yesterday's Shopping* (1969) 368/3 *Book ends.* Mahogany, inlaid Marqueterie, heavily weighted, for keeping books in position, pair 23/0. 1932 E. BOWEN *To North* xxi. 228 'What did she sell?'..'Oh, paraphernalia—lampshades, book-ends.' 1961 *Lebende Sprachen* VI. 39/2 *Book ends*, die bücherstützen. 1945 in *Amer. Speech* (1946) XXI. 66/2 Pasadena Junior College has a booketeria. 1947 *Ibid.* XXII. 306/1 The first *Booketeria*..is located in a popular grocery... Borrowers choose their own reading matter. They select and charge books and the store cashiers record the charges. 1825 LOUDON *Encycl. Agric.* IV. 1133/1 *Book farmers*..are those who know agriculture only by reading about it. 1920 R. FROST *Let.* 21 Mar. (1964) 102 'Amatoor!' all the leaves began to murmur. 'Book-farmer!' 1823 R. B. THOMAS *Farmer's Almanack 1824* 18–24 Dec., Be not stubborn and unreasonable in your prejudices against what is called *book-farming*. 1923 D. H. LAWRENCE *Stud. Classic Amer. Lit.* viii. 154 And that's why the idealists left off brook-farming, and took to book-farming. 1872 C. L. BRACE *Dangerous Classes N.Y.* 166 She went to be a book-folder downtown. 1903 *Daily Chron.* 24 Feb. 8/5 (Advt.), Book-folder. Apply..Printing Dept. 1925 *Public Opinion* 5 June 538/3 Blurbs, those interesting little paragraphs which appeared on bookfolders. 1849 ROSSETTI *Let.* 18 Oct. (1965) I. 82 They will be bound..that they may go in the book-form. 1856 *Chambers's Jrnl.* V. 322/2 M. Dumas had previously published it in the book-form. 1893 *Photogr. Ann.* 333 A light camera, with..book-form double dark slides. 1902 'MONKSHOOD' & GAMBLE *Kipling* 161 This story passed from 'Lippincott's Magazine' to the pomp and pride of a book-form Edition. 1964 P. F. ANSON *Bishops at Large* vi. 196 They appeared in book-form in 1913. 1897 PARKER & HASWELL *Textbk. Zool.* I. xi. 621 External appendages or gills (book-gills). 1885 *Encycl. Brit.* XVIII. 143/2 Down to the time of the introduction of printing, writing ran in two lines—the set book-hand and the cursive. 1893 E. M. THOMPSON *Handbk. Gk. & Lat. Palæogr.* xix. 301 We find it convenient to treat the cursive or charter hand as a separate branch of mediaeval English writing apart from the literary or book hand. 1928 *Daily Tel.* 19 July 15/5 A fifteenth century English manuscript..with others written in a vernacular book-hand. 1880 LANG *XXII Ballades in Blue China* 23 He book-hunts, though December freeze. 1905 *Daily Chron.* 19 Dec. 6/2 An interesting copy of the works of Horace, having John Kemble's leather book-label on both covers. 1938 *Times Lit. Suppl.* 3 Dec. vi/2 His new work is a book-length lyric in prose. 1953 *Encounter* Nov. 49/1 Vigolo..has prefaced his edition with a book-length essay. 1867 *Amer. Naturalist* I. 312 The little wingless book-louse (*Atropos*) scampering irreverently over the musty pages of his *Systema Naturae*. 1897 PARKER & HASWELL *Zool.* I. 604 The organs of respiration are sometimes tracheæ, similar to those of Insects, sometimes book-lungs or sacs containing numerous book-leaf-like plates. 1862 GEO. ELIOT *Let.* 24 Dec. in J. W. Cross *Life* (1886) II. 295, I have been discontented with the Coventry book-marks. 1838 C. M. YONGE *Let.* 6 Aug. in C. Coleridge *C. M. Yonge* (1903) iv. 135 Your W.H.W.B.W. bookmarker. 1939 E. AUGUST *Black-Out Book* 30/1 Hold two 'book' matches side by side between your finger and thumb. 1883 *Encycl. Brit.* XVI. 528/2 *Cheyletidæ*, the so-called book mites,..quite unconnected with books. 1937 *Amer. Speech* XII. 30 State newspapers of Friday, September 4, 1936, made announcement that the Nebraska public library commission had purchased a half-ton panel truck to be used as a *bookmobile*. 1941 *Ibid.* XVI. 311/1 A bookmobile is an automobile fitted with shelves and other necessary equipment for serving rural districts and patrons who cannot visit a library. 1969 *Telegraph* (Brisbane) 30 May

4/1 This year the long-awaited bookmobile service would serve some outlying areas. 1759 Newport Mercury 10 Apr. 4/2 Book Muslin, Cambricks, silk Ferrets. c 1793 Jane Austen Volume the First (1954) 72 She lies wrapped in a book muslin bedgown. 1878 Britten & Holland Dict. Eng. Plant-n., Aconite, a common book-name for Aconitum Napellus. 1885 Lisbon (Dakota) Star 27 Mar. 5 A China-man..gets a book-name when he goes to school. 1930 Wyndham Lewis Lett. (1963) 197 The literary editors of the London book-pages. 1932 N. & Q. CLXII. 84/1 In recent years the same thing was done with photostat prints, in copying book-pages. 1969 Daily Tel. 11 Aug. 18 Book page turners for people who have lost the use of their hands. 1902 Encycl. Brit. XXVI. 305/1 'Book-piles', ex-emplified by the ex-libris of W. Hewer (Samuel Pepys's secretary). 1922 Joyce Ulysses 430 Bloom pats with par-celled hands..bookpocket. 1955 J. E. Liberty Pract. Tailoring v. 64 The Hare Pocket..is like a long welt pocket and is sometimes called a book pocket. It is made in the lining, with the welt 1½ in. wide, lined with linen, and with a hole and button. 1861 Rossetti Let. 18 Jan. (1965) II. 389, I will send it you by book-post. 1868 Geo. Eliot Let. 31 Mar. (1955) IV. 426, I send by book post all the printed sheets of the poem. 1934 H. G. Wells Exper. Autobiogr. I. v. 286 Like so many people who have had the benefit of a simple English education she was book-shy. 1941 V. Woolf Between Acts 26 Book-shy she was..and gun-shy too. 1929 Times 27 May 12/2 The Book Society's choice for May is Valentin Kataev's 'The Embezzlers'. 1938 Times 5 Jan. 12/3 He was not conscious of any bad effect on his own business of book clubs and book societies. 1967 E. Grierson Crime of One's Own i. 16 'Sold over five thousand and a Book Society Recommend,' its creator declared with pride. 1900 Library Jrnl. Nov. 679/2 Convenient elevators for passengers and freight are provided in the book-stacks. 1968 Bodl. Libr. Rec. VIII. 60 Mobile shelving in the bookstack. 1893 Funk's Standard Dict., Bookstamp. 1909 Cyril Davenport (title) English Heraldic Bookstamps. 1807 W. Scott Let. 13 Jan. (1932) I. 346 The great genius who invented the gilded inlaid or Japan bookstands for boudoirs & drawing rooms. 1891 Kipling Light that Failed xiv. 303 A bookstand that supported a pile of sketch-books. 1895 G. Stikeman Adjustable Book Shelving 4 Book supports, for partially filled shelves. 1829 M. Hare Let. 12 Sept. in A. J. C. Hare Mem. Quiet Life (1872) I. vi. 275, I ordered a book table according to my own fancy, having two shelves above, a bureau part, and shelves below, with a cupboard at each end. 1905 Daily Chron. 23 May 4/6 A lovely inlaid book-table. 1932 Book Tokens (Nat. Book Council) 5 In order to give readers some idea of the appearance of the Book Token this leaflet has been made similar in size and format. 1938 Times 5 Jan. 12/3 They had a 7s. 6d. book token which they exchanged for one 5s. book and five 6d. books. 1875 T. Seaton Fret Cutting iv. 42 It is a book-tray end; the full size is six inches by five. 1916 E. F. Benson David Blaize vii. 132, I love looking through old book-trays. 1907 Yesterday's Shopping (1969) 131/1 Book Troughs ..Fumed oak, length 16 in...with repoussé plates..4/3. 1929 E. Bowen Last Sept. I. ii. 17 She glanced intently along the books in the book-trough. 1961 T. Landau Encycl. Librarianship (ed. 2) 52/1 Book trough, a V-shaped wooden shelf or rack placed on library tables for the display of books in such a manner that the titles are clearly visible. 1888 Encycl. Brit. XXIII. 699/1 Types are divided into two classes—book type, including Roman and Italic, and job type. 1899 Westm. Gaz. 14 June 6/1 Eight years ago the book value of the Stella was £60,000. 1952 Economist 6 Sept. 575 The five main companies have fixed assets totalling more than £6 million in book value, but worth probably four times that at replacement values. 1889 Cent. Dict., Book-work. 1926 W. H. Slater What a Compositor should Know III. 1 The work of the composing department..is divided roughly into two sections: 'Bookwork' and 'Job-work'. 1858 Bagehot Coll. Wks. (1965) I. 332 In the book-world they [sc. the Liberal party] enjoyed a domination. 1906 Daily Chron. 7 May 3/5 The book-world, which is the edited reflection of life, brings the great facts of contrast into added prominence.

book, *v.* Add: **2. c.** To make an entry of or against a person's name; *esp.* to enter (a name) in a police register for an alleged offence; see also quot. 1846.

1841 Fistiana 58 The names of individuals of distinction were 'booked' for indictment, should the prosecution of the principal..end in a conviction. 1846 Snowden Magistrate's Assistant 344 Caught, taken, or disposed of: booked. 1902 Wodehouse Pothunters iii. 49 If he books a chap out of bounds it keeps him happy for a week. 1935 Steinbeck Tortilla Flat viii. 122 The police sergeant said he hadn't booked them for a long time. 1961 P. Barry Unwillingly to School xv. 204 If you hadn't been a learner driver..I'd have booked you for that!

d. To put (tackle) in a fishing-book.

1892 Field 18 June 922/3 We therefore book our cast, and wind up for the day.

3. b. To enter (the arrival or departure of an employee, hotel guest, etc.) in a book; so *to book in, out.* Also intr. *to book off,* to sign an attendance book on going off duty.

1902 Daily Chron. 13 May 10/5 Baker's..Bookkeeper.—Young lady required, with good experience, to book men and keep books. Ibid., Wanted young lady,..one able to book in. Ibid., Shopwoman wanted, capable of booking out men. 1928 Observer 3 Feb. 12/7 To-night he drives his engine for the last time. To-morrow he 'books off'..and the Line knows him no more. 1958 Spectator 22 Aug. 251/1 Booking in immediately before the flight. 1958 Times 3 Sept. 13/4 O'Brien-Greer booked in at the hotel on August 20.

4. b. Also intr. *to book through*: to obtain a ticket to cover a whole journey.

1844 Dickens Mart. Chuz. li. 592 The other [man], seating himself on the steps of the coach, remained in conversation with Slyme...'He's booked,' observed the man. 'Through,' said Slyme. 1858 Penny Cycl. 2nd Suppl. 565/2 A man may now 'book through' from London to so many continental cities. 1884 [in Dict.].

c. (Earlier example.)

1807 Lamb Let. 29 Jan. (1868) I. 251 Dear Wordsworth—We have book'd off from Swan and Two Necks, Lad Lane, this day (per coach) the Tales from Shakspear.

5. (Further examples.) Also with *up*.

1906 Daily Chron. 20 Oct. 7/2 The defendant..wrote: 'Thanks for calling. I am pleased to book you for four—or it may be five Sundays.'..Mr. —..said he was engaged at £2. 2. 0 a day. 1923 Wodehouse Inimit. Jeeves iii. 35 Before I went I had been booked up to take brother and the girl for a nice drive that afternoon.

bookable (bu·kăb'l), *a.* [f. Book *v.* + -able.] That may be booked.

1903 Daily Chron. 30 Mar. 6/6, I think some [seats] should be bookable, and some should be unreserved. 1926 Music & Letters Apr. 102 The house..would be designed..to accommodate the largest possible audience..with all seats bookable.

bookbindery. (Earlier examples.)

1815 Niles' Reg. VIII. 141/2 There are [sc. in Pittsburgh] ..4 book binderies. 1842 [see *bakery 2].

bookbinding, *vbl. sb.* (Earlier example.)

1771 B. Franklin Autobiogr. in Life & Writings (1905) I. 290 Meredith was to work at press, Potts at book-binding.

bookcase. (Earlier example.)

1726 in N. & Q. (1942) CLXXXII. 47/1 [A London shop-sign] Desk & Bookcase. James Field, cabinet-maker.

booked, *ppl. a.* Add: **4.** Entered in an official book or list; scheduled.

1892 Daily News 3 June 5/4 That the Board of Trade shall have compulsory powers..to order a railway company to revise the booked time of the men. 1898 Ibid. 15 Dec. 7/4 Certain booked trains will be discontinued.

5. Having (a specified amount of) orders or engagements in one's book or books. Also with *up*; and in transf. sense: having engagements, engaged (cf. 3).

1905 Westm. Gaz. 3 Oct. 9/1 The iron, steel, and allied trades are heavily booked. 1911 John Vincent in Rep. Labour & Social Cond. Germany III. Nos. vi and vii. 52 We found..the trades were busy and well booked up with orders. 1952 Granville Dict. Theatr. Terms 30 Booked up, all seats reserved for a performance.

bookful, *a.* Restrict †*Obs.* to sense in Dict. and add: **2.** Full or stored with books.

1896 Literary World 7 Aug. 115/2 Coleridge, in this bookful age, is..getting to be more and more nominis umbra. 1904 Daily Chron. 3 Oct. 3/4 In bookful loneliness.

bookie[2] (bu·ki). Also **booky.** Colloquial modification [see -y[6]] of Book-maker 3.

1885 [see -y[6]]. 1887 Pall Mall Gaz. 15 Sept. 5/1 Both played their parts so well that they collared both the 'bookies' without any trouble at all. 1934 Sarovan Daring Young Man (1935) 219 Red, the bookie-clerk..had been stabbed by a crazy Russian who had lost twenty dollars on the ponies. 1949 N. Algren in Penguin New Writing XXXVI. 95 A tavern with a bookie in the back. 1968 Times 15 Nov. 16/2 One of his thirties-style bookie check suits.

booking, *vbl. sb.* Add: **4. booking-office,** also, the place where tickets are sold for a theatre or other place of entertainment (1889 *Cent. Dict.*).

1948 Mencken Amer. Lang. Suppl. I. 453 Punch's theatre article used to be headed 'Our Booking Office'. Today everybody speaks and writes of the box-office of a theatre. Only a railway ticket-office is a booking-office. 1952 Granville Dict. Theatr. Terms 30 Booking office, a theatre-booking agency in the city. Patrons are able to book seats through the agency, which communicates with the theatre concerned.

boo·k-keep, *v. intr.* [Back-formation f. Book-keeping.] To do book-keeping, keep books.

1886 G. B. Shaw Cashel Byron's Profession Prologue 12, I could book-keep by double entry. 1917 Dialect Notes 408 Book-keep, to act as bookkeeper. 'He book-kept for the camp.'

book-making. Add: **1.** (Later examples.)

1899 T. Veblen Theory of Leisure Class vi. 162 Artistic book-making. 1930 Publishers' Weekly 5 Apr. 1892/1 The increasing attention that the book-trade is giving to the art of book-making.

2. (Examples of *attrib.* use.)

1807 Cabinet I. 113 This is a fine book-making age. a 1856 in K. H. Digby Lover's Seat (1856) II. xviii. 222 Of all the books in this book-making world the philosophical books are the least intelligible.

Hence (as back-formation) **boo·k-make** *v. intr.* (*rare*).

1819 Byron Let. 6 Apr. (1900) IV. xvii. 284, I could have spun the thoughts of the four cantos of that poem into twenty, had I wanted to book-make. 1845 R. Browning Let. 16 Apr. (1899) 48 Mrs Norton has gone and book-made at a great rate about the Prince of Wales.

3. (Earlier examples.)

1824 Sporting Mag. XV. N.S. 51/2 Betting at present proceeds but slowly..what is done consists merely in book-making and speculation. 1836 R. S. Surtees in Mrs. Mathews Mem. C. Mathews (1839) IV. ix. 184 He entered into the spirit and excitement of the thing with the true ardour of a turfite, without any knowledge however of the science of book-making.

book-store. Chiefly *U.S.* [Book *sb.* 17 a.] A bookshop.

1763 Boston Even. Post 3/3. 1785 A. Ellicott in C. V. Mathews Life & Lett. (1908) 51 Examined several book-stores. 1810, 1884 [see Book sb. 17 a]. 1959 Times Lit. Suppl. 6 Nov. p. xx/1 The paperback revolution is at work everywhere, but especially in college towns and in college bookstores.

booksy (bu·ksi), *a. colloq.* [f. books pl. of Book *sb.* + -y[1].] Having literary or bookish pretensions; usu. in jocular or derisory use.

1934 A. Woollcott While Rome Burns III. iv. 185 The library of this booksy Aspasia. 1945 'M. Innes' Appleby's End ii. 11 In Mr. Raven's presence everything turned booksy. 1958 Woman's Own 3 Sept. 55/1 I'm going to borrow Jack and snitch him away from your booksy conversations.

booky, *a.* (Earlier and later examples.)

1832 Paulding Westward Ho! I. xiii. 194 You're one of the booky fellers, that think on one thing while they are talking about another. 1941 Masefield Gautama the Enlightened 13 Basil Blackwell's shop..with above Some booky house Half England's scholars nibble books or browse.

booky. Slang representation of Bouquet. Cf. *bokay.

1848 Thackeray Van. Fair i. 2 'We have made her a bow-pot.' 'Say a bouquet, sister Jemima, 'tis more genteel.' 'Well, a booky as big almost as a hay-stack.'

Boolean (bū·liăn), *a.* Also **Boolian.** [f. the proper name *Boole* (see below) + -an, -ian.] Of or pertaining to the work of George Boole (1815–64), English mathematician and logician; *Boolean algebra*, an abstract system of postulates and symbols applicable to problems in logic and the manipulation of sets; a Boolean ring; *Boolean expansion*, an expansion of a Boolean expression involving 'or' in terms of a logically equivalent series of expressions each involving only 'and'; *Boolean operation* (see quot. 1962); *Boolean ring*, a ring with unity in which every element is idempotent.

1851 Cambr. & Dublin Math. Jrnl. VI. 192 The Hessian, or as it ought to be termed, the first Boolian Determinant. 1889 Cent. Dict., Boolian algebra. 1913 Trans. Amer. Math. Soc. XIV. 481 A set of five independent postulates for Boolean algebras. 1924 Ibid. XXVI. 175 A commutative boolean operation which always has an inverse is also associative. 1936 M. H. Stone in Ibid. XL. 38 We shall..take as the central theme of this paper not merely Boolean algebras, but, more generally, rings in which every element is idempotent, designating the latter systems as Boolean rings or generalized Boolean algebras. 1948 A. Ambrose & Lazerowitz Fund. Symbolic Logic v. 78 Such a list of the conjunctive conditions for the truth of a disjunction will be called a Boolean expansion, because it expands a function into a form which exhibits all the conjunctive possibilities for its being true. 1955 J. L. Austin How to do Things with Words (1962) ii. 17 Despite the name, you are not when bigamous marry twice. (In short, the algebra of marriage is Boolean.) 1962 Gloss. Autom. Data Processing (B.S.I.) 29 Boolean operation, an operation depending on the application of the rules of Boolean algebra. By extension, any operation in which the operands and results take either one of two values or states, i.e. any logical operation on single binary digits. 1964 Language XL. 366 The designatum of the whole expression is a Boolean sum of the designata of the members.

boom, *sb.*[2] Add: **1. c.** *Aeronaut.* (See quot.)

1916 H. Barber Aeroplane Speaks 135 Boom, a term usually applied to the long spars joining the tail of a 'pusher' aeroplane to its main lifting surface.

d. A movable bar supporting a microphone or camera. Also *attrib.*

1931 [see *microphone 3]. 1932 Wireless World 10 Feb. 146/1 A microphone suspended on a boom. 1958 Manch. Guardian 10 June 6/2 The narrative swings as if on a boom-arm from character to character. 1959 Viewpoint July 34 The boom operator has found his first boom, microphone, shadow.

3. b. *pl.* The floating timbers placed between portions of the lines of piles marking the regatta course at Henley-on-Thames, to prevent the encroachment of boats during a race.

1899 Daily News 5 July 4/7 It will be impossible to pronounce definitely on the success of the booms until the regatta is in full swing. 1902 Encycl. Brit. XXXII. 306/2.

4. b. *fender, glancing,* or *sheer boom*, a boom erected to guide logs in the desired direction. *U.S.*

1896 U.S. Monthly Weather Rev. Nov. 407 (Cent. Dict. Suppl.), The driving of piles..to hold a sheer boom for the purpose of running the logs.

5. (sense 4) *boom boat, -chain, fence, log, -man, -stick* (earlier example), *timber* (all *N. Amer.*); **boom end** *v. trans.*, to run out a boom or spar at the end of a yard so as to extend the foot of a sail; **boom mat** (cf. Mat *sb.*[1] 4); **boom net,** a fishing-net connected with a boom.

1966 Canad. Geogr. Jrnl. June 214/3 Log booms, loose logs, and boom boats were swept out of the bay. 1969 Islander (Victoria, B.C.) 7 Sept. 10/2 Mainland Point had been en route up coast to pick up a tow, with a boom boat tied astern. 1883 J. Fraser Shanty, Forest, & River Life 281 The first business of the drive is to..con[illegible] there by long half-square logs called 'boom timbers' fastened at the ends by 'boom chains'. 1969 Islander (Victoria, B.C.)

31 Aug. 16/3 He was an excellent swimmer but was carrying boom chains around his shoulders as he went out on the booms. 1840 R. H. Dana Bef. Mast xxvii. 90 The studding-sail halyards were let go, and the yards boom-ended. 1890 Clark Russell Ocean Trag. I. viii. 164 Then boom-ending her fore-topmast studdingsail she backed her main topsail. 1848 Thoreau Maine Woods (1864) 42 Surrounded by a boom fence of floating logs. 1945 Reader's Digest Aug. 86 He and Dad are looking for boom logs. 1908 M. A. Grainger Woodsmen of West 28 The work for a practised boom-man, was now to take a long, light pole, and jumping upon a floating log, to stand upon the log and pole it into the boom-stick enclosure. 1883 Man. Seamanship Boys' Training Ships 182 In making a heavy boom mat a fiddle is used instead of a loom. 1925 Glasgow Herald 25 Aug. 5 The relationships of the young year classes caught by boom-net and seine-net to the mature herring in the drift-net fishery. 1850 S. Judd R. Edney xvi. 207 They found Chuk in trouble; his guys had parted, and his boom-sticks were broken. 1883 Boom timber [see boom chain above].

boom, *sb.*[3] For *U.S.* read 'orig. *U.S.*', and add: **1.** (Later examples.)

1911 E. M. Clowes On Wallaby ii. 31 The Land Boom—'the Boom', as it is always called..had a most potentially humanizing effect on the people. 1936 M. Plowman Faith called Pacifism 28 The people of this country were enjoying a post-war boom. 1955 Bull. Atomic Sci. Mar. 88/2 Thus the uranium boom began. 1966 Economist 19 Nov. 778 The country is in boom and therefore deficit.

b. Phr. *boom and (or) bust*: a period of great prosperity followed by a severe depression. orig. *U.S.*

1943 H. S. Canby Walt Whitman iii. 18 The building trade, as usual, suffered from boom-and-bust. 1947 D. Riesman in Yale Law Jrnl. Dec. 194 The luxury market would be..entitled..to its privilege of boom and bust. 1962 Times Lit. Suppl. 13 July 502/1 Cataclysmic alternatives—destruction or utopia, boom or bust.

4. *attrib.*, as **boom city, town,** one that owes its origin, growth, or prosperity to a boom in a particular trade, mining, etc.

1886 Leslie's Pop. Monthly XXI. 306/1 Cities..whose inhabitants had yet to be gathered in from the four corners of the earth by boom magic. 1891 Anthony's Photogr. Bull. IV. 23 The old town of San Juan Capistrano I found had been divided in the 'boom' times. 1896 Chautauquan XXIII. 219/1 Who first began to talk about..'boom' towns? 1904 Philad. Public Ledger 14 June 6 No 'boom city' of the West can boast such a record of amazing..growth. 1944 J. S. Huxley On Living in Revol. xiii. 138 The boom period in 1928. 1955 Times 20 May 14/1 Some call Wolverhampton a 'boom town'. There are full employment, high wages, and no industrial unrest. 1969 Scotsman 25 Apr. 9/3 The country's two boom areas—the South-east of England and West Midlands.

boom, *v.*[2] Add: **2. b.** To protect (a regatta course) from encroachment by pleasure boats during a race, by placing floating booms (*boom *sb.*[2] 3 b) between the piles. Also with *off.* Hence **boo·ming** *vbl. sb.*

1899 Daily News 29 Apr. 7/3 It is proposed to 'boom' the whole length of the course along the Bucks side and for some distance on the Berks side. 1900 Ibid. 6 July 7/4 Owing to the booming of the course, the expenses of the regatta are much heavier. 1902 Encycl. Brit. XXXII. 306/2 This proposal to boom off the [regatta] course was very severely criticized.

boom, *v.*[3] For *U.S.* read 'orig. *U.S.*' and add: **1.** (Earlier examples.)

1871 'Mark Twain' Lett. to Publishers (1967) 55 My popularity is booming, now. 1875 Scribner's Mag. July 272 Stocks may 'boom' to-day, but droop to-morrow.

2. (Further examples.)

1891 Confectioners' Union 15 Aug. 451/1 Messrs. J. S. Fry..are booming their manufactures in a novel way. 1894 Country Gentlemen's Catal. 15/1 If..cider [were] properly 'boomed', the consumption of it might be vastly increased. 1901 Daily Colonist (Victoria, B.C.) 27 Oct. 10/1 He says the country was boomed out of all reason, and is now suffering from an inevitable reaction. 1915 W. Owen Let. 5 Mar. (1967) 325 My prime object is not..to boom a monster business. 1926 Spectator 30 Oct. 752/2 Dreyer's 'diaplytes', most deplorably boomed here, with official support. 1966 New Statesman 25 Mar. 414/2 One minor political figure in Alabama, a certain Shorty Price, decided to boom his own wife for governor.

boom-boat. (Also **bomboat.**) [Boom *sb.*[2]] Any of the boats stowed in the booms of a vessel.

1867 Smyth Sailor's Word-bk. 1874 Bedford Sailor's Pocket Bk. vii. 226 By placing boats inside her present boom boats. 1893 Daily News 3 July 5/6 Seeing the critical condition of the Victoria,..boats were immediately prepared and boom boats got ready to be hoisted out.

boomer.[2] Add: Also formerly **boomah.** (Earlier and later examples.)

1830 Ross Hobart Town Almanack 110 (Morris), Snapped the boomah's haunches, and he turned round to offer battle. 1852 Mrs. Meredith My Home in Tasmania I. 244 (Morris), The oldest and heaviest male of the herd was called a 'Boomer'. 1925 H. Graham Last of Biffins ii. 28 A man whom even the older kangaroos ('boomers' as they are locally called) looked up to. 1967 Sunday Mail Mag. (Brisbane) 16 Apr. 2 Catching boomers isn't easy because they are big kangaroos.

b. *transf.* Something very large or notable of its kind. *Austral. slang.*

1885 Australasian Printers' Keepsake 76 (Morris), When the shades of evening come, I choose a boomer of a gum. 1928 'Brent of Bin Bin' Up Country xiii. 135 Old Healey is always telling some boomer of a scandal behind people's backs. 1941 Baker Dict. Austral. Slang 12 Boomer, a particularly ambitious lie. Used loosely to denote anything large or noteworthy. 1962 Austral. Women's Wkly. Suppl. 24 Oct. 3/1 Boomer, big wave.

boomerang, *sb.* Add: **2.** *attrib.* and *Comb.*; esp. *fig.* (with reference to its action in returning to the thrower).

1892 Daily News 30 June 5/5 It is..quite a boomerang business. Tories built the fort in anticipation of the battle of 1880, and to-day the Liberals hold it. 1898 Ibid. 21 Apr. 5/2 Rawson, the Australian boomerang thrower, will take part. 1949 C. Fry Lady's not for Burning II. 58 Boomerang rages and lunacies. 1955 Times 20 May 10/6 The..agitation ..may ultimately have a boomerang effect on the party itself. 1968 Daily Tel. 16 Dec. 1/8 Russia's unmanned Zond 5 and 6 made successful 'boomerang' flights round the moon.

boomerang (bū·mərӕŋ), *v.* [f. the sb.] *intr.* To throw a boomerang; to fly back to the starting-point, after the manner of a boomerang when thrown; also *fig.* Also with advbs. *off, on.* So **boo·meranging** *vbl. sb.*

1880 J. B. Stephens Misc. Poems 26 War shouts and universal boomeranging. 1899 Longman's Mag. XXXIII. 475 Boomeranging is dangerous for on-lookers, till the thrower is a perfect master of his weapon. 1900 H. Lawson On Track 59 The horse..boomeranged off again and broke away through the scrub. 1936 Discovery Apr. 125/2 It is easily possible for a deflecting wind to cause a shot to boomerang. 1943 H. Pearson Conan Doyle iii. 39 If your magnet is so strong as all that, you would have your own broadside boomeranging back upon you. 1965 Listener 24 June 928/1 But the psychologically dangerous stratagem soon boomeranged.

booming, *ppl. a.* **2.** For '*U.S.*' read 'orig. *U.S.*' and add further examples.

1896 J. A. Hobson Probl. of Unempl. 50 Where improvements of machinery occur during periods of 'booming' trade they do not occasion any large quantity of unemployment or distress. 1964 Sun-Herald (Sydney) 21 June 136/4 The railways have been doing booming business this year. 1968 Times 9 Dec. 7/2 The booming new science of infra-red astronomy.

boomlet (bū·mlet). orig. *U.S.* [f. Boom *sb.*[3] + -let.] A small boom, esp. on the Stock Exchange.

1880 Puck (U.S.) 21 Apr. 106/3 All the other booms and alleged boomlets. 1897 H. Bottomley in J. R. Ware Passing Eng. (1909) s.v., I may mention that during the recent West Australian boom—or, as some of my Stock Exchange friends prefer to call it, 'boomlet'—we succeeded in realising, etc. 1898 Westm. Gaz. 23 Apr. 9/2 A very favourable feature has been a recovery in Westralian stocks. Something like a boomlet prevailed at first. 1900 London Stock Market Rep. 18 Aug. 5 What may be called a Republican boomlet has been started in the Yankee corner. 1961 New Statesman 27 Oct. 626/2 The boomlet in shipping shares collapsed at once. 1966 Economist 5 Mar. 919/1 A little boomlet in consumer demand towards the end of last year.

‖**boomslang** (bū·mslӕŋ, bō·mslaŋ). *S.* and *E. Afr.* [Afrikaans, f. *boom* tree + *slang* snake.] The tree-snake *Dispholidus typus.*

1793 tr. Thunberg's Trav. I. 156 A serpent..was found here, called the Boomslang, or (tree serpent) on account of its being frequently found in trees. 1849 Andrew Smith Illustr. Zool. S. Afr., Reptilia Plate xiii, The natives of South Africa regard the Boom-slange as poisonous; but in their opinion we cannot concur. The Boom-slange [Bucephalus capensis] is generally found upon trees. 1912 Fitzsimons Snakes S. Afr. 136. 1952 Cape Argus 7 June 2/7 Cobras, puff-adders, boomslangs and other deadly snakes. 1970 Cape Times 22 Jan. 6/3 The ban applies to black mambas as well as green..and also to cobras, rinkhalse, boomslange..and the..harmless mole snake.

boomy (bū·mi), *a.* [f. Boom *sb.*[1] + -y[1].] Having the noise or quality of a boom; = Booming *ppl. a.* 1.

1909 Westm. Gaz. 31 Mar. 2/3 The motor-omnibus..was too bulky, boomy, bumpy. 1939 Amat. Radio Handbk. viii. 116 If the speaker is too close to the microphone the quality will tend to be 'boomy'. 1956 E. Delaney in S. Traill Play that Music 55 Tune it properly: side-drum crisp, bass-drum nice and deep—not boomy but thick.

boondock (bū·ndǫk). *U.S. slang.* [ad. Tagalog *bundok* mountain.] Rough country; jungle; an isolated or wild region. Usu. in *pl.* Also *attrib.*

1944 C. Wynn in C. Metcalf Marine Corps Reader III. 139 The sand and boondocks of Paris Island. 1950 H. L. Miller in Word Study Oct. 7/1 Today Marines use boondock clothes and boondock shoes for hikes and maneuvers. 1951 N.Y. Times Mag. 10 June 39/1 Sculthorpe is known to its temporary guests as 'the boondocks', as much for its isolation as anything else. 1965 Spectator 12 Mar. 317/1 Those who have been feeling the public pulse out in the boondocks report a good deal of unrest.

Hence **boo·ndockers,** shoes suitable for rough outdoor use.

1953 L. M. Uris Battle Cry II. iii. 127 Andy Hookans was dumping a can of footpowder into his boondockers. 1957 New Yorker 16 Nov. 193/1 My hacking jacket and pair of old Marine Corps boondockers.

boondoggle (bū·ndǫg'l), *sb.* and *v. U.S. slang.* [Origin unknown.] **A.** *sb.* **a.** (See quots. 1935.) **b.** A trivial, useless, or unnecessary undertaking; wasteful expenditure. **B.** *v. intr.* To engage in trifling or frivolous work. Hence **boo·ndoggler, boo·ndoggling.**

1935 R. Marshall in N.Y. Times 4 Apr. 2 'Boon-doggles' is simply a term applied back in the pioneer days to what we call gadgets today. 1935 Word Study Sept. 2 Boondoggle was coined for another purpose by Robert H. Link of Rochester. Through his connection with scouting the word later came into general use as a name given to the braided leather lanyard made and worn by Boy Scouts. 1935 Chicago Tribune 4 Oct., To the cowboy it meant the making of saddle trappings out of odds and ends of leather, and they boondoggled when there was nothing else to do on the ranch. 1935 H. L. Ickes Secret Diary (1953) I. 435 I am for substantial, worth-while, and socially desirable public works, while Hopkins is for what has come to be known as boondoggling. 1937 Amer. Speech XII. 6 [In the 1936 American election] boondoggling became the current term for describing the waste assertedly evident in..government agencies and bureaus. Administrators of relief became boondogglers to the Republican press and orators. 1947 Chicago Tribune 8 June I. 22/2 The cost of this boondoggle has been estimated at perhaps 50 million dollars. 1949 R. K. Merton Social Theory (1951) vi. 178 This eliminates the very rationale of the intellectual's work and dissipates his interest in his work, leading to the 'boon doggling neurosis'. 1969 N.Y. Rev. Books 2 Jan. 5 (heading) Nixon and the arms race: the bomber boondoggle.

boong (bŭŋ). *Austral. slang.* Also **boang.** [Aboriginal word.] An (Australian) Aboriginal; a native of New Guinea; also, any coloured person.

1941 K. Tennant Battlers ix. 110 Those boangs are all too matey with the police. 1943 Baker Dict. Austral. Slang (ed. 3) 13 Boong, an aboriginal; a native of New Guinea. 1945 R. J. Oakes in Coast to Coast 1944 99 He had six wounded men...and four boongs to help him. 1950 'N. Shute' Town like Alice 75 'I thought you were a lot of boongs,' he said. 'You say you're English.' 1962 Economist 10 Nov. 577/1 In Australian eyes, Indonesia was a nation of poor struggling 'boongs' (a slang word which in its narrowest sense is applied to the natives of New Guinea). 1969 Times 1 Apr. p. ii/5 He is trying to lead Australians away from what he calls the 'poor old bloody boong' mentality.

boongary (bŭngēə·ri). [Native name: *bangaray* in the Port Jackson dialect.] The tree-kangaroo of North Queensland, *Dendrolagus lumholtzii.*

1889 R. B. Anderson tr. Lumholtz's Among Cannibals xviii. 227 Upon the whole, the boongary is the most beautiful mammal I have seen in Australia. 1965 E. Troughton Furred Animals Austral. (ed. 8) 175 The blacks concluded the Boongary never drank.

boorka, var. *burka[1, 2].

boost, *sb.*[2] For '*U.S. colloq.*' read '*colloq.* (orig. *U.S.*)'. Add to def.: help, encouragement (by means of publicity, etc.), increase (in value, reputation, etc.). Also, the action of *boost *v.* 2. Add earlier and later examples.

1825 J. Neal Bro. Jonathan II. 101 (Th.), Shall I give him a boost? or no? 1830 Seba Smith Major Jack Downing (1860) 114 (Th.), I got a pretty good boost in Boston, by the editors giving me recommendations. 1919 Wodehouse My Man Jeeves 24 You see. A boost for the uncle right away. 1958 Spectator 6 June 722/2 After this initial boost, recruiting will fall off again. 1964 Daily Tel. (Colour Suppl.) 2 Oct. 7/1 A piece of national prestige-building and a boost for exports too. 1967 Listener 6 Apr. 473/3 The recording needs treble boost, but is otherwise excellent. 1969 Gramophone Apr. 1502/3 A small amount of bass boost gives a well-balanced performance. 1969 Scotsman 26 Apr. 6/6 [Scottish manufacturers] have expressed the opinion that they are doing all right..with their exports and need no additional boosts.

2. In an internal-combustion or jet engine: = *supercharging *vbl. sb.*; supercharger pressure. Also, = *booster 2 c. Hence *attrib.* and *Comb.*

1931 Air Ann. Brit. Empire 88 It has been necessary to evolve a practical automatic boost-controlling mechanism. 1931 Handbk. Aeronaut. viii. 500 Boost Gauge, Mark II A. 1934 Aircraft Engin. Apr. 110/3 Boost pressure is the difference between induction pipe pressure and 760 mm. of mercury. 1941 P. Richey Fighter Pilot 57, I immediately started clambering after them, with my 'plug' (boost override) pulled. 1943 L. Cheshire Bomber Pilot i. 13 He tested..the airscrew pitch controls and the boost. 1950 Jrnl. Brit. Interplanet. Soc. IX. 174 The thrust of the boost motors. 1962 F. I. Ordway et al. Basic Astronautics ix. 385 The boost phase of flight.

boost, *v.* For '*U.S. colloq.*' read '*colloq.* (orig. *U.S.*)'. Add to def.: To assist over obstacles, to advance the progress of; to support, encourage; to increase (in value, reputation, etc.); to praise *up*, to extol; also *absol.* Add earlier and later examples.

1815 D. Humphreys Yankey in England 103 Boost, raise up, lift up, exalt. 1834 Seba Smith Major Jack Downing 139 You..give me a lift into public life, and you've been a boosting me along ever since. 1845 Yale Lit. Mag. XI. 34 (Th.), There is one poor fellow getting his comrade to boost him, while he hangs to the skirts of the one above. 1887 F. Francis Jr. Saddle & Mocassin 121 If you think that I'm trying to boost the place up because it belongs to us. 1909 Daily Chron. 21 Oct. 3/5 In times like these, when trade and other factors are bad, it is the duty of the Press to 'boost' in the interests of the nation, for when trade is good it will 'boost' itself. 1923 Wodehouse Inimit. Jeeves ii. 22 Young Bingo must have boosted me for some purpose. 1926 Publishers' Weekly 16 Jan. 147 Perhaps advertising might help boost their sales.

2. *Electr.* To increase or otherwise regulate the electromotive force in (a circuit, battery, etc.).

1906 A. Russell Theory Altern. Currents II. x. 282 The pressure can be boosted positively or negatively. 1911 Engineer 10 Mar. 237/2 It is far more profitable to boost the pressure at the generating station. 1959 H. Burstein Stereo, How it Works ii. 43 With stereo program material it is no longer necessary to boost the intensity above live concert level to hear fine details.

Hence **boo·sting** *vbl. sb.* and *ppl. a.*

1906 A. Russell Theory Altern. Currents II. x. 282 When a transformer is used in this fashion it is called an auto-transformer or a boosting transformer, or simply a booster. 1911 A. B. Smith Mod. Amer. Telephony v. 76 When the current is too small for effective transmission of signaling the necessity of 'boosting'..arises. 1912 Baltimore Amer. 5 Nov., The quiet citizen who has not indulged in boosting will decide the result. 1920 Chambers's Jrnl. Aug. 557/2 These arrangements are supplemented by 'boosting' fans at intervals of about 2000 feet. 1962 Nisbett Technique Sound Studio i. 17 Switching centres, boosting amplifiers and frequency correction networks on landlines.

booster (bū·stəɹ). *colloq.* (orig. *U.S.*). [f. Boost *v.* + -er[1].]

1. One who boosts.

1890 Stock Grower 1 Feb. 3/2 The..company, of which Col. Partial Smyth is chief booster, seems to be worried. 1920 Glasgow Herald 23 Aug. 7 It is not suggested that we should become 'blatant-boosters'. 1928 Observer 1 July 11/2 President Coolidge, a true 'booster' of his country. 1937 G. B. Shaw in Malvern Fest. Bk. 8, I am a booster of festivals because they are markets for my plays. 1968 Globe & Mail (Toronto) 17 Feb. 1/5 An ardent Trudeau booster.

2. *Electr.* A machine interposed in a circuit for the purpose of increasing or otherwise regulating the electromotive force acting in the circuit.

1894 Jrnl. Inst. Electr. Engin. 25 Jan. 79, I found in use in Chicago a system of raising the voltage on the feeders when the pressure falls, owing to the increase of load, by switching in motor dynamos. They have given the name 'boosters' to these instruments. 1902 Encycl. Brit. XXV. 34/1 Taking the current from the mains to the battery through a 'booster', that is, a dynamo arranged so that its E.M.F. is added to that of the mains. 1905 R. E. B. Crompton Unsolved Problems in Electr. Engin. 16 We have obtained regulation by means of automatic motor-generators, to which the name of 'boosters' has been generally given. 1906 Times (Engineering Suppl.) 22 Aug. 267/3 The use either of a large full-voltage battery for regulating purposes, a set of cells as back E.M.F.'s for the near feeders, or as a third alternative, the provision of extra boosters. 1911 Engineer 10 Mar. 237/3 A booster in use for boosting the pressure of an overhead conductor of an electric tramway system. 1911 A. B. Smith Mod. Amer. Telephony v. 77 Each line must be provided with its own booster battery. 1924 Roget Dict. Electr. Terms 24/2 Booster Transformer (or Boosting Transformer), a transformer with its secondary in series with the line to compensate for line drop on an A.C. feeder.

b. A radio-frequency amplifier for intensifying signals.

1935 Nilson & Hornung Pract. Radio Communic. viii. 383 If the lines are long, it is necessary to insert booster amplifiers... For high-quality transmission such boosters should be inserted every 10 to 20 or more miles. 1954 Times 10 Nov. 2/6 The Truleigh Hill booster, installed to improve Sussex reception. 1959 Economist 7 Mar. 878/2 Hundreds of remote communities have owed their television fare to 'booster' stations that are built on high places and that relay and amplify the original broadcast signal.

c. *Aeronaut.* An auxiliary engine or rocket, esp. one used to give initial speed to a rocket or missile which is afterwards left to continue under its own power.

1944 A. L. Murphy Rockets 72 Some being catapulted by booster rockets, while other craft will take off as rockets. 1950 Jrnl. Brit. Interplanet. Soc. IX. 175 A single annular-shaped tank, the bottom of which is attached to the booster motors. 1961 Spectator 21 Apr. 546 The huge boosters that launched Major Gagarin's capsule. 1968 Times 16 Dec. 7/4 To begin with, Apollo 8, still attached to the third stage of the Saturn 5 booster, will be parked in a 119-mile earth orbit.

d. *Med.* A dose or injection of a substance that increases or prolongs the effectiveness of an earlier dose or injection. Also *attrib.* and *fig.*

1950 in Webster Add. 1953 Fishbein Med. Progress 66 The use of 'booster doses' will usually bring the process back under control. 1956 Collier's Year Bk. 563/1 The effect of the vaccine, once a child has had both primary shots and a booster, seven or more months later, will last for a long time. 1957 I. Asimov Earth is Room Enough (1960) 13 Additional work as a postdoctorate teaching fellow acted as a booster shot. 1957 Times 9 Dec. 11/5 In the case of diphtheria inoculations, it is usually considered necessary to give what is technically known as a 'booster' injection when the child starts school.

e. Other *techn.* uses.

1917 A. Marshall Explosives (ed. 2) II. xxxi. 527 Recently brass tubes 4 inches long containing a charge of trinitro-toluene have been introduced under the name of renforts or boosters. 1927 Engineering 11 Nov. 631 (caption) Booster pumps and compression tank. 1938 Fuller & Snow Air Conditioning v. 196 One of the standard types of booster heaters. 1945 Reader's Digest Apr. 97/2 The 74 booster stations in France's long-distance telephone system. 1958 Observer 15 June 10/2 Manufacturers [of shampoos] can meet these difficulties by..adding a foam booster.

3. A shoplifter. *U.S. slang.*

1912 A. H. Lewis Apaches of N.Y. (Partridge, Dict. Underworld), A gifted booster..of the feminine gender. 1925 Collier's Mag. 8 Aug. 30/1 The lowest type of thief is the 'booster'.

boosy. Add: *attrib.*, as **boosey close,** the close in which the cow-sheds stand; **boosey** or **boozy pasture,** pasture land lying near the cowsheds.

1794 T. Brown Gen. View Agric. Derby 45 A specified close, which the way-going tenant has for foddering his cattle in, under the name of a Boosey pasture. 1844 Jrnl. R. Agric. Soc. V. I. 86 Tenancy commences..on the 2nd of February as to all the other lands, except the boosey pasture. 1862 Catal. Internat. Exhib. II. x. 27 Boosey or manger bricks. 1905 Oswestry & Border Counties Advertizer (Advt.), 6 acres of Boozy Pasture. 1922 S. J. Weyman Ovington's Bank xxxvi. 423 A countryman, whom the news had only just reached in his boosey-close or his rickyard.

boot, *sb.*[3] Add: **1. b.** *the boot is on the other leg* (delete † and add examples); also *the boot (is) on the wrong leg* or *foot*; *to put* (or *sink*) *in the boot* or *to put the boot in* (esp. Austral. and N.Z.): to kick (in a brutal manner); also *fig.*; *boots and all* (Austral. and N.Z. colloq.): with no holds barred, wholeheartedly; also *attrib.*

1834 Carruthers Kentuckian in N.Y. I. 97 He's got the boot on the wrong leg. 1854 Whyte-Melville Gen. Bounce (1855) II. xvi. 47 The young woman as owns that house has got the boot on the other leg. 1861 'F. G. Trafford' City & Suburb 385 That's what I call putting the boot on the other leg with a vengeance. 1863 (title) The Boot on the Other Leg; or, Loyalty above Party. 1866 F. Moore Women of War 173 'Ah,' replied the jolly rebel, 'the boot is on the other foot now.' 1899 [see Leg sb. 2 a]. 1916 C. J. Dennis Songs Sentim. Bloke 42 Plunks Tyball through the gizzard wiv 'is sword, 'Ow I ongcored! 'Put in the boot!' I sez. 'Put in the boot.' c 1926 Transport Workers' Songbook (N.Z.) 107 All of them helping the worker down, by putting in the 'boot'. 1936 G. B. Shaw Simpleton iii. 29, I should say..that the boot is on the other leg. 1942 V. Palmer in Coast to Coast 26 Wait till he gets his opening, Charlie will, and then sink in the boot. 1947 D. M. Davin Rest of our Lives xix. 96 The next thing he'll do is counter-attack, boots and all. 1949 Economist 10 Sept. 566 [N.Z. correspondent] Longer political experience, a greater tactical sense and a 'boots-and-all' ruthlessness. 1955 Times 13 Aug. 7/2 When the boot was on the other foot and his own Democratic Party was in opposition the People's Party, then in power, deplored their rivals' use of the boycott weapon. 1955 V. Palmer Let Birds Fly 128 This could be no light affair for either of them. 'It's boots and all...Boots and all for both of us.' 1964 Guardian 2 Mar. 7/6 When he's lying there some cow in the front row puts the boot in.

c. *to give* (a person) *the boot* or *the order of the boot*: to 'kick out', dismiss, 'sack'. So *to get the boot.*

1888 Rider Haggard Col. Quaritch xii. 215 There'll be the money to take over the Moat Farm and give that varmint Janter the boot. 1904 Minister's Gaz. Fashion Dec. 219/2 His vivacious accounts of 'padding the hoof', getting the 'boot', [etc.]. 1917 'Taffrail' Sub ii. 62 An habitual slacker ..generally got the Order of the Boot at the end of his third term. 1927 R. A. Freeman Certain Dr. Thorndyke I. ii. 22 If you hadn't come I should have got the order of the boot to a certainty.

d. *U.S. slang.* A recruit at a boot camp (see sense 8).

1915 Recruiter's Bull. (U.S.) Apr. 11/1 One of the 'boots' transferred to the Recruit Depot recently. 1944 [see sense 8 below]. 1963 Amer. Speech XXVIII. 78 It is taught to the 'boot' before he leaves boot camp.

e. *slang.* A Negro.

1957 C. MacInnes City of Spades I. iii. 17 In England some foolish man may call me sambo, darkie, boot or munt or nigger. 1962 H. Simmons Man walking on Eggshells II. xxii. 162 A lot of paddy studs still didn't know that boots were human.

4. c. (Later examples.) Now the ordinary name for the luggage compartment usu. at the rear of a motor vehicle. Also *attrib.*

1868 W. E. Waters Life among Mormons 41 The mail is carried in the boots of stage-coaches. 1886 Leslie's Pop. Monthly XXI. 66/1 The great boot was securely strapped down over the baggage. 1933 Boy's Mag. XLVII. 35/1 The spare wheel is carried in an enclosed luggage boot at the rear. 1955 Times 29 June 12/6 The tailboard formed by the lowered boot-door.

d. *U.S.* (See quots.)

1828–32 Webster, Boot, 3...An apron or leathern cover for a gig or chair, to defend persons from rain and mud. This..application is local and improper. 1911 J. C. Lincoln Cap'n Warren's Wards i. 9 The 'boot' was a rubber curtain buttoned across the front of the buggy, extending from the dashboard to just below the level of the driver's eyes.

6. e. The feathered legs of some varieties of pigeons and poultry.

1855 Poultry Chron. III. 348/2 The boots, or as Shanghai fanciers would style it, the vulture hock, must be white. 1875 Contemp. Rev. XXVI. 949 Instances..in which the feet of pigeons or fowls are abnormally feathered, or, as it is termed, furnished with 'boots'.

f. *Ornith.* An entire tarsal envelope characteristic of the legs of some birds.

1864 E. Coues in Proc. Philad. Acad. Nat. Sci. 82 The very long tarsi present the remarkable feature of having their anterior and lateral aspects covered with one smooth unbroken podotheca or 'boot'.

7. *boot-heel* (examples), *-sponge* (example), *-upper* (Upper *sb.* 1).

1870 'Mark Twain' Sk. New & Old (1875) 99 The rims of his boot-heels. 1891 Kipling Light that Failed vii. 117 You're a work-woman, darling, to your boot-heels. 1920 D. H. Lawrence Lost Girl vii. 139 Striking a match on his boot-heel. 1863 Dickens Mrs. Lirriper's Lodgings in All Year Round (Christmas no.) 6/1 My boot-sponge was in my hand. a 1877 [see boot-crimp in 8 below]. 1879 Birmingham Weekly Post 21 June 5/3 Charged with..stealing a quantity of boot-uppers. 1906 Times 13 Dec. 4/2 The [stolen] boot uppers were found at the shop.

8. boot-black (earlier U.S. example); **boot-blacking,** (*a*) the polishing of boots and shoes; also *attrib.*; (*b*) blacking for polishing boots and shoes; **boot-boy,** a boy employed to clean boots and shoes; **boot camp** *U.S. slang*, a centre for the initial training of American naval or Marine recruits; **boot-clamp, -crimp** (see quots.); **boot-closer** (example); **boot-faced** *a. colloq.* [f. phr. *to have a sea boot face*, see quot. 1925], grim-faced, sad-faced; with an expressionless face; **boot-grain,** a cowhide leather used for heavy boots; **boot-hole,** the place where boots are cleaned in a large establishment; **boot-hook** (example); **boot machine operator, boot machinist,** any person engaged in any machine operation in the manufacture of boots; **bootman,** a dealer in boots and shoes; **boot powder,** a powder, as of soapstone, used for dusting the inside of a boot or shoe; **boot-rack** (examples).

1817 in Essex Inst. Hist. Coll. VIII. 246 They have a boot-black and barber. 1866 J. C. Gregg Life in Army 189 Here are..boot-blacking establishments. 1948 C. Day Lewis Otterbury Incident iv. 45 Ted had smears of boot-blacking on his face and hands. 1860 Trollope Tales of all Countries 58 'He, he, he,' laughed the boot boy as he turned the boot for me to look at. 1925 W. Deeping Sorrell & Son x. 84, I don't share out,—with the boot-boy. 1944 G. P. Bailey Boot; a Marine in the Making Foreword, Marine recruits are called 'Boots' and it is Marine Corps custom to send them all through a grim process called 'boot camp'. 1955 C. S. Forester Good Shepherd 45 That boy was one of the new draft, fresh out of boot-camp. a 1877 Knight Dict. Mech., Boot-clamp, a device for holding a boot while being sewed. 1824 J. Constable Let. 12 July (1964) II. 366 Mary is married to her cousin..a boot closer. a 1877 Knight Dict. Mech., Boot-crimp, a tool or a machine for giving shape to the pieces of leather designed for boot uppers. 1925 Fraser & Gibbons Soldier & Sailor Words 253 To have a sea boot face, to look gloomy. 1942 M. Dickens One Pair of Feet viii. 180 Everyone wore a face like a boot. 1958 W. Graham Foreign Affair ix. 111 Laughing nervously at the boot-faced British. 1961 Bookseller 12 Aug. 1087 Commenting on this remarkable achievement..the author nicely bootfaced. 1965 J. Porter Dover Two ix. 113 Dover came down to breakfast more boot-faced than ever and lost no time in burying himself in the morning paper. 1870 Little Folks II. 162/2 It was Grandmother who had to discover him sitting in the boot-hole under the stairs. 1840 'A Connoisseur' Fashionable Biogr. 83 When the Romans used boot-hooks, and of what kind and shape, we have not been able to ascertain. 1891 Daily News 11 Feb. 3/2 Boot machinist. 1927 Daily Express 27 May 9 [Obtainable] from all Bootmen. Fixing Extra. 1837 Dickens Pickw. xlix. 538 A boot-rack and boot-jack. 1907 Yesterday's Shopping (1969) 277/2 Boot Racks..18 in—.

boot, *v.*[3] Add: **4.** To kick (a person).

1877 Bartlett Dict. Amer. (ed. 4) 59 To 'boot' a man is to kick him. 1883 D. C. Murray By Gate of Sea II. xiii. 43, I have felt..an electric sensation in the right foot, indicative..of a desire to boot a noble swell or two who hold up her train. 1891 Daily News 11 Feb. 7/3 At him, lads..., and kick him! Kill him! 1892 Stevenson & Osbourne Wrecker xii. 195, I saw a big hulking beast of a Dutchman booting the ship's boy. 1913 'Ian Hay' Happy-go-Lucky ii. You will be booted for that afterwards, my lad.

b. To eject (a person); = Kick *v.*[1] 8 a. Also with *out.* Also *fig.*

1880 Harper's Mag. Dec. 160/2 He angrily bade the 'boot' to leave..and never show his face there again; if he did he would be booted out. 1902 Westm. Gaz. 13 Nov. 2/2 Who flocked together in the House of Lords..to 'boot' the Home Rule Bill. 1907 Daily Chron. 5 Nov. 10/4 That German scrub wants me to boot [him]. 1909 Westm. Gaz. 1 Dec. Burgess literally booted him out of his office. 1941 'R. West' Black Lamb I. 317 Were not the Turks booted out of here in 1878? 1966 B. Kimenye Kalasanda Revisited 45 Though he led the ladies with charming politeness to the office, his baser instincts were at the same time urging him to boot the whole lot of them down the steps.

5. To kick (the ball) with more than the usual vigour. *Football colloq.*

1914 Morning Post 2 Mar. 4/1 The ball was booted too hard and the defence got the touch down. 1932 A. J. Worrall English Idioms 14 The right-back booted the ball far up the field.

booted, *ppl. a.* Add: **2. b.** (Earlier and later examples.) Also applied to other birds.

1781 Latham Gen. Synop. Birds I. i. 75 Falcon... F... legs feathered to the toes. 1814 Lewis & Clarke Exped. to Sources of Missouri II. 182 The large black and white pheasant..is also booted to the toes. 1954 R. Peterson et al. Field Guide to Birds 72 Booted Eagle. Hieraëtus pennatus..Tarsi heavily feathered. 1966 Van Tyne & Berger Fund. Ornith. ii. 48 Several types of passerine tarsal scutellation have been described. Booted ('ocreate'): Scutella fused into a single smooth sheath or 'boot', except in some cases, at the very lower end.

bootee. (Examples.)

1799 Aurora (Philad.) 15 Nov. (Th.), For sale, 180 pairs of bootees. 1801 Spirit of Farmer's Museum 262 He is, in fine, made up of marechal powder, cravat, and bootees. 1844 Congress. Globe 11 Mar. 361/3 Men's boots or bootees of leather,..women's boots or bootees,..children's boots, bootees, and shoes. 1848 Bartlett Dict. Amer., Bootee,

dimin. of boot, a boot without a top, or a shoe made like a boot without a leg. 1848 Boag Eng. Dict. 155/1 Bootee, a word sometimes used for a half or short boot. 1861 Revised Regulations U.S. Army 238 Mounted men may receive one pair of 'boots' and two pair of 'bootees', instead of four pair of bootees. 1929 Sears Catal. Fall 76 These bootees make a warm and cozy place to tuck little feet.

bootery (bū·təri). *U.S.* [f. Boot *sb.*[3] + -ery.] A shop where boots and shoes are sold.

1920 N.Y. Times 25 Jan., [In San Francisco] You can buy your groceries in grocerterias..and your shoes in booteries. 1945 Steinbeck Cannery Row xi. 47 The final climax came with the front of Holman's bootery broken out and the party trying on shoes in the display window.

booth, *sb.* **2. b.** = *telephone booth.*

1930 Bell System Techn. Jrnl. IX. I. 12 A form of booth furnished by the telephone companies, provided with a seat and with lighting. 1952 A. Baron With Hope, Farewell 95 He stood in the booth, fumbling with his notebook to find the telephone number.

boot-lace (bū·tlēᵊs). [Boot *sb.*[3] 7.] **a.** A lace for fastening a boot or shoe (see Lace *sb.* 3 a).

b. (See quot. 1934.) Hence as *v. N.Z.* and *Austral.*

1934 Press (Christchurch, N.Z.) 13 Jan. 13/7 Boot laces, narrow strips of skin cut off by rough shearers, generally when opening up the neck. Hence the verb, to b. l. 1955 G. Bowen Wool Away! iii. 41 You are much more likely to bootlace the sheep—i.e. cut the wrinkles off in long thin strips of skin. 1965 J. S. Gunn Terminol. Shearing Industry I. 9 Good shearers could easily 'bootlace' wrinkly sheep so it is not the same as 'tomahawking'.

c. A piece of liquorice made in the shape of a boot-lace.

1935 W. de la Mare Early One Morning xviii. 230 Such casual bounties as brandy balls, pink sherbet, roast sparrow, 'boot laces', [etc.].

boot-leg, bootleg (bū·tleg). [Boot *sb.*[3] 7.]

1. The leg of a tall boot, or the leather, etc., cut out for this (see also quot. 1875).

1634 [see Boot sb.[3] 7]. 1843 Knickerbocker XXI. 523 A pair of linsey-woolsey breeches plunged into his boot-legs. 1855 M. M. Thomson Doesticks xxi. 181 The man who would..hopefully essay the concoction of a satisfactory stew from jack-knife-handles and bootlegs. 1875 W. D. Parish Dict. Sussex Dial. 20 Boot-legs, short gaiters, not reaching to the knee. 1889 'C. E. Craddock' Broomsedge Cove iv. 65 He..paused only to slip into his long boot-leg a 'shootin' iron'.

2. *attrib.* **a.** With reference to illicit trading in liquor. Also *ellipt.* Cf. *boot-legger. orig. *U.S.*

a 1889 Omaha Herald (Barrère & Leland), There is as much whisky consumed in Iowa now as there was before,..'for medical purposes only', and on the boot-leg plan. 1921 W. D. Newton in Chambers's Jrnl. Mar. 154/2 Joe left him apparently sleeping the solid sleep of 'bootleg' whisky in his shack. 1928 H. Crane Let. 31 Jan. (1965) 315 Gradually I'm becoming acquainted with all the brands of bootleg that the Westcoast offers. 1929 Morn. Post 7 June 11/2 Alleged bootleg ring. 1931 E. Linklater Juan in America II. ii. 70 She'll get nothing but bootleg rye and bath-tub gin..after this.

b. Of other commodities, persons, etc.

1928 Daily Express 5 Mar. 11/6 Bootleg [i.e. smuggled] baby. 1929 Variety 10 Apr. 1/2 There is almost as big a market for bootleg disk records as there is for bootlegged books. 1931 Daily Express 15 Oct. 11/2 A 'bootleg' house is one which has been erected in defiance of the Building Act. 1944 M. Laski Love on Supertax vii. 72 The occasional bootleg lemon. Ibid. ix. 84, I can't eat, said Clarissa, pushing away her bootleg egg.

Hence **boo·t-legger,** one who carries liquor in his boot-legs; hence, an illicit trader in liquor; **boo·t-legging,** illicit trade in liquor; also *attrib.* and *ppl. a.*; whence **boo·tleg** *v.*, to traffic illicitly in (liquor); **boo·tlegged** *ppl. a.*, illicit, smuggled; also *transf.*

1889 Sanger Rep. in J. B. Thoburn Hist. Oklahoma (1916) I. 223 Liquor dealers (or as they are called here 'boot-leggers'). 1890 Voice (N.Y.) July 17 The 'boot-legger' is a grim spectre to the anti-Prohibitionist... He is a man who wears boots in whose tops are concealed a flask or two of liquor. 1903 Cincinnati Enquirer 3 Jan. 1/3 In the pulpit was a minister when arrested for 'boot-legging'. 1906 in Dialect Notes III. 127 William Castell, charged with bootlegging whiskey, was tried..this morning. 1919 T. K. Holmes Man fr. Tall Timber vi. 59 Who's got the forty-rod, Steve?..There's a bootlegging place somewhere, I'll be bound. 1922 S. Lewis Babbitt v. 66 The bright lights and the bootlegged cocktails. 1927 'Sax Rohmer' Moon of Madness 14 A deck load of Dagos..that would have frightened a Chicago bootlegger. 1928 Observer 5 Feb. 18/2 Negroes who carried bootlegged liquor. Ibid., The result is that books are bootlegged in Boston as liquor is bootlegged in other cities. 1928 Daily Express 5 Mar. 11/6 She 'bootlegged' a baby into her home and..pretended to her husband that it was hers. 1936 Wodehouse Laughing Gas vii. 76 Somebody is bootlegging it [sc. candy] to him, and I mean to find out who it is. 1948 Chicago Tribune 28 Nov. 41/5 The sale of liquor is banned here on the Sabbath, and that's when the bootlegger bootlegs. 1958 Manch. Guardian Weekly 1 May 6/1 Customs officers.. regarded it as no business of theirs to prevent the 'boot-legging' of the long-play records imported privately in thousands.

boot-lick, bootlick (bū·tlik), *v.* and *sb. slang.* [f. Boot *sb.*[3] + Lick *v.*] **A.** *v. trans.* and *intr.* To curry favour (with); to toady to). **B.** *sb.* A toady, a sycophant; one who curries favour; also *attrib.* *U.S.*

1846 J. J. Hooper Adv. Simon Suggs (1851) v. 58 A young man who was inclined to boot-lick any body suspected of having money. 1849 Yale Banger 6 Nov. 6 When Boot-lick hypocrites upraised their might. 1850 Yale Battery 14 Feb., Then he arose, and offered himself as a 'boot lick', to the Faculty. 1885 Milnor (Dakota) Teller 1 May 2/3 They must drink, truckle, and bootlick to keep their greatness uppermost. 1932 J. Farrell Young Lonigan (1936) ii. 46 He boot-licked around until he became a ward committeeman.

Hence **boo·t-licker,** = sense B above; **boo·t-licking** *vbl. sb.* and *ppl. a.* orig. *U.S.*

1846–47 Mrs. Whitcher Widow Bedott P. (1856) xxvii. 331 Sweezer's very intimit with the squire's folks—a kind o' bootlicker tew 'em. 1849 Gallinipper Dec., Those crouching, fawning, boot-licking hypocrites. 1851 Hall College Words 24 Some [students are]..very apt to linger after recitation to get a clearer knowledge of some passage. They are Bootlicks, and that is known as Bootlicking. 1890 Farmer Slang, Bootlick, subs. (American), a flunkey, hanger-on... In England such a one is called a 'bootlicker'. Verb. To toady; to hang on; to undertake 'dirty' work. 1894 Current Hist. (Buffalo, N.Y.) IV. 472 Working his way by scheming and bootlicking into the good graces of.. a young idiot of a lord. 1906 Westm. Gaz. 27 June 2/1 If boot-licking became permanently fashionable, we should in course of time have various grades of boot-lickers. 1938 'G. Orwell' Homage to Catalonia viii. 140 No privilege and no boot-licking. 1961 L. Mumford City in History v. 154 The Greek eventually turned the citizen into..the cadger and bootlicker. 1964 Daily Mail 4 Aug. 3/8 That reverential boot-licking tone which we reserve for the heights of human achievement.

boot-strap, bootstrap (bū·tstrӕp). [f. Boot *sb.*[3] + Strap *sb.*] **1.** A strap sewn on to a boot to help in pulling it on or looped round a boot to hold down the skirt of a lady's riding habit; a boot-lace.

1891 R. P. Chope Dial. of Hartland, Devonsh. 29 Boot-strap, a boot-lace. 1908 Daily Chron. 11 Sept. 7/5 He put her up, adjusted boot-strap and skirt.

2. Colloq. phr. *to pull* (*lift, raise,* etc.) *oneself* (*up*) *by one's* (*own*) *boot-straps*: to raise or better oneself by one's own unaided efforts; hence *allusively.*

1922 Joyce Ulysses 630 There were..others who had forced their way to the top from the lowest rung by the aid of their bootstraps. 1936 Kunitz & Haycraft Brit. Authors 19th Cent. 213/1 A poet who lifted himself by his own boot-straps from an obscure versifier to the ranks of real poetry. 1937 V. D. Scudder On Journey III. ii. 306 Humanity could never pull itself up by its own bootstraps. 1960 D. Lessing In Pursuit of English 35, I had no money, I could have got some by writing to my family, of course, but it had to be the bootstraps or nothing. 1962 Listener 23 Aug. 271/2 A rather naive faith in humanity's ability to pull itself up by its own bootstraps.

3. Chiefly *attrib.* in various technical usages (see quots.). Hence **boo·tstrapping** *vbl. sb.*

1946 Jrnl. Inst. Electr. Engin. XCIII. III A. 308/1 The 'Bootstrap' Circuit..is much used for generating a linear rise of voltage with time, for time-base and other purposes. ..It is called a 'bootstrap' circuit because the potential at A is apparently being 'pulled up by its own bootstraps'. 1949 Electronic Engin. XXI. 198 The bootstrap valve may be anode loaded to obtain the push-pull output. 1949 R. Kelner in B. Chance et al. Electronic Time Measurements v. 125 (heading) Bootstrap Triangle Generator with Diode Comparator. 1960 Aeroplane XCVIII. 147/2 The particular process is not, strictly speaking, 'bootstrapping', which in rocket parlance merely means that the engine runs as a self-contained unit, using its own propellents in one way or another to drive its pumps. 1962 Simpson & Richards Junction Transistors xiii. 317 The process of multiplying the apparent value of a resistance by applying nearly equal signal voltages to each end is known as 'bootstrapping'. 1962 Gloss. Terms Automatic Data Processing (B.S.I.) 42 Bootstrap, 1. A form of program input in which simple preset computer operations are used to read in initial instructions which in turn cause further instructions to be read until the complete program is assembled. 2. The process of using parts of a compiler to construct the remainder of the same compiler. 1965 K. Nicol Elem. Programming iii. 11 When the computer was constructed, a small set of basic instructions were permanently wired into the store and the control unit is made to obey the first of these by pressing an external button. They constitute a small program (known as the 'bootstrap' or 'initial input' routine) whose task is to read characters from the paper tape reader and place them in successive storage locations.

boot-top. Add: **2. c.** = *boot-topping c. Also *attrib.*

1788 Earl Dundonald Direct. for using Coal Tar 24 Painting vessels boot-tops with white lead and oil, has been discontinued..in the Navy. 1937 H. A. Calahan Ship's Husband xviii. 229 Between the topsides and the bottom, it is frequently customary to paint a boot-top. The boot-top ..separates the topsides from the bottom. 1949 P. Heaton Sailing 247 When painting..mark in the boot top with tacks. 1958 Times 22 Sept. 14/1 The red boot-top line on the yacht's hull was more clearly visible forward than aft.

So as *v.*, to subject (a vessel) to 'boot-topping'.

1724 C. Johnson Gen. Hist. of Pyrates (ed. 2) xi. 211 Here they watered, boot-topp'd their Ship, and made ready for the designed Cruise. 1768 James Cook Jrnl. 15 Nov. (1893) i. 19 Got all the Empty Casks on shore, and set the Coopers to Work to repair them; Heeld and Boot Topt the Starboard side. 1908 Chambers's Jrnl. Nov. 829/1 The system is in operation upon the Clyde with great success, boot-topping the fast pleasure-craft plying on the river.

boot-topping. Add: **c.** (See quot. 1894.)

1788 Earl Dundonald Direct. for using Coal Tar Proofs, p. 22 The worm had not touched the bottom lower than the boot-topping. 1894 H. Paasch From Keel to Truck (ed. 2) 46/1 Boot-topping, is the term applied to that portion of the outside-plating or planking of a vessel between the light-water-line and the load-line. The coating (also the wood-sheathing of wooden vessels) on this part of a ship is likewise so named. 1927 Bradford Gloss. Sea Terms 18/2 Boot-topping, a band of paint at the water-line; usually red; the particular kind of paint used is called boat-topping. 1961 Times 2 Aug. 4/1 The paint of the boot-topping gleamed in the fitful sun.

booze, *sb.* Add: **1. b.** Alcoholic drink, chiefly beer; *U.S.* esp. spirits.

1859 Hotten Dict. Slang. 1895 Daily Tel. 2 Dec. 5/1 She heard some men shout that they wanted some more booze. Mr. Justice Wright: 'What?' Mr. Willis: 'Booze, my lord, drink.' Mr. Justice Wright: 'Ah!' 1896 Voice (N.Y.) 16 July 1/6 The Duckworth club..consumed large quantities of booze. 1900 Daily News 8 Dec. 6/2 On his way home he drops into a pub, and gets some 'booze'. 1932 T. S. Eliot Sweeney Agonistes 30 We're gona sit here and drink this booze.

c. *attrib.* and *Comb.*

1895 Voice (N.Y.) 9 May 5/4 Lost souls—lost through the booze traffic. 1922 H. L. Foster Adv. Trop. Tramp v. 60 After that the police had a private booze-party of their own on the proceeds. 1934, 1941 [see *artist 10].

2. b. Phr. *on the booze*: having a spell of hard drinking.

1850 H. Halse Let. 26 Nov. in McLean Papers (MS., Wellington, N.Z.) VIII. 271 The sawyers who were to have got the board out of the bush, had been on the booze. 1888 E. W. Payton Round about N.Z. xxiii. 306 The man..often does not know..how long he has been 'on the booze'. 1889 'R. Boldrewood' Robbery under Arms I. ii. 20 Poor old Mr. Howard wasn't always on the booze. 1892 Heslop Northumberland Words I. 82 'He's on the booze'—that is, he has a drinking fit. 1946 F. Sargeson That Summer 31 He was a bit of a one for going on the booze, Bill was. 1959 J. Cary Captive & Free 96 If I didn't you'd go on the booze and say it was all my fault.

c. booze-hound, a drunkard; **booze-up,** a drinking bout. *slang.*

1926 Wood & Goddard Dict. Amer. Slang 23 Hound (as in booze-hound, smut-hound, etc.), one fond of a thing. 1947 K. Tennant Lost Haven i. 19 This ruddy old booze-hound. 1897 Session Paper 26 Oct. 860 We..had a booze up together. 1939 I. Brown Down on Farm in Best One-Act Plays of 1939 131 We're celebrating...So we're going to have what the vulgar call..a bit of a booze-up. 1957 J. Braine Room at Top x. 91 The traditional lunchtime booze-up.

boozed, *ppl. a.* Add: (Further examples.) Also with *up.* Also *transf.*

1886 Lantern (New Orleans) 22 Sept. 3/3 This fortune-teller gets boozed up. 1897 W. S. Maugham Liza of Lambeth v. 77, I believe I'm boozed. 1936 C. Day Lewis Friendly Tree xiv. 209 One looks like that when one is a bit boozed-up. 1954 Encounter May 19/1 Camino Real..depends too much on boozed writing and aureate diction. 1969 Daily Tel. 12 Nov. 13/4 A boozed and dilapidated Don.

boozer. Add to def.: See Bouser.

2. A public house. *slang.*

1895 Daily News 4 Jan. 3/7 (E.D.D.), I pops around [stays] at the boozer. 1929 'H. Green' Living iii. 29 He'd be glad when he was dead: 'glad, more'n glad, I'll go straight into the boozer and 'ave one'. 1966 P. Moloney Plea for Mersey 38 The boozer on the corner.

boozeroo (bū:zərū·). *N.Z. slang.* [f. Booz(e) *sb.* + *-eroo.] A drinking spree.

1943 J. A. W. Bennett in Amer. Speech XVIII. 89 A 'good spree' would be described as a 'proper old boozeroo', this word being of the same pattern as various American words in -eroo. 1952 Landfall VI. 224 The Saturday night boozeroo in the Sydenham side-street with the keg in the kitchen-sink. 1963 N. Hilliard Piece of Land 28 He played Beethoven at boozeroos. 1970 N. Z. Listener 12 Oct. It's going to be a real boozeroo.

boozily (bū·zili), *adv.* [f. Boozy *a.*[1] + -ly[2].] In a boozy or intoxicated manner.

1893 in Funk's Standard Dict. 1912 English Review Feb. 493 To claim that women would vote more..'boozily' than men do is mere blarney. 1957 V. Palmer Seedtime 140 'Conscience!' repeated Donovan a little boozily.

bop (bǫp), *sb.*[1] *colloq.* [Echoic.] = Pop *sb.*[1] 2 a.

1937 Hemingway To Have & Have Not III. x. 149 Harry ..heard the bop-bop-bop-bop, small and hollow sounding in the wail of the siren. 1942 'B. J. Ellan' Spitfire! xv. 89 Shells seemed to be exploding very close to us although the familiar 'bop' was inaudible.

bop (bǫp), *sb.*[2] orig. *U.S.* = *bebop. Also *attrib.* Hence **bo·pper, bo·pster** = *bebopper.

1948 Language Mar. 132 People in Chicago use rebop and bop for the same kind of music. 1948, 1955 [see *bebop]. 1948 Sat. Rev. 25 Dec. 48/2 It is, in the words of bop hepsters, real crazy. 1950 J. Vedey Band Leaders p. xi, The BBC has not been lacking in fostering its development, even to the point of broadcasting Bop—most advanced of all jazz forms. Ibid. 27 He [Harry Roy] is an acknowledged past master at the art of Bop-Scat singing. 1951 M. L. Wolf Dict. Arts 573/1 Be-bop and bop, its adherents..are known as bopsters. 1957 New Yorker 26 Oct. 174/2 One of the leaders of an increasingly fashionable school of modern jazz called..'hard bop' or 'funky' is Horace Silver. Ibid., The hard-boppers employ a heavy or sharply accented beat, a florid, staccato attack, and a hardness of timbre that is in direct contrast to the soft, inky sounds of the cool school. 1962 Radio Times 10 May 42 Hard bop, aggressive, neurotic brand of bop, with a delinquent, flick-knife cutting edge.

1965 G. Melly *Owning-Up* vi. 66 A whole row of the audience raised, during Bruce Turner's first alto chorus, a long banner reading 'go home dirty bopper'.

bop (bɒp), *v.* [Echoic; cf. *bop *sb.*[1]] **1.** *trans.* **a.** (See quot.) *dial.*
1887 Parish & Shaw *Dict. Kentish Dial.* 16 *Bop*, to throw anything down with a resounding noise.
b. *colloq.* To hit, strike, punch. *orig. U.S.*
1931 D. Runyon *Guys & Dolls* (1932) vi. 117 Dave the Dude reaches across the table and bops One-eyed Solly right in the mouth. 1937 'E. Queen' *Door Between* iii. xv. 158 If the doc hadn't been on a ship..when she was bopped, I'd say he did it himself. 1938 Wodehouse *Summer Moonshine* xv. 181 I'll bop you over the head with this chair. 1948 C. Day Lewis *Otterbury Incid.* 8, I can use it [sc. a football] to bop them on the head.
2. *intr.* To fight. So **bo·pping** *vbl. sb.* and *ppl. a.* *U.S. slang.*
1958 *Daily Tel.* 8 Apr., Bop, to fight. Bopping club, a fighting gang. 1958 *Life* 28 Apr. 78 You gotta go on bopping (gang fighting) and hanging around street corners all your life? 1959 *Listener* 29 Jan. 201/1 The 'turf' of a well-known 'bopping club'—and that means the small area of pavement and street called their own by a well-known fighting gang. 1959 H. Salisbury *Shook-up Generation* iii. 49 Arms and armament..are not a problem for bopping gangs.

bo-peep. Add: **2.** A look; = Peep *sb.*[3] 1 a. *Austral.* and *N.Z. colloq.*
1941 *Coast to Coast 1941* 67 'I'll 'ave a bo-peep,' he said. 'You gotter watch 'er. We don't want no dead pups.' 1944 L. Glassop *We were Rats* vi. 79 Let's take a bo-peep at what they got in the canteen. 1946 F. Sargeson *That Summer* 61 A smart-looking piece of goods, having a bo-peep out of the bathroom window. 1961 B. Crump *Hang on a Minute, Mate* (1963) 91 Have a bo-peep at this little lot. 1969 *Landfall* XXIII. 323 Take a bo-peep at old Lionel.

Bora[1] (bǭ·rä). Also **bor(r)ah**. [Hindi *bohrā*.] A Muslim trader. Also *attrib.*
1698 Fryer *Acc. E. India & P.* 93 Schisms they have made, as *Bilhim, Jemottee*, and the lowest of all is *Borrah*. 1813 M. Graham *Jrnl. Res. India* 33 The inside of a Borah's box is like that of an English country shop. 1863 G. P. Badger in J. W. Jones tr. *Varthema's Travels* p. xlix, A considerable trade is still carried on, chiefly by Borah merchants of Guzerat and Cutch. 1876 *Encycl. Brit.* IV. 537/1 Borás, a class of commercial Mahometans.

bora[2] (bǭ·rä). Also **borah**. [Aboriginal Australian.] A rite amongst the Aborigines of eastern Australia, constituting the admission of a young person to the rights of manhood.
1866 W. Ridley *Kamilaroi* 17 Kubura, young man who has attended a bora. 1883 J. Fraser *Aborigines of N.S. Wales* 12 This assembly—the most solemn and unique in the tribal life—is called the Bora and sometimes the Kobbora. *Ibid.* 17 The inner Bora customs. 1885 Mrs. C. Praed *Austral. Life* 14 The great mystery of the Blacks is the Bora—a ceremony at which the young men found worthy receive the rank of warriors. 1948 V. Palmer *Golconda* xiii. 104 The white-haired maker of corroborees..holding forth to the young men about the need to bring back the borah and all the ancient rites of the tribe. 1953 J. McLaren in I. Bevan *Sunburnt Country* ii. 72 The 'Bora' ceremonies of initiation are held in some secluded part of the bush, and have rarely been seen by whites.

borak (bǭ·ræk). *Austral.* and *N.Z. slang.* Also **borac(k), borax.** [Aboriginal Australian. Cf. *barrack *v.*[2]] Nonsense, humbug; chaff, banter; esp. in *to poke (the) borak*, to make or poke fun.
1845 T. M'Combie *Arabin* 273 *Borack*, gammon, nonsense. 1882 *Sala Ballarat* (Sydney) 5 Sept. 9 A smart fellow was 'poking borak' at them, and asked, 'Is the snow in Japan the same as it is in Tasmania?' 1898 in M. Davitt *Life & Progr. Australasia* xxxv. 192 A jest is 'poking borak'. 1904 *Blackw. Mag.* June 833/1 One of the crowd was poking borak and said something pretty bad to him at the beginning. 1916 J. B. Cooper *Coo-oo-ee* ix. 113 At the same time he wondered whether Nipper was not 'pokin' borak' at him. 1944 J. M. Fullarton *Troop Target* iv. 34 You wouldn't be poking the borax, would you?

Boran (bǭ·rän). [Native name.] A breed of cattle native to drier parts of East Africa.
1935 *East Afr. Agric. Jrnl.* I. iii. 251/2 Two types of cattle are employed, namely, Boran and Masai. The Boran native type were obtained from a European farmer under whose care a certain amount of selection had taken place. 1953 Faulkner & Brown *Improvement of Cattle in Brit. Colonial Territories in Africa* 5 The Boran and Nandi in Kenya have shown conclusively what can be achieved by the selection of pure indigenous stock. 1959 R. B. Kelley *Native & Adapted Cattle* iii. 35 The second kind of cattle from the desert-grass area are known as Boran, after the nomadic tribes that own them in the Lake Rudolf region. 1969 *Times* (Somali Republic Suppl.) 16 Sept. p. vi/3 Somali cattle (boran variety) were driven 200 miles out of the republic.

borane (bǭ·rē[i]n). *Chem.* [ad. G. *boran* (A. Stock 1916, in *Ber. d. Deut. Chem. Ges.* XLIX. 111), f. Bor(on + -ane (after *methane*, etc.).] The systematic name for a boron hydride; *spec.* the (hypothetical) compound BH_3; also, any substituted derivative of a boron hydride.
1916 *Jrnl. Chem. Soc. Abstr.* CX. ii. 319 BH_4 (unknown).. would be described as borane. 1923 *Chem. Abstr.* XVII 2242 All attempts to obtain a mono- and a triborane failed. 1933 A. Stock *Hydrides of Boron & Silicon* i. 17 Our present knowledge of the chemistry of boron leads us now to recommend..that..the stable, so to say normal, hydrides ..be called 'boranes'. The less stable hydrides..should be called 'hydroboranes'. 1965 *New Scientist* 4 Nov. 341 Thought towards solution of the fuel problem was aimed at substituting hydrogen-boron compounds (the boranes) for the hydrocarbon constituents of the classical fuels.

borax. Add: **3. borax carmine** (see quot. 1890).
1887 *Amer. Naturalist* XXI. 596 For staining, borax-carmine was used. 1890 Billings *Nat. Med. Dict.*, *Borax carmine*, an aqueous solution of borax and carmine... A pure and intense nuclear stain if bleached with acid.
4. Cheap, inferior, or ostentatious goods, esp. furniture; inferior and tasteless design, 'so called fr. the reputed custom of a producer of borax soap of giving cheap furniture as a coupon premium' (Webster 1961). Also *attrib.* *orig. U.S.*
1942 R. Chandler *High Window* (1943) xxv. 171 A standing lamp from the basement of some borax emporium. 1948 *Archit. Rev.* CIV. 92 (*title*) Borax, or the chromium-plated calf. *Ibid.* 92/2 'Borax'..is generally restricted to consumer goods where obviously heavy forms and elaborate jazzy ornament are used in order to add spurious eye-appeal. The term originated in the furniture industry and by analogy is sometimes applied to kitchen appliances and more rarely to automobile design. 1963 *Times* 11 Feb. 13/3 There..appears to be far too much unnecessary design or 'borax', as useless embellishment has come to be known in financial and trade circles.

borax, var. *borak.

borazon (bǭ·räzɒn). [f. Bor(on + Az(o- + -on.] A crystalline form of boron nitride, extremely hard and resistant to oxidation at high temperatures.
1957 R. H. Wentorf in *Jrnl. Chem. Physics* XXVI. 956/1 The name 'Borazon' is proposed for this cubic form of boron nitride. 1960 N. Kneale *Quatermass & Pit* iii. 78 We could try borazon. That's several times harder still [than diamond]. Can you get hold of a drill with a borazon bit?

borborygm. Add: (Later *fig.* example.) Hence **borborygmic** *a.*, characterized by borborygms or rumblings.
1927 H. G. Wells in *Sunday Express* 1 May 12/4 Elephant hunters say that they can tell the proximity of a herd by the borborygmy (see dictionary) noises the poor brutes emit. 1928 A. Huxley *Point Counter Point* xi. 170 The stertorous borborygms of the dyspeptic Carlyle! 1938 *Times* 24 Nov. 12/3 The borborygmic note of the Arabian camel. 1959 E. Fenwick *Long Way Down* vii. 54 The room was very quiet, except for its borborygmic old radiator.

bord. Add: Still current in *bord-and-pillar* (see *board *sb.* 16).

Bordeaux. Add: **2.** A shade of red produced by any of several red azo-dyes derived from beta naphthol.
1904 *Westm. Gaz.* 10 Mar. 4/2 Bordeaux-coloured straw. 1909 *Ibid.* 11 Jan. 5/1 The various shades of violet, blue, and what we have called Bordeaux.
3. *Bordeaux mixture*: a mixture composed of blue vitriol, lime, and water, used for the destruction of fungi.
1891 *Potato Dis.* (R. Veitch & Son) 4 The Bordeaux Mixture, or 'Bouillie Bordelaise'. 1921 *Discovery* May 130/1 Bordelaise, or Bordeaux mixture, was first used to cover grapes growing alongside paths and roads to prevent 'finger blight'. 1959 *Sci. News* LI. 50 The latter causes foot rot and can be dealt with by spraying with Bordeaux mixture.

bordello. Delete †*Obs.* and add later examples.
1920 E. Pound *XXX Cantos* xxviii. 130 And was lodged in a bordello (promptly). 1961 J. Heller *Catch-22* (1962) xxiii. 236 The distant resources of the strange and marvellous bordello.

border, *sb.* Add: **5. c.** The upper edge of a basket.
1907 T. Okey in *Jrnl. Soc. Arts* 12 Jan. 190/2 Besides common borders, many other forms, such as plaited, roped, tracked borders, are used [in basket-making]. 1960 E. Lucas *Country Baskets* 37 At the top, the side stakes are turned down to form the border..thus completing the basket.
8. Substitute for def.: Usu. *pl.* (See quot. 1957.) (Add earlier and later examples.)
1824 J. Decastro *Memoirs* 16 It is that [part] of the scenic department from whence the borders of chambers or clouds drop. 1831 J. Boaden in *Corresp. Garrick* I. p. xlviii, A splendid show for his theatre; where a temperate sky always hangs from the borders. 1957 *Oxf. Compan. Theatre* (ed. 2) 882 *Border*, a narrow strip of painted cloth, battened at the top edge only, used to mask-in, or hide, the top of the stage as seen from the auditorium.
11. *border-country, -state, -war.*
c 1885 in *Westm. Gaz.* (1904) 24 Sept. 3/1 In the gay border-country of youth. 1945 E. Step *Wayside & Woodland Ferns* (ed. 2) 76 In the Border country it is known as Dead-man's Hands. 1842 J. Sturge *Visit to U.S.* 166 Many planters, with their slaves, have emigrated thither [sc. to Texas] to escape their creditors from the border States. 1809 'D. Knickerbocker' *Hist. N.Y.* (1820) v. ii. 301 Heart-rending cruelties that disgraced these border wars. 1965 *New Statesman* 30 Apr. 671/1 Ayub Khan..describes this border war as 'a useless quarrel'.

12. border ballad = *riding ballad* (see Riding *vbl. sb.* 5 d); **Border Leicester**, a variety of sheep originating from the cross-breeding of Cheviot and Leicester sheep; **Border terrier**, a small rough-haired terrier originating in the Cheviot Hills.
1863 E. Dickinson *Poems* (1955) II. 569 Better entertain Than could Border Ballad—or Biscayan Hymn. 1941 L. MacNeice *Poetry of Yeats* iv. 79 The best and glitter of Housman's verse, partly derived from Heine and the Border ballads. 1873 *Country Gentleman's Mag.* 2 Nov. 308/1 The distinguishing features of the Yorkshire and Border Leicesters, though springing from the same source, have diverged. 1874 W. C. Spooner *Hist. of Sheep* (ed. 3) 70 The sheep which prevail mostly in the lowlands of Scotland and the good land of the Border Counties are called the Border Leicesters. 1894 R. B. Lee *Mod. Dogs* (*Terriers*) i. 21 The 'Border terriers'..have been for a long time indigenous to the Border counties, and so far south as Westmorland, Lancashire, and Yorkshire. 1928 F. T. Barton *Kennel Encycl.* 51 Border terriers are very hardy and the puppies easy to rear as a rule.

borderland. Add: **2. b.** *attrib.*, esp. in the senses (*a*) of or pertaining to the 'land' between this world and the next; (*b*) = *border-line 2.
1886 F. Podmore *Apparitions & Thought-Transference* xi. 249 The voice in this case awoke the percipient, but it was of the 'borderland' type. *Ibid.* 257 The hallucination should be classed as a 'borderland' case. 1896 *Brit. Med. Jrnl.* 25 Jan. 61/1 Mental or Borderland case. 1898 *Daily News* 30 July 5/5 One thing about the deceased deputy may interest 'Borderland' readers. 1904 F. G. Younger *Insanity in Every-Day Pract.* 99 Borderland states..are cases of nervous weakness which do not quite fit in with the recognised forms of insanity. *Ibid.* 102 Dr Walsh has told me of a borderland case under his care which benefited greatly under his treatment. 1904 *Westm. Gaz.* 1 Dec. 6/1 He regarded the deceased's as a 'borderland' case. 1909 *Times Lit. Suppl.* 14 Jan. 9/1 It is a vague, borderland word.

border-line. [Border *sb.*] **1.** The strip of land along the border between two countries or districts; a frontier-line; often *fig.*, the boundary between areas, classes, etc.
1863 *Cassell's Mag.* Dec. 31/1 On the very border-line of the Black Country. 1889 Kipling *Barrack-R. Ball.* (1892) 81 Thou must harry thy father's hold for the peace of the Border-line. 1890 W. James *Princ. Psychol.* II. xvii. 22 Simultaneous contrast is always strongest at the border-line of the two fields. 1917 'Contact' *Airman's Outings* 143 Weather conditions place such duties near the border-line of possible accomplishment. 1965 M. Spark *Mandelbaum Gate* vii. 213 The armed border patrol..returned along the border-line.
2. *attrib.* or as *adj.* Occupying a border-line; esp. in phr. *border-line case*; *spec.* (*a*) verging on the indecent or obscene; (*b*) verging on insanity.
1907 F. J. Poynton in *Edin. Med. Jrnl.* Sept. 232 There are border-line cases which are well worthy of consideration by those who are interested in this form of arthritis. 1913 R. H. Cole *Mental Dis.* 23 Many come within the range of the so-called 'Border-line' cases. 1917 E. F. Ballard *Epit. Mental Dis.* 20 Therapeutic suggestive conversation with these early or border-line types of disorder is never wasted. 1928 *Sunday Express* 29 Apr. 3/4 Men are fond of 'thrillers'. Girls prefer rather shocking 'border-line' books. 1928 *Daily Tel.* 11 May 7 [He] described the cases as border-line cases, and bound them over under probation for two years. 1937 *Mind* XLVI. 478 This clear central nucleus is surrounded by a penumbra of borderline cases, such as the credibility of witnesses. 1965 J. E. Floud *Social Class* iv. 54 To aid the assessment of borderline candidates, a Composition test was administered. 1965 F. G. Cassidy in Bessinger & Creed *Mediæval & Linguistic Stud.* 78 This has left uncertain the status of the 'borderline cases'.
Hence **border-liner**, a border-line case.
1953 E. Partridge *'Shaggy Dog' Story* ii. 48, I finally reduced the candidates to five that can fairly be adjudged eligible for the description 'border-liners'.

bore, *sb.*[1] **3.** Australian examples: used esp. in sense 'water-hole for cattle'.
1906 J. W. Gregory *Dead Heart of Australia* iv. 331 Geysers... With regard to the bore at Westland, there were some very fine features in the evolution of firedamp. 1933 *Bulletin* (Sydney) 23 Aug. 25/1 He cleaned bore-drains at Myall Creek. 1936 I. L. Idriess *Cattle King* xxxviii. 318 The bore-drain carrying the water from the bore-head would meander for many miles. 1938 F. Clune *Millionaire in Memories* iii. 73 'Good,' I thought, 'I have time to..bathe in the "bore"'—the Government had bored for water at Barcaldine, and..had struck an inexhaustible supply. 1955 J. Cleary *Justin Bayard* xix. 285 That night they camped by a bore. The cattle took their fill of the mineral-tasting water. 1963 V. B. Crawley *27,000 Miles through Australia* x. 62 We..filled our water reserves at the bore along the way... Lilly Vale Bore... Wild Dog Bore... Tickalara Bore, and others.

boreal, *a.* Add: **4.** Applied by Blytt to the next period of vegetation in Scandinavia after the arctic period, and later by others to the climate of other areas.
1876 [see *Arctic *a.* 3]. 1939 V. G. Childe *Dawn Europ. Civ.* (ed. 3) i. 2 A continental Boreal phase, characterized by summers longer and warmer than to-day, but severe snowy winters. 1946 L. D. Stamp *Britain's Struct. & Scenery* xiv. 157 The abrupt change from a cold or 'Boreal' climate to the milder, damper 'Atlantic' climate. 1963 *Field Archaeol.* (Ordnance Survey) (ed. 4) 23 The application of the techniques of pollen analysis and Carbon 14

investigations have shown conclusively that the site belongs to the Boreal climatic phase which succeeded the last Ice Age, and that it may be dated with some confidence to 8,000–7,500 B.C.

boree[1] (bǭri·). [f. Bore *v.*[2] + -ee[1] 2.] A person who is bored.
1841 J. S. Mill *Let.* 1 Mar. in *Wks.* (1963) XIII. 466 Instead of being the bore on the subject of an unfinished article I have a strong vocation for being the *boree* in respect of it. 1872 Butler *Erewhon* ix. 79 They would be consigned to the Hospital for Incurable Bores, and made to work at being bored..; in fact..they would be kept as professional borees.

boree[2] (bǭ·ri). [Aboriginal *booreah* fire.] A variety of myall, *Acacia pendula*, found in Eastern Australia.
1878 R. Boldrewood *Ups & Downs* iii. 29 Myall and boree belts of timber. 1889 J. H. Maiden *Usef. Native Plants* 365 'Weeping', or 'True myall'... Called 'Boree' by aboriginals, and often 'Boree' or 'Silver-leaf Boree', by the colonists of Western New South Wales. 1934 *Bulletin* (Sydney) 7 Nov. 24/1 Gidya and boree are good in most places [as fence posts].

boresome (bǭ·rsəm), *a.* [f. Bore *sb.*[2] + -some.] Tending to be a bore, boring. So **bo·resomeness.**
1868 R. Gower *Rec. & Remin.* (1903) 150 So little real enjoyment to make up for so much loss of time and boresomeness. 1895 *Nineteenth Cent.* Sept. 474, I spent a boresome fortnight at Aden. 1905 E. Glyn *Visits Elizabeth* 13 They were all casual and indifferent to their poor wives! and boresome, and bored!! 1925 P. A. Scholes *Second Bk. Gramophone Rec.* p. xii, There is..a degree of boresomeness in some of Beethoven's compositions.

boring, *ppl. a.*[1] Add: **1. boring sponge**, a saltwater sponge of the genus *Cliona*, which bores into shells.
1881 R. Hunter et al. *Encycl. Dict.* 1885 *Encycl. Brit.* XVIII. 107/2 *Cliona*, the boring sponge, destroys the shells and so injures the oyster.

bo·ringness. [f. Boring *ppl. a.*[1] + -ness.] The quality of being boring or annoying.
1893 *Sketch* 31/6 His boringness is a quality inherent in him. Boringness springs, surely, in some measure from want of tact. 1971 *Daily Tel.* 29 Apr. 6/6 The brutishness and boringness of..Eskimo village life.

born, *pple.* and *a.* Add: **A. 3.** Also *born and bred*: see Breed *v.* 1 c. *born in the purple*: see Purple *sb.* 2 d; *born yesterday*: see Yesterday *adv.* 2.

borne, *ppl. a.* Add: **2. b.** With prefixed sb., as *air-, carrier-, chair-, glider-borne*: see the sbs. (See 1944 *Amer. Speech* XIX. 222 f.)

‖**borné** (borne), *a.* [Fr., pa. pple. of *borner* to limit.] Limited in scope, outlook, mental equipment, etc.
1795 J. Moore *Diary* 17 May (1904) I. v. 153 His mind is borné and illiberal to a degree. 1850 *Househ. Words* 3 Aug. 434/1 The Rockvilles remained high, proud, bigoted, and borné. 1873 J. S. Mill *Autobiogr.* vi. 193 A jobbing and borné local oligarchy. 1936 T. Watts-Dunton in G. K. Chesterton *Autobiogr.* xiii. 282 Anyone so borné as poor Browning was. 1961 *Partisan Rev.* 199 Certain critics, certain characters, in their impudence or awfulness, have the power of making us feel borné.

Bornean (bǭ·rniăn), *a.* and *sb.* Also 9 **Borneon.** [f. *Borneo* (see below) + -an.] **A.** *adj.* Of or pertaining to Borneo, the largest island in the Indonesian Archipelago. **B.** *sb.* A native of Borneo.
1813 I. Hunt in H. Keppel *Exped. Borneo* (1846) II. xxxi. The fruit-bearing trees which enrich and adorn the Indian continent offer, on the Bornean shore, all their kindred varieties. 1839 J. Brooke *Jrnl.* 19 Sept. in *Ibid.* I. iv. 28 Let me speak it to the honour of the Borneons, that they neither cock-fight nor smoke opium. 1850 Kingsley *Alton Locke* II. xv. 122 A baby-ape in Bornean forests. 1863 *Boys' Own Mag.* II. 396 Such a course is quite impracticable with a Bornean. *Ibid.* 511 And here I may say I am an end of my Bornean adventures. 1920 *Q. Rev.* July 175 Their ancient Indian forefathers..were clearly once at much the same grade of social development as the Bornean tribes. 1936 A. Huxley *Jesting Pilate* vi. 228 Genuine and aboriginal Borneans. 1965 P. Wayre *Wind in Reeds* xv. 215 A trio of Bornean Great Argus and several Greater Bornean Crested Fireback Pheasants as well as a number of the rare Chestnut Flanked Bornean Tree Partridges.

Bornholm (bǭ·rnhǫlm). [The name of a Danish island in the Baltic Sea.] Used *attrib.* in **Bornholm disease**, an epidemic virus disease characterized by acute pain in the intercostal muscles; epidemic pleurodynia.
[1932 E. Sylvest in *Bull. de l'Office intern. d'hygiène publique* XXIV. 1431 La Maladie de Bornholm.] 1933 *Brit. Med. Jrnl.* 4 Nov. 817/2 'Bornholm' disease. 1962 *Lancet* 12 May 1016/1 Quite sharp pains, similar to those in Bornholm disease.

bornite (bǭ·rnəit). *Min.* [ad. G. *bornit* (1845), f. the name of Ignatius von *Born*, an Austrian mineralogist (1742–91): see -ite[1].] A brittle reddish-brown sulphide of copper and iron; also called *purple copper ore* or *erubescite*.
1854 Dana *Syst. Min.* (ed. 4) 58 Erubescite. Bornite, Haid. Variegated Copper. 1896 *Times* 22 Aug. 10/3 The great bornite mine at Nelson [British Columbia]. 1946 J. R. Partington *Gen. & Inorg. Chem.* xiii. 323 The commonest ores of copper are sulphides of copper and iron: chalcopyrite ..and bornite or erubescite or variegated copper ore. 1965 G. J. Williams *Econ. Geol. N.Z.* viii. 98/2 The Bonanza section of Maria mine—where covellite was reported with a little bornite.

borohydride (bǭrohəi·draid). *Chem.* [f. Boro- + Hydride.] A compound containing the radical $-BH_4$, or a substituted form of this radical. Hence **borohydride reduction**, selective reduction of a compound by a borohydride.
1940 *Jrnl. Amer. Chem. Soc.* LXII. 3422 Its probable structure has led us to adopt the name aluminium borohydride. *Ibid.* 3430 The basic structures of the aluminium, beryllium and lithium borohydrides are closely related. 1961 W. Gerrard *Org. Chem. Boron* x. 136 Selectivity of reduction can be achieved by employing sodium borohydride in conjunction with a particular solvent. 1965 H. O. House *Mod. Synthetic Reactions* ii. 40 Sodium borohydride reductions of α,β-unsaturated compounds are only rarely complicated by reduction of the carbon-carbon double bond.

boron. Add: **boron hydride**, any of a series of compounds of boron and hydrogen; at one time considered as potential rocket fuel, and among the earliest examples of electron-deficient compounds; = *borane.
1879 F. Jones in *Jrnl. Chem. Soc.* XXXV. 41, I am not aware that any attempt has been made since 1809 to obtain boron hydride. 1933 A. Stock *Hydrides of Boron & Silicon* ii. 30 The yield of boron hydride..was better if they used phosphoric instead of hydrochloric acid for decomposing the boride. 1959 *Q. Rev.* (*Chem. Soc.*) XIII. 176 The boron hydrides and some of their derivatives are examples of electron-deficient compounds. 1961 *New Scientist* IX. 464 These boron hydrides appeared to be of academic interest only until..the need for rocket fuels became apparent.
b. Examples in *Nucl. Physics.*
1938 *Proc. Camb. Phil. Soc.* XXXIV. 290 (*title*) The Disintegration of Boron by Slow Neutrons. *Ibid.* 291 By such measurements on the image of the cloud track resulting from the boron disintegration we have been able to determine the..origin of the He and Li particles. 1949 Rossi & Staub *Ionization Chambers & Counters* viii. 189 The construction of a highly sensitive boron chamber that can be used for detecting neutrons of all energies. *Ibid.* 194 These detectors have been named long boron counters. 1957 *Encycl. Brit.* XVI. 315/1 Because of its high absorption of neutrons, boron may be used (e.g., in the form of boron steel) as a control absorber in thermal nuclear reactors. 1963 B. Fozard *Instrumentation Nucl. Reactors* iii. 30 Typical sensitivities for a boron-coated reactor control ionisation chamber.

boronia (borō̆u·niă). [f. name of Francesco *Borone*, an Italian botanist (1769–94).] A plant of the rutaceous genus so named, a sweet-scented Australian shrub.
1838 G. C. Mundy *Antipodes* I. 36 The Boronias shoot up their slender stems..towards the sun. 1906 *Melbourne* 28 Syd. 35 (Morris), The scent of boronia was the colour of wattle-blossom greet us from baskets of the flower-girl. 1926 K. S. Prichard *Working Bullocks* v. 55 Ti-tree, boronia bushes, and spray of purple and yellow. 1966 (W. Aust. Suppl.) p. iv/2 The scent of the boronia's monk-brown-and-golden bells.

borosilicate (bǭ·rosi·likėt). [f. Boro- + Silicate.] A silicate incorporating some boric oxide. *borosilicate glass*, silicate glass containing boric oxide, used esp. in heat-resistant glassware.
1817 T. Thomson *Syst. Chem.* (ed. 5) III. 409 Family of Borosilicates. 1933 Rutherford *Coll. Papers* (1965) III. 311 The..glass tube..a tube of borosilicate glass. 1947 *Endeavour* July 117/1 Potash glasses give a purer blue than the corresponding sodium glasses, and..borosilicate glasses..strengthen the superposition that the whole effect might be due to boron. 1947 *Endeavour* July 117/1 Potash glasses give a purer blue than the corresponding sodium glasses... 1963 Phillips & Williams *Inorg. Chem.* I. xiv. 553 Soda glass is conveniently worked at a lower temperature than borosilicate glass, and the latter at a lower temperature than quartz glass.

borough. Add: **3. d.** An incorporated town or village; a town having a warden or chief burgess as its official head. *U.S.*
1718 in *Pennsylvania Col. Rec.* III. 58 The said town might be erected into a Borough by a Charter of Incorporation. 1864 *Stat. Gen. State Connecticut* 309 The warden and a majority of the burgesses of any such borough, may, in like manner, authorize such an alley to be kept at any place in any such city or borough. 1925 G. P. Krapp *Eng. Lang. in Amer.* I. iii. 178 Now that the word Manhattan has been legalized as the name of the Borough of Manhattan, it is possible it may become colorlessly official and lose some of its romantic glamour.
e. (*a*) In New Zealand, a village, township, or town having a special governing body called a borough council. (*b*) In New South Wales, a municipal corporation of not less than 1,000 inhabitants and not more than 9 square miles in area. In Victoria, such a municipality of not less than 300 inhabitants.
1865 R. P. Whitworth *Baillière's Vict. Gazetteer* 59 Brunswick is a borough township,..in the..electoral district of E. Bourke boroughs. 1867 *Acts N.Z.* 31 *Vict.* No. 24 § 29 There shall be in and for each single borough a council consisting of nine councillors. 1874 *Silver's Handbk. Australia & N.Z.* (ed. 2) 131 At the end of 1871 there were sixty-four corporate towns and boroughs, containing within their municipal limits about one-half the population. A borough must not have an area of more than nine square miles.
7. c. borough court, a court of limited jurisdiction held in a borough by special privilege; **borough sessions**, a court held by the recorder of a borough, usually quarterly, established under the Municipal Corporations Act of 1835.
1769 Blackstone *Comm.* Index, Borough courts. 1959 Jowitt *Dict. Eng. Law* I. 266/2 Borough courts..are now of little importance, most business having been transferred ..to the county court. 1835 *Act* 5 & 6 *Will. IV* c. 76 § 110 *marg.*, Offenders committed to Borough Sessions whose Jurisdiction is taken away to be tried in the adjoining County.

Borrovian (bǫrō̆u·viăn), *sb.* and *a.* [f. the name of George *Borrow* (1803–81) English writer + -ian, after *Harrovian*, etc.] **A.** *sb.* **1.** A student or admirer of George Borrow. **2.** The diction characteristic of Borrow's works. **B.** *adj.* Of, relating to, or characteristic of Borrow or his writings.
1888 *Reflector* 8 Jan. 26/1 The born Borrovian—for men are born Borrovians, not made. 1900 W. I. Knapp in *Borrow's Lavengro* 568/2 *Engro* (mere *ending*), Borrovian for 'master', 'fellow', 'chap'. 1904 *Daily Chron.* 21 July 3/6 Bampfylde Moore Carew, who, born with a genius for roving truly Borrovian, ran away from school to become King of the Gypsies. 1956 F. Swinnerton *Background with Chorus* vi. 49 Edmund Gosse, no Borrovian, anonymously and poohpoohingly reviewed..*George Borrow and his Circle*.

borrow, *v.*[1] Add: **2. b.** *to borrow trouble*: to go out of one's way to meet trouble. *U.S. colloq.*
1854 H. H. Riley *Puddleford & its People* ix. 119 Bird, you're allers bor'ring trouble. 1934 J. H. Wallis *House of Murder* xiii. 128 Don't borrow trouble, Mr. Gundlach...If the hand doesn't pain any more, it will probably be all right. 1963 D. B. Hughes *Expendable Man* (1964) vi. 182 Let's not borrow trouble.
c. In organ-building: see *borrowed *ppl. a.* 2 b.
1894 T. Elliston *Organs & Tuning* 63 Borrowing from one stop to form part of another is undesirable; the wind coming sometimes from one source, and at other times from another, or from both sources, make it impossible for the note to be always in tune...The Pedal Organ is somewhat exempt from the faults of borrowing, inasmuch as usually only one note is used at a time. 1902 J. W. Hinton *Organ Constr.* (ed. 2) 64 *Borrowing*, arranging a certain number of pipes so as to be common to two or more stops. 1905 T. Casson *Pedal Organ* 24, I once tried the experiment of borrowing the Quint from a manual Bourdon. *Ibid.* 27 The borrowing must be economical; that is, it must cost less in room or money, or both, than actual independent pipes. 1927 *Organ* VII. 75 While he was about it, the builder might have borrowed this 'creamy' bourdon into the pedal.
d. *trans.* and *intr. Golf.* (See quots.)
1897 *Encycl. Sport* I. 472/1 *Borrow*, to play a ball up a hill or slope, instead of straight across it, so that the slope will cause the ball to return towards the hole. 1909 Vaile *Mod. Golf* 55 You allow for the slope..and the curve..will be the result if you have correctly estimated length and the amount you ought to 'borrow'.... In golfing language, 'always borrow enough'. *Ibid.* 64 Both these cuts may be used when putting across a slope to hold the ball up against the natural tendency to run down a hill. If they are used it will obviously be unnecessary to 'borrow' so much in allowing for the slope.

borrowed, *ppl. a.* Add: **1.** *borrowed time*: an unexpected extension of time, esp. of a person's life.
1898 *E.D.D.* s.v., A man who lives on borrowed time lives on trespass-ground. Ay, all mine is borrowed time, noo. 1939 R. Chandler *Big Sleep* xviii. 148 Brody was living on borrowed time. 1961 R. Jeffries *Evidence of Accused* iii. 26 After the age of forty-five one's living on borrowed time.
2. *borrowed light*, (*a*) reflected light (see quot. 1834 in Dict.); also *fig.*; (*b*) see quot. 1963.
1657 Bp. H. King *Poems* 139 Even such is man, whose borrow'd light Is streight call'd in, and paid to night. 1934 H. G. Wells *Exper. Autobiogr.* I. ii. 39 Behind the shop was an extremely small room, the 'parlour', with a fireplace, a borrowed light and glass-door upon the shop and a larger window upon the yard behind. 1963 *Gloss. Build. Terms* (*B.S.I.*) 13 *Borrowed light*, a glazed..opening in an internal wall or partition designed to admit light.
b. In organ-building, said of a pipe, a stop, or a set of them which is sounded at the expense of another or is incomplete of itself and is eked out by the use of pipes of another stop or set.
1840 in Grove *Dict. Mus.* (1880) II. 600/2 'Borrowed' Solo Organ. 1880 *Ibid.* 595/1 Choir Organ. 2 real stops; 4 borrowed...Borrowed by communication from the Great Organ. *Ibid.* 595/2 The extra department consisted of a complete borrowed organ of 13 stops derived from the Great Organ. *Ibid.*, Second Great Organ. 13 borrowed stops.

bo·rrow-pit. [app. f. Borrow *v.*[1]] In civil engineering, an excavation formed by the removal of material to be used in filling or embanking. Also **bo·rrow-hole.**
1893 Kipling *Day's Work* (1898) 2 Tiny asses climbing out of the yawning borrow-pit below with sackfuls of stuff. 1901 *Practitioner* Mar. 258 'Borrow-holes' in railway embankments. 1907 *Notes on Books* June 267/2 An easily applied check on borrow-pit measurements.

borsch (bǭɪʃ, borʃtʃ). Also **borscht, borshch, bortsch.** [Russ. *borshch*.] A Russian soup of several ingredients, esp. beetroot and cabbage.
1884 J. Paget *Let.* 2 Sept. in *Mem. & Lett.* (1901) ii. vi. 346 A real Russian dinner—first there was a strange thing called Borsch. 1927 *Blackw. Mag.* Sept. 294/1 Caviare, crimson bowls of bortsch, with thick seasoned cream. 1929 *Daily Tel.* 22 Jan. 7/6 The borscht is a delicious consommé of beef and duck, ham, beetroot. 1939 *Collier's* 21 Jan. 40/1 The other new words came in a steady flood: Big time..the borscht circuit (Catskill Mt. summer resorts booking life-of-the-party m.c.'s). 1963 V. Nabokov *Gift* iii. 153 She was slowly mixing a white exclamation mark of sour cream into her borshch.

borsella (bǫɪse·lä). [Perversion of It. Procello.] In glass-making, an instrument for modifying the form of vessels.
[1699 tr. *H. Blancourt's Art of Glass* 31 The Instruments mark'd E. serve to finish the Work, which the Italians call *Ponteglo, Passago, Procello, Spiei*, and also *Borsello*, whereof we want the Figure.] 1823 G. Crabb *Technol. Dict.*, *Borsella* (Mech.), an instrument with which glass makers contract or extend their glasses at pleasure.

Borstal, borstal (bǭ·ɪstăl). [Name of a village near Rochester in Kent.] In full *Borstal institution*: a reformatory for 'juvenile adults', conducted according to the method first put into practice at the reformatory at Borstal and adopted afterwards elsewhere. Also *attrib.* So *Borstal system*: a system established in 1908 whereby young persons convicted of criminal offences between the ages of 16 and 23 may be sent to a Borstal institution for a period of reformative training, usu. 3 years, after which they are released subject to further supervision by the *Borstal Association*. Hence **Borsta·lian**, an inmate of a Borstal.
[1902 *Encycl. Brit.* XXXII. 8/1 In 1901..a 'juvenile-adult reformatory' was opened at Borstal, near Rochester, by the conversion of a part of the existing convict prison.] 1907 *Daily Chron.* 6 June 5/4 The 'Borstal prisoner'. 1907 *Borstal System* 2 Every lad who is imprisoned in Dartmoor or Borstal passes into the care of the Borstal Association on his discharge. 1917 *Times* 7 Feb. 5/6 Four youths have escaped from the Borstal Institution, Rochester. 1921 *Glasgow Herald* 15 Nov. 6 The Borstal boys have five meals a day at Portland. 1923 (*title*) The Borstalian. 1951 *Lancet* 3 Mar. 519/1 The only open borstal for girls is at East Sutton. *Ibid.* 519/2 A survey of 300 borstal girls. 1957 *Encycl. Brit.* III. 924/1 There are now in England four Borstal institutions—at Chatham, Feltham and Portland for youths, and that for girls at Aylesbury. 1958 *Observer* 19 Oct. 21/6 There are sharp sketches of fellow Borstalians.
Also *transf.* as *adj.*
1936 'J. Tey' *Shilling for Candles* x. 112 Grant thought how Borstal she was in spite of her soignée exterior. That air of resentment against the world in general and her own fate in particular was very familiar to him.

bort. (Further examples.)
1698 Fryer *Acc. E. India & P.* iii. iii. 113 A flat Press, where under Steel-wheels the Diamonds are fastned; and with its own Bort are worn into what Cut the Artist pleases. 1959 *Times* 18 Nov. 21/1 The diamonds produced in the laboratory..are similar to so called boart or abrasive grit.

Borussian (bo[u]rɒ·ʃăn), *sb.* and *a.* [f. med. L. *Borussi* pl. or *Borussia* (app. etymologizing perversion of stem *Prūs-*, as if f. Slavonic *po* by, alongside + *Russia*): see -ian.] = Prussian.
1607 Topsell *Four-F. Beasts* 211 They [elks] are..found ..in the wood Hercynia, and among the Borussian-Scythians. 1608 —— *Serpents* 23 The auncient Borussians, worshipped a naturall Serpent of the earth. 1718 T. Purney *Chevalier de St. George* 49 As regal Beast From Dwolm Borussian bursts. 1882 *Encycl. Brit.* XIV. 702/1 The Lithuanian stem was divided into three main branches:— the Borussians or Prussians; the Letts..and the Lithuanians. 1921 *Trans. Scott. Eccles. Soc.* 126 So far as they are free from Wendish or Borussian admixture. 1925 *Contemp. Rev.* Jan. 72 The country was occupied by aboriginal tribes of Finns, Letts, Lithuanians, Borussians, and Poles or kindred Slavs.

borzoi (bǭ·ɪzoi). Also 9 **barzoi.** [a. Russ. *borzói*, a male dog of the breed called *borzáya*, f. *bórzyi* swift.] A breed of dog, also called the Russian or Siberian wolf-hound.
1887 *Field* 14 May 679/1 The Russian barzoi, or Siberian wolfhound, is one of the noblest of all dogs. 1892 *Pall Mall Gaz.* 11 Feb. 7/2 The Grand Duke owns seventy of these barzois or Russian wolfhounds. 1892 *Field* 5 Mar. 325/1 There seems to be a general feeling amongst owners of Borzois that the time has now arrived for the successful starting of a special club. 1895 *Westm. Gaz.* 9 Oct. 7/2 The Borzois Club is going to be remodelled, but I am glad to hear..that there is no immediate intention of altering the existing scale of Borzoi points. 1945 C. L. B. Hubbard *Observer's Bk. Dogs* 30 The Borzoi arrived in England about 1875. 1969 *Times Lit. Suppl.* 6 Mar. 236/4 One inevitably comes up with Gollancz's yellow jackets,..Alfred A. Knopf's borzoi colophon, and so on.

Bosch(e: see *Boche.

bose (bō[u]s). Colloq. abbrev. of *bosun.
1912 J. Masefield in *English Rev.* XII. 376 The bose half blind, Spat. 1927 J. Sampson *Seven Seas Shanty Book* 47 A handy Bose and a handy Sails.

bose (bō[u]s), *v.* [cf. E.D.D. *Boss* v.[3] to bang, *Boss* adj. hollow, empty (= Boss *a.*).] To test (ground) for the presence of buried structures by noting the sound of percussion from a weighted rammer. So **bo·ser**, an instrument used for this purpose; **bo·sing** *vbl. sb.*, the action or process of testing ground in this way.
1929 *Antiquity* III. 231 The plan made by Dr. Curwen, based upon 'bosing' and excavation, is full of interest. 1930 *Ibid.* IV. 30 The 'boser' can easily be made out of a narrow cylindrical tin filled with about 8 lbs. of lead. 1953 R. J. C. Atkinson *Field Archæol.* (ed. 2) i. 31 The first of these [methods], known as '*bosing*', consists in percussing the surface of the ground with a weighted rammer and listening to the sound thus produced. Over undisturbed ground the sound is dull; ..over a filled-up ditch or pit it changes to a more resonant note...A heavy pick-axe may be used as a boser.

Bose (bō[u]z), *sb.* The name of S. N. *Bose* (see *boson) used *attrib.* in place of *Bose-Einstein, as *Bose condensation, statistics*.
1931 *Sci. Abstr.* A. XXXIV. 187 The Bose statistics. 1932 *Physical Rev.* XL. 1029 One can as yet not decide from isotherm measurements alone whether or not real gases obey the Bose statistics. 1967 J. Wilks *Liquid & Solid Helium* xi. 309 A perfect gas of even atomic mass should obey Bose statistics, and exhibit a Bose condensation below a certain temperature. 1968 C. G. Kuper *Introd. Theory Superconductivity* ix. 156 This formalism is a generalization of the quantum mechanics of a single particle to a system of many non-interacting particles obeying Bose statistics.

Bose-Einstein (bō[u]:z,əi·nstəin). *Physics.* [The names of S. N. *Bose* (see *boson) and Albert *Einstein* (1879–1955), German-born American physicist.] *Bose-Einstein condensation*, in a system of bosons, the existence of a proportion of the particles in a zero-energy state when the temperature is below a certain value; *Bose-Einstein particle* = *boson; *Bose-Einstein statistics*, a type of quantum statistics used with systems of indistinguishable particles which have the property that any number can occupy the same quantum state; cf. **Fermi-Dirac statistics*.
1928 *Proc. Physical Soc.* XL. 329 In the Bose-Einstein statistics λ is essentially positive. 1938 *Physical Rev.* LIV. 947 (*heading*) On the Bose-Einstein condensation. 1948 Mott & Sneddon *Wave Mech. & its Applications* xi. 354 Field theory for Bose-Einstein particles. 1955 H. B. G. Casimir in W. Pauli *Niels Bohr* 131 The remarkable feature of Bose-Einstein condensation is that a sizable fraction of the particles is forced by the statistics into..a state which should have curious and essentially non-classical properties. 1968 C. G. Kuper *Introd. Theory Superconductivity* ix. 156 A collection of particles obeying Bose-Einstein statistics has a wave function which is symmetrical under the interchange of any two particles.

bosey: see *bosie.

bosh, *sb.*[3] **2.** (Earlier example.)
1850 C. Kingsley *Let.* 31 May in *Life & Works* (1902) VII. 28 Theirs is now discovered not to be a necessary trade. Bosh! The question is this—[etc.].

bosh, *sb.*[4] *slang.* [ad. Romany *bosh-* to crow, fiddle, etc., a. Skr. *vāś-* to low, bellow.] A fiddle. *Comb.*: **bosh-faker, -killer, -man**, one who plays a fiddle.
1846 *Swell's Night Guide* 47 A boshman every Tuesday night for hopping and chaunting. 1859 Hotten *Dict. Slang*, *Bosh*, a fiddle. *Bosh-Faker*, a violin player. 1865 F. H. Nixon *Peter Perfume* 102 'Boshman' in the old-hand vernacular signifies a fiddler. 1876 W. Green *Cheap Jack* 231 Can you rocker Romanie, Can you fake a bosh? 1935 X. Petulengro *Romany Life* xxxiii. 119 Rudy and Adolphus were the *bosh*-killers (we spoke English Romany now); I played the melodeon.

bosher (bǫ·ʃəɪ). *slang.* [f. Bosh *sb.*[3] + -er.[1]] One who talks 'bosh' or nonsense.
1913 D. H. Lawrence *Let.* 17 Jan. (1932) 94 Don't ever mind what I say. I am a great bosher, and full of fancies that interest me. 1939 H. G. Wells *Holy Terror* i. iii. 89 'The man's a blatant bosher.' 'He's not. He knows exactly what he is up to.'

bosh-shot: see *boss-shot.

boshy (bǫ·ʃi), *a.* [f. Bosh *sb.*[3] + -y[1].] Of the nature of bosh; contemptible, worthless.
1860 J. M. Atkinson *Let.* 6 May in *Richmond-Atkinson Papers* (1960) I. x. 577 The concentration of 'boshy' gossip in town is wearisome. 1864 'Mark Twain' *Sk. Sixties* (1927) ii. 132, I read your boshy criticisms on the opera with the most exquisite anguish. 1882 'F. Anstey' *Vice Versâ* iv, There was no dancing, only boshy games and a conjuror.

bosie (bō[u]·zi). *Austral.* Also **bosey.** [Hypocoristic f. the name of B. J. T. Bosanquet, an English cricketer: see -ie.] = *googly *sb.* Also *attrib.*, as *bosie ball, bowler, bowling*.
1912 *Australasian* 2 Mar. 481/2 Then he lifted the 'Bosie' bowler high to the on. 1920 E. R. Wilson in P. F. Warner *Cricket* ii. 74 The 'googly' or 'Bosie ball' as it was afterwards christened in Australia. 1927 *Observer* 13 Feb. 23/6 The finest exponent of 'Bosie' bowling in the world. 1930 C. V. Grimmett *Getting Wickets* i. 22 It was at this time that I learned to bowl the 'bosie' or 'googly'—an off-break with a leg-break action. 1954 A. G. Moyes *Austral. Batsmen* xiv. 186 The changes in bowling style and tactics..from off-spin to speed, to 'boseys'.

bosker (bǫ·skəɪ), *a.* *Austral.* and *N.Z. slang.* Now *obsolescent.* Also formerly **bo·scar, bo·shter.** [Origin unknown.] Good, excellent, delightful. Cf. *bonzer *a.* Hence as *sb.*
1906 E. Dyson *Fact'ry 'Ands* i. 1 'She's er little boshter!' he said vehemently. 'Y' orter seen 'er.' *Ibid.* xii. 151 He promised to show Feathers a 'boshter knack for passing out gazobs'! 1909 A. H. Adams *Galahad Jones* i. 8 It's a 'bosker' castle. 1911 C. E. W. Bean *'Dreadnought' of Darling* xxxv. 314 'That's it—my word, it's a bosker, that is' the driver whispered to the Sydney passenger. 1911 'Kiwi' *On Swag* vii. 14 We gave him a boscar funeral. 1916 *Anzac Book* 36 'A boshter night for a walk,' I remarked, buttoning my coat about me. 1922 A. E. Mulgan *Three Plays N.Z.* 36 That's a boscar song. 1926 I. M. Peacocke *His Kid Brother* iv. 54 What bosker fun. 1943 F. Sargeson in *Penguin New Writing* XVIII. 63 It turned out a bosker day. 1952 A. Grimble *Pattern of Islands* 68 'Cripes!' he said. 'She's a fair bosker, ain't she, son!'

Boskop (bǫ·skǫp). [The name of a place in the Transvaal, South Africa.] In *attrib.* use with *man, race*, etc.: of or belonging to the early type of man indicated by the skull of the late Pleistocene period found at Boskop. Hence **Bo·skopoid** *a.* [-oid], characterized by the type of skull found at Boskop; also as *sb.*
1915 *Nature* 5 Aug. 615/2 The Boskop man was of the Neanderthal race, but more advanced in intelligence. 1926 *Bantu Studies* II. 219 Comparison has been made mainly with the Boskopoid remains from Zitzikama reported upon..during the last two years, and with the descriptions of the original Boskop remains. 1948 A. L. Kroeber *Anthropol.* (ed. 2) iii. 108 Undoubtedly Neanthropic is the Boskop type...The Boskop type skulls are massive and large. 1959 J. D. Clark *Prehist. S. Afr.* iv. 97 The physical types belonging with the earlier occupation levels at this site were in fact 'Bush-Boskopoids'. 1966 E. Palmer *Plains of Camdeboo* vii. 118 The Middle Stone Age people—the Boskop race—of South Africa were not one race but several.

†**bosman.** *U.S. Obs.* Also **bosseman.** [ad. Fr. *bosseman* (cf. G. *bootsmann* boatswain).] = *bowsman. *Obs.*
[1807 C. C. Robin *Voyages* II. 212 Les rameurs sont distribués également de chaque côté; à l'arrière est le patron qui gouverne, et en avant un homme, nommé bosman, une perche à la main, sonde les lieux où l'on craint de toucher.] 1876 *Scribner's Monthly* Jan. 404/2 A man in the bow called the *bosman*, who generally wielded a sort of a boathook, watched the course. 1906 *Trans. Kansas State Hist. Soc.* 1905–6 IX. 272 The captain of the boat, called the 'patron', did the steering, and his assistant, called the 'bosseman', stood on the bow, pole in hand, and gave directions to the men at the cordelle.

Bosniac (bǫ·zniæk). Also **-ak.** [ad. F. *Bosniaque*, or G. *Bosniake*, ad. Russ. *Bosnyak*.] = *Bosnian *sb.*
1836 *Penny Cycl.* V. 231/1 The inhabitants of Bosnia are composed of Bosniaks, a race of Sclavonian origin. 1848 E. Lear *Jrnl. of Landscape Painter in Albania* (1851) 17 The packed Wallachians, and Bosniacs, and Jews started crampfully from the deck. 1920 *Edin. Rev.* Oct. 218 The ruling race absorbed large numbers of Christians, Greeks, Slavs, and later on Albanians and Bosniacs.

Bosnian (bǫ·zniăn), *a.* and *sb.* [f. *Bosnia*: see -ian. Cf. F. *bosnien*.] **A.** *adj.* Of or pertaining to Bosnia, a province of Yugoslavia. **B.** *sb.* A native of Bosnia.
1788 [see Servian *sb.*]. 1830 C. Fraser *Hist. War in Bosnia* 17 A number of females who, like the ancient Bosnian women, acquired the courage of heroes. *Ibid.* 63 Mohammed set off with five or six thousand horsemen, Bosnians. 1836 *Penny Cycl.* V. 230/1 The Verbas, another Bosnian river, rises in the heart of the country. 1847 [see Turk[1] 2]. 1847 Mrs. A. Kerr tr. *Ranke's Hist. Servia* 167 He opposed to the great Bosnian army about 1500 men. 1924 *Contemp. Rev.* Nov. 620 'The Bosnians are never satisfied', we were told. 1949 L. Durrell *Spirit of Place* (1969) 103 The Bosnian peasants in their dramatic costume.

bosom, *sb.* Add: **3. d.** The front of a shirt. *U.S.*
1863 *Horticulturist* (Albany, N.Y.) Dec. 4 Shirt and bosom makers. 1872 *N.Y. Times* 24 Apr. 8 Advt. (Hoppe), Shirts made to order, with beautifully embroidered Bosoms. 1945 B. A. Botkin *Lay My Burden Down* 51 Please don't let my gal see under my coat, 'cause I got on a bosom and no shirt.
7. b. *ellipt.* for Bosom friend.

1913 A. N. Lyons *Simple Simon* i. vi. 90 He's a darling. He and me are bosoms. 1959 B. Ruck *Romantic Afterthought* xxvii. 138, I don't go in for buddies and 'bosoms'.

bosomy, *a.* Add: **2.** Of a woman: having a prominent bosom.
1928 Blunden *Undertones of War* xiv. 156 A most formidable woman of bosomy immenseness. 1952 *Landfall* VI. 114 A rare young creature, straight and slender yet bosomy. 1967 *Coast to Coast 1965–6* 250 Glossy photos of bosomy girls.

boson (bō[u]·zǫn). *Nuclear Physics.* [f. the name of the Indian physicist S. N. *Bose* (1894–1974) + *-on.] Any particle which has a symmetric wave-function and which therefore obeys Bose–Einstein statistics; also *attrib.* Cf. *fermion. Also *intermediate boson*, a boson postulated to exist as a quantum or intermediary of weak interactions.
1947 P. A. M. Dirac *Princ. Quantum Mech.* (ed. 3) ix. 210 The new statistics was first studied by Bose, so we shall call particles for which only symmetrical states occur in nature *bosons*. 1953 M. Schönberg in *Nuovo Cimento* X. 429 When the coordinates and momenta are given at the time t = 0, we must take for bosons (28a)..and for fermions (28b). 1955 W. Pauli *N. Bohr* 33 The coupling between one Boson field and one Dirac field. 1955 D. ter Haar in *Rev. Mod. Physics* XXVII. 312/2 Fermi–Dirac particles. Such particles are sometimes called *fermions* and the Bose-Einstein particles, *bosons*. 1958 *Physical Rev.* CX. 1482/2 The existence of such a heavy boson..will itself lead to the occurrence of decays which are not found in nature, and which would not occur in any detectable amount if there are no intermediate bosons. 1961 Powell & Crasemann *Quantum Mech.* xii. 453 Particles whose spin is an integral multiple of ħ are bosons. 1965 *New Statesman* 17 Sept. 393/3 One of these tracks may well be the first sign of the existence of the long-sought particle of matter called intermediate boson.

boss, *sb.*[1] Add: **3. g.** A soft pad used in ceramics and glass-manufacture for smoothing and making uniform the colours applied with oil to a glass or porcelain surface, and for cleaning gilded surfaces.
1860 Ure *Dict. Arts* (ed. 5) III. 506 The 'boss' is made of soft leather. a 1877 Knight *Dict. Mech.* s.v. *Bossing*, The bossing is laid on with a hair-pencil, and leveled with a boss of soft leather. 1879 E. C. Hancock *Amat. Pott. & Glass Painter* 49 The boss consists of a lump of cotton wool, screwed up, as it were, in two or three thicknesses of fine soft linen. 1961 M. Jones *Potbank* xii. 44 Bert..polished it [sc. a plate] with a boss—a piece of cloth stuffed with wool.
h. The central portion of the propeller of an aeroplane.
1916 H. Barber *Aeroplane Speaks* iv. 121 If a weight.. placed in a bolt-hole on one side of the boss fails to disturb the balance, then the propeller is usually regarded as unfit for use.

boss, *sb.*[6] Add: **a.** (Earlier and later examples.) Now in general use in Britain.
[a 1649 J. Winthrop *Hist. New England* (1908) I. 166 Here arrived a small Norsey bark..with one Gardiner, an expert engineer or work base [= Du. *werk-baas*], and provisions. 1653 F. Newman et al. *Let.* May in E. Hazard *Hist. Collections* (1794) II. 236 From our Place of Residence at the Basses house in the Monhatoes.] 1806 W. Irving *Let.* 26 May in P. M. Irving *Life & Lett.* (1862) I. xi. 138, I had to return, make an awkward apology to boss, and look like a nincompoop. 1813 Ld. Yarmouth *Let.* 12 Dec. in E. Taylor *Taylor Papers* (1913) vii. 98 There are some peasants watching, one of whom has frightened the boss with an alarm of a sortie. 1936 G. B. Shaw *Millionairess* Pref. 106 A born boss is one who rides roughshod over us by some mysterious power that separates him from our species and makes us fear him. *Ibid.* 128 Clearly we shall be boss-ridden in one form or another. 1937 'G. Orwell' *Road to Wigan Pier* viii. 157 The accent and manners which stamp you as one of the boss class. 1962 *Listener* 16 Aug. 238/1 Most people make mistakes when they start a new job, and it is irritating for the boss.
c. (Earlier examples.)
1836 in J. R. Commons *Doc. Hist. Amer. Industr. Soc.* (1910) IV. 287, I am a boss shoemaker. 1840 J. P. Kennedy *Quodlibet* 221 Charley Moggs, long known as the boss loafer of Bickerbray. 1848 W. E. Burton *Waggeries* 63 (Th.), 'How d'ye do, folks?'..'is the boss devil to hum?'
d. *Comb.* **boss-boy** *S. Afr.* (see *boy *sb.*[1] 3 e); **boss-cocky, -cockie** *Austral. slang* [Cocky *sb.*[2] 2], a farmer who employs labour and works himself; hence in extended use, a person in authority; **boss-man** (orig. *U.S.*) = Boss *sb.*[6] a; **boss of (or over) the board** [*board *sb.* 2d] *Austral.* and *N.Z.*, the overseer of a shearing-shed.
1898 Morris *Austral English* 46/1 Boss-cockie. 1916 J. B. Cooper *Coo-oo-ee* i. 16 Mrs. Muller, before she married the German, was Kate Hardley, the daughter of a boss cockie farmer fifty miles away. 1928 'Brent of Bin Bin' *Up Country* xiii. 218 Rab was the boss cocky of it [sc. an orchestra]. 1969 *Coast to Coast 1967–8* 5 But what was her place now? She was a kind of boss cocky. 1934 C. Carmer *Stars fell on Alabama* (1935) iv. iv. 212 When old man Huckaby died a year ago Wade came up to the house and stayed night and day, waiting on the old boss-man. 1965 H. Gold *Man who was not with It* xii. 103 Bossman on the biggest power-generator Sunday school show in the Southland. 1969 J. Wainwright *Big Tickle* 25 Divisional Chief Superintendent Sullivan..was boss-man of North End Division. 1896 H. Lawson *While Billy Boils* xiv. 89 There are tally lies..and lies about getting the best of squatters and bosses-over-the-board. *Ibid.* xliii. 253 The third shearer was telling a yarn...'So I said to the boss-over-the-board, "you're a nice sort of a thing," I sez.' 1901 M. Franklin *My Brilliant Career* xxii. 186 A big strike among the shearers when the narrator had been boss-of-the-board out beyond Bourke. 1948 V. Palmer *Golconda* vii. 51 Macy the Battler, they had called him at his last shearing-shed after he had tackled a nagging boss-of-the-board.

boss (bǫs), *sb.*[7] *U.S.* [app. the same word as Eng. dial. (south-western) *borse, boss, buss* six-months-old or half-grown calf (1790 Grose *Prov. Dict.*, etc.).] A word used in addressing a cow. (Cf. *bossy *sb.*) Also, the American bison.
1800 in *Wash. Hist. Quart.* (1928) XIX. 268 The Indians traded..Tongues and Bosses. 1848 Bartlett *Dict. Amer.*, *Boss*, among the hunters of the prairies, a name for the buffalo. 1874 *Rep. Vermont Board Agric.* II. 706 So-o-o boss! There, you've kicked it over—All that milk, now, I declare! 1901 *Nation* 18 Apr. 314/2 The call 'Co' boss is familiar to most of the inhabitants of our Northern States and Canada.

boss, *sb.*[8] = *boss-shot.
1898 *Eng. Dial. Dict.* s.v., He then tried to jump the ditch to the big stone, but in his hurry he made a boss and fell into the water.

boss, *v.*[1] Add: **2. b.** In ceramics, to smooth a surface of boiled oil on pottery by means of a boss (*boss *sb.*[1] 3 g).
1860 Ure *Dict. Arts* (ed. 5) III. 506 A coat of boiled oil adapted to the purpose being laid upon the ware with a pencil, and afterwards levelled, or as it is technically termed 'bossed', until the surface is perfectly uniform. 1879 [see *bossing *vbl. sb.*[1] d]. 1881 [see Boss *v.*[1] 2, where this quot. is wrongly placed].

boss, *v.*[2] (Further examples. In Britain now in general colloquial use, not only humorous.)
1882 W. D. Hay *Brighter Britain* I. vi. 166 The way that Old Colonial 'bossed' them round was an edifying sight to see. 1933 C. A. Macdonald *Pages from Past* v. 62 He 'bossed' a timber-cutting camp. 1944 R. Lehmann *Ballad & Source* i. iv. 35 'Well, I won't be bossed by her,' she said gruffly.

boss (bǫs), *v.*[3] *dial.* and *slang.* [Cf. *boss-eyed, *boss-shot.] *trans.* To miss or bungle (a shot); *gen.* to bungle, make a mess of. Also *absol.*
1887 *N. & Q.* III. 236/2 To boss is schoolboy slang for 'to miss'. 1889 Barrère & Leland *Dict. Slang* s.v., To *boss* anything, to make a mess of it, to spoil it. 1898 *Eng. Dial. Dict.* s.v., He had six shies at the cocoa-nuts, and he bossed every time. 1903 'Marjoribanks' *Fluff-Hunters* 74 You're simply bossing up the whole show by philandering with a widow.

bossa nova (bǫ:să nō[u]·vă). [Pg. *bossa* tendency + *nova* fem. sing. of *novo* new.] A style of Brazilian music related to the samba; a dance performed to this music.
1962 *Daily Mail* 2 Nov. 14/4 In a hesitant way the bossa nova is already starting. At Edmundo's club..his band plays bossa nova while the customers perform their own version, a kind of lazy samba. 1962 *New Statesman* 21 Dec. 911/1 In Brazil *bossa nova* is in no sense a dance. It is a way of playing and singing. 1963 *New Yorker* 8 June 99 (Advt.), And shop for Brazilian antiques and bossa nova records. 1967 *Guardian* 22 Apr. 5/6 The same band handles bossa nova, rock-and-roll, and rhythm-and-blues with equal panache.

bossdom (bǫ·sdəm). *orig. U.S.* [f. Boss *sb.*[6] + -dom.] The condition of being a boss; that which constitutes the sphere of influence of a political boss; the control of politics by bosses.
1888 J. B. Bryce *Amer. Commonw.* II. lxiii. 462 The extinction of the Boss himself and of bossdom. 1893 *Black & White* 1 Apr. 383/1 He was going to make John Bull realise that his days of bossdom were over. 1894 *Citizen* (Albion, Mich.) 293 It is not healthy for a party, if the few are allowed to do all and say all. That way Rings and Bossdom lie.

boss-eyed, *a.* *dial.* and *slang.* [Cf. *boss *v.*[3], *boss-shot.] Having only one good eye; squint-eyed, cross-eyed. Also *fig.* oblique, crooked, one-sided.
1860 Hotten *Dict. Slang* (ed. 2). 1882 Chamberlain *W. Worc. Gloss.*, *Boss-eyed*, squinting. 1890 [see *boss-shot]. 1898 *Eng. Dial. Dict.* s.v., The horse shied and we ran up against the gate-post, and knocked the step of the cart all boss-eyed. 1959 I. & P. Opie *Lore & Lang. Schoolchildren* xi. 214 It is unlucky for a cross-eyed woman to look at you... When somebody who is boss-eyed goes by you spit on the ground.

bossiness[2] (bǫ·sinės). [f. Bossy *a.*[2] + -ness.] The quality of being bossy or overbearing.
1928 J. Devanny *Dawn Beloved* i. ii. 19 She savagely resented any 'bossiness'. 1958 *Listener* 30 Oct. 699/1 The bossiness and love of power so often characteristic of dwarfish persons of insecure social status.

bossing, *vbl. sb.*[1] Add: **d.** In ceramics, the process of smoothing the surface of colour applied to pottery over a coating of boiled oil; also, the coating of oil used for this purpose.
a 1877 Knight *Dict. Mech.*, *Bossing*, ground-laying the surface of porcelain in an unfinished state, to form a basis of adherence for the color. ... The bossing is a coat of boiled oil, to hold the color. 1879 E. C. Hancock *Amat. Pott. & Glass Painter* 49 When the oil has become somewhat set, so as to be 'tacky' to the finger, it is ready for the second process, called 'bossing'.

boss-ship, bossship (bǫ·sʃip). *orig. U.S.* [f. Boss *sb.*[6] + -ship.] The rule or position of a boss or bosses, esp. in politics.
1882 *Nation* 2 Nov. 371/3 To enable Mahone to build up a new..boss-ship in Virginia. 1889 'Mark Twain' *Conn. Yankee* xiii. 160 The thing that would have best suited..my nature would have been to resign the Boss-ship and get up an insurrection. 1894 *Voice* (N.Y.) 6 Sept., It was thought to be an auspicious time to shake off the 'bossship' exercised by Mr. Platt for many years. 1928 D. H. Lawrence *Lady Chatt.* xiii. 217 'But who is boss of the show?' she asked. 'The men who own and run the industries'...'They don't take their boss-ship seriously enough,' she said.

boss-shot. *dial.* and *slang.* [See *boss *v.*[3]] A bad shot or aim; *fig.* an unsuccessful attempt.
1890 Farmer *Slang*, *Boss*, to miss one's aim; to make such a shot as a boss-eyed person would be expected to make. Boss-shot is a common phrase. 1898 *Eng. Dial. Dict.* s.v., A bad shot with a stone is called a boss-shot. 1912 W. Deeping *Sincerity* xviii. 141 Harkness made a clutch at the bow, but Master Brandon was too quick for him. He retreated three paces, grinning. 'Boss shot, that!' 1919 E. H. Jones *Road to En-dor* (1920) xxviii. 300 The student ..made two boss shots before he hit the bull.
Also in corrupt form **bosh-shot.**
1939 'G. Orwell' *Coming up for Air* iii. i. 181 The Nazis chop people's heads off..and sometimes the executioner makes a bosh shot. 1959 'O. Mills' *Stairway to Murder* vii. 80 You haven't been having another bosh shot at the Crown Jewels, have you?

bossy, *a.*[2] For '*U.S. colloq.*' read: '*colloq.* (orig. *U.S.*).' (Add later U.K. examples.)
1902 *Westm. Gaz.* 16 June 3/1 They cannot forget his often irritating ways and his decided tendency to be 'bossy'. 1933 N. Coward *Design for Living* i. i. 7 Personally, I never cared for her very much. A bossy woman.

bossy (bǫ·si), *sb.* *U.S.* [Eng. dial. (south-western), dim. of *boss*, *boss *sb.*[7]] = *boss *sb.*[7]; a calf or cow. Also *bossy-calf, -cow.*
1844 'J. Slick' *High Life N.Y.* II. 181 She only bust out in a new spot; and, like a great bossy calf, I had to jine in again. 1848 Bartlett *Dict. Amer.*, *Bossy*, a familiar name applied to a calf. 1863 'G. Hamilton' *Gala-Days* 95 Bossy starts from the post, tail up in a hand gallop. 1907 *N.Y. Even. Post* 25 Feb. 3 He..will go out to interview a bossy [cow] who has eaten her last wisp of hay. 1911 H. Quick *Yellowstone N.* xii. 314 A notion o' what it means to incorporate the fruit of the nest [sc. eggs] with the bossy. 1927 H. Crane *Let.* 19 Mar. (1965) 291, I had just landed in town after three months with the bossy cows.

bost, var. *boast *v.*[2]

bosthoon (bǫstū·n). *Irish.* Also **bostoon.** [ad. Ir. *bastún* whip made of green rods, soft or spiritless fellow.] An awkward fellow; a tactless, senseless person.
1833 W. Carleton *Traits* I. 65 Sure only for this, I say, you bosthoon..where 'ud the purty colleen be? 1841 S. C. & A. M. Hall *Ireland* I. 75 Ye grate Bosthoon. 1914 Joyce *Dubliners* 197 Is this what we pay rates for?..To feed and clothe these ignorant bostooms [*sic*]. 1932 D. Johnston *Moon in Yellow River* i. 20 Are you the ignorant bosthoon that's banging and hammering away at my knocker? 1933 Davies & Thomson tr. *O'Sullivan's 20 Yrs. A-Growing* xvii. 227 Remember your ancestors! Strike the bostoon! 1970 M. Kenyon *100,000 Welcomes* xiii. 105 If it hadn't been for that seethin' bosthoon Dempsey we'd have had it wrapped up.

Boston[2]. Add: (Earlier examples.)
1800 *Sporting Mag.* XVI. 185/2 (*heading*) Rules for the Game of Cards called Boston. *Ibid.* 186/1 The eldest hand ..if he thinks..that he can get five tricks or more, played as at Whist, says Boston.
2. A variation of the waltz. Also *Boston dip* (*waltz*) (see quot. 1885); *Boston two-step*, a variation of the two-step.
1879 *Amer. Punch* Oct. 116/1 The Hardshell Baptists will not dance even the Boston Dip Waltzes. 1885 A. Dodworth *Dancing* 73 Boston...When stepping with the right foot, the left knee is slightly bent, producing the dip, from which the name Boston Dip was derived. 1887 *Courier-Journal* (Louisville) 8 May 12/8 The young English baron..[was] dancing the Boston with Miss Bazaine. 1913 E. Scott *All about the Boston* 21 Mr. Henry Zay wrote: 'I have danced the "Boston" for twenty years (I am an American) and say emphatically that there is no set figure that can be called the "Boston". It is a series of steps or figures—such as the ordinary waltz-step, the "dip", the "run", the "reverse",' etc. *Ibid.* 24 At the present time the term 'Boston' is applied to the kind of movement that in its best and most graceful form would be far more consistently described as Rectilineal or Diagonal waltzing. 1918 A. Bennett *Pretty Lady* iii. 11 Do play for me. Play a boston—a two-step.
3. Boston (bull) terrier, a small smooth-coated terrier originating in Massachusetts from a crossing of the bulldog and terrier. Also *ellipt.* and *attrib.*
1894 *Outing* (U.S.) Mar. 465/1 After much discussion the name of Boston Terrier was finally selected because all other names indicating the origin of the dog were more or less in conflict with those of older breeds. 1907 F. T. Barton *Terriers* ix. 90 The so-called *Boston terriers* appear to be but indifferent specimens of the bull-terrier cropped. 1945 C. L. B. Hubbard *Observer's Bk. Dogs* 31 Boston Bull-

Terrier... The breed was recognised by the American Kennel Club in 1893. **1948** *Chicago Tribune* 4 Apr. (Grafic Mag.) 20/4 The Boston originated from a cross between the English bulldog and the white English terrier. **1955** W. W. Denlinger *Complete Boston* i. iv. 112 The correct Boston neck fits neatly into and between long shoulders.

Hence **Bo·ston** *v. intr.*, to dance the Boston. *colloq.*

1913 A. N. Lyons *Simple Simon* iii. vi. 328 He was saying to her 'Do you Boston?'. **1920** S. Lewis *Main St.* i. 2 Scores recited more accurately and dozens Bostoned more smoothly.

Bostonese (bǫstǝnī·z). *U.S.* [f. *Boston*, Mass. + -ese.] **a.** *collect.* Natives or inhabitants of Boston (see also quot. 1785). **b.** (See quot. 1888.)

1785 J. Hadfield *Englishman in Amer.* 29 June (1933) 61 He asserted he had tomahawked and scalped 50 Bostonese, a name they [sc. Indians in Canada] give generally to the Americans of the States. **1876** *Cincinnati Enquirer* 17 June 3/2 His reply was not negro at all, it was purest Bostonese. **1888** *N.Y. Herald* 29 July (Farmer), There were a number of people present, principally Bostonese. **1888** Farmer *Americanisms*, *Bostonese*, a method of speech or manners supposed to be specially affected by the residents of that city. **1925** *N.Y. Times* 8 Sept. 24/3 Squire Sam Jones..in the second innings surrendered four runs to the Bostonese. **1948** Mencken *Amer. Lang.* Suppl. II. 179 The educated speech of the State..is 'untainted by Bostonese'.

Bostonian (bǫstōu·niăn), *sb.* and *a.* [f. as prec.] **A.** *sb.* A native or inhabitant of Boston.

1682 J. W. *Let. from N.-Eng.* 7 These Bostonians enrich themselves by the ruine of Strangers. **1698** C. Mather *Bostonian Ebenezer* Title-p., The Bostonians. **1774** H. Walpole *Let.* 2 Feb. (1904) VIII. 418 The Bostonians have canted three hundred chests of tea into the ocean. **1847** W. I. Paulding *Noble Exile* in *Amer. Comedies* 107 Nay.. we Bostonians have the reputation of being Bostonians wherever we go. **1897** C. M. Flandrau *Harvard Episodes* 18 When I say 'a Bostonian'..I mean of course a Bostonian that one knows. **1965** *Listener* 3 June 815/1 To be a poet in the age of Eliot, to grow up in the shadow of a frosty Bostonian.

B. *adj.* Belonging or native to Boston.

1698 C. Mather (*title*) The Bostonian Ebenezer. **1794** Humphreys *Industry* 15 Where Bostonian maids, with songs, prepare The canvass wings. **1920** *Chambers's Jrnl.* 15 May 374/1 His carefully nuanced Bostonian accent.

bosun, bo'sun, var. Boatswain, representing the common pronunciation (bōu·s'n).

1868 W. S. Gilbert *Bab Ballads* 85 And a bo'sun tight, and a midshipmite, And the crew of the captain's gig. **1894** Stevenson & Osbourne *Ebb-Tide* i. v. 77 He had a bo'sun's chair rigged over the rail..and went overboard with a pot of paint. **1930** 'Gregorian' *Tinker, Tailor* vi. 134 We were swung alongside in a bosun's chair to scrape the dirt off the masts and spars. **1963** *Times* 26 Feb. 12/6 The bo'sun's call is the object one sees..hanging on a chain round the neck of the sailor on duty at the gangway of a British warship. Some people..call it a pipe.

Boswell (bǫ·zwel). The name of James *Boswell* (see Boswellian *a.*), used allusively for: a constant companion or attendant who witnesses and records what a person does.

1858 O. W. Holmes *Autocrat Breakf.-Table* (sub-title), Every man his own Boswell. **1892** A. Conan Doyle *Adv. Sherlock Holmes* (1892) 6, 'I think that I had better go, Holmes.' 'Not a bit, Doctor. Stay where you are. I am lost without my Boswell.' **1932** N. Mitford *Christmas Pudding* iii. 41, I never thought of biography, but of course that's the very thing for me... May I be your Boswell, darling?

Boswellian, *a.* Add: Also **Boswellean**. (Earlier examples.)

1843 Carlyle *Past & Present* II. i. 54 The 'Chronicle', or private Boswellean Notebook, of Jocelin. **1855** *Rambler* III. 485 The average Englishman..has been bored by the Boswellian adoration of the Corsican Paoli.

B. *sb.* A student or admirer of Boswell.

1908 *Daily Chron.* 11 Nov. 3/4 Their appearance in print was hailed with pleasure by Johnsonians and Boswellians. **1921** *Glasgow Herald* 29 Jan. 13 Johnsonians and Boswellians alike will therefore receive with joyful gratitude Dr. J. T. T. Brown's paper.

Bo·swellize, *v.* **1.** *intr.* (See Boswellian *a.*)

2. *trans.* To treat in the Boswellian manner; to observe and record the actions, etc. of (someone).

1837 J. G. Lockhart *Mem. Sir W. Scott* IV. v. 151 Private lucubrations..designed to Boswellize Scott. **1855** *Tait's Mag.* XXII. 444/2 We had rather the many-sided man should remain to us the mystery he is, than be Boswellised after the fashion which is now current. **1912** A. Lang *Shakespeare, Bacon & Gt. Unkn.* ii. 29 There was no enthusiastic curiosity about him; nobody Boswellised any playwright of his time.

bot, bott. Add: **1. a.** (Further examples.) Also *fig.*

1918 E. Pound *Pavannes* 36 Mosher's propagandas That are the nation's botts, collicks and glanders. **1922** Wodehouse *Clicking of Cuthbert* v. 109 Like a sheep with the botts. **1951** Colyer & Hammond *Flies Brit. Isles* xxiii. 294 Bots do not appear to be the cause of permanent damage to their hosts.

b. (Earlier and later examples.)

1721 Ramsay *Poems* 30 She never ran sour Jute [liquor], because It gee's the Batts. **1844** H. Stephens *Bk. Farm* II. 178 Going out to work..with a full stomach, may bring on an attack of batts or colic.

c. *Austral.* and *N.Z. slang.* (See quots. 1919 and 1941.)

1919 Downing *Digger Dial.* 13 *Bott*, (1) a cadger; (2) a useless person; (3) a hanger-on. **1941** Baker *N.Z. Slang* vi. 51 In early uses *bot* is rendered as a germ, doubtless from *bot-fly*. From this comes the phrase of greeting, *How are the bots biting?* By the 1920's a *bot* is being used extensively for a troublesome person, for a persistent borrower, a first-class parasite...Of fairly recent development in New Zealand is the phrase *to have the bot*, to be sick or out of sorts, moody or disagreeable. **1960** J. Fingleton *Four Chukkas to Australia* 63 One of..the officials was berating Pressmen..as a lot of bots who wanted everything for nothing'.

3. *bot-worm.*

1877 *Rep. Vermont Dairym. Assoc.* VIII. 105 Grub-in-the-head is a bot-worm,..cousin to the bots in horses.

bot, bott (bǫt). Colloq. abbrev. of Bottom *sb.* 1 b.

1922 Joyce *Ulysses* 527 Spank your bare bot right well, miss, with the hair-brush. **1951** Auden *Nones* (1952) 54 The cute little botts of the sailors. **1959** I. & P. Opie *Lore & Lang. Schoolchildren* xiii. 301 A kick up the bot for being a clot.

bot (bǫt), *v.* *Austral.* and *N.Z. slang.* [f. *bot, bott 1 c.] *intr.* (See quot. 1941.) So **bo·tting** *vbl. sb.*

1934 *Bulletin* (Sydney) 7 Nov. 46/2 Settle up when I see me next picture... Never did like botting on a bloke. **1941** Baker *N.Z. Slang* vi. 52 *To bot*, to borrow money, to impose on others, and *botting*, the practice. *To bot on* a person is widely used..*cold botting* is a straight-out request for food at house-doors.

‖ **bota** (bo·tă). [Sp.] **1.** A small leather bottle for holding wine.

1832 W. Irving *Alhambra* 20 His bota, or leathern bottle, which was of portly dimensions, filled to the neck with choice Valdepeñas wine. **1923** *Blackw. Mag.* Aug. 180/2 The bota was unslung, and we all toasted one another in tarry wine. **1940** E. Hemingway *For whom Bell Tolls* xxi. 267 Give me the *bota* with wine.

2. A boot, esp. for riding. *U.S.*

1834 A. Pike *Prose Sk. & Poems* 138 The men with their pantalones of cloth,..the botas of striped and embroidered leather. **1901** *Out West* June 481 She..looked up ..at the pretty red-stained *botas* studded with silver buttons. **1967** 'La Meri' *Sp. Dancing* (ed. 2) 149 *Botas*, Spanish dancing boots, male.

botallackite (botæ·lăkait). *Min.* [f. the *Botallack* mine, Cornwall: see -ite¹.] A name given by A. H. Church in 1865 to a variety of atacamite containing a large proportion of water.

1865 A. H. Church in *Jrnl. Chem. Soc.* III. 213. **1868** Dana *Syst. Min.* (ed. 5) 121 Botallackite occurs at the Botallack mine, Cornwall, in thin crusts of minute interlacing crystals, closely investing killas.

Botany. Add: **2. b.** Short for *Botany wool.*

1884 W. S. B. McLaren *Spinning* vi. 116 When the material is very short Botany, the advantage is not so great.. ..It is confined to commission combers and Botany spinners. **1900** *Daily News* 16 Oct. 2/6 In the botany trade it is difficult for spinners to obtain orders.

botch, *sb.*² Add: **4. b.** = Botcher¹ 3. *dial.* and *colloq.*

1829 J. Kenney *Illust. Stranger* II. i. 24 Some botch of an embalmer, who had not done justice to Your precious remains. **1868** J. C. Atkinson *Gloss. Cleveland Dial.* 59 He's nobbut an aud *botch*. He's mair lahk t'mar an t'mend.

botchy, *a.*² (Example.)

1879 G. M. Hopkins *Let.* 10 Mar. (1935) 24 Each verse is a work of art, no botchy places.

botel (bōu·te·l). orig. *U.S.* Also **boatel.** [Blend of Boat *sb.* and Hotel *sb.*] A hotel catering for boat-owners (see quot. 1960).

1956 *N.Y. Times* 3 June 25 A 'boatel' in the park for the overnight accommodation of yachtsmen and outboard boat operators. **1959** *Woman's Day* July 28/2 The area is accessible only by boat or sea-plane, and there are no automobiles on the island. Hence the Inn at Wild Harbor and the Lodge at Rock Harbor are referred to as botels rather than motels. **1960** *Daily Tel.* 20 Aug. 7/1 Botel. It means what you would expect: a waterfront hotel where you can bring your own boat, or hire one, and have the best of both worlds by spending your holidays sailing yet being able to live ashore in warm, dry comfort. **1962** *Ocean Times* (R.M.S. 'Queen Mary') 20 Oct. 8/1 (Advt.), Boulters Inn Botel..situated on the River Thames. **1969** *Daily Tel.* 11 Jan. 19/3 It will include two hotels, 'boatels', yacht clubs.

2. A ship with the facilities of a hotel. Also *attrib.*

1959 *Listener* 31 Dec. 1162/1 (Advt.), A new concept of river cruising by the botels 'Amsterdam' and 'Arnhem'. **1962** *Times* 3 Jan. 2/3 'Botel cruises' through the heart of Europe.

bother, *v.* **2. b.** (Earlier examples.)

1840 Dickens *Old C. Shop* xxxiii. 281 To this amorous address Miss Brass briefly responded 'Bother!' **1844** — *Mart. Chuz.* xlix. 564 'Bother Mrs. Harris!' said Betsey Prig...'I don't believe there's no sich a person!'

botheration. (Earlier examples as sb. and exclamation.)

1797 *Sporting Mag.* Mar. 342/1 After a long *botheration* ..Mr. Lucius Concannon was convicted. **1841** C. J. Lever *C. O'Malley* xv. 81 'It's for Counsellor Kinshella, below-stairs'..'Botheration.' **1847** C. M. Yonge *Scenes & Char.* vii. 67 'Rachel is going away.'..'Rachel! Rachel! botheration!' roared Reginald.

Bothrodendron (bǫþrǫde·ndrǫn). *Bot.* [mod. L., f. Gr. βόθρος pit + δένδρον tree.] A genus of fossil plants, found in coal measures (cf. Lepidodendron); also (with lower-case initial), a plant of this genus.

1835 Lindley & Hutton *Fossil Flora Gt. Brit.* II. 97 Bothrodendron. Stem not furrowed, covered with dots. Scars of cones, obliquely oval. **1862** Buxton *Bk. Humbers.* v. Your other hard names—your ..lepidodendron, and bothrodendron. **1902** *Encycl. Brit.* XXXI. 423/2 The cuticles of a Bothrodendron.

bott, see Bot in Dict. and Suppl.

bott (bǫt). [f. Bat *sb.*⁴]

1. The name given by lace-makers to the cushion on which lace is woven.

1849 in Craig.

2. In founding, a clay plug used to close a hole against molten iron (Cent. Dict. Suppl. 1909). Also *attrib.* So **bo·tting** *vbl. sb.* (see quot. a 1877).

a 1877 Knight *Dict. Mech.*, *Botting*, restopping the tapping-hole of a furnace after a part of its charge has been allowed to flow therefrom. **1888** *Lockwood's Dict. Mech. Engin.*, *Bott Stick.* **1900** J. Simpson *Mod. Foundry Pract.* 14 A round ball of this [loam] is placed on a disc of iron at the end of an iron rod, and is forced into the tap-hole..when it is wished to stop a tapping out with the bott or bod stick.

‖ **bottega** (bǫtte·gă). [It. *bottega* small shop, studio, f. L. *apotheca*: see Apothec.] An artist's workshop or studio, esp. in Italy.

1900 F. Litchfield *Pott. & Porc.* vii. 198 The group of ateliers or bottegas in ..Italy..where work of this kind was carried on by individual artist potters, and their assistants or pupils. **1934** *Burlington Mag.* July 30/1 A frieze of blue terra-cotta from the bottega of Andrea della Robbia.

Böttger (bö·tgǝɹ). The name of J. G. Böttger (1682–1719), a German maker of porcelain, used *attrib.* to designate a type of red stoneware.

[**1850** J. Marryat *Hist. Pott. & Porc.* 233 *Böttcher Ware*, a fine red stoneware which was made by a discovery of that name, at Dresden, and which led to the discovery of porcelain in Europe.] **1869** Lady C. Schreiber *Jrnl.* (1911) I. 32 Böttger tankard. *Ibid.* 43 Two white Dresden teapots ..and a red Böttger one. **1925** W. W. Worster tr. *Hannover's Pott. & Porc.* III. 11. 45 The Plaue ware is thicker in the body and coarser in its shape than Böttger ware.

bott-hammer (bǫ·t-hæ:mǝɹ). [ad. G. *bott-hammer*, f. *botten* to break flax + *hammer* Hammer *sb.*] A wooden hammer used to break the stalks of flax.

1858 Simmonds *Dict. Trade*, *Bott-hammer*, a wooden block with a long bent helve or handle, and having channels or flutings under its face used to beat flax. **a 1877** Knight *Dict. Mech.*, *Bott-hammer*, a wooden mallet with a fluted face, used in breaking flax upon the floor to remove the boon.

Botticellian (bǫtitʃe·liăn), *a.* [f. the name of Sandro *Botticelli* (1444–1510), Florentine painter: see -an.] Having the characteristics of Botticelli's work. Also **Botticel-l(i)esque** *a.* and **Botticelli** *attrib.*

1890 G. B. Shaw in *Star* 28 Apr. 2/1 Her slim figure, Tuscan school profile, and Botticellian grace. **1927** *Observer* 8 May 7 Mr. Berenson traces the original Botticellian type of the head. **1930** P. Armstrong in *Oxf. Poetry* 1 Quivering Botticelli leaves Tantalize the water-plants. **1934** M. Beerbohm *Mere Pricks against Kicks* 156 Enormous breasts Botticelli thighs. **1939** *Burlington Mag.* Feb. 100/1 All the Botticelliesque gestures. *Ibid.* June 291 Pollaiuolo-esque-Botticellesque forms. **1966** Cox *Illustr. Dict. Hairdressing & Wigmaking* 21/1 *Botticelli style*, from the hair style depicted on women by his painting 'Spring'. **1967** *Coast to Coast 1965–6* 175 The..lacquer-red of their lipsticks had caused..even Botticellian schoolgirls to feel fashionable.

bottine. 2. (Earlier example.)

c 1845 C. Brontë *Professor* (1857) I. xii. 194 Large feet tortured into small bottines.

bottle, *sb.*² **1. e.** Add to def.: a baby's feeding-bottle.

1848 Thackeray *Pendennis* I. iii. 25 His first socks..his bottle, and other interesting relics of his infancy. **1858** — *Virginians* I. xviii. 141 Baby..is bawling over his bottle. **1966** *Child Care* (Brit. Med. Assoc.) 21 An expectant mother who..has decided..to put her baby on the bottle is probably influenced by one or more of the following objections.

f. A hot-water bottle.

1857 Mrs. Gaskell *Let.* 9 Oct. (1966) 483 We got two great bottles & slept together & heaped shawls & kept warm. **1967** R. Mackay *House & Day* 142 'I'll put a bottle in your bed. I put a wee flask on the table.' **1968** K. V. Beste *Repeat Instructions* xiv. 147 I've put a kettle on for my bottle.

g. In various *slang* uses. (*a*) Phr. *no bottle*: no good; bad(ly), useless(ly).

1846 *Swell's Night Guide* 76 She thought it would be no bottle, cos her first would go in a buster. **1931** W. F. Brown in *Police Jrnl.* Oct. 501 When he got up the steps, he had a mouthpiece who was no bottle.

(*b*) A collection or share of money.

1893 P. H. Emerson *Signor Lippo* v. 12 We never count the denarly on the pitch, but put each man's bottle into the sack as it is till sharing time. **1928** *Radio Times* 2 Nov. 302/1 His [sc. a busker's] show ended, he passes along the line with his hat and proceeds to investigate the contents, or 'bottle'. **1939** J. B. Priestley *Let People Sing* x. 256 'Knocker brought out some money... 'Not much bottle. A nicker, half a dollar.'

(*c*) A reprimand. *Naval.*

1938 'Giraldus' *Merry Matter Again* 177 A 'bottle' from the captain of the quarter-deck who is usually the ugliest P.O. in the ship. **1962** G. H. Jones *Word Enemy* 220 Others came in to see me over-anxious to please, full of 'yes, sirs' expecting always to be given what is called a 'bottle'.

3. b. *colloq.* A thermionic valve.

1940 in *Chambers's Techn. Dict.* **1945** *Electronic Engin.* XVII. 424 Vacuum bottles..had to be produced on an ever-increasing scale.

4. *bottle-cork* (earlier and later examples); *bottle-shaped* adj.

1746 W. Dunkin in P. Francis tr. *Horace's Ep.* II. ii. 134 The Felon's Fork Detach'd the Square of a Bottle-Cork. **1940** Dylan Thomas in *Life & Lett.* 2/2 A bursting sea with bottlecork boats. **1733** P. Miller *Gard. Dict.* (ed. 2) s.v. *Cucurbita*, The Fruit of some Species is long, of others round ..bottle-shaped. **1951** A. G. L. Hellyer *Sanders' Encycl. Gardening* (ed. 22) 426 Bottle-shaped fruits.

5. bottle-age, the length of time that a wine, etc., has remained in the bottle; **bottle-arsed** *a.* (*Printers' slang*), of type: wider at one end than at the other; **bottle-baby**, a baby reared by means of a feeding-bottle; **bottle-end**, a round of glass resembling the bottom of a bottle, used in windows; **bottle-fed** *a.*, (of an infant or young animal) brought up on the bottle (see sense 1 e); cf. **breast-fed*; hence (as back-formation) **bottle-feed** *v. trans.*; **bottle-grass** *U.S.*, a variety of fox-tail grass, esp. *Setaria viridis*; **bottle-imp**, also, a Cartesian devil, a hollow figure suspended in a bottle of water; **bottle-jack** (earlier example); (*b*) applied to an escapement in a clock or watch resembling that of a bottle-jack; (*c*) a kind of lifting-jack (Knight *Dict. Mech.*, a 1877); **bottle-opener**, an implement for opening bottles; **bottle-party**, a party to which each guest contributes a bottle (of wine, etc.); also an establishment, usu. a night-club, where drinks ordered in advance are served after licensed hours; **bottle-screw** (delete † and add later examples); **bottle-shaker**, an apparatus used in centrifugation; **bottle-shop**, a shop licensed to sell wines and spirits only in the bottle; **bottle store** (*a*) *S. Afr.* = prec.; (*b*) a place where bottles are stored; **bottle-swallow**, an Australian bird, a species of martin; **bottle-washer** (earlier and later examples); **bottle-windowed** *a.*, having windows made up with bottle-ends (see above).

1959 *Spectator* 28 Aug. 255/1 It..will be better still with a little more bottle-age. **1770** Luckombe *Hist. Printing* 433 It [sc. the type] drives out, or gets in, either at the head, or at the foot, as the Printers call it, Bottle-arsed. **1838** Timperley *Printers' Man.* 64. **1890** Farmer *Slang*, *Bottle-arsed*, type thicker at one end than the other—a result of wear and tear. **1893** *Daily News* 9 Mar. 2/7 Was it what you call a bottle-baby? **1905** *Westm. Gaz.* 23 Oct. 4/1 Wanted, nurse for bottle baby; only one thoroughly accustomed to bottle babies. **1907** W. De Morgan *Alice-for-Short* ix. 92 A..window..filled with what some called bottle-ends, and some bull's-eyes. **1847** ... This might be good because of the..infantile mortality, especially amongst bottle-fed children. **1960** *Farmer & Stockbreeder* 29 Mar. 65/3 She has built up an enterprise which last year bottle-fed 82 lambs. ... **1850** E. B. Denison *Clock & Watchm.* 50 The bottle-jack or 'vertical' pallets. **1860** *Ibid.* (ed. 4) 35 The bottle-jack escapement is precisely the same as in De Vick's clock. **1931** *Kansas City Times* 17 Dec. 20 There were a few keys and a corkscrew and bottle-opener in the bunch. **a 1953** Dylan Thomas *Quite Early One Morning* (1954) 31 Why should the bottle-opener be under the hall-stand? **1926** C. Beaton *Diary* 9 Dec. in *Wand. Years* (1961) 151, I was invited to Madge Garland's bottle-party. **1931** A. Powell *Afternoon Men* I. i. 23 'Is it a bottle party?' 'You'd better bring a bottle of something,' said Barlow, 'in case there isn't anything to drink at all.' **1937** *Daily Herald* 26 Jan. 4/5 There may also be provisions to deal with bottle parties. **1775** J. Granger *Biogr. Hist. Eng.* (ed. 2) III. 148 Her hair is dressed in many formal curls, which nearly resemble bottle-screws. **1938** Masefield *Dead Ned* 133 A clasp-knife having at its back a corkscrew. **1969** E. H. Pinto *Treen* 50 Corkscrews, also known as bottle screws, screws, and steel worms, were being made in Tudor times. **1913** *Oxford Univ. Gaz.* 4 June 943/2 Motor driven centrifuge and bottle-shaker. **1929** *Times* 30 Jan. 9/7 These were what were known as 'bottle shops', and could not sell less than a bottle of spirits and a half-bottle of wine at any one time. **1862** G. H. Mason *Zululand* ii. 17 Another..formerly kept a small bottle store. **1944** J. A. Lee in D. M. Davin *N.Z. Short Stories* (1953) 104 There are rats in the bottle store. **1950** *Cape Times* 30 June (W.-e. Mag.) 5 As soon as the bottlestore opens, the Native is there. He gets his requisition from the town and takes this to the shebeen. **1898** Morris *Austral Engl.* 47/2 *Bottle-Swallow*, a popular name for the bird *Lagenoplastis ariel*, otherwise called the Fairy Martin... The name refers to the bird's peculiar retort-shaped nest. **1837** *Southern Lit. Messenger* III. 656 They have yet founded no city to themselves ..but are willing to remain the boot-cleaners and the bottle-washers of the whites. **1894** *Daily News* 8 Mar. 8/6 (Advt.), Handy man as Bottlewasher or Kitchen Porter. **1928** *East End Star* May 3/3 She is Superintendent, Treasurer, cook and bottlewasher. **1899** Kipling *Stalky* 224 A little bottle-windowed..half-dairy, half-restaurant, a dark-browed, two hundred-year-old house.

bottle, *v.*¹ Add: **3.** *Printing.* To make bottle-arsed.

1877 *Design & Work* 15 Sept. 341 The letters stand fair and square upon their shanks, 'not bottled', as we say in the trade. 'Bottling' arises from the following cause—imperfect locking up, or lines badly spaced out... The risk of 'bottled' letter is, however, not very great... Amateurs are in danger of 'bottling' their own letter.

4. *intr.* To collect money. So **bo·ttling** *vbl. sb. slang.* Cf. *bottle *sb.*² 3 g (*b*).

1933 P. Allingham in *Evening News* 9 July 11/3 He is an expert at that delicate part of the business [street entertaining] known as 'bottling', which means the art of persuading people to put money into your hat. **1936** W. A. Gape *Half Million Tramps* xi. 159, 'I only sing the old favourite songs. You can "bottle" until your heart's content.'..'To bottle' is the slang term for collecting. **1939** Adeler & West *Remember Fred Karno!* iii. 47 They commenced operations, performing as often as they could draw a crowd, and collecting, or 'bottling', before the crowd dispersed.

5. *trans.* To admonish. *Naval slang.*

1946 J. Irving *Royal Navalese* 37 To bottle someone is to dress them down very thoroughly.

bottle-brush. Add: **2.** (Further examples.) Also applied to various other plants or flowers (see quots.); *bottle-brush grass*: *Asperella hystrix.*

1843 J. Torrey *N.Y. Nat. Hist. Surv.: Flora* II. 478 Bottle-brush Grass... Moist, rocky woods, and along shady ravines: not rare. **1852** G. C. Mundy *Our Antipodes* I. ii. 26 The Bottle-brush, one of the most characteristic plants of the bush..has rough, twisted branches, and a leaf something like the holly. Sir Joseph Banks gave it the botanical name of Banksia. **1883** A. Easther *Gloss. Dial. Almondbury* 15 *Bottlebrush*, a plant otherwise called Common Spurry, or Farmer's Ruin: *Spergula arvensis*... Another plant bears the same name—the Mare's Tail, or *Hippuris vulgaris*. **1883** *Encycl. Brit.* XV. 174/1 Bottlebrush (in Queensland), the *Callistemon lanceolatus*. **1923** D. H. Lawrence *Kangaroo* xviii. 397 Gold red brushes of the bottle-brush tree. **1957** P. White *Voss* ii. 30 Belle had a spray of the crimson bottlebrush that she had torn off recklessly. **1961** *Amat. Gardening* 30 Sept. Suppl. 8/1 *Callistemon*..the unusual flowers..have a mass of prominent stamens giving the appearance of a bottle brush. **1967** *Southerly* XXVII. 199 The flame of bottle-brush.

bottled, *ppl. a.* Add: **2.** *bottled gas*, gas stored in liquid form in portable containers.

1930 *Engineering* 4 July 28/1 In the United States, 'bottled' gas, in the form of liquified propane and butane..is employed. **1960** *News Chron.* 20 July 6/6 The little bottled gas cookers are just like ordinary ones.

3. b. Stored up, concentrated.

1872 W. W. Reade *Martyrdom of Man* 399 Life is bottled sunshine.

c. *bottled lightning*: (*a*) = Lightning 2; (*b*) concentrated vigour or energy; also *attrib.*

1838 Dickens *Nickleby* xlix. 489 Bring in the bottled lightning, a clean tumbler, and a corkscrew. **1899** W. James *Talks* 209 To all who looked upon her an impression as of 'bottled lightning' was irresistibly conveyed. *Ibid.* 210 There are plenty of bottled-lightning temperaments in other countries, and plenty of phlegmatic temperaments here.

d. *slang.* Drunk, intoxicated.

1927 *N.Y. Times* 9 Jan. (heading) British pilot tells police, 'If you say I am bottled, I will agree'. **1932** A. Huxley *Brave New World* v. 93 Drunk as she was... Lenina did not forget to take all the contraceptive precautions prescribed by the regulations. **1939** J. B. Priestley *Let People Sing* iii. 64 The gov'nor must be good an' bottled.

bottle-neck, bottleneck (bǫ·t'lnek). [f. Bottle *sb.*² + Neck *sb.*¹] **1.** The neck of a bottle (see Neck *sb.*¹ 10 a).

1922 Joyce *Ulysses* 423 A room lit by a candle stuck in a bottleneck.

2. A narrow entrance to or stretch in a road, comparable to the neck of a bottle in shape; *gen.* a narrow or confined space where traffic may become congested.

1896 [see sense 4 below]. **1907** *Westm. Gaz.* 21 Aug. 5/2 The bottle-neck, known as London-road, at the Elephant and Castle. **1915** W. J. Locke *Jaffery* x. 123 Through the bottle-neck of Brentford,..we crawled as fast as we were able. **1928** *Britain's Industr. Future* (*Liberal Ind. Inquiry*) IV. xxiii. 314 Any failure to maintain dock and harbour facilities .. results in delays. .. Ports then become the 'bottle-necks' of ocean traffic and congestion results.

3. *fig.* Anything obstructing an even flow of production, etc., or impeding activity, etc.

1928 *Observer* 15 July 10/3 It is hoped to make one side of the higher science forms of the school a bottle-neck through which boys of special intelligence..may pass. **1936** *Economist* 14 Mar. 581/1 Frequent complaints of deliveries falling into arrears..reveal the existence of numerous 'bottlenecks'. **1947** *Ann. Reg. 1946* 13 Import programmes, the bottleneck of which was no longer shipping, but finance. **1968** *Brit. Med. Bull.* XXIV. 192/2 The technology, therefore, exists; the bottle-neck..is in the education and training of those whose activities can benefit from its effective use.

4. *attrib.* and *Comb.* Also quasi-*adj.*, = *bottle-necked *ppl. a.*

1896 *Daily News* 26 Dec. 3/1 The widened portions at Holloway and elsewhere are rendered useless by narrow bottle-neck approaches to Finsbury-park. **1898** *Ibid.* 19 Oct. 3/1 [He] called Old Jewry 'a bottle-neck-shaped street'. **1908** *Daily Graphic* 21 Mar. 13/3 Our desire at present is to look so feminine that bottle-neck shoulders are praised. **1938** L. MacNeice *Earth Compels* 7 The bottle-neck harbour collects the mud.

bottle-neck, bottleneck (bǫ·t'lnek), *v.* [f. prec.] *trans.* To confine or impede in a bottle-neck; to pass (something) through a bottle-neck.

1928 *Daily Express* 12 June 3/4 He is 'bottle-necked' between Hungerford Bridge and the Hotel Cecil. **1933** *Planning* I. viii. 8 It is easier to organise an export trade which is necessarily bottlenecked through one channel. **1954** *Information Please Almanac 1955* 35 *Bottleneck*, to delay progress; to hold up a process, especially at a critical point.

So **bo·ttle-necked** *ppl. a.*, shaped like the neck of a bottle.

1932 H. Simpson *Boomerang* vii. 140 No bottle-necked alleys to delay them.

bottle-o(h (bǫ·t'lōu). *Austral.* and *N.Z. colloq.* Also **bottle-o-er.** [f. Bottle *sb.*² + O *int.*] A collector of empty bottles.

1906 E. Dyson *Fact'ry 'Ands* xvi. 217 Half-a-dozen of them would have died for the bibulous comp. despite the bottle-o's stock garnered in the trouser fringe. **1914** *Bulletin* (Sydney) 14 May 36/2 Bottle-o Bill and his tart Could join in the catchy refrain. **1915** E. N. G. Poulton *Countess of Liverpool's Gift Bk.* 95 He was the great panjandrum of his 'profession', the King of Bottle-O-ers. **1943** *Coast to Coast 1942* 105 A bottle-o followed him crying cheerily, 'Bottles, bottles, any empty bottles.' **1954** *Ibid. 1953–4* 161 They must pay for what they took; no thieving like other bottle-ohs. **1967** D. Whittington *In Search of Australian* 90 'What do you do for a living?'..'I'm the local bottle-O'.

bottler (bǫ·tlǝɹ), *sb.*² and *a.* *Austral.* and *N.Z. slang.* [Cf. *bonzer *a.*] (Something or somebody) excellent.

1890 'R. Boldrewood' *Col. Reformer* x. 154 He's a bottler, that's what he [sc. a horse] is. **1941** Baker *N.Z. Slang* vi. 51 Of children's terms..we may note..*bottler* and *snozzler*, descriptive of something superlative or excellent, both as nouns and adjectives. **1959** G. Slatter *Gun in Hand* 239 Congratulations boy, a glorious try, a real bottler, you won the game.

bottle-tree (bǫ·t'ltrī). [f. Bottle *sb.*² + Tree *sb.*] An Australian tree of the sterculia family, either the Queensland tree *Sterculia rupestris* or the similar *Sterculia diversifolia* of Victoria, so called from the bottle-like shape of its trunk.

1846 C. P. Hodgson *Remin. Australia* 264 The sterculia or bottle-tree is a very singular curiosity. It generally varies in shape between a soda water and port wine bottle. **1880** Mrs. C. Praed *Head Stat.* 179 In dense scrub, where the bottle-trees rose weird and white. **1889** J. H. Maiden *Useful Native Plants* 60 A 'Kurrajong'. The 'Bottle-tree' of N.E. Australia, and also called 'Gouty-stem'. **1898** *'Coo-ee'* 284 A great white bottle tree, its trunk perfectly bare. **1931** F. D. Davison *Man-Shy* (1934) vii. 107 Through open bottle-tree country. **1958** R. Stow *To Islands* 52 Not far from the aerodrome strip there, under the bottle trees.

bottom, *sb.* Add: **1. c.** *Bottoms up!*: a call or toast to finish one's drink to the last drop. Cf. Bottomer c. Hence as *adv. phr.*

1917 G. J. Nathan (*title*) Bottoms up. **1928** *Vanity Fair* Dec. 79 Bottoms up to Vanity Fair! **1934** S. Kingsley *Men in White* i. iii. 48 Come on! Bottoms up! *She smiles back at him, and drains the glass.* **1934** J. O'Hara *Appt. in Samarra* (1935) iv. 119 The old priest..drank his highball almost bottoms up. **1964** L. Nkosi *Rhythm of Violence* 51, I say bottoms up both to women and to glasses! [He raises his glass.]

4. b. (Later examples.) Now esp. *U.S.*

1837 Peck *Gaz. Illinois* i. 3 The term 'bottom' is used throughout the west to denote the alluvial soil on the margin of rivers, usually called 'intervals' in New England. **1851** C. Cist *Cincinnati* 18 Cincinnati itself is built on an ancient alluvial plain, lying in two levels called the 'upper and lower bottoms'. **1907** Mulford *Bar-20* xx. 200 They crawled to the last line of brush and looked out over the extensive bottom. **1942** W. Faulkner *Go down, Moses* 33 Messing around up yonder in the bottom all last night.

c. In gold-mining, the channel of an old river (also called the gutter) containing rich deposits of gold; also, the layer below it. *Austral.*

1855 W. Howitt *Land, Labour & Gold* I. xiii. 223 We have a hole within a few feet of the bottom, which I am confident will turn out well. **1887** Hayter *Christmas Adv.* 5 (Morris), We reached the bottom, but did not find gold.

5. d. *Mining.* Usually *pl.* The lowest workings in a mine. Also *attrib.*, as *bottom captain, coal, worker.*

1778 Pryce *Min. Cornub.* 174 The Bottom-Captains, whose business is to see that the common men perform due labour down in the mine. **1860** *Eng. & For. Mining Gloss.* (ed. 2) 5 *Bottoms*, the lowest workings either in a stope, level, or elsewhere. **1883** Gresley *Gloss. Coal-mining*, *Bottom*, the bottom of the shafts and roadways, &c., near the shafts. **1892** *Daily News* 26 Feb. 5/6 It comprises about 280,000 miners, of whom 200,000 are 'bottom workers'. **1900** *Daily Express* 28 June 7/3 There is an immense quantity of coal known locally as 'bottom coal' practically intact. **1967** *Gloss. Mining Terms* (*B.S.I.*) viii. 7 *Bottom coal* (bottoms, floor coal), the lowest part of a seam, which may or may not be extracted.

e. The part of a boot or shoe below the uppers; the sole, heel, and shank.

1841 *Penny Cycl.* XXI. 410/2 The employing master.. prepares and sorts the sole or bottom-stuff for the maker. **1881** *Instr. Census Clerks* (1885) 76 Bottom Finisher. **1886** *Encycl. Brit.* XXI. 831/1 He then pares off inequalities and 'levels the bottoms'. **1911, 1921** [see *bottom-scourer* in *19].

11. c. In fig. phrases: *the bottom falls* (or *drops*) *out of*: there is a collapse of; *to knock the bottom out of*: see Knock *v.* 6 b.

1637 Rutherford *Let.* 9 Sept. (1664) 144 The bottom hath fallen out of both their wit and conscience at once. **1868** *Iowa Agric. Soc. Rep. 1867* 64 The bottom has at length dropped out of this humbug. **1872** 'Mark Twain' *Roughing It* (1873) lviii. 420 Gould and Curry soared to six thousand three hundred dollars a foot! And then—all of a sudden, out went the bottom and everything..went to ruin and destruction! **1923** Wodehouse *Inimit. Jeeves* iv. 45, I felt as if the bottom had dropped out of things with a jerk. **1926** E. M. Dell *Black Knight* i. x, 'I try to take things as they come.'..'And when the bottom falls out of everything —what do you do then?' **1957** M. Banton *W. Afr. City* vi. 103 In the 1930s the bottom fell out of the market in ginger and coffee.

12. *at bottom* (earlier examples).

1780 Sheridan *Sch. Scand.* i. i. 8 Every body was sure there was some reason for it at bottom. **1838** Dickens *Nickleby* xxiii. 215 He's a good pony at bottom.

18. a. *bottom-discharge*; *bottom rock*, also *fig.*; *bottom-heavy* adj. **b.** *bottom-liver*. **f.** (sense 1 b) *bottom-pincher*, *-pinching*.

a 1877 Knight *Dict. Mech.*, *Bottom-discharge Water-wheel*, a turbine from which the water is discharged at the bottom instead of at the sides. **1900** *Daily News* 24 Oct. 7/7 Orders for 160 bottom-discharge trucks have been placed in America. **1927** *Sunday Times* 6 Mar. 7/3 The double basses are generally too plentiful at these concerts, and..they too often made the music sound bottom-heavy. **1927** Haldane & Huxley *Animal Biol.* xii. 302 They are all marine and essentially bottom-livers. **1959** J. Blish *Case of Conscience* xii. 119 He was fundamentally nothing more complicated than a bottom-pincher. **1939** Auden & Isherwood *Journey to War* 53 Their horse-play, bottom-pinching..and endless jokes about *les poules*. **1955** Auden *Shield of Achilles* ii. 41 The honking bottom-pinching clown. **1887** C. B. George *40 Yrs. on Rail* v. 93 About the time I had reached bottom rock in my financial troubles,..I met A. B. Pullman.

19. bottom-bed, the lowest stratum of a formation of rocks; **bottom-boarding**, the bottom-planks of a boat; **bottom-boards**, boards at the bottom of a boat serving to protect the outer planking; **bottom-dish** (earlier example); **bottom dog** = Underdog (cf. *top dog* s.v. Top *sb.*¹ 32); also *attrib.*; so **bottom-doggy** *a.*, pertaining to or characteristic of a bottom dog; **bottom dollar** *U.S.*, (one's) last dollar, usu. in collocations with *bet*; **bottom drawer**, lit. the lowest drawer of a chest of drawers, etc., in which a woman stores clothes, linen, etc., in preparation for her marriage; **bottom facts** *U.S.*, the fundamental facts; **bottom fermentation**, that during which the yeast cells collect at the bottom of the liquid; also *attrib.*; **bottom gear**, the lowest-speed gear (see Gear *sb.* 7) in a motor; **bottom-land, bottomland** *U.S.*, low-lying land, esp. a stretch of level land near a river; = Bottom *sb.* 4 b; also *attrib.*; **bottom-line**, the lower part of a fishing-line; **bottom-planks** the outer planking of the bottom of a boat; **bottom-plate**, (*a*) an iron plate in a printing-press; (*b*) the set of knives forming the bed of a pulping machine in paper-making; **bottom prairie** *U.S.*, a prairie lying along the bank of a river; **bottom-sampler**, a grab for dredging samples from the sea-bottom; hence *bottom-sampling* ppl. a. and vbl. sb.; **bottom-scourer**, an operative who smooths the 'bottoms' of boots and shoes; **bottom-set bed** *Geol.* (see quots.); **bottom-side** = sense 1 a; **bottom timber** *U.S.*, timber growing in bottom-lands; **bottom-tool**, a tool used in wood-turning; **bottom wool** (see quot.); **bottom yeast** = **bottom fermentation* yeast.

1845 P. B. Brodie *Hist. Fossil Insects* 58 Hard blue limestone, ('bottom bed') with Ostrea, Modiola minima, and other shells. **1900** G. Swift *Somerley* 110 At the risk of tearing out what was left of the bottom-boarding, we hauled her on to the beach. **1861** T. Hughes *Tom Brown at Oxford* I. xi. 192 I've larded the bottom boards under my seat so that not a drop of water will..come through. **1883** *Man. Seamanship for Boys' Training Ships* 84 Q. What are bottom boards? A. Long pieces of wood nailed together, which lay from the stern sheets to the bow. **1747** H. Glasse *Art of Cookery* ix. 94 A collar of Fish in Ragoo...This is a fine Bottom-dish. **1884** Bottom dog [see sense 17 in Dict.] **1926** D. H. Lawrence *Plumed Serpent* vi. 119 There was a touch of bottom-dog insolence about her. **1927** *Daily Express* 12 Aug. 9/5 The award will be received with disappointment by..the 'bottom dogs' of the service. **1925** D. H. Lawrence *Let.* 2 Apr. (1962) II. 832 *Canaille* of the most bottom-doggy order. **1857** *San Francisco Call* 24 Jan. 4/1 Sometimes, however, luck will run against him, and..he 'slips up for his bottom dollar'. **1866** *Congress. Globe* Mar. 1474/1 His opinion is that a State can go out of the Union and he is willing to bet his bottom dollar on his judgment. **1904** Harben *Georgians* v. 43 You bet yore bottom dollar I'm open to criticism myself. **1958** *Dissent* V. i. 80 And I'd bet my bottom dollar that Negro hipsters, among themselves, often put down the whites. **1886** R. Holland *Gloss. Cty. Cheshire* 407 If a young woman were to buy a set of teathings, or a tablecloth, or what not, and were asked what use she had for such things, she would answer, 'Oh! they're to put in my *bottom drawer*.' **1902** A. Bennett *Anna of Five Towns* xiii. 343 The bride took all the house-linen to her husband...As soon as a girl had passed her fifteenth birthday, she began to sew for the 'bottom drawer'. **1959** *Woman's Own* 14 Feb. 58/3 She had been saving furiously for her 'bottom drawer' ever since she became engaged. **1877** *N.Y. Tribune* 17 Mar. (Bartlett), Curiosity has been on the tiptoe these many weeks to know the bottom facts. **1883** 'Mark Twain' *Life Mississippi* xlii. 393 Though there ain't only one or two ways when you come down to the bottom facts of it. **1902** *Encycl. Brit.* XXVI. 367/1 In the Continental *bottom-fermentation* system, the pitching and fermentation take place at a very low temperature. **1905** J. L. Baker *Brewing Industry* 100 Brewers' yeast (*Saccharomyces cerevisiæ*) is divided into two classes, top fermentation and bottom fermentation yeast. **1923** *Westm. Gaz.* 2 Feb., A stiff climb on bottom gear brings Mosul..in sight. **1968** *Listener* 1 Aug. 140/3 Still our red-hot old bus engines ploughed on uphill in bottom gear. **1728** *Boston News-Letter* 23–30 May 2/2 Fifty Acres of...Meadows and Meadow Bottom Land. **1841** C. Cist *Cincinnati* 66 The larger streams are now found meandering through alluvial plains called 'bottom lands'. **1903** *N.Y. Even. Post* 12 Sept., The tract consists of a bottom land along the Ohio River, and a plateau elevated 300 feet above the bottom land. *Ibid.* 19 Sept., To complete the maturity of the bottomland crops. **1926** *Chambers's Jrnl.* 1 May 345/1 The lakes and sloughs ran in a vast network over the bottom lands. **1961** L. Mumford *City in History* xiii. 405 A discouraging site: bottomland, bordered by a swamp on the Potomac side. **1837** Kirkbride *Northern Angler* 91 The bottom or casting-line must consist of three lengths of twisted gut. **1891** Kipling *Light that Failed* ii. 27 The whale-boat..chose to hit a hidden rock and rip out half her bottom-planks. **a 1877** Knight *Dict. Mech.*, *Bottom-plate* (printing), a plate of iron belonging to the mold of a printing-press, on which the carriage is fixed. **a 1884** *Ibid.* Suppl., *Bottom plate* (paper-making), the gang of knives forming the concave or bed beneath the cylinder of a rag-grinding machine or pulping engine. **1804** J. Ordway in *Wis. Hist. Coll.* (1916) XXII. 95 A beautiful Bottom Prarie..about 2000 acres of Land covered with wild rye and wild potatoes. **1882** Worthen *Econ. Geol. Illinois* II. 73 The latter are the so-called 'ridge prairies', while the former are sometimes designated as 'bottom prairies'. **1911** Petersen & Jensen *Danish Biol. Station Rep.* XX. 73 By means of bottom-samplers..it is shown that the uppermost brown layer of the sea-bottom must be regarded as dust-fine detritus. **1959** A. Hardy *Fish & Fisheries* v. 104 Petersen's quantitative bottom-sampling grab. **1911** F. Sellers in *Rep. Labour & Social Conditions in Germany* III. 95 Bottom-scourers 24s. (Frankfurt per week). 12s. to 16s. (Leeds per week). **1921** *Dict. Occup. Terms* (1927) § 429 *Scourer*,.. designated according to parts upon which he works, *e.g.* bottom or naumkeag scourer, heel scourer. **1905** Chamberlin & Salisbury *Geol.* iii. 191 The sediment rolled at the bottom of the current is dumped on reaching the steep slope, and constitutes the inclined *fore-set* beds...The material in suspension is carried farther, settles more gradually, and constitutes the *bottom-set* beds. **1942** C. A. Cotton *Geomorphology* (ed. 3) xv. 206 In a delta some of the silt layers are covered over by advancing fore-set beds, and then become the *bottom-set beds* of the delta. **1856** Cozzens *Sparrowgr. Papers* vii. 88 It was vexatious enough to see our lawn bottom-side up on a festive occasion. **1869** *Rep. U.S. Commissioner Agric. 1868* 327 Put the box in a pan of water and turn it nearly bottom-side up. **1964** *New Scientist* 12 Mar. 686/1 The regions above and below the height of maximum density are generally referred to as the 'topside' and 'bottomside'. **1834** Peck *Gaz. Illinois* ii. 150 The bottom timber consists of oaks [etc.]. **1874** J. W. Long *Amer. Wild-fowl Shooting* ix. 150 How much better walking it is in this bottom-timber than in the woods of New England. **a 1877** Knight *Dict. Mech.*, *Bottom-tool*.., a turning-tool having a bent-over end, for cutting out the bottoms of cylindrical hollow work. **1848** H. W. Haygarth *Bush Life Australia* v. 47 The wool nearest the skin, or, as it is called, the 'bottom wool', which is the hardest to cut, but the most weighty and valuable. **1910** *Encycl. Brit.* X. 278/1 It has not..been possible to transform a typical top yeast into a permanent typical bottom yeast.

bottom, *v.* Add: **4. c.** *intr.* Of prices, trade, etc.: to reach the lowest level. Also with *out.*

1892 *Daily News* 17 Nov. 7/1 Discount rates appear to have bottomed for the time. **1920** *Glasgow Herald* 6 Sept. 9 Others with shallower purses are content to wait until prices have bottomed. **1958** *Times* 14 July 13/3 With the recession apparently having bottomed out there is now much less insistence..that the Government take some vigorous action. **1969** *Daily Tel.* 21 Apr. 2/1 This is not the time to go liquid. If the index bottoms at 420 unless your timing is absolutely spot on it will pay to sit tight and ride out the squalls. **1970** *Ibid.* 10 Feb. 19/4 Analysts saw the advance as part of the market's 'bottoming out' pattern.

5. *trans.* In mining, to reach the bottom of (a mine); to reach a point (in a mine) beyond which further mining is useless; also *absol.*, to get down to the bed-rock or clay; to reach earth which contains gold. So *to bottom on*: to strike or reach (gold); also *fig.* **b.** *intr.* Said of a claim, etc.: to be worked to the bottom. *Austral.* and *N.Z.*

1853 E. Clacy *Lady's Visit to Gold Diggings of Australia* v. 73 Their hole had been fairly 'bottomed', but not a nest of nuggets discovered. **1858** McCombie *Hist. Victoria* xv. 219 In their anxiety to bottom their claims, they not seldom threw away the richest stuff. **1890** 'R. Boldrewood' *Miner's Right* I. viii. 214 As soon as the majority of the claims began to bottom, gold flowed in with a tumultuous profusion. *Ibid.* II. xxii. 25 Though they had bottomed a few of us, we bottomed on the same day, and the whole of the field was aware that Olivera's half-share men had struck another duffer. **1892** R. Wardon *McPhee* ... They shifted their pegs to fresh ground and again and again bottomed on gold! **1900** H. Lawson ... 143 One day Peter..told us that his party would 'bottom' during the following week. *Ibid.*, ... news that 'McKenzie and party' had bottomed on payable gold. **1926** J. Doone *Timely Tips for New Australians* Gloss., To 'bottom on to gold', to strike gold. **1963** A. Lubbock *Austral. Roundabout* 79 Shafts have been sunk 'blind',..on the chance of bottoming on gold.

6. *intr.* In mechanics, to strike against the bottom or far end (see quot. a 1877).

a 1877 Knight *Dict. Mech.* s.v., Cogs are said to bottom when their tops impinge upon the periphery of the co-acting wheel. A piston which strikes or touches the end of the cylinder is said to bottom. **1959** *Motor Manual* (ed. 36) v. 123 The spring deflections with a full load may be excessive and the vehicle will 'bottom' if a bad bump is encountered.

7. *Electr.* (See quot. 1960.) So *bottoming vbl. sb.*

1946 *Electronic Engin.* XVIII. 143 A master oscillator produces a sine wave...This is amplified and squared by 'bottoming' a valve anode. **1948** *Ibid.* XX. 63 ..the screen grid on its negative resistance portion..the valve must 'bottom'. **1954** K. W. Gatland *Devel. Guided Missile* (ed. 2) 237 When a valve 'bottoms' its anode will pass no more current. **1960** H. Carter *Dict. Electronics* 32 *Bottoming*, a thermionic valve is said to 'bottom' when, by reason of the potential applied to one or other of its grids, the anode current falls to zero.

bottoming, *vbl. sb.* Add: **1. d.** *attrib.*, as **bottoming-hole**, the open mouth of a glass-making furnace; **bottoming-tap**, a square-edged tap for cutting an internal thread uniformly to the bottom of a hole.

1839 Ure *Dict. Arts* 582 This spheroid having become cool and somewhat stiff, is next carried to the bottoming hole. **a 1877** Knight *Dict. Mech.*, *Bottoming-hole*, the open mouth of a furnace at which a globe of crown glass is exposed during the progress of its manufacture, in order to soften and allow it to assume an oblate form. **a 1884** *Ibid.* Suppl., *Bottoming tap*, one for carrying the thread of a tap to the bottom of the hole.

2. *Austral.* In mining, the process of reaching bed-rock. Cf. *bottom *v.* 5.

1890 'R. Boldrewood' *Miner's Right* I. ix. 189 The bottoming of three or more shafts on the supposed line of lead shall be a sufficient test.

botty (bǫ·ti). *slang.* [f. Bottom *sb.* 1 b + -y⁴.] (See quot. 1874.)

1874 Hotten *Slang Dict.* 94 *Botty*,..an infant's posteriors. —Nursery. **1922** Joyce *Ulysses* 759 Like a willwhipped childs botty. **1924** R. Firbank *Prancing Nigger* v. 36 De time de scorpion bit her botty.

botulin (bǫ·tiulin). *Med. rare.* [f. L. *botul-us* sausage + -in¹.] Botulinus toxin (see below). So **botuli·nic** *a.*, derived from, pertaining to botulin; **botulinum** (bǫtiulai·nǝm) [mod.L.], specific name of the bacillus of botulism, used *attrib.*; **botuli·nus** [mod.L. (E. van Ermengem 1896, in *Centralbl. f. Bakteriol.* XIX. 443)], former specific name of the bacillus of botulism, used *attrib.*, as **botulinus toxin**, an exceptionally powerful neurotoxin produced by the bacillus *Clostridium botulinum*, usu. in preserved foods that have been imperfectly sterilized.

1890 Billings *Med. Dict.* 178/2 *Botulinic acid*, Buchner's name for a substance found in poisonous sausages. **1899** *Clinical Jrnl.* 7 June 112/2 An efficient antitoxin for the botulinic poison. **1900** Dorland *Med. Dict.* 117/1 *Botulin*, a poisonous ptomain sometimes found in putrefying sausage; it is produced by *Bacillus botulinus*. **1902** Muir & Ritchie *Man. Bacteriol.* (ed. 3) xvi. 382 The properties of the botulinus toxin..have been found to correspond closely..with the toxins of diphtheria and tetanus. **1910** T. C. Allbutt's *Syst. Med.* (ed. 2) VII. 185, I have concluded from experimental poisoning by botulin, absinthe, phosphorus, and other poisons,..that the neurons are not equally affected. **1928** W. Giltner *Elem. Text Bk. Gen. Microbiol.* xviii. 243 The bacteria either split the protein of the food into a toxic molecule..or they secrete a toxin, e.g., botulinum toxin, while growing in the food. **1929** W. Bulloch in *Syst. Bacteriology* III. 380 In precipitin tables..antigenic substances are formed capable of fixing complement in the presence of ordinary botulinus antitoxin. *Ibid.* 386 Botulinus toxin is unaffected by peptic digestion. *Ibid.* 394 Botulism can hardly arise from the

ingestion of botulinus spores. *Ibid.* 397 The botulinus serum should be a polyvalent one. **1962** *Jrnl. Neuropath. & Exper. Neurol.* XXI. 610 Botulinus toxin in its purified form is the most toxic biological substance known. **1970** *Daily Tel.* (Colour Suppl.) 20 Feb. 21 Six ounces of botulinus toxin is enough to kill everybody in Britain (or North America) if it is distributed in the right way.

botulism (bǫ·tiuliz'm). *Med.* [ad. G. *botulismus* (also in Engl. use), f. L. *botul-us* sausage: see -ism.] Poisoning caused by eating food, usu. imperfectly preserved, that contains botulinus toxin.

1878 W. B. Woodman tr. *von Ziemssen's Cycl. Med.* XVII. 539 In the larger towns, botulismus occurs extremely seldom. **1887** A. M. Brown *Anim. Alkaloids* 156 In Germany it is known by the term *botulism*, a form of poisoning of the organism observed to result from the ingestion of putrefying meats. **1899** *Clinical Jrnl.* 7 June 112/1 This disease or botulism is quite different in its nature. **1922** H. Zinsser *Textbk. Bacteriol.* (ed. 5) xxxvi. 742 Botulismus toxin is produced under conditions of strict anaërobiosis. **1922** *Times* 6 Sept. 7/3 He had no doubt the deaths in this case were due to botulism, for which disease antitoxins.. had proved disappointing. **1959** *Science* 25 Sept. 769/2 In botulism the immediate cause of death is usually a paralysis of the skeletal musculature.

bouchées. Add: (Examples.) Also *sing.* and *attrib.*

1846 Soyer *Cookery* 153 The bouchées must be cut with a fluted cutter not larger than half-a-crown piece. **1928** Blunden *Undertones of War* iii. 27 Their front windows.. exhibited..chocolate bouchées in silver paper. **1960** *Good Housek. Cookery Bk.* 48/2 *Bouchées*, very small pastry cases filled with a savoury mixture, and served as after-dinner savouries, or with cocktails. Special tins, like miniature patty pans (called bouchée moulds), are used to bake the pastry cases.

‖ **bouclé** (bukle), *a.* [Fr., = buckled, curled.] (See quot. 1909.) Hence as *sb.*, a yarn of looped or curled ply; fabric made from this.

1895 *Montgomery Ward Catal.* 30/3 Boucle Cloth..plain, smooth weave, over the surface of which is the boucle effects, produced by curly mohair noils of same colors, woven irregularly through the cloth, and having the appearance of silk tufts. **1909** Webster *Bouclé*, woven so as to have a knotted and curled appearance, by using a two-ply yarn one thread of which is partly drawn out into a loop. **1930** *News Chron.* 6 Oct. 13/2 Another good fabric.. is bouclé coating, an all-wool material with a bouclé surface. **1960** *Which?* Feb. 39/2 *Bouclé*. Fabric named from its fancy looped yarn, which gives characteristic, uneven surface. Often in wool, silk and viscose rayon. Used for dresses and coats. **1960** *Times* 14 Mar. 15/1 Dior's two-piece of navy bouclé wool. **1964** *Which?* Sept. 284/2 *Loop* or *bouclé*, a compound yarn with loops at regular intervals.

‖ **boudin** (budæ̃n). [Fr.] A blood-sausage, a black pudding; also, force-meat shaped like a sausage. Also *white boudin* [Fr. *boudin blanc*], a white pudding.

1845 E. Acton *Mod. Cookery* xiii. 279 Small mushrooms.. stewed quite tender in butter may be mixed with the boudin after it is taken from the mortar. **1861** Mrs. Beeton *Bk. Househ. Managem.* 472 *Boudins* of a long shape, the size of the dish they are intended to be served on. **1947** M. Lowry *Under Volcano* i. 25 Puddings known as black or blood puddings..boudin, don't you know, Jacques. **1967** C. Durrell tr. *R. Oliver's French at Table* viii. 316 The hors d'œuvre.. were..solid stuff: sausage, white boudin, truffled pasties.

boudoir. Add: **c.** *attrib.*

1803 *Lett. Miss Riversdale* III. 92 The duke had fitted up these pretty little *boudoir* recesses. **1858** Bagehot *Coll. Works* (1965) II. 50 The *Lady of the Lake* is a sort of boudoir ballad. **1903** *Daily Chron.* 17 Jan. 8/4 A dressing gown..is sometimes glorified ..into a compromise between a dressing gown and a teagown, and then is known as a boudoir gown. **1912** in A. Adburgham *Shops & Shopping* (1964) xxii. 258, 2 boudoir caps, at 18/9. **1914** Wyndham Lewis *Let.* 2 Apr. (1963) 59 The boudoir suggestiveness and Yellow Book Gallicisms. **1937** M. Sharp *Nutmeg Tree* vii. 78 Mrs. Packett was sitting up in bed wearing a very smart boudoir-cap. **1966** Cox *Illustr. Dict. Hairdressing & Wigmaking* 23/2 *Boudoir helmet*, a decorated hair net of silk or other suitable material for wear by a woman in her boudoir to keep her hair in position during her toilet activity.

bouffant, *a.* Add: **b.** Of a hair-style: puffed out; arranged in a swelling or fluffy style. Hence as *sb.*, such a hair-style.

1955 *Vanity Fair's Guide for Bride* 52 René..designed a simple hairstyle...Watch that smooth but bouffant silhouette. **1959** S. Gibbons *Pink Front Door* xx. 232 Her eyes glistened dreamily within their little burrow of white wool and *bouffant* fair hair. **1959** *Sunday Times* 27 Sept. 21/1 The bouffant, hair-styler's joy and milliner's grief. **1966** *Economist* 9 Apr. 130/1 Since lofty bouffant hairdos have long been favoured here, Spain had a head start in the manufacture of..switches, chignons, postiches, wiglets and wigs.

bough, *sb.* Add: **5. bough-house**, (*a*) *U.S.*, a temporary structure made of boughs; (*b*) *dial.*, see quot. 1852.

1811 Wilson *Amer. Ornith.* III. 111 Their destroyers construct for themselves lurking holes made of pine branches, called *bough-houses*. **1852** *N. & Q.* 17 Apr. 371/2 Witnesses spoke of a 'bough-house', and the explanation given was, that certain houses where beer, &c. was sold at fair-time only had boughs outside to indicate their character. **1882** Mrs. Chamberlain *Gloss. West Worcs. Words* 4 *Bough-house*, house opened at fair-time only, for the sale of liquor. (Pershore.) Suppressed 1863. **1894** *Outing* (U.S.) July 281/1 Down in the bough house the campers reclined. **1946** Blunden *Shelley* i. 20 To the July Fair..folk came in from the country by hundreds and thousands, and 'bough houses' for their refreshment crowded the roadways.

bought, *sb.*¹ **2. b.** Delete †*Obs.* and substitute: Now *poet.* (revived by Tennyson).

1872 Tennyson *Gareth & Lynette* p. 16 The dragon-boughts and elvish emblemings Began to move, seethe, twine and curl. **1906** C. M. Doughty *Dawn in Britain* IV. xv. 148 Like as some serpent, which..in the sun, displays her glittering boughts.

bought, *ppl. a.* Add: **a.** *bought-in, bought-out*: purchased from an outside source (*i.e.* not raised or produced on one's own premises). (See also Buy *v.* 6, 8.)

1894 *Daily News* 8 June 9/6 Several parcels have been cleared since the late auctions, at the bought-in prices. **1947** N. Balchin *Aircraft Builders* 26 Joint committees on ..materials, sub-contracting, and bought-out parts and spares. **1960** *Farmer & Stockbreeder* 16 Feb. 109 Bought-in store cattle.

bouillabaisse (buyabē·s, -bę·s). [Fr., ad. mod. Prov. *bouiabaisso.*] A dish of Provençal origin, composed of fish stewed in water or spiced white wine. Also *fig.*

1855 Thackeray *Miscell.* I. 44 This Bouillabaisse a noble dish is—A sort of soup or broth, or brew, Or hotchpotch of all sorts of fishes. **1902** *Westm. Gaz.* 9 July 2/1 Let it be noted that the greater variety of fish employed to fashion a *bouillabaisse*, the greater the grace and excellence of the same. **1911** R. W. Chambers *Common Law* x. 305 Only art is one delicious bouillabaisse to him. **1932** H. Simpson *Boomerang* viii. 175 His wife had skill in the preparation of bouillabaisse. **1966** [see *boutique b].

bouillon. Add: **1. bouillon cube**, a cube of beef (or other meat) extract with seasonings.

1934 in Webster. **1954** E. David *Ital. Food* 39 Chicken bouillon cubes..are quite excellent. *Ibid.*, The same firm also make good beef bouillon cubes. **1960** *Sunday Express* 10 Apr. 15/4 A little light stock made from a bouillon cube.

b. A broth used as a medium for the culture of bacteria. Also *attrib.*

1887 T. M'C. Anderson *Treat. Dis. Skin* 289 Injections made under the skin with Bouillon of recent cultivation.. set up gangrene of the skin. **1897** *Jrnl. Path. & Bacteriol.* IV. 183 Peptone Bouillon at 37° C.—A cloudiness appears in the broth in 12–18 hours. **1903** *Daily Chron.* 21 Oct. 10/1 Five cubic centimetres of a highly poisonous bouillon culture of certain bacteria (streptococci) was injected under the skin of the ears of rabbits. **1909** *Practitioner* Nov. 591 Five decigrams of a bouillon culture of streptococcus of erysipelas.

Boulangism (bula·ńʒiz'm). [ad. F. *Boulangisme*, f. the name of Georges Ernest Jean Marie *Boulanger* (1837–91), French general and politician.] The principles and methods of Boulanger and his party who, from about 1886 to 1889, advocated a policy of militarism and revenge against Germany. So **Boula·ngist** *a.* and *sb.*

1888 *Times* 16 Mar. 5/4 Nothing proved that Boulangism had anything in common with Bonapartism. **1888** *Appletons' Ann. Cycl. 1887* 292/1 Boulangist Demonstrations... Crowds gathered around the Opéra cheering Boulanger. **1889** F. Turner *General Boulanger* 197 Another method invented by the Government to crush out Boulangism was the employment of numbers of police spies to watch the General, Count Dillon, and all the principal members of the National party. **1889** *St. James's Gaz.* 16 Jan. 10/2 The source of the Boulangist election expenditure is a burning question in France. *Ibid.* 3 Aug. 9/2 Fifteen telegraph clerks, reported to the Postmaster-General as Boulangists, **1954** A. J. P. Taylor *Struggle for Mastery* 315 If there had ever been a Boulangist crisis, it was now over.

boulder, *sb.*¹ Add: **4.** *boulder-strewed, -strewn* adjs.; **boulder-belt**, a belt of boulders deposited by a glacier on melting; **boulder-clay** (earlier example); **boulder-pavement**, a bed of boulders naturally arranged; **boulder-train**, boulders deposited by the melting of a glacier.

1894 J. Geikie *Gt. Ice Age* (ed. 3) 742 In some instances these boulder-belts are all the marginal morainic representatives that can be detected over considerable tracts. **1859** H. Miller *Sketch-Bk. Pop. Geol.* i. 30 The boulder-clay exhibits certain unique appearences. **1894** J. Geikie *Gt. Ice Age* (ed. 3) 523 Here and there 'boulder-pavements' occur. **1897** *Outing* (U.S.) XXIX. 339/2 A wearisome tramp over the boulder-strewed mountain side. **1897** *Daily News* 5 May 3/1 Arta Hill, bleak and boulder strewn. **1899** *Westm. Gaz.* 11 Dec. 8/1 The precipitous boulder-strewn heights. **1967** *Oceanogr. & Marine Biol.* V. 135 Boulder trains between drumlins along the New England coast are effective means of such protection.

bouldering, *vbl. sb.* Add: **2.** *Mountaineering.* Practice climbing on large boulders.

1920 G. W. Young *Mountain Craft* iv. 152 The introduction to climbing customarily..is practice upon single rocks, low cliffs, quarries and erratic boulders, with or without the aid of a rope held from above. This 'bouldering', or problem climbing,..is of little use as commencing practice. **1954** W. Noyce *South Col* iv. 63 John spotted a pointed boulder some thirty feet high by the wayside. Daring routes were made up it; and henceforward bouldering.. became a popular pastime.

Boule²: see *Boulle.

boule³ (bu·li, bū·le). [a. Gr. βουλή senate.] A legislative council of ancient Greece, originally aristocratic and consisting of the heads of families, later consisting of representatives chosen by lot.

1846 Grote *Hist. Greece* II. i. xx. 89 The Boulê, or council of chiefs, and the Agora, or general assembly of freemen. **1905** *Spectator* 4 Mar. 318/2 The Boulê, which answers practically to the House of Commons.

‖ **boule⁴** (bul). [Fr., = Bowl *sb.*²] **1.** A game resembling roulette (see quot. 1911). Also *attrib.*

1911 *Encycl. Brit.* XXI. 308/2 In recent years there has been a tendency to supplant the *petits chevaux* at French resorts by the *boule* or ball game..; in this a ball is rolled in a basin-shaped table so that it may eventually settle in one of a number of shallow cups, each marked with a number. **1928** *Daily Tel.* 10 July 15/5 Rooms for boule and baccarat. **1937** M. Allingham *Dancers in Mourning* i. 3 A shower of counters on a *boule* table.

2. A French form of bowls, played on rough ground, usu. with metal balls. Also in pl. form.

1924 W. J. Locke *Coming of Amos* xvii. 220, I..asked her to play a childish game of boule. **1942** W. Simpson *One of our Pilots* iv. 107 Groups of soldiers were playing 'boules', a crude form of bowls played on rough ground which was very popular in France. **1953** F. Stark *Coast of Incense* ii. 96 The game of boule—bowls without the bowling-green.

boulevard. Add: Now freq. (esp. in *U.S.*), a wide or well laid-out street or avenue. (Earlier and later examples.)

1769 H. Walpole *Let.* 30 Aug. (1857) V. 183 She and I went to the Boulevard last night after supper. **1873** *Scribner's Monthly* Sept. 541/2 The boulevard which starts from Lincoln Park, connects the Central and Douglas Parks, and then continuing [etc.]. **1903** A. B. ... *Actual Govt.* 328 Hence have grown up systems of boulevards, broad, winding, and well-surfaced, reaching from park to park and often from city to city. **1938** J. Cary *Castle Corner* 65 A head and face that might have belonged to any senior military club or Cheltenham boulevard. **1958** A. Sillitoe *Sat. Night & Sun. Morning* xi. 153 They took a long walk back to her house, by the boulevard that bordered the estate.

b. *attrib.* and *Comb.* **boulevard theatre**: see quot. 1961; so *boulevard farce*, etc.

1838 *Times* 24 Feb. 5/2 The scribblers of the small Boulevard-theatres are its real masters. **1918** W. Hutchinson *Doctor in War* (1919) xviii. 260 The superb, boulevard-wide..military roads of the Italian engineers. **1927** D. H. Lawrence *Lett.* (1938) 613 I'm always reading the ..men I like: none of them boulevard idols. **1929** *Observer* 17 Nov. 11/3 The piece is not only amusing. It deserves to be judged by a higher standard than the mere boulevard farce. **1941** Koestler *Scum of Earth* 48 The boulevard press..tried to prove that France was fighting a war for democracy. **1961** *Times* 17 Jan. 4/1 The expression 'boulevard theatre', which up to a few years ago was in common use to describe that part of French theatrical production whose principal aim was to amuse..becomes nowadays less and less useful.

boulevarded (bu·lvaɹdĕd), *a.* [f. Boulevard + -ed².] Provided with boulevards.

1891 *Pall Mall Gaz.* 21 Dec. 1/3 An ideal of a boulevarded London. **1897** *19th Cent.* June 931 It has no boulevarded Parisianised streets. **1951** Dylan Thomas *Let.* Jan. (1966) 352 O evergreen..boulevarded..cradle of Persian..

boulevardeer, var. Boulevardier.

1916 H. L. Wilson *Somewhere in Red Gap* ix. 374 Roystering blades from Pittsburgh or boulevardeers from Bucyrus—not a New Yorker in sight.

Boulle (bul). [f. the name of André-Charles *Boulle* (1642–1732), French cabinet-maker.] The correct form of the word commonly spelt Buhl. Cf. Boule².

1876 *Encycl. Brit.* IV. 446/1 Boulle-work is a kind of inlaying and ornamentation of cabinet-work. **1910** *Ibid.* 322/2 The most splendid and sumptuous specimens of Boulle are diminishing in number. **1955** R. Fastnedge *Eng. Furn. Styles* xi. 257 Boulle furniture was favoured in some quarters.

Boulton (bōu·ltǝn), *a.* [The name of a firm of glove-manufacturers.] Of the thumb of a glove: cut with a shaped piece extending upwards into the palm.

1922 *Daily Mail* 12 Dec. 1 Real Kid Nappa Gauntlet Gloves,..boulton cut thumb. *Ibid.*, Mocha Gloves.. Pique sewn, boulton thumb. **1929** *Penberthy's Sale Catal.* Mar. 24 An ideal Glove for hard service. Boulton cut. *Ibid.* 26 Men's English Tan Cape,..Boulton cut.

bounce, *sb.*¹ Add: **3. b.** An act of bouncing or ejecting. Also *fig. colloq.* (orig. *U.S.*).

1876 *N.Y. Times* 23 June 1/5 Tilden is no stronger than he was, although his friends are already playing the 'bounce' game that was so successful at Cincinnati. **1877** Bartlett *Dict. Amer.* (ed. 4) Add., *Bounce*. To get the *grand bounce* is to be dismissed from service; particularly from an office under government. **1906** E. Dyson *Fact'ry 'Ands* ix. 119 Pee give him er bounce off their land. **1910** 'O. Henry' *Rolling Stones* (1916) 125 'Had you ever thought,' I asks,..'of giving her the bounce yourself?' **1966** 'E. Lathen' *Murder makes Wheels go Round* i. 8 At these companies..the big boys who went to jail got the bounce.

5. A buoyant rhythm. Also *attrib. colloq.* (orig. *U.S.*).

1937 *Amer. Speech* XII. 45/1 *Bounce*, a light medium-fast tempo with a light accent on the first and third beats. **1956** G. Chisholm in S. Traill *Play that Music* iii. 41 Medium Bounce Tempo. **1958** B. Ulanov *Hist. Jazz in Amer.* xxv. 350 *Bounce*, used by some musicians, especially Duke Ellington, to describe a particularly buoyant beat.

6. *Comb.* **bounce-flash** *Photogr.*, reflected flash-light.

1952 J. F. W. Frerk *All about Flash Photogr.* 25 Splash or Bounce Flash. This is a way of using diffused flash light only, by directing the flash against the ceiling. **1953** A. Matheson *Leica Way* 176 For very soft and even illumination..fire the flash at a light wall or ceiling, utilizing the reflected light... This 'bounce-flash' illumination needs either a stronger flash bulb or a larger aperture than direct light.

bounce, *v.* Add: **6. b.** *trans.* To cause to rebound. Also *transf.*, to cause to be reflected.

1876 'Mark Twain' *Tom Sawyer* i. 9 I'll take and bounce a rock off'n your head. **1929** Wodehouse *Summer Lightning* iii. 76 He poised the tennis-ball and..bounced it on the silver medallist's back. **1950** *Sci. News* XV. 67 A beam of electrons is 'bounced off' the surface (in a vacuum) and collected on a photographic plate. **1959** *Daily Tel.* 16 May 1/1 An experiment was now in hand for bouncing radio waves from the planet Venus. **1959** I. & P. Opie *Lore & Lang. Schoolch.* vii. 114 Repeated while bouncing two balls against a wall.

c. *intr.* Of a cheque: to be returned to the drawer because there are insufficient funds to meet it. Occas. *trans.*, to present (such a cheque).

1927 *New Republic* 26 Jan. 277/2 'Bouncer'..may be either (1) a rubber check returned by the bank as no good, or (2) the person who passes (bounces) the rubber check. **1928** *Sunday Express* 2 Dec. 2/6 'Rubber checks', i.e., the type that comes bouncing back from the bank. **1943** Hunt & Pringle *Service Slang* 16 *Bounce*, to be returned by the Bank [of a dud cheque]. **1951** *News Chron.* 12 Dec. 4/4 If a customer draws a cheque for $25 when there is only $20 in his balance, the cheque will 'bounce'—it will be returned to the customer who paid it in with the uncomplimentary remark, 'Insufficient funds'. **1955** *Times* 25 Aug. 9/4 If.. you..then bounce a cheque, you will be in trouble.

8. b. *trans.* To eject summarily. Chiefly *U.S. colloq.* Cf. *Bouncer 5.

1877 Bartlett *Dict. Amer.* (ed. 4) 62, I daresn't go in there; the bar-tender's drunk, and I might get bounced. **1883** [see *bouncer 5]. **1891** C. Roberts *Adrift Amer.* 128 Here I jumped another train and got 'bounced' at Bernalillo.

c. To throw over (as a suitor). *U.S. colloq.*

1893 'O. Thanet' *Stories Western Town* 213 You don't suppose it would be any use to offer Esther a cool hundred thousand to promise to bounce this young fellow?

bouncer. Add: **4.** For reference in first quot. substitute:

1842 De Quincey in *Blackw. Mag.* July 127/2.

5. One engaged to eject undesirable or unruly persons from a saloon, ballroom, etc.; a 'chucker-out'. *colloq.* (orig. *U.S.*).

1865 *Nat. Police Gaz.* (U.S.) 29 Apr. 4/2 Old Moyamensing is almost as famous for its lawless gangs of boys and young men, as it was in the days of the 'killers' and 'bouncers'. **1883** *Daily News* 26 July 4/8 The Bouncer..is merely the English 'chucker out'. When liberty verges on licence and gaiety on wanton delirium, the Bouncer selects the gayest of the gay, and—bounces him. **1888** A. C. Gunter *Mr. Potter* xx, Several of the fighting brigade of the establishment, that in American slang would be termed 'bouncers'. **1903** A. Adams *Log Cowboy* xiii. 204 The bouncer of the dance hall of course had his eye on our crowd. **1938** Wodehouse *Summer Moonshine* i. 19 He held down a job for a time as bouncer at some bar. **1961** *Evening Standard* 21 Aug. 12/6 Bouncers required for dance Sat. evenings.

6. A ball that bounces high; *spec.* in *Cricket*, one that rises sharply off the pitch.

1913 *Cricket* 305/2 Every bowler pitches short sometimes, and when..he does so the resultant..'bouncer'..is no more than an ordinary risk. **1955** *Times* 24 June 14/1 Heine's first ball, a bouncer, was as the drawing of a sword.

bouncing, *vbl. sb.* Add: **5.** *Firework-manuf.* (See quot.)

1888 W. H. Browne *Firework Making* 15 Furnishing the squib with its report is called bouncing. *Ibid.* 43 Bouncing, or giving to the saucisson its report.

6. *Comb.* **bouncing-pin,** an apparatus for measuring 'knocking' in an internal combustion engine; also *attrib.*

1930 *Engineering* 10 Jan. 44/1 This sleeve..carries a bouncing-pin mechanism. **1934** *Jrnl. R. Aeronaut. Soc.* XXXVIII. 354 Instructions for setting of the bouncing-pin contacts. *Ibid.*, Then remove the diaphragm and bouncing pin.

bouncing, *ppl. a.* Add: *bouncing putty*: a soft elastic silicone polymer (see quot. 1950).

1944 J. G. E. Wright *U.S. Patent 2,541,851* (1951), Novel compositions which because of their unusual properties may best be described as 'bouncing putties'. **1950** *Jrnl. Brit. Interplan. Soc.* IX. 56 Another contribution by General Electric is 'bouncing putty', a viscous and highly resilient silicon material previously used as a core for golf balls. The idea is that certain delicate equipment is buried in the material and shocks thereby distributed evenly over the entire surface of the instruments.

bouncy (bau·nsi), *a.* [f. Bounce *v.* + -y[1].]

1. That bounces; having a buoyant manner.

1921 *Chambers's Jrnl.* Jan. 1/2 The little, bouncy man waved his umbrella in excitement. **1928** *Sunday Dispatch* 2 Sept. 15 Once more George..forestalled the bouncy Irishman. **1962** C. Watson *Hopjoy was Here* ix. 105 He sounded jolly bouncy over the phone.

2. Resilient, springy.

1926 *Sunday at Home* Jan. 220/2 This big, old house with the bouncy sofa and springy chairs.

bound, *sb.*[1] Add: **2. c.** *pl.* The limit or boundary beyond which soldiers, sailors, students, schoolchildren, etc., resident in a particular building, quarters, or area, may not pass. Now chiefly in *out of bounds*, outside or beyond this boundary.

1681 R. Knox *Hist. Ceylon* iv. ix. 156 Plain reason would tell him, that we being prisoners were without our bounds. *c* **1805** *Regul. Sherborne School*, Every other part of the town is out of School bounds, except the Church-Yard. **1857** Hughes *Tom Brown* i. ix, The chief offenders..were flogged and kept in bounds. **1865** *Blackw. Mag.* Apr. 472/1 The reason of putting the river out of bounds was the danger incurred by boys who could not swim. **1890** A. Conan Doyle *Firm of Girdlestone* v. 32 A lad coming up to an English University..must be within bounds at a fixed time. **1895** [see Out of *prep. phr.* III]. **1909** D. Sladen *Trag. Pyramids* xiii. 201 The decree of the General, which made the Considines out of bounds for the Army, like a Mohammedan festival. **1965** M. Spark *Mandelbaum Gate* ii. 43 She got into the car and made him drive out of school bounds, miles away.

5. **bound(s)-beater,** one who takes part in the ceremony of beating the bounds; **bound(s)-beating,** the ceremony of beating the bounds.

1909 *Essex Rev.* XVIII. 188 After the perambulation there was a dinner, jointly with the Little Leighs bounds-beaters, at the inn. **1927** *Sunday at Home* June 553/2 The religious part of the bounds-beating was prohibited.

bound, *ppl. a.*[2] Add: **4. b.** *Philol.* Designating a grammatical element, such as the present participial ending *-ing*, the adverbial ending *-ly*, etc., which occurs only in combination with another ('bound' or 'free') form.

1926 L. Bloomfield in *Language* II. 155 A form which may be an utterance is *free*. A form which is not free is *bound*. Thus, *book, the man* are free forms; —*ing*.., —*er*..are bound forms. **1957** S. Potter *Mod. Ling.* iv. 78 *Book-s*.. consists of two morphemes, the free form *book* and the bound form *-s* (bound because it cannot be used independently).

boundary. Add: **1. b.** *boundary-ditch, fence, -keeper, -line* (earlier U.S. example); **boundary dog** *Austral.* and *N.Z.* (see quot. 1945); **boundary layer,** the layer of fluid adjacent to a moving body; esp. in *Aeronaut.*, the layer of air adjacent to an aircraft in motion; **boundary light** (see quot. 1951); **boundary-rider** (examples); also *N.Z.*; hence *boundary-riding*.

1941 *Oxoniensia* VI. 87 One of the field-systems attached to the..settlement may probably be identified in the boundary-ditches on the lower half of the air-photograph. **1876** D. Kennedy *Colonial Trav.* xv. 196 The first [dog] seen was a boundary dog, chained to a break in a fence, to prevent sheep straying from one run to another. **1945** Baker *Austral. Lang.* iii. 72 A sheep dog chained to a kennel at a gateway..to prevent sheep passing through is called a *boundary-dog*. **1857** R. B. Paul *Lett. fr. Canterbury* ii. 27 The boundary fence which separates the poor man from the rich. **1926** M. L. Skinner in *Adelphi* May 794 Somebody's boundary fence beyond the railways. **1933** L. Acland in *Press* (N.Z.) 16 Sept. 15/7 *Boundary keeper*, a shepherd who keeps sheep from passing an unfenced boundary. **1922** J. M. Burgers in *Proc. K. Akad. Wetensch.* (*Amsterdam*) XXIII. 1097 We can calculate the distribution of the vorticity and the current in the boundary layer, when we suppose the velocity outside the boundary layer to be known. **1924** *Flight* 20 Nov. 737/1 The deductions from the boundary layer theory gave a rather poor approximation to the truth. **1949** O. G. Sutton *Science of Flight* ii. 48 The air, because of its viscosity, sticks to the surface of a moving body, and forms a boundary layer. **1937** *Reports & Mem., Aeronaut. Res. Committee* No. 1793 p. 1, Light signals used in aviation..at the aerodrome..boundary lights. **1951** *Gloss. Aeronaut. Terms* (*B.S.I.*) iii. 23 *Boundary lights*, lights defining the boundary of a landing area. **1705** in *Rec. Col. Rhode Island* (1858) III. 528 The boundary lines between this her Majesty's Collony..and her Majesty's Province of the Massachusetts Bay. **1865** *Australasian* 15 July 13/4 The shepherds and boundary-riders of the past and present. **1890** E. W. Hornung *Bride from Bush* xviii. 279 A boundary-rider..sees that the sheep in his paddock 'draw' to the water, that there is water for them to draw to, and that the fences and gates are in order. **1944** *Living off Land* iv. 94 There are no boundary riders on these cattle holdings. **1890** 'R. Boldrewood' *Col. Reformer* I. x. 149 You'll have messages to carry, boundary riding to do.

2. *Cricket.* The bounds or limit-line of a cricket field.

1867 *J. Lillywhite's Cricketers' Compan.* 7 Always agree at starting..what are the boundaries. **1899** W. G. Grace *Cricket. Remin.* ii. 23 There were no fixed boundaries at Lord's when I first played there. **1905** H. A. Vachell *Hill* xii. 248 [He] cut the sixth ball to the boundary. **1920** E. R. Wilson in P. F. Warner *Cricket* 64 Boundaries ..were adopted at Lord's for the Eton *v.* Harrow match of 1866.

b. A hit to the boundary; also, the number of runs allowed for the hit.

1896 *Westm. Gaz.* 24 July 5/2 His placing on the leg side, especially off Giffen, was extremely fine, and gave him several boundaries. **1910** *Ibid.* 26 Feb. 16/2 A four boundary is often the result. **1955** *Times* 13 July 8/6 He then hit Goddard almost for 6 to long-on and hooked him for another vivid boundary.

c. *attrib.*, as *boundary-bye, -hit, -play, -stroke*.

1887 F. Gale *Game Cricket* ix. xvi. 270 'Boundary byes' ought to be abolished..the batsmen being entitled to run as many as they could. *Ibid.* 268 There were no boundary hits. **1896** *Westm. Gaz.* 7 Feb. 3/1 A splendid display of true cricket relieved by free 'boundary' play. **1905** *Ibid.* 15 June 7/1 It was not until the third over of the match that he opened the scoring with a boundary-stroke.

bounder (bau·ndəɹ), *sb.*[2] [f. Bound *v.*[2] + -er[1].]

†**1.** *slang.* A four-wheeled cab or trap, so called from the bounding motion of the vehicle in passing over rough roads. *Obs.*

1842 *Hints to Freshmen* (Hotten, 1865), The man who drives has a well-appointed 'bounder' of his own. **1859** Hotten *Dict. Slang*, *Bounder*, a four wheel cab. **1865** *Ibid.*, *Bounder*,..a University term for a trap.

2. A person of objectionable manners or anti-social behaviour; a cad. Also in milder use as a term of playful abuse. (Occas. applied to a woman.) *colloq.*

a **1889** in Barrère & Leland *Dict. Slang* (1889) s.v., If I ordered the particular hat I desired I should be taken for a bounder. **1889** *Ibid.*, *Bounder* (university), a student whose manners are despised by the *soi-disant élite*, or who is beyond the boundary of good fellowship.. (society), a swell, a stylish fellow, but of a very vulgar type. **1890** *Times* 2 May 13/5 To speak of a man as a bounder is to allude to him as an outsider or cad. **1899** W. Archer *Study & Stage* 48 That is an anti-social proceeding, the conduct of a 'bounder'. **1912** A. Brazil *New Girl at St. Chad's* viii. 126 Flossie is a bounder! **1917** J. Adams *Student's Guide* 27 A prig is one who has too much self-respect, a bounder one who has too little. **1919** Wodehouse *Damsel in Distress* vi. 70 He had been marched up the Haymarket in the full sight of all London by a bounder of a policeman. **1930** W. S. Maugham *Cakes & Ale* xvii. 195 Women..adore a bounder.

bounderish (bau·ndəriʃ), *a.* [f. *Bounder *sb.*[2] 2 + -ish[1].] Having the character or characteristics of a bounder. So **bou·nderishly** *adv.*, **bou·nderishness.**

1899 E. Phillpotts *Human Boy* 35 With all his bounderishness, he was awfully clever. **1921** D. H. Lawrence *Sea & Sardinia* 265 He smoked his cigarette bounderishly. **1928** *Sunday Dispatch* 23 Dec. 12 An awfully bounderish and unbearable sort of young man. **1959** *Guardian* 30 Sept. 7/1 The hero's Nietzschean vitality and young bounderishness.

bounding, *ppl. a.*[1] Add: **b.** Of the pulse.

1879 *St. George's Hosp. Rep.* IX. 783 A very bad illness, marked by headache, bounding, hardly compressible pulse.

c. *fig.* Increasing by leaps and bounds.

1887 *Charity Organis. Rev.* III. 230 And yet, those.. bounding vagrancy returns!

2. = *Bounderish *a.* *colloq.*

1904 *Spectator* 31 Dec. 1089/1 A certain repulsion excited by his 'bounding' behaviour. **1924** H. A. Vachell *Quinney's Adv.* 169, I caught him looking at that bounding Tommy Loring.

boundless. Add: **B.** *sb.* That which has no bounds, the illimitable.

1909 *Westm. Gaz.* 10 July 3/2 We, a handful of men, concerning ourselves deeply in small matters, are brought face to face with the boundless.

bounty. Add: **5. e.** *King's* (or *Queen's*) *bounty*: a sum of money given from the royal purse to a mother who has given birth to three or more children at once.

1872 *Porcupine* 8 June 156/1 Frenchwomen are seldom recipients of the Queen's bounty. **1910** *Encycl. Brit.* IV. 324/2 King's Bounty. **1939** *Times Lit. Suppl.* 18 Mar. 162/3 Here, too, are..'quins' and the King's Bounty.

bouquet. Add: **1. b.** *fig.* A compliment, praise; phr. *to throw bouquets*, to pay compliments.

1904 *Journalist* 28 May 89 We do not wish to say 'I told you so', or to 'throw any bouquets' in our own direction. **1907** N. Munro *Daft Days* xxiii. 203 I've never done throwing bouquets at myself about it ever since. **1955** *Times* 4 May 9/6 The union would inevitably receive their share of the brickbats and bouquets.

3. a. Also **bouquet garni** [F., 'garnished bouquet'] (see quots.).

1852 F. Bishop *Illustr. London Cookery Bk.* 408 *Bouquet*, a bunch of parsley and scallions tied up to put in soups, &c. *Bouquet garni*, or *Assaisonné*, the same, with the addition of cloves or aromatic herbs. **1960** *Woman* 3 Dec. 42 Classic ingredients of a bouquet garni are half a bay leaf, three or four long-stemmed sprigs of parsley, and a sprig of thyme.

‖**bouquetier** (bukətiə·ɹ, Fr. buktye). Also erron. **-ière.** [Fr., f. Bouquet.] A small holder for a bunch of flowers, esp. one carried in the hand.

1786 J. Wedgwood *Let.* 22 July (1965) 300 It consisted of upwards of £100 worth, of vases, bouquetiers, [etc.].

1871 A. Nesbitt in W. Chaffers *Catal. Slade Coll. Glass* 80 Large Blue Bouquetière. **1875** Meteyard *Wedgwood Handbk.* 392 Bouquetier. A flower-pot,..often with a pierced cover for the insertion of the stems of flowers.

Bourbon. Add: **1.** After 'Naples' read: and until 1931 that of Spain.

3. b. A rose belonging to a group of hybrids (*Rosa × borboniensis*) produced by *R. chinensis* and *R. damascena*, believed to have originated in Réunion (Isle de Bourbon) *c* 1825, and used as a parent of many modern varieties.

1829 Loudon *Encycl. Plants* 446 *Rosa centifolia*...Garden Varieties...Bourbon. **1841** *Penny Cycl.* XX. 158/1 Bourbon rose..is a natural hybrid between *R. Indica* and a variety called red four-seasons. This hybrid was found among a number of the latter plants in a hedge in the Isle of Bourbon. **1852** S. Warner *Queechy* liv, The air was full of the sweetness of damask and Bourbon varieties. **1869** S. R. Hole *Book about Roses* 173 Of the Bourbons.. there are several which are valuable additions to a general collection of Roses. *Ibid.* 174 Catherine Guillot, with Louise Odier,..are..two winsome maids of honour in waiting upon the Bourbon Queen. **1962** *Amat. Gardening* 24 Mar. 4/3 The species and older shrub roses such as the albas, Bourbons, damasks and gallicas.

d. *Bourbon tea*: see Tea *sb.* 6.

1861 Bentley *Man. Bot.* 667 *Angræcum fragrans*.—The dried leaves of this fragrant species are used as a kind of tea in the Mauritius. This is commonly known as *Faham* or *Bourbon tea*.

4. Whisky of a kind originally made in Bourbon County, Kentucky; a glass of this whisky. Also *attrib.* orig. *U.S.*

1846 in *Amer. Speech* (1963) XXXVIII. 158 Old Bourbon. **1850** 'M. Tensas' *Odd Leaves from Louisiana 'Swamp Doctor'* 37, I would have sworn it was good old Bourbon whiskey. **1862** *Congress. Globe* May 2288/3 One barrel of genuine Bourbon liquor. **1957** J. Braine *Room at Top* vii. 75 George poured me another Bourbon. **1958** G. Greene *Our Man in Havana* v. v. 238 Bourbon is stronger than Scotch.

5. *Bourbon biscuit*: a chocolate-flavoured biscuit with a chocolate cream filling.

1932 L. Golding *Magnolia Street* v. 103 Mrs. Poyser sent in some smoked salmon and a bag of Bourbon biscuits. **1960** R. Collier *House called Memory* iii. 44 The chocolate biscuits, the bourbon biscuits, and the bridge rolls.

bourdon[1]. Add: **2. b.** *Bell-ringing.* (See quot. 1938.)

1927 R. Church *Dream*, etc. 10 Loud peal and pæan, bourdon and burden Swinging, one voice, ringing 'Rejoice'. **1938** *Oxf. Compan. Mus.* 111/2 The lowest string of the lute and violin used to be called the bourdon, as is still the lowest bell in a ring of bells.

Bourdon[3] (bŭə·ɹdǫn). The name of Eugène *Bourdon* (1808–84), French hydraulic engineer, used *attrib.* and in the possessive to designate his inventions: **Bourdon('s) barometer, gauge, manometer,** a pressure gauge employing a Bourdon tube; **Bourdon coil, spiral, tube,** a coiled metallic tube which tends to straighten out when pressure is exerted within it.

1859 *Negretti & Zambra's Catal. Opt. & Meteorol. Instruments* 23 Bourdon's Barometers, card dial. metal dial and plate glass front. **1864** *Negretti & Zambra's Treat. Meteorol. Instruments* 7 Bourdon's Pressure Gauge, with metal taps, adapted for all pressure below nine atmospheres. *a* **1877** Knight *Dict. Mech.* s.v. *Bourdon Barometer*, The Bourdon is commonly known as the metallic barometer. **1886** *Q. Jrnl. R. Meteorol. Soc.* XII. 124 Each thermometer consists of a very thin curved metal case (a Bourdon tube). **1901** W. W. F. Pullen *Steam Engin.* iv. 143 The Bourdon Gauge consists of an oval tube bent nearly into a complete circle, one end being fixed and the other free to move. **1923** Glazebrook *Dict. Appl. Physics* III. 494/2 The Bourdon tube thermograph. *Ibid.*, The Bourdon tube thermometer is much less sensitive to change of temperature than the bimetallic thermometer. **1930** *Engineering* 14 Feb. 214/2 Any fall in pressure in the Bourdon tube will tend to open the pilot valve. **1959** H. Barnes *Oceanogr. & Marine Biol.* 121 [In the bathythermograph] the temperature-responsive unit is not a wire but a liquid in metal thermometer with a Bourdon spiral.

‖**bourdonnée** (buəɹdǫ·ne), *a.* *Her.* [Fr.] Terminating in knobs or balls.

1612 Peacham *Gentl. Exerc.* 153 *Cosm.* Be there no more crosses then one borne in Armes? *Eud.* Yes...The most ordenary are these...Croiseè...Composeè...Bourdonneè. **1722** Nisbet *Her.* I. 119 A Cross Bourdonee, as if it were made of Pilgrims Staves, which use to have a round Ball at the Top. **1780** Edmondson *Heraldry* II. Gloss. **1880** *Encycl. Brit.* XI. 696/1 Originally..the arms of the cross ended in knobs like the handles of a pilgrim's staff, thence called 'bourdonnée'.

bouree, var. *Bourrée.

bourgeois, *sb.*[1] and *a.* Add: **A.** *sb.* Also fem. **bourgeoise,** a Frenchwoman of the middle class.

1794 J. Courtenay *Pres. State Manners of France & Italy* ii. 25 Here the pretty *Bourgeoise*, drest in smiles and in charms. **1807** J. Beresford *Miseries Hum. Life* II. 49 A *Bourgeoise*, who is privileged by wealth..to issue cards, and lose her money, to her superiors. **1842** [in Dict.]. **1959** *Listener* 10 Dec. 1051/2 A garrulous *bourgeoise*.

2. Used disparagingly. **a.** In communist or socialist writings: a capitalist; anyone judged to be an exploiter of the proletariat.

1883 [see Communist 1]. **1886** Marx & Engels *Manifesto of Communists* ii. 17 The Bourgeois, not content with having the women and daughters of their wage-slaves at command, ..find it a capital amusement to seduce each other's wives. **1886** F. K. Wischnewetzky tr. *Engels's Condition of Working-Class* (1892) 277 It is utterly indifferent to the English bourgeois whether his working-men starve or not, if only he makes money.

b. A socially or æsthetically conventional person; = Philistine *sb.* 4. Hence *to shock the bourgeois* [F. *épater le(s) bourgeois*], to behave unconventionally, to utter novel opinions; so *bourgeois-shocker*. Cf. *Épater.

1930 A. Huxley *Brief Candles* 190 It's better to be a good ordinary bourgeois than a bad ordinary bohemian..or a second-rate intellectual. **1934** R. Campbell *Broken Record* ii. 48 This tame and cowardly gang of bourgeois-shockers. **1960** C. P. Snow *Affair* v. 49 It was mildly ironic, when one thought how, as a young woman, she had shocked the bourgeois, to find her set on seeing him a cosy, bourgeois success.

B. *adj.* **2.** (Earlier and later examples.) Also used disparagingly: selfishly materialistic or conventionally respectable and unimaginative; = Philistine *a.* 2. (See also quot. 1960.)

1764 S. Foote *Lyar* ii. 11 Decency is..a mere bourgeois plebeian quality. **1775** H. Walpole *Let.* 10 Aug. (1904) IX. 231 Consider how *bourgeois* it would be in me to talk of her Highness my niece. **1894** 'A. Hope' *Dolly Dialogues* xvi. 84 '*Bourgeois*,' I observed, 'is an epithet which the riff-raff apply to what is respectable, and the aristocracy to what is decent.' **1921** Shaw *Back to Methuselah* ii. 43 You and I were brought up in the old bourgeois morality... Savvy..is a Bolshevist. **1930** A. Huxley *Brief Candles* 189 How can you be so disgustingly *bourgeoise*, Pamela? So crass, so crawling? **1949** F. Maclean *Eastern Approaches* i. iii. 32 Up to a year or two before [1937] jazz, or 'dzhaz', as it was called, had been frowned on [in Russia] as bourgeois stuff. **1960** C. S. Lewis *Studies in Words* i. 21 When I was a boy *bourgeois* meant 'not aristocratic, therefore vulgar'. When I was in my twenties this changed...*Bourgeois* began to mean 'not proletarian, therefore parasitic, reactionary'.

3. (Cf. sense A. 2 a above.) Capitalistic, non-communist; esp. used disparagingly.

1850 H. Macfarlane tr. *Marx & Engels's Manifesto of German Communist Party* in *Red Republican* 23 Nov. 183/2 The old Bourgeois Society with its classes, and class antagonisms, will be replaced by an association, wherein the free development of each is the condition of the free development of all. **1886** J. L. Joynes tr. *Marx's Wage-Labour & Capital* 11 Increase of capital..cannot abolish the opposition between his *sc.* [the labourer's] interests and those of the bourgeois or capitalist class. **1928** G. B. Shaw *Intelligent Woman's Guide to Socialism* lxxiv. 369 When the Russian Communist and his western imitators speak of the proprietors and their political supporters as 'bourgeois'. **1946** Koestler *Thieves in Night* 90 Powder and cosmetics are banned from our stores as attributes of 'bourgeois decay'.

4. Of French wine: next in quality to wines classified as the best (see quot. 1908).

[**1846** C. Cocks *Bordeaux* ii. 197 After the Second–Fourth growths, comes a list of wines known by the name of *Bons Bourgeois*.] **1908** E. & A. Vizetelly *Wines of France* iii. 93 Below the classified growths, throughout the Bordelais, come..those which are known locally as (1) 'bourgeois', (2) 'artisan', and (3) 'peasant' wines. *Ibid.*, A superior bourgeois Médoc of 1904. **1920** G. Saintsbury *Notes on Cellar-bk.* iv. 49 A sound *bourgeois* wine, much above *ordinaire*. **1967** C. Ray *Compleat Imbiber* IX. 66 Some lesser classed growths, both great and *bourgeois*.

5. *Comb.*, as *bourgeois-capitalistic, -democratic, -liberal* adjs.; *bourgeois-mindedness*.

1936 Wirth & Shils tr. *Mannheim's Ideology & Utopia* i. 28 The liberal beginnings of the bourgeois-capitalistic era. **1937** E. Snow *Red Star over China* iii. ii. 90 The Communists, of course, regarded the successful fulfilment of the 'bourgeois-democratic' revolution as a necessary preliminary for any Socialist society. **1948** J. Towster *Political Power in U.S.S.R.* i. 10 The Soviet state arose on the ruins of the bourgeois-democratic state form. **1936** Wirth & Shils tr. *Mannheim's Ideology & Utopia* v. 249 The bourgeois-liberal mode of thought. **1955** H. Hodgkinson *Doubletalk* 18 The rise of 'bourgeois-mindedness'.

Hence **bourgeoi·sdom** [-dom], the political ascendancy of the bourgeoisie; bourgeois people collectively.

1887 Moore & Aveling tr. *Marx's Capital* I. p. xxx, In its rational form it [*sc.* Hegelian dialectic] is a scandal and abomination to bourgeoisdom. **1937** 'C. Caudwell' *Illusion & Reality* iv. 87 In its early stages bourgeoisdom requires the shattering of all feudal forms.

bourgeoisie. Add: **b.** In Communist writings: the capitalist class. Cf. *Bourgeois A. 2 a and B. 3.

1888 S. Moore tr. *Marx & Engels's Manifesto* 7 By bourgeoisie is meant the class of modern Capitalists, owners of the means of social production and employers of wage-labour. **1886** F. K. Wischnewetzky tr. *Engels's Condition of Working-Class* (1892) 281 We have seen..how the bourgeoisie exploits the proletariat in every conceivable way for its own benefit. **1957** R. N. Carew-Hunt *Guide to Communist Jargon* iii. 6 In present communist usage the term bourgeoisie is given, however, such an extension as to include all those who, whether they possess capital or not, have an interest in preserving the capitalist system.

bourgeoisify (buɹʒwā·zifəi), *v.* [f. Bourgeois *a.* + -ify.] To convert to a bourgeois outlook or mode of life. (Used esp. in Communist writings.) So **bourgeoi·sified** *ppl. a.*

1930 tr. *Lenin's Coll. Wks.* XVIII. 44 Marx shows how.. the British proletariat is becoming 'bourgeoisified'. **1937** 'C. Caudwell' *Illusion & Reality* iv. 75 The bourgeoisie and the bourgeoisified nobility. **1938** 'G. Orwell' *Homage to Catalonia* v. 78 The workers' militias..must be preserved.. and every effort to 'bourgeoisify' them must be resisted. **1960** *New Left Rev.* Sept.–Oct. 13/2 Temporary industrial monopoly enabled the capitalists to 'bribe' and 'bourgeoisify' the working class.

So **bourgeoisifica·tion,** conversion to a bourgeois outlook or mode of life.

1937 J. M. Murry *Necessity of Pacifism* ii. 32 'The bourgeoisification of the English proletariat', which he [*sc.* Marx] deplored. **1938** 'G. Orwell' *Homage to Catalonia* v. 71 A general 'bourgeoisification', a deliberate destruction of the equalitarian spirit of the first few months of the revolution, was taking place.

‖**bourrée** (bure). Also 8 **bouree.** [Fr.; see Boree.] **a.** A lively dance, of French origin, in common time (two beats in a bar). **b.** A musical composition, written usually in duple rhythm and in two-measure phrases, in which the character of the lively dance so named is represented.

1706 J. Weaver tr. *Feuillet's Orchesography* 57 Suppose, for example, the *Step* to be a *Bouree*. **1776** Hawkins *Hist. Music* IV. 390 The Bouree is supposed to come from Auvergne in France; it seldom occurs but in compositions of French masters. **1879** Grove *Dict. Mus.* I. 264/1 The bourrée is often to be found in the older suites, especially in those of Bach. **1891** *Queen* 3 Jan. 18/3 Minuet in D (No. 1) is fairly interesting, Bourrée in G (No. 2) is rather wild. **1963** *Times* 12 Jan. 4/1 An excellent set of French dancers became available. Their *bourrée* was a great delight.

2. *bourrée step* or *pas de bourrée*, a sideways step in dancing in which one foot crosses behind or in front of the other.

1706 P. Siris *Art of Dancing* 41 (*heading*) A Table of Bourrée-Steps or Fleurets. **1710** J. Essex tr. *Feuillet's Treatis of Chorography* 15 One may make little hopps or Bouree steps but little hopps are more in fashion. **1786** S. J. Gardiner *Definition Minuet-Dancing* 25 The lady must then make a pá tombéz with her left Foot, and the Gentleman a pá de bouréz with his left Foot. **1830** R. Barton tr. *Blasis's Code of Terpsichore* ii. 102 Afterwards come..the *pas de bourrée* and the various movements of different kinds of pirouettes. **1957** G. B. L. Wilson *Dict. Ballet* 206 *Pas de bourrée*, lit. a bourrée step; a step probably deriving its name from the Bourrée..with which, however, it has now no apparent connexion.

Boursault (bŭə·ɹsǫlt). [The name of a Parisian rose-grower, Henri *Boursault* fl. *c* 1810.] A species of climbing rose originating from the Alpine rose (*Rosa alpina*), introduced in 1683; also, a rose of this species.

1826 R. Sweet *Hortus Brit.* I. 138 Rosae..Boursoulti. Boursoult's. Hybrid. 1821. **1837** T. Rivers *Rose Amateur's Guide* 50 The Boursault Rose. (Rose Alpina.) This is a most distinct group of roses, with long, reddish flexible shoots. *Ibid.* 51 The Red Boursault is our oldest variety. **1864** S. Hibberd *Rose Bk.* 29 The Boursault roses...The Old Red Boursault..was the first double variety of *Rosa alpina*. **1956** *Dict. Gardening* (*R.H.S.*) (ed. 2) IV. 1824/2 The Boursault Roses (hybrids of *R. alpina*, pillar roses with reddish smooth stems, sometimes without prickles).

bout, *sb.*[2] Add: **5.** The inward curve of a rib in a violin or similar instrument, by which the waist is formed.

1889 *Cent. Dict.* **1893** *Fiddler's Handbk.* 4 Bouts,..the sides of the fiddle, divided into the lower, middle, and upper bouts. **1898** H. R. Haweis *Old Violins* iii. 35 The curve of the bouts tilted, contracted, or elongated. *Ibid.* 38 Observe the improved purfling, the bouts and mitres cut with clear intention.

bout (baut), *v.* [f. Bout *sb.*[2] 1 b.] *trans.* To plough in such a manner as to make bouts. Hence **bou·ted** *ppl. a.*, ploughed in this manner.

1733 W. Ellis *Chiltern & Vale Farm.* ii. 36 Bout it up at Allhollantide. *Ibid.* 37 In June harrow it down and bout it the same way; in July hack it overthwart, or bout it up across. **1844** *Jrnl. R. Agric. Soc.* V. i. 16 Two of these harrows cover a single ridge of four furrows; four cover an eight-furrow stretch, consisting of two ridges bouted into one. **1864** *Ibid.* XXV. ii. 291 The fallows are broken up in the autumn.., ploughed back in spring, then twice across, and bouted in 27-inch ridges.

boutade. Delete † *Obs.* and add later examples.

1865 'Ouida' *Strathmore* I. x. 164 One of those tantalising *boutades* that were her most cruel and certain witcheries. **1905** *Spectator* 28 Jan. 141/2 Miss Burney had the good fortune to see only the better side of the Doctor...She was seldom witness of his boutades. **1924** *Blackw. Mag.* June 783/2 A certain notoriety for boutades among my associates.

bouting, *vbl. sb.* (Earlier examples.) Also *attrib.*

1733 W. Ellis *Chiltern & Vale Farm.* xxxviii. 311 Bouting, or Bouting-up, is a Half-ploughing of the Ground. *Ibid.* 312 Bouting-down, is done by making a shallow Thorough on each Side of the Ridge of the Bout. **1786** Washington *Diaries* (1925) III. 43 The field..is divided into 3 parts, by bouting Rows running crossways.

boutique (buti·k). [F. (14th c. in Littré), f. OPr. *botica* (It. *bottega*), ad. L. *apothēca*, a. Gr. ἀποθήκη (see Apothec).] A small shop.

1767 J. Long *Sel. Unpubl. Rec. Govt.* (*Fort William*) (1869) 501 The street..has been greatly encroached upon by a number of golahs, little straw huts, and boutiques. **1780** *India Gaz.* 9 Dec. (Y.), Mrs. Henpeck..is a great buyer of

Bargains, so that she will often go out to the Europe Shops and the Boutiques, and lay out 5 or 600 Rupees in articles that we have not the least occasion for. **1854** *Househ. Words* IX. 43/1 A collection..presided over by a very solemn man ..calling aloud at intervals to the passers-by to patronise the *boutique* at six sous. The attractions of his booth include soaps of all colours and patterns. **1859** Sala *Twice round Clock* 185 The merchants who have here [Burlington Arcade] their tiny *boutiques*. **1926** *Glasgow Herald* 18 Sept. 4 A small Sinhalese child..disappears into one of the 'boutiques' (small native shops) on the other side.

b. *spec.* A small fashion-shop or department that sells ready-to-wear clothes designed by a couturier; a small shop selling 'trend-setting' clothes or other articles, esp. for young or fashionable people. Also *attrib.*

1953 *N.Y. Times* 26 Jan. 12/5 The usual boutique sports clothes were not so evident as swirls of summer cocktail and evening dresses went by. **1954** *New Yorker* 27 Nov. 143/1 On the first floor is still another boutique, this one awave with ostrich-feather fans on new or antique mountings. **1957** *Times* 21 Oct. 13/1 Boutique departments in the big stores, designed to fill the gap between custom-made couture clothes and those made by wholesale houses are now well established. **1957** *Observer* 17 Nov. 11/3 The idea of 'Boutiques', those small shops set inside couture establishments to sell ready-to-wear. **1964** *Queen* 1 Jan. 57/2 [In] the third of the..boutiques..is a pot-pourri of pretty rococo and sometimes gilded objects, handsome ranges of household glass and china, and a special men's section. **1966** *Vanity Fair* May 116/2, I..love the look of boutique clothes. **1966** M. Quant *Quant by Quant* 35 It was agreed that if we could find the right premises for a boutique..we would open a shop. It was to be a *bouillabaisse* of clothes and accessories..sweaters, scarves, shifts, hats, jewellery, and peculiar odds and ends.

bouton (butoṅ). [Fr., = button.]

1. In *pearl bouton, bouton pearl*, anglicization of *perle bouton*, a round pearl with a flat back marking the place where it was attached to the shell.

1851 *Illustr. Catal. Gt. Exhib.* iii. 689 Brilliant tiara, ornamented with fine oriental pearl drops and boutons. **1907** *Daily Chron.* 5 Apr. 4/5 An exquisite pearl and brilliant necklace, formed as fifteen graduated drops, each composed of one bouton pearl, one brilliant, and one pear-shaped pearl drop. **1927** *Daily Express* 9 May 5/3 If it is a round pearl, with a flat back showing where it was attached to the shell, it will be called a 'bouton' pearl.

2. The button-like end of a honey-bee's tongue.

1886 F. R. Cheshire *Bees & Bee-Keeping* 95 The central and side ducts run down to that part of the tongue where the spoon, or bouton is placed.

‖**boutonnière** (butonyę̄r). [Fr.]

1. A spray of flowers worn in the button-hole.

1877 B. Harte *Story of a Mine* xi, She had distributed it to make *boutonnières* for other gentlemen. **1883** *Standard* 10 Nov. 3/2 Sir John Bennett came into court..carrying a huge bouquet, as well as a scarlet *boutonnière*. **1919** *Hist. Amer. Lit.* II. 268 He had worn a high hat, cane, and boutonnière. **1970** *Daily Tel.* 26 Feb. 17 Mr Bronson cooks and receives diners alternately..wearing a *boutonnière* in the lapel of his dark lounge suit.

2. *Surg.* An incision made in the urethra in order to extract a calculus or remove a tumour.

1884 H. Thompson *Tumours of Bladder* 76 The high operation is a much more formidable and hazardous proceeding than the simple boutonnière. **1908** *Practitioner* Feb. 180 A projecting intravesical lobe was wrenched out through a boutonnière incision.

bouvardia (buvā·ɹdiă). *Bot.* [mod.L., f. the name of Dr. Charles *Bouvard* (died 1658), superintendent of the Jardin du Roi, Paris: see -ia.] Any plant of the rubiaceous genus of this name, which was introduced from Mexico and Central America, bearing handsome red, yellow, or white flowers.

1846 Lindley *Veg. Kingd.* 764 The fragrance or beauty.. of the Gardenias,..Bouvardias, Catesbæas, &c. is unsurpassed in the vegetable kingdom. **1873** A. D. Whitney *Other Girls* vi. 64 Scarlet bouvardias and snowy deutzias. **1907** *Westm. Gaz.* 9 Oct. 2/1 The garden, with odorous bouvardias all awake.

bouvier (bū·vie). [F., lit. 'cowherd'.] A breed of dog (see quots. 1934 and 1947).

1934 Webster, *Bouvier des Flandres*, a powerfully built, rough-coated dog of a breed originating in Belgium. It has a slightly tousled appearance, with definite eyebrows, a mustache, and a beard. In color it ranges from fawn to black, through pepper-and-salt, and gray and brindle. **1947** G. L. Hubbard *Working Dogs* ii. 58 The Bouvier des Ardennes is the Cattle Dog of the Belgian Ardennes...It has many rivals among other local Bouviers or Bouvier types...It appears to be a descendant of the Bouvier de Flandres. **1963** *Guardian* 5 Jan. 5/3 Among the dogs successfully trained for police work are the Rottweiler, the Reisenschnauzer, and the Bouvier.

bouzouki (bŭzū·ki). [ad. mod.Gr. μπουζούκι; cf. Turkish *bozuk*.] In Greece, a sort of mandoline. Also *attrib.*

1952 L. MacNeice *Ten Burnt Offerings* ii. 22 Not all The gadarene jeeps nor all the taverna bouzoukis Can utterly drown the pack that yelps on the scent. **1959** 'N. Blake' *Widow's Cruise* 90 The three Greek musicians were warming up with *bouzouki* songs. **1964** *Punch* 4 Nov. 700/3 A bouzouki, the most haunting sounding of all Greek musical instruments.

bovarism (bōu·vāriz'm). Also **bovarysm(e.** [ad. F. *bovarysme*, f. the name of the principal character in Flaubert's novel *Madame Bovary* (1857) + -ism.] (Domination by) a romantic or unreal conception of oneself. Hence **bovaric, bovaristic** *adjs.*; **bovarize, bovaryze** *v. trans.* and *intr.*

[**1902** J. de Gaultier (*title*) Le Bovarysme.] **1929** A. Huxley *Do what you Will* 273 By a process of what Jules de Gaultier has called 'Bovarysm'..we impose upon ourselves a more or less fictitious personality. *Ibid.* Our earnest efforts to bovaryze ourselves into imaginary unity. *Ibid.* The bovaric personage..is firmly established. **1934** T. S. Eliot *Eliz. Essays* iii. 40, I do not believe that any writer has ever exposed this *bovarysme*, the human will to see things as they are not, more clearly than Shakespeare. **1936** A. Huxley *Olive Tree* 30 The French philosopher, Jules de Gaultier, has said that one of the essential faculties of the human being is 'the power granted to man to conceive himself as other than he is'. He calls this power 'bovarism' after the heroine of Flaubert's novel Madame Bovary. *Ibid.* 31 People have bovarized themselves into the likeness of every kind of real or imaginary being. *Ibid.* 32 Realizing, if only in words, his bovaristic dreams. **1952** H. Levin in *Ess. in Crit.* II. 3 If to Bovarize is simply to daydream. *Ibid.* 16 An all-pervasive state of mind: Bovarism.

bovinely (bōu·vəinli), *adv.* [f. Bovine + -ly[2].] In a bovine, dull, or inert manner.

1900 Crockett *Joan of Sword Hand* xxxiv. 227 Seydelmann..simpered bovinely down upon the girls. **1902** *Daily Chron.* 7 Nov. 5/2 The slow-witted politician, bovinely chewing the cud of his correspondent's schemes.

Bovril (bǫ·vril). Also **bovril.** [f. L. *bōs, bovis*, ox, cow.] **1.** The proprietary name of a concentrated essence of beef, invented in 1889 by J. Lawson Johnston. Also *fig.*

1889 *Athenæum* 18 Aug. 187/1 He ought to have drunk nothing but water, or at most a comforting cup of bovril. **1924** T. E. Lawrence *Lett.* (1938) 452 Your work..is the essence of thought, a variety of mental Bovril. **1935** Auden & Isherwood *Dog beneath Skin* 12 Tourists, to whom the Tudor cafés Offer Bovril and buns upon Breton ware.

2. Facetious alteration of Brothel *sb.* 3.

1936 R. Ackland *After October* i. 49, I had to take a job as a sort of dancing partner, and the place was practically a bovril. **1948** S. P. B. Mais *Caper Sauce* xii. 124 I'm not having my house turned into a Bovril.

bovrilize (bǫ·vrilaiz), *v.* [f. prec. + -ize.] *trans.* To concentrate the essence of; to epitomize, condense.

1901 *Daily News* 11 Jan. 5/1 Here..is one of these 'bovrilised' announcements..: Wanted, Sit. as Comp. by Eng. lady, 35; good Fr.; good refs.; would trav. R.R., 100. **1902** *Daily Chron.* 22 Aug. 8/7 He must give us not life, but the essence of it—a bovrilised version, as it were. **1902** *Academy* 13 Dec. 655/2 His fondness for bovrilising thought ..into so small a compass of words that the sentences are really too 'meaty'. **1928** *Musical Times* Nov. 1003 The exercises..are short enough not to fatigue the attention, and they bovrilise the difficulties or virtues admirably.

bow, *sb.*[1] Add: **1. b.** *Calligraphy.* A curved stroke forming part of a letter.

1914 E. A. Loew *Beneventan Script* vii. 127 The important elements of which the letters are composed are..the bow, the tall upright stem, [etc.]. **1957** N. R. Ker *Catal. MSS. Anglo-Saxon* p. xxx, The bow of p is regularly open in the early manuscripts.

4. c. Proverb, *a bow too long bent waxes dull*: relaxation is desirable; hence in other allusive phrases.

c **1532** Sir Adrian Fortescue in *Oxf. Dict. Proverbs* (1948) 59/1 A bowe that is longe bent, will waxe dulle. **1721** J. Kelly *Scot. Proverbs* 33 A Bow o'er bent will weaken. *Eng.* All Work and no Play makes *Jack* a dull Boy. **1817** Keats *Let.* 5 Sept. (1931) I. 38 But let us refresh ourself from this depth of thinking, and turn to some innocent jocularity —the Bow cannot always be bent. **1876** C. M. Yonge *Womankind* xi. 80 A strain which makes it very desirable. to unbend the bow, by a journey abroad, a sea-side sojourn.

e. *Cupid's bow*: see *Cupid 1.

11. d. A curved piece of metal used to make contact with an overhead wire in electric traction; = *bow trolley* (see 19 below). Also *attrib.*

1909 *Cent. Dict.* Suppl., *Bow-spring*, in electr. traction, the spring which keeps the bow spread out, when a bow is used to take current from a trolley-wire.

e. A leg of a pair of spectacles; also, the frame of either of the lenses. *U.S.*

1711 in *Essex Inst. Hist. Coll.* (Salem, Mass.) IV. 187/1 To Madam Rebekah Brown, I give my spectacles with gold bows. **1847** Longf. *Evang.* i. iii. 4 Glasses with horn bows Sat astride on his nose. **1890** *Harper's Mag.* Oct. 720/1 A pair of ancient silver-rimmed spectacles from which the bows were lost.

16. (Earlier and later examples of *bow-knot*.)

1547 Salesbury *Welsh Dict.*, *Kwlwm dalen*, a bowe knot. **1861** Mrs. Stowe *Pearl Orr's Isl.* I. v. 30 When he had once seated himself in a double bow-knot at a neighbor's evening fireside. **1896** *Godey's Mag.* Apr. 446/2 A woman with her back doubled into a bow-knot.

19. **bow-and-arrow** *attrib.*, belonging to or characteristic of the period when the bow and arrow was the chief weapon of war; **bow-iron** (see quot.); **bow-key** *U.S.* = *bow-pin*; **bow-pin** (example); **bow-spring,** a bow-shaped spring; **bow-(neck)tie,** a necktie in the shape of a bow (sense 16 b); **bow trolley,** a device for collecting the current from an overhead wire in electric traction; **bow-whip** *U.S.* (see quot.).

1899 *Westm. Gaz.* 27 July 2/2 We may yet work back to the bow-and-arrow period if modern inventions make war with their aid too grotesquely horrible and difficult. **1907** *Macm. Mag.* Oct. 935 Bow-and-arrow men. *a* **1877** Knight *Dict. Mech.*, *Bow-iron*, the staple on the side of a wagon-bed which receives the bows of the tilt or cover. **1859** J. Young in B. Young et al. *Jrnl. Discourses* VI. 230/2 You that have on such a yoke had better pull out the bow-keys. **1913** M. Stacpoole *Monte Carlo* vi. 63 He was wearing a rather exaggerated bow necktie. **1856** *Mich. State Agric. Soc. Trans.* 1855 VII. 55 Sample bow pins. **1840** *Civil Engin. & Arch. Jrnl.* III. 56/2 A vertebrated carriage..with bow-spring bearers and buffers. **1897** *Sears, Roebuck Catal.* 223/2 Gentlemen's silk bow ties. For turn-down collar. **1910** *Westm. Gaz.* 24 Jan. 3/1 He wore a check suit..and a pink cotton bow-tie. **1901** *Trans. Amer. Inst. Elect. Engin.* 111 (Cent. Dict. *Suppl.* s.v. *Trolley*), Bow trolley. **1902** *Encycl. Brit.* XXVIII. 98 On the continent of Europe considerable use has been made of bow trolleys, which consist of light metallic bow-shaped structures..running along on the under side of the wire against which they rub. **1890** *Harper's Mag.* Oct. 718/1 His whip was the fashionable 'bow whip' of the period, common enough now, to be sure, with a long lash, tapering down to a fine silk 'snapper' on the end.

bow, *sb.*[3] Add: **1. d.** *bow(s on*, with the bow of the vessel turned towards the object considered or in view.

[**1858** T. Williams *Fiji & Fijians* I. vi. 205 The commander ordered it to be run with its bows on the shore.] **1877** *Design & Work* 218/2 To hit a craft coming bows on. **1893** Kipling *Many Invent.* 104 A galley coming up bow-on. **1903** *Westm. Gaz.* 15 June 3/1 The 'standard type' has equal offensive strength in all directions—whether bows-on or broadside. **1967** *Jane's Surface Skimmer Systems 1967–68* 49 The flat bottom hull..permits the craft to run bow-on to any flat, sloping bank to embark passengers.

3. **bow-fast** (examples); **bow-oar** (examples); also, in a whale-boat, the foremost oar but one; **bow-wave,** (*a*) *Naut.* the wave set up at the bows of a ship under way; (*b*) *transf.* a shock wave produced in front of a body passing through the air.

1822 *Western M. Rev.* III. 354 His bow-fast (a grape vine) parted, and his frail bark put to sea without a pilot. **1913** Masefield *Daffodil Fields* 30 The bowfast was cast off, the screw revolved. **1851** Melville *Moby Dick* II. xxx. 205 Being the savage's bowsman—that is, the person who pulled the bow-oar in his boat (the second one from forward). **1867** *Harper's Mag.* Oct. 655/1 That man, the smallest of the lot, is the 'Bow Oar'. **1877** W. H. White *Man. Naval Archit.* 450 The position of the crest of the last wave in the train of waves that follow the bow wave.. exercises a very sensible effect on the resistance. **1949** S. P. Llewellyn *Troopships* 32 Porpoises..leaping and plunging in the bow-wave. **1959** J. L. Nayler *Dict. Aeronaut. Engin.* 238 For a sharp nosed wedge of semi-angle θ at zero incidence with the bow wave attached. **1962** *Listener* 1 Mar. 370/1 It was most probably the supersonic bow wave from a large meteorite falling through the atmosphere.

Bow (bōu). The name of a make of china originally manufactured at the Bow China Manufactory at Stratford-le-Bow in Essex.

1753 *Derby Mercury* 9 Mar. in Jewitt *Ceramic Art* (1878) I. vii. 200 Bow China Warehouse was opened on Wednesday, the 7th of February, near the Royal Exchange, in Cornhill, London,..where it will continue to be sold in the same manner as formerly, at Bow. **1863** W. Chaffers *Marks & Monograms* 138 A punch bowl of Bow china. **1869** Lady C. Schreiber *Jrnl.* (1911) I. 2 Small broken Bow figure. *Ibid.* 31 An imperfect..'Dovecote' of Chelsea or Bow. **1961** *Connoisseur* Dec. 310 Three dated specimens of Bow porcelain..have been purchased by the British Museum.

bowenite (bōu·ėnəit). *Min.* [f. the name of George T. *Bowen*, 19th-cent. American mineralogist: see -ite[1].] A hard, compact, translucent form of serpentine, resembling nephrite.

1850 Dana *Syst. Min.* (ed. 3) 265 Bowenite..as Bowen states, differs from nephrite in containing a large percentage of water. **1911** *Encycl. Brit.* XV. 123/1 The green jade-like stone known in New Zealand as *tangiwai* is bowenite, a translucent serpentine with enclosures of magnesite.

bower, *sb.*[3] **2.** Delete † *Obs.* and add later example.

1921 *Dict. Occup. Terms* (1927) § 278 *Bower, fish hook bower*,..sets, feeds and operates small fly press which bends fish hooks and forms ring at end.

Bower-Barff (bau·əɹ bāɹf). *Metallurgy.* [The names of G. and A. S. *Bower* and F. S. *Barff*, 19th-cent. American engineers.] **Bower-Barff process**: a development of the Barff process (see quots. and *Barff). Hence **Bower-Barffed** *ppl. a.*, treated by this process; **Bower-Barffing** *vbl. sb.*, treatment by this process. (Also with lower-case initials.)

1884 G. Bower in *Trans. Soc. Engin. 1883* 59 The coating consists of a film of magnetic oxide, and is produced directly by the Barff process, and indirectly by the joint process of the author and his son. So long as these were worked independently, the one was called the Barff, the other the Bower process; but now that the various English patents

are in the hands of one company, they are worked together and the process is called the Bower-Barff process. **1940** *Chambers's Techn. Dict.* 106/2 *Bower-Barffing*, an anti-corrosion process applied to sanitary ironwork. **1944** *Jrnl. Iron & Steel Inst.* CL. 83 The Bower-Barff process in which the steel is heated to about 800° C. for 20 min. first in air, then in superheated steam and finally in a producer-gas atmosphere. **1952** *Archit. Rev.* CXI. 138/2 A pan of twenty gallons working capacity in bower-barffed cast-iron.

bower-bird. Add: **2.** *fig.* A person who collects ornaments, odds and ends, etc. (see also quot. 1943). Also *attrib.* Hence as *v. intr.*, to pick up odds and ends.

1926 Fowler *Mod. Eng. Usage* 193/2 Such bower-birds' treasures as *au pied de la lettre, à merveille*. *Ibid.* 194/1 Every writer..who suspects himself of the bower-bird instinct should make and use some such classification system. **1941** K. Tennant *Battlers* xxvii. 301 George the Bower-bird..was..prowling around deserted camps, swooping on rubbish. *Ibid.*, I don't want him bower-birding round this camp. **1943** Baker *Austral. Slang* (ed. 3) 13 *Bower bird*, a petty thief.

bowery, *sb.* Restrict † *Obs.* to the senses 'A farm; a "plantation"'. Add examples in sense 'the Bowery' in New York City, an area of a squalid and wretched character noted for its cheap places of amusement and frequented by homeless vagrants.

1787 M. Cutler in W. P. & J. P. Cutler *Life & Corr.* (1888) I. 305, I..left the city by way of the Bowery. *c* **1844** R. H. Collyer *Lights & Shadows Amer. Life* 7 The crowd and bustle of business in Chatham Street, and the Bowery. **1930** E. Pound *XXX Cantos* xxviii. 130 Stiff as a cigar-store Indian from the Bowery.

b. *attrib.*, in sense 'of, pertaining to, or characteristic of the Bowery'; **Bowery boy,** a rough or rowdy of a type at one time characteristic of the Bowery.

1840 *Daily Picayune* (New Orleans) 28 Aug. 2/1 The Bowery boys of New York have..eclipsed the nice young men of Baltimore. **1852** C. A. Bristed *Upper Ten Thousand* 29 Its occupants are of not-to-be-mistaken Bowery cut—veritable b'hoys. **1856** *Spirit of Times* 1 Nov. 149/1 Star-flower of the blooming Bower-y girls. **1882** J. D. McCabe *New York* 642 The original 'Bowery Girl' must have been made of a rib of the original 'Bowery Boy', so exactly was she his counterpart. **1884** *Thompson St. Poker Club* 14 Mr. Tooter Williams and the odor of a Bowery cigar entered together. **1924** J. Buchan *Three Hostages* xviii. 263 'Hell!' he cried, with a torrent of Bowery oaths. *a* **1952** E. J. Brady in R. Ward *Penguin Bk. Austral. Ballads* (1964) 185 The Bowery gal she knows 'er know; The Frisco gal is silly.

bow-head (bōu·hed). [f. Bow *sb.*[1] + Head *sb.*] In full *bow-head whale*. The Arctic right whale, *Balæna mysticetus*.

1887 *Pall Mall Gaz.* 8 Sept. 2/2 A bow-head whale preparing to 'sound'. **1888** *Encycl. Brit.* XXIV. 523/2 The 'bowhead' of the Okhotsk Sea and Behring Strait. **1896** Kipling *Seven Seas* 129 Hath he..Spoke on the ice with the Bow-head—followed the Sabre-tooth home? **1938** *Geogr. Jrnl.* XCII. 371 The mammoth bowhead whale.

bowing, *vbl. sb.*[1] **3.** *bowing acquaintance* (earlier example).

1807 J. Beresford *Miseries Hum. Life* II. xviii. 164 A horse with an unceasing trick of nodding..his head up and down, as if he had a bowing acquaintance with every thing he meets.

bowl, *sb.*[1] Add: **7.** **bowl-barrow** (earlier example).

1812 R. Colt-Hoare *Anc. Hist. S. Wilts.* 21 Bowl Barrow. This is, I think, the most ordinary shaped barrow, and more frequently met with than any of the others.

bowl, *sb.*[2] Add: **2. e.** A delivery of the ball in cricket: now usually *ball*.

1862 *Chambers's Encycl.* III. 320/1 At the end of every four bowls, the bowler, wicket-keeper, long-stop, and fielders, change places.

f. A turn in the game of bowls; also, the delivery of the ball.

1889 in *Cent. Dict.* **1894** E. T. Ayers *Bowls* (ed. 2) 80 'Driver' on such occasions often comes in for reproach after an unsuccessful bowl.

bowl, *v.*[1] Add: **4. b.** *to bowl with one's head*: to bowl intelligently; *to bowl short*: to pitch the ball short of a good length; *to bowl over the wicket*: to bowl with the bowling-arm nearest to the bowler's wicket; opp. *to bowl round* (or formerly *outside*) *the wicket*: to bowl with the bowling-arm away from the bowler's wicket.

1851 J. Pycroft *Cricket Field* ii. 20 How is it that Clarke's slow bowling is so successful?..'You see, sir, they bowl with their heads.' *Ibid.* viii. 161 Playing him back all day if he bowls short, and hitting him hard along the ground whenever he overpitches. **1854** *Ibid.* (ed. 2) xi. 265 Any round-armed bowler (who does not bowl 'over the wicket'). **1887** F. Gale *Game Cricket* 154 Learning to play a round-arm bowler, bowling round the wicket. **1893** R. Daft *Kings of Cricket* xiii. 230 Harrison..seems to me to bowl better 'with his head' than he used to formerly. **1955** *Times* 9 May 15/1 On Saturday he bowled over the wicket.

5. (Earlier examples.) Also, **d.** To bowl *an over*. **e.** To put on (a player) to bowl in a cricket match.

1719 in H. T. Waghorn *Dawn of Cricket* (1906) 5 The Kentish men were bowled out. **1736** in —— *Cricket Scores* (1899) 14 When there were but two, and one wicket to bowl down. **1744** *Laws of Cricket* in *Dict. Arts & Sci.* (1755) IV. 3459/1 The bowler..when he has bowled one ball, or more, shall bowl to the number four before he changes wickets. *Ibid.*, If..the player is bowled out. *Ibid.*, If the wicket is bowled down, it is out. **1833** *New Sporting Mag.* Sept. 325 Floyer bowls the over. **1862** *F. Lillywhite's Cricket Scores* i. 415 His underhand bowling..was so fast that it was not always safe to bowl him. **1882** Pardon *Austral. in Eng.* 111, I think he might have bowled Boyle more. **1888** R. H. Lyttelton *Cricket* xi. 343 Mr. Ward bowled thirty-two overs for 29 runs. **1955** A. Ross *Australia 55* 96 Hutton bowled Tyson and Statham for an hour.

6. (Earlier examples.)

1793 *Sporting Mag.* 29 Sept. 371 Field-tennis threatens ere long to bowl out cricket. **1805** Capt. Crumby *Let.* in *19th Cent.* No. 273. 721 He wished me to be made acquainted with it [*sc.* Ld. Nelson's memorandum], that in the event of his being 'bowl'd out' I might know how to conduct the ship. **1867** Trollope *Claverings* II. xii. 147 You certainly did bowl her over uncommon well.

bowler[1]. Add: **2.** (Earlier example.) Also *bowler's* (or *bowlers'*) *match*: a cricket match in which the bowling is superior to the batting; *bowler's wicket*: one more favourable to the bowler than to the batsman.

1722 *Weekly Jrnl.* 21 July 2296/1 The Taylor was a good bowler. **1863** *Baily's Mag.* Aug. 360 It was a bowlers' match all through. **1876** *Ibid.* June 415 Much..will depend on..whether it is a batsman's or a bowler's wicket on the day of the match. **1888** *Sat. Rev.* 2 Sept. 312/2 It was clearly to be a bowler's match. **1895** *Badminton Mag.* Aug. 132 Any one can bowl on a bowler's wicket.

bowler[2]. Add: **b.** *fig.* As a symbol of civilian life (as opposed to service in the armed forces), or of the process of demobilization. Hence *bowler-hatting* vbl. sb. Occas. in extended use, with reference to dismissal.

1925 Fraser & Gibbons *Soldier & Sailor Words* 34 *To be given one's bowler*, to be demobilised and returned to civil life. **1928** 'Ian Hay' *Poor Gent.* iv. 74 Nelson and Trafalgar put the bowler hat on that scheme. **1931** Brophy & Partridge *Songs and Slang, 1914–18* (ed. 3) 287 *To be given one's bowler*, to be demobilized. *To get a bowler hat*, to be sent home, to be 'sacked'. Officers' slang; applied only to major and higher officers. **1953** *Economist* 8 Aug. 372/1 Little has been said.. to suggest that the settlers have any agreed programme to offer—beyond the 'bowler-hatting' of colonial civil servants of whom they have made scapegoats. **1959** *Observer* 4 Jan. 11/3 He [*sc.* Lord Mountbatten] did not ask for his bowler hat when Mr. Sandys reorganised defence a short time ago.

Hence **bowler-hatted** *adj.*, (*a*) wearing a bowler-hat; (*b*) civilian, demobilized.

1909 M. Beerbohm *Yet Again* 47, I picture him frock-coated, bowler-hatted, and evidently nervous. **1950** E. Hyams *From Waste Land* 16 The bowler-hatted City gent. **1959** *Economist* 30 May 841/2 Those warriors who want to spend vastly more on defence are mostly either bowler-hatted or arm-chairborne.

bowless, *a.* (Examples.)

1845 Eliza Cook *Poems* Ser. ii. 235 The bowless blind boy [*viz.* Cupid]. **1861** *Temple Bar* I. 282 His bow-less white neckcloth. **1905** *Daily Chron.* 25 Mar. 3/2 Her slippers were even bowless.

bowling, *vbl. sb.* Add: **2. b.** The strength or resources of the bowlers in a cricket team.

1845 W. Denison *Cricketer's Compan.* (ed. 2) p. viii, Their great deficiency..is in bowling. **1893** R. Daft *Kings of Cricket* xiii. 234 Surrey..possesses plenty of bowling and of quite the best quality.

c. *fig.*

1959 *Listener* 6 Aug. 222/2 A Minister at question time trying to stand up to accurate opposition bowling.

3. **bowling saloon** *U.S.*, a building equipped with bowling-alleys.

1842 Dickens *Amer. Notes* I. vi. 208 A painted lamp directs you to the Bowling Saloon, or Ten-Pin alley.

Bowman[2] (bōu·mæ̆n). *Anat.* [The name of Sir William *Bowman* (1816–1892), English surgeon.] Used in the possessive in *Bowman's capsule*: a Malpighian capsule; *Bowman's glands*: tubular glands in the olfactory mucous membrane; *Bowman's membrane*: the anterior elastic lamina (see quot. 1900).

1882 D. B. Delavan in T. E. Satterthwaite *Man. Histol.* (ed. 2) xxii. 372 Bowman's glands, peculiar to the olfactory mucous membrane, are found in it in large numbers. A. Mayer in *Ibid.* xiv. 206 Bowman's capsule is composed of a structureless basement-membrane surrounding each glomerulus. **1898** W. H. H. Jessop *Man. Ophthalmic Surgery* v. 69 The anterior homogeneous membrane (Bowman's). **1900** E. H. Starling *Elem. Hum. Physiol.* (ed. 4) x. 445 The glomerular epithelium acts simply as a filter allowing the water and salts of the blood-plasma to pass into Bowman's capsule. **1900** C. H. May *Man. Diseases of Eye* viii. 112 Bowman's membrane is a thin, homogenous membrane which separates the corneal epithelium from the proper substance of the cornea. **1927** Haldane & Huxley *Animal Biol.* iv. 115 The end of the kidney tubule..is dilated to form a capsule (Bowman's Capsule).

bowser[2] (bau·zəɹ). [Trade name.] A petrol pump (chiefly *Austral.* and *N.Z.*); also, a petrol tanker used for fuelling aircraft, tanks, etc. Also *transf.*

1934 *Bulletin* (Sydney) 18 July 11/4 In those days there was but one [petrol] pump storage system—the Yankee line, manufactured by one Bowser... 'Bowser' had found general acceptance, at least in Australia. **1935** 'H. Adair' *Wanted a Son* ii. viii. 228 They came to a petrol pump—a 'bowser' Alec called it. **1942** W. Simpson *One of our Pilots* 18 The hundred odd vehicles of our M.T. columns—petrol bowsers, tractors,..etc. **1943** *Amer. Speech* XVIII. 87 [In New Zealand] a pump at a filling-station is a *bowser*. **1963** D. Irving *Destruction of Dresden* iii. iii. 139 The bowsers were waiting to top up the tanks once again. **1968** P. Sharp *Railway Stations* 16 A water bowser (or tank on wheels) refills the cisterns in the roof of the coaches for the kitchens and toilets.

bowsman (bau·zmæ̆n). [ad. Fr. *bosseman* *Bosman, misunderstood as 'bow's-man'.] = Bowman[2]: used of a man positioned at the bow without an oar, having certain specified duties.

1776 M. Cocking *Jrnl.* 10 Mar. in *Publ. Hudson's Bay Rec. Soc.* (1951) XIV. 37 He..will make a proposal to them for employing Canadians to be engaged by him for a term of years at Montreal to serve as Bowsmen. **1804** Lewis in *Jrnls. Lewis & Clark Exped.* (1904) I. 34 The one not engaged at the oar [of the batteau] will attend as the Bowsman. **1840** *Knickerbocker* XVI. 270 The delicious white-fish are so numerous, that the bow's-man takes his scoop-net and literally dips them into the boat. **1851** Melville *Moby Dick* II. xii. 107 The mutineer was the bowsman of the mate. **1935** *Geogr. Jrnl.* LXXXV. 271 Not only the steersman but the bowsman. **1968** R. M. Patterson *Finlay's River* 63 The six members of Black's crew..consisted of bowsman and steersman, and four middlemen.

bow-wow, *int.* and *sb.* **2. a.** Delete quot. *a* 1832 (see 2 b below) and insert:

1785 [see *barking *ppl. a.*[1] 2 b]. **1826** Galt *Last of Lairds* xviii. 165 It's a sore thing for a man to be frightened into his first marriage by the bow wow o' a kirk session.

b. (Earlier example.)

1826 Scott *Jrnl.* 14 Mar. (1939) I. 135 The Big Bow-wow strain I can do myself like any now going.

c. Delete quot. 1838 and insert:

1785 Ld. Pembroke in Boswell *Jrnl. Tour Hebrides* 8 Dr. Johnson's sayings would not appear so extraordinary, were it not for his *bow-wow way*.

3. (Earlier and later examples.) Also *to go to the bow-wows*: to go 'to the dogs'. *jocular colloq.*

1785 Grose *Dict. Vulgar T.*, *Bow-wow*, the childish name for a dog. **1839** Dickens *Nickleby* lxiv. 617 It is all up with its handsome friend, he has gone to the demnition bow-wows. **1893** W. K. Post *Harvard Stories* 114 Everything was going to the bow-wows. **1917** H. A. Vachell *Fishpingle* xiii. 263 He was going fast to the bowwows before I went to India. **1931** R. Campbell *Georgiad* i. 20 All the bow-wows, poodles, tykes and curs.

bowyang (bōu·yæŋ). *Austral.* and *N.Z.* Also **boyang.** [f. dial. *bowy-yanks* (see *E.D.D.*), *bow-yankees* (see Halliwell), leather leggings. Cf. also *Sc. Nat. Dict.* s.v. *Booyangs, Bonanks*.] A band or strap worn about the trousers below the knee, esp. by labourers.

1893 *Warracknabeal Herald* 22 Sept., To those not in the cult of 'boyang worship', it may be necessary to explain that the two straps used to hitch the lower part of labourers' trousers are 'boyangs'. **1927** J. Devanny *Old Savage* 121 He missed his bowyangs dreadfully. **1948** A. P. Gaskell in *Coast to Coast 1947* 259 Progressive all right... I was the first in my district to appear at a dance without bowyangs. **1956** G. Bowen *Wool Away!* (ed. 2) ii. 15 Without bowyangs, and with sheep working on the legs all the time, trousers would soon work down. **1965** *N.Z. Listener* 26 Feb. 15/1 *Bow-yangs*, laces tied around the trouser legs below the knee to prevent drag when shearers bend over.

box, *sb.*[1] Add: **3. b.** **box-edged** *a.*, having a border of box plants; so *box-edge* (cf. quot. 1884 in Dict. under 3 a); **box-gum** *Austral.*, one of various species of *Eucalyptus* (cf. c below).

1932 Auden *Orators* ii. 63 Between box-edges, past the weathering urns. **1909** *Cent. Dict.* Suppl., Box-edged. **1945** E. Waugh *Brideshead Revisited* 71 The box-edged walks of the kitchen gardens. **1887** D. Macdonald *Gum Boughs* 7 The clumps of box-gums clinging together for sympathy.

c. Applied to many Australasian species of *Eucalyptus* (see quots.), *Tristania conferta* (Bastard, Brisbane, Brush, Red, or White Box), and some other trees: see Morris *Austral English*.

1820 J. Oxley *Jrnl. Two Exped. into N. S. Wales* 15 The timber, dwarf box, and gum trees (all eucalypti), with a few cypresses and camarinas. *Ibid.* 227 The country.. thickly timbered, chiefly with the species of eucalyptus called box. **1866** Lindley & Moore *Treas. Bot.*, Box.. White, of Australia, *Eucalyptus albens*. Yellow, of Australia, *Eucalyptus melliodora*. **1889** J. H. Maiden *Usef. Native Pl. Australia* 121 Native box..is greedily eaten by sheep,.. usually a small scrub, in congenial localities it develops into a small tree. *Ibid.* 468 *Eucalyptus hemiphloia*...This is a common 'Box' of New South Wales and Queensland. **1964** R. Holt in R. Ward *Penguin Bk. Austral. Ballads* 198 Gidgee, myall, box and jarrah.

box, *sb.*[2] Add: **1. d.** *Austral.* and *N.Z.* A mixing up of different flocks of sheep; also *transf.*; also with *up*. Cf. sense 21 and *Box *v.*[1] 5 b.

1872 C. H. Eden *My Wife & I in Queensland* iii. 67 Great care must of course be taken that no two flocks come into collision, for a 'box', as it is technically called, causes an infinity of trouble. **1941** Baker *N.Z. Slang* v. 39 A *box-up*, a state of confusion, and *to be in a box*, to be in a confused state of mind, in a quandary.

BOX

3. g. A receptacle or pigeon-hole at a post office in which letters to a subscriber are placed; hence, a similar receptacle or the like at a newspaper office in which replies to an advertiser are placed. orig. *U.S.* Cf. *box-letter, box number, box-rent* (sense *24).

1832 [see *box-letter]. **1833** B. F. Hallett *Trial E. K. Avery* 43 E. K. Avery had a private box at my office. **1897** E. W. Brodhead *Bound in Shallows* 242 The following evening Dillon found in his post-office box a letter of one line. **1919** Wodehouse *My Man Jeeves* 119 My address will be Box 341, *London Morning News*. **1971** *Times Lit. Suppl.* 23 Apr. 487/2 The Times Canadian Service Division, Box 490, King City, Ontario.

h. (See quot.)

1889 *Atalanta* June 597/1 For flower-painting never use what is technically termed 'box', viz.: the muddy colour.. that is left on the sides of the colour-box from former usage.

i. = Safe *sb.* 1 b. *slang* (orig. *U.S.*).

1904 'No. 1500' *Life in Sing Sing* xiii. 261 We got a country jug on our first touch, but the box wasn't heavy enough for five. **1904** H. Hapgood *Autobiogr. of Thief* vi. 120 He was one of the most successful box-men (safe-blowers) in the city. **1926** J. Black *You can't Win* viii. 89 We've got the combination of that box, kid. *Ibid.* ix. 104 Shorty was one of the patricians of the prison, a 'box man' doing time for bank burglary.

j. *colloq.* A gramophone, wireless set, or television set; spec. *the box*: television; a television set. Cf. *magic box.

1924 T. E. Lawrence *Lett.* (1938) 436, I.. play Beethoven & Mozart to myself on the box. **1930** Kipling *Limits & Renewals* (1932) 208 It was one of his prerogatives to announce what the Man in the Box [*i.e.* wireless announcer] said about the sick Padishah. **1950** G. Marx *Let.* 6 Dec. (1967) 168, I have solved the television problem by having a remote control installed on the ugly box. **1958** *Observer* 18 May 14/2 'Contradiction is the very moving principle of the world!' The Box, with its abrupt switches from the trivial to the profound, is always providing illustrations to Hegel's pet dictum. **1963** E. Humphreys *Task* I. xi. 120, I saw one of your plays, Dicky. On the old box.

k. A coffin. *slang.*

Cf. quot. 1674 s.v. Black *a.* 1 a.

1925 Fraser & Gibbons *Soldier & Sailor Words* 34 *The box*, a coffin. **1957** 'W. Henry' *Jesse James* 66 Personally, I'll believe he's dead when the box is shut and covered up.

10. b. = Confessional *sb.* 2 a.

1922 Joyce *Ulysses* 726, I wonder did he know me in the box.

11. b. *U.S.* The station occupied by various players in baseball; esp. either of the spaces in which the pitcher or the batter stands.

1881 *Detroit Free Press* 26 Sept. 1/5 Weidman..will have to go into the box for the remaining four games. **1909** *Collier's* 15 May 14/1 The pitcher was now contained in a box six feet square. **1944** Runyon *R. à la Carte* (1946) vi. 95 Putting the zing on Bill Terry for not taking Walker out of the box when Walker is getting a pasting from the other club.

c. *Cricket.* = *gully *sb.*[1] 2 d.

1913 *Daily Mail* 7 July 9/2 Splendid catch anywhere near the wicket, especially in 'the box'. **1926** *Ibid.* 29 June 10/2 Hobbs was caught in that nondescript position which is variously known as 'the box' and 'the gully'.

13. c. Short for *telephone box.

1935 G. Greene *England Made Me* I. 16, I rang up.. from a box in the Circus. **1961** 'T. Hinde' *For Good of Company* i. 16 Tony seized the phone and held it out of the box towards him.

15. d. A group of aircraft flying in close formation. Also *attrib.*

1941 *Battle of Britain Aug.–Oct. 1940* 13 Enemy bomber formations were also protected by a box of fighters. **1946** G. Gibson *Enemy Coast Ahead* ix. 121 They looped and rolled in perfect box formation. **1951** O. Berthoud tr. *Clostermann's Big Show* I. 45 The Luftwaffe would not in fact have time to concentrate on the first box.

e. *Mil.* An enclosed area heavily defended in all directions.

1942 *Hutchinson's Pict. Hist. War* 10 June–1 Sept. 125 The best defended 'box' will not hold out very long against greatly superior armoured strength. **1944** *Return to Attack* (Army Board, N.Z.) 9/1 'Baggush by the sea' was the fortress 'box' on the Mediterranean shore originally constructed in 1940 by the first contingent to leave New Zealand.

f. A light shield worn by cricketers to protect the genitals.

1950 N. Cardus *Second Innings* iv. 90 Not every player in those days used a 'box'. **1964** I. Fleming *You only live Twice* xi. 139 'What is a box?' 'It is what our cricketers wear to protect those parts when they go out to bat. It is a light padded shield of aluminium.'

19. b. *Printing.* A space enclosed within borders or rules, esp. one to draw attention to a heading, an announcement, etc. Cf. *box *v.*[1] 3 d.

1929 F. A. Pottle *Stretchers* 317 Its scare heads and leaded sub-heads, its boxes and rules. **1933** P. Macdonald *Myst. Dead Police* v. 36 In the centre of the page, inset in a black-edged 'box' and different type, was a condensed biography of the arrested man. **1957** Steinbeck *Pippin IV* 58 Louella Parsons had a front-page box headed: *Will Clotilde come to Hollywood?*

21. So *to be in the same box*: to be in a similar (unhappy) predicament.

1865 H. Sedley *M. Rooke* 143 If I weren't in something like the same box as his'n. **1884** Rider Haggard *Dawn* xlvii, Well, we are in the same box. **1911** H. Walpole *Mr. Perrin* v. 92 He always told himself that all the members of the staff were in the same box.

b. *Austral.* and *N.Z.* colloq. phrases: (*one*) *out of the box*: an excellent person or thing; (*to be*) *a box of birds*: (to be) fine, excellent.

1931 F. D. Davison *Man-Shy* (1934) v. 74 She's one out of the box, all right. **1943** F. Sargeson in *Penguin New Writing* XVII. 73 'Hello Terry,' he said, 'how's things?' 'A box of birds,' Terry said. **1947** D. M. Davin *For Rest of Lives* xxx. 146 Everyone is a box of birds, still celebrating being alive. **1949** H. Wadman *Life Sentence* 11 Well, anything very good of its kind is 'one out of the box'.

23. b. (*c*) *box-book* (hence *box-book-keeper*), *-lobby* (examples), *-ticket*.

1783 *London Chron.* 22 Nov. 501/2 A box lobby Puppy comes in at half price, and immediately goes to the box-book to see *who's there*. **1812** *Dramatic Censor 1811* 294 For the benefit of Mr. Spring, Box book-keeper. **1849** *Theatrical Programme* 4 June No. 1, p. 12 Mr. Massingham the obliging and gentlemanly box-book keeper of the Princess's Theatre, takes his benefit on Friday next. **1842** Box-lobby [see Lobby *sb.* 2 a]. **1768** *Pennsylvania Gaz.* in G. O. Seilhamer *Hist. Amer. Theatre* (1888) I. xxii. 247/1 After the tea equipage was removed, one of the gentlemen produced some box tickets for the play. **1866** M. Mackintosh *Stage Reminisc.* 175 For the evening performance there were so many 'box' tickets taken that we had to remove the barrier and enlarge our reserved seat accommodation.

24. box-annealing *vbl. sb.*, a process of annealing in which the metal is enclosed in a metal box or pot to prevent oxidation; also *attrib.*; hence **box-anneal** *v.*; **box-back** *a.*, designating a coat or jacket of which the back has a squared, box-like appearance; **box barrage,** an artillery barrage concentrated on a particular 'box' or area; **box-board** (orig. *U.S.*), board suitable for making boxes; also *attrib.*; **box-camera,** (*a*) (see quot. 1842); (*b*) a hand camera of the form of a box; **box-cañon, -canyon** *U.S.*, a narrow canyon having a comparatively flat bottom and vertical walls; **box-car** *U.S.*, a large closed-in railway goods wagon; **box-cart** *U.S.*, a cart having a box-shaped body; **box-churn,** a churn resembling a box in shape; **box-cloth,** a thick coarse cloth material, usually of a buff colour, from which riding garments are made; also applied to the colour; **box-coil,** a heating apparatus consisting of a coil of straight tubes joined at the ends, and occupying a cubical space; **box-coloured** *a.*, coloured by immersion in a box or tray of dye; **box-cutter,** a person employed in cutting out the material for boxes; **box-desk,** a desk of a box-like shape; **box-fitter,** a worker in an iron and steel foundry who attaches fittings and adjusts the parts of the moulding boxes; **box-food,** food which is given to animals in a box; **box-frame,** (*a*) the enclosed space in a window-frame for sash windows, in which the balance-weights are hung; (*b*) a frame or framework shaped like a box; also *attrib.*; see also quot. 1931; **box-girder** (see quot. 1865); **box-grain,** a grain given to leather in which lines are crossed in rectangular fashion; **box gutter,** a gutter of rectangular cross-section; **box-hand** (earlier example); (*b*) a person engaged in the manufacture or packing of boxes; (*c*) the compositor who sets up the type for stop-press news; **box-hat** *colloq.*, a tall (silk) hat; = *boxer[4] 1; **box-head,** (*a*) an indented heading in a printed article; (*b*) the fresh-water squaw-fish, *Ptychocheilus oregonensis*; **box-hook,** a hook used to handle, close, or raise boxes; **box-house** *U.S.*, a square-built house suggestive of a box; **box junction,** a road junction with a grid of yellow lines painted on the road forbidding the road-user to enter the junction area until his exit is clear; also *ellipt.*; **box-key** = *box-spanner*; **box-kite,** (*a*) a toy kite having the form of a box; (*b*) a kite invented by Lawrence Hargrave, of Sydney, Australia, consisting of two light rectangular boxes secured together horizontally, formerly used in meteorological experiments; cf. *Hargrave; (*c*) = **box-kite aeroplane,** an early form of biplane in which the arrangement of the planes resembled a box-kite; **box-letter** *U.S.*, a letter placed in a private box at a post office instead of being sent out and delivered to the addressee; **box-level,** a surveyor's level consisting of a glass-covered box instead of a level and tube; **box-loom,** a loom with more than one shuttle-box at either end of the lathe; **box lunch** *U.S.*, a packed lunch; **box-master** *Sc.*, a treasurer; **box-mattress** = *box-spring*; **box-meat,** meat packed in boxes for transport; **box-motion,** the machinery for operating the shuttle-boxes of a loom; **box number,** the number of a 'box' (sense *3 g) at a post office or newspaper office; **box-nut,** a screw nut with a closed end; **box-ottoman,** an ottoman (Ottoman *sb.*[2] 1) with a hinged upholstered lid forming the seat, with a receptacle below; **box-oyster** *local U.S.*, a fine large oyster, formerly packed in boxes instead of barrels; **box-plan,** a plan of the boxes or seats in a theatre; **box-rent** *U.S.*, the charge for a private post office box; **box-room,** a room for storing boxes, trunks, etc.; **box-seat,** the driver's seat on the box of a coach (see sense 6); (in quot. 1838, a seat on the roof of an early type of railway-carriage); **box-set,** a theatrical scene closed in with walls and ceiling; **box-shutter,** a shutter that folds back into a box, also called *boxing-shutter*; **box-spanner,** a spanner with a socket-head at one or both ends which fits over the nut, etc., to be turned; **box-spring,** one of a set of spiral springs contained in a box-like mattress frame; **box-square,** a metal-working tool used for marking parallel lines on round shafts; **box-stair** (see quot.); **box-staircase** *U.S.*, a closed staircase; **box-staple,** the staple on a door-post into which the bolt of a lock is shot, when the staple is so shaped that it covers the end of the bolt; **box-stone** *Geol.*, a rounded piece of brown sandstone containing a fossil; **box-strap,** a flat bar bent at right angles to confine a square bolt or projection; **box-string,** a string-board of a staircase in which the ends of the steps are entirely boxed in, also called *close string*; **box-swivel,** a swivel designed to prevent a fishing-line from tangling; **box-tail,** a box-shaped stabilizer of a biplane; **box-tappet,** a cam for working the shuttle-boxes of a loom; **box-tenon,** a tenon at an angle; **box-toe,** in boots and shoes, a toe with a stiff, strong lining; **box-tool(s,** an attachment to a lathe consisting of tools secured in a box-shaped holder (Lockwood); **box-top,** the top of a box; *spec.* a voucher attached to the packaging of groceries, etc., which offers a free gift or comprises part of a special offer; **box-trap,** a trap, shaped like a box, used for capturing animals; **box-tricycle,** a tricycle with a box in which articles can be carried; **box-valve,** a short rectangular section of a pipe, containing a valve; **box-van,** a van with a flat roof; **box-wagon,** (*a*) *U.S.* = *box-car*; (*b*) an open wagon with a box-shaped body; **box-wallah,** (*b*) a shop-keeper, retailer, or business-man; **box-wrench** = *box-spanner*.

1929 *Jrnl. Iron & Steel Inst.* CXX. 483 Normalised sheets command a much higher price in the United States than material which has been box-annealed. **1928** *Ibid.* CXVIII. 356 The American Rolling Mill Co... has built a series of twelve box-annealing furnaces. **1930** *Engineering* 2 May 583/1 (*heading*) Normalising and box-annealing of sheets. **1946** R. Blesh *Shining Trumpets* (1949) viii. 196 In their tight 'box-back' coats, the players blasted away. **1918** in F. A. Pottle *Stretchers* (1929) 116 Germans put over a box barrage entirely around the wood, hemming our boys in. **1941** *Hutchinson's Pict. Hist. War* 19 Mar.–13 May 16 Our airmen..had to make their way through an intense box barrage. **1941** *Illustr. London News* CXCVIII. 207 (*caption*) British artillery, responsible for the most severe 'box barrage' of the campaign, pounding away at Tobruk just before our troops entered. **1841** *Amer. Almanac for 1842* 116/2 Paper. Sheathing, binders, wrapping, and box boards. **1910** R. W. Sindall *Paper Technology* (ed. 2) 240 *Box boards*, used for box-making of all kinds, and manufactured from mechanical wood pulp, old waste papers, hemp, &c. **1932** F. Lloyd Wright *Autobiogr.* (1945) IV. 273 The box-board cabins..are to be connected by a low staggered box-board wall. **1842** Francis *Dict. Arts* s.v. *Camera Obscura*, The box camera is constructed as follows: procure a box, about 14 inches long E, having another box sliding in it F; the inner box having but one end, and in the centre of that end a double convex lens. **1902** *Encycl. Brit.* XXXI. 690/1 Single-magazine Box Camera. **1873** J. G. Bourke *Jrnl.* 21 Mar. (D.A.E.), We descended into a box cañon and made camp. **1947** *Canadian Alpine Jrnl.* June 91 We were unable to follow this route (with a pack-horse) beyond a box canyon. **1856** *Mich. State Agric. Soc. Trans. 1855* VII. 334 There are on the road..11 four-wheeled box cars. **1898** *Engineering Mag.* XVI. 69 The Illinois Central equipment was of the standard box-car type. **1890** *Harper's Mag.* Mar. 569/2 Jim..returned with the box-cart and horse. **1844** H. Stephens *Bk. Farm* II. 341 For a box-churn, whether horizontal or vertical, the plunger should make about 60 revolutions per minute. **1853** *Jrnl. R. Agric. Soc.* XIV. I. 70 In the box-churn the whey often escapes through the spindle-hole. [**1748** Lady Sherard *Let.* 17 Oct. in M. M. Verney *Verney Lett.* (1930) II. xxxiii. 244 He had two box clothes and no harness.] **1890** *Peel City Guardian* 29 Mar. 3/5 We saw it in emerald green velvet to stone box cloth. **1898** H. Graves et al. *Cycling* 20 We are not aware of the existence of all-wool box-cloth. *a* **1884** Knight *Dict. Mech.* Suppl., *Box Coil*, a steam or hot-water coil of many members, occupying a cubical space comparable in its proportions to a box. **1903** L. A. Flemming *Pract.*

Tanning 80 Sumac-tanned skins..are usually box-colored, that is, dyed in trays or dye boxes. **1901** *Daily Chron.* 10 Sept. 10/7 Box cutter..wanted, for box factory at Watford. **1905** *Ibid.* 21 Nov. 5/6 Two 'paltry pieces' of silk and a box-desk were among the gifts. **1920** *Glasgow Herald* 18 Sept. 8 An agreement under which fitters, pattern-filers, and box-fitters in the Glasgow area are to receive an advance of 5s. 6d. per week. **1886** C. Scott *Sheep-farming* 129 If it be intended to use box food, the sheep require to be trained to eat from the boxes when hoggs. *a* **1877** Knight *Dict. Mech.*, *Box-frame*, a casing behind the window-jamb for counter-balance-weights. **1931** Roget *Dict. Electr. Terms* (ed. 2) 34/2 *Box-frame motor*, an ironclad motor without a split frame, as in some traction motors. **1948** *Archit. Rev.* CIV. 187 Concrete box-frame construction. **1865** Brande & Cox *Dict. Sci.*, etc., *Box Girder*, a form of girder resembling a box, made out of boiler plate, fastened together by means of angle irons, which are riveted respectively to the top and bottom plates. **1874** Thearle *Naval Archit.* 73 The complex and varied systems of box-girder keels and keelsons. **1914** H. R. Procter *Making of Leather* 133 If the drawing down is repeated across the skin, a square or 'box' grain is formed. **1876** *Encycl. Brit.* IV. 503/1 Parallel or box-gutters are necessary next parapets where a curb roof is formed. **1954** *Highway Engin. Terms* (B.S.I.) 23 *Box gutter*, a drainage channel or gutter of rectangular cross section. **1796** *Sporting Mag.* IX. 126/2 The profits of this trade were immense. There was what they called the box-hands, that is, if a person threw three times together successfully, he gave one silver piece to the use of the house. **1921** *Dict. Occup. Terms* (1927) § 449 *Box hand, box man*,..fills monorail bogies with *masse cuite*. *Ibid.* § 522 *Box hand, fudge box hand*, a compositor who sets up type, for 'latest news' or 'stop press', which is put into a box and inserted in stereo cylinder. *Ibid.* § 554 *Box hand, box maker*,..turns up, to form box side, pieces of cardboard already cut to shape and creased. **1886** F. T. Elworthy *Dial. W. Somerset* 86 He had on a box-hat too! **1919** J. C. Snaith *Love Lane* xvi. 73 Broadcloth trousers allied to a prehistoric box-hat. **1909** Jacobus et al. *Standard Bible Dict.* p. xxii, The larger articles [in the Standard Bible Dictionary] will be found to be divided into sections by box-heads... This is done to facilitate easy cross-reference. *a* **1884** Knight *Dict. Mech.* Suppl., *Box Hook*. 1. A hook used in handling boxes; somewhat like a cotton-hook, which see. 2. A hook made on the plan of a cant-hook; used in closing boxes packed full of fish. 3. Hooks used in pairs in swinging boxes from a lifting-tackle. **1881** *Rep. Indian Affairs* (U.S.) 83 Some box-houses constructed for the purpose. **1966** *Guardian* 30 Dec. 14/5 So successful have London's yellow box junctions been in easing the traffic flow, that the Greater London Council is proposing 33 more. ..The Ministry of Transport first introduced the boxes with their warning 'Do not enter the box until your exit is clear' three years ago. *a* **1877** Knight *Dict. Mech.*, *Box-key*, an upright key, used for turning the nuts of large bolts, or where the common spanner cannot be applied. **1898** *Science Siftings* 17 Dec. 137/1 A few of the principal shops are now offering box kites. **1908** *Westm. Gaz.* 11 May 4/1 The double box-kite aeroplane with which Mr. Farman won the Archdeacon Prize in Paris recently. **1918** H. Barber *Aeroplane Speaks* Pl. vi, In 1909 came the semi-Wright biplane... Then the first box-kite flown by Mr. Grace at Wolverhampton. *Ibid.* Pl. xii, The Curtiss bi-planes..the box-kite type, 1909, on which Mr. Curtiss won the Gordon-Bennett Race at Reims. **1928** C. F. S. Gamble *North Sea Air Station* xx. 351 By means of a 5-foot linen box-kite and spare aerial with which I had been experimenting at Yarmouth we were able to get into and keep into [*sic*] constant wireless communication with Yarmouth. **1832** *U.S. Postal Regul.* 43 Box-letters. **1909** *Century Dict.* Suppl., Box-loom. **1954** J. L. Morse *Unicorn Bk. 1953* 179 It went into such details as allowing troops twenty minutes for their box lunches. **1960** G. Sanders *Mem. Professional Cad* II. vii. 157 The crowd outside the Chapel of the Psalms had brought box lunches, bawling babies, and hula hoops. **1873** *Contemp. Rev.* XXI. 741 The treasurer of an ordinary Friendly Society, in Scotland, is sometimes its 'box-master'. **1885** D. Beveridge *Culross* II. 155 The privilege of having a deacon and box master of their own. **1928** *Daily Mail* 7 Aug. 1/2 (Advt.), You can enjoy *all* the luxury and comfort of the most expensive Box Mattress. **1906** *Daily Chron.* 7 June 5/4 The import of 'box' meat. **1894** T. W. Fox *Mech. Weaving* 393 The ever-increasing number of different box motions. **1923** D. L. Sayers *Whose Body?* iv. 67 Here's our advertisement. ..Perhaps it would have been safer to put a box number. **1888** *Lockwood's Dict. Mech. Engin.*, Box Nut. **1912** 'K. Mansfield' *Scrapbk.* (1939) 7 Box-ottomans and beds. **1811** C. Mathews *Let.* 1 July in Mrs. Mathews *Mem. C. M.* (1838) II. vi. 126, I have an immense box-plan already; and I expect a good house. **1912** A. Bennett *Matador of 5 Towns* 246 The box-plan could be consulted at the principal stationers. The Alexandra Hall contained no boxes whatever, but 'box-plan' was the phrase sacred to the occasion. **1841** *Congress. Globe* 20 Feb., App. 343/2 [I proposed] to cure the abuse growing out of box-rents. **1881** *Congress. Rec.* Mar. 2283/1 Postal funds are such funds as arise from box rents and from the sale of postage-stamps. **1927** Conan Doyle *Case-Bk. S. Holmes* ix. 248 My mind is like a crowded box-room with packets of all sorts stowed away therein. **1838** *Osborne's Guide to Grand Junction Railway* 107, I will suppose you mounted on the box seat. **1849** De Quincey in *Blackw. Mag.* Oct. 488/2 The public took to bribing, giving fees to horse-keepers, &c., who hired out their persons as warming-pans on the box-seat. **1853** 'C. Bede' *Verdant Green* xii. 116 Mr. Verdant Green tipped for the box-seat. **1889** *Cent. Dict.*, Box-set. **1933** P. Godfrey *Back-Stage* xi. 143 Ceiling-pieces were introduced and box-sets for domestic interiors with real doors and windows. **1888** *Lockwood's Dict. Mech. Engin.*, Box Spanner. **1962** *Which?* (Suppl.) Apr. 67/1 The Austin A60 had a grease gun, nave plate remover and box spanner. **1895** *Montgomery Ward Catal.* 621/1 Wire Mattresses... Box Spring: has 72 spiral steel japanned springs covered with..tow. **1927** *Daily Tel.* 11 May 17/7 (Advt.), All the beds are fitted with boxsprings and hair overlays of the finest quality. **1902** P. Marshall *Metal Tools* 18 A box-square is an elongated form of square, the chief use of which is for marking parallel lines on round shafts or spindles. **1901** R. Sturgis *Dict. Arch.*, *Box Stair*, one made with two closed strings, so that it has a boxlike form of construction. **1907** M. H. Norris *Veil* i. 5 Returning to the hall he opened the door of a box stair-case, ascending unconsciously on tiptoe a broad flight of shallow stairs to an immense attic. *a* **1877** Knight *Dict. Mech.*, *Box-staple* (Carpentry), the box or keeper on a door-post, into which is shot the bolt of a lock. **1870** E. R. Lankester in *Q. Jrnl. Geol. Soc.* XXVI. 499, I have..spent a good deal of time in working at the nodules, which I propose to call 'Box-stones', since the name of 'boxes' has been applied to those which exhibit the remains of a shell on being broken open by the phosphate-diggers of Suffolk. **1893** Geikie *Geol.* (ed. 3) 1009 Rounded pieces of brown sandstone, known as 'box-stones', evidently derived from the denudation of a single horizon, and enclosing casts of marine shells. *a* **1877** Knight *Dict. Mech.*, *Box-strap*, a flat bar, bent at the middle, to confine a square bolt or similar object. **1847** Stoddart *Angler's Comp.* 138 The box-swivel is a very necessary part of the minnow-tackle. **1909** *Westm. Gaz.* 23 Oct. 9/1 His Voisin biplane,..with its box-tail wagging high in the air. **1894** T. W. Fox *Mech. Weaving* 287 A lateral motion..is given to the short shafts by box tappets and lever connections. **1909** Webster, Box toe. **1913** W. H. Dooley *Man. Shoemaking* 181 Box Toe, used to hold up the toe of the shoe so as to retain the shape. It is generally of sole leather, but often made of canvas or other material and stiffened with shellac or gum. **1937** *Amer. Speech* XII. 101 *Hook* is that part of the commercial which urges you to send in the box tops. **1967** I. Hamilton *Man with Brown Paper Face* ii. 27 He used to send in all the box-tops for the do-it-yourself handcuff kit. **1770** G. Cartwright *Diary* 26 Sept. (1792) I. 39, I made a box-trap for martens, and set it on the opposite side of the river. **1785** T. B. Hazard *Jrnl.* 2 Apr. (1930) 78/2 Made door for Box Trap for Biger Babcock Junr. **1936** *Discovery* July 227/2 Shrews..escaped between the wires of box-traps. **1896** *Daily News* 8 Oct. 9/2 The father of the deceased said his son was in the habit of riding a box tricycle. *a* **1884** Knight *Dict. Mech.* Suppl., *Box Valve*, a box section in a pipe containing a valve, and having a cover for access. **1908** *Westm. Gaz.* 30 Mar. 5/2 A light box-van. **1847** H. Bushnell *Address Hartford Co. Agric. Soc.* 14 To..live on the coarsest fare, to ride in a box waggon or cart. **1874** *Congress. Rec.* Apr. 3377/2 A small..covered carriage, sufficient to enable a Bureau officer to come to the Capitol.. upon a rainy day like this, and not be soaked in a 'box-wagon'. **1886** *Encycl. Brit.* XX. 247/1 Open or box waggon. **1889** Kipling *From Sea to Sea* (1899) I. i. 7 This Young Man must have been a delight to the Delhi boxwallahs. **1956** W. Slim *Defeat into Victory* 133 It was the 'Box Wallahs', the commercial community, who in those hot, anxious months..turned Eastern India into a base and workshop.. for Burma.

box, *v.*[1] Add: **3. c.** *to box in* or *up*: to enclose in a box or casing.

1864 *Jrnl. R. Agric. Soc.* XXV. II. 354 Provision should be made, when practicable, to fence in or 'box up' the moving parts... It would be very easy to box up the gearing of a fixed thrasher... It would also be impossible to completely box-in a chaff-cutter. **1919** *Autocar Handbk.* (ed. 9) 129 In other cars the motor is boxed in nearly air-tight by a shield underneath and a closed bonnet above.

d. *Printing.* To enclose within rules; to print with a border. Also with *in*.

1904 Goodchild & Tweney *Technol. & Sci. Dict.* 65/1 *Box in*, a term indicating that rules should be placed round as a border. **1924** W. M. Raine *Troubled Waters* xxii. 229 A leaded advertisement..boxed to draw more attention.

5. b. To mix up or allow to be mixed up (different flocks of sheep). Also const. *up*, and *absol.* *Austral.* and *N.Z.*

1864 *Puketoi Diary* 19 Apr. (MS.), Lambs boxed. **1871** Lady Barker *Christmas Cake* IV. iii. 278 Tom's sheep were always coming to grief..and never a week passed without his getting boxed. That's mixed-up, ma'am. **1881** A. C. Grant *Bush Life in Queensland* I. 253 All the mobs of different aged lambs which had been hitherto kept apart were boxed up together. **1888** 'R. Boldrewood' *Robbery under Arms* III. xii. 168 After they'd got out 20 or 30 they'd get boxed, like a new hand counting sheep, and have to begin all over again. *a* **1889** in Barrère & Leland *Dict. Slang* s.v., Now, mind yourselves, for if you box, You'll play the mischief with the flocks. **1959** H. P. Tritton *Time means Tucker* 16/2 In the event of two mobs getting 'boxed' they would have to be taken to the nearest yard and drafted out.

box, *v.*[2] Add: **2. c.** *to box on*: to continue boxing or fighting; also *fig.* *colloq.*

1919 Downing *Digger Dial.* 13 *Box on*, (1) continue; (2) fight. **1954** F. C. Avis *Boxing Ref. Dict.* 14 *Box*, an order of the referee to the contestants to carry on with the contest; also..Box On. **1959** 'D. Buckingham' *Wind Tunnel* vii. 61, I would face her with Paddy's story and then 'box-on' from there. **1965** *New Statesman* 7 May 725/1 However clever and facile I was, I lost friends and failed to influence people. But I boxed on.

d. *to box clever*: to behave cleverly; to use one's wits. *slang.*

1936 J. Curtis *Gilt Kid* vi. 62 He knew, however, that he would have to box clever. **1950** A. Baron *There's no Home* 210 If you box clever and keep your mouth shut,.. you ought to be able to count on a suspended sentence.

3. b. colloq. or dial. phr. *to box Harry*: to go without a meal; to have a poor meal so as to save expense.

1823 'Jon Bee' *Slang* 16 *Box-harry*, to go without victuals. Confined truants, at school, without fire, fought or boxed an old figure nicknamed 'Harry', which hung up in their prison—to keep heat. **1862** G. Borrow *Wild Wales* II. i. 3 Those [commercial travellers] whose employers were in a small way of business, or allowed them insufficient salaries, frequently used to 'box Harry', that is have a beef-steak, or mutton-chop, or perhaps bacon and eggs..instead of the regular dinner of a commercial gentleman. *Ibid.* Having made arrangements for 'boxing Harry' I went into the tap room. **1902** *N. & Q.* 7 June 450/1 An old woman..was telling me that she had only by her a very poor supply of seed [potatoes], and finished up by ejaculating, 'Never mind, I must box Harry..' When questioned..she said.. she must needs do without. *Ibid.* 5 July 13/2 '*Box Harry*'. This is well known in the Northern Counties in the sense of doing things 'on the cheap'... In Mr. Page's instance the woman seems to have meant she must beg some potatoes to make up the deficiency.

Box and Cox. The name of a farce written by J. M. Morton (1811–91) in 1847, in which two characters, John Box and James Cox, occupy the same apartment (the one by day and the other by night); hence applied allusively to an arrangement in which two persons take turns in sustaining a part, occupying a position, or the like.

1881 *Punch* 3 Sept. 100/1 This sort of Box and Cox life in a cabin. **1927** C. E. Raven *Creator Spirit* iii. 78 Representing mind and body as playing a perpetual game of Box and Cox. **1959** *Spectator* 14 Aug. 180/2 The French Community.. shares, Box-and-Coxwise, the Luxembourg Palace with the French Senate.

box-calf. [Named about 1890 by Edward L. White, of White Bros. & Co., Boston, Mass., U.S.A., after Joseph *Box*, bootmaker, of London. (The picture of a calf in a box was adopted as an advertising device.)] A calf-skin tanned with chrome salts and having a grain of rectangularly crossed lines. Cf. *box-grain*.

[**1899** *Moniteur de la Cordonnerie* 437 (Bonnaffé *Dict.*).] **1904** P. N. Hasluck *Leather Working* 15 Box Calf.—The grain side is the face of this leather. It is somewhat like firm ooze calf, only black. **1906** A. Watt *Leather Manuf.* (ed. 5) 356 In box-calf it is usual to grain two ways only, once from the head to the tail, and then from belly to belly. **1908** Trotman *Leather Trades Chem.* 143 Ordinary chromed box-calf shavings containing 5·7 per cent. of chromic oxide.

boxed, *ppl. a.* Add: (Further examples.) Also with *in*.

1865 J. D. Burn *Three Years among Working-classes in U.S.* 254 It is often dangerous for a person with boxed-up notions to try the experiment of letting them loose in company. **1900** *Times* 15 Mar. 8/1 The boxed machinery. **1905** *Westm. Gaz.* 29 Aug. 10/1 The reception and despatch of the boxed fruit. **1939** N. Coward *Play Parade* II. p. x, *Boxed in.* Part of Stage surrounded by a three or fourfold Scene set down to the Proscenium or False Proscenium. **1960** *Times* 16 Sept. 16/7 The acoustics of the theatre.. giving a boxed-in sound to the score.

boxer[2]. Add: **2.** (With capital initial.) A member of a Chinese secret society organized in the late 19th century, whose primary aim was to save the Chinese empire (and was consequently anti-dynastic); the movement inspired the attack on foreigners in Pekin in 1900. Also *attrib.*

The name represents Chinese *i ho chuan* or *chuen*, lit. righteous harmony boxers (fists).

1900 *Westm. Gaz.* 26 Apr. 8/4 A number of Boxers attacked a village where some Chinese Catholics live. *Ibid.* 25 May 5/2 The 'Boxer' menace in China continues to excite anxiety. *Ibid.* 7 June 5/7 A Boxer force was for several weeks drilling..within sight of the foreign concession. **1902** *Encycl. Brit.* XXIX. 228/2 The faith which he [*sc.* Sir Robert Hart] put in the Chinese made him turn a deaf ear to the warnings of the threatening Boxer movement in 1900. **1933** *Granta* 19 Apr. 358/2 No one could have believed that the brutal sadism to which for a moment it gave rein had survived the Boxer rising. **1959** *Chambers's Encycl.* II. 477/2 The Boxers indulged in a cult of magic.

Hence **Boxerism** (bǫ·ksəriz'm), the conduct and practices of the Boxers in China.

1901 *Westm. Gaz.* 11 Apr. 1/2 It is scarcely surprising.. that Boxerism is still lurking beneath the surface. **1907** *Daily Chron.* 28 Sept. 1/7 An outbreak of Boxerism has taken place in China.

boxer[3]. [f. Box *sb.*[1] + -er[1].] A boxwood peg-top.

1840 *Peter Parley's Ann.* 85 I'll give you my two pegs for your boxer. **1853** *N. & Q.* 1st Ser. VIII. 63/1 Schoolboys call tops made of boxwood, *boxers*. **1881** F. Y. *Ev. Man his own Mech.* II. ii. 267 'Boxers', as they [*sc.* spinning tops] are called from the material of which they are made; are considered to be the best.

bo·xer[4]. [? f. Box *sb.*[2] + -er[1].] **1.** *dial.* A tall hat.

1877 *Legends in Dial. of Gloucestershire* 51 Twur genelmen wi' boxers on. **1884** H. D. Rawnsley in *Trans. Wordsworth Soc.* VI. 163 Wearing a big wideawake, or a bit of an old boxer.

2. A bowler or billycock hat. *Austral.*

1897 *Argus* 9 Jan. 14/2 (Morris), And will you wear a boxer that is in a battered state? **1904** *Westm. Gaz.* 9 June 12/1 In Melbourne..a..man in the ordinary Sunday suit of a worker, a boxer hat, and a blue shirt with a white collar.

boxer[5]. [G., a. Eng. Boxer[2].] A smooth-coated, square-built, fawn or brindle breed of dog of the bulldog type, originating in Germany.

1934 *Hutchinson's Dog Encycl.* I. 184 Boxer. This dog, which appears to derive from the 'Dogues' that had been made use of, during past centuries, for animal-baiting, is very little known in Britain. **1952** Koestler *Arrow in Blue* 200 My dog Jessy—the first of a long series of Alsatians, Welsh sheep dogs, Boxers and St. Bernards. **1958** F. King *Man on Rock* vi. 170 They cost you a fortune, big dogs. A pal of mine has a boxer, a German boxer, that's what they call it.

boxing, *vbl. sb.*[1] Add: **I.** Various technical uses: see quots.

a **1877** Knight *Dict. Mech.*, *Boxing*,..the fitting of the shoulder of a tenon in the surface of the timber, which is mortised for the reception of the tenon. *a* **1884** *Ibid.* Suppl., *Boxing Machine*, a machine for boring out the boxes of hubs. **1900** *Coal & Metal Miners' Pocketbk.* (ed. 6) 574 Boxing, a method of securing shafts solely by slabs and wooden pegs. **1903** *Daily Chron.* 10 Mar. 10/5 Wheeler..one used to rivetting and boxing. **1913** W. H. Dooley *Man. Shoemaking* 180 *Boxing*, a term used to designate the stiffening material placed in the toe of a shoe to support it and retain the shape; such as leather, composition of leather and paper, wire net..etc.

4. b. A wooden casing, conduit, etc., constructed after the manner of a box; the lining of a well.

1867 *Trans. Ill. Agric. Soc. 1865–6* VI. 319 A water pipe or boxing eighteen inches in the clear is laid under ground to the centre of the yard. **1907** M. H. Norris *Veil* v. 44 Its [*sc.* a well's] slimy, moss-covered stone boxing. **1950** *N.Z. Jrnl. Agric.* May 483/2 Material should be rammed well as the walls are being built... The boxing may be removed the day following placing.

7. *Austral.* and *N.Z.* (See quots. and *box *v.*[1] 5 b.)

1871 Lady Barker *Christmas Cake* IV. iii. 278 We calls it boxing when your sheep go and join another mob feeding close by, and you can't tell one from another except by the brand or the ear-mark. **1959** H. P. Tritton *Time means Tucker* 16/2 Getting his sheep mixed-up with the travelling mob..was known as 'boxing-up'.

boxing-day, boxing-night. (Earlier examples.)

1833 in A. Mathews *Mem. C. Mathews* (1839) IV. viii. 173 To the completion of his dismay, he arrives in London on *boxing*-day. **1837** Dickens *Pickw.* xxxii. 343 No man ever talked in poetry 'cept a beadle on boxin' day. **1837** —— in *Bentley's Misc.* Mar. 296 The most turbulent sixpenny gallery that ever yelled through a boxing-night.

box office, box-office. [Box *sb.*[2] 8 a.]

1. The office at which seats may be booked for a theatrical performance or other entertainment (orig. for the hiring of a box). Also *attrib.*

1786 Betsy Sheridan *Jrnl.* (1960) 77 After we had nearly reach'd the Box Office a cry of Pick-pocket raised a general confusion. **1812** [see Box *sb.*[2] 23]. **1812** J. & H. Smith *Rej. Addresses* 74 Close to Mr. Spring's Box office door, I've stood and eye'd the builders. **1929** J. B. Priestley *Good Compan.* II. iv. 327 The box office was at the entrance to the Pier.

2. The financial element of such a performance; *ellipt.* or *attrib.* (and *transf.*), success(ful) in attracting or appealing to an audience; = Draw *sb.* 3; by extension, the paying audience.

1904 G. B. Shaw *Let.* 6 Dec. (1956) 44, I dont want to sacrifice my aim to the box office. **1907** *Daily Chron.* 16 May 3/5 The box-office value of every artist of prominence. **1923** G. B. Shaw in *Shaw on Theatre* (1958) 163 All that Mr Bridges Adams can do is to pretend that the front of the existing stage is a forestage, and make the box office pretend. **1926** Galsworthy *Silver Spoon* II. viii. 183 'Don't know why he keeps on an amateur like that.' 'Box office, dear boy; she brings the smart people.' **1932** A. Huxley *T. H. Huxley* in *Huxley Mem. Lectures 1925–32* 17 This kind of popular science is thoroughly popular in the other, the box-office sense of the word. **1935** Wodehouse *Luck of Bodkins* xxiv. 304 Exactly the touch the treatment needed to make it box-office. **1936** *Amer. Speech* XI. 221 Our play may have turned out to be good ticket or good b[ox] o[ffice]. **1960** *Observer* 11 Dec. 40/7 Mr. Khrushchev's box office performance at the United Nations. **1962** *Listener* 8 Mar. 446/2 His canny equation of box-office and genuine artistic ambition.

box-tree. Add: **b.** Applied to many Australasian species of *Eucalyptus*. Cf. *box *sb.*[1] 3 c.

1827 Hellyer in Bischoff *Van Diemen's Land* (1832) 168, I found the stump of a withered box tree. **1849** Sturt *Narr. Exped. Central Australia* I. 116 The flats of the Darling..are..lightly wooded, having low and useless box-trees (the *Gobero* of Sir Thomas Mitchell), growing on them. *a* **1882** Kendall *Poems* (1886) 19 Keen, fitful gusts,.. Fleet down by whistling box-tree butt. **1940** A. W. Upfield *Bushranger of Skies* xx. 193 I've put in half a handful of box-tree seeds. They'll act like a double dose of painkiller.

boxty (bǫ·ksti). *Irish.* (See quot. 1880.)

1880 W. H. Patterson *Gloss. Antrim & Down* 11 *Boxty*, or *Boxty-bread*, a kind of bread made of grated raw potatoes and flour; it differs from 'potato bread', or 'potato cake', of which cold boiled potatoes form the principal part. **1949** 'M. Innes' *Journeying Boy* xiii. 162 Gracie's boxty is very eatable, very eatable indeed. **1969** *Observer* 12 Jan. 33 Boxty Bread—otherwise Boxty-in-the-Pan, Boxty Pancakes and Boxty-on-the-griddle..are all made with potatoes. *Boxty* is eaten on Hallow-e'en, mostly in the north.

boxy (bǫ·ksi), *a.*[2] [f. Box *sb.*[2] + -y[1].]

1. Resembling a box in shape; comparable to a box.

a **1861** T. Winthrop *John Brent* (1883) xxvi. 222 A frowzy county town, with a dusty public square, a boxy church, and a spittley court-house. **1883** W. Harcourt in A. G. Gardiner *Life* (1923) I. 492 A tall boxy house will not look well there. **1928** *Observer* 26 Feb. 20/2 A very covered and boxy effect. **1959** *Landfall* XIII. 122 Her nose was flat, with square boxy nostrils.

b. *spec.* Of clothes: having a squared appearance.

1936 *Times* 10 June 19/6 It may be of pique or linen in a short boxy swagger style. **1960** *Woman* 23 Apr. 15/1 A good, firm binding that gives a tailored look to a boxy jacket.

2. Of the feet of a horse or mule: high and narrow.

1908 *Animal Managem.* 224 'Mule', 'boxy', 'narrow', or 'club' feet are the opposite of 'flat'. *Ibid.* 271 The feet [of the mule] are narrow and boxy, as compared with the horse.

3. Of recorded or reproduced sound: lacking a full range of tone; sounding as if enclosed in a box.

1941 *Time* 14 Apr. 55/1 The recording—taken from a radio performance—sounds boxy. **1952** *Times Lit. Suppl.* 12 Dec. 812/2 The recording is unthinkably boxy. **1962** A. Nisbett *Technique Sound Studio* ii. 50 The results seem to sound more boxy than they should.

boy, *sb.*[1] Add: **2. b.** (Further examples of use in Ireland.) Cf. *b'hoy.

1847 *Paddiana* I. 263 Judge Moore having decided in my hearing, that in Ireland the word 'Boy' has no reference to age. **1908** *Westm. Gaz.* 16 Oct. 11/2 In Ireland anyone who is not married is called a boy... John Gillan, the 'boy', a sturdy young man, then gave evidence.

c. In expressions of encouragement or admiration, etc., esp. *that's the boy!* (see *attaboy).

[**1843** T. C. Haliburton *Attaché* xxv. 61/2 It's a great advantage havin' the minister with you. He'll fell the big stiff trees for you; and I'm the boy for the saplin's.] **1902** J. J. Bell *Wee Macgreegor* ii. 13 'If a beast wis gaun fur to pu' ma heid aff,' remarked Macgregor, who had grown suddenly bold, 'I-I-I wud—I wud gi'e't a kick!' 'Ye're the boy!' said his father. **1932** E. Wallace *When Gangs came to London* ii. 26 'Ain't you the boy!' he said. **1936** J. Curtis *Gilt Kid* xvi. 166 'Got it?' echoed Scaley. 'That's the boy.' **1962** H. Hood in R. Weaver *Canad. Short Stories* (1968) 2nd Ser. 205 John's the boy. Oh, he's a sharp lad is John.

3. e. In S. Africa, a Coloured or Native labourer or servant of any age. So *boss boy*, a Coloured or Native overseer; *Cape boy*, a Cape Coloured man or boy.

1812 A. Plumptre tr. *Lichtenstein's Trav. S. Afr.* I. i. viii. 119 A Hottentot..expects to be called by his name if addressed by any one who knows it; and by those to whom it is not known he expects to be called Hottentot..or boy. **1833** J. Philip in *Lett. Amer. Missionaries* (1950) 40 An example of a native boy at Pacaltsdorp conducting a small school. *Ibid.* 42 A school on the British system taught by a Hottentot boy. **1896** *Spectator* 2 May 629 A Cape 'boy' fighting at Bulawayo is..a coloured native enlisted and drilled within the Colony. **1906** *Daily Chron.* 11 Apr. 3/6 One white man in the mine is expected to 'boss' forty blacks or Chinese, which he cannot do with safety, in fact the black 'boss-boy' is left to do much of the blasting. **1963** *Times* 2 May 12/7 The Johannesburg city council, which employs thousands of Africans in jobs ranging from welfare officers to dustmen, has banned the use of the word 'boy' to describe any of them.

f. (*the*) *boy*: champagne. *slang.*

The derivation is uncertain: see Barrère & Leland *Dict. Slang*, Farmer *Slang*.

1882 *Punch* 11 Feb. 69/1 He'll nothing drink but 'B. & S.' and big magnums of 'the Boy'. **1882** O. Wilde *Let.* 12 May (1962) 117 You and I will sit and drink 'Boy' in our room. **1929** *Melody Maker* Jan. 20/2 Lord Delamere came up to them with a foaming magnum of champagne and said, 'Well, boys! you've given us a glorious time! What do you say to a beaker of "the boy"?'

g. From 1918 to 1924, the lowest rank in the Royal Air Force.

1918 *Army Orders* Nov. 35 The following additional column will be *inserted* in the table of corresponding ranks.. Royal Air Force Air Mechanic, 3rd Class. Private, 2nd Class. Clerk, 3rd Class. Boy. **1924** *R.A.F. King's Regulations* I. p. vii, Airman, or Airmen. These words, wherever they occur, will be held to include a warrant officer, a N.C.O., an aircraftman, and a boy.

5. *old boy* (see also Old *a.* 8 a).

6. *the old boy* (*U.S.*) (earlier and later examples).

1802 *Balance* (Hudson, N.Y.) 14 Oct. 317 (Th.), The devil has been nick-named the old boy. **1854** M. J. Holmes *Tempest & Sunshine* xv. 203 Where the old boy is your mistress? **1953** A. Miller *Crucible* (1956) I. 32 The Church's enemies relied no less upon the Old Boy to keep the human mind enthralled.

b. A rough or rowdy (e.g. of the streets); freq. *pl.* in *the boys* = criminals; *spec.* the thieves and swindlers who frequent race-courses.

1834 *Knickerbocker* III. 34 The landlord after telling me not to mind the *boys*, went about his business. **1840** *Daily Picayune* (New Orleans) 28 Aug. 2/1 The Bowery boys of New York have..eclipsed the nice young men of Baltimore. **1843** *Punch* 29 Apr. 179/2 The comments and cheers of those very important members of street society, the boys. *a* **1889** in Barrère & Leland *Dict. Slang*, Cleansing the rings from..those criminal scoundrels known as the boys. **1938** F. D. Sharpe *S. of Flying Squad* i. 13 Down goes the Squad the night before to greet 'the boys' at the turn-stiles. *Ibid.* v. 64 Word went round among the 'boys' that the takings had been hefty.

c. *pl.* Men of the armed forces; soldiers. So *boys in blue*: see *blue *sb.* 3 b.

1861 O. W. Norton *Army Lett.* (1903) 19 It was then announced that the whole thing was a canard, started just to keep the boys quiet. **1881** F. E. Weatherly *Old Brigade* 2 Where are the boys of the Old Brigade? *c* **1915** L. G. Ford (*song*) *Keep the Home Fires Burning*, Turn the dark cloud inside out, Till the boys come home. **1959** M. Shadbolt *New Zealanders* 36 The time when we would watch the boys swing, bayonets flashing,..towards the grey waiting ships.

d. *pl.* Members of a group sharing common interests; one's fellows or habitual companions; esp. in colloq. phr. *one of the boys*: one who belongs to such a group; *spec.* one who conforms to its interests or practices, 'a good sport'. *colloq.* So *jobs for the boys*, appointments for one's supporters or favourites.

1886 *Lantern* (New Orleans) 8 Sept. 3/1 When he happens in with the boys, he can enjoy himself. **1889** W. Skey *Pirate Chief* 195 He goes on Sundays to the 'pub.' And sits among 'the boys'. **1893** *Ladies' Home Jrnl.* Nov. 20/3 She doesn't want to be treated like a lady because she wants to be 'one of the boys'. **1905** M. Beerbohm in *Sat. Rev.* 11 Nov. 620/1 On him, somehow, the blight of the theatre has not fallen... He is not, and may never become, 'one o' the boys'. **1930** Wodehouse *Very Good, Jeeves!* vii. 192 A chummy lion-tamer—a tamer who, after tucking the lions in for the night, relaxes in the society of the boys. **1939** I. Baird *Waste Heritage* v. 63 He..stopped to watch a half-dozen of the boys playing blackjack. **1950** D. & C. Christie *His Excellency* I. i, It's just a political racket—Jobs for the Boys. **1955** M. Gilbert *Sky High* i. 19 'It wasn't exactly a popular appointment, was it?' 'It certainly wasn't,' said the General..'Jobs for the Boys.' **1969** *New Yorker* 3 May 64/3 He doesn't do it by being one of the boys. That's not his nature. He's a lone wolf.

e. With defining word or phrase prefixed = men (of the kind indicated by the defining element). (Cf. *backroom boy.) *colloq.*

1941 F. & R. Lockridge *Murder out of Turn* iv. 39 The B.C.I. boys will be along. **1945** A. Huxley *Time must have Stop* xxx. 288 Just a little bit of Wordsworth, say the blue-dome-of-nature boys. *Ibid.* 291 Adler and Freud, the Dialectical Materialism boys and the Behaviourists. **1958** *Spectator* 7 Feb. 167/2 The public relations boys could really go to town. **1963** R. Parker tr. *Solzhenitsyn's One Day in Life of Denisovich* 36 Oh no, he wasn't ill, the security boys were keeping him back.

7. *boy-baby* (example); **boy-crazy** *a.*, (of a girl) eager to associate with boys; **boy-farm** *slang*, a school (cf. Farm *sb.*[2] 7); so *boy-farmer*; **boy friend, boy-friend** *colloq.*, a male friend; *spec.* a woman's favourite male escort or companion; also with implication of an illicit relationship, a paramour; occas., the associate of a homosexual.

1881 *Atlantic Monthly* Jan. 89/1 They are baby socks. They are always blue. And so they all have to be given to the boy babies. **1923** *Cosmopolitan* Sept. 72/2 Going boy-crazy at fifty. **1955** 'C. H. Rolph' *Women of Streets* 151 Ena became boy-crazy early. **1890** W. Morris *News from Nowhere* vi. 43, I had best say nothing about the boy-farms which I had been used to call schools. **1901** *Daily Chron.* 16 Sept. 2/6 The professional boy-farmers..are naturally trying to supply what is desired. **1896** E. Turner *Little Larrikin* iv. 41 He went on ahead with his boy-friend. **1909** G. W. E. Russell *Coll. & Recoll.* 330 The young ladies.. meet their boy-friends at all hours and places. **1929** Wodehouse *Summer Lightning* ii. 53 'Who is this Pilbeam?' he demanded. 'Pretty much the Boy Friend, I take it, what?' **1953** E. Taylor *Sleeping Beauty* ix. 156 Boy-friend of my mother's just come in. **1957** J. Braine *Room at Top* xxx. 242 Pansies only use pubs for picking up boy friends.

8. Phrases. *boy and girl* (also with hyphens), *attrib. phr.*, pertaining to or involving a boy and a girl; applied esp. to a juvenile attachment; *boy-meets-girl* (*colloq.* esp. *attrib.*), referring to a copybook romance; *boy next door* (*colloq.*), the young man in a conventional romance; a simple, unsophisticated young man; *boys will be boys*: an expression of resignation towards childish ways.

1841 Dickens *Barn. Rudge* xxvi. 84, I have found it necessary to take some active steps towards setting this boy and girl attachment quite at rest. **1870** L. M. Alcott *Old-Fashioned Girl* viii. 132 It's only a boy-and-girl fancy, that will soon die a natural death. **1934** P. Bottome *Private Worlds* v. 54 It was what people called 'a boy and girl affair', but they can go quite deep. **1945** 'A. Gilbert' *Don't Open Door!* ix. 80 The murder..had more of those features that make a crime fascinating to the general public. There was no boy-meets-girl element. **1947** *Landfall* I. 45 This immediately..reduces the story to the familiar Hollywood formula of boy-meets-girl. **1955** T. Sterling *Evil of Day* viii. 80 The boy-next-door parody was meant to amuse her. **1958** *Photoplay* Oct. 54/1 His whole build-up is based on The Boy Next Door—the boy who's within reach of every girl fan. **1848** Thackeray *Van. Fair* xiii. 112 And as for the pink bonnets..why boys will be boys. **1858** C. M. Yonge *Christmas Mummy* ix. 131 All he had ever known against Asaph Harper was keeping company with the like of them: but boys would be boys. **1964** Wodehouse *Frozen Assets* iii. 50, I tried to tell him that boys will be boys and you're only young once.

9. (*oh*) *boy!* a colloq. (orig. *U.S.*) exclamation of shock, surprise, excitement, etc.; freq. used to give emphasis to a statement that follows it. Cf. *attaboy.

1917 *Amer. Mag.* Mar. 13/1 'I told that dame I was Kid Hanlon.'..'Oh, boy!' I yells. **1927** *Punch* 7 Sept. 263 Oh, boy, I feel good! **1930** D. H. Lawrence *Nettles* 17 And they blushed, they giggled, they sniggered, they leered.. and said: Oh boy!..that's pretty hot! **1934** M. Hodge *Wind & Rain* I. i. 18 Boy! They don't wear a damned thing! **1942** L. D. Rich *We took to Woods* (1944) ii. 34 Maine guides have a legend of quaintness to uphold, and boy! do they uphold it. **1958** 'N. Shute' *Rainbow & Rose* i. 2, I slithered in over the fence and put her [*sc.* the aeroplane] down and boy! was I glad to be on the ground!

boyang, var. *bowyang.

boyey (boi·i), *a.* [f. Boy *sb.*[1] + -y[1].] Having the characteristic qualities of a boy.

1885 Mrs. Molesworth *Carrots* i, He was a very boy-ey boy, very much inclined to look down upon girls in general. **1902** *Harmsworth Lond. Mag.* June 494/2 A delightfully boyey boy.

boyla (boi·lā). [Native name.] An aboriginal Australian sorcerer.

1865 W. Howitt *Discov. Australia* I. 386 A superstitious notion that by eating them [*sc.* mussels] he should subject himself to the absolute power of boylas, or evil sorcerers. **1966** W. S. Ramson *Austral. Eng.* vi. 129 Several of these words, notably..*boyla* and *koradji*, 'an aboriginal medicine-man or witchdoctor', are used only in their original and specific senses.

boyo (boi·o). *colloq.* and *dial.* (chiefly *Anglo-Irish*). Also **boyoh.** [f. Boy *sb.*[1]] Boy, lad; esp. as a jovial form of address.

1870 G. M. Hopkins *Jrnl.* (1937) 133 Irish expressions.. *Boyo, Lado* = Boy and a half, [etc.]. **1898** J. Macmanus *Bend of Road* 258 While there's a whack in the belt, boyo, I'll lay-on ye. **1907** G. B. Shaw *John Bull's Other Island* 80 There was..me bould English boyoh in front at the machinery. **1914** Joyce *Dubliners* 201 If you want a thing well done..go to a Jesuit. They're the boyos have influence. **1938** J. Cary *Castle Corner* 46 The wee boyo grows like the Corners. **1943** P. Cheyney *You can always Duck* i. 11 There was nothin' too bad to say about that boyo. **1953** Dylan Thomas *Under Milk Wood* (1954) 30 Nogood Boyo. Me, Nogood Boyo, up to no good in the wash-house. **1955** K. Amis *That Uncertain Feeling* xvii. 235 Get along home, there's good boyos. **1960** G. Avery *Elephant War* iii. 49 Well, my boyo, you're in a hurry this fine morning.

boy scout: see Scout *sb.*[4] 2 c.

boysenberry (boi·zĕn-, boi·sĕnberi). [f. the name of Rudolph *Boysen*, 20th-cent. American horticulturist who developed it + Berry *sb.*[1]] The fruit of a bramble, a hybrid of several plants of the genus *Rubus*; also, the bramble that bears this fruit.

1935 *Market Growers Jrnl.* 1 Dec., The Boysenberry is a cross between Cuthbert raspberry, Loganberry and blackberry. **1941** *Reader's Digest* Oct. 89/1 Boysenberry pies baked by Mrs. Knott. **1945** *Greeley* (Colorado) *Daily Tribune* 15 Mar. 10/7 Eldorado [blackberry] is plenty hardy, but Boysenberries both thorny and thornless must be covered thru the winter. **1956** *Dict. Gardening* (R.H.S.) Suppl. 98/1 *Boysenberry*, July–Aug. Large, purplish-black when ripe. Juicy but acid.

bozo (bōu·zo). Orig. and chiefly *U.S. slang.* [Origin unknown.] A person, fellow.

1920 *Collier's* 11 Dec. 5/3 The weaknesses of the other bozo. **1921** *Ibid.* 11 June 5/1 Joe is the bozo which I write all them letters to from France. **1924** Wodehouse *Bill the Conqueror* vi. 134 If there's one bozo in this world I got no use for it's a little squirt that double-crosses his pals. **1935** S. Lewis *It can't happen Here* xi. 104 Come on, you bozos! **1961** *Encounter* Apr. 26 Frank, the grey bozo behind the counter.

‖**bozzetto** (bǫtse·to, -z-). Pl. **bozzetti.** [It., dim. of *bozzo* rough stone, sketch, alt. of *bozza*: see Boss *sb.*[1]] A small rough model for a larger sculpture; also, a sketch for a larger painting.

1935 *Burlington Mag.* Oct. 144/2 A bozzetto..which was shown at the Exhibition of Venetian pictures. **1936** *Ibid.* Apr. 169/1 A plastic figure, a bozzetto, a workshop model. **1944** *Ibid.* Nov. 280 The terracotta is a bozzetto for the statue. **1959** *Times* 23 Mar. 3/3 Drawings and prints, marble busts and terracotta *bozzetti*.

bra (brā). Formerly also (as *sing.*) **bras.** Colloq. abbrev. of *brassière. Also *attrib.*

1936 W. B. M. Ferguson *Somewhere off Borneo* i. 26 She wore nothing but a 'bras', the briefest of French knickers, and the sheerest of white silk hose. **1936** *New Yorker* 28 Mar. 7 (Advt.), Bras..50 c. **1937** *Night & Day* 30 Sept. 27 (Advt.), The Bra..is one of the famous 'Alphabet' Bra's. **1948** *Punch* 2 June iii/1 (Advt.), The two-piece is Rayon-Lastex with a bra adjustable in three.. ways. **1959** *Manch. Guardian* 3 July 5/6 Elasticated sides and reinforced bra top. **1968** J. Ironside *Fashion Alphabet* 66 *Bra-slip*, petticoat or slip with top part shaped like a brassière and giving the support of a brassière.

‖**braaivleis** (brāi·flēⁱs). *S. Afr.* [Afrikaans, lit. 'grilled meat', f. *braai* to grill + *vleis* meat, Flesh.] An out-door meal, esp. a picnic, at which meat is grilled; a barbecue.

1939 A. W. Wells *S. Afr.* 410 Some Afrikaans words are used in English conversation, e.g...*braaivleis*. **1942** *Cape Times* 6 Nov. 3 A braaivleis in aid of war funds will be held at Alto, Helderberg. **1947** *Ibid.* 22 Dec. 9/7 Braaivleis Grillers. **1955** J. Packer *Valley of Vines* ii. 30 A braaivleis is a sort of barbecue, Hal. We cook our steaks and chops out of doors on the embers.

Hence **braai** *sb.* and *v. trans.* and *intr.*

1959 *Cape Argus* 10 Mar. 9/9 Buy and braai your own meat. **1959** *Cape Times* 11 Apr. 4/2 Chicken Braai at Somerset West. **1961** *Cape Argus* 11 Mar. 8/8 We picnic and we braai.

brace, *sb.*[2] Add: **7. b.** *Dentistry.* A wire device for straightening the teeth.

1952 *News Chron.* 8 July, If she wears corrective glasses or braces on her teeth, pretty hair makes her less conscious of her handicaps. **1952** M. McCarthy *Groves of Academe* (1953) iv. 56 They had been routed to the dentist for braces.

9. b. Delete 'App. not before 19th c.' and add earlier examples.

1798 Jane Austen *Let.* 27 Oct. (1952) 23 There were no narrow braces for children and scarcely any notting silk. **1806** J. Beresford *Miseries Hum. Life* I. x. 261 When dressing in violent haste—your braces becoming suddenly.. entangled.

15. c. (Further examples.) *a brace of shakes*: see Shake *sb.*[1] 2 h. Hence in *Cricket*: *a brace* (*of ducks*), a score of nought in both innings of a match; *to bag a brace*, to score nought in both innings.

1719 Defoe *Crusoe* 200 The two Muskets I loaded with a Brace of Slugs each. **1755** Mrs. C. Charke *Life* 45 A heavy Blunderbuss, a Muscatoon, and two Brace of Pistols. **1867** *J. Lillywhite's Cricketers' Compan.* 57 Tom Humphrey achieved the feat of 'bagging a brace'. **1891** W. G. Grace *Cricket* xi. 329 In 1868, he got me out for a brace of 'ducks' at Neath. **1903** P. F. Warner in H. G. Hutchinson *Cricket* xiv. 398 More than one well-known cricketer has 'bagged a brace' there. **1912** A. A. Lilley *Twenty-four Years Cricket* v. 65 Noble and Gregory..were thus dismissed for a 'brace'. **1929** *Chicago Sunday Trib.* 25 Aug. I. 3/7 They found Edward Barnett..and his brother..sleeping with a brace of automatics under their pillows.

17. b. In the theatre, a rod or length of timber used to brace a flat.

1866 W. Davidge *Footlight Flashes* xv. 151 They are called braces, and are used for sustaining the weight of cottages, trees, and set pieces of all kinds. **1941** N. Coward *Australia Visited* vii. 46 The actors stumble around.. making believe that a few chairs, braces, and empty sugar-boxes are the palace of a King.

20. brace-rod, a connecting rod used to support or give rigidity to any part of a structure; **brace-shot** *U.S.* = *bracket *sb.* 5 b.

1844 H. Stephens *Bk. Farm* I. 428 In this plough..there is usually applied a brace-rod V. **1907** *Westm. Gaz.* 27 Dec. 4/2 On a special collar of the enclosed propeller-shaft are two brace-rods, extending triangular fashion to the sleeves of the live axle. **1914** R. H. Davis *With the Allies* (1915) 134 To find the range the artillery sends what in the American army are called brace shots.

brace, *v.*[1] Add: **5. b.** Also freq. with *up*.

1816 Jane Austen *Emma* I. vii. 113 It was..necessary to brace her up with a few decisive expressions. **1887** M. Corelli *Thelma* I. xvii, He paused—then suddenly bracing himself up, added [etc.]. **1891** Hardy *Group of Noble Dames* ix. 230 It gave him an opportunity to brace himself up. **1903** 'No. 7' *25 Years in 17 Prisons* xiii. 125 When I heard the words 'sad news'..I braced myself up, clenched my teeth..and prepared to hear the worst.

c. *intr. to brace up*: to brace oneself; to pull oneself together for an effort; also, to take a drink for this purpose. *orig. U.S.*

1809 *Deb. Congress* 20 Jan. 1148 We have been..bracing up; we have had plenty of good wine. **1817** S. R. Brown *Western Gaz.* 354 If the stomach be foul..take an emetic, and then brace up with bark. **1845** C. M. Kirkland *Western Clearings* 62 He braces up for the occasion. *a* **1910** 'Mark Twain' *Myst. Stranger* (1916) 130 All that was needed..was that Wilhelm should brace up and do something that should cause favorable talk. **1958** M. Dickens *Man Overboard* ix. 145 Don't make a scene, Mum... Brace up.

brace, *a.* *U.S. slang.* [? attrib. use of Brace *sb.*[2]] *brace game*, a game in which there is concerted cheating; *brace box*, in faro, a dealing box designed to facilitate cheating.

1875 *Chicago Tribune* 25 Aug. 8/1 The brace game flourishes..to cheat the gambling fraternity. **1908** S. E. White *Riverman* vi. 58 'I tell you, you can't win!' cried Newmark disgustedly. 'It's a brace game pure and simple.' **1908** G. H. Lorimer *J. Spurlock* vi. 116 Life's not even a gamble in this age of commercialism, fo' Fo'tune deals from a brace box.

bracelet. Add: **1. b.** *Palmistry.* A wrinkle crossing the wrist at its junction with the hand.

1883 Frith & Allen *Chiromancy* 125 The Rascette or 'bracelets'..will be found in two or three distinctly marked lines underneath the hand. **1891** *Chambers's Encycl.* VII. 727 These last (the bracelets), if well marked, strengthen the effect of the line of life, each bracelet indicating thirty years of life.

c. A narrow band of hair left on the legs of an animal after it has been clipped.

1904 *Windsor Mag.* Jan. 290/1 The only sign of hair being on the ears and the 'bracelets' around the legs.

bracer[1]. Add: **3.** *spec.* A drink taken to brace one up. *colloq.* (orig. *U.S.*).

1829 *Savannah* (Ga.) *Mercury* 1 July (Th.), If I take..a cooler at nine, a bracer at ten, a whetter at eleven,..who has a right to complain? **1877** J. Habberton *Jericho Road* 12 Treat him to whisky; he needs a bracer. **1924** Wodehouse *Bill the Conq.* xx. 288 He took some worms and slipped them a stiff bracer. **1951** M. Kennedy *Lucy Carmichael* II. iii. 113 We're all going into the Swan for a bracer...Do come..and have a drink before your drive back.

brachial, *sb.* [f. the adj.] **1.** (See Brachial *a.* 1 b.)

2. *Zool.* **a.** One of a series of bones in fishes to which the pectoral fins are attached.

1873 Mivart *Elem. Anat.* 162 The fourth or lowest of the four *brachials* which together may represent the humerus, and to which the fin-rays are attached.

b. One of the calcareous plates in the branches of a crinoid.

1888 Rolleston & Jackson *Anim. Life* 572 The joints of the arms [in Crinoids] are termed *brachials*. **1962** D. Nichols *Echinoderms* ii. 24 The skeletal pieces supporting the arms are called *brachials*.

brachiate (bræ·kiˌēⁱt), *v.* [As from Brachiate *a.*; see -ate[3].] *intr.* (See quot. 1948.) So **bra·chiating** *ppl. a.* and *vbl. sb.*; **brachia·tion,** the act of brachiating; **bra·chiator,** an animal that brachiates.

1899 *Proc. Zool. Soc.* 7 Mar. 306 The hand of the Chimpanzee is adapted for brachiation. *Ibid.* 305 The arm of the Chimpanzee is that of the brachiators, anthropoids like the Orang and Gibbon, which use the arms as one of the main organs of locomotion. **1932** J. S. Huxley *Probl. Rel. Growth* vii. 238 Man..is undoubtedly descended from brachiating ancestors with relatively long arms. **1934** Webster *Brachiate* v.i. **1948** Weiner in *New Biol.* V. 70 Their [*sc.* apes'] ability to brachiate, that is to swing their way from branch to branch by their arms. **1957** *Antiquity* XXXI. 191 They lacked the brachiating specializations of modern apes. **1962** D. Morris *Biology of Art* v. 143 A hanging and swinging form of locomotion termed brachiation.

Brachiosaurus (bræːkiosǭ·rǝs). *Palæont.* [mod.L., f. Gr. βραχίω-ν arm + σαῦρος lizard.] A genus of huge dinosaurs, with the forelegs longer than the hind legs; also, an animal of this genus.

1903 E. S. Riggs in *Amer. Jrnl. Sci.* XV. 299 The term *Brachiosaurus altithorax* is therefore proposed in recognition of the great size and unusually long humerus of this specimen. *Ibid.* 306 The thoracic centra..of *Brachiosaurus*.. range from 39 to 43 centimeters in length. **1931** *Discovery* May 142/2 A large dinosaur, which is perhaps allied to the brachiosaurus. **1934** W. E. Swinton *Dinosaurs* vii. 90 There is little doubt that *Brachiosaurus* was an immense creature.

brachy-. Add: **bra·chyblast** [Gr. βλαστός sprout, shoot] = Spur *sb.*[1] 7 a; **brachycra·nial** *a.*, having a short cranium or head; **brachypterous** *a.* (examples); also applied to insects; hence **brachy·pterism,** the state or condition of being brachypterous.

1895 W. R. Fisher *Schlich's Man. Forestry* IV. III. ii. 402 The fungus [*sc.* larch-blister] can gain admission only through wounds, frequently of brachyblasts eaten by *Coleophora laricella*. **1902** *Biometrika* I. 462 In both races platycephaly is associated with brachycranial characters. **1937** *Discovery* Mar. 91/2 This brachypterism, as it is termed, occurs also in the Orthoptera and Rhynchota. **1939** *Nature* 15 Apr. 645/2 Ruwenzori provides numerous examples of species of insects which are either short winged (brachypterous) in the female or in both sexes... Apart from this phenomenon of 'brachypterism' no obvious cases of adaptation to mountain conditions among insects were noted. **1842** Brande *Dict. Sci.* 163/1 *Brachypterous*, in ornithology, when the folded wings of a bird do not reach to the base of the tail.

brachycephal (bræ·kisifæl). [Back-formation from Brachycephalic *a.*] A brachycephalic person.

1901 G. Sergi *Mediterranean Race* i. 12 The present populations of southern Germany are in great part brachycephals, among whom mesocephals are rare and dolichocephals quite isolated. **1928** V. G. Childe *Most Anc. East* x. 234 The European beakers were made and used by brachycephals. **1930** C. G. Seligman *Races of Africa* vi. 132 Short to medium brachycephals, with a stature rather under 65 inches.

brachydactyly (bræːkiˌdæ·ktili). *Path.* [f. Brachy- + Gr. δάκτυλ-ος finger + -y[3].] Abnormal shortness of the digits. So **bra:chyda·ctylism**; **bra:chyda·ctylous** *a.*

1881 *Syd. Soc. Lex.* I, *Brachydactylous*, short-fingered. **1886** Stedman in *Buck's Handbk. Med. Sci.* III. 498/2 A loss of one or two phalanges only (brachydactylism). **1906** *Green's Encycl. & Dict. Med. & Surg.* I. 460/1 *Brachydactyly* (shortness of the digits). **1908** *Lancet* 25 Apr. 1212/1 Dr. H. Drinkwater read a paper on Brachydactyly—a Study in Heredity, giving an account of a brachydactylous family. **1944** K. Mather in Darlington & Mather *Genes, Plants & People* (1958) xii. 131 Traits, like brachydactyly, inherited in this way, could virtually be wiped out in one generation.

brachyodont (bræ·kiŏdǫnt), *a.* Also **brachydont.** [f. Brachy- + ὀδούς (ὀδοντ-) tooth.] Designating teeth with short or low crowns and well-developed roots.

1883 W. H. Flower in *Encycl. Brit.* XV. 430/1 Modification of [the selenodont form] from a brachyodont to a hypsodont type. **1884** *Geol. Mag.* 548 Detached upper molars of a smaller form, with a brachydont..structure. *Ibid.*, Several lower molars and a fragment of a mandible from Hempstead indicate a brachydont anthracotheroid. **1891** [see *bilophodont *a.*]. **1968** R. Zangerl tr. *Peyer's Compar. Odontology* 191 The teeth of the earliest monodelphous mammals..are at the same time brachyodont—low-crowned and provided with well-developed roots.

brack, *sb.*[7] [Abbrev. of *barnbrack; cf. Ir. *breac* speckled.] An Irish cake or loaf containing seeds or fruit; = *barnbrack.

Quots. 1855 prob. represent attempts to render Irish *bairghean breac* *BARNBRACK.

1855 E. ACTON *Mod. Cookery* (rev. ed.) xxvi. 546 Rich Brawn Brack, or Borrow Brack. *Ibid.*, To convert the above into the popular Irish 'speckled bread', or *Brawn Brack*.. add to it three ounces of carraway-seeds. *Ibid.* xxvi. 555 Common *brown brack*, or Irish seed-cake. **1959** J. O'DONOVAN *Fiddle* xxix. 182 She..made tea and sliced a home-made brack. **1960** *Guardian* 10 Dec. 12/6 Breads, bracks, baps, scones.

bracket, *sb.* Add: **5. b.** *transf.* The (specified) distance between a pair of shots fired, one beyond the target and one short of it, in order to find the range for artillery; chiefly in the phrase *to establish a bracket*.

1899 *Daily News* 6 Dec. 5/7 At first I fire at 100 yards, and if I find that my shell is short I fire my second, say at 3300, in order to go beyond the object. If I see that my shot does go over I am satisfied that I have established what is called 'a long bracket', that is to say, I have found two ranges, 200 yards apart, between which the object must lie. ..I..fire another shot to shorten the distance within which I can then know that the target must be. This we call, on the same principle as the other, 'a short bracket'. **1916** 'BOYD CABLE' *Action Front* 42 The German gun had got its bracket. **1927** *Blackw. Mag.* Apr. 476/2 The shell passed over the ship, to be followed by a second one which fell short, establishing a 'bracket', which..is all that a gunner desires.

c. A group bracketed together as of equal standing in some graded system, as *income bracket*: a class of persons grouped according to income.

1890 [see sense 5 in Dict.]. **1932** *N.Y. Times* 1 May III. 1/3 The most striking fact in vital statistics is the increase in the upper-age brackets. **1940** *Jrnl. R. Aeronaut. Soc.* XLIV. 506 The general bracket of utilisation within which the aeroplanes will fall. **1940** F. SCOTT FITZGERALD *Let.* 18 Mar. (1964) 66 Competent people with a little pull have no trouble finding places in the same income brackets. **1943** *Gen* 2 Jan. 29/1 Once he got in the upper brackets, fistically speaking. **1952** *News Chron.* 15 July 6/8 'At my age,' he [sc. an athlete] added, 'I have only two years left in the top bracket'. **1956** 'M. INNES' *Appleby plays Chicken* v. 43 They were both from the same social bracket.

d. *Skating.* A series of turns resembling a bracket or 'brace' (see sense 5 in Dict.). Also *attrib.*

1892 [see *COUNTER *sb.*⁴ 6]. **1901** *Encycl. Sport* IV. 366/1 Three turns and bracket turns are accomplished upon two edges. *Ibid.* 370/1 Turn loops are turns worked into the form of a loop, and *bracket loops* are brackets skated in a similar way. **1935** *Times* 14 Nov. 6/7 The great stumbling block in this test for the average skater is the bracket figure. **1967** *Daily Tel.* 3 Mar. 14/7 She coolly executed a smooth change-edge loop figure and an equally accurate bracket-change bracket.

6. bracket clock, a clock designed to stand on a shelf or wall-bracket; **bracket fungus, mushroom,** any fungus which grows on trunks of trees forming a bracket-like projection; **bracket system** (see quot.).

1894 F. J. BRITTEN *Former Clock & Watchmakers* 185 Bracket or pedestal clocks..were in favour before..the long-case variety. The earliest English wooden bracket clocks have cases of the square pattern. **1958** 'W. HAGGARD' *Slow Burner* i. 13 The handsome bracket clock on the table. **1909** *Cent. Dict. Suppl.*, Bracket fungus. **1910** *Encycl. Brit.* IV. 366/2 *Bracket-fungi.* The term 'bracket' has been given to those hard, woody fungi that grow out like a bracket..in the form of semicircular brackets. **1927** *Observer* 28 Aug. 18/1 The so-called bracket-mushrooms that chiefly flourish in rotting trunks have been platforms rather than brackets. **1874** THEARLE *Naval Archit.* 86 The Bracket System is a development..of the transverse and longitudinal systems combined, by which iron-clad ships have been built since their introduction.

bracket, *v.* Add: **3.** *intr.* To project like a bracket.

1855 FERGUSSON *Handbk. Archit.* IX. iv. 428 A number of small imitations of arches, bracketing one beyond the other.

4. To find the range for artillery by means of a bracket or series of brackets (*BRACKET *sb.* 5 b). Also *trans.* Hence **bra·cketing** *vbl. sb.*

1909 in WEBSTER. **1914** *Times* 12 Oct. 7/4 They [sc. the Germans] dispense to a great extent with the method of ranging known by us as 'bracketing', especially when acting on the defensive, and direct fire by means of squared maps and telephone. **1916** J. BUCHAN *Greenmantle* xxi. 293 The shell dropped ten yards to our right. A second later another fell behind us... 'They know their business. They're bracketing.' **1926** 'J. CONNINGTON' *Death of Swaythling Court* xvi, I took the liberty of bracketing the Lethal Ray machine..on Swaythling Court. **1957** M. K. JOSEPH *I'll soldier no More* (1958) 187 In the distance, a puff of smoke suddenly appeared... Almost looks as if we're being bracketed.

Braconid (bræ·konid), *a.* and *sb.* [f. mod.L. *Bracon* + -ID³.] **A.** *adj.* Of or belonging to the Braconidæ, a family of small ichneumon flies. **B.** *sb.* A fly of this family.

1893 in *Funk's Stand. Dict.* **1913** *Oxf. Univ. Gaz.* 4 June 961/1 The Braconids *Caelinus niger* and *Pezomachus fasciatus*. **1916** *Canadian Entomologist* XLVIII. 89 *Dinocampus americanus*..a common Braconid parasite of many of the larger species of Coccinellids. **1957** E. B. FORD *Butterflies* (ed. 3) vi. 109 [The Ichneumons are] again divided between the true Ichneumons (family Ichneumonidae) and the closely related Braconids (family Braconidae).

bract. Add: **1. b.** *bract-like* adj.

1847 W. E. STEELE *Field Bot.* 167 *S[piranthes] autumnalis*, root-leaves oblong, those of stem bract-like. **1911** F. O. BOWER *Plant-Life on Land* iv. 64 The calyx sprang from bract-like leaves.

Bradbury (bræ·dbəri). [The name of John Swanwick *Bradbury*, Secretary to the Treasury 1913–19.] Former colloquial name for a currency note of £1. (Cf. *FISHER².)

[**1914** *Punch* 21 Oct. 341/3 Not a burglar on our books for the last six weeks..not a coiner, not a note expert. And they had the opportunity of their lives with the John Bradbury notes!] **1917** *Punch* 22 Aug. 128 Parting with three Bradburys and three shillings. **1920** *Oxf. Tracts Econ. Subj.* No. 27. 4 The 'Bradbury', nominally a pound, and by statute legal tender for a pound of debt, has lost, from its abundance, a part of its purchasing power. **1926** G. FRANKAU *Masterson* xvi, Cynthia had decided to 'risk a couple of Bradbury's each way'.

Bradenham (bræ·d'nəm). [Trade-name.] Applied *attrib.* to a dark, sweet-cured ham. Also *ellipt.*

1906 *Daily Chron.* 24 Sept. 7/2 Next door the Wiltshire Bacon Curing Company shows a decided novelty in 'Bradenham' hams and chaps. They are perfectly black, a result achieved, it is said, by curing with treacle. **1927** D. L. SAYERS *Unnatural Death* I. vi. 68 Tell me how it comes that your little waitress and her railway clerk come down to Epping Forest to regale themselves on sandwiches made from coal-black, treacle-cured Bradenham ham? **1959** *Hist. Wiltshire* (V.C.H.) IV. 223/1 In 1897 the company [sc. the Wiltshire Bacon Curing Co. Ltd.] took over the Bradenham Ham Co. and since it was wound up in 1921 'Bradenham' hams, cured by a secret recipe, have been manufactured by the Wiltshire Bacon Co. **1960** *House & Garden* Apr. 111/2 There are counters of hams including Bradenhams (the coal-black York).

Bradford (bræ·dfəd). Shortened form of *Bradford-on-Avon*, the name of a town in Wiltshire, used *attrib.* in *Bradford clay* (see quots.). Hence **Bradfordian** *a.*

1858 [see *BATHONIAN *a.*]. **1903** GEIKIE *Textbk. Geol.* (ed. 4) II. VI. 1142 Its [sc. Forest Marble's] lower portion near Bradford-on-Avon is a grey marly clay with thin layers of tough limestone and calcareous sandstone about 10 feet thick, and this argillaceous band has been separately designated the Bradford clay. **1910** *Encycl. Brit.* IV. 372/1 *Bradford clay*, a thin, rather inconstant bed of clay or marl situated in England, at the base of the Forest Marble, the two together constituting the Bradfordian group in the Bathonian series of Jurassic rocks.

Bradleian (bræ·dliən, -lī·ən), *a.* Also **Bradleyan**. [f. *Bradley* (see below) + -AN.] Of, pertaining to, or characteristic of: (*a*) the English idealist philosopher Francis Herbert Bradley (1846–1924) or his writings; or (*b*) his brother, the Shakespearean critic Andrew Cecil Bradley (1851–1935).

1909 W. JAMES in *Jrnl. Philos.* 2 Mar. 115 It gets rid of any need for an Absolute of the Bradleyan type. **1909** — *Pluralistic Universe* ii. 71 The functions of the Bradleian absolute are in this particular identical with those of the theistic God. **1943** E. L. MASCALL *He who Is* vi. 77 Idealistic philosophies of the Bradleian type. **1961** *Listener* 21 Dec. 1083/2 He does not try to put Othello back on the Bradleian pedestal.

bradoon (brădū·n), var. BRIDOON.

1803 *Sporting Mag.* XXI. 220/2 A Weymouth bridle, with bit and bradoon, is in my opinion, preferable to any other sort for the road. **1862** *Catal. Internat. Exhib.* II. No. 4593 Safety Springs, for riding and driving reins, to both bits, and one rein in hand, to act on the bradoon. **1968** *Encycl. Brit.* XI. 708/2 The snaffle (also bridoon or bradoon)..may be jointed, twisted or straight.

Bradshaw (bræ·dʃǭ). Colloquial designation of 'Bradshaw's Railway Guide', a time-table of all railway trains running in Great Britain, the earliest form of which was first issued at Manchester in 1839 by George *Bradshaw* (1801–53), printer and engraver. (Ceased publication in 1961.) Also *transf.* and *fig.*

1847 F. A. KEMBLE *Let.* in *Records of Later Life* (1882) III. 289 You ask me what book I read now to put me to sleep—why, Murray's 'Handbook for France', and the foreign 'Bradshaw'. **1851** *Fraser's Mag.* Jan. 126/1 It is always dangerous to travel upon the faith of an old Bradshaw. **1914** 'BARTIMEUS' *Naval Occasions* xi. 77 The young Doctor looked up from the year-old 'Bradshaw' with which he was wont to enliven moments of depression by arranging mythical week-ends at friends' houses in various parts of England. **1923** J. M. MURRY *Pencillings* 201 Reading reviews is a minor and fascinating science like the study of the continental Bradshaw. **1954** J. MASTERS *Bhowani Junction* xxx. 257 The girl's a walking Bradshaw. **1959** *Times* 20 May 11/7 It would be the easiest thing in the world for a new D.N.B. to become the Tom Tiddler's ground of historical specialists and so to cease to count as more than a 'Bradshaw' for students. **1961** *Economist* 18 Mar. 1047/1 Poor Bradshaw is, no doubt, irreplaceable. His approaching demise, in May, is one that countless thousands [illegible] mourn.

Hence **Bra·dshaw** *v.*, (*a*) *trans.* to make a 'Bradshaw' for; (*b*) *intr.* (*R.A.F. slang*) to follow a railway in flying. Hence **Bra·dshawing** *vbl. sb.*

1887 MRS. D. DALY *Digging & Squatting S. Australia* 153 Mr. Westgarth says: 'If the through route of Central Australia is some day to be Bradshawed.' **1946** E. C. CHEESMAN *Brief Glory* ii. 21, I knew the only thing to do was to 'Bradshaw': in other words to follow the railways. **1946** A. PHELPS *I couldn't care Less* iii. 25 Bradshawing can sometimes lead into trouble... I dislike following a railway except in extreme emergency when forced to fly low.

brady-, comb. form of Gr. βραδύ-ς slow, as in **bradycardia** [Gr. καρδία heart], slowness of the pulse; **bradykinin** (-kai·nin) [Gr. κίν-ησις motion + -IN¹], a polypeptide stimulating the action of the visceral muscles; **bradyphrasia** [Gr. φράσις speech], slowness of speech due to mental defect or disease; **bra·dyseism** [SEISM], a slow rise and fall of the earth's crust.

1890 BILLINGS *Med. Dict.* I. 182/1 Bradycardia. **1908** J. MACKENZIE *Dis. Heart* 133 True bradycardia..only occurs when all the chambers of the heart participate in the slow action. **1965** J. POLLITT *Depression & its Treatment* iii. 13 Bradycardia. **1949** M. ROCHA E SILVA et al. in *Amer. Jrnl. Physiol.* CLVI. 261 (*title*) Bradykinin, a hypotensive and smooth muscle stimulating factor released from plasma globulin by snake venom and by trypsin. **1878** J. A. MCCARTHY tr. *von Ziemssen's Cycl. Med.* XIV. 816 We find bradyphrasia and paraphrasia developed in a similar manner in conditions of morbid hebetude and intellectual weakness. **1896** *Daily News* 11 Feb. 6/2 The slow movements, or bradyseisms..were those which had resulted in the formation of continents. **1905** A. A. G. SCHIEFERDECKER *Geol. Nomencl.* 106/2 *Bradyseism*, slow movement of the ground [illegible] upheaval, subsidence or a shifting of shore line. *Ibid.* 268/1 *Bradyseism*, slow vertical movements of the earth's crust, caused by volcanic action.

brag, *a.* Add: **5.** Prime, first-rate, surpassingly good. *U.S.*

1836 *Jeffersonian* 3 May 96 (Th.), The Moselle was a new brag boat, and had made several quick trips. **1857** *Knickerbocker* Sept. 292 Isaac had been the 'brag hand' of the plantation. **1893** *Outing* (U.S.) XXII. 92/1 His [sc. the pup's] sire won the brag runner of the country. **1904** W. N. HARBEN *Georgians* xvi. 169 A boy that blowed any army bugle come, an' the brag singer, a young man that sung 'What is my wanderin' boy to-night?'

braga-beaker (brā·gă,bī:kəɹ). Also **brage-goblet**, etc. [f. ON. *braga-* in *braga(r)full* the cup drunk at funeral feasts.] A cup from which a toast is drunk.

Quots. 1847 and 1851 show that some writers, mistaking the etymology, thought the word was connected with Bragi, the god of poetry in Norse mythology.

1847 G. STEPHENS tr. *Tegner's Frithiof's Saga* 270 Thus should be stand opposite the Brage-goblet, first make a vow, and then drink out the Bumper. **1847** I. A. BLACKWELL in T. Percy tr. *Mallet's Northern Antiq.* 549 Bragi's bumper—the Bragafull. **1851** B. THORPE *Northern Mythol.* I. 190 At gilds the Bragafull, or Bragi-cup was drunk. **1854** W. B. JERROLD (*title*) A Brage beaker with the Swedes; or, notes from the North in 1852. **1921** *Public Opin.* 15 July 61/3 Wielding Thor's hammer and drinking from the braga-beaker in the hall.

Bragg (bræg). The name of Sir William Henry *Bragg* (1862–1942) and of his son Sir William Lawrence *Bragg* (born 1890), English physicists, used *attrib.* or in the possessive to designate certain laws, effects, etc., in *Physics* (see quots.).

1913 *Phil. Mag.* XXV. 628 The air equivalent of the carbon was calculated from Bragg's law to be about 78 per cent. of the whole stopping power. **1920** RUTHERFORD in *Proc. R. Soc.* A. XCVII. 387 It can readily be calculated from Bragg's rule. **1930** *Ibid.* CXXIX. 214 [illegible] **1923** *Phil. Mag.* XLV. 11 The fact that Bragg's law cannot be strictly true seems to have been pointed out first by Darwin. **1931** *Physical Rev.* XXXVIII. 1410 In this investigation, regular Bragg reflections from the crystal [illegible]. **1934** [illegible] Bragg's law [illegible]. **1940** *Chambers's Techn. Dict.* [illegible] **1957** [illegible] the average number of ions per unit distance along a beam of initially monoenergetic ionizing particles.

bragger. *sb.* **2.** *Card-playing.* In the game of brag, a nine or knave (see also BULLET *sb.*¹ 6 in Dict. and Suppl.). *U.S.*

1807 [see BULLET *sb.*¹ 6]. **1938** H. ASBURY *Sucker's Prog.* 21 In American Brag there were eight 'braggers'—the jacks and nines of each suit.

braggite (bræ·gait). *Min.* [f. *Bragg* + -ITE¹.] (See quot.)

1932 BANNISTER in *Min. Mag.* XXIII. 188 This being the first new mineral to be isolated and determined by X-ray methods, it may very appropriately be named *braggite* in honour of Sir William H. Bragg and his son Prof. W. L. Bragg, the pioneers of the new method for the investigation of crystals. **1940** C. M. RICE *Dict. Geol. Terms* 52/1 *Braggite*, sulphide of platinum, palladium, and nickel, (Pt, Pd, Ni) S, tetragonal, as minute grains in the concentrates of the Bushveld norite, Transvaal.

bragite (bræ·gait). *Min.* [ad. Norw. *bragit* (D. Forbes and T. Dahll 1854, in *Nyt Mag. Naturvidenskab.* VIII. 227), f. *Brage*, a Scandinavian deity: see -ITE¹.] A variety of fergusonite found in Norway and Sweden.

1868 DANA *Syst. Min.* (ed. 5) 276, 525. **1896** CHESTER *Dict. Min.* 38 *Bragite*, a mineral occurring in brown, tetragonal crystals, not fully examined, but referred both to zircon and fergusonite.

braguette (brăge·t). Also **brayette.** [OFr.] A piece of armour of the fifteenth century corresponding to the cod-piece.

1867 W. MCDOWALL *Hist. Dumfries* xxvii. 349 His bulky dagger-proof coat of green velvet and scarlet braguette. **1895** *Oracle Encycl.* I. 180/1 Brayette and loin-guard.

‖ **brahmacharya** (brāmătʃa·riă). [ad. Skr. *brahmacarya*, f. *brahman* prayer, worship + *carya* conduct.] Purity of life, esp. regarding sexual matters; celibacy; sexual self-restraint; freq. used with reference to the life and teachings of M. K. Gandhi. Hence **brahmacha·ri**, one who practises brahmacharya.

1920 M. K. GANDHI *Self-Restraint v. Self-Indulgence* (1947) I. ix. 70 The word in Sanskrit corresponding to celibacy is *Brahmacharya*, and the latter means much more than celibacy. *Brahmacharya* means perfect control over all the senses and organs. For the perfect *Brahmachari* nothing is impossible. **1925** *Ibid.* ii. 39 There can be no two opinions about the necessity of birth control. But the only method handed down from ages past is self-control or *brahmacharya*. **1958** J. V. BONDURANT *Conquest of Violence* i. 12 The opinion that a satyagrahi must be a vegetarian, must observe *brahmacharya* (continence), must develop *aparigraha* (non-possession) and must manifest other ideal Gandhian attributes. **1962** A. HUXLEY *Island* v. 56 He had been brought up to reverence the Ideal of Purity. '*Brahmacharya*, if you know what that means.' **1969** R. SHANKAR *My Music* ii. 58 He strongly insists that the students follow *brahmacharya*—for the disciple, a traditional Hindu way of life that includes only the absolute essentials of material needs.

Brahmi (brā·mi). [Skr.] The name of one of the oldest alphabets of India, probably of Semitic origin.

1895 G. BÜHLER *Orig. Indian Brāhma Alphabet* 33 The Brāhmī lipi is the real old Indian alphabet, which was popularly used in the third century B.C. all over India. **1902** *Encycl. Brit.* XXXIII. 901/2 The alphabets of India all spring from two sources: (*a*) the Kharoṣṭhī, (*b*) the Brāhmī alphabet. **1920** *Blackw. Mag.* May 621/2 A long inscription in Brahmi characters of about 150 B.C. **1963** A. H. DANI *Indian Palaeogr.* iii. 28 This similarity hardly leaves any doubt that Indian Brāhmī was created on the basis of the North Semitic letters.

Brahmic, *a.* For date of first example read: 1852.

Brahmin. Add: **b.** *spec.* A member of the upper class of Boston, Mass., U.S.A.

1881 *Homes & Haunts of our Elder Poets* 155 To be a missionary of Boston culture..must have pleased the anxious thought of this medical Brahmin. **1931** J. T. ADAMS *Epic of Amer.* viii. 219 The West..was dominating the American outlook, in spite of the smug Boston Brahmins. **1963** M. MCCARTHY *Group* ii. 45 Their exact social position in Boston, which Kay greatly over estimated—they were not 'Brahmins', horrid word, at all.

Brahmoism (brā·moiz'm). [f. Bengali *Brāhmo* in *Brāhmo* (or *Brāhma*) *Samāj* society worshipping the supreme spirit (BRAHMA) + -ISM.] A reformed Hinduism of a theistic character founded by Ram Mohun Roy in 1830. So **Brah·mo, Brah·moist,** a follower of Brahmoism; also *attrib.*

1857 DASS *Supreme Being of Brahmo Theol.* 7 There is an impassable gulf which separates unitarian Christianity from Brahmoism. **1862** C. H. MANUEL *Destiny of Human Life* 17 Brahmoism—pure Theism—is essentially anti-idolatrous. **1870** *Contemp. Rev.* XIII. 232 Which has raised Brahmoism from a small Hindoo sect into a comprehensive religion. *Ibid.* 239 A congregation of Brahmos. *Ibid.* XV. 131 The Brahmoists of both sections may already be numbered by thousands. **1885** [see BRAHMISM]. **1927** P. C. RAY *Life of C. R. Das* iii. 25 He married according to Brahmo rites (Act 3 of 1873) Basanti Devi.

Brahmsian (brā·mziăn), *a.* and *sb.* [f. *Brahms* (see def.) + -IAN.] **A.** *adj.* Of, relating to, or characteristic of the German composer Johannes Brahms (1833–97) or his music. **B.** *sb.* A follower or admirer of Brahms. So **Brah·msite** *a.* and *sb.*

1894 G. B. SHAW *Music in London 1890–94* (1956) III. 145 A German Brahmsite critic proclaims them the latest products of the great school in chamber music. *Ibid.* 146 It was at one of these recent Brahmsian Populars that Mr Oswald sang some vocal pieces. **1929** C. A. PHILLIPS tr. *Niemann's Brahms* xxviii. 355 A few typically Brahmsian technical formulas. **1947** C. GRAY *Contingencies* ii. 59 An exclusive and esoteric cult such as that of the Brahmsians. **1949** *Penguin Mus. Mag.* July 85 The early 'renaissancers' were Brahmsites. **1963** *Listener* 7 Feb. 264/3 It used dated features like Brahmsian minor-key themes.

Brahui (brāhū·i), *sb.* and *a.* Also 9 **Braho(o)e.** **A.** *sb.* A pastoral people of Baluchistan; a member of this people; their language. **B.** *adj.* Of or pertaining to the Brahui or their language.

1816 H. POTTINGER *Trav. Beloochistan* iv. 54 The Belooches..are divided into two great classes, severally known by the appellations of Belooche and Brahooe. *Ibid.*, The Brahooe tongue..is so dissimilar [from Persian] in its sound and formation, that I never recollect to have remarked in it a single expression in any way approaching the idiom of Persian. **1826** J. C. PRICHARD *Physical Hist. Mankind* (ed. 2) II. v. iii. 11 These nations are the Belooches..and the Brahooés. *Ibid.* 12 The Brahooés are divided into a great number of petty tribes. **1844** *Ibid.* (ed. 3) IV. x. 63 We must mention the Brahui, who are divided into a number of different tribes. **1848**, **1875** [see *BALUCHI]. **1873** H. W. BELLEW (*title*) From the Indus to the Tigris..Together with a synoptical grammar and vocabulary of the Brahoe language. **1924** *Blackw. Mag.* Jan. 113/1 Luckily we were dealing with Brahuis not Mahsuds. **1932** W. L. GRAFF *Language* 414 The Brahui dialect..in Baluchistan. **1943** S. K. CHATTERJI *Lang. & Linguistic Problem* 8 Brahui (207,000) in Baluchistan, a remnant of the great Dravidian bloc of western India. **1953** W. J. ENTWISTLE *Aspects of Lang.* iii. 347 The Dravidian languages of the Deccan have left behind them a solitary outpost in the Brahui of Baluchistan.

braided, *ppl. a.* Add: **c.** Applied to a stream that divides, esp. at low water, into several channels.

1901 *Science* 15 Nov. 778/1 This extensive deposit..is the product of aggradation by braided or laced streams. **1940** *Geogr. Jrnl.* XCVI. 352 Three major types of rivers, actively downcutting, balanced meandering, and aggrading or braided.

brailed, *ppl. a.* Add: **c.** Hauled (*up*, etc.) by means of brails. So **brai·ling** *vbl. sb.* (also of a tent and *fig.*).

1856 G. CUPPLES *Green Hand* xiv. 138 The brailed foresail. **1902** C. J. CORNISH *Naturalist Thames* 8 A furling and brailing-up of the rain-clouds. **1910** F. H. LAWRENCE *Let.* 19 Aug. in *Home Lett. T. E. Lawrence* (1954) 600 They had left a piece of the tarpaulin outside the brailing of the tent. **1921** *Spectator* 7 May 589/1 The brailing up of the loose-footed sail. **1924** *Weekly Westm. Gaz.* 13 Sept. 568/3 The brailed brown mainsail of a barge. **1939** *War Illustr.* 16 Dec. 440/2 A bell-tent with sandbagged brailing. **1964** C. WILLOCK *Enormous Zoo* v. 96, I looked out under the brailing of the tent.

Braille (brē¹l). Also **braille.** The name of the French teacher of the blind Louis *Braille* (1809–52) used *attrib.* to designate a system of embossed printing for the blind, perfected by him in 1834. Also *ellipt.*

In this system the symbols for the letters, etc., are composed of raised dots arranged in different ways. It was adopted as the standard system in England by the British and Foreign Blind Association, 1869–70.

1853 E. C. JOHNSON *Tangible Typogr.* 22 System Braille.. a system of raised dots..this new arbitrary system of dots [has been] adopted throughout the French schools. **1871** T. R. ARMITAGE *Educ. Blind* i. 8 A modification of the Braille system has been suggested by Dr. Russ, of New York. *Ibid.* iii. 21 The reasons which have induced the Council to adhere to the original Braille, are briefly these:—1st. The Braille type is very generally diffused [etc.]. **1887** A. M. SULLIVAN *Let.* 2 June in H. Keller *Story of my Life* (1903) III. ii. 324, I gave her my braille slate to play with, thinking that the mechanical pricking of holes in the paper would amuse her. **1891** *Daily News* 21 Feb. 5/4 The assistant transcribed her answers, written in Braille, into the ordinary characters. **1930** J. M. RITCHIE *Concerning the Blind* vii. 126 Braille has reached the proud position of being the one and only system now in use. *Ibid.* 127 There was a dearth of music in Braille.

Hence as *v. trans.*, to transcribe or print in Braille; to mark in Braille. So **Brai·ller** (see quot. 1951); **Brai·llist,** a person who transcribes Braille.

1909 *Daily Chron.* 15 July 4/4 A worker who is both a skilled musician and a skilled Braillist. **1944** A. ROHRBACK *Volunteer Braille Transcribing* 5 Qualified braillists began the transcription of books for libraries and schools for the blind. **1951** *Sci. News Let.* 1 Dec. 352/1 Brailler, a typewriter for the blind... The only moving part is the Brailling unit which carries a small embossing head across the page. **1956** *Bookseller* 20 Oct., A select list of new titles Brailled and added to stock. **1964** *Listener* 5 Mar. 394/1 Their [sc. blind people's] cookers can be brailled for temperature markings.

brain, *sb.* Add: **3. b.** *to have* (something) *on the brain* (examples); also, *to have got* (something) *on the brain.*

1862 MRS. GASKELL *Lett.* (1966) 698 Our poor people would get work, and..we should not be killed with 'Poor on the Brain', as I expect we shall before the winter is over. **1869** *Congress. Globe* Jan. 182/2 The Gazette seems to have the franking privilege 'on the brain'. **1884** CROMER *Let.* 11 Mar. in Marq. of Zetland *Lord Cromer* (1932) ix. 100, I have got it on the brain that much writing is not a good thing for the moment. **1900** C. M. YONGE *Modern Broods* i. 12 Child, I believe you have bicycles on the brain. **1911** A. BENNETT *Hilda Lessways* II. iv. 173 Tom..had for the moment got Beethoven on the brain. **1957** M. SPARK *Comforters* v. 108 Caroline had it very much on the brain that her phantom should be outwitted in this one particular.

c. An electronic device that performs complicated operations comparable to those of the human brain; *spec.* an electronic computer. Cf. *electronic brain.*

1934 *Scoops* 10 Feb. 15/2 An attempt to construct a mechanical man who can think..is to be made by the Massachusetts Institute of Technology. The 'Brain' of the new robot is an accomplished fact. **1945** *War Illustr.* 7 Dec. 492/1 The 'brain' of the shell is a fuse, a tiny radio set—transmitter, receiver and aerial all combined—in the nose of the shell. **1951** R. KNOX *Stimuli* xii. 129 Recently the Press recorded the invention of a calculating machine.. They called it a 'brain'. **1952** *Economist* 1 Nov. 305 Univac, the 'Giant Brain', an electronic automatic computer. **1954** K. W. GATLAND *Devel. Guided Missile* (ed. 2) iv. 109 In the self-homing device we have a 'brain' capable of making high-*g* turns and of operating..control entirely beyond human capabilities.

4. c. *colloq.* A clever person; so *the brains*: the cleverest person (in a group, etc.); a 'master-mind'. (Cf. 1844 *fig.* quot. for sense 3 a in Dict.)

1914 W. OWEN *Let.* 21 Dec. (1967) 309 This gentleman is, all round, an 'interesting' pupil, and what the French call a 'Brain'. **1933** E. WALLACE *Capt. Souls* xxxv. 197, I felt like a fourth form boy listening to a 'brain', and found myself being respectful! [illegible] **1954** 'N. BLAKE' *Whisper in Gloom* I. v. 75 The Brain's got what it takes. **1958** *Times* 20 Jan. 5/5 Admiral Sir William Wynter, 'the brains' of the victory.

5. a. brain-capacity, -condition, -cortex, -work (earlier and later examples).

1936 *Discovery* Nov. 351/2 The largest [skull on record] is that of Turgenev, the Russian novelist, [who] had a brain capacity of 2,030 c. cm. **1890** W. JAMES *Princ. Psychol.* I. vi. 162 Certain brain-conditions occur together which, if they occurred separately, would produce a lot of lower states. *Ibid.* x. 399 We speculate on the brain-condition during all these different perversions of personality. *Ibid.* xiv. 567 The amount of activity at any given point in the brain-cortex is the sum of the tendencies of all other points to discharge into it. **1845** GEO. ELIOT *Let.* 13 June (1954) I. 195 It always gives one satisfaction to see any evidence of brain-work. **1931** J. S. HUXLEY *What dare I Think?* ii. 69 [illegible]

6. brain-axis = *brain-stem; **brain-ball,** the brain of an enemy slain in combat made into a ball by mixing it with lime and preserved as a trophy; **brain-cap,** the upper part of the skull; **brain-centre,** any nerve-centre in the brain, esp. one of those supposed to be the controlling centres of particular functions; also *fig.*; **brain-child** *colloq.*, the product of a person's mind; an invention; **brain-coral** (examples); **brain drain,** phrase used *colloq.* of the 'loss' of highly trained or qualified people by emigration, particularly to the U.S.; **brain-dressed** *a.*, of skins, dressed with a liquor prepared by boiling deer brains; **brain-fag,** exhaustion of the brain by prolonged mental strain; **brain-fagged** *a.*, suffering from brain-fag; **brain-fever bird,** an Indian hawk-cuckoo (*Cuculus varius*) having a persistent cry; **brain-mantle,** the upper part of the brain; **brain-mass,** (*a*) material quantity of brain; (*b*) the brain, regarded as a material object; **brain-path,** one of a number of supposed lines of conduction in the brain, along which nervous impulses come to travel more readily than by other routes; also *attrib.*; **brain-picker,** one who 'picks' the brains of another (cf. PICK *v.*¹ 9); **brain-racking** *vbl. sb.*, racking of the brain; mental torture or anguish; also *ppl. a.*; **brain-sand,** minute particles of gritty substance found chiefly in the pineal gland; **brain-stem,** the central trunk of the brain upon which the cerebrum and cerebellum are set, and which continues downwards to form the spinal cord; **brain-stone** (examples); **brain-storm,** (*a*) 'a succession of sudden and severe phenomena, due to some cerebral disturbance' (Gould 1894); (*b*) *U.S. colloq.* = *brain-wave (*c*); (*c*) *U.S.*, a concerted 'attack' on a problem, usu. by amassing a number of spontaneous ideas which are then discussed; also *attrib.*; so as *v.*, to make such an attack; hence **brain-storming** *vbl. sb.* and *ppl. a.*; **brain-stuff,** (*a*) medicine for the brain; (*b*) the product of thinking; ideas; **brain-sugar** (see quot.); **brain-teaser** or **-twister** (orig. *U.S.*), a difficult problem; a puzzle; **brain-vibration,** an excitation or nervous discharge in the brain; **brain-wave,** (*a*) a hypothetical telepathic vibration; (*b*) (usu. in *pl.*) a measurable electrical impulse in the brain; (*c*) *colloq.*, a sudden inspiration or bright idea.

1894 GOULD *Dict. Med.* 233 *B[rain] axis*, that portion of the brain-substance including the island of Reil, the basal ganglia, the crura, pons, medulla, and cerebellum. [**1809** D. O'CONNOR tr. *Keating's Gen. Hist. Ireland* I. 279 Ceat.. placed his ball of brains in a sling.] **1904** *Westm. Gaz.* 9 July 5/1 The old king, who sat with the brain-ball on his head. **1907** *Folk-Lore* June 228 Mesgegra's brain-ball, an object that could be slung from a sling. **1895** A. H. KEANE *Ethnology* iii. 46 The brain-cap of many savages has been found to be larger and heavier than that of some higher races. **1887** W. JAMES *Coll. Ess. & Rev.* (1920) 290 Even in the case of congenital defect of the extremities, the brain-centres might feel in the usual ancestral way. **1902** —— *Var.*

Relig. Exper. ix. 212 Our emotional brain-centres strike work, and we lapse into a temporary apathy. **1904** *Westm. Gaz.* 3 Mar. 1/3 From its brain-centre in the middle of the floor of the New York Stock Exchange the recording instruments of the Exchange Company print out continuously the news of the transactions. [illegible] The brain-children [i.e. novels] of the illustrious dead. **1882** SWINBURNE *Lett.* (1960) IV. 284 The most unlucky and despised of all my brain-children. **1921** WODEHOUSE *Jill the Reckless* xvii. 254 The almost maternal yearning to see his brain-child once more, which can never be wholly crushed out of a young dramatist. **1952** J. CANNAN *And be a Villain* iv. 81 How well I know that filing system. It was my own brain-child. **1790–1** PETIVER *Gazophyl.* Decas Sept. 63 Tab. 28 Common American Brain Coral... So call'd for its likeness to humane Brains. **1936** RUSSELL & YONGE *Seas* (ed. 2) vi. 182 The well-known Meandrina or Brain-coral. **1963** F. FAIRLEY in *Evening Standard* 7 Jan. 1/2 Nearly one quarter of Britain's best young scientists and technologists are being magnetised to jobs in North America. About 10 per cent are settling there. This is the shock finding of experts who have spent months investigating the 'brain drain' across the Atlantic. **1964** *News Front* Apr. 15 The United Kingdom is deeply concerned about its 'brain drain'. **1889** *Harper's Mag.* June 63/2 These [deerskin leggings] were prepared of brain-dressed skins that perfectly turned the rain and dew. **1851** DUNGLISON *Dict. Med. Sci.* (ed. 8) 396 A hypochondriacal condition, termed by some *cerebropathy*; by others, *brain-fag*. **1884** W. JAMES in *Mind* IX. 17 In states of extreme brain-fag the horizon is narrowed almost to the passing word. **1903** MCFAUL *Ian Golden* iii. 13 He continued in this brain-fagged and mentally-deranged condition for several weeks. **1889** MARCHIONESS OF DUFFERIN in *Vice-regal Life in India* (1890) ii. 30 The 'brainfever' bird repeats his name over and over again until he nearly gives you the malady itself. **1901** *Blackwood's Mag.* Mar. 243 The ceaseless, irritating cry of the brain-fever bird. **1879** G. H. LEWES *Problems* III. iii. 416 The corona (what the Germans call the *Brain-mantle* because it covers the other parts). **1894** H. ELLIS *Man & Woman* v. 109 The cerebrum, or the brain-mantle. **1879** G. H. LEWES *Problems* III. ii. 67 The popular argument..showing more brain-mass to be accompanied by greater intelligence, breaks down. **1915** E. R. LANKESTER *Diversions of Naturalist* 270 The mere shape of the brain-mass. **1890** W. JAMES *Princ. Psychol.* II. xviii. 81 [illegible] **1892** — *Coll. Ess. & Rev.* (1920) 279 The brain-path theory [illegible]. **1925** C. FOX *Educ. Psychol.* 143 [illegible] brain-paths to account for the phenomena of memory. **1810** LADY LYTTELTON *Let.* 11 Feb. (1912) 94 He will meet a formidable body of brain-pickers. **1963** *Times* 17 Mar. 15/1 Very successful farmers are apt to be inveterate brain-pickers. **1892** 'MARK TWAIN' *Amer. Claimant* i. 4 My very footfalls time themselves to the brain-racking rhythm. **1897** *Daily News* 10 Mar. 8/1 It is this that causes the English in South Africa so much brain-racking. **1905** *Westm. Gaz.* 21 Mar. 1/3 It was a noisy..monster and began its torturing, brain-racking persecution at eight in the morning. **1905** *Daily Chron.* 29 Sept. 3/5 [illegible] brain-racking railway guide. **1857** DUNGLISON *Med. Lex.* 718/1 *Pineal Gland*.. It almost always contains sabulous particles, *Sabulum conarii*, Brain Sand, Pineal Sand. **1873** T. H. GREEN *Introd. Pathol.* (ed. 2) 75 The corpora amylacea..are very liable to become calcified, and they then constitute one form of 'brain sand'. **1879** G. H. LEWES *Probl. Life & Mind* 3rd Ser. III. xiv. 416 That large portion of the neural axis which Germans call the Brain-stem. **1927** HALDANE & HUXLEY *Anim. Biol.* vi. 142 When we are 'doing nothing' the cortex is all the time [illegible] the postural centres in the brain-stem from producing rigidity. **1681** GREW *Museum* II. i. iv. 2 Brain-Stone. **1758** BORLASE *Nat. Hist. Cornw.* 240 The fossil corals..such as brain-stone. **1855** KINGSLEY *Glaucus* 34 A beautiful madrepore or brainstone on your mantelpiece, brought home from some Pacific coral-reef. **1871** T. R. JONES *Anim. Kingd.* (ed. 4) 112 The beautiful structures known as *Meandrina*, or 'brain-stones'. **1907** *Daily Chron.* 13 Feb. 7/5 Ordeals of mind which formed a brain-storm or mental explosion. **1908** *Westm. Gaz.* 8 Aug. 4/3 In the closing years of his active life Ruskin had suffered from recurrent brain-storms. **1922** *Daily Mail* 2 May 5/4 If he were crossed he had brain storms which took the form of violent temper and depression. **1925** *College Humor* Feb. 43/2 He had a brainstorm. **1932** *Amer. Speech* VII. 329 *Brain storm*, a sudden and usually fortunate thought. **1947** I. ASIMOV in E. Crispin *Best SF* II (1962) 103 When our missing robot failed of location anywhere..we brain-stormed ourselves into counting the robots left. **1953** *Manch. Guardian Weekly* 12 Feb. 3 They carry behind their very brain-storm the authority of the White House. **1957** *San Francisco Exam.* 15 Mar. [Pict. Rev. Sect.] 4/2 Alex Osborn..has contributed to American business a new technique and terminology—the 'brain storm session'. **1959** *N.Y. Times* 8 Nov. 83/5 They are being taught..how to brainstorm their way to conclusions concerning subjects ranging from world affairs to specific engineering puzzles. [illegible] C₆H₁₂O₆: a principle derivable from the brain-substance, and having a resemblance to diabetic sugar. **1923** H. C. WITWER *Fighting Blood* iv. 105 While you're puzzling over that brain teaser, I'll get back to Judy. **1966** OGILVY & ANDERSON *Excurs. Numb. Theory* p. xi Here are some of the super brain-teasers that Sierpinski asks us to ponder. [illegible] brain twisters never so hard as to discourage nor so easy as to fall of sustained interest. **1942** R. G. COLLINGWOOD *New Leviathan* xvi. 116 This is a famous brain-twister planted upon the world by Kant and Fichte. **1890** W. JAMES *Princ. Psychol.* I. v. 119 Then the last brain-vibration would discharge downward into the motor tracts. **1905** A. R. WALLACE *My Life* II. 309 Thought or brain-vibrations, may be carried by the ether to other brains. **1869** C. READE *Put Yourself in his Place* (1870) II. 61 Now that very afternoon..as if by the influence of what they call a brain-wave, Grace Carden..was moved to ask [etc.]. **1871** A. T. RITCHIE *Lett.* (1924) 147 You must have sent a brain wave, for oddly enough we had all just read the book..when your letter came saying you too had been reading it. **1886** *Proc. Soc. Psych. Research* Oct. 178 Such expressions as 'brain-waves' (Knowles), 'mentiferous ether' (Maudsley)..testify to the natural, though premature, desire to ticket or identify a force which..cannot at present be correlated with nerve-force. **1890** *Harper's Mag.* Apr. 744/1 Lucilla, with what she was fond of terming a brain wave, comprehended the situation. **1926** *Blackw. Mag.* Aug. 184/1 Then the wires got brain waves, saw the folly of their first orders, and began to panic terribly. [illegible] **1926** GALSWORTHY *Escape* ii. 72 Look here! I've got a brain wave. Let's all go into Widecombe in the car! **1935** *Discovery* Jan. 31 By means of electrical records made through the skull various changes in the brain can be recognised; but the 'brain waves' thus recorded do not appear to be the result of thought-action. **1959** *Listener* 1 Oct. 534/3 Study of the activity of the brain of puppies with the electro-encephalograph for 'brain waves'..reveals a juvenile type of wave form.

brainily (brē·nili), *adv.* [f. BRAINY *a.* + -LY².] In a brainy fashion; with clever use of the wits. So **brai·niness**.

1930 *Daily Mail* 21 Nov. 10 Lewis's tactics..were extremely bad. There was no suggestion of braininess. **1931** *Daily Exp.* 28 Sept. 10 Against Lewis,..he certainly fought brainily.

brain trust, brains trust. [transf. sense of TRUST *sb.* 7.] **a.** The name (usu. in form *Brain Trust*) given to a group of experts appointed in 1933 to advise the American President F. D. Roosevelt on political and economic matters. **b.** (usu. *brains trust*) A group of persons assembled (orig. in *Broadcasting*) to give their impromptu views on topics of current or general interest. Both forms (in Britain usu. *brains trust*) also used *transf.* of any group of experts.

Quot. 1910 is a chance occurrence of the expression *brain trust*; the main use began in 1933.

1910 in *Amer. Speech* (1957) XXXII. 57 Brain trust. **1929** *Papers Michigan Acad. Sci., Arts & Lett.* X. 181/2 *Brain Trust* [Am.], the General Staff. **1933** *Sun* (Baltimore) 27 Aug. 8/6 The 'brain trust' of the gang chooses a city for the 'operations' located within a hundred miles of a State boundary line. **1933** *Newsweek* 1 Sept. 4/1 The President's Brain Trust, a little band of intellectuals. **1934** H. G. WELLS *Exper. Autobiogr.* II. ix. 791 The new President..was using these new brains into consultation. Some journalist..invented the phrase the 'Brains Trust'. *Ibid.* 792 This Brains Trust movement. [illegible] The 'brain trust' of the [illegible] are men who have read widely, thought deeply on industrial question, and might be characterized as 'intellectuals'. **1937** C. ODETS *Golden Boy* I. iii. 45 [illegible] 'brain trust' You could fit in a telephone booth! **1940** *Amer. Speech* XIV. 246 The term *brains trust*..was originally used by James Kieran during the campaign of 1932. He used the plural, *brains*. **1940** *War Illustr.* 1 Nov. 425/2 The 'brain trust' of the R.A.F. in France. **1941** *New Statesman* 27 Sept. 300/3 A question recently discussed by the Brains Trust. [illegible] **1941** *Manch. Guardian Weekly* 7 Dec. 365/1 In one unit the commanding officer, a captain, and a private form the Brains Trust. **1948** W. CHURCHILL [illegible] the 'Brains Trust' movement..so rankly started by the B.B.C., was sweeping England. Every town, every village, every parish, every street had its Brains Trust, at whose meetings earnest seekers after knowledge discussed the scientific, religious or political problems of the day. **1962** *Lancet* 1 Dec. 1180/1 A brains trust discussed practical health education of village parents. **1966** A. HAILEY *Airport* i. viii. 83 There were other sessions, some of them 'brain trust' affairs involving Kennedy aides.

Hence **brain(s)·truster**, a member of a brain(s) trust.

1934 *Sun* (Baltimore) 28 Mar. 3/3 The Roosevelt 'brain trusters'. **1953** *Manchester Guardian Weekly* 23 Apr. 2 Foreign Ministers, advisers, and anonymous brain trusters. **1958** *New Statesman* 1 Feb. 127/2 He believes the Brains-trusters really are equipped to pronounce themselves upon, virtually, anything.

brai·nwashing. orig. *U.S.* [f. BRAIN *sb.* + WASHING *vbl. sb.* 1 b.] The systematic and often forcible elimination from a person's mind of all established ideas, esp. political ones, so that another set of ideas may take their place; this process regarded as the kind of coercive conversion practised by certain totalitarian states on political dissidents. Also *attrib.* and *transf.* Hence, by back-formation, **brainwash** *v. trans.*, to practise brainwashing on; **brainwashed** *ppl. a.*; **brainwasher**, one who practises it.

1950 E. HUNTER in *New Leader* 7 Oct. 7/2 'Brain-reform' is the objective, popularly referred to as 'brain-washing'. **1953** *Time* 11 May 42/1 [illegible] was Red China's chief indoctrinator or, as he was generally called, Brainwasher No. 1. **1953** *Manch. Guardian Weekly* 21 May 15 You refer to brainwashing. You feel..that you have been brainwashed. **1953** *Sat. Even. Post* 31 Oct. 20/1 The anticommunist soldiers..may be blackmailed or brain-washed or third-degreed. **1955** *Times* 10 June 7/4 Snake-charmers..have so far failed in their attempts to brain-wash the sentries. **1959** *Harper's Mag.* July 16 On this subject—and others—the Texans have brain-washed themselves so thoroughly. **1961** *Times* 22 Aug. 8/5 Realistic 'brainwashing' procedures for those who are 'captured'. **1957** J. D. KNIGHT *Apothecary's Shop* 225 (*title*) The Brain-Washed Muse. Some Second Thoughts on Tradition. **1968** M. WOODHOUSE *Rock Baby* viii. 83 [illegible] idea of psycho-analysing a brain-washed robot.

brainy, *a.* Delete 'Chiefly in U.S.' and add earlier and later examples.

1845 LEIGH HUNT *Let.* 19 July (1862) II. 104 That was a good, hearty, brainy, valuable bit of existence you gave us the other night. **1951** J. CANNAN *And All I Learned* x. 274 All of them had passed except poor old Timothy and she had to confess that he wasn't at all brainy. **1960** M. DICKENS *Angel in Corner* xi. 228 Betty's fiancé was an undersized but brainy boy.

brak (brak), *a.* and *sb.* *S. Afr.* Also (rarely) **brack.** [Afrikaans, ad. Du. *brak*.] **A.** *adj.* Brackish; alkaline. **B.** *sb.* Brackishness; alkalinity; alkaline soil. Cf. BRACK *a.* and *sb.*⁴

1793 tr. *Thunberg's Trav.* II. 51 *Brak-water* is water stagnating in valleys and low places; it contains a kind of brine, and tastes very salt and bitter. [illegible]

C. *Comb.* **brakbush,** name of various salt-loving plants, esp. species of *Atriplex, Salicornia* and *Salsola*.

[illegible]

brake, *sb.*³ Add: **2.** (Later examples.) Hence **brakesman, break(s)man,** a man who operates a baker's kneading-machine; **brake-** (or **break-**) **staff** (see quots.).

1836 P. BARLOW in *Encycl. Metrop.* (1845) VIII. 801/2 [illegible] **1968** [illegible]

brake, break, *sb.*⁷ (The former is now the usual spelling.) Add: **3. brake-lever, -pedal, brake-cylinder** (see *brake-pipe*); **brake-drum,** a cylinder attached to a wheel or hub, upon which the brake shoe presses; **brake-gear,** the whole braking apparatus of a motor-car or train; **brake-handle,** a hand lever controlling a brake, taking the place of the usual brake-wheel; **brake-horse-power** (see *HORSE-POWER); **brake lining,** a strip of fabric attached to the face of a brake-shoe to increase friction and provide a renewable surface; **brake mean effective pressure** (see quot.); **brake parachute** *Aeronaut.*, a parachute attached to the tail of an aeroplane and opened to serve as a brake; **brake-pipe,** the pipe of an automatic air-brake, which conveys compressed air to the cylinders operating the brakes of a railway train; **brake-strap,** a strap which surrounds the pulley of a brake worked by friction.

a1877 KNIGHT *Dict. Mech.* I. 336/2 The air-pipe under each car-body..called the brake-pipe. [illegible] A brake-drum of peculiar construction placed under the seat. [illegible] **1968** [illegible] Suppl. Oct. 134/1 The figures for brake lining life..are calculated.

calculated. **1927** V. W. PAGÉ *Mod. Aircraft* (1928) vii. 302 Brake mean effective pressure—The net unit pressure which, if applied during the power strokes to the pistons of an engine having no mechanical losses, would produce the given brake horsepower at the stated speed. **1942** H. S. ZIM *Parachutes* 251/1 (*index*) Brake parachute. **1903** R. J. MECREDY *Dict. Motoring* 33 The motion of the brake pedal causes mechanism..to close the two halves inward and grip the brake drum. **1906** D. LEECHMAN *Autocar Handbk.* xv. 117 In many cars the brake pedal is coupled up to the clutch, so that applying the brake automatically disengages the motor from the driving gear. **1886** *Encycl. Brit.* XX. 249/1 The brake can be at once released by restoring the pressure in the brake-pipe. **1835** URE *Philos. Manuf.* 53 The frame..acts, of course, upon the slide-pieces..which hold the brake-strap. **1897** *Outing* (U.S.) June 262/1 Right thumb on spool of reel, left thumb in leather brake-strap.

brake, *v.*⁵ Add: **2.** *intr.* To be checked by a brake. Also with *up*.

1891 E. S. ELLIS *Check 2134* ii. 13 He felt the train braking up for the Station. **1937** J. SQUIRE *Honeysuckle & Bee* 149 A car suddenly braked to a standstill outside the door.

brakesman. 2. Delete 'in U.S. (*brakeman*) the guard', and add earlier and later examples.

1833 *Amer. Railroad Jrnl.* II. 738/1 Two brakemen $450 00. **1838** *Civil Engineer & Architect's Jrnl.* I. 388/1 The engine was thrown off the line..and the 'breakman' so dreadfully injured that his life is despaired of. **1921** *Daily Colonist* (Victoria, B.C.) 22 Mar. 1/5 Mr. Murdock asserted that $5.12 was not too much for a railway brakeman for an eight-hour day or a run of 100 miles. **1967** *Boston Herald* 1 Apr. 1/3 Railroad conductors and brakemen were barred Friday from proceeding with a nationwide railroad strike.

braking (brē¹·kiŋ), *vbl. sb.* [f. BRAKE *v.*⁵ + -ING¹.] The action of applying a brake (to a wheel); also *attrib.*, as *braking distance* (see quot. 1950).

1904 *Daily Chron.* 15 June 6/7 A clever braking arrangement secures a representation of the difficulties encountered in running on the metals. **1905** *Ibid.* 13 Feb. 3/7 Frictional braking is jarring, uncertain, difficult of gradual application, and most destructive to the parts concerned. **1909** *Westm. Gaz.* 16 Sept. 5/1 Fixed to these chains is a braking band. **1950** *U.S. Bur. Publ. Roads Highway Capacity Manual* 18/1 *Braking distance*, the total distance traversed by a vehicle while it is being brought to rest, measured from the position of the vehicle at the instant the brake shoe touches the brake drum. **1964** *Times* 11 Feb. 11/7 Drivers moving at speeds above their vision-and-braking distances.

Bramantesque (brămănte·sk), *a.* [ad. It. *Bramantesco*, f. the name of *Bramante* d'Urbino (1444–1514), a celebrated Italian architect.] Designating the style of architecture now known as Renaissance.

1883 C. C. PERKINS *Handbk. Ital. Sculp.* 182 *note*, The artist who introduced Renaissance architecture, then called Bramantesque, into Lombardy. **1908** E. WHARTON *Hermit* 162 The Library cupola assumed a Bramantesque grace.

Bramley (bræ·mli). [f. the name of Matthew *Bramley*, an English butcher, in whose garden at Southwell, Notts., this apple is said to have been first grown *c.* 1850 (see 1931 *N. & Q.* 14 Nov. 351/2).] A type of cooking apple; in full *Bramley('s) seedling*.

1900 *Gardening* 3 Mar. 1 Vigorous-growing varieties, especially Bramley's..will make strong wood. **1902** *Encycl. Brit.* XXVIII. 530/1 For storing, the cooking sorts favoured now are..Bramley's Seedling, Golden Noble, and Lane's Prince Albert. **1957** E. HYAMS *Speaking Garden* 34 No matter how sour and sharp the apples used—for example, Bramleys. **1959** *Listener* 19 Nov. 907/1, 4 medium-sized Bramley cooking apples.

Bramling (bræ·mliŋ). *local.* Also **Brambling.** [The name of *Bramling* Farm, Ickham, Kent, where the first sets were raised.] A species of hop.

1899 *Westm. Gaz.* 15 Sept. 9/3 In many of the important parishes the Bramlings are all picked. **1902** *Times* 22 Sept. 11/6 Both the Bramlings and the Goldings have grown out fairly satisfactorily.

bran, *sb.*¹ Add: **3.** *bran-mash* (earlier example), *-tea* (example); † **bran-dance** *U.S.* (see quot. 1833); **bran-drench,** a bath of bran and water in which leather is placed to remove the lime used in liming; **bran-duster** (example); **bran-pie** (see PIE *sb.*² 2); **bran-tub** = *bran-pie*; also *fig.*

1833 *Sk. D. Crockett* 148 This is the famous bran-dance of the west, and derives its name from the fact that the ground is generally sprinkled with the husk of Indian meal. **1887** *Harper's Mag.* Dec. 61/2 It all kem about from that thar bran dance. **1883** HALDANE *Workshop Rec.* Ser. II. 373 Bran-drench..is prepared by soaking wheaten-bran in cold water..and straining the extract. **1849** *Rep. Comm. Patents* (U.S.) 374 No. 6952.—Improvement in Bran Dusters. **1838** *Penny Cycl.* XII. 309/1 A bran-mash is given after a day of more than usual fatigue. **1849** C. BRONTË *Let.* in Mrs. Gaskell *Life* (1857) II. 91 The pains..return..but I combat them steadily with pitch plasters and bran tea. **1858** C. PARRY in E. Parry *Mem.* vii. (1870) 173 It quite reminded me of the bran-tub itself as I unpacked each separate article. **1909** *Westm. Gaz.* 22 Apr. 8/2 Sideshows will contain the ever-popular phrenologist's tent and bran-tub. **1963** *Times Lit. Suppl.* 26 Apr. 313/3 This is a mathematical bran-tub.

branch, *sb.* Add: **2. a.** (Earlier examples referring to railways.)

1791 *Act 31 Geo. III* c. 65 (Preamble), To make and maintain One other Rail or Waggon Way, or Stone Road, with Branches therefrom. **1831** in *N. & Q.* (1868) II. 102/1 The railroad is not supposed to answer vastly well, but they are making a branch to Warrington. **1840, 1846** [see *branch-railway, -line* s.v. sense 12 in Dict.].

b. (Earlier and later examples.)

[**1624** JOHN SMITH *Hist. Virginia* II. 23 Here doth the river divide it selfe into 3 or 4 convenient branches.] **1663** in *North Carolina Col. Rec.* (1886) I. 20 That Parcell of land..Beginning at a small creek or Branch. **1796** B. HAWKINS *Lett.* 25, I came to a branch covered with reeds. **1832** J. HALL *Leg. West* (1833) 27 He proceeded cautiously towards a rivulet, or in the vernacular of the country, a branch, that meandered along the foot of the hill. **1917** MATHEWSON *Sec. Base Sloan* v. 67 An' there's a branch close by it too, mighty nice tastin water, Mas Wayne.

11. (Examples referring to the metal piece on the end of a fireman's hose.) Also, the hose itself.

1897 *Daily News* 22 Nov. 7/3 Engineer Heather..ordered his men to drop their 'branches' and run. **1945** S. SPENDER *Citizens in War* 46 Two firemen may then have to stand for hours holding the 'branch'..from which the water proceeds. **1959** R. COLLIER *City that wouldn't Die* viii. 132 Swing your branch to the left, son—give the Shoe Lane corner a drink.

12. b. *branch-road*; **c.** *branch bank* (examples).

1796 *Boston Directory* 302 Branch Bank, State street. **1832** *Chambers's Edin. Jrnl.* I. 162/3 The conductors of the branch banks give no loans..without consulting with their masters. **1831** *Deb. Congress* 1 Mar. 830 The opening of branch roads. **1842** DICKENS *Amer. Notes* I. iv. 149 Except when a branch-road joins the main one, there is seldom more than one track of rails.

13. branch house, an offshoot of a religious community, business firm, etc.; **branch-island,** an island beside a river formed by an *ANA-BRANCH; **branch library,** a library other than the main one in an area; hence *branch librarian*; **branch-point** *Math.*, a point in the complex plane at which two or more branches of a function of a complex variable coincide.

1840 K. H. DIGBY *Mores Cath.* X. i. 9 Cisteaux, the mother house of the order, [was] founded..in 1098... La Ferté was the first branch house. **1872** GEO. ELIOT *Middlem.* VI. lxi. 346 There was a branch house at the west end. **1834** J. R. JACKSON in *Jrnl. R. Geogr. Soc.* IV. 79 Thus, such branches of a river as after separation re-unite, I would term anastomosing-branches; or, if a word might be coined, ana-branches, and the islands they form, branch-islands. **1938** H. A. SHARP *Branch Libraries* iii. 62 A superintendent of branches..is necessarily a mediary between the chief and the branch librarian. **1862** MRS. GASKELL *Let.* 16 June (1966) 689 Would it be much trouble for you to enquire from him [sc. Mr. Mudie] whether he has any Branch-Library in Paris..? **1927** *Cmd. 2868* (*Publ. Libr. Rep.*) 66 The branch library is the most important form of library extension. **1878** CAYLEY in *Proc. Lond. Math. Soc.* IX. 32 There are certain points *V* called branch-points (Verzweigungspuncte), such that to each point *V* there correspond two united points..and *n*-2 other distinct points. **1966** E. G. PHILLIPS *Topics Complex Analysis* ii. 20 On traversing loop *A* in the positive sense, the straight pieces have no effect on *f(z)* since they do not contain a branch point.

branched, *ppl. a.* Add: **4.** *Chem. branched chain*: an open chain (*CHAIN *sb.* 5 g) of atoms having one or more side chains. Freq. *attrib.*

1889 G. M'GOWAN tr. *Bernthsen's Text-bk. Org. Chem.* i. 43 The boiling point becomes lowered continuously the more the carbon atom chain is branched. **1903** A. J. WALKER tr. *Holleman's Text-bk. Org. Chem.* I. 46 In branched chains there are carbon atoms which are directly linked to three or four others. *Ibid.*, Branched-chain compounds are often distinguished by the prefix *iso*. **1939** *Jrnl. R. Aeronaut. Soc.* XLIII. 42 Straight-run petrols..and branched-chain paraffins are the outstandingly useful aviation fuels. **1946** *Nature* 14 Dec. 863/1 One fundamental approach to this problem was based on fatty acids with branched-chains.

branchial, *a.* Add: **branchial arch,** one of a series of bony or cartilaginous structures supporting the gills of fishes and amphibians, and found embryonically in mammoths.

1875 *Encycl. Brit.* I. 757/2 There is no trace of any fourth branchial arch, such as exists in tadpoles and in young salamanders. **1936** J. T. JENKINS *Fishes Brit. Is.* (ed. 2) 7 The variation in the structure of the gills and their supporting branchial arches.

branchite (bræ·ŋkəit). *Min.* [ad. G. *branchit* (P. Savi 1842, in *Neues Jahrb. f. Mineralogie* 459), f. the name of J. *Branchi* of Pisa: see -ITE¹.] A mineral resin found in fossil pine wood.

1851 DANA *Man. Min.* 97 *Branchite*, found with coal, especially brown coal, and resembling wax or tallow. **1883** *Encycl. Brit.* XVI. 429/1 Branchite,..is similar [to Scheererite].

brand, *sb.* Add: **4. d.** *spec.* A mark of ownership impressed on cattle, horses, etc., by branding. Also *attrib.* and *Comb.*, as **brand blotter,** one who steals cattle and obliterates the ownership-marks. Chiefly *U.S.*

1665 in *Conn. Public Rec.* (1852) II. 28 They shal enter such saile..in the said brand booke. **1667** *Ibid.* 58 For yr brand of horses they shal have ye letter V on ye near buttock. **1707** *Boston News-Let.* 10 Feb. 2/2 A pretty large bright bay Gelding, having a Brand on his near Buttock of this form d. **1869** C. L. BRACE *New West* xxii. 288 The brands, both of cattle and horses,..are controlled by law in California. **1888** *Century Mag.* Apr. 860 A man must have natural gifts, as well as great experience, before he becomes a good brand-reader. **1910** W. M. RAINE *B. O'Connor* 107 You red-haided [= headed] son of a brand blotter. **1930** L. G. D. ACLAND *Early Canterbury Runs* iv. 77 He was probably D.A. or J.D. Brittan, who registered a brand between them in 1854. **1963** A. LUBBOCK *Austral. Roundabout* 147 A brand-forger with a dozen aliases.

e. A steer or other animal bearing a particular brand-mark. *U.S.*

1881 G. W. ROMSPERT *Western Echo* 186 It is seldom they kill their own brands. **1903** A. ADAMS *Log Cowboy* vii. 90, I must have inspection papers before I can move a brand out of the county in which it is bred.

9. *attrib.* (sense 6) and *Comb.*, as **brand-image,** the impression of a product in the minds of potential users or consumers; also *transf.* and *fig.*, the general or popular conception of some person or thing; **brand-name,** a trade or proprietary name; also *transf.*

1958 M. MAYER *Madison Avenue U.S.A.* iii. 59 David Ogilvy, of Ogilvy, Benson & Mather, apostle of the 'brand image'. *Ibid.* iii. 63 Ogilvy's brand-image advertising.. works essentially on the consumer's conscious mind in an effort to convince him that brand A, technically identical with brand B, is somehow a better product. **1959** *News Chron.* 14 July 4/6 In the jargon of the P.R. trade, there is as yet no 'brand image' for the Prime Minister of Japan. **1960** K. AMIS *New Maps of Hell* ii. 45 Space-opera with a full complement of BEMs and a small staff of mad scientists attended by scantily clad daughters constitutes..the main brand-image of science fiction. **1961** *Guardian* 27 May 6/2 Methods..used to determine brand loyalty among smokers. **1922** *Hotel World* 25 Mar. 9 (*heading*) Brand names on menus? **1950** A. H. SAYER et al. *Economics in our Democracy* IV. xiii. 99 Each package has a brand name, which identifies the product as having been made by a certain manufacturer. **1952** E. PARTRIDGE *From Sanskrit* 49 The association existed long before any Scotch whisky received the brand-name *the real Mackay*. **1958** *Spectator* 27 June 836/2 The brand-name wrapped loaf. **1960** *20th Cent.* Sept. 234 Mr. Wesker's enemies dismiss him as a mere brand-name oversold by the theatrical Left.

brand, *v.* Add: **2. a.** *spec.* To mark (cattle or horses) with a brand.

1644 in *Conn. Public Rec.* (1850) I. 118 Owners..shall earemarke or brand all their Cattle and swyne that are aboue halfe a yeare old. **1652** [see BRANDED *ppl.a.*¹ 1 a]. **1669** in *Springfield* (Mass.) *Rec.* (1899) II. 101 There are many horses to be branded. **1765** WASHINGTON *Diaries* (1925) I. 216 Calves and Yearl[in]gs branded on the Right Shoulder GW. **1869** C. L. BRACE *New West* xxii. 288 Each owner..lassoed the cattle which were his, branding the calves and those whose marks were somewhat obliterated.

‖ **brandade** (brandad'). [Fr., ad. Pr. *brandado*, lit. 'thing which has been moved, shaken'.] In full *brandade de morue*. A Provençal dish made from salt cod.

1825 *Fr. Domestic Cookery* 195 (*heading*) Brandade of salt cod. **1866** H. TOOGOOD *Treas. Fr. Cookery* 46 (*heading*) Cod.—brandade de morue. **1877** E. S. DALLAS *Kettner's Bk. Table* 84 The first recipe for a brandade..was written by Grimod de la Reynière. **1951** R. CAMPBELL *Light on Dark Horse* xxi. 294 Nothing is better, says a Provençal proverb, for reviving and remaking a man than living in the clear sunlight, speaking Provençal, eating 'brandade', and drinking the good red wine of Provence. **1966** 'K. NICHOLSON' *Hook, Line & Sinker* v. 64 Brandade de Morue..cod fillets cooked with garlic, cream, olive oil, lemon juice and a teeny suggestion of fennel.

branded, *ppl. a.*² Add: **1. c.** Labelled with a trade or proprietary name; having a brand-name.

1897 *Daily News* 16 Dec. 7/2 A customer asked for a certain branded ham. **1946** D. L. SAYERS *Unpopular Opinions* 128 Any proposal to control the marketing of branded goods..will be violently opposed, on the loftiest hygienic grounds, by the papers that carry the branded advertising.

brandenburgs. Add: **b.** (Also **brande(n)bourgs.**) Ornamental trimmings (see quot. 1882) on a woman's dress, in fashion from about 1870 to 1910.

1873 *Young Englishwoman* Mar. 130/2 One [mantle].. is fastened in front with brandenbourgs, finished off with macarons and round balls of the same style. **1882** CAULFEILD & SAWARD *Dict. Needlew.* 44/2 *Brandenbourgs.* Synonymous with 'Frogs'. A button formed somewhat in the shape of a long and narrow barrel, smaller at the ends than the middle, and made of silk on a wooden foundation. **1891** *Queen* 17 Jan. 103/1 Costumes of Hussar-blue cloth with black braid brandebourgs. **1902** *Daily Chron.* 8 Nov. 8/3 The fulness of the fronts is finished with brandebourgs of silk cords. **1909** *Westm. Gaz.* 17 July 15/1 It hangs rather full from the waist, is open in front, or caught with Brandenburgs of braid or tinsel.

brandified, *ppl. a.* Add: **2.** Mixed or treated with brandy.

1841 THACKERAY *Mem. Gormandising* in *Wks.* (1900) XIII. 576 A brandyfied liquor called sherry. **1877** G. EASTON *Wine of Cana*, Brandified and whiskeyfied wines.

branding, *vbl. sb.* Add: **2. branding-chute** *U.S.*, a gradually narrowing enclosure into which cattle are driven to be branded.

1882 *Lippincott's Mag.* May 429/2 Several small fenced enclosures..and a branding-chute are soon completed. **1911** Mulford *Bar-20 Days* xx. 197 Johnny Nelson waited.. on the platform of the branding chute.

brandisite (bræ·ndisəit). *Min.* [ad. G. *brandisit* (1846), f. the title of Clement, Count of *Brandis*, after whom the mineral was named: see -ite¹.] A variety of seybertite.

1868 Dana *Syst. Min.* (ed. 5) 508. **1883** *Encycl. Brit.* XVI. 413/2 Brandisite is similar [to Clintonite].

brand-new, bran-. For the last sentence in etym. read: The commoner form is now *brand-new*.

brandtite (bræ·ndtəit). *Min.* [ad. Sw. *brandtit* (1888), f. the name of Georg *Brandt*, Master of the Swedish Mint: see -ite¹.] Hydrated arsenate of calcium and manganese occurring in crystal form near Pajsberg in Sweden.

1896 Chester *Dict. Min.*

brandy, *sb.* Add: **1. b.** A drink of brandy. Similarly *brandy-and-soda* (cf. *B. and S.* s.v. B. III in Dict. and Suppl.).

1884 G. Moore *Mummer's Wife* (1887) 193 The brandies and sodas supped in the dressing room. **1900** E. Glyn *Visits Elizabeth* 253 Charlie had two brandies-and-sodas instead of his usual glass of milk. **1903** *Daily Chron.* 11 Nov. 5/2 Took a brandy before going to kirk, lest I should smell of whisky in the house of the Lord! **1932** E. Bowen *To North* xvii. 174 They would want two brandies.

c. *fig.* Something that stimulates or excites.

1903 G. B. Shaw *Man & Superman* III. 99 Hell is full of musical amateurs: music is the brandy of the damned. **1930** M. Enthoven tr. *Janni's Machiavelli* 80 Machiavelli is the most glorious distiller of political brandy who has ever existed.

2. *brandy-punch* (earlier example); *brandy-sodden* adj.; **brandy-ball** (earlier examples); **brandy-butter,** a hard sauce made of brandy and butter; **brandy-cherry** (examples of sense 'a cherry preserved in brandy'); **brandy-smash** *U.S.* (see quot. 1909 and Smash *sb.*¹ 5).

1825, 1849 Brandy-ball [see Hard-bake]. **1939** 'Vivette' in A. L. Simon *Concise Encycl. Gastronomy* I. 29/2 In U.S.A., a *Hard Sauce* is made with one measure of fresh butter to two of castor sugar...In England, a similar sauce is called *Brandy Butter* or *Rum Butter*. **1853** *Harper's Mag.* VII. 275/2 A swine drunk on brandy-cherry stones is disgusting. **1885** C. M. Yonge *Two Sides of Shield* II. v. 101 And there was Miss Hacket getting brandy cherries and strong coffee. **1689** Sewall *Diary* (1878) I. 306 In the Lord's Hall Guy's Pot was filled with Brandy Punch. **1850** A. W. Thaxter *Poem before Iadma* 7 Or didst thou at the Pemberton absorb a brandy-smash? **1909** *Cent. Dict.* Suppl., *Brandy-smash*, a drink made by mixing brandy with crushed ice and putting a few sprigs of mint in the glass. **1854** Whyte Melville *Gen. Bounce* (1855) I. ix. 203 Sufficient to destroy the ideal in the most brandy-sodden brain.

branner (bræ·nəɹ). [f. Bran *sb.*¹ + -er¹.] **1.** An operative who cleans tinned plates with bran.

1881 *Instr. Census Clerks* (1885) 94 White Branner. **1921** *Dict. Occup. Terms* (1927) § 279 *Branner*, a tinman's helper attending branning machine,..sometimes specifically designated according to whether black or white plates are being cleaned, e.g., black branner, white branner.

2. A machine for removing the oil from tinned plates by means of bran and slaked lime.

1902 *Sci. Amer.* 1 Nov. 290/2 As the plates leave the tin pot, they have upon them a thin coating of oil which has to be removed...They are put into a branner which is located conveniently at the side of the tinning machine.

brash, *a.*² Add: Latterly in general colloq. use, in imitation of U.S. currency: impulsive, assertive, impudent; crude, insensitive; flashy. Also as *adv.*

(Quots. 1837 and 1889 are U.S.)

1837 R. M. Bird *Nick of Woods* I. viii. 120 Strannger thar's as brash as a new hound in a b'ar fight. **1889** 'C. E. Craddock' *Despot of Broomsedge Cove* ii. 27 Ye notice how turrible brash Josiah Preen be,—can't wait fur pa'son ter summons him. **1928** *Punch* 4 Apr. 374/2 He was going out in his brash street-suit with the flash tie. **1946** J. B. Priestley *Bright Day* viii. 248 That feeling of inevitable national superiority..which makes decent people seem brash and insensitive. **1948** W. Sansom *South* 114 The dogs of Italy do not chase cats like their brash northern brothers. **1950** R. Macdougall *To Dorothy, A Son* in *Plays of Yr.* 418 The room has the lurid, brash appearance of an American cocktail bar. **1956** A. Wilson *Anglo-Saxon Att.* I. i. 7 This brash young American little knew what sore places he was invading with his clumsy fingers. **1958** *Oxf. Mag.* 1 May 390/2 One of those amiable institutions which survive in an alien age until they stand directly challenged by some brash newcomer.

b. Hard, harsh, rough. Also as *adv. U.S.*

1868 *Putnam's Mag.* Dec. 675/1 See here,..you are playing this a little too brash. **1872** Schele de Vere *Americanisms* 446. **1880** A. W. Tourgée *Bricks without Straw* 116 He was pretty brash wid me, an 'llowed ter hit me wid a stick. **1896** G. Ade *Artie* iii. 23, I swore I'd get next, no matter what kind of a brash play I had to make. **1904** W. H. Smith *Promoters* 51 We won't rob 'em entirely; there isn't any use in being altogether too brash.

2. Active, quick. Also as *adv. U.S.*

1884 'Mark Twain' *Huck. Finn* 62 When I got to camp I warn't feeling very brash, there warn't much sand in my craw. **1888** 'C. E. Craddock' *Keedon Bluffs* 63 Whar's that buckeye tree ye war a-goin ter cut down fur me so brash? **1891** M. E. Ryan *Pagan of Alleghanies* 118, I ain't so brash in the timber as I'd like to be.

brash (bræʃ), *v.*² [See Brash *sb.*³ and cf. *brush *v.*² 7.] *trans.* To remove the lower branches of (a tree). Also with *up*. Hence **bra·shing** *vbl. sb.*

1950 *Q. Jrnl. Forestry* XLIV. 72 As a fire precaution..a belt 30 ft. wide should be completely brashed up as soon as possible. *Ibid.* 75 'Brushing up or brashing' means the removal of the lower boughs up to 5 to 7 ft. **1959** W. K. Richmond *Brit. Birds of Prey* xiii. 150 Forestry Commission plantations before they have been 'brashed'.

brashly (bræ·ʃli), *adv.* orig. *U.S.* [f. Brash *a.*²] In a brash manner.

1865 'Mark Twain' *Sk. Sixties* (1926) 179, I mixed into this business a little too brashly—so to speak—and without due reflection. **1884** [see Brash *a.*²]. **1949** *Here & Now* Oct. 11/3 A brashly self-confident member of an Opposition which had had the confidence largely knocked out of it. **1952** C. Day Lewis tr. *Virgil's Aeneid* v. 104 What that exhibitionist yonder So brashly relies on. **1957** 'E. Ferrars' *Furnished for Murder* iii. 25 An energetic, rather brashly handsome woman.

bra·shness¹. orig. *U.S.* [f. Brash *a.*¹] Brittleness.

1864 *Rep. Agric. Soc. Maine* 54 They [sc. sheep] become very fat in the summer, which increases the brashness of the wool. **1929** *Ann. Appl. Biol.* XVI. 41 The term 'brashness' arose in America, and is now in common use in the timber trade of both this country and the U.S.A. In this paper its use will be restricted to that condition of timber.. which produces a peculiar 'carroty', short, fracture.

bra·shness². orig. *U.S.* [f. Brash *a.*²] The state or quality of being brash, impetuous, crude, etc.

1883 [see Brash *a.*²]. **1894** P. L. Ford *Honorable P. Stirling* (1898) xliii. 258 'That is a bargain,' he said, with a brashness simply disgraceful in a good business man. **1957** *Guardian* 9 Sept. 4/4 One needs..a maximum of brashness in out-staring curious sightseers. **1958** J. Carew *Wild Coast* v. 62 His front of brashness was his way of finding out how far he could go.

brasque (brask), *v.* [f. Brasque *sb.*] *trans.* To line with brasque. Hence **brasqued** *ppl. a.*

1880 *Encycl. Brit.* XIII. 319/2 The bed of this latter [finery] is 'brasqued' or lined with charcoal powder moistened and rammed in. **1885** *Spons' Mechanics' Own Bk.* 17 If brass be heated in a brasqued crucible.

brass, *sb.* Add: **1. e.** (Earlier example.)

1829 S. Glover *Hist. Derby* I. v. 234 Many of the coal-seams..have considerable quantities of brasses or drosses in them, which are lumps of iron pyrites.

2. d. (Further examples.)

1876 G. B. Shaw *How to become Mus. Critic* (1960) 10 A good deal of piccolo, drum, and cymbal, relieved by an effective melody for the brass. **1926** Whiteman & McBride *Jazz* ix. 195 Musicians recognize four general classes of instruments in speaking of the orchestra—strings, wood winds, brasses, and the battery of traps.

e. *slang* (orig. *U.S.*). Senior officers in the armed forces (in allusion to their brass or gold insignia); 'brass-hats' collectively; esp. *the big* (or *top*) *brass*. Also *transf.* and *attrib.*

1899 *Boston Herald* 26 July 4/8 It was not a big brass general that came; But a man in khaki kit. **1945** *Life* 2 July 13/3, I don't suppose that Congress and the Big Brass would ever agree to that. **1949** *Bulletin* (Philad.) 14 Sept. 4/1 The top police brass spreads out a hot carpet for the local cops. **1951** E. Ambler *Judgment on Deltchev* xvii. 204 Some of his revelations..were deeply shocking to the Anglo-American brass. **1951** *Economist* 15 Dec. 1463/1 The 'high brass' of American business was also well represented at the meeting. **1952** *Newsweek* 19 May 21 Top Pentagon brass are taking the aviation-gasoline restriction seriously. **1959** A. C. Clarke *Across Sea of Stars* 123 The general was unaware of his *faux pas*. The assembled brass thought for a while.

f. *Typogr.* A brass block or die, esp. one used for a design or lettering on the cover of a book. Often with defining word.

1930 M. Sadleir *Evol. Publishers' Binding Styles 1770–1900* 90 Such lettering was printed from a specially cut binding-brass...This brass was sometimes discarded and plain type lettering employed. **1951** S. Jennett *Making of Books* 175 The brass from which the lettering and decoration is blocked on a modern book.

g. *slang.* = *brass nail.

1934 [see sense *7]. **1952** N. Streatfeild *Aunt Clara* 161 If Mr. Willis thought she was a brass, he had got another think coming...If you looked at those brown eyes you could see she was innocent as a baby. **1958** F. Norman *Bang to Rights* 10 His old woman who was a brass on the game.

4. b. Colloq. phr. *as bold as brass*: very bold(ly) or impudent(ly); brazen-faced(ly).

1789 G. Parker *Life's Painter* 162 He died damn'd hard and as bold as brass. An expression commonly used among the vulgar after returning from an execution. **1849** Lytton *Caxtons* I. I. iv. 27 Master Sisty (coming out of the house as bold as brass) continued rapidly [etc.]. **1922** S. J. Weyman *Ovington's Bank* xvii. 188 Seeing as he hung back I up to him bold as brass.

5. b. Colloq. phr. *to come* (or *get*) *down to brass tacks* (or *nails*): to concern oneself with basic facts or realities. orig. *U.S.*

1897 H. A. Jones *Liars* I. 23 Come down to brass tacks. What's going to be the end of this? **1903** *N.Y. Sun* 28 Nov. 3 This bold sister was the first..to get down to brass tacks in a discussion of the scandal. **1911** H. Quick *Yellowstone N.* xi. 288 When you come down to brass nails. **1927** *Daily Express* 20 June 2 (Advt.), Let's get down to Brass Tacks. **1932** T. S. Eliot *Sweeney Agonistes* 25 That's all the facts when you come to brass tacks: Birth, and copulation, and death. **1953** L. A. G. Strong *Personal Remarks* 10 When we put theories aside, and come down to brass tacks.

6. b. *brass-bound* (examples); also *fig.*; **c.** *brass-bold* (later example), *-bright*; **d.** *brass-studded.*

1880 G. M. Hopkins *Poems* (1918) 50 Now the other was brass-bold. **1867** W. Morris *Jason* VI. 110 The brass-bound tiller. **1901** *Daily Chron.* 29 Aug. 3/1 The soldierly old brass-bound General. **1913** W. de la Mare *Peacock Pie* 88 His brass-bound cart. **1908** Hardy *Dynasts* III. v. vi. 451 The flames making the faces of the crowd brass-bright. **1858** Longfellow *M. Standish* IV. 53 A Bible, Ponderous, bound in leather, brass-studded, printed in Holland.

7. brass band (earlier examples); **brass-blacking,** a dead black surface given to brass ornaments by treatment with chemicals; **brass-bounder,** (*a*) see quot. 1890; (*b*) an apprentice on board ships of English companies, so called from the brass on his uniform; **brass-colour,** (*a*) a preparation used to colour objects to resemble brass; (*b*) a preparation of oxidized brass used to stain glass to various tints of blue and green; **brass-hat** *slang*, an officer of high rank in the British army (or other Service), so called from the gilt insignia on his cap; hence *brass-hatted* adj.; **brass-helmeted** *a.*, wearing a brass helmet; **brass-man,** one who plays a brass musical instrument; **brass nail** [rhyming slang for *tail *sb.*¹], a prostitute; **brass plate, brass-plate** (further examples); **brass-plater** *slang*, a man of the professional class; **brass-powder,** a powder consisting of copper or one of its alloys used in varnish; **brass-rag,** in slang (orig. *Naut.*) phr. *to part brass-rags*, to quarrel; **brass-rubbing,** the process of rubbing a brass (see Rub *v.*¹ 1 e); also, the impression thus obtained; so *brass-rubber*.

1834 C. Brontë in W. Gérin *C. Brontë* (1967) vi. 84 There are to be five brass bands each consisting of two trumpets, three bombardones, [etc.]. **1837** Hawthorne in *Democratic Rev.* Oct. 35 A company of summer soldiers,..attended by the 'brass band'. **1849** *Theatrical Programme* 9 July 44 The Brass Band on the Lawn will perform an admired Selection of Popular Overtures. *a* **1884** Knight *Dict. Mech.* Suppl., *Brass-blacking*, a dead black color; used freely with French optical instruments. **1890** Farmer *Slang*, *Brass-Bounder* (nautical), a midshipman. **1927** *Blackw. Mag.* Sept. 374/2 There were none but sailors, brass-bounders, stewards. **1797** *Encycl. Brit.* III. 519/2 The finest brass-colour is made with powder brass.., diluted into a varnish. **1839** Ure *Dict. Arts* 169 Brass Colour..is prepared by exposing for several days thin plates of brass upon tiles in the *leer* or annealing arch of the glass-house, till it be oxidized into a black powder. **1893** Kipling *Many Invent.* 210, I tell you der big brass-hat pizness does not make der trees grow. **1940** *War Illustr.* 5 Jan. 546/1 No one today hears sarcastic reference to 'brass hats', the traditional term for the staff officers of the High Command. **1903** Kipling *Traffics & Discov.* (1904) 73 There's a crowd of brass-'atted blighters there which will say I've been absent without leaf. **1917** 'Taffrail' *Off Shore* 82 The brass-hatted potentate who regards our quarterly list of breakages with a horny and unsympathetic eye. **1897** *Westm. Gaz.* 22 Nov. 9/1 The brass-helmeted firemen. **1872** T. Hardy *Under Greenwood Tree* I. I. iv. 50 They should ha' stuck to strings. Your brass-man, is brass—well and good; your reed-man, is reed—well and good. **1958** B. Rust in P. Gammond *Decca Bk. of Jazz* iv. 60 These two fine brassmen made good jazz. **1934** P. Allingham *Cheapjack* 317 Brass (Brassnail), prostitute. **1938** F. D. Sharpe *S. of Flying Squad* i. 15 Ladies whom the Underworld calls 'brass nails'. **1771** *Encycl. Brit.* III. 511/1 The compositor..puts..this slip of brass-plate..in the composing-stick. **1836** Dickens *Sk. Boz* I. 96 A brass-plate on the private door with 'Ladies School' legibly engraved thereon. **1894** Brass plate [see Plate *sb.* 5 b]. **1921** *Glasgow Herald* 9 Apr. 6 Steelworkers and bankers, ship-platers and 'brass-platers', 'workers' and 'parasites', we shall all have to take off our coats. **1839** Ure *Dict. Arts* 169 Only so much of the brass powder and varnish should be mixed at a time as is wanted for immediate use. **1898** W. P. Drury *Tadpole of Archangel* 141 The graceful figures stiffened, passing each other with..eyes fixed on futurity...'Don't you *know* that we've parted brassrags?' *Ibid.* 142 When 'Pincher' Martin, Ordinary, and 'Nobby' Clarke, A.B., desire to prove the brotherly love..with which each inspires the other, it is their..custom to keep their brasswork cleaning rags in a joint ragbag. But, should relations..become strained between them, the bag owner casts forth upon the deck.. his sometime brother's rags; and with the parting of the brassrags hostilities begin. **1903** Kipling *Traffics & Discov.* (1904) 49 You'll shut your mouth..or you an' me'll part brass rags. **1922** Mrs. A. Sidgwick *Victorian* vii, If you dare to use that word flapper in my hearing..we part brass rags. **1959** *Economist* 14 Mar. 979/2 He seems to have finally parted brass rags with the Arab nationalists and

President Nasser. **1856** *Athenæum* 17 May 626 The quiet haunts of the brass-rubber. **1886** C. M. Yonge *Chantry House* II. xiv. 135 Her greatest achievement in brass-rubbing, a severe and sable knight. **1890** H. W. Macklin *Monumental Brasses* ii. 27 Brass-rubbings are greatly improved by being mounted. **1897** (*title*) The Oxford Journal of Monumental Brasses, being the Journal of the Oxford University Brass-Rubbing Society. **1952** E. Coxhead *Play Toward* ii. 68 He'd cycled over with one of your girls, and they were taking brass rubbings in the church.

brass, *v.*¹ Add: **2.** *intr.* To pay up. Also *trans.* *slang.*

1898 J. D. Brayshaw *Slum Silhouettes* 147 Now..p'raps you'll pay the man. Go on—brass up! *Ibid.* 150 Along comes Mister Internashonal, an' brasses up every stiver o' that twenty-eight quid. **1939** F. Thompson *Lark Rise* vi. 119 Husbands and sons at work 'brassed up' on Friday nights. **1949** Wodehouse *Mating Season* viii. 79 What did he soak him? Five quid? And Gussie brassed up and was free?

3. *slang* (orig. *Services*'). **a.** (See quot. 1925.) **b.** (See quot. 1943.)

1925 Fraser & Gibbons *Soldier & Sailor Words* 35 *To brass off*: to grumble. **1943** Hunt & Pringle *Service Slang* 16 The verb *brass off* means to tell off severely. **1964** V. Canning *Scorpio Lett.* iv. 75 After I'd brassed you off for pinching my parking space.

brassed, *ppl. a.* Add: **2.** With *off*: disgruntled, 'fed up', 'browned off'. *slang* (orig. *Services*').

1941 J. Sommerfield in *Penguin New Writing* VIII. 46 I'm brassed off waiting. **1943** P. Brennan et al. *Spitfires over Malta* 12 Very tired and brassed off, we bundled our kit on our shoulders. **1959** J. Jefferies *13 Days* x. 158, I thought you was brassed off!

‖**brasserie** (brasəri). [Fr., orig. = brewery, f. *brasser* to brew.] A beer saloon, usually one in which food is served.

1864 *Realm* 15 June 8/3 May his monument outlive all other brasseries! **1887** *Athenæum* 1 Jan. 10/3 [They] are delighted to earn a cheap reputation at the *café* or the *brasserie*. **1888** *Pall Mall Gaz.* 6 Mar. 4 The employment of girls in brasseries, which have so evil a name. **1926** *Sunday at Home* June 536/1 You are impressed at once by the contrast of the innumerable *brasseries* and restaurants with our wretched public-houses.

brassey, brassie: see *brassy *sb.*²

brassière (bræ·siēə, -ia, -ɛ·). Also **brassiere.** [Fr., orig. (17c.) bodice; now chiefly in pl. 'leading strings (of infant)'.] A woman's undergarment worn to support the breasts.

1911 *Daily Colonist* (Victoria, B.C.) 5 July 3/7 (Advt.), Brassieres of fine cambric, lace and embroidery trimmed. **1912** *Queen* 27 July 10 (Advt.), The Stylish Figure of To-Day requires a Brassière. **1936** W. Holtby *South Riding* IV. vi. 262 Her young body, partially covered by pink brassière, knickers, [etc.]. **1957** F. Hilles *Catch* x. 20 No unhooked brassiere was as close as you ever hoped to get to Paradise.

brassily (bra·sili), *adv.* [f. Brassy *a.* + -ly².] **1.** With impudence or brazen confidence.

1880 in *Cent. Dict.* **1952** S. Kauffmann *Philanderer* iv. 61 He doesn't mind sticking his bespectacled face into things and grinning broadly and brassily.

2. With a brassy noise.

1898 Kipling *Day's Work* 84 Its band playing clashily and brassily a popular but impolite air.

brassiness. (Later examples.)

1921 H. Crane *Let.* 17 Oct. (1965) 67, I have the apparent brassiness to call myself a person of rather catholic admirations. **1952** S. Kauffmann *Philanderer* iv. 61 But there was more to Percy than brassiness. **1956** H. Gold *Man who was not with It* iii. 41 The brassiness of Baldwin..he acquired, the characteristic mellow tone free from any disagreeable brassiness.

Brasso (brɑ·so). [f. Brass *sb.*] The proprietary name of a preparation for polishing brass and other metals.

1905 *Trade Marks Jrnl.* 24 May 670 *Brasso*...Preparations for polishing Metals and other articles, included in Class 50. Reckitt and Sons, Limited,..Yorkshire; Manufacturers. **1925** *Glasgow Herald* 14 Mar. 8 A tin of Brasso, oily rag. **1936** E. Levett Yeats *Friendly Tree* iii. Anna had once seen a duck dive its head and emerge with the top of a Brasso tin in its beak and swallow it. **1942** A. L. Rowse *Cornish Childhood* 83 Friday..was the day on which the scales and weights in the shop were cleaned: whether it was..the smell of the brasso, I don't know, but I disliked Friday.

brassy, *a.* Add: **3. c.** In medical use, describing a cough.

1880 Barwell *Aneurism* 91 Severe brassy cough. **1895** *Oracle Encycl.* II. 221/2 The patient..awakens..with a peculiar cough, called by physicians 'brassy'.

brassy (bræ·si), *sb.*² *Golf.* Also **brassey, brassie.** [f. Brass *sb.* + -y⁶.] A wooden club shod with brass.

1888 *Daily News* 2 July 5/1 The golfer will hunt for his 'brassey' in vain. **1889** Linskill *Golf* iii. 20 A brassey is very similar to a wooden niblick, but..the sole of the head is shod with a plate of brass as a protection to the wood and bone. **1890** [see *cleek *sb.* 1 b]. **1929** Wodehouse *Mr. Mulliner Speaking* vi. 203 John Gooch, smiting vigorously with his brassie.

b. *ellipt.* for *brassy shot.

1906 *Daily Chron.* 29 May 9/5 He sliced his drive badly, but played a perfect brassey to within four yards of the pin. **1909** *Ibid.* 22 Apr. 8/4 His tee shots and brassies bring of fine length and direction.

c. *attrib.*, as *brassy player, shot, stroke.*

1894 *Westm. Gaz.* 21 Dec. 7/2 To the third hole in, he got away a fine tee shot, which he followed by a good brassy stroke to the green. **1897** *Ibid.* 30 Dec. 7/3 Mr. Ramsay [Hunter] a seagull on the wing with a brassey shot. **1904** *Ibid.* 1 Jan. 3/1 He is a good brassy player.

brat, *sb.*¹ Add: **1. b.** (Later examples.) (See also quot. 1962.)

1885 Mrs. Lynn Linton *Chr. Kirkland* I. i. 15 The women held their aprons ('brats' we called them). **1959** *Times* 15 Oct. 14/6 The picking basket or pail was rejected in favour of the 'brat', a long apron made from a meal sack into which the potatoes were first picked. **1962** J. B. Priestley *Margin Released* I. iii. 24, I..had put on the chequered overall known locally as a 'brat'.

‖**bratwurst** (brā·tvŭrst). Pl. **bratwürste.** [G.] A type of German sausage.

1911 *Cosmopolitan* Feb. 314/1 See that I have blutwurst, bratwurst, and mettwurst every day for breakfast. **1932** L. Golding *Magnolia Street* I. viii. 128 The gentile guests drank beer and ate Bratwurst. It was a real German party. **1962** *Punch* 11 July 51/2 We ate two delicious *Bratwurste* apiece.

brave, *a.* Add: **3. b.** *brave new world* (also with capital initials): the title of a satirical novel (1932) by Aldous Huxley (after Shakespeare's *Tempest* v. i. 183) portraying a society in which 'progress' has produced a nightmarish 'utopia'; freq. used allusively.

1933 *Ann. Reg. 1932* 25 The driving force that sweeps Mr. Huxley on to generating every nook and cranny of his Brave New World to the hereditary light of inquiry is the modern world, much to be brought to a lot. **1940** *Tablet* 21 Sept. 16 The practice and particularly the theory of agriculture were the subject of many of these treatises on post-war planning—'blueprints'..of a brave new world.

c. *brave west winds*, the strong prevailing westerly winds in 'the Roaring Forties'.

1858 [see Forty *sb.* 4].

brave, *v.* **7.** Delete †*Obs.* and add: Now in *to brave it out*. (Perh. rather sense 3?)

1855 Tennyson *Maud* iv. v. 17 However we brave it out, we men are a little breed.

braveness. Delete ?*Obs.* and add later example.

1927 *Daily News* 14 May 4/5, I was much moved by the braveness and sincerity of the words I read.

bray, *sb.*¹ Add: **4.** *fig.*

1929 E. Sitwell *Gold Coast Customs* 17 The bray Of the eyeless mud. **1953** C. Day Lewis *Ital. Visit* I. 21 A great Elgarian clash and bray of sunshine.

bray (brēi), *sb.*³ *Her.* Also **brey.** [a. OF. *braie*, pl. *breie, now *broie*.] **1.** A semicircular figure representing a barnacle or bit to restrain a restive horse.

1863 C. Boutell *Man. Her.* 45 Breys, barnacles or for a horse's nose, used in breaking the animal. **1864** —— *Her. Hist. & Pop.* (ed. 3) xv. 175 Three breys or barnacles in pale or.

2. (Corresponding to Fr.) A tool used for breaking hemp, used as a bearing.

1880 Cussans *Handbk. Her.* 109 A Hemp-brey is really the same instrument as a Horse-brey, except that they were used for different purposes, and that the former is in Armory always represented as being open whereas the latter.

bray (brēi), *v.*⁴ *S. Afr.* Also ‖**brei, brey.** [ad. Afrikaans *brei*, f. Du. *breien* Braid *v.*¹] *trans.* To prepare, dress (the skin of an animal).

[**1801** W. J. Burchell *Trav.* I. xv. 351 The trunk of a tree is fixed up near the hut, for the purpose of preparing (or, as they call it, *breyen*) leathern *reims*.] *a* **1835** J. Goldswain *Chron.* (1946) 90 Six or eight sheep skins which they bray or rub them in thear hands. **1840** W. C. Harris *Game & Wild Animals S. Afr.* i. 4 The hide, when brayed, is employed by the Colonists for *reims*, or thongs. **1897** 'F. Macnab' *On Veldt & Farm* 14 *a* **1900** ...always busy braying a skin, sewing a karross [etc.]. **1925** S. C. Cronwright-Schreiner in *Centenary Bk.* I. *Verse* 61 The hardy Boer..cut the strip And brei'd and rolled and hammered it round to make the Wagon-whip. **1934** K. Campbell *Broken Record* vii. 170, I shot two seals,..brayed their skins and sold them to Corporal. **1940** *Nat. Geog. Mag.* LXXVII. *Makes* ... **1955** E. A. Ritter *Shaka Zulu* xvii. 203 Everyone lay down to sleep—the maidens on their mats with well brayed skins as blankets.

brayette: see *braguette.

braying, *ppl. a.* Add: **2.** *fig.*

1922 E. Sitwell *Façade* 15 The hard and braying light.

braze (brēⁱz), *sb.* [f. Braze *v.*²] The process of brazing; a brazed joint. So **bra·zeless** *a.*

1897 *Westm. Gaz.* 6 Dec. 9/1 Among the other novelties are the brazeless but fixed joints shrunk together. **1898** *Ibid.* 16 Apr. 6/1 The other process, which is described as brazeless. **1934** Webster, Braze. **1955** *Brazing Man.* (Amer. Welding Soc.) iii. 21 The American Welding Society defines a brazing filler metal as a metal to be added in making a braze.

Brazilian, *a.* and *sb.* (Earlier and later examples.)

1607 Topsell *Four-f. Beasts* 19 By Brasilians..they are.. called Sagoines. *Ibid.* 705, I take it to be a Brasilian Hedghog. **1628** Drake *World Encompassed* 14 It is not likely that many doe use it to that end, which the Brasilians doe. **1825** C. Waterton *Wanderings in S. Amer.* ii. 95 The Brazilians were told, that..education would go on. **1851** K. H. Digby *Compitum* V. viii. 313 Accompanying him to the Brazilian frontier. **1860** [see *Argentine *sb.*² 1]. *a* **1910** W. James *Memories & Stud.* (1911) i. 9 Certain officials of the Brazilian empire smiled. **1961** G. Durrell *Whispering Land* vi. 153 Two pigmy Brazilian rabbits, with ginger paws and white spectacles of fur round their eyes.

brazilianite (brăzi·liănəit). *Min.* [f. Brazil¹ 3 (where it was first discovered) + -ian + -ite¹ 2 b.] †**1.** = Wavellite. *Obs.*

1818 J. Mawe *Descr. Catal. Min.* (ed. 3) 54 Wavellite.. *Brazilianite.* Occurs massive; botroidal; and crystallized in flat rhombic prisms.

2. A mineral consisting of a basic phosphate of sodium and aluminium.

1945 Pough & Henderson in *Amer. Mineralogist* XXX. 574 The authors decided that such an important mineral should bear the name of its country of origin...Our.. choice was *brazilianite*. **1947** *Times* 9 Jan. 7/4 A new mineral of gem quality named Brazilianite.

brazilite (brăzi·ləit). *Min.* [ad. G. *Brazilit* (Hussak 1892, in *Neues Jahrb. für Min.* II. 141), f. Brazil¹ 3 (where it was first discovered) + -ite¹ 2 b.] = *baddeleyite.

1893 L. Fletcher in *Min. Mag.* X. 158 Since the results of the above investigation [on Baddeleyite]..a description of a new mineral, *Brazilite*, has appeared in the *Neues Jahrbuch für Mineralogie*..as part of a letter sent by Dr. Hussak from Brazil. *Ibid.* 160 Dr. Hussak, with characteristic courtesy, offers to simplify mineralogical nomenclature by withdrawing the name Brazilite. **1955** [see *baddeleyite].

breach, *sb.* Add: **3. b.** breach of promise (examples).

1590 Shakes. *Com. Err.* iv. i. 49 You vse this dalliance to excuse Your breach of promise to the Porpentine. **1613** [see Promise *sb.* 1]. **1786** Paley *Mor. Philos.* (ed. 2) III. I. xvi. 164 God will punish false swearing with more severity than a simple lie, or breach of promise. **1895** G. B. Shaw *Shaw on Theatre* (1958) 61 Trial by Jury is..unintelligible except as a satire on..the breach-of-promise suit. **1949** M. Mead *Male & Female* xv. 299 Breach-of-promise cases are a silly excrescence in a world in which women do half the proposing.

c. In colloq. use, short for *breach of promise*.

1840 Dickens *Old Cur. Shop* viii, There's the chance of an action for breach. **1905** *Daily Chron.* 16 Aug. 6/7 'The breach action was not brought by her,' said Mr. Burnett, opening the present proceedings on behalf of the major. *Ibid.*, At the breach trial.

breachy, *a.*² **1.** (Earlier and later U.S. examples.)

1780 E. Parkman *Diary* (1899) 275 To my sorrow, my Oxen have been breachy at Mr. Isaac Parker's and let in Cattle with them into his Cornfield. **1899** A. Brown *Tiverton Tales* 133 That cussed breachy cow..hooked it [*sc.* a fence] down.

bread, *sb.* Add: **2. d.** bread and cheese (further examples). See also Water *sb.* 2 a.

1875, etc. [see E.D.D.]. **1952** F. White *Good Eng. Food* I. vi. 69 Hawthorn..the budding leaves of which..are called by the children 'bread and cheese'.

g. *bread and circuses* [tr. L. (Juv. *Sat.* x. 80 Duas tantum res anxius optat, Panem et circenses.)], used allusively for food and entertainment, esp. when provided by a government to assuage the populace.

1914 H. P. Eden (*title*) Bread and circuses. **1924** Kipling *Debits & Credits* (1926) 217 Rome has always debauched her beloved Provincia with bread and circuses. **1937** 'N. Blake' *There's Trouble Brewing* i. 23 Bread and circuses, with me figuring as the circus. **1967** *Listener* 16 Mar. 373/1 The almost Roman outpouring of stale bread and dull circuses in what is supposed to be the more 'entertaining' side of television.

3. b. In full *altar-bread*. A sacramental wafer. Usu. *pl.*

1849 Altar-breads [see Altar B. II]. **1877** J. D. Chambers *Div. Worship* 352 The Breads being now on the Corporal. **1899** W. J. S. Simpson *Mem. W. S. Simpson* 154 An iron instrument for stamping the altar breads.

5. b. *in good bread*: in a good living or position (?*obs.*); *in bad bread*: in a bad state, in difficulties; in disfavour *with* a person. *dial.* and *U.S.*

1763 in *Essex Inst. Hist. Coll.* (1913) XLIX. 139 Mr. Barnard..is now in good bread, and seems loth to affront his people by telling them plainly of these public sins. **1778** *Ibid.* (1907) XLIII. 11 Old England I beleve is got into Bad Bread. *Ibid.* 16 Hope it is the French Fleet, if not we shall be in Bad Bread, but we must see it out with them. **1785** Grose *Dict. Vulgar T.* s.v. *Bread*, In bad bread; in a disagreeable scrape, or situation. **1825** Jamieson *Suppl.*

s.v. *Bread, To be in bad bread*, to be in a dilemma, or in an evil taking. **1894** P. H. Hunter *J. Inwick* xviii. 223, I saw fine I was gaun to be in bad breid wi' baith sides.

c. Money. *slang* (orig. *U.S.*).

[**1939** Ramsey & Smith *Jazzmen* (1940) iii. 63 Inside the low, smoky room, the musicians sweated for their bread.] **1952** *Down Beat* 18 June 15 If I had bread (Dizzy's basic synonym for loot) I'd certainly start a big band again. **1965** L. W. Holt in J. H. Clarke *Harlem* 207 There won't be much 'cake' unless the brothers should happen to get some *bread* to buy it with. **1967** C. Drummond *Death at Furlong Post* iii. 28 So me with all that bread..maybe a week, and then I get the plane.

8. (Earlier and later examples of *bread pudding*.)

1723 J. Nott *Cook's & Confectioner's Dict.* P. § 235 To make a Bread-pudding. **1968** *Listener* 26 Sept. 423/2 The secret of a good bread pudding is adding enough spices and fruit, especially currants.

9. a. *bread-bag, -knife* (earlier examples); *bread-pill, -sauce* (example); **b.** *bread-earner* (example).

1729 Byrd *Field Jrnl.* in *Colonial Rec. N. Carolina* (1886) II. 795 The woods..tore the very deer skins that guarded the Bread bags. **1882** A. Bain *J. S. Mill* v. 193 Mill has done more than any single person for the bread-earners of the sex. **1833** *Chambers's Jrnl.* II. 285/3 With a bread-knife arrested in his surprised hand. **1892** G. & W. Grossmith *Diary of Nobody* xiii. 172 Daisy and Lupin..began throwing bread-pills at each other. **1747** H. Glasse *Art of Cookery* i. 4 Some love Bread-Sauce in a Bason.

10. bread-board, breadboard, (*a*) a board on which bread is cut, dough kneaded, etc.; (*b*) *slang*, a name given to a board on which an experimental electric circuit, etc., is set out; hence as *v. trans.*, to make a flat model of (such a circuit, etc.); so **brea·dboarding** *vbl. sb.*; **bread-crust bomb** *Geol.* (see quot. 1909); **bread-jelly** (see quot. 1868); **bread-line** (orig. *U.S.*), (*a*) a queue of poor people waiting to receive bread or other food given as charity; also *fig.*; (*b*) = **subsistence level*; **bread-stick,** a long slender roll of crisp bread; **bread-ticket** (orig. *U.S.*), (*a*) a ticket entitling the possessor to bread; (*b*) *fig.* = **meal ticket* b; **bread-unit,** a unit of value in rationing bread.

1857 *Mrs. Putnam's Receipt Book* 3 Mix..the dough; roll it on a bread-board about an inch thick. **1948** *Gloss. Computer Terms* (U.S. Office of Naval Res. Special Devices Center: M.I.T. Servomechanisms Lab. Rep. R—138) 4 *Breadboard*, a roughly constructed experimental model of a circuit. **1952** R. F. Jones in W. Sloane *Stories for Tomorrow* (1955) ii. 75 Breadboard lay-outs were assembled with maximum care. **1953** *Electronic Engin.* XXV. 143 Drift was encountered in the initial 'breadboard' model. *Ibid.* 417 During the late 1920's the supporting panel or 'breadboard' for the components tended to become horizontal in the box. **1956** R. A. Heinlein *Door into Summer* (1960) iii. 44, I could bread-board a rig for that. **1959** *New Scientist* 26 Mar. 695/1 Years would be lost by breadboarding, since it required..the redesign of an operating breadboard model into a submarine hull. **1967** H. Harrison *Technicolor Time Machine* (1968) ix. 86 I'm afraid a good deal of the circuitry was breadboarded. **1909** *Cent. Dict.* Suppl., *Bread-crust bomb*, a somewhat rounded mass of igneous rock, ejected in a volcanic eruption, whose surface presents an appearance similar to that of baked bread, as if the interior had expanded beyond the stretching capacity of the surface crust. **1938** *Nature* 18 June 1106/1 Origin of breadcrust bombs of the peléan type. **1853** Mrs. Gaskell *Cranford* xi. 208 Mrs. Forrester made some of the bread-jelly, for which she was so famous, to have ready as a refreshment. **1868** M. Jewry *Warne's Model Cookery* 661/1 *Bread Jelly*... Take the crumb of a penny roll; cut it into thin slices, and toast them...Put them into a quart of spring water. Let it simmer over the fire till it has become a jelly. Strain..and flavour..with a little lemon juice and sugar. **1900** *Lippincott's Mag.* LXV. 3 [Story by A. B. Paine entitled] The Bread Line. *Ibid.* 12 That's the bread line. They get a cup of coffee and a loaf of bread every night at twelve o'clock. **1909** H. N. Casson *Life C. H. McCormick* 12 This.. republic could not develop beyond the struggle for food. It was chained to the bread-line. **1929** *Dundee Courier* 7 Nov., My life has been spent among people..close to the bread line. **1959** *New Statesman* 28 Feb. 294/3 The average African family in the urban area lived calamitously below the bread-line. **1909** *Cent. Dict.* Suppl., *Bread-sticks*, bread-dough rolled into sticks and baked: served with bouillon, soup, or tea. **1936** J. dos Passos *Big Money* 470 Picking up a breadstick and snapping it into his mouth. **1855** *Chicago Times* 21 Mar. 3/1 Entering a bake-shop and stealing therefrom bread tickets to the value of $16. **1956** I. Bromige *Enchanted Garden* III. iv. 164, I wanted so much more...Not just a bread ticket. I wouldn't care how poor we were. **1946** *Lancet* 6 July 29/2 The ration will be measured in bread units. A 1 lb. 12 oz. loaf will cost 4 bread units, 1 lb. of flour 3 bread units, and 1 lb. of flour confectionery 2 bread units.

bread, *v.* (Earlier examples.)

1629 Parkinson *Parad.* II. iii. 476 Some doe use the pouder of the herbe dryed..to mixe with grated bread, to breade their meate. **1797** E. Chambers *Let.* 29 Nov. in *Papers of J. Steele* (1924) I. 152 What [corn] I have..will Scarcely Bread the Negroes.

bread and butter. Add: **1.** (Earlier example of *bread-and-butter pudding*.) Also *Comb.*, as *bread-and-butterless* adj.

1729 E. Smith *Compl. Housewife* (ed. 3) 81 A Bread and Butter Pudding...Take a two-penny Loaf..spread it in very thin slices [etc.]. **1849** [see Tealess *a.*]. **1853** Mrs. Gaskell *Cranford* i. 14 He..lessened the pretty maidservant's labour by waiting on empty cups, and bread-and-butterless ladies.

b. *dial.* A slice of bread and butter.

1853 W. D. Cooper *Gloss. of Provincialisms in Sussex*, *Bread and butters*, slices of bread buttered. **1927** W. E. Collinson *Contemp. Eng.* 54, I well remember the disgust we children felt at a lady..who always said a bread and butter, where we used *a piece of bread and butter*.

2. (Earlier and later examples.) See also Quarrel *v.* 1 a.

1732 Swift *Let.* 12 Aug. (1965) IV. 60 Your quarrelling with each other upon the subject of bread and butter is the most usual thing in the world. **1738** —— *Polite Conv.* I. 17, I won't quarrel with my Bread and Butter for all that: I know when I'm well. **1929** *Publishers' Weekly* 30 Nov. 2588/1 The old stand-bys, the bread-and-butter books in every department. **1939** A. Christie *Murder is Easy* xii. 128 One mustn't quarrel with one's bread and butter. **1955** *Times* 11 May 6/1 Providing furniture for new houses was the bread and butter of the industry.

4. b. bread-and-butter letter orig. *U.S.*, a letter of thanks for hospitality; a Collins. Also *ellipt.*

1901 Howells *Pair of Patient Lovers* 82 His prompt bread-and-butter letter. **1933** N. Streatfeild *Tops & Bottoms* xxiv. 308 Please *never* write me bread-and-butter letters. **1964** E. Bowen *Little Girls* III. i. 164 Rude? Should have written a bread-and-butter?

bread-basket. Add: **3.** *slang.* A large bomb containing smaller bombs; esp. in *Molotov *bread-basket*.

1940 [see *Molotov]. **1941** *Word Study* Nov. 1/1 A few radio commentators and newspapermen have applied the phrase *Goering*, or *Goering's*, *breadbasket* to a similar type of bomb dropped sometimes by the Germans in raids over London. **1944** J. H. Fullarton *Troop Target* xxiv. 179 Lone bombers dropping 'bread-baskets' which spew scores of tiny bombs over a wide area.

bread-crumb, *v.* [f. the sb.] *trans.* To dip in bread-crumbs in preparation for cooking. Hence **bread-crumbed** *ppl.a.*; **bread-crumbing** *vbl. sb.*

1846 Soyer *Cookery* 103 Egg and bread-crumb each piece. **1854** —— *Shilling Cookery* 35 The heads when cooked may be egged and bread-crumbed over. **1877** E. S. Dallas *Kettner's Bk. Table* 179 The plain cutlets and the bread-crumbed ones. *Ibid.* 228 For breadcrumbing, the French plan is to dip the cutlet into oil or butter and then to roll it in crumbs. **1927** *Daily Express* 17 Nov. 5/2 The chicken was 'egged and breadcrumbed' in readiness for the dish.

bread-kind (bre·d,kəind). [f. Bread *sb.* + Kind *sb.*] Food like bread; esp. a West Indian name for yams, sweet potatoes, and similar food-stuffs.

1697 Dampier *Voy.* 311 The Mindanao People live 3 or 4 months of the year on this food for their Bread kind. **1712** W. Rogers *Voy. round World* 376 At which Allowance we have not above 12 Days at most, being all the Bread Kind we have in the Ship. **1891** *Wesleyan Meth. Mag.* May 362 A brown woman..with a load of 'bread-kind' on her head. **1899** W. P. Livingstone *Black Jamaica* v, The principal food consisted of what is locally called 'bread-kind'; yams, sweet potatoes, cocos, plantains.

breadth. Add: **4. b.** Undue freedom or lack of decorum in dealing with indelicate matters; grossness or licence of expression. (Cf. Broad *a.* 6 c, Broadness 2.)

1873 W. C. Hazlitt *Feudal Period* Pref. p. ix, A few of them exhibit a breadth which is scarcely consonant with modern ideas of decorum.

break, *v.* Add: **2. e.** *spec.* To change (a banknote or the like). (Cf. quot. 1741 in Dict.)

1844 Dickens *Mart. Chuz.* xliii. 494 It was the same note; he hadn't broken it. **1880** W. H. Patterson *Gloss. Antrim & Down* 12 Can you break that pound note for me? **1959** N. Mailer *Advts. for Myself* (1961) 81 He stretched himself out,..thinking..of the thrill of breaking a five-dollar bill.

i. In leather manufacture, to scrape a skin smooth and clean on the flesh side.

1842 *Penny Mag.* XI. 215/2 The lamb-skins having been steeped in water, 'broken' on the flesh side, and drained. **1845** Dodd *Brit. Manuf.* V. 187 The goat-skins are..soaked in water..to soften them, and then undergo the process of 'breaking'.

j. *Phonology.* To cause breaking (*breaking *vbl. sb.* 1 e) of (a vowel). Also *intr.*

1845 J. M. Kemble in *Proc. Philol. Soc.* II. 135 Not satisfied with transforming *i* into *ĕ*, before *h, l, m*, it [sc. Anglo-Saxon] broke the vowel into *ĕo*. **1871** F. A. March *Compar. Gram. Anglo-Saxon* I. 11 Before a consonant combination beginning with *l, r, h*, it [sc. *a*] breaks to *ea*. *Ibid.* 20 *l, r, h*, oftenest before a consonant, break foregoing *a* to *ea*. **1959** A. Campbell *Old Eng. Gram.* v. 56 *æ* was broken, and appears as *ea*..before *r* followed by a consonant.

k. *to break the wicket* (Cricket): to dislodge a bail or the bails in stumping or running out a batsman.

1875 F. Gale in *Baily's Mag.* Sept. 274 He took her [*sc.* the ball] close to the bails and just broke the wicket. **1901** *Strand Mag.* June 616/1 The ball was thrown in from the field, the bowler took it, and broke the wicket, so as to run the batsman out.

l. (See quot. 1889.) orig. *U.S.*

1889 *Cent. Dict.* s.v., *To break a gun*, to open it by the action. **1956** M. Procter *Pub Crawler* 125 With the casual ease of long practise he 'broke' the gun and ejected the six rounds.

5. c. *intr.* Of the surface of water: to present a broken appearance, caused by waterbloom (see quots. and *breaking *vbl. sb.* 2 c). *dial.*

1873 G. C. Davies *Mountain, Meadow, and Mere* 16 The Ellesmere water..*breaks*. Every summer..the water becomes full of some matter held in suspension...The other meres do not *break* to such an extent. **1887** T. Darlington *Folk-Speech of S. Cheshire* 128 Bar-mere's bin *breekin'* this afternoon.

7. d. *Cricket.* *to break one's duck('s egg)*: to score one's first run in an innings, thus avoiding a 'duck' (Duck *sb.*¹ 7).

1867 G. H. Selkirk *Guide to Cr. Ground* ii. 26 If he makes one run he has 'broken his duck's egg'. **1900** W. A. Bettesworth *Walkers of Southgate* 19 Parr broke his duck, but could get no further, being bowled by Atkinson for one run. **1912** A. Brazil *New Girl at St. Chad's* vii. 112 Her first ball, being a wide, served to increase the confidence Honor had felt in breaking her duck.

8. b. Also of weather: to change suddenly, esp. after a long settled period. Cf. sense *56 h.

1887 Yeats *Lett.* (1954) I. 51 The weather breaking might send me off any time, as my uncle stops here only so long as it is fine. **1930** W. S. Maugham *Cakes & Ale* viii. 91 The weather broke suddenly.

c. Of prices of commodities, stocks, etc.: to fall suddenly or sharply. orig. *U.S.*

1870 W. W. Fowler *Ten Yrs. in Wall St.* 435 Gold had broken to 87, and then..ran up to 194. **1899** *Daily News* 15 May 2/6 Under the influence of Mr. Flower's death, what are known as Flower stocks broke in overwhelming volume. **1929** *Times* 30 Oct. 14/1 Prices broke far below the previous low levels of the year.

9. *esp.* To defeat the object of (a strike) by engaging other workers.

1905 [implied in *strike-breaker*, Strike *sb.* 20]. **1914** *Round Table* Mar. 367 The farmers contributed the bulk of the power that..broke the strike.

b. To nullify or set aside (a will) by legal methods.

1891 *Melbourne Argus* 12 Dec. 11/8 [New York.] Under the law [she] would be entitled to one-half of the estate, should the will be broken.

c. To better (a record, a score, etc.). (See Record *sb.* 5 d.)

1886 [see Record *sb.* 13 a]. **1909** Webster s.v., He broke the record for the high jump. **1955** F. Brown *Angels & Spaceships* 171 Up to that hole he [*sc.* a golfer] had an excellent chance to break a hundred.

d. To win against (an opponent's service) in lawn tennis or a similar game. Also *intr.* or with *through*.

1959 *Times* 2 July 3/1 Mackay saved his next game, broke to 8–7 in an uproar, and served out heroically for the set. *Ibid.* 4 July 3/7 He broke service in the first game. **1961** *Ibid.* 4 July 4/1 True, Wilson..did break for 2–4 and then move to 3–4. **1964** *Observer* 1 Nov. 19/1 Sangster broke service in the ninth game and went on to win 6–4. *Ibid.*, Sangster broke through Bungert's service in the fifth game.

15. f. *to break ship*: to fail to rejoin a ship on the expiration of leave.

1905 'Q' *Shining Ferry* III. xviii, I brought across a sailor-looking chap...Thinks I, 'You've broken ship, my friend.' **1907** *Daily Chron.* 3 Apr. 1/7 The serious offence of 'breaking ship'. **1909** *Ibid.* 28 June 8/7 In the afternoon he broke ship, but was undiscovered.

16. b. To solve (a code or cipher); to decipher.

1928 P. Buranelli et al. *Cryptogram Bk.* p. ii, We were amazed at the ease with which anyone could break a coded message. **1931** *N. & Q.* 30 May 379/2 Their centre in New York receives messages by the thousand in a code that has not yet been broken. **1956** C. Simak *Time & Again* xxxv. 168 No one else could break the language in which his notes were written.

17. a. Delete †*Obs.* and add: In modern use, only in phr. *to break and enter*: see *breaking *vbl. sb.* 1 c.

1797 Tomlins *Jacob's Law-Dict.* I. Bb3/3 To break and enter a *shop*..is not burglary, but only larceny. **1817** [see Enter *v.* 10 d]. **1959** A. Sillitoe *Loneliness Long-Dist. Runner* 11 There's a shop to break and enter. **1961** J. Maclaren-Ross *Doomsday Bk.* I. iv. 56 He broke-and-entered through a back window.

19. *to break bounds* (earlier examples).

1816 Jane Austen *Emma* III. vii. 116 You had..broken bounds yesterday, and run away from your own management. **1856** [see Bound *sb.*¹ 4].

22. c. To publish or reveal (an item of news); to make available for publication. (Cf. sense *39 b.) *Journalists' colloq.*

1906 G. W. Peck *Bad Boy with Circus* 21 (Weingarten). **1935** M. M. Atwater *Murder in Midsummer* xxviii. 262 Are you breaking the story in the morning papers? **1961** 'B. Wells' *Day Earth caught Fire* vii. 108 But she didn't break the story.

25. c. *Naut. trans.* To free and shake out (a flag or sail which has been furled); also with *out*.

1889 *Times* 6 Aug. 8/3 The Royal Standard was broken on board the Victoria and Albert, and immediately H.M.S. Valorous..began to fire a salute. **1899** *Daily News* 9 Oct. 6/2 The Columbia broke out her spinnaker. **1901** *N. & Q.* 9th Ser. VII. 176/2 When a standard is 'broken' it is unfurled after being hoisted. **1928** *Daily Tel.* 20 Mar. 13/7 The Afghan standard was broken from the Majestic's mainmast. **1945** C. S. Forester *Commodore* 36 A black ball was soaring up the mast, and as it reached the block a twist of the seaman's wrist broke it out.

32. b. (Earlier examples.)

1847 W. Denison *Cricketer's Companion* p. xix, The tendency of his bowling is to make the ball break back from the 'off', to the 'leg'. **1866** 'Capt. Crawley' *Cricket* 36 A.. ball breaking in from the leg-side.

34. b. In boxing or wrestling, to separate from one's opponent after a clinch; esp. as an order from the referee. Also with *away* (see sense *49 d).

1932 Wodehouse *Hot Water* vi. 114 Next thing you know they're rolling on the floor, and me acting as referee and telling them to break. **1966** 'J. Hackston' *Father clears Out* 199 Tom, who was also referee, told them to break, and both girls still kept hanging on.

37. c. (Earlier and later examples.)

1693 Dryden *Ovid's Met.* I. l. 483 in *Examen Poeticum* 31 A second Deluge, o're our heads may break. **1961** L. van der Post *Heart of Hunter* I. i. 37 The country where the rains had broken. **1965** *Listener* 3 June 828/1 The monsoon, heaviest in the world in that area, had broken.

e. *Athletics.* To get off the mark prematurely at the start of a race. Also with *away*.

1897 *Encycl. Sport* I. 55/2 While the runner has his hands on the ground he cannot 'break away' from the mark, and if a runner 'breaks away' he is, under the A.A.A. rules, put back a yard. **1959** *Observer* 26 July 24/6 H. Smit.. broke twice in the 100 yards and should by international rules have been disqualified.

38. b. To make a dash; to set off at a run. So *to break back*: to set off running in a reverse direction. *U.S.*, *Austral.*, and *N.Z.*

1834 Crockett *Narr. Life* ii. 11 He gathered about a two year old hickory, and broke after me. *Ibid.* xiv. 96 When my lead dog..raised his yell, all the rest broke to him. **1835** A. B. Longstreet *Georgia Scenes* 125 The way she [*sc.* a horse] now broke for Springfield 'is nothing to nobody'. **1889** A. Reischek *Story of Wonderful Dog* vi. 34 If the sheep broke he ordered a second dog to assist. **1907** S. E. White *Arizona Nights* v. 98 The cattle would attempt to 'break' past the end and up the valley. **1933** E. Jones *Autobiogr. Early Settler* xii. 59 If a mob [of sheep]..broke back,..the rest of us would have to wait while the shepherd on whose beat the sheep had broken back, went for them. **1946** F. Davison *Dusty* ix. 96 With the dogs so far forward, the rearmost sheep..had a chance to break back.

c. Of a horse, esp. in trotting or pacing: to change gait; to lose a level stride. orig. *U.S.*

1839 *Spirit of Times* 13 July 222/3 While Awful was ahead, and his backers were counting the spoils in advance, he broke! **1852** C. A. Bristed *Upper Ten Th.* 26 Suddenly the pacer stops short and capers. He is used up and has 'broken'. **1908** *Springfield Weekly Republican* 8 Oct. 2 A trotter in a race breaks, that is, loses his level stride and reverts to an impossible kind of gallop which is not permitted by the rules. **1969** *Sydney Morning Herald* 24 May 27/3 Yogi Hall, who started 10–9 on favourite in the Trotters' Handicap, was barred because he broke badly during the running.

39. b. Of an item of news: to become public or available for publication. *Journalists' colloq.*

1934 F. Baldwin *Innoc. Bystander* (1935) ix. 188 The papers have left me alone for a long time...I was sick with worry when..Merry acted the fool. But nothing broke. **1936** E. Ambler *Dark Frontier* xv. 163 You'll give us an exclusive when the story does break, I hope. **1938** E. Waugh *Scoop* II. i. 111 A big story is going to break.

c. To happen, occur. *slang* (orig. *U.S.*).

1914 *Sat. Even. Post* 15 Aug. 8/1 They say my homer was lucky..but, believe me, it was time things broke for me. They been breakin' for him all his life. **1934** Webster s.v., Things are breaking right for me. **1936** J. Curtis *Gilt Kid* xi. 118 Everything'll break good.

42. b. To get *into*, or make a sudden appearance in, some occupation, activity, etc. *colloq.* (orig. *U.S.*).

1899 *Chicago Daily News* 10 May 6/1 Nichols will be anxious to break into the game pretty soon. **1907** *Collier's* 5 Oct. 11/2 Mr. F. C. Wheeler..decided recently for the first time in his life to break into print. **1939** J. Dell *Nobody ordered Wolves* i. 7 Phillip was one of the countless thousands whose consuming ambition was to 'break' into the motion picture industry.

44. e. *Boxing.* (See quot. 1897.)

1889 E. B. Michell *Boxing* iii, in W. H. Pollock et al. *Fencing*, etc. 154 The proper style of breaking ground or shifting, or slipping, is by movements to the right. **1897** *Encycl. Sport* I. 139/1 *Break ground*, to take up a fresh position to the right or left.

45. See also Ice *sb.* 2b.

47*. *to break even*: **a.** *intr.* In *Faro* (see quot.). *U.S.*

1909 *Cent. Dict.* Suppl. s.v., *To break even*, in *faro*, to bet that each card will win or lose an even number of times on the deal.

b. To emerge from a transaction, enterprise, etc., with balancing gains and losses. orig. *U.S.*

1914 S. Lewis *Our Mr. Wrenn* 33 To go bumming around like you do and never have to worry about how the firm's going to break even. **1920** *Flight* XII. 508/2 What must we earn to 'break even'? **1950** J. Dempsey *Champ. Fighting* ii. 12 He and I broke even in our three four-rounds bouts. **1958** *Listener* 2 Oct. 498/1 Today a novel needs to sell 5,000 copies for a publisher to break even.

49. break away. d. *Boxing.* (See quot.)

1897 *Encycl. Sport* I. 139/1 *Break away*, to get away from an opponent. **1904** A. J. Newton *Boxing* viii. 67 So clinched, they remain..on the lookout for an opportunity to break away in the most favourable manner. **1923** [see *clinch *v.*¹ 2 e].

50. break down. d. Also of an engine, a machine, vehicle, or the like: to cease to function, esp. through the fracture or dislocation of a part.

1837 *United Service Jrnl.* May 112 One of her engines 'broke-down', as it is technically called. **1904** Kipling *Traffics & Discoveries* 314 They knew my car had broken down. **1958** *Times* 9 Sept. 4/6 Congestion that might be caused when a vehicle breaks down.

f. *trans.* To divide (logs) into timber (see quot. 1922).

1878 *Technol. Dict.* (ed. 3) II. 84/1 *To break down timber*,.. débiter le bois..en planches, etc. **1883** M. P. Bale *Saw-Mills* iii. 35 The band-sawing machine can..be used for breaking down heavy logs. **1922** R. C. Bryant *Lumber* 451 *Break down*, v. 1. To reduce large logs to a size which can be sawed on the main log saws in a sawmill. 2. To cut a log into cants. (Pacific Coast Forests.) **1927** J. F. Stewart *Man. Forest Engineering & Extraction* xiii. 171 The logs.. are broken down to suitable sizes by this saw.

g. To analyse or classify (figures, statistics, etc.). orig. *U.S.*

1934 Webster, *Break down*,..to separate (an account or a budget) into its component parts or subdivisions. **1941** *Amer. Speech* XVI. 45 A vast amount of raw material has been broken down for classification. **1948** *Hansard Commons* CDXLVIII. 1663 The programme account..is not broken down as between the Home, Light, and Third Programmes.

h. To stop (something objectionable); to 'give over'; esp. in *break it down*: stop it, 'come off it'. *Austral.* and *N.Z. colloq.*

1941 *Coast to Coast 1941* 127 Ah, break it down, feller. Everybody knew you had her on the town. **1943** *N.Z.E.F. Times* 15 Mar. 6/5 Break down the swearing. **1944** L. Glassop *We were Rats* I. v. 35 'Break it down,' I said, 'nothing would interest me less.' **1949** J. R. Cole *It was so Late* 13 'Break it down!' Wood shouted from the telephone. 'I can't hear a thing.'

52. break in. g. In paper-manufacture, to subject (rags) to a process of washing and pulping.

1865 *Chambers's Encycl.* VII. 243 They are thoroughly washed and partly pulped; or, as it is technically called, *broken in*.

h. To bring (virgin land) into cultivation. Also **breaking-in** *vbl. sb.* *Austral.* and *N.Z.* (orig. *dial.*: see *E.D.D.* and *Sc. Nat. Dict.*)

1891 R. Wallace *Rural Econ. Austral. & N.Z.* i. 24 The single-furrow plough is employed to break in the land. *Ibid.* vi. 109 Such results are only expected for a few years after breaking in until the wood ashes left after burning the bush are exhausted. **1939** E. E. Vaile *Pioneering Pumice* v. 99, I was breaking in some country about two and a half miles from the homestead. **1950** *N.Z. Jrnl. Agric.* Feb. 141/1 In many parts of the North Island, farm country still offers a stubborn resistance to breaking-in.

54. break out. f. (N.Z. examples referring to a drinking bout.) See also *E.D.D.*

1898 J. Bell *In Shadow of Bush* xvii. 101 Davie..had thought it best on the one occasion in which Dan had broken out, to give him a wide berth. **1904** *N.Z. Illustr. Mag.* IX. 429 Bill did 'break out'.

g. *trans.* To open up (a receptacle or the like) and remove its contents; to get (articles) out of a place of storage; hence, to prepare (food or drink) for consumption. *colloq.* (chiefly *U.S.*).

[**1832** Marryat *P. Simple* (1834) v, He was breaking Casks out of the hold.] **1840** R. H. Dana *Bef. Mast* xiv, There is always a good deal to be done in the hold: goods to be broken out. **1849** N. Kingsley *Diary* (1914) 22 Broke out our chests to-day, found all our things in good order. *Ibid.* 73 They broke out the baggage room to-day to get iron for various purposes. **1877** *Fraser's Mag.* XV. 221 Afterwards the fish are *broken out* and washed, and then packed in wooden hogshead casks. **1962** K. Orvis *Damned & Destroyed* vi. 47, I went home and broke out a fresh bottle of Scotch. **1968** C. Burke *Elephant across Border* vi. 236 'Break out some more coffee.' Lori made more coffee.

h. *intr.* Of a goldfield: to come into operation. Also *transf.* *Austral.* and *N.Z.*

1862 E. Hodder *Memories of N.Z. Life* 180 The Mammoth Caverns, which have been discovered since the diggings broke out. **1873** Trollope *Australia & N.Z.* II. xxiii. 380 When Gold 'broke out', as the phrase goes, on the western side of the Middle Island. **1894** C. J. O'Regan *Voices of Wave & Tree* 14 When the Coast broke out, we roughed it thro' the bush. **1901** 'M. Franklin' *My Brilliant Career* ii. 11 Ere the diggings had broken out on Bruggabrong, our nearest neighbour..was seventeen miles distant.

56. break up. d. Colloq. phr. *break it up*: (*a*) *imp.* disperse; stop (a fight); (*b*) *U.S.* (see quot. 1946).

1936 S. Kingsley *Dead End* II. 113 Break it up!..Come on, break it up!..go on home! Go on, break it up! **1946** Mezzrow & Wolfe *Really the Blues* 371 *Break it up*, bring the house down. **1947** 'N. Blake' *Minute for Murder* vii. 149 A policeman..forced his way..through the crowd, shouting '..Stand aside! Break it up!' **1959** *Encounter* Aug. 28/2 If someone had stepped in and said, 'break it up', ..all would have gone well.

e. (Earlier examples referring to a school.)

1707 C. Mordaunt *Let.* 5 May in E. Hamilton *Mordaunts* (1965) iv. 74 Wee break up the Saturday after next and I desiare if you be in town to send for me. **1740** Richardson *Pamela* II. 364 When you break up next, my Dear, said he, if you're a good Girl, you shall make your new Aunt a Visit. *a* **1855** C. Brontë in *Cornh. Mag.* (1860) Apr. 495, I wrote..to the friends of my pupils, notifying the day when we break up.

h. Of any kind of weather: to change.

1544 *Late Exped. Scot.* (1798) 10 And for asmoch as the myst yet contynued, and dyd not breake..we concluded, if the wether did not breake vp, to haue encamped our selues vpon the same ground. *a* **1586** Sidney *Arcadia* (1912) II. ix. 202 The weather breaking up, they were brought to the maine lande of Pontus.

o. *trans.* To disconcert, upset, disturb; to excite; *spec.* (orig. *U.S.*) to convulse with laughter. Also *intr. colloq.*

1825 J. Constable *Let.* 23 Oct. (1964) II. 404 She says, her sister is going to be married—& that she fears it will break her up. **1860** O. W. Holmes *Prof. Breakf.-t.* i. 11 This episode broke me up, as the jockeys say, out of my square conversational trot; but I settled down to it again. **1895** 'Mark Twain' in *N. Amer. Rev.* Jan. 61 Well, humour is the great thing, the saving thing,..so, when M. Bourget said that bright thing about our grandfathers, I broke all up. **1895** *Harper's Mag.* Sept. 545/2 A most pathetic stream of arguments and blasphemy, which broke Joan all up, and made her laugh as she had not laughed since she played in the Domremy pastures. **1902** L. Bell *Hope Loring* 240 What language you use!..If you knew how it breaks me up when you use slang! **1959** H. Gardner *So What Else is New?* 2 The remark broke up the other people in the elevator, but the diminutive culprit continued to stare defiance. **1967** *New Yorker* 21 Jan. 52 The number broke the place up, and Marsala invited me back to play that night. **1968** *Listener* 4 Jan. 27/3 The camera had only to turn to Tommy Cooper for the audience to break up with laughter.

break-. Add: **I. 1. break-circuit** a device for opening and closing an electric circuit; **break-wind** (*b*) a screen or protection against the wind.

a **1884** Knight *Dict. Mech.* Suppl., *Break-circuit*, an arrangement on an electro-magnetic or magneto-electric instrument, by which an operator can open or close the circuit at pleasure. **1833** in W. S. Ramson *Austral. Eng.* (1966) 95 Breakwind. **1862** J. S. Dobie *S. Afr. Jrnl.* 26 Sept. (1945) 32 A tarpaulin hung on weather side for a break-wind. **1863** *Fraser's Mag.* Mar. 282/2 What the Australians call a 'breakwind', *i.e.*, a pent roof, looking like the falling flap of a large bird-trap. **1875** *Encycl. Brit.* II. 317/2 The Norway maple..is a hardy tree, used as a breakwind in exposed situations on the east coast. **1890** *Athenæum* 18 Oct. 516/1 [Tasmanians] were frequently content with a mere break-wind in lieu of any covered structure. **1934** A. Russell *Tramp-Royal* vii. 54 The only form of shelter I needed was a small breakwind.

2. break-bone fever, also *ellipt.*; **break-teeth** or **-tooth,** difficult to pronounce.

1862 *N.Y. Tribune* 16 May, Another fever, to which the natives [of the south-western United States] give the name ..of Breakbone. **1788** Grose *Dict. Vulgar T.* (ed. 2), *Break-teeth Words*, hard words, difficult to pronounce. **1825** H. Wilson *Mem.* I. 48 Not to put in any break-teeth long words. **1827** Scott *Jrnl.* 11 Feb. (1939) II. 21 The Admiral with the break-tooth name.

II. break-piece = *break *sb.*¹ 17 a; **break-roll,** one of a pair of rollers between which wheat-grains are split; **break-signal,** a signal used to separate distinct parts of a telegraphic message.

1842 Francis *Dict. Arts* Q 2 b/1 The fore part of the spindle is terminated by a wire, and a break piece at the end of it. **1879** Breakpiece [in Dict.]. **1910** *Encycl. Brit.* X. 551/2 The first pair of break-rolls used to be called the splitting rolls, because their function was supposed to be to split the [wheat] berry longitudinally down its crease. **1876** Preece *Telegraphy* 287 These parts are separated from each other by a distinct signal, called the break signal.

break, *sb.*¹ Add: **1. b. break-back,** a sudden backward movement (see also sense 5 in Dict. and Suppl., and *break-back *a.* 2); **break-in** (further examples); **break-out** (further examples); also *Austral.* and *N.Z. slang*, a drinking bout.

1920 *Blackw. Mag.* Feb. 196/2 This 'break-back' of his had certainly been a brilliant achievement. **1960** E. S. & W. J. Higham *High Speed Rugby* xvii. 239 If the scrum-half tries a blindside run, the flank will follow him round just far enough to make a break-back impossible. **1903** *Daily Chron.* 10 Feb. 6/4 There was a further break-in of the river bank. **1944** *Times* 22 July 4/4 A successful break-in by the British..is never exploited by pursuit. **1888** 'R. Boldrewood' *Robbery under Arms* I. xi. 128 He saw him once in one of his break-outs, and heard him boast of something he'd done. **1908** W. H. Koebel *Anchorage* 49 A break-out doesn't seem to oil your tongue to run any more'n usual. **1947** *Ann. Reg. 1946* 24 The Russian break-out from the Baranovo bridgehead. **1958** *Economist* 29 Nov. 764/1 Nothing is more important than a British breakout from the rigid positions of the cold war.

4. b. An act of breaking out or away (see Break *v.* 49 c and 54 b); a rush or dash; an escape; freq. with *to make*. orig. *U.S.*

1833 *Sk. & Eccentr. D. Crockett* 82 Just before I got there, the old bear made a break and got loose. **1846** J. J. Hooper *Adv. Simon Suggs* (1851) xii. 143, I maid a brake on a bee line for Urwinton. **1888** T. Roosevelt in *Cent. Mag.* May 49 Our three men..understood perfectly that the slightest attempt at a break would result in their being shot down. **1910** [see *jail *sb.* 2]. **1929** 'G. Daviot' *Man in Queue* vi. 65 The man had gone to ground instead of making a break for it.

c. *Hort.* A bud or shoot that sprouts from a plant-stem. Also *attrib.*, as *break bud* (see quot. 1954).

1933 *Jrnl. R. Hort. Soc.* LVIII. 99 There are varieties too that on natural break buds are not good. *Ibid.*, Secure the plants well at the top break. 1954 A. G. L. Hellyer *Encycl. Garden Work* 30/2 A break is a branch or fork. *Ibid.* 31/1. If a rooted chrysanthemum cutting is left to its own devices, it will after a time, produce an abortive flower bud at the top of the stem, which will prevent further lengthening of this particular stem and force it to produce side shoots or breaks. In consequence, this abortive flower bud is often known as the 'break bud'. 1959 *Listener* 22 Oct. 706/1 From these [shoots], new breaks will appear which will produce the blooms for next season.

5. (Earlier examples.)
1851 Pycroft *Cricket Field* vii. 137 Look hard for the twist, or a 'break' will be fatal. 1855 F. Lillywhite's *Guide to Cricketers* 21 Without a 'break-back', the thing is impossible with any but an over-pitched ball.

8. b. *spec.* Of a trotter or pacer, the act of breaking away from a steady stride. (Cf. *Break v. 38 c.) orig. *U.S.*
1839 *Spirit of Times* 13 July 222/3 It was as bad a break as we ever saw. a 1867 H. Woodruff *Trotting Horse of Amer.* (1868) xli. 41 The penalty of a break was such that the rider..would be afraid to push his horse up to the top of his speed. 1903 A. D. McFaul *The Gidden* xiii. 290 When rounding into the home stretch his horse broke, and suddenly went to a wild swerving break that carried him to the complete outside of the track. 1968 *Wanganui Chron.* 13 Nov. 6/3 Stylish Major, the beaten favourite on Tuesday after going into a break trying to match the early pace.

e. A short interval between lessons, usu. in the middle of morning or afternoon school. Also *transf.* Cf. *coffee-break, *tea-break.
1861 H. Spencer *Educ.* ii. 65 Short breaks during school-hours, excursions into the country,..in these and many like traits, the change may be discerned. 1913 C. Mackenzie *Sinister Street* I. i. vi. 94 Well, see you to-morrow in the break, young Fane. 1921 S. Thompson *Rough Crossing* ii. 3 'At break' Elizabeth met Lilian again. 1933 Mrs. C. S. Peel *Life's Enchanted Cup* xi. 133 In many workrooms no morning break was permitted. We finally decided that..the girls should be allowed a ten minutes' break at 11 o'clock.

f. On the Stock Exchange, a sudden decline in prices. *U.S.* (Cf. *Break v. 8 c.)
1870 J. K. Medbery *Men & Myst. Wall St.* 203 To endure an occasional 'break' in stocks. 1931 *Economist* 23 May 1110/2 The trend continued downwards, with particularly sharp breaks among high-priced stocks.

g. *slang.* A collection taken in aid of a prisoner awaiting trial or recently discharged.
1879 *Macmillan's Mag.* Oct. 502/1 The mob got me up a break (collection). 1896 A. Morrison *Child of Jago* xxv. 252 Get up a 'break' or subscription to pay for his defence.

h. The angle between the brim and crown of a hat.
1882 in Ogilvie.

i. A mistake, blunder; esp. in phr. *a bad break*: a serious mistake. *colloq.* (orig. *U.S.*).
1884 E. W. Nye *Baled Hay* 200 Possibly source may be wrong. We have known science to make bad little breaks. 1887 F. Francis *Saddle & Mocassin* 146 You've made one or two bad breaks since you've been in town. 1902 G. H. Lorimer *Lett. Merchant* 31 When a clerk makes a fool break, I don't want to beg his pardon for calling his attention to it. 1905 Kipling *Actions & Reactions* (1909) 96 We're..moving in worlds not realised, and we shall make some bad breaks. 1931 Wodehouse *If I were You* vii. 82 He'd always be worrying..for fear he was going to make a break of some kind.

j. A freak or abnormal development from the parent stock.
1921 *Conquest* Sept. 491/3 These 'mutations', 'sports' or 'breaks', as they are variously called. 1933 *Jrnl. R. Hort. Soc.* LVIII. 388 We are always looking out for natural breaks of variations.

k. *Broadcasting.* (See quot. 1941); *spec.* in phr. *natural break* (see quot. 1962).
1941 *B.B.C. Gloss. Broadc. Terms* 5 Break, interruption, either momentary or prolonged, in the transmission of a programme. 1959 *Manch. Guardian* 11 Aug. 4/5 [illegible] 1962 *Rep. Comm. Broadc.* [illegible] natural breaks.

9. c. In jazz, a short solo or improvised phrase; a passage of a few bars during which an instrumentalist plays unaccompanied. orig. *U.S.*
1926 A. Niles in W. C. Handy *Blues* 8 The notes..which follow this rest, fill in the following break, and themselves are called 'the break', or 'the jazz'. 1927 *Melody Maker* Apr. 377/1 Now try a two-bar break composed of Type A and Type B. 1958 P. Oliver in P. Gammond *Decca Bk. of Jazz* i. 21 Of each line of four bars, he may sing only two or three, allowing room for an instrumental or vocal 'break'.

10. *spec.* Rough, irregular country; broken country. *local U.S.*
1820 J. C. Gilleland *Ohio & Miss. Pilot* 171 Some of the breaks rise in deep circular glens called coves. 1902 Webster Suppl., *Break*,..a line of cliffs, and associated spurs and small valleys, at the edge of a mesa. (*Western U.S.*) 1903 Clapin *Dict. Amer.* 74 *Break*, a rough, irregular piece of ground. (Neb.) 1918 S. S. Visher *Geogr. S. Dakota* 117 Badlands or 'breaks' afforded protection from winter storms.

e. A broken or disturbed portion on the surface of water. *U.S.*
1852 *Trans. Mich. Agric. Soc.* III. 231 They will make a break in the water near the shore with their tail. 1897 'Mark Twain' *Following Equator* ix. 109 With..the 'break' spreading away from its head, and the wake following behind its tail.

14. In type-founding, a surplus piece of metal remaining on the shank of a newly cast type.
1683 Moxon *Mech. Exerc., Printing* 370 *Break*,..the Mettle that is contiguous to the Shank of a New Cast Letter: This Break is formed in the Mouth-piece of the Letter-mould, and is called a Break, because it is always broke from the Shank of a Letter. 1843 Holtzapffel *Turning* I. 374 The breaks, or the runners, of the types are first broken off. a1877 Knight *Dict. Mech.*

15. The quantity of hemp which is prepared or sold in one year. *U.S.*
1796 *Mass. Mercury* 29 Apr. (Cent. Dict.), Best St. Petersburg clean Hemp of the break of the year 1796. 1907 *Daily Chron.* 7 Mar. 6/6 A 'break' of hemp, which in America means the quantity sold in a year.

16. A portion of a crop of turnips, etc., set aside for sheep to feed on.
1856 R. W. Dickson *Pract. Agric.* II. 671 Removing them [sc. sheep] to fresh portions or breaks every eight or ten days. 1866 C. Scott *Sheep-farming* 48 A certain breadth or portion of the turnips, called a 'break', is given to be eaten off. 1923 L. Ackland in *Press* (N.Z.) 13/7 Break, a portion of land..divisions made in a paddock so that stock shall feed off the turnips..in sections. 1950 *N.Z. Jrnl. Agric.* May 461/1 Breaks of winter forage crops such as turnips or chou moellier.

17. *Electr.* and *Telegraphy.* An apparatus for interrupting or changing the direction of an electric current; a commutator.
1834 *Tomlinson's Cycl. Useful Arts* I. 560/1 The other pole..communicates..with the little wheel, called the break, the circumference of which is partly of metal and partly of wood or ivory, so as to interrupt and renew..the metallic connexion. a1877 Knight *Dict. Mech.*

b. *Electr.* The action of breaking contact in an electric circuit; the position in which contact is broken (in phr. *at break*). See also Make *sb.*[1] 9.
1875 F. Guthrie *Magn. & Electr.* 206 The automatic make and break. Fig. 181 shows the 'hammer break'. 1876 *Nature* XIV. 62/2 The increase of excitability was manifested towards make, and scarcely at all towards break.

18. *colloq.* A chance, an opportunity; a piece of good luck; freq. with defining word, as *an even break*: an equal or fair chance. orig. *U.S.*
1911 H. Quick *Yellowstone N.* v. 126 It's allus an even break whether they'll stan' and freeze in their tracks, or chase after some bunch of..natives. 1921 Mulford *Bar-20 Days* xxiii. 231 Now he wanted an 'even break', where once he would have called all his wits into play to avoid it. 1928 *Daily Express* 11 July 11 The chances in the 'quarter-mile' seem to give the Americans only an even break for a first place. 1930 *Publishers' Weekly* 8 Feb. 705/2 These buyers and their stores get what are known as 'the breaks'. 1938 G. Greene *Brighton Rock* II. i. 62 A break like that's too good. 1946 L. A. G. Strong *Trevannion* 196 Give the boy a break, they thought indulgently.

19. *Boxing.* The act of separating after the contestants have been in a clinch.
1928 *Daily Express* 2 Aug. 13/5 Lewis was disqualified for hitting on the break.

20. *attrib.* **break-lathe**, a lathe having a portion of its bed open or removable so as to admit work of larger diameter; **break-line** *Typogr.*, the last line of a paragraph.
1869 *Encycl. Brit.* XV. 134/1 Break lathes..were made by Mr. (now Sir Joseph) Whitworth as long ago as 1840. 1683 Moxon *Mech. Exerc., Printing* 226 Nor do good Compositors account it good Workmanship to begin a Page with a Break-line. 1868 Stower *Printer's Gram.* 163 Part of a word, or a complete word in a break line, if it contain no more than three or four letters, is improper. 1884 J. Johnson *Typogr.* II. 90 To take a comprehensive view of the copy..to notice..the number of break lines. 1967 *Hart's Rules for Compositors* (ed. 37) 65 Break-lines should consist of more than five letters, except in narrow measures.

break, *sb.*[2] **2.** (Earlier example.)
1831 C. M. Yonge *Daisy Chain* I. xxvi. 285 Norman's fate conveyed him to the exalted seat beside the driver of the break.

breakable. Add: **B.** *sb. pl.* Things which are capable of being broken.
1820 Byron *Let.* 12 June in R. E. Prothero *Lett. & Jrnl. Byron* (1901) V. 44 Pack up all the breakables. 1904 H. G. Wells *Food of Gods* I. iv. 117 The child was born with good intentions. 'Fudda be good, good,' he used to say as the breakables flew before him. 1909 —— *Tono-Bungay* I. ii. §4 There was a plaster of Paris horse to indicate veterinary medicines among these breakables.

break-away, breakaway (brēi·kăwēi). Pl. **break-aways, breaks-away.** [f. phr. *to break away*: see Break *v.* 49.]
1. The action of breaking away; severance.
1897 *Badminton Mag.* IV. 421 A break-away occurs in the ranks [of flying rooks]. 1909 *Daily Chron.* 13 July 1/4 The owners report that there is no 'breakaway' from the agreement. 1923 *Glasgow Herald* 1 Feb. 6 A challenging breakaway from rhythm. 1923 *Daily Mail* 31 May 13 There is a natural breakaway of the water on one side into a bog. 1950 H. Read *Educ. for Peace* iv. 55 A complete break-away from a pedagogic tradition which had its origins in the Revival of Learning.

2. *Austral.* **a.** A panic rush of animals, usually at the sight or smell of water; a stampede.
1891 'The Breakaway', title of picture by Tom Roberts at Victorian Artists' Exhibition (Morris *Austral Eng.*).
b. An animal that leaves the herd.
1893 *Argus* 29 Apr. 4/4 (Morris), The smartest stock horse that ever brought his rider up within whip distance of a breakaway.

3. In various sports, the act of breaking away or getting free. **a.** *Athletics, Racing.* A false start to a race. **b.** *Boxing.* The getting away from one's opponent or the separating of the contestants after a spell of in-fighting. **c.** *Association Football.* A sudden rush of a player or players with the ball towards the opponents' goal (esp. after a period of pressure); in *Rugby Football* used esp. of the action of a player moving quickly away from the scrummage. **d.** *Cycling.* (See quots. 1961); also, a cyclist who is leading in a race.
1882 *Times* 4 June 10/3 After several breaks away the 12 competitors were despatched to an excellent start. 1906 *Daily Chron.* 7 Sept. 9/4 They scored from a breakaway. 1909 *Westm. Gaz.* 19 Nov. 12/2 The visitors fully deserved their win, for, save for a few spasmodic breaks-away by the home team, they were pressing continually. 1928 *Daily Mail* 9 Aug. 14/1 A bad preliminary breakaway, which delayed the start. 1930 I. M. M. Stuart *Theory Mod. Rugby Practice* vi. 121 The wing forwards would be well advised to..hold themselves in readiness for a quick break-away. 1961 F. C. Avis *Sportsman's Gloss.* 149/1 *Breakaway*, a sudden and significant opening up of a gap in advance of the main group of riders in a cycle race. 1961 Partridge *Dict. Slang Suppl.* 1013/1 *The breakaway*, those competitors who have established a substantial lead in racing cyclists' coll.: since ca. 1925. 1961 *Times* 7 June 5/6 The exception to the general massing of the riders were the early breakaways of Jacobs..and Tarr.

4. *attrib.* or as *adj.* That breaks away or has broken away; seceding.
1934 in Webster. 1949 Koestler *Promise & Fulf.* III. i. 310 The so-called 'General Zionists' and the 'Progressive Party', a break-away group of the former. 1951 *Engineering* 13 July 56/2 Breakaway unions were condemned by..the Minister of Labour. 1961 *Listener* 28 Dec. 1116/1 The 'breakaway' province of Katanga.
b. *spec.* in *Rugby Football.* Applied to a forward in the side row of the scrummage. Also *ellipt.*
1954 J. B. G. Thomas *On Tour* 114 A twenty-eight-year-old salesman and tall breakaway forward. 1955 *Times* 23 Aug. 9/1 The breakaway men, Fry, Retief, and Ackermann, covered a vast amount of ground and showed great speed in the open. 1966 *Australian* 24 May 36/7 Other NSW Country forwards who could force their way into the State side tomorrow are breakaway Dick Cocks, and prop Ross Turnbull.

break-back, *a.* Add: **2.** Of a roof: having the lower portion at a different angle from the upper. Also *ellipt.* as *sb. U.S.* ? *Obs.*
1856 S. G. Goodrich *Recoll. of Lifetime* (1857) I. 78 The house..was a low edifice, two stories in front, the rear being called a *breakback*, that is, sloping down to a height of ten feet. 1859 Bartlett *Dict. Amer.* (ed. 2), *Break-back*, a term applied to a peculiar roof, common in the country, where the rear portion is extended beyond the line of the opposite side, and at a different angle. The addition thus acquired is used as a wash-room, a storehouse, or for farming implements.

break-back, *sb.*: see *Break *sb.*[1] 1 b.

break-down. Add: **1. b.** (Earlier and later examples.)
1838 *Civil Engineer* I. No. 11. 296/1 Railway Casualties—obstruction from cattle..various break-downs. 1893 *Funk's Standard Dict.*, *Breakdown van* (Gt. Brit.), a wrecking-car. 1933 in *Amer. Speech* (1942) XVII. 4/2 Morton, my car's outside Savarin's..Get your breakdown lorry round for it. 1953 A. Smith *Blind White Fish in Persia* i. 22 One of its hands..prepared the breakdown lorry.
c. Add to def.: esp. of the mental powers; *spec. nervous breakdown*: (a case of) neurasthenia; a vague term for any severe or incapacitating emotional disorder.
[1904 J. London *Sea-Wolf* xxxvi. 300 There had been his terrible headaches, and I was convinced that they were the result of a brain break-down.] 1905 A. Bennett *Sacred & Profane Love* III. i. 213, I read in the papers that you were suffering from neurasthenia and nervous breakdown. 1907 J. London *Iron Heel* x. 141 Hints were made of mental breakdown on his part. 1927 J. S. Huxley *Relig. without Rev.* iv. 125 The phase of conflict ended with that crash known generally as a 'nervous breakdown'. 1930 B. Russell *Conquest of Happiness* 75 One of the symptoms of approaching nervous break-down is the belief that one's work is terribly important. 1959 J. Braine *Vodi* v. 82 He was never seen at the school again. The official explanation was a complete nervous breakdown.

d. *Electr.* The sudden passage of electric current through an insulating medium. Also *attrib.*, as **breakdown voltage,** the voltage required to cause a break-down.
1915 F. W. Peek *Dielectric Phenomena* vi. 154 The moisture may even be removed from the space between the electrodes by the action of the field, in which case its presence would not be detected by low-voltage breakdowns. *Ibid.* 155 The relative breakdown voltages of gaps in oil, at 60 cycles, and for impulse voltages..are given. 1962 Simpson & Richards *Junction Transistors* iv. 67 The voltage V_B at which this sudden decrease in resistance occurs is called the breakdown voltage.
e. Chemical or physical decomposition. Also *attrib.*, as **breakdown product,** a product resulting from the disintegration of a substance.
1928 A. B. Callow *Food & Health* 24 Certain chemical compounds when eaten cause a flow of gastric juice. The chief of these substances are the break-down products of proteins (peptones, etc.). 1959 *Listener* 2 July 38/2 Organic materials of a fibrous nature..encourage green water and algae during the inevitable breakdown processes. 1961 *Lancet* 19 Aug. 395/2 Hæmoglobin breakdown was completely inhibited.
f. An analysis or classification (of figures, statistics, etc.). Cf. *Break *v.* 50 g.
1936 Harrison & Mitchell *Home Market* 140 Such families were extremely few and their breakdown according to density was impossible. 1948 *Observer* 18 Apr. 4/5 The latest threat to clarity is the use of 'breakdown' to mean 'analysis' or 'classification'. 1957 *Times* (Canada Suppl.) 12 Nov. p. viii/1 Figures rose..to 73,578 for the corresponding period this year. The breakdown of this..figure was English 47,240, Scots 15,124, Irish 9,646, Welsh 1,568.
3. *Sawmilling.* (See quot. 1957.) Also (*N.Z.*) applied to the building in which the initial cutting of timber from logs is done. Also *attrib.* Cf. *breaking-down.
1923 C. M. Malfroy *Small Sawmills* 17 The laying of the foundations of the breakdown should first be proceeded with. *Ibid.*, The logging delivery-tram, mill log-skids, engine, breakdown bench. 1943 *Amer. Speech* XVIII. 85 The cleared tracks..carry logs to the mill, where it is handled by a *break-down man.* 1957 *Brit. Commonw. Forest Terminol.* II. 33 *Breakdown*, the initial operation in converting from the round, by sawing a log longitudinally into cants, and, by extension, cants into large timber, preparatory to further manufacture.

breaker[1]. Add: **4. c.** In paper-manufacture, a machine in which rags, etc., are washed and partly pulped. Also *attrib.*, as *breaker-plate.*
1880 J. Dunbar *Pract. Papermaker* 71 It may be mentioned that the breaker-plate ought to be sharp when starting to blottings. 1902 *Encycl. Brit.* XXXI. 456/1 The next step is that of washing and 'breaking in', which takes place in an engine called the 'breaker'. 1963 R. R. A. Higham *Handbk. Papermaking* ii. 23 Different types of breakers are available for use with rag, wood pulp sheets, waste paper and broke.
d. In anthracite mining, an apparatus for breaking, sizing, and cleaning coal for the market.
1885 *Encycl. Brit.* XVIII. 501/2 The 'breaker', an anthracite invention and a monster of destruction, is an edifice of wood and iron 100 feet high,..with rollers set with teeth to crush the larger lumps, with bolting screens to separate the sizes. 1900 *Coal & Metal Miners' Pocket-bk.* (ed. 6) 574 *Breaker Boy*, a boy who works in a coal breaker.
e. In cheese-making, an implement for breaking the curd into small pieces.
1844 *Jrnl. R. Agric. Soc.* V. I. 88 The first process of breaking down the curd in the cheese-tub is..performed by a breaker or curd-cutter.
7. A horse that breaks (*Break *v.* 38 c). orig. *U.S.*
a1867 H. Woodruff *Trotting Horse in Amer.* (1868) xxiii. 201 Although a trotter of remarkably fine speed and power, he was such a bad breaker. 1965 *Weekly News* (N.Z.) 8 Dec. 59/2 Breaking horses have always been a problem... There is an alarmingly high percentage of breakers.

break-even (brēi·ki·v'n). orig. *U.S.* [f. *to break even* (see *Break *v.* 47* b).] Usu. *attrib.*: designating or pertaining to the point at which one 'breaks even' (see *Break *v.* 47* b); of or pertaining to a balance of expenditure and revenue, profit and loss, etc.
1938 *Barron's National Financial Weekly* 14 Nov. 27/3 Tire companies in the first half of the year were around the break-even point on the average. 1948 *Jrnl. R. Aeronaut. Soc.* LII. 688/2 By matching the right size of aeroplane to the stage length, 'break-even' fares of between fourpence and sixpence per passenger mile could be achieved. 1949 *Ibid.* LIII. 199/2 Sales of 400 aircraft would be required before the constructor reached 'break-even' point and recovered his outlay. 1952 A. E. Benn *Management Dict.* 66 *Break-even point*, that point at which the level of balance is equal, as the level of production at which there will be no profit or loss. 1958 *Times* 12 Dec. 12/1 (*heading*) Earliest possible 'break-even'. *Ibid.*, Steps necessary to achieve the earliest possible break-even date. 1965 *Listener* 1 July 9/2 Half the tenants..can afford a break-even rent.

breakfall (brēi·kfǭl). [f. Break *v.* 28 b + Fall *sb.*[1]] In judo, a movement whereby the impact of a fall is diminished. Hence as *v. intr.*, to execute a breakfall.
1906 Miyake & Tani *Game of Ju-Jitsu* v. 35 The sideways break-fall..may nearly always be used when you are pitching forward on to your head. 1954 E. Dominy *Teach yourself Judo* ii. 20 Spend 5 minutes on breakfalls. *Ibid.* iii. 36 As he begins to fall you must let go of his jacket with your right hand so that he can breakfall.

breakfast, *sb.* Add: **2. b.** = Wedding-*breakfast.*
1848 Thackeray *Van. Fair* xv. 132 We'll have a wedding, Briggs, and you shall make the breakfast, and be a brides' maid. 1903 G. B. Shaw *Man & Superman* iv. 167 I'm off for my honeymoon...Oh don't say that, Violet. And no wedding, no breakfast.
3. *breakfast-china, -cloth, -cup, -food, -hour, -nook, -parlour* (earlier example), *-party* (earlier example), *-room* (earlier example), *-service, -stall* (example), *-table* (earlier examples), *things, -time* (earlier example), *-tray.*
1811 Jane Austen *Sense & Sens.* I. ii. 26 The set of breakfast china is twice as handsome as what belongs to this house. 1929 E. Bowen *Joining Charles* 124 Breakfast-china..with a red rim. 1778 G. L. Way *Learning at Loss* I. 107 Dipping the Corner of the Breakfast Cloth into his Tea-cup. 1850 Dickens *Dav. Copp.* xxv. 270 A clean breakfast-cloth. 1762 Boswell *London Jrnl.* 12 Dec. (1950) 81, I ordered some large breakfast cups. 1911 A. Bennett *Hilda Lessways* I. v. 47 Her big breakfast-cup full of steaming tea. 1898 *Ladies' Home Jrnl.* Feb. 28/1 (Advt.), Ralston Health Club Breakfast Food. 1941 M. Allingham *Traitor's Purse* vii. 75 The familiar cartons of breakfast food. 1811 L. M. Hawkins *Countess & Gertrude* I. v. 80 An unforeseen indisposition..kept her ladyship in almost hopeless doubt during her breakfast-hour. 1876 C. M. Yonge *Womankind* xiii. 92 It is very unfortunate that most people's breakfast hour coincides with this only period permitted for religious teaching. 1931 *Kansas City Times* 9 Oct., A cute little room called the breakfast nook. 1958 J. Cannan *And be a Villain* iii. 50 You can come into my kitchen..there's a breakfast nook, as they calls it in the books. 1802 C. Wilmot *Let.* 14 Feb. in *Irish Peer* (1920) 43 If 'Italy is the Garden of Europe',..Scotland the breakfast Parlour, and so on, I really think one may say France is the 'Drawing-room'. 1814 Jane Austen *Mansf. Park* II. x. 232 Sir Thomas asked Crawford to join the early breakfast party. 1732 Sarah, Duchess of Marlborough *Let.* 4 July (1943) 36 The rooms I mean, is that where the red velvet bed stands and the room next to it, which they call the breakfast room. 1838 J. Romilly *Camb. Diary* 16 Oct. (1967) 156 She ordered a beautiful breakf. service for him. 1855 Trollope *Warden* viii. 122 The breakfast-service on the table was equally costly and equally plain. 1939 A. Thirkell *Before Lunch* ix. 249, I shall give you a breakfast service and a very large Persian cat. 1853 Dickens *Bleak Ho.* xlvii. 449 A breakfast-stall at a street corner. 1762 T. Chippendale *Gentleman & Cabinet-Maker's Director* (ed. 3) 8 Two Designs of Breakfast-Tables. One hath a Stretching-Rail... The other hath a Shelf. a 1817 Jane Austen *Northanger Abbey* (1818) II. v. 67 Never in her life before had she beheld half such variety on a breakfast-table. 1841 'M. A. Titmarsh' in *Britannia* 15 May 315/2 All the breakfast things begin to clatter. 1960 'J. Bell' *Well-known Face* iv. 37 She had moved both cups off the little tray to make room for her breakfast things. 1599 Lady M. Hoby *Diary* 25 Sept. (1930) 74 After I had praied I walked in the garden tell break fast time. 1865 Mrs. Gaskell *Wives & Daughters* (1866) II. xxi. 210 Mrs. Gibson breakfasted in bed...Her own private letters always went up on her breakfast-tray. 1893 E. F. Benson *Dodo* (ed. 3) I. iv. 82 He met a footman carrying a breakfast-tray.

break-front (brēi·k,frɒ·nt). [f. Break- + Front *sb.*] Used esp. *attrib.*: a piece of furniture having a front of which the line is broken by a curve or angle.
1928 *Daily Tel.* 24 July 12/2 A large break-front sideboard. 1955 R. Fastnedge *Eng. Furn. Styles* vii. 168 Some prominence was given to the library breakfront bookcase. 1962 *Times* 19 Apr. 15/3 The very stately Chippendale break-front had glazed locked doors.

break-in: see Break *sb.*[1] 1 b in Dict. and Suppl.

breaking, *vbl. sb.* Add: **1. b.** (Examples with *down.*)
1842 Dickens *Amer. Notes* I. viii. 307 Blowings-up in steam-boats and breakings down in coaches. 1957 J. S. Huxley *Relig. without Rev.* viii. 198 The breaking-down of other substance.
c. The act of forcing a passage into another person's house or other building; freq. in phr. *breaking and entering*, = Housebreaking.
1617 [see Housebreaking]. 1729 G. Jacob *Law-Dict.* O1/2 If a Thief unlocks a Door, or draws the Latch of a Room, to rob..these are a Breaking. 1797 Tomlins *Jacob's Law-Dict.* I. Bb3/1 A Felony at common law, in (1) *breaking and entering* (2) the *mansion house* of another..to the intent to commit some *felony.* 1855 [see Burglary[1] a]. 1939 N. Marsh *Overture to Death* viii. 93 A breaking and entering job at Moorton Park with..her ladyship's jewellery gone.
d. In woollen manufacture, the operation by which short combed slivers are combined and made into continuous lengths. Also *breaking-in.*
1843 *Penny Cycl.* XXVII. 554/1 The breaking being thus effected, the sliver of wool proceeds to a large bobbin or cylinder. 1915 R. Beaumont *Woollen & Worsted* 631 The piece, having been scoured, milled, dried, and tentered, is evenly damped, raised, or raised across before being passed on to the teazle machine. This is called 'breaking-in'.
e. *Phonology.* [After G. *brechung.*] = Fracture *sb.* 5. Also applied to different sound-changes in Old Norse and other Germanic languages.
1871 F. A. March *Compar. Gram. Anglo-Saxon* I. 20 *Breaking* is the change of one vowel to two by a consonant. 1874 A. J. Ellis *E. E. Pronunc.* IV. xi. 1270 Grimm considers breaking mainly due to the action of a following *r, h.* 1937 *Language* XIII. 123 (*title*) Breaking in Old Norse and Old English. *Ibid.*, Under breaking I include here the change of a front vowel in a stressed syllable to a diphthong by the influence of following velar elements whether these be consonants in the same syllable [in O.E.] or vowels in the next [in O.N.].
2. c. *breaking* (*of the meres*) = *water-bloom. Cf. *Break *v.* 5 c.
1884 W. Phillips in *Trans. Shrops. Archæol. Soc.* VII. 285 The breaking is called in German 'Wasserblüthe' (water-blossom). 1927 West & Fritsch *Treat. Brit. Freshwater Algae* 451 The phenomena of 'water-bloom' and the 'breaking of the meres' are due to the sudden and often periodical appearance of large quantities of a few species of Myxophyceae. 1948 *New Biol.* V. 21 This 'water bloom' or 'breaking of the meres' may be quite sudden.
8. *breaking(-up) plough*; **breaking-down,** the action of converting a log into sawn timber; also *attrib.*; cf. *Break *v.* 50 f and *break-down 3; **breaking-down rollers** (see quot.); **breaking-engine,** (*a*) in paper-manufacture, a machine for washing and pulping rags, a breaker; (*b*) in woollen-manufacture, a carding-machine; **breaking-off,** the removal of the piece of surplus metal from newly-cast type; also *attrib.*; **breaking-point,** (*a*) the point, or degree of stress, at which a particular material breaks; (*b*) *fig.* the point at which a person's strength or endurance fails, or at which a situation becomes critical; **breaking-rollers,** an apparatus for the mechanical kneading of dough; **breaking-strain, -strength, -stress,** the strain or stress required to break a particular material or object.
1883 M. P. Bale *Saw-Mills* xxxviii. 331 *Breaking down*, in sawing, is dividing the baulk into boards or planks. 1913 A. I. Carr *Country Work & Life in N.Z.* xxiv. 40 'Breast' benches (where the flitches from the breaking-down saw are cut into commercial sizes) are still used in many mills. 1922 R. C. Bryant *Lumber* iv. 81 The breaking down of the log continued until it was reduced to a size which could be worked by the saws C and D. 1949 E. de Mauny *Huntsman in Career* 162 The scream of the bandsaws on the breaking-down bench. 1839 Ure *Dict. Arts* 860 Two pairs of rollers, which, from being used to consolidate the metal by rolling whilst hot, are termed breaking-down rollers. a1877 Knight *Dict Mech.*, *Breaking-engine*, the first of a series of carding-machines, to receive and act on the lap from the lapper; it has usually coarser clothing than the finishing-cards. 1880 J. Dunbar *Pract. Papermaker* 71 When furnished in the breaking-engine, wash thoroughly before letting down the roll. 1683 Moxon *Mech. Exerc., Printing* 176 Breaking off is commonly Boys-work: It is only to Break the Break from the Shanck of the Letter. 1839 Ure *Dict. Arts* 1261 From the breaking-off boy the types are taken to the rubber. 1921 *Dict. Occup. Terms* (1927) § 279 *Breaker*.., *breaker-off*, *breaker-of boy* [also *breaking-off boy*], breaks off superfluous wedge-shaped piece of metal, which adheres to lower surface of type when type leaves casting machine, by pressing lower surface of type against table. 1853 *Knickerbocker* XLII. 593 The great 'breaking-plough'..goes tearing..through the roots and grubs. 1899 Pitman *Key to Business Corresp.* 23 The breaking point of the yarn is guaranteed to be not less than 36 pounds weight for 120 yards. 1908 H. G. Wells *War in Air* iv. 102 Elaborating the apparatus of war, until the accumulating tensions should reach the breaking-point. *Ibid.* xi. 353 Under the stresses of the war their endurance reached the breaking point. 1922 Joyce *Ulysses* 365 Transparent stockings, stretched to breaking point. 1845 Dodd *Brit. Manuf.* V. 24 The dough is..placed under the breaking-rollers..which perform the office of kneading. 1886 S. W. Mitchell *R. Blake* xix. 181 The engineer speaks of the breaking-strain in material; the breaking-strain in morals was near for Octopia. 1888 [see Strain *sb.*[2] 9]. 1902 *Encycl. Brit.* XXXIII. 11/2 A medium steel..showed a breaking strength of 39 tons per square inch. 1960 *B.S.I. News* Jan. 2 The belt has a breaking strength of 2 tons and weighs only a little over 2 lb. 1940 *Chambers's Techn. Dict.* 111/1 Breaking stress. 1959 *Chambers's Encycl.* XIII. 224/2 Once the breaking stress has been measured it is easy to calculate the permitted working stress by dividing by the appropriate factor of safety. 1781 M. Patten *Diary* (1903) 438 Our 4 oxen and breaking up plow helped james Walker break up.

break-out: see Break *sb.*[1] 1 b in Dict. and Suppl.

break-through, breakthrough (brēi·kþrū). [f. phr. *to break through*: see Break *v.* 55; cf. G. *durchbruch.*] An act of breaking through (a barrier of any kind); spec. *Mil.*, an advance penetrating a defensive line or the like; also *fig.*, esp. (*a*) *U.S.* a sudden increase in prices or values; (*b*) a significant advance in knowledge, achievement, etc.; a development or discovery that removes an obstacle to progress. Also *attrib.*
1918 *Daily Express* 5 Nov. 1/2 The attempted break-through for which the English and French have again been striving on a front of over thirty-seven miles was frustrated. 1921 A. Lunn *Alpine Ski-ing* vii. 98 A break-through into a crevasse can be instantly checked. 1936 *Mind* XLV. 248 Barriers have to be erected having negative valence if

a break-through be attempted. 1937 A. L. Rowse *Sir R. Grenville* vii. 154 Drake had captured 200,000 ducats of the King's property..in his break-through into the South Sea. 1942 *Times* 8 Feb. 4/4 British tanks..have been forced to renounce their break-through attempt. 1949 *Amer. Speech* XXIV. 174 A strong and sudden increase [in the market] is a *breakthrough.* 1957 *Economist* 23 Nov. 736/1 The break-through to industrial prosperity began with the First Industrial Revolution. 1958 *Listener* 11 Sept. 376/2 The technological break-through which allowed both the United States and the U.S.S.R. to produce H-bombs within a year of each other.

break-up. Add: **break-up price** or **value,** the price or value of assets at the break-up of a concern.
1899 *Westm. Gaz.* 4 Dec. 10/2 A trade valuer was examined to show that he had advised the Grices to sell their business..at a break-up price. 1902 *Ibid.* 13 Nov. 7/1 At break-up values the assets of the company would pay 10s. in the pound to preference shareholders. 1930 *Economist* 1 Feb. 233/1 Shares of most of them are now selling below or close to their 'break-up' asset value. 1968 *Globe & Mail* (Toronto) 17 Feb. B 8 The breakup value is no more than 10 cents a share.

bream, *sb.* Add: **2. b.** Any of various fishes of the family Centrarchidæ, or sun-fishes, resembling the common European bream. *U.S.*
1634 W. Wood *New Eng. Prospect* II. xvi. 90 Catching of Pikes, Pearches, Breames, and other sorts of fresh water fish. 1791 W. Bartram *Trav. Caroline* 176 The golden bream or sun-fish, the red bellied bream..also abound here. 1884 Goode *Nat. Hist. Aquatic Anim.* 406 The Blue Sun-fish—*Lepomis pallidus*..is known as the 'Blue Bream'. 1965 A. J. McClane *Standard Fishing Encycl.* 143/2 Bream. A regional colloquialism (Southern United States) for various sunfishes, usually pronounced as 'brim'.

breast, *sb.* Add: **9. h.** A large roller or cylinder in a carding-machine. Also *attrib.*, as *breast cylinder.*
a1884 T. Lister in W. S. B. McLaren *Spinning* (1884) x. 229 A breast large enough to entitle it to the name of a cylinder. 1888 *Encycl. Brit.* XXIV. 659/1 Against the licker-in revolves the 'angle-stripper', the function of which is to remove the wool..and deliver it over to the great breast cylinder.
11. breast-beating, an exaggerated and ostentatious demonstration of woe, remorse, etc.; **breast-fed** *a.*, (of infants) fed at the mother's breast; so *breast-feeding* (opposed to *bottle-feeding*); *breast-feed* vb.; **breast-girth,** (*a*) measurement around the breast; (*b*) (see quot. 1958); **breast-height,** (*b*) the height of a man's chest above ground-level, usually taken as 4 ft. 3 in. (in some countries 4 ft. 6 in.), the standard height used for measuring the 'girth, diameter and basal area of standing trees'; **breast-pump** (examples); **breast-shore,** each of a line of props transversely supporting a vessel in dry dock; **breast-stroke** *Swimming*, a stroke in which the breast is squarely opposed to the water, the arms are pushed forward and outwards in a wide arc, and the legs perform a frog-like action; also as *v. intr.*; so *breast-swimming.*
1952 *Granta* 15 Nov. 10/1 There are times when the noise of breast-beating and intellectual self-analysis sounds like a drum and fife band. 1969 *Nature* 9 Aug. 550/2 A proper and urgent concern for the environment is overtaken by self-indulgent breast-beating. 1903 *Daily Chron.* 15 Jan. 5/2 The death rate was thirty times as high in children fed on cow's milk as in those breast-fed. 1909 *Daily Chron.* 3 June 6/4 This figure including those breast-fed by the mothers. 1928 *Daily Express* 29 Feb. 3 Taken regularly—both before and after the birth—[it] enables nearly every mother to breast-feed her baby. 1904 *Fabian News* XIV. 25/1 The decline of breast-feeding. 1909 *Daily Chron.* 3 June 3/3 'Weak heart', or 'insufficient breast-girth', or other physical defect. 1958 J. Hislop *From Start to Finish* 167 *Breast-girth*, strap made of webbing which goes from one side of the saddle to the other (being attached to the girths) and stops the saddle from slipping back. 1932 *Forestry* VI. 53 The common breast-height form-factor, which represents the ratio between the volume of the tree and that of a cylinder with the same height and the same girth or diameter as that of the tree at breast height. 1861 Mrs. Beeton *Bk. Househ. Managem.* 1036 Every mother..should be provided with a breast-pump, or glass tube, to draw off the super-abundance. 1924 J. S. Fairbairn *Gynæcol. with Obstetrics* v. xxi. 406 The use of a breast-pump. 1894 W. H. White *Man. Naval Archit.* (ed. 3) 322 The shores under the bilges and bottom take part of the weight, and the 'breast shores' assist in maintaining form. 1867 C. Steedman *Man. Swimming* 93 There are four distinct kinds of motions for the arms, and the same number and kind for the legs, used in the common plain or breast-stroke. 1922 E. Raymond *Tell England* II. vii. 237 Other flies fell into my tea, and did the breast-stroke for the side of the mug. 1928 *Daily Express* 18 Feb. 5/1 He..breast-stroked away toward the setting moon. 1867 C. Steedman *Man. Swimming* 106 Its [sc. side swimming's] superiority over breast swimming.

breast-high, *a.* (Later example, *spec.* in *Forestry.*)
1905 *Terms used in Forestry & Logging* 7 *Breasthigh*, at or having a height of 4½ feet above the ground.

breastwork. Add: **3.** The brickwork or masonry forming the breast of a fire-place.
1806 *Massachusetts Spy* 23 July (Th.), On the breastwork over the fireplace was the distinct impression of a bloody hand. 1833 Loudon *Encycl. Archit.* § 79 The fire-places to have each a strong iron chimney-bar (bar for supporting the breast-work, or front side of the flues).

breasty (bre·sti), *a. colloq.* [f. Breast *sb.* 1 a + -y[1].] = *bosomy *a.* 2.
1944 H. Hunt *Limit of Darkness* 23 The others kidded him about being breasty because of the twin bulges made by the pipe and his tobacco pouch. 1959 *Numbers* (N.Z.) IX. 38 A pretty breasty, curly haired girl in blue. 1960 *20th Cent.* Dec. 582 An Amis hero pouncing on a breasty girl.

breath. Add: **3. a.** *to keep* (*save, spare*) *one's breath to cool one's* (*own*) *porridge*: see Porridge *sb.* 4.
5. c. *to take a person's breath* (*away*): to cause him to hold his breath owing to sudden emotion; hence, to dumbfound, flabbergast.
1864 Browning *Likeness* in *Dram. Personæ*, He never saw..What was able to take his breath away. 1905 T. Dixon *Clansman* 351 The daring campaign these men were waging took his breath. a1910 'Mark Twain' *Myst. Stranger* (1916) 14 He said it placidly, but it took our breath for a moment and made our hearts beat. 1965 *Listener* 3 June 826/1 It is the really bold planting that takes one's breath away.
10. (Earlier example.)
1842 *Penny Cycl.* XXII. 429/2 It will be observed..that these consonants have no voice throughout their duration; that they each have breath-sound.
11. *breath-control, -force, -holding* sb. and adj., *-stream*; **breath-catching** *a.* = *breath-taking; **breath-group** *Phonetics*, the succession of words, whether a sentence or part of a sentence, uttered without pause, in a single breath; **breath-taking** *a.*, surprising, thrilling, dumbfounding (cf. sense 5 c, above); hence *breath-takingly* adv.; **breath-test,** measurement of the amount of alcohol in the blood by means of a breathalyser; also **breath-testing** *vbl. sb.* and *ppl. a.*
1928 *Daily Tel.* 6 Nov. 14/5 Green, Wood and Violet..do some breath-catching tumbles. 1958 L. W. Tancock in *Aspects of Translation* 50 His personal endowments of voice and breath-control. 1935 M. Schubiger *Role of Intonation* 1 The amount of breathforce (stress). 1963 *English Studies* XLIV. 60 Amplitude is not only a function of breath-force but also of vowel quality. 1877 H. Sweet *Handbk. Phonetics* 86–7 The only division actually made in language is that into 'breath-groups'...Within each breath-group there is no pause whatever. 1909 D. Jones *Pronunc. of Eng.* 58 The following are examples of breath-groups: *Yes; Good morning; Shall we go out for a walk?* 1964 E. A. Nida *Toward Sci. Transl.* viii. 178 In this type of translating there are several important factors (1) timing, both of syllables and breath groups. 1890 W. James *Princ. Psychol.* II. xx. 174 The breath-holding stillness of the boy playing 'I spy', to whom the seeker is near. 1937 J. R. Firth *Tongues of Men* iii. 40 The larynx or Adam's apple—the breath-holding, whisper-, and voice-making instrument. 1968 *Brit. Med. Bull.* XXIV. 250/1 CO_2 excretion is abolished..by breath-holding or by rebreathing. 1934 Priebsch & Collinson *German Lang.* p. xv, A total stoppage of the outgoing breath-stream. 1880 'Mark Twain' *Tramp Abroad* xxxiii. 324 It was a sort of breath-taking surprise. 1908 *Westm. Gaz.* 1 Apr. 7/3 This breath-taking assertion was made to a 'Westminster' representative. 1966 *Illustr. London News* 30 July 35 (Advt.), It tastes like a view from the top of the Eiffel Tower. Breathtaking. 1928 *Manch. Guardian Weekly* 31 Aug. 175/3 Breathtakingly beautiful. 1966 *Economist* 12 Nov. 650/3 She has dropped her proposal for random road-side breath-tests for drivers. 1968 *Listener* 21 Mar. 391/3 It has been forecast from recent casualty figures that the breath-test should cut deaths by 1,600 a year. 1960 *Daily Tel.* 27 Jan. 11/4 He subjected a colleague and myself to analyses of alcohol in the blood on his breath-testing machine. 1967 *Spectator* 17 Nov. 605/2 The main object of breath testing is, of course, deterrence.

breathalyser (bre·þăläizəɹ). Also **-zer.** [f. Breath + (An)alyser, -zer.] (See quot. 1960[1].) Also *attrib.*
1960 *Times* 19 Jan. 7/4 The Breathalyser, an American instrument for measuring the percentage of alcohol in the blood from a breath sample, was put on view..yesterday. 1960 *Daily Mail* 19 Jan. 5/1 The Breathalyzer, the snap-test machine..for picking out drunken drivers. 1960 *News Chron.* 16 June 6/1 The breathalyser test is favoured by many. 1967 *Spectator* 20 Oct. 450/2 It may well be that the breathalyser test works and one cause of death on the roads has been banished from the scene.

breather. Add: **3. b.** A short rest in which to recover breath; a breathing-space. (Cf. Breathe *v.* 13.)
1901 *Daily Chron.* 15 Oct. 5/4 But the horses have had their breather, and we must on. 1902 *Ibid.* 21 May 3/3 A field battery on the left had a hot time of it just at this moment, and drew out of action for a breather quite close to our guns. 1940 L. A. G. Strong *Sun on Water* 178 He worked fast for an hour and a half, then sat back for a breather. 1959 *Daily Tel.* 9 Nov. 1/1 They stopped for a 'breather' on the hard shoulder, which is out of bounds, except for vehicles in distress.
5. In full, *breather pipe*: a vent in the crank-case, etc., of an internal-combustion engine, for the release of excessive pressure or gases.
1929 W. Faulkner *Sartoris* II. vi. 145 He lifted the hood and removed the cap from the breather-pipe. 1935 *Times* 22 Oct. 9/1 Here also is the oil filler with a breather. 1962 *Which?* (Car Suppl.) Oct. 136/1 Cleaning the crankcase breather.

breathiness, see Breathy *a.* 2.

breathing, *vbl. sb.* Add: **10. breathing-room** = *breathing-space*; **breathing-space,** also, a rest; a period of inactivity; **breathing-tube,** a tube through which to breathe.
1678 R. L'Estrange tr. *Seneca's Morals* II. xv. 188 But they had open Ayr, and Breathing-Room. 1891 W. James *Will to Believe* (1897) 203 That howling mob of desires, each struggling to get breathing-room for the ideal to which it clings. 1922 D. H. Lawrence et al. tr. *Bunin's Gentleman from San Francisco* 2 He had almost reached the level of those whom he had taken as his ideals, so he made up his mind to pause for a breathing space. 1889 *Cent. Dict.*, *Breathing-tube*, in *entom.*, the respiratory tube of certain aquatic larvæ and dipterous puparia. 1946 *Jane's Fighting Ships 1944–5* 229/2 The 'Schnorkel', or breathing tube,..enabled submarines to remain submerged for much longer periods.

breccia. Add: **c.** *fig.*
1873 L. Stephen *Freethinking* vi. 203 His prose, at excited moments, becomes a kind of breccia of blank verse. 1944 H. G. Wells *'42 to '44* 149 With this sort of breccia in their heads, these gentlemen seem able to sleep of nights without ordinary narcotics.

Brechtian (bre·ҳtiăn), *a.* and *sb.* [f. the name *Brecht* (see def.) + -ian.] **A.** *adj.* Of, pertaining to, or characteristic of, the German playwright and poet Bertolt Brecht (1898–1956) or his plays or dramatic technique. **B.** *sb.* An admirer or follower of Brecht.
1935 *Life & Letters Today* Sept. 66 The film..lies halfway between the Brechtian ideal and commercial materialism. 1948 E. R. Bentley *Brecht's Private Life of Master Race* 92 The Brechtian theatre has few technical demands to make. *Ibid.* 93 The Brechtian characters are without hate, without love. 1958 *Times* 10 Nov. 14/2 His delight in Brechtian visual effects. 1959 *Observer* 15 Mar. 11/3 Her stagecraft is boldly Brechtian. 1959 *Times Lit. Suppl.* 6 Nov. 637/1 'Relaxed' is a word which Brechtians are especially fond of applying to the Master's style of acting.

breck. Add: **2.** (Further examples.) Also *attrib.*; **Breckland,** a name given to the region of brecks in Norfolk.
1837 *Penny Cycl.* VIII. 282/1 The naked brecks (or undulating downs) of Norfolk. 1840 *Jrnl. R. Agric. Soc.* I. iv. 360 The first damside breck meadow on the plan. 1879 Lubbock *Fauna of Norfolk* Introd. p. viii, On the 'Breck' district the lordly Bustard roamed. 1894 *Naturalists' Jrnl.* Oct. 90 The 'brecks'..are the upland breaths and huge fields of this district [sc. south-west Norfolk]. *Ibid.*, We at length reach a typical breckland wild—Roudham Heath. 1897 W. Rye *Songs of Norfolk* 124 Such cramped wild country, half rough breck land and half marsh. 1956 O. Cook *Breckland* 59 Many of the Breckland cottages..have no gardens...This is typical of Breckland.

bredbergite (bre·dbəɹgait). *Min.* [f. the name of B. G. Bredberg, who first described it: see -ite[1].] A name given by J. D. Dana in 1868 to a variety of andradite found at Sala, Sweden, which contains a large amount of magnesium.
1868 Dana *Syst. Min.* (ed. 5) 270 Lime-Magnesia Iron-garnet; Bredbergite.

bredie (brī·di). *S. Afr.* Also **breedi.** [Afrikaans, from Malagasy.] A stew of meat and vegetables.
1815 A. Plumptre tr. *Lichtenstein's Trav. S. Afr.* II. xxxiii. 82 *Breedi* signifies in the Madagascar tongue *Spinage*; the word is brought hither by the slaves, and at present, throughout the whole colony, every sort of vegetable which, like cabbage, spinage, or sorrel, is cut to pieces and dressed with Cayenne pepper, is included under the general term *Breedi.* 1870 *Cape Monthly Mag.* Oct. 224 All sorts of vegetable 'breedies', are importations from India. 1947 L. G. Green *Tavern of Seas* (1952) viii. 59 She made every variety of bredie, the stews that the Malays brought to the Cape; especially the tamatie bredie.

bredigite (bri·digait). *Min.* [f. the name of M. A. Bredig (b. 1902), German-born American chemist: see -ite[1].] A mineral consisting of the metastable orthorhombic phase of dicalcium silicate.
1948 Tilley & Vincent in *Min. Mag.* XXVIII. 257 It is in recognition of Bredig's important contribution to the solution of the problem of calcium orthosilicate polymorphism that it is proposed to designate the new mineral phase bredigite. 1966 *Amer. Min.* LI. 1766 The α' (bredigite), β (larnite) and γ (unnamed) forms of dicalcium silicate (Ca_2SiO_4) occur together in the contact zone around a syenite-monzonite intrusion in Marble Canyon.

breech, *sb.* Add: **2. b.** Phr. *too big for one's breeches*: see *Big *a.* 3 e.
4. (Examples of use in *Obstetr.*)
1673 H. Chamberlen tr. F. Mauriceau's *Accomplisht Midwife* II. xxiv. 201 The Chirurgeon perceiving the Child to come with the Breech foremost, ought to put it back, if he can. 1752 W. Smellie *Treat. Theory & Pract. Midwifery* III. iv. 323 When the legs are delivered, let him wrap a cloath round the breech of the child. 1840 T. Castle *Blundell's Princ. & Pract. Obstetricy* IV. vi. 292 Let then, the natural

efforts bring the breech to the outlet of the pelvis, then lay hold of the hips. 1962 D. E. Reid *Textbk. Obstetr.* xx. 532/2 During labor, as the breech descends, the shoulders must also descend.
(*b*) *ellipt.* for *breech delivery, position, presentation.*
1781 A. Hamilton *Treat. Midwifery* 244 The varieties of the *breech* are, 1st, The fore parts of the child placed to the *pubes* of the mother; 2dly, To the *sacrum*; 3dly, To either side. 1903 J. W. Williams *Obstetr.* xiii. 262 If interference becomes necessary, the complete breech offers more satisfactory conditions for immediate delivery. 1958 H. Speert *Essays on Eponymy* lxv. 567 The 'assisted breech' has long been taught as the properly conservative method of managing most breech births. 1963 D. G. W. Clyne *Textbk. Gynaecol. & Obstetr.* xxviii. 666 Maternal death from an uncomplicated breech is almost unknown nowadays.
7. (sense *4) *breech case, delivery, labour*; *breech-clout* (examples).
1774 W. Smellie *Treat. Theory & Pract. Midwifery* (ed. 5) III. 52 A Breech case, from Dr. Tathwell. 1924 J. S. Fairbairn *Gynæcol. with Obstetr.* III. xv. 283 In a doubtful breech case the finger should be passed along the cleft backwards and forwards. 1757 R. Putnam *Memoirs* 9 July (1903) 12 Having nothing to cover us from the Natts & Musketoes..but a Shirt and Breech Clout. 1897 *Outing* (U.S.) XXX. 246/1 A breech-clout for the men, and a short skirt for the women. 1947 J. Bertram *Shadow of War* VII. iv. 235 We stripped down to our *fundoshi*—a kind of Japanese breech-clout. 1882 W. T. Lusk *Sci. & Art Midwifery* x. 200 (*heading*) The configuration of the foetus in breech deliveries. 1964 J. M. Brudenell *Obstetr.* xi. 94 Because of the risks to the foetus breech deliveries should only be performed by an experienced obstetrician in hospital. 1885 W. S. Playfair *Treat. Sci. & Pract. Midwifery* (ed. 5) II. vi. 365 After a difficult breech labour is completed the child should be carefully examined.
8. (sense 4) **breeches-part** (earlier example); (sense *4 in *Obstetr.*) **breech baby** *colloq.*, a fœtus in the womb in a breech presentation; **breech position,** in a breech presentation, any of the four possible orientations of the fœtus relative to the mother's pelvis; **breech presentation,** a presentation (Presentation 8 b) in which the buttocks are the nearest part of the fœtus to the *os uteri.*
1969 *Woman* 11 Oct. 15/1 There were two of us booked for Caesarian operations. They had decided on that for me because they thought she was a breech baby. 1779 T. Holcroft *Let.* 30 Oct. in Hazlitt *Mem. T. Holcroft* (1816) III. 250 Who on the stage has considerable merit in breeches' parts, coquets, &c. 1876 W. S. Playfair *Treat. Sci. & Pract. Midwifery* I. III. v. 352 The phenomena of delivery in the first and third breech positions. 1811 *Lond. Pract. Midwifery* (ed. 3) x. 190 In breech presentations the parts are gradually and well dilated. 1962 D. E. Reid *Textbk. Obstetr.* xx. 527/1 In rare instances the diagnosis of breech presentation may be established only by use of the x-ray.

breed, *sb.* Add: **2. c.** (Later examples.) Now *colloq.*
1881 *Folk-Lore Record* IV. 106 The word breed is peculiarly used [in Irish folk-lore], as they speak of 'breeds of cabbages', 'breeds of potatoes', &c. 1889 'Mark Twain' *Connecticut Yankee* xxiii. 228 All the different breeds of rockets. 1964 *Ann. N.Y. Acad. Sci.* CXV. 569 The purpose of this paper is not to propose the design and construction of new breeds of computers. 1967 *Boston Sunday Herald* 26 Mar. I. 9/3 Development Administrator Edward J. Logue—a new breed of planner.
e. A person of mixed descent; a half-breed. Also *attrib.* orig. *N. Amer.*
1870 *Canadian Illustr. News* 26 Feb. 271/3 The 'breeds' in their ire said on him they'd fire For him 'twas a regular sell, sell, sell. 1892 *Outing* (U.S.) Jan. 287/1 A tall, wiry 'Breed' hunter, called Dave, or 'Injun Dave', according to taste. 1892 *Harper's Mag.* Feb. 387/2 One-quarter of the number of 'breeds' could read and write. 1905 D. Wallace *Labrador Wild* iii. 47 Eskimos and 'breeds', the latter being a comprehensive name for persons whose origin is a mixture in various combinations and proportions of Eskimo, Indian, and European. 1926 J. Black *You can't Win* xvi. 229, I soon mastered Chinook, practicing on the two 'breed' boys.
4. breed-cup, breed-prize, a prize at a show, etc., given to the best animal of a particular breed; **breed-society,** a society which is concerned with the production of a particular breed of animal.
1888 *Pall Mall Gaz.* 10 Dec. 8/2 There are seven silver breed cups for the cattle classes. 1896 *Daily News* 8 Dec. 5/1 The Breed Cup for shorthorns was taken by the Earl of Rosebery's 'Proud Madam'. *Ibid.* 4 Mar. 8/6 Mr. G. Jackson, of Birmingham, has the breed prize. 1892 *Ibid.* 14 Sept. 5/5 The future of stock fairs will be in some degree affected by the breed-societies.

breed, *v.* Add: **6. c.** *Nuclear Engin.* To produce (fissile material) in a breeder reactor.
1948 *Nature* 28 Aug. 318/1 Uranium 235..must be made to breed 'secondary' fuel from more abundant materials. 1955 *Times* 5 Aug. 9/7 The more spectacular third-stage reactor, in which nuclear fuel is 'bred'.
9. c. To put (an animal) *to* (another) for breeding.
1886 C. Scott *Sheep-farming* 161 The ewes to which he is bred. 1955 W. W. Denlinger *Complete Boston* 57 Half of the top bitches in the entire country may have been bred to him upon the strength of his winning record.
III. *to breed out*: restrict † *Obs.* to sense in Dict. and add: Also, to eliminate (a characteristic) by (controlled) breeding.
1922 R. Leighton *Complete Bk. of Dog* xii. 178 Most Irish water spaniels have bad, straight shoulders, a defect which should be bred out. 1941 J. S. Huxley *Uniqueness of Man* xv. 299 We could theoretically breed out much of human variety.

breeder. Add: **4.** *Nuclear Engin.* In full *breeder reactor, pile*: an apparatus that can create more fissile material than it consumes in the chain reaction.
1948 Lapp & Andrews *Nucl. Radiation Physics* xvi. 401 The development of a breeder-type reactor. *Ibid.* 481 Breeder pile. 1948 *U.S. Atomic Energy Comm. 3rd Semi-Ann. Rep.* 12 The 'breeder' type of reactor, a 'nuclear chain reactor which over a period of time will actually create more fissionable material than is put into the reactor as fuel to sustain the reaction'. 1949 *Sci. Amer.* July 36 A 'breeder' reactor that would produce new fissionable atoms as fast as it used the old ones up. 1950 *Ann. Reg. 1949* 423 An engineering design for an experimental 'breeder' atomic pile had been completed. 1950 *Chemical Industries* Jan. 28/2 Of the four reactors, two are breeders and two are non-breeders. 1963 *Ann. Reg. 1962* 380 The successful operation of so-called breeder reactors, which made new fuel as they burnt their own. 1969 *Sci. Jrnl.* July 41/3 Eight or nine countries..are hurrying to develop practical reactors that will produce more fissile material than they consume; that is, breeder reactors. *Ibid.*, Breeders are seen as opening the door to virtually unlimited energy.

breeding, *vbl. sb.* Add: **1. c.** *Nuclear Engin.* The production of fissile material. Cf. *Breed *v.* 6 c.
1947 G. Gamow *One Two Three* vii. 184 The favorable and unfavorable conditions for neutron breeding. 1956 A. H. Compton *Atomic Quest* 324 A breeding reactor. *Ibid.*, The process by which all the uranium and thorium is made available as fission fuel is known as 'breeding'.
5. *breeding-dress, -ground* (also *fig.*), *-habit, plumage, range, station, -stock, storm, territory.*
1936 *British Birds* XXIX. 234 In May numbers appear to be in full breeding dress. 1931 V. A. Demant *This Unemployment* i. 15 The atmosphere of acquiescence..is the most fruitful breeding-ground of practical atheism. 1937 *British Birds* XXXI. 84 Observations concerning breeding-habits. 1938 *Ibid.* XXXII. 19 Mr. Blake caught it and noticed the breeding plumage and oiled underparts. 1890 *Stock Grower* 1 Feb. 11/2 Breeding range, on the Pecos River, New Mexico. 1920 E. Howard *Territory in Bird Life* ii. 44 The district..lies well within the limits of the breeding range of most of our common species. *Ibid.* 59 Guillemots and Razorbills return at intervals to the breeding-stations early in the season. 1840 J. Morton in *Rep. Sel. Farms* (Libr. Usef. Knowl., Husb. III) 17 The healthiness..of the stock, upon this farm..is a great inducement to keep a large breeding-stock. 1961 *Guardian* 21 Nov. 8/5 Man..has started a breeding storm in his own species. 1920 E. Howard *Territory in Bird Life* 26 The evolution of the breeding territory may have been influenced by relationships in the inorganic world.

breeze, *sb.*[2] Add: **3. b.** Slang phrases: *to hit, split* or *take the breeze*: to depart; *to get* (*have*) or *put the breeze up*: to get or put the wind up (see Wind *sb.*[1] 10 b).
1910 'O. Henry' *Whirligigs* xix. 166 We got to be hittin' the breeze. 1925 Fraser & Gibbons *Soldier & Sailor Words* 35 *Breeze up, to have the*: to be nervous, to have the 'wind up'. 1931 Runyon *Guys & Dolls* (1932) 29 And with this she takes the breeze and I return to the other room. 1934 D. L. Sayers *Nine Tailors* iii. 279 He got a vertical breeze up. 1948 D. Ballantyne *Cunninghams* 89 She was only making out she hadn't seen you so's you wouldn't get the breeze up. 1951 J. B. Priestley *Fest. Farbridge* 196 Put the breeze up the— 1959 I. & P. Opie *Lore & Lang. Schoolch.* x. 193 Expressions inviting a person's departure, for instance:..sling your hook, split the breeze, [etc.].
4. c. *slang.* Something easy to achieve, handle, etc. orig. *U.S.*
1928 G. H. Ruth *Babe Ruth's Own Bk. Baseball* 299 *Breeze*, an easy chance. 1958 M. Dickens *Man Overboard* ix. 136 This will be a breeze for you. 1962 S. Carpenter in *Into Orbit* 75 All in all, the test was a breeze.
5. *breeze-swept* adj.
1872 Calverley *Fly Leaves* 4 Lingers on, till stars unnumber'd Tremble in the breeze-swept tarn. 1894 G. Bell *Safar Nameh* 48 On the threshold of his breeze-swept dwelling.

breeze, *sb.*[3] Add: **b.** *attrib.*, as **breeze-block,** a building-block made of 'breeze' (or some similar material) (see quot. 1923); so *breeze-concrete.*
a1877 Knight *Dict. Mech.* I. 367/1 *Breeze-oven*, a furnace adapted for burning coal-dust or breeze. 1923 *Harmsworth's Househ. Encycl.* I. 521/2 *Breeze block.* Fine cinders, crushed coke, furnace clinker and burnt brick are used under the name of breeze in the construction of blocks for housebuilding...Breeze-blocks are fire-resisting, light and porous, suited for partitions and other internal walling. 1930 E. F. Benson *Account Rendered* ii. 10 We put up breeze-block boxes to look like houses. 1930 *Engineering* 11 July 41/3 Breeze and clinker concrete.

breeze, *v.*[2] Add: **1. b.** *colloq.* To move or proceed briskly; to depart. So *to breeze in*: to arrive or enter briskly. orig. *U.S.*
1907 *Chicago Even. Post* 4 May 9 He breezed through the Louvre at such a pace that he broke all the rapid sight-seeing records. 1922 J. A. Dunn *Man Trap* ix. 129 If they show, breeze right along and forget me. 1923 Wodehouse *Inimit. Jeeves* iv. 51 'Ten o'clock, a clear night, and all's well, Jeeves,' I said, breezing back into the good old suite. *Ibid.* xv. 204 To..warn young Bingo to turn his coat-collar up and breeze off snakily by some side exit. 1930 W. S. Maugham *Bread-Winner* I. 45 When did you breeze in, old bean? 1951 R. Campbell *Light on Dark Horse* 246 My favourite brother George..breezed in to look for me.

breeze-way, breezeway (brī·zwēi). orig. *U.S.* [f. Breeze *sb.*[1] + Way *sb.*[1]] A roofed passage, usu. open at the sides, connecting two buildings or parts of a house.
1931 K. N. Burt *Man's Own Country* 39 A small log building attached to the end of his own ranch-house by means of what is known to the Far West as a breeze-way. This construction is a floored and roofed-over passage, open at the sides. 1952 M. McCarthy *Groves of Academe* (1953) ix. 170 From the other end of the breeze-way, through the open door of the dark garage, they heard Cathy's voice thinly calling. 1957 *New Yorker* 21 Sept. 39/1 Now if you don't have a floodlit crèche in the breezeway and your roof outlined with colored lights you're nobody. 1963 J. N. Harris *Weird World Wes Beattie* (1964) ii. 18 A modern split-level house complete with car port, breezeway and Chinese elm hedge. 1968 *Globe & Mail* (Toronto) 17 Feb. 46/9 (Advt.), Two acres of trees and lawn,..breezeway and double garage. 1969 *Sydney Morning Herald* 24 May 37/3 (Advt.), Large breezeway..connects Triple Garage to Rumpus Room.

breezily, *adv.* (Later examples.)
1935 F. W. Crofts *Crime at Guildford* xi. 146 'Good morning, gentlemen,' he said breezily. 1963 *New Statesman* 20 Dec. 902/3 She..refers quite breezily to her lack of relatives.

breezy, *a.* Add: **2.** *fig.* Also, characterized by brisk vigour or activity; lively, jovial.
1873 'Mark Twain' & Warner *Gilded Age* lii. 470 Said in almost his breezy old-time way. 1884 *American* VIII. 87 The chapter on 'Value' is particularly fresh and breezy. 1896 Mrs. Caffyn *Quaker Grandmother* 112 A strong, vigorous, breezy, old woman. 1930 W. S. Maugham *Cakes & Ale* ii. 21 Roy smiled at the steward with breezy cordiality. 1942 *New Statesman* 11 July 25/1 American Jews are just like any other Americans, complete with breezy manners and deep voice. 1955 *Times* 27 Aug. 7/5 Some breezy remarks on Dutch or Flemish pictures.

Breguet (brəge). Also erron. **Brèguet.** The name of Abraham Louis Breguet, French watchmaker of Swiss origin (1747–1823), used attrib. in *Breguet* (*hair*)*spring*, the overcoil balance spring of a watch.
1881 F. J. Britten *Watch & Clockmakers' Handbk.* (ed. 4) 63 Breguet springs are now often used for pocket chronometers instead of the helical form. 1908 *Sears Roebuck Catal.* 259/1 This movement..has 21 fine ruby jewels in gold settings, cut expansion balance, Breguet hairspring, [etc.]. 1929 G. H. Baillie *Watches* vii. 168 Properly, a Breguet spring should have no regulating curb pins, because the curve of the overcoil is correct only when it is of one definite length, and any movement of the curb pins alters its length and partially destroys its effect. 1952 Flood & West *Dict. Sci. & Techn. Words* (1953) 42/1 *Breguet spring*, special form of hair-spring..in a watch.

brei, var. *Bray *v.*[4]

breitschwanz (brai·tʃvants). Also **breitschwantz, breitswanz.** [G., = broad tail.] = *broadtail.
1923 *Daily Mail* 16 Apr. 15 The frilled coat had a sailor collar of dark blue breitschwanz. 1927 *Daily News* 26 Sept. 2/4 Especially good is the imitation or broadtail known as breitswanz. 1928 *Daily Express* 24 Jan. 5/4 Shaved lamb, breitschwantz, caracul, and astrakhan are seen in beige and light shades of golden brown. 1960 B. Marshall *Divided Lady* 119 Bice looked lovely in a Breitschwanz coat and a black hat.

brekker (bre·kəɹ). (*University*) *slang.* [f. Break(fast *sb.* + *-er[6].] Breakfast.
1889 in Barrère & Leland *Dict. Slang.* 1900 G. Swift *Somerley* 66 Have you had any brekker? 1908 *Daily Chron.* 20 Apr. 4/7 A young man who looked like an undergraduate wondered when his 'brekker' would be ready. 1928 D. L. Sayers *Unpleasantness at Bellona Club* xi. 129 He'd only that moment swallowed his brekker. 1965 J. P. Carstairs *Concrete Kimono* i. 9, I complained of an undesirable brekker kipper.

brekkie, brekky (bre·ki). *slang* (orig. *children's*). [f. Break(fast *sb.* + -y[6].] Breakfast.
1904 E. Nesbit *Phoenix & Carpet* xii. 218, I've brought your brekkie, and I've put the little cloth with clover-leaves on it. 1926 E. Bowen *Ann Lee's* 66 Do call poor Bingo in..and give him his brekky. 1969 *Private Eye* 25 Apr. 12, I don't reckon I feel like brekkie!

bremsstrahlung (bre·mʃtrālŭŋ). *Nuclear Physics.* [G., lit. 'braking radiation', f. *bremsen* to brake + *strahlung* radiation.] The electromagnetic radiation produced by the retardation of a charged particle, usu. an electron passing through the field of a nucleus.
1944 W. Heitler *Quantum Theory Radiation* (ed. 2) iii. 160 This is..the case for the Bremsstrahlung (continuous X-ray emission). The Bremsstrahlung is one of the most important processes affecting the penetrating power of high-energy electrons by passing through matter. 1948 Mott & Sneddon *Wave Mech.* x. 275 Among [radiation] processes of the third order the most important is 'Bremsstrahlung'—the emission of radiation by an electron passing through the field of a nucleus. 1958 *New Scientist* 16 Jan. 14/1 A radiative heat loss due to this process (technically known as *bremsstrahlung* r diation). 1962 *Newnes Conc. Encycl. Nucl.*

Energy 90/2 *Bremsstrahlung*. .The name (braking radiation) originates in the radiation produced when electrons are stopped in an absorber. **1965** *New Scientist* 30 Dec. 905/1 The third possibility is that the X-rays from the Crab are produced by bremsstrahlung or the braking of electrons in a hot gas cloud at a temperature of 10–100 million °K.

Bren (bren). [f. *Brno*, a town in Czechoslovakia, where the gun was originally made + first syllable of *Enfield*, Middlesex, England, seat of a small-arms factory.] In full, *Bren gun*. A type of light, quick-firing machine-gun. Also *attrib.*, as **Bren (gun) carrier,** a small bullet-proof tracked vehicle, designed to carry the Bren gun and its crew.

1937 *Times* 29 Sept. 6/3 Use of the Bren Gun. **1939** *War Illustr.* 28 Oct. 200 A long train of Bren gun carriers is passing towards the front. **1940** *Times Weekly* 7 Aug. 8/4 A motor-cyclist detachment of the Lancashire Fusiliers followed by their Bren carriers. **1946** Koestler *Thieves in Night* 67 The steady rattle of an automatic, a Lewis or a Bren.

brer (brē̆ɹ). *colloq.* (orig. *U.S.*). [Representing an American Negro or southern U.S. pronunc. of Brother *sb.*] Brother.

1881 J. C. Harris *Uncle Remus* i. 17 Arter Brer Fox bin doin' all dat he could fer ter ketch Brer Rabbit. **1889** G. B. Shaw *London Music 1888–89* (1937) 145 Last night we had Brer Jean [de Reszke] as Raoul and Brer Edouard as Marcel. **1895** Kipling *Lett. of Travel* (1920) 107 No one seems to know particularly where Brer Coon lives.

bretelle (brĕte·l). [Fr., strap, sling; in *pl.* braces.] Each of the ornamental shoulder-straps extending from the waist-belt in front to the belt behind of a woman's dress. Chiefly *pl.* Also *attrib.* and *Comb.*

1857 *Godey's Lady's Book* July 68/2 This bretelle makes a sufficient trimming for any simple evening-dress. **1890** *Daily News* 21 Oct. 2/1 Bretelle-shaped lines of fine passementerie in green and gold. **1896** *Ibid.* 4 July 6/3 The bodice. .has frills of black chiffon passing over the shoulders in bretelle fashion. **1909** *Daily Chron.* 22 Feb. 7/5 These are arranged back and front bretelle-wise to fall over the deep, folded taffetas belt. **1932** *Daily Express* 20 Sept. 19/1 Two bretelles in front that are knotted at the centre front of the corsage.

Breton (bre·tǫn), *sb.* and *a.* [ad. OF. *Breton* (see Briton).] **A.** *sb.* A native of Brittany; the Celtic language of Brittany. **B.** *adj.* Belonging to or characteristic of Brittany, its inhabitants, or their language. *Spec.*, designating a type of hat with a round crown and an upward-curved brim; also *ellipt.*

† *Breton tackle* (Naut.), app. the earlier form of Burton *tackle*.

c **1386** Chaucer *Prol. Franklin's T.* (1894) 709 Thise olde gentil Britons. . .Of diuerse auentures maden layes. *a* **1400** *Kyng Alisaunder* (1952) 5881 Þe kynges oost. . maugre Picard and Bretoun, Breken þere þe wal adoun. **1495** *Naval Acc. Hen. VII* (1896) 188 Brytton takles. *Ibid.* 198 Bretton takles. *Ibid.* 210 Breton takles. *a* **1500** *Sir Orfeo* (Harley MS.) 16 In Brytayn þis layes arne ywryte. . Of aventures þat fillen by dayes, Wherof Brytouns made her layes. **1525** [see See *v.* 15 b]. [**1592** *Surv. or Topogr. Descr. France* 42 The other three dioceses do speake a mixed language, sometimes French sometimes Britton. *Ibid.* 44 The Brittains are generally tractable, but those that are neerer the sea coast are not so courteous as the rest. **1652** Heylin *Cosmogr.* i. 168/1 The Britains. .were questionless one of the first Nations that possessed any part of Gaul. See also Britain *sb.*[2] 2, *a.* 3.] **1653** Urquhart tr. *Rabelais' Gargantua* xxxi. 139 The very barbarous Nations of the Poictevins, Bretons, Manceaux. **1818** Mrs. Stothard *Lett. Tour Normandy* (1820) xxiv. 253 The Breton language appears to me. .far more corrupted than the Welsh. *Ibid.* 254 The Bretons do not resemble in countenance either the Normans or French, nor have they much of the Welsh character. **1830** *Cambrian Q. Mag.* II. 192 The Breton-speaking Brittany. **1855** Tennyson *Maud* xxiii. 78 Here on the Breton strand! Breton, not Briton. **1861** *Chambers's Encycl.* II. 332/1 The Breton has generally a tinge of melancholy in his disposition. **1903** [see Sailor 4]. **1913** J. Morris Jones *Welsh Gram.* Introd. 1 Keltic. .The P division, consisting of Gaulish, and the British group, comprising Welsh, Cornish and Breton. **1941** *Amer. Speech* XVI. 67/2 Later *Bretons* and *Sailors* were pushed, and the standard *Snap-Brim*. **1952** *Vogue* June 61 Simone Mirman's huge Breton hat in black straw. **1968** K. O'Hara *Bird-Cage* xvi. 129 A string of Breton onions. **1968** J. Fleming *Hell's Belle* i. 49 They. .ate Breton pancakes filled with cheese and mushrooms. **1969** *Listener* 6 Feb. 165/3 Local schools used to display a notice which read: 'No spitting, and no talking Breton.' And even today there is a celebrated Breton couple six of whose 13 children have never been officially registered because they insisted on giving them Breton names.

breunnerite (broi·nərait). *Min.* [f. the name of Count *Breunner*, a 19th-cent. Austrian nobleman: see -ite[1].] The name given by Haidinger to a variety of magnesite in which some of the magnesium is replaced by iron.

1825 Haidinger tr. *Mohs' Treat. Min.* I. 411 Brachytypous. Rhombohedral. . .Breunnerite. **1857** J. D. Dana *Man. Min.* (new ed.) vi. 248 *Mesitine spar*, (Breunnerite.) A carbonate of iron and manganese occurring in yellowish rhombohedrons. **1929** H. A. Miers *Mineralogy* (ed. 2) I. iii. 255 The anhydrous rhombohedral carbonate known by the name of breunnerite. **1960** W. H. Dennen *Princ. Min.* vii. 288 *Magnesite*. . .Chemical varieties: Ferroan (breunnerite), manganoan, calcian.

breviary. Add: **2. c.** fig. phr. *matter of breviary* (= *matiere de breviaire*, Rabelais, Pantagruel IV. viii): a thing that admits of no question or doubt.

1694 Motteux tr. *Rabelais* iv. viii. 35 It is written *Mihi vindictam*, &c. matter of breviary, Mark ye me; that's holy stuffe. **1889** *Sat. Rev.* 12 Jan. 29/1 It is a matter of breviary with Gladstonians that Unionist journals are not to be trusted. **1894** *Ibid.* 27 Oct. 463 Is it not. .rather matter of controversy than matter of breviary whether the distinction applies?

brevicite (bre·visəit). *Min.* [ad. Sw. *brevicit* (Berzelius 1834, in *Årsberättelse om Framstegen i Fysik och Kemi* (1834) 179), f. *Brevig* (now *Brevik*), a town in Norway: see -ite[1].] Natrolite, as found in certain parts of Norway.

1836 T. Thomson *Outl. Min., Geol.*, etc. (ed. 7) I. 680 Brevicite. This name has been given by Berzelius to a mineral sent him by M. Strom, from Brevig, in Norway, which appears to have filled up an amygdaloidal cavity in a trachyte-looking rock. **1892** *Dana's Syst. Min.* (ed. 6) 602 Brevicite is the same as spreustein, though the name has also been used as synonymous with radiolite. *Ibid.*, Radiolite, bergmannite, spreustein, brevicite, palæo-natrolite, are names which have been given to the natrolite from the augite-syenite of southern Norway,. .where it occurs fibrous, massive, and in long prismatic crystallizations.

brew, *v.* Add: **3. b.** Colloq. phr. *to brew up*: to make tea.

1916 *Daily Mail* 1 Nov. 4/4 'Brew up' or 'drung up' (to make tea, over-seas expressions). **1943** A. Clifford *Three against Rommel* xxxiii. 389, I thought we might brew up.

brew, *sb.*[1] Add: **b.** *brew-up*: (a pause for) the making of tea; cf. prec. *colloq.*

1944 A. Jacob *Traveller's War* vi. 123 The crews halted for a 'brew up' near us: some drank tea,. .others took a nap. **1963** *Times* 26 Jan. 9/7 The. .petrol tins which the Desert Rats found equally handy for washing in or for a 'brew-up'.

brew (brū̆), *sb.*[2] Local var. Brow *sb.*[1] 6 b.

1887 Hall Caine *Deemster* xxvii. 176 Nearer the cliff I found this, and this; and then down the brew itself. .I saw this other one. **1891** 'L. Keith' *Halletts* II. v. 107 He'll stand quiet enough;. .it's the grass on the brew he's after. **1927** *Chambers's Jrnl.* Feb. 126/1 The brews of the ditches or hedge-bottoms.

brewer. Add: **1.** *brewers' grains* (see Grain *sb.*[1] 4 b).

brewing, *vbl. sb.* **4. brewing-up,** (*a*) gradual formation or production; (*b*) (see quot. 1961).

1953 R. Lehmann *Echoing Grove* 103 She began to speak. . of the first signs. .of the brewing-up of Corrigan's psychological collapse. **1961** *Times* 7 Nov. 19/1 In Britain a thriving business has grown up in tuning and modifying the engines of existing models [of cars] to give more performance. . .Referred to by enthusiasts as 'brewing up',. .or merely 'hotting up'.

brewis. Add: Also (brūz). (*Canad.* examples.)

c **1850** in E. Fowke et al. *Canada's Story in Song* (1960) 164 Tho' Newfoundland is changing fast, some things we must not lose: May we always have our flipper pie, and codfish for our brewis. **1906** J. Lumsden *Skipper Parson* vi. 87 A popular dish in Newfoundland is 'brewis', pronounced *broose*. **1964** *Canad. Geogr. Jrnl.* Apr. 135 Only in Newfoundland were we served. .fish-and-brewis.

brewsterlinite (brū·stəɹlinəit). *Min.* [f. *brewsterlin*(*e* + -ite[1]; Dana orig. used *brewsterline*, f. the name of the Scottish physicist Sir David *Brewster* (1781–1868), who first described the liquid in 1823 + *-line*, prob. repr. L. *oleum* oil + -in[1] (as in Cryptolin); this he changed to *brewstoline* and then to *brewsterlinite*.] A transparent liquid of uncertain composition found in inclusions in certain minerals.

[**1850** J. D. Dana *Syst. Min.* (ed. 3) v. vii. 559 Brewsterline. Fluid. Colorless. . .We have thought proper to recognize this fluid among minerals and name it, although yet imperfectly described. **1854** *Ibid.* (ed. 4) I. v. 471 Brewstoline.] **1864** *Ibid.* (ed. 5) 761 Brewsterlinite. A new fluid in the cavities of minerals. **1902** H. A. Miers *Mineralogy* ii. xviii. 443 'Brewsterlinite'. .was investigated by Brewster, who found it to have the low refractive index 1·2. **1966** W. A. Deer et al. *Introd. Rock-Forming Min.* i. 47 Microscopic cavities are sometimes found in topaz, usually filled with a liquid: one such liquid, brewsterlinite, has a refractive index of 1·13 and has been supposed to be liquid CO_2.

brey, var. *bray *sb.*[3]

brey, var. *bray *v.*[4]

Briard (briā·ɹ(d)). [Fr., adj. of *Brie* (region of France).] A sheep-dog of a breed of French origin.

1934 in Webster. **1938** A. Woollcott *Let.* (1946) vii. 168 The livestock there has been augmented by a Briard. He. .is known as Mugwump and is roughly the size of the National Gallery. **1948** C. L. B. Hubbard *Dogs in Britain* 191 The Briard (Chien de Brie), a French Sheepdog taken from Corsica to Egypt during the Napoleonic expedition. **1967** H. Kemelman *Nine Mile Walk* (1968) 156 Duke was no ordinary dog. He was a briard, a Belgian sheepdog, a huge creature with a long rough iron-grey coat, the shaggy fur covering even his face.

brick, *sb.*[1] Add: **2.** Now extended to blocks or slabs made of sand and lime, concrete, and other materials.

1875, 1879 Slag-brick [see Slag *sb.* 5 a]. **1922** D. M. Liddell *Handbk. Chem. Engin.* II. 948 At present bricks made from sand and lime are extensively used, while they have been used in Europe for 50 years.

4. Examples of (*a*) a wooden block for a child to play with; (*b*) a brick-shaped block of ice-cream. Also, further *transf.* examples.

1832 F. Trollope *Dom. Manners* I. xix. 301 A numerous collection of large wooden bricks. **1861** C. M. Yonge *Young Step-Mother* xxix. 434 He. .built up a tower with her wooden bricks. **1922** S. Lewis *Babbitt* ix. 123 He gulped down a chill and glutinous slice of the ice-cream brick. **1938** R. W. Lawson tr. *Hevesy & Paneth's Man. Radioactivity* (ed. 2) ix. 94 The α-particles. .are not the ultimate bricks of which the nucleus is composed. **1939** L. MacNeice *Autumn Jrnl.* xix. 76 Baby Croesus crawls in a pen With alphabetical bricks. **1959** I. & P. Opie *Lore & Lang. Schoolch.* ix. 165 Cornet, Brick, and Lollipop Taste very nice when bought from the shop. **1964** F. L. Westwater *Electronic Computers* iii. 49 Only a limited number of electronic 'bricks', as they are called, are used. These bricks, which are circuit elements, have been thoroughly tested beforehand.

b. *brick couching*, in embroidery, couching in which the laid threads or cords are secured by cross stitches resembling, in their arrangement, the vertical joints of brickwork. Hence *in bricks*, in divisions resembling bricks. Cf. *brick-stitch* (sense 10 below).

1882 Caulfeild & Saward *Dict. Needlew.* 180/1 The chief varieties of Flat Couching are Brick, Broad, Burden, Diagonal, and Diamond. **1911** A. Dryden *Church Embroid.* 112 The commonest form of stitching the gold is in bricks, each couching-stitch being in between the two stitches of the preceding line.

c. The colour of brick; brick-red.

[**1856** M. N. Thomson *Elephant Club* 163 A [red] head of hair which the youth of America are accustomed to designate as a 'brick-top'.] **1897** *Sears, Roebuck Catal.* 21/3 Send for illustrated color card. .Pink—Milwaukee Brick—Quaker Drab. [**1912** *Dialect Notes* III. 572 *Brick-topped*, red-headed.] **1922** Joyce *Ulysses* 48 A dull brick muffler strangling his unshaven neck. **1923** *Daily Mail* 16 Jan. 1 Blanket Cloth Magyar Wrap Coats. . . Cream, Beaver, Mole, Nigger, Brick.

5. Also (orig *U.S.*), *like a thousand* (occas. *hundred*) *of brick*(*s*); *like a ton of bricks*: see *come *v.* 56 g.

1836 Hill *Yankee Stories* 32 (Weingarten *Suppl. Notes to D.A.E.*), If I don't be into him with a thousand of brick. **1840** 'Samuel Slick' *Clockmaker* Ser. iii. 50 If I don't pitch into Ben Parsons' ribs like a tousand of bricks. **1841** *Picayune* (New Orleans) 16 Mar. 2/2 They. .rounded the first turn pretty much in a heap like a thousand of brick. **1860** *Ibid.* 27 Apr. (De Vere), He fell upon us like a thousand of bricks. **1896** C. James *Yoke of Freedom* 161 Once let a man play me false, I'm down on him like a hundred of bricks. **1911** A. Bennett *Hilda Lessways* iii. ii. 226, I had the whole gang down on me instantly like a thousand of bricks. **1929** W. S. Maugham *Writer's Notebk.* (1949) 228 I'd have had the whole community down on me like a ton o' bricks. **1967** E. Grierson *Crime of one's Own* xii. 98 The . .gentleman. .made his living by selling books, and not *those* sort of books either, or the Super would have been down on the place long since like a ton of bricks.

b. Slang phr. (orig. *U.S.*), *to have* (or *wear*) *a brick in one's hat*: to be under the influence of liquor.

1847 *Knickerbocker* XXIX. 569 A youth who came home one night. .having 'a brick in his hat'. **1848** Durivage & Burnham *Stray Subjects* 61 He wore a 'brick' within that hat. **1849** [see Hat *sb.* 5 c]. **1889** Barrère & Leland *Dict. Slang* I. 180/1 *Brick in the hat* (common), intoxicated, top-heavy.

c. *to drop a brick*: to commit a verbal indiscretion, make a 'bloomer'. *colloq.*

1923 *Punch* 3 Oct. 334 It was hinted to me pretty plainly that I had dropped a brick, as you say. **1924** Galsworthy *White Monkey* iii. xii, I've got to keep my head shut, or I shall be dropping a brick. **1928** 'Sapper' *Female of Species* xvii. 307 The stones of Stonehenge are little pebbles compared to the bricks you dropped, but I forgive you.

8. a. *brick-and-mortar*: see also Mortar *sb.*[2] a (quots. 1863 and 1895).

9. a. *brick-colour* (examples), *-stack*. **c.** *brick-floored*.

1598 Florio *Worlde of Wordes* 149/2 *Gioggiolino*, a kinde of colour which we call flesh-colour, or brickcolour. **1818** W. Tucker *Fam. Dyer & Scourer* (ed. 2) 53 To make a Brown inclining to a Brick Colour. **1898** Countess of Arnim *Eliz. & German Garden* 10, I used to. .go. .slowly across the brick-floored hall. **1899** *Westm. Gaz.* 7 Jan. 5/1 A brick-stack, 60 ft. in height, fell, crashing on to the sheds. **1915** *Times* 15 Apr. 4/1 A severe bombardment of the 'brickstacks' and the enemy's trenches.

10. brick-bread (example); **brick-loaf** (earlier example); **brick-on-edge** *a.*, designating a construction built of bricks laid on their sides; **brick-pond** *U.S.*, a pond in a brickfield; **brick-stitch** = *brick couching* (see 4 b above).

1762 *Boston Selectmen's Minutes* (1887) 29 Nov., A 4*d.* loaf of brick bread is 3 oz. less than a 4*d.* white loaf. **1723** J. Nott *Cook's & Confectioner's Dict.* No. 7 E, 'Till you have made it in the form of a Brick-loaf. **1851** *B'ham & Midl. Gardeners' Mag.* Dec. 220, I should recommend a brick-on-edge wall being run up. **1811** *Massachusetts Spy* 9 Jan. 3/3 Two boys. .were. .drowned in a brick pond in the vicinity of the city [*sc.* Philadelphia]. **1894** *Naturalist* Feb. 56 Myriophyllum alterniflorum. Blythe Nook brickponds. **1872** Mrs. A. Mercier *Our Mother Church* xvi. 348 The edge is done in 'brick' stitch, which can be easily worked by looking at the pattern. **1882** Caulfeild & Saward *Dict. Needlework* s.v., Brick stitch was largely used as backgrounds in ancient embroideries. **1960** B. L. Snook *English Hist. Embroidery* 32 Split stitch was still used. .accompanied by laid-work, couching, brick stitch, satin stitch and long and short stitch.

brickbat. Add: **b.** Also *attrib.*

1890 W. James *Princ. Psychol.* I. vii. 196 The continuous flow of the mental stream is sacrificed, and in its place an atomism, a brickbat plan of construction, is preached.

c. *fig.* An uncomplimentary remark; adverse criticism (cf. quot. 1642 in Dict.).

1929 *Daily Express* 7 Nov. 17/5 And now for the brickbats. **1955** [see *bouquet 1 b]. **1966** *Listener* 30 June 960/3 There were some much-needed brickbats thrown at our hero's wife.

2. *Astr.* (See quots.) *colloq.*

1892 Ranyard *Proctor's Old & New Astr.* 640 Clerk Maxwell used to describe the matter of the rings [of Saturn] as a shower of brickbats, amongst which there would inevitably be continual collisions taking place. **1898** A. Clerke et al. *Astr.* 340 It may be that collisions are infrequent in this conglomeration of 'brickbats'. **1926** H. C. Macpherson *Mod. Astr.* 78.

brickette (brike·t). Also **bricket.** [f. Brick *sb.*[1] + -et, -ette.] A small brick, esp. of ice-cream. Cf. *brick *sb.*[1] 4 and *briquette 3.

1934 *Archit. Rev.* LXXVI. p. xliii, The paths, constructed of herringbone brickets, are both colourful and durable. **1959** *Manch. Guardian* 4 Aug. 3/6, 180 brickettes [of ice-cream] a minute. **1962** *Engineering* 4 May 597/1 Brickettes destined to become choc-bars.

brickfielder. (Earlier example.) Add to def.: Now applied to a hot northerly wind in various regions of Australia.

1843 J. Backhouse *Narr. Visit Austral. Col.* 3 Jan. 236 This kind of wind. .is frequent in the summer, and coming upon the town from the direction of some brick-fields, has obtained the name of a Brick-fielder.

bricking (bri·kiŋ). [f. Brick *sb.*[1] + -ing[1].] **1.** Building with brick; brickwork. Also *attrib.*

1770 in *Maryland Hist. Mag.* (1917) XII. 368 Pray defer yr Stable, Bricking in yr Garden or any other Jobs untill you have compleated the stone Wall. **1924** *Glasgow Herald* 20 Mar. 12 But between him and his coal, before ever he had sunk his bricking ring, intervened the war.

2. An imitation of brickwork, as on a plastered or stuccoed surface; in embroidery, brick-stitches collectively.

a **1877** Knight *Dict. Mech.*, *Bricking*, the imitation of brickwork on a plastered or stuccoed surface. **1911** A. Dryden *Church Embroid.* 112 For ordinary bricking use about ten stitches to the inch.

brickish, *a.* Add: **2.** = Bricky *a.*

1879 G. M. Hopkins *Poems* (1918) 41 Thou [*sc.* Oxford] hast a base and brickish skirt there. **1900** E. Glyn *Visits Eliz.* 257 She has quite a different coloured chest to the top bit that shows above her pearl collar, which is brickish-red from hunting. **1960** Betjeman *Summoned by Bells* i. 1 Brickish Kentish Town seen through the leaves of Highgate.

So **bri·ckishness** *slang*, the quality of being good-hearted or 'brickish' (sense in Dict.).

1906 *Daily Chron.* 26 July 3/2 Janet's sheer 'brickishness' held her faithful to her organist. **1924** F. M. Ford *Some do Not* i. vi. 163 They had talked. .about the brickishness of the parson in taking her in.

brick wall, *sb.*[1] Add: (Earlier example.)

c **1440** [see Brick *sb.*[1] 8].

2. *fig.* An impenetrable barrier. Phr. *to talk to a brick wall*, to fail to elicit any response from one's interlocutor; *to see through a brick wall*: see Wall *sb.*[1] 18.

1886 H. Baumann *Londinismen* 16/2 *Brickwall*: he can see thro a brickwall *er kann durch eine Mauer sehen*. **1898** *Westm. Gaz.* 19 July 1/2 We have been putting this dilemma to Liberals and Irish and the answers which we have obtained from both have brought us to what we have called the 'Irish brick-wall'. **1909** Jerome *They and I* vii, We mustn't have to tell 'em the same thing over and over again, like we was talking to brick walls. **1936** L. A. G. Strong *Last Enemy* 309 You're a bit second-sighted, aren't you? I mean you see farther through a brick wall than most? **1963** A. Smith *Throw out Two Hands* iii. 36 A chain reaction followed of other similarly frustrating brick walls. **1968** G. Butler *Coffin Following* ii. 58 You still look as if you could see through a brick wall.

bricky, *sb.* Also **brickie.** (Earlier and later examples.)

1880 D. W. Barrett *Life & Work among Navvies* ii. ii. 39 'Navvy' and 'bricky' seem to them amusing specimens of wit. **1964** *Economist* 19 Dec. 1363/1 Something to help the brickie with corners.

bricky, *a.* Add: **2.** Like a 'brick' or good fellow. *slang.*

1863 J. C. Atkinson *Stanton Grange* i, Old Milburn was the 'brickiest' master. .at Elmdon.

bridal, *a.* Add: **2.** *bridal suite*, in a hotel, a suite of rooms designed especially for use by a newly-married couple.

1925 N. Coward *Fallen Angels* ii, The room decorated like a Bridal Suite. **1938** H. Asbury *Sucker's Progr.* 265 The bridal suite was called 'Paradise'. **1970** *Woman's Own* 24 Jan. 3/2 London's Royal Garden Hotel has offered the couple a bridal suite.

c. *bridal wreath*: a Chilean shrub of the family *Saxifragaceæ* (see quot. 1962); see also quot. 1961.

1889 *Cent. Dict.*, *Bridal wreath*, the *Francoa ramosa*. **1961** F. Perry *Shrubs & Trees for Smaller Garden* v. 139 *Spiraea x arguta*, Bridal Wreath. The most widely grown species, with very slender twigs festooned with dainty white flowers. **1962** *Amat. Gardening* 27 Jan. 23/1 The plant known as Bridal Wreath is Francoa ramosa and comes from Chile. It produces long sprays of white flowers in July and August.

bride, *sb.*[1] Add: **1. b.** A girl, woman, esp. a girl-friend. *slang.*

1935 G. Ingram *Cockney Cavalcade* ix. 146 'I must git a bride too,' announced Patsy. . .Patsy, like most of his class, was eager to get a girl. **1964** *Listener* 31 Dec. 1053/2 This load of squaddies. .ain't got any brides with them.

5. *bride-money, -people, -sheet*; also in Social Anthropology, as *bride-abduction, -capture, -purchase, -service, -wealth.*

1937 R. H. Lowie *Hist. Ethnol. Theory* v. 46 There is nothing to show that it regularly accompanies bride-abduction. **1878** E. B. Tylor *Early Hist. Mankind* (ed. 3) x. 287 The import of the Spartan marriage, which he calls the 'form of capture', as indicating previous habit of bride-capture. **1937** C. M. Arensberg *Irish Countryman* iii. 106 A last remnant of a primeval bride-capture. **1926** D. H. Lawrence *David* x. 71 Is Merab not worth the bride-money? **1816** Jane Austen *Emma* I. i. 3 The wedding over and the bride-people gone, her father and herself were left to dine together. **1920** R. H. Lowie *Primitive Society* vii. 178 The functionally related customs of clear-cut bride-purchase and patrilocal residence. **1937** —— *Hist. Ethnol. Theory* xiii. 240 Kirchhoff has unearthed significant South American relations: matrilocal residence, bride-service. **1930** T. S. Eliot tr. *St.-J. Perse's Anabasis* 63 Loud acclamations. .for the publication of the bride-sheets! **1931** E. E. Evans-Pritchard in *Man* Mar. 36 There has been a considerable amount of discussion. .about an alternative expression for 'bride-price'. . I may be excused for putting forward the term 'bride-wealth'. **1951** *R. Anthrop. Inst. Notes & Queries* ii. iv. 116 Bride-price, bride-wealth, marriage payment, are terms used for the goods, gifts or payments transferred by the bridegroom or his family to the bride or her family on the occasion of marriage.

bridge, *sb.*[1] Add: **1. b.** (Later examples.) Also *attrib.*, as *bridge church, passage* (see *11 b).

1931 *Ann. Reg. 1930* ii. 238 The replacement of the Marqués de Estella's Cabinet by a 'bridge' Government. **1950** *Time* 4 Sept. 33/1 Radio dramas have always depended on music to get listeners from one mood or scene into another, fast. In the trade, such six- or eight-bar snatches of music are called cues or 'bridges'. **1962** A. Nisbett *Technique Sound Studio* ix. 158 Effects can often be peaked as a bridge between two parts of the same scene. *Ibid.* 159 As an alternative to the simple fade-out/fade-in technique. . music bridges can be used.

e. Proverbial phr. *to cross a* (or *the*) *bridge when one comes to it*: to deal with a problem when and if it arises; *to burn one's bridges behind one*: to burn one's boats (see Burn *v.*[1] 9 c).

1850 Longfellow *Jrnl.* 29 Apr. in *Life* (1886) II. 165 Remember the proverb, 'Do not cross the bridge till you come to it.' **1892** 'Mark Twain' *Amer. Claim.* 94 It might be pardonable to burn his bridges behind him. **1914** E. R. Burroughs *Tarzan of Apes* xxviii. 399 Because she had been afraid she might succumb to the pleas of this giant, she had burned her bridges behind her. **1930** J. C. Ransom *God without Thunder* (1931) vii. 166 If Plato had had the opportunity, he might have become a Christian, and burned his bridges behind him. **1943** E. S. Gardner *Case of Empty Tin* i. 13 We'll cross that bridge when we come to it.

9. h. Substitute for def.: An electrical circuit for measuring resistance or other properties by equalizing the potential at two points. (Add earlier and later examples.) Also *bridge circuit*.

1872 [see Wheatstone]. **1933** A. Hund *High-Frequency Measurements* iii. 97 The system shown in Fig. 57 is a bridge circuit balanced by a direct current *I*. **1942** J. G. Brainerd *Ultra-High-Frequency Techniques* iii. 164 Either of these circuits is equivalent to a balanced bridge circuit. **1964** *Times* 31 Aug. 7/4 The Zwislocki acoustic bridge, which measures the acoustic impedance of the ear.

i. *Dentistry.* A false tooth or teeth usually connected to natural teeth on each side. Also *attrib.*; cf. *bridge-work* (s.v. *11 b).

1883 J. L. Williams in *Dental Cosmos* Dec. 629 A bridge of the six superior front teeth. **1891** *Brit. Jrnl. Dental Sci.* XXXIV. 65 Dr. Melotte. .made a small bridge of one tooth. **1963** C. R. Cowell et al. *Inlays, Crowns & Bridges* ii. 4 A gold inlay is the most suitable type of bridge retainer. **1965** H. Gold *Man who was not with It* xxx. 281 Mrs. Nancy. .gave me a smile with her new bridge.

10. *spec.* in *Euchre* (see quot.).

1891 'L. Hoffman' *Cycl. Card & Table Games* 95 (*Euchre*) If one side has scored four, and the other side only one, such position is known as the 'bridge'. . . The elder hand is the only one who should order up at the bridge.

11. a. *bridge-builder* (example), *-building.*

1772 C. Hutton *Bridges* 6 A Bridge-builder should be employed. **1751** (*title*) Gephyralogia. .including. .an abstract of the rules of bridge-building. **1931** J. S. Huxley *What dare I Think?* vi. 199 There is no such instinct, any more than there is. .a bridge-building instinct.

b. bridge church, a faith in which elements of two systems of religious belief are found together; **bridge circuit** (see *9 h); **bridge-head,** also, any military position established in the face of the enemy, e.g. by a landing force; also *fig.*; **bridge passage,** a transitional passage in a musical or literary composition; **bridge-rail,** (*b*) a rail around the bridge of a ship; **bridge spectacles** *U.S.* = Pince-nez; **bridge-work** *Dentistry* = sense *9 i; also, the insertion of such a bridge.

1927 *Church Times* 2 Sept. 266/1 Dr. Gore accepts for it [*sc.* the Anglican Church] its description by a Lutheran as a brücke Kirche—a bridge Church between Catholicism and Protestantism. **1930** *Economist* 22 Mar. 632/2 The Channel tunnel. .would be a hopelessly vulnerable 'bridge-head'. **1938** *Star* (Kansas City) 22 Apr. 1/1 Counterattacks had reduced insurgent bridgeheads. **1940** R. W. B. Clarke *Britain's Blockade* 6 The Nazis aim at using these territories as bridge-heads for the attack on Britain. **1944** *Ann. Reg. 1943* 25 British troops established a bridgehead of considerable width. **1959** *Listener* 2 Apr. 583/2 Once we have secured these export bridgeheads we can of course expect greater benefits. **1927** *Melody Maker* Sept. 926/2 Rhythmic construction of introductions, bridge passages, modulations, interludes [etc.]. **1959** *Times Lit. Suppl.* 5 June 330/5 There are no bridge passages in this book. **1962** N. Del Mar *R. Strauss* ii. 41 There is a bridge passage based on the rising octave figure. **1915** 'Bartimeus' *Tall Ship* i. 14 The Captain, clinging to the bridge rail to maintain his balance. **1830** J. F. Watson *Annals Philadelphia* 180 The only spectacles she ever saw were called 'bridge spectacles', without any side supporters, and held on the nose solely by nipping the bridge of the nose. **1883** J. L. Williams in *Dental Cosmos* Dec. 624 So far as the 'bridge-work' is concerned. **1885** I. E. & R. E. Clifford (*title*) Crown, Bar & Bridge-work. **1935** Wodehouse *Luck of Bodkins* xxi. 267 Causing George to double up like a pocket-rule and nearly swallow his bridge-work. **1963** C. R. Cowell et al. *Inlays, Crowns & Bridges* vii. 65 Very occasionally, jacket crowns may be used as retainers in anterior bridgework.

bridge (bridʒ), *sb.*[2] Also 9 **biritch, britch,** 9– **Bridge.** [Etym. unascertained; prob. of Levantine origin, since some form of the game appears to have been long known in the Near East; the origin of the seemingly Russian forms *biritch, britch*, is unknown; an unrecorded Turkish form **bir-üç* 'one-three' (since one hand is exposed and three concealed) was postulated as the source in *N. & Q.* (1969), 430–1.] A card-game based upon whist. In the original form of the game the dealer or his partner (dummy) named trumps, dummy's hand was exposed after the lead, and the odd tricks varied in value according to the suit named as trumps. Now = *auction* or *contract bridge*.

The game is said to have been played in Constantinople and the Near East about 1870. Formerly also called *Bridge Whist* (see sense c below). The sense in quot. 1843 is uncertain; *biritch* in quots. 1886 is applied to the call of 'no trumps'.

[**1843** J. Paget *Let.* 18 Apr. in *Mem. & Lett.* (1901) i. vi. 144 We improved our minds in the intellectual games of Bagatelle and Bridge.] **1886** *Biritch, or Russian Whist* 2 The one declaring may, instead of declaring trumps, say 'Biritch', which means that the hands shall be played *without trumps*. *Ibid.* 3 The odd tricks count as follows:—If 'Biritch' is declared each [odd trick counts] 10 points. *Ibid.* 4 There are four honours if 'Biritch' is declared, which are the four aces. **1898** 'Boaz' (*title*) The Pocket Guide to Bridge. **1898** *Nat. Rev.* Aug. 809 At a game of *wint* or *bridge*. **1901** 'Slam' *Mod. Bridge* Introd., 'Bridge', known in Turkey as 'Britch'. **1963** G. F. Hervey *Handbk. Card Games* 131 The modern game of Bridge, more correctly Contract Bridge, to distinguish it from its now-defunct predecessors, was developed by Harold S. Vanderbilt.

b. *auction bridge*: a variety of the game which superseded the original form. The right to name trumps and to play with the dummy goes, for each deal, to the player who undertakes to make the highest score. See also *contract bridge* *contract *sb.*[1] 1 g).

1903 O. Crawford in *Times* 16 Jan. 5/6 'Auction bridge' . .is more lively than dummy bridge. **1908** Dalton *Auction Bridge* p. iii, Auction Bridge is really a clever combination of the two games of Poker and Bridge. **1959** T. Reese *Bridge Player's Dict.* 18 Auction bridge, the predecessor of contract, held sway from about 1911. .to 1928, when contract began to be popular. **1960** Betjeman *Summoned by Bells* i. 5 Happy and tense they played at Auction Bridge.

c. *attrib.* and *Comb.*, as *bridge-coat, -drive, -four, -party, -player, -playing* adj., *-scorer, -table, -whist*; **bridge-marker,** a pad on which bridge scores are noted; **bridge roll** [or ?Bridge *sb.*[1]], a soft, oval, bread roll.

1905 *Daily Chron.* 29 May 8/4 A dainty bridge-coat in white, soft glacé silk. **1928** R. Macaulay *Keeping up Appearances* viii. 78 She had no bridge coat, a mysterious garment apparently donned by bridge-players. **1927** *Auction Bridge Mag.* May 19/2 'Bridge Drives'. .make you go 'all out' when there is a chance of a big killing. **1914** Compton Mackenzie *Sinister St.* II. iii. v. 576 Michael and a large party of freshmen. .had sustained themselves with dressed crab and sleepy bridge-fours. **1953** D. Parry *Going up—Going Down* vi. 315 Afterwards there was a bridge four and Clive played billiards. **1907** *Yesterday's Shopping* (1969) 386/3 Bridge Scoring Block. .Bridge Scoring Book. .Bridge Book. Ruled for Engagements. Scoring and Total. .Bridge Marker.

1914 'Saki' *Beasts & Super-Beasts* 258 'What did one send them?'. .'Bridge-markers.' **1910** —— *Reg. in Russia & Other Sk.* 96 Occasionally she went to bridge parties. **1949** M. Laski *Little Boy Lost* i. i. 9 The bridge-parties and the brittle gossip. **1899** A. G. Hulme-Beaman *Pons Asinorum* 50 No amount of rule and precept will suffice to make a first-class Bridge player. **1967** E. Lemarchand *Death of Old Girl* xvii. 194 His brilliance as a bridge player. **1907** E. Wharton *Fruit of Tree* ii. xvi. 244 A cluster of young bridge-playing couples. **1967** J. Symons *Man who killed Himself* i. i. 10 Half a dozen bridge-playing couples. **1926** D. D. C. Taylor *Good Housekeeping Menu* 106 Social Tea. . Bridge Rolls and Cress, White and Brown Bread and Butter. **1951** *Good Housek. Home Encycl.* 371/2 Bridge rolls are split, buttered and filled with a variety of sweet and savoury fillings. **1908** A. Bennett *Buried Alive* vi. 139 Manilla cigars, bridge scorers, [etc.]. .seemed to be the principal objects offered for sale. **1951** R. Senhouse tr. *Colette's Chéri* 111 Let's. .buy playing-cards, good wine, bridge-scorers. **1905** *Daily Chron.* 27 Dec. 4/5 Are we, as they say at the bridge-table, 'content'? **1899** A. G. Hulme-Beaman *Pons Asinorum* 46 As in Bridge Whist everybody plays his own game.

d. *ellipt.* = **bridge-party*.

1907 R. Brooke *Let.* 7 Apr. (1968) 81 The ugly friend of the Simpson's, who won a prize at our bridge last winter. **1966** H. Kemelman *Saturday the Rabbi went Hungry* (1967) xx. 121 My wife gave a bridge and invited his wife.

Hence **bridge** *v.*[3] *intr.*, to play bridge; **bri·dger** (cf. F. *bridgeur*, 1893), a bridge-player.

1907 Mrs. H. De La Pasture *Lonely Lady* xvi. 279 Miss de Courset, come and play billiard-fives,. .unless you are a bridger. Are you a bridger? **1908** *Daily Chron.* 14 Nov. 6/4 We must dine and we must 'bridge'. **1928** *Sunday Express* 27 May 15 Shall she Charleston, Blues or Bridge that evening?

bridgeable (bri·dʒăb'l), *a.* [f. Bridge *v.*[1] + -able.] That can be bridged or spanned by a bridge. Also *fig.*

1865 Carlyle *Fredk. Gt.* (1873) VII. xviii. iii. 135 Wadeable, bridgeable. **1890** *Advance* (Chicago) 13 Mar., Intervening islands break this distance [60 miles] into bridgeable lengths. **1898** *Daily News* 27 July 6/2 From the young officer to the preacher of non-resistance, there is a great gulf, but bridgable.

Bridgetin. Add: (Further examples.) Also **Bridgettine.** Also *attrib.* and as *adj.*

1756 A. Butler *Lives Saints* I. 514 March 22. .St. Catherine of Sweden. . .See her life written by Ulpho, a Brigittine friar. **1779** —— *Ibid.* (ed. 2) X. 148 The Brigittin nuns of Sion also had leave to retire abroad. **1873** J. H. Blunt *Pref.* to *Myr. Oure Ladye* p. xii, A Brigittine community of nuns. **1902** F. M. Steele *Convents Gt. Brit.* 62 The Bridgettine Rosary consists of seven *Paters* and sixty-three *Aves*. **1969** *Observer* (Colour Suppl.) 13 Apr. 54/2 *Bridgettines* . .run guesthouses at Iver Heath, Bucks, Rome and Lugano.

bridging, *vbl. sb.* Add: **1. c.** *Mountaineering.* (See quots.)

1941 C. Kirkus *Let's go Climbing* iv. 56 You are now standing astride the chimney. .left hand and foot on the left wall and the right hand and foot on the right wall. You can now climb straight up in the more natural position. This is known as bridging. **1957** R. G. Collomb *Dict. Mountaineering* 35 *Bridging*, a method of climbing chimneys and corners; it can also be any series of upward movements on a rock face when the legs are astride and the feet are being used on pressure holds.

bri·dging, *ppl. a.* [f. Bridge *v.*[1] + -ing[2].] **1.** Constructing or appointed to construct a bridge or bridges. (Perhaps orig. attrib. use of the vbl. sb.)

1891 *Daily News* 26 May 3/7 The bridging battalion of Royal Engineers. **1901** 'Linesman' *Words by Eyewitness* (1902) 94 Losing eleven men of the bridging party, and having every pontoon and plank struck by the Boer marksmen.

2. bridging species *Bot.*, one of a series of plant-species on which a parasitic fungus may be trained or adapted to infect species otherwise immune. Also *bridging host*.

1903 H. M. Ward in *Phil. Trans.* B. CXCVI. 34 A certain species of grass (A) may be capable of infection by means of spores from two other host-plants (B and C), neither of which is predisposed to reciprocal infection, though both may be infected from such a 'bridging' species (A) as is referred to above. **1922** *Encycl. Brit.* XXX. 478/2 'Bridging species'. *Ibid.*, 'Bridging hosts'.

bridle, *sb.* Add: **5. g.** (See quot. 1906.)

1899 C. F. Marvin in *Yearbk. U.S. Dept. Agric. 1898* 211 The one-point attachment of bridle. .is better suited to strong than light winds. **1906** A. F. Collins *Man. Wireless Telegr.* 209 *Bridle*, a cord attached to a kite that holds the latter at the proper angle in the wind; the kite-cord is attached to the bridle.

bridle, *v.* Add: **1. c.** (See *bridle *sb.* 5 g.)

1899 C. F. Marvin in *Yearbk. U.S. Dept. Agric. 1898* 210 Two methods of bridling the kite.

3. a. (Later example.)

1848 A. Brontë *Tenant Wildf. Hall* I. iv. 71 She bridled her long neck and smiled.

4. *intr.* of a horse: to rise to or answer the bridle.

1929 *Daily Express* 5 Jan. 7/5 Mr. Wroughton's horse never bridled well at the fence. . .It slipped and brushed through the fence, hardly rising.

bridled, *ppl. a.* Add: **2.** In names of birds, having bridle-shaped markings.

1869 *Amer. Naturalist* III. 340 Dr. Gambel. .bestowed upon it the title of the Bridled Tern (*S. frenata*). **1934** *Discovery* Oct. 293/1 The recent addition of six new birds to the British Isles,. .and the Bridled or Lesser Sooty Tern (*Sterna anaethetus*)—by the Council of the British Ornithologists' Union. **1935** Huxley & Haddon *We Europeans* iii. 98 The normal and so-called bridled variety of the guillemot, which latter has a white spectacle mark round the eye.

bri·dle-wise, *a.* *U.S.* [Bridle *sb.* 1.] Of a horse: readily guided by a touch of the bridle. Also *transf.*

1840 *Picayune* (New Orleans) 6 Oct. 2 The horse was not fit for a lady to ride; he was not *bridle wise*. **1898** H. S. Canfield *Maid of Frontier* 100 Just like a woman. You can't never make 'em bridle-wise. **1921** *Chambers's Jrnl.* Oct. 629/1 Quick-step. .was the smartest and most bridle-wise pony that ever went on shoes.

Hence **bri·dle-wi:sdom.**

1895 *Outing* (U.S.) XXVI. 477/1 Not that in the heat of play one relies upon this bridle-wisdom.

Brie (brī). Also **brie.** A kind of soft cheese made in *Brie*, an agricultural district in the north of France.

[**1848** Thackeray *Van. Fair* li. 453 An assembly. . attended by the Duchess (Dowager) of Stilton, Duc de la Gruyère,. .Marchese Alessandro Strachino, Comte de Brie.] **1876** *Encycl. Brit.* V. 456 The principal kinds of cheese at present known in commerce. . .Camembert, Parmesan, Gruyère, Brie, Roquefort [etc.]. **1951** M. Kennedy *Lucy Carmichael* vi. i. 276 Wives. .took the bus to Fenswick in search of *brie*, because Mrs. Beauclerc had discovered a shop there which stocked imported cheeses.

brief, *sb.* Add: **4. c.** In various slang uses (see quots.).

1860 Hotten *Dict. Slang* (ed. 2) 105 *Brief*, a pawnbroker's duplicate. **1874** *Ibid.* (ed. 4) 97 *Brief*,. .a raffle card, or a ticket of any kind. **1879** *Macmillan's Mag.* XL. 501/2, I. .took a brief (ticket) to London Bridge. **1889** Barrère & Leland *Dict. Slang* I. 180/1 *Brief* (prison), a note or letter. . .(Thieves), a ticket, pocket-book, pawnbroker's duplicate. **1939** H. Hodge *Cab, Sir?* xv. 217 It [*sc.* a cab-driver's licence] is also called the 'brief'. **1962** *John o' London's* 25 Jan. 82/2 A policeman's warrant card is his *brief*.

7. Also *fig.* in phr. *to hold a brief for* (a person): to express oneself like an advocate rather than an unbiased and critical appraiser. Freq. in neg. *to hold no brief for*: to be no advocate or supporter of.

1888 M. Arnold in *19th Cent.* Jan. 24 Professor Dowden holds a brief for Shelley; he pleads for Shelley as an advocate pleads for his client. **1918** R. A. Knox *Spiritual Aeneid* 215 When I was at Balliol, we used to adapt the phrase 'I hold no brief for So-and-so'.

b. = *briefing *vbl. sb.* 2.

1856 W. Bagehot in *National Rev.* III. 164 The calling of a constitutional statesman is very much that of a political advocate; he receives a new brief with the changing circumstances of each successive day. **1940** *Sphere* 10 Aug. 164 The pilots and navigators receive their 'briefs'. **1940** *Times* 22 Aug. 5/6 (*heading*) The 'Brief' For A Long Journey and Distant Target. **1949** *Economist* 27 Aug. 465/1 The last touches have been put to the brief which Sir Stafford Cripps and Mr Ernest Bevin will be taking with them for the Washington financial talks.

c. (See quot.) Cf. *brief-paper* (sense *11).

1923 H. A. Maddox *Dict. Stationery* 15 *Brief*, (1) A standard size of legal bags or envelopes measuring 14 in. by 5 in. (2) A legal pattern of ruled or watermarked foolscap comprising 36 or 42 feint lines and a marginal line.

11. *brief-paper*; **brief-bag,** the blue or red bag in which a barrister carries his briefs to and from court; **brief-case,** a small case made of leather, etc., for carrying papers, documents, and the like.

1848 *Punch* XV. 190/1 If my brief-bag is clear, so is my conscience. **1910** *Encycl. Brit.* IV. 562/1 English brief-bags are now either blue or red. Blue bags are those with which barristers provide themselves when first called, and it is a breach of etiquette to let this bag be visible in court. The only brief-bag allowed to be placed on the desks is the red bag, which by English legal etiquette is given by a leading counsel to a junior who has been useful to him in some important case. **1926** *Amer. Speech* I. 444/1 'Why don't you rest to-night?'. .'The why is in my briefcase.' **1954** T. S. Eliot *Confid. Clerk* i. 15 Enter Colby Simpkins with brief case. **1967** S. Beckett *Film* 36 He drops briefcase. **1796** *Archæologia* XII. 116 Brief paper, even and thin, but yellow with age.

12. *pl.* Very short knickers (see *knicker[2]), or trunks.

1934 *Books of To-Day* Nov. 10/1 I'm bored to tears with 'scanties', I'm sick to death of 'briefs'. **1959** 'M. Neville' *Sweet Night for Murder* iii. 40 Cathy, in minute briefs and bra, struggling into her dress. **1968** J. Ironside *Fashion Alphabet* 66 *Briefs*, snugly fitting panties with straight leg and crutch.

brief, *v.*[2] Add: **3. b.** To give instructions or information to. Cf. *briefing *vbl. sb.* 2.

1866 Lever *Sir B. Fossbrooke* xliv, They had not been well 'briefed', as lawyers say, or they had not mastered their instructions. **1940** *Times* 22 Aug. 5/6 These calm young men are being 'briefed' for a dangerous task. **1955** *Bull. Atomic Sci.* Apr. 119/3 Dr. McNair went to the Strategic Air Command Headquarters in Omaha to brief selected top staff officers on the results of his study. **1959** A. Lejeune *Crowded & Dangerous* x. 113 He handed them over to the secret police and was briefed for another job.

briefing, *vbl. sb.* Add: **2.** The action of giving information or instructions relating to a particular situation; information of this kind. Also *attrib.*

Applied esp., in the war of 1939–45 and since, to pre-flight conferences.

1910 H. Belloc *Pongo & Bull* ix. 133 The House'll be full. . .You don't want Eddie to give you any briefing, do you? **1940** *Times* 22 Aug. 5/6 Everything is ready for the 'briefing' which forms an indispensable part of every bombing raid. **1942** *Penguin New Writing* XIII. 27 His name, and the names of his crew, were quietly erased from the blackboard in the briefing-room. **1942** T. Rattigan *Flare Path* i. 113 Take-off 22.00. Briefing 19.45. **1942** R. Storrs in E. Kennington *Drawing the R.A.F.* 31 The precise but detached tones of the Briefing Officer. **1949** 'J. Tey' *Brat Farrar* vi. 47 It had been a wonderful briefing. For a whole fortnight. .they had. .rehearsed. .the lie of a land he had never seen.

brier, briar, *sb.*[2] (Examples of sense 'a pipe of this wood'.)

1882 *Graphic* 16 Dec. 683/2 Nowadays, every third man you meet has a cigarette or a 'briar' in his mouth. **1885** M. E. Braddon *Wyllard's Weird* i, He sat. .lazily puffing at his black briarwood. **1891** 'L. Keith' *The Halletts* xxviii, Pass me the brier-root. **1909** *Daily Chron.* 23 July 3/3 A briar being his constant companion while writing.

brig. Add: **1. d.** A place of detention, orig. on board a ship; a military or naval prison. *slang* (orig. *U.S.*).

1852 *Knickerbocker* XXXIX. 404 In less than a minute I was in the 'brig', in double irons. *Ibid.*, They call the place where prisoners are confined, 'the brig'. **1934** A. Woollcott *While Rome Burns* 13 Our dreamy old mess sergeant was even then languishing in the brig, awaiting trial. **1956** J. Masters *Bugles & Tiger* xxii. 293 My friends were hauled off to spend a day in the brig.

brigade, *sb.* Add: **4. brigade group** (see quot. 1953).

1945 *Diamond Track* (Army Board, N.Z.) 6/1 A brigade group with vehicles moving at 100 yards intervals. **1948** Lindsell & Benoy *Lindsell's Mil. Organization* (ed. 27) 26 The Territorial Army is in future therefore to include. . Four independent infantry brigade groups. **1953** E. Smith *Guide to Eng. Traditions* 9 There may even be a composite 'brigade group', in which three infantry battalions are supported by a regiment of artillery, a squadron of Royal Engineers, and signal and transport units, &c.

brigalow (bri·gălōu). *Austral.* Also **bricklow.** [ad. native name *būriagalah*.] Any one of several species of acacia, esp. *A. harpophylla*. Also *attrib.*

1847 Leichhardt *Jrnl.* i. 4 The Bricklow Acacia, which seems to be identical with the Rose-wood Acacia of Moreton Bay. **1862** Kendall *Poems* 79 Good-bye to the Barwan and brigalow scrubs. **1885** Mrs. C. Praed *Head Stat.* xvii, We are not fit for anything but store-cattle, we are all blady grass and brigalow scrub. **1901** F. Campbell *Love* 263 Long shafts of silvery moonlight creep in between the waving branches of blue gum and brigalow. **1935** *Bulletin* (Sydney) 5 June 46/4 Some clumpy brigalow country. **1936** A. Russell *Gone Nomad* i. 4, I am cattle-banging in western Queensland, the scent of the brigalow and mulga in my nostrils. **1957** P. White *Voss* viii. 202 After the midday halt, which was spent in the shade of some brigalow scrub.

bright, *a.* Add: **1. e.** Also, hopeful, encouraging, cheering.

1839 J. Romilly *Cambridge Diary* 14 Jan. (1967) 162 She is. .very talkative & disposed to look on the bright side of every thing. **1917** C. Mathewson *Sec. Base Sloan* xiii. 174 Toonalta's chance to pull the game up high and dry looked bright.

4. b. *spec.* Denoting tobacco of a light shade or colour. Also *bright-leaf*. orig. *U.S.*

1765 in *Amer. Hist. Rev.* (1921) XXVII. 71 Saw some of the bright couloured tobaco which sels So Dear in foreign markets. It is of a light yelow Coulour. . .The Inhabitants call it bright tobacco. **1933** *Discovery* Aug. 250/2 The [Cacao] moths have a preference for bright-leaf kiln-cured tobacco. *Ibid.*, Enormous stocks of Rhodesian bright tobacco.

7. b. *bright young thing* (etc.), a member of the younger generation in fashionable society (esp. in the 1920s and 1930s), noted for exuberant and outrageous behaviour (cf. Thing *sb.*[1] 10 a).

1927 *Punch's Almanack* 7 Nov. [8] Since a section of the 'Bright Young People' literally 'set the Thames on fire', things have been a little quiet. **1928** D. H. Lawrence *Phoenix II* (1968) 526 *Show* me somebody, then! And she shows me some guy, or some bright young thing. **1929** G. K. Chesterton *The Thing* iv. 38 If the bright young thing cannot be asked to tolerate her grandmother. .why should the grandmother. .have tolerated the bright young thing? **1931** R. Aldington *Colonel's Daughter* iii. 142 The Bright Young Idiots, who seem determined to queer the whole pitch to the puritans, by being as vicious as they can. **1936** *Morning Post* 15 July 14/5 There is a section of the community. .whose life seems to consist of cocktail and sherry parties, cabarets and midnight revelries . . . These are decadent 'bright young things'. **1953** S. Spender *Creative Element* 159 Dreams of the Old England and the Bright Young Things.

10. *bright-eyed* (later examples); also *transf.*

1881 O. Wilde *Burden of Itys* in *Wks.* (1948) 724 Light-winged and bright-eyed miracle of the wood! **1920** A. Huxley *Leda* 13 The bright-eyed bliss Perished. **1959** *Times* 24 Nov. 3/6 Arthur Jacobs's bright-eyed, allusive libretto.

b. bright emitter (see quot. 1931); so *bright-emitting* adj.; **bright lights,** the city, as a place of entertainment; urban gaiety; **bright-line,** (*a*) *Physics*, applied to a discontinuous spectrum consisting of bright lines resulting from radiation from an incandescent vapour or gas; (*b*) *Photogr.*, applied to a view-finder in which the area of the picture appears framed by a white line; **bright work,** polished metal-work on ships, etc.

1923 *Pop. Wireless Weekly* 24 Nov. 463/3 A new low-consumption valve, of the bright emitter type. **1931** *B.B.C. Year-Bk.* 437/2 *Bright Emitter*, a thermionic valve in which the filament gives its normal emission only when heated to a high temperature so that it glows brightly. **1943** *Gloss. Terms Telecomm.* (*B.S.I.*) 29 *Bright-emitting cathode*, a cathode, usually of pure metal, designed to be used at a relatively high temperature (e.g., above red heat). **1922** Sinclair Lewis *Babbitt* v. 66 A nice expensive vacation in New York. .with the bright lights and the bootlegged cocktails. **1938** F. D. Sharpe *S. of Flying Squad* iv. 43 She went back to live with her mother and promised not to come up again near the bright lights. **1890** G. F. Chambers *Handbk. Astron.* (ed. 4) II. 371 The spectra of bright-line stars. *Ibid.* 372 The meteorites in nebulæ giving a bright-line spectrum. **1954** R. H. Bomback *Basic Leica Technique* iii. 39 The most compelling new feature of the Leica M.3 is. .the Measuring Bright Line Viewfinder. *Ibid.* 40 Superimposed on the field of view is the 'Bright Line' contour which shows clearly the extent of the image covered by any one of the three standard lenses. **1959** *Which?* Nov. 158/2 One camera had a 'brightline' viewfinder—the area of the picture was framed by a white line when seen through the viewfinder. **1841** *South. Lit. Mess.* VII. 769/2 It was a part of my duty. .to superintend the cleaning of the 'bright work'. **1912** 'Aurora' *Jock Scott, Midshipman* i. 10 The bright-work on the quarter-deck. .shone like silver. **1933** Masefield *Conway* iii. 92 The lower deck was the show deck. . .There was. .much brasswork and bright-work upon it. **1962** *Times* 25 May 18/5 The external brightwork, including the bumpers [of a motor car].

bright, *adv.* Add: **1. c.** Phr. *bright and early*: very early in the morning. orig. *U.S.*

1837 W. Irving *Capt. Bonneville* II. ii. 17 Captain Bonneville and his three companions set out, bright and early, to rejoin the main party. **1900** *Daily News* 18 July, Persons desirous of being present at this gathering should be there bright and early. **1926** 'J. J. Connington' *Death at Swaythling Court* xvi. 286 Next morning I got up bright and early.

brightener. Add: *spec.* in *Electro-plating* (see quots.).

1924 Blum & Hogaboom *Princ. Electroplating* xviii. 293 It is possible to produce bright silver deposits from solutions containing very small amounts of carbon disulphide (CS_2), which serves as a 'brightener'. **1930** Field & Weill *Electro-Plating* vii. 94 A number of materials are not infrequently added to the cyanide copper bath to improve the quality of the deposit. These are commonly called 'brighteners'.

brilliant, *a.* Add: **1. b.** *Photogr.* Applied to a type of reflecting view-finder (see quot. 1958).

1894 *Photogram* Oct. 247/1 The 'Adams Brilliant' finder is a new departure of considerable importance. **1896** *Photography Ann. 1895* 429 Brilliant view finders are now becoming the rule. **1937** *Miniature Camera Mag.* Jan. 92/1 With the additional aid of a 'brilliant' viewfinder. **1958** M. L. Hall *Newnes Compl. Amat. Photogr.* iii. 40 Many simple cameras are fitted with so-called brilliant finders—simply a metal container with a lens in front, a mirror set at 45°, and another which receives the image and into which the user looks.

c. *brilliant cut*: in *Glass-cutting*. (See quot. 1962.)

1933 *Archit. Rev.* LXIV. 22/1 The glass panel with the brilliant-cut (that is engraved with emery wheel) figure had real beauty. **1953** *Glass for Glazing* (*B.S.I.*) 21 (*caption*) Types of brilliant cut. **1962** *Gloss. Terms Glass Ind.* (*B.S.I.*) 26 *Brilliant cut*, a decorative process in which designs are produced on flat glass with abrasive and polishing wheels.

brilliantined (brilyănti·nd), *a.* Dressed with Brilliantine.

1906 *Daily Chron.* 6 Apr. 9/5 Their adipose cheeks and brilliantined hair. **1926** *Chambers's Jrnl.* May 315/2 A sleek and pear-shaped gentleman heavily brilliantined and scented. **1967** A. Wilson *No Laughing Matter* i. 11 Thickly brilliantined heads.

brim, *sb.*[2] Add: **4.** spec. *brim of the pelvis* or *pelvic brim*: the part of the pelvis that forms the boundary of the superior pelvic aperture and separates the false or greater pelvis above from the true or lesser pelvis below; also, the aperture itself. Also *ellipt.*

1754–64 [see Pelvis 1]. **1810** [in Dict.]. **1885** R. & F. Barnes *Syst. Obstetr. Med.* II. xiii. 627 The filling up of the brim, and even of a part of the pelvic cavity sometimes, by the breech. **1957** D. Sinclair *Introd. Functional Anat.* xxi. 275 This basin is very incomplete, and ends centrally in a fairly abrupt rim of bone called the pelvic brim.

brim (brim), *sb.*[5] *U.S.* and *Austral.* var. of Bream *sb.* In U.S. *spec.* the long-eared sunfish (*Lepomis auritus*); cf. *bream *sb.* 2 b.

1795 J. Scott *U.S. Gazetteer* Hv, A great variety of fish, as rock, mullet,. .brim, and sturgeon. **1887** *Harper's Mag.* July 270/1 If they could slip away. . there would be a diminished number of 'brim' and 'goggle-eye' in the ditch. **1894** *Outing* (U.S.) XXIII. 403/2 The brim, a small, red fish, which is excellent fried. **1898** Morris *Austral Eng.* 52/2 The popular pronunciation is *Brim*, and the fishes are all different from the various fishes called *Bream* in the northern hemisphere. **1936** *Amer. Speech* XI. 314/1 *Brim*, black perch. **1945** Baker *Austral. Lang.* xii. 214 Popular fish-names peculiar to the Australian include: *brim* for bream. **1954** *Sydney Morning Herald* 24 May 8 Cod up to nine pound and brim up to seven.

brimfulness. (Later example.)

1891 Hardy *Tess* xxvii, The brimfulness of her nature breathed from her.

brimmer, *sb.* **3.** Delete †*Obs.* and add to def.: Now chiefly *local*; spec. a straw hat.

1888 Lowsley *Gloss. Berks. Words*, *Brimmer*, a hat. **1893** Dartnell & Goddard *Gloss. Words Wilts.*, *Brimmer*, a broad-brimmed hat. **1902** *Pall Mall Gaz.* 14 Apr. 2/3 The early brimmer marked a conspicuously daring spirit here and there against the crowd of silk hats.

Cf. *Local rhyme* (*Birmingham*). Straw brimmer, Ate the donkey's dinner.

brimmy (bri·mi), *a.* [f. Brim *sb.*[2] + -y[1].] Of a hat: having a wide brim; broad-brimmed.

1896 A. Morrison *Child of Jago* 112 The brimmy tall hat. **1908** *Westm. Gaz.* 28 Mar. 2/2 The cult flew. .to big buttons on parachute-skirted coats,. .to brimmy hats, brimmier hats, brimmiest hats, brimless hats. **1970** *Daily Tel.* 9 Nov. 10/2 She still had. .a brimmy black hat.

brimstony, *a.* Add: **b.** *transf.* and *fig.* Sulphurous, fiery.

1885 *Chambers's Jrnl.* Aug. 22 That brimstony old reprobate next door. **1924** Scholes *1st Bk. Gramophone Rec.* 93 The betrayer. .came to a bad and brimstony end.

Brinell (brine·l). The name of J. A. *Brinell* (1849–1925), Swedish engineer, used *attrib.* to designate his method (introduced *c.* 1900) of measuring the hardness of substances. (See quot. 1954.)

1915 *Proc. Amer. Soc. Test. Materials* XV. 47 (*title*) Relation between Brinell Hardness and Scleroscope Hardness. *Ibid.*, There was very little difference noted in the hardness at the center, or the outside, and a Brinell test was then made. . .The diameter of the depression was measured under a microscope and the corresponding Brinell number calculated. **1916** *Metal Industry* 25 Aug. 236 (*title*) Testing non-ferrous metals for hardness: application of the Brinell method. *Ibid.*, A series of combined microscopic and Brinell hardness tests. **1954** *Gloss. Terms Iron & Steel* (*B.S.I.*) i. 7 *Brinell hardness test*, a test to determine hardness by pressing a hard steel ball of known diameter under a standard load into the surface of the material and measuring the diameter of the indentation produced.

bring, *v.* Add: **1. d.** Colloq. phr. *to bring home the bacon* (fig.): to succeed in an undertaking; to achieve success.

1924 Wodehouse *Ukridge* viii, It may be that my bit will turn out to be just the trifle that brings home the bacon. **1928** *Daily Express* 10 Aug. 3/5 If I fail to 'bring home the bacon' I will give £10 to any charity selected by your ladyship. **1946** P. Larkin *Jill* 33 The College takes a number of fellows like him to keep up the tone. .but they look to us to bring home t' bacon.

7. a and **b.** (Later examples.)

1831 J. M. Peck *Guide for Emigrants* 47 The bottoms. . will bring three or four crops of corn without manure. **1834** *Encycl. Metrop.* XXII. 365/2 They [*sc.* ferrets] breed twice a year, bringing five or six at a time. **1843** *Amer. Pioneer* II. 172 The moose is an animal similar to the deer. . . They usually bring two young at a time.

15. bring down. a. (Further examples; also of aircraft.)

1768 S. Bentley *River Dove* 9 The Partridge, here oft as it flies, The Sportsman brings down with his Gun. **1917** 'Contact' *Airman's Outings* 23 Perhaps a German machine had been brought down. *Ibid.* 178 To 'bring down'. .an enemy was extremely difficult.

18. bring in. j. (Examples.) Also *ellipt.* and *transf.*

1684 [see Verdict *sb.* 2]. **1804** *Sporting Mag.* XXV. 127/2 The jury. .brought in a verdict for the plaintiff. **1841** T. Hood in *New Monthly Mag.* LXI. 272 The Jury debated from twelve till three What the verdict ought to be And they brought it in as Felo de Se, 'Because her own Leg had killed her!' **1865** Meredith *Rhoda Fleming* xviii, He's mad. . .There ain't a doubt as t'what the doctors 'd bring him in. . .Lunatic's the word! **1905** Conan Doyle *Return of Sherlock Holmes* 385 The coroner's jury brought in the obvious 'Wilful murder'. **1931** D. L. Sayers *Five Red Herrings* xxix. 351 'If the jury are sensible people, they'll bring it in self-defence or justifiable homicide.'. .They brought it in manslaughter. **1938** 'N. Blake' *Beast must Die* i. 12 A verdict of manslaughter was brought in against some person or persons unknown. **1960** 'J. Bell' *Well-Known Face* xiv. 149 And Mrs. Prentice, too? The jury brought that in suicide.

k. = *to bring to* (see *25 f).

1860 J. Cargill *Otago, N.Z.* 29 Open land is covered with either fern or grass,. .and is easily brought in. **1860** Ruskin in *Cornhill Mag.* Nov. 561 Bringing in of waste lands. **1950** *N.Z. Jrnl. Agric.* Oct. 371/1 Swedes and turnips. .their usefulness as feed for dairy cows and in the bringing in of new land.

19. bring off. d. To carry to a successful issue; to achieve.

1928 *Sat. Rev.* 27 Oct. 550/2 Theorizing about anything so personal. .seems a forlorn endeavour; but Mr. Beresford has brought it off. **1936** *Discovery* Aug. 241/1 He strains forward. .and. .brings off one of his special 'stunts' of marksmanship. **1952** M. Laski *Village* ii. 40 They each hoped to goodness Daisy could bring it off.

20. bring on. a. (Further examples.) Still current in the sense 'to advance the growth or development of'.

1629 Parkinson *Paradisus* 25 According to. .the temper of the climate. .to bring them on earlier or later, as it doth with all other fruits. **1847** Dickens *Dombey* xiv. 132 But he said. .that study would do much. . .'Bring him on, Cornelia! Bring him on!' **1848** Thackeray *Van. Fair* lii. 464 Briggs was a capital mistress for him, and had brought him on. . famously in English. **1932** A. J. Worrall *Eng. Idioms* ix. 61 His trainer brought on the horse in fine style.

e. To produce (a play, etc.) on the stage. (Quot. 1768 is perh. sense *20 a.)

1768 A. Dow *Let.* 16 July in *Corresp. D. Garrick* (1831) I. 306, I think it very unnecessary to submit the tragedy to any man's judgment but yours. Take it with you to the country; make your objections: if they can with facility be removed, I shall request the favour of you to bring it on. **1932** A. J. Worrall *Eng. Idioms* ix. 61 Mr. Blank is bringing on his show at His Majesty's [Theatre].

f. To bring forward or into action; *spec.* in *Cricket*, to put (someone) on to bowl. So in *U.S. colloq.* phr. *bring on your bears*, a defiant challenge to an adversary to do his worst.

1860 *Baily's Monthly Mag.* Sept. 429 Hayward and Parr were then brought on [as bowlers]. **1886** *Chicago Tribune* 13 Sept. 4/3 Bring On Your Bears. What with offensive Ministers and erratic Consuls,. .burden after burden of trouble has been laid upon Secretary Bayard's shoulders. *Ibid.*, He can request England or Canada. .to bring on their bears. **1904** P. F. Warner *Recov. Ashes* xiii. 261 Arnold was brought on, and in his first over clean bowled Duff. **1954** Wodehouse & Bolton *Bring on the Girls* i. 11 He says: 'Bring on the girls!' It is the panacea that never fails. . .The impresario has his way. The girls are brought on.

21. bring out. e. (Earlier example.)

1790 *Loiterer* 23 Jan. 12 A young woman seldom did well who was *brought out* before she was eighteen.

h. *to bring out one's bat* (in *Cricket*) = to carry it (out) (see Carry *v.* 53 c).

1833 J. Mitford in *Gentl. Mag.* Sept. 236/1 Tom [Walker] scored the amazing number of 95 runs in his first innings, and brought his bat out with him. **1870** *Times* 15 July 12/5 Mr. Green. .brought out his bat with the total at 198.

23. bring round. a. (Earlier example.)

1834 'A Virginian' *Kentuckian in New York* I. vi. 94 You want something to make your blood circulate: a small taste or two would soon bring you round.

c. To persuade; to convert *to* an opinion.

1862 Trollope *Rachel Ray* I. xiii. 259 Don't you think you could say something civil to Mr. Tappitt, so as to—to bring him round again? **1889** Bridges *Feast of Bacchus* iv. 1055 You've got. .your father Brought nicely round: and all through my good management. **1932** A. J. Worrall *Eng. Idioms* ix. 62 With some difficulty I brought him round to my way of thinking.

25. bring to. f. To bring (land) into good condition. *U.S.*

1814 in *Amer. Speech* (1947) XXII. 273 To *bring to* a piece of land—to bring it into a state of cultivation, or rather perhaps into a state fit for cultivation. **1838** H. Colman *Rep. Agric. Mass.* 77 One of these gentlemen. . has found this sort of land after it was thus 'brought to' extremely favorable to the growth of rye.

27. bring up. j. (Examples.)

1719 Defoe *Further Adventures of Robinson Crusoe* 200 Then I grew sick, and reached to vomit, but could not; for I had nothing in my Stomach to bring up. *Ibid.* 201 My Stomach loathed the Sugar, and brought it all up again. **1746** T. Tomkyns tr. *Lamotte's Gen. Treat. Midwifry* ii. xxi. 213 A little cough, which kept encreasing till it brought up a little purulent spittle. **1945** M. Dickens *Thursday Afternoons* viii. 234 Hoping that Ugly [*sc.* a dog] would manage to swallow whatever it was he had taken before he brought it up.

m. To lead (troops, etc.) to the scene of action.

1885 U. S. Grant *Pers. Mem.* I. 415 The troops from Corinth were brought up in time to repel the threatened movement without a battle.

bring-and-buy, *adj. phr.* Descriptive of a charity bazaar, stall, etc., to which people bring objects for sale, and buy those brought by others. Also *ellipt.*

1932 *N.Z. Home & Country* 1 Dec. 20/1 A 'Bring and Buy' stall was held, the proceeds to be donated to the Plunket Society. **1950** S. Ertz *Prodigal Heart* xii. 210 A recent 'Bring and Buy' bazaar in aid of the Crippled Boys' Home. **1952** A. Christie *Mrs. McGinty's Dead* xiv. 115 'I got it at the B. and B. at Christmas.' 'B. and B.?'. .'Bring and Buy. . .You bring things you don't want, and you buy something.' **1967** *Span* July 4 A bring and buy sale and lunch was held at Mrs. Miriam Stevens' house on Boars Hill on June 8th.

bringing, *vbl. sb.* Add: **2. bringing up** (*Naut.*): see Bring *v.* 27 f.

1908 *Westm. Gaz.* 1 July 7/4 The 'Patrol' lost her anchor and cable in bringing up.

4. *attrib.* **bringing money,** money formerly paid by a recruiting officer to a recruit on his joining the forces.

1892 *Rep. Comm. Terms & Cond. Service in Army* 63 The Committee have found a strong consensus of opinion that the reintroduction of bringing-money would probably give a great stimulus to recruiting.

brink. Add: **5. c.** *spec.* the verge of war. Hence **bri·nkmanship** [*-MANSHIP], the art of advancing to the very brink of war but not engaging in it; also *transf.* and *fig.*; hence (as a back-formation), **bri·nkman,** one who practises brinkmanship; **bri·nkmanlike** *a.*

[1840 MILL *Let.* 30 Dec. in *Wks.* (1963) XIII. 459 They had been brought to the brink of a war.] 1956 *Life* 16 Jan. 78 Says Dulles '..Of course we were brought to the verge of war...If you try to run away from it, if you are scared to go to the brink, you are lost...We walked to the brink and we looked it in the face.' 1956 *N.Y. Times* 26 Feb. 1/5 He [*sc.* Adlai Stevenson] derided the Secretary [*sc.* J. F. Dulles] for 'boasting of his brinkmanship—the art of bringing us to the edge of the nuclear abyss'. 1958 *Ann. Reg. 1957* 183 Anglo-French 'brinkmanship' over Suez had failed to stop at the brink. 1958 S. POTTER *Supermanship* 127 Brinkmanship is a clever way of describing the Dulles attitude. *Ibid.*, Khrushchev is the true Brinkman: his existence depending..on enemy-at-the-gatemanship. 1958 *Times* 11 Nov., Jackson Pollock..was..one to whom every new painting was..almost an act of spiritual brinkmanship. 1958 *Economist* 27 Dec. 1134/1 He [*sc.* Potter] is about to plunge, brinkmanlike, into International Lifemanship. 1967 *Spectator* 18 Aug. 177/3 A policy of muddling through, of economic brinkmanship.

briny (brəi·ni), *sb. colloq.* or *joc.* Also **briney.** [f. BRINY *a.*[1]] The sea, the ocean.

1831 J. BANIM *Smuggler* I. xii. 276 What is he to do without a sharp 'un to chaffer with the Parleys across the briney? *Ibid.* III. 40 He was seen at t'other side of the briny. 1880 H. EVANS *Brighton Beach Loafer* (1888) 6 Tales of unutterable woe and adventures undergone on the 'briny'.

brio. (Earlier example.)

1734 J. CONSTABLE *Refl. Accuracy Style* iv. 140 Their hardy and venturesome expressions..does often give a kind of vigour and *Brio* (if the word is sufficiently naturaliz'd) to their lines.

‖ **briquet**[2] (bri·ke). [Fr.] A steel for striking light from a flint; a representation of this; esp. *Her.*, one of the ornaments used to form the collar of the order of the Golden Fleece.

1824 SCOTT *St. Ronan's* III. iv. 90 Lighting one [*sc.* a cigar] with his *briquet.* 1913 W. OWEN *Let.* 27 Dec. (1967) 223 However the *Briquet* is precisely the thing I want. They are very common in France. 1920 *Edin. Rev.* Oct. 314 The cross and briquet of Burgundy. 1952 J. B. OLDHAM *Eng. Blind-Stamped Bindings* 63 *Briquet*, a conventional representation of a steel for striking a light.

briquetage (brikĕtā·ʒ). *Archæol.* Also **bricquetage, briquettage.** [F. *briquetage.*] Objects fashioned of burnt clay (see also quot. 1960).

1902 *Amer. Anthropol.* IV. 162/2 The name 'bricquetage' is given to masses of oven-baked clay found in heaps in the region of Vic and elsewhere. 1908 *Athenæum* 28 Mar. 391/2 The clay objects mentioned above, to which the Committee have provisionally given the non-committal name of 'briquettage'. 1960 *Lincs. Archit. & Archæol. Soc.* VIII. 70 The word 'briquetage'..can be used to include debris produced by so many different activities, from potting and salting to corndrying, and it can also include material which may be purely domestic, such as the remains of hearths and chimneys, or the clay walls of buildings which have been burnt.

briquette, briquet[1]. **2.** Substitute for def.: A block of compressed coal-dust, usu. with addition of a binding substance such as pitch. (Later example.)

1963 *Ann. Reg. 1962* 509 Exports of coal, coke, and briquettes decreased.

3. A small block of ice-cream. Cf. *BRICKETTE.

1927 W. C. COLLINSON *Contemp. Eng.* 16 Cyclists selling ice-cream briquettes.

briquette (brike·t), *v.* [f. the sb.] *trans.* To form (coal-dust, etc.) into briquettes. So **briquetted** *ppl. a.*, **briquetting** *vbl. sb.*

1898 *Eng. Mechanic* 7 Oct. 194/1 Briquetting Sawdust. 1916 LANTSBERRY tr. *Franke's Handbk. Briquetting* I. 621 Briquetting of Sawdust, and other Waste Wood. These materials can generally be briquetted fairly readily without a binding material. 1928 *Daily Tel.* 25 Sept. 11/5 The Yallourn briquetting plant of the State Electricity Commission of Victoria. 1962 *Times* 26 Mar. 5/3 They also intended to develop a briquetted fuel.

brisance (bri·zaňs). [f. F. *brisant*, pres. pple. of *briser* to break.] The shattering effect of such high explosives as nitroglycerine and guncotton. (See also quot. 1935.) Also *attrib.*

1915 A. MARSHALL *Explosives* 390 It has been proposed by Bichel to use the expression $\frac{mv^2}{2}$ for the 'brisance' or violence of the blow given by an explosion, *m* being the mass of gas evolved and *v* the velocity of the detonation. *Ibid.* 320 For the brisance test quantities of 10 g. were used. *Ibid.* 407 *marg.*, Brisance meter. 1935 *Discovery* Feb. 43/2 Little is yet known as to the exact nature of brisance, or the property of detonating at the slightest shock, which these explosives possess.

brisant (bri·zaň), *a.* [Fr.: see prec.] Of explosives: shattering, smashing, breaking.

1905 LARSEN tr. *Bichel's New Meth. Test. Explosives* 18 In this translation the word 'brisant' (explosives) has sometimes been used in preference to the rather inadequate expression 'high'. 1906 *Times* 14 Mar. Although these brisant substances belong directly to chemistry. 1918 COLVER *High Explosives* 594 The genuine brisant explosives can only be detonated by means of the energy of a primary explosive.

‖**brisé** (bri·ze). *Dancing.* [Fr., pa. pple. of *briser* to break.] Any movement in which the feet or legs are beaten together in the air.

1786 S. J. GARDINER *Def. Minuet-Dancing* 54 The Steps generally made use of in Cotillions are..Assembléz, Brizéz, [etc.]. 1830 R. BARTON tr. *Blasis's Code of Terpsichore* II. vi. 78 An attitude upon one leg, as in..the *cabriole*, *brisés* and the *ronde-de-jambe en l'air.* 1952 KERSLEY & SINCLAIR *Dict. Ballet Terms* 19 A single *brisé* can begin with one foot on the other ankle.

‖**brise-bise** (bri·zbiz). Also erron. **brise-à-bise, bris-à-bris**; colloq. **brisby.** [Fr., lit. 'wind-breaker'.] A curtain of net or lace for the lower part of a window. Also *attrib.*

1912 *Queen* 10 Aug. 266/1 Brise-bise curtains of striped net. 1923 *Daily Mail* 14 Feb. 1 (Advt.), Cream Hemstitched Lace and Insertion Casements and Brise Bise. *Ibid.* 17 Apr. 14 Insertion Brise a bise. *Ibid.* 29 May 15 The *Brise-bise* Lace Curtain (familiarly called a 'Brisby'). *Ibid.* 14 Aug. 1/3 Cotton net..suitable for gowns, bris-à-bris.

‖**brise-soleil** (briz,solē̆ⁱ·). [Fr., lit. 'sun-breaker'.] A device (whether a perforated screen, louvers, or projections) for shutting out direct or excessive sunlight.

1944 *Archit. Rev.* XCV. 69 Two main types of sun-baffle have been devised..: the pierced concrete screen or 'cambogé'; and the quebra-sol or brise-soleil. 1962 *Listener* 15 Nov. 805/1 As in a modern *brise-soleil*, the two horizontal projections—balcony and roof—give shelter from the high midday sun.

brish, *v.* Dial. var. *BRUSH *v.*[2] 7. Also *intr.* Hence **brishing** *vbl. sb.*

1636 in PARISH & SHAW *Dict. Kentish Dial.* (1887) 19 For shredinge of the ashes and brishinge of the quicksettes.. vj[d]. 1914 KIPLING *Diversity of Creatures* (1917) 377 The first wood down is sere and small,..The brishings off the hills. 1916 BLUNDEN *Harbingers* 59 The morning hedger with his brishing-hook. 1920 —— *Waggoner* 25 No hedger brished nor scythesman swung.

brisk, *v.* Add: **2.** *to brisk up*: to become brisk, to behave or move in a brisk manner. (Also without *up.*)

1830 in WEBSTER, *Brisk up*,..to come up with life and speed; to take an erect, or bold attitude. 1884 L. B. WALFORD *Nan* II. 76 After this adventure, we seemed..to brisk up afresh. 1915 H. L. WILSON *Ruggles of Red Gap* (1917) iii. 40 As I brisked out of bed the following morning at half-after six. 1916 'BOYD CABLE' *Action Front* 170 The rattle of rifle fire dwindled away at times to separate and scattered shots, brisked up again and rose to a long roll. 1938 L. MACNEICE *Earth Compels* 40 You should see her in jodhpurs Brisking in to breakfast from a morning canter. 1966 *Listener* 24 Nov. 764/1 Claire..brisks up the even flow of strictly grammatical dialogue.

brisken (bri·sk'n), *v.* [f. BRISK *a.* + -EN[5].] **1.** *trans.* To make brisk or lively. Also with *up.*

1799 A. YOUNG *Agric. Linc.* 451 Let your fuel be coke, which bears the bellows to brisken the fire, without raising a flame. 1838 *Eclectic Rev.* I. 551 Briskening the then dull tenor of periodical criticism. 1901 *Daily Chron.* 27 Dec. 3/3 The Coronation, they think, may brisken up trade generally. 1905 E. F. BENSON *Image in Sand* i. 8 Sir Henry briskened his pace a little.

2. *intr.* To become brisk, to speed up.

1876 W. MATHEWS *Getting on in the World* (1884) Ser. II. iii. 68, I heartily wish that business may brisken a little. a1909 *Scribner's Mag.* (Webster), Meanwhile our artillery fire briskened. 1965 *Guardian* 27 Oct. 8/7 Leduc describes it with awe, then briskens.

Hence **bri·skening** *vbl. sb.* and *ppl. a.*

1901 E. F. BENSON *Luck of Vails* I. i. 4 These agreeable influences..seemed to produce a briskening effect on the two. 1907 *Athenæum* 23 Mar. 364/1 This piece, with briskening of the action, will prove diverting. 1927 *Scots Observer* 4 June 17/2 A briskening of the air.

brisky, *a.* Delete †*Obs. rare*[−1] and add later and *fig.* examples.

1894 *Cornh. Mag.* Jan. 43 As we proceed, stopping regularly at every wayside public-house, many of the passengers betray symptoms of becoming 'brisky'. 1905 *Daily Chron.* 14 Mar. 6/7 The lamb..gambols about the farm in as brisky a manner as any of its companions.

brisling (bri·sliŋ, bri·z-). Also **bristling.** [Norw., Da. *brisling* sprat.] A small, sardine-like fish of the herring family, which is widely distributed in the north-east Atlantic and the North Sea, and is cured and tinned for use as food.

1902 WEBSTER Add., Bristlings. 1913 *Times* (Weekly ed.) 14 Nov., 'Skipper Sardines' were bristling, or sprats. 1915 *Law Rep.* K. B. Div. III. 742 The brisling is the nearest approach to the sardine which the Norwegian seas provide. 1926 *Spectator* 2 Jan. 11/2 If the brisling fisheries fail the Stavanger Theatre shows a loss. 1969 A. WHEELER *Fishes Brit. Isles & N.-W. Europe* 132 In Norway..the resultant sprats, seasoned and canned in oil, are sold as 'brislings'.

brisque (brisk, brisk). [Fr.] In bézique and other card games, a privileged card, such as the aces and tens in bézique; also the name of a card game.

1870 'CAVENDISH' *Bézique* 8 Brisque (queen of spades and knave of diamonds, now called bézique), scored 40. Curiously enough, brisque did not exist in the game of *Brisque* itself. 1890 'BERKELEY' *Bézique & Cribbage* 10 If the loser's score, with his brisques, is less than 1,000, he is said to be rubiconed. 1963 G. F. HERVEY *Handbk. Card Games* 20 Every Ace and 10 that a player holds in the tricks that he has won is scored for at the rate of 10 points each. They are known as brisques.

bristle, *sb.* Add: **6. bristle-bird,** a name given to certain Australian reed-warblers; **bristle-grass** (examples); **bristle-worm,** a chætopod.

1827 VIGORS & HORSFIELD in *Trans. Linnæan Soc.* XV. 232 He [*sc.* Mr. Caley] calls it in his notes 'Bristle Bird'. 1865 GOULD *Birds Australia* I. 343 Sphenura longirostris.. Long-billed Bristle-bird. 1911 LUCAS & LE SOUËF *Birds Australia* 330 The Bristle-birds have a shy disposition, and live in reed-beds and thickets. 1967 M. SHARLAND *Birds of Sun* 177 Such species as emu, bristle-bird and two or three other 'historical' kinds. 1841 *Penny Cycl.* XXI. 299/1 Setaria...Two [species] are indigenous in England, *S. verticellata* and *S. viridis*, and called bristle-grass. 1961 R. W. BUTCHER *Brit. Flora* II. 1025 The Green Bristle-grass is a loosely tufted annual with bent or erect stems 4–24 in. (10–60 cm.) high. 1908 *Westm. Gaz.* 8 Aug. 16/3 There is..a similarity in the eyes of the bristle-worm to those of the fly. 1941 J. S. HUXLEY *Uniqueness of Man* ix. 192 Certain marine bristle-worms (Polychaetes).

bristling, var. *BRISLING.

Bristol. Add: **2. b.** Applied to a type of porcelain or pottery similar to Delft ware manufactured in Bristol.

1776 J. WEDGWOOD *Let.* 24 Jan. (1965) 190, I believe this to be true of nearly all our present compositions in England and France, except the Bristol. 1784 H. WALPOLE *Descr. Strawberry-Hill* 13 A cup and saucer, white with green festoons, of Bristol porcelaine. 1787 *Bristol Gaz.* Jan. in W. Chaffers *Marks & Monogr.* (1863) 138 (Advt.), Bristol Pottery, Temple Backs. 1869 LADY C. SCHREIBER *Jrnl.* (1911) I. 66 There is no Bristol, no Plymouth, scarce any characteristic Bow. 1873 H. OWEN *Two Cent. Ceramic Art Bristol* xiii. 328 Previous to the year 1859, Bristol delftware was unknown to collectors. 1900 F. LITCHFIELD *Pott. & Porc.* vii. 91 Many specimens of Bristol porcelain are unmarked. 1969 V. C. CLINTON-BADDELEY *Only Matter of Time* 28 Four milk white cups and saucers with a blue and gold band, which Mrs. Bazeley was prepared to describe as late Bristol, though..they might be early New Hall.

c. Applied to an opaque coloured (esp. blue) or white glass manufactured in Bristol.

1880 LADY C. SCHREIBER *Jrnl.* (1911) II. 292 An exquisite pair of Bristol glass candlesticks. 1882 *Ibid.* 393 Smelling bottles of fine Bristol glass. 1939 O. LANCASTER *Homes Sweet Homes* 36 Bristol glass candlesticks. 1948 M. ALLINGHAM *More Work for Undertaker* xxiv. 283 Oh, is that the glass Lawrence drank from?..They're old Bristol. 1966 J. GLOAG *Sentence of Life* vii. 73 He glanced back at the bristol blue lamp.

3. Bristol cream, the proprietary name of a type of sherry (see quot. 1959); cf. *Bristol milk.*

1886 *Trade Marks Jrnl.* 23 June 580 *Bristol Cream.* John Harvey & Sons, Denmark Street, Bristol.: Sherry. 1924 J. BUCHAN *Three Hostages* iv. 57, I had a glass of the Bristol Cream for which the club was famous. 1959 W. JAMES *Word-Bk. of Wine* 32 *Bristol Cream*, an oloroso sherry blended by Harvey's of Bristol from wines imported from several sources in Spain.

5. *pl.* (with lower-case initial). *Rhyming slang* [ellipt. for *Bristol Cities* = titties (TITTY[3])]. The breasts.

1961 in PARTRIDGE *Dict. Slang* Suppl. 1962 R. COOK *Crust on its Uppers* ii. 30 These slag birds used to go trotting upstairs .., arses wagging and bristols going. 1969 J. LEASOR *They don't make them like that any More* vii. 194 She was in her early twenties, with a wonderful pair of bristols; I could see their outline, firm and round, against her dress. 1969 *Observer* 2 Feb. 28/3 The main point (or should it be points?) of this programme is Miss Barbara Windsor's bristols which are..well-developed.

Brit, *sb.*[3] Colloq. shortening of BRITON *sb.* 1 c or BRITISH(ER. Also *attrib.* or as *adj.*

1901 'LINESMAN' *Words by Eyewitness* (1902) 62 The Brit is at his old game. 1948 P. WYNDHAM LEWIS *Let.* 3 June (1963) 443 Are you trying to fool the Brit Public..? 1961 S. PRICE *Just for Record* viii. 69 Your working-class Brit is a glutton for celebrities. 1969 R. PETRIE *Despatch of Dove* II. 106 Goddam Limey! You're a Brit.

Britain, *sb.* Add: **1. b.** *pl.* with reference to the several dominions and dependencies of Great Britain; cf. *Britt.* (*Omn.*) (= of (all) the Britains) in the legend on coins.

1897 EARL OF ROSEBERY in *Daily News* 5 July 4/5 'Regina Britanniarum'—the Queen of the Britains...She is sovereign, not of one or two, but of numberless Britains, all self-supporting. 1901 *Westm. Gaz.* 11 Dec. 2/2 Lord Rosebery has succeeded with his cry of 'All the Britains', as the three letters 'Omn' on the new coins are to testify... Our King henceforth is to be King of All the Britains.

Britannia. Add: **3. b.** *ellipt.* = *Britannia metal*, or an article made of this. Also *attrib.*, made of Britannia metal.

1814 W. BENTLEY *Diary* Feb. (1914) IV. 237 He had begun the work of the Britania Ware. 1840 *Picayune* (New Orleans) 28 July 1 Britannia Table and Tea Spoons. 1852 A. CARY *Clovernook* 91 The cupboard.., where the china and britannia were wisely set for show. 1887 M. E. WILKINS

Humble Romance 107 Pouring the tea from the shiny britannia teapot into the best pink china cups. 1970 *Globe & Mail* (Toronto) 28 Sept. 26/6 What is Britannia ware? It is a very fine type of pewter made by a spinning process instead of being cast.

britannicize (britæ·nisəiz), *v.* [f. BRITANNIC *a.* + -IZE.] *trans.* To make Britannic or British in form or character.

[1811 *Ann. Reg. 1809* Pref. p. iv, Such measures as may *Britannize*, as it were, all the other Grecian islands.] 1887 W. F. RAE *Miss Bayle's Romance* xvii, American gentlemen are 'britannicized Indians'. 1921 *Spectator* 16 Apr. 487/2 St. Paul did not fight a lifelong battle against judaizing the Christian Church in order that we should britannicize it nowadays.

britch. Still current = BREECH *sb.* 4 b. (See also quot. 1940.) Also in phr. *too big for one's britches*: see *BIG *a.* 3 e.

1885 [see BREECH *sb.* 4 b]. 1940 *Chambers's Techn. Dict.* 114/1 *Britch* or *breech*, a wool-sorter's term for wool obtained from the thighs and root of the tail, estimated as the lowest quality in a fleece. 1963 *Times* 6 Feb. (N.Z. Suppl.) p. v/3 Britch wool.

Briticism. (Earlier U.S. example.)

1868 R. G. WHITE in *Galaxy* Mar. 335 This use of the word is a widespread Briticism.

Briticization (bri:tisəizē̆ⁱ·ʃən). Also **Britticization.** [f. as BRITIC(ISM + -IZE + -ATION after *Anglicization*, etc.] The process of making British (esp. with reference to the ancient Britons), or the result of this.

1953 K. JACKSON *Lang. & Hist. Early Brit.* 702 The *Bern-* and *Birn-* of Nennius are not British but are Britticisations of Latin *Bernicia.* 1955 *Antiquity* XXIX. 86 The consequent re-Britticization of Dumfriesshire.

British, *a.* Add: **2. b. British Empire**: the empire consisting of Great Britain and the other British possessions, dominions, and dependencies; now replaced by the British Commonwealth (see *COMMONWEALTH 4 c). Cf. *EMPIRE *sb.* 5 b.

1604 J. DEE *Pet. to King*, God..make your Maiestie to be the most blessed and Triumphant Monarch, that euer this *Brytish* Empire, enioyed. 1768 GOLDSMITH (*title*) The present state of the British Empire in Europe, America, Africa and Asia, containing a concise account of our possessions in every part of the globe, [etc.]. 1813 HECTOR CAMPBELL (*title*) The Impending Ruin of the British Empire, its Cause and Remedy considered. 1884, etc. [see *COMMONWEALTH 4 c]. 1887 [see EMPIRE *sb.* 5 a]. 1902 *Encycl. Brit.* XXXIII. 393/1 The British Empire League and the Imperial Trade Defence League endeavour to promote inter-Imperial trade.

c. British English: the English language as spoken or written in the British Isles; *esp.* the forms of English usual in Great Britain, as contrasted with those characteristic of the U.S.A. or other English-speaking countries.

[1869 ELLIS *E. E. Pronunc.* I. 20 Practically the speech of the American English is archaic with respect to that of the British English.] 1892 H. SWEET *New Eng. Gram.* I. 224 The influence of the vulgar London or 'Cockney' dialect is stronger in Australasian than in British English. 1926 H. W. FOWLER *Mod. Eng. Usage* 52/2 These words do not mean in American..use what they mean in British English. 1932 R. W. CHAPMAN *'Oxford' English* 540 This expression may be current in America, but it is not British English. 1967 'J. CROSS' *To Hell for Half-a-crown* x. 130 The English that Neumann was babbling was not..British English.. but American.

5. British-Israelite = *ANGLO-ISRAELITE; so **British-Israel** (*attrib.*), **British-Israelism**; **British Restaurant,** a government-subsidized restaurant opened during the war of 1939–45; **British Standard Time,** the time system introduced in Britain on 18 Feb. 1968, which is the same as Central European Time and is equivalent to the extension of Summer Time throughout the year; **British Summer Time** = SUMMER-TIME 2; **British thermal unit** (see THERMAL *a.* 2); **British warm** (see WARM *sb.*[2] 2).

1907 (*title*) The British-Israel Ecclesia. 1920 S. C. (*title*) British-Israelism. 1934 *Times Lit. Suppl.* 1 Feb. 77/2 Are all the present millions of British-Israelites mad? 1948 'N. SHUTE' *No Highway* v. 133 He has been in trouble with the police arising out of his activities with the British Israelites. 1941 W. S. CHURCHILL *Minute* 21 Mar. in *Grand Alliance* (1950) 663, I hope the term 'Communal Feeding Centres' is not going to be adopted...I suggest you call them 'British Restaurants'. 1950 A. WILSON *Such Darling Dodos* 152 Returning from lunch at the British Restaurant. 1967 *Times* 24 Oct. 11/3 The Secretary of State for Home Affairs announced on June 23 that a bill will be introduced early next session to extend 'summer time' permanently throughout the year...We understand that the name British Standard Time is being considered; this is ambiguous, is likely to lead to confusion and misunderstanding and should be rejected. 1970 A. P. HERBERT *In Dark* 11 By the beginning of February some real advantages will at last be seen to flow from British Standard Time. 1930 [see *B.S.T. *s.v.* B III]. 1958 L. R. MUIRHEAD *Blue Guide to Northern Spain* (ed. 2) p. cxxxvi, Railway time is now 1 hour later than Greenwich time, and coincides with British Summer Time when the latter is in force. 1970 A. P. HERBERT *In Dark* 90 We arrived punctually at one o'clock Local Time, two o'clock Newfoundland Government Time, 1730 British Summer Time and 1530 Greenwich Mean Time.

6. *the best of British luck*, an expression of encouragement, often with the ironical implication that good luck will not be forthcoming. Also ellipt. *best of British.*

1963 L. MEYNELL *Virgin Luck* ii. 39 As soon as they [*sc.* kittens and young birds] can fend for themselves they get shoved out into the world and the best of British luck to them. 1965 EVA-LIS WUORIO *Z for Zaborra* 190 The best of British luck to you. 1967 W. KEENAN *Lonely Beat* vii. 74 'Best of British.' With that the reporter vanished.

Britishism (bri·tiʃiz'm). [f. BRITISH *a.* + -ISM.] The characteristic qualities of the British; with *a* and *pl.*, any of these qualities; a British peculiarity, form of expression, or the like.

1894 *Harper's Mag.* Jan. 315 Doubtless he could use 'Britishisms' if he chose. *Ibid.*, Nor should we advise an American statesman to attempt a 'Britishism'. 1906 *Daily Chron.* 20 Oct. 4/4 Americanism yet Britishism, cosmopolitanism but Imperialism. 1915 SANDAY *Meaning of War* 80 For the Englishman, *Britishism* (Britentum) and civilization..are one and the same. 1927 *Observer* 20 Nov. 11/1 These are a few of the concrete Britishisms which engage the vigilance of Chicago's mayor. 1959 *Encounter* Dec. 54 Almost all his characteristics were British..sometimes they were Britishisms exaggerated.

Britishly (bri·tiʃli), *adv.* [f. BRITISH *a.* + -LY[2].] †**1.** In the ancient British tongue. *Obs. rare.*

1654 VILVAIN *Enchir. Epigr.* VI. xxiii, Glassenbury, Brytishly cald Inis Avalon.

2. After the manner of the British, in British fashion.

1892 STEVENSON & OSBOURNE *Wrecker* xxiii. 362 Britishly chuckle-headed. 1892 F. E. TROLLOPE *That Wild Wheel* xiv, I shall come home..to grumble, Britishly, on my half-pay. 1900 *Daily News* 3 Mar. 6/1 As he took up his position, very Britishly in front of the fire. 1904 *Windsor Mag.* Jan. 294/1 A certain sedate and Britishly respectable old codger. 1968 *Economist* 13 Apr. 64/2 Rioting very Britishly in front of the City Hall.

Britishness (bri·tiʃnės). [f. BRITISH *a.* + -NESS.] †**a.** The conditions of the ancient Britons. **b.** The quality or character of the British.

1682 W. RICHARDS *Wallogr.* 88 Primitive Brittishness was never acquainted with the habiliment of a Shirt. 1872 W. CLARK RUSSELL *Perplexity* xxvi, There were Frank's pure eyes, his clear-cut nose,..his manliness, his high-bred air, and, if I may coin a word, his thorough *Britishness.* 1904 *Daily Chron.* 9 Sept. 3/2 As Napoleon pulverised ancient kingdoms..so Hazlitt hurled himself courageously at the solid mass of complacent Britishness.

Brito- (bri·to), comb. form f. L. *Brit(t)o* BRITON, used:

1. In sense 'belonging to the ancient Britons and ..', as *Brito-Pictish, Brito-Roman* adjs.

1654 VILVAIN (*title*) Enchiridium Epigrammatum Latino-Anglicum; an epitome of essais, Englished out of Latin.. Containing six Classes or Centuries of 1. Theologicals. 2. Historicals. 3. Heterogeneals. 4. Bryto-Anglicals. 5. Miscellaneals. 6. Mutuatitials. 1860 *Chambers's Encycl.* I. 376/2 Of Roman or Brito-Roman manufacture. 1926 *Glasgow Herald* 25 Oct. 6 A long list of British or Brito-Pictish ministers.

2. In sense 'British and ..', as *Brito-arctic, Brito-Canadian, Brito-Japanese.* Also **Brito-centric** *a.*, having Britain as the centre.

1866 A. BELL *Hist. Canada* II. 321 In brief, what is denounced as a crime in a Gallo-Canadian shall pass for public virtue in a Brito-Canadian. 1898 *Geogr. Jrnl.* XI. 134 It is evident, therefore, that the basalts of Cape Flora and Hooker island are similar to types widely distributed in the Brito-arctic volcanic province. 1905 *Westm. Gaz.* 11 Sept. 2/3 'Anglo-Japanese' certainly is an easier form than 'Brito-Japanese'. 1924 R. MACAULAY *Orphan Island* xvi. § 3 The world, as viewed from Orphan Island, wore a curious, Brito-Centric aspect.

Britonic, var. *BRITTONIC *a.* and *sb.*

brittle, *a.* Add: **1. c.** *Metallurgy.* Applied to a type of fracture of material (see quot. 1946).

1930 *Engineering* 14 Feb. 231/1 A typical brittle-material fracture. 1946 *Gloss. Metallurg. Terms* (Firth Brown) 27 In a brittle fracture, metal fails by cleavage because its cohesion is exceeded. 1965 *New Scientist* 15 Apr. 173/1 The important characteristic of a brittle fracture is that a crack can propagate under a stress which is less than the design stress..and can cause complete failure of a structure.

4. *brittle-minded* adj.; **brittle heart** (see quot. 1938).

1934 DADSWELL & LANGLANDS in *Jrnl. Council Sci. & Ind. Res., Australia* VII. 190 (*heading*) Brittle heart in Australian timbers. 1938 —— in *Empire For. Jrnl.* XVII. 58 Brittle heart has been defined as that central portion of the tree which is extremely brittle and of comparatively low strength. 1887 W. JAMES *Let.* 2 July in R. B. Perry *Thought & Char. of W. J.* (1935) I. 376 My narrow and brittle-minded bachelor state.

brittlely, *adv.* Delete '*rare ? Obs.*' and add later examples.

1923 D. H. LAWRENCE *Birds, Beasts & Flowers* 44 Almond tree..Grey, lavender sensitive steel, curving thinly and brittly up in a parabola. 1929 W. FAULKNER *Sartoris* IV. 304 Ridges skeletoned brittlely with frost. 1949 M. MEAD *Male & Female* v. 120 The Mundugumor have faced new events brittlely. 1958 *Spectator* 18 July 86/1 This facile Neapolitan, whose music and plot are brittly innocent of modulation.

brittly, *adv.* See BRITTLELY *adv.*

Brittonic (britǫ·nik), *a.* and *sb.* Also **Britonic.** [f. Old Celtic *Britton- BRITON + -IC.] **A.** *adj.* Of or pertaining to the ancient Britons; *spec.* = BRYTHONIC *a.* **B.** *sb.* The Brythonic language.

1923 *Glasgow Herald* 16 Feb. 11 The extensive Britonic and Bangor missions. 1926 *Ibid.* 25 Oct. 6 The Brittonic, or British, missionaries completed the Christianisation of the Northern Britons. 1953 K. JACKSON *Lang. & Hist. Early Brit.* 3 Until fairly recently, the term *Brythonic*, coined by Rhys, was regularly used to describe the language brought to Britain by the bearers of that variety of primitive Celtic speech known as *P-Celtic*...Of late there has been an increasing tendency to use *Brittonic* instead. *Ibid.* 154 A Brittonic-speaking population. 1960 REANEY *Orig. Eng. Place-Names* v. 87 Brittonic was still spoken in Somerset and Dorset at the end of the seventh century. *Ibid.* 88 Here Brittonic names are rare.

Brix (briks). The name of A. F. W. *Brix* (1798–1890), a German scientist, used *attrib.* of a hydrometer calibrated according to the *Brix scale*, a scale for the measurement of the specific gravity of a sugar solution. Also used *absol.*

1897 G. L. SPENCER *Handbk. Chemists Beet-Sugar Houses* 55 The Brix and Baumé Scales. The degree Brix is the percentage by weight of sucrose in a pure sugar solution. 1910 *Chem. Abstr.* IV. 684 In basing conclusions upon results by apparent Brix the relationship of apparent to actual solids must always be borne in mind. 1912 C. A. BROWNE *Handbk. Sugar Analysis* iii. 45 The Brix hydrometer or spindle is supplied in a variety of forms. 1959 *Encycl. Brit.* XI. 996/1 The Brix saccharometer simply indicates percentage of sugar by weight when read in sugar solutions at 17·5° C.

bro (brōᵘ). A written or colloq. abbrev. of BROTHER *sb.* Pl. *bros.* (joc. pronounced brǫs), in the title of a firm.

a1666 EVELYN *Diary* an. 1638 (1955) II. 20, I accompanyd my Eldest Bro (who then quitted Oxford) into the Country. 1860 E. WALFORD *County Families* 49 (Advt.), Bass's East India Pale Ale..Berry, Bros., & Co. 1905 W. M. GALLICHAN *Cheshire* 161 Notice the old timbered house now occupied by Messrs George Bros., ironmongers. 1937 PARTRIDGE *Dict. Slang* 94/2 *Bro.* Brother: Charterhouse: C. 20. 1940 'N. BLAKE' *Malice in Wonderland* ix. 132, I must..report to my bro on the sinister bloke we've just chased off.

broach, *sb.*[1] Add: **4. b.** A shuttle used in weaving tapestry.

1783 *Encycl. Brit.* X. 8536/1 They serve to keep the warp open for the passage of broaches wound with silks, woollens, or other matters used in the piece of tapestry. 1878 MRS. SKETCHLEY tr. *A. de Champeaux's Tapestry* 2 The material for the woof is wound on a wooden shuttle, called a 'broach' or 'flute'. 1888 *Encycl. Brit.* XXIII. 212/1 The design [of tapestry and pile carpets] is formed by short stitches knotted across the warp with a wooden needle called a broach.

14. Also *in broach.*

1826 DISRAELI *Vivian Grey* II. xii. 169 As fine a barrel of ale in broach as you ever tasted.

III. (sense 5) *broach-river*; *broach-splitting.*

1899 RIDER HAGGARD in *Longman's Mag.* Mar. 410, I found..Rough Jimmy..employed in splitting broaches to be used for thatching stacks. This is the process of broach-splitting. 1921 *Dict. Occup. Terms* (1927) § 499 *Broach river*; rives timber with a cleaver, and shapes the pieces of timber with a hand knife to form broaches, *i.e.* pointed implements used in thatching.

broach (brōᵘtʃ), *sb.*[2] In 7 **baroche.** [Place-name.] A Surat cotton grown in the *Broach* district, Gujarat State, India.

1617 R. COCKS *Diary* (Hakl. Soc. 1883) I. 330 We gave our host..a peece of backar baroche to his children to make them 2 coates. 1877 *Encycl. Brit.* VI. 482/2 The principal sorts [of Surats] are Hingunghât, Oomrawuttee, Broach, Dhollera, and Dharwar. 1959 *Chambers's Encycl.* II. 587/1 Raw 'Broach' cotton.

broach, *v.*[1] Add: **9.** To enlarge and finish (a drilled hole) with a 'broach' or boring-bit. Also with adv. *out.*

1846 HOLTZAPFFEL *Turning* II. 572 Flutes and clarionets are first perforated with the nose-bit, and then broached with taper holes. 1859 *Ibid.* IV. 363 The work being removed from the chuck, the hole is broached out to size. 1889 HASLUCK *Model Engin. Handybk.* viii. 91 The hole in cross-head for piston-rod will have to be broached out.

broacher. Add: **3.** One who broaches holes.

1921 *Dict. Occup. Terms* (1927) § 200 *Broacher; broaching machinist*; a driller who enlarges or smooths out core hole, a hole previously drilled, using a broach or reamer in a drilling machine.

broaching, *vbl. sb.* Add: **4. b.** The action or operation of enlarging and finishing a drilled hole. (Cf. *BROACH *v.*[1] 9.) Also *attrib.*, as *broaching machine, machinist.*

1846 HOLTZAPFFEL *Turning* II. 575 For large works, broaching machines are employed. 1888 *Lockwood's Dict. Terms Mech. Engin.* s.v., Broaching is chiefly done in a drilling machine or a lathe. 1921 Broaching machinist [see *BROACHER 3].

broad, *a.* Add: **1. e.** Of bran: consisting of large particles.

1908 *Animal Managem.* 98 Two varieties are distinguished as 'broad' or 'fine'; in 'broad' bran the wheat husk is more or less whole and gives the article a flaky appearance.

5. c. Denoting a type of phonetic transcription in which separate symbols are used only to denote distinctive sound units (phonemes); opp. *narrow.*

1877 H. SWEET *Handbk. Phonetics* 105 The different values of each of the vowel signs in this system, which I will call 'Broad Romic', in apposition to the scientific 'Narrow Romic', as indicating only *broad* distinctions of sound. 1908 —— *Sounds of English* 9 A broad notation is one which makes only the practically necessary distinctions of sound in each language, and makes them in the simplest manner possible, omitting all that is superfluous. 1912 *Princ. Internat. Phonetic Assoc.* 14 A *broad* transcription may be more accurately defined as a transcription obtained by using the minimum number of symbols requisite for representing without ambiguity the sounds of one language independently of other languages. 1932 D. JONES *Outl. Eng. Phonetics* (ed. 3) x. 50 A transcription of the type 'one letter per phoneme' is called a Broad Transcription. *Ibid.* 305 The transcription used in this book is not the broadest possible.

B. *sb.* **5.** (Earlier example.)

1659 SIR T. BROWNE *Let.* 16 Nov. (1946) 351 Such deluges..might..settle lakes and broades.

6. (Earlier and later examples.) So **broa·dsman,** a card-sharper. *slang.*

1789 G. PARKER *Life's Painter* xv. 129 Who are continually looking out for flats, in order to do them upon the broads, that is, cards. 1860 HOTTEN *Dict. Slang* (ed. 2) 105 *Broads*, cards. *Broadsman*, a card sharper. 1938 SHARPE *S. of Flying Squad* xviii. 193 Broadsmen, or three-card sharpers, kept the Flying Squad busy in its early days. *Ibid.* xx. 219 They..were also playing the Broads on the trains.

7. *Turning.* A tool having a disc or angular end with a sharpened edge used for turning the insides and bottoms of cylinders.

1846 HOLTZAPFFEL *Turning* II. 515 The broad..requires to be held downwards or underhand.

8. A woman; *spec.* a prostitute. *slang* (orig. and chiefly *U.S.*).

1914 JACKSON & HELLYER *Vocab. Criminal Slang* 19 *Broad*, ..A female confederate; a female companion; a woman of loose morals. 1915 G. BRONSON-HOWARD *God's Man* II. ii. 131 Listen, broad,..you got your roasting clothes on to-day and you better take 'em off quick or I'll slam you one in the kisser. 1927 HEMINGWAY *Men without Women* 114 There were a couple of broads sitting at the next table. 1928 *Amer. Speech* III. 218 *Broad*, a plump, shapely girl; the words [*sic*] sometimes carries a disparaging moral significance. 1931 E. LINKLATER *Juan in Amer.* II. xvi. 177 Slummock..had got into a jam with a broad; no ordinary broad, but a Coastguard's broad. 1962 *John o' London's* 25 Jan. 82/3 Prostitutes are variously termed *tarts, toms, broads.*

D. 1. a. *broad-beamed* (transf.; cf. *BEAM *sb.*[1] 16 b).

1945 M. DICKENS *Thursday Afternoons* ii. 77 George in his broad-beamed blue shorts and striped fisherman's jersey.

2. broad aisle or † **alley** *U.S.*, the main aisle or passage in a church or meeting-house; **broad band,** (*b*) *Electr.* a band (see *BAND *sb.*[2] 14) with a wide range of frequencies; also *transf.* and *attrib.*; **broad-banded** *a.*, having broad bands of colour as a distinctive marking, esp. defining a species of armadillo; **broad-bill,** (*b*) in full *broadbill swordfish*: the swordfish *Xiphias gladius*; **broad-billed** *a.*, having a broad bill, esp. defining a species of sandpiper; **broad-brow, broadbrow** [f. as *HIGHBROW, etc.] *colloq.*, a person of broad tastes or interests, not a highbrow or a lowbrow; **broad-headedness** *Ethnol.*, brachycephaly; **broad jump** = *long jump* (see JUMP *sb.*[1] 1 b); **broad-lace,** a woollen material about four inches wide, used as an ornamental border in carriage upholstery; **broad-leaf,** (also) a settlers' name for a New Zealand tree *Griselinia littoralis*; **broadloom** *a.*, applied to a carpet woven in broad widths; also *absol.*; **broad-mindedness,** the condition of being liberal or tolerant in thought or opinion; **broad-spectrum** *a.* (orig. *U.S.*), of a drug, effective over a wide range of diseases or micro-organisms; **broad stone** = ASHLAR 1, FREESTONE[1] 1.

1807 *Massachusetts Spy* 25 Mar. (Th.), Another pew at the right hand of the broad aisle, esteemed the pleasantest in said house. 1887 *Harper's Mag.* Dec. 161/2 Miss Flint shall have pew No. 40 in the broad aisle. 1731 in H. S. Sheldon *Doc. Hist. Suffield* 250 Whether the Broad alley in the meeting House should be fil'd up. 1806 *Intelligencer* (Lancaster, Pa.) 21 Oct. (Th.), Mr. Deming was sitting in the Pew east of the broad Alley. 1956 HEFLIN *U.S.A.F. Dict.* 92/2 *Broadband*, a band having a wide range of frequencies. 1960 H. CARTER *Dict. Electronics* 32 Broad-band radio systems. 1962 H. C. WESTON *Sight, Light & Work* (ed. 2) vi. 186 By accepting this 'broad-band' objective it is possible to simplify the application of the method. 1904 *Westm. Gaz.* 8 Sept. 10/1 The broad-banded species (*Xenurus unicinctus*) is a rare creature [*sc.* armadillo] from Surinam. 1927 Z. GREY *Tales of Swordfish* iii. 107 Few novices at the game ever held a broadbill longer than a few moments. *Ibid.* iv. 125 There were never any flukes in broadbill swordfishing. *Ibid.* vii. 157 The broadbill swordfish was the one fish in the world which could not be caught through luck. 1969 *Guardian* 8 Mar. 7/5 If you are dedicated to catching marlin swordfish and broadbills, it is an obsession. 1839 PEABODY in *Zool. & Bot. Surv. Mass.: Birds* 367 The Broad billed Sandpiper..is very rare in the United States. 1927 A. P. HERBERT in *Punch* 12 Jan. 51/1 (*title*) Ballads for Broad-brows. 1927 *Observer* 11 Sept. 21/1 It is no longer highbrow versus lowbrow. We are all broadbrows, and what we want to listen to depends on our mood. 1929 H. G. WELLS *King who was King* i. § 2. 22 The Broadbrow is as anxious not to be 'arty' as the Low-brow and as terrified of the cheap and obvious as the High-brow. 1890 T. H. HUXLEY in *20th Cent.* Nov. 758 In the extreme north..marked broad-headedness is combined with low stature..in the Lapps. 1846 DODD *Brit. Manuf.* VI. 132 The lace employed..is used as a binding or edging for various parts of the interior [of a coach]; the finest is called 'broad-lace'. 1879 W. N. BLAIR *Building Materials of Otago* 155 There are few trees in the bush so conspicuous, or so well known as the broadleaf. 1908 W. P. REEVES *N.Z.* iv. 94 The foliage which the broadleaf puts forth quite eclipses the leaves of most of the trees upon which it rides. 1956 *N.Z. Timber Jrnl.* July 54/1 *Broadleaf*, Griselinia littoralis Raoul. Cornac. Papaumu. A small tree of N.Z. with wide distribution from sea-level to beyond timber line. 1925 *Carpet & Rug World* Apr. 50/3 Belgrade Broadloom Carpet. 1954 *Archit. Rev.* CXV. 138/1 Broadloom carpets are woven generally at widths of 10/4, 12/4, 14/4 and 16/4. 1963 *Which?* Mar. 70/2 Many carpets are sold as 'broadloom'. This is not a type, but a width. 1893 *Athenæum* 2 Dec. 770/3 There was a rare combination in him of bigotry and broad-mindedness. 1952 *Sci. Amer.* Apr. 49/3 They are known as the broad-spectrum drugs, because each of them attacks a wide range of infections. 1954 S. DUKE-ELDER *Parsons' Dis. Eye* (ed. 12) x. 121 The 'broad spectrum' antibiotics..are clinically effective against both Gram-positive and Gram-negative organisms. 1969 *Times* 3 Mar. 5/8 The new concept of integrated pest control was discussed...It means avoiding the broad spectrum chemicals and changing lines of attack. 1703 T. N. *City & C. Purchaser* 56 Broad-stone..the same with Free-stone,..so called, because they are raised broad and thin out of the Quarries. 1842 GWILT *Archit.* Gloss., *Broad Stone*, the same as free-stone.

broad acres. In phr. (*the land of the*) *broad acres*: Yorkshire. So **broad-acred** *a.*, of or characteristic of Yorkshire. Cf. ACRE 1 b.

1898 *Windsor Mag.* VII. 696/2 Yorkshire has in recent years..played its matches in various parts of the broad-acred shire. 1907 *Minister's Gaz. Fashion* Oct. 182/1 'There's a trip in fra' Leeds, my lad,' explained the man from broad acres. 1908 A. N. COOPER (*title*) Across the Broad Acres. 1928 *Sunday Express* 17 June 12/6 A red-faced..man of the broad-acred type. 1928 *Sunday Dispatch* 8 July 22/3 The youth of the Broad-acres.

broad-brim. b. (Earlier example.)

1749 FIELDING *Tom Jones* III. vii. x. 78 This the Quaker had observed, and this..inspired honest *Broadbrim* with a Conceit.

broadcast. Add: **A.** *adj.* **3.** Disseminated by means of radio or television.

1922 *Times* 6 Apr. 17/7 Request to the Postmaster-General to consider the desirability of including weather forecasts in any broadcast distribution of information by wireless telephony. 1924 *Wireless World* 3 Sept. 664/1 Broadcast Announcements in Two Languages. 1936 *Discovery* Apr. 124/2 Mr Thornton's broadcast talks on exotic music are widely known.

C. *sb.* **2.** The action or an act of broadcasting by radio or television. Also *attrib.*

1922 *Daily Mail* 8 Aug. 7/3 (heading) World Broadcast. 1924 REITH (*title*) Broadcast over Britain. 1924 *Westm. Gaz.* 4 Dec., The Postmaster-General's power to include control over broadcast receivers. 1926 *Glasgow Herald* 16 Nov. 9 The ban on the broadcast of controversial topics. 1955 *Times* 10 May 10/3 Whether Sir Winston Churchill will make a sound or television broadcast for the Conservative Party.

broadcast, *v.* Add: **3.** To disseminate (a message, news, a musical or dramatic performance, or any audible or visible matter) from a radio or television transmitting station to the receiving sets of listeners and viewers; said also of a speaker or performer. Also *absol.*

Inflected pa. t. and pa. pple. *broadcast* (cf. *BROADCAST *a.* 3); occas. *broadcasted.*

1921 *Discovery* Apr. 92/1 The [wireless] station at Poldhu is used partly for broadcasting Press and other messages to ships, that is, sending out messages without receiving replies. 1922 *Daily Mail* 8 Aug. 7/3 The largest and most powerful wireless station that can broadcast to the world. *Ibid.* 11 Nov. 7 Government arrangements for broadcasting. 1924 *Daily News* 13 Dec. 6/7 The speech broadcast to our homes to-day. 1956 M. W. STEARNS *Story of Jazz* (1957) xvi. 188 Coon-Sanders..and Henry Halstead broadcasted over the radio.

Hence **broa·dcasted** *ppl. a.*, = *BROADCAST *a.* 3. Now *rare.*

1923 *Glasgow Herald* 3 Feb. 8 In cities the 'broadcasted' entertainment can never prove a serious rival to the theatre and concert hall. 1924 *Ibid.* 28 Aug. 7 The crowning event was the broadcasted evening service at Spurgeon's Tabernacle on Sunday.

broadcaster (brǭ·dkɑstəɹ). [f. BROADCAST *v.* + -ER[1].] **1.** A person whose speech, performance, etc., is broadcast on radio or television; also, a broadcasting company, station, or instrument.

1922 *Daily Mail* 30 Nov. 8 The Prince of Wales..made a great hit as a 'broadcaster'..when he delivered a message by wireless to the Boy Scouts. 1923 *Weekly Dispatch* 28 Jan. 2 When such music is the fare the broadcasters offer. 1926 C. R. BOYD FREEMAN *Towards Answer* i. 35 Imagine all 'wireless' installations being not only receivers but broadcasters. 1955 *Times* 27 Aug. 6/1 A frequent broadcaster and pamphleteer in favour of the all-German talks.

2. A machine for sowing seeds broadcast.

1934 in WEBSTER. 1960 *Times* 15 Feb. 19/2 The main implements are, in fact, the broadcaster, the forage harvester, [etc.].

broadcasting (brǭ·dkɑstiŋ), *vbl. sb.* [f. BROADCAST *v.* + -ING[1].] The action of *BROADCAST *v.* 3. Also *attrib.* or as *ppl. a.*, as *British Broadcasting Company* or *Corporation* (abbrev. *B.B.C.*, q.v. as separate entry).

1922 *Glasgow Herald* 21 Apr. 10 The 'broadcasting' of pre-determined material of public interest from central stations. *Ibid.* 1 May 10 Whether the 'broadcasting' station itself shall be under or independent of official control. 1922 *Westm. Gaz.* 19 Oct., The British Broadcasting Company will broadcast news, information, concerts, lectures, educational matter, speeches, weather reports, and theatrical entertainments. 1926 *Encycl. Brit.* III. 1046/2, Feb. 23 1920 saw the opening of the first wireless telephone broadcasting service in the world, when a programme of vocal and instrumental music..was transmitted. 1962 (*title*) Report of the Committee on Broadcasting, 1960.

broaden, *v.* Add: **1.** Also with *out.*

1888 MRS. H. WARD *R. Elsmere* I. iii. 79 Her round comfortable face brightened and broadened out into a beaming smile. 1894 B. HARRADEN *In Varying Moods* vii, The narrow, dull, everyday existence broadened out into many interesting possibilities.

broad gauge. (Earlier examples.) (In quot. 1858, *fig.*)

1844 [see NARROW GAUGE]. 1858 BRANNAN in F. J. Meine *Tall Tales of S.W.* (1930) 255 The Universalists.. get on the broad gage and goes the hull hog-ah!

Broadland (brǭ·dlænd). [f. BROAD *sb.* 5 + LAND *sb.*] The district of the Broads; East Anglia, or a section of it. Also *attrib.* Hence **Broa·dlander.**

1889 H. M. DOUGHTY (*title*) Summer in Broadland. 1899 *Daily News* 27 Dec. 6/4 The sluggish streams and low meadows of the Norfolk Broadland. 1902 N. EVERITT (*title*) Broadland Sport. 1903 W. A. DUTT et al. *Norfolk Broads* vi. 78 Among the characteristic Broadlanders.. the wherrymen are the most numerous and conspicuous. 1926 *Daily Tel.* 6 Aug. 5/2 Broadland bream at this season are particularly partial to stewed wheat.

broad-leaved, *a.* Add: **c.** spec. in *Forestry* (see quot. 1957).

1905 *Terms Forestry & Logging* 39 *Hardwood*, a broad-leafed, or dicotyledonous, tree. 1927 *Forestry* I. 14 Scots Pine..in natural forests was almost always associated with a broad-leaved species, generally birch. 1953 H. L. EDLIN *Forester's Handbook* viii. 122 The current rate of planting of the broadleaved trees, or hardwoods is much lower than that of the conifers. 1957 *Brit. Commonw. Forest Terminol.* II. 33 *Broadleaved tree*, one belonging to the Dicotyledonous group of Angiosperms and producing timber known commercially as hardwood...The use of this term in certain African countries to distinguish trees having relatively large leaves as compared with *Acacias* is deprecated.

broadside, *sb.* Add: **4.** *attrib.* **broadside array** *Radio*, an aerial array having its directional effect perpendicular to the elements of the array.

1932 F. E. TERMAN *Radio Engin.* xiv. 511 The Broadside Array. The simple broadside antenna array consists of a number of antennas spaced at uniform distances along a horizontal line and connected so as to be excited in phase. 1959 DAVIES & PALMER *Radio Studies* iii. 31 Aerials of large size can be constructed by connecting together large numbers of dipole or yagi elements. Such aerials are known as broadside arrays.

broadside, *adv.* Add: **2.** *Printing.* Of letterpress, illustrations, tables, etc.: set sideways. Also as *adj.*, of a page so set (see quot. 1948); and *sb.* Chiefly *U.S.*

1925 *Univ. Chicago Press Man. Style* (ed. 8) 143 Set long box-headings broadside (i.e., vertically) so as to read up from the bottom of the text page. 1948 *Words into Type* (Appleton-Century-Crofts) 539 A broadside table or illustration is one printed with the top at the left side of the page, requiring a quarter turn of the book to the right to be in position for reading. 1957 R. A. HEWITT *Style for Print* iii. 30 All broadside illustrations, whether printed with the text or inset, should have their legends on the right-hand side of the page, i.e. in the inner margin on a left-hand page and the outer margin on a right-hand page.

broa·dside, *v.* [f. the sb.] *intr.* Of a motor cycle in dirt-track racing: to skid and slide in the cinders as part of a deliberate manœuvre when the rider is 'hugging' a corner. Also in extended use of a car, pedal cycle, etc., esp.

in a controlled sideways skid. So **broadsiding** *vbl. sb.*

1930 *Aberdeen Press & Jrnl.* 28 Mar. 1 (*caption*) A dirt-track rider broadsiding at Crystal Palace, London. 1932 *Daily Express* 25 June 17/5 Forty-six miles an hour does not seem much to the uninitiated, but it is the broadsiding that makes dirt-track cycling what it is. 1967 *Autocar* 28 Dec. 24/3 The car broadsided for 100 yards before turning over. 1969 *N.Z. Listener* 16 May 16/1 In the early days of 'dirt-track' the people came to see the extreme slides made by the machines as they were raced into the cinder-strewn corners. Later, specially-designed machines enabled the riders to hug the corners without 'broadsiding'.

broadtail (brǭ·dtē̆ⁱl). The skin or pelt of a young Persian lamb, having a lustrous moiré appearance. Also *attrib.* Cf. *BREITSCHWANZ.

1892 H. POLAND *Fur-Bearing Animals in Commerce* 330 Persian Lamb takes a brilliant dye...The *Broadtail*, or *Breitschwanze*, is probably..the unborn Persian Lamb; it is undoubtedly the skin about which so much..has been written, as to slaying the parent sheep for the sake of its skin. 1898 *Westm. Gaz.* 13 Jan. 3/2 A broadtail cycling suit is the latest whim in the Bois. *Ibid.* 3 Nov. 3/2, I can as strongly commend black broadtail. It is a fur very easy to manipulate. 1899 *Ibid.* 21 Sept. 3/2 The coat of black broadtail, or caracal, or Persian lamb or astrachan, all varieties of a species, is prophesied as the pet garment this winter. 1927 *Observer* 20 Nov. 21 Long waistcoats of ermine or squirrel or broad-tail. 1963 *Retail Trading-Standards Assoc. Bull.* Aug./Sept. 3/1 The following names [for furs] may be used notwithstanding the general rule:—*Animal*, Lamb (Persian). *Accepted Trade Name*, Broadtail Persian or Persian Broadtail.

Broadway. Add: Used allusively with reference to the theatres for which Broadway in New York is famous. Also *attrib.* and in *Comb.*

1881 *Harper's Mag.* Apr. 712/2 Another suggests some Broadway idea. 1907 ZANGWILL *Ghetto Comedies* 269 He had heard vaguely of 'Hamlet'—as a great play that was acted on Broadway. *Ibid.* 271 The 'happy endings' of Broadway. 1925 O. SITWELL *Poor Young People* 35 A widow-theme That is..Ancien-Régime More than Broadway-American.

Broadwood (brǭ·dwud). A piano made by John Broadwood (1732–1812) or by the London firm founded by him. Also *attrib.*

1832 W. GASKELL *Let.* 17 Sept. in A. B. Hopkins *Eliz. Gaskell* (1952) iii. 52 First to buy the piano (a Broadwood). 1888 G. B. SHAW *London Music 1888–89* (1937) 46 The ensemble of the three Broadwood grands was not so dreadful as might have been expected. 1895 KIPLING *Song of Banjo* in *New Rev.* June 602 You couldn't pack a Broadwood half a mile—You mustn't leave a fiddle in the damp. 1969 *Listener* 14 Aug. 225/1 A Broadwood of 1787.

Broca (brōᵘ·kä). The name of the French surgeon and anthropologist Pierre Paul *Broca* (1824–80), applied (chiefly in the possessive case) to anatomical features, etc., discovered by or named after him (see quots.). *Broca's aphasia*, motor aphasia; *Broca('s) area*, part of the premotor cortex in the inferior frontal convolution of the brain, concerned with the movements required for speech; *Broca's point*, the centre of the external auditory meatus; *Broca's table*, a graduated table of skin-colours.

1875 *Encycl. Brit.* II. 111/1 The colour of the eyes and hair is also to be defined accurately by Broca's table. 1877 *Jrnl. Mental Sci.* XXII. 406 Case of Sudden and Complete 'Aphasia' and Partial 'Right Hemiplegia', Lesion of 'Broca's Convolution'. 1884 T. G. STEWART *Dis. Nervous Syst.* II. 22 This includes what has been named Broca's convolution, from his having made out its relation to aphasia. 1890 W. JAMES *Princ. Psychol.* I. ii. 40 Broca's region is homologous with the parts ascertained to produce movements of the lips, tongue, and larynx. 1898 BEEVOR *Dis. Nervous Syst.* xix. 283 The speech centre of Broca, situated in the posterior part of the left third frontal convolution. *Ibid.* 285 Motor aphasia..is caused by a lesion of ..Broca's centre. 1905 W. D. HALLIBURTON *Handbk. Physiol.* (ed. 7) xlviii. 689 In the higher apes..faradisation of the Broca area does not evoke vocalisation. 1911 *Ibid.* (ed. 10) xlviii. 729 That Broca's area is the chief speech centre has for long been a matter of doubt among physiologists and pathologists. 1959 *Psychol. Rev.* LXVI. 46/1 Broca's aphasia, indistinguishable from that caused by damage of Broca's area.

brocade, *sb.* Add: **3. brocade-matting,** a floor matting of Japanese manufacture consisting of a texture of reeds and cotton yarn with a coloured design woven upon it.

1902 *Encycl. Brit.* XXIX. 696/2 The 'brocade-matting' industry of Okayama.

‖**broché** (brōᵘ·ʃe, Fr. broʃe), *a.* [Fr., pa. pple. of *brocher* to stitch: see BROCHE *v.*] **1.** Of a material, esp. silk: woven with a pattern on the surface. Also *sb.*, a material of such a texture.

a 1877 KNIGHT *Dict. Mech.*, *Broché-goods*, goods embroidered or embossed. 1880 *Drapers' Jrnl.* 27 May 5/2 Bodice of broché front breadths. *Ibid.*, A chip hat to match the dark shade of broché. 1882 CAULFEILD & SAWARD *Dict. Needlework*, *Broché*, a French term denoting a velvet or silk textile, with a satin figure thrown up on the face. 1898 *Queen* 1 Jan. 31/2 Some silk crépon gauzes, rich brochés, and fancy velveteen.

2. Of an unbound or paper-covered book: stitched.

1889 in *Cent. Dict.* 1952 J. CARTER *ABC for Bk. Collectors* 127 A continental book in its original wrappers (*broché*).

brochette. Add: **2.** A pin or bar used to fasten medals, orders, etc., to the coat or uniform of the wearer.

1849 THACKERAY *Pendennis* I. xxvi. 255 He wore three little gold crosses in a brochette on the portly breast of his blue coat. 1896 *Daily News* 5 Mar. 7/5 M. Faure wore his brochette of Russian Orders.

brochure. Add pronunc. (brōᵘ·ʃiūᵊɹ). Also, *spec.* a small pamphlet or booklet describing the amenities of a place, etc.

1925 *Brit. Hotels* (E. J. Burrow & Co.) 136/2 Torquay, Allerdale Hotel. Situated in own spacious grounds...Write for illustrated brochure (inclusive terms). 1935 *Punch* 10 July 36/2 Innumerable hotels in England offer to send me an 'Illustrated Brochure on request'. 1963 V. NABOKOV *Gift* v. 316 The sun's impact restores the deficiency..and the brazen body no longer experiences shame. All this sounds like a nudist brochure.

Brocken (brǫ·k'n). The name of the highest of the Harz Mountains in Saxony, reputed to be the scene of witches' Walpurgis-night revels. Applied *attrib.* to a magnified shadow of the spectator thrown on a bank of cloud in high mountains when the sun is low and often encircled by rainbow-like bands (first observed on the Brocken). Also applied to a dramatic representation of the Walpurgis-night revels, and *transf.*

1801 [see SPECTRE *sb.* 3]. 1844 E. B. BROWNING *Poems* II. 45 Waved backwards (as a wind might wave a Brocken mist, and with as brave Wild roaring). 1860 [see SPIRIT *sb.* 3 *transf.*]. 1888 G. B. SHAW *How to become Mus. Critic* (1960) 136 The Brocken business, which is musically and scenically childish. 1924 —— *St. Joan* Pref. p. xvii, Brocken spectres, echoes and the like. 1951 L. MACNEICE tr. *Goethe's Faust* II. II. 202, I could essay many a Brocken-stunt But heathendom seems barred on every front. 1962 *Listener* 11 Jan. 68/1 We may conceive the divine either as an abstract entity..or else as a Brocken-spectre projection of humanity.

Brock's benefit. *colloq.* [f. the name of C. T. *Brock*, maker of fireworks + BENEFIT *sb.* 4 a.] From the spectacular display of fireworks held annually at the Crystal Palace from 1865 to 1936: (*a*) fireworks; (*b*) a brilliant illumination at night, esp. in war, from searchlights, flares, artillery, etc.; (*c*) *fig.* any comparable outburst (of excitement, etc.).

1920 P. GIBBS *Realities of War* IV. vii. 188 They ..opened such a Brock's benefit that the enemy must have been shocked with surprise. 1921 'IAN HAY' *Willing Horse* ix. 159 The Germans were furnished with bombs which exploded on impact; ours were of the Brock's Benefit type, and had to be lit with a match. 1929 *Melody Maker* Jan. 19/1 The engines puff out..a regular Brock's Benefit of sparks. 1930 BLUNDEN *Poems* 173 Thence *Brock's Benefit* commanded endless fireworks by two nations. 1942 *New Statesman* 19 Sept. 186/2 The searchlights and anti-aircraft fire and bursting bombs are collectively referred to as.. *Brock's benefit.* 1948 *Economist* 26 June 1060 The third reading of the Representation of the People Bill produced a regular Brock's benefit of insults and expletives. 1957 *Sunday Times* 20 Oct. 13/3 The whole of the southern flank ..opened up and it looked like a real Brock's Benefit.

brodder (brǫ·dəɹ). *Sc.* and *north.* [f. BROD *v.* + -ER[1].] One who brods or uses a brod; *rug-brodder* (see quot. 1921).

1877 *Covenant Times* 200 This torture consisted in the free use of the 'brodder's' needles. 1885 W. ROSS *Aberdour & Inchcolme* xi. 328 'Who,' you say, 'was the Brodder?'.. His office was to settle the question, whether those accused of being witches were so or not..by searching for the 'devil's mark' on their bodies, by 'brodding' or pricking it with a sharp needle. 1921 *Dict. Occup. Terms* (1927) § 398 *Rug brodder* (Yorks. term)..makes rugs, by hand, on a canvas ground, by piercing canvas with an awl and inserting cloth cuttings.

brodekin, brodkin. (Later examples in the form brodequin.)

c 1845 C. BRONTË *Professor* (1857) I. x. 159 Trim Parisian brodequins showed her..feet, to complete advantage. 1886 M. F. SHELDON tr. *Flaubert's Salammbô* vii. 184 He soiled his purple brodequins.

‖**broderie anglaise** (brodri aṅglē̆z). [Fr., = English embroidery.] Open embroidery on linen, cambric, etc. Also ellipt., *broderie.*

1852 *Illustr. Lond. News* 26 June 504/2 The *pardessus*,.. trimmed with *broderie Anglaise.* 1882 CAULFEILD & SAWARD *Dict. Needlework* s.v., True broderie Anglaise patterns are outlines of various sized holes, arranged so as to make floral or geometrical devices. 1905 *Westm. Gaz.* 11 May 4/2, I am perfectly charmed with these embroidered lawn and Broderie Anglaise petticoats. 1907 *Daily Chron.* 25 June 8/2 Coats in Broderie Anglaise...A plain full skirt.. finished at the hem with a band of the broderie. 1955 *Times* 8 July 10/4 The bride..wore a gown of organdie and *broderie anglaise* and a tulle veil. 1963 *Times* 17 Jan. 12/2 Long white cotton dresses with *broderie* frills. *Ibid.*, They had *broderie anglaise* headdresses.

Brodrick (brǫ·drik). The name of the Secretary of State for War (1900–3), the Rt. Hon. W. St. J. *Brodrick*, afterwards Viscount Midleton, applied facetiously to: **a.** a soldier enlisted under the lower standard of physique introduced under his régime; more explicitly *little Brodrick*; **b.** an army cap invented and introduced by him.

1903 *Westm. Gaz.* 19 Mar. 8/1 'Brodricks' in South Africa. 1903 *Daily Chron.* 17 Sept. 7/2 There were comparatively few so-called 'Little Brodricks' in these sturdy line battalions of the 1st Army Corps. 1905 *Westm. Gaz.* 25 May 8/2 The other half in undress uniform, with..its fatigue jackets, its 'Brodrick' caps, and riding whips. 1915 A. D. GILLESPIE *Lett. from Flanders* (1916) 74 The soft 'Gor' bli'me' hat..does not give such a mark as the flat-topped 'Brodrick'. 1960 E. WAUGH in *Spectator* 15 July 95/3 The officials..wore khaki service dress and Brodrick caps.

bröggerite (brö·gərəit). *Min.* [ad. Sw. *bröggerit* (C. W. Blomstrand 1884, in *Geol. För. i Stockholm Förh.* VII. II. 66), f. the name of Waldemar Christopher *Brögger* (1851–1940), Norwegian mineralogist and geologist + -ITE[1].] A thorian variety of uraninite.

1884 *Jrnl. Chem. Soc.* XLVI. 1102 The mineral, which has been named bröggerite by the author [*sc.* Blomstrand], is closely related to cleveite. 1901 *Ibid.* LXXX. II. 396 The specimen of bröggerite..from which radioactive lead was obtained..had a hardness 5·5 and a sp. gr. 9·06 at 15°. 1951 J. R. PARTINGTON *Gen. & Inorg. Chem.* (ed. 2) xxvi. 759 Cleveite and bröggerite are varieties richer in thorium and rare earths.

brogue, *sb.*[2] Add: **1. b.** In full *brogue shoe.* A strong shoe for country and sports wear, having characteristic bands of ornamental perforations.

1906 HASLUCK *Boot & Shoe Pattern Cutting* 57 Brogue Shoes.—The gentleman's brogue is always a strong shoe for shooting, golf, fishing, etc. The wing of the vamp and cap are longer..than for a lady's brogue. 1917 *Mod. Boot & Shoe Maker* III. 234 Highland Brogue Shoe...The general design is similar to a very heavy golfing brogue. 1925 *Blackw. Mag.* Jan. 35/1 An enormous pair of unlaced black brogue shoes.

4. brogue heel, a low heel like that of a brogue shoe; **brogue vamp,** a stout vamp made like that of a brogue shoe.

1927 *Chambers's Jrnl.* XVII. 286/2 She'd only brogue heels on her feet; and her..shoulders were just about on a level with his. 1889 T. BROPHY *Pattern-Cutting made Easy* iv. 86 We have had boots with all sorts of..variations of design...We have had Derby and brogue vamps fitted to elastic patterns.

brogued, *a.* Add: **b.** Of a boot or shoe: made with a strong vamp like that of a brogue. Hence **broguing** (brōᵘ·giŋ) *vbl. sb.*, the manufacture of brogued shoes or brogue vamps; the vamp itself.

1894 *Standard* 11 Apr. 3/1 Strong brogued walking boots. 1894 *Queen* 17 Nov. 870/1 White kid laid under the black broguing. 1906 W. GREENFIELD in P. N. Hasluck *Boot & Shoe Pattern Cutting* i. 22 In all cases where outside fittings are to be brogued.

broguish (brōᵘ·giʃ), *a.* [f. BROGUE *sb.*[3] + -ISH[1].] Inclined or tending to a brogue.

1899 *Echo* 6 July 1/5 A bright and intermittently brogue-ish Irishman. 1921 *Glasgow Herald* 25 Apr. 8 Fine big, broguish fellows, mostly from Ireland. 1928 *Daily Express* 11 Aug. 8/5 He has no ear at all and we all laughed at his broguish cacophonies.

broil, *sb.*[3], **bryle** (broil, brəil). *Min.* (Cornwall.) Loose fragments, often of a metallic nature, found lying on the surface above a vein or lode.

1778 PRYCE *Min. Cornub.* 125 Upon the top of most Tin Lodes..is that mineralized substance, which is called the Broil or Bryle of the Lode. 1818 W. PHILLIPS *Min. & Geol.* (ed. 3) 210 Loose fragments or portions of earthy or stony substances, having generally more or less of an ochreous tinge..called the 'bryle of the load'. 1839 DE LA BECHE *Rep. Geol. Cornwall* xv. 528 The upper part of a lode is usually now termed the broil, or bryle. 1859 FORFAR *Pentowan* v, Sometimes we do discover the lode by a broil.

broiler[1]. Add: **1. b.** A gridiron or similar utensil used in broiling. Now *U.S.*

1393 in *M.E.D.* 1632 SHERWOOD, A broiler, *gril.* 1828– WEBSTER. 1907 'O. HENRY' *Trimmed Lamp* 64 Two minutes longer on the broiler would have made this steak fit to be eaten by a gentleman. 1907 *Yesterday's Shopping* (1969) 182/2 Capt. Warren's Batchelor's Broiler, 10½ in. 2/9. 1967 *Boston Sunday Herald* 26 Mar. (Advt.), Easy dinners with non-drip cook top, smokeless broiler, and built-in aluminium griddle.

2. (Earlier and later examples.) Now normally reared in close confinement in a broiler house. Also *transf.* and *attrib.*

1876 *Rep. Vermont Board Agric.* III. 244 The cockerels should be sold for broilers when large enough. 1952 *Daily Tel.* 4 Dec., A broiler..is a young bird, weighing about 3 lb dead weight, which has been raised on a new system. It is plump and tender enough to grill or to fry in the traditional American manner. *Ibid.*, Broiler production has been flourishing in America for the past six years. **1959** *Ibid.* 5 Mar. 11/3 The continued spread of broiler houses, the buildings for rearing chickens. **1960** *Farmer & Stockbreeder* 16 Feb. 91/3 Five batches of broilers would go through the broiler-house in the life of one steer. **1966** *New Statesman* 25 Feb. 260/1 A state of affairs which seems to be treating these old *people* [in hospitals] as if they were broiler fowls. **1970** *Times* 14 Feb. 9/6 The individual birds live a natural life in wild conditions, very unlike those of a broiler house.

broke, *ppl. a.* Add: **3.** *slang.* In predicative use = BROKEN *ppl. a.* 7; penniless; also *broke to the wide* (see *WIDE *sb.*) or *broke to the world.* Freq. with qualifying word, as *clean, dead, flat broke, stone-broke* (see STONE *sb* 20), *stony broke* (see STONY *a.* 6).

[Cf. the following, which are properly instances of BREAK *v.* 11: **1665** PEPYS *Diary* 6 July (1895) V. 6 It seems some of his creditors have taken notice of it, and he was like to be broke yesterday in his absence. **1669** *Ibid.* 12 Mar. (1896) VIII. 258 Being newly broke by running in debt.] **1716** J. STEUART *Let.* 28 Dec. (1915) 38 Alexr. Mackpherson ..is much in arear and quit broke. **1821** in *N. Carolina Hist. Comm.* I. 220 I have been broke now twelve months,.. yet I move on in the old way. **1842** *Spirit of Times* 2 Apr. 58/1 Barrett, poor fellow, is dead broke. *Ibid.* 21 May 138/1 very friend of Old Whitenose would have been flat broke! **1843** *Ibid.* 14 Jan. 544/3, I was clean broke in less than four hours. **1846** *Ibid.* 25 Apr. 101/2, I unfortunately am short of funds, flat broke, busted, collapsed. **1851** N. KINGSLEY *Diary* (1914) 173 To day men have come along 'dead broke' and have gone to work for 4 dollars pr. day. **1886** H. SMART *Outsider* vii, Well, sir, I was broke—so broke as I hope I never shall be again—'dead stoney', barely expresses it. **a 1889** in Barrère & Leland *Dict. Slang.* s.v., Then came the *fiasco*, And Ben cried 'Carrasco! I'm bested, broke, busted—or partly!' **1889** *Pall Mall Gaz.* 14 Aug. (Farmer), I see that Sullivan made 21,000 dols. out of his fight, but as he was 'dead broke' before the battle, there won't be much of it left. **1898** 'O. THANET' *Heart of Toil* 141 Think of them boys, who are all stone-broke.., wanting to lend *me* money. **1918** W. J. LOCKE *Rough Road* iii, I believe you good people think I've come back broke to the world. **1926** J. BLACK *You can't Win* v. 53 [The landlady] wanted the rent. I told her I was broke.

4. Of animals: broken to harness; = BROKEN *ppl. a.* 8. Chiefly *U.S.*

1800 *Sporting Mag.* XVI. 117/2 The grand manege consists in teaching a horse, already perfectly broke in the common way, certain artificial motions. **1833** M. A. HOLLEY *Texas* v. 97 This brutal process is repeated until the animal is thoroughly broke and rendered docile. **1850** W. MILES *Jrnl.* (1916) 12, 500 broke mules were to be in readiness. **1856** *Trans. Mich. Agric. Soc.* VII. 275 Oxen exhibited as working cattle, for their being the best broke, must be hitched to either a wagon or cart. **1893** T. ROOSEVELT *Wilderness Hunter* xx. 426 The light-hearted belief..that any animal which by main force has been saddled and ridden ..is a 'broke horse'.

5. = *BROKEN *ppl. a.* 1 e. Also *ellipt.*

1888 CROSS & BEVAN *Paper-making* vi. 104 *'Broke' Paper.*—Under this head may be included all the partially formed paper which is always obtained..when a paper-machine is started, or such portions as are occasionally unavoidably damaged in its passage over the drying cylinders. *Ibid.* 105 'Broke' paper may be advantageously disintegrated by means of an edge-runner. **1954** SOUTHWARD *Mod. Printing* (ed. 7) II. xlii. 449 Broke. The third grade of imperfect paper.

broke, *v.* **3.** (Later examples.)

1926 G. K. CHESTERTON *Outl. Sanity* v. ii. 203 If men were not brokers, it was because they were not able to broke. **1965** 'W. HAGGARD' *Hard Sell* v. 54 I'm a stock-broker...*I* broked for Franchin.

broken, *ppl. a.* Add: **1. c. broken letter** *Typogr.*, distributed type.

1683 MOXON *Mech. Exerc., Printing* 371 By broken Letter is..meant..the breaking the orderly Succession the Letters stood in..and mingling the Letters together.

e. In paper-making, seriously damaged, denoting a quality of defective paper inferior to retree.

1807 [see RETREE]. **1880** J. DUNBAR *Pract. Paper-maker* 48 This method..saves broken [paper], and can be worked so near the edge that the impression is taken off at the cutter. **1907** CROSS & BEVAN *Paper-making* (ed. 3) v. 150 The fibres of the broken paper are..separated.

f. *Phonology.* Subjected to breaking (see *BREAKING *vbl. sb.* 1 e).

1845 J. M. KEMBLE in *Proc. Philol. Soc.* II. 135 A tendency in the vowel to become dulled or broken when placed in particular positions. **1887** SKEAT *Princ. Etym.* 45 The symbol *ea* denotes that the vowel was, to speak technically, 'broken', i.e. was resolved into the diphthong *e-a.*

g. *broken (over)*: in bookbinding, applied to the creasing down of a small part of an inserted print near the binding margin, for the purpose of giving support to the binding thread.

1879 ZAEHNSDORF *Art Book-binding* 169 Broken over. When plates are turned over or folded a short distance from the back edge, before they are placed in the volume, so as to facilitate their being turned easily or laid flat, they are said to be broken over. When a leaf has been turned down the paper is broken.

10. b. *broken home*, a home from which either the father or the mother of the children is absent, usu. through legal separation or divorce.

a 1846 B. R. HAYDON *Autobiogr.* (1853) I. v. 80, I left the vault, and returned to our broken home. **1919** J. C. COLCORD (*title*) Broken homes, a study of family desertion and its social treatment. **1952** M. MCCARTHY *Groves of Academe* (1953) iv. 56 The wild-haired progressive-school rejects, offspring of broken homes. **1958** *Listener* 25 Sept. 480/3 There can be no doubt that a broken home can badly affect a child. **1959** B. WOOTTON et al. *Social Sci.* x. 313 The broken home..normally means homes broken by death, desertion, separation or divorce, and often also by long absence on account of illness.

12. c. *Hort.* Of a breeder tulip: that has developed into a striped or variegated flower. (See BREAK *v.* 32 c.)

[**1724** P. MILLER *Gardeners & Florists Dict.* II. TU 32 Some Tulips that have been already broke, or have come to stripe, do one Year abound in the dark Colours, and come finely mark'd the next Year.] **1731** —— *Gard. Dict.* 8 D/1 If one of these Flowers [*sc.* tulips] is quite broken.. it will never lose its Stripes. **1824** LOUDON *Encycl. Gard.* (ed. 2) 832 Save seed from these in preference to the finest of the variegated or broken sorts [of tulips]. **1852** G. W. JOHNSON *Cottage Gardeners' Dict.* 894/2 After some years the petals of these [*sc.* tulips] become striped, and they are then said to be *broken.* **1956** *Dict. Gardening* (R. Hort. Soc.) (ed. 2) IV. 2162/1 Each group of broken Tulips..is subdivided according to the colouring and ground.

f. *broken time*, time lost from regular employment.

1872 *Porcupine* 18 May 102/2, I..was in the category of 'broken time'. **1895** *Westm. Gaz.* 8 Nov. 1/3 The formation of a new Union on the basis of payment for 'broken-time', as it is called by courtesy. **1912** R. W. POULTON in *Life* (1919) 214 A carefully arranged payment for 'broken time' for men who are paid weekly or monthly for the hours they work.

17. broken-up (earlier U.S. examples).

1637 in *Cambridge Reg. Bk. Lands* (1896) 42, 20 ac[res] of broken upp grounde..& 25 ac[res] unbroken upp lying by it. **1684** in *Essex Inst. Hist. Coll.* (1862) IV. 68/2 He should have liberty to make use of part of ye improved & broken up ground upon ye sd ffarme.

b. broken-down, (*c*) having ceased to function (cf. *BREAK *v.* 50 d).

1914 MILLICENT, DUCHESS OF SUTHERLAND *Six Weeks at War* vi. 77 We saw a number of transport trains carrying broken-down auto-wagons on the trucks. **1944** AUDEN *For Time Being* (1945) 23 Where a crown Has the status of a broken-down Sofa. **1958** *Times* 9 Sept. 4/6 On three-lane highways..the problem of broken-down vehicles is less acute.

brokeress. Delete †*Obs. rare*⁻¹ and for def. read: (*a*) A procuress; (*b*) a female stock-broker. Add later examples.

1749 J. CLELAND *Mem. Woman Pleasure* II. 106 Three hundred guineas to myself, and an hundred to the brokeress. **1827** CARLYLE *Germ. Rom.* I. 32 The talking brokeress.. was far from giving him a true disclosure of her blabbing. **1865** SWINBURNE *Chastelard* II. i, Yea, and she said, the Italian brokeress, She said such men were good for great queens' love. **1872** SCHELE DE VERE *Americanisms* 655 A couple of ladies having established their 'Exchange Office' in..Wall-street, they were at once spoken of in the New York papers as bankeresses or brokeresses.

brolga (brǫ·lgă). [Native name.] The Australian native companion crane, *Grus rubicunda.*

1896 *Westm. Gaz.* 6 Oct. 2/1 The native companion crane, otherwise known as the brolga. **1911** BEAN *'Dreadnought' of Darling* xvii. 169 Far up one of the Darling tributaries, the brolga (native companion) and crane and ibis seemed fairly thick. **1944** A. RUSSELL *Bush Ways* xv. 75 It was a flock of native companions, or brolgas—immense birds, nearly five feet tall. **1959** J. CLEARY *Back of Sunset* 214 They..saw two brolgas, tall long-legged birds, going through their weird dance.

brolly (brǫ·li). *colloq.* **1.** Clipped and altered form of UMBRELLA 2.

1874 HOTTEN *Slang Dict.*, *Brolly*, an umbrella. Term used at both Oxford and Cambridge Universities. **1885** *Punch* 6 June 273/2 Pair o' pattens and brolly are more in your line. **1899** KIPLING *Stalky* 209 What are you stealin' the gentleman's brolly for? **1925** *Sunday at Home* July 595/1 The lass was a clever thief who had hooked his brolly.

2. *slang.* A parachute. Also *attrib.* and *Comb.*, as **brolly-hop**, a jump made with a parachute; so *brolly-hopping* vbl. sb.

1934 *Daily Express* 27 June 10/4 A brolly hop is a hop into space—with a parachute! Brolly-hopping will become a commonplace to the younger generation. **1934** G. P. OLLEY *Million Miles in Air* ii. 72 Never having made what pilots refer to..as a 'brolly hop'. **1940** J. M. B. BEARD in Michie & Graebner *Their Finest Hour* xii. 180, I was floating still and peacefully with my 'brolly' canopy billowing above my head.

bromelia (bromī·liă). *Bot.* [Named by Linnæus after Olaus *Bromel*, a Swede.] A plant of the genus *Bromelia*, or of the family *Bromeliaceæ*, consisting of plants indigenous to S. America and W. Indies, the species of which have short stems, and generally lance-shaped leaves with spiny margins.

1822 W. J. HOOKER *Exotic Flora* I. 41 *Bromelia Pallida*, pale-flowered Bromelia. **1833** *Penny Cycl.* I. 447/1 Orchideous plants and bromelias overrun their limbs. **1963** R. P. DALES *Annelids* ix. 179 Such specialized habitats as the reservoirs of bromelias.

b. *attrib.*, as **bromelia water**, water contained in the rosette of leaves of a bromeliad.

1860 MAYNE REID *Odd People* 83 The thorny point of the *bromelia* leaf. **1908** *Smithsonian Misc. Coll.* V. 73 The species was also bred from bromelia water near Tabernilla.

bromeliad (bromī·liæ̆d). *Bot.* [f. mod.L. *Bromeliaceæ*, f. *BROMELIA: see -AD.] Any plant belonging to the family *Bromeliaceæ.*

1866 LINDLEY & MOORE *Treas. Bot.* 170/1 Some of the Bromeliads grow attached to the branches of trees, and are called Air-plants. **1902** *Encycl. Brit.* XXV. 438/1 In some aerophilous Bromeliads the rosette of leaves forms a cup in which water collects. **1927** *Glasgow Herald* 23 July 4 The 'vegetable horse hair' (Tillandsia Usneoides) is a rarely flowering Bromeliad that hangs in grey festoons from the branches of the trees. **1962** *Amat. Gardening* 17 Feb. 7 Until recent years almost all bromeliads have been expensive.

bromelin (brōᵘ·mĭlin). *Chem.* [f. *BROMEL(IA + -IN¹.] A proteolytic enzyme obtained from the juice of the pineapple.

1894 *Jrnl. Chem. Soc.* LXVI. II. 63 Bromelin..appears to be associated with a peculiar proteose-like substance. **1901** *Ibid.* LXXX. I. 355 Bromelin..readily forms leucine, tyrosine, and tryptophan, even in acid media. **1928** *Daily Express* 13 Dec. 5/5 Pineapple..contains a peculiar digestive enzyme known as bromelin.

bromellite (brǫ·mĕləit). *Min.* [ad. G. *bromellit* (G. Aminoff 1925, in *Zeitschr. f. Krystallogr.* LXII. 122), f. the name of Magnus von *Bromell* (1679–1731), Swedish physician and mineralogist + -ITE¹.] A mineral of the zincite group, consisting of beryllium oxide and occurring in white hexagonal prisms.

1926 *Chem. Abstr.* XX. 29 An analysis of an unknown mineral from Långban gave 98% BeO... It is named bromellite. **1965** *Amer. Min.* L. 22 Synthetic bromellite (BeO) is finding wide application not only in high-temperature refractory ware but also as a moderator in gas-cooled nuclear reactors and as a dielectric in electronic applications.

bromide. Add: **2.** A dose of potassium bromide taken as a sedative.

1903 *Smart Set* IX. 14/1 I'll give you a bromide when you're ready for bed.

3. *fig.* A person whose thoughts and conversation are conventional and commonplace. Also, a commonplace saying, trite remark, conventionalism; a soothing statement. *slang* (orig. *U.S.*).

[**1903** *Daily Chron.* 9 May 4/5 Literature is resentful at being mistaken for bromide.] **1906** G. BURGESS (*title*) Are you a Bromide? **1909** W. RALEIGH *Lett.* (1926) II. 340 Bromides are dull partly because everyone pretends to understand them. **1924** HICHENS *After the Verdict* II. xvii, For once Mrs. Baratrie gave way to a bromide. She said: 'How good little Clive was!' **1925** *Contemp. Rev.* Oct. 469 There is the rise of slums which 'ought not to be in a new country', but which, in spite of this oft-quoted bromide, certainly existed in still earlier days. **1926** *Publishers' Weekly* 20 Feb. 563 The old bromide that poetry never sells is once again proved to be wrong. **1950** *Manch. Guardian Weekly* 29 June 2/3 The Republicans would have to fall back on the old bromide about the incurable quarrelsomeness of 'old, sick Europe'. **1961** B. FERGUSSON *Watery Maze* i. 15 These two bromides..were quoted by the faithful..until they were worn as thin as a Queen Victoria bun penny.

4. *Photogr. bromide developer*, a developer suitable for bromide paper; *bromide emulsion*, a gelatine emulsion impregnated with a bromide, esp. silver bromide; *bromide paper*, a paper coated with gelatino-bromide emulsion, used for contact printing and enlargements; also *bromide print, printer, printing* (of or with reference to bromide paper).

1885 *Amateur Photographer* 27 Mar. 409 Britannia Bromide Paper, specially for enlargements. **1892** A. BROTHERS *Photogr.* 78 Opal glass and paper are coated with silver bromide emulsion. *Ibid.*, Bromide-Printing Process. **1902** *Bromide Monthly* Jan. 10 One well-known Bromide printer we know of makes his exposures in contact printing to the light of an ordinary candle from preference. **1904** GOODCHILD & TWENEY *Technol. & Sci. Dict.* 71/1 *Bromide Prints*..are developed and fixed like dry plates. **1923** S. E. SHEPPARD in *Photography* 165 Characteristic Curves for Bromide Papers. **1971** *Ann. Rep. Curators Bodl. Libr.* 1969–70 46 Photography from Library material..consisted of..3,544 bromide prints.

bromidic (bromi·dik), *a.* *slang* (orig. *U.S.*). [f. *BROMIDE 3 + -IC.] Of the nature of a bromide; commonplace or conventional.

1906 G. BURGESS *Are you a Bromide?* 19 The Bromide can't possibly help being bromidic. **1911** H. S. HARRISON *Queed* xix, 'Did genius fail to burn?' he asked, employing a bromidic phrase. **1927** WILLEY *Early Ch. Portraits* 98 Athanasius did not desire the office of Bishop, in spite of the bromidic and venerable injunction of the Apostle.

bromo-. Add: (Earlier and later examples.) Also **bromophe·nol blue,** a dye used as an acid-base indicator; also *bromphenol blue*; **bro·mo-seltzer** (orig. *U.S.*), a proprietary effervescent mixture used as a sedative, etc.; also, a dose of this; **bro:msulphtha·lein** [f. BROM- + SULPH- + *phenol*)*phthalein* (see

PHENOL d)], a dye, $C_{20}H_{10}Br_4O_{10}S_2$, derived from phenolphthalein and used in a test of liver function; also *bromsulphalein.*

1837 R. D. THOMSON *Brit. Annual* 345 Bromobenzoic acid. **1838** T. THOMSON *Chem. Org. Bodies* 314 (*heading*) Of Bromoform. **1862** *Catal. Internat. Exhib., Brit.* II. No. 3150, Hardwich's bromo-iodized collodion. **1885** REMSEN *Org. Chem.* ix. 131 Two brom-propionic acids. **1896** *Illustr. London News* 11 Jan. 63 (Advt.), Emerson's Bromo-seltzer, the most successful American Remedy... Three doses..cure any headache. **1913** *Jrnl. Chem. Soc.* CIV. I. 1023 Two Bromo-substituted Acidyl-carbamides: Bromural and Adaline. **1916** LUBS & CLARK in *Jrnl. Washington Acad. Sci.* VI. 485 *Chemical Name*, Tetrabromphenolsulphone-phthalein. *Short Name*, Bromphenol blue. **1920** A. B. BAXTER *Parts Men Play* x. § 1 Lord Durwent read the *Morning Post* as a sort of 'prairie-oyster', or 'bromoseltzer'. **1924** ROSENTHAL & WHITE in *Jrnl. Pharmac. & Exper. Therap.* Nov. 287 Dyes..indicating hepatic injury... Bromsulphalein appears to be the most suitable for this purpose. **1937** A. J. CRONIN *Citadel* IV. xii. 353 By introducing a small bromoform manometer close to the needle we avoid rarefaction. **1949** *New Gould Med. Dict.* 155/1 Bromophenol blue. **1955** *Sci. Amer.* Aug. 98/2 The albumin can be made more strikingly visible by labeling it with a few crystals of bromphenol blue. **1961** *Brit. Med. Dict.* 222/2 *Bromsulphthalein*,..a non-toxic, non-irritant dye used as an indicator of liver function. **1961** *Lancet* 16 Sept. 631/2 Other flocculation tests..and bromsulphthalein retention were normal.

bromo (brōᵘ·mo). *colloq.* [f. BROMO-.] (A dose of) one of the proprietary sedatives containing a bromide mixture; a bromo-seltzer. Also *bromo-tablet.*

1916 W. OWEN *Let.* 3 Feb. (1967) 378 Would I had taken 'Bromo-Tablets', and never revived at all! **1934** J. T. FARRELL *Young Manhood* (1936) xiii. 320 You had to dope yourself with bromos, bicarbonate of soda, [etc.]. **1957** M. MILLAR *Soft Talkers* v. 46 There's some bromo in the bathroom. **1961** *Encounter* Apr. 20/2 For God's sake a Bromo!

bromoil (brōᵘ·moil). *Photogr.* [f. BROM- + OIL *sb.*¹] In full *bromoil process*, a process in which pigment is applied to a bleached bromide print; so *bromoil transfer*, a picture in reverse taken from a freshly pigmented print.

1909 *Westm. Gaz.* 9 Jan. 14/2 How many..camera-users ever attempt work in gum-bichromate or in bromoil? **1909** *Amateur Photographer* 3 Aug. 108/2 Most workers in the bromoil process not only use, but advocate, smooth, platino-matt. **1952** E. O'DION *All about Photogr.* xxix. 154 The Bromoil Process. Bromoil is a 'high art' process used largely for exhibition work, mainly because of the amount of control it allows.

Brompton (brǫ·mptǫ̆n, brɒ·-). [Name of a former hamlet, now part of London, where the Brompton Park Nursery was founded in 1681.] In full *Brompton stock*, a biennial variety of the stock (see STOCK *sb.*¹ 43).

1724 P. MILLER *Gardeners & Florists Dict.* I. GI 22 There are five Sorts of..Stocks..the strip'd Sort, and the large red Brompton-Stock...The Brompton Kind is esteem'd the best. **1798** C. MARSHALL *Gardening* xix. 337 Stock, Brompton, scarlet, blush, and white. **1852** G. W. JOHNSON *Cottage Gard. Dict.* 583 For the latter purpose [*sc.* spring-flowering] none [of the stocks] beats the intermediate, Queens, Bromptons, and other biennials. **1880** *Encycl. Brit.* XII. 249/2 The Brompton Stock (Matthiola incana simplicicaulis) is a robust plant, growing 3 feet high, with a long central flower stem bearing very large flowers, which are crimson, purple, or white. **1956** *Dict. Gardening* (R. Hort. Soc.) (ed. 2) IV. 2034/1 The Brompton Stocks are robust, vigorous plants.

bronc (brǫŋk). *colloq.* (orig. and chiefly *U.S.*). Also **bronch, bronk.** [Abbrev. of BRONCO.] = BRONCO; a horse. Also *attrib.* and *Comb.*

1893 T. ROOSEVELT *Wilderness Hunter* xx. 418, I saddled up the bronc' and lit out for home. **1908** MULFORD *Orphan* v, Keep it up, bronchs!.. We'll win! **1910** W. M. RAINE *B. O'Connor* 72 You're going to..learn to stick to your saddle when the bronc and you disagree. **1911** H. QUICK *Yellowstone N.* x. 245 Old Jim Bridger was obliged to go up f'r three days on his bronk one time. **1913** MULFORD *Coming of Cassidy* 153 That's shore a long time to ride this bronc train. **1924** *Glasgow Herald* 12 June 9 The champion bronk riders, steer wrestlers, and trick horsewomen of America. *Ibid.* 4 July 9 The international bronk riding contest..took place yesterday afternoon at the Wembley Stadium.

bronchioge·nic, *a.* [f. BRONCHIO- + -GENIC.] = **bronchogenic* adj.

a 1909 in *Buck's Handbk. Med. Sci.* IV. 356 (Cent. Dict. Suppl.). **1927** *Archives Internal Med.* XL. 347 Primary bronchiogenic carcinoma of the right lung. **1947** *Radiology* XLIX. 279/1 There is no clear-cut evidence that tumors of bronchiogenic origin occur in mice of this strain.

bronchiolitis (brǫŋki͵olǝi·tis). *Path.* [f. BRONCHIOLE + -ITIS.] Inflammation of the bronchioles.

1887 VICKERY & KNAPP tr. *Strümpell's Textbk. Med.* IV. 157 Curschmann..claims that the anatomical basis of these cases is an exudative bronchiolitis. **1961** *Lancet* 16 Sept. 660/2 When these areas are infected, bronchiolitis, or broncho-pneumonia may result.

bronchitis. For second sentence in etym. read: First brought into use by C. Badham *Observations on the Inflammatory Affections of the Mucous Membrane of the Bronchiae* (1808).

Add: **b.** *attrib.*, as **bronchitis kettle**, a kettle with a long tube and a detachable medicator used for keeping the atmosphere of a room humid and for giving a medicated vapour inhalation to a patient in a case of bronchitis.

1886 *Cassell's Family Mag.* Jan. 127/2 (*heading*) New Bronchitis Kettle. **1896** *Lancet* 18 Apr. 1056/1 The unfortunate patient is kept in a room with the atmosphere saturated with moisture from a bronchitis kettle.

broncho, var. BRONCO.

broncho-. Add: **broncho-cavernous** *a.* (see quots.); **broncho-constriction**, stricture of the bronchi; **bronchodilator**, that which dilates the bronchi, *esp.* a drug; **bronchogenic** *a.*, of bronchial origin; **broncho-pneumonic** *a.*, pertaining to broncho-pneumonia or inflammation of the bronchi and the lungs; **broncho-pulmonary** *a.*, pertaining to the bronchi and the lungs (cf. *bronchio-pulmonary*, s.v. BRONCHIO-); **bronchospasm**, a spasm of the bronchi; **broncho-vesicular** *a.*, bronchial and vesicular.

1890 BILLINGS *Med. Dict.*, *Broncho-cavernous respiration*, sound heard from consolidated lung surrounding a cavity. **1922** F. W. PRICE *Textbk. Pract. Med.* 905 Broncho-cavernous breathing is incomplete cavernous breathing, inspiration being bronchial, while expiration is cavernous. **1910** *Practitioner* June 859 An amount of air may be sucked through the broncho-constriction. **1903** *Jrnl. Physiol.* XIX. 164 Broncho-dilators: atropine, hyoscyamine, hyoscine, [etc.]. **1961** *Lancet* 12 Aug. 341/1 An increase in airflow resistance..is relieved by bronchodilator substances. **1966** *Ibid.* 5 Feb. 307/2 Aerosols hold their own against oral bronchodilators. **1934** WEBSTER, Bronchogenic. **1962** *Lancet* 8 Dec. 1193/2 A bronchogenic carcinoma which was confirmed at necropsy. **1883** G. S. WOODHEAD *Pract. Pathol.* 249 It [*sc.* capillary bronchitis] only occurs in connection with the broncho-pneumonic process. **a 1909** *Buck's Handbk. Med. Sci.* V. 848 (Cent. D. Suppl.), Broncho-pulmonary. **1912** OSLER & MACRAE *Princ. & Pract. Med.* (ed. 8) 636 Broncho-pulmonary hæmorrhage. **1963** *Brit. Med. Jrnl.* 7 Dec. 1425 (*heading*) Meaning of diagnostic terms in broncho-pulmonary disease. **1901** DORLAND *Med. Dict.* (ed. 2) 121/2 Bronchospasm. **1929** H. L. ALEXANDER *Bronchial Asthma* ii. 29 Bronchospasm .. stimulates the mucous gland. **1966** *Lancet* 31 Dec. 1436/1 There was evidence of bronchospasm or a clinical history of bronchial asthma. **1887** *Buck's Handbk. Med. Sci.* V. 732/2 The broncho-vesicular breathing..becomes..tubular in dyspnœa.

bronchoscope (brǫ·ŋkǒskōᵘp). [f. BRONCHO- + -SCOPE.] An instrument for inspecting the interior of the bronchi.

1899 *Westm. Gaz.* 20 Dec. 10/2 Kilian's (Berlin) bronchoscope was..introduced into the trachea. **1904** *Electr. World & Engin.* 16 Jan. 140 (Cent. D. Suppl.), A bronchoscope, which consists of a tube, the inner surface of which is highly polished to serve as a reflector, with an electric lamp arranged so as to throw a strong light on the tube. **1962** H. C. WESTON *Sight, Light & Work* (ed. 2) vi. 193 The miniature lamp..is used..in cystoscopes and bronchoscopes.

So **bro:nchosco·pic, bro:nchosco·pical** *adjs.*, of or pertaining to a bronchoscope or to bronchoscopy; **bronchoscopy** (brǫŋkǫ·skǒpi) [ad. G. *bronchoskopie* (G. Killian 1898, in *Münch. med. Wochenschr.* 5 Juli 844/2)], (the science or study of) the examination of the bronchi with a bronchoscope.

1908 *N.Y. Med. Jrnl.* 15 Aug. 293/2 Two bronchoscopical tubes of 5 and 7 mm. in diameter for upper bronchoscopy. **1910** F. W. F. ROSS tr. *Bruck's Dis. Nose, etc., & Larynx* IV. 381 The bronchoscopic tubes are perforated at the side, so as not to hinder respiration. **1962** *Lancet* 8 Dec. 1205/1 Prevent..lung infection, treating this (if it arises) by bronchoscopic suction and antibiotics. **1903** *Therapeutic Gaz.* Jan. 60 (Cent. D. Suppl.), Bronchoscopy allows the whole bronchial tree to be searched. **1927** *Daily Express* 6 July 9/2 The chair of bronchoscopy in the University of Pennsylvania.

bronco. Add: Also **broncho.** (Earlier examples.) Also *gen.*, a horse.

1869 S. BOWLES *Our New West* v. 101 A well-broken Indian pony or a 'broncho' (a California half-breed horse). **1878** J. H. BEADLE *Western Wilds* 454 Our bronchos carried us with ease and safety. **1924** R. CAMPBELL *Flaming Terrapin* I. 18 Bellerophon, the primal cowboy..slewed his white-winged broncho out to sea.

b. *attrib.* and *Comb.*, as *bronco-horse, -mule, -pony, -team*; **bronco-buster** *colloq.* (orig. *U.S.*), a breaker-in of broncos; so *bronco-busting* vbl. sb. and ppl. a.

1887 *Century Mag.* in Farmer *Americanisms* (1888) 89/2 An Eastern or English horse-breaker and Western broncho-buster have so little in common with each other. **1888** T. ROOSEVELT in *Century Mag.* Feb. 507 The flash riders, or horse-breakers, always called 'bronco busters'. **1913** R. BROOKE *Let.* in *Coll. Poems* (1918) p. lxxix, A bold, bad, bearded broncho-buster. **1889** *Regina* (Sask.) *Jrnl.* 18 July 1/6 Broncho 'busting' and base-ball are the sports most indulged in [in] these times. **1931** L. STEFFENS *Autobiogr.* I. i. 9 A gun-playing, broncho-busting vaquero. **1963** *Times* 7 June 5/7 Attend a bronco-bustin' rodeo. **1883** Broncho horse [in Dict.]. **1963** W. E. HARNEY *To Ayers Rock & Beyond* v. 48 It was a simple matter to draft the animals they required into an adjacent yard, brand them on the bronco-panel where the motor displaced the bronco-horse. **1895** *Outing* (U.S.) XXVII. 244/2 Their pack train composed of hardy little broncho-mules. **1869** S. BOWLES *Our New West* v. 101 The mule and the Indian and 'broncho' ponies will live on the rich grasses of the country. **1892** GUNTER *Miss Dividends* 163 You can drive down in a day with a good tough broncho-team.

B. Hence as *adj.* Wild, uncontrollable, rough. *U.S. colloq.*

1866 *Weekly New American* 21 July 1/4 The Territory did not keep fast horses and other things, and go to bronco bailes and play whiskey poker. **1887** F. FRANCIS *Saddle & Mocassin* 146 Sam's too broncho: he gets all-fired mean sometimes when he's full. **1947** *Westerners' Brand Book* 75 A man who made a false step wasn't necessarily all bad or 'broncho' as he expressed it.

bronk, var. *BRONC.

Brontëan (brǫ·ntĭăn), *a.* [See -AN.] Of, pertaining to, or characteristic of, the Brontës (esp. the sisters Charlotte (1816–55), Emily (1818–48), and Anne (1820–49)), or their literary works. So **Brontëa·na,** literature concerning the Brontës; **Bro·ntëism,** a quality characteristic of the Brontës (see also -ISM *suffix* 2); **Bronte·sque** *a.* = *BRONTËAN *a.*

1887 *Pall Mall Gaz.* 31 Aug. 3/1 The blusterous Brontëism which..is an affliction hardly less ludicrous than Whitmania. **1900** *Daily News* 17 Aug. 6/1 'Charlotte Brontë and her Circle' seems..to have finally exhausted the vast and well-explored mines of Brontëana. **1905** *Daily Chron.* 4 Nov. 4/7 Almost Emily-Brontesque in its concentrated austerity. **1945** A. HUXLEY *Time must have Stop* vi. 69 'You mean, that I should be acting as Mrs. Gamble's companion?' The final word..took on its fullest, Brontëan significance. **1950** BLUNDEN *Reprinted Papers* 186 Some Brontesque force.. in the work of the young painter-author.

bronto-, comb. form of Gr. βροντή thunder, as in **brontogram** (brǫ·ntogræm), the record made by a brontometer or brontograph; **bro·ntograph,** a recording brontometer; also, a chart of the phenomena recorded by that instrument; **brontolo·gical** *a.*, pertaining to BRONTOLOGY; **bronto·meter,** an instrument for recording the phenomena associated with thunderstorms; **brontopho·bia,** intense dread of thunder and thunderstorms.

1888 *Symons's Monthly Meteorol. Mag.* June 71 Our German friends..were clearly the first as to thunderstorms, or rather Brontological research. *Ibid.* May 50 Messrs. Richard Frères, of Paris, are constructing for Mr. Symons a very complicated apparatus (Brontometer).. for recording the details of thunderstorms. **1905** W. G. HOLMES *Justinian & Theodora* I. iii. 298 He was affected with brontophobia in his later years.

Brontosaurus (brǫntǒsǭ·rǝs). *Palæont.* [mod.L., f. Gr. βροντή thunder + -O + σαῦρος lizard.] An extinct genus of dinosaurian reptiles existing on all the continents during the Jurassic and Cretaceous periods; also (with lower-case initial), a reptile belonging to this genus.

1879 O. C. MARSH in *Amer. Jrnl. Sci. & Arts* XVIII. 503 *Brontosaurus excelsus*, gen. et sp. nov. **1905** *Harper's Mag.* Dec. 73 The mighty brontosaurus came striding into camp. **1907** *Westm. Gaz.* 3 Apr. 8/1 A brontosaurus, or a chondrosteosaurus. **1924** *Public Opinion* 28 Mar. 300/3 [The Anglican Church] is going the way of the dinosaurus and the brontosaurus.

Bronx (brǫŋks). [f. the name of Jonas *Bronck*, a 17th-cent. landowner in New York.] **1.** The name of a borough of New York City, used *attrib.* or *absol.* of a kind of cocktail.

1906 'H. MCHUGH' *Skiddoo!* 39 Every time there was a lull in the conversation Charlie Swayne kept yelling for a Bronx cocktail. **1909** *Sat. Even. Post* 10 Apr. 37/2 The first-night death-watch..could recognize a block away a Bronx, Martini or Manhattan cocktail by its color. **1917** WYNDHAM LEWIS *Let.* 24 July (1963) 90 Then you would go to the American bar and have a Bronx or two. **1930** H. CRADDOCK *Savoy Cocktail Book* I. 37 *Bronx cocktail.* The Juice of ¼ Orange, ¼ French Vermouth, ¼ Italian Vermouth, ¼ Dry Gin.

2. *Bronx cheer*: a sound of contempt or derision made by blowing through closed lips, usually with the tongue between; = *RASPBERRY 5. orig. *U.S.*

1929 *Collier's* 23 Feb. 10/4 Maxim give him a Bronx cheer. **1932** WODEHOUSE *Hot Water* i. 21 She told me..that she was through... No explanations. Just gave me the Bronx Cheer and beat it. **1955** E. HYAMS *Slaughterhouse Informer* 179 That rasping sound variously known as the raspberry or the Bronx cheer.

bronze, *sb.* Add: **7.** *bronze-faced* adj.; **bronze-backer** *U.S.*, angler's name for the black bass; **bronze diabetes,** a disorder of iron metabolism in which hæmosiderin is deposited in the tissues and the skin becomes bronzed; also called *hæmochromatosis*; **bronze disease,** a form of corrosion affecting the surface of bronze; **bronze-founder,** one who founds or casts bronze, or fashions articles of bronze; so

bronze founding; **bronze powder**: substitute for def. = BRONZE *sb.* 4; **bronze-wing** (further examples); also *bronze-wing pigeon*; **bronze-winged** (also *bronzed-winged*) **pigeon** = *bronze-wing.*

1888 GOODE *Amer. Fishes* 56 'Bronze-backer' is one of its pet names among the anglers. **1894** *Outing* (U.S.) XXIV. 452/1 This old bronze-backer [*sc.* small-mouth bass] weighed ..six pounds and five ounces. [**1900** E. KLEEN *On Diabetes Mellitus* iv. 137 Bronze-colored diabetes generally appears in men between forty and sixty years old.] **1901** DUNGLISON *Dict. Med. Sci.* (ed. 22) 1214/2 *Bronze diabetes*, diabetes accompanied with pigmentation of various secretory organs and with sclerosis of the liver and pancreas. **1966** P. E. LACY in W. A. D. Anderson *Pathology* (ed. 5) II. xxix. 958/2 The term 'bronze diabetes' is sometimes used since increased pigmentation of the skin, diabetes mellitus, and cirrhosis of the liver may be present in hemochromatosis. **1925** FINK & ELDRIDGE *Restoration Anc. Bronzes* 42 The usual immediate cause of the bronze disease is the presence of a trace of chloride, and the action is..electrolytic. **1961** *Antiquaries Jrnl.* XLI. 31 The *bucranium*..has a few spots of active copper chloride ('bronze disease'). **1896** *Godey's Mag.* Apr. 404/1 That gentle figure of contentment, bronze-faced and white-apparelled. **1839** URE *Dict. Arts* 333 The bronze-founder should study to obtain a rapid fusion. **1869** *Van Nostrand's Eclectic Engin. Mag.* I. 834/1 Brass and bronze-founding is much more of a speciality in France.. than it has (hitherto at least) ever been in England. **1897** *Daily News* 18 Jan. 6/4 Two works which Benvenuto Cellini wrote ..on bronze founding. **1832** in Bischoff *Van Diemen's Land* ii. 31 The pigeons are by far the most beautiful birds in the island; they are called bronze winged pigeons. **1838** T. L. MITCHELL *Jrnl.* 10 Aug. in *3 Exped. E. Australia* (1838) I. 305 The bronze-wing pigeon was .. the most numerous of that kind of bird. **1852** MUNDY *Antipodes* xviii, I killed..a few bronzed-winged pigeons. **1884** Bronze-wing pigeon [see PIGEON *sb.* 2 b]. **1897** *Daily News* 24 May 9/4 A pair of Smith's partridge bronze-winged pigeons (*Geophaps Smithi*) from Northern Queensland. **1927** M. M. BENNETT *Christison* vii. 72 A bronzewinged pigeon flew past him. **1936** F. CLUNE *Roaming round Darling* xvii. 162 There seemed an unlimited supply of bronze-wing and top-knot pigeons. **1961** *Coast to Coast 1959–60* 60 A bronze-wing pigeon, lurking in the fine sand, whirred away from almost under my feet.

bronzed, *ppl. a.* Add: **4. b.** **bronzed diabetes** = **bronze diabetes.*

[**1886** HANOT & SCHACHMANN in *Arch. de Physiol. Normale et Pathologique* VII. 62 C'était là comme un *diabète bronzé.*] **1898** R. T. WILLIAMSON *Diabetes Mellitus* 411/2 (Index) Bronzed diabetes. **1909** *Practitioner* Feb. 194 That very rare disease called bronzed diabetes. **1966** W. A. D. ANDERSON *Path.* (ed. 5) I. iii. 70/2 Hemochromatosis is a rare disease of iron-pigment metabolism..characterized clinically by cirrhosis of the liver, diabetes mellitus, and skin pigmentation (bronzed diabetes).

bronzer (brǫ·nzǝɹ). [f. BRONZE *v.* + -ER¹.] One who coats with, or colours like, bronze.

1865 *Congress. Globe* Feb. 682/1 The bronzer then [in the U.S. Treasury Department] puts the paper through his bronzing machine, and when it is all bronzed it is counted again by the man bronzing it.

bronzing, *vbl. sb.* Add: **b.** (See quots.)

1868 LEA *Photography* 42 By this time the dark shadows ought to show the greenish, almost metallic look, known as 'bronzing'. **1885** W. K. BURTON *Mod. Photogr.* (ed. 5) Index 127 Bronzing of prints. **1889** E. J. WALL *Dict. Photogr.* 23 *Bronzing*, a peculiar metallic lustre seen on looking at the shadows of some prints at a certain angle. **1903** H. R. PROCTER *Leather Manuf.* 404 'Bronzing', the dichroic effect produced by light reflected from the surface of many colouring matters, complementary to that transmitted by them. **1968** *Gloss. Terms Offset Lithogr. Printing* (*B.S.I.*) 30 *Bronzing*, the process of dusting a metallic powder upon a wet printed base.

(Examples of *bronzing machine.*)

1865 [see *BRONZER]. **a 1877** KNIGHT *Dict. Mech.*, *Bronzing-machine*, a machine for bronzing wall-papers or printed sheets. **a 1884** *Ibid.* Suppl. **1889** *Century Dict.*, *Bronzing machine*, a machine for decorating wall-papers, fabrics, labels, etc. with bronze-powder.

brood, *sb.* Add: **1. f.** Of bees and wasps: the larvæ while in the brood-cells; *foul brood*: see FOUL *a.* 1 b.

1754 P. TEMPLEMAN *Remarks Physics* II. 82 Till all the brood have sallied forth in the form of bees. **1806** J. G. DALYELL tr. *Huber's Nat. Hist. Bees* v. 102 Another piece of comb, containing the brood of workers. **1869** *Good Words for Young* 1 Sept. 515/1 At length the first brood [of wasps] is hatched.

2. d. A state of brooding or mental contemplation, esp. *in a brood.*

1895 HARDY *Jude* VI. vi. 469 You seem all in a brood, old man. **1941** N. MARSH *Surfeit of Lampreys* xvi. 251 We'll all have a brood over the beastly thing. **1959** *Guardian* 16 Oct. 9/1 Bill Mortlock is apt to go into a sudden brood about ethics or the rightness of Proust.

6. *brood-mare* (earlier U.S. example); **brood-box** = *body-box* (*BODY *sb.* 30); **brood-cell,** (*a*) a cell in a honeycomb, made for the reception of a larva, as distinguished from a honey-cell; (*b*) *Bot.*, an asexually produced reproductive cell (*Funk's Stand. Dict.* 1893); **brood-chamber,** (*a*) a chamber for holding the eggs or brood of an animal, etc.; (*b*) a chamber folded off from the uterus and closed around the embryo in some species of *Peripatus* (*Cent. Dict.* Suppl.); **brood-food,** a prepared food for young bees; a substance derived from pollen by digestion, and serving as a pap for a brood of bees; **brood-lamella,** 'in crustaceans, a part of an appendage modified to form a protective cover for the eggs or young' (*Cent. Dict.* Suppl.); **brood nest,** the space inside a hive occupied by the queen and brood; **brood-space** *Anat.*, a cavity in the body of an animal, in which eggs or young are received and remain for a time; **brood spot** (see quot.).

1888 F. R. CHESHIRE *Bees & Bee-Keeping* II. 99 The section-racks..are constructed on the general plan of the brood-chambers..; their edges..abut accurately upon the brood-boxes. **1884** T. W. COWAN *Brit. Bee-Keeper's Guide Book* (ed. 4) ii. 11 A small portion [of pollen] is used by the mature bees..for..capping brood-cells. **1914** GEDDES & THOMSON *Sex* iv. 90 A unique shell..which is used as a brood-chamber for the developing ova. **1792** *Sporting Mag.* I. 153/2 Stallions...Brood mares. **1875** *Encycl. Brit.* III. 494/1 In the early spring, if a clean empty piece of drone comb be put into the centre of the brood nest, the queen will usually fill it with drone eggs. **1878** BELL & LANKESTER tr. *Gegenbaur's Comp. Anat.* 268 An Egg in the brood-space formed between the body and the mantle. **1896** KIRKALDY & POLLARD tr. *Boas' Text Bk. Zool.* 450 Usually the sitting Bird is provided with brood spots, regions from which the feathers have fallen off, so that the eggs may come into direct contact with the warm skin.

brooder. Add: **2.** A hen, etc., that broods or hatches eggs.

1599 T. M[OUFET] *Silkewormes* 26 Now what are seedes and egges of wormes or foule, But recrements of pre-existing things..? Yea, from themselues corruption onely springs, Vnlesse by brooders heate..They changed be to belly, feete, or wings. **1854** *Poultry Chron.* I. 153 An anxious Brooder.

3. An apparatus for the artificial rearing of young chickens and other birds. Also *attrib.* orig. *U.S.*

1880 H. TOMLINSON *Artificial Incubation* vii. 32 [Chickens] may..be transferred to an artificial mother, or *brooder* as the Yankees call it. **1935** AUDEN & ISHERWOOD *Dog beneath Skin* I. 19 The brooder's really excellent. I rear now 98 per cent. **1945** BETTY MACDONALD *Egg & I* (1946) 44 The brooder house was finished and the seven hundred and fifty yeeping chicks and the brooder were installed. **1959** *Times* 9 Mar. 19/4 A new heat storage chick brooder..has been developed.

broody, *a.* Add: **B.** *sb.* A broody hen.

1904 *Daily Chron.* 30 July 4/7 He would..defer putting the 'broody' on them [*sc.* the eggs] till dusk. **1923** *Daily Mail* 20 Jan. 11 Broodies of these highly developed strains are..unreliable sitters. **1967** *Guardian* 17 May 2/5 We've no broodies left in the farmyard.

Brooklynese (bru:klinī·z). [f. *Brooklyn* (see below) + -ESE; cf. LONDONESE *sb.*] The form of speech generally found in uncultured use in New York City, but esp. associated with the borough of Brooklyn.

1948 *Commonweal* 25 June 249/2 Brooklynese and American. **1963** *Economist* 5 Oct. 39/2 His rich Brooklynese.

broom, *sb.* Add: **6. broom-croft,** a croft or field in which broom grows; **broom horse** [HORSE *sb.* 6 e] (see quot. 1921); **broom-straw** *U.S.*, the straw of broom corn; also the plant itself.

1871 KINGSLEY *At Last* x, Grand masses of colour..are supplied by a heather moor, a furze or broom-croft. **1921** K. S. WOODS *Rural Industries round Oxford* II. i. 95 The broom-maker sits on a 'broom horse' which has a grip to hold one end of the band while binding the twigs. **1785** WASHINGTON *Diaries* (1925) II. 365 Tussics of broom Straw. **1846** J. J. HOOPER *Simon Sugg's Adv.* (1851) iii. 38 He wont be able to step over the butt cut of a broom straw.

broomie (brū·mi). *Austral.* and *N.Z.* [f. BROOM *sb.* + -IE.] A broom-hand who keeps the floor clean in a shearing shed.

1933 L. G. D. ACLAND in *Press* (Christchurch) 16 Sept. 15/7 *Broomie*, a boy who keeps the board swept of locks, etc., at shearing. **1966** G. W. TURNER *Eng. Lang. Austral. & N.Z.* vii. 147 The *broomie*, the broom-hand on the board (*on the board* is the shearer's way of saying 'in the woolshed').

Broseley, the name of a town in Shropshire, used in *Broseley porcelain*, a soft-paste porcelain made by Thomas Turner (1749–1809) of Caughley, near Broseley, esp. the blue transfer-printed china of Oriental design, in which the *Broseley dragon* was extensively used.

1836 P. BARLOW in *Encycl. Metrop.* (1845) VIII. 464/2 The blue colours are supplied by the oxide of cobalt... Blue... Weak, (Broseley.)—Blue calc 20, flux 80. (Flint 70, nitre 9, borax 21;) or, calc 15, petuntse 85. **1878** L. JEWITT *Ceramic Art* I. 43 The white..is made of what is commonly called Brosely clay. *Ibid.* 268 The 'Broseley Blue Dragon' pattern.

brosier, brozier (brōᵘ·ziǝɹ, brōᵘ·ʒ¹ǝɹ), *v.* [Etym. unknown.]

1. *pass.* To be bankrupt. *dial.*

1796 T. MORTON *Way to get Married* I. i, I am completely brozier'd, cut down to a sixpence, and have left town.

2. *trans.* In Eton College phraseology: to attempt to exhaust the supply of food at a meal, as an expression of dissatisfaction with the fare provided; esp. in the phrase *to brosier my dame* or *my tutor.*

1850 *N. & Q.* 1st Ser. II. 44/1, I well remember the phrase, 'brozier-my-dame', signifying to 'eat her out of house and home'. **1888** W. ROGERS *Remin.* 15, I joined a conspiracy to 'brozier' him. There were ten or twelve of us [at breakfast], and we devoured everything within reach. **1899** C. K. PAUL *Mem.* III, If a tutor or a dame was suspected of being niggardly, it was determined to 'brosier' him or her

brosier, brozier (brōᵘ·ziǝɹ, brōᵘ·ʒ¹ǝɹ), *sb.* [Cf. prec.] **a.** A bankrupt. *dial.* **b.** A boy who has spent all his pocket-money. *Eton slang.* **c.** The custom of brosiering: see *BROSIER *v.* 2.

1826 WILBRAHAM *Gloss. Cheshire* (ed. 2), *Brosier*, a bankrupt. It is often used by boys at play, when one of them has nothing further to stake. **1850** *N. & Q.* 1st Ser. II. 44/1 A boy at Eton was a 'brosier', when he had spent all his pocket-money. **1907** *Daily Chron.* 17 Dec. 3/4 You have heard of the Eton custom of a 'brozier'? The attempt to eat out of house and home. **1926** *Glasgow Herald* 27 Dec. 6 It was reserved for Eton..to invent the..'brozier', where eating even to beyond repletion is indulged in to gratify a sense of injury.

broth, *sb.* Add: **1. b.** *Bacteriol.* A liquid (as a sterilized infusion of meat) prepared or used as a culture medium.

1885 WOODHEAD & HARE *Pathological Mycol.* IV. 107 To carry on cultivations with the sterile broth thus prepared, it is convenient to have it divided into small quantities. **1897** PEARMAIN & MOOR *Appl. Bacteriol.* II. 57 Glycerin-broth is used for the cultivation of the tubercle bacillus. **1899** G. NEWMAN *Bacteria* I. 21 To provide peptone beef-broth, ten grains of peptone and five grains of common salt are added to every litre of acid beef-broth. **1930** *Syst. Bacteriol.* (Med. Res. Council) I. ix. 354 Old cultures in ordinary broth often yield a considerable variety of colonies when plated on agar or gelatin. **1956** R. HARE *Outl. Bacteriol. & Immunity* II. 37 Infusion broth..consists of the extractives which go into solution when minced meat is allowed to steep in water. **1959** [see **broth culture*].

4. *Comb.*: **broth culture,** (the micro-organisms in) a sample of broth used as a culture medium.

1897 *Jrnl. Path. & Bacteriol.* IV. 196 To test the virulence of our bacilli we injected guinea-pigs subcutaneously with broth-cultures. **1909** *Practitioner* Nov. 596 Over the surface of the agar..pour a two-days old broth culture of bacillus prodigiosus. **1959** F. S. STEWART *Bigger's Handbk. Bacteriol.* (ed. 7) x. 190 Serial dilutions of the drug are made in broth and inoculated with the organism—the inoculum usually being a small amount of a broth culture.

brothel, *sb.* Add: **4. b. brothel-creeper (shoe)** *slang*, (in pl.) suède or soft-soled shoes.

1954 G. SMITH *Flaw in Crystal* ix. 81 'Poncing about the place in those brothel-creepers of his!'.. He always wore plush suede shoes. **1959** 'A. FRASER' *High Tension* ii. 24 He had immaculately creased grey flannel trousers and brothel-creeper shoes. **1959** H. HOBSON *Mission House Murder* iv. 31 Those ghastly suède shoes with two-inch crêpe soles and corrugated edges, known to the Edwardian *élite*..as brothel-creepers. **1969** J. FREDMAN *Fourth Agency* ix. 85 Suède brothel-creeper boots.

brotherhood. Add: **5. c.** A railway trade union. *U.S.*

1883 J. D. FULTON *Sam Hobart* 83 The Brotherhood of Locomotive Engineers..was formed in the city of Detroit in August, 1863, as the Brotherhood of the Footboard. **1959** *Economist* 6 June 947/2 The political influence of the railway brotherhoods, as the unions are called.

Brother Jonathan: see JONATHAN 1.

brotherliwise (brɒ·ðǝɹliwəiz), *adv.* [f. BROTHERLY *a.* + -WISE.] In brotherly fashion.

1890 KIPLING *Life's Handicap* (1891) 117 Who brotherliwise had followed his kinsman's fortune.

brouhaha (brū·hāhā). [a. F. *brouhaha* (15th c. in Littré).] A commotion, a to-do, a 'sensation'; hubbub, uproar.

1890 O. W. HOLMES *Over Teacups* v. 94, I enjoy the *brouhaha*..of all this quarrelsome menagerie of noise-making machines. **1931** C. MORLEY *John Mistletoe* ii. 95 He was immediately captivated by the jargon and brouhaha of the sales department. **1937** WYNDHAM LEWIS in L. Russell *Press Gang!* 276 The peculiar esoteric *brouhaha* of the New York underworld. **1946** 'BRAHMS' & SIMON *Trottie True* vii. 186, I shall never forget the brou-ha-ha.. when Cousin Geraldine married into Trade. **1964** *Times* 24 July 16/2 Whenever there is a City brouhaha of this kind, all sorts of voices are raised for all kinds of official inquiries.

‖**brouillon** (bruyoṅ). [Fr.] A rough draft.

1678 W. TEMPLE *Let.* 19 Aug. in *Wks.* (1731) II. 492 What was signed at first was rather a *brouillon*, than any fair and formal Draught. **1735** POPE *Let.* 17 June (1956) III. 469 Your lordship has still in your custody the Brouillons of verses. **1735** *Pope's Lett.* I. 37 The first Brouillon of those Verses, and the second Copy with Corrections. **1921** *Blackw. Mag.* Feb. 251/2 'Papa Crémieux' was called in to furnish *brouillons* for letters to duchesses and authors.

brow, *sb.*¹ Add: **5. g.** [= the second element in *HIGHBROW, *LOW-BROW, etc.] *colloq.* Level of intellectual attainment or interest. Also *attrib.*

1923 J. AGATE in *Sunday Times* 9 Sept. 4/2 There is

nothing quite so abysmally boring in the theatre as your author who has got no brow at all. **1931** H. G. WELLS *Work, Wealth & Happiness* (1932) 20 That final title..had to present a candid attractive brow to the world, broad rather than high. **1954** F. CORNFORD *Epitaph for Reviewer* in *Coll. Poems* 112, I hope to meet my Maker brow to brow And find my own the higher. **1959** *Guardian* 19 Dec. 4/4 The only differences worth noting among playwrights are their brow-levels.

10. brow-band (examples); **brow-ridge,** a superciliary ridge; **brow-spot,** the interocular gland of a frog or toad.

1844 H. STEPHENS *Bk. Farm* I. 620 Brass or plated buckles and brow-bands..serve only to load and cover the horses when at work. **1958** J. HISLOP *From Start to Finish* 167 *Browband*: Usually called forehead-piece in racing stables; the piece of the bridle which goes round the horse's forehead, below the ears. **1898** *Guide Mammalia Brit. Mus.* 15 Enormous brow-ridges give them a ferocious and savage appearance. **1927** PEAKE & FLEURE *Hunters & Artists* 60 The great frontal torus or overgrown brow-ridges of Neanderthal Man. **1964** G. B. SCHALLER *Year of Gorilla* (1965) vi. 135 The gorillas..looked thoroughly miserable with the water dripping off their brow ridges. **1875** HUXLEY & MARTIN *Elem. Biol.* xiii. 187 The browspot or inter-ocular gland.

Browallia (browǭ·liă). *Bot.* [Named by Linnæus after J. *Browall* (1707–55), Swedish botanist.] A plant of the genus of South American annual plants so named, bearing violet, blue, or white flowers.

1782 ABERCROMBIE *Ev. Man his own Gardener* 84 Tender Annual Flowers. The choicest kinds are..browallia, &c. **1866** LINDLEY & MOORE *Treas. Bot.* 172/2 *Browallia*, the name of certain plants belonging to the order of linariads. **1960** *Farmer & Stockbreeder* 29 Mar. (Suppl.) 6/2 The little blue-flowered browallia with its glossy leaves.

browed, *a.* (Later examples.)

1888 'L. MALET' *Counsel of Perfection* x. 233 The tall, browed, black figure in the doorway. **1899** A. WERNER *Capt. Locusts* 98 Heavily browed eyes.

browless, *a.* Add: **2.** Without eyebrows.

1823 *New Monthly Mag.* VIII. 531 A composition which ..utterly destroyed the hair, and left him literally browless! **1905** E. PHILLPOTTS *Secret Woman* II. ix. 179 Browless eyes. **1906** B'NESS VON HUTTEN *What became of Pam* I. iv, The large..eyes,..so almost browless and lashless.

brown, *a.* Add: **6. a.** *brown-black, -gold, -rosy*; **b.** *brown-skinned* (later example).

1849 D. CAMPBELL *Inorg. Chem.* 246 The brown-black bisulphide. **1942** S. SPENDER *Life & Poet* 28 The field is green in summer and brown-black in winter. **1909** *Daily Chron.* 7 May 4/4 The sun..glittered on her brown-gold hair. **1926** D. H. LAWRENCE *Sun* v. 19 Ripe now, and brown-rosy all over with the sun. **1904** E. F. BENSON *Challoners* viii. 162 Helen's soft brown-skinned hand.

7. brown algæ, the algæ of seaweeds constituting the class *Phæophyceæ*, which contain a brown colouring matter in addition to chlorophyll; **brown-back** *U.S.*, (*a*) the red-breasted snipe or dowitcher in summer plumage; (*b*) the great marbled godwit, *Limosa fedoa* (Cent. Dict., 1889); **brown-banded snake** (see quots.); **Brown Betty** chiefly *U.S.*, a baked pudding containing apples and breadcrumbs; **brown earth,** a type of soil developed in temperate humid regions under deciduous forests, having a brown surface-layer rich in humus; **brown-footed rat** (see quot.); **brown heart,** (*a*) a disease of stored apples and pears characterized by internal brown decay; (*b*) a disease of turnips and related plants in which a deficiency of boron in the soil causes internal brown decay; **brown house** *U.S.* = **brown-stone* house; **brown job** *slang*, a soldier; also *collect.*, the Army; **brown moth,** an artificial fly used in angling by night; **brown-nose** ['from the implication that servility is tantamount to having one's nose in the anus of the person from whom advancement is sought' (Webster 1961)], chiefly *U.S. slang*, a sycophant; hence as *v. trans.* and *intr.*, to curry favour (with), to flatter; so *brown-noser*; **brown rice,** unpolished rice, with only the husk of the grain removed; **brown rot,** any of various diseases of plants and trees characterized by browning and decay; **brown sherry,** sweet sherry; **Brown-shirt,** a Nazi; so called from the brown shirt worn as part of the uniform; so *brown-shirted* ppl. a.; **brown snake,** one of several poisonous Australian snakes of the genus *Pseudonaja* (formerly included in *Demansia*); **brown-stone,** (*b*) *U.S.*, one or other variety of a dark-brown sandstone used for building; also *ellipt.*, a house built of brown-stone, and *attrib.* and *transf.*, designating the well-to-do; **brown-tail(ed) (moth)** (see quots.); **brown top,** any of various pasture grasses in Australia and New Zealand (see quots.); **brown tubes** *pl.* (see quot.).

1902 *Encycl. Brit.* XXV. 272/1 A white efflorescence which appears on certain Brown Algæ..when they are dried in the air. **1844** J. E. DEKAY *Zool. N.Y.* II. 255 The Dowitchee..or Brown-back. **1872** COUES *N. Amer. Birds* 252 *Macrorhamphus. Red-breasted Snipe.* Gray Snipe. Brown-back. **1893** *Outing* (U.S.) XXII. 96/2 'Brown-back' snipe. **1869** KREFFT *Snakes Australia* 55 The large-scaled snake..is confounded, in Victoria in particular, with the Tiger or Brown-banded Snake (*Hoplocephalus curtus*). **1887** *Encycl. Brit.* XXII. 197/1 The Brown-banded Snake (*Hoplocephalus curtus*), with a similar distribution [throughout Australia], and also common in Tasmania, from 5 to 6 feet long. **1864** *Yale Lit. Mag.* XXIX. 187 (Th.), [In training,] tea, coffee, pies, and 'brown Betty' must next be sacrificed. **1911** S. E. WHITE *Bobby Orde* (1916) x. 126 It was the season of..apple-tapioca and Brown Betty. **1948** 'J. TEY' *Franchise Aff.* xv. 164 Brown-betty with thick cream. **1932** *Forestry* VI. 27 The soil type of the dales is the 'brown earth', characterized in the vertical section (profile) by the dark brown to black colour of the top layers of the mineral soil. **1953** *Proc. Prehist. Soc.* XIX. 131 Deforestation.. would in time change such a forest soil by decalcification into an ordinary brownearth. **1884** *Encycl. Brit.* XVII. 5/2 *Mus fuscipes*, Waterh., the Brown-footed Rat of western and southern Australia. **1923** F. KIDD & C. WEST in *Dept. Sci. & Ind. Res., Food Invest. Board, Spec. Rep.* XII. (*title*), Brown-heart—a functional disease of apples and pears. **1928** F. T. BROOKS *Plant Dis.* ii. 16 *Brown Heart* in apples has been shown..to be caused by asphyxiation of the fleshy part of the fruit when there is a large increase of carbon dioxide and a decrease of oxygen in the surrounding atmosphere. **1950** *N.Z. Jrnl. Agric.* May 476/3 Although the size and outward appearance of turnips or swedes affected with brown heart are quite good, when opened up they may have internal greyish or brownish markings in the flesh. **1845** S. JUDD *Margaret* II. viii. 324 The Deacon's..was a small, one story brown house. **1861** H. B. STOWE *Pearl Orr's Isl.* I. ii. 10 Down near the end of Orr's Island..stands a brown house. **1943** HUNT & PRINGLE *Service Slang* 18 *Brown job*, the Army (R.A.F. name). **1952** SHERBROOKE-WALKER *Khaki & Blue* i. 4 Another aspect of R.A.F. dress which struck the 'brown job', as they called their brothers in the Army. **1954** 'N. BLAKE' *Whisper in Gloom* II. xvi. 230 Can you hear me, bluebottles and brown job? **1963** *Economist* 29 June 1391/3 General Delacombe was a pretty undiplomatic brown-job. **1837** KIRKBRIDE *Northern Angler* 48 The Brown Moth..is made with a feather from the wing of the dark-brown owl. **1939** *Amer. Speech* XIV. 25/1 *Brown nose, v.*, to curry favor, especially for rank; *n.*, a cadet who curries favor. **1950** HEMINGWAY *Across River* xxxiii. 208 The brown-nosers..and all the jerks. *Ibid.* vii. 57, I have erected the defence against brown-nosing my superiors and brown-nosing the world. **1960** J. SYMONS *Progress of Crime* xv. 91 If you don't.. get cracking on a few little jobs for this paper instead of spending your time brown-nosing Mr. Fairfield, you [etc.]. **1969** M. PUGH *Last Place Left* v. 34 It was part of the tradition to hate a Highland laird or be a brown-nose. **1916** *Daily Colonist* (Victoria, B.C.) 20 July 11/5 The Vancouver consignment included sake, brown rice, white rice, shoyu. **1969** *Listener* 17 July 79/3 Macrobiotics is a life science discovered in Japan by Georges Ohsawa and it's based on the two pillars of philosophy and diet—the staple food being brown rice. **1894** C. M. WEED *Fungi & Fungicides* 56 The brown rot of stone fruit is probably familiar to every grower of plums, peaches or cherries in..the United States. **1899** G. NEWMAN *Bacteria* I. 37 Bacterial brown-rot of potatoes and tomatoes is another plant disease probably due to a bacillus. **1936** J. PERCIVAL *Agric. Bot.* (ed. 8) 816 One of the best examples of bacterial plant diseases is that known as the 'Brown Rot'..of the cabbage... The disease is caused by a bacterium named *Pseudomonas campestris.* **1968** *Gloss. Terms Timber Preservation* (*B.S.I.*) 9 *Brown rot*, a type of decay caused by fungi that attack chiefly cellulose, leaving a brown friable residue of unattacked lignin. **1849** THACKERAY *Pendennis* I. xxxi. 306 A pint bottle of pale ale, and some brown sherry. **1854** E. TWISLETON *Let.* 3 Apr. (1928) x. 171 A Golden Sherry, then true Brown Sherry... The Golden was £70 the butt, the Brown.. £100. **1965** *Guardian* 24 Oct. 5 Sherry..these are the main types..fino, amontillado, oloroso, cream brown. **1932** *Observer* 26 June, The concession made to Herr Hitler's brown shirts. **1939** R. C. K. ENSOR *Who Hitler Is* 7 The heavy jack-booted hero afterwards idolized by the Brown-shirts. **1934** *Times* 16 Jan., A brown-shirted Fascist organization which manifests strong anti-Jewish proclivities. **1896** F. G. AFLALO *Nat. Hist. Australia* 172 The brown snake is found all over Australia and New Guinea. **1931** R. L. DITMARS *Snakes of World* xiv. 192 The Brown Snake, *D[emansia] textilis*, broadly distributed and usually four to five feet long, is rated as highly dangerous. **1963** E. WORRELL *Reptiles of Australia* 139 Brown Snakes are fast-moving and easily aroused. **1966** 'J. HACKSTON' *Father clears Out* 163 It was the longest tiger or brown snake ever killed in the district. [**1836** *Knickerbocker* VIII. 390 His poor remains.. in one corner..—a brown stone at his head and foot.] **1858** *Spirit of Times* 13 Feb. 377/2 A solid substantial brown-stone front house. **1865** 'G. HAMILTON' *New Atmosphere* 32 The brown-stone friends are shocked and scandalized. **1909** 'O. HENRY' *Options* 22 Two old-fashioned, brownstone front residences. **1948** *Time* 8 Mar. 25/1 Nightclubs in sorry brownstones. **1957** *New Yorker* 21 Sept. 132/3 The part of Brooklyn we were riding through was.. lined with old and ugly brownstones. **1782** W. CURTIS *Brown-tail Moth* 10 The Brown-tail Moth..is about two-thirds of the size of the Moth produced from the Silk-worm. **1815** KIRBY & SPENCE *Entomol.* I. vi. 202 The brown-tailed moth (*Bombyx phæorhæa*, F.). **1832** J. RENNIE *Butterfl. & Moths* 42 The Brown Tail (*Porthesia Auriflua*, Stephens) appears in August. **1891** R. WALLACE *Rural Econ. Austral. & N.Z.* xxii. 294 *Erianthus fulvus*, Kunth.— Sugar grass or brown top. **1902** *Encycl. Brit.* XXXII. 108/2 Some [Queensland] stockholders consider that the 'Sugar grass' or 'Brown Top' (*Pollinia fulva*) surpasses them [*sc.* the 'Mitchell grasses'] in its quickness of bursting into leaf. **1928** R. G. STAPLEDON *Tour Australia & N.Z.* viii. 53 New Zealand Chewing's fescue and New Zealand Brown top (*Agrostis vulgaris*) are at present in great demand in America. **1950** *N.Z. Jrnl. Agric.* Jan. 19/3 The dominant pasture species on this store sheep country are browntop and danthonia. **1902** *Encycl. Brit.* XXXII. 637/1 The excretory organs [of the Sipunculoidea]..serve as channels by which the reproductive cells leave the body, and they are sometimes spoken of as 'brown tubes'.

brown, *sb.* Add: **3. a.** Also, (a drink of) brown ale (in quot. 1862, brown sherry).

1820 'PETER CORCORAN' *Fancy* 109 *Brown*, porter.— *Heavy Brown*, stout. **1862** DICKENS *Somebody's Luggage* in *All Yr. Round* (Christmas) 5/1 Bottle old East India Brown £0 8 0. *Ibid.* 47/1 He ordered a bottle of Old Brown. **1934** T. S. ELIOT *Rock* I. 12 Only if I 'ad a pint o' brown, I'd show you 'ow your money goes. **1959** A. SINCLAIR *Breaking of Bumbo* iii. 53 Let's go and have a brown in the NAAFI.

c. In later use, a mulatto.

1862 *Independent* 10 Apr., The jealousy between the blacks and browns, which has done so much mischief in the West Indies, is not fostered by American people of color. **1890** H. M. FIELD *Bright Skies* 147 In Jamaica..they are distinguished as the whites and blacks and *the browns.* **1968** *Listener* 31 Oct. 57 53 A white girl found herself the only white in a class of blacks and browns.

4. *the brown*: lit. the brown-coloured mass of a flock of game-birds; in phr. *to fire into the brown (of them)*: to fire into the midst of a covey instead of singling out a bird; also *transf.*, to fire, or launch a missile, indiscriminately into a mass.

1845 J. MAYER *Sportsman's Directory* (ed. 7) 21 Always aim at one particular bird, not firing at random at the whole covey, or 'into the brown of them'. **1871** *Punch* 16 Dec. 256/2 Sportsmen, whose sport must mainly consist in 'firing into the brown'. **1885** *Ibid.* 31 Jan. 53 'Pop! Bang! Whose bird?' That's the French notion of a tag, as the husband looses off 'into the brown' of his wife's adorers. **1888** KIPLING *Phantom 'Rickshaw* (1889) 87 We three Martinis firing into the brown of the enemy. **1899** *Daily News* 4 May 8/2 The sportsman..not firing into the brown, but taking careful aim at some particular bird. **1955** C. S. FORESTER *Good Shepherd* 23 Three square miles of sea, an immense target for any torpedo fired 'into the brown'.

brown, *v.* Add: **3.** [from *BROWN *sb.* 4.] To fire indiscriminately at (a covey of birds or a mass of men).

1873 *Sat. Rev.* Sept. 310 We seldom did ourselves anything like justice until the coveys were out on the feed towards evening, when we were apt to brown them as they rose in little clouds. **1885** RIDER HAGGARD *K. Solomon's Mines* xiii, Good..began to fire,..industriously 'browning' the dense mass before him with a Winchester repeater. **1921** *Blackw. Mag.* Jan. 122/2 He waited till the troop was some hundred and fifty yards away, and then he 'browned' it. **1962** B. SCOTT JAMES *Asking for Trouble* xix. 145 We all shot a little wildly..one guest, a German, browned the beaters.

brown bread. 2. brown-bread ice, ice cream made with brown-bread crumbs, popular in Victorian and Edwardian times.

1846 'A LADY' *Jewish Manual* vii. 148 The ice should be prepared in a round mould—brown-bread ice is particularly well adapted to a Juditha. **1875** L. TROUBRIDGE *Life among Troubridges* (1966) 117, I had a brown bread ice, which was one tiny drop of consolation. **1965** T. FITZGIBBON *Art Brit. Cooking* 201 *Brown Bread Ice Cream.* This was a most popular Edwardian ice-cream... The brown breadcrumbs give the ice a nutty flavour.

browned, *ppl. a.* Add: **2.** *browned off*, bored, 'fed-up', disgusted. Hence *browned-offness* [-NESS], the condition of being 'browned off'. *slang.*

Partridge *Dict. Slang* (1961) states 'Regular Army since ca. 1915; adopted by the R.A.F. ca. 1929'; but pre-1938 printed evidence is lacking.

1938 J. CURTIS *They drive by Night* ii. 28 What the hell had he got to be so browned off about? He ought to be feeling proper chirpy. **1941** *MenOnly* May 8 We have worked hard for success, and now that it has come we are robbed of it by this paper shortage. Believe us, we are properly browned off. **1942** W. SIMPSON *One of our Pilots* ii. 19 Most of our time was spent moping about the aerodrome getting thoroughly 'browned-off', and hoping against hope for some real action. **1943** W. ROBSON *Let.* 15 Aug. (1960) 53 Splendid things for combating the deadly browned-offness. **1958** *Observer* 9 Nov. 4/4 Medical boards were always being begged by browned-off invalids to pass them fit for active service. **1966** *Ibid.* 27 Feb. 28/7 This [programme] has got over an initial stage of browned-offness, mutiny on the knees, dumb insolence.

Brownie¹. Add: **2.** [Named from the colour of their uniform.] A member of the junior section of the organization known as the 'Girl Guides'. Also *attrib.*

1916 *Home Chat* 30 Sept. 558 For the younger girls who are not eleven, and therefore not old enough to become Girl Guides, the Brownie movement has been started. **1918** BADEN-POWELL *Girl Guiding* 23 In the Brownie Pack every Brownie obeys the wishes of the leader. *Ibid.* 24 When a girl has passed her test as a recruit she is admitted into her Six as a Brownie. **1959** I. & P. OPIE *Lore & Lang. Schoolch.* viii. 123 Scout's, Cub's, Guide's and Brownie's honour, are the only pledges deliberately sown by adults to have taken root.

brownie², **browny** (brau·ni). [Subst. use of BROWNY *a.*] **1.** A sweet bread made with brown sugar and currants. *Austral.* and *N.Z.*

1883 J. EDGE-PARTINGTON *Random Rot* vii. 312 Each with a huge hunch of 'browny' (bread sweetened with brown sugar and currants) in one hand. **1900** H. LAWSON *On Track* 129 Pile of 'brownie' on the bare black boards at the end of the table. 1908 A. Gunn *We of Never-Never* xxiii. 331 Half the Vealer, another huge pudding, several yard of sweet currant 'brownie'. 1959 H. P. Tritton *Time means Tucker* v. 39/1 He called us over to have a mug of tea snd a hunk of brownie.

2. A small square of rich, usu. chocolate, cake containing nuts. *U.S.*

1897 *Sears, Roebuck Catal.* 17/3 *Fancy Crackers, Biscuits, Etc.*..Brownies, in 1 lb. papers. 1954 J. Steinbeck *Sweet Thursday* xii. 80 Do you like brownies? 1968 L. J. Braun *Cat who turned on & off* (1969) x. 96 On her tray were chocolate brownies..frosted chocolate squares topped with walnut halves.

3. An angler's name for the trout.

1914 *Chambers's Jrnl.* July 432/2, I got him into the net at last, a thick, deep-bodied brownie of pounds weight. 1928 *Observer* 22 July 28/2 It is a difficult matter to creel a brace or so of brownies. The trout are there, but they stir not.

Brownie³ (brau·ni). The proprietary name of a simple type of camera.

1902 *Westm. Gaz.* 24 Mar. 3/2 Five shillings..is the price of the Brownie Kodak. 1906 A. Bennett *Whom God hath Joined* viii. 285 Gater..had decided to sell his high-geared bicycle in order that he might become the possessor of a Brownie camera. 1952 P. Bonner *SPQR* (1953) xxvi. 234 Clicking our Brownies and scribbling postcards to the folks back home.

Browning (brau·niŋ). The name of John M. *Browning* (1855–1926) of Ogden, Utah, U.S.A., applied *attrib.* to various weapons ranging from pistols to machine-guns, esp. automatic ones, designed by him. Also *ellipt.*

1905 *Daily Chron.* 9 Feb. 5/2 Hohental fired all the seven chambers of a Browning revolver at Herr Johnsson. 1906 *Ibid.* 6 Jan. 8/1 The party of freedom have to depend on revolvers, especially the 'Brownings'. 1906 *Westm. Gaz.* 6 Apr. 6/3 They were armed with Mauser pistols and Brownings. 1964 H. L. Peterson *Encycl. Firearms* 68/1 Successful Browning automatic-weapon designs include the ..Browning Automatic rifle M1918, the recoil operated Browning machine guns Model 1917 and Model 1919, [etc.].

Browningesque (brauniŋe·sk), *a.* [f. the surname *Browning* + -esque.] Of, pertaining to, or characteristic of Robert Browning or his style.

1880 G. M. Hopkins *Let.* 23 Mar. (1935) 101 Your brother appears to admire Browning..the *Elder Brother* is quite Browningesque. 1901 *Daily Chron.* 29 June 3/1 Monologues ..which trace their lineage to *Lippo Lippi* and the rest of that most Browningesque series. 1953 H. Read *True Voice of Feeling* i. vii. 128 A good deal of Browningesque swagger.

So **Browningese** (brauniŋī·z), *a.* = prec.; *sb.* the language or style of Browning; **Brow·ningite**, an admirer of Browning; also *adj.* = *Browningesque *a.*

1880 G. M. Hopkins *Let.* 26 Oct. (1935) 111 That piece is worse, indeed *it is* Browningese. *c* 1882 R. Browning *Lett.* (1933) 212, I am quite other than a Browningite. 1889 *Spectator* 14 Dec., [He] translated them all into Browningese forms. *Ibid.*, You never lose the Browningite manner of deliverance. 1910 H. Walker *Lit. Vict. Era* 320 Thorold, Luria, Djabal, Valence, all speak Browningese.

bro·wn-out. Chiefly *Austral.* and *N. Amer.* [f. Brown *v.*, after *black-out *sb.*] A partial black-out. Also *transf.* and *fig.*

1942 in *Amer. Speech* (1945) XX. 143/1 Brown-out..used in Australia to denote semidarkening a city as distinguished from the complete darkening of a blackout. 1943 *Ibid.* (1944) XIX. 149/1 The suggested conservation measures for electricity involve a national brownout, the extinguishing of all ornamental and display lighting and signs after 10 p.m. 1950 *N.Y. Herald* 18 Feb., New York Brownout is Ordered as Coal Dwindles. 1955 *Times* 4 May 10/4 The new 'brown-out' on information in Washington. 1969 *Daily Colonist* (Victoria, B.C.) 6 Dec. 23/1 Premier Bennett..was asked why the government found it necessary to route the transmission [line] through the park. He said:..'It is to prevent brown outs.'

browny, *a.* Add: **2.** *Comb.*: with the name of another colour, as *browny-green, -grey*.

1905 *Westm. Gaz.* 5 Aug. 10/2 It is a very becoming tone, that bronze browny green. 1906 *Daily Chron.* 10 Dec. 4/4 The browny-grey soldiers of Russia. 1907 *Ibid.* 12 Apr. 9/3 The browny-grey back of her sharply-pointed wings. 1945 G. Millar *Maquis* vii. 147 He wore a very old browny-green tweed coat and trousers.

browse, *v.* Add: **1. b.** (Later examples.)

1927 J. S. Huxley *Religion without Revelation* iv. 127 Browsing in the public library at Colorado Springs,..I came across some essays of Lord Morley. 1965 *Crescendo* Sept. 1/1 While browsing through a local bookshop recently I came across your excellent publication. 1968 *Listener* 22 Aug. 242/3 *Hobson-Jobson* is not of course a book to read right through but to browse in—something like, but on a vaster scale, Fowler's *Modern English Usage*.

3. (Later U.S. example.)

1876 *Rep. Vermont Board Agric.* III. 74 It was customary, in years past, when farmers were short of hay, to browse their cattle, as it was called.

browser. Add: **2. b.** *fig.*; *spec.* a person who browses among books.

1863 Geo. Eliot *Romola* I. vii. 126 Friends who were ready to praise his writings:..amiable browsers in the Medicean park along with himself. 1937 *Atlantic Monthly* Apr. 502/1 But the perfect browser must have one possession ..a *ladder*; a library ladder. 1955 *N.Y. Times* 5 May 32/3 Conrad Library, where often the casual browser has stayed to become a serious reader. 1965 *Listener* 2 Dec. 926/2 Bookshop browsers.

browsing, *vbl. sb.* Add: **1. b.** *fig.*

1899 S. A. Barnett *Let.* in H. O. Barnett *Canon S. A. B.* (1918) II. 127 Many men owe as much to browsing as to the running...They laid hold of books to their tastes. 1902 W. Raleigh *Let.* Aug. (1926) I. 241, I will not say that I have read it..all through, but I found browsings to interest me. 1907 E. Godfrey *Eng. Children* xix. 309 Those early browsings in the fields of romance by no means unfitted him to become an explorer in the paths of science. 1966 *Listener* 20 Jan. 108/2 The book makes incomparable browsing.

browsy (brau·zi), *a.* Also **brousy.** [f. Browse *v.* + -y¹.] †**1.** Of vegetation: scanty; twiggy. *Obs.*

1738 W. Ellis *Timber-tree* i. 6 The knotty, mossy Bodies and brousy Heads of Oaks, Ashes, and Beeches. *Ibid.* 42 These [beech sets] being of the Tree Kind..will get a brousy, bushy Head, like the shrubby Sort.

2. Characterized by or suitable for browsing or casual reading.

1910 A. Bennett *Jrnl.* 20 Oct. (1932) I. 388 Miscellaneous browsy reading in the evening. 1966 *Guardian* 6 May 9/3 A very quiet, wise, browsy sort of book about country life.

brozier, var. *brosier.

brrr (br-r-r), *int.* Also **b'rrh, brrrr.** [Imitative.] An interjection expressive of shivering with cold or apprehension.

1898 R. Hughes *Lakerim Athletic Club* vi. 109 The whirling snow hid the banks of the lake...'Brrr!' said Tug. 1905 H. G. Wells *Kipps* ii. vii. 286 B-r-r-r. It didn't do to think of his Aunt and Uncle. 1938 O. Nash *I'm Stranger here Myself* 38 They regretfully hoist themselves up and shiver and say Brrrr! 1961 'N. Blake' *Worm of Death* ii. 22 B'rrh, it's cold outside. I've swallowed about a hundred cubic feet of fog walking here.

brucellosis (brūselōᵘ·sis). *Path.* [mod.L., f. *Brucell(a*, f. the name of Sir David *Bruce*, a Scottish physician (1855–1931) + -osis.] A disease caused by bacteria of the genus *Brucella*; as applied to the disease in men, also called Malta fever, undulant fever, etc., and, in cattle, contagious abortion, etc.

1930 *Jrnl. Amer. Med. Assoc.* XCIV. 1905 (*title*) Brucellosis...The significance of brucella agglutinins in the blood of vegetarians. 1933 *Chem. Abstr.* XXVII. 5114 The problem of eradication of cattle brucellosis. 1936 Topley & Wilson *Princ. Bacteriol.* (ed. 2) lxxii. 1333 The term brucellosis has been suggested on analogy with tuberculosis, to apply to all types of Brucella infections...We prefer ourselves to use instead the term 'Brucella infection'. 1955 *Sci. News Let.* 9 July 25/3 Brucellosis is a bacterial disease contracted in a mild but persistent form by the drinking of raw milk. 1969 *Times* 27 Jan. 10/8 Brucellosis accredited cattle only will be shown at the Surrey county show.

Brückner (brü·knəɪ). *Meteorol.* The name of Professor E. *Brückner* (1862–1927), German meteorologist, used *attrib.* to designate a climatic cycle, lasting about 35 years, of alternating cold, wet periods and warm, dry periods.

1905 H. W. Clough in *Astrophysical Jrnl.* XXII. 56 W. J. S. Lockyer pointed out that a cycle of about 35 years exists in the variations of the interval from one sun-spot minimum to the succeeding maximum...He considers this as the source of the Brückner cycle. 1910 *Encycl. Brit.* VI. 525/2 A somewhat longer period of slight fluctuations or oscillations of climate, known as the Brückner cycle. 1931 A. A. Miller *Climatology* xv. 270 The best known of these is the 35-year or Brückner Cycle, found by Brückner to recur..in a variety of phenomena.

brugmansia (brugmæ·nziă). *Bot.* [mod.L., f. the name of S. J. *Brugmans* (1763–1819), Dutch botanist: see -ia¹.] **1.** A plant of the solanaceous genus so named, native to S. America (formerly included under the genus *Datura*), the species of which have white, orange, or red tubular blossoms.

1884 Lady Brassey *The Trades* 352 The garden contains some..pretty 'lily-trees', as they call them here; although I should describe them as a sort of datura or brugmansia. 1961 *Amat. Gardening* 14 Oct. Suppl. 17/1 *Datura*..half-hardy plants..(also known as Brugmansias and Angels' Trumpets)..striking evergreen subjects for the cool greenhouse.

2. A plant of a genus of parasitic plants of the family Cytinaceæ, found in the Malay islands, each plant of which consists of little more than a flower.

1885 *Encycl. Brit.* XVIII. 265/2 *Rafflesia* and *Brugmansia* consists one may say of a single flower.

bruise, *v.* Add: **8.** To become bruised.

1912 W. Deeping *Sincerity* xxii. 175 The huge mouth seemed to bruise like an over-ripe love-apple.

bruiser. Add: **1.** Also, an operative who pulverizes materials.

1921 *Dict. Occup. Terms* (1927) § 056 *Sample bruiser*; pulverises average sample of ore with flat-headed hammer, ready for sampler.

2. Also *gen.*, a brawny, muscular man. (Later examples.)

1907 *Daily Chron.* 11 Oct. 3/3 A 'hero' of a sufficiently 'bruiser' type to please the most athletic-minded youth. 1934 J. T. Farrell *Young Manhood* (1936) iv. 205 Two of the bruisers were drawing close to him. He started to run.

4. Also, any machine, tool, or other instrument used for bruising or crushing.

In quot. 1828 applied to a tooth.

1828 Fleming *Brit. Anim.* 9 In the lower jaw [of the badger], the bruiser is small, the chewer large. 1881 Ogilvie (Annandale), *Bruiser*.., the name of various machines for bruising grain, &c., for feeding cattle.

brumby, brumbie (brɒ·mbi). *Austral.* Also **brumbee.** [Origin unknown.] A wild or unbroken horse.

1880 *Australasian* 4 Dec. 712/3 (Morris), These our guide pronounced to be 'brumbies', the bush name here [Queensland] for wild horses. 1887 Kipling *Plain Tales fr. Hills* (1890) 153 People who lost money on him [*sc.* a race-horse] called him a 'brumby'. 1899 Semon *In Austral. Bush* 95 When one of the brumbies begins to move all the others follow. 1916 *Anzac Book* 143 A lanky, sawny bushman who ..saddled his brumby, and rode for the nearest town. 1950 'N. Shute' *Town like Alice* 90, 'I got thrown once,' he said, 'breaking in a brumby to the saddle.'

b. *attrib.*

1895 *Chambers's Jrnl.* 702/1 The Brumbie Horse of Australia..is the descendant of runaways of imported stock. 1897 *Pall Mall Mag.* Feb. 190 And so to the unknown interior, past the points where sheep-tracks become brumby-trails. 1925 *Chambers's Jrnl.* 810/1 Wild 'brumbie' mobs, which consisted of mares and one horse, the harem. 1926 *Brit. Weekly* 19 Aug. 412/5 An unbroken, raging devil of a brumby colt. 1967 *Sunday Mail Mag.* (Brisbane) 16 Apr. 2 Brumby hunters if they're successful, finish up with a batch of..wild horses.

Brummellian (brɒme·liăn), *a.* [f. the proper name *Brummel* + -ian.] Of, pertaining to, or characteristic of 'Beau Brummel' (G. R. Brummell, 1778–1840), leader of fashion in London. So **Bru·mmelism.**

1831 Carlyle *Misc. Ess.* (1872) IV. 26 Werterism, Byronism, even Brummelism, each has its day. 1832 *Ibid.* 127 Pharisaical Brummellean Politeness. 1908 'Lewis Melville' *Beaux of Regency* II. 70 The last three carried on the Brummellian traditions.

brummer fly. [Obs. var. Afrikaans, Du. *brommer* blowfly, bluebottle; cf. Du. *brommen* to hum, buzz, *bromvlieg* bluebottle.] A species of blowfly generally known as the locust-fly (genus *Wohlfahrtia*).

1913 Pettman *Africanderisms* 92 Brummer fly, *Cynomyia pictifacies*...An insect somewhat like the common house-fly, but considerably larger. It is useful in the destruction of locusts. 1924 *Chambers's Jrnl.* XIV. 314/2 The brummer fly (*Wolfahrtia brunis palpis*) lays its eggs in the neck of the locust.

Brummy (brɒ·mi). *colloq.* Also **Brummie.** [dim. of Brummagem.] A local name for a native or inhabitant of Birmingham. Also *attrib.*

1941 G. Kersh *They die with their Boots Clean* ii. 65 You're a Brummy Boy. I can tell by your accent. 1956 *Archit. Rev.* CXX. 124/3 Now it tries hard to avoid being swamped by Birmingham—permanently, in the way of urban sprawl; temporarily, by the ebb and flow of the country-bound Brummies at week-ends. 1965 *New Statesman* 30 Apr. 670/2 He proclaims proudly, in a modulated Birmingham accent that makes him sound like a well-bred Australian: 'I'm a natural born Brummie.'

brunch (brɒntʃ). orig. *University slang.* [A 'portmanteau' word f. Br(eakfast and L)unch.] A single meal taken late in the morning and intended to combine breakfast with lunch.

1896 *Punch* 1 Aug. 58/2 To be fashionable nowadays we must 'brunch'. Truly an excellent portmanteau word, introduced, by the way, last year, by Mr. Guy Beringer, in the now defunct *Hunter's Weekly*, and indicating a combined breakfast and lunch. 1900 G. Swift *Somerley* 93 Brunches (*i.e.* breakfast-cum-lunch). 1900 *Westm. Gaz.* 19 Dec. 2/3 Perish Scrambling breakfast, formal lunch! Hardened night-birds fondly cherish All the subtle charms of 'brunch'. 1924 *Blackw. Mag.* Apr. 465/1 We proposed to have a substantial 'brunch' at eleven. 1941 in Mencken *Amer. Lang.* (1963) vi. iv. 316 Sunday strollers' brunch, $1 per person, served from 11 a.m. to 3 p.m. 1967 *Boston Sunday Herald Mag.* 26 Mar. 8/2 Easter is the day when more fashion marches forth..to brunches and to egg hunts.

b. *attrib.*, as *brunch-bar*; **brunch coat,** a woman's short house-coat.

1942 *N.Y. Times* 24 May 31(Advt.), A brunch coat that's indispensable in the Summertime. 1959 *Vogue Pattern Bk.* June–July 50 Good companion to the nightgown: short brunchcoat or dressing gown to make in a plain poplin. 1960 *New Statesman* 16 Jan. 75/3 A 'split-level' house on Long Island becomes a 'Georgian split, with a bi-level brunch bar in a maître d'kitchen'. 1966 *Observer* 13 Feb. (Suppl.) 28/1 Brunchbars (Have a bean-burger?) have made their appearance.

brune (brūn). [a. F. *brune*, fem. of *brun* brown.] A dark-complexioned girl or woman, a brunette.

1865 'Ouida' *Strathmore* I. ii. 41 Now with a blonde, and now with a brune. 1870 R. Broughton *Red as Rose is She* xiii, Priscilla, a beady-eyed, brisk brune. 1894 Meredith *Ld. Ormont* x, Brunes are exceptional in England.

brunet (brune·t), *a.* and *sb.* [a. F. *brunet*, dim. of *brun* brown.] **A.** *adj.* Dark-complexioned. **B.** *sb.* A dark-complexioned person. Hence **brune·tness.**

1887 Dole tr. *Tolstoi's Russ. Proprietor* 247 But I should like to see what sort of a man this hussar is,—whether he is brunet or blondin. 1890 T. H. Huxley in *19th Cent.* Nov. 757 The brunet broad-heads now met with in central France. *Ibid.* 767 The present contrast of blonds and brunets existed among them. 1899 Ripley *Races Eur.* 147 Our map of the distribution of brunetness. 1932 *Times Lit. Suppl.* 1 Dec. 915/1 For what reason the union of brunet-white with blackish should bring about something blacker still, is here wisely left to the biologist. 1967 T. Wells *Dead by Light of Moon* (1968) v. 52 The boy looks like the mother, the father is brunet.

brunette, *sb.* Add to def.: Also, a brown-haired girl or woman.

1908 G. Wodnil *Brunette or Blonde* 4 What shade her hair?..Brunettes they are witty, I have heard say. 1965 T. Wolfe *Kandy-Kolored Tangerine-Flake Streamline Baby* (1966) i. 9 A pregnant brunette walks in off the street wearing black shorts.

b. A variety of the satinette pigeon.

1879 L. Wright *Pract. Pigeon Keeper* 174 Brunettes are lighter Satinettes. 1891 R. Woods *Pigeon-Culture* 138.

brung (brɒŋ), dial. pa. t. and pa. pple. of Bring *v.*

Bruno (brū·no). In **St. Bruno's lily,** a bulbous plant of the species *Anthericum* (*Paradisea*) *liliastrum*, cultivated for its white sweet-scented flowers, which resemble small lilies.

1760 J. Lee *Introd. Bot.* App. 325 Saint Bruno's Lily, *Hemerocallis.* 1882 *Garden* 3 June 391/3 St. Bruno's Lily is beautiful in a glass by itself.

brunsvigite (brɒ·nzvigəit). *Min.* [ad. G. *brunsvigit* (J. Fromme 1902, in *Tschermak's min. und petrogr. Mittheil.* XXI. 171), f. Da. *Brunsvig* Brunswick + -ite¹.] A type of oxidized chlorite occurring in gabbro in the Radauthal, Germany; *spec.* one with 5·6–6·0 per cent of silicon and six to ten per cent of iron.

1902 *Jrnl. Chem. Soc.* LXXXII. ii. 512 A chlorite mineral of wide distribution in the gabbro of the Radauthal, Harz, is described under the new name *brunsvigite*. 1954 *Min. Mag.* XXX. 280 The manganese-bearing chlorites include the remarkable species pennantite..; grängesite, a manganiferous brunsvigite; and manganese-pennine.

brush, *sb.*¹ Add: **2.** (U.S. examples.)

1791 in *Amer. Speech* XV. 161/2 To a white Oak & red Oak near a hollow in the Edge of Brush. 1801 *Massachusetts Spy* 23 Dec. 3/4 The imprudence of a person who set on fire a quantity of brush, &c. near Cambridge. 1887 I. R. *Ranche Life Montana* 8 The bright red of the brush by the river-side.

4. *brush house, stable, tent, whisky U.S.*; **brush-fire** orig. *U.S.*, a fire in brush; also *transf.*; *attrib.*, (of a war) arising suddenly and limited in scale or area; **brush wallaby** *Austral.*, several species of the genus *Wallabia*, esp. *W. rufogrisea*, found esp. in coastal brushes.

1850 L. H. Garrard *Wah-to-yah* xix. 238 The spiral smoke..rose from the brushfire. 1947 *Chicago Daily News* 15 May 1/3 The family outcast is stirring up a brush fire of liberal resentment against the Truman administration. 1955 *Times* 14 May 7/5 He opposed any reduction in man-power because of the risk of 'brush fire' wars. 1854 B. Young in *Jrnl. Discourses* I. 166 Families went there and lived in wagons and brush houses. 1835 *Southern Lit. Messenger* I. 581 The pony..moves homeward with accelerated velocity, leaping every obstacle in his way to his brush stable. 1862 *Harper's Mag.* June 16/1 In the yard.. were several chapadens or brush-tents in which whisky, gin,..and other refreshments..were for sale. 1878 J. H. Beadle *Western Wilds* xix. 310 John A. Lee..had his wife living there in a sort of brush tent. [1841 G. R. Waterhouse *Marsupialia* 221 The Walabee of New South Wales somewhat resembles the Brush Kangaroo of Van Diemen's Land.] 1896 F. G. Aflalo *Nat. Hist. Australia* ii. 40 Into the specific descriptions of the rock, swamp, brush, scrub and other wallabies I shall not enter. 1926 Le Souef & Burrell *Wild Animals of Australasia* 189 The red-necked wallaby commonly known as the scrub and in places as the brush wallaby, is found in the drier forest country of Eastern Australia. 1966 V. Serventy *Continent in Danger* iii. 66 The fleetness of foot of these brush wallabies led them to be hunted for sport. 1885 'C. E. Craddock' *Prophet Gt. Smoky Mts.* xv. 275 The constable's heart was warmed by the brush whiskey. 1913 M. W. Morley *Carolina Mts.* 66 That important beverage, variously known as..'blockade', 'brush whiskey', and..'corn whiskey'.

brush, *sb.*² Add: **1. b.** One of a pair of thin sticks set with long wire bristles with which to make a soft hissing sound on drums, cymbals, etc.; in full *wire brush*. So *brush-work*.

1927 *Melody Maker* Aug. 807/3 In this article I have discussed brush-work in conjunction with the side drum stick. 1955 in M. Stearns *Story of Jazz* (1957) xxiii. 288 Snare drum, sticks, brushes, [etc.]. 1961 A. Berkman *Singers' Gloss.* 11 *Brushes*, wire brushes used to play the drums.

8. b. Short for *brush-off* (see *brush *v.*² 5 b). So *brusheroo* [*-eroo].

1941 in *Amer. Speech* (1942) XVII. 12/1 That's why I'm getting the brusheroo. 1947 B. Schulberg *Harder they Fall* i. 27 The ones who had already made up their minds almost always got the brush. 1953 'S. Ransome' *Drag the Dark* (1954) ii. 25 So far I had found no chance to give Goodlee the brush. 1962 E. Lacy *Freeloaders* viii. 175, I told Daniele what a crawling punk her boyfriend was, and she gave him the brush.

11. a. *brush-stroke.*

1898 *Westm. Gaz.* 17 Nov. 3/1 [Gainsborough's] brush-strokes are scarcely due to separate acts of volition. 1963 *Times* 17 Jan. 4/4 The canvas becomes a web of shimmering, delicate brushstrokes.

12. **brush borer** = **brush driller*; **brush-colour** (see quot.); **brush discharge** *Electr.* = Brush *sb.*² 6 a; **brush drawer,** an operative who puts in the bristles in 'drawn brushes'; **brush driller,** an operative who drills the holes in the stocks of brushes and brooms; **brush-grain,** a grain produced in painting woodwork by drawing the brush over a wet coat of paint so that the under-coat is seen through the brush-marks; **brush-hat** (see quot.); **brush-holder** (see quot. 1904); **brush-tail(ed) porcupine** (see quot. 1885); **brush-varnish** (see quot.); **brushware,** goods consisting of all kinds of brushes; **brush-work,** (*a*) painting, as distinguished from drawing; *spec.* the characteristic method (of a painter) of laying on the colours; (*b*) (see sense *1 b).

1921 *Dict. Occup. Terms* (1927) § 688 Brush borer. 1845 G. Dodd *Brit. Manuf.* IV. 130 Floor-cloth manufacture... A second coating of paint is laid on..wholly with a brush... Hence it is called the 'brush-colour', to distinguish it from the first or 'trowel-colour'. 1849 Noad *Electricity* (ed. 3) 42 The difference between the brush discharge and the spark is, that in the former discharge begins at the root [etc.]. 1923 *Popular Wireless* 13 Oct. 11 *Brush discharge*, a discharge of high-tension electricity, which takes the form of a luminous glow. 1900 *Daily News* 6 Nov. 9/1 Brush drawer. 1921 *Dict. Occup. Terms* (1927) § 688 Brush driller. 1901 *N. & Q.* 9th Ser. VIII. 310/1 In the painting of wood-work, when the second coat, say of vermilion, is made to show through the third, say of brown, by passing the brush over it while the last coat is still wet, the result is spoken of as 'brush-grain'. 1968 *Gloss. Terms Offset Lithogr. Printing* (*B.S.I.*) 21 *Brush grain*, a fine grain produced by the action of abrasive brushes. *a* 1877 Knight *Dict. Mech.*, *Brush-hat*, one in which the surface is continually brushed by a hand-brush, during the process of sizing. 1894 W. P. Maycock *Electric Lighting* (ed. 2) i. vi. § 114. 179 Construction of Brush-holders. 1904 Goodchild & Tweney *Technol. & Sci. Dict.* 72/1 *Brush Holder*, the support or frame carrying the copper (or carbon) strips by which the current enters or leaves a motor or dynamo. 1885 *Encycl. Brit.* XIX. 518/2 The second genus of Old-World porcupines is *Atherura*, the Brush-tailed Porcupines,..with long tails tipped with bundles of peculiar flattened spines. 1953 G. Durrell *Overloaded Ark* iv. 84 A Brush-tailed Porcupine..about the size of a cat...He was mostly black in colour. 1962 *Times* 20 Dec. 9/7 The Borneo Brushtail Porcupine. 1875 T. Seaton *Fret Cutting* 31 Should you wish to varnish the work that has been fret cut or carved, you must do it with brush-varnish, made with spirits of wine.. laid on with a camel-hair brush. 1923 *Glasgow Herald* 9 July 9 Brushware and pottery. 1960 *Times* 13 Jan. 17/4 Manufacture of household and toilet brushware is generally highly mechanized. 1868 Brushwork [see Dict., sense 11]. 1886 *Encycl. Brit.* XX. 218/2 Works..wanting in the trenchant brush-work..of subsequent productions. 1893 *Daily News* 8 Apr. 3/6 The brush work of this incomparable painter. 1935 *Discovery* Sept. 261/1 Flaws in the brushwork of the eyelids.

brush, *sb.*³ Add: **1. b.** *at* or *after the first brush*: at or after the first encounter or meeting.

1815 Scott *Guy M.* lii, So you intend to give up this poor young fellow at the first brush? 1857 Hughes *Tom Brown* ii. ii, The people were..civil to you if you were civil to them, after the first brush.

c. A rapid run or race; a contest in speed. *dial.* and *U.S.*

1841 *Spirit of Times* 16 Oct. 390/3 The third mile was a 'brush' throughout. 1860 Trollope *Framley P.* xiv, Mark..would enjoy a brush across the country quite as well as he himself. *a* 1867 H. Woodruff *Trotting Horse of Amer.* (1868) ix. 105 He may have a couple of brushes of a quarter of a mile each. 1906 *N.Y. Even. Post* 16 June, Apart from the annual regatta, there are endless minor 'brushes' for the 'fresh-water sailormen'.

brush (brɒʃ), *sb.*⁵ *Austral.* and *N.Z. slang.* [Of uncertain origin.] A girl, a young woman; freq. *derog.* Also *collect.*

1941 in Baker *Dict. Austral. Slang.* 1945 F. Sargeson *When Wind Blows* vii. 55, I don't go looking for trouble with brushes that are under age. 1947 D. M. Davin *Gorse blooms Pale* 200 What comes along but an Iti bint, a real grouse brush she was. 1953 K. Tennant *Joyful Condemned* iii. 26 To him all girls were collectively 'the brush'. 1960 N. Hilliard *Maori Girl* iii. ix. 239 It's the good-looking brush that give a man all the trouble.

brush, *v.*¹ Add: **2. b.** To force on (*fig.*); to drive hard. *U.S.*

1755 *Connecticut Gaz.* 29 Nov. (Schele de Vere), As tending to beget ill will, and brushing a disunion in the several governments in America. 1827 J. F. Cooper *Prairie* x I have at this moment a dog brushing a deer. *a* 1867 H. Woodruff *Trotting Horse of Amer.* (1868) v. 70 Eight or ten days prior to the race..brush him a half mile. 1904 *N.Y. Times* 28 Nov. 5 The drivers..spent a couple of hours before dusk brushing their fast steppers on the upper stretch.

brush, *v.*² Add: **2.** Also *absol.* (*rare*) and *intr.*

1811 *London Pract. Midwifery* (ed. 3) vi. 140 The practitioner should always be cheerful...Whenever he perceives his patient looking at him, he should brush up, and appear as cheerful as he can. 1818 Keats *Let.* 27 Apr. (1958) I. 275 Don't you think I am brushing up in the letter way? 1835 Dickens *Lett.* (1965) I. 66, I felt rather tired this morning when I got up; but as I did not do so until *past eleven*, I soon brushed up again. 1840 *Knickerbocker* XVI. 162, I thought I must brush up for the occasion. 1903 *Dial* 1 Sept. (Advt.), If you wish to brush up on your English, you will find nothing better. 1904 *Hartford Courant* 5 Oct. 8 The ex-governor must brush up a bit on his ecclesiastical studies.

Hence **brush-up,** the action or process of 'brushing up'.

1897 E. Terry *Let.* 3 July in *Ellen Terry & Bernard Shaw* (1931) 224 She looked quite nice when she'd had a nice 'wash and a brush up'. 1912 [see Wash *sb.* 1 b]. 1951 A. Christie *They came to Baghdad* xix. 192, I left her to have a wash and brush up. 1951 in M. McLuhan *Mech. Bride* 127 If your knowledge is hazy, rusty, in need of a brush-up.

5. b. *to brush off*: *fig.*, to rebuff, dismiss (a person, etc.). So *brush-off*, a rebuff, dismissal. orig. *U.S.*

1941 J. R. Parker *Attorneys at Law* i. 10 I'd have given my eye teeth to hear Forbes getting the brush-off. 1941 B. Schulberg *What makes Sammy Run?* vi. 132 Since Sammy was waiting for Julian the chances are he'd only brush me off. 1943 Lewis Browne *See What I Mean?* i. 8 No matter where I turned, I was given the brush-off. 1944 *Amer. Speech* XIX. 310/2 The organizer of a *Brush-off-club* 'made up of mournful soldiers who were given the hemlock cup by femmes back home'. 1947 J. Steinbeck *Wayward Bus* 71 Casual kindness in a man she had found to be the preliminary to a brush-off. 1958 M. Dickens *Man Overboard* v. 68 The bleakly familiar: 'The post has been filled', or the more courteous brush-off: 'We will keep your letter on record in case a suitable post arises'. 1969 *Listener* 31 July 131/2 The problem of the future of British sovereignty can no longer be brushed off with humorous references to accepting foreign referees' decisions in international football matches.

7. To trim (a hedge or tree, the sides of a ditch or path). *local.*

1513 [implied in Brushing *vbl. sb.* 1]. 1809 *Warehorne Highway-Bk.* 29 June (E.D.D.), For brushing the footpath, 1s. od. 1845 *Jrnl. R. Agric. Soc.* VI. ii. 479 They [*sc.* hedges] are carefully brushed, or clipped, twice a year. 1886 R. Holland *Gloss. Chester* 48 *Brush*, to trim a hedge.

8. To beat (a covert). Also *intr.* in *vbl. sb.*

1876 *Coursing Cal.* 223 Our long dragging beats taking us..round the far side by Fliskoe Forest, in proximity to which the ranges were brushed, but with no good results. 1895 W. Rye *Gloss. E. Anglia* 26 *Brush*, to beat a covert; 'a day's brushing with the governor'.

9. *Hunting.* To take the 'brush' from (a killed fox) as a trophy of the chase.

1879 *Tinsley's Mag.* XXIV. 334 As they then rode in the master 'brushed' him [*sc.* a fox], while the hounds were baying. 1893 *Field* 11 Feb. 191/3 Some twenty minutes later he had the satisfaction of brushing his fox.

10. Painting. *to brush (in)*: to put in with the brush, to paint in.

1897 *Daily News* 16 Jan. 6/3 For flesh painting, the torso ..is so firm, so luminous; the draperies, too, are decisively brushed in. 1901 *Ibid.* 7 Mar. 6/6 These are vivid, quickly brushed impressions by an artist who has an eye for.. Italian landscape.

brushed, *ppl. a.* Add: **2.** *spec.* Designating knitted or woven fabrics that have been finished with a nap.

1926 G. G. Denny *Fabrics* (ed. 2) i. 26 *Brushed wool*, knit fabrics for sweaters, scarfs, trimmings which have been napped. Usually contain mohair fibres which make long, silky nap. 1957 M. B. Picken *Fashion Dict.* 37/1 *Brushed rayon*, napped rayon fabric, used for sleeping garments, sportswear, shirts. 1966 J. Aiken *Trouble with Product X* iv. 69, I took off my brushed-nylon waist slip. 1967 *Observer* 14 May 29/6 French brushed cotton in black with pink roses.

brushed (brɒʃt), *a.* [f. Brush *sb.*¹ + -ed².] Overgrown with brushwood.

1666 in *Duxbury Rec.* (Mass.) (1893) 16 A certain parcel of land,..until you come to a low brushed swamp. 1832 *Jrnl. R. Geogr. Soc.* II. 110 The line of sight extended over a great extent of densely wooded, or brushed, land. 1888 *Century Mag.* Jan. 453/1 A cañon, liberally wooded or 'brushed' with wild plums. 1938 J. Steinbeck *Long Valley* 63 There was a draw below him..brushed with dry struggling sage and chaparral.

brusheroo: see *brush *sb.*² 8 b.

brushing, *vbl. sb.* Delete †*Obs.* and add later examples.

1886 R. Holland *Gloss. Chester* 48 *Brushings*, the trimmings of hedges. 1887 Darlington *Folk Speech S. Cheshire* 131 *Brushin' hook*, the hook used in *brushing* a hedge. 1969 *Gloss. Landscape Work* (*B.S.I.*) v. 19 *Brushing*.., clearance of light, woody growth ('brush') from rough ground by slashing.

4. *Coal-mining.* The action of, or the work involved in, cutting or blasting down the roof, or building supporting and blocking walls in a coal mine. Also *attrib.* and *Comb.*, as

brushing contractor; **brushing-bed** (see quot. 1883).

1883 Gresley *Gloss. Coal-m.*, *Brushers* (Scotland), men who brush the roof, build packs and stoppings, which work is called brushing. *Brushing-bed* (Scotland), the stratum brushed or ripped. 1923 *Glasgow Herald* 16 Oct. 9 The 'brushing' contractor.

5. The action or process of applying the enamel in the manufacture of enamel-ware. Also *attrib.*

1893 *Daily News* 15 Dec. 5/4 The brushing department of enamelled plate works.

6. Brush-work, as distinguished from drawing.

1896 *Daily News* 6 Apr. 6/5 No incompleteness of drawing, of brushing, or of line.

brushless, *a.* Add: **2.** Of shaving-cream: made for use without a brush.

1933 *Drug & Cosmetic Industry* July 40/2 Introducing a new shaving cream of both the brushless and lather types. 1941 Jenkins & Hartung *Chem. Org. Med. Products* 7 Sun-tan oils, brushless shaving creams, and hair tonics. 1960 H. L. Lawrence *Children of Light* i. 9 He..took his Gillette and brushless shaving cream.

brush-off, brush-up: see *brush *v.*² 5 b, 2.

Brussels. Add: **1.** (Earlier and later examples.)

1813 D. Wordsworth *Let.* in M. Moorman *W. Wordsworth* (1965) II. vii. 230 We are going to have a *Turkey*!!! carpet in the dining-room, and a Brussels in William's study. 1882 W. D. Howells *Woman's Reason* viii, The reception-room ..was respectable in threadbare brussels.

2. Brussels carpet (earlier example).

1799 *Times* 1 June 4/2 Brussells and Wilton carpets.

‖ **brut** (brüt), *a.* [Fr.] Of wines: unsweetened.

1891 *Longman's Mag.* Aug. 417 An especial brand of brut champagne. *Ibid.*, Brut wines. 1896 *Pall Mall Mag.* Mar. 399 Tell my man to bring me a quail, broiled, and a pint of Piper Heidsieck, *brut*. 1932 Wodehouse *Louder & Funnier* 12 Washing it down with a *brut* champagne of a vintage year.

brutalism. Add: **2.** A style of art or architecture characterized by deliberate crudity of design (see quot. 1953).

1953 A. & P. Smithson in *Archit. Design* Dec. 342/2 House in Soho..bare concrete, brickwork and wood.. would have been the first exponent of the 'new brutalism' in England, as the preamble to the specification shows: 'It is our intention in this building to have the structure exposed entirely, without internal finishes.' 1954 *Archit. Rev.* CXV. 274/2 The attitude taken by certain younger English architects and artists, and known, half satirically, as the New Brutalism. 1957 *New Yorker* 2 Nov. 95/1 This sculpture is the fulfillment of a lifetime dedicated to Constructivism.. unshaken by the more popular fashions of..Brutalism, and Disintegrationism. 1962 *Observer* 1 Apr. 13/8 A style which Lady Cargill describes as 'a mild version of the "new brutalism"'.

brutalist (brū·tălist). [f. Brutal(ism + -ist.] One who exhibits brutalism; *spec.* an exponent of brutalism in art, architecture, or literature. Also *attrib.* or as *adj.*

1934 in Webster. 1937 *John o' London's* 12 Mar. 984/3 It is a good example of the modern Brutalist literature that rejoices in tearing every veil aside. 1954 *Archit. Rev.* CXV. 343/3, I understand that Paolozzi, in common with some young architects, now calls himself a 'brutalist'. 1958 *Ibid.* CXXIII. 1/3 Objects of gross physical plasticity, such as Brutalist sculptures and American cars. 1959 *Manch. Guardian* 7 Feb. 4/5 Churchill College at Cambridge will be built by a modern architect—perhaps even by a 'new brutalist'. 1960 *Observer* 17 Apr. 15/2 Furniture designs.. which include..a very simple desk cum storage-cabinet on brutalist lines.

brutalitarian (brutælitēᵊ·riăn). [f. Brutality, after *humanitarian*.] One who practises or advocates the practice of brutality. Also as *adj.*

1904 (*title*) The Brutalitarian, a Journal for the Sane and Strong. 1909 G. K. Chesterton *Tremendous Trifles* 215 It is only these two types, the sentimental humanitarian and the sentimental brutalitarian, whom one hears in the modern babel. 1910 —— *Shaw* 83 And in this the brutalitarians hate him not because he is soft, but..because he is not to be softened by conventional excuses. 1960 *Spectator* 5 Feb. 171 There is obviously no point, even on the most brutalitarian of grounds, in producing weapons of more than a certain amount of destructiveness.

brute. Add: **A.** *adj.* **4.** (Later example.)

a 1861 T. Winthrop *John Brent* (1883) i. 6 The precious metal was to the brute mineral in the proportion of perhaps a hundred pin-heads to the ton.

b. *brute fact*, a crude, isolated, or unexplained fact.

1874 G. H. Lewes *Probl. Life & Mind* I. v. 296 Science looks *through* the brute fact, to contemplate the Abstraction which gives it connection, significance. 1879 W. James in *Mind* IV. 329 We discriminate between brute fact and explained fact. 1926 A. E. Taylor *Plato* xvii. 455 Science.. is always forced to retain *some* element of brute fact, the merely given, in its account of things. 1957 *Economist* 21 Sept. 931/1 He refuses to accommodate the brute facts of historical development within a convenient but artificial pattern.

brutedom (brū·tdəm). [f. Brute *sb.*¹ + -dom.] Brutish nature.

1890 A. R. Wallace in *Fortn. Rev.* Sept. 331 In proportion as men leave brutedom behind and enter into the fulness of their human heritage. 1904 *Westm. Gaz.* 29 June 2/1 The paths that lead to..the depths to which sheer brutedom may descend.

bruting (brū·tiŋ), *vbl. sb.* [Rendering F. *brutage*, f. *brut* rough, unworked, unformed (cf. *brut *a.*, Brute *a.* 4) as in *diamant brut*.] The roughing-out of a diamond (see quots.). So **brute** *v. trans.*; **bru·ter.**

1903 L. J. Spencer tr. *M. Bauer's Precious Stones* 82 Bruting..consists in rubbing together two diamonds, each being cemented at the end of a stick or holder, until the desired form is obtained. *Ibid.* 243 The stone to be bruted is fixed to a handle. *Ibid.* 244 The three operations..are entrusted to..skilled workmen, namely cleavers, bruters, and grinders or polishers. 1906 L. Claremont *Gem-cutter's Craft* iv. 41 The bruting of diamonds consists of rubbing two diamonds together in such a way that by continual friction each can be made to assume the desired shape. *Ibid.* 42 Upon the principle of 'diamond cut diamond' the stones are roughly fashioned by the bruter into whatever symmetrical form he has designed them to be when finished.

‖ **brutum fulmen** (brū·tŏm fɒ·lmen), *phr.* Pl. **bruta fulmina.** [L., = 'senseless thunderbolt' (Pliny *N.H.* ii. xliii).] A mere noise; an ineffective act or empty threat.

1603 C. Heydon *Def. Iudic. Astrol.* 55 The Councells and decrees of the Church..prooue but *bruta fulmina*, making vaine cracks without any touch of that which I defend. *a* 1680 T. Goodwin *Works* (1685) XI. 131 It hath been *brutum fulmen* to us, a thunderbolt of no force. 1767 J. Wesley *Let.* 5 Mar. in *Jrnl.* (1827) III. 266 Till this is done, all you add (bitterly enough) is mere *brutum fulmen*. 1835 J. W. Croker in *Q. Rev.* Sept. 550 This rhodomontade..was..a mere *brutum fulmen*. 1963 *Times* 8 Mar. 5/4 No legal aid certificate in a limited form could be issued for such a limitation would be *brutum fulmen*.

Brutus. (Earlier examples.)

1798 H. L. Piozzi *Let.* 27 Mar. (1914) 152, I wonder if the pretty Misses go in *self* coloured drawers..and Brutus Heads with you as they do here. 1804 W. Irving *Jrnls. & Notebks.* (1969) I. 464 She caught hold of his hat & it came off with his brutus wig in it. 1807 Southey *Lett. from Eng.* lxxi. 304 During one period of the French Revolution the Brutus head-dress was the mode.

bruvver (brɒ·vəɪ). Representation of a Cockney pronunciation of Brother *sb.*

1898 J. D. Brayshaw *Slum Silhouettes* 120 Her 'big bruvver' told her..'this was the path up which the angels ran to God.' 1958 J. Townsend *Young Devils* xv. 143 'My bruvver will get you,' he promised. 1966 'S. Harvester' *Treacherous Road* ii. 23 We know what our 'Arold and 'is bruvvers think of us.

bryle, var. *broil *sb.*³

bryophyte (brəi·ŏfəit). *Bot.* [f. mod.L. *Bryophyta* (A. Braun 1864, in P. Ascherson *Flora d. Brandenburg* I. 22), f. Gr. βρύον moss + φυτόν plant.] One of the *Bryophyta*, a group of plants comprising the liverworts and mosses. Hence **bry:ophy·tic** *a.*

1878 W. R. McNab *Bot. Classif. Plants* i. 2 The transition from the Thallophyta to the next sub-kingdom, the Bryophyta, is not an abrupt one, hence..in the lower bryophytes the thallus is still to be met with. *Ibid.* ii. 71 The bryophytes can be divided into two classes, the Liverworts and the Mosses. 1928 *Bryologist* XXXI. 123 The hot arid conditions ..are little favorable to bryophytic growths. 1958 *New Biol.* XXVII. 87 The phylum or class of bryophytes..forms the lowest group of land plants.

Bual (biū·ăl). [ad. Pg. *boal.*] A variety of wine-making grape; the madeira made from this.

1882 [see Sercial]. 1883 *Encycl. Brit.* XV. 178/1 (*Madeira*) Other high-class wines, known as Bual, Sercial and Malmsey. 1927 J. B. Priestley *Adam in Moonshine* vii. 134 'Old Madeira. ..Bual, they call it.' 'What!' cried the Baron. .. 'An old Bual!'

bub, *sb.*³ Read: A contraction of Bubby¹. Add examples.

1860 Hotten *Slang Dict.* (ed. 2) 105 *Bub*, a teat, woman's breast. 1955 *Essays in Criticism* V. 123 'Breasts' are always 'bubs'. 1957 E. Pound tr. *Rimbaud* 11 The gal with the big bubs And lively eyes.

bub, *sb.*⁴ *U.S. colloq.* [See Bubby².] A form of familiar address to boys or men. (Cf. *bud *sb.*³)

1839 C. F. Briggs *Adv. H. Franco* II. 189 'Speak louder, Bub,' said one of the vice presidents, encouragingly. 1872 'Mark Twain' *Roughing It* v. 51 Well, I shall have to tear myself away from you, bub. 1896 *McClure's Mag.* VI. 485/2 She always called him 'bub' when she wanted to vex him. 1902 S. E. White *Blazed Trail* xxviii, Well, bub, .. blowed that stake you made out of Radway, yet? 1948 *Chicago Star* 24 Apr. 3/4 Hey bub—can I get a squint at yer uppers?

bubaline (biū·bălǝin), *a.* [ad. mod.L. *Bubalinæ*, f. *būbalus* Bubal.] Of or pertaining to antelopes formerly grouped in a sub-family Bubalinæ or buffaloes of the genus *Bubalus*.

1827 Griffith tr. *Cuvier's Anim. Kingd.* IV. 378 The Bubaline Group. 1907 *Westm. Gaz.* 2 Jan. 10/2 The Bubaline antelope is uniformly coloured bright bay.

bubble, *sb.* Add: **2. c.** A motor-car. *U.S. colloq.* Also (in U.K.) short for **bubble car*.

1918 Wodehouse *Piccadilly Jim* xxiii. 209 From the direction of the street, came the roar of a starting automobile...'Gee! He's beat it in my bubble—and it was a hired one!' 1920 —— *Coming of Bill* ii. xiii. 231 I'll take you out in the bubble—the automobile, the car, the chug-chug wagon. 1958 *Economist* (Suppl.) 25 Oct. 5/1 Entirely different kinds of car: first, a standard large-sized model for the export markets..; second, some kind of 'people's car' to tap the new market of our own masses; more recently motorised bubbles to float their way through growing traffic congestion. 1966 P. Moloney *Plea for Mersey* 18 Here rows of meters guard from crowding troubles, Rolls, Bentleys, Daimlers, Jaguars and Bubbles. 1967 J. B. Priestley *It's Old Country* vi. 65 Half a dozen cars, ranging from a gigantic old Rolls to a three-wheel bubble, were already parked.

d. The transparent domed canopy over the cockpit of an aeroplane. Freq. *attrib.*

1945 *Jrnl. R. Aeronaut. Soc.* XLIX. 315/2 You..showed great foresight..[with] such developments as the gyro stabilised gun sight, bubble canopy,..gas turbine and jet power plant. 1949 *Aeronautics* Jan. 35/2 The awkward cockpit canopy has been redesigned as a 'bubble' type. 1955 *Amer. Speech* XXX. 117 *Bubble*, n., the plexiglass canopy covering the cockpit.

6. **bubble bath,** a bath in which the water has been made to foam by a perfumed toilet preparation; such a preparation, in a liquid or crystal form; also *fig.*; **bubble car,** a miniature motor-car with a transparent domed top; **bubble chamber,** a container of superheated liquid for the detection of ionizing particles; **bubble-dancer** *U.S.*, a woman who dances as if in the nude, covered by one or more balloons; **bubble-gum,** chewing-gum which can be blown into large bubbles; **bubble sextant** (see quot. 1935); **bubble-trier,** an instrument used for testing the accuracy of the tubes of spirit-levels; **bubble-tube,** the glass tube of a spirit-level containing spirit and enclosing an air-bubble.

1949 L. Charteris in *Queen's Awards* (1951) IV. 299 'I was having a bubble bath,' said Pauline Stone. 1960 Koestler *Lotus & Robot* ii. vi. 166 Then the lights go up, the town changes into a bubble-bath of coloured neon. 1969 *Woman* 10 May 22 Bath essence, bath oil, bubble bath, bath cubes. 1957 *Observer* 20 Oct. 3/3 The B.M.C...are not interested in bubble cars as now known, but only in properly engineered vehicles. 1958 *Spectator* 13 June 762/1 The tiniest bubble car I ever set eyes on. 1953 *Physical Rev.* XCI. 496/2 (*title*) A possible 'bubble chamber' for the study of ionizing events. *Ibid.* 762/1 Bubble-chamber tracks of penetrating cosmic-ray particles. 1969 *Times* 5 Feb. 13/6 The tracks are recorded by taking high-speed photographs of the tank, or bubble chamber as it is called. 1936 *Time* 11 May 28/3 *Pre-honeymoon* is concerned with the love of a U.S. Senator for a bubble dancer. 1939 A. Huxley *After many a Summer* i. iii. 28 Candies, bubble gum, balloons. 1958 P. Mortimer *Daddy's gone A-Hunting* i. 7 The steady, pungent smell of bubble gum. 1920 *Flight* XII. 375/2 Bubble sextant. 1935 C. G. Burge *Compl. Bk. Aviation* 194/1 *Bubble Sextant*, a sextant employing a bubble device to provide an artificial horizon. Normally employed for determining the altitude of a celestial body. *a* 1877 Knight *Dict. Mech.*, *Bubble-trier*, an instrument for testing the delicacy and accuracy of the tubes for holding the spirit in leveling-instruments. 1890 W. F. Stanley *Surveying Instr.* 88 The Bubble Trier is a bar or bed 12 to 20 inches long, with two extended feet ending in points at one end, and a micrometer screw, the point of which forms a resting foot, at the other end, thereby forming a tripod. 1888 *Lockwood's Dict. Mech. Engin.*, *Bubble Tube, or Spirit Glass*, the tube of a spirit-level which contains the enclosed spirit. 1890 W. F. Stanley *Surveying Instr.* 86 Level Tubes, or Bubble Tubes as they are technically termed, are used in nearly all important surveying instruments.

bubble, *v.* Add: **7.** *trans.* To make (a baby) bring up wind; = *burp *v.* 2. Also *intr. U.S.*

1943 L. J. Halpern *How to raise a Healthy Baby* i. 19 During the process of belching or 'bubbling' small amounts of milk come up with the swallowed air. 1946 B. Spock *Common Sense Bk. Baby & Child Care* 83 You need to 'bubble' your baby in the middle of a feeding only if he swallows so much air that it stops his nursing. 1963 M. McCarthy *Group* x. 221 The baby belched...'Someone should have come in to bubble him,' she said. 'He swallowed a lot of air.'

bubble-and-squeak. Add: (Earlier and later examples.) Nowadays potatoes and other vegetables are often used instead of meat.

1772 T. Bridges *Burlesque Trans. of Homer: Iliad* xi. 507 We therefore cooked him up a dish Of lean bull-beef, with cabbage fry'd,..Bubble, they call this dish, and squeak. 1951 *Good Househ. Home Encycl.* 373/2 In the modern version of bubble and squeak the meat is usually omitted.

b. *transf.* and *fig.*

1927 T. E. Lawrence *Lett.* (1938) 509, I can make the most lovely bubble and squeak of a life for myself. 1960 *Times* 17 Sept. 9/4 The battle of the organs—British boom versus baroque bubble-and-squeak.

2. *Rhyming slang* for (*a*) Beak *sb.*³; (*b*) Greek *sb.* 1 a; also ellipt. *bubble*.

1935 Curtis & Wallace *Mouthpiece* i. 16 You ought to have heard him talking to the old bubble and squeak. 1937 Partridge *Dict. Slang* Add. 983/1 *Bubble and squeak*, a Greek. 1962 R. Cook *Crust on its Uppers* i. 20 All the best Anglo-Saxon grafters come from mine [*sc.* my school], and the Bubbles and the Indians from the other. *Ibid.* Gloss., Bubble = bubble-and-squeak = Greek. 1968 L. Berg

Risinghill 106 'Why do they call Greek children "Bubbles"?' said Mr. Colinides to me...Later it dawned on me that it was short for 'bubble-and-squeak'; rhyming slang.

bubblement (bɒ·b'lmĕnt). [f. Bubble *v.* + -ment.] Effervescence (*lit.* and *fig.*).

1890 *Pall Mall Gaz.* 24 Mar. 1/2 Berlin is in a state of bubblement. 1893 *Field* 17 June 895/2 A flash, and swirl, a bubblement. 1902 J. H. Skrine *Pastor Agnorum* 200 The wicked man..was an etymologist and knew that zeal means bubblement, effervescence.

bubbly, *a.* Add: *bubbly water* (slang), champagne; also *sb.*, short for this.

1910 *Daily Chron.* 9 Apr. 9/3 'Too much bubbly water', so he explained, was responsible for the breach of the rules. 1920 *Chambers's Jrnl.* 346/2 It goes to the head like bubbly. 1927 *Blackw. Mag.* Feb. 231/2 [He] had finished up at dinner with some capital oysters and a bottle of bubbly. 1951 J. B. Priestley *Festival at Farbridge* iii. iii. 566 Bubbly for you...Bubbly for me. George, two bubblies.

bubonic, *a.*¹ Add: Also as *sb.*, short for *bubonic plague*.

1901 *Daily News* 11 Feb. 5/3 A native child died last night, and the post-mortem proved the existence of bubonic. 1908 Kipling *Lett. Trav.* (1920) 121 A rot at what is called 'headquarters' may spread like bubonic, with every steamer.

bubu(c)kled (biū·bɒk'ld), *a. pseudo-arch.* [f. Bubukle (only Shakes.) + -ed².] Covered with 'bubukles' or large pimples.

1829 Cooper *Good's Study Med.* (ed. 3) II. 357 A welky or bubukled face. 1888 Stevenson *Across the Plains* (1892) iv. 163 Poor bubuckled Commissary!

bucchero (bukēᵊ·ro). *Archæol.* [It.] A type of pottery, characteristic of ancient Tuscany, of a uniform black, neither glazed nor painted, but decorated with figures in low relief, or, in later times, with figures moulded separately and applied to the pottery. Also *attrib.*

1889 in *Cent. Dict.* 1905 H. B. Walters *Anc. Pottery* II. 295 In the earlier chamber-tombs no *bucchero* is found. 1910 *Encycl. Brit.* V. 722/1 *Bucchero* ware—the national pottery of Etruria. 1922 *Edin. Rev.* July 54 Some goblets of finely modelled and burnished black bucchero. 1950 *Proc. Prehist. Soc.* XVI. 56 There is much to be said for correlating the Aeolic dialect with the areas that produced that grey bucchero pottery known as Grey Minyan ware.

bucco- (bɒ·ko), used as comb. form (for the correct bucci-) of L. *bucca* cheek, with the meaning 'relating to the cheek (or mouth)', as *buccocervical, -facial, -labial, -lingual, -nasal, -pharyngeal* adjs.

1892 *Phil. Trans.* B. CLXXXIII. 98 Dastre and Morat find..that the vaso-dilator fibres for the bucco-facial region of the Dog issue by the IInd., IIIrd., IVth., and Vth. thoracic nerves. 1903 Dorland *Med. Dict.* (ed. 3), *Buccocervical*, pertaining to the neck and the buccal surface of a tooth. *Buccolabial*, pertaining to the cheek and lip. *Buccopharyngeal*, pertaining to the mouth and pharynx. 1911 St. Clair Thomson *Dis. Nose & Throat* ii. vii. 111 The bucco-nasal membrane. 1967 G. M. Wyburn et al. *Conc. Anat.* 137/2 The overlapping constrictor muscles in their sheath of buccopharyngeal fascia.

Buchan (bɒ·kăn). The name of a Scottish meteorologist, Alexander *Buchan* (1829–1907), used to designate certain specified periods of cold weather foretold by him as of annual occurrence.

1923 *Daily Mail* 11 May 7 The period from May 9 to 14 is known as 'Buchan's Winter'. Buchan, an old meteorologist,..50 years ago laid down six periods of which May 9–14 is the coldest. 1929 *Daily News* 17 Apr. 8/7 Is it not a fact that the proposal to fix the date of Easter will mean that this holiday will in future coincide with 'Buchan's Cold Spell' which we are at present enjoying?

Buchanite (bɒ·kănəit). *Hist.* [f. the name of Mrs. Elspeth *Buchan* (1738–91), commonly known as Lucky Buchan + -ite¹.] A member of a sect which arose in 1783, in the Relief Congregation, Irvine, Ayrshire, Scotland, and became extinct in 1848 on the death of the only surviving member.

Mrs. Buchan believed herself to be the woman of Rev. xii.

1846 J. Train (*title*) The Buchanites from first to last. 1910 *Encycl. Brit.* IV. 714/1 Elspeth Buchan..founder of a Scottish religious sect known as the Buchanites.

Buchmanism (bu·kmăniz'm, bɒ·k-). [f. the name of Frank *Buchman* (1878–1961), founder of the Oxford Group + -ism.] The theories or practice of the Oxford Group Movement. So **Bu·chmanite** *a.* and *sb.*

1928 *Daily Express* 29 Feb. 1/4 The disclosure in the 'Daily Express' this morning of a revival here [*sc.* in Oxford] of a movement known as Buchmanism has led to a storm of comment in the University...The Buchmanites believe that if they conceive a sudden thought, without the exercise of a process of reasoning, this is a divine revelation, and should be acted on at once. 1928 *Isis* 16 May 1/2 Buchmanism is not widespread; probably in Oxford it never will be. 1933 H. H. Henson *Oxford Groups* 48 The Buchmanite Group reminds us irresistibly of the Russian Soviet. 1936 A. Huxley *Eyeless in Gaza* ii. 13 Under Guidance, as the Buchmanites would say. 1936 *New Statesman* 11 July 46/1 They tell us that Buchmanism is good business. 1961 *Times* 9 Aug. 10/3 A religious movement variously known as the Oxford Group, Moral Re-Armament, or, less politely, as Buchmanism.

bucholzite (bɒ·kǫlzəit). *Min.* [ad. G. *bucholzit* (R. Brandes 1819, in *Jrnl. Chem. und Physik* XXV. 125), f. the name of C. F. *Bucholz*, a German chemist: see -ite¹.] A variety of fibrolite.

1831 *Trans. R. Soc. Edinburgh* XI. 263 The colour of Bucholzite is a greyish-white, with a very slight tinge of yellow, not recognizable in the purest specimens. 1892 *Dana's Syst. Min.* (ed. 6) 498 *Fibrolite*. Fibrous or fine columnar...Bucholzite and monrolite are here included.

buck, *sb.*¹ Add: **1. a.** (Later example.)

1869 C. L. Brace *New West* xviii. 237 In the fall of 1861, W. Landrum obtained two bucks from a grower in the State of Georgia.

c. Also, the male of the ferret.

1904 *Westm. Gaz.* 11 Oct. 3/1 An old buck broken out of bounds is selfishly disposed.

d. A ram. Also *attrib. U.S.*

1812 *Niles' Reg.* II. 240/1 The product [of wool] was as follows: a Buck (*Judas*) 12 lbs. 4 oz. 1852 *Trans. Mich. Agric. Soc. 1851* III. 95 Sheep...Best buck over 2 years old. *Ibid.* 96 A lot consisting of 1 buck, 3 ewes, 10 ewe lambs and 2 buck lambs. 1948 *Daily Ardmoreite* (Okla.) 25 June 2/2 Choice native spring lambs 29.00; bucks out at 28.00.

2. d. Also applied to any male Indian, Negro, or Aboriginal. (Earlier and additional examples.) So *buck Aborigine, Indian, Maori, Negro, nigger*. Also (illogically) *buck-woman*. Chiefly *U.S.*

1800 J. McKenzie *Jrnl.* 9 Apr. in L. R. Masson *Les Bourgeois* (1890) II. 385, I..kept the woman to be disposed of in the season when the Peace River bucks look out for women. 1806 Pinckard *Notes West Indies* II. 405 The accuracy of the Bucks, in shooting with the long arrow, and in blowing the short poisoned one. 1835 *Gentleman's Vade-Mecum* (Phila.) 17 Jan. 3/6 A buck nigger is worth the slack of two or three hundred dollars. 1840 C. F. Hoffmann *Greyslaer* II. xii. 54 There they lay on the grass, six big buck Injuns, likely fellows all. 1853 *Southern Lit. Messenger* XIX. 221/2 A big buck negro. 1857 W. Chandless *Visit Salt Lake* i. 98 You could not mistake a squaw for a buck Indian. 1879 H. R. Michels *Sage Brush Leaves* 167 The buck aborigine takes more solid comfort than the female of his tribe. 1883 *Congress. Rec.* June 4147/1, I told the boys that we wanted 20,000 'bucks', buck niggers, in Indiana this year. 1933 F. E. Baume *Half-Caste* 23 Four bucks from Raglan came in drunk and there was a fight. 1941 Baker *N.Z. Slang* vi. 55 A *buck Maori*, a large well-built native (a somewhat unwarranted construction on *buck nigger*). 1951 E. Mittelhölzer *Shadows move among Them* i. iii. 20 His black smooth Buck hair seemed to gleam. 1958 J. Carew *Wild Coast* xix. 234 Don't talk to me about that Buck-woman. 1964 *People* (Australia) 16 Dec. 2/1 The bucks..strike large boulders said to be the petrified forms of the dingo slayers.

3. *buck-rabbit* (examples); **buck-ague** (also *buck-fever*) *N. Amer.*: see quot. 1872; **buck-brush** *N. Amer.*, one or other variety of brush on which deer feed; **buck-bush** *U.S.*, 'a species of *Symphoricarpos*, also a buckbrush' (D.A.); also applied to various Australian shrubs; **buck-pot** (cf. sense 2 d), an earthenware pot found in parts of Guyana; **buck-rake,** a large rake for farm use, freq. fixed to a vehicle (orig. *U.S.*); **buck rarebit,** a Welsh rabbit served with a poached egg on top.

1844 G. W. Kendall *Santa Fé Exped.* I. 172 There is a very common disease prevalent among young and inexperienced hunters in Texas, which is known as the 'buck ague'. 1872 Schele de Vere *Americanisms* 208 The *buck*.. makes him speak of *buck ague*, or *buck fever* when he wishes to describe the nervous agitation of the inexperienced sportsman. 1894 *Outing* (U.S.) XXIV. 344/1 My confusion ..was the direct result of buck-ague. 1933 F. B. Willoughby *Alaskans All* 146 An instant's 'buck ague', a single quiver of the arm, and there would have been a wounded beast in the boat. 1874 J. W. Long *Amer. Wild-fowl Shooting* ix. 152 Hang up your cartridge bag on a branch of the buck-brush. 1969 *Islander* (Victoria, B.C.) 22 June 2/3 Leather facing on both [trouser] legs keeps thorns, buck brush and thistle from drawing blood. 1911 C. E. W. Bean *'Dreadnought' of Darling* xxix. 246 We knew it was a lake because of the line of buck-bush—low tufted shrubs the colour and shape of big cabbages in a Chinaman's garden—that grew along the bed of it. 1918 S. S. Visher *Geogr. S. Dakota* 93 The buck-bush, is a transition stage between grassland and woodland. 1940 A. Upfield *Bushranger of Skies* xi. 125 The growth of buckbush, cotton- and flannel-bush. 1841 *S. Lit. Messenger* VII. 224/2 If you see a deer.. you'll be sure to git the buck fever. 1955 C. S. Forester *Good Shepherd* 57 He shared the tenseness of the others... He knew that hand would tremble if he allowed it to; this was buck-fever, unmistakably. 1851 *Illustr. Catal. Gt. Exhib.* iv. i. 986/2 Buck pot, used in preparing pepper pot. 1838 *Knickerbocker* XI. 447 Your land is so poor that a single buck-rabbit would make a famine in your whole country. 1915 D. H. Lawrence *Rainbow* (1916) ii. 66 'Now my young buck-rabbit,' he said. 'Slippy!' 1893 *Funk's Stand. Dict.*, *Buck-rake*, a two-horse hay-rake having horn-like teeth projecting 6 or 8 feet in front for gathering and transferring hay to a stacker. 1958 *Times* 1 July i/(*caption*) Tractor and buckrake cut out the heavy labour. 1927 Lindsay & Mottram *Man. Mod. Cookery* 83 *Buck rarebit*. Heat the cheese, milk, and seasoning; pour on to the hot toast. Poach the egg and serve on top of the cheese. 1959 P. Bull *I know the Face* vi. 102 Black coffee with the Buck Rarebits after that ordeal.

4. *ellipt.* = Buck-shot 2. *U.S.*

1845 W. G. Simms *Wigwam & Cabin* Ser. ii. 107 On using big buck, he numbered two sevens for a load; the small buck, three. 1876 *Fur, Fin & Feather* Sept. 135 The doctor soon drew a bird charge from his gun and loaded it with buck and fired. 1889 *Century Dict.*, *Buck-and-ball*, a cartridge for smooth-bore firearms containing a spherical bullet and three buckshot: now little used.

buck, *sb.*⁴ Add: (Earlier examples.) Also, on the river Thames, a wooden framework at a weir, supporting eel-baskets. Cf. *eel-buck* (Eel 6). Also *buck gate*.

1694 *Act* 6 & 7 *Will. & Mary* c. 16 *Preamble*, For the.. convenience of the Navigation [of the Thames] there.. are diverse Lockes Weares Buckes Winches..and other Engines. 1791 *Rep. Committees Ho. Comm.* (1803) XIV. 263/2 (*Thames & Isis*) The difficulties made by the millers and fishermen, of drawing their bucks and sluices. 1791 R. Mylne *2nd Rep. Thames Navig.* 12 Nuneham Wear.. is now used as a Fishery only, having a Tumbling, solid Wear, and two sets of Buck Gates. 1798 *Sporting Mag.* XII. 7 Having laid down our bucks in the river Loddon..we missed one. 1902 *Thames Conserv. Bye-Laws Regul. Thames* 32 No person shall without the previous consent of the Conservators erect any new buck or weir.

buck, *sb.*⁵ Add: **3.** **buck-wagon** (U.S. and S. Afr. examples). Also *S. Afr.* in other *Combs.*, as **buck-rail,** the rail of a buck-wagon; **buck-sail** [partial tr. Afrikaans *bokseil*], a large canvas or tarpaulin, esp. one used to cover a buck-wagon.

1896 H. A. Bryden *Tales S. Afr.* viii. 182 The tent I've fastened on to the buck-rail. 1955 W. Robertson *Blue Wagon* iv. 35 The wagon canted over at an angle with the buckrails below the surface [of the water]. 1882 Mrs. Heckford *Lady Trader in Transvaal* i. 11 An open buck-waggon..with a tarpaulin, or what is here called 'a buck-sail', thrown over it to protect the goods. 1961 *Argosy* Mar. 20 Caught in a summer storm with only a buck sail thrown over the cart to protect him from the rain. 1864 Webster, Buckwagon. 1870 in S. H. Pellissier *J. P. Pellissier van Bethulie* (1956) 662 Buckwagons, (empty,) drawn by 16 Oxen. 1877 G. McKiernan *Jrnl.* 27 June (1954) 121 James' buck wagon broke the king bolt before we got off the place. 1947 H. Vaughan-Williams *Visit to Lobengula 1889* iv. 13 We bought two full-sized buck-wagons, half tented for sleeping in.

buck, *sb.*⁷ (Earlier examples.)

1817 J. K. Paulding *Lett. from South* I. 189 He bought himself a buck and saw, and became a redoubtable sawyer. 1839 C. F. Briggs *Harry Franco* II. i. 2 There were also wood sawyers sitting listlessly on their bucks.

buck (bɒk), *sb.*⁸ *slang* (orig. and chiefly *U.S.*). [Origin obscure.] A dollar.

1856 *Dem. State Jrnl.* (Sacramento) 3 July 3/2 Bernard, assault and battery upon Wm. Croft, mulcted in the sum of twenty bucks. 1896 Ade *Artie* xii. 106 Jimmy can afford to buy wine at four bucks a throw when he's only getting three a week out o' the job. 1921 *Blackw. Mag.* Aug. 264/2, I wonder if I've done right forking out five bucks. 1927 M. de la Roche *Jalna* xi. 116 He's afraid some of us will want to borrow a few bucks. 1953 A. Baron *Human Kind* 183 'What did you do before the war?'..'Anythin' fer a buck.' 1968 *Globe & Mail* (Toronto) (Mag.) 17 Feb. 5 If you are a supporter of the profit motive in your own life and good times, blame no one. This is strictly a matter of bucks, like selling meat.

buck (bɒk), *sb.*⁹ [Origin obscure.] An article used in the game of poker; *to pass the buck* (see quot. 1887). *U.S.*

1865 *Weekly New Mexican* 14 July 1/3 They draw at the commissary, and at poker after they have passed the 'buck'. 1872 'Mark Twain' *Roughing It* xlvii. 332, I reckon I can't call that hand. Ante and pass the buck. 1887 J. W. Keller *Draw Poker* 38 They resort to the bold and ludicrous experiment of 'passing the buck'. The 'buck' is any inanimate object, usually knife or pencil, which is thrown into a jack pot and temporarily taken by the winner of the pot. Whenever the deal reaches the holder of the 'buck', a new jack pot must be made.

b. fig. *to pass the buck (to)*, to shift responsibility (to another). *colloq.* (orig. *U.S.*). Hence *buck-passing* vbl. sb.; *buck-passer*, one who passes the buck.

1912 W. Irwin *Red Button* 341 The Big Commissioner will get roasted by the papers and hand it to the Deputy Comish, and the Deputy will pass the buck down to me, and I'll have to report how it happened. 1932 E. Wilson *Devil take Hindmost* viii. 79 He invariably passes the buck to his subordinates. 1933 E. Cunningham *Buckaroo* 110 'Why, you lying buck-passer!' cried Dud, indignantly. 1933 *New Republic* 22 Nov. 37/1 (*heading*) Some Expert Buck-Passing. 1955 A. L. Rowse *Expansion Eliz. Eng.* x. 404 He [*sc.* Sir Francis Vere] is very good at passing the buck. 1959 *Listener* 15 Jan. 92/1 No one is less a buck-passer than is President de Gaulle. 1963 *Times Lit. Suppl.* 24 May 365/2 A human element of buck-passing and self-exculpation.

buck (bɒk), *sb.*¹⁰ [f. Buck *v.*³] An act of bucking; the power to buck. Cf. Buck-jump.

a 1877 in *Chicago Tribune* (Bartlett), The *buck* consists of the mustang's springing forward with quick, short, plunging leaps, and coming down stiff-legged [etc.]. 1883 Sweet & Knox *Through Texas* 69 If there had been any buck in them, it would have developed itself at an early stage in the journey. 1890 L. C. D'Oyle *Notches* 34 In two months from now the worst 'buckers' amongst them will not have a 'buck' left in them. 1908 Somerville & 'Ross' *Further Exp. Irish R.M.* v, The white horse got over the ground in bucks like a rabbit. 1942 Baker *Dict. Austral. Slang* 14 *Buck*, the plunge of a horse in 'bucking'.

b. A try, an attempt. *Austral.* and *N.Z. colloq.*

1913 A. Bathgate *Sodger Sandy's Bairn* ix. 71 I've a mind to have a buck at this new rush myself. 1941 Allen & Hall *Farmer in N.Z.* iv. 104 Readiness to 'try anything once', or to 'give it a buck' when some innovation swims into its ken. 1942 Baker *Dict. Austral. Slang* 14 *Give it a buck*, make an attempt at (something); to try. Also, 'have a buck at'.

buck (bʊk), *sb.*¹¹ *slang.* Also **bukh.** [a. Hind. *bak*, Hindi *buk buk*.] Talk, conversation; *spec.* boastful, bragging talk; insolence; esp. in phr. *old buck*.

1895 Mrs. Croker *Village Tales* (1896) 20 Having a 'bukh' with the elders. 1912 A. Neil Lyons *Clara* x. 99 We've 'ad enough of your old buck. 1941 *Penguin New Writing* III. 64 Nah then, none o' yer ol' buck, Ernie.

buck (bʊk), *a.*² *U.S. slang.* [Prob. f. Buck *sb.*¹ 2.] Belonging to the lowest grade of a specific military rank.

1918 H. C. Witwer *Baseball to Boches* iii. 109 Here I am nothin' but a buck private, and I been in the army goin' on four months! 1928 Fraser & Gibbons *Soldier & Sailor Words* 37 *Buck*,..a U.S. Army term for a Private, a smart soldier—a Buck Private. 1955 H. Roth *Sleeper* iii. 22 He had acquired the rank of buck sergeant. 1962 *Times* 14 June 15/4 From general officer to buck private.

buck, *v.*³ Add: (Earlier Australian example.) Also *refl.*

1848 H. W. Haygarth *Bush Life in Australia* vii. 78 Australian horses have a vicious habit known as 'buck-jumping', or as it is more familiarly called, 'bucking'. 1923 *Outward Bound* Mar. 60/2 Many ambitious young horses have bucked themselves to a finish without dislodging my cargo.

buck, *v.*⁴ Add: **2.** *intr.* Of persons. Chiefly *fig.* with *against* or *at*. orig. *U.S.*

1857 *San Francisco Call* 21 May 3/1 They think it hardly worth while to 'buck' against the present law prohibiting the pursuit of their 'science'. 1859 Bartlett *Dict. Amer.* (ed. 2), *To buck*, used instead of *butt*,—metaphorically of players at football and such games, pugilists, etc. 1870 in Schele de Vere *Amer.* (1872) 327 You'll have to buck at it like a whole team, gentlemen, or you won't hear the whistle near your diggings for many a year. 1900 G. Bonner *Hard-Pan* ii. 41 There's no good bucking against bad luck. 1906 E. Dyson *Fact'ry 'Ands* 76 Well, if we're mostly earth t' begin with, where's the sense in buckin' at er bit extry on the outside. 1920 J. Devanny *Butcher Shop* xvi. 190 The horses were terrible snacks whose shelter..a dog would 'buck at'.

3. *trans.* To butt into or against. Freq. *fig.*, to come up against, find oneself opposed to, oppose. orig. *U.S.*

1861 *Harper's Mag.* July 276/1 Mr. Fusilbury..was in a dream of philosophy, bucking a lamp-post. 1891 C. Roberts *Adrift Amer.* 114 There was a snow plough with two engines to it 'bucking the snow' (as the expression goes here). 1904 C. J. Steedman *Bucking the Sagebrush* ii. 14 If any convert..bucked the authority of the Church, he disappeared from his 'sphere of influence'. 1929 *Publishers' Weekly* 14 May 2436/2 It is wasted time and energy to try to buck any legitimate movement, and I think it is time and energy wasted for you men to attempt to buck the book clubs. 1947 *Time* 3 Feb. 68/1 In winter months they buck four to ten foot drifts. 1959 R. Collier *City that wouldn't Die* vii. 102 Duty won; you didn't buck an order from the chief of the Luftwaffe.

buck, *v.*⁵ Add: **2.** *to buck up.* **a.** *intr.* To cheer up, be encouraged. Also *trans.* in causal sense.

1844 *Graham's Mag.* Jan. 38 'I don't see the trouble,' said Mrs Fitzgig, 'why can't a man buck up?' 1859 Barrère & Leland *Dict. Slang* s.v., (Winchester College) to 'buck up' is to be glad. 1890 Farmer *Slang*, *Buck up*,..(Winchester College), to be glad; pleased... The usual expression is 'Oh buck up', a phrase which at Winchester School would have a very different meaning, namely, 'exert yourself'. 1894 *Punch* 27 Oct. 193/1 Buck up, mate; you're no call to be yaller, nor a permanent hoo, neither! 1901 W. W. Lawson et al. *Winchester Coll.* 14, Buck up, Hurrah! The original meaning, which is still used. Hence later:—Cheer up, hurry up. 1906 D'Urban von Hutten *What became of Pam* ix. 113 In..Don't spoil it all by being weepy... Come, buck up, like a dear, and wish me joy. 1909 H. G. Wells *Tono-Bungay* ii. i. § 1 Never saw her so larky. This has bucked her up something wonderful. 1920 W. J. Locke *Simon the Jester* xviii, Now and again one does help a lame dog over a stile which bucks one up, you know. 1926 *Lay Thoughts of a Dean* 233, I asked the medical members..in particular whether it was impossible that microbic diseases..might be benefited by 'bucking up' the patient. 1966 'J. Hackston' *Father clears Out* 37 As if to buck us up after our recent loss, he promised us poultry on the table.

b. *intr.* To make an effort, to 'brace up'; to hurry up.

1890 [see a]. 1909 H. G. Wells *Tono-Bungay* i. i. § 8 It was equally impossible for him to either buck-up and beat me, or give in. 1910 W. J. Locke *Simon the Jester* ii, You must buck up a bit, Simon, and get your name better known about the country. 1913 'Ian Hay' *Happy-go-lucky* i, 'Hallo, you fellows—finished?' 'Yes, buck up!' commanded Rumbold.

buck (bʊk), *v.*⁸ *U.S.* [Origin obscure, but cf. *buck *sb.*⁴] **1.** *intr.* To play at a game of chance. Usu. with *against* or *at*. Also *fig.* Hence *to buck the tiger* (see Tiger *sb.* 9 a in Dict. and Suppl.).

1849 *New Orleans Picayune* 18 May 1/6, I left them 'bucking' away, desiring only one more to 'get even'. 1869 Bret Harte *Luck of Roaring Camp* (1871) 95 Why don't you say you want to buck agin' faro? 1872 Schele de Vere *Americanisms* 327 The fact that players at 'Three-Card Monte'..are said to buck at monte, causes the familiar phrase of *bucking at* anything. 1898 H. S. Canfield *Maid of Frontier* 109 The man who bucks against that kind of faro bank.

2. *trans.* To bet or lose (money) in gambling.

1851 *Alta California* 8 July (D.A.), The money Percy took to the El Dorado, where he duly bucked it off against a faro bank. 1851 Mrs. Clappe *Let. from Calif.* (1922) 121 Little John was..betting, or, to speak technically, 'bucking' away large sums at monte.

buck (bʊk), *v.*⁹ *N. Amer.* [f. Buck *sb.*⁷] To cut (wood) with a buck-saw.

1870 *Phila. Press* 1 Jan., [The] Pennsylvanian does not saw wood; he 'bucks' it. 1905 *Terms Forestry & Logging* (U.S. Dept. Agric.) 31 *Buck*, to saw felled trees into logs (Pacific Coast Forest). 1923 *Jrnl. Commonw. Forest Terms* I. 21 *To buck*, to cut felled trees into any required lengths (Canada). 1964 *Times* 18 Sept. 13/7 One lumber executive..can still buck a log. 1969 *Islander* (Victoria, B.C.) 16 Nov. 3/3 The tree had to be bucked into seven lengths varying in size from 10 to 40 feet.

buck (bʊk), *v.*¹⁰ *slang.* Also **bukh, bukk.** [Cf. *buck *sb.*¹¹] *intr.* To swagger, talk big or bumptiously, brag.

1880 G. Aberigh-Mackay *21 Days in India* 164 He bucks with a quiet, stubborn determination that would fill an American editor..with despair. 1890 Kipling *Many Invent.* (1893) 37 You're ordered to Aden. 1895 Mrs. Croker *Village Tales* (1896) 157 Those who were wont to assemble..of an evening to smoke, argue, and bukh. 1899 *Daily News* 8 June 6/4 There is not much in this for Etonians to 'buck' about. 1921 *Blackw. Mag.* Apr. 517/2 Arthur and I have 'bukked' till we're hoarse.

buck-and-wing. *U.S.* [? f. Buck *v.*² + Wing *n.*] A dance of a lively character, usually performed by one person. Also *attrib.*

1895 *N.Y. Dram. News* 23 Nov. 13/4 Burt Jordon, an exceedingly agile buck and wing dancer. 1907 Mulford *Bar-20* x. 110 In the centre of the room was a large man dancing a fair buck-and-wing to the time so vigorously set by his companions. *Ibid.* 112 Up by the door Bigfoot Baker, elated at winning the buck-and-wing contest, was endeavouring to learn a new step. a1910 'O. Henry' *Rolling Stones* (1916) 191 He heard..the light, stirring staccato of a buck-and-wing dance. 1964 J. J. Mathews *Talking to the Moon* 45 He would get up from his chair..and do a buck-and-wing on the pine flooring.

buckaroo, buckayro. *U.S.* Also **bakhara, buckeroo, buckhara,** etc. [Corruption of Vaquero.] A cowboy. Also *attrib.*

1827 W. B. Dewees *Let.* 16 Jan. in *Lett. fr. Texas* (1852) 66 These [rancheros] are surrounded by..peons and bakharas, or herdsmen. 1889 *Century Dict.*, *Buckayro*, (Western U.S.). 1890 Farmer *Slang*, *Buckhara* (American), a name given in California to a cattle-driver or cowboy. 1904 *N.Y. Tribune* 17 July, He was herding a big bunch of cattle there with the help of half a dozen buckaroos. 1905 S. E. White *Arizona Nights* ii, If you were going to be a buckeroo, you couldn't go into harder training. 1909 J. Hart *Vigilante Girl* iv. 60, I can talk what they call 'buckayro' Spanish. 1916 H. L. Wilson *Somewhere in Red Gap* iv. 129 Nothing ever did worry that buckaroo as long as his fingers wasn't too cold to roll a cigarette. 1946 Mencken *Amer. Lang.* (ed. 4) iv. 152 Buckaroo seems to have dropped out.

bu·ckboard, *sb.* orig. and chiefly *U.S.* [See Buck *sb.*⁸ 3.]

1. A plank slung upon wheels, forming the body of a light vehicle.

1839 C. F. Hoffman *Wild Scenes* I. 10 Did he ever see a teamster riding upon a buckboard? a stout, springy plank, laid upon the bare bolsters of a waggon. 1861 [see Buck *sb.*⁸ 3]. 1947 H. Vaughan-Williams *Visit to Lobengula* 1889 35 When on the trek all cooking pots were slung on the rails of the buckboard of the wagons.

2. A vehicle of this description. Also *Comb.*

1874 'H. Churton' *Toinette* (1881) iv. 45 The shivering splattered figure on the 'buck-board'. 1883 [see Buck *sb.*⁸ 3]. 1886 *Outing* (U.S.) Oct. 57/2 For some reason or other, the buckboard driver, who usually brings us our mail, did not bring it yesterday. 1893 *Daily News* 5 May 5/4 There is a buck-board seen across Wokingham Creek..as far as Bourke in New South Wales. 1904 C. Johnstone *Winter & Summer in Canada* 64, I hired a 'rig' or 'buckboard' for 3 dollars to take me as far as Carlton. 1935 J. Steinbeck *To God Unknown* 175 He hitched up the buckboard and drove to town. 1953 A. Upfield *Murder must Wait* xxv. 217 A buckboard drawn by two horses.

Hence **bu·ckboard** *v. intr.*, to ride on a buckboard.

1904 F. Lynde *Grafters* i. 2 Two men who buckboarded in ahead of the track-layers.

buckbuck, see *boobook.

buck dance. *U.S.* = *buck-and-wing. Hence *buck dancer, dancing.*

1840 *Spirit of Times* 2 May 103/1 The extra clearing in Turn for the finishing buck dance. 1896 *N.Y. Dram. News* 29 Aug. 8/3 Conwell and O'Day, buck dancers, made a gigantic hit. 1897 Ade *Pink Marsh* 73, I use' to know cullud boy..'at got job on 'e stage doin' buck-dancin'. 1933 P. Godfrey *Back-Stage* xv. 187 The girls in these troupes have to undergo a hard training in widely different types of dancing: buck, ballet, tap, and so on.

bucked (bʊkt), *ppl. a. colloq.* [f. Buck *v.*⁵ + -ed¹.] Cheered, encouraged, elated.

1907 *Varsity* 31 Jan. 194/3 We are very 'bucked' that he fitted in a visit this Term. 1920 *Pyramids* i. 35 So bucked at being with you again. 1922 A. A. Milne *Red House Myst.* vii. 81 He was unusually bucked with himself this morning. 1928 *Punch* 23 May 580/1 I am so bucked that you have asked me what to wear when you are accompanying at the concert next month. 1960 *Listener* 19 May 896/3 I'm pleased and bucked and I feel pretty good.

buckeen² (bʊki·n). [ad. Guiana Du. **bokin*, fem. of *bok* goat, buck (see Buck *sb.*¹ 2 d).] A female aboriginal Indian in Guiana.

1806 Pinckard *Notes West Indies* II. 405 The morning was..spent..in hearing M. Heyneman relate his adventures among the Indians; whom he termed Bucks and Buckeens. 1868 W. H. Brett *Ind. Tribes Guiana* 94 *note*, The Indian men and women were called by the Dutch 'Bucks' and 'Buckeens'. 1899 J. Rodway *In Guiana Wilds* 39 They called her a 'Buckeen'.

Buckelkeramik (bʊ·k'lkɛræ:mik). *Archæol.* [G., lit. 'knobbed ware'.] A type of late Bronze Age pottery with protruding decorative knobs.

1902 W. Dörpfeld et al. *Troja und Ilion* i. vi. 201 Für die Feststellung der oberen Grenze ist die Zeit der VII. Schicht maassgebend, weil die..Topfwaaren und ebenso die Buckelkeramik keine genaue Datirung gestatten. 1925 V. G. Childe *Danube in Prehist.* xiii. 261 *Buckelkeramik*. The commonest [Hungarian Bronze Age] pottery is the distinctive 'Toszeg ware' (*Buckelkeramik*, Figs. 160–3). 1937 *Amer. Jrnl. Archaeol.* XLI. 49 Debris..dated by pottery fragments of Buckelkeramik to the period VII b (i.e., at Troy). 1940 *Ann. Brit. Sch. Athens* 1936–7 XXXVII. 21 The sudden appearance of Buckelkeramik at Troy now indicates an abrupt change in foreign contacts. 1950 H. L. Lorimer *Homer & Monum.* i. 36 Then came the intrusion from Europe of a new people..with the characteristic Buckelkeramik.

bu·cker². *N. Amer.* [f. *buck *v.*⁹] One who saws trees into logs.

1900 *Treasurer's Bur. Statistics* Nov. 1116, A logging crew consists of a foreman, fallers,..buckers. 1956 *Amer. Speech* XXXI. 149 *Bucker*, a logger who saws felled trees into log lengths. 1963 *Times* 24 Jan. 11/6 The attendants, the fallers and buckers, often make £15 a day.

bucket, *sb.*¹ Add: **1.** (Buckets are now made of various materials, esp. metal, and used as containers for many things. Add further examples.)

1895 *Army & Navy Co-op. Soc. Price List* 427/1 India Rubber Bucket, Rope Handle...Canvas Bucket. 1907 *Ibid.* 181/1 Fire Buckets...tinned steel, with iron handles, painted red and lettered. 1922 Joyce *Ulysses* 100 A servant, bare-headed, came out with a bucket. *Ibid.* 348 The rain falling on the rusty bucket. 1969 *Sears, Roebuck Catal.* 1260/1 Ice bucket..made of easy-clean vinyl with a plastic liner.

3. b. A scoop operated by power, used for hoisting coal, grain, etc., and in dredging and excavating.

1877 Knight *Dict. Mech.*, *Bucket*, (3) The scoop of a dredging machine. 1881 [see Grab *sb.*¹ 3]. 1902 *Encycl. Brit.* XXVII. 532/2 Grabs or buckets for dredging purposes.

4*. *Rowing.* [f. Bucket *v.* 5.] 'A plunge forward with the body when the stroke is concluded before the arms have been straightened out, and the hands at least passed the knees' (*Encycl. Sport*).

1898 W. B. Woodgate *Boating* iv. 70 Lateness in swing may arise *per se*, and so may a 'bucket'. 1898 R. P. P. Rowe et al. *Rowing* 75 The swing forward should be kept as slow as is reasonably possible, and a 'bucket' avoided.

5. bucket bag, a woman's bag, resembling a bucket in shape; **bucket chain,** (*a*) a chain linked to a bucket; (*b*) a line of persons formed to pass buckets of water or sand to a conflagration; **bucket-door,** the cover of an opening which gives access to the buckets of a pump; **bucket dredge, dredger,** a dredge that excavates by means of scoops or buckets; so **bucket-dredging** *vbl. sb.*; **bucket-elevator** [Elevator 3 a], a chain of buckets for raising material or liquids to a higher level; **bucket-ladder** (see quot. 1877); **bucket-seat,** in a motor-car, aeroplane, etc., a seat partly resembling a bucket in shape; **bucket-wheel,** also, an overshot or breast wheel operated by the impulse and weight of the water falling into the buckets or receptacles on its rim.

1956 'J. Bell' *Death in Retirement* iv. 49 Mrs. Weaver produced a fat envelope from her bucket bag. 1959 *Woman* 31 Oct. 74/3 Her hand had crept down to take a purse from the open bucket bag. 1911 *Daily Colonist* (Victoria, B.C.) 21 Apr. 14/3 It was expected the dredger would be ready about a week ago but the bucket chain slipped and some time was occupied in replacing this. 1932 *Daily Express* 28 June 1/2 When one fire brigade failed, the village girls.. assisted the men in a 'bucket chain' until another brigade arrived. 1797 J. Curr *Coal Viewer* 74 Cross Bars and Bolts for Bucket and Clack Doors. 1907 *Jrnl. Soc. Arts* LV. 1009/1 The gold may rest in crevices, from which the bucket dredge could not collect it. 1902 *Encycl. Brit.* XXVII. 530/2 Where it is necessary to finish off the dredging work to a uniform flat bottom..bucket dredgers are better adapted. 1930 *Engineering* 15 Aug. 195/3 The bucket dredger is the only type which has been uniformly successful from the time of its introduction. 1898 *Engineering Mag.* XVI. 116/1 The bucket-dredging-apparatus. 1903 *Chambers's Jrnl.* 30 May 415/2 The grain is then lifted by a bucket-elevator to the upper deck. 1930 *Engineering* 25 July 102/3 The function of the bucket elevators..on the main washing plant..will thus be understood. 1877 *Encycl. Brit.* VII. 464/2 The main feature of the machine [*sc.* steam dredger] is the bucket-ladder...Along this ladder a series of buckets traverse which cut into the bottom..and return loaded with the excavated material. 1902 *Ibid.* XXVII. 529/1 Bucket-Ladder Dredgers. 1908 *Westm. Gaz.* 29 Oct. 4/1 One is fitted with a luxurious body and the other left bare with two bucket-seats. 1943 *Jrnl. R. Aeronaut. Soc.* XLVII. Abstr. 212 The prone position is less tiring [to the pilot] than the normal bucket seat. 1958 *Vogue* June 129 Individual bucket seats are ideal since they allow drivers to adjust them to their own requirements. 1797 *Encycl. Brit.* XVIII. 904/1 A bucket-wheel has been executed lately..of a construction entirely new. 1880 *Ibid.* XII. 523/1 When..the supply is from 3 to 25 cubic feet per second, it is possible to construct a bucket wheel on which the water acts chiefly by its weight. 1970 *Times* 23 Feb. (Canada Suppl.) p. vi/3 (*caption*) A giant bucket-wheel excavator scooping up oil-bearing tar sands.

bucket, *v.* Add: **2. b.** *intr.* Of rain, etc.: to pour down heavily.

1926 *Chambers's Jrnl.* Dec. 844/1 Tramps don't walk about a marsh in bucketing rain in the midnight hours. 1970 *Woman* 21 Feb. 45/1 She stood, umbrella-less..not even a headscarf between her and the bucketing sky.

4. Also, to move or drive (a vehicle, etc.) roughly or jerkily. Also *intr.*

1879 L. S. Wingfield *My Lords of Strogue* iv, He..was painfully alive to the possibility of finding his daughter stolen one day,..to be bucketed about the country without a change of linen till his reluctant consent was wrung. 1904 *Westm. Gaz.* 13 Dec. 10/2 He..was bucketed about the garrisons from the Canadas and Cape Breton to the Mediterranean. 1911 C. E. W. Bean *'Dreadnought' of Darling* i. 10 The coach bumped and bucketed over boulders. 1921 *Blackw. Mag.* Feb. 233/1 Poor preparation..for bucketing about the Baltic all the following day. 1922 *Chambers's Jrnl.* 437/2 The envious beggar..buckets his old boat along for nothing. 1944 *Jrnl. R. Aeronaut. Soc.* XLVIII. 276 There is little tendency to swing or bucket. *Ibid.* 280 The aircraft can be taxied fast without danger of bucketing.

bu·cketed, *a.* [f. Bucket *sb.*¹] Having the form of a bucket.

1886 Bret Harte *Snow-bound* 124 A vast niche or bucketed shelf in the hollow flank of the mountain.

bucketful. Add: (Additional examples.) Also *transf.* and quasi-*adv.*

1843 *Spirit of Times* 30 Sept. 366/3 The rain..came pouring down in buckets-full. 1892 C. M. Yonge *That Stick* II. xxxviii. 181 She cried bucketsful. 1936 W. R. Titterton *G. K. Chesterton* ii. iv. 145 Turning out books in buckets-full.

bucket-shop. Delete ? in etym. Add examples of obs. *U.S.* sense 'gin-mill', 'low-class liquor-shop', and further examples of sense in Dict.

1875 Mrs. Stowe *We & Neighbors* 380 The lowest, the most dreadful of all, was what they called the bucket shops. There the vilest of liquors are mixed in buckets and sold to the wretched, crazed people. 1881 *N.Y. Even. Post* Oct. (Th.), A 'bucket-shop' in New York is a low 'gin-mill', or 'distillery', where small quantities of spirits are dispensed in pitchers and pails (buckets). When the shops for dealing in one-share or five-share lots of stocks were opened, these dispensaries of smaller lots than could be got from regular dealers were at once named 'bucket-shops'. 1880 *Bradstreet's* 1 Dec. 1/4 The failure of the 'Produce Exchanges', or bucket shops..caused little excitement. 1910 *N.Y. Even. Post* 4 Apr. 4 What..our people call a bucket-shop, is not only a gambling establishment pure and simple but is in most cases a gambling establishment which pretends to be something else. 1957 *Encycl. Brit.* IX. 703/2 'Bucket shop' means the business of a sham share broker who gambles with his customers on the rise or fall in the prices of stock exchange securities without acquiring these securities at all. 1960 *Times* 10 June 7/2 It might be said that he was conducting a bucket-shop..except that his stock in trade was not securities but British Colonial stamps.

buck-eye. *U.S.* Add: **1.** (Earlier and later examples.) Also, a tree of this species.

1763 in R. T. Durrett *Louisville* (1893) 132 Beginning at a hoop-ashe and buckeye. 1784 J. Filson *Kentucke* 23 Here also is the buck-eye, an exceeding soft wood, bearing a remarkable black fruit. 1878 Bret Harte *Man on Beach* 109 Looking down upon me through the buckeye bushes.

b. The nut or fruit of the buck-eye.

1797 in *Filson Club Hist. Q.* (1928) II. 166 Buck eye.. resembles the Chestnut, but is as large as a hickory nut of the largest Size. 1892 *Jrnl. Amer. Folk-Lore* V. 20 The negroes believe if one carries buckeyes in the pocket he will have no chills through the year. 1948 *Chicago Tribune* 28 Mar. vii. 1/3 You carry a buckeye to ward off rheumatism.

2. (Examples.)

1822 E. James *Exp. Rocky Mts.* I. 20 In allusion to this circumstance, the indigenous backwoodsman is sometimes called buck-eye. 1882 *Congress. Rec.* July 5811/1 The regenerated statesmanship of the modern Buckeye.

attrib. 1840 *Buffalo Commerc. Advertiser* 12 Sept. (Th.), Queer carryalls did these Buckeye boys construct. 1894 *Congress. Rec.* Jan., App. 79/1 The very author of the tariff plank in the Chicago platform was last fall buried beneath more than 80,000 votes in that grand old 'Buckeye' State.

3. A type of canoe or schooner (see quots.).

1885 C. P. Kunhardt *Small Yachts* 234 The Buckeyes.. are an exaggeration of the dugout canoe...The primitive builder bored two holes, one on each side of the stem, through which to pay out his cables. These were simply two round holes, bored with a large auger, and, when the boat was coming head on, resembled to the fancy of the negroes, the eyes of a buck. 1889 *Cent. Dict.*, *Buckeye*... A flat-bottomed centerboard schooner of small size.. decked over, and with a cabin aft, used in oyster-fishing in Chesapeake Bay. Also called *bugeye*. 1923 *Outing* (U.S.) Jan. 187/1 The craft he suggests is one of 'those Chesapeake Bay buckeyes'.

buck-horn. Add: **4.** In full *buck-horn sight*: a branched form of sight of a rifle or sporting gun. *U.S.*

1877 R. I. Dodge *Hunting Grounds Gt. West* vii. 105 The very-best sight, and the one almost universally in use by sportsmen and professional hunters on the plains, is the plain 'buckhorn'. *Ibid.*, Sportsmen who use the 'buck-horn' must learn to sight 'on the barrel'. 1901 S. E. White *Westerners* xi. 78 He had viewed..the scout through the buck-horn sights of his long rifle.

Buck House (bʊk haus). A jocular appellation of *Buckingham Palace*, the London residence of the Sovereign.

1922 David, Earl Beatty *Let.* 28 Feb. in W. S. Chalmers *Life & Lett.* (1951) xvii. 390 A succession of very heavy solemn collections of functions at Buck House. 1954 J. B. Priestley *Magicians* i. 11 I'm told Karney plays with millions now. Pops up all over the place. Including Buck House for a Knighthood. 1962 R. Jeffries *Exhibit No. Thirteen* vi. 58 Come on in—the door's wide open and this isn't Buck House.

buckish (bʊ·kiʃ), *a.*² [f. Buck *v.*³ + -ish¹.] Inclined to buck; hence, high-spirited, in good fettle (*slang*, of persons).

1911 *Scott's Last Exped.* (1913) I. xiii. 392 The ponies are very buckish and can scarcely be held in.

buck-jump, -jumper, -jumping. (Earlier examples.)

1848 H. W. Haygarth *Bush Life in Australia* vii. 78 An expert 'buck-jumper' usually begins when his rider is in some degree off his guard. 1848 Buck-jumping [see *buck *v.*³]. 1861 *Harper's Mag.* June 8/1 The Captain.. compels his animal to slide or make 'buck jumps' over the worst places.

bucklandite (bʊ·klændait). *Min.* [f. the name of Dr. William *Buckland* (1784–1856), English geologist: see -ite¹.] A variety of allanite or the related mineral epidote.

1824 M. Levy in *Ann. Philos.* VII. 134 At the suggestion of Mr. Heuland, I propose to give the name of Bucklandite (in honour of the celebrated Professor of Oxford), to a mineral substance, the crystallographical characters of which I find to differ from any hitherto described. 1854 Dana *Syst. Min.* (ed. 4) II. 208 Bucklandite is from Achmatowsk, Arendal, and Lake Laach. 1910 *Encycl. Brit.* IX. 689/1 Several other names (achmatite, bucklandite, escherite,..&c.) have been applied to this species [*sc.* epidote].

buckle, *sb.* Add: **1. b.** † *to hold* or *bring* (*bare*) *buckle and thong together* (in U.S., *to make buckle and tongue meet*): to make both ends meet. *Obs.*

1600 *Weakest goeth to the Wall* sig E2, My benefice doth bring me in no more But what will hold bare buckle and thong together. 1675 *Pepys Ballads* (1930) III. 12 Hold Buckle and Thong together. 1732 T. Fuller *Gnomologia* 100 He'll bring buckle and thong together. 1859 *Fisher's River* 249 All they cared for was 'to make buckle and tongue meet' by raising stock..and a little corn for bread. 1888 *Harper's Mag.* Apr. 703/1 Beginning without money, he had as much as he could do to make 'buckle and tongue meet', as the phrase goes.

d. An iron loop for fastening the blade to the frame of a wood-saw.

1846 Holtzapffel *Turning* II. 725 The chairmakers' saw is in general a diminutive of the ordinary pit saw, and has a central blade strained by buckles and wedges.

e. [app. f. Buckle *v.* 2 c.] A vigorous struggle or conflict.

1845 J. Keegan *Leg. & Poems* (1907) 268 Master Tom and the girls running out to..welcome me home, as if I was just returning from a buckle with the Terry Alts, or a rencontre with a scalping party of Huron Indians. 1876 *Coursing Cal.* 6 Wheatear and The Pet had a tremendous tight buckle for the lead. *Ibid.* 181 Liberator and Controversy made a tight buckle for pace.

buckle, *v.* Add: **2. c.** Also *to buckle down to.*

1865 *Atlantic Monthly* XV. 301 If he would only buckle down to serious study. 1871 Bret Harte *Poems* 55 Chiquita Buckled right down to her work. 1934 Wodehouse *Right Ho, Jeeves* xxii. 280 The recollection of what this bell could do when it buckled down to it gave me pause.

6. a. Also *fig.* in *pass.*

1921 *Contemp. Rev.* Mar. 291 It [*sc.* the French budget] may..for practical purposes, be considered to be 'buckled'.

c. *to buckle up*: to become warped and bent, to collapse. Also *fig.* (cf. 7).

1866 *Jrnl. R. Agric. Soc.* II. i. 140 Its expansion and contraction..rendering it [*sc.* plane iron] liable to 'buckle up'. 1927 *Daily Tel.* 1 Nov. 12/2 Such a man as Biquet is not given to buckling up.

Buckley's (bʊ·kliz). *Austral.* and *N.Z. colloq.* [Of uncertain origin.] In full *Buckley's chance* (or *hope*, etc.): a forlorn hope, no chance at all.

1898 *Bulletin* (Sydney) 17 Dec. 31/2 'Devil shoot me!' muttered Tim..'if I see Buckley's chance of a shindy tonight.' 1903 'T. Collins' *Such is Life* 339 His show is little better than Buckley's. 1916 C. J. Dennis *Songs Sentim. Bloke* (1916) 118 *Buckley's* (*chance*), a forlorn hope. 1918 —— *Digger Smith* 81 Doreen an' ole Mar Flood 'as got a scheme. ..But they've got Buckley's, as I tells me wife. 1934 *Bulletin* (Sydney) 2 May 25/1 He was especially sure..that the tourists had Buckley's hope of winning a Test. 1948 V. Palmer *Golconda* xiii. 103 Buckley's chance we have of getting our price if we're left to face the companies alone. 1955 D. Niland *Shiralee* 216 You reckon I haven't got Buckley's?

buckling (bʊ·kliŋ), *sb.*² Also **bückling** (pl. *bücklinge*). [ad. G. *bückling* bloater.] A smoked herring; a bloater (see quots.).

1909 *Cent. Dict. Suppl.*, *Buckling*, a large smoked and salted herring. 1924 G. B. Stern *Tents of Israel* iv. 54 Bücklinge.., which were herrings smoked by herself according to old Maria's Viennese recipe. 1934 'Taffrail' *Seventy North* iii. 66 The 'buckling'..was a herring smoked in a special oven and cooked at the same time. 1967 M. Gilbert *Dust & Heat* i. 182 This isn't a smoked trout at all. It's a buckling.

bucko (bʊ·ko). *Naut. slang.* [f. Buck *sb.*¹ + *-o.] **a.** A blustering, swaggering, or domineering fellow; sometimes used as a term of address. Also *attrib.* or as *adj.* = blustering, swaggering, bullying; esp. in phr. *bucko mate.* **b.** Swagger, bluster.

1883 J. F. Keane *On Blue-Water* xv. 190 After that, no sailor will deny that a 'bucko mate' is not sometimes useful. 1899 M. Robertson *Where Angels Fear* 107 Stand by here, mates. These buckoes'll kill some one yet. 1909 *Times Lit. Suppl.* 9 Sept. 325/2 Hudson was included among the victims because he was given to 'bucko' (to use a piece of sea slang). 1924 *Ibid.* 6 Nov. 712/3 Bucko skippers. 1926 *Spectator* 6 Feb. 229/2 A youngster who went to sea as apprentice and finished bucko mate of a Yankee packet. 1927 *Blackw. Mag.* Jan. 5/1 A great big bucko of a man. 1939 T. S. Eliot *Old Possum's Practical Cats* 16 His bucko mate, Grumbuskin, long since had disappeared. 1969 *Courier-Mail* (Brisbane) 18 Aug. 5/2 The old buckos in the peaked caps..never thought they would live to see the day when Irishmen applauded the arrival of British troops.

buckshee (bʊ·kʃi, bʊkʃi·), *sb.* and *a. slang* (orig. *Army*). [Alteration of Baksheesh; cf. Bukshi.] **A.** *sb.* Something extra, free, or to spare; an allowance above the usual amount. **B.** *adj.* Free; spare, extra. Hence as *adv.*

1916 *Daily Mail* 1 Nov. 4/4 'Buckshee' (probably derived from back-sheesh, meaning extra rations, or anything over after an issue has been made—buckshee loaf, buckshee 'fags', etc.). 1919 *Athenæum* 1 Aug. 695/2 'Buckshee' deserves attention as probably, with 'scrounge', the most popular slang towards the end of the war. 1920 *Outward Bound* Nov. 66/2 The police help themselves freely to buckshee refreshments. 1921 *Glasgow Herald* 1 Nov. 5 To..give the goods almost buckshee to our cut-throat friends, the tribesmen of the N.-W. Frontier of India. 1942 C. Barrett *On the Wallaby* iii. 48 The Chief of Staff..snapped, 'Want a buckshee trip, eh?'

buckskin. Add: **5.** A kind of strong twill cloth.

1894 *Tailoring* 3 Feb. 1/1 Coatings.—Black and Blue Serges, cut and covered in every size of twill,..Buckskins, 10s. 1895 *Oracle Encycl.* I. 528/2 *Buckskin*, a strong twilled woollen fabric for trouserings. The web is usually about 27 inches wide, and when finished the pile or nap is so shorn that the texture is seen through it. 1968 J. Ironside *Fashion Alphabet* 216 Buckskin cloth..is a closely woven woollen cloth simulating buckskin.

buckwheat. Add: **1. c.** *ellipt.* A buckwheat cake. *U.S.*

1830 *Collegian* 41 (D.A.E.), Six muffins, a dozen buckwheats. 1904 G. H. Lorimer *Old Gorgon Graham* 227 A fellow'll load up with pie and buck-wheats for breakfast, and go around wondering about his stomach-ache.

bu·ckwood. *U.S.* [Buck *sb.*¹ 1.] = Buck-eye 1.

1787 M. Cutler in W. P. & J. P. Cutler *Life* (1888) II. 397 The more useful trees..butternut,..buckwood. 1810 Morse *Amer. Gazetteer* s.v. *Ohio*, Hickory, cherry, buckwood or horse chestnut, [etc.].

bucolic, *a.* Add: **3.** *bucolic cæsura*, a cæsura after the fourth foot in a dactylic hexameter.

1887 G. M. Hopkins *Let.* 20 Feb. (1938) 130 The 'bucolic caesura' (between fourth and fifth foot, pause or no pause in sense). *Ibid.*, The rarity of spondees before the bucolic caesura. 1957 *Encycl. Brit.* XXII. 56/1 A feature in his [*sc.* Theocritus'] versification .. is the so-called bucolic caesura. The rule is that, if there is a pause at the end of the fourth foot, this foot must be a dactyl.

bucolicism (biukǫ·lisiz'm). [f. Bucolic + -ism.] Bucolic qualities or characteristics; the bucolic style.

1879 M. Pattison *Milton* iii. 41 This factitious bucolicism is pervaded by a pathos, which, like volcanic heat, has fused into a new compound the dilapidated débris of the Theocritean world. 1893 *Temple Bar* Jan. 63 Beer, brandy and bucolicism notwithstanding.

bud, *sb.*¹ Add: **3. d.** A girl who is just 'coming out'; a débutante. Also more fully *bud of promise*. Chiefly *U.S. colloq.*

1880 R. Grant *Confessions of Frivolous Girl* i. 39 'This is your first party, I believe, Miss Palmer?'..'Yes, I am what is called a "bud".' 1888 *Charlestown Enterprise* (Farmer), The young, unmarried girl, in sport, Is called a bud of promise. 1889 *Harper's Mag.* Sept. 571/1 As dashing a belle as there was in the rooms—not a bud—a belle of some six or seven years standing. 1903 *N.Y. Tribune* 4 Oct., Among the latter [*sc.* the débutantes] will be Miss Enid Shaw, the only bud of the Cabinet circle. 1919 *Sphere* 1 Nov. 120/2 One of those dream-like young girls that only seem to happen in London during the height of the summer and winter seasons...This gilded 'bud' drove away in a cloak of warm pink panne.

4. (See also Nip *v.*¹ 5 b.)

5. bud-graft *v. trans.* = Bud *v.*¹ 5; also as *sb.*, a shrub or tree grown by this process; so *bud-grafted* ppl. adj.; **bud-rot,** rotting of the buds of a plant or tree; a disease characterized by this; **bud-sport** [Sport *sb.*¹ 6 b], an abnormal variation produced from a bud; **bud-worm,** a larva that feeds upon the buds of corn, tobacco, fir-trees, etc.

1930 *Times* 26 Mar. 25/6 The yields from bud-grafted trees will exceed those from selected seedlings. Of the 4,343 acres planted, 2,200 acres are either bud-grafted or alternate rows of bud-grafts and selected seedlings. 1936 *Economist* 18 Apr. 152/1 Those young areas..had been budgrafted. 1906 E. J. Butler in *Agric. Jrnl. India* I. 310 A cocoanut palm disease known as 'bud rot'. 1946 *Nature* 14 Dec. 882/1 Bud-rot in the eastern tropics, hitherto fortunately never confused with eelworm attack, is due to *Phytophthora*. 1900 B. D. Jackson *Gloss. Bot. Terms* 38/1 Bud sport. 1905 H. de Vries *Species & Varieties* xiv. 427 Bud-sports of variegated hollyhock, elms, chestnuts, beeches and others. 1956 C. Auerbach *Genetics in Atomic Age* vi. 41 If it [*sc.* a mutated gene] is of a kind to produce a visible effect, a so-called 'bud-sport' will result, for example, a branch with white berries on a red-currant bush. 1850 *Rep. Comm. Patents 1849* (U.S.) 459 With the bud-worm you must be more particular. 1922 W. Schlich *Man. Forestry* (ed. 4) I. ii. 188 Financial assistance..for the balsam bud-worm investigation.

bud (bʊd), *sb.*³ *U.S. colloq.* [Childish or colloq. pronunc. of Brother *sb.*, or abbrev. of *buddy *sb.* Cf. *bub *sb.*⁶] Brother; used chiefly as a form of address.

1851 *Polly Peasblossom's Wedd.* 19 (Th.), 'An't you joking, bud?' asked Polly of her boy brother. 1889 *Harper's Mag.* Aug. 459/1 He said that his name was 'Bud' Lightwood...'It's *brother*,' he said..'"bud" and "sis", you know.' 1920 Mulford *J. Nelson* iv. 46 But, say, bud, we don't have to go to Highbank at all. 1953 W. R. Burnett *Vanity Row* i. 9 Gamblers..would often hand him a quarter ..and say: 'Keep it, bud.'

budda (bʊ·dā). Also **buddah, buddha,** etc. [Native name.] An Australian myoporaceous plant, *Eremophila mitchelli*.

1890 *Sydney Mail* 14 June 1300/1, I would state that the tree known as sandalwood on the Darling and in the West generally, one of the 'Eremophylla', called 'Butha' by the natives, is not eaten by sheep. 1911 C. E. W. Bean *'Dreadnought' of Darling* xv. 141 A feathery little bunch of light green..which we often saw along the Darling banks; the stumpy budda. 1923 *Chambers's Jrnl.* 24/2 The air was rich with the scent of sandalwood and buddah. 1968 K. Weatherly *Roo Shooter* 18 The old doe and her joey camped under a thick buddha bush.

buddawong, var. *burrawang.

Buddhistically (budi·stikăli), *adv.* Also **buddhistically.** [f. Buddhistical *a.* + -ly².] In a Buddhistical manner.

1920 *19th Cent.* July 59 'The Bull' makes us sympathise almost Buddhistically with the huge beast. 1921 D. H. Lawrence *Let.* 8 May (1962) II. 653, I suppose I buddhistically removed her beyond sorrow.

buddleia (bʊ·dliă, bʊdli·ă). *Bot.* Also **buddlea, buddleja.** [mod. L. (Linnæus *Hortus Cliffortianus* (1737) 35), f. the name of Adam *Buddle* (died 1715), botanist: see -ia¹.] Any plant of the genus of shrubs and trees of this name of the family Loganiaceæ, natives of America, Asia, and South Africa; esp. the large deciduous shrub *B. davidii*, with mauve flowers in panicles.

1791 *Curtis's Bot. Mag.* 174 Round-headed Buddlea. 1864 Grindon *Brit. & Garden Bot.* 415 In genial situations the Buddlea will often attain the height of 12–15 feet. 1885 *Encycl. Brit.* XVIII. 673/2 The *Buddleia*..flourishes at a height of 12,000 feet round the shores of Lake Titicaca [Peru]. 1924 *Blackw. Mag.* Apr. 446/1 Red admirals in August sunning themselves on the buddleia. 1927 *Times* (Weekly ed.) 22 Sept. 327/3 Purple buddleia. 1964 F. A. Bush *Trees & Shrubs* viii. 109 Buddleias flourish in a deep well-drained soil in sunny but sheltered situations.

buddy (bʊ·di), *sb. colloq.* (orig. *U.S.*). [Possibly an alteration of Brother *sb.* (cf. *bud *sb.*³) or a variant of Butty.] Brother; companion, friend; freq. as form a of address. So **buddy-buddy** *sb.*, a friend; also *adj.*, friendly.

1850 'Barton Premium' *Eight Yrs. Brit. Guiana* 218 Buddy (brother) how you can fink me sha' talk so to you? 1852 Whitmore *Diary* 25 Dec., Wrote to my folks and took a Christmas dinner with my buddy. 1858 *Harper's Mag.* Jan. 284/2 'Look, sister, see; the sky's got the measles!' 'No, buddy,' said she,..'it's only freckled.' 1927 J. Barbican *Confessions of Rum-Runner* ii. 24 He had a buddy [*f.n.* Chum], and they went into a liquor deal together. 1932 N. Mitford *Christmas Pudding* iv. 51 Little Bobby Bobbin..is a great buddy of mine. 1937 *Daily Express* 17 Mar. 12/2 When I went into the night nursery to get the boys up I was greeted with a shout of 'Stick 'em up, buddy'. 1952 M. Laski *Village* vi. 102 Those two are getting to be such buddies they can't bear to be parted. 1960 *Woman* 26 Mar. 12/3 He rather enjoyed Miss Hopkins's naïve assumption that now...he and she were buddies. 1962 L. Deighton *Ipcress File* xviii. 115 This way they stopper up the information without offence to old buddy buddies. 1962 K. Orvis *Damned & Destroyed* xxv. 184 Those two got real buddy-buddy.

Hence as *v. intr.* (U.S. colloq.), to become friendly. Usu. with *up*.

1931 G. Irwin *Amer. Tramp & Underworld Slang* 38 'Me and Slim buddy up and take a trip'—Slim and I struck up a friendship and went for a trip together. 1948 N. Algren in *Penguin New Writing* XXXIV. 137 My cot was next to his, and we started buddying up.

budgeree (bʊ·dʒəri), *a. Austral. colloq.* Also **boojery, budger(r)y.** [Native word: cf. next.] Good, excellent.

1793 J. Hunter *Port Jackson* viii. 213 They very frequently, at the conclusion of the dance, would apply to us..for marks of our approbation..which we never failed to give by often repeating the word *boojery*, which signifies good; or *boojery caribberie*, a good dance. 1832 T. L. Mitchell *Jrnl.* 3 Jan. in *Three Expeds. E. Australia* (1838) I. 63 In vain did Dawkins address them thus, 'What for you jerran budgerry whitefellow?' 1850 *Household Words* I. 476/2 They ..spoke in a friendly strain; Budgery Master always gibit bullock along im Black fellow. 1918 *Chambers's Jrnl.* Apr. 268/1 Instead of being 'bong', he was 'budgerie' (all right). 1966 W. S. Ramson *Austral. Eng.* vi. 105 Except, perhaps, in the north and west of Australia, *budgeree* has passed out of use.

budgerigar (bʊ·dʒərigā:ɪ). Also **betcherrygah, betshiregah, bougirigard, budgeragar, budgereg(h)ar, budgery garr, budgeregore.** [Native Australian ('Port Jackson dialect', Morris *Austral English*), f. *budgeri, boodgeri* good + *gar* cockatoo.] A small Australian parrot, the grass or zebra parakeet, *Melopsittacus undulatus*, a popular cage-bird in Britain and elsewhere. Cf. *budgie.

1847 Leichhardt *Overland Exped.* 297 The Betshiregah (Melopsittacus undulatus, *Gould*) were very numerous. 1848 H. W. Haygarth *Bush Life in Australia* xii. 139 A most brilliant little parrot..about the size of a bullfinch,.. called the budgery garr. 1857 W. Howitt *Tallangetta* I. ii. 48 Young paroquets, the green leeks, and the lovely speckled budgeregores. 1857 F. J. A. Hort in A. F. Hort *Life & Lett.* (1896) I. 388 A small green creature like a miniature cockatoo, called a Budgeragar. 1889 *Times* 16 Feb. 4/3 Crystal Palace Cagebird Show 1889...Two Australian budgerigars. 1922 E. V. Lucas *Genevra's Money* xix. 132 Little foreign birds for the most part, avadavats, Java sparrows, budgerigars. 1968 K. Weatherly *Roo Shooter* 118 Budgerigars came in thousands, wheeling in aerial manœuvres of unbelievable intricacy.

budget, *sb.* Add: **1. b.** (Later examples.)

1847 A. Brontë *Agnes Grey* III. xiv. 219 There's Matilda ..and I must go and open my budget to her. 1861 Trollope *Tales of all Countries* 133 At length Miss Jack was allowed to open her budget, and to make her proposition.

3. a. Esp. of news; *spec.* a long letter full of news.

1807 C. Wilmot *Let.* 15 May in *Russ. Jrnls.* (1934) ii. 241 Months have intervened since your delightful Budget reach'd these Realms. 1852 E. Ruskin *Let.* 16 Jan. in M. Lutyens *Effie in Venice* (1965) ii. 246, I am going out to tea ..but have time to begin my weekly budget before I go. 1855 Trollope *Warden* xii. 185 The budget of news which was prepared for her father. 1868 C. M. Yonge *Chaplet of Pearls* I. xiv. 190 He gathered up the sense of the letters.. and said, 'This is a woful budget, my poor son.' 1960 C. Day Lewis *Buried Day* ii. 30, I had a budget from her last week.

4. Now freq. (and not merely humorously) applied to the domestic accounts (of income and its manageable expenditure) of a family or individual; also, the money available for domestic spending; so *on a budget*, with a restricted amount of money.

1854 C. M. Yonge *Heartsease* I. i. vi. 92 Your budget? Are you good at arithmetic? 1899 R. Whiteing *No. 5 John St.* iii. 18 To the Budget, then. Rent, 2s. 6d. a week; coal and candle, 6d. 1901 B. S. Rowntree *Poverty* p. ix, Chapter viii. deals with workmen's budgets, and especially the *diet* of the working classes. 1909 C. F. G. Masterman *Condition of England* iv. 98 The Blue Book..analyses over a thousand 'family budgets', each giving details of how much is spent weekly on butter, tapioca, or treacle. 1932 *Listener* 4 May 630/1 Wheat occupies a much smaller place in the housewife's budget than it once did. 1955 *Oxf. Jun. Encycl.* XI. 154/1 By 1951, 11% of the weekly budget, almost as much as the rent, was being spent on milk. 1959 *Economist* 4 Apr. 46/1 Those on a budget go to Florida in spring or late autumn, the 'off seasons' when charges there are reduced.

b. *attrib.* or quasi-*adj.* Designed or suitable for someone of limited means; cheap.

1958 *Woman* 29 Nov. 6/2 This is just the drink to give party guests a glow—at a budget price. 1960 *Housewife* May 31/1 There are two restaurants catering for both luxury and budget tastes. 1969 *Woman's Own* 12 Apr. 27 Budget meals for the family.

7. budget-trimmer, a man who prepares and fixes in position the leather fittings on coaches and carriages; **budget-wise,** (*a*) *adv.* (orig. *U.S.*), with reference to a budget; (*b*) *adj.*, making full use of limited resources.

1881 *Instr. Census Clerks* (1885) 56 Budget trimmer. 1909 *Daily Chron.* 4 Mar. 4/7 Wanted..Budget Trimmers, accustomed to head work. 1952 T. Pyles *Words & Ways of Amer. English* vii. 189 Combinations with..*-wise*,.. *stylewise, budgetwise*. 1956 A. H. Compton *Atomic Quest* 196 The Metallurgical Project had by now grown until it was budget-wise the major part of the University's activity. 1958 *Woman* 22 Nov. 31/1 Budget-wise dishes. Family recipes..that are easy on the purse.

budget, *v.* **b.** Substitute for def.: *intr.* To draw up or prepare a budget (Budget *sb.* 4); esp. *for* a certain supply or establishment, or *for* a particular financial result.

1884 *Daily News* 9 Oct. 4/6 An army of 6,000 men and a force of 7,757 police were budgeted for in 1883. 1893 *Ibid.* 24 Mar. 5/6 Although the Government of India are most unwilling to budget for a deficit. 1900 *Westm. Gaz.* 23 Mar. 2/2 Every Chancellor of the Exchequer budgets with the fear of the Irish members before his [eyes]. 1901 *Ibid.* 4 June 2/2 When Sir Michael Hicks-Beach Budgeted for 1901–1902. 1922 G. A. Greenwood *England to-day* 28 There..is the inevitable wear and tear of the home to be budgetted for. 1957 C. Morgan *Challenge to Venus* iv. i. 197 The small professional class do worry. .. They budget. They keep personal accounts.

c. *trans.* To arrange (for) in a budget.

1890 *Sat. Rev.* 16 Aug. 191/1 General revenue, as budgeted for the years 1890–91, does not maintain the improvement of the previous year. 1909 D. Lloyd George in *Daily Chron.* 23 Oct. 1/1 The increment duty, which I budgetted to yield £50,000 this year. 1944 *Bath Wkly. Chron. & Herald* 24 June 3/1 (Advt.), By budgeting my points to work in with the rest of the rations, we make out very well.

Hence **bu·dgeting** *vbl. sb.*, the preparation of a budget; financial planning.

1945 G. Williams *Women & Work* iii. 88 Budgeting and shopping on a small income. 1951 *Good Housek. Home Encycl.* 30/2 The first essentials for budgeting are to keep weekly or monthly accounts.

budgie (bʊ·dʒi), colloq. abbrev. of *budgerigar.

1936 W. Watmough *Cult of Budgerigar* xv. 207 Although Budgies are so hardy..reasonable care should..be exercised to protect them from chills. 1959 'A. Gilbert' *Death takes Wife* xv. 186 We've got a budgie..that Maureen's teaching to talk.

‖**Buen Retiro** (bwen reti³·ro). [Sp., lit. 'good retreat'.] The name of a palace near Madrid, used *attrib.* or *absol.* to designate a soft-paste porcelain made there during the reign of Charles III.

1863 W. Chaffers *Marks & Monogr.* 173 This monogram of Charles III is said to be found on the Buen Retiro porcelain, without the crown. 1869 Lady C. Schreiber *Jrnl.* (1911) I. 11 A set of Buen Retiro white (moulded) china. .. Figure of a man in white porcelain, .. marked Buen Retiro.

buer (biū³ɪ). *north. dial.* and *tramps' slang.* Also 9 **bure, buor, bewer.** [Orig. unknown.] A woman, spec. one of loose character.

1807 J. Stagg *Misc. Poems* (1808) 144 A bure her neame was Meg, A winsome weel far'd body. 1886 W. Newton *Secrets of Tramp Life Revealed* 8 *Buor*, a woman. 1889 Barrère & Leland *Dict. Slang* I. 110/2 *Bewer*, (tinkers' slang), a woman...Young *bew'r*, a girl. 1938 G. Greene *Brighton Rock* i. i. 8 'Christ,' the boy said, 'won't anybody stop that buer's mouth?' *Ibid.* vi. i. 228 That was what always happened if you took up with anything but a buer; they gave you the air.

Buerger (bɜ̄·ɪgəɪ). [The surname of L. *Buerger* (1879–1943), an American physician and surgeon.] *Buerger's disease*, inflammation and thrombosis in the small and medium-sized blood-vessels of the extremities, freq. leading to gangrene; also called *thrombo-angiitis obliterans*.

1914 *Trans. Philad. Acad. Surgery* XVI. 319 (*heading*) Thrombo-angiitis obliterans. (Buerger's disease). .. The disease usually attacks Polish and Russian Jews between the ages of twenty and..forty years. *Ibid.* 321 A man upon whom he had operated at different periods for Buerger's disease was stricken with a cerebral embolus. 1946 J. M. Murry *Let.* 27 May (1961) 278 The condition was diagnosed eventually as a rare disease—sometimes called Buerger's disease. 1968 *Brit. Jrnl. Surgery* LV. 452/1 Many surgeons do not recognize a disease entity, Buerger's disease, and prefer to ascribe the condition to atherosclerosis.

buff, *sb.*² **3.** Delete (somewhat *arch.*) and add later examples.

1956 V. Jenkins *Lions Rampant* i. 17 They went swimming, sunbathed, did their training stripped to the buff. 1965 G. McInnes *Road to Gundagai* ix. 153 The wizened fellow..observed us undressing down to the buff. 1969 *Rolling Stone* 28 June 4/1 The girls call themselves the Groupies and claim they recorded their song in the buff.

5. b. In full *Buff Cochin*, a variety of the Cochin fowl, in which both cock and hen are of a uniform buff colour.

1855 *Poultry Chron.* III. 173 Our old friends, the Cochins, mustered pretty strong,..the buffs..were very good. 1873 L. Wright *Bk. Poultry* 210 We have several shades in the Buff Cochin cock. *Ibid.* 213 The earliest and greatest breeders of Buff Cochins. 1899 Norris-Elye *Brahmas & Cochins* 61 Evenness of colour is perhaps the greatest difficulty in breeding..buff Cochins.

6. (Royal assent was given to the restoration of buff facings to the East Kent Regiment on 19 August 1890.)

b. 'An enthusiast about going to fires' (Webster 1934); so called from the buff uniforms worn by volunteer firemen in New York City in former times. Hence *gen.*, an enthusiast or specialist. Chiefly *N. Amer. colloq.*

1903 *N.Y. Sun* 4 Feb. iv. 2/1 The Buffs are men and boys whose love of fires, fire-fighting and firemen is a predominant characteristic. 1907 A. M. Downes *Fire Fighters & Pets* xiii. 159 The 'buff' is a private citizen who is a follower, friend, and devoted admirer of the firemen. 1931 Lavine *Third Degree* vi. 62 A dentist, known to many cops as a police buff (a person who likes to associate with members of the department and in exchange for having the run of the station house does various courtesies for the police). 1955 *Sci. Amer.* Aug. 88/3 No choo-choo buff can be without Sampson. 1962 *Listener* 1 Nov. 704/2 A neighbour of mine who is a hi-fi buff. 1963 *Economist* 20 July 244/2 The 'Pugwash' meetings between western and Russian scientists and other disarmament buffs. 1966 *New Yorker* 17 Sept. 130 For ballet buffs, Tuesday evening of last week was a great occasion. 1968 *Globe & Mail* (Toronto) 17 Feb. 37 Sports buffs will enjoy many diversions, with bicycling and camping..heading the list.

buff, *sb.*⁷ **1.** (With capital initial.) Short for *buffalo *sb.* 1 e. *colloq.*

1879 *The Buffalo* 16 Jan. 3/3 The great scheme of a technical university now being taken up by the City companies, was first started by Buffs. 1888 [see *buffalo *sb.* 1 e]. 1897 *Buffalo World* Sept. 3/2 It should..be the aim of every loyal Buff..to show his desire to help on the cause for which we are fighting, viz. Progress and the Brotherhood of man. 1909 *Daily Chron.* 31 July 4/5 A..belief..that the 'Buffs', as it is generally called, originated at the Harp Tavern, in Russell-street, Drury-lane, in 1822.

2. Short for Buffalo *sb.* senses 1 a to d. *colloq.*

1583 G. Peckham *True Reporte* in Hakluyt *Voy.* (1600) III. 175 He and his company did finde in one cottage aboue two hundred and fortie hides..and with this agreeth Dauid Ingram, and describeth that beast at large, supposing it to be a certaine kinde of Buffe. 1665 P. E. Radisson *Voyages* (1885) 212 They have very handsome shoose laced very thick all over wth a peece sowen att the side of ye heele, wch was of a haire of Buff. 1884 *Bismarck Tribune* Aug., The ball struck the unsuspecting animal...But the old 'buff' took the fling as an insult. 1935 Hemingway *Green Hills of Africa* (1936) ii. iii. 98 I'd rather get another buff than rhino. *Ibid.* 113 Where the rhinos and the buff had come out of the reeds. 1964 C. Willock *Enormous Zoo* ii. 24 When the buff was nearly up with him, the boy took off his hat and put it over the animal's eyes.

buff, *v.*³ Add: **2.** To impart a buff colour to.

1897 Rothwell *Textile Fabrics* 237 The pieces are to be 'buffed' or 'slop-padded' with substantive dyestuffs in solutions containing soap.

buffalo, *sb.* Add: **1. c.** (Earlier examples.)

1635 *Relat. Maryland* iii. 23 In the upper parts of the countrey there are Bufeloes, Elkes, Lions, Beares, Wolues, and Deare there are in great Store. 1705 R. Beverley *Virginia* ii. 39 The Elks, Buffaloes, Deer and greater Game.

d. *collect.*

1765 G. Croghan *Jrnl.* (1875) 132 The country hereabouts abounds with buffalo, bears, [etc.]. 1817 S. R. Brown *Western Gaz.* 30 The buffaloe..have lately disappeared. 1895 C. King *Fort Frayne* xviii. 260 A deep cleft in the foothills through which the buffalo in bygone days had made their way.

e. (With capital initial.) A member of the Royal Antediluvian Order of Buffaloes, founded in 1822 for sociable and benevolent purposes. Hence **Bu·ffaloism.**

1869 P. Egan *Finish Adv. Tom, Jerry, & Logic* v. 120 At the Harp, in Great Russel Street, opposite Drury Lane Theatre, the Buffalo Society was first established, in August, 1822. 1879 *The Buffalo* 16 Jan. 3/3 Bro. Barrett, the Buffalo Bootmaker of Walworth. 1881 (*title*) The Buffalo Review and Lodge Reporter. 1888 C. Hindley *True Hist. Tom & Jerry* 162 *Buffs—Buffaloes*—and *Buffaloism*. —A society..established in August, 1822, by an eccentric young man of the name of Joseph Lisle, an artist, in conjunction with Mr. W. Sinnett, a comedian, to perpetuate, according to their ideas upon the subject, of that hitherto neglected ballad of *We'll chase the Buffalo!* 1897 *Daily News* 16 Mar. 8/3 A room in which certain 'Buffaloes' were holding a lodge meeting. 1897 *Buffalo World* Sept. 3/1 Buffaloism can boast an existence of 300 years at least. 1970 *Sunday Times* 18 Jan. 52/3 We used to have a branch of the Buffaloes at the pub. *Ibid.*, Her family were very high up in the Buffaloes.

f. An amphibious tank.

1944 *Hutchinson's Pict. Hist. War* Oct. 139 (*caption*) British troops were carried across the Scheldt in assault craft...This 'Buffalo' assault craft is carrying back some of the prisoners taken. 1945 *Times* 1 Mar. 4/3 The generals.. rode in a 'buffalo' and inspected it minutely. 1959 *Times Lit. Suppl.* 13 Feb. 86/2 This division was also the British Army's nursery of the amphibious 'Buffaloes' of American origin.

5. (Earlier and additional examples.) Also **buffalo-bean,** a milk vetch of the western United States, *Astragalus crassicarpus*; **buffalo fly, gnat,** a small biting insect of the genus *Simulium*; **buffalo grass,** for *Sesleria dactyloides* read *Buchloë dactyloides*; also used generally (see quot. 1950); (*b*) any of various African grasses used for pasture and fodder; (*c*) *Austral.* and *N.Z.*, the grass *Stenotaphrum americanum*, introduced from the United States, and first noticed near Buffalo Creek in New South Wales (Webster 1911); **buffalo-horn,** (*a*) the horn of a buffalo; (*b*) an African tree, *Zizyphus mucronata*; (*c*) *U.S.* (see quot. 1887²).

1906 P. A. Rydberg *Flora of Colorado* 202 Geoprumnon... Buffalo Beans, Ground Plums. 1922 *Chambers's Jrnl.* 219/1, I..became acquainted with a creeping plant that grows a bean...I have since heard it called buffalo-bean. 1805 *Massachusetts Spy* 17 July 2/3 (Th.), Scions of a newly discovered berry, called the buffaloe berry. 1840 *Picayune* 11 Oct. 2 We raised an extensive cloud of smoke from burning 'buffalo chips' to keep off the musquitos. 1767 in *N. Carolina Col. Rec.* VII. 1007 Buffalow Clover was extremely thick here. 1774 D. Jones *Jrnl.* (1865) 111 Another kind of fish called buffaloe fish, many of which are larger than our sheepshead. 1849 Charles Lyell *Second Visit N. Amer.* II. 89 There were swarms of buffalo flies to torment his horses, and sand flies to sting him and his family. 1932 *Discovery* July 210/2 The buffalo fly is another extremely serious pest, in this case of the cattle industry. 1822 J. Woods *Eng. Prairie* 278 We had no buffalo gnats. 1959 J. Clegg *Freshwater Life Brit. Isles* (ed. 2) xiv. 235 The Black-flies..under such names as Buffalo Gnats.. are only too well known as pests of cattle. 1784 J. Filson *Discovery of Kentucke* 24 Where no cane grows there is abundance of wild-rye, clover, and buffalo-grass,..affording excellent food for cattle. 1868 J. Chapman *Trav. S. Afr.* II. 457 (Pettman), The Buffalo grass has a large, broad, corrugated leaf and is greedily eaten by horses and cattle. 1876 F. M. Bailey in *Papers & Proc. R. Soc. Tasmania 1875* 132 The Buffalo grass, *Stenotaphrum Americanum*..is a very fine and desirable species. 1950 *Amer. Speech* XXV. 164 The ground is covered with 'buffalo grass', which once designated a specific type but now means any tough grass that grows on the poor soil of the plains. 1783 W. Fleming in N. D. Mereness *Trav. Amer. Col.* (1916) 665 We picked up..a petrified Buffalo horn. 1887 Moloney *Forestry W. Afr.* 300 'Buffalo-Horn' (*Zizyphus mucronata*, Willd.). 1887 *Scribner's Mag.* II. 507 The latter fixes his attention on the saw-like, serrated crowns, or summits, which are.. typical..of true mountainous form. There are plenty of such features in the Rocky Mountains, and natives call them 'buffalo-horns'. 1810 Z. M. Pike *Exped. Sources Mississippi* App. ii. 34 Restricting (by edicts) the buffalo hunts to certain seasons. 1824 A. Ross *Jrnl.* 28 Mar. in *Oregon Hist. Soc. Quart.* XIV. 376 The buffalo hunters came back today, buffalo in plenty; thirty killed. 1775 W. Calk *Jrnl.* 10 Apr. in *Filson Club Pubn.* II. 36 Some of the company went over the River a bufelo hunting but found none. 1857 Gray *Botany* 382 Buffalo-nut..[is] a low straggling shrub, with..small greenish flowers. 1775 Adair *Hist. Amer. Indians* 118 Living very scantily, even in a buffalo range, under a strict rule, lest by luxury their hearts should grow evil. 1804 Clark in *Lewis & C. Exped.* (1904) I. 130 The [Sioux] Squars wore Peticoats & a white Buffalow roabe. 1732 *S.C. Gazette* 25/1 Preparing Buffelo, Deer, Sheep, Goat or Kid-Skins in Oil.

bu·ffalo, *v. N. Amer. slang.* [f. the sb.] *trans.* To overpower, overawe, or constrain by superior force or influence; to outwit, perplex. So **bu·ffaloed** *ppl. a.*

1903 *Cincinnati Enquirer* 9 May 13/1 Buffaloed—Bluffed. 1904 *N.Y. Even. Post* 25 Oct. 10 All the rest [of the newspapers] were what we used to term in the Southwest 'buffaloed' by the McKinley myth—that is, silenced by the fear of incurring the resentment of a people taught to regard McKinley as a saint. 1910 W. M. Raine *B. O'Connor* 77 O'Connor admitted that he was 'buffaloed' when he attempted an analysis of his unusual feeling. 1947 E. A. McCourt *Flaming Hour* 118 Jerry Potts himself would have been buffaloed.

buffe (buf). [See Buff *sb.*²] In ancient armour, a chin piece pierced with breathing-holes, worn with the burgonet.

1600 [see Buff *sb.*²]. 1885 H. A. Dillon *Fairholt's Costume* (ed. 3) II. 102 A separate and distinct chin-piece called a *buffe* is so often associated with it [*sc.* the burgonet]. 1909 C. ffoulkes *Arm. & Weapons* iv. 83 The face-guard, when used with the burgonet, is called the Buffe. 1922 *Daily Tel.* 12 June 20/3 The 'Buffe' of French 16th century work, which belongs to the 'Colbert' casque.

bu·ffel. †**1. buffel duck** = *bufflehead 2. Also *buffel's head duck. Obs. U.S.*

1731 Catesby *Nat. Hist. Carolina* I. 95/1 The Buffel's Head Duck. *Ibid.*, The length and looseness of these Feathers make the Head appear bigger than it is, which seems to have given it the Name of Buffel's Head, that Animal's Head appearing very big by it's being covered with very thick long Hair. 1785 Pennant *Arctic Zool.* II. 559 Pied and buffel duck. 1831 Wilson & Bonaparte *Amer. Ornith.* IV. 351 Red headed buffel duck.

2. buffel grass [Du. *buffel gras*] = *buffalo grass* (b).

[1915 R. Marloth *Flora S. Afr.* IV. 22 Buffelgras.] 1955 J. Cleary *Justin Bayard* xii. 182 Between the patch of buffel grass and the grey-blue slate was a small strip of fine sand. 1958 *Austral. Encycl.* IV. 126/1 Introduced species [of]..fodder plants..from Africa, veldt..buffel (Cenchrus ciliaris)..grasses.

buffer¹. Add: **1. b.** A substance or a mixture of substances, usu. of a weak acid or base and its salt, which stabilizes the degree of acidity or alkalinity of a solution; also, a buffer solution. Also *attrib.* in *buffer action, base, salt*; **buffer solution** (see *3).

1914 *Chem. Abstr.* 3561 The action of moderators (buffers) in the shifting of the acid-base equilibrium in biological liquids...A monobasic acid shows its max. buffer reaction at a H-ion conc...The max. buffer action of all weak acids in equiv. concs. is the same. 1922 *Encycl. Brit.* XXXI. 59/1 Death will be due to acid poisoning, and the administration of what are called 'buffer salts'..is instrumental in postponing the fatal issue. 1926 Tansley & Chipp *Stud. Vegetation* vii. 124 This substance acts as an acid 'buffer' to the soil solution. 1961 *Lancet* 22 July 176/1 The solvent contains 45 ml. of phosphate buffer (pH = 7·4). 1965 [see *alkalosis].

c. *Biol.* (See quot. 1939 and cf. *buffer *v.* c.)

1939 H. J. Muller in *Biol. Rev.* XIV. 271 Some 'buffers' may eventually arise—that is, genes that compensate for a change in one or more of the secondary effects of the primary gene in question. 1943 *Nature* 16 Jan. 68/2 The distinction between switch genes and buffer genes.

d. A 'memory' device in a computer (see quot. 1962); a buffer memory.

1948 *Math. Tables & other Aids to Computation* III. 289 Two 32-word delay lines, or reservoirs..are used as buffers between the main electronic part of the machine and the tape. 1956 *Control Engineering* Nov. 117/1 Input Information..is transferred to the internal memory of the computer either directly or through an external storage device or buffer. 1958 M. Phister *Logical Design Digital Computers* viii. 216 The buffer memory must be independent of the operation of the rest of the main memory. 1962 S. Handel *Dict. Electronics* 40 *Buffer*,..a memory device used to compensate for a difference in the rate of flow of information in the different parts of a computing system.

2. *spec.* (chiefly *attrib.* or quasi-*adj.*). (Designating) a state, zone, etc., lying between two others, usu. owing allegiance to neither, and serving as a means of preventing hostilities between them.

[1876 B. Frere in W. B. Worsfold *Sir B. Frere* (1923) 43 Using the Afghans as a 'buffer' to avoid immediate contact between our frontier and the Russian.] 1883 [in Dict.]. 1908 *Westm. Gaz.* 7 May 1/3 A 'buffer' zone, inhabited by independent tribes, lies between the boundaries of British India and Afghanistan. 1920 *Discovery* May 133/2 That treaty settlement of neutrality which regulated the status of Luxemburg as a buffer-duchy. 1931 A. P. Herbert *Derby Day* 27 Belgium's a buffer nation.

3. *buffer-bank*; **buffer amplifier** *Electr.* (see quot. 1938); **buffer-block,** (*a*) a framework of timber set up at the end of a railway line or siding; (*b*) a block on the end of a coach, which acts as a buffer; (*c*) the flat head of a buffer; **buffer-box, -case,** the case which encloses the buffer-rod and -spring; **buffer-knot,** an arrangement of two knots joining two parts of a fisherman's line in which the strain is taken by a piece of waxed silk which acts as a buffer; **buffer-plunger,** the portion of a buffer which slides in the buffer-case and carries the shock to the spring; **buffer pool** = **buffer stock*; **buffer solution** *Chem.*, a solution containing a 'buffer' (see *1 b); **buffer stock,** a stock of a commodity held in reserve so as to reduce fluctuation in prices when supplies are low; **buffer-stop** = *buffer-block* (*a*); **buffer store,** (in a computer) = sense *1 d above.

1933 J. H. Morecroft et al. *Princ. Radio Communic.* (ed. 3) vii. 727 The crystal oscillates at 2500 kc. and gives in its tank circuit about 50 volts of this frequency, and the buffer amplifier raises this to about 400 volts at the same frequency. 1938 G. P. Harnwell *Princ. Electricity* vii. 215 Multielectrode tubes not only amplify but also isolate the plate and grid circuits, avoiding the many undesired effects of interaction; these are frequently known as *buffer amplifiers*. 1900 *Westm. Gaz.* 21 Aug. 6/1 The accident at Kent House Station, near Penge (in which a train ran into a buffer-bank, injuring eleven people). 1889 *Century Dict.*, Buffer-block. 1892 *Daily News* 14 Jan. 3/4 The effect was to cause the waggon furthest from the engine to mount the fixed buffer block. a1877 Knight *Dict. Mech.* s.v. *Buffer*, The buffer bar or beam is attached to the framework of the car, and carries the buffer-box, in which is the buffer-rod. 1880 *Engineering* XXX. 294 Projections h' on the buffer case. 1885 Treherne in Pennell et al. *Fishing* 41 Although I can lay no claim to be the inventor of the 'buffer knot', I can honestly say that I had never seen or heard of it before. 1895 'J. Bickerdyke' et al. *Sea Fishing* 69 The buffer knot is another good one;..I learnt the way of tying it, shown in the illustration, from an Irish friend, who has greatly improved on the old original buffer. 1880 *Engineering* XXX. 294 Tijou's mode of securing buffer plungers. 1940 *Economist* 24 Feb. 338/1 The price of the metal..had been previously pegged at the upper limit of £230 per ton under the buffer pool scheme. 1921 *Jrnl. Biol. Chem.* XLIX. 183 The use of standard buffer solutions for colorimetric comparison. 1951 A. I. Vogel *Text-Bk. Quantitative Inorg. Analysis* i. 31 Buffer solutions usually consist of solutions containing a mixture of a weak acid or base and its salt. 1935 *Economist* 23 Feb. 456/1 At the end of February 7,476 tons of tin, being part of the buffer stock held by the committee, will appear in 'visible supplies'. 1878 *Engineer* XLVI. 7 The standard buffer stop on the London and South-Western Railway. 1962 *Gloss. Terms Automatic Data Processing*

(*B.S.I.*) 64 *Buffer store*, a store used to compensate for a difference in rate of flow of data, or time of occurrence of events, when transmitting data from one device to another. 1964 F. L. Westwater *Electronic Computers* iv. 80 Buffer stores are essentially a device for gaining time or for reconciling the different time scales inherent in a computing system.

buffer³. Add: **d.** A farrier's shoeing tool having a blunt chisel at one end to remove clinch nails and a point at the other to punch out nails embedded in the hoof.
1902 V. Jacob *Sheep-Stealers* xiv, A smith's buffer. 1907 *Yesterday's Shopping* (1969) 706/2 Farriers' tools, set of,.. containing shoeing hammer, pointing hammer, rasp, whetstone, buffer, [etc.].

buffer⁴. Add: **4.** *Naut.* A chief boatswain's mate. Also, a petty officer.
1864 Hotten *Slang Dict.* (ed. 3) 87 *Buffer*, a navy term for a boatswain's mate, part of whose duties is to administer the 'cat'. 1916 'Taffrail' *Stand By!* 30 It wus all I could do to stop meself larfin', 'specially when Number One sings art fur the chief buffer. 1941 *Weekly Telegraph* 25 Oct. 6/3 The 'Buffer'..is a petty officer 'go-between', his duty being that of passer-on of orders from officers to seamen.

buffer (bʊ·fəɹ), *v.* [f. Buffer².] *trans.* To lessen the impact of, with or as with a buffer; to act as a buffer to; freq. *fig.*
1894 *Speaker* 16 June 658/1 The crude..opinionativeness of the permanent official—which is used to being discounted and buffered by a lay chief. 1958 *Listener* 2 Jan. 32/3 Continental statesmen..saw, with a clarity which has so far eluded our more buffered countrymen. 1958 *House & Garden* Apr. 73/2 The kitchen..buffers the children's area from the main reception rooms.
b. *Chem.* To treat with a buffer (see *buffer² 1 b). Also *intr.*, to act as or use a buffer.
1923 W. M. Clark *Determ. Hydrogen Ions* (ed. 2) ii. 44 Unless a solution is buffered..it is almost impossible to make an accurate electrometric determination of the pH... The failure to buffer against the effect of so-called neutral salts..may lead to gross error. 1955 *Sci. Amer.* Feb. 60/3 A thin layer of protein solution..is inserted between a pair of liquids, one acid and one alkaline, which are 'buffered' to maintain a constant pH. 1956 *Nature* 10 Mar. 478/1 The salivary glands..were incubated on a welled microscope slide in buffered substrate (pH 6·8).
c. *Biol.* To limit the effect of.
1936 J. S. Huxley in *Rep. Brit. Assoc. Advancem. Sci.* 82 Each such step is immediately *buffered* by ancillary changes in genes and gene-combinations. 1944 —— *On Living in Revol.* vi. 71 Slightly deleterious genes have been rendered harmless or even beneficial by being 'buffered'.. by new combinations of other genes.

buffering (bʊ·fəriŋ), *vbl. sb.* [f. Buffer² or prec. + -ing¹.] The action of bringing buffers into play; also, buffers collectively.
1898 *Times* 6 Jan. 7 The buffering up of the wagons caused one of them to tilt over. 1928 *Daily Express* 29 Dec. 9/4 The use of side buffering with screw couplings.
b. The action of *buffer *v.* sense b. Also *attrib.* or as *ppl. a.*
1927 *Forestry* I. 122 Plants in which the buffering agent has a reaction of a more favourable type..counteract this acid-buffering. 1932 Fuller & Conard tr. *Braun-Blanquet's Plant Sociol.* vi. 175 Each soil has its characteristic pH value..which normally is little changed by external conditions... This property of tenaciously maintaining a 'reaction constancy' is called buffering.
c. The action of *buffer *v.* sense c. Also *attrib.* or as *ppl. a.*
1936 J. S. Huxley in *Rep. Brit. Assoc. Advancem. Sci.* 82 The adjustment of such mutations..may occur entirely through recombination of existing modifiers, or, after a preliminary and partial buffering by this means, the final adjustment may have to wait. 1943 *Nature* 16 Jan. 68/2 In genetical language, the integrated genotype acts as a buffering system..to limit the variation of the organism's response to environmental fluctuations.

buffet, *sb.*¹ Add: **c.** *Aeronaut.* = *buffeting *vbl. sb.* 2. Also *attrib.*
1951 *Jrnl. R. Aeronaut. Soc.* Oct. 629/2 With very few exceptions buffet comes from the tail. 1958 *Chambers's Techn. Dict.* 962/1 *Buffet boundary*, the maximum Mach number at which a subsonic aeroplane may be safely flown without risk of uncontrollability due to compressibility drag.

buffet, *sb.*³ Add: **1. b.** In various collocations, *buffet meal, party, supper, table*, etc., with sense 1 extended to cover the refreshments set out on the sideboard, table, etc., and where guests or customers are usually served standing. Also *ellipt.*
1888 Mrs. Beeton *Bk. Househ. Managem.* 1443 (*caption*) Buffet Tea-Table Arranged For From Forty Guests. 1906 *Daily Colonist* (Victoria, B.C.) 30 Jan. 6/6 For the first time in Victoria, the buffet supper will be introduced, following the plans of the suppers now usually given at swell balls in Eastern cities. 1922 *Hotel World* 22 Apr. 13/1 At the close of the meeting those in attendance were tendered a buffet luncheon in the main restaurant. 1930 R. F. Wilson *How to wine & dine in Paris* v. 112 The café serves..a cold buffet lunch. 1933 E. Shanks *Enchanted Village* x. 138 The girls and young men round the buffet-table. 1937 M. Hillis *Orchids on your Budget* vi. 102 Buffet suppers are a triumphant solution of the no-maid-and-little-money party. 1951 *Good Housek. Cookery Bk.* 45 Arrange your buffet table in the most convenient place. *Ibid.*, A fair number of people will need to reach the buffet at the same time. 1951 *Good Housek. Home Encycl.* 193/2 Paper serviettes..are inexpensive and convenient..for buffet meals. *Ibid.* 582/1 Suitable food for a buffet party includes sandwiches and bridge rolls.
3. (Earlier examples.)
1792 *Observer* 19 Feb. 3/3 At two o'clock, the buffets were opened, and the company regaled with a cold collation. *c* 1810 W. Hickey *Mem.* (1913) I. xi. 129 The Buffets, which were numerous, were abundantly supplied with refreshments of every kind.
b. buffet-car orig. *U.S.*, a railway carriage containing a refreshment bar.
1887 C. B. George *40 Yrs. on Rail* 248 Buffet,..dining and sleeping cars have all been added to meet the needs and tastes of this enterprising age. 1895 J. C. Wait *Car-Builder's Dict.* (ed. 3) 21 *Buffet-car*, a term..applied to a style of sleeping-car or parlor-car which has an ornamental buffet where light lunches can be prepared for the passengers. 1899 *Railway Times* 11 Mar. 340/1 Five express trains will be run.., two having first and third class dining-cars attached, and one a buffet car. 1969 L. Meynell *Of Malicious Intent* iv. 42 A fast train back to Liverpool Street ..with a buffet car attached to it.

buffeting, *vbl. sb.* Add: **2.** *Aeronaut.* Irregular oscillation, caused by air eddies, of any part of an aircraft.
1931 Frazer & Duncan in *Aeronaut. Res. Committee Rep. & Mem.* No. 1360 Jan. 84 The term 'buffeting' is here used to denote an irregular, and more or less severe, oscillatory movement of the organs of the tail unit. 1961 C. B. Smith *Testing Time* x. 168 The 'buffeting' was related to a basic aerodynamic problem, the turbulence which was bound to occur when an aircraft not suitably streamlined attained to a speed approaching the speed of sound.

buffing, *vbl. sb.*² Add: **c.** The operation of reducing the thickness of a hide by shaving off the grain surface with a currier's knife or splitting-machine; also, the thin pieces of leather so removed. Also *Comb.*, as *buffing-machine, buffing-slicker.*
a 1884 Knight *Dict. Mech.* Suppl., *Buffing* (leather), taking off thin shavings from the grain side with a buffing-slicker until the skin is very thin; the object being to make cowhide imitate calfskin. The operation is finished by whitening. 1897 C. T. Davis *Leather* (ed. 2) 444 The buffings which are not required for japanning are sold in russet for making stained buffings. 1920 *Chambers's Jrnl.* Sept. 622/1 Leather buffings and shavings.

buffish (bʊ·fiʃ), *a.* [f. Buff *a.* + -ish¹.] Somewhat buff, approaching buff in colour; also in comb. with other adjs. of colour.
1802 D. Wordsworth *Grasmere Jrnl.* 14 Feb. in *Jrnls.* (1941) I. 113 The snow-covered mountains were spotted with rich sunlight, a palish buffish colour. 1896 *Brit. Birds, Their Nests & Eggs* I. 3 The cheeks..are buffish white. 1902 E. Glyn *Refl. Ambrosine* 83 Terra-cotta and buffish brocade. 1910 *Westm. Gaz.* 29 Jan. 11/1 The white and buffish brown or ashy streaks..of his plumage.

bufflehead. Add: **2.** A North American duck (*Bucephala albeola*), the head of which appears to be disproportionately large. *U.S.* Cf. next and *buffel 1.
1858 S. F. Baird *Birds Pacific* in *Rep. Explor. Railroad to Pacific* (U.S. War Dept.) IX. 798 The name buffle head is a corruption of buffalo head. 1870 *Amer. Naturalist* IV. 49 Buffel Head (*Bucephala albeola* Baird). 1874 J. W. Long *Amer. Wild-fowl Shooting* Introd. 16 In the deep water varieties,.I shall treat of the..buffle-head or butter-ball.

buffle-headed, *a.* Add: **3.** *U.S.* (See quots. and prec.)
1831 Wilson & Bonaparte *Amer. Ornith.* III. 232 *Anas albeola*..Buffel-headed Duck. 1840 E. Blyth et al. tr. *Cuvier's Anim. Kingd.* 264 The little Buffel-headed Garrot (*Anas albeola* Lin.), common in North America, is nearly allied [to the Golden-eyed Garrot]. 1874 J. W. Long *Amer. Wild-fowl Shooting* 279 Buffle-headed Duck (*Fuligula alvesta*).

buffy (bʊ·fi), *a.*² *slang.* [Origin obscure.] Intoxicated, 'squiffy'.
1858–9 C. W. S. Brooks *Gordian Knot* (1860) viii. 57, I must have conducted myself with extreme propriety, and not as you did at the Clissolds', when you came in buffy. 1866 E. H. Yates *Land at Last* I. vi, Flexor was fine and buffy when he came home last night. 1924 A. Huxley *Little Mexican* 225 She did like boasting about the amount of champagne she could put away without getting buffy.

bufotenine (biūfote·nin, -in). *Biochem.* and *Pharm.* Also **-in.** [ad. F. *bufoténine* (Phisalix & Bertrand 1902, in *Compt. Rend.* 7 July 48), f. L. *būfo* toad + *ten-* (perh. repr. L. *ten-āx* holding fast (f. *ten-ēre* to hold, cf. F. *tenir*), in allusion to its 'paralysing influence') + -ine⁵.] A hallucinogenic tryptamine alkaloid that produces vasoconstriction and hypertension when injected and is present in various amphibia (esp. toads), mushrooms, and tropical shrubs.
1902 *Jrnl. Chem. Soc.* LXXXII. ii. 576 The toxic action [of the venom of the common toad] is due to two principal substances; one..is of resinoid nature...The other, termed bufotenin, is very soluble in both water and alcohol; it has a paralysing influence. 1936 *Jrnl. Biol. Chem.* CXVI. 91 Bufotenine has been obtained from the secretions of *Bufo bufo bufo* (*Bufo vulgaris*) and *Bufo viridis viridis.* 1966 *New Scientist* 25 Aug. 414/3 Bufotenine (5-HDMT) is the active hallucinogen in potions prepared from plants by witch-doctors in the West Indies.

bug, *sb.*¹ Add: **1. b.** *big bug* (earlier and later examples).
1827 *Harvard Reg.* Oct. 247 He who desires to be a big-Bug, rattling in a natty gig. 1892 J. Nie *Rob. Crusoe* (MS.) 21 That you're a big bug here is understood. 1932 E. Waugh *Black Mischief* viii. 300 He seems to have been quite a big bug under the Emperor. Ran the army for him. 1939 'G. Orwell' *Coming Up* ii. ix. 156, I saw..a big bug. You know how it is with these big business men.

bug, *sb.*² Add: **3. bug-hunter,** in various *slang* uses (see quots.); *spec.* an entomologist, a naturalist; so *bug-hunt* v. intr., *bug-hunting*; **bug-juice** *U.S. slang*, bad whisky; **bug-trap,** *Naut. slang* (see quot.).
1796 Grose *Dict. Vulgar T.* (ed. 3), *Bug-hunter*, an upholsterer. 1862 Bug-hunter [see Moocher 3]. 1889 M. E. Bamford *Up & down Brooks* 46 It seems sometimes to the bug-hunter as though there would be but very few vacant rooms to rent in Nature's house. 1962 A. Wise *Death's-Head* vii. 70 Was she one of them? I thought—a passionate bughunter? 1855 Kingsley *Glaucus* 7 The naturalist was looked on as a harmless enthusiast, who went 'bug-hunting', simply because he had not spirit to follow a fox. 1905 *Westm. Gaz.* 19 May 4/2 The pursuit that in schoolboy days of irreverence we used to call 'bug-hunting'. 1869 *New No. West* (Deer Lodge, Mont.) 22 Oct. 1/5 Citizens glad to see us—freedom of the city—'bug juice', *ad lib.* 1888 Farmer *Americanisms, Bug-juice*, the Schlechter whiskey of the Pennsylvania Dutch—a very inferior spirit. Also called *bug-poison*. These terms are now applied to bad whiskey of all kinds. 1895 *Boy's Own Paper* XVII. 427/1 Small vessels are..commonly called 'bug-traps', because they soon get filled up with cockroaches.
4. In various slang uses. **a.** A person obsessed by an idea; an enthusiast. Freq. with defining word, as *jitterbug, litterbug*; cf. *fire-bug* under sense 1 and *s.v.* Fire *sb.* B 5 a. Also, an obsession, a craze. orig. *U.S.*
1841 *Congress. Globe* June 133 Mr. Alford of Georgia warned the 'tariff bugs' of the South that..he would read them out of church. 1909 'O. Henry' *Roads of Destiny* xiii. 208 He's got bugs. Sitting on ice and calling his best friends pseudonyms. 1911 *Daily Colonist* (Victoria, B.C.) (Magazine Section) 18 Apr. 1/2 There are no more critical people than what are generally classified as baseball 'bugs'. 1937 N. Coward *Present Indicative* ix. 377 Bushell and Guerrier, having firmly inoculated me with the naval bug, obtained permission from their captain..for us both to travel..with them. 1946 *Penguin New Writing* XXVIII. 26 The boy's got the religious bug. 1948 'N. Shute' *No Highway* xii. 303, I love being on aerodromes and seeing aeroplanes. It's a sort of bug that gets in you. 1959 *Which?* Dec. 171 A boy bitten by the railway bug.
b. A defect or fault in a machine, plan, or the like. orig. *U.S.*
1889 *Pall Mall Gaz.* 11 Mar. 1/1 Mr. Edison, I was informed, had been up the two previous nights discovering 'a bug' in his phonograph—an expression for solving a difficulty, and implying that some imaginary insect has secreted itself inside and is causing all the trouble. 1935 *Jrnl. R. Aeronaut. Soc.* XXXIX. 43 Casting, forging and riveting are processes hundreds of years old, and, to use an Americanism, 'have the bugs ironed out of them'. 1956 'N. Shute' *Beyond Black Stump* v. 138 They worked..until the rig had settled down and all the bugs had been ironed out. 1958 *Engineering* 14 Mar. 336/2 The seven-and-a-half years..was not an excessive time to..get the 'bugs' out of a new system of that kind.
c. Schoolboys' slang for 'boy'; usu. with defining word, as *day-bug.
1909 [see *day *sb.* 24]. 1927 W. E. Collinson *Contemp. Eng.* 29 Day-bugs and boarder-bugs. 1936 A. Huxley *Eyeless in Gaza* vi. 63 It really wasn't right to treat New Bugs the way he did—as though they were equals. 1960 J. Rae *Custard Boys* i. i. 17 You're new, Curlew, and new bugs should be seen and not heard.
d. A microbe or germ; also, a disease. Hence in *pl.*, bacteriology or biology.
1919 W. A. Fraser *Bulldog Carney* iii. 129 'Gee! now I will get well,' he said; 'I'll beat the bug out now—I'll have heart.' 1927 W. E. Collinson *Contemp. Eng.* 58 Disease-germs are sometimes referred to as bugs. 1932 Kipling *Limits & Renewals* 350 He's bugs—agar-agar—guinea-pigs—slides—slices. The microbe game. *Ibid.* 362 He used to talk bugs to me, too. 1933 Partridge *Slang To-day & Yesterday* iii. 191 *Bugs*, according to the context, bacteria or bacteriology. 1941 J. Cary *House of Children* lxvii. 279 May I get into your bed, Harry?—I'm freezing. I won't breathe any of my bugs on you. 1943 Koestler *Arrival & Departure* iv. i. 128 The whole thing had probably been due to some new kind of influenza, an unknown variety of the bug. 1963 *New Society* 22 Aug. 5/2 'Bugs' may still be used for biology.
e. A burglar-alarm system. *U.S.*
1925 in Partridge *Dict. Underworld*. 1930 J. P. Burke in *Amer. Mercury* Dec. 454/2 *Bug*, a burglar-alarm. 'The casa's bugged.' 1950 H. E. Goldin *Dict. Amer. Underworld* 35/2 There ain't no bug on this joint...Let's charge out (go to work).
f. A concealed microphone (see quots.). orig. *U.S.*
1946 W. L. Gresham *Nightmare Alley* (1947) xi. 171 That would have been a beautiful place to plant a bug if you wanted to work the waiting room gab angle. 1948 F. Brown *Murder can be Fun* (1951) xiv. 215 There's been a bug on your phone line for three days. Man on duty in the basement. 1955 *N.Y. Times* 4 May 1/3 A telephone 'bug', or tiny microphone and wire, attached to Mr. Celler's own phone in the hearing room. 1961 A. Christie *Pale Horse* xvi. 164 Perhaps you have some idea that this office of mine might have a bug in it?

bug (bʊg), *v.*¹ [f. Bug *sb.*²] **a.** *trans.* To clear (plants, etc.) of insects. **b.** *intr.* (See quot. 1889.)
1869 *Champaign Co.* (Ill.) *Gazette* 26 May 2/1 If every tree in the township was 'bugged' daily..the destruction of this little pest would be certain. 1889 *Cent. Dict.*, *Bug*.., to hunt for bugs; collect or destroy insects: chiefly in the present participle: as, to go *bugging*. 1895 *Voice* (N.Y.) 8 Aug. 7/6 While 'bugging' potatoes this season I came across a number of beetles..that I have never seen any mention of.
2. *trans.* To equip with an alarm system or a concealed microphone. (Cf. *bug *sb.*² 4 e, f.) Also *intr.* and in extended use. *slang* (orig. *U.S.*). So **bugged** *ppl. a.*, **bugging** *vbl. sb.* and *ppl. a.*
1919 M. Acklom in *Bookman* Apr. 209/1 The possibilities of the joint being bugged. 1931 *Amer. Speech* VII. 104 *Bugged*, wired to a burglar alarm. 1949 *Times* (Los Angeles) 16 Aug. i. 4/6 Cohen seemed well aware that his house was 'bugged' and that his conversations were tapped. 1955 *Newsweek* 24 Oct. 32/1 'I'll guarantee you that you're not going to do any more bugging!'..Kalven's group had hidden microphones in Wichita, Kans., district court jury rooms and recorded the jurors' traditionally secret deliberations. 1958 J. D. MacDonald *Executioners* (1959) v. 81 We bugged both suites. 1958 *Times* 9 July 8/6 The police..explained that 'bugging' as distinct from wire-tapping, was no crime. 1960 *News Chron.* 13 Sept. 4/8 The 'bugged' conversations of a foreign Government's leaders.
3. To annoy, irritate. *slang* (orig. and chiefly *U.S.*).
1949 *Music Library Assoc. Notes* Dec. 40/2 *Bug*, popularized by swing musicians and now much used by be-boppers: to be annoying. 1952 B. Ulanov *Hist. Jazz in Amer.* 350 *Bug*, to bewilder or irritate. 1959 J. Osborne *Paul Slickey* ii. ix. 71 It will surely bug you when there is no man to hug...You will be bugged for ever. 1959 *Times* 31 Oct. 7/3 The heroine..inquires picturesquely of the hero 'What's bugging you?' and he replies, succinctly, 'Life.'

bug (bʊg), *v.*² *intr.* Of the eyes: to bulge *out*. *U.S. colloq.* Also *trans.* (*rare*).
1877 'Mark Twain' in *Atlantic Monthly* XL. 446 His dead-lights were bugged out like tompions; and his mouth stood..wide open. 1883 —— *Life on Mississippi* xxxvi. 346 Wouldn't their eyes bug out, to see 'em handled like that? 1929 W. Faulkner *Sartoris* iii. 225 They was.. buggin' their eyes at him. 1961 D. McN. Douglass *Saba's Treasure* (1963) vi. 97 Her mouth dropped open and her eyes bugged.

bug (bʊg), *v.*³ *slang* (chiefly *U.S.*). [Origin uncertain; perh. connected with *bug *v.*¹ or *bug *v.*²] *intr.* To get *out*; to leave quickly; to 'scram'.
1953 in Partridge *Dict. Slang* Suppl. (1961) 1304/2 If one were to 'swan' southward with the purpose of moving on from the enemy, the act would be called 'bugging out'. 1959 J. Christopher *Scent of White Poppies* vii. 114 There was no sign of movement...'Give it five minutes. If there's nothing showing by then, either he's bugged out or he's asleep.' 1969 *Daily Colonist* (Victoria, B.C.) 21 Oct. 1/7 He also said that Canada is not 'bugging out' of NATO.

bugaboo. Add: **2.** Loud or empty talk, nonsense, rubbish.
1897 [see Guff¹ 2]. 1959 *Listener* 15 Jan. 121/3 So straightforward an inquiry can produce so rich a harvest of pure bugaboo.

bugan. *dial.* Also **buggan(e), buggin.** [See Bug *sb.*¹] An evil spirit, hobgoblin, bogy. So *to play the bugan*: to play the devil (with).
1839 G. C. Lewis *Gloss. Herefordshire* 16 *To play the bugan*, to play the devil. 1872 G. M. Hopkins *Jrnl.* 12 Aug. in *Note-Books* (1937) 165 A foolish legend of a tailor and a goblin called a Buggane. 1879 G. F. Jackson *Shropshire Word-Bk.* 55 If yo' dunna be qweet I'll let bugan tak' yo. 1887 T. Darlington *Folk-Speech S. Cheshire* 132 Ah darna go a-milkin', The buggin's in the bush. 1894 Hall Caine *Manxman* vi. i. 359 You'd have a buggane riding on your breast the night through.

bugbear. Add: **2. a.** In weakened senses: an annoyance, bane, thorn in the flesh.
1880 Geo. Eliot *Let.* 14 Sept. (1956) VII. 322 Our only bugbear—it is a very little one—is the having to make preliminary arrangements towards settling ourselves in the new house. 1955 *Sci. Amer.* Jan. 90/1 Richness of context was their bugbear. 1966 *Observer* 10 Apr. 12/3 The great bugbear of economic management is the near impossibility of devising policies with a particular objective in view without..making it harder to attain other ..desirable ends.
b. (Later examples.)
1853 Mrs. Gaskell *Cranford* xii. 223 Indiscretion was my bugbear fault. 1930 E. Sitwell *Coll. Poems* 252 A bugbear bone that bellows white.

bugeye. *U.S.* [?f. Bug *sb.*² 1.] = *buck-eye 3.
1881 E. Ingersoll *Oyster Industry* 242 A bugeye is always decked over and has a cabin aft. 1889 [see *buck-eye 3]. 1938 *Times Lit. Suppl.* 3 Sept. 574/3 She was the type of craft known as a Chesapeake 'bugeye'.

bug-eyed, *a.* orig. *U.S.* [f. *bug *v.*²] Having bulging eyes; esp. in phr. *bug-eyed monster*, an extra-terrestrial monster with bulging eyes; abbrev. *B.E.M.*
1922 H. L. Wilson *Merton of Movies* xi. 193 Kind of innocent and bug-eyed the way he'd rubber at things. 1943 R. Chandler *Lady in Lake* (1944) viii. 47 An angular bug-eyed man with a sad sick face. 1953 Koestler *Trail of Dinosaur* (1955) ii. 143 Young space cadets, for instance, dislike meeting Bems—for bug-eyed Monsters. 1957 P. Moore *Science & Fiction* 45 He was no dabbler in fiction of the bug-eyed monster and ray-gun type. 1958 *New Scientist* 18 Sept. 861/3 'Space opera' and 'B E M stories' (Bug-Eyed Monsters). 1960 K. Amis *New Maps of Hell* ii. 44 In space-opera,..Indians turn up in the revised form of what are technically known as bug-eyed monsters, a phrase often abbreviated to BEMs. 1960 C. Day Lewis *Buried Day* viii. 165 The bug-eyed, frantic immobility of a rabbit confronted by a stoat.

bugger, *sb.* Add (all *coarse slang*): **2. b.** (Later examples.) Delete 'in Eng. dial. and in U.S.'
1929 F. Manning *Middle Parts of Fortune* I. v. 103 Not when there are two poor buggers dead, and five more not much better. 1934 R. Blaker *Night-Shift* xx. 261 The words that had voiced the shock at the bottom of a startled heart..were 'pore bugger'. 1955 *Times* 27 Jan., A remark of the policeman to him was: 'Don't argue, get those buggers out of here.' 1969 *Listener* 9 Jan. 53/3 Come and sit on my other side. Otherwise they will put me beside that bugger Oparin.
c. Something unpleasant or undesirable; a great nuisance.
1936 'G. Orwell' *Let.* ?16 Apr. (1968) I. 216 This business of class-breaking is a bugger. 1940 Harrisson & Madge *War begins at Home* viii. 189 Heard one old lady say, 'It's a bugger this dark!' 1942 *Penguin New Writing* XV. 28 Drilling before breakfast's a bugger, believe me.
d. = Damn *sb.* 2.
1922 Joyce *Ulysses* 576, I don't give a bugger who he is. 1939 Dylan Thomas *Let.* 2 Nov. (1966) 243, I don't care a bugger whether you won't or will. 1960 F. Raphael *Limits of Love* i. iii. 32 It'd be a wonderful thing to have a magazine that just didn't give a bugger what it said about anyone.
e. *bugger-all*, nothing. (See *all A. 8 f.)
1937 in Partridge *Dict. Slang* 103/1. 1961 I. Jefferies *It wasn't Me!* v. 63 'What did they offer to give you?' 'Bugger-all.' 1967 J. Cleary *Long Pursuit* ii. 39 Way out here in the middle of bugger-all, and you turn up!

bugger, *v.* Add: (Earlier and later examples.)
1598 Florio *Worlde of Wordes* 39/1 *Bardascia*, a bardash, a buggering boy, an ingle. 1868 *Index Expurgatorius of Martial* 49 You open your doors and bugger tall youths, Amillus. 1930 E. E. Cummings *Let.* 12 Apr. (1969) 116 Hats off to thea orthodox flea who attempted to bugger a bee. 1965 *New Statesman* 30 Apr. 687/1 Her German maid, whom he buggers. 1968 *Peace News* 16 Feb. 8/3 Some were in..for homosexuality, which is still an offence in the Army...Usually they'd have done it to get out: it was a saying, you had to bugger your way out of the Army. 1968 *Observer* 29 Sept. (Colour Mag.) 24/3 The thought of actually buggering a little boy is repulsive to me.
2. *coarse slang.* **a.** = Damn *v.* 5.
1794 *Sessions Paper* 86/1 She said, b–st and b–gg–r your eyes, I have got none of your money. 1862 A. M'Gilvray *Poems* (ed. 2) 21 Tho' Henderson, the sheriff-clerk, May damn, and swear, and buggar. 1867 Rossetti *Let.* 14 Aug. (1965) II. 628 I'm on the right side of the hedge this time, and b–g–r his drawings! 1886 Baumann *Londinismen* 17/2 *Bugger* (*you*)! geh' zum henker! 1897 *Shetland News* 22 May (E.D.D. Suppl.), 'Dat be bugger'd!' I said, 'Doo's shurly leein', Sibbie.' 1923 [see *2 b, below]. 1929 F. Manning *Middle Parts of Fortune* I. v. 92 'Cushy be buggered,' said Minton angrily. 1936 Dylan Thomas *Let.* 9 Mar. (1966) 168 Whatever I do now, bugger me it's literary. 1939 Joyce *Finnegans Wake* 522 Buggered if I know! 1953 S. Beckett *Watt* i. 42 Bugger these buttons! *Ibid.* 45 I'll be buggered if I can understand how it could have been anything else. 1967 D. Pinner *Ritual* xix. 189 Bugger me, he thought, looking at the grin on his watch, it's three o'clock!
b. To mess *up*; to ruin, spoil. In *pass.*, to be tired out.
1923 Manchon *Le Slang* 74 *Bugger!* or *bugger you!* merde! Salaud! *Ibid.*, *Buggered off* or *buggered up*, foutu. 1929 F. Manning *Middle Parts of Fortune* I. ii. 31 Buggered-up by a joy-ride in the train from Rouen to Méricourt. 1934 J. O'Hara *Appointment in Samarra* (1935) iii. 77 He's probably forgotten about it, and my going there will bugger things up proper. 1942 F. Warnick *Dialect of Garrett County, Maryland* 4 Booger up, v. phr., to treat roughly (always used in past tense) 'all buggered up'. 1947 *Coast to Coast 1946* 216 We're buggered, too—been burnin' the floor boards orf the dunny. 1953 Dylan Thomas *Let.* 31 Mar. (1966) 401, I hope I haven't boomily buggered the poems up. 1961 A. Wilson *Old Men at Zoo* i. 32 No hippos in their natural lovely setting of the Severn or beavers buggering up the Broads. 1968 H. C. Rae *Few Small Bones* ii. i. 72 He was so utterly buggered that he had no hunger left.
c. *intr.* With *off*: to go away, depart.
1922 Joyce *Ulysses* 586 Here bugger off, Harry. There's the cops! 1945 J. B. Priestley *Three Men in New Suits* iv. 63, I didn't come back for their bloody leavings. Bugger off! 1958 P. Scott *Mark of Warrior* ii. 202 They sort of stared at the river a bit, then buggered off. 1969 *Private Eye* 12 Sept. 14 Let's get up to palace, pick up O.B.E.'s and bugger off 'ome, like.
d. With *about*: (see quot. 1937); also with *around*. Also *trans.*, to hound from pillar to post; to mess about with.
1929 F. Manning *Middle Parts of Fortune* I. iii. 87 We're just 'umped an' bumped an' buggered about all over..France. 1937 Partridge *Dict. Slang* 103/1 *Bugger about*, potter about; fuss; act ineffectually; waste time on a thing, with a person. Hence *bugger about with*, to caress intimately; interfere with (person or thing). 1946 B. Marshall *George Brown's Schooldays* 43 Pious Aeneas went buggering about the Aegean. 1957 M. Kennedy *Heroes of Clone* iii. iii. 174 Do I then have to be buggered about by a lot of professors and critics..asking how I rate? 1961 'T. Hinde' *For Good of Company* v. 56 It was properly buggered around when it came back. 1968 J. Wainwright *Web of Silence* 12 Let's not bugger around being polite.

buggery. Add: **b.** Also used of unnatural intercourse of a man and a woman.
1729 G. Jacob *Law-Dict.*, *Buggery*..is defined to be *carnalis copula contra Naturam, & hoc vel per confusionem Specierum, sc.* a Man or Woman with a brute Beast; *vel Sexuum*, a Man with a Man, or Man with a Woman. 1966 J. Sparrow *Controversial Ess.* 41 Lawrence weaves into his story not merely a defence but a panegyric of this practice, making *Lady Chatterley's Lover* a vehicle for conveying his belief that it is a proper, if not a necessary, element in a full sexual relationship between man and woman...The practice approved by Lawrence is that known in English law as buggery.
c. In various slang uses (see quots.) = Hell *sb.* 9; *to play buggery*: to play havoc. (Cf. 1851 *attrib.* quot. in Dict.)
1898 *Shetland News* 11 June (E.D.D. Suppl.), You wye 'at dey geng an' buy at private bargains ootside da ring plays buggery. 1923 Manchon *Le Slang* 74 All to buggery, foutu. 1929 F. Manning *Middle Parts of Fortune* I. iii. 39 I saw 'im, sir; 'e were just blown to buggery. 1937 Partridge *Dict. Slang* 103/2 *Like buggery*, either vigorously, cruelly, vindictively; or, as an expletive, certainly not! 1939 Dylan Thomas *Let.* Mar. (1966) 228 Old stories mostly, but cut and pruned to buggery or sense. 1961 *Coast to Coast 1959/60* 83 'Pipe down, Rymill!' 'Go to buggery, Rymill!' 1966 'E. Lindall' *Time too Soon* (1967) xiii. 142 'Sah. You sick.' 'Go to buggery.' Minogue snarled. 'Yes, sah,' Basikas said, and stood aside.

bugging, *vbl. sb.* and *ppl. a.*: see *bug *v.*¹ 2.

Buggins (bʊ·ginz). [A 'typical' proper name used generically.] **Buggins'(s turn,** the principle of assigning an appointment to persons in rotation rather than according to merit.
1901 Ld. Fisher *Let.* 13 Jan. in A. J. Marder *Fear God & Dread Nought* (1952) I. 181 Favouritism was the secret of our efficiency in the old days...Going by seniority saves so much trouble. '*Buggins's turn*' has been our ruin and will be disastrous hereafter! 1917 *Let.* 8 Jan. in Ld. Fisher *Memories* (1919) iii. 39, I was sorry for Jellicoe superseding Callaghan when the war broke out, but I remembered your old saying, 'Some day the Empire will go down because it is Buggins's turn'! 1944 W. S. Churchill in *Second World War* (1952) V. 616 The departmental view is no doubt opposed to long tenures and the doctrine of 'Buggins's turn' is very powerful. 1962 *Guardian* 18 Oct. 1/6 Merit, not 'Buggins's turn', was the standard. 1970 *Observer* 8 Feb. 44/4 A sort of Buggins's-turn job, with the vice-chairman automatically moving into his boss's seat.

buggy, *sb.* Add: **1. b.** Extended to various other vehicles or the like, as: (*a*) a perambulator (see *baby buggy*); (*b*) = Caboose 2; (*c*) an automobile. *slang* (orig. *U.S.*).
1890 [see *baby *sb.* B. 2]. 1894 *St. Paul's* 11 Aug. p. ii (Advt.), Safety perambulators...Coach-panel 'Buggy' mail cart,..33/- to 5 guineas. 1899 *Boston Globe* 18 Mar. 12/2 Estimating the total length of these [freight] cars, with the engines and buggies. 1911 *Jrnl. Amer. Med. Assoc.* LVI. xiv. p. vi/2 My use of this motor buggy has shown me that the auto is a good thing for either city or country. 1929 *Nebraska Alumnus* June 166 A doll bed and dresser and buggy with several dolls and their wardrobes. 1931 G. Irwin *Amer. Tramp & Underworld Slang* 39 *Buggy*, an automobile; a contraction of the term 'gasoline buggy', by which the earlier automobiles were known. 1949 J. Creasey *Battle for Insp. West* xiii. 119 'Come on,' he said, 'you're changing buggies.' Grant hesitated; rain splashed into the car. 1961 *Economist* 27 May 918/1 The green fees, he said, were a bit stiff, and the electric buggies cost $14 a day.
3. buggy-ride orig. *U.S.*, a ride on a buggy; esp. in colloq. fig. expression of gratitude *thanks for the buggy-ride.*
1862 G. W. Wilder *MS. Diary* 16 Jan. (D.A.E.), The officers took a buggy ride; joy go with them. 1865 *Trans. Ill. Agric. Soc. 1863* V. 255 The one on the gang plow would have a fine buggy ride. 1926–7 in *Shopper's View of Canada's Past* (1969) 261 Records..Thanks for the Buggy Ride—Fox Trot. 1938 N. Marsh *Artists in Crime* xv. 231 When the spot of trouble comes along it's 'Thanks for the buggy ride, it was OK while it lasted'.

bu·ghouse, *sb.* and *a.* *slang.* **A.** *sb.* **1.** [Cf. *bug *sb.*² 4 a.] A lunatic asylum. orig. *U.S.*
1902 O. V. Limerick *Billy Burgundy's Opinions* 44 Place his name upon the list of permanent guests of the county bughouse. 1907 'O. Henry' *Heart of West* 226, I thought he was in the bughouse. 1919 F. Hurst *Humoresque* 316 It's too bad a nut from the bug-house bought the Brooklyn Bridge to-day or I'd try to sell it to you. 1940 N. Marsh *Surfeit of Lampreys* (1941) viii. 113 You're bigger bloody fools than anybody outside a bughouse.
2. [f. Bug *sb.*² 1.] An opprobrious term for a theatre or cinema. Cf. *flea pit* (*flea *sb.* 6).
1946 in Partridge *Dict. Slang* Suppl. (1961) 1018/1. 1952 'N. Shute' *Far Country* ix. 326 There's a little picture theatre...It's a bit of a bug-house. 1957 J. Osborne *Entertainer* i. iii, If there's nothing else on, I still go..to the bug house round the corner.
B. *adj.* [Cf. sense *A. 1.] Crazy; very eccentric. Chiefly *U.S.*
[1891 *Contemp. Rev.* Aug. 255 Begging is called 'battering for chewing';..insanity, 'bug-house'.] 1895 *Century Mag.* June 291/2 How's that for bein' bughouse, eh? 1896 Ade *Artie* xvii. 161 If I don't get mine inside of a week I'll go bug-house. 1917 Conan Doyle *His Last Bow* viii. 292 It's enough to make a man bughouse when he has to play a part from morning to night. 1930 'Sapper' *Finger of Fate* 187 For a moment I thought he'd gone bughouse.

bugle, *v.* Add: **a.** Also, to make a sound like a bugle. **b.** Also, to make known with or as with a bugle.
1876 G. M. Hopkins *Wr. Deutschland* (1918) xi, And storms bugle his fame. 1890 Meredith *One of our Conq.* (1891) I. iv. 58 He blinked, bugled in his throat,..and smiled. 1911 W. Owen *Let.* 7 May (1967) 72, I almost bugled forth the fact that she was addressing a B.Sc.
So **bu·gling** *vbl. sb.*
1847 *Infantry Man.* (1854) 93 Too much bugling..is to be avoided. 1895 *Chambers's Jrnl.* XII. 648/2 The eternal bugling on the Canea walls still continued. 1948 A. L. Rand *Mammals Eastern Rockies* 207 The bugling or whistling of the bull elk in the early autumn is the signal.. that the rut is starting.

bugology (bʊgǫ·lŏdʒi). *U.S. humorous.* [f. Bug *sb.*² 1 + -ology.] The science of 'bugs' or insects; entomology. Hence **bugo·logist,** an entomologist.
1843 'R. Carlton' *New Purchase* II. l. 171 Chemistry, botany, anatomy, conchology, bugology. 1881 *Nat. Republican* (U.S.) 24 Feb. 2/4 Mr Riley, the eminent bugologist. 1898 *Congress. Rec.* Apr., App. 455/2 Those.. acquainted with bugology know there is rather a disreputable bug that looks one way and rolls the other. 1910 *Sat. Even. Post* 2 July 49/2 Government bugologists.. studied his habits.

bugong (bū·gǫŋ). Also **bogong, bougong.** [Native name.] An Australian noctuid moth, *Danais limniace* or *Agrotis spina*, highly prized by the Aborigines as an article of food.
1834 G. Bennett *Wand. N.S.W.* I. 265 It is named the 'Bugong Mountain', from the circumstance of multitudes of small moths, called Bugong by the aborigines, congregating at certain months of the year about masses of granite on this and other parts of the range. 1859 H. Kingsley *G. Hamlyn* xxxix, To collect and feed on the great grey moths (Bougongs) which are found on the rocks. 1878 R. B. Smyth *Aborig. Victoria* I. 207 The Bugong moths..are greedily devoured by the natives. 1919 *Nature* CIII. 345/2 In Australia at certain seasons a 'cutworm' moth, known as the 'bogong' or 'bugong' (Agrotis infusa), swarms in myriads in many places. 1961 R. Park *Hole in Hill* (1962) ii. 13 The big bogong moths would be coming in.

buhr, var. Burr *sb.*⁵

build, *v.* Add: **2. b.** *to build in*: also, to construct or insert (something, esp. furniture) as an integral part of a larger unit; also *fig.*; chiefly as pa. pple. (see *built *ppl. a.* 1 b).
1933 *Telegr. & Teleph. Jrnl.* XIX. 151/1 In New York telephones are 'built-in' and when you become a tenant.. you 'phone as often as you like. 1956 A. H. Compton *Atomic Quest* 326 Better control of the amount of rubber that is built into tires. 1965 *Listener* 4 Nov. 687/1 The legacy of those years has been built in to the domestic and foreign policies of both countries.
3. *built like a castle*: said of a horse having a strong and sturdy frame.
1835 G. Stephen *Adv. Search of Horse* i, He [*sc.* a cob] was, to use the accepted phrase, 'built like a castle!' 1882 *Illustr. Sporting & Dram. News* 4 Feb. 502/1 Miss Bell's colt is built like a castle, and full of massive strength from head to heels.
b. *trans.* and *intr.* *Tailoring.* To make (clothes).
1840 Barham *Ingol. Leg.* 22 [The trousers] were cleverly 'built', of a light-grey mixture. 1897 *Globe* 11 Mar. 3/4 A tailor would..have had his work cut out for him to build that..chubby creature a costume. 1897 G. Du Maurier *Martian* iv. 183 Is it still Skinner who builds for you?
c. *to be built* (*that way*, etc.): to be (so) constituted or naturally disposed. *colloq.* (orig. *U.S.*).
1882 *Amer. Humorist* 12 May (Farmer), Even womankind is not built as she was a few brief years ago. 1888 *Missouri Republ.* 25 Jan. (Farmer), 'Why didn't you roll down?' 'I wasn't built that way.' 1912 A. Bennett *Matador of Five Towns* 75 I'm not built the same way myself.
d. In card-playing: (see quot. 1901). Also *transf.* *U.S.*
1901 *Munsey's Mag.* XXIV. 871/2 To build down..is to place a card upon one of the next higher denomination. To build up..is to do just the opposite—that is, to place an eight on a seven. 1903 A. Adams *Log Cowboy* vi. 76, I built right up to him.
4. b. *to build up*: to establish or enhance the reputation or prominence of (a person, nation, etc.); to 'boost'. orig. *U.S.*
1935 S. Lewis *It can't happen Here* ix. 86 Sarason had, as it was scientifically called, been 'building up' Senator Windrip for seven years before his nomination as President. 1939 *Ann. Reg. 1938* 262 The desire to 'build-up' the figure of the Leader (*El Caudillo*) in the approved Fascist style. 1944 A. Jacob *Traveller's War* xii. 212 Rommel had been 'built up' by the British press into a great figure. *Ibid.* 213 He set out to build himself up in the eyes of an army that had tasted defeat.
c. *to build up. trans.* and *intr.* Of an electric current, volume of sound, etc.: to increase.
1936 *Discovery* July 222/2 The sound builds up from silence to strength. *Ibid.* 223/1 The amplifier building up the sound they [*sc.* the oscillating valves] produce before it reaches the loud-speaker. 1949 *Chr. Sci. Monitor* 30 Apr. Mag. Sect. 2/4 Five relay stations..build up the signal.
d. *to build up. intr.*: to accumulate, collect; to grow.
1956 A. L. Rowse *Early Churchills* ii. 22 Such was the spirit that was building up on either side in this deplorable war. 1962 *Which? Car Suppl.* Oct. 139/2 Some rust had built up behind the chrome strips on both front wings.

buildable (bi·ldăb'l), *a.* [f. Build *v.* + -able.] Capable of being built (on).
1927 *Times Weekly Ed.* 10 Nov. 527/3 [A design] compact, orderly and simple—and eminently buildable. 1963 W. Holford in *Oxf. Univ. Gaz.* 23 May 1257/2 By adding a little more than a quarter of Merton Field..an extremely buildable site could be obtained. 1967 *Listener* 6 July 10/2 The megalomaniac diagrams of the late eighteenth-century French neo-classical architects were made buildable by pragmatic nineteenth-century English architects.

builder. Add: **a.** *spec.* One who 'builds' clothes; a tailor (see *build *v.* 3 b).
1845 *Ainsworth's Mag.* VIII. 214 In former days our builders of shooting jackets used to perch the buttons of the waist underneath our shoulder blades.
b. *builder's paper*, a strong paper sheeting used in building. *U.S.*
1917 Mathewson *Sec. Base Sloan* vii. 88 Red builder's paper superseded the boards across the window frames.
c. *builder-upper*, one who builds up (something or someone). (Cf. quots. 1825 and 1827 in Dict.) *colloq.*
1936 *Esquire* Sept. 59/3 You, Mr. Sports Fan—it is you who plays the role of 'builder-upper' [of athletes' morale]. 1944 D. Wecter *When Johnny comes marching Home* iv. ii. 493 A strength 'that cannot be pooh-poohed even by the most agile tearer-downers nor cheapened by the most blatant builder-uppers'. 1945 L. Shelly *Hepcats Jive Talk Dict.* 22 *Builder upper*, publicity man. 1951 A. H. Cherry *Yankee R.N.* iii. 83 'Off reefers, up shirt-sleeves!' barked the Royal Navy's 'fitness and morale' builder-upper.
2. That part of a spinning-machine which regulates the yarn as it is wound on the bobbins.
1884 W. S. B. McLaren *Spinning* vii. 136 The lifter, or builder as it is called, moves more slowly as each layer is placed on the bobbin.
b. A substance added to soap or to a detergent to increase its efficiency.
1931 *Soap* Nov. 29/2 The various bases and salts showing alkaline reaction..are also called fillers in soap, builders, and detergents. 1958 *Spectator* 4 July 16/1 Cotton.. cannot be washed so easily and it is necessary to use slightly alkaline 'builders' (sodium tripolyphosphate, etc.). 1963 A. J. Hall *Textile Sci.* vi. 284 Some soap-containing proprietary washing powders also contain mild alkalis and silicates and phosphates which are termed 'builders'.

building, *vbl. sb.* Add: **1. a.** Freq. with *up* (also *attrib.*).
1604, 1852 [in Dict.]. 1901 *Daily Chron.* 5 Dec. 3/4 The synthetic or building-up chemistry. 1942 W. S. Churchill *End of Beginning* (1943) 36 The steady building-up of very powerful forces and bases there.
4. *building-board, -land, -lot, -material, -site, -stone*; **building-block,** (*a*) = Block *sb.* 12 b; also *fig.* and *attrib.*; (*b*) one of the temporary supports for a ship's keel while the ship is being built (Knight *Dict. Mech. a* 1877); **building motion,** in *Cotton-spinning*, apparatus for winding and shaping uniformly the roving on the bobbins of a fly frame or the yarn of a cop on a mule; **building paper** orig. *U.S.*, a heavy paper used by builders as a covering or lining material; **building-slip,** a slip (see Slip *sb.*³ 1 b) on which vessels are built; **building-society** (earlier examples); (see also quot. 1965).
1846 *Boston Herald* 14 Oct. 3/1 Jewsharps, Games, Building Blocks, Harmonicas. 1857 *Mich. Agric. Soc. Trans.* IX. 316 A dozen Patent Building Blocks. 1915 J. Webster *Dear Enemy* (1917) 152 Punch was occupying a rug..engaged with building-blocks. 1936 J. Kantor *Objective Psychol. Gram.* iii. 39 To such a building-block comparison of speech, psychological grammar is strongly opposed. 1949 *Rev. Eng. Studies* XXV. 369 Four..pitch levels..are the 'basic building blocks' of the tunes. 1962 F. I. Ordway et al. *Basic Astronautics* iv. 153 Some hope to learn from gravitational research more about the microstructure of the building blocks of nature, the subatomic particle. 1969 *Times* 5 Feb. 13/7 The most successful symmetry scheme is the one in which the elusive particle, the quark, is the fundamental building block from which all the heavier particles are made. 1917 *U.S. Pat. 1,227,767* 29 May 1/2 Building board is put in place on the outside of a house..or secured on the inside. 1959 M. S. Briggs *Conc. Encycl. Archit.* 61 *Building board*, a term covering a wide range of artificial products used internally in modern building. 1905 *Westm. Gaz.* 12 May 7/1 The L.C.C. purchased squares at building-land price. 1701 in *Conn. Col. Rec.* IV. 357 Pasture, building lot, and long lott. 1881 W. O. Stoddard *E. Hardery* 15 The high prices of all building lots. 1835 C. F. Hoffman *Winter in West* I. 69 A species of yellow freestone..which, for elegance as a building material, is not surpassed by marble itself. 1904 Goodchild & Tweney *Technol. & Sci. Dict.* 73/1 Building motion. 1873 *Newton Kansan* 20 Feb. 3/4 Building paper, the best substitute for plastering. 1955 G. Bowen *Wool Away!* x. 115 Building paper should be used under the iron above the shearing board. 1871 Geo. Eliot *Middlemarch* i. iv. 52 They were driving home from an inspection of the new building-site. 1966 D. Jenkins *Educated Society* ii. 50 Constructive forces..have no more attraction than building sites. 1846 Dodd *Brit. Manuf.* VI. 147 Every ship-yard has got one or more 'building-slips'. 1894 Building slip [see Slip *sb.*³ 1 b]. 1848 H. C. Robinson *Diary* 31 Dec. (1967) 250 Miss Martineau.. is now full of a prospect of forming here building-societies for the benefit of the poor in imitation of the Birmingham societies. 1852 Geo. Eliot *Let.* 21 Oct. (1954) 62 There has been an intelligent gentleman visitor today who is interested in Miss Martineau's Building Society. 1965 J. L. Hanson *Dict. Econ.* 51/2 *Building societies*, institutions which accept 'deposits and then use their funds to lend on mortgage to people who wish to buy their own houses. 1790 *Pennsylv. Packet* 2 Jan. 4/4 Bourdeaux rough hewn building stone. 1924 R. M. Ogden tr. *Koffka's Growth of Mind* v. 325 Building-stone of a peculiar shape.

build-up (bi·ldʊp). [f. Build *v.* 4 and 5.]
a. An accumulation of favourable publicity designed to popularize a person, product, etc. Also, simply, preparatory work, preparation. Cf. *build *v.* 4 b. *colloq.* (orig. *U.S. slang*).
1927 *Collier's* 3 Dec. 10/4 That's the old build-up for the Patsys. 1935 *Time* 24 June 26 *One Night of Love* had a build-up unrivaled in cinema history. 1935 *Evening News* 29 June 3/2 The swindler's talk..arouses the interest and acquisitive instincts of the mug...The '*build-up*' has been made. 1936 Wodehouse *Laughing Gas* x. 103, I thought it might soften her a little if you gave the old boy a build-up. 1942 Berrey & Van den Bark *Amer. Thes. Slang* § 241/1 Preparation,..build-up. *Ibid.* § 489/1 *Build-up*, preparatory work for a crime. 1950 R. Chandler *Trouble is my Business* 228 The threats were a build-up for a killing. 1953 J. Pudney *Ring for Luck* 16 A big build-up for the indispensable Miss Motting, 'without whom nobody ticks in Fragos'. 1955 R. M. Lumiansky *Of Sondry Folk* viii. 85 The scene is to be understood as his careful preparation—the present-day term 'build-up' is apt here—for playing a joke on the Host. 1957 *Economist* 12 Oct. 120/2 Mr Howard gives more space for her early life—the perfect build-up for her subsequent adventures.
b. (Usu. gradual) accumulation, increase; spec. *Mil.* an accumulation of troops, weapons, etc. Also *attrib.*
1943 *Daily Tel.* 22 Apr. 1/2 The last four months' patient build up of Allied air power. 1944 *Times* 15 May 4/3 The build-up in the Eighth Army bridgehead across the River Rapido continued during the night. *Ibid.* 25 July 4/4 The build-up period lacks the excitement of the initial assaults. 1955 *Times* 14 July 4/7 The build-up of case law by the tribunal. 1958 I. Brown *Words in our Time* 29 There is now evidence of some spread in the population locally, but no evidence at present of any large build-up. 1964 *Ann. Reg. 1963* 150 On the basis of one of these reports, the Security Council noted with alarm the arms build-up in South Africa. 1969 *Times* 2 Jan. 16/1 Professor Semple's team detected a massive build-up of the bacteria in several kinds of food on the premises.
c. The gradual development or exposition of a theme, argument, work of art, etc.; artistic construction.
1942 Berrey & Van den Bark *Amer. Thes. Slang* § 522/10 *Build-up*, initial plot development. 1953 A. Upfield *Murder must Wait* ix. 78 The build-up of the background against which five infants had been stolen. 1958 *Spectator* 8 Aug. 191/3 The slow build-up of *All My Sons*..is deliberate. 1959 H. Read *Concise Hist. Mod. Painting* vi. 193 The whole build-up, or orchestration, of form and colour is purposively expressive.

built, *ppl. a.* Add: **1.** Also *built-over*.
1935 E. Bowen *House in Paris* ii. i. 88 A smoky built-over hill. 1954 L. MacNeice *Autumn Sequel* 131 The Roman slabs lay snug below In that built-over darkness.
b. built-in: (*a*) constructed to form an integral part of a larger unit, esp. of the fittings or appurtenances of a house; (*b*) *fig.* inherent, integral, innate.
1898 *Electrical Engin.* July 2 The first and simplest is the 'solid' or 'built-in' system, where wires, which are insulated thoroughly and thoroughly protected from mechanical disturbances, are buried in the ground. 1902 *Harper's Monthly Mag.* Jan. 302/2 A built-in refrigerator. 1929 M. Skipper *Meeting-Pool* 6 'I shall be waiting then,' murmured the Crocodile, snuggling his long face, with the built-in smile on it, into the warm sand. 1930 *Engineering* 7 Mar. 309/2 The employment of built-in and flange motors is also increasing. 1933 *Discovery* July 219/1 Bedrooms are small, adequate space for the storing of clothes, for a desk and so on, being found in built-in furniture. 1946 Koestler in *New Writing & Daylight* 82 Archetypes are..inherited, built-in patterns of instinct-conflicts. 1951 *Good Housek. Home Encycl.* 162/1 *Counter Unit* with .. built-in storage space underneath. 1962 *Listener* 26 July 131/1 The phrase 'built-in obsolescence'..was very fashionable, especially among cynics, about ten years ago.
Hence as *sb.*, a built-in piece of furniture, etc. Chiefly *N. Amer.*
1930 *San Antonio* (Texas) *Light* 31 Jan. 14/6 (Advt.), Houses for sale...large screened porch, cabinets, built-ins. 1963 *House & Garden* Feb. 7 (Advt.), A Moffat Built-In is a Moffat *plus*. 1970 *Globe & Mail* (Toronto) 28 Sept. 26/4 (Advt.), 3 large bedrooms, 3 washrooms, many built-ins.
c. built-up: (*a*) (see sense 1 in Dict.); (*b*) constructed of parts, esp. of parts that are separately prepared and afterwards joined or welded together; so *built-up gun*, a gun whose parts are constructed separately and united in such a way that the elastic quality of the metal is fully utilized; (*c*) designating a locality where buildings abound, esp. in *built-up area.* Also *built-up butt*, a raised as distinguished from a sunken shooting-butt; *built-up rope* (see quot. 1908). Of a shoe: heightened.
1829 [see sense 1 in Dict.]. 1853 C. Brontë *Villette* I. x. 188 This demi-convent, secluded in the built-up core of a capital. 1865 P. Barry *Shoeburyness & the Guns* vi. 86 The built-up system demands the same care..as the Krupp blocks. 1886 *Harper's Mag.* Oct. 786/2 The means of providing against this successive rupture of overstrained parts is found in the 'built-up gun'. 1887 *Cassell's Fam. Mag.* 509/1 A Built-up File. 1904 *Westm. Gaz.* 26 Aug. 3/1 A built-up butt. 1908 *Animal Managem.* (War Office) 143 The 'built-up' [picketing] rope, a portion of which is carried by each man. 1935 *Ann. Reg. 1934* ii. 71 The recently passed Road Traffic Act..was extended shortly

afterwards to all 'built-up' areas throughout the country. 1937 *Sunday Times* 10 Jan., The perils of built-up by-passes. 1946 E. Johnston *Writing* xv. 255 Built-up letters are composed of compound strokes. 1950 A. L. Rowse *England of Eliz.* v. 162 The actual built-up area grew in extent in Elizabeth's reign. 1969 T. Parker *Twisting Lane* 19, I have to wear this special built-up shoe.
d. The simple *built* is used in the same sense as (*b*) above.
1860 [see sense 1 in Dict.]. *a* 1877 Knight *Dict. Mech.*, *Built-beam*, a compound beam made up of a number of planks, or thin, deep beams, laid parallel and secured together. *Ibid.*, *Built-rib*, an arched beam made of parallel plank laid edgewise and bolted together. 1908 *Westm. Gaz.* 14 Nov. 19/2 The built all-in-one frock of lace net. 1909 *Ibid.* 20 Mar. 8/3 Three-bladed propellers of the 'built' type.

buisson (bwi·soṅ). *Hort.* [Fr.] (See quots.)
1832 J. Lindley *Introd. Bot.* i. ii. 48 If a shrub is low, and very much branched, it is often called *dumosus*..: this kind of shrub is what the French understand by their word *buisson*. 1852 G. W. Johnson *Cottage Gardeners' Dict.* 159/2 Buisson, is a fruit tree on a very low stem, and with a head closely pruned. 1951 *Dict. Gardening* (R. Hort. Soc.) I. 327/2 Buisson, name given at one time to a dwarf fruit tree on a low stem, closely pruned annually.

bukh, bukk, varr. *buck *sb.*¹¹, *v.*¹⁰

Bukhara, see *Bokhara.

Bukharinism (bŭkā·riniz'm). [f. the name of N. I. *Bukharin* (1888–1938), a Russian leader and editor + -ism.] The political principles of Bukharin. So **Bukha·rinist** *a.* and *sb.*
1937 *Times* 1 Oct. 15/6 Responsible Communist officials who are held up for public abuse as Trotskyists, Bukharinists,..and enemies of the people. 1949 I. Deutscher *Stalin* ix. 384 Conspiratorial activities of the Trotskyist and Bukharinist leaders. *Ibid.* xi. 486 The evils of Trotskyism, Bukharinism, and other deviations.

bulb, *sb.* Add: **4.** In full (*electric*) *light bulb*. The glass bulb-shaped container of the incandescent filament used for producing electric light.
1856 *National Rev.* July 88 The ray of the electric bulb, so sharply defined that all beyond its pencil falls into depth of darkness. 1882 *Electric Light* 1 Sept. 70/1 Volatilized carbon being deposited on the inside of the bulb. 1884 [in Dict.]. 1890 J. W. Urquhart *Electric Light Fitting* 170 It becomes a question whether it is economical to run such blackened bulbs longer after a certain percentage of light has been so cut off. 1964 Naipaul *Area of Darkness* v. 110 A large bulb..was attached to a stunted flexible arm..: this was the lamp.
b. *bulb of percussion*, the convex protuberance on the fractured surface of a flint.
1872 J. Evans *Anc. Stone Impl.* xii. 247 Where a splinter of flint is struck off by a blow, there will be a bulb or projection, of a more or less conical form, at the end where the blow was administered...This projection is usually known as the 'bulb of percussion'. 1923 *Discovery* Dec. 316/2 Even the bulb of percussion which arises when a flint is broken by a violent blow, owing to the elasticity of its substance, can be produced by [natural] forces.
c. A pneumatic rubber bulb-shaped device on syringes, camera-mechanisms, etc.
1885 *Army & Navy Co-op. Soc. Price List* 874 The bulb can be disconnected, and fitted to any of the pipes as an Injection Bottle. 1911 *Encycl. Brit.* XXI. 514/2 J. Cadett's system of pneumatic pressure, applied by means of a compressible rubber bulb and tube, which may drive a piston acting on the lever holding the shutter. 1966 *McGraw-Hill Encycl. Sci. & Technol.* II. 431/2 Arrangements are made for keeping the shutter open for long times if needed (bulb and time or B and T). 1968 *Listener* 19 Dec. 811/3 He took from his pocket..a small glass instrument with a rubber bulb at the end. With this bulb he sucked some of the whisky into the tube.
5. bulb-fin, bulb-keel, a keel of a yacht having a cigar-shaped attachment which in section presents a bulb-like appearance; also *ellipt.* a yacht having such an attachment.
1893 [see Fin *sb.* 6]. 1894 E. R. Sullivan et al. *Yachting* I. 91 She was very deficient in stability when the lead slab forming the keel was recast in the form of a bulb on the bottom of the plate, the completed design simply forming one of our modern bulb fin keels. 1895 *Outing* (U.S.) Sept. 481/1 The great bulb-fins *Jubilee* and *Pilgrim*. 1927 R. A. Freeman *Certain Dr. Thorndyke* i. ix. 136 Spoon-bows and bulb keels were things as yet undreamed of.

bulberry (bu·lbĕri). *U.S.* Also **bullberry, bull berry.** [? f. Bull *sb.*¹] The buffalo-berry. Also *attrib.*
1841 T. J. Farnham *Trav. West. Prairies* 121 A fruit called bullberry. 1880 *Libr. Univ. Knowl.* (N.Y.) X. 159 Bulberries, which resemble red currants, grow by the rivers [in Montana]. 1888 Roosevelt in *Century Mag.* XXXVI. 45/2 A great patch of bulberry bushes. 1946 E. B. Thompson *Amer. Daughter* 190, I..went bullberry-picking down in the sharp ravines.

bulblet. Add: **b.** = Bulbil a.
1890 in Webster. 1904 *Daily Chron.* 7 Oct. 8/1 By-and-by little bulblets begin to form—sometimes as many as thirty ..growing from the one source. 1933 *Jrnl. R. Hort. Soc.* LVIII. 16 Clusters of bulblets formed round the base of the parent bulb.

bulbosity (bʊlbǫ·sĭti). [f. Bulbous *a.*: see -ity, -osity.] The condition or quality of being bulbous or bulb-shaped.
1901 'G. Douglas' *House with Green Shutters* 209 He had ..a body of such bulging bulbosity, that all the droppings of his spoon..were caught on his..waistcoat. 1963 G. F. Kantorowicz et al. *Inlays, Crowns & Bridges* ix. 107 Other areas..present bulbosity.

Bulgar (bʊ·lgaɹ), *sb.* [ad. med.L. *Bulgarus* (F. *Bulgare*, G. *Bulgar*), ad. OBulg. *Blŭgarinŭ* (Bulg. *Bălgarin*, Russ. *Bolgáry* pl., *Bolgárin* sing.).] Any member of an ancient Finnish tribe who conquered the Slavs of Mœsia in the seventh century A.D. and settled what is now Bulgaria, becoming Slavonic in language; a native or inhabitant of Bulgaria. Also *attrib.* or as *adj.*, Bulgarian.
1759 *Mod. Part Universal Hist.* IV. iv. i. 498 Bulgars. 1886 *Encycl. Brit.* XXI. 78/2 The Bulgars, whose origin still remains doubtful. 1902 *Ibid.* XXX. 395/1 The Bulgar *millet* comprises the Bulgarians who accept the rule of the exarchate. 1920 *Q. Rev.* Jan. 63 Basil II, whom the Greeks of to-day still admire as the 'Bulgar-slayer'. 1921 S. Graham *Europe—Whither Bound?* 27 The Bulgars have certainly hit on a novelty. *Ibid.* 33 Refugee Bulgars from the lost Bulgar territories. 1965 H. M. Smyser in Bessinger & Creed *Medieval & Linguistic Stud.* 93 The Bulgars and.. the so-called Jewish Khazars, about whom Ibn Faḍlān learned from his Bulgar hosts.
Hence **Bu·lgarize** *v. trans.*, to make Bulgarian in character; so **Bu:lgariza·tion.**
1869 Tozer *Highl. Turkey* I. 183 They become denationalized and Bulgarized. *Ibid.*, The Bulgarians are endeavouring the ecclesiastical superiority over them, and ..will bring about their Bulgarisation. 1903 *Daily Chron.* 9 Nov., The Bulgarisation of race-divided Macedonia. 1925 *Glasgow Herald* 2 Oct. 5 Fully Bulgarised Macedonians.

Bulgarian (bʊlgē°·riăn), *a.* and *sb.* [f. med.L. *Bulgaria*, f. *Bulgarus*: see prec. and -ian.] **A.** *adj.* Of or pertaining to the ancient Bulgars or to Bulgaria, a country in the Balkans. **B.** *sb.* **a.** Any member of the Bulgarian people; a native of Bulgaria; a Bulgar. **b.** The language of Bulgaria. *Old Bulgarian*: the oldest extant form of the Slavonic group of languages, also called Old Slavonic, Church Slavic or Slavonic (G. *kirchslavisch*).
1555 R. Eden tr. *P. Martyr's Decades* f. 289ᵛ The Slauon tounge..vsed of..the Mysians, Seruians, Bulgarians. 1797 *Encycl. Brit.* III. 769/1 The Bulgarians anciently inhabited the plains of Sarmatia that extended along the banks of the Volga. *Ibid.*, Bulgarian Language, the same with the Sclavonic. 1861 *Chambers's Encycl.* II. 418/2 The Bulgarian Language is divided into two dialects—Old Bulgarian and New Bulgarian; the former, the richest and best of the Slavonic dialects. 1869 Tozer *Highl. Turkey* I. 182 Even the priest, a Bulgarian, did not know a word of Greek. *Ibid.* 199 Children are taught to read and write both Greek and Bulgarian. 1959 J. Chapin tr. *Giovannetti's We Have a Pope* ii. 33 He delivered part of his Christmas sermon in Bulgarian. 1964 P. F. Anson *Bishops at Large* x. 469 Brother Michael, Bulgarian-born superior of the White Brotherhood.
So **Bulgaric** (bʊlgæ·rik), *a.*, of or pertaining to the ancient Bulgars or their modern representatives; *sb.*, the language of the ancient Bulgars. **Bu·lgaro-,** comb. form of *Bulgar, *Bulgaria*, or *Bulgarian, as in *Bulgaro-Wallachian* adj., **Bulgarophil(e, -philism, -phobe.**
1880 *Encycl. Brit.* XI. 119/1 The Bulgaro-Wallachian kingdom. 1888 *Ibid.* XXIV. 269/2 The Bulgaro-Vlach Empire. 1917 *Observer* 15 Apr. 4/2 The war of 1912 made us Bulgarophil; the war of today makes us—with much better reason—Bulgarophobe. 1921 *Glasgow Herald* 30 Dec. 6 His passionate Bulgarophilism. 1931 *Times Lit. Suppl.* 19 Mar. 218/3 Bulgarophil, pro-Turk or Serbophil. 1949 I. Deutscher *Stalin* vii. 243 An aristocratic revolutionary Bulgaro-Rumanian family.

bulge, *sb.* Add: **3. b.** *fig.* Usu. with *the*: the advantage or upper hand; the superior position; esp. in phr. *to have the bulge on*: to have the advantage over. *slang* (orig. *U.S.*).
1841 *Spirit of Times* 18 Dec. 498/3 Kate got the bulge on her at the start. 1872 'Mark Twain' *Innoc. at Home* in *Roughing It* (1873) xlvii. 332 Well, you've ruther got the bulge on me. Or maybe we've both got the bulge, somehow. 1899 E. W. Hornung *Amat. Cracksman* 197 We had the bulge before; he has it now. 1963 Wodehouse *Stiff Upper Lip, Jeeves* xv. 120 The Assyrians had the bulge on him.
c. A slight swell made on the surface by a fish moving through water as it feeds on flies, etc.
1878 C. Hallock *Hallock's Amer. Club List* p. ii, *Bulge*, the swirl made by a salmon rising to the surface. 1889 F. M. Halford *Dry-fly Fishing* vi. 122 [The angler] knows only too well that the apparent rises are bulges. 1892 *Field* 16 July 104/2 The rises, splashes, and bulges of burly brown fish.
d. A rise in prices or shares. *U.S. colloq.*
1890 Biff Hall *Turnover Club* 208 There is quite a bulge on June cocktails, and I fear a corner. 1908 G. H. Lorimer *J. Spurlock* xi. 272 The city house which the Bonsalls had occupied just before the big bulge in Southern Pacific landed them on top.
e. *Mil.* A bulging part of a military front; a salient.
1927 W. S. Churchill *World Crisis 1916–18* ii. xviii. 433 The German line formed a salient or bulge fifteen kilometers deep and sixty-four wide in the original British positions. 1942 *Daily Tel.* 28 Jan., The Russian officers I met around the Smolensk bulge.
f. *colloq.* A temporary increase in volume or numbers; *spec.* the increased number of children of school age resulting from the rise in the birth-rate at the end of the 1914–18 and 1939–45 wars. Also *attrib.*
1930 *Times Educ. Suppl.* 26 Apr. 186/2 Accommodation would have to be provided which would not be required when the 'bulge' years had passed. 1933 *Planning* I. xiii. 9 The impending flood of excess juvenile labour (the post-war bulge). 1935 *Economist* 23 Feb. 426/2 A temporary phenomenon, which corrected a 'bulge' in the curve without affecting its main upward trend. 1956 *Times* 2 June 7/7 The school population 'bulge' is moving up toward the 11–15 ages.
4. b. A protuberance on the hull of a ship to increase stability or to protect against under-water attack (cf. *blister *sb.* 2 c).
1919 *Chambers's Jrnl.* Aug. 543/1 Immunity from the evil effects of torpedoes and mines is sought by the provision of a swelling, commonly called a 'bulge' or a 'blister', below the water-line on each side. 1920 *Glasgow Herald* 10 Sept. 8 The modified 'bulge' for the protection against under-water attack.

bulge, *v.* Add: **3. b.** Of a fish: to make a bulge (see prec. 3 c). Hence **bu·lger, bu·lging** *vbl. sb.* and *ppl. a.*
1889 F. M. Halford *Dry-fly Fishing* vi. 116 A fish taking caddis, shrimp, or snails is said to be tailing, from its tail appearing at intervals above water,..when feeding on larvæ or nymphæ it is described as bulging, from its motion through the water. *Ibid.* 123 A bulging fish is feeding and looking out for food. *Ibid.* 124 At times trout bulge at shrimp, snail, and caddis. 1889 *Sat. Rev.* 18 May 612/2 Mr. Halford..mentions a short way with bulgers, which he condemns as unsportsmanlike. 1892 *Field* 4 June 838/2 The fish were smutting or bulging on the shallows. 1904 *Westm. Gaz.* 13 Dec. 4/2 The question of herrings 'bulging', as it is called.
5. *intr.* To rush *in*, make a rush *for*. Also *trans.* (causative). To cause to rush *off*. *U.S. colloq.*
1834 Crockett *Narr. Life* xiv. 96 My dogs..bulged in, and in an instant the bear followed them out. *Ibid.* xvi. 105 As soon as we struck, I bulged for my hatchway. *a* 1867 H. Woodruff *Trotting Horse Amer.* (1868) xxii. 207 Brooks and Harry Jones bulged them [*sc.* the horses] off in the lead at such a rate I was forced to let them take the pole on the turn. 1884 'Mark Twain' *Huck. Finn* 372 Here comes a couple of the hounds bulging in from under Jim's bed.

bulger (bʊ·ldʒəɹ). [f. Bulge *v.* + -er¹.] **1.** (See quot. 1872.) *U.S. slang.*
1835 Crockett *Tour* 37 We..soon came in sight of the great city of New York, and a bulger of a place it is. 1872 Schele de Vere *Americanisms* 587 *Bulger*,..in the United States generally designates anything very large. 'That's a bulger of a story.'
2. *Golf.* A wooden club with a convex face. (Disused.)
1889 *Sat. Rev.* 25 May 622/2 In golf there is little to note beyond the apparition of the Bulger. This club is the invention of a fresh and scientific mind. 1890 W. Park, Jun. in *Golf* 17 Oct. 70/1, I know for a fact, which I can prove, that I invented the bulger..; and I am undoubtedly entitled to the honour of producing the first bulger cleek [*read* club]. 1907 *Westm. Gaz.* 15 Feb. 4/2 The 'bulger'.. that club with the round face which was designed as an aid to straightness of hitting, but has now..been quite discarded.
3. (See *bulge *v.* 3 b.)

bulgily (bʊ·ldʒili), *adv.* [f. Bulgy *a.* + -ly².] In a bulgy manner.
1891 V. C. Cotes *Two Girls on Barge* 156 Settling down bulgily, much as the robins do before they go to sleep.

bulgingly (bʊ·ldʒiŋli), *adv.* [f. Bulging *ppl. a.* + -ly².] In a bulging manner; *fig.*, prominently.
1896 *Daily News* 25 Feb. 4/7 They only offend persons in whom the moral sense is so bulgingly hypertrophied as to prompt them to yell 'Plagiarist!'

bulk, *sb.*¹ **1. a.** Delete †*Obs.* and add: A pile of tobacco made up to undergo sweating. *U.S.*
1784 J. Smyth *Tour U.S.* II. 135 When the tobacco house is quite full,..all that is within the house is..carefully placed in bulks, or regular rows one upon another. 1850 *Rep. Comm. Patents 1849* (U.S.) 322 Two rows or bundles are put in a bulk. 1902 *U.S. Dept. Agric., Farmers' Bulletin* No. 60. 14 Before the sweat is completed the bulk is pulled down and built up eight or ten times.
attrib. 1693 in *Cal. Virginia St. Papers* (1875) I. 48 An answer to a former message of yours relating to the Act of Ports & Bulke Tobacco.
c. *in bulk* (earlier and later examples); also, in large quantities.
1678 *New Castle* (Del.) *Court Rec.* (1904) 253 Tobacco which was struck & Lay in bulke. 1908 *Mod. Business* II. ii. 165/1 It is possible for traders to effect a considerable saving by buying in bulk. 1928 [see *coupon 2].
4. c. *Paper-making.* The thickness of paper (see quots.).
1903 C. Beadle in C. F. Cross et al. *Paper Testing* i. 9 The simplest numerical expression of 'bulk', (*i.e.*, the bulking qualities of the fibres composing the paper). 1906 R. W. Sindall *Paper Technology* ix. 100 The bulk of a paper may be expressed in terms of the thickness of a single sheet or the thickness of a ream. 1920 H. A. Bromley *Paper & its Constituents* iii. ii. 161 Bulk in its most correct

sense may be defined as the ratio of fibre volume to total volume. 1969 *Brit. Printer* June 65/2 In the field of book papers..one can still obtain a ton or two, tailor-made to a particular requirement of shade, bulk and finish.

d. *spec.* The thickness of a book without its covers.

1906 L. L. WALTON in F. H. Hitchcock *Building of Book* 27 The bulk or thickness that the book must be, to make a volume of proper proportions, is determined. 1960 G. A. GLAISTER *Gloss. Book* 48/1 *Bulk*, the thickness of a book without its covers. The bulk will be less after binding than before.

e. = *ROUGHAGE 2.

1940 G. BOURNE *Nutrition & War* i. 8 This necessity for bulk in food is one reason why we are not likely to have all our food requirements reduced to one small pill. 1950 *N.Z. Jrnl. Agric.* May 485/2 There are three groups of crops suitable for feeding to pigs: Concentrates, semi-bulk foods, and bulk foods. 1962 *Which?* Jan. 25/1 These are all harmless laxatives, useful if your normal diet produces insufficient bulk. *Ibid.* 26/2 All preparations used as laxatives are effective by acting as bulk-suppliers, or irritants, or lubricants.

7. *attrib.* = in bulk, as *bulk-buying, -purchasing, supply*, etc.

1693 [see sense *1 a.] 1848 *Rep. Comm. Patents 1847* (U.S.) 527 Bulk pork is that which is intended for immediate use or smoking. 1892 *Daily News* 13 May 5/8 Bulk transport threatens 'danger to the security, safety, and freedom from interruption of the Suez Canal'. 1906 *Daily Chron.* 3 Mar. 4/4 Bulk power generation. *Ibid.* 13 June 4/6 A monopoly of the bulk supply [of electricity]. 1930 *Economist* 19 July 107/1 The question of import boards for bulk purchases. *Ibid.* 9 Aug. 272/1 The project of setting up bulk-purchasing import boards. 1932 *Times Lit. Suppl.* 10 Nov. 824/3 Not only is it impossible for the potentialities of bulk-buying.. to be exploited, [etc.]. 1940 'M. INNES' *Secret Vanguard* xix. 209 Let us worry rather about petrol and bulk wheat. 1949 *Hansard Commons* CDLXV. 1453, 50 per cent..of the total imports of the country are bulk-purchased by the Government. *Ibid.* 1558 When we talk about bulk buying, we obviously mean three or four different things. We mean an ordinary large single purchase; we mean centralised buying; and, of course, we mean State trading.

b. bulk barrel, a barrel of 36 gallons of wort or beer without regard to specific gravity (as distinguished from *standard barrel*); so *bulk gallon*; **bulk eraser** (see quot. 1959); **bulk modulus** *Math.* (see quots.).

1905 *Daily Chron.* 29 July 4/5 The discrepancy between the number of 'standard' barrels of beer upon which duty is paid and the number of 'bulk' barrels actually brewed. 1909 *Ibid.* 12 May 1/4 The bulk barrel may be of any specific gravity. The average is, I should say, about 1·053. 1956 R. E. B. HICKMAN *Magnetic Recording Handbk.* v. 120 A tape may be cleaned more rapidly by the use of a bulk eraser. 1959 W. S. SHARPS *Dict. Cinematogr.* 82/1 *Bulk eraser*, a device designed to pass a high intensity alternating current through wound reels of magnetic tape such that the recorded magnetic patterns are completely erased in a matter of seconds. 1889 in G. Birch *Handbk. Gauging* (1894) 67 When the deduction for tenths reduces bulk gallons to less than those of next lower inch. 1908 E. S. ANDREWS *Theory & Design of Structures* i. 9 There is an additional modulus called the *bulk* or *volume modulus* (K), which represents the ratio between the unital change in volume and the intensity of pressure or tension on a cube of material subjected to pressure or tension on all faces. 1935 C. G. BURGE *Compl. Bk. Aviation* 194/1 The bulk modulus is that which expresses the relation between stress and change in unit volume when a body is subjected to equal stresses on all faces, such as when a body is under pressure.

bulk, *v.*[1] Add: **4.** Also, to pile (tobacco) in the course of preparing it for use. *U.S.*

1850 *Rep. Comm. Patents 1849* (U.S.) 322 Stripping should never be done in drying..weather, unless the tobacco is bulked up almost as fast as it is stript. 1863 *Rep. Agric. Soc. Maine* 163 When the weather again becomes moist,..take it [*sc.* tobacco] down and carefully bulk it away as before directed. *Ibid.*, Care must be taken that the tobacco does not..get too high in case before it is bulked. 1902 *U.S. Dept. Agric., Farmers' Bulletin* No. 60. 17 These are tied up into hands and bulked down for a short time.

5. b. (See quot.)

1931 C. MAUGHAN *Markets of London* xxix. 104 Most of the descriptions of coffee..are poured out from the bags on to special floors, where they are 'bulked', or mixed, in order to ensure that the contents of all the bags are of uniform quality, and they are then rebagged.

6. *trans.* To put together (two or more consignments of goods) for transport as one. Also *absol.*

1908 *Modern Business* Sept. 164/2 Had they been 'bulked'—*i.e.* sent as one consignment, from one consignee to an agent to deliver—the company would have had no alternative but to charge the lower rate. *Ibid.* 165/1 If a merchant can, by bulking several parcels, get them through at a much lower rate. 1928 *Daily Express* 10 Mar. 9/4 Bulking means.. that two or more consignments for different consignees in one town, forwarded at one and the same time, can be charged together as one lot.

7. To enlarge a book by adding to the number or thickness of its leaves; esp. to make a book look big by printing it on paper of abnormally loose texture. Also *intr.*, to have a specified bulk (see *BULK *sb.*[1] 4 c).

1920 CROSS & BEVAN *Textbk. Paper-Making* (ed. 5) xiv. 403 A paper either 'bulks high' or 'bulks low' in relation to its 'substance'. 1932 B. BLACKWELL *World of Books* 41 Setting their faces against the artificial bulking of books. 1937 E. J. LABARRE *Dict. Paper* 114/1 Light and 'bulky' papers, the opposite of which are heavy or dense, have 'bulking' qualities, and papers are said to 'bulk well'.

1957 *Times Lit. Suppl.* 27 Dec. 787/2 For this deterioration publishers are in part to blame in their demand for 'bulking' properties.

bulked, *ppl. a.* Add: **b.** Having its bulk increased; *spec.* of books (cf. *BULK *v.*[1] 7) and of yarn (see quot. 1957).

1954 J. H. KENNEDY in *Textile World* Oct. 80/2 Pioneering in the field of bulked-Orlon yarns from tow. 1957 *Times Lit. Suppl.* 27 Dec. 787/2 Max Reinhardt should ask public librarians for their view of 'bulked' books. 1957 *Textile Terms & Defs.* (ed. 3) 30 *Bulked yarns*, yarns which have been treated physically or chemically so as to have a notably greater 'apparent volume' or bulk.

bulker[2]. Add: **2.** One who makes up tobacco into piles for curing. *U.S.*

1863 *Rep. Agric. Soc. Maine* 163 This [bulking] is done by lapping the bundles over each course..the bulker having his knees upon the bulk. *Ibid.*, Under the pressure of the bulker's knee.

bulkhead. Add: **1. a.** (Earlier example.)

1496 *Naval Accts. Henry VII* (1896) 167 Amendyng of the Bulke hede for Couching of Bere.

c. A similar partition in an airship, an aeroplane, or a train.

1921 *Discovery* II. 97/1 The fuselage [*sc.* of the airship].. consists of riveted sheet duralumin, kept rigid by a series of internal duralumin girder bulkheads. 1935 C. G. BURGE *Compl. Bk. Aviation* 306/2 The main precaution against fire is the inclusion of a fireproof bulkhead between the engine compartment and the rest of the aeroplane. 1948 'N. SHUTE' *No Highway* iii. 80 You must..sit down on the floor facing to the tail, with your back against the forward bulkhead. 1952 *Railway Age* 24 Nov. 10 (Advt.), Bulkheads and partitions in this luxury car are 73% lighter than standard steel construction.

d. In a motor-car: the partition dividing the engine from the body. Cf. *SCUTTLE *sb.*[2] 1 d.

1958 F. J. CAMM *Practical Motorist's Encycl.* (ed. 9) 473/2 Holes are drilled in the metal bulkhead or scuttle, and the holder fixed. 1962 *Which?* (*Car Suppl.*) Oct. 138/2 The mounting bracket for the accelerator linkage came adrift from the bulkhead.

bull, *sb.*[1] Add: **1. c.** *to take the bull by the horns* (earlier examples); (*like a*) *bull at a* (*five-barred*) *gate*: with direct violence or impetuosity; so *bull-at-a-gate*, used attrib. to describe a direct and vigorous attack.

1711 SWIFT *Conduct of Allies* 58 To engage with France, was to take a Bull by the Horns. 1816 SCOTT *Old Mort.* in *Tales of my Landlord* III. xii. 258 He had not, as the phrase goes, taken the bull by the horns, or advanced in front of the enemie's fire. 1896 *Westm. Gaz.* 14 Oct. 5/3 [He] cross-examined in his usual blusterous bull-at-a-gate sort of fashion. 1957 *Times* 11 May 7/3 If the whole scheme is to be rushed through, bull at a gate, there is bound to be some initial chaos. 1963 *Times* 2 Mar. 3/4 Blair has always possessed cover and straight drives of the purest quality, but has usually adopted bull-at-a-gate tactics.

7. b. A locomotive. *U.S. slang.*

a 1889 *On the Trail* (Barrère & Leland), Had just touched a bloke's leather as the bull bellowed for the last time. 1889 FARMER *Americanisms*, *Bull* (Cant), a locomotive; the word is sometimes lengthened into Bullgine.

c. A policeman. *U.S. slang.*

[1859 G. W. MATSELL *Vocabulum* 15 *Bull-traps*, rogues who personate [police] officers for the purposes of extorting money.] 1893 J. FLYNT in *Century Mag.* Nov. 103/2, I have seldom met a hobo who was very angry with a New York 'bull'. 1909 J. LONDON in *Contemp. Rev.* June 699, I noticed the bull, a strapping policeman in a grey suit...I never dreamed that bull was after me. 1959 'A. GILBERT' *Death takes Wife* xvii. 217 Sam putting the bulls on you.

8. b. bull point *colloq.*, a point of advantage or superiority, a great 'score'.

1900 *Westm. Gaz.* 27 Sept. 9/3, I am afraid that Lord Lansdowne has proved anything but a bull point to the House. 1923 *Daily Mail* 12 Oct. 7/2 The great bull point of our manufactures is their reputation for quality. 1961 *Times* 14 Apr. 5/4 It is a bull point in his favour that the visitors..found him eminently satisfactory.

10. a. *bull-meat, -team.* **b.** *bull-mouthed, -throated* adjs.

1812 R. STUART *Narratives* (1935) 160 Poor Bull meat or Buck Antelope. 1859 R. B. SAGE *Rocky Mountain Life* 64 Bull-meat at this time..is unprecedentedly tough, strong-tasted, and poor. 1893 KIPLING *Seven Seas* (1896) 22 When the bull-mouthed breakers flee. 1855 *Golden Era* (San Francisco) 1 Apr. 4/2 The humbler occupation of swaying a bull-team. 1888 *San Francisco Weekly Exam.* 23 Feb. (Farmer), I gave instructions to the wagon boss, and the long bull-team moved away. 1888 KIPLING *Departm. Ditties*, etc. (ed. 4, 1890) 69 Hans the blue-eyed Dane, Bull-throated, bare of arm. 1928 *Daily Express* 12 Sept. 8 Bull-throated foremen bawled for more and yet more labour.

11. bull and cow, rhyming slang for Row *sb.*[3]; **bull-ant** = BULL-DOG 4 b; **bull-bat** (earlier example); **bull dust** *Austral.*, (*a*) a coarse dust; (*b*) nonsense, rubbish (*slang*); **bull-fiddle** *U.S. colloq.*, a bass-viol or double-bass; hence *bull-fiddler*; **bull-holder** (see quot. 1940); **bull-horn,** a megaphone; **bull-kelp,** any of several varieties of large seaweed found in Pacific and Antarctic waters; cf. *bull-head kelp* (KELP[1] 1 a); **bull-nose,** (*a*) = *bull's nose* (see 11 b in Dict.); (*b*) see quots.; also *attrib.*; so **bull-nosed** *a.* (see quots.); **bull oak** *Austral.*, a name given to several species of *Casuarina*, esp. *C. luehmanni* (cf. OAK 3 b); **bull-puncher,** (*a*) *Austral.*, a bullock-driver; (*b*) *U.S.*, a cowboy, cow-puncher; so *bull-punching*; **bull-ring,** also *transf.* (*slang*) a military training-ground; **bull-roarer,** *spec.* used by Australian Aborigines in certain (religious) ceremonies; **bull-snake** *U.S.*, a large North American snake (*Pityophis melanoleucus*); the pine-snake; **bull-sticker** (see quots.); **bull-strong** *a. U.S.*, strong enough to resist a bull; **bull's wool, bullswool,** (*a*) *Army slang*, coarse woollen cloth or yarn; (*b*) *Austral.* and *N.Z. colloq.* = *BULL *sb.*[4] 3; see also quot. 1898; **bull-tongue** (**plough**) *U.S.*, a simple form of plough; **bull-whack** *sb. U.S.* (see quot.); **bull-whack** *v. trans.* and *intr.* (*U.S.*), to drive (cattle); **bull-whacker** (earlier example); (*b*) = *bull-whack* sb.; **bull-whip** *U.S.*, a whip with a long heavy lash, used for driving cattle.

1859 HOTTEN *Slang Dict.* 142 *Bull and cow*, a row. 1962 'A. GILBERT' *No Dust in Attic* xi. 138 The murder might have been the result of a private bull-and-cow. 1900 *Daily News* 26 May 3/5 As eager for fight as a bull ant on a hot plate. 1908 *Westm. Gaz.* 10 Dec. 1/3 His first bite from an Australian 'bull-ant'. 1948 B. JAMES in *Coast to Coast 1947* 162 But Tommy could fight—game as a bull-ant. 1838 P. H. GOSSE *Lett.* (1859) 62 The common people here generally call these birds by the name of bull-bats. 1943 BAKER *Dict. Austral. Slang* 16 *Bull dust*.. bullsh. 1943 W. E. HARNEY *Taboo* 127 The gypsum crumbles into copi or bulldust. 1954 B. MILES *Stars my Blanket* xvi. 105 We found the track..to be inches deep in bull dust—a soft, powdery dust that seeps through into everything. 1967 J. HAMILTON *Man with Brown Paper Face* iv. 50 I'm not in the mood for any of your bulldust. Where have you been all night? 1880 G. A. SALA *Amer. Revisited* 209/1 A 'bull-fiddle' which is American for violoncello. 1941 STEINBECK & RICKETTS *Sea of Cortez* (1951) iv. 30 A deep and yet penetrating tone like the lowest string of an incredible bull-fiddle. 1957 W. C. HANDY *Father of Blues* xvii. 237 As usual the bull-fiddler sawed away in G. 1940 *Chambers's Techn. Dict.* 119/2 *Bull-holder*, forceps for grasping the nasal septum of cattle as a means of restraint. 1955 C. S. FORESTER *Good Shepherd* ii. 38 Her captain shouting himself hoarse through his bull-horn at the laggards. 1959 *Ottawa Citizen* 24 Sept. 48/6 Mr. Garst tried to explain something to reporters through an electric bull-horn. *c* 1929 W. MARTIN *N.Z. Nature Book* II. i, The Giant Bull Kelp (*Durvillea utilis*) is a truly Antarctic sea-weed. 1954 *New Biol.* XVII. 96 The genera vary in size from the great bull-kelps..to a small parasite. 1858 *Skyring's Builders' Prices* 41 Circular styles to bull-nose corners. *a* 1884 KNIGHT *Dict. Mech.* Suppl., *Bull-nose Rabbet Plane*, a plane with the bit at the end, in order to enable it to work up close into corners. 1895 *Funk's Standard Dict.* s.v. *Nose*, *Bull-nose*, a front coupler on a locomotive: named from its shape; a bull-nose coupler. 1904 GOODCHILD & TWENEY *Technol. & Sci. Dict.* 75/2 *Bullnosed bricks*, bricks having one angle on the end rounded off. *Bullnosed step*, a step with a rounded end. 1933 *Archit. Rev.* LXXIII. 198/2 It could be chiselled with a bull-nosed chisel. 1952 'W. COOPER' *Struggles of Albert Woods* II. ii. 83 Albert pointed his bull-nosed Morris Cowley in the direction of Islip. 1884 A. NILSON *Timber Trees N.S.W.* 7 The most widely-distributed and best-known species are..'Pine' or 'Cypress Pine'..'She Oak', 'He Oak', 'Bull Oak'. 1963 W. E. HARNEY *To Ayers Rock & Beyond* ii. 23 The famous 'Drunk's-seat', beneath a shady bull-oak tree. 1872 C. H. EDEN *My Wife & I in Queensland* ii. 49 The 'bull-puncher', as bullock-drivers are familiarly called. 1874 *Chambers's Jrnl.* 543/2 Commissariat beeves, guarded by the commissariat 'bull-punchers'. 1887 M. ROBERTS *Western Avernus* ii. 19 He followed the profession of a 'bull-puncher', that is, he went in charge of the cattle destined for slaughter and canning in the distant North, and made money at it. *Ibid.* 20, I found this bull-punching a very wearisome and dangerous business. 1917 'H. H. RICHARDSON' *Fortunes R. Mahony* I. v. 39 Stock-riders and bull-punchers rubbed shoulders with elegants in skirted coats. 1928 BLUNDEN *Undertones of War* i. 3, I associate it [Étaples], as millions do, with 'The Bull-Ring', that thirsty, savage, interminable training-ground. 1949 E. DE MAUNY *Huntsman in Career* I. i. 13 Drawing equipment at the Q.M., drilling on the bull-ring. 1898 *Daily News* 23 July 6/1 Among Australian blacks, the bullroarer is associated with tales of a deluge. 1943 W. E. HARNEY *Taboo* 43 In this ritual the lads are taken away from the women, initiated into the secrets of the 'bullroarer' and..live as outcasts. 1784 J. FILSON *Kentucke* 27 The bull, the horned and the mockason snakes. 1791 W. BARTRAM *Travels* 276 The pine or bull snake is very large. 1878 J. H. BEADLE *Western Wilds* ix. 133 The 'bull-snake'..an immense thing of four or five feet in length, which gets its name from its blunt head and thick clumsy body. 1933 *Discovery* Oct. 319/2 Numerous varieties of burins called by such pleasant names as spitstickers, bullstickers, [etc.]. 1958 J. R. BIGGS *Woodcuts* 49 The bull-sticker..is virtually a spitstick with bulging sides. The line it makes will therefore have a very rapid 'spread'. 1859 *Harper's Mag.* Oct. 712/2 A fence that is bull strong, horse high, and pig tight! 1850 'TWO MOUNTED SENTRIES' *Horse Guards* 70 The 'sealed pattern' [of cloth] provided for the British soldier, and familiarly known among the men by the euphonious cognomen of 'bull's wool'. 1893 KIPLING *Many Inventions* 268 If iver you feel that you've got a felt sole in your boot instid av a Government bull's-wool, come to me. 1898 MORRIS *Austral Eng.* 64/1 *Bulls-wool*, colloquial name for the inner portion of the covering of the Stringybark-tree. 1911 E. M. CLOWES *On Wallaby* ix. 247 To build one's fire, kindling it surely and quickly with what is called 'bull's-wool', the thick, dry fibre, like fine cocoa-nut matting, which forms the hair shirt of the gum-tree between the white skin and the cream and green and madder-tinted bark. 1933 *Bulletin* (Sydney) 13 Dec. 40/1 'And I'm dropping fifteen!' 'Bullswool!' declared Tommy. 'What you'd drop it'd take a bloke with a micro-

scope to find.' 1950 G. MEEK in A. E. Woodhouse *N.Z. Farm & Station Verse* 153 Don't think that it's all bullswool. 1957 I. CROSS *God Boy* (1958) ix. 67 That last bit was bullswool of course, but I had to be careful. 1831 M. A. HOLLEY *Texas* (1833) 139 Many farmers use the coulter and bull-tongue plough. 1837 in N. E. Eliason *Tarheel Talk* (1956) 262, 25¢ for making a bul tung. 1886 *Harper's Mag.* June 58/2 Ploughing is commonly done with a 'bull-tongue', an implement hardly more than a sharpened stick with a metal rim. 1869 A. K. MCCLURE *Rocky Mts.* 102 You will often find some graduate of Yale 'bull-whacking' his own team from the river to his mines. 1885 C. L. NORTON in *Mag. Amer. Hist.* XIII. 98 In Texas and western Louisiana the 'bull-whack' is a terrible whip with a long and very heavy lash and a short handle. It is used by drovers to intimidate refractory animals. 1906 *Dialect Notes* III. 129 He's a bull-whackin'. 1858 *Valley Tan* (Salt Lake City) 17 Dec. 2/2 An example that will make the blush of shame mantle upon the cheek of the bull-whacker. 1887 E. B. CUSTER *Tenting on Plains* (1888) vii. 229 There is no sound like the snap of the lash of a 'bull-whacker'. 1852 *Southern Lit. Messenger* XVIII. 749/2 An overseer..omitted..laying down at once his bull-whip for the yard-stick. 1935 J. STEINBECK *To God Unknown* xxii. 205 Romas snapped his bull-whip and the popper spat up the dirt like an explosion.

b. bull's blood, a full-bodied red wine made in and around the village of Eger in Hungary.

1926 P. M. SHAND *Bk. Wine* ix. 242 An almost black wine called..'Bulls' Blood Wine' used to be made here [*sc.* Eger] by a very protracted process of fermentation. 1967 A. LICHINE *Encycl. Wines* 294/1 Bull's blood of Eger is a blend of three kinds of grapes.

bull, *sb.*[4] Add: **2. b.** A bad blunder. *U.S.*

1846 D. CORCORAN *Pickings* 13 When we speak of 'Irish Evenings' in New Orleans, we are guilty of neither bull nor blunder. 1855 'P. PAXTON' *Capt. Priest* 226, I had committed a bull myself, by intruding where I evidently was *de trop*. 1904 *N.Y. Times* 4 July 3 They are going to nominate Parker, and they are going to make a bull by doing it. 1934 J. T. FARRELL *Young Manhood* (1936) xviii. 379 It was bull number one for him, [a] bad way to start the evening off.

3. Trivial, insincere, or untruthful talk or writing; nonsense. *slang* (orig. *U.S.*).

Popularly associated with *BULLSHIT 1.

1914 *Dialect Notes* IV. 162 *Bull*, talk which is not to the purpose; 'hot air'. 1925 WODEHOUSE *Sam the Sudden* xx. 160 You threw a lot of bull about being the brains of the concern. 1932 *Times Lit. Suppl.* 8 Dec. 933/3 'Bull' is the slang term for a combination of bluff, bravado, 'hot-air', and what we used to call in the Army 'Kidding the troops'. 1946 G. GIBSON *Enemy Coast Ahead* xii. 159, I have never heard such a line of bull in all my life. 1952 M. MCCARTHY *Groves of Academe* (1953) vi. 116, I never thought I'd be listening to that old bull slung at Jocelyn.

b. *attrib.*, esp. in **bull session** (orig. and chiefly *U.S.*), an informal conversation or discussion, esp. of a group of males.

1920 T. WOLFE *Let.* 26 Nov. (1956) 11 With no more delightful 'bull sessions', I have turned to work. 1924 P. MARKS *Plastic Age* x. 77 Religion and sex, the favorite topics for 'bull sessions'. *Ibid.* xxiv. 286 The monthly meetings were nothing but 'bull fests'. 1931 'DEAN STIFF' *Milk & Honey Route* 201 *Bull artist*, a hobo with the gift of gab. 1960 *Guardian* 8 Dec. 12/3 The kind of college 'bull session' that is common among English students.

4. Unnecessary routine tasks or ceremonial; excessive discipline or 'spit-and-polish'; = RED-TAPE b. Cf. *BULLSHIT 2. orig. *Services' slang.*

1941 *New Statesman* 30 Aug. 218/3 *Bull*, discipline. 1942 I. GLEED *Arise to Conquer* vi. 51 The Squadron..felt very bolshie about all the bull that was flying around the station. 1953 A. BARON *Human Kind* xxiv. 178 Them turning out the guard for us, us marching past eyes right, all that sort of bull. 1958 *Economist* 8 Feb. 470/1 The drudgery and 'bull' in an MP's life.

bull, *sb.*[7] Ellipt. for BULL-DOG 1 a.

1827 PIERCE EGAN *Anecdotes of Turf* 107 Turk was a thorough-bred bull, and the other two were half-bred between a bull and a mastiff. 1939 T. WOLFE *Web & Rock* (1947) 17 The little bull..had his fierce teeth buried..in the great throat of the larger dog.

Hence **bull-mastiff, bullmastiff,** a dog of a cross-breed between a bull-dog and a mastiff.

1871 *Field* 29 Apr. 343/2 The fight which took place in the earlier part of this century between Wombwell's two lions and bull-mastiffs. 1948 D. J. NASH in B. Vesey-Fitzgerald *Book of Dog* 384 In 1795 an advertisement appeared for a lost Bullmastiff. 1959 *Country Life* 10 Dec. 1139/1 When the bull-mastiff or bull mastiff graduated from the gamekeeper's kennel..to the show-ring and a place in the Kennel Club register, the breed became known as the bullmastiff.

bull (bul), *sb.*[8] [Origin obscure.] A game resembling quoits. Also *attrib.*

1864 TREVELYAN *Compet. Wallah* 16 In search of sport these join the circle full That smokes and lounges round the game of 'Bull'. 1889 *Pall Mall Gaz.* 22 June 3/2 We tried to help on the dreary time with..a game called 'bull'—a kind of sea-quoits. 1928 *Blackw. Mag.* Mar. 418/2 Indifferent to the call of the 'Bull-board' or the deck-quoit. 1963 M. MALIM *Pagoda Tree* xiii. 77 She was good at bull-board—a matter of lobbing little sacks of sand into numbered squares upon an inclined board.

bull, *sb.*[9] Short for BULL'S-EYE 7; also, a shot hitting the bull's-eye.

1900 *Westm. Gaz.* 13 June 5/2 Able to hit a two-foot bull five times out of ten at 500 yards. 1932 A. J. WORRALL *Eng. Idioms* 3 He scored seven bulls in eight shots. 1955 *Times* 11 Aug. 8/4 The uninitiated..soon learn to refer to a 'gold', and not to a 'bull' or an 'inner'.

bull, *v.*[1] Add: **2. b.** *intr.* To advance in price; *fig.* to be in the ascendant.

1928 S. VINES *Humours Unreconciled* 252 Music was 'bulling' in Japan and the Conservatory crammed to bursting point.

3. To behave or move like a bull; to act with violence in the manner of a bull. Also *refl.* *U.S. slang.*

1884 'MARK TWAIN' *Huck. Finn* xvi. 144 Up-stream boats..bull right up the channel. *Ibid.* xxvii. 276 The old fool he bulled right along. 1947 A. MILLER *All my Sons* II, Don't come bulling in here. If you've got something to say, be civilized about it. *Ibid.* III, You can't bull yourself through this one, Joe, you better be smart now. 1956 *Time* 10 Sept. 30/1 A mob of about 400 Texans bulled about the school grounds.

bull, *v.*[3] Delete †*Obs.* and add later example.

1927 J. BARBICAN *Confessions of Rum-Runner* xxiii. 256, I thought he was trying to bull me.

2. *intr.* To talk emptily or boastfully (cf. *BULL *sb.*[4] 3).

1850 T. M. GARRETT *Diary* in *Amer. Speech* (1951) XXVI. 182 Elaborate bulling about a point that has been exploded for years. 1941 BAKER *Dict. Austral. Slang* 14 *To bull*, to brag, talk nonsense.

3. *trans.* To polish (equipment, etc.) in order to meet excessive standards of neatness. Hence **bulled** *ppl. a.*[3] Cf. *BULL *sb.*[4] 4. *Services' slang.*

1957 *Times* 4 Oct. 13/5 Those army recruits who spend so much time 'bulling' their boots with a hot spoon. 1967 'M. HUNTER' *Cambridgeshire Disaster* v. 32 The bed.. collapsed, spilling equipment haberdashery over the bulled-up floor. 1969 D. CLARK *Nobody's Perfect* ii. 35 His shoes were bulled so that the toecaps gleamed like patent leather.

bullamacow (bu·lāmăkau). Also bulli-. [Fiji combination of *bull* and *cow*.] **a.** Cattle. **b.** Corned beef; bully beef.

1887 W. B. CHURCHWARD *My Consulate in Samoa* xxiii. 277 'Bulli-ma-cou', applicable both to the meat and the animal...On the first animals being landed the natives.. were told that the beasts they saw before them were a bull and a cow; so combining the two English words they made one of their own. 1902 *Westm. Gaz.* 30 July 2/1 Their supper of rice and bullmacow [*sic*]. 1925 *Chambers's Jrnl.* Feb. 87/2 If they wanted *bullamacow*..he would readily provide that food. 1927 *Observer* 6 Feb. 19/5 The Fijians had no words for bull or cow and now have only one portmanteau word, 'bulumacau'. 1952 A. GRIMBLE *Pattern of Islands* 80 We never entertained one whose face brightened perceptibly when bullimacow and beetroot were placed before him.

bullberry, bull berry: see *BULBERRY.

bull-dike (bu·ldaik). *slang.* Also **bull-diker, -dyke(r.** [Origin unknown. Cf. *DIKE *sb.*[3]] A lesbian with masculine tendencies.

1926 C. VAN VECHTEN *Nigger Heaven* 285 *Bulldiker*, Lesbian. 1942 BERREY & VAN DEN BARK *Amer. Thes. Slang* § 405/3 Masculine woman, *boy, bulldike, bulldyke, bulldiker*. 1964 J. RECHY *City of Night* 174 On the dance-floor, too, lesbians—the masculine ones, the bulldikes—dance with hugely effeminate queens.

bull-dog. Add: **1. b.** *bull-dog breed.*

1857 C. KINGSLEY *Two Years Ago* II. iv. 67 Knowing him to be of the original British bull-dog breed, which, once stroked against the hair, shows his teeth at you for ever afterwards. 1897 F. MCGLENNON *Sons of Sea* 4 Sons of the Sea! All British born!..boys of the bull-dog breed Who made old England's name. 1940 'G. ORWELL' *Inside Whale* 40 The typical English boasting, the..'bulldog breed'..style of talk.

4. a. (Earlier example.) In full *bull-dog fly*. **b.** (Earlier examples.)

1792 P. TURNOR *Jrnl.* (1934) 488 A kind of fly..called bull dogs. 1848 R. M. BALLANTYNE *Hudson's Bay* (ed. 2) vii. 147 The whole room was filled with mosquitoes and bull-dog flies. 1853 E. CLACY *Gold Diggings Austr.* xvi. 249 The largest [ants] are called by the old colonists, 'bull-dogs', and formidable creatures they are. 1860 W. B. CLARKE *Res. S. Gold Fields N.S.W.* vii. 120 It was difficult to find a spot on which to lay our blankets, on account of the 'Bull Dog Ants'.

7. b. Used (freq. *attrib.*) of a clip or other fastening having a strong closure.

1908 *Sears, Roebuck Catal.* 204/3 Bull Dog Wire Grip. The more you pull the tighter it grips. 1923 *Man. Seamanship* II. 307 Special screw clamps, after the fashion of bull-dog grips. 1943 *Trade Marks Jrnl.* 26 May No. 3400 224/2 (*heading*) Bulldog, B 620,897 metal clips being articles of stationery .. 6th Nov. 1942 [date of registration]. 1944 'N. SHUTE' *Pastoral* ii. 28 Messages and signal forms in bulldog clips. 1969 D. CLARK *Nobody's Perfect* v. 153 Her invoices hung in large bulldog clips from hooks on the walls.

9. In full *bull-dog pipe*. A short tobacco-pipe of brier-wood.

1885 A. EDWARDES *Girton Girl* I. iv. 84 The pipe was a black, ferociously Bohemian-looking 'bulldog'. 1900 W. C. MORROW *Bohemian Paris* 306 Three-days-in-Paris English tourists wearing knickerbockers and golf-caps, and always smoking bulldog pipes. 1966 L. J. BRAUN *Cat who could read Backwards* (1967) v. 56 'I use a pipe.' Qwilleran searched for his quarter-bend bulldog and his tobacco pouch.

10. *bull-dog edition*, the earliest edition of a daily or Sunday newspaper. *U.S.*

1926 *Nation* 13 Oct. 342/2 This story got into the bull-dog edition of one of the papers before he could finish his midnight rounds. 1934 D. WILKIE *Amer. Secret Service Agent* iii. 33 That night I took the bulldog edition with my spread to a restaurant where the newspaper men congregated.

bull-dog, *v.* Chiefly *U.S.* [f. the sb.] *trans.* To attack like a bull-dog; to assail or treat roughly; *spec.* to wrestle with and throw (a steer or other animal). Hence **bu·ll-do:gger, bu·ll-do:gging.**

1842 *Congress. Globe* 29 Apr. 457/3 Mr. Whitney had not been 'dogged' to the door of the committee-room, but, when inside, he had been 'bull-dogged' with a vengeance. *Ibid.* 4 May 478/1, I made the reply about bull-dogging for the gentleman from Virginia. 1883 MEREDITH *Let.* 20 July (1970) II. 706 Greenwood..is bull-dogging France. 1907 *Outing* (U.S.) Dec. 329/1 'No more necked calves,' they announced, 'catch 'em by the hind legs, or bull-dog 'em yourself.' 1907 S. E. WHITE *Arizona Nights* I. viii. 148 The two 'bull-doggers' immediately pounced upon the victim. *Ibid.* 151 One of the men..reached well over the animal's [*sc.* a calf's] back to get a slack of the loose hide next the belly, lifted strongly, and tripped. This is called 'bull-dogging'. 1953 R. CAMPBELL *Mamba's Precipice* viii. 94 He wasn't as good as Antonio..in bulldogging them by the horns and throwing them on their sides. 1963 *Times* 7 Mar. 9/4 A wonderful exhibition of..bulldogging (leaping from a horse and throwing a running bull to the ground).

bull-dogged (bu·ldǫgd), *a.* [f. BULL-DOG *sb.*] Characteristic of or like that of a bull-dog.

1904 *Daily Chron.* 4 Jan. 9/1 The face has become pushed back, or 'bull-dogged'. 1921 GALSWORTHY *To Let* III. II. xi, All that was bull-dogged in Soames stared for a moment out of his grey eyes.

bull-dose, -doze, *sb.* and *v.* For *U.S. colloq.* read: orig. *U.S. colloq.*

Add: **B.** *vb.* (Later examples.) (The usual spelling, influenced by *BULLDOZER 2 a, is now *bulldoze*.)

1897 E. A. BARTLETT *Battlefields of Thessaly* iii. 53 There is a remarkable resemblance..between the way in which English public opinion has been 'bulldozed' and misled in both cases. 1916 J. B. COOPER *Coo-oo-ee!* viii. 104 Debenham backed Danvers up by..pointing out to Hawley the folly of handing a loaded revolver to Boder to examine. They simply bull-dozed Hawley. 1941 G. G. SCHOLEM *Major Trends in Jewish Mysticism* viii. 320 The philosophers who tried to bulldoze us into accepting the God of Aristotle as the God of Religion. 1954 *Encounter* July 31/1 The men..were..trying to bulldoze the creative Czech artist..into conformity with the precepts of Socialist Realism.

2. *intr.* To use a bulldozer (*BULLDOZER 2 a); also, to push *one's way* by means of a bulldozer. Also *trans.*, to move, clear, or level by means of a bulldozer. So **bu·lldozed** *ppl. a.*, **bu·lldozing** *vbl. sb.* and *ppl. a.*

1937 *Geogr. Jrnl.* XC. 369 'Bull-dozing', in which the vertical walls at the head of a gully are destroyed and a sloping surface constructed. 1942 *Interpretation Aerial Photographs* (U.S. War Dept., Techn. Man. TM5–246) 188 The road was constructed by bulldozing the earth and painting the edges with asphalt paint. 1944 *Reader's Digest* Aug. 93 Men were coming out of the sea continually and starting to work—digging, hammering, bulldozing. 1946 *Spectator* 12 Apr. 372/1 Americans had to bulldoze their way through the rubble. 1948 *Time* 5 July 19/3 Trucks were lumbering..up a goat path, newly bulldozed. 1953 *Proc. Prehist. Soc.* XIX. 232 The destruction of many of the down-land earthworks by bulldozing and deep ploughing.

b. *transf.* and *fig.*

1948 *Sat. Rev. Lit.* 21 Feb. 26 She bulldozed her way through her songs. 1949 *Good Housekeeping* June 76/2 Every panacea-pamphlet that pours from the bull-dozing pens of doctrinaire Utopians. 1950 G. BARKER *News of World* 37 The juggernauts Go bulldozing through my thoughts. 1963 *Rev. Eng. Studies* XIV. 319 The second edition of 1934..has been bulldozed away and a new edifice constructed.

bulldozer (bu·ldōᵘzəɹ). orig. *U.S.* Also **bull-doser, bull-dozer.** [f. BULL-DOSE, -DOZE *v.*] **1.** (See BULL-DOSE, -DOZE *v.*)

1876 in *Congress. Rec.* (1877) 9 Jan. 500/1 A band of bull-dozers came into Saint Francisville. 1878, etc. [in Dict.]. 1899 M. KINGSLEY *Let.* 19 Mar. in S. Gwynn *Life of M.K.* (1933) 210 They leave that to the bulldozers, and the present system mistakes these bulldozers for representative men.

2. A heavy caterpillar tractor fitted with a broad steel blade in front, used for removing obstacles, levelling uneven surfaces, etc.

1930 *Water Works & Sewerage* (U.S.) June 262/3 The bulldozer is built for heavy duty. 1941 *N. & Q.* CLXXXI. 119/1 If they can bring in American machinery, why can't they bring in bull-dozers? 1942 *Times* 9 Oct. 2/2 There are.. machines for levelling—motor-propelled scrapers—tractors, dumpers, angle-dozers and bull-dozers.

b. *fig.* (Also *attrib.*)

1945 R. J. OAKES in *Coast to Coast 1944* 100 The fourth man was a gunner, a bulldozer of a man. 1952 *Sat. Rev.* 9 Aug. 12 The bulldozer determination with which he plowed through confused happenings. 1955 *Times* 21 June 9/4 Such spotters can..obtain 'bulldozer' rights for a patrol on a hot scent to pass through the areas of other units without being shot in error. 1959 *Times Lit. Suppl.* 20 Feb. 102/1 The bulldozer detective tactics of Inspector Evans.

buller (bu·ləɹ), *sb.*[3] [f. BULL *sb.*[1] + -ER[1].] (See quots.)

1858 *Jrnl. R. Agric. Soc.* XIX. 151 The cow..became what is called a 'perpetual buller', that is, always in a state to take the bull. 1901–3 *Rep. Kansas State Board Agric.* 253 (*Cent. Dict.* Suppl.), *Buller*, in *stock-raising*, a cow of irregular reproductive habit. 1953 S. J. BAKER *Australia Speaks* II. iii. 76 From dairy farm folk have come..*buller*, a cow on heat.

buller (bu·ləɹ), *sb.*[4] *University slang.* [See *-ER.[6]] = BULL-DOG 2.

1906 *'Varsity* 1 Feb. 165/3 There's a buller over there. Wonder if he's waiting to spot anyone. 1919 *Isis* 5 Nov. 7/2 The Proctor..on a motor-scooter, accompanied by a couple of attendant 'bullers' on a push-bike. 1937 *Evening News* 5 May 10/3 A 'buller' was sent round to their house to collect the fines of 13/4 each.

bullet, *sb.*[1] Add: **1. b.** *pl.* Beans or peas. *slang.*

1929 *Papers Michigan Acad. Sci., Arts & Lett.* X. 282/2 *Bullets*, beans. 1943 HUNT & PRINGLE *Service Slang* 18 *Bullets*, peas. 1963 *Daily Mail* 26 Aug. 4/3 He [*sc.* a schoolboy] calls peas 'bullets'.

3. b. *fig.* Notice to quit, the 'sack'. *slang.*

1841 SAVAGE *Dict. Art Printing* s.v., When a workman, at case or press, either for neglect, want of punctuality, or for gross misconduct, is discharged *instanter*, and the usual notice of 'a fortnight' is not given, it is said, *He has got the Bullet*. 1868 T. WRIGHT *Great Unwashed* 254 [One] who has ..got 'the bullet', as the formal note intimating that, 'owing to a reduction of our establishment your services will no longer be required', is called among working men. 1874 HOTTEN *Slang Dict.* s.v., To shake the bullet at any one, is to threaten him with 'the sack', but not to give him actual notice to leave. 1887 *Punch* 17 Sept. 126/1, I have just got the 'bullet', Mate—sacked without notice. 1929 *Melody Maker* Jan. 9/1 You and your band are getting the bullet, Bert Ambrose is taking over the Savoy in your place. 1960 H. PINTER *Caretaker* I. 10 The guvnor give me the bullet. Making too much commotion, he says. 1967 *Crescendo* Oct. 14/1 It was only the boss's inherent good nature that saved me from the bullet.

6. Substitute for def.: An ace in the game of brag or poker; esp in phr. *two bullets and a bragger* (see *BRAGGER 2): a winning hand; also *fig.* *U.S.* (Add further examples.)

1841 *Spirit of Times* (N.Y.) 23 Oct. 402 Zenith, Magnate, and Jim Bell are 'two bullets and a bragger' that Kentucky can 'travel on' and 'pay expenses'. 1878 F. H. HART *Sazerac Lying Club* 151 'Here's four bullets,' said Brown, as he reached for the pot. 1889 BARRÈRE & LELAND *Dict. Slang* I. 196/1 The highest hand in the game is three white (or real) aces, the next highest is 'two bullets and a bragger'.

7. *to bite* (*on*) *the bullet*: see *BITE *v.* 16.

8. *bullet-hole* (examples), *-pouch*; *bullet-riddled, -swept* adjs.; **bullet-head** *colloq.*, a bullet-headed person; **bullet-headed** *a.*, having a head that is round or shaped like a bullet; obstinate; so *bullet-headedness*.

1848 LOWELL *Biglow P.* ix. 129 He ain't No more 'n a tough ole bullethead. 1857 E. BEADLE *To Nebraska* (1923) 19 A clever bullet-headed Kentuckyan. 1947 AUDEN *Age of Anxiety* (1948) ii. 54 Bullet-headed bandit. *a* 1849 POE *Marginalia* lxxiv, The disgusting sternness, captiousness, and bulletheadedness of her husband. 1869 'MARK TWAIN' *Sk. New & Old* (1875) 49 Freckles me with bullet-holes. 1917 'CONTACT' *Airman's Outings* 246 We land..and count the bullet-holes on the machines. 1757 LOUDOUN & LYMAN *Gen. Orders* (1899) 65 The Deficiency of Powder Horns & Bullet Pouches. 1849 PARKMAN *Calif. & Oregon Trail* 20 His bullet-pouch and powder-horn hung at his side. 1905 *Daily Chron.* 10 Jan. 4/5 The head of the drum was so bullet-riddled that it resembled a sieve. 1897 *Westm. Gaz.* 3 June 2/2 The bullet-swept square. 1901 'LINESMAN' *Words by Eyewitness* (1902) 33 The bullet-swept hills above Ladysmith.

bulletin. Add: **2. b.** A broadcast report of news, weather, etc. Also *fig.*

1925 *Times* 23 July 8/3 The news given out as a bulletin on a very recent Sunday from the London Station must have made many listeners pause. *Ibid.* 3 Aug. 5/5 The news bulletins are broadcast in the exact terms in which they are received from the news agencies. 1938 AUDEN & ISHERWOOD *On Frontier* 7 The drums tap out sensational bulletins; Frantic the efforts of the violins. 1938 WODEHOUSE *Summer Moonshine* i. 7 The weather-bulletin announcer of the British Broadcasting Corporation.

4. *attrib.*, esp. in **bulletin-board** *U.S.*, a notice-board.

1831 *Boston Transcript* 5 July 2/4 From the City Hall Bulletin Board. 1869 'MARK TWAIN' *Innoc. Abr.* 333 A great public bulletin-board in Pompeii. 1897 —— *Following Equator* iv. 75 To-day per the bulletin-board at the head of the companionway, it is September 10. 1949 *Lisle* (Ill.) *Eagle* 10 Mar. 5/1 With the new scroll placed on the bulletin board all may see who made the honor roll this time.

bullet-wood. [Cf. F. *bois de balle* and *boulet de canon*.] The wood of the bully tree (BULLY *sb.*[4]).

1843 HOLTZAPFFEL *Turning* I. 77 Bullet-wood, from the Virgin Isles, West Indies. *Ibid.*, Bullet-wood, another species so called, is supposed to come from Berbice; its colour is hazel brown..; it is..well adapted to general and to eccentric turning. 1858 [see *BALATA 1].

bullgine (bu·ldʒəin). *colloq.* (orig. *U.S.*). Also **bulgine, bulljine.** [f. BULL *sb.*[1] + EN)GINE *sb.*] A locomotive or steam-engine.

1848 F. A. DURIVAGE *Stray Subjects* 38 [He made] himself agreeable to his officers by..imitating the 'bull-gine'. 1848 in *Amer. Speech* (1946) XXI. 116/1 Going ober to Hobuc, in de steamboat, De bulgine busted and we all got afloat! *Ibid.*, Clar de track, de bullgine's coming. 1889 [see *BULL *sb.*[1] 7 b]. 1922 JOYCE *Ulysses* 395 The men of the island..let the bullgine run..and put to sea. 1927 in J. Sampson *Seven Seas Shanty Book* 32 So clear the track let the Bullgine run.

bull-head. Add: **1. b.** Any of various North American fresh-water fish of the genus *Amiurus* or allied genera, esp. the bull-pout or horned pout (*Amiurus nebulosus*).

1674 J. JOSSELYN *Two Voy.* 113 Blew-fish, Bull-head, Bur-fish. 1758 J. WILLIAMS *Redeemed Captive* (ed. 3) 18 There seven of us supped on the Fish, called Bull-head or Pout. 1814 MITCHILL *Fishes N.Y.* 380 Eighteen-spined Bullhead, *Cottus octodecem-spinosus*. 1947 *Sports Afield* (U.S.) Feb. 21/2 Our lake is full of bullheads.

4. (Later U.S. example.)

1840 W. G. SIMMS *Border Beagles* 487 We've time enough to scud and run to-night, and to-morrow we can turn upon that bullhead, Rawling.

5. A brick wider at one end than the other, used for arches.

1862 *Catal. Internat. Exhib. Industr. Dept. Brit. Div.* II. x. 27/1 Common fire-bricks, of various forms, arch bricks, bull-heads, pin bricks, soap bricks &c.

bullhood (bu·lhud). [f. BULL *sb.*[1] + -HOOD.] The condition of being a bull.

1845 FORD *Hand-bk. Spain* I. 290 The priest..selected a bull and christened him;..but..on the morrow he relapsed into his former bullhood and brutality. 1885 *Fortnight in Waggonette* 92 He was destined never to reach the full maturity of bullhood.

Bulli (bu·ləi). *Austral.* The name of *Bulli*, a town south of Sydney, New South Wales, used (chiefly *attrib.*) to designate a type of soil used esp. for cricket pitches.

1904 *Westm. Gaz.* 18 Apr. 3/1 This Bulli soil is wonderful in its resistance to wet...The wet does not run through the Bulli. 1907 *Daily Chron.* 19 Dec. 4/7 The famous rain-resisting 'Bulli soil' of which the pitch on the Sydney Cricket Ground is composed. 1929 CONAN DOYLE *Maracot Deep* 244 The perfect Bulli-earth wicket, so far as England could supply that commodity, reminded our visitors of their native conditions. 1963 *Guardian* 4 May 4/6 The flawless bulli soil and merry creek pitches which were the terror of all bowlers half a century ago.

bulling, *vbl. sb.*[1] (Later example.) Also *attrib.*

1950 J. G. DAVIS *Dict. Dairying* 84 *Bulling heifer*, a heifer which is the right age and size for being served. 1960 *Times* 9 May 3/3 The calfhood vaccination scheme allowed animals to be vaccinated up to the age of bulling.

bullion[2]. Transfer †*Obs.* to 1 and 3, and add to 2:

1881 *Spons' Encycl. Industr. Arts* 1064 Pressing this lump upon an iron point, so as to give it the form of a little cup, he fits it, when thus shaped, on to the bullion-point, to which it soon becomes firmly attached. The lump thus formed is called the 'bull's-eye' or 'bullion' of the developed plate. 1885 *Spons' Mech. Own Bk.* 630 'Roundels' and 'bullions' are small discs of glass, some made with a knob in the centre, and used in fretwork with cathedral glass.

4. *Comb.*: **bullion-bar,** the bar on or against which the end of the sphere of glass is pressed in blowing crown glass; **bullion embroidery** (see quot. 1968); **bullion knot** = *bullion stitch*; **bullion-point,** the point or end of a bulb that is being worked on a blow-pipe; also, the thick centre of a disc of blown glass, the bull's-eye; **bullion-rod** = *bullion-bar*; **bullion stitch** (see quot. 1968).

a 1854 TOMLINSON *Cycl. Usef. Arts* I. 773/2 In again blowing out the bulb, the man supports it on a horizontal smooth iron rod, called the bullion-bar. 1882 CAULFEILD & SAWARD *Dict. Needlework* 55 Bullion embroidery, when used for letters and large pieces, is applied to the material, as in appliqué. 1968 J. IRONSIDE *Fashion Alphabet* iii. 82 Bullion embroidery is an ancient embroidery done with gold wires instead of threads. 1882 CAULFEILD & SAWARD *Dict. Needlework* 55 *Bullion knot*, useful in crewel and silk embroideries, and largely employed in ancient embroideries for the foliage of trees and shrubs, and the hair of figures. 1881 Bullion-point [see above]. *a* 1884 KNIGHT *Dict. Mech.* Suppl., *Bullion Point* (glass), the thick portion at the center of a disk of crown glass. 1890 W. J. GORDON *Foundry* 143 The globe is heated and again blown, and becomes a Florence flask, the 'bullion-point', the apex of the old cone, being still conspicuous. 1862 *Chambers's Encycl.* IV. 780/1 The workman..next ma[r]vers it, without, however, using the bullion-rod. *c* 1890 tr. *T. de Dillmont's Encycl. Needlework* 231 For bullion stitch, select a needle, a little thicker towards the handle, and finer than you would use for any other crochet stitch. 1968 J. IRONSIDE *Fashion Alphabet* iii. 82 *Bullion-stitch*, a decorative stitch formed by twisting the thread several times round the needle before inserting it.

bullioned (bu·liənd), *a.* [f. BULLION[4] 2 + -ED[2].] Ornamented with bullion.

1902 *Daily Chron.* 1 Feb. 8/3 Of clear white muslin.. heavily bullioned up the centre breadth. 1905 *Westm. Gaz.* 13 July 5/1 The well-known ball dress of the British Navy, with its heavily bullioned epaulettes.

bullishness (bu·liʃnės). *Stock Exchange.* [f. BULLISH *a.*[1] + -NESS.] The quality or condition of being bullish.

1895 *Daily News* 9 Feb. 8/4 'Bullishness' is a thing almost unknown. 1903 *Westm. Gaz.* 3 Jan. 7/1 The Kaffir Circus has come to a pause in its rakish career of bullishness. 1922 *Daily Mail* 1 Nov. 10 American Futures closed steady after recent bullishness.

bullock, *sb.* Add: **6. a.** *bullock-bell, -car, -carriage, -gear*; **b. bullock-puncher** (examples) (also *N.Z.*); hence *bullock-punching*.

1911 C. E. W. BEAN *'Dreadnought' of Darling* v. 48, I fancy there is a bullock-bell somewhere, Joe. 1928 'BRENT OF BIN BIN' *Up Country* viii. 126 The bullock bells added to the orchestra of frogs and the millions of crickets. 1830 W. S. MOORSOM *Lett. fr. Nova Scotia* 241 The Portuguese bullock-car of Peninsular memory. 1903 *Daily Chron.* 11 Mar. 8/2 Mr. Chamberlain..subsequently proceeded to Mount Church, partly by municipal bullock-car. 1839 J. C. MAITLAND *Lett. fr. Madras* (1843) 299 All the poor widows, with their palanquins and bullock-carriages covered with black cloth. 1848 HAYGARTH *Bush Life Australia* ii. 19 Every sort of saddlery, bullock-gear, and harness of every description. 1856 W. H. S. ROBERTS *Diary* 19 Sept. in H. Beattie *Early Runholding in Otago* (1947) I. vi. 43 The whip..was a powerful flagellator in the hands of an experienced 'bullock-puncher'. 1921 H. GUTHRIE-SMITH *Tutira* xxxviii. 382 Owners and employees had worked shoulder to shoulder as..bullock-punchers. 1891 G. CHAMIER *Philosopher Dick* II. xv. 411 He soon got charge of a team, and was loud in expatiating on the art of bullock-punching.

bullock, *v.* Restrict †*Obs.* exc. *dial.* to sense in Dict. and add: **2.** *intr.* To work like a bullock, *i.e.* strenuously without intermission. So **bu·llocking** *vbl. sb.* and *ppl. a.* *Austral.* and *N.Z. colloq.*

1888 'R. BOLDREWOOD' *Robbery under Arms* vii, It would have paid us better if we'd read a little more and put the 'bullocking' on one side. 1900 H. LAWSON *Over Sliprails* 12 You'll never make money, except by hard graft—hard, bullocking nigger-driving graft. *Ibid.* 19 We've..come down to have a bit of a holiday before going back to bullock for another six months or a year. 1946 K. TENNANT *Lost Haven* (1947) 52 If he had not 'bullocked' and sweated and driven himself.

3. *to bullock one's way*: to make one's way with heavy violence.

1909 *Westm. Gaz.* 1 June 1/2 Sir William has frequently been referred to as 'bullocking' his way through Parliament. 1921 *Public Opinion* 10 June 538/2 To bullock his way through a British Amateur Championship. 1930 V. PALMER *Passage* i. 15 Fred was tough as tarred canvas, able to bullock his way anywhere. 1965 *Economist* 31 July 416/2 They picked, by a narrow majority, the man they reckoned most likely to bullock their way back into power.

bullocker (bu·ləkəɹ). [f. BULLOCK *sb.* + -ER[1].]

1. A bullock-driver. *Austral.* and *N.Z.*

1889 BARRÈRE & LELAND *Dict. Slang* s.v. *Bullocky*, Bullock[e]rs in Australia are as proverbial as bargees or Billingsgate fishwives in England for the forcibleness of their language. 1894 J. K. ARTHUR *Kangaroo & Kauri* II. ii. 91, I once saw..a bullocker who was working along with his team.

2. A ship used for carrying cattle.

1915 J. SIBREE *Naturalist in Madagascar* ii. 20 'Bullockers' ..were still considered good enough to convey from two to three hundred oxen from Tamatave to Port Louis or Réunion.

bullocky, *a.* Delete *nonce-wd.* and add: **2.** Having to do with driving bullocks or managing cattle. *Austral.*

1885 T. L. WORK in *Australasian Printers' Keepsake* 16 'When you *make* Mokepilly,' quoth one of the sunburnt bullocky men, 'keep on the brush fence...' 1890 'R. BOLDREWOOD' *Col. Reformer* xii. 121 By George, Jack, you're a regular bullocky boy.

3. Resembling a bullock.

1890 *Temple Bar* Jan. 118 With more enthusiasm than persons of his bullocky conformation commonly exhibit. 1892 *Illustr. Sporting & Dram. News* 27 Aug. 862/3 Deer get bullocky if fed on ordinary cattle rations. 1954 E. C. STUDHOLME *Te Waimate* (ed. 2) xx. 179 The swine made a bullocky rush for home.

B. *sb.* A bullock-driver. *Austral.* and *N.Z. colloq.*

1889 in BARRÈRE & LELAND *Dict. Slang*. 1891 G. CHAMIER *Philosopher Dick* II. xii. 335 'And as for them bullockies,' said she, [etc.]. 1900 H. LAWSON *On Track* 49 The party had been increased by Jimmy Nowlett, the bullocky. 1934 T. WOOD *Cobbers* vi. 82 He was a bullocky, one of a race which speaks straight and spares none, like the bargees. 1966 'J. HACKSTON' *Father clears Out* 92 A red-bearded, six-foot bullocky, known everywhere for his cleverness with his whip.

b. Language of the sort used by bullock-drivers; swearing. *Austral.* and *N.Z. slang.*

1916 *Anzac Bk.* 103/2 We heard fragments of language, of the category known in Australia as 'bullocky'. 1941 BAKER *N.Z. Slang* v. 47 The noun *bullocky* has even been evolved to describe the language used by such men.

c. *bullocky's* (or *bullockys'*) *joy*: treacle or golden syrup. *Austral.* and *N.Z. slang.*

1918 R. H. KNYVETT *Over There* IV. xvii. 158 This last is merely molasses or 'golden syrup' called 'bullocky's joy'.. because it is the chief covering for slices of bread with the bullock-driver. 1929 A. RUSSELL *Diary* 6 Mar. in *Tramp-Royal* (1934) xxxi. 202 Damper and 'bullockys' joy' (treacle) take the place of porridge. 1946 L. R. C. MACFARLANE *Amuri* III. 109 Out in camp he lived on hard bread, meat and bullockey's joy (treacle).

bull-pen. [f. BULL *sb.*[1] + PEN *sb.*[1]]

1. A pen or enclosure for bulls; *spec.* a square enclosure made of logs, freq. used in early times as a place of confinement; hence, a lock-up or prison. Also *transf.*, any enclosure or enclosed area. orig. and chiefly *U.S.*

1809 M. L. WEEMS *Life F. Marion* xxv. 214 The tories were all..confined together under a centinel, in what was called a *bull pen*, made of pine trees. 1820 in *Pubn. Col. Soc. Mass.* (1906) VIII. 391 A circumscribed valley within the mountains, called the Bull-pen. 1865 *Nation* I. 15 A guardhouse..which appears to have gone by the name of the 'Bull-pen'. 1912 *N.Y. Even. Post* 5 Sept. 2/5 Rioting convicts who broke out of 'bull pens' here are gathered in the prison yard, considering further steps. 1926 J. BLACK *You can't Win* vii. 76, I was taken downstairs and locked in a cell; I saw no more of the 'bull pen' where I spent the night. 1931 *Amer. Speech* VII. Oct. 49 [Lumberjacks'] 'free-for-all' when the 'shanty men' join the 'bull pen boys' around the box stove. 1933 *Jrnl. R. Aeronaut. Soc.* XXXVII. 14 To erect in front of the terminal building a semi-circular fence enclosing an area colloquially known as the 'bull pen'.

b. In baseball, a place outside the playing limits where pitchers are exercised for possible use during a game. *N.Amer.*

1924 *Chicago Tribune* 5 Oct. II. 1/1 Blake was rushed from the battlefield and Rip Wheeler from the bullpen. 1957 O. NASH *You can't get there from Here* 93 The pitcher too unrenowned, Who is always in the bull pen, But never reaches the mound.

2. A schoolboys' ball game, played by two groups, one group outlining the sides of a square enclosure, called the 'bull-pen', within which are the opposing players. *U.S.*

1857 *Spirit of Times* 19 Dec. 241/1 Some playing—the boys at bull-pen, the girls at jumping the rope. 1944 DUNCAN & NICKOLS *Mentor Graham* 99 Sometimes he joined the boys at 'bull pen'.

bull's-eye. Add: **7. b.** A shot that hits the bull's-eye of a target; *fig.* a 'shot' that hits the mark.

1857 J. BLACKWOOD *Let.* 30 Jan. in G. S. Haight *G. Eliot Lett.* (1954) II. 290 The public is a very curious animal and.. how difficult it is to tell what will hit the bull's eye. 1887, 1891 [see INNER B. *sb.*]. 1894 [see MAGPIE 7]. 1951 *Essays in Criticism* I. 1. 6 Some of his observations..seem also to be very just—especially his opinion of Amiel, which is a bulls-eye.

13. An old-fashioned type of watch, with the case partly enclosing the glass. Also *attrib.* *U.S.*

1833 J. NEAL *Down-Easters* I. 78 Lugging out a heavy silver watch,..a genuine bull's eye. 1854 B. P. SHILLABER *Mrs. Partington* 26 The old bull's-eye watch on the nail over the mantel-piece.

14. A fish of New South Wales, *Priacanthus macracanthus*, excellent for the table.

1883 E. P. RAMSAY *Food Fishes N.S.W.* 9 The 'bull's-eye', a beautiful salmon-red fish with small scales. 1966 T. C. MARSHALL *Trop. Fishes Gt. Barrier Reef* 181 Family Priacanthidae: Bulls-eyes or Big-eyes.

bullsh (bulʃ). *coarse slang.* Abbrev. of next.

1919 DOWNING *Digger Dialects* 14 *Bullsh*, (1) insincerity; (2) an incorrect or insincere thing; (3) flattery; (4) praise. 1948 J. B. PRIESTLEY *Linden Tree* 33 Look—I'm talking too much—and most of it bullsh, I suppose.

bullshit (bu·lʃit). *coarse slang.* [f. BULL *sb.*[1] + SHIT *sb.*] **1.** Rubbish, nonsense; = *BULL *sb.*[4] 3.

c 1915 WYNDHAM LEWIS *Let.* (1963) 66 Eliot has sent me Bullshit and the Ballad for Big Louise. They are excellent bits of scholarly ribaldry. 1928 E. E. CUMMINGS *Enormous Room* vii. 194 When we asked him once what he thought about the war, he replied, 'I t'ink lotta bullsh–t.' 1953 G. LAMMING *In Castle of Skin* vi. 135 Some say they had no time for all that bullshit. 1961 R. KEE *Refugee World* xii. 130 'What do you think of the criticisms..?' 'Sheer bullshit, frankly.' 1969 P. ROTH *Portnoy's Complaint* 97, I swear to you, this is not bullshit or a screen memory, these are the very words these women use.

2. = *BULL *sb.*[4] 4.

1930 T. E. LAWRENCE *Let.* 8 Jan. (1938) 677, I would like to know (1) Guards or Fire Pickets? (2) Drill Parades? (3) Equipment..? When worn? (4) Bullshit generally? 1957 R. HOGGART *Uses of Literacy* iv. 93 The world of special parades in the Services, of 'blanco and bullshit'.

Hence as *v. trans.* and *intr.*, to talk nonsense (to); = BULL *v.*[3]; also, to bluff *one's way through* (something) by talking nonsense.

1942 BERREY & VAN DEN BARK *Amer. Thes. Slang* § 151/6 Talk nonsense,..*bull-shit*. 1949 E. POUND *Pisan Cantos* lxxiv. 12 Wot are the books ov the bible? Name 'em, don't bullshit ME. 1967 E. AMBLER *Dirty Story* I. iii. 25 Never tell a lie when you can bullshit your way through. 1969 P. ROTH *Portnoy's Complaint* 105 Please, let us not bullshit one another about 'love' and its duration.

bully, *sb.*[1] Add: **5. bully-boy,** (*a*) (see sense 1 b in Dict.); (*b*) a young ruffian; a 'tough'; esp. = sense 3 b in Dict.

1932 B. DE VOTO *Mark Twain's America* iii. 60 Or pulled at the end of ropes which the bully-boys lugged through swamp and mire. 1952 S. KAUFFMANN *Philanderer* (1953) xii. 197 Typical of this whole generation of bully-boy, movie-tough-guy, Hemingway idolaters. 1963 *New Statesman* 30 Aug. 245/1 Tenants..told one how Rachman's bully-boys had beaten them up.

bully, *sb.*[2] Add: **2.** *Hockey.* The procedure of putting the ball in play by two players, one from each side, who strike with their sticks, first the ground then their opponent's stick, three times, after which the ball is in play. Also *bully-off*.

1883 *Boy's Own Paper* 13 Oct. 30/1 The game shall be commenced and renewed by a bully in the centre of the ground. 1895 BATTERSBY *Hockey* 92 From the bully off until his opponents' 'twenty-five' is invaded. 1897 *Encycl. Sport* I. 516/2 A penalty bully is given for deliberately unfair play by the defending side in their own circle. 1901 *Daily News* 28 Feb. 7/5 The Oxford centre-forward got away directly after the bully-off. 1963 *Times* 18 Feb. 3/7 West's centre forward put the ball into the net almost before South had moved from the bully.

bully, *sb.*[5] (Earlier examples.)

1753 SMOLLETT *Ct. Fathom* I. xxiv. 160, I could get no eatables upon the ruoad, but what they called Bully, which looks like the flesh of Pharaoh's lean kine stewed into rags and tatters. 1788 J. WOODFORDE *Diary* 18 Sept. (1927) III. 47 We had for Dinner some Hare Soup, a Couple of Chicken boiled and Ham—Some Beef Bulley, Stewed Pork—Partridges,..&c.

bully, *sb.*[7] Add: **b.** Abbrev. of *COCKABULLY. *N.Z.*

1912 B. E. BAUGHAN *Brown Bread fr. Colonial Oven* 2 The beloved creek where bullies wait the hook. 1943 G. E. MANNERING *80 Years in N.Z.* xxi. 158 In the North Island the bully run follows the whitebait.

bully, *a.*[1] Add: **2.** Also *spec.* in the earlier *bully-boat*. *U.S.*

1844 *Scribblings & Sk.* 181 (Th.), A two days' race with bully-boats combines every sort of pleasing excitement. It were well to inform you that a bully-boat means a boat that beats everything on those [Mississippi] waters. 1847 W. T. PORTER *Quarter Race Kentucky* 126 (Th.), Our 'bully' boat sped away like a bird.

bully (bu·li), *v.*[2] *Hockey.* [f. *BULLY *sb.*[2] 2.] *trans.* To put (the ball) in play by a bully. Also *intr.*, usu. with *off*, to start play in this manner.

1886 *Rules of Game of Hockey* 11 The game shall be started by one player of each side bullying the ball in the centre of the ground. 1909 *Westm. Gaz.* 30 Mar. 12/2 When two players are bullying-off they stand perfectly square. 1967 J. POTTER *Foul Play* vi. 77 The two centre forwards bullied. *Ibid.* ix. 110 Julian bent over the ball to bully-off with a stocky, close-cropped French student.

bullyrag, *v.* Delete †*obs.* from sense a, and add earlier and later examples. Also, to scold, harass, badger. Hence **bullyragging** *ppl. a.*

a 1790 T. WARTON *Poet. Wks.* (1802) II. 210 You vainly thought to ballarag us With your fine squadron off Cape Lagos. 1820 M. WILMOT *Let.* 5 Aug. (1935) 76 What a Ballyragging foulmouthed son of a gun art thou. 1888 G. M. HOPKINS *Let.* 30 Nov. (1938) 50, I am afraid this ballyragging will make you gloomy. 1889 BARRÈRE & LELAND *Dict. Slang* s.v., To *ballyrag* a man [at Oxford] is to mob him and play practical jokes upon him, to hustle him. To *ballyrág* a man's rooms is to turn them upside down, to make 'hay' of them. 1935 WODEHOUSE *Blandings Castle* vi. 148, I won't have her ballyragged. Understand that! 1958 B. HAMILTON *Too Much of Water* v. 121, I like *Samson* too...Don't be bullyragged out of it.

buln-buln (buln buln). *Austral.* Also **bullen-bullen.** [Imitative.] An Aboriginal name for the lyre-bird.

1857 D. BUNCE *Australasiatic Remin.* xi. 70 The ground had been torn or scratched up...This was the work of the Bullen Bullen, or lyre bird, in its search for large worms, its favorite food. 1871 *Athenæum* 27 May 660/3 The Gipps Land and Murray districts have been divided into the following counties:..Buln Buln (name of Lyre bird), [etc.]. 1965 *Austral. Encycl.* V. 393 Very few aboriginal names for the lyrebird have been recorded. The best-known one is *bullen-bullen*, or *buln-buln*, a term based on one of the bird's calls.

bult (bult, ‖bōlt). *S. Afr.* [Afrikaans, a. Du. *bult* hump, hunch.] A hillock, ridge.

1852 C. BARTER *Dorp & Veld* vii. 96 They [*sc.* the lions] retreated slowly up the *bult*. 1864 T. BAINES *Explor. S.W. Afr.* xii. 371 We crossed the seringa bult at a narrower place. 1901 TURNBULL *Tales from Natal* 149 Hidden from view by a slight bult he was then ascending. 1926 *Blackw. Mag.* May 638/1 Dairy cows grazing on a rise or 'bult' (as these are called locally). 1946 M. WILMAN *Check List Fl. Plants Griqualand W.* 320 A sturdy, decumbent plant on lime bults.

bultong, var. BILTONG.

1883 OLIVE SCHREINER *Afr. Farm* I. ii, Did not Tant' Sannie keep in the loft 'bultongs', and nice smoked sausages?

bum, *sb.*[1] Add: **4. bum-freezer, -perisher, -shaver, -starver** (*slang*), a short jacket, coat, etc.; **bum-sucker** *slang* = SYCOPHANT *sb.* 3; hence **bum-suck** *v. intr.*, to toady; **bum-sucking** *vbl. sb.*, toadying.

1932 L. GOLDING *Magnolia St.* I. iii. 58 He wore..an Eton coat, or 'bum-freezer', as they termed it in Magnolia Street. 1943 G. GREENE *Ministry of Fear* I. vi. 98 The tight little blue trousers and the bum-freezer jacket. 1955 H. SPRING *These Lovers fled Away* 73 A nice little Eton suit—what Greg inevitably called my bum-freezer. 1889 BARRÈRE & LELAND *Dict. Slang* s.v. *Bum perisher*, or *shaver*, a short-tailed coat. 1936 E. WAUGH *Waugh in Abyssinia* 121 A gloomy, uniform row in white bum-shavers. 1930 O. ONIONS *Open Secret* i. 15 'Is it at Eton they wear toppers?' 'Yes.' 'And bum-starvers?' 1930 in PARTRIDGE *Dict. Slang* (1937) 107/1 Bum-suck v. 1953 D. PARRY *Going up—Going Down* iv. 132 He crawls round bum-sucking to the Press. 1960 L. COOPER *Accomplices* III. i. 151 He bumsucked to all the rich men. 1877 SWINBURNE *Let.* 7 Sept. (1960) IV. 18 Mr. Bumsucker Walford (excuse the Rabelaisian term current at our universities to define a sycophant). 1943 'G. ORWELL' in Comfort & Bayliss *New Road* 156 The lords of property and their hired liars and bumsuckers. 1949 PARTRIDGE *Dict. Slang* (ed. 3) 1003/2 *Bum-sucking*, arse-crawling. 1963 C. MACKENZIE *My Life & Times* II. 21 Being accused of sucking up, or even of bum-sucking.

bum (bʌm), *sb.*[4] *slang* (orig. and chiefly *U.S.*). [Prob. short for BUMMER[3]; cf. BUM *sb.*[1] 2.] **1.** A lazy and dissolute person; an habitual loafer or tramp; = BUMMER[3]. See also quot. 1933.

1864 *Gold Hill* (Nevada) *News* 15 Apr. 5/1 The policemen say that even their old, regular and reliable 'bums' appear to have reformed. 1891 C. ROBERTS *Adrift Amer.* 68, I don't believe in feeding professional bums. 1926 J. BLACK *You can't Win* ix. 104 A thoroughgoing bum from the road. The term 'bum' is not used here in any cheap or disparaging sense. In those days it meant any kind of a traveling thief. 1931 E. LINKLATER *Juan in Amer.* II. iii. 75, I'll fix the dirty bum that framed me! 1933 *Observer* 2 Apr., 'Bum', a term of affectionate obloquy which young American friends have applied to me..means not merely a fool, but a droning fool. 1941 A. L. ROWSE *Poems of Decade* 103 Lord, how he pontificates, Lays down the law to these poor bums. 1958 *Punch* 5 Feb. 218/2 The bums in the dosshouse have reached bottom.

b. *bum's rush*: forcible ejection.

1925 L. O'FLAHERTY *Informer* iii. 46 They might give him 'the bum's rush', breaking his neck silently like a rabbit's neck. 1931 E. LINKLATER *Juan in Amer.* II. xv. 167, I told him I'd give him the bum's rush if he tried to pull that stuff on me. 1959 M. CUMBERLAND *Murmurs in Rue Morgue* xxxi. 176 Chotin was being given what the vulgar term the 'bum's rush'. He was down the steps..through the gate and flat on his back on the pavement.

2. A debauch or spree.

1871 L. H. BAGG *At Yale* 153 Aside from the annual convention on Commencement night, there are two other 'bums' held during the year. 1885 E. CUSTER *Boots & Saddles* xx. 193, I intend to celebrate their return by going on a tremendous 'bum'.

3. *on the bum*: (*a*) vagrant; begging (cf. *BUMMEL *sb.* and *v.*); (*b*) in a state of disorder.

1895 *Century Mag.* Oct. 941/2 Plans are made also for going 'on the bum' the moment they are free. 1896 ADE *Artie* iii. 28, I sized it up that the house was on the bum and she didn't want me to see it. 1931 D. RUNYON *Guys & Dolls* (1932) ix. 185 Trade is strictly on the bum. 1932 J. T. FARRELL *Young Lonigan* (1936) vi. 141 He vowed he'd blow the place, and go on the bum.

bum, *v.*[6] *slang* (orig. and chiefly *U.S.*). [? back-formation from BUMMER[3]. Cf. *BUM *sb.*[4]] **1.** *intr.* To wander *around*, to loaf; to go 'on the bum'; to act as a 'bum'. Also with *it*.

1863 *Boston Herald* 2 Aug. 2/5 They are just fit to..read the News and Express, bum round rum-shops [etc.]. 1876 *Wheatland* (California) *Free Press* 4 Mar. 2/2 The Professor is readier..to 'flunk' the student, who spent his time 'bumming' the night before. 1883 C. S. KEENE in Layard *Life & Lett.* (1892) 350 I've been bumming around all day.. and haven't caught a darned fish. 1890 L. C. D'OYLE *Notches* 168 Qualifications which eminently fitted a man to 'bum it' on such a community. 1897 KIPLING *Capt. Cour.* 72 You ought to hev more sense than to bum araound on deck this weather. 1942 P. ABRAHAMS *Dark Testament* I. xiv. 75 Dinnie was the fellow with whom I went bumming in Vrededorp when I was a kid. 1950 *Manch. Guardian Weekly* 12 Jan. 7 The unshaven months he spent bumming around New York.

2. *trans.* To beg; to obtain by begging; to cadge.

1863 *Unionville* (Nevada) *Humboldt Reg.* 4 July 2/1 He offered to pay, and didn't undertake to bum a puff out. 1931 W. FAULKNER *Sanctuary* ii. 14, I have been walking and bumming rides ever since. 1931 'DEAN STIFF' *Milk & Honey Route* p. v, Nobody knows where the hobo..bums his feeds. 1941 L. A. G. STRONG *Bay* 279 An odd sort of bloke..bummed a light and a fill of tobacco off me.

b. To travel on (a train) without a ticket.

1896 *Pop. Sci. Monthly* L. 254 Several of the 'lads' had been 'pulled' at the Rapids for 'bumming the freights'.

c. To beg (a person) *for* (something); to cadge from.

1923 H. L. FOSTER *Beachcomber in Orient* i. 2 Then he bummed me for the price of a 'square meal'. 1931 'DEAN STIFF' *Milk & Honey Route* 191 He had bummed every guy up in Portland, And they all came across with the goods.

d. *to bum one's way*: to make one's way by begging; to hitch-hike.

1925 F. SCOTT FITZGERALD *Great Gatsby* (1926) vii. 154 He was probably bumming his way home. 1932 E. WILSON *Devil Take Hindmost* ii. 8 Some will bum their way—others will have their transportation provided.

bum, *a. slang* (orig. *U.S.*). [Cf. *BUM *sb.*[4]] Of poor, wretched, or miserable quality; spec. *bum steer*, false or poor information or advice.

1859 in *Pacific N.W. Quart.* (1940) XXXI. 292 Bum River Ferry. 1888 *Nation* (N.Y.) 31 May 439/2 One of them.. heard B. called a 'bum actor'. 1896 ADE *Artie* xii. 109 He didn't have a sou markee except what was tied up in a bum little grocery store. 1911 H. QUICK *Yellowstone N.* vii. 190 A stranger that had seen better days and had a bum lung. 1924 G. C. HENDERSON *Keys to Crookdom* 399 Bum steer, poor advice. 1931 A. POWELL *Afternoon Men* I. iii. 40 This is a bum party. 1934 J. M. CAIN *Postman always rings Twice* ix. 87 If I told a bum story first. 1957 W. H. WHYTE *Organization Man* 137 The muddy-headed way so many of us do [= talk] gives young men a bum steer.

bumble, *sb.*[1] Delete †*Obs.* exc. *dial.* and add: **2. c.** An angler's artificial fly.

1873 H. St. J. Dick *Flies & Fly Fishing* ix. 138 *The Bumble*. This is a Derbyshire grayling fly used in that part of the country nearly all the season. *Ibid.* 39 It is, I suppose, meant for some water insect, but the local fishermen have spring, summer, and autumn bumbles, all dressed differently. 1880 F. M. Halford *Dry-Fly Fishing* vi. 123 An *orange bumble*, floated occasionally over the feeding fish, may be successful.

bumble, *sb.*² Add: **3.** A beadle; a member of a municipal corporation, parish council, or the like, to whom official pomposity and fussy stupidity are attributed; a consequential jack-in-office.

1856 [in Dict.]. 1865 Hotten *Slang Dict.*, *Bumble*, a beadle. a 1889 *Punch* (Barrère & Leland), The apish antics of a bumble crew. 1890 Farmer *Slang* s.v., *Bumble-Crew*, a collective name for corporations, vestries, and other official bodies. 1899 Morris in Mackail *W.M.* (1899) II. 308, I hope we shall beat our Bumbles.

4. bumble-foot, (also) a disease of the feet of domestic fowls, etc.

1854 *Poultry Chron.* I. 105 Bumble-foot comes from the ball of the foot. 1886 J. W. Hill *Dis. Poultry* 87 Occasionally the sole of the foot becomes the seat of a thick corn-like growth, which ultimately festers and exposes a ragged ulcerated wound. Such a condition is commonly termed 'Bumble-Foot'.

bumble, *v.*¹ Transfer †*Obs.* to sense 2. **1.** (Add later examples.) Also *transf.*

1908 'Ian Hay' *Right Stuff* iii. 198 The bees were bumbling in the heather. 1925 C. Dodd *Farthing Spinster* 240 Grasshoppers sang, bees bumbled. 1941 Wyndham Lewis *Let.* 10 Aug. (1963) 296 How can people read books with war-planes incessantly bumbling away over their heads.

b. To speak ramblingly, to drone *on* (in some examples influenced by Bumble *v.*²).

1958 *Listener* 2 Jan. 36/1 To doze contentedly over my set, bumbling on about the good fortune of my colleague. 1968 *Punch* 29 Jan. 181/2 His style of oratory in peculiar, as he bumbles along like a metaphysical farmer. 1969 *Sunday Times* (Colour Suppl.) 9 Nov. 80/3 You can quite happily bumble on without too much trouble if that's what you want.

bumble, *v.*² Delete †*Obs.* exc. *Sc.* and add later examples. Cf. *Bumbling *ppl. a.*

1876 *Covatonga Cal.* 212 Merry Girl beat Unknown in good style, the latter bumbling very much at his fences. 1926 *Chambers's Jrnl.* 87/1 Ploughmen of thirty learning to hold a pen and 'bumbled' through the Single Carritch'. 1959 E. Pound *Thrones* xcvii. 32 The artigianato bumbles into technology.

bu·mblepuppet. [f. Bumble-puppy + -er¹.] One who plays unscientific whist. So **bu·mblepuppist.**

1880 'Pembridge' *Whist*, etc. 2 The Bumblepuppist only admires his own eccentricities. 1891 *Daily News* 10 Sept. 5/1 The careless Bumblepuppet dreads the expert.

bumble-puppy. Add: **b.** Also of bridge. Also *attrib.*

1936 E. Culbertson *Contract Bridge Complete* i. 34 Persons who claim they 'play no conventions' either play bumble-puppy Bridge or do play conventions that are tacitly understood. 1947 W. S. Maugham *Creatures of Circumstance* 104 Templeton isn't the sort of chap to play bumble-puppy bridge with a girl like that unless he's getting something out of it.

c. A game in which a ball slung to a post is struck with a racket by each player in opposite directions, the object being to wind the string entirely round the post; also, the post so used.

1900 L. B. Walford *One of Ourselves* xiv. They had had a great game of 'bumble-puppy'. a 1918 J. T. B. McCudden *Five Yrs. R.F.C.* (1919) xii. 227 We had a wonderful game called 'Bumble-puppy', which one played with tennis rackets. 1940 M. Sadleir *Fanny by Gaslight* i. 43 One of the boys seized a chance to occupy the bumble-puppy...It was great fun hitting the ball in its string-bag so that it wound tightly round the pole.

bumbling, *vbl. sb.* Delete †*Obs.* and add:

1952 *Essays in Criticism* II. 11 The incongruity between Emma's high-flown sentiments and Charles's pedestrian bumblings.

bu·mbling, *ppl. a.* [f. Bumble *v.*² + -ing².] Awkward, blundering.

1950 M. Lynn Linton *Paston Carew* ix. The rector's only son, a big bumbling young fellow. 1954 N. Balchin *Last Recoll. Uncle Charles* iv. Of these people there are people who'd like to have a really nice and not some bumbling old cab. 1955 *Amer. Dial. Soc. Publ.* XXIV. 40 Every pickpocket has to believe that he is good,..even if he is the most bumbling of operators.

bumbo² (bʌ·mbo). Also **bombo, bumboo, bungo.** [Native name.] A fabaceous tree, *Daniellia thurifera*, of Sierra Leone, yielding a fragrant resin; also *bumbo-* or *bungo-tree*. Also, the gum or resin obtainable from this tree.

1874 Lindley & Moore *Treas. Bot.* Suppl. s.v. *Daniellia*. 1916 C. E. Lane-Poole *Trees of Sierra Leone* 32 *Daniellia* [sic] *thurifera*, Bennett...The Frankincense Tree...Creole: Bungbo. 1965 G. Kunkel *Trees of Liberia* 88 *Daniellia thurifera* Benn...Names: Daniellia, Copal Tree, Bungo, Bumbo, Faro, Gum Copal.

bumby(e, *adv. dial.* [var. *Bimeby *adv.*] By and by; presently.

1786 *Boston Exchange Advertiser* 19 Oct. (Th.), [Negro talk.] Oh! he say, land dear now, bumbye buy him five dollars' an acre. 1830 C. Clark J. *Noakes* 15 John Noakes, bum-by, come to be dead, When Many seem'd more settled. 1872 'The Village Schoolmaster' *Giles's Trip to London* v. 51 But bum-by I woke up.

bumf (bʌmf). *slang.* Also **bumph.** [Short for *bum-fodder* (see Bum *sb.*¹ 4).] Toilet-paper; hence, paper (esp. with contemptuous implication), documents collectively. Also *attrib.*

1889 Barrère & Leland *Dict. Slang*, *Bumf* (schoolboys), paper...A *bumf-hunt* is a paper-chase. 1912 V. Woolf *Let.* 16 Nov. in *Woolf & Strachey Lett.* (1956) 46 Is this letter written upon Bumf? It looks like it. 1930 Wyndham Lewis *Apes of God* (1932) vi. 161 Low-lid fodder or high-brow bumph! 1930 E. Raymond *Jesting Army* i. vi. 90 The Brigadier pushed back the mess of papers on his desk and said, 'You'll keep all that bumf till next time, won't you, padre?' 1938 E. Waugh *Scoop* ii. iv. 211, I shall get a daily pile of bumf from the Ministry of Mines. 1957 M. K. Joseph *I'll soldier no More* (1958) 21 Matthews is bringing the bumf...He says to be sure and type it on Army Form A2.

bumfiddle. (Later examples.)

1940 H. W. Hickey *Mem.* (1960) xix. 311, I, of course, shall pay, and like it, must pay my bum fiddle. 1825 H. Wilson *Mem.* I. 91, I am puzzled to guess..how, you came to shew me, an utter stranger, your bum-fiddle!

bummaree. Add: **2.** Also **bummeree.** A licensed porter at Smithfield meat-market in London. Also *attrib.*

1954 *Meat Marketing* 29 Sept. 44/1 There are adequate bummerees or mobile porters already. 1955 *Times* 1 Dec. 7/6 A 'test case' as to whether or not a retail butcher was entitled to remove meat purchased at Smithfield Market by himself or his full-time employees, or whether he had to employ porters known as 'bummarees'. 1968 M. C. Borer *England's Markets* 19 There are at present about 200 of these bummarees, and thirty 'bummarees' at Smithfield, and they are all self-employed.

bummel (bʌ·mĕl, bʊ·məl), *sb.* and *v.* [a. G. *bummel* a stroll, *bummeln* to stroll; cf. Bummer².] **A.** *sb.* A leisurely stroll or journey. **B.** *v. intr.* To stroll or wander in a leisurely fashion. Hence **bu·mmelling** *vbl. sb.*, wandering, sauntering.

[1891 *Pall Mall Gaz.* 29 Aug. 3/2 The verb to 'bummeln', apparently equivalent of the French 'flâner'.] 1900 J. K. Jerome (title) *Three Men on the Bummel.* *Ibid.* xiv. 284 He..lays out his time bummelling, beer drinking, and fighting. *Ibid.* xiv. 327 A 'Bummel'..I should describe as a journey, long or short, without an end. 1909 *Daily Chron.* 24 July 6/4 Hitherto it has been the proud prerogative of males [in Berlin] to 'bummel' [loaf]. 1947 F. Smythe *Again Switzerland* x. 187 It is an easy mountain...a ski tourer's 'bummel'. 1952 H. W. Tilman *Nepal Himalaya* ii. xviii. 212, I had already been 'bummelling' about Nepal for five months.

bummer². (Earlier examples.)

1855 *Oregonian* (Portland) 27 Jan. 1/4 Come, clear out, you trunken loafer! We don't want no *bummers* here! 1856 *San Francisco Call* 25 Dec. (Th.), Put my word I'm no bummer. I never ate a lunch in all my life without taking a square drink.

bummer³ (bʌ·mǝɪ). *Austral.* (See quot.)

1905 *Terms Forestry & Logging* 32 *Bummer*, a small truck with two low wheels and a long pole, used in skidding logs. Syn.: *bum*.

bu·mming, *vbl. sb.* *U.S. slang.* The action of *Bum *v.*⁴

1857 *San Francisco Call* 9 Jan. 1/2 The 'Bumming and Gassing Company' were out in full strength. 1860 *Yale Lit. Mag.* XXV. 196 (Th.), Another great shame connected with our social life is that of ignoring or 'bumming'. 1891 C. Roberts *Adrift Amer.* 66 The idea of begging or 'bumming' as it is popularly called out there, went strongly against my stomach.

bump, *sb.*¹ Add: **1.** So *with a bump* (*fig.*), abruptly, with a shock.

1920 G. W. Holmes *Lark Fr. in Holmes-Laski Lett.* (1953) I. 259, I must go in 5 minutes to a conference of the [J] and therefore run down with a bump. 1935 W. G. Hardy *Father Abraham* i. viii. 111 His mood of exaltation fell with a bump into the trough of melancholy. 1955 E. Hillary *High Adventure* ix. 171, I came back to full consciousness with a bump.

b. (See quots.)

1883 Gresley *Gloss. Coal-m.*, *Bump*, a very sudden breaking, sometimes accompanied by a settling down, or upheaval of the strata, during the working away of the mineral, accompanied by a loud report or bumping noise heard in the mine. 1893 *Trans. Fed. Instit. Mining Eng.* V. 381 A bump (or earth explosion) occurred on November 1th, 1891. 1960 *Times* 16 Apr. Suppl. p. i/2 For 'bump' and 'bump-hole'...What was a 'bump', a subterranean movement caused mainly by the settling of strata disturbed by mining activity.

c. *Cricket.* The rise of a ball from the pitch to a greater height than is usual. Cf. *Bump *v.*¹ 2 b.

1901 R. H. Lyttelton *Out-door Games* i. 31 A man who plays fairly straight,..and can meet the ball with the bat when it comes on straight with no hang or bump.

d. *Aeronaut.* An air-pocket.

1914 Rosher *In R.N.A.S.* (1916) 15 While flying at 200 feet, the machine suddenly bumped [*note*, met an air-wave]....These bumps are due to the sun's action on the air and are called 'sun bumps'. 1918 E. M. Roberts *Flying Fighter* 279 When correcting bumps or small erratic air currents one has often to resort to his inclinometer.

e. *slang.* Usu. in *pl.* An uneven landing of an aeroplane.

1943 Hunt & Pringle *Service Slang* 19 *Bumps*, the touching down of the aircraft during landing due to uneven ground or bad handling. 1958 N. Shute' *Rainbow & Rose* v. 207 Rather than keep him at the dreary round of circuits and bumps I had been teaching him aerobatics.

f. The action of thrusting forward the abdomen or hips in a dance or the like. (Cf. *Grind *sb.*¹ 5.) *slang* (orig. *U.S.*).

1946 Mezzrow & Wolfe *Really the Blues* vi. 75 She [a dancer] went through her whole routine, the bumps and grinds and shakes and breaks. 1964 *Punch* 26 Aug. 295/2 Sing a song, and do a bump-and-grind routine.

5. *Phr. like a bump on a log*: stupidly silent or inarticulate. *colloq.* (orig. and chiefly *U.S.*).

1863 Mark Twain' *Mark Twain of Enterprise* (1957) ii. xvi. 103 You have been sitting there for thirty days like a bump on a log, and you never rightly understand anything. 1922 S. Lewis *Babbitt* xvii. 226 With that he drove on and left the fellow standing there in the road like a bump on a log! 1959 N. L. McCluing *Clustering* 19 He went 1st. 280 You couldn't expect her to sit there like a bump on a log, Mrs. Mossery.

III. bump-ball *Cricket*, a ball hit hard upon the ground close to the bat; coming with a long hop to the fieldsman, and having the specious appearance of a catch; also (erron.) *bum(-ball)*; **bump-car** = *Dodgem; **bump-supper** (examples); **bump-up**, a sudden increase (cf. *Bump *v.*¹ 1 d).

1867 *Australasian* 9 Mar. 300/4 The apparent sincerity of a..wicket-keeper, when appealing for a 'leg before', or a 'bum'. 1870 *Marlburian* 8 June 58/2 Woollcombe...was caught off what appeared to be a 'bum ball'. 1877 C. Box *Eng. Game Cricket* 444 *Bump ball*, a ball caught after it has bounded from the ground. 1963 *Times* 18 Feb. 10/1 When an appeal was made the umpire ruled that it was a bump ball. 1937 Hull & Whitlock *Far-distant Oxus* xx. 276 She was wondered off to pay for bump-car. 1853 'C. Bede' *Adv. Verdant Green* x. 94 A Bump-supper...that is, a supper to commemorate the fact of the boat of one college having, in the annual races, bumped, or touched the boat of another college. 1940 J. Buchan *Memory Hold-the-Door* iii. 65 Raymond [Asquith] wrote the poem...On a Viscount who died on the Morrow of a Bump Supper. 1958 *Economist* 13 Sept. 801/1 They are excited because of the bump-up in their support and by-election votes this year.

bump, *sb.*³ Add attrib. uses of *bump-mill*.

1881 *Instr. Census Clerks* (1885) 68 Bump Mill Worker. 1921 *Dict. Occup. Terms* (1927) § 365 *Spinner, bump mill*...spins candle wick yarn from coarse cotton waste.

bump, *v.*¹ Add: **1. b.** *spec.* in *Services' slang*: to explode (a mine or mine-field); to shell.

1915 'Bartimeus' *Tall Ship* ix. 166 We haven't bumped a mine-field. *Ibid.* 168 The chance of 'bumping' a mine'. 1919 *Athenæum* 1 Aug. 695/2 An artilleryman speaks of having 'bumped' a town or spot, meaning shelled.

c. *to bump off*: to remove by violence; to kill. Also ellipt., *to bump*. *slang* (orig. *U.S.*).

1910 *Sat. Even. Post* 8 Oct. 13/1 To bump off is... 1914 Jackson & Hellyer *Vocab. Criminal Slang* 21 He copped a cuter and got bumped making a get-away. 1927 *Cleveland Press* 29 Jan.. Senator Thomas J. Heflin...informed his colleagues that a 'drag rat' threatened to 'bump him off'. 1927 F. C. Coe *Me—Gangster* iii. 54 Who bumped that poor chump that was driven out there? 1930 *Punch* 16 Apr. 442 Jimmy is duly bumped off. 1938 E. Waugh *Black Mischief* vii. 266 They had two shots at bumping me off yesterday. 1943 P. Cheyney *You can always Duck* xii. Glessy was going to bump Lillian. 1958 Hayward & Harari tr. *Pasternak's Dr. Zhivago* v. 211 A few were bumped off by way of example.

d. *to bump up*: to increase or raise (prices, etc.) suddenly. *colloq.*

1940 N. Mitford *Pigeon Pie* xii. 192 Olga bumps up his allowance every time he horsewhips anybody for making a pass at her. 1958 *Spectator* 10 Jan. 42/2 It is wise at night to look out for places which bump up their prices.

2. b. *Cricket.* Of a ball: to rise abruptly to an unusual height.

1863 *Daily's Mag.* June 257 At one time the ball would hit a batsman on the ribs, another time bump up and fly yards over his head. 1871 *Ibid.* June 172 There was an ugly place where the ball bumped near the north wicket. 1882 Pardon *Australians in England* 173 Bates was caught, a bumping ball from Spofforth going off the shoulder of his bat. 1888 W. G. Grace in Steel & Lyttelton *Cricket* 310 Emmett was in his glory, his bowling bumping and kicking up as I have never seen it since. 1891 —— *Cricket* 130 The first ball...bumped and hit him on the head. 1920 *Times* 28 Feb. 15/2 A pitch which, apart from the fact that two or three balls bumped in the afternoon, was beautifully easy.

c. *Chem.* To give off vapour intermittently and with almost explosive violence. (Cent. Dict. 1889). Cf. Bumping *vbl. sb.* 2.

1883 *Encycl. Brit.* XVI. 52/2 Mercury 'bumps' badly on boiling. 1942 *Industr. & Engin. Chem.* 34 Yet gives a characteristic tendency to 'bump' badly on distillation. 1950 P. J. Durrant *Org. Chem.* iii. 64 As the pressure of a boiling liquid is lowered, the tendency to 'bump' is greatly increased.

d. *Aeronaut.* To move irregularly owing to an inequality of air pressure.

1914 [see *Bump *sb.*¹ 1 d].

e. *to bump into*: to meet (a person) by chance, to run into (a person). *colloq.*

[1886 'P. Perkins' *Fam. Lett.* 22 Went down those old stairs lickety-wallop, and bumped into that old party who was evidently running for the...station.] 1953 W. P. McGivern *Big Heat* ii. 24 We just bumped into each other on Market Street. 1958 E. Dundy *Dud Avocado* i. v. 82 What a *mad* coincidence bumping into John.

bumper, *sb.*¹ Add: **2.** (Earlier example.) Esp. freq. in attrib. use = exceptionally abundant or good (see also quot. 1864). Cf. sense 3.

1759 *Gentl. Mag.* XXIX. 171/2 In some of the midland counties, anything large is called a *bumper*, as a large apple or pear...1864 Hotten *Slang Dict.* 89 A match at quoits, bowls, &c., may end in a 'bumper game', if the play and score in all on one side. 1885 [in Dict.]. 1898 *Westm. Gaz.* 18 Nov. 8/1 The past fortnight's bumper trams. 1908 *Daily Chron.* 8 Jan. 4/4 So far as the foreign trade of this country is concerned, 1907 was a 'bumper year'. 1955 *Times* 22 June 11/7 Instead of an expected crop of 600,000 bags there was a bumper crop of 1,400,000 bags.

3. (Earlier examples.)

1789 J. B. Watson *Let.* 17 Aug. in L. Sumbel *Mem.* (1811) III. 144 Her benefit, at Gloster, which, if a bumper, is every sixth of which the receipts were...far more than I most cordially wish it. 1798 Tate Wilkinson *Wand. Patentee* IV. 36 A bumper of a house.

5. b. (See quot.)

1868 Fairley *Gloss. Coal-Mining* ii. 5 A massive piece of iron, so fixed by the side of the cage as it at the bottom of the cut, it will draw it empty to the top, and when the cage at the top is laden, it will act as a balance as the cage descends; this piece of iron is called the bumper.

c. A log, bar, etc., serving as a fender or shock-absorber; *spec.* a metal bar attached to either end of a motor vehicle to lessen the shock in a collision. Also *attrib.* and *Comb.*, as **bumper-to-bumper** *adv.* and *a.*, (of cars) travelling very close together.

1867 Smyth *Sailor's Word-bk.* 144 *Bumpers*, logs of wood placed over a ship's side to keep off ice. 1883 Gresley *Gloss. Coal-Mining* 38 *Bumpers*, projecting blocks of wood attached to pump spears for preventing damage in case of a break. 1889 *Cent. Dict.*, *Bumper-timber*, in some locomotives, a timber to which the cow-catcher or pilot is fastened, designed to receive the shock or blow of a collision. 1902 *Law Times* 31 May 29/2 An elevator car...passed downward until it struck the bumpers at the bottom of the shaft. 1926 *Morris Owner* Feb. 1600/2 The front face of the bumper bar is attractively finished in bright nickel plate. 1928 *Punch* 25 Apr. p. xxxiii (Advt.), 'The Bumper with the Leaf Spring Buffer.' This unique and ingenious feature evenly distributes and reduces the shock of an impact. 1938 'Ellery Queen' *Four of Hearts* (1939) v. 75 Los Feliz Boulevard was jammed with cars crawling bumper to bumper. 1959 *Manch. Guardian* 18 May 1/1 A bumper-to-bumper traffic jam. 1959 *Motor* 21 Oct. 346/2 Lights...repositioned behind the front bumper. 1967 G. Legman *Fake Revolt* 16 The bumpersticker approach to hallucinatory drugs and sex technique.

6. One who or a thing that bumps; *spec.*

Bookbinding = *Smasher¹ 3.

1881 *Instr. Census Clerks* (1885) 43 Rocker or Bumper. 1881 C. C. Rhys *Minora Carmina* 267 Up at Oxford by eights on the Isis, The glorious of bumpers and of bumpers the glow. 1921 *Dict. Occup. Terms* (1927) § 279 *Bumper* (tin boxes, etc.), packs into bundles, strap tinplate tett from stamping or cutting, by shovelling it into iron box, ramming or bumping it into compact shape with long-handled iron ram. *Ibid.* § 409 *Bumper*, a plankner who operates a bumping machine; places felt form in a sort of trough, sets machine in motion, so that forms are bumped about against arms of machine. 1951 S. Jennett *Making Books* xi. 171 The machine appropriately called the smasher or bumper is...in effect an automatic clamp.

bumper, *sb.*² *Cricket.* A bumping ball (see *Bump *v.*¹ 2 b).

1855 *Bell's Life in London* 19 Aug. 8/3 From the fact of the ground not being a good wearing one, the 'bumpers' of Lillywhite could not be mastered. 1904 *Westm. Gaz.* 24 Aug. 3/2 With the likelihood of...an occasional 'bumper' even such great batsmen..might have failed. 1955 *Times* 30 Aug. 3/2 Heine bowled a number of rude, honest bumpers.

c. *bumper car* = *Dodgem.

1949 M. Laski *Little Boy Lost* iii. xiv. 195 'We'll go in the bumpy-cars,' he said, and climbed into a bright blue car. 1958 R. Brown *As far as Singapore* vi. 124 They are big amusement parks...They have bumper cars. 1966 *New Left Rev.* Jan.–Feb. 52/1 Going to a fair to ride on the bumper cars.

bu·mper, *sb.*³ *colloq.* [*-er⁶.] A bumping-race.

1899 *Daily News* 1 Apr. 5/1 The Bumpers, to give them their familiar name, are split in such a way that a day of rest is allowed in the middle of the four days' racing. 1920 H. W. Chadwick *Rec. Rowing Club S. Philip & S. James*', *Oxf.* 15 Each of its two crews secured four bumps up the City 'Bumpers'.

bu·mper, *sb.*⁴ *Austral.* and *N.Z. slang.* [App. telescoped form of Butt *sb.*² and Stump *sb.*¹ 3 + *-er¹ 3.] A cigarette end.

1916 *Anzac Book* 47/1 While we was standin' to arms 'e lights up a bumper. 1945 *Salt* 2 July 43/2 Then he busily engaged in searching in the dust for enough bumpers to roll a smoke. 1958 S. Stow *To Islands* ii. 42 Gabumbe, engaging it, requested 'Bumper, body,' and Heriot, after stubbing it placed the butt in the open mouth. 1967 Southerly XXVII. 122 He patted the bare mattress...where a bumper had burned a hole sometime in the past.

bumpety, bumpity (bʌ·mpɪtɪ), *adv.* A childish form of Bump *v.*¹ 8; *bumpety-bump*, with repeated bumps. Also *attrib.* and as *v. intr.*

1874 Hardy *Far fr. Madding Crowd* lvii, Souls alive, what news! It makes my heart go quite bumpity-bump! 1904 B. Pain *Kindn. Celestial* 190 You could almost fancy that you heard the man going bumpety-bump down the stairs. 1902 W. de la Mare *Songs of Childh.* to a bumpety ride in a wagon of hay. 1958 *Spectator* 14 Feb. 194/3 His plane bumpety-bumps to a standstill. 1967 K. Giles *Death in Diamonds* vi. 99 My heart went bumpity because it seemed she looked right at me.

bumph, var. *Bumf.

bumping, *vbl. sb.* Add: **3.** **bumping-post**, **bumping-table** (see quots.).

a 1877 Knight *Dict. Mech.*, *Bumping-post* (Railway Engineering), a timber or set of timbers at the termination of a railroad track, to limit the motion of the train in that direction. 1880 P. Wilford *Dict. Mining Terms* ed. 2) 13 A bumping-table is an appliance used in a stamp-mill for treating tailings. It consists of an inclined table, which is given a bumping or jerking motion which serves to force upwards by each successive bump the material or heavier portion of the slimes, while the lighter portion is washed off the lower end of the table. 1902 *Encycl. Brit.* XXXI. 371/2 *Bumping Tables*—Rittinger's table is a rectangular bumping-plane surface which by a bumping motion throws the heavy particles to one side while the current of water washes down the quartz to another.

bumpo·logist. *humorous.* [f. Bump *sb.*¹ 4 + -ology.] One who is learned in bumpology. So **bumpologi·cally** *adv.*

1824 *Blackw. Mag.* XVI. 137 He for you recollect is one, not of the Bumpologists, but of the Fistologists. 1837 *Southern Lit. Messenger* III. 107, I once laid my head examined—bumpologically—in a regular way. 1848 J. Richardson *Trav. Sahara* I. vi. 166 This afternoon examined phrenologically, bumpologically, the heads of many children. 1899 *Westm. Gaz.* 3 Jan. 8/2 It is not only porters who patronise the bumpologist.

bumpy, *a.* Add: **b.** *Cricket.* Of a ball: that rises abruptly from the pitch; of bowling: using or characterized by 'bumpers' (see *Bumper *sb.*²).

1897 *Australasian* 9 Mar. 300/3 The batsmen seemed afraid to look at him, especially after the first bumpy over. 1871 'Incognito' *Cricketers in Council* 28 Balls too high to strike the wicket, and too 'bumpy' to be hit down. 1906 *Westm. Gaz.* 8 May 2/1 He delivers a very fast bumpy ball.

c. *Aeronaut.* Full of bumps (see *Bump *sb.*¹ 1 d); uneven because of bumps.

1911 H. R. P. Reynolds in A. E. Berriman *Aviation* (1913) 166, I scarcely moved my control lever until I got to Bletchley, where it [the air] began to get rather 'bumpy'. 1918 *Punch* 3 Apr. 222/3 Weather looks dud—you're going to have it bumpy in the morning, if you're on a gap. 1959 *Times* 15 June 95 It was a nightmare journey for pilot and navigator alike. Conditions were very bumpy. 1963 V. Gielgud *Goggle-box Affair* xxii. 149 He was...suffering from the after-effects of a bumpy flight.

bun, *sb.*³ Add: **1.** Slang phrases: *to take the bun*, to take the cake (see Cake *sb.* 7); *a bun in the oven*, a child conceived; '*to do one's bun*' (N.Z. slang), to lose one's temper.

1889 in *Amer. Speech* (1950) XXV. 33/1 But the pale and yellow joke of her where sister 'takes the bun'. 1896 *Dialect Notes* I. 414 'That takes the bun', that's very good. 1906 'Jacko's the bun'. 1901 Kipling *Kim* xii. 315 You take the bally bun, by Jove! It was splendid. 1934 L. van der Post *In a Province* ii. 108, I've seen many dressed-up niggers, but that one takes the bun. 1941 *Korero* II. no. xix. The most important of Army slang expressions, however, has been 'doing the scone' with its variant 'doing the bun', used for losing one's temper. 1949 'The Sarge' *Excuse my Feet* 72. xix 'O.K. O.K. Don't do your bun!' 1951 N. Monsarrat *Cruel Sea* II. vi. 105. 'I bet you left a bun in the oven, both of you', said Bennett thickly. Lockhart explained...the reference to pregnancy. 1958 *Spectator* 10 ... 1949 *Church Times* 19 Apr. 400/3 Victorian fashion, with hats perched on the head, permitted a free view of chignon, 'bun', or curls.

b. *transf.* Hair coiled at the back of the head in a shape suggesting a bun. Also *attrib.*

1894 *Daily News* 26 May 6/4 The fashionable ladies to be seen in the Park with their [buns]. 1899 *Cassell's Fam. Mag.* Dec. 73/1 The days of the 'bun' coiffures are over. 1921 *Chambers's Jrnl.* Dec. 781/1 His chief glory was, however, his head, the hair of which was dressed in a large 'bun'. 1949 *Church Times* 19 Apr. 400/3 Victorian fashion, with hats perched on the head, permitted a free view of chignon, 'bun', or curls.

c. In full **bun hat.** A bowler hat. *N.Z. slang.*

1941 Baker *N.Z. Slang* vi. 52 A bowler hat has become a *bun* in colloquial speech. 1950 *Landfall* IV. 21 He was thin ..and was wearing an old bun hat.

3. Add to def.: *attrib.*, *bun-shop*; **bun-face**, a face resembling a bun; hence **bun-faced** *a.* (also *fig.*); **bun-fight**, a jocular expression for a tea-party (cf. *tea-fight*); **bun foot**, of furniture (see quot. 1952); **bun-penny** [sense *1 b], a penny showing the head of Queen Victoria with her hair worn in a bun; **bun-struggle**, = *bun-fight*.

1906 B'ness von Hutten *What became of Pam* i. v. The bun-and-milk establishments. 1913 R. Brooke *Let.* 1 Dec.

(1968) 540 Fijians swinging along half naked with bun-faces. 1927 W. Deeping *Kitty* xx. 254 A local decorator, a little bun-faced man with bright eyes. 1966 *Listener* 2 June 806/3 The novelist-farcer is a serious man; and the bun-faced academic is a frivolous one. 1928 K. Campbell *Wayzgoose* 7 If the 'wayzgoose' contrives the functions of a bun-fight, an Eisteddfod and an Olympic contest. 1904 P. Macquoid *Hist. Eng. Furniture* iv. 115 The solid back is a modern addition, as are the bun feet on which the buffet stands. 1952 J. Gloag *Dict. Furniture* 154 *Bun foot*, a foot resembling in form of a bun-like, flattened sphere, used on chairs, tables, etc...during the latter part of the 17th century. 1958 S. Hyland *Who goes Hang?* xxxii. 187 The Victorian 'bun-penny' was first minted in the year 1860. 1884 *Dramatic Notes* 1888 145 An innocent flirtation in a bun-shop. 1903 *Daily Chron.* 25 May 3/3 Bun-shop waitress. 1889 Barrère & Leland *Dict. Slang*, *Bun-struggle* or worry (army), a tea meeting; an entertainment for soldiers in a garrison. 1899 R. Whiting *No. 5 John St.* vi. 53 He wants yer to show up at a sort o' bun-struggle in 'er room...kind of a tea-fight. 1911 W. De Morgan *Likely Story* 224 Madeline...had been going to the bun-fight.

bun, *sb.*⁴ *slang.* [Origin unknown.] A drunken condition, esp. in *to get, have, tie a bun on*: to become drunk.

1901 'H. McHugh' *John Henry* i. 16 You've got another bun on! How dare you trail into my flat with your tile high enough to float a battleship! 1914 W. L. George *Making of Englishman* ii. v. 302 You've had your bun, and I'm not stayin' here for you to get a bun on. 1954 J. van Druten *I am a Camera* i. ii, We'll celebrate tonight, if you do. We'll get out, then we'll tie a bun on anyway, dear.

buna, Buna (biū·nǝ). [a. G. *Buna*, f. *Bu(tadiene + Na(trium).] A synthetic rubber first developed in Germany, made by the polymerization of butadiene.

1936 *Chem. Abstr.* XXX. 7374 The new German synthetic (Buna) rubber. 1938 *Encycl. Brit. Bk. of Yr.* 19/8 282/2 Buna, or synthetic rubber, was awarded the Grand Prix at the Paris Exhibition for 1937. 1940 Graves & Hodge *Long Week-end* xviii. 302 German Buna car tyres were said to last half as long again. 1943 H. M. Langton in R. S. Morrell *Synthetic Resins* (ed. 2) i. 40 Buna 85 is a pure butadiene polymer, Buna S a mixed polymer of butadiene and styrene.

Bunbury (bʌ·nbǝrɪ). The name of an imaginary person used as a fictitious excuse for visiting a place or avoiding obligations (see quot. 1899). Hence used allusively in various formations (see quots.).

1899 Wilde *Importance of being Earnest* i. 14, I have Bunburyed all over Shropshire on two separate occasions. *Ibid.* 16, I have invented an invaluable permanent invalid called Bunbury, in order that I may be able to go down into the country whenever I choose. *Ibid.* 17 Now that I know you to be a confirmed Bunburyist I naturally want to talk to you about Bunburying. 1959 *Listener* 12 Feb. 300/1 He may even be able to kill the faint hope in many hearts that the former has merely gone Bunburying. 1960 *Times* 27 Apr. 10/1 The perils of Bunburying like the Wellerdom were to be the creation of a spurious alibi—increase in proportion to the complexity of the story told. 1965 P. Moyes *Johnny under Ground* ix. 117 I've evolved this rather attractive *alter ego*—Mr. Reginald Derbyshire-Bennett. Quite Bunburyish, in his own little way. 1969 *Listener* 5 June 794/3 For he who lives more lives than one *More deaths* than one must die...at least the words are an apt motto for a Bunburyist.

bunce. (Later examples.)

1962 *Guardian* 19 Nov. 7/5 You make your 'bunce' (big profit) in the summer. 1968 C. Drummond *Death & Leaping Ladies* viii. 194 They take the place for a fee and pocket any bunce.

bunch, *sb.*¹ **1. a.** Delete †*Obs.*, and add later U.S. example.

1874 *Rep. Vermont Board Agric.* II. 428 Their bite is poisonous to a certain extent, as bunches can be felt around their bites.

2. b. A certain quantity fastened together for sale, as a bundle of flax.

a 1877 Knight *Dict. Mech.*, *Bunch*...(Flax-manufacture), three bundles, or 180,000 yards, of linen yarn. 1882 P. Sharp *Flax, Tow & Jute* 143 The bunch generally weighs about 40 lbs., the number of banks depending on the size of the yarn.

4. (Later examples.) Also, a company or group of persons.

1840 C. F. Hoffman *Greyslaer* I. v. 92 You'll find them pretty much here and there, in bunches, helping one another. 1893 W. S. Gilbert *Utopia* I. 1 Because we are, By furlongs far The best of all the bunch. 1909 G. B. Shaw *Lett. to Granville Barker* (1956) 156 He will be the best of the bunch, like all new converts. 1936 D. Powell *Turn, Magic Wheel* i. 37 He liked knowing the 'Greenwich Village Bunch'.

5. b. *bunch of fives*: see Five *sb.* 3 b.

7. bunch-bean *U.S.*, a dwarf kidney-bean, also called *bush-bean*; **bunch-berry**, the dwarf cornel, *Cornus canadensis*, or its fruit; **bunch flower**, a liliaceous plant of N. America, *Melanthium virginicum*; **bunch-grass**, any of various grasses, chiefly of western North America, characterized by growing in clumps; **bunch greens**, greens sold by the bunch; **bunch-knot**, a knot joining broken ends of yarn; **bunch-oyster**, a wild oyster which grows in clusters; **bunch pink**, a name for the sweet-william.

1787 Washington *Diaries* (1925) III. 212 The bunch Nonnery bean. 1869 R. Parkinson *Tour in Amer.* 341 The bunch bean...produces abundantly. 1860 Darlington *Amer. Weeds* 107 The...Dwarf or Bunch Bean. 1845 S. Judd *Margaret* I. xiv. 106 She got boxberry flowers and fruit, bunch-berry and star-of-Bethlehem flowers. 1960 *Islander* (Victoria, B.C.) 31 Aug. 11/1 The bunch berries, I thought, were loveliest of all. 1818 A. Eaton *Man. Bot.* (ed. 2) 316 *Melanthium racemosum* bunch flower. 1840 T. W. Sanders *Encycl. Gardening* (ed. 3), *Melanthium* (American Bunch-flower). 1837 W. Irving *Capt. Bonneville* I. xii. 203 The upland bunch grass...retained its nutritious properties...in the autumn. 1878 J. H. Beadle *Western Wilds* xiii. 125 One may ride all day through...bunch-grass pasture. 1959 McLintock *Descr. Atlas N.Z.* 24 Tussock (bunch grass) grasslands, generally with species of *Poa* and *Festuca* dominant. 1884 W. S. B. McLaren *Spinning* xi. 238 When one end of yarn breaks or runs out, the other must be broken too, and what is called a bunch-knot tied. 1881 E. Ingersoll *Oyster Industry* 242 *Bunch oysters*, those growing in clusters (South). 1857 Gray *Botany* 54 Sweet William or Bunch Pink. 1877 *Rep. Vermont Board Agric.* IV. 99 Bunch pinks and candytufts coming from self-sown seeds.

bunch, *v.*² Add: **1. c.** To crowd together in a body. Also with *up*.

1873 Beadle *Undevel. West* 60 Buffalo grass and grama grass...show a tendency to bunch together, leaving large portions of the surface bare. 1887 F. Francis jr. *Saddle & Mocassin* vii. 124 They had got scared, and had bunched up like a bevy of quail. 1909 *Westm. Gaz.* 23 Dec. 12/3 The result...is to force the opposing defences back to their own goal and so bunch. 1924 Galsworthy *White Monkey* i. ii, The really big people don't talk—and don't bunch—they paddle their own canoes in what seem backwaters. 1959 I. & P. Opie *Lore & Lang. Schoolch.* xii. 269 Lookers usually 'bunch up in' groups of three or four'.

d. *Mining.* Of a vein or lode of ore: to form an irregular mass. Cf. Bunch *sb.*¹ 6.

1883 Baring-Gould *J. Herring* III. xlviii. 102 The vein 'bunched', and the bunch of nearly pure metal was before him. 1889 *Temple Bar* LXXXV. 26 The lode ran under Orleigh gardens and promised freely to 'bunch' under the mansion.

2. (Earlier U.S. examples.) Also *absol.* (see quot. 1887). Also with *up*.

1828 Webster, *Bunch*, v.t., to form or tie in a bunch or bunches. 1869 A. K. McClure *Rocky Mountains* 99 The beeves..have been 'bunched' at either end of the hostile country. 1873 'Mark Twain' & Warner *Gilded Age* (1903) I. 16 The speaker bunched his fingers like the claws... stem end of a tomato. 1887 *Overland Monthly* (Farmer), Two men often bunch on the market, thus unite their herds for convenience in driving. 1893 G. B. Shaw *Let.* 27 Apr. (1965) 392 The way you...bunch up your back.

b. In technical use (see quots.). In *Baseball*, to secure (hits) in close succession. Chiefly *U.S.*

1883 *Chicago Tribune* 5 July 6/5 Detroit played a wretched muffing game today and failed to bunch hits. 1905 *Terms Forestry & Logging* 32 To *bunch load*, to encircle several logs with a chain and haul them at once, by steam or horse-power. To *bunch logs*, to collect logs in one place for loading. 1909 Webster s.v., To bunch...shots (in target practice).

3. To present (a woman) with a bunch of flowers. Also *transf.*

1901 *Daily Chron.* 7 Dec. 8/3 The King's gift of a bouquet to Miss Brodrick upon her wedding morning, in that it revives the charming old fashion of 'bunching' young beauties, a very modish practice, that only languished a few years ago. 1905 N. Marsh *Singing in Shrouds* iv. 59 Captain Bannerman felt that in a way he would be bunching Mrs. Dillington-Blick by presenting her with a No. 1 Personality. 1962 A. Cooke *Art of Economics* iv. 95 Send flowers next day with a brief thank-you note. The older hostess, the more she will like being 'bunched'.

bunched, *ppl. a.* Add: **b.** *bunched up* (examples); also of other things than a dress.

1917 D. H. Lawrence *Look! We have come Through!* 62 A bunched-up dress, its four little feet Clustered dead. 1934 *Burlington Mag.* Mar. 118/2 A wild movement of bunched-up draperies. 1959 D. Davie *Forests of Lithuania* iv. 42 I heard a knot Of bunched-up moses.

d. *Bot.* Having convex protuberances.

1776 J. Lee *Introd. Bot.* (ed. 3) 378 *Fasciculatus*, bunched. 1900 B. D. Jackson *Gloss. Bot. Terms*, *Bunched*, gibbous.

e. Clustered, gathered into a bunch or bunches. Cf. Bunch *v.*² 2.

1904 R. J. Farrer *Garden Asia* 42 The ground is thick with the bunched blue stars of a wee-iris. 1964 *Economist* 14 Nov. 737/2 'Bunched gains'—heavy realisations coming together in one tax year.

buncher (bʌ·ntʃǝɪ). [f. Bunch *sb.*¹ or *v.*² + -er¹.]

1. One who or that which bunches; *spec.* a machine for forming bunches or collecting things in bunches.

1881 *Instr. Census Clerks* (1885) 74 Straw Plait Buncher. 1909 Webster s.v., An asparagus buncher. 1921 *Dict. Occup. Terms* (1927) § 309 *Buncher, yarn buncher*; (i) ties several hanks of yarn into bunches, in readiness for succeeding processes, bleaching, etc.; (ii) [cigar] = roller. 1930 C. S. Johnson *Negro in Amer. Civilisation* (1931) v. 64 The lowest median weekly earnings were...Tobacco bunchers $1.85. 1939 M. Spring *Rick Working-Class Wives* vii. 164 She worked as a 'buncher' in flower gardens.

2. A device that modulates a beam of charged particles and causes them to collect together in 'bunches' or groups. Also *attrib.*

1939 R. H. & S. F. Varian in *Jrnl. Appl. Physics* X. 324/1 The rhumbatron in which the electrons are given their first modulation of speed is called the 'buncher', and the one into which the bunches 'break' is called the 'catcher'. 1948 J. B. H. Kuper in D. R. Hamilton et al. *Klystrons & Microwave Triodes* i. 17 The velocity of each electron depends on the instant at which it crossed the buncher gap. 1955 *Science* CXXII. 1133/3 The ion beam passes through the buncher, which applies a 24-megacycle ripple of 5 to 7 kev to the beam. 1962 Livingston & Blewett *Particle Accelerators* x. 333 The buncher...is merely a section of accelerator in which the phase velocity is matched to the particle velocity.

bunchiness. Delete †*Obs.* and add later example.

1922 Joyce *Ulysses* 501 Those pannier pockets of the skirt and slightly pegtop effect are devised to suggest bunchiness of hip.

bunching, *vbl. sb.* Add: **c.** (Further examples.)

1901 L. M. Waterhouse *Conduit Wiring* 51 The system of bunching which, with wood casing, is objectionable from a fire insurance point of view. 1906 Dewar *Fairy Year* 27 Bunching up of the cress for local retail business. 1967 *Economist* 7 Jan. 41/1 The negative attitude of London bus drivers to their work, e.g., bunching, largely stems from their low wages.

d. The action of a buncher (see *Buncher 2). Also *attrib.*

1939 R. H. & S. F. Varian in *Jrnl. Appl. Physics* X. 326/2 A large bunching distance increases the difficulty of getting a large fraction of the beam into the catcher. 1946 *Electronic Engin.* XVIII. 153 Another method...makes use of the klystron bunching principle.

bunchy, *a.* Add: **2. b.** *Mining.* (See quots.)

1778 Pryce *Min. Cornub.* 88 The Ore in this vein is bunchy and uncertain. 1796 Kirwan *Elem. Min.* (ed. 2) II. 173 The surface often uneven and bunchy. 1849 J. Wealе *Dict. Terms Archit.* s.v. *Bunch*, A mine that is termed bunchy when the ore is found in irregular and sparsely distributed masses. 1867 *Ure's Dict. Arts* IV. 151/2 *Bunch*, A lode is said to be bunchy when the ore is found in irregular and sparsely distributed masses.

c. *bunchy top*, a virus disease of plants (esp. bananas) in which the leaves are crowded at the tip of the stem.

1919 *Agric. Gaz. N.S. Wales* XXX. 809, I lately visited the Tweed River district..to investigate the disease known as 'Bunchy Top' in bananas. 1930 *Discovery* June 186/1 'Bunchy top' of bananas, which has done so much damage..[etc.]. 1935 *Crown S. Afr. Dept. Agric. Sci. Bull.* CXXXIX. 46 (title) Further investigations on the bunchy-top disease of tomato. 1967 *Times* 14 Nov. 14/6 Bunchy Top Disease, a virus disease, transmitted by the banana aphid, *Pentalonia nigronervosa*.

bunco (bʌ·ŋkǝ), *sb.* *U.S. slang.* Also **banco, bunko.** [Said to be ad. Sp. *banca*, a card-game similar to monte.] A swindle perpetrated by means of card-sharping or some form of confidence trick. Freq. *attrib.* or as *adj.*; esp. **bunco-steerer**, a swindler; so **bunco-steering** *vbl. sb.* and *ppl. a.*

1872 *Chicago Tribune* 18 Oct. 8/4 A quintet of bunco thieves were 'taken in'. 1878 *Ibid.* 30 Sept. 2/2 The criminal classes...proceeded to introduce the business of 'bunko-steering'. *Ibid.* 2 Dec. 2/2 The 'bunko-steerer' seems to be a subordinate confidence-man. *Ibid.* 8 Dec. 12/3 This man was 'nonchalant' merely a 'confidence' or 'bunko' game on both sides. 1883 [see Bunko]. 1891 S. E. White *Westerners* xiii. 94 Bunco men can clean him out in a gambling joint. 1945 *Chicago Daily News* 5 Feb., I've come unpleasantly to talk about 'working from the bottom up'. 1949 *Atlantic Monthly* 28/1 I'm glad we got Aggie by a bunco-steering trick. 1963 *Spectator* 5 July 9 The bunco-artists from the lunatic fringe of the Democratic party.

bunco, *v.* *U.S. slang.* Also **bunko.** [f. prec.] *trans.* To swindle, to cheat.

1875 *Chicago Tribune* 6 July 8/1 The fugitive is the same person who bunkoed a stranger out of $75 recently. 1883 [see Bunko *v.*]. 1892 *Congress. Rec.* Mar. 2651/1 The farmer is always buncoed by the three-card monte, green-goods protection sharper. 1958 J. Black *You Can't Win* xiii. 174 The way to sell a brass brick is to bunko yourself first into the belief that your brick is solid gold.

Bund, bund (bʊnt), *sb.*⁴ [G.; related to Band *sb.*², Bind *v.*] A league, confederacy, or association; *spec.* (*a*) the confederation of German states; (*b*) a Jewish Social Democratic workers' organization in Eastern Europe, founded in 1897; (*c*) *U.S.*, an American pro-Nazi organization founded in 1936. Hence **Bu·ndist** *sb.*

1850 *Ann. Reg.* 1849 363/2 The Governments of Prussia, Saxony, and Hanover have therefore agreed, according to the 11th Article of the Act of Confederation, to enter into a union (*bund*) that has for its object the mutual protection of Germany. 1861 *Times* 16 Oct., The Government is required to join this *bund* called into existence by the danger of the moment. 1861 *Fraser's Mag.* Feb. 144/2 The new 'Bund', with all the motley crew of his fifteen nationalities. 1903 *Forest & Stream* 24 Jan. 78/3 The institution of revolver shooting is regarded likely to create much interest, as many members of the bunds favor this sort of work. 1905 *Westm. Gaz.* 4 July 4/1 After the massacre at Kischineff, after the bloodshed at Homel, the idea of self-defence took root...the Jewish 'Bund'. 1905 I. Zangwill *Ghetto Comedies* 257 Our Bund is the soul of the Russian revolution; our self-defence bands are bringing the heroic days of Judas Maccabæus. *Ibid.* 398 'A Bundist!'...From the bravest revolutionary party in Russia he could surely call a recruit or two. 1938 C. G. Robertson *Bismarck* ii. 81 German Liberalism called for the abolition of the Bund and the Diet. 1939 *Life* 27 Nov. 69 German American Bund. 1940 *G. Mann Lat.* 12 June (1967) 18 I'm not a Bundist in doubt, so let me sleep... I see Bund members dropping down my chimney. *Cambridge* ... 1951 *Sat. Even. Post* ... the Bund meetings in Germany...; the Bundists was not Communist? This Bundist... 1967 ... an anti-Bundist... [etc.]

bund (bʌnd), *v.* [f. Bund *sb.*²] *trans.* To embank.

1883 F. Day *Indian Fish* 41 Rivers which can be easily bunded. 1897 D. Dollar *Mem.* xiv. 141 The river front had been substantially *bunded*.

So **bu·nding** *vbl. sb.*, embanking; embankment.

1939 *Geogr. Jrnl.* XCIII. 136 The more common forms [of soil conservation] include silt pits, contour trenching [or bunding], and contour drains. (1967) E. M. Hough *Cooperative Movement in India* (ed. 3) iii. 170 The bunding of fields to prevent erosion.

bunder, var. *Bandar.

Bundesrat (bu·ndǝsrāt). Formerly **-rath.** [G., f. gen. of *bund* *Bund *sb.*⁴ + *rat(h* council.] A federal council; *spec.* (*a*) the upper house of the German or Austrian parliament; (*b*) the federal council of Switzerland.

1872 *Ann. Reg.* 1871 I. 239 This Bill was adopted by the Reichstag; but the *Bundesrath*, or Federal Council, refused to agree to it. 1879 *Encycl. Brit.* IX. 52/1 Delegates of the various confederated Governments form the Bundesrath. *Ibid.* 62/2 The emperor...has power...with the consent of the Bundesrath, to declare war in name of the empire. 1887 *Ibid.* XXII. 799/2 The [Swiss] federal council or executive (Bundesrath) consisted of seven members elected by the federal assembly. 1950 *Ann. Reg.* 1949 233 The...Constitution for the 'Federal Republic of Germany'...provided for two Chambers, the *Bundestag* (or Lower House), elected for four years by a mixed electoral system,..and the *Bundesrat* (or Upper House), representing the constituent confederate Länder and consisting of members of their Governments.

Bundestag (bu·ndǝstāk). [f. as prec. + *-tag*, prob. f. G. *tagen* to meet, sit in conference, f. *tag* (see Day).] An assembly of representatives of a league, confederacy, etc.; *spec.* (since 1949) the lower house of parliament of the Federal Republic of Germany.

1879 *Encycl. Brit.* X. 458/1 At a later era we may..were sold, the German bundestag [federal council] not being in sympathy with and separated from the...states. 1950 [see *Bundesrat].

bundle, *sb.* Add: **2. f.** Two reams of printing or brown paper, a quantity fixed by statute.

1724 *Act* 11 *Geo. I* c. 7 p. 367/1 Brown Paper, the Bundle containing 40 Quires. *Ibid.* 10 *Geo.* 1V. 4.

g. *slang.* Also **bundle**. A woman, esp. a fat one. Cf. Baggage 6 and 7.

1830 R. Forby *Vocab. E. Anglia* I. 46 *Bundle*, an opprobrious term applied to females, equivalent to baggage, when perhaps means strictly, a follower of the camp. a 1841 T. Sharp *Gloss. Words Warwicksh.* (1865) 11 *Bundle*, a large fat woman. 1922 Joyce *Ulysses* 751 The bundle of Nationalists. 1941 G. Heyer *Envious Casca* vi. 109 ...

h. (See quot. 1922.) So in phr. *to go a* (or *the*) *bundle on*: to bet much money on; *fig.* to be very fond of. *slang* (orig. *U.S.*).

1903 J. Flynt *Tramping with Tramps* (1900) IV. 392 *Bundle*, plunder, swag. 1922 'M. Hough' *You can't win* 119 He brought back with the bundle...1922 N. & Q. 12th Ser. XI. 206/2 *Pocket, parcel, and bundle*, ... 1945 Baker *Austral. Lang.* ... 1950 ... go a bundle on ... I don't go a bundle on you,' he said. 1958 C. McInnes *Absolute Beginners* ... I'd go a bundle if he knows his mate's gone to get shot...

i. *slang.* A fight or scrap.

1958 J. Curtis *Gilt Kid* xii. 124 If there was going to be a bundle he did not want to be bashed sitting down. 1963 'A. Garve' *Sea Monks* iii. 88 None o'them ain't goin' to start a bundle if he knows he's going to get shot for it.

3. b. *to drop one's bundle*: to give up hope, surrender, resist or compete no further. *Austral.* and *N.Z. slang.*

1915 C. J. Dennis *Songs Sentim. Bloke* (1916) 119 *To drop the bundle*, to surrender; to give up hope. 1928 *Bulletin* (Sydney) 7 Mar. 39/1 'E [a dog] wouldn't chase the 'are...'E dropped 'is bundle. 1947 P. Newton *Wayleggo* (1949) x. 125 My confidence immediately disappeared...I could not drop my bundle', so into the jungle I went.

c. *Colloq. phr.* (*to be*) *a bundle of nerves*: (to be) very nervous, timid, or highly-strung (see quot. 1940). (Cf. quot. 1802 for sense 2 b in Dict.)

1940 Partridge *Dict. Clichés* 35 *Bundle of nerves, to be a*, to be in an extremely nervous condition; to be ready to start or to be startled, to flinch, to jump at every trifle. 1948 ... a bundle of nerves. 1965 J. Fleming *Nothing is the Number* ... when you are a bundle of nerves...

4. bundle-man *Naut. slang*, a married seaman (see quot. 1925); **bundle-wood**, fire-wood made up into bundles.

1895 *United Service Mag.* 518 If one of the officers is on the sick list it is well to have a spare room in the bundle-man's house. 1925 Fraser & Gibbons *Soldier & Sailor Words* 39 *Bundle-man*, a Lower Deck phrase for a married man, apparently suggested by the small bundle tied up with a blue handkerchief which married seamen.. usually take ashore with them when going on leave. 1879 *Good Words* 781/2 The wood-yard gates are the storage ground of the 'bundle' fire-wood trade. *Ibid.* 781/2 The bundle wood coming into his own household.

bundler. Add: **1.** (Examples of technical uses.)

1879 *Good Words* 781/2 Children, working shifts as half-timers, 'pile' for the bundler. 1881 *Instr. Census Clerks* (1885) 70 Flax, Linen Mill...Drying and Bundling; Dryer. 1921 *Dict. Occup. Terms* (1927) § 399 *Bundler*, ..made waste from various departments of mill up into bundles.

bundy (bʌ·ndɪ), *a.* [f. Bundle *sb.* + -y¹.] Like a bundle.

1898 *Westm. Gaz.* 21 Mar. 7/3 The clothes were all loose. Were they bundly? Perhaps. 1925 D. H. Lawrence *Let.* 21 Apr. (1962) II. 838 We look such a bundly Mexican outfit.

‖**bundobust, bandobast** (bʌ·ndǝbʌst). *India.* Also **banda-, bando-, bunda-, bunder-, bundo-, bundoo-; -bast, -bust, -busst.** [Hind. (Pers.) *band-o-bast* tying and binding.] An arrangement, organization; preparations.

1776 *Trial Jos. Fowke, Depositions* 17/1 In the year 1180, when the Bundobast of the farms took place. 1813 *Gloss. 5th Rep. Sel. Comm. E. India Co.* 10/1 Bundobust... 1866 ... 1886 C. Arbuthnot *Twenty-one Days India* iii. This unhappy creature whose mind is a perfect blank regarding *Faujdari* and *Bandobast*...will...actually presume to discuss Indian affairs with you. 1896 A. Forbes *Camps, Quarters*, & c. *Casual Places* 170 George's sacred not make quite so bad a *bundobust* after all. 1920 *Blackw. Mag.* Sept. 357/1 In bandobast matters, good bandobast. 1923 E. F. Norton *Fight for Everest* 1924 163 [We] set off with a minimum bundobast to penetrate as far down into Nepal as we could. 1927 *Blackw. Mag.* 403 The bandobast...was excellent. 1938 ... 1958 M. Mais *Pagoda Tree* xviii. 157 A great deal of trouble attached with a tremendous lot of determination 'making *bundobusts*' right, left and centre.

bundook (bʌ·ndūk). *India.* [Hind. *bandūq*, a. Pers. *bunduq* filbert nut, musket or cannon ball, firearm, ad. Gr. Ποντικόν (sc. κάρυον hazel nut) Pontic *a.*] A musket or matchlock; a rifle.

1886 in Yule & Burnell *Hobson-Jobson*. 1916 'Sapper' *Men, Women & Guns* v. 156, I had words with a sniper at the frontier, but I put it across him with his own bundook. 1923 *Blackw. Mag.* Apr. 505/1 Only, a quick gun and a bundook were none too. 1926 *Ibid.* Nov. 664/2 Fortunately bundooks...do not shoot very straight.

bundu (bu·ndū). *Southern Afr.* [Bantu.] An uncivilized rural region; wilds. Also *attrib.*

1946 *Cape Times* 23 Oct. 8/7 The King is expected to wear a bush shirt and shorts and, while he is in Rhodesia, is to go to the bundu. 1950 *Cape Argus* 27 May 4/6 He...finds life in the bundu tough. 1962 *Listener* 4 Jan. 29 ... 1963 ... *World of Strangers* ii. 54 It's absolutely in the bundu, of course, forty-three miles from Neksburg.

bung, *sb.*¹ Add: **3, b.** A brewer, or landlord of a public house. Also, the brewing interest (as in politics); hence *attrib.* or as *adj.*, favouring the brewers or their interests in politics.

1860 Hotten *Dict. Slang* (ed. 2), *Bung*, the landlord of a public-house. 1884 *Graphic* 23 Feb. 170/1 That Sir Wilfrid Lawson has carried 'bung' can easily be imagined. 1887 ... the Lord High Bung. a 1889 *Truth* (Barrère & Leland), Within the last few years several bungs have been made Peers. 1891 *Pall Mall Gaz.* 13 June 1/3 There must be no further compensation for injured ...the local bung ... 1893 ... that the 'bung' party fail. 1897 ... 1909 ... 1903 *Daily Chron.* 5 Feb. 1/5 They've a pub in North London which might be the 'bung' is named Scoot. 1960 ...

bung, *sb.*² *Austral.* and *N.Z. slang.* Also formerly **bong.** [Austral. (Aboriginal word.)]

A. Dead. **B.** Bankrupt. **C.** Ruined, useless. Also in phr. *to go bung*, to die; to fail; to go bankrupt.

1841 D. Lang *Cooksland* x. 430 A place called *Umpie bung*, or the dead houses. 1885 A. W. & J. D. Boyd *Old Colonials* (1882) 23 ... 'gone bung'. 1885 H. Finch-Hatton *Adv. Australia* x. 142 Directly the bung ... 1893 ... 1898 ... 1902 ... 1905 ... 1910 ... 1932 ... 1958 Gillespie ... 1964 ...

bung, *v.*¹ *slang* (orig. *dial.*). [Echoic.] *trans.* To throw (violently); to send; to put forcibly. Also with *up*.

1825 Jamieson, To *bung*, to throw with violence. Aberd. 1835 *Sessions' Paper Cent. Criminal Court* May 37, I heard one say to the other, 'Bung them down'... 1894 ... 1899 ... 1904 ... 1910 ... 1913 ... 1921 ... 1929 ... 1933 ... 1944 R. Lehmann *Ballad & Source* ... 1950 ... 1952 ... 1963 ...

bung, *v.*² *Criminals' slang.* [Origin unknown; perh. f. prec.] *trans.* To bribe; to pay; to tip. Cf. *Bung *sb.*⁴

1950 P. Tempest *Lag's Lexicon* 27 'Did he bung you?' — did he give you a tip?' 'I will let you know (you give) a few quid... 1958 ... 1963 T. & P. Morris *Pentonville* 378 To 'bung' a person is to give him a fixed amount of tobacco each week to keep quiet. 1966 J. Ashford *Consider Evidence* i. 13 ...

bung, *adv.* [Cf. prec. and Bang *v.*¹ 8.] In or into the very middle of things; = Smack *v.*² 7 b.

1899 Kipling *Stalky & Co.* 132 They have bables and terrifying and measles and all that sort of thing right bung in the middle of the school. 1901 ... 1940 N. Shute' *Landfall* 25 'Where was the ship, anyway?' 'Bung in the middle of Matheson's zone.'

bungaloid (bʌŋgǝlɔɪd), *a.* [f. Bungalo(w + -oid after *fungoid*.] Having the appearance or style of a bungalow or bungalows; characterized by the presence of bungalows. Also *sb.*, a bungaloid building.

1927 *Daily Express* 21 Nov. 17/3 Hideous allotments and bungaloid growths make the approaches to any city repulsive. 1928 *Sunday Dispatch* 30 Sept. 11/2 Mr. Shaw designates our modern urban communities 'bungaloid promiscuities', and denounces them as the ugliest blight of new civilizations. 1929 *Morning Post* 4 May 1/3 Will somebody ever manage a bungaloid rash on the freehold and preserve its open tracts... 1947 ... 1968 ...

bungalow. Add: In modern use, any one-storied house. Also *attrib.* and *Comb.*

1886 W. Raleigh *Let.* May (1926) I. 72 Ten marches with a bungalow rest house of some description to stop at. 1903 *Arch. & Contr. Reporter* 24 Apr. 272/2 The buildings have been designed in a bungalow type. 1906 *Daily Chron.* 9 Mar. 8/3 The bungalow village. 1907 *Westm. Gaz.* 28 Apr. 9/1 The meeting of bungalow-holders. 1936 *Discovery* Feb. 42/1 The permitting of extensive uncontrolled camping and of inferior-type bungalow-towns.

bungarum (bʌŋgǝrǝm). Also **bungarus, bungar, bongar.** [Bengali (Skr., neut. sing. of *bangura* bent, curled (sb. bend of a river), f. *bhanj* to break, bend).] Any of the venomous snakes of the genus *Bungarus*, esp. the krait of India.

1835 *Encycl. Metrop.* XXIII. 640/1 *P[seudoboa] Fasciata*. A native of India. In Bengal it is called *Bungarum Pamah* and Sankeener. *Ibid.* 1869 ... 1870 ... 1887 *Encycl. Brit.* XXII. 196/2 The fangs of the Bungarums are shorter than those of the cobras. 1917 F. Wall *Snakes of Ceylon* 435 The Telugu word 'bungarum',..gold...is applied by the Telugus to the banded krait.

bungee. Add: = *Pongee.

bunger, bungie, bungy, varr. *Punga.

bung-ho (bʌŋhōᵘ), *int.* Also **bung-o, -oh.** An exclamation used at parting or as a drinking toast.

1925 *Springfield Union* 20 Nov., They now say 'Bungo!' instead of 'Here's how!' over cocktails. 1926 Hemingway *Fiesta* (1927) iv. 118 Good night...'Bung-o,' Mike said. 'Get along there old Jake.' 1928 D. Sayers *Unpl. Bellona Club* ii. 8 'Dry Martini,' said Wimsey...'Bung-ho!' 1935 N. Marsh *Enter a Murderer* xi. 124 'Bung-ho,' said Nigel 'Bung-ho,' said Nigel... 1965 L. Payne *Nose on my Face* xxi. 257 'Bung ho!' I...had a long and refreshing drink.

bungie, bungy (bʌ·ndʒɪ). *slang.* Also **bunjie** (e, **bunjie, bunjy.** [Origin unknown.]

a. India-rubber; a rubber. **b.** (A nickname for) a Naval physical-training instructor.

1915 'Bartimeus' *Tall Ship* iv. 75 'Bungie,' said the First Lieutenant, 'come to the club and have tea and play "pills" afterwards!' The Indiarubber Man shook his head. 1927 E. Scott *War among Ladies* i. 11. 24 Where's my rubber? John, have you pinched my bunjy? 1934 N. & Q. CLXVII. 442/2 In the schoolroom fifty years ago, the piece of india-rubber for erasure purposes was referred to as 'india-bungie' or 'bungie'...bungie in the middle of Matheson's zone. 1946 ... going to be quite...

bu·nglesome, *a.* *U.S.* [f. Bungle *v.*] Awkward to handle, bungling and troublesome.

1889 in *Century Dict.* 1897 *Congress. Rec.* Feb., App. 55/1 An unexcusable and bunglesome fraud or mistake on the part of some one of the paper has occurred... 1915 Mrs. Stratton-Porter *M. O'Halloran* xv. 337 But this sheet is going to be rather bunglesome. You bungle...

bungo, var. *Bumbo².

Bu·ngtown. *U.S.* Also with lower-case initial. [App. a fictitious local name, associated with Rehoboth in Massachusetts (Schele de Vere *Americanisms* 587).] *Bungtown copper* or *cent*, a counterfeit English halfpenny or other spurious copper coin. Also *ellipt.*

1787 *Newport Mercury* 13 Aug. in *Mag. Amer. Hist.* (1885) XIII. 206 We are informed that all Coppers by law, in New York, except Bungtowns, are fixed at 180 for a Dollar. 1840 *Knickerbocker* XV. 183 'He took a five cent piece and two bungtown coppers out of the till. 1853 ... 1890 ... Coppers, bogus Bungtown, Bungtowns, etc.

bungy (bʌ·ŋgɪ). *sb.* *India.* Also **bunghea.** [Hind. *bhangī*.] In Bombay, a member of the low caste of sweepers.

1823 *Trans. Lit. Soc. Bombay* III. 374 *Bhunghea*, the sweeper, particularly of filth. 1886 Yule & Burnell *Hobson-Jobson* s.v., The colloquial application of this term to the Bombay people is, we believe, ...

bungy, bunje, etc. See *Bungie.

bunk, *sb.*¹ Add: **1.** (Earlier examples.) Freq. of one of two or more beds arranged in a tier.

1758 L. Lyon in A. Tomlinson *Mil. Jrnls.* (1855) 35 We made us up 2 straw bunks for 4 of us to lay on. 1780 W. Heath *Let.* 2 Feb. in *Mass. Hist. Soc. Coll.* (1905) V. 28 The bunks and bedding at the bomb proof were taken out. 1810 *Deb. Congress* (1853) XXI. App. 2448, 20 bunks at $1 per bunk. 1817 Harper in *Alvine Mag.* (1926) 1. 83 The bunks..in New Orleans.

b. *attrib.*, as *bunk-car, -room*; **bunk-bed** = bunk (sense *1); **bunk-house**, a house where workmen, etc., are lodged.

1951 *Catal. Exhibit. Festival of Britain* 127/2 Bunk beds. 1967 *Observer* 21 May 30/7 Bunk beds that splits into single beds. 1804 P. L. Ford *P. Stirling* xi. 256 By the light, one of the superintendents found the bunk cars. 1877 R. W. Raymond *8th Rep. Mines* 332 Bunk-house. 1891 E. S. White *Claim Jumpers* xxxi. 274 The old 'bunk house' was accommodated a good-sized gang of miners. 1952 H. Innes *Campbell's Kingdom* i. iii. 50 This is to the bunk-house. 1958 J. R. P. Barker... *Barker Glance* at N.Y. 24 He pointed to the bunk-room.

c. *spec.* The lodging of a student at St. Andrews university. Hence *attrib.* in *bunk-wife*, the (or a) keeper of a 'bunk'; landlady.

1928 ... *Scottish N.Y.* 2nd s. ... 1936 *St. Andrews Univ.* 8/1 ... 1937 *Ibid.* 27 Feb. 7/3 Those bunk-wives who refuse students any comforts.

2. (Examples.)

1770 M. Patten *Diary* (1903) 238, I brewed a bunk and a Slat for my boys lodging Sled. 1901 S. E. White *Blazed Trail* 72 These sleighs, with...bunks nine feet in width for the reception of logs.

BUNK

bunk, *sb.*[3] [f. BUNK *v.*[2]] In slang phr. *to do a bunk*: to make an escape; to depart hurriedly.

c **1870** *Broadside Ball.*, 'Peck's Bad Boy' (Farmer), The keeper tried to catch him, but the bad boy did a bunk. **1905** E. CANDLER *Unveiling of Lhasa* xiii. 256 The old bloke's done a bunk. **1919** J. B. MORTON *Barber of Putney* ix, 'All right, son,' said Curly. 'They [*sc.* Germans] done a bunk.' **1921** G. B. SHAW *Back to Methuselah* IV. III. 199 If my legs would support me I'd just do a bunk straight for the ship.

bunk, *sb.*[4] *slang* (orig. *U.S.*). [Abbrev. of BUNKUM.] Humbug, nonsense.

1900 ADE *More Fables* 15 He surmised that the Bunk was about to be Handed to him. **1916** H. FORD in *Chicago Tribune* 25 May 10/1 History is more or less bunk. **1930** *Punch* 5 Mar. 265/3 No tempting blurb, no critics' bunk Can animate this mass of junk. **1932** W. S. MAUGHAM *For Services Rendered* III. 70 It's all bunk what they're saying to you, about honour and patriotism and glory, bunk, bunk, bunk.

bunk, *v.*[1] Add: (Earlier examples.) Also *to bunk down*: to go to bed, retire to bed.

1840 R. H. DANA *Bef. Mast* viii, We turned in to bunk and mess with the crew forward. **1850** H. C. WATSON *Camp-Fires Revol.* 250 It's about time for us to bunk. **1940** *War Illustr.* 2 Feb. 59/2, I..was snugly bunked down when the explosion happened. **1965** M. SHADBOLT *Among Cinders* ii. 9, I was wondering if this might be a decent place to bunk down for the night.

bunk, *v.*[2] Add: Also const. *about*.

1949 R. ALDINGTON *Strange Life of C. Waterton* xiii. 160 A schoolboy prefers to bunk about the fields..rather than work in class.

2. *trans.* To expel from school. So **bunked** *ppl. a.*, expelled.

1890 FARMER *Slang* I. 385/2 *Bunk*, (Wellington College)—To expel (from the school). *c* **1898** WYNDHAM LEWIS *Lett.* (1963) 6 A sixth (form boy) has been bunked (expelled) for stealing out of the shop in town. **1902** W. S. MAUGHAM *Mrs. Craddock* xxviii. 283, I was bunked from Rugby. Well, that wasn't my fault...I'm blowed if I was worse than anybody else. **1923** H. WALPOLE *Jeremy & Hamlet* i. 21 A chap called Bates got bunked for stealing.

bunked (bʌŋkt), *a.* [f. BUNK *sb.*[1] + -ED[2].] Furnished with or having a bunk or bunks.

1892 STEVENSON & OSBOURNE *Wrecker* xiii. 209 A double bunked division for the cook..and second mate. **1940** *Economist* 9 Nov. 575/2 A survey is being taken of private basement shelters, much of which will be bunked.

bunker, *sb.*[1] **3.** Delete (*Sc.*) in def., and add:

1921 *Dict. Occup. Terms* (1927) § 279 *Bunkerman*..tips wagons to discharge coal or iron into bunkers at ironworks. **1930** *Engineering* 16 May 629/3 This conveyor distributes the coal to three bunkers.

b. *pl.* = *bunker coal* (see 5 below).

1883 GRESLEY *Gloss. Coal-m.*, *Bunkers*, steam coal consumed on board ship. **1898** *Daily News* 18 Apr. 11/5 More money has to be paid for steam, bunkers, and gas coals. **1935** *Economist* 5 Jan. 48/2 Prime unscreened bunkers, 13s. 6d.

c. = *bunker-man* (see 5 below).

1921 *Dict. Occup. Terms* (1927) § 744 *Bunker*, tips coal from baskets or trucks into coal bins or bunkers as they are lowered by hoist...*Collier*..includes bunker, trimmer, loader.

4. Now, an artificial sand-hole with a built-up face; also, any natural obstruction (as water, long grass, etc.) on a golf-course.

1857 H. B. FARNIE in J. L. Stewart *Golfiana Miscellanea* (1887) 119 The surface is dotted over at frequent intervals with sandy holes, technically called bunkers, from two to six feet deep, of irregular forms and sizes. **1893** H. G. HUTCHINSON *Golfing* 60 Sometimes you may be driven to invent hazards, by throwing up banks, cutting bunkers or planting bushes. **1897** *Encycl. Sport* I. 458/1 Between the teeing-ground and the putting-green should be found, whether they be natural or artificially formed, various 'hazards' in the shape of sand-pits or 'bunkers'. *Ibid.* 472/1 *Bunker*, originally a natural sand hole on the golf course. Now used also of artificially made hazards with built-up faces. **1955** *Times* 17 May 5/6 Mrs. Smith cut her second shot into the rough and put her recovery into a bunker on the right.

b. *fig.*

1900 A. BIRRELL in *Cornhill Mag.* Mar. 313 If you want to find the natural man at work you must look for him in the bunkers of life. **1905** *Westm. Gaz.* 28 Oct. 16/1 The Princess frock is a bunker not to be cleared by any economies or adaptations.

c. A military dug-out; a reinforced concrete shelter.

1939 *War Pictorial* 13 Oct. 29/2 A Nazi field gun hidden in a cemented 'bunker' on the Western front. **1945** *Over-All Report* (U.S. Strategic Bombing Survey, Europ. War) 30 Sept. 104/1 Germany promised its people 'bombproof' shelters for all, and planned the construction of extensive above-ground concrete structures known as 'bunkers'. **1947** H. TREVOR-ROPER *Last Days of Hitler* iv. 117 A curved stair led downwards to a still deeper and slightly larger bunker. This was..Hitler's own bunker, the stage on which the last act of the Nazi melodrama was played out. **1949** F. MACLEAN *Eastern Approaches* III. iv. 354 The turf-covered 'bunkers' in which the Germans and Ustaše had made their last stand.

5. *attrib.* and *Comb.* (See senses 1 and 3 in Dict.) Also (sense 4) *bunker-iron*; **bunker coal,** *spec.* steam coal; also, coal carried by a steamer for its own use and not as cargo; **bunker-man** (see quot. 1921).

1885 Bunker coal [see sense 3 in Dict.]. **1888** *Daily News* 10 July 5/2 The exports of 'bunker' coal..show an improvement. **1857** H. B. FARNIE in J. L. Stewart *Golfiana Miscellanea* (1887) 142 Some bunker irons of the old make are round bottomed. **1886** A. LANG in *Longman's Mag.* July 332 The iron head makes it more like a bunker iron than a play club. **1921** *Dict. Occup. Terms* (1927) § 089 *Bunkerman*, in charge of bunkers at coke ovens where coal is stored. *Ibid.* § 279 [see sense 3 a above].

bu·nker, *sb.*[2] *local U.S.* Short for MOSSBUNKER.

1842 J. E. DEKAY *Zool. N.Y.* IV. 260 The Mossbonker... At the east end of the island, they are called Skippangs or Bunkers. **1888** [see MOSSBUNKER].

bu·nker, *v.* [f. BUNKER *sb.*[1]]

1. *trans.* To fill the bunkers of (a steamer) with coal or oil for its own consumption. Also with the coal or oil as object.

1891 *Pall Mall Gaz.* 19 Jan. 4/1 Many..of the great steamship companies of Liverpool were simply begging for coal to either bunker or cargo their steamers. **1893** *Times* 11 July 3/6 The ordinary rate of bunkering coal by manual labour. **1925** *Blackw. Mag.* June 836/1 She was bunkered for the twenty-nine days' run to Batavia.

b. *intr.* To take in a supply of coal or oil for consumption on a voyage.

1893 *Whitby Gaz.* 3 Feb. 2/6 The foreign coaling clause in the outward coal charter bound them to bunker with the agents of the charterers. **1895** *Ibid.* 11 Apr. 3/2 We bunkered at Malta at four o'clock in the morning. **1925** *Chambers's Jrnl.* Dec. 778/2 There was some delay at Durban, where she bunkered.

2. *Golf. pass.* **a.** Of the ball: to be hit into a bunker. Of a player: to have one's ball in a bunker.

1886 H. G. HUTCHINSON *Hints Game Golf* 40 Your adversary is badly bunkered. **1891** —— *Famous Golf Links* v. 56 His ball lofted on Mr. Maitland's, knocked it out of the hazard, and lay bunkered in its place. **1903** *Punch* 22 Apr. 283 On..the Golf Links..watching the Colonel, who has been bunkered for the last ten minutes. **1955** *Times* 19 May 4/7 She was bunkered at the 18th.

b. To be furnished with a bunker or bunkers.

1907 *Daily Chron.* 17 Aug. 7/2 The point of controversy..is as to how such a hole should be bunkered, or rather as to whether it should be bunkered at all.

c. *fig.* (*colloq.*) To be placed in a situation from which it is difficult to extricate oneself. Also, to place in such a situation.

1894 BARON RIBBLESDALE in *Westm. Gaz.* 6 Sept. 4/2 The Liberal peers were powerless. To use a golfing simile, they were bunkered. **1899** *Geogr. Jrnl.* May 474 In the long narrow ice-bound valleys which lead up from the Hindu Kush to the Pamirs..we were once, to use a familiar term, fairly bunkered. **1905** *Westm. Gaz.* 27 Dec. 1/3 The truth is that Mr. Balfour is bunkered by his own record. **1917** H. A. VACHELL *Fishpingle* ix, 'Perhaps you regard golf as a sort of epitome of life?'..'I suppose I do.' 'If you found yourself "bunkered", you would not lose heart?' **1930** G. B. SHAW *Apple Cart* I. 18 *Balbus.* Thatll do it. He couldnt face that. *Crassus.* Yes: thatll bunker him.

bunkering (bʌ·ŋkəriŋ), *vbl. sb.* [f. *BUNKER *v.*] **1.** The action of filling a ship's bunkers with coal or oil. Also *attrib.*

1899 *Westm. Gaz.* 7 Nov. 9/3 It is thought that the North country coal will be used by transports for bunkering. **1920** *Act* 10 & 11 *Geo. V* c. 50 § 3 Regulating the export of coal and the supply of coal for the bunkering of vessels. **1922** *Daily Mail* 9 Nov. 3 A group of foreign bunkering depots. **1958** *Times Rev. Industry* May 51/3 The bunkering trade has increased in proportion to output and Kuwait now delivers well over one million tons of bunker oils a year to ocean-going vessels.

2. The action of furnishing (a golf course) with bunkers.

1909 *Westm. Gaz.* 8 Mar. 12/2 The final bunkering of the course.

bunkery (bʌ·ŋkəri), *a.* [f. BUNKER *sb.*[1] 4 + -Y[1].] Full of or abounding in bunkers.

1890 *Sat. Rev.* 31 May 665/2 That mountainous range of bunkery sandhills. **1894** *Blackw. Mag.* Oct. 561 The sand was..strewed in a thick bunkery mattress over all the ladies' links.

bu·nkie, bu·nky. *U.S. colloq.* [f. BUNK *sb.*[1] 1 + -Y[4].] One who shares a bunk with another.

1858 VIELÉ *Following the Drum* 218 Which triumph over paternal love I rewarded by giving him Jack for his 'bunkie'. **1903** *N.Y. Even. Post* 10 Oct., In a logging camp in Maine a lanky Yankee had been unburdening himself of his past life for the entertainment of his 'bunky'. **1910** W. M. RAINE *B. O'Connor* 58 If it is certain that my old bunkie came to his death foully. **1948** *Southwest Rev.* Winter 29/2 The next morning Bud's 'bunkie' sat watching him shave.

bunko. See also *BUNCO *sb.* and *v.*

bunk-up, *sb.* [Cf. BUNT *sb.*[8] and E.D.D. 'a bunt up' (1888).] A lifting-up.

1919 DOWNING *Digger Dialects* 14 *Bunk-up*, a lifting up. **1938** F. D. SHARPE *S. of Flying Squad* v. 51, I was given a bunk up, and..got through the wire. **1951** A. BARON *Rosie Hogarth* II. iii. 103 'Here,' he appealed, 'give us a bunk-up, someone.'

bunned (bʌnd), *a.* [f. *BUN *sb.*[2] 1 b + -ED[2].] Of hair: coiled into a bun.

1944 *Penguin New Writing* XXI. 77 Hair straggled and bunned. **1951** W. SANSOM *Face of Innocence* iv. 59 Her tweeds and her efficiently bunned hair.

bunnia (bʌ·nyā). *India.* Also **buneeya, bunia, -ar; bunneah, bunniah, bunnya; bania, -(i)ya.** [Hind. *banyā*, a. Gujarati *vāṇiya* (see BANIAN).] A trader or merchant. Also *attrib.*

The usual form now is *bania*.

1794 MORSE *Amer. Geogr.* 687 The tribe of Beise, who are chiefly merchants, bankers, and banias or shopkeepers. **1829** *Encycl. Metrop.* XX. 33/2 Some of the Banyás are Awaks, or seceders from the Brahmanical faith. **1879** MRS. A. E. JAMES *Ind. Househ. Managem.* 77 The native *bunniahs*..do not give long credit. **1880** *Encycl. Brit.* XII. 760/1 In the interior of the Bombay Presidency, business is mainly divided between two classes, the Baniyas of Guzerat and the Marwaris from Rajputana. **1926** *Blackw. Mag.* Dec. 799/1 The Hindu Bunia in remote villages is still bullied by the wealthy zamindar of the neighbourhood. **1936** J. NEHRU *Autobiogr.* liii. 432 The *bania* civilisation of the capitalist West. **1945** 'P. WOODRUFF' *Call Next Witness* I. 30 Banias reckoning the day's gains.

bunny[2]. Add: **1. c.** Rabbit-fur. *colloq.*

1950 H. MCCLOY *Through Glass Darkly* (1951) v. 44 Girls in wolf or bunny that looked almost like fox or ermine. **1955** M. LASKI *Apologies* 59 You know bunny wears just as well.

d. In full *bunny girl*. A night-club hostess, or the like, dressed in a costume which is partly imitative of a rabbit. Also *attrib.*

1960 *Playboy* Aug. 42/1 The girls are called Bunnies and they're invitingly attired in brightly coloured rabbit costumes, complete to the ears and white cotton tails. *Ibid.*, Proportions..as remarkable as those of the Bunny Girls. **1963** *Listener* 7 Feb. 260/3 American Bunny Clubs, with their Freudian fantasy-tease hostesses. **1963** *Daily Mail* 16 Feb. 6/6 These bunnies are the newest import to London night-club life from America. *Ibid.* 6/7 The bunny costumes with their stylised unreality somehow defuse the provocation of the dress. **1966** 'E. E. SUMNER' *Second-hand Death* vii. 131 The girl..worked as a night club Bunny. **1967** J. GARDNER *Madrigal* i. 12 A fallen Bunny girl, with strange fetishes.

bu·nny hu:g. [f. BUNNY[2] 1 a + HUG *sb.*] A dance in ragtime rhythm, esp. popular in the early part of the 20th century. Hence **bu·nny-hu:gger, -hu:gging.**

1912 *Daily Mirror* 6 Jan. 7/1 Fashionable society in New York..is fighting upon the great question as to whether..'The Bunny Hug'..shall be..allowed. **1914** *Art* (U.S.) 290 It is not in the souls of the bunny-huggers that the new ferment is potent. **1916** *Daily Colonist* (Victoria, B.C.) 11 July 3/7 Such dances should not be classed with fox-trotting, bunny-hugging, Wilson gliding, [etc.]. **1926** GALSWORTHY *Silver Spoon* II. vi. 158 What is there to give up—bunny-hugging? **1937** *Evening News* 26 Jan. 9/4 Some time ago many people were dancing the bunny hug. **1954** DANNETT & RACHEL *Down Memory Lane* 70 The bunny hug survived its contemporaries and made the transition to jazz.

bunodont (biū·nŏdǫnt), *a.* and *sb.* [f. Gr. βουνός mound + ὀδοντ-, ὀδούς tooth.] **A.** *adj.* Designating molar teeth whose crowns are elevated into tubercles; having tuberculate molars. **B.** *sb.* A mammal with teeth of this pattern.

1874 E. D. COPE in *Jrnl. Acad. Nat. Sci.* VIII. 72 Bunodont type; crown supporting tubercles. *Ibid.* 73 The Bunodonts, except some *Carnivora*, are all of the former or isognathous type. **1883** *Encycl. Brit.* XV. 429/2 The grinding surfaces of the molar teeth either of a distinctly tubercular (bunodont) or of a crescentic ridged (selenodont) form. **1902** *Nature* 25 Sept. 523 A series of six cheek-teeth, distinct and bunodont in type. **1969** *Ibid.* 22 Nov. 821/2 The Nagri suids..are low-skulled, with bunodont dentitions.

‖ **bunraku** (bu·nrāku). [Jap.] A kind of traditional Japanese puppetry practised at the Bunraku-za marionette theatre by its company; also, a contracted name for this company; hence, the Japanese puppet theatre in general. Also *attrib.*

1920 B. KURE *Hist. Devel. Marionette Theatre Japan* 30 There are two schools in the marionette art: Yuki School, which manipulates the dolls by strings and Bunraku School, which makes the dolls act by holding them. **1936** *Travel in Japan* II. 24 Even on a stage of Bunraku's pretention, only puppets in important roles are handled by three manipulators. **1955** A. C. SCOTT *Kabuki Theatre of Japan* iv. 56 If ever there was a case for State support it is that of the Bunraku. *Ibid.*, The title Bunraku has now come to mean the doll theatre as a whole. **1959** *Listener* 12 Feb. 299/1 Like puppets in the Bunraku doll-theatre.

bunt, *sb.*[8] Add: (Earlier U.S. example.)

1767 *Boston Gaz.* 19 Jan. (Th.), [The black ram] will sometimes come behind a great Weather..and give him a paultry Bunt at unawares.

2. *Baseball.* An act of stopping the ball with the bat without striking. Also *bunt-hit*. *U.S.*

1889 *Chicago Tribune* 8 Aug. 6/1 Connor got around on bases on balls given himself and Richardson, Ward's bunt toward third..and a wild pitch. **1896** KNOWLES & MORTON *Baseball* 114 A bunt hit is a deliberate attempt on the part of the batsmen to hit a ball slowly within the infield so that it cannot be fielded by any fielder in time to retire the batsman. **1917** MATHEWSON *Sec. Base Sloan* ii. 169 Despaigne started out poorly enough, trickling a bunt to third. **1968** *Washington Post* 4 July C 1/8 With Rich Reese looking for a bunt and charging in from first base, Azcue punched his game-winning bit into right field.

bunt, *sb.*[9] *Aeronaut.* [perh. f. BUNT *v.*[2]] A manœuvre in aerobatics involving half an outside loop followed by a half roll.

1932 *Techn. Rep. Aeronaut. Research Comm.* 1930–31 I. 47 Load factors one half that of normal flight would be adequate to cover the manoeuvre known as the bunt. **1935** C. G. BURGE *Compl. Bk. Aviation* 89/2 Like most aerobatic manoeuvres the bunt was first done by Pégoud.

bunt, *v.*[3] Add: **2.** *Baseball.* To stop (the ball) with the bat without swinging the latter. Also *absol.* So **bu·nted** *ppl. a.*, **bu·nting** *vbl. sb.*

1889 *Reach's Base Ball Guide* 144 Bunted Ball. **1892** *Courier-Journal* 2 Oct. 13/5 There is not a man in his team that can bunt. **1896** *Spalding's Base Ball Guide* 77 The 'bunting' of the ball, so as to cause it to drop to the ground almost dead. **1912** MATHEWSON *Pitching* ii. 23 Doyle bunted and was safe, filling the bases. *Ibid.* xiii. 298 Once,..McGraw planned a bunting game against Overall. **1967** *Boston Sunday Herald* (This Week Mag.) 9 Apr. 5/2 Don't wait till after a pitch is thrown to decide whether the batter will bunt or try for the hit-and-run.

buntal (bʌ·ntāl). [Philippine name. See also *BALIBUNTAL.] A straw prepared from the fibres of the petioles, or leaf-stems, of the talipot or buri palm (*Corypha umbraculifera*), used for the manufacture of hats in the Philippine Islands. Also *attrib.*

1910 H. H. MILLER *Philippine Hats* 38 Very lately, buntal straw has been imported into Baliwag and there woven into very beautiful hats in the Baliwag weave. **1911** C. B. ROBINSON in *Philippine Jrnl. Science* June 114 Until the beginning of 1909, all buntal hats were made in one weave, that of Lucban, Tayabas. **1926** *Commerce & Industry Jrnl.* (Manila, P.I.) Nov. 11 The hat manufacturers of China..are making hats out of buntal fibers from the Philippines.

bunter[1]. (Later examples.) (Also see quots.)

a **1852** MAYHEW *Lond. Labour* (1861) II. 142/1 Old women alone gathered the substance [dogs' dung], and they were known by the name of 'bunters', which signifies properly gatherers of rags. **1861** *Ibid.* Extra vol. (1862) 223/1 There is a class of women technically known as 'bunters', who take lodgings, and after staying some time run away without paying their rent. **1891** C. WORDSWORTH *Rutland Words* 5 *Bunter*, a disreputable woman. 'She stood at the gate and called me a *bunter*.'

bunting, *sb.*[1] Add: See also CIRL, ORTOLAN 2 a, REED-BUNTING, SNOW-BUNTING.

b. Applied by extension to any bird of the bunting subfamily, and to similar birds of other families. *U.S.*

1831 WILSON & BONAPARTE *Amer. Ornith.* II. 242 Black-throated Bunting...In their shape and manners they very much resemble the yellow-hammer of Britain. *Ibid.* 245 *Fringilla Graminea*,..*Emberiza Graminea*..Bay-winged Bunting. **1893** NEWTON *Dict. Birds* 459 Indigo-bird,..a well-known North-American species,..American ornithologists give full accounts of the habits of this bird, together with those of..the still more gaudy Painted Bunting. **1964** A. L. THOMSON *New Dict. Birds* 112/1 The buntings (Emberizinae) are predominantly terrestrial.

4. *attrib.*, esp. **bunting lark,** the corn bunting; also *bunting-lark fly*, an angler's fly.

1802 MONTAGU *Ornith. Dict.* I, Bunting-lark. **1837** KIRKBRIDE *Northern Angler* 25 The Bunting Lark Fly. **1876** *Encycl. Brit.* IV. 525/2 The true Bunting (or Corn-Bunting, or Bunting-Lark, as it is called in some districts). **1884** COUES *N. Amer. Birds* (ed. 2) 356 Bill very small and truly conic, well exhibiting 'emberizine' or 'bunting' characters.

bunting, *sb.*[2] Add: **b.** *attrib.* **bunting-tosser** *Naval slang*, a signaller.

1905 *Daily Chron.* 23 Jan. 3/1 From which breathless catalogue it will be gathered that the path of the 'bunting-tosser' is not an easy one! **1909** in J. R. WARE *Passing Eng.*

buntons (bʌ·ntənz). [orig. in sing. *bunting*, a piece of squared timber, of obscure etym.

1633 *Gateshead Church Bks.* in Heslop *Northumb. Words* (1892) s.v., It. p[d] for one bunting and two sparres to a yeat and the makeing it, 4s. 4d.]

Strong balks of timber placed crosswise in the shaft of a mine serving to divide it into compartments and to support the guides on which the cages run, etc. (Rarely *sing.*)

1839 URE *Dict. Arts* 971 In quadrant shafts the buntons cross each other towards the middle of the pit. *Ibid.* 986 The compartment intended for the upcast is made air-tight at top, by placing strong buntons or beams across it. **1860** *Eng. & For. Mining Gloss.* (ed. 2) 51 *Buntons*, strong balks of timber. **1883** *Encycl. Brit.* XVI. 450/2 Pieces of timber parallel to the end-pieces (buntons or dividings) are fixed across the shaft, and serve to stay the wall-plates. **1888** GREENWELL *Gloss. Coal Trade Northumb. & Durh.* (ed. 3) 81 A cistern which stands on a strong oak bunton. **1967** *Gloss. Mining Terms* (B.S.I.) IX. 6 *Bunton*, one of a series of horizontal beams set at intervals across a shaft to support rigid guides, cables and pipes and, in rectangular shafts, to act as struts.

bunty, *a.* [f. BUNT *sb.*[2] + -Y[1].] Of grain: infected with bunt.

1855 MORTON *Cycl. Agric.* I. 363/2 A year or two since, we saw a very bunty crop in the fields of a first rate farmer. **1896** *Chron.* 19 Dec. (E.D.D.), If he sowed bunty seed he should expect to reap bunty crop.

bunya (bʌ·nyă). Also **bunya-bunya.** [Native name.] An Australian tree, *Araucaria bidwillii*, which bears a cone of great size yielding a nutritious vegetable pulp. Also *attrib.*

1843 L. LEICHHARDT *Let.* in J. D. Lang *Cooksland* (1847) 82 The bunya-bunya tree is noble and gigantic. **1844** *Ibid.* 89 The kernel of the Bunya fruit. **1844** *Port Phillip Patriot* 25 July 4/4 The Bunya Bunya or *Araucaria* on the seeds of which numerous tribes of blacks are accustomed to feed. **1875** *Encycl. Brit.* II. 310/2 The Bunya-Bunya pine, found on the mountains of Eastern Australia. **1885** R. C. PRAED *Head Station* 179 In dense scrub..where the stately bunya-branches drooped. **1887** J. MATHEW in Curr *Austral. Race* III. 161 In laying up a store of bunyas, the Blacks exhibited an unusual foresight. **1934** T. WOOD *Cobbers* xvi. 191 Orange groves and bunya nuts. **1957** P. WHITE *Voss* iii. 60 They had gone..as far as the elbow and the bunya bunya.

Bunyanesque (bʌnyăne·sk), *a.* [f. *Bunyan* (see below) + -ESQUE.] Of, pertaining to, or characteristic of either (*a*) John Bunyan (1628–88), the English writer, or (*b*) Paul Bunyan, the legendary American hero.

a. 1888 F. THOMPSON in *Merry England* Nov. 6 A rustic alehouse bench, upholstered with cushions by a daring flight of Bunyanesque fancy. **1965** M. DRABBLE *Millstone* 19 Their eye-for-an-eye and Bunyanesque attention to the detail of offence.

b. 1952 D. HOFFMAN *Paul Bunyan* v. 149 Mr. Steven's Bunyanesque maxims for business executives. **1961** B. MALAMUD *New Life* (1962) 59 Enormous leaves,..he had thought, until the old woman told him..that they were from some Bunyanesque species of maple tree.

bunyip (bʌ·nyip). Also **bunyup. 1.** The Aboriginal name of a fabulous monster inhabiting the rushy swamps and lagoons in the interior of Australia. Also *attrib.*

1848 W. WESTGARTH *Australia Felix* 391 Certain large fossil bones..have been referred by the natives..to a huge animal of extraordinary appearance, called in some districts the Bunyup, in others the Kianpraty, which they assert to be still alive. **1888** *Athenæum* 14 Jan. 47/2 There are plenty of sea-gods, little better than salt-water kelpies or marine bunyips. **1891** *Coo-ee* 275 When a black fellow disappears, it is generally understood that the Bunyip has got hold of him. **1894** A. ROBERTSON *Nuggets*, etc. 61 A weird boom, from bittern or bunyip, came from the swamp. **1916** J. B. COOPER *Coo-oo-ee!* xiv. 199 Were not chivalrous men in these venal days as extinct as the bunyip? **1936** M. FRANKLIN *All that Swagger* i. 7 The wild Murrumbidgee sinking into the Bunyip Hole..would dissolve.

2. An impostor. Hence *attrib.* *Obs.*

1852 MUNDY *Antipodes* (1858) ix. 215 Bunyip became, and remains a Sydney synonyme for *impostor*, *pretender*, *humbug*, and the like. **1853** W. C. WENTWORTH in H. Parkes *Fifty Years Austral. Hist.* (1892) I. 41 A mushroom, a Brummagem, a bunyip aristocracy.

‖ **buon fresco** (bwǫn fre·sko). [It., lit. 'good fresco'.] = FRESCO *sb.* 2 a.

1874 *Amer. Cycl.* VII. 483/1 Painting designs in colors ground in water and mixed with lime upon the freshly laid plaster..was called by the Italians *buon fresco*, or the true fresco. **1957** *Encycl. Brit.* IX. 835/2 In the 15th century, Italian painters distinguished between *fresco secco*—the pigments being mixed with egg as a medium, and *buon fresco*—the pigments being mixed with water and laid on the fresh wet plaster.

buoy, *sb.* **3.** (Further examples.)

1870 *Chambers's Jrnl.* 15 Jan. 34/2 This 'buoy-shifting' is a duty which calls forth all the skill and energy of the officers and men. *a* **1877** KNIGHT *Dict. Mech.*, *Buoy-safe*, a metallic body divided into compartments, by which it is braced, and having water-tight doors opening to the inside. **1921** *Dict. Occup. Terms* (1927) § 731 Buoy master.

buoyage. Add to def.: Also, buoys collectively, a series of buoys which mark out a channel.

1855 in OGILVIE Suppl.

buoyancy. Add: **4.** *attrib.*, as *buoyancy chamber*, *garment*, *tank*.

1930 *Engineering* 17 Jan. 69/1 The buoyancy chamber..is divided by a longitudinal watertight bulkhead along the centre line. **1959** A. HARDY *Fish & Fisheries* i. 1 Many of the fish are swollen..as their air-bladders (buoyancy chambers) are violently distended. **1962** *B.S.I. News* May 9/2 Buoyancy garments are being used in increasing numbers by yachtsmen, canoeists and other 'weekend' sailors. **1928** C. F. S. GAMBLE *Story of North Sea Air Station* xviii. 309 In order to dive a submarine her commander fills the buoyancy tanks with sea-water.

bur, burr, *sb.* Add: **5. b.** An ornamental veneering wood or veneer, esp. of walnut, containing knots. Also *attrib.*, as *burr-walnut*. Cf. *BURL *sb.*[1] 4 b.

1885 *Spons' Mech. Own Bk.* 357 Walnut burrs are best cut with scissors. **1901** *Tradesman's Catal.* 1 Bedroom Suite, in solid American Walnut and Burr. **1908** *Daily Report* 5 Sept. 8/2 A burr-walnut armoire. **1923** *Daily Mail* 23 Jan. 1 Sideboard in oak..with finely figured panels of burr walnut. **1938** *Times* 17 Oct. 8/5 The cabinet work is in two shades of burr walnut.

8. bur-oak (examples).

1815 D. DRAKE *Cincinnati* ii. 82 The most valuable timber trees are the..bur oaks. **1876** *Encycl. Brit.* IV. 704/2 The bur oak (*Q. lobata*). **1888** *Ibid.* XXIII. 808/1 The burr oak (*Q. macrocarpa*) has almost as wide a range as the white oak. **1962** *Victoria & Albert Museum Internat. Art Treasures Exhib.* 55/1 A Louis XV Tortoiseshell Casket on burr oak.

buran (bū·rān). [a. Russ. *burán*, ad. Turki *boran*.] In the steppes, a snowstorm, esp. one accompanied by high winds; a blizzard.

1886 *Encycl. Brit.* XXI. 76 Gales with snow (*burans*, *myatels*), lasting from two to three days..are especially dangerous to man and beast. **1898** J. T. BEALBY tr. *Hedin's Through Asia* I. viii. 99 These burans or snow-hurricanes come on with startling suddenness. **1936** P. FLEMING *News from Tartary* VI. xi. 305 When we went on the evening sky was overcast and presently the *buran* hit us.

Burano (burā·no). The name of an island near Venice, used attrib. in *Burano (point) lace*, a needle-made lace having a net ground, resembling Alençon and Brussels lace.

1865 MRS. B. PALLISER *Hist. Lace* iv. 54 (*caption*) Burano. **1869** *Ibid.* (ed. 2) iv. 50 (*caption*) Burano Point. **1874** *Queen Lace Book* viii. 22 In the latest specimens of Burano Point, the ground is a heavy fabric formed of square meshes. **1934** *Discovery* Sept. 255/2 Her sleeve ruffles are of the finest Burano point lace.

Burberry (bʌ̄·ɹbəri). The registered trade mark (in form *Burberrys*) distinctive of cloth or clothing made by the firm of Burberrys Ltd.; *spec.* a raincoat made by this firm.

1903 *Burberry Yarn Proofs* 3 Burberry Yarn Proof Coatings. **1909** *Trade Marks Jrnl.* 13 Oct. 1717 *Burberrys*. Cloths and Stuffs of Wool, Worsted and Hair. The Firm trading as Burberrys, 30 to 33 Haymarket..Outfitters. **1915** T. CASSELS *Men Knotted Heart* iii. 37 Somebody presented to him once a Burberry. **1919** C. ORR *Glorious Thing* iii, She was clad in a drooping burberry coat. **1944** A. CHRISTIE *Towards Zero* 89 'It's raining, you know.' 'I know, I've got a Burberry.'

burble, *sb.*[1] Transfer †*Obs.* to senses in Dict. and add: **3.** A murmurous flow of words.

1898 G. W. STEEVENS *With Kitchener to Khartum* 33 An inarticulate burble more like the sound of a distant railway train than any known form of human speech. **1909** J. H. SKRINE *Pastor Ovium* 140 What I listened to was a burble of platitudes in a honeyed voice. **1923** *Blackw. Mag.* Dec. 767/1 The low burble of petition-reading and the murmurous flow of false evidence were still proceeding.

4. *attrib.*, as **burble point,** the point at which the smooth flow of air over an airfoil is broken up.

1918 COWLEY & LEVY *Aeronautics in Theory & Experiment* ii. 25 At the burble point the lifting force drops sharply and just as quickly rises again. **1918** W. E. DOMMETT *Dict. Aircraft* 11 *Burble point*, that point on the lift curve of a wing which is reached when the angle of incidence has become so great that the stream lines change from a steady to a fluctuating and eddying state, causing the lift to fade and the drag to increase. **1920** *Conquest* I. 439/2 The angle at which the loss of lift is first noticed is called the 'critical angle' or 'burble point'.

burble, *v.*[1] Add: **2. a.** To speak murmurously; to 'ramble' *on*. **b.** *trans.* To say (something) murmurously or in a rambling manner. Also *transf.*

[**1871** 'LEWIS CARROLL' *Through Looking-Glass* i. 22 The Jabberwock..Came whiffling through the tulgey wood, And burbled as it came!] **1891** KIPLING *Light that Failed* viii, You only burble and call me names. **1906** B'NESS VON HUTTEN *What became of Pam* III. iv, Miss Wantage..began to burble, and then to roar. **1920** MULFORD *J. Nelson* vii. 67 'Forty feet of rope an' a sycamore tree,' burbled Smitty. **1921** *Blackw. Mag.* July 31/2 A sleepy dinner it was. We burbled a few plans for next day, and fell asleep by the fire. **1934** T. E. LAWRENCE *Let.* 6 Aug. (1938) 813 You send me a sensible working-man of a letter..and I burble back in this unconscionable way. **1965** *Parade* 15 May, 'I think they just called our flight number,' burbled Carter.

Hence **bu·rbling** *vbl. sb.*; **bu·rbler** *sb.*

1920 *Blackw. Mag.* July 44/2 They..hold his answering burblings to be the divine voice of Kali. **1923** KIPLING *Land & Sea Tales* 139 Now are you satisfied, you burbler? **1934** *Punch* 7 Mar. 280/2 *Lady Placidia* was a confirmed burbler, and if at times she is in danger of exceeding her burbling allowance, she remains entirely lovable and amusing.

Burdekin (bʌ̄·ɹdekin). The name of a river in the eastern part of Queensland, Australia; used attrib. in *Burdekin plum*, an Australian timber tree (see quots.); *Burdekin vine*, an Australian vine, *Vitis (Cissus) opaca*, bearing large edible tubers, also called *round yam*.

1889 J. H. MAIDEN *Useful Native Plants Austral.* 67 *Vitis Opaca*, 'Burdekin Vine', 'Round Yam'. *Ibid.* 599 *Spondias pleiogyna*, 'Sweet Plum', or 'Burdekin Plum'. **1902** *Encycl. Brit.* XXXII. 108/1 Other orders..furnish equally serviceable, large-sized timber, particularly the following:..'Burdekin Plum' (*Pleiogynium Solandri*), [etc.]. **1963** *Austral. Encycl.* I. 174/2 Burdekin plum (*Pleiogynium cerasiferum*) is a very common tree in coastal Queensland from about the Burnett River northwards.

burden, *sb.* Add: **2. a.** *the white man's burden*: a rhetorical expression for the responsibility of the white for the coloured races.

1899 KIPLING *White Man's Burden* vi, Take up the White Man's burden—Ye dare not stoop to less. **1911** H. G. WELLS *New Machiavelli* I. iv. 128 We were all..Imperialists also, and professed a vivid sense of the 'White Man's Burden'. **1922** JOAD *Common-Sense Theology* 135 Little nationalised Jingoes who are ready enough to adopt any parrot cry such

as 'The White Man's Burden', or 'The Kultur of the Fatherland'. **1966** *Observer* 17 Apr. 10/6 In the seventies we can and should lay down the White Man's Burden with a clear conscience.

6. (Earlier and later examples.)

1825 J. NICHOLSON *Operat. Mechanic* 329 In proportion to the quantity of lime and ore that is added to the standard quantity of the coke, the furnace is said to carry a greater or less burthen. **1944** *Jrnl. Iron & Steel Inst.* CL. 419 We are operating at present on a burden of 60% brown Northampton ore and 30% carbonate ore. **1952** *Gloss. Welding & Cutting Metals* (B.S.I.) 43 *Burden*, the layer of melt and fused metal above the welding zone in submerged-arc welding.

VI. 13. *pl.* The floor boards of a rowing boat; *side burdens*, the side seats in a rowing boat.

1857 P. COLQUHOUN *Comp. 'Oarsman's Guide'* 29 The flooring is termed *burthens*. *Ibid.* 31 *Side burthens* are extra thwarts laid in provisionally to carry sitters: *burthens* are the bottom boards. **1898** ANSTED *Dict. Sea Terms* 37 In boats the burdens are the footwalings.

burdensome, *a.* Add: **2.** Capable of carrying a good burden. *U.S. Obs.*

1763 *Boston Even. Post* 9 May (Th.), A very good and burthensome Schooner for sale. **1834** M. SCOTT *Cruise Midge* (1836) II. vi. 210 'Pull under the stern of that large ship.'..'A fine burdensome craft that, sir.' **1834** HOFFMAN *Winter in West* (1835) II. xxvii. 74 The burthensome steamboats from New-Orleans reach here at the lowest stage of the river.

burdensomely, burthen-, *adv.* For *a* **1873** J. S. MILL (O.) read: **1848** J. S. MILL *Pol. Econ.* II. v. vi. § 3. 425.

bure[1]. Delete †*Obs.* and add earlier and later examples.

1585 *Edinb. Test.* XIV. 230 b (D.O.S.T.), Ane steik of buir. **1591** *Ibid.* XXIII. 298 Of braid claytht for cloiking callit bure. **1853** C. BRONTË *Villette* III. xxx. 23 He wished to counsel me, whenever I wore it [*sc.* a coloured dress] to do so..as if its material were 'bure'. **1909** *Westm. Gaz.* 4 Sept. 15/2 A favourite material with some of the best French designers is the coarse sacking of which nuns make their dresses...It is called bure; the finer makes of it are combined with velvet. *Ibid.* 9 Oct. 15/1 The fabric of the season is undoubtedly the coarse bure which takes its name from the stuff used for the costumes of monks in the Middle Ages.

‖ **bure**[2] (biū·re). [Fijian.] A Fijian house.

1843 T. WILLIAMS *Jrnl.* 21 Jan. (1931) I. 140 He was sitting in a *bure* with some of the 'old ones' whilst his yanggona was being prepared. **1963** A. BURNS *Fiji* 6 There are typical Fijian houses, the *bure*, with thatched roofs and sides. **1970** *Observer* (Colour Suppl.) 15 Feb. 24/1 When you've gotten your bearings in Nadi Town head out to native *bures* (thatched huts).

bure, var. *BUER.

bureau. Add: **1. b.** A chest of drawers. *U.S.*

1819 *St. Louis Enquirer* 15 Sept. (Th.), Look in the bureaus and trunks of modern men of fashion and see the number of coats, waistcoats, pantaloons, &c. **1837** *Southern Lit. Messenger* III. 333 The only bureau and the only rocking-chair were in Charlotte's room. **1890** MRS. CUSTER *Following Guidon* 253 Our *bureaus* were always called *bureaus*; but they were in fact packing boxes.

c. *attrib.*

1843 'R. CARLTON' *New Purchase* I. 199 The bureau-top was consecrated to Bibles and Hymn-Books. **1854** M. J. HOLMES *Tempest & Sunshine* xxiii. 328 Safely stowed away at the bottom of her bureau-drawer..was a big daguerreotype. **1867** MRS. WHITNEY *L. Goldthwaite* vi, Bits of most delicate ferns..filled a great shallow Indian china dish upon her bureau-top. **1939** T. S. ELIOT *Old Possum's Pract. Cats* 18 He likes to lie in the bureau drawer. **1969** R. JAFFE *Fame Game* xii. 184 When she was out of the house her mother went through her bureau drawers.

2. *spec.* In modern use, an office or business with a specified function; an agency for the co-ordination of related activities, the distribution of information, etc.

1926 Information bureau [see *INFORMATION 8]. **1932** L. GOLDING *Magnolia St.* II. xi. 417 The family was to make all arrangements at Garden's Travel Bureau. **1962** WODEHOUSE *Service with a Smile* v. 70 A typewriting bureau, eh? ..One of those places full of machines and girls hammering away at them like a lot of dashed riveters.

bureaucracy. Add: (Earlier examples.) Also †*bureaucratie*, *bureau-ocracy*.

1818 LADY MORGAN *Florence Macarthy* II. i. 35 Mr. Commisioner..represented the *Bureaucratie*, or office tyranny, by which Ireland has been so long governed. **1834** *Tait's Mag.* I. 180 The trade-ocracy and bureau-ocracy must now..prepare themselves. **1837** J. S. MILL in *Westm. Rev.* XXVIII. 71 That vast net-work of administrative tyranny..that system of *bureaucracy*, which leaves no free agent in all France, except the man at Paris who pulls the wires. **1843** R. R. MADDEN *United Irishmen* II. xvii. 367 This 'bureau-cracy' was an inveterate evil of Ireland, in the early part of Earl Grey's administration.

bureaucrat. (Earlier example.)

1842 MOTLEY *Let.* 10 Jan. (1889) I. 94, I don't know whether I have ever described to you the great bureaucrat [Count Nesselrode] of the great autocrat [Emperor of Russia].

bureaucratize (biu°rǫ·krătəiz), *v.* [f. BUREAUCRATIC + -IZE.] *trans.* To govern by, or transform into, a bureaucratic administration or system. So **bureau:cratiza·tion**; **bureau·cratized** *ppl. a.*; **bureau·cratizing** *vbl. sb.*

1892 A. C. MORANT tr. *Schäffle's Impossibility Soc. Democr.* 39, I do not by any means contemplate the bureaucratizing and nationalization of industry. **1916** *Trans. Soc. Engineers* 112 It is essential to guard against the bureaucratisation of science. **1920** *Times Lit. Suppl.* 2 Sept. 559/2 The centralization of power..was..accompanied by a progressive system of bureaucratization. **1920** *Contemp. Rev.* Dec. 882 It proved itself to be a mere bureaucratization of distribution. **1945** 'G. ORWELL' in *New Saxon Pamphlets* III. 39 The radio was bureaucratised so early in its career that the relationship between broadcasting and literature has never been thought out. **1961** B. R. WILSON *Sects & Society* I. ii. 39 The bureaucratisation of the movement had begun. **1966** *Economist* 22 Oct. 377/2 American life has become too bureaucratised, too centralised.

burette. Add: **2.** (Later examples of the form *buret*.)

1945 E. A. BRECHT in R. A. Lyman *Amer. Pharmacy* I. i. 20/2 Burets are used for the very accurate measurement of volumes, as in volumetric analysis. **1964** R. J. HENRY *Clinical Chem.* vi. 115 Micro burets fall into two general classes. The first is the Rehberg Buret.

3. A cruet used for the wine or the water at the Eucharist.

1856 J. C. ROBINSON *Invent. Objects Mus. Ornamental Art* 57 Plated Flagon or Burette. *Ibid.* App. 11 Pair of Sacramental Cruets, or 'Burettes'. **1871** O. SHIPLEY *Gloss. Eccl. Terms* 79/1.

burg. Add: **2.** A town or city. (Also *berg*, *burgh*.) *U.S. colloq.*

1843 *Spirit of Times* 25 Mar. 43/1 Two 'individs' in this 'burg' will give our friend Greer 'the run of his teeth' whenever he visits New York. **1846** S. SMITH *Theatr. Apprent.* 151 The stranger..also disembarked at the same burgh. **1888** *Battle Creek Weekly Jrnl.* 8 Feb., If successful, it will make a lively burg in the next few years. **1901** 'J. FLYNT' *World of Graft* ii. 30 All cities where fakirs show up strong in the streets and tramps have a quarter of their own, are recognized by grafters in general as 'open bergs'. **1903** A. ADAMS *Log Cowboy* vi. 71 The wagon and a number of the boys went into the burg. **1944** D. RUNYON *Runyon à la Carte* (1946) vi. 92 There are no more opportunities for you in this burg.

burger (bʌ̄·ɹgəɹ). A familiar shortening of *HAMBURGER. Also used as a terminal element, e.g. in *beefburger*, *porkburger*, etc., usu. denoting a roll, sandwich, etc., containing the foodstuff specified in the first element. orig. *U.S.*

1939 *Amer. Speech* XIV. 154. **1941** *Word Study* Nov. 7/1 A favorite broth of the word-brewers..forms like *-burger*, *-krieg*, *-teria*. **1946** *Amer. Speech* XXI. 88 *Burger*, hamburger sandwich. 'Burger steak' is hamburger steak. **1960** *Observer* 28 Feb. 13/4 Recently the Hamburger has become just a 'burger', and there are 'beefburgers', 'chefburgers', 'cheeseburgers', 'eggburgers' and even 'kingburgers'. **1960** *Guardian* 10 Dec. 6/6 Lines of frozen food, such as fish, vegetables, steakburgers. **1966** T. CAPOTE *In Cold Blood* ii. 72 'Perry, baby,' Dick said, 'you don't want that burger. I'll take it.' Perry shoved the plate across the table.

burgh, var. *BURG 2.

burgher, *sb.* Add: **4.** *S. Afr.* A citizen of the Cape Colony, the Natal or Transvaal Republics, or the Orange Free State before the advent of British rule. Also *attrib.*

1879 B. FRERE et al. *Speech Cape Town* 24 To encourage the brave Burghers and Southey's Volunteers to do their best to put an end to the war on the northern border of the Colony. **1879** (*title*) Interview between..Sir Bartle Frere, and the Deputation from the Boer Committee, At the Burgher Camp, April, 1879. **1881** F. R. STATHAM *Blacks, Boers, & British* iii. 43 To save them from annihilation at the hands of the Free State burghers, the Basutos were, in 1868, taken under British protection. **1898** KRUGER in *South Africa* 1 Jan. 11/1 Burghers and fellow-countrymen, the times are such that a wise and judicious development of our sources of aid requires the most earnest consideration. **1958** L. VAN DER POST *Lost World of Kalahari* iii. 56 Every burgher was permitted, if not actually enjoined, to shoot a Bushman on sight.

burgheress (bʌ̄·ɹgərès). [f. BURGHER *sb.* + -ESS[1].] A female burgher.

1901 *Westm. Gaz.* 7 May 2/2 Stout German burgheresses patrolling the streets in whaleboned uniforms. **1912** R. BROUGHTON *Between Two Stools* ix, A tender-hearted burgher-ess of Catterwell.

burgherly (bʌ̄·ɹgəɹli), *a.* [f. BURGHER *sb.* + -LY[1].] Of, belonging to, or characteristic of a burgher.

1762 P. MURDOCH tr. *Büsching's Syst. Geogr.* IV. 445 The town consists of what are called free and burgherly houses, as also of free and burgherly inhabitants, and of the garrison. **1906** *Daily Chron.* 19 Feb. 3 There met in Bruges..a princely and burgherly splendour. **1969** *Daily Tel.* (Colour Suppl.) 16 May 40/2 The quietly smiling creators of such burgherly tosh.

burglar, *sb.* Add: **2.** *burglar-alarm* (earlier examples); also *attrib.*; *burglar-proof* (earlier example).

1840 in M. D. Leggett *Index Pat. Inventions* (1874) 173 Burglar-alarm, L. E. Denison, Saybrook, Conn., Oct. 22, 1840, 1,835. **1877** *Telegraphic Jrnl.* V. 19/1 The application of the magneto-electric current for..burglar alarms. **1889** *Cent. Dict.*, *Burglar-alarm lock*, a lock having an attachment which when set will sound an alarm if the bolt is improperly moved. **1963** *B.S.I. News* Apr. 10/1 Bad news for burglars is BSI's decision to give burglar alarm systems some close attention. **1856** *Spirit of Times* 13 Dec. 247/3 Manufacturers of..Burglar Proof Safes for stores and dwelling houses.

burglar (bʌ̄·ɹgləɹ), *v.* [f. the sb.] **a.** *trans.* To steal (goods) or rob (a place) as a burglar. **b.** *intr.* To commit a burglary. Cf. BURGLE *v.* Hence **bu·rglared, bu·rglaring** *ppl. adjs.*

1890 MRS. HUNGERFORD *Born Coquette* II. xiii. 128 He certainly burglared Nan. He broke into the house..and stole her away. **1890** *Mercury* (Tasmania) 27 Dec., A news agency ..was burglared yesterday morning. **1896** 'MARK TWAIN' in *Harper's Mag.* Aug. 345/2 They used to hear about him robbing and burglaring now and then. **1909** *Daily Chron.* 31 Aug. 1/2 'Raffles' remains a more endeared and far more possible character than the burglaring 'Duke'. **1919** W. T. GRENFELL *Labrador Doctor* (1920) iv. 76 He..got alarmed when busy burglaring. **1928** *Sunday Dispatch* 22 July 12/4 The burglared shop.

burglarize, *v.* Add: (Earlier and later examples.) Also *intr.*

1871 *Southern Mag.* Apr. (Schele de Vere), The Yankeeisms donated, collided, and burglarized, have been badly used up by an English magazine-writer. **1876** *Congress. Rec.* July 4419/2, I found that the house of a lady moving in good society had been burglarized. **1926** J. BLACK *You can't Win* xi. 142 It was built to be burglarized. **1947** *Jrnl. Crim. Law & Criminol.* Nov.–Dec. 319, I tried to resist the urge to get outside and burglarize.

Hence **bu·rglarizing** *vbl. sb.*

1872 SCHELE DE VERE *Americanisms* 655 In like manner the burglar's occupation has been designated as burglarizing. **1888** *Merchant Traveler* (Farmer), 'What have you been doing for a living lately?'..'Burglarizing.'

burgle (bʌ̄·ɹg'l), *sb.*[2] [f. BURGLE *v.*] An act of burgling; a burglary.

1889 *Punch* 26 Jan. 37/2 The Burgle Song. **1898** *Tit-Bits* 26 Feb. 418/1 This afternoon we did a little burgle on our own account.

bu·rgoma:stership. [-SHIP.] The office of burgomaster.

1896 *Daily News* 28 Apr. 3/5 Dr. Lüger..proceeded to the Town Hall and communicated his renunciation of the Burgomastership. **1909** *Daily Chron.* 25 May 6/7 The burgomastership of Breslau.

burgoo. Also **burgou(t).** Add: **2.** A soup or stew made with a variety of meat and vegetables, used especially at outdoor feasts. *N. Amer.*

1743 J. ISHAM in *Publ. Hudson's Bay Rec. Soc.* (1949) XII. 132 There is a Sort of mawse which grows upon the Rocks, which is of a Brownish Colour, which the Indians Eats frequent, they wash itt clean, then Boil itt for a considerable time till itts tender, then mixing it with Ruhiggan Burgoe or other Victuals, and Reckon itt Good Eating. **1853** J. L. MCCONNEL *Western Char.* 363 Around a burgou pot..he shone in meridian splendor. *Note.* A kind of soup, made by boiling all sorts of game with corn, onions, tomatoes, and a variety of other vegetables. **1906** *Washington Post* 16 June 6 Who, excepting Kentuckians and their favored Southern friends and kinsmen, has ever really known the bliss of genuine burgoo? **1906** H. D. PITTMAN *Belle of Bluegrass C.* xix. 272 The old timers smacked their lips as they,..in fancy,..tasted the burgout. **1966** *Times Lit. Suppl.* 22 Sept. 879/4 Burgoo..is traditionally served at a barbecue for thousands of Kentucky colonels during the weekend of the Kentucky Derby.

Burgundy. Add: **2. c.** A red wine resembling the Burgundy of France, made in other wine-growing districts, esp. Australian and Californian 'Burgundy'.

1886 *Catal. Colonial & Ind. Exhibition, W. Australian Court* 53 One dozen Burgundy, vintage 1882. *Ibid.*, *S. Australian Court* 36/2 Number of gallons manufactured last vintage 18,000; of Burgundy type. **1888** *W. & A. Gilbey's Wine List* 11 Australian Wines—..Burgundy Red. **1894** C. F. OLDHAM *California Wines* 8 Of natural dry red wines, the Burgundy types are perhaps the best. **1907** *Daily Chron.* 12 Oct. 9/4 You find 'Burgundy' imported into this country from Australia, California, Spain, Italy and Hungary.

d. A shade of red of the colour of Burgundy wine.

1881 *Household Words* 9 July 215/2 New colours in gloves are steel and burgundy. **1891** *Daily News* 30 Dec. 2/2 Every tone of red, from brilliant military scarlet to the deep claret and burgundy tones. **1942** R. CHANDLER *High Window* (1943) i. 5 A big solid cool-looking house with burgundy brick walls.

6. *Burgundy mixture*, a preparation of soda and copper sulphate used for spraying potato-tops to destroy fungi.

1894 *Board of Agric. & Fisheries, Leaflet No. 23* By the use of Bordeaux or Burgundy mixture, the spores of the fungus are prevented from germinating. **1951** *Dict. Gardening* (R. Hort. Soc.) II. 849/2 Burgundy mixture is made like Bordeaux mixture but washing soda is used instead of lime.

burhel, var. *BHARAL.

buri (bū°·ri). [Tagalog.] The talipot palm. Also *attrib.*

1890 J. FOREMAN *Philippine Isl.* xxii. 404 Tall *buri* palms. *Ibid.* 423 The canoe..was gaily decked out with festoons of *buri*. **1911** C. B. ROBINSON in *Philippine Jrnl. Science*

June 106 Three different parts of the leaves of the buri palm are used for hats. **1926** *Commerce & Industry Jrnl.* (Manila, P.I.) Nov. 6 In the trade, 'buri hats' refer only to those made of the bleached buri leaf, as distinguished from those made from the unbleached leaves which are known as the 'Arayat hats'.

burial. Add: **5. burial-case** *U.S.* (examples); **burial club** = *burial-society*; **burial permit** *U.S.*, a certificate authorizing the burial of a deceased person; **burial-society** (examples).

1851 C. CIST *Cincinnati* 191 Every possible variety in which iron can be cast, from a butt hinge to a burial case. **1870** 'MARK TWAIN' in *Wks.* (1900) XIX. 259, I am talking about your high-toned, silver mounted burial-case. **1848** MRS. GASKELL *Mary Barton* I. vi. 109 The town must bury him; he had paid to a burial club..but by a few weeks' omission, he had forfeited his claim. **1910** A. BENNETT *Clayhanger* I. x. 76 To be an established subscriber to the Burial Club was evidence of good character and social spirit. **1888** *St. Louis Globe Democrat* (Farmer), Yesterday's Burial Permits. **1908** *Daily Chron.* 22 May 1/3 The case..has developed sensational features through the refusal of the authorities to grant a burial permit. **1812** in C. S. Davies *Hist. Macclesfield* (1961) IV. 180 Sick and burial societies which are the germ of revolution. **1850** *Rep. Extramural Sepulture* 63 in *Parl. Papers* XXI. 575 Mr. George Downing, a mechanic and secretary to a burial society,..represents the sentiments prevalent among persons of his own class on this subject.

6. As second element in comb. with sbs., as *house-*, *ship-*, *tree-*, *urn-burial*, etc.: see under the first element.

Buriat (bū·riăt). Also **Buryat.** A member of a Mongolian people inhabiting the borders of Lake Baikal, Siberia; their language. Also *attrib.* or *adj.* = **Buria·tic** *a.*, belonging to the Buriat people.

1836 *Penny Cycl.* VI. 29 The Buriates constitute one of the three great divisions of nations into which the Mongols are divided. **1888** *Encycl. Brit.* XXIV. 1/2 *Ural-Altaic languages*..*Mongolic*..*Buriat* or *Siberian Mongolian*, east and west of Lake Baikal. **1895** *Westm. Gaz.* 23 Nov. 7/1 The thirty Mongolian Buriats who, on the Emperor's initiative, have been brought to St. Petersburg for their education. **1902** *Ibid.* 28 Feb. 2/1 Disguised as a Buryat, and accompanied by one Buryatic Cossack..I set off. **1903** *Daily Chron.* 18 Nov. 3/2 He made the experiment..in disguise as a Buriat Lama. **1947** D. DIRINGER *Alphabet* II. v. 318 The three principal [Mongolian] dialects, Khalkha, Kalmuck and Buriat, do not differ much. **1957** *Encycl. Brit.* IV. 410/2 *Buriat* (*Buryat*) *Mongol Republic*, an autonomous Soviet Socialist Republic in Asiatic Russia, created as a region in Jan. 1922, and as a republic in Sept. 1923.

buried, *ppl. a.* Add: **1. b.** *buried treasure* (later example); also *fig.* and *attrib.*

1851 (*title*) Buried treasures. Part I. The law of liberty: a letter on toleration, by John Locke. Part II. Milton 'on the civil power in ecclesiastical causes'. **1886** A. SYMONS *Introd. Browning* Pref., Criticism is..a hazel-switch for the discovery of buried treasure. **1896** H. JAMES in *Cosmopolis* I. 390 Vereker's secret..the string the pearls were strung on, the buried treasure. **1904** E. NESBIT *Phoenix & Carpet* ii. 42 'I believe it is a buried treasure,' he cried...I wonder what you would say if you suddenly came upon a buried treasure? **1964** *Times* 11 Feb. 12/7 Catalogues..have a buried-treasure charm.

3. *Comb.*, as *buried-alive* (also *fig.*).

1851 H. MELVILLE *Moby Dick* II. xxxvi. 244 Poor, buried-alive Tashtego. **1871** GEO. ELIOT *Let.* 6 June in J. W. Cross *Life* (1885) III. xvi. 131 A resurrection of a buried-alive friendship.

burin. Add: **3.** *Archæol.* A flint tool with a point like that of a chisel.

1916 H. F. OSBORN *Men of Old Stone Age* iv. 308 The primitive graving-tool, or *burin*..which we know was designed by the Crô-Magnon artists for their early engravings on stone. **1921** M. C. BURKITT *Prehist.* iv. 69 Gravers or *burins* fall into two main groups, the difference being determined by an observation of the working edge. In one variety this is straight like a screw-driver,..the other variety, being curved like a gouge, is known as the gouge type. **1939** W. B. WRIGHT *Tools & Man* ii. 15 The burin or graver of flint. **1957** *Encycl. Brit.* II. 239/2 *Aurignacian*..The tool types include various kinds of steep-ended scrapers,..busked gravers or burins and split-base bone points.

burka[1] (bʌ̄·ɹkă). Also **burqa, boorka, bourkha, burko.** [Hind. (from Arabic) *burqa'*.] A long enveloping garment worn in public places by Muslim women to screen them from the view of men and strangers.

1836 E. W. LANE *Manners Mod. Egyptians* I. i. 51 The *boor'ko'*, or face-veil, which is a long strip of white muslin, concealing the whole of the face except the eyes, and reaching nearly to the feet. **1884** KIPLING *Plain Tales* (1888) 149 He went..clad in a *boorka*, which cloaks a man as well as a woman. **1905** HOLMAN HUNT *Pre-Raphaelitism* I. 386 His pleadings to be allowed to satisfy his eyes as to the features hidden under the black *burko*. **1905** *Daily Chron.* 16 Nov. 5/4 The Begum was clothed in strictest purdah costume, and wore a burka. **1927** *Blackw. Mag.* May 755/1 A burka with netted eyeholes. **1929** *Daily Express* 15 Jan. 1/1 The Queen [of Afghanistan] is wearing the boorka—a heavy shapeless garment which effectually hides her beauty. **1955** G. BAND *Road to Rakaposhi* x. 119 Their gay red and yellow burqas, which cover the wearer from head to foot and permit only a latticed window before the eyes, made a brilliant patch of colour. **1959** *Times* 21 Sept. 13/4 The bourkhas which veiled the students from top to toe.

burka[2] (bʌ̄·ɹkă). Also **burkha, boorka, burqa.** [Russ.] A long Caucasian cloak of felt or goat's hair.

1898 J. T. BEALBY tr. *Hedin's Through Asia* I. viii. 103 A Caucasian *burkha* (cloak or mantle). **1916** J. Y. SIMPSON *Self-Discov. Russia* 138 Caucasian Cossack officers with..black burka sweeping behind them. **1924** *Blackw. Mag.* Feb. 180/2 The ample folds of the ankle-long burqa.. **1927** *Ibid.* Sept. 299/1 Broad-built men, like giants in their hairy boorkas and astrakhan caps.

burke, *v.* Add: **2.** (Earlier example.) Also, to evade, to shirk, to avoid.

1835 J. A. ROEBUCK *Dorchester Labourers* 6/1 (*note*), The reporters *left it out*...Those who spoke in favour of the poor men, were what the reporters call *burked*. **1921** *Daily Colonist* (Victoria, B.C.) 3 Apr. 32/1 He had before him a clear issue—and he burked it. **1931** *Economist* 4 July 25/1 The problem, as it concerns the investor, of the holding company and its accounts is one which it is not wise to burke. **1953** R. GRAVES *Poems* 4 Socrates and Plato burked the issue.

Burke (bʌ̄ɹk), *sb.* A colloquial designation of 'A General and Heraldic Dictionary of the Peerage and Baronetage of the United Kingdom', the first edition of which, issued in 1826, was compiled by John Burke (1787–1848). Also *transf.*

1848 THACKERAY *Van. Fair* li. 451 Her ladyship is of the Kingstreet family, see Debrett and Burke. **1901** F. H. BURNETT *Making of Marchioness* I. iii. 41 'If we were not in Debrett and Burke, one might be reserved about such matters,' poor Lady Claraway wrote. **1966** D. FRANCIS *Flying Finish* i. 7 Bastard I was not: not with parents joined by bishop with half Debrett and Burke in the pews.

burk(e, var. *BERK.

burl, *sb.*[1] Add: **4. b.** An overgrown knot or excrescence in walnut and other woods, used in veneering; also, a log or piece of timber containing such a knot; also, a veneer made with this wood. Cf. *BUR, BURR, *sb.* 5 b.

1885 *Spons' Mech. Own Bk.* 356 French walnut burls. *Ibid.*, The ash burls..avoid the necessity of a stay log by having a sufficient part of the trunk on which the burl grew left to serve for this purpose.

burl (bʌ̄ɹl), *sb.*[2] [App. var. north. dial. *birl* a whirring sound, a rapid twist or turn, etc. (see *E.D.D.* and *Sc. Nat. Dict.*); cf. BIRL *v.* to revolve, rotate.]

1. ? Roundness, fullness. *poet.*

1876 G. M. HOPKINS *Wr. Deutschland* (1918) st. 16 The burl of the fountains of air. **1883** —— *Serm. & Writings* (1959) 154 Therefore in that 'cleave' of being which each of his creatures shews to God's eyes alone (or in its 'burl' of being[uncloven) God can choose countless points in the strain (or countless cleaves of the 'burl') where the creature has consented, does consent, to God's will in the way above shewn.

2. Also **birl.** An attempt, a try, a test; esp. in phr. *to give it a burl*, to make an attempt. *Austral.* and *N.Z. colloq.*

1917 *N.Z.E.F. Chrons.* 16 May 137/2 So up they [pennies] went and spinning well And betters cried, 'Fair "burl"!' **1933** *Bulletin* (Sydney) 8 Nov. 20/4 He [a whale] decided to give the launch a birl. He butted it frequently with his nose. **1939** K. TENNANT *Foveaux* IV. i. 348 How about giving it a burl? **1947** D. M. DAVIN *For Rest of Lives* xlv. 227, I thought I'd give it a burl. And I made it, got clean away. **1955** D. NILAND *Shiralee* 108, I'm going to give Eucla a birl. Want to get there as soon as I can. **1969** *Private Eye* 23 May 14 Might as well give this mead another burl.

bu·rlap, *v.* [f. the sb.] *trans.* To wrap round with burlap. So **bu·rlapped** *ppl. a.*

1908 *Sears, Roebuck Catal.* 415 Each article..is securely wrapped, burlapped or crated. **1927** *Ladies' Home Jrnl.* Dec. 109/2 Evergreens should be balled and burlapped, if they are to be successfully transplanted. *Ibid.*, Insist on carefully balled and burlapped stock.

burlesque, *sb.* Add: **B. 3. a.** The concluding portion of a Negro minstrel entertainment, containing dialogue and sketches. *U.S.*

1857 *Porter's Spirit of Times* I. 344/3 The nightly concert which precedes the burlesque. *c* **1864** *Corsican Twins* t-p. (D.A.E.), An Ethiopian Burlesque..As performed by Griffin & Christy's Minstrels. **1957** *Oxf. Compan. Theatre* (ed. 2) 556/1 The third part [*sc.* of a Negro minstrel show] consisted of after-pieces—farces, comic opera, or burlesque.

b. A variety show, frequently featuring strip-tease. Also *attrib.* orig. and chiefly *U.S.*

1870 T. A. BROWN *Hist. Amer. Stage* 66/2 She..reappeared during the winter of 1868, at the Fifth Avenue Opera House, New York, in burlesque. **1870** O. LOGAN *Bef. Footlights* 563 There are numberless people..who are utterly unable to see any difference between the dancing of a ballet-girl and the caperings of a jigging burlesque woman. **1887** 'M. CORELLI' *Thelma* II. II. iv. 217 You'd better not go to the Brilliant Theatre to-night—it's only a burlesque, and is sure to be vulgar and noisy. **1895** *N.Y. Dramatic News* 12 Oct. 5/3 The only burlesque show in town [*sc.* Chicago]. **1930** C. WITTKE *Tambo & Bones* iv. 158 Cheap burlesque 'girl' shows. **1956** *N.Y. Times* 10 Dec. 33/6 [The License Commissioner] felt burlesque was 'synonymous with the strip tease and the dialogue of unvarnished salaciousness'.

Burley, burley[3] (bʌ̄·ɹli). *U.S.* [? Personal name.] An American tobacco, of which there are two varieties, red and white. Also *attrib.*

1881 *Southern Planter* in B. W. Arnold *Tobacco Ind.* (1897) 35 The White Burley produced in the west has now thoroughly substituted our dark grades. **1900** *Yearbk. U.S. Dept. Agric.* 435 The White Burley is entirely air cured, except in exceedingly damp weather, when wood fires may be used. **1909** *Westm. Gaz.* 10 Feb. 5/2 From the Blue Grass also comes most of what is called the Burley, another fine quality, which is made into chewing tobacco for the American and German. **1952** *Economist* 29 Nov., Over 155,000 growers of burley tobacco voted on whether they wished to continue..price support for their 1953 crop.

burley, var. *BERLEY.

burling (bʌ̄·ɹliŋ), *vbl. sb.*[2] and *ppl. a. poet.* [f. *burl* vb., var. BIRL *v.* to revolve, rotate.] Whirling, rotating.

1874 G. M. HOPKINS *Jrnls. & Papers* 11 Aug. (1959) 251, I marked the bole, the burling and roundness of the world. **1886** —— *Poems* (1918) 72 Ringlet-race on burling Barrow brown.

burly, *sb.* (Later example in sense 'bluster'.)

1876 G. M. HOPKINS *Wr. Deutschland* (1918) st. 27 In wind's burly and beat of endragoned seas.

Burma. In full *Burma cheroot*. A kind of cheroot manufactured in Burma and with a peculiar aroma.

1891 KIPLING *City Dreadf. Night* iv. 26 'Honest Bombay Jack' supplies nothing but Burma cheroots and whisky. **1899** —— *From Sea to Sea* I. xxi. 422 He should bring with him thousands of cheroots..Singapur is the last place on the line where you can buy Burmas. **1969** *Fine Cigars* (Harrods Ltd.) 7 Burma cheroots.

Burman (bʌ̄·ɹmæ̆n), *a.* and *sb.* Also †**Birman, Burmhan.** [f. *Burma* + -AN.] Of or belonging to, a native of, Burma; also, (of) the Burman language.

1800 M. SYMES *Embassy to Ava* 7 The Birmans..became masters of Ava. **1814** F. CAREY *Gram. Burman Lang.* 1 The Burman alphabet consists of forty-five letters. *Ibid.* 39 This mark (') is the only stop the Burmans have. **1821** HIRAM COX (*title*) Journal of a Residence in the Burmhan Empire. **1826** *Mod. Traveller: Birmah* 1 The vast region of Asia..comprehending the Birman empire, the kingdoms of Siam and Anam. *Ibid.* 10 The Birman, which is spoken in Ava and Arracan. **1835** *Penny Cycl.* IV. 439 Among the vegetable productions of the Birman forests the teak holds the first place. **1883** A. P. PHAYRE *Hist. Burma* 1 The Burman people..were formed into a nation by the union of Mongoloid tribes. **1965** B. SWEET-ESCOTT *Baker St. Irreg.* viii. 243 The Burmans, or the peoples of the Irrawaddy delta and the great central plain of Burma. *Ibid.*, The Burma Defence Army under the command of a number of young Burman officers trained by the Japanese in Japan.

Burmese (bʌ̄ɹmī·z), *a.* and *sb.* Also †**Birmese.** [f. *Burma* + -ESE.] **A.** *adj.* Of or pertaining to Burma or its inhabitants or their language.

1823 A. JUDSON in F. Wayland *Mem.* (1853) I. 256 Translated from the Burmese original. **1827** H. G. BELL in *Constable's Misc.* IX. (*title*) A Narrative of the Late Military and Political Operations in the Birmese Empire. **1842** A. JUDSON *Burmese Lang.* 5 The Burmese alphabet consists of ten vowels..and thirty two consonants. **1876** *Encycl. Brit.* IV. 551 The Burmese empire with its present limits contains no maritime districts. **1920** *Blackw. Mag.* Oct. 519/2 Adorable Burmese babies. **1965** [see sense B. 2].

b. In the specific names of minerals, fauna, etc., found in Burma, as *Burmese naphtha*; **Burmese cat,** a breed of domestic cat (see quots.); also *ellipt.*; **Burmese squirrel,** a squirrel native to Burma and Tenasserim, closely allied to *Sciurus caniceps*; **Burmese worm,** a mulberry-feeding silkworm found domesticated in India.

1939 I. M. MELLEN *Pract. Cat Book* i. 3 Various long-haired cats of the Orient, Persian, Angora, Russian, Burmese, etc. *Ibid.* i. 26 The Burmese Cat..is of medium size, muscular, with slim legs, the hind legs being long and tilting the body slightly downward. **1962** *House & Garden* Dec. 6/3 The fashionable Burmese, which..combine the fluffy coat of the Persian with Siamese colouring and eyes. **1858** FOWNES *Man. Chem.* (ed. 7) 655 The Burmese naphtha (Rangoon tar)..consists principally of liquid homologues of marsh gas, associated with small quantities of hydrocarbons of the benzol-series, [etc.]. **1887** *Encycl. Brit.* XXII. 60 *Bombyx arracanensis*, the Burmese worm. *Ibid.* 438 Burmese squirrel.

B. *sb.* **1.** A native of Burma; also *collect.*

1824 T. EVANS *Disp.* 11 Oct. in W. James *Naval Hist. Gt. Brit.* (1837) VI. 465 The Burmese jumping overboard to save themselves. **1842** A. JUDSON *Burmese Lang.* 9 The character °..is reckoned among the consonants, by the Burmese. **1876** *Encycl. Brit.* IV. 551 The sugar cane appears to have been long known to the Burmese. *Ibid.* 552 The Burmese in person have the Mongoloid characteristic.. **1876** 'J. BRADLEY' *Trav. & Sport* ii. 29 These people, like all other Burmese, were vivacious and noisy. **1885** G. C. WHITWORTH *Anglo-Ind. Dict.* 54/2 *Burman*, a native of Burma, a Burmese. **1937** C. V. WARREN *Burmese Interlude* xxi. 141 The 'victories' claimed by the Burmese in their riots during 1930 against the Chinese and Indians.

2. The language of Burma.

1842 A. JUDSON *Burmese Lang.* 6 The pronunciation of the cerebrals and the dentals, though different in the Sungskrit, is the same in the Burmese. **1894** R. F. ST. JOHN *Burmese Reader* Pref., It is a great mistake to suppose that Burmese ..can be learnt properly from a phrase-book. **1965** B.

Sweet-Escott *Baker St. Irreg.* viii. 246 None of us could speak Burmese, but..he proposed to read us the Burmese text of a letter he had received.

burmite (bʊ̄·ɹmait). [f. *Burma* + -ITE¹.] A variety of amber found in Burma, used by the Chinese in the manufacture of objects of art.

1893 Bruhl tr. *Helm's Note on Burmite* in *Rec. Geol. Surv. India* XXVI. 62 My researches..have rendered it evident, that we have to deal with a remarkable fossil resin, which differs essentially from all fossil resins known hitherto, and to which I have given the name of Burmite. **1893** F. Noetling *Ibid.* 31 The Burmese amber is totally different from..amber (Succinite)...In concordance with Dr Helm's suggestion, the name of Burmite is adopted for the new fossil resin. **1932** *Times Lit. Suppl.* 12 May 343/2 The deep-red opalescent Burmite from Burmah.

burn, *sb.*³ Add: **1. b.** *spec.* An instance of burning the vegetation on land as a means of clearing it for cultivation. (Cf. Burning *vbl. sb.* 8 a and *e.) **c.** A place where the trees or brush have been burned; a clearing in the woods made in this way. *N. Amer.*, *Austral.*, and *N.Z.*

1792 J. Belknap *Hist. New Hampshire* III. 132 Much depends on getting what is called a *good burn*, to prepare the ground for planting. **1834** S. Moodie *Jrnl.* 28 Apr. in *Six Years in Bush* (1838) 92 [Canada] A great burn, during which the wind rose so high as to endanger my shanty. **1839** E. Holmes *Rep. Explor. Aroostook River* 69 Very little ploughing is as yet done, since most of the crops are raised on a 'burn'. **1854** J. M. Richmond *Let.* 19 Feb. in *Richmond–Atkinson Papers* (1960) I. 143 He wants to preserve a magnificent red pine..that a good burn would probably have killed. **1868** *Amer. Naturalist* Oct. 468 They [*sc.* deer] resort always to a recent burn, when grass and weeds are just shooting again and are soft. **1905** W. Baucke in D. M. Davin *N.Z. Short Stories* (1953) 85 Suddenly..where one had looked for fresh wonders of forest beauty, spread a settler's recent burn! **1950** *N.Z. Jrnl. Agric.* Sept. 215/3 In places where a good primary burn was not obtained stumps and tree trunks [still] litter the ground.

d. A manifestation of anger or frustration; usu. in phr. *slow burn*, a display of slowly-mounting anger; the act or state of gradually becoming enraged. *colloq.* (orig. *U.S.*).

1938 *Daily News* (Los Angeles) 2 Aug. 18 The saga of the Slow Burn Man, or 'Killer' [Edgar] Kennedy. **1951** 'J. Tey' *Daughter of Time* xiv. 184 Just a nice polite reasonable Act for him to swallow and like it. I bet he did a slow burn about that one. **1969** H. Carvic *Miss Seeton draws Line* v. 93 A slow burn began inside Bob.

e. A 'smoke'; tobacco, esp. a cigarette; *to have a burn*: to smoke a cigarette. *slang.*

1941 G. Jeffery in *Penguin New Writing* II. 89, I just sat there, having a burn, dressed to go home. **1943** Baker *Dict. Austral. Slang* (ed. 3) 17 *Burn*, a cigarette. 'To twist a burn', to roll a cigarette. **1956** A. Thorne *Baby & Battleship* II. 89 Rolling cigarettes for 'a quiet burn'. **1964** *Times* 26 June 14/6 To a non-smoker [in prison] it is..an advantage to acquire a steady income of 'burn'.

f. The provision of thrust by the engine of a spacecraft. Also *attrib.*

1965 K. W. Gatland *Spacecraft & Boosters* II. 266/2 First stage thrust, 172,000 lb, burn time 146 sec; second stage thrust, 7,500 lb, burn time, 165 sec. **1968** *Radio Times* 19 Dec. 41/1 The rocket burn that takes Apollo away from Earth orbit. **1969** *Guardian* 8 Mar. 1/1 Relieved ground controllers heard that the burn had gone as planned.

burn, *v.*¹ Add: **2. c.** *to burn out*, spec. (*a*) of an electrical valve, fuse, etc.; (*b*) of a space rocket. Also, *to burn oneself out*: to exhaust one's strength (by over-exertion) (1937 in Partridge *Dict. Slang* s.v.).

1931 *Boys' Mag.* XIV. 171/1 To operate the valve with increase of voltage does not solve the problem, for this means that the valve (which is already partially ruined) will soon cease to function, or in other words 'burn out'. **1955** R. Bannister *First Four Minutes* iv. 48, 7½ miles is much too far for me—I want to be a miler and I shall burn myself out. **1958** *Listener* 11 Dec. 992/1 The American rocket..burns out over the Sahara Desert.

d. Of nuclear fuel: to undergo fission or fusion. Also const. *up*.

1958 *Times Lit. Suppl.* 20 June 337/2 When a uranium pile burns—physicists use the word habitually and correctly—useful energy is released under control. **1959** *New Scientist* 29 Jan. 242/2 So that the element will not buckle as it burns up, a metal frame is welded around it to brace it in the middle.

3. a. (Examples of usages referring to the ears.)

c **1563** *Jack Juggler* s. B2v., I feel a vengeable burning in my left ere. **1607** W. S. *Puritaine* I. s. B4v., I warrant my Kinsman's talking of me, for my left eare burnes most tyrannically. **1738** Swift *Polite Conv.* i. 70 Miss, didn't your Left Ear burn last Night?..Because..you were extoll'd to the Skies. **1868** Dickens *Let.* 25 May (1938) III. 651, I dine with Dolby..and if your ears do not burn from six to nine this evening, then the Atlantic is a non-conductor. **1967** G. Fallon *Rendezvous in Rio* xvii. 146 'Richard! We were just talking about you!'..'Is that so? No wonder my ears were burning!'

4. d. Of the engine of a spacecraft: to consume fuel and provide thrust. (Cf. sense *2 c.)

1964 K. W. Gatland *Spacecraft & Boosters* I. 278/2 The Thor first stage burned for approximately 160 sec, propelling the vehicle to an altitude of 41 miles. **1969** *Daily Tel.* 8 Mar. 1/2 The ascent stage's engine will burn and place the two returning astronauts..back into orbit round the moon.

8. b. *to burn up*: also *fig.*, to irritate, to upset, to enrage. *U.S. slang.*

1923 H. C. Witwer *Fighting Blood* i. 23, I certainly burnt Ajariah Stubbs up that day..I fell asleep..and I give a guy pepsin bismuth and a stiff argument, when all he says he asked for was a plain chocolate soda. **1931** D. Runyon *Guys & Dolls* (1932) ii. 34 Naturally this crack burns Handsome Jack up quite some. **1934** J. O'Hara *Appointment in Samarra* (1935) vii. 211 Ed is plenty burned up, and, my God, I don't blame him. **1935** S. Lewis *It can't happen Here* xi. 124 What burns me up is the fact that..7 per cent of all the families in the country earned $500 a year or less. **1945** F. Cheyney *You can always Duck* ii. 39, I told him the story, an' was he burned up!

d. To 'spend' or use freely; esp. *to have* (money, etc.) *to burn*, to have in abundance or to spare. orig. *U.S.*

1896 Ade *Artie* 106 Two years ago he was on his uppers and now he's got money to burn. **1897** *Congress. Rec.* Mar. 400/1 Mr. Simpson..No; I have not time to burn. **1904** *Louisville Courier Jrnl.* 2 July 5 She has..already had literary experience to burn. **1909** 'O. Henry' *Options* (1916) 76 The gentleman of the family had owned plantations and had slaves to burn. **1910** W. M. Raine *B. O'Connor* 30 We'll keep an eye on the gambling hells and see who is burning up money. **1911** H. Quick *Yellowstone N.* ix. 240 The gall of my swearing against these boys to burn. **1915** 'Ian Hay' *First Hundred Thous.* xvi. 220 You will get baccy and cigarettes to burn out there. **1917** H. A. Vachell *Fishpingle* x, Why not? I have money to burn. **1928** *Sunday Express* 6 May 6 People in the States have 'money to burn'.

9. b. *to burn the earth* or *wind*: to go at full speed. *U.S.*

[**1881** G. W. Romspert *Western Echo* 164 The first day the mustangs will burn the prairie.] **1891** 'O. Thanet' *Otto the Knight* 219 An' we all ayfter 'im...Didn't he burn the wind, though! **1903** A. Adams *Log Cowboy* iii. 37, I was half a mile in the lead burning the earth like a canned dog. **1910** W. M. Raine *B. O'Connor* 20 So burn the wind, and go through the car on the jump. *Ibid.* 57 When he finds out how the horse he's after is burning the wind, his suspicions grow stronger.

c. *to burn one's bridges*: see *Bridge *sb.*¹ 1 e.

13. c. *to burn on*: to add (a part) to an injured or incomplete casting by running in a stream of molten metal.

1888 *Lockwood's Dict. Terms Mech. Engin.* 53 A sand mould of the portion to be burned on is made and placed in proper juxtaposition to the old casting in the bed of the foundry floor.

f. *to burn off*: to clear (land) for cultivation by burning the vegetation; to burn dry or rank vegetation (tussock, etc.). Also *absol.* (Cf. *Burn *sb.*³ 1 b and Burning *vbl. sb.* 8 a and *e.) *N. Amer.*, *Austral.*, and *N.Z.*

1843 C. A. Dillon *Let.* 16 Jan. (1954) 17 The fern was all burnt off by the surveyors. **1852** S. Moodie *Roughing It* 90 Moodie and Jacob had chopped eight acres during the winter, but these had to be burnt off and logged-up before we could put in a crop of wheat. **1860** G. Duppa in S. S. Crawford *Sheep & Sheepmen of Canterbury* (1949) v. 46 Burn off portions of the run for winter feed to destroy tuft grass. **1932** K. S. Prichard in Murdoch & Drake-Brockman *Austral. Short Stories* (1951) 189 He cleared and grubbed, burnt off, and cultivated his land. **1959** A. McLintock *Descr. Atlas N.Z.* 39/1 The Maoris had long been aware that much of the bush country was more fertile [than the open country], for they burned off patches for cultivation. **1966** 'J. Hackston' *Father clears Out* 87 The men started burning off opposite and the heavy clouds of smoke hid the tree for a while.

g. To vulcanize (india-rubber) by mixing it with sulphur or metallic sulphides and heating it.

1900 Sadtler *Handbk. Indust. Chem.* (ed. 3) 106 In vulcanizing by the first process, that of 'burning', as it is termed, the crude caoutchouc is mixed with varying amounts of sulphur.

h. To utilize the nuclear energy of (uranium, etc.).

1946 *Rep. Internat. Control Atomic Energy* (Dept. of State, U.S.) III. i. 35 Such power reactors would 'burn' the active materials and require replenishing from time to time. **1949** *Britannica Bk. Year 1948* 686/2 *Burn*, to utilize the atomic energy of (nuclear fuel). **1957** *Encycl. Brit.* II. 649/2 Some losses could be tolerated and still leave a neutron to initiate another fission of uranium, thus producing a chain reaction which would continue to burn uranium. *Ibid.* 651/2 It is also possible to burn plutonium in the presence of U-238 and make more plutonium. **1962** *Newnes Concise Encycl. Nuclear Energy* 278/1 To achieve power balance, it is necessary also that at least about 1 per cent of the fuel is 'burned' before being lost from the system.

14. f. To swindle. (See *Sc. Nat. Dict.*)

In quot. 1655 the sense may be 'to suffer'.

1655 R. Baillie *Let.* (1842) III. 290 Our people were so ill-burnt, that they had no stomach for any farder medling. [*a* **1700** B. E. *Dict. Cant. Crew*, Burnt the Town, when the Soldiers leave the Place without paying their Quarters.] **1808** Jamieson, *Burn*, to deceive, to cheat in a bargain. **1844** *Philad. Spirit of Times* 19 Aug. (Th.), Two negro burners were arrested in the act of trying to burn two Pottsville boatmen with a plated chain worth about fifteen cents. **1926** J. Black *You can't Win* ix. 106 If you'd burnt Shorty for his end of that coin, you'd have been here just the same. **1969** *Sunday Truth* (Brisbane) 16 Mar. 39/2, I figured I'd burn the guy for a thousand.

15. (Later examples.)

1851 C. Cist *Cincinnati* 250 Henry Albro, who was burnt out some months since, on Front street, has recently put up new veneer and saw mills. *a* **1861** T. Winthrop *John Brent* (1883) i. 7 They had been burnt out, they had been cleaned out, they had been drowned out.

16. *to burn a hole in one's pocket* (earlier example).

1857 Trollope *Three Clerks* II. ix. 198 How was she to give him the purse? It was burning a hole in her pocket till she could do so.

18. *to burn out*: to fuse by means of an electric current. Cf. *burn-out* b, s.v. *Burn-³.

1924 *Discovery* June 83/2 The Germans were using some objectionable form of frightfulness that burnt out the magneto of French planes flying across German territory.

19. *slang.* To smoke (tobacco). Cf. *Burn *sb.*³ 1 e.

1958 F. C. Bowen *Sea Slang* 20 *Burning*, smoking in the training ships. **1958** F. Norman *Bang to Rights* iii. 82 The more [tobacco] we got the more we used to burn.

burn-. Add: **3.** With the verb+adv., as **burn-off** (cf. *Burn *v.*¹ 13 f); **burn-out**, (*a*) a complete destruction by fire; also = *Burn *sb.*³ 1 c; (*b*) *Electr.*, the fusing of a wire or other electric conductor by excess of electric current; also *attrib.*, as *burn-out fuse*, *alloy*, one that melts at a comparatively low temperature and serves as a safeguard against damage by excess of current; (*c*) (the moment of) final consumption of fuel by a space rocket, etc., also *attrib.*; **burn-up**, (*a*) the consumption of fuel in a nuclear reactor; (*b*) *slang*, a ride on a motor-cycle, etc., at an extremely high speed (cf. Scorch *v.*¹ 3).

1861 W. Morgan *Jrnl.* 27 Feb. (1963) iii. 28 Not an over excellent burn off—there having been of late a good deal of wet. **1869** J. May *May's Guide to Farming in N.Z.* 18 In due time we fired it [the bush], and had the satisfaction of having a clean burn-off. **1933** *Bulletin* (Sydney) 6 Dec. 28/2 A good burn-off gets rid of a lot of rubbish in the form of insects and other pests, but it also destroys the organic matter (*i.e.*, humus) without which no soil can be cropped successfully. **1903** *Daily Chron.* 29 June 7/5 It has been a burn-out of three floors and roof destroyed. **1907** *Installation News* Mar. 4/1 Incipient fires and burnouts, due to the earthing of high voltage systems on building fronts, etc. **1940** W. Stegner in *Atlantic Monthly* June 774/1 Even without shoes he would have run across burnouts, over stretches so undermined with gopher holes that sometimes he broke through to the ankle. **1941** —— in *Harper's Mag.* Jan. 160/1 The topless Ford lurched, one wheel at a time, through the deep burnout. **1952** *Jrnl. Brit. Interplan. Soc.* XI. 10 Ideal performance of multi-stage vehicles is secured if..the burnout weights of each stage form a geometric progression. **1953** *Time* 14 Sept. 89/2 The three tons of fuel lasted less than three minutes. At 'burnout', Carl was at 75,000 ft. **1957** *Spaceflight* I. 64/2 Four wings provide the lift necessary for controlled flight and four small fins at the rear are used for steering after burn-out. **1954** R. Stephenson *Introd. Nucl. Engineering* vii. 276 As a reactor continues to operate, the fissionable material is gradually used up and the reactivity may decrease accordingly. This is known as fuel depletion, or burnup. **1959** *New Scientist* 29 Jan. 239/1 A major aim is to obtain a large 'burn up'—in other words to use up as large a proportion of the fuel as possible between refuelling operations. **1961** *Guardian* 18 Mar. 2/3 If I was going for a real burn-up, you wouldn't have caught me. **1963** A. Prior *Z Cars Again* iii. 26 Ton-up boys were doing early morning burn-ups at the Turntable Roundabout.

Burne-Jones (bɜ̄·ɹn,dʒōᵘ·nz). The name of E. C. *Burne-Jones* (1833–98), English artist and designer, used *attrib.* to designate art or a type of beauty suggestive of or characteristic of the work of Burne-Jones.

1892 E. Lytton *Let.* 15 Nov. in E. Lutyens *Blessed Girl* (1953) ix. 172 Lady Grandby is a beauty of the Burne-Jones type. **1958** J. Cannan *And be a Villain* i. 25 Her Burne-Jones beauty. **1967** *Vogue* June 78 The Burne-Jones type was pale and lissom and romantic, hollow-cheeked and with haunted eyes.

Hence **Burne-Jo·nesian** *a.* and *sb.*

1895 G. B. Shaw *Theatres in Nineties* (1932) I. 17 He has a beautiful costume, mostly of plate-armor of Burne-Jonesian design. **1908** D. H. Lawrence *Let.* 26 Oct. (1962) I. 32 Shall I make her longer or shorter, fatter or frailer, a Burne-Jonesian or a Moore? **1925** W. Deeping *Sorrell & Son* xxv. 241 His sisters, pale, sweet, Burne-Jonesian. **1965** G. McInnes *Road to Gundagai* ix. 139 The injection of a few rich commercial corpuscles into our pale Burne-Jonesian fluid.

burner. Add: **1. c.** A swindler. *U.S.* ? *Obs.* (Cf. *Burn *v.*¹ 14 f.)

1838 *Lexington Observer & Reporter* 3 Nov., He pulls out his pocket book, it is seized by the burner who makes off with it. **1842** *Philad. Spirit of Times* 15 Jan. (Th.), The burners make better plots than most of our dramatists. **1844** [see *Burn *v.*¹ 14 f]. **1845** *Congress. Globe* 6 Jan. App. 118/1 The Empire Club [of New York]..consisted of gamblers, pickpockets, droppers, burners, thimble-riggers and the like.

4. b. *Welsbach incandescent gas burner*, a burner devised by Auer von Welsbach for producing an incandescent light by means of a mantle (see Mantle *sb.* 5 g) and Bunsen burner. Also called the *Auer*, *incandescent*, or *Welsbach burner*.

1894 [see Incandescent *a.* 1 d]. **1902** *Encycl. Brit.* XXX. 260/2 Incandescent Burners. The invention of the Welsbach mantle places at the disposal of lighthouse authorities the means of producing a light of high intensity.

c. In a gas cooker, the part containing the hole or holes through which the gas passes before combustion. See also *Cook *v.*¹ 1 b.

1885 *Army & Navy Co-op. Soc. Price List* Jan. 368 Gas stoves..lever taps to hot-plate, and extra simmering burner in centre of ring burner, with brass handle tap. **1963** *Good Housek. Setting Up Home* viii. 113 A cooker with four burners and a grill is the usual choice for family cooking. *Ibid.*, Oven burners may be arranged at the sides or back.

burnet, *sb.*³ Add: **3.** In the names of species of moths belonging to the genus *Zygæna*; cf. *burnet-moth* in 2.

1775 M. Harris *Engl. Lepidoptera*..*Catal. Moths & Butterflies* 15 English Names..Burnet, sphinx..six spots of scarlet, inferiors scarlet. Burnet, companion..Brown clouded, inferior yellow wing'd with brown. **1832** J. Rennie *Butterfl. & Moths* 22 The Melilot Burnet..The Trefoil Burnet..The Five Spot Burnet. **1921** *19th Cent.* Feb. 275 The Zygaenidae, or Burnets. **1961** R. South *Moths* 334 The Six spot Burnet..is the most generally common of our Burnets.

Burnham (bɜ·nəm). The name of Harry Lawson Webster Levy-Lawson, first Viscount *Burnham* (1862–1933), who was chairman of the Standing Joint Committee of Education Authorities and Teachers, set up on the 12th Sept. 1919; hence applied to the scale of salaries, etc., recommended by this committee and periodically revised.

1920 *Jrnl. Education* Jan. 18 The notable modesty of the proposals of the Burnham Committee on Salaries. **1921** *Ibid.* July 426 The Burnham scales of salaries. **1929** *Daily Mail* 7 June 7 The National Union of Teachers..at South Shields is demanding the Burnham scale 3 in place of the lower scale paid by the local education committee. **1945** *Educ. Syst. Eng. & W.* (H.M.S.O.) 37 *Burnham scales*, scales of salaries payable to all teachers in publicly maintained schools and institutions, the scales having been fixed by the Burnham committees, consisting of representatives of L.E.A.s and teachers.

burning, *vbl. sb.* Add: **8. b.** Also, the quantity of bricks burnt at one operation.

1901 *Pall Mall Gaz.* 9 Dec. 8/2 No two burnings (a burning is a kiln full of bricks) have exactly the same shade of colour.

e. **burning off.** *Austral.* and *N.Z.* See *Burn *v.*¹ 13 f.

1800 P. G. King in *Hist. Rec. N.S.W.* (1896) IV. 188 For falling, burning off, and breaking up an acre of ground. **1844** C. Chapman *Let.* 30 Nov. in A. Drummond *Married & Gone to N.Z.* (1960) iv. 69 Xmas is the time for burning off. **1860** in A. F. Ridgway *6 Years from Auckland* 75 The burning off should be driven as late in the season as it is safe to do so. **1891** R. Wallace *Rural Econ. Austral. & N.Z.* xv. 232 Falling [of the bush] is done in winter, and burning off in the middle or end of summer, when everything is withered and the weather dry. **1909** D. Walker *Wren of Austral.* 281 This was the hot season and all 'burning off' strictly forbidden.

10. *Comb.* **burning-ghat**: see *Ghaut, Ghat 4.

Burnsian (bɜ·ɹnziən), *a.* and *sb.* [f. the name of Robert *Burns* (1759–96), Scottish poet + -IAN.] **A.** *adj.* Of or relating to Burns, his works, or his style. **B.** *sb.* An admirer of Burns or his works. So **Burnsia·na** [-ANA], things connected with Burns; **Bu·rnsite** = *Burnsian *sb.*

[**1866** J. Mackie (*title*) Bibliotheca Burnsiana. Life and works of Burns: title pages and imprints of the various editions in the private library of James M'Kie, Kilmarnock.] **1874** J. Gibson (*title*) The Burns calendar: a manual of Burnsiana; relating events in the poet's history, names associated with his life and writings, a concise bibliography, and a record of Burns relics. **1904** *Daily Chron.* 21 Aug. 3/1 The accumulative literature of Burnsian topography. **1909** *Westm. Gaz.* 12 Jan. 12/1 Collectors of Burnsiana. **1909** *Daily Chron.* 14 July 6/4 All Burnsites—common or peculiar. *Ibid.*, Devotees of the Burnsian cult. **1920** *Glasgow Herald* 28 July 6 A representative gathering of Scottish Burnsians.

Burnside, burnside (bɜ·ɹnsaid). *U.S.* [f. the proper name.] A style of beard such as that affected by General Burnside (1824–81), consisting of a mustache, whiskers, and a clean-shaven chin (*Cent. Dict.* Suppl. 1909). Freq. *pl.* Also *attrib.* Cf. *side-burn* (Side *sb.*¹ 27).

1875 *Cincinnati Enquirer* 6 July 2/1 His whisker was of the Burnside type, consisting of mustache and 'mutton-chop', the chin being perfectly clean. **1881** I. M. Rittenhouse *Jrnl.* in *Maud* (1939) i. 36 The older one has lovely burn-sides. **1907** *Outing* (U.S.) L. 279 Such various patterns of ornamental whiskers as the 'Piccadilly Weeper' (No. 2), the 'Burnside' [etc.]. **1930** *Publishers' Weekly* 8 Feb. 679 In the days of copper-toed boots and burnsides..our grandfathers were buying this book.

burnt, burned, *ppl. a.* Add: **2. a.** (Further examples of *fig.* use and of spelling *burned-out*.)

1887 *Manch. Guardian* 31 May 5 The burned-out company of the Opéra Comique. **1908** Hardy *Dynasts* III. vii. ix. 347 This is my burnt-out hour. **1917** T. S. Eliot *Prufrock* 24 The burnt-out ends of smoky days. **1919** F. Hurst *Humoresque* 256 She looked up at him with a tired, a burned-out, an ashamed smile. **1926** E. Bowen *Ann Lee's* 93 Mrs Pym was a fair, burnt-out young woman of twenty-five. **1969** *Listener* 3 Apr. 472/1 Joss Ackland as Danton, the not entirely burnt-out volcano.

c. Of a leper: cured (see quot. 1959), esp. in *burnt-out case*. Also *fig.* (freq. with influence of sense 2 a).

1959 G. Greene *Congo Jrnl.* 10 Feb. (1961) 42 Leprosy cures where disease has been arrested and cured only after the loss of fingers or toes are known as burnt-out cases. **1961** —— (*title*) A burnt-out case. **1961** *New Statesman* 24 Feb. 318/1 Yet had this priest not burned at the stake in 1634, he could be discovered in the Congo as a nobler but still walking 'burnt-out case'. **1961** *Encounter* XVI. 79 The burnt-out leper is not too obtrusive. **1961** *Times Lit. Suppl.* 2 June 340/2 His construction 'seemed to have broken down, and no career in life lay open' to him. A burnt-out case. **1968** R. Harris *Nice Girl's Story* ii. 18 He was probably a burnt-out case—perhaps his wife had perished in the gas chambers. His interest in life, except to get through it somehow, might be dead.

4. c. **burnt sienna** (examples).

1844 Thackeray *May Gambols* in *Wks.* (1899) XIII. 441, I have so often wandered before with burnt-sienna ploughboys. **1846** Dickens *Pict. from Italy* 41 Two burnt-sienna … **1946** R. Lehmann *Gipsy's Baby* 78 The expanses of burnt sienna mud.

e. **burnt cream** = **crème brûlée*.

1723 J. Nott *Cook's & Confectioner's Dict.* C. 209 Burnt Cream. Take Yolks of four or five Eggs, beat them well in a Stew-pan..Set the Cream on the Furnace [etc.]. **1969** *Observer* 12 Jan. 32/8 Burnt Cream (layers of custard and clotted cream under caramelised sugar).

6. c. (See quot.)

1909 *Daily Chron.* 23 Feb. 7/2 The buyer should beware of..burnt furs. Such furs have been over-dyed, and the hair will soon become rusty and fall out.

7. **burnt almond**, an almond enclosed in burnt sugar; hence, a flashing shade of brown; **burnt cork**, cork that has been burnt so that it can be used for blackening the face, hands, etc.; freq. *attrib.*, as *burnt-cork artist*, a Negro minstrel (see Negro 3); **burnt-cork** *v. trans.*, to blacken with burnt cork; **burnt feed** *Austral.* (see quot.); **burnt stuff** *Austral.* (see quot. 1945).

1859 *Family Friend* III. 327/1 Put a burnt almond..in the centre. **1893** *Encycl. Pract. Cookery* 143/1 Pound the Burnt Almonds..in a mortar. **1895** *Cassell's Fam. Mag.* June 554/1 [A bonnet] of burnt-almond straw. **1913** 'Ascott R. Hope' *Half & Half Tragedy* 121 Treating me to twopence worth of 'burned almonds'. *c* **1800** C. Mathews in Mrs. Mathews *Mem. C. M.* (1838) I. xv. 305 Camel's hair pencils, hare's feet, whiting, burnt corks. **1840** Burnt cork [see Cork *sb.*¹ 1]. **1869** *Porcupine* 3 July 123/1 The Theatre Royal has within the last week or two been patronized by a company of 'burnt cork' performers. **1873** 'Ascott R. Hope' *Night before Holidays* (1874) 98 We had one very fine false beard, and … **1888** E. James *Amat. Negro Minstrel's Guide* 10 A pair of legs such as Nelse Seymour had … burnt-cork artist. **1893** P. H. Emerson (*title*) Signor Lippo, burnt-cork artiste. **1885** … *Stage* 3 We..burnt-corked our hands and faces. **1905** *Daily News* 6 May 2/6 For money making it's necessary to wear masks or to burnt-cork the face. **1848** H. W. Haygarth *Bush Life Australia* vii. 73 A patch of 'burnt feed' (as the young herbage is called which springs up..where the old grass has been set on fire). **1854** J. Bonwick *Wild Gold Digger* 9 Some neighbouring bearded digger turns round and condescendingly remarks, that it is only the 'burnt stuff'. **1945** Baker *Austral. Lang.* v. 94 *Burnt stuff*, a stratum of iron-hard rock or compacted clay at rock encountered during digging.

8. **burnt** (colour), a deep shade of yellowish brown; so **burnt-coloured** adj.

1896 *Daily News* 1 July 8/7 Hats of 'burnt' straw, this being the technical name of a deep shade of yellowish brown. **1899** *Westm. Gaz.* 29 Apr. 3/3 Burnt coloured straw. **1933** *Weekly Dispatch* 25 Feb. 14 Colours: Navy..Nut, Mastic and Burnt.

b. Of a colour or shade of colour: having the appearance of darkening by scorching.

1897 *Sears, Roebuck Catal.* 360/1 Colors for Artists..Burnt Roman Ochre—Burnt Sienna—Burnt Terre Verte. **1923** *Daily Mail* 19 Feb. 5 Coloured Shantung..In a full range of new colourings, including..Rose, Burnt Orange, Almond, [etc.]. *Ibid.* 19 Mar. 1 Burnt Gold.

burnwood (bɜ·ɹnwud). A species of sumac, *Rhus metopium*, found in the West Indies and southern Florida.

1874 Lindley & Moore *Treas. Bot.* Suppl., Burnwood of the West Indian negroes. *Rhus Metopium.* **1926** Fawcett & Rendle *Flora of Jamaica* V. 9 R[hus] Metopium..Jamaica Sumach. Burn Wood.

burp (bɜːp), *v. slang* (orig. *U.S.*). [Imitative.] **1.** *intr.* To belch. Hence **bu·rping** *vbl. sb.* and *ppl. a.*, belching or making a similar sound.

1932 *Amer. Speech* VII. 330 Burp, to belch. **1934** *Etude* Aug. 456/3 Saxophonists also go in for this slapping effect; … burping springs. **1953** W. R. Burnett *Vanity Row* xvii. 131 He belched, 'It's an old Arab custom...You no like food—no burp—host insulted.' **1958** *Spectator* 6 June 726/3 At the hot springs the mud bubbles and burps.

2. *trans.* To cause to belch or bring up wind.

1940 in *Amer. Speech* (1941) XVI. 145/2 Chronic air swallowers should be 'burped' three or four times..during each feeding. **1950** *Lancet* 4 Feb. 218/1 In the U.S.A. babies are 'burped' during and at the end of feeds. **1968** 'G. Bagby' *Another Day* ii. 28 'What would you do with a baby?' I asked. 'Feed it, diaper it, and burp it,' she said.

burp (bɜɹp), *sb. slang* (orig. *U.S.*). [f. the vb.] **1.** A belch. Also *transf.* and *fig.*

1932 *Amer. Speech* VII. 330 *Burp*, sound made when belching. **1946** P. Kesten *Radio Alphabet* 18 *Burp*, an interloping noise on transmitting or receiving stations. **1951** J. Masters *Nightrunners of Bengal* v. 64 He forced a small musical burp and giggled. **1957** *Observer* 3 Nov. 19/7 Hollywood gives vent to one of its periodic burps of sentiment over an entertainer whose screws drove him to the bottle. **1962** V. Nabokov *Pale Fire* 22 A comfortable burp told me he had a flask of brandy concealed about his warmly coated person.

2. **burp gun** (see quot. 1946). *Army slang.*

1946 *Finito! Po Valley Campaign* 41 The whirr of burp guns. **1946** *Amer. Speech* XXI. 246 Practically any type of German automatic or semi-automatic small arm was apt to be described as a *burp gun*, presumably from the resemblance of shots or short bursts of fire to hiccoughs; but I have been told by a combat infantryman that the name is properly applicable only to the Schmeisser machine pistol.

burqa, var. *Burka¹ and ².

burr, *sb.*³ Delete †*Obs.* and add later examples.

1857 P. Cartwright *Autob.* (1858) viii. 46, I struck a sudden blow in the burr of the ear and dropped him to the earth. **1928** P. Green *In Valley* 121, I whammed him in the burr of the ear and piled him. **1954** L. L. Hubbard *Compl. Dog Breeders' Man.* xxxvi. 318 *Burr*, the irregular formation inside the ear.

burr, bur, *sb.*⁴ Add: **2. a.** Short for *burr-drill* (examples); also, a similar instrument used for surgical operations on the bones. Hence **burr-hole**, a hole made by such an instrument.

1859 J. Taft *Pract. Treat. Operative Dentistry* iv. 99 The burs and drills may be made of pieces of wire..and fitted to a socket handle. **1881** A. C. Harris *Princ. & Pract. Dentistry* (ed. 10) III. iii. 305 Dr. Forbes has adapted to enamel burrs, chisels and gouges an ingenious handle. **1899** C. Truax *Mech. Surgery* xviii. 352 Surgeons' burrs..may be either olive shaped or in cylindrical form, the former being generally preferred. **1939** Parritt & Herbert *Oper. Dental Surg.* (ed. 4) ii. 115 The introduction of diamond burs in recent years has almost revolutionized cavity cutting. **1948** E. H. Botterell in *Brit. Surg. Practice* II. 379 Bilateral burr holes are made in the mid-temporal region.

3. **burr-drill** (examples).

1850 C. A. Harris *Dental Surg.* (ed. 4) III. iii. 290 The flat and burr-headed drills are very useful for enlarging the opening into the cavity. **1859** J. Taft *Pract. Treat. Operative Dentistry* iv. 99 Bur Drills..should be manufactured of the best steel, and wrought with the greatest care.

burr, *sb.*⁵ Add: **5. b.** **burr millstone** (U.S.).

1771 Washington *Writings* (1931) III. 65 A pair of French Burr Mill Stones. **1855** Cummings *Western Pilot* 23 This..is famous for its quarries of stone; from which are manufactured burr mill stones. **1857** C. Cist *Cincinnati* 182 James Bradford & Co...manufacture yearly seventy-five pairs burr millstones.

burr, *v.*¹ **3.** (Examples.)

1838 Audubon *Ornith. Biogr.* IV. 355 We..saw the males [*sc.* humming birds] in numbers, darting, burring, and squeaking. **1886** [see Burring *ppl. a.*]. **1906** E. Tennant *Lost Haven* (1908) xxiii. 321 She hated moths. Let it stay there burring and fluttering. **1959** C. Usher *Death in Bag* xvi. 167 The telephone clacked and burred at length.

burr (bɜː), *v.*⁴ [f. Burr *sb.*⁴] **a.** *intr.* To use a burr. **b.** *trans.* To excavate (a tooth) with a burr. Hence **bu·rring** *vbl. sb.* used *attrib.*, as **burring-engine**, a dentist's machine for driving a burr-drill, etc.; **burring tool** (see Burr *sb.*⁴ 2 b).

1875 *Dental Cosmos* XVII. 310 With the burring-engine I ground off enough of the cusp.

burramundi, -munda, see *Barramundi.

burra sahib (bʌ·rə,sɑ·ib). *India.* Also with capital initials. [Hind. *barā* great, (*ṣāḥib* master (see Sahib).] A title of respect used by Indians in referring to the head of a family, the chief officer in a station, the head of a department, etc. Also *transf.* So **burra beebee** or **mem(-sahib)**, a similar title for a lady.

1807 G. Elliott *Let.* 20 Sept. in *Lord Minto in India* (1880) ii. 29 The Burro Bebee, or lady of the highest rank. *c* **1810** W. Hickey *Memoirs* (1925) II. x. 136 The Governor-General, Mr. Wheler, General Stibbert, Mr. Barwell, and in fact all the Burra Sahibs (great men) of Calcutta. **1848** J. H. Stocqueler *Oriental Interpreter* 44 *Burra-beebee*, or *burra-mem*, a great lady; the appellation bestowed upon the female head of a house, or the wife of the principal personage at a station or presidency of India. **1863** Trevelyan *Dawk Bungalow* 1, Chota Sahib one rupee give. Burra Sahib two rupee. **1885** Lady Dufferin *Viceregal Life India* (1889) I. 57 The great lords and ladies (Burra Sahib and Burra Mem Sahib). **1922** *Blackw. Mag.* Sept. 283/2, I made my way to the burra Police Sahib. **1922** *Outward Bound* Nov. 137/2 Part of the headdress or person of a *burra-mem*. **1928** *Chambers's Jrnl.* Feb. 180/2 The Hindu..announced it to be the habitation of a very *burra* (high in station) sahib. **1934** 'G. Orwell' *Burmese Days* xxv. 371 The position for which Nature had designed her from the first, that of a burra memsahib. **1936** R. W. Chapman *Names, Designations & Appellations* 240 The societies of industry, commerce..are commonly divisible into three grades...Members of A say '*Smith*' to each other (..but a Burra Sahib may be '*Sir*'). **1963** M. Malim *Pagoda Tree* xxii. 145 The gathering throng of eminently decorous knights, ladies, *burra sahibs* and *burra mems* gathering on the lamplit lawn below.

burrass. (Earlier example.) Cf. Barras¹, and *barras* sb.² in Eng. Dial. Dict.

1770 C. Carroll *Let.* 24 Oct. in *Maryland Hist. Mag.* (1918) XIII. 66 Pray write for 6 strong matrasses..strongly Quilted & Covered with Burras or a Coarse strong Canvass.

burrawang (bʊ·rāwæŋ). *Austral.* Also **buddawong, burrawong, burrowan, burwan.** [f. the name of Mt. *Budawang*, New South Wales.] An Australian palm-like tree, *Macrozamia spiralis*; the nut produced by this tree.

1826 J. Atkinson *Agric. & Grazing N.S.W.* 19 The burwan is a plant with leaves very much like the cocoa nut. **1851** J. Henderson *Excurs. N.S. Wales* II. 238 The Burrowan, which grows in a sandy soil. **1877** E. A. Heron *Balance Pain* 108 A Buddawong seed-nut fell to earth. **1889** J. H. Maiden *Usef. Native Plants Australia* 41 'Burrawang Nut', so called because they used to be, and are to some extent now, very common about Burrawang, N.S.W. **1934** *Bulletin* (Sydney) 3 Oct. 21/1 The burrawang is another slow grower. **1956** *Landfall* X. 122 Straight-trunked burrawong trees.

burred (bɜ̄ɹd), *a.* [f. Bur *sb.* + -ED².] Rough and prickly like a burr.

1906 *Westm. Gaz.* 8 Sept. 2/3 He falls furiously on the ball..until it is hopelessly burred and gashed. **1924** *Glasgow Herald* 4 Apr. 8 The burred fruits are accounted for by their clinging to the feathers of birds.

burrel-fly. Delete †*Obs.* and add earlier and later examples.

1658, 1829 [see Whame *sb.*]. **1951** Colyer & Hammond *Flies Brit. Isles* vi. 102 Various names have been bestowed upon Tabanidae, including burrel-flies, gad flies..horse flies [etc.].

burrell, burrhal, burrhel, varr. *Bharal.

burring, *vbl. sb.*¹ Add: **burring rollers** *pl.*, an apparatus for removing the burrs from wool in preparing it for carding.

1884 W. S. B. McLaren *Spinning* v. 83 The burrs..lie along the fibres of wool...To clear them off, burring rollers are fixed on the top of two of the lickers-in. **1888** *Encycl. Brit.* XXIV. 658 The swift as it travels round is met by a series of three burring rollers rotating in an opposite direction.

burro. Add to def.: Now esp. *U.S.* (common in Western states).

1845 T. J. Green *Jrnl. Texian Exped.* xii. 166 The sick were permitted to ride upon 'burros' (jack-asses). **1882** *Rep. Prec. Met.* (U.S. Bureau of Mint) 569 With these attached a burro or horse runs the machine. **1932** E. Wilson *Devil take Hindmost* xviii. 197 One of the burros is laden with wooden casks.

bursary. Add: **3.** In modern use, no longer restricted to Scotland. Also, in extended use, an endowment to persons other than students.

1907 *Act 7 Edward VII* c. 43 § 11 The powers and duties of a local education authority..shall include a power to aid by scholarships or bursaries the instruction in public elementary schools of scholars from the age of twelve up to the limit of age fixed. **1910** *Encycl. Brit.* IV. 863/1 *Bursar*..is also applied to the holder of 'a bursary'..in England a scholarship or exhibition enabling a pupil of an elementary school to continue his education at a secondary school. **1943** *Hansard* (Canada) 1 Mar. 787/1 A third field to which the committee should give some consideration is that of providing federal bursaries and scholarships at both secondary school and university levels. **1955** *Times* 10 May 12/4 This bursary has been awarded to a playwright or producer in order to enable him to make a study of the theatre.

bursiculate (bʊɹsi·kiŭlĕt), *a.* *Bot.* and *Anat.* [ad. mod.L. *bursiculatus*, f. *bursicula*, dim. of L. *bursa* purse.] Resembling a purse or pouch; bursiform.

1880 in Webster Suppl. **1957** Snell & Dick *Gloss. Mycology* 23/1 Bursiculate, purselike; pouchlike.

bursitis (bʊɹsəi·tis). *Path.* [mod.L., f. Bursa + -ITIS.] Inflammation of a bursa.

1857 Dunglison *Med. Lex.*, *Synovitis*...When it affects the bursæ it is sometimes called Bursitis. **1908** *Practitioner* June 865 Myositis, phlebitis, bursitis. **1950** R. A. Willis *Princ. Path.* xii. 138 The traumatic pre-patellar bursitis known as 'housemaid's knee'.

burst, *v.* Add: **2. d.** With *up*. To become 'broken' or bankrupt. Cf. *Bust *v.*² c.

1848 W. Armstrong *Stocks* 9 If any firm or individual does not fulfill his obligations..he is unable to do so, or,..he has 'burst up'. **1865** Dickens *Mut. Fr.* III. xii, Then you think, Mrs. Lammle, that if Lammle got time, he wouldn't burst up?—To use an expression..which is adopted in the Money Market.

6. b. *to burst out crying* (example).

1863 H. E. P. Spofford *Amber Gods* 126 The other girls burst out crying at the sight of the coffin.

8. e. To spend (money) extravagantly; *esp.* to spend it 'on the burst' or 'on the spree'. *slang*.

1892 J. Murdoch *From Australia & Japan* 151 It sounds odd to be told that a fellow's conscientious scruples are lax enough to permit him to 'burst' $6 50c. on the very much off-chance of ever seeing a cent of his investment again. **1908** *Westm. Gaz.* 31 Mar. 10/3 Two natives..returned to visit their old relatives at home, and burst a little money on the spree.

burst, *sb.* Add: **2. c.** House-breaking, burglary. *slang*.

1857 'Ducange Anglicus' *Vulg. Tongue* 3 *Burst*, burglary. **1863** *Sessions' Paper* Apr. 786, I asked Simpson where they had done the *burst*—that is what is commonly called house-breaking.

4. b. *spec.* Of gunfire, esp. from an automatic weapon.

1893 in *Funk's Stand. Dict.* **1902** *Times Hist. War in S. Afr. 1899–1902* II. 159 The effect of this sudden burst—the first experience of the massed fire of modern rifles in the war—did not stay the advance. **1916** *King's Roy. Rifle Corps Chron. 1915* 41 The Companies in the trenches fired short bursts of rapid fire. **1942** T. Rattigan *Flare Path* III. 158 O.K., Kyle. Shoot. Give us the five-second burst. **1964** *Times* 21 Aug. 8/1 A Soviet sentry fired..a second burst as the car was leaving the area.

c. *Physics.* (See quot. 1960.)

1933 *Physical Rev.* XLIV. 779/1 The time intervals between the occurrence of bursts of ionization..were measured. **1942** J. D. Stranathan *Particles* xii. 509 The cosmic ray phenomenon responsible for..excessive momentary ionizations is referred to as a burst. *Ibid.* 519 It has been remarked that the cosmic ray burst is probably identical with the shower. **1949** *Austral. Jrnl. Sci. Res.* A. II. 214 (*title*) Bursts of solar radiation at metre wavelengths. **1960** Cooke & Markus *Electronics & Nucleonics Dict.* 59/1 *Burst.* 1. A sudden increase in the strength of a signal being received from beyond line-of-sight range. It is believed due to meteors passing through the upper atmosphere and momentarily affecting the ionized layers that reflect radio waves back to earth. 2. An exceptionally large pulse observed in an ionization chamber, signifying the arrival of several ionizing particles simultaneously. It may be caused by a cosmic-ray shower.

6. b. (Earlier example.)

1789 *Loiterer* 14 Feb. 6 Pompous accounts of *sharp bursts*, and long chaces.

c. Hence, in other sports; *spec.* a short spurt, etc., at greater speed than that employed for the remainder of the course.

1824 *Mirror* III. 290/2 The dogs..pursue it..with great impetuosity, which sportsmen term a burst. **1925** 'Ian Hay' *Paid with Thanks* xvi. 211 They..proceeded to row courses, half courses..and short bursts. **1957** Duncan & Bone *Oxf. Pkt. Bk. Athletic Training* (ed. 2) iii. 22 *Bursts*, a sprint put into the middle of a period of striding or jogging.

7. (Earlier example. See also *Bust *sb.*³)

1849 T. T. Johnson *Sights in Gold Region* xix. 183 Bill and Gus had come over from the Middle Fork for a particular, general and universal 'burst'.

burst, *ppl. a.* Add: **1.** Also with advs., as *burst-out*, *burst-up*.

1824 [in Dict.] **1900** *Engineering Mag.* XIX. 746/1 The burst-up condition of the decks..showing..the behaviour of the lyddite shells.

burster. Add: **2.** Also *N.Z.* (Earlier and later examples.)

1851 *Austral. & N.Z. Gaz.* 29 Nov. 483 The order of the day has been for some time past, rain, rainbows, and southerly bursters. **1922** W. G. Kendrew *Climates of Continents* xlvii. 364 On the east coast of the continent..cold winds from the south, known as Southerly Bursters. *Ibid.* 365 A striking roll of cumulus cloud may accompany the Burster, and there is usually heavy rain.

bursting, *vbl. sb.* Add: **6.** **bursting point**, the internal pressure at which an enclosed vessel will explode; usu. *transf.*

1902 W. James *Varieties Relig. Exp.* x. 236 The tension of subliminal memories reaching the bursting-point. **1931** E. Linklater *Juan in Amer.* II. iv. 81 Their columns had been stuffed to bursting-point with superlatives. **1933** D. L. Sayers *Murder must Advertise* i. 9 A small, inconvenient cubicle, crowded at the moment to bursting-point. **1944** R. Lehmann *Ballad & Source* III. 168 There came the bursting point of a complexity of hideous fears and pressures.

burton. Add: **b.** A term denoting the thwartship stowage of casks, bags, etc. Also **a-burton** (see quots. 1846, 1867).

1846 A. Young *Naut. Dict.* s.v. *A-burton*, Casks are said to be stowed a-burton, when placed athwartships in the hold. **1867** Smyth *Sailor's Word-bk.*, *A-burton*, the situation of casks when they are stowed in the hold athwart ship, or in a line with the beam. **1928** *Observer* 22 July 10/7 Terms as used by stevedores..'Burton' to describe the thwart-ship stowage of bags, as against the fore and aft method. *Ibid.* 29 July 7/3, I have..both stowed and worked out 'Burtons' many scores of times during the twenty-five years I spent along the East End waterside.

Burton² (bʊ̄·ɹt'n). The name of a town in Staffordshire (in full *Burton-on-Trent*) used *attrib.* and *absol.* to designate a type of ale: (*a*) = *Bass *sb.*⁶; (*b*) (see quot. 1953).

1738 in *Westm. Gaz.* (1902) 24 Feb. 2/3 Genuine Burton Ale, Brew'd to the Greatest Perfection for Keeping by Sea and Land. **1843** Tizard *Brewing* 88 The Burton ales principally owe their superior qualities and uniform permanency to the nature of the water there used. **1861** *Let.* 3 Aug. in C. Tovey *Wines & Wine Countries* (1862) iv. 171 The workman takes..'Bitter Beer', instead of his Burton. **1882** E. G. Hooper *Man. Brewing* (ed. 2) 118 The Burton beer has been long noted for its high character. **1896** H. G. Wells *Wheels of Chance* vi, Burton and biscuit and cheese he had, which indeed, is Burton in its proper company. **1953** *Word for Word* (*Whitbread & Co.*) 15/1 *Burton*, a strong ale, dark in colour, made with a proportion of highly dried or roasted malts. It is not necessarily brewed in Burton, and the term is in general use for all varieties of 'strong' or 'old' ales.

burton³ (bʊ̄·ɹt'n). Also **Burton.** [Origin unknown; perh. connected with prec.] In slang phr. *to go for a burton*, (of an airman) to be killed; (of a person or thing) to be missing, ruined, destroyed.

None of the several colourful explanations of the origin of the expression is authenticated by contemporary printed evidence.

1941 *New Statesman* 30 Aug. 218/3 *Go for a Burton*, crash. **1943** C. H. Ward-Jackson *Piece of Cake* 32 *Gone for a Burton*, killed, dead. **1946** E. Roberts in Raymond & Langdon *Slipstream* 38, I can see those flowers going for a burton. **1947** 'N. Shute' *Chequer Board* iii. 49 He went for a Burton over France last year. **1957** J. Braine *Room at Top* xx. 176 We noncoms used to say *got the chopper*. Going for a Burton was journalist's talk.

Burtonize (bʊ̄·ɹt'nəiz), *v.* [f. *Burton² + -IZE.] *trans.* To harden (water for brewing) by treatment with the sulphate and chloride of magnesium and calcium or other salts.

1882 E. G. Hooper *Man. Brewing* (ed. 2) 120 Burtonising...A hardening mixture giving solids similar to those of the Burton water may be made as follows:—Four parts of gypsum [etc.]. **1892** H. E. Wright *Handy Bk. Brewers* 116 The presence of bicarbonates or carbonates may explain certain negative..results which sometimes attend the so-called 'Burtonising' of brewing water. **1940** H. L. Hind *Brewing* II. xxiv. 589 The fall in p_H from its original value of 5·6–5·8, which are normal figures with liquors that contain 10–12 parts per 100,000 of carbonates and little or no gypsum, or rather lower values if 'Burtonizing' is adopted, is very slight.

Hence **Burtonizer** (bʊ̄·ɹt'nəizəɹ): see quot.

1908 *Daily Chron.* 12 Nov. 1/2 Burtonisers..are chiefly sulphates and chlorides of calcium and magnesium.

burwan, var. *Burrawang.

bury, *v.* Add: **2. a.** *to bury the hatchet* (earlier and later examples).

[**1680** S. Sewall in *New England Hist. & Genealogical Reg.* (1870) XXIV. 121 Meeting wth ye Sachem the[y] came to an agreemt and buried two Axes in ye Ground;..which ceremony to them is more significant & binding than all Articles of Peace the Hatchet being a principal weapon wth ym.] **1754** in *Mass. Hist. Soc. Coll.* (1836) 3rd Ser. V. 10 We have ordered..our Governor of New York to hold an interview with them [*sc.* the Six Nations] for delivering those presents, [and] for burying the hatchet. **1796** 'P. Pindar' *Wks.* IV. 485 Gentle Reader, Wouldst thou not have imagined that the war hatchet was buried for ever? **1794, 1837** [see Hatchet *sb.* 2]. **1954** I. Murdoch *Under Net* xiv. 189 It is possible to break the ice without burying the hatchet.

4. b. Of a vessel: to plunge the bows under water. Also *trans.*

1866 'Argonaut' *Arts Rowing & Training* I. iv. 28 The oar will be feathered under water, and thus the boat will be buried. **1886** *Outing* (U.S.) Nov. 117/1 It was asserted that she [*sc.* the keel schooner] was too fine forward,..that she would bury in driving hard. **1888** W. B. Woodgate *Boating* x. 128 The aspirant to sculling honours had better..take his first lesson in a gig...A dingey buries too much on the stroke, and spoils style.

Buryat: see *Buriat.

burying, *vbl. sb.* Add: **3.** *burying-party*.

1887 [see Dict.]. *a* **1918** W. Owen *Poems* (1920) 19 The burying-party, picks and shovels in their shaking grasp.

burying, *ppl. a.* Add: **burying-beetle** (earlier example).

1818 Kirby & Spence *Entomol.* I. ix. 258 The burying-beetle (*Necrophorus Vespillo*, F.) inters the bodies of small animals.

bus, *sb.*² Add: **1. b.** Phr. *to miss the bus* (fig.): to lose an opportunity; to fail in an undertaking. *slang*.

[**1886** J. Morley *Miscell.* III. 147 Though he [*sc.* Mark Pattison] appeared..as much a Catholic at heart as Newman..it was probably his constitutional incapacity for heroic and decisive courses that made him, according to the Oxford legend, miss the omnibus.] **1915** C. J. Dennis *Sentimental Bloke* 118 The deeds and words of some un'appy bloke Who's missed the bus. **1922** *Daily Mail* 28 Oct. 8/3 The Prime Minister has 'missed the 'bus'...He has thrown away the greatest opportunity ever offered..to any statesman. **1940** *Manch. Guardian Weekly* 10 May 337 He [*sc.* Mr. Chamberlain]..boasted that Hitler has 'missed the bus'.

2. *colloq.* **a.** Short for *bus-bar* (see *3) (*Cent. Dict.* Suppl. 1909). **b.** An aeroplane. **c.** A motor car. **d.** A motor cycle.

1910 *Flight* 13 Aug. 648 (*caption*) A Blériot 'Bus' being Built. *Ibid.*, M. *Blériot*..has nearly completed a four-seater monoplane...the day of the aerial 'bus will soon be with us. **1916** C. Winchester *Flying Men* 213 We were about 2500 feet above the 'buses'. **1921** *Times Lit. Suppl.* 24 Feb. 113/4 Cadets..marching in their companies and battalions or gadgeting with 'buses on Port Meadow. **1924** J. Buchan *Three Hostages* xii. 179 Got here last night after a clinkin' journey, with the bus [*sc.* an aeroplane] behavin' like a lamb. **1927** D. L. Sayers *Unnat. Death* xi. 131 'Excuse me,' said ..a youth in motor-cycling kit, 'but could you give me a hand with my 'bus?' **1930** *Proc. Inst. Radio Engin.* XVIII. 438 A panel voltmeter is connected to the filament bus to facilitate the adjustment of the bus voltage.

3. *attrib.* and *Comb.*, as *bus company*, *conductor* (Conductor 7), *conductress*, *crew*, *driver* (examples), *load*, *queue* (Queue *sb.* 3), *ride*, *route*, *station*, *terminal*, *ticket*, *time-table*, *top*; *bus-riding* adj.; **bus-bar, -conductor** *Electr.*, a system of conductors in a generating station on which all the power of all the generators is collected for distribution or, in a receiving station, on which the power from the generating station is received for distribution; also *attrib.* (cf. Omnibus *a.* 2 c); **bus-boy** orig. *U.S.* = Omnibus *sb.* 4; **busman**, the driver of a bus; so **busman's holiday**, leisure time spent in occupations of the same nature as those in which one engages for a living; **bus-rod** = **bus-bar*; **bus-shelter**, a roadside structure affording protection from the weather to passengers intending to travel by bus; **bus-stop**, a place at which a bus makes a regular halt.

1893 Sloane *Stand. Elect. Dict.*, *Bus Rod*...*Synonyms*—Omnibus Rod, Wire or Bar—Bus Bar, or Wire. **1897** *Daily News* 9 Dec. 10/3 The currents of electricity generated at these dynamos are led to one common conductor called 'the bus-bar'. **1911** *Engineer* 10 Mar. 238/3 The busbar voltage suddenly dropped. **1948** *Electronic Engin.* XX. 38 Two cores of magnetic material linked the busbar so that the current caused them to be magnetically saturated. **1913** *Industrial Worker* (Spokane, Washington) 12 June 4/2 They are cooks, bus boys, dishwashers. **1947** Auden *Age of Anxiety* (1948) ii. 36 A bus-boy brushing a table. **1965** R. Howard tr. *S. de Beauvoir's Force of Circumstance* x. 477 After a difficult adolescence, he had been a sailor, then a busboy in a London restaurant, and I don't know what else. **1899** R. Whiteing *No. 5 John St.* xi. 107 My Samaritan gave me the letter to the 'bus company. **1905** *Times* 9 Mar. 13/4 The London and District Motor Bus Company. **1846** *Chambers's Jrnl.* V. 28/2 The only cry heard would be that of the 'bus-conductor for Paddington and Holloway. **1886** *Punch* 27 Nov. 254/1 Bus Conductor (shouting from the Foot-board), Wes'minister! **1902** Webster Suppl. s.v. *Bus*, Bus..conductor, etc. (*Elec.*). **1919** *Manch. Guardian* 22 Oct. 8/6 Women 'bus conductors. **1935** *Discovery* Feb. 58/2 London bus-conductors on the new 'Q' type motor-buses are having a busy time. **1916** *Sphere* 30 Dec. 237 His [*sc.* a soldier's] sister in the garb of a 'bus conductress, an employment which she has taken up in her brother's absence. **1939** *Daily Mail* 13 Sept. 3/3 The first women bus conductresses appeared in Manchester yesterday. **1958** *New Statesman* 7 June 713/2 The admittedly underpaid bus crews. **1870** D. J. Kirwan *Palace & Hovel* (1963) xi. 98 The cabbies are to the buss drivers a sort of gypsies. **1905** G. B. Shaw in *Shaw on Shakespeare* (1962) 149 Beatrice..saying things that a flower-girl would spare a busdriver as if they were gems of delicate intuition. **1888** G. & W. Grossmith *Diary of Nobody* (1892) iii. 42 All our 'bus-load..seemed to be going in. **1959** I. & P. Opie *Lore & Lang. Schoolchildren* xvi. 359 Bus-loads of youngsters come in from the surrounding villages. **1851, 1887** Busman [in Dict.]. **1893** *Eng. Illustr. Mag.* 488/2, I shall indeed take a holiday soon,..but it will be a 'Busman's Holiday'. **1921** *Times* (weekly ed.) 19 Aug., The proverbial 'busman's holiday' is nothing to that of the man who retires from business. **1927** *Observer* 21 Aug. 13/2 The U.S.A. Secretary for War..said..'No, I did not go to see the military manœuvres. Busmen's holidays do not give me any delight.' **1938** R. Finlayson *Brown Man's Burden* 31 He was lucky to get safely back to the village at all! A friendly bus-man put him off at the right place. **1950** 'R. Crompton' *William—the Bold* ii. 58 You'd think it was a treat to stand in a bus queue. **1908** Kipling *Lett. Travel* (1920) 199 A 'bus-ride down the Strand. **1952** M. Laski *Village* xii. 175 It would save the bus-ride. **1898** *Daily News* 14 Apr. 7/2 The 'bus riding public. **1898** A. Bennett *Man fr. North* vii. 44 Cab-fares, bus-routes, and local railways. **1936** *Discovery* Sept. 299/2 *Birds of the Green Belt* contains a useful transport guide of train and bus routes. **1945** *City of Oxf. Council Rep.* 16 Feb. 240 It is recommended that..the principle of the provision by Council of roadside bus shelters be accepted. **1960** C. Wilson *Ritual in Dark* II. ii. 212 An old man, crouched in a bus shelter. **1952** M. Laski *Village* vii. 120 The big mock-marble cinema opposite the bus-station. **1916** E. Pound *Lustra* 25 You loiter at the corners and bus-stops. **1930** *City of Oxf. Council Rep.* 20 Feb. 241 That notices 'Bus Stop' be marked on the road at the usual stopping places at Carfax. **1947** M. Lowry *Under Volcano* ix. 254 Had not Geoffrey met her at the Bus Terminal? **1949** *Granta* Christmas No. 43 Punctopapyrists (bus ticket collectors to you). **1928** R. Knox *Footsteps at Lock* xiv. 138 He asked for a railway guide and a 'bus time-table. **1905** *Westm. Gaz.* 31 May 10/2 'The Delights of London', introducing various amusing 'imitations'—in a tea-shop, on a 'bus-top, and so forth. **1927** *Glasgow Herald* 18 Aug. 9 The most interesting 'bus-top ride which London can offer.

bus (biz), *sb.*³ Colloq. abbrev. of Business 20 in Dict. and Suppl.

1864 Hotten *Slang Dict.* (ed. 3) 90 *Bus*, business (of which it is a contraction) or action, on the stage. **1933** Auden *Dance of Death* 10 *A.* Shall I show you? *B.* Please. [*Bus.*] **1949** Wodehouse *Mating Season* ix. 90 'Throughout the script the word "*bus*" in brackets occurs. ..Can you explain it?' 'It's short for "*business*". That's where you hit Mike with your umbrella. To show the audience that there has been a joke.'

bus, *v.* Add: (Later example.) Also without *it*.

1889 *Punch* 2 Mar. 107/1 'Bus'-ing on the cheap. **1936** J. B. Priestley *They walk in City* viii. 220 He had plenty of time to walk, bus, walk again. **1963** *Times* 11 Mar. 13/3 Mother now wins the car for termtime and the foreseeable future, and father has to bus it or bicycle.

2. *trans.* To transport (people) by bus from one place to another, esp. in order to encourage or achieve racial integration. Hence **bus(s)ed** *ppl. a.* *U.S.*

1961 Webster s.v., The children were bussed to school. **1969** *New Yorker* 31 May 86/3 Nine thousand children are now bused—about half as many as would have to be bused in the final stage of integration. **1969** *Ibid.* 29 Nov. 151/1 What had been a restless gathering of bused-in teenagers turned into a rapt adoring cult as soon as Dayan began to speak.

Hence **bu·s(s)ing** *vbl. sb.*, travelling or transportation by bus; (*Mil.*) boarding a bus.

1888 G. & W. Grossmith *Diary of Nobody* (1892) v. 76 'We had better get into this blue 'bus.'..'No blue-bussing for me.' **1889** [see above]. **1923** Kipling *Irish Guards* I. 108 They were instructed in march-discipline..as well as in bussing and debussing against time into motor-buses. **1965** *Economist* 19 June 1405/1 Local authorities should adopt the policy known in the United States as 'bussing'—that is to say, spreading immigrants' children around over a wide area. **1969** *Guardian* 18 Dec. 2/1 Southern Senators made a strong attempt today to prevent the integration of schools in the South by the bringing of black children from Negro areas to..schools in predominantly white areas. This controversial practice has become known simply as busing.

‖**busaun** (bŭzɑu·n). [G.] = Posaune.

1776 [see Posaune 1]. **1876** Stainer & Barrett *Dict. Mus. Terms* 65/2 *Busaun. Busain. Buzain.* A reed-stop on the organ.

busbied (bʊ·zbid), *a.* [f. Busby 2 + -ED².] Wearing a busby.

1914 G. Lloyd *Englishwoman's Adventures* 13 Four busbied officers. **1960** D. Potter *Glittering Coffin* ii. 24 The busbied guards are on parade.

buser (bʊ·səɹ). (Disused.) Also **busser.** [f. Bus *sb.*² + -ER¹.] A bus horse.

1894 *Daily Tel.* 10 Nov. 10/5 [For sale] By order of the Star Omnibus Company, six useful busers. **1900** *Ibid.* 5 Sept. 7/2 The English 'bussers'..performed their part admirably as heavy artillery horses. **1901** *Morn. Post* 9 Apr. 8/1 Active short-legged bussers. All warranted quiet in harness.

bush, *sb.*¹ Add: **6. b.** A signalling instrument used in Cornish pilchard fishing. *local*.

1880 in M. A. Courtney *W. Cornw. Gloss.* **1892** *Graphic* 13 Aug. 194/1 The huers on land making signals to the boatmen with two instruments called 'bushes', which are hoops crossing each other, and covered with a white bag, and fastened to the end of short rods.

c. The cat-o'-nine-tails. *slang*.

1895 *Daily News* 13 Sept. 7/6 They might give him twenty years, and he should not care, so long as they did not order him the bush.

9. a. (Further examples.) For U.S. examples see *D.A.*, *D.A.E.*

1826 J. Atkinson *Agric. & Grazing N.S.W.* iv. 64 When any person finds himself overstocked..they go into the interior, or *bush*, as it is termed, beyond the occupied parts of the country. **1836** W. B. Marshall *Two Visits N.Z.* 152 They [*sc.* the interpreters] took to the bush for shelter by day. **1851** N. J. Merriman *Jrnl.* 21 Sept. in *Kafir, Hottentot, & Frontier Farmer* (1854) 121 His mother and sister had escaped into the bush for refuge. **1857** R. B. Paul *Lett. fr. Canterbury* iv. 65 A pleasant walk of three hours through the bush. **1888** *Castle Line Guide to S. Afr.* 69 (Pettman), The soil..having been covered to a large extent by a thick forest of trees (usually termed *bush*). **1953** A. Paton *Too Late the Phalarope* vii. 52 The kloof was wooded, not with forest, but with what we in South Africa call the bush. **1968** K. Weatherly *Roo Shooter* 27 He's been in the bush so long he's started to look like a roo.

†**c.** A clump of trees. *N.Z. Obs.*

1856 T. Tancred in *Edin. New Phil. Jrnl.* III. 7 Over these tracts are scattered some small 'bushes', or woods. **1857** R. B. Paul *Lett. fr. Canterbury* vi. 94 You should try to have a bush on or near your section. **1867** Lady Barker *Station Life N.Z.* (1870) x. 62 Most stations have a bush near the homestead.

d. *pl.* = sense 9 a. *U.S.*

1879 Tourgée *Fool's Err.* xxii. 130 That refuge of free thought at the South, the woods (or 'the bushes', as the scraggly growth is more generally termed). **1912** C. Mathewson *Pitching* x. 210 The youngsters, who have come from the bushes and realize that this is their..chance to make good.

e. *to go bush*, to go into the country; to leave the city; to disappear from one's usual surroundings. Also *transf.*, to run wild, to go berserk. orig. and chiefly *Austral.*

1908 A. Gunn *We of Never-Never* 8 *She* went bush with me when I'd nothing but a skeeto net and a quart-pot to share with her. **1933** *Bulletin* (Sydney) 8 Feb. 20 It is rare for this fellow [*sc.* a dog] to go bush, but it has been recorded. **1934** A. Russell *Tramp-Royal* x. 78 Milbuka had fled, 'gone bush' that morning, and could not be found. **1946** F. Davison *Dusty* xiv. 156 From the fact that separate killings [of sheep] were reported it was natural to conclude that they were the work of a sheep dog gone bush. **1953** V. Bartlett *Struggle for Africa* i. 24 An African may do the same job day after day for months or years, and then suddenly 'go bush', omit some vital part of his routine, even say or do things which he himself cannot remember or explain when the mood is over. **1958** R. Stow *To Islands* 118 Brother Heriot has disappeared. He's gone bush somewhere and he might never come back. **1964** *Economist* 30 May 1000/1 To take refuge from [Australian] urban problems by 'going bush'.

10. a. *bush-clad*, *-covered*, *-fringed* adjs.

1909 *Westm. Gaz.* 4 June 5/3 The bush-clad plains of the North-Eastern Transvaal. **1873** J. H. H. St. John *Pakeha Rambles through Maori Lands* v. 67 The dark bush-covered hills of the Hunua. **1901** 'Linesman' *Words Eye-witness* (1902) 81 The rocky bush-covered foot of Schwartz Kop. **1891** 'J. Evelyn' *Baffled Vengeance* 47 The river between its bush-fringed banks.

c. *bush-country*, *-fire* (earlier example), *-flat*, *-girl*, *-hand*, *-hut*, *-land* (earlier example), *-line*, *-range*, *-school*, *-shanty*, *-tea*, *-track* (examples), *-walking* (so *-walk*, *-walker*), *-worker*.

1855 W. G. Simms *Forayers* 544 Who would have thought of any fellow being such a..booby as to bring a bathing-tub ..into..bush country? **1859** J. Rochfort in *Jrnl. R. Geogr. Soc.* (1862) XXXII. 297 In looking across this lake you perceive a flat bush-country. **1954** J. Collin-Smith *Scorpion on Stone* xx. 291 The lightening landscape of uninhabited scrub and bush-country. **1845** L. Leichhardt *Jrnl. Exped. Austral.* 12 Feb. (1847) v. 147 The smoke of extensive bush-fires was observed under Lord's Table Range. **1847** *N.Z. Jrnl.* VII. cxc. 90 The Ma-Whera river, another very considerable stream running through fine bush flats. **1960** B. Crump *Good Keen Man* 136 Four deer and three hours later we came to a long bush-flat where the trees were enormous. **1822** *New Monthly Mag.* Sept. 414/2 She was, in fact, the prettiest young Bush-girl I had yet seen. **1963** W. Soyinka *Lion & Jewel* 9 Bush-girl you are, bush-girl you'll always be; Uncivilized and primitive—bush-girl! **1863** S. Butler *First Year Canterbury Settlement* x. 147 How many hands shall you want? We will say a couple of good bush hands, who will put up your hut and yards and wool-shed. **1867** Lady Barker *Station Life N.Z.* (1870) xxi. 183 First came two of the most experienced 'bush-hands'. **1775** S. Thayer *Jrnl.* (1867) 12 Our troops..had not the satisfaction..to build..a Bush hut to pass the tedious night in. **1867** Lady Barker *Station Life N.Z.* (1870) xxi. 181 The mistress of this charming bush-hut insisted on our having some hot coffee. **1885** Mrs. Praed *Australian Life* iii. 73 The usual bush hut of slabs and bark. **1842** C. Heaphy *Narr. Residence N.Z.* viii. 103 The bush land will be cleared with less..expense than at Wellington. **1889** R. Paulin *Wild West Coast N.Z.* xiv. 119 Fresh snow ..came down to within 2000 feet of sea level—*i.e.*, considerably below the bush-line. **1955** J. K. Baxter *Fire & Anvil* iii. 58 A hut above the bush-line of the Southern Alps. **1879** T. W. Gudgeon *Reminisc. War N.Z.* xii. 70 Allowing the Hauhaus to erect a strong pah in the bush-ranges. **1852** G. C. Mundy *Our Antipodes* III. ii. 61 The humble hedge-school—or rather bush-school..and a crowd of flaxen..children rushing from its porch. **1896** H. Lawson *While Billy Boils* 3 You remember when we hurried home from the old bush school. **1936** G. Greene *Journey without Maps* II. 97 Even in the Sierra Leone Protectorate..most natives ..will attend a bush school. **1942** J. S. Huxley in *Political Q.* XIII. 395 Over most of Africa..not 10 per cent of the schools are anything but the most primitive sub-elementary bush-schools, confining themselves to hymn-singing, the catechism, and the rudiments of the three R's. **1857** S. H. Hammond *Wild Northern Scenes* 169 Crop crept close alongside of me, in our bush-shanty. **1888** 'R. Boldrewood' *Robbery under Arms* III. xi. 161 Like a man in..a bush shanty, not likely to wake before sunrise. **1924** H. T. Gibson *That Gibbie Galoot* 66 Most bush shanties possess such a butt. **1891** R. Wallace *Rural Econ. Austral. & N.Z.* ii. 43 Bush tea is..boiled in a can. **1832** *N.S.W. Cal. & G.P.O. Directory* 51 Several bush tracks lead to the farms. **1916** J. B. Cooper *Coo-oo-ee* i. 17 He..catches his father's favourite mare..and drives like mad over the five miles of rough bush track. **1968** K. Weatherly *Roo Shooter* 12 It was the shooter going to work, following a bush track that skirted the box flat. **1956** S. Hope *Diggers' Paradise* 196 Nothing deters young hikers from going bare-legged on bushwalks. **1955** *Times* 21 June 9/5 Four hundred soldiers, police and bush-walkers..are searching 5,000 ft Mount Baw Baw for Mihran Haig. **1853** in *Richmond–Atkinson Papers* (1960) I. 133 Beyond it there are two miles of bush walking along what is called 'a line'. **1957** *Times* 2 Dec. 9/6 Four members of a bushwalking club who were burnt to death when trapped by a bush fire in the Blue Mountains yesterday. **1959** *Times* 23 Sept. 1/4 Experience in out-door life such as skiing..bushwalking. **1936** I. L. Idriess *Cattle King* v. 41 He found a fair sale from travelling bush-workers.

d. (Extended use of sense 9, passing into *adj.*) Crude; rough and ready; without the formal training or qualifications usually considered necessary for an occupation.

1851 E. J. Wakefield *Let. to Sir G. Grey* 31 The stock-owner, though brought up as a gentleman, if he lives long in the 'bush', learns first to be proud of the 'bush' manners, and then becomes unfit for any but 'bush' society. **1870** R. P. Whitworth *Martin's Bay Settlement* 48/2, I found.. tied to a stick (a bush candlestick), about two inches of candle. **1873** Lady Barker *Station Amusem. N.Z.* vi. 101 He was what is called a bush-carpenter: *i.e.* a wandering carpenter, who travels from station to station, doing any little odd jobs wanted. *Ibid.* ix. 150 A bush doctor..was likely to be round by Simmons', cos o' his missus. **1891** G. Chamier *Philosopher Dick* I. vi. 141 The table was laid in regular bush style, with tin plates and pannikins, iron forks and spoons. **1916** J. B. Cooper *Coo-oo-ee!* ix. 115 Pilkins' gate, a skilful piece of rough bush carpentry, swinging on a wooden pivot. **1933** E. Jones *Autobiogr. Early Settler* xi. 52 As it was too far to get a tradesman up for any repairs, we were all, what was called, 'Bush Carpenters'. **1944** *Return to Attack* (Army Board, N.Z.) 9/2 Any available timber was turned into ingenious bush furniture. **1960** B. Crump *Good Keen Man* 114, I..put nine stitches in his [*sc.* the dog's] side with a pack-needle and string...He recovered..from both the wound and the bush surgery.

11. **bush-baby**, an African lemur of the species *Galago senegalensis*; **bush baptist** *slang* (chiefly *Austral.* and *N.Z.*) (see quot. 1959); **bush-bean** (examples); **bushboy**, a native Australian or South African bushman; **Bush Brotherhood**, a society of missionaries, clerical and lay, established to evangelize the inhabitants of the Australian bush; hence *Bush Brother*; **bush burn** *N.Z.*, the burning of bush on cultivable land; land so cleared; so *bush-*

BUSH

burning vbl. sb.; **bush canary,** the popular name of various birds in Australia and New Zealand (see quots.); **bush-car** (see quot. 1926); **bush cow,** (*a*) a wild cow of the bush; (*b*) the tapir; **bushcraft,** skill in matters pertaining to life in the bush; **bush dassie,** a S. African hyrax, *Dendrohyrax arboreus arboreus*; also *attrib.*; **bush deer,** in W. Africa, a gazelle; **bush dog** (see quot.); **bush-dray** *Austral.* (see quot.); **bush-drive,** a drive of game in the South African bush; **bush eel** (see quots.); **bush-faller**: delete '?'; **bush-falling,** the felling of trees in the bush; **bush flea,** a variety of *Pulex* so called in Natal; **bush-fly,** an Australian blow-fly of the family *Calliphoridæ*; **bush-goat,** a S. African warbler, *Camaroptera brachyura*; **bush gourd,** the squash gourd, *Cucurbita melopepo*; **bush-hawk,** the New Zealand falcon, *Falco novæseelandiæ*; **bush-hen** *N.Z.*, the weka, *Gallirallus australis*; **bush-house,** a house or hut in the bush; in Australia, also one in a (suburban) garden; **bush jacket,** a belted cotton jacket; **bush-lawyer,** (*b*) *Austral.* and *N.Z.*, a layman who fancies he has a knowledge of law; an argumentative person; see also quot. 1874; **bush-magpie,** an Australian crow-shrike of the genus *Gymnorhina*; **bush nurse,** a qualified nurse who is 'on call' in the remote districts of Australia; **bush-pig,** (*a*) a species of S. African swine, *Potamochœrus porcus koiropotamus*; (*b*) *N.Z.*, a wild pig; **bush pilot,** the pilot of an aeroplane which flies over sparsely inhabited country (chiefly *N. Amer.*); **bush poppy,** an evergreen Californian shrub, *Dendromecon rigida*; **bush-quail,** (*a*) a HEMIPOD; (*b*) an Indian bird of the genus *Perdicula* (Funk's Stand. Dict. 1893); **bush-rat,** a popular name for many small rodents; **bush road** (earlier and later examples); **bush robin,** the popular name esp. in Africa of various small birds; **bush-scythe** (later U.S. examples); **bush-sheep** *U.S.* (see quot.); **bush shirt,** a loose-fitting light shirt worn by men in hot climates; **bush-shrike,** any of various birds belonging to the African subfamily Malaconotinæ; **bush-sickness** (see quots.); so *bush-sick* adj.; **bush tea,** tea made from the leaves of certain shrubs, esp. those of species of *Cyclopia*, *Borbonia*, etc., dried and used medicinally in S. Africa; **bush telegraph** orig. *Austral.*, bush-rangers' confederates who disseminated information as to the movements of the police; *transf.*, rapid spreading of information, or of a rumour, etc.; the 'grapevine'; **bush-tick** (see quot. 1886); also *attrib.* in **bush-tick berry,** the fruit of *Osteospermum moniliferum*; **bush-tit,** a bird of the genus *Psaltriparus* (Cent. Dict. 1889); **bush-titmouse** *U.S.* (see quot.); **bush-warbler,** any of several genera of warblers belonging to the family Sylviidæ, found in Asia, Australasia, and Africa; **bush-water,** rain water that collects in the low-lying parts of tropical forests; **bush willow,** in S. Africa, a plant of either of the species *Combretum erythrophyllum* or *C. salicifolium*; **bushwoman,** a woman living in the Australian or African bush; **bush-wren,** the New Zealand name for a bird of the species *Xenicus longipes*.

1901 A. R. R. TURNBULL *Tales fr. Natal* 81 The occasional cry of a bush-baby alone broke the awful silence. **1928** *Daily Tel.* 15 May 14/3 Two bush babies, the pets of Baroness de Tuyll. **1902** J. MILNE *Epistles of Atkins* i. 18 Nothing is left to the imagination by the corporal who ranks himself among 'Bush Baptists and other fancy religions', in order to evade Sunday Service. **1959** BAKER *Drum* 97 *Bush Baptist*, a person of dubious religious persuasion or one who has no religious persuasion at all. Rare. **1821** *Plough Boy* II. 358/3 An opinion prevails here (Columbus, Ohio) that our soil is *too rich*, for the profitable culture of the *bush bean*, (called, I believe, at the eastward, the *fisher bean*). **1865** *Trans. Ill. Agric. Soc. 1862* V. 518 We usually plant bush beans in garden drills. *a* **1834** T. PRINGLE *Poet. Wks.* (1839) 8 Afar in the Desert I love to ride, With the silent Bush-boy alone by my side. **1850** R. G. CUMMING *Hunter's Life S. Afr.* (1902) 32/1, I found a funny little fellow in the shape of the Bushboy before alluded to, awaiting my arrival. **1899** J. MILNE *Romance Proconsul* xiv. 149 He would.. shoulder his rifle, and start off, with a couple of bush-boys for gillies. **1930** *Bulletin* (Sydney) 26 Mar. 31/2 The Bush Brother turned out to be a woman, born in Queensland. **1950** 'N. SHUTE' *Town like Alice* 315 They were married by a travelling Church of England priest, one of the Bush Brothers. **1903** *Daily Chron.* 7 Nov. 5/5 Founding a Bush Brotherhood under the direction of the Bishop. **1861** W. MORGAN *Jrnl.* 27 Apr. (1963) iii. 29 Last week sowed some grass seed on bush burn. **1900** J. G. WILSON in *Rep. Agric. Societies N.Z.* 132 On bush-burns, if sown at the rate of a few pounds per acre [cocksfoot] rapidly takes possession. **1950** *N.Z. Jrnl. Agric.* Dec. 508/3 Thousands of acres were ploughed (much of it after a bush burn). **1964** *Weekly News* (Auckland) 22 Jan. 39/2 To burn everything cleanly but the heavy trunks is so important to the success of a bush burn. **1898** J. BELL *In Shadow of Bush* xxx. 198 The season promised to be a good one for bush-burning. **1904** HUTTON & DRUMMOND *Animals N.Z.* II. 91 *The Bush Canary. Mohua ochrocephala*...The New Zealand Canary has a sharp, strident call, and its movements are quick and active. **1918** *Bulletin* (Sydney) 14 Feb. Red Page 4 *White-throated Flyeater* (Bush Canary) and other members of the genus *Gerygone*. **1936** H. GUTHRIE-SMITH *Sorrows & Joys N.Z. Naturalist* viii. 67 The rain-forests beloved of the Bush Canary. **1946** J. C. ANDERSEN in *Jrnl. Polynesian Soc.* June 154 *Mohua*, a gregarious bird (*Mohoua ochrocephala*), South Island form of the bush-canary, the North Island form (*Mohoua albicilla*) being upokotea. **1921** *United Free Ch. Mission Rec.* Apr. 116/1 She left at midday in a bush-car. **1926** *Ibid.* May 227/1 The bush-car is an arm-chair placed high above one wheel, excellent for the narrow paths and high grass of Africa. **1847** 'A. HARRIS' *Settlers & Convicts* xiv. 287 On a new farm..the stockyard..is necessary for milking bush cows. **1851** W. H. BRETT *Indian Missions Guiana* 37 The tapir or maipuri, called the bush-cow by the settlers. **1897** M. KINGSLEY *W. Africa* 734 The bush cow came on, and drove its horns through his thigh. **1871** C. L. MONEY *Knocking about in N.Z.* iii. 29 He gave me my first lessons in bushcraft such as a knowledge of edible roots, modes of crossing rivers, snaring birds. **1897** *Westm. Gaz.* 22 July 4/2 Leichardt perished because he lacked the rudiments of bushcraft. **1911** C. E. W. BEAN *'Dreadnought' of Darling* xxxvi. 324 Real scouting and bush-craft will always be part of the station life. **1911** *East London Dispatch* 20 Dec. 5 (Pettman), Bush dassie flesh was the staple food for the Hottentots. **1897** M. KINGSLEY *W. Africa* 734 He..put his net into the forest, and caught bush deer (gazelles). **1883** *Encycl. Brit.* XV. 438/2 *Icticyon*, with one small species, *Canis venaticus*, the Bush Dog, from Guiana and Brazil. **1848** HAYGARTH *Bush Life Australia* v. 48 The bush-dray, the only vehicle used in New South Wales for the conveyance of wool and other produce, is open and low, more resembling a brewer's dray than any other description of dray known in England. **1899** *Proc. Zool. Soc. Lond.* 831 At the bush-drives so common in the [Cape] Colony, Blue-buck are seldom turned out. **1828** HONE *Table-book* II. 224 At this season when persons, at inns in Lincolnshire, ask for 'eel pie', they are presently provided with 'bush eels'; namely snakes, caught for that purpose in the bushes. **1965** R. & D. MORRIS *Men & Snakes* vii. 160 In hard times the grass snake and other species were eaten as 'hedge eels' or 'bush eels'. **1882** W. D. HAY *Brighter Britain!* I. 184 We worked steadily at bush-falling. **1921** H. GUTHRIE-SMITH *Tutira* xxii. 203 Bush-falling had barely been started. **1899** G. RUSSELL *Old Durban* 503 The Bush Flea..is quite content to share your camp blanket if you do not resent his liberty of action. **1934** A. RUSSELL *Tramp-Royal* xxvii. 172 No sooner were the bush-flies at rest than plagues of sand-flies would rise up to take their places. **1952** M. BOYD *Cardboard Crown* v. 85 The room..was thick with the buzzing of bush-flies, a species of blow-fly. **1908** HAAGNER & IVY *Sk. S. Afr. Bird-life* 79 This Warbler is called the Bush-goat on account of the plaintive goat-like call to which the bird gives utterance. **1842** Bush gourd [see *squash-melon pumpkin* s.v. SQUASH *sb.*[2] 4]. **1882** W. BULLER *Man. Birds N.Z.* 2 The Bush-Hawk is generally met with on the outskirts of the woods. **1939** J. MULGAN *Man Alone* 80 Bush-hens were calling across the valley. **1834** C. A. DAVIS *Lett. J. Downing* 367 Saratogue, for politicians, is jist like the bush-houses for killing pigeons. **1901** F. CAMPBELL *Love, the Atonement* ix. 133 Let us have ices in the bush-house. **1902** W. SATCHELL *Land of Lost* xviii. 164 'Come around to the bush-house.'.. They seated themselves together on a rustic seat among the ferns. **1939** J. CARY *Mister Johnson* 11 The station..consists of six old bush houses, with blackened thatch reaching almost to the ground. **1959** M. NEVILLE *Sweet Night for Murder* xviii. 175 The bush-house..was made of brushwood and housed seed boxes..indoor plants, and bulbs under sacks waiting for planting. **1939** M. B. PICKEN *Lang. of Fashion* 16/1 *Bush jacket* or *coat*, belted, hip-length jacket. **1959** 'M. DERBY' *Tigress* ii. 62 He took the plastic flask from his bush-jacket pocket. **1961** *Listener* 24 Aug. 292/2 The familiar picture of Fidel Castro with beard, bush-jacket and peak-cap. **1874** A. BATHGATE *Colonial Exper.* xvi. 225 The bush lawyers, or mining agents, which is the name they accept,..are of great use to the diggers. **1908** E. J. BANFIELD *Confessions of Beachcomber* I. i. 13 A 'bush' carpenter is a very admirable person, when he is not also a bush lawyer. **1926** J. DOONE *Timely Tips New Australians* Gloss., *Bush-lawyer*, a man who gratuitously voices legal opinions although possessing no qualifications for doing so. **1933** N. SCANLAN *Tides of Youth* viii. 91 You're a regular bush-lawyer. **1948** B. JAMES in *Coast to Coast 1947* 168 Mrs. Bolton loved the touch of legality, being a good deal of a bush lawyer. **1948** V. PALMER *Golconda* iv. 29 It's easier to find a bush-lawyer than a man who's bent his back at all sorts of jobs. **1890** E. E. MORRIS *Cassell's Picturesque Australasia* II. 235 The College precincts are sacred to the classic muse and the omnipresent bush-magpie. **1933** *Bulletin* (Sydney) 28 June 20/4 He was under the bush nurse for a fortnight. **1840** J. S. POLACK *Manners & Customs N.Z.* II. 270 The narrator had been hunting the *cochon maron* or bush-pig. **1844** J. BACKHOUSE *Narr. Mauritius & S. Afr.* 213 The Bosch Vark, *Bush Pig*. **1907** W. H. KOEBEL *Return of Joe* 20 Mutton down to the price of bush-pig. **1910** J. BUCHAN *Prester John* xiv, I was inclined to think him a very large bush-pig. **1936** *Beaver* Mar. 52/2 The northern bush pilot is dependent solely on his own good judgement, resourcefulness and initiative. **1948** 'N. SHUTE' *No Highway* ix. 243 The pilot was to be a civilian bush-pilot called Hennessey, a thick-set tough who knew that country [*sc.* Labrador] intimately. **1948** *Shell Aviation News* CXXV. 2/1 The small aircraft, usually single-engined, flown by the bush pilot. **1869** J. MUIR *First Summer in Sierra* (1911) 51 A marked plant is the bush poppy (*Dendromecon rigidum*). **1900** W. D. DRURY *Bk. Gardening* 281 Californian Bush Poppy. **1893** H. A. BRYDEN *Gun & Camera in S. Afr.* 158 Tiny bush-quail (*Turnix lepurana*), dainty creatures, scarcely bigger than sparrows. **1964** A. L. THOMSON *New Dict. Birds* 625/1 The so-called 'bush quails' of India (*Perdicula*) are dwarf partridges. **1867** *Amer. Naturalist* I. 399 The Bush Rat (*Neotoma Mexicana*) is abundant throughout the territory [*sc.* Arizona]. **1889** *Trans. N.Z. Inst.* XXII. 301 On the habits of the New Zealand Bush-rat (*Mus maorium*). **1947** I. L. IDRIESS *Isles of Despair* xxxii. 213 There were bush rats and water rats. **1966** *Southerly* XXVI. 93 You, long-dead entomologist Sidney, like this bush rat, have found the crumbling edge cave in. **1827** P. CUNNINGHAM *Two Years N.S.W.* I. 123 A *made* bush-road is one where the brushes have been cleared, banks of rivers and gullies levelled, [etc.]..; while a *natural* bush-road signifies one to which nothing has been done except notching the trees, the carts simply following each other's track. **1916** J. B. COOPER *Coo-oo-ee* i. 2 Along the side of the Ironbark ranges was a bush road, leading to the hamlet. **1966** *Weekly News* (N.Z.) 1 June 43/2 In the area that we visited, which is already served by the bush road, we saw isolated kauri trees more than five feet in diameter. **1901** A. C. STARK *Birds S. Afr.* II. 217 Tarsiger stellatus. White-starred Bush Robin. *Ibid.* 219 Tarsiger silens. Silent Bush Robin. **1932** *Discovery* July 231/2 No bird calls but the bush-robin with chrome-yellow underparts and silver stars on his slate-blue forehead. **1856** *Trans. Mich. Agric. Soc.* VII. 54 D. O. and W. S. Perry..[exhibited] three bush scythes. **1874** *Rep. Vermont Board Agric.* II. 194 It is possible to subdue them..by cutting them off near the ground with a bush-scythe. **1869** *Trans. Ill. Agric. Soc. 1867–8* VII. 457 Sheep shipped in for sale because they 'didn't flourish' on prairie grass. These 'bush sheep', as they are called, have been in abundant supply. **1909** M. S. KISCH *Let. Nigeria* (1910) xii. 169 This is the kind of costume I go about in; a bush shirt..and long native-made boots. **1953** D. LESSING *Five* iii. 144 Now he strode fast over the ground, his loose bush-shirt flying around him. **1893** NEWTON *Dict. Birds* 21 The large genus *Thamnophilus*, containing upwards of 50 species, is one of the most important of the so-called 'Bush-Shrikes'. **1932** *Discovery* Jan. 25/2 A wonderful new black-breasted bush-shrike..an entirely new species..christened *Chlorophoneus nigrescens*. **1953** R. CAMPBELL *Mamba's Precipice* iv. 40 A pretty bush-robin came out..followed by a lovely bush-shrike with a bright green back, a yellow stomach, and a brilliant crimson chest. **1950** *N.Z. Jrnl. Agric.* Jan. 31/1 The amount of cobaltised fertiliser used..in 1949..would be sufficient to correct cobalt deficiency in 940,000 acres of bush-sick pasture. **1929** *Times* 1 July 15/6 Lack of minerals in pastures causes innumerable diseases, such as..'bush-sickness'..in New Zealand. *Ibid.* 16/2 Able to..identify 'Nakuruitis' as similar to 'bush-sickness' in New Zealand. **1950** *N.Z. Jrnl. Agric.* July 67/3 As a result of research in both New Zealand and Australia, the cause of bush sickness was found in 1935 to be a deficiency of cobalt. **1768** *Holyoke Diaries* (1911) 30 Began to take Bush Tea. **1838** J. E. ALEXANDER *Exped. Interior Afr.* I. 141 He regaled Mr. Schmelen and myself on boiled salt beef and bush tea. **1858** SIMMONDS *Dict. Trade, Bush Tea*, the leaflets of a species of *Cyclopia*, probably *C. latifolia*,..supposed to possess expectorant and restorative properties. **1902** 'X. C.' *Everyday Life in Cape Colony* ix. 122 In most of these Colonial stores 'bush' tea can be bought. **1946** *Cape Argus* 27 Feb. 6/9 In the country districts bushtea has many names—honey tea, boer tea, rooibos and so on. **1878** *Australian* I. 507 (Morris), The police are baffled by..the number and activity of the bush telegraphs. **1893** J. A. K. MACKAY *Out Back* v, A hint dropped in this town set the bush telegraphs riding in all directions. **1934** *Bulletin* (Sydney) 21 Nov. 21/1 The bush telegraph sends tidings to and fro. **1946** U. KRIGE *Way Out* x. 121 We had heard too many 'latrinograms' now to be unduly impressed by the particular form of bush-telegraph..practised by the Italian peasants. **1951** 'N. SHUTE' *Round the Bend* 208 'How did the Imam get to know about it?'..'The bush telegraph works very well, here in Bahrein.' **1954** L. P. HARTLEY *White Wand* i. 12 'How did he know that I was here?' I asked—a silly question from someone who knew the workings of the Venetian bush-telegraph as well as I did. **1856** C. J. ANDERSSON *Lake Ngami* ii. 20 Besides myriads of fleas, our encampment swarmed with a species of bush-tick. **1865** HARVEY & SONDER *Flora Capensis* III. 436 A large bush.. the Colonial name is Bush-tick Berry. **1886** *Pall Mall Gaz.* 22 July 4/1 The *carrapato*, or bush-tick..is a degenerate spider. **1893** NEWTON *Dict. Birds* 83 *Chamæa*..'Bush-Tit' and 'Ground-Wren'. **1881** *Amer. Naturalist* XV. 213 That diminutive little bird, the least bush titmouse (*Psaltriparus minimus*). **1898** MORRIS *Austral Eng.*, *Gerygone*...In New Zealand they are called Bush-warblers. **1808** *Ann. Reg. 1806* 856 Some were cutting wood for firing—Some collecting bush-water with a calabash. **1871** J. E. JENKINS *Coolie* ix. 120 That strange ebonised 'bush-water', which ..anon curls and eddies round us like the smiles on a Negro's face. **1891** *Eng. Illustr. Mag.* Feb. 383 The plantations..were surrounded by four dams or embankments;..one behind to exclude the 'bush water', the accumulated rain of the interior. **1917** R. MARLOTH *Dict. Common Names of Plants* 89 The so-called 'Bushveld [Willow]' or 'Bush [Willow]' is *Combretum salicifolium*. **1863** LYELL *Antiq. Man* 484 The human brain here given.. is that of an African bushwoman. **1874** W. M. BAINES *Narr. E. Crewe* viii. 192 The white bushwomen—creatures of a mature age, hideous to look upon. **1905** *Daily Chron.* 16 Dec. 4/7 Bushmen and bushwomen within a radius of forty or fifty miles ride to these functions. **1887** W. L. BULLER *Birds N.Z.* I. 115 Bush-wren [*Xenicus longipes*]..is generally met with singly or in pairs.

bush, *v.*[1] Add: †**9.** To camp in the bush. Usu. with *it*. *Austral.* and *N.Z.* *Obs.*

1827 W. J. DUMARESQ *Let.* in G. Mackaness *Fourteen Journeys* (1950–1) 99 Not being provided for *bushing* it, in these early frosts, we made up our minds to return. **1846** *N.Z. Jrnl.* VI. 166/1, I passed the night under a pine-tree..and awoke, after my first experience of 'bushing it', exceedingly refreshed. **1853** MRS. C. CLACY *Visit to Gold Diggings Australia* 245 If this fails, you must just bush it for the night. **1862** J. GOLDIE *Jrnl.* 5 May in H. Beattie *Pioneers explore Otago* (1947) 98, I resolved to scramble along the side of the lake..even although we had to 'bush it' for a night or two. **1868** *People's Mag.* II. 365/2, I have 'bushed' it many a rough night in Australia.

bushed, *ppl. a.*[1] Add: **5. b.** *transf.* and *fig.* Lost as in the bush. *Austral.* and *N.Z. colloq.*

1885 MRS. PRAED *Australian Life* 29, I get quite bushed in these streets. **1898** *Westm. Gaz.* 29 Sept. 3/2 He tangled himself up and got 'bushed', and frantically implored.. everybody..to help him with his contract. **1900** H. LAWSON *Over Sliprails* 1 The deeper you read..about things that end in ism..the more likely you are to get bushed. **1916** *Anzac Book* 144/1 To be 'bushed' in the heart of London became a common experience with him. **1944** J. H. FULLARTON *Troop Target* v. 45 We're bushed behind the enemy lines about a hundred miles from nowhere. **1953** 'N. SHUTE' *In Wet* ii. 39 It is a very easy country to get bushed in; the sense of direction can be easily lost.

c. Tired, exhausted. *N. Amer.*

1870 *Nation* July 57/1 To be 'bushed' was to be tired. **1910** W. A. FRASER *Red Meekins* 266, I was that danged near bushed, toward the last that I was feared I might go right on sleepin'. **1958** 'CASTLE' & HAILEY *Flight into Danger* x. 132 You thought you'd reached the end then—completely bushed, with not another ounce left in you. **1966** *Oxford Mail* 4 June 1/1 Astronaut Eugene Cernan's.. spacewalk was postponed..because he and the Gemini-9 command pilot..were 'pretty well bushed' from their exertions in space.

d. Suffering from the effects of isolation (see quots.). *Canada.*

1952 J. MARSHALL in R. Weaver *Canadian Short Stories* (1960) 289 'You had three years here alone,' she began. 'I have never been bushed,' Toddy interrupted. **1959** *Maclean's Mag.* 14 Feb. 40/2 It was geographically isolated, and its inhabitants were cut off in separate buildings by the cold and by storms, and often..psychologically isolated—that is, bushed.

bushed (buʃt), *ppl. a.*[2] [f. BUSH *v.*[3] + -ED[1].] Fitted with a bush or lining; lined.

1907 *Installation News* May 11/1 Bushed outlets. **1909** *Ibid.* III. 121 These..boxes are provided with bushed holes.

bushel (bu·ʃĕl), *v.*[2] *U.S.* [perh. f. G. *bosseln* to do odd jobs, to do poor work.] *trans.* and *intr.* To repair (garments). So **bu·shelman,-woman,** a man or woman employed in repair tailoring.

1864 WEBSTER 177/3 Bushelman. **1877** BARTLETT *Dict. Amer.* 777 *To bushel*,..to repair garments. **1889** *Cent. Dict.*, *Bushelwoman*, a woman who assists a tailor in repairing garments. **1909** 'O. HENRY' *Options* (1916) 92 You would say he had been brought up a bushelman in Essex Street.

bushido (bū·ʃido). [Jap.: see quot. 1900.] In feudal Japan, the ethical code of the Samurai or military knighthood.

1898 *Trans. Asiatic Soc. Japan* Dec. 149 The knowledge of *Bushidō*, or 'Way of *Samurai*', is absolutely necessary for any one desirous of knowing something about the Japanese people. **1900** I. NITOBÉ *Bushido* 3 *Bu-shi-do* means literally Military-knight-ways—the ways which fighting nobles should observe in their daily life as well as in their vocation. **1902** *Encycl. Brit.* XXIX. 709 It is essential to know something of the ethical code of the *samurai*, the *bushi-do* ('way of the warrior') as it was called. **1923** *19th Cent.* Jan. 133 The old samurai spirit of *bushido* will prove equal to stemming the tide of radicalism. **1961** R. SETH *Anat. Spying* iv. 70 They [*sc.* the Japanese] brought espionage within the scope of *bushido*, their extremely strict and elevated code of morals and conduct.

bushie, var. ***BUSHY** *sb.*

bushing, *vbl. sb.*[2] Add: **1.** (Further examples.) *spec.* in *Electr.*, an insulating device.

1896 R. ROBB *Electric Wiring* v. 151 Where the cord enters the socket there is gradual wear of the insulation by abrasion unless the hole is bushed with an 'insulating bushing'. **1934** *Times* 28 Feb. 8/6 The failure was caused by a flash-over on a transformer bushing at Croydon substation. **1943** *Electronic Engin.* XV. 410 Components like bushings..where metal parts have to be affixed to the porcelain insulator. **1957** *New Scientist* 23 May 24/1 The glass thus formed flows into the forehearth and then into platinum bushings, which are heated electrically to keep the glass fluid.

bushman, Bushman. Add: **1. b.** *Bushman grass*, in S. Africa any of various grasses, esp. species of *Aristida* and *Stipa*.

[**1789** W. PATERSON *Narr. Journeys Country of Hottentots* 63 Here I found many new species of Gramina, particularly that which the Dutch call Boshman's Grass, from the use made of it by that people, who eat the seed of it.] **1857** A. WYLEY *Rep. Min. Struct. Namaqualand* App. 44 There the various kinds of Bushman grass prevail, almost to the exclusion of every other plant. **1886** G. A. FARINI *Through Kalahari Desert* 448 Bushman grass, the best that grows on the Kalahari. **1915** R. MARLOTH *Flora S. Afr.* IV. 19 *Stipa* (bushman grass).

2. b. (N.Z. examples.) Also, one who fells timber.

1846 *N.Z. Jrnl.* VI. 274/1 E Kehu, our guide, is thus a perfect bushman, and is of very great service on an expedition. **1848** T. BRUNNER *Jrnl. Exped. Middle Isl.* 360, I have now acquired the two greatest requisites for bushmen in New Zealand, viz., the capability of walking barefoot, and the proper method of cooking and eating fern root. **1849** S. C. BREES *Guide & Descr. N.Z.* 29 This tree [*sc.* Titoki] is tall..and the wood..is prized by the bushmen for axe-handles. **1867** LADY BARKER *Station Life N.Z.* (1870) xxi. 180 The 'bushmen'—as the men who had bought 20 acre sections and settled in the bush are called—had scattered English grass-seed. **1916** G. THORNTON *Wowser* xii. 189 These bushmen..lead a very different life from that of the woodcutters and sawyers (although they too are called bushmen) employed at a saw-mill. **1961** B. CRUMP *Hang on a Minute* (1963) 32 Just find out if he wants a couple of experienced bushmen.

3. The language of the aboriginal bushmen of South Africa.

1869 BLEEK in R. Noble *Cape & its People* 277 Many nouns in Bushman vary in their terminations according to their position or use. *Ibid.* 278 The Bushman nouns do not appear to possess any representative parts. **1874** J. M. ORPEN in *Folklore* (1919) XXX. 146 Then he sent another bird, the tinktinki..—qinqininyq in Bushman. **1960** *Times* (S. Afr. Suppl.) 31 May p. xv/4 Afrikaans also borrowed from..the Hottentot and Bushman tongues.

Bushmanoid (bu·ʃmănoid), *a.* [f. BUSHMAN 1 + -OID.] Of a Bushman type.

1940 *Jrnl. R. Anthrop. Inst.* 15 The Bushmanoid races are not negroid stock. **1959** J. D. CLARK *Prehist. S. Afr.* iv. 99 Or else both represent parallel specializations from an ancestral proto-Bushmanoid stock.

bushmanship. (Further examples.)

1922 *Times Lit. Suppl.* 28 Dec. 869/3 He made his way back to South Africa and..found himself serving in the Intelligence Department, for which the knowledge of bushmanship and native customs obtained in lion-hunting specially recommended him. **1931** T. A. HARPER *Windy Island* II. v. 146 A friendship based on bushmanship. **1968** K. WEATHERLY *Roo Shooter* 56 The first rule of bushmanship is 'Don't cross a fence or go through a gate when you're bushed'.

bushveld (bu·ʃfelt, -velt). Also **bush veldt.** [ad. Afrikaans *bosveld*: see BUSH *sb.*[1] and VELD.] **a.** Veld composed largely of bush. **b.** (Usu. with capital initial.) The wooded region of north-western, northern, and eastern Transvaal, the low-lying portion of which is called the Lowveld or Low Country.

1879 *Chambers's Jrnl.* 1 Mar. 134/2 For big game, the low country and Bushveld is that part of the Transvaal which the hunter must seek. **1887** *Atalanta* Nov. 80/1 The roar of a lion..in the solemn bush veldt. **1887** A. A. ANDERSON *Twenty-five Yrs. in Waggon* (1888) 153 The Notuane River, in what is termed the Bush Veldt. **1901** *Contemp. Rev.* Mar. 333 An efficient guide, whose knowledge of the dense bushveld proved of great value. **1903** KIPLING *Five Nations* 205 The Low Bush-veldt that sends men stragglin' unaware. **1903** 'INDICUS' *Labour S. Africa* 19 Their farm is of about 5,000 acres in extent, on the lower or Bush Veldt. **1907** P. FITZPATRICK *Jock of Bushveld* 14 Between the goldfields and the nearest port lay the Bushveld. **1940** V. POHL *Bushveld Adv.* i. 41 My family had moved to the Bushveld of the Northern Transvaal. **1958** L. VAN DER POST *Lost World of Kalahari* (1961) vii. 126 Soon the main stream carried us away from the bush-veld banks.

bushwa, -wah (bu·ʃwā). *N. Amer. slang.* Also **booshwa(h), bushwha.** [app. a euphemism for ***BULLSHIT.**] Rubbish, nonsense. Also *attrib.*

1920 in WENTWORTH *Amer. Dial. Dict.* **1921** J. DOS PASSOS *Three Soldiers* (1922) IV. 215 Can't do anything without getting a general order about it. Looks to me like it's all bushwa. **1924** *Chicago Tribune* 1 Oct. 25/8 The Bull, the Glad Hand, the Old Oil, and Il Bushwa. **1932** *Amer. Speech* VII. 329 *Booshwah*, nonsense. **1932** J. DOS PASSOS *1919* 162 They said this war-talk was a lot of bushwa propaganda. **1934** J. O'HARA *Appt. Samarra* (1935) vi. 164 'Oh, bushwah on you,' said Irma. **1936** MENCKEN *Amer. Lang.* (ed. 4) vi. 301 The college boys and girls launched *bushwah* ..and a number of other thinly disguised shockers. **1949** E. F. RUSSELL in E. Crispin *Best SF* (1962) 205 'Certain of them may have secret knowledge..' 'Bushwa!' defined Queth, unhesitatingly. **1959** 'J. R. MACDONALD' *Galton Case* (1960) x. 83 If you're a detective, what was all that bushwa about Hollywood and Sunset Boulevard?

bushwhacker. Add: **4.** One who clears the land of bush, esp. an axeman engaged in cutting timber. *N.Z.*

1898 J. BELL *In Shadow of Bush* iv. 18 Davie isn't qualified as a bush whacker yet. **1907** W. H. KOEBEL *Return of Joe* 257 The most skilful bush-whacker in the district. **1948** R. FINLAYSON *Tidal Creek* II. v. 149 How nice after Uncle Ted's bushwhacker style to see a table with a crisp white cloth.

bushwhacking, *vbl. sb.* Add: **3.** Felling or clearing bush (with an axe). *N.Z.*

1906 E. W. ELKINGTON *Adrift in N.Z.* xvi. 262 Bush-felling, or, as it is termed, bush-whacking was a favourite pastime of mine. **1907** W. H. KOEBEL *Return of Joe* 287 You new chums cuttin' good terbacker as if you was bush-wackin'. **1930** W. SMYTH *Wooden Rails* iii. 39 Don't you like saw-mills and bush whacking and all that?

bushy, *a.* Add: **3. c.** *Ent.* Of antennæ: covered with long, erect hairs (Cent. Dict. 1889).

6. *bushy-browed, -tailed* adjs.; **bushy stunt,** a virus disease of tomato plants (see quot. 1956).

1912 W. OWEN *Let.* 2 July (1967) 148 A bushy-browed and horny-fisted *blacksmith's assistant*. **1965** G. MCINNES *Road to Gundagai* x. 176 The crinkly lines round his bushy browed eyes were thoughtful. **1936** G. C. AINSWORTH in *Jrnl. Min. Agric.* XLIII. 266 It is proposed to call this disease of tomato, 'Bushy Stunt'. **1939** *Ann. Reg. 1938* 376 The virus of bushy stunt of tomato was obtained in a fully crystalline state. **1956** *Dict. Gardening* (*R.H.S.*) IV. 2124/1 Another virus disease [of tomato plants] is Bushy Stunt, in which there is enormous production of secondary shoots with a resulting bushy appearance in the plant. **1868** *Amer. Naturalist* II. 535 It seems widely separated..in habits from its nearest relative *Nycteris occidentalis*, or bushy-tailed Bat. **1947** J. STEVENSON-HAMILTON *Wild Life S. Afr.* XXV. 207 The bushy-tailed meercat (*Bdeogale crassicaude*).

7. Concerned with the (Australian) bush.

1900 H. LAWSON *On Track* 37 The foreman was a bushman; his sympathies were bushy. **1904** *Daily Chron.* 19 Apr. 3/5 Her stories are of the bush bushy.

bushy (bu·ʃi), *sb.* *Austral.* and *N.Z.* Also **bushie.** [f. BUSH *sb.*[1] + -Y[6].] A dweller in the bush; a bushman as distinguished from a townsman.

1896 H. LAWSON *While Billy Boils* 144 Bushies don't generally carry their swags out of pubs in their sleep. **1899** *Bulletin* (Sydney) 7 Jan. 14/1 The usual summer query—Why won't the bushy wear straw hats? **1924** H. T. GIBSON *That Gibbie Galoot* xvii. 66 The unlucky 'bushie' whose mannerisms or objectionable traits attract overmuch attention from his mates. **1934** *Bulletin* (Sydney) 19 Dec. 20/1 The sweet test of a bushie! He looks round for a rail to lean his elbows on, while the townie negligently leans up against a post. **1968** K. WEATHERLY *Roo Shooter* 21 The bushie spoke for the first time.

business. Add: **11.** Phr. *to make it one's business*: to undertake as a self-appointed task (*to* do something).

1642 [in Dict.]. **1723** G. P[ARRY] tr. *Cicero's De Orat.* II. 134 When, as you said, I had first made it my Business not to examine but to aggravate. **1735** BERKELEY *Defence Free-Thinking in Maths.* 54 And since the publication thereof, I have myself freely conversed with Mathematicians of all ranks, and some of the ablest Professors, as well as made it my business to be informed of the Opinions of others. **1946** 'P. WENTWORTH' *Clock strikes Twelve* ix. 43 'I don't know how she knew.' 'She's the sort of woman who makes it her business to know.'

12. c. Phr. *business as usual*: things proceeding normally in spite of disturbing circumstances.

1884 *Punch* 12 Apr. 178/2 The true way she could show respect to Her Majesty was by letting her shopmen carry on 'business as usual' for the benefit of Her Majesty's subjects. **1914** WILSON & HAMMERTON *Great War* I. 84 'Business as usual' was the motto of London. **1944** J. S. HUXLEY *On Living in Rev.* iii. 38 The subordination of the profit motive and all ideas of 'business as usual' to the non-economic motive of success in war. **1969** *Times* 25 Nov. 21/7 We can never expect it to be a case of, after the squeeze, business as usual.

13. d. *to do* (a person's) *business* (later examples). Also *fig.*

1773 GOLDSMITH *Stoops to Conq.* v, Oh, Tony, I'm killed! ..That last jolt, that laid us against the quickset hedge, has done my business. **1816** JANE AUSTEN *Emma* I. viii. 122 Her visit to Abbey Mill..seems to have done his business. He is desperately in love. **1883** J. GREENWOOD *Odd People in Odd Places* 7 It was the bricks and mortar that did *his* business, poor chap. **1891** J. M. DIXON *Dict. Idiom. Eng. Phr.* 47 His last imprudent exposure of himself to the night air did the business for him.

15. d. *letters of business*: a royal letter authorizing Convocation to transact business.

[**1839** CARDWELL *Doc. Ann. Ch. Eng.* II. 359 No business can be undertaken in convocation, unless it has been specially proposed to them by royal license. **1842** LATHBURY *Hist. Convocation* 350 Parliament was summoned in February, 1713: and the convocation met on the 16th... On the 17th, the convocation was authorized, by a royal letter, to proceed to business.] **1873** PHILLIMORE *Eccles. Law* 1934 In 1713, convocation had royal letters of business, and considered various subjects,—penance, excommunication, forms for the visitation of prisoners. **1906** *Convocations Cant. & York* in *Parl. Papers* LXXXIV. 805 You may see your way to advise His Majesty the King to direct that Letters of Business be issued.

16. f. *like nobody's business*, beyond the normal range (of a person's capacity); in no ordinary way; 'like anything'. Hence also *nobody's business*, an extraordinary affair. *colloq.*

1931 E. LINKLATER *Juan in America* 242 'How I love you is just nobody's business,' she said. **1938** WODEHOUSE *Code of Woosters* vii. 163 The fount of memory spouting like nobody's business. **1941** 'N. BLAKE' *Abominable Snowman* xii. 132 Plays the piano like nobody's business. **1957** H. CROOME *Forgotten Place* vii. 92 My head this morning is nobody's business.

20. (Later examples.) Also in phr. *business of the stage*.

1763 GARRICK *Let.* 10 Aug. (1831) I. 163 If you mean by the *warmth of temper* you have accused me of to Mr. Johnson, a certain anxiety for the business of the stage, your accusation was well founded. **1849** *Theatrical Programme* 4 June 13/1 Mr Hurlstone is not sufficiently alive to the business of the stage to make a figure among professionals. **1893** I. ZANGWILL *Children of Ghetto* (ed. 3) xiii. 123 An actor who knows all the 'business' elaborated by his predecessors. **1923** WODEHOUSE *Adv. Sally* vi. 78 'Bit o' business,' she announced, at length. 'What do you mean, a bit of business?' 'Character stuff,' explained Miss Winch.. 'Thought it out myself. Maids chew gum, you know.' **1949** [see *BUS *sb.*[3]].

21. c. The audience or attendance at a theatre; a 'house'. Also, the total of box-office receipts.

1755 MRS. C. CHARKE *Life* 130 Business continuing very shocking. **1811** C. MATHEWS *Let.* 5 Dec. in Mrs. Mathews *Mem. C. M.* (1838) II. viii. 173 They may promise a salary, and I am sure they would pay it; but can they promise business? **1837** in W. R. Alger *Life E. Forrest* (1877) I. 324 Will conclude with her benefit on Friday evening when she will probably have between $900 and $1,000...This is considered a very handsome business. **1895** *N.Y. Dramatic News* 12 Oct. 5/2 Hanlon brothers' Superba has played to 'banner' business.

d. *Bridge.* Calling for the purpose of gaining a penalty. Freq. *attrib.*

1925 A. E. M. FOSTER *Auction Bridge* 46 The two players with the better cards are going to get the contract, or they are going to force the others into a bid when a real 'business' double for penalties can be made...A double of four or more is always a 'business' double. *Ibid.* 51 The Business Redouble is seldom sound business. **1927** *Observer* 6 Mar. 25 This Business Pass is one of the most formidable weapons. It converts the Informatory Double into a Business Double. **1959** *Listener* 23 July 154/3 It is standard practice to regard a double as primarily for business.

24. *business agent, centre, college, committee, efficiency, girl, hours* (earlier example), *letter* (examples), *life* (examples), *proposition, school, suit, woman*; **business card** (earlier and later examples); **business doctor** (see quot. 1909); **business edge,** cf. *business end*; **business end** *colloq.*, the operative part (further examples); **business lunch(eon),** a luncheon at which commercial transactions are discussed; **business man** (earlier examples); **business manager,** a manager of the business or commercial side of an enterprise; hence *business-manage* vb. trans.; **business part,** the sphere of business (also *concr.* = *business end*).

1849 C. LANMAN *Alleghany Mts.* xi. 85 The 'guide, counsellor, and friend' of the Indians, as well as their business agent. **1901** MERWIN & WEBSTER *Calumet 'K'* i. 15 All that remained was to wait until the business agent made the next move. **1840** *Boston* (Mass.) *Almanac* 119 Business cards printed in the most expeditious manner. **1959** T. S. ELIOT *Elder Statesman* III. 102 Here's my business card With the full address. **1851** C. CIST *Cincinnati* 278 Business centre. **1888** J. KIRKLAND *McVeys* 4 In the 'business centre' one might see an occasional tall, narrow, straight-sided brick structure. **1865** *Indianapolis Daily Jrnl.* 14 Sept. 2/4 Students in Bryant's Business College. **1903** A. D. MCFAUL *Ike Glidden* xvi. 124 He had just graduated from a business college, and claimed to know how to do business 'in a business like manner'. **1838** W. L. GARRISON in Garrison & Jackson *Life* (1885) II. 227 A business committee was then appointed. **1901** *Daily Express* 6 Aug. 6/2 A very novel profession has been lately started in the City. It may be called that of the business doctor. **1909** *Modern Business* Jan. 606/1 In America.. there exists a body of men who are known as 'Business Doctors', men who are called in to give advice upon the proper conduct of business. **1935** *Antiquity* IX. 211 The business edge of the chisel-ended arrow. **1926** A. HUXLEY *Jesting Pilate* iv. 316 Reduced to an Indian diet, Americans would be a good deal less interested than they actually are in business efficiency, uplift and the Charleston. **1953** D. PARRY *Going Up* iv. 154, I was supposed to be a business-efficiency expert, one of those menaces who crawl round doing time-and-motion studies. **1936** 'R. CROMPTON' *Sweet William* ix. 227 The business end of a geometrical compass was jabbed into Douglas's arm. **1955** *Sci. Amer.* Sept. 197/1 The business end of the coronagraph is the quartz polarizing monochromator. **1962** *Sunday Express* 25 Feb. 16/2 The business end of a rifle barrel. **1888** C. M. YONGE *Beechcroft at Rockstone* I. ix. 167, I..mixed her up with the ordinary class of business girls. **1958** BETJEMAN *Coll. Poems* 215 (*title*) Business girls. **1839** DICKENS *Nickleby* xl. 390 You will be surprised..to witness this, in business hours. **1766** J. ROSE *Let.* 8 Apr. in A. & H. Tayler *Lord Fife* (1925) ii. 31, I will have business letters also to write. **1914** W. OWEN *Let.* 6 Mar. (1967) 237, I make money for this by doing a few translations..of business letters. **1941** T. S. ELIOT *Dry Salvages* iii. 12 The passengers are settled To fruit, periodicals and business letters. **1868** W. COLLINS *Moonstone* II. v. 148 Female Boards..drew the breath of their business-life through the nostrils of Mr. Godfrey. **1951** M. MCLUHAN *Mech. Bride* 137/2 People..could maintain an intimate link with ordinary social and business life. **1926** S. LEWIS *Mantrap* xxv. 289, I don't really know a soul.. except for meeting them at business lunches. **1954** L. MACNEICE *Autumn Sequel* XII. 76 The foregone Conclusion of a business lunch. **1963** P. MOYES *Murder à la Mode* v. 83 He knew enough of the protocol of business luncheons.. not to be surprised..that Goring studiously avoided all reference to the matter in hand until the coffee arrived. **1826** H. C. ROBINSON *Diary* 9 June (1967) 91 Watts is a business man and is editor and publisher on his own account. **1832** *Congress. Globe* 30 Jan. 1511 Having been in the practice of the law..and somewhat conversant with business men. **1843** DICKENS *Chr. Carol* iv. 124 One little group of business men. **1860** O. W. HOLMES *Prof. Breakf.-t.* i. 16 People of cultivation, of pure character,..shrewd business-men, men of science [etc.]. **1901** R. LORAINE in W. Loraine *R. L.* (1938) I. iv. 79 Mr Frohman would finance the enterprise and business manage it entirely. **1852** *Chambers's Edin. Jrnl.* XVII. 306/2 Clerks, book-keepers, foremen, business-managers. **1906** B. STOKER *Pers. Remin. H. Irving* II. lxxiii. § 3. 319, I was Sir Henry Irving's business manager. **1838** J. S. MILL in *Westm. Rev.* Apr.–Aug. XXIX. 490 He [*sc.* Bentham] committed the mistake of supposing that the *business* part of human affairs was the whole of them. **1910** T. E. LAWRENCE *Let.* 29 Aug. (1938) 86 The business part of the log with which you are going to block your staircase. **1901** S. E. WHITE *Claim Jumpers* v. 70, I have a plain business proposition to make. You and I are going to be great friends. **1909** 'O. HENRY' *Options* (1916) 11 He had been used to having his business propositions heard of. **1916** *Nat. Educ. Assoc. U.S. Addresses & Proc. 1915* 325 (*heading*) The service of business schools at the close of the Great War. **1966** 'N. BLAKE' *Morning after Death* i. 19 'What do you actually do in the Business School?' 'There are courses in economics, management, salesmanship, commercial history, theory of exchange, the ethical aspect of business—all that kind of thing.' **1870** *Harper's Bazaar* 5 Nov. 707 Business suits. **1882** *Advt.* in *W. Burnot's Mother Goose* (Elephant & Castle Theatre) 30 Business Suits—21s. **1932** E. WILSON *Devil take Hindmost* iv. 30 A prosaic, gray business suit. **1844** *Southern Lit. Messenger* X. 486 Reputation of being a 'business woman'. **1958** BETJEMAN *Coll. Poems* 215 A thousand business women Having baths in Camden Town.

busk, *v.*[2] Add: **3.** (Further examples.) Now usu., to play music or entertain in the streets, etc.

1897 *Daily News* 21 Sept. 8/3 A highly-accomplished lady..begs for a dress in which to go busking. Busking is the jargon for wandering minstrels—folk who play the perambulating pianos we see in the streets or on the sands—folk who sing from morning till midnight. **1905** *Evening News* 12 Aug. 5/3 We are all 'busking' this year. It would surprise the public if they knew who constituted many of the troupes of pierrots and mysterious minstrels that are performing at the various holiday resorts. **1934** P. ALLINGHAM *Cheapjack* 318 Busk, to perform in the street.

b. *trans.* and *intr.* To improvise (jazz or similar music). *Musicians' slang.*

1934 S. R. NELSON *All about Jazz* ii. 51 The drummer can still busk his part, and except for roughly glancing at the score, that is what the best drummers do today. **1966** *Crescendo* Feb. 35/1 Many drummers are with small groups or busking outfits and therefore never see a drum part.

busker[2] (bʊ·skəɹ). [f. BUSK *v.*[2] + -ER[1].] One who 'busks'; an itinerant entertainer or musician.

1857 *National Mag.* II. 167/1 (*heading*) The Busker.. His avocation is strictly peripatetic; and hence he takes his title from the short boot, or 'buskin', which has been a common article of stage-apparel..ever since the earliest days of the drama. **1859** HOTTEN *Dict. Slang*, *Busker*, a man who sings or performs in a public house. *Scotch.* **1908** *Daily Chron.* 1 Sept. 7/4 'Buskers'..can be counted as belonging to the most genuine of the professional vagrant fraternity. **1951** L. MACNEICE tr. *Goethe's Faust* I. 33 Leave me not here a hopeless busker!

busser, var. ***BUSER.**

bust, *sb.*[1] Add: **2. b.** The measurement around a woman's body at the level of her bust, usually measured in inches. So *bust measure, measurement, size.*

1895 *Montgomery Ward Catal.* 36/3 Unlaundered Waists. Sizes 32 to 42 inches bust measure. **1961** M. DICKENS *Heart of London* i. 14 Beautiful model offers her services. Anything considered. Bust 40. Waist 20. Hips 37. **1969** *Woman* 4 Oct. 34/2 Her bust measurement had never been more than 38 in. **1969** *Woman's Own* 11 Jan. 26/2 *Flower-printed red dress*...In bust sizes 32–36 ins. **1970** *Woman* 7 Mar. 19/1 Dress and coat...Sizes 32 in. to 42 in. bust.

4. *bust bodice, -costume, -improver, -line.*

1903 in C. W. Cunnington *Eng. Women's Clothing* (1952) ii. 59 Patent bust bodice worn above the corset. **1961** M. SPARK *Prime of Miss Jean Brodie* iii. 73 Her chest was a slight bulge flattened by a bust bodice. **1910** *Maggs' Catal.* No. 255, 85 Etchings, Illustrating the Head-Dress and Bust-Costume Worn by Ladies during the Early 17th Century. **1849** in C. W. Cunnington *Fem. Attitudes* (1935) 312 The Registered Bust Improver. For the purpose of.. giving an elegant figure and appearance to the wearer. **1905** *Daily Chron.* 1 July 4/5 Many artificial aids have been devised, such as hip pads, bust improvers. **1939** *Ottawa Jrnl.* 24 June 10/6 The brief bra top is an uplift and is webbed to cling below the bustline. **1969** *Woman* 4 Oct. 34/3 Slim, happy and confident with a 36 in. bustline.

bust, *sb.*[3] *Vulgar* and *U.S.* = BURST *sb.* Phrs. *to go on the bust, to go a bust.*

1764 in *Essex Inst. Hist. Coll.* XLIX. 284 Stray'd or stolen..a Bay mare, with a cut main, and a Bust on the near Side of the Hind Flank. **1840** *American Joe Miller* 116 Away with the expense..when a fellow is on a bust. *a* **1860** [see BUST]. **1890** 'R. BOLDREWOOD' *Col. Reformer* vii, There would be a slight probability of some of the party going 'on the bust' after three or four months' teetotalism. **1909** *Westm. Gaz.* 4 Sept. 2/3 We wish to go on the bust mildly. **1912** BEERBOHM *Christmas Garland* 17 Wot'll 'e do then poor devil? Go a bust on 'is conduc' money. **1963** L. MEYNELL *Virgin Luck* vi. 127, I was in pocket now. So I decided to go a bust and back a number.

b. A sudden failure or collapse of trade, etc. (cf. *burst-up* s.v. BURST *sb.*[2] 2 b); *spec.* opp. BOOM *sb.*[3] 1.

1842 *Knickerbocker* XX. 99 'A mistake!' exclaimed the other; 'not a bit of it! It's a reg'lar built bu'st!' **1942** H. A. WALLACE *Cent. of Common Man* 28 Dec. (1944) 41 We cannot afford either a speculative boom or its inevitable bust. **1947** *Chr. Sci. Monitor* 24 Jan. 1/1 We had an agricultural 'boom and bust' after the other World War. **1969** *Times* (Suppl.) 5 May p. ii/2 The general reasoning is that it is better to have a slowdown now than a bust later. At least that is the thinking of many economists and politicos.

c. A person, etc., who is a failure; *spec.* in *Poker*, a busted flush or straight (see ***BUSTED** *ppl. a.*); hence, a bad hand at cards.

1928 *Amer. Speech* III. 218 *Bust*...Sometimes the word is used to mean failure...'The Theta dance sure was a bust'. **1931** D. RUNYON *Guys & Dolls* (1932) vi. 134 As Cupid I am a total bust. **1932** *Amer. Speech* VII. 330 *Bust*, a very poor bridge hand. *Ibid.* 435 A worthless hand [at poker] is called..a 'bust'. **1934** M. ELLINGER *Poker* 166 *Bust*, a hand which the player has endeavoured and failed to make into a straight or a flush. **1957** WODEHOUSE *Over Seventy* iv. 54 At the age of ten I was a social bust. **1959** H. O. YARDLEY *Education of Poker Player* II. i. 86 A bust is a useless hand.

d. = ***BURST** *sb.* 2 c.

1859 in G. W. MATSELL *Vocabulum* 16. **1887** J. W. HORSLEY *Jottings fr. Jail* i. 23 'Rem for a bust'..means..remanded for a burglary. **1947** *Sci. News* IV. 50 The back of a pub where you and a 'screwer'..had decided to 'do a bust'.

e. A blow with the fist.

1925 H. LEVERAGE in *Flynn's* III. 693/2 *Bust*, a blow; a stroke (in the face). **1962** J. F. POWERS *Morte d'Urban* viii. 176 How'd you like a bust in the nose?

f. A police raid or arrest. *slang* (orig. *U.S.*). Cf. ***BUST** *v.*[2] e.

1938 *New Yorker* 12 Mar. 38/3 'One whiff [of marijuana]' said Chappy, 'and we get a bust.' ('Bust' is Harlem for a police raid.) **1959** W. BURROUGHS *Naked Lunch* 15 Provident junkies..keep stashes against a bust. **1969** *It* 10–23 Oct. 10/1 At the moment, there are over a hundred of our kids in nick as a result of the busts at 144 Piccadilly & Endell Street.

bust, *v.*[2] *Vulgar* and *U.S.* = BURST *v.* Add: (Examples.)

1806 M. LEWIS in *Lewis & Clark Exped.* (1905) V. 137 Windsor busted his rifle near the muzzle. **1839** DICKENS *Nickleby* lvii, His genius would have busted all bonds. **1843** —— *Mart. Chuz.* xvi, Keep cool, Jefferson...Don't bust! **1844** *Ibid.* xxxiv, If the biler of this vessel was Toe bust, Sir. **1891** G. CHAMIER *Philosopher Dick* II. 249 We must have a spell anyhow. Nobody has ever *bust* himself as I have for the old man. **1915** A. D. GILLESPIE *Let.* 3 Mar. (1916) 30 The shells make a scraping sound...When they 'bust' in the distance, I see a bright flash. **1918** H. C. WITWER *Baseball to Boches* vii. 249, I oughta bust you in the nose! **1919** WODEHOUSE *Damsel in Distress* iii, I shall infallibly bust you one on the jaw. **1930** *Diary Public Sch. Girl* (ed. 2) 25 Quite a decent game. Bust my crosse though. **1942** T. RATTIGAN *Flare Path* I, Caught us bending proper this time. Group must be fair busting their stays with laughter. **1953** M. SCOTT *Breakfast at Six* xxii. 181 It's yours. Have a good time with it [*sc.* £15]. Bust it any way you like. **1963** *Listener* 28 Mar. 568/3 Protocol would hardly permit him to..bust his interviewer one on that earnest Canadian snoot of his.

b. *spec.* To break (a horse). Cf. ***BUSTER** 4.

1891 *Harper's Mag.* July 208/2 The whole secret of 'busting'..lies in completely exhausting the bronco at the first lesson. *Ibid.* 210/1 Two rides will usually bust a bronco so that the average cow-puncher can use him. **1941** *Nat. Geogr. Mag.* Mar. 300/1 He's too old now to..bust a bronco.

c. *trans.* To reduce to insolvency (cf. BURST *v.* 8 d). Also *intr.*, to go bankrupt.

1829 R. C. SANDS *Writings* (1834) II. 153 The Aigle Bank was *bussted*. **1837** *Knickerbocker* X. 329 Can it truly be That 'Providence, which oft afflicts the just', Has fore-ordain'd that all the banks should *bu'st*? *a* **1860** [see BUST in Dict.]. **1880** *Harper's Mag.* Oct. 729/1 After the fate of the town, be it to 'boom' or 'bust', has been decided. **1909** H. G. WELLS *Ann Veronica* vi. 124 Gives you a fight to hang on to the old man until he busts—practically. **1923** H. CRANE *Let.* 15 Feb. (1965) 123 *Broom*, by the way, has busted; N.Y. office closed last Saturday. **1923** T. E. LAWRENCE *Let.* 22 Sept. (1938) 431, 300 copies could be produced, with the fifty or sixty portraits I've bust myself upon, for £10 a copy.

d. To break into (a house, etc.). Cf. ***BUST** *sb.*[3] d.

1859 in G. W. MATSELL *Vocabulum* 16. **1879** *Sessions' Papers* 29 May 184 *Busting* means burglary or house-breaking. **1927** E. WALLACE *Feathered Serpent* ix. 110 There's a little house just outside of Thatcham..me and Harry..thought we might 'bust' it and get a few warm clothes. **1948** WODEHOUSE *Spring Fever* xiv. 143 What! You're asking me to bust a pete?

e. Usu. in pa. pple. *busted*: dismissed, demoted; arrested, jailed.

1918 R. W. LARDNER *Treat 'Em Rough* 81 His captain.. busted him and I don't mean he cracked him in the jaw but when a man gets busted in the army it means you get reduced to a private. **1930** *Amer. Speech* V. 382 *Busted, to be*, to be demoted. **1943** HUNT & PRINGLE *Service Slang* 19 *Bust*, to reduce to the ranks or a lower rank; to deprive an N.C.O. of his stripes. **1953** W. R. BURNETT *Vanity Row* v. 40 Roy showed his [police] badge. 'You'll get busted for this,' shouted the man. **1958** *Landfall* XII. 115 The little man came out of his cell...'This your first time busted?'

f. *intr.* In vingt-et-un, to exceed the score of twenty-one, and so lose one's stake. Also *to go bust.*

1939 PHILLIPS & WESTALL *Complete Bk. Card Games* 195 When each of the players has had the cards he wants (those who are 'busted' having disclosed the fact), the Dealer turns up his two cards. *Ibid.*, Players who have gone 'bust', i.e. exceeded 21, will have forfeited their stakes automatically. **1945** A. A. OSTROW *Complete Card Player* 39 If a player's total count passes 21..he has 'gone over', 'gone overboard' or 'busted'. Banker collects the bet. **1950** *Chambers's Encycl.* XIV. 326/2 Each of the players may.. have cards 'twisted'..until such time as he either 'sticks' on the hand he has, or 'busts' (i.e. exceeds 21).

bust, *ppl. a.* = ***BUSTED** *ppl. a.* Also *bust-up*.

1913 R. BROOKE *Let.* Aug. (1968) 499 *The Blue Review* has gone bust, through lack of support. **1915** J. BUCHAN *39 Steps* vi. 140 A bust-up motor-car. **1921** WODEHOUSE *Jill the Reckless* viii. 121 Jill Mariner had gone completely bust. **1928** D. L. SAYERS *Unpl. Bellona Club* xiv. 165 Well, it's all gone bust—but it was a darn' good stunt while it lasted. **1952** KOESTLER *Arrow in Blue* xxv. 350, I woke up in the morning..financially broke, and with a bust-up car. **1964** *Observer* 26 July 8/2 Companies do go bust. *Ibid.* 8/5 The firm..is noted for its history of winding up bust companies.

bu·sted, *ppl. a. colloq.* (orig. *U.S.*). [f. BUST *v.*[2] to burst.] Burst, broken; bankrupt or ruined; *spec.* in *Poker* denoting a flush or straight that one fails to complete (also *fig.*). Also *busted-up*.

1837 *Knickerbocker* X. 170 He was a busted man. **1860** *Harper's Mag.* July 278/2 Having lost all he had..on a 'busted' bank at a game of poker. **1868** M. M. POMEROY *Nonsense* xx. 185 Now I'm a gone nutmeg, a busted what-do-you-call-it. **1909** 'O. HENRY' *Roads of Destiny* vi. 97 The precarious busted flush. *Ibid.* xiii. 212 There was a buck coon who had drifted down there from a busted-up colored colony. **1919** W. S. MAUGHAM *Moon & Sixpence* xlvii. 198 Busted?..I'll get you some breakfast. **1920** MULFORD *J. Nelson* vi. 58 They took him over to th' SV to set Ol' Arnold's busted laig. **1953** A. UPFIELD *Murder must Wait* xviii. 161 Damn this busted leg. **1956** 'N. SHUTE' *Beyond Black Stump* 205 There's no goddam oil here. This is a dry hole, a busted flush.

bustee. Add: Also **busti.** (Earlier, later, and *attrib.* examples.)

1883 *Sat. Rev.* 8 Sept. 301/2 Collections of huts, which are known as 'Bustees'. **1907** *Westm. Gaz.* 16 Sept. 8/3 A large tiger..lurking among the 'bustis'. **1960** 'S. HARVESTER' *Chinese Hammer* XXV. 202 *Bustee* hovels and sugar-cake temples..Calcutta. **1963** *Times* 30 May 13/6 In *bustees*—as the shanty towns where 25 per cent of the population live are called—..fresh, clean water is never available.

buster. Add: **2. b.** Also used as a slang form of address, usu. friendly or slightly disrespectful; 'mate', fellow; *old buster*: an affectionate or disrespectful designation for an elderly man.

1905 H. A. VACHELL *Hill* vi. 137 You funny old buster. **1919** WODEHOUSE *My Man Jeeves* 79 An extremely wealthy old buster. **1931** M. A. VON ARNIM *Father* ix. 213 Shall you be all right, shut up alone with that old buster? **1948** A. SEAGER *Inheritance* 174 'Hi-ya, buster. What's new?' he heard a woman's coarse voice say. **1962** A. SHEPARD in *Into Orbit* 101 'OK, Buster,' I said to myself, 'you volunteered for this thing.' **1965** P. ARROWSMITH *Jericho* xix. 199 If you go on accusing me of attacking you lot, buster, you'll have the police to answer to.

3. a. (Earlier and later examples.) Also, any violent gale. (Usage not confined to Australia.)

1848 BARTLETT *Dict. Amer.* 57 'This is a buster,' i.e. a powerful or heavy wind. **1852** G. C. MUNDY *Our Antipodes* I. ii. 83 The Brickfielder, or, as the Port Jackson boatmen call it, the Sútherly Būster! **1888** *Longman's Mag.* Apr. 630 In New Zealand there are the *Southerly Busters*, following the dry hot 'nor'-wester'. **1935** 'L. LUARD' *Conquering Seas* xvii. 231 Flamers in sky. Northerly buster, I'm thinking. **1952** A. GRIMBLE *Pattern of Islands* v. 107 If the northerly buster takes your mainsail aback close-hauled to the south-easter. **1953** F. ROBB *Sea Hunters* vi. 59 Whatever was on the way would be no normal line squall—no summer 'buster'! **1956** S. HOPE *Diggers' Paradise* viii. 71 Sometimes the south wind or a 'south-westerly buster' puts a chill on the city [*sc.* Sydney].

b. (Examples.) = BURSTER 1 c. Also *fig.*

1928 'BRENT OF BIN BIN' *Up Country* xi. 184 The fancy riding resulted in 'busters' galore to skilled and unskilled. **1952** D. NILAND in *Coast to Coast 1951–2* 198 She doesn't look like having a buster. **1968** *Sunday Truth* (Brisbane) 30 June 20 The Australian Government has come an incredible double buster on the design for its Vietnam campaign medal.

4. A horse-breaker. (Cf. *bronco-buster* s.v. *BRONCO.) *U.S.*

1891 *Harper's Mag.* July 208/2 The buster must be careful to keep well away from sheds and timber. **1903** *Wide World Mag.* Apr. 545 On a large ranch which employs many cowboys, there is much rivalry among them as to who is the best rider or buster.

-buster. As the second element of an objective compound, as in **bronco-buster*; in familiar designations of guns, bombs, etc., as in **block-buster*.

bustle, *v.*[1] Add: **2. d.** Of a place: to be full of activity or bustle; to be alive *with*.

1880 *All Year Round* 9 Oct. 514 Transports..bustling with sailors. **1905** F. TREVES *Other Side of Lantern* x. 85 Those who walk along its empty terraces..see it as it was when it bustled with men and women three centuries ago.

bust-up (bʊ·st,ʊp). [See *BUST *sb.*[3]] = *burst-up* (see BURST *sb.* 2 b): an explosion (*lit.* and *fig.*); a flare-up, an altercation; excitement.

1846 *Knickerbocker* XXVIII. 313 The houdaciousest bust-up I ever seed. **1899** KIPLING *Stalky & Co.* 133 Then there's a big bust-up and a row that gets into the papers. **1908** W. DE MORGAN *Somehow Good* xi, There are hundreds of English equivalents for *éclaircissement*. There's bust-up. **1914** *English Rev.* Sept. 250 It [*sc.* this war] is the 'bust-up' of materialism. **1945** A. L. ROWSE *West-Country Stories* 148 They were having a tremendous bust-up with the railway porters about their belongings.

busty (bʊ·sti), *a.* [f. BUST *sb.*[1] + -Y[1].] Having a prominent bust.

1944 S. J. PERELMAN *Crazy like a Fox* (1945) 117 The leitmotiv of the campaign was a busty Polynesian hussy. **1954** K. AMIS *Lucky Jim* viii. 91 Bertrand's blonde and busty Callaghan piece.

busy (bi·zi), *sb.*[2] *slang.* [f. BUSY *a.*] A detective.

1904 *Daily Chron.* 17 Sept. 6/6 We had better slide; he looks like a 'busy'. **1928** *Sunday Dispatch* 2 Sept. 3/1 A porter..replied: 'Not to-night, sir...You see, we've had the "blow" that the "Busies" are coming.' **1948** M. ALLINGHAM *More Work Undertaker* (1949) xiv. 179, I don't know 'ow long we've got before the busies come trampin' in.

busy, *a.* Add: **1.** *to get busy*: to become active; to begin to act. *colloq.* (orig. *U.S.*).

1904 *Louisville Courier-Jrnl.* 27 Sept. 3 It was necessary to call upon the sergeant-at-arms...When that functionary got busy there came near being a riot. **1906** 'O. HENRY' *Four Million* 121 'Ikey,' said he,..'get busy with your ear. It's drugs for me if you've got the line I need.'

2. b. Of a telephone line: engaged; *busy tone*: the signal indicating that a line is engaged.

1893 KEENAN & RILEY *Transmitted Word* iv. 69 In this way..the operator..knows if the line is 'busy'. **1894** *Jrnl. Electr. Engin.* XXIII. 58 The operator has rarely to say, 'Wire engaged', or 'Busy', which is the common expression. *Ibid.* 63 False busy signals are frequently received. Now a third wire and an automatic restoring indicator has made the busy test reliable. **1914** W. ATKINS *Princ. Automatic Teleph.* 27 If all the lines are engaged the wipers will rotate to this busy-tone circuit. **1921** *Conquest* II. 127/2 The 'busy tone' is sent back to the calling subscriber if the line he wants is busy.

8. b. Of ornamentation: full of detail. Of a picture, photograph, etc.: having much or excessive background detail.

1903 *Burlington Mag.* III. 86/1 The decoration, in dealers' language, being too 'busy' for broad effects. **1909** *Athenæum* 1 May 535/3 His pictures..look a little 'busy', and ask to be displayed..with..reasonable relief of bare space. **1909** MRS. H. WARD *Daphne* v. 99 The ceilings spoilt, the decorations 'busy', pretentious, overdone.

11. busy Lizzie, any of various house plants of the family Balsamaceæ, *Impatiens sultani* or *I. holstii* or hybrids of these two species.

1956 X. FIELD *House Plants* 79 *Impatiens*..gathered nicknames, and among them were Patient Lucy, Patience Plant or just Busy Lizzie. **1970** *Woman's Own* 21 Mar. 23/1 The table where the variegated busy lizzie fought for living-space with the telephone.

busybody. Add: **b.** *transf.* A mirror attached to a building, reflecting a view of the street, etc. *U.S.*

1892 *Outing* (U.S.) Mar. 487/1 Only a 'tell-tale mirror'—otherwise 'a busybody'—set at a tantalizing angle outside the window. **1942** C. MORLEY *Thorofare* xxxv. 159 One of those telltale mirrors called 'busybodies' projecting from a three-sided bay upstairs.

busyness. Delete *rare* and add earlier and later examples.

1849 THOREAU *Week* 380 Behind every man's busy-ness there should be a level of undisturbed serenity. **1929** J. BUCHAN *Courts of Morning* iii. 351 It was a week of desperate busyness. **1961** A. WILSON *Old Men at Zoo* vii. 323 Nothing could chill the warmth I generated in my cocoon of busyness.

but, *conj.* Add: **27. b.** Also with preceding exclamation occasionally omitted, esp. as a gallicism (cf. F. *mais oui, mais énormément,* etc.), used to give emphasis to a following word or statement, and with the sense of 'indeed'.

1887 M. CORELLI *Thelma* II. II. ii. 156 'I believe you would do it if I asked you!' he said. 'But, of course!' **1941** B. SCHULBERG *What makes Sammy Run?* viii. 153 Now you've really bitched yourself up. But good. **1960** L. COOPER *Accomplices* II. ii. 84 If the other..looks like getting in the way it must be turned out, but quick. **1965** I. FLEMING *Man with Golden Gun* vi. 81 I'm goin' fix that man, but good.

c. Introducing an emphatic repetition.

1920 R. MACAULAY *Potterism* VI. iv. 239 A weekly, and it promised to sell better than any other weekly on the market, but far better. **1921** 'K. MANSFIELD' *Coll. Stories* (1945) 422 She knew what it was to be in love, but-in-love! **1928** D. H. LAWRENCE *Lady Chatt.* xiv. 240 And about *everything* I talked to her: but everything. **1950** S. ERTZ *Prodigal Heart* iv. 68 They're very gay, and entertain a lot and I feel years older than they are. But years older. **1958** B. NICHOLS *Sweet & Twenties* iv. 59 He sold enormously—but enormously. *Ibid.* x. 132 There is nothing, but nothing, to be said for the female knee.

butadiene (biū·tādəi·īn). *Chem.* [f. *BUTA(NE + DI-[2] + -ENE.] Either of the two isomeric hydrocarbons $CH_2{\cdot}CH{:}CH{\cdot}CH_3$ (also called *methylallene*) and $CH_2{:}CH{\cdot}CH{:}CH_2$ (also called *erythrene*), the latter being used as a constituent of various artificial rubbers (cf. *BUNA). Also *attrib.* and *Comb.*, as *butadiene-acrylonitrile, -styrene.*

1900 *Jrnl. Chem. Soc.* LXXVIII. I. 2 According to the author's theory of unsaturated compounds, butadiene.. should yield an αδ-dibromide on direct addition of bromine. **1911** *Chem. Abstr.* V. 3519 Hydrocarbons of the Butadiene Series and Artificial Caoutchoucs. **1937** *Discovery* Aug. 247/1 The real raw material of synthetic rubber—a product known as butadiene. **1959** *Times* 13 Mar. 10/1 (Advt.), Cariflex general purpose butadiene-styrene rubber. **1959** *Times* (Suppl.) 27 Apr. vi/7 For oil and fuel resistance the butadiene-acrylonitrile type rubbers are in common use.

butane (biū·tēⁱn). *Chem.* [f. BUT(YL + -ANE 2 b.] Butyl hydride, C_4H_{10}; = TETRANE. Also *attrib.* and *Comb.*

1875 *Jrnl. Chem. Soc.* XXVIII. 561 The simplest saturated hydrocarbons which afford substitution-products belonging to all the three groups of primary, secondary, and tertiary nitro-compounds, are the two isomeric butanes. **1877** *Fownes' Chem.* (ed. 12) 47 Normal Butane, Diethyl, or Methyl-propyl. **1918** COLVER *High Explosives* 60 The propane, butane, and pentane hydrocarbons in the form of gases and vapours which can only be condensed with difficulty. **1935** *Discovery* Dec. 361/1 The allied gas, butane, compressed similarly in cylinders, is now in frequent use for gas-cookers etc. **1953** *Ann. Reg. 1952* 74 The experimental use of butane–air mixture [for gas supplies] at Whitland. **1958** *Engineering* 7 Feb. 192/3 A butane burning blowlamp is now marketed in Britain.

butanol (biū·tānǫl). *Chem.* [G., f. *BUTAN(E + -OL.] Butyl alcohol, esp. the normal isomer.

1894 G. M'GOWAN tr. *Bernthsen's Text-bk. Org. Chem.* (ed. 2) 93 Normal butyl alcohol. [*Note*] 1-Butanol. *Ibid.*, Secondary butyl alcohol. [*Note*] 2-Butanol. **1957** *Encycl. Brit.* IV. 472/2 Normal butyl alcohol (butanol) is made commercially by the fermentation of corn or molasses and also from acetylene. **1964** BODEN & KAMPA in *Oceanogr. & Marine Biol.* II. 351 Considerable purification has been obtained with butanol extraction.

butanone (biū·tānōᵘn). *Chem.* [f. *BUTAN(E + -ONE.] Methyl ethyl ketone, $CH_3{\cdot}CH_2{\cdot}CO{\cdot}CH_3$, an ethereal liquid used industrially.

1905 *Jrnl. Chem. Soc.* LXXXVIII. I. 355 *cyclo* Butanone ..prepared by brominating tetramethylenecarboxylic acid and treating the bromo-anhydride thus obtained with ammonia. **1964** N. G. CLARK *Mod. Org. Chem.* x. 194 *Butanone*..also known as methyl ethyl ketone..is largely used as a solvent for cellulose nitrate..and vinyl resins.

butch (butʃ), *sb.*[1] *slang* (orig. *U.S.*). [Origin unknown, but perh. = next.] A tough youth or man; a lesbian of masculine appearance or behaviour. Also *attrib.* or as *adj.* In the U.S. also applied to a type of short haircut, crew-cut.

[**1902** *Pinkerton's Nat. Detective Agency Circular* 3 Feb., George Parker, alias 'Butch' Cassidy, alias George Cassidy. **1930** D. RUNYON *Guys & Dolls* (1932) 71 (*title*) Butch minds the baby.] **1941** AUDEN *New Year Let.* III. 72 And culture on all fours to greet A butch and criminal élite. **1947** *Cromwellian* May 35 *Butch*, a strongly-built person. **1949** W. R. BURNETT *Asphalt Jungle* (1950) v. 30 His white-blond hair disfigured by a butch haircut. **1950** *Neurotica* Spring 37, I no longer remotely resembled a 'butch' fairy. **1954** *News* (San Francisco) 10 Sept. 1 Then some of the girls began wearing mannish clothing. They called themselves 'Butches'. **1954** *Observer* 26 Sept. 9/2 The homosexual Stoics with their crew-cuts and 'butch' behaviour. **1965** *New Statesman* 26 Mar. 493/1 This rejection of the female role is very common among the 'butch' type of lesbian. **1966** *Ibid.* 13 May 696/3 One of the femmes, secure in the loving protection of her butch. **1967** W. MURRAY *Sweet Ride* ix. 148 Three burly men with butch haircuts.

butch, *sb.*[2] Colloq. abbrev. of BUTCHER *sb.* (see quots.).

1919 DOWNING *Digger Dialects* 14 *Butch*, doctor. **1942** W. FAULKNER *Go Down, Moses* 164 Boon bought..a bottle of beer from the news butch. **1960** WENTWORTH & FLEXNER *Dict. Amer. Slang* 80/2 *Butch*,..a vendor.

butcher, *sb.* Add: **3.** (Earlier example.)

1860 C. M. YONGE *Hopes & Fears* I. II. iv. 229 The doctor the butcher, the duchess, and all her other fabrications of.. feathers.

b. A vendor of sweets, fruit, etc., in a railway train, a theatre, etc. *U.S. colloq.*

1882 J. J. JENNINGS *Theatrical & Circus Life* 513 In the spring ye 'candy-butcher' shows confections old and tough. **1883** G. W. PECK *Peck's Bad Boy* 54 They never prayed in circus, 'cept the lemonade butchers. *a* **1889** *Detroit Free Press* (Barrère & Leland), On a Michigan central train the other day as the butcher came into the car with a basket of oranges [etc.]. **1924** W. M. RAINE *Troubled Waters* vii. 70 From the train butcher he bought a magazine and settled himself for a long ride.

c. Short for *butcher('s) blue*.

1922 *Daily Mail* 13 Dec. 1 (Advt.), Strong Cotton *Dresses*. In plain Butcher, Navy, Brown, [etc.]. **1923** *Ibid.* 30 May 4 (Advt.), Mauve, Pink or Butcher on White ground. **1952** 'P. WENTWORTH' *Brading Coll.* xv. 93 Most of the time she wore blue...Navy, or butcher.

d. A glass or measure of beer (see quots.). *Austral. slang.*

1898 MORRIS *Austral Eng.* 73/1 *Butcher*, South Australian slang for a long drink of beer, so-called (it is said) because the men of a certain butchery in Adelaide used this refreshment regularly. **1934** *Bulletin* (Sydney) 7 Mar. 10/3 A 'butcher'..is identical in volume with the fourpenny glass of the other capitals. **1945** BAKER *Austral. Lang.* ix. 168 *Butcher* is Adelaide slang; in the early days it was used for a glass containing about two-thirds of a pint. In modern times the size has dropped to about half a pint.

5. butcher blue = *butcher's blue*; **butcher-boots** *pl.*, high boots without tops (see TOP *sb.*[1] 10); **butcher-crow,** a crow-shrike (Funk's Stand. Dict. 1893); **butcher's knife, butcher-knife** (examples); also, any large, strong-bladed knife of many uses; **butcher's sleeves,** short sleeves covering the forearm from elbow to wrist, worn by butchers as a protection against soiling the sleeves of their ordinary clothes.

1909 D. LEVITT *Woman & Car* ii. 28 Indispensable to the motoriste..is the over-all. This should be made of butcher blue or brown linen. **1923** R. MACAULAY *Told by an Idiot* III. vi. 199 The girl in a short butcher-blue cotton frock. **1960** *Times* 18 Jan. 15/2 Dresses in linens, such as one in butcher blue. **1861** WHYTE MELVILLE *Market Harb.* v, My friend sharing with me a strong prejudice against what have been termed 'Butcher-boots'. **1886** *Eng. Illustr. Mag.* Mar. 414/1 A man in a round hat and butcher-boots is as out of place at a hunt as a man in a tweed suit at a ball. **1897** *Badminton Mag.* IV. 397 Men in cords and butcher boots, tweeds and gaiters. **1941** *Manch. Guardian Weekly* 17 Jan. 45 Butcher boots are de luxe knee-boots used by officers. **1714** *Boston News-Let.* 1–8 Mar. 2/2 Butchers knives. **1766** H. BROOKE *Fool of Qual.* II. 591 Pulling out his butcher's knife from a sheath in his side-pocket, he..made a stab at

my heart. **1822** *Massachusetts Spy* 25 Dec. (Th.), Her foot slipt, and she fell upon a large butcher-knife which she had in her hand. **1878** J. H. Beadle *Western Wilds* xviii. 294 We fell to with our..butcher-knives and dug several holes. **1856** *Household Words* XIII. 206/2 After a long delay the doctor came in, with scientific butcher's sleeves on his arms, and an apron tied round his portly waist.
6. *butcher's*, short for *butcher's hook*, rhyming slang for 'look'.
1936 J. Curtis *Gilt Kid* ii. 23 And while he's there he takes a butchers. **1961** N. Amis *Take Girl like You* xxvii. 311 Have a butcher's at the *News of the World*.
b. *Austral.* and *N.Z. rhyming slang.* (See quots.)
1941 Baker *Dict. Austral. Slang* 16 *To be butcher's*, to be angry, annoyed (about something). Often 'go butchers at' (i.e., 'go butcher's hook' or crook). **1943** J. A. W. Bennett in *Amer. Speech* XVIII. 90 *To go crook* is to show anger or annoyance, to 'sling off at'; and *to go butcher's hook* is presumably a development of this in rhyming slang [in New Zealand]. **1951** D. Stivens *Jimmy Brockett* 126 As soon as Sadie came in I went butcher's hook. 'What's this bloody nonsense about a studio, Sadie?' I said, going straight to the point.

butcher, *v.* Add: **4.** To cut up or divide (an animal or flesh) after the manner of a butcher; to cut *off* or *from* in this fashion. *U.S.*
1822 J. Fowler *Jrnl.* (1898) 122 The former killed two Elk, and left the latter to butcher them. *a* **1848** Ruxton *Life Far West* (1849) iv. 128 The..body of one of the Indian squaws, with a large portion of the flesh butchered from it. *Ibid.* 160 Bill..called to him..to butcher off a piece of meat and put it in the pot. **1855** Mayne Reid *Hunters' Feast* xxxiv, The fat cows only were 'butchered'. The bulls were left where they had fallen, to become the food of wolves.
5. *intr.* To do butchering. *U.S.*
1865 *Atlantic Monthly* XV. 454 If it isn't about time to butcher; we butchered last year [etc.]. **1896** *Scribner's Mag.* VI. 484/1 'Don't butcher next week. Friday is Christmas day.'..'Well, we always butcher Christmas week, don't we?'

butcherdom (bu·tʃədəm). [f. Butcher *sb.* + -dom.] Butchers collectively, or their trade.
1889 in *Cent. Dict.* **1904** H. Hawkins *Remin.* II. 109 The butcher's slander was one that seemed to shake the very foundations of butcherdom throughout the world.

butchering, *vbl. sb.* Add: **3.** The slaughtering of cattle. Also *attrib.*, as *butchering cow*.
1773 in *Maryland Hist. Mag.* (1920) XV. 63, I expect one more steer from the Island; the last we had from thence was..miserably mangled in the butchering. **1900** *Daily News* 10 Apr. 8/6 Fat butchering cows.

bute, var. *beaut. *U.S.*
1903 A. H. Lewis *The Boss* 272, I know that gang of card sharps,..an' they're a bunch of butes at that!

butea (biū·tiă). [mod.L. (W. Roxburgh 1792, in *Asiatick Researches* III. 469), named after John Stuart, Earl of *Bute* (1713–92).] A member of a genus of Indian or Chinese trees or climbers so named, belonging to the family Leguminosæ; esp. *Butea frondosa*, the dhak or palas of India. Also the resin of these trees (in full *butea gum*).
1792 W. Roxburgh in *Asiatick Researches* III. 469 (*title*) A Description of the Plant Butea. *a* **1815** —— *Flora Indica* III. 236 Here it [*sc.* the gum of *Pterocarpus marsupium*] differs from Butea gum. **1866** Lindley & Moore *Treas. Bot.* 183 Butea kino or gum butea. **1879** Watts *Dict. Chem.* I. 686 Butea gum. Bengal kino.—The juice of *Butea frondosa*, Roxb., often sent into the market instead of genuine kino. **1921** R. S. Troup *Silvicult. Indian Trees* I. 261 The *Butea* seedlings suffered greatly from the attacks of porcupines.

butenyl (biū·tinil). *Chem.* [f. Buten(e + -yl.] Any of the three isomeric forms of the radical C_4H_7—, of which the two normal butylenes are hydrides.
1881 *Jrnl. Chem. Soc.* XL. 711 A liquid is obtained, which, on evaporating, leaves butenylglycerol as a thick yellow liquid. **1884** Roscoe & Schorlemmer *Treat. Chem.* III. II. 377 Butenyl Alcohol, $C_4H_7(OH)_3$. **1957** T. S. Stevens in E. H. Rodd *Chem. Carbon Compounds* IVa. iii. 228, 2-Butenylthiophen yields 95% of thionaphthen when passed over kieselguhr-P_2O_5 at 600°.

butine (biū·tain, -in). *Chem.* [ad. G. *butin*, f. L. *būt-yrum* Butter *sb.*¹ + -ine⁵.] †**1.** The name given by W. Heintz (*Ann. der Physik und Chem.* (1853) XC. 151) to a substance present in butter. *Obs.*
1867 Bloxam *Chem.* 582 Butter..contains also butine, which yields glycerine and butic acid ($HO.C_{40}H_{39}O_3$) when saponified.
2. Orig., any of the four unsaturated unbranched hydrocarbons with the formula C_4H_6; later restricted to the two isomers that have a triple bond, those with two double bonds being called *butadiene.
The word is now obsolescent, being replaced by *butyne.
1884 Roscoe & Schorlemmer *Treat. Chem.* III. II. 463 Butine, $C = H_3CH.CH = CH_2$. When erythrite, $C_4H_6(OH)_4$, is heated with concentrated formic acid to 230°, butine is formed. **1888** *Amer. Chem. Jrnl.* X. 431 The fourth butine, methylisoallylene, was the subject of this investigation. **1889** G. M'Gowan tr. *Bernthsen's Text-bk. Org. Chem.* 57 Crotonylene or Butine, $CH_2=CH—CH=CH_2$, is contained in illuminating gas. *Ibid.*, Pyrrolylene, C_4H_6, from pyrrolidine, is probably identical with the butine from erythrite. **1895** Thomson & Bloxam *Bloxam's Chem.* (ed. 8) 524 The hydrocarbon C_4H_6 (butine) can exist in two forms, each of which will have a pair of trebly-linked carbon atoms. **1934** *Chem. Abstr.* XXV.II. 2321 EtI and NaC:CH in liquid NH_3 at −45° give 78% of 1-butine.

butler, *sb.* **3. butler's pantry** (earlier example).
1816 Jane Austen *Emma* II. vi. 113 A bad butler's pantry.

butlerish (bʌ·tlərɪʃ), *a.* [f. Butler *sb.* + -ish¹.] Belonging to or characteristic of a butler.
1923 A. Huxley *Antic Hay* iii. 45 He moved with a certain pomp, a butlerish gravity.

butling: see *buttle *v.*

butt, *sb.*² Add: **3. butt-beaker** *Archæol.*, a butt-shaped beaker; **butt-howel,** a howelling-adze used by coopers (Knight *Dict. Mech.* 1874); **butt-shaped** *a.*, shaped like a butt or cask; spec. *Archæol.* applied to a type of Belgic pottery.
1933 *Antiquity* VII. 29 There are no butt-beakers, no imported Italic wares. **1941** *Proc. Prehist. Soc.* VII. 140 Butt-beaker of sandy, biscuit-coloured ware. **1916** T. May *Pottery at Silchester* 168 Tall 'butt-shaped vase'. *Ibid.*, This is a still later development of the 'butt-shaped' beaker. **1930** —— *Catal. Rom. Pottery in Colchester & Essex Museum* 12 'Butt-shaped' beaker.

butt, *sb.*³ Add: **6. butt-hole,** a blind hole, a cul-de-sac.
1905 *Westm. Gaz.* 3 Mar. 3/2 The old dog's got him [*sc.* a badger] in a butt hole.
7. The piece of the inner margin of a single leaf of a book, which projects as a narrow strip beyond the sewing or other fastening when the book is bound.
1921 A. Esdaile in *Library* Dec. 185 The last leaf of B. is a single leaf, whose butt is visible after B. 1.
8. The fag-end of a cigar or of a cigarette.
1847 W. H. Gregory *Paddiana* I. 235 Will yer honor give me the butt? **1888** Kipling *Departm. Ditties* (1890) 106 Like the butt of a dead cigar. **1918** Wodehouse *Piccadilly Jim* i. 5 'Smoking cigarettes.'..'There are two butts in the ash-tray.' **1940** Dylan Thomas *Portr. Artist* 117, I cupped a match..I puffed my last butt. **1958** S. Ellin *Eighth Circle* (1959) II. v. 62 A litter of used paper cups and cigarette butts.

butt, *sb.*⁴ **1.** Delete †*Obs.* and substitute: *Obs.* exc. *dial.* and *U.S.*
1838 W. Holloway *Dict. Provinc.* 23 Butts and Bounds, the borders of a person's estate. *E. Sussex.* **1903** McFaul *Ike Glidden* vii. 44 Have you any documents for reference in order to fix the butts and bounds?
2. c. In grouse-shooting, a position either sunken or on the level ground, protected by a wall or bank of earth behind which the sportsman may stand and fire unobserved by the game.
1885 W. S. Stanhope *Let.* 28 Nov. in Walsingham & Payne-Gallwey *Shooting: Moor & Marsh* (1886) I. 11, I began to shoot grouse in 1841; we had our regular drives then, but without butts. **1897** *Encycl. Sport* I. 489/2 The butts, or batteries, as they are indifferently called. **1935** *Encycl. Sports* 322/2 A semi-circular, or..almost circular wall of peat or turf, covered with heather..is the usual form of butt. **1955** *Times* 13 May 12/6 The butts all ready set up for the grouse-shooting.

butt, *sb.*⁷ Add: **2. butt-chain** (see quot.); **butt-joint** *v. trans.*, to join with a butt-joint; **butt-riveting,** riveting in which a butt-strap is used; **butt-strip** = *butt-strap* sb.; **butt-weld** *sb.*, a butt-joint made by welding; *v. trans.*, to join with a butt-weld (Webster 1909); **butt-welded** *a.*, joined with a butt-weld; so *butt-welding*.
a **1877** Knight *Dict. Mech.*, *Butt-chain* (Saddlery), a short chain which reaches from the leather tug to the single-tree, to each of which it is hooked. **1885** *Spons' Mechanics' Own Bk.* 361 There are 3 or 4 ways of butt-jointing curbs. **1904** Goodchild & Tweney *Technol. & Sci. Dict.* 77/2 *Butt riveting*, a riveted joint where the plates touch at the edge only, and a strip overlaps and is riveted to both of them. *Ibid.*, *Butt strip*, the strip of plate used to cover a butt joint. **1864** Webster s.v. *Weld*, Butt-weld, or jump-weld. **1944** *Jrnl. Iron & Steel Inst.* CL. 50A The abrupt change in section at the edge of the reinforcement of a butt weld is a serious stress raiser. **1927** *Glasgow Herald* 27 Aug. 12 Butt-welded tubes. **1962** *B.S.I. News* Feb. 10/1 Storage tanks (vertical mild steel welded with butt-welded shells). **1925** *Jrnl. Iron & Steel Inst.* CXII. 463 Butt welding and its application to joining wires.
3. b. A place where the stratum of rock to be quarried is cut off by other rock.
1900 *Coal & Metal Miners' Pocket Bk.* (ed. 6) 576 The butt of a slate quarry is where the overlying rock comes into contact with an inclined stratum of slate rock.

butt (bʌt), *sb.*¹⁴ *dial.* Also **but.** A shoemaker's knife. In full *butt-knife*.
1847 Halliwell, *But*, a shoemaker's knife. *North.* **1905** *Daily Chron.* 7 Feb. 3/1 Butt-knives..of French and Swedish makes.

butt, *v.*¹ Add: **1. d.** *to butt in*: to thrust oneself unceremoniously and uninvited into an affair, discussion, etc.; to intrude, interfere without good reason. orig. *U.S.*
1900 G. Ade *Fables in Slang* 106 One Student..whose people butt into the Society Column with Sickening Regularity. **1904** *Philad. Even. Telegraph* 8 June … **1915** W. J. Locke *Jaffery* viii, If a man loves a woman, he ought to know what to do with the guy that butted in, without being told. **1920** R. Macaulay *Potterism* iv. i, I've not gone there or written, or anything yet, because I didn't want to butt in. **1922** … **1928** R. Campbell *Wayzgoose* ii. 53 And what if total strangers butted in? **1957** E. Eager *Magic by Lake* 69 'I'm sorry,' he said, 'butting in like this, but I've got to tell you something.'

butt, *sb.*⁵ Add: **1.** (Later U.S. examples.)
1816 U. Brown *Jrnl.* in *Maryland Hist. Mag.* (1915) X. 366 This 100 acres is butted and bounded and described directly as the grant is from the Commonwealth of Virginia. *Ibid.* 367 This John Hall Butted and Bounded of the 2 first lines with our 70,000 Acre tract.
3. (Later U.S. example.)
1853 P. P. Kennedy *Blackwater Chron.* i. 6 A large spur..apparently the Backbone itself—keeps straight to the south, and butts down on the Cheat.

butt, *v.*² **2.** Delete '?' and add earlier examples.
1774 M. Patten *Diary* (1903) 331, I cut of 9 Rail cuts and butted 8. **1850** S. Judd *R. Edney* 41 Richard took an axe and very neatly proceeded to 'butt' a log; that is, to cut the end of it square off.

bu·tt-cut. *U.S.* [Butt *sb.*³ 2.] **a.** The first portion of a tree cut off above the stump. Also *fig.*, an important or a heavy man. **b.** The section of tan-bark taken from the butt of a tree before felling it for further peeling (*Terms Forestry & Logging* 1905).
1830 *Northern Watchman* (Troy, N.Y.) 19 Oct. (Th.), [He] weighs little short of 450 lbs. and is familiarly known as the But-cut. **1840** J. P. Kennedy *Quodlibet* 172 Nebuchadnezzar couldn't beat him at a speech. He's the Butt cut of democracy. **1878** J. H. Beadle *Western Wilds* x. 143 The 'butt cut' of the tree lies as it fell, the top reached by means of a ladder. **1948** *Country Gentleman* May 177/1, I rived enough boards to kiver it from the butt-cuts.

butt-end, *sb.* Add: **4.** *Hockey* and *Lacrosse*. A jab or thrust with the handle end of a hockey-stick. Chiefly *Canadian*.
1963 A. O'Brien *Headline Hockey* 79 Blair told Ted O'Connor to hand No. 11 of the Swedish national team a butt-end. **1965** *Globe & Mail* (Toronto) 9 Apr. 38/1 Nobody ever gave you a better butt-end.

butt-end, *v.* Add: **2.** *Hockey* and *Lacrosse*. To jab or thrust at (an opponent's body) with the handle end of the stick. So **butt-ending** *vbl. sb.* Chiefly *Canadian*.
1955 *Telegram* (Toronto) 21 Apr. 6/8 A hockey coach says even youngsters playing hockey are taught kneeing and butt-ending. **1965** *Globe & Mail* (Toronto) 9 Apr. 38/1 The art of butt-ending is almost an obsolete illegality in hockey.

bu·tt-ended, *a.* [f. Butt-end *sb.* + -ed².] Having a blunt end; having ends that butt or come flat, the one against the other.
1897 Kipling *Capt. Cour.* v. 124 *Standez awayez*, you butt-ended *mucho bono*! **1898** *Cycling* 72 A hopeless single-tube tyre may be slit open at the valve and a butt-ended air-tube inserted.

butter, *sb.*¹ Add: **3. c.** A perfumed fat obtained by inflowering or maceration with a heated fat.
1885 *Encycl. Brit.* XVIII. 526 For the manufacture of perfumes for the handkerchief the greases now known as pomades, butters, or philocomes are treated with rectified spirit of wine..which practically completely abstracts the odour.
4. c. *butter-bright*, *-coloured* (example), *-like* (earlier and later examples), *-smooth* adjs.
1868 G. M. Hopkins 17 July in *Jrnls. & Papers* (1959) 176 The sun coming out..with a butter-bright lustre. **1894** *Daily News* 20 Mar. 3/2 A deep frill of butter-coloured lace. **1700** W. Salmon tr. *Bate's Pharm. Bat.* (ed. 2) 380/2 A butter-like oil. **1964** M. Hynes *Med. Bacteriol.* (ed. 8) xi. 157 To give a butter-like flavour to margarine. **1920** Galsworthy *In Chancery* II. v. 170 His grandfather's first gold hunter watch, butter-smooth with age. **1969** *Observer* (Colour Suppl.) 23 Mar. 27/2 (Advt.), There isn't even a clutch pedal. Just a Selective-Automatic transmission. With butter-smooth gears.
5. butter-and-egg man *U.S. slang*, a wealthy, unsophisticated man who spends money freely; **butter-basher, butter-boy** *slang*, a new driver of a taxi-cab; **butter-bush,** an Australian tree, *Pittosporum phyllirœoides*, of which the wood is used for turnery and the leaves as fodder; **butter-cake,** a rich cake usu. containing butter, sugar, flour, and eggs;

butter cloth, a thin loosely-woven cloth with a fine mesh used primarily as a wrapping for butter; **butter colour,** (*a*) the colour of butter; (*b*) a preparation used to give a good colour to butter and butter substitutes (Cent. Dict. 1889); **butter-cooler** (earlier examples); **butter-duck** *U.S.* (see quot.); **butter fat,** the essential fats of pure butter; also *attrib.*; **butter-knife** (example); **butter letter,** a letter issued on ecclesiastical authority giving permission to eat butter in Lent; **butter muslin** = *butter cloth*; **butter oil,** that part of refined cotton-seed oil which is used in making oleomargarine; **butter paper,** a semi-transparent waterproof wrapping-paper for butter, cream cheese, etc.; **butter-pat** (examples); **butter salt,** fine common salt in small crystals obtained by rapid evaporation of brine, used in salting butter; **butter scoop** (see quot. 1902); **butter-slide,** a slide (Slide *sb.* 9) made of butter or ice; also *fig.*; **butter spade,** a wooden spatula used in cutting butter from a firkin or other vessel, or used (as one of a pair) for making up butter; **butter stamp** = Butter-print 1; **butter-stick,** a wooden implement used in working butter; **butter substitute,** a substance used as a substitute for butter in food; **butter tongs** (see quot.); **butter trier** *U.S.*, a segment of a tube used to pierce a firkin of butter for sample; **butter week** (see quots.); **butter-worker** (examples); **butter-working,** the moulding of butter into rolls, prints, pats, etc., for sale; **butter yellow,** a coal-tar dye formerly used for colouring butter, oils, etc.
1926 H. C. Witwer *Roughly Speaking* 229 A couple of big butter and egg men from Verona, New Jersey. **1927** *Daily Express* 31 Aug. 8/7 'Butter and egg man' is an American slang expression practically equal to our term 'greenhorn', that is, a man of money who spends lavishly and is an easy prey of the gold-digger and other unscrupulous persons. **1948** *Antioch Rev.* Spring 105 The 'butter-and-egg' man who startles the foreign lecturer with blunt questions. **1939** H. Hodge *Cab, Sir?* xv. 216 Contemptuous cabmen, therefore, called these blacklegs 'Butter-bashers'. *Ibid.* x. 134 During my 'butter-boy' period. *Ibid.* xv. 215 The new driver is called a 'Butter-Boy'. **1960** C. Ray *Merry England* 26 [The] owner-driver..is called a 'butter-boy' when he first appears on the rank, taking the butter from the older hands' bread, they say. **1885** *Outing* (U.S.) Nov. 180/2 A thick hedge of butter-bush. **1936** I. L. Idriess *Cattle King* xxviii. 252 The rabbits had killed all the white wood, apple-bush and butter-bush. **1747** H. Glasse *Art of Cookery* xv. 139 (*heading*) To make a Butter Cake. **1897** G. du Maurier *Martian* II. 53 Scotch butter-cake. **1963** *Times* 11 Mar. 13/5 Irish butter cakes...Serve hot or cold. **1885** Wilde *Lett.* (1962) 172 My wife has a huge bill against you—for your meat-safe and the buttercloth. **1894** *Daily News* 20 Mar. 3/1 Yellowish cream-colour, called butter-colour by the modistes. **1895** *Montgomery Ward Catal.* 573/1 Improved Butter Color...1 Gallon Cans. **1790** *Pennsylvania Packet* 7 Dec. 3/3 Butter coolers. **1875** G. H. Lewes *Problems Life & Mind* II. 135 The china service and glass butter-cooler. **1857** J. G. Swan *Northwest Coast* 357 The Colonel saw a 'butter-duck' in a shallow creek...These ducks are the black surf-duck (*Fuligula perspicillata*). **1889** in A. Davis *Package & Print* (1967) Pl. 69 Powdered milk without butter fat. **1899** *Daily News* 17 Feb. 8/3 The sample..afforded no evidence of the presence of fat other than butter fat. **1906** *Macmillan's Mag.* June 612 If wanting in butter-fat, it [*sc.* milk] was not fit for the purpose for which it had been sold. **1946** *Nature* 12 Oct. 522/1 The obvious technique for assessing levels of performance is milk and butter-fat recording. **1850** Dickens *Dav. Copp.* lxi. 602 Fish-slices, butter-knives, and sugar-tongs. **1893** *Westm. Gaz.* 25 Feb. 5/3 In Italy, butter is prohibited [in Lent]...The Northerners, however,..would have none of this, and special 'butter-letters' were consequently dispatched to them from the obliging Vatican. **1902** *Connoisseur* II. No. 8. p. xvii, Frilled Butter Muslin. **1903** Tisdale & Robinson *Soft Cheesemaking* 34 Instead of paper, the cheese is done up in butter muslin. **1906** —— *Buttermaking* 55 Place a damp butter-muslin over the roller and butter-board. **1951** *Good Housek. Home Encycl.* 193/1 The coarser muslins, such as butter muslin, are used for household purposes, for example, straining liquids. **1894** *Dairy Rev.* Aug. 46/2 Some makers used to prepare the annatto in butter oil. **1895** *Montgomery Ward Catal.* 573/1 Waxed Butter Paper, grease proof. **1844** *Ainsworth's Mag.* VI. 547 The two oars appeared to be playing the parts of butter-pats with him. **1953** Dylan Thomas *Under Milk Wood* (1954) 30 My crisp toast-fingers, my home-made plum and butterpat. **1884** R. Holland *Gloss. Wds. Chester*, *Butter salt*, salt-making term. A fine boiled salt, not stoved, used specially for making up butter. **1892** *Cornhill Mag.* Sept. 264 The unmoulded salt—locally termed 'butter-salt'—is sent away in trucks. **1872** O. W. Holmes *Poet Breakf.-t.* i. 2 As the market people run a butter-scoop through a firkin. **1902** *Encycl. Brit.* XXVII. 358 The butter-scoop is of wood, and is sometimes perforated; it is used for taking the butter out of the churn. **1887** Wilde in *Court & Soc. Rev.* 2 Mar. 207/2 He met with a severe fall, through treading on a butter-slide, which the twins had constructed. **1927** W. E. Collinson *Contemp. Eng.* 20 Ice to make slides (if very slippery sometimes called a *butterslide*). **1928** *Observer* 1 Apr. 22 It will do us all the good in the world to slip on a mental butter-slide now and again. **1895** *Montgomery Ward Catal.* 572/2 Wooden Butter Spades and Ladles. **1906** *Chambers's Jrnl.* Jan. 119/1 An old Dublin butter-spade with ivory handle. **1881** Ogilvie (Annandale) *Butter-stamp*, a piece of carved wood used to mark cakes of butter. **1836** *Southern Lit. Messenger* II. 480 To beat the collected ends of the fingers with an implement..made like a butterstick. **1888** Butter substitute [see Substitute *sb.* 6 b]. **1906** *Macmillan's Mag.* June 607 What are termed 'butter-substitutes',—in other words, fraudulent adulterants. **1908** *Westm. Gaz.* 5 Aug. 2/2 Vegetarians use very extensively a butter substitute derived from the fat of nuts. *a* **1877** Knight *Dict. Mech.*, *Butter-tongs*, an implement for cutting and transferring pieces of butter. **1868** *Rep. Comm. Patents 1867* (U.S.) II. 1218/1 *Butter Tryer*...The scraper fits the trough of the gouge to remove the butter therefrom. **1762** P. Murdoch tr. *Busching's Syst. Geogr.* I. 384 The Butter-week..when eating of flesh is forbidden and butter is allowed, is the week immediately preceding the great Fast of Lent. **1923** *Daily Mail* 3 Mar. 10 Maslenitza, or Butter Week, as the Russians call the fortnight preceding Lent, is always celebrated with feasting and drinking in Russia. **1854** *Rep. Comm. Patents 1853* (U.S.) 247 Improvement in Butter Workers. **1908** *Daily Chron.* 16 Nov. 7/1 Churns, cream separators, and butter-workers are turned out by the million in Stockholm. **1906** *Ibid.* 25 Sept. 2/6 One is reluctantly obliged to conclude that butter-working is a lost art amongst grocers' assistants. **1909** *Cent. Dict.* Suppl., Butter Yellow. **1956** *Nature* 24 Mar. 576/2 Rat liver tumours induced by butter yellow.

butter, *v.* Add: **2. b.** Now usu. with *up*. Hence **buttering-up** *vbl. sb.*
1819 T. Moore *Tom Crib's Memorial* 40 This *buttering-up*, against the grain, We thought was *curs'd* genteel in Bob. **1832** Lytton *Eugene A.* II. II. viii. 42 Your honour should see how they fawns and flatters, and butters up a man. **1924** E. M. Forster *Passage to India* ix. 106 'This is a great relief to us, it is very good of you to call, Doctor Sahib,' said Hamidullah, buttering him up a bit. **1943** H. Pearson *Conan Doyle* iii. 42 The little country practitioner who had been buttering them up for a quarter of a century found that he might as well put up his shutters.

bu·tter-ball. 1. A ball of butter; butter moulded into a ball.
1892 J. C. Harris *Plantation Printer* 58 She wur a fine cow too, ez fat ez a butter-ball. **1901** Greenough & Kittredge *Words* (1902) 177 Thus we have butterball, a ball that consists of butter. **1952** L. MacNeice *Ten Burnt Offerings* IX. 85 One minute quicksilver, next minute butter-balls.
2. The buffle-head or buffle-headed duck, *Clangula albeola*, of North America, so called from its exceeding fatness in autumn. *U.S.*
a **1813** Wilson *Amer. Ornith.* (1814) VIII. 51 Buffel-headed Duck...usually known by the name of the Butter-box, or Butter-ball. **1874** [see *bufflehead 2]. **1895** *Outing* (U.S.) XXVII. 212/1 Over me they went, so close that I could tell that they were butter balls. **1902** S. E. White *Blazed Trail* xiii. 85 Butterballs and scoters paddled up at his approach.

bu·tter-bean. A variety of the French bean, *Phaseolus vulgaris*, with almost white pods, or the dried seeds of the Lima bean, *P. limensis*.
1819 W. Cobbett *Amer. Gardener* (1821) iv. § 197 The *Lima*-bean..is sometimes called the *butter*-bean. **1884** Miller *Plant-n.* s.v. *Bean*, Butter Bean, a tender-podded variety of *Phaseolus vulgaris*. **1906** *Macmillan's Mag.* July 676 Butterbeans somewhat resemble white fish in colour and the task of turning out a butter-bean fritter..to look like a fried sole would not be a difficult one. **1951** *Dict. Gardening* (R. Hort. Soc.) I. 244/1 Pods without green colouring..are known as waxpod or Butter Beans, but they are not to be confused with the Butter Beans of the grocer. *Ibid.* 244/2 The seeds [of Lima Beans] vary considerably in colour and size, the white varieties being best for use as dry seeds, constituting the Butter Beans of the grocer.

butter-box. Add: **1.** (Later example.)
1874 *Contemp. Rev.* XXIV. 698 The wits declared that Denmark had sent a hundred thousand herrings, Holland a hundred thousand butter-boxes.
b. *transf.* A vessel or vehicle resembling a butter-box. Also *attrib.*
1840 Dana *Bef. Mast* ix, She was the *Loriotte*,..and was engaged in the hide and tallow trade. She was a lump of a thing, what the sailors call a butter-box. **1851** H. Melville *Moby Dick* I. xvi. 110 You may have seen many a quaint craft in your day,..butter-box galliots, and what not. **1893** Clark Russell *List, ye Landsmen* ix, Why the deuce don't the shipwrights ease off when they come aft, instead of holding on with the square run of the butter-box to the very lap of the taffrail? **1909** *N.Y. Even. Post* 28 Jan. (Th.), What New York youngster ever heard of a butterbox? This is the name applied [in the country] to the spring wagons of farmer and grocer, divested of wheels and set up on runners for the winter season.
2. Delete †*Obs.* and substitute: †*Obs.* exc. in Naut. slang. Also, a Dutch ship.
1867 Smyth *Sailor's Word-Bk.* 148 *Butter-box*..A cant term for a Dutchman. **1929** F. Bowen *Sea Slang* 21 *Butter Box*..Dutch ship or seaman.
3. = *butter-ball 2. *U.S.*
1806 W. Clark in *Lewis & Clark Exped.* (1905) IV. 50 The same species of duck..called the butterbox. *a* **1813** [see *butter-ball 2]. **1874** J. W. Long *Amer. Wild-fowl Shooting* xxix. 281 Local names [of the buffle-headed duck]: butter-box, butter-ball, and little whistler.

buttercupped (bʌ·təɪkʌpt), *a.* [f. Buttercup + -ed².] Abounding in or covered with buttercups.
1872 C. S. Calverley *Fly Leaves* 89 Looking far over buttercupped leas. **1888** Mrs. H. Ward *R. Elsmere* xliv, A wavering rainy light played..over the buttercupped river meadows. **1924** *Public Opinion* 9 May 400/1 Banks all buttercup'd and burning.
So **buttercuppy** (bʌ·təɪkʌpi), *a.*
1871 W. Morris in Mackail *Life* (1899) I. 237 The fields are all butter-cuppy. **1958** Betjeman *Coll. Poems* 259 That burning buttercuppy day.

buttered, *ppl. a.* Add: **2. c.** *buttered eggs*: eggs beaten up and cooked with butter; now applied to the dish otherwise called *scrambled eggs*.
c **1410** *Master of Game* (MS. Digby 182) xii, Buttured egges doth hem moche goode. **1865** Mrs. Beeton *Dict. Everyday Cookery* 113/1 Buttered Eggs. Ingredients.—4 new-laid eggs, 2 oz. of butter.

bu·tter-fish. Any of several fishes having a slippery coating of mucus, esp. the Gunnel found in British waters; the Murray perch, *Oligorus mitchelli*, a fresh-water fish of Australia; the kelp fish of New Zealand (see quot. 1880); the dollar-fish, *Stromateus* or *Poronotus triacanthus*, a food-fish of the eastern U.S.
1674–1883 [see Butter *sb.*¹ 5]. **1842** J. E. DeKay *Zool. N.Y.* IV. 153 The American Butter-fish, *Gunnellus mucronatus*. **1850** J. B. Clutterbuck *Port Phillip* iii. 44 In the bay are large quantities of..butter-fish. **1880** Günther *Fishes* 533 The 'butter-fish', or 'Kelp-fish' of the colonists of New Zealand (*C[oridodax] pullus*), is prized as food. **1883** E. P. Ramsay *Food Fishes N.S. Wales* 12 *H[aplodactylus] obscurus*,..known to our southern fishermen as the 'butter-fish', is highly esteemed. **1888** G. B. Goode *Amer. Fishes* 221 The 'Butter-fish' of Massachusetts and New York, *Stromateus triacanthus*. *Ibid.* 232 [*Selene setipinnis*] is a frequent summer visitor all along the coast as far north as Woods Holl, Mass., where it has a peculiar name,..the 'Hump-backed Butterfish'. **1946** K. Tennant *Lost Haven* (1968) xxii. 393 One of them little butter-fish, all fins and flaps. **1962** K. F. Lagler et al. *Ichthyology* xiv. 443 Among the many other fishes of the neritic zone are the..butter-fishes (Stromateidae).

butterfly, *sb.* Add: **2. c.** Phr. *to break a butterfly on a wheel*, to use unnecessary force in destroying something fragile. Also *ellipt.*
1735 [see Break *v.* 7 b]. **1874** Trollope *Way we live Now* I. ix. 70 One doesn't want to break a butterfly on the wheel. **1931** W. Holtby *Poor Caroline* vi. 213, I can't bear to see a woman in the dock—butterfly on the wheel. **1951** N. Annan *L. Stephen* 292 Why break a butterfly on the wheel of scholarship?
d. Applied to persons whose periods of work or occupation of a place are transitory or seasonal.
1890 *Chambers's Jrnl.* 10 May 289/2 A 'butterfly' man rests for a moment to wipe his streaming brow, when the warder's stern voice bids him proceed with his work. **1891** *Daily News* 29 Dec. 6/4 The 'butterfly man', who is given cabs by the proprietors in the height of the season. **1895** *Westm. Gaz.* 8 Mar. 3/1 Those cabbies who come upon the streets in the fine days and disappear with the autumn leaves are called 'butterflies'. **1902** *Daily Chron.* 2 June 7/1 Chelsea will welcome the return of the truant 'butterfly' to a region always to be associated with his artistic fame. **1912** *Standard* 15 Apr. 6/7 It was stated..that the word 'butterflies' was a 'technical term' for painters and decorators who worked upon bank holidays.
e. A fanciful name (usu. *pl.*) used of the fluttering sensations felt before any formidable venture, esp. in phr. *butterflies in the stomach, tummy*, etc. Also *attrib.*
1908 F. Converse *House of Prayer* iv. 43 The three o'clock train going down the valley..gave him a sad feeling, as if he had a butterfly in his stomach. **1943** *Word Study* Oct. 6/1 The expression some aviators use to describe their condition before taking off. They have 'butterfly stomach', they say, so marked is the fluttering in the Department of the Interior. **1944** H. Croome *You've gone Astray* v. 52 There was no electrical response to the movement of that firmly gentle hand, no butterflies on the backbone. **1955** J. Cannan *Long Shadows* viii. 132 With butterflies in her stomach..she ascended the pretentious flight of dirty marble steps. **1958** *Woman* 20 Sept. 69/3, I still have 'butterflies' even now when I hear the Tiger Moth plane throttling back, which is my signal to prepare for the jump. **1959** *Sunday Times* 25 Jan. 15/5 'I always have butterflies when I open Parliament,' she [*sc.* Queen Elizabeth II] remarked.
4. b. In full *butterfly bow*, a bow made up or tied with the loop and end on each side spread apart like the expanded wings of a butterfly. So *butterfly tie*.
1870 *Young Ladies' Jrnl.* 1 Mar. 138/2 The butterfly bow..is of black or coloured velvet. **1888** *Cassell's Fam. Mag.* Feb. 182/1 A bonnet à la Folle, with a tricoloured butterfly bow at the top. **1887** E. B. Custer *Tenting on Plains* (1889) xv. 502 It was then the fashion for men to wear a tiny neck-bow, called a butterfly tie. **1914** G. K. Chesterton *Wisdom Fr. Brown* xi. 264 A very young gentleman with..a black butterfly tie. **1920** *Punch* 4 Aug. 97/2 The wearing of a butterfly bow with a double event collar was a solecism past forgiveness.
d. *Swimming.* (See quot. 1957.) Also *attrib.*
1936 *N.Y. Times* 15 Aug. 8/2 The men's 200-metre breast-stroke..seemed to prove that the butterfly stroke is not all that some..think it is. *Ibid.* 15 Aug. 8/2 Higgins..used the butterfly only at the beginning and end [of the race]. **1937** *Off. Rep. XIth Olympiad 1936* 168 The breast stroke swimmers used the butterfly style, which was a failure. **1938** *Times* 10 Aug. 6 The butterfly, it may be explained, is made by recovering both arms at once out of the water. *Ibid.* 15 Aug. 5 A proposal will be made that the Butterfly stroke be abolished from international competitions. **1957** *Encycl. Brit.* XXI. 665/2 The butterfly is a competitive style only...In the butterfly the arms pull under the body all the way to the thighs, then emerge and fling forward above surface in circular motion.

5. Also *attrib.*, as *butterfly apparatus, catch*.
1887 *Daily News* 11 Jan. 2/7 The butterfly apparatus..had acted, but..the bolts..were torn away. **1909** *Daily Chron.* 8 Jan. 5/3 When the winding rope was detached the safety 'butterfly' catches failed to act.
7. *butterfly-brained*.
1961 *Times* 6 Dec. 17/3 The butterfly-brained society hostess.
8. butterfly blenny = *butterfly-fish*; **butterfly bomb** (see quot.); **butterfly kiss** (see quots.); **butterfly lily** = Mariposa lily; **butterfly lobster,** a marine crustacean, *Ibacus incisus*, found in Tasmanian waters; **butterfly lupus,** lupus of the nose and cheeks; **butterfly-net,** a net used for catching butterflies; **butterfly nose,** a dog's nose when spotted or mottled; **butterfly orchis** (examples); **butterfly-pea**: see Pea¹ 3; **butterfly ray,** an Australian sting-ray, *Gymnura tentaculata*; **butterfly snail,** a mollusc of the sub-class *Pteropoda*, a sea-butterfly; **butterfly tulip** = Mariposa lily.
1897 McIntosh & Masterman *Life-Hist. Brit. Marine Food-Fishes* vii. 205 (*heading*) Butterfly Blenny. **1959** A. Hardy *Fish & Fisheries* x. 213 The beautiful little butterfly blenny..which is not uncommon to the south west. **1944** *Sci. News Let.* 14 Oct. 247/2 Recently they developed a 'butterfly bomb', with wings that open up as soon as the bomb is released, and act like a parachute to slow its descent. **1871** Geo. Eliot *Middlem.* I. i. v. 73 Celia knelt down..and gave her little butterfly kiss. **1932** E. Waugh *Black Mischief* ii. 58 'I've invented a new way of kissing. You do it with your eye-lashes.' 'I've known that for years. It's called a butterfly kiss.' **1945** G. Endore *Methinks the Lady* (1947) iii. 47 Don't you know what a butterfly kiss is?..You flutter your eyelashes against his cheek, and then he flutters his against yours. **1902** V. K. Chesnut *Plants used by Indians Calif.* 323 Calochortus venustus..the commonest species of the Mariposa or butterfly lilies. **1880** L. A. Meredith *Tasmanian Friends & Foes* 248 'Butterfly lobsters'..the shell of the head and body..expands into something like wing-forms. **1913** Dorland *Med. Dict.* (ed. 7) s.v. *Lupus*, Butterfly lupus. **1827** M. Wilmot *Jrnl.* 25 July in *More Lett.* (1935) 278 Edmund and Wilmot amused themselves with their butterfly nets. **1939** T. S. Eliot *Fam. Reunion* II. i. 77 The day I lost my butterfly net. **1883** W. G. Stables *Our Friend Dog* vii. 59 *Butterfly nose*, a nostril with white spots in it. **1597** Gerarde *Herball* 165 (*caption*) *Ornithophora Candida*, Butter-flie Orchis. **1629** Parkinson *Parad.* xxii. 192 *Orchis Hermaphroditica candida*, the white Butter-flie Orchis. **1898** C. M. Yonge *J. Keble's Parishes* xiv. 155 Butterfly or honey-suckle orchis, *Habenaria*. **1963** *Times* 25 Apr. 14/6 A solitary exquisite butterfly-orchis. **1931** J. R. Norman *Hist. Fishes* xvi. 325 The Butterfly Rays (*Pteroplatea*). **1890** *Chambers's Encycl.* V. 110/2 The yet more closely allied 'butterfly-snails' or Pteropods. **1886** J. M. Hutchings *In Heart Sierras* 92 The charming Mariposa, or 'Butterfly Tulip'. **1896** T. W. Sanders *Encycl. Gardening* (ed. 2) 47 *Calochortus* (Butterfly Tulip).

butterfly, *v.* Add: **b.** *fig.* To flirt or philander.
1893 *Chambers's Jrnl.* 12 Aug. 504/2 The young graduate was only butterflying after all. **1906** B'ness von Hutten *What became of Pam* III. ix, 'What about Wantage?'..'He is still butterflying.'

buttermilk. Add: **c.** (Later example.) Also quasi-*adj.* in *buttermilk land* U.S.
1843 'R. Carlton' *New Purchase* ix. 58 They had been sufficiently fortunate as to get a taste of 'buttermilk land'—'spouty land'. **1918** J. Galsworthy *Let.* 9 Aug. in E. V. Lucas *Post-Bag Div.* (1934) 70 Do you think..that this shows..that I ate too much of Mrs. Endacott's butter-milk bread?

butter-print. Add: **3.** The Indian mallow, *Abutilon theophrasti*, bearing a round seed-capsule marked with radiating furrows.
1872 *Illinois Dept. Agric. Trans. 1871* p. ix, The Indian Mallow (*Abutilon Avicennae*) is..known as..'butter print'. **1899** *Mem. Amer. Folk-Lore Soc.* VII. 120 *Abutilon Avicennae* is called 'butter-print'..because its pods are used to stamp butter.

butt-in. *U.S.* One who butts in (see *butt *v.*¹ 1 d); an intruder. So **butt-iner.**
1906 H. Green *Actors' Boarding House* 97 Gettee 'way...No want flesh butt-ins round! *a* **1910** 'O. Henry' *Rolling Stones* (1916) 198 Any of the Flat bush or Hackensack Meadow kind of butt-iners. **1911** J. C. Lincoln *Cap'n Warren's Wards* viii. 124 If I had my way the old butt-in should understand exactly what I think of him. **1930** *Liberty Mag.* 1 Nov. 43/1 She [*sc.* a gangster's moll] may be just a butt-in.

buttinsky (bʌti·nski). *slang* (orig. *U.S.*). Also **buttinski.** [Jocular, f. *butt in* (see *butt *v.*¹ 1 d) + *-sky*, final element in many Slavonic names.] = prec. So (nonce-wd.) **butteri·nsky.**
1902 Ade *Girl Proposition* 70 The Friend belonged to the Buttinsky Family and refused to stay on the Far Side of the Room. **1903** *Cincinnati Enquirer* 9 May 13/2 Piker..a cheap grafter; a noser; a butterinsky. **1922** S. Lewis *Babbitt* xx. 252 If you think I'm a buttinsky, then I'll just butt in. **1933** D. L. Sayers *Murder must Advertise* iv. 69, I never..met with such a bunch of buttinskis...Nothing is sacred to you. **1960** Wodehouse *Jeeves in Offing* v. 50 It is never pleasant for a man of sensibility to find himself regarded as a buttinski and a trailing arbutus.

buttle, *v.* Add: **b.** To do a butler's work. *jocular*.
1918 Mrs. H. Ward *War & Elizabeth* iii, The under-housemaid 'buttles' for him like a lamb. **1923** F. H. Kitchin *Diversions of Dawson* 292 Nobody could buttle like James who had not been born in a pantry and taken pap out of silver spoons. **1929** W. J. Locke *Ancestor Jorico* xvi, Peters—the head steward—is a fat fool...Seems he buttled for decaying noble families. **1954** R. Fuller *Fantasy & Fugue* iv. 91 He valeted him as well as buttled. **1968** J. C. Holmes *Nothing More to Declare* 98 Eric Blore spluttered and buttled like a paranoid chipmunk.
Hence **bu·ttling, bu·tling** *vbl. sbs.*
1918 Wodehouse *Piccadilly Jim* xix. 185 How on earth did you come to be here? What's the idea? Why the buttling? **1945** A. L. Rowse *West-Country Stories* 37 The baronet did his own butling. **1968** A. Diment *Gt. Spy Race* x. 177 Petite was there, dressed in his butling clothes.

buttock, *sb.* Add: **8.** *Coal-mining*. The portion of the working-face of coal to be broken out next. Hence **buttocker,** a man who works at the buttock.
1883 in Gresley *Gloss. Coal Mining*. **1912** Bulman & Redmayne *Colliery Working* (ed. 3) 294 *Buttocker*, one who breaks down the coal which has been undercut by the 'holers'. A 'getter'. **1967** *Gloss. Mining Terms* (*B.S.I.*) VIII. 8 *Buttock*, in some longwall faces, a short step in the line of face, and substantially at right angles to it, from which coal can be more conveniently worked.

button, *sb.* Add: **1. e.** In phrases expressing weakness of intellect, as: *not to have* (*got*) *all his buttons on*, *to be a button short*. Similarly *he has all his buttons* (*on*): he is sound in intellect, 'all there'.
1860 Hotten *Dict. Slang* (ed. 2) 109 *Not to have all one's buttons*, to be deficient in intellect. **1890** *Daily News* 21 May 6/3 He is 83 years of age, but as we say hereabouts, has all his buttons on. **1892** *Leeds Merc. Suppl.* 23 Jan. (E.D.D.), In Wilsden, one lacking full mental capacities has 'some of his buttons off'. **1893** Mrs. A. Kennard *Diogenes' Sandals* xi, They said..he had not 'got all his buttons', meaning that he was not 'all there'. **1961** *Times* 16 Jan. 4/5 But East Midlands, too, had all their buttons on.
h. A button of a particular colour, or bearing a distinctive design, worn as a party badge.
1895 *Montgomery Ward Catal.* 180 Emblem Pins and Buttons..Masonic..Odd Fellows..Eastern Star. **1900** *Daily News* 5 Nov. 7/1 Another feature of an American Presidential campaign is the lavish display of political 'buttons'. **1915** V. Woolf *Voyage Out* xix. 314 The particoloured button of a suffrage society. **1970** *Times* 14 Mar. 5/6 One [demonstrator] wore a Mao button.
4. b. Phr. *to press the button*: to push back a disc, pin, knob, or the like and thus produce the required result by completing an electric circuit, operating the shutter of a camera, etc. Often *fig.* in colloquial use, to perform an action that automatically brings about the required state of affairs.
1860 *Nat. Mag.* VIII. 278/2 By pressing a button..the light can be given in flashes. **1865** *Mech. Mag.* 10 Mar. 155/1 On pressing one of these buttons with the finger, the bell..is rung loudly. **1893** *Cassell's Fam. Mag.* Mar. 318/1 Then he pressed the button of the camera. **1893** *Jrnl. Soc. Arts* XLI. 629/2 President Cleveland..pressed the button which started all the..machinery...The button..was of ivory. **1893** [see Press *v.*¹ 1 a]. **1905** *Minister's Gaz. Fashion* July 138/1 Pressing the Button. A Plea for Modern Methods...When the art of cutting will be reduced to a mechanical science. **1914** E. Grey in *Europ. Crisis, Corr.* (Parlt. Papers CI) 46 Mediation was ready to come into operation..if only Germany would 'press the button' in the interests of peace. E. Goschen *Ibid.* 59 The Chancellor told me last night that he was 'pressing the button' as hard as he could. **1962** Wodehouse *Service with Smile* i. 12 Thinking..that the ideal way of opening Parliament would be to put a bomb under it and press the button.
Also *to press button A* or *B*: of a former type of telephone coin-box; *button A* completed the connection after the called subscriber had answered; *button B* actuated a mechanism to return the coins if there was no answer or if the required number was engaged.
1934 *Discovery* Mar. 58/2 Distinguishing between buttons A and B, deciding when to press either or both. **1935** G. Greene *England made Me* I. 17, I rang up four times from a box..and I remembered three times to press Button B and get my money back. **1942** J. D. Carr *Seat of Scornful* xiii. 188 Unless you put the money in, you couldn't press Button A and the connection wouldn't work. **1961** 'T. Hinde' *For Good of Company* i. 16 Tony seized the phone and held it out of the box towards him. 'Press Button A,' he shouted.
c. The leather projection on an oar, by which it is kept in position in the rowlock.
1866 'Argonaut' *Arts Rowing & Training* ii. 11 Both [oars and sculls] are kept in their proper place in the row-lock by a circular button of a peculiar shape, on the leather, which plays against the inner side of the thowl. **1888** W. B. Woodgate *Boating* iii. 63 If the body swings true, the oar will keep home to the rowlock; there should be just sufficient fraction of weight pressed against the button to keep it home. **1963** *Times* 18 Feb. 3/7 Mead was pulling his button away from the rowlock, almost clear of the leather.
d. In an organ, a round piece of leather that keeps the tracker from jumping out of place.
1876 Stainer & Barrett *Dict. Mus. Terms* s.v. *Organ*, A tracker, a strip of light wood provided with a tap wire and leather button at the end. **1877** Stainer *Organ* 17 The little wire passing from the end of the tracker into the hole in the backfall is made like a screw,..so, where it appears below the backfall, a little leather button can be screwed on to it.
e. Each of the keys of an accordion.
1876 Stainer & Barrett *Dict. Mus. Terms* s.v. *Accordion*, The first instruments had only four buttons, or keys, each of which acted on two reeds.
5. g. The point of the chin. *U.S. slang.*
1921 H. C. Witwer *Leather Pushers* 48 The Kid..floored him with a right cross to the button of the jaw. **1931** D. Runyon *Guys & Dolls* (1932) 278, I never saw a more accurate puncher than Rusty Charley, because he always connects with that old button. **1936** Wodehouse *Laughing Gas* ii. 27 He soaked him on the button, don't you know.
11. b. *Aleppo button*: cutaneous leishmaniasis (oriental sore), a form of the disease endemic in parts of Asia and Africa and caused by infection with *Leishmania tropica*; also, the local lesion characteristic of this infection. So also *Baghdad*, *Biskra button* (also *boil*), etc.
The F. form *bouton* was formerly also used.
1874 *Med. Times & Gaz.* 24 Jan. 94/2 It is generally believed that the Aleppo Button takes about a year before it heals: hence the name of the *year pimple*, given to it by the natives. **1876** *Brit. Med. Jrnl.* 19 Feb. 225/2 (*heading*) Aleppo boil or 'Biskra bouton'. **1897** *Daily News* 18 Sept. 6/2 The 'Bagdad button' (a painful species of boil). **1911** T. E. Lawrence *Home Lett.* (1954) 138 The Aleppo button is the effect of a fly. **1969** Kurban & Chaglassian in Simons & Marshall *Ess. Trop. Dermatol.* 187 Various synonyms are used for oriental sore, depending on the region in which it is found, *e.g.*, Aleppo boil, Baghdad boil, Delhi boil, Biskra button, Jericho boil and many others.
c. *Med.* Any small, rounded elevation on the cutaneous or mucous surface. (Cf. prec. sense.)
[**1876** *Brit. Med. Jrnl.* 19 Feb. 226/1 Villemin mentions that the 'boutons on the feet are sometimes multiple, simulating syphilitic ulcers'.] **1888** C. M. Doughty *Trav. Arabia Deserta* II. xv. 452, I washed the wound..but a red button remained. **1900** *Jrnl. Exper. Med.* 15 Dec. 259 More characteristic lesions..are the so-called 'buttons'.
12. button brass, (*a*) (see quot. 1884); (*b*) a strip of brass slipped under a metal button to shield the garment while the button is being cleaned; **button bud,** a bud resembling a button; **button-bush** (earlier example); **button-down** *a.* (orig. *U.S.*), applied to a collar the points of which are buttoned to the shirt; **button ear,** an ear, of a dog, that laps over and hides the inside; hence *button-eared* adj.; **button fastener,** (*a*) a spring loop, the free ends of which are passed through the shank of the button to keep it in place; (*b*) (see quot. 1884); **button-grass** *Austral.* (see quot. 1898); **button key** = *button fastener (a); **button-nosed** *a.*, having a small roundish nose; **button-quail** (see quots.); **button-scar,** a scar drawn up into button-shape, used for ornamentation of the body by some African peoples; **button shell,** a small marine univalve of the genus *Rotella*, with a lenticular polished shell (*Funk's Stand. Dict.* 1893); **button-stick,** a soldier's appliance for use in button-polishing (= *button brass b, but usu. made of wood).
1884 Lock *Workshop Rec.* Ser. III. 16/2 For button brass, an alloy of 8 parts of copper and 5 of zinc is commonly used by the Birmingham makers, under the name of 'platin'. **1899** *Daily News* 27 Dec. 8/3 The 'button brass'...This little plate tucked under the button with its shank in the slit enables the button to be well rubbed without mischief. **1869** Blackmore *Lorna D.* xvii, The opening cones were struck with brown, in between the button buds. **1754** J. Eliot *Ess. Field-Husb.* (1760) v. 124 There was not the same Success attending the cutting these Button Bushes. **1934** J. O'Hara *Appt. Samarra* (1935) v. 151 Soft white shirt with button-down collar. **1960** *Guardian* 9 June 1/1 The egg-head in the button-down collar..an American type. **1883** W. G. Stables *Our Friend Dog* vii. 59 *Button-ear*, an ear that falls in front, entirely concealing the inside. **1952** R. Leighton *Complete Bk. Dog* 73 Unfortunately, within the last few years the 'button' and 'semi-tulip' ear have been rather prevalent. **1867** Mrs. Whitney *L. Goldthwaite* x, She had hooks and eyes, and button-fasteners, when these gave out. *a* **1877** [see *button key]. *a* **1884** Knight *Dict. Mech.* Suppl., *Button Fastener*, a clasp which hooks over the eye of a shoe button and is then clinched to the shoe. **1898** Morris *Austral Eng.*, Button-grass, *Schœnus sphærocephalus*, Poiret, N.O. *Cyperaceæ*...So called from the round shaped flower (capitate inflorescence), on a thin stalk four or five feet long, like a button on the end of a foil. **1927** *Blackw. Mag.* Oct. 470/1 A steep razor-backed hill, covered with ragged clumps of button-grass and dwarfed ti-tree. **1958** 'N. Shute' *Rainbow & Rose* i. 6 Button-grass plains where no feed grows that will sustain a horse. *a* **1877** Knight *Dict. Mech.*, *Button Key* or *Fastener*, a spring loop..to..keep the button in place. **1909** H. G. Wells *Tono-Bungay* III. ii, Button-nosed, pink-and-white Aunt Susan. **1885** *Encycl. Brit.* XVIII. 46/1 The Hemipodes or Button-Quails. **1893** Button quail [see Hemipod]. **1947** J. Stevenson-Hamilton *Wild Life S. Afr.* xxxii. 276 Button quail or riet kwartel (*Turnicidæ*).—Three species of Hemipodes, or button quail, exist in South Africa. They are smaller than true quail and lack a hind toe. **1897** A. J. Butler tr. *Ratzel's Hist. Mank.* II. 394 In the so-called 'button-scars', a row of button-shaped warty scars runs from the edge of the forehead to the tip of the nose; this is found both on the Congo and on the Zambesi. **1890** Kipling *Barrack-r. Ballads* (1892) 20 I've a head like a concertina: I've a tongue like a button-stick. **1969** E. H. Pinto *Treen* 150 Button sticks..were used for cleaning uniform buttons.

button, *v.* Add: **3.** *to button* (*a person's*) *mouth*: to silence (a person); *to button* (*up*) *one's lip, face* (slang): to be silent; also *ellipt.*
1647 J. Trapp *Comment. Four Evang.* 526 How easily can God button up the mouths of our busiest adversaries.

1868 *N. & Q.* I. 603 At school, it was thought quite an accomplishment in the young gentlemen who were fast of tongue to be able to silence a talkative comrade with the phrase 'button your lip'. **1936** A. Huxley *Eyeless in Gaza* xxv. 352 Mr. Beavis..began to describe his researches into modern American slang..Horse feathers, dish the dope, button up your face—delicious! **1945** L. Shelly *Hepcats Jive Talk Dict.* 22/2 *Button up your lip*, don't talk. **1962** K. Orvis *Damned & Destroyed* ix. 61 If the little pusher chose to button up now, I would be left helpless.
4. Hence **button-through** *a.*, applied to a garment fastened with buttons from top to bottom; **button-up** *a.*, that buttons up; also *absol.* as *sb.*
1920 *Ware-Pratt Co. Stylebook* Nov., It is a double-breasted, button-through model with long rolling lapels, narrow gorge and velvet collar. **1952** C. W. Cunnington *Eng. Women's Clothing* viii. 275 Wool housecoat in button-through style. **1962** L. Deighton *Ipcress File* xxxi. 198 A new round-necked, sleeveless, button-through dress. **1943** K. Tennant *Ride on Stranger* ii. 13 Anyway there would be trouble..over the soggy state of her button-up boots. **1951** J. Frame *Lagoon* 62 To have button-up shoes instead of lace-ups. *Ibid.* 67 O for button-ups! O for a dress with a cape collar! **1956** A. Miller *Memory Two Mondays* (1958) 360 He wears a blue button-up sweater.

buttoned, *ppl. a.* Add: **2. b.** *fig.* With *up*, spec. of persons: reserved, uncommunicative. Cf. Button *v.* 3.
1936 R. Lehmann *Weather in Streets* I. v. 92 Plump, buttoned-up face. **1946** M. Dickens *Happy Prisoner* viii. 158 Why is she so quiet and buttoned up? **1957** *Economist* 2 Nov. 393/1 The British are notoriously bad..at human relationships; buttoned-up, standoffish.
c. *fig.* With *up*. Of a plan, etc.: successfully arranged. *slang.*
1940 'N. Shute' *Old Captivity* vi. 197, I was thinking about this photography. But I believe we've got that buttoned up now. **1942** *Gen* 1 Sept. 12/2 When army things are under control they are 'buttoned up'.

button-hole, *sb.* Add: **1. c.** *spec.* one in the lapel of a coat.
1842 *Knickerbocker* XIX. 44 A bud and two leaves..at his buttonhole. **1847** Dickens *Dombey* xxxi, The Major..wears a whole geranium in his button-hole. **1863** [in Dict., sense 1]. **1929** Edington *Studio Murder Myst.* xvii, The white gardenia in his buttonhole.
2. b. *Surg.* A small straight opening in an organ or part.
1884 B. Bramwell *Dis. Heart & Thoracic Aorta* v. 478 In cases of this description, the orifice, when seen from above, looks like a narrow slit, hence the term button-hole mitral which has been applied to it. **1907** *Practitioner* Oct. 526 Without making a 'button-hole' through the muco-perichondrium. **1946** H. T. Hyman *Integrated Pract. Med.* III. xix. cxxxvii. 2971 Metacarpal shaft driven through joint capsule between flexor tendons (button-hole dislocation).
4. *button-hole cutter, gimp, hand, scissors* (examples), *twist*; **button-hole globe, light,** an electric bulb to be fastened in one's button-hole as a reading light.
1869 *Rep. U.S. Commissioner Agric. 1868* 289 Button-hole twist is the same, with a tighter twist. **1873** *Young Englishwoman* Mar. 147/2 Cutting-out scissors, small scissors, and button-hole scissors. *a* **1877** Knight *Dict. Mech.*, Button-hole Cutter. **1890** *Daily News* 27 Jan. 3/1 The apparatus is about the same as that of the buttonhole light, only that instead of a buttonhole globe at the end of a thread there is a fine, minute drill. **1899** in A. Adburgham *Shops & Shopping* (1964) xxii. 261 Button hole gimp and twist. **1908** *Daily Chron.* 13 June 9/6 Tailoring.—Wanted button-hole hand for coats and vests. **1964** *McCall's Sewing* v. 62/1 *Buttonhole scissors*, a handy item if you make a lot of garments with machine-made buttonholes, these are designed to cut open buttonholes accurately.

button-hole, *v.* Add: **1. c.** To make button-hole openings in.
1908 *Practitioner* Oct. 522 The hand is kept in that position ..by long strips of adhesive plaster..; these are button-holed over the knuckles to adapt themselves to them.
d. *Surg.* To make (esp. accidentally) a button-hole incision in.
1907 *Practitioner* Oct. 530 It may seem impossible to strip off the muco-chondrium without buttonholing it. **1964** S. Duke-Elder *Parsons' Dis. Eye* (ed. 14) xxvii. 423 The dissection is made by sharp-pointed scissors in the subconjunctival tissue..care being taken to avoid button-holing the flap.
2. (Later examples.)
1951 'A. Garve' *Murder in Moscow* iii. 48 Jeff..had buttonholed a Russian who spoke a few words of English. **1953** *Encounter* Oct. 58/2 Scientists were free to..publish their findings, talk about them to colleagues (or to anyone else they could buttonhole).
button-holing *vbl. sb.*: also **button-holeing,** and as *ppl. a.*
1862 in Sperber & Trittschuh *Amer. Pol. Terms* (1962) 68/2 Quarreling, fighting, button-holeing. **1894** *Athenæum* 10 Feb. 176/1 The reticence of Mr. Maartens..contrasts vividly enough with the buttonholing familiarity of his English model. **1909** H. G. Wells *Tono-Bungay* II. iii. 179 That alluring, button-holeing..style of newspaper advertisement. **1963** *Times* 13 June 15/5 His book is a button-holing chronicle of daring theft and boudoir hanky-panky.

buttress, *sb.*¹ Add: **4. c.** *Bot.* = buttress-root. (See quots.) Cf. *buttress-root.
1900 B. D. Jackson *Gloss. Bot. Terms* 39/1 *Buttress*, the knee like growths of trunk or roots in certain trees. **1969** *Gloss. Landscape Work* (*B.S.I.*) v. 27 *Buttress*, a projecting growth linking the trunk of a tree to a major root.
6. buttress-root, = *plank-buttress* (see *plank *sb.* 7); **buttress thread,** a screw-thread having one face at right angles to the axis of the bolt or shaft.
1914 M. Drummond tr. *Haberlandt's Physiol. Plant Anat.* iv. 188 This complex condition is exemplified by the stilt- or buttress-roots which occur in the genus *Pandanus* [etc.]. **1930** *Discovery* Nov. 381/1 Buttress roots [of trees] resemble thick planks and are often four or five feet in height at the trunk. **1964** J. P. Clark *Three Plays* 8 The burden was voided and buried In the crook of twin buttress roots. **1887** D. A. Low *Introd. Machine Drawing* iii. 15 The Buttress thread..is designed to combine the advantages of the V and square threads. **1930** *Engineering* 6 June 721/2 In asymmetrical threads, such as..certain buttress threads.

butty. Add: **4.** In full *butty-boat*, a second barge or freight-boat in tow by a first.
1909 Quiller Couch *True Tilda* ix, With two horses hauling at the first [barge], and the second (which Sam called a butty-boat) towed astern. **1923** *Blackw. Mag.* Nov. 663/2, I overhauled two barges, the foremost with its little steam-engine towing the second or 'butty'. **1946** *Archit. Rev.* C. 161/1 They [*sc.* barges] go in pairs, a diesel-engined motor boat and a butty boat towed behind...The cabin of the butty is down into the bottom of the boat.

bu·tty². *north. dial.* Also **buttie.** [f. Butt(er *sb.*¹ + -y⁶.] A slice of bread and butter.
1855 Mrs. Gaskell *North & South* II. xi. 142 He's always mithering me for 'daddy' and 'butty'; and I ha' no butties to give him, and daddy's away. **1927** W. E. Collinson *Contemp. Eng.* 121 Buttie, general in North for a piece of bread and butter (together with jam-buttie). **1959** I. & P. Opie *Lore & Lang. Schoolchildren* ix. 162 Spread it on the butty nice and thick. **1965** *Oxford Mail* 17 Nov. 11/5 The biggest jam butty in the world.

butyl. Add: **2.** *butyl rubber*, a type of synthetic rubber made from isobutylene; also *ellipt.*
1940 *Economist* 20 July 100/1 The Standard Oil Company of New Jersey has evolved an entirely new product, which it calls 'butyl'. **1940** R. M. Thomas et al. in *Ind. & Engin. Chem.* XXXIII. 1283 (*title*) Butyl Rubber: A new hydrocarbon Product. *Ibid.* 1284/2 The designation 'butyl rubber' must be considered a generic term. **1957** *Encycl. Brit.* IV. 459/2 Isobutylene..polymerizes to a spongelike material called butyl rubber, which is used for the production of high-grade tire tubings. **1962** *Times* 6 Mar. 15/7 Butyl-rubber tyres.
3. butylamine (examples).
1866 H. E. Roscoe *Elem. Chem.* xxxi. 279 Primary Monoamines..Butylamine. **1964** N. G. Clark *Mod. Org. Chem.* xii. 245 The *butylamines* and *pentylamines* are obtained by the action of ammonia on the corresponding chlorides.

butyne (biū·tain). *Chem.* [altered form of *butine: see *-yne.] Either of the two isomers $CH_3{\cdot}CH_2{\cdot}C{:}CH$ and $CH_3{\cdot}C{:}C{\cdot}CH_3$ (also called *crotonylene*) of the hydrocarbon C_4H_6. (Formerly called *butine.)
1935 *Proc. Indiana Acad. Sci.* XLIV. 148 Ethyl sulphate reacts vigorously with calcium acetylide in ammonia and gives a theoretical yield of pure butyne-1. **1951** A. W. Johnson in E. H. Rodd *Chem. Carbon Compounds* Ia. ii. 264 The acetylenic hydrocarbons are isomeric with the corresponding diolefines, e.g. propyne (allylene)..is isomeric with allene and butyne with butadiene.

butyrometer (biutirǫ·mitəɪ). [f. Butyro- + -meter.] An instrument for estimating the percentage of butter-fat in milk. Cf. *lacto-butyrometer* s.v. Lacto-.
[**1863** *Jrnl. R. Agric. Soc.* XXIV. II. 315 Marchand's lacto-butyrometer.] **1902** *Encycl. Brit.* XXVII. 360/1 The butyrometer is extremely useful, alike for measuring periodically the fat-producing capacity of individual cows in a herd, for rapidly ascertaining the percentage of fat in milk delivered to factories [etc.]. **1961** *B.S.I. News* Sept. 13/1 Milk hydrometers, butyrometers, [etc.].

buvette (buve·t). Add: Also, a refreshment bar or room.
1885 *Cornhill Mag.* June 597 We..went into the stove-stifling heat of the little buvette of the station to keep ourselves warm.

buy, *v.* Add: **15.** *to buy money*: see quot. 1922. *Racing slang.*
1906 Fox-Davies *Dangerville Inheritance* vii. 99 The public had left off buying money, and the wagering had become slack. **1922** *N. & Q.* 12th Ser. XI. 206/2 *Buying money*, laying heavy odds on a favourite. **1928** *Daily Express* 12 July 12/2 Backers..had to buy money over On Avon and Rainbow Bridge.
16. In slang use.
a. To suffer some mishap or reverse; *spec.* to be wounded; to get killed, to die; (of an airman) to be shot down. Freq. with *it*.
1825 W. N. Glascock *Naval Sketch-Bk.* (1826) I. 30 Never mind, in closing with *Crappo*, if we didn't buy it with his raking broadsides. **1920** W. Noble *With Bristol Fighter Squadron* v. 70 The wings and fuselage, with fifty-three bullet holes, caused us to realize on our return how near we had been to 'buying it'. **1925** Fraser & Gibbons *Soldier & Sailor Words* 41 *To buy*, to have something not desired, such as a job, thrust on one unexpectedly, *e.g.*, 'Just as he was going out, he ran into the Corporal and bought a fatigue.'..Another meaning: to be scored off or victimized. Of a man getting an answer to a question which made him ridiculous: 'He bought it that time.' **1943** Hunt & Pringle *Service Slang* 39 *He bought it*, he was shot down. **1943** C. H. Ward-Jackson *Piece of Cake* 16 *He's bought it*, he is dead—that is, he has paid with his life. **1944** J. E. Morpurgo in *Penguin New Writing* XXII. 11 I'm afraid we want you elsewhere...Jim Barton bought it, and you'll have to take on his troop. **1953** R. Lehmann *Echoing Grove* 261 He'd lived in London before the war, but the whole street where he'd hung out had bought it in the blitz.
b. To believe; to accept, to approve. Chiefly *U.S.*
1926 E. Wallace *More Educ. Evans* vi. 139 'It's rather early in the day for fairy-tales,' he said, 'but I'll buy this one.' **1944** *Amer. Speech* XIX. 72/1 If the work is perfect, the inspector *buys* it...In the drilling departments, one might hear a worker say, 'I am waiting for the company to buy this hole.' **1949** *Time* 2 May 8/1 After talking it over with the President..Secretary Johnson bought the Air Force point of view. **1951** I. Shaw *Troubled Air* xiii. 213 People feel that the best way to prove how loyal they are is to be as nasty..as they know how, and I'm not buying any of that. **1952** M. McCarthy *Groves of Academe* (1953) ix. 182 It doesn't seem to me likely that they cooked it up between them...More likely she half guessed and he told her. I'm willing to buy that for what it's worth.
c. *I'll buy it* (in reply to a question or riddle): I give it up (as an invitation to reveal the answer), I don't know, I'll accept your answer.
1930 E. Wallace *White Face* xvi. 248 'You thinking, too?' growled Mason. 'All right, I'll buy it.' **1932** D. L. Sayers *Have his Carcase* xi. 128, I'll buy it, Inspector. *What* did he do with it? **1957** P. Frankau *Bridge* 136 'Confession coming,' he said. 'I'll buy it. Something that happened last night?'

buy (bai), *sb.* orig. *U.S.* [f. Buy *v.*] A purchase; *best buy*, the most worth-while purchase or bargain. Also *fig.* Phr. *on the buy*: actively buying.
1879 F. A. Buck *Yankee Trader* (1930) 274, I believe the Mammoth Mine here to be the best buy in the lot. **1890** Van Dyke *Millionaires of Day* 134 Biggest buy in town. **1903** *Longman's Mag.* Mar. 444 What do you think of my new buy? **1911** H. Quick *Yellowstone N.* vii. 191, I believe it's a good buy! **1929** *Star* 21 Aug. 18/2 His clients are 'on the buy'. **1952** J. Pinckney *My Son & Foe* ii. 14 Knowing the intrinsic quality of the goods, what the best buys are that life puts out on the counter. **1957** *Times* 21 Oct. 13/1 Among the best 'buys' from these departments are the deceptively simple models that..bear an unmistakable *chic*. **1964** *Which?* Feb. 43/2 Because each of these prams had some drawbacks, we do not choose a Best Buy.

buyer. Add: **3.** *buyers' market*: one in which goods are plentiful and low prices favour buyers.
1926 *Textile World* 11 Dec. 91 (*heading*) Buyers' market. **1930** *Economist* 13 Dec. 1105/1 The problem..is the marketing of about 300 million bushels of Canadian grain at adequate prices in what is obviously a buyers' market. **1959** *Times Rev. Industry* Feb. 18/3 The Board now has to face heavy losses in fighting back on a buyers' market.

buying, *vbl. sb.* Add: **3.** The purchasing of shares on the stock exchange. *buying-in day*, the day on which, owing to non-delivery within the appointed time of shares bought, the buyer may purchase the shares on the market; *buying-in rule*, the rule with regard to buying-in day.
1900 *Westm. Gaz.* 24 Mar. 7/1 'Buying-in Day.'..He immediately delivers the shares, usually on the day after the buying-in takes place. **1902** *Encycl. Brit.* XXXII. 865 The 'corner' in Northern Pacific common shares produced..the suspension for two or three weeks of the 'buying in' rule.

buzz, *sb.*¹ Add: **1. b.** *Phonetics.* A voiced hiss (see Hiss *sb.* 1 b).
1877 Sweet *Handbk. Phonetics* 79 The voiced buzzes admit of more variety than the voiced stops. **1887** *Encycl. Brit.* XXII. 383 A hiss (s), followed without a positional glide by the buzz (z). **1888** Sweet *Hist. Eng. Sounds* 24 Some consonants..are pronounced with..a complete absence of buzz. **1908** —— *Sounds of English* 43 The English ɔ· is vowellike in sound, being quite free from buzz.
c. A round game in which each player in turn utters a number in numerical order, with the exception that 'buz(z)' must be substituted for 7 and multiples of 7.
1864 Hotten *Slang Dict.* 91 *Buz*, a well-known flash game. **1868** L. M. Alcott *Little Women* iii, They..were in the midst of a quiet game of 'buzz'.
d. *spec.* The buzzing sound made by a telephone. Hence *slang*, a telephone call.
1913 G. B. Shaw *Let.* 14 July (1952) 132, I rang a second time, but the answer was buzz, buzz. **1930** *Punch* 26 Feb. 236 One of the cops..directed another to beat it to his desk and give headquarters a buzz. **1938** E. Bowen *Death of Heart* I. ii. 38 The Quaynes had a room-to-room telephone, which, instead of ringing, let out a piercing buzz. **1959** G. Usher *Death in Bag* xii. 128 Shall I give him a buzz?
3. b. (Later examples.)
1892 Lugard *Diary* 8 June (1959) III. viii. 290 Juma..asked him if he had heard of this. He said *Yes*, but not from any responsible chief, merely a buzz through the country. **1919** 'Etienne' *Strange Tales fr. Fleet* 135 There's a buzz floating round that we are slipping off at 8 p.m. **1962** B. Knox *Little Drops of Blood* iii. 62 There's a strong buzz on the go that his team are building some new engine.
5. buzz-bomb *colloq.* = *flying bomb; **buzz-box** *slang* = *buzz-wagon; **buzz-planer,** a small wood-planing machine (Knight *Dict. Mech.* Suppl. *a* 1884); **buzz-saw** (earlier and

attrib. examples); **buzz-wagon** *slang*, a motor-car.

1944 in *News Rev.* 10 May (1945) 7/1 The Germans sent over 11 pilotless planes or 'Buzz-bombs'. 1946 *Amer. Speech* XXI. 246 The V-1 was in the U.S. and British official parlance a *flying bomb*...The troops, however, generally called it a *buzz bomb*, probably because it normally came in at rather low altitude, reminiscent of an airplane buzzing the ground. 1920 'Sapper' *Bull-dog Drummond* x. 264 How long will it take me to get the old buzz-box to Laidley Towers? 1934 *Passing Show* 12 May 10/1 Ring up Mason's yard..and ask 'em to send round the old buzz-box. 1858 *Varieties* (San Francisco) 17 July 3/1 'Any taste for music?' 'Strong. Buzz and buck saws in the day time, and-wolf howling and cat fighting nights.' 1897 'Mark Twain' *Following Equator* xxxiv. 313 Whizzing green Ballarat flies..with..stunning buzz-saw noise. 1914 *Dial. Notes* IV. 104 *Buzz-wagon*, automobile. 1918 Wodehouse *Piccadilly Jim* xxi. 197 Dere's a buzz-waggon outside, waitin'. 1923 'Ian Hay' *Lucky Number* ix. 253 Let's go to the stable and start up your little friend's buzz-wagon.

buzz, *sb.*² Add: **2. b.** Also quasi-*adj.* of an artificial fly (see quot. 1877).

1877 Hallock *Sportsman's Gazetteer* 599 A fly is said to be buz when the hackle is wrapped on thick and it looks 'bushy' as we Americans would term it. 1889 F. M. Halford *Dry-Fly Fishing* ix. 205 Arguments in favour of dressing spinners hackle or buzz fashion.

buzz, *v.*¹ Add: **2.** (Later example.) Also with (*a*)*round*, *along*.

1923 Wodehouse *Inimit. Jeeves* ii. 20 Anyhow, things seemed to be buzzing along quite satisfactorily. 1924 —— *Leave it to Psmith* i. 36 Greatest mistake go buzzing about to different dentists. 1925 T. Dreiser *Amer. Tragedy* (1926) II. i. 165 Too many youths and men were already buzzing around. 1959 T. S. Eliot *Elder Statesman* I. 10 To have the waiters All buzzing round you.

b. *slang*. To go (quickly). *to buzz off*: to go off or away quickly. Also *to buzz in*: to come in (quickly), to enter.

1914 E. Pugh *Cockney at Home* 144 'Here you!' to the Cub, 'you'd better buzz off—quick!' 1925 A. Huxley *Those Barren Leaves* IV. v. 313 So I buzzed after you till I saw old Ernest wiv ve car. 1931 L. Robinson *Far-Off Hills* I, Are you buzzing too? You're very short and sweet. 1938 E. Bowen *Death of Heart* II. iii. 232, I asked Daphne who you were the moment you buzzed in. *Ibid.* iv. 241 We may buzz back here for tea. 1948 C. Day Lewis *Otterbury Incident* ii. 19 We were just about to buzz, when the yard door opened.

6. b. *Phonetics*. To pronounce as or with a buzz. Cf. *buzz *sb.*¹ 1 b.

1877 Sweet *Handbk. Phonetics* 37 (j) in N[orth] G[erman] is often distinctly buzzed. *Ibid.*, Buzzed (j) is the ordinary G. *g* in 'liegen', 'regen'.

8. a. Recently in extended use: to fly (an aircraft) fast and close to. Also *buzzing* vbl. sb. Also *transf.*

1941 *Amer. Speech* XVI. 164/1 *Buzzing a town*..in Air Corps, to fly over it. 1942 *Time* 14 Dec. 82/2 They said he could buzz the camouflage off the top of a hangar without touching it. 1948 in Berg *Dict. New Words* (1953) 49/2 Two fighters buzzed a Bristol Wayfarer. 1958 *Daily Mail* 18 July 1/4 The reported 'buzzing' of British air transports by Israeli fighters as they crossed the coast. 1959 *Times* 23 May 6/4 The commander said that the Chaplet..circled and 'buzzed' the Odinn before hitting her in the stern. 1969 *Daily Tel.* 17 Dec. 10/6 It can be a frightening experience to be shadowed, or 'buzzed' by a heavy lorry in fog.

9. To telephone or signal (a call or message) by the 'buzzer'. *to buzz off*: to ring off on the telephone. Also *intr.* of a message: to come *in* by the 'buzzer'. Hence (*slang*) to telephone (a person) (cf. *buzz *sb.*¹ 1 d).

1914 *Pears' Christmas Ann.* 20/2 Are you the *Bainbridge*? Then buzz off!..You there—have you had a call from the *Bainbridge*? 1916 'Boyd Cable' *Action Front* 173 The telephonists..'buzzed' even more monotonous strings of longs and shorts on the buzzer. *Ibid.*, The messages that had just 'buzzed' in over their wires. *Ibid.* 183 It's bad enough..to get all these messages through by voice. I haven't a dog's chance of doing it if I have to buzz each one. 1929 'E. Queen' *Roman Hat Myst.* viii. 117, I wouldn't have buzzed you so early in the morning except that Ritter just phoned. 1956 R. Heinlein *Double Star* (1958) vi. 101 He's gone to his room. I'm buzzing him.

10. To cut (wood) with a buzz-saw. *U.S.*

1925 *British Weekly* 5 Mar. 554/5 His home-built contraption for 'buzzin'' wood.

11. To throw swiftly or forcibly. *colloq.*

1890 Kipling *Many Invent.* (1893) 35 Dennis buzzed his carbine after him, and it caught him on the back of his head. *a* 1917 E. A. Mackintosh *War, the Liberator* (1918) 113 If we cannae throw a live We can aye buzz a dud. 1948 C. Day Lewis *Otterbury Incident* i. 10 The Prune buzzed a half-brick at Ted.

buzzard, *sb.*¹ **1. b.** Delete † and add later examples.

1775 N. Cresswell *Jrnl.* (1924) 147 We are between Hawk and Buzzard. 1832 [see Hawk sb.¹ 1 c]. 1895 E. C. Brewer *Dict. Phr. & Fable* 193 *Between hawk and buzzard*. Not quite a lady or gentleman, nor quite a servant. Applied to tutors in private houses [etc.]. 1904 *Courier-Jrnl.* (Louisville, Ky.) 12 July 4 The intelligence of the Commonwealth found itself literally between hawk and buzzard. It hovered in the balance.

2. Delete '†*Obs.* exc. *dial.*' and add later examples; also used *euphem.* = *bastard *sb.* 1 c.

1889 Barrère & Leland *Dict. Slang*, *Buzzard* (American), an oppressive, arrogant person, jealous of rivalry, and vindictive. 1918 Mulford *Man fr. Bar-20* viii. 77 'Pop', he said, sharply, 'who is this buzzard?' *Ibid.* xi. 108 You two buzzards are about as cheerful an' pleasant as a rattler in August. 1939 A. Huxley *After Many a Summer* I. v. 57 Not that he was doing anything spectacular with the old buzzard at the moment. 1960 J. Wain *Nuncle* 163, I could never have accused the old buzzard of caring excessively for me as a person.

buzzer¹. Add: **4.** An electric mechanism for producing an intermittent current and a buzzing sound or series of sounds; used chiefly as a call or signal. Also *attrib.*

a 1884 Knight *Dict. Mech.* Suppl., *Buzzer*, a telegraphic call in which a vibrating hammer strikes a sounding piece and gives out a buzzing sound, which, in certain cases, is preferable to a bell. 1901 'Linesman' *Words Eyewitness* (1902) 203 The little station, with its brave air of business, its stationmaster, and its electric 'buzzer'. 1916 'Boyd Cable' *Action Front* 183 He could hear the morse signals on the buzzer plain enough. *a* 1917 E. A. Mackintosh *War, the Liberator* (1918) 99 If..his bloody barrage-fire's Broken all your buzzer wires Don't get flurried. 1920 *Conquest* June 404/1 There is a local buzzer-circuit in the call box. 1943 Hunt & Pringle *Service Slang* 20 *The Buzzer*, another name for the telephone, and particularly the modern 'buzz' boxes or house-phone systems. 1968 *Globe & Mail* (Toronto) 5 Feb. 17/5 Oakland's Ted Hampson and Minnesota's Mike McMahon..scuffled after the final buzzer [in ice hockey].

fig. 1930 H. Nicolson *Swinburne* i. 11 Although, so to speak, this obstruction exists only on one line of communication, yet it acts as a buzzer which disturbs the rest.

b. Hence (*Services' slang*), a signaller.

1915 'Ian Hay' *First Hundred Thousand* vii. 55 One the Battalion signallers—or 'buzzers', as the vernacular it, in imitation of the buzzing of the Morse instrume 1917 'Taffrail' *Off Shore* 1 His friends..refer to him 'Buzzer'..because the instruments of which he is custodian..emit buzzing and humming sounds.

c. A door-bell. *colloq.* (orig. *U.S.*).

1934 R. Stout *Fer-de-lance* xvii. 295 Fritz, the buzzer, attend the front door, please. 1959 'J. Welcome' *Stop at Nothing* viii. 132 The door buzzer sounded in the hall...She ..pressed the button that freed the lock.

5. *Electr.* The trembler of an induction coil.

1882 W. H. Preece in J. J. Fahie *Hist. Wireless Telegr.* (1899) 138 *Buzzers*, little instruments that make and break the current very rapidly with a buzzing sound. 1888 *Chambers's Jrnl.* 14 Jan. 25 It is called a 'buzzer'...It is a rapid current-breaker.

bwana (bwā·na). Also **Bwana.** [Swahili.] A term of respectful address or reverence (formerly used) in East Africa, equivalent to '(the) master', 'Mr.', or 'Sir'.

1878 H. M. Stanley *Dark Cont.* I. iii. 59 Bwana, you see these scars. *Ibid.* II. xvii. 478 It is Bwana Stanley's expedition that has returned. 1887 E. C. Dawson *James Hannington* 212 The cries of 'Run, bwana, run!' were accentuated by a double roar. 1921 *Blackw. Mag.* Jan. 119/1 He had not been able to tell his Bwana about the bustard. 1963 *Punch* 24 July 113/1 An African moving..on to his ex-*bwana's* farm.

by, *sb.*¹ Add: Also, an instance of a place-name in *-by*.

1884 *Pall Mall Gaz.* 20 Feb. 5/1 Dr. Taylor..had already taught us to recognize the general tokens of Scandinavian settlement in the..*bys* where they [*sc.* the pirates] made their solitary..homesteads. 1908 W. G. Collingwood *Scandinavian Brit.* 113 'Thorpes' indicating villages as opposed to 'byes' or isolated farmsteads..are found.

by, *prep.* Add: **26.** Phr. *do as you would be done by*: see Do *v.* 37.

29. Also with omission of the verb. Phr. *by the name of*: see Name *sb.* 13.

1869 J. T. Coleridge *Mem. Keble* 217 Dialogues, in which a mason by that name [*sc.* Richard Nelson] bears a principal part.

33. c. Used for: written, painted, executed by (an author, painter, sculptor, etc.).

1570 (*title*) The Scholemaster..¶By Roger Ascham. 1595 (*title*) Colin Clovts Come home againe. By Ed. Spencer. 1673 (*title*) Poems, &c. upon Several Occasions. By Mr. John Milton: Both English and Latin, &c. Composed at several times. 1779 *Mirror* No. 24 Can the representations of moon-light, even by Homer, Milton, and Shakespeare, be more exquisitely finished? 1832 Disraeli *Cont. Fleming* II. xiv, I must get 'Manstein' directly, if it be by young Moskoffsky. 1901 *Lincoln City & Cathedral* 154 The latter [window], by a Nuremburg executant, is poor and feeble. 1966 *Observer* 23 Oct. 22/2 (Advt.), Lady Windermere's Fan. Directed by Anthony Quayle. Scenery & costumes by Cecil Beaton.

d. Of a public house, etc.: kept or managed by (as licensee).

1840 Dickens *Old C. Shop* xxix, This is the Valiant Soldier, by James Groves. 1885 Henley & Stevenson *Macaire* I. iii, *Auberge des Adrets*, by John Paul Dumont. 1919 Masefield *Reynard the Fox* 2 The meet was at 'The Cock and Pye By Charles and Martha Enderby'.

e. Followed by a personal pronoun or personal name in expressions indicating agreement: with, as far as I (etc.) am concerned. *colloq.* (orig. *U.S.*).

1930 *Amer. Mercury* Dec. 456 Five skins is jake by me. 1940 'N. Shute' *Old Captivity* i. 37 If it pleases you to think like that, it's O.K. by me. 1956 —— *Beyond Black Stump* vi. 167 I'd like to go on..if that's all right by you. 1960 *New York Times Bk. Rev.* 30 Apr. 8 He is regarded as a youngish-type people's critic..and this is fine by Mr. Fiedler.

34. b. From, after, according to (a model).

a 1650 E. Norgate *Miniatura* (1919) 84 When the Italians have not the Life to draw by, they make use off Models. *Ibid.* 86 By these and such others they draw. 1654 H. Vaughan tr. *Nieremberg's Discourses* 88 Pictures that have not so much as an ayre of those faces they were drawn by.

B. *adv.* **5.** Over in duration, finished, at an end. Of time: past, gone by. Also *by with*. *Sc.* and *north*.

1784 Burns *Ep. Rankine* x, As soon 's the clocking-time is buy, And the wee pouts begun to cry. 1846 Alex. Laing *Wayside Flowers* 20 Whan the buryin' was bye, an' relations a' gane. 1896 Crockett *Grey Man* xii, The days of curses are by with.

b. Of a person: done for, ruined, dead: esp. in *to be by with it*. *Sc.* and *north*.

1890 Service *Notandums* vi. 34 When the dykes are broken you're bye, ye ken,—ouay! fairly bye! 1893 Stevenson *Catriona* xxx, You're by with it, James More. You can never show your face again. 1900 Kipling in *Daily Express* 26 June 4/6 I'll not call it farmin'—up yonder, but ye're by with that even.

by-. Add: **2. b. by(e)-child, -son,** a bastard. *dial.*

Cf. By-blow 3.

1873 *Lett. from Jamaica* 81 Concubinage is a universal institution, and bears with it no disgrace. The offspring of such connexions—'bye-children', 'out-children', or 'love-children', as they are called,—generally follow the mother. 1894 Hall Caine *Manxman* III. xvi, You'd be hearing of the by-child, it's like? 1887 A. E. Barr *Border Shepherdess* xii, 'That play-acting by-son o' the Graeme.' 'He was no by-son.'

3. d. *by-practice*.

1913 C. Read in *Eng. Hist. Rev.* Jan. 48 Walsingham believed that his [*sc.* Burghley's] by-practice through Hunsdon with Arran was influenced by these considerations.

4. *bye-betting*, *-wager*.

1886 H. Smart *Outsider* II. i. 2 On no race of late years had there been so much bye-betting, that is to say, wagers in which one horse was backed against one other. *Ibid.*, Others..who had laid these heavy bye-wagers, looking upon the horse as having no possible chance, had never taken the trouble to secure themselves.

5. by-form, a collateral and sometimes less frequent form; spec. in *Philology*; **by-stake,** each of the short intermediate stakes used in basket-making; hence **by-stake** *v.*, to furnish with by-stakes; **by-staking** *vbl. sb.*

1887 By-form [see Dict.]. 1905 *Science* 26 May 802/2 The English masque is a by-form of the English drama. 1934 Priebsch & Collinson *German Lang.* II. ii. 172 The plural *Dinge* has a pejorative bye-form in *Dinger*. 1912 T. Okey *Art of Basket-making* viii. 84 It is usual to supplement the stakes with bye-stakes..which are inserted in the waling of the upsett between the pairs of stakes. This is termed half bye-staking. Square work is full bye-staked. 1960 E. Legg *Country Baskets* 43 Double stakes—this is side stakes and long bye-stakes.

by and by. Add: **B. 2.** The name by which cannon were known to some African tribes.

1857 J. Shooter *Kafirs of Natal* 112 They believe that the fearful *by-and-bye* eats up everything. 1893 B. Mitford *Gun-Runner* xxiv, We laugh at their *bai-nbai*. What are guns, big or small, against the broad shields and devouring spears of the ever-conquering Amazulu? 1894 C. H. W. Donovan *With Wilson in Matabeleland* x. 234 They used to call common shells 'by-and-byes', because they could see the smoke, and by and by a shell would explode in their midst.

by and large, *advb. phr.* **1.** *Naut.* See By *adv.* 1 d.

1669, 1833 [see By *adv.* 1 d]. 1961 F. H. Burgess *Dict. Sailing* 42 *By and large*, with the wind near the beam.

2. In one direction and another, all ways; now esp., in a general aspect, without entering into details, on the whole.

1706 [see Large *adv.* 7 b]. 1769 in *Southern Lit. Mess.* XVII. 183/2 Miss Betsey, a charming frigate, that will do honour to our country, if you take her by and large. 1833 J. Neal *Down-Easters* I. 23 A man who feels rather perplexed on the whole, take it by and large. 1869 'Mark Twain' *Innoc. Abr.* v. 47 Taking it 'by and large', as the sailors say, we had a pleasant..run. 1929 Wodehouse *Mr. Mulliner Speaking* viii. 248 Taking it by and large—Australia seemed to him a pretty good egg. 1955 *Times* 23 May 9/2 The virtue of sound broadcasting was that, by and large, the content mattered more than anything else.

bye, *sb.* Add: **1. c.** In *Assoc. Football* and *Lacrosse*. Also *bye-line*, the line extending the alignment of the goal-posts.

1928 *Sunday Express* 16 Dec. 21/1 There was scarcely one who was not positive that the ball had passed the bye-line. 1959 *Times* 23 Feb. 3/2 Viollet pulled his pass back from the by-line. 1962 *Times* 6 Feb. 4/6 As is usual at the universities nowadays, the game [lacrosse] was played with side-lines and bye-lines.

d. esp. in *Cockfighting*: in full *bye-battle* (also *transf.* and *fig.*), as distinguished from the 'main': see Main *sb.*³ 3.

1716 [see Main *sb.*³ 3]. 1754 *Connoisseur* 22 Aug. I. 178 Our present race of spindle-shanked beaux had rather close with an orange wench at the playhouse, than engage in a bye battle at Tottenham Court. 1859 Lennox *Picts. Sporting Life* I. 175 Eleven a-head on the main and byes seven. *Ibid.*, On the usual fighting night..at the same pit, for bye-battles. 1882 R. Caldecott *'Graphic' Picts.* I. 13/1 There were carpet-dances on off-nights by way of byes.

e. *Golf.* The hole or holes of the stipulated course that are unplayed when the match is finished.

1887 *Golfing* 92 *Bye*. Any hole or holes that remain to be played after the match is finished, are played for singly; unless the sides agree to make another match of them. 1890 *Sat. Rev.* 31 May 666/1 Prestwick golfers of to-day do not play for such stakes as a soul on the round and a nose on the bye.

3. = *bywater.

1928 J. R. Sutton *Diamond* 34 Yellow Diamond...Fine White. White. First Cape...First Bye.

by-election. [f. By- 4 + Election *sb.* 1 b.] The choice of a parliamentary representative at a time other than that of a General Election. Also *attrib.*

1880 [see By- 4]. 1897 *Daily News* 6 Jan. 4/5 Very actively engaged in by-election work. 1957 *Times* 11 May 7/2 In the borough elections the leftward movement seen in national by-elections has been repeated.

Hence **by-electioneering** *vbl. sb.* and *ppl. a.*, (the) contesting (of) a by-election. Also **by-electioneer**, a (successful) candidate at a by-election.

1897 *Westm. Gaz.* 25 Oct. 2/2 By-electioneering exigencies. 1938 *Times* 28 Apr. 15/4 By-electioneering will not be added to the work of Ministers with special responsibilities. 1962 *Punch* 4 July 25/1 Three by-electioneers had to take their seats.

So **by-elec·toral** *a.*, of, belonging to, or characteristic of a by-election.

1892 *Pall Mall Gaz.* 6 July 2/2 The Liberals are deprived of their by-electoral successes.

Byelorussian, var. *Belorussian.

bygone. Add: **B.** *sb.* **1. d.** *sing.* A person or thing of the past. Also, *spec.* a domestic, industrial, etc., artefact of a disused kind.

1857 Melville *Confidence-Man* xxii. 161 The working or serving man, shall be a buried by-gone, a superseded fossil. 1891 Hardy *Tess* III. xlv. 89 Bygones would never be complete bygones until she was a bygone herself. 1897 *Westm. Gaz.* 24 July 4/2 We would let the bygone of the Rae Mine, whatever its case, be a bygone. 1935 *Burlington Mag.* Feb. p. xiii/2 These diverse dispersals should give a real spurt to the collecting of so-called 'by-gones'. 1940 *Ibid.* Dec. 173/2 Italian Maiolica, pottery and porcelain of other lands, curious and entertaining 'bygones' of many kinds. 1960 H. Hayward *Antique Coll.* 52/1 *Bygones*, any objects no longer in use, such as obsolete agricultural implements, old-fashioned spits [etc.].

‖**bylina** (bəli·nā). Pl. **byliny, bylinas.** [Russ.] A Russian traditional heroic poem.

1886 I. F. Hapgood *Epic Songs Russia* 1 Amid the vast swamps and forests of Northern Russia the *bylinas* are sung to-day by scores of peasants. 1902 L. Wiener *Anthol. Russ. Lit.* I. 10 The bylinas of the Vladimir cycle, the last and most numerous of all that are preserved, point to an old poetic tradition. 1932 N. K. Chadwick *Russ. Heroic Poetry* 2 Even in the province of Olonets on Lake Onega, the district in which the great majority of the *byliny* have been collected, the practice of singing *byliny* has for long been restricted to a limited area.

by-line (bai·lain). [f. By- + Line *sb.*²]

1. A line giving the name of the writer of an article in a newspaper or magazine. orig. *U.S.*

1926 Hemingway *Fiesta* (1927) ii. 19, I sorted out the carbons, stamped on a by-line. 1928 *Sat. Even. Post* 12 May 36/2 The term 'by-line' means the signature of the writer. 1938 I. Kuhn *Assigned to Adv.* iii. 26 The thing I wanted most was a by-line—that magic inch of print above a story I had written which would identify me as the author of the gem. 1963 *Punch* 14 Aug. 238/3 To win a by-line is.. the rising newspaperman's dream.

2. A secondary line; a side-line.

1936 *Scrutiny* V. June 16 But when we think of Johnson and Crabbe..then Gray, Thomson..and the rest belong plainly to a by-line.

Hence as *v.*, to print a by-line. So *by-lined* ppl. a.; **by-liner,** one who writes under a by-line.

1944 *Birmingham* (U.S.) *News* 7 Apr. 9 He discovered a young writer and launched him in the newspaper business, where he soon became a popular sports by-liner. 1958 'E. A. Robertson' *Justice of Heart* iii. 35 It's my business to decide whether I by-line the stuff on my page or not! 1959 I. Ross *Image Merchants* (1960) viii. 126 He..soon had a by-lined column on science. 1961 *Guardian* 30 Sept. 12/6 The ace byliners found their stories on the back page. 1970 G. Lord *Marshmallow* xv. 129 He would swap two years of by-lined cuttings for one book on the shelf with his name on the spine.

bymeby, var. *bimeby *adv.*

byon (byōᵘn). [ad. Burmese *brun* refuse, as of grain, peas, etc., the matrix earth of rubies and the rejected stones; app. related to *prun*, *phrun* to be worn out or exhausted.] The ruby-bearing clay of the ruby mines district of Upper Burma; also *attrib.*

1892 E. W. Streeter *Prec. Stones* (ed. 5) 153 A brown or yellowish clay, known locally as *Byon*, seems to be the typical Ruby-bearing earth. 1895 *Standard* 24 July, The valley byon beds of Mogok, Tagoungnandine, and Loodah. 1922 *Chambers's Jrnl.* July 447/1 Coolies will still..be required to take the 'byon' to the sluice.

by-pass (bai·pɑs), *sb.* Also formerly **bye-pass.** [f. By- B. 3 b + Pass *sb.*¹]

1. A secondary pipe issuing from the main or service pipe below a stop-tap or cock, allowing the free passage of a small supply of gas, steam, etc., when the main supply is shut off; *esp.* the small tube and pilot light of a gas-jet, which remains alight when the jet is turned off. Also *attrib.*

1848 W. Pole tr. *Alban's Steam Engine* 264 It is a sort of by-pass to allow the steam to travel freely from the upper into the lower box. 1876 *Amer. Gas-L. Jrnl.* 3 July 20 (Knight *a* 1884), Farmer's hydraulic main, with dip-pipe and bye-pass. 1888 *Morning Post* 12 Jan. 2/3 The only service from the stage supply that was open being one half inch bye-pass for the pilot light of auditorium sun burner. 1895 *Daily News* 10 Oct. 6/4 The innovation..consists of a little 'by-pass' arrangement by which a tiny flame is always kept going. 1901 [see *accelerator c]. 1959 *Times* 25 July 9/4 An external by-pass valve.

2. *Electr.* A circuit or element providing an alternative path for the flow of current. Also *attrib.*

1914 R. A. Philip in H. Pender *Amer. Handbk. for Electr. Engineers* 364 To avoid the excessively high voltage which would occur when the circuit opened..an automatic by-pass is provided in multiple with the lamps. 1930 *Engineering* 9 May 600/1 The sides of these two switches, which are remote from the transformer, being connected together to form a 'by-pass' connection. 1931 *Answers* 10 Oct. 36/2 It might be found advantageous to connect a by-pass condenser across the telephones in the case of short-wave sets. 1964 R. F. Ficchi *Electrical Interference* v. 49 A typical bypass filter made up of a capacitor component.

3. A road diverging from and re-entering a main road, *esp.* one constructed as an alternative route to relieve congestion of traffic in a town. Also *attrib.*

1922 *Daily Mail* 2 Dec. 5 New roads and by-passes, which should remove some of these danger spots. 1923 *Times* (*Weekly Ed.*) 25 Jan., The Kingston by-pass will begin at the Robin Hood gate. 1929 *Times* 13 Nov. 9/4 It was recommended that the proposal to make a by-pass road be dropped. 1937 [see *by-pass *v.* 1 b]. 1955 *Times* 13 June 5/3 Before they considered secondary roads..the trust should urge..the completion of the by-passes. 1966 J. Betjeman *High & Low* 70 This had to happen at the corner where the by-pass Comes into Egham out of Staines.

4. *transf.* and *fig.*

1928 T. E. Lawrence *Let.* 15 Mar. (1938) 579 They will come to you, round about through the parcel mail...I sent them by an official by-pass, for safety. 1957 *Listener* 18 July 102/3 To overstress Crashaw's capacity for writing hypnotic poetry would be a dangerous simplification, a crude bypass for those who want to dodge the implications of his belief.

5. *Aeronaut.* Applied to a type of jet engine (see quot. 1955).

1948 *Jrnl. R. Aeronaut. Soc.* LII. 714/1 One stage on towards eliminating the propeller, the ducted fan or 'by-pass' engine offers interesting possibilities. 1955 *Times* 26 Aug. 4/5 In a conventional jet engine all the air is compressed and then heated by the injection of burning fuel, expanded through the turbine, and finally ejected at high velocity. In a by-pass engine only a proportion of the air is compressed and heated; the remainder by-passes the combustion system and turbine and rejoins the heated gases in the jet pipe, to mix with them and lower their temperature before the whole mixture is ejected at a lower speed than that in the 'simple' jet engine.

6. An alternative passage for the circulation of blood during a surgical operation (on the heart). Also *attrib.*

1957 *Times* 1 Nov. 11/7 The circulating blood may have to be diverted and oxygenated through a by-pass circuit while the heart is opened, so that a defect within it may be rectified. 1961 *Lancet* 22 July 182/2 The patient was on bypass for eighty-one minutes. *Ibid.*, The first open repair on the heart, using a cardiopulmonary bypass, was carried out by Cooley et al. (1958).

by-pass (bai·pɑs), *v.* [f. prec.]

1. *trans.* To furnish with a by-pass.

1886 *Sci. Amer.* Suppl. 4 Dec. 9099/3, I next by-passed the outlet valve with a one inch pipe. 1929 *Times* 28 May 17/4 Schemes are on foot for by-passing both Leatherhead and Dorking.

b. To take an indirect route around, to avoid (a locality, military position, etc.).

1928 *Even. Standard* 12 Mar. 6/2 King's Langley and Tring by passing Bushey and Watford. 1937 *Times* 13 July 10/4 The scheme has been referred to as 'the Selborne by-pass', but it does not by-pass Selborne at all. 1942 *Hutchinson's Pict. Hist. War* 10 June–1 Sept. 72/1 German armoured divisions which had by-passed our positions west of Mersa Matruh. 1959 *Listener* 15 Jan. 115/2 We had to by-pass Tippaburra.

c. *fig.*

1941 *Spectator* 10 Oct. 358/1 Congress is only one party in India, but by general consent it is the largest one, and it's no good trying to by-pass it. 1949 Koestler *Promise & Fulf.* vii. 70 Conditioned during the course of a whole generation to bypass the law. 1959 *Listener* 4 June 971/1 Whenever he was at odds with his Secretary of State he simply by-passed him.

2. To conduct (liquid, gas, etc.) by means of a by-pass.

1909 in *Cent. Dict.* Suppl. 1922 *Encycl. Brit.* XXX. 41/1 A valve for bypassing the whole or part of the exhaust gas directly into the atmosphere is provided to enable the output from the blower to be regulated. 1924 *Glasgow Herald* 6 Nov. 9 A small portion of the gas..is by-passed to a boiler.

by-product. Add: **b.** *attrib.* and *transf.*

1905 *Daily Report* 9 Dec. iv/1 By-product coke ovens. 1926 Fowler *Mod. Eng. Usage* 304/1 But a regrettable by-product of their activities has been a relapse into primitive methods of soliciting attention. 1952 S. Kauffmann *Philanderer* (1953) ii. 23, I was going to say the principal thing is that you're here. But that's a by-product. The principal thing..is that you know me. 1969 *Nature* 27 Dec. 1297/2 By-product gypsum formed during the wet process manufacture of phosphoric acid contains various impurities.

bysmalith (bi·zmăliþ). *Geol.* [f. Gr. βύσμα plug + λίθος stone: see -lith.] A large roughly cylindrical body of igneous rock which, when it was forced upward, lifted up the overlying rock.

1898 J. P. Iddings in *Jrnl. Geol.* VI. 706 By this mode of intrusion, the vertical dimension of the intruded mass becomes still greater as compared with the lateral dimensions, so that its shape is more that of a plug or core. Such an intruded plug of igneous rock may be termed a *bysmalith*. 1905 Chamberlin & Salisbury *Geol.* I. 477 Between the bysmalith and the laccolith there are various gradations.

Hence **bysmali·thic** *a.*, characteristic of a bysmalith.

1933 R. A. Daly *Igneous Rocks & Depths of Earth* vii. 132 We have no direct evidence that the intracrustal space occupied by a typical batholith was gained principally through a laccolithic or bysmalithic lifting of the roof.

byssinosis (bisinōᵘ·sis). *Path.* [mod.L., f. Gr. βύσσινος made of byssus (see Byssine) + -osis.] A chronic disease of the lungs caused by the inhalation of fine particles of textile fibres, esp. cotton dust, over a long period.

1890 Billings *Med. Dict.* I. 203/2 *Byssinosis*, production of lung disease by inhalation of cotton-fibres. 1948 *Lancet* 14 Feb. 253/1 Byssinosis develops after long and continuous exposure to cotton dust. 1958 *Times* 20 June 11/7 A worker [at Calder Hall] is apparently a good deal healthier than one in a Lancashire cotton mill where byssinosis has not yet been eradicated.

Hence **byssino·tic** *a.*, affected with, characteristic of byssinosis; also as *sb.*, a byssinotic person.

1952 *Brit. Jrnl. Industr. Med.* IX. 138/2 Prausnitz.. stated that byssinotics were hypersensitive..to a substance which could be extracted from cotton dust. *Ibid.* 195/1 The skin testing of byssinotic persons with cotton dust extracts fails to distinguish them from their symptomless fellow workers. 1964 *Lancet* 19 Sept. 609/2 The most severe attacks of byssinotic symptoms are after the annual holidays.

byssogenous (bisǫ·dʒinəs), *a.* [f. Byss-us + -gen 2 + -ous.] That (normally) produces a byssus.

1886 *Ann. Rep. Smithsonian Inst. 1885* I. 777 Lamellibranchs generally exhibit more or less well marked traces of this byssogenous apparatus. 1895 *Camb. Nat. Hist.* III. 453 Foot with byssogenous slit, but no byssus.

byte (bəit). *Computers.* [Arbitrary, prob. influenced by *bit *sb.*⁴ and Bite *sb.*] A group of eight consecutive bits operated on as a unit in a computer.

1964 Blaauw & Brooks in *IBM Systems Jrnl.* III. 122 An 8-bit unit of information is fundamental to most of the formats [of the System/360]. A consecutive group of *n* such units constitutes a field of length *n*. Fixed-length fields of length one, two, four, and eight are termed bytes, halfwords, words, and double words respectively. 1964 *IBM Jrnl. Res. & Developm.* VIII. 97/1 When a byte of data appears from an I/O device, the CPU is seized, dumped, used and restored. 1967 P. A. Stark *Digital Computer Programming* xix. 351 The normal operations in fixed point are done on four bytes at a time. 1968 *Dataweek* 24 Jan. 1/1 Tape reading and writing is at from 34,160 to 192,000 bytes per second.

by-the-way, *sb.* [f. phr. *by the way*: see By *prep.* 12 b.] An incidental remark.

1896 *Punch* 1 Feb. 52/1, I think I may indulge myself with a short by-the-way on the subject of hampers. 1907 *Daily Chron.* 2 Oct. 3/3 Here are two 'by-the-ways' from her pages of observation.

bywater (bai·wǭtəɹ). [f. By- B. 5 + Water *sb.* 20.] A diamond of inferior water, yellowish in colour; also *attrib.*

1878 [see Off colour 2]. 1895 *Westm. Gaz.* 1 Oct. 2/1, I..woke up to consider two fresh lots of treasure, 'whites' and 'by-waters', so termed from their colour. 1916 F. B. Wade *Diamonds* 25 Yellows, or by-waters.

bywoner (bai·wōnəɹ, ‖ bēⁱ·vōnər). *S.Afr.* Also (corruptly) **beiwoner, bywonner.** [Afrikaans, f. *by* with + *woon* live + *-er*, pers. suff.] A poor tenant farmer who lives on the farm of another man to whom he renders certain services (with or without payment in produce or money), being allowed to carry on some farming on his own account. Also *attrib.*

1886 J. J. Aubertin *Six Months in Cape Colony & Natal* ix. 235 Then there is what are called the Baywhoner [*sic*] tribe among the Dutch..that live upon the landowners, their friends. 1889 H. A. Bryden *Kloof & Karroo* 253 A beiwoner (a sort of sub-farmer on the estate of a richer farmer, who is expected to perform certain duties for the privilege of running his stock). 1901 *Daily News* 22 Mar. 4/7 Steyn and De Wet and their following of 'bywoners'. 1916 F. E. Mills Young (*title*) The Bywonner. 1955 L. G. Green *Karoo* i. 15 Farmers lived on biltong and brak water, and their bywoners ate dead donkeys.

by your leave, *sb.* [f. phr. *by your leave*: see Leave *sb.* 1.] An expression of apology for not having asked permission; the asking of permission.

1914 H. R. Martin *Barnabetta* xii. 106 She was sportively handling huge sums like this without a By-Your-Leave. 1924 W. M. Raine *Troubled Waters* i. 13 'With not even a by your leave. You're a claim jumper,' she said. 1936 *Scrutiny* V. 38 Jung, indeed, insists that he is a bold man nowadays who would roundly declare, without so much as by-your-leave from the doctor, that he is altogether free of it [*sc.* neurosis]. 1948 'J. Tey' *Franchise Affair* viii. 90 He picked Robert's glass out of his hand without a by-your-leave and rose to fill it.

Byzantine, *a.* and *sb.* Add: **A.** *adj.* Also, reminiscent of the manner, style, or spirit of Byzantine politics. Hence, intricate, complicated; inflexible, rigid, unyielding.

1937 Koestler *Spanish Testament* iv. 75 In the old days people often smiled at the Byzantine structure of the Spanish Army. 1965 *Economist* 25 Dec. 1404/3 From the byzantine procedural caution of the approach work [to the Common Market] on both sides, it seems that substantive issues are still beyond the diplomats' grasp. 1966 G. Steiner *Lang. Silence* (1967) 399 It was precisely on this occasion that Stalin struck the new ominous note of the cult of personality, of the Byzantine homage to the leader. 1966 *Listener* 26 May 765/1 To hint that one does not quite catch the drift of their byzantine prose..pierces to the heart of their intellectual pride.

Byzantinism. Also *fig.*

1945 Koestler *Yogi & Commissar* III. iii. 225 The antidote to Eastern Byzantinism is Western revolutionary humanism. 1963 *Times* 21 Feb. 15/2 The Byzantinism of much academic criticism in America. 1967 *Economist* 18 Mar. p. xxxix/3 The only way to cut through the dead wood is often to peddle influence in Rome...In the end it only aggravates the tendency to confuse political byzantinism with energetic administration.

Byzantinist (bəi-, bizæ·ntinist). [f. Byzantine + -ist.] A student of or an expert in Byzantine matters.

1892 *Nation* (N.Y.) 21 Apr. 306/1 The collaboration of Greek, English, and Italian 'Byzantinists' is already pledged to support it. 1927 *Observer* 24 Apr. 20/3 The Congress of Byzantinists has concluded its work. 1958 *Times Lit. Suppl.* 21 Mar. 148/3 Viewed through the eyes of a distinguished and enthusiastic Byzantinist.

C

C. Add: **I. 2.** *C-scroll*: a decorative scroll shaped like the letter C.

1904 P. Macquoid *Hist. Eng. Furniture* iii. 63 A strapwork of C scrolls and cocksheaded arabesques.

II. 1. b. C 3: the lowest grade in the scale of physical fitness for military service employed in the classification of recruits conscripted under the Military Service Act, 1916; hence *fig.* of the lowest grade, of grossly inferior status or quality.

1918 D. Lloyd George in *Times* 13 Sept. 8/2 You cannot maintain an A1 Empire with a C3 population. 1923 *Daily Mail* 1 Mar. 7 He would agree prisoner's left arm would be a C3 left arm. *Ibid.* 11 July 13 Sunshine all the way, no C3 affair but a magnificent blaze of light. 1924 Galsworthy *White Monkey* I. viii, Eight years her senior and C3 during the war!

2. Applied to a tenor saxophone in C. Also *ellipt.*

1932 B. Davis *Saxophone* xxx. 154 Because it is not necessary to change the key when reading for the C Melody, this instrument has come to be regarded as non-transposing, in order to differentiate it from those for which it is necessary to change key. 1955 L. Feather *Encycl. Jazz* i. 20 Frank Trumbauer with his C-Melody saxophone.

III. 3. c., c. and b. (*Cricket*): earlier examples; c. = cubic, as in c.c. = cubic centimetre; C (American), $ 100; C, cocaine; ©, copyright (followed by the name or other indication of the owner of the copyright); C.A.T., College of Advanced Technology; C.B., confined to barracks, as a punishment in the army; C.B.(W.), chemical and biological (warfare); C.D., corps diplomatique; C.E., C.Æ., Common Era; occas., Christian Era; C.E.M.A., Council for the Encouragement of Music and the Arts; C.F., Chaplain to the Forces; C.G.S., centimetre-gramme-second; C.I.A. (American), Central Intelligence Agency; C.I.D., Committee of Imperial Defence; Criminal Investigation Department; C.I.E., Companion of the Order of the Indian Empire; C.I.F., c.i.f., Cost, Insurance, plus Freight; C. in C., Commander in Chief; C.K.D., completely knocked down; C.Litt., Companion of Literature; C.N.D., Campaign for Nuclear Disarmament; C.O., Commanding Officer; conscientious objector; C.O.D., see separate entry; C. of A., Certificate of Airworthiness; C. of E., Church of England; C.O.R.E. *U.S.*, Congress of Racial Equality; C.O.S., Charity Organization Society (*Cent. Dict.* Suppl. 1909); C.P., Communist Party; C.P.R., Canadian Pacific Railway; C.Q.D., in wireless telegraphy, the signal formerly used by ships in distress, consisting of C.Q., the international sign for 'all stations', followed by D indicating 'urgent'; after 1908 superseded by S.O.S.; C.R.T., c.r.t., cathode-ray tube; CS [the initials of B. B. Corson (b. 1896) and R. W. Stoughton (1906–57), American chemists who discovered its properties in 1928], a designation of *o*-chlorobenzalmalononitrile, $ClC_6H_4CH{:}C(CN)_2$, a substance that causes irritation of the skin, lachrymation, coughing, etc., and is used in the form of a finely divided solid as a quick-acting irritant for riot control and other purposes; so *CS gas*, etc.; C.S.E., Certificate of Secondary Education.

1744 in J. Nyren *Young Cricketer's Tutor* (1833) 111 Smith o C by Bartrum. 1810 in Alverstone & Alcock *Surrey Cricket* (1902) ii. 54 Lord F. Beauclerck..c. & b. Lambert 21. [1866 J. J. Griffin *Chem. Handicraft* 298 Graduated for Centimetre Cubes..1 cc. 6d.] 1899 Edser *Heat for Adv. Students* iv. 64, 1 c.c. will possess a mass of m/v grams. 1955 *Times* 18 July 12/6 The meeting opened with a 17-lap race for 500 c.c. cars. 1839 *Spirit of Times* 13 Apr. 66/3, I had no idea of betting more than an 'L', or a 'C'. 1930 *Liberty* 11 Oct. 30/3 We gave him five C notes and two tens, or 10 per cent [of $5,200] to make the payoff. 1946 *Science Digest* Aug. 23/2 A goodly supply of crisp C-notes. 1955 'H. Robbins' *Stone for Danny Fisher* II. xiii. 159 My biggest worry was somebody's clipping the five C's from my trousers back there in the dressing room. 1922 E. F. Murphy *Black Candle* II. xi. 212, I cried and made a fuss when I could not get enough 'M' or 'C', so we moved to a house where no one would hear me. 1959 W. Burroughs *Naked Lunch* 29 The craving for C lasts only a few hours, as long as the C channels are stimulated. 1947 *U.S. Congress, Statutes* c. 391 § 19 The notice of copyright..shall consist either of the word 'Copyright' or the abbreviation 'Copr.', accompanied by the name of the copyright proprietor...In the case..of copies of works specified in subsections (f) to (k)..the notice may consist of the letter C enclosed within a circle, thus ©, accompanied by the initials, monogram, mark, or symbol of the copyright proprietor. 1957 *Encycl. Brit.* VI. 429/1 The form of notice, which is required for literary, musical and dramatic works and may be used on any other works, consists of the word 'Copyright' or the abbreviation 'Copr.' or the symbol ©, accompanied by the name of the copyright owner and the year in which the copyright was secured; e.g. '© 19— by John Doe'. 1957 *Technology* July 167/2 (*caption*) CAT for North. 1964 *Economist* 27 June 1485/1 Universities, CATs and the professional social work institutions. 1888 Kipling *Soldiers Three* 11 Now, I put ut to you, Sorr, is ten days' C.B. a fit an' a proper tratement for a man who has behaved as me? 1892 —— *Barrack-room Ball.* 20 O it's pack-drill for me and a fortnight's C.B. For 'drunk and resisting the Guard'! 1919 *War Slang* in *Athenæum* 8 Aug. 787/2 When doing C.B...he [*sc.* the soldier] was doing 'jankers' or 'Paddy Doyle'. 1928 W. Empson in *Granta* 2 Nov. 74/2 We would have put the cooks in C.B...if they'd served up this cat's food. [1949 T. Rosebury *Peace or Pestilence* ix. 99 In its modern form BW has never been used in a military operation. 1960 M. Stubbs in *Advances in Chem.* XXVI. 36 A growing awareness of the potential threat of a CW–BW attack.] 1964 *Bull. Atomic Sci.* XX. Oct. 35/1 More attention to CB weapons is required. *Ibid.*, In 1961, the budget for Army expenditures for CBW research, development, and procurement was somewhat over 100 million dollars per year. 1966 *New Scientist* 29 Sept. 717/1 The employment of any one CB weapon weakens the barriers to the use of others. 1967 *Times* 29 May 4/4 John H. Hoskins..denied all charges that Yale was in any way engaged in classified research on C.B.W. 1942 Partridge *Dict. Abbrev.* 23/1 *CD*, *Corps Diplomatique*..On, e.g., motor-cars and letters. 1955 G. Greene *Quiet American* II. ii. 104 Along the route to Tanyin flowed a fast stream of staff and C.D. cars. 1961 *Guardian* 4 May 2/5 Lord Lansdowne..said in the Lords..that the 'CD' plate did not.. afford the occupant of a car any privilege or immunity. 1838 E. H. Lindo *Jewish Calendar* (title-p.), Tables for continuing the calendar to A.M. 6000–2240 C.Æ. *Ibid.* 111 (*heading*) 3760 C.Æ. Commencement of the Christian Æra. 1886 K. Magnus (*title*) Outlines of Jewish History from B.C. 586 to C.E. 1885. 1940 *Times* 15 June 7/4 The C.E.M.A. has one expert representative of each art on its Council. 1958 *Times* 4 June 11/3 In 1946 C.E.M.A. was incorporated by royal charter and given the name Arts Council. 1909 *Cent. Dict.* Suppl., *C.F.* Chaplain to the Forces. 1873 C.G.S. [see *absolute unit]. 1875 J. D. Everett (*title*) Illustrations of the Centimetre-Gramme-Second (C.G.S.) System of Units. 1962 Corson & Lorrain *Introd. Electromagn. Fields* ii. 29 In the c.g.s. system of units, K is made unity by choosing appropriate units for these quantities. 1951 *Sat. Even. Post* 17 Mar. 71 The CIA, or Central Intelligence Agency, which deals with the war capabilities of an enemy. 1910 E. H. Richardson *War, Police & Watch Dogs* iii. 56 Detective F. H. Carr of the C.I.D. 1914 W. S. Churchill *World Crisis* (1923) xii. 267 The situation..is entirely different from those which have been discussed in the Invasion Committee of the C.I.D. 1960 *Times* 3 Oct. 13/6 Every C.I.D. man must start as a uniform constable. 1886 Kipling *Departm. Ditties* (ed. 2) 8 Then the Birthday Honours came...Stood against the Rajah's name nothing more than *C.I.E.* 1937 *Discovery* Jan. 6/1 Dr. J. H. Hutton, C.I.E. 1902 *Times* 7 July 3/3, 90 per cent. f.o.b. invoice on the basis of 52s. 6d. c.i.f. sawn pitch pine 35 cubic feet average Blaenavon. 1907 *Westm. Gaz.* 7 May 2/2 The United Kingdom figures are c.i.f. at the ports of arrival. 1958 *Times Rev. Industry* June 63/1 Redwood deals and battens have dropped from £88 to £84 c.i.f. 1889 Barrère & Leland *Dict. Slang* I. 3/2 *C. in C.*, Commander-in-Chief. 1951 L. MacNeice tr. *Goethe's Faust* II. iv. 256 No, you shall win it, believe you me. It's you to-day are C.-in-C. 1937 *Times* 13 Apr. iv/1 The successful working at Wellington, New Zealand, of a C.K.D.—'Completely Knocked Down'—plant for the manufacture of the same make of car. 1961 *Times* 19 Apr. 13/3 The new 'literary honour'..is to be distributed by the Royal Society of Literature to not more than ten holders at any one time and the first batch of five living British writers..become 'Companions of Literature' and free to put 'C.Litt.' after their names from May 10. 1958 C. Judd in C. Driver *Disarmers* (1964) ii. 47 There are major points in the policy of the CND which UNA cannot support. 1961 *Times* 8 May 17/3 The C.N.D. has ceased to be a movement of moral protest..and has become a political organization. 1889 Barrère & Leland *Dict. Slang* I. 3/2 *C.-O.*, Commanding-Officer. 1890 Kipling *Many Inv.* (1893) 29 'Who was your C.O.?' said Boileau. 1915 F. H. Lawrence in *Home Lett. T. E. Lawrence* (1954) 701 One officer was C.O., and the other four company commanders. 1916 *Tribunal* 23 Mar. 4/1 (*heading*) Treatment of C.O.'s. 1919 J. Buchan *Mr. Standfast* i. 30 'Launcelot's a C.O., you know,' said Miss Doria...I remembered that the letters stood..for 'Conscientious Objector'. 1968 *War Resistance* II. xxiv. 27 Nazarene leaders have made petitions to Marshal Tito for the release of the COs. 1932 *Flight* 15 Apr. 318/1 This is in excess of the normal C. of A. gross weight of 5,400 lb. 1913 W. T. Rogers *Dict. Abbrev.* 41/2 *C. of E.*, Church of England. 1954 J. Betjeman *Poems in Porch*, Still it gives the chance to me To praise our dear old C. of E. 1962 in *Amer. Speech* (1963) XXXVIII. 229 An official of the Congress of Racial Equality (CORE). 1968 *Chicago Tribune* 9 July I. 21/1 (*heading*) Chapters in 3 cities drop out of C.O.R.E. 1910 Chesterton *What's Wrong with World* IV. xiii. 248, I do not expect the schoolmaster to hate hospitals and C.O.S. centres so much as the schoolboy's father. 1936 J. Curtis *Gilt Kid* iii. 29, 'I want to talk to you about joining.' 'About joining the C.P.?' 1969 *Listener* 9 Jan. 54/2 One of the few commendable features of the CP is that, at that time, its outlook on many issues was seldom more than a year behind enlightened opinion. 1892 Kipling *Lett. Travel* (1920) 80 The traveller is on the C.P.R. train at Vancouver. 1968 *Globe & Mail* (Toronto) 3 Feb. 1/8 A CPR right-of-way in Eastview. 1909 *Daily Chron.* 17 Feb. 3/5 Among the ships responding to the 'C.Q.D.' message were the Lucania, [etc.]. 1928 *Manchester Guardian Weekly* 7 Dec. 450/3 That..when his ship took a list..[he] should have sent out a peremptory C.Q. call. 1941 *Rev. Sci. Instruments* XII. 298/2 The connection of the lower deflecting plate of the *CRT* is incorrect. 1946 *Electronic Engin.* May 149/1 The c.r.t. is an essential part of radar equipment. 1969 *Computers & Humanities* IV. 79 A crt screen with only ten lines of 40 characters each cannot completely replace a printed page. 1960 *Armed Forces Chem. Jrnl.* Nov.–Dec. 27/1 CS is an Army chemical symbol for an agent that causes burning and watering of the eyes. 1961 *Techn. Man.* (U.S. Dept. of Army) 3-215, change 2, 25 Sept. 4 CS..a white crystalline solid..has a pungent, pepperlike odor. 1969 *Listener* 4 Sept. 297/1 The effects of the CS chemical contained in this gas are now the subject of an inquiry. 1970 *Daily Tel.* 7 Apr. 2/7 CS gas has been used by British troops during peace-keeping operations in Ulster, and by American forces in Vietnam. 1963 *Daily Tel.* 1 Oct. 1/7 In three or four years' time the CSE examination may be taken by as many as 300,000 candidates. 1966 *New Statesman* 10 June 859/3 Mathematics on sound traditional lines is taken to CSE, 'O' and 'A' levels.

caa'ing whale, var. Ca'ing-whale.

1924 *Nature* CXIII. 532/1 Among the true dolphins (Delphinidae) the caa'ing whale (Globicephala melaena) and Risso's dolphin (Grampus griseus)..are known to subsist, partly at least, on cuttlefish. 1959 A. Hardy *Fish & Fisheries* xv. 287 Yet another northern species..is the pilot whale, blackfish or caa'ing whale.

caama, var. Kaama.

caatinga (kā·tiŋgă). Also **catinga.** [Tupi, f. *caa* natural vegetation, forest + *tinga* white.] In Brazil, a forest consisting of thorny shrubs and stunted trees.

1846 G. Gardner *Trav. Brazil* 166 Woods..consisting of low trees and shrubs..called by the inhabitants *Catingas*. 1869 Burton *Explor. Highl. Brazil* I. 61 Low woods known in Brazil as Caatingas and Carrascos. 1927 Kendrew *Climates of Continents* (ed. 2) 328 A region of caatinga or dry thornwood. 1944 S. Putnam tr. *da Cunha's Rebellion in Backlands* II. i. 30 The caatinga..repulses him with its thorns and prickly leaves, its twigs sharp as lances.

cab, *sb.*³ Add: **1. a.** Applied also to motor-driven vehicles (see Taxi-cab).

1899 *Westm. Gaz.* 15 Nov. 4/3 The cab-without-a-horse. 1954 [see *bumbling *ppl. a.*].

3. *cab-body*, *-door*, *-minder*, *-user*, *-washer*, *wheel*; **cab-rank,** also *transf.* (see quots.), esp. a line of aircraft waiting in readiness; also *attrib.*; **cab-shelter,** a shelter for cabmen; **cab-stand** (earlier example).

1908 *Westm. Gaz.* 30 Mar. 5/2 All Humber cab-bodies are interchangeable. 1888 A. C. Gunter *Mr. Potter of Texas* xxii. 261 He..opens the cab door. 1898 *Daily News* 16 Nov. 5/4 C. B...described as a cab minder. 1919 Downing *Digger Dial.* 15 *Cab-rank*, transport lines. 1939 H. Hodge *Cab, Sir?* 268 Cab-ranks are appointed by the Commissioner of Police. 1945 *Times* 1 Mar. 5/7 A 'cab-rank' has been maintained over the bridgehead. *Ibid.* 6 Mar. 5/6 'Cab-rank' fighter bombers have been overhead continuously. 1946 L. E. O. Charlton *Britain at War* IV. 275 R.A.F. Typhoons ..came in time to develop..the 'cab rank' technique. Based on their own special air strips,..they there awaited a wireless summons. 1936 E. M. Forster *Abinger Harvest* I. 17, I heard profanity in the cab-shelter. 1848 Thackeray *Van. Fair* liv. 484 The people joked at the cabstand about his appearance, as he took a carriage there. 1894 *Westm. Gaz.* 22 May 5/3 The ins and outs of the matter are but imperfectly understood by the mass of cab-users. 1897 *Daily News* 19 Oct. 2/3 The death of C. W...a cabwasher. 1862 Geo. Eliot *Let.* 12 Mar. (1956) IV. 21 The sound of cab wheels.

cab, *v.*¹ (Earlier examples.)

1835 Dickens *Let.* ?29 Oct. (1965) I. 84 Worth your while to walk or Cab so far East. 1845 W. C. Macready *Diary* 1 Aug. (1912) II. 300 Walked and cabbed down to Forster.

2. To drive a cab. Cf. Cabbing *vbl. sb.*

1887 A. Conan Doyle *Study in Scarlet* (1888) II. vi. 160, I went on cabbing it for a day or so. 1939 H. Hodge *Cab, Sir?* 19 When a man first starts cabbing, he usually rubs his neck sore with this constant turning of his head.

caba. Add: (Earlier examples, and see quot. 1865.) Also earlier in U.K. in Fr. form *cabas*.

1833 Lady Morgan *Dram. Scenes* II. 74 Bring me down my *cabas*. I must have something to toss these things in. *c* 1845 C. Brontë *Professor* (1857) I. x. 171 Day-pupils, tearing down their cloaks, bonnets, and cabas from the wooden pegs. *Ibid.* xiii. 229 She proceeded..to open her cabas, to take out her books. 1865 Webster, *Cabas*, a flat basket..for figs, &c.; hence, a lady's flat work-basket or reticule; — often written *caba*. 1877 H. Ruede *Sod-House Days* (1937) 27 The pump..furnished the liquor and the caba furnished the solids for the meal.

‖**cabaletta** (kæbăle·tă). *Mus.* Pl. **-ette, -ettas.** [It., prob. ad. *coboletta*, dim. of *cobola* stanza, couplet, L. *copula*: see Couple *sb.*]

a. A short aria in simple style with a repetitive rhythm. **b.** The final section of an aria, marked by a quick, uniform rhythm. Also *transf.*

1842 J. F. Warner *Dict. Mus. Terms* 16/2 Caballetta, (frequently also *Cabaletta*, better, perhaps *Cavaletta*). This term is applied to the fashionable musical language of the Italians to a very particularly musical, agreeable and exciting passage occurring in an air (aria). It not unfrequently happens that an Italian audience..listen to a whole air simply for the purpose of being entertained by the *cavaletta*. 1896 G. B. Shaw *Our Theatres in Nineties* (1932) II. 13 Old-fashioned arias with florid cabalettas at the ends of them. 1947 A. Einstein *Music in Romantic Era* xvi. 276 The lyric *cabalette* or shortened arias introduced by Rossini. 1947 Auden *Age of Anxiety* (1948) iii. 82 And a chronic chorus of cascades and birds Cuts loose in a wild cabaletta. 1970 *Times* 29 Aug. 11/6 In the closing cabaletta ..*fioritura* was most perfectly dissolved into the vocal line itself.

caballada (kæbælä·dä). *U.S.* Also **caballado, cavallado.** [Sp., f. *caballo* horse. Cf. Cavallard.] A herd or train of horses (or mules, etc.).

1841 W. L. McCalla *Adventures in Texas* 57 (Bentley), Fifty or sixty horses..forty of which were from one caballado. 1844 J. Gregg *Commerce of Prairies* I. 27 [The Indians] drove off the entire *caballada* of near five hundred head of horses, mules, and asses. 1844 G. W. Kendall *Santa Fé Exped.* (1855) I. 97 Nothing can exceed the grandeur of the scene when a large cavallada, or drove of horses, takes a 'scare'. 1900 Smithwick *Evol. State* 22 He had his caballada driven in for us to choose from.

caballero. Add: (Earlier examples.) See also Cavalier *sb.*

1835 R. M. Bird *Infidel* I. 217 He has invited all my caballeros on this paper. 1868 Geo. Eliot *Sp. Gypsy* I. 11 A goodly knight, A noble caballero.

cabana. (Earlier U.S. example.)

1840 W. G. Simms *Border Beagles* 65 I'll go a quart and a dozen cabanas upon it.

cabana² (kăbä·nă, -nyă). Chiefly *U.S.* [Sp., see Cabin *sb.*] A cabin; esp. a hut or shelter at a beach or swimming-pool.

1898 *Land of Sunshine* (Los Angeles) Jan. 61 Though not lacking a certain picturesqueness, what the *cabana* means, and what it stands for, is a large thing. 1932 N. Coward *Coll. Sk. & Lyrics* 191 The scene is the Lido Beach...In the foreground a row of cabanas with coloured striped awnings. 1953 W. R. Burnett *Vanity Row* x. 74 A swimming-pool with cabanas. 1957 F. Rickards' *Practise to Deceive* (1959) xiii. 185 He asked why Lane had gone into the bath-house cabana. 1959 *Encounter* Sept. 50/2 The physical stage..looks like a beach cabana. 1967 *London Opinion* Free Press 21 June 7/2 Cabana sets with matching beach jacket & bathing suit.

cabane (kăbā·n). *Aeronaut.* [a. F. *cabane* Cabin.] A pyramidal structure supporting the wings of an aircraft.

1913 A. E. Berriman *Aviation* iii. 33 Each spar is supported at intervals by wires that run overhead to a mast or cabane above the pilot's seat. 1924 *Aeronaut. Jrnl.* Oct. 315 Cabane, a French word to denote the mast structure projecting above the body to which the top load wires of a monoplane are attached. 1930 *Flight* 12 Sept. 1015/2 The framework of the wing cellule comprises a central 'cabane' supported by four streamlined struts.

cabaret¹. For pronunc. read: (kæ·bărē¹). Add: **2. b.** A restaurant or night-club in which entertainment is provided as an accompaniment to a meal; also, the entertainment so provided, a floor-show. Also *attrib.*

1922 W. Irwin *Red Button* 42 He spends most of his extra money at post parties, Austrian villages and cabaret shows. 1915 T. Burke *Nights in Town* 254 Those melancholy places, the night clubs and cabarets. 1920 F. Scott Fitzgerald *This Side of Paradise* (1921) ii. ii. 211 A famous cabaret star was at the next table. 1922 *Daily Mail* 13 Nov. 7 Cabaret, as exemplified by the Midnight Follies at the Hotel Metropole, also broke all records...This latest form of entertainment is very popular just now. *Ibid.* 25 Nov. 13 The new cabaret show which was to have been started on Tuesday has been rehearsed for a fortnight. 1932 *Westm. Gaz.* 22 May, There is to be a super cabaret at the Hotel Metropole on Friday evening. 1938 G. Greene *Brighton Rock* ii. i. 68 The floor was cleared for the last cabaret of the evening.

3. A porcelain tea, coffee, etc., service sometimes accompanied by a small table or tray.

1856 J. C. Robinson *Invent. Mus. Ornamental Art* 32 Old Sèvres porcelain...'Cabaret', rose du Barry ground...the set consisting of four pieces. 1904 *Daily Chron.* 9 June 3/2 A cabaret, painted with flowers, fruit, and trophies..by Lene père. 1968 R. H. R. Smithies *Shopfitter* (1969) v. 98 A cabaret tray of Swansea porcelain.

cabas, see *caba.

cabbage, *sb.¹* Add: **1. b.** As a term of endearment: *my cabbage* [tr. F. *mon chou*], my dear, darling.

1840 Thackeray *Miscellanies* (1855) I. 488 Oui, mon chou, mon ange; yes, my cabbage, my angel, my darling, quite right. 1896 C. M. Yonge *Release* v. vii. 71 'Ah, my dear little Cabbage,' she began, 'I fear they will never forgive you!' 1924 J. M. Barrie *Mary Rose* I, This wasn't the drawing-room, my cabbage; at least not in my time. 1968 'J. T. McIntosh' *Take a Pair of Private Eyes* iii. 41 Ambrose drew her close and murmured meaningly: 'But I'm completely merciless, my little French cabbage.'

c. = *chou 2. Also *attrib.*

1888 *Daily News* 21 Sept. 5/6 A large 'cabbage' bow of surah in the same colour. 1895 *Ibid.* 19 Oct. 6/3 The toque worn with this had a large green velvet 'cabbage' at either side. 1896 *Ibid.* 14 Nov. 6/5 Folds of black satin, held down by 'cabbage' knots. 1899 *Ibid.* 3 June 8/3 The inevitable chou, or cabbage-bow, of black or dark green velvet.

d. *fig.* = *cabbage-head.* So *cabbage-looking* adj., stupid, 'green'.

1870 Mark Twain' *Writings* (1922) VII. 1 All this human cabbage could see was that the watch was four minutes slow. 1898 *Westm. Gaz.* 3 Nov. 4/1, I said I knew 'ow many beans make five..., and I 'l wer cabbage-looking I weren't green. 1924 Joyce *Ulysses* 307 Gob, he's not as green as he's cabbagelooking. 1938 R. Campbell *Wayzgoose* i. 9 What wonder if..any cabbage wins the critics' praise Who wears his own green leaves instead of bays! 1969 *Guardian* 5 Mar. 7/3, I stayed at home for nearly a year. It was awful. I became a cabbage.

e. Money. *slang* (chiefly *N. Amer.*).

1926 J. Black *You can't Win* xv. 213 'You carry this bundle of cabbage, Kid,' passing me a pack of greenbacks. 1960 *Observer* 14 Jan. 5/1 The White Cross, cabbage, poppy, lolly, in other words cash.

5. *cabbage-bed, -patch* (also *fig.*), *-seed.* **b. cabbage-head** *fig.* (later U.S. examples); **cabbage-leaf hat** = *cabbage-tree hat* (see Cabbage-tree 1 b); **cabbage-palm, palm-tree** *Austral.* = Cabbage-tree 1 e; **cabbage-palmetto,** the West Indian cabbage-tree.

1816 Jane Austen *Emma* II. vi. 84 Cabbage-beds would have been enough to tempt the lady. 1865 *Nation* (N.Y.) I. 369 We hear persons whose talents are rather of the solid than the brilliant order familiarly spoken of as 'cabbage heads'. 1880 J. R. Lowell *Biglow Papers* ii. 164 When all's come to past, The kebbage-heads'll cure the thing at last. 1899 Lytton *Caxtons* III. xvii. 199 A cabbage-leaf hat shading a face rarely seen in the Bush. 1847 Leichhardt *Jrnl.* iii. 72 My companions suffered by eating too much of the cabbage palm. 1853 Mundy *Antipodes* III. ii. 46 The cabbage palm-tree...[is] becoming scarce. 1802 J. Drayton *S. Carolina* 66 Cabbage palmetto. 1890 *Rep. Comm. Patents* 1849 (U.S.) 250 Pulp made from the cabbage palmetto tree to the manufacture of brooms. 1862 R. H. Newell *Orpheus C. Kerr Papers* I. 217 How often does man, after making something his particular forte, discover at last that it is only a cabbage-patch. 1898 Daily Chron....

[illegible]

cabbage-tree. 2. (N.Z. examples.)

1848 T. Arnold *Let.* 26 Apr. (1966) 41 There is a certain stiffness in the appearance of a New Zealand forest..but some of the trees and shrubs are very beautiful...There is the cabbage-tree [etc.]. 1905 W. B. *Where White Man Treads* 145 Then the engine would..show off its paces to an admiring cluster of ancient cabbage trees. 1930 L. G. D. Acland *Early Canterbury Runs* (1951) 65 Cabbage trees only grow..in swampy land. 1935 'J. Guthrie' *Little Country* xxviii. 390 In the clump of cabbage trees..one palm, alas, had been blown down by the fierce southerly gale. 1967 *N.Z. Listener* 29 Sept. 10/1 A New Zealand play without sentimentality, bracken or cabbage trees.

3. *cabbage-tree hat* (further examples).

1905 W. B. *Where White Man Treads* 197 A hurrying woman in a man's soiled cabbage-tree hat. 1908 E. J. Banfield *Confessions of Beachcomber* ii. iii. 314 A stock-rider..in a..cabbage-tree hat. 1928 'Brent of Bin Bin' *Up Country* i. 2 He flung off his streaming oilskin and shook the water from his wide-brimmed cabbage-tree hat, with its bobbing corklets attached as fly chasers.

cabbagy, *a.* Also cabbagey. Delete *rare* and add earlier and later examples.

1855 C. Kingsley *Lett.* (1877) I. 432 ...quite a delicious flavour to a very cabbagey German citizen. 1957 C. MacInnes *City of Spades* ii. iii. 123 Muriel was cooking something cabbagey. 1966 A. Burgess *Right to Answer* xi. 141 The cabbagy, beery breathing smells of winter England.

cabbing, *vbl. sb.* (Further examples.)

1939 H. Hodge *Cab, Sir?* 13 During my cabbing life I have worked at all hours. *Ibid.* 219 'Cabology', of course, is the art or science of cabbing. 1939 *Listener* 16 Aug. 299/1 Let no one laugh at cabmen until he's tried cabbing.

‖**cabildo** (kăbi·ldo). [Sp., f. late L. *capitulum* chapter-house: see Chapter *sb.*] **1.** A town hall or town council.

1824 J. R. Poinsett *Notes on Mexico* (1825) iv. 53 Opposite to it stands the cabildo, or town house. *Ibid.* x. 156 The *ayuntamientos*, or *cabildos*, were, and still are, an important branch of the government of Spanish America. 1934 A. Huxley *Beyond Mexique Bay* 160 In the little plaza between the church and the cabildo, or municipality. 1960 R. M. Fyson *Britain in Argentina* 104 Cabildo...All purchases..would be at moderate prices fixed by the cabildo or municipal government.

2. (Usu. with capital initial.) The chapter-house of a cathedral or collegiate church, or the chapter itself.

1880 G. W. Cable *Grandissimes* xii. 320 He saw his clerks..standing idle and shaky in the arcade of the Cabildo. 1924 L. Hearn *Amer. Miscellany* xviii, I wish to thank those who did this service for me..at the Cabildo, in New Orleans.

cabin, *sb.* Add: **5. a.** Also in an aircraft or spacecraft.

1908 H. G. Wells *War in Air* vii. 214 There followed upon these things a long, deep swaying of the airship, and then Bert began a struggle to get back to his cabin. 1913 C. Grahame-White *Aviation* 274 The pilot and his passenger are provided with a completely covered body, which they enter through a small door. This cabin, which has a roof, walls, and floor, is equipped with celluloid windows. 1931 M. Corelli *Secret Power* viii. 87 The steering cabin and accommodation for the pilot and observer. 1962 J. Glenn in *Into Orbit* 41 You have one large handle for repressurizing the cabin with oxygen in case of a bad leak.

8. Cabin class, the designation of a type of accommodation in a passenger ship (cf. *first, tourist class*); also *attrib.*; **cabin cruiser,** a cruiser with a cabin for living in (in quot. 1921, a flying-boat); **cabin ship,** a vessel carrying only one class of cabin passengers.

1929 *Evening News* 18 Nov. 5/5 The plans for the cabin-class vessels. 1936 *Times* 3 Feb. 17/5 There are four agreements in the North Atlantic passenger route—i.e., first-class, cabin, tourist, and third-class. 1921 *Aircraft Year Book* 187 Practical experimental work..was also carried on with the Dayton Wright R.T. 'Cabin Cruiser'. 1928 *Daily Express* 6 Aug. 73/1 Among the standard craft..is the 38-foot single cabin cruiser. 1959 *Manchester Guardian* 25 Aug. 5/2 Everywhere in the United States the outboard motor, the cabin cruiser..are to be seen. 1926 *Daily Colonist* (Victoria, B.C.) 3 July 7/6 (Advt.), Fine Scotch cabin ships to the old country, and real Scotch hospitality. 1958 B. Hamilton *Too much of Water* i. 9 What used to be called a cabin ship, with accommodation in a single class.

cabinet, *sb.* Add: **5. a.** Also, one containing a radio or television receiver or the like.

1934 Webster, *Console*..a cabinet, often decorated, for a radio receiving set, and designed to be placed against a wall. 1962 A. Nisbett *Technique Sound Studio* 245 Many people..'like a loudspeaker to sound like a loudspeaker'.. to exhibit cone and cabinet resonances. *Ibid.* 251 *Enclosure*, a loudspeaker cabinet.

13. (Further examples.)

1932 etc. [see sense 8 in Dict.] 1806 *Deb. Congress* 5 Mar. (1852) 562 My answer was (and from a Cabinet Minister too) 'There is no longer any cabinet.' *Ibid.* 13 Mar. 765 The gentleman's [sr. Mirabeau's] fondness for Cabinet talk and Utopian glory. 1825 *Ann. Reg.* 1812 (Chron.) 133 List of His Majesty's Ministers, January 1812. Cabinet Ministers. Earl Camden..Lord President of the Council. 1867 W. Bagehot *Eng. Constitution* i. 12 This critical operation is the very source..of cabinet government. 1938 *Ann. Reg.* 1937 31 The second reading of the Ministers of the Crown Bill..was moved on April 12 by Sir John J. Simon, who pointed out that it was the first time placing on the Statute-book the terms 'Cabinet' and 'Cabinet Minister'.

14. cabinet pudding, a pudding made of bread or cake, dried fruit, eggs and milk, usually served hot with a sauce.

1821 Kitchiner *Cook's Oracle* (ed. 3) 430 Newcastle or Cabinet Pudding. Butter a half melon mould, or quart basin, and stick all round with dried cherries, or fine raisins, and fill up with bread and butter. 1834 *Ure French Cook* (ed. 7) 348 Cabinet Pudding or Chancellor's Pudding. 1958 B. Hamilton *Too much of Water* i. 18 Picking with extreme caution at the last vestiges of a sort of cabinet pudding.

cabinetable (kæ·binĕtăb'l), *a. colloq.* [f. Cabinet *sb.* + -able.] That is fit to be a member of a political cabinet.

1896 *Daily News* 18 Nov. 4/7 The Prime Minister is chosen..practically by public opinion, and a small list of men whom they may call 'Cabinetable' men. 1905 *Westm. Gaz.* 5 Dec. 4/1 There is rather an unusually strong reserve of 'Cabinet-able' men.

ca·binless, *a.* [f. Cabin *sb.* 5 + -less.] Not furnished with a cabin.

1890 S. Allen in D. J. Browne *Amer. Poultry Yd.* 320 Small dirty, cabinless shallops, of a few tons burthen. 1890 *Harper's Mag.* Mar. 558/1 The tops..are deckless and cabinless.

cable, *sb.* Add: **1. d.** Short for *cable-stitch* (see *7).

1943 *Mary Thomas's Bk. Knitting Patt.* 74 The effect is a vertical Waved Rib, not a Cable.

2. b. *to cut (the) cable(s)*: to depart; to make a break.

1870 Geo. Eliot *Let.* 8 June (1954) III. 79, I want to get rid of this house—cut cables and drift about. 1895 G. B. Shaw *Let.* 29 Jan. (1965) 480 By cutting the cable before the supplies are exhausted you will prove that you are not merely making the best of a bad job. 1917 — *Matter with Ireland* (1962) 156 Any nation less sheepish than the English would have cut the cable long ago and insisted on having a Parliament of its own for its own affairs.

6. (senses 1 and 2) *cable-traction*; (sense 3) *cable-layer, -laying.*

1908 *Daily Chron.* 25 Aug. 3/3 There is only room in the hearts of men for one cable-layer. 1901 *Westm. Gaz.* 12 Sept. 3/2 The primary object of cable-laying is to facilitate communications. 1887 Cable traction (see *cable-grip* below).

7. cable-car, a carriage moved by a chain or cable, e.g. on an overhead cable-way; **cable-carrier,** a system of tubs or buckets slung from an overhead cable for the purpose of carrying heavy materials across a space; **cable-grip** (see Grip *sb.²* 5 b); so *cable-gripper, -gripping*; **cable pattern** (see quot.); **cable-railroad, -railway, -road,** one along which carriages are drawn by an endless cable; **cable-ship,** a ship used to lay a submarine cable; **cable-station,** a station from which a cable message may be sent; **cable-stitch,** any of various twisted rope-like stitches in knitting and embroidery; **cable system,** a system of traction by cable, or of telegraphy by submarine cables; **cable-tier** (earlier U.S. example); **cable tramway** (see quot. 1887); **cable-way,** (*a*) = *cable-railroad*; (*b*) an overhead cable and apparatus for the transport of materials or passengers.

1908 J. B. Smith *Cable Traction* 42 The excellent control of the cable cars is..admirably demonstrated upon this line. 1888 S. Hale *Let.* (1919) 200 Two rival processions which encumbered the streets..almost prevented our getting there in the cable-car. 1867 'Mark Twain' *Following Equator* xvi. 161 It has an elaborate system of cable car service. 1902 E. Banks *Newspaper Girl* 254, I..took a cable car for Capitol Hill. a1884 Knight *Dict. Mech. Suppl.*, *Cable Carrier*, a means of transporting rough materials; stone, sand, lime, coal, earth, by a suspended bucket traveling on a wire cable. 1904 *Westm. Gaz.* 21 Nov. 8/2 Mr. Wall has discontinued the use of the overhead cable-carrier. 1887 J. B. Smith *Cable Traction* 17 Elevated cable traction systems, for which he devised an ingenious cable grip or catch. 1947 *Life* 24 Feb. 16 Cable grip fits into the slot between the tracks. With it the gripman made the car grasp or release the cable, which moves at a constant 9.5 mph. a1877 Knight *Dict. Mech.* I. 419/1 *Cable-gripper*, a lever compressor over the cable-well, and by which the cable is stopped from running out. 1887 J. B. Smith *Cable Traction* 30 The intermediate slot through which the cable gripper passed into the interior of the tube. *Ibid.* 16 A cable-gripping apparatus fitted with vertically acting gripping jaws. 1882 Caulfeild & Saward *Dict. Needlework* 387/1 Cable Pattern.—This is also known as Chain Stitch. 1887 J. B. Smith *Cable Traction* 16 The Claystreet cable railroad was opened for public traffic in August, 1877. *Ibid.* 15 The success that attended the working performances of early cable railways. 1882 *Chambers's Jrnl.* ...Francisco) 25 July 2/3 The cable-or 'grip' road, as they term it in Cincinnati. 1882 *Electrician* XIV. 184/1 An additional cable..has been laid to Jersey by the Post Office's new telegraph cable ship 'Monarch'. 1907 *Daily Chron.* 18 Jan. 5/1 The cable-ship Henry Holmes left here for Jamaica yesterday...She goes to repair the Cuban cable. 1901 *Daily Colonist* (Victoria, B.C.) 8 Sept. 21/1 A staff of forty men is maintained at the cable station, mostly operators. 1920 Blackw. Mag. Aug. 181/2 Chatham Bay was ...had a cable-stitch. 1959 T. de Dillmont's *Encycl. Needlework* 180 Cable or chain stitch used for strengthening the edge. 1899 W. G. P. Townsend *Embroidery* 95 Cable-stitch... The first stitch of all is to make a small link. 1950 J. Cannan *Murder Included* ii. 19 His cable-stitch stockings. 1887 J. B. Smith *Cable Traction* Introd. 3 From the capabilities of the 'cable system', it is now quite feasible to effect tramway communication in the most hilly districts. 1902 *Encycl. Brit.* XXXIII. 223/2 This spread of the cable system has naturally followed trade routes. 1778 in *Essex Inst. Hist. Coll.* XXVI. 195 They drove us..towards the frigate, and down in the cable tier we must go. 1887 J. B. Smith *Cable Traction* 2 A 'cable tramway', or in other words a tramway on which the cars are drawn or hauled by means of a cable or rope receiving its motion from a stationary and distant source of power. 1899 *Daily News* 11 Mar. 3/2 A cable-way from London to Brighton. 1904 *Alpine Jrnl.* Nov. 356 The road which connects the Wetterhorn Hotel with the starting-point of the cable-way is finished. 1904 Goodchild & Tweney *Technol. & Sci. Dict.* 78/2 Cable-way...Overhead cables, suitable for haulage of materials. 1962 *Times* 3 Jan. 18/5 Erecting a pylon..for the 45 mile cableway.

cabler (kēⁱ·b'lə). [f. Cable *v.* + -er¹.] One who sends a cable message.

1890 *Daily News* 12 Dec. 6/3 The next I heard was that one of the cablers was among those who forced the offices. 1920 *Discovery* June 173/2 Those who held the view that the majority of cablers require speed without regard to cost were entirely mistaken.

cablese (kēⁱb'lī·z). [Contracted form of *cable-ese* (also used), f. Cable *sb.* 3 c + -ese.] The contracted or cryptic jargon used in cablegrams, esp. by journalists.

[1895 A. Roberts *Adventures* viii. 104 The cold language of cablegrams fails to do justice to my prowess.] 1952 Koestler *Arrow in Blue* xxv. 205 And, thus translated into French cablese, pasts them on to telegram forms. 1952 H. Kurnitz *Invasion of Privacy* (1956) ii. 14 Messages were now framed in cablese..combining and contracting words until the output..bordered on gibberish. 1958 E. A. Robertson *Justice of Heart* iv. 45 One of his correspondents..was said to have wired back a long despatch in plain English, and been told to use cable-ese, to save telegraph bills.

cableless, *a.* Add: Also of railway engines.

1887 C. B. George *40 Yrs. on Rail* vii. 129 The cableless engines gave no shelter for engineer or fireman.

cabling, *vbl. sb.* Add: **2.** The transmitting of a message by electric cable.

1901 *Westm. Gaz.* 3 Jan. 2/2 Cable Chess: Great Britain *v.* America...The cabling will be done in the room.

3. *collect.* Lengths of cable.

1957 *Daily Tel.* 1 Nov. 10/7 The mechanical apparatus consists of a generator, an air-screw..., cabling, [etc.].

cabman. (Earlier examples.)

1834 C. Mathews *Let.* 22 May in Mrs. Mathews *Mem. C. M.* (1839) IV. xiii. 164 This city of horrors—this collection of ruffianly Cab-men, and Buss-es. 1837 [see Five *sb.* 3 b].

Caboclo (kabō·klo). [Brazilian Pg., derived by some from Tupi *caboculo* dipilated.] A civilized Amerindian descended from aboriginals of Brazil. Also applied to others of mixed Indian and Negro or Indian and white parentage.

1816 Koster *Trav. Brazil* 367 Indians in a domesticated state, who are called generally Caboclos. 1825 Caldcleugh *Trav. S. Amer.* I. 80 Caboclos. Note, Mixture of the Indian races. 1874 R. F. Burton in *Captivity of H. Stade* 45 Mamalucos, meaning the offspring of a white man by an 'Indian' woman, is obsolete in S. Paulo, where Caboclo... has taken its place. 1958 *Times Lit. Suppl.* 28 Mar. 168/4 The Portuguese-speaking natives [in Brazil]...Caboclos in the south or Mulattos further north. 1956 R. Redfield *Peasant Soc.* 32 The more remote frontiersmen call for attention. Such are the *caboclo* of the Brazilian Amazon and coastal selva.

caboodle (kăbū·d'l). *slang* (orig. *U.S.*). [Supposed to be a corruption of the phrase *kit and boodle* (see *kit *sb.*¹ 1).] *the whole caboodle*: the whole lot (of persons or things).

a1848 *Ohio State Jrnl.* (Bartlett, Add.), The whole caboodle will act upon the recommendation of the Ohio State Journal. 1888 *Lowell Times*...1873 B. Harte *Fiddletown* 5 She had more sense than the whole caboodle of them put together. 1901 H. Pain *De Omnibus* 79, I was forced for give 'im eleven coppers, which 'e took and then dropped the 'ole caboodle. 1904 *Daily Chron.* 6 May 8/4 'Give them a week's notice..the whole caboodle..' 1919 J. B. Morton *Barber of Putney* xvi. 236 If we knew they're comin', you can bet your sweet life there was the 'ole caboodle fr. war] knows it too. 1923 *Strand Mag.* Oct. 351 Actually, the whole caboodle, sold, not pawned, produced seventy, not fifty—hundred and twenty in all.

cabook (kăbū·k). Also **kabook.** ['Perhaps the Port. *cabouco* or *cavouco* a quarry. It is not in Singh. Dictionaries' (Yule).] The name given in Ceylon to a reddish gneissoid building-stone, soft when quarried but hardening by exposure to the air; laterite.

1834 S. C. Chitty *Ceylon Gazetteer* 75 The houses are built of cabook, and neatly whitewashed with chunam. 1859 Tennent *Ceylon* I. 17 The most characteristic formation..the decomposition of clay iron-stone called *cabook*. 1858 Simmonds *Dict. Trade*, Kabook. 1941 *Archit. Rev.* XC. 75/1 In more expensive types of building solid walls—of cabook, stone or mud, or occasionally brick—replace the bamboo framework.

caboose. Add: **2.** (Earlier example.) Also *attrib.*

1861 H. Dawson *Remin. Life Locomotive Engineer* 90 Another midnight ride in the 'Caboose' of a freight train. 1880 Ashcroft's *Railway Directory* 1862 *Index of Cabooses*...Cars [on the Central Railroad of New Jersey]. 6. 1903 *New York Even. Post* 25 Aug., The rest of the crew..saw from the caboose windows the bodies..lying along the tracks.

3. A hut or poor dwelling. Chiefly *N. Amer.*

1859 *Congress. Globe* 15 Feb. 1035/1 She was seated in her caboose in our village...and in his little caboose of a post office I found electioneering interferences. 1874 V. Pyke *Life* ... W. Pratt (1940) iii. 31 It was the most wrong thing to allow in your caboose, mister. 1926 G. Parker *World for Sale* xxiv. 301 He's been lying drunk at Gautry's caboose ever since yesterday morning'...'Gautry's tavern—that isn't a caboose.'

b. A mobile hut or bunk-house, moved on wheels or runners. *Canad.*

1908 E. Ferguson *Open Trails* 106 In the later winter one sees..a caboose on a sled, heading for the Peace River district or for some point up north. This is a comfortable way of travelling, and on arriving at his claim the home-steader shifts the caboose to the ground and uses it for a house. 1954 A. M. Bezanson *Sodbusters invade Peace* xiv. 90 Our honeymoon trip was to be made in a caboose—a part-lumber part-canvas house built on a farm sleigh.

c. = Calaboose. *slang* (orig. *U.S.*).

1865 *Republican Banner* (Nashville, Tenn.) 12 Oct. 3/2 The 'caboose' is neatly packed with 'picked' offenders of municipal law. 1929 F. Bowen *Sea Slang* 21 *Caboose*...On shore it meant prison. 1939 *They was in Love* (U.S.) 146 If they put me in the caboose for cruelty to roaches and water bugs.

cabotage. Add: **2.** *Aeronaut.* (The reservation to a country of) the air-traffic within its territory. Also *attrib.*

1933 *Aeroplane* 24 May 962/1 As far as the progress of international air transport is concerned cabotage is about the finest form of 'sabotage'. 1958 *Sunday Times* 15 June 3/5 India..though possessed of an enormous area protected by cabotage, has neither the resources to exploit this potential nor a population rich enough to indulge in large scale air transport. *Ibid.* Members of the Commonwealth do not form a cabotage unit. 1960 *Times* 11 Feb. 9/4 B.O.A.C.'s possession of end-to-end cabotage rights on these routes.

‖**cabotin** (kabotæ̃). Fem. **cabotine** (-tīn). [Fr., see next.] A low-class actor. Also *attrib.* Hence **cabotinism,** the state or quality of being a cabotin; = next.

1903 A. Bennett *Truth about Author* xv. 196 He is terrible against cabotins, not the slightest where he finds them. 1909 H. James *Tragic Muse* I. p. xvii, The mountebank, the mummer and the cabotin. 1922 R. Fry *Transformations* 120 The incalculable difference between Velasquez's distinction, detachment and scrupulous reserve as compared with Caravaggio's blustering cabotinism. 1930 J. Agate *Red Letter Nights* (1944) 129 There remains those impudences ...which fell from Diaz like sour benedictions, from Sarah with the cabotine's natural, slightly vulgar good nature. 1951 L. P. Hartley *My Fellow Devil* i. 8 He was a cabotin sort of character, always playing a lonely part. *Ibid.* xxi. 218 'Cabotin' character...seems to mean someone who enjoys playing..a rather second-rate part, who has no core of personality. 1966 *Punch* 30 Nov. 826/1 Sarah Bernhardt was a *cabotine*; she excelled in the 'sort of tasteless.

‖**cabotinage** (kabotinaʒ). [Fr., f. *cabotin* strolling player, perh. ultimately f. *caboter* to coast, because of the resemblance between players who travel from town to town and coasting vessels.] The life or behaviour characteristic of low-class actors, with implication of 'playing to the gallery'.

1894 *Nation* (N.Y.) 22 Mar. 211 We can find the spirit of 'cabotinage'..even in the Church. 1895 *Daily News* 8 May 8/2 Dickens, yielding..to his native cabotinage, descended so low as to give readings from his own books! *Ibid.*, We cannot..blame the undesirable cabotinage of the great Napoleon. 1921 *Times Lit. Suppl.* 29 Sept. 626/2 In the narrow field of *cabotinage* there is little scope for originality.

cabré, *a.* Add: †**2.** *Aeronaut.* = *tail-down* adj. (see *tail *sb.*¹ 14). Also as *sb. Obs.*

1910 R. Loraine in *W. Loraine Far Lorraine* (1938) vi. 118, I was alarmed lest my biplane should suddenly do a cabré (fall backwards), if I kept her climbing. 1913 A. E. Berriman *Aviation* x. 225 So to tilt the machine nose upwards into a position that is called *cabré*. 1914 H. M. Buist *Aircraft in German War* i. 35 (caption) The craft flying cabré, or tail down, in a steep climb.

cabriole (kæ·briōʊl). Restrict †*Obs.* to senses in Dict. and add: **1. b.** = Capriole *sb.* 1; spec. in *Ballet*, a springing step in which one leg is extended and the second leg is brought up to the first.

1805 F. Peacock *Sk. Dancing* 110 The steps may be accompanied with bendings, risings, leaps, cabrioles. 1830 R. Barton tr. *Blasis's Code of Terpsichore* ii. vi. 78 An attitude upon one leg, as in, *cabriole*, *brisés* and the *rond-de-jambe en l'air*. 1952 *Ballet Ann.* VI. 126/2 Eglevsky's special leaps for which he is famed; the high kicks and the cabrioles, the long suspension of the body in the air for an unbelievable moment during his *cabrioles*.

2. In full *cabriole chair*.

1781 *Eng. Chron.* 27–30 Jan. 1/2 The..original Manufactory for all sorts of Cabriole Chairs. 1848 H. Forster *Stowe Catal.* [Terminal Advt.] 13 Cabriole Drawing Chairs.

4. A form of curved leg, frequent in Queen Anne and Chippendale furniture, so called from its resemblance to a quadruped's fore-leg making a leap or caper.

1883 J. Marshall in *Catal. Exhib. Decor. Handiwork, Edin.* 59 Settees and chairs with their cabriole legs and claw-feet. 1904 I. W. Lockwood *Col. Furniture Amer.* 56 Walnut and Inlay Cabriole-legged Dressing-table. 1907 H. C. Candee *Decor. Styles* 201 The cabriole leg is the one great point of this decorative period with which collectors..must arm themselves. 1940 A. W. Lewis *Gloss. Woodworking Terms* 11 *Cabriole leg*, furniture leg which curves outwards at the top and inwards at the bottom.

cabriolet. Add: (Earlier examples.) Also, the top or open section of a carriage (see quot. 1815). Also *attrib.*

1766 Smollett *Trav.* i. v. 73 He goes in a one-horse chaise, which is here called a *cabriolet*. 1770 H. Walpole *Let.* 7 July (1941) X. 323 We walked in the garden or drove about it in cabriolets, till it was time to dress. 1815 J. Scott *Visit to Paris* iii. 39 A conductor is attached to each Diligence, whose..chief business, according to his practice, is to sleep, closely shut up in the Cabriolet (which is a covered seat in front). 1820 J. S. Mill *Jrnl.* 27 May in *Boyhood Visit to France* (1960) 9, I..set off at 2 P.M. in the cabriolet of the diligence. We had chosen the cabriolet, thinking the interior would be too hot. 1846 Dickens *Pictures from Italy* ii. 9 The Diligence..its cabriolet head on the roof, nodding and shaking, like an idiot's head. 1907 *Westm. Gaz.* 18 Jan. 3/1 Cabriolet-fares were one-third less than those for hackney coaches.

b. A motor car with fixed sides and a folding top.

1930 *Westm. Gaz.* 4 Feb. 4/3 The illustration given is that of a special type of cabriolet. *Ibid.* 17 Nov. 5/ The exhibit [sc. a Lancia] is comprised of a 20-h.p. chassis fitted with a cabriolet body of admirable proportions.

2. A bonnet or hat shaped like a cabriolet.

1771 H. Walpole *Let.* 31 July (1904) 65, I have bespoken two cabriolets for her, instead of one, twelve in it made them very dear. 1923 *Daily Mail* 22 June 11 Cabriolet hats are in fashion again...With a cabriolet you must have ribbon streamers falling over one shoulder.

ca'canny (ka,kæ·ni). Also (after northern dialects) -conny. [See Call *v.* 15, Canny *a.* 10.] Moderation, caution; *spec.* the practice of 'going slow' at work; a deliberate policy of limiting output of work. Freq. *attrib.*

1896 *Westm. Gaz.* 25 Mar. 1/3 He was the first English Labour leader to refuse to have any part in the 'ca' canny' policy. 1896 *Seamen's Chron.* 24 Oct. In *Times* 18 Nov. (1901) 10/1 What is 'ca' canny'? It is a simple and handy phrase which is used to describe a new instrument or policy which may be used by the workers in place of a strike. 1902 *Westm. Gaz.* 25 Apr. 8/1 A 'ca'canny' policy said to have been deliberately adopted by certain trade union officials. 1958 *Economist* 29 Nov. 784/1 The teaching staff defends itself skilfully from battle to battle with cynicism and ca'canny.

Hence **ca'ca·nnyism, ca'ca·nnyness** (or **-i-ness**), ca'canny policy or behaviour; caution.

1917 *Glasgow Herald* 9 Aug. 3/7 The letter..unintentionally detracts, by its extreme 'ca'cannyness', from the merits of the proposed scheme. 1921 *Glasgow Herald* 17 June 6 Mr. Ramsay Macdonald recommends 'a ingeniously organised system of passive resistance', which, plainly interpreted by his followers, means Ca'cannyism. 1926 W. S. Bruce *Salt & Sense* 186 To-day we have a new word for a novel vice, which began first on the Clyde. It is called 'Ca'cannyism'. 1963 *Economist* 16 Mar. 976/2 The entrenched forces of ca'canniness.

cacao. Add: **4.** *cacao-bush, -farm, -planter, -powder*; **cacao-bean,** the seed of the cacao-tree; **cacao-mother,** a tree used to protect the delicate cacao-tree.

1839 Ure *Dict. Arts* 292 The cacao-beans lie in a fruit somewhat like a cucumber. 1871 Kingsley *At Last* vii. Lombardy poplar[s]..the beauty of these 'Madres de Cacao', Cacao-mothers as they call them here because their shade is supposed to shelter the Cacao-trees. *Ibid.* xiii. The cacao-bush which produces chocolate. *Ibid.* xvi. Can nothing be done to increase the yield of the cacao-farms? 1903 W. H. Smith *Cacao Planting* 18 As long as cacao powder is so much in vogue. *Ibid.* 21 The Jamaica cacao planters. 1936 *Discovery* Feb. 41/2 In 1939 *Ephestia elutella*, formerly destructive to cacao beans and then to cacao-beans storage, was found in London tobacco warehouses.

cache, *sb.* (Now usu. pronounced kæʃ.) **1.** (Delete quot. 1595, which belongs to Cash *sb.¹* 1.)

b. (Earlier N. Amer. examples.)

1797 C. Chaboillez *Jrnl.* in B. C. Payette *Northwest* (1964) 152 [He] had a large Cash of Provisions near that river. 1805 Lewis & Clark *Jrnls.* (1904) II. 134 These holes in the ground or deposits are called by the engages caches. 1806 *Ibid.* VII. 59 We put in the cash or hole..the bellows. 1817 J. Bradbury *Trav.* 118 The Arikaras could not spare any provisions, as the excessive rains had penetrated into their caches, and spoiled the whole of their reserved stock.

cache, *v.* (Now usu. pronounced kæʃ.) Add to def.: orig. *U.S.* (Earlier examples.)

1805 Lewis & Clark *Jrnls.* (1904) VII. 139 We carried our baggage we concluded to cash in the place of caching. 1823 E. James *Exped. Rocky Mts.* I. x. 193 They then proceeded to cache, or conceal in the earth these acquisitions.

Hence **cached** (kæʃt), *ppl. a.*

1901 S. E. White *Westerners* vii. 47 Lone Wolf's band took up quarters within striking distance of the cached schooners. a1910 'O. Henry' *Trimmed Lamp* (1916) 230 The man from Nome, loyal to her who had resurrected his long cached heart.

‖**cache-peigne** (kaʃpeɲ). [Fr., f. *cacher* to hide + *peigne* comb.] A bow or hat ornament, usually worn at the back of a woman's hat. Also *attrib.*

1873 *Young Englishwoman* Feb. 79/2 The cache peigne is quite gone out of fashion. 1896 *Westm. Gaz.* 7 May 3/1 pretty Parisian toque..trimmed..with a cache-peigne of peach-coloured roses. 1905 *Daily Chron.* 8 May 8/4 In very many instances..that which has a wreath of roses..finishing in the form of a cache-peigne at the back. 1936 *Burlington Mag.* June 261/1 The heavy cache-peigne cap helps to pull the reflection downwards.

‖**cache-pot** (kaʃpo, kæ·ʃpɒt). [Fr., f. *cacher* to hide + *pot* Pot *sb.*¹] An ornamental holder for a flower-pot.

1872 Lady C. Schreiber *Jrnl.* (1911) I. 162 Two fine Chantilly cachepots. 1898 *Daily News* 12 Mar. 6/6 The multi-coloured vases, jars, plaques, cache-pots, &c. 1968 *Times* 27 July 21/2 Staffordshire cache pot, decorated with pictures of HRH the Duke and a Blériot monoplane about 1914.

‖**cache-sexe** (kaʃ-, kæʃ,seks). [Fr., f. *cacher* to hide + *sexe* sex, genitals.] A covering for the genitals.

1928 *Spectator* 13 Feb. 270 Revues the feature of which was two *grandes ensembles* of women naked except for the minute *cache-sexe*. 1930 H. Lawrence *Nettles* 15 English artists might finally learn When they painted a nude, to put a *cache sexe* on. 1967 P. Ustinov *Loser* xi. 246 She oscillated listlessly in nothing but a spangled brassière and a sequin-covered *cache sexe* the shape of a heart.

cachet. Add: **4.** A covering of paste, gelatine, or other digestible material, enclosing (nauseous) medicine; = Capsule 5.

1884 *Pharmac. Jrnl.* XV. 42/2 Cachets are..sheets of unleavened bread cut to a round or oval shape with a..concave centre to receive the drug. A powder or medicine to be taken. 1898 Q. Hogg in Ethel M. Hogg *Biography* (1904) 349 My experience and quick cachets not to use to him. 1901 *Contemp. Rev.* Mar. 405 One cachet..to be taken with the midday meal and one in the evening.

cachou. 2. (Examples.)

1880 in Webster *Suppl.* 1895 *Montgomery Ward Catal.* 261/2 Aromatic cachou lozenges for perfuming the breath. 1898 *Army & Navy Co-op. Soc. Price List*, Cachous, Atkinson's White Rose, Violet, Heliotrope, Citron..per box o/8. 1909 *Daily Chron.* 13 Feb. 5/6 The tabloids might easily be taken for 'cachous'. 1922 *Glasgow Herald* 3 Apr. 4 He carried a little round, flat gilt box of Prince Albert cachous.

cacique. Add: **2.** In Spain or Latin America, a man who owes his ascendancy to his power or influence; a political 'boss'; also used *attrib.* of a political system in which the power is in the hands of such a man or men. Also *attrib.* Hence **caci·qu(e)ism,** the cacique system.

1872 Schele De Vere *Americanisms* 71 The West India term Cacique..is the more usually applied..to mayors of New Mexican towns, and any somewhat pompous and self-sufficient man is apt to be nicknamed the Cacique of his town. 1903 *Times* 3 June 5/3 Caciquism, the Spanish equivalent for the methods of Tammany. 1923 *Glasgow Herald* 23 Oct. 7 This decree, of great importance for the destruction of 'caciquism', will be applied to about 100 towns. 1929 *Contemp. Rev.* Nov. 615 The cacique whether an employer of labour or a moneylender has the majority of the constituency under his thumb. 1933 *Glasgow Herald* 13 Dec. 12 The whole cacique system is..an immense satire on local authority in Spain. 1964 *Economist* 4 July 37/2 The provincial *caciques* and city bosses.

cack-handed (kæ·k,hæ·ndĕd), *a. dial.* or *colloq.* Also **cag-, keck-handed.** Left-handed; ham-handed, clumsy, awkward. Hence **ca·ck-ha·ndedness.**

1854 A. E. Baker *Gloss. Northants.* I. 365 *Keck-handed*, awkward, left-handed. 1859 N. & Q. VIII. 483 If a man, at hay time or harvest, holds his fork with his left hand lowest, they say, 'Ah! he's no good: he's keck-handed'. 1893 Darlington *S. Cheshire Gloss.*, *Kei-handed*...ham-handed. 1938 E. Waugh *Scoop* 19 Cack-handed, cag-handed. 1952 Dylan ... Sprat' would you ever? How is..kack-handed; 1955 *Speaker* 21 Sept. 1927 Doesn't seem cack-handed; he feels that as a very cack-handed...1962 C. Hanley ... 'Back to Methuselah'. 1967 J. Potter *Foul Play* x. 125 When he saw the ball going the wrong side of him, he lunged his [hockey] stick out behind him...'Just thought I'd try a bit of your cack-handed stuff.'

cackle, *sb.* Add: **3. a.** Colloq. phr. *cut the cackle (and come to the horses)*: stop talking (and get to the heart of the matter, the real business). Hence *cackle-cutting* vbl. sb. and ppl. a.

1889 Barrère & Leland *Dict. Slang* I. 216/1 The great Ducrow..was wont to apostrophise the performers in his equestrian dramas after this fashion: 'Come, I say, you mummers, cut your cackle, and come to the 'osses'. 1899 *Westm. Gaz.* 27 June 2/1 Nine-tenths would be delighted if the business place were confined to the 'osses', and so on upon. 1919 G. B. Shaw in *Shaw on Theatre* (1958) 124 Out with the lot of them, then: let us cut the cackle and come to the 'osses. 1921 — *Back to Methus.* v. 239 Cut the cackle; and come to the synthetic couple. 1930 Wyndham Lewis *Apes of God* xi. 5 Cut the cackle, Arthur—I'm pressed for time! 1957 *Economist* 28 Dec. 1106/1 Nor ..will delegates reassemble in September after parting only in March—a decidedly cackle-cutting feature of this last session. 1958 J. Wain *Contenders* ix. 193 You must be wondering what all this is about, and I'll cut the cackle.

4. cackle-berry *slang* (orig. *U.S.*), an egg.

1916 *Dial. Notes* IV. 272 Pass the *cackleberries*. 1925 G. Y. Knapp *Eng. Lang. in Amer.* I. 272 The language is complicated in its suggestiveness, like *cackleberry*, meaning egg. 1960 *John o' London's* 14 June 571/1 In cackle-berry is an egg (in hen slang).

cacody·late. A salt of Cacodylic acid.

1908 *Practitioner* Aug. 338 Arseniate of soda is more brought, as it does not, as does cacodylate, the action of which is prolonged and deliberate.

cacogenesis. Add: Also *transf.*

1895 A. H. Keane *Ethnology* ix. 200 From this 'kakogenesis' or spread of races the deeper degeneration of races, there follow some important inferences.

cacogenic (kækədʒe·nik), *a.* [f. Caco- after Eugenic *a.*; cf. Cacogenesis.] The reverse of *eugenic*; = *dysgenic *a.* Hence **cacoge·nics** *sb. pl.*, the breeding of a weak race (opp. *eugenics*).

1917 W. R. Inge in *Edin. Rev.* Jan. 80 To How will it escape the cacogenic effects of family restriction in the better classes combined with reckless multiplication among the refuse? 1920 — in *Outsp. Ess.* II. iii. 265 The whole policy of subsidising the unsuccessful by taxes extorted from the industrious was cacogenics erected into a principle. 1932 R. C. Kettle *Heredity in Man* 180 ...1962 ... cacogenic marriages and (3) a cacogenic strain.

cacomistle (kæ·komis'l). Also **cacomixl,** etc. [Amer. Sp. *cacomixtle* (also used), f. Nahuatl *tlacomitli*.] A raccoon-like animal of the south-western United States and Mexico, *Bassariscus astutus*.

1869 W. H. Flower in *Proc. Zool. Soc.* 31 The interesting little American Carnivore the *Bassaris*...The animal was mentioned by Hernandez under the name of Tepe-Maxtlaton or Cacamitztli, meaning..the 'Rush-Cat'. 1889 *Cent. Dict.*, Cacomixl. 1931 *News Chron.* 12 Feb. 7/2 Cacomistles..With the paws of a cat, the face of a fox and a ringed tail. A cacomistle, the nocturnal prowler of the Mexican desert. 1958 Rydberg & Mannix *Wildest Game* (1959) xvi. 200, I managed to pick up some cacomistles, or ring-tailed cats, in the Southwest.

caconym (kæ·konim). [f. Caco- + Gr. ὄνυμα name.] An example of bad nomenclature or terminology, esp. in biology and botany.

1891 *Science* 20 Feb. 109/2 O. F. Cook in *Amer. Naturalist* XLVIII. 301 A name rejected for linguistic reasons [is] a caconym. 1936 *Nat. Cactus & Succulent Jrnl.* II. 3/1 A name may qualify as a caconym in different ways. First, from sheer length...Second, from the clash of consonants making it difficult (for a European at least) to articulate.

cactus. Add: **3. cactus wren** *U.S.*, a North American wren of the genus *Campylorhynchus*, frequenting cactus plants.

1869 *Amer. Naturalist* III. 183 The Rock Wren..and Cactus Wren (*Campylorhynchus brunneicapillus*) change but little..1962 *Sat. Eve. Post* ... The cactus wren, so called from its habit of nesting in the cactus wherever available.

cacuminal (kækiū·minăl), *a. Phonetics.* [f. L. *cacūmināre* (see Cacuminate *v.*) + -al.] Of sounds: produced with the tip of the tongue 'inverted' or curled upwards towards the hard palate; retroflex. Hence as *sb.*

1862 *Jrnl. Amer. Oriental Soc.* VII. 354 Müller proposes 'cacuminal' as a name for the class [of consonants]. 1902 E. W. Scripture *Elem. Exper. Phonetics* xvii. 397 The term 'cacuminal' (or 'cerebral', or 'inverted') is applied to a frontal articulation, in which the point of the tongue is turned up and back. 1913 G. Noël-Armfield *Gen. Phonetics* xvii. 99 Indian grammarians class these consonants as *cerebrals*. The name for them in the terminology of phonetics is *retroflex* or *cacuminal*. *Ibid.* 100 The cacuminals. 1933 [see *coronal a.* 5].

caddie, *sb.* Add: **2. b.** Also *attrib.*, as *caddie-car, -cart.*

1961 F. C. Avis *Sportsman's Gloss.* 202/2 *Caddie-Car*, a light two-wheeled rack for holding golf-clubs, drawn by the golfer himself. 1962 *Punch* 22 Nov. 747/1 A moment's weakness, and I began to tow a caddy-car.

caddy, *sb.¹* Read 1 ½ for 1 ¼, and add: **1. b.** *attrib.* and *Comb.*, as *caddy-spoon*, a short-handled spoon of a special shape used for measuring tea out of the caddy.

1929 *Daily Express* 31 Aug. 14 Old silver caddy-spoons that have survived from Georgian days.

caddy (kæ·di), *v.* [f. *caddy*, var. Caddie *sb.*] *intr.* To act as caddy for a golfer. Also *transf.*

1908 *Daily Chron.* 26 Aug. 4/7 You've caddied for me before...I'll caddy for you. 1933 Wodehouse *Heavy Weather* xix. 236 I should go off and caddy for Honoria on a shopping tour down Regent Street. 1948 *Daily Mail* 7 Sept. 3/4 The Prince had a local caddie named William Everett..who caddied for him.

-cade, *suffix.* Taken by a false division of Cavalcade *sb.* and used in various *Combs.*, as *aquacade, *motorcade, etc., in the sense 'a procession, a show'. Chiefly *U.S.*

1936 Mencken *Amer. Lang.* (ed. 4) v. ii. 180 In the case of *motorcade, autocade, camelcade* and *aerocade*,..a new suffix, *-cade*, seems to have come in.

cadeau. (Earlier examples.)

1808 Wynne *Diaries* 10 Dec. (1940) III. 321 He has brought..a very handsome *joint Cadeau* to us of a Silver Tea Urn. 1826 M. Kelly *Reminisc.* (ed. 2) II. 320 Sheridan ..selected certainly not the worst for *cadeaux*.

cadelle (kăde·l). [Fr., ad. Pr. *cadello*.] The larva or adult of a beetle (*Trogosita mauritanica*) that is destructive to grain.

1861 *Chambers's Encycl.* II. 464/1 Cadelle (*Trogosita Mauritanica*...), an insect sometimes found in granaries in Britain.

cadence, *v.* Add: **2.** *intr.* To flow in rhythm; to move in a cadence. So **ca·denced** *ppl. a.*, **ca·dencing** *vbl. sb.* and *ppl. a.*

1907 P. L. Falon *Love's Re-Awakening* 96 The cadencing majestic beat. Of poesy's most noble strains. 1928 Quiller-Couch *Studies in Lit.* 10 The verse cadences to the feeling. 1930 N. S. Colby *Remembering* vii. 159 You could see her sight-reading cadences among the tracks. 1959 *Scot. Mag.* ...

cadential, *a.* Add: (Earlier and later examples.) Also, of or pertaining to a cadenza.

1880 [see Thirteenth *sb.* 2]. 1954 L. R. Palmer *Latin Lang.* viii. 214 A return to concord in the cadential part of the line. 1955 E. Dent in H. van Thal *Fanfare for E. Newman* 103 Apart from a few cadential flourishes, the part of Leonore is entirely devoid of *coloratura*. 1958 *Times* 22 Jan. 37 Such a common feature of eighteenth-century recitative as a cadential appoggiatura. 1969 *Language* XLV. 253 Pāda-initial and pada-final (cadential) occurrences account for 43 of 71 instances of *hāra*-.

cadet. Add: **3. b.** A boy in an ordinary school who receives military training with or without a view to entering the army. Also *attrib.*, as *cadet corps*, a company of schoolboys who receive such training.

1873 *Programme of Review at Charterhouse School* 6 Aug., The young gentlemen [of Charterhouse Cadets] will be drawn up in Line at Open Order on the Cricket Ground. 1901 *Public School Mag.* Dec. 432 The Cadet Corps paraded in front of the School and marched to the field. 1942 *Punch* ...played The Land o' the Leal'. 1957 *Times* 3 Dec. 12/3 Cadet corps in Birmingham schools came under fire from many members of the city education committee to-day.

4. *N.Z.* A young man learning sheep-farming on a sheep-station. Hence **cade·t(t)ing** *vbl. sb.*, **cadetship,** the position or status of such a person.

1842 R. G. Jameson *N.Z., S. Aust. & N.S.W.* xxiv. 337 We [are] in want of a new kind of good, steady Cadet. Colonial cadetships. 1853 J. Rochfort *Adv. Surveyor in N.Z.* ii. 11 Ont they had just finished their 'cadetship', that is, they had been learning sheep-farming under a settler. 1862 H. R. Chudleigh *Diary* 21 Mar. (1950). 1.29 There were four Cadets learning sheepfarming. 1868 H. H. Vogel *Maori Maid* xix. 147 A cadet is a young man, generally from England, who is paying a run-holder so much a year for the honour and privilege of working for him. *Ibid.* xii. 148 Otherwise called cadeting..is a swindle. 1930 L. G. D. Acland *Early Canterbury Runs* ii. 23 Reginald Wade managed for the Chamberlains, and at one time had no fewer than ten cadets on the station.

Cadet¹ (kădĕ·t). Also **Kadet.** [Russ. *Kadét*, f. the names (*Ka de*) of the initials of *Konstitutsiónnyĭ demokrat* Constitutional Democrat, with ending assimilated to that of Cadet¹.] In Russian politics, a member of the Constitutional Democratic Party.

This party was formed in 1905 by a group of the group favouring autonomy for Poland and a federal constitution for the Russian empire with the [so-called] Independence Party formed by political exiles at Paris in 1905.

1906 *Daily Chron.* 21 Jan. 5/2 The 'Cadets' (Constitutional Democrats)..have decided..to wait until the agrarian question comes on for discussion. 1908 *Westm. Gaz.* 19 Aug. 4/1 The moderate Liberals, the so-called Cadets, who led the Octobrists..took no hand. 1917 *Daily Mail* 16 Nov. Other Cadets joined. 1938 W. H. Chamberlin *Russ. Revolution* I. 237 The Cadets' advanced their claims. 1965 ...the Kadets must properly be described as the 'Kadets'. 1951 R. H. ... The moderate brilliant Zemstvo Liberals, who did so much to found the Cadet (or Liberal) Party. 1958 [see *Bolshevik sb.*]. 1961 [Guardian 23 Mar. 11/1 The Kadets must properly be described as radicals. *Ibid.* 17 Jan. 31/4 The Cadets advanced the claim..to a sovereign and democratically elected legislature.

cadmium. Add: **b. cadmium (mercury) cell,** a type of voltaic cell used as a standard of electromotive force; **cadmium green, orange, red,** pigments obtained from cadmium compounds.

1893 *Electrician* 13 Oct. 645/2 Notes on the E.M.F. and Temperature Coefficients of the Cadmium Mercury Cell. 1908 *Phil. Trans.* A. CCVII. 393 When the solution is saturated at all temperatures, i.e. when solid cadmium sulphate is always present in the cell, the name 'Cadmium Cell' has been frequently assigned to it in order to distinguish it from the original form. 1940 W. Garner *Ch.* in *Then War* xvii. 215 Jagger switched on the small light..from the light been threading through the dark like a sword blade. 1934 H. Hiler *Technique of Painting* ii. 113 *Cadmium green*...are mixtures of cadmium yellows and ultramarine. 1936 A. Huxley *Eyeless in Gaza* iv. 49 His grey trees have shadows of cadmium green. 1962 *Montgomery Ward Catal.* 253/1 Water colors..Pale cadmium yellow—cadmium orange—French blue. 1934 H. Hiler *Technique of Painting* ii. 112 *Cadmium orange*...of the same chemical nature and composition as the cadmium yellows, and the most permanent of all the cadmium colors. 1866 H. C. Standage *Artists' Man. Pigments* 47 Cadmium red, a rather recently introduced colour. 1962 *Listener* 28 Jan. 126/2, I love cadmium red or orange.

cadre. Add: **2.** Also *attrib.* Also of an R.A.F. squadron.

1908 *Daily Chron.* 26 Nov. 6/4 The German Army..is a 'cadre' army, which can only be set on a war footing by drawing large reserves. 1929 *Air Ann. Brit. Empire* 54 The Service..is made up of three parts—the Regular Air Force, the Cadet Squadrons and the Auxiliary Air Force. 1931 *Flight* 1 May 374/2 The explanation is that the lower part of number indicates a Cadre Squadron, which is mainly composed of Special Reserve personnel.

3. In Communist countries, a group of workers, etc., acting to promote the interests of the Communist Party; also, a member of such a group; = *cell *sb.*¹ 12 b.

1930 *Economist* 1 Nov. (Russian Suppl.) 10/2 The number, quality and devotion of these *cadres*, as the industrial army of the leaders in the Soviet are called. 1932 [see *Rabfak]. 1937 *Times Lit. Suppl.* 9 July 536/4 The six 'cadres'..chosen as typical—Communist, Young Communist, shock worker, cultural worker, collectivized peasant and Red Army man—represent the sum total of the Soviet Government's support. 1957 E. Snow *Red Star over China* ii. 61 Military training was secretly given to 2,000 cadres. 1950 D. Hyde *I Believed* 64 92 In communist jargon to be a *cadre* meant to be someone trained and prepared to take leadership for communism. 1963 *Ann. Reg.* 1962 215 The specialized technical cadres needed for a further economic development [in Poland]. 1967 G. Steiner *Lang. Silence* 341 Assured through the hard core of French Stalinism, a harsh and disciplined *cadre*..then the German Stalinist, Brecht's..fought in the *hauts fascia* and flourished a large and animated world of intellectual Marxism.

cady (kē¹·di). *local.* Also **cadey, -i(e).** [Origin unknown.] A hat or cap.

1846 *Spirit of Times* 6 June 170/2, I may be able to discover my lost 'Cady'. 1869 *N. & Q.* III. 406 In Lancashire a straw hat [is vulgarly called] a cady or cadey. 1887 *Walford's Antiquarian Mag.* XI. 251 Cadey found its way into a music-hall song some years ago. The chorus was something like this: 'He wore a penny cady, a penny cady'. A penny I gave for my stick. 1887 *Jamieson's Scot. Dict. Suppl.*, *Cadie*..a boy's cap; generally applied to a square cap, as a Glengarry cadie. 1908 R. Mavor *Cat.* Doc. Dec. 25/4 In which I was doomed to a cadie, for shoe-cotton has cost a shilling..for which wad hang up the like in the fringes. 1909 *Daily Chron.* 8 Feb. 6/7 The British navy is never so happy as when..the sinister rat..a spray of the fateful [peacocks'] feathers 1959 *Times*

28 Sept. 11/4 Isn't it time we relieved our modern rozzer of his incongruous cady?

cænogenesis (sīnodʒe·nĕsis). More regular form of Kenogenesis (1879). So **cænogene·tic** *a.*

1900 *Cent. Dict. Suppl.*, Cænogenesis. 1930 G. R. de Beer *Embryol. & Evol.* i. 6 Haeckel concluded that a new or cænogenetic stage had been interpolated in the ontogeny. *Ibid.* xv. 109 The appearance of characters in the early stages of development is cænogenesis. 1937 *Jrnl. R. Microsc. Soc.* 122 It would clearly be a cænogenetic character.

Caen-stone. Add: (Examples.) Also *attrib.*

1421 in *Lett. Q. Margaret of Anjou* (1863) 20, I have purveyed xiij. tons tight [weight?] of Caen stone. 1508 *Stow Surv.* 362 Part of the ruines of the old Temple were seene to remaine, being builded of Caen stone. 1848 C. M. Yonge *J. Keble's Parishes* ix. 102 The material chiefly used in the cathedral was Caen stone. 1906 *Westm. Gaz.* 26 Apr. 12/1 The beautiful Caen-stone staircase. 1954 *Architect* VIII. 50 The tomb ..built up of Caen stone. 1961 *Punch* 8 Nov. 686 'Ere, my ...ivory. 99 They've just got..a facing of Caen stone this time.

Caerphilly (kē³ɑ·fi·li). [The name of a town in S. Wales.] *Caerphilly cheese*: a mild cheese (originally) made in Caerphilly; also *ellipt.*

1901 *Jrnl. R. Agric. Soc.* LXII. 151 The local variety of *Caerphilly* cheese was divided into two classes, according to size. 1908 Both the classes of Caerphilly cheese contained various samples of. 1908 *Westm. Gaz.* 24 Dec. 10/1 They both..of Caerphilly cheese. 1960 *Good Housek. Home Encycl.* 376/1 Caerphilly. Originally a Welsh cheese, this is now made in Somerset, Wiltshire, Devon and Dorset. 1958 M. Dickens *Man Overboard* xv. 243 His face remained the colour of Caerphilly cheese.

Cæsar. Add: **1. a.** *Cæsar's wife*: In various uses with allusion to Plutarch *Cæsar* x. 6 ('I thought my wife ought not even to be under suspicion').

1579 Lyly *Euphues* (1868) 329 Al women shal be as Cæsar would have his, that not onelie free from sinne, but from suspition. 1748 Richardson *Clarissa* III. xxxi. 108 The wife of Cæsar must not be suspected. 1866 J. C. & A. W. Hare *Guesses at Truth* 187 Cæsar's wife ought to be above suspicion. 1938 E. Waugh *Handful of Dust* ... Bell 'Death in Retirement' iii. 35 'I had to do a Cæsar,' answered Mr. Clayton. 1963 *Economist* 20 Apr. 239/1 That 'Cæsar's' in the maternity ward. 1964 *Guardian* 8 Feb. 8/5 One Roman Catholic voter..will awaken this convenient custom of Cæsar's wife.

†c. A baby delivered by Cæsarean section. *Obs. rare.*

1940 R. Jonas tr. *Roesslin's Byrth of Mankynde* i. ix. fol. 20, Suche as be borne after this fashion be called cesares, for because they be cut out of theyr mothers belly, wherevpon also the noble Romane cesar the .j. of that name in Rome toke his name.

d. (A case of) Cæsarean section. *Med. slang.*

1958 'K. Gordon' *Doctor in House* ix. 97 'How many Cæsars have you had?'..'Forty-nine. That includes a Cæsar.' 1962 *Observer* ...'Bell' *Death in Retirement* iii. 35 'I had to do a Cæsar,' answered Dr. Clayton. 1963 *Economist* 20 Apr. 239/1 That 'Cæsar' in the maternity ward.

2. *c. to appeal (un)to Cæsar* (with allusion to *Acts* xxv. 11): to appeal to the highest authority.

1855 *Wesleyan Methodist Mag.* Apr. 305 The conduct of our agent..is condemned by the public. There, they appealed to Cæsar. 1864 T. B. Reed *Dog with Bad Name* xv. 156 'Can't she come, Father' said Percy..adroitly appealing to Cæsar. 1929 Kipling *Stalky & Co.* 27 'I appeal to the Head, sir.'..'Thou hast appealed unto Cæsar: unto Cæsar shalt thou go.' 1951 J. Chamberlain in *Westm. Gaz.* 22 Oct. 5/1 If this policy..were not accepted by the heads of the Government. I should have to appeal to Cæsar. 1958 P. Gudalla *Palmerston* iii. 51 Obviously the matter might go to the Emperor. Repent the indignant soldier appealed to Cæsar.

Cæsaro-papism (sī:zăro,pē¹·piz'm). [f. Cæsar + -o- + Papism.] The supremacy of the civil power in the control of ecclesiastical matters. So **Cæsaro-pa·pal** *a.*, **-pa·palism, -pa·pist** *a.*

1890 *Edin. Rev.* Apr. 349 Such a *régime* as the Cæsaro-papism of Justinian. 1903 *Camb. Mod. Hist.* II. xvi. 567 A hill which were the full Henrician length in its Cæsaro-papalism and was made it right, as in St. Paul's case, for him to appeal to Cæsar. 1905 Fowler-Jackson *Studies in Life Early Ch.* 240 Another effect of the triumph of the Church in which Christianity won the respect and subservience of the Roman state was what is called Cæsaro-papalism. 1929 A. J. Toynbee *Study Hist.* IV. 592 The abortive renaissance of the Roman Empire ...in 'orthodox Christendom. *Ibid.*, The 'Cæsaro-papal' authority. 1941 'R. West' *Black Lamb* (1942) II. 249 A Cæsaropapist empire whose emperor was the Vicar of Christ.

cæsium. Add: Also (*U.S.*) **cesium.** (Later examples.) **cæsium cell** (see quot. 1940).

1931 *Engineering* 16 Jan. 87/1 The new Osram monatomic cæsium cell has been specially suitable for white light. 1940 *Chambers's Techn. Dict.* 125/1 Cæsium cell, a photo-electric cell having a cathode consisting of a film of cæsium deposited on minute globules of silver; it is particularly sensitive to red radiation. 1955 *Sci. News Let.* 12 Aug. 123/1 He developed a cesium oxide layer, which is sensitive to red radiation, the luminous alkali metals, sodium and potassium. 1958 C. C. Adams et al. *Space Flight* xix. 348 Both rubidium and cesium have been considered as the propellant for the ion drive. 1958 *Times Rev. Industry* July

25/2 Machines now in service are stated to be equivalent in power to cobalt 60 sources of well over 100,000 curies, and, in the isotope Cæsium 137, of about 300,000 curies. 1960 *Gloss. Atomic Terms* (H.M.S.O.) 5 *Cæsium* (137)..a long-lived (30 yr. half life) gamma-emitting radioactive isotope of cæsium 137 is found in fission products, the 'ash' of nuclear reactors, and is used for cancer therapy in some hospitals..and for industrial radiography.

cafard. Restrict *Obs.* to sense in Dict. and add: **2.** Melancholia; depression; 'the blues'.

1924 P. C. Wren *Beau Geste* I. i. 14, I say my handful of cafard-stricken men in my best mood. 1926 Gould *Med. Dict.* (ed. 8) 963/2 Cafard, a subacute melancholia, characterized by attacks of the 'blues'...[prevalent] in soldiers. 1929 *Times Lit. Suppl.* 28 May 1/2 Miss Jesse was stricken with the superb *cafard* of the Romantics. 1951 W. Sansom *Face of Innoc.* ix. 131 The itch of over-eating and the *cafard* that follows every feast.

café. Substitute for pronunc.: (kæ·fe). Also vulgarly or jocularly pronounced (kæf) or (kef), and written in the form *caf*; cf. *caff. (Earlier and later examples.)

1802 M. Starke *Lett. from Italy* I. 225, I breakfasted at the *Café d'Angleterre*. 1852 C. Wilmot *Let.* 19 June in *Irish Peer* (1920) 73 All the Cafés are out of doors. 1870 D. J. Kirwan *Palace & Hovel* (1963) xv. 151 In the corners of the streets and along the sides, cafés and other refreshment bars. 1906 A. Bennett *Whom God hath Joined* ...5 Workshops, theatres, concerts, cafes, pawnshops. 1929 *S.P.E. Tract* XXXII. 37/4 It popular tea-shops paint their title of café over their doors the word will not be degraded like *chafe* and *calf*. 1938 [see *Friday* sb. c]. 1969 J. Fleming *Man with Golden Gun* v. 71 'There's the café' (she pronounced it caff).

2. *attrib.* and *Comb.*, as *café-bar, -habit, -haunter, -restaurant, -window; café-haunting* adj.

1938 *Times Lit. Suppl.* 8 Oct. 641/3 In the other wing is a café-bar. 1910 *Daily Chron.* 5 Mar. 4/4 Any slight modification in the national temperament which the café-habit might..bring. 1931 *Mind* LX. 133 It is the café-haunters, the preachers, the metaphysicians, and the calendar makers who talk of beauty. 1866 M. Arnold *Friendship's Garland* (1871) 167, I do not wish them to grow into a café-haunting, dominoes-playing Frenchmen. 1926 'C. Barry' *Detective's Holiday* iv. 40 The café-restaurant which the forester had called the canteen. 1907 W. O. Lilliefield *Where Trail Divides* 56 A café-window..spell stretched across a café window.

b. café chantant (ʃantā) [lit. 'singing café'], a café in which the customers are entertained by singers or other music; **café concert,** a musical or variety concert given in a café; also = *café chantant*; **café society** (orig. *U.S.*), a group of people who frequent fashionable restaurants, night-clubs, and resorts; also *attrib.*

1854 Bayle St. John *Purple Tints Paris* II. iii. 67 Go out to the Luxembourg, to a *café chantant*, or to the country. 1859 Sala *Twice round Clock* 164 Leicester Square..with the monster *cafés chantants*. 1958 *Times* 16 Feb. 13/5 The tradition of cafe-chantant artist. 1968 *Times* 13 Nov. 10/4 Downstairs..in the same new establishment. There is a café-chantant. 1891 *Harper's Mag.* Dec. 49/2 A café concert over in the Bowery. 1934 C. Lambert *Music Ho!* iii. 197 Chabrier's..tunes, though evocative of the café-concert are in no way pastiches of café-concert tunes. 1937 *Fortune* Dec. 123 A blending of old socialites and new rich, and the like, under the so-called cafe society. 1952 *Time* 8 Sept. 4/2 All the newspapers call Café Society. 1959 *Times Lit. Suppl.* 9 May 320/5 His friends among the café society set. ...the admission so that it was seen in the society which ...

3. In French phrases, with the sense 'coffee', as **café au lait,** coffee taken with milk; white coffee; also, the colour of café au lait, a brownish cream colour; **café complet** (see quot. 1966); **café crème,** coffee with cream; **café-filtre,** (a cup of) coffee made by filtering boiling water through coffee; cf. *filtre; **café noir,** black coffee, i.e. coffee without milk.

1763 H. Walpole *Let.* 18 Oct. (1904) V. 382 Pray send me some *café au lait*: the Duc de Richmond has a jar to carry out. 1823 J. Griscom *Year in Europe* II. 32 We..refreshed..with a cup of café au lait. 1824 *Ann. Reg.* ...1839 Ure *Dict. Arts* 430 Red with yellow, produces ...with in café au lait. 1907 *Westm. Gaz.* 13 Aug. 9/1 Milk-women..selling the chief material of the Parisian *café-au-lait*. 1910 *Westm. Gaz.* 15 July 2/2 In a café-au-lait..marigold. 1949 Blackwood's *Wand.* ...Those trays of frippery called *café complet*, which, for the coppery taint of chicory, one..perhaps complementary to a teacup. 1937 M. V. Hughes *London Home in Nineties* viii. 137 Our breakfast consisted of *café complet*. 1966 *Guardian* 24 Dec. 5 ...*café complet* means coffee with rolls, butter and jam, and ...Sunday Times ...1955 C. Connolly *Rock Pool* xiii. 203 A hundred café crèmes steamed on the marble tables. 1906 C. Brent *Case of Good Employer* x. 99, I asked for a large *café crème*. 1959 *Listener* 25 Feb. 360/2 Cheeses and coffee, with ...where *café filtré*, or coffee produced by the filtration method, is a speciality. 1956 D. Vine *Leaves* ...1958 ...The strong *café noir*, usually drunk in Paris, served in a ... The Chairman ...1966 P. O'Donnell *Modesty Blaise* vii. 79 He set out two large cups and the perforated tins containing the *café filtre*. 1846 Dickens *Lett.* ...(café noir). 1848 *Cornhill Mag.* Aug. 255 The widow brought our *café noir* to us after dinner. 1956 *Daily Express* 29 Sept. 2/7 Men the colour of ebony, *café noir*, *café au lait*.

cafeteria (kæfĭtɪ³·riă). orig. *U.S.* [a. Amer.-Sp. *cafetería* coffee-shop.] A coffee-house; a restaurant, esp. now a self-service restaurant.
1839 J. L. Stephens *Trav. Russian & Turkish Emp.* I. 157 Every third shop, almost, being a cafeteria [*sic*] where a parcel of huge turbanded fellows were at their daily labours of smoking pipes and drinking coffee. **1894** *Lakeside Directory Chicago* 2188 Cafeteria Catering Co. 44 Lake. **1895** *Ibid.* 2231 'Cafeteria', 46 Lake, 80 Adams, 108 Quincy and 93 Vanburen. **1896** *Chicago Tribune* 28 June 41 Gerbach used to be a waiter in a West Side restaurant subsequent to his employment by the cafeteria company. **1912** *Jrnl. Home Economics* IV. 245 Exactly the same menu was served in a large college dining room and at the cafeteria. **1916** H. Newmark *Sixty Years in S. Calif.* x. 133 Then came the *cafeteria*... It was rather a place for drinking than for eating, and in this respect the name had little of the meaning current in parts of Mexico to-day, where a *cafeteria* is a small restaurant serving ordinary alcoholic drinks and plain meals. **1923** *Mod. Lang. Notes* Mar. 188 Every one knows by this time that a cafeteria is a 'help yourself' restaurant. **1925** *Glasgow Herald* 30 July, Cafeterias, although a commonplace in America, are just beginning to have a hold in Paris. **1958** *Oxf. Mag.* 8 May 410/2 Breakfast and lunch are served cafeteria style in Hall... How many colleges eat cafeteria-wise?

‖ **cafetière** (kaftyɛr). Also erron. **cafétiere, cafetiere, cafétière.** [Fr.] A coffee-pot; a coffee percolator.
1846 A. Soyer *Gastronomic Regenerator* 711 Pour either into the silver cafetière or the cups. **1888** *Queen* 17 Nov. in L. de Vries *Victorian Advertisements* (1968) 108/2 Royle's patent self-pouring cafetières and kettles. **1906** Mrs. Beeton *Housek. Managem.* xlix. 1475 Place the coffee in the coffee chamber of a cafetière, and pour the boiling water through the distributor on to the coffee. **1907** *Yesterday's Shopping* (1969) 169/1 Cafetiere, tin... Simplex Circulating Cafetiere. **1927** Francis Brett Young *Portrait of Clare* v. iv. 459 She..let her emotion flow into a laughing, triumphant recitation of her lavish purchases; the lordly Uffdown presents, Steven's bicycle, Aunt Cathie's cafetière.

caff (kæf). Vulgar or jocular slang for Café. Also *attrib.*
1931 D. Morse-Boycott *We do see Life!* i. 15, I was sitting in a caff in Compton Street. **1950** P. Tempest *Lag's Lexicon* 28 *Caff boys*, the youths and men, the 'spivs' and 'wide boys', who hang around certain cafés. **1959** C. Freeman *Jack would be a Gentleman* ix. 191 We'll find a caff somewhere an' 'ave our food out. **1959** *Daily Mail* 3 Mar. 7/1 The news reached Chrimes, 30-year-old boss of the low-life 'caff society' with his headquarters in a juke-box cafe in Uxbridge.

caffeism (kæ·fiiz'm). [f. Caffe(ine + -ism.] A morbid condition arising from the prolonged or excessive use of beverages containing caffeine. So **caffeinism**.
1886 *American* XII. 169 That class of diseases in which morphinism, caffeism and vanillism are found. **1889** *Cent. Dict.*, Caffeinism. **1906** *Westm. Gaz.* 27 Sept. 4/2 The beverage—I even hate its name: Guess it who can!—That makes the direful caffeism rage.

caffle (kæ·f'l), *v. dial.* [f. Cavil *v.*] *intr.* To cavil, argue; to prevaricate. Hence **ca·ffler, ca·ffling.** Cf. Caffling *ppl. a.*
1851 T. Sternberg *Dial. Northants.* 16 *Caffle*,..to quarrel. Probably a corruption of *cavil*. **1877** E. Peacock *Gloss. Manley* 45/1 Noo none o' your caffiin', tell us all aboot it straight out. **1883** A. Easther *Gloss. Almondbury* 21 *Caffler*, a shuffler, excuse-maker, &c. **1913** D. H. Lawrence *Love Poems* 50 To think I should have to haffle an' caffle Wi' a woman an' pay 'er a price. **1932** S. O'Faolain *Midsummer Night Madness* 66 Are you going to be stopped by a city caffler?

Caffrarian (kæfrɛ³·riăn), *a.* Also **Kaffrarian.** [See -an.] Of, pertaining to, or characteristic of Caffraria (see Caffre 2).
*a*1828 J. Bernard *Retrosp. Amer.* (1887) 80 A certain German duchess had lately given Mrs. Siddons a Caffrarian [illegible] fully equalled the magnificence of her other displays. **1874** Geo. Eliot *Legend of Jubal* 153 No lions then shall lap Caffrarian pools. **1882** C. F. G. Cumming *Fire Fountains* I. 258 From Crimean winters to Kaffrarian summers.

cafoufle, cafuffle, var. Curfuffle *sb.*
1953 *John o' London's* 3 July 602/3 The word cafuffle is still in general use in her part of Scotland... as a noun meaning a state of confusion. **1960** K. Martin *Matter of Time* 187 The post went on and her boy friend are having a wee cafouffle in the garden. **1961** *Radio Times* 14 Dec. 3/2 You remember the cafuffle there was when the Ministry of Transport introduced their ten-year test for cars.

caftan. Add: **2.** A wide-sleeved, loose-fitting shirt or dress worn in Western countries, resembling the original garment worn in the East.
1966 *Vogue* July 43 Kaftan, 39 gns. **1966** *Daily Tel.* 17 Oct. 10/6 Caftans, the season's fashion talking point, won murmurs of delight in lightweight silk, hand-painted, in versions both short and long from artist Noel Dyrenforth. **1968** *Guardian* 6 June 5/5 [He] wore an astrakhan cap and a blue caftan shirt.

cag (kæg), *sb.*³ *Naut. slang.* Also **kagg.** [Cf. Cag *v.* 2.] An argument.
1916 M. T. Hainsselin *In Northern Mists* xviii. 69 We had a right-down regular genuine old-fashioned Ward-room Cag about it. **1918** 'Bartimeus' *Navy Eternal* 330 This..is developing into a 'Branch-kagg'. **1932** C. Morgan *Fountain* 150 He was one with... a passion for argument on remote and unprofessional subjects. He would sit down to what he called a 'cag' as eagerly and patiently as a dog before a rabbit bone.
So as *vb.*, to argue, to nag.
1919 'Etienne' *Strange Tales fr. Fleet* 23, I've never met such a crowd for 'kagging'. **1932** 'N. Shute' *Lonely Road* xi. 233 I'm not going to worry you, or cag about this any more.

cage, *sb.* Add: **2. b.** A (barbed-wire) camp enclosure for prisoners of war. *colloq.*
1919 Downing *Digger Dial.* 15 *Cage*, a prisoner of war compound. **1939** *War Illustr.* 7 Oct. 103 Polish prisoners are seen in a 'cage' to which they have been marched immediately after capture. **1956** A. Crawley *Escape from Germany* xxvi. 280 Having seen that captives were inevitable..he had..packed three large suitcases..to take..to whatever 'cage' he was sent.
10. *cage-mate*; **cage aerial** (see quot. 1926).
1926 S. O. Pearson *Dict. Wireless Techn. Terms*, *Cage Aerial*, an aerial in which a number of component wires are held in position round small star-shaped spreaders or round small hoops in such a manner as to form a 'cage'. This is done to reduce the high-frequency resistance of the aerial. **1904** *Westm. Gaz.* 26 Sept. 10/1 Till he becomes acquainted through the bars with the animals that are in future to be his cage-mates. **1935** 'J. Doyle' *Marmosite's Misc.* 10 Little David Garnett, a cage-mate of mine.

caged, *ppl. a.* Add: **1.** (Later examples.) Also with *in*.
1949 R. Graham *Dream Days* 15 'What's that got to do with it?' retorted his sister, resuming her caged-lion promenade. **1937** W. de la Mare *This Year, Next Year*, A bookworm... Timid, half-blind, caged in, afraid.

cager (kē¹·dʒəɪ). [f. Cage *v.* or *sb.* + -er¹.] **1.** One who encloses in a cage. *rare.*
1889 Browning *Asolando* 5 Boy-Cupid's exemplary catcher and cager.
2. An operative who attends to a cage (in various trades).
1908 *Daily Chron.* 24 Apr. 7/5 The cager was..engaged in another part of the mine. **1921** *Dict. Occup. Terms* (1927) §§ 043, 056, 339, 399.

cagey (kē¹·dʒi), *a. colloq.* (orig. *U.S.*). Also **cagy.** [Etym. unknown.] Not forthcoming, reticent, wary, non-committal. Hence **ca·gily** *adv.* Also **ca·geyness, ca·giness,** the state or quality of being cagey.
1909 *Sat. Even. Post* 1 May 5/3 See? He's cagey about going to 'em, but when a good medium gets him in front of her he swallows it all, hook, stock and barrel. **1922** *Short Stories* Feb. 258/2 The Battler was cagy and covered up for the greater part of the round. **1926** *New Yorker* 11 Sept. [illegible] **1927** [J. Barbican] *Confessions of Rum-Runner* xxii. 259 We hoped they would come out and pick us off, but they were too cagey for that. **1948** *Time* 25 Oct. 23/1 The Yankee management had turned cagey. **1950** A. Lunn *Revolt against Reason* x. 112 The caginess of a Darwinist when cross-examined on his claims. **1953** G. Heyer *Detection Unlimited* ii. 31 Aunt Miriam's always a bit cagey about it. What happened? **1955** *Archit. Rev.* CXVII. 178/3 This is not the first London exhibition to reflect the conventions of 'cageyness'. [illegible] **1966** *Listener* 10 Nov. 681/2 My own is a cagey mathematician where rent, food, and expenses are concerned.

Cagian, var. Cajan.

cagmag, *sb.* and *a.* **1. b.** Add to def.: Anything worthless or rubbishy. (Further examples.)
1811 *Lex. Balatronicum*, *Cag Magg*, bits and scraps of provisions. Bad meat. **1875** Twitchell *Cleveland Dial.* 37 Any ill-conditioned stuff. **1948** J. H. Johnson *Railway Pattern* 80 Maudie is the best beloved woman in London, with the grandest manner. She makes Royalty look like cag-mag.

cagmag, *v. dial.* [f. the sb.] **a.** *intr.* To quarrel. **b.** *trans.* To nag.
1882 Mrs. Chamberlain *Gloss. W. Worc. Words* 6 It's only them two now 'cause they quarrels a cagmaggin' like they allays be. **1932** H. J. Massingham *Wold without End* 296 Cotswold possesses a number of these dramatic words... The farm labourer..will 'grizb',..his wife 'cagmags', not henpecks, him.

‖**cagnotte** (kanyɔt). [F.] Money reserved from the stakes for the bank at certain gambling games (see quots.). Also *attrib.*
1928 *Daily Express* 29 Aug. 1/5 The club has adopted the equivalent of the Cagnotte system... A member winning a sum such as twenty-five pounds at a session is assessed approximately ten shillings. If he wins fifty pounds or over he pays the club one pound. **1959** *Times* 24 Feb. 11/7 The bank's cut, or cagnotte. **1961** I. Fleming *Thunderball* xv. 156 One table of chemin de fer, whose cagnotte yields a modest five per cent.

‖**Cagoulard** (kagulär). [F., lit. 'wearer of a monk's cowl', f. *cagoule* a sleeveless hooded garment + -ard.] A member of a secret right-wing organization in France in the 1930s.
1937 *Times* 17 Sept. 11/7 A secret para-military organization of the extreme Right known as the *Cagoulards* (hooded ones). **1958** *Listener* 21 Aug. 277/2 When Raoul Dautry took over the Ministry of Armaments in September 1939, he was staggered to discover the chief Cagoulard in an important post. **1966** R. R. D. Foot *SOE in France* vi. 137 We have his own word for it that he was not a *cagoulard*—that is, did not belong to the rough French equivalent of the Ku Klux Klan.

cagy, var. *Cagey *a.*

cahier (kaye). Add to def.: An exercise-book, pamphlet, or fascicle.
*c*1845 C. Brontë *Professor* (1857) I. xvi. 271 Her eyes were fastened on the cahier open before her. **1909** E. Nesbit *Daphne* iii. 37 You can divide my school books and cahiers among you. **1958** *N. & Q.* Nov. 488/1 A cahier is being prepared in commemoration of the centenary of Eleanora Duse's birth. **1964** E. Bowen *Little Girls* i. 24 He returned his cahier to it contentedly.
2. *Hist.* The instructions prepared by each of the three representative bodies as a guide for their policy at the National Assembly of 1789 in France.
1792 A. Young *Travels* 27 June (1929) i. 162 These cahiers being instructions given to their deputies, I have now gone through them all. **1805** *Edin. Rev.* VI. 152 Busily employed in preparing the *cahiers* or instructions for the direction of their deputies. **1879** *Encycl. Brit.* IX. 597/1 Early in 1789 all France was busy with the elections to the States-General, and in drawing up the *cahiers*, or papers of grievances. **1887** Lecky *Eng. in 18th Cent.* V. xx. 450 The cahiers of the clergy showed a frank willingness to surrender all privileges.

cahoot. Add to def.: Freq. in pl., esp. in phr. *in cahoot(s)* (*with*): in league or partnership (with). (Earlier and later examples.) For 'U.S.' read 'orig. U.S.'.
1829 S. Kirkham *Engl. Gram.* 207 Hese in cohoot with me. **1834** S. S. Prentiss in J. D. Shields *Life* (1884) 239, I will splice the member from North Carolina to vote, and for a short time will consider you in cahoot. **1862** G. K. Wilder *MS. Diary* 14 Jan. (D.A.E.), [illegible] in cahoots in a store. **1862** [S. Bowers] *Golden Days* 26 Are you willing to work in cahoots with yours truly? **1892** *Congress. Rec.* 16 Mar. 2113/1 Let's go into cahoots on this [illegible]. **1899** G. Ade *Doc' Horne* xv. 280, I have good reasons for saying that these two were in cahoots. **1953** A. Upfield *Murder must Wait* xii. 108 She was in cahoots with a doctor.

cahoun (kăhū·n). Also **cahoon.** Frequent variant of Cohune.
1861 Bentley *Man. Bot.* 685 From the seeds of *A[ttalea] Cohune*, the Cahoun Palm, a fatty oil may be obtained. **1866** [see Cohune]. **1880** *Encycl. Brit.* XII. 133/2 The cahoon or coyol palm..producing clusters of nuts, from which is extracted a valuable oil.

cahow, var. Cohow.

caid: see Kaid (in Dict. and Suppl.).

cailcedra (koil·se·drā). [Origin unknown.] A lofty tree of West Africa, *Khaya senegalensis*, of which the wood is specially adapted for joinery and the bark furnishes a bitter tonic.
1866 Lindley & Moore *Treas. Bot.*, *Cailcedra-wood*, the timber of *Flindersia australis*. **1886** *Encycl. Brit.* XXI. 662/1 The wood of the cailcedra.. is used for ship-building. **1887** Moloney *Forestry W. Afr.* 297 Cailcedra or Mahogany Tree of the Gambia.

Cain². Add: **1. b.** *to raise Cain*: see Raise *v.*¹ 20 b. *colloq.* (orig. *U.S.*).
1840 *Daily Pennant* (St. Louis) 2 May (Th.), Why have we ever reason to believe that Adam and Eve were both rowdies? Because..they both raised Cain. **1841** 'Dow, Jr.' *Short Patent Sermons* xxix. 73 They will feel that they have been raising Cain and breaking things. **1852** [see Raise *v.*¹ 20 b]. **1881** R. L. Stevenson *Treasure Island* (1883) iii. 20 If I get the horrors, I'm a man that has lived rough, and I'll raise Cain. **1931** J. B. Priestley *Angel Pavement* vii. 345 If we stand here talking another minute the mistress'll be raising Cain the way she'll say she's destroyed with the noise.

ca'ing-whale: see also *Caa'ing whale.

cainosite, var. *Cenosite.

Cairene (kai³·ri·n), *sb.* and *a.* Also **Caireen.** [f. *Cairo* + *-ene*, after *Nazarene*, etc.] **A.** *sb.* A native or inhabitant of Cairo, the capital of Egypt. **B.** *adj.* Of or pertaining to Cairo.
1844 Sophia Poole *Englishwoman in Egypt* I. xi. 156 If we were conducted by a Caireen, no Turkish ladies were likely to address us. **1851** *Illustr. London News* 8 Feb. 102/1 An interesting specimen of Caireen house-building. **1854** B. Taylor *Life fr. Egypt* xv. 205 A harem of Cairene ladies. **1855** R. F. Burton *El-Medinah* I. 270 The people of Suez are a finer and a fairer race than the Cairenes. **1893** [see Nilot]. **1902** S. Lane-Poole *Story of Cairo* i. 14 The Cairene tucks his legs up under him on the divan. **1936** E. M. Forster *Abinger Harvest* iv. 254 The joint authors..are Egyptian—one of them a Cairene Jew. **1945** R. Hargreaves *Enemy at Gate* 107 A Cairene rebellion which was only suppressed at the cost of considerable bloodshed.

cairn. Add: **2.** *Cairn terrier* [said to be so named from being used to hunt among

cairns], the smallest breed of terrier in Great Britain, somewhat long in the body and deep in the ribs, with short straight legs and a shaggy coat; also *ellipt.*
1910 *Kennel Christmas No.* 461/1 Cairn terriers promise to rank high in the near future, and already they are being bred to a fairly uniform standard. **1922** R. Leighton *Complete Bk. Dog* xvii. 271 It is now rare to find a Cairn with a bad mouth. **1924** *Westm. Gaz.* 31 Oct., The handy little Sealyhams and Cairns.

cairn, *v.* [f. the sb.] *trans.* To mark with a cairn. So **cai·rned** *ppl. a.*, **cai·rning** *vbl. sb.*
1859 [see Cairned *a.*]. **1937** *Geogr. Jrnl.* XC. 309 Points were fixed with the latter, and where possible cairned. **1960** *Guardian* 2 Nov. 6/6 If any route marking is done in mountain districts, I hope it will be done by cairning.

cairngorm. Add: *attrib.* and *Comb.*
1870 Meredith *H. Richmond* (1871) I. viii. 136 In satisfying Janet's wishes for riding-whips, knives, pencil-cases, Cairngorm buttons, and dogs. **1881** *Instr. Census Clerks* (1885) 42 Lapidary, Cairngorm, Cameo, Cutter, Worker. **1921** *Dict. Occup. Terms* (1927) § 658 Cairngorm grinder.

caisson. Add: **4. caisson sickness** = *caisson disease.*
1911 *Engineer* 10 Mar. 243 Caisson Sickness and Compressed Air.
Hence **caissonier** (kē³sani·ə), one who works in a caisson; so **cai·ssoning.**
1903 *Strand Mag.* Jan. 98/2 The lives of 'the men in the shaft', the caissoniers, are in the greatest danger. [illegible] 10/2 After all these terrors it is perhaps astonishing to be told..that caissoning would be sought by any considerable number of men.

Cajan, Cajen, Cajian, Cajun, etc., colloq. corruptions of Acadian *a.* and *sb. N. Amer.*
1868 *Putnam's Mag.* I. 54 Among them were..Cagians, the descendants of the old Acadians. **1878** *Hallberger's Illustr. Mag.* 577 (Farmer), The Native Louisianian and the Acadian, more universally known..as the Cajen. **1880** *Scribner's Monthly* Jan. 383 Acadian—or rather its corruption 'Cajun', as they pronounce it—is regarded..as implying contempt. **1885** *Outing* (U.S.) Feb. 337/1 The Cajun fisherman will gladly teach you his art of catching trout. **1942** T. H. Randall *His Majesty's Yankees* 26 We'd pluck the French off their farms—off the face of the 'Cajun lands around Fundy Bay. **1959** P. Capon *Amongst those Missing* 43 My father's a Cajun from New Orleans.

cake, *sb.* Add: **4. c.** A substance (such as cotton-seed, linseed, etc.) compressed in a flat form and used for feeding cattle, etc. Freq. with defining word, as Cotton-*cake*, Linseed *cake*. Also *attrib.* and *Comb.*
1757 [see Oilcake]. **1833** *Niles' Reg.* XLV. 222/1 The cake is the very best food for stock. *a*1884 Knight *Dict. Mech. Suppl.* 152 *Cake grinder*, a machine for breaking linseed oil cake for food for stock. **1886** R. E. C. Cole *Gloss. S.W. Lincs.* 24 *Cake*, usual term for the Linseed Cake, used for fattening cattle. Some men run up a great cake bill their last year. **1894** *Country Gentleman's Catal.* 14/2 A few acres of autumn cabbage will maintain and fatten, with the aid of corn or cake, of course, a large flock of sheep. **1907** *Yesterday's Shopping* (1969) 57 Patent Meat Fibrine Vegetable Dog Cakes... Patent Cod Liver Oil Dog Cakes. **1915** E. Blunden *Barn* 20 The smell of apples..And homely cattle-cake. **1960** *Farmer & Stockbreeder* 26 Jan. 93/3 Those ewes whose body weights were maintained..by cake feeding. *Ibid.*, The cake-fed ewes.
d. In the manufacture of artificial silk: (see quots.).
1927 T. Woodhouse *Artificial Silk* 42 An annular package of yarn is gradually built up by the succeeding layers of yarn. This annular package is called a 'cake'. **1963** A. J. Hall *Textile Sci.* ii. 56 Fibres..are collected into the form of a thread..to be drawn over rotating Godet wheels B and fall down into the rotating centrifugal Topham pot T to be built up inside as a ball or 'cake'.
7. *to take the cake* (†*U.S. cakes*) (earlier and later examples); often used ironically or as an expression of surprise. Cf. *biscuit 1 d.
1847 W. T. Porter *Quarter Race Kentucky* 120 [illegible] **1884** *Lisbon* (Dakota) *Star* 25 July, Sherrill Mease takes the cake for the first wheat-harvesting in Ransom county. **1900** T. Dreiser *Sister Carrie* xxiii. 249 Pack up and pull out, eh? You take the cake. **1904** A. Bennett *Great Man* xxv. 275 My bold buccaneer, you take the cake... There is something about you that is colossal, immense, and magnificent. **1938** G. Heyer *Blunt Instr.* ix. 158 I've met some kill-joys in my time, but you fairly take the cake.
b. In phrases, as *the national cake*, or, allusively, (*the*) *cake*, the assets or proceeds of a national, etc., economy, regarded collectively as something to be shared out.
1750 [see Dict.]. **1949** *New Statesman* 22 Oct. 443/2 A general free fight between capital and labour for their respective shares of the national cake. **1957** *Listener* 8 Aug. 181/1 German labour may be about to demand a larger slice of the recovery cake. **1958** *Engineering* 28 Mar. 391/1 They [sc. trade unions] have for long thought that their job of getting more of the cake out of employers included advising on how to increase the cake.
c. Colloq. phr. *a piece of cake*: something easy or pleasant.
1936 O. Nash *Primrose Path* 172 Her picture's in the papers now, And life's a piece of cake. **1942** T. Rattigan *Flare Path* I, Special. Very hush-hush. Not exactly a piece of cake, I believe. **1943** P. Brennan et al. *Spitfires over Malta* i. 31 The mass raids promised to be a piece of cake, and we anticipated taking heavy toll of the raiders. **1960** T. McLean *Kings of Rugby* 205 They took the field against Canterbury as if the match were 'a piece of cake'.
8. *one's cake is dough*: delete †*obs.* and add earlier and later examples.
1563 Becon *Displ. Popish Mass* in *Wks.* III. f. xlviii, Or els your Cake is dough & al your fatte lye in the fyre. **1860** [see Dough *sb.* 1 b]. **1934** J. Farnol *Winds of Fortune* v. 31 Thy cake's dough, eh, Japhet; art cheated o' thy dear vengeance, lad?
9. a. *cake-basket* (examples), *-cutter*, *-plate*, *-stand*; **d. cake-eater** *U.S. slang*, a self-indulgent or effeminate man; a playboy; **cake-hole** *slang*, a person's mouth; **cake-mix,** the prepared ingredients of a cake sold ready for cooking; **cake-mixer** (see quot. *a* 1877); **cake-tin,** (*a*) a tin in which cakes are baked; (*b*) a tin in which cakes are stored.
1860 in Mrs. E. S. Bowne *Life* (1888) 205, I plated Cake Basket silver rims. **1956** G. Taylor *Silver* viii. 145 The many elaborately pierced cake-baskets. **1845** E. Acton *Mod. Cookery* xi. 263 Croutons..stamped out..with a round or fluted paste or cake cutter. **1922** *Daily Ardmoreite* (Ardmore, Okla.) 6 Jan. 10/4 He calls on 'lounge lizards, tea drinkers, cake eaters and all that'. **1943** Hunt & Pringle *Service Slang* 20 *Cake hole*, the airman's name for his or anyone else's mouth. **1959** I. & P. Opie *Lore & Lang. Schoolchildren* x. 174 Shut your cake-hole. [**1938** Fiene & Blumenthal *Handbook Food Manuf.* 332 Sponge cake mix.] **1950** *Canad. Home Jrnl.* Jan. 24 The cake mix... The cake mixes should yield you the principal makings for a good cake. **1957** *B'ham Jrnl.* I. 15/1 Packaged cake-mixes are undoubtedly time-and-labour-saving. *a* **1877** Knight *Dict. Mech.* I. 422/2 *Cake-mixer*, a device for incorporating together the ingredients of cake. **1906** *Daily Colonist* (Victoria, B.C.) 1 Jan. 5/5 Come down to No Douglas street and get..a Cake Mixer (it whips cream, beats eggs and mixes cake). **1867** A. D. T. Whitney *L. Goldthwaite* x. Cake plates were garnished with wreathed oak-leaves. **1851** J. J. Hooper *Widow Rugby's Husb.* 100 Here he went..dancing all our woman and her cake stand at a jump. **1895** *Montgomery Ward Catal.* 331/1 China Fruit or Cake Stand..measure 12 inches across and 5¼ inches high. **1903** A. Bennett *Leonora* iii. 80 A scratched cake-tin. **1953** G. Greene *Living Room* I. i. A cake-stand with bread-and-butter on one level and a plain cake on another. **1963** *Times* 25 May 6/1 He described the 'cake stand' of ministerial hierarchy—chocolate biscuits on top, cream buns underneath, and buttered scones at the bottom. **1846** 'A Lady' *Jewish Manual Cookery* xii. 119 Put the apples into an oval cake tin. **1906** E. Nesbit *Railway Children* i. 15 In the pantry there was only a rusty cake-tin and a broken plate. **1965** 'T. Howard' *Games of Chance* ii. i. 153 I'd..been relieved to find in the cake-tin that yesterday's Swiss roll had only been reduced by three inches.

cake, *v.* Add: **3.** *trans.* To feed (cattle, etc.) on cake (see *Cake *sb.* 4 c). Also *absol.*
1911 *Jrnl. R. Agric. Soc.* XII. ii. 314 Many farmers cake their hogs on the turnips. **1889** E. Peacock *Gloss. Manley* (ed. 2) I. 626 *Cake*, to feed cattle with linseed or cotton cake. I also cake my yoes o' winter as well as th' hogs. **1904** Kipling *They* (1905) 71 You've sixty-seven [bullocks] and you don't cake.

cakelet (kē¹·klĕt). [f. Cake *sb.* + -let.] A small cake.
1830 Ure *Dict. Arts* 458 These cakelets must be dried upon tiles. **1908** M. & J. H. Findlater *Crossriggs* xix. 134 [illegible] **1928** *Daily Express* 13 June 3/6 These elusive cakelets [sc. cookies] are evolved by mixing [illegible].

cake-walk (kē¹·k,wǫk), *sb.* orig. *U.S.* [f. Cake *sb.* + Walk *sb.*]
1. a. 'A walking competition among negroes, in which the couple who put on most style "take the cake"' (Thornton). **b.** A dance modelled on this.
It originated among the Negroes of the southern United States.
1879 *Harper's Mag.* Oct. 799/1 Reader, didst ever attend a cake walk? **1887** E. Farmer *Americanisms* s.v. *Cake*, In certain sections of the country, cake-walks are in vogue among the colored people. It is a walking contest, not in the matter of speed, but in style and elegance. **1897** *Blackw. Mag.* Mar. 341/2 'Cake-walks' and frolics and preachings filled the cabins with sound and merriment. **1902** Harben *Abner Daniel* 53, I could have done the cake-walk with that fat Howard girl from Rome. **1947** *Penguin Music Mag.* May 25 Ragtime was most certainly responsible for Debussy's 'Golliwog's Cake Walk'.
c. *transf.* and *fig.* In quots. 1916, 1966 = 'something easy'.
1863 H. Eggar *Jrnl.* in *Montana Hist. Soc. Contrib.* (1900) III. 133 Around and around that bush we went..We had a good high old cake-walk. **1894** 'M. Twain' in *Critic* 7 July 8/1 This Shelley biography is a literary cake-walk. **1916** B. Cooper *Coo-oo-ee* xi. 155 Whether they would give him victory or fight-to-a-finish it was a cake-walk; he did not know. **1966** J. M. Brett *Cargo of Spent Evil* x. 87 This should be a cakewalk for you.
2. A form of entertainment consisting of a promenade moved by machinery on which people walk to the accompaniment of music.
1909 *Oxford Times* 11 Sept. 9/5 In dealing with the fair itself there were really no new features..except that of the Brooklyn cake-walk, an ingenious rocking platform which gave those who patronised it the sensation of a cake-walk dance...The novelty was in operation at the White City last year. **1914** *Ibid.* 12 Sept. 10/3 The absence of the popular joy-wheel, the cake-walk [etc.]. **1968** D. Braithwaite *Fairground Archit.* p. ix, The boneshaking old Cake-walk changes its name to suit the fashion of the day, becoming at one time the Jolly Jersey Bounce and more recently the Rock an' Roll.
Hence **ca·ke-walk** *v. intr.*, to walk or dance in the manner of a cake-walk (sense 1); also *transf.* and *fig.* So **ca·ke-walker**; **ca·ke-walking** *vbl. sb.* and *ppl. a.*
1898 Williams & Walker *Let.* 16 Jan. in J. W. Johnson *Black Manhattan* (1930) x. 105 We, the undersigned world-renowned cake-walkers..hereby challenge you to compete with us in a cake-walking match. **1898** *Daily Tel.* 14 Mar., Cake-walking is, in fact, a graceful motion, conducted upon the toes and ball of the foot. **1898** *Westm. Gaz.* 3 Dec. 7/7 The cake walkers at Covent Garden. **1904** *Daily Chron.* 22 Mar. 4/7 The genuinely tip-top men Were those who never cake-walked. **1904** 'Saki' *Reginald* 90 A mouse used to cake-walk about my room. **1905** *Westm. Gaz.* 17 Aug. 8/1 French singers, cake-walking coons, and fifth-rate English dancers. **1927** *Melody Maker* Sept. 931/2 The syndicate..cake-walks to prosperity. **1958** Blesh & Janis *They all played Ragtime* 3 Soon the French were cakewalking in the streets of Paris to *le temps du chiffon*. *Ibid.* v. 99 Cakewalking developed into a real art. **1967** V. Nabokov *Speak, Memory* (ed. 2) xv. 309 Pale-blue and pink underwear cakewalking on a clothesline.

Calabrese (kælăbrē¹·se). [It., = Calabrian.] A variety of sprouting broccoli.
1930 L. H. & F. Z. Bailey *Hortus* 130 The Asparagus or Sprouting broccoli, *Calabrese*, is a different plant. **1939** *Times* 11 Jan. 15/4 Calabrese..is one of the most quickly grown and delicately flavoured of all greens. **1957** E. Hyams *Speaking Garden* 58 *Calabrese*, perhaps the most delicious of all cabbages.

Calabrian (kălē¹·briăn), *a.* and *sb.* [f. *Calabria* (see below) + -an.] **A.** *adj.* Of or pertaining to Calabria, a region of Italy (anciently the south-eastern, now the south-western, projection of the Italian peninsula). **B.** *sb.* A native or inhabitant of Calabria (see also quot. 1905).
1594 Kyd *Cornelia* v. 239 As in the faire Calabrian fields. **1607** Topsell *Four-f. Beasts* 599 The Calabrian, Milesian, and Arentinean sheepe. **1615** Sandys *Travels* iv. 250 There are not that professe Christ, a more vnciuill people then the vulgar Calabrians. *Ibid.*, A certaine Calabrian hearing that I was an Englishman, came to me. **1783** H. More *Let.* (1925) 88 The miseries which have visited the devoted Calabrians. *a* **1821** Keats *Otho* (1848) v. v, What wine? The strong Iberian juice, or mellow Greek? Or pale Calabrian? **1833** [see *en règle*, s.v. *en *prep.*]. **1905** A. Hopwood *Old Eng. Sheep Dog* i. 1 The mighty Calabrian, or sheep dog of the Pyrenees,..stands over thirty inches high. **1924** tr. *Pastor's Hist. Popes* XIII. viii. 223 A Calabrian, 33 years of age, was sent..to the Jesuits. **1963** *Times* 22 Feb. 6/4 These people are Calabrian peasants.

calamistrum (kælămi·strŏm). Pl. **-a.** [L., curling-iron.] A comb-like structure on the metatarsi of the fourth pair of legs of certain spiders, used to card and curl the silk as it issues from the spinnerets.
1866 E. F. Staveley *Brit. Spiders* 14 These [spines] are called calamistra, and are used in the construction of the web. **1875** *Encycl. Brit.* II. 292 The function of the calamistrum has been proved..to be the carding, or teasing and curling, of a peculiar kind of silk, secreted and emitted from the fourth pair of spinners. **1958** W. S. Bristowe *World of Spiders* viii. 79 Three families of British spiders have a cribellum and a calamistrum.

calamitean (kælămɔi·tiăn), *a.* [f. Calamite 1 + -an.] Belonging or relating to calamites.
1895 *Naturalist* Aug. 237 The histology of calamitean leaves. **1904** *Amer. Nat.* Apr. 250 Thus such transitions are well known, though of a relatively simplified form in the structure of the calamitean stem.

calamitously, *adv.* (Later example.)
1896 *Home Missionary* (N.Y.) July 145 Churches..brought suddenly and calamitously into missionary conditions.

calamity. Add: **3.** *attrib.* and *Comb.*, as *calamity-howler*, *-howling*, *-prophet*, *-shouting* (U.S. colloq.); **Calamity Jane,** the nickname of Martha Jane Burke (*née* Canary) (? 1852–1903), a famous American horse-rider and markswoman, applied to a prophet of disaster.
1892 *Congress. Rec.* 2 Mar. 1654/1 We had some 'calamity howlers' here in Washington. **1905** D. G. Phillips *Plum Tree* 264, I..sent Woodruff East to direct a campaign of calamity-howling in the eastern press. **1876** *Cheyenne* (Wyoming) *Daily Leader* 23 Nov. in N. Mumey *Calamity Jane* (1950) 55 Calamity Jane..now slingeth hash as a waiter in a Custer City Hotel. **1882** *Street & Smith's New York Weekly* 16 Jan. 1 Calamity Jane, the Queen of the Plains. A Tale of Daring Deeds by a Brave Woman's Hands. **1885** E. L. Wheeler (*title*) Deadwood Dick on deck; or, Calamity Jane the heroine of Whoop-up. **1930** *N. & Q.* 27 Sept. 232/1 A crepe-hanger is the ultimate in depressing persons; 'wet-blankets','gloomy Gus's', 'calamity Janes', are all a degree milder. **1960** *TV Times* 8 Jan. 11/1 I'm a real Calamity Jane. **1894** *Republican Campaign Text-bk. for 1894* 229 This is going to be a bad, sad year for the calamity prophets of both parties. **1911** J. C. Lincoln *Cap'n Warren's Wards* i. 3 The pair of calamity prophets broke off their lament. **1892** *Congress. Rec.* 17 Mar. 2160/2 Calamity-shouters whose occupation is gone unless they can prove that calamity stalks abroad.

calandra, var. Calender.
1803 [see Calander]. **1906** *Westm. Gaz.* 19 Apr. 12/1 In Milan, and also in Florence,..he saw enormous masses of small birds..field and calandra larks, and robin-redbreasts. **1953** Bannerman *Birds Brit. Isles* II. 8 The calandra is a Mediterranean species. *Ibid.*, The calandra lark is found almost exclusively in cultivated areas.

calandria (kălæ·ndriă). [Sp., lit. 'lark (bird); calander'.] A closed cylindrical vessel with a number of tubes passing through it, used as a heat exchanger in an evaporator and, in some nuclear reactors, to separate a liquid moderator from the fuel rods and coolant.
1929 B. Heastie in E. Hausbrand *Evap. App.* (ed. 4) xxvii. 445 The Kestner Salting Type Evaporator consists essentially of..the separator and the calandria. The separator is a large cylindrical vessel around which the calandrias are grouped. *Ibid.* 446 Steam is admitted to the calandria, and the liquor then passes..down into the bottom box of the calandria, up the calandria tubes, and back into the separator. **1955** *Bull. Atomic Sci.* Feb. 72/2 Working chiefly by remote control, the intensely 'hot' calandria, the heart of the reactor, was removed for safe burial. **1963** *New Scientist* 7 Nov. 305 In a tank (or 'calandria') which contains the heavy-water moderator there are fixed vertical tubes to contain the nuclear fuel.

calcar². Add: **1.** (Earlier example.)
1832 J. Lindley *Introd. Bot.* I. 120 Sometimes a petal is lengthened at the base into a hollow tube, as in Orchis, &c.: this is called the spur or calcar, and by some nectarotheca.
2. *Anat.* Any of various spur-like bones of vertebrates, or the tibial spur of some insects.
1895 *Camb. Nat. Hist.* V. iii. 104 The spines at the top of the tibia [of insects], projecting beyond it, are called spurs, or calcares. **1898** Parker & Haswell *Textbk. Zool.* II. 254 On the tibial side of the first [digit] is a spur-like structure or calcar..: such a rudimentary digit is called a præhallux. **1951** C. K. Weichert *Anat. Chordates* x. 484 A small additional bone, the prehallux, or calcar, occurs on the tibial side of the tarsus in most salientians.

calcicole (kæ·lsikŏl), *a. Bot.* [f. L. *calc(i)-* lime, Calx + *colĕre* to inhabit.] That grows best in calcareous soil. Hence as *sb.*, a calcicole plant. So **calci·colous** *a.*
1882 *Encycl. Brit.* XIV. 562/1 As to saxicole lichens..they may be divided into two sections, viz. *calcicole* and *calcifugous*. To the former belong such as are found on calcareous and cretaceous rocks. **1886** J. E. Bagnall *Handbk. Mosses* 33 Another very characteristic calcicolous moss is Eucladium verticillatum. **1932** Fuller & Conard tr. *Braun-Blanquet's Plant Sociol.* vi. 183 Mere traces of lime..enable calcicoles to survive. *Ibid.* 184 The occurrence of calcicolous (lime-constant) communities. **1952** P. W. Richards *Tropical Rain Forest* ix. 223 Calcicole and calcifuge species. *Ibid.*, True calcicoles..dependent on the chemical rather than the physical properties of the limestone.

calcicrete (kæ·lsikrĭt), see *calcrete.

calciferol (kælsi·fĕrǭl). *Chem.* [f. Calciferous *a.* + -ol, as in *ergosterol.] Vitamin D_2.
1931 T. C. Angus et al. in *Proc. R. Soc.* B. CVIII. 340 We think the provisional adoption of a name is justified..and suggest 'Calciferol' in view of the high antirachitic activity of the substance. **1942** *Endeavour* I. 31/1 The first few almost microscopic crystals of calciferol were inspected by enthusiastic chemists. **1964** S. Duke-Elder *Parsons' Dis. Eye* (ed. 14) xv. 174 Cases of lupus frequently respond to calciferol.

calcifuge (kæ·lsifiūdʒ), *a. Bot.* [f. *calci-* (see *calcicole *a.*) + -fuge.] Not suited by calcareous soil; that grows best in acid soil. Hence as *sb.*, a calcifuge plant. So **calci·fugous** *a.*
1882 [see *calcicole *a.*]. **1909** *Cent. Dict.* Suppl., Calcifuge *a.* **1926** Tansley & Chipp *Study of Veget.* vii. 122 When the acidity is well marked the soil bears special plant communities (often called *calcifuge*) excluding many species of plants altogether. **1946** *Nature* 17 Aug. 240/1 Several other blue-fruiting calcifuges. **1957** P. Greig-Smith *Quant. Plant Ecology* v. 109 Compare, for example, the occurrence of *Calluna vulgaris*, normally calcifuge, on calcareous soils under extreme climatic conditions.

calcimine. Substitute for etym.: Later modification of Kalsomine, after L. *calci-*, *calx* lime + inorganic *-m-* + -ine⁴. Add further examples.
1885 *Spons' Mech. Own Bk.* 612 The wash or calcimine can be used for ordinary purposes. **1897** *Sears, Roebuck Catal.* 299/2 Should the ceilings and walls be calcimined or whitewashed, wash off old calcimine or whitewash. **1903** *Westm. Gaz.* 23 Jan. 4/3 The old-gold calcimine ..that covers the wall of the drawing-room. **1911** H. S. Harrison *Queed* xi. 134 Sharlee tapped the calcimine with her pointed finger-nails. **1919** *Chambers's Jrnl.* May 327/1 The quarters were..somewhat garishly calcimined within. **1930** T. S. Eliot tr. *St.-J. Perse's Anabasis* 27 My heart twittered with joy under the glare of the calcimine.

calciphilous (kælsi·filəs), *a. Bot.* [f. *calci-* (see *calcicole *a.*) + Gr. φίλος (see -phil).] Of a plant: well suited to calcareous soil. So **ca·lciphile** *a.* (see also quot. 1952).
1909 *Cent. Dict.* Suppl., Calciphilous. **1934** Webster, Calciphile. **1938** [see *calciphobous *a.*]. **1952** J. Clegg *Freshwater Life* xvi. 255 It is possible to group the fresh-water Molluscs into those that need and are found only in 'hard' waters—the so-called *Calciphile* species..—and those that can thrive in quite soft waters.

calciphobous (kælsi·fŏbəs), *a. Bot.* [f. *calci-* (see *calcicole *a.*) + Gr. -φόβος (see -phobe).] = *calcifuge *a.*
1907 W. R. Fisher *Schlich's Man. Forestry* (ed. 2) IV. 408 Holly makes excellent hedges, but requires plenty of humus. It is calciphobous like sweet-chestnut. **1938** Weaver & Clements *Plant Ecology* (ed. 2) viii. 233 Certain species of plants have been grouped into lime-loving (*calciphilous*) and lime-fearing (*calciphobous*) species according to whether they grow on limestone or siliceous soils.

calcrete (kæ·lkrĭt). Also **calcicrete.** [f. L. *calc(i)-* lime, Calx + Con)crete *sb.*] A breccia or conglomerate formed by the cementation of rock debris, etc., by calcareous material.
1902 *Irish Naturalist* Oct. 231 Over the Boulder-clay and the calcrete is a bed of limestone sand and gravel. **1903** *Geol. Mag.* Mar. 139 No one would be likely to quarrel with 'calcicrete' and 'silicicrete', of which one would be two, the other three, letters longer [than 'calcrete', 'silcrete']. **1903** *Nature* 22 Oct. 614/1 In the Gulf of Manaar, calcareous masses ('calcretes') of great extent are formed *in situ* on the sea-bottom by the cementing of sand and other loose material by calcareous incrusting Polyzoa. **1959** J. D. Clark *Prehist. S. Afr.* ii. 45 The tools of those times are found in calcretes and ferricretes.

calculated, *ppl. a.* **1.** (Further examples.)
1930 *Economist* 12 July 59/1 This important statement had been communicated to the Press as a result either of a misunderstanding or of a 'calculated indiscretion'. **1956** R. Heinlein *Double Star* (1958) i. 18 'You haven't any right to jeopardise everybody else by telling him. You don't know a thing about him.' 'It's a calculated risk.' **1959** *Listener* 22 Oct. 672/2 Obviously, the Soviet Union is taking a calculated risk.

calculating, *vbl. sb.* Add: **calculating machine,** any machine designed to carry out calculations, esp. one that performs arithmetical operations mechanically.
1832 D. Brewster *Lett. Natural Magic* xi. 292 The calculating-machine now constructing under the superintendence of the inventor [*sc.* Babbage]. **1855** *Proc. R.Soc.* VII. 499 Report of a Committee appointed by the Council to examine the Calculating Machine of M. Scheutz. **1890** Conan Doyle *Sign of Four* ii, You really are an automaton—a calculating machine. **1901** *Nature* 11 July 268/2 The advantages of the calculating machines..are so great, and they are in so many ways preferable to logarithms where they can be used. **1955** Koestler *Trail of Dinosaur* 184 The calculating machines called electronic brains.

calculating, *ppl. a.*¹ Add: **calculating boy,** a child prodigy in arithmetic.
1866 *North Brit. Rev.* XLV. 39 Colburn, the American 'calculating boy', who was then being exhibited as a curiosity in Dublin. **1937** H. G. Wells *Star Begotten* vi. 91 The proportion of children of the calculating-boy and musical-prodigy type seemed to be increasing quite markedly.

calculiform (kæ·lkiŭlifǫɹm), *a.* [f. L. *calculus* pebble + -(i)form.] Pebble-shaped.
1900 *Spectator* 24 Mar. 417/2 The people are still known as the Mayas, and the writing is called calculiform. **1930** *Antiquity* IV. 289 The calculiform signs..have proved no Rosetta stone for the decipherment of..surviving New Empire codices.

Caledonian. Add: **A.** *adj.* **2.** *Geol.* Designating or pertaining to the great mountain-building episode of late Silurian and early Devonian times in Britain and Scandinavia; also applied to contemporary mountain-building elsewhere. Hence **Ca·ledonoid** *a.*, of or pertaining to the direction assumed by the Caledonian mountain-folds, generally from north-east to south-west.
1903 Geikie *Text-Bk. Geol.* (ed. 4) I. iii. i. iii. 394 In north-western Europe, the prevalent line along which terrestrial plications took place during the earlier half of Palæozoic time was from S.W. or S.S.W. to N.E. or N.N.E. —the Caledonian chain of Professor Suess. **1906** Sollas tr. *Suess's Face of Earth* II. iii. ii. 82 These pre-Devonian mountains, which proceed from Norway and form the whole of Scotland..we call the *Caledonian mountains*. **1913** C. Lapworth in G. A. Auden *Handbk. for Birmingham* 598 The trend..N.E., S.W...which suggests that trend characteristic..of the 'Caledonian Movement' becomes..naturally referred to as the *Caledonoid Trend*. **1929** L. J. Wills *Physiogr. Evol. Brit.* ii. vi. 76 'Caledonian' implies a fold produced during the post-Silurian orogeny, but 'Caledonoid' refers to a south-west to north-east fold of any age.

calendarist. Delete †*Obs.*, and add example.
1875 O'Hanlon *Irish Saints* I. 379 All our calendarists agree in assigning St. Fechin's feast to the 20th of January.

calendric, -ical, *a.* Delete *rare* and add further examples. Also, occurring on a special day or days indicated in a calendar.
1949 G. Bateson in M. Fortes *Social Structure* 44 You made a beautiful structure of flowers and fruit for the calendric feast in his temple. **1959** I. & P. Opie *Lore & Lang. Schoolchildren* xiii. 293 The calendrical framework in which folklorists commonly arrange their collectings. **1961** Y. Olsson *Syntax Eng. Verb* vii. 202 A calendrical back-reference.

calf². Add: **3. calf-length** *a.* (of a garment, boots, etc.) reaching down to, or up to, the calf of the leg.
1965 G. McInnes *Road to Gundagai* ix. 143 Turning up..in a calf-length white motoring coat. **1967** *Harper's Bazaar* Sept. 60/1 New calf-length skirt for the country. **1968** *Ottawa Jrnl.* 24 June 17/5 Sue Ellen..started down the aisle in white calf-length boots. **1969** J. Gardner *Founder Member* vii. 115 A pair of heavy calf-length stockings.

Caliban. Add: Also *attrib.* and *Comb.*, as *Caliban-like* adj. So **Ca·libanish** *a.*
1839 Kemble *Jrnl. Georgian Plantation* (1863) 222 The Calibanish wonderment of all my visitors..is very droll. **1872** Du Chaillu *Country of Dwarfs* 62 The fiendish countenances of the living calibanish trio. **1909** *Lady's Realm* Feb. 465/2 He was a Caliban-like creature, primitively ugly. **1921** *Chambers's Jrnl.* 22/2 A lunatic..indulges in Caliban-like gambols, unheeded. **1965** F. Sargeson *Memoirs of Peon* iv. 90 Mr Gower-Johnson with his Caliban fish-eye.

calibrate, *v.* Add: Also, to determine the correct position, value, capacity, etc., of; to set an instrument so that readings taken from it are absolute rather than relative; *spec.* to mark (a radio) with indications of the position of various wavelengths or stations. Also *transf.* Hence **ca·librated** *ppl. a.*, **ca·librating** *vbl. sb.* (usu. *attrib.*).
1897 *Westm. Gaz.* 14 Jan. 6/3 The tube was handed on to a calibrating machine, which accurately 'shaped' it. **1909** *Installation News* III. 95/1 A very accurately calibrated check meter. *Ibid.* 160 A complete Testing and Calibrating Plant. **1930** *Daily Mail* 4 Jan. 7/1 To get the [wireless] set to do its best it must be calibrated. **1959** Halas & Manvell *Film Animation* xix. 228 All scene movements which he has to calibrate in terms of camera and rostrum movements. **1960** M. Sharp *Something Light* viii. 74 Calibrating a cup of char with Rossy as one extreme, and tea at Gladstone Mansions as the other, tea with the Meares..came about halfway.

calibration. Add: (Further examples.) Also, a set of graduations or markings; a classification. Also *attrib.*
1922 F. W. Aston *Isotopes* v. 55 This is our first *calibration curve*—of necessity inaccurate owing to the gaps between the points. **1930** *Daily Mail* 4 Jan. 7/1 With calibration you can tune in at will to any foreign programme that is within the receptive powers of the set. **1959** Halas & Manvell *Film Animation* xix. 228 He may also have to readjust the camera-lens..for other effects of movement in the scene according to the calibrations he has previously worked out. *Ibid.* 337 *Calibrations*, markings to indicate the movement of a background in panning shots, the movement of the camera in tracking shots, or as a guide to the position of in-between drawings. **1965** *Language* XLI. 173 As an example of special calibration, consider 400 indirect object prefixes.

calibrator (kæ·librē¹:təɹ). [f. Calibrate *v.* + -or.] One who, or that which, calibrates; *spec.* in *Med.*, an instrument for measuring the calibre of a passage, etc.
1900 Dorland *Med. Dict.* 127/2 *Calibrator*, an instrument for dilating the urethra or for measuring the caliber of a passage. **1932** *News Chron.* 23 Sept. 7/4 A novel wireless calibrator, called the Easy Station Finder..enables owners of non-calibrated tuning arrangements to tune-in to any desired station with accuracy. **1958** *Spectator* 8 Aug. 201/2 The aisles and transepts are piled high with sheaves of crankshafts and bunches of calibrators.

caliche (kăli·tʃe). *Min.* [Amer. Sp., f. Sp. *caliche* pebble in a brick, flake of lime.] In arid areas of North and South America, any of various mineral deposits (esp. native Chile saltpetre) containing sodium nitrate, calcium carbonate, or other salts.
1858 Simmonds *Dict. Trade*, *Caliche*, a name for nitrate of soda found in Peru. **1883** R. Haldane *Workshop Rec.* 349/2 Iodine occurs in caliche or raw nitrate deposit, as iodate of sodium. **1892** E. S. Dana *J. D. Dana's Syst. Min.* (ed. 6) 871 In the district of Tarapaca, northern Chili,..the dry pampa..is covered with beds of this salt (caliche) several feet in thickness. **1939** W. H. Twenhofel *Princ. Sedimentation* ix. 342 Caliche..is a deposit of calcium carbonate and other salts made in semiarid regions.

caliciform, *a.* Add: **b.** *Archæol.* Resembling a calyx.
1902 [see *bell-beaker* (*bell *sb.*¹ 11 c)]. **1933** W. F. Albright in *Ann. Amer. Sch. Oriental Res.* 1931–32 XIII. 66 The caliciform vessels have a rudimentary stem. **1949** —— *Archaeol. Palestine* v. 80 This ceramic culture came down from Syria, where is it known as 'caliciform' because of the tendency of potters to prefer the calyx form of vases to any other.

calico. Add: **3. b.** Coloured in a way suggestive of printed calico; variegated, piebald. Chiefly of horses. Also as *sb.*, a calico horse. *U.S.*
1807 W. Irving *Salmag.* 24 Nov. 372 Bantering nature fairly out of countenance—representing her tricked out in all the tawdry finery of copper skies, purple rivers, calico rocks, red grass, [etc.]. **1812** —— *Hist. N.Y.* (ed. 2) II. vii. iii. 182 Behold..Van Corlear, mounted on a..calico mare. *a* **1861** T. Winthrop *Canoe & Saddle* (1883) x. 144 A hundred horses, roans, calicos..blacks and whites. **1878** B. F. Taylor *Between Gates* 207 There would be scant room for the calico horses to canter. **1954** J. Potts *Go, Lovely Rose* (1955) ix. 60 Havelka's calico cat..was taking a fastidious stroll.
4. calico-bush, the American mountain laurel (*Kalmia latifolia*).
1814 Pursh *Flora Amer. Sept.* I. 297 *Kalmia latifolia*..called Laurel or in the mountains Callico-bush. **1829** Loudon *Encycl. Plants* 356 *Kalmia latifolia*, Calico-bush. **1914** L. H. Bailey *Standard Cycl. Hort.* II. 627/2 Calico bush: *Kalmia*. **1969** Hay & Synge *Dict. Garden Plants* 314/1 *Kalmia* (*Ericaceae*) *latifolia*. Mountain Laurel, Calico Bush.

California (kælifǭ·ɹniă). [The name of the state on the Pacific Coast of North America.]
†**1.** Money. *Obs. slang.*
1851 *London at Table* 1. 5 Some 'California', as the fast young men of the day term 'money', is necessary for these houses. **1852** C. M. Yonge *2 Guardians* xi. 183 You had plenty of money... I know you keep California in your pocket.
2. Used *attrib.*, esp. in the names of various species of animals and plants; = *Californian *a.*
1831 F. W. Beechey *Voyage to Pacific* II. 81 The California quail (*tetrao virginianus*). **1840** R. H. Dana *Bef. Mast* xiv, A few hides were brought down, which we carried off in the California style. *Ibid.*, Telling us that it was 'California fashion' to carry two on the head at a time. **1846** in *Calif. Hist. Soc. Q.* V. 380, I begin to conclude that californea Horses are not a hardy race of animals. **1874** Coues *Birds of Northwest* 363 Ferrugineous Buzzard, or California Squirrel Hawk. **1881** *Appleton's Ann. Cycl.* XII. 312/2 *Ceanothus thyrsiflorus* is a small tree producing an abundance of light blue flowers, and known as the California lilac. **1891** F. F. Victor *Atlantis Arisen* 225 The California poppy, *Eschscholtzia*, is found in Southern Oregon. **1911** *Daily Colonist* (Victoria, B.C.) 15 Apr. 6/3 A California grey whale, the first to be captured off the Island coast, has been brought in to the Sechart whaling station. **1926** *Ibid.* 16 July 4/4 My attention has been drawn recently to some California poppies, growing on the roadside along the Malahat [Road]. **1927** C. M. Russell *Trails plowed Under* 3 A high-forked, full-stamped California saddle. **1967** 'E. Queen' *Face to Face* xxi. 96 A bottle of undistinguished California burgundy.

Californian (kælifǭ·ɹniăn), *a.* and *sb.* [f. prec.]
A. *adj.* Of or belonging to, native or peculiar to, California; esp. in the names of species of birds, beasts, and plants.
1785 Cutler in *Mem. Amer. Acad. Arts & Sci.* I. 400 The Spaniards are said to have procured from the Californian Indians, the art of dyeing the best black. **1801** Latham *Gen. Synop. Birds* Suppl. II. 281 Californian Quail. **1839** A. Forbes *California* 192 Many of the [Indian] baskets are ornamented with..the black crest feathers of the Californian partridge. *Ibid.* 251 The wheels of the Californian ox-cart.. are of a most singular construction. **1845** J. C. Frémont *Exped.* 245 Some of the banks being absolutely golden with the Californian poppy, (*eschscholtzia crocea*). **1884** R. L. Stevenson *Silverado Squatters* iii. 34, I was interested in Californian wine.., still in the experimental stage. *c* **1900** in A. Davis *Package & Print* (1967) pl. 185 (*label*) Extract Eonia Californian Poppy. J. & E. Atkinson London. **1907** *Yesterday's Shopping* (1969) 34/1 Soaps..Californian Poppy —box of 3 tablets 3/3. **1926** A. Huxley *Jesting Pilate* iv. 260 Californian figs and oranges. **1938** G. Greene *Brighton Rock* i. iii. 46 A handkerchief scented with Californian Poppy. **1956** *N.Z. Timber Jrnl.* July 54/2 Californian redwood, *Sequoia sempervirens*.
B. *sb.* A native or inhabitant of California.
1789 Morse *Amer. Geogr.* 479 The characteristics of the Californian, are stupidity and insensibility. **1840** R. H. Dana *Bef. Mast* xiv, We saw three men..dressed partly like sailors and partly like Californians. **1893** K. A. Sanborn *Truthful Woman in S. Calif.* ii. 18 Some one says that Californians 'irrigate, cultivate, and exaggerate'. **1969** I. Kemp *Brit. G.I. in Vietnam* iii. 48 A soft-spoken Californian of medium build with hair already going a little grey.

californite (kælifǭ·ɹnəit). *Min.* [f. *California + -ite¹ 2 b.] A compact form of green vesuvianite found in California.
1903 G. F. Kunz in *Amer. Jrnl. Sci.* XVI. 397 (heading) *Californite* (*Vesuvianite*),—a new ornamental stone,.. recently [discovered] in California. *Ibid.* 398 This interesting mineral..is a form of vesuvianite distinctive enough to warrant giving it a special variety name,..I therefore propose the name 'Californite' for this massive, translucent mineral. **1955** Brown & Dey *India's Min. Wealth* (ed. 3) 633 A compact green variety [*sc.* of idocrase] known as californite resembles jade.

californium (kælifǭ·ɹniŏm). [f. the University of *California*, where it was discovered, + -ium.] A transuranic radioactive element; symbol Cf; atomic number 98.
1950 in *Amer. Speech* (1951) XXVI. 291/2 Scientists have created a new element—Californium—carrying the heaviest atom ever known, it was announced Friday. **1954** *Sci. News* XXXII. 109 Through the absorption of 13 successive neutrons and 4 emissions of electrons (beta decay), it [*sc.* plutonium-239] was converted into californium-252, which was separated chemically. **1969** *Times* 22 Apr. 6/6 Element 104 has successfully been synthesized by bombarding the artificial element californium.

calina (kăli·na). [Sp.] (See quot. 1887.)
1887 *Encycl. Brit.* XXII. 296/2 In July and August the plains of New Castile..are sunburnt wastes;..the atmosphere is filled with a fine dust, producing a haze known as *calina*. **1927** Kendrew *Climates of Continents* (ed. 2) 244.

calinda (kăli·ndă). [American Sp.] An American Negro dance, found in Latin America and the southern United States.
1763 tr. *Le Page du Pratz's Hist. Louisiana* I. iv. iv. 271 Under pretence of *Calinda* or the dance, they [*sc.* the Negroes] sometimes get together to the number of three or four hundred. **1880** G. W. Cable *Grandissimes* 121 There our lately met *marchande*..led the ancient Calinda dance. **1958** P. Oliver in P. Gammond *Decca Bk. Jazz* i. 22 Dances such as the bamboula and the calinda, which are still to be found in the West Indies.

caliology (kæliǫ·lŏdʒi). [f. Gr. καλιά wooden dwelling, hut, nest + -ology.] That department of ornithology which is concerned with birds' nests. Hence **caliolo·gical** *a.*
1875 *Encycl. Brit.* III. 772 There are not many works on nidification, for 'Caliology' or the study of nests has hardly been deemed a distinct branch of the science. **1884** Coues *N. Amer. Birds* (ed. 2) 227 One of the most delightful departments of ornithology, called caliology. **1902** C. Dixon *Birds' Nests* Introd. 4 The late J. G. Woods' popular treatment of birds' nests..practically exhausts the special literature of caliology.

calisthenic, -ics, variants of Callisthenic *a.*, Callisthenics *sb. pl.*

calix. Add: *Gr. Antiq.* = Cylix.
1849 A. Rich *Illustr. Comp. Lat. Dict.* **1912** H. B. Walters in *Catal. Gr. Vases Brit. Mus.* I. ii. 228.

call, *v.* Add: **1. e.** (*d*) in *Bridge* (*trans.* and *intr.*), to bid.
1906 A. Sutro *Walls of Jericho* ii, *Duchess.* I call no trumps. *Tiny.* Shall I play to no trumps, partner? **1923** *Harmsworth's Househ. Encycl.* I. 532/1 The best that you can do is to call one of the suit you want led in case B goes no trumps. **1928** A. Waugh *Nor many Waters* ii. 74, I called, 'Three No Trumps.' And the man on my left doubled. **1958** *Listener* 11 Dec. 1012/1 West was the dealer and the opponents did not call.
f. To make a telephone call. (Cf. sense *4 n.)
1882 J. E. K. *The Telephone* 19 The means by which the Exchange operator knows which subscriber is calling is very ingenious and very simple. *Ibid.* 38 An anxious mother.. called through the Exchange for the doctor. **1928** Hecht & MacArthur *Front Page* ii. 72 *Endicott* (*into phone*): Endicott calling. Gimme a rewrite man.
4. b. *spec.* To summon to another world.
1526 [in Dict.]. **1830** [see sense 4 a in Dict.]. **1886** *F. Leslie's Pop. Monthly* XXI. 611/2 All the doctors in Christendom..can't save him. He's called.
m. *Cricket.* Of an umpire: to declare (a bowler) to have bowled a 'no ball'; to declare (a delivery) illegal.
1850 *Nottingham Rev.* 11 Oct. 6/5 Regarding height in delivery as to warrant the umpire in calling them. **1862** *Baily's Mag.* Oct. 201 He was getting high, and if he did not keep it down he should have to 'call' him. **1902** A. Shaw *Reminisc.* xvi. 168 Will Oscroft..bowled a wicket from outside the return crease, and as the umpire did not 'call' him he went down that path again..every wicket taken being illegal. **1959** *Oxford Mail* 2 Feb. 8/6 His drag of the back foot was most noticeable but umpire McInnes, though watching him closely, did not 'call' him.
n. To communicate with (a person) by radio or telephone. (Cf. 1 f above.)
1889 'Mark Twain' *Connecticut Yankee* 477 In the telegraphic line.. I said..'Lively, now, call Camelot'. **1921** [see *call-sign* s.v. *call *sb.* 15]. **1932** E. Wallace *When Gangs came to London* xxii. 190 The operator..said she would 'call her back'. **1936** N. Coward *Hands across Sea* 23 *Clare* (at telephone):..All right, darling—call me in the morning. **1968** *Globe & Mail* (Toronto) 17 Feb. 45 (Advt.), For more information, call Jud Newell 485-9191.
o. To summon (an actor, etc.) to be ready to appear on the stage; to inform an actor, etc., of (the part of a performance that is about to commence). Cf. *call *sb.* 6 h.
1938 N. Coward *Operette* ii. vii. 127 *Duggie.* Overture beginners, please... *Dora.* He's calling the overture. **1962** *Listener* 25 Jan. 162/1 He was called for the last act.
17. e. *to call out of one's name*, to address by a name other than the true one.
1848 Dickens *Dombey* ii. 12 Perhaps if she was to be called out of her name, it would be considered in the wages. **1885** C. M. Yonge *Two Sides of Shield* I. iii. 38 She had rather not be called out of her name.
f. *to call it a day*: see *day *sb.* 20 b; so *to call it a night.*
1934 'J. Spenser' *Limey breaks In* xi. 180 There were at least sixty pounds there, and I quickly collared the lot and called it a night. **1968** K. Weatherly *Roo Shooter* 55 At length, when he had about half a ton of meat on the Rover, he decided to call it a night.
g. *to call one's* (or *the*) *bluff*: see *bluff *sb.*² 3.
23. call on or upon. **c.** (*b*) To require or urge (a horse) to exert itself further. Cf. *ask *v.* 2 b.
1850 'H. Hieover' *Pract. Horsemanship* viii. 163 In the last few strides [of a race], where sudden and increased exertion is called for, and the horse is, in technical phrase, 'called upon'. **1886** Ld. Suffolk & W. G. Craven *Racing* v. 86 Romanus is seen to..lose his pace. Wood calls on him without mending matters. **1894** Custance *Riding Recoll.* xi. 162 When I called on the gallant animal for the final effort, he got up and won.
26. call back. e. *intr.* To revert to type; = *throw back*, Throw *v.* 38 d.
1853 *Jrnl. R. Agric. Soc.* XIV. i. 112 Isolated individuals appear, which, in the phraseology of breeders, 'call back' to their more remote progenitors. **1855** *Ibid.* XVI. i. 22 The offspring are said..to call back to their grand parents.
27. call down. e. To rate or reprove; to challenge sharply. *colloq.*
1896 Ade *Artie* iii. 27, I didn't want to call her down. **1897** Kipling *Capt. Courageous* ix. 196 An unsatisfied dough-faced youth who took delight in 'calling down the old man' and reducing his mother to tears. **1904** F. Lynde *Grafters* v. 58 He..so far lost his temper as to get himself called down by the judge. *a* **1910** 'O. Henry' *Trimmed Lamp* (1916) 209 When Fernando wanted to give me several thousand dollars for my trousseau he called him down something awful. **1940** H. G. Wells *Babes in Darkling Wood* i. i. 35 It's all very well for you to call it down, young lady, and criticise it.
30. call off. c. *trans.* To cancel (an engagement, etc.), draw back from (an undertaking). Also *intr.*
1888 Mrs. Oliphant *Second Son* v, Why, in the name of all that's idiotic, do you call off now, and disappoint her.. and defy me? **1900** Ade *More Fables* 158 He was about to Call Off the Vestry Meeting, the Dinner, and all other Engagements for a Week to come. **1902** *Daily Chron.* 17 Oct. 5/3 The delegates of the Miners' Convention must first pass a vote upon the question of calling off the strike. **1927** *Observer* 14 Aug. 6 That he would have been profoundly relieved if the whole expedition had been called off. **1952** V. Gollancz *My Dear Timothy* 388 But I am almost certain that, war, or no war, I should have called it off.
32. call out. b. *spec.* To summon to active or permanent service in a campaign or in a state of emergency.
1779 *Digest of Militia Laws* 112 Every such person, having served in the Militia when called out into actual service. **1849** Macaulay *Hist. Eng.* I. iii. 291 When the trainbands were called out against an enemy. **1853** Bunn *Old Eng.* II. 53 Shot by the military, who had been called out for the occasion. **1890** *Chambers's Jrnl.* 5 July 423/1 The fog-signalmen..are often called out for a night's 'fogging' just as they have finished a hard day's work. **1921** *Act* 11 & 12 *Geo. V* c. 15 § 9 Where..a man of the Naval Reserves..is called into actual service or called out for permanent service ..on an occasion of great emergency.
e. To summon (workers) to strike. orig. *U.S.*
1895 H. P. Robinson *Men born Equal* 284 Ugly threats, moreover, were being made by the strikers that the members of other labor organizations would be 'called out'. **1947** *Times* 1 May 5/2 The chairman..talked of launching a national strike and of calling out the seamen, the road transport workers, and the engineers. **1957** *Screen Printer & Display Producer* July 1/1 It was never the intention of the Union to call all its members out.
35. call up. f. To call to battle; *spec.* to summon to military service. Cf. *call-up.
[? **1684** in *Roxburghe Ballads* (1897) VIII. 453 Come fill up my cup, come fill up my Can; come saddle my horse and call up my man.] **1827** Scott *Bonnie Dundee* in *Lit. Gaz.* 8 Dec. 786 Come saddle my horses and call up my men. **1857** *Blackw. Mag.* LXXXII. 281/2 The landwehr of the first band are liable..in the event of war, to be called up. **1899** Atteridge *Wars of Nineties* 550/2 Thus Japan had an army of nearly 70,000 men on a peace footing, which by calling up the reserves could be expanded into a war force of more than a quarter of a million. **1914** *Eng. Rev.* Sept. 258 We saw young Belgians crowded in trains *en route* for the front, men who were 'called up' against the enemy.
g. To summon up (Summon *v.* 7).
1889 *Illustrations, a Pict. Rev.* 143 Calling up whatever remnants of valour were left to me,..I advanced.
h. To summon (a person) on the telephone.
1898 [implied in *caller-up]. **1900** [see Phone *sb.*² and *v.*] *a* **1910** 'O. Henry' *Strictly Business* (1917) ii. 29 Kelley went to the nearest telephone booth and called up McCrary's café. **1921** G. B. Shaw *Back to Methuselah* iii. 137 Engaged! Who is she calling up now?

call, *sb.* Add: **1. d.** A summons or communication by telephone; a telephone conversation. (See also attrib. uses.)
1878 *Design & Work* IV. 306/3 Apparatus..to enable the sound of the voice while singing to be heard all over a room, and which I use as a 'call', instead of an electric bell. **1879** G. B. Prescott *Speaking Telephone* i. 23 It being necessary to keep the vibratory bells at each station in circuits, in order that the calls may be heard. **1882** J. E. K. *The Telephone* 19 The number of calls made upon the Exchange clerks. **1884** *Routledge's Every Boy's Ann.* 199/1 Before we follow the series of operations forming a complete call, let us examine the system of telephones used in the Broadway Office. This..allows these batteries to be used for the calls to the subscribers by means of ordinary electric bells. **1899** *Post Office Guide* July 533 This deposit is refunded if the call is not extended. **1944** 'N. Shute' *Pastoral* ii. 22 Give me twopence for the call, and I'll give him a tinkle in the morning. **1953** R. Lehmann *Echoing Grove* 290 There was a call for you about a quarter of an hour ago. From London.
6. c. (Earlier and later examples.)
1825 *News* 4 Sept. 286/1 Mr. Kean came forward, and addressed the audience...'It is impossible to withstand so gratifying a call.' **1921** 'Ian Hay' *Willing Horse* viii. 129 Seven *legitimate* calls after the first act.
h. Substitute for def.: A notification, either verbal or written, summoning theatrical, etc., performers to attend (at a rehearsal or during a performance). (Earlier and later examples.) Cf. *call *v.* 4 o.

1780 G. Colman *Manager in Distress* 22 They were all drest, sitting ready for the call in the Green Room. **1933** P. Godfrey *Back-Stage* i. 16 The call-boy leaves the stage to cry: 'Overture beginners.'..As the call reaches them they take one last approving glance at their reflections. *Ibid.* 17 Each 'call' is plainly marked in the prompt book sufficiently ahead of the actual entrance to allow the actor concerned to receive an individual visit from the call-boy. **1955** *Times* 19 Aug. 4/1 The schedule for to-morrow's work arrived with all its facts and figures about the crew call, the set, [etc.].

i. Also in poker (orig. *U.S.*), 'a demand for a show-down; the show-down itself' (*Cent. Dict.*, 1889); in Bridge, a bid (*sb. 2), pass, double, or redouble.

1850 'M. Tensas' *Louisiana Swamp Doctor* 123 A gambler who has staked his whole pile, and found at the call that he has been bluffing up against a greenhorn with 'three white aces'. **1853** J. G. Baldwin *Flush Times Alabama* 8 A negro *ante* and twenty on the call, was moderate playing. **1885** *Encycl. Brit.* XIX. 283/1 When all the stakes are thus equal, it becomes a *call*. **1923** *Harmsworth's Househ. Encycl.* I. 531/1 Doubling..reopens the bidding and gives each player the chance of a fresh call. **1928** A. Waugh *Nor many Waters* ii. 74 The man on my right suddenly cut into the bidding. 'Well, it's my call,' he said. 'Three spades.'

l. *Cricket.* A shouted direction by a batsman to his partner either to run or to remain in his crease.

1854 J. Pycroft *Cricket Field* (ed. 2) x. 212 Let men run by some call; mere beckoning..leads to fatal errors...'Yes', 'no', or 'run', 'stop' are the words. **1898** G. Giffen *With Bat & Ball* viii. 103 Murdoch..was run out through a bad call of Tom Garrett's. **1955** *Times* 15 July 3/3 Fellows Smith failed to scramble home after being sent back for a ridiculous call.

m. A signification of impending death; *spec.* in phr. *to get one's* (or *the*) *call*, to die, be about to die. *dial.*

1884 D. Grant *Lays & Legends* 172 His wife, wi' his grainin' sae weary, Was fain to have seen him awa'...Hersel' was the first gat the ca'. **1889** Tennyson *Crossing the Bar* in *Compl. Wks.* (1896) 894/1 Sunset and evening star, And one clear call for me! **1892** C. M. Yonge *Old Woman's Outlook* ii. 39 There came a little robin about the door. We knowed it was a call, and we thought it must be for granfer; so we put 'im to bed. **1915** J. Buchan *Salute to Adv.* xxiii. 328 His breath laboured, and there was pain in his eyes. 'I've got my call,' he said faintly.

7. b. A need to defecate or urinate; freq. in phr. *call of nature.* Colloq. phr. *to pay a call*, to go to the lavatory.

1761 Sterne *Tr. Shandy* IV. 43 A city..who neither eat, or drank,..or hearkned to the calls either of religion or nature for seven and twenty days. **1852** in *Tailor & Cutter* (1966) 14 Oct. 111/1 The calls of Nature are permitted and Clerical Staff may use the garden below the second gate. **1858** G. Rawlinson tr. *Herodotus's Hist.* I. cxxxiii. 274 To vomit or obey natural calls in the presence of another, is forbidden among them. **1926** Galsworthy *Escape* I. i, *Warder*...Where's your mate? *Fellow Convict.* 'Ad a call, sir...Went over to that wall. **1926** A. Huxley *Two or Three Graces* 249 He would suggest dropping in at the Monico, pretext a call of nature. **1939** F. Thompson *Lark Rise* 56 A type of complaint that could always be countered by pleading, 'Call o' Nature, please, sir.' **1959** 'O. Mills' *Stairway to Murder* xv. 155 He'd had quite a bit of beer, and I'd an idea he might have to get up and pay a call. **1961** J. McCabe *Laurel & Hardy* (1962) i. 33 On the way there we became aware of Nature's urgent call.

15. call-bird, (*b*) *transf.*, a lure used by tradesmen to promote sales; **call-board** *Theatr.*, a notice board; **call-box,** a telephone-booth; **call-boy,** add: **d.** a messenger boy, *spec.* = *call-man*; **e.** a device for transmitting messages and orders; **call-button,** a push-button or other device for ringing a call-bell or alarm; **call-disc,** a disc indicating the provenance of a telephone call-signal; **call-down** *U.S.*, a scathing rebuke, an abusive tirade (cf. *Call *v.* 27 e); **call-girl** (orig. *U.S.*), a prostitute who makes appointments by telephone; also *attrib.*; **call-house** *U.S.*, a brothel; *spec.* a house occupied or used by call-girls; **call-letter,** a letter forming part of a call-sign; **call-loan** (earlier *U.S.* example); **call-man,** a man who calls up the members of a fire-brigade for duty; **call number** (orig. *U.S.*), the press-mark or shelf-mark of a book; **call-office,** (*a*) a telephone office or central station where the call signals are received and where the connections necessary to intercommunication are made; (*b*) = *call-box*; **call-on** (see quot. 1958); also *attrib.*; **call-over,** (also) in betting parlance, a calling over or reading aloud of a list of prices; **call-room,** (*a*) = *call-box*; (*b*) *U.S.*, the room in a Stock Exchange where calls are announced; **call-sheet** *Theatr.* and *Cinemat.* (see quot. 1959); **call-sign,** a conventional sign used to indicate the identity of the sender of a wireless message, a transmitting station, etc.; **call signal,** (*a*) a signal for calling up on the telephone; (*b*) = *call-sign*; **call slip** *U.S.*, the slip on which a reader or borrower in a library enters details of the book or books required.

1901 *Farm, Field & Fireside* 13 Dec. 358/1 Frequently a man buys two or three couples of English spring chickens, and then has a case of *eighty* 'Russians' sent in. He uses the English as 'call birds' on his shop front and those few Surreys are the means of disposing of his whole case of Russians. **1958** *Observer* 5 Jan. 7/4 Certain shops are studying..women's response to the 'call-bird'. That is the name they give to extraordinary bargains placed in the windows as lures for the sales-minded. **1886** J. B. Howe *Cosmopolitan Actor* 134, I saw the cast of 'Colleen Bawn' still on the call-board. **1901** C. Morris *Life on Stage* xxiii. 277 On the wall..there hung that shallow, glass-covered frame or cabinet called, variously, 'the call-board', 'the call-case', or even the 'call-box'. **1885** *Electrician* 13 Nov. 10/2 At intervals along the streets..we shall have 'call-boxes', by means of which conversation, by telephone, may be carried on with *any* desired person. **1893** *Cassell's Fam. Mag.* Apr. 338/2 Thus it was that I first became introduced to the call-box system. **1951** 'N. Shute' *Round Bend* 26 Her dad went down the road to the call-box and telephoned the police. **1848** Thackeray *Van. Fair* li. 453 The gigantic footmen..were billeted off in the neighbouring public-houses, whence, when they were wanted, call-boys summoned them. **1868** *Cassell's Mag.* 15 Aug. 256/1 Some of our great iron-clads have been furnished with electrical call-boys. **1887** *16th Ann. Rep. Oxf. Volunteer Fire Brigade* 9 The improved method of calling by electric bells, instead of by call boys. **1878** *Telegr. Jrnl.* VI. 113/2 When the attention of either station has been called on its bells by pressing the call-button. **1966** J. Chamier *Cannonball* vi. 52, I..pressed the call button for the lift. **1884** *Routledge's Every Boy's Ann.* 124/2 The circles in the figure are the little windows at which the call-disc makes its appearance. **1901** 'H. McHugh' *John Henry* i. 11 The four-flush call-down makes you back-pedal. **1917** Wodehouse *Man with Two Left Feet* 121 The feller that tries to get gay with me is going to get a call-down that'll make him holler for his winter overcoat. **1940** in *Amer. Speech* (1942) XVII. 204/1 Call Girls Die Young. **1951** E. Kefauver *Crime in America* (1952) xvii. 193 The hotel..has shown up in some of the call-girl operations. **1957** *Economist* 7 Sept. 736/2 If the streets are cleared, however, there is certain to be an extension of prostitution off them—a growth of the 'call-girl' system. **1929** W. R. Burnett *Little Caesar* vii. v. 270 Sometimes at night he would go to one of the call-houses on a nearby street and spend a couple of hours with one of the women. **1936** J. Dos Passos *Big Money* 340 It wasn't a hotel or a callhouse. *a* **1940** F. Scott Fitzgerald *Last Tycoon* (1949) v. 106 She went to what you call call-houses. **1913** *Year-bk. Wireless Telegr.* 286 Combinations of 'call letters' which are in turn allotted to ship and land stations. **1852** *N.Y. Wkly. Tribune* 10 Apr. 8/4 The rates are 4 @ 4½ on Government Stocks, 5 @ 6 on general call loans. **1905** *Westm. Gaz.* 19 Apr. 8/1 When instructions were sent to call-man Hills to order out all the fogmen. **1876** *Public Libraries U.S.* (Bureau of Educ.) I. 626 Determining the character of any book simply from its call number. **1964** N. R. Ker *Medieval Libraries* (ed. 2) 225 No entry follows the call number of the book if the inscription is in the common *ex dono* or *dedit* form. **1885** *Electrician* XIV. 102/1 Giving increased facilities to the public in the shape of 'call' offices. **1895** *Daily News* 13 Sept. 5/3 The charge for conversations..within a distance of 25 kilometres from the call-office..has so far been 5d. **1899** *Post Office Guide* July 533 The following additional charges..also apply to conversations between call offices. **1955** *Times* 28 June 8/3 There was a steady trickle of men arriving for the call-on [at the docks]. *Ibid.* 4 July 8/2 It was decided to hold another meeting..before the normal docks call-on time to-day. **1958** *Ibid.* 7 Mar. 7/1 Dockers..having to attend a 'call-on' twice a day...They present themselves for work at 7.45 a.m. and again at 12.45 p.m. **1927** *Daily Tel.* 8 Mar. 15/1 That all such bets transacted at the club's 'calls-over' would be free of tax to the backer. **1882** J. E. K. *The Telephone* 33 Call-rooms have been established in different parts of London, in which are placed telephones connected with the Exchanges. **1886** *Harper's Mag.* July 213/1 The Call Room daily presents an impressive spectacle of the traffic in grain. **1955** *Times* 25 Aug. 12/4 The rehearsal call-sheet lists animals side by side with ladies and gentlemen of the chorus. **1959** W. S. Sharps *Dict. Cinemat.* 82/2 *Call sheet*, the itemised schedule of calls for performers and production personnel. **1919** *Times* (weekly ed.) 17 Jan., She signalled the letters of her 'call-sign' and our wireless picked up her message. **1921** *Glasgow Herald* 4 Feb. 9 Shortly after two o'clock she was called, but did not answer her call-sign. **1884** *Routledge's Every Boy's Ann.* 120/2 This system can make a Bell telephone..speak loud enough to be heard throughout a room; and it comprises besides a call-signal within itself. **1912** in *Year-bk. Wireless Telegr.* (1913) 52 The call signals must be differentiated from one another, and each one must consist of a group of three letters. **1881** *Amer. Library Assoc. Papers* 4/1 All books are to be asked for on call-slips made out from the catalog.

calla (kæ·lă). *Bot.* [mod.L. *calla* (C. Linnæus *Hortus Cliffortianus* (1737) 435).] Substitute for def.: **1.** A monotypic genus of aquatic plants, native to northern Europe and eastern North America, and belonging to the family Araceæ.

1816 *Curtis's Bot. Mag.* 1831 Marsh Calla...Linnæus applied that [*sc.* the name] of Calla, a name borrowed from Pliny. **1845** A. Wood *Class-bk. Bot.* 364 *C. Palustris.* Northern Calla. **1885** *Outing* (U.S.) Nov. 178/2 An interesting plant is the wild Calla, growing in cold, wet places. **1968** R. Hay *Gardening Year* 174/2 Calla (bog arum)..bears white lily flowers in spring.

2. A popular name for some plants of the genera *Zantedeschia* (esp. *Z. æthiopica*) and *Arum*, both members of the family Araceæ.

1805 *Curtis's Bot. Mag.* 832 Ethiopian Calla. **1870, 1884** [in Dict.]. **1949** L. H. Bailey *Man. Cult. Plants* (rev. ed.) 187 Arum, L. A dozen species now recognized..a few occasionally grown for curiosity, often under the name Calla. **1951** *Dict. Gardening* (R. Hort. Soc.) IV. 2302/2 Zantedeschia...In gardens these plants are often known as Callas or Richardias. Calla is a different genus, typified by *Calla palustris.*

callable (kǭ·lăb'l), *a.* [f. Call *v.* + -able.] That may be called (or called in, called upon, etc.).

1826 *Examiner* 673/2 Judges called, or callable, rural. **1889** *Sat. Rev.* 16 Mar. 313/2 Without a cash reserve, or without callable loans with bill-brokers. **1892** *Ibid.* 11 June 678/2 As there is no callable capital, the shareholders cannot suffer more than the loss of the capital paid up. **1959** *Economist* 18 Apr. 237/2 With $450 million of the amount for hard loans remaining callable as backing for ordinary bond issues.

‖**callalloo, callaloo, callalou.** Also cullaloo. = Calalu; also, a soup or stew made with this.

1810 F. Cuming *Tour Western Country* 297 Mr. Green made me observe..the cullaloo or Indian Kail [near Natchez, Mississippi]. **1892** J. C. Harris *On Plantation* 122 There was Callalou—a mixture of collards, poke salad, and turnip greens boiled for dinner and fried over for supper. **1929** W. J. Locke *Ancestor Jorico* viii. 107 We were given..callaloo, which is an apotheosis of the American Okra soup. **1953** G. Lamming *In Castle of my Skin* xiv. 273 A vegetable muddle called callalloo cooked with crab. **1969** *Daily Tel.* 11 Jan. 14/1 Calalu (sometimes spelt 'callaloo'), a green vegetable like spinach, makes wonderful soup—to bring tears to ex-patriate Jamaican eyes just mention calalu.

callee (kǭlī·). [f. Call *v.* + -ee.] One who is called or called upon.

1872 Butler *Erewhon* xxi. 197 The callee would have been deaf to the caller. **1883** *Cambridge Staircase* v. 73 Our callee suggested that there were more comfortable seats. **1959** M. Chamberlin *Dear Friends* (1960) iv. 29 It is the common practice of the caller to demand the number of the callee immediately.

caller, *sb.* Add: **1. e.** One who announces the changes of steps during a dance. Also *caller-out.* orig. and chiefly *U.S.*

1882 *Century Mag.* Oct. 878/2 The 'caller-out'..not only calls out the figures, but explains them at length to the ignorant, sometimes accompanying them through the performance. **1931** *Amer. Speech* VII. 50 To the 'fiddlin'' for the square dances the 'caller' sing-songs: All to your places, straighten up your faces, Join eight hands and circle left. **1938** *Times* 10 Jan. 10/4 The 'caller'..is a figure still prominent in American 'play-parties'. **1966** *Melody Maker* 7 May 18/1 (Advt.), Beckenham Ballrooms are interested in commencing a Country Style evening every Tuesday. They need a caller and musicians.

f. One who makes a call on a telephone. Also *caller-up.*

1898 *Daily News* 6 June 6/2 The caller-up knowing whether he is through or not by the ringing or not ringing of his bell. **1899** *Post Office Guide* July 533 If a deposit has been made by the caller in the first instance to cover the fee for the second period of three minutes. **1954** K. Amis *Lucky Jim* 95 Your second three minutes are up, caller. **1959** [see *Callee].

g. A man or boy employed to knock up men for duty, esp. the members of a railway staff. Also *caller-up.*

1898 H. E. Hamblen *Gen. Manager's Story* 72 The caller made his rounds with orders to call the first man he found off duty. **1899** R. Whiteing *No. 5 John St.* vi. 47 The caller-up begins his rounds with the dawn. **1935** A. J. Cronin *Stars look Down* I. viii. 65 Next morning the caller woke him, at two o'clock he was in the pit working the early fore shift.

calliard (kæ·liăɹd). *local.* [Of doubtful origin; perh. connected with F. *caillou* pebble, and so ultimately with L. *calculus.* But cf. the later Galliard *sb.*²] A hard, smooth, flinty gritstone.

1781 J. Hutton *Tour to Caves* (ed. 2) Gloss. (1873), *Callierd*, an hard stone. *a* **1835** J. Phillips *Geol.* in *Encycl. Metrop.* (1845) VI. 587/2 Some less regular sandstone beds, called 'Cankstone', approach very nearly to the nature of the ganister or calliard rocks of the coal strata. **1865** Page *Handbk. Geol. Terms*, *Calliard*..., a local name for any hard siliceous stone; often applied by English miners and quarrymen to beds of cherty or siliceous limestone. **1876** Woodward *Geol. Eng. & Wales* 81 Some of the beds contain gritstone or greywacke, provincially called 'calliard'.

calligraphic, *a.* Add: **b.** Of or pertaining to calligraphy (sense *4); having beauty of line.

1928 *Observer* 1 July 14/1 Swift calligraphic strokes which invest the picture with..vitality. **1930** *Times Lit. Suppl.* 9 Jan. 25/3 Leonardo displays his extraordinary gift of enclosing a volume by a flowing and calligraphic line. *Ibid.*, The calligraphic beauty of line. **1962** *Listener* 29 Nov. 909/2 These brilliantly coloured, calligraphic paintings.

calligraphy. Add: **4.** The brush-work (of a painting); beauty of line (of a painting, drawing, etc.).

1928 *Observer* 1 July 14/1 Gainsborough adopted the elegant calligraphy, the loose, flickering touch. **1962** *Listener* 2 Aug. 178/3 The three studies of 'Woman in Rocking Chair' show Kline's characteristic bold calligraphy in the process of being formed.

calling, *vbl. sb.* Add: **8.** *calling-down* (orig. *U.S.*), a rebuke, upbraiding; *calling-up* (see Call *v.* 35).

1877 *Design & Work* III. 713/3 For calling-up purposes I have a bell in the bedroom of a very drowsy domestic. **1890** *Peel City Guardian* 12 July 6/1 The calling up of constables on probation. **1901** G. Ade *40 Mod. Fables* 55 He would give the National Administration a sharp Calling Down every few Days. **1902** *Electr. & Magn.* xviii. 226

(Govt. Milit. Bk.) Calling-up may be done by means of an ordinary bell and battery. **1918** *Act 8 Geo. V* c. 5 § 4 (2) With a view to preventing..the calling up of himself..for any form of military service. **1930** Pertwee *Pursuit* I. xiii. 63 N.C.O.'s and other ranks were within hearing of the calling-down Frost was receiving. **1936** M. Plowman *Faith called Pacifism* 94 When the calling-up paper comes along again, under the Conscription Act.

III. (Further examples.) Freq. in the sense 'visiting'.

1814 T. Creevey *Let.* 14 June (1963) 109, I called on her this morning, and saw some very different names in her calling book from what I had ever seen before. **1877** E. S. Phelps *Story of Avis* 397 Lest society strike him from her calling-list. **1888** Kipling *Plain Tales from Hills* 150 [He] put on his calling-clothes and called on the ladies of the Station. **1893** *Westm. Gaz.* 19 Apr. 6/2 There will be four calling stations. **1896** *Daily News* 24 Oct. 5/6 A note from him, written by himself on his calling card, placed the matter beyond all doubt. **1908** *Westm. Gaz.* 27 Apr. 10/2 To ensure regularity in arriving at the various calling ports.

calliope. Add: (Earlier and later examples.)

1858 J. Cook *Let.* 19 Sept. (1946) 45 On board the *Armenia*..is a Calliope, or an instrument resembling an organ & played in connection with the engine. **1908** E. C. Booth *Cliff End* v. 35 The calliope at the Crystal Palace would just about begin to realise our musical aspirations—with steam drum attachment. **1923** V. Lindsay *Coll. Poems* 118 Proud men..Call me the 'Calliope'...I am the Gutter Dream, Tune-maker, born of steam...I am the Kallyope. **1955** R. Blesh *Shining Trumpets* (ed. 3) v. 100 A calliope outside the circus tent.

2. *attrib.* **calliope hummingbird,** a humming-bird, *Stellula calliope*, of the Western United States and Mexico. *U.S.*

1878 *U.S. Nat. Museum Proc.* I. 426 *Stellula calliope.*—Calliope Humming-bird. **1962** *Amer. Speech* XXXVII. 28 There was a December-May marriage, conducted with all the calm and sobriety of a calliope concert.

Hence **calli·opeist**, one who plays the calliope.

1932 Hemingway *Death in Afternoon* xiii. 150 An organist..or a calliopeist.

calliper, caliper. Add: **4.** A metal support for a broken or diseased leg. Also *calliper splint*.

1886 H. O. Thomas *Contrib. Surg. & Med.* vi. 85 We decided to fix the whole extremity in a Calliper splint, which enabled him to resume his personal attention to business. **1890** *Ibid.* vii. 50 A most efficient mechanical means of assistance..I found to be the 'calliper-knee-splint'...Its defects are, that it (the 'calliper') has to be altered in shape [etc.]. **1959** *B.S.I. News* Mar. 32 Thigh corsets and cuff tops for orthopaedic calipers.

callipygous (kælipi·dʒəs, -pəi·gəs), *a.* = Callipygian *a.*

1923 A. Huxley *Antic Hay* iv. 64 One reality..and that is callipygous. **1928** —— *Point Counter Point* vii. 122 One does not fall..in love with a loud speaker..however attractively callipygous. **1949** Wodehouse *Uncle Dynamite* vi. 95 He was unwilling to relinquish his memories of a callipygous Plank. **1967** J. Rathbone *Diamonds Bid* ii. 22 The callipygous matrons with their shopping.

Hence **callipy·gously** *adv.*

1939 A. Huxley *After Many a Summer* ii. 14 Young ladies stretching, writhing, callipygously stooping to tie their sandals.

callisthenics, *sb. pl.* Add: Also **calisthenics.** Also *sing.*, *transf.* and *fig.*

1839 F. A. Kemble *Jrnl. Residence Georgian Pl.* (1863) 234 To follow me through half a day with any species of lively participation in my feelings would be a severe breathless moral calisthenic to most of my friends. **1946** Mezzrow & Wolfe *Really the Blues* viii. 117 We'd all raise our arms high, like a calisthenics class. **1956** M. Stearns *Story of Jazz* (1957) xxiv. 301 *The Jazz Dance*..captures unerringly the stereotype notion..of frantic, abandoned calisthenics.

callithump (kæ·lipʌmp), *sb.* and *v.* *U.S. colloq.* [App. back-formation from next.] **A.** *sb.* A callithumpian band; = Charivari. **B.** *v. intr.* (See quot. 1888.)

1856 Hall *College Words* (ed. 2) 342 The band corresponds to the *Calliathump* of Yale. **1872** Schele de Vere *Americanisms* 589 Callithump seems..to be of American origin. It represents the French *charivari*, the German *Katzenmusik*. **1888** Farmer *Americanisms*, *To callithump*, to caterwaul; to produce discordant 'musical' sounds by means of instruments, either incongruous in themselves or in conjunction—such as tin kettles, bells, rattles, etc.

callithumpian (kælipʌ·mpiăn), *a.* and *sb.* *U.S. colloq.* [? fanciful formation; but cf. E.D.D. *Gallithumpians.* 1. A society of social reformers. 2. Disturbers of order at Parliamentary elections.] **A.** *adj.* Designating, pertaining to, or resembling, a band of discordant instruments. **B.** *sb.* A member of a callithumpian band.

1836 Hill *Yankee Stories* 9 (Weingarten), He said it was the *callathumpians*..that always went round New Year to kick up a sort of jollification. *c* **1845** G. Furman in *N.-Y. Hist. Soc. Q. Bull.* (1939) XXIII. 16 A celebrated detachment of these Rioters has long assumed the name of the 'Callithumpian Band', and has been distinguished for being more noisy and uproarious than the others. *Ibid.* 17 It was impossible for the Callithumpians to effect any meeting. **1848** Bartlett *Dict. Amer.* s.v., This party is called the *Callithumpians* or the *Callithumpian Band.* **1886** *Harper's Mag.* July 213/2 The call [on the exchange] lasts ten or fifteen minutes, and occasionally has the accompaniment of callithumpian discord. **1946** R. Blesh *Shining Trumpets* (1949) vii. 155 Then the swift notes poured out, a black, callithumpian music.

callop (kæ·lǫp). [Aboriginal name.] An Australian freshwater fish, *Plectroplites ambiguus.*

1921 *Rec. S. Austral. Mus.* II, No. 1 (Index), *Plectroplites ambiguus* Richardson (Callop, Tarki). **1935** *R. Comm. Fishing Industry S. Austral., 2nd Progress Rep.* 23 Larger quantities of Callop are marketed than of any other river fish. **1949** Vesey-Fitzgerald & Lamonte *Game Fish of World* 306 The callop is preferred to the cod by many people. **1966** *S. Austral. Yearbk.* 369 Approximately 210 part-time commercial fishermen using drum nets and set lines are engaged in fishing on the River Murray for Murray cod and callop.

callosal (kălōᵘ·săl), *a.* *Anat.* [f. L. *callōs-us* + -al.] Of or belonging to the corpus callosum.

1868 R. Owen *Anat. Vertebrates* III. xxviii. 101 Its [sc. that of the corpus callosum] hind part is embraced by the 'callosal convolution'. **1890** W. Turner in *Jrnl. Anat.* XXV. 1. 116 Marginal, callosal and hippocampal gyri are all differentiated. **1966** W. H. McMenemey in Wright & Symmers *Systemic Pathol.* II. xii. 1275/1 The symptoms that might result from a callosal defect..are obscured by the mild dementia and loss of coordination that characterize the drunkard.

calloso-marginal (kælōᵘ:so,mā·ɪdʒinăl), *a.* [cf. *Callosal *a.*, Marginal *a.*] *Anat.* Pertaining to the callosal and marginal gyri of the brain.

1876 *Quain's Elem. Anat.* (ed. 8) II. 531 The calloso-marginal fissure (cmf) commences beneath the anterior extremity of the corpus callosum. **1890** W. James *Princ. Psychol.* I. 31 The calloso-marginal convolution. **1961** *Lancet* 30 Sept. 746/1 An inter-hemispheric clot may be recognised..by separation of the callosomarginal from the pericallosal artery.

callousing (kæ·ləsiŋ), *ppl. a.* [f. Callous *a.* or *v.* + -ing².] That makes callous.

1921 *Sunday at Home* Jan. 214/1 My fellow-citizens are engaged on labour which is hard and callousing. **1928** *Sunday Express* 8 Jan. 9/1 The hardships and callousing brutalities of a mercenary force.

Callovian (kælōᵘ·viăn), *a.* *Geol.* Also *Kellovian. [ad. F. *callovien* (A. d'Orbigny *Paléont. française. Terrains jurassiques* (1842) I. 608), f. mod. L. *Callovien-sis*, f. *Kellaways* (given by d'Orbigny as *Kelloway*), a village in Wiltshire.] Epithet of a stage of the Jurassic between the Oxfordian in the Upper Jurassic and the Bathonian in the Middle Jurassic. Also *absol.*

1881 *Q. Jrnl. Geol. Soc.* XXXVII. 560 Near Bazinghen there is nothing but a few feet of clay between the Callovian grit with *Terebratula humeralis* (true) and *Rhynchonella varians* and the Nerinæan or so-called Astartian Oolite. **1885** A. Geikie *Text-bk. Geol.* (ed. 2) 801 Oxfordian, divisible into (a) Callovian, with zones of *Amm. macrocephalus*, and *A. anceps*, and (b) Oxfordian, with zones *A. Lamberti, A. Mariæ, A. cordatus.* **1946** W. J. Arkell in *Bull. Geol. Soc. Amer.* LVII. 8 Formerly, (1933) the present author maintained that the Callovian should comprise only the Kellaways beds of Kellaways, Wiltshire, and that the whole of the Oxford Clay of Oxford should be included in the Oxfordian; however, he has come to the conclusion that this would be wrong...The line of division should..be drawn between the *Lamberti* and the *Mariae* zones. **1955** G. G. Woodford tr. *Gignoux's Stratigr. Geol.* vii. 338 The Callovian is a more or less calcareous or clayey sandstone.

Calluna (kælū·nă). *Bot.* [mod.L. (R. A. Salisbury 1802, in *Trans. Linn. Soc.* VI. 317), f. Gr. καλλύνειν to beautify, sweep clean, f. κάλλος beauty.] A monotypic genus of plants of the family Ericaceæ, the only known species being *C. vulgaris*, the common heather; also (with lower-case initial), a plant of this species. Also *attrib.*

1824 J. E. Smith *Eng. Flora* II. 224 *Calluna.* Ling...Mr. Salisbury..has judiciously called our common Ling, *Calluna*, from καλλυνω; which is doubly suitable, whether..we take it to express a *cleansing* property, brooms being made of Ling; or whether we adopt the more common sense of the word, to *ornament* or *adorn*, which is very applicable to the flowers. **1920** G. Jekyll *Wall & Water Gardens* (ed. 6) xxiv. 195 The Callunas should begin at the back..and then come forward to the path. **1946** Zeuner *Dating Past* iii. 56 Peats growing above the water level:..*Calluna* peat (heather peat). **1962** *Amat. Gardening* 17 Feb. 2 The ericas and callunas have very much the same symmetrical habit of growth.

call-up. [*Call *v.* 35 f.] = Conscription 4. Also *attrib.*

1940 *Economist* 10 Aug. 175/2 A continuous system of call-up and training for the Industrial Army. **1941** D. Reed *Prophet at Home* 264 The rate of the call-up was too slow. **1945** *News Chron.* 1 June 4/2 We still propose to go on calling up young men under 30 as they reach the call-up age. **1958** *Spectator* 14 Feb. 191/2 The call-up cannot now be abandoned unless the bait for volunteers is made much tastier.

callus (kæ·lǝs), *v.* [f. the sb.] *intr.* To form a callus.

1864 J. S. Hibberd *Rose-bk.* 284 In all cases keep cuttings and eyes alive and fresh by sprinkling their tops frequently rather than making the soil they are in very wet, they will in fact callus quicker if the soil is nearly dry. **1958** *Times* 22 Nov. 9/4 They allow the natural healing processes to carry on underneath so that in time the wound will callus over.

calm, *sb.*² **2.** Delete † and add later examples.

1885 P. J. Davies *Standard Pract. Plumbing* 31 Put the sharpened end of the calme in between the cutters and turn the handle. **1885** *Spons' Mech. Own Bk.* 630 The use of lead 'calmes' for fixing window panes is of venerable antiquity. **1955** *Antiquity* XXIX. 217 No evidence was found of calms with horn panels. **1970** H. Braun *Parish Churches* viii. 111 A network of delicate grooved strips..known as 'calms'—pronounced 'cames'.

calmingly (kā·miŋli), *adv.* [f. Calming *ppl. a.* + -ly².] In a calming manner.

1908 A. Bennett *Buried Alive* viii, 'Of course you haven't,' she said calmingly. **1965** M. Frayn *Tin Men* xiii. 68 A paper whose language was..calmingly incomprehensible.

Calmuc(k, var. *Kalmuck.

Calor (kæ·lǭɪ). [a. L. *calor* heat.] The proprietary name of liquefied gas (largely butane) supplied under pressure in containers for domestic use, etc.; usu. *Calor gas.* Also *attrib.*

1936 *Nature* 23 May 862/2 The manufacture of butane and isobutane in Great Britain has been undertaken by Imperial Chemical Industries, Ltd.,..and the gas is being marketed as 'Calor' gas. **1938** *Trade Marks Jrnl.* 8 June 699/1 *Calor*..Liquefied fuel gas. Calor-Gas (Distributing) Co. Ltd. **1938** 'G. Orwell' *Let.* 5 July (1968) I. 339 There is a Calor Gas Stove,..but there is also a little oil oven. **1948** *Times* 9 Jan. 2/3 Many people have to rely on calor gas for cooking and other essential purposes, particularly in rural areas. **1959** C. Fremlin *Uncle Paul* ii. 16 The Calor gas stove.

caloric (kălǫ·rik), *a.* [f. L. *calor* heat + -ic.] Of or pertaining to heat.

[**1853** (see Caloric 3).] **1865** J. R. Mayer in W. R. Grove *Correl. Physical Forces* 275 The velocity of an asteroid when it strikes the sun measures from 445,750 to 630,400 metres; the caloric effect of the percussion is consequently equal to from 27½ to 55 millions of degrees of heat. **1896** F. W. Hume *Myst. Hansom Cab* (ed. 2) xxi. 198 To look at them merely was to increase one's caloric condition. **1925** T. Dreiser *Amer. Tragedy* I. viii. 69 Having acclimated himself to this caloric atmosphere. **1969** *Sci. Jrnl.* May 28/2 The animals were then dried to constant weight and combusted in a calorimeter to give caloric values.

calorie. Substitute for def.: A unit of heat or energy based on the specific heat of water; *esp.*: **a.** The amount of heat required to raise the temperature of 1 kilogramme of water 1 degree centigrade; used also as a measure of the heat- or energy-producing value of food or for a quantity of food having this value; more fully *great*, *kilogramme*, *large* or *major calorie*; also called *kilocalorie*. **b.** The amount of heat required to raise the temperature of 1 gramme of water 1 degree centigrade (see quot. 1963); more fully: *gramme*, *lesser* or *small calorie*. Also *attrib.* (Earlier and later examples.)

In quot. 1821 corresponds in both sense and chronology to *caloric*, and is prob. due to a misreading or a misprint; the word has not been found in French before 1813 and was not in general English use before 1880.

1821 M. Wilmot *Let.* 21 June (1935) 116, I am writing to you, wrapped up in my rug cloak, with four lined shoes to keep in an atom of calorie. **1863** E. Atkinson tr. *Ganot's Elem. Treat. Physics* iv. viii. 279 The unit chosen for comparison, and called the thermal unit, is not everywhere the same. In France it is the quantity of heat necessary to raise the temperature of one kilogramme of water one degree centigrade: this is called a calorie. [*corrected to* calorie *in ed.* 2 (1866).] **1889** M. Foster *Text-bk. Physiol.* (ed. 5) II. 802 The following results expressed in calories, that is in gramme-degree units of heat. **1892** *Pall Mall Gaz.* 22 June 6/1 A pound of beefsteak contains..870 calories of energy. **1926** *Public Opinion* 13 Aug. 158/3 When the customary measure of caloric value is applied to cake and bread, it is easily found that there is very little difference between these two staple foods. **1951** Wodehouse *Old Reliable* i. 9 A large, stout, elderly gentleman..who looked like a Roman Emperor who had been doing himself too well on starchy foods and forgetting to watch his calories. **1957** J. I. M. Stewart *Use of Riches* 112 He had to admire an energy that didn't seem very substantially based on an intake of calories. **1963** Jerrard & McNeill *Dict. Scientific Units* 29 The energy represented by a calorie varies according to the temperature of the water, thus there was the International steam calorie (4·1868 joules), the 15° C calorie (4·1855 joules), the 4° C calorie (4·2045 joules), and the mean 0–100° C calorie (4·1897 joules). These differences were overcome when the calorie was replaced by the joule as the primary unit of heat in the metric system in 1950.

calorific, *a.* Add: **1.** Esp. in *calorific value* (see quot. 1904).

1904 Goodchild & Tweney *Technol. & Sci. Dict.* 81/2 *Calorific value*, the measure of the amount of heat obtainable from a given weight of fuel. **1950** *Ann. Reg. 1949* 465 Adjustments were made to coal prices in pursuance of the policy of making prices reflect relative calorific values.

calorizing (kæ·lǒrəiziŋ), *vbl. sb.* [f. L. *calor* heat: see -ize.] The coating (of a metal) with aluminium by a process involving heat.

1930 *Engineering* 27 June 819/2 In cases where surface treatment is applied, e.g. calorising, whereby concentric zones of modified metal are produced. **1940** *Chambers's Techn. Dict.* 128/2 *Calorising*, a process of rendering the surface of steel or iron resistant to oxidation by spraying the surface with aluminium and heating to a temperature of 800° to 1000° C.

calotte. Add: **3.** (Further example.)

1945 *Electronic Engin.* XVII. 326/2 The sensitiveness of this arrangement at low pressures may be materially improved by attaching a quartz calotte of a few μ thicknesses to the two quartz fibres where they are fused together.

6. An ice- or snow-cap.

1894 J. W. Gregory in *Q. Jrnl. Geol. Soc.* L. 515 A 'calotte' or snow-cap, similar to those on Kibo..and Chimborazo. **1957** Collomb *Dict. Mountaineering* 38 *Calotte*, a crest, tip or pyramid of snow which caps the summit of a mountain. The *calotte* of Mont Blanc is best known.

calque (kælk). *Philol.* [Fr., lit. 'copy', f. *calquer* to trace (a design, etc.), ad. It. *calcare*, ad. L. *calcāre* to tread.] A loan-translation (see *Loan *sb.*¹). Also as vb.

1937 *Amer. Speech* XII. 44 The speaker begins saying things in one form only, instead of using two forms which are almost exact *calques* of each other. **1941** C. T. Onions in *Medium Ævum* X. 160 The possibility of a direct 'calque' on a Norse *wrabba ok wrægja should perhaps not be ruled out. **1957** G. V. Smithers *Kyng Alisaunder* II. 81 *Fecche mood*..is evidently a calque on OF. *porter ire*, as in *Chanson de Roland.* **1958** A. S. C. Ross *Etym.* 34 MnE *that goes without saying* is a translation-loan of (better, is calqued on) MnFrench *cela va sans dire.*

calsomine, var. Kalsomine, Calcimine (q.v. in Dict. and Suppl.).

1930 J. Dos Passos *42nd Parallel* 300 He was in a small room calsomined bright yellow. **1953** A. Upfield *Murder must Wait* iii. 19 The faint smear on the cream calsomine.

caltrop. Add: **4.** In the nomenclature of the spicular elements of sponges, a tetraxial spicule with four equal arms radiating from a central point, so called from its resemblance to a caltrop (sense 2).

1887 Sollas in *Encycl. Brit.* XXII. 416/2 Tetraxon Quadriradiate Type (Calthrops). **1940** L. H. Hyman *Invertebrates* I. vi. 337 The typical spicule of the simpler tetractinellids is the calthrops, or tetraxon with four approximately equal rays.

calumny, *v.* Delete †*Obs.* and add later examples.

1895 *Pall Mall Gaz.* 19 Jan. 7/1 The President has not been in office twelve hours..and is already calumnied. **1963** *Times* 10 May 13/5 We therefore consider it especially shameful that..the work of Pope Pius XII should be wrongly portrayed and his thoughts calumnied.

calutron (kæ·liutrǫn). *Physics.* orig. *U.S.* [f. *Ca*lifornia *U*niversity cyclo*tron*.] (See quots.)

1945 H. de W. Smyth *Atomic Energy* xi. 187 The 'calutron' mass separator. *Ibid.* 189 The 37-inch cyclotron was dismantled..and its magnet was used to produce the magnetic field required in what came to be called a 'calutron'. **1946** *Ann. Reg. 1945* 355 An electromagnetic method [of separating 2 isotopes] using a magnetic separator, called a calutron, first constructed from the California University cyclotron. **1956** *Nature* 28 Jan. 157/1 The calutrons which are now used for enrichment of the isotopes of all the elements.

calvados (kæ·lvādǫs). Also **Calvados.** [f. *Calvados*, a department of Normandy, France.] A spirit distilled from cider, named after the region in which it is traditionally made; apple-jack.

1906 C. Scudamore *Normandy* viii. 39 Calvados produces much cider and perry as well as a kind of brandy termed 'Calvados'. **1930** R. F. Wilson *How to wine & dine in Paris* v. 102 Cider, good ordinary wine, and fine old apple brandy (Calvados). **1942** E. Waugh *Put out more Flags* 139 The principal ingredients were vodka and calvados.

calve, *v.*¹ Add: **2. b.** *to calve down*: to breed from (a cow). Also *intr.* = *pass.*

1858 *Jrnl. R. Agric. Soc.* XIX. I. 27 These stock are generally calved down when little more than two years old, or else sold. *Ibid.* 28, I have myself known stock costing 6 *l.* per head worth at the end of the same year 13 *l.* or 14 *l.*, and the increase is just as great when they calve down.

calycate (kæ·likeᶦt), *a.* *Bot.* Also **calicate.** [ad. mod.L. *calycātus*, f. L. Calyx: see -ate².] Provided with a calyx.

1866 Lindley & Moore *Treas. Bot.*, *Calicate*, furnished with a calyx.

calypso (kăli·pso). Pl. **-os.** [Origin unknown.] A West Indian ballad or song in African rhythm, usually improvised to comment on a topic of current interest. Also *attrib.* Hence **calypsonian** (kælipsōᵘ·niăn), an entertainer who composes and sings calypsos.

1934 A. Huxley *Beyond Mexique Bay* 18 A Calypso Tent..is..a tin roof on posts—in which..the local talent assembles to rehearse certain songs composed against the coming of Carnival. *Ibid.* 19 The tunes to which these songs are sung is always some variant of an old Spanish air called Calypso; the words are home-made and topical. *Ibid.*, The singers..call themselves 'Calypsonians'. **1938** *Times* 26 May 17/6 'Calypso' songs (as they are curiously called), to judge from examples perhaps too discreetly chosen, are mere doggerel, and the music, if it does more than beat out the rhythm, is the most commonplace jazz. **1955** *Times* 28 Jan. 9/6 Calypso is a form of doggerel lampoon composed and sung to a guitar, often extempore...A Calypsonian both composes and accompanies his songs. **1957** *Observer* 24 Nov. 10/6 The calypso is a commentary. It seeks to tell of an event which has made an impression on the singer.

calyx. Add: **3. calyx-bursting,** bursting of the calyx, a defect in carnations; **calyx(-form) crater** or **krater** *Gr. Antiq.*, a crater (sense 1) of the shape of a calyx.

1900 *Westm. Gaz.* 7 May 4/2 Calyx-bursting is a fatal objection to many varieties otherwise superb. **1896** C. H. Smith in *Catal. Gr. Vases Brit. Mus.* III. 280 Calyx-form Craters. **1912** J. D. Beazley in *Ann. Brit. Sch. Athens* XVIII. 225 The pattern No. 15 does not occur on any other rf. kalyx-krater. **1915** *Oxf. Univ. Gaz.* 3 Feb. 378/2 The other vases include two *hydriæ*,..two amphoræ,..and a calyx-crater.

cam, *sb.*¹ Add: **b. cam-box,** a frame surrounding a cam and designed to compel the rod which the cam drives to follow the return motion of the projecting lobe; also, a casing enclosing the cam and its rollers in order that copious lubrication may be secured by having the cams revolve in a bath of oil (*Cent. Dict.* Suppl. 1909); **cam-cutter,** a machine-tool specially adapted for cutting and finishing cams; **cam-pump,** a pump in which the valve motion is given by a cam; **cam-shaft,** a shaft bearing a cam or cams; also *attrib.*; **cam-yoke,** a frame attached to a valve-stem or other reciprocating piece to which it gives intermittent straight-line motion from a cam on the face of a rotating disc; used in steam-engine valve-gears (*Funk's Standard Dict.* 1893).

a **1884** Knight *Dict. Mech.* Suppl. 156 Cam Cutter. **1922** Jacobs *Cam Design* 74 The hand-made master cam is now placed in position on the cam cutter head spindle. *a* **1884** Knight *Dict. Mech.* Suppl. 157 Drayton Cam Pump. *a* **1877** —— *Dict. Mech.* I. 435/2 *Cam-shaft*, a shaft having cams or wipers, for raising the pestles of stamping-mills. **1908** *Westm. Gaz.* 9 Jan. 4/1 Depressing the cam-shaft pedal. **1955** *Times* 20 Aug. 4/3 The new B.R.M. has a 2½-litre short stroke 4-cylinder engine with overhead cam-shafts.

CAM (kæm). Also **C.A.M. ship, ca·m-ship.** [f. the initials of *C*atapult *A*ircraft *M*erchant-ship.] A merchant-ship equipped with a fighter plane launched by catapult.

1943 *Cosmopolitan* Aug. 14/1 Every eye was turned instinctively to the C.A.M., the convoy's Catapult Aircraft Merchantship, the defiant reply of the Royal Air Force to Nazi air assault on British merchant shipping. **1944** A. M. Taylor *Lang. World War II* 18 CAM: Catapult Aircraft Merchantship. Adopted by the RAF, to accompany and afford protection to convoys. **1945** L. R. Gribble *Battle Stories RAF* vii. 18 Volunteered to fly with convoys as a catafighter on a Cam-ship. **1954** P. K. Kemp *Fleet Air Arm* 152 These catapult ships were known as Camships. **1956** 'Taffrail' *Arctic Convoy* xix. 201 A 'C.A.M.' ship..an ordinary cargo-carrier fitted with a catapult forward with a single Hurricane fighter.

Camaldolese (kămæ·ldōlīz). Also **Camaldulese.** Pl. **-ese.** = Camaldolite. Also *attrib.* or as *adj.*

1828 in B. Ward *Eve Cath. Emanc.* (1912) III. xlv. 198 Two have embraced the Order of Camaldolese Hermits. **1850** in *Ushaw Mag.* (1907) Mar. 62 In Rome the *Candela dell' Elevazione* is used by the Camaldolese only. **1873** Newman *Mission Bened. Order* (1908) 75 Monte Cassino excelled in illumination and in mosaic, the Camaldolese in painting, and the Olivetans in wood-inlaying. **1880** H. Collins *Heaven Opened* II. xix. 247 The Carthusians and Camaldulese have but one fixed recreation in the week. **1929** D. Gwynn *Cdl. Wiseman* i. 16 Clothed in his white robes as a Camaldolese monk. **1952** S. Rigby tr. *Bessières' Wife Mother & Mystic* III. ii. 164 The saintly Camaldolese used a capacious snuff-box.

‖**camalote** (kæ·mălōᵘte). Also **cameloté, -a.** [Amer. Sp.] An aquatic plant of the genus *Pontederia*, native to North and South America; also, floating islands of vegetation including this plant.

1881 E. W. White *Cameos fr. Silver-Land* I. 243 When the river is high, it is no very uncommon circumstance to see the whole surface of this archipelago covered with the debris (*camelotas*) of these mud-formed islands, buoyed up by matted roots. **1882** *Ibid.* II. 3 Blue-flowered camalotes, consisting chiefly of *Pontederia*, which spread their broad leaves as sails. **1905** R. B. Cunninghame Graham *Progress* 63 Camelotes brought down by the flood were wreathed about them like gigantic eels. *Note*, The camelote is a very thick-growing water-lily, which sometimes chokes small streams. **1918** W. H. Hudson *Far Away & Long Ago* xx. 261 The fourth lakelet..was..covered with a luxuriant growth of the floating *camaloté*, a plant which at a distance resembles the wild musk. **1950** J. G. Kerr *Naturalist in Gran Chaco* i. 3 We passed on up the wide estuary with its muddy fresh water dotted with green floating islands of camelote (*Pontederia*). **1956** G. Durrell *Drunken Forest* vii. 146 Then, round the bend of the river, appeared the vanguard of the camelotes..islands of lilies, convolvulus, and grass.

caman (kæ·măn). Also **camman.** [Gaelic.] The stick or club used in shinty.

1891 *Daily News* 12 Oct. 5/6 The *camman* is the club..with which the Gaelic athletes play their favourite game. **1900** *19th Cent.* Aug. 307 Each man has a *caman* or hurley—a stick, about three feet long, with a bend or curve at the end. **1905** P. A. Sheehan *Glenanaar* i, The crack of the camans as they crossed in the air above or on the grass beneath.

camanchaca (kamăntʃā·kă). [Amer. Sp.] A heavy mist on the Peruvian coast.

1907 C. R. Enock *Andes & Amazon* i. 5 The heavy mist upon the coast, known as *Camanchaca.* **1922** *Daily Mail* 17 Nov. 8 In the Northern rainless zone, the camanchacas, or heavy, overhanging mists, became much denser.

camaron (kæmărōᵘ·n, kæ·mărǫn). Also **cammaron.** [ad. Sp. *camarón* shrimp, f. L. *cam(m)arus* sea-crab.] A freshwater shrimp or prawn resembling the crayfish.

1880 T. H. Huxley *Crayfish* 329 These fluviatile prawns (known in many places by the name of '*Cammarons*') are not unfrequently confounded with true crayfishes. **1920** *Glasgow Herald* 31 Dec. 12 Camarons are excellent when boiled. **1947** M. Lowry *Under Volcano* vii. 229 The boy brought them camarones, red shrimps in a saucer. **1967** *Oceanogr. & Marine Biol.* V. 371 Nephrops norvegicus...is known by the following common names:..Norway lobster, Dublin Bay prawn, North Shields prawn, Beardog (Britain)..Langostino, Camaron, Cigala, Maganto, Escamarlanch (Spain).

camata (kămā·tă, -ēᶦ·tă). [It.] The commercial name for the half-grown acorns of *Quercus ægilops*, dried and used for tanning. Cf. next.

1858 Simmonds *Dict. Trade.* **1866** Lindley & Moore *Treas. Bot.* 950/1.

camatina (kæmăti·nă, -əi·nă). [dim. of prec.] The commercial name for the incipient acorns of *Quercus ægilops*.

1858 Simmonds *Dict. Trade.* **1866** Lindley & Moore *Treas. Bot.* 950/1. **1884** *Encycl. Brit.* XVII. 694/2 The valonia of commerce, one of the richest of tanning materials, is the acorn of *Q[uercus] Ægilops*...Immature acorns are sometimes exported under the name of camatina.

camber, *sb.* Add: **1. b.** The transverse arch of the surface of a road.

1905 *Westm. Gaz.* 13 Oct. 2/7 Another suggestion is that the 'camber' (i.e., the upward curve) of roadways should be lessened. **1907** *Ibid.* 30 July 8/1 The heavy camber of the sides..is the cause of very many mishaps. **1925** *Public Opinion* 11 Dec. 588/2 Hodge has always allowed his horse to take the top of the camber.

c. The curvature of the wings of an aeroplane. Also *attrib.*

1910 R. Ferris *How it Flies* xx. 456 *Camber*, the distance from the chord of the curve of a surface to the highest point of that curve, measured at right angles to the chord. **1918** H. Barber *Aeroplane Speaks* (ed. 6) 5, I must have a certain chord to make it possible for my Camber (that's curvature) to be just right for the Angle of Incidence. **1935** K. D. Wood *Techn. Aerodynamics* ii. 52 The shape of the median camber line of the variable-density-tunnel series is nearly the same as for other good airfoils. *Ibid.* 53 The maximum median camber must next be located relative to the line through the leading and trailing edges.

d. *Automobile Engin.* (See quot. 1959².)

1936 *Motor Man.* (ed. 29) vii. 121 How a steering head is set to give castor action and camber. *Ibid.*, The point..is thrust forward by the slope or castor angle of the king pin and slightly to one side according to the camber chosen for the wheel and pin. **1959** *Ibid.* (ed. 36) v. 102 Each steering swivel..is also given..a camber angle so that its axis, when viewed from the front, meets the road at a point..close to the centre line of the wheel. **1959** *Motor* 2 Sept. 92/1 Camber (the sideways inclination of the wheel).

cambered, *ppl. a.* (Later examples.)

1909 *Flight* 20 Feb. 104/1 *Cambered*, this term denotes that the plane or wing has a curved transverse section. **1919** *Autocar Handbk.* (ed. 9) 223 When driving a car on a much-cambered or arched road. **1932** G. M. Boumphrey *Story of Wheel* 57 His roads were cambered (raised slightly in the middle to throw off water). **1951** W. F. Hilton *High-Speed Aerodynamics* ii. x. 241 The 8% cambered wing would be outside the limits of accuracy at M = 10. **1967** *Gloss. Mining Terms* (*B.S.I.*) xi. 6 *Cambered girder*, a roof bar projecting forward to support the roof beyond the propped area.

Cambodian (kæmbōᵘ·diăn), *a.* and *sb.* [f. *Cambodia*, a country in S.E. Asia + -ian.] **A.** *adj.* Of or pertaining to Cambodia or its inhabitants. **B.** *sb.* An inhabitant of Cambodia; also, the language of its people.

1770 W. Guthrie *New Geogr., Hist., & Comm. Gram.* 481 The betel..is the highest luxury of the Cambodians, from the King to the peasant. **1853** *Jrnl. Indian Archipelago & E. Asia* VII. 292 The most approved forms of Cambodian correspondence. *Ibid.*, The Cambodians always use Indian ink. *Ibid.*, The letter was then inserted into an envelope addressed in Cambodian. **1862** *Proc. R. Geogr. Soc.* VI. 82 The Cambodians had invented a written phonetic character, which they used at the present time; therefore there could be no difficulty in understanding a Cambodian manuscript. **1877** [see *Annamite *a.* and *sb.*] **1937** R. H. Lowie *Hist. Ethnol. Theory* x. 162 Their first pyramids were erected five or six centuries later as copies of Cambodian..models. **1966**

H. L. Shorto in C. E. Bazell *In Memory of J. R. Firth* 398 In Cambodian, for example, words may be distinguished by a complex of features which includes both elements ascribed to 'register' and differences in vowel quality. **1966** *Guardian* 31 Aug. 11/3 Armed Cambodian terrorists..said to belong to the 'Free Cambodians'.

‖**cambré** (kaṅbre), *a.* [Fr., pa. pple. of *cambrer* to arch, to bend: see Camber *v.*] Curved, arched; spec. in *Dancing* (see quots.). Also as *sb.*

1913 C. d'Albert *Dict. Dancing* 33 *Pas cambré*..Open the heels and pivot on the toes; the points are inward, the feet in the form of an arch, toes nearly touching in front, heels open in the rear. Raise the shoulders..then pivot on the heels, opening the toes..and assemblé, letting the hands fall. **1929** D. H. Lawrence *Lovely Lady* (1932) 97 Her figure, though not stout, was full, strong, and *cambré.* **1952** Kersley & Sinclair *Dict. Ballet Terms* 22 *Cambré*, a bend of the body from the waist in any direction.

cambric. Add: **c. cambric tea** *U.S.*, a drink composed mainly of hot milk and water, given to children in place of tea.

1888 *Union Signal* (Chicago) 21 Jan. 3 [She] gave me a vast easy chair to sit in..and offered me tea, cambric tea to be sure, but in a beautiful cup. **1944** *Greeley* (Colo.) *Daily Tribune* 28 Sept. 6/4 Many children dearly love cambric tea, which is made by pouring about two tablespoons of weak tea into a cup of hot milk and adding a dash of sugar.

Cambridge (kēᶦ·mbridʒ), the name of a university town in England, used *attrib.*, as **Cambridge blue,** a light blue; **Cambridge calf** (see quot. 1895); **Cambridge chimes,** the composition of Joseph Jowett and William Crotch, first employed in 1793 at the Church of St. Mary the Great, Cambridge; **Cambridge coprolite, greensand** *Geol.* (see quots. 1881, 1882); **Cambridge roller** *Agric.* (see quot. 1954); **Cambridge sausage,** a variety of sausage.

1580 in *Cath. Rec. Soc. Publ.* (1961) LIII. 194 Robert Kent prest, a cambridge man borne. **1830** M. R. Mitford *Our Village* IV. 108 Your thoroughbred Londoner..grumbling over..his thin milk and his Cambridge butter. **1840** Dickens *Let.* 1 Feb. (1969) 17 The Cambridge sassages of right down English Manafacter. **1850** Denison *Clock & Watch-m.* 226 A clock made for a nobleman a few years ago, who intended to have the Cambridge chimes. **1877** E. S. Dallas *Kettner's Book of Table* 94 It is notorious that what has been called Cambridge butter is a mixture of foreign butters. **1881** *Spon's Encycl. Industr. Arts* IV. 1260 The most valuable beds of the mineral in this country are in the Upper Greensand formation, lying chiefly in Cambridgeshire, and merging into Buckinghamshire. These are known as 'Cambridge' coprolites. **1882** Geikie *Text-bk. Geol.* 809 The so-called 'Cambridge Greensand'—a bed about 1 foot thick lying at the base of the Chalk of Cambridge, and largely worked for phosphate of lime derived from coprolites and bones. **1883** *Cassell's Fam. Mag.* Mar. 206/2 A huge wooden spoon, ornamented with bows of Cambridge blue. **1891** R. Wallace *Rural Econ. Austral. & N.Z.* xviii. 262 About one pound of seed is sown per acre..distributed from a sowing-box placed behind a Cambridge roller. **1895** *Windsor Mag.* I. 403/2 Knickerbockers of dark blue velveteen trimmed with Cambridge blue. **1895** Zaehnsdorf *Short Hist. Bookbinding* 20 Cambridge Calf.—Fine and dark sprinkled calf of two tints, a square panel being left in centre of sides. **1898** *Westm. Gaz.* 3 Mar. 9/2 The bulky little volume..in its Cambridge-blue cover. **1909** *Daily Chron.* 16 Feb. 4/6 As the 'Cambridge Chimes' at St. Stephen's strike 2 p.m. the King will arrive to open Parliament. **1947** G. Greene *19 Stories* 227 A tin of Cambridge sausages. **1954** *Gloss. Terms Agric. Mach.* (*B.S.I.*) 12 *Cambridge roller*, a roller consisting of loosely mounted ring segments each usually about 3 in. wide and tapering at the periphery to a narrow rim. **1970** *Sunday Times* (Colour Suppl.) 1 Feb. 40/1 The finest breakfast I have ever had..finnan haddock, Cambridge sausages, York ham.

Cambridgeshire (kēᶦ·mbridʒʃəɪ). The name of the eastern county in which Newmarket (q.v.) is situated, used attrib. in *The Cambridgeshire Handicap* (*Stakes*) as the name of a horse-race originated in 1839 and run annually at Newmarket: usually ellipt. *The Cambridgeshire.*

1840 J. C. Whyte *Hist. Brit. Turf* II. 482 The Cambridgeshire stakes. **1856** 'Stonehenge' *Brit. Sports* 373/1 Cambridgeshire Course..1 mile 240 yards. **1891** G. Chetwynd *Racing Remin.* I. 10 The next day Vestminster won the Cambridgeshire. **1930** Wodehouse *Very Good, Jeeves!* iv. 99 She married old Tom Travers the year Bluebottle won the Cambridgeshire.

Cambro- (kæ·mbro), mod.L. *Cambro-*, as in *Cambro-Britannicus* (1592), used as combining form in the sense 'pertaining to Cambria, Welsh'. Also in sense 'Cambrian' (Geol.)

1612 Drayton *Poly-Olb.* Pref., To my friends, the Cambro-Britans. **1748** Smollett *Rod. Rand.* I. xxv. 233 A prescription..which..the Welchman..got up to prepare. . . This Cambro Briton..ordered the tar to run to his mess-mate. **1853** W. J. Rees (*title*) Lives of the Cambro British Saints. **1871** Lowell *Study Windows* 164 Tennyson in the Cambro-Breton cyclus of Arthur. **1925** J. Joly *Surface-Hist. Earth* iii. 57 Almost complete submergence of North America in Cambro-Ordovician times. **1967** *Oceanogr. & Marine Biol.* V. 317 Permanent ocean basins which have not changed significantly in their area or position since Cambro-Ordovician times, some 500 million years ago.

camel, *sb.* Add: **1. d.** The characteristic colour of a camel, a shade of fawn.

1881 C. C. Harrison *Woman's Handiwork* I. 50 Camel color is the most recent variety of écru shades, coming to us from England. **1922** *Daily Mail* 11 Dec. 14 Brushed Wool Scarf-wrap..Can be supplied..in plain colours—White,..Camel, Beige, Grey. **1923** *Ibid.* 26 Feb. 1 In good shades of Grey, Camel, Fawn. **1924** *Tourist* Winter Sports No. 20 Sports hats..of camel check Tweed with scarves to match.

4. *camel-cart*, *-dung* (also *camel's dung*), *-guide*, *-skin*.

1900 *Daily News* 25 Sept. 3/4 Our caravan..included..six camel carts for the ladies and children. **1907** *Westm. Gaz.* 25 Nov. 2/1 The most striking..conveyance is a camel-cart. **1903** W. C. Russell *Overdue* vi, Recollection reeks of the flavour of the camel-dung cigarettes of Alexandria. **1886** R. F. Burton *Arab. Nts.* X. 193 'Take care of the glass-phials!' cried the Prophet to a camel-guide. **1827** *Every Night Book* 87 We strongly recommend you..to illumine the butt-end of your cigar with camel's dung. **1879** W. J. Loftie *Ride in Egypt* xii. 261 There is an all-pervading smell..caused..by the use of camel's dung for fuel. *a* **1425** in *M.E.D.*, Camel skyn. *c* **1450** Capgrave *Life St. Aug.* 38 þe haue girdilis lich knytys; and þei with þongis of chamel skynnys, as Hely and Ion, go girt in her lendes. **1497** Bp. Alcock *Mons Perfeccionis* E i b/1 Clothed in a camell skynne. **1903** *Month* Aug. 165 St. John in his camel-skin robe.

5. camel-back orig. *U.S.*, an inferior or synthetic rubber used to retread tyres; **camel-gun,** a gun, as a machine gun, made light and short so as to be transportable by camels; **camel-swallower, -swallowing** (see sense 1 c); **camel-thorn,** (*b*) *S. Afr.*, the tree *Acacia giraffæ* or *A. hirtella*; camel-tree; **camel-trot, camel-walk,** a ball-room dance resembling the walk of a camel.

1942 in *Amer. Speech* (1943) XVIII. 302/2 The term 'camelback', broadly used, refers to the uncured rubber applied to the worn tire to make the new tread. **1959** *Times* 27 Apr. (Suppl. Rubber Ind.) p. vii/6 A general-purpose cold rubber for tyre stocks, camelbacks. **1891** Kipling *Light that Failed* ii. 29 Aren't the camel-guns ever going to begin? **1802** C. Wilmot *Let.* 15 Nov. in T. U. Sadleir *Irish Peer* (1920) 114 The Friar..was a delightful sketch of a wholesale camel swallower. He believed in the most extravagent miracles. **1840** C. H. Townshend *Facts in Mesmerism* 332 The gnat-strainers and camel-swallowers may be content to accept this story. **1858** Dickens *Lett.* (1879) II. 82 All manner of camel-swallowing and of gnat-straining. **1824** W. J. Burchell *Trav. S. Afr.* II. 292 Some scattered trees of Camel-thorn, or *Mokaala.* **1896** H. A. Bryden *Tales S. Afr.* 75 The camel-thorn trees [*f.n.* giraffe acacias] grew pretty thickly all around. **1947** L. Hastings *Dragons are Extra* i. 14 Among the scrub and camelthorn trees. **1961** L. van der Post *Heart of Hunter* i. 21 Camel-thorn trees in leaf..growing in a part of the desert which was not typical camel-thorn country at all. **1923** *Weekly Dispatch* 8 Apr. 8 They call the modern dances camel-trots. **1921** *Frontier* May 16 The morbid minded may read them as openly as they danced the shimmy and the camel-walk a year ago. **1969** *New York* 15 Feb. 29/3 Rubbery-legged dances, like the Camel Walk.

camelious (kămī·liəs), *a.* [f. Camel *sb.* + -ious.] Jocular word invented by Kipling (in form *cameelious*) to describe the hump given to the lazy camel in *Just So Stories.* Hence *allusively* (with ref. to Hump *sb.* 3).

1902 Kipling *Just So Stories* 27 Kiddies and grown-ups too-oo-oo, If we haven't enough to do-oo-oo, We get the hump—Cameelious hump—The hump that is black and blue! **1909** 'Ian Hay' *Man's Man* xvi, The men have both got camelious hump.

camellia. Add: **camellia-red,** a bright red, the colour of red camellias.

1890 *Daily News* 21 Oct. 2/1 Camellia red felt.

cameloid (kæ·mĕloid), *a.* and *sb.* *Zool.* [ad. mod.L. *Cameloidea*: see Camel *sb.* and -oid.] **A.** *adj.* Of or pertaining to the Camelidæ or camel family of ruminants. **B.** *sb.* An animal of this family.

1885 tr. *Schmidt's Mammalia* 156 The cameloid type of Ruminant. **1888** *Longman's Mag.* July 298 That the existing cameloids should be so strangely distributed. **1924** *Glasgow Herald* 8 Nov. 4 Herds of camels..trekked across..to Europe, leaving North America..with a glorious graveyard of cameloid progenitors.

camelote, var. *Camalote.

camel's hair. Add: **1. b.** Used *attrib.* of a cloth or garment, etc., made from the hair of the camel. Also *ellipt.*

1678 *New Castle Court Rec.* 361 A Camell haire Rugg. **1854** M. Cummins *Lamplighter* 178 That boy..has sent you home a camel's hair scarf. **1860** [see sense 1 in Dict.]. **1865** *Atlantic Monthly* XV. 10 Cotton-bales, bound in striped camel's-hair cloth. **1918** W. Owen *Let.* 10 Sept. (1967) 575 I'm sitting in my Camel Hair Coat at my bed-table. **1964** *McCall's Sewing* iv. 56/1 *Camel hair*, a coating fabric made from the natural-coloured hair of the camel, frequently mixed with wool. **1968** 'E. Trevor' *Place for Wicked* v. 70 Well I'll bring my camel-hair. It's got pockets like rucksacks.

Camembert (kæ·maṅbēᵊɪ). [Name of a village near Argentan, France.] In full *Camembert cheese*: a rich soft cheese made in the vicinity of Camembert; also, any cheese of the same type, wherever made.

1878 *Cassell's Fam. Mag.* Aug. 535/1 Gruyère, Camembert, and Gorgonzola are all delicious cheeses. **1910** [see *Gorgonzola]. **1951** *Good Housek. Home Encycl.* 381/1 The best Camembert cheeses are obtained during the summer months.

cameo. Add: **b.** Special Combs.: **cameo-embossing** (see quot.); **cameo glass,** a decorative glass consisting of layers of different colours, the outermost being cut away so as to leave the design or designs in relief, an example being the Portland vase; **cameo-incrustation,** the art of producing bas-relief casts within a coating of flint-glass; **cameo-type,** in photography, a name formerly given to a small daguerreotype which could be mounted in a jewelled setting; **cameo ware,** pottery with figures in relief on a background of a different colour, as in Wedgwood ware.

1878 *Encycl. Brit.* VIII. 160/2 In the second variety [of colour embossing]—called cameo embossing—the colour is applied to the flat parts of the design by means of a small printing roller, and the letters or design in relief is left uncoloured. **1879** *Ibid.* X. 649/2 The first place among those processes in which one colour was superimposed on another may be given to that by which the cameo glass was produced. **1961** E. M. Elville *Collector's Dict. Glass* (1967) 32 Cameo glass first made its appearance shortly after the Great Exhibition of 1851. **1849** A. Pellatt *Curiosities of Glass Making* 119 Cameo Incrustation was unknown to the ancients, and was first introduced by the Bohemians, probably about a century since. *a* **1877** Knight *Dict. Mech.* I. 433/2 *Cameo-type*, a fanciful name given to a small vignette daguerreotype for mounting in a jeweled frame like a cameo.

2. *transf.* and *fig.* Esp. a short literary sketch or portrait; a small character part that stands out from the other minor parts. Freq. *attrib.*

1851 *Monthly Packet* I. 5 Cameos from English history. **1881** E. W. White (*title*) Cameos from the Silver-Land. **1901** *Daily News* 19 Jan. 6/1 This volume is mainly composed of biographical sketches... Altogether there are here about ninety of these cameo-biographies. **1917** A. Waugh *Loom of Youth* iv. vii. 314 He could give a clean-cut cameo impression of that monarch in two lines..: 'A dreamer who unfortunately allowed his dreams to encroach on his waking moments.' **1928** *Daily Mail* 6 Aug. 10/7 A daring act on motor-cycles..was followed by a cameo of the war. **1950** 'E. Crispin' *Frequent Hearses* i. 34 A cameo part..the film equivalent of a bit part on the stage. **1967** D. Pinner *Ritual* x. 108, I was an actor. Beloved for my cameos in the Classics.

camera. Add: **3. c.** *Television.* That part of the equipment which forms the image and converts it into electrical impulses.

1928 *Television* July 26 (*caption*) What the television 'camera' of the future may look like. **1936** *Nature* 19 Sept. 514/1 The system employs..scanning cameras..by means of which the scenes to be transmitted are directly and continuously transformed into electrical impulses without the intermediary of a film device. **1957** *Encycl. Brit.* IV. 218/1 The function of the television camera is analogous to that of the microphone in sound broadcasting.

d. *attrib.* and *Comb.*, as *camera-angle*, *-crew*, *-work*; **camera booth** (see quot.); **camera-eye,** an eye that records detailed impressions, a camera-like eye; also, a person capable of unusually detailed or detached observation or memory; so *camera-eyed* adj.; **camera gun** *Aeronaut.* (see quots.); **camera-man,** a man who uses or operates a camera professionally; **camera rehearsal,** a dress rehearsal for a film or for a television programme; **camera shake,** unintentional movement of the camera during photography or filming; **camera-shy** *a.*, fearful or nervous of cameras; not liking to be photographed or filmed.

1928 Camera angle [see *Angle *sb.*² 1 e]. **1930** J. B. Priestley *Angel Pavement* iv. ii. 168 He knew nothing about camera angles and 'cutting' and all the intricacies of crowd work. **1968** A. Diment *Gt. Spy Race* viii. 151 You could see what was happening from two camera angles. **1929** *Photoplay* Apr., *Camera booth*, the movable sound-proof box with a glass front, in which cameras are enclosed in a talking picture studio in order that the sound of the camera may not intrude in the picture. **1940** *Chambers's Techn. Dict.* 129/2 Camera crew. **1950** 'E. Crispin' *Frequent Hearses* i. 8 Camera crews; continuity girls; youngish directors. **1968** *Listener* 8 Feb. 164/2 A BBC camera crew went round Washington collecting vox pops. **1930** J. P. Burke in *Amer. Mercury* Dec. 455/1 *Camera eye*, a retentive memory for faces. 'Put him on the door. He's camera eye.' **1946** J. B. Priestley *Bright Day* v. 143 Staring at her with a cold camera eye. **1960** *Guardian* 17 Nov. 8/2 The 'I' will be speaking, not an uninvolved camera eye, but a very much involved eye. **1935** S. Lewis *It can't happen Here* xxxv. 377 You were supposed to be the camera-eyed gink that kept up on everything that goes on. **1918** F. H. Colvin *Aircraft Mechanics Handbk.* xxii. 316 For machine-gun practice at enemy airplanes, the camera gun is..used. **1939** *War Illustr.* 2 Dec. 371 The first training of the gunners of fighting 'planes is given with a camera gun which registers on a photographic film the hits made by the gun. **1908** *Westm. Gaz.* 21 Apr. 7/1 After both had posed to the inevitable camera-men. **1920** *Q. Rev.* July 183 The camera-man will film you anything. **1957** *Times* 9 Oct. 18 (*caption*) Wild camels of central Asia..dash to the safety of the foothills..as cameramen made what is believed to be the first photographic record of these animals. **1959** W. S. Sharps *Dict. Cinemat.* 83/1 *Camera rehearsal.* Otherwise *run through.* **1963** E. Humphreys *Gift* II. v. 239 During camera rehearsals ..I always bought her a packet of glucose tablets. **1940** *Wall's Dict. Photogr.* (ed. 15) 364 *Camera shake.* Movement of the camera during the exposure is a very common cause of poor definition. **1962** L. Deighton *Ipcress File* xxiv. 154 A slow motion movie would be less subject to camera shake. **1922** E. J. Kimble *Kimble's Vocational Vocab.* 189 Camera-shy. **1937** H. G. Wells *Brynhild* i. 9 He had always been a little aloof and camera-shy in his publicity. **1958** *Manchester Guardian* 30 June 6/4 It is not difficult to sympathise with the camera-shy M.P. who fears that his daily life is to become too public with the arrival of television in the House. **1908** *Daily Chron.* 14 Mar. 3/3 They are most excellent examples of camera-work.

camerata (kæmĕrā·tă). [mod.L., f. *camera* Chamber.] Each of the groups into which students of English theological colleges at Rome are divided.

1846 J. H. Newman *Let.* in W. Ward *Life* (1912) I. iv. 132 The whole body of students is divided into eight classes or portions (cameratas?)—who are never allowed to speak to each other. If you and Christie and Penny went, they would of course put you into three separate cameratas. **1912** B. Ward *Eve Cath. Emanc.* III. xxxiv. 13 The students..likewise felt aggrieved at having to walk out in 'Camerata'.

camerist[2] (kæ·mərist). Chiefly *U.S.* [f. Camera 3 b + -ist.] One who uses a camera, a photographer.

1890 *Anthony's Photogr. Bull.* III. 19 Theoretically, all camerists believe in a good negative. **1891** *Ibid.* IV. 85 A steady head is often needed to keep an enthusiastic camerist from being unduly depressed or exalted. **1906** *Westm. Gaz.* 15 May 12/3 Many of the pictures..are excellent examples of artistic photography...They have been taken by camerists..whose names are thoroughly well known. **1921** *Chambers's Jrnl.* Aug. 547/1 The camerist determines on some other makeshift foreground. **1936** *Discovery* June 191/2 The trick manœuvres of the 'camerist' in contracting or expanding the dimensions of time and space.

camerlingo (kæməɹli·ŋgo). Also **camarlingo, camerlengo.** [It. *camerlingo*: see Chamberlain.] A chamberlain or treasurer. **a.** In a secular state. **b.** The Pope's chamberlain and financial secretary; the highest officer in the papal household. **c.** The cardinals' chamberlain, the treasurer of the sacred college. Hence **camerli·ngate**, the office of camerlingo.

1625 Purchas *Pilgrims* II. x. 1834 The Inhabitants are governed by a *Camarlingo*, in the behalfe of Venice. **1753** Chambers *Cycl. Suppl.* s.v., The *camerlingo* is the most conspicuous officer in the court of Rome...The cardinals have also their *camerlingo*, or treasurer of their college. **1759** A. Butler *Lives Saints* IV. i. 524 St. Charles..absolutely refused the Camerlingate, the second and most lucrative dignity in the Roman court. **1923** *Glasgow Herald* 10 May 9 He was all confused when taken to the presence of Her Majesty by a camerlengo.

camerostome (kæ·mĕrostōᵘm). *Zool.* [ad. mod.L. *camerostoma*, irreg. f. L. *camera* vault + Gr. στόμα mouth.] The anterior part of the body of arachnids which forms an arch over the mouth.

1888 Rolleston & Jackson *Anim. Life* 523 A well-developed chitinoid piece—the camerostome..overhangs the mouth anteriorly. **1949** *Trans. Connecticut Acad. Arts & Sci.* XXXVII. 260 The camarostome is formed by the upper lip and the pedipalpal coxae.

Camford (kæ·mfǒɹd). [f. Cam(bridge + Ox)ford.] = *Oxbridge.

1850 Thackeray *Pendennis* II. xiv. 135 He was a Camford man and very nearly got the English Prize Poem. **1924** [see *Oxbridge]. **1939** H. G. Wells *Holy Terror* I. ii. 37 Camford..had never made the slightest attempt to give any coherent picture of the universe to the new generation that came to it for instruction. **1961** *Observer* 26 Mar. 29/4 The Camford-style Sermon on the Mount.

cami- (kæ·mi), abbreviated form of *camisole 2 b, used in *Comb.*, as **cami-bocker** [f. (*knicker*)*bocker*], **cami-knickers** or **-knicks**, an undergarment which combines camisole and knickers; so **cami-petticoat**, etc.

1908 in C. W. Cunnington *English Women's Clothes* (1952) ii. 84 The cami-skirt, worn over the corsets, has its skirt usually divided into two. **1915** *Home Chat* 30 Oct. 229 My new cami-knickers. **1923** *Daily Mail* 2 May 1 Cami-petticoats made of..heavy Artificial Silk Stockinette. **1926** *Good Housekeeping* July 58 Cami-bocker in cotton crêpon. **1930** T. S. Eliot tr. *St.-J. Perse's Anabasis* 37 The girls' camiknickers hanging at the windows. **1937** Partridge *Dict. Slang* 122/2 Cami-knicks. **1952** B. Hamilton *So Sad, so Fresh* iv. 35 How are the cami-knicks coming along, Miss Demarest? **1966** *Punch* 5 Jan. 20/3 Her dress falls off to reveal eau-de-nil camiknickers.

camion (kæ·miǫn). Restrict † *Obs.* to sense in Dict. and add: Also, a large dray; a lorry; a bus.

1885 Warren & Cleverly *Wand. Beetle* 102 The ponderous camion thundered over the uneven pitching of the streets. **1922** *Blackw. Mag.* Jan. 23/2 The French..had in their turn a splash of khaki on each of their *camions*. **1942** 'N. Shute' *Pied Piper* 46 There's lots and lots of camions and motors at the station. **1967** *Punch* 15 Feb. 234/1 Following the kerb does enable one to see past the camion along straight roads.

camisole. Add: **2. b.** An underbodice, often embroidered and trimmed with lace. Also *attrib.*

1866 Advt. in A. Adburgham *Shops & Shopping* (1964) xii. 133, 2 Long-cloth petticoats..3 camisoles. **1895** *Army & Navy Co-op. Soc. Price List* 1062 Tucked Camisoles. **1903** *Daily Chron.* 14 Mar. 8/4 There are many camisole patterns. **1906** *Ibid.* 7 May 11/6 Camisole machinists. **1916** *Sphere* 19 Aug. p. vi/1 Cambric and nainsook combinations are..becoming obsolete, the camisole-knicker or camisole-chemise taking their place.

camman, var. *Caman.

camomile. 2. camomile-tea (later examples).

1965 *New Statesman* 30 Apr. 674/3 At my daughter's kindergarten, the children drink camomile tea, not milk, at break. **1968** G. Butler *Coffin Following* ix. 200 She sipped her camomile tea delicately.

camote (kāmōᵘ·te). [Mexican Sp., ad. Nahuatl *camotli.*] A name in Mexico and other Spanish-speaking countries for any one of several tuber-bearing plants, e.g. the sweet potato and yam.

1842 'A. T. Myrthe' *Ambrosio de Letinez* I. 143 This is not a good country for praties, no how—save for the swate ones, that the Mexican people call *camotes*. **1885** *Encycl. Brit.* XVIII. 751/1 Sweet potatoes (camote)..are pretty generally grown [in the Philippines]. **1909** *Teachers' Assembly Herald* II. 44/2 [To] relieve the Igorot girls from their laborious toil in the camote patch.

camoudie (kāmū·di). Also **-di, camoodie, -oedi, -udi.** [Native name.] A boa constrictor of tropical America.

1825 Waterton *Wand. S. Amer.* i. 11 The Camoudi snake ..from thirty to forty feet long. **1851** W. H. Brett *Ind. Missions in Guiana* 35 The reptile, a large camudi, sprang upon him. **1866** R. Duff *Brit. Guiana* 42 The camoedi was killed with a cutlass. **1887** *Timehri* Dec. 202 A very fine skin of a large Land-camoodie. **1898** H. Kirke *25 Yrs. Brit. Guiana* 90 A camoudie shot over him and wrapped itself round his body. **1904** W. H. Hudson *Green Mansions* ii. 31 If dangerous creatures had existed there—tigers, or camoodis, or solitary murderous savages.

camouflage (kæ·mŭflāʒ), *sb.* [Fr., f. *camouflet* Camouflet.] The disguising of any objects used in war, such as camps, guns, ships, by means of paint, smoke-screens, shrubbery, etc., in such a way as to conceal it from the enemy; also, the disguise used in this way; freq. *attrib.*

1917 *Daily Mail* 25 May 4/4 The act of hiding anything from your enemy is termed 'camouflage'. **1917** *Ibid.* 16 July 5/3 The King paid a visit to what is called a camouflage factory. **1922** C. E. Montague *Disenchantment* viii. 108 A French aerodrome across which the French camouflage painters had simply painted a great white high-road. **1948** *Sci. News* VII. 84 Features..which in a normal colour photograph might escape detection could often be clearly differentiated in this 'camouflage detection' film. **1957** *Granta* 9 Mar., We took home a few pieces of camouflage-painted aluminium.

b. *transf.* and *fig.*

1918 G. B. Shaw *Pen Portraits & Reviews* (1932) 35 The first necessity of such souls when truth is about, as it always is, is camouflage, or, better still, complete cover. **1920** R. Macaulay *Potterism* i. iii, It's a very laudable object, and needs no camouflage. **1963** V. Nabokov *Gift* iv. 251 Striped and spotted with words, dressed in verbal camouflage, the important idea he wished to convey would slip through.

camouflage (kæ·mŭflāʒ), *v.* [f. the sb.] To conceal by or as by camouflage. So **ca·mouflaged** *ppl. a.*, **ca·mouflaging** *vbl. sb.*

1917 *Daily Mail* 16 July 5/3 The King saw all the latest Protean tricks for concealing or, as we all say now, for 'camouflaging' guns, snipers, observers. **1918** W. Owen *Let.* Apr. (1967) 545 Your portrait is certainly slightly camouflaged. **1919** Downing *Digger Dialects* 15 *Camouflaged Aussy*, an Englishman serving with the A.I.F. **1920** *Blackw. Mag.* Mar. 332/1 Number One, ensconced in the little camouflaged control. **1921** *Spectator* 23 Apr. 518/2 The house telephone.., its extremely ugly box 'camouflaged' with the pattern of the paper. **1922** W. J. Locke *Tale of Triona* ii. 21 These are real eggs, although they're camouflaged in a Chinese scramble. **1924** Galsworthy *White Monkey* ii. xi, Queer how Nature camouflaged her schemes. **1970** *Daily Mail* 24 Mar. 4/9 Television needs a lot more money to improve quality and variety. Colour merely camouflages the same old material.

camouflet. Add: **2.** A subterranean cavity formed by a bomb exploding beneath the surface of the earth.

1941 *Nature* 17 May 596/2 If the bomb is too deep, the breaking of the surface does not occur; it is simply heaved up and drops back more or less into place. This is called a 'camouflet'.

camp, *sb.*[2] Add: **4. c.** *Austral.* and *N.Z.* A resting or assembly place (of sheep or cattle). Also *attrib.*

1891 D. Ferguson *Bush Life* xxiv. 170 A long string of lambs and ewes..all making off as fast as they could for their camp. **1946** F. Davison *Dusty* x. 107 [The dog].. startled a few sheep huddled in camp. **1950** *N.Z. Jrnl. Agric.* Aug. 141 Variegated thistle established on a stock camp site under a tree.

d. Quarters for the accommodation of detained or interned persons, as *concentration camp.

1917 *Sphere* 10 Feb. 128 Several copies have reached England of *The Ruhleben Camp Magazine* issued by the prisoners. *Ibid.*, The Lancashire and Cheshire civilians—who number over 500—in the camp [in Germany]. *Ibid.*, A parody of *The Mikado*..which jovial play seems to have been performed in the internment camp.

6. b. A local division or lodge of a society or league. *U.S.*

1880 Tourgée *Fool's Err.* ii. v. 415 Sometimes several 'camps' or 'dens' [of the Ku-Klux] would, independently of each other, direct a warning to be sent to the same individual. **1904** Harben *Georgians* 132 The general is invited to address nearly all the veteran camps over the State when the badges of honor are presented once a year.

8. a. (Further examples.) So *to have a foot in both camps*, to belong to or sympathize with two opposite groups, factions, etc.

1933 J. G. Cozzens *Cure of Flesh* ii. iii. 157 You never know when they may pull a fast one on you. I think you're in the wrong camp, George. **1935** W. Empson *Versions of Pastoral* vi. 217 The divine Polly has a foot in both camps. **1958** *Listener* 6 Nov. 715/1 The world is in fact divided into two camps, Communist and anti-Communist, with a number of uncommitted nations standing on the sidelines.

15. *S. Afr.* [ad. Afrikaans *kamp.*] A fenced-in portion of a farm.

1877 *Queenstown Free Press* 25 Sept. (Pettman), He purchased three birds to establish a camp at Somerset East in 1853. **1883** O. Schreiner *Story Afr. Farm* I. i. 5 The.. Englishman, whose grave lay away beyond the ostrich-camps. **1896** R. Wallace *Farm. Ind. Cape Col.* xi. 223 Ostriches require to be enclosed in camps. **1947** H. C. Bosman *Mafeking Road* 60 The wire he had borrowed from me for his new sheep-camp.

VII. b. camp-ground *U.S.*, (*a*) the site of a camp-meeting; (*b*) a camping-ground; **camp-muster** *Austral.* (see quot.); **camp-oven** *Austral.* and *N.Z.* (see quot. 1933); **camp-stool** (earlier examples).

1806 L. Dow *Travels* II. 94, I viewed the Camp-ground, and preparations making for the meeting. **1816** U. Brown *Jrnl.* in *Maryland Hist. Mag.* XI. 360 Their Pilot..never could find their camp ground. **1895** *Outing* (U.S.) Dec. 254/2 The soil of the camp-ground is light and sandy. **1968** *North Carolina Travelbk.* 1968–69 20/2 Easily accessible public camp grounds, picnic areas, small lakes, fishing streams. **1933** *Bulletin* (Sydney) 9 Aug. 21/3 The camp muster was an annual event in the old days before general fencing, when every station had a general muster on the main cattle camps, and men from all the stations came along to identify and cut out their own cattle. **1846** H. Weekes *Jrnl.* 25 Feb. in Rutherford & Skinner *New Plymouth Settlement* (1940) i. vii. 118 Our cooking was now an open-air affair..with a camp-oven and gypsy kettles. **1900** H. Lawson *On Track* (1945) 62 There was a camp-oven with a leg of mutton and potatoes sizzling in it on the hearth. **1933** L. G. D. Acland in *Press* (Christchurch, N.Z.) 23 Sept. 13/7 *Camp oven*, an iron pot with three short legs and a flat top, so that it can be used to boil, bake, or fry in. **1968** K. Weatherly *Roo Shooter* 21 There's a roast leg in the camp oven. **1794** R. Wellford *Diary* 29 Sept. in *William & Mary College Q.* (1903) XI. 5 To have three Canvass Camp stools made directly, which will serve for seats in the day, & a bedstead at Night. **1817** Jane Austen *Sanditon* (1954) iv. 383 Two Females..with their books & camp stools.

camp, *v.*[1] Add: **2. c.** *trans.* To excel or surpass in a contest. *Austral.*

*a***1882** H. C. Kendall *Poems* (1886) 207 At punching oxen, you may guess There's nothing out can 'camp' him.

camp, *v.*[2] Add: **1. b.** With *down.* *lit.* and *fig.* *U.S.*

1781 T. Johnson *Jrnl.* 8 Mar. in G. Powers *Hist. Sketches Coos* (1841) 197 Camped down on the River Lamoille this night. **1850** W. Colton *3 Years Calif.* 310, I have seen this *savan* camp down and snore soundly through the night. **1869** L. M. Alcott *Lit. Women* (1870) II. vii. 100 I'll be hanged if I don't make them camp down before her table afterward. *a* **1888** *Spirit of Times* (Farmer), They..camped down a smart piece off the trail.

2. b. (Earlier examples.) Also *transf.* and *fig.*

1748 Washington *Jrnl.* 18 Mar. in *Writ.* (1889) I. 3 We camped out in ye field this night. **1835** A. B. Longstreet *Georgia Scenes* 9 The old gentleman and his lady had consented to *camp out* for a day. **1853** Mrs. C. Clacy *Lady's Visit to Gold Diggings* iii. 33 We..determined to 'camp out' as much as possible. **1901** 'L. Malet' *Hist. Sir R. Calmady* ii. ii. 96 He..took to camping-out on one of the broad window-seats of the Long Gallery.

3. (Later examples.)

1920 J. M. Hunter *Trail Drivers of Texas* 63, I told the cook..to take the wagon and camp it up the river. **1930** L. G. D. Acland *Early Canterbury Runs* v. 98 An old circular manuka yard which had been put up to camp the sheep in at night.

4. *intr.* Of birds: to flock together, gyrate in the air (*dial.*). Of sheep or cattle: to flock together, usu. for rest or at night (*Austral.* and *N.Z.*).

1847 A. Harris *Settlers & Convicts* xii. 234 A flock of sheep 'camping', as the shepherds call it, under the shade of a tree from the noon-tide heat. **1879** *Norfolk Archæol.* VIII. 168 'The rooks are camping' is an expression often heard in the autumn when those birds assemble together and gyrate in the air. **1933** L. G. D. Acland in *Press* (Christchurch, N.Z.) 16 Sept. 15/7 Sheep which camp together are also said to 'break camp' when they move off to feed at dawn. **1938** F. S. Anthony in D. M. Davin *N.Z. Short Stories* (1953) 220 The first thing we saw was our twenty cows, camped alongside the fence, chewing their cuds.

camp (kæmp), *v.*[3] *slang.* [Etym. obscure.] **a.** *trans.* To make (something) 'camp' (see *camp *a.*); esp. in phr. *to camp it up*, to use exaggerated movements, gestures, etc., to over-act. **b.** *intr.* To be 'camp'; to be or behave like a homosexual.

1931 *New Broadway Brevities* (N.Y.) ii. 10/1 Boys and men with painted faces and dyed hair flaunt themselves camping and whooping for hours each night. *Ibid.* 10/2 His greatest triumph had been the management of a drag of his own at Bryant Hall, overlooking Bryant Park, so that most of the customers merely crossed the street to camp in a hall instead of a park. **1959** *Spectator* 13 Nov. 667/3 Most of the time he camps it up for sniggers with manly gestures fading into womanly wriggles. **1960** *Encounter* Feb. 23/2 He..had taken to an open, defiant 'camping'. **1962** R. Cook *Crust on its Uppers* Foreword, You all love to shiver and say 'Ooh!' and camp about at the mention of the word 'crime'. **1965** G. Melly *Owning-Up* xv. 188 We were all very impressed by the thought of being used [in a film] as the basis for characters and camped it up like mad.

camp (kæmp), *a.* (and *sb.*[5]) *slang.* [Etym. obscure.] Ostentatious, exaggerated, affected, theatrical; effeminate or homosexual; pertaining to or characteristic of homosexuals. So as *sb.*, 'camp' behaviour, mannerisms, etc. (see quot. 1909); a man exhibiting such behaviour.

1909 Ware *Passing Eng.* 61/2 *Camp* (*Street*), actions and gestures of exaggerated emphasis. Probably from the French. Used chiefly by persons of exceptional want of character. 'How very camp he is.' **1931** *New Broadway Brevities* (N.Y.) ii. 7/1 (*heading*) Drags, camps, flaunting hip-twisters and reefer peddlers run afoul of cops on the lam. **1933** M. Lincoln *Oh! Definitely* vi. 62 Dennis, slightly more 'camp' than usual, opened the front door. **1941** S. J. Baker *Dict. Austral. Slang* 16 *Camp* (adj.), homosexual. **1952** A. Wilson *Hemlock & After* i. v. 101 The..gossip of the golden spiv group...the 'camp' end of the room. *Ibid.* ii. i. 112 The incoherence of his speech, the..absence of the customary 'camp'. *Ibid.* iii. i. 191 Whether Terence was really 'queer'..how much happier he was when he was not being 'camp'. **1954** C. Beaton *Glass of Fashion* viii. 153 Hearty naval commanders or jolly colonels acquired the 'camp' manners of calling everything from Joan of Arc to Merlin 'lots of fun', and the adjective 'terrible' peppered every sentence. **1954** C. Isherwood *World in Evening* ii. iii. 125 High Camp is the whole emotional basis of the Ballet.. and of course of Baroque art. **1956** L. McIntosh *Oxford Folly* vii. 103 'He was—you know—one of *those*'..'What, a pansy?' 'That's right,' said Julian, 'he was camp.' **1959** *Observer* 1 Feb. 17/1 The cute little dirty chuckle and the well-timed 'camp' gesture have made stage and audience indistinguishable from any would-be-smart cocktail-party. **1964** S. Sontag in *Partisan Rev.* XXXI. 515 (*title*) Notes on 'Camp'.

campaign, *sb.* Add: **5. b.** (Earlier U.S. examples.) Also *attrib.*

1809 J. Steele *Papers* (1924) II. 601 The electioneering campaign having become much warmer than I had anticipated. **1844** *Talladega* (Ala.) *Dem. Watchtower* 12 June 2/6 We issue our Campaign Paper to meet the wants of numerous Associations. **1857** S. Bowles *Let.* 16 Feb. in G. S. Merriam *Life & Times S. Bowles* (1885) I. xxv. 291 We should get those amendments out of the way before we strike out for the summer campaign. **1880** *Daily Union* (San Diego) 1 Sept. 1/1 He will take care of the Indiana campaign expenses, with the understanding that he be not called on to contribute to the campaign fund for other states. **1882** *Nation* 21 Dec. 522/3 Garfield seems to have reposed great confidence in Dorsey as a campaign manager. **1907** *Daily Mail* 14 Jan. 7/3 A born campaign-manager. **1968** W. Safire *New Lang. Politics* 62/2 Just as there is an unwritten law that takes for granted campaign exaggerations, there is also an unwritten law that a candidate may not afterward disown his campaign statements.

campaigned (kæmpē[i]·nd), *ppl. a.* [f. Campaign *v.* + -ed[1].] Of estates in Ireland which came under the Plan of Campaign (see Campaign *sb.* 5 c).

1889 *Daily News* 24 July 3/3 Much light has been thrown on the grievous state of affairs..on the 'campaigned' estate of Mr. Leader in county Cork. **1892** *Sat. Rev.* 1 Oct. 381/2 Neither Mr. Dillon..nor Mr. Redmond..has as yet made it his business to stump the 'Campaigned' districts.

campan (kæ·mpæn, ‖kaṅpaṅ). [Name of a town in the French Pyrenees.] A marble of which there are several varieties.

1795 Kirwan *Elem. Min.* (ed. 2) I. 115 The sp. gr. of green campan is 2,741. **1839** *Civil Eng. & Arch. Jrnl.* II. 452/2 Campan marble proper unites all these three varieties by very large stripes.

campana. Add: **4.** A bell-shaped vase (see quot. 1957).

1802 *Christie's Catal. Etruscan Vases J. Clark Collection* 4 Two Vases, of the form called *Campana. Ibid.*, A campana Vase. *Ibid.* 5 A Vase of the campana form. **1957** Mankowitz & Haggar *Encycl. Eng. Pott. & Porc.* 41/2 *Campana vase*, a vase made in the neo-classic style, during the first decade of the nineteenth century.

campanularian (kæmpæniulē[ə]·riăn), *a.* and *sb.* *Zool.* [f. Campanularia + -an.] Pertaining to, a hydroid of, the genus *Campanularia* or to calyptoblast hydroids generally. So **campanula·ridan.**

1868 T. Hincks *Brit. Hydroid Zoophytes* p. iv, Some very skilfully mounted specimens of Campanularian and other zoophytes. **1883** *Science* I. 197/1 The Leptomedusæ..from the campanularian hydroids. **1888** Rolleston & Jackson *Anim. Life* 247 In addition to the possession of hydrothecae, a Campanularian differs from a Tubularian in three important respects. *Ibid.* 766 The destruction of Campanularidan colonies by parasitic Protophytes. **1898** *Naturalist* 218 The empty campanularian cups. **1940** L. H. Hyman *Invertebrates* I. vii. 370 Calyptoblastea or Leptomedusae or Thecophora, the campanularian hydroids.

Campari (kæmpā·ri). The name of an Italian firm, used to designate an aperitif. Also *attrib.*

1923 A. Huxley *Antic Hay* vi. 95 Cinzano, Bonomelli, Campari—illustrious names. **1930** H. Craddock *Savoy Cocktail Book* i. 115 'Old Pal' Cocktail. ⅓ Canadian Club Whisky. ⅓ French Vermouth. ⅓ Campari. **1959** 'M. Ainsworth' *Murder is Catching* xv. 171 He poured himself a half-glass of Campari and put an olive in it. **1959** P. Moyes *Dead Men don't Ski* iii. 28 A dark, smiling girl served them with Campari-sodas. **1960** *Spectator* 23 Dec. 1022 An account of their campari-drinking and conversations.

Campbellite. Add: (Earlier examples.) Also *attrib.*

1830 *Massachusetts Spy* 22 Dec. (Th.), Elder Rigdon, one of the early Mormons, is described as having been 'a Campbelite leader of some notoriety'. **1834** J. M. Peck *Gaz. Illinois* i. 91 The Cambellites, or 'Reformers'..have several traveling, and a number of stationary preachers. **1847** J. Palmer *Jrnl. Trav. Rocky Mountains* 23 A Campbellite preacher, named Foster, was reading a hymn.

2. *transf.* A sunfish of the Mississippi region.

1872 *Harper's Mag.* July 315/2 'What do you call those fish?' 'Campbellites,' promptly responded the boy. **1884** Goode *Nat. Hist. Aquatic Anim.* 407 The Crappie—*Pomoxys annularis*...Other names are..'New Light' and 'Campbellite' in Kentucky and Indiana.

camper[2]. **3.** (Earlier example.) *U.S.*

1806 L. Dow *Trav.* II. 61 We held Quarterly-meeting on Clarke's creek; some supposed I would get no campers... This prepared the way for the Camp-meeting.

‖ **campesino** (kæmpəsī·no). [Sp.] A peasant farmer.

1937 *New Statesman* 20 Nov. 834/1 He stopped in the next village and..took on board..an aged *campesino* bearing two live cockerels. **1961** *Times* 10 Oct. 10/5 Fried banana, which is the staple of the Colombian *campesino*. **1965** C. D. Eby *Siege of Alcázar* (1966) iii. 70 A ragged *campesino* had smashed an image of the Virgin in San Vicente's Church. **1969** J. Mander *Static Soc.* vii. 214 The Church has..by distributing transistor radios to the *campesinos* of the backlands,..ensured herself a captive audience.

camph-. Add: **ca·mphane** [-ane], an inert, saturated bicyclic terpene, $C_{10}H_{18}$; now usu. called *bornane*; **ca·mphoid** [-oid], a preparation of pyroxylin and camphor in alcohol; **ca·mpholide** (also †*-id*) [*-olid* (see quot. 1896)], either of two isomers of the lactone $C_{10}H_{16}O_2$, a crystalline substance resembling camphor.

1895 *Jrnl. Chem. Soc.* LXVIII. i. 426 (*heading*) Action of Nitrous acid on Oximes of the Camphor (Camphane) Series. **1932** J. L. Simonsen *Terpenes* II. v. 221 The saturated dicyclic hydrocarbon, camphane, $C_{10}H_{18}$, is of outstanding importance, since it is the parent hydrocarbon of the camphor group. **1952** *Chem. & Engin. News* 3 Mar. 930/3 The parent of borneol and camphor..is called bornane rather than camphane because of conflicting names beginning with camph-. **1897** F. P. Foster *Reference-bk. Pract. Therapeutics* I. 203/2 *Camphoid*, a solution of 1 part of soluble gun-cotton and 20 parts of camphor in 20 parts of absolute alcohol, is used like collodion. **1929** Woolley & Forrester *Pharmaceutical Formulas* (ed. 10) I. 54/2 Camphoid. Collodion Substitute.... Used as a vehicle for iodoform, salicylic acid,..and ichthyol. **1896** M. O. Forster in *Jrnl. Chem. Soc.* LXIX. i. 39 The isomeric lactone will be called *campholid*, the termination *olid* indicating, according to systematic nomenclature, a compound of lactonic character. **1938** *Thorpe's Dict. Appl. Chem.* (ed. 4) II. 247/2 Reduction of camphoric anhydride with sodium amalgam..or catalytically..gives α-campholide, whilst reduction of β-methyl hydrogen camphorate gives β-campholide, m.p. 218°–220°.

camphor, *sb.* Add: **3.** *camphor julep* (examples); **camphor-chest, -trunk** *U.S.*, a clothes-chest containing camphor as a protection against moths; **camphor ice** *U.S.*, a solid preparation of camphor; **camphor laurel** *Austral.* (see quots.); **camphor-wood**, the popular name for several trees (including prec.) of the families Dipterocarpaceæ and Lauraceæ, or the wood from these trees, which is fragrant.

1861 Mrs. Stowe *Pearl of Orr's Isl.* i. viii. 59 That ar shawl your mother keeps in her camfire chist. **1889** R. T. Cooke *Steadfast* v. 59 Mrs. Dennis was packing away blankets in the camphor chest upstairs. **1880** A. A. Hayes *New Colorado* xv. 197 In the alkali regions, glycerine, or what is called 'camphor ice', should be used on face and hands. **1788** F. Burney *Diary* 5 Nov. (1891) III. 61, I gave her some camphor julep. **1835** Dickens *Let.* (1965) I. 73, I hope you are well and have taken some Camphor Julep. **1894** *Proc. R. Soc. Queensland* XI. 23 The identity of the camphor of *Cinnamomum oliveri* with that of Camphor Laurel was proved by me. **1897** R. T. Baker in *Proc. Linn. Soc. N.S.W.* XXII. 282 *Cinnamomum virens*, sp. nov. 'Wild Camphor Laurel'. **1932** R. H. Anderson *Trees N.S.W.* 50 Camphor Laurel (*Cinnamomum camphora*). A handsome, dense-topped tree. *Ibid.* 127 *Cinnamomum virens* is sometimes known as the Native Camphor Laurel. **1869** Mrs. Stowe *Oldtown Folks* 34 Mrs. Major had a real *Ingy* shawl up in her 'camphire' trunk. **1895** *Century Mag.* July 323/2 Cedar-chest and camphor-trunk and flowered bandbox have been called upon to disgorge their treasures. **1923** Dallimore & Jackson *Handbk. Coniferæ* ii. 173 *Callitris verrucosa*, R. Brown. Turpentine Pine..Camphor Wood; Rock Pine. **1950** C. W. Bond *Colonial Timbers* 55 In spite of its hardness Borneo camphorwood is pleasant to work. *Ibid.* 63 East African camphorwood is interesting. **1955** *World Timbers* (Timber Development Assoc.) II. 9 Borneo Camphorwood [Dryobalanops aromatica] should not be confused with East African Camphorwood (*Ocotea usambarensis*)..nor with true camphorwood (Cinnamomum camphora). **1965** *Austral. Encycl.* II. 247/2 Camphorwood is the name standardized in the timber trade for the indigenous C[*innamomum*] *oliveri.*

Campignian (kæmpi·nyăn), *sb.* and *a.* *Archæol.* [ad. F. *campignien* (P. Salmon in *Dict. Sci. Anthrop.* (1886)), f. Le *Campigny* (Seine-Maritime, France), the type site.] **A.** *sb.* A Stone Age culture of Europe. **B.** *adj.* Of or pertaining to this culture.

1921 M. C. Burkitt *Prehistory* xii. 156 The early Neolithic people (Campignians) owe nothing to their predecessors. *Ibid.* xiii. 164 The early Campignian industries also occur in Ireland. **1927** Peake & Fleure *Hunters & Artists* 114 Campignian picks have been found in most parts of Europe except the extreme south. **1942** H. L. Movius *Irish Stone Age* iv. 239 Although the main roots of the Campignian are Mesolithic, the real development began in Neolithic times.

campimeter (kæmpi·mĭtəɹ). [f. L. *campus* field + -meter.] An apparatus designed to measure the field of vision, or the sensitivity of the retina to colour and space in indirect vision. Hence **campi·metry**, the use of the campimeter, the measuring of retinal areas; **campime·trical** *a.*

1889 *Buck's Handbk. Med. Sci.* VII. 660/1 The campimeter of De Wecker. **1901** Titchener *Exper. Psychol.* I. 9 The Distribution of Colour Sensitivity over the Retina: Campimetry. *Ibid.* 10 The campimeter must be set so high above the mixer that no shadow is cast by it upon the spot of colour seen through the circular opening. *Ibid.* 12 The experiment may be repeated with as many combinations of the colour discs as the time allotted to campimetrical work permits. **1964** S. Duke-Elder *Parsons' Dis. Eye* (ed. 14) xiii. 138 For more accurate investigation of details campimetry must be employed.

Campine (kæmpī·n). [The name of a district of Belgium.] A breed of domestic fowl, resembling the Hamburgs.

1892 Simmonds *Dict. Trade* Suppl., *Campine*, a Belgian breed of fowls. **1902** *Encycl. Brit.* XXXI. 876/1 The hardy Campine or Braekel, resembling the pencilled Hamburgh in plumage, but larger and with a single comb, and laying a large egg in great numbers. **1924** *Glasgow Herald* 26 Apr. 5 Few birds make a prettier show than the still uncommon Silver Campine, a gallant little Belgian.

‖**campo**[2] (kæ·mpo). [American Sp., f. Pg. or Sp. *campo* field, open country (see Camp *sb.*[2]).] In Brazil, a grass plain with occasional stunted trees, a savannah. Also *attrib.*

1863 Bates *Nat. Amazons* II. i, The country around Santarem..is a *campo* region; a slightly elevated and undulating tract of land, wooded only in patches, or with single scattered trees. **1931** L. D. Stamp in W. Rose *Outl. Mod. Knowledge* 831 The campos of Brazil. **1944** S. Putnam tr. *E. da Cunha's Rebellion in Backlands* ii. 79 Leaving the Alpine-like regions,..he will cross the great campos, a huge arena made to the measure of..the vaqueiros.

campodean (kæmpōᵘ·diăn), *a.* and *sb.* *Ent.* [f. mod.L. *Campodea*, f. Gr. κάμπη caterpillar + -ode: see -an.] (An elongated wingless insect) of the genus *Campodea.*

1895 *Naturalist* 62 The rediscovery of the blind campodean *Lipura stillicidii.*

campodeiform (kæmpōᵘ·dĭ,ifǭɹm), *a.* *Ent.* [f. *Campodea* (see prec.) + -form.] Resembling insects of the genus *Campodea*, used esp. of larvæ.

1888 Rolleston & Jackson *Anim. Life* 150 There are two types of larval (or young) Insecta: one known as Campodeiform, from a more or less close resemblance to the genus *Campodea* among *Thysanura*; the other as eruciform, of which a caterpillar may be taken as a good example. **1895** *Nature* 19 Dec. 155 Lubbock's far bolder attempt to derive his Campodeiform larva from a Rotifer-like ancestor. **1959** E. F. Linssen *Beetles* I. 28 The campodeiform type has the head, thorax, and abdomen well defined, and possesses three pairs of well-developed legs.

campoi (kæmpoi·). [Cantonese pronunc. of Chinese *kien* picked, selected + *pei* fire.] A fired variety of Congou tea.

1842 *Penny Cycl.* XXIV. 291/1 An Account of the Quantity and Prices of several sorts of Tea sold in England..: Bohea..Congou..Campoi..Souchong..Pekoe. **1860** Ure *Dict. Arts* (ed. 5) III. 855 To the black tea belongs [*sic*] the varieties known as Bohea, Congou, Campoi, Souchong, Caper, and Pekoe. **1875** [see Congou].

campshed (kæ·mpʃĕd). Also 5 **camshide** 6 **cambshide**, 6–7 **camshed(d**, 7, 9 **campshead.** Early variant (the second element = Shide *sb.*) of Camp-shot, whence Camp-shed *v.*, Camp-shedding (*camp-sheading*), q.v.

1471 in P. E. Jones *Calendar of Plea & Memoranda Rolls 1458–82* (1961) 70 A tenement..at ende of vj fote of Assise from the Camshide and north-west Corner of the said wharf. **1531** in *Lett. & Pap. Hen. VIII* (1880) V. 183, 2 sawiers strangers, sawing with the sawers of the ordinary waigis, as nedylles, bynders, anckers, camp shedes, grete postes, planckes, and other necesares for the new frame of the est juttye. **1570** in *Crt. Min. Surrey & Kent Sewers Comm.* (1909) 85 To fill and to planke iiij roddes of the Cambshide againste the Thames. **1622** *Admir. Crt. Misc.* 1420, lf. 16 (MS.), The end of three piles at the topp of yt [*sc.* the wharf] are out of the campshead. **1632** in E. B. Jupp *Carpenters' Co.* (1887) 301 The making of Wharfes Camshedds Cranes and bridges of timber. **1841** *Civil Eng. & Arch. Jrnl.* IV. 106/1 These piles are placed from 3 to 4 feet apart from centre to centre, and have a capping or campshead 7 inches square. **1888** *Times* 26 Mar. 4 The starting boats were moored in mid-stream at Putney opposite the end of the campshed on the Fulham side.

camptonite (kæ·mptənəit). *Petrogr.* [a. G. *camptonite* (H. Rosenbusch *Mikrosk. Physiogr. d. Min. und Gesteine* (ed. 2, 1887) II. 333), f. *Campton*, a town in New Hampshire, U.S.A.: see -ite[1].] Any of a group of dark-coloured, usu. porphyritic dike-rocks resembling basalt and typically consisting of phenocrysts of hornblende or barkevikite in a matrix of plagioclase.

1895 Dana *Man. Geol.* (ed. 4) 87 Camptonyte.—Rock resembling diabase and dolerytе. Consisting of hornblende (as an original mineral of the rock) and probably anorthite. **1901** *Nature* 19 Sept. 513/1 The sills of camptonite and felsite intrusive in the Cambrian rocks. **1961** F. H. Hatch et al. *Petrol. Igneous Rocks* (ed. 12) iv. ii. 475 Localized intrusions of basic lamprophyric rocks including camptonites and monchiquites occur in or near the Ayrshire vents.

campus (kæ·mpǿs). orig. *U.S.* [a. L. *campus* field: see Camp *sb.*[2] First used at Princeton, New Jersey.] The grounds of a college or university; the open space between or around the buildings; a separate part of a university. Hence allusively, university or college life or people. Also *attrib.*

1774 in J. F. Hageman *Hist. Princeton* (1879) I. 102 Having made a fire in the Campus, we there burnt near a dozen pounds [of tea]. **1826** R. Mills *Statistics S. Carolina* 701 The whole disposed so as to form a hollow square containing about ten acres which is called the Campus. **1833** J. Finch *Trav. U.S. & Canada* 282 In front of the College is a fine campus ornamented with trees. **1879** H. J. Vandyke Jr. in *Princeton Bk.* 382 The central point of the Campus, the hub of the college world, is undoubtedly the big cannon. **1904** H. N. Snyder in *Sewanee Rev.* Jan. 87, I am almost willing to shut my eyes to the excesses of the noisy strenuosity of the athletic mood if it bring into the campus life a warm, vital sense of college unity. **1939** *Nature* 26 Aug. 392/1 Frome Avenue, on the opposite side of which is the [Adelaide] University campus. **1958** *Times* 10 Mar. 12/7 Not only in the cloistered courts of Cambridge but also on college campuses in America. **1958** *Sunday Times* 27 Apr. 20/7 The first walls are rising of Sir Hugh Casson's new arts faculty campus. **1959** *Listener* 19 Feb. 326/2 As in this country, some of the best of this kind of history is written off campus. **1968** *Brit. Univ. Ann.* 30 To my eye, the Birmingham campus has now developed into one of the most attractive in the country.

Hence as *v. trans.*, to confine to the campus (see also quot. 1928). *U.S. colloq.*

1928 *Amer. Speech* III. 218 When a Freshman is campussed..he is not actually confined to the campus but merely deprived of some privilege or other. **1949** 'N. R. Nash' *Young & Fair* 79 You've campused Dru Eldridge?

campy (kæ·mpi), *a.* *slang.* = *camp *a.*

1959 *Spectator* 2 Jan. 11/2 Kenneth Williams and Ted Durante as the campy sisters [in *Cinderella*] manage to seem five times the size of the two stars. **1961** *Harper's Bazaar* Apr. 138/2 Noël Coward (with his campy diaeresis above the 'e'). **1965** *New Statesman* 20 Aug. 263/1 The movie has been called 'campy', and yet I have never seen comedy direction of such conviction.

can, *sb.*[1] Add: **1. c.** A revolving cylinder open at the top to receive the sliver from a carding-machine. Also *attrib.* and *Comb.*

1825 J. Nicholson *Oper. Mech.* 382 The roving-frame.. used..where mule-spinning is carried on..is termed the *can roving-frame.* **1844** G. Dodd *Textile Manuf.* i. 30 In the 'can-roving frame',..the cardings coming from two cans, and passing between the pairs of rollers, become elongated and fall into the can. **1853** Ure *Dict. Arts* (ed. 4) II. 697 The missing band or sliver was supplied out of a can, being the produce of a single carding-engine working into cans. **1861** [see Sliver *sb.*[1] 2 *attrib.*]. **1882** *Spon's Encycl. Industr. Arts* V. 2091 The several drawings are passed into a coiling-can, by which they are loosely twisted into one. **1912** Barker & Priestley *Wool Carding* 201 A 'can' delivery apparatus. *Ibid.*, If the box is a 'can' box the sliver passes through a funnel and press rollers into a cylindrical can. *Ibid.* 228 The slivers are fed into the comb just as if they were fed out of the can-coiler from the carder.

d. A large, usu. cylindrical, metal container or bin, as in **ash-can*, **garbage can*. Chiefly *U.S.*

e. A lavatory, water-closet. *U.S. slang.*

1900 *Dialect Notes* II. 26 *Can*, water-closet. **1933** A. Woollcott *Let.* 22 Feb. (1946) 97 He was always organizing quartets and being fired for practicing in the can during working hours. **1951** J. D. Salinger *Catcher in Rye* x. 90 She kept saying..corny..things, like calling the can the 'little girls' room'.

f. *to carry* (or *take*) *the can* (*back*) [origin of use unkn.; said to be the 'beer-can' which one (soldier) carries for all his companions]: to bear the responsibility, take the blame (also see quots.). *slang* (orig. Services').

1929 Bowen *Sea Slang* 22 *A Can*, a reprimand. *Ibid.* 23 *To Carry the Can*, to be reprimanded (Navy). *Ibid.* 137 *To Take the Can Back*, to be reprimanded. **1936** *Daily Herald* 11 Aug. 8/5 Railwaymen and road transport workers use the phrase, 'Taking the can back' for (respectively) being held responsible for a mishap and being imposed upon. **1938** F. D. Sharpe *S. of Flying Squad* 333 '*Taking the Can Back*', being left to do the dirty work. **1943** Hunt & Pringle *Service Slang* 20 *Carrying the can back*, accepting the blame for your own or another's error. **1957** *Times* 23 Feb. 7/4 Senior officers who were forced to 'carry the can' because of the misdeeds of others. **1959** D. Barton *Loving Cup* i. 11 Officially you have to take the can. **1959** J. Braine *Vodi* x. 148 It's always my fault, everything's my fault. I always carry the bloody can back. **1959** *Even. Standard* 6 July 5/2 He has enough political nous not to wish to carry the can for people like Aneurin Bevan. **1967** *Spectator* 30 June 763/3 No Department wanted to carry the can for cable-vetting—quite irrespective of security considerations.

3. (Earlier and later examples.)

1867 A. D. Richardson *Beyond Miss.* 147 Mitchell..was fined two cans of oysters for contempt. **1941** *Manch. Guardian Weekly* 26 Sept. 206/1 Cans of food are kept in the factory for a period before distribution. *Ibid.*, The can of carrots also contained gases at high pressure. **1968** *Gloss. Terms Mechanized & Hand Sheet Metal Work* (*B.S.I.*) 26 *Can*, lightweight container (usually for processed foods) fabricated from separate components, i.e. bottom, body and lid (or cover). (Usually known as an open-top can.)

b. A prison. *slang* (orig. *U.S.*).

1912 D. Lowrie *My Life in Prison* xi. 125, I was in th' can ag'in, up against it f'r robbery. **1926** J. Black *You can't Win* xv. 216 Those two..never allowed any of 'their people' to languish in the 'can' overnight. **1961** *20th Cent.* Mar. 236 I'll stand by my man Though he's in the can.

c. *in the can*: of a cinema film or sequence, completed. *colloq.* Cf. *can *v.*[3] b.

1930 C. Beaton *Diary* in *Wand. Years* (1961) 193 The scene was 'in the can' after twenty-five retakes. **1934** *Tit-Bits* 31 Mar. 12/2 When a film is completed it is 'in the can'. **1946** 'Brahms' & Simon *Trottie True* i. vi. 158 John Bunny..was stirring a bucket of whitewash...When this shot was in the can Trottie..was due to come along..and push him into it. **1968** *Observer* (Colour Suppl.) 17 Nov. 48/4 Godard got the film into the can on schedule.

d. A protective jacket covering the fuel element in a nuclear reactor (see quot. 1962).

1945 H. D. Smyth *Devel. Methods of using Atomic Energy for Mil. Purposes* 106 Mechanical jackets or cans of thin aluminum were feasible from the nuclear point of view. **1962** *Gloss. Terms Nuclear Sci.* (*B.S.I.*) 19 *Can*, a container used for a fuel element in a reactor to prevent the escape of fission products and possibly corrosion of the fuel, and sometimes to increase the mechanical strength of the fuel rod.

4. can-opener orig. *U.S.* (see quot. *a*1877); = *tin-opener.*

a **1877** Knight *Dict. Mech.* I. 452/1 *Can-opener*, a domestic implement for opening cans containing fruit, oysters, and what not. **1936** *Economist* 4 Apr. 4/2 It is now possible to obtain complete meals in tinned form. The can-opener must surely be regarded as a sign of the times.

can, *v.*[1] Add: **A. 7.** *used to could*, a common phrase in certain dialects of England and in the United States for: used to be able to.

1823–1940 dialectal examples in Wentworth's *Amer. Dial. Dict.* (1944).

1827 J. F. Cooper *Prairie* II. xvi. 257 A small and trifling matter is it, to what I used-to-could offer in the way of bargains. **1848** Bartlett *Dict. Amer.* 372 *Used to could*, a vulgarism used in the Southern States for could formerly. **1848** A. B. Evans *Leicestersh. Words.* **1872** Schele de Vere *Americanisms* 646. **1892** in *Eng. Dial. Dict.* s.v. *Can*, Ah can't walk five mile i' t' hahr nah, but ah used to could! **1899**—in *Ibid.* Suppl. s.v. *Use.* **1940** *Sat. Even. Post* 6 Jan. 15/2 She used to could smell that old goat of yours from here to yonder.

B. 6. b. To be allowed to, to be given permission to; = May *v.*[1] 4 a. *colloq.*

1879 Tennyson *Falcon* 12 Can I speak with the Count? **1894** T. B. Reed *Dog with Bad Name* xv. 156 Father says you can come. **1905** *Ch. Times* 3 Feb. 136/3 No one can play the organ during service time without the consent of the Vicar.

8. c. Colloq. phr. *can do*: it is possible, it is within the power of (the speaker). So *no can do.*

1903 Kipling in *Collier's Weekly* 3 Oct. 16/3 'Four hundred and twenty knots'...'Can do,' said Moorshed. **1923** H. C. Witwer *Fighting Blood* v. 141 I've tried everything *I* know to get this gil to fight us and no can do! **1951** 'E. Crispin' *Long Divorce* iii. 29 'Can do, sir,' said Mogridge with watery affability. **1958** L. A. G. Strong *Treason in Egg* i. 11 'No. Sorry, old boy...No can do.'

IV. (Further example.) Also *can-doing*; *can't-doing*; *can't-eat*, *can't-work*: one who cannot eat, work, etc.

1842 J. Aiton *Cler. Econ.* iv. 204 Let me have a first-rate goer, a good 'can do',—not that I mean in general to ride fast. **1900** *Westm. Gaz.* 15 Dec. 1/3 Achievement..is open to fewer persons than can't-doing. *Ibid.* 2/1 That the art of can't-doing has sprung from (1) the difficulty; (2) the barrenness of can-doing. *Ibid.*, Can't-eat, when applied, say, to lobster, is one thing; can't-eat, when relating to bread, is quite another. **1904** *Daily Chron.* 24 Feb. 6/5 We have then the 'Can't Works'.

can, *v.*[3] Add: (Earlier U.S. example.)

1861 *Trans. Ill. Agric. Soc.* IV. 511 Good fruit..is always marketable in large cities..and much will be dried, or canned, for export.

b. *transf.* and *fig.* Esp., to record or preserve on film or as a recording. Cf. *can *sb.*[1] 3 c.

1865 *Atlantic Monthly* XV. 395/1 The copper vessel wherein Solomon had so cunningly 'canned' the rebellious Afrit. **1914** *Wireless World* July 246/1 In their efforts to 'can' colloquial expressions the students have issued a manifesto. **1935** *Punch* 23 Oct. 456/2 *Anna Karenina*..has been rapidly canned for Greta Garbo. **1936** [see *ad-lib *v.*]. **1940** H. G. Wells *Babes in Darkling Wood* iv. i. 307 Ten or twelve [discourses] on fundamental ideas which have to be embalmed or canned or potted or whatever you like to call it, upon steel gramophone records. **1958** *Times* 28 July 9/6 The progress towards 'canning' television programmes.

c. *spec.* To cover (the fuel element in a nuclear reactor) with a protective jacket. (Cf. *can *sb.*[1] 3 d.)

1945 H. D. Smyth *Devel. Methods of using Atomic Energy for Mil. Purposes* 87 No one..is likely to forget the 'canning' problem, i.e., the problem of sealing the uranium slugs in protective metal jackets. **1955** *Times* 18 Aug. 6/2 Zirconium is used in atomic plants for canning nuclear fuel elements. **1957** *Listener* 9 May 738/1 There is the metal used to enclose the uranium rods, to 'can' them as they say. At Windscale the canning was done with aluminium.

2. To discharge or suspend from a situation; to expel from school or college. *U.S. slang.*

1905 *Dialect Notes* III. 73 Jim..got canned for two weeks. **1911** H. Quick *Yellowstone N.* ii. 37 Did you get canned for letting me in? **1914** G. Atherton *Perch of Devil* ii. 269 They would merely be..canned—I beg pardon, fired. **1937** J. Steinbeck *Of Mice & Men* 41 Won't ever get canned 'cause his old man's the boss.

b. To stop, leave off (something); to 'cut out'. *slang* (orig. *U.S.*).

1906 H. Green *At Actors' Boarding House* 187 'Now, see here; can that line o' comedy!' shouted Miss Gray. **1912** A. H. Lewis *Apaches N.Y.* 20 'Can that black-jack guff,' he retorted. **1920** Wodehouse *Coming of Bill* i. ix. 99 Can the rough stuff, Colonel. **1934** G. B. Shaw *Too True to be Good* iii. 79 Can all that stuff, Sergeant. **1953** 'E. Ferrars' *Murder in Time* ix. 78 Carver winced at the noise. 'Can that bloody row, can't you?' he grunted.

caña (kæ·nyă). [Sp., = cane.] A spirit resembling rum made from sugar cane.

1881 E. W. White *Cameos fr. Silver-Land* I. 40 Caña (White Rum). **1904** Gallichan *Fishing Spain* 211 A poor substitute for whisky is the spirit called caña. **1918** *Blackw. Mag.* Apr. 446/1 Supper and the tot of caña sacred to such occasions. **1967** A. Lichine *Encycl. Wines* 536/1 The Uruguayans distil quantities of brandy...Caña—a type of rum—is popular.

Canader (kănē[i]·dəɹ). *University slang.* [f. Canad(ian + *-er[6].] A Canadian canoe.

1893 H. W. Greene in *Oxford Mag.* 18 May, Beneath the Magdalen shadows, We'll drift in a 'Canader' When afternoons are warm. **1899** A. D. Godley *Lyra Frivola* 24 O it's Youth in a Canader with the willow boughs to shade her. **1919** R. B. Dawson *Light & Shade in Sarawak* iv. 33 'Get in carefully, old man,' said he; 'it's like climbing into a *Canader* at Oxford.'

Canadian, *a.* and *sb.* Add in etym. [after F. *canadien*]. (Earlier and later examples.)

1568 Hacket tr. *Thevet's New found World* lxxvi. 123 *marg.*, The amiable maners of these Canadiens. *Ibid.* 124 *marg.*, How these Canadians doe chase the dere & other wilde beastes. **1697** C. Mather *Pietas* 103 How beneficial an Undertaking it would have been for them to have pursued the Canadien Business, for which the New-Englanders were now grown too Feeble. **1746** Shirley *Let.* 21 Nov. in E. Richard *Acadia* (1895) I. xii. 221 Some allowance may likewise be made for their bad situation between Canadians, Indians and English. **1757** in *General Orders of 1757 issued by the Earl of Loudoun & P. Lyman* (1899) 121 Two Canadian Prisoners..confind in yᵉ Fort at Albany. **1774** J. Langhorne *Country Justice* i. 18 Cold on Canadian Hills, or Minden's Plain, Perhaps that Parent mourn'd her Soldier slain. *c* **1789** *Encycl. Brit.* (1797) IV. 76/2 The commodities required by the Canadians from Europe are, wine, or rather rum;..linen; and wrought iron. *Ibid.* 77/2 The many specimens of profit to be made by the Canadian trade, at last induced the public to think favourably of it. **1836** *Penny Cycl.* VI. 215/1 The Canadians further urge that the province contains no aristocracy. **1862** *Brit. Canad. Rev.* Dec. 3 Shall we become a united people, truly Canadian in principle? **1887** *Trans. Hist. & Sci. Soc. Manitoba* XXVIII. 14/2 Thirty years ago, we, who speak French, were called by every one purely and simply 'Canadians'; others were known as English, Scotch or Irish. Lately the fashion has grown up of calling others Canadians and distinguishing us as French. **1907** *Yesterday's Shopping* (1969) 15/1 *Cheese*.. Canadian Cheddar, first quality, white or coloured—lb 0/8½. **1925** J. Joly *Surface-Hist. Earth* viii. 131 The mountains themselves have long ago been base-levelled by denudation to the existing peneplanes of the Canadian Shield, where they extend over an area of at least a million square miles. **1964** *English Studies* XLV. (Suppl.). 11 He has grouped languages of Canadian Indians together. **1969** *Nature* 6 Sept. 994/2 The Canadian mayflower (*Maianthemum canadense*)..is usually only found in cold, high altitudes.

Canadiana (kănē[i]diā·nă). [f. *Canada* + *-iana *suff.*] Things relating to or characteristic of Canada.

1837 W. B. Wells (*title*) Canadiana. **1921** *Daily Colonist* (Victoria, B.C.) 27 Mar. 24/3 Many have kept all the series for future reference, indexing them in a scrap book and thus making a unique piece of Canadiana. **1966** *Times* (Canada Suppl.) 28 Feb. p. ix/3 Until a few years ago it was mostly

Canadiana—paintings, artifacts..from the Old West. **1968** *Globe & Mail* (Toronto) 3 Feb. 28/4 The now famous Apple Rose which is pure Canadiana with apple.

Canadianism (kănē[i]·diăniz'm). [f. Canadian *a.* and *sb.* + -ism.] The condition of being Canadianized; the spirit of the Canadians; a Canadian idiom or word.

1875 *Canadian Monthly* Nov. 429/2 In other words, the feeling of Canadianism is not yet sufficiently strong to override all conflicting local feelings and interests. **1899** *Daily News* 17 Nov. 6/6 He declares that Afrikanderdom is the very opposite to Canadianism. **1906** *Westm. Gaz.* 20 May 12/1 What is called Canadianism may grow more pronounced with the growth of population. **1926** *Brit. Weekly* 20 May 130/5 A flat in Charing Cross-road, London, from which he cultivated his virile Canadianism. **1928** *Daily Express* 4 Dec. 10/3 Perhaps one of the most descriptive Canadianisms is the word 'kick' instead of thrill. **1958** *Times* 24 Nov. p. iii/2 The feeling of unity and national identity and..other less tangible aspects of Canadianism.

Canadianize (kănē[i]·diănəiz), *v.* [f. Canadian + -ize.] *trans.* To render Canadian in character. Also *intr.* Hence **Cana:dianiza·tion.**

1829 J. Mactaggart *Three Years in Canada* I. 37 Some of the unthinking Scotch ape the manners of the latter [English], and are termed *Canadianized Scotchmen.* **1902** *Monthly Rev.* Oct. 55 When he calls himself 'French-Canadian', he simply wants to differentiate his racial origin from that of his English, Scotch, or Irish fellow citizens, who, in his mind, are but partially *Canadianised.* **1925** *Contemp. Rev.* Nov. 605 They are in process of rapid 'Canadianisation'. **1964** *Economist* 21 Mar. 1134/3 Corporations ..will..have their own incentive to 'Canadianise'. *Ibid.*, Firms that do meet the requirements of Canadianisation.

‖**Canadien** (kanadyaṅ), *sb.* and *a.* fem. **Canadienne** (-yęn). [Fr., = Canadian.] **A.** *sb.* A French Canadian. (Cf. quot. 1568 s.v. *Canadian *a.* and *sb.*) **B.** *adj.* French-Canadian.

1832 *Vindicator* (Montreal) 9 Mar. 2/5 [He] justly remarks that the Canadiens have good grounds for complaining. *Ibid.*, No man who feels as a Briton ought to feel would have thus prostituted a Canadien press. **1863** E. H. Walshe *Cedar Creek* 82 The aged Canadienne arose, with the politeness so natural to her Gallic descent, and bade them welcome. **1906** *Daily Colonist* (Victoria, B.C.) 20 Jan. 8/1 (Advt.), This pretty Canadienne saved from terrible kidney disease by 'Fruit-a-tives'. **1952** D. Hoffman *Paul Bunyan* iv. 97 The idea persists that the origin of Paul Bunyan must be in *Canadien* folk tradition. **1958** *Times Lit. Suppl.* 31 Jan. 55/3 Some of the most important married into important 'Canadien' families. **1960** *Guardian* 5 Nov. 7/2 The 'Canadiens', the most self-centred, tenacious, clannish of minorities. **1963** *Western Weekly Suppl.* 13 Mar. 6/1 Louis loved his horses..almost as much as he did his petite Canadienne wife.

canaigre (kănē[i]·gəɹ). [Mexican Sp.] A species of dock, *Rumex hymenosepalus*, which is grown on sandy soils from Texas to Lower California, and whose roots are rich in tannin; also, the tannin obtained from this plant.

1878 *Rep. Comm. Agric.* (U.S.) 119 In many respects cañaigre root resembled rhubarb. **1884** *Encycl. Brit.* XVII. 401/2 The cañaigre has long been known to possess powerful tanning properties. **1901** *Westm. Gaz.* 21 Oct. 5/1 A plant indigenous to New Mexico, Arizona, and Lower California, known as Rumex hymenosepalus, or 'canaigre'. **1902** *Encycl. Brit.* XXV. 629/1 Considerable promise attends.. the experiments upon canaigre as a source of tannin. **1959** P. A. Munz *California Flora* 358 *R*[*umex*] *hymenosepalus*... Canaigre. Wild-Rhubarb...The dry roots contain as much as 35% tannin.

canal, *sb.* Add: **6. b.** Any of the faint seasonal markings of doubtful nature observed on the planet Mars.

A misleading rendering of It. *canali* (Schiaparelli) = channels.

1888 *Edin. Rev.* Jan. 26 Indications derived as to the nature of the mysterious Martian canals. **1891** E. Dunkin *Midnight Sky* 253 Networks of dark lines, to which Schiaparelli has given the name of 'canals', were noticed by him, in 1877 and 1882, to overspread the continents..of the planet. **1926** H. Macpherson *Mod. Astr.* iv. 56 These lines he [*sc.* Schiaparelli] designated by the Italian word 'canali', which actually means 'channels', but was translated into English as 'canals'. **1969** *Times* 19 Feb. 13/4 The close-ups will pick out detail down to 900 ft. across and may resolve the question of the curious linear markings nicknamed canals.

10. canal incline, canal-lift, an incline or elevator used instead of a lock for transferring canal-boats from one level to another; **canal rays** [tr. G. *kanalstrahlen* (Goldstein 1886, in *Berl. Ber.* XXXIX. 691)]: from the openings in the cathode through which the ions pass, = **positive rays.*

1882 L. F. Vernon-Harcourt *Rivers & Canals* I. 102 Canal inclines are similar to inclines so common in mines. **1902** *Encycl. Brit.* XXVI. 555/1 Canal inclines were early adopted on canals where loss of water in lockage was of importance. **1858** Simmonds *Dict. Trade*, Canal-lift. **1876** *Min. Proc. Instit. Civ. Eng.* XLV. 107 Hydraulic Canal Lift at Anderton, on the River Weaver. **1904** Rutherford in *Technics* July 12/2 The rays from radium are very similar to those produced when a strong electric discharge is sent through a vacuum tube...The α rays are very analogous to the 'canal' rays discovered by Goldstein. **1957** *Encycl. Brit.* XVIII. 882C/1 Goldstein first observed..streams of luminous gas back of a perforated cathode..as if..ionizing particles were coming through the holes..and ionizing the gas...These radiations..were first called *kanalstrahlen*, or canal rays.

canal, *v.* Add: Also, to make (a river, etc.) navigable by furnishing it with locks like a canal.

1819 E. Dana *Geogr. Sketches* 20 The operation of canalling and locking the falls. **1905** *Daily Chron.* 29 July 9/2 The river..has now been canalled, and is controlled by fourteen locks.

canaliform (kănæ·lifǭɹm), *a.* [f. L. *canalis* Canal *sb.* + -form.] Having the form of, or resembling, a canal.

1826 Kirby & Spence *Entomol.* IV. xlvi. 349 *Canaliform*.., when it [*sc.* the postscutellum] is a deepish elongate channel running from the postdorsolum to the abdomen. **1830** R. Knox tr. *Béclard's Anat.* 132 The villosities do not appear conical, or cylindrical, or canaliform. **1926** H. Macpherson *Mod. Astr.* v. 75 Canaliform appearances.

canalization. Add: **3.** *transf.* A furnishing with (underground) ducts or channels for the conveyance of cables, etc.

1889 *Daily News* 7 Oct. 3/1 This is what Mr. Crompton, borrowing a French expression, calls his system of undersurface 'canalization'.

4. *fig.*

1929 *Atlantic Monthly* Sept. 387/1 He [*sc.* William James] went on to say that the 'canalization of pity was an engineering feat'. **1950** *Brit. Birds* XLIII. 344 On the West Coast a 'canalisation' of Gannets from various stations.

canalize, *v.* Add: **3.** *transf.* To furnish with (underground) ducts or channels for the conveyance of cables, etc.; to convey through cables so furnished.

1886 *Jrnl. Soc. Telegr. Engineers* XV. 547 Electricity, to use a French phrase, is so easily 'canalised', or conveyed through insulated conductors. **1889** *Daily News* 7 Oct. 3/1 He has 'canalized' nearly the whole length of the principal streets in the South Kensington–Knightsbridge region.

4. *fig.* To lead in a desired direction, so as to control or regulate.

1922 J. M. Muir *Short Hist. Brit. Commonw.* II. ix. v. 359 The development of means for 'canalising' the nation's unspent wealth. **1943** H. Read *Politics of Unpolitical* ii. 23 Organizations designed to canalize the national spirit. **1945** R. Knox *God & Atom* x. 133 The urge to which it gave rise in the patient's nature may be canalized in a new direction. **1965** W. Lamb *Posture & Gesture* viii. 107 It is a pity that the new thinking on physical behaviour has become canalised to such an extent into the teaching of physical education as a subject.

b. *intr.* To take a certain direction, to form a channel.

1927 *Observer* 13 Nov. 8/5 A latent crisis in civilisation, a crisis which should have canalised into a religious revolution.

canapé (kæ·năpi). [Fr. (both senses).]

1. A piece of bread or toast, etc., on which small savouries are served.

1890 *Mrs. Beeton's Cookery Bk.* 759/1 Anchovy Canapés. **1908** *Daily Chron.* 10 Apr. 7/5 As an appetiser nothing is better than a canape of chicken. **1955** J. Cannan *Long Shadows* i. 11 She had sipped, smoked, examined pictures, nibbled canapés, moved here and there.

2. A sofa.

1892 F. Litchfield *Illustr. Hist. Furn.* vi. 167 (*caption*) Carved and Gilt Canapé or sofa. **1955** R. Fastnedge *Eng. Furn. Styles* vi. 144 Settees..were produced in emulation of Louis XV *canapés.*

canard. Add: **2.** A smaller surface on an aeroplane or hydrofoil providing stability or a means of control and placed forward of the main lifting surface; also (and orig.) an aeroplane with its wings so placed. Also *attrib.*

1916 H. Barber *Aeroplane Speaks* 137 *Canard*, literally 'duck', the name which was given to a type of aeroplane of which the longitudinal stabilizing surface (*empennage*) was mounted in front of the main lifting surface. **1928** C. F. S. Gamble *North Sea Air Station* Introd. 11 These monoplanes were of the 'Canard' (or 'tail first') type. **1931** *Flight* 2 Jan. 4/1 His brother experimented with canard models. **1961** *New Scientist* 16 Nov. 416/3 Most tentative designs for a Mach 3 liner provide for a canard form with the main wing at the rear. **1964** *Sci. Amer.* June 27/3 SCAT 17 is a delta-wing design with a canard, or balancing surface, at the nose. **1967** *Jane's Surface Skimmer Systems 1967–68* 95/2 The foils have been arranged..in a canard configuration, with one foil forward and two foils aft.

3. A bright, deep blue, like the colour which is found on a duck's wing.

[**1902** *Daily Chron.* 13 Dec. 8/4 The peculiar bright, yet deep, blue known in Paris as 'canard'.] **1908** *Westm. Gaz.* 22 Feb. 13/2 *Canard*—a new shade of blue inspired by the lovely patch of iridescent greeny blue that occurs on a duck's wing. **1923** *Daily Mail* 21 June 1 Over 40 shades including Ivory,..Apricot, Canard.

Canarese, var. *Kanarese.

Canarian (kănē[ə]·riăn), *a.* and *sb.* [f. Canary *sb.* + -ian.] **A.** *adj.* Of or pertaining to the Canary Islands. **B.** *sb.* A Canary Islander.

1793 *Sporting Mag.* II. 23/2 His name, Ardagoma, signifies, in the Canarian language, shoulders of rock. *Ibid.* 24/1 These Canarians have given such remarkable proofs of their athletic powers. **1871** F. Locock tr. *Pégot-Ogier's Fortunate Isles* I. vii. 97 The Canarians burning and plundering the ships. *Ibid.* xix. 304 Canarian ladies are not fond of walking. **1922** D. A. Bannerman *Canary Islands* i. 13 Carrying with him a number of the Canarians as captives. *Ibid.* iii. 57 The commoner forms of Canarian bird-life.

canariensis (kănē[ə]rie·nsis). [app. popular alteration of the specific name *canariense* in *Tropæolum canariense*, a former systematic name of *T. aduncum*. (*Canariensis* occurs as the specific name of canary seed, *Phalaris canariensis*, and canary wood, *Persea canariensis.*)] Canary-creeper, *Tropæolum aduncum.*

1897 *Hearth & Home* 3 June 174/2, I have them [*sc.* tubs] filled with multi-coloured dwarf Nasturtiums, round which at the edges trailed Canariensis. **1908** *Daily News* 21 Sept. 4/2 They frequently attack both the nasturtium..and the canariensis. **1929** W. Deeping *Roper's Row* xxiv. 260 She bought canariensis and climbing nasturtiums.

canary, *sb.* Add: **5. b.** A gold coin, so called from its colour. Also *transf. slang.*

[**1785** [see Canary-bird 2].] **1860** Hotten *Dict. Slang* 110 *Canary*, a sovereign. **1890** *Pall Mall Gaz.* 8 May 6/2 'Canaries', or promissory notes, were returned for considerably over £2,000. **1928** 'Brent of Bin Bin' *Up Country* xiii. 203 The 'Sweep Stakes', for which every entrant had to pay a 'canary' and the winner scoop the pool.

c. A convict. *Australian slang.*

1827 P. Cunningham *N.S. Wales* II. 117 Convicts of but recent migration are facetiously known by the name of *canaries*, by reason of the yellow plumage in which they are fledged at the period of landing. **1890** 'R. Boldrewood' *Col. Reformer* vi. 49 Can't you get your canaries off the track here for about a quarter of an hour and let my mob of cattle pass?

7. Canary banana, the dwarf banana, *Musa nana*; **canary pudding**, a lemon-flavoured sponge pudding.

1951 *New Biol.* XI. 67 The dwarf or Canary banana (formerly *M. Cavendishii*) which flourishes in Mediterranean lands is also widespread in the Far East, and hence the synonym *M. Chinensis.* **1861** Mrs. Beeton *Bk. Househ. Managem.* 636 *Canary pudding.*..The weight of 3 eggs in sugar and butter, the weight of 2 eggs in flour, the rind of 1 small lemon, 3 eggs. **1951** *Good Housek. Home Encycl.* 382/2 Many different sponge puddings can be made, using the canary pudding mixture as a basis.

canary-bird. Add: **2.** (Later example in the sense of a gold coin.)

1842 *Ainsworth's Mag.* II. 74 Leaving fifty good canary birds with their landlord when they go away.

canasta (kănæ·stă). [Sp., lit. 'basket', ult. f. L. *canistrum* Canister.] A card game of Uruguayan origin, in which two packs are used with four jokers, combining features of rummy and pinochle; *canasta* is also the name of a meld of seven cards.

1948 *Harper's Bazaar* July 47 The Uruguayan rummy game, *canasta*,..is rapidly winning over even the most fervent Oklahoma devotees. *Ibid.*, Here the Baroness counts up the canastas at the end of a hand. **1949** *Bookseller* 12 Nov. 1171 (Advt.), A new card game—*Canasta*—is said to be sweeping the world. **1958** 'J. Welcome' *Run for Cover* iv. 82, I resolved to go for my canastas as soon as possible.

canaveg, var. Candavaig.

1847 Stoddart *Angler's Comp.* 366 A variety of the salar, termed canavegs.

cancan. Add: *Comb.*, as *cancan-dancing* ppl. a.

1894 G. du Maurier *Trilby* II. vi. 201 This long-legged, cancan-dancing, Quartier Latin grisette.

cancel, *sb.* Add: **3.** Sense b (in full, *cancel-leaf*) is now the prevailing use.

1861 D. G. Rossetti *Let.* May (1965) II. 401 There are five *cancel* leaves already in the book. **1908** F. Madan in *Trans. Bibliogr. Soc.* IX. 62 Cancels are newly-printed leaves intended to take the place of cancelled leaves. **1908** Pollard & Greg *Ibid.* 44 When we speak of a cancel nobody else ever knows whether we mean the leaf cut out and destroyed or the leaf inserted as a substitute. If we are careful we distinguish between a cancelled leaf and a cancel-leaf, but the person to whom we are speaking probably does not catch the subtle distinction. A change of terminology might be desirable. **1914** R. B. McKerrow *Ibid.* XII. 299 In one case the leaf has been simply cut out, but in all the others it is replaced by a cancel-leaf. **1924** R. W. Chapman in *Library* V. 249 Notes on Cancel Leaves. **1927** [see *cancellandum]. **1964** F. Bowers *Bibliogr. & Text. Crit.* v. iv. 152 Two versions of a cancel leaf.

cancel, *v.* Add: **2.** Also with *out.*

1924 Hichens *After Verdict* ii. xix, Her will, so it seemed to her, had been cancelled out by little Clive's death.

b. Delete † *rare* and add later examples, const. *out.*

1925 *Wireless Weekly* July 449/1 There will be a position where the effects of the two field coils cancel out. **1965** *Listener* 16 Sept. 421/1 The personal preferences of your contributors are..likely to cancel out.

cancellandum (kænsĕlæ·ndǿm). *Printing* and *Bibliography*. Also anglicized **ca·ncelland.**

[neut. gerundive (sc. *folium* leaf) of L. *cancellāre* CANCEL *v.*] In full, *cancellandum leaf*: a leaf, or portion thereof, for which another is substituted. So **ca·ncellans** [pres. pple.], the substituted leaf, etc.

1923 R. W. CHAPMAN in *Library* IV. 173 The function of a signature on a cancel is not to distinguish the *cancellans* from the *cancellandum*, but to tell the binder where the cancel is to be placed. 1927 MCKERROW *Introd. Bibliogr.* 223 It is convenient to have a means of distinguishing clearly between the original sheet or portion of a sheet which is intended to be cancelled and what is intended to replace it. We may call the former the *cancellandum* or 'cancelland', the latter the *cancellans* or simply 'cancel'. 1938 *Times Lit. Suppl.* 6 Aug. 524/2 There is also a presentation copy of Boswell's 'Life of Johnson', a first edition in boards, and Sheridan's copy with a rare cancelland. 1964 F. BOWERS *Bibliogr. & Text. Crit.* IV. v. 131 We know very little about the process of mutilating a cancellandum leaf. *Ibid.*, If the cancellans leaf had been machined before the printer collated the copies.

cancellation. Add: **1. b.** The action or fact of cancelling a seat, room, place, etc., that has been reserved; a seat or room cancelled thus.

1953 E. S. GARDNER *Case Green-Eyed Sister* (1959) viii. 105 Luckily I managed to pick up a cancellation and came right through.

cancer, *sb.* Add: **5. cancer bush** *S. Afr.* [ad. Afrikaans *kankerbos*] (see quot. 1895); **cancer stick** *jocular* or *colloq.*, a cigarette.

1895 A. SMITH *S. Afr. Materia Medica* (ed. 3) xxi. 138 *Sutherlandia frutescens*, R. Br.—Dutch, *Cancer bush*. This shrub has been brought forward recently as a remedy for cancer. 1949 *Cape Argus* 9 July (Mag. Section) 1/3 This is where the..cancer bushes and chincherinchees for his rockeries are raised. 1966 E. PALMER *Plains of Camdeboo* xvii. 277 The early settlers..used..the pretty little *Sutherlandia humilis*—the cancer bush—not only for cancer but for flu. 1959 J. BRAINE *Vodi* xxii. 242 There was a packet of cigarettes on the locker. She took one out and lit it. 'First cancer stick today,' she said. 1967 *New Statesman* 20 Jan. 77/2 To be able to..tempt fate by a debonair pull on a cancer stick is a way of asserting the individual's right to choose his own end.

cancrizans (kæ·ŋkrizænz), *a. Mus.* [med.L., pres. pple. of *cancrizāre* to walk backwards, f. *cancr-*, *cancer* crab (CANCER *sb.*): see -IZE.] Applied to a canon in which the theme or subject is repeated backward in the second part. Hence as *sb.*, such a canon.

1782 C. BURNEY *Hist. Mus.* II. 494 The same part sings *in retro*, or, as it is called in the musical technica of the times, cancrizans. 1876 STAINER & BARRETT *Dict. Mus. Terms* 72/1 The following is a canon cancrizans with a bass part *per recte et retro*. 1879 GROVE *Dict. Mus.* I. 302/2 Sometimes a canon is both cancrizans and by contrary motion. 1926 A. B. SMITH *Studies & Caprices* 53 Inversions, diminutions, cancrizans exist only on paper. 1938 SCHOLES *Oxf. Compan. Music* 134/2 A highly artificial form in which the imitating voice gives out the melody backwards...This has been called the *Canon Cancrizans*..but crabs move *sideways*. 1942 *Scrutiny* XI. 10 This composer's elaborate *cancrizans*. 1969 *Listener* 24 Apr. 585/2, I strongly suspect that the obvious musical reversals were subtilised by the technical device of *cancrizans*.

candela (kændī·lă). [L., see CANDLE *sb.*] A unit of luminous intensity (see quot. 1968).

1950 *Commission Internat. de l'Éclairage, 1948* 14 It is recommended that the new unit for luminous intensity (which is such that the luminance or photometric brightness of a black body at the temperature of freezing platinum equals 60 units of intensity per square centimetre) shall be called in all countries by the Latin name 'candela', with the symbol '*cd*'. 1957 R. W. G. HUNT *Reprod. Colour* viii. 89 A purely physical or photometric quantity, measured, for instance,..in candelas or lumens per square foot if light units are used. 1960 *Aerodrome Lighting* (*B.S.I.*) 13 It may be said that in clear conditions the intensity of the runway lights should not exceed about 50 candelas. 1968 *Nature* 16 Nov. 651/2 The candela is the luminous intensity, in the perpendicular direction, of a surface of 1/600 000 square metre of a black body at the temperature of freezing platinum under a pressure of 101 325 newtons per square metre.

candela·braed, *a.* [f. *candelabra* (see CANDELABRUM) + -ED.] Furnished with or as with a candelabrum. Also used as pa. pple. of a hypothetical verb **candelabra*.

1923 *Daily Mail* 19 Feb. 8 Throngs gather round the cheap lottery booths established under the candelabra'd lamp-posts. 1929 HEMINGWAY *Farewell to Arms* viii. 47 We passed stone farmhouses with pear trees candelabraed against their south walls.

candelabrum. Add: **3.** Simple *attrib.* Also prefixed (in form *candelabrum* or *candelabra*) to the names of trees with foliage shaped like a candelabrum, esp. a tropical African tree of the genus *Euphorbia*.

1834 [see EUPHORBIA]. 1878 R. J. HINTON *Handbk. Ariz.* 343 To what age the candelabra cactus attains is a matter of mere conjecture, their growth being exceedingly slow. 1892 J. G. BOURKE *On Border* 53 The majestic 'pitahaya', or candelabrum cactus, whose ruby fruit had long since been raided upon. 1909 WEBSTER, Candelabrum tree. 1927 D. H. LAWRENCE *Mornings in Mexico* 80 Even the organ cactus,..and the candelabrum cactus, seem to be slowly wheeling and pivoting upon a centre. 1946 J. P. R. WALLIS *North. Goldfields Diaries T. Baines* I. 170 *Euphorbia ingenus* or candelabra tree. 1957 *Timber Technol.* LXV. 467 *Botanical name*: *Euphorbia* spp...*Trade name*: Candelabra tree...All species resemble each other in their giant cactus or candelabra formation. 1960 *Guardian* 7 Nov. 1/3 They had chased the Masai beyond the candelabra tree.

candelilla wax (kændēli·lyă). [Amer. Sp. *candelilla*, f. Sp., little candle.] A vegetable product obtained from various American shrubs, esp. of the genus *Euphorbia*, prepared for some special purpose in the arts, dentistry, etc.

1909 *Jrnl. R. Soc. Arts* 25 June 664/2 Moulded into phonograph records, the candelilla wax will register the sounds perfectly. 1910 *Jrnl. Ind. & Engin. Chem.* II. 203/2 Candelilla wax..is found coating the entire surface of a plant that grows wild in the semi-arid regions of northern Mexico and southern Texas. 1951 R. MAYER *Artist's Handbk.* ix. 285 *Candelilla Wax*..obtained from a weed native to Texas and Mexico..is next in hardness to carnauba, and finds a use in industry as a cheaper substitute for it. 1961 J. N. ANDERSON *Appl. Dent. Mat.* (ed. 2) xv. 134 Candelilla wax is..light brown in colour.

candid, *a.* Add: **5. c.** Of a photograph or photography: unposed, informal. So *candid camera*, a small camera for taking informal photographs of persons, freq. without their knowledge; also *attrib.* Also as *sb.*, an unposed photograph.

1929 *Graphic* 11 May 265 At the foreground table, unaware of the proximity of the candid camera, will be seen Dean Inge enjoying his cigar. 1935 *N.Y. Times* 8 Dec. Sect. xi. 14/3 Small enough to be manipulated without attracting attention, the miniature camera as used by 'candid camera' hobbyists is..becoming a highly revealing adjunct of the contemporary scene. 1937 *Miniat. Camera World* Dec. 5/2 'Candid' photography was at one time synonymous with 'miniature' photography. 1939 A. HUXLEY *After many a Summer* I. iii. 31 Candid Camera portrait of the President of Consol Oil. 1946 T. GODSEY *Free Lance Photogr.* iii. 36 He liked these 'candids' so well that a contract to do his new publication entirely with the miniature was the result. 1957 P. MANSFIELD *Final Exposure* i. 11 Those delightfully candid shots. *Ibid.* ii. 33 Candid photographs. *Ibid.* iv. 51 Can't you..take real candids—without people knowing?

candidate, *v.*[2] *U.S.* (Earlier and later examples.)

1884 *Century Mag.* June 308/1 Let him put the question to some [choir-singers] who every spring have to candidate for a situation. 1909 *Springfield Weekly Republ.* 2 Sept. 14 Mr. Seccombe candidated in the Goschen church last spring.

candiru (kændirū·). [a. Pg. *candirú*, a. Tupi *candirú*, *candérú*.] A tiny, bloodsucking catfish, *Vandellia cirrhosa*, of the family Pygidiidæ, found in the Amazon river, where it attacks other animals, including man.

1841 R. H. SCHOMBURGK in *Ann. Mag. Nat. Hist.* VI. 395, I was frequently warned by the inhabitants to be cautious while bathing of a small fish called *Cancliru* [sic]. 1897 J. BACH in *Proc. Zool. Soc.* 901 The 'Candyrú', as the fish is called, is much dreaded by the natives of the Jurua. 1930 E. W. GUDGER *Candirú* p. vii, 'Candirú' is the collective name given to certain small catfishes of the Amazon River and its tributaries to which are attributed the evil habit of entering the urethra of men and the vulva of women bathers. 1962 K. F. LAGLER et al. *Ichthyol.* xiv. 438 One fish that can perhaps be regarded as a true parasite is the small candiru, a South American catfish.

candle, *sb.* Add: **1. d.** Freq. called *international candle*. Also, with prefixed numeral, = *candle power* (see *7). Replaced as a unit of luminous intensity by the *new candle* (see quot. 1937) = *CANDELA.

[1860 *Act* 23 & 24 *Vict.* c. 125 § 25 The Quality of the Common Gas..shall be..such as to produce..a Light equal in Intensity to the Light produced by not less than Twelve Sperm Candles.] 1880 J. W. URQUHART *Electr. Lt.* 279 A 2,000-candle light. *Ibid.* 282 A hundred-candle Sugg gas-burner. 1917 *Trans. Illum. Engin. Soc.* XII. 440 *Candle*, the unit of luminous intensity maintained by the national laboratories of France, Great Britain and the United States. ..This unit, which is used also by many other countries, has..been referred to as the international candle. [1937 *Com. Internat. Poids & Mesures* XVIII. 216 L'adoption d'une nouvelle définition de l'unité d'intensité lumineuse... Le Comité se rallie à la proposition d'appeler 'bougie nouvelle' l'unité qui vient d'être définie.] 1938 *Light & Lighting* XXXI. 186/1 The initiation of the new international candle, based upon..a primary standard of light (one-sixtieth of the luminous intensity of one square centimetre of a black body maintained at the temperature of solidification of platinum)... Laboratories are preparing groups of lamps to be measured in terms of the new candle, which is to come into operation in 1940.

6. *candle-dish, -shade*; *candle-lit* adj.

1899 *Pall Mall Mag.* Jan. 77 Two silver candledishes. 1868 WHITTIER *Meeting* in *Compl. Wks.* (1898) 486/1 No altar candle-lit by day. 1890 *Pall Mall Gaz.* 22 Oct. 2/3 This candlelit old hall. 1916 H. G. WELLS *Mr. Britling* ii. § 9 Equally unexpected was the supper on a long candle-lit table without a cloth. 1780 *Hickey's Bengal Gaz.* 8 Apr. (Y.), Borrowed last Month by a Person or Persons unknown ..a very elegant Pair of Candle Shades. *a* 1916 H. JAMES *Middle Years* (1917) 76 Milford Cottage, with its innumerable red candles and candle-shades. 1967 J. RATHBONE *Diamonds Bid* xvii. 147 The glass candle shade on my own table.

7. candle-bush, a South African plant, *Sarcocaulon patersoni*, so called from the readiness with which it burns; **candle-foot** = **foot-candle*; **candle-metre,** the illumination of a standard candle at a distance of one metre; **candle power,** the illuminating power of an electric lamp, etc., reckoned in terms of the light of a standard candle.

1890 A. MARTIN *Home Life Ostrich Farm* 60 The *kersbosch*, or candle-bush, a stunted, thorny plant, if lighted at one end when in the green state, will burn steadily just like a wax candle. 1966 E. PALMER *Plains of Camdeboo* xvi. 258 The men cut long forked sticks, spiked on the fork a Candlebush which burns like paraffin, and..moved quickly from clump to clump, burning off the noors spines. 1892 A. P. TROTTER in *Min. Proc. Inst. Civil Engineers* CX. 71 As the heights of lamp-posts and the width of streets are measured in feet, the candle-foot, that is, the illumination produced by one standard candle at a distance of 1 foot, will be taken in the present Paper as the unit of illumination. 1902 *Encycl. Brit.* XXVIII. 83/2 In order that street surfaces may be well lighted, the minimum illumination should not fall below 0·1 candle-foot. 1928 *Sunday Dispatch* 30 Dec. 7/5 (L.C.C. Regulation) The lighting maintained in no part of the auditorium shall be less than ·025 candle-foot. 1908 *Westm. Gaz.* 23 Mar. 4/2 The values are expressed in 'candle-mètres'. 1877 *Engineering* XXIV. 333 In measuring the candle power of the light produced by each machine. 1880 J. W. URQUHART *Electr. Lt.* 278 Six electric lights of 480-candle power each. 1883 Candle power [see CANDLE *sb.* 1 d]. 1894 SALOMONS *Electr. Lt. Install.* (ed. 7) II. 248 The amount of energy required to incandesce the lamp determines its candle-power efficiency. 1934 *Discovery* June 155/2 The official physical standards of the country, e.g., length, mass, temperature, candle-power.

candleless (kæ·nd'l,lės), *a.* [f. CANDLE *sb.* + -LESS.] Without a candle or candles.

1906 *Westm. Gaz.* 20 Jan. 5/1, I am left soapless, candleless, and forlorn. 1920 SANTAYANA *Char. & Opin. U.S.* i. 6 Like those candlesticks, probably candleless, sometimes displayed as a seemly ornament in a room blazing with electric light.

ca·ndle-li:ghting. (Later U.S. examples.)

1784 *Maryland Jrnl.* 17 Sept. Advt. (Th.), Said School to begin at Candle-lighting, and continue till Nine o'Clock, P.M. 1869 H. B. STOWE *Oldtown Folks* xxxiv. 449 Designing to bring up at the red school-house..at 'early candle-lighting'. 1896 *Dialect Notes* I. 385 Evenin' meetin' took up at early candle-lightin'.

ca·ndler. [f. CANDLE *v.* 2.] One who tests eggs by the light of a candle or an electric bulb.

1906 *N.Y. Even. Post* 5 June 9 The candlers say ten hours' work on a stretch is harmful to their eyesight. 1921 *Dict. Occup. Terms* (1927) § 449 Egg candler.

candlewick. Add: **2.** A soft material, usually cotton yarn, used to produce a tufted surface, also called *candlewicking*; material embroidered with tufts of this yarn. Also *attrib.*

1930 *Sears Catal.* Fall 357 Candlewick Embroidery is a popular vogue in needlework. A candlewick tufted bedspread. 1934 *Archit. Rev.* LXXVI. 183 Candlewick work.. is the apotheosis of the spot [design], consisting of blobs of fluffy wool stuck on to a material. 1939 M. B. PICKEN *Lang. of Fashion* 18/2 *Candlewicking*, tuftings of threads to give a napped surface to fabric. 1951 *Good Housek. Home Encycl.* 152/2 Tweeds, folk weave, and candlewick..provide a wide range of surface-texture. 1958 P. MORTIMER *Daddy's gone A-Hunting* xx. 119 Her candlewick dressing-gown.

candy, *sb.*[1] Add: **2.** *candy-girl, -merchant, pink, -shop*; *candy-coloured, -pale* adjs.; **candy butcher** (see *BUTCHER *sb.* 3 b); **candy-floss** [FLOSS[2]], a sweet confection, usually pink, of fluffy spun sugar; also in *transf.* and *attrib.* use as a type of meretriciousness; **candy-stripe(d),** pattern(ed) in alternate stripes of white and colour, as in a popular kind of candy.

1888 G. M. HOPKINS *Poems* (1918) 89 A candycoloured.. river. 1951 *Springfield Sunday Republican* 6 May 29A (Advt.), Salt water taffy..pop corn—candy floss. 1952 *Times* 2 Oct. 6/2 They could not solve problems of foreign policy on a diet of rhetorical candy floss. 1957 J. FRAME *Owls do Cry* II. xxii. 100 You won't get any ice creams or.. candy-floss. 1957 R. HOGGART *Uses of Literacy* vii. 171 (*heading*) Invitations to a candy-floss world: the newer mass art. 1855 M. M. THOMPSON *Doesticks* xxxiii. 299 With what an affectionate air couldst thou..box the ears of the little candy-girl. 1870 'FANNY FERN' *Ginger-Snaps* 61 To the delight of these youngsters and the candy-merchants. 1920 E. SITWELL *Wooden Pegasus* 48 As they Shelter the children, candy-pale. 1937 L. BROMFIELD *Rains Came* I. i. 15 He wore an enormous Rajput turban in shades of poison green and violet and candy pink. 1845 *Knickerbocker* XXV. 424 Candy-shop keepers. 1886 *Harper's Mag.* June 93/2 A considerable portion of the refined sugars find their way to the candy shops. 1939 M. B. PICKEN *Lang. of Fashion* 18/2 *Candy stripe*, stripe like those in stick candy. 1894 'MARK TWAIN' *Pudd'nhead Wilson* 234 The candy-striped pole..indicated..the barber shop. 1941 'R. WEST' *Black Lamb* (1942) I. 497 His candy-striped pyjamas. 1959 *Housewife* June 100/1 Candy-striped sheets and pillowcases.

cane, *sb.*[1] Add: **1. d.** *U.S.* (*a*) Canes collectively; (*b*) a field of cane; (*c*) = *cane-brake* (*a*) (see Dict.).

1784 J. FILSON *Discovery Kentucke* 18 This great tract is..covered with cane, wild rye, and clover. 1796 B. HAWKINS *Let.* in *Georgia Hist. Soc. Coll.* IX. 14 There is plenty of young cane and provisions. 1836 J. HALL *Statistics of West* ii. 27 The inhabitants drive their cattle to the cane in the autumn. 1847 in D. Drake *Life Kentucky* (1870) i. 14 Their practice was..to..lodge separately among the cane, which flourished in great luxuriance. 1925 Z. A. TILGHMAN *Dugout* 91 George secured men to..put in a crop of kafir and cane.

9. a. *cane-piece* (earlier examples = *cane-field*); **b.** *cane-carrying*.

1852 MUNDY *Antipodes* I. iv. 137 They laughed at the cane-carrying soldiers. 1924 *Glasgow Herald* 16 Apr. 10, I had not imagined..that cane-carrying was peculiar to some nations and not others. 1774 J. SCHAW *Let.* 12 Dec. in *Jrnl. of Lady of Quality* (1921) ii. 84 We walked thro' many cane pieces, as they term the fields of Sugar-canes. 1861 TROLLOPE *Tales of all Countries* 134 He took Mr. Leslie through his mills and over his cane-pieces.

10. cane-brake (earlier and later examples); **cane colour,** the colour of cane as applied to pottery ware; pottery of this colour; also as adj.; so **cane-coloured** *a.* (also *transf.*); **cane-fruit,** a commercial name for such fruit as raspberries and blackberries which grow on canes (*Cent. Dict. Suppl.* 1909); **cane grass,** (*a*) *U.S.*, the plant *Arundinaria macrosperma* forming the cane-brakes of the southern United States; (*b*) *Austral.*, *Glyceria ramigera*; **cane knife** *U.S.*, a large knife used in cutting cane; **cane-mill** (U.S. example); **cane-rat,** any of several large African rodents, as *Thryonomys swinderianus* and *Aulacodus* s.; **cane-top** *U.S.* (see quot. 1833); **cane trash,** (*b*) (see quot.); **cane-work,** strips of cane interwoven and used to form the backs of chairs and other articles of furniture; also *attrib.*; **cane-worker,** one who makes articles of cane.

1770 *South-Carolina Gaz.* 18 Oct., There is a large Neck, or Island, of Swamp or Cane-Brake Land. *a* 1930 D. H. LAWRENCE *Last Poems* (1932) 14 So in the cane-brake he clasped his hands in delight. 1866 METEYARD *Jos. Wedgwood* II. p. xxiv, Cane-colour Inkstand. 1875 —— *Wedgwood Handbk.* Gloss. 393 *Cane-colour*, ware the colour of cane... Cane-colour was applied both to ornamental and to useful purposes. 1787 Jos. WEDGWOOD *Catalogue* (ed. 6) 2 Bamboo, or cane-coloured bisqué porcelain. 1865 L. JEWITT *Wedgwoods* 311 The 'bamboo, or cane-coloured' ware. 1910 W. DE LA MARE *3 Mulla-Mulgars* xiv. 196 Short, fleecy, and cane-coloured whiskers. 1827 *Western Monthly Rev.* (Cincinnati, U.S.) I. 209 The cane grass of the vast swamps and savannahs on the Gulf of Mexico. 1898 MORRIS *Austral Eng.* 78/2 Cane-grass. 1953 A. UPFIELD *Murder must Wait* xxii. 195 A clump of low cane-grass. 1798 A. ELLICOTT in *Life & Lett.* (1908) 159 [The country] could only be explored by using the cane knife and hatchet. 1887 *Harper's Mag.* July 272/1 The children..squabbling for the possession of one cane-knife to split kindlers. 1833 B. SILLIMAN *Man. Sugar Cane* 30 The cane mill consists of three cast iron cylinders. 1876 H. BROOKS *Natal* 116 The cane-rat or ground-rat, that feeds upon the sugar-canes, is properly more of a porcupine than a rat. 1934 *Nature* 7 Apr. 524/2 Another addition to the Zoo worthy of note is three young cane-rats (*Aulacodus swinderianus*) from West Africa. 1954 G. DURRELL *Bafut Beagles* ii. 39, I could see we had caught a very large Cane Rat...It measured about two and a half feet in length, and was covered with a coarse brownish fur. It had a chubby, rather beaver-like face, small ears set close to the head, a thick naked tail and large naked feet. 1826 J. BRADFORD *Hist. Notes Kentucky* (1932) 11 Cane tops. 1833 B. SILLIMAN *Man. Sugar Cane* 12 But a part of the planting is done with cane tops, or that portion of the Cane which is rejected in cutting it for the mill. 1790 CASTLES in *Phil. Trans.* LXXX. 349 Burning the cane trash (or straw of the cane). 1858 T. VIELÉ *Following the Drum* 53 Divans of cane-work. 1887 A. FORBES *Insulinde* 25 The backs of the open canework chairs. 1934 *Burlington Mag.* Nov. 201/2 A back formed of a single panel of cane-work is something new. 1858 SIMMONDS *Dict. Trade*, *Cane-worker*, a maker of articles in rattans, Spanish and other canes; a basket-maker. 1901 *Daily Chron.* 24 Aug. 5/6 W.Y. ..cane-worker, pleaded guilty. 1921 *Dict. Occup. Terms* (1927) § 472 *Caner, cane worker*,..fills in framework of baskets, wicker furniture, and other basket ware by interweaving cane.

caney, var. CANY *a.*

canezou (ka·nzū). *Hist.* [Fr., of unknown origin.] A woman's blouse-like garment of muslin or cambric. Also *attrib.*

1827 *Lady's Mag.* Sept. 510/2 A canezou spencer of embroidered muslin. *Ibid.* 511/1 Muslin canezous over high dresses. 1893 G. HILL *Hist. Engl. Dress* II. 241 A cambric canezou..with sleeves full to the elbow. 1898 *Daily News* 26 Sept. 6/4 When the Restoration came in 1815,..Fleur-de-lys appeared on everything...The canezou replaced the hideous spencer.

Canfield (kæ·nfīld). *U.S.* [The name of R. A. *Canfield* (1855–1914), an American gambler.] A form of the game of patience similar to Klondyke.

1912 *Sat. Even. Post* 6 July 3/1 Playing double Canfield with my husband. 1932 N. MITFORD *Christmas Pudding* xx. 299 She would..play canfield on the schoolroom table. 1957 *Encycl. Brit.* XVII. 379/2 The most popular solitaire game of the U.S. is generally called Canfield..but..the game..should be called by its original name, Klondike... Canfield is a different though similar game.

canfieldite (kæ·nfīldəit). *Min.* [f. the name of F. A. *Canfield* (1849-1926), Amer. mining engineer: see -ITE[1].] A rare sulphide of silver and tin.

1893 S. L. PENFIELD in *Amer. Jrnl. Sci.* XLVI. 113 There can be no doubt that canfieldite and argyrodite have the same chemical composition. 1894 —— in *Ibid.* XLVII. 454 As the Freiberg argyrodite has been shown to be isometric, and the name canfieldite cannot therefore be applied to the germanium compound, it is proposed now to transfer the name to the new isomorphous tin compound. 1957 *Encycl. Brit.* II. 340/1 Isomorphous with argyrodite is the corresponding tin compound Ag_8SnS_6, also found in Bolivia as cubic crystals, and known by the name canfieldite.

canid (kæ·nid). *Zool.* [f. mod.L. *Canidæ* (see def.), f. L. *canis* dog + -ID[3].] A member of the Canidæ, a family of digitigrade carnivorous mammals including dogs, wolves, and related animals.

1889 in *Cent. Dict.* 1945 G. G. SIMPSON *Princ. Classif. Mammals* III. 222/2 Fossil canids are very abundant. *Ibid.* 224/1 Borophaginae are large, later Tertiary canids with heavy jaws. 1950 *Proc. Prehist. Soc.* XVI. 128 An incomplete tibia of a canid.

canine, *a.* (*sb.*) Add: **1. c.** *canine letter* = DOG'S LETTER.

1886 PHELPS & FORBES-ROBERTSON *Life S. Phelps* 15 He attached considerable importance to the canine letter, and often said that the rolling of the *r* [etc.]. 1890 HOWELLS *Shadow of Dream* 24 'Hermia' said Faulkner, sounding the canine letter in her name with a Western strength.

B. *sb.* For 'Jocosely used for dog' substitute: A dog (sometimes *joc.*). (Later examples.)

1947 J. STEVENSON-HAMILTON *Wild Life S. Afr.* xii. 83 The little dogs rush in barking, as only such small canines can. 1969 *Nature* 18 Oct. 244/1 Rabies is believed to have originated as a disease of wild canines.

canities (kăni·ʃiiz). *Path.* [L.] Whiteness or greyness of the hair.

1807 MORRIS & KENDRICK *Edin. Med. Dict.* 1890 *Science* 3 Oct. 186/1 The sudden change in canities, when due to violent emotions. 1968 A. ROOK et al. *Textbk. Dermatol.* II. xlvi. 1413/1 Canities is a physiological manifestation of the ageing process.

cank (kæŋk), *sb. local.* The name in the Midland coalfields for a hard ferruginous sandstone. Also *cankstone*.

a 1835 J. PHILLIPS *Geol.* in *Encycl. Metrop.* (1845) VI. 587/2 Some less regular sandstone beds, called 'Cankstone', approach very nearly to the nature of the ganister or calliard rocks of the coal strata. 1860 *Engl. & For. Mining Gloss.* (ed. 2) 34 *Cank*, whinstone. 1877 [see GALLIARD *sb.*[1]]. 1964 *Gloss. Mining Terms* (*B.S.I.*) v. 6 *Cank*, a hard, dark-grey massive rock consisting largely of ankerite.

cankery, *a.* **3.** (Later example.)

1864 TROLLOPE *Can you forgive Her?* I. xxxviii. 296 One of them cankery chiels as never have a kindly word for man nor beast.

cannabidiol (kæ:năbidəi·ǫl). *Chem.* [f. *CANNABI(S + DI-[2] + -ŌL 1.] (See quot. 1949.)

1940 R. ADAMS et al. in *Jrnl. Amer. Chem. Soc.* LXII. 196/2 The isolation of a pure compound was accomplished. It proved to have one of the two empirical formulas suggested above [*viz.* $C_{21}H_{30}O_2$] and has been given the name cannabidiol. 1949 *Blakiston's New Gould Med. Dict.* 175/1 *Cannabidiol*, a constituent of cannabis which, on isomerization to a tetrahydrocannabinol, exhibits to a great degree the activity of cannabis. 1967 *New Scientist* 31 Aug. 436/1 The classical analysis of hashish..yielded three types of related compounds as the characteristic components of the drug. These were cannabidiol, cannabinol and tetrahydrocannabinol.

cannabinol (kæ·năbinǫl, kănæ·b-). *Chem.* [f. as prec. + -OL.] A crystalline phenol, $C_{21}H_{26}O_2$, obtained from cannabis resin.

1896 T. B. WOOD et al. in *Jrnl. Chem. Soc.* LXIX. 546 The name, 'cannabindon', which Leib Lapin proposes for his substance..does not recommend itself to the authors who propose the name *cannabinol* as the compound is undoubtedly a hydroxyl derivative. 1940 *Jrnl. Amer. Chem. Soc.* LXII. 196/1 Cannabinol is very toxic but has no marihuana activity. 1963 *Tetrahedron* XIX. 2073 Up till now the structure of only one compound of the cannabis group, namely the physiologically inactive cannabinol (I), has been fully elucidated. 1967 [see prec.].

cannabis (kæ·năbis). [mod. L. (Linnæus *Species Plantarum* (1753) II. 1027), f. earlier L. *cannabis*, Gr. κάνναβις hemp.] **1.** Common hemp, *Cannabis sativa*, a tall erect herb of the family Moraceæ having long dentate leaves on long petioles and common in central Asia and other warm regions; different regional varieties, occas. distinguished as *Cannabis americana*, *Cannabis indica* (Indian hemp), etc., are cultivated for their fibre, their intoxicating properties, or the oil obtained from their seeds.

[1728 CHAMBERS *Cycl.* I, *Hemp*, by Naturalists call'd *Cannabis*, bears a near Analogy to Flax, *Linum*.] 1798 W. ROXBURGH *Plants Coast of Coromandel* II. 50/1 They have no idea of the superior quality of the bark of the common hemp-plant (Cannabis), which is indigenous in all parts of India. 1851 *Monthly Jrnl. Med. Sci.* XIII. 27 Two species of Cannabis have been described by many botanists —viz., *C. Sativa*, and *C. Indica*; but repeated comparison has not detected any material difference. 1904 *Sci. Mem. Officers Med. & San. Depts. Govt. India* XII. 1 (*heading*) On the Morphology, Teratology and Diclinism of the Flowers of Cannabis. 1937 A. F. HILL *Econ. Bot.* xii. 293 Cannabis has been of little importance in the United States, but recently the use of marijuana cigarettes, illicitly made from hemp, seems to be increasing, in spite of all attempts to stamp out the habit. 1962 *Times* 7 May 11/7 Indian hemp, or cannabis as it is now generally termed, grows freely over great stretches of the earth.

2. (Orig. *ellipt.* for *Cannabis sativa* or (esp.) *indica*.) Any of various preparations of different parts of the hemp-plant which are smoked, chewed, or drunk for their intoxicating or hallucinogenic properties and were formerly used medicinally; bhang (marijuana), ganja, and charas (hashish) are different forms of these preparations and there are many other names.

1848 HEMPEL & QUIN tr. *Jahr's New Manual* I. 386 According to Noack and Trinks, Cannabis may be employed in the following affections: Hysterical conditions [etc.]. 1860 A. STILLÉ *Therapeutics & Mat. Med.* II. VI. 82 The mental effects of cannabis when fully developed are exceedingly curious. 1888 *Practitioner* Feb. 97 Even in more advanced cases of tropical diarrhœa cannabis will sometimes prove very useful. 1935 *Lancet* 6 Apr. 811/2 Frequent indulgence in cannabis..results in loss of judgment and restraint. 1939 *Times Lit. Suppl.* 9 Sept. 531/3 Marihuana, otherwise cannabis, hemp, hashish, nepenthe, bhang, ganja, dagga, and about a hundred other names, has not yet become in this country a popular way of escape from the worries of normal life. 1952 *Martindale's Extra Pharmacopœia* (ed. 23) I. 321 Cannabis was formerly employed in mania and nervous disorders as a cerebral sedative or narcotic, but owing to the uncertainty of its action it is now seldom employed in this manner. 1962 *Times* 7 May 11/7 The taking of cannabis does not at the moment present the Home Office with anything like a serious problem. 1967 *Guardian* 31 July 2 The general effects of cannabis seem to be the liberation of certain inhibitions, accompanied by mild hallucinations. 1970 *Times* 12 Mar. 1/2 The Bill also halves the maximum prison sentence for possessing cannabis.

3. *attrib.* and *Comb.*, as *cannabis plant*, *-smoker*; **cannabis resin,** the sticky resin produced by the hemp-plant, esp. the flowering tops of the female, and containing the active principles.

1896 K. L. DEY *Indigenous Drugs India* (ed. 2) 60 The *Charas* or cannabis Resin exudes naturally on the leaves, stem and fruits..on plants growing on the mountain tracts at an altitude of 6,000 to 8,000 feet. 1957 *Bull. Narcotics* IX. 4/1 In Europe, the cannabis plant produces a valuable fibre while showing little or no tendency to produce the narcotic principle. 1962 *Times* 7 May 11/7 Cannabis abuse is very likely to be the forerunner of addiction to more dangerous addicting drugs. 1967 *Listener* 3 Aug. 131/3 Only a tiny percentage of cannabis-smokers escalate to heroin. 1970 *Times* 10 Mar. 2/6 Three boys have been expelled..for smoking cannabis resin. 1970 *Oxford Mail* 8 Apr. 1/8 Flying Squad officers..charged them early today with possessing about 60 lb. of pure cannabis resin, worth about £30,000 on the black market.

canned, *ppl. a.* Add: (Earlier and later examples.)

1859 R. B. MARCY *Prairie Traveler* 31 Canned vegetables are very good for campaigning. 1888 E. H. D'AVIGDOR *Antipodean Notes* xxiv. 173 Diggers all use immense quantities of jam and 'canned' fruits. 1937 *Daily Express* 24 Feb. 16/5 Canned beer has..flopped in Great Britain. 1954 *Sun* (Baltimore) 11 Dec. B. 15/2 The park area...where gather 'boys who like canned heat'.

b. *fig.* Mechanically or artificially reproduced, esp. of music.

1904 'O. HENRY' *Cabbages & Kings* vi. 93 'The Latin races ..are peculiarly adapted to be victims of the phonograph.' 'Then,' says I, 'we'll export canned music to the Latins.' 1908 *Westm. Gaz.* 5 Aug. 2/2 The latest invention is the 'canned speech' delivered by a gramophone. 1930 *Punch* 2 July 21/1 'Canned music', by which is meant music transmitted by mechanism along with the film instead of being played by the theatre band. 1934 *B.B.C. Year-Book* 117 Broadcasting has been derided as 'canned' entertainment. 1959 *Camb. Rev.* 7 Feb. 321/1 The film as *pictures* (not merely as canned drama).

c. Intoxicated, 'tight'. *slang.*

1914 C. MACKENZIE *Sinister Street* II. IV. ii. 882 She was a bit canned that night, and I suppose I'd had one or two myself. 1926 'J. J. CONNINGTON' *Death at Swaythling Court* iv. 70 Being rather canned, he sticks the candle on the table, and forgets all about it.

cannella, var. CANELLA.

1836 J. M. GULLY tr. *Magendie's Formul.* 91 Cannella powder. 1887 A. M. BROWN *Anim. Alkaloids* 53 A precipitate of a cannella colour.

cannelloni (kænelōu·ni), *sb. pl.* Also formerly **cannelons.** (Both occas. used in *sing.*) [It. *cannellone*, pl. *cannelloni*, f. *cannello* stalk.] **1.** (Only in form *cannelons.*) Rolls of pastry stuffed with a savoury filling or cream and eaten as a desert. **2.** Rolls of pasta filled with seasoned meat.

1845 E. ACTON *Mod. Cookery* xvi. 337 Cannelons. 1892 *Encycl. Pract. Cookery* I. 281/1 Cannelons..are composed chiefly of nouille paste in the shape of small pipes about 3 in. long and ½ in. in diameter. They are generally made by rolling the paste out very thin, cutting into strips 3 in. by 2 in., and rolling up round small pieces of wood, which are removed after baking. 1906 MRS. BEETON *Househ. Managem.* xvi. 511 (*heading*) Cannelon of beef. *Ibid.* xxiii. 669 (*heading*) Cannelons of chicken. 1937 C'TESS MORPHY *Good Food from Italy* 55 Toss the *cannelloni* in the *sugo*. 1952 P. BONNER *SPQR* (1953) vi. 52 A large slice of *cannellone*. 1960 *Guardian* 29 Aug. 5/6, I ate scampi and cannelloni with fresh Parmesan.

canner. Add: **b.** A beast fit only for canning. Chiefly *U.S.*

1890 *Stock Grower* (Las Vegas, New Mexico) 11 Jan. 5/2 Fit only for canners and feeders. 1892 *Pall Mall Gaz.* 8 Dec. 2/1 'Canners', which is the designation of all animals collected at the Chicago and other markets, that are refused by the butchers as unfit for their trade. 1906 *Daily Chron.* 13 June 7/3 A 'canner'..is an animal with little, if any, fat on it, in fact often nothing more than skin and bone. 1964 *Weekly News* (Auckland) 22 Jan. 40/3 The canner muttons and ewes were not so much in demand.

c. A machine for canning food.

1928 *Daily Express* 6 July 5/5 Until now canning has been beyond me for want of a canner.

ca·nnibal[2]. [? Corruption of *Camdeboo*: see Pettman *Africanderisms*.] Used *attrib.* in *cannibal stink-wood*, a South African name for *Celtis kraussiana*.

1859 R. J. MANN *Natal* 156 (Pettman), There is a variety of this wood known under the name of the Cannibal stinkwood. 1877 LADY BARKER *Yr.'s Housekpg. S.Afr.* 325 What rhyme or reason, what sense or satire can there be in such a name as 'Cannibal Stink-wood'?—applied..to a graceful, handsome tree whose bark gives out an aromatic..perfume.

cannibalize (kæ·nibăləiz), *v.* [f. CANNIBAL[1] + -IZE.] *trans.* To take parts from one unit for incorporation in, and completion of, another (of a similar kind). Hence **ca:nnibaliza·tion,** the removal of a part (of something) for incorporation in something else; **ca·nnibalized** *ppl. a.*

1943 REDDING & LEYSHON *Skyways to Berlin* (1944) xi. 173 One by one the other aeroplanes have gone, some destroyed by enemy action, others cannibalized for parts and still others retired because of age. 1947 *Hansard Commons* CCCCXLV. 973 It would be necessary to dismantle or, as they say, cannibalise half a dozen to provide one good Nissen hut. 1947 *Jrnl. R. Aeronaut. Soc.* LI. 981/2 The necessity for removing serviceable parts from one aircraft to service another. Cannibalisation, often of a new aircraft, was frequently the only method of maintaining reasonable serviceability. 1957 *Times Lit. Suppl.* 20 Dec. 771/1 The *Abwehr's* so-called Brandenburg Division was cannibalized, and five further Commando Battalions were formed. 1960 *How TV Works* 11/1 The whole process of cannibalising a travelling eye..began again in London. 1964 'M. UNDERWOOD' *Crime of C. Wise* ii. 24 Wise had subsequently made him one [radio] out of cannibalised parts.

Cannizzaro (kæniză·ro). The name of Stanislao Cannizzaro (1826–1910), Italian chemist, used *attrib.* or in the possessive of a reaction of aldehydes with caustic alkali (see quot. 1964).

1912 *Chem. Abstr.* 1431 The conversion of aromatic aldehydes by alcoholates into acids and alcs. (Cannizaro's reaction) is found to be untenable. 1964 N. G. CLARK *Mod. Organic Chem.* x. 187 Cannizzaro reaction. When aldehydes containing no hydrogen atoms on the α-carbon atom (adjacent to the functional group) are treated in the cold with concentrated aqueous or alcoholic sodium hydroxide, a dismutation involving two molecules takes place. One molecule is reduced to the alcohol, while the other is oxidized to a salt of the corresponding carboxylic acid.

cannon, *sb.*[1] Add: **2. d.** A pistol, a revolver. *U.S. slang.*

1901 'J. FLYNT' *World of Graft* 137 The thief had him covered with his 'cannon'. 1915 [see sense *10 below]. 1926 J. BLACK *You can't Win* xiii. 185 One of them.. started to lug out his 'cannon'.

e. A shell-firing gun in an aircraft.

1919 *Radio Pamphlet No. 30* (*Signal Corps, U.S. Army*) 3 These are..biplanes, mounting two to six or more machine guns and often a cannon. 1921 *Flight* XIII. 294/1 The machine carries one pilot and two gunners, operating in all eight machine-guns and one cannon. 1949 F. MACLEAN *Eastern Approaches* III. viii. 394 He showed me how the rear cannon fired, and how the intercom. worked.

7. b. Substitute for def.: *transf.* An act of cannoning (see CANNON *v.* 3 b); a collision between persons or animals (as in racing).

1806–7 [in Dict.]. 1876 *Coursing Cal.* 35 Bellini came best out of a cannon, and never let Hawkseye have a look in, except to kill.

9. cannon-fodder [tr. G. *kanonenfutter*; cf. Shakespeare's *food for powder* (1 Hen. IV, IV. ii. 72)]: men regarded merely as material to be consumed in war.

1891 [in Dict.] 1898 G. B. SHAW *Our Theatres in Nineties* (1932) III. 287 Peter [the Great], regarding children as future cannon-fodder, was as terribly severe on infanticide as he was infinitely indulgent to illegitimacy. 1950 PARTRIDGE *Here, There & Everywhere* 65 *Cannon-fodder* was never very popular with British troops.

10. A thief, spec. a clever pickpocket. *U.S. slang.*

1915 JACKSON & HELLYER *Vocab. Criminal Slang* 22 *Cannon*, General currency. A revolver. In pickpocket parlance it signifies a pickpocket of indefinite order. 1936 *Evening News* 9 Dec. 8/5 The art of the 'dip' or 'cannon', as these light-fingered gentry are known in the underworld, probably reached its climax in the person of 'Diamond Dick'. 1955 D. W. MAURER in *Amer. Dial. Soc.* XXIV. 89 *Cannon*, an intensification..of *gun* was, and still is, used with some sense of indicating a better-than-ordinary pickpocket.

cannonade, *v.* **1.** (Later example.)

1937 W. S. CHURCHILL *Great Contemp.* 104 The possibility of the German fleet..cannonading Calais.

cannon-ball. Add: **2.** Also simple *attrib.*

1920 TILDEN *Lawn Tennis* xiv. 146 A fast cannon-ball smash.

ca·nnulate, *v.* [f. the adj.] *trans.* To introduce a cannula into (a cavity). So **cannula·tion,** the action of the verb.

1926 *Jrnl. Amer. Med. Assoc.* 28 Aug. 641/2 The duct of Santorini was then cannulated, no attempt being made to cannulate or ligate the duct of Wirsung. 1944 *Science* 9 June 476/1 The 10 to 15 minutes used for the cannulation. 1956 *Nature* 18 Feb. 339/2 The blood-flow..has been measured by cannulating the sagittal sinus. 1961 *Lancet* 19 Aug. 410/1 Repeated hæmodialysis by means of conventional cannulation techniques.

canoe, *sb.* Add: **2.** See also PADDLE *v.*[2] 2 b.

3. *canoe load*; *canoe-man* (earlier U.S. examples); **canoe-shell,** a shell shaped like a canoe, spec. *Scaphander lignarius*.

1691 in J. Munsell *Annals of Albany* (1850) II. 115 Ye def't accused him of stealing 1/2 canoe load of water millions. 1753 in Washington *Writ.* (1889) I. 15 They were sent from New-Orleans with..8 Canoe-Loads of Provisions. 1890 A. HENRY *Trav.* 63 Several canoe-loads of fish were exported. 1936 J. C. BEAGLEHOLE *New Zealand* i. 14 Canoe-loads of savages. 1698 L. HENNEPIN *New Discovery Amer.* xlv. 188 One of our Canow-Men look'd after our Things. 1755 L. EVANS *Geogr. Essays* 17 The Canoe Men are often obliged to carry over Land. 1711 PETIVER *Gasophyl.* VI. Tab. liii, Small Indian Canoe-shell...It's of an odd Shape, and sticks to larger Shells. 1850 MISS PRATT *Common Things Sea-side* 216 The Tufted Canoe-shell. 1945 E. STEP *Shell Life* xvi. 272 The Canoe-shell (*Scaphander lignarius*) is a many-whorled spiral of reddish hue.

canon[1]. Add: **4.** Also, those writings of a secular author accepted as authentic.

1885 *Encycl. Brit.* XIX. 211/1 The dialogues forming part of the 'Platonic canon'. 1953 C. J. SISSON *Shakespeare: Compl. Works* p. xviii (*heading*) The canon and the text.

canon (kæ·nən), *v.* [f. CANON *sb.*[1] 7.] *trans.* To treat (a musical theme) in canon fashion. Also *absol.* or *intr.*

1894 G. DU MAURIER *Trilby* I. I. 41 They..fugued and canoned and counterpointed [that simple melody]. 1927 *Music & Lett.* July 347 The phrases [of bird singing] were more often alike and several times canoned one into the other in a most delightful way.

cañon, *sb.* (Earlier U.S. examples.)

1834 A. PIKE *Sketches* 20 Two cañons ran up into the bosom of the ridge (by which word *cañon* the Spaniards express a deep, narrow hollow among the mountains). 1846 R. B. SAGE *Scenes Rocky Mts.* 111 The Platte forces its way through a barrier of table lands, forming one of those striking peculiarities incident to mountain streams, called a 'cañon'.

cañon (kæ·nyən), *v.* [f. the sb.]

1. *intr.* To form, or flow in, a cañon.

1851 MAYNE REID *Scalp-hunters* I. v. 58 The stream, after running parallel to the ridge, swept round and cañoned through it. 1911 J. S. CHASE *Yosemite Trails* 208 Above it 'cañons' to the long gorge that is known as the Grand Cañon.

2. *trans.* To pierce with cañons.

1889 J. S. DILLER in *8th Ann. Rep. U.S. Geol. Surv. 1886–87* I. 426 Deeply cañoned by numerous streams.

canonizable (kæ:nŏnəi·zăb'l), *a.* [f. CANONIZE *v.* + -ABLE.] Worthy of canonization.

1913 *Nation* (N.Y.) 16 Jan. 58 A fictitious Crispi—..wise, righteous, canonizable. 1933 *Times Lit. Suppl.* 26 Oct. 736/1 Ampère, the scientist, whose laboratory was the home of 'a canonizable saint'.

canonry. Add: **2.** An establishment of canons or canonesses.

1877 SKENE *Celtic Scot.* II. 243 Boys and youths who are educated in the canonry. 1925 C. S. DURRANT *Flem. Mystics* I. ix. 130 John Busch himself travelled across Germany with three nuns of Bronope to reform an Austin Canonry.

canoodle (kănū·d'l), *v. slang* (orig. *U.S.*). Also 9 **cannoodle, conoodle.** [Origin obscure.] *intr.* To indulge in caresses and fondling endearments. Also formerly *trans.*, to persuade by endearments or deception. Hence **canoo·dler, canoo·dling** *vbl. sb.* and *ppl. a.*

1859 SALA *Tw. round Clock* 11 a.m., A sly kiss, and a squeeze, and a pressure of the foot or so, and a variety of harmless endearing blandishments, known to our American cousins..under the generic name of 'conoodling'. 1864 —— in *Temple Bar* Dec. 40 He is an adept in that branch of persuasive dialectics known as 'conoodling'. He will 'conoodle' the ladies..into the acquisition of whole packages of gimcrack merchandise. 1879 *Punch* 15 Mar. 117/2 Then he and the Matchless one struggle, snuggle, and generally conoodle together rapturously. 1894 *Mexborough & Swinton Times* 2 Feb. 2/7 The money which he had lent in return for the privilege of canoodling his daughter. 1894 *Idler* Sept. 207 She is not to be overcome by courting or *canoodling*. 1903 'MARJORIBANKS' *Fluff-Hunters* 172 Two canoodlers tied up in knots and fastened with sealing-wax at the lips. 1919 J. M. BARRIE *Alice Sit-by-the-Fire* I. 31 She's not fond of a canoodlin' way. 1921 H. WALPOLE *Young Enchanted* III. iii, She's in there...I'm off on some business of my own for an hour or two, so you can canoodle as much as you damned well please.

canophilist (kănǫ·filist). [irreg. f. L. *canis* dog + -PHIL + -IST.] A lover of dogs. So **canophi·lia,** affection for dogs.

1879 W. L. LINDSAY *Mind in Lower Animals* I. I. iii. 24 A distinguished author, well known as a canophilist. 1889 *Macm. Mag.* Apr. 463/1 A great comfort to those who devote themselves to canine pets, and to canophilists generally (a pretty new word). 1935 AUDEN & ISHERWOOD *Dog beneath Skin* II. i, *2nd M.O.* Travelling with a dog. *1st M.O.* Hm. Canophilia. 1960 *Spectator* 7 Oct. 532 His canophilia..is chronic, sad and private.

Canopic. Add: *Canopic jar*, = *Canopic vase*.

1893 BUDGE *Mummy* 171 At each of the four corners or sides of the bier, is placed one of the so-called Canopic jars. 1957 *Encycl. Brit.* XV. 954/2 The internal organs were.. placed in four vases known as the 'Canopic Jars'.

canopy, *sb.* Add: **2. a.** *spec.* The uppermost layer of branches in a forest.

1905 *Terms Forestry & Logging* 8 *Crown cover*, the canopy formed by the crowns of all the trees in a forest. 1952 P. W. RICHARDS *Tropical Rain Forest* ii. 23 A canopy means a more or less continuous layer of tree crowns of approximately even height.

4. *Naut.* (See quots.)

1867 SMYTH *Sailor's Word-Bk.*, *Canopy*, a light awning over the stern-sheets of a boat. 1961 F. H. BURGESS *Dict. Sailing* 44 *Canopy*, a canvas covering on a metal frame or supported by stanchions, for protective coverings over hatchways, etc.

5. A hood over a carriage or motor car. Also *attrib.*

1895 *Montgomery Ward Catal.* 580/1 Canopy Top Surrey ...A very nice carriage. 1906 W. W. BEAUMONT *Motor Vehicles* (ed. 2) II. 150 A..car with a closed tonneau body or limousine with canopy top and wind guard. 1930 *Motor Body Building* LI. 105/1 If the longitudinal framing is carried over the driver's seat the extension is called the canopy rail.

6. The 'umbrella' of a parachute, which fills with air when released from its packing.

1930 C. DIXON *Parachuting* vi. 53 The moment the silk canopy meets the full shock of the air currents—which spread it to its maximum extension—it pulls the airman off the wing and swings him into space. 1940 [see *BROLLY 2]. 1959 *Times* 18 May 8/7, I heard the crack of the canopy opening above me.

7. The cover of the cockpit in an aircraft.

1939 *Jane's All World's Aircraft* 48c/1 Enclosed pilot's cockpit over wing. Sliding canopy with quick-release for emergency exit. 1944 *Aircraft of Fighting Powers* V. 29/2 The introduction of the 'tear-drop' cockpit canopy on the P-51D has resulted in a decrease in the keel area ahead of the tail assembly.

‖ **canot** (kano). The French word for 'canoe', used in certain French phrases relating to the Canadian fur trade: **canot du maître** (kano dü mę̄tr'), the largest canoe of the fur trade, up to 40 feet long and carrying a cargo of 4 to 5 tons, formerly used esp. on the Great Lakes and St. Lawrence; **canot du nord** (kano dü nor), a canoe about 25 feet long and carrying over a ton of cargo, formerly used by the fur trade on the rivers of North-West Canada.

1828 A. MCDONALD *Peace River* (1872) 41 The largest kind of canoes used in the trade, viz., those which used to be dispatched from Lachine, on first open water, to Fort William, Lake Superior, and which were called '*Canots du Maître*'. *Ibid.*, The Canot du M. was of six fathoms, measured within, and the C. du Nord about four, more or less. 1961 H. MACLENNAN *Rivers of Canada* 23 The Nor'Westers used two types of canoe which they called the *canot du maître* and the *canot du nord*, the former for the run out of Montreal, the latter, which was lighter and carried less than a ton and a half of cargo, for the run west of Fort William where the streams were shallower. 1963 *Canad. Geogr. Jrnl.* Dec. 210/3 The bateaux were concealed, and the whole party of twenty-five embarked in a canot-du-maître for Fort Mackinac.

canroy (kæ·nroi). Also **candroy.** A machine, used in calico-printing establishments, through which cotton cloth is passed before printing. Hence **ca·nroyer,** one in charge of a canroy.

1836 *Penny Cycl.* VI. 151/2 If they are not calendered, they are run through a machine called in Lancashire the *candroy*, which spreads them smoothly in the act of rolling them upon a cylinder. 1858 [see CANDROY]. 1896 DUERR *Bleaching & Calico-printing* 26 After being sewn, each batch is taken to the shearing and winding-on machines, generally called a 'canroy'. Sometimes the shearing machine and canroy are two separate machines. 1921 *Dict. Occup. Terms* (1927) § 384 *Canroyer*, a calenderer minding a canroy calender, having two iron rollers through which cloth is passed to break down stiffness after starching.

cant, *sb.*[1] Add: **4. b.** One segment of the rim of a wooden cog-wheel.

a 1877 in KNIGHT *Dict. Mech.*

12. cant-file, a file with cutting faces at an obtuse angle to each other; **cant-line** (see quots.); **cant-rail,** a timber or other stiffening member which supports the roof of a railway carriage either at an angle or longitudinally; also *transf.*

a 1877 KNIGHT *Dict. Mech.*, *Cant-file*, a file having the shape of an obtuse-angled triangle in its transverse section; used in filing the inner angles of spanners and wrenches for bolts with hexagonal and octagonal heads. 1867 SMYTH *Sailor's Word-bk.*, *Cant-line*, synonymous with *girt-line*, as to cant the top over the lowermast-head. 1909 *Cent. Dict. Suppl.*, *Cantline*, the space between the sides or ends of barrels. 1961 F. H. BURGESS *Dict. Sailing* 44 The 'cant line' is the groove between strands of a rope, rows of casks or drums, etc. 1871 *Saddl. Harn. & Carriage Builder's Gaz.* 1 Dec. 12/2 Levers and links, aided by springs, to throw open the 'cant rails' and 'uprights' of carriage heads by inside pressure. 1930 *Motor Body Building* LI. 105/1 *Cant Rail*, the longitudinal framing of the Roof. 1951 *Engineering* 8 June 705/3 The roof cantrail is an interesting design. 1958 *Ibid.* 14 Mar. 344/1 Continuous longitudinal stiffening members such as cantrails. 1969 *Jane's Freight Containers 1968–69* 505/2 End frames: Fabricated hollow section corner pillars with 6·35 mm (¼ in) fixed end cantrail and hollow section.

13. cant-fall, the tackle connected with the cant-blocks of a whaling ship.

1867 SMYTH *Sailor's Word-bk.*, *Spike-tackle and cant-falls*, the ropes and blocks used in whalers to sling their prey to the side of the ship.

cant, *sb.*[2] (Examples of use in *Forestry*.)

1928 *Forestry* II. 82 The season's *coupe* having been marked out previously into lots or 'cants'. 1953 H. L. EDLIN *Forester's Handbk.* xv. 255 The portion of the wood that is due for felling is marked out on the ground and, if it is large enough, divided into several parcels, called *cants* in Kent. 1965 *Punch* 15 Dec. 890/2 Our neighbour sold his at auction at the end of October in five cants—that is, slices, in this instance of roughly three acres apiece.

Cantabrian (kæntēi·briăn), *a.* and *sb.* [f. L. *Cantabria*: see -IAN.] **A.** *adj.* Pertaining to the Cantabri, an ancient warlike tribe of northern Spain, or to Cantabria, the region formerly occupied by them. **B.** *sb.* **a.** One of the ancient Cantabri. **b.** The language of the Cantabrians.

1642 J. HOWELL *For. Trav.* x. 124 The Cantabrian tongue in Biscaie. 1746 AKENSIDE *Hymn to Naiads* in Dodsley *Poems* (1758) VI. 8 The rough Cantabrian coast. 1747 [see SEPT *sb.*[1]]. 1797 *Encycl. Brit.* IV. 122/1 The Cantabrians had waged war with the Romans for upwards of 200 years. *Ibid.* 122/2 The Cantabrian does not appear to have any affinity with any other known language. 1813 *Q. Rev.* Oct. 256 Classes and Families of Languages...Indoeuropean... Cantabrian. Sclavic. 1861 *Chambers's Encycl.* II. 572/1 Their bravery was evinced in the Cantabrian war, a six years' contest with the Romans. *Ibid.*, *Cantabrian mountains*, the general name of the several ranges of coast and boundary mountains. 1904 T. R. GLOVER *Virgil* vii. 145 In the course of the Cantabrian war he wrote to the poet from Spain letters full of playful entreaties. 1957 P. KEMP *Mine were of Trouble* v. 97 These two victories forced the gateway through the Cantabrian Cordillera.

Cantabrigian, *a.* and *sb.* Add: (Earlier examples.)

c 1540 [see OXONIAN *sb.* a]. 1616 J. CHAMBERLAIN *Let.* 27 Mar. (1939) I. 618 The King..had a play at Roiston acted by some of the younger sort of our Cantabrigians.

2. Belonging to Cambridge, Mass., or to Harvard University.

1887 *Harper's Mag.* Mar. 589/1 Mrs. Sainsbury was Boston-born, as well as Mrs. Pasmer, and was Cantabrigian by marriage. 1893 W. K. POST *Harvard Stories* 26 The New Haven men struggled to the Cantabrigian twenty-yard line.

Cantal (kæ·ntăl). [Name of a district in south central France.] In full *Cantal cheese*. A hard cheese made chiefly in the Auvergne, France.

1890 J. MACDONALD *Stephens's Bk. Farm* (ed. 4) IV. 517/2 The Cantal cheese..is an extremely important one upon the Continent. 1918 [see *BLEU]. 1961 *Listener* 20 Apr. 704/1 The delicious cheeses Cantal, Bleu d'Auvergne, and Saint-Nectaire. 1966 K. WHITE *Lett. from Gourgounel* xxii. 103 Last year in Auvergne..I was to be met on the hills with a bit of Cantal cheese.

cantaloup. The spelling *cantaloupe* is now common. Delete 'Chiefly *U.S.*' and add earlier and later examples.

1739 MILLER *Gard. Dict.* II. s.v. *Melo*, The Cantaleupt [*sic*] Melon:..the Flesh..is of a rich vinous Flavour. 1763 MILLS *Pract. Husb.* IV. 169 The Cantaleupe..is held in the greatest esteem by all the curious in Europe. 1777 COWPER *Let.* 23 Oct. (1904) I. 143 The melon is a crimson Cantalupe. 1786, 1813 [see ROCK *sb.*[1] 9 e]. 1890 *Daily News* 18 Sept. 3/1 The very first item on the list is 'iced cantaloupes'. 1929 *Times* 2 Aug. 14/1 Melon or bailer shells..are almost exactly the same shape, size and colour as rock melons, or canteloups. 1970 *Natural History* Feb. 18/3 Cantaloupes.. produce a crop only occasionally.

cantarist (kæ·ntărist). *Hist.* [ad. med.L. *cantarista*, f. *cantaria* CHANTRY: see -IST.] A chantry priest.

1800 *Gentl. Mag.* Mar. 201 The almshouse..for one chaplain, or cantarist, and three almsmen. 1894 H. FISHWICK *Lancashire* ix. 213 Many of these cantarists, as they were called, were now pensioned off for life.

cantate (kæntēi·ti, kæntā·te). [a. L. *cantate* sing ye, the first word of the psalm.] The ninety-eighth psalm (ninety-seventh in the Vulgate) used as a canticle (e.g. as an alternative to the Magnificat at Evening Prayer in the Church of England).

Cantate Sunday, the fourth Sunday after Easter, so called because the introit for that day is taken from this psalm.

c 1550 BALE *K. Johan* (Camden) 65 Now may we synge Cantate, And crowe Confitebor with a joyfull Jubilate. 1880 *Grove's Dict. Mus.* I. 305/2 The 17th canon of the council of Laodicea appointed lessons and psalms to be read alternately; and on this principle the 'Cantate' is to be considered as a 'responsory psalm' coming between the lessons. *Ibid.*, 'Cantate Services' are..rare. 1957 *Oxf. Dict. Chr. Church* 231/2 In the American Prayer Book of 1789, the 'Magnificat' was..replaced by the 'Cantate'.

cantatrice. (Earlier example.)

1803 SCOTT *Let.* 6 Mar. (1932) I. 178 Immense parties where none dare tell his mind to his neighbour should it involve anything more important than an opinion on the merit of the newest *cantatrice* or *figurante*.

cant-dog. Delete 'in *U.S.*', and add earlier and later examples.

1850 S. JUDD *R. Edney* 272 Leaning on a cant-dog, he could talk with Melicent and Barbara. 1942 *Amer. Speech* XVII. 220/2 *Cant dog*, a peavey; the riverman's tool. *a* 1951 B. CRONIN in *Austral. Short Stories* (1951) 167 The duty of a cant-dog man is to orient the logs.

canteen. Add: **1.** (S. Afr. examples.)

1837 F. OWEN *Diary* (1926) 7 The annoyance I had been put to, in one of the Hottentots being found on the floor at a canteen, at the time when he ought to have been at his place at the waggon. 1852 C. BARTER *Dorp & Veld* 9 (Pettman), But the inns sadly need reformation—they are in fact little better than canteens. 1950 *Cape Times* 20 Sept. 16/2 Men in uniform have been encouraged by their officers to use their own canteens rather than public bars.

b. In extended use. Now usu. a refreshment-room at a factory, school, or the like.

1870 D. J. KIRWAN *Palace & Hovel* (1963) xv. 143 The Canteen of the Alhambra is..nothing more than a subterranean bar-room. 1908 *Parish Councils* (Fabian Tract No. 137) 13 At Limpsfield (Surrey) the school manager appointed by the parish council personally started a canteen, which fed 60 children who came from a distance. *a* 1930 D. H. LAWRENCE *Phoenix II* (1968) 18 Dinners are given in the canteen in one of the mean streets, where the children feed in a Church Mission room. 1937 *Discovery* Sept. 268/2 A works canteen, a matter on which employees are sensitive, especially if..the canteen is a virtual monopoly. 1955 *Times* 12 May 18/5 Every consideration will also be given to facilities for canteens and both indoor and outdoor recreation.

3. b. A chest or case containing a set of table knives, forks, spoons, etc.

1895 *Army & Navy Co-op. Soc. Price List* 756 Canteens for wedding presents, yachting, college use, &c. 1963 *Times* 30 Jan. 4/3 Generous expenses and a canteen of cutlery!

5. *canteen-keeper* (earlier example); (sense *1 b) *canteen-manager, -work*.

1832 *Cape Gd. Hope Lit. Gaz.* 1 Feb. 243 (Pettman), We would call for certificates of character from every canteen-keeper. 1918 A. BENNETT *Pretty Lady* xxvi. 171 Concepcion..had decided that she ought to take up canteen work. 1937 *Discovery* Sept. 268/2 A director who occasionally goes round dinners with his canteen manager.

‖ **cante hondo, jondo** (ka·nte χo·ndo). [Sp., lit. 'deep song'.] A popular type of Spanish song, often mournful.

1932 E. HEMINGWAY *Death in Afternoon* iv. 41 Do not look for beautiful women..in the brothels or the canta honda [*sic*] places. 1944 W. APEL *Harvard Dict. Mus.* 268/1 The gypsy style is supposed to have originated in the early 19th century from the *cante hondo* or *jondo*..of Andalusia, a highly emotional and tragic type of song probably influenced by the Sephardic Jews. 1961 *New Statesman* 22 Dec. 954/1 The agonised wail of *cante hondo* from beer-and-prawn cafés. 1969 D. CORY *Night Hawk* 48 Her voice had the real old cante jondo rasp.

canter, *sb.*[1] Add: **1. b.** In a sawmill, a machine placed over the carriage and used to cant or roll over the log on the carriage in making the first cuts; a canting-machine (*Cent. Dict.* Suppl. 1909).

canter, *v.*[2] Add: **4.** (Earlier example.) Also *fig.*

1845 'TITMARSH' *Leg. of Rhine* xii, in *Cruikshank's Table-Bk.* 243 The knight gracefully cantering an elegant cream-coloured Arabian. 1930 V. WOOLF *Diary* 20 Feb. (1953) 156, I must canter my wits if I can.

Canterbury, *sb.* Add: **A. 3.** *Canterbury hoe* (see quot. 1954).

a 1887 JEFFERIES *Field & Hedgerow* (1889) 79 A two-spean spud, or Canterbury hoe, with points instead of a broad blade. 1954 A. G. L. HELLYER *Encycl. Garden Work & Terms* 124/2 What is virtually a modification of the draw hoe, is a type known as the Canterbury hoe in which the blade consists of three fork-like prongs attached to the handle at right angles.

4. [Name of province in South Island, N.Z.] *Canterbury lamb*, term used in Great Britain for lamb or mutton imported from New Zealand; in N.Z., for certain grades of such meat. Also *absol.*

1898 *N.Z. Farmer, Bee & Poultry Jrnl.* July 232/1 Does he mean to say that intrinsically as food..English mutton is worth about three times as much as prime Canterbury frozen? 1903 *Cyclopedia N.Z.* III. 80 We are basking in prosperity, now that we can turn off a large quantity of 'Prime Canterbury' from the plains. 1928 R. G. STAPLEDON *Tour Australia & N.Z.* viii. 58 A brief reference to the New Zealand 'Canterbury lamb'..may not be out of place. 1959 J. PASCOE *N.Z. Sheep-Station* 3 Every time we buy Canterbury lamb from the butcher we are reminded that the meat we are going to eat has come from a New Zealand sheep-farm. 1966 'K. NICHOLSON' *Hook, Line & Sinker* ii. 25 We'll call at the butcher first; Len has saved a nice piece of Canterbury for us.

B. 2. (Earlier and later examples.)

1803 T. SHERATON *Cabinet Dict.* 127 *Canterbury*..has of late years been applied to some pieces of cabinet work, because, as the story goes, the bishop of that see first gave orders for these pieces. One piece is a small music stand... The other piece which receives this epithet, is a supper tray, made to stand by a table at supper. 1904 A. BENNETT *Great Man* i. 2 The Canterbury with its spiral columns. 1962 *Times* 17 Nov. 11/7 In England the supper canterbury was essentially a specialized form of dumb waiter.

cantharus, kantharos (kæ·nþărŏs, -ǫs). Pl. **canthari, kantharoi.** Also **cantharos.** [L. *cantharus*, Gr. κάνθαρος.]

1. *Antiq.* A large, two-handled drinking-cup.

1853 *Dict. Archit.*, *Cantharus*, the name of a peculiarly shaped drinking vessel particularly consecrated to the personifications of Bacchus. 1875 E. METEYARD *Wedgwood Handbk.* 393 Cantharos. 1885 *Encycl. Brit.* XIX. 180/1 Silver Cantharus from Rhodes, with gold mounts. 1909 *Daily Chron.* 18 Mar. 3/1 Twenty-six drinking cups, one kantharos. 1957 CHILDE *Dawn Europ. Civilization* (ed. 6) ii. 33 A two-handled tankard or cantharos with quatrefoil lip.

2. A fountain or laver placed in the court-yard of an ancient church for the use of worshippers.

1842 GWILT *Archit.* Gloss. 943. 1902 *Encycl. Brit.* XXXII. 626/2 A court enclosed the whole; near the porch was a laver (*cantharus*) for the ablutions of intending worshippers.

‖ **cantiga** (kæntī·gă). [Sp. and Pg.: cf. CANTICA.] A Spanish or Portuguese poem or folk-song.

1915 A. F. G. BELL in *Mod. Lang. Rev.* X. 338 Written in the Galician–Portuguese language..there is not a page of the *Cantigas* which does not throw some curious light on the life of the thirteenth [century]...Some of these poems express beautiful and widely-extended legends. 1959 *Chambers's Encycl.* III. 56/2 *Cantiga* or *cantar* was the designation of all forms of poetry...*cantiga de amor*..*cantiga de amigo*..*cantiga de escárneo* (satirical). The religious songs are represented by the *Cantigas de Santa Maria*. 1963 *Times* 18 May 11/3 The single-voiced *cantigas* of Alfonso the Wise (1221–1284), who turned the art of the *trobadors* to spiritual ends.

cantilena. Add: (Earlier examples.) Also *attrib.*

1740 GRASSINEAU *Mus. Dict.* 19 *Cantilenæ* are no more than songs, and signifies in general pieces of melody well composed. 1775 'J. COLLIER' *Mus. Trav.* (ed. 3) 97 My *cantilena* was often rude. 1832 C. MACFARLANE *Lives Banditti* I. i. 38 They are sung in a sort of recitative, monotonous *cantilena* style. 1964 *Listener* 8 Oct. 558/1 A singer's cantilena style.

ca·ntilevered, *ppl. a.* [f. CANTILEVER + -ED[2].] Projecting like a cantilever; supported by a cantilever.

1910 H. BECHER *Reinforced Concrete* 229/1 Cantilevered beam. 1943 J. S. HUXLEY *TVA* xi. 85 The cantilevered overhead gantry crane. 1962 E. GODFREY *Retail Selling & Organ.* ii. 10 New stores..have made extensive use of prestressed concrete and cantilevered construction.

cantillation. Add: Also **cantila·tion.**

1901 W. B. YEATS *Let.* 16 Mar. (1954) IV. 348 Mr. Runciman said that I called a method of speaking verse.. 'Cantilation'...You will perhaps permit me to say that Mr. Runciman invented the word. I never used it, and I don't mean to. 1919 E. POUND *Quia Pauper Amavi* 39 Up, up my soul, from your lowly cantilation.

Hence (*rare*) **ca·ntilate** *v.*, **cantila·tionary** *a.*

1902 G. B. SHAW *Let.* 6 June in C. Bax *Farr, Shaw & Yeats* (1941) 19, I should do no good by entering into cantilationary polemics...Half the curates in the kingdom cantilate like mad all the time.

‖ **cantina** (kæntī·nă). [Sp. and It.] **a.** A (Spanish) canteen (sense 1, *1 b). **b.** A bar-room, a saloon (in Central and South America and south-west U.S.). **c.** An (Italian) wine-shop.

1892 *Dialect Notes* I. v. 245 *Cantina*: bar-room; of frequent use [in Texas]. Often found on signs of Mexican bar-rooms. 1925 HEMINGWAY *In our Time* (1926) xi. 143 At the cantina near the bridge they trusted him for three more grappas. 1939 G. GREENE *Lawless Roads* i. 36 Life..went on behind the swing doors of the cantinas and billiard saloons. 1959 *Listener* 15 Jan. 125/3 When a station was reached the passengers swarmed down to the platform and round the tiny *cantina*. 1969 A. MARIN *Rise with Wind* xv. 190 Clay found a cantina on a side street. 1970 *Daily Tel. Mag.* 3 Apr. 43/1 Every hotel, every restaurant, every cantina paid him [*sc.* a Cuban chauffeur] his commission.

cantino (kantī·no). [It.] The treble string of a violin or similar instrument.

1876 in STAINER & BARRETT *Dict. Mus. Terms*. 1944 W. APEL *Harvard Dict. Mus.* 117/1 *Cantino*, the highest string of lutes, viols, etc.

ca·ntly, *adv.*² [Cant *sb.*² 4 e.] In canting phraseology; in slang.

1828 *Niles' Reg.* 27 Sept. XXXV. 66/2 This is cantly called 'talking to Bunkum'. 1890 *Field* 1 Nov. 655/2 They become open professionals, and 'go for the pieces', as it is cantly termed. 1906 Chesterton *Dickens* xi. 288 In attempting to decide whether an author will, as it is cantly expressed, live, it is necessary to have [etc.].

Canto·n, *sb.*³ The name of the city in southern China used *attrib.* to denote various manufactured articles, as *Canton china, crape, enamel, flannel, matting.*

1860 *Texas Almanac* 244 Shirts, Carpets, Canton-Matting, shoes. 1865 M. Eyre *Lady's Walks* xvii. 195 A lady's worked Canton crape shawl. 1881 C. C. Harrison *Woman's Handiwork* i. 47 Canton flannel...a soft, downy fabric, the same on both sides. *Ibid.* III. 231 Blue Canton china of the willow pattern. 1889 *Anthony's Photogr. Bull.* II. 98 A good quality of canton flannel. 1910 S. W. Bushell *Chinese Art* (ed. 2) II. 82 Painted enamels on copper.. often known as 'Canton enamels'. 1968 J. Ironside *Fashion Alphabet* 222 *Canton crêpe.* Originally made of silk from Canton, this crêpe is widely used for dresses; it has a very pebbly surface and drapes and hangs beautifully.

Cantonese (kæntōnī·z), *a.* and *sb.* [f. *Canton*, China + -ese.] **A.** *adj.* Of or pertaining to Canton or its inhabitants. **B.** *sb.* **a.** An inhabitant of Canton. **b.** The dialect of Canton.

1857 *Bombardment of Canton* 29 Between you and ourselves, the Cantonese, there have been relations of friendship. 1857 'An Asiatic' *China Question* 10 The Cantonese stir will be for the benefit of humanity. 1861 *Chambers's Encycl.* II. 573/2 The Cantonese are notorious for their turbulence and hatred of foreigners. 1872 Eitel (*title*) A Chinese dictionary in the Cantonese dialect. 1883 J. D. Ball (*title*) Cantonese made Easy. 1893 *Athenæum* 27 May 668/1 The book of English phrases compiled by a Cantonese for the use of those of his countrymen who aspire to become Hong Kong 'boys'. 1957 *Encycl. Brit.* IV. 772/2 The principal parts were played by Cantonese. *Ibid.*, The Cantonese delegates.

cantoniza·tion. [f. Cantonize *v.* + -ation.] The process of making cantonal; a division into cantons.

1936 *Times Lit. Suppl.* 23 May 439/1 A cantonization after suppression of a serious revolt. 1949 Koestler *Promise & Fulf.* xii. 132 A scheme of Cantonization prepared by the Colonial Office.

cantor. Add: **3.** = *chazzan.

1893 I. Zangwill *Ghetto Tragedies* 3 The quaint monotonous sing-song of the Cantor reading the Law. [illegible] 1958 *Times* 23 Sept. 2/7 A wandering synagogue-cantor.

‖ **cantoris** (kæntō·ris). [L., genitive of *cantor* Cantor.] Of or belonging to the cantor or precentor; *cantoris side, stall*, the side occupied by the cantor, the north (exceptionally, the south) side. In *Music* used to indicate that side of the choir in antiphonal singing. (Correlative to Decani.)

1641 J. Barnard *First Book Sel. Ch. Mus.*, Tenor Cantoris. [illegible] 1760 Boyce *Cath. Mus.* I. 67. 1880 *Grove's Dict. Mus.* I. 306/1. 1894 [see Decani]. 1955 M. Gilbert *Sky High* i. 8 The Cantoris tenor..bent across to say something to..the Cantoris alto.

cantus (kæ·ntŏs). *Mus.* Pl. **cantus** (-tūs). [L.] A song or melody, especially ecclesiastical melody; also, the principal voice. Also *attrib.*

[1481 Caxton *Reynard* (1895) xvi. 69 What was it, prose or ryme, metre or verse..? I knowe it was cantum, for I herde you synge.] 1590 Whythorne (*title*) Cantus. [Bassus.] Of Duos, or Songs for two voices. 1597 Morley *Introd. Mus.* 70 In this Cantus there is no difficultie if you sing your Semibreves three Minyms a peece. 1662 T. Davidson (*title*) Cantus, Songs and Fancies, To Three, Four, or Five Parts, both apt for Voices or Viols. With a brief Introduction to Musick. 1887 *Athenæum* 25 June 842/1 The work is written for cantus, altus, and tenor—a rather unusual combination of voices. 1962 *English Studies* XLVI. 169 The only extant cantus part in Hanover.

b. *cantus firmus*: see Canto fermo.

1847 A. L. Phillips *Little Gradual* p. xiv. The choir be directed by persons well skilled in the ecclesiastical chant (which is called *Cantus Planus* or *Firmus*). 1956 A. Huxley *Adonis & Alphabet* 234 Those composers who wrote their masses around a *cantus firmus*—a melody borrowed, almost invariably, from the closed, symmetrical music of popular songs. 1962 *Listener* 12 Apr. 652/2 The *cantus firmus* of Palestrina's early Mass, '*Ecce sacerdos magnus*'.

Canuck (kănʌ·k). *colloq.* Also **Kanuck,** etc. [App. f. the first syllable of *Canada*.] **A.** *sb.* **1.** A Canadian; *spec.* a French Canadian. **2.** A Canadian horse or pony. **3.** The French-Canadian patois. **B.** *adj.* Of or pertaining to Canada or its inhabitants.

In *U.S.* usage, *gen.* derogatory.

1835 H. C. Todd *Notes upon Canada* 92 Jonathan distinguishes a Dutch or French Canadian, by the term *Kanuk*. 1849 J. E. Alexander *L'Acadie* I. xvi. 273 'Come boys and have some grog, I'm what you call a canuck:' a (Canadian). 1855 *Knickerbocker* XLV. 341 [We gave] our donkey into the keeping of a lively Canuck. 1860 Holland *Miss Gilbert's Career* ii. 29, I'll hang on the tail of it and try legs with that little Kanuck of his. 1862 *Congress. Globe* 29 Apr. 1867/3 To Canada to buy the little Canuck ponies. 1884 *Harper's Mag.* June 125/1 A 'Kanuck' or French Canadian. 1895 *Century Mag.* Sept. 674/1 That would be convenient over the line among the Canucks. 1904 H. F. Day *Kin o' Ktaadn* 145 'Roule, roulant, maboule roulant,' it's all Canuck but a good song. 1910 T. E. Lawrence *Let.* 17 Dec. (1954) 121 The three Canuck priests. 1964 *Calgary Herald* 19 Mar. 18/6 The Scottish skip missed a wide open [illegible] leaving the Canucks another single. 1965 M. Gold *Man who was not with It* xxvi. 249 Bon jour, Grack, tu sueux enjin?' That's Canuck for you ain't been a son to your ma.

canvas-back. 2. (Earlier U.S. examples.)

1785 Jefferson *Notes Virginia* vi. 130 Besides these [birds] we have..The Widgeon, Sheldrach, or Canvas back, Black head. 1809 W. Irving *Hist. N.Y.* II. 282 The gastronomical merits of..canvas-back ducks.

canvass, *sb.* Add: **5. b.** A scrutiny of votes in an election. *U.S.*

1778 *Laws State N.Y.* c. 16 § 9 To determine upon such Canvas and Estimate. 1888 Bryce *Amer. Commw.* II. App. 667 If all the returns have not been received, the canvass must be postponed...Upon the completion of said canvass and returns, the said Board shall immediately certify the same. 1903 *N.Y. Sun* 20 Nov. 5 The official canvass of the vote in the borough of Brooklyn was practically finished yesterday.

canvass, *v.* **4. c.** Delete ? and add: *Obs.* exc. *U.S.* Add examples.

1778 *Laws State N.Y.* c. 16 § 9 The said joint Committee ..[shall] canvass and estimate the votes. 1792 Low *Corr.* (1891) II. 428 Tioga will most probably not be canvassed, as the box was delivered by a person deputed by a deputy. 1888 Bryce *Amer. Commw.* II. App. 682 The officers...whose duty it is..to receive and canvass the returns from the several precincts of their respective counties..must then and there proceed to canvass the returns. 1947 *Chicago Daily News* 17 Jan. 14/1 When the popular votes for governor are canvassed by the legislature only a candidate with a majority of all the votes shall be qualified.

canvasser. 1. d. (Later U.S. example.)

1904 *Newark Even. News* 15 Nov. 5 Governor Murphy has appointed..as the State Board of Canvassers. The board will..officially determine the number of votes which were cast for the different candidates.

canvassy (kæ·nvăsi), *a.* Also **canvasy.** [f. Canvas *sb.* + -y¹.] Made of canvas, resembling canvas.

1892 *Dram. Opinions* 13 Jan. 1/1 The scenery..does not come up to the Lyceum standard, being more palpably canvassy than usual. 1903 *Westm. Gaz.* 12 Feb. 4/2 That nice-crisp softness in the way they hang peculiar to things canvassy. 1923 *Blackw. Mag.* July 4/1 The material was quite loosely woven and canvassy.

canyon, *sb.* (Earlier and later examples.)

1837 O. Russell *Jrnl.* (1955) 50 A deep narrow kanyon of rock. 1841 Farnham *Trav. Gt. Western Prairies* (1843) I. xv. 167 About midway from the Great Gap and the Kenyon of the south Fork of the Platte. *Ibid.* 268 This Kenyon terminates thirty miles above the Gulf. 1946 *Nat. Geogr. Mag.* Jan. 33/2 High pine-studded canyons.

ca·nyon, *v.* [f. the sb.] **a.** *intr.* To flow in(to) a canyon. **b.** *trans.* To cut into canyons.

1869 S. Bowles *Summer Vac. Colorado* 25 They 'canyon', as, by making a verb of the Spanish noun, the people here describe the streams as performing the feat of cutting such rock passages. 1870 J. H. Beadle *Utah* 441 Bear River..forms a great U in Idaho, then turning southward 'canyons' downward three miles. 1878 I. L. Bird *Lady's Life in Rocky Mts.* (1879) xi. 195 Rocks, cleft and canyoned by the river. 1904 *Christian Sci. Monitor* 26 Dec., The great inland rivers of America, as wide as the Missouri and as canyoned as the Colorado.

canzon. Delete †*Obs.* and substitute: *arch.* (Later example.)

1958 E. Pound *Personae* 258 Safe may'st thou go my canzon whither thee pleaseth.

‖ **canzonetta** (kænzone·tă). Pl. **canzonettas, canzonette.** [It., lit. 'little song'.] = Canzonet; Canzone.

1724 *Short Explic. For. Words in Mus. Bks.* 19 Canzonetta, is a little Song, or Tune, Cantata, or Sonata. 1811 Byron *Let.* 7 Sept. in *Wks.* (1832) II. 61 Spin canzonettas for Vauxhall. 1947 C. Gray *Contingencies* ii. 149 Several of his madrigals and canzonette. 1958 *Listener* 14 Aug. 250/3 The canzonetta, villotta, carnival songs, and so on which were all off-shoots of the madrigal.

Caodaism (kā,odai·iz'm). [Vietnamese *Cao Dai*(also used), lit. 'great palace' + -ism.] A syncretistic religion founded in Cochin China in 1926. Hence **Caodaist** (-dəi·ist), an adherent of Caodaism.

1937 V. Thompson *French Indo-China* vi. 474 The birth of a new politico-religious sect in Cochin-China called Caodaism in 1926. *Ibid.*, Nationalists and Caodaists use to good effect the plea for liberty of conscience. 1953 *Americana Annual* 332/2 The Caodaist sect fuses elements of all the great religions. 1953 *Newsweek* 4 May 48/3 Private armies of politico-religious sects such as the Caodaists (a hodgepodge of Buddhism, Confucianism, Spiritualism, and Christianity). *Ibid.* 21 Sept. 62/3 Among the most interesting religions..is Caodaism in Indo-China. 1955 *Times* 2 May 12/2 The Caodai 'pope', spiritual leader of the Caodai sect. 1958 A. J. Toynbee *East to West* xx. 62 A spiritual journey leading up to Caodai-ism as its goal. 1969 *New Yorker* 20 Sept. 110/3 The scattered religious elements in the South—the Buddhists, the Catholics, the Cao Dai sect.

‖ **caoine** (ki·nə). [Ir.] = Keen *sb.*

1797 E. Ledwich *Antiq. Irel.* 309 Caoine is a sort of Verse used in Elegies or Funeral Poems, and sometimes also in Panegyricks and Satyrs. Each versicle of a Caoine consists of only four Feet, and each Foot most commonly of two Syllables. 1844 T. C. Croker in *Percy Soc. Early Eng. Poetry* XIII. p. xix, Keen, which is here written according to its sound to the English ear, is, in its correct orthography, *Caoine*. 1926 Stanford & Forsyth *Hist. Music* x. 211 The principal styles [of Irish folk-music] are the songs, reels, jigs, caoines. 1954 *Grove's Dict. Mus.* (ed. 5) II. 50/1 The caoine is identical with the *ghnak* of the Old Testament, where it is a lamentation for the death by professional wailing-women.

caoline, var. Kaolin.

1865 Baring-Gould *Werew.* x. 140 On the tops of the plateau..lies a film of caoline.

cap, *sb.*¹ Add: **4. e.** Also *spec.* the cap, of a special form or colour, denoting selection as a member of a representative team, crew, etc.; hence, one who is awarded such a cap. Cf. *cap *v.*¹ 1 c.

1879 *Boy's Own Paper* 18 Jan. 2/3, I..was despairing of ever rising to win my 'first fifteen cap'. 1889 Barrère & Leland *Dict. Slang* I. 367/1 At Rugby when the school played football in white ducks, the ordinary 'caps' were allowed to wear flannels. 1895 *Cassell's Family Mag.* June 518/2 He may possibly be one of the crew, and the vision of the Light Blue cap flits before his enraptured eyes. 1902 *Football Who's Who 1901–2* 133 Cowan, James.. Earned caps and medals galore. 1906 Gibson & Pickford *Assoc. Football* IV. 28 He was eight seasons with the grand old club, and won his English cap. 1955 *Times* 10 Aug. 4/4 A brilliant first wood by Coulson, who gained his England cap this season.

g. *cap-in-hand* attrib. phr.

1889 C. M. Yonge in *Monthly Packet* Xmas 33, I don't care for those cap-in-hand ways of your people here. 1960 *Farmer & Stockbreeder* 29 Mar. 109/1 A more militant approach is called for and an end to this cap-in-hand begging for fair play.

6. Delete †*Obs.* and add later examples.

1902 *Westm. Gaz.* 2 Dec. 3/2 Masters of Hounds are naturally averse to that method of enriching their treasuries which consists in taking a 'cap' from strangers. 1928 *Daily Mail* 3 Nov. 13/2 [illegible] the Quorn he will have to pay a cap of £3 3s. per day. 1959 *Daily Tel.* 30 Jan. 19/8 The annual subscription to ride with the Exmoor Foxhounds is going up from £25 to £30 and the cap for visitors will be £3 instead of £2.

9. *to pull caps* (earlier examples).

1749 Cleland *Mem. Woman of Pleasure* i. 99 Oh! he was such a beauty!..they would pull caps for him! 1754 Richardson *Let.* 15 Sept. (1804) V. 27 She scrupled not..to pull caps in good-humoured roguery.

10. d. *Ornith.* The pileum or top of a bird's head, especially when distinctively coloured. Cf. Black cap 3.

1889 in *Cent. Dict.* 1890 Coues *Ornith.* 142 'Top of the head' is a collective term for all the upper surface, from base of bill to nape, and laterally to about the level of the upper border of the eyes. This is the pileum or 'cap'...; it is divided into three portions.

e. The calyptra of mosses.

1861 Oliver *Elem. Bot.* 181 The sporange is covered at first by a cap (calyptra). 1900 B. D. Jackson *Gloss. Bot. Terms.*

11. c. (Also *blue cap*.) In coal-mining, a circle of bluish flame appearing above and around the wick of a safety lamp when a dangerous quantity of fire-damp is present.

1877 *Encycl. Brit.* VI. 72/2 When a lighted candle is exposed in a non-explosive mixture of this gas [*i.e.* fire-damp], the flame gradually elongates, forming a conical cap, floating above the wick. 1883 Gresley *Gloss. Coal-m.* 21 Blue Cap, To carry on work in an atmosphere which shows a cap is unsafe. 1887 *Good Words* 99/1 The marsh gas is revealed..by the appearance of what is known as a 'cap' upon the flame. 1893 Lupton *Mining* 248 If there is 1 per cent. of fire-damp in the atmosphere, a small blue-cap may be observed.

12, and **13.** Further examples of various tech. uses.

1884 Knight *Dict. Mech.* Suppl., *Cap*, a short terminal section of a pipe, with a removable stopper called a plug. 1886 *Cassell's Fam. Mag.* 638/1 An outer cap protecting the point [of a fountain pen]. 1895 Burns *Gloss. Techn. Terms* 18 *Cap and lining*, in gasfitters' work a joint used for connecting a composition pipe to an iron pipe. 1902 *Westm. Gaz.* 4 June 5/1 The effect of fitting shot or shell with a forged steel cap. 1921 *Dict. Occup. Terms* (1927) § 254 *Cap maker* (electric lamps); a draw press hand making brass caps for electric lamps in a die press. 1958 M. L. Hall *Newnes Compl. Amat. Photogr.* 110 Large flashbulbs are fitted with an ES(E27) cap and smaller types with an ASCC(B15s) cap.

13. c. *Mining: in (the) cap*: see quots.

[1871 R. W. Raymond *Mines of Pacific States* 313 The 'cap', a term usually employed to express the impoverished condition of the vein, may be due either to the pinching together of the walls of the fissure, or..to the filling of the vein with barren rock.] *a* 1884 Knight *Dict. Mech.* Suppl. s.v., A vein is 'in the cap' when it is much contracted. 1889 P. Milford *Dict. Mining Terms* (ed. 2) 14 A mine when the vein matter is barren or when the vein is pinched, or contracted, is said to be 'in cap'. *c* 1931 G. F. Willison *Here they dug Gold* vii. 122 Most of the hard-rock mines are in cap.

d. A contraceptive device, usu. made of rubber, covering the neck of the womb. Cf. *Dutch cap.

1916 *Declining Birth-Rate* ii. 247 A woman..will sometimes wear a cap over the neck of the womb, which takes the place, in the female, of the 'letters' that men wear. 1918 M. C. Stopes *Wise Parenthood* iii. 18 The great advantage of this cap is that once it is in and properly fitted it can be entirely forgotten. 1935 E. F. Griffith *Mod. Marriage* 166 There are numberless caps on the market, most of which are made of rubber and vary considerably in thickness. 1970 *Sunday Times* 15 Mar. 50/4 Like the pill, the diaphragm, or cap..are used almost entirely by the more prosperous sections of the community.

14. b. The paper percussion cap of a toy pistol; = *amorce.

1872 C. M. Yonge *P's & Q's* ii. 13 He was only just exploding a few caps to teach the child to stand fire. 1877 [illegible] III. 521/3 Those small caps..called French amorces. 1883 [see *amorce]. 1921 *Dict. Occup. Terms* (1927) § 149 *Snap cap maker*, tends machine which makes snap caps..for toy pistols.

18. a. *cap-badge, -border* (example).

1897 *Sears, Roebuck Catal.* 236/1 Nickel Plated Cap Badges. [illegible] Men of divers varieties of cap badges. 1960 D. Lambert *Angels in Snow* xii. 192 His Army cap badge, a couple of Service ribbons. 1878 Mrs. Stowe *Poganuc P.* xxii. 246 A thin old delicate face, with its aureole of white hair and its transparent cap-border.

19. cap-bar *Spinning*, an attachment to a drawing-frame supporting the bearings of draft rolls; **cap-cell** *Bot.* (see quot. 1900); **cap-frame**, a type of spinning-frame in which the guide for the yarn takes the form of a cap; **cap-gun** = *cap-pistol; **cap-man**, a man who inspects the lamps attached to miners' caps; **cap-pistol**, a toy pistol which fires caps (sense *14 b); **cap-ribbon**, a band round a sailor's cap bearing the name of his ship; **cap rock** *Geol.*, an overlying rock or stratum; see also quot. 1956; **cap-screw** = *tap-bolt* (s.v. Tap *sb.*¹ 6); **cap-sheaf** (earlier and later examples); **cap-tally** *Naut. slang*, = *cap-ribbon; see also quot. 1946.

1867 W. S. Taggart *Cotton Spinning* II. 91 It was formerly the practice to make the cap bars of cast-iron. 1898 *Ibid.* III. 241 The cap bar, for keeping the top rollers in position, is provided, so that it can readily be moved over out of the way when the rollers require attention. 1882 Capcell [in Dict.]. 1900 B. D. Jackson *Gloss. Bot. Terms*, *Cap Cells*, the upper sister-cells of the embryo-sac in the ovule [illegible]

cap (kæp), *sb.*⁴ Also **cap, cap'.** Colloq. abbrev. of Captain.

1759 S. Merriman *Diary* 21 June in G. Sheldon *Hist. Deerfield* (1895) I. 662 A covering party, consisting of wone cap, 3 subbs 4 sargents 100 rancks & file. 1840 *Porter's Spirit of Times* (N.Y.) 17 Oct. 391/3 The old cap. wanted to kill one of them varmints. 1902 Kipling *Traffics & Discoveries* (1904) 29 'Well, Cap,' I says. 1909 Ware *Passing Eng.* 63/2 *Cap*, equivalent to 'Sir'—but really abbreviation of 'Captain'. Common in America—gaining ground in England. 1933 'L. Luard' *All Hands* 252 Where's the cap'?

cap (kæp), *sb.*⁵ Colloq. abbrev. of Capsule. Chiefly *U.S.*

1942 Berrey & Van den Bark *Amer. Thes. Slang* § 509. 8 *Cap*, a capsule of narcotics. 1962 J. Glenn in *Into Orbit* 144 The Cap Sep or 'Capsule Separation' event. 1963 *New Society* 7 Nov. 11/2 If they were big dope pedlars, they [illegible] [heroin] for delivering something.

cap, *v.*¹ Add: **1. c.** To award (a player) his cap (*cap *sb.*¹ 4 e); to select a representative player for a country, etc.

1902 *Football Who's Who 1901–2* 131 He was first capped as far back as 1893 against Ireland. 1963 *Times* 6 Feb. 4/2 Baker..capped five times for England in 1959–60.

4. c. (Earlier and later U.S. examples.)

1804 *Lancaster (Pa.) Intelligencer* 21 Feb. (Th.), Your correspondent caps the climax of Misrepresentation. 1891 *Fur, Fin & Feather* Mar. 158 A section of country..that caps the climax for quail, especially along the little creeks.

10. *intr.* To take cap-money (see Cap *sb.*¹ 19).

1854 R. S. Surtees *Handley Cross* ix. xxxiv. 269 Mr. Jorrocks allowed Pigg [his huntsman] to cap when they were [illegible]. 1896 Mordaunt & Verney *Ann. Warwicks. Hunt* I. xiv. 188 They used to cap for us then.

‖ **capable de tout** (kapabl də tu), *adj. phr.* [Fr.] Capable of anything.

['Habacuc était capable de tout.' Attributed to Voltaire (N. & Q. (1941) CLXXXI. 46/2).] 1908 M. Beerbohm *More* 166 He is *capable de tout*. 1953 'N. Blake' *Dreadful Hollow* vii. 93 You consider Célandine a vindictive character, *capable de tout*. 1962 N. Marsh *Hand in Glove* iv. 125 He..would certainly be capable of going too far—*capable de tout*.

capacitance (kăpæ·sităns). *Electr.* [f. Capacity + -ance.] The ratio of the change in an electric charge to the corresponding change in potential; also, the ability to store a charge of electricity, capacity.

1909 in *Cent. Dict.* Suppl. 1916 *Standardiz. Rules Amer. Inst. Electr. Engineers* 18 It is..recommended that..the term 'Capacitance' be used when referring to the electrostatic capacity of a device. *Ibid.* 53 Machinery of Low Capacitance. 1939 *Nature* 9 Sept. 458/2 The mica capacitor was compared by alternating current with an air capacitor, the capacitance of which can be evaluated in terms of resistance and time. 1940 *Chambers's Techn. Dict.* 132/1 *Capacitance grading*, grading of the properties of a dielectric in order to equalize the stress from conductor to sheath is [illegible]. 1956 Amos & Birkinshaw *Telev. Engin.* II. vii. [illegible] The input capacitance of a cathode follower is small only if the output load is high. 1962 A. Nisbett *Technique Sound Studio* iv. [illegible] In building the size of a broadcasting studio centre there is a danger..of high frequency losses due to capacitance.

capacitate, *v.* Add: **1. b.** *Physiol.* To cause (a spermatozoon) to undergo capacitation (see next).

1957 *Nature* 2 Feb. 258/2 'Capacitated' sperm were obtained from rabbits that had either been mated normally or inseminated with epididymal sperm. *Ibid.* 259/2 Epididymal sperm has to be capacitated. 1969 *Ibid.* 6 Sept. 1042/1 The Fallopian tube fluid is still effective in capacitating sperm even when heavily diluted with a simple chemical medium.

capacitation. Add: **b.** *Physiol.* The process or change that a spermatozoon undergoes in the female reproductive tract rendering it capable of penetrating the *zona pellucida* of an ovum and so fertilizing it.

1951 C. R. Austin in *Austral. Jrnl. Sci. Res.* B. IV. 594 The principal conclusion..is that, at least in the rat and rabbit, the sperm must undergo some form of preparation or capacitation before it can penetrate the zona. 1969 *Nature* 6 Sept. 1042/1 The acrosome, the cap-like structure at the head of the spermatozoon, is punctured in the course of capacitation and releases enzymes which dissolve materials such as compose the outer coverings of the egg.

capacitive (kăpæ·sitiv), *a.* *Electr.* [f. Capacity + -ive.] Pertaining to electrostatic capacity. Also **capa·citative.**

1916 *Standardiz. Rules Amer. Inst. Electr. Engineers* 97 *Capacitive Coupler.* An apparatus which, by electric fields, couples two or more radio frequency circuits. 1934 Webster, s.v. Capacitive. 1943 *Gloss. Terms Telecomm.* (*B.S.I.*) 74 *Common-impedance coupling*, the coupling between two oscillating circuits by means of a current path common to both. (This coupling may be *inductive*, *capacitive* or *resistive*.) *Ibid.*, *Capacitive tuning*, a form of tuning in which a capacitance component is varied. 1946 *Electronic Engin.* XVIII. 71 There will be two forms of parasitic coupling present: capacitive and resistive. 1962 A. Nisbett *Technique Sound Studio* 256 An inductive impedance increases with frequency; a capacitive impedance decreases with frequency. 1964 R. F. Ficchi *Electrical Interference* 214 Capacitive coupling is the association of two circuits with one another by means of capacitance mutual to the circuits.

capacitor (kăpæ·sitǫı). *Electr.* [f. Capacity + -or.] A device which stores electricity during part of an operation; a condenser.

1926 *Gloss. Terms Electr. Engin.* (*B.S.I.*) 43 Condenser.. It is suggested that the new term capacitor shall be used for this device in order to avoid confusion with a steam 'Condenser'. 1939 [see *capacitance]. 1946 *Electronic Engin.* XVIII. 13 A range of electrolytic capacitors of plain foil construction. 1962 A. Nisbett *Technique Sound Studio* 243 Capacitors are used in electronics for coupling, smoothing and tuning purposes.

capacity. Add: **1. d.** The power of an apparatus to store static electricity; also = *capacitance. Also *attrib.*, denoting an apparatus which gives additional capacity, as *capacity cage, earth.*

1777 T. Cavallo *Compl. Treat. Electr.* II. i. 103 The celebrated Father Beccaria supposes that the action of rubbing increaseth the capacity of the Electric, *i.e.* renders that part of the electric, which is actually under the rubber, capable of containing a greater quantity of electric fluid. 1903 *Westm. Gaz.* 25 Feb. 5/1 The oscillatory circuit consists of a light wire cage, or 'capacity'. 1906 A. F. Collins *Manual Wireless Telegr.* 209 *Capacity cage*, a cylindrical cage made of wire and placed at the top of the aerial wire to give it greater capacity. 1923 *Daily Mail* 3 Mar. 5 If you are particularly bothered by the proximity of wires it is worth while to try the effect of substituting a 'capacity earth' for the ordinary earth connection. The capacity is a wire or wires stretched out below and parallel to the aerial above but insulated from the earth. 1932 C. J. Smith *Intermediate Physics* xxxix. 507 The capacity of a sphere expressed in E.S.U. is numerically equal to its radius in centimetres. 1948 A. C. Coulson *Electr.* 43 If the potential difference between the plates is *V*, we define the capacity *C* by the formula *C* = *Q/V*. 1959 *Chambers's Encycl.* III. 826/1 The capacity of a charged conductor is..increased, i.e. it can hold a bigger charge for a given potential.

e. Ability to provide accommodation (for the discharge, transport, etc.) of a certain amount or volume; also *spec.* in *Physical Geogr.*, ability to carry away detritus, measured by the quantity which passes a given point in a given time.

1885 *Eng. Illustr. Mag.* Sept. 813/2 The [illegible] does not discharge its waters so rapidly as it ought. [illegible] bends if you want a more efficient discharging capacity. 1892 *Jrnl. Soc. Arts* XLI. 96/2 The capacity of the present tunnel is sufficient to provide 100,000 horse-power applied to the turbines. 1900 *Ibid.* XLVIII. 835/1 The capacity of the larger canal would be over 20,000,000 tons a year.

f. The largest audience that a place of entertainment can hold; a situation in which a theatre, cinema, etc., is full; *spec.* in phr. *to capacity*: with all available room occupied; hence in general use, full; to the limit.

1908 L. Merrick *Man who understood Women* 152 On the second Saturday night they played to 'the capacity of the house'. 1919 *Honey Pot* July 5 Why there is hardly a theatre in the West End playing to capacity, people are turned out. 1929 J. B. Priestley *Good Compan.* II. [illegible] We've got going in the town... Another week and it'll be capacity every night. *Ibid.* 329 Turning money away, we are, capacity to the roof, ol' man. 1937 D. Marlowe *Coming, Sir!* viii. 123 The other stows were filled to capacity. 1949 Wodehouse *Mating Season* xxii. 182, I took my place among the standees at the back, noting that..I should be playing to absolute capacity. 1958 *Spectator* 20 June 819/1 The bank-note paper mills worked to capacity throughout the year.

g. In industry: the ability to produce; equivalent to 'full capacity'.

1931 *Times Trade & Engin. Suppl.* 24 Jan. p. x/3 The production of bituminous coal [in U.S.] by 'mechanical mining' increased to 37·8 million tons in 1929, but this does not represent capacity, for short time and intermittent work reduced the total below the maximum. 1940 *Economist* 8 June 999/1 The existing capacity must not only be worked to the limit, it must be expanded. 1959 *Listener* 12 Feb. 273/1 The British economy, so far from being inflationary, was suffering from a marked amount of excess capacity.

11. *attrib.* and *Comb.*, as *capacity-catching.*

1904 *Westm. Gaz.* 17 Sept. 7/1 A certain grounding of school knowledge, enough..to serve for what Professor Huxley called 'capacity catching'. 1905 *Daily Chron.* 31 Jan. 4/5 If you are out for what Mr. Sidney Webb would style 'capacity-catching'.

b. passing into *adj.* That reaches or fills the utmost capacity.

1920 *Times* (weekly ed.) 17 Sept., Profits are not unduly large in view of the fact that it is necessary to play almost to capacity business. 1925 *Times Lit. Suppl.* 27 Aug. 554/1 Both the play and film are now drawing capacity houses in London. 1928 *Morning Post* 20 Oct. 10/6 A 'capacity' audience greeted the appearance of M. Ravel. 1929 *Ibid.* 28 Aug., Swimming baths, of course, attracted capacity crowds throughout the day. 1965 *Listener* 16 Sept. 416/3 *A Hard Day's Night* is running in six cinemas...The Beatles do capacity business.

cape, *sb.*² Add: **3. b.** *cape and sword* (also *cape and cloak*): phr. used to characterize romantic fiction or drama with a more or less historical background. Cf. *cloak *sb.* 6.

Cf. F. *roman de cape et d'épée.*

1898 *Westm. Gaz.* 3 Nov. 2/3 Plays of poetry, romance.. Cape and cloak, are all the fashion. 1898 *Daily News* 4 Nov. 3/5 The drama of the 'Cape and Sword'. 1920 *Westm. Gaz.* 2 Feb. 6/4 The cape-and-sword romanticism of Anthony Hope.

c. *transf.* The short feathers on a fowl's back falling below the hackle.

1899 A. H. Evans *Birds* 548 *Xanthomelus aureus*..has a ..black throat, tail, and part of wings and back, and a cape of hackled plumes falling over the last.

4. *cape-bonnet* (earlier example); **cape-work,** 'work' done by a bull-fighter in exciting and engaging the bull with his cape.

1838 C. Gilman *Recoll. Southern Matron* xix. 131 A young girl..dressed in homespun, with a *cracker* or cape-bonnet of the same material. 1926 Hemingway *Fiesta* (1927) xv. 193 She liked Romero's cape-work. 1962 *Guardian* 29 Oct. 5/6 Instead of practising their capework during idle moments, Spanish boys nowadays play..football.

cape, *sb.*³ Add: **2.** *Cape* (or *cape*) *hide*; also ellipt. for *Cape leather* or *capeskin, and *attrib.*

1897 *Sears, Roebuck Catal.* 228/2 Men's Genuine Medium Weight Cape Goat Driving Gloves. 1915 K. J. Adcock *Leather* ii. 11 Large quantities of Cape hides are sent to England. 1921 B. E. Ellis *Gloves & Glove Trade* iv. 58 Real Cape gloves are usually bark-tanned..but many gloves sold as 'Capes' are tawed and dyed by the dipping process. 1929 [see *Boulton *a.*]. 1956 *Gloss. Leather Terms* (*B.S.I.*) 7 *Cape*, originally a soft, grain gloving or clothing leather made from South African hair sheep skin; now any similar leather made from hair sheep skin, but not finished leather made from E.I. native vegetable tanned hair sheep skin.

4. Cape boy (see *boy *sb.*¹ 3 e); **Cape cart,** a two-wheeled, horse-drawn hooded cart peculiar to South Africa; **Cape cobra,** a cobra (*Naja nivea*) of southern Africa, variable in colouring; **Cape Coloured** *a.*, of or designating the Coloured or brown population group of the Cape Province, especially of the Western Province of the Cape; *sb.*, a person (or the people) of this group; **Cape doctor,** a strong south-east wind in S. Africa (cf. *doctor *sb.* 6 b); **Cape Dutch,** (*a*) South Africans of Dutch extraction; (*b*) the Dutch spoken in South Africa, Afrikaans; also as adj. phr.; **Cape gooseberry,** delete hyphen and read, *Physalis peruviana*, a herb of the family Solanaceæ, native to South America, or its fruit; (add earlier examples); **Cape jessamine,** any of various flowers of the genus *Gardenia*, esp. *G. jasminoides*; **Cape lobster,** the Cape crawfish (see *crawfish *sb.* 1 b); **Cape pigeon,** read 'a pigeon-sized petrel, *Daption capensis*'; (earlier example); **Cape robin,** a species of chat-like thrush, *Cossypha caffra*; **Cape salmon,** name given to various fishes having a resemblance to the European salmon, esp. the *geelbek and the *kabeljou; **Cape smoke** *slang*, South African brandy; **Cape sparrow,** the S. African bird *Passer melanurus*; *mossie; **Cape wagon** (see quot. 1850); **Cape weed,** (*b*) a common yellow-flowered herb, *Cryptostemma calendulaceum*, now a troublesome weed in Australia and N.Z.; (*c*) (see quot. 1933). (See also Pettman *Africanderisms* for many other specific names.)

1892 J. R. Couper *Mixed Humanity* ii. 20 Cape carts, driven by Malays and Cape boys, rattled up and down the streets in numbers. 1896 F. C. Selous *Sunshine & Storm* 59 This force was, however, augmented by about 150 Cape boys, chiefly Amaxosa Kafirs and Zulus. 1877 Lady Barker *Year's Housek. S. Afr.* i. 17 It was decided that I ought to take a drive in a Cape cart. 1881 Statham *Blacks, Boers, & Brit.* 53 Do you know what a Cape cart is? It is a peculiar, but pleasant, institution—something like what was once in England called a 'Whitechapel'. 1910 J. Buchan *Prester John* vii, The half-caste who called him 'Sir' and drove his Cape-cart. 1910 F. W. Fitzsimons *Snakes S. Africa* vi. 74 The Cape Cobra (*Naia flava*). Geel Slang, Bruin Slang, Spung Slang. The Cape Cobra is by far the commonest species of Cobra inhabiting South Africa. 1959 *Cape Argus* 31 Oct. 9/7 Cape cobras and puffadders provide most of the venom. 1897 *Milner Papers* (1931) iii. 89 The better treatment of Cape Coloured people. 1927 W. M. Macmillan *Cape Colour Question* iii. 29 The great mixed mass, descended from Hottentots, slaves, and Europeans, but forming now one distinct class, whom we describe as 'Cape Coloured'. 1938 N. Devitt *Spell of S. Afr.* xxi. 176 She jerks her body in a half-contemptuous shake, a common gesture among Cape coloureds of her class. 1943 *Cape Argus* 30 Jan. 7 Complaints and suggestions affecting the welfare of the Cape Coloured. 1960 *20th Cent.* Sept. 284 Kenneth Makeer..is a Cape Coloured, light enough in complexion to pass as white. 1861 Lady Duff Gordon *Let.* 19 Oct. (1875) 213 It portends a 'south-easter'...This wind..is the Cape doctor, and keeps away cholera, fevers [etc.]. 1890 A. Martin *Home Life on Ostrich F.* 15 That rough but benevolent south-east wind, which, owing to its kindly property of sweeping away the germs of disease, is called 'the Cape doctor'. 1966 *Listener* 18 Aug. 237/2 The Cape south-easter was blowing—the wind they call the Cape doctor because it blows the rubbish from the streets. 1826 *New Monthly Mag.* II. 488 The Cape Dutch..possess many estimable qualities. 1850 Appleyard *Kafir Lang.* 11 A grammar, [with] 'Proeve van Kaapsch Taaleigen', where the peculiarities of Cape-Dutch usage are exposed. 1852 *Punch* 3 July 19/1 They've christened me Zekoe—that's Cape Dutch for Sea-Cow! 1944 *Archit. Rev.* XCVI. 97/1 The 'Cape Dutch' style in town and farm building. 1833 W. F. W. Owen *Voyages* II. xix. 238 The physalis (Cape gooseberry, or winter cherry) is here..a most delicious fruit. 1840 [see Winter cherry 1]. 1870 *Cape Monthly Mag.* Oct. 218, I prefer the preserved *Cape* gooseberry to everything I have tasted. *a* 1776 J. Schaw *Jrnl. Lady of Quality* (1921) 246 While we were admiring a row of cape jessamine, which even now is covered with flowers. 1804 J. Barrow *Trav. S. Afr. 1797–8* II. 82 The *Gardenia Thunbergia*, or the wild Cape Jessamine. 1858 T. Vielé *Following the Drum* 58 Cape jessamine hedges. 1793 tr. *Thunberg's Trav.* I. 240 The Cape lobster (*Cancer arctos*) ..has no large claws, and is craggy all over. 1902 H. J. Duckitt *Hilda's Diary Cape Housekeeper* 47 'Crayfish', or 'Kreeft', is also plentiful all through the summer. We also call it 'Cape lobster'. 1913 W. W. Thompson *Sea Fisheries Cape Col.* ii. 51 The 'Cape lobster', as it [*sc.* the crawfish] is sometimes called. 1798 S. H. Wilcocke tr. *Stavorinus' Voyages* II. 31 We saw..the birds called 'cape-pigeons'. 1867 E. L. Layard *Birds S. Afr.* 132 Bessonornis Phoenicurus..is the Cape 'robin'. 1913 Pettman *Africanderisms* 114 *Cape robin.*—*Cossypha caffra*, a bird that resembles the English robin somewhat, but is without the red breast. 1846 H. H. Methuen *Life in Wilderness* i. 17 The cape salmon, a heavy fish, in size and in external aspect somewhat resembling its British namesake. 1865 *Hardwicke's Science-Gossip* 64/2 *Cape Salmon.*—Under this name the 'Geelbeck'..has been eulogized... Why call it Cape Salmon? 1906 [see *kabeljou]. 1846 H. H. Methuen *Life in Wilderness* viii. 232 Revelling in the luxuries of *Cape smoke*, or brandy, and sheep-tail fat. *a* 1871 J. Goldswain *Chron.* (1946) I. 36, I did not stop at this place Long for thear was to much Cape Smoke. 1954 'D. Divine' *Golden Fool* v. 47 It was a better brandy than the Cape Smoke most of them drank. 1936 E. L. Gill *First Guide S. Afr. Bird* 21 *Cape Sparrow*, Mossie, a very common bird about Cape Town and up the west coast region, and thence across the Karroo and High Veld. 1952 *Cape Times* 8 Nov. (Week-end Mag.) 6/5 What a handsome fellow the cock Cape sparrow is. 1798 Lady Anne Barnard *Let.* 6 June in *S. Africa a Century Ago* (1901) viii. 157 Of course, it was a Cape waggon; any other sort..is impossible..for such an excursion. 1837 J. E. Alexander *Narr. Voy. Observ.* xiv. 348 It is quite astonishing to a stranger what severe work Cape wagons undergo without injury. 1850 R. G. Cumming *Hunter's Life S. Afr.* I. ii. 22 The Cape waggon is a large and powerful, yet loosely-constructed vehicle, running on four wheels. 1877 *Trans. N.Z. Inst.* X. 367 The Cape weed, which is plentiful in Auckland. 1884 W. R. Guilfoyle *Austral. Bot.* (ed. 2) 107 Cape Weed, *Cryptostemma calendulacea*. (Natural Order, *Compositæ*). This weed, which has proved such a pest in many parts of Victoria, was introduced from the Cape of Good Hope, as a fodder plant. 1933 L. G. D. Acland in *Press* (Christchurch, N.Z.) 23 Sept. 13/7 *Cape Weed*, Hypochaeris radicator. The English catsear is always so called in Canterbury. 1965 *Austral. Encycl.* II. 261/2 Capeweed is now so common on good pasture land..that the whole countryside may become a sea of yellow during late spring.

capeador (kæpiădō̆ᵊ·ɹ). [Sp., f. *capear*, to trick a bull with a cape, f. *capa* Cape *sb.*²] One who aids a bull-fighter by distracting and tricking the bull with his cloak.

1909 in *Cent. Dict.* Suppl. 1924 Kipling *Debits & Credits* (1926) 220 The capeadors—the men with the cloaks—were advancing to play Apis. 1932 R. Campbell *Taurine Provence* iii. 73 Since Gordito's time..bandilleros have had most recourse to the aid of capeadors.

Cape Cod (kē¹p kǫd). *N. Amer.* The name of a promontory in Massachusetts, U.S.A., used *attrib.* or *absol.* to designate a type of compact rectangular house, usually with a steep gable roof, reminiscent of dwellings on Cape Cod. Also *ellipt.*, as *Cape*.

1916 J. E. Chandler *Colonial House* ii. 60 The remembrance of a most charming Cape Cod cottage. 1926 *House Beautiful* Sept. 255/1 The Cape houses..were carefully constructed, with pitch roofs and walls hung on a solid frame of hewn oak. 1945 *Geogr. Rev* XXXV. 445 Contemporary 'Cape Cod' invades the hills of the Los Angeles suburban zone to be neighbor to..'English half-timbered'. 1966 H. Kemelman *Saturday the Rabbi went Hungry* (1967) iii. 23 There was no confusing the Moderne with its flush door and three small diagonal panes of glass with the Cape Cod, which had a white panelled door flanked by two long narrow windows. 1967 *Boston Herald* 1 Apr. 8/1 (Advt.), A charming Cape with exquisite authentic detail. 1968 *Globe & Mail* (Toronto) 13 Feb. 30/1 (Advt.), Attractive Cape Cod, storey and a half, located on a well landscaped lot.

capelet (kē¹·plėt). [f. Cape *sb.*² + -let.] A small cape.

1912 *Home Chat* 1 June 444/2 The new Puritan capelet of white lawn. 1968 *Globe & Mail* (Toronto) 3 Feb. 11/5 The top is covered up by a broad-shouldered capelet that looks like a giant, beautiful bib.

capeline. Add: **3.** Later, a hat for a girl or woman, having a wide brim often consisting of many folds of muslin, or the like.

1899 *Westm. Gaz.* 13 July 3/2 A girl..in a black muslin *capeline* with a band round the crown and a big front bow of vivid cerise. 1901 *Daily News* 11 Feb. 3/2 The black tulle capeline made with a brim of perhaps twenty to thirty double folds of tulle. 1928 *Observer* 12 Feb. 23 Later on, brims will widen, and capelines of a particularly picturesque type will accompany the dainty dresses worn in the afternoon.

caper, *sb.*² Add: **1. c.** *transf.* Any activity or pursuit, spec. a fashionable occupation. Also, a 'game', dodge, racket. (There are many shades of meaning in U.S., N.Z., and elsewhere.)

1839 *Spirit of Times* 9 Nov. 423/2 When they are short on't for cradles, a feedin trof is jest the caper. 1840 C. F. Hoffman *Greyslaer* I. 84 The bizness is a bad one, any how you can fix it, capting; but I think I understand the caper on't. 1851 Mayhew *Lond. Labour* I. 416/1, I used to dress tidy and very clean for the 'respectable broken-down tradesman or reduced gentleman' caper. 1867 *London Herald* 23 Mar. 221 'He'll get five years penal for this little caper,' said the policeman. *a* 1889 *Boston Herald* (D.A.E.), Mind-reading is now the proper caper. 1897 Conrad *Nigger Narcissus* iv. 81 'I know his caper,' he said, in a low voice. 1926 J. Black *You can't Win* x. 131 If anything had gone wrong with this caper and we had to take a pinch. 1944 J. A. Lee *Shining with Shiner* 75 It's getting worse for this caper every year. 1959 *N.Z. Listener* 12 June 5/2 Marxism was the caper. 1959 *Times* 26 May 12/7 Now let's see if the car stops at all after that little caper. 1964 J. Burke *Hard Day's Night* v. 104, I know your caper. The kidney punch and the rabbit clout.

capeskin (kē¹·pskin). [f. Cape *sb.*³ 2 + Skin *sb.*] A soft leather made from South African hair sheepskin.

[1915 K. J. Adcock *Leather* ii. 13 Of the imported varieties, the..Cape sheep skins are the best.] 1934 in Webster. 1938 *Times* 11 Mar. 19/3 The Gay Nineties type of flat crown... Of particular interest are those of patent leather or capeskin. 1959 *Observer* 5 Apr. 13/1 The supplest and lightest weight of capeskin leathers, silky to the touch, printed in delicate marbled designs.

Capetian (kăpī·ʃ¹ăn), *a.* [ad. F. *Capétien.*] Pertaining to the third dynasty of French kings, founded by Hugh Capet in A.D. 987. Also as *sb.*

1836 *Penny Cycl.* VI. 264/1 The third, or,..Capetian dynasty of French princes. 1839 K. H. Digby *Mores Cath.* IX. xi. 333 The Capetians, whose title dates from Robert-le-fort. 1929 Belloc *Joan of Arc* ii. 39 From him all the Kings were descended, and that Crown Capetian which was the centre and prop of the world and the eye of Christendom.

capillary, *a.* Add: **4. b.** (See quot. 1962.)

1895 F. H. King *Soil* v. 176 The slower rise of capillary water in a dry soil. 1929 Weaver & Clements *Plant Ecol.* ix. 182 The water-retaining capacity..includes the hygroscopic water as well as the much larger quantity that the soil holds besides, commonly called capillary water. 1932 Fuller & Conard tr. *Braun-Blanquet's Plant Sociol.* vii. 212 The capillary water of the soil forms a thin, coherent water film around each solid particle. 1962 Hanson *Dict. Ecol.* 66 *Capillary water*, the portion of soil water which is held by cohesion as a continuous film around particles and in spaces; most of it is available to plants.

capilotade. Delete †*Obs.* and add later examples.

1889 *Mrs. Marshall's Cookery Bk.* 111 Capilotade of sweetbread. 1892 T. F. Garrett *Encycl. Pract. Cookery* I. 284/1 Capilotade..is literally the French term for a hash or ragoût, consisting usually of giblets and pieces of cold meat or poultry. 1950 *Mrs. Beeton's Househ. Managem.* xxv. 594 Giblets, capilotade of.

capital, *a.* and *sb.*² Add: **A. 2. b.** The distinction (from 1957 to 1965) of *capital murder* (see quot.).

1957 *Act* 5 & 6 *Eliz. II* c. 11 § 5 The following murders shall be capital murders..(*a*) any murder done in the course or furtherance of theft; (*b*) any murder done by shooting [etc.]. *Ibid.*, Where it is alleged that a person accused of murder is guilty of capital murder, the offence shall be charged as capital murder in the indictment.

5. b. Introducing (a repetition of) the initial letter of a word, for emphasis, etc.

1863 Dickens in *All Yr. Round* Christmas 3/2 'Capital D her!' bursts out Caroline. 1892 I. Zangwill *Childr. Ghetto* (1893) II. xv. 381 'Oh, Leon, Leon, you'll turn Catholic soon!' said Strelitski reprovingly. 'Not with a capital C,' said Raphael, laughing. 1902 H. James *Wings of Dove* iv. 66 Kate had mentioned..that her aunt was Passionate.. uttering it as with a capital P. *a* 1930 D. H. Lawrence *Phoenix* (1936) iv. 534 Life with a capital L is only man alive. 1956 A. J. Toynbee *Historian's Appr. Relig.* xi. 147 He assumed that this was the mature and perfect form of Civilized Society: Civilization with a capital 'C'. 1966 *Sunday Times* 13 Nov. 10/4 A subtle attack upon the stability of our very own Monarchy, with a capital M.

6. c. Delete †*Obs.* and add earlier and later examples.

1652 in *Mariner's Mirror* (1926) XII. 399 They are between the I. of Wight and Portland with 45 sail, 12 of which are capital ships. 1793 Ld. Howe in Barrow *Life* (1838) 214 Under a repeated fire from three or four of their capital ships. 1805 D. Macpherson *Ann. Commerce* III. 250 They ..took from our English East-India company their most important fort..; they also took one of that company's capital ships. 1909 *Hansard Commons* I. 1110/2 Mr. Robert Harcourt asked the First Lord of the Admiralty if he was prepared to give an official definition of the term capital ship? Mr. M'Kenna: The Board of Admiralty have never sanctioned the official use of the term, and they do not deem it expedient to do so. 1919 *Daily Tel.* 11 Aug. 9/6 The battle-fleet force became a vast assemblage of capital vessels, cruisers, light cruisers, destroyers and submarines. 1928 *Britain's Industr. Future* v. xxx. 426 A capital-ship base at Singapore.

B. 3. Also, the general body of capitalists or employers of labour, esp. with regard to its political interests and claims. (Cf. Labour *sb.* 2 b.)

1869 *Eng. Mechanic* 4 June 237/3 We might feel inclined to despair over the chances of Giant Capital and Dwarf Labour ever working harmoniously. 1929 D. H. Lawrence in *Star Rev.* Nov. 626 The Soviet hates the real physical body far more deeply even than it hates Capital. 1940 W. Temple *Thoughts in War-time* iv. 26 When we reach the stage of justice in the relations between capital and labour.

6. (attrib. use of 3) *capital account, stock*; *capital bonus*, a pro rata bonus distributed in shares; *capital expenditure* (see quot. 1959); *capital gain* (orig. *U.S.*), a profit from the sale of investments or property; freq. used *attrib.* in *capital gains tax*; *capital goods*, commodities forming capital; economic goods (e.g. railways, ships, machinery, buildings) destined for use in production, opp. to *consumers' goods*; *capital levy*, the confiscation by the state of a proportion of privately-owned wealth or property.

1895 *Min. Proc. Inst. Civil Engin.* CXXII. 224 When the North-Eastern Railway Company..includes the cost of between three and four hundred locomotives, which had presumably been charged to revenue, in the capital stock without any increase in capital account. *Ibid.* 234 The engine-miles run (excluding those by 'capital' engines) were 58,202,648. 1898 S. S. Dawson *Accountant's Compendium* 47/2 *Capital expenditure*..In its more restricted sense the term implies the expenditure of the capital receipts of a company or other body upon the construction of a particular work, *e.g.*, a railway. 1899 J. B. Clark *Distrib. Wealth* ix. 116 The differences that science must recognize between 'capital' and 'capital-goods'. 1919 *Hansard Commons* CXVI. 235, I do not see how you can have a capital levy under war conditions. 1921 *Washington Post* 16 Aug. 4/2 The bill redefines capital gain and capital loss, declaring the former to mean 'taxable gain from the sale or exchange of capital assets'. 1928 *Daily Chron.* 9 Aug. 8/4 The directors.. have distributed a number of capital bonuses among their fortunate Ordinary shareholders. 1931 *Times Trade & Engin. Suppl.* 24 Jan. 430/4 The production of 'capital goods' which are not 'consumed' immediately the money is spent upon them, but contribute a quota to the national wealth over many future years. 1951 L. H. Seltzer (*title*) The nature and tax treatment of capital gains and losses. 1959 Jowitt *Dict. Eng. Law* I. 310/1 *Capital expenditure*, expenditure from which benefits may be expected over a relatively long period, as opposed to revenue expenditure;..expenditure on capital or fixed assets. 1962 *Times* 10 Apr. 13/2 The Chancellor of the Exchequer..proposed an uncommonly large number of tax changes...Several are controversial: the pledge on ending Schedule A; the capital gains tax, [etc.]. 1970 *Money Which?* Mar. 4/1 Capital Gains Tax. You may have to pay this tax on most capital gains—for instance, the gain you make when you sell shares at a higher price than you bought them for.

capitalist. Add: (Earlier example of *attrib.* use.) Also *Comb.*, as *capitalist-imperialist*.

1845 J. S. Mill in *Edin. Rev.* LXXXI. 525 This is true of capitalist farmers..but not of labourer-farmers. 1937 'G. Orwell' *Road to Wigan Pier* xii. 247 The capitalist-imperialist governments..will not fight with any conviction against Fascism.

capitalistic, *a.* (Earlier and later examples.)

1873 in *Doc. Hist. Amer. Industr. Soc.* (1910) IX. 371 The growth of capitalistic association and monetary institution. 1935 *Discovery* Mar. 81/1 These two systems 'communistic' or 'capitalistic'.

capitalistically (kæ:pităli·stikăli), *adv.* [f. Capitalistic: see -ically.] In a capitalistic manner.

1889 W. Wilson *State* (1893) xvi. 666 Churches are spiritually convenient; joint-stock companies are capitalistically convenient. 1921 *Glasgow Herald* 24 June 9 When the concern was capitalistically owned. 1931 *Time & Tide* 3 Oct. 1130 The *rentier* has been unfeeling enough to practice what is capitalistically called virtue.

capitalization. Add: **1. b.** The sum or figure resulting from the action of converting into capital.

1906 *Daily Chron.* 13 Feb. 5/1 The companies have a capitalisation of 80,000,000 dols. 1970 *Daily Tel.* 22 Apr. 21/3 IBM—whose market capitalisation is larger than that of any other United States company—is the post-war growth stock *par excellence*.

3. (Later examples.)

1906 R. L. Ramsay in *Skelton's Magnyf.* p. xx, The orthography is that of the original; punctuation and capitalization are modern. 1908 M. W. Sampson *Milton's Lyr. & Dram. Poems* p. v, The text follows the first editions as closely as modern spelling, capitalization and punctuation permit. 1964 F. Bowers *Bibliogr. & Text. Crit.* vi. iii. 177 The details of spelling, capitalization, punctuation, and word-division.

capitalize, *v.* Add: **1. b.** To invest with capital. *U.S.*

1870 J. K. Medbery *Men & Myst. Wall St.* 11 The variations of its [*sc.* Wall St.'s] share market affect the whole volume of capitalized indebtedness the country through. 1897 *Daily News* 13 Mar. 2/2 This Company is very moderately capitalised at £80,000 in Shares of £1 each. 1948 *Chicago Tribune* 18 Aug. III. 5/5 The Canadian unit..will be capitalized at 5,000 shares of $100 par value stock. 1970 *Sci. Amer.* Mar. 35/2 A conglomerate can..apply the debt capacity of safe, mature businesses to capitalize rapidly growing but unstable ventures.

c. To make capital out of, turn to account. Also *absol.* orig. *U.S.*

1869 in *Trans. Kansas Hist. Soc.* (1900) VI. 64 To capitalize this land in the hands of speculators. 1926 *Publishers' Weekly* 22 May 1701 The book-stores and the libraries can capitalize on that same interest. *Ibid.* 29 May 1795 The publishers..do not intend to capitalize the publicity. 1935 *Punch* 15 May 578/3, I can only 'capitalize' ..something which already belongs to me...And therefore I condemn the advertisement..which asserts that the makers..have 'capitalised this mother-instinct'. 1953 *Encounter* Nov. 9/1 The *Tokyo Evening News*..capitalises on the time-difference between Japan, America, and Europe to be up-to-the-minute with the news. 1958 *Times* 18 Nov. 13/5 His material was indifferent..but..he was able to capitalize brilliantly on every disadvantage.

3. Delete *nonce-use*. Also, to print an initial letter in capitals. (Add earlier and later examples.) Also in *vbl. sb.* = next.

1764 *Acc. Coll. New-Jersey* 25 All these compositions.. are critically examined with respect to the..pointing, capitalizing, with the other minutiæ. 1809 W. Cunningham Jr. *Let.* 23 Sept. (1823) xlix. 165, I capitalized the prophetic parts of the letter..and italicized the Latin. 1890 *Phonetic Jrnl.* 26 Some peculiarity in capitalising. 1944 [see *Afro-]. 1964 F. Bowers *Bibliogr. & Text. Crit.* vi. iv. 186 The single Q capitalization is also capitalized in F.

capitalling (kæ·pitălıŋ), *vbl. sb.* [f. Capital *sb.*² + -ing¹.] The furnishing of a word with a capital letter.

1683 Moxon *Mech. Exerc., Printing* 261 The Correcter.. examines the Proof, and considers the Pointing, Italicking, Capitalling, or any error. 1770 Luckombe *Hist. Printing* 249 Some [authors] give themselves no concern about capitaling. 1904 *Athenæum* 13 Feb. 210/2 The first variation, apart from detail of pointing and capitalling, is in line 17.

‖ **capitao** (kapitā·o). Also **capitow, capito.** [Pg. *capitão*, = Captain.] A head-man, leader of a gang, etc., in Africa.

1899 *Daily News* 21 Mar. 7/7 Parties of villagers who elect to work on the lines are in charge of the 'Capitows', or gang-leaders. 1925 *Chambers's Jrnl.* Oct. 655/2 Accompanied by my 'capitao', I reached the river bank. 1926 *Blackw. Mag.* Apr. 550/1 Our Safari consisted of A. and myself, two capitos (headmen), a few personal boys and gun-bearers, and forty-six porters. 1957 V. W. Turner *Schism in Afr. Soc.* iv. 95 A capitao in charge of road maintenance.

capitol. 2. (Earlier, later, and *attrib.* examples.)

1699 *Acts Assembly Virginia* (1727) I. 205 An Act directing the Building the Capitol and the City of Williamsburgh, &c. 1706 *Cal. Virginia St. Papers* (1875) I. 109 The main street..extending from the Capitol to ye utmost Limits of the City Westward. 1752 J. Blair in *Official Records of R. Dinwiddie* (1883) I. 146, I laid the last top Brick on the Capitol Wall. 1793 Jefferson *Let.* 1 Feb. in *Writ.* (1854) III. 508 Doctor Thornton's plan of a capitol. 1795 J. Scott *U.S. Gazetteer* s.v. *Washington* (D.C.), The capitol is now building and is situated upon a beautiful eminence. 1834 C. D. Arfwedson *United States* I. 318 The Capitol, as it is called, or Statehouse occupied by the legislature of the State. 1888 *Harper's Mag.* June 50/1 The part of the Capitol which is completed (Senate Chamber, House of Representatives, Library, Historical Society, etc.). 1928 S. V. Benét *John Brown's Body* 155 Muddy Washington, with its still-unfinished Capitol. 1943 *Chicago Daily News* 12 June 6/6 To provide the State of Texas with a new capitol.

‖ **capitonné** (kapitone), *a.* [Fr., pa.pple. of *capitonner* to upholster, quilt.] (See quots. 1882 and 1957.)

1873 *Young Englishwoman* Mar. 131/2 A dress of tulle over white tafettas, capitonné into bouillons with small clusters of Parmese violets. 1882 Caulfeild & Saward *Dict. Needlework* 61/1 *Capitonné.* This is a French term, signifying drawn in at intervals, as a stuffed sofa, chair, or pincushion, which is buttoned down at each attachment of the double material, at the front and back. 1939 C. Beaton *My Royal Past* ii. 17 She graciously patted the *capitonné* sofa. 1957 M. B. Picken *Fashion Dict.* 53/1 *Capitonné embroidery*, decorative tufting similar to that used on furniture, but adapted and modernized for use on hats, yokes, panels of dresses, etc., giving quilted appearance. 1968 *Home & Garden* Nov. 86/2 Deep capitonné Chesterfield...In black hide it costs £237 10s.

capitulationism (kăpi:tiulē¹·ʃəniz'm). [f. Capitulation + -ism.] The act or principle of capitulation; *spec.* in Communist usage (see quots.). Hence **capi:tula·tionist** *sb.* and *a.*, one who advocates capitulationism; characterized by capitulationism.

1957 R. N. Carew Hunt *Guide to Communist Jargon* xlv. 152 Wilhelm Zeisser and Rudolf Herrnstadt were..expelled ..on account of the attitude they were said to have adopted at the time of the Berlin rising of the previous June, the specific charges against them being that they had advocated 'capitulationism', and had exalted the spontaneous role of the masses at the expense of the Party. 1962 [see *adventurist *sb.*]. 1963 *Economist* 5 Jan. 18/3 Peking's campaign against such 'capitulationists' as Mr. Khrushchev. 1963 *Times* 8 Jan. 7/1 Scarcely veiled accusations of Russian 'capitulationism' and 'Trotskyism', and the abandonment of peaceful coexistence by the Chinese, were exchanged between Peking and Moscow. 1963 *New Statesman* 1 Feb. 145/1 The reported Chinese memorandum..[refers to] Krushchev's 'capitulationist' attitudes towards the 'imperialists'.

capitulum. Add: **3. b.** *Bot.* In characeous plants, a head-cell borne by each manubrium.

1875 Bennett & Dyer tr. *Sachs's Bot.* 520. 1877 Huxley & Martin *Elem. Biol.* (ed. 4) 45 At the free end of the manubrium is a rounded body, the *capitulum*, which bears six smaller, *secondary capitula*. 1898 H. C. Porter tr. *Strasburger's Bot.* 339 Each manubrium terminates in a knob-like cell or capitulun, from which a large number of short cells grow out into the cavity of the antheridium.

4. b. *Zool.* (See quots.)

1920 I. F. & W. D. Henderson *Dict. Sci. Terms* 42/1 *Capitulum*..the exsert part of the head in ticks. 1964 M. Hynes *Med. Bacteriol.* (ed. 8) xxx. 460 Ticks are eight-legged arthropods with no division between head, thorax or abdomen. Their mouth-parts are carried on a separate movable *capitulum*.

capless (kæ·plės), *a.* [f. Cap *sb.*¹ + -less.] Without a cap; having no cap.

a 1839 Galt *Demon Destiny* (1840) viii. 53 His hoary hair streamed capless. 1891 *Daily News* 30 July 2/7 He could not have a capless maid to answer the door. 1904 H. G. Wells *Food of Gods* I. ii. 22 A very little old woman, capless, with dirty white hair. 1958 M. L. Hall *Newnes Compl. Amat. Photogr.* 110 A range of capless fitting flashbulbs.

Capo di Monte (ka·po di mǫ·nte). Also **Capodimonte.** The name of a palace near Naples applied *attrib.* to a type of porcelain first produced there in the mid-eighteenth century. Also used *absol.* and *ellipt.*

1850 Marryat *Hist. Pott. & Porc.* xi. 217 The Capo di Monte tea and coffee services are perhaps the most beautiful description of Porcelain which has ever been manufactured in Europe. 1869 Lady C. Schreiber *Jrnl.* (1911) I. 34 One small Mennecy figure (marked), and a white Capodimonte (unmarked). *Ibid.* 49 Saw Mr. Gladstone's collection..the Capo di Monte I do not understand. 1926 V. Trees *Castles in Air* iii. 123 Beardsley drawings or better, a glass of Capo di Monte china. 1942 A. E. W. Mason *Musk & Amber* vi. 84 Candlesticks of the Capodimonte porcelain. 1969 *Times* 25 Mar. 12/4 The Antique Porcelain Company gave 19,500 gns. for a very rare Capodimonte Italian comedy group.

capo tasto (ka·po ta·sto). *Mus.* Also **capodastro.** [It., lit. 'head stop'.] A device consisting of a bar or movable nut attached to the fingerboard of a stringed instrument for the purpose of raising the pitch of all the strings at once.

1876 Stainer & Barrett *Dict. Mus. Terms* 75 The capo tasto, or capodastro as it is sometimes called, is screwed over the strings on to the finger-board and forms a temporary nut. 1879 Grove *Dict. Mus.* I. 306/2 The position of a capo tasto varies according to the stringing and the length of the neck of the instrument it is to be applied to. 1907 *Army & Navy Co-op. Soc. Price List*, Capodastros. 1967 C. Bunting in A. Baines *Mus. Instr.* vi. iii. 141 The use of the thumb as a movable 'nut' (like the guitarist's capodastro).

cappa (kæ·pă). *Eccl.* [It.: see Cape *sb.*¹] A cloak (cf. Capa); a cloak forming part of a religious habit; a cope.

1859 J. Morris *S. Thomas Becket* vii. 64 [S. Thomas of Canterbury] put on a black cappa, which was closed all round and reached his feet...The black cappa with lambs-wool..was not the monastic habit of his monks of Christ Church. It was that of the Black Canons Regular, to which order Merton belonged. 1867 R. Palmer *Philip Howard* 15 In tropical climates where the cappa is less used. 1925 *Glasgow Herald* 10 May 9 Camerlengos and attendants with the spada and cappa.

Cappadocian (kæpădōᵘ·ʃ¹ăn), *a.* and *sb.* Also †**Capadocian.** [f. *Cappadocia*: see -an.] **A.** *adj.* Of or pertaining to Cappadocia or its inhabitants. **B.** *sb.* **a.** An inhabitant of Cappadocia, an ancient kingdom of Asia Minor, now part of Turkey. **b.** The language of the Cappadocians.

1607 Topsell *Four-f. Beasts* 285 The Cappadocian is also a breed of the Parthian horsses. *Ibid.* 286 The Cappadocian horsses are swift and lusty in their old age. 1762 Sterne *Tr. Shandy* V. xxvii. 98 The Phoenicians,—the Arabians, the Capadocians..did it. 1836 *Penny Cycl.* VI. 271/1 The Cappadocians were very generally known during the Roman occupation of their country for their unprincipled and vicious character. 1845 *Encycl. Metrop.* XIV. 255/1 The Cappadocian horses have been the theme of much eloquent panegyric from the pens of several historians. 1876 *Encycl. Brit.* V. 74/2 In the time of Herodotus, the Cappadocians occupied the whole region from the chain of Mount Taurus on the south to the shores of the Euxine. 1934 A. J. Toynbee *Study of Hist.* II. 80 The Cappadocian Fathers of the Church. 1954 Pei & Gaynor *Dict. Linguistics* 35 *Cappadocian*, an extinct language..of undetermined linguistic affinities..classified as Asianic.

capped, capt, *ppl.a.* Add: **5.** (Examples.) Also, astonished, surprised. Cf. Cap *v.*¹ 4 b.

1862 C. C. Robinson *Dial. Leeds* 262 Capt—'surprised.' 'I wur capt to see him thear'. 'Fair capt'—clean astonished; as at the tricks of a juggler. 1866 J. Scholes in J. Harland *Lancashire Lyrics* 133 A'm capt' 'at folk wantin' to wed. 1876 F. K. Robinson *Gloss. Whitby* 32/1, I was sair capp'd te tell. 1959 J. Ashworth *Murder in Borough Library* xvii. 135 Just saw her cat appearing into the wall. I was fair capped for a minute, because I didn't know there was a door. 1967 S. Woods *And shame Devil* ii. 41 'Why should she have been?' 'Beats me. I was fair capped when they called her.'

capper, *sb.*¹ Add: **4.** One who or that which caps: in various technical senses.

a 1884 Knight *Dict. Mech.* Suppl., *Capper*, a tool used in placing the priming cap on its seat in the rear end of the metallic shell. 1890 *Pall Mall Gaz.* 28 June 4/3 Employed by the prosecuting firm [of perfumers] as 'cappers'. 1921 *Dict. Occup. Terms* (1927) § 138 *Glass capper*, cuts off glass as it leaves tubes of glass blowing machine. *Ibid.* § 459 *Capper, bottle*,..fixes a small paper, tin, or lead foil cover over corks of bottles. 1961 J. G. Davis *Dict. Dairying* (ed. 2) 90 A combined unit which fills the bottle and seals it with its own aluminium cap. .. A 20-head filler and 8-head capper.

5. An accomplice, esp. a confederate in a gambling game; a dummy bidder at an auction. *slang* (chiefly *U.S.*).

1753 J. Poulter *Discoveries* (ed. 2) 27 There are generally four Persons concerned, that is, the Sailor, call'd a Cull, to pinch the Nobb; the next is the Capper, who always keeps with the Sailor. 1853 *Alta California* (San Francisco) 25 Apr. 1/7 A glib-mouthed auctioneer and a set of cappers, puffers, or decoy-ducks. 1872 Schele de Vere *Americanisms* 319 Cappers they ['strikers'] are called when the game is the famous Three-Card Monte. 1926 J. Black *You can't Win* xiii. 184 His 'cappers', 'boosters', and 'shills' ..always beat them to the pieces of soap containing the money.

capping, *vbl. sb.* Add: **1. c.** Also *N.Z.*

1905 *Westm. Gaz.* 28 Aug. 2/3 Sir Robert Stout..

announced at the last capping at Wellington that..if the students persisted in their senseless conduct there would be no more capping ceremonies in public. **1966** *Weekly News* (N.Z.) 27 Apr. 10/4 Invading students from Massey University did not have permission from the Auckland City Council to sell their capping magazine in Queen St.

d. The practice of taking a definite sum of money for a day's hunting from a non-subscriber to the hunt. (See *Cap *sb.*¹ 6.)

1890 *Pall Mall Gaz.* 26 Feb. 3/1 The proceeds of the capping should go to a damage fund. **1897** *Daily News* 16 Jan. 7/6 It has just been decided by the Hunt Committee to introduce the 'capping' system into the Quorn country next season.

3. *spec.* Of honey or cells in a honeycomb.

1934 in Webster. **1950** *N.Z. Jrnl. Agric.* Jan. 27/2 When drained free of surplus honey, wax cappings [of cells] should not be thrown away.

‖**cappuccino** (kaputʃī·no). [It., see Capuchin.] Coffee with milk; white coffee, esp. as served in espresso coffee-bars, topped with white foam. Also, a cup of such coffee.

1948 R. O'Brien *This is San Francisco* 84 You drink a *cappuccino*, gray, like the robe of a capuchin monk, and made of chocolate that is laced with brandy or rum, and heated by steam forced through coffee. **1955** *Archit. Rev.* Sept. 166 Do you think I dare say cappucino? **1959** C. MacInnes *Absolute Beginners* 17 She blew a little brown nest in the white froth of her *cappuccino*. **1960** *Guardian* 25 Mar. 16/6 The first meal in a café should be a cappuccino or two and a dry roll or bun.

Capri (kă·prī·). [Name of an island in the Bay of Naples.] **1.** Any of various wines produced originally in Capri. Cf. Caprik(e.

1877 E. S. Dallas *Kettner's Bk. of Table* 482 Some of the Italian white wines—as White Capri, or White Lachryma Christi—would do as well. **1890** Kipling *Life's Handicap* (1891) 209 Raw, rasping Capri with all the strength of whisky. **1961** B. Malamud *New Life* (1962) 137 George, the fine waiter, served..manchiego cheese and some dry white capri, iced in buckets.

2. *Capri pants*: women's close-fitting trousers with tapering legs, reaching to just above the ankle. Also *ellipt.* as *Capris*. orig. *U.S.*

1956 *Amer. Speech* XXXI. 307 Capri pants, pants tapered to just above the ankle. **1966** H. Nielsen *After Midnight* (1967) i. 12 The shrieking pink Capris and the wild silver blonde hair-do. **1967** P. McGerr *Murder is Absurd* ix. 113 He appraised the high-necked sweater and capri pants. **1969** R. Airth *Snatch!* iv. 31 She was wearing capris, green and tight.

capriped, *a.* Add: Also as *sb.*

1916 E. Pound *Lustra* 46 And what, pray, do you know about horticulture, you capriped? **1925** A. Huxley *Those Barren Leaves* i. iv. 46 I'm nothing but an old capripede.

caprolactam (kæprolæ·ktăm). *Chem.* [f. Capro(ic *a.* + *Lactam.] A crystalline amide, $C_6H_{11}NO$, used esp. in the manufacture of a type of nylon.

1944 *Chem. Abstr.* XXXVIII. 122/2 Production of N-(ethylmercaptomethyl)caprolactam. **1946** L. H. Smith *Synthetic Fiber Dev.* iii. 452 Perlon..Caprolactam was the main nylon-type raw material used during the war. **1960** *Times* 2 Nov. 19/5 An arrangement with Courtaulds to produce the chemical caprolactam in Britain. **1961** *Economist* 10 June 1144/3 Caprolactam, the raw material for the nylon 6 polymer.

caps. (Further examples.)

1856 Geo. Eliot in *Westm. Rev.* X. 450 That particular view of Christianity which..condenses itself into a sentence of small caps. **1960** K. Hopkins *Dead agst. Principles* ii. 15 'Original designs TATTOOED'—large caps, Manciple.

capsa (kæ·psă). [L., see Case *sb.*²] (See quot. 1959.)

1932 B. L. Ullman *Anc. Writing* xvii. 204 The grouping of the books of Livy in decades is due to the fact that ten books were kept in each *capsa*. **1954** C. H. Roberts in *Proc. Brit. Acad.* XL. 193 The hypogaeum of Trebius Justus where codices, roll, and capsa are all represented. **1959** L. M. Harrod *Librarians' Gloss.* (ed. 2) 58 *Capsa*, a cylindrical box used in Roman libraries to hold one or more rolls standing upright.

Capsian (kæ·psiăn), *a.* and *sb.* *Archæol.* [ad. F. *capsien* (J. de Morgan et al. 1910, in *Rev. de l'École d'Anthrop.* XX. 110), f. L. *Capsa*, substituted for Gafsa (see def.) + -ian.] **A.** *adj.* Belonging to the palæolithic period represented by remains found at Gafsa in central Tunisia. **B.** *sb.* **a.** The culture of Gafsa. **b.** A person of this period.

1915 W. J. Sollas *Anc. Hunters* (ed. 2) viii. 290 In the greater part of Spain, almost the whole of Italy, and across the Mediterranean, as in Tunisia..this industry..persists throughout the Solutrian and Magdalenian ages...It is known as the Capsian or Getulian. **1921** R. A. S. Macalister *Textbk. Europ. Archæol.* I. 537 The Capsian thus corresponds chronologically to the European Upper Palæolithic, and culturally to the Aurignacian stage. **1931** *Times Lit. Suppl.* 1 Jan. 6/2 The so-called Capsians of Eastern Spain. **1951** *Proc. Prehist. Soc.* XVII. 91 The large backed blades, so characteristic of the Capsian, suggest that comparison should be made with the north rather than Africa. **1958** A. R. Radcliffe-Brown *Method in Soc. Anthropol.* i. iv. 107 The affinities of the Solutrean or Capsian culture.

capsid (kæ·psid), *a.* and *sb.*¹ [f. mod.L. *Caps-us* (see def.) + -id³.] **A.** *adj.* Belonging to the Capsidæ, a family of heteropterous insects including the genus *Capsus*. **B.** *sb.* An insect of this family.

1889 in *Cent. Dict.* **1932** *London Gaz.* 28 Oct. 6796 The Apple Capsid (Essex) Order of 1932...On being satisfied that the Apple Capsid exists on any such trees or bushes [etc.]. **1934** *Discovery* Sept. 252/2 In Kenya..it [*sc.* pyrethrum] is effective against capsid and antestia bugs, which are harmful to coffee plants. **1959** Southwood & Leston *Land & Water Bugs* viii. 201 Great care should be taken when collecting capsid bugs. **1960** *Times* 25 July 11/6 Scales, capsids, tsetses.

capsid (kæ·psid), *sb.*² [ad. F. *capside* (A. Lwoff et al. 1959, in *Ann. de l'Inst. Pasteur* XCVII. 286), f. L. *capsa* box (see *Capsa) + -id².] (See quots.)

1964 *Inst. Biol. Jrnl.* XI. 93 *Capsid*, the quasi-crystalline shell, built up by the regular arrangement of capsomeres, to produce the structure that is the stable and transmissible form of some viruses. **1964** *New Scientist* 23 July 210 Some viruses consist of a single protein coat or 'capsid' enveloping the infective nucleic acid. **1964** G. H. Haggis *Introd. Molecular Biol.* iv. 83 A number of other small spherical viruses have been shown to contain nucleic acid within similar protein envelopes, now called *capsids*.

capsomere (kæ·psomīᵊɹ). [ad. F. *capsomère* (A. Lwoff et al. 1959, in *Ann. de l'Inst. Pasteur* XCVII. 286), f. as *Caps(id *sb.*² + -o + *-mere*, f. Gr. μέρος part.] (See quot. 1964².)

1962 *Lancet* 5 May 965/1 Preliminary evidence suggested that the antigens A and C were in fact the two kinds of capsomeres believed to form the outer shell of the virus particle. **1964** M. Hynes *Med. Bacteriol.* (ed. 8) xxiv. 349 A typical virus consists of a core of either DNA or RNA surrounded by numerous identical subunits of protein, termed capsomeres. **1964** *Inst. Biol. Jrnl.* XI. 93 *Capsomere*, one of the units from which the closed shell or tube of some viruses is built up.

capstan. Add: **1. b.** (See quot. 1959.)

1948 *Sat. Rev.* 28 Feb. 56 Tape..is usually capstan driven at constant linear speed. **1949** Frayne & Wolfe *Sound Recording* xxix. 580 The recording material is pulled by a capstan against which it is held by a roller. The capstan may be driven directly by a motor or by means of a belt drive. **1959** W. S. Sharps *Dict. Cinemat.* 84 *Capstan*, in magnetic tape recording, this is the motor-controlled spindle surmounted by a roller which grips the tape and draws it past the sound head at constant speed.

2. Add: *capstan labourer, -man*; **capstan table,** a round table on a column, often with a revolving top.

1881 *Instr. Census Clerks* (1885) 94 Capstan labourer. **1889** P. H. Emerson *Eng. Idyls* 67 On deck, a countryman who had shipped as capstan-man, was standing for'ard with the master. **1921** *Dict. Occup. Terms* (1927) § 706 *Capstan man*.., moves trucks in goods depôt by depressing foot lever actuating automatic capstan. *Ibid.* § 741 *Capstan man*, operates capstan,..allows capstan to revolve,..etc. **1921** *Glasgow Herald* 1 June 9 Cranemen, capstanmen, and others. **1927** *Daily Tel.* 31 May 4/1 An old English 'capstan' writing table.

capsule. Add: **2. capsule of Tenon,** Tenon's capsule (Tenonian *a.*).

1867 *Amer. Jrnl. Med. Sci.* July 241 The capsule of Tenon,..which incloses the whole eyeball, with the exception of the cornea, consists of two very different portions. **1884** [see *Capsulitis]. **1964** S. Duke-Elder *Parsons' Dis. Eye* (ed. 14) x. 112 At the time of an operation after the capsule of Tenon has been opened.

8. (*a*) A pressurized compartment in an aircraft, also used for an emergency escape. (*b*) The detachable nose-cone of a rocket or space missile for carrying an astronaut, instruments for recording and transmitting scientific data, etc. orig. *U.S.*

1954 M. Caidin *Worlds in Space* v. 105 The ship could be controlled from within the sealed capsules. **1955** *Sci. News Let.* 11 June 377/3 An escape capsule for pilots flying supersonic aircraft..has been patented. **1958** *Wall St. Jrnl.* 9 Oct. 1/3 Later efforts would be made to send a man, again encased in a pressurized capsule, circling the earth for 24 hours. **1959** *Daily Tel.* 23 Feb. 11/6 For the space pilot..there must be the ability, in emergency, to eject the whole capsule in which one sits. **1959** *Listener* 17 Sept. 440/1 The first American space capsule. **1964** *Ann. Reg. 1963* 185 Because of a fault in some of the automatic control equipment he had had to control the Mercury capsule manually during the critical period of re-entry into the earth's atmosphere. **1968** *Times* 10 Dec. 6/8 Solar telescopes which can be linked up in space to an Apollo capsule.

B. As *adj.* Brief, condensed, compressed; small and compact. orig. *U.S.*

1938 in J. Beatty *Saturday Rev. Gallery* (1959) 279 One of those capsule biographies of the author that sometimes appear on the jackets of detective stories. **1944** *Amer. N. & Q.* June 42/2 (*heading*) Capsule language for fighting men. **1953** R. Stout *Golden Spiders* 55 An estimated million and a quarter New Yorkers got an impressive capsule demonstration of the might of American armed forces. **1955** Keepnews & Grauer *Pict. Hist. Jazz* xiv. 149 It's not so much a Kansas City 'style' as a capsule history of jazz in one town. **1958** *Vogue* May 199 Blue and white blazer..slim skirt, really a capsule wardrobe in itself.

capsulitis (kæpsiuləi·tis). *Path.* [f. Capsule + -itis.] Inflammation of a capsule (e.g. of the eyeball); tenonitis.

1830 W. Mackenzie *Pract. Treat. Dis. Eye* x. 470 By the term aquo-capsulitis is meant inflammation of the cartilaginous membrane..which lines the internal surface of the cornea. **1884** H. R. Swanzy *Handbk. Dis. Eye* xxii. 414 Capsulitis, or Inflammation of the Capsule of Tenon, is an affection concerning the occurrence of which there is some doubt. **1951** W. S. C. Copeman in *Brit. Encycl. Med. Pract.* (ed. 2) V. 437 A number of these cases appear to progress to the formation of intra-capsular adhesions, resulting ultimately in what the Americans term the 'frozen shoulder', or 'fibrosing capsulitis'.

capsulize (kæ·psiuləiz), *v.* orig. *U.S.* [f. Capsul(e + -ize.] *trans.* To compress or condense (information, news, etc.) into a brief and compact form. (Cf. *Capsule B.) Hence **ca·psulized** *ppl. a.*

1950 *Sat. Rev. Lit.* 11 Nov. 17/1 This is a brief, almost capsulized account of postwar USSR. **1958** *New Yorker* 11 Oct. 39/1, I wanted..a slogan that would capsulize the lip-smacking flavor of the soups. **1959** *Economist* 7 Feb. 497/2 The Readers' Digest supplements its capsulised journalism.

capsulotomy (kæpsiulǫ·tǒmi). *Surg.* [f. Capsuli-, capsulo- + -tomy.] The incision of a capsule, *spec.* that of the lens of the eye in an operation for cataract.

1877 *Arch. Ophthalmol. & Otology* VI. 550 The peripheric capsulotomy may lessen the danger still attending Graefe's method of extraction. **1907** *Practitioner* Dec. 778 The operation of capsulotomy for some cases of acute nephritis with threatened suppression of urine. **1969** S. Duke-Elder *Syst. Ophthalmol.* XI. i. iii. 272 The subsequent capsulotomy which is required if the posterior capsule remains intact should be undertaken soon after the eye has become quiet.

captain, *sb.* Add: **1. b.** *captain of industry*, an industrial magnate, a merchant prince.

1843 Carlyle *Past & Pr.* iv. iv. 361 (*heading*) Captains of Industry. **1896** G. B. Shaw *Our Theatres in Nineties* (1932) II. 179 Attentive observers of great 'captains of industry' know that their success often comes to them in spite of themselves. **1933** J. M. Keynes *Essays in Biogr.* i. iii. 42 Mr. Bonar Law has liked to think of himself as..handling wars and empires and revolutions with the coolness and limited purpose of a first-class captain of industry. **1931** *Economist* 10 Jan. 51/2 A representative conference of trade union leaders and captains of industry.

5. b. An officer in the Salvation Army.

1878 *Christian Mission Mag.* Dec. 320 These stray remarks added to the interest we felt in our contemplated visit to the command of Captain Louise Agar. **1884** W. Booth *General's Lett.* 6 Dec. (1890) 18 Go to the crowd of sinners, or spot them individually..because out of them you may make Lieutenants, and Captains, and Majors, and Generals. **1965** R. Collier *General next to God* iii. 75 Not all in those pioneer years could afford..a guinea for a captain's outfit.

7. b. The pilot of a civil aircraft.

1929 *Proc. Internat. Civil Aeronaut. Conf.* 253 It will be indispensable..that a uniform juridical system govern..the aircraft, its owner..its captain and crew, as well as the passenger. **1955** *Times* 13 May 7/1 Having decided to retire from commercial flying after being an airline captain continuously since 1922 [etc.].

9. b. One in charge of a group of waiters or bell-boys in a hotel. Chiefly *U.S.*

1942 L. Bemelmans *Hotel Splendide* iii. 43 These two have been nothing but waiter and captain for twenty years. *Ibid.* 44 Captains, ah! Maîtres d'hôtel they call themselves. **1952** M. Steen *Phoenix Rising* iii. 59 He indicated his wants to the waiter...The captain went to his table..and nodded sharply to the waiter. **1968** *Globe & Mail* (Toronto) 3 Feb. 23/1 The Imperial Room opened when the hotel was built in 1929 (one captain and one waiter have been there from the beginning).

11. *Cricket.* (Earlier example.)

1823 Miss Mitford in *Lady's Mag.* July 387/1 Our captain applied to him instantly to play.

Captain Cooker (kæ·ptėn ku·kəɹ). *N.Z.* Also **Captain Cook.** [f. Captain James *Cook* (1728–79), navigator and explorer.] A wild boar, *Sus scrofa* (supposedly descended from the domesticated pigs introduced into New Zealand by Cook).

1879 *N.Z. Country Jrnl.* III. 55 The immense tusks..attest..the size of the wild boars or 'Captain Cooks' as the patriarchs are generally termed. **1892** W. E. Swanton *Notes on N.Z.* 116 The old boars among them are called 'Captain Cooks', after the esteemed founder of their order. **1902** W. S. Walker *Zealandia's Guerdon* iv. 42 [The dog] got..'bailed up' a regular 'Captain Cooker', long, lean and very muscular. **1957** D. Glover in J. Reid *Kiwi Laughs* (1961) 206 He has taken the liberty of bringing along a complete history of the district from the sighting of the first Captain Cooker by a white man.

captaincy. Add: **2. b.** spec. in *Cricket.*

1853 *Bell's Life* 22 May 6/2 The England side was under the captaincy of Mr. Nicholson. **1873** *Baily's Mag.* Aug. 44 The good captaincy added quite a new attraction.

4. captaincy-general = *Captain-generalcy.

1846 Worcester (citing Murray). **1876** *Gentl. Mag.* XVI. 433 We want..changes of supreme importance:—Abolition of the Captaincies-General, [etc.].

captain-generalcy. The office or province of a captain-general. So **captain-generalship.**

1846 Worcester (citing *Sat. Mag.*), Captain-generalcy. **1896** *Cosmopolitan* XX. 412/1 The captain-generalcy of Venezuela had then been under the Spanish crown two hundred and forty years. **1898** *Daily News* 28 Nov. 7/4

Before the colonies had secured their independence four of them were rated as Viceroyalties, and five as Captain-Generalships. *Ibid.* 10 Dec. 6/1 Promotion to the Captain-Generalship of Cuba. **1900** *Ibid.* 9 May 9/3 A headquarter [cyclist] corps in London, of which Mr. A. J. Balfour had accepted the Captain Generalcy.

ca·ptaining, *vbl. sb.* [f. Captain *v.* + -ing¹.] In Cricket, etc.: acting as the captain of a side.

1903 *Westm. Gaz.* 28 May 8/2 On fielding and on captaining. **1904** *Daily Chron.* 10 Aug. 7/3 Their captaining capacities.

captan (kæ·ptăn). [Shortened from Mercaptan.] The name of a fungicide and pesticide produced from mercaptan.

1953 A. R. Kittleson in *Jrnl. Agric. & Food Chem.* I. 677/1 SR-406 or captan, the common name..is a member of a new class of highly active organic fungicides, which is currently undergoing commercial development as an agricultural pesticide. *Ibid.* 678/1 Captan is synthesized by the reaction of sodium..with perchloromethyl mercaptan. **1955** *Rose Annual* 193 (Advt.), Orthocide Wettable 50% Captan for black spot on roses.

caption, *sb.* **2.** Delete †*Obs.* and add later example.

1922 *Public Opinion* 14 July 36/1 Genius like hers stands above envy and caption.

4. Earlier U.S. examples and examples of later (esp. British) use. Also used (orig. *U.S.*) for the title below an illustration; in cinematography and television, a sub-title. Also *attrib.* and *Comb.*

1789 J. Madison *Writ.* (1904) V. 355 You will see in the caption of the address that we have pruned the ordinary stile of the degrading appendages of Excellency, Esqrs. &c. **1821** *Massachusetts Spy* 24 Oct. (Th.), [The statute] is under the caption of 'Fees in the Secretary's office'. **1894** H. Frederic *Copperhead* 83 Spreading eagles in front, over the printed captions. **1919** H. L. Wilson *Ma Pettengill* ii. 43 The caption says of Vida Sommers: 'Her love has turned to hate.' **1920** *Glasgow Herald* 23 July 7 The Speaker said that this part of the bill—the caption, he believed was its proper title—was not submitted to the House. **1923** *Yorkshire Post* 17 Dec. 6/8 A continuous alternation of pictures and those pieces of text that are, one believes, known as captions. **1924** D. McCarthy *Drama* (1940) 360 It is true that the expression on a film actor's face may occasionally suggest that he, or she, is saying something worth hearing, but the audience cannot supply it from their imaginations; neither can 'the caption' writer. **1936** *Punch* 10 June 654/2 The film *The Emperor's Candlesticks*..is in German, and..those who do not know German..have to collect the sense through captions in our own tongue. **1938** [see *Cut *sb.*² 35]. **1964** T. Rattigan *Heart to Heart* in *Coll. Plays* III. 426 In on David. Super his caption. *Ibid.*, Take out caption. Cue David. **1968** *Listener* 20 June 814/2, I have seen one of these pictures used in a Chinese magazine over the caption 'US aggressor flees'.

caption, *v.* [f. the sb.] *trans.* To provide with a caption, heading, or title; to entitle. Hence **ca·ptioning** *vbl. sb.*

1901 *Science* 22 Nov. 808 An effective poem..captioned 'The Song of the Innuit'. **1912** J. London *Son of Sun* vii. ii, It means the feathers of the sun. Thus does this base interloper caption himself. **1927** *Observer* 27 Nov. 14/5 [His] article captioned 'Cecil Sharp'. **1957** *Listener* 12 Dec. 1001/2 The captioning sometimes strays awkwardly on to the picture.

ca·ptionless, *a.* Of a cartoon: without a caption.

1944 *Newsweek* 25 Sept. 102 This captionless effort..is from a collection of cartoons. **1961** *Times* 23 Nov. 17/4 The captionless jokes in the *New Yorker.*

captivate, *v.* **1.** (Later U.S. examples.)

1840 C. F. Hoffman *Greyslaer* I. x. 114 We can captivate those chaps complete,..if they only move a little further down stream. **1873** J. M. Bailey *Life in Danbury* 137 Captivating coons is not a very easy task.

captive, *a.* and *sb.* Add: **A. 1. b.** (Examples.)

1855 *Househ. Words* XI. 150/2 Pilâtre de Rosier..had already the honour of being the first man who ascended..in a captive balloon. **1897** *Aeronaut. Jrnl.* Jan. 5/2 My own apparatus..is simply a system of large kites only for captive use. *Ibid.*, My present object is to get an apparatus to serve instead of a captive balloon. **1939** G. Greene *Lawless Roads* x. 251 The twenty-four churches rose like captive balloons.

d. *captive audience*: an audience of captives; esp. in extended use, an audience that cannot escape a particular form of entertainment or instruction. orig. *U.S.*

1902 W. D. Howells *Lit. & Life* 57 It was on a Sunday that we crouched in an irregular semicircle..within the prison pale, and faced the captive audience in another semicircle. **1949** in *Amer. Speech* (1951) XXVI. 208/1 It [tram radio] gives a radio station what is referred to in less polite circles as a 'captive' audience. **1956** A. Huxley *Adonis & Alphabet* 119 If they [*sc.* publishers]..possess a large enough captive audience for their text-books to be able to implement their good intentions. **1956** *Kenyon Rev.* XVIII. 463 The captive audience of the class-room. **1959** Thurber *Years with Ross* v. 82 The captive audiences in Grand Central Station, where passengers had been forced to listen to broadcast commercials. **1966** *Listener* 2 June 785/2 We must be especially careful with a captive audience. No airline would inflict a stream of dirty stories..on the helpless passengers on a long-distance journey. **1970** B. Aldiss *Hand-reared Boy* 42, I cannot remember a word she said, being merely a captive audience and bored with the whole visit.

e. Applied to a mine that supplies its products only to commercial concerns under the same ownership, i.e. not to the open market. *U.S.*

1924 *Min. Resources U.S. 1922* II. 546 It was not possible, however, to distinguish..between the output of commercial mines and the output of the 'captive mines' owned or controlled by consumers. **1948** *Economist* 10 July 59/2 The steel companies which operate 'captive' coal mines.

capture, *sb.* Add: **1. b.** *Physical Geogr.* The process by which a stream by headwater erosion encroaches on the basin of a stream at a higher level, and diverts the upper waters of the latter into its own channel; also the point of such diversion; said also of a glacier.

1898 *Geogr. Jrnl.* XI. 441 M. Meunier explains the distribution of erratics in the neighbourhood of the Alps by the 'capture' of one glacier by another, the head of which has eaten back through the dividing wall, and thus tapped the ice-supply. **1908** J. Lomas in *Nature Book* I. 165 Standing at the elbow of the capture of the Hodder, and looking seawards. **1965** A. Holmes *Princ. Phys. Geol.* (ed. 2) xix. 560 Eventually C_2 is intercepted, its headwaters are diverted into S_1 and its lower course is beheaded. This process is called river capture. The rectangular bend *e* at the point of diversion is known as the elbow of capture.

c. *Astr.* The process whereby a star or planet brings an object within its gravitational field. Also *attrib.*

1910 T. J. J. See *Res. Evolution Stellar Systems* II. xxiv. 731 The Capture Theory is so overwhelmingly indicated by the most diverse phenomena of the Starry Heavens, that I cannot doubt that it represents an ultimate truth...Moons and planets..always drift towards the powerful centres of attraction, and are finally captured or absorbed. **1940** *Monthly Not. R. Astron. Soc.* C. 552 The decisive moment for capture is when the parent stars have only just separated after the collision. **1958** *Listener* 20 Nov. 821/2 A few years ago Hoyle also believed in the capture theory but in a different form.

d. *Physics.* The process in which an atomic particle is absorbed by another particle, an atom, or a nucleus. (Formerly believed to be a process by which one particle cohered to another.)

1923 Rutherford in *Proc. Camb. Phil. Soc.* XXI. 510 The capture of electrons by flying α particles has thus opened up a new and interesting field of enquiry. **1929** J. Chadwick in *Proc. Roy. Soc.* A. CXXIII. 383 The disintegration [of the nucleus] is supposed to occur when an α-particle penetrates into a nucleus and is captured by it, the result of the capture being the emission of a proton. **1938** R. W. Lawson tr. *Hevesy & Paneth's Man. Radioactivity* (ed. 2) ii. 33 In the course of the flight of an α-particle this capture and loss of electrons is repeated several thousand times. **1955** Friedman & Weisskopf in W. Pauli *Niels Bohr* 136 The actual predominance of capture over scattering in low energy resonances is easily described.

capture, *v.* Add: **b.** *Physical Geogr.* Of a river or glacier: to divert by capture (see prec. 1 b).

1900 *Geogr. Jrnl.* XVI. 36 It may then have happened that whole basins, like that of Lake Mascardi, drained formerly towards Nahuelhuapi, have been captured towards the Pacific slopes. **1908** J. Lomas in *Nature Book* I. 166 Along the Vale of York, where the rocks are softer than those of the Pennines or East Yorkshire, the Don sent up a tributary and successively captured the rivers to form what is now the Ouse. **1965** A. Holmes *Princ. Phys. Geol.* (ed. 2) xix. 562 The headwaters of the Aire..were captured by the Ribble.

c. *Astr.* Of a star or planet: to bring an object within its gravitational field (see prec. 1 c).

1910 [see *Capture *sb.* 1 c]. **1958** tr. *J. Verne's Round Moon* 11 It had however travelled so near to the moon that it had been 'captured' by the lunar gravitation. **1959** *Listener* 23 Apr. 708/1 Mercury has a captured rotation; that is to say, its sidereal period is equal to its axial rotation period, eighty-eight Earth days in each case. **1961** *Ibid.* 9 Nov. 766/1 According to one theory, minor satellites of this sort are really in the nature of captured asteroids. **1969** *Times* 13 June 7/7 It seems that the moon was either a wandering planet that was captured by the earth or else was formed from a cloud of matter.

d. *Physics.* To bring about the capture of (a particle) (see prec. 1 d).

1923 Rutherford in *Proc. Camb. Philos. Soc.* XXI. 506 The α particle in passing through matter occasionally captures an electron. **1929** [see *Capture *sb.* 1 d]. **1956** A. H. Compton *Atomic Quest* 18 The uranium atom when it captured the neutron, became unstable and divided into roughly two equal parts.

caput. **4.** Add: *caput lupinum* (lit., wolf's head), an outlaw: see Wolf's-head 2.

[**1797** *Encycl. Brit.* IV. 156/1 Anciently an outlawed felon was said to have *caput lupinum*.] **1837** Macaulay *Crit. & Hist. Essays* (1843) II. 393 That a valetudinarian..should be treated as a *caput lupinum* because he could not read the Timæus without a headache, was a notion which the humane spirit of the English school of wisdom altogether rejected. **1888** *Guardian* 4 Apr. 488/2 The National League, if it did not formally decree the death of Fitzmaurice for disobedience to its orders, at least proclaimed him as a *Caput Lupinum*.

car, *sb.*¹ Add: **1. e.** = Motor car.

1896 L. Serraillier tr. *Farman's Auto-Cars* 132 The latter drove with a daring which may have been dangerous to himself, but which never affected his car. *Ibid.* 135 The three cars which came in next after Mr. Levassor's were all Peugeot cars. **1900** W. W. Beaumont *Motor Vehicles* I. 615 Hill-climbing trials alone would not of course be sufficient as a test of the wearing power or durability of a car. **1902** Harmsworth *Motors & Motor Driving* 23 The first car built by the Daimler Company at Coventry. **1948** A. Huxley *Ape & Essence* (1949) i. 12 He threw the car into gear and we were off.

2. (Earlier U.S. example.) Also used in Great Britain, usu. as the second element in combinations.

1826 *Laws Commonw. Mass.* 4 Mar. c. 183 § 5, p. 331 The conveyance of stone and other property, in their cars and vehicles on said railways. **1873** C. M. Yonge *Pillars of House* IV. xlv. 274 Even Pullman's cars shook him. **1875** [see Pullman]. **1899** [see *Buffet *sb.*² 3 b]. **1901** M. Carmichael *In Tuscany* 284 The restaurant car has come in since. *a* **1930** D. H. Lawrence *Phoenix II* (1968) 252 One sits in the breakfast car in the train, coming to London... In America the Pullmans..don't shake like our cars. **1939** T. S. Eliot *Old Possum's Practical Cats* 40 It is Skimble who's in charge Of the Sleeping Car Express. **1967** *Jane's World Railways 1967–68* p. iii, Computers..are being extended into every possible field of operation.., from sending messages to locating freight cars.

b. *ellipt.* A car-load. *U.S.*

1869 *Trans. Ill. Agric. Soc. 1867–9* VII. 446 A farmer had far better send..one car of good sheep in the twelve months, than six cars of..bad breed. **1902** G. H. Lorimer *Lett. Merchant* 133 The last car of lard was so strong that it came back of its own accord from every retailer they shipped it to. **1948** *Green Bay* (Wis.) *Press-Gaz.* 30 June 23/4 Receipts were: wheat 38 cars, corn 83, oats 34 and soybeans 9.

4. b. The cage of a lift. Chiefly *U.S.*

1870 *Gold Hill* (Nevada) *News* 25 Aug., The cable fell..upon the car and cage. **1904** *N.Y. Even. Post* 2 Feb. 2 The elevator man seemed to lose control of the elevator at the fifth floor. The car made a sheer drop from the fourth story to the basement. **1943** D. Baker *Trio* III. 171 When leaving car, close car gate and landing door, otherwise elevator becomes inoperative.

6. *car-body, -door, -drive, -fare* (U.S.), *-ferry, -hire, -horse, -key, -load, -ride, -window* (examples); *car-carrying* adj.; *car-owner; car-owning* vbl. sb. and ppl. a.; see also *Car-less *a.*; **car-coat,** a short coat designed especially for the use of car-drivers; **car-hop** *U.S. colloq.*, a waiter or waitress who serves customers in their parked cars; cf. **bell-hop*; hence as *v. intr.*, to act as a car-hop; **car-loading,** the amount of freight carried in goods trains within a given period; freq. in *pl.*; also *fig.*; **car-park,** a building, open space, etc., for the parking of motor vehicles; also *attrib.*; **car-port** (orig. *U.S.*), a motor-car shelter, open at the side or front, adjoining or built into a house; **car-sick** *a.* [after *sea-sick*], sick from the motion of a motor car; so *car-sickness*.

1838 *Civil Eng. & Archit. Jrnl.* I. 390/1 The truck is guided by the carbody with..mathematical precision. **1959** *New Statesman* 1 Aug. 141/1 A car-body firm. **1959** *Times* 10 June 9/7 The Boulogne–Lyons car-carrying trains. **1963** *Punch* 10 Apr. p. xiv, Men's wear..includes foam-backed car coats. **1965** *Guardian* 31 Mar. 16/3 The now ubiquitous car-coat..is usually made from a cotton suedette or brushed nylon shell fabric and lined with an acrylic pile liner. **1863** 'G. Hamilton' *Gala-Days* 22 The heads of the two columns collide near the car-door. **1924** Geikie *Long Life's Wk.* 220 We had long walks and still longer car-drives. **1870** 'Fanny Fern' *Ginger-Snaps* 182 What troubles me most is, whether I am to pay six cents for car-fare. **1900** *Dial. Notes* II. 16 A fish-scale..is the nickel with which a [Yale] student pays his car-fare. **1958** S. Ellin *Eighth Circle* (1959) ii. x. 104 She said she needed carfare to get back to school. **1884** H. Hall in *Rep. U.S. Census of 1880* VIII. 221 Side-wheel car-ferry Transport..with three railroad tracks on deck. **1898** *Strand Mag.* Apr. 443/2 The..necessity for appeasing the ire of passengers, who were compelled in the old days to change from train to ferry-boat, and then to train again, is the real reason for the existence of the car ferry. **1955** 'Lory' *Brit. Railways' Sands across Sea* 264 A Car Ferry Terminal has been constructed at..Dover. **1959** *Listener* 31 Dec. 1156/1 Even car-hire services possess a long ancestry. **1937** *Amer. Speech* XII. 320/1 At roadside drive-ins..the waitresses are now being called *curbies* or *car-hops*. **1960** J. McManus in *Introduction* 188, I had another job..car-hopping six days a week at a little beef-and-burger kerbstand. **1800** M. Edgeworth *Parent's Assistant* V. 75 He had been used..to lead car-horses. **1907** M. H. Norris *Veil* v. 38 A car-horse, sore-footed and generally weather-beaten..completed this part of his purchase. **1962** M. Kelly *Due to a Death* vii. 123, I dropped the car key in the table drawer. **1854** Thoreau *Walden* 40, I do not see in my mind any retinue at their heels, any car-load of fashionable furniture. **1958** R. Godden *Greengage Summer* vi. 72 The car-loads and char-à-bancs of visitors. **1959** N. Mailer *Advts. for Myself* (1961) 210 A carloading of homogenized words. **1963** *Economist* 21 Dec. 1292/2 Railway shares have benefited especially from such things as consistent increases in carloadings. **1903** *Motoring Ann.* 91 Car-owners always consider it a privilege to have Miss Wilson on their front seat. **1925** A. Huxley *Along Road* i. 16, I can talk for hours about motors with other car-owners. *Ibid.* 17 Car-owning may have the worst effect on the character. **1959** *Daily Tel.* 15 Oct. 12 The traffic problems of a car-owning democracy. **1926** *Daily Mail* 1 Dec. 9/6 Glastonbury Car Park. Indignation has been aroused..by a proposal..to purchase part of the land..as an extra parking space for motor cars. **1935** *Archit. Rev.* LXXVIII. p. xlvi, The car-park problem. **1955** *Times* 9 May 18/2 A large construction programme [in Canada] of multi-deck car parks with direct access to the stores. *Ibid.* 25 Aug. 8/6 The car park attendant. **1939** *Life* 8 May 17/1 At the left is the 'carport' which opens on main entrance of building. **1947** *Archit.*

Rev. CII. 117 Above the living-room–kitchen part is an open roof-deck, below the bedrooms a car-port. **1963** *Sunday Times* 21 Apr. 45/6 Terrace houses overlooking Torbay which incorporate tiled carports under the houses. **1902** Harben *Abner Daniel* 25 Now I have a long car-ride before me, and it's growing late. **1908** *Lancet* 12 Sept. 828/1 Individuals who are car-sick may be good sailors and *vice versâ*. *Ibid.*, By car-sickness is meant a series of unpleasant symptoms, as giddiness, staggering, nausea, and vomiting due to riding in a vehicle or even in a baby-carriage. **1941** J. P. Marquand *H. M. Pulham* xxxi. 343 You always get car-sick if you eat too much. **1947** C. P. Stewart *Sword Between* xv. 80 Maria suffered from car-sickness, and frequent stops had to be made. **1854** M. J. Holmes *Tempest & Sunshine* xvii. 239 From the car windows Fanny watched the long blue line of hills. **1903** Kipling *5 Nations* 115 Where the white car-windows shine.

car, *v.* Add: Also *intr.*: to go by car.

1907 *Daily Chron.* 23 Sept. 4/4 You just get out of your train, walk or 'car' to your hotel.

carabid (kæ·răbid), *a.* [f. the sb.] Of or pertaining to the Carabidæ.

1901 *Trans. Entomol. Soc.* 193 (*title*) The Carabid genus Pheropsophus. **1966** C. Sweeney *Scurrying Bush* iv. 52 Our search disclosed a number of tenebrionid and carabid beetles, the former feeding on the dung, the latter carnivorous.

carabideous (kæräbi·dīəs), *a.* *Ent.* [f. Carabid + -eous.] = prec.

1835 J. Duncan *Nat. Hist. Beetles* 119 The majority of carabideous insects secrete an acrid and caustic fluid. **1835** [see Square *a.* 1]. **1889** A. R. Wallace *Darwinism* 44 The curious little carabideous beetles of the genus Notiophilus.

‖**carabinero** (kæräbineᵊ·ro). [Sp.] A customs or revenue officer; a frontier guard.

1845 R. Ford *Handbk. Spain* I. iii. 324 The expensive preventive service of *Resguardos, Carabineros*, etc. which is every where established to put down the smuggler. **1883** Ld. Saltoun *Scraps* I. ii. 216 The *carabineros*, or revenue officers, might be so disposed of as to leave the coast clear. **1955** J. Thomas *No Banners* viii. 67 They knew they could never hope to reach Barcelona with a pack of irate *carabineros* on their trail. **1957** P. Kemp *Mine were of Trouble* i. 13 *Carabineros*, or frontier guards, in light green and flat peaked caps.

‖**carabiniere** (kæräbinyēᵊ·ri). Pl. -ieri. [It.] A member of an Italian Army Corps which serves as a police force.

1847 J. H. Newman *Let.* 25 July (1962) XII. 104 Minardi..the head of the Carabinieri, was found to be implicated. **1924** D. H. Lawrence in M. Magnus *Memoirs of Foreign Legion* 52 The carabinieri wanted to arrest him at the monastery, so he has escaped here. **1946** G. Millar *Horned Pigeon* ix. 129 The four of us squeezed into the middle of a first-class compartment with the officer and the *carabinieri* occupying all the corners and another *carabiniere* in the corridor. **1969** P. Loraine *Mafia Kiss* i. i. 13 What if the carabinieri had taken the child, questioned him?

caracul(e), var. *Karakul.

Caradoc (kăræ·dǫk). *Geol.* [f. Caer *Caradoc*, name of sandstone hills in Shropshire.] Applied *attrib.* to a stage or series or to the rocks of the Ordovician System in Britain. Hence **Carado·cian** *a.*

1835 R. I. Murchison in *Phil. Mag.* vii. 47 *Caradoc sandstones.* This name..has been derived from the..ridge of Caer Caradoc. **1855** J. Phillips *Man. Geol.* 192 There is also a narrow tract of Caradoc sandstone running N.N.W. in the Lickey Hills. **1910** *Encycl. Brit.* V. 300/1 In the Lake district the Coniston limestone series represents the Upper Caradocian. **1960** L. D. Stamp *Britain's Struct.* (ed. 5) xii. 109 Caradocian and Ashgillian..being the lower and upper parts respectively of the old Bala Group.

carafe. Add: Also, one used at table, etc., for wine; *attrib.*, designating an ordinary wine served in a carafe.

1939 E. Ambler *Mask of Dimitrios* xiii. 274 The carafe burgundy which his host had recommended so warmly at dinner. **1950** J. Cannan *Murder Included* iii. 58 A *carafe* of white wine, biscuits, butter and cheese. **1957** R. Postgate *Good Food Guide 1957–58* 263 A small carafe of white wine. **1959** *Ibid. 1959–60* 260 Carafe wines, white and red. **1962** *Punch* 26 Dec. p. x, The 'Tent'..serves simple English dishes and carafe wines for the hurried or the hard-up. **1966** *Guardian* 15 Dec. 6/6 An admirable carafe Sancerre.

caramba (kăræ·mbă), *int.* Also **carramba.** [Sp.] An exclamation of surprise or dismay: gracious! strange! confound it! damn!

1835 J. H. Ingraham *South-West* I. 139 'Sacré diable!' 'Carramba!'—fell upon the ear. **1845** R. Ford *Handbk. Trav. Spain* i. i. 36 More becoming will it be to the English gentleman to swear not at all; a reasonable indulgence in *Caramba* is all that can be permitted. **1851** H. Melville *Moby Dick* II. i. 2 Caramba! have done, shipmate, will ye? **1906** *Daily Chron.* 3 Jan. 3/4 An obvious villain had snapped 'caramba!' at her.

caramel, *sb.* Add: **d.** The colour of caramel; brown. Also *attrib.*

1909 *Daily Mirror* 4 Oct. 10/3 Caramel is the name for various new shades of brown. **1927** *Daily Express* 12 Mar. 3/5 Caramel, a useful light brown, suitable for all kinds of walking costumes. **1970** C. Drummond *Stab in Back* vi. 135 She was wearing the caramel suit she disappeared in.

caramel, *v.* Add: Also *intr.*

1887 *Century Mag.* Nov. 114/1 He seeks to keep the temperature down to 130°. If it is too high some of the sucrose will 'invert' or 'caramel' into glucose. **1897** *Yearbk. U.S. Dept. Agric.* 515 The sugar in the milk caramels in baking and browns the crust.

ca:rameliza·tion. [f. Caramelize *v.* + -ation.] The production of caramel; conversion into caramel or into the colour of caramel.

1889 in *Cent. Dict.* **1955** J. G. Davis *Dict. Dairying* (ed. 2) 163 *Caramelisation*, the browning of sugar solutions by heat. **1960** *Times Rev. Industry* July 19/3 Care must be taken to prevent browning caused by caramelization of the sugars.

carancha, carancho, varr. *Carrancha.

carangid (kăræ·ŋgid), *a.* and *sb.* [f. mod.L. *Carangidæ*.] **A.** *adj.* Pertaining to or resembling the Carangidæ, a family of perciform fish. **B.** *sb.* A member of this family.

1889 *Cent. Dict.*, Carangid *n.* **1931** J. R. Norman *Hist. Fishes* v. 94 The Scad or Horse Mackerel (*Trachurus*), a member of a large group of fishes known as Pampanos or Carangids (Carangidae). **1931** J. R. Norman in W. P. Pycraft *Standard Nat. Hist.* xi. 461 (*caption*) A Carangid fish known to the ancients as Pompilus. **1962** K. F. Lagler et al. *Ichthyology* xi. 337 Certain carangids such as *Trachurus* have prominent paired swellings designated the cristae cerebelli.

carangoid (kăræ·ŋgoid), *a.* and *sb.* [f. as prec. + -oid.] = prec.

1862 *Proc. Acad. Nat. Sci. Philadelphia* ix. 430 My attention having been..attracted especially to the Carangoids, it has been discovered that the nomenclature of several was quite erroneous. **1926** H. M. Kyle *Biol. Fishes* v. 113 Many normal fishes among the Carangoids show this stage.

carapacial (kæräpēˡ·ʃᵊăl), *a.* [f. Carapace + -ial.] Of or pertaining to a carapace.

1880 T. H. Huxley *Crayfish* iv. 217 The lateral portions of the carapacial ridge.

So **carapacic** (kæräpēˡ·sik) *a.*

1903 *Ann. & Mag. Nat. Hist.* XI. 442 The carapacic portion of the dorsal scute.

carapato, var. *Carrapato.

Caravaggesque, Caravaggiesque (kæ:ravædʒe·sk, -dʒie·sk), *a.* Also **Caravaggioesque.** [f. the name *Caravaggio* (see below) + -esque.] Of, resembling, or characteristic of the Italian painter Michelangelo Merisi Caravaggio (1573–1610), or his works.

1925 A. Huxley *Along Road* iii. 184 The most spectacular Caravaggioesque light and shade. **1936** *Burlington Mag.* Mar. 132/1 A Caravaggiesque *Madonna*. **1938** *Ibid.* Jan. 4/1 The third great style of the century is the 'Caravaggesque'. **1958** *Spectator* 24 Jan. 106/2 As a painter of religious or genre subjects in artificial light he has obvious associations with the Caravaggesque.

caravan, *sb.* Add: **3. b.** A company of travellers, traders, or emigrants, with their wagons, mules, or packhorses, esp. in the Western States; a train (of wagons, etc.). *U.S.*

1748 Catesby *Carolina* App. p. iv, Indian Traders, whose Caravans travel these uninhabited Countries. **1791** W. Bartram *Trav. Carolina* (1792) 376 Our caravan consisting of about twenty men and sixty horses, we made a formidable appearance. **1817** S. R. Brown *Western Gaz.* 77 General Harrison..was accompanied in his march through the wilderness by a caravan of waggons. **1897** J. L. Allen *Choir Invisible* ii. 14 A company of travellers with pack-horses—one of the caravans across the desert of the Western woods.

4. Now freq. used as a stationary dwelling. Also *attrib.*, as **caravan park, site,** a place where caravans are parked and used as dwellings.

1935 *Caravan Ann.* (*sub-title*), A list of over 1,200 caravan sites in Britain. **1937** 'G. Orwell' *Road to Wigan Pier* iv. 61 The dreadful caravan-dwellings that exist in numbers in many of the northern towns. *Ibid.*, Parts of the population have overflowed into supposedly temporary quarters in fixed caravans. *Ibid.*, Some of the caravan-dwellers. **1959** M. Steen *Tower* i. iii. 44 The back streets, the caravan sites and the pre-fabs. **1963** *Times* 18 Feb. 16/5 The difficult problem of finding enough caravan parks on which to put the end product.

caravan, *v.* [f. the sb.] **1.** *intr.* To travel or live in a caravan.

1885 [see Caravaning *vbl. sb.*]. **1909** R. Brooke *Let.* Sept. (1968) 184 Will you Caravan with me in the Spring, or Summer? **1930** W. de la Mare *Desert Islands* 39 In the 'second part' of his adventures Crusoe is transformed into a mere globe-trotter, caravanning from Nanking to Tobolsk. **1963** *Times* 12 Mar. p. iv/1 Having caravanned all over this continent, he admitted that there is an awful lot of it to love.

2. *trans.* To convey by caravan. *rare.*

1898 G. W. Steevens *With Kitchener to Khartum* 118 The Greek gets his stuff up everywhere:..he caravans it with a double-barrelled rifle on his shoulder.

caravaneer. Add: **b.** One who lives or travels in a caravan.

1893 *Cornhill Mag.* Mar. 288 The joy of the caravaneer was ours.

caravanner (kæ·răvænəɹ). Also **-aner.** [f. Caravan + -er¹.] **a.** One who lives in or travels in a caravan (esp. on a pleasure tour). **b.** The conductor of a caravan in the East (Cf. prec.)

1909 *Westm. Gaz.* 5 Apr. 4 The [Caravan] club decided to spell 'caravanner', 'caravanning', with two n's—not one. **1916** C'tess von Arnim (*title*) The Caravaners. **1923** *Weekly Dispatch* 21 Jan. 5 Victor Maclaglen, who gives a magnificent performance as a gipsy king in the story, was urged to prolong his stay by the caravanners. **1929** *Daily Tel.* 22 Jan. 11/3 There is little doubt that information which caravanners gave me yesterday has now reached Kabul. **1958** *Economist* 25 Oct. 310 The shanty towns into which caravanners have been herded.

caravanning (kæ·răvæniŋ), *vbl. sb.* Also **caravaning.** [f. *Caravan *v.* + -ing¹.] Travelling or living in a caravan. Also *attrib.*

1885 [see Caravaning *vbl. sb.*]. **1892** *Pall Mall Gaz.* 5 May 3/1 The rage of this season is to be caravanning. **1914** *Lancet* 9 May 1369/1 Caravanning offers freedom, movement, variety, open-air life. **1957** *Listener* 24 Oct. 643/2 A caravanning holiday. **1958** *Times* 19 Mar. 6/6 Caravanning, he said, was 'not a question of poverty or squalor'. It was a 'new way of life'. **1962** *Listener* 5 July 14/2 Facilities are provided for camping and caravanning in special fields.

caraway. **1.** For *Carum Carui* read *Carum carvi*.

carbanion (kāɹbæ·naiǫn). *Chem.* [f. Carb- + Anion.] An organic ion containing negatively charged carbon. Cf. *Carbonium (*b*).

1933 Wallis & Adams in *Jrnl. Amer. Chem. Soc.* LV. 3838 For the sake of clearness in discussion all tricovalent carbon compounds may be divided into three classes: (1) 'carbanions'..(2) free radicals..(3) carbonium ions... The above word [carbanion] is here proposed for a negatively charged carbon ion in contrast to the term 'carbonium', which indicates a positively charged ion. **1956** J. E. Leffler *Reactive Interm. of Org. Chem.* ix. 187 Many of the reactions of the weak carbon acids are reactions of the carbanion, the rate being the rate of ionization.

carbide. Add: *spec.* Short for 'calcium carbide'.

1898 W. E. Gibbs *Lighting by Acetylene* 70 Water from an elevated reservoir enters T until the carbide is reached. **1904** A. B. F. Young *Compl. Motorist* 257 Carbide lamps give a beautiful light when they are working properly. **1957** P. Kemp *Mine were of Trouble* iv. 59 A bare room..lit by a single smelly carbide lamp.

carbo-. Add: *carbo-hydrogen; -hydrous, -methylic* adjs.; **carbo-cyclic** *a.*, containing a ring of carbon atoms only; **carbo-dynamite,** a patented explosive, in which fine charcoal is used as the absorbent.

1899 *Jrnl. Chem. Soc.* LXXVI. i. 59 Carbocyclic Compounds. **1904** *Nature* 11 Aug. 341/1 The stability of carbocyclic and heterocyclic chains. **1888** *Times* 2 Oct. 3/1 Some carbo-dynamite, which had lain for eight months in water. **1866** Odling *Anim. Chem.* 75 The carbo-hydrogen of vegetable tissue. **1881** *Ann. Rep. Smithsonian Inst.* 393 The carbohydrous material which is present in the plant. **1838** R. D. Thomson in *British Annual* 323 Carbo-methylic acid.

carbohydrate. Add: (Now usu. written without hyphen.) Also *attrib.* and *Comb.*

1886 *Jrnl. Franklin Inst.* CXXII. 274 (*title*) Carbohydrate and fatty foods. *Ibid.* 285 The glycogen of the liver greatly diminishes in amount in the absence of carbohydrate food stuffs. **1888** *Jrnl. Chem. Soc.* LIV. 972 Acid in Healthy and Disordered Stomachs during a Carbo-hydrate Diet. **1900** *Westm. Gaz.* 21 Dec. 3/2 Concerning malt liquors, we are told that the large quantity of carbohydrate matter in them [etc.]. **1908** *Carnegie Trust Rep.* 22 The carbohydrate material and ferments of blood. **1910** *Practitioner* Jan. 129, I want to insist that pyorrhoea alveolaris is a local disease due to germ, or carbohydrate-germ. **1964** M. Hynes *Med. Bacteriol.* (ed. 8) i. 7 Carbohydrate-fermenting bacteria may produce enough acid in culture to kill themselves.

carbol-fuchsin, -ine (kā:ɹbǫl,fū·ksin). [f. Carbol(ic *a.* + Fuchsine.] A mixture of fuchsine and carbolic acid, used as a staining-solution.

1897 Pearmain & Moor *Appl. Bacteriol.* iii. 84 (*heading*) Ziehl's Carbol-Fuchsine. **1909** *Practitioner* Feb. 304 Leprosy bacilli..can be stained with carbol-fuchsin in the ordinary way. **1956** *Nature* 3 Mar. 430/2 Sections were..stained with..dilute carbol fuchsin.

carbolic, *a.* Add: Also *ellipt.* for *carbolic acid* or *soap*.

1884 *Medical Uses Carbolic Acid* (F. C. Calvert & Co.) 35 The Carbolic will be readily given off in vapour sufficient to fill an ordinary room. *Ibid.* 47, I used 1 pint of Carbolic to each room. **1907** M. C. Harris *Tents of Wickedness* i. iv. 58 The stone floors are so cold, and there is such a smell of carbolic.

carbomycin (kā:ɹbomai·sin). *Med.* [f. Carbo- + Gr. μύκης fungus (see Myco-) + -in¹.] An antibiotic obtained from the fungus *Streptomyces halstedii*.

1952 *Antibiotics & Chemotherapy* II. 441 Magnamycin is the trade-mark name of Chas. Pfizer & Co.'s antibiotic carbomycin. **1960** M. E. Florey *Clinical Appl. Antibiotics* IV. ii. 36 Like erythromycin, the antibiotic carbomycin..was isolated in 1952 and has similar antibacterial properties.

carbon, *sb.* Add: **1. b.** A form of diamond, the black diamond or carbonado.

1869 *Eng. Mechanic* 20 Aug. 475/1 Diamond or carbon (the latter name is preferred) stands the severest tests for mechanical purposes without apparent wear. **1903** *Westm. Gaz.* 6 June 9/1 'Black diamonds'..technically known as 'carbons'.

c. Short for **carbon-paper* (*b*).

1895 Collyns *Typists' Man.* 55 Errors must not be erased while the carbons are in the machine. **1913** E. W. Sargent *Technique of Photoplay* (ed. 2) 25 For carbon copies get the carbon second sheets. **1920** R. Macaulay *Potterism* i. iv. 46 Jane extracted carbons from a drawer and fitted them to her paper.

d. Short for **carbon-copy*.

1937 in Partridge *Dict. Slang* 127/1. **1940** H. Innes *Trojan Horse* iii. 62, I signed the carbon and placed it in a foolscap envelope. **1964** R. Petrie *Murder by Precedent* iii. 46 There has to be something in the file first. The carbon of the finished letter.

e. carbon 14: a long-lived radioactive isotope of carbon, of mass-number 14, the regular decay-rate of which makes possible the dating of organic materials from ancient deposits; also used as a tracer element in biochemistry. Symbol: C^{14} or ^{14}C. Cf. *Radio-carbon.

1936 *Physical Rev.* XLIX. 778/1 McMillan has obtained evidence of a radio-active C^{14}. **1941** *Ibid.* LIX. 349 There are 5 known isotopes of carbon, stable C^{12}..C^{13}..C^{10}..C^{11} and C^{14} (half-life 10^3–10^5 years). *Ibid.* 351/2 Production of C^{14} is achieved as a by-product of cyclotron operation. **1946** *Lancet* 12 Oct. 535/2 A so-called unit of carbon 14 (C^{14}). **1950** *Amer. Speech* XXV. 25 Carbon 14..is being extensively used as a tracer. **1951** *Proc. Prehist. Soc.* XVII. 7 There can be age-determination..by the Carbon 14 method. **1958** *Listener* 6 Nov. 720/1 Carbon 14 dating has become vital to the study of archaeology. **1958** *Times* 12 Dec. 17/6 Carbon 14..breaks down only slowly—half of it in about 5,600 years. **1969** E. H. Pinto *Treen* 5 The scientific system of radio carbon dating, or carbon-14 dating, as it is also called, has come into being.

3. a. carbon-black = Lamp-black; also *attrib.*; **carbon-copy,** a copy made by using carbon-paper; also *fig.* and *attrib.*; **carbon cycle,** (*a*) *Biol.*, the cycle in which carbon dioxide is absorbed from the atmosphere by plants and replaced mainly by the respiration of plants and animals and the decay of organic matter; (*b*) a cycle of thermonuclear reactions in stellar regions, in which carbon acts as a catalyst in the conversion of hydrogen into helium, the consequent energy released being held to be the source of the energy radiated by the sun and stars; **carbon fibre,** a fibre of carbon; *spec.* a very thin, polycrystalline filament of carbon, freq. very strong, obtained by the pyrolysis of organic textile fibres and consisting of crystallites that are preferentially aligned parallel to the axis of the filament; used esp. to increase the strength-to-weight ratio of plastic or other material in which filaments are incorporated as in a matrix; also, such filaments collectively; **carbon-paper,** (*a*) *Photogr.*, paper used in carbon printing (*disused*); (*b*) thin paper coated on one side with a preparation of carbon-black or some other pigmented material, used between two papers to make a duplicate copy of what is written, typed, etc., on the upper sheet (earlier called **carbonic paper*); **carbon steel,** a steel the properties of which are determined mainly by the percentage of carbon present.

1889 *Cent. Dict.* s.v., Carbon-black..is almost pure carbon in a finely divided form. **1930** *Engineering* 9 May 605/3 The carbon-black industry in the United States. **1934** *Discovery* May 138/2 Indian ink contains carbon black. **1895** Carbon copy [see **carbon-paper*]. **1944** Auden *For Time Being* (1945) 58 Everything, the massacres, the whippings, the lies, the twaddle, and all their carbon copies are still present. **1959** *Observer* 15 Mar. 11/5 A group of indolent, shiftless, carbon-copy 'West Side' youths. **1961** *Daily Tel.* 3 Apr. 7/4 Victim of the week-old 'carbon copy' murder. **1912** E. J. Russell *Soil Conditions* iv, The carbon and nitrogen cycles in the soil. **1940** H. A. Bethe in *Astrophysical Jrnl.* XCII. 118 The nuclear reactions involved in the carbon cycle. **1946** *Ann. Reg. 1945* 356 It is generally accepted that the Bethe carbon cycle..provides the main source of stellar energy. **1960** *Chem. Abstr.* LIV. 1966 (*heading*) The morphology of carbon fibers. **1966** *Engineer* 27 May 816/1 Work on the application of these new carbon fibres is directed toward the manufacture of carbon-fibre composites having resinous matrices. **1969** *Guardian* 10 June 4/1 The Lotus car company is examining ways in which carbon fibre can be used in car production. **1878** *Jrnl. Chem. Soc.* XXXIV. 919 (*heading*) Carbon-paper rendered sensitive without a chrome-bath. **1895** Collyns *Typists' Man.* 55 In correcting the carbon copies, a narrow strip of carbon paper should be placed upon the ribbon to prevent the corrections being conspicuous. **1903** A. H. Hiorns *Steel & Iron* xix. 336 The mild blister steel is used for springs, and the higher carbon steel after fagoting and welding is termed 'shear' steel. **1925** *Jrnl. Iron & Steel Inst.* CXI. 305 Plain carbon steels high in carbon have a partially austenitic structure when sufficiently intensely quenched.

b. carbon-dioxide snow, carbon dioxide solidified by cooling (cf. *Carbonic *a.* 1 a); **carbon disulphide,** see Disulphide; **carbon tetrachloride,** see Tetra- 2 a.

1910 *Lancet* 14 May 1350/1 The beneficial effects of treatment by carbon dioxide snow in Xeroderma pigmentosum. **1934** *Discovery* Aug. 223/2 In its solid state it [*sc.* carbon dioxide] is used by hospitals for cauterisation, in the form of 'carbon dioxide snow'.

c. carbon-arc, an arc between electrodes that consist wholly or mainly of carbon; also *attrib.*; **carbon microphone,** a microphone depending for its action on the varying electrical resistance of carbon granules when subjected to sound waves of varying pressure (see quot. 1962); **carbon transmitter,** = **carbon microphone*.

1908 *Jrnl. & Proc. R. Soc. N.S.W.* XLII. 201 An investigation of the phenomena associated with the relighting of the carbon arc. **1922** Glazebrook *Dict. Appl. Physics* II. 145/2 The cored carbons which are used in ordinary open type or enclosed type lamps, for the so-called pure carbon arc, have the core canal filled with a mixture of carbon and potassium silicate. **1927** *Lancet* 1 Jan. 15 The therapeutical use of the carbon-arc light. **1879** *Telegraphic Jrnl.* VII. 1/1 The carbon microphone..has already proved a practical instrument for sending articulate sounds. **1962** A. Nisbett *Technique Sound Studio* 244 *Carbon microphone*, microphone in which the audio signal is produced by varying the resistance of a button of granular carbon to which a polarizing voltage is applied (a diaphragm presses on the carbon button). **1879** Carbon transmitter [in Dict.]. **1889** *Telephone* I. 248/1 A carbon transmitter is used in connection with this receiver.

carbon, *v.* [f. the sb.] *intr.* or *pass.* Of the cylinders in the engine of a motor car: to become coated with carbon deposit. Const. *up*.

1922 *Daily Mail* 28 Nov. 6 [A motor engine] longer to carbon up and easier to decarbonise than its rivals. **1925** *Morris Owner's Man.* 70 Thanks to good carburation, the Morris engine carbons up very slowly. **1928** *Daily Express* 3 Aug. 9 Even if the engine is carboned up, you get wonderful pulling.

carbona (kaɹbōᵘ·na). *Cornish Mining.* [Origin unknown.] An irregular mass of ore, usually found in the vicinity of a lode.

1843 W. J. Henwood in *Trans. R. Geol. Soc. Cornwall* V. 21 The extraordinary deposit of tin ore which..has obtained the name of the *Carbona*. **1855** Leifchild *Cornwall* 40 The formation of the tin ore in this mine is very singular, and is provincially termed *carbona*.

carbonade, *sb.* Restrict †*Obs.* to sense in Dict. and add: **2.** (Also in Fr. form *carbonnade*.) A beef stew (see quots.).

1877 E. S. Dallas *Kettner's Bk. of Table* 101 The carbonade has degenerated in France into a stew, having meant originally a grill; and attempts are made to introduce the word into England as corrupted by the French cooks. **1935** Morphy *Recipes of all Nations* 550 (*heading*) Les Carbonnades Flamandes (Flemish carbonades). This is one of the most well known of all Belgian dishes...The word 'carbonnades'..is now used for a method of slow stewing. *Ingredients*..boned neck of beef..onions..1 bottle of beer. **1959** R. Postgate *Good Food Guide 1959–60* 73 Occasionally a more unusual dish appears, such as carbonade of beef (braised with onions in beer). **1970** *Times* 7 Mar. (Suppl.) p. v/2 Carbonnade of beef is a rich stew, combining beef, onions and beer.

carbonado (kāɹbonēˡ·dǫ), *sb.*² [Pg.] A dark, opaque variety of diamond, found near Bahia in Brazil, used in rock-drilling and stone-polishing.

1853 J. Tennant *Gems & Prec. Stones* 94 A considerable quantity of a black substance was found, of specific gravity like the Diamond, but lamellar...It was termed 'Carbonado' by the discoverers from its charcoal like appearance. **1879** *Encycl. Brit.* IX. 345/1 The dressing and grooving of millstones is generally done by hand-picking, but sometimes black amorphous diamonds (carbonado) are used. **1951** J. R. Partington *Gen. & Inorg. Chem.* (ed. 2) xvii. 439 Black or dark-coloured..diamonds, carbonado and bort (or boart), of no value as gems, are used for rock-drills.

carbonatation (kā:ɹbǝnēˡtēˡ·ʃǝn). [f. Carbonate *v.*¹ + -ation.] = Carbonation.

1887 *American* XIV. 24 The process of carbonatation, or removing the excess of lime used in defecating the juice by carbonic acid gas. **1888** *Harper's Mag.* June 47 The best method of 'carbonatation' of the saccharine juice.

carbonate, *sb.* Add: **2.** In full *carbonate ore*.

1890 W. J. Gordon *Foundry* 96 The hæmatite,..unlike the carbonate ores,..requires no roasting. **1892** *Daily News* 29 Oct. 6/4 The new bodies of carbonate ore.

3. = *Carbon *sb.* 1 b.

1883 Gresley *Gloss. Coal-m., Carbonates*, black imperfectly crystallised form of diamond used for rock boring; the abrasion of the diamond removes the rock in an annular form, producing cores.

carbonator (kā·ɹbǝnēˡtǝɹ). [f. Carbonate *v.*¹ + -or.] One who or that which carbonates.

1896 *Voice* (N.Y.) 21 May 4/5 (*title*) American Bottler and Carbonator. **1921** *Dict. Occup. Terms* (1927) § 459 *Carbonator*..runs a specified quantity of carbonic acid gas, in solution, into beer. **1935** *Discovery* Dec. 363/1 The juice with the milk of lime then passes into the first carbonators where it is treated with carbon dioxide gas.

carbonic, *a.* Add: **1. a. carbonic-acid snow** or **carbonic snow**: carbon-dioxide snow; **carbonic anhydrase** (see quot. 1949).

1841 [see Snow *sb.*¹ 4 b]. **1932** Meldrum & Roughton in *Jrnl. Physiol.* LXXV. 4 P, Recently we have succeeded in separating the catalytic system from hæmoglobin... For this new enzyme system we suggest the name Carbonic Anhydrase. **1940** *Nature* 3 Aug. 164/1 (*heading*) Sulphanilamide as a specific inhibitor of carbonic anhydrase. **1949** S. & L. M. Miall *New Dict. Chem.* (ed. 2) 109/2 Carbonic anhydrase is an enzyme which serves to accelerate the formation of carbon dioxide from bicarbonate in the blood vessels of the lung. **1964** S. Duke-Elder *Parsons' Dis. Eye* (ed. 14) xxi. 309 Further lowering of the tension may be obtained by the systemic administration of carbonic anhydrase inhibitors. **1905** Gould *Dict. New Med. Terms* 197/2 A bag filled with carbonic snow at a temperature 176° F. below zero is applied daily for half an hour to the pit of the stomach.

c. *carbonic paper*, early name of **carbon-paper* (*b*).

1808 S. Napier *Let.* 14 Nov. in H. A. Bruce *Life Gen. Sir William Napier* (1864) I. iii. 66, I am enabled to write this with my own hand upon carbonic paper, invented by Wedgwood for taking copies. **1876** Preece & Sivewright *Telegraphy* 289 The rest of the message..is transcribed on the public copy by carbonic paper. **1881** *Instr. Census Clerks* (1885) 82 Carbonic Paper Maker. **1884** W. Lynd *Pract. Telegraphist* 26 The carbonic paper must be changed sufficiently often to admit of a clear and distinct impression being made. **1895** *Army & Navy Co-op. Soc. Price List* 548 Carbonic Paper. For Tracing and Transfers, Black both sides.

carbonite (kā·ɹbǝnəit). [f. Carbon *sb.* + -ite¹.] (See quots.)

1891 Thorpe *Dict. Applied Chem.* II. 83/2 Carbonite is an explosive introduced by Hellhoff, of Berlin; stated to be composed of nitrobenzene, potassium nitrate, sulphur, and Kieselguhr. **1892** H. W. Hughes *Coal-mining* 79 Carbonite is another explosive of late introduction. It is said to be composed of nitro-glycerine, sulphur, and nitro-benzol. **1898** A. W. & Z. W. Daw *Blasting of Rock* xix. 204 H.M. Inspectors of Mines strongly condemn the use of blasting powder in coal mining..and they recommend that a high explosive practically free from flame, which property is claimed for carbonite, should be substituted. **1910** *Encycl. Brit.* VIII. 764/1 Carbonite contains 25% of nitroglycerin. **1922** *Ibid.* XXX. 86/2 Bombs of a type known as 'Carbonite'. **1931** Payman & Statham *Firedamp Explosions* viii. 123 The latter regulation cuts out the old 'Carbonite' class of explosives, which..appears to be the one class of explosive capable of causing the ignition of coal dust more rapidly than firedamp.

carbonium (kāɹbōᵘ·niǒm). *Chem.* [f. Carbo- + *-onium* as in Ammonium.] (*a*) Formerly, 'the univalent radical CH_3 in the capacity..of a base former analogous to ammonium' (Webster 1909); (*b*) used *attrib.* of an organic ion containing positively charged carbon. Cf. *Carbanion.

1902 *Jrnl. Chem. Soc.* LXXXII. i. 600 Triphenylmethyl... Carbonium Salts. **1905** *Ibid.* LXXXVIII. i. 282 Distinction is drawn between an ionisable and a non-ionisable valency; the former is termed a 'carbonium valency'. *Ibid.* 282 Carboxonium salts..are regarded as carbonium salts, corresponding with the salts of triphenylmethyl. **1933** [see *Carbanion]. **1936** *Nature* 8 Aug. 224/1 Ionic reactions..generally involve carbonium anions and cations. **1962** J. Hine *Physical Org. Chem.* (ed. 2) vii. 163 The carbonium ion is a very reactive intermediate that is rapidly transformed into the final product.

carbonizable (kā·ɹbǝnəi:zăb'l), *a.* [f. Carbonize *v.* + -able.] That can be carbonized.

1883 *Brit. Pat.* 5978, Carbonisable material in a plastic or semi-liquid state.

carbonize, *v.* Add: **4.** To destroy vegetable impurities in (wool, etc.) by treatment with an acid and heat, which reduces the vegetable matter to carbon dust. Hence **ca·rbonizer,** an operative who performs or assists in the operation.

1892 *Sci. Amer.* 6 Feb. 84/3 Washing..follows, by which the chloride is removed and the residue of carbonized matter washed away. **1920** *Glasgow Herald* 6 July 7 Australian wool...There will be a considerable surplus of low and medium cross-breds and carbonising wool. **1921** *Dict. Occup. Terms* (1927) § 361 *Piece carboniser*; carbonises woollen fabrics in the piece. **1964** *Times Rev. Industry* Mar. 45/1 Carbonising has become the almost universally recognised method of burr removal.

carbonizer (kā·ɹbǝnəizǝɹ). [f. Carbonize *v.* + -er¹.] One who or that which carbonizes. See also prec.

1908 *Daily Chron.* 24 Aug. 9/6 (Advt.), Foreman (Working) wanted for gas works; must be a good carboniser. **1912** G. Martin *Industr. & Manuf. Chem., Org.* 392 Continuous carbonisers, such as Woodall-Duckham and Glover-West, have also been tried. **1921** *Dict. Occup. Terms* (1927) § 693 *Carboniser, carbonising man*..general terms for men charging retorts or furnaces either by hand or machine.

carbonizing, *vbl. sb.* and *ppl. a.* (Further examples.)

1898 *Westm. Gaz.* 17 Feb. 8/3 The company..pays 3s. 6d. per ton of coal for carbonising wages. **1920** [see *Carbonize *v.* 4]. **1921** [see prec.]. **1955** *Times* 5 July 5/1 The carbonizing industries had been allocated 3,000,000 tons more than dispatches in the summer of last year. This should produce nearly 1,500,000 tons more coke. **1964** [see *Carbonize *v.* 4].

carbonnade: see *CARBONADE *sb.* 2.

carborundum (kābǫrʌ·ndǝm). orig. *U.S.* [Trade-name, f. CARBO(N *sb.* + CO)RUNDUM.] A compound of carbon and silicon, SiC, a very hard crystalline substance, used either as a powder or in blocks for polishing and scouring, for grinding tools and as a refractory lining in furnaces. Also *attrib.*

1892 *Official Gaz. U.S. Patents* LIX. 1014/1 [Trade-mark registered by] The Carborundum Company, Monongahela City, Pa. *Ibid.*, Trade-marks registered June 21, 1892 [include] Carborundum. 1893 *Amer. Jrnl. Sci.* XLVI. 472 While examining the hardness of 'carborundum', a carbide of silicon, made by Mr. Acheson of Pittsburg, it was found that it readily scratched red, blue, white, pink, and yellow corundum in the form of fine grains. [illegible] 1917 *Mod. Shoe & Shoe Maker* [illegible] 1924 *Chambers's Jrnl.* 39/1 The cutting edge is bestowed by grinding on a carborundum wheel. 1963 J. OSBORNE *Dental Mechanics* (ed. 5) xv. 343 The sprues must next be removed with either a diamond disc or a carborundum-type 'cut-off' disc.

carboxy- (kāɹbǫ·ksi), comb. form of CARBOXYL; as in **carboxy-hæmoglobin**, a combination of carbon monoxide and hæmoglobin formed in the blood when carbon monoxide is inhaled; **carboxy-(poly)peptidase**, an enzyme (e.g. in the pancreas) that hydrolyses peptides and polypeptides by removing the amino acids containing free carboxyl groups.

1891 *Jrnl. Chem. Soc.* LX. 1512 Conversion of carboxy-hæmoglobin into methæmoglobin, and detection of carbonic oxide in blood. 1908 *Practitioner* Dec. 843 Carbon-monoxide has a marked affinity for the colouring matter of the blood, and forms carboxy-hæmoglobin, the chief characteristic of which is stability. 1962 *Lancet* 13 Jan. 68/1 Each animal was gassed until the blood-level of carboxyhaemoglobin was 70%. 1930 *Chem. Abstr.* 1129 Proteinase and carboxy-polypeptidase of the pancreas. 1935 *Science* LXXXI. 467/1 Carboxypolypeptidase splits the amide linkages of certain amino-acid compounds..with the liberation in each case of an amino-acid which in the intact compound has a free carboxyl group. 1940 *Chem. Abstr.* 1397 Ext[racts] of pancreas contg. carboxypeptidase were subjected to adsorption. 1955 *Sci. Amer.* July 78/2 The enzyme carboxypeptidase..removes amino acids from the ends of protein chains. 1956 *Nature* 17 Mar. 527/2 Treatment with carboxypeptidase..revealed leucine as the C-terminal amino-acid.

carboxylase (kāɹbǫ·ksilēˈs, -ēˈz). *Chem.* [G. (Neuberg and Karczag 1911, in *Biochem. Zeitschr.* XXXVI. 73), f. CARBOXYL + *-ASE.] An enzyme that breaks down ketonic acids into the corresponding aldehydes and carbon dioxide.

1911 *Jrnl. Chem. Soc.* C. II. 1020 These decompositions are not brought about when yeast which has previously been heated is employed... It is therefore due to an enzyme to which the name carboxylase is given. 1955 *Biochem. Jrnl.* LIV. 378/1 Carboxylase is a thiol enzyme. *Ibid.* 389/2 Highly purified carboxylase is sensitive towards thiol reagents such as heavy metals, trivalent arsenicals, [etc.].

carbuncly (kāɹbʌŋkli), *a.* [f. CARBUNCLE + -Y¹.] Of or resembling a carbuncle; bearing carbuncles.

1873 LELAND *Egypt. Sketch-Bk.* 74 Their combs and wattles became of a deep carbuncly red. 1922 JOYCE *Ulysses* 348 His carbuncly nose with the pimples on it.

carburant (kāɹbiurǎnt). [f. CARBUR(ET *v.* + -ANT¹.] A liquid or vaporized hydrocarbon used to carburet air or gas for the production of light or mechanical energy. Also **ca·rburant**.

1893 *Funk's Standard Dict.*, Carburetant. 1909 *Cent. Dict. Suppl.*, Carburant. 1928 *Daily Tel.* 4 Sept. 9/3 To find a carburant less inflammable than petrol.

carburation (kāɹbiurēˈʃǝn). [f. CARBUR(ET *v.* + -ATION.] The process of charging air with hydrocarbon in a finely divided liquid form, the resulting gas being burnt for the production of energy.

1896 B. DONKIN *Gas, Oil, & Air Engines* (ed. 2) II. xxi. 303 The carburation of the air takes place in the middle division. 1902 HARMSWORTH *Motors & Motor-Driving* 166 There may be a slight flare up, but this will assist the carburation. 1920 *Westm. Gaz.* 1 Apr. 4/1 Few improvements have been made to the various carburation systems. 1959 *Motor* 1 Apr. 319/2 Carburation is by means of a pair of..carburetters.

carburetted. Add: *carburetted air*, air which has been impregnated with fine particles of hydrocarbon, and which provides the power by which petrol engines are driven; *carburetted water-gas*, water-gas which has been enriched by mixing with a hydrocarbon gas.

1879 *Encycl. Brit.* X. 101/2 The efforts to introduce carburetted water-gas have been numerous and persistent. 1896 B. DONKIN *Gas, Oil, & Air Engines* (ed. 2) II. xxi. 298 Inflammable petroleum essence..is perhaps best distinguished by the term usually applied to it abroad—'carburetted air'. 1899 *Motor-Car World* Oct. 9/1 A 'carburetter'..[illegible] 'carburetted air'. [illegible]

carburettor, -etter. Add: **b.** (Also formerly **carburator**.) In petrol engines, the apparatus for impregnating air with fine particles of fuel and thus preparing the explosive mixture for the cylinders.

1896 *Jrnl. Soc. Arts* 27 Nov. 22/1 Where a vaporiser is employed, or a carburetter. 1898 tr. *L. Lockert's Petroleum Motor-cars* 145 The carburator, in which the petroleum essence..is vaporized. 1900 *Motor-Car World* Aug. 254 Supplemental exhaust tube for heating carburettor tank. 1902 HARMSWORTH *Motors & Motor-Driving* 166 Too rich a mixture may be caused by the pressure of too much petrol in the carburator. 1912 *Motor Man.* 12 To obtain an explosive mixture of paraffin and air a special form of carburetter is required. 1959 [see *CARBURATION].

carbyl (kāɹbil). *Chem.* Also **carbyle.** [f. CARB- + -YL.] **1.** [a. G. *carbyl*, in *carbylsulphat* (G. Magnus 1839, in *Ann. d. Physik und Chem.* XLVII. 512).] A name for the basic hydrocarbon radical derived from ethylene; little used except in **carbyl sulphate**, ethionic anhydride, $(—CH_2SO_3—)_2$, a white, deliquescent, crystalline substance.

1844 FOWNES *Chem.* 395 Carbon, hydrogen, and the elements of sulphuric acid. Sulphate of carbyle. 1890 *Watts' Dict. Chem.* I. 714 The compound formed by SO_3 with ethylene..is termed carbyl sulphate or ethionic anhydride. 1950 N. V. SIDGWICK *Chem. Elements* II. 903 With alcohol it [sc. sulphur trioxide] forms 'carbyl sulphate', which is..the anhydride of ethionic acid.

2. [a. F. *carbyl*, in *carbylamine* (A. É. J. Gautier 1867, in *Compt. Rend.* LXV. 902).] A divalent radical of carbon; little used except in **carbylamine** (kāɹbilæ·min, kāɹbilǎmīn), the isocyanide radical, —NC; also, any of the compounds (also called carbamines, isocyanides, isonitriles) consisting of this radical attached to a hydrocarbon radical, which are foul-smelling poisonous liquids; *spec.* phenyl isocyanide, C_6H_5NC; **carbylamine test**, a test for detecting primary amines by heating the substance with chloroform in a basic solution, the presence of an amine being indicated by the characteristic smell of an isocyanide.

1870 WATTS *Dict. Chem.* (ed. 2) VI. 522 In the one class..the alcohol-radicle is in direct union with the carbon; in the other, discovered independently by Hofmann and by Gautier in 1867, and designated as isocyanides, carbylamines, or carbamines, it is combined directly with the nitrogen. 1886 E. F. SMITH tr. *V. von Richter's Org. Chem.* 216 The carbylamine test of Hofmann for detection of primary amines. 1887 A. M. BROWN *Contrib. Anim. Alkaloids* 30 Heated with potash and chloroform, cadaverine does not give carbylamine. 1946 *Nature* 31 Aug. 295/1 [illegible] 1963 W. E. FLOOD *Orig. Chem. Names* 59/1 A comparison of the structure of a carbylamine with that of a simple amine shows that the carbylamine may be regarded as an amine in which the two hydrogen atoms of the amino-group are replaced by carbyl (i.e. a carbon 'radical').

carcajou. **1.** (Earlier examples.)

1703 tr. *La Hontan's Voy. N. Amer.* I. 62 The Holes or Dens of the Carcaious. *Ibid.* 132 Carcaious, an Animal not unlike a Badger. 1744 A. DOBBS *Hudson's Bay* 40 The Beavers have three enemies, man, otters, and the Carcajou [*sic*].

carcass, carcase, *sb.* Add: **6.** The framework of a cabinet or other piece of furniture. Also *attrib.*

1803 *Edinb. Cabinet Makers' Bk. of Prices* (1821) 10 A Library Book Case. Six feet six inches long, seven feet nine inches high, in six carcases. 1940 *Burlington Mag.* Sept. 91/2 The carcase-work and the doors are of soft pine. [illegible] Veneers are thin sheets of wood..which are used to cover a 'carcase' of commoner and cheaper wood.

carcass, *v.* Add: **2.** To make a carcass of.

1906 *Daily Chron.* 7 June 5/4 These animals realise only one-third as much as those fit for carcassing.

carcel. Add: **2.** A French unit of illumination, equal to the light emitted by a standard carcel lamp burning 42 grammes of colza-oil an hour and with a flame 40 millimetres in height. (*Disused.*)

1884 KNIGHT *Dict. Mech. Suppl.*, *Carcel-bec*, the French unit of illumination. Given the preference over candle-power..at the Electrical Congress, in Paris, 1881. 1884 F. KROHN tr. *Glaser de Cew's Magn. & Dyn.-Electr. Mach.* 119 The expenditure of work per second per Carcel-burner (7·4 candle power)...An illuminating power of 1·40 Carcels (10·36 candles). 1902 *Encycl. Brit.* XXX. 260/2 The values are expressed in carcels (9·5 candles) per square centimetre of the mean horizontal focal plane of the luminous source.

carceral, *a.* Delete †*Obs.* and add later examples.

1909 *Westm. Gaz.* 5 Jan. 2/1 Any punishment, carceral or otherwise. 1960 V. NABOKOV *Invit. to Beheading* xii. 191 The [prison] door opened, whining, rattling and groaning in keeping with the rules of carceral counterpoint.

carcinogen (kāɹsi·nǝdʒěn). *Path.* [f. CARCINO(MA + -GEN 1; or back-formation from *CARCINOGENIC *a.*] A substance or agent that produces cancer.

1936 *Nature* 10 Oct. 651/2 This result affords some support for the suggestion that the 'substrates' upon which carcinogens act are protein in nature. [illegible] Carcinogens can excite oestrus and oestrin can have carcinogenic effects. 1955 *Sci. News* XXXVI. 14 Continued treatment with a true carcinogen..speeds up tumour formation.

carcinogenesis (kāːsinodʒe·nėsis). *Path.* [f. CARCINO(MA + -GENESIS.] The production of cancer in living tissues. Also, (the study of) the changes that occur in tissues or cells during the origin of cancer.

1926 in *Gould's Med. Dict.* (ed. 8) 279/1. 1930 *Lancet* 21 June 1334/1 (*heading*) Experimental carcinogenesis. 1962 R. J. LUDFORD in G. H. Bourne *Cytology* (ed. 2) ix. 407 Such cases render it improbable that cellular differentiation can be an essential factor in carcinogenesis.

carcinogenic (kāːsinodʒe·nik), *a.* [f. CARCINO(MA + *-GENIC.] Cancer-producing.

1926 *Jrnl. Cancer Res.* X. 229 Separation of the crude coal tar into its various substances has disproved the presence of a specific carcinogenic agent. [illegible] 1957 *Times* (Agric. Suppl.) 2 Dec. p. vi/4 Fear of the carcinogenic effects of oestrogenic hormones.

Hence **carcinogeni·city**, carcinogenic quality.

1936 *Jrnl. Path. & Bacteriol.* XXXIII. 134 We had practically detoxicated the tar as far as its carcinogenicity is concerned. 1955 *Times* 13 July 4/5 Cigarette tobacco extracts from the butts of naturally smoked cigarettes have also proved inactive in tests for carcinogenicity. 1962 *Lancet* 13 Jan. 101/1 Benzidine, because of its carcinogenicity, must not be used as a routine reagent.

carcinoma. **1.** Now usually restricted to a malignant tumour of epithelial origin (the general term being CANCER).

carcinomatosis (kāːsinǫ·mǝtōˈ·sis). *Path.* [mod.L., f. L. *carcinomat-* (see CARCINOMATOUS *a.*) + -OSIS.] Widespread dissemination of carcinoma throughout the body; applied also to a more limited spread when it is substantial in amount but diffuse, without the usual separate nodules.

1903 GOULD *Med. Dict. Suppl.* 1531. 1964 S. DUKE-ELDER *Parsons' Dis. Eye* (ed. 14) xxv. 372 The patient is generally in the stage of general carcinomatosis.

card, *sb.*² Add: **2. b.** *sure card* (earlier and later examples); *to have a card* (or *cards*) *up one's sleeve*: to have a plan, resources, etc., in reserve.

c1560 *Enterlude Called Thersytes* (1848) 87 Nowe thys is a sure carde, nowe I maye well saye. 1836 DICKENS *Let.* [illegible] Harley wrote, when he had read the whole of the opera, saying 'It's a sure card—nothing wrong there'. 1872 J. P. KENNEDY *Quodlibet* (ed. 3) i. 40 Consider me a sure card in that line. 1898 *Tit-Bits* 23 July 328/3, I took the precaution of having what, I believe, you sporting men call a card up my sleeve. 1907 E. PHILLIPS OPPENHEIM *Secret* xxxvi. 213 If Guest has yet another card up his sleeve, he has kept it secret from me. 1933 'G. ORWELL' *Down & Out* vi. 38, I have got a few cards up my sleeve. There are people who owe me money, for instance. 1953 R. LEHMANN *Echoing Grove* 206 [illegible]

c. Also without adjective: a 'character', an 'original'; a clever, audacious, etc., person.

1853 DICKENS *Bleak Ho.* xiii. 196 You know what a card Krook was for buying all manner of old things. 1905 A. BENNETT *Tales of Five Towns* 1 It would be a topic for years, the crown of his reputation as a card. 1911 — (*title*) The Card—a Story of Adventure in the Five Towns. 1929 W. DEEPING *Roper's Row* xxi. [illegible] 1947 W. B. JOHNSON *Widening Stain* (1943) iii. 34 That old Witch-Hammer was really quite a card.

d. (Later examples.) *to lay, put* (or *play with*) (*all*) *one's cards on the table*: 'to show one's cards', to reveal one's resources.

1753 [see PLAY *v.* 22 b]. 1801 M. EDGEWORTH *Belinda* I. i. 24 A man of gallantry..who was known to play his cards well, and to have good luck whenever hearts were trumps. 1848 MRS. GASKELL *M. Barton* I. vi. 156 Come, cards on table; was it true or false? 1871 W. S. GILBERT *Palace of Truth* II. p. 29 One who shows her cards so candidly. 1907 E. PHILLIPS OPPENHEIM *Secret* viii. 55, I began to think that I had been rash to lay my cards upon the table. 1914 SHAW *Fanny's First Play* III, Tramps are often shameless; but theyre never sincere. Swells—if I may use that convenient name for the upper classes—play much more with their cards on the table. 1923 J. M. MURRY *Pencillings* 195 He seems to put his cards on the table and to be saying in the friendliest way: 'That's my opinion. What's yours?' 1925

W. DEEPING *Sorrell & Son* xvi. 155 There is no reason why we should not put all our cards on the table. 1930 W. S. MAUGHAM *Cakes & Ale* vii. 85 If you play your cards right you ought to marry well. [illegible]

5. b. *transf.* Anything having the form of a card. *U.S.*

1823 J. F. COOPER *Pioneers* ix. With 'cards of gingerbread'. 1853 J. G. BALDWIN *Flush Times Alabama* 103 He distributed..a plug of tobacco there, or a card of town gingerbread to the little snow-balls. 1884 *Phin Dict. Apiculture* 20 Card, a frame filled with honeycomb. A sheet of honeycomb.

6. c. *spec.* in *Cricket*, a score-card.

1849 *Punch* XVII. 12/1 Paid..6*d.* for a Card of the Innings. 1903 A. D. TAYLOR *Annals of Lord's* 111 The printing tent was introduced at Lord's, on June 26th, 1848, and the public, for the first time, could secure a 'correct card' of the game.

f. A card held by a delegate at a trade union meeting or congress and representing a certain number of the members of his union. Cf. *card vote.

1909 *35th Ann. Rep. Trades Union Congress* 69 A vote by card was then demanded, and resulted in the Committee being supported by 790,000 to 375,000. *Ibid.* 85 The method of voting shall be by card, to be issued to the delegates of trade societies according to their membership, and paid for on the principle of one card for every 1,000 members or fractional part thereof represented.

g. *pl.* An employee's documents (e.g. national-insurance card, etc.) held by an employer and returned when employment ceases; hence in various phrases alluding to dismissal or resignation from employment. *colloq.*

1950 H. GREEN *Living* vii. 103 Arthur Bridges was sending for his cards after 34 years' work. 1958 M. DICKENS *Man Overboard* xi. 176 You're going to go and ask for your cards this morning, I suppose. 1958 'A. GILBERT' *Death against Clock* 93 Wouldn't surprise me to know he'd helped himself from the till, and that's why they gave him his cards.

7. (Earlier and later examples.)

1769 *Boston News-Letter* 2 Feb. (Th.), A Card from the London and British Merchants to the American Merchants. 1788 *Gen. Even. Post* (London) 1–3 Jan. 2/3 A Card. Dr. Norris..desires to inform the public [etc.]. 1931 G. T. CLARK *Leland Stanford* viii. 305 That there might be no misunderstanding as to where they stood they published the following card over the signature of James Anthony and Company. 1946 *Bristol* (New Hampshire) *Enterprise* 15 Feb. 4/1 (Advt.), Births, marriage and death notices inserted free. Card of thanks, $1.00.

13. (sense 1 a) *card-dealer, -game, -play, -room* (earlier examples), *-sense*; (sense 6 c) *card-basket* (examples), *-exchanging* adj., *-leaving, -plate.*

1841 A. BACHE *Fire-Screen* viii. To this treasury does the tasteful disposer of drawing-room decorations consign her accumulated card-baskets. 1872 C. M. YONGE *P's & Q's* vi. 58 She began to search in the card basket. 1902 O. WISTER *Virginian* iii. 31 Into my thoughts broke the voice of that card-dealer. 1899 *Westm. Gaz.* 26 May 10/1 The Hague, in the opening days of the Peace Conference, was simply a mass of card-exchanging foreigners. [illegible] Other principal card-games of the period were hombre, noddy, bankerout, and post-and-pair. 1853 E. RUSKIN *Let.* 10 Nov. in M. Lutyens *Millais & Ruskins* (1967) 113 Continual letter-writing and note-sending and card-leaving on every body in and around Edinburgh. 1900 *Manners & Rules of Good Society* iii. 16 The etiquette of card-leaving is a privilege which society places in the hands of ladies. [illegible] 1931 *Essays & Studies* XVI. 185 The terms of card-play. [illegible] I'll go this Moment into the Card-Room, and watch whom she whispers with. 1930 W. S. MAUGHAM *Cakes & Ale* ix. 106 I did not play well, but Mrs. Driffield had a natural card sense.

14. **card-carrying** *a.*, having a membership card of a specified organization, esp. of the Communist Party; so **card-carrier**; **card-edge gilder**, a man who, or machine which, gilds the edges of cards; **card-holder**, one who possesses a membership-card of a certain organization; a member; **card-index** (orig. *U.S.*), an index in which each item is entered on a separate card; a card-catalogue; also *transf.* and *fig.*; hence **card-index** *v. trans.* and *intr.*, to make a card-index (of); also *fig.*; so **card-indexed** *ppl. adj.*, **card-indexing** *vbl. sb.*; **card-rack** (earlier examples); **card-sharp** (orig. *U.S.*), a card-sharper; **card vote**, in trade union meetings, etc., a method of voting by which the vote of each delegate counts for the number of his constituents (cf. sense 6 f above).

1966 'W. HAGGARD' *Power House* iii. 27 His wife was a communist...She wasn't a card-carrier. 1948 B. ANDREWS *Washington Witch Hunt* ii. 96 The most dangerous Communists in the nation today are not the open, avowed, card-carrying party members. 1953 *Wall St. Jrnl.* 18 Aug. 2/3 'Millions of workers who are just card-carrying members' and who take no interest in union affairs. 1902 *Westm. Gaz.* 1 Jan. 1/7 Book Card-edge Gilders. 1921 *Dict. Occup. Terms* (1927) § 538 *Card edge gilder*;..places cards in press after they are trimmed..; brushes them over with size, lays on gold leaf and allows it to dry; burnishes gold leaf by hand with agate or bloodstone burnishing tool. 1934 WEBSTER, *Cardholder*, a person who has a card issued by a library entitling him to borrow books. 1955 *Times* 17 June 7/3 A fellow traveller is not a card-holder. 1970 *Daily Tel.* 25 Apr. 17/3 Individual card holders may find their credit ceiling upgraded and find Barclaycard less pressing in future. 1850 *Rep. Comm.* [illegible] the card index formed [illegible] the card in the slit of the [illegible] *Engineering Mag.* XIV. [illegible] items for card-index purposes. [illegible] vi. 67 She had touched [illegible] index system, known as his [illegible] 130/1 All the vast store [illegible] up in that marvelous [illegible] almost as if his mind were [illegible] *Purchasing & Storing* 16 Card [illegible] will make them available [illegible] 1920 ROSE MACAULAY *Potterism* I. iv, [illegible] office, card-indexing, [illegible] *Sunday Dispatch* 13 Jan. [illegible] possible to 'card-index' my [illegible] 1927 *Daily Express* [illegible] example of the 'card-[illegible] 1924 KIPLING *Debits & Credits* [illegible] drawer of card-indexed [illegible] *Business* Dec. 544/2 Card [illegible] an essential requirement [illegible] *Pennsylvania Packet* 11 Dec. 3/2 [illegible] *Lett. Miss Riversdale* III. [illegible] her card rack, I don't [illegible] HARTE *On Frontier* 273 [illegible] 1963 *Punch* 3 July 33/3 [illegible] thief and cardsharp. [illegible] *Congress* 74 Should a [illegible] course be taken. 1924 [illegible] Should its [*sc.* the [illegible] a card vote after the [illegible] Congress? 1966 *Listener* [illegible] different forms of voting [illegible] the card vote—and [illegible] down.

card, *v.*¹ Add: **3. b.** To write (something) on a card, e.g. for use in a card-index. Also const. *up*.

1925 MAWER & STENTON *Place-Names Bucks.* p. v, Mrs. Stenton..[has given invaluable] help..in the work of carding up and identifying the place-names. 1934 J. E. [illegible] I. p. vi, To supplement these we must collect and 'card' our own material. 1968 *Times Lit. Suppl.* 30 May 559/4 This isn't to say..that other readers may not have found and carded them [*sc.* words] earlier.

cardan (kā·ɹdǎn). The name of *Cardan* (Geronimo Cardano, 1501–1576, Italian mathematician), used *attrib.* in *cardan joint*, a universal joint, a joint permitting free motion of the different parts of the mechanism; *cardan shaft*, a shaft having a universal joint at one end or at both ends for transmitting motion from one shaft to another not in a direct line with it. Also *absol.*

1902 HARMSWORTH *Motors & Motor-Driving* 191 In order to permit of the free vertical movement of the wheels under the springs, two universal or 'Cardan' joints..are fitted within the length of the shaft. 1906 *Daily Chron.* 14 July 6/4 The cardan shaft, which connects the gear-box with the differential gear that works the driving wheel. 1906 *Westm. Gaz.* 2 Oct. 7/1 While oiling his motor as it was running, M. Santos-Dumont caught the sleeve of his leather jacket in the cardan. 1907 *Ibid.* 13 Nov. 9/2 The 15-h.p. is the only model having a cardan drive. 1908 *Ibid.* 16 Apr. 4/2 A very short cardan-jointed shaft. 1951 *Engineering* 22 June 752/2 A twin-motor cardan-shaft drive.

cardanic, *a.* Delete †*Obs.* and add: *cardanic suspension*, a form of support in which an instrument is hung on gimbals, so as to allow free movement in all directions.

1892 *Review of Reviews* July 718/2 A highly sensitive vibrator, with cardanic suspension.

cardboard. Add: (Earlier example.)

1848 A. BRONTË *Tenant Wildf. Hall* I. xviii. 325 The pencil..leaves an impression upon card-board that no amount of rubbing can efface.

b. *fig.* (*attrib.* or as *adj.*) Unsubstantial, unreal, 'pasteboard'.

1893 *Jrnl. Soc. Arts* XLI. 476/1 When his cardboard empire of the East fell to pieces. 1927 J. DEVAL (*title*) Her Cardboard Lover. 1955 *Illustrated* 6 Dec. 8 The cardboard family that has become larger than life.

cardboardy (kā·ɹdbōˈɹdi), *a.* [f. CARDBOARD + -Y¹.] Resembling cardboard; also *fig.*

1925 D. H. LAWRENCE *Refl. Death of Porcupine* 233 All you can do now is to twiddle-twaddle about golden boughs, because you are empty..hollow, deficient, and cardboardy. [illegible] *Country* 37 A huge piece of pastry..that had gone slightly cardboardy from being kept in the fridge. 1966 H. R. F. KEATING *Inspector Ghote hunts Peacock* i. 8 The big, light brown, rather cardboardy—no, very cardboardy—suitcase. 1969 *Listener* 6 Mar. 324/1 Mrs Spark probably intended Miss Brodie to be cardboardy in her passions.

carded. Add: **4.** Entered on a programme card, score-card, filing-card, etc.

1900 *Daily News* 18 Aug. 7/3 The next race carded was for the Rear-Commodore's Cup. 1927 *Daily Express* 19 Oct. 3/4 Nine times out of ten the carded distance is wrong.

cardinal, *a.* Add: **1. b.** *cardinal vowel*: one of a series of vowel-sounds proposed as a standard for phonetics.

1922 D. JONES *Outl. Eng. Phonetics* (ed. 2) p. ii, By referring the vowels of all languages to a set of Cardinal Vowels, the relations between them are made clearer. 1932 *Ibid.* (ed. 3) vii. 28 There is only one way of making written descriptions of vowels intelligible to a large circle of foreign readers, and that is to describe the sounds with reference to a scale of 'Cardinal Vowels', i.e. a set of fixed vowel-sounds having known acoustic qualities and known tongue and lip positions. 1943 K. L. PIKE *Phonetics* vii. 137 Jones's 'cardinal vowel' scheme is perhaps the most convenient objective classification.

cardinal, *sb.* Add: **1. d.** Either of two (Senior and Junior) of the minor canons of St. Paul's Cathedral (see quot. 1868).

1748 CHAMBERLAYNE *Pres. St. Gt. Brit.* (ed. 37) II. III. 231 12 Minor Canons [of St. Paul's Cathedral]...The Rev. Mr. William Rayner, Subdean, sen. The Rev. Mr. Dechair sen. Cardinal, The Rev. Mr. Hilman jun. Cardinal, [etc.]. 1868 WALCOTT *Sacred Archæol.* 113 Cardinal. The word, when applied to an altar, means the high or principal altar, and from their attendance upon it two minor canons at St. Paul's are still called the senior and junior cardinals. 1877 J. D. CHAMBERS *Divine Worship* 90 At S. Paul's, London..there were four Vicars called Cardinals.

cardinalatial (kāːɹdinǎlēˈ·ʃiǎl), *a.* = CARDINALITIAL *a.*

1888 *Scottish Leader* 1 June 3/7 Two names..are being considered by Leo XIII for elevation to the Cardinalatial dignity. 1961 *Times* 31 July 17/3 The Pope..gave them [*sc.* two cardinals] the cardinalatial privileges of preaching or hearing confessions anywhere in the world.

cardinality. Restrict †*Obs.* to sense in Dict. and add: **2.** *Math.* The property of having a certain cardinal number.

1935 *Mind* XLIV. 343 We may imagine set out before us all the examples, of whatever cardinality, involving a single dyadic relation. 1959 K. R. POPPER *Logic Sci. Discovery* vi. 114 The concept of the cardinality (or power) of a class. 1966 *McGraw-Hill Encycl. Sci. & Technol.* XII. 205 An ordering is one way of setting up a 1–1 correspondence between two sets of the same cardinality.

cardinalize (kā·ɹdinǎlaiz), *v.* Restrict †*Obs.* to sense 2 in Dict. and add later example of sense 1.

1921 S. LESLIE *Manning* xiii. 196 Meantime Cullen had been Cardinalised.

carding, *vbl. sb.*¹ **3.** **carding-machine** (examples).

1788 *Salem Mercury* 25 Nov., A Providence paper informs that the Carding and Spinning Machines used in England in manufacturing cotton stuffs, are introducing into that town by some publick spirited gentlemen. 1831 J. KENNEDY in *Mem. Lit. & Phil. Soc. Manchester* V. 321 In 1784 or 5 he [*sc.* Samuel Crompton] made a carding machine, the working of which was a little different from those in common use.

cardio-. Add: **ca:rdioblast**, in insect embryology, one of a row of mesodermal cells from which the heart develops; **ca:rdio-diaphragma·tic** *a.*, pertaining to the heart and the diaphragm; **ca·rdiogram**, the tracing made by a cardiograph or electrocardiograph; **ca:rdio·grapher**, one who uses a cardiograph or electrocardiograph; esp. a technician who has received training in (electro)cardiography; **ca:rdiogra·phic** *a.*, of or pertaining to a cardiograph; hence **ca:rdiogra·phically** adv.; **ca:rdio-inhi·bitory** *a.*, checking or arresting the heart's action; **ca:rdioli·pin**, a substance extracted from beef hearts, capable of acting as an artificial antigen in serum tests for syphilis; **ca:rdiolo·gical** *a.*, of or pertaining to cardiology; **cardio·logist**, one who specializes in the study or treatment of the heart and its diseases; **cardio·lysis**, resection of portions of the cardiac area in order to free the heart from surrounding structures to which it has become adherent; **ca:rdiomo·tor** *a.*, pertaining to the action of the heart; **ca:rdiopto·sis**, displacement of the heart downwards; **ca:rdio-pu·lmonary** *a.*, pertaining to the heart and the lungs; **ca:rdio-re·nal** *a.*, pertaining to the heart and the kidneys; **ca:rdio-respi·ratory** *a.*, relating to the action of both heart and lungs; **ca:rdio-sclero·sis**, induration of the tissues of the heart; **ca·rdioscope** (see quot. 1890); **ca:rdio-spasm**, spasmodic contraction of the cardiac sphincter of the stomach; **ca:rdiotacho·meter**, an instrument for measuring the rate at which the heart beats; **cardio·tomy**, dissection or incision of the heart; **ca:rdioto·nic** *a.*, serving to invigorate the heart; **ca:rdio-va·scular** *a.*, relating to both the heart and the blood-vessels.

1898 A. S. PACKARD *Text-Bk. Ent.* II. 572 A long string or row of cells (*cardioblasts*), which on each side border the mesodermal layer of the primitive band. 1907 *Practitioner* Apr. 528 A smaller movement of the right cardio-diaphragmatic angle. 1876 *Nature* 13 Apr. 471/2 More than one

physiologist has obtained far more satisfactory 'cardiograms' by applying the sphygmograph. 1888 *Buck's Handbk. Med. Sci.* VI. 95/2 The Cardiogram.—A graphic record of the impulse of the heart against the chest-wall can be obtained by means of sphygmographs constructed on the principle of Marey's; or by modifications of the same for their better application to the thorax, as Galabin's cardiograph. 1892 A. E. SANSOM *Dis. Heart* 493 Dr. Hercules Macdonald obtained cardiograms from the posterior aspect of the heart. 1964 *Whitley Councils for Health Services P.T.B. Circular* No. 174, 2 A Cardiographer is an officer who has completed..appropriate..training..and is employed on electrocardiography. 1884 B. BRAMWELL *Dis. Heart* 751 Cardiographic tracings are usually obtained from the pulsations of the left apex-beat. 1907 *Practitioner* Nov. 610 Both cardiographically and clinically, its duration is well defined. 1884 B. BRAMWELL *Dis. Heart* 31 The cardio-inhibitory centre in the medulla. 1907 *Practitioner* Nov. 693 The conservative influence of cardio-inhibitory action. 1942 M. C. PANGBORN in *Jrnl. Biol. Chem.* CXLIII. 247 It is proposed to designate this substance 'cardiolipin'. 1946 *Nature* 27 July 119/2 Almost as active preparations could be obtained with cardiolipin. 1922 *Encycl. Brit.* XXXI. 350/2 Another aspect of the cardiological problem..is the so-called 'nervous or irritable heart'. 1927 *Daily Tel.* 21 June 15 The cardiological department contains an electrocardiograph. 1885 *Lancet* 26 Sept. 576/1 Being near the great cardiologist, he thought he would hear what he had to say. 1965 *Math. in Biol. & Med.* (*Med. Res. Council*) I. 34 He claims that the computer can diagnose correctly as often as three experienced cardiologists using the same clinical data. 1903 *Lancet* 18 July 188/2 Dr. Brauer performed an operation which has been called cardiolysis and which consists in resecting some costal cartilages and a part of the sternum so that the heart would have more freedom of movement. 1918 F. W. PRICE *Dis. Heart* 346 The advisability of performing the operation of 'cardiolysis' in some cases of severe adhesions between the pericardium and the chest wall. 1910 J. MACKENZIE *Dis. Heart* (ed. 2) 399/1 Cardio-motor centres. 1924 R. W. G. HINGSTON in E. F. Norton *Fight for Everest: 1924* (1925) 250 The cardiomotor mechanism of the vigorous man at a height of 16,000 feet closely resembles that of the physically inefficient person at sea-level. 1905 *Medical Annual* 185 Palpitations, attacks of vertigo, and occasionally inability to lie on the left side, are the principal symptoms...Cardioptosis does not endanger life. 1918 F. W. PRICE *Dis. Heart* 24 The apex-beat..may be displaced downwards in cardioptosis, emphysema, and pneumothorax. 1908 *Practitioner* Mar. 324 Such murmurs are louder during inspiration, and are literally cardio-pulmonary, the pulmonary artery being compressed by the flattened chest-wall through the medium of the lung. 1959 *Times* 14 Oct. 14/5 A cardio-pulmonary research centre. 1913 *Jrnl. Chem. Soc.* CIV. I. 1022 Respiration and Metabolism in Cardio-renal Disease. 1892 OSLER *Princ. Med.* 227 The so-called cardio-respiratory murmur. 1908 J. MACKENZIE *Dis. Heart* 26 A cardio-respiratory reflex. 1964 L. MARTIN *Clinical Endocrinol.* (ed. 4) ii. 61 A cardio-respiratory syndrome. 1894 GOULD *Dict. Med.*, Cardio-sclerosis. 1908 J. MACKENZIE *Dis. Heart* p. xviii, The term cardio-sclerosis, unless qualified, will always refer to the group with arterial degeneration. 1890 BILLINGS *Nat. Med. Dict.* I. 230/2 *Cardioscope*, instrument for examining the movements of the heart. 1900 DORLAND *Med. Dict.* 135/1 Cardiospasm. 1914, 1939 Cardiospasm [see *ACHALASIA]. 1928 *Arch. Internal Med.* XLI. 403 The cardiotachometer, an instrument to count the totality of heart beats over long periods of time. 1956 *Newsweek* 26 Nov. 104/2 The cardiotachometer..detects heart stoppage on the operating table seconds before it is recognizable to the surgeon. 1922 *Encycl. Brit.* XXXI. 348/1 Doyen attempted cardiotomy on a patient believed to be suffering from mitral stenosis. 1936 *Nature* 19 Sept. 484/1 None of the recognized cardiatonic [*sic*] drugs seems to have been used in China at the date, A.D. 1596, when this list was compiled. 1949 *Blakiston's New Gould Med. Dict.* 182/1 Cardiotonic. 1879 *St. George's Hosp. Rep.* IX. 798 Cardio-vascular pulsation. 1909 *Practitioner* Nov. 617 The cardio-vascular hypertrophy. 1918 F. W. PRICE *Dis. Heart* 79 Those articles of food which especially stimulate the cardio-vascular system. 1962 *Lancet* 29 Dec. 1386/2 Deaths should be assigned to the cardiovascular or cerebrovascular class of disease.

cardol (kā·ɹdǫl, -ōᵘl). *Chem.* Also 9 **-ole.** [a. G. *cardol* (G. Städeler 1847, in *Ann. d. Chem. und Pharm.* LXIII. 141), f. mod.L. *anacard-ium*, generic name of the cashew tree + -OL 3.] A vesicatory oil, $C_{15}H_{27}\cdot C_6H_3(OH)_2$, obtained from the pericarp of the cashew-nut.

1848 G. STÄDELER in *Chem. Gaz.* 15 Jan. 30, I propose for this fatty acid the name of anacardic acid, and for the active constituent that of cardol. *Ibid.* 1 Feb. 61 The medicinal effects of pure cardol far exceed those of the mixture extracted from the pericarp by æther. 1884 H. G. GREENISH tr. *Dragendorff's Plant Analysis* 146 Cardol is a colourless oil accompanying anacardic acid in the cashew nut. 1954 *Current Science* Aug. 265/1 A sample of cardol, isolated from Indian cashew nut shell liquid.

care, *sb.*¹ Add: **4. a.** *in care of* (U.S.): = *care of*.

1917 S. MERWIN *Temperamental Henry* 256 If you care to write a good-bye, address me in care of the ship. 1928 *N. & Q.* 4 Aug. 90 If owners of such letters will write to me in care of the Oxford University Press. 1967 *Boston Sunday Globe* 23 Apr. B. 63/7 Questions on home improvement and repair may be addressed to The House Doctor in care of this newspaper.

b. *to take care of*: to look after (see LOOK *v.* 12 f); to deal with, provide for, dispose of. (Further examples.)

a1724, etc. [see PENNY 9 h]. 1747 LD. CHESTERFIELD *Let.* 6 Nov. (1932) III. 1051, I recommend to you to take care of minutes; for hours will take care of themselves. 1762 BOSWELL *London Jrnl.* 31 Dec. (1950) 113 She mentioned one consequence that in an affair of gallantry might be troublesome...'Why, to be sure, if such a person should appear, he must be taken care of.' 1801 T. JEFFERSON *Let.* 18 Sept. in *Writings* (1897) VIII. 94 We are bound to take care of them. Could we not procure them as good berths as their former at least, in some of the custom-houses? 1816 JANE AUSTEN *Emma* II. xvi. 309 Young ladies should take care of themselves.—Young ladies are delicate plants. They should take care of their health and their complexion. 1861 MILL *Repr. Govt.* ii. 22 Things left to take care of themselves inevitably decay. 1925 *Ladies' Home Jrnl.* May 26/2 'Never mind, ma. I'll take care of them,' put in Edd. 1926 *Publishers' Weekly* 22 May 1709/2 Such an information desk should take care of all inquiries regarding books and reading. 1930 *Ibid.* 31 May 2728 The budget..was not sufficient to take care of the increasing expenses. 1932 *Atlantic Monthly* Apr. 421/2 Money..which would enable him to take care of all arrears on the property. 1931 H. F. PRINGLE *Theodore Roosevelt* I. xiv. 198 He..found himself besieged with petitions for jobs...It was impossible to take care of all of them. 1959 *Listener* 9 Apr. 628/1 That little problem had been taken care of.

c. (*in need of*) *care or* (also *and*) *protection*, a legal formula (see quot. 1933) used of a destitute or dangerously circumstanced child who is judged fit for official guardianship; also *attrib.*, as *care or* (*and*) *protection order*, one by which a magistrate effects such guardianship; *in care*, under official guardianship.

1932 *Act* 22 & 23 *Geo. V* c. 46 § 9 (*heading*) Juveniles in need of Care or Protection. 1933 *Act* 23 *Geo. V* c. 12 § 61 For the purposes of this Act a child or young person in need of care or protection means a person who is (a) a child..who, having no parent or guardian.., is either falling into bad associations, or exposed to moral danger, or beyond control; or (b) a child..in respect of whom any of the offences mentioned in..this Act has been committed; [etc.]. 1944 A. E. IKIN *Education Act 1944* 102 Children and young persons in need of care and protection. 1948 *Act* 11 & 12 *Geo. VI* c. 43 § 10 (*heading*) Duty of parents to maintain contact with local authorities having their children in care. 1969 *Listener* 15 May 666/1 The family..ends with the father in lodgings and the mother and children in a hostel (and frequently some children in care).

6. **care-and-maintenance** *attrib.*, describing a building, area, etc., maintained in good condition though not in present use; **care-committee**, a committee which charges itself with the care of the poor; **care-labelling**, the securing of labels on clothes and fabrics, giving advice about cleaning and ironing processes; so *care-label*.

1930 *Economist* 13 Sept. 475/1 Once estates are put on to a 'care and maintenance' basis, it will take a considerable increase in price to bring them back into production. 1958 *Times Rev. Industry* Aug. 40/2 Two former..sites (one of which had been kept on a partial care-and-maintenance basis) were reopened. 1908 *Times* 3 Feb. 11/5 He had been told to devote the whole of his time to assisting the Care Committee. 1909 M. FRERE *Children's Care Committees* 35 In 1902 the name [*sc.* Charitable Funds Committee] was changed to Children's Relief Committee, and finally, in 1908, to Children's Care Committee. 1931 W. HOLTBY *Poor Caroline* v. 157 He schooled his..sense of humour to docility in the face of care committees and church workers. 1967 *Spectator* 1 Dec. 682/1 The most recent spot check by *Which?* found that only two thirds of garments had care labels of any kind. 1962 *Rep. Comm. Consumer Protection* 39 in *Parl. Papers 1961–62* (Cmnd. 1781) XII. 317 The next topic we explored in relation to clothing and textiles was the very important subject of care-labelling.

care, *v.* **2. b.** Delete 'Now only with *for*', and add: Also in colloq. phrases expressing or implying lack of interest or concern: *for all I care, see if I care, who cares?*

1844 'J. SLICK' *High Life N.Y.* I. vi. 68, I didn't exactly like the feel of it, but 'Who cares', sez I to myself. 1856 *Harper's Mag.* Sept. 561/2 Who cares? 1931 *Maclean's Mag.* 15 Dec. 24/2 'I'll be killed,' the man gasped. 'Who cares?' was the brutal reply. 1934 J. HILTON *Goodbye Mr. Chips!* vii. 55 You can go to blazes for all I care. 1947 V. M. AXLINE *Play Therapy* 356 Fall on the floor, damn you! See if I care.

4. a. (*c*) Colloq. phr. (*I*, etc.) *couldn't care less*: (I am, etc.) completely uninterested, utterly indifferent; freq. as phr. used *attrib.* Hence *couldn't-care-less-ness.*

1946 A. PHELPS (*title*) 'I couldn't care less.' 1947 B. MARSHALL *Red Danube* vi. 53 The couldn't-care-less boys, the chaps who imagined that now that the war was over there was no need for further effort. 1947 *People* 22 June 2/4 If I suggest that it should be good because the book was by a top-line author she simply couldn't care less. 1955 *Essays in Criticism* V. 76 Exhibiting a vulgar couldn't-carelessness. 1957 F. H. KING *Man on Rock* iv. 120 The phrase he most used was 'I couldn't care less': which seemed to sum up his character. 1965 *Times Lit. Suppl.* 25 Nov. 1083/1 The couldn't-care-less attitude of people with little to lose.

cared (kēᵊɹd), *ppl. a.* [f. CARE *v.* + -ED¹.] In comb. with prep., as *cared-for*, attended to, looked after; liked.

1901 *Westm. Gaz.* 7 May 2/1 Never may we have to see any cared-for play therein produced. 1911 *Rep. Labour & Social Condit. Germany* VI–VII. 190 The..cared for appearance of the children. 1924 R. MACAULAY *Orphan Island* xviii, He..had a cared-for looking white skin.

careen, *v.* Add: **4. b.** *transf.* To lean over; to tilt.

1883 G. MEREDITH *Poems* 157 Now his huge bulk o'er Africa careened. 1895 CONRAD *Almayer's Folly* xii. 260 The big office desk, with one of its legs broken, careened over like the hull of a stranded ship. 1920 C. H. STAGG *High Speed* (ed. 2) ix. 152 A hundred times their throats choked as the car careened on a bank. *Ibid.* xiv. 251 The car should have careened until it almost turtled. *Ibid.* xv. 272 A sickening skid,..a careening lurch that brought a cry from Dan. 1938 *British Birds* XXXI. 333 The bird was careening from side to side as though there were waves.

5. [Influenced by CAREER *v.* 2.] To rush headlong, to hurtle, esp. with an unsteady motion. Chiefly *U.S.*

Quot. 1925 may belong to sense *4 b.

1923 [see *CRUISER 2]. 1925 T. DREISER *Amer. Tragedy* (1926) II. xxiii. 312 There came a contact..which set his thoughts careening in an entirely different direction. 1928 F. HURST *President is Born* xiii. 146 With terrible, terrifying, careening strides, that zigzagged crazily. *Ibid.* xxxi. 315 The tears jetted and careened down to her lips. 1940 *Amer. Speech* XV. 72 *Careen* of recent years has come to mean 'to rush headlong', or 'hurtle', doubtless because of its resemblance to *career*. 1957 H. ROOSENBURG *Walls came tumbling Down* iv. 91 A lot of Russians careening along the road on liberated bicycles. 1965 H. GOLD *Man who was not with It* (1965) v. 46 A shrill cry careened down the street.

career, *sb.* Add: **5. b.** Freq. *attrib.* (orig. *U.S.*), esp. (*a*) designating one who works permanently in the diplomatic service or other profession, opp. one who enters it at a high level from elsewhere; (*b*) *career girl, woman*, etc., one who works permanently in a profession, opp. one who ceases full-time work on marrying. Also, **careers master, mistress**, a schoolteacher who advises and helps pupils in choosing careers.

1927 *Lit. Digest* 25 June 14/2 The foundation of any sound Foreign Service must consist of 'career men' who have become expert. 1931 F. J. STIMSON *My U.S.* xviii. 190 The career professors look somewhat askance at one who comes in from the outside world—just as career secretaries in diplomacy do upon a chief who has not gone through all the grades. 1936 *Yale Rev.* XXV. 288 Other steps essential to a well-rounded career service remain to be taken... The prospect of permanent undersecretaryships for career men needs to be realized. 1937 *Collier's Wkly.* 26 June 20 (*heading*) Career girl. 1937 *Sat. Rev. Lit.* 9 Oct. 16/3 Most career women are different. 1943 *Assistant Masters' Year Book* 24 The Committee enables members who are careers masters to exchange information. 1947 *Daily Mail* 25 Aug. 2/3 (*heading*) Should the career woman make dates? 1951 R. HOGGART *Auden* vi. 200 The career-girl Rosetta yearns for her lush English landscape. 1954 F. P. KEYES *Royal Box* viii. 101 He might well have expected the offer of an embassy...It's only occasionally that they go to career diplomats like me. 1959 *Times* 15 Apr. 13/4 As careers mistress in a grammar school it is certainly not my habit. 1970 *New Yorker* 17 Oct. 167/1 Philip Habib, a competent, if unusually brusque, career diplomat.

careerism (kǎrīᵊ·riz'm). [f. CAREER *sb.* + -ISM.] The practice or policy of a careerist.

1933 RAHMAT ALI *Now or Never* (1934) 4 They [*sc.* Muslim leaders] have, time and again, sacrificed their own political principles and our national patrimony for the sake of sheer opportunism and sordid careerism. 1946 E. BENTLEY *Playwright as Thinker* 58 The striving of modern careerism with all its vast implications.

careerist (kǎrīᵊ·rist). [f. CAREER *sb.* + -IST.] A person (esp. a holder of a public or responsible position) who is mainly intent on the furtherance of his career, often in an unscrupulous manner. Also *attrib.* or as *adj.*

[1910 H. G. WELLS *Mr. Polly* vii. 225 He called him the 'chequered Careerist'.] 1917 *Times* 5 June 7/2 Half the present unpopularity of the 'lawyer-politician'..is due to the fact that he is too often a 'carpet-bagger' and a 'careerist'. 1926 S. JAMESON *Three Kingdoms* v. 153 I'm one of those damned careerist women. 1929 G. B. SHAW in *Times* 6 Aug., There were already..members of it [*sc.* the Labour party] who were 'careerists'—men who wanted to have a political career and joined the party they thought would give them the best prospects. 1934 *Punch* 21 Mar. 336/2 He states plainly that he was a 'careerist', power and money were in his hands and it is no wonder that he was dazzled by them. 1940 G. BARKER *Lament & Triumph* 15 The careerist politician and the vague thinker. 1969 *Daily Tel.* 4 Jan. 23/2 Accused..of being a 'double-dealer' and 'political careerist'.

carefree, care-free (kēᵊ·ɹfrī), *a.* [f. CARE *sb.*¹ + FREE *a.*] Free from care or anxiety. Hence **ca·refreedom, ca·refreeness**, the state or quality of being carefree.

1795 S. J. PRATT *Gleanings through Wales* III. lxiii. 121 As heart-whole, and care-free, as any man in the circle of Westphalia. 1854 B. TAYLOR *Life & Landscapes fr. Egypt* vii. 96 We..wander..as happy and care-free as two Adams in a Paradise without Eves. 1919 E. GRUBB *Relig. Exper.* xi. 133 This joyous, care-free, self-abandoning life. 1924 *Public Opinion* 21 Mar. 292/2 It makes for joyous indolence and carefreeness of heart. 1928 G. COOKE *Theory of Music* 75 The apparently carefree joyousness of individual performance. 1960 *House & Garden* May 79/1 A special kind of carefreedom in the furnishings.

careless, *a.* **4. c.** Add: Esp. in phr. *careless talk*, applied during the war of 1939–45 to talk which, if overheard, might assist the enemy.

1940 *War Illustr.* 16 Feb. 98/3 A Finnish soldier is reading a Government poster warning the people that careless talk may help the enemy. 1942 T. RATTIGAN *Flare Path* I. 95 Talk about careless talk! *Ibid.* 100 There must be some news, or are you going all careless talk on me?

caressable, *a.* Delete *rare*⁻¹ and add later examples. Also **caressible.**

1890 W. CORY *Let.* 21 Aug. (1897) 554 Anglican curates, invitable to garden-parties at Lambeth, caressible for ladies. 1900 M. CORELLI *Master-Christian* 343 A small, slight gossamer thing of beauty,..soft and caressable as a dove. 1969 R. WOLLHEIM *Family Romance* 148 The lovely caressable girl.

caressively (kǎre·sivli), *adv.* [f. CARESSIVE *a.* + -LY².] Caressingly.

1913 D. H. LAWRENCE *Sons & Lovers* i. 10 'Now do come and have this one wi' me,' he said caressively. 1922 — *England, my England* (1924) 113 Her gloomy black eyes softened caressively to me for a moment.

caretake (kēᵊ·ɹtēˈk), *v.* [Back-formation f. CARE-TAKER.] *trans.* To take charge of, watch over, and keep in order (a house, estate, business premises, etc.) in the absence of the owner or customary occupants. Also *absol.*

1893 *Cornhill Mag.* Nov. 507 Martha caretakes a decrepit City warehouse. 1904 *Daily Chron.* 14 Apr. 9/5 (Advt.), Housekeeper (Working): reliable; occasionally to caretake. 1921 *Glasgow Herald* 6 Aug. 6 They have to be electrical and mechanical engineers, and to 'care-take' the whole establishment, as well as to operate. 1960 H. PINTER *Caretaker* III. 81 I'll keep an eye on the place for you.., I'll caretake for you.

care-taker. Also **caretaker.** Add: **c.** *attrib.*, esp. designating a government, administration, etc., in office temporarily; = STOP-GAP 5.

[1885 J. CHAMBERLAIN in *Cobden Club Dinner, Special Rep.* 11 It is only upon those terms that what will be known in history as the 'Stop-gap' Government can invite the toleration of its opponents...I see no reason why they should not remain as caretakers on the premises—(great laughter and cheering)—until the new tenants are ready in November for a prolonged..occupation.] 1945 *Times* 26 May 5/2 This is evidently not 'a caretaker administration', but a carefully balanced team. 1945 *New Statesman* 2 June 346/3 There are already plenty of signs that the advent of the 'caretaker' Government means a serious setback to reconstruction. 1946 *News* (Birmingham, Ala.) 13 Nov. 20/4 A stopgap or caretaker operation will be established to carry on at the state's biggest college. 1963 *Ann. Reg. 1962* 370 Far from being a 'caretaker' Pope,..he was proving one of the most popular Popes.

care-taking, *vbl. sb.* and *ppl. a.* (Also written as one word.) Add later examples, esp. in the sense of taking charge of a house, etc.

1879 'E. GARRETT' *House by Works* I. 174 There will be none to send to market but that caretaking woman. 1883 A. THOMAS *Mod. Housewife* 139 To rely solely upon..'trusty' servants, or care-taking models. 1895 *Atlantic Monthly* July 61 The feeding and care-taking inseparable from life in the nest.

Carey Street (kēᵊ·ri,strīt). A street in London, formerly the location of the Bankruptcy Department of the Supreme Court, used allusively to indicate a state of bankruptcy.

1922 J. AGATE in *Sat. Rev.* 29 July 172/2 The melancholy gentleman in direful Carey Street. 1936 N. COLLINS *Trinity Town* v. 80 A man like Mr. Broster who might find himself in Carey Street at any moment.

cargador (kāɹgǎdōᵊ·ɹ). Pl. **cargadores.** [Sp.] In Spanish America: a porter (see also quot. 1892).

1811 J. BLACK tr. *Humboldt's Polit. Essay* I. 39 The existence of these valleys prevents the inhabitants from travelling except on horseback, afoot, or carried on the shoulders of Indians, (called *cargadores*). 1844 G. W. KENDALL *Santa Fé Exped.* II. xix. 362 Cargadores, with their leather trousers rolled up. 1889 K. MUNROE *Golden Days* iv. 36 Nearly naked *cargadores*, or native porters [bent] beneath heavy burdens. 1892 *Dialect Notes* I. 245 *Cargador*, the man in charge of the packs, in a pack train. 1904 CONRAD *Nostromo* viii. 81 A cargador would fly out head first and hands abroad. 1926 D. H. LAWRENCE *Plumed Serp.* xvi. 245 The cargadores were busy at the charcoal boats, carrying out the rough sacks. 1958 *Alaska Sportsman* Dec. 34/3 The 'cargodore' is Chinese, two of the packers are Tahitian natives and the cook..is Chinese.

cargo, *sb.*¹ Add: **3.** *cargo-fall, -hold, -man, -rice, -winch*; **cargo-cult** (see quots.).

1949 *Sydney Morning Herald* 18 June 2/4 The fantastic 'cargo cult' among New Guinea natives has become a serious problem for Australian administrators...Natives influenced by it believe that the spirits of their ancestors will arrive shortly in ships and aeroplanes carrying cargoes of food, tobacco, axes, and other goods. 1951 R. FIRTH *Elem. Soc. Organization* iii. 112 The extraordinary economic operations of some groups, particularly in what has been termed the Cargo cults of some of the New Guinea natives. 1966 'E. LINDALL' *Time too Soon* (1967) iii. 26 The cargo cults, in which the cultists sat waiting for ships to come sailing from America laden with all the luxuries and geegaws of white civilisation. 1924 R. CLEMENTS *Gipsy of Horn* viii. 151 Making the harpoon-line fast to the cargo-fall. 1920 *Blackw. Mag.* Mar. 316/2 The cargo-holds were refurbished. 1922 *Glasgow Herald* 14 June, A homeward bound cargoman. 1832 J. R. MCCULLOCH *Dict. Commerce* 908 There are an immense number of different sorts of rice. The rice exported from Bengal is chiefly of the species denominated *cargo* rice. 1907 *Motor Boat* 19 Sept. 190/2 A good, simple, and cheap cargo winch.

cargo (kā·ɹgo), *v.* [f. CARGO¹.] *trans.* To load. Hence **ca·rgoed** *ppl. a.*

1889 *Guardian* 10 July 1050/3 In the race for literary immortality the heavy Spanish galleon may be cargoed with golden ingots. 1892 *Congress. Rec.* 18 Mar. 2188/1 Pelts and peltries, and anything else the teamster may be cargoed with. 1909 *Westm. Gaz.* 5 Nov. 2/3 Our ship with store of cargoed memories. 1924 R. MACAULAY *Orphan Island* i. 15 They were for leaving the island with the two boats, cargoed with fruit and cocoa-nuts.

cariama: see SERIEMA.

Carian (kēᵊ·riǎn), *sb.* and *a.* [f. L. *Caria*, Gr. Καρία Caria + -AN.] **A.** *sb.* A native or inhabitant of Caria, an ancient province of Asia Minor; the language of Caria. **B.** *adj.* Of or belonging to Caria.

1607 TOPSELL *Four-f. Beasts* 142 Among the Carians..they sacrifized a Dog in stead of a Goate. 1680 OLDHAM in *Rochester's Poems* 122 Thus the vain glorious Carian, I'll out do. 1818 KEATS *Endym.* II. 374 This Dew-dropping melody, in the Carian's ear. *Ibid.* IV. 52 Thou, Carian lord, hadst better have been tost Into a whirlpool. 1839 *Civil Engin. & Arch. Jrnl.* II. 435/1 Jassos..was named also Carian marble. 1876 *Encycl. Brit.* V. 102/2 The Carians are spoken of by all ancient writers as a distinct nation from their neighbours. 1891 W. CORY *Ionica* 7 My dear old Carian guest. 1902 *Encycl. Brit.* XXXIII. 896/2 Symbols exactly like φ, χ, ψ..are found in the Carian alphabet. 1933 BLOOMFIELD *Language* iv. 65 We have copious inscriptions in Lycian..and less extensive ones in Carian, from the seventh century B.C. 1948 DIRINGER *Alphabet* 466 Names of Carian mercenaries scratched upon Egyptian monuments.

caribe (karī·be, kæribēˈ). Also **cariba.** [Sp.: see CARIB.] A characinoid fish of the genus *Serrasalmus*, in the rivers of tropical S. America, noted for its voracity and sharp bite; the piranha.

1868 R. PAEZ *Trav. S. & Cent. Amer.* v. 62 Several dangerous fish..such as..the *caribe*, whose ravenous and bloodthirsty propensities have caused it to be likened to the cannibal tribe of Indians. 1880 GÜNTHER *Fishes* 613 The fish hooked is immediately attacked by the 'Caribe' (as these fishes are called). 1904 BOULENGER in *Cambr. Nat. Hist.* VII. 578 The 'Piranha' or 'Cariba' (*Serrasalmo*)..whose bite has been compared to the cut of a razor. 1904 W. H. HUDSON *Green Mansions* xiii. 181 Some little caribe fishes had made their appearance. 1931 J. H. NORMAN *Hist. Fishes* vii. 130 Another fish which is renowned for its ferocity is the Caribe or Piraya (*Serrasalmus*).

caribou. Add: (Earlier examples.)

c1665 P. E. RADISSON *Voyages* (1885) 202 We killed severall..Carriboucks. 1672 JOSSELYN *New Eng. Rarities* 20 The Maccarib, Caribo, or Pohano, a kind of Deer. 1703 tr. *La Hontan's Voy. N.-Amer.* I. 59 Harts and Caribous are kill'd both in Summer and Winter.

b. *attrib.* and *Comb.* **Caribou Eskimo,** one of a group of Eskimos living in the Barren Grounds of N. Canada and depending on caribou; **caribou moss,** reindeer moss.

1779 in *Essex Inst. Hist. Coll.* XLIX. 112 Sold..82 caribou skins, at £6 10. 0. 1860 H. Y. HIND *Narr. Canad. Red River Exped.* I. v. 110 The general surface was either bare..or else thickly covered with cariboo moss and tripe de roche. 1887 *Harper's Mag.* Feb. 458/2 The hunter..covers his feet with..a pair of moose or caribou shanks, with the hair outside. 1903 S. E. WHITE *Conjuror's House* viii. 94 She felt..the hunger that yet could not stomach the..hairy, black caribou meat. 1904 — *Silent Places* xiii. 137 The foundation he made of caribou moss. 1929 K. B. SMITH *Caribou Eskimos* I. 9 'Caribou Eskimo' is the name which the Fifth Thule Expedition has attached to a group of Eskimo tribes in the southern part of the extensive Barren Grounds west of Hudson Bay. 1936 *Discovery* July 200/2 According to his view the eastern or Caribou Eskimo are neither original nor degenerate, but represent a fusion of Eskimo and Indian cultures. 1963 *Times* 25 Feb. (Canada Suppl.) p. xvi/5 The Chipewyan Indians—the greatest of the caribou-eating tribes.

Caridea (kǎri·diǎ), *sb. pl. Zool.* [mod.L. (neut. pl.), f. Gr. καρίς, καριδ- shrimp.] A group of macrurous decapod crustaceans, containing the shrimps and prawns. Hence **carid** (kæ·rid), a member of this tribe; also *attrib.*; **cari·dean, ca·ridid** *adjs.*, of or belonging to the Caridea; **ca·ridoid** *a.*, resembling the Caridea.

1852 DANA *Crustacea* I. 528 In arranging the Caridea into groups, much stress is usually laid upon external form and length of beak. 1877 THOMSON *Voy. Challenger* II. iii. 193 Scarlet caridid and peneid shrimps. 1904 *Ann. & Mag. Nat. Hist.* XIII. 147 This combination of characters [in Malacostraca] goes to make up what might be called the caridoid 'facies'. 1907 *Ibid.* XIX. 460 The absence [in the Penæidea] of the Caridean bend in the abdomen. *Ibid.* 461 The gill-series in the lower Reptantia are fuller than in either Penæids or Carids. 1909 G. SMITH in *Cambr. Nat. Hist.* IV. 158 Two of the Caridea, viz. the Shrimp..and the Prawn. 1964 BODEN & KAMPA in *Oceanogr. & Marine Biol.* II. 354 The penaeid and carid decapod crustaceans.

caries. Add: **a.** Also *attrib.* and *Comb.*

1937 T. W. & E. V. B. WIDDOWSON *Dental Surgery* (ed. 3) ix. 228 Many teeth are caries-free until cementum is exposed late in life. 1944 MARZELL & BRONNER *Dental Surgeon's Handbk.* 10 Caries prophylaxis should begin in early infancy. 1963 J. OSBORNE *Dental Mech.* (ed. 5) x. 193 These may form a caries-prone area.

carillon. Add anglicized pronunciation: (kǎri·lyǝn). **1.** (Earlier examples.)

1775 J. COLLIER *Mus. Trav.* (ed. 2) 39 Dr. Burney treats all *Carillons* with sovereign contempt, I confess I was much pleased with these. 1776 *Ibid.* (ed. 4) App. 15 An accurate history of the carillons and church-clock.

carilloner (kǎri·lyǝnǝɹ), anglicization of CARILLONNEUR.

1930 *Aberdeen Press & Jrnl.* 20 Jan. 6/6 Again and again the carilloner had to give encore pieces.

carinated, *ppl. a.* Add: Also of pottery.

1936 *Oxoniensia* I. 49 This contained pottery of the first century, including a carinated bowl with cordons. 1937 *Proc. Prehist. Soc.* III. 338 The commonest grave-vessel is a plain carinated bowl with pronounced shoulder. 1963 H. N. SAVORY in Foster & Alcock *Culture & Environment* iii. 35 The fine, carinated bowls of Clegyr-boia.

carination. Add: Also of pottery.

1936 *Oxoniensia* I. 60 Bowl with heavy moulded rim and slight carination at base.

carioca (kæriōᵘ·kǎ). [Pg.] **1.** A native of Rio de Janeiro. Also *attrib.*

1830 R. WALSH *Notices of Brazil* I. 499 A fountain in the largo or square of the Carioca. *Ibid.* 500 The term Carioca is applied to a citizen of Rio in the same sense as cockney to a citizen of London. 1925 L. E. ELLIOTT in *Countries of World* V. XXXIII. 3364/2 Every one born in the capital [Rio de Janeiro] is a 'carioca'; the true native is supposed to be born within sight of the historic Largo de Carioca (a fine square with a great fountain). *Ibid.* 3372/1 The freedom of the carioca pencil and the pungency of carioca wit. 1965 *New Statesman* 10 Dec. 922/2 The Cariocas have left the cool and attractive hillsides to the shanty-dwellers.

2. A type of dance related to the samba, originating in Brazil; also, the music for this dance.

1934 *Washington Post* 13 Mar. 3/5 The polka has gone out of fashion, but why don't they learn the carioca? 1935 WODEHOUSE *Luck of Bodkins* xviii. 216 A certain brand of cigarette—one puff of which..will make a week-old corpse spring from its bier and dance the Carioca. 1962 — *Service with Smile* iii. 43 A pal of mine who had sprained an ankle while trying to teach choirboys the carioca.

‖ **carissima** (kǎri·simǎ). [It., superl. of *cara* beloved, dear.] A term of affection addressed to a woman: darling, best-loved.

1857 W. C. YELVERTON *Let.* in D. Crow *Theresa* (1966) ix. 109 How are you getting on in health, carissima, and how do the dreams progress? 1909 KIPLING *Actions & Reactions* 87 Don't be provincial, *carissima*. 1969 M. GILBERT *Etruscan Net* II. iv. 121 The old woman put an arm round her waist and said, 'Nothing is wrong, carissima, and nothing shall be wrong.'

caritas (kæ·ritās). [a. L. *cāritās*: see CHARITY.] = CHARITY 1. Also *attrib.*

1862 GEO. ELIOT *Let.* 29 Dec. (1956) IV. 72 What charity meant in the elder English and what the translators of the Bible meant in their rendering of the XIIIᵗʰ chapter of I Corinthians—*Caritas*, the highest love or fellowship. 19[illegible] (*title*) Caritas. The quarterly review of the Guild of the Love of God. 1932 A. G. HEBERT tr. *Nygren's Agape & Eros* I. I. i. § 5. 36 Mediæval Caritas is a complex conception containing elements both of Agape and of Eros. 19[illegible] H. G. WELLS *Anat. Frustration* xix. (*b*). 241 A free wide generous, unpossessive caritas for the individuals about us. 1939 P. S. WATSON tr. *Nygren's Agape & Eros* II. II. iv. [illegible] Mediæval Christianity as a whole is Caritas-religion and Caritas-ethics.

caritative, *a.* Add: **2.** *Gram.* Endearing, hypocoristic.

1906 DRIVER *Jeremiah* 368 It seems more probable that [illegible] has a diminutive, caritative force.

caritive (kæ·ritiv), *a. Gram.* [f. L. *carit-*, ppl. stem of *carēre* to lack + -IVE.] Applied to the case used (in Caucasian languages, etc.) to express the lack of something. Also *absol.*

1860 [see *ADESSIVE *a.*]. 1939 L. H. GRAY *Found. Lang.* vii. 201 The caritive denotes 'lack of'.

car-less, carless (kā·ɹlės), *a.* [-LESS.] Not possessing, or unprovided with, a (motor) car; without cars.

1927 *Blackw. Mag.* Nov. 696/1 The transport problem—for of course I was carless—was solved by the arrival of Bosworth, one of whose cars I instantly seized. 1955 T. H. PEAR *Eng. Social Diff.* 147 The car-less middle-class man. 1970 *Daily Tel.* 26 Nov. 16/6 Car-less shopping areas.

Carley (kā·ɹli). The name of an American, Horace S. *Carley*, used *attrib.* (chiefly as *Carley float*) to designate a type of large raft carried on board ships for use in emergency (see quot. 1922). Also *ellipt.*

[1903 *U.S. Pat.* 734,118 Horace S. Carley of Hydepark, Massachusets, Assignor to Carley Life Float Company, of Philadelphia...I, Horace S. Carley,..have invented a new and useful Improvement in Life-Rafts.] 1915 *Illustr. War News* 26 May 23/2 Our illustration shows a valuable life-saving device in the form of the Calley [*sic*] Life-Buoys. 1922 *Encycl. Brit.* XXXII. 451/1 The Carley Life Raft..is made in various sizes. A large copper pipe is bent into the form of an O, brazed up to be airtight, surrounded by cork and canvas, provided with a strong rope netting to form a floor within the O, and fitted with hand ropes, etc. 1925 FRASER & GIBBONS *Soldier & Sailor Words* 82 'Carley Floats', life-saving rafts of circular shape. 1942 *Ann. Reg. 1941* 131 A search..produced no more evidence of the catastrophe than two empty life-boats and a Carley float. 1951 N. MONSARRAT *Cruel Sea* III. vi. 161 Alongside the boats were the Carley floats. *Ibid.*, There were two Carleys.

carlicue, var. CURLICUE. *U.S.*
1840 C. F. HOFFMAN *Greyslaer* II. x. 27, I soon saw, by the way in which the white man's track doubled and doubled again..that the fellow could not be cutting such carlicues for nothing. 1848 BARTLETT *Dict. Amer.* Add., *Carlacue*, a caper or boyish trick. 'To cut up carlacues', is a common expression, equivalent to 'cutting up didoes'. Used in New York. 1898 *Literature* 17 Sept. 263/2 In Canada..the English-speaking country people..often used the word..to signify a trifle or a thing of little value—e.g. 'I don't care a carlicue.' 'It is not worth a carlicue.'

Carlowitz (kā·ɹlǒvits, -wits). Also **Karlowitz(er).** [a. G. *karlowitzer.*] A red wine of Carlowitz on the Danube (above Belgrade).
1858 *Murray's Handbk. Trav. S. Germany* 519 A good full-bodied red wine, known under the name of Karlowitzer. 1888 *Catal. Cellar Wines* (Christie, Manson & Woods) 24 July 7 One Bottle of Carlowitz. 1892 W. & A. GILBEY *Price List Wines* 9 Castle Hungarian Claret Karlowitz. 1920 G. SAINTSBURY *Notes Cellar-bk.* 93 The commoner vintages were not intolerable; you *could* drink Carlowitz if you tried, and the Austrian Vöslauer was not to be despised.

Carlsbad (kā·ɹlzbæd). The German form of the name of Karlovy Vary, a town in Czechoslovakia, used *attrib.* in *Carlsbad plum*, a blue-black dessert plum, usu. crystallized.
1885 *Army & Navy Co-op. Soc. Price List* Jan. 72 Carlsbad plums..per lb. 1/7. *a* 1916 'SAKI' *Toys of Peace* (1919) 130 Trying to decide between the merits of Carlsbad plums and confected figs as a winter dessert. 1960 R. COLLIER *House called Memory* vi. 92 Sweet purple-black Carlsbad plums that would not [keep].

Carmelite, *sb.* and *a.* Add: **1. b.** Belonging to, or a member of, an order of nuns organized on the model of the Carmelite or White Friars.
1611 T. CORYAT *Crudities* 14, I was at the Nunnery of the Carmelite Nunnes. 1670 CRASHAW *Steps to Temple* 61 The Admirable Sainte Teresa Foundresse of the Reformation of the Discalced Carmelites, both Men and Women. 1739 GRAY *Lett.* 1 Apr. (1900) I. 17 We saw the chapel of Minims and the Carmelite Nuns. 1888 H. J. COLERIDGE *St. Teresa* III. 9 St. Mary Magdalene of Pazzi was a 'Mitigated' Carmelite. 1909 *Dublin Rev.* Jan. 61 We have..Carmelites at Lanherne, Darlington, and Chichester.

carmined (kā·ɹmind), *a.* [f. CARMINE *sb.* + -ED², after *rouged.*] Reddened with carmine.
1893 *Scribner's Mag.* June 702/2 These coarse carmined Delilahs. 1902 G. MEREDITH in *Westm. Gaz.* 4 Mar. 8/2 Warfare with carmined savages. 1920 *Chambers's Jrnl.* 344/1 Two parted, carmined lips.

carminophilous (kāɹminǫ·filəs), *a. Biol.* [f. CARMINE *sb.* + Gr. φίλος loving + -OUS.] Epithet of those cytoplasmic granules which are readily stained by carmine.
1901 G. N. CALKINS *Protozoa* 143 Carminophilous granules ..are easily stained by carmine and many aniline colours. 1903 E. A. MINCHIN in E. R. Lankester *Zool.* I. ii. 182 The so-called carminophilous granules,..composed apparently of an albuminoid substance which is stained red by picro-carmine and acetocarmine.

Carnaby Street (kā·ɹnābi strīːt). The name of a street in central London, used allusively to refer to fashionable clothing designed for young people. Also *ellipt.* as *Carnaby*.
[1964 *New Statesman* 29 May 853/1 Window dressers and Carnaby street pants-peddlers.] 1965 *Observer* (Colour Suppl.) 29 Aug. 21/1 A mother's-boy spy, still living at home to save all his income for Carnaby Street. 1966 K. GILES *Provenance of Death* iii. 86 Leicestershire wore..his Carnaby Street gear. 1966 *Guardian* 25 Apr. 7/1 Discothèque-crazed tele personalities jerking in Carnaby plumage. 1969 C. BOOKER *Neophiliacs* iii. 80 Why was it going to be Britain that would produce the Beatles and Carnaby Street and television satire..?

carnaptious (kāɹnæ·pʃəs, kăn-), *a. Sc.* and *Irish dial.* Also **carnapshus, curnaptious.** Bad-tempered; quarrelsome; cantankerous.
1858 *Ulster Jrnl. Archæol.* VI. 44 A nivver seen wan so curnaptious. 1879 W. G. LYTTLE *Readings by Robin* 48 (S.N.D.), He's a cross carnapshus wee brat, so he is! 1931 H. S. ROBERTSON *Curdies* 114 (S.N.D.), That belangt to ane they ca'd Rab Frew, a carnaptious auld deevil he was. 1963 *Times Lit. Suppl.* 21 June 459/3 The carnaptious old Irish hotel-keeper.

carnauba (kāɹnǭ·bă, -nau·-, kāɹnăˌū·bă). [Pg.] The Brazilian wax-palm, *Copernicia cerifera*; so *carnauba wax*, a wax exuded from the leaves of the carnauba and used in the manufacture of polishes, candles, etc.
1854 R. D. THOMSON *Dict. Chem.* 521/2 *Wax, Carnauba*... From the leaves of a palm in Brazil; soluble in hot alcohol and ether. 1866 LINDLEY & MOORE *Treas. Bot.* I. 225 *Carnaüba*, a Brazilian palm, *Corypha cerifera*, the leaves of which yield a wax, which is used for making candles. 1919 *Chambers's Jrnl.* Apr. 271/1 It is a question whether there is a more wonderful or more useful tree to man than the carnauba (or carnahuba) tree of Brazil. 1958 *Listener* 4 Dec. 967/1 The hard, most efficient polishing waxes (carnauba wax, for instance) had often to be mixed with less efficient, softer wax.

carnet (kā·ɹne). [F., lit. 'note-book, booklet'.] **1.** A note-book.
1897 *Daily News* 17 Sept. 6/2 Little scenes..that might have been taken bodily from the jottings of the artist's 'carnet' of subjects for illustration.

2. (*a*) A permit issued to an aviator (see quot. 1926¹). (*b*) A permit allowing a motor-vehicle, etc., to be taken across a frontier. (*c*) A permit giving admission to some camping-sites.
1926 *Spectator* 20 Mar. 556/2 The A.A. issues Triptyques or Carnets to its members. 1926 *Flight* 2 Dec. 774/1 A 'carnet'..serves as a Customs pass on foreign aerodromes and relieves the holder of all troublesome formalities. 1955 *Times* 13 Aug. 7/7 One in eight of the motoring parties ..took camping carnets which give them the run of camping sites recognized by the Alliance Internationale de Tourisme.

carney, carny (kā·ɹni), *a. slang.* [f. CARNY, CARNEY *v.* (and *sb.*).] Artful, sly.
1881 *N. & Q.* 6th Ser. III. 318 Eh, she *was* carny when she was a-sayin' that. 1925 in FRASER & GIBBONS *Soldier & Sailor Words* 47. 1955 E. BLISHEN *Roaring Boys* iv. 203 Macbeth was pretty carney in the way he handled Banquo.

carney: see *CARNY *sb.*¹ and *sb.*²

carnival. Add: **2.** Now usu. = FESTIVAL *sb.* 1 a (see also quots. 1950).
1916 JOYCE *Portrait of Artist* v. 257 She passed now dancing lightly across his memory as she had been that night at the carnival ball, her white dress a little lifted, a white spray nodding in her hair. 1921 *Daily Colonist* (Victoria B.C.) 19 Mar. 11/5 The Arena was the scene of a brilliant assemblage of people last night when over eighteen hundred guests attended the ice carnival, and well over one thousand guests joined in the skating. 1950 *Oxf. Jun. Encycl.* IX. 114/2 We now use the word 'carnival' in a general sense to describe a particular kind of public celebration or entertainment which includes a fancy dress procession through the streets. *Ibid.* 115/1 During the summer holiday season,..seaside resorts sometimes arrange imitations of the continental carnival, the main feature being a beauty competition for the election of the Carnival Queen. 1966 *Oxf. Mail* 1 Oct. 1/5 It was not a serious demonstration. Grinning youngsters..baited the police and the crowd of 3,000 turned the occasion into a boisterous carnival. 1968 M. ALLINGHAM *Cargo of Eagles* xiv. 158 Travels for a small firm..who make carnival novelties, streamers, funny hats and so on.
b. A fun-fair; circus. *N.Amer.*
1931 J. LITTELL *Carnival Girl* (1933) ii. 41 An oblong of trucks surrounding a village of tents. A carnival company! *Ibid.* iii. 45 The man was offering her a job—with the carnival! 1939 [see *CARNY *sb.*²]. 1955 D. W. MAURER in *Amer. Dial. Soc.* XXIV. 31 A bank robber would be likely to distrust a carnival grifter on principle. 1968 *Globe & Mail* (Toronto) 17 Feb. 6/5 Nobody would be allowed to set up a permanent carnival ride on the grounds of any church in the City of Toronto.
Hence **ca·rnivaling, ca·rnivalizing** *vbl. sbs.*; also **ca·rnivalite, ca·rnivaller,** one who takes part in a carnival.
1841 THACKERAY *Shrove Tuesday in Paris* in *Wks.* (1900) XIII. 570 When they grow old, perhaps, they leave off gallantry and carnivalising. 1881 *Pop. Sci. Monthly* XIX. 151 All shouting and cheering, merry as carnivallers. 1893 *Westm. Gaz.* 25 Feb. 5/3 Feasting on wonderful and uncanny—but very dainty—dishes is another important part of carnivaling. 1896 *Daily News* 19 Feb. 3/4 These fashionable Shrove-tide carnivalites.

carnotite (kā·ɹnǒtait). *Min.* [ad. F. *carnotit* (Friedel & Cumenge 1899, in *Compt. Rend. Acad. Sci.* CXXVIII. 534), f. the name of Marie Adolphe *Carnot* (1839–1920), French inspector-general of mines: see -ITE¹.] A yellow earthy hydrated vanadate of potassium and uranium, found in south-western Colorado, and worked as a source of vanadium, uranium, and radium. Also *attrib.*
1899 *Jrnl. Chem. Soc.* LXXVI. ii. 434 A new mineral containing uranium and vanadium, to which the authors give the name *carnotite*, is found in yellow, friable masses, mixed with very variable quantities of silica, together with malachite and chessylite, in pockets at the surface of a grit in Montrose Co., Colorado. 1920 *Discovery* May 143/1 The carnotite beds on the borders of Utah and Colorado. 1945 *Times* 8 Aug. 4/2 In 1938 the United States produced 4,290 short tons of carnotite ore.

carny, *sb.*¹ Now usually written **carney.** Add: (Earlier and later examples.) Also, a smooth talker, a flatterer.
1818 *London Guide & Stranger's Safeguard* p. x, *Carney*, softening talk. *Ibid.* 205 The thoroughbred sycophant may be known by his *carney* or small talk. 1869 *Good Words* 1 Oct. 561/1 He's a little bit too much of a carney. 1923 J. MANCHON *Le Slang* 79 *To come the carney*, flatter, flagorner.

carny (kā·ɹni), *sb.*² *U.S. slang.* Also **carney, carni(e).** [f. *CARNIVAL 2 b + -Y⁶.] = *CARNIVAL 2 b; also, a person who works at a carnival. Also *attrib.* or as *adj.*
1931 *Amer. Speech* VI. 330 *Carnifolks*, persons who engage in the carnival business. 1933 B. J. CHIPMAN *Hey Rube* 193/2 *Carny*, carnival. 1939 *New Yorker* 12 Aug. 22/2 Sixty thousand outdoor show people, the 'carnies', who travel from town to town with carnivals. 1948 F. BROWN *Dead Ringer* i. 5 All around the midway carneys were letting down banners and running rope. *Ibid.* 10 A town kid might possibly be around the sideshow top this late at night, but not without his clothes on. For a carney kid that wasn't too strange. *Ibid.* 11 One of the new girls with the posing show. She'd been with the carney only a week. 1955 R. BRADBURY *October Country* (1956) 10 Life fixed him so he's good for nothing but carny shows. 1956 H. GOLD *Man who was not with It* (1965) i. 4 Joy..stopped in her roaming through this carnie where she dwelled and listened. *Ibid.*, He was a sallow stooped carnie with vapors fuming in his eyes. 1961 *Times Lit. Suppl.* 27 Jan. 62/2 Ernie, the carni man, is violent, stupid, and wholly untouched by any kind of decency.

Carolean (kærǒlī·ăn), *a.* and *sb.* [f. L. *Carolus* Charles + *-ean* as in JACOBEAN *a.* (*sb.*).] **A.** *adj.* = CAROLINE *a.* 1 b. **B.** *sb.* One who lived in the reign of Charles I or II.
1911 *Chambers's Jrnl.* 15 July 513/1 The rapacity and meanness of the Ministers in the Carolean era. 1927 *Observer* 8 May 15/2 The spirit that animated the restored Caroleans in their exhumatory operations against the regicides. 1927 *Daily Tel.* 12 July 5/5 A Carolean Poet [*sc.* Marvell]. 1939 *Archit. Rev.* LXXXV. 58/1 This seeming late 18th century window is in strange company, surrounded as it is by what appear the familiar swags of essentially Carolean aspect.

Carolina. Add: **Carolina rice,** a variety of rice, the ripe husk of which is yellowish.
1787 JEFFERSON *Let.* 30 July in *Writings* (1853) II. 195 The Carolina rice..crumbles in certain forms of preparation. 1845 E. ACTON *Mod. Cookery* i. 44 The Patna..rice..is not so good as the Carolina for the general purposes of cookery. *Ibid.* 45 The Carolina rice even answers, well dressed, in this way. 1962 *Guardian* 12 Jan. 8/5 Soak a breakfast-cupful of Carolina rice.

Caroline, *a.* Add: **1. a.** *spec.* Designating a style of minuscule handwriting developed in France at the time of Charlemagne.
1850 F. MADDEN tr. *J. B. Silvestre's Univ. Palæogr.* I. cxxix. 346 The text is in clear, well proportioned, Caroline minuscules, with the words not divided, the tails and tops of the letters of proper length, and the strokes of the *m* and *n* inclined towards the left; graphic characters which indicate the ninth century, and the kind of writing termed Caroline. 1883 I. TAYLOR *Alphabet* II. viii. 181 Owing to its manifold excellencies..the rapidity with which it could be written, the ease with which it could be read, and economy of parchment, the Caroline minuscule, as it is usually called, grew rapidly in favour. 1897 [see next]. 1957 N. R. KER *Catal. MSS. containing Anglo-Saxon* p. xxii, It should be assumed that..writing dated s. xi or later is caroline minuscule. 1962 D'ARDENNE in Davis & Wrenn *Eng. & Medieval Studies* 85 The manuscript is written in Latin on parchment in a smooth round English form of the caroline minuscule.

Carolingian, *a.* Add: *spec.* = prec.
1897 H. W. JOHNSTON *Latin MSS.* ii. 72 Of the better forms the Caroline (Carolingian) may be regarded as the type, as it finally became the literary hand of all Western Europe. 1912 E. M. THOMPSON *Introd. Gr. & Latin Palaeogr.* xvi. 367 At Tours, where, under the rule of Alcuin of York, who was abbot of St. Martin's from 796 to 804, was specially developed the exact hand which has received the name of the Carolingian Minuscule. 1962 D. B. UPDIKE *Printing Types* (ed. 3) I. iii. 50 This Carolingian minuscule..became the dominant handwriting of western Europe.

Carolinian, *a.* (and *sb.*) Add: **b.** (Earlier examples.)
1705 PENN *Let.* 9 Dec. in *Penn–Logan Corr.* (1872) II. 105 The Carolinian Lords. 1707 J. ARCHDALE *New Descr. Carolina* 15 By the Encouragements of several Carolinians than in England my Going was concluded on. 1818 *Mass. Agric. Repository & Jrnl.* V. 60 Populus Angulata—Carolinian Poplar, name given to it in Europe, because first brought from Carolina.
c. = CAROLINE *a.* 1 b. Also as *sb.*, a poet of the time of Charles I or II.
1949 M. BEWLEY in *Scrutiny* Mar. 16 The tone of the colloquialism is Carolinian. *Ibid.* 19 The Carolinians exploit the fashion for the grace and elegance of conceit it allows them.

carom, carrom. B. *v. intr.* Add def.: To strike or glance and rebound. Also *fig.* Chiefly *U.S.* (Further examples.)
1911 MULFORD *Bar-20 Days* iv. 45 The table skidded through the door on one leg and carromed off the bar at a graceful angle. 1943 K. TENNANT *Ride on Stranger* (1968) xvii. 219 It was here that Mrs. Brewster made the mistake.. of caroming off a telegraph post into the ditch. 1946 H. CROOME *Faithless Mirror* xviii. 192 The car lurched crazily..up the gully, caroming from side to side. 1952 B. WOLFE *Limbo '90* xix. 284 The phrase caromed through his mind. 1967 *Boston Globe* 18 May 36/8 Rockets carom to the moon.

carotene, carotin (kæ·rǒtīn, -in). [a. G. *carotin* (H. W. F. Wackenroder 1831, in *Mag. Pharmacie* XXXIII. 148), f. L. *carota* carrot + -ENE, -IN¹.] An orange or red hydrocarbon, $C_{40}H_{56}$, synthesized in several isomeric forms by carrots and many other plants, and an important source of vitamin A. Also *attrib.*
1861 MRS. BEETON *Bk. Househ. Managem.* 563 A peculiar crystallizable ruby-red neuter principle, without odour or taste, called carotin. 1895 *Naturalist* 24 [Berries of mountain ash.] Their colouring matter is due to carotin. 1897 *Jrnl. Chem. Soc.* LXXII. ii. 225 The author does not attempt to decide whether the crystals formed in this reaction are all xanthophyll or all carotin crystals, or whether these are identical or whether they consist in part of colouring matters nearly related to carotene. 1951 *Science News* XXII. 75 Vitamin A..is one half of a carotene molecule and is formed from carotene in the intestinal wall of animals. 1960 [see *CRYPTOXANTHIN].

Hence **ca:rotenæ·mia** (U.S. -nem-), **ca:rotinæ·mia,** the presence of carotene in the circulating blood.
1919 A. F. HESS & V. C. MYERS in *Jrnl. Amer. Med. Assoc.* LXXIII. 1745/1 In cases of carotinemia the urine was colored yellow as well as the serum. 1959 *Listener* 11 June 1017/1 Occasional cases of a condition known as carotinaemia were encountered...It arose from an over-indulgence in carrots.

carotenoid, carotinoid (kæ·rǒtinoid). *Biochem.* [ad. G. *Carotinoïde* (M. Tswett 1911, in *Ber. d. Deut. Bot. Ges.* XXIX. 630); see prec. + -OID.] Any one of a group of pigments including the carotenes, the xanthophylls, and fucoxanthin, found in many plants and animals. Also *attrib.* or as *adj.*
1913 *Chem. Abstr.* 2956 (*title*) The Occurrence of Carotinoids in Plants. 1917 HAAS & HILL *Introd. Chem. Plant Products* (ed. 2) v. 229 Carotin, Xanthophyll, and Fucoxanthin..are known collectively as the Carotinoids. 1922 L. S. PALMER (*title*) Carotenoids and related pigments. 1930 *Brit. Jrnl. Exper. Path.* Apr. 81 The relation between the carotenoid pigments and vitamin A. 1952 *New Biol.* XIII. 40 Doubling the number of chromosomes in pure yellow corn caused a 40% increase in the carotenoid pigment content, including the active provitamin A fraction of the carotenoids. 1968 *Times* 16 Nov. 9/7 Sporopollenin is a polymer formed of sub-units known as carotenoids.

carousel. Add: **2.** A merry-go-round, a roundabout. Also *attrib.* Chiefly *U.S.* (where freq. written *carrousel*).
1673 R. FOLVARTE *Let.* in D. Braithwaite *Fairground Archit.* (1968) iii. 34 A new and rare invencon knowne by the name of the royalle carousell or tournament being framed and contrived with such engines as will not only afford great pleasure to us and our nobility in the sight thereof, but sufficient instruction to all such ingenious young gentlemen as desire to learne the art of perfect horsemanshipp. 1899 *N.Y. Times Illustr. Mag.* in F. Fried *Pict. Hist. Carousel* (1964) iii. 82 A carousel costs from $300 to $10,000 according to the decoration and finish...A carousel that will seat 60 riders measures 40 feet in diameter and costs $2,200. 1909 *Sat. Even. Post* 13 Mar. 64/1 We make everything..from a hand-power Merry-Go-Round to the highest grade Carousselles. 1951 J. D. SALINGER *Catcher in Rye* xxv. 250 She sat down on this big..horse. Then the carousel started, and I watched her go round and round. 1956 E. AMBLER *Night-Comers* iii. 58 There was even a small fair in progress. A carousel had been set up. 1958 S. ELLIN *Eighth Circle* (1959) II. xi. 122 A faraway sound of carrousel music. 1968 R. PETRIE *MacLurg goes West* x. 89 Bracketed on, so that they hung semi-rampant..were the two carousel horses.
3. (See quot. 1961.) Now esp. one at airports for the delivery of passengers' luggage.
1961 WEBSTER, *Carrousel*, a conveyor (as for assembly-line work) on which objects are placed and carried round a complete circuit on a horizontal plane. 1970 *Which?* Nov. 352/1 Older people..found it hard to tug their cases off the carousels.

carp, *sb.*¹ Add: **3. carp-louse,** a name for small crustaceans of the family Argulidæ, parasitic on fishes; cf. FISH-*louse*.
1889 in *Cent. Dict.* 1909 J. J. LISTER in A. Sedgwick *Textbk. Zool.* III. 410 Carp-lice. Copepods with large compound eyes. 1931 J. R. NORMAN *Hist. Fishes* xx. 417 Mention may also be made of the so-called Carp Louse (*Argulus*), another Crustacean parasite.

carp, *sb.*² Restrict †*Obs. rare* to senses *a* and *b*, and add further examples for sense *c*.
1904 *Westm. Gaz.* 9 May 4/1 Criticism—what a lady I know calls 'the carpers carping with their carps'. 1922 F. SCOTT FITZGERALD *Let.* Jan. (1964) 331 But one more carp before I close. 1967 *Observer* 24 Oct. 25/2, I have one carp, however. His own performance..tends to become confusing.

Carpano (kāɹpā·no). [Name of Antonio *Carpano*, who made the first commercial preparation of vermouth in Italy in 1786.] The proprietary name of an Italian vermouth; a drink of this.
1921 *Chambers's Jrnl.* 6 Aug. 572/2 The little waiter had that moment emerged..with a wet tray and a cool glass of *carpano*. 1950 D. AMES *Corpse Diplomatique* xii. 89 Dagobert ordered a *Carpano* for me and a *Campari* for himself. 1961 *Guardian* 21 Nov. 13/2 Carpano is..a dark, bitter-sweet vermouth. 1968 HURD & OSMOND *Send him Victorious* 20 Raikes began to fumble in his pockets for money to pay for his Carpano.

Carpathian (kaɹpē¹·piăn), *a.* [f. *Carpathos*: see -IAN.] Of or pertaining to Carpathos or Karpathos, an island in the Ægean Sea.
1634 MILTON *Comus* 872 By hoary Nereus wrincled look, And the Carpathian wisards hook. 1797 *Encycl. Brit.* IV. 179/1 The sea which, from this island, is called the Carpathian Sea. 1925 *Glasgow Herald* 30 July 6 In a spell of Carpathian silence.

‖**carpe diem** (kā·ɹpī dai·ˌem). [L., 'enjoy the day, pluck the day when it is ripe'.] An aphorism quoted from Horace (*Odes* I. xi) affirming the need to make the most of the present time. Also *attrib.*
1817 BYRON *Let.* 2 Jan. in Moore *Life* (1830) II. 68, I never anticipate,—*carpe diem*—the past at least is one's own, which is one reason for making sure of the present. 1853 'C. BEDE' *Verdant Gr.* x. 89, I daresay old Horace gives very good advice when he says, 'carpe diem'. 1867 'OUIDA' *Under Two Flags* II. ii. 41 The reckless life of Algeria..with..its gay, careless carpe diem camp-philosophy. 1901 *Daily News* 7 Mar. 6/1 The 'Carpe diem' philosophy is not the philosophy of happy people. 1914 T. A. BAGGS *Back from Front* xix. 91 'Carpe diem' is their motto, and indeed, they enjoy life while they may. 1957 N. FRYE *Anat. Criticism* 299 The *carpe diem* poem based on a moment of pleasure in experience.

carpenter, *sb.* Add: **5. carpenter-work,** carpentry. Also *fig.*
1553 UDALL *Respublica* (1952) v. vi, I woulde ere long of yowe [haue] made suche carpenter weorke, That ye shoulde haue saide Policie had been a clerke. 1720 in *Jrnl. Derbysh. Archæol. Soc.* (1905) XXVII. 216 Carpenter work 1 6 0. 1844 H. STEPHENS *Bk. Farm* I. 168 They embrace the particulars of mason-work, carpenter-work, slater-work,..smith-work. 1909 *Daily Chron.* 2 June 5/2 The play is at best a piece of very crude carpenter-work.

carpenter, *v.* Add: Also *fig.*
1844 [see CARPENTERED *ppl. a.*]. 1885 A. BRERETON *Dram. Notes* 50 Mr. Paul Meritt and Mr. Henry Pettitt..know how to carpenter a play for the stage. 1908 *Daily Chron.* 23 Oct. 6/1 The acting may be bad, the play cribbed and carpentered, but if people are genuinely moved the essence is there. 1909 MRS. STRATTON-PORTER *Girl of Limberlost* xi. 212 When I think of how you are carpentered, I'm adoring the result.

carpet, *sb.* Add: **1. b.** (Later example.) Also *colloq.* (orig. *U.S.*) (with admixture of sense 2 a): undergoing, or summoned to receive, a reprimand. Cf. sense 2 d in Dict. and *MAT *sb.*¹
1900 G. ADE *Fables in Slang* 120 Next Morning she had him up on the Carpet and wanted to know How About It. 1902 LORIMER *Lett. Merchant* 134 The boss of the canning-room [will be] called up on the carpet and made to promise that it will never happen again. 1921 LD. DUNSANY *If* I. i. 9 You been on the carpet, Bill? 1936 *Sketch* 22 Jan. 148/1 His manager had just had him on the carpet, pointing out that his work had been getting steadily bad for the last few months. 1944 'N. SHUTE' *Pastoral* v. 136 After the first shock of realising that she herself was on the carpet. 1961 *Times* 8 May 13/2 Such a solution is not on the carpet now.
2. e. *to sweep* (or *push*) (something) *under the carpet*: to conceal (something embarrassing or unpleasant), in the hope that it will go unnoticed or be forgotten.
1963 *Times* 24 Jan. 6/1 It would be self-deception to think that unemployment could be dealt with by emergency measures and pushed under the carpet. 1966 *Listener* 17 Mar. 390/2 You cannot sweep these problems under the carpet as the Labour Government is doing.
3. b. *spec.* in *Cricket*, the surface of the field, the ground. Also *attrib.*, as *carpet drive.*
1882 *Cricket* 22 June 93/3 His hits..are never high; on the contrary they are mostly, to use the slang of the cricket field, 'on the carpet'. 1896 *Westm. Gaz.* 23 July 7/2 A grand carpet drive to the off-boundary. 1899 SNAITH *Willow the King* v. 74 It flew like a streak to mid-off all along the carpet. 1908 W. E. W. COLLINS *Country Cricketer's Diary* xiv. 233 Leaving us in a parlous bad condition by putting catch after catch..upon the carpet. 1927 M. A. NOBLE *Those 'Ashes'* 201 Bowley tried a carpet drive through the covers. 1950 W. HAMMOND *Cricketers' School* x. 98 The batsman's art..is to learn how to turn the ball..so that.. it shall stay 'on the carpet'.
c. *Aeronaut. slang.* The ground; *on the carpet*, at or near ground level.
1918 J. M. GRIDER *War Birds* (1926) 258, I was right on the carpet and over a little ruined village. 1934 V. M. YEATES *Winged Victory* xxi. 165 He could not go down on to the carpet again point blank into the mouths of the machine guns.
d. *Civil Engin.* (See quot. 1954.)
1920 H. E. GOLDSMITH *Mod. Road Constr. & Maintenance* App. I 86 *Carpet*, a bituminous surface of appreciable thickness, generally formed on top of a roadway by the application of one or more coats of bituminous material with gravel, sand, or stone chippings added. 1930 *Engineering* 15 Aug. 210/1 It consisted of a 2 in. stone-filled asphalt carpet laid on a reinforced-concrete foundation. 1954 *Gloss. Highway Engin. Terms* (B.S.I.) 31 *Carpet*, a wearing course containing a road tar or bitumen binder and having a compacted thickness not greater than 1½ inches.
e. A large number of bombs dropped to damage an area intensively, esp. to clear the way for advancing ground troops. Also *attrib.* and *Comb.*
1944 T. H. WISDOM *Triumph over Tunisia* xxiii. 181 Our massed bombers..set off to lay Tedder's bomb carpet; to blast a passage for the armies right through to Tunis. 1945 *Baltimore Sun* 23 Feb. 3/5 Several ministries were laid in ruins that Saturday as a result of the American carpet bombing. 1945 *Life* 4 June 25 A carpet raid is an effort to pulverize an area in front of ground troops to facilitate a breakthrough. 1946 *Aviation Ann.* 16 Carpet bombing.. consists simply of dropping such a heavy concentration of bombs in a small area that the defenders are stunned and demoralized for a short time. *Ibid.*, A carpet attack in a certain sector. 1968 L. VAN DER POST *Portrait Japan* i. 29 Carpet bombing by the United States Air Forces flattened the capital.
5. carpet beetle, any of various beetles or their larvæ, especially *Dermestidæ*, destructive to carpets and other fabrics; **carpet dance** (earlier example); **carpet shark,** a shark of the genus *Orectolobus* having spots on the back suggesting the pattern of a carpet; **carpet slipper,** a slipper the uppers of which were orig. made of carpet-like material; also used *attrib.* to express an informal or slip-shod method of work; **carpet sweeper,** a household implement with a revolving brush or brushes for sweeping carpets.
1889 *Cent. Dict.*, *Carpet-beetle*, a popular name of *Anthrenus scrophulariæ*, a beetle of the family *Dermestidæ*. 1961 *Which?* July 157/1 The varied carpet beetle, often known in its larval phase as the woolly bear (*Anthrenus verbasci*). 1846 R. FORD *Gatherings from Spain* xxiii. 325 An extempore ball, which might be called a *carpet*-dance, if there were any. 1929 S. ELLIOTT NAPIER in *Times* 2 Aug. 14/1 The 'Leopard' or 'Carpet' shark. 1956 *Coast to Coast 1955–56* 128 The two spearsmen came in with a five-foot carpet shark. 1851 MAYHEW *Lond. Labour* I. 157/1 The large carpet slippers that served her for shoes. 1902 *Chambers's Jrnl.* 10 May 353 Etymologists relieve the tedium of a hard day's work by a relapse into the carpet-slipper side of their science. *Ibid.*, The old monks were adepts at carpet-slipper derivation. 1905 H. A. VACHELL *Hill* i. 17 He slimes about in carpet slippers—the beast! 1967 S. BECKETT *Eh Joe* 15 Joe, late fifties, grey hair, old dressing-gown, carpet slippers, in his room. 1859 *Rep. Comm. Patents 1858* (U.S.) II. 444 That description of carpet or floor sweepers in which a revolving brush..is made to take up and deposit the sweepings in a case covering the brush. 1892 *Ladies' Home Jrnl.* Dec. 19/4 Bissell carpet sweeper. 1960 *Which?* Apr. 85/1 The ideal carpet sweeper is one which picks up a lot of dirt, but only a little carpet pile, and which is easy to use and empty.
7. *Criminals' slang. Ellipt.* for *CARPET-BAG 2.
[1903 'No. 7' *Mark of Broad Arrow* xi. 166 Your 'Autoleyne' cares little about a 'drag' (three months), a sixer (a 'carpet' it is generally called), or a 'stretch'.] 1917 E. WALLACE *Just Men of Cordova* x. 133 'Carpets' (three months' hard labour) almost innumerable had fallen to his share. 1936 J. CURTIS *Gilt Kid* 17 Long enough to've been in Wandsworth and done a carpet. 1963 T. & P. MORRIS *Pentonville* xi. 227 A *carpet* (3 months).

carpet-bag. Add: **a.** (Earlier examples.)
1830 *Boston Directory* 28 (Advt.). 1836 DICKENS *Early Coaches* in *Sk. Boz* II. 173 The new carpet-bag you bought.
2. *Rhyming slang.* = DRAG *sb.* 8 b.
1938 SHARPE *S. of Flying Squad* 330 *Carpet* (or *Carpet Bag*, or *Drag*), three months' imprisonment.
3. *Comb.* **carpet-bag steak,** a steak stuffed with oysters.
1958 *Times* 28 Aug. 7/7 At the Gas Council stand she can see ground nut stew, carpet bag steaks, biryani..being prepared. 1969 *Times* 22 Dec. 9/8 Many 'foreign' dishes are really British. Like Australian 'carpetbag steak'—rare fillet stuffed with oysters.

ca·rpet-ba:g, *v. colloq.* (orig. and chiefly *U.S.*). [f. the sb.] *intr.* To travel with very little luggage (see also quot. 1889). Also *transf.*
1872 *Trans. Ill. Dept. Agric. 1871* IX. 266 Almost the entire force of our common schools tends to..send [our young men]..'carpet bagging', down South. 1889 *Century Dict.*, *Carpet-bag*, to act or live in the manner of a carpet-bagger. (U.S. slang.) 1890 *Congress. Rec.* 4 June 5598/2 It has been stated that Mr. McDuffie carpet bagged from somewhere down into Alabama...Now I will tell the House how Judge McDuffie happened to carpet-bag down into that district. 1968 A. WILLIAMS *Brotherhood* I. iv. 51, I don't have to carpet-bag to your kind of publicity, Owen. Now piss-off.

ca·rpet-ba:gging. orig. and chiefly *U.S.* Travelling with only a carpet-bag to contain one's effects; *spec.* the practice of a carpet-bagger. Also *attrib.*
1869 *Atlantic Monthly* June 747 After three weeks' delightful Carpet-Bagging. 1888 BRYCE *Amer. Commw.* III. lxxv. II. 621 Negro suffrage produced, during the few years of 'carpet-bagging' and military government which followed the war, incredible mischief. 1904 *N.Y. Even. Post.* 1 Feb., The infernal 'carpet-bagging' system at present in vogue should be done away with. 1955 *Times* 26 May 4/3 In this carpet-bagging age, a home-grown politician is as rare as a home-grown professional footballer.

carpeting, *sb.* **2.** (Earlier U.S. example.)
1758 B. FRANKLIN *Corr.* in *Wks.* (1840) VII. 165 Some carpeting for a best room floor.

carpincho (kaɹpi·ntʃo). [Native name.] A local name in South America for the capybara.
1839 DARWIN *Jrnl.* iii. 57 These great Rodents are generally called 'carpinchos'. 1869 W. H. HUDSON *Let.* 14 Dec. in *Ornithol. Buenos Ayres* (1951) i. 2, I have known the *ciervo jaguar* aguará and carpincho with other large mammels. 1904 *Westm. Gaz.* 6 Sept. 10/1 When alarmed the carpincho utters a kind of grunting cry. 1949 *Oxf. Jun. Encycl.* (1955) II. 81 The Capybaras, or Carpinchos, are 4 feet long and weigh about 98 pounds.

carrageen, -gheen. In def., for *Chrondus* read *Chondrus*.

carrancha (kāræ·ntʃā). Also **carancha, car(r)ancho.** [Native name.] A South American caracara or carrion-hawk, *Polyborus tharus*.
1839 DARWIN *Jrnl.* iii. 64 *Polyborus Braziliensis*..is most numerous on the grassy savannahs of La Plata (where it goes by the name of Carrancha). 1869 W. H. HUDSON *Let.* 22 Dec. in *Ornithol. Buenos Ayres* (1951) ii. 10 This *sayno* swamp is a great breeding place for the Carranchos, and for hawks. 1889 SCLATER & HUDSON *Argent. Ornithol.* II. 81 *Polyborus tharus*..Carancho Carrion-Hawk. 1911 *Encycl. Brit.* XX. 901/1 The carrancha or carrion-hawk.. is one of the characteristic features of a Patagonian landscape. 1920 W. H. HUDSON *Birds of La Plata* II. 76 South of Paraguay the Spanish name is *Carancho*, possibly a corruption of *Keanché*, the Puelche name for the allied *Milvago chimango*, in imitation of its peevish cry. *Ibid.* 77 The Caranchos pair for life, and may therefore be called social birds.

carrapato (karāpā·to). Also **carapato**; and in Sp. form **garrapata.** [Pg.] A tick of the genus *Ixodes*. Also *attrib.*, as *carrapato disease* (quot. 1903).
1886 [see *bush-tick*]. 1892 G. FLEMING *Neumann's Treat. Parasites* 103 These are popularly designated by such names as Ticks..and Carapatos. 1903 *Jrnl. Tropical Med.* 2 Nov. 341/1 The carapato disease of the Zambesi basin is certainly produced by the bite of a tick—*Argas moubata*. 1906 F. C. GLASS *Through Heart Brazil* 20 The carrapato is a kind of minute tick. 1930 *Discovery* July 229/2 Such parasites as the garrapato and pinalia abound.

Carrara (kărā·ră). The name of a town in N. Italy, used *attrib.* to designate a variety of marble quarried there; *spec.* in *Pottery* (see quot. 1957).
a 1728 [see MARBLE *sb.* 1 b]. 1795 R. KIRWAN *Elem. Min.* (ed. 2) I. vi. 85 Carrara marble is next in purity. 1883 *Encycl. Brit.* XV. 528/2 It is in Carrara marble that the finest works of Michelangelo and of Canova are executed. 1907 E. WHARTON *Fruit of Tree* I. iii. 32 The numerous gas-jets..shed their lights..on Carrara busts and bronze Indians. 1928 D. H. LAWRENCE *Lady Chatt.* vii. 87 She began to be afraid of the ghastly white tombstones, that peculiar loathsome whiteness of Carrara marble. 1957 MANKOWITZ & HAGGAR *Encycl. Eng. Pott.* 43/1 *Carrara*, name given to the parian paste by Josiah Wedgwood & Sons. 1967 J. RICHARDSON *Courtesans* i. 22 In Blanche d'Antigny's bathroom, the bath was made of the finest Carrara marble.

carrel(l (kæ·rəl), var. CAROL *sb.* 5 a. *Obs.* exc. *Hist.*
1593, 1721 [see CAROL *sb.* 5 a]. 1890 J. W. CLARK *Cambr.* 274 They..retired thither for study after dinner, each sitting in his 'carrell' as it was called. 1893 F. MADAN *Bks. in MS.* 34 These carrels were fully open on one side of the cloister walk. 1911 L. WEAVER *House & Equipment* 39 In monastic houses reading was generally done in carrells, which were little bays in one or more of the Cloister walks.
b. A private cubicle provided in a library for use by a reader.
1919 *Library Jrnl.* XLIV. 635 The provision of carrels for research workers. 1933 *Times* 11 Dec. 19/5 Experiments in library administration, including the use of carrels and research rooms adjoining appropriate sections of the stack. 1957 V. NABOKOV *Pnin* iii. 76 He then returned to his carrell for his own research. 1960 *Medium Ævum* XXIX. 109 The study cubicles in the college library at Ampleforth are still called 'carrels'. 1970 *Libr. Assoc. Rec.* Apr. 161/2 In the Reference Library there are ten carrels and six semi-carrels, as well as open study-desks.

‖**carretera** (karetē·ra). [Sp.] A main road.
[1845 R. FORD *Handbk. Spain* I. i. 12 The cross roads and minor roads of Spain are bad...They are divided into those which are practicable for wheel-carriages, '*camino carretero*', '*de carruage*', '*carretera*', and those which are only bridle-roads.] 1900 *Engineering Mag.* XIX. 683 A *carretera*, or macadamized road is the 'best road' or route of the 'fast mail' in Porto Rico. 1923 *Blackw. Mag.* Aug. 178/1 The extreme provincialism of the Spanish *carratera*. 1924 *Ibid.* June 804/2 Very pretty they look sailing slowly along the dusty *carretara* to do the day's marketing. 1958 L. R. MUIRHEAD *Blue Guide to Northern Spain* (ed. 2) p. cxlii, Apart from the Carreteras Nacionales, however, many of the roads are stony and often under repair.

carriage. Add: **25.** *spec.* ellipt. for *railway-carriage.*
1830 E. GROSVENOR *Jrnl.* Aug. (1965) vi. 127 There were 3 steam engines and carriages attached to each. 1859 *Carriage Builders' & Harness Makers' Art Jrnl.* I. 66/1 A train started from Worcester at half-past three, p.m., and consisted of six carriages with an engine and tender. 1921 A. HUXLEY *Crome Yellow* i. 2 What right had he to sit in the sunshine, to occupy corner seats in third-class carriages, to be alive? 1958 J. BETJEMAN *Coll. Poems* 265 The train at Pershore station was waiting that Sunday night Gas light on the platform, in my carriage electric light.
33. (Earlier examples.) Also *attrib.*
1690 H. KELSEY *Papers* (1929) 2 Through Rivers wth run strong with falls [there were] thirty three Carriages five lakes in all. 1744 A. DOBBS *Hudson's Bay* 34 The River..having about 30 Falls..where they must carry their Canoes. Two or three of them are Carriages of a League or two. 1775 J. MELVIN *Jrnl.* (1857) 10 The carriage-place is about a mile in length.
35. carriage-cradle, a luggage-rack; **carriage-folk** = *carriage-company*; **carriage free,** used esp. as *advb. phr.* (examples); **carriage-house** (earlier U.S. example); **carriage paid** *a.* (with hyphen) and *advb. phr.*, for which the cost of carrying has been paid; **carriage trade,** (*b*) trade or custom from those who use carriages, i.e. from the wealthy (Webster 1909); *transf.* (*colloq.*), wealthy people collectively; **carriage-way** (earlier example and later examples; cf. *dual carriageway).
1866 *Chambers's Jrnl.* 28 July 466, I saw [him] clutch at the carriage-cradle above his head, as though it would have relieved his mind to climb. 1819 C. BROWN *Let.* 24 Jan. in *Lett. of Keats* (1931) II. 302 The whole of the family are *shuffling* to carriage folks for acquaintances, *cutting* their old friends. 1952 W. DE LA MARE in *N.Y. Times* 7 Sept. VII. 30/2 They were carriage-folk, and thoroughly genteel. 1769 J. WEDGWOOD *Let.* 11 Nov. (1965) iii. 83 He delivers them safe, and carriage free to London. 1873 *Young Englishwoman* IV. 131/2 The Little Yankee Sewing Machine..is delivered carriage free. 1761 in F. H. Stewart *Notes on Old Gloucester County, N.J.* (1917) I. 316 Their fine carriage house is finished and painted. 1771 SMOLLETT *Humph. Cl.* I. p. xv, I have sent an Almanack..directed for him at Mr. Sutton's, bookseller, in Gloucester, carriage paid. 1869 TROLLOPE *He knew*, etc. II. lviii. 68 A parcel..sent by the railway, carriage paid. 1894 *Daily News* 13 Nov. 5/4 Carriage-paid tickets, with which the carriage of letters for any place within the capital could be prepaid. 1970 *Observer* 29 Mar. 31/6 Tubular steel fruit cages...Immediate despatch. Carriage paid. 1958 *Spectator* 25 July 136/3 For the carriage trade there is still polished mahogany. 1800 *Act* 39 & 40 *Geo. III* c. 47 § 1 Where there is a regular Continuation of Carriage Way Pavement from either of the said Cities. 1957 *Times* 3 Dec. 7/7 Although the 16ft. carriageway is marked as two 8ft. lanes, in practice many of the larger lorries occupy more than one lane, so that the traffic at this point is often reduced to a single line.

Ca:rrickmacro·ss. The name of a town in Co. Monaghan, Ireland, used to designate certain forms of lace made there (see quots.).
1861 QUEEN VICTORIA *Let.* 23 Aug. in *Dearest Child* (1964) 346 A very pretty flounce...But it is not Carrick Mackross. 1864 *Morning Post* in B. Palliser *Hist. Lace* (1875) XXXV. 394 A lady's dress is described as having the train, and petticoat 'trimmed with Carrickmacross point lace'. 1888 A. S. COLE *Irish Art Lace-making* 6 Carrickmacross Appliqué Lace..stitched on to net (*appliqué*)..has an appearance somewhat similar to Brussels *appliqué* lace. 1919 R. FIRBANK *Valmouth* viii. 141 A rich muff of Carrickmacross lace. 1947 J. E. PETHEBRIDGE *Man. Lace* 11 Carrickmacross lace is the oldest lace made in Ireland. 1970 *Sunday Tel.* 22 Mar. 9/2 An industry that 10 years ago was..based on Irish tweeds, hand-knitted sweaters, and hand-made Carrickmacross lace.

carrier. Add: **1. e.** A small low detached cloud, betokening rain. *local.*
1844 [see *water-wagon*, WATER *sb.* 29]. 1884 R. LAWSON *Upton-on-Severn Words* 23 *Messenger*, a small detached cloud (*cumulus*) floating low, and supposed to betoken rain. Sometimes called a Carrier.
f. A case in which letters, etc., are enclosed for dispatch by pneumatic tube. Also, a small light capsule for carrying messages, attached to a homing pigeon.
1872 *Min. Proc. Inst. C.E.* XXXIII. 7 The carriers for the reception of telegrams, letters, or light parcels, consist of small cylinders made of gutta percha [etc.]. 1876 *Ibid.* XLIII. 60. 1892 *Pall Mall Gaz.* 18 Aug. 1/3 The actual form on which the message was written is put into a little cloth box, called a carrier, and blown through a tube to the central telegraph office. 1908 H. R. KEMPE *Engineer's Year-Bk.* 253 The carriers in which the messages are placed consist of a cylindrical tube of gutta-percha covered with felt. 1920 *Blackw. Mag.* Dec. 762/1 He took a message form, wrote a few words on it, and taking a pigeon from the basket, fixed a carrier to its leg. 1969 C. R. HILL *Pet Library's Pigeon Guide* i. 16 In a small waterproof envelope were message carriers, message pad and pencil.
g. A vessel which conveys fish from the fishing-ground to a port or market.
1883 R. F. WALSH *Irish Fisheries* 16 Many plans of steam carriers have been devised and proposed...Amongst [them] is the vessel with false hold or bottom, which allows the water to pass through with a view to bringing the fish alive to market. 1896 *Daily News* 27 Feb. 8/7 The fish..arrived in eight carriers. 1907 *Westm. Gaz.* 25 Feb. 6/3 The 'Speedwell', a steam carrier.
h. A device for filling the magazine of a gun or rifle with a group of cartridges; a charger.
1885 *Marine Engineer* 1 July 95/2 The carrier which draws the cartridges out of the belt, and deposits them in the feed wheel. 1901 *Westm. Gaz.* 21 Mar. 4/3 By means of the carrier the cartridges are dropped into the magazine receptacle and the empty carrier thrown away. 1903 C. B. MAYNE *Infantry Weapon* 139 This..can only be done by rapidly loading the rifle by means of groups of cartridges temporarily held together by 'clips' or by 'carriers', also called 'chargers'. [*Note*] The 'carrier', or 'charger', is a strip of tin that grips the cartridges by their projecting bases only; the 'clip' is a tin framework that, more or less, grips the whole cartridge-case.
i. A box, metal plate, or other contrivance attached to a bicycle or motor-cycle for carrying parcels, luggage, or a pillion-rider.
1885 *Naturalist's World* Jan. 6 A 'carrier' can be fixed on to the rod supporting the seat. 1887 BURY & HILLIER *Cycling* 201 Luggage..should be carefully affixed [to the machine] by means of one or other of the carriers described elsewhere. *Ibid.* 429 The Carrier Cycle. 1911 C. S. LAKE *Motor Cyclist's Handbk.* 253 Luggage Carrier and Stand. It is common practice to make the carrier of tubular material. 1915 B. E. JONES *Motor Cycles* 137 A very light carrier can be fitted, as this machine is not strong..enough for passenger-carrier riding. 1967 N. FITZGERALD *Affairs of Death* ii. 20 The bicycle..was a new one;..to the carrier had been affixed a small bundle.
j. A substance used as a base or supporting medium.
(i) An insoluble substance (e.g. barytes, china clay, gypsum) used as a base to receive the colouring matter in the preparation of certain pigments.
1892 G. H. HURST *Painters' Colours* 268 The base or carrier exerts a most important influence on the value of the lake as a pigment. 1915 J. C. SMITH *Manuf. Paint* (ed. 2) 170 Many so-called lakes consist simply of a carrier.. saturated with a soluble dye.
(ii) *Chem.* A supporting medium for a catalyst.
1923 E. E. REID tr. *Sabatier's Catalysis in Org. Chem.* § 946 It is best to use the palladium precipitated on an inert carrier. 1943 H. GILMAN *Org. Chem.* (ed. 2) I. ix. 785 Asbestos has also been employed as a carrier for platinum black.
(iii) An inert gas used to convey volatile material, esp. in chromatographic analysis. Also *carrier gas.*
1940 *Nature* 27 July 129/1 The vapour of the iodides at low pressure..and diluted with a carrier gas..was passed through a tube. 1955 L. B. LOEB *Basic Processes of Gaseous Electronics* ii. 185 Since carriers in gases are in constant contact with the ambient gas molecules, they partake of the random chaotic heat motions of the molecules. 1956 *Nature* 14 Jan. 84/2 Low-boiling materials prove difficult to remove from the carrier-gas stream. 1959 R. L. PECSOK *Gas Chromatogr.* i. 5 A carrier gas containing the vapor of a substance..is passed through the column.
(iv) A substance (usu. non-radioactive) that will carry a trace of a substance of similar chemical properties (but usu. radioactive) with it through a chemical or physical reaction.
1947 M. D. KAMEN *Radioactive Tracers* ii. 38 It is usual to add small quantities..of the element to be purified so that ordinary chemical manipulation is possible...Such material is called 'carrier'. 1950 *Thorpe's Dict. Appl. Chem.* X. 432/1 The carrier is simply a macroscopic concentration of an ion of similar properties to the tracer component. 1953 A. F. RUPP in J. R. Bradford *Radioisotopes* ix. 170 It is..desirable to produce as many radioisotopes as possible by other processes in which they are obtained carrier-free, or nearly so. *Ibid.* 183 Carrier-free separation of radioisotopes by adsorption of radiocolloids. 1959 G. E. FRANCIS et al. *Isotopic Tracers* (ed. 2) iii. 42 If this radioactive isotope is purified and separated from any 'carrier' or non-radioactive isotope of the same element, it is described as carrier-free.
(v) A substance that transports an ion across a cell membrane.
1957 H. H. USSING in Q. R. Murphy *Metab. Aspects of Transport across Cell Membranes* iii. 52 The principal property of the hypothetical membrane-carrier is to bind specifically the ion to be transported. 1964 G. H. HAGGIS *Introd. Molecular Biol.* vi. 181 The number of carrier molecules required to account for known membrane effects is so small that it may be impossible to detect their presence among membrane lipids by current chemical methods.
k. (i) *Chem., Biochem.* A substance that effects a transference of an element or property, etc. (See quots.)
1892 BEDSON & WILLIAMS tr. *Meyer's Theoret. Chem.* 214 Certain bodies act as carriers of chlorine in a similar way to the oxygen carriers. 1902 J. B. COHEN *Theoret. Org. Chem.* 368 If chlorine or bromine acts upon benzene in presence of a 'carrier', substitution occurs. 1921 D. LL. HAMMICK *Org. Chem.* 156 [The iron] probably acts as a halogen 'carrier', ferric bromide being first formed, and then 'handing on' bromine to the benzene molecules. 1929 *Biochem. Jrnl.* XXIII. 128 Will a hydrogen donator react with a hydrogen acceptor in the presence of their specific enzymes but in the absence of any intermediate hydrogen carrier? 1934 *Ibid.* XXVIII. 1819 These substances might catalyse carrier-linked reactions.
(ii) *Physics.* An electron, atom, or group carrying an electric charge.
1901 RUTHERFORD in *Phil. Mag.* II. 221 The velocity of the negative carrier produced by ultra-violet light is about the same as that of an ion produced by *x*-rays; ..it behaves at low pressures as if it were identical with the cathode-ray carrier. 1902 —— & BROOKS in *Phil. Mag.* IV. 20 Excited radio-activity is due to the conveyance of radioactive matter of some kind on positively charged carriers, which travel through the air in an electric field.
spec. In a semi-conducting material: any of the mobile electrons or holes which carry charges.
1939 A. H. WILSON *Semi-conductors & Metals* iv. 47 Since the 'holes' behave in many ways like positive electrons, this question can be settled by measuring any quantity which depends on the first power of the charge of the carrier. 1945 *Jrnl. Appl. Physics* XVI. 564/2 Such 'holes' will give rise to a conductivity described by Eqs. (1)–(3) but the sign of the charge carriers will appear to be positive. 1946 *Chem. Abstr.* XL. 6907 (*title*) Measurement of mobilities of charge carriers in real semiconductors. 1964 K. W. CATTERMOLE *Transistor Circuits* (ed. 2) ii. 10 The main charge carriers, holes and electrons respectively, are called majority carriers.
l. (i) A person, animal, or plant that carries a pathogenic agent and acts as a potential source of infection without suffering from clinically obvious disease; sometimes restricted to organisms that are hosts of the pathogenic agent (cf. *VECTOR).
[1899 *Leisure Hour* 176/2 Two-winged flies..act as carriers of disease.] 1906 *Medical Annual* 182 Carriers [of diphtheria] without symptoms or demonstrable contact. 1908 [see TYPHOID B. c]. 1910 R. ROSS et al. *Prevent. Malaria* 195 If the local carrier belongs to a species..which feeds almost entirely upon man. 1930 *Discovery* June 198/2 The human carrier of typhoid fever and other diseases... A person, himself immune, bears the germ of the disease and is capable of infecting other healthy but more susceptible persons. 1937 F. D. HEALD *Introd. Plant Path.* xvi. 319 A virus may exist in a plant without causing any external evidence of its presence, and such symptomless 'carriers' may yield the virus to insects feeding upon them. 1955 GAIGER & DAVIES *Vet. Path. & Bacteriol.* (ed. 4) vii. 152 Convalescent animals are often carriers...In some diseases..the carrier state may remain for years and the animal becomes a danger to other susceptible animals.

(ii) *Genetics.* An organism that has a recessive gene for some genetic characteristic (as a hereditary disease) and so can pass the characteristic on to a descendant although not itself showing it.
1933 *Brit. Jrnl. Psychol.* July 7 An intermediate type of inheritance of mental defect in which the heterozygote or 'carrier' is recognizable as a dull person. 1944 DARLINGTON in *Nature* 5 Aug. 167/1 Viruses..[are] suppressed by some host genotypes and permitted by others...There are, therefore, 'susceptible' and 'carrier' genotypes...Infection of one susceptible species can take place from another through an immune carrier species. 1970 *Observer* 12 Apr. 25/4 It's estimated that every human being is a carrier for about half a dozen lethal and perhaps a dozen crippling recessive genes.
m. = *aircraft carrier. Also *attrib.*
1917 W. L. WADE *Flying Book* 12 The big cruising seaplane, operating either from port or from a seaplane carrier. 1919 L. R. FREEMAN *To Kiel* 114 Aeroplanes launched from the 'carrier' *Furious*. 1922 *Advis. Committee for Aeron., Techn. Rep. 1918–19* I. 29 A large amount of wind channel work has been done on models of different aeroplane carrier ships. 1958 *New Statesman* 28 June 822/2 Fighters and bombers operating from the fleet's three carriers.
n. *Telecomm.* An electromagnetic wave or an alternating current or voltage that is modulated by a wave, etc., of lower frequency representing the signal to be transmitted.
1921 *Jrnl. Inst. Electr. Engin.* Apr. 402/2 The most valuable feature of the high-frequency carrier-wave system is its adaptability to multiplex telephony. 1921 COLPITTS & BLACKWELL in *Ibid.* 412/2 Carrier-current Telephony and Telegraphy. 1922 *Encycl. Brit.* XXXII. 712/1 The high frequency current serves as a 'carrier' for the telephone current over the line. 1923 *Electr. Communication* I. iv. 16 In a carrier current system a number of telephone or telegraph messages are simultaneously superposed on a single pair of wires by means of high frequency currents of different frequencies on which individual messages are impressed. 1929 E. MALLETT *Telegr. & Teleph.* ix. 249 The frequency ω_1 is called the *carrier* frequency. 1931 *B.B.C. Year-Bk.* 437/2 *Carrier Wave*, the high frequency oscillations emitted by a wireless telephone transmitter. These are modulated during telephony. The analogy is that the telephony (music, speech, etc.) is 'carried' by the high-frequency oscillations from the transmitter to the receiver. 1933 *Ibid.* 443 The television signals are made to modulate the carrier-wave sent out by the broadcasting station. 1962 A. NISBETT *Technique Sound Studio* 254 Unless two carriers on the same wavelength have almost the same strength, the stronger 'captures' the area. *Ibid.* 267 Systems of carrier modulation in use are amplitude modulation and frequency modulation.
o. A carrier-bag.
1936 W. HOLTBY *South Riding* I. v. 58 Mothers with laden paper carriers. 1959 S. DELANEY *Taste of Honey* I. i. 7 Pass me that bottle—it's in the carrier.
2. b. (With capital initial.) [See quot. 1906.] A people of Athapascan Indians inhabiting British Columbia; a member of this people; also, the language of this people. Also *attrib.*
1793 A. MACKENZIE *Voy. fr. Montreal* 3 July (1801) II. vii. 284, I found no difference in their language from that of the Nagailas or Carriers. 1820 D. W. HARMON *Jrnl. Voy. & Trav.* 403 (*heading*) A specimen of the Tacully or Carrier tongue. 1845 C. WILKES *U.S. Exploring Exped.* IV. xiii. 480 This part..is inhabited by the two great nations of the north, the Takali, and Atnaks or Shouswaps: the former are also known by the name of the Carriers. 1846 [see *ATHAPASCAN *sb.* 1]. 1890 FRAZER *Golden Bough* I. iii. 239 Amongst the Takilis or Carrier Indians of North-West America. 1906 A. G. MORICE *Hist. N. Int. Brit. Columbia* 6 Among the Carriers, the widow of a deceased warrior used to pick up from among the ashes of the funeral pyre the few charred bones..and *carry* them on her back in a leathern satchel—hence the name of the tribe—until the co-clansmen had amassed a sufficient quantity of eatables and dress skins to be publicly distributed..in the course of an ostentatious ceremony called 'potlatch'. 1921 E. SAPIR *Language* iv. 71 Athabaskan languages..including..Navaho..Carrier, Chipewyan. 1957 *Encycl. Canadiana* II. 253 The Carrier have not made a successful adjustment to European civilization.
3. a. (Later example.)
1903 E. R. JOHNSON *Amer. Railway Transp.* 124 A copy of the bill must be sent to each of the railroads..and to any freight association of which the carrier may be a member.
7. *carrier-rocket*; **carrier-bag,** a strong (usu. paper) bag fitted with a handle; **carrier-based** or **-borne** *a.*, of aircraft: operating from or accommodated on an aircraft-carrier.
1907 *Yesterday's Shopping* (1969) 339/3 The 'Sensible' carrier bag. With strings..is the only paper Bag with a firm bottom, and capable of carrying wet fruit, pastry, &c., without..bursting the bag. 1956 *Times* 7 Aug. 6/7 Most of the new arrivals stepped out of the train with nothing more than one small suitcase or a paper carrier-bag. 1935 *Flight* 9 May 498 American carrier-based aircraft. 1968 *New Scientist* 1 Feb. 230/1 United States bases in the Pacific were warned to expect a carrier-based air strike.. some days before the attack upon Pearl Harbour. 1939 K. EDWARDS *Navy of Today* iv. 81 The number of carrier-borne aircraft accounted for by the ships already building. 1946 H. HARPER *Dawn of Space Age* p. vii, The large 'carrier' rocket would have a flat nose, and to this would be attached the small rocket. 1959 *Listener* 18 June 1057/1 The carrier rocket of the Russian sputnik.

carrion. Add: **C. carrion-beetle,** any beetle of the family *Silphidæ*, which feed on carrion.
1817 KIRBY & SPENCE *Entomol.* II. xxi. 242 Those unclean feeders, the carrion beetles (*Silphæ*, L.)..are at the same time very fetid. 1959 E. F. LINSSEN *Beetles* I. 159 Burying beetles, carrion beetles, rove beetles, etc.

‖**carro** (kā·ro). [Pg.] In Madeira, a sledge usu. drawn by bullocks.
1882 E. M. TAYLOR *Madeira* 198 Those Mount sledges, or *carros*, in which people are impelled down the steep Mount road at a very rapid rate. 1900 A. J. D. BIDDLE *Madeira Isl.* I. 118 The *carros* (bullock-cars mounted on runners). 1928 J. E. HUTCHEON *Things seen in Madeira* iii. 64 Madeira is indebted to Englishmen for two of these unique modes of conveyance, the bullock-carro, and the running-carro of the mountains. *Ibid.* 66 There are ox-carros and mule-carros. 1953 E. NICHOLAS *Madeira & Canaries* viii. 59 In the old days horses, mules, and bullock carros were used as a means of transport.

Carrollese (kæ·rǒlīz), **Carrollian** (kærǫ·liăn), **Carrolline** (kæ·rǒlain), *adjs.* [See -ESE, -IAN, -INE¹.] Resembling, or characteristic of, the style of 'Lewis Carroll' (C. L. Dodgson (1832–98), author of 'Alice's Adventures in Wonderland'). So **Carrollia·na** [see *-IANA].
[1899 S. D. COLLINGWOOD *L. Carroll Picture Bk.* vii. 321 (*heading*) Miscellanea Carrolliana.] 1907 *Westm. Gaz.* 25 July 2/3 There is nothing Carrollese about the lines. 1924 S. H. WILLIAMS (*title*) Some rare Carrolliana. 1927 *Observer* 3 Apr. 8 The book has the Gilbertian or, rather, Carrolline, title of 'Ships and Sealing Wax'. 1932 *Times Lit. Suppl.* 28 Jan. 49/2 The Carrollian heroine. 1963 *Listener* 14 Mar. 474/3 The King Shag, which, with Carrollian panache, makes its nest from guano, scurvy-grass, and mesembryanthemum. 1969 J. BROWN *Rhapsody of Words* 70 The Carrolline adjective galumphing. 1970 *Guardian* 10 Apr. 10/4 Membership of the society..includes most of the important collectors of Carrolliana in this country and abroad.

carrot, *sb.* Add: **2. a.** *fig.* (With allusion to the proverbial method of tempting a donkey to move by dangling a carrot before it.) An enticement, a promised or expected reward; freq. contrasted with 'stick' (= punishment) as the alternative.
1895 *Westm. Gaz.* 24 Aug. 2/2 Among other carrots dangled before the electors last month was Bimetallism. 1916 E. W. GREGORY *Furnit. Collector* 228 The spectacle of an otherwise intellectual individual engaged in trying to plumb the depths of duplicity to which dealers can descend in faking old furniture is like that of the donkey pressing eagerly forward after the dangling carrot. It would..be very pleasant to possess the carrot of complete knowledge, but the conditions render it impossible. 1948 *Economist* 11 Dec. 957/2 The material shrinkage of rewards and the lightening of penalties, the whittling away of stick and carrot. 1954 J. A. C. BROWN *Social Psychol. of Industry* i. 15 The tacit implication that..most men..are..solely motivated by fear or greed (a motive now described as 'the carrot or the stick'). 1963 *Listener* 21 Feb. 321/2 Once Gomulka had thrown away the stick of collectivization, he was compelled to rely on the carrot of a price system favourable to the peasant.
b. (Later U.S. examples.)
1857 *Trans. Ill. Agric. Soc.* II. 360 The creoles manufactured the tobacco into carrots, as they were called. A carrot is a roll of tobacco twelve or fifteen inches long, and three or four inches in diameter at the middle of the roll, and tapered towards each end. 1890 *Congress. Rec.* 27 Aug. 9213/2, I have here some carots [*sic*] of Cuban tobacco.
5. carrot (rust) fly, a small fly (*Psila rosæ*) whose larva feeds mainly on carrots.
1840 J. & M. LOUDON tr. *Kollar's Treat. Insects* II. 161 The larva of the carrot-fly is cylindrical. 1882 [in Dict.]. 1951 COLYER & HAMMOND *Flies Brit. Isles* xv. 197 The common Carrot Fly or Carrot-rust Fly, an agricultural pest of some significance.

carrot (kæ·rət), *v.* [f. the sb., from the yellow colour imparted to the fur.] *trans.* To treat (fur) with nitrate of mercury (see quot. 1906). Cf. CAROTING *vbl. sb.* and SECRETAGE.
1862 *Chambers's Encycl.* IV. 560/1 Furs have their felting property sometimes increased by the process of *carroting*, in which the action of heat is combined with that of sulphuric acid. 1879 *Encycl. Brit.* IX. 837/1 Furs..of the hare and the rabbit,..dressed, carroted, and cut from the skin. 1906 WATSON SMITH *Chem. Hat Manuf.* 17 The *sécretage* or 'carrotting' process..consists in a treatment with a solution of mercuric nitrate in nitric acid, in order to improve the felting qualities of the fur.

carroty, *a.* **a.** (Later examples.)
1887 *Daily News* 9 Dec. 8/5 He was a tall, carroty man. 1912 C. N. & A. M. WILLIAMSON *Heather Moon* III. 297 The plainest, oldest, and carrotiest of the three red-headed maids. 1947 W. S. MAUGHAM *Creatures of Circumstance* 16 A shock of untidy carroty hair.

carrousel, var. CAROUSEL (in Dict. and Suppl.).

carry, *v.* Add: **1. d.** *intr.* (for *pass.*). Of soil: to stick to the feet, or to horses' hoofs.
1892 *Field* 30 Jan. 155/2 The frost..caused the fallows and seeds to 'carry' a good deal, and they could only hunt very slowly. *Ibid.* 156/3 An expanse of ploughed soil which 'carries' considerably.
9. b. *Golf* and *Cricket.* Of the ball, or the player hitting it, or the club, etc.: to cover (a distance) or pass (a point) at a single stroke. Also *absol.* Cf. *CARRY *sb.* 4 b.
1875 R. CLARK *Golf* 213 The balls carried considerably higher than the weather-cock. 1887 W. G. SIMPSON *Golf* 155 Many prefer it [*sc.* a dragging shot] to an ordinary loft at shortish distances, the latter being more difficult within, say, thirty yards than when the player has further to carry, round, or play short of, a bunker, it ought to be allowed to try to carry. 1903 J. BRAID in Benson & Miles *Bk. Golf* 30 At a greater distance where the mashie will not carry I should very often use an iron. 1929 *Morning Post* 17 June 16/2 Off the next ball—a full-pitch—he only just failed to carry the ring. 1953 B. LOCKE *On Golf* II. xvi. 117 It is important at times to be able to hit very high iron shots to carry formidable obstacles, such as big bunkers or even trees.
c. *intr.* Of sound: to travel or be heard at a distance. Cf. *CARRYING *ppl. a.* b.
1896 M. CORELLI *Mighty Atom* iii. 54 Lionel's voice could not now 'carry' far enough to echo the farewell. 1932 E. V. LUCAS *Reading, Writing* ii. 45 He [*sc.* Andrew Lang] had a voice that did not carry—'roupy' he himself called it. 1934 *Discovery* Dec. 354/2 The sound..carries remarkably well.
16. c. To gain (a district, etc.) in an election. (Cf. 17.) *U.S.*
1848 LOWELL *Biglow P.* ix. 124, I thought our ticket would ha' caird the country with a resh. 1905 D. G. PHILLIPS *Plum Tree* 122 We, our party, carried the state, as usual. Our legislative majority was increased by eleven. 1965 T. C. SORENSEN *Kennedy* viii. 212, I told him [*sc.* J. F. Kennedy]—mistakenly as it turned out—that he had carried California.
19. b. *to carry the war into the enemy's camp*, etc.: to move the scene of fighting to the enemy's camp, to take up the attack; freq. *fig.*; *to carry the war into Africa* (*U.S. colloq.*): to act aggressively, to go over to the attack.
1828 *Reg. Deb. Congress* IV. I. 1315/1, I shall not..act in mere self-defence. I shall carry the war into Africa. 1835 R. J. MACKINTOSH in J. Mackintosh *Life* I. 81 Nor was this production altogether defensive; the war was now and then carried into the enemy's quarters. 1845 *Q. Rev.* II. 162 Having in so far attempted to vindicate Condorcet, we carry the war into the enemy's camp by asserting that Lord Brougham's biography is obnoxious to all the charges. 1855 J. B. JONES *Winkles* 202 But the way to be redressed.. is to carry the war into Africa. 1927 L. P. HARTLEY in C. Asquith *Black Cap* 47 'I don't think Rollo is slow,' remarked Jimmy, hoping to carry the war into her country. 1938 H. MCCLOY *Dance of Death* xx. 237 Her smile annoyed Basil. He carried the war into Africa. 'Are you aware that M. Pasquale takes morphine?'
38. c. To hold or keep on hand (securities, merchandise, a stock, etc.). orig. *U.S.*
1848 W. ARMSTRONG *Stocks* 10 It is nominally considered that the stock is meanwhile 'carried' or possessed by the seller. 1869 J. H. BROWNE *Gt. Metrop.* iii. 48 (Funk), Operators can..carry such an amount of stocks as astounds the weaker ones of the street. 1870 MEDBERY *Men & Myst. Wall St.* 77 When a broker agrees to 'carry' stock, he says, Seven per cent, unless the market tightens. 1898 *Westm. Gaz.* 14 Jan. 8/3 The only remedy we see..is for the sufferers to carry smaller stocks. 1917 TWYFORD *Purchasing & Storing* 325 It is not economical to carry in stock several variations of articles of a similar nature. 1930 *Publishers' Weekly* 11 Jan. 214/1 Mr. Brady explained that he did not carry the book. 1963 J. MITFORD *Amer. Way Death* 233 *Casket and Sunnyside* is carried in only six eastern libraries.
d. To maintain or keep up with financial (or other) support. orig. *U.S.*
1879 *Bradstreet's* 8 Oct. 4/4 He is forced to pay on loans necessary to 'carry' the farmer. 1883 *Harper's Mag.* Nov. 877/2 The men of business..have for years carried the New York Academy of Music. 1917 ATWOOD *Exch. & Specul.* 51 In common parlance the customer trades on a 'margin'. Expressed in another way the broker 'carries' the customer for all except a small part of the cost. 1931 *Times Lit. Suppl.* 19 Feb. 124/3 When the production period of capital goods ends, therefore, there will be no savings to 'carry' them, and the boom must end. 1944 *Ann. Reg. 1943* 157 The Central Government could not continue indefinitely to 'carry' a province to which Nature had given so generous a crop. 1947 J. BERTRAM *Shadow of War* VII. i. 211 If one man 'swings the lead' in a coal gang of four, the other three must 'carry him' by working all the harder. 1959 G. SLATTER *Gun in Hand* ii. 23, I been carryin ya all mornin.
40. Also, to maintain (a population).
1905 *19th Cent.* Nov. 816 Mackay, with back country carrying about 15,000 people.
b. Of a journal, newspaper, etc.: to print (an article) in its pages. Also *transf.*, to broadcast. orig. *U.S.*
1926 *Publishers' Weekly* 22 May 1676/1 There are many towns in which the newspapers do not carry book reviews. 1927 *Ibid.* 12 Feb. 609 *Publishers' Weekly* of January 8th carried a letter signed by a committee of the American Booksellers' Association..which commented adversely on the Literary Guild. 1929 E. WALLACE *Kennedy the Con Man* iv. in *Red Aces* 178 We carry big ads. in all the papers. 1946 D. L. SAYERS *Unpopular Opinions* 128 Any proposal to control the branded goods..will be violently opposed (on the loftiest hygienic grounds) by the papers that carry the branded advertising. 1957 *BBC Handbk.* 73 The Light Programme carries the People's Service in the morning and the 'Sunday Half-hour' of community hymn singing in the evening. *Ibid.* 170 The German Service carried programmes and discussions on this subject.
51. carry off. e. To take away, abduct, steal.
1817 JANE AUSTEN *Sanditon* (1954) viii. 405 If she could not be won by affection, he must carry her off. 1829 PEACOCK *Misf. Elphin* xi. 141 The mountain sheep are sweeter, But the valley sheep are fatter; We therefore deemed it meeter To carry off the latter. 1855 [see sense 5 b in Dict.]. 1969 G. PAYTON *Payton's Proper Names* 266/1 *Lochinvar*, the hero of a song in Scott's *Marmion* who carries off the fair Ellen at her wedding feast.
52. carry on. d. Also, in military use, to continue as before, resume the former situation or

occupation; to proceed to carry out instructions, to 'go ahead'. Hence *gen.*, to 'keep going', to persist; to make the best of things.

1909 *Daily Chron.* 24 July 4/4 'Carry on!' is a word they have in the Navy. It is the 'great word' of the Service... To-morrow the workaday life of the Fleet begins again, and the word will be, 'Carry on!' **1915** 'Bartimeus' *Tall Ship* i. 14 The ship..began to heel slowly over. The Captain..raised the megaphone to his mouth. 'Carry on!' he shouted. 'Every man for himself.' **1915** A. D. Gillespie *Lett. from Flanders* (1916) 183 All except the officers were just carrying on as usual, which meant that, except for the sentries, you could see nothing except boots sticking out from the dug-outs. **1915** 'Ian Hay' *First Hundred Thou.* xiv. 190 'Do you understand my order?' thundered the Colonel.. 'I do, sir,' replied Blaikie politely, 'but—' 'Then, for heaven's sake, carry on!' **1919** —— *Last Million* vii. 97 I'm not one to ask for sympathy when there's others needs it more... Carry on—that's my motto! **1927** C. E. Montague *Right off the Map* 156 'We'll let the men carry on resting,' said Willan. **1932** 'F. Iles' *Before Fact* xv. 261 Linda said..'I want to talk to you.' 'Carry on, sergeant,' said Mr. Thwaite amiably. **1940** *War Illustr.* 19 Jan. 623 But War caught these essential transport toilers at their job, and in the good British spirit they—not perhaps without an excusable grumble—felt they could but 'carry on'.

e. Also *spec.* (*a*) To behave, esp. to speak, in a rowdy, excited, or bad-tempered way; (*b*) to engage in flirtation, esp. of a dishonourable nature, to have an affair (with).

1828 Mrs. Royall *Black Bk.* II. 27 They romped and squalled, and to use a Yankee phrase, 'carried on' at such a rate that he and Mrs. C. were greatly annoyed. **1876** Besant & Rice *Golden Butterfly* xxxv, She and I carried on for a whole season. People talked. *a* **1876** E. Leigh *Gloss. Cheshire* (1877) 37 *Carry on, v.* 'She carried on shameful', *i.e.* she used very unladylike language, or she shewed bad conduct. **1892** R. L. Stevenson *Uma* i, in *Illustr. London News* 2 July 11/1 What's she carrying on about? *Ibid.* ii, 9 July 42/1 There was Adams in the middle, gone luny again, and carrying on about copra like a born fool. **1930** W. S. Maugham *Cakes & Ale* viii. 92 It was impossible that she could be 'carrying on' with Lord George. **1947** 'N. Shute' *Chequer Board* 4 She don't half carry on about the beer I drink.

53. carry out. c. (Earlier and later examples.) Also freq. with omission of *out*. So *to carry one's bat through*: to go in first and remain undismissed at the end of the innings.

[**1833** J. Mitford in *Gentl. Mag.* Sept. 236/1 Tom scored the amazing number of 95 runs in his first innings, and brought his bat out with him.] **1833** *New Sporting Mag.* Sept. 325 Take care..or through the game Your bat you will not carry. **1846** W. Denison *Cricket: Sk. Players* 18 Brown carried his bat out with 112 runs marked against his name. **1867** G. H. Selkirk *Guide Cr.-Ground* ii. 23 If the player carrying his bat out was one of the two who first went in, he is said to have carried his bat through. **1882** *Cliftonian* June 228 The former carried his bat for a lucky 14. **1895** *Westm. Gaz.* 18 May 7/1 Grace has carried his bat twenty-two times when scoring centuries. **1933** D. L. Sayers *Murder must Advertise* xviii. 311 The satisfaction of carrying out his bat for 14.

54. carry over. c. To transfer.

1889 E. Carpenter *Civilis.* iv. 105 The ideal passion of that period..was that of comradeship, or male friendship carried over into the region of love.

carry, *sb.* Add: **4. b.** *Golf.* The distance between the spot from which a ball is struck and that where it first lands; also, the trajectory of the ball. Cf. *carry *v.* 9 b.

1887 W. G. Simpson *Golf* 112 Getting both hands well under the club also produces a low carry. **1890** H. G. Hutchinson *Golf* xvii. 445 *Carry*, the distance from the place where the ball is struck to the place where it pitches. Hence a *long carry*, and a *short carry*. **1896** W. Park *Golf* 104 A well hit drive should be almost all carry; the ball should not run any distance after it falls. *Ibid.* 262 A long carry or a short carry are used to signify the distance a ball must be lofted usually over a hazard. **1899** *Golf Illustrated* 29 Dec. 319/2 The carry alone is estimated to have been close on two hundred and fifty yards. **1953** B. Locke *On Golf* II. xvi. 118 You must not take a divot with this shot, otherwise you will not get the loft and the carry needed.

carry-all, carryall. a. (Earlier example.)

1714 J. Stoddard *Jrnl.* in *N.-E. Hist. & Gen. Reg.* V. 27 Mr. Longuille sent a carryall for us.

ca·rry-away. [f. *carry away*, Carry *v.* 46 c.] In *Yachting*, the breakage of a spar, rope, etc.

1901 *Westm. Gaz.* 23 May 5/1 The 'carry-away'..was the most serious that could have happened. First bowsprit snapped, and topmast bent to leeward. **1928** *Daily Tel.* 11 Sept. 15/6 Mrs. Percy Sabel's Widgeon had a carry-away, and was forced to give up when well placed.

ca·rry-cot. [Cot *sb.*[4]] A portable cot for a baby.

The proprietary name *Karri-Kot* and variants also occur. **1943** H. Croome *O Western Wind* xvi. 131 He slept all day long in his carry-cot. **1951** A. Baron *R. Hogarth* 140 Young couples..carrying two-handled carry-cots between them from which..babies bellow.

carry-forward. *Comm.* [f. *carry forward*, Carry *v.* 49.] A balance of money carried forward, esp. after providing for a dividend, reserves, etc.

1898 *Westm. Gaz.* 17 May 8/1 The last carry-forward was £132. **1901** *Ibid.* 7 Aug. 7/1 This is one of the few companies which does not announce its carry forward with the dividend. **1959** *Economist* 28 Feb. 817/2 The ordinary dividend..is paid by drawing upon the carry-forward.

carrying, *vbl. sb.* Add: **2. carrying-over** = *carry-over a.

1907 Poley & Gould *Stock Exch.* 168 The General Contango Day (these days are also known as Continuation or Carrying-over Days). *Ibid.* 175 Where the broker is himself the taker-in rendering a carrying-over note, he is not entitled to charge both commission and contango. **1910** H. Withers *Stocks & Shares* 277 If no charge is made for carrying over, a full commission is usually paid when the bargain is finally closed.

3. *carryings-on* (later examples); also *sing.*

1866 Brierley *Fratchingtons* i. 5 Theau'd weary th' patience ov a jackass wi' thi carryins on. **1890** J. Service *Notandums* xi. 80 What carryin's on have I no seen there! **1909** Masefield *Trag. Nan* i. 13 You'll let 'er marry 'im, after 'is carrying on along o' Jenny?

4. carrying-capacity, the number of people or animals (esp. sheep or cattle) that a given area of land will support; also *transf.*; **carrying-chair,** a chair in which a person is carried; **carrying-place** (earlier Amer. example).

1883 'Mark Twain' *Life on Mississippi* 570 Demands made upon their [boats'] carrying capacity. **1930** L. Acland *Early Canterbury Runs* vii. 151 Hoare.., by ploughing, had raised the carrying capacity [of the station] to over 30,000 sheep in 1890. **1958** *Geogr. Rev.* XLVIII. 5 By relating actual to potential use the capability of the land to support population—that is, its 'carrying capacity'—can be measured. **1880** *Queen* 13 Mar. (Advt.), Invalid furniture.. carrying chairs, £2. 16s. 6d. **1894** *Outing* (U.S.) XXIV. 129/2 The Chinese mandarin..when seated proudly in his fancy carrying-chair. **1905** *Westm. Gaz.* 25 Feb. 16/3 The carrying-chair used by Leo XIII. **1689** in *Mass. Hist. Soc. Coll.* (1861) 4th Ser. V. 221 Then we marched down to.. several of the carrying-places.

carrying, *ppl. a.* Add: **b.** Of sound: far-reaching, penetrating. Cf. *carry *v.* 9 c.

1893 Yonge & Coleridge *Strolling Players* ii. 9 A small, slight girl..sending her clear, sweet 'carrying' voice before her. **1932** *Punch* 6 Jan. 14/1 My offices are just above and you have a carrying voice. I am not known as Towser to the staff. **1940** Wodehouse *Eggs, Beans & Crumpets* 156 Beamish's voice is of a robust and carrying timbre.

2. Special combinations: **carrying comb** *Spinning* (see quots.); **carrying party** *Mil.*, a party detailed to carry or bring up supplies.

1866 *Brit. Pat.* 1123 5 In Figure 6 an arrangement is shown for obtaining motion to the parts from a column in the centre of the main circle of carrying combs. **1884** W. S. B. McLaren *Spinning* (ed. 2) 94 The carrying-comb carries off the wool from between its two plates. **1889** Burnley *Wool & Woolcombing* 269 The carrying comb advances in as near as possible a perpendicular position close to the nipper mouth and takes off the tuft of fibres. **1884** *Instr. Mil. Engin.* (*H.M.S.O.*) (ed. 3) I. II. 109 A certain number..will be told off as carrying party. They will be provided with bags of hay, shavings, wool, &c., boards, fascines. **1928** Blunden *Undertones of War* ix. 95 A carrying-party from another battalion was to meet me in Hamel.

carry-on (kæ·ri,ǫn). *dial.* and *slang.* [f. *carry on*, Carry *v.* 52 e in Dict. and Suppl.] Fuss, to-do, excitement; carryings-on.

1890 J. Service *Notandums* v. 29 Byla [bailie] Stech, who had been cheated out of his dinner by a' the carry on. **1927** *Scots Observer* 23 Apr. 2/3 Ladies of the Hannah More type lamented the 'voluptuous' carry-ons of miners. **1948** L. Walmsley *Master Mariner* i. vii. 123 There's a different carry on up here than what we've had at the other bazaar concerts. **1959** P. Bull *I know the Face* vi. 111 We were all engaged for a radio version of Hamlet...I had never realized the incredible carry-on connected with these productions. **1963** V. Gielgud *Goggle-Box Affair* iii. 29 Hargest was completely sold on Eurovision and all that carry-on. **1964** *Guardian* 2 Mar. 7/6 They want madmen and blood and wild carry-ons and all that.

carry-over. [f. *carry over*, Carry *v.* 54 b.] On the Stock Exchange: postponement of payment of an account until the next settling-day; the amount so kept over. Also *attrib.* Cf. Contango.

1894 *Daily News* 29 Jan. 2/5 Grand Trunk stocks are from 2 to 4 per cent. higher than at the last 'carry over'. **1895** *Ibid.* 19 Oct. 2 The carry-over price was fixed at 2¼. **1922** *Daily Mail* 12 Dec. 3 It was Carry-over day in the Mining market, and 'new-time' dealing there was on a small scale. In view of to-day being the general Carry-over, there was a certain amount of realising.

b. *transf.* Something remaining or transferred from one period, process, etc., to the next; a balance carried forward; a transference.

1925 *London Teacher & Lond. School Rev.* 13 Nov. 393/2 The Board in 1921 announced suddenly that they would not pay grants on the full carry-over of the three Burnham scales. **1927** *Observer* 27 Nov. 21/2 A very heavy carry-over of work from the adjournment on December 21 is inevitable, as hundreds of cases are awaiting trial in the King's Bench Division. **1931** *Economist* 17 Jan. 104/1 To make room for the absorption of the present unprecedentedly huge carry-over by restricting further production or export during the coming five years. **1934** Balmer & Wylie *After Worlds Collide* 43 'He was indubitably wrong about hundreds of other theories.' Tony grinned at this carry-over of this curious man's prejudices and attitudes. **1940** H. G. Wells *All aboard for Ararat* iii. 85 Will they evaporate with the rest of their class? Will there be no carry-over from them for the Ark? **1952** *Scrutiny* XVIII. III. 197 *Women in Love* is wholly self-contained, and, for all the carry-over of names of characters, has no organic connexion with *The Rainbow*.

1955 J. G. Davis *Dict. Dairying* (ed. 2) 107 Carry-over in many [milk-bottle-washing] machines of the ordinary detergent solution is inevitable.

Carshuni, var. *Garshuni.

carstone (kā·ɹstōᵘn). *Geol.* [app. = *carnstone*, for *quarn stone*, local form of Quernstone.] A species of the Lower Greensand.

1815 W. Smith *Mem. Map Strata Eng. & Wales* 43 Iron Sand, or Carstone. **1822** Conybeare & Phillips *Outl. Geol. Eng. & Wales* 136 Iron Sand...We are able to trace and identify the present series throughout the island. In different counties it has received the name of Carstone and Quern stone. **1840** Spurdens *Suppl. Forby's Voc. E. Anglia* (1858), *Car-stone*, a peculiar stone, found principally near Swaffham. **1876** Woodward *Geol. Eng. & Wales* 230 The hard beds, locally termed Carstone (or Quern stone), are worked for building-purposes. **1940** *Nature* 8 June 902/1 A carstone rubble bed. **1961** M. W. Barley *Eng. Farmhouse & Cottage* II. iv. 103 Chalk, like carstone, was little used before the end of the eighteenth century.

cart, *sb.* Add: **4. b.** The carapace or upper shell of a crab. *colloq.* or *dial.*

1860 Hotten *Dict. Slang* (ed. 2) s.v. *Carts*, In Norfolk the carapace of a crab is called a *crab cart*. **1887** *Girl's Own Paper* 16 July 661/2 Pick the meat from the cart, the breastplate, and the claws.

5. (Later examples.) Also *in the cart*: in an awkward, false, or losing position; in serious difficulties (cf. sense 2 c in Dict.). *slang.*

1705 E. Ward *Hud. Rediv.* II. iii. p. 26 Excuse me, that the Muses force The Cart to stand before the Horse. **1863** Kingsley *Water-Babies* iv. 176 They..having, as usual, set the cart before the horse, and taken the effect for the cause. **1889** *Evening Standard* 25 June 1/3 In two races..Sir George Chetwynd—to use a vulgarism—had been 'put in the cart' by his jockey. **1908** *Punch* 12 Feb. 110 We were simply all over 'em, and had 'em in the cart in no time. *Ibid.* 4 Nov. 334 Something..tells me I am fairly in the cart. **1909** A. Bennett *Literary Taste* 27 The passionate few do not read 'the right things' because they are right. That is to put the cart before the horse. **1914** W. S. Maugham *Smith* 13 Emily was left in the cart? **1924** J. B. Hobbs *Cricket Mem.* 158 We made 238, which was enough practically to put South Africa hopelessly in the cart. **1945** R. Hargreaves *Enemy at Gate* 152 There is a glib saying..that 'trade follows the flag'; an apophthegm that succeeds in putting the cart before the horse with greater aplomb than almost any other cant phrase in common use. **1966** G. W. Turner *Eng. Lang. Austral. & N.Z.* vi. 137 Sargeson's French translator..had some slang shared with English to translate, and job (boulot), tight (noir), spree (nouba), in the cart (dans le pétrin)..are neatly managed.

6. cart-body (examples); **cart-lodge** [Lodge *sb.* 1 c] *local*, a shed or out-house where carts are kept; **cart-road** (earlier example).

c **1325** *Gloss. W. de Bibbesw.* in Wright *Voc.* (1857) 167 *Le chartil*, the carte body. **1407** in Kennett *Par. Antiq.* (1818) II. 213 Et pro sarratione et dolatione unius Cartbody..vjdd. **1779** in *Narrag. Hist. Reg.* I. 93 Went to Tower hill for a cart body. **1834** W. G. Simms *Guy Rivers* 414 The conflagration..destroyed his cart-body and calicoes. **1888** [see Lodge *sb.* 1 c]. **1915** Kipling *Div. Creatures* (1917) 432 A little cart-lodge was sliding sideways amid a clatter of tiles. **1933** *Brit. Birds* XXVII. 25 A Blackbird..built a nest on the horizontal beam..above the open front of a cart-lodge in south-west Kent. **1607** Topsell *Foure-Footed Beasts* 536 The Shrew, which falling by chaunce into a Cart-roade or tracke doth die vpon the same.

cart, *v.* Add: **1. a.** *to cart off* or *away*: to carry off or away in a cart; hence *gen.* to carry off, take away, remove.

1881 *Punch* 8 Oct. 166/2 Napier and Havelock might be carted off to join the poor Duke of Kent at the top of Portland Place. **1891** Farmer *Slang.*

c. *Anglo-Indian.* (See quot.)

1837 T. Bacon *First Impr. Hindostan* I. 137 Carting a girl, or riding out with her, is considered in India as a regular publication of the banns.

d. *Cricket.* *trans.* and *intr.* To hit (a ball) hard and high. So with the bowler as obj.

1903 Wodehouse *Tales St. Austin's* 35 Half-way through the over I carted a half-volley into the pav. **1903** —— *Prefect's Uncle* xiii. 188 'When have you seen him?' 'In a scratch game between his form and another. He was carting all over the shop. Made thirty something.' **1961** *Times* 30 May 4/1 He carted Halfyard to square leg for six.

4. *trans.* To put (someone) 'in the cart' (see prec. 5); to cause to feel let down. *slang.*

1889 Barrère & Leland *Dict. Slang* I. 228/2 An owner is said to be 'in the *cart*', or *carted*, when his horse is prevented winning by some fraud on the part of those in his employment. **1923** T. E. Lawrence *Let.* 13 June (1938) 425 A temporary job at a high salary would only cart me worse than ever in the end. **1934** —— *Let.* 16 Nov. 825 Poets are always..savagely political: and the real politician..always carts them properly. Poets hope too much. **1948** *Hansard* CDXLVIII. 1054 Many had lived in camp according to the promised scale of pay, with the result that they lived beyond their means. To use an Army expression, 'They were properly carted'.

‖**carte d'identité** (kart didantite). [Fr.] An identity card; a document which gives personal particulars.

1923 *Michelin Guide to Gt. Brit.* (ed. 7) III. 861 Passports are indispensable...It is also advisable to obtain a 'Carte d'Identité' at the Passport Office. **1933** 'G. Orwell' *Down & Out* vii. 56, I had not sufficient papers of identification; my *carte d'identité* was not enough. **1947** *Horizon* XVI. Oct. 100 As indispensable to the common man as a birth

certificate, a passport, a *carte d'identité*. **1965** G. Lyall *Midnight plus One* xxiv. 183, I wanted him on the defensive in the hope he'd forget to ask for my *carte d'identité*.

‖**carte du pays** (kart,dü,pēⁱ). [Fr., lit. 'map of the country'.] The lie of the land; also *fig.*

1744 H. Walpole *Let.* 22 July (1955) XVIII. 483 Not being quite perfect in the *carte du pays*, told my Lady, [&c.]. **1840** J. B. Fraser *Trav. Koordistan* II. xviii. 436 My companion, who knew the *carte-du-pays* well, had been prowling about to discover..the means of getting on. **1853** Mrs. Gaskell *Ruth* II. ix. 228 [He] knew, from the *carte du pays* which the scouts of the parliamentary agent had given him, that Mr. Benson was a person to be conciliated. **1905** R. Broughton *Waif's Progress* ii. 11, I want, before you go, to give you a tiny *carte du pays*.

cartel, *sb.* Add: **3. c.** [After G. *kartell.*] (Orig. in Germany) an agreement or association between two or more business houses for regulating output, fixing prices, etc.; also, the businesses thus combined; a trust or syndicate. Also *attrib.* and *transf.*

1902 *Daily Chron.* 24 May 6/3 He laid stress on the injury which would be done to the Indian industry if the country were flooded with 'Cartel' sugar. **1902** *Pol. Sci. Q.* XVII. 381 The cartel, or producers' syndicate. **1902** *Westm. Gaz.* 24 Nov. 10/1 The new cartel includes practically every important iron and steel interest in the Dual Monarchy. **1934** *Punch* 4 Apr. 377/1 The cartel of American lyrists has drawn up regulations forbidding the use of any adjective other than 'swell', 'red-hot' or 'blue'. **1935** *Economist* 12 Jan. 68/2 The Vienna coal trade has arranged a cartel for house coal, and a 4 per cent. price rise has occurred. **1958** *Times Rev. Industry* Apr. 12/1 A powerful world cartel.., controlling about 95 per cent of world production, was formed to restore balance.

d. *Hist.* The coalition formed in 1887 between the Conservatives and the National Liberals in Germany to support each other's candidates, for the furtherance of Bismarck's military and imperial policy. Hence as a name for similar coalitions in other countries.

1889 *Ann. Reg. 1888* 270 On this occasion the German Unionists, or 'Cartell party', composed of the Conservatives and National Liberals, again spoke for the Government measure. **1918** C. G. Robertson *Bismarck* 453 The Chancellor's political cartel was complete; it consisted of the Conservatives, the old National Liberals, and the Centre; and the union gave him a decisive and obedient majority. **1919** A. W. Ward *Germany* III. 125 The entire Cartel, this time with the Centre, agreed to proceed with the loan. **1926** *Encycl. Brit.* II. 94/1 The Radicals, Radical-Socialists and Socialists formed a Cartel [France, 1924]. **1927** *Contemp. Rev.* Aug. 154 Hence an electoral cartel was inevitable, and to this the Transylvanians and Tsaranists consented. **1928** *Daily Tel.* 21 Aug. 8/3 The 'cartel' of Republican parties formed by him [*sc.* M. Veniselos]. **1959** B. & R. North tr. *Duverger's Political Parties* (ed. 2) II. i. 231 The Left-wing Cartel, the Popular Front, and so on. *Ibid.* ii. 326 The Cartel..which won the 1887 elections and lost the 1890 one.

5. cartel clock (see quots.).

1899 F. J. Britten *Old Clocks & Watches* 275 Hanging or 'Cartel' Clocks of Louis XV period were usually of metal thickly gilt and graceful in form. **1960** H. Hayward *Antique Coll.* 57/1 *Cartel clock*, a mural clock, usually of somewhat flamboyant design. .. English are usually of carved wood, whereas the French are usually of cast brass or bronze and gilt.

Hence **ca·rtel(l)ism,** the system of cartels, cartelization; **ca·rtel(l)ist,** a member or supporter of a cartel; **cartel(l)iza·tion,** the formation of or into cartels; **ca·rtel(l)ize** *v. trans.* and *intr.*, to form (into) a cartel; hence *cartel(l)ized* ppl. adj., *cartel(l)izing* vbl. sb.

1923 *Contemp. Rev.* June 716 Hitherto textiles were not regarded as favourable fields for cartelisation. **1925** *Glasgow Herald* 16 Apr. 9 Passionate Cartelist though he is. **1926** *Ibid.* 4 Jan. 7 The leading exponent of Cartellism. **1927** F. S. Browne tr. *R. Liefmann's Internat. Cartels* 98 Baron Tibbaut (Belgium) believed that cartellization spelled the death of 'protectionism'. *Ibid.* 101 In the face of a cartellized industry, strong organization of the consumers..was necessary. **1929** *Times* 5 June 15/3 In the late Council Anti-Cartellists and Cartellists were exactly balanced. **1935** *Economist* 6 Apr. 779/1 The Conservative advocates of the cartelising policy known euphemistically as 'Self-government for Industry'. **1955** *Times* 4 July 9/7 Whereas only lately, at the request of the Allies, decartelization in Germany was the order of the day, it is now again cartelization. **1961** *Economist* 25 Nov. 819/2 This [air travel], the most highly-cartelised industry in the world.

cartel, cartle, varr. Kartel.

1880, 1885 [see Kartel]. **1910** J. Buchan *Prester John* viii. 151, I..flung my kaross on the cartel which did duty as bed.

carter.[1] Add: **4. b.** In full **carter spider**: an arachnid of the order Opiliones; = Harvest-man 2.

1665 Hooke *Microgr.* 198 The Carter, Shepherd Spider, or long-legg'd Spider. **1736** Hooke in E. Albin *Nat. Hist. of Spiders* 57 Microscopical observations on the Carter Spider and Jumping Spider. **1746** B. Martin *Ess. Electricity* 22 That Sort [of spider] we call a Carter, or Father Long-legs.

Cartesian, *a.* Add: **Cartesian co-ordinates**: see Co-ordinate *sb.* 2.

Carthaginian (kāɹþădʒi·niăn), *a.* and *sb.* Also 6–9 **Carthagenian.** [f. L. *Carthago, -inis* Carthage + -ian.] **A.** *adj.* Of, pertaining to, or characteristic of Carthage, an ancient city of north Africa, or its people; *Carthaginian faith*: see Faith *sb.* 11 b; *Carthaginian peace*, a peace settlement that is very severe to the defeated side. **B.** *sb.* A native or inhabitant of Carthage. Cf. Punic *a.* and *sb.*

1592 [see Perfidy]. **1594** Kyd *Cornelia* II. 484 Fought, before the fearefull Carthagenian walls. **1596** Spenser *View Irel.* in *Wks.* (1893) 627/2 The Carthagenians in all the long Punicke Warres. **1601** [see Punic *a.* 1]. **1618** [see Trick *sb.* 3]. **1652** [see Controversy *sb.* 1 a]. **1671** Milton *P.R.* III. 35 Young Scipio had brought down The Carthaginian pride. **1711** [see Faith *sb.* 11 b]. **1846** R. Ford *Gatherings from Spain* ii. 11 The lower classes partake of the Greek and Carthaginian character, being false, cruel, and treacherous. **1869** 'Mark Twain' *Innoc. Abr.* viii. 79 The Phoenicians, the Carthagenians, the English, Moors, Romans, all have battled for Tangier. **1940** *Economist* 3 Feb. 191/2 Intending to annex slices of German territory, or to impose a Carthaginian peace. **1948** Kroeber *Anthropology* xvii. 729 The Etruscans and the Carthaginians..were slow in taking up coinage. **1967** *Guardian* 11 July 6/5 The idea that the Israelis are anxious to..dictate a Carthaginian peace is ludicrously wide of the mark.

cartogram (kā·ɹtogræm). [ad. Fr. *cartogramme.*] (See quot. 1965.)

1890 *Rep. Insane, Feeble-Minded, &c.* (11th Census U.S.) 9 Cartogram 1 indicates by depths of shading the ratios of the insane to each 100,000 of the population in the different states. **1909** Webster, *Cartogram*, a map showing geographically, by shades or curves, statistics of various kinds; a statistical map. **1934** E. Raisz in *Geogr. Rev.* XXIV. 292 The statistical cartogram is not a map. Although it has roughly the proportions of the country..the cartogram is purely a geometrical design to visualize certain statistical facts and to work out certain problems of distribution. **1937** *Geogr. Jrnl.* LXXXIX. 349 A cartogram is arithmetically accurate but topographically untrue. **1938** E. Raisz *General Cartography* xxiii. 257 How 'diagrammatic' a map must be in order to be called a cartogram depends largely on individual judgment. **1965** F. J. Monkhouse *Dict. Geogr.* 55/2 *Cartogram*, a map on which statistical information is presented in diagrammatic form.

carton[2] (kā·ɹtən). [a. F. *carton*: see Carton in Dict.] **a.** Light pasteboard or cardboard used for making boxes, etc. Also *attrib.*

[**1786** T. Blaikie *Diary Scotch Gardener* (1931) 202 The Marquis de Cray..proposed to make buildings of Paper or as he called it Carton. *Ibid.* 203 The Marquis still persisted in his carton projects.] **1891** Mahaffy *Flinders Petrie Papyri* 9 The coffins..were made of layers of papyrus,.. stuck together so as to form a thick carton, painted within and without with designs and religious emblems. These carton-cases were made to fit the swathed body. **1908** J. H. Judd (*title*) The 'A.L.' Carton Work. **1910** —— Carton Designing for Juniors. **1913** *Bk. School Handwork* I. 188 Domestic Handicraft for Girls in Paper, Carton, Cardboard, and Woodwork. *Ibid.* 195 Paper folding and carton work.. in the making of toys.

b. A box or case made of light pasteboard, cardboard, or plastic, for holding goods. Hence **cartoned** (kā·ɹtənd) *a.*, packed in a carton.

1906 *Daily Chron.* 24 Oct. 6/6 A pound carton [of soap]. **1921** C. C. Martin *Export Packing* 559 Sundry Cartoned Goods. **1924** *Public Opinion* 12 Sept. 258/3 Motorists' sandwich papers and cigarette cartons. **1926** *Spectator* 18 Sept. 412/1 The fish..are packed in specially designed cartons. **1957** *Encycl. Brit.* III. 989/1 Cartons form an important branch of paper box manufacture. **1971** *Garden News* 25 June 14/4 Transparent yoghurt cartons.

cartonnage (kā·ɹtənĕdʒ). *Archæol.* Also **cartonage.** [Fr., f. *carton*: see prec. and -age.] An Egyptian mummy-case made of layers of linen or papyrus tightly pressed and glued together and fitting closely to the embalmed body; also, the material thus used. Also *attrib.*

[**1834** T. J. Pettigrew *Egypt. Mummies* 116 The case I am describing is called by the French the 'Cartonage', from the resemblance of its composition to pasteboard.] **1841** J. G. Wilkinson *Anc. Egyptians* II. 477 The innermost covering of the body..was the *cartonage*. This was a pasteboard case fitting exactly to its shape. **1881** Rawlinson *Hist. Anc. Egypt.* I. 512 The swathed body was covered by a 'cartonnage', consisting of twenty or forty layers of linen tightly pressed and glued together, so as to form a sort of pasteboard envelope, which then received a thin coating of stucco, and was painted in bright colours with hieroglyphics and figures of deities. **1889** Petrie *Hawara* 15 Mummies with gilt cartonnage heads. **1957** G. Clark *Archaeol. & Society* (ed. 3) iii. 78 A number of papyri..were recovered from the cartonnage of a mummy excavated by Flinders Petrie.

‖**carton-pierre** (kartoñ pyęr). [Fr., = 'cardboard (of) stone'.] A kind of papier mâché made to imitate stone or bronze.

1850 Thackeray *Pendennis* I. xxxviii. 372 Old family portraits from Wardour Street..and battle-axes made of *carton-pierre*. **1851** *Illustr. Catal. Gt. Exhib.* III. 747/1 Bracket-table,..the figures supporting it of carton-pierre. **1858** Simmonds *Dict. Trade, Carton-pierre*..has been used for roofing, and is composed of the pulp of paper mixed with whiting and glue. **1961** *Connoisseur Guide to Antique Eng. Furn.* 115 A great deal of Early Victorian furniture made use of patent materials such as gutta-percha, stamped leather, carton-pierre, [etc.].

cartoon, *sb.* Add: **2.** (Earlier and later examples.) Now, a humorous or topical drawing (of any size) in a newspaper, etc. Cf. *strip-cartoon.*

1843 *Punch* 24 June 258/2 *Punch* has the benevolence to announce, that in an early number of his ensuing Volume he will astonish the Parliamentary Committee by the publication of several exquisite designs, to be called Punch's Cartoons! *Ibid.* 15 July 22/2 Substance and Shadow. Cartoon No. I. **1900** M. Beerbohm *Lett. to R. Turner* (1964) 298 His [*sc.* Pellegrini's] famous *Vanity Fair* cartoons. *Ibid.*, A pair of original cartoon-drawings. **1910** *Daily Chron.* 1 Mar. 6/6 The cartoon nowadays..is merely party-political. You find a scratchy sketch in the corner of a paper devoted to the week-end football matches or the spring handicaps, or Mr. Asquith's breakfast with the King. 'Cartoon!' **1959** *Chambers's Encycl.* III. 117/1 Daumier's example can still be traced in the modern newspaper cartoon. *Ibid.*, Caricatures..began to appear as political cartoons.

b. *Cinematography.* An animated cartoon (see *animated *ppl. a.* 1 d). Also *attrib.*

1916 *Sci. Amer.* 14 Oct. 354/1 Knowledge of motion is a requisite in properly animating a drawing, in spite of the seeming simplicity of the cartoons when viewed on the screen. *Ibid.*, Here may be learned..the production of cartoon films, from the inception of the idea to the projection of the film on the screen. **1923** *Lit. Digest* 5 May 27/2 The moving cartoons now familiar at the picture shows are made with the aid of transparent sheets of pyralin. **1936** *Discovery* Apr. 122/2 The making of motion pictures, cartoons, and sound recording. **1956** *Ann. Reg. 1955* 385 The first feature-length cartoon film, *Animal Farm.* **1959** I. & P. Opie *Lore & Lang. Schoolchildren* vii. 112 Popeye.. another cartoon character originating in America.

cartoonery (kaɹtū·nəri). [f. Cartoon *sb.* + -ery.] The making of cartoons; cartoons in general.

1902 *Strand Mag.* Oct., The cartoon which we reproduce.. appealed particularly to the intellectually-minded, who hold that cartoonery should be something else than buffoonery. **1914** *N.Y. Herald* 16 June 10/7 Typical of a vast amount of cartoonery.

cartoo·ning, *vbl. sb.* [f. Cartoon *v.* + -ing[1].] The drawing or execution of cartoons; representation in a cartoon.

1846 Browning *Lett.* 9 July (1899) II. 319 Then comes the rest from cartooning and exhibiting. **1887** [see Cartoon *v.* 1]. **1927** *Sunday at Home* June 540/1 The cartooning of nature has been aided by some touching up from a tricksy human hand. **1967** *Punch* 27 Dec. 954 Within the limits of cartooning.

cartophily (kāɹtǫ·fili). [f. F. *carte*, It. *carta* (see Card *sb.*[2]) + -o + -phil + -y[3].] The pursuit of collecting, arranging, and studying cigarette-cards and similar items. So **carto·philist,** a person devoted to cartophily; **carto·phi·lic** *a.*

1936 C. L. Bagnall in *Cigarette Card News* Apr. 87, 30,000 people in the British Isles..collect cigarette cards... I have coined a new word for my clients and call them cartophilists. **1936** *Morning Post* 11 Aug. 12/6 There is a magazine entirely devoted to 'cartophily'...For every one serious 'cartophilist' in 1930 there are 25 now. **1937** *Cigarette Card News* Apr. 120 In a mansion..the new cartophilic headquarters is to be found.

cartridge. Add: **1. d.** (i) *Photogr.* A spool of film in a (cylindrical) light-proof container designed for daylight loading; *esp.* one that requires mere insertion into a suitable camera.

1918 *Photo-Miniature* Mar. 9 *Film Cartridge*, roll of sensitive film (for negatives) wound on a spool and enclosed within a longer band of black paper for daylight changing. **1940** A. L. M. Sowerby *Wall's Dict. Photogr.* (ed. 15) 111 The term cartridge is sometimes applied to roll films in spools. **1969** *New Yorker* 11 Oct. 17/2 (Advt.), This is it... The most automatic automatic. Just drop in the film cartridge. **1970** *Amat. Photographer* 11 Mar. 32/3 The 126 cartridge consists of two film chambers joined by a double 'bridge' which forms a channel for the film to pass through.

(ii) A small container for chemicals.

1920 Webster, *Cartridge*, a small case or capsule of chemicals. **1940** A. L. M. Sowerby *Wall's Dict. Photogr.* (ed. 15) 111 Dry photographic chemicals are often put up in cartridge form for the use of tourists.

(iii) = *cassette d.

1960 *Tape Recording & Hi-Fi Mag.* 20 Apr. 18/2 Sufficient tape is contained within the three-inch wide cartridge to provide 65 minutes' playing time without interruption. **1966** *Tape Recording Year Bk.* 7 For some years tape recorder manufacturers have been trying to come to terms with the tape cassette (alternatively called a cartridge, or sometimes a magazine). **1970** *Daily Tel.* 16 Mar. 24/2 The '8 track' cartridge..gives four stereo tracks of about 20 minutes each on a closed loop of tape.

e. Part of the pick-up arm of a gramophone or record-player (see quots. 1961, 1962).

1941 J. F. Rider *Automatic Record Changers & Recorders* ii. 43/1 In the case of crystal pickups..it is inadvisable to make any attempt to repair the crystal, it being preferable to replace the defective crystal with a new cartridge. **1946** *Electronics* Sept. 164/2 A standard type of crystal pickup arm with the carbon button assembly mounted in place of the crystal cartridge. **1961** G. A. Briggs *A to Z in Audio* 28 *Cartridge*, the head of a pickup containing the voltage generating mechanisms and stylus assembly, which clips on or plugs in to the tone arm. **1962** A. Nisbett *Technique Sound Studio* 243 Cartridge can mean removable pick-up head or capsule. A tape capsule (or cartridge) is a double spool of tape in a single container (comprising both feed and take up) which does not need to be laced up before replay. **1963** *Which?* Jan. 8 The head contains the cartridge, and set into the cartridge is the stylus.

4. cartridge starter, a device for starting an internal-combustion engine, esp. in an aeroplane, by means of an explosive charge.

1922 *Flight* XIV. 65/1 For use when no accumulators are carried there is the 'cartridge' starter, originated and, we believe, patented by Farmans. **1956** J. Johnson *Wing Leader* 68 Which provided extra power and incorporated the Coffman cartridge starter.

cart-wheel, *sb.* Add: **3.** Also without 'to turn' and in *sing.* Also *transf.* and *attrib.*

1895 [see Split *sb.*[1] 4 b]. **1917** 'Contact' *Airman's Outings* 246 A medley of swift darts, dives, and cart-wheel turns. **1933** P. Godfrey *Back-Stage* xv. 187 The girls in these troupes..must be expert in high kicks, cartwheels, and splits. **1955** *Simple Gymnastics* ('Know the Game' Series) 27/1 *Cartwheel.* Start from standing aside...Swing your legs up as high as you can and look over your shoulder with the head well back. **1960** B. Keaton *Wonderful World Slapstick* (1967) 149, I could do..a series of cartwheels in a circle, without touching hands to the floor.

4. A kind of fire-work. Cf. Catherine wheel 3.

1840 R. H. Dana *Bef. Mast* xiii. 72 We had..everything ..from Chinese fire-works to English cart-wheels.

5. A hat with a wide circular brim. Also *attrib.*

1884 L. Troubridge *Life amongst Troubridges* (1966) 169 Mr. and Mrs. Oscar Wilde to tea, she dressed for the part in ..a huge cartwheel Gainsborough hat. **1895** *Westm. Gaz.* 29 Jan. 3/1 Complaints in respect to the cart-wheel hat. *Ibid.*, The indignant playgoer who happens to be seated directly behind a superb cart-wheel of the latest Parisian make. **1946** Koestler *Thieves in Night* 99 Four boys in khaki shorts and cartwheel hats of straw. **1959** B. Goolden *For Richer, for Poorer* iv. 61 They come in to buy a cart-wheel..and they happen to see a completely entrancing cloche.

cart-wheel, *v.* [f. the sb.] *intr.* To move like a rotating wheel; said esp. of an aeroplane which makes a crash landing on one wing-tip.

1920 *Blackw. Mag.* July 85/1 He cart-wheeled round right-handed to get behind her again. **1925** E. F. Norton *Fight for Everest, 1924* 114 Somervell's axe slipped from his numb fingers and went cart-wheeling down the slopes below. **1928** *Daily Express* 24 Sept. 7/4 The machine continued to lose height, and suddenly cartwheeled vertically to the ground. **1962** *Daily Tel.* 29 Dec. 1/6 A Carvair aircraft..hit a snow-covered dyke and 'cartwheeled' four times as it came in to land.

ca·rucated, *a. Hist.* Measured and assessed by carucates.

1908 W. G. Collingwood *Scandin. Britain* 111 Leicestershire..was a carucated district.

carvacrol (kā·ɹvăkrǫl). Also 9 **carvacrole.** [a. G. *carvacrol* (E. Schweizer 1841, in *Jrnl. f. prakt. Chem.* XXIV. 263), f. mod.L. *carv-i* in *Carum carvi*, botanical name of Caraway, + L. *ācr-is* (see Acrid *a.*) + -ol 2.] A liquid phenol, $C_{10}H_{14}O$, obtainable from camphor, origanum, oil of caraway, savory, etc., and used in perfumery and as a fungicide, etc.

1854 R. D. Thomson *Cycl. Chem.* 153/1 *Carvacrole*, a colourless oil, heavier than water. **1860** H. Watts tr. *Gmelin's Chem.* XIV. 414 Carvol when *heated* appears to be transformed into carvacrol. **1911** *Brit. Pharm. Codex* 266 An iodide of carvacrol, in the form of a light yellow or brownish powder, has been used as a substitute for iodoform. **1963** D. L. Wedderburn in H. W. Hibbott *Handbk. Cosmetic Sci.* xxi. 452 The isomeric methyl isopropyl phenols, thymol and carvacrol, are about 28 times as effective as phenol against *Salmonella typhosa.*

carval (kā·ɹvăl). *Isle of Man.* Also **carvel.** [Manx *carval.*] A carol, a ballad on a sacred subject.

1873 W. Harrison in *Mona Miscellany* 2nd Ser. p. x, A specimen of a Manx *carval* is given...It would have been easy to have given many of these *carvals*, which may be termed a literature entirely peculiar to the Manx people, consisting chiefly of ballads on sacred subjects which have been handed down in writing...Most of these carvals are from 50 to 150 years old, and amongst the favourites may be mentioned 'Joseph's History', 'Susannah's History', 'The Nativity', 'The Holy War', 'David and Goliath', 'Samson's History', 'Birth of Christ'. **1887** Hall Caine *Deemster* III. xxxiii. 60 Sometimes he crooned a Manx carval. **1910** *Encycl. Brit.* V. 639/2 Most of the existing literature..consists of ballads and carols, locally called carvels. **1956** *Archivum Linguisticum* VIII. I. 88 The carval manuscripts with their rhymes and abundant non-literary spellings.

carve, *v.* Add: **1. c.** *trans.* (*slang*). To slash (a person) with a knife or razor; esp. *to carve* (*a person*) *up*. Hence **carved** *ppl. a.*, **carving** *vbl. sb.*

1929 D. Hammett *Dane Curse* xi. 113 The man had.. stood..waiting to carve me when I came out; and my fall had saved me, making him miss me with the blade. **1938** G. Greene *Brighton Rock* II. i. 73 How do we make you safe, Spicer?...If carving'd do it... *Ibid.* III. i. 98 They just meant to carve him up, but a razor slipped. **1938** Sharpe *S. of Flying Squad* v. 65 He feared that they would 'carve him up'. *Ibid.* xiv. 157 Carve him with a razor. **1960** M. Spark *Ballad of Peckham Rye* ix. 188 We got to carve up that boy one of these days.

10. b. *to carve up*: to cheat, swindle. *slang.*

[**1909** Ware *Passing Eng.* 65 *Carve up* (Amer.), to annihilate completely. That dear grave holds a disappointed chap who cum out here from Reno to carve me up.] **1933** C. E. Leach *On Top of Underworld* x. 138 *Carve up*, swindle accomplice out of share. **1936** J. Curtis *Gilt Kid* ix. 96 I'm going to trust you to play square with me at the end. No carving-up or else. **1959** H. Pinter *Birthday Party* I. 10 Then after that, you know what they did? They carved me up. Carved me up. It was all arranged, it was all worked out.

Carver (kā·ɹvəɹ). *U.S.* The name of John *Carver* (1576–1621), the first governor of Plymouth Colony in America, used *attrib.* to designate a chair of a type owned by him, having a rush seat, arms, and a back usually consisting of three horizontal and three vertical spindles.

1902 L. V. Lockwood *Colonial Furnit. Amer.* v. 129 Figures 85 and 86 are Elder Brewster's and Governor Carver's chairs respectively, and were, according to tradition, brought over in the Mayflower...Chairs in the fashion of Figure 86 are commonly known as 'Carver chairs', and are more frequently met with than any other pattern of the turned chair. **1913** —— *Furnit. Collectors' Gloss.* 10 *Carver*, an Americanism for a turned chair of the early seventeenth century, the back of which has three horizontal turnings and three vertical spindles, between the two lower horizontal turnings. So called because of one owned by Gov. Carver. **1923** *Furnit. Jrnl.* 3 Feb. 69 (*caption*) No. 1461 Carver. **1924** Nutting *Furniture Pilgrim Cent.* 299 This great Carver chair has a back superior to any other that we have seen, in its massiveness, and the character of its turnings. **1937** Langdon *Everyday Things* 32 The Carver Chair and the Brewster Chair.

carve-up (kā·ɹvɒp). *slang.* [f. Carve *v.* 10 and *10 b.] (The situation resulting from) a sharing of spoils; a division, sharing-out, cutting-up (with derogatory contextual overtones).

In quot. 1935 the reference is to money bequeathed in a will.

1935 M. Harrison *Spring in Tartarus* I. 27 The de Launes hadn't a tosser between them. Oh my Lord, what a carve up! **1936** J. Curtis *Gilt Kid* ix. 92 There would probably be a carve-up at the end when it came to sharing out the dough. Scaley was certainly the sort of man to stick to more than his fair share. **1943** 'R. Llewellyn' *None but Lonely Heart* xxxiv. 251 'How much can we make a week?'..'Depends on the carve up,' Slush says. **1959** *News Chron.* 3 July 6/6 In practice it is a carve-up among the Big Four. **1961** *Guardian* 12 Jan. 8/2 Public monopoly is preferable to commercial carve-up. **1963** *Times* 24 Jan. 5/1 Is the selection of justices of the peace in Britain..a 'political carve-up', as alleged by some of the more vociferous of the system's opponents? *Ibid.* 29 Apr. 7/6 Apart from one farcical carve-up of a passage from *Coriolanus*, the exercises owed nothing to written drama.

caryo-: see Karyo-.

Casanova (kæzănōᵘ·vă, kæs-). Used as *sb.* allusively of a man whose amorous activities resemble those of the Italian adventurer, Giovanni Jacopo *Casanova* de Seingalt (1725–98), author of memoirs, in French, describing his escapades and amours in most countries of Europe. So **Casano·va-ish, Casanove·sque, Casano·vian** *adjs.*; also **Casano·vism.**

[**1888** *Encycl. Brit.* XXIII. 215/1 There is also no doubt a touch of Casanova in Barry Lyndon's character.] **1903** G. B. Shaw *Man & Superman* p. xiv, Don Juanism is no longer misunderstood as mere Casanovism. **1925** A. Huxley *Those Barren Leaves* I. vii. 68 Boasting..of one's Casanovesque capacities. **1928** E. Cantor *My Life is in your Hands* v. 52 My grandmother..came to box my ears and take the infant Casanova home. **1932** W. Faulkner *Light in Aug.* (1933) i. 4 Young bachelors, or sawdust Casanovas. **1940** *Time* 4 Mar. 74/3 It concerns the adventures of a Casanovian Irishman, Gideon Ouseley, among the English. **1958** J. Kerouac *On Road* 170 Decadent, Casanova-ish men. **1959** J. Braine *Vodi* vi. 94 A resplendent Casanova in a royal blue sports jacket.

‖**casaque** (kazak). [Fr. *casaque* Cassock.] A woman's blouse or jumper.

1872 *Young Englishwoman* Dec. 646/2 Walking costumes are still made with either polonaise or casaque tunic. **1894** *Daily News* 30 Oct. 6/6 The 'casaque', or Louis XVI jacket, in brocade or very handsome fancy cloth, is another revival of this season. **1923** *Daily Mail* 19 Feb. (Advt.), Knitted Wool Casaques.

‖**casaquin** (kazakæ̃n). [Fr.: see Cassakin.] A kind of bodice or jumper.

1879 J. R. Planché *Cycl. Costume* II. 181 A *casaquin* (a loose jacket) of grey velvet laced with silver..and lined with cloth of silver. **1922** *Daily Mail* 4 Dec. 15 The casaquin bodice was joined to a full ankle-length skirt. **1923** *Ibid.* 18 Jan. 11 Another frock from Paris has a casaquin..over a plain skirt. *Ibid.* 12 Feb. 15 The casaquin jumper of this three-piece suit.

cascade, *sb.* Add: **2. b.** (Examples.)

1749 in A. St. H. Brock *Pyrotechnics* (1922) iv. 30 A large vertical Sun moved by double Fires, Cascades, Pyramids. **1875** W. H. Browne *Art Pyrotechny* xii. 118 The chapter will contain instructions for the construction of..brilliant suns, cascades [etc.]. **1922** A. St. H. Brock *Pyrotechnics* v. 128 Cascade..a feature of the Crystal Palace displays.

d. (Earlier example.) Also applied to other electrical devices connected in such a manner that each operates the next one in turn; freq. in phr. *in cascade.*

1868 *Q. Jrnl. Sci.* V. 117 We have here a compact form of Leyden battery, arranged for 'cascade'. **1915** *Proc. Inst. Radio Engin.* III. 230 Where a greater amplification than can be obtained with one audion is required, cascade working of the radio frequency systems may be resorted to by coupling two or more audion systems. *Ibid.* 286 The 'exponential' method of tuning, involving the use of radio frequency pliotron amplifiers in cascade, is shown to have given remarkable selectivity. **1930** *Engineering* 7 Mar. 312/3 The two 500,000-volt transformers are connected in cascade to give 1,000,000 volts. **1940** *Jrnl. R. Aeronaut. Soc.* XLIV. 373 The current control can be carried out.. mechanically by means of a special cascade transformer. **1946** *Electronic Engin.* XVIII. 151 The required high voltage may be obtained..by the use of cascade transformers and rectifiers.

e. *spec.* Applied to a succession of stages or processes in some operation or event in Physics, Chemistry, etc. Freq. *attrib.*

1902 *Encycl. Brit.* XXVII. 187/1 The method by which we try to obtain successively lower temperatures by making use of successive gases is called the 'cascade method'. **1937** *Nature* 13 Nov. 837/2 In the cascade process, showers are built up from the successive conversion of electron energy into radiation, radiation into pair production and then further loss of electron energy to radiation. **1940** *Rep. Progr. Physics* VI. 65 Hertz has..arranged these in a cascade similar to the cascade used for the diffusion of isotopes through porous walls. Figure 3 shows cascades of three units. **1945** *Ann. Reg. 1944* 383 Cascade showers are largely initiated by electrons. **1958** *Chambers's Techn. Dict.* 963/2 *Cascade*, separation of isotopes by similar successive stages in a process, each stage increasing the concentration of each. **1960** *Economist* 15 Oct. 268/3 The production method..is to turn the uranium into a gas and filter it through porous membranes through which the lighter uranium 235 passes first. This process works on a 'cascade' principle. **1964** N. G. Clark *Mod. Org. Chem.* v. 86 The necessary low temperatures are obtained by a 'cascade' process involving the pressure liquefaction and evaporation of propane.

cascade, *v.* Add: **3.** *Electr.* *trans.* and *intr.* To link or connect (valves, etc.) in stages to form a cascade (cf. prec. 2 d). Hence **casca·ded** *ppl. a.*, **casca·ding** *vbl. sb.*

1919 *Wireless World* July 187/1 These types of alternator ..may be classified as follows:—1. Machines in cascade. 2. Internally-cascaded machines. *Ibid.* 189/2 In effect the machine Figure 2 represents in a single machine the summation of the four cascaded machines of Figure 1. **1930** *Jrnl. Inst. Radio Engin.* XVIII. 994 Cascading must give the same type of transmission curve as coupled circuits with very weak coupling...Two staggered single-circuit curves of the same shape cascade into a coupled-circuit transmission curve...The amount of transmission is improved in the cascaded case. *Ibid.* 1007 Consideration..must be given to characteristics of any other selective devices of the amplifier system..if single or coupled circuits are cascaded with the coupled circuit under consideration. **1949** *Electronic Engin.* XXI. 61 Simple cascading (i.e. direct coupling between anode and grid) requires a relatively high potential source for the later stages. **1956** Amos & Birkinshaw *Telev. Engin.* II. vii. 114 Decrease in stage gain to give constant passband when cascading. *Ibid.* viii. 122 A conventional cascaded amplifier has an upper frequency limit beyond which it is impossible to amplify. **1971** *Physics Bull.* Apr. 208/3 It is possible to cascade either germanium-silicon or lead telluride type modules thermally with bismuth telluride.

casca·ding, *ppl. a.* [f. Cascade *v.* + -ing[2].] Falling in or as in a cascade.

1889 C. Edwardes *Sardinia* 199 The cascading stream. **1916** 'Boyd Cable' *Action Front* 179 From overhead..there came..the clatter and rattle of cascading bricks and tiles. **1923** *Glasgow Herald* 9 May 11 Climbing and cascading plants. **1957** *Encycl. Brit.* XXIII. 412/2 Another cause for hanging tributaries and consequently a sheer or cascading plunge..is simply more rapid deepening of that valley.

‖**cascalho** (kaska·lyo). Also **-hao.** [Pg.] In Brazil, a deposit of pebbles, gravel, and ferruginous sand, containing diamonds or gold.

1812 J. Mawe *Trav. Brazil* v. 77 The gold lies..in a stratum of rounded pebbles and gravel, called *cascalhão*. **1823** —— *Diamonds* (ed. 2) 35 In one part of the Diamond District the *cascalho* forms a solid breccia. **1867** C. W. King *Precious Stones* (ed. 2) 59 The gravel—*cascalhao*—filling their beds is dug out. **1910** *Encycl. Brit.* VIII. 160/1 The diamond is found..in the gravels of the present rivers, embedded in a ferruginous clay-cemented conglomerate known as *cascalho*. **1957** *Ibid.* IV. 838/2 Carbonado..is found almost exclusively in the state of Bahia in Brazil, where it occurs in the *cascalho* or diamond-bearing gravel.

cascara. Add: **2.** *Med.* (An extract of) the bark of a Californian buckthorn, *Rhamnus purshiana*, used as a laxative or cathartic; in full *cascara sagrada*, 'sacred bark' (see Sacred *a.* 7). Popularly pronounced (kæskā·rā).

1879 A. O'Neill in *New Preparations* (Detroit) III. 140. **1885** *Brit. Pharmacopœia* 151 Extract of Cascara Sagrada. *Ibid.* 152 Mix the cascara with two pints of the spirit. **1922** Joyce *Ulysses* 68 One tabloid of cascara sagrada. **1951** A. Grollman *Pharmacol. & Therapeutics* xix. 390 Cascara sagrada is a very popular remedy in habitual constipation. **1966** J. Gardner *Amber Nine* ii. 35 If you are given any reason to think you've been detected, move like an overdose of cascara.

cascode (kæ·skōᵘd). *Electr.* [f. *casc(ade *sb.* 2 d + cath)ode.] (See quot. 1958.) Also *attrib.*

Quot. 1939 does not illustrate the current sense: see quot. 1948.

1939 Hunt & Hickman in *Rev. Scient. Instruments* X. 16/2 This tube connection we have called the 'cascode'... It is generally useful for d.c. amplifier work in which it is inconvenient to supply the additional bias voltages for a pentode. The dual triode, as connected, may be shown to

be equivalent to a single triode. **1948** H. Wallman et al. in *Proc. Inst. Radio Engin.* XXXVI. 701/1 Search for a concise name for the grounded-cathode, grounded-grid combination led to the designation 'cascode', after a somewhat similar arrangement employed by Hunt and Hickman. *Ibid.*, The low-noise property that forms the feature of the present circuit is, however, entirely unconnected with the original use of the cascode as a d.c. amplifier in a voltage stabilizer. **1955** *Electronic Engin.* XXVII. 323 The features of the cascode type of amplifier. **1958** *Chambers's Techn. Dict.* 964/1 *Cascode*, thermionic valve circuit in which a grounded-cathode triode followed by a grounded-grid triode provides a low noise amplifier for very high radio frequencies. **1960** *Electronic Engin.* XXXII. 44 A transistor cascode amplifier does not require neutralization.

case, *sb.*[1] Add: **3. b.** An infatuation; a situation in which two people fall in love. So *to have a case on*: to be infatuated with or enamoured of. *slang* (orig. *U.S.*).

1852 *Harper's Mag.* V. 338/2 Young America..voted it 'a case'. The elderly ladies thought it a 'shocking flirtation'. **1860** Hotten *Slang Dict.* 112 Among young ladies at boarding-schools a *case* means a love affair. **1872** M. E. Braddon *To Bitter End* III. xvii. 285 They have only been engaged three weeks; but from the day we first met Lord Stanmore..the business was settled. It was a 'case', as you fast young men say. **1893** 'S. Grand' *Heavenly Twins* I. II. iv. 237 Edith..blushed. She could not reason about him...'That's a case, I think,' said Mrs. Guthrie Brimston. **1928** F. N. Hart *Bellamy Trial* iii. 73 Everybody knew they had a terrible case on each other. **1931** *Story-Teller* Aug. 598/2 By the end of the second year the girls were saying that Salesby *had* quite *a case on* Chips. **1951** M. Kennedy *Lucy Carmichael* VII. iii. 372 'I think it's a case all right.'.. 'Yes,' said Melissa. 'She's lost to the world.'

4. b. *in the case of*: as regards (a specified thing or person); *in that case*: if that is true; if that should happen; that being so; similarly, *in the first case*, etc. Also *in any case*: whatever may happen, whatever the fact is (cf. 13); *in many cases*: in a number of instances; similarly, *in some cases.*

1791 J. Walker *Pron. Dict.* s.v. *Medicinal*, Poets..who have, and, perhaps, in some cases, ought to have, a language different from prose. **1848** *Sporting Life* 11 Mar. 23/1 'In that case,' replied the guard, 'I will go and fetch the steward myself.' **1872** Thudichum & Dupré *Treatise on Wine* xiii. 435 When he comes to his vineyard he finds his crop either eaten by the flies, or rotten, or over-ripe, and in any case yields him a lesser quantity or quality of wine. **1881** *Spons' Encycl. Industr. Arts* IV. 1252 In many cases, flax-spinning establishments have weaving branches in connection with them. **1894** Sweet *Anglo-Saxon Reader* (ed. 7) 227 The parts of speech are not marked in the case of adjectives, pronouns and weak verbs. **1896** L. Fletcher *Introd. Study Meteorites* 36 In some cases the chondrules consist wholly or in great part of glass. **1909** Beynon *Drapery* 76 'And if I can show you cheaper lines than you are buying, you won't buy?' 'No, I will not buy even in that case.' **1912** C. N. & A. M. Williamson *Heather Moon* II. iii. 128 'It didn't seem to me there was anything romantic about Mr. Douglas, except his name.' 'In that case, you are a little flirt,' said he. **1916** E. W. Gregory *Furnit. Collector* 227 But in any case worm-eaten furniture is not at all desirable. **1921** A. L. Simon *Wine* 59 In the first case, the wines are sent over here, as a rule two years after vintage. **1929** Pinero in *Eighteen-Seventies* 133 The difficulties of great men are intensified in the case of little ones.

5. b. *in good case* (later U.S. example). Also, esp. in U.S., *spec.* of tobacco.

1640 in *Archives of Maryland* (1883) I. 98 Bad Tobacco shall be judged..[what is] sooty, wett, or in too high Case. **1800** W. Tatham *Cult. & Com. Tobacco* 37 It must be stretched gently over the ends of the fingers and knuckles, and if it is in good case, *i.e.* plight, or condition, it will discover an elastic capacity. **1845** Dodd *Brit. Manuf.* 132 An exposure to the air for..about five weeks makes the leaves of tobacco elastic and tough, and slightly covered with a glossy kind of moisture. The tobacco is then said to be *in case*. **1852** Mrs. Stowe *Uncle Tom's Cabin* xii. 72 What he would sell for, if he was kept fat and in good case. **1865** *Trans. Ill. Agric. Soc. 1863* V. 667 The fires should be suffered to go out, and the tobacco be suffered to come in case, or get soft again. **1944** *Dial. Notes* Nov. 65 *In case*: adj. phr., in proper condition—cured and having the correct amount of moisture to ensure handling without injury or loss.

8. b. Add to def.: Also (*colloq.*), a patient.

1851 [in Dict.]. **1940** N. Mitford *Pigeon Pie* xiii. 197 Heatherley and Winthrop..would not be able to leave the Treatment Room for a moment, except to carry 'cases' upstairs to the Hospital.

10. b. *in case*, esp. in *just in case*, orig. with aposiopesis, *in case* —, to indicate an unspecified apprehension of accident.

1898 Kipling *Fleet in Being* 28 One leg over the edge of the bunk—in case. **1929** Wodehouse *Mr. Mulliner Speaking* ix. 301 She rather thought she wouldn't be able to, but she said leave her ticket at the box-office in case. **1951** *Teazle's News-Let.* 24 Apr. [5] A picture..of a London policeman directing the traffic at a busy point in Paris, with a French traffic constable standing by, just in case.

14. (sense 9) **case-distinction, -function, -system.** Also, **case-book,** also *transf.*, *fig.* and *attrib.*; **case-form** *Gram.*, a morphological variant (of a noun, adjective, pronoun, etc.) that indicates its case by its form; **case-history,** the record of a person's origins, personal history, etc., orig. *Med.*, compiled for diagnostic or other purposes and comprising all matters relevant to an episode of illness in a patient; in extended use, any brief conspectus of individual development, cf. *life-history*; **case-load,** the total number of cases that a doctor, social worker, probation officer, or other professional person or agency, is concerned with at any one time; **case-phrase** *Gram.*, a phrase expressing a relation by means of a preposition and a noun instead of the bare case-form of the noun; **case-sheet** *Med.*, a document giving a patient's clinical history; also *transf.*; **case-study,** the attempt to understand a particular person, institution, society, etc., by assembling information about his or its development; the record of such an attempt.

1934 Webster, *Casebook*, a compilation of detailed information concerning actual cases, legal, medical, psychological, economic, etc., for reference and study. **1949** E. Coxhead *Wind in West* iv. 102 They sat discussing the literature of the passions, but let somebody present them with..a case-book page and..they all thought fit to take offence. **1953** *Proc. Prehist. Soc.* XIX. 129 This paper is a small selection from our 'case-book' in the Department of Environmental Archaeology. **1958** *Times Lit. Suppl.* 15 Aug. 457/3 The first pages are casebook reportage; then there appears to be an attempt at a symbolist dimension. **1894** Jespersen *Progress in Lang.* 31 The doing away with the old case distinctions in English has facilitated many extremely convenient idioms. **1934** Priebsch & Collinson *German Lang.* 153 The loss of case-distinctions..becomes still more marked in M.H.G. **1875** Case-form [in Dict.]. **1922** Jespersen *Lang.* xviii. 345 English shows that a small number of case forms is not incompatible with perfect clearness and perspicuity. **1951** R. H. Robins *Anc. & Med. Gram. Theory* i. 34 The most systematic semantic analysis of case-forms in antiquity is by the Byzantine scholar, Maximus Planudes. **1894** Jespersen *Progress in Lang.* 161 The apostrophe was at that time used (without any regard to case-function) where a syllable was added in pronunciation. **1912** *Lancet* 2 Mar. 589/2 (*title*) Case Histories in Neurology. **1931** E. Wilson *Axel's Castle* v. 185 Is he telling us his own case-history with symbols? **1934** C. Day Lewis *Hope for Poetry* iv. 18 A very instructive entry in the case-history of poetic feeling. **1935** Woodworth *Psychol.* (ed. 10) ii. 23 Reconstructing the individual's history..from his memory, the memories of his acquaintances. .. This case history method has obvious disadvantages. **1945** 'L. Lewis' *Birthday Murder* (1951) iii. 40 A good many of our friends were case histories. They belonged in a textbook on mildly abnormal psychology. **1958** *Engineering* 14 Mar. 346/3 This practice of giving applications in the form of case histories is now common in the industrial engineering field. **1950** C. Morris *Social Case-work in Gt. Brit.* ii. 51 Pseudo-cases are soon dropped from the caseload. **1961** R. Kee *Refugee World* 1 The..human beings who are the refugees are lumped together into the 'case-load'. **1961** 'C. H. Rolph' *Crime & Punishment* x. 158 It used to be thought that the maximum 'case-load' a good probation officer could carry was about fifty. **1911** *Rep. Joint Comm. Gram. Terminol.* 27 Recommendations. .. That in English and French the combination of a Preposition with a Noun or Pronoun may be called a *Case-phrase*. **1927** E. A. Sonnenschein *Soul of Gram.* 44 In Indo-European the case-system was supplemented..by..the case-phrase system. **1925** W. Deeping *Sorrell & Son* xxviii. 273 A few words had been written in haste on the back of a case-sheet. **1958** *Listener* 16 Oct. 621/2 What we saw was the case sheet of Sinforth. **1933** *Proc. Brit. Acad.* XVIII. 371 His [*sc.* Kenny's] ideas as to the value of case-study fitted neatly into his general conception of English legal education. **1940** Woodworth *Psychol.* (ed. 12) v. 162 Those who are professionally concerned in judging personality..depend largely on the interview and the inclusive procedure known as the case study. **1956** *Planning* XXII. 148 It is intended to illustrate the findings by brief case-studies of industries, firms, or products. **1959** *Listener* 29 Oct. 744/3 A shrewd and stimulating case-study of the Lancashire election of 1868. **1875** Whitney *Life & Growth of Lang.* vi. 107 The downfall of the case-system was accompanied by the uprise of the class of prepositions. **1953** W. J. Entwistle *Aspects of Lang.* xi. 349 The Caucasian languages develop the case system up to twenty-three places.

case, *sb.*[2] Add: **1. d.** *U.S.* In the game of faro, the fourth card of a denomination, when the other three have been taken from the dealing-box. So *to keep cases*: to note the cards as they come from the dealing-box; also *transf.*, to keep a close watch (*on*). Also, *to come* (or *get*) *down to cases* (chiefly *U.S.*): to come to the point.

1856 *Harper's Mag.* Dec. 69/1 He has no great faith in 'cases', but believes in betting on three cards at a time. [**1856** *San Francisco Bull.* 4 Dec. 2/2 He was sitting in front, keeping the 'cases'.] **1896** G. Ade *Artie* iii. 24 A Johnny-on-the-spot..was tryin' to keep cases on her. *Ibid.* xi. 103 When it comes down to cases they're just as good as a lot of people that make a bigger front. **1902** O. V. Limerick *B. Burgundy's Opinions* 58 To get right down to cases, she was a human-four-card-flush. **1903** A. Adams *Log of Cowboy* xiii. 199 We found Quince Forrest and Wyatt Roundtree playing the faro bank, the former keeping cases. **1918** Mulford *Man fr. Bar-20* ii. 21 Comin' down to cases, you ain't really a cow-puncher. **1920** —— J. *Nelson* xiv. 144 I'm keeping' cases on these cattle. **1937** Runyon *More than Somewhat* i. 7 It takes him some time to get down to cases and tell me what is eating him.

6. c. (Earlier and later examples.) Also, a brothel.

1536 R. Copland *Hye Way to Spyttell Hous* Eiii, Toure the patryng coue in the darkman cace. **1699** B. E. *Dict. Cant. Crew, Case*,..also a Bawdy-house. **1821** P. Egan *Life in London* II. ii. 177 In the motley group are several *Coves* of *Cases* and procuresses...The proprietors of houses of ill-fame. **1905** A. M. Binstead *Mop Fair* viii. 135 They arranges to stop 'private' in Brighton, at a little case in Black Lion Street where Tom Reeder annually took his old woman every August. **1931** *Police Jrnl.* Oct. 501 Whilst he was sleeping (kipping), she told a detective (bogey) she knew that Jack was in the brothel (case). **1942** A. Gardner *Lower Underworld* xx. 58 It's used as a case, you see, by some of the girls.

12. case-house *slang*, a brothel; **case-keeper** (*a*) *U.S.*, one who 'keeps cases' in faro (see *1 d); the device used for this purpose; (*b*) a brothel-keeper; **case-oil,** oil or other petroleum products transported in containers packed in wooden cases.

1912 C. Mackenzie *Carnival* xxx. 322 You don't suppose I'd go on living in what's no better than a common case-house. **1924** M. Kennedy *Constant Nymph* viii. 118, I wonder at you for bringing the young lady here, for it's nothing better than a dirty case-house. **1867** *Terr. Enterprise* (Virginia, Nevada) 18 Aug. 3/1 A 'case keeper' ..at a game kept on C Street. **1890** J. P. Quinn *Fools of Fortune* 201 A record of the game is kept by means of an instrument known as a 'case-keeper'. **1909** *Cent. Dict. Suppl.*, *Case-keeper*, in *faro*, the player who marks the cards as they come from the dealing-box...Small buttons are pushed along wires for this purpose. **1914** C. Mackenzie *Sinister St.* II. iv. ii. 879 You call me a case-keeper? What men have I ever let you bring back here? **1938** H. Asbury *Sucker's Progr.* 14 Case-keeper, a device for keeping a record of the cards as they were drawn. Also, the man who operated this device. **1892** *Daily News* 13 May 5/8 A steamer's carrying her oil in small hermetically sealed tin cases.. would not necessarily secure her..from fire...But..no instance is on record of a fire having occurred on board a steamship laden with case-oil while in transit through the Canal. **1930** Masefield *Wand. Liverpool* 49 The ship discharged her chalk..and loaded case-oil. **1953** tr. F. C. *Gerretson's Hist. R. Dutch* I. v. 226 In order to sell the remaining supplies of Russian case oil, the cases had to be removed first.

case, *sb.*[4] (Earlier U.S. examples.)

1851 Melville *Moby Dick* II. xxxv. 238 The upper part, known as the Case. *Ibid.* III. viii. 64, I am ready to squeeze case eternally.

case, *v.*[1] **4.** Delete †*Obs.* and substitute *Obs.* exc. *U.S.* (Later U.S. examples.)

1821 J. Fowler *Jrnl.* (1898) 44 The Hunters killed two deer, [and] Cased the Skins for Baggs. **1900** N. Smithwick *Evol. State* 178 The vessels for carrying water were made of deer skins cased—stripped off whole—the legs and necks tied tightly with sinews.

5. To examine, inspect, size up beforehand. Phr. *to case the joint* (*gaff*, *job*), to study the layout of premises before robbing them. *slang* (orig. *U.S.*).

1915 Jackson & Hellyer *Vocab. Criminal Slang* 23 *Case*,..to watch; to observe; to scrutinize. **1928** *Amer. Mercury* May 81/1 Has it [*sc.* a building] been *cased*? '*Casing* a mark without getting a rank' is the most difficult part of a robbery. **1929** *Sat. Even. Post* 13 Apr. 54/3 If he [a crook] intends to prowl a place, he first cases the joint. **1934** 'D. Hume' *Too Dangerous to Live* x. 108 We've cased a job, and it looks good. **1940** J. O'Hara *Pal Joey* (1952) 36, I..cased the mouse and got a look at her kisser. **1946** *Amer. Speech* XXI. 69/2 *To case*..to criminals..means simply to look at, to watch, or to look over. **1947** M. Gilbert *They never looked Inside* vii. 106 He was one of the curious specialists who flourish on the fringe of the kingdom of crime; he was infinitely patient at 'casing' a job and infinitely crafty at finding or making a way in. **1954** J. Steinbeck *Sweet Thursday* ii. 15 He was casing the field for a career. **1957** M. Gair *Sapphires* iii. 42 What he was doing was casing the gaff; or, in police terms, 'loitering with intent to commit a felony'.

caseate (kē^i·si,ē^it), *v. Path.* [f. L. *cāseus* cheese + -ate[3].] *intr.* To undergo caseous degeneration. See *caseation b.

1871 T. H. Green *Introd. Pathol.* 76 In..fatty degeneration..subsequent changes invariably take place; the part either softening, caseating, or becoming the seat of calcification itself. **1907** *Practitioner* Dec. 737 If the appendix is subsequently removed, you may find a caseating centre near it. **1910** *Ibid.* June 743 Caseated glands. **1964** S. Duke-Elder *Parsons' Dis. Eye* (ed. 14) xvii. 226 *Tuberculosis of the Sclera*..may be primary, forming a localized nodule which caseates and ulcerates.

caseation. b. Substitute for def.: A type of necrosis characteristic of tuberculosis, in which the infected tissue becomes a firm, amorphous, yellowish-white material resembling cheese in appearance; also, caseated material. (Add later examples.)

1954 E. P. Abraham in H. W. Florey *Lect. Gen. Path.* viii. 165 Caseation is a form of coagulation necrosis in which the dead tissue looks like cheese and contains a mixture of coagulated protein and fat. **1965** Walter & Israel *Gen. Path.* (ed. 2) xx. 348 There is now produced the tubercle follicle.., which consists of a central mass of caseation surrounded by epithelioid and giant cells.

cased, *ppl. a.* Add: **2.** Slang phr. *to be cased up*: (*a*) to cohabit (*with* someone); (*b*) to be in a brothel. Cf. Case *sb.*[2] 6 c.

1936 J. Curtis *Gilt Kid* ii. 23 She was cased up with a bloke...She's got a bloke, a regular customer, get me, who pays the rent of the flat. **1939** H. Hodge *Cab, Sir?* xv. 219 A 'Case', or 'Knocking shop', is what the police call a 'disorderly house'. To be 'cased up' is to be in one.

casein. Now commonly pronounced (kēⁱ·sin). Add to def.: also, the product precipitated as curd and applied to various commercial uses.

On the applications of *casein* and *caseinogen* see next.

1891 [see *CASEINOGEN]. **1919** E. E. SLOSSON *Creative Chem.* 142 A mixture of casein and celluloid has something of the merits of both. **1939** *Nature* 29 Apr. 734/1 A wool-like fibre is made from casein by dissolving in alkali solution and forcing under pressure into a solution containing acid, formaldehyde and either salts or sugars. **1966** A. W. LEWIS *Gloss. Woodworking Terms* 37 *Casein* or *cold-water glue*. Curd obtained from skimmed milk is dried and ground to powder and then mixed with lime and sodium carbonate.

2. *attrib.*, as **casein plastic**, a plastic material prepared by a reaction of casein with formaldehyde.

1889 *Cent. Dict.*, *Casein glue*, a glue made by dissolving casein in a strong solution of borax, used as a substitute for ordinary glue by bookbinders and joiners. **1934** *Thorpe's Dict. Appl. Chem.* Suppl. I. 255 (*heading*) Casein plastics. **1937** *Jrnl. R. Aeronaut. Soc.* XLI. 526 Unlike casein cements, they are proof against the attacks of mould and fungus. **1943** V. E. YARSLEY in R. S. Morrell *Synth. Resins* (ed. 2) ii. 59 Casein plastic works in all respects like horn and ivory, and it can be turned, drilled, polished, stamped, moulded, and engraved. **1951** R. MAYER *Artist's Handbk.* iv. 179 Emulsions of casein solutions with oils are easily made.

caseinogen (kēⁱsi,i·nŏdʒĕn). [f. CASEIN + -O + -GEN.] A protein of milk producing casein when acted upon by rennin; CASEIN (see quots. 1938 and 1961).

1891 W. D. HALLIBURTON *Textbk. Chem. Physiol.* xxvi. 580 It will be more convenient..to reserve the term casein for the clotted proteid, and to use the word *caseinogen* for its precursor in the milk. **1906** *Jrnl. Chem. Soc.* XC. I. 56 A comparison of solutions of casein and paracasein (or, as they are usually called in English, caseinogen and casein respectively). **1914** *Chem. Abstr.* VIII. 3059 The conversion of caseinogen into casein by enzyme action is accompanied by the cleavage of N, P, and Ca. **1934** *Times Lit. Suppl.* 1 Nov. 758/3 Caseinogen hydrolysates completely freed from tyrosine are still able to support growth in young rats. **1938** *Thorpe's Dict. Appl. Chem.* II. 411/1 Many American and some English authors have preferred to use the term *casein* for the commercial product produced by precipitation from milk, and the word *caseinogen* for the protein.. of milk in its natural condition...In view of the doubt whether the latter term refers to a single specific chemical compound, and of the fact that there are two quite distinct types of the commercial product, the term *casein* is used in both senses throughout this article. **1961** *Brit. Med. Dict.* 262/1 *Caseinogen*, a phosphoprotein present in milk and convertible by the enzyme chymosin (rennin) into casein. (In the U.S.A. caseinogen is called *casein* and casein *paracasein*.) It is the substance which is precipitated when milk goes sour by the formation of lactic acid, and the precursor of the casein formed by rennin in the coagulation of milk

casemate. Add: **1. d.** *Naut.* An armoured enclosure for guns in a warship.

1888 *Engineering* 17 Feb. 159/2 Italian Ironclad 'Italia'... The barbettes are contained in an armoured casemate, which is supported by the unarmoured structure of the ship. **1899** *Daily News* 21 July 10/4 Twelve out of the sixteen 6-inch guns are in casemates, a term borrowed, I fancy, from the land gunner. It is a neat little apartment, containing one gun, with the hoist from the magazine into it, and all complete.

casement. Add: **1. b.** The matrix cut in stone to receive a monumental brass.

1454 in Dugdale *Antiq. Warwicksh.* (1656) 354/1 Either of the said long plates for writing shall be in bredth to fill justly the casements provided therefore. **1890** J. T. FOWLER in *Proc. Soc. Antiq.* XIII. 34 It has been proposed to revive 'casement', an equally good word at one time in use. **1891** W. H. ST. JOHN HOPE *Ibid.* 213 Beneath the figure is the casement of the inscription. **1903** *Yorks. Archaeol. Jrnl.* 307 The top slab with the casement of the brass taken up and inserted in the north wall.

5. casement cloth, a fabric (*spec.* a cotton or mock-linen cloth) used primarily for casement curtains, but also as a dress material, etc.; so **casement fabric**, which is of wider application; **casement curtain**, a curtain made to fit a casement window; **casement flux**, the orig. name of casement cloth (in the *spec.* sense).

1903 *Popular Fallacy* (*Heal & Son*) (back cover), Pattern books of..printed cottons, printed linens and woollen casement cloths. **1908** *Home Chat* 6 June 572/2 Linen, or casement cloth, employed as concluding note to a skirt of the same. **1932** *Manchester Guardian* 28 Jan. 15/7 Furnishing and casement cloths have been in rather better demand. **1957** *Encycl. Brit.* VI. 572/1 *Casement cloth*, flat, lustrous, British curtain fabric. **1830** TENNYSON *Mariana* ii, She drew her casement-curtain by. **1910** *Daily Chron.* 5 Apr. 9/4 Our windows are furnished with casement curtains. **1912** *Queen* 27 July 180/2 Patterns of Sundour casement fabrics and washing materials. **1920** *Home Chat* 4 Dec. 428 Casement Fabrics made to measure. **1903** *Popular Fallacy* (*Heal & Son*) (back cover), Casement flux: 36 ins. wide.

caser² (kēⁱ·səɹ). *slang.* [Yiddish; f. Heb. *kesef* silver.] A crown; five shillings; in the *U.S.*, a dollar.

1849 'A. HARRIS' *Emigrant Fam.* I. ix. 212 A caser (dollar) if you fine him a night of it; and four if he gets what'll make him quiet. **1860** HOTTEN *Slang Dict.* 111. *Caser* is the Hebrew word for a crown. **1879** J. W. HORSLEY in *Macm. Mag.* Oct. 501/2 One morning I found I did not have more than a caser (5s.). **1899** *Bulletin* (Sydney) 14 Jan., *Caser*, a five-shilling piece [cited as N.Z. slang]. **1900** H. LAWSON *Over Sliprails* (1945) 145, I borrowed half a caser off a murderer once. **1939** J. B. PRIESTLEY *Let People Sing* x. 256 Knocker brought some money and examined it... 'A nicker, half a bar, a caser an' a hole.' **1967** *Sunday Mirror* 19 Mar. 24/4, 5s.—Oxford scholar or caser.

case-work (kēⁱ·swɜ̄ɹk). [f. CASE *sb.*¹ 8 b + WORK *sb.* 2.] Social work carried out by the study of individual persons or groups. Hence **ca:se-wo·rker**, one engaged in such study.

1896 *Westm. Gaz.* 23 Mar. 8/2 There are two kinds of criticism urged frequently against this body—one, from outside, that it is all organisation and no charity; the other, from among its own members, that the officials are buried in 'case' work and do no organising. **1935** C. F. WARE *Greenwich Vill. 1920–1930* xiii. 374 The presence of 'begging tendency' as a possible problem to be checked on a case worker's record. **1940** *Economist* 30 Mar. 573/2 The weakness of Miss Hill's book is her lack of interest in the 'case work' side of the problem. **1942** *Soc. Insurance* (App. G to Beveridge Rep.) 36 The Health Centre could be the focus for all social work among the incapacitated, and its social workers could act as case workers for the Ministry. **1959** *Times Lit. Suppl.* 24 Apr. 244/4 She is a magistrate herself, and a social worker, apparently gifted with a lawyer's appreciation of the one function and a caseworker's knowledge of the other. **1959** B. WOOTTON *Soc. Sci. & Path.* App. II. 360 The difference between caseworker and the intrained or untrained worker lies in the skill required in assessing the effect of environment.

cash, *sb.*¹ Add: **2. b.** *cash on delivery*: applied to the forwarding of goods to order, payment being made to the carrier or postman when the goods are delivered. Abbreviated *C.O.D.

1851 *Illustr. London News* 11 Oct. 442/2 One Sydney merchant has sold one ton of flour..for £70, cash on delivery. **1904** *Daily Chron.* 13 Apr. 6/3 The cash-on-delivery system of transmitting goods by parcel post.

f. *cash down* (DOWN *adv.* 12): ready money. orig. *U.S.*

[**1722** P. LLOYD *Let.* 28 July in *Maryland Hist. Soc. Publication* (1894) XXXIV. 31 A Reserve was made of Allmost all the Lands upon the Western shore, for the Value of £120 Cash pᵈ downe.] **1800** *Green's Impartial Observer* I. 29 Nov. 4 (Advt.), I have for sale..a few Negroes, for Cotton or Cash down. **1817** CUMMINGS & HILLIARD *Let.* 22 July in *Proc. Amer. Antiq. Soc. 1938* (1939) XLVIII. 38 We now address you to ascertain on what terms you would sell us six terrestrial, & one celestial globe, that is—for what each, cash down. **1855** HALIBURTON *Nat. & Hum. Nat.* II. 111 What's the price..cash down on the nail? **1907** I. ZANGWILL *Ghetto Comedies* 238 You should have made it a rule—cash down.

g. *cash and carry*, a system whereby the purchaser pays cash for goods and takes them away himself. Usu. *attrib.* Orig. U.S. *Spec.* used with reference to purchases of arms from the U.S. in the period immediately before 1941. Also, *cash and carry away*.

1917 *Ladies' Home Jrnl.* July 27/3, I would recommend to every woman that you follow the 'cash and carry' plan of buying in preference to the 'credit and delivery' plan. **1922** S. LEWIS *Babbitt* iv. 54 One of those cash-and-carry chain stores. **1930** *Economist* 24 May 1178/2 Marks and Spencer, being a 'cash and carry' concern, is liquid in every respect. **1937** *Ann. Reg. 1936* 294 The President should be given some measure of discretion to permit, say, the victims of aggression to buy, pay for, and transport at their own risk such supplies, not actually munitions of war, which they might need. This policy was described by its proponents as the 'cash and carry' policy. **1940** *Ann. Reg. 1939* 308 It [*sc.* a Bill of U.S. Senate] permitted the country to sell arms to belligerents on a 'cash-and-carry' basis. **1962** H. E. BEECHENO *Introd. Bus. Stud.* xi. 101 These [discount] shops restrict themselves to selling goods on a cash-and-carry-away basis.

3. *cash-box* (earlier example), *-girl*; **cash-account** (earlier and later examples); **cash carrier** *U.S.*, a device employed in shops by which money is carried in a receptacle running on a line between the cash-desk and the several counters; **cash-credit** (earlier example); **cash-crop** (orig. *U.S.*), a crop cultivated primarily for its commercial value (opp. to one for subsistence, etc.); hence **cash-cropping** *vbl. sb.* and (as back-formation) **cash-crop** *v.*; **cash-nexus**, a relationship constituted by, and usu. consisting solely in, monetary transactions; also *attrib.*; **cash-price** (example); **cash register**, orig. *U.S.*, a till for recording and adding the amounts put into it; **cash-sale** (examples); **cash-store** *U.S.*, a store in which credit is not given; **cash-value**, the value in cash; spec. in *Insurance* (in full *cash surrender value*), the value of a policy, etc., cashed before it matures; *fig.* (Philos.), the empirical content of a concept, word, or proposition.

1768 J. WEDGWOOD *Let.* 13 June (1965) 65 Your Cash Account is much wanted. **1786** BURNS *Poems* 88, I might, bythis, hae..strutted in a Bank and clarket My Cash-Account. **1954** T. S. ELIOT *Confid. Clerk* II. 59 Claude has just accepted me like a debit item Always in his cash account. **1834** *Chambers's Edin. Jrnl.* III. 149/1 Tills, or cash-boxes in counters, are now..banished from the higher class of the trading community in London. **1889** *Century Dict.*, Cash-carrier. **1903** G. ADE *In Babel* 18 He had thought out an overhead cash-carrier of the kind used in retail stores. **1832** *Chambers's Edin. Jrnl.* I. 186/1 It is now a hundred and three years since the first cash credit was instituted. **1869** *Rep. U.S. Dept. Agric. 1868* 18 Wheat is a cash crop, and demands a small outlay of labor. **1934** F. R. IRVINE *Text-Bk. W. Afr. Agric.* p. vii, (*heading*) Cash Crops. **1937** *Geogr. Jrnl.* XC. 75 This grafting of cash-crop production on subsistence agriculture in tribal communities. **1942** *Rep. Comm. Land Utilisation in Rural Areas* 5 Cash crops are crops sold directly off the farm, as opposed to those used on the farm. **1950** *N.Z. Jrnl. Agric.* Apr. 365/2 Peas and potatoes also play an important part in the cash-cropping programme. **1957** M. GLUCKMAN in V. W. Turner *Schism & Continuity* p. xiii, The development of wage-earning and cash-cropping. **1960** *Farmer & Stockbreeder* 8 Mar. 91/1 Try to reduce the acreage per cow to 1¼ and cash-crop what you save. **1880** *Harper's Mag.* June 37/1 The cash-girls are paid a dollar and a half a week. *a* **1910** 'O. HENRY' *Trimmed Lamp* (1916) 78, I was a cash-girl and a wrapper and then a shop girl. [**1839** CARLYLE *Chartism* vi. 149 Cash Payment had not then grown to be the universal sole nexus of man to man.] **1855** MRS. GASKELL *North & South* II. xxvi. 353 My only wish is to have the opportunity of cultivating some intercourse with the hands, beyond the mere 'cash nexus'. **1904** *Society in New Reign* iv. 93 London is a European suburb, united by a cash nexus to New York. **1936** C. DAY LEWIS *Friendly Tree* II. xiii. 193 I'm bound to them by a cash-nexus. They paid for my ticket down here. **1781** in *Cal. Virginia St. Papers* I. 438 The articles were furnished at cash prices. **1879** *U.S. Pat. Off. Gaz.* XVI. 847/1 [Patent No.] 221,360. Cash Register and Indicator. **1886** *Cassell's Family Mag.* 123/1 The cash register which is represented in the woodcut is only twelve inches in height. **1938** S. BECKETT *Murphy* ix. 178 If his mind had been on the correct cash-register lines. **1808** J. STEELE *Let.* 28 Aug. in *Papers* (1924) II. 558 A cash sale at present I found to be totally impracticable. **1816** U. BROWN *Jrnl.* in *Maryland Hist. Mag.* XI. 350 [He] advises me to sell the whole of Clement Brooks property for cash, or at a Cash Sale. **1838** H. COLMAN *Rep. Agric. Mass.* 90 At the same time we are always sure of a cash sale. **1879** TOURGÉE *Fool's Err.* viii. 36 The plantation would never have brought that price at a cash sale. **1811** *Raleigh* (N.C.) *Star* 7 Mar. 1/2 Cash Store. S. Bond having taken in a partner, the business in future will be conducted under the firm of *Bond & Jones*. **1830** PAULDING *Chron. Gotham* 156 The Honourable Peleg Peshell, cash-store keeper at Peshellville. *Ibid.* 159 Passing a unanimous resolution, not to buy anything at his cash-store. **1849** N. P. WILLIS *Rural Lett.* xviii. 156 You do injustice to the 'cash stores' of Oswego. **1898** W. JAMES *Coll. Ess. & Rev.* (1920) 434 The great English way of investigating a conception is to ask yourself right off, '..What is its *cash-value*, in terms of particular experience?' **1902** —— *Var. Relig. Exper.* xviii. 443 So Berkeley with his 'matter'. The cash-value of matter is our physical sensations. **1911** A. E. SPRAGUE *Treat. Insurance Companies' Accounts* iii. 26 The cash value of bonuses surrendered when the policy itself remains in force. **1915** S. S. HUEBNER *Life Insurance* xviii. 234 Some companies allow cash values equal to the full reserve at the end of the second or third year. **1929** C. I. LEWIS *Mind & World-Order* i. 32 These empirical criteria..are the 'cash-value' of the category. **1930** *Pitman's Dict. Life Assurance* 506/1 Every policy-holder who fails to pay his premiums..when they fall due renders his policy liable to lapse...However, the life office allows him a cash surrender value if he discontinues. This cash value is based on the reserve held against his policy...The minimum cash surrender allowed is one-third of the total payments the policy-holder has made, plus the cash value of any bonuses which have been allotted to the policy. **1935** *Mind* XLIV. 143 But what is the cash-value of this slogan 'Essence involves existence'? **1966** *Performing Right* Oct. 9/1 Twice a year, the total number of points logged is divided into the total distributable revenue to establish the cash value of each point.

b. Applied adjectively to (*a*) commodities purchasable for cash, (*b*) tradesmen or commercial houses doing business for ready money only. Cf. *cash-sale*, etc., above.

1875 *Chicago Tribune* 13 Sept. 6/1 A large Premium on Cash Pork, Wheat and Corn. **1898** *Daily News* 15 June 6/2 Mr. Armour to-day bought all Mr. Leiter's cash wheat in the north-west. **1903** *Daily Chron.* 7 Apr. 5/2 Cash Dispensing Chemists. **1929** D. H. LAWRENCE *Pansies* 80 Turned my modest penny Over on Boot's Cash Chemist's counter. **1958** *Economist* 18 Oct. 265/1 The continued recovery in copper has now brought cash metal in London to £241 a ton.

cash, *v.*² Add: **1. b.** *Bridge.* To lead (a winning card); to win (a trick) by leading a winning card.

1934 E. CULBERTSON in *Amer. Speech* IX. 11/1 *To cash* a card is simply to take it while the taking is good. **1936** —— *Contract Bridge Complete* xlii. 479 Cash all idle top cards in trump or plain suits. **1959** *Times* 14 Jan. 10/4 Suppose that he cashes four spades and two hearts and can safely assume that East..has nothing left but clubs. **1963** *Listener* 14 Feb. 314/1 The best line of play is to cash the top winners.

2. cash in, to settle accounts in the game of poker; hence in general use, to clear accounts; to close up a matter. (Sometimes *trans.* with *checks* as object.) *U.S. colloq.*

1888 [see below]. **1889** *Kansas Times & Star* 20 Mar., The market value now is about $1,700 a front foot, and many members favor 'cashing in' at such a fancy price, and building elsewhere. **1896** G. ADE *Artie* v. 46 If you're struck on him I'll cash in right here and drop out of the game. **1899** —— *Doc Horne* xxi. 232, I lost back the $2,500 and cashed in. **1904** S. E. WHITE *Blazed Trail Stories* xii. 224 By all the rules of the game, Peter should have failed long since, should have 'cashed in and quit' some five years back.

b. *fig.* To die. (Also without *in*.) Also with *checks* as object.

1884 H. DOUGHERTY *Oratorical Stump Speaker* 14 When Bob cashes in his checks and is toasted like a sirloin steak.. on the top of Old Nick's pitchfork. **1888** *Amer. Humorist* 11 Aug. (F.), Till death calls upon you to cash in your earthly checks. **1908** MULFORD *Orphan* xix. 250 The Orphan not only saved me but also some of them, for I'd a gotten some of them before I cashed. **1920** —— *J. Nelson* xx. 220 He's been follerin' me around steady since Wolf cashed in. **1948** *Sat. Even. Post* 10 July 88/2 Cashing in or shipping out, it made no difference as long as you didn't watch them die. **1966** D. VARADAY *Gara-Yaka's Domain* vii. 75 Because of the size of the dead animal, at first I thought it to be buffalo. 'Poor Bill or Phyl, cashed in?'

c. To 'get in' *with*; now usu., to make a profit *on*, (fig.) to take advantage of (an opportunity, etc.). orig. *U.S.*

1904 S. E. WHITE *Blazed Trail Stories* viii. 146, I don't stack very high in the blue chips when it comes to cashin' in with th' gentle sex. **1927** *Daily Express* 12 Sept. 11 An enterprising American company..'cashed in' on Mr. Arlen by acquiring the screen-rights of one of his earliest stories. **1928** *Sunday Express* 16 Dec. 4/3 She is appearing in too many films, even for a star who would be justified in 'cashing in' on her popularity while the popularity is good. **1930** *Publishers' Weekly* 1 Mar. 1040 Cash in on this tremendous wave of interest and enthusiasm! A large national advertising campaign will start the novel toward a big sale! **1934** WODEHOUSE *Right Ho, Jeeves* ii. 26 With a thing like that to give you a send-off, why didn't you cash in immediately? **1935** *Economist* 8 June 1295/1 Japan's diplomats are now trying to 'cash in on' the opportunities which its soldiers have created. **1955** A. L. ROWSE *Expansion of Eliz. Engl.* ix. 368 That rather unattractive journalist, Barnabe Rich, cashes in on the rising interest in military matters with a series of tracts. **1958** *Spectator* 1 Aug. 156/2 A possible autumn election, designed to cash in on what the Conservatives hope will be the flood tide of their popularity. **1966** *Listener* 2 June 794/2 Are not some of them..cashing in quite shamelessly on the current debased fascination with evil?

d. *trans.* To pay *in* to a bank; to earn, gain.

1904 'MARK TWAIN' *$30,000 Bequest* (1906) 10 I'm going to cash-in a whole three hundred on the missionaries. *a* **1910** 'O. HENRY' *Trimmed Lamp* (1916) 229 With his gold dust cashed in to the merry air of a hundred thousand..the Man from Nome sighed to set foot again in Chilkoot. **1910** W. M. RAINE *B. O'Connor* 21, I know your kind—hell-bent to spend what you cash in. **1933** D. L. SAYERS *Murder must Advertise* xvi. 278 If all these vouchers were cashed in at once, it would send up the cost per packet.

Hence **cash-in** *sb.*, an instance of 'cashing in'.

1940 HARRISSON & MADGE *War begins at Home* iv. 95 Apart from this commercial cash-in, all the comments.. were satirical. **1950** 'G. ORWELL' in *World Rev.* June 35 Margesson's entry into the Cabinet is..a deliberate cash-in on Wavell's victory.

3. To pay *over* or *up*. Now chiefly *U.S.*

1818 KEATS *Let.* 10 Jan. (1958) I. 203, I will..ask Kingston and Cᵒ to cash up. **1825** *New Monthly Mag.* XIV. 193 When it came to 'cashing-up', affairs assumed a soberer complexion. **1831** *Examiner* 296/2 A certain Alderman..did not cash up to his supporters on the former election. **1842** BARHAM *Ingol. Leg.* II. 54 He could not cash up, spite of all he could do. **1854** M. J. HOLMES *Tempest & Sunshine* xxi. 227 Tempest is in a desput hurry to know whether I'm goin to cash over and send her to market in New Orleans. **1924** WODEHOUSE *Leave it to Psmith* I. 26 I'm game to spill it and leave it to your honesty to cash up if the thing looks good to you. **1969** D. CLARK *Death after Evensong* ii. 37 Don't parsons cash up the takings after the service with the church wardens?

b. In *pass.*: to be supplied with money. *Austral.* and *N.Z. colloq.*

1940 F. D. DAVISON *Woman at Mill* 151, I..went to Sydney well cashed up. **1961** B. CRUMP *Hang on a Minute* (1963) 191, I couldn't even go on the bash when we were cashed-up.

cashable (kæ·ʃăb'l), *a.* [f. CASH *v.*² + -ABLE.] Capable of being cashed. Also *fig.* (cf. *cash-value).

1891 *Pall Mall Gaz.* 21 Sept. 2/1 The money card would not be cashable. **1928** *Daily Express* 2 July 10/2 Cashable orders on retail or manufacturing firms. **1953** H. H. PRICE *Thinking & Exper.* ii. 36 It is part of the nature of a concept to be 'cashable' by instances, whether or not it is actually cashed. *Ibid.* vii. 229 There can be no symbols unless *some* symbols are empirically cashable.

cashaw, var. *CUSHAW.

cashless, *a.* (Earlier U.S. example.)

1833 *Niles' Register* 20 July XLIV. 347 They with rich land, and staple productions..are becoming poor and cashless.

Casinese, var. *CASSINESE.

casing, *vbl. sb.* Add: **1. b.** The action of *CASE *v.*² 5; inspection, planning, etc., esp. in preparation for a robbery (see quots.). *slang* (orig. *U.S.*).

1928 *Amer. Mercury* May 80/2 Laying out the route of escape before consummating a robbery comes under *casing*. **1942** *Amer. Speech* XVII. 91/1 *Casing*, looking over the prospective customers [by street vendors], so that they will be able to judge to some extent the amount of business they will be able to do. **1946** *Ibid.* XXI. 70/1 Casing and 'planning a crime' can..hardly be synonymous, for the casing of an individual, a place of business, the movements of payrolls or registered mail, etc., only supplies certain information which enables the mob to plan the crime. **1960** *Argosy* May 53 The casing will be done by the Thomases. They can go anywhere.

2. c. *U.S. Mining.* (Earlier example.)

1869 J. R. BROWNE *Adv. Apache Country* 525 They are all true fissure veins, with well-defined casings.

d. The framing round a door or window. orig. *U.S.*

1873 J. H. BEADLE *Undevel. West* xxii. 449 Grasping the casing I got in at the door. **1913** MRS. STRATTON-PORTER *Laddie* ii. 65 All the casings were oiled wood, and the walls had just a little yellow. *Ibid.* xviii. 584 He carefully loaded his gun, and leaned it against the front casing. **1940** *Chambers's Techn. Dict.* 140/2 *Casing*, the frame enclosing the sash-weights...The frame within which a door hangs.

casino. Add: **4.** A building for gambling, often with other amenities. (Now the usual sense.)

1851 E. RUSKIN *Let.* 15 Aug. in M. Lutyens *Effie in Venice* (1965) II. 180 He lost in gambling at Chamouni to the Master of the Casino 25,000 francs. **1950** *Oxf. Jun. Encycl.* IX. 115/1 *Casinos*..were originally public buildings with music and dancing rooms...Gambling games..were soon introduced, and gradually the word began to be accepted as meaning a gambling house. **1966** *Oxford Mail* 23 Sept. 1/7 A conspiracy to defraud a Plymouth casino by using illegal dice in a game of craps, and obtaining gaming chips by false pretences.

cask, *sb.* Add: **5.** *cask-body, -head, -steamer, -washer.*

1874 *Spons' Dict. Engin.* VIII. 2919 Having thus far followed the shaping of the staves, and the conversion of the same into cask bodies, it will be necessary to direct our attention to the formation of cask-heads. **1905** *Daily Chron.* 10 Oct. 2/5 The old-fashioned cask-heads with the familiar legend of 'Fine old Port'. **1881** *Instr. Census Clerks* (1885) 61 Cask Washer, Steamer. **1891** *Daily News* 12 Feb. 2/8 A cask-washer, employed at the Berkshire Brewery. **1921** *Dict. Occup. Terms* (1927) § 459 *Washer, barrel; cask washer*..(i) *cask steamer*; rolls barrels to feed end of washing machine..(ii) cleans barrels by pouring water and stone into them.

casket, *sb.*¹ **3.** (Earlier and later examples.)

1849 C. SPENCER *Let.* 20 Mar. in N. E. Eliason *Tarheel Talk* (1956) 264 The casket, which held this jewel [*sc.* her dead friend], was worthy of it. **1863** HAWTHORNE *Our Old Home* 102 'Caskets'! a vile modern phrase, which compels a person..to shrink..from the idea of being buried at all. **1895** *Daily News* 29 June 2/1 The strange-looking mahogany coffin (it is called a casket in the United States). **1948** *Atlantic Monthly* Sept. 87/2 Shrouds have long since become slumber robes, and coffins caskets. **1963** J. MITFORD *Amer. Way Death* v. 180 The only difference between an English 'casket' and a 'coffin' is in the shape, the former being rectangular, and the latter, tapered or 'kite-shaped'.

Caslon (kæ·zlən). *Typography.* The name of William *Caslon*, father (1692–1766) and son (1720–1778), applied to the type foundry established by the father, and to the old-face type cut there, or (later) one cut in imitation of this. Also *Comb.*, as *Caslon-shaped* adj.

1825 T. C. HANSARD *Typographia* 353 The Caslon foundry is still upheld. **1836** *Gentl. Mag.* May 557/2 The printers, whose offices were generally stored with the Caslon founts. *Ibid.* 558/1 The true Caslon-shaped Elzevir types. **1898** J. SOUTHWARD *Mod. Printing* I. xxii. 136 The Elzevir types prevailed for a long time all over Europe; and the first William Caslon adopted them as his models—hence his types are known as Caslon Elzevirs. **1922** UPDIKE *Printing Types* II. 106 Caslon types are..beautiful in mass, and above all..legible. **1927** MᶜKERROW *Introd. Bibliogr.* 300 It may perhaps be well to warn the reader that 'Caslon' as applied to type has not an absolutely precise connotation. Though, for example, the founts of different sizes cut by Caslon himself are in the same general style, the forms of the letters do not correspond with mathematical exactness...When it is added that 'Caslon' founts have been repeatedly cut in modern times, it will be evident that we must be prepared for a certain range of variety. *Ibid.* 301 The Caslon italic retained something of the angularity of the current script. *Ibid.* 305 The thick strokes of 'old style' are on the whole thinner than in Caslon.

Caspian (kæ·spiăn), *a.* and *sb.* [f. L. *Caspius* (Gr. Κάσπιος): see -AN.] **A.** *adj.* Of or pertaining to the *Caspian Sea*, an inland sea of central Asia. **B.** *sb.* **a.** *ellipt.* The Caspian Sea. **b.** *Anthropol.* (A member of) the easternmost branch of the Mediterranean race.

1586 MARLOWE *1st Pt. Tamburl.* I. i, Fled to the Caspean or the Ocean maine. **1590** SPENSER *F.Q.* II. VII. xiv. 3 Who swelling sayles in Caspian sea doth crosse. **1607** TOPSELL *Four-f. Beasts* 96 The haire of the caspian camels is so softe. *Ibid.* 221 Amongst the Caspians they [*sc.* foxes] abound. **1757** J. DYER *Fleece* IV, And borderers of the Caspian, who renew That ancient path to India's climes. *Ibid.* 174 On the right The Caspian lake. **1759** A. BUTLER *Lives Saints* IV. II. 982 The coming of the numerous swarm of the Turci from the coasts of the Caspian sea. **1785** PENNANT *Arctic Zool.* II. 526 *Caspian Tern*..*Terna Caspia*, inhabits the Caspian Sea. **1788** [see *ALBANIAN *a.*² and *sb.*²]. **1854** J. H. NEWMAN *Hist. Turks* I. 9 This is the country, forming the two basins of the Aral and the Caspian. **1876** *Encycl. Brit.* V. 179/1 The temperature of the Caspian area is remarkable for its wide range, both geographical and seasonal. **1923** R. B. DIXON *Racial Hist. Man* 21 Names have been given to the eight fundamental types as follows ..Caspian. *Ibid.* V. I. 409 It is impossible to say whether the dolichocephalic element in the Micmac is the Caspian, which would ally them with the Eskimo, or the Mediterranean. **1954** R. PETERSON et al. *Birds Brit. & Eur.* 153 Caspian tern, *Hydroprogne caspia*. .. Almost as big as Herring Gull.

cassab (kăsā·b). Also **cussab.** [Hind., a. Arab. *kaṣṣāb* butcher.] A seaman of Asian origin employed in the merchant service.

1881 *Instr. Census Clerks* (1885) 35 Cussab or Cassab..in P. and O. Service. **1921** *Dict. Occup. Terms* (1927) § 735 *Cassub, cussab*; a member of Asiatic crew employed, either as lamp trimmer (on deck) or as storekeeper in steward's department. **1924** *Glasgow Herald* 17 May 4 There is usually an elder, not necessarily a Haji, who carries a copy of the Quran. On board ship he is, as a rule, the cassab. **1960** *Times* 16 Mar. p. xiv/4 There will be the specialist posts of cook, store-keeper and lampman (bhandary, cassab and battiwallah).

cas(s)alty, -elty, dial. ff. CASUALTY.

1886 *Cornhill Mag.* June 582 A Cas'alty Corner is a feature of every district of outcast London, is to be found wherever the poor of the great city most do congregate. *Ibid.*, Cas'alty labourers. **1892** TENNYSON *Church-w. & Curate* i, Nasty, casselty weather!

Cassandra (kăsæ·ndră). [L. *Cassandra*, Gr. Κασ(σ)άνδρα.] The name of a daughter of Priam, sought in love by Apollo, who gave her the gift of prophecy; when she deceived him he ordained that no one should believe her prophecies, though true; used allusively, *attrib.*, and *Comb.*

a **1668** LASSELS *Voy. Italy* (1670) Pref. a x, Other Governours (Cassandra like) telling their Pupils many excellent truths, are not believed by them. **1711** ADDISON *Spect.* No. 130 ¶ 2 A Cassandra of the [Gypsy] Crew..told me, That I loved a pretty Maid in a Corner. **1837** CARLYLE *Fr. Rev.* II. I. ii, A Cassandra-Marat cannot do it. **1863** LONGFELLOW *Tales of Wayside Inn* 191 The cawing of the crow,..Cassandra-like, prognosticating woe. **1901** *N. Amer. Rev.* Feb. 236 Far be from me the Cassandra task of attempting to persuade my countrymen that an army of any given size is a necessity for the Republic. **1926** CHESTERTON *Incredulity of Fr. Brown* v. 177 Lady Diana had recovered a little from her trance of Cassandra. **1941** KOESTLER *Scum of Earth* 155 The censorship continued to suppress Kerillis's Cassandra-cries against the traitors in the Ministries. **1959** *Times* 18 June 13/3 Dr. Bertram gets off to a slow start because he is a very zestful Cassandra.

Hence **Cassa·ndrian** *a.*

a **1876** EADIE *Comm. Thess.* (1877) 347 Baxter..accused Grotius of a design to reconcile Papists and Protestants in a Cassandrian Popery. **1903** H. BEGBIE in *Daily Chron.* 28 May 3/7 Remembering the Cassandrian comparison which has been made between the Transvaal and Ireland.

cassata (kăsā·tă). [It.] Ice-gâteau; Neapolitan ice-cream containing candied fruit and nuts.

1927 *Harper's Mag.* Feb. 319, I was sitting in front of Gilli's one May afternoon eating *cassata*. **1953** tr. *Bonhoeffer's Lett. fr. Prison* (1963) 114 We..consumed vast quantities of *granitos* and *cassatas* on the way. **1958** *New Scientist* 2 Jan. 9/1 More exotic concoctions like cassata. **1959** C. MACINNES *Absolute Beginners* 16, I ordered my striped *cassata*.

‖**casse** (kæs, kas). [Fr., vbl. n. f. *casser* to break.] Incipient souring of certain wines, accompanied by loss of colour and a deposit of sediment.

1883 J. GARDNER *Brewer, Distiller* 226 If the breakage, or *casse*, as it is termed, has not exceeded 7 or 8 per cent. by the time August is reached, he..lets the wine remain. **1911** *Encycl. Brit.* XXVIII. 720/1 The disease known as *tourne* or *casse* is generally caused by the wine having been made..from grapes affected by mildew. **1959** W. JAMES *Word-Bk. Wine* 43 *Casse*, a common disease of wines in which an excess of metallic salts causes cloudiness and an off-taste.

Cassegrainian, *a.* Also **Cassegrain**, used *attrib.* and *absol.*

1888 *Encycl. Brit.* XXIII. 145/1 The Cassegrain telescope...In comparatively recent years the Cassegrain has acquired importance from the fact of its adoption for the great Melbourne telescope. **1961** *Listener* 7 Sept. 353/3 The Cassegrain mirror is swung down into a compartment let into the telescope tube.

Cassel (kæ·sĕl). The name of Cassel (now Kassel), a town in Germany, used attrib. to designate various pigments, as **Cassel brown, Cassel earth**, a brown prepared from impure lignite, Vandyke brown; **Cassel green**, a green consisting chiefly of barium manganese, Manganese green; **Cassel yellow**, a patent yellow pigment, Chinese yellow, mineral yellow.

1860 URE *Dict. Arts* (ed. 5) I. 805 Vandyke, Cappah, Rubens, Cassel, and Cologne Browns. **1875** T. SEATON *Fret-Cutting* viii. 105 The brown pigment is well known to artist's colourmen as Cassel earth. **1882** *Encycl. Brit.* XIV. 379/1 Another oxychloride, $PbCl_2.7PbO$, known as 'Cassel yellow'. **1885** *Ibid.* XIX. 88/1 Cassel Green, called also Rosenstiehl's Green, is a fine innocuous pigment made by melting together sulphate of baryta and oxide of manganese. ..Vandyke brown and Cologne or Cassel brown, peaty ochres...Real Vandyke Brown..ought to be a kind of bituminous peaty earth..allied to which are Cologne and Cassel Earth.

casserole. Delete ‖. Now usu. with pronunc. *ca·sserole*. Add: **1.** (Later example.) Now, a dish cooked and served in a casserole. Also *à la casserole*, now usually *en casserole*. Also *attrib.*, as *casserole cookery, dish, pan*.

1870 U. DUBOIS *Cosmopol. Cookery* § 727 Capon of Caux, roasted 'à la casserole'...Braised or roasted in a stewpan. **1904** C. H. SENN (*title*) Chafing Dish and Casserole Cookery. **1905** A. KENNEY-HERBERT *Common-sense Cookery* 292 A poulet, or pheasant *à la casserole*. **1906** FILIPPINI *Internat. Cook Bk.* 378 Heat one tablespoon butter in an earthen casserole pan. *Ibid.* 565 Sweetbreads en Casserole. Blanch and trim six heart sweetbreads. Place in an earthen casserole dish [etc.]. **1928** A. CHRISTIE *Myst. Blue Train* xxxii. 264 The waiter deftly served them with chicken *en casserole*. **1958** A. WHITE tr. *Colette's Claudine in Paris* i. 9 The yellow chanterelles that go so well with creamy sauces and casserole of veal. **1960** *Observer* 17 Jan. 14/4 A casserole, once upon a time, was a pan with a handle, made in tinned copper, earthenware, enamelled iron or some other heat-resistant material... In Britain to-day the casserole has become a portmanteau word for any receptacle in a fire-resistant material, with or without a lid, that goes both into the oven and on to the dinner-table.

3. *fig.*

1930 W. J. LOCKE *Town of Tombarel* iv. 127 The affairs of women..are equally important in this stew, this vast casserole of male and female existence. **1962** *Economist* 17 Nov. 653/1 The..Bill is one of those casserole measures into which bits and pieces..from various reports are dropped.

So **ca·sserole** *v. trans.*, to cook in a casserole.

1930 D. L. SAYERS *Strong Poison* ix. 114 A casseroled chicken with turnips and carrots done in the gravy. *Ibid.* 115 It seemed a shame to casserole it [*sc.* a chicken], for it would 'ave roasted beautiful. **1960** *News Chron.* 11 Oct. 8/6 Instructions for casseroling pigeons in home-made wine.

cassette. Add: **b.** Also, a light-proof (cylindrical) container for a spool of film; a container for an X-ray plate or film.

1934 ST. JOHN & ISENBURGER *Industr. Radiogr.* xi. 81 To secure the proper contact it is customary in medical practice to mount two intensifying screens permanently in a metal 'cassette'. This consists of a cast-aluminium frame to which is attached a sheet of Bakelite or aluminium, forming a window or front transparent to x-rays. **1934** WEBSTER, *Cassette*, a film or plate holder, esp. one for the intensifying screen in X-ray photography. **1940** SOWERBY *Dict. Photogr.* (ed. 15) 111 *Cassette*, container in which 35-mm. film, which has no backing paper, is put up for daylight loading into the camera...The name 'cassette' is also applied to special containers designed for X-ray duplitised film technique...They ensure even pressure over the whole area of the film when intensifying screens are used. **1962** A. GÜNTHER *Microphotogr.* 18 A copying cassette taking 10 metres (30 ft.) or 30 metres (100 ft.) of 35 mm. film. **1964** 'E. PETERS' *Flight of Witch* iii. 44 Miles Mallindine and Dominic Felse were devotedly disentangling finished cassettes from cameras, and securing them in their little yellow bags ready for the post. **1970** *Amat. Photogr.* 11 Mar. 61/3 This is another 'first' for 9·5 mm, the gauge which first introduced the amateur to the joys of cassette-loading cameras and projectors.

c. *Ceramics.* = SAGGAR *sb.* 1.

1909 in WEBSTER.

d. A closed container of magnetic tape with both supply and take-up spools, so designed that it needs merely to be inserted into a suitable tape recorder to be ready for use.

1960 *Tape Recording & Hi-Fi Yearbk. 1959–60* 5 One of the new decks..is the first British product designed to operate with cassettes (or magazines), with the tape enclosed so that there is no need to thread it in front of the magnetic heads on the deck, nor to anchor it in the spools. **1967** J. GARDNER *Madrigal* viii. 199 Mostyn had his briefcase open and a cassette tape-recorder ready to run. **1969** *Tape Recording Yearbk.* (ed. 9) 6 Insertion of a cassette automatically brings the tape transport system into operation.

cassie² (kæ·si). = CASSE PAPER; also see quots.

1688 [see CASSE PAPER]. **1770** LUCKOMBE *Concise Hist. Printing* 496 Cassie paper, broken paper. **1889** BARRÈRE & LELAND *Dict. Slang* I. 218/2 *Cassie* (printers), wrinkled, or outside sheets of paper. **1940** *Chambers's Techn. Dict.* 140/2 *Cassie*, the damaged tops and bottoms of a ream of paper.

cassie³ (kæ·si). *U.S.* [Fr., ad. Pr. *cacio* acacia.] A leguminous shrub with yellow flowers, *Acacia farnesiana*.

1876 C. E. HOBBS *Bot. Hand-bk.* 20 Cassie, the flowers yield a perfume, Acacia Farnesiana. **1905** *Yng. Woman* Apr. 230/1 Mignonette, and cassie. **1952** A. G. L. HELLYER *Sanders' Encycl. Gardening* (ed. 22) 3 *A*[*cacia*]..*Farnesiana*, 'Popinac', 'Cassie'..Tropics.

Cassinese (kæsini·z), *a.* and *sb.* Also **Casinese.** [a. It. *Cassinese*, f. Monte *Cassino*, on which the earliest Benedictine monastery was founded *c* 529: see -ESE.] **A.** *adj.* Of or pertaining to the Benedictine monks of Monte Cassino, or to a congregation of Benedictine abbeys. **B.** *sb.* A monk of the monastery of Monte Cassino, or of the Cassinese congregation; an inhabitant of the town of Cassino below Monte Cassino.

[*c* **1643** in *Cath. Rec. Soc. Publ.* (1933) XXXIII. 82 Some Bennedictine fathers, who were of the Italian Congregation, otherwise called the Cassine or the Congregation of S. Justina.] **1878** J. L. PATTERSON *Maguire's Pius the Ninth* xviii. 357 Supposing..the libraries were not open to the public, what right had the State to confiscate them... The Cassinese, the Roman College, and the Angelica libraries..have been massed together. **1881** *Weldon's Chronol. Notes* Pref. p. x, The reigning Pope Julius II gave the name of the Cassinese Congregation to the whole body of the reformed Benedictines of Italy. *Ibid.* Index p. iv, Cassinese Monks on English Mission. **1893** ADDIS & ARNOLD *Cath. Dict.* 82/2 There is a monastery at Ramsgate belonging to the Cassinese branch of the order. **1910** M. HAILE *Life R. Pole* 127 The Benedictines of St. Giustina at Padua—who were beginning to call themselves Casinese to denote their union with Monte Casino. **1954** *Biogr. Studies* II. 225 Preston was the Mission Superior of the Cassinese in England. **1957** F. MAJDALANY *Cassino* I. viii. 53 The few thousand Cassinese who, citizens of Cassino, had stubbornly insisted on their right to stay there. *Ibid.* III. iii. 117 The bitterness and bewilderment of the Cassinese monks.

cassiopeium (kæ:siŏpi·ŏm). [mod. L. (C. A. von Welsbach in *Sitzungsberichte d. Akad. d. Wissenschaften Wien, Math.-naturw. Klasse*, sect. 2 b, CXVI. 1435), f. *Cassiopeia* (see CASSIOPEIAN *a.*).] An alternative name (now *disused*) for a rare-earth element now known as lutetium or lutecium.

1908 *Chem. Abstr.* IV. 1393 Following the publication of the first article by the author [*sc.* G. Urbain] on this subject, Auer von Welsbach..described two substances obtained from ytterbium...He gave them the names aldebaranium and cassiopeium...The author claims priority in this work. **1966** *McGraw-Hill Encycl. Sci. & Technol.* VII. 623 C. F. Auer von Welsbach..named the element cassiopeium, but lutetium is the accepted form at present.

cassiri (kæsīᵊ·ri). Also **casserie, cassiree, kasiri.** [a. Carib *cassiri*: see CASSAREEP.] An intoxicating liquor made in Guyana from fermented sweet potatoes.

1796 STEDMAN *Surinam* I. 392 Another drink called *cassiree* is also much used by these Indians. **1851** W. H. BRETT *Indian Missions Guiana* 103 There is another drink made in a more cleanly manner from potatoes, called Kasiri. **1854** H. G. DALTON *Hist. Brit. Guiana* I. 82 They have also another intoxicating beverage, called cassiri. **1879** BODDAM-WHETHAM *Roraima* 248 The favourite beverage among the inhabitants was a disagreeable-looking compound called cassiree. **1904** W. H. HUDSON *Green Mansions* I. 26 More cups of casserie followed this outburst. **1934** E. WAUGH *Handful of Dust* v. 272 Cassiri..the local drink made of fermented cassava.

cassis (kæ·sis). [Fr., black currant; black-currant liquor, app. f. L. *cassia* (see CASSIA¹).] A syrupy liquor made from black currants and used chiefly to flavour vermouth, spirits, etc. Hence *vermouth cassis*.

[**1877** E. S. DALLAS *Kettner's Bk. Table* 106 Cassis, the French name for black currants and for the syrup made from them. The cassis of Dijon has a great reputation..as a cooling drink.] **1907** *Yesterday's Shopping* (1969) 100/2 Cassis, made from Black Currants grown in Burgundy. **1911** M. BEERBOHM *Z. Dobson* xxiv. 346 Vermouths secs, vermouths cassis. **1913** C. MACKENZIE *Sinister St.* I. II. x. 312 We always play billiards in the evening, and drink cassis. **1948** A. WAUGH *Unclouded Summer* iii. 25 A *vermouth cassis* and a cigarette. **1954** W. FAULKNER *Fable* (1955) 47 The bottle of wine..and one of *cassis*.

cassolette. Add: **3.** A small casserole; a dish cooked in such a receptacle.

1813 L. E. UDE *French Cook* 418 Mind that the *cassolettes* are to be made quite cold before you take them out of the mould. **1898** MRS. ROUNDELL *Pract. Cookery Bk.* 195 Scoop out the insides and fill them with mince and sauce, after heating the Cassolettes in the oven. **1905** A. KENNEY-HERBERT *Common-sense Cookery* 177 Potato cassolettes. **1951** *Good Housek. Home Encycl.* 393/1 The cassolettes may be served as a hot hors d'œuvre, an entrée or an after-dinner savoury.

‖**cassone** (kassō·ne). Pl. **-ni** (nĭ), **-es.** [It., augmentative of *cassa* chest.] A large Italian coffer, esp. one made to hold the trousseau of a bride, often elaborately carved and decorated. Also *attrib.*

1882 J. W. MOLLETT *Illustr. Dict. Art & Archæol.* 60 *Cassone*, an Italian chest, richly carved and gilt, and often decorated with paintings, which frequently held the *trousseau* of a bride. **1886** *Athenæum* 22 May 687/2 The first and second [pictures] are decorative panels from *cassoni*, and represent scenes at tournaments. **1904** *Studio* Sept. 303/2 The collecting of wedding-coffers, or 'cassones' as they are sometimes called. **1922** *19th Cent.* May 803 The fourteenth century round-backed chair and the *cassone* attributed to Ammannati, which are shown in the same room. **1932** *Times Lit. Suppl.* 21 Apr. 285/2 The cassone panels representing Satyrs hunting. **1949** E. POUND *Pisan Cantos* lxxvi. 47 And the gilded cassoni.

‖**cassoulet** (kasulę). [Fr. (19th c.), dim. of dial. *cassolo* tureen, stew-pan.] A stew-pan; hence, a ragout, orig. a regional speciality of Languedoc, consisting of meat (esp. duck, goose, or pork) and haricot beans.

[**1939** M. MOSSER *Good Maine Food* 93 The only satisfactory way to cook an old wild goose is to..place the strips in a bean pot, cover them with a quart of pea beans and a pound of salt pork, and bake 8 hours in a slow oven. This dish, in the south of France, would be called a *cassoulet*.] **1940** G. WESCOTT *Pilgrim Hawk* 65 They always telephoned a certain small unsuccessful restaurant to prepare a supreme *cassoulet* which took two days. **1942** B. BREUER in *55 Short Stories from 'New Yorker'* (1949) 328 A casserole, the veritable French *cassoulet*, and inside..squabs and tiny vegetables. **1959** *Times* 16 Feb. 11/3 *Cassoulet* is one of the great dishes of rural French cooking. **1963** N. FREELING *Because of Cats* xi. 168 Good lunch... Cassoulet. Beans with goose.

cast, *sb.* Add: **41. c.** *Austral.* and *N.Z.* The sweep a trained dog makes when mustering sheep. Cf. *CAST *v.* 60 c.

1933 L. G. D. ACLAND in *Press* (Christchurch) 30 Sept. 15/7 A heading dog..goes wide round sheep so as not to disturb them and make them go faster. This curve or sweep is called a cast. **1946** F. D. DAVISON *Dusty* (1947) 115 The trial had four phases; the cast, when the owner sent the dog forward by himself to find the sheep. **1947** P. NEWTON *Wayleggo* (1949) ii. 28 [The dog] had a tremendous cast, and I used to marvel at the unerring way he would land out at the head of sheep in country where his cast took him far out of sight of them.

42. cast-net, a net which is cast and drawn immediately, as distinguished from a net which is set and left.

1647 HEXHAM *Dutch Dict.* s.v. *Net*, A cast-net, *een werpnet*. **1767** FAWKES *Theocritus* XXI. 13 The seine, the cast-net, and the wicker maze. **1883** DAY *Indian Fish* 46 In some places several cast-nets are joined together, to stop up all passage of fish along a stream. **1952** HEMINGWAY *Old Man & Sea* 13 I'll get the cast net and go for sardines. **1963** *Times* 8 Feb. 12/6 Most of his days are spent plying his cast-net in the shallow waters of the south Arabian coast.

cast, *v.* Add: **6. d.** *trans.* To throw the line over (a piece of water). Hence **ca·stable** *a.*

1892 *Field* 16 July 104/2 A fairly strong stream of only about 2 ft. in depth, and just a nice 'castable' width. *Ibid.* 1 Oct. 522/2, I therefore began to cast the lower portion of the pool. **1953** J. MASTERS *Lotus & Wind* xviii. 229 She.. took her rod and..began listlessly to cast the pool.

23. b. Of crops: to yield, bear fruit, produce. *dial.*

1890 GISSING *Village Hampden* vii, They tell me as the Lammas wheat be a-casting badly. **1893** *Field* 8 Apr. 530/1 A dry March..is of universally good omen for the coming corn crops. These never 'cast' so well as they do when a warm soil causes the under ground progress to be earlier and quicker than is that above the surface.

60. c. *N.Z.* Of a trained dog: to make a wide sweep when mustering sheep. Also, to direct a dog to make such a sweep. Cf. *CAST *sb.* 41 c.

1911 W. H. KOEBEL *In Maoriland Bush* v. 77 He must acquire the art of 'casting' a sheep-dog. **1947** P. NEWTON *Wayleggo* (1949) 153 It is instinctive for a heading dog to cast when running out i.e. to make a wide detour so as to get round his sheep without disturbing them. **1966** *Weekly News* (N.Z.) 6 Apr. 45/5 She cast out very wide and then she spotted the sheep on the huntaway course.

79. cast off. j. *Printing.* To estimate how much printed matter will correspond to (a piece of MS. copy). Also *absol.* (Cf. *CAST-OFF *sb.*¹ 2.)

1683 MOXON *Mech. Exerc., Printing* xxii. ¶ 9 Casting off Copy..is to examine..how much..of Printed Copy will Come-in into any intended number of Sheets..; or how much Written Copy will make an intended number of Sheets..: Therefore if I shew you how the Compositer Casts off Written Copy, I do at the same time inform you how to Count off Printed Copy. **1784** B. FRANKLIN in *Ann. Reg.* (1817) *Chron.* 389 The compositors in your chapel do not cast off their copy well. **1808** STOWER *Printer's Gramm.* 135 To cast off manuscript with accuracy and precision, is an essential object. **1824** J. JOHNSON *Typogr.* II. 89 To cast off manuscript..is a task of a disagreeable nature. **1892** OLDFIELD *Man. Typogr.* xii. 97 Sometimes copy is so badly arranged that it is almost impossible to cast-off accurately...In casting-up allowance must be made for chapter-heads, &c.

cast, *ppl. a.* Add: **1.** *cast shadow*, in painting, a shadow cast by an object within the picture, and serving to bring it out against the objects behind it.

1849 J. D. HARDING *Less. Art* liii, In shading this Lesson care must be taken to make the cast shadow GHF darker at the points G and F, and also sharp on the edge. **1890** *Adeline's Art Dict.* 362/1 The cast shadow is always darker than the shadow, properly so called, if the body casting the shadow and the surface receiving it are of the same tonality.

b. Cf. CAST *v.* 12. (Further examples.)

1877 *Trans. N.Z. Institute* 310 Sheep that were 'cast' were soon attacked by the blow fly. **1947** R. B. KELLEY *Sheep Dogs* v. 69 She [*sc.* a ewe]..was 'cast', that is, lying on her back in a depression, unable to rise. **1953** BANNERMAN *Birds Brit. Isles* I. 15 A 'cast' sheep, whether healthy or sick, is always in danger, the eyes being torn from the sockets [by crows].

c. *Austral* and *N.Z.* *cast-for-age*, see quot. 1965.

1933 L. G. D. ACLAND in *Press* (Christchurch) 30 Sept. 15/7 *Cast for age*, etc., means 'culled' for, etc. This term is used for stud sheep rather than flock sheep. **1950** *N.Z. Jrnl. Agric.* Sept. 217/3 Cast-for-age ewes..are sold in late summer or autumn. **1965** J. S. GUNN *Terminol. Shearing Indust.* I. 13 *Cast for age*, a term to describe sheep that are got rid of..because they are too old.

8. b. *cast stone*: a manufactured substance resembling stone.

1925 *Pop. Mechanics Mag.* May 743 (*title*) Making cast stone; a new building material from marble. **1938** *Archit. Rev.* LXXXIII. 223 The photograph shows the rough texture of the facing bricks which contrasts with the smooth white cast-stone trim of the fenestration. **1956** *Gloss. Terms Concrete* (*B.S.I.*) 9 *Cast stone*, a building material manufactured from cement and natural aggregate, for use in a manner similar to and for the same purpose as natural building stone.

Castalian, *a.* (Later examples.) Also *absol.*

1863 C. READE *Hard Cash* III. ii. 41 He had better have the courage to plunge into the Castalian stream, like Virgil and Lucan. **1882** G. M. HOPKINS *Let.* 4 Nov. in *Lett. to R. Bridges* (1955) 159 What I call Castalian or Parnassian, that is the language of poetry draping prose thought.

castanite (kæ·stănəit). *Min.* [ad. G. *castanit* (L. Darapsky 1890, in *Neues Jahrb. Min.* II. 267), f. Gr. κάσταν-α chestnut, in allusion to its colour: see -ITE¹.] = *HOHMANNITE.

1892 DANA *Syst. Min.* (ed. 6) 964 Castanite..Monoclinic, with a prismatic angle of 82°. **1951** PALACHE et al. *Dana's Syst. Min.* (ed. 7) II. 614 Castanite has been considered to be identical with amarantite but was later shown to be in all probability the same as hohmannite.

caste. Add: **4.** *caste-mark*; *caste-bound* adj.

1955 T. H. PEAR *Eng. Social Diff.* 133 The mass of Americans..took it for granted that the typical Briton is essentially a caste-bound snob. **1901** KIPLING *Kim* xi. 290 Kim splashed in a noble caste-mark on the ash-smeared brow. **1941** C. KIRKUS *Let's go Climbing* xiv. 177 At the temple at Gangotri red caste-marks were put on our foreheads.

Castel Durante (kæ:stel,duræ·nti). The name of a small Italian town, now called Urbania, used to designate a kind of tin-glazed ware of which the finest examples were produced in the 16th cent.

1857 MARRYAT *Hist. Pott. & Porc.* (ed. 2) iv. 58 The manufacture of Castel Durante..flourished under Duke Francesco Maria II (1574 to 1631). *Ibid.*, The characteristics of the Castel Durante majolica are very uniform. **1951** 'M. INNES' *Operation Pax* III. iii. 91 The Castel Durante dishes that were a relic of her childhood in Rome. **1962** *Vict. & Albert Mus. Internat. Art Treasures Exhib.* 76/1 A Castel Durante plate by Giovanni Maria, with sunk centre, painted with a seated putto within a narrow white border.

casteless (kā·stlės), *a.* [f. CASTE *sb.* + -LESS.] Having no caste or castes.

1886 [in Dict. s.v. CASTE *sb.*]. **1898** *Daily News* 15 Aug. 6/2 Reformed casteless Hindoos. **1906** *Macm. Mag.* July 714 While he [*sc.* the Maharajah of Travancore] journeyed over the black water to distant, and to them casteless and almost godless, Britain. **1949** G. BATESON in M. Fortes *Social Struct.* 46 It is wrong for a casteless person to address a prince in other than the 'polished language'.

castellated, *ppl. a.* Add: **1. c.** *transf.* Of a nut or disc: having grooves or recesses on its upper face.

1904 A. B. F. YOUNG *Complete Motorist* iv. 74 Castellated nuts are used throughout, with split pins. **1922** *Times* 20 June 8/5 The wheel and consequently the castellated shaft will be rotated.

Castelli (kæste·li). The name of an Italian town used to designate a kind of majolica formerly made there.

1868 MARRYAT *Hist. Pott. & Porc.* (ed. 3) 540/2 Castelli (Abruzzo) majolica. **1877** LADY C. SCHREIBER *Jrnl.* (1911) II. 34 Only rewarded by one Castelli plate. **1960** H. HAYWARD *Antique Coll.* 60/1 Castelli wares of late 17th cent. date are often decorated with mythological or historical scenes in a subdued range of colours, dominated by a pale, greyish blue.

caster. Add: **2. b.** (Earlier U.S. example.)

1793 *Ship Owner's Manual* (1795) 141 Many seamen, keelmen, casters.

c. One who takes or makes a model by running some liquid or forcing a plastic substance into a mould.

1921 *Dict. Occup. Terms* (1927) § 105 Potters; ware-makers, casters and finishers. *Ibid.* § 414 *Caster*, takes plaster cast of foot where any special form of boot is required, as in case of malformation, etc.

d. (See quot.)

1921 *Dict. Occup. Terms* (1927) § 049 *Caster*, examines coals sent from screens, and removes splints, *i.e.* slaty coal, in readiness for sale as house coal.

casting, *vbl. sb.* Add: **1.** *esp.* The action or process of founding (metal or glass).

1428–1474–5 in *M.E.D.* **1540, 1542** [see Dict.]. **1832** G. R. PORTER *Porc. & Glass* II. i. 139 The first English establishment of magnitude for the casting of plate glass was undertaken in 1773. **1962** *Gloss. Terms Glass Ind.* (*B.S.I.*) 21 *Casting*, a process of shaping glass by pouring it into a mould or on to a table or passing it between rollers.

c. *Theatr.* and *Cinemat.* The assigning of parts to suitable actors and actresses.

1814 JANE AUSTEN *Mansf. Park* I. xiii. 253 From the first casting of the parts, to the epilogue, it was all bewitching. **1926** *Contemp. Rev.* June 757 The initial failure of Ivanov in a private theatre..was accidental and due mainly to wrong casting. **1952** ELIOT & HOELLERING *Film of Murder in Cathedral* 8 In the theatre, the first problem to present itself is likely to be that of casting.

d. In ploughing, the method and operation of turning all the furrow-slices of a ridge in one direction, and those of the adjoining ridge in the opposite direction.

1825 LOUDON *Encycl. Agric.* II. v. 471 The form of the old ridges, and the situation of the inter-furrows, are preserved by what is called *casting*, that is, the furrows of each ridge are all laid in one direction, while those of the next adjoining ridges are turned the contrary way. **1837** J. F. BURKE *Brit. Husb.* II. 46 It is sometimes desirable to throw two ridges into one...This operation is called casting. **1855** MORTON *Cycl. Agric.* II. 646/1 The mysteries of 'gathering up', 'rown and furrow' ploughing, 'casting', 'yoking or coupling' ridges [etc.].

3. a. (Earlier examples.)

1788 J. FITCH *Original Steam-boat Supported* 10 His application..for castings for a steam-engine. **1851** *Crystal Palace & Great Exhib.* xi. 156 In the castings, for which Germany is deservedly famous, there is much to admire.

4. casting-couch *colloq.* (orig. *U.S.*), see quots.; **casting director**, one responsible for casting (sense *1 c); **casting-ladle**, an iron ladle used for conveying the molten metal into the mould in casting; **casting-machine**, (*a*) in a blast furnace a machine used in casting metal; (*b*) a machine for founding type; **casting-pit**, that part of a foundry where the moulds are placed and the castings made; **casting-plate** = *casting-table*; **casting-pot**, (*a*) a box in which a stereotype plate is cast; (*b*) a crucible; **casting-press**, an apparatus for subjecting metal to pressure during the process of casting; **casting-reel**, the reel of a casting-line; **casting sand**, black moulding sand that has been used for castings; **casting-shop**, the place where the operation of casting metal, etc., is carried out; **casting-slab, -table**, a table of polished metal with raised edges which serves as a mould for plate-glass.

1948 MENCKEN *Amer. Lang.* Suppl. II. XI. ii. 704 Terms emanating from Hollywood wits..*casting-couch* for the divan in a casting-director's office. **1963** *Sunday Express* 27 Jan. 22/6 In the old days..the only way anyone got anywhere in this business was by way of the casting couch. **1966** C. FENN *Pyramid of Night* ix. 183 Get a load of that casting couch. What girl wouldn't want to be laid on velvet? **1924** G. S. DOUGHERTY *Criminal as Human Being* ii. 48 You have all heard of the casting director who selects actors of suitable type for the movies. **1950** T. S. ELIOT *Cocktail Party* III. 150 We've got the casting director: He's looking for some typical English faces. **1861** FAIRBAIRN *Iron* 157 No sooner is the mixture of the metals effected than the casting-ladle is brought under the mouth of the vessel. **1880** *Encycl. Brit.* XIII. 345/1 The casting ladle into which the contents of the converters are emptied. **1899** *Chambers's Jrnl.* 25 Nov. 829/2 Pig-iron casting-machines. **1902** *Census Bulletin 216*, 28 June 58 (Cent. Dict. Suppl.), The monotype ..consists of two machines—a perforating device operated by a keyboard, and a casting-machine. **1921** *Dict. Occup. Terms* (1927) § 524 *Monotype caster attendant*..operates and adjusts casting machine. **1884** W. H. GREENWOOD *Steel & Iron* xxi. 469 In the centre of the [Bessemer] casting-pit is fixed a hydraulic crane. **1921** *Dict. Occup. Terms* (1927) § 279 *Moulder, casting pit*..shapes moulds in sand, in casting pit. **1881** *Spons' Encycl. Industr. Arts* III. 1061 The impressions are given by projections on the casting-plate, which acts as a mould. **1846** DODD *Brit. Manuf.* VI. 60 The casting-pot, with the mould,..is gradually forced down into the molten mass. **1858** SIMMONDS *Dict. Trade* 75/1 *Casting-pot*, a pot adapted for melting metals. *Casting-pot and Crucible maker*, a special trade in the iron districts. *a* **1877** KNIGHT *Dict. Mech.*, *Casting-press*, one in which metal is cast under pressure, as in the car-wheel press. **1892** NIVEN *Brit. Angler's Lexicon* 192 The 'Malloch' casting reel is used for spinning only. **1849** N. KINGSLEY *Diary* (1914) 30 Our sugar is black enough for casting sand. **1871** *Daily News* 2 Jan. 3/5 He was carrying some lead from one part of the yard to the casting shop. *a* **1877** KNIGHT *Dict. Mech.*, Casting-slab. **1728** CHAMBERS *Cycl.* s.v. *Glass*, When the first Annealing Furnace is full, the Casting-table is to be carried to another. **1838** *Penny Cycl.* XI. 256/1 The cuvette is withdrawn from the furnace and taken to the casting-table.

castle, *sb.* Add: **3. f.** A heap of brushwood or sticks under which rabbits hide when being hunted.

1898 *Encycl. Sport* Mar. 175/1 The whole space is laid out in lines of some 20 to 30 yards in breadth, marked out by heaps of sticks or brushwood euphemistically termed 'castles'. *Ibid.* 175/2 When three or more residents turn out together from the same 'castle', and perchance bolt backwards, then complications ensue.

g. *Cricket.* The wicket a batsman defends.

1959 I. PEEBLES in *Sunday Times* 31 May 38/3 In support came..Alfie Hall..with a low action which ensured his hitting the castle. **1960** —— *Bowler's Turn* 59, I..knocked.. Robinson's castle down first ball.

12. castle-nut, a castellated nut (see *CASTELLATED *ppl. a.*); **castle pudding**, a pudding steamed or baked in a dariole mould.

1902 *Automobile Rev.* July 1, *Castle nut*, a nut having three grooves cut across its top face to receive a split pin. **1940** *Chambers's Techn. Dict.* 141/2 *Castle nut*, a six-sided nut in the top of which six radial slots are cut. Two of these line up with a hole drilled in the bolt or screw, a split pin being inserted to prevent loosening. **1845** E. ACTON *Mod. Cookery* xviii. 472 *Sutherland or Castle Puddings*..equal weight of eggs..butter..flour..sugar... Pour the mixture..into well-buttered cups, and bake. **1965** E. J. HOWARD *After Julius* xv. 250 Hanwell was having some castle puddings in the kitchen. **1965** A. CHRISTIE *At Bertram's Hotel* xi. 107 Veal cutlets..succeeded by some small castle puddings with a blackberry sauce.

castle, *v.* Add: **3. b.** (Earlier example.)

1764 R. LAMBE *Hist. Chess* 97 After you have opened your Game a little, and castled your King, bring out your Pieces.

Castleford (kɑ·s'lfɔ̈ɹd). The name of a town in the West Riding of Yorkshire used to designate a type of pottery made there at the beginning of the 19th century.

1863 W. CHAFFERS *Marks & Monogr.* 133 *Castleford Pottery*. Castleford, in the West Riding of Yorkshire, a pottery about 1800. **1871** LADY C. SCHREIBER *Jrnl.* (1911) I. 114 Found a small piece of Castleford pottery. **1961** *Connoisseur New Guide to Antique Eng. Pott.* 74 The comparatively shallow relief of Castleford and other earlier analogous wares.

castlet. Delete †*Obs.* and add later examples.

1813 'ÆDITUUS' *Metrical Remarks* 22 These walls so thick, ..With castlets by their sides, squat shapeless things. **1907** E. H. COLERIDGE in S. T. Coleridge *Christabel* 24 Triermain Castle must have been but a 'castlet' compared with Naworth.

cast-off, *sb.*¹ Add: **2.** *Printing.* A calculation of the amount of space which will be required by a given amount of copy. (Cf. *CAST *v.* 79 j.)

1898 J. SOUTHWARD *Mod. Printing* I. xlii. 263 These two lines must be reckoned for in the cast off. **1917** F. S. HENRY *Printing for School & Shop* iii. 32 If the cast-off leaves but two or three lines on the last page, it is better to have the few previous pages each a line long. **1934** *Proc. Brit. Acad.* XIX. 388 In February 1903 fifty Letters of Erasmus were dispatched to the Press for a 'cast-off'.

castor.¹ Add: **6.** Glove leather made from goat-skins; it is given a very soft finish of a grey colour.

1897 C. T. DAVIS *Manuf. Leather* xxxviii. 519 When finished, [they] bear a close resemblance in texture and quality, to deer-skin or castor. **1910** FLEMMING *Tanning* (ed. 2) 292 The tanner who wants to make mocha castor glove leather from kid and goat skins. **1923** *Daily Mail* 14 Feb. 11 Ladies' good quality Washable Castor Gloves.

7. A light drab colour.

1904 *Daily Chron.* 24 Oct. 8/4 Castor-coloured cloth (a soft beaver shade). **1923** *Daily Mail* 5 June 1 Colours: Nude, Fawn,..Mouse, Castor, Dark Tan.

castor.² Add: **3.** *attrib.* and *Comb.*, as *castor-stand*; **castor action** (see quot. 1940); **castor angle**, the angle at which the steering-head of the front wheels of a motor vehicle is set; **castor-sugar**, powdered sugar, so called from its suitability for use in a castor; **castor-wheel**, a small wheel which turns on its own and a vertical axis, used to support or steer an agricultural machine, or enable it to be turned short round.

1926 *Motor Man.* (ed. 26) xix. 193 *Caster action*, an action tending to maintain the front wheels in the course they are following, obtained by tilting the steering pivots as viewed from the side. **1936** Castor action [see *CAMBER *sb.* 1 d]. **1940** *Chambers's Techn. Dict.* 141/1 *Caster action*, the use of inclined king-pins by which the steerable front wheels of a motor vehicle are given fore-and-aft stability.. on the principle of the domestic caster. **1936** Castor angle [see *CAMBER *sb.* 1 d]. **1959** *Motor* 2 Sept. 92/1 Caster angle (the fore-and-aft inclination of the swivel axis). **1963** BIRD & HUTTON-STOTT *Veteran Motor Car* 151 There was no rake or castor-angle to the steering-heads. **1867** F. S. COZZENS *Sayings* III. 13 The reflected sunshine from those cut bottles in the castor-stand. **1855** E. ACTON *Mod. Cookery* xx. 405 Morella cherries..simmered..with three quarters of a pound of castor-sugar. **1894** T. F. GARRETT *Encycl. Pract. Cookery* II. 533/1 White [sugar]..being sometimes sold in the form of a powder as 'caster Sugar'. **1901** *Daily Chron.* 30 Nov. 8/4, 1½ oz. of castor sugar. **1951** MRS. BEETON *Househ. Managem.* xxxiv. 870 *Sago snow*..2 ozs. of sago, 2 pints of milk, 2 ozs. of castor sugar, 2 or 3 eggs, vanilla. **1805** DICKSON *Pract. Agric.* I. Plate ix. to face p. 40 The profile of a six-shared horse-hoe on the principle of the pentagraph, improved by Mr. Amos by the addition of two castor-wheels to regulate depth. *a* **1877** KNIGHT *Dict. Mech.* s.v., The caster-wheel is used as a support to the front parts of machines, such as harvesters..and plowing machines, to enable them to be steered or to turn short around at the end of the row.

Castor⁶ (kɑ·stǫɹ). Applied to ancient Romano-British pottery made in the neighbourhood of Castor in Northamptonshire.

[**1828** E. T. ARTIS *Durobrivae of Antoninus* Pl. 48 Fine red Ware in relief, collected in excavating the remains of a Roman Pottery in the parish of Castor.] **1857** C. R. SMITH *Coll. Antiqua* IV. 81 Ancient pottery discovered in this country..a particular kind..belongs to a class..termed Castor or Northamptonshire ware, on account of its having occurred frequently in immediate connection with potters' kilns discovered near Castor, by the late Mr. E. T. Artis. **1923** COLLINGWOOD *Roman Britain* 73 The imported Samian ware..provoked the British manufacturer to..put on the market a style of pottery which resembled it in having ornament in relief and a highly glazed surface. Thus was produced what is known as Castor ware. **1945** S. E. WINBOLT *Britain under Romans* ix. 123 Somewhat more fragile, but more interesting in their designs, were the slate-coloured vessels of 'Castor' ware, decorated in raised creamy slip with hunting-scenes. **1963** *Times* 12 Mar. 5/2 A Roman beaker of about A.D. 200 decorated with a hound chasing a hare, possibly Castor ware.

ca·stor-bean. *U.S.* [See CASTOR OIL.] The castor-oil bean. Cf. *CASTOR OIL b.

1819 *Baltimore Morn. Chron.* 23 Apr. 4/5 Physical herb seed..palma christi or castor bean. **1835** J. MARTIN *Gaz. Virginia* 250 There is here [*sc.* Northampton, Va.] an article of culture which is not much met with in other parts of the state—it is the palma christi, called castor bean. **1865** *Trans. Ill. Agric. Soc. 1862* V. 508 The cotton plant in its habits of growth is similar to the castor bean. **1901** C. MOHR *Plant Life Alabama* 56.

castoring (kɑ·stəriŋ), *ppl. a.* [f. CASTOR *sb.*² + -ING².] Acting as a castor (cf. CASTOR *sb.*² 2).

CASTOR OIL

1948 *Jrnl. R. Aeronaut. Soc.* LII. 32/1 The castoring undercarriage makes possible the general construction at low cost of many single strip private landing grounds. 1950 *Engineering* 29 Sept. 280/2 The triangular base is mounted on three rubber tyred castoring wheels.

castor oil. Add: **b. castor-oil bean,** the bean or seed from which the oil is obtained; also, the castor-oil plant. Cf. *CASTOR-BEAN.
1814 F. Pursh *Flora Amer. Septentr.* II. 603 Ricinus communis...Frequent in old plantations in Virginia and Carolina...Known by the name of Castor-oil Bean. 1845 A. Wood *Class-bk. Bot.* II. 336 *R[icinus] communis*...From its seeds is expressed the well known castor-oil of the shops ..Castor-oil Bean. 1901 C. Mohr *Plant Life Alabama* 594 *Ricinus communis*. Castor Oil Bean.

castor pomace.
1877, 1878 [see Pomace 2 b].

castrate, *sb.* Delete †*Obs.* and substitute *arch.* (Later examples.)
1782 Elphinston *Martial* I. iii. 3 No castrate or suborner shall there be: Erewhile the castrate was the debauchee. 1905 W. G. Holmes *Justinian & Theod.* I. 134 *note*, The emperor cannot even uncover his head without the castrates closing round him to intercept the gaze of rude mankind.

castrate, *v.* **3.** *transf.* and *fig.* Delete †*Obs.* and add later example.
a 1930 D. H. Lawrence *Last Poems* (1932) 163 The Victorians..Successfully castrating the body politic.

castration. Add: **1. castration-complex** *Psycho-analysis*, a group of repressed ideas based on a feared potential loss of the genitals in childhood, and resulting in anxiety.
1914 tr. *Freud's Psychopath. Everyday Life* ix. 223 A 'castration-complex'—namely, a childhood fear, often continued in a disguised form into adult life. 1922 J. Riviere tr. *Freud's Introd. Lect. Psycho-anal.* xiii. 175 The *castration complex*, the reaction to that intimidation in the field of sex or to that restraint of early infantile sexual activity which is ascribed to the father. 1929 P. Mairet *Adler's Problems of Neurosis* v. 67 The Freudians have interpreted this fact as the so-called 'castration complex', because girls frequently have the fantasy that the male organs have been surgically removed from them. 1962 T. Kaye *David, from where he was Lying* xvii. 147 No doubt, I've suffered from a castration complex like every-one else.

castrative (kæstrēⁱ·tiv), *a.* [f. L. *castrāt-* ppl. stem of *castrāre* (see Castrate *v.*) + -ive.] Of, or relating to, or tending to produce the same effect as, castration.
1943 T. M. French in S. S. Tomkins *Contemp. Psychopath.* xiii. 208 A little girl will ward off the dangerous temptation to father incest by castrative impulses toward the father. 1949 M. Mead *Male & Female* iv. 85 Societies in which .. the initiatory rites are not accompanied by any such overwhelming castrative feelings. 1960 J. B. Broadbent *Some Graver Subject* i. 33 It seems likely that, powerful, widowed, son-ambitious, a jealous god, he had a 'castrative' effect on Milton.

castrensic (kæstre·nzik), *a.* [f. L. *castrensis* (f. *castra* camp) + -ic.] Of or pertaining to a camp; military; = Castrensian *a.*
1840 Keightley *Rom. Emp.* III. iv. 348 The emperor distributed civic, naval, and castrensic crowns to those who had most distinguished themselves.

Castroism (kæ·stroˌiz'm). [See -ism.] The political principles or actions of Fidel *Castro* Ruz (1927–), Cuban statesman, or of his adherents or imitators. So **Ca·stroist,** an adherent of Castroism; also **Ca·stroite** *a.* and *sb.* Cf. *Fidelism.
1960 *U.S. News & World Rep.* 14 Nov. 60/3 Three members of the International Staff of 'U.S. News and World Report' have just completed an on-the-spot survey of Castroism in the Caribbean. 1961 W. Benton *Voice Latin Amer.* iv. 132 More than 400 members of the..faculty have been fired, many of them..because they lacked sufficient enthusiasm for Castroism. 1961 *Daily Tel.* 28 Aug. 10/2 They must be, it seems, either Castroist and pro-Communist, or reactionary and fostered by Washington. 1962 *Times* 12 Jan. 11/6 The bugbear..of Mexican Castroism. 1963 *Economist* 4 May 425/2 Some form of Castroite uprising. 1964 *New Statesman* 17 Apr. 587/3 The [Mexican] left came to regard him [*sc.* de Gaulle] almost as a Castroite.

castrum (kæ·strɒm). Pl. **castra.** [L.] A Roman encampment or fortress.
1836 *Penny Cycl.* VI. 353/2 Roman Castra were probably sometimes formed on the sites of British works. 1850 C. R. Smith *Antiq. Richborough* 31 The remains of the castrum at Richborough. 1886 *Athenæum* 13 Mar. 365/2 Another characteristic of a Roman castrum,..a huge wine cellar filled with thousands of amphoræ.

cast-steel. [Cast *ppl. a.* 8.] A hard steel made from broken-up blistered steel melted in a crucible and run into ingot-moulds; crucible-steel. Also *attrib.*
1778 P. Thicknesse *Year's Journey* (ed. 2) I. xvi. 149 An English cast steel razor. 1792 *Brit. Pat. 1869* 2 The greatest part of the metal will be found to be steel, and sufficiently converted for making into cast steel. 1800 *Repertory of Arts* XIV. 177 The general principles of my process or processes are the fusion of malleable iron or of iron ore in such manner and by such means as immediately to convert them into cast-steel. 1812, 1858 [see Steel *sb.*¹ 1 c]. 1824 [see Cast *ppl. a.* 8]. 1848 W. H. Emory *Notes Mil. Reconn.* 132 The wooden plough, the harrow, and the cast-steel axe. 1868 *Jrnl. Chem. Soc.* XXI. 281 Cast-steel containing less than 0·3 per cent. of carbon is no longer capable of being hardened. 1880 *Encycl. Brit.* XIII. 356/2 Cast steel products. 1923 *Man. Seamanship (H.M.S.O.)* II. 257 Cast Steel is employed [in ship-building] in several qualities.

casual, *a.* (*sb.*) Add: **1. d.** Golf. *casual water*: see quot. 1899.
1899 *Rules of Golf, St. Andrews* 3 'Casual water' shall mean any temporary accumulation of water (whether caused by rainfall or otherwise) which is not one of the ordinary and recognised hazards of the course. 1957 *Encycl. Brit.* X. 506A/1 Paths, bare patches of ground and temporary accumulations of water (called casual water) are not hazards.
6. b. Showing (real or assumed) unconcern or lack of interest.
1916 'Boyd Cable' *Action Front* 105 'Don't feel particularly hungry,' answered Toffee, with an attempt to appear as off-handed and casual and at ease as his questioner. 1924 R. Macaulay *Orphan Island* xvii. 230 She was casual and indifferent, but Rosamond..stood up and said 'Yes'.
9. *casual labour*: see quot. 1923. So *casual hand, labourer*.
a 1852 Mayhew *Lond. Labour* (1861) II. 220/2 This mixture of constant and casual hands is..a necessary consequence of all trades which depend upon the seasons...Those who have paid attention to the subject of dock labour and the subject of casual labour in general. *Ibid.* 336/1 Greenwich may be looked upon as the first stage or halt for casual labourers, on their way to London. 1876 [in Dict.]. 1905 *Daily Chron.* 25 July 5/4 He would not admit that the policy of the guardians had increased the casual labour system. 1908 H. W. Caslon *Letterpress Printing* 64 Casual labour is, and must be, a disturbing element in the costing of work. 1923 J. D. Hackett *Labor Terms* in *Management Engineering* May, *Casual Labor*. 'Unskilled help, employed and discharged at frequent intervals, and dependent upon the varying demand of the labor market from day to day, without any prospect of continuous employment.' (*What's What in Labour Movement*, Waldo R. Brown, p. 59.)
11. Of clothes: suitable for informal wear. (Cf. *B. 4.)
1939 M. B. Picken *Lang. of Fashion* 23/3 Casual, designed for easy, informal wear; of sports or semi-sports type; as, a casual coat. 1952 *Evening News* 5 Jan. (Advt.), 560 Prs. casual shoes. 1958 *Times* 20 Jan. 11/1 It is sport and casual clothes that visitors come especially to see.
B. *sb.* **3.** (Further examples.)
a 1852 Mayhew *Lond. Labour* (1861) II. 218/1 The 'casuals' being mostly paid by the day. *Ibid.* 220/2 Of the scavagers proper there are..two distinct orders of workmen, 'the regulars and casuals' to adopt the trade terms. 1887 *19th Cent.* Oct. 488 The true casual is seldom employed. 1889 C. Booth *Life & Labour* I. 202 The work of the casuals was a dead loss to the contractor. 1930 L. G. D. Acland *Early Canterbury Runs* v. 94 He did not of course pay the casuals very big wages.
c. *Bot.* and *Zool.* A plant, animal, etc., found away from its normal area or habitat; an alien, casual immigrant.
1899 *Westm. Gaz.* 24 Aug. 3/1 The wanton slaughter of our rarer-winged 'casuals' is positively exasperating. 1926 Tansley & Chipp *Study of Veget.* ii. 11 A species which is a mere alien intruder into the association..may be called a *casual*. 1959 A. R. Clapham et al. *Excursion Flora Brit. Isles* 550 *Casual*, an introduced plant which has not become established though it occurs in places where it is not cultivated. 1959 E. F. Linssen *Beetles* I. 12 'Casuals', i.e. casual immigrants, accidental introductions, etc.
4. Esp. in *pl.* Casual clothes (cf. *A. 11); *spec.*, a style of shoe with a low heel and shallow vamp.
1941 *Amer. Speech* XVI. 96 They are all right for *casuals* or *spectator* wear, but I'd get that formal. 1958 *Woman's Jrnl.* Mar. 3 (Advt.), Classic casual in pebbleweave. 1958 *Observer* 20 Apr. 11/4 Into the series of fashion shows will go..casuals from Italy and France. 1959 *Vogue* June 71 Lastly, casuals—not all necessarily flat but on rounded cotton-reel and thimble-heels.

ca:sualiza·tion. [f. Casual + -ization.] Conversion to a system of employing casual labour; the system itself. Hence (as back-formation) **ca·sualize** *v. trans.*, to convert (regular employees) into casual labourers. Cf. Decasualization s.v. De- II. 1.
1920 *Rep. Ld. Shaw's Ct. of Inquiry on Dock Lab. (H.M.S.O.)* 9 The system of casualisation must..be torn up by the roots. 1934 *Times* 20 July 12/5 The employment of those engaged in theatres..would be reduced to such a state of casualization that they would have to go back to the state of casual employment among the dockers in the eighties for a parallel. 1950 *Hansard Commons* CDLXXVII. 825 The major reason..is the decision..to refrain from casualising the waiters and waitresses.

casualness. Delete *rare* and add later examples.
1924 R. Macaulay *Orphan Island* xvi. 222 There's a prodigious deal too much casualness about the business... I'm all for marriage or nothing. 1958 *Times* 20 Jan. 11/2 These three models are typical of spring styles...The casualness is both disciplined and chic.

casualty. Add: **2. c.** Used of an individual killed, wounded, or injured. Also *fig.*
1844 A. W. Kinglake *Eothen* xvi. 226 Although there may have been some 'casualties' in the way of eyes black.. and women 'missing', there was no return of 'killed'. 1900 W. S. Churchill *London to Ladysmith* 393 In spite of more than a hundred casualties, the advance never checked for an instant. 1916 'Boyd Cable' *Action Front* 144 Such casualties as could walk back walked. 1925 E. F. Norton *Fight for Everest, 1924* 87 With half a dozen porters to wait for our return in case there should be casualties to carry down the glacier. 1956 A. H. Compton *Atomic Quest* 198 The education of students at the university of Chicago was a war casualty.
6. Similarly *casualty nurse, sister*; **casualty insurance** (chiefly *U.S.*) = *accident insurance* (*Accident *sb.* 10); **casualty list,** a list of the dead, wounded, etc. in an engagement or campaign; so *casualty returns*; **casualty man** = *casual *sb.* 3.
1846 *Colburn's United Service Mag.* May 127 Casualty Return. 1864 *Standing Orders, Dress Regul. Artillery* 73 The casualty returns being the documents on which the correctness of Brigade Records mainly depends. *Ibid.* 173 (in List of Periodical Returns) Casualty list. 1887 Beatrice Potter in *19th Cent.* Oct. 489 A considerable number of men, possessing a preferred right to employment, act as an intermediate class between the permanent staff and the 'casualty' men. 1896 *Daily News* 17 June 8/3 The casualty nurse. 1902 *Encycl. Brit.* XXIX. 509/2 Another large class of casualty insurance applies to various forms of damage to property. 1919 G. B. Shaw *Heartbreak House* p. xxii, The emotional strain, complicated by the offended economic sense, produced by the casualty lists. 1925 W. Deeping *Sorrell & Son* xxix. 294 He performed three minor operations, with the..casualty-sister assisting him with critical and voiceless composure. 1958 L. van der Post *Lost World of Kalahari* i. 22, I have forgotten the precise extent of the casualty list but I remember there were.. several horses among the dead.

‖ **casus belli** (kēⁱ·sɒs be·lǝi). [f. L. *casus* Case *sb.*¹ + *belli*, gen. of *bellum* war.] An act justifying, or regarded as a reason for, war. Also *transf.*
1849 J. S. Mill in *Westm. Rev.* LI. 28 To assist a people struggling for liberty..is not a *casus belli* set down in Vattel. 1853 H. Greville *Diary* 22 June (1884) 60 He thought the Russians would soon occupy the principalities, which, however, the other Powers would not pronounce to be a *casus belli*. 1868 Trollope *He knew*, etc. I. xlviii. 374 The Juno from the Close had come quite prepared to declare her casus belli..and to fling down her gauntlet. 1878 *Times* 2 Feb. 6/3 He did not say what was to be the *casus belli* or the *casus armandi*. 1920 D. H. Lawrence *Touch & Go* 11 Some men might acknowledge the bone to be merely a pretext, another hollow *casus belli*.

‖ **casus fœderis** (kēⁱ·sɒs fi·dĕris). [f. L. *casus* Case *sb.*¹ + *fœderis*, gen. of *fœdus* treaty.] A situation or occurrence covered by the provisions of a treaty or compact, and so requiring the action of the parties thereto. Also *attrib.* and *fig.*
1780 J. Adams *Let.* 31 Dec. in *Wks.* (1852) VII. 348 These powers will not be duped by the artifice of the British Court, and adjudge this war not a *casus fœderis*. 1834 H. Greville *Diary* 30 July (1883) 16 A quadripartite treaty, of which your person is the *casus fœderis*. 1882 *Standard* 20 Dec. 5/7 The Treaty of Alliance with the Austrian Empire, including those paragraphs..which define the special obligations of each contracting party under a casus fœderis. 1941 Auden *New Year Let.* III. 51 Which of these calls to conscience is For mine the casus foederis. 1962 *Times* 4 Jan. 9/3 Dr. Salazar admits the *casus foederis* argument.

caswellite (kæ·zwĕlǝit). *Min.* [Named after John H. *Caswell*, 19th-c. Amer. mineralogist: see -ite¹.] A micaceous aluminosilicate of manganese of a copper-red colour, probably an altered form of biotite.
1894 A. H. Chester in *Trans. N.Y. Acad. Sci.* XIII. 183 Inasmuch as these results seem to justify the view that it is a new mineral species, I propose for it the name Caswellite. 1896 *Jrnl. Chem. Soc.* LXX. II. 309 It is supposed that the caswellite has been derived by the local action of water containing manganese and calcium on this biotite. 1966 *Amer. Min.* LI. 1120 The name caswellite has since been applied more specifically..to the garnet pseudomorphs themselves...The name caswellite lacks species or varietal significance and should be abandoned.

cat, *sb.*¹ Add: **2.** *fig.* An itinerant worker. *U.S. slang.*
1926 J. Black *You can't Win* vi. 67 Buy nothin'...It's you kind of cats that make it tough on us, buyin' chuck. *Ibid.* xvi. 220 Harvest workers were called blanket stiffs or gay cats. *Ibid.* 221 They stuck up the cats, took their money, [etc.].
c. An expert in, or one expertly appreciative of, jazz. *slang* (orig. *U.S.*). Cf. *hep-cat.
[1922 J. A. Carpenter (*title of ballet*) Krazy Kat.] 1932 *Melody Maker* Oct. 836/1 [citing L. Armstrong] All the cats were there. 1935 *Down Beat* 1 Nov. 8 The slanguage of swing-terms that 'cats' use. 1936 *Delineator* Nov. 49/2 *Cats*, the musicians of a swing orchestra. 1937 *Amer. Speech* XII. 183/1 *Cats*, those members of the audience who are receptive to jazz music or who understand it. 1937 L. Armstrong *Swing that Music* xiii. 111, I wanted to give 'em a load of how we swing that music at home. My 'cats' understood it the same way and began lickin' their chops, as we say it. 1955 Shapiro & Hentoff *Hear Me talkin' to Ya* xix. 299 Minton's was just a place for cats to jam... When you went in you'd see cats half-stewed who weren't paying much mind to what was happening on stage. 1958 *Woman's Own* 19 Feb. 22/1 'It's got beat and a lot of excitement,' said one teenage 'cat' I talked to.
d. *slang.* A 'regular guy', fellow, man.
1957 Mezzrow & Wolfe *Really the Blues* 372 *Cat*, regular fellow, guy. 1959 A. Anderson *Lover Man* 116 'At-*dam*, man, youre the selfishest kat I seen yet. 1959

C. MacInnes in *Encounter* Aug. 35/2 The coloured cats saw I had an ally, and melted.
4. b. (Earlier Amer. example.)
1705 R. Beverly *Hist. Virginia* (1722) 129 Conger-Eels, Perch, and Cats.
11. b. *pl.* The salt which crystallizes round the edge of the pan or beneath the holes in the bottom of the trough in which salt is put to drain. Cf. *cat *v.* 5.
1886 R. Holland *Gloss. Chester*, *Cats*, salt-making term. Masses of salt formed under a pan when it leaks. 1892 *Cornhill Mag.* Sept. 265 The 'cats', or salt that has become encrusted round the edges of the pan, is sent to the pottery works for glazing pipes and pitchers.
13. h. *like a cat on hot bricks*: see *hot *a.* 12 c. **i.** *not a cat (in hell)'s chance*: no chance whatever. **j.** *to make a cat laugh*: said of something excruciatingly funny. **k.** *that cat won't jump* (orig. *U.S.*): that suggestion is implausible or impracticable. **l.** *the cat's pyjamas, whiskers* (*slang*, orig. *U.S.*): the acme of excellence. **m.** *to look (feel) like something the cat has brought in*: to appear, or to feel, exhausted or bedraggled.
i. 1796 Grose *Dict. Vulg. Tongue* (ed. 3), No more chance than a cat in hell without claws; said of one who enters into a dispute or quarrel with one greatly above his match. 1902 C. Hyne *Mr. Horrocks, Purser* 100 Crutches by themselves wouldn't have stood a cat's chance. 1927 *Daily Express* 13 Dec. 16/6 There did not seem a cat's chance for Oxford on comparative form. 1958 M. Procter *Man in Ambush* ii. 19 He hasn't a cat-in-hell chance. 1962 'H. Howard' *Double Finesse* xvi. 168 You haven't a cat-in-hell's chance of getting away. 1966 *Guardian* 22 June 3/6 One seaman said the union had not 'a cat in hell's chance' of beating the Government as well as the shipowners.
j. [*c* 1598 Haughton *Englishm. for my Money* (1616) E 4, Oh maister Mouse,..it would make any Mouse, Ratte, Catte, or Dogge, laugh to thinke, what sport we shall have.] 1838 J. R. Planché *Puss in Boots* 24 Allow us just applause to win Enough to make a cat laugh. 1907 W. W. Jacobs *Short Cruises* 230 It would ha' made a cat laugh. 1929 J. B. Priestley *Good Compan.* III. iv. 551 Make a cat laugh, the way she takes people off.
k. 1838 Haliburton *Clockmaker* xvi. 242 Them Yankee villains would break up our laws, language, and customs; that cat wouldn't jump at all, would it? 1906 Galsworthy *Man of Property* III. ix. 372 'They talk of suicide here,' he said at last. '*That* cat won't jump.' 1934 F. W. Crofts *12.30 from Croydon* xiii. 178 They know I was hard up...No, Charles, that cat won't jump. 1965 'S. Troy' *No More a-Roving* ii. 57 If you're telling me she fell in, just like that—oh no! That cat won't jump.
l. 1923 W. A. Roberts in *Saucy Stories* 1 Mar. 121/1 It would have landed us in jail, if we had published it... But as literature, it was the cat's whiskers. 1924 Wodehouse *Leave it to Psmith* ix. 181 'Well, if this ain't the cat's whiskers!' said Miss Peavey. 1925 S. Lewis *M. Arrowsmith* xxxix. 460 This kid used to think Pa Gottlieb was the cat's pyjamas. 1930 *Times Lit. Suppl.* 4 Sept. 693/2 We were the absolute and utter cat's pyjamas. 1958 *Times* 15 Aug. 9/4 Lord Montgomery..holds that to label anything the 'cat's whiskers' is to confer on it the highest honour.
m. 1928 R. Knox *Footsteps at Lock* viii. 79 Bredon felt, in an expressive modern phrase, like something the cat had brought in. 1940 R. Postgate *Verdict of Twelve* II. ii. 112 Rosalie realized that for the family she was something the cat brought in. 1967 A. Wilson *No Laughing Matter* III. 170 The sweet toothy smile froze to a 'what's that the cat's brought in' disdain.
17. a. *cat-food.* **c.** *cat-headed* adj.
1907 *Yesterday's Shopping* (1969) 58 Cat food, doz. packets, 0/9. 1966 A. Prior *Operators* vii. 87 The food cupboard..contained at least three dozen tins of cat food. 1905 E. F. Benson *Image in Sand* i, Close to them stood the great cat-headed statue. 1905 *Daily Chron.* 4 Sept. 3/1 The power was divinised in the cat-headed Pasht.
18. and **19.** **Cat-and-mouse Act,** nickname for the Prisoners (Temporary Discharge for Ill-health) Act of 1913 to enable hunger-strikers to be released temporarily; used chiefly *attrib.* (now without capital initials) of (*esp.* official) action taken (repeatedly or for a prolonged period) against a weaker party; **cat-bear,** the red bear-cat or lesser panda; **cat-burglar,** a burglar who enters by extraordinarily skilful feats of climbing; hence **cat-burgling** *vbl. sb.*; **cat-castle** (see Cat *sb.*¹ 6 and quot. 1907); **cat chain** (see quot.); **cat-door,** a small door, usually swinging, which can be opened by a cat for its own ingress and egress; **cat-footed** *a.* (*a*) stealthy in movement; (*b*) (see quot. 1883); hence **cat-footedness,** surefootedness; **cat-heather,** the name given to various kinds of heather in Scotland (Jamieson, 1825); see *Sc. Nat. Dict.* s.v. *Cat.* n.¹ I. 2; **cat-house,** (*b*) a house for cats; (*c*) *slang*, a brothel; **cat-lick,** colloq. expression for a perfunctory manner of washing; also as vb.; **cat('s)-nap** (earlier example); hence as *v. intr.*, to take a cat-nap; **cat-squirrel** (earlier and later U.S. examples); **cat-steps,** (*b*) *Geol.* (see quot.) *U.S.*; **cat-suit,** an all-in-one garment reaching from neck to feet, generally tight-fitting, and with trouser legs; **cat('s) whisker,** a fine adjustable copper or gold wire in a crystal wireless receiver or in certain types of electronic circuit (see also sense *13 l).
1913 *Punch* 23 July 81/1 Plural Residence, which will still be permitted after the abolition of Plural Voting, is being encouraged by the Cat-and-Mouse Act. 1926 G. B. Shaw *Transl. & Tomfooleries* 225 The Cat and Mouse principle.. is a part of the law of England. 1937 F. P. Crozier *Men I Killed* viii. 156 The Absolutists [conscientious objectors] went to jail, again and again and again, as men 'deemed to have been enlisted'...During this cat-and-mouse performance every conceivable kind of inducement, temptation, and privation were used by the Government..in order to make the objectors surrender. 1949 Koestler *Promise & Fulf.* ii. 18 The Administration played a curious cat-and-mouse game with the Jewish self-defence organization. 1965 G. McInnes *Road to Gundagai* ii. 28 He was..prepared to enjoy a game of cat-and-mouse with the superior little English boy. 1888 W. T. Blanford *Mammalia* 190 *Aelurus fulgens*. The red Cat-bear...South-eastern Himalayas..lives in holes of trees, or perhaps amongst rocks. 1931 H. L. Myers *Prince Jali* vi. 61 His parents had given him a pair of cat-bears, charming creatures that were exceedingly tame. 1907 *Daily Chron.* 18 Apr. 1/7 Owing to his skill in climbing he was known as 'the cat burglar'. 1927 *Daily Express* 24 Mar. 2/6 A 'cat' burglar broke into the house.. by climbing a stackpipe. 1958 D. Emmet *Function, Purpose & Powers* ix. 253 What about a man seized with an irresistible urge for cat-burgling. 1861 *Chambers's Encycl.* II. 668/1 *Cat*, or *cat-castle*, in the military engineering of the middle ages, was a kind of movable tower to cover the sappers as they advanced to a besieged place. 1907 Collingwood in *Trans. Cumbld. & Westmld. Archæol. Soc.* (1908) VIII. 100 'Catcastle' in local [Kendal] dialect is the second figure in the game of Cat's-cradle. 1883 *Man. Seamanship Boys' Training Ships* 195 Q. What is a cat chain? A. A chain which is rove through the cat block, and shackled on to the upper end of ground chain to bring the anchor to cat head. 1959 M. Summerton *Small Wilderness* i. 16 Bella jumped down, and went through her private cat-door. 1598, 1847 Cat-footed [see Cat *sb.*¹ 17 in Dict.]. 1883 C. Stables *Our Friend the Dog* vii. 59 *Cat-footed*—Having the toes well knuckled up, making the foot short and round. 1929 W. J. Locke *Jorico* 94 He had the peculiar, sure cat-footedness of those who follow the sea. 1864 T. Guthrie *Let.* 9 July in *Autobiogr.* (1877) xii. 645 The hum of bees who are all on the *qui vive*, as the heather (the 'cat-heather' as it is called) is now coming out. 1886 Britten & Holland *Dict. Eng. Plant-Names* 92 *Cat-heather*, more than one kind of heath seems to be so called in Scotland..(*Calluna vulgaris*),..*Erica cinerea*,..or possibly *E. Tetralix*. 1922 D. H. Lawrence *England, my England* 33 Bits of cat-heather were coming pink in tufts. 1895 C. J. Cornish *Life at Zoo* 236 It is obvious that so active and beautiful an animal could not be seen with advantage..in the cramped little cages of the present Cat House. 1931 'D. Stiff' *Milk & Honey Route* 202 *Cat house*, a brothel. 1934 Thornton Wilder *Heaven's My Destination* vi. 113 On Sunday you raped a whole cat-house. 1935 'G. Orwell' *Clergyman's Daughter* ii. 113 He's took her abroad an' sold her to one o' dem flash cat-houses in Parrus. 1968 D. Francis *Forfeit* x. 120 He walked straight out of the Cat House. [*c* 1450 *Jacob's Well* 187/25 Þe feend seyde: 'come hedyr, freend, þou art catlycked & qwyt of þi synnes in þi schryfte.'] 1859 Geo. Eliot *A. Bede* xi, Th' men ne'er know whether that floor's cleaned or cat-licked. 1892 *Leeds Merc. Suppl.* 12 Mar. (E.D.D.), That's nobbut gien thisen a cat-lick asteead ov a reight wesh. 1898 *Eng. Dial. Dict.* s.v. *Cat* sb.¹, Yer may ev catlicked the flooer; yer hevn't weshed it. 1906 W. De Morgan *Joseph Vance* vii. 65 Anne..soaped me with a vigour far beyond any experience of washing I had had up to that date. My method had been Cat-licking, she said. 1915 D. H. Lawrence *Rainbow* ii. 61 Let's finish wiping your face—it'll pass wi' a cat-lick. *a* 1953 Dylan Thomas *Quite early one Morning* (1954) 30 They catlicked their hands and faces, but never forgot to run the water loud and long as though they washed like colliers. 1823 J. F. Cooper *Pioneers* xxxii, I just closed my eyes in order to think the better with myself...It was only some such matter as a cat's nap. 1938 *Amer. Speech* XIII. 182/2 To cat-nap. 1958 *Times* 19 May 3/7 Mr. Hugh Bentley, who cat-naps every afternoon between 3 and 3.30. 1826 J. Godman *Amer. Nat. Hist.* II. 129 The Cat-Squirrel, *Sciurus Cinereus*. 1834 McMurtrie tr. *Cuvier's Anim. Kingd.* 80 The Cat Squirrel (*Sciurus cinereus*, Lin.) of America is cinerous above, white beneath. 1943 W. J. Hamilton *Mammals East. U.S.* 225 Gray Squirrel. Cat Squirrel. *Sciurus carolinensis*. 1939 A. K. Lobeck *Geomorphol.* iii. 93 Small, backward-tilted terraces due to slumping, features which are sometimes called *catsteps*. 1960 *Guardian* 16 Nov. 5/3 Miss Odell wears a 'cat-suit' which is sleeveless, low-cut, tight round the shanks and everywhere else. 1970 *Oxf. Mail* 27 Jan. 1/9 Courreges' best inspiration for ready-to-wear were his flare-legged cat suits in P.V.C. or cotton. 1915 A. F. Collins *Bk. of Wireless* 205 Adjust the wire until the pointed end presses on the crystal and you will have what is called a cat-whisker detector. 1923 *Daily Mail* 28 June 13 A crystal called 'Radiocite'..used with a 'cats-whisker' contact, and the pressure necessary is extremely light. 1948 *Electronic Engin.* XX. 354 The catswhiskers can be mounted in slots at the top of the copper base pins, using a dummy crystal..to align the tips of the wires. 1949 *Ibid.* XXI. 448 Two IN34's will be required to provide the two cats-whiskers, leaving a spare crystal.

cat, *sb.*³, var. Kat.
1877 *Encycl. Brit.* VI. 110/2 In Arabia the beverage [*sc.* coffee]..only supplanted a preparation from the leaves of the cat, *Celastrus edulis*. 1904 *U.S. Consular Rep.* No. 285. 549 The cat is a plant containing a medicinal principle which acts as a tonic upon the muscles of the heart. 1934 Dylan Thomas *Let.* 15 Apr. (1966) 103 If I..could see them as a Yucatan people, call them to a cat-drinking ceremony.

cat, *sb.*⁴ Colloq. abbrev. of Catalytic *a.*, in *cat cracker, cracking*, etc. (cf. *catalytic *a.*).
1943 *Fortune* Sept. 50/1 'Cat cracker' is the oil industry's nickname for the new catalytic cracking processes now producing high-octane gas. 1952 *Economist* 6 Sept. 581/1 (*heading*) Premium Petrol and Cat Crackers. 1957 *New Scientist* 18 July 36/2 'Cat-cracking'..show[s] the power of catalysis in large-scale chemical industry.

cat, *v.* Add: **1.** Also *absol.*
1890 W. Clark Russell *Ocean Trag.* iii, They had catted and were fishing the anchor forwards.
5. *intr.* To be deposited in the manner of salt, etc., round objects, in crevices, or the like. (Cf. *cat *sb.*¹ 11 b.)
a 1909 *Buck's Handbk. Med. Sci.* VII. 901 (Cent. Dict. Suppl.), The material which cats here is in a state not capable of ready absorption, and must act locally.

cata-: see also Kata-.

catabolic (kætăbǫ·lik), *a. Biol.* = Katabolic.
1957 *Encycl. Brit.* XV. 303/1 The manifold chemical transformations involved in the building up (anabolism) of more complex materials and the breaking down (catabolism) of these and other substances...exchange and storage of energy occur—storage of energy in anabolic and release of energy in catabolic change.

catabolism (kătæ·bǫliz'm). *Biol.* = Katabolism.
1889 *Nature* 26 Sept. 525/1 The words in question, 'anabolism', which being interpreted means winding up, and 'catabolism', running down, are the creation of Dr. Gaskell. Prof. Hering's equivalents for these are 'assimilation', which, of course, means storage of oxygen and oxidizable material, and 'disassimilation', discharge of these in the altered form of carbon dioxide and water. 1957 [see *catabolic].

catabothron: see Katabothron.

cataclastic (kætăklæ·stik), *a. Geol.* [ad. Norw. *kataklas(-struktur)* (T. Kjerulf 1885, in *Nyt Mag. Naturvidensk.* XXIX. 268), f. Gr. κατά down + κλαστός broken (κλᾶν to break).] Designating a structural character due to intense crushing.
1887 J. J. H. Teall in *Geol. Mag.* IV. 493 The structures are of the kind for which Prof. Kjerulf has proposed the term cataclastic. I venture to suggest..that we should distinguish between the three types of clastic rocks at present recognized by using the terms epiclastic, cataclastic and pyroclastic...*Cataclastic*—Rocks largely composed of fragments produced during the deformation of older rocks by the earth stresses. 1896 J. W. Judd *Student's Lyell* 560 Many metamorphic rocks exhibit a similar 'cataclastic' structure. 1903 Geikie *Text-bk. Geol.* (ed. 4) 421 When the temperature is 300° C. or 400° C. no cataclastic structure is observable. 1958 *Van Nostrand's Sci. Encycl.* (ed. 3) 287/1 *Cataclastic*..has the same meaning as crush-breccias. This term is also applied to the deformation and granulation of minerals such as may take place during dynamic metamorphism.

cataclinal (kætăklǝi·năl), *a. Geol.* Now *rare*. [f. Gr. κατακλινής sloping, f. κατά down + κλίνειν to bend + -al. Cf. Monoclinal *a.* and *sb.*] (See quot. 1875.) Cf. *anaclinal *a.*
1875 J. W. Powell *Expl. Colorado River* II. xi. 160, I have ..classified these valleys..in the following manner.. *cataclinal*, valleys that run in the direction of the dip. 1960 B. W. Sparks *Geomorphol.* vi. 104 Terms originally proposed by Powell, such as cataclinal..have found no wide usage, as they are unnecessarily obscure.

cataclysmically (kætăkli·zmikăli), *adv.* [f. Cataclysmic: see -ically.] By a cataclysm.
1889 I. Taylor *Aryans* iii. 132 The civilization..was not introduced cataclysmically, by the immigration of a new race.

cat-a-cornered: see *cater *adv.*

catagenesis (kætădʒe·nėsis). [See Cata- and Genesis.] Retrogressive evolution. Cf. *anagenesis.
1884 E. D. Cope in *Amer. Assoc. Advancem. Sci.* XXXIII. 468 The process of creation by the retrograde metamorphosis of energy, or, what is the same thing, by the specialization of energy, may be called *catagenesis*. 1893 [see *anagenesis]. 1900 B. D. Jackson *Gloss. Bot. Terms*.
Hence **catagenetic** (kæ:tădʒėne·tik) *a.*, of the nature of retrogressive evolution.
1896 E. D. Cope *Org. Evol.* 482 If the tendency of the catagenetic energies is away from vital phenomena.

cataglottism. Delete †*Obs.*—⁰ and add later example.
1905 H. Ellis *Stud. Psychol. Sex* IV. 5 The kiss is not only an expression of feeling; it is a means of provoking it. Cataglottism is by no means confined to pigeons.

Catalan. Add: *sb.* The language of Catalonia (examples). *Catalan forge* (earlier example); so *Catalan furnace, hearth*, etc.
1792 A. Young *Travels* 25 Region of the Pyrenees..the language of the country is a mixture of Catalan, Provençal, and French. 1807 Southey *Lett. fr. England* III. lxiii. 335 Scotch..differs far more from English than Portugueze from Castilian, nearly as much as the Catalan. 1839 Ure *Dict. Arts* 709 The Catalan forge can be profitably employed only where wood is exceedingly cheap and abundant. 1861 Fairbairn *Iron* 42 The disposition of the Catalan hearth during the process of reduction. 1867 Bloxam *Chem.* 321 In the Pyrenees, where the Catalan process is employed. *a* 1877 Knight *Dict. Mech.*, *Catalan-furnace*, a blast-furnace for reducing iron ores, extensively used in the North of Spain. 1906 Belloc *Hills & Sea* 261 Hundreds of men had spoken to me in Catalan. 1933 H. J. Chaytor

Hist. Aragon iv. 62 Catalan presents a different problem; the question whether it belongs to the Hispanic or to the Gallo-Roman branches of the Romance languages is even yet in dispute. 1960 W. D. Elcock *Romance Lang.* v. 435 The question whether Catalan should be considered as a language has at times given rise to debate, much of it ill-formed and coloured by political partisanship...During the later medieval period it came to rank as one of the 'great languages' of western Europe.

Catalanist (kæ·tălænist). [f. Catalan + -ist.] One who favours the independence of Catalonia; usually *attrib.* or as *adj.* So **Ca·talanism,** the favouring of independence for Catalonia; also, an idiom or mode of expression belonging to the Catalan language.
1905 *Daily Chron.* 23 Sept. 5/3 The Catalanist Agitation. 1907 *Westm. Gaz.* 25 May 2/1 The 'Catalanist' Home Rulers. 1923 *Glasgow Herald* 8 Oct. 9 The suspension of..more or less seditious Catalanist journals. 1930 S. de Madariaga *Spain* xviii. 304 Catalanism is, above all, a Barcelona affair. 1949 —— *C. Columbus* 433 Ulloa..led by his strong Catalanist bias, paints this struggle as an effort of the Catalans to liberate themselves. 1958 *Archivum Linguisticum* X. 31 Gallicisms (including Provençalisms and Catalanisms).

catalase (kæ·tălēⁱs, -z). *Chem.* Also **katalase.** [f. Catal(ysis + *-ase.] Any of the hæm-containing enzymes that catalyse the reduction of hydrogen peroxide.
1901 O. Loew in *Rep. U.S. Dept. Agric.* LXVIII. 12 Since it is clear that the power of catalyzing hydrogen peroxid is not due to any of the known enzyms, it appears justifiable to ascribe this power to a special enzym. The writer proposes to call this catalase. 1904 *Brit. Med. Jrnl.* 20 Aug. 28 Yeast contains..nuclein, zymase, endotrypsin, and katalase. 1911 *Encycl. Brit.* XIX. 921/1 Certain oxydases, catalases and de-amidizing enzymes..play an important part in the various metabolic processes. 1958 *New Biol.* XXVII. 118 Hydrogen peroxide, a powerful poison against which most organisms—but not *Ascaris*—are protected by an enzyme, catalase, which destroys it.

catalectic, *a.* Add: Often in postposition in imitation of Latin. **b.** *sb.* A catalectic line or verse.
1631 B. Jonson *Staple of News* iv. iv, Pentameters, Hexameters, Catalecticks. 1842 *Penny Cycl.* XXIV. 228/1 The Iambic Tetrameter Catalectic. 1903 Saintsbury *Short Hist. Eng. Lit.* (ed. 3) II. ii. 52 A strict iambic form of tetrameter catalectic, or alternate dimeters acatalectic and catalectic.

catalecticant (kætăle·ktikănt). *Math.* (See quots. 1852, 1860.) So **catale·ctic** *a.*, **catale·ctically** *adv.*
1851 J. J. Sylvester *Canonical Forms* in *Coll. Math. Papers* (1904) I. 211 The theory of the catalectic forms of functions of the higher degrees of two variables. *Ibid.*, If, however, certain further relations obtain between the coefficients of *F*, the canonical form reappears catalectically, the function becoming in fact representable as a single cube. 1852 *Ibid.* 293, I shall hereafter refer to a determinant formed in this manner from the coefficients of *f* as its catalecticant. 1860 Cayley *Coll. Math. Papers* (1891) IV. 606 The name catalecticant denotes a certain invariant of a binary quantic of an even order.

catalexis (kætăle·ksis). [f. Gr. κατάληξις termination, f. καταλήγειν to leave off.]
1. Absence of a syllable in the last foot of a verse.
1830 Seager tr. *Hermann's Metres* II. xix. 46 Cretic verses are for the most part terminated by that same foot, and have no other catalexis. 1898 Saintsbury *Short Hist. Eng. Lit.* II. iii. 77 Halidon is told [by Minot] in octave eights admitting catalexis. 1957 W. Beare *Latin Verse* vii. 82 From iambic trimeters we pass freely to dimeters, monometers, iambics with catalexis and syncopation.
2. A catalectic verse or line.
1850 Mure *Lit. Greece* III. 55 The combination of a single short verse or 'catalexis' with one or more longer verses.

catalo (kæ·tălo). *U.S.* Also **cattalo, cattelo.** [f. Cat(tle + Buff)alo.] A cross between the male buffalo and the domesticated cow.
1889 *Kansas Times & Star* 20 May, Colonel [Charles] Goodnight was the first man to experiment with crossing buffalo and cattle. A big herd of the hybrids, called 'cattalo', is now on his Texas ranch. 1894 *San Francisco Midwinter Appeal* 10 Feb. 3/3 Grand Exhibit of Buffaloes and Cattleos. 1899 C. J. Jones in H. Inman *Buffalo Jones' Forty Years of Adv.* 243 To these cross-breeds I have given the name 'Catalo', from the first syllable of *cattle* and the last three letters of *buffalo*. 1906 *Harper's Mag.* Apr. 798 [The buffaloes] sad captives sinking to slow extinction in the hybrid cattelo with his mongrel name. 1923 *Chambers's Jrnl.* July 454/1 The crossing of buffalo with domestic cattle. From the mating..has been evolved the 'cattalo'. 1958 *Irish Times* 23 Aug., The Cattalo, which was bred in Alberta, is an animal with fine beef qualities.

catalogue, *sb.* Add: **3. a.** A list of college or university graduates, alumni, or teachers. **b.** A university or college calendar. *U.S.*
1682 J. Bishop in *Mass. Hist. Coll.* (1868) 4th Ser. VIII. 311, I lately received..a Catalogue of Harvard's sons. 1786 in J. Maclean *Hist. Coll. N.J.* (1877) I. 344 Ordered, That a complete catalogue of the graduates of this College be prepared and published at the expense of the present Senior class. 1812 (*title*) Catalogue of the Officers and Students[of Harvard]. 1823 (*title*) Catalogue of officers and students in Yale College November 1823. 1842 Z. Thompson *Hist. Vermont* II. 155 Middlebury College..Catalogue of Alumni and Honorary Graduates. 1873 J. H. Beadle *Undevel. West* xxxi. 686 The 'University of Deseret' puts forth a pretentious catalogue, with a lengthy list of professors. 1945 G. Santayana *Middle Span* viii. 159 My name had figured in the Harvard Catalogue..for 30 years.
4. Miscellaneous attrib. uses.
1892 *Photogr. Ann.* II. 127 Each of the catalogue plates will have two exposures. 1899 *Daily News* 29 Apr. 7/3 A Piccadilly firm have offered to buy at catalogue price the picture by Policeman Jones, of Leeds, accepted for the Royal Academy. 1908 *Amer. Libr. Assoc. Catalog Rules* p. v, The size and quality of catalog cards. 1961 *Oxf. Mail* 6 Oct. 2/3 (Advt.), Catalogue paster required.

‖ **catalogue raisonné** (katalog rezone). Also erron. **raisonnée** [Fr., = carefully studied or methodical catalogue.] A descriptive catalogue arranged according to subjects, or branches of subjects; hence *gen.* or loosely, a classified or methodical list.
1784 H. Walpole *Descr. Strawberry-Hill* p. i, Catalogues raisonnés of collections are very frequent in France and Holland. 1792 C. Smith *Desmond* III. 169 He had glided away on a descriptive tour to his own seat near Bath; and was giving a *catalogue raisonnée* of its conveniences. 1803 *Edin. Rev.* Oct. 79 A *catalogue raisonnée*, if executed with judgment and impartiality, would be a very useful appendage to every work. 1818 Hazlitt *Eng. Poets* viii. 324 There is no Gay in the present time to give a *Catalogue Raisonné* of the performances of the living undertaker of epics. 1835 *Penny Cycl.* IV. 380/2 The best specimen of a *Catalogue Raisonné* that we know of any of the more considerable collections of this country, is that of the library of the writers to the Signet in Edinburgh, published in one volume quarto in 1805. 1903 F. L. Gardner (*title*) A Catalogue raisonné of works on the occult sciences. 1948 *Ann. Reg.* 1947 227 An expenditure of $37,528 million. It was a huge figure, but the *catalogue raisonné* of its component items challenged its critics at their weakest point. 1960 *Guardian* 16 June 8/5 The erudition embodied in the catalogue raisonné is quite exceptional.

Catalonian (kætălōᵘ·niăn), *a.* and *sb.* [f. *Catalonia*, the Spanish province (*Cataluña*, Cat. *Catalunya*): see -ian.] = Catalan.
1707 Mortimer *Husb.* xviii. 496 The Catalonian or Spanish jessamine..is not so high as the former. 1781 Dillon *Lett. Eng. Trav. Spain 1778* v. 63 The Aragonians and Catalonians adopted the Castilian dialect. 1823 T. Ross tr. *Bouterwek's Hist. Span. & Pg. Lit.* I. 6 In the kingdom of Arragon, the language in general use was the Catalonian. 1829 K. H. Digby *Broad Stone, Orlandus* xxi. 531 He spoke Greek, Hebrew, Latin, Catalonian, and Italian. 1836 *Penny Cycl.* VI. 361/2 During the war of the Spanish succession, the Catalonians..took the part of the Archduke Charles of Austria against Philip of France. 1875 *Encycl. Brit.* II. 433/2 This Catalonian work of the 14th and 15th centuries. 1876 *Ibid.* V. 217/2 The Catalonians are a frugal, sharp-witted and industrious people. 1958 A. R. Radcliffe-Brown *Method in Social Anthropol.* II. v. 185 From the many Latin dialects there came into existence the present Romance languages—French, Provençal, Italian, Catalonian, Castilian, Portuguese. 1963 *Times* 25 Feb. 9/2 The accused, all Catalonians, included six lawyers.

catalyse (kæ·tălǝiz), *v.* Also †**kata-, -lyze.** [f. Catalysis after *analyse, analysis*.] *trans.* To increase the rate of (a reaction or process) by catalytic action; to produce by means of catalysis. Also *fig.* Hence **ca·talyser, -or,** (*rare*) **ca·talysator,** a catalyst; **ca·talysing** *vbl. sb.* and *ppl. a.*
1890 *Nature* 13 Nov. 25/1 Every micro-organism produces, from the substances which it katalyzes..a material or materials, which, on accumulation, inhibit its growth. 1893 *Ann. Rep. Smithsonian Inst.* 237 Numberless specific catalysators exist which act only upon certain phenomena. 1901 *Jrnl. Amer. Chem. Soc.* Apr. 236 The ratio of the velocity of the reaction to the concentration of the catalysor is nearly constant in dilute solutions of strong acids. 1904 *Jrnl. Phys. Chem.* May 373 Measurements were made of the effect of catalyzers on the formation of C_6H_5Cl and $C_6H_4Cl_2$ from benzene and chlorine. 1904 J. McCabe tr. *Haeckel's Wonders of Life* ii. 47 Many recent chemists and physiologists are of opinion that plasm is a colloid catalysator. 1917 *Chambers's Jrnl.* Nov. 726/2 Some of these natural catalysers can be manufactured. 1926 *Spectator* 21 Aug. 280/1 [Advertising] is..the great mover of merchandise, the catalyser of commerce. 1937 *Discovery* June 181/1 Traces of impurities that catalyse reactions may lead to erroneous inferences. 1943 Koestler in *Horizon* Apr. 242 But tradition might act on a man in two ways: either as a sterilizing, or as a catalysing agent. 1944 J. S. Huxley *Living in Rev.* xi. 115 They do not impose their own plans, but they catalyse planning jointly with others. 1955 *Sci. News Let.* 29 Oct. 274/1 Peroxidase..catalyzes, or sparks, the transfer of oxygen from hydrogen peroxide..to another substance. 1966 *McGraw-Hill Encycl. Sci. & Technol.* V. 19/2 Most of the chemical reactions that take place in living organisms are catalyzed by specific enzymes, without which such reactions would proceed at a negligible rate, if at all. 1969 *Nature* 27 Dec. 1250/1 The technology bred of science has catalysed stupendous economic growth.

catalyst (kæ·tălist). *Chem.* [f. Catalysis, on the analogy of *analyst*.] A substance which when present in small amounts increases the rate of a chemical reaction or process but which is chemically unchanged by the reaction; a catalytic agent. (A substance which similarly slows down a reaction is occas. called a *negative catalyst*.) Also *fig.*
1902 *Nature* 3 Apr. 523/1 No reactions are possible under the influence of catalysts that could not take place in their absence without a breach of one of the laws of energy. 1920 *Chambers's Jrnl.* 284/1 Nickel is used as a 'catalyst', or a carrier of hydrogen in the hydrogenation..of oils for use in the manufacture of margarine. 1927 Haldane & Huxley *Anim. Biol.* viii. 169 We have begun to isolate many of the intermediate products of metabolism and the catalysts that govern the course of the reactions by which they are formed. 1943 H. Read *Politics of Unpolitical* xiv. 160 Shelley called poets the *unacknowledged* legislators of the world, and the epithet was well chosen. The catalyst is unchanged, unabsorbed; its activity therefore not acknowledged. 1948 Glasstone *Physical Chem.* (ed. 2) xiii. 1126 If the reaction is hindered by the added substance it is said to be a negative catalyst, and the word catalyst, when used alone, is almost invariably taken to imply acceleration of the chemical process. 1953 *Sci. News* XXVIII. 117 There is no reason why a catalyst..should be rendered totally inactive by a change in the geometry of the substance acted upon. 1954 J. I. M. Stewart *Mark Lambert's Supper* 139 In the intricate chemistry that gives motive-power to the machine he has himself acted as an obscure catalyst. 1968 A. White et al. *Princ. Biochem.* (ed. 4) x. 209 Enzymes are catalysts peculiar to living matter.

catalytic, *a.* Add: Also *fig.* Also *catalytic cracker*: the device in which catalytic cracking is carried out; *catalytic cracking* [*crack *v.* 23]: the cracking of petroleum oils by a process using a catalyst.
1927 *Chem. Abstr.* 644 (*title*) Catalytic cracking of heavy fractions of petroleum...A special exptl. app. for catalytic cracking, in which a continuous influx of oil and recovery of the gas, oil and volatile products is carried out, is described. 1943 in *Amer. Speech* (1944) XIX. 150/1 Supersonics..may usher in a new age of chemistry with radio being used as a catalytic agent. 1945 Koestler *Yogi & Commissar* 32 It has to act as a catalytic agent, as the saliva in the process of creative assimilation. 1951 *Economist* 11 Aug. 358 The modern catalytic cracker..may use as many as 350 instruments measuring and controlling such variables as flow, pressure and temperature.

catalytical (kætăli·tikăl), *a.* = Catalytic *a.*
1889 in *Cent. Dict.* 1923 *Glasgow Herald* 30 May 8 Catalitical salts.

catamaran. Add: **1.** In recent use, a sailing boat with twin hulls placed side by side, widely used as pleasure craft and in sailing contests.
1957 *Times* 13 Dec. 15/2 The catamaran has strongly caught the fancy of those to whom speed is the prime satisfaction to be had from sailing. There were races for catamarans at Cowes last summer.

catananche (kætănæ·ŋkɪ). *Bot.* [mod.L. (Linnæus 1735), f. L. *catanancē* plant used in love potions, Gr. καταναγκη, f. κατά down + ἀνάγκη compulsion.] The name of a genus (*Catananche*) of herbs of the family Compositæ with blue or yellow flowers; a plant of this genus.
1798 tr. *J. F. Gmelin's Linnæus' Syst. Nat.* XIV. 291 Compositae...The genera are..Catananche, Candia lion's-foot. 1836 Loudon *Encycl. Plants* 678 *Catananche*...The modern genus, which contains two or three species of ornamental border annuals, can have no reference to that of the ancients. 1868 S. Hereman *Paxton's Bot. Dict.* 116/2 *Catananche*..a pretty genus, that succeeds well in common soil. 1961 *Times* 25 Nov. 11/4 The artemisias, catananches, [etc.]. 1962 *Amat. Gardening* 19 May 21 A packet of seed of catananche which gave me a lovely group of lavender-blue daisy-like flowers in August.

cat and dog. **2.** (Earlier and later examples.)
[*a* 1652 R. Brome *City Wit* (1653) IV. i, It shall raine.. Dogs and Polecats.] 1949 A. Wilson *Wrong Set* 188 It always 'rained cats and dogs'.

cataphatic (kætăfæ·tik), *a. Theol.* [ad. Gr. καταφατικ-ός affirmative (καταφῆναι to assent).] Defining God positively or by positive statements, opp. *apophatic *a.*
1869 [see *apophatic *a.*]. 1937 Wall & Adamson tr. *Maritain's Degrees Knowl.* iv. 291 It is clear..that *apophatic* theology, which knows God by the mode of negation or ignorance, knows him better than *cataphatic* theology, which proceeds by that of affirmation and science. 1951 *Theology* LIV. 29 In an end age he [*sc.* Berdyaev] turned..from a cataphatic to an apophatic theology.
Hence **catapha·tically** *adv.*
1937 Wall & Adamson tr. *Maritain's Degrees Knowl.* iv. 297 Theological faith..must first advance cataphatically, making known the mysteries of the Godhead to us in communicable enunciations.

cataphoresis (kæ:tăfŏri·sis). Also **kata-.** [f. Gr. κατά down + φόρησις being carried.] **a.** *Med.* The action of causing medicinal substances to pass through the skin into living tissue by the use of an electric current. **b.** = *electrophoresis. Hence **ca:taphore·tic** *a.*
1889 *N.Y. Med. Jrnl.* 27 Apr. 449/1 (*heading*) Electric cataphoresis as a therapeutic measure. *Ibid.*, The cataphoretic action of electricity has often been made use of experimentally to introduce drugs into the system through the skin. 1895 *Buck's Handbk. Med. Sci.* Suppl. 267/1 The use of iodine by cataphoresis has been attempted in cases of goitre. 1908 *Times* 9 Oct. 10/1 The possibility of using cataphoresis as a means of conveying foods and drugs to the system. 1944 *Electronic Engin.* XVI. 341 When a solid

particle becomes suspended in a liquid medium of higher dielectric constant it becomes, in general, negatively charged relative to the dispersion medium and will therefore be attracted to the anode of an electrode system placed in the solution. This phenomenon is known as cataphoresis. *Ibid.*, It is logical to regard electrolytic conduction as the limiting case of cataphoretic conduction. 1944 S. Field *Princ. Electrodeposition* xx. 279 Since colloidal suspensions are electrically charged they migrate to one electrode or the other under the influence of an applied P.D. This motion is known as cataphoresis or electrophoresis. 1949 *Electronic Engin.* XXI. 405 One application of these aspect ratio resistors is to coat a sheet of insulating material by a capillary process, by spraying or by cataphoresis with a uniform resistance film of constant value of ohms per square. 1954 *Ibid.* XXVI. 404 It has been found that the cataphoretic coating process has some advantages.

cataphoric (kætăfǫ·rik), *a.* = Kataphoric; cataphoretic.
1890 M. A. Starr in *Electricity in Daily Life* (1891) 271 The second action of a continuous galvanic current is to move along with it the fluids which lie in its path. This is called its cataphoric action. 1895 *Buck's Handbk. Med. Sci.* Suppl. 267/1 The cataphoric transfer of molecules of protoplasm and liquid from one cell to another. 1903 *Med. Record* 7 Mar. 363/1 The cataphoric electrode..is connected with the positive pole of the battery.

cataphractic, *a.* Add to def.: Covered with or as with armour.
1890 Meredith *One of our Conq.* (1891) I. v. 67 Not even the flower..would hold constant, as they, to the constantly unseen—a trebly cataphractic Invisible.

catapult, *sb.* Add: **3.** A mechanical contrivance by which aircraft are launched at a high speed; also *attrib.* So *catapult launching*, etc.
1927 *Daily Express* 12 Dec. 11/4 Béarn, new aircraft carrier, will carry forty airplanes to be launched by a compressed air catapult at forty-seven miles per hour after a run of sixty-five feet along the deck. 1928 *Times* 28 Aug. 12/6 Catapult Air Mail. 1929 *Air Ann. Brit. Empire* 56 Catapult launching which has been successfully developed and can be used from ships. 1934 *Flight* 15 Feb. 148/1 For use as a catapult ship on the South Atlantic route. 1940 E. C. Shepherd *Britain's Air Power* 11 The aeroplane is set on the catapult carriage while its engine is started and run up. 1941 W. S. Churchill *Secret Session Speeches* (1946) 39 A number of special vessels carrying catapult Hurricane aircraft are employed on patrolling duties. 1959 *Times* 28 Apr. 11/7 Eagle..equipped to take the new generation of aircraft, made possible by the installation of steam catapults.

catapult, *v.* Add: **a.** *trans.* (Further example.) Also *fig.*
1929 C. C. Martindale *Risen Sun* 202 Like a fresh breeze, thus to get the pure truth catapulted at you! 1944 Blunden *Cricket Country* iv. 56 The second ball which T. catapulted down, swung out. 1959 *Times* 13 Jan. 9/6 Modern technology is catapulting some aspects of Asian life into the twentieth century.
d. To discharge by means of a catapult contrivance. Cf. prec. 3.
1912 *Sci. Amer.* 14 Dec. 512 (*heading*) Catapulting a Hydro-aeroplane from a Fighting Ship. 1928 *Daily Express* 29 Aug. 8/2 The Post Office has just announced that for a special fee it will accept British letters and postcards for the United States on the understanding that before arriving in New York they are to be catapulted from the ship by seaplane. 1928 *Scotsman* 30 Aug. 9/6 The aeroplane will be catapulted into the air on approaching the American coast. 1931 *Air Ann. Brit. Empire* 318 The machine is a tractor biplane..capable of being catapulted from the deck. 1938 *Jrnl. R. Aeronaut. Soc.* XLII. 864 The German catapulting ships stationed in mid-ocean and forming seaplane stations. 1959 *Daily Tel.* 23 Feb. 11/6 The rocket-powered North American X-15..will be catapulted from a B-36 bomber at 35–40,000 ft. and..may reach a height of 300 miles and a speed of 4,000 m.p.h.
e. *intr.* To fling oneself as though hurled by a catapult.
1928 *Daily Express* 10 Apr. 5/2 Langley catapulted into the room and drove his axe through the window. 1962 K. Orvis *Damned & Destroyed* xx. 148, I catapulted to my feet.

cataractal (kæ·tăræktăl), *a.* [f. Cataract + -al.] Of the nature of a cataract (*lit.* and *fig.*).
1888 W. Clark Russell *Death Ship* xxxviii, A cataractal roaring of water. 1891 *Illustr. Lond. News* 17 Jan. 74/2 The sea swept..with a cataractal fury. 1926 C. L. Warr *Principal Caird* 237 The outburst of popular indignation.. was simply terrific in its power of cataractal denunciation.

catasetum (kætăsi·tɒm). *Bot.* [mod.L. (L. C. Richard 1822, in K. S. Kunth *Synopsis Plantarum* I. 330) f. Gr. κατά down + L. *seta* bristle.] A plant of the genus *Catasetum*, a genus of epiphytic orchids native to tropical America, in which the rostellum of the male flower is developed into two slender horns or bristles.
1829 Loudon *Encycl. Plants* I. xx. 756. 1843 *Florist's Jrnl.* IV. 184 A beautiful dwarf orchideous plant, resembling a catasetum in habit. 1914 *Chambers's Jrnl.* Oct. 631/1 Catasetums have established themselves, some with huge pseudo-bulbs. 1956 *Dict. Gardening* (*R. Hort. Soc.*) (ed. 2) I. 412/2 When growing, all Catasetums benefit by a tropical atmosphere. 1962 *Amat. Gardening* 7 Apr. 6/3 In most of the catasetums and mormodes the male and female flowers are carried on separate spikes or separate plants.

catathymia (kætăþǝi·miă). *Psychiatry.* [mod.L., ad. G. *katathymie* (H. W. Maier 1912, in *Zeitschr. f. Neurol. & Psychiat.* XIII. 555), f. Gr. κατά according to + θυμός spirit, temper.] A condition in which the mind falls under the control of the emotions. Hence **catathy·mic** *a.*, **catathy·mically** *adv.*
1934 E. B. Strauss tr. *Kretschmer's Text-bk. Med. Psychol.* II. vi. 71 By *catathymia*..we mean the transformation of the psychic content by affective influences. The primitive world-picture is much more catathymic than our own...This is sufficient to bring about the catathymically determined conviction. *Ibid.* vii. 101 In such cases of schizophrenia the type of thought is again almost entirely catathymic, even to the extent of the loss of all contact with the realities. 1949 *Brit. Jrnl. Psychol.* XL. 13 In catathymia we are dealing with changes to which the psychic content is subjected, and transformed, by affective influences.

catatonia (kætătōᵘ·niă). Now the usual form of Katatonia.
1917 C. R. Payne tr. *Pfister's Psychoanal. Method* I. vi. 132 Probably there was catatonia; her brother suffered from a severe form of this disease and was cared for in an insane asylum. 1937 'C. Caudwell' *Illusion & Reality* 258 In catatonia the affects are repressed. 1956 A. Huxley *Heaven & Hell* App. 84 In psychological terms..from catatonia and feelings of unreality to a sense of heightened reality in vision and, finally, in mystical experience. 1963 *New Scientist* 18 July 146 The inert state called catatonia which is characteristic of some kinds of schizophrenia.

catatonic (kætătǫ·nik), *a.* [f. as prec.: see -ic 1.] Characterized by *catatonia. Hence as *sb.*, one affected by catatonia.
1908 *Practitioner* Jan. 12 There may be catatonic stupor, automatic obedience, and occasionally impulsive automatism. 1917 C. R. Payne tr. *Pfister's Psychoanal. Method* II. xix. 499 The overcoming of the resistance is impossible in catatonics of an advanced stage. 1948 *Brit. Jrnl. Psychol.* Dec. 89 Catatonic schizophrenia is characterized by negativism, catalepsy, suggestibility, stupor, excitement, mannerisms, stereotypy. 1957 P. Lafitte *Person in Psychol.* 7 The scale or scope of activity may vary, down to the limit of the catatonic schizophrenic, whose endless wars and reconciliations are conducted in strict privacy.

catavothron. = Katabothron.
1869 H. F. Tozer *Highl. Turkey* I. vii. 160 The Mediævals ..had the idea of there being a catavothra from the lake. 1878 *Encycl. Brit.* VIII. 685/2 The subterranean course of the streams is frequently indicated by peculiar vents or pits caused by the subsidence of the soil; they are popularly known in Greece as *catavothra*.

catawampous, *a.* Add: (Earlier and later examples.) Also, askew, awry.
1840 *Spirit of Times* 25 Jan. 561/2 Him is done up—used up catawampous—kicked up into eberlasting hoki! 1844 [see Chaw *v.* 3]. 1885 'C. E. Craddock' *Prophet Gt. Smoky Mts.* ix. 153 She got me plumb catawampus. 1889 —— *Broomsedge Cove* iii. 44 But it's a powerful differ ter please this man an' not git that one set catawampus. 1917 L. M. Montgomery *Anne's House of Dreams* xxxvi. 308 Dear me, everything has gone catawampus with me this week.

catawampus, *sb.* For def. read: A bogy, a fierce imaginary animal. (Earlier and later examples.)
1843 'R. Carlton' *New Purchase* I. xxviii. 265 The tother one what got most sker'd is a sort of catawampus (spiteful). 1893 Yonge & Coleridge *Strolling Players* xvii. 145 Classes had better..swallow each other, like the crocodile and the catawampus.

catbird. Add: **b.** The name given to several species of Australian birds whose cry resembles the mewing of a cat.
1848 J. Gould *Birds Australia* IV. pl. 11 Ptilonorhynchus Smithii...Cat Bird. *Ibid.*, Situations suitable to the Regent and Satin Birds are equally adapted to the habits of the Cat Bird. 1887 D. Macdonald *Gum Boughs* 36 One of the most peculiar of birds' eggs found about the Murray is that of the locally-termed 'cat-bird', the shell of which is veined thickly with dark thin threads as though covered with a spider's web. 1889 R. B. Anderson tr. *Lumholtz's Among Cannibals* vii. 96 The cat-bird (*Ælurædus maculosus*), which makes its appearance towards evening, and has a voice strikingly like the mewing of a cat. 1957 *Encycl. Brit.* V. 24/2 *Catbird*... In Australia, a name given to any of several bowerbirds, especially to *Ailuroedus crassirostris*, which builds no bower.
2. Phr. *the catbird seat*: a superior or advantageous position. *U.S. slang.*
1942 J. Thurber in *55 Short Stories from New Yorker* (1949) 61 'Sitting in the catbird seat' meant sitting pretty, like a batter with three balls and no strikes on him. 1958 Wodehouse *Cocktail Time* xiii. 114 'I get you. If we swing it, we'll be sitting pretty.' 'In the catbird seat.'

cat-boat: see Cat *sb.*²

catch, *sb.*¹ Add: **1. c.** In Rugby football and baseball (see quots.).
1867 [see *fair *a.* 10 d]. 1868 H. Chadwick *Base Ball* 45 A running catch is made when the ball is caught on the fly while the fielder is on the run. 1896 R. G. Knowles & Morton *Baseball* 14 He..must be as proficient in making running catches as in bringing off standing ones. 1897 *Encycl. Sport* I. 431/1 If a player makes a fair catch he shall be awarded a free kick. 1960 E. S. & W. J. Higham *High Speed Rugby* xiii. 183 The method of making a fair-catch is to make a mark on the ground with the heel as the ball is caught, and to call: 'Mark!'
d. *Rowing.* The grip of the water taken with the oar at the beginning of a stroke.
1881 *Standard* 30 Mar. 3/7 The shallow waters of the Cam ..make it very difficult for a crew to imitate the catch at the beginning of the stroke. 1898 *Encycl. Sport* II. 296/2 *Catch*, the instant application of the weight and muscles of legs and body to the oar at the moment it enters the water.
e. In full *glottal catch* (see *glottal *a.*). Often used as a synonym of the more frequent *glottal stop*.
1925 P. Radin tr. *J. Vendryès's Lang.* I. i. 30 It is directly after the vowels, when the emission of air is complete, that this catch or 'stop' occurs. 1964 Crystal & Quirk *Prosodic Features in Eng.* iii. 43 One flap of tremulousness..is equivalent to a 'catch' in the voice (i.e. one flap or brief roll of glottal trill).
2. c. A crop, esp. one sufficient to render further sowing unnecessary. *U.S. colloq.*
1868 G. Brackett *Farm Talk* 128 That's one reason why I sowed the field to barley—so as to get a good catch. 1941 *Harper's Mag.* Aug. 329/1 My newly laid down field, where I didn't get a very good catch of grass.
3. Add after def.: Also, a ball so hit that a fielder may catch it. *catch-and-bowl*, a catch made by the bowler.
1816 W. Lambert *Instr. & Rules Cricket* 15 Strikers are generally cautious at first, which will frequently cause a catch. 1888 Steel & Lyttelton *Cricket* iii. 110 A catch in the slips or at point, or a catch and bowl, is not infrequently the result. 1906 A. E. Knight *Compl. Cricketer* iv. 146 Rightly judged by the eye, a catch should drop into the hands. 1907 *Westm. Gaz.* 20 July 15/1 Hence the number of catches-and-bowls he used to bring off.
b. (Earlier example.)
1854 F. Lillywhite *Guide to Cricketers* 58 [He is a] beautiful field at long-leg, being a sure catch.
7. b. A hidden element (in a proposal, etc.) designed to take advantage of another person; something concealed with the intention of catching or tripping one up; hence, an unforeseen difficulty or awkwardness, a 'snag'.
1855 Barnum *Life* 120 The old farmer, who was pretty 'cute, was sure that there was some 'catch' in this offer. 1913 *Punch* 18 June 474/1, I say, I'm a burglar. There is no catch in it. 1914 *Daily Express* 22 Dec. 2 It is not a toy or a 'catch', but a genuine game. 1919 Wodehouse *Damsel in Distress* x, There's nobody I think a more corking sportsman than Maud, if you know what I mean, but—this is where the catch comes in—I'm most frightfully in love with somebody else. 1951 C. V. Wedgwood *Last of Radicals* iv. 83 I've just been reading your Henry George..and I..can't see the catch in it.
11. b. *colloq.* A person matrimonially desirable on account of wealth or position.
[1722 Defoe *Col. Jack* (1723) 240 She intended to have me, if she could catch, and it was indeed a kind of catch.] 1749 Cleland *Mem. Woman Pleasure* II. 215 Your surprize that one of my blood and relish of life, should count a gallant of threescore such a catch. 1835 T. Moore *Fudge Fam. in Eng.* i. 2 Sole encumbrance, Miss Fudge to be taken therewith. Think, my boy, for a Curate how glorious a catch! 1837 Marryat *Snarleyyow* xx, She..fully proved to his satisfaction that, independent of her beauty, she would be a much greater catch than Frau Vandersloosh. 1917 H. A. Vachell *Fishpingle* x, Matrimonially he was no great 'catch' for an heiress of quality.
c. *Austral.* and *N.Z.* (See quot. 1933.)
1933 L. G. D. Acland in *Press* (Christchurch) 23 Sept. 13/7 *Catch*. Just before stopping time in a wool shed, a shearer tries to finish the sheep he is on and catch another which he can finish at ease after knock-off. This is called *getting a catch*. E.g., 'How many more can you do this run?'—'Two and a catch.' 1956 G. Bowen *Wool Away!* (ed. 2) 155 *Catch*, the last sheep of a mob. Very often this sheep produces one extra for the tally of the shearer who shears well enough to catch it first. 1965 D. S. Gunn *Terminol. Shearing Ind.* I. 7 The 'bell sheep', or 'the catch' as it is often called, may be an easy one.

catch, *v.* Add: **9. b.** *pass.* (Always in pa. t. or as pa. pple.) To become pregnant. Also *caught out*. *colloq.*
1858 Queen Victoria *Let.* 15 June in *Dearest Child* (1964) 115 The pride of giving life to an immortal soul is very fine..perfectly furious as I was to be caught. 1919 M. Stopes *Let. to Working Mothers* 4 Very often it happens that you get 'caught', and you know that the baby that you feared might come has really begun. 1935 E. Bowen *House in Paris* II. ix. 202 Being caught is the word for having a child, sometimes. 1955 'C. Brown' *Lost Girls* xiii. 142 She's been caught..she's about five months gone. 1957 J. Osborne *Look Back in Anger* I. 29 I'm pregnant. After three years of married life, I have to get caught out now.
12. b. *Rowing.* To grip the water with the oar at (the beginning of a stroke).
1886 *Encycl. Brit.* XXI. 32/1 The oar should be struck down firmly and decisively into the water.., and the weight of the body be thrown entirely upon it, by which the beginning of the stroke is caught. 1898 *Encycl. Sport* II. 273/2 The beginning must be caught with full power. 1902 *Encycl. Brit.* XXXII. 307/2 Swivel rowlocks..are considered unsuitable to eight-oared rowing, where the beginning of the stroke has to be firmly and smartly caught.
c. Of a crop: to germinate and grow. *U.S.* (Cf. *catch *sb.*¹ 2 c.)
1843 W. Oliver *Eight Months in Illinois* 95 Putting in hayseed without ploughing the ground...The seed catches quick and grows well.
14. Also in wider application.
1925 G. L. Mallory in E. F. Norton *Fight for Everest, 1924* 213 The wind and sun between them have fairly caught us all these last three days.

16. b. To fasten, attach (some object) *back* or *up*.

1893 *Funk's Stand. Dict.*, To *catch up*, to raise by attaching something; festoon; loop up; as, her dress was caught up with ribbons. **1898** *Daily News* 11 May 4/4 [Her] rose-tinted brocaded train was caught back at one side.

24. c. (Earlier examples.) Also *absol.*, to make a catch. *to catch and bowl*: usu. in pa. t. *caught and bowled*, caught by the bowler; also as *ppl. a.* and *sb.*

1712 *Devil & Peers* (Broadside) in W. J. Lewis *Lang. Cr.* (1934) 44 I'll catch them both out in three or four stroaks. **1744** *Laws* [of Cricket] in *New Dict. Arts & Sci.* (1755) IV. 3459/2 So as to hinder the bowler from catching her. **1873** R. A. Fitzgerald *Wickets in West* 287 Gilbert..was also caught and bowled by Eastwood. **1883** [in Dict.]. **1897** *Encycl. Sport* I. 245/2 *Caught and Bowled*, caught by the bowler who delivered the ball. **1904** P. F. Warner *Recov. Ashes* v. 78 Rhodes missed catching and bowling Gregory. **1924** J. B. Hobbs *Cricket Mem.* 169 Then Mr. Simms got rid of Woolley with a magnificent 'caught and bowled'. **1950** W. Hammond *Cricketers' School* v. 52, I have seen Larwood take some of the speediest single-handed caught-and-bowled catches ever put up on any cricket field. **1956** R. Alston *Test Commentary* xvi. 136 A 'caught-and-bowled' off a hard hit from May.

d. *Baseball.*

1858 F. Pigeon in A. G. Spalding *Amer. Nat. Game* (1911) iv. 61 Next man got scared; caught out. **1874** Chadwick *Base Ball Man.* 88 There he was when Mills was caught out on the fly. **1902** *Encycl. Brit.* XXVI. 160/2 In base-ball if the ball is knocked in a certain direction it is called a foul, and the player who knocked it has not the privilege of making a run, but may be caught out.

e. In the game of baseball: *trans.* To catch the pitcher's deliveries. *intr.* To act as catcher.

1865 *Wilkes' Spirit of Times* (N.Y.) 8 July 301/1 Wansley caught behind in a handsome manner. **1887** *Courier-Jrnl.* (Louisville, Ky.) 26 May 2/6 Young Love Cross caught Ramsey in fine style, and Greer also handled Porter's delivery as well as could be desired. **1890** Will Carleton *City Legends* 39 'An' will you pitch or catch?' Says I, 'I'll catch, if so desired'.

30. *to catch one's death* (*of cold*): later examples.

1904 'G. B. Lancaster' *Sons o' Men* xvi. 204 Don't you dare take him till I get some more clothes on him! He'd catch his death! **1951** G. Greene *End of Affair* v. vii. 215 She had walked in the rain seeking a refuge and 'catching her death' instead.

35. b. To watch (a theatrical performance or television programme); to listen to (a concert, etc.).

1906 H. Green *At Actors' Boarding House* 150 Where are you on the bill at Moctor's this week? I must come in an' ketch you guys. **1937** *Amer. Speech* XII. 45/2, I caught Wright's band last night and are they tough. **1965** *Listener* 7 Oct. 551/2 This modest programme..was the kind of late-evening music item that is so often well worth catching. **1969** *Oxf. Times* 15 Aug. 13/1 You can have a cigarette or a drink, read the newspaper or catch the television news.

39. *catch as catch can*: also as *sb.* and *attrib.*

1764 K. O'Hara *Midas* III. 62 There's catch as catch can, hit or miss Luck is all, And Luck's the best tune of life's Toll lol de roll. *a* **1777** S. Foote in M. Edgeworth *Harry & Lucy Concluded* (1825) II. 153 They all fell to playing the game of catch as catch can, till the gun powder ran out at the heels of their boots. **1936** *World Film News* Sept. 4/1 The present catch-as-catch can method of entry. **1949** R. Harvey *Curtain Time* 130 The production was usually a hurried, catch-as-catch-can affair. **1958** G. Barker *Two Plays* 23 Davy Jones and his daughter at catch-as-catch-can. **1959** *Manchester Guardian* 14 July 7/1 To-night's papers are full of catch-as-catch-can interviews with the..survivors.

b. *catch-as-catch-can*, the Lancashire style of wrestling. Also *attrib.* and *transf.*

[**1617** Middleton & Rowley *Fair Quarrel* II, I'le wrastle with any man for a good supper...I'le take your part there, catch that catch may.] **1889** W. Armstrong *Wrestling* Introd. p. xiv, In 1871, the late Mr. J. G. Chambers ..endeavoured to introduce and promote a new system of wrestling at the Lillie Bridge Grounds, West Brompton, which he denominated, 'The Catch-as-catch-can Style; first down to lose'. **1898** *Encycl. Sport* II. 548/2 The principal chips associated with catch as catch can wrestling are the double Nelson, the half Nelson, the heave [etc.]. *Ibid.* 549/2 Turkish wrestling is principally carried out in catch as catch can style. **1905** *Daily Chron.* 21 Dec. 9/5 A catch-as-catch-can wrestler needs to be wonderfully active. **1913, 1934** [see *all-in 2]. **1935** C. Isherwood *Mr. Norris changes Trains* xv. 243 Arthur's orientally sensitive spirit shrank from the rough, healthy, modern catch-as-catch-can of home-truths and confessions. **1957** *Encycl. Brit.* XXIII. 806/2 The Lancashire style, generally known as 'catch-as-catch-can', is practised in Lancashire, throughout Great Britain generally, and is the most popular style in the United States, Canada, Australia, Switzerland. **1958** *Times* 13 Aug. 9/4 A bull..decided that here was a catch-as-catch-can event in which he was free to join.

40. Further examples with pronouns other than *me*.

1890 G. B. Shaw in *Time* Feb. 201 'Does Christine ever lecture them?'..'Catch her at it!' said Krogstad..'They would soon show her the door.' **1936** N. Coward *Fumed Oak* 41 *Mrs. Rockett*: I can always go to a boarding-house or a private hotel. *Doris*: Catch you!

46*. *to catch* (someone) *bending*: see *bend *v.* 9 d.

50. d. Hence *catch-on* sb., a success. *rare.*

1895 G. B. Shaw *Our Theatres in Nineties* (1932) I. 274 The ordinary commercial west end theatre, with its ignoble gambling for 'a catch-on'. **1897** —— *Ibid.* III. 28 Commercial enterprise, always dreaming of 'catches-on', long runs, and 'silver mines', attempted to exploit the occasion in the usual way.

51. catch out. c. *fig.* To catch in a mistake, catch napping or in the act.

1816 Jane Austen *Emma* II. xiv. 275 Ah! there I am—thinking of him directly. Always the first person to be thought of! How I catch myself out! **1881** Mrs. Lynn Linton *My Love* xvii, Randolph caught himself out in the vileness of wishing that she was just a trifle less superior. **1956** A. L. Rowse *Early Churchills* xii. 230 His methods were distinctly unorthodox: that was what alarmed the Dutch text-book generals and caught out the French.

53. catch up. e. Also in non-physical senses, and *intr.* esp. in *to catch up on, to, with*.

1886 *Calcutta Englishman* in J. M. Dixon *Dict. Idiomatic Eng. Phrases* (1891) 58 He has not caught up (overtaken) his rival by the time earlier educational honours are distributed. **1923** H. Crane *Let.* 5 Dec. (1965) 159 Getting things straightened around again and catching up on our supply of wood. **1925** E. F. Norton *Fight for Everest, 1924* 114, I had to wait quite half an hour for him to catch up. **1925** *Times* (weekly ed.) 26 Nov., The police caught up on the men just as they entered a dark archway. **1926** *Ibid.* 5 Aug., Its sanitation can never catch up to its needs. **1927** Jespersen *Mod. Eng. Gram.* (1928) III. 271 Where an Englishman says 'I shall *catch* you *up*', or 'I'll *catch up with* you', Americans know only the latter phrase. **1941** A. L. Rowse *Tudor Cornwall* xv. 412 Killigrew found himself in prison; his own misdemeanours..had at last..caught up with him. **1967** 'La Meri' *Sp. Dancing* (ed. 2) vii. 89 Toward the middle of the [nineteenth] century the *Seguidillas Manchegas* caught up the waning popularity of the Bolero.

So **catching-up** *vbl. sb.*

1942 *19th Cent.* Feb. 90 This rapid catching-up on Western ideas.

catch-. Add: **1. catch-water** (earlier and later examples); (*b*) a vessel designed to catch water.

1799 A. Young *Agric. Lincoln* xii. 275 The catch-water drain runs all winter. **1838** *Civil Engin. & Archit. Jrnl.* I. 256/2, I shall now proceed to describe the mode of discharging, by catch-water courses or drains, all the brooks and rivers which flow into it. **1842** G. Francis *Dict. Arts*, *Catch-water drains*, drains, or channels, cut in a slanting direction across and down embankments, therefore catching and carrying off the water which falls upon them. *a* **1877** Knight *Dict. Mech.* I. 503/2 *Catch-water Drain*, a drain to intercept waters from high lands, to prevent their accumulation upon lower levels. **1888** *Lockwood's Dict. Terms Mech. Engin.*, *Interceptor*, a T-shaped cylindrical vessel employed in connection with marine engines to prevent particles of water from being carried over with the steam into the cylinders...Called also catch water. **1901** M. W. Travers *Exper. Study Gases* 33 The water runs into a catch-water below the bulb, and is conducted away. **1963** *Times* 7 Mar. 10/6 Fourteen miles of catchwaters have been built around the valley, and it is hoped they will catch about 90 per cent of the rain that falls.

3. a. *catch-bar, -bolt, -boom, -pot, ratline* (see Ratlin(e 2), *-tank*; **catch-basin** (example); (*b*) a reservoir for catching and retaining surface-drainage over large areas; **catch-box,** a box-like clutch of a spinning machine; **catch-fake** (see quot.); **catch-point,** a switch or point intended to derail a train, wagon, etc. (e.g. to prevent it from running on to a main line); **catch-stitch,** (*a*) *Bookbinding* = Kettle-stitch; (*b*) (see quot. 1968); also as vb.; **catch-wheel,** a wheel capable of motion in one direction only, a ratchet-wheel; **catch-work,** also *attrib.*

1850 *Rep. U.S. Comm. Patents 1849* 170 The second crank to slide the catch bar. *a* **1877** Knight *Dict. Mech.* 503/1 *Catch-bar* (Knitting-machine), a bar employed to depress the jacks. *Ibid.*, *Catch-basin*, a cistern at the point of discharge into a sewer. **1884** *Science* III. 372/1 Whether any..system of catch-basins or reservoirs, could..mitigate ..such..floods. **1859** *U.S. Patent Off., Ann. Rep. 1858* I. 537 The levers or arms are designed to force back the catch-bolt and lock-bolt. *a* **1877** Knight *Dict. Mech.* 503/1 *Catch-bolt*, a cupboard or door-bolt which yields to the pressure in closing and then springs into the keeper in the jamb. Usually retracted by a small knob. **1905** *Terms Forestry & Logging* 33 *Catch boom*, a boom fastened across stream to catch and hold floating logs. **1825** J. Nicholson *Oper. Mech.* 426 When the catch-box 14 is in contact with the sheeve *s*. **1892** Nasmith *Students' Cotton Spinning* 320 The wheel..is provided with a catch box..and by its means drives the shaft. **1867** Smyth *Sailor's Word-bk.*, *Catch-fake*, an unseemly doubling in a badly coiled rope. **1883** *Peel City Guardian* 27 Jan. 6/1 The mineral train came on at considerable speed, passed the signals, and through the catch-points. **1895** *Daily News* 7 Dec. 7/7 That catch points should be more clearly indicated. **1902** *Encycl. Brit.* XXXII. 144/2 A throw-off or derailing switch ('catch-points'). **1936** *Gloss. Terms Rly. Signalling* (*B.S.I.*) 12 *Catch points*, trailing points provided on a rising gradient for the purpose of derailing a vehicle running back after breaking away. **1909** *Westm. Gaz.* 26 Oct. 11/4 Tin..runs through a channel ..into a catch-pot, whence it is ladled by small gourds.. and poured. **1962** *New Scientist* 12 July 82/3 The catchpot was designed to hold the cathode in the event of its failure by fracture or melting. **1846** Dodd *Brit. Manuf.* VI. iv. 87 'Kettle-stitch'..is supposed by some to be a corruption of 'catch' or 'ketch' stitch, while others refer it to 'chain' stitch. **1906** D. Cockerell in E. Johnston *Writing & Illum.* xvi. 347 At about ½ inch from either end make an additional line across the back for the 'kettle' or catch stitch. **1932** D. C. Minter *Mod. Needlecraft* 123/1 Herring-bone Stitch..used for the same purposes as catch-stitch. *Ibid.* 208/2 Catch-stitch..all round on to the linen—no stitches showing through. **1964** *McCall's Sewing* xiv. 259/1 Lap front shoulder darts to stitching line and catch-stitch halfway down from the top. **1968** J. Ironside *Fashion Alphabet* iii. 83 *Catch-stitch*, large cross-stitch used to hold heavy hems, invisible on right side. **1920** *Blackw. Mag.* May 706/2 The glen stream was never again diverted from its course, nor the catch-tank drained. **1845** *Encycl. Metrop.* VIII. 619/1 This cylinder carries a catch wheel..the teeth of which engage the click..attached to the wheel C by a screw. **1799** T. Wright *Art Floating Land* 82, I have seen the common plan of catch-work watering resorted to. **1844** H. Stephens *Bk. Farm* III. 1022 Catch-work irrigation.

b. *catch-cry*; **catch-line,** a short line of type that catches the eye; *spec.* in *Typogr.* (see quot. 1938); **catch title,** an abbreviated title sufficiently expressive of the full title to identify the book.

1901 *Daily Chron.* 20 Nov. 4/5 Some very sound remarks ..on certain catch-cries of the day. **1916** W. B. Yeats *Eight Poems*, The clever man who cries The catch cries of the clown. **1866** Catch-line [see Dict.]. **1909** Webster, *Catch line*,..(*a*) A line containing the catchword at the foot of a page. (*b*) A short line in displayed matter. **1938** L. M. Harrod *Librarians' Gloss.* 35 *Catch line*, a line [of type] inserted at the top of matter by the compositor in order to identify it. **1958** T. Landau *Encycl. Librarianship* 66/1 *Catchline*, a temporary descriptive headline on galley proofs. Also a short line of type in between two large displayed lines. **1909** Webster, *Catch title*, a short expressive title used for abbreviated book lists, etc. **1959** L. M. Harrod *Librarians' Gloss.* (ed. 2) 62 *Catchword title*. ..Also called 'Catch title'.

c. catch question, a question that catches one out or has a catch in it; also as vb.

a **1860** Alb. Smith *Med. Student* (1861) 14 Legendary 'catch questions'. *Ibid.* 116 The inquisitors..are willing to help a student out of a scrape, rather than 'catch question' him into one. **1905** *Westm. Gaz.* 22 Apr. 3/1 The critics and commentators for centuries have been, like the philosophers, fooled by the catch-question of the Stuart King concerning the weight of a live fish in a bucket full of water.

d. That is or may be taken or 'caught' to one's advantage.

1895 *Westm. Gaz.* 4 Dec. 7/1 Until the end of President Cleveland's term Great Britain has a statesman, and not a catch-vote politician, to deal with. **1905** *Spectator* 7 Jan. 5/2 To put the policy of development at the mercy of a catch vote. **1907** *Daily Chron.* 15 Aug. 7/5 There was a strong catch tide in favour of the swimmers.

4. catch-weight, also in *Boxing* and *Wrestling* (see quot. 1897).

1863 *Punch* XLV. 86 The Archimandrite Nilos has offered to fight the Bishop of London for 20 pound a side, catch-weight. **1887** G. B. Shaw *Let.* 8 Feb. (1965) 164 The Socialist League have been challenged by C. Bradlaugh to pick a man to fight him at catch weight. **1897** *Encycl. Sport* I. 139/1 *Catchweight* (*To box at*)—Boxing without restrictions as to weight. **1907** *Daily Chron.* 18 Oct. 9/3 To wrestle the best of three falls for £50 a side at catch-weight. **1971** *Wrestling Rev.* XVI. v. 4/1 A catchweight contest.

ca·tch-all, *sb.* orig. *U.S.* [Catch- 1.] A general receptacle. Also *fig.* and *attrib.*

1838 *Congress. Globe* 16 Apr. App. 275/1 [The party includes] old Federalists,..Antimasons, and Abolitionists. They have, sir, been a kind of catch-all, or *omnium gatherum*. **1866** Mrs. Stowe *Lit. Foxes* ii. 27 The general catch-all and menagerie..for all the family litter. **1875** [see Catch- 1]. **1892** *Harper's Mag.* June 29/1 A shrewd spider..had spread his gossamer catch-all beneath the bramble. **1897** 'Mark Twain' *Following Equator* xli. 383 It seemed to have been designed as a catch-all for every thing that can damage it. **1923** L. J. Vance *Baroque* xiv. 86 The tenement yard was a simple black hole, for generations a common catch-all. **1945** 'L. Lewis' *Birthday Murder* (1951) vi. 88 That cupboard..has been a catchall for odds and ends for a long time. **1957** N. Frye *Anat. Criticism* 304 The word novel..has since expanded into a catchall term which can be applied to practically any prose book that is not 'on' something. **1961** *Times* 10 Feb. 17/4 At Cambridge..history had become a 'catch-all' for students who found classics too difficult and science too serious. **1963** *Lancet* 19 Jan. 125/2 There is danger that it may become a catch-all term for maladies with puzzling clinical and anatomical features.

catch-hold. *Wrestling.* [f. phr. *to catch hold*: see Catch *v.* 45.] (See quot. 1889.)

1883 *Encycl. Brit.* XXIV. 690/2 The 'catch hold, first down to lose' style of wrestling. **1889** W. Armstrong *Wrestling* 34 The catch-hold fashion is a most rational style of wrestling, as the competitors are..at liberty to catch each other as they please..provided they do not hold by the legs or clothes. **1898** *Encycl. Sport* II. 546/1 A style on the 'catch hold' principle.

ca·tchiness. [f. Catchy *a.*] The quality of being catchy.

1898 *Century Mag.* Mar. 773 In spite of its dangerously instantaneous catchiness, it [*sc.* an étude] expresses just the sentiment of flattery in all its shades.

catching, *vbl. sb.* Add: (Examples of use in *Cricket*.)

1767 R. Cotton in F. S. Ashley-Cooper *Hambledon Cricket Chron. 1772–96* (1924) 183 Here's guarding, and catching, and throwing, and tossing. **1904** P. F. Warner *Recov. Ashes* viii. 160 Rhodes bowled splendidly, as did Hirst, but our catching was wretched.

c. *attrib.*, as *catching-hook, -pen, -season.*

a **1877** Knight *Dict. Mech.*, *Catching-hook*, a crochet-hook. A crook or animal-catching hook. **1874** Hardy *Madding Crowd* xxii, In one angle a catching-pen was formed, in which three or four sheep were continuously kept ready for the shearers to seize without loss of time. **1894** A. Robertson *Nuggets* 4 He dashed into the catching pen, and seized the smaller of two sheep that remained. **1904** H. Buckland in M. Cradock *Sport in N.Z.* II. vi. 172 The 'wool shed' where all shearing is done has a row of pens ('catching pens', they are called) in it..each pen holding some twenty-five to thirty sheep. **1914** H. B. Smith *Sheep & Wool Ind. Australasia* vi. 34 Two shearers generally have one pen to catch their sheep from. They are called catching pens. **1891** *Scribner's Mag.* X. 470 During the planting-time [for oysters] as well as in the catching-season.

catching, *ppl. a.* **3.** (Later U.S. example.)

1876 *Rep. Vermont Board Agric.* III. 481 The 'catching' rains of harvest time..will always fetch a lugubrious wail from any farmer.

catchment. Add: *catchment area*: also *transf.* and *fig.*

1959 *Times* 28 May 13/7 The catchment area for Leeds [prison] is dense—reputedly four and a half million people. **1960** *Library Assoc. Record* Aug. 261/2 *Catchment Area*, in library planning denotes the area from which readers may be expected to be drawn to a given library service point. **1961** *Lancet* 12 Aug. 357/2 It [*sc.* the hospital] has been used for patients of good prognosis from all parts of the Metropolitan area, instead of being more strictly attached to a catchment area. **1962** A. Nisbett *Technique Sound Studio* 248 It [*sc.* the diaphragm of a microphone] should be..big enough to present a sufficiently large catchment area to the pressure of the sound wave. **1970** *Daily Tel.* (Colour Suppl.) 3 Apr. 21/1 For the children's sake, they are going to buy a house in the catchment area of a modern primary school which has a nursery class.

catchword. Add: **1.** Also in *Manuscripts*.

1885 E. M. Thompson in *Encycl. Brit.* XVIII. 144/2 Catch-words to connect the quires date back to the 12th century. **1957** N. R. Ker *Catal. MSS. Anglo-Saxon* p. xl, Some quires of..f. 245^v have catchwords.

2. c. (Example.)

1814 Jane Austen *Mansf. Park* I. xviii. 346 The others aspired at nothing beyond his remembering the catchword, and the first line of his speech.

4. catchword entry (see quots.).

1893 in *Funk's Stand. Dict.* **1938** L. M. Harrod *Librarians' Gloss.* 35 *Catchword entry*, entry by some striking word in a book's title, other than the first, and likely to be remembered by an enquirer. **1956** F. C. Avis *Bookman's Concise Dict.* 54/1 *Catchword Entry*, an entry of the name of a book in a catalogue under its most important word, as Pickwick Papers.

catchy, *a.* Add: **2.** Also, liable to 'trip one up', difficult to manage or execute.

1877 *Coursing Calendar* 285 Mr. Hedley, for the second time, pleased everyone with his judging; whilst Johnston, who slipped for the first time on such catchy ground, performed his duties well. **1882** R. Hunter et al. *Encycl. Dict.* II. 1, *Catchy*,..difficult, not easy to learn or to execute. (*Colloquial*.)

catdom (kæ·tdəm). [f. Cat *sb.*¹ + -dom.] The condition or quality of cats; the world of cats.

1888 *Pall Mall Gaz.* 25 Oct. 3/1 A charming specimen of catdom in one 'Jimmy'. **1890** *Illustr. Lond. News* 10 May 599/3 The winning ways of 'catdom' and 'kittenhood'.

catechol (kæ·titʃǫl). *Chem.* [f. Catech(u + -ol.] = *pyro-catechin* (s.v. Pyro- 2 b); Catechin.

1880 *Jrnl. Chem. Soc.* XXXVIII. 417 The author denies the presence of catechol in the leaves of *Ampelopsis hederacea*. **1892** *Photogr. Ann.* II. 87 Catechol in dilute solution develops slowly, but gives excellent detail and opacity. **1964** N. G. Clark *Mod. Org. Chem.* xxi. 439 The three dihydroxybenzenes are called catechol, resorcinol and hydroquinone (quinol). **1964** D. Nichols in *Oceanogr. & Marine Biol.* II. 408 The coelomic fluid will slowly oxidize catechol and 'dopa' *in vitro* at pH7·4 and 30° C.

catechu. Add: Also *attrib.* or *Comb.*: **catechu brown,** a brown colour produced by the use of catechu as a dye.

1860 Ure *Dict. Arts* (ed. 5) I. 805 Catechu Brown, rich and transparent. **1880** J. Dunbar *Pract. Papermaker* 41 Catechu dyed papers. **1902** *Encycl. Brit.* XXVII. 564/1 Catechu browns are fast to a variety of influences.

catechumenal (kætĭkiū·mĕnăl), *a.* [f. Catechumen + -al.] = Catechumenical *a.*

1883 C. C. Perkins *Italian Sculpture* Introd. p. liv, He had laid aside his white catechumenal robes.

categorial (kætigō*·riăl), *a.* Also **categoreal.** [ad. G. *kategorial* (W. Wundt *Logik* (1880) I. 107), f. Category + -al.] Relating to, or involving, categories, *spec.* in Logic and Linguistics. Hence **catego·rially** *adv.*

1912 S. Alexander in *Mind* XXI. 18 In material things we contemplate their categorial characters. **1920** —— *Space, Time & Deity* I. II. i. 185 The pervasive characters of existents are what are known from Kant's usage as the categories of experience, and I shall call them, in distinction from the empirical ones or qualities, categorial characters. **1925** C. D. Broad *Mind & its Place* iv. 216 (*title*) The Categorial Factor in Sense-Perception. **1929** A. N. Whitehead *Process & Reality* 10 The gradual elaboration of categoreal schemes. **1959** P. F. Strawson *Individuals* v. 172 The categorial criterion. *Ibid.* 175 The categorially predicable. **1965** N. Chomsky *Theory of Syntax* ii. 68 Functional notions like 'Subject'..are to be..distinguished from categorial notions such as 'Noun Phrase'. **1965** *Language* XLI. 490 Categorial and phrase-structure grammars.

category. Add: **1. c.** Hence in more general use (see quot. 1901). Also *attrib.*

1901 Baldwin *Dict. Philos.* I. 161/2 The term category, in post-Kantian philosophy, comes to mean any relatively fundamental philosophical conception. **1938** G. Ryle in *Proc. Arist. Soc.* XXXVIII. 206 'Quality', 'state', 'substance', [etc.]..we could call..'category-words'. **1949** —— *Concept of Mind* i. 16 A category-mistake..represents the facts..as if they belonged to one logical type or category.. when they actually belong to another. **1960** J. O. Urmson *Encycl. Western Philos.* 79/2 Today the word 'category' is used by philosophers, if at all, for any supposedly ultimate type, without any settled convention about what it is a type of.

2. b. spec. in *Linguistics.* (See quots.)

1933 Bloomfield *Language* xvi. 270 Large form-classes which completely subdivide either the whole lexicon or some important form-class into form-classes of approximately equal size, are called *categories*. Thus, the English parts of speech (substantive, verb, adjective, and so on) are categories of our language. **1964** Halliday et al. *Ling. Sciences* ii. 23 Grammar deals with closed system choices, which may be between items ('this/that'..) or between categories (singular/plural, past/present/future). *Ibid.* 24 The four theoretical categories that are required if we want to account fully for the kind of patterning in language that we recognize as the level of grammar..class and system..unit and structure. **1965** N. Chomsky *Theory of Syntax* ii. 68 The notion 'Subject'..designates a grammatical function rather than a grammatical category.

catena. Add: **c.** In full *soil catena* (see quots.).

1935 G. Milne in *Soil Research* IV. 194, I propose the word catena (Latin, = a chain). This term will help to indicate that the soils so grouped are linked by their topographic relationship. *Ibid.*, The Uganda soils might be spoken of as the Bukalasa catena. **1954** W. D. Thornbury *Princ. Geomorphology* iv. 78 A soil catena consists of a group of soils within a particular soil region which developed from similar parent material but differ in the characteristics of their profiles because of the varying topographic and drainage conditions under which they formed.

catenist (kăti·nist). [f. Catena + -ist.] A maker of a catena of authorities or evidence.

1880 Swete *Theodore of Mopsuestia on St. Paul's Ep.* I. 240 Theodoret followed his master, without, however, condescending to the level of the mere catenist.

catenoid (kæ·tĭnoid), *a.* and *sb.* [f. Catena + -oid.] **A.** *adj.* Catenary, chain-like. **B.** *sb. Math.* The surface formed by the revolution of a catenary about its axis.

1876 *Encycl. Brit.* V. 68/2 This catenoid..is in stable equilibrium only when the portion considered is such that the tangents to the catenary at its extremities intersect before they reach the directrix. **1901** Calkins *Protozoa* 156 Such pseudoconjugation frequently leads to the formation of catenoid colonies.

cater, *adv.* Now chiefly *U.S.* Add: **cater-cornered** (later examples); also as *adj.* Also **catacornered, catercorner, catty-cornered,** etc., used as *adv.* and *adj.* So **cater-cornering, catty-cornering** *a.* and *sb.*

1838 J. C. Neal *Charcoal Sk.* 196 One of that class..who, when compelled to share their bed with another, lie in that engrossing posture called 'catty-cornered'. **1843** 'R. Carlton' *New Purchase* xxvii. 261 With directions how.. to secure by two strings diagonally fastened, or as he better understood it—'katterkorner'd-like'. **1841** 'Dow, Jr.' *Short Patent Sermons* xxi. 52 Unless you..took too many catty-cornered tracks, Making but little headway with the tacks, While calling on the ladies. **1885** *Century Mag.* Nov. 64/1 He just takes them records..and brings his side line down catercornerin'—that way. **1888** J. Kirkland *McVeys* 59 Now, suppose the railroad runs diagonally across a field, 'cater-cornering', as he says. **1893** *Harper's Mag.* June 112/1 He leaned against the wall, eying me catacornered, but innocent. **1902** S. E. White *Blazed Trail* vii. 53 When the log had been cat-a-cornered from its bed, the chain was fastened [etc.]. **1906** *Harper's Mag.* July 252 'You do leave things so catacornered,' Martha observed. **1919** F. Hurst *Humoresque* 86 The optical department, catercorner from the children's shoes. **1945** J. Steinbeck *Cannery Row* i. 4 If you walk catty-cornered across the grass-grown lot. *Ibid.* v. 16 Lee Chong's grocery was on its catty-corner right and Dora's Bear Flag Restaurant was on its catty-corner left. **1957** *New Yorker* 29 June 22/1 There was a girl sitting down one row, catty-corner from me. **1959** *Listener* 22 Jan. 174/2 A square cat, going cater-cornered among the leeks. **1962** R. Deming *Careful Man* ii. 21 She took a catty-corner shortcut. **1964** *New Yorker* 21 Nov. 222/2 Temple Square..is catercorner to the site of the Kennecott Copper Corporation's new eighteen-story building.

catering (kē*·təriŋ), *ppl. a.* [f. Cater *v.*¹ + -ing².] That caters.

1923 *Glasgow Herald* 21 Mar. 11 A large London catering firm. **1951** *Oxf. Jun. Encycl.* VII. 88/1 The catering industry in this country is specially interesting, for the way in which it reflects social and economic changes.

caterpillar. Add: **1. b.** In full *caterpillar tractor* [*Caterpillar*, proprietary term]: a type of tractor which travels upon two endless steel bands, one on each side of the machine, to facilitate travel over very rough ground. Also *caterpillar lorry, tank, wheel*, etc.; *caterpillar-wheeled* adj.

1908 *Sci. Amer.* 16 May 348/1 The 'Caterpillar' Tractor. For some months past the British military authorities have been experimenting with a new type of tractor for the haulage of heavy vehicles over rough and unstable ground. ..The soldiers at the Aldershot military center, where it is in operation, promptly christened it the 'caterpillar'. **1911** *Official Gaz. U.S. Pat. Off.* 28 Nov. 1079/2 The Holt Manufacturing Company, Stockton, Cal. *Caterpillar*. Gasolene, Steam, and Traction Engines, Harvesters, and Road-Working Machines. **1914** *Illustr. London News* 5 Sept. 369 A 21-centimetre siege-mortar—with 'caterpillar' wheels. **1915** W. S. Churchill *Let.* 5 Jan. in *World Crisis* (1923) II. 74 The caterpillar system would enable trenches to be crossed quite easily. —— *Memo.* 3 Dec. *Ibid.* 87 The Caterpillars will be so close to the enemy's line that they will be immune from his artillery. **1915** *Lit. Digest* 4 Sept. 467/1 Government road-building throughout the interior has paved the way for automobiles, caterpillars and traction-engines. **1915** *Trade Marks Jrnl.* 8 Dec. 1224 Class 6. *Caterpillar*. Tractors and Traction Engines, being Machinery included in Class 6. Caterpillar Tractors, Limited,..Queen Victoria Street, London, E.C. **1922** *Westm. Gaz.* 28 Dec., From In-Salah the caterpillar-wheeled cars will cross the plateau of Tidikelt. **1923** *Contemp. Rev.* Oct. 487 The arrangement of 'caterpillar' traction, with which they were fitted. **1925** Fraser & Gibbons *Soldier & Sailor Words* 50 *Caterpillar*, a familiar name for the tractor fitted with 'girdles', or flat slabs of metal, round their wheels, employed to haul heavy guns and large vehicles on soft ground. **1935** H. G. Wells *Things to Come* v. 33 Long lines of tanks and caterpillar lorries. **1935** *Times* 21 Dec. 9/3 The main body of the Italian troops, with light caterpillar tanks, advanced southwards. **1940** *War Illustr.* 16 Feb. 108 Caterpillar tractors built in the U.S.A. are awaiting shipment to France. **1958** *Listener* 19 June 1004/1 We were shown a huge iron machine on caterpillar tracks.

Also applied to the undercarriage of an aeroplane equipped with a similar device.

1931 *Jrnl. R. Aeronaut. Soc.* XXXV. 492 (*heading*) Safety in flying and caterpillar undercarriages. *Ibid.*, The caterpillar landing wheel has an equivalent wheel diameter ..with small resistance in flight. **1943** in *Jrnl. R. Aeronaut. Soc.* (1944) XLVIII. (Abstr.) 73 (*title*) Caterpillar track landing gear.

6. A member of the *Caterpillar Club* founded by Leslie Leroy Irvin in 1922 (see quots. 1930).

1925 *Literary Digest* 26 Sept. 50 (*title*) Are you eligible for the Caterpillar club? **1930** C. Dixon *Parachuting* vii. 61 Each caterpillar has saved his life with a parachute. **1930** *Engineering* 26 Dec. 811/3 The Caterpillar Club (this being the name given to persons who have saved their lives by the agency of a parachute).

ca·terpillar, *v.* [f. the sb.] To move like a caterpillar or on caterpillar tracks (see prec. *1 b).

1916 *Daily Sketch* 23 Nov. 3/1 Three tanks started...One of the machines moved..caterpillaring its laborious way up the slope. **1928** *Daily Express* 2 July 3 These 'tank' drivers have developed enormous calf muscles, due to caterpillaring over rough country.

So **ca·terpillared** *ppl. a.*, (*a*) (see Dict.); (*b*) fitted with caterpillar tracks.

1608 [in Dict.]. **1917** *Blackw. Mag.* Mar. 379/2 New armoured cars, caterpillared and powerfully armed, would make their bow to Brother Boche.

cat-fish. Add: Also applied to various species of fish in Australia, New Zealand, and Africa.

1834 G. Bennett *Wanderings N.S.W.* I. 343 The '*Cat-fish*', (*Silurus*,) said to have the power of stinging with the tentaculae or feelers, which pend from about the external part of the mouth,..and several species of bream, were caught in..the splendid harbour of Port Jackson. **1851** J. Henderson *Excurs. N.S.W.* II. 207 The Cat-fish, which I have frequently caught in the McLeay, is a large and very ugly animal. **1864** J. T. Baines *Explor. S.W. Africa* i. 3 Beyond was a broad flat, covered with cat, dog and other mud-frequenting fish. **1871** *Cape Monthly Mag.* II. 135 A most horrible creature called a 'catfish', but which ought more properly to have been named 'a sea devil',..as it was all arms and legs, and huge goggle-eyed head. **1897** Parker & Haswell *Text-bk. Zool.* II. 212 The Cat-fishes (*Siluridæ*). **1900** H. A. Bryden *Anim. Africa* 196 When..African rivers dry up, there is nothing left for the Cat-fish but to burrow in the mud. **1947** K. H. Barnard *Pict. Guide S. Afr. Fishes* iii. 59 The best known South African river Cat-fish is the Mud-Barbel or Platkop Barger (*Clarias mossambicus*). **1963** P. H. Greenwood *Norman's Hist. Fishes* (ed. 2) xvi. 284 The only fishes of the order Ostariophysi in the Australian region are certain freshwater genera and species of Cat-fishes. These belong to two families (Ariidae; Plotosidae)...The characteristic African families of..Cat-fishes (Clariidae, Mochocidae, Amphiliidae, etc.).

ca·t-foot, *v.* Chiefly *U.S.* [Cat *sb.*¹ + Foot *sb.*] *intr.* To walk stealthily or noiselessly. Also with *it*.

1916 H. L. Wilson *Somewhere in Red Gap* iii. 119 Mebbe it's a Blackhander's camp, I think; so I didn't yell any more. I cat-footed. And in a minute I was up close. **1928** *Collier's Mag.* 10 Nov. 42/2 Tichenor arose and nonchalantly cat-footed down the field for a deceitful touchdown. **1960** 'R. Simons' *Frame for Murder* viii. 103 Crow began to cat foot it up the stairs.

Cathar (kæ·þăɹ). Also **-are.** Pl. **Cathars,** ‖**Cathari.** [ad. med.L. *Cathari*: see Catharan.] = Catharan. Also used *attrib.* or as *adj.*

1637 [see Catharan]. **1806** M. Pembridge *Roman Cath. Ch. Vind.* III. 560 There was another sect, of *Cathari*, or Catharists, so named from a Greek word, signifying a cleansing or purging, from a certain execrable manner of cleansing from their execrable uncleanness, that they made use of. **1907** *Cath. Encycl.* I. 557/1 The Cathares and the Patarines. **1910** *Encycl. Brit.* V. 515/2 Cathars (Cathari or Catharists). **1920** H. G. Wells *Outl. Hist.* II. xxxiii. 465/2 In the south of France the people..were called the Cathars or Albigenses. **1927** F. J. E. Raby *Hist. Chr.-Lat. Poetry* xiii. 416 Based like the Cathar and Waldensian [religions], on poverty and renunciation. **1947** S. Runciman *Med. Manichee* vi. 125 The Dualism of the Cathars. *Ibid.* 126 In England Cathar heretics were found as late as 1210.

catharsis. Add: Also **katharsis.**

b. The purification of the emotions by vicarious experience, esp. through the drama (in reference to Aristotle's *Poetics* 6). Also more widely.

[**1867** J. A. Symonds *Let.* 22 Aug. (1967) I. 751 The world desiderates now..a trilogy, whereof the whole third part shall exhibit 'the height, the space, the gloom, the glory', of ultimate final and perfect κάθαρσις.] **1872** G. S. Morris tr. *Ueberweg's Hist. Phil.* I. 1. 179 Aristotle can not have meant..to exclude from among the effect of the Tragedy, its effect as..ethical discipline. With the 'Catharsis'.. are..joined..the other effects of the same,—the latter effects flow from the 'Catharsis'. **1897** Costelloe & Muir tr. *Zeller's Philos. Greeks* II. xv. 311/2 According to Aristotle there is a kind of music which produces a catharsis, although it possesses no ethical value..—namely, exciting music. **1904** Dowden *Browning* 289 Balaustion, stricken at heart, yet feels that this tragedy of Athens brings the tragic katharsis. **1920** D. H. Lawrence *Touch & Go* III. i. 72 It's a cleansing process—like Aristotle's Katharsis. We shall hate ourselves clean at last, I suppose. **1924** L. Cooper *Aristotelian Theory Com.* 180 Aristotle..would recognize some sort of catharsis, and the resultant pleasure, to be the proper end of comedy. **1924** Selbie *Psychol. Relig.* 159 There may..be cases where experiences of this kind produce a moral catharsis which has good results. **1959** *Chambers's Encycl.* I. 592/1 The word *catharsis* (purgation), in which he [*sc.* Aristotle] summed up the emotional effect of tragedy, has also received much fanciful interpretation; in reality it is a medical term, with no directly moral or spiritual implications.

c. *Psychotherapy.* The process of relieving an abnormal excitement by re-establishing the association of the emotion with the memory or idea of the event which was the first cause of it, and of eliminating it by abreaction.

1909 A. A. Brill in *Freud's Sel. Papers Hysteria* 6 The German abreagiren..has different shades of meaning, from defense reaction to emotional catharsis. **1951** J. C. Flugel *Hundred Years Psychol.* (ed. 2) viii. 280 The mere bringing back and discussing of memories..which Freud and Breuer called subsequently 'abreaction' or 'catharsis'.

ca·t-haul, *sb.* *U.S.* [Cat *sb.*¹ 17.] (See quot.) Also **ca·t-haul** *v. trans.*, to subject to this punishment; *fig.* to examine stringently; hence **cat-hauling** *vbl. sb.*

1824 'A. Singleton' *Lett. fr. South & West* 79 The cat-haul; that is, to fasten a slave down flatwise..and then to take a huge fierce tom-cat by the tail backward, and haul him down along the..bare back, with his claws clinging into the quick all the way. **1840** *Congress. Globe* 12 Jan., App. 99/2 White people of the South..hunting slaves with dogs and guns,—cat-hauling slaves, &c. **1844–7** [see Cat *sb.*¹ 17]. **1881** *Congress. Rec.* 28 Feb. 2202/2 You begin to ransack and examine and cat-haul the whole navy, big and little. **1950** R. Starnes *And when she was Bad* xvii. 109 Brafferton's cathauling by the Harford Committee commanded a banner headline.

Cathayan (kăþē*·ăn), *a.* and *sb.* Also **-aian.** [f. med.L. *Cat(h)aya* + -an. Cf. Cataian.] Chinese.

1667 Milton *P.L.* x. 293 Mountains of Ice, that stop th'imagin'd way Beyond Petsora Eastward, to the rich Cathaian Coast. *Ibid.* xi. 388 Cambalu, seat of Cathaian Can. **1757** J. Dyer *Fleece* in *Poems* (1761) III. 140 A double wealth; more rich than Belgium's boast, Or the Cathayan's, whose ignobler care Nurses the silkworm. **1876** *Encycl. Brit.* V. 628/1 The identity of these Cathayans with the Seres of classic fame. **1928** *Blackw. Mag.* Jan. 1/1 To push back a Cathayan encroachment. **1928** H. Lamb *Genghis Khan* xiii. 126 The Cathayans must have suffered that winter.

cat-head, *sb.* Add: **3. c.** 'An attachment to a lathe to assist in supporting long bars when they are being turned' (*Cent. Dict.* Suppl. 1909).

1940 *Chambers's Techn. Dict.* 141/2 *Cathead* or *spider*, a lathe accessory consisting of a turned sleeve.

4. *attrib.*, as *cat-head stopper* (see small-type note in Dict.).

1829 *Patents* in *Ann. Reg.* 551/2 Improvements in the construction of cat-head stoppers. **1883** *Man. Seamanship for Boys' Training Ships* 10 For lifting the anchor from the water's edge to the cat head in order to pass the cat head stopper.

cat-head, *v.* (Earlier U.S. example.)

1840 R. H. Dana *Bef. Mast* xxv, Everything was sheeted home and hoisted up, the anchor tripped and cat-headed.

cathect (kăþe·kt), *v.* *Psycho-analysis.* [Back-formation f. *cathectic *a.*] To charge with mental energy; to give (ideas, etc.) an emotional loading.

1936 A. Strachey tr. *Freud's Inhibitions* xi. 145 A repressed instinctual impulse can be activated (newly cathected) from two directions. **1951** C. Kluckhohn et al. in Parsons & Shils *Toward Gen. Theory of Action* IV. ii. 399 Disvalued activities are cathected.

cathectic (kăþe·ktik), *a.* *Psychol.* [ad. Gr. καθεκτικός capable of holding.] Of or relating to cathexis.

1927 J. Riviere tr. *Freud's Ego & Id* iv. 63 We know this trait; it is characteristic of the cathectic processes in the id. **1936** A. Strachey tr. *Freud's Inhibitions* v. 58 The commonest symptoms of conversion hysteria..are cathectic processes. **1951** C. Kluckhohn et al. in Parsons & Shils *Toward Gen. Theory of Action* IV. ii. 394 Values synthesize cognitive and cathectic elements in orientations to an object world.

cathedral, *sb.* Add: **3. cathedral glass,** coloured glass leaded after the fashion of the stained windows of churches, used (e.g.) in the panels of the vestibule doors of houses.

1850 *Archaeol. Inst. Gt. Brit.: Mem. Lincoln 1848* 122 Many modern windows in which stain is used, especially those composed of the yellow tinted 'Cathedral glass', appear at a little distance as if they were wholly yellow. **1885** *Spon's Mechanic's Own Bk.* 630 'Roundels' and 'bullions' are small discs of glass..used in fretwork with cathedral glass. **1905** H. A. Evans *Highways & Byways Oxf. & Cotswolds* 329 The exquisite pale green transparent glass of the windows,..displaced to make room for the vulgar abomination known as 'cathedral glass'. **1960** Willmott & Young *Family & Class* i. 11 Stained 'Cathedral' glass..is used for the top halves of doors.

cathepsin (kăþe·psin). *Biochem.* [ad. G. *kathepsin* (Willstätter & Bamann 1929, in *Zeitschr. f. physiol. Chem.* CLXXX. 130), f. Gr. καθέψ-ειν to boil down (ἕψειν to boil) + -in.¹] Any of various proteolytic enzymes, present in most animal tissues, which aid in autolysis of cells after death or in diseased cells.

1929 *Chem. Abstr.* 1653 The enzymes of leucocytes... For the proteinase which is active at slightly acid reaction the name cathepsin is proposed. **1963** *Lancet* 19 Jan. 153/1 In..the concentration of enzymes within the diseased muscle-cell..the acid phosphatases and cathepsins, both acid and alkaline, were increased.

catheterize, *v.* Add: *trans.* To employ a catheter on (a patient); to introduce a catheter into.

1887 R. Harrison *Surg. Disorders Urinary Organs* (ed. 3) v. 60 If you are only accustomed to catheterise patients standing with their backs against a wall how can you expect to be dexterous with the lithotrite. **1908** *Practitioner* Jan. 44 A true stricture,..forming an irregular channel, through which urine passes with difficulty, and which it is almost impossible to catheterise. **1954** M. Campbell *Urology* II. xiii. 1472 Having catheterized the ureters, the specimens are collected. **1967** S. Beckett *No's Knife* 83, I catheterize myself, unaided, with trembling hand.

cathexis (kăþe·ksis). *Psychol.* [a. Gr. κάθεξις holding, retention; intended as a rendering of G. (*libido*)*besetzung* (Freud).] The concentration or accumulation of mental energy in a particular channel.

1922 J. Strachey tr. *Freud's Group Psychol.* iii. 48 Dread in an individual is provoked either by the greatness of a danger or by the cessation of emotional ties (libidinal cathexes [orig. *Libidobesetzungen*]). **1923** E. Jones in *Internat. Jrnl. Psycho-Analysis* IV. 299 In hetero-suggestion ..a hypercathexis of the idea of the operator is correlated with a hypo-cathexis of all ideas in conflict with his. **1948** *Brit. Jrnl. Psychol.* XXXVIII. 172 There are a number of very specific factors responsible for any particular process of 'cathexis'. **1951** C. Kluckhohn et al. in Parsons & Shils *Toward Gen. Theory of Action* IV. ii. 398 Since value always involves affect, cathexis and value are inevitably somehow interrelated.

‖**cathisma** (kăþi·zmă). Pl. **cathismata.** [a. Gr. κάθισμα seat, f. κατά down + ἵζειν to sit.] In the Greek Church: a portion of the psalter, containing from three to eleven psalms. Also, a short hymn used as a response.

1850 J. M. Neale *Holy Eastern Ch.* I. II. 844 The Greeks rarely sit in church: the cathismata are therefore pauses for rest; and are longer than the usual troparia. **1880** *Encycl. Brit.* XII. 580/1 In various parts of the services solitary troparia are sung, under various names, 'contacion', 'œcos', 'cathisma', &c.

cathode. Add: **c.** *attrib.* and *Comb.* **cathode dark space** = *Crookes (*dark*) *space*; **cathode follower,** a type of electric circuit (see quots.); also *attrib.*; **cathode ray,** a beam of electrons issuing from the cathode of a high-vacuum tube under the action of an electric field; hence **cathode-ray oscillograph**: see *oscillograph and quot. 1922; **cathode-ray oscilloscope** (see quot.); **cathode-ray tube,** a vacuum tube in which cathode rays are projected upon a fluorescent screen.

1914 *R. Soc. Catal. Sci. Papers 1800–1900, Subject Index* III. II. 685/2 Cathode dark space, origin. **1920** *Discovery* July 217/1 The dark space around the negative pole, which has since been referred to as the Crookes or Cathode Dark Space. **1939** *Electronics* Mar. 35 Connections of the cathode-coupled ('cathode-follower') stage, used for low-impedance output applications. **1941** *Wireless World* July 176/1 In essence, the cathode-follower stage is a device to avoid mismatching. *Ibid.* 177/1 The cathode-follower gives an output voltage equal to its input voltage (the cathode potential *follows* the grid potential—hence the name 'cathode-follower'). **1942** *Electronic Engin.* XV. 287 The 'cathode follower' type of circuit has the distinction of being one of the most versatile circuits used by the Radio Engineer.. and derives its name from the fact that an input signal that makes the grid go positive with respect to HT. negative will also make the cathode go positive with respect to HT. negative. **1905** *Jrnl. Chem. Soc.* LXXXVIII. II. 224 With copper, nickel, gold, and bright platinum, the cathode potential for a given current density goes on increasing for hours. *Ibid.*, Cathode process and anode liquid are opposed to each other. [**1879** *Phil. Trans. R. Soc.* CLXX. 144 It excludes such rays of the cathode as impinge upon it from reaching the side.] **1880** *Phil. Mag.* X. 411 It was further examined whether the negative discharges which excite the green light, the so-called kathode-rays, do really propagate themselves only in straight lines. **1898** [see *corpuscle 2 c]. **1900** *Jrnl. Chem. Soc.* LXXVIII. II. 587 Electrical Conductivity in Gases traversed by Cathode Rays. **1912** J. W. Mellor *Mod. Inorg. Chem.* xliii. 825 The cathode rays is [*sic*] a stream of negatively charged electrons sent from the cathode with a high velocity. **1951** J. R. Partington *Gen. & Inorg. Chem.* (ed. 2) viii. 184 These cathode rays were discovered in 1859 by Plücker, who showed that they are easily deflected by a magnet. **1957** *Encycl. Brit.* V. 41/1 Cathode rays have many applications, one of the chief being the excitation of X-rays by the impinging of swift electrons against a hard anticathode. **1922** Glazebrook *Dict. Appl. Physics* II. 146/2 *Cathode ray oscillograph*, an apparatus for delineating the instantaneous values of the current or voltage in a circuit by the deflection of a fine cathode stream. **1937** *Discovery* Jan. 4/1 A cathode ray oscillograph..is an instrument in which a fast beam of electrons passes between deflecting plates and impinges on a screen, where it produces a spot of light. **1951** Uvarov & Chapman *Dict. Science* (ed. 2) 44 *Cathode ray oscilloscope*, apparatus consisting essentially of an 'electron gun' producing a beam of electrons which passes through horizontal and vertical deflecting plates, to fall upon a fluorescent screen. **1905** Varley & Murdoch in *Electrician* 16 June 335/1 (*title*) Some Applications of the Braun Cathode-Ray Tube. **1934** *Discovery* Apr. 106/1 The advances made in the use of short waves and the introduction of cathode-ray tubes have enabled us to transmit images on screens 3 or 4 feet square. **1957** Amos & Birkinshaw *Television Engin.* I. ix. 210 In the majority of television receivers the picture is produced on the face of a cathode-ray tube which is viewed directly or by reflection at a plane mirror.

cathodic, *a.* Add: **3.** *Electr.* Of or pertaining to a cathode.

1837 [see *anodic *a.* 1 a]. **1896** *Daily News* 18 Jan. 5/4 Taking photographs with cathodic rays. **1951** *Engineering* 13 Apr. 443/1 [In corrosion] If the water contains a salt.. the products formed..at the cathodic (unattacked) part.. are freely soluble.

b. *cathodic protection*: protection of an underground or underwater structure from corrosion by a technique that causes the structure to act as the cathode of an electrolytic cell.

1931 *U.S. Bur. Standards Techn. News Bull.* May 52/1 A number of soils..are severely corrosive with respect to the commonly used pipe materials. Four methods of preventing corrosion in these soils have been considered: (1) the surrounding of the pipes by noncorrosive soils transported from other regions; (2) electrical or cathodic protection; [etc.]. **1964** *Electronics Weekly* 18 Nov. 24/3 The cathodic protection equipment..has been installed in a number of naval vessels. Designed to protect the ship's hull against corrosion..the system..sets up a current flow between a set of anodes and the ship's hull, which acts as the cathode.

cathodo- (kæ·þŏdo), used as combining form of Cathode.

1909 *Physical Rev.* XXVIII. 349 With regard to the dependence of kathodo-luminescence upon discharge potential and current strength. **1914** *Ibid.* 2nd Ser. IV. 21 The brightness of the crests of the red and blue bands of the white and pink kunzite under kathodo-excitation. **1923** *Amer. Jrnl. Sci.* Oct. 373 The minimum energy required to excite cathodo-luminescence.

cathodograph (kăþǫ·dograf). Also **cathode-graph, kathodograph.** [f. *cathodo- + -graph.] A photograph of normally invisible objects taken by means of cathode rays, an X-ray photograph. (*Disused.*) Hence **catho·dograph** *v. trans.*, to take an X-ray photograph of; **cathodography** (kæþodǫ·grăfi), photography by cathode rays.

1896 *Century Mag.* May 120/1 No school or college has considered the day well spent in which..it has not taken 'cathodographs' of hands and coins. *Ibid.* 121/1 Key cathodographed through a book of 526 pages. *Ibid.* 126/2 A cathodograph picture. *Ibid.*, Cathodography of corked bottles would uncork them. **1897** *Daily News* 2 Feb. 5/2 Its transparent honesty was recognised by the 'cathode-graph' showing only the setting, the diamond being represented by a white space.

Catholic. Add: **A.** *adj.* **11.** *Comb.*, as *Catholic-minded* adj.

1879 *Dublin Rev.* Jan. 95 The learned, irresolute, yet pious and Catholic-minded men at the head of whom was Fisher's friend, Cuthbert Tunstal. **1964** P. F. Anson *Bishops at Large* xi. 534 An alternative to Roman Catholicism to a catholic-minded people.

B. *sb.* **2.** Examples of *English Catholic* = English Roman Catholic.

1584 in Foley *Rec. Eng. Prov. S.J.* (1880) VI. 740 He said..that all English Catholics were bound to pray for the King of Spain. **1641** J. Louth in A. H. Mathew *Convers. Sir T. Matthew* (1904) 176 The innocency and loyalty of English Catholics towards others. **1715** in Estcourt & Payne *Eng. Cath. Nonjurors of 1715* (1885) 8, I, Henry Englefield, do declare that I am, by the grace of God, an English Catholic. **1800** C. Butler *Life Alban Butler* xvi, A person would deserve well of the English Catholics who should translate it into English. **1876** [in Dict.] **1889** J. O. Payne (*title*) Records of the English Catholics of 1715.

b. *old Catholic*, a term introduced after the secession of John Henry Newman and others to distinguish members of Catholic families in England since the Reformation from Catholic immigrants and converts.

1846 J. H. Newman *Let.* 14 July in Gasquet *Ld. Acton* (1906) p. xiii, It will be one of your collisions with old Catholics. **1909** *Dublin Rev.* Jan. 56 The friction between converts and old Catholics..was inevitable. **1918** L. Strachey *Emin. Victorians* I. v. 56 It seemed as if the harvest was to be gathered in by a crowd of converts, who were proclaiming on every side as something new and wonderful the truths which the Old Catholics..had not only known, but for which they had suffered, for generations. **1962** V. A. McClelland *Cardinal Manning* i. 3 If one is to understand the opposition to Cardinal Manning and to the Oxford converts, one has to appreciate the feelings and position of the 'Old Catholics'.

3. b. *German Catholic* (examples); *Old Catholic* (later examples); also applied to members of other churches separated from Rome, and united by acceptance of the Declaration of Utrecht of 1889.

1871 *Union Rev.* 273 For German Catholics to succumb to the Vatican decrees, would be an act of moral suicide. **1931** W. Temple *Thoughts on Probl. of Day* iv. 92 The Conference..was concerned with advances towards union in two directions—on the one hand towards union with the Orthodox and the Old Catholic Churches, and on the other hand with the non-episcopal Churches. **1948** C. B. Moss *Old Catholic Movement* i. 1 The Old Catholic Churches are a group of self-governing national churches, united by their acceptance of the Declaration of Utrecht (1889) as their dogmatic basis. *Ibid.* xxviii. 348 Eight Dutch Old Catholic priests came to England to see the English Church for themselves. **1969** D. W. D. Shaw tr. *Heyer's Catholic Church* vi. 149 Political factors had produced an initial wave of interest in the 'German Catholic' movement.

catholicization (kăþǫ·lisaizē*·ʃən). [f. Catholicize *v.* + -ation.] The action of making Catholic.

1905 *Nation* (U.S.) 28 Dec. 522 The catholicization of France by Clovis. **1945** R. Hargreaves *Enemy at Gate* 309 Italy's conquest of Abyssinia was the occasion of the peremptory Catholicisation of the old Coptic Church of Ethiopia.

catholyte (kæ·þŏlait). [f. Cath(ode + Electr)olyte.] That part of the electrolyte which adjoins the cathode.

1890–1949 [see *anolyte].

Catilinarian, *sb.* and *a.* (Earlier and later examples.)

1824 J. H. Newman *Cicero* in *Hist. Sketches* (1872) II. 291 The eloquence of his Catilinarians and Philippics. *Ibid.* 292 The Catilinarian conspirators. **1889** Skrine *Mem. Thring* 155 The senate which quelled the Catilinarians met in the Temple of Concord. **1925** *Glasgow Herald* 16 Apr. 9 He is still a sort of Catilinarian doomed to reckless experiments.

cation. Add modern pronunc. (kæ·taiǫn) and substitute for def.: An ion carrying a positive charge which moves towards the cathode (negative electrode) during electrolysis. Opp. *anion*. (Add example.) **cation exchange,** ion exchange involving cations; also *attrib.*; so **cation exchanger,** a substance capable of cation exchange. **cation-active** *a.*, having an active cation.

1834, 1931 [see *anion]. **1946** *Nature* 26 Oct. 585/2 The addition of divalent metallic ions greatly enhances the surface-active properties of anion-active detergents, and it was suggested..that a similar effect should occur on adding divalent anions to cation-active substances. **1931** *Soil Science* XXXI. 320 If the cation exchange capacity depends upon the number of free silicate or other acidoid valences, [etc.]. **1950** *Amer. Jrnl. Med. Sci.* CCXX. 547 (*title*) Treatment of Potassium Retention in Anuria with Cation Exchange Resins. **1957** *Encycl. Brit.* XII. 573/2 In water softening..calcium and magnesium ions..are removed by a granular bed of cation exchanger and replaced by sodium ions from the cation exchanger. **1963** [see *anion].

cationic (kætai,ǫ·nik), *a.* [f. Cation + -ic.] **a.** Of, relating to, or consisting of cations. **b.** Characterized by an active cation.

1928 *Jrnl. Chem. Soc.* I. 905 The former may repel electrons..but it cannot without disruption provide a seat for the cationic charge. **1946** *Nature* 3 Aug. 155/1 Following the relatively rapid achievement of cationic equilibrium between perfusate and soil, manganese usually disappears from the perfusate. **1957** *Textile Terms & Defs.* (ed. 3) 23 *Cationic Dye*, a dye which dissociates in aqueous solution to give a coloured ion which is positively charged. **1964** M. Hynes *Med. Bacteriol.* (ed. 8) ii. 22 Cationic detergents are more bactericidal than the anionic types used for domestic cleaning.

cationoid (kæ·taiǒnoid, kætai·ǒnoid), *a.* Also **kationoid.** [f. Cation + -oid.] Resembling a cation.

1925–1953 [see *anionoid *a.*].

catnip. Add: **b.** *fig.*

1934 A. Woollcott *While Rome Burns* 183 It was so much catnip to 'that ill-looking, squinting man'. **1962** *Observer* 25 Mar. 24/3 She is catnip to failures.

catocalid (kætǒkē*·lid), *a.* and *sb.* *Ent.* [f. mod.L. *Catocala*, f. Gr. κάτω below + καλός beautiful: see -id³.] **A.** *adj.* Belonging to the genus *Catocala* of noctuid moths. **B.** *sb.* A moth of this genus.

1896 J. W. Tutt *Brit. Moths* x. 224 It is possible that the Geometers..went a short way with the Catocalids. *Ibid.* 225 We have only four species of Catocalid Moths inhabiting Britain, one of which has blue-banded hindwings, the others red. **1913** *Oxf. Univ. Gaz.* 4 June 960/2 A fine set of 28 Catocalid moths. **1963** V. Nabokov *Gift* ii. 95 Your blue stripe, Catocalid, shows from under its grey lid.

catostomid (kătǫ·stŏmid), *sb.* and *a.* [ad. mod.L. *Catostomidæ*, f. *Catostomus*, the typical genus, f. Gr. κάτω down + στόμα mouth.] **A.** *sb.* A freshwater fish of the family Catostomidæ, found chiefly in North America, and including the suckers and buffalo fishes. **B.** *adj.* Of or pertaining to this family.

1889 in *Cent. Dict.* **1908** *Smithsonian Misc. Coll.* V. 97 This character suggests a cyprinid or a Catostomid. **1957** P. J. Darlington *Zoogeography* i. 32 This is consistent with its [*sc.* *Myxocyprinus*] being a relict of an earlier Chinese catostomid fauna...The closest relatives of the catostomids are the cyprinids, the carps and their allies.

catpiece (kæ·t,pīs). *U.S.* [f. Cat *sb.*¹ + Piece *sb.*] In logging, a perforated bar of wood fitting on the tops of uprights in floating booms, so that the space between the booms may be narrowed when necessary.

1905 in *Terms Forestry & Logging* 33.

cat's-eye. Add: **2.** Also *attrib.*

1859 [in Dict.]. **1874** *Jrnl. Chem. Soc.* XXVII. 555 The so-called Cat's-eye and Fibrous Quartz. **1899** T. M. Ellis (*title*) The three Cat's-eye Rings. **1903** *Daily Mail* 8 Sept. 7/4 A..narrow edging of cat's-eye green.

4. The calcareous operculum of turban shells, esp. the south Pacific *Turbo petholatus*.

1898 Morris *Austral Eng.* 83/1 *Cat's Eyes*..the name given in Australia to the opercula of *Turbo smaragdus*. **1949** E. de Mauny *Huntsman in Career* ii. 70 He collected a pocketful of cats' eyes, and sat on a shelf of rock to watch the sea. **1962** Abbott & Zim *Sea Shells of World* 32 Its [*sc.* *Turbo petholatus*] operculum is the famous blue-green 'cat's eye'.

5. One of the chain of light-reflecting studs used to demarcate traffic lanes on roads at night. Also *attrib.*

1940 *New Statesman* 20 Jan. 63/1 Few motorists..in Oxfordshire will deny that the 'cats-eyes', with which the County Council has studded its main roads, are an improvement for night driving on the old white line. **1948** *Times* 4 Dec. 3/2 Drivers of vehicles..had reason to challenge the decision of the Ministry of Transport..to dispense with 'cat's eye' reflector studs in roads provided with street lighting. **1958** E. Hyams *Taking it Easy* iv. 200 The endless chaplet of cat's eye reflectors..was conjured out of the road's long darkness.

catskin. Add: **1. b.** *catskin earl*, each of the three senior earls in the House of Lords, viz. the Earls of Shrewsbury, Derby, and Huntingdon.

1869 W. Hook *Lives Abps.* 2nd Ser. III. iv. 264 The Earl of Huntingdon is one of the three catskin earls of the present day,—one of the first three earls in the House of Lords. **1926** G. E. Cokayne *Complete Peerage* VI. 655 The robes of an Earl consist now of three rows of ermine, but in some early representations an Earl is depicted with four rows...It seems..probable that the four rows..may have given the name of *Catskin* (or Quatreskin) to the Earl of ancient creation whose robes were thus described.

cat-stairs. *dial.* and *U.S.* Also **cat's-.** [Cat *sb.*¹ 19.] (See quots.)

1825 Jamieson *Suppl.*, *Cat's-stairs*, a plaything for children, made of thread, small cord, or tape, which is so disposed by the hands as to fall down like steps of a stair, Dumfr., Gall. **1828** Carr *Dial. Craven* (ed. 2) 64 *Cat-stairs*, Tape, &c. so twisted, that by its alternate hollows and projections, it resembles stairs. **1886** *Leslie's Pop. Monthly* XXI. 150/1 'Twere vain to tell what sylvan treasures accumulated in our back yard that summer. The flights of catstairs [etc.].

cat-tail: see Cat's tail.

cattalo, var. *catalo.

catted (kæ·tĕd), *ppl. a.* See Cat *v.* 1. So **ca·tting** *vbl. sb.*

1867 Smyth *Sailor's Word-bk.*, *Catting*, the act of heaving the anchor by the cat-tackle. **1883** *Man. Seamanship for Boys' Training Ships* 210 When stowing the anchor for sea after the second catting. **1901** *Daily Chron.* 12 June 7/2 The 'catted' port anchor of the schooner.

cattily (kæ·tili), *adv.* [f. *catty *a.* + -ly².] In a catty manner.

1924 N. Coward *Rat Trap* III. 59 *Naomi*: I'm sure her Cockney humour is inimitable (cattily). **1946** R. Macaulay *They went to Portugal* 307 Shaw..added, rather cattily, that though the Lancers were in very good order and looked smart, they were seldom in contact with the enemy.

cattimandoo (kætimæ·ndū). Also **cattemundoo, kattimundoo, callemundoo.** [Telugu *kattimandu*, the name of the plant.] A resinous substance obtained from the milky juice of *Euphorbia cattimandoo* or other species of *Euphorbia*; used as a cement and as a medicine for rheumatism. **b.** The plant from which the gum is obtained.

1851 *Catal. Gt. Exhib.* IV. I. 877/2 Cutteemundoo, or Kattimundoo gum. **1858** Simmonds *Dict. Trade*, Cattemundoo, Callemundoo. **1880** *Encycl. Brit.* XI. 339/2 Many of the euphorbias yield milky juices...The chief among these are the cattimandoo..and the Indian spurge tree.

ca·ttiness. [f. *catty *a.* + -ness.] = *cattishness.

1920 *19th Cent.* Nov. 752 Her antagonism to her own sex —known colloquially as 'cattiness'. **1928** *Sunday Express* 6 May 4 Manx folk are wonderfully good-natured,..and cattiness..is practically non-existent. **1951** R. Knox *Stimuli* III. v. 98 It is not an impossible Lenten resolution to give up..the delights of saying 'I told you so': to cut from our conversation..the spirit which I think St Paul.. would probably have labelled 'cattiness'.

cattish. Add: **b.** Of a sly or spiteful character. (Cf. Cat *sb.*¹ 2.)

1883 W. Black *Yolande* III. vii. 127 The cattish temper of an old woman. **1893** E. E. Saltus *Madam Sapphira* 15 The stealthy air and cattish smile of the mother.

Hence **ca·ttishly** *adv.*, in a cattish manner or spirit; **ca·ttishness,** the quality of being 'cattish', spitefulness.

1894 Meredith *Ld. Ormont* III. iv. 69 Thither he walked.. prepared for cattishness. **1907** *Daily Chron.* 8 July 3/2 Thus ..was Mrs. Steel's cattishness brought home to her. **1909** *Westm. Gaz.* 8 June 3/2 Elizabeth, we are afraid, is developing into a little 'cat', as her female friends would probably say...That is where the 'cattishness' of Elizabeth shows itself. **1913** W. Owen *Let.* 7 Nov. (1967) 207, I drank milk so doggedly, (or cattishly) even with raw egg in it, that strength has returned. **1923** M. Arlen *These Charming People* 180 'Society Hostess Robbed.' It's almost worth it for her, he thought cattishly, to be called a Society Hostess. **1938** R. G. Collingwood *Princ. Art* v. 87 The biographies of cattishness, whose aim is to release the reader from the irksome reverence..for persons who were important in their day. **1967** J. Gardner *Madrigal* iii. 58 The sergeant turned out to be cattishly agile.

cattle. Add: **8. a.** *cattle-maiming, -raiding, -raiser, -raising, -rustler, -rustling, -thief.*

1907 *Westm. Gaz.* 6 Sept. 5/1 The renewed outbreak of cattle-maiming in this parish [*sc.* Great Wyrley]. **1965** K. H. Connell in *Pop. in Hist.* xvii. 433 Arson and murder, the boycott and cattle-maiming were some of their weapons. **1899** *Daily News* 13 Nov. 7/4 The real object of this cattle-raiding expedition. **1853** 'P. Paxton' *Yankee in Texas* 122 He lived on the frontier amid the Ingens, and cattle-raisers. **1896** *Daily News* 16 Jan. 5/6 All the victims were well-known cattle-raisers. **1878** I. L. Bird *Lady's Life in Rocky Mts.* (1879) x. 170 Perry's Park is one of the great cattle-raising ranches in Colorado. **1883** *Athenæum* 2 June 693 In Galicia cattle-raising is rapidly superseding tillage. **1923** *Daily Mail* 15 Feb. 8 A great crisis has fallen upon the cattle-raising industry of this Republic. **1953** E. Smith *Guide to Eng. Trad.* 3 Some livestock farmers..specialize in sheep-rearing rather than in cattle-raising. **1903** A. Adams *Log Cowboy* vii. 86 The stampede..was the work of cattle rustlers. **1907** S. E. White *Arizona Nights* I. iii. 60 We ..saw the beginning of the cattle rustling. **1960** *Farmer & Stockbreeder* 15 Mar. 75/1 The alleged case of cattle-rustling. **1862** T. E. DeVoe *Market Bk.* I. 172 A foraging party, under..the city's former governor..extensively known as a 'Cattle Thief'. **1903** A. Adams *Log Cowboy* vii. 101 The biggest cattle-thief ever born in Medinah County.

b. *cattle-camp* (*camp *sb.*² 4 c), *-country, -kraal, -path, -show* (examples), *-track, -yard*; *cattle-boat, -car, -ship*; **c.** *cattle-proof* adj.

1860 Sala *Make your Game* 14 Not a cattle-boat luckily, though, in some pens forward, there were a few sheep. **1889** C. Edwardes *Sardinia* 375 This Black Hole of a cattle-boat. **1945** Wyndham Lewis *Let.* 13 Mar. (1963) 381 There still is no alternative: a cattle-boat..or stop here. **1900** H. Lawson *Verses Pop. & Humorous* 219 The old bark-school..Is a cattle-camp in winter. **1931** V. Palmer *Separate Lives* 122 Its wild horns and glossy red coat had become a familiar figure on the cattle-camp when he mustered to cut out the half-yearly mob of fats. **1864** C. H. Cooke *Let.* 4 May in *Wisconsin Mag. Hist.* (1921) V. 65 We took the train for Chattanooga. Our cars were cattle cars. **1951** M. McLuhan *Mech. Bride* 23/1 Why doesn't somebody write of a last-minute gamble for happiness in a cattle car headed for Buchenwald? **1886** T. Roosevelt in *Cent. Mag.* July 340/1 The cattle country of western Dakota. **1943** *Collier's* 28 Aug. 11/1 This stirring tale of the cattle country. **1897** J. Bryce *Impressions S. Afr.* ix. 86 All round on the lower ground are large inclosures rudely built of rough stones, and probably intended for cattle-kraals. **1932** C. Fuller *L. Trigardt's Trek* xi. 142 When this area was first settled by Europeans, old cattle-kraal sites..were found in the fly country. **1838** Hawthorne *Amer. Note-Bks.* 9 Sept. (1868) I. 257 Followed a cattle-path till I came to a cottage. **1887** *Outing* (U.S.) May 117/2 The bank was worn away on the other side by a cattle-path just wide enough for one. **1882** Armstrong & Campbell *Austral. Sheep Husbandry* xvii. 187 This fence possesses the advantage over the ordinary wire fence of being cattle-proof, or nearly so. **1908** *Daily Chron.* 17 Sept. 7/2 A twelve-foot barrier of cattle-proof wire. **1915** *N.Z. Jrnl. Agric.* 20 Feb. 190 Can you advise me as to the best hedge to grow..to make a good cattle-proof fence? **1630** J. Winthrop *Hist. New Eng.* (1908) I. 29 Mr. Weatherell, whose father was master of one of the cattle ships. **1891** *Scribner's Mag.* X. 610 The loading of cattle-ships. **1815** *N. Amer. Rev.* II. 136 The Cattle show..at Pittsfield. **1844** *Ainsworth's Mag.* VI. 534 Farmers,..who had been in town enjoying the spectacle of the 'cattle-show'. **1877**

HARRIS' *Settlers & Convicts* 294 A Cattle-racket. The term at the head of this chapter was originally applied in New South Wales to the agitation of society which took place when some wholesale system of plunder in cattle was brought to light. It is now commonly applied to any circumstance of this sort, whether greater or less, and whether really springing from a felonious intent or accidental. **1857** OLMSTED *Journ. Texas* 160 Some live upon the produce of farms and cattle-ranches owned in the neighborhood. **1879** BODDAM-WHETHAM *Roraima* 114 Two boatmen who once rowed us over..to visit a cattle ranch, were both generals. **1946** *Illinois State Arch. Soc. Jrnl.* July 32/1 She owns and operates a large cattle ranch. **1866** 'MARK TWAIN' *Lett. fr. Hawaii* (1967) 289 The whole country is given up to cattle ranching. **1888** T. ROOSEVELT in *Century Mag.* Feb. 500 Cattle-ranching can only be carried on in its present form while the population is scanty. **1640** in *Essex Inst. Hist. Coll.* V. 170I Ordered that none of the land within the cattle range shall be granted..to any man. **1835** C. F. HOFFMAN *Winter West* II. 130 We entered at once upon a large and beautiful park or chase (*Note*, called a cattle-range, if I mistake not, in Kentucky). **1948** *Southwestern Rev.* Summer 272 They..had passed across a forbidden cattle range. **1905** *Spectator* 18 Feb. 248/1 The important feature which the cattle-roads make even to-day in modern embankments. **1853** C. B. HALL *Let.* 6 Sept. in T. F. Bride *Lett. Vict. Pioneers* (1898) 218 Various cows and bullocks, on a cattle run. **1944** W. E. HARNEY *Taboo* (ed. 3) 86 Roy Johnstone was boss of a cattle-run. **1903** *Pall Mall Gaz.* 7 July 7/1 Cattle-sickness is alarmingly on the increase in Rhodesia. **1950** *N.Z. Jrnl. Agric.* July 74/1 It was in this period that cobalt was introduced as a cure for cattle sickness. **1851** *Lyttleton Times* (N.Z.) 1 Mar. 8/3 A Canterbury colonist..would undertake the entire management of a cattle Station or Farm. **1949** D. M. DAVIN *Roads fr. Home* 250 John worked his way round the cattlestops by the..glare of the engine's fire-box. **1953** M. C. SCOTT *Breakfast at Six* i. 12 Presently we came to a cattle-stop...Paul bumped noisily across. **1869** *Amer. Naturalist* III. 51 The Cattle Tick[s]..drop from the cattle..along the cattle paths. **1950** *N.Z. Jrnl. Agric.* Sept. 196/1 Where cattle ticks are plentiful spraying or dipping should be carried out. **1905** HUBBARD *Neolithic Dew-Ponds & Cattle-Ways* 38 The length of the ascending cattle-way is a quarter of a mile or more.

cattleya (kæ·tliă). *Bot.* [mod.L., f. name of William *Cattley*, an English patron of botany: see -IA.] An epiphytal plant belonging to the orchidaceous genus *Cattleya*, native to Central America and Brazil, bearing handsome violet, rose-coloured, or yellow flowers.

1828 *Bot. Reg.* XIV. 1172 Curled-petaled Cattleya. **1838** *Paxton's Mag. Bot.* V. 6 Cattleyas thrive best in a degree of heat below that required for the major part of orchidæous plants. **1927** *Times* (weekly ed.) 2 June 603/1 An absolutely clear yellow cattleya. **1938** S. BECKETT *Murphy* iv. 50 Neary arrived with a superb bunch of cattleyas.

catty (kæ·ti), *a.* [f. CAT *sb.*[1] + -Y[1].] Of or pertaining to cats; concerned with the breeding or exhibiting of cats.

1903 F. SIMPSON *Bk. Cat* v. 84 The question of open judging at cat shows has frequently been discussed in catty circles.

b. = *CATTISH b.

1886 MRS. ARGLES *Lady Branksmere* ix, There is a sly, *catty* look about her. **1909** *Westm. Gaz.* 11 Dec. 16/2 A noted prelate..said recently in addressing a community composed of the fair sex that all such communities had the temptation to be 'catty'. **1936** A. CHRISTIE *ABC Murders* xxiii. 173 Women..are a bit catty about other women.

catty-corner(ed), -cornering: see *CATER *adv.*

catwalk (kæ·twǭk). [f. CAT *sb.*[1] + WALK *sb.*[1]] A narrow footway or platform.

[**1885** B. POTTER *Jrnl.* 20–22 Mar. (1966) 136 Fine rooms and such a slip of a garden at the back, a cat-walk.] **1917** H. J. STEPHENS *Gloss. Aeronaut. Words* 4 *Cat walk*, narrow passage in the interior of an airship. **1925** FRASER & GIBBONS *Soldier & Sailor Words* 50 *Cat walk*, a Western Front name for the brick-paved pathway, usually one brick (nine inches) wide, laid down across farm fields in Flanders. **1952** W. GRANVILLE *Dict. Theatr. Terms* 39 *Cat-walk*, a narrow bridge communicating with the two fly-galleries above and below the proscenium arch. **1959** C. S. FORESTER *Hunting the Bismarck* 63 On the cat-walk two dark figures of Fleet Air Arm officers watched them circle. **1961** F. H. BURGESS *Dict. Sailing* 47 *Cat walk*, any long narrow gangway, especially when erected high over anything or above a deck that is awash in heavy weather. **1970** J. YARDLEY *Kiss the Boys* ii. 38 Halfway up the catwalk she stopped and the cape fell clear of her body. *Ibid.* 39 Lucie began..her..mannequin perambulations...the spotlight followed her down the catwalk and back again.

catydid, var. KATYDID.

1825 PAULDING *J. Bull in Amer.* iii. 35 The frogs croaked, the caty-dids caty-didded it, the crickets chirped. **1838** E. FLAGG *Far West* II. 214 Even until the morning dawned did a concert of whippoorwills and catydids keep up their infernal oratorio. **1859** A. CARY *Pict. Country Life* iv. 92 The caty-dids..were noisily welcoming the early autumn.

Caucasian, *a.* and *sb.* Delete '(Now practically discarded.)' and add later examples, spec. (of or belonging to) a member of the 'white race', opp. one of other ethnic descent.

1937 C. W. FERGUSON *Fifty Million Brothers* v. 76 At first it was said that the candidates must be 'Caucasian' but this evidently was aimed at Negroes, for Jews were later admitted. **1939** A. HUXLEY *After many a Summer* I. i. 5 The first thing to present itself was a slum of Africans and Filipinos, Japanese and Mexicans...The population took on a more Caucasian tinge. **1958** LEDERER & BURDICK *Ugly*

cavayard of twenty [mares]. **1851** MAYNE REID *Scalp-Hunters* xxvi, How are 'ee gwine to get at it with this cavayard 'ithout makin' sign?

β. **1821** S. F. AUSTIN in *Texas Hist. Quart.* VII. 288 Erasmo had captured a Caviard of mules & horses. **1824** W. B. DEWEES *Lett. fr. Texas* (1852) viii. 53 Corasco..was driving a large cavyyard of horses and mules to Louisiana. **1836** EDWARD *Hist. Texas* 107 When this powerful tribe wishes to raise the wind as the saying is, they will cavy back a Mexican cavy-yard. **1853** 'P. PAXTON' *Stray Yankee in Texas* 97 Two or three more [men] were mounted, and sent into the prairie in search of the 'caviarde' of horses—and we went in to dinner. **1874** J. C. MCCOY *Hist. Sk. Cattle Trade* 11 The extra horses not under the saddle are called the cavvie-yard, and are driven behind the camp wagon. **1901** *Munsey's Mag.* XXV. 404/2 Each man roped a fresh horse from the cavviyard. **1942** E. E. DALE *Cow Country* 47 The band of horses..known as..the *caballado*, sometimes corrupted into 'cavvy yard', 'cavalry yard', or 'cavvy'.

γ. **1908** C. E. MULFORD *Orphan* ii. 24 He was soon able to count seven warriors who were driving another 'cavvieyeh' of horses. **1920** —— *J. Nelson* xxiv. 267 The caviya of a hundred and thirty saddle horses. **1940** E. FERGUSSON *Our Southwest* 32 Unskilled boys wrangled the horse herd, called the 'caviya'.

cave, *sb.*[1] Add: **5.** *cave-mouth, -phantom, -pool*; **cave-art**, depiction of animals and figures, etc., on the interiors of caves by prehistoric or primitive peoples; hence *cave-artist*; **cave-dweller** *fig.*, one who is uncivilized in behaviour like a prehistoric cave-man; similarly **cave-man, -woman**; *cave-man* also = *CAVER 2; **cave-painting** = **cave-art*; also, such a painting or drawing; hence *cave-painter*.

1921 M. C. BURKITT *Prehistory* xv. 193 The authenticity of the cave art. **1930** *Times Lit. Suppl.* 3 July 547/1 The meaning of Pleistocene cave-art. **1923** H. H. WILDER *Man's Prehist. Past* 187 In modern art and in imaginative literature the early cave artists are commonly represented as working upon, or admiring their work. **1906** B'NESS VON HUTTEN *What became of Pam* II. xiii, You will always be a cave-dweller,..for you always were a little savage. **1897** MARY KINGSLEY *W. Africa* x. 208 These pots have a cave-man look about them; they are unglazed unlidded bowls. **1924** W. M. RAINE *Troubled Waters* xxviii. 280 He was a throw-back to the cave man. **1926** *Westm. Gaz.* 7 Aug., The devotion of the modern girl to the 'cave man' of fiction. **1928** A. HUXLEY *Point Counter Point* xxi. 405 'That passionateness of his, that violence —.' Philip laughed. 'Quite the irresistible cave-man.' **1932** E. A. BAKER *Caving* i. 3 The caveman must be prepared physically and mentally for any emergency. *Ibid.* 4 The caveman will stuff his pockets with all sorts of gadgets. **1949** A. CHRISTIE *Crooked House* ix. 67 Why do men always think that a caveman must..be..attractive to the opposite sex? **1906** *Westm. Gaz.* 11 Apr. 2/3 High rocks above my cavemouth stand. **1937** H. READ *Art & Society* i. 31 The cave-painter at Altamira. **1882** *Mag. of Art* V. 249 The cave paintings of the Australians and the bushmen in South Africa. **1942** *Burlington Mag.* June 140/2 The old cave paintings of India. **1930** S. BECKETT *Whoroscope* 3 Shall I swallow cave-phantoms? **1939** DYLAN THOMAS *Map of Love* 16 Curl-locked and animal cavepools of spells and bone. **1903** E. T. FOWLER *Place & Power* v, Miss St. Just..belonged to the cave-woman species. **1951** M. MCLUHAN *Mech. Bride* 23/2 The reader is treated as the sluggish male is treated by the sex-hungry cave woman in the shirt ads.

cave, *sb.*[4] [f. CAVE *v.*[3]] A fall of earth, a cave-in.

1876 B. HARTE *G. Conroy* VI. viii, Gabriel was amazed to find that during the earthquake a 'cave' had taken place in the drift. **1877** RAYMOND *Statist. Mines & Mining* 319 A very serious cave occurred about 170 feet from the entrance of the drift. **1887** FARRELL *How He Died* 164 A 'cave' had happened in a mine.

cave, *v.*[3] Add: **1.** (Earlier U.S. examples.)

1707 S. SEWALL *Diary* (1879) II. 186 Grave was caved in. **1728** J. COMER *Diary* 5 Nov. (1893) 57 This day a man..digging a well, after he had dug 20 feet deep, it cav'd in upon him.

b. Without *in*.

1848 THOREAU in *Atlantic Monthly* (1892) LXIX. 744 His cellar..has caved and let one end of the house down. **1873** J. H. BEADLE *Undevel. West* iii. 69 He had dug two wells,..but struck sand which 'caved so he could not curb'. **1882** *U.S. Rep. Prec. Met.* 639 Wherever the rock in the tunnel has a tendency to cave.

c. *transf.* To yield to outward pressure.

1898 HAMBLEN *Gen. Manager's Story* 32, I was caught between the corners of the cars..and heard my ribs cave in.

2. b. (Earlier and later U.S. examples.)

1844 *Spirit of Times* 23 Mar. 42/2 List 'till I tell you, and if you do not agree with me, why, I'll *cave*. **1855** 'P. PAXTON' *Capt. Priest* 64 The one who 'caves' first shall pay the shot. **1860** HOLLAND *Miss Gilbert's Career* xxii. 390, I tell you when a man gets in front of him Sunday, he catches it—no use dodging—might as well cave. **1961** 'A. A. FAIR' *Stop at Red Light* (1962) xi. 169 The guy caved... The guy broke down and admitted the whole damned business.

3. Delete *rare*. Add earlier and later examples.

1857 *Knickerbocker* Mar. XLIX. 278 He would feel like caving my head in, if he knew. **1873** B. HARTE *Mrs. Skagg's Husbands* 61 Reckons she's caved in his head the first lick! **1916** 'BOYD CABLE' *Action Front* 56 'If we can plant a bomb or two in the right spot, it will bottle up any Germans working inside?' 'Sure to!' said Ainsley. 'It will cave in the entrance completely.' **1957** E. EAGER *Magic by Lake* 96 The ship turned in craven flight and hurried away, fearful of being rammed and caved in.

b. *to cave down*: to bring down by an excavation caving in. *U.S.*

a **1762** S. NILES *Wars* in *Mass. Hist. Soc. Coll.* (1861) V. 340 Providence prevented them by sending a great rain, and caved down the sides of their trench. **1851** C. CIST *Cincinnati* 244, I obtained permission to open a sand-pit, which had long been closed for fear of caving down a house, by further excavation.

Hence **caved** *ppl. a.* Freq. with *in*.

1862 C. F. BROWNE *Artemus Ward, his Book* 92 A old kaved in hat. **1882** 'MARK TWAIN' *Innoc. at Home* vii. 309 An Hour in the Caved Mines. **1953** R. GRAVES *Poems* 19 A caved-in chest, hairy black mandibles.

cave (kēⁱ·vi), *int. School slang.* [L., imper. of *cavēre* to take care, beware.] Beware! A signal of warning, e.g. of the approach of a master. Also used subst. in phr. *to keep cave*.

[**1584** R. GREENE *Carde of Fancie* 7 Nowe thou wilte crye Caue when thy coyne is consumed, and beware when thy wealth is wracked.] **1868** *Cassell's Mag.* 17 Oct. 390/1 [Title of Poem] Cave! **1873** 'A. R. HOPE' *Night before Holidays* 110 There was a heavy footstep sounding along the passage... '*Cave!*' '*Canem*,' responded Lessing, burying himself under the bedclothes again. **1906** E. NESBIT *Railway Children* xiv. 295 He won't keep *cave*, shirks his turn And says he came to school to learn! **1922** *Blackw. Mag.* May 557/2 One of their number doing sentry-go gives the native equivalent for the schoolboy's 'Cave' on the reappearance of their employer. **1959** I. & P. OPIE *Lore & Lang. Schoolchildren* xvii. 373 The term 'keeping cave'..only rarely extends to boys who do not possess any Latin...The look-out in a grammar school..may call just 'Cave!' (pronounced *kave* or *kay-ve*).

cavea (kēⁱ·viă). *Rom. Antiq.* Pl. **-æ**. [L., = a hollow.] The auditorium of an ancient theatre, so called from its concave shape; also the whole theatre.

1611 CORYAT *Crudities* 314 The seats or benches..incompassing the Cavea. **1842** L. SCHMITZ in W. Smith *Dict. Gr. & Rom. Antiquities* 959/1 The whole of the cavea in the Attic theatre must have contained about 50,000 spectators. **1886** *Athenæum* 4 Dec. 751/2 The peculiarity of this theatre [at Laurium] is the strange form of the cavea, which sweeps inward in a loop to the right as viewed from the proscenium. **1955** *Times* 19 Aug. 2/5 Extending 263 ft. in diameter, the hollow of the cavea is divided into five blocks of white marble seats, separated by narrow stairways.

caveat, *sb.* Add: **1. b.** *caveat emptor* [lit., let the purchaser beware], let the purchaser examine the article he is buying before the bargain is completed, so that in case of disappointment after purchase he may not blame the seller.

1523 FITZHERBERT *Husb.* f. xxxvi, He [*sc.* the horse] is no chapmans ware yf he be wylde: but and he be tame and haue ben rydden vpon than caueat emptor be ware thou byer. **1629** T. ADAMS *Pol. Hunting* in *Wks.* 118 We compell none to buy our Ware; *Caueat emptor*. **1809** H. MORE *Let.* 14 Aug. (1925) 139 Mr. C. in his last Review..feels it his duty to say, '*Caveat Emptor*'. **1902** *Economic Jrnl.* XII. 12 *Caveat emptor*. It is the employer on whom the responsibility rests of testing the quality of the article he buys. **1950** T. H. MARSHALL *Citizenship & Social Class* iv. 133 The principle of *caveat emptor* is at least plausible when you are buying a horse.

caver. Restrict ?*Obs.* to sense in Dict. and add: **2.** An explorer of caves.

1932 *Times Lit. Suppl.* 17 Mar. 191/3 The reader is expected to be something of a caver. **1958** *Spectator* 1 Aug. 172/3 A number of local archæologists and speleologists..declared the discoveries to be false, and the paintings and engravings to be forgeries done by the *Maquisards* and by cavers in the post-war years.

cavernicolous (kævəmi·kŏləs), *a.* [f. L. *caverna* CAVERN + *-colus* inhabiting.] Cave-dwelling.

1889 in *Cent. Dict.* **1924** *Glasgow Herald* 13 Dec. 4 Another feature of many cavernicolous animals is the delicacy of their integument. **1947** *Sci. News* V. 51 The 'Proteus', that extraordinary amphibious newt belonging to the Cave of Istria, which became the most celebrated cavernicolous animal. **1966** C. SWEENEY *Scurrying Bush* iv. 61, I hoped that it might lead to a further system of caves, perhaps to water, for, if so, I anticipated finding some truly cavernicolous animals.

cavernous, *a.* Add: **3. b.** *Med.* Applied to respiration marked by a prolonged hollow resonance. (Cf. *broncho-cavernous* adj.)

1853 W. O. MARKHAM tr. *Skoda's Auscult.* 94 'By the term cavernous respiration,' says Laennec, 'I understand the murmur which occurs during inspiration and expiration in a cavity formed in pulmonary tissue.' **1890** F. TAYLOR *Man. Pract. Med.* 341 Breath-sounds which deserve the name of cavernous are often heard over solid lung. **1944** R. COOPE *Dis. Chest* v. 71 All cavities will not necessarily act in this way as resonating chambers, and..cavities may be present without cavernous breathing.

caviare. Add: **2.** *slang.* A passage blacked out by a censor (orig. a Russian censor) by the use of a stamp which when inked and applied to the paper leaves a close network of white lines and black diamonds, resembling to some extent the appearance of caviare spread upon bread and butter. So **caviare** *v. trans.*, to black out or censor in this way.

celebrate, *v.* **3.** Add: Also *absol.* (see quot. 1937).
1929 *Randolph Enterprise* (Elkins, W.Va.) 26 Sept. 3/2 [He] came over..Sunday night to celebrate a little. **1937** PARTRIDGE *Dict. Slang* 136/1 *Celebrate*, v.i., to drink in honour of an event or a person; hence, to drink joyously. **1963** J. T. STORY *Something for Nothing* i. 40 It's Treasure's wedding day. Somebody's got to celebrate.

celebrational (selĭbrēⁱ·ʃən'l), *a.* [f. CELEBRATION + -AL.] = CELEBRATIVE *a.*
1925 P. B. BALLARD *Changing School* xiii. 201 New methods, of a ceremonial or celebrational kind, which will supply this stronger appeal. **1937** F. SWINNERTON *Autobiogr.* xi, He was going to give a celebrational dinner to his fifty best friends.

celebrative, *a.* Delete *rare*⁻¹ and add further examples.
1959 *Listener* 8 Jan. 80/1 There is little twentieth-century art in any medium that is celebrative in this sense. **1961** D. HOLBROOK *Eng. for Maturity* 45 Our essential cultural experiences are not celebrative, they are, rather, onanistic.

celebratory (se·lĭbrēⁱtəri), *a.* [f. CELEBRATE *v.* + -ORY².] Serving to celebrate; used in or designed for the celebration of an event, etc.
1926 *Glasgow Herald* 1 June 7 Without a celebratory or consolatory bite and sup. **1955** *Times* 23 May 9/3 Celebratory parades at home and oversea.

‖ **celebret** (se·lĭbret). [L., = 'let him celebrate', 3rd pers. sing. pres. subj. of *celebrāre* to CELEBRATE.] A document, signed and sealed by a bishop, giving a priest permission to say mass in a certain parish.
1844 IGNATIUS SPENCER in Fr. Pius *Life Fr. Ignatius* (1866) xiii. 282, I had forgotten to get credentials from Dr. Wiseman, and so he hesitated, but gave the *celebret*. **1907** *Daily Chron.* 25 Sept. 4/6, I am told that the Pope will give me a celebret. **1908** *Cath. Encycl.* III. 477/1 The absence of the celebret does not suffice for the refusal of permission to say Mass, if persons worthy of belief bear positive testimony to the good standing of the priest. **1918** M. D. PETRE *Modernism* 237 The subsequent refusal, on the part of his bishop, to grant him a *celebret*.

celery. Add: **2.** *Comb.* **celery pine, celery-leaved, -topped,** or **-top pine,** any Australasian tree of the genus *Phyllocladus*, in which the upper part of the branchlets resembles the foliage of the celery; **celery salt,** a mixture of ground celery seed and salt used for seasoning; **celery seed** (see quot. 1964).
1827, etc. [see PINE *sb.*² 2 a]. **1851** *Illustr. Catal. Gt. Exhib.* IV. i. 994 Celery-topped pine (*Phyllocladus asplenifolia*) of Tasmania. *Ibid.*, Celery pine slab (*Phyllocladus aspleniifolia*), squared. **1883** [see TANEKAHA]. **1889** T. KIRK *Forest Flora N.Z.* 9 The tanekaha is one of the remarkable 'celery-topped pines'. **1927** *Blackw. Mag.* Oct. 464/1 Graceful featherwoods and celery-top pines. **1897** *Sears, Roebuck Catal.* 13/2 Celery Salt..4 oz. sifting top bottles. **1964** R. HEMPHILL *Spice & Savour* (1965) 99 Commercial celery salt is flavoured with ground celery seed, or with ground dried celery stems. **1861** MRS. BEETON *Househ. Managem.* 189 *Celery Vinegar*..½ oz. of celery seed, 1 pint of vinegar. **1964** R. HEMPHILL *Spice & Savour* (1965) 99 Celery seed is the dried fruit of the celery plant.

celesta (sĭle·stă). [app. pseudo-latinization of F. *céleste* (cf. CELESTE).] A keyboard instrument with piano-like action, having hammers that strike upon steel plates placed over wooden resonators, invented by Auguste Mustel of Paris in 1886.
1899 E. PROUT *Orchestra* II. vi. 158 Only a few of the strings are employed to accompanying the *celesta*. **1927** *Grove's Dict. Mus.* I. 591 The tone of the celesta itself is of exquisite purity. **1966** *Listener* 10 Mar. 364 The celesta player on both occasions was Richard Rodney Bennett.

celeste. Add: **2. c.** = prec.
1934 S. R. NELSON *All about Jazz* v. 93 A couple of bars by solo violin, which is taken the second time by celeste. **1958** B. ULANOV *Hist. Jazz* xiv. 164 He tried out every sort of scoring,..celeste interludes, [etc.].

celestial, *a.* and *sb.* Add: **A.** *adj.* **1.** *celestial equator* = EQUATOR 1 (cf. EQUINOCTIAL *sb.* 1); *celestial mechanics*, gravitational astronomy: see quot. 1959; *celestial pole*: see POLE *sb.*² 1; *celestial triangle* = *ASTRONOMICAL *triangle*.
1875 *Encycl. Brit.* II. 764/2 The celestial equator..is the great circle having for its poles..the poles of the heavens. **1950** *Chambers's Encycl.* V. 379/1 The celestial equator is the great circle in the sky corresponding to the extension of the equator of the earth. **1821** *New Monthly Mag.* III. 184/2 A simple lunar theory, and a rigorous system of celestial mechanics. **1936** *Discovery* Jan. 31/1 Important residual effects like the advance of the perihelion of Mercury in celestial mechanics. **1959** *Encycl. Brit.* V. 93/1 Celestial Mechanics is the branch of astronomy that deals with the mathematical theory of the motions of celestial bodies. **1956** W. A. HEFLIN *U.S.A.F. Dict.* 106/2 Celestial triangle.
4. (Further examples.)
1840 *Knickerbocker* XVI. 447 We have seen a Chinese map of the world, in which the celestial country occupies the entire space, with the exception of a few island-like circles. **1872** 'M. TWAIN' *Roughing It* 396 We ate chow-chow with chop-sticks in the celestial restaurants. **1878** H. JAMES *Europeans* I. vi. 246 He possessed the most delightful *chinoiseries*—trophies of his sojourn in the Celestial Empire.
B. *sb.* **2.** (Earlier example.)
1842 W. DYOTT *Diary* 24 Nov. (1907) II. 365 Accounts arrived with glorious news of peace with China...The terms of peace with the celestials highly favourable.

cell, *sb.*¹ Add: **4. b.** *pl.* Imprisonment, esp. in solitary confinement as a punishment for offences against military law. *colloq.*
1891 KIPLING *Life's Handicap* 23 You 'ave been absent without leave an' you'll go into cells for that. **1910** GALSWORTHY *Justice* III. ii, Now I'll get cells, I suppose, or seven days' bread and water. **1918** W. J. LOCKE *Rough Road* xvii, At the worst they might give him cells when he recovered. **1929** GALSWORTHY *Exiled* III. i, In the Court yesterday, you give me a night cells for sleepin' out.
9. c. *Math.* See PEAUCELLIER CELL.
12. b. *fig.* A small group of people (occas. a single person) working within a larger organization as a nucleus of political, esp. revolutionary, activity; also, the headquarters of such a group.
1925 *Glasgow Herald* 25 Apr. 10 They were Communists and were carrying membership cards of Communist cells. **1931** *Morning Post* 6 Aug. 12/4 The Viennese police have discovered a Communist 'cell' in..a district of the city. **1937** A. HUXLEY *Ends & Means* viii. 73 Ten is the number of individuals constituting a Communist cell. **1939** G. B. SHAW *Geneva* I. 20 My butler..tells me that my footman..is a cell...A communist cell. Like a bee in a hive. Planted on me by the Communists to make their dreadful propaganda in my household! **1958** L. A. G. STRONG *Treason in Egg* v. 93 What if the thing was organised like an underground cell in war-time, when each member of the resistance knew only his own little bit of the chain of activity.
15. *cell-free* adj.; **cell-count,** a count or estimate of the number of cells in a given amount of blood, etc.; **cell-layer** = GERMINAL *layer*; **cell line,** a population of cells *in vitro* which are descended through one or more sub-cultures from a single primary culture; also, a group of cells *in vivo* which share a common characteristic because of a common descent; **cell-lineage,** the stages in the development of a cell or blastomere; **cell-spot,** a spot of colour occurring in the spaces between the nerves in the wings of certain lepidopterous insects; **cell strain,** a population of cells sharing some common characteristic that persists to a greater or lesser extent when they are propagated in cultures.
1910 J. TURNER in *Jrnl. Mental Sci.* LVI. 487 The value of a cell-count for diagnostic purposes. **1962** *Lancet* 5 May 929/1 The cell-count reached 2000 cocci per ml. **1946** Cell-free [see *ANTIBIOTIC *a.* 2]. **1967** *Listener* 3 Aug. 142/3 The solution of the genetic code has come mainly from biochemical studies of protein synthesis in cell-free extracts. **1893** TUCKEY *Hatschek's Amphioxus* 55 All changes from the blastula onwards can be traced to these primitive cell-layers. **1951** *Trans. N.Y. Acad. Sci.* XIII. 324/2 A study of the virus of equine encephalomyelitis on different normal and tumor cell lines. **1961** M. HYNES *Med. Bacteriol.* (ed. 7) xxv. 379 Cell lines such as the HeLa strain are generally used for routine culture and antibody titration. **1961** *Lancet* 19 Aug. 435/2 If these cells are XXY then the presence of such abnormal cell lines is presumably restricted to a few tissues. **1892** E. B. WILSON in *Jrnl. Morphology* VI. 361 The cell-lineage of Nereis. **1911** J. A. THOMSON *Biol. of Seasons* ii. 182 The primordium..is continuous through unspecialised cell-lineage with the fertilised egg from which the parent arose. **1946** *Nature* 5 Oct. 461/2 Tissue culture made it possible to prove that the cell-lineages of the ordinary somatic cells of the body are indefinite or indeterminate. **1902** *Proc. Zool. Soc.* I. 49 The black cell-spots on the primaries. **1936** *Amer. Jrnl. Cancer* XXVII. 69 Although the diagnoses were different, the cell strains established from D.Au., H.Da. and A.Ha. I and II were strikingly similar in cytological appearance and cultural behavior. **1958** *Jrnl. Immunol.* LXXXI. 426/1 Advances in tissue culture techniques have made available numerous stable cell strains..which have contributed to recent developments in virology.

cell, *sb.*³ *Cinemat.* Also **cel.** [Abbrev. of CELLULOID.] A transparent sheet of celluloid or similar film material; esp. such a sheet bearing opaque matter (e.g. a drawing), used in combination with other sheets to produce a composite picture. Also *attrib.*
1933 G. H. SEWELL *Commercial Cinemat.* x. 160 The two drawings are..replaced on the desk, a cell placed on top of them, and the drawing of the ram in its first position is made on the cell. The subsequent cells receive images of the successive positions of the ram. **1938** S. W. BOWLER in S. G. B. Stubbs *Mod. Encycl. Photogr.* I. 253/2 The next stage is the substitution of a transparent celluloid 'overlay' for the papers on which the figures are drawn. With these celluloid overlays, or 'cels' as they are called, there are almost unlimited possibilities of action available, since at least two or three may be used..over the same background. *Ibid.* 253/2 The drawings are usually made on these 'cels' in non-waterproof ink. **1959** HALAS & MANVELL *Technique Film Anim.* xix. 223 The head of the colouring department ..is responsible for handing out the traced cells to the individual colourists. **1959** *Listener* 23 July 126/2 A new pencil called the cell-graph, invented by the London film cartoonist John Halas, enables the key artist to draw his own work directly on to the celluloid sheets. **1966** *New Scientist* 10 Feb. 346/1 The cels have to be individually photographed.

cell, *v.* Transfer †*Obs. rare* to sense a and add: **b.** *spec.* To share a prison cell *with* another person. *slang.*
1903 J. FLYNT *Rise of Ruderick Clowd* iii. 108, I celled for a couple o' years with old Darbsey—he was doin' life. **1965** T. CAPOTE *In Cold Blood* iii. 130 He was the first fellow I celled with. We celled together I guess a month.
c. *trans.* To store in cells.
1819 WIFFEN *Aonian Hours* 75 Honey, which the bee Cells beneath briery boughs. **1927** *Chambers's Jrnl.* Feb. 90/1 Last autumn one of my stocks celled twelve pounds in twelve hours.

cellar, *sb.* Add: **6. cellar-book** (delete quot. in Dict. and insert earlier and later examples); **cellar-way** (earlier Amer. example).
1769 G. SELWYN *Let.* 4 July in *Hist. MSS. Comm.* (1897) 15th Rep., App. VI. 248 According to my cellar book you will have had in all ten dozen. **1848** THACKERAY *Van. Fair* xxiv. 202 He..overhauled the butler's cellar-book. **1920** G. SAINTSBURY (*title*) Notes on a cellar-book. *a* **1762** S. NILES in *Mass. Hist. Soc. Coll.* (1861) V. 512 Two or three were found lying..in the cellar-way.

cellarless (se·ləɹlės), *a.* [f. CELLAR *sb.* + -LESS.] Having no cellar.
1853 R. S. SURTEES *Sponge's Sp. Tour* ii. 7 Little cellarless wine-merchants. **1864** [see CELLAR *sb.* 6]. **1900** *Westm. Gaz.* 9 May 2/2 All this legislation is passed by the cellared population and not by the cellarless population.

cellifugal (selĭ·fiŭgăl), *a.* Also **cellulifugal.** [f. *celli-*, comb. f. L. *cella* CELL *sb.*¹ + *fugere* to flee: see -AL.] Of the nerve-impulses in a ganglion-cell: passing from the body of a cell. Also **cellipetal** (selĭ·pĭtăl) *a.* [L. *petere* to seek], moving towards the body of a cell.
1900 W. S. HALL *Text-bk. Physiol.* xiii. 538 In monaxons the impulse passes always away from the cell-body, hence is cellulifugal. **1904** TITCHENER tr. *Wundt's Physiol. Psychol.* I. 42 The dendrites are devoted exclusively to cellipetal, the neurites to cellifugal conduction.

'cellist, cellist (tʃe·list). Shortened form of VIOLONCELLIST, after 'CELLO. Also **'celloist, celloist** (tʃe·loist).
1888 MRS. H. WARD *R. Elsmere* v. xxxi, The 'cellist with the hair. **1897** *Daily News* 25 Mar. 9/2 The pianist came, but the 'celloist did not turn up at all. **1954** *Grove's Dict. Mus.* VIII. 823/1 The many fine performances by women cellists.

'cello. Add: (Earlier example.) (See also quot. 1954.)
1876 STAINER & BARRETT *Dict. Mus. Terms* 82/1. **1954** *Grove's Dict. Mus.* II. 138/2 Cello..may be considered as having become an English word by adoption, and there is thus no necessity to spell it with an apostrophe ('cello), as is still frequently done, nor is the Italian plural (*celli*) desirable.

cellobiose (selobəi·ōᵘs). *Chem.* [f. CELL(ULOSE + -O + *BIOSE.] A disaccharide, $C_{12}H_{22}O_{11}$, that with dextrose forms the basis of the chemical structure of cellulose.
1902 *Jrnl. Chem. Soc.* LXXXII. 135 The name cellobiose is now used for the substance previously termed cellose. **1929** *Birm. Post* 22 Feb. 5/4 Over 50 % of starch can be converted into maltose, and..cellulose can be transformed into the related disaccharide cellobiose.

celloidin (seloi·din). [a. G. *celloidin*, f. CELL(ULOSE + -OID + -IN¹.] A pure form of pyroxylin, soluble in ether, used chiefly in microscopy for embedding specimens of tissues so that sections may be prepared.
1883 *Jrnl. R. Microsc. Soc.* III. 305 Very elegant results may also be obtained by an imbedding mass originally invented by Duval, and recently much improved by Merkel and Schiefferdecker. This is collodion, or, preferably a solution of so-called celloidin. **1892** *Photogr. Ann.* II. 107 Making an enlarged transparency..on a rapid celloidin film. **1908** *Practitioner* Feb. 201 Three healthy foetuses.. were hardened in formol, two of them being cut in celloidin. **1932** *Discovery* May 158/1 The methods of fixing, section-cutting, embedding in paraffin wax or celloidin, [etc.]. **1969** *Jrnl. Anat.* CIV. 212 A large central block, containing the sphenoid bone, pituitary and brain, was then cut out, embedded in celloidin, and cut serially in either the sagittal or the coronal plane.

Cellon (se·lɒn). [f. CELL(ULOSE *sb.*] Proprietary name of a composition of cellulose acetate, used as an insulating material, etc. Also *attrib.*
1913 *Chem. Abstr.* VII. 2113 'Cellon' is non-combustible celluloid. **1934** *Archit. Rev.* LXXV. 12/2 As a constructivist painter Moholy-Nagy was the first to use new synthetic materials like galalith, trolith and cellon. **1944** 'N. SHUTE' *Pastoral* iv. 95 Gunnar..scrambled back down the fuselage to the wireless position, where there was a cellon window.

Cellophane, cellophane (se·lŏfēⁱn). [f. CELL(ULOSE + -O + -*phane* as in DIAPHANE *a.* and *sb.*] Proprietary name of a glossy transparent material made from regenerated cellulose, used chiefly for wrapping goods, food, etc. Also *fig.* and *attrib.*
1912 CROSS & BEVAN *Res. Cellulose 1905–10* III. v. 162 The 'viscose film' (cellulose) under the powerful auspices

of the Société Industrielle de Thaon is at length a *fait accompli*, and is an article of commerce under the descriptive term 'Cellophane'. *Ibid.*, Specimens of 'Cellophane' film. **1921** *Spectator* 23 Apr. 522/2 Ciré, raffia,..tinsels, cellophane, and other ornaments. **1933** R. E. SHERWOOD *Reunion in Vienna* p. xii, To hell with them..and their neuroses, all tidily encased in cellophane. **1935** *Times* 24 Oct. 11/4 A frock in a mixture of cellophane and wool. **1957** *Listener* 26 Sept. 484/1 English itself comes to him as something sparkling Cellophane-fresh instead of the shopworn, finger-soiled article it inevitably seems to ourselves.
Hence **ce·llophaned** *a.*, wrapped in cellophane; also *fig.*
1952 *Landfall* VI. 185 The galaxy of cellophaned packets. **1953** C. DAY LEWIS *Italian Visit* i. 14 The white-faced addicts of a patent, cellophaned future.

cellose. An earlier name for *CELLOBIOSE.
1901 *Jrnl. Chem. Soc.* LXX. 370 Cellose, a Biose from Cellulose.

cellotape, var. *SELLOTAPE, SELLOTAPE.

cellular, *a.* (and *sb.*) Add: **2. b.** Of open texture, as *cellular linen*; also *sb.*, a material of open texture.
1888 *Cassell's Fam. Mag.* Dec. 60/1 Cellular Clothing. Under-clothing of every kind is now being made of a cellular cloth. **1889** *Pall Mall Gaz.* 5 Jan. 7/1 Cellular linen. ..Gentlemen have tennis-shirts of real silk cellular now. **1964** *Which?* Apr. 123/1 Cellular, mesh or eyelet fabrics have a regular pattern of small holes over the surface.

cellulase (se·liulēⁱs, -ēⁱz). *Chem.* [f. CELLUL(OSE + *-ASE.] An enzyme which brings about the decomposition of cellulose.
1903 *Jrnl. Chem. Soc.* LXXXIV. II. 503 Most moulds attack cellulose, and the action is due to a specific enzyme for which the author [*sc.* G. van Iterson, jun.] supports the name *cellulase*. **1950** *New Biol.* VIII. 99 All enzymes which break down cellulose are called cellulases.

cellule. Add: **1. b.** A small room or cell.
1894 A. MORRISON *Tales Mean Streets* 222 The family should take a whole house..instead of the two rooms and a cellule upstairs now rented. **1908** *Westm. Gaz.* 12 June 8/1 'Each go into a cell and shut the door!' ordered a wardress, indicating a row of tiny cellules.

cellulitic (seliuli·tik), *a.* [f. CELLULITIS + -IC.] Pertaining to cellulitis.
1906 FERGUSON in *Trans. Edinb. Obstet. Soc.* XXXI. 127 Where there is no cellulitic abscess which can be evacuated.

celluloid. Substitute for def.: A solid inflammable material consisting essentially of soluble cellulose nitrate and camphor, used in the manufacture of many articles, esp. photographic film. orig. *U.S.*, as a trade-name. (Further examples.)
1871 *Amer. Dental Assoc. Trans.* XI. 152 We have many so-called cheap materials..rubber, celluloid, pyroxyline, [etc.]. **1892** W. E. WOODBURY *Encyl. Photogr.* 110 *Celluloid Films.* Films of celluloid..may be coated with a gelatine sensitive emulsion, the celluloid taking the place of the glass of the photographic dry plate, the advantage being in the reduction of bulk and weight. **1951** YARSLEY & KITCHEN in H. M. Langton *Synthetic Resins* (ed. 3) ii. 116 The bulk of the..cine film used throughout the world to-day is still celluloid.
b. A piece or sheet of celluloid; *spec.* = *CELL *sb.*³
1936 A. JENKINSON *America came my Way* xvii, When each picture comes to be photographed it may actually consist of five different pictures—the background and four celluloids placed one on top of the other. **1959** HALAS & MANVELL *Technique Film Anim.* xix. 223 He is also responsible for seeing that the dye of the celluloids is correct.
c. *transf.* Films; the 'cinema'.
1934 A. G. MACDONELL *How like Angel* II. viii. 147 The British Board of Film Censors..who draw the tenuous line between the decorous and libidinous in the world of celluloid. **1937** A. CALDER-MARSHALL in C. Day Lewis *Mind in Chains* 65 Celluloid now gives them..relief from present trouble. **1938** C. MORGAN *Flashing Stream* 32 There are no heroes but in celluloid.
2. *attrib.* or as *adj.* Made of celluloid.
1871 [in Dict.]. **1882** SWEET & KNOX *Texas Siftings* 9 He is usually swung to a satchel containing a comb and brush,..a clean celluloid collar, and a newspaper. **1892** [see above]. **1923** *Manual of Seamanship* II. 21 Celluloid films should not be allowed on board unless stowed in well-fitting iron boxes. **1962** *Unesco Bull. for Libraries* XVI. 1 In 1887 H. Goodwin produced the first photographic film on celluloid base.
b. Of or pertaining to films; appearing in a film; photographed. Also *fig.*, synthetic; lifeless, unreal.
1922 *Frontier* Nov. 18 The celluloid hero flashed his impartial smile across the screen. **1926** E. WALLACE *More Educ. Evans* (1928) ii. 50 Harold Lloyd favoured them with a celluloid smile as they slipped into the dark interior. **1928** D. H. LAWRENCE *Lady Chatt.* 141 Something that has gone out of the celluloid women of today. **1948** WODEHOUSE *Spring Fever* i. 12 His offspring was contemplating marrying into celluloid circles. **1958** *Times Lit. Suppl.* 15 Aug. p. xxviii/2 A confected thriller plot, complete with flat celluloid characters. **1959** I. & P. OPIE *Lore & Lang. Schoolch.* vii. 118 The celluloid dream-world really begins to have an effect on their thought.

cellulose. (Now freq. stressed on first syllable and with final -ōᵘz.) Add: **B.** *sb.* Also *Comb.*, as *cellulose-digesting* adj.
1924 J. A. THOMSON *Science Old & New* xxix. 161 Sheep and cattle live very largely on grass, and yet they are also without cellulose-digesting ferments.
b. Now *esp.* as the basis of important commercial products. In popular use the word commonly designates compounds of cellulose, particularly cellulose acetate and cellulose nitrate, solutions of which form the 'cellulose' finishes used in varnishing metal, woodwork, etc. Also *attrib.*, as *cellulose lacquer, paint.*
1876 *Jrnl. Chem. Soc.* XXX. 231 (*title*) Cellulose manufacture. **1898** *Daily News* 9 Nov. 4/6 A sawmill and a cellulose factory have just been set up at Mitrovitza. **1902** *Westm. Gaz.* 6 Jan. 2/1 The sulphite cellulose process is due to the labours in the laboratory of Al. Mitscherlich and Tilghmann. **1927** *Observer* 16 Oct. 27 Bodies finished in cellulose. **1930** *Motor Body Building* LI. 105/2 *Cellulose Lacquer*, a finishing material containing nitro-cellulose. **1935** *Discovery* Nov. 326/2 Cellulose lacquers, paints, distempers, etc. **1951** *Good Housek. Home Encycl.* 321/1 Cellulose paint is not very suitable for outdoor work.
c. cellulose acetate, any of several compounds formed by acetylating (usu. with acetic anhydride) some form of cellulose, esp. cotton; used in the manufacture of synthetic fibres, plastics, etc.; **cellulose nitrate,** a compound formed by the action of nitric acid on a cellulose; nitrocellulose.
1880 *Jrnl. Chem. Soc.* XXXVIII. 372 The author describes the properties of five cellulose nitrates. **1895** C. F. CROSS et al. *Cellulose* 35 The cellulose acetates about to be described are of undetermined molecular weight. *Ibid.* 39 Although gun cottons, or pyroxylines, are generally spoken of as nitro-celluloses, they are perhaps more correctly described as cellulose nitrates. **1909** KEANE *Mod. Org. Chem.* 199 Conditions..favourable to explosibility are present in gun-cotton, a mixture of cellulose nitrates, and in blasting gelatine. **1927** T. WOODHOUSE *Artificial Silk* 27 Cellulose Acetate artificial silk, also known as 'Celanese' artificial silk. **1958** *Economist* 20 Dec. 1051/1 Cellulose Acetate makes tough, water-resistant wire and cable coatings. Cellulose Acetate Butyrate produces lacquers, adhesives, airplane dopes and melt and peelable coatings possessing high strength, flexibility and excellent weathering properties.
Hence as *v. trans.*, to treat or coat with a compound of cellulose, esp. with a cellulose lacquer or paint. So **ce·llulosed** *ppl. a.*
1934 *Punch* 29 Aug. 248/2 Dingy Skipper may need re-cellulosing but is fine defensive navigator. **1935** *Times* 23 Oct. 4/5 This last is cellulosed in two shades of metallic blue. **1937** *Times* 23 Dec. 4/7 A less elaborate affair is a cellulosed bed-table. **1958** W. SANSOM *Cautious Heart* 6 Small refectory tables, carved and terribly cellulosed.

cellulosic (seliŭlōᵘ·zik), *a.* and *sb.* [f. CELLULOSE *sb.* + -IC.] **A.** *adj.* **a.** (See CELLULOSE *a.* and *sb.*). **b.** Made from (a compound of) cellulose.
1881 [in Dict.]. **1928** *U.S. Pat.* 1,817,963 3 A film sheet for natural color photography comprising a light transmitting support of a cellulosic compound. **1932** *Kodak Abstr. Bull.* XVIII. 508 Process of making cellulosic films. **1949** H. R. SIMONDS et al. *Handbk. Plastics* (ed. 2) i. 14 Synthetic resins..account for about 90% of all the plastics materials now produced. The cellulosic plastics are the second most important group. **1968** J. IRONSIDE *Fashion Alphabet* 248 A cellulosic rayon fibre.
B. *sb.* A substance made from (a compound of) cellulose.
1946 J. M. DEBELL et al. *Germ. Plastics Practice* ix. 203 (*heading*) Water-Soluble Cellulosics and Cellophane. **1949** H. R. SIMONDS et al. *Handbk. Plastics* (ed. 2) xii. 683 Cellulosics are used for bonding paper, thermoplastics, and leather.

celosia (sĭlōᵘ·siă, -ʃ¹ă). *Bot.* [mod.L., f. Gr. κήλεος burning, κηλός dry, so called from the burnt appearance of the flowers of some species.] A plant of the amaranthaceous genus so named, esp. the cockscomb, *Celosia cristata.*
1807 T. MARTYN *Miller's Gard. & Bot. Dict.* I. 7 D/8 Celosias or Cock's-combs, are all herbaceous plants. **1899** *Westm. Gaz.* 16 Aug. 4/2 A very fine collection of celosias, a handsome feathery flower, in shades of gold and dark crimson. **1962** *Amateur Gardening* 17 Feb. 4 For an unusually showy plant..nothing can be more effective than celosia or cockscomb.

Celsius (se·lsiŭs). The name of a Swedish astronomer, Anders *Celsius* (1701–44), used to designate the centigrade type of thermometer and temperature-scale invented by him in 1742 (see CENTIGRADE *a.*).
1797 *Encycl. Brit.* XVIII. 497/2 Celsius's thermometer. **1850** WEALE *Dict. Terms Archit.* 451/2 The Celsius, or Centigrade thermometer. **1863** E. ATKINSON tr. *Ganot's Physics* § 235 On the continent..this division [into 100 parts] is called the centigrade or Celsius scale. *a* **1928** in Gamble *Story North Sea Air Station* 410 My altitude was.. about 5,800 metres, the temperature 28° Celsius. **1961** *Times* 23 Dec. 7/4 The Meteorological Office proposal to use the centigrade (Celsius) scale for temperature reports and forecasts to the public.

Celtdom (se·ltdǫm, k-). [f. CELT¹ + -DOM.] Celtic peoples collectively; the Celtic spirit or genius.
1895 A. NUTT *Voyage Bran* I. 206 That natural magic which we seek in vain outside Celtdom. **1920** *Glasgow Herald* 25 Sept. 6 An Comunn Gaidhealach has done much to unveil the heart of Celtdom.

Celtiberian (seltibī³·riăn), *a.* and *sb.* [f. L. *Celtiberia*: see CELT¹ and IBERIAN.] **A.** *adj.* Of or pertaining to Celtiberia, an ancient province of Spain lying between the Tagus and the Ebro, or to its inhabitants the Celtiberi, a union of Celts with Iberians. **B.** *sb.* An inhabitant of Celtiberia.
1607 TOPSELL *Four-f. Beasts* 288 The Horsses of the Celtiberans [are] somewhat a dusty colour. **1622** PEACHAM *Compl. Gent.* 48 How Titus Sempronius Gracchus subdued the Celtiberian Spaniards. **1797** *Encycl. Brit.* IV. 284/1 The Celtiberians were very cruel towards their enemies. **1845** *Encycl. Metrop.* XXV. 1345/1 The Celtiberian alphabet, of which there are several specimens in inscriptions and on coins, has not as yet been fully deciphered. **1924** *Glasgow Herald* 28 July 5 Numantia, the capital of the powerful Celtiberian tribe, the Arevaci. **1936** *Antiquity* X. 274 The Celtiberians had a two-piece plough.

Celtic, *a.* Add: **2.** *Celtic cross*: a Latin cross, the centre of which is surrounded by a circle; *Celtic fringe* (or *edge*): (the land of) the Scots, Welsh, Irish and Cornish, regarded as occupying the fringe or outlying edge of the British Isles (freq. *derogatory*); *Celtic twilight*: W. B. Yeats's name for his collection of stories, etc., based on Irish folk-tales; hence *gen.*, (sometimes disparagingly) the atmosphere of, or artistic tendencies associated with, the folklore and legends of Celtic Britain, esp. of Ireland.
1873 QUEEN VICTORIA *Jrnl.* 9 Sept. (1968) 281 We..saw the Celtic cross at Logierait put up to the late Duke of Athole. **1890** MARQUIS OF SALISBURY *Mr. Parnell & Irish Question* 9 The great defect of our present representation is that the Celtic edges of the country on both islands are represented enormously out of proportion to the rest of the Anglo-Saxon population. **1893** W. B. YEATS (*title*) The Celtic twilight. **1894** G. ALLEN *Post-Prandial Philos.* xviii. 155 If Lord Salisbury thinks we are a Celtic fringe he is vastly mistaken. **1899** A. H. KEANE *Man Past & Present* xiv. 523 The 'Keltic fringe', that is, the strips of territory on the skirts of the Teutonic and Neo-Latin domains in the extreme west. **1907** A. S. T. GRIFFITH-BOSCAWEN *Fourteen Yrs. in Parlt.* ii. 11 Their majority [in 1892]..came entirely from Scotland, Wales, and Ireland, or, as Mr. Balfour aptly said, 'the Celtic fringe'. **1908** *Westm. Gaz.* 2 June 2/2 Apparently he has now got tired of his Celtic-fringe seat. **1923** A. HUXLEY *On Margin* 30 If Mr. Yeats understood the Einstein theory..he too could give us, out of the Celtic twilight, his lyrics of relativity. **1936** R. LEHMANN *Weather in Streets* IV. ii. 398 The mixture of Celtic twilight and Aubrey Beardsley décor which .. enshrouded her. **1938** L. MACNEICE *I crossed Minch* I. i. 3 That natural.. culture which..is only found on the Celtic or backward fringes. **1969** C. CARFAX *Silence with Voices* x. 67 All the tombstones had their backs to the water except one. This, a lichened Celtic cross, faced the others.

Celticist (se·ltisist, k-). [f. CELTIC *a.*] = CELTIST.
1912 *Irish World* 12 June 7/2 The celebrated philologist and Celticist. **1964** *Punch* 15 July 87/2 A Belgian Celticist.

celtiform (se·ltĭfǭɹm), *a.* *Archæol.* [f. CELT² + -FORM.] Shaped like a celt.
1932 J. G. D. CLARK *Mesolithic Age* iv. 65 An almost celtiform shape. **1937** *Proc. Prehist. Soc.* III. 283 Celtiform axes or adzes of Large type have been reported from the Solway.

Celtish (se·ltiʃ, k-), *a.* [f. CELT¹.] Celt-like, somewhat Celtic; = CELTIC *a.*
1889 in *Cent. Dict.* **1906** *Daily Chron.* 6 Jan. 6/5 The Celtish melancholy and the Celtish gleam in his eyes. **1915** R. BROOKE *Let.* 24 Jan. in Marsh *Mem.* (1918) 136 Where our huts are was an Iberian fort against the Celts—and Celtish against Romans—and Roman against Saxons.

celtium (se·ltiŭm). *Chem.* [a. F. *celtium* (Urbain 1911, in *Compt. Rend.* CLII. 142), f. L. *Celtæ* (see CELT¹) + -IUM.] An element now generally known as *HAFNIUM.
1911 *Jrnl. Chem. Soc.* C. II. 115 The name celtium is given to the corresponding element, and the symbol Ct assigned to it. **1922** *Nature* 17 June 781/1 Very different conclusions have been reached as to the identity of celtium with the missing element of number 72 on the Moseley classification.

‖ **cembalo** (tʃe·mbălo). *Mus.* [It.; see CEMBALIST.] Shortened form of *CLAVICEMBALO; the harpsichord.
[**1801** BUSBY *Dict. Mus.*, *Cembalo*, the Italian name for a harpsichord.] **1865** WEBSTER, *Cembalo*, the harpsichord. [*Rare.*] **1928** *Radio Times* 23 Nov. 520/1 A cembalo, the delicate-toned ancestor of the pianoforte. **1947** C. GRAY *Contingencies* xi. 161 Scipione Dentice, a writer on music and player of the *cembalo*.
b. = CYMBALO.
1879 [see CYMBALO].

cembra (se·mbră). Also **-o.** [mod.L., f. G. dial. *zember, zimber*, var. of *zimmer* TIMBER *sb.*¹] In full *cembra pine*: the Swiss stone-pine, *Pinus cembra*. Also *attrib.*, as *cembra nut.*
1785 W. MARSHALL *Planting* 283 *Pinus Cembra*: the Cembro, or Cembro-Pine. **1807** T. MARTYN *Miller's Gard. &*

Bot. Dict. II. 8 G/1 The Cembra Pine grows higher up the Alps than any other Pine. **1848** DUNGLISON *Med. Lex.* (ed. 7) s.v. *Pinus Cembra*, The nuts, Cembro nuts, have an eatable kernel. **1909** *Westm. Gaz.* 9 Oct. 14/2 Two bent and stunted cembras. **1920** C. COLTMAN-ROGERS *Conifers* i. 19 The cone of the Cembra is a little round football-shaped specimen.

cement, *sb.* Add: **5. b.** (See quots.) Cf. *cement-gold.*
a **1877** KNIGHT *Dict. Mech.*, *Cement*,..3... *a.* The brown deposit in the precipitation tank, wherein the soluble chloride of gold, obtained by the chlorination process, is deposited by the addition of sulphate of iron to the solution. *b.* The material in which the metal is imbedded in the cementing-furnace. **1889** [see next].

cementation. Add: **2. c.** *attrib.*
1884 W. H. GREENWOOD *Steel & Iron* xviii. 406 The Cementation process for the conversion of bar-iron into cementation or blister steel...The process is conducted in a converting or cementation furnace. **1887** D. A. LOW *Introd. Machine Draw.* xvii. 79 In the cementation process, bars of wrought iron are imbedded in powdered charcoal in a fireclay trough. **1889** *Q. Rev.* July 137 This is known as the 'converting' or 'cementation' process, and the charcoal employed as the recarbonizing agent is termed 'cement'. **1967** *Gloss. Mining Terms (B.S.I.)* IX. 7 *Cementation process*, a method of sealing off water-bearing strata by the injection of cement grout through boreholes.

cemented (sĭme·ntėd), *ppl. a.* [f. CEMENT *sb.* or *v.* + -ED.] Treated with cement; united with or as with cement; *cemented carbide*, powdered carbide of one or more heavy metals compressed into a solid mass and used in cutting-tools, etc.
a **1877** KNIGHT *Dict. Mech.* 509/1 Cemented-back carpet. **1903** *Daily Chron.* 16 June 7/4 To realise the fair dream of a cemented Empire. **1909** *Ibid.* 10 Apr. 3/4 Dry cemented roads. **1909** *Westm. Gaz.* 20 Sept. 4/1 Krupp cemented armour. **1932** *Jrnl. Iron & Steel Inst.* CXXVI. 581 The steels are arranged under the following types: Carbon, high-speed, special alloy, non-deforming, cemented carbides and unclassified.

cementer. Add examples illustrating technical uses.
1903 *Daily Chron.* 24 Feb. 8/5 Envelope Cementers.. wanted. *Ibid.* 9 Dec. 9/6 Tortoiseshell Worker.—Good cementer, used to repairs. **1921** *Dict. Occup. Terms* (1927) § 429 *Solutioner* (boots); *cementer*; uses a rubber solution or cement, on lining of upper.

cementite (sĭme·ntəit). [f. CEMENT *sb.* + -ITE¹.] A hard and brittle carbide of iron, Fe_3C.
1888 H. M. HOWE in *Engin. & Mining Jrnl.* 18 Aug. 132 Minerals which compose iron. Name suggested here.. Cementite. Iron with cement carbon. *Ibid.* 25 Aug. 154/1 Cementite..appears to be intensely hard, brilliant, homogeneous, structureless carbide of iron, containing the whole of the cement carbon. **1889** [see PEARLITE 2]. **1922** C. H. DESCH *Metallography* (ed. 3) xvii. 373 The carbide, cementite, Fe_3C, is the only compound of carbon and iron which has yet been recognized by microscopical means.

cementoma (sĭmentōᵘ·mă). Pl. **-omata.** Also **ceme·ntome.** [f. L. *cementum* CEMENT *sb.* + *-OMA.] (See quot. 1893.)
1893 J. B. SUTTON *Tumours* 35 When the capsule of a tooth becomes enlarged,..and these thick capsules ossify, the tooth will become embedded in a mass of Cementum. To this form of odontome the name cementoma may be applied. *Ibid.*, The largest cementome from a horse known to me..weighs seventy ounces. **1910** *Practitioner* Jan. 124 The curious odontomata known as cementomata.

cenacle. Add: **b.** A place in which a group of people meet for the discussion of common interests; also, the group of people so meeting, *spec.* a literary clique. Also in Fr. form *cénacle* (senakl).
[**1879** *Encycl. Brit.* IX. 676/1 The famous *cénacle* or clique in which Hugo was chief poet, Sainte-Beuve chief critic.] **1889** H. F. WOOD *Englishman Rue Cain* xiii, The pundit of some cenacle in Garrick Street. **1899** A. C. LYALL *Asiatic Studies* 2nd Ser. 61 Within the philosophic cenacle I should entirely concur. **1926** *Contemp. Rev.* Jan. 85 In England and France schools and 'cénacles' of literature sprang up round the great writer of the moment.

Cenomanian (sĭnomēⁱ·niăn, sen-), *a.* *Geol.* [ad. F. *cénomanien*, f. med.L. *Cenomania*, now Le Mans, France, f. L. *Cenomani*, an ancient Gallic tribe of northern Italy: see -IAN.] Designating a subdivision of the Upper Cretaceous period, corresponding to the Lower Chalk and Upper Greensand of British geologists; of flora, etc., belonging to or found in these strata.
1882 GEIKIE *Text-bk. Geol.* 829 The Cenomanian formation consists in Hanover of earthy limestones and marls. **1902** *Encycl. Brit.* XXXI. 434/2 The Cenomanian flora of central Europe. **1914** *Brit. Mus. Return* 206 The Cenomanian Chalk Marl of Cambridge. **1960** L. D. STAMP *Britain's Struct. & Scenery* (ed. 5) xii. 138 A great marine transgression..referred to as the 'Cenomanian Transgression'.

cenosite (se·nosəit). *Min.* Also **kainosite, cainosite.** [ad. *kainosit* (A. E. Nordenskiöld 1886, in *Geol. Föreningens Förhandl.* VIII. 143), f. Gr. καινός novel, in reference to its unusual composition, + -ITE¹.] A hydrous silicate and carbonate of yttrium, erbium, and calcium, occurring in Norway and Sweden in the form of greasy, yellowish-brown, prismatic crystals.
1888 *Jrnl. Chem. Soc.* LIV. 234 Kainosite. **1892** E. S. DANA *Dana's Syst. Min.* (ed. 6) 698 Cenosite. **1930** *Amer. Min.* XV. 207 The Nordmark cenosite occurs in druses, which have apparently originated through the solution of calcite.

cenotaph, *v.* [f. the sb.] *trans.* To honour or commemorate with a cenotaph. So **ce·notaphed** *ppl. a.*
1845 PLANCHÉ *Golden Fleece* II. 29 Lying, dying, Cenotaphed, and paragraphed. **1891** KIPLING *Light that Failed* iv, The oblivion that is preceded by toleration and cenotaphed with contempt. *a* **1893** J. H. BONER *Poe's Cottage* vii. (Funk), And fate that then denied him, And envy that decried him, And malice that belied him, Have cenotaphed his fame.

cenote (senōᵘ·te). [Yucatan Sp., f. Maya *conot.*] A natural underground reservoir of water, such as occurs in the limestone of Yucatan.
1841 J. L. STEPHENS *Incidents Travel Cent. Amer.* II. xxiii. 408 The servant urged us to go immediately and see a cenote...It was a large cavern or grotto, with a roof of broken, overhanging rock..and at the bottom water pure as crystal..resting upon a bed of white limestone rock. **1902** *Encycl. Brit.* XXV. 379/2 The cenotes or underground reservoirs were the important factors in locating the ruins of Northern Yucatan. **1902** *Amer. Anthropologist* IV. 128 Offerings..were cast into..the deep *cenotes*, or natural wells, to appease the gods believed to dwell therein. **1940** *Geogr. Jrnl.* XCV. 123 All of the ancient cities were located close to sink holes or cenotes. **1963** D. W. & E. E. HUMPHRIES tr. *Termier's Erosion & Sedimentation* xiv. 308 The forms most typical of karst are the *cenotes*.

cense, *v.*¹ **2.** Delete †*Obs.* Cf. CENSING *vbl. sb.*¹, *CENSING *ppl. a.*

censing, *vbl. sb.*¹ (Later example.)
1873 J. B. DYKES in Fowler *Life* (1897) 306 The prevalence of the practice of censing in the diocese.

ce·nsing, *ppl. a.* [f. CENSE *v.*¹ + -ING².] That censes or offers incense.
1893 *Athenæum* 26 Aug. 296/2 A censing adult angel.

censor, *sb.* Add: **2. b.** More explicitly *dramatic censor, film censor.*
1839 E. BULWER *Let.* 20 Dec. in C. H. Shattuck *Bulwer & Macready* (1958) 148 Wd the office of censor (Dramatic) be one..agreeable to yourself? **1888** *Encycl. Brit.* XXIII. 227/2 The master of the revels appears to have been the dramatic censor from 1545 to 1624. **1912** *Times* 30 Nov. 4/6 Mr. G. A. Redford..now Censor of Cinematograph Films, is engaged in organizing the British Board of Film Censors, of which he is the director. **1930** G. B. SHAW in *Times* 17 Feb. 15/5 Mr. Edward Shortt, who lately succeeded the late T. P. O'Connor as Film Censor.
e. An official whose duty it is to censor private correspondence (as in time of war: cf. *CENSOR *v.*).
1914 (*Stamped on envelope of soldier's letter from the Front*) Passed by Censor. **1915** 'IAN HAY' *First Hundred Thou.* xix. 290 These [letters]..are stamped with the familiar red triangle and forwarded to the Base, where they are supposed to be scrutinised by the real Censor. **1918** B. MIALL tr. *A. Hamon's Lessons of World-war* 146 In a letter written from England to a French non-commissioned officer, the censor cut out all that concerned the miners' strike in Wales! *Ibid.* 147 Correspondence from neutrals was examined and bore the paper band now well known in the United Kingdom: 'Opened by the censor'.
4. *Psychol.* [Mistranslation of G. *zensur* censorship.] A mental power or force which represses certain elements in the unconscious and prevents them from emerging into the conscious mind. Also *attrib.* Cf. *CENSORSHIP 3.
1912 *Maclean's Mag.* Nov. 152/1 But the 'censor', to use the Freudian terminology, is easily deceived. **1913** BRILL tr. *Freud's Interpr. Dreams* vi. 287 We may state, as a second condition which the elements must satisfy in order to get into the dream, that they must be withdrawn from the censor of resistance. **1920** B. LOW *Psycho-Analysis* 63 This force of the Unconscious..at times so powerful that the Censor-barrier..is swept aside. **1926** G. COSTER *Psycho-Analysis* 200 The endopsychic censor, the judge within the soul.

censor, *v.* Delete *rare* and add examples having special reference to the control of news and the departmental supervision of naval and military private correspondence (as in time of war) or to the censorship of dramatic or cinematographic productions. Often in *ppl. a.*
1895 *Blackw. Mag.* Feb. 320/1 The severely censored columns of the Russian daily press. **1897** *Daily News* 13 Sept. 5/2 All news is being rigorously censored. **1899** *Westm. Gaz.* 9 Mar. 5/3 Everything read to him is carefully censored. *Ibid.* 22 Nov. 2/1 Finally the correspondent is handed back his censored despatch, probably reduced to a mere sequence of words conveying little intelligence. **1904** *Daily Chron.* 13 May 4/4 Censoring is very strict during the war. No news is allowed to be published which has anything to do with the movements of the army. **1907** *Westm. Gaz.* 21 Oct. 9/2 Ibsen's censored play entitled 'Ghosts'. **1915** R. BROOKE *Let.* Apr. (1968) 681, I must go & censor my platoon's letters.

censorable (se·nsərăb'l), *a.* [f. CENSOR *v.* + -ABLE.] Subject to censoring; in need of censoring.
1906 *Westm. Gaz.* 27 Jan. 2/2 To avoid Censorable references to prominent politicians. **1926** *Spectator* 30 Oct. 767/2 Mr. Gilhooley is not more censorable, and not less suitable for what used to be called family reading, than psycho-analytic text-books. **1966** D. VARADAY *Gara-Yaka's Domain* xvii. 184 My remarks to, and concerning Freddie in particular and the masonja in general..were highly censorable.

censorial, *a.* **2.** Delete †*Obs.* and add later example.
1866 BP. GOTT *Lett.* (1918) 145 The very nature of a creed is personal, never censorial of others.

censorship. Add: **2. a.** *spec.* (see *CENSOR *sb.* 2 b, e).
1879 *Encycl. Brit.* IX. 143/1 A notable incident in the history of the stage, inasmuch as it led to the institution of the dramatic censorship. **1912** *Times* 4 July 6/2 (*heading*) Cinematograph film censorship. **1918** B. MIALL tr. *A. Hamon's Lessons of World-War* 146 The censorship of all correspondence was a stupendous task. **1930** *Times* 17 Feb. 15/5 Mr. Shaw on Film Censorship.
3. *Psychol.* = *CENSOR *sb.* 4; the function of a mental censor.
1924 W. B. SELBIE *Psych. Relig.* 80 Unpleasant experiences..driven out of consciousness, and kept there by means of what Freud calls a censorship. *Ibid.* 90 The repressions and censorships of which Freud and others make so much are connected with changes in the nervous system. **1925** tr. *Freud's Coll. Papers* IV. iii. 54 Dream-formation takes place under the sway of a censorship which compels distortion of the dream-thoughts.

census, *sb.* Add: **3. b.** *census-table, -taker.*
1831 J. M. PECK *Guide for Emigrants* 199 The census table of last autumn. **1840** *Picayune* (New Orleans) 25 Aug. 2/2 The following took place between a census taker and a married lady. **1936** *British Birds* XXIX. 264 To guide future census-takers through some of the difficulties.

cent¹. **4. a.** (Further examples.) See also RED *a.* 3 c.
1803 J. DAVIS *Trav. in U.S.A.* 389 But I never wronged Master of a cent. **1830** 'MAJOR J. DOWNING' *Let.* 17 Feb. in *Life & Writings* (ed. 2, 1834) 68 They don't seem to rip up worth a cent since the first night they begun. **1872** E. EGGLESTON *End of World* 11, I don't believe that you'd care a cent if she did marry a Dutchman! **1902** HARBEN *Abner Daniel* 59 He'd cut her off without a cent. **1947** K. TENNANT *Lost Haven* (1968) 12 If you marry that Alec, you don't get a cent. *Ibid.* i. 23 He left with my schnapper boat that he promised to pay for and never paid a damn cent. **1968** *Globe & Mail* (Toronto) 17 Feb. 23/6 If Vancouver just had to pay for an orchestra it wouldn't make a cent.

centaurea (sentǭ·riă, sentǫ·riă). [med.L. *centaurea*: see CENTAURY. Adopted by Linnæus in his *Species Plantarum* (1753) II. 909 as the name of a genus.] A plant of a large genus of herbs, belonging to the family Compositæ, and including the cornflower.
1829 J. C. LOUDON *Encycl. Plants* 735 *Centaurea.* It is said, that with this plant, the *Centaur* Chiron cured the wound in his foot made by the arrow of Hercules. **1835–6** MAUND *Botanic Garden* VI. No. 506 Chiron, the wise centaur, having been represented as using the original Centaurea in the cure of a wound. **1864** OLIVER *Elem. Bot.* 190 Corn Centaurea, or Bluebottle. **1900** *Daily News* 5 May 4/5 The centaurea has for long been regarded as only half-hardy. **1907** *Westm. Gaz.* 18 July 6/3 An exceedingly fine show of roses and centaureas. **1955** A. G. L. HELLYER *Flowers* 30 The centaureas may be considered in three distinct groups.

centavo (sentā·vo, ‖þentā·vo). [Sp., f. L. *centum* a hundred.] A small coin of Spain and Portugal, and of Central and South America.
1883 *Encycl. Brit.* XVI. 733/1 Chili—100 Centavos = 1 Peso. **1896** *Westm. Gaz.* 18 Mar. 8/1 The 1-centavo stamp is orange-yellow,..the 20-centavos black. **1920** *Chambers's Jrnl.* 465/2 Two centavos to the escudo is the 'assistance' tax in hotels and restaurants.

centennial, *a.* (*sb.*). (Further examples.)
1940 E. H. MCCORMICK *Lett. & Art in N.Z.* p. viii, Some readers will have seen..the Centennial Exhibition of New Zealand Art. *Ibid.* vii. 194 Government patronage of artists has more recently marked the celebration of New Zealand's Centennial. **1961** *Spectator* 17 Mar. 382/1 Many English newspapers..are now referring to the Civil War 'centennial'.

‖ **centesimo** (tʃente·zimo). Pl. **centesimi.** [It.] An Italian coin of the value of one hundredth of a lira.
1851 E. RUSKIN *Let.* 2 Nov. in M. Lutyens *Effie in Venice* (1965) II. 210 Gondoliers carrying people across..for which they were paid a centesimo. **1922** D. H. LAWRENCE et al. tr. *Bunin's Gent. from San Francisco* 33 He gambled away to the last farthing all those copper coins.., in all..forty centesimi. **1959** E. POUND *Thrones* civ. 90 And a $ to look like a franc, the franc worth a centesimo.

centibar (se·ntibāɹ). *Meteorol.* One hundredth of a bar (see *BAR *sb.*⁸ 2).

centimetre. Add: **b.** *attrib.* and *Comb.*, as **centimetre-gramme-second,** used attrib. to designate a system of measurement introduced in 1874 in which the unit of length is the centimetre, the unit of mass the gramme, and the unit of time the second; commonly abbreviated *C.G.S.*; **centimetre wave,** an electromagnetic wave of wavelength between 1 and 10 centimetres.
1875 J. D. EVERETT (*title*) Illustrations of the Centimetre-Gramme-Second System of Units. **1883** *Encycl. Brit.* XV. 699/1 In the Centimetre-Gramme-Second system of units, the absolute unit of force produces in one second, in a mass of one gramme, a velocity of one centimetre per second. **1884** F. KROHN tr. *Glaser de Cew's Magn. & Dyn.-Electr. Mach.* p. xiii, The absolute or C. G. S. (centimetre-gramme-second) units. **1940** *Chambers's Techn. Dict.* 147/1 Centimetre waves. **1947** CROWTHER & WHIDDINGTON *Sci. at War* Pl. xxiii, The aerial of a centimetre-wave radar.

centimetric (se:ntime·trik), *a.* [f. CENTIMETRE + -IC.] A centimetre in length; *spec.* designating or employing centimetre waves (see prec.).
1941 S. R. ROGET *Dict. Electr. Terms* (ed. 4) 54/1 *Centimetric waves*, electro-magnetic waves of a wave-length between 0·1 and 0·01 metre. **1945** *Times* 15 Aug. 2/1 The result was a centimetric ground Radar set. **1946** *Nature* 6 July 33/1 The numerous war-time applications of centimetric waves.

centimo (s-, ‖ þe·ntimo). [Sp.] A Spanish coin of the value of one hundredth of a peseta.
1899 R. ST. BARBE *Mod. Spain* 64 Countless five-centimo cigars. **1926** *British Weekly* 20 May 125/3 Dr. Irwin exhibited an indulgence which he bought for 70 centimos (7*d.*) in Madrid.

centipede. Add: **c.** *Naut.* A device consisting of a long piece of wood pierced with holes through which ropes are rove, used for suspending an awning. Also, a strong piece of rope running the length of the boom, with short cross-pieces used in stowing jibs.
1883 *Man. Seamanship for Boys' Training Ships* 138 A Centipede, or, as it is sometimes called, a Euphroe..is used as a crowfoot, fitted with a number of legs, for ridge ropes of awnings.

centipoise (se·ntipoiz). *Physics.* [f. CENTI- + *POISE *sb.*²] A unit of dynamic viscosity, one hundredth of a *POISE.
1916 P. C. MCILHINEY in *Jrnl. Ind. & Engin. Chem.* VIII. 434/1 It would therefore seem to be a rational procedure to use the name 'poise' in the way suggested by Deeley and Parr, to use the 'centipoise' which is one-hundredth the size of this as a unit for practical measurements. **1935** *Jrnl. R. Aeronaut. Soc.* XXXIX. 916 Gasoline viscosities range from about 0·40 centipoises to 0·50 centipoises at 30° C. **1962** J. T. MARSH *Self-Smoothing Fabrics* xi. 160 The final value of the viscosity is expressed in centipoises, against the value of unity for water.

centisecond (se·ntisekənd). [f. CENTI- + SECOND *sb.*¹] One hundredth of a second.
1950 J. R. FIRTH *Papers in Linguistics 1934–51* (1957) xiii. 174 When we measure any features of a kymogram in centiseconds, we must realize what the machine itself abstracts from the single instance and what is abstracted by the employment of the kymogram. **1964** *Amer. Speech* XXXIX. 230 Eight Swedish vowels spoken between three voiced plosives /b/, /d/, /g/ at five syllabic rates produced vowels of approximately 5, 10, 15, 20, and 25 centiseconds. **1969** *English Studies* L. 327 As regards [bdg], the average duration of this element turned out to be 3 centi-seconds.

centistoke(s (se·ntistōᵘk, -ōᵘks). *Physics.* [f. CENTI- + *STOKE(S *sb.*⁴] A unit of kinematic viscosity, one hundredth of a *STOKE(S.
1934 *Proc. World Petroleum Congress 1933* II. 508/2 As the unit of kinematic viscosity..the stokes and the derived centistokes have been adopted. **1937** *Jrnl. R. Aeronaut. Soc.* XLI. 735 The two oils had nearly the same viscosity at 50° C., viz., 78 and 76 centistokes respectively. **1956** *Electronic Engin.* XXVIII. 59 Except for the very low viscosity members (0·65 centistokes to about 50 centistokes) the dielectric properties vary very little with viscosity.

central, *a.* Add: **1. d.** *Phonetics.* Of a vowel: formed with the tongue in a middle position between front and back. Cf. *BACK *a.* 1 c; *FRONT *sb.* 13 b; *MIXED *ppl. a.* 6 b.
1932 D. JONES *Outl. Eng. Phonetics* (ed. 3) xiv. 86, a: is a central vowel; in other words the central part of the tongue is raised in order to make it. **1933** JESPERSEN *Ess. Eng. Gram.* 24 The vowels..are either front, central, or back, according to the part of the tongue that is highest. **1965** W. S. ALLEN *Vox Latina* 4 Vowels intermediate between front and back are referred to as central.
2. a. Also const. *to.*
1902 *Daily Chron.* 10 June 3/2 The older views of the authority and inspiration of Scripture are central to his whole system. **1937** *Tablet* 23 Oct. 541/1 The Christian religion..is in no sense central to their conception of the African's future. **1957** N. FRYE *Anat. Crit.* 243 Literature seems in a way to be central to the arts.
4. *Central American* adj. and sb., (an inhabitant or native) of Central America, a region extending roughly from Mexico to Colombia; *Central European* adj. and sb., (an inhabitant or native) of Central Europe; cf. **middle-European*; *central heating*, a system of heating a building by hot water, steam, or air conveyed through pipes or ducts from a central source; also applied to any form of heating which simultaneously warms several rooms in a building; so *central-heated* adj., supplied with central heating; *central school*, a secondary school of a type which by the Education Act of 1944 usually became 'secondary modern' schools (see quot. 1919).
1857 LD. NAPIER *Let.* 12 Mar. in *Parl. Pap. 1860* LXVIII. 695 A settlement of the Central American question in conformity with the spirit of the Treaty. **1934** A. HUXLEY *Beyond Mexique Bay* 46, I have seen a fair amount of Central American art. *Ibid.*, The art of the Central Americans. **1959** T. S. ELIOT *Elder Statesman* III. 86 He's Federico Gomez, the Central American. **1931** P. MACDONALD *Crime Conductor* II. iv. 176 Donna Sigsbee..has tired of the central European charms of Mr. Vanesco. **1940** N. MARSH *Surfeit Lampreys* (1941) ii. 34 Aunt V. is..all Central-European. **1941** A. HUXLEY *Grey Eminence* viii. 187 Large numbers of Central Europeans were in process of being starved and slaughtered. **1967** J. B. PRIESTLEY *It's Old Country* iii. 26 Oldish women with Central European accents. **1922** D. H. LAWRENCE *Aaron's Rod* xiii. 160 The fresh morning air comes startling after a central-heated house. **1906** *Internat. Libr. Technol.* LXXIV. § 38. 5 For use in connection with central heating plants, such as are frequently installed for heating the various buildings of a university or state institution, the vacuum system of heating is well adapted. **1913** F. W. RAYNES *Heating Systems* i. 14 A good method of reducing the production of smoke is by district and large central heating systems in which either water or steam is circulated through the pipes. **1921** R. HICHENS *Spirit of Time* iii, There was no central heating on. **1951** *Good Housek. Home Encycl.* 35/2 The efficiency of a central-heating system naturally depends on that of the boiler. **1919** A. W. NEWTON *Eng. Elementary School* ix. 167 Central schools.. are intended to attract the best scholars from the upper classes of neighbouring [elementary] schools, and to provide a higher course than the ordinary school can offer. **1956** J. E. FLOUD *Social Class* 15 The central schools..are officially designated 'secondary modern schools'.

central, *sb.* *U.S.* [f. the adj.] A central telephone exchange; hence, any telephone exchange.
1889 'MARK TWAIN' *Conn. Yankee* xv. 184, I used to wake ..and say 'Hello, Central!' just to hear her dear voice. **1908** *Busy Man's Mag.* Mar. 113/1 You may establish a direct, permanent connection with the doctor, the fire department, or the police and call them immediately without the services of 'central'. **1948** *Chicago Sun-Times* 7 Sept. 47/3 He thought all those other rings were somebody calling 'central'.

Centralia (sentrēⁱ·liă). [f. CENTR(AL *a.* + AUSTR)ALIA.] A name orig. proposed for what was then South Australia, but now used to designate the remote central area of Australia. Hence **Centra·lian** *a.* and *sb.*
1896 J. S. LAURIE *Story of Australasia* xxxiii. 299 The name 'South' Australia, embracing as it does both the central and the northern territory, is anomalous...For telegraphic, postal, and general purposes *one* word is desirable for a name—*e.g.* why not *Centralia*; for West Australia, *Westralia*; for New South Wales, *Eastralia*? **1931** I. L. IDRIESS *Lasseter's Last Ride* i. 3 Through Centralia runs a continuation of the famous Kalgoorlie ore belt. **1936** —— *Cattle King* xix. 179 The utter stillness of a Centralian night. **1944** F. CLUNE *Red Heart* 1 Although his title [*sc.* 'Dead Heart'] applied more particularly to the sunken Lake Eyre region, it was gradually extended to the whole Centralian Area. *Ibid.* 15 Many a camp-fire in Centralia. **1964** *Panorama* (Brisbane) Sept. 7/3 As far as Centralians are concerned, manana is good enough for them!

centralism. Add: (Earlier example.) Also, *democratic centralism* (see quot. 1951).
1831 *Deb. Congress* 2 Feb. 51 A system of centralism, hostile to the federative principle of our Union. **1951** *Britannica Bk. of Yr.* 686/1 *Democratic centralism*, a Communist system of government in which recommendations are passed upwards through successive stages of popular assemblies. **1963** *Ann. Reg. 1962* 22 Mr Michael Foot..charged the Gaitskellites with Soviet-style 'democratic centralism' in running the party.

centralist. (Earlier examples.)
1826 E. CRAVEN *Mem. Margravine of Anspach* II. i. 15 Masons of every denomination,—Rosicrucians, Centralists, Illuminati,—had all, under his reign, the liberty of establishing lodges. **1836** D. B. EDWARD *Hist. Texas* 125 The second faction of this party, called the Centralists, wish to..form a central government. **1849** *Tait's Mag.* XVI. 241/2 A federalist must in Texas be a welcome man, simply because General Santa Anna was a centralist.

centralization. Add: **3.** *Phonetics.* The pronunciation of a vowel in the central (*CENTRAL *a.* 1 d) position.
1962 *Amer. Speech* XXXVII. 169 A third distinctive characteristic of Ocracoke pronunciation..is the centralization of the onsets of the diphthongal allophones of /i/ and /e/. **1964** A. C. GIMSON in D. Abercrombie et al. *Daniel Jones* 134 The current complete centralization of the first element in all situations.

centralized, *ppl. a.* Add: **2.** *Phonetics.* Of a vowel: made central (see *CENTRAL *a.* 1 d).
1964 J. C. CATFORD in D. Abercrombie et al. *Daniel Jones* 35 A 'muffled' or 'centralized' vowel-quality. **1965** W. S. ALLEN *Vox Latina* ii. 57 The sound in question must..have been a more centralized, i.e. fronted, variety of *u* than the inherited short *u*.

centre, *sb.* Add: **6. a.** *spec.* A place or a collection of buildings forming a central point in a town, district, etc., or the main area for a particular activity, interest, or the like; freq. with defining word, as *city centre, civic centre, shopping centre, training centre* (see *CITY, *CIVIC *a.*, etc.). orig. U.S.
1843 J. HAYWARD *Gaz. U.S.* 48 Buxton Centre. *Ibid.* 49 Cumberland Centre. **1855** E. A. CHARLTON *New Hampshire* 295 At Meredith Centre are a meeting house..and three stores. **1896** *Punch* LXI. 143 But here we are at Market Street...This is the spot our parsons call 'The city's pulsing centre'. **1898** E. HOWARD *To-morrow* vii. 75 The Arcade is ..designed to be..the great shopping centre of the town. **1928** [see *TRAINING *vbl. sb.* 5]. **1959** *Cambridge Rev.* 9 May 493/1 The planners of Plymouth have..built a 'centre' which is tolerable only to a man in a hot tin box who has already become a part of the box. **1961** *Deb. Senate Canada* 25 Sept. 1170/2, I am delighted to know that the Prime Minister is interested in having a centre in Ottawa for the performing arts. **1970** *Sunday Times* 19 Apr. 27/5 The new arts centre would operate all year round with summer schools and special courses.
11. d. In various games, a player on each side whose position is the middle of a line or field of players; esp. in rugby football = centre-threequarter; in other games, the player in the middle of the field.
1866 *Spirit of Times* (N.Y.) 23 June 262/3 Edwards, as centre, is decidedly the best fielder in the nine. **1868** *Chambers's Encycl.* X. 597/1 La Crosse... In the arrangement of the men on each side..centre is in the centre of the field. **1892** *Young England* Sept. 442/1 The big 'centre', taking in the situation at a glance, slips round. **1897** *Encycl. Sport* I. 411/2 [Rugby football.] Three Three-quarters.—The Centre—A centre is the correlative of the half-back. *Ibid.*, The centre will, of course, be on the look out for a drop at goal. **1922** *Daily Mail* 8 Dec. 12 M. A. Ap-Rhys Price, Marlborough's best centre, was unable to play.
e. Of chocolates and other sweets: the central portion which is enclosed in chocolate or other covering.
1877 [see *PAN-DROP]. **1929** *Sears, Roebuck Catal., Spring & Summer 1929* 469 Extra fine..filled and hard candies. A tempting assortment with fruit and nut centers. **1935** *Discovery* Nov. 322/1 The centres [of the chocolates] are passed by an automatic feeding attachment through a curtain of temperature-controlled chocolate. **1959** *Economist* 13 June 1015/2 A hard covered chocolate with a soft centre.
f. *Cricket.* A guard (GUARD *sb.* 3 b) covering the middle stump.
1887 *Cricket* 24 Nov. 460/1 Miss Rogers..fell when taking 'centre'. **1923** E. W. HORNUNG *Old Offenders* 250 He waited by the umpire while Chrystal took centre.
g. *Football* and *Hockey.* A kick or pass from either of the wings towards the centre of the pitch. Cf. *CENTRE *v.* 4 b.
1900 *Westm. Gaz.* 23 Apr. 8/3 The forwards..were never where they should have been when centres..were put across. **1927** *Daily Express* 20 Apr. 13/2 Hill met with his head a glorious centre by Williams. **1967** J. POTTER *Foul Play* vi. 79 Bob enjoyed two dramatic races down the right wing, but they came to nothing when Julian wasn't fast enough to pick up the centres.
h. *the Centre*, the remote interior of Australia, Central Australia. See also **Red Centre.*
1912 SPENCER & GILLEN *Across Austral.* I. i. 5 Central Australia and the Northern Territory..consists of four distinct parts...Of these four areas the second and the third, which together occupy a large part of the Centre, may be known as the Australian Steppe lands. **1934** A. RUSSELL *Tramp-Royal in Wild Austral.* ix. 71 For years he had spent his time drifting about 'the Centre', which he probably knew as well as any man living. **1954** B. MILES *Stars my Blanket* xxiii. 198, I asked him if he had much trouble with scrub bulls as they do in The Centre.
15. (Further examples.)
1907 [see *HISTORY *sb.* 4 c]. **1962** *Listener* 3 May 774/2 Those socialists and others orientated left (or right) of centre who are automatically biased against the United States. **1962** *Ibid.* 19 July 87/2 All three parties in Bonn are now parties of the centre, and there is virtually no opposition.
16. *centre of gravity*: also *fig.*, the point or object of greatest importance or interest.
1903 'VIGILANS SED AEQUUS' *German Ambitions* 48 Loss of millions of Germans..has moved the world's centre of gravity in a sense unfavourable to them [*sc.* the German people]. **1937** *Discovery* Sept. 266/2 The wage-earner is now the centre of political gravity. **1963** *Higher Educ.* (Cmnd. 2154) x. § 384 The centre of gravity should be in science and technology.
19. centre-bully, a bully taken in the middle of the field at the start or re-start of play (see *BULLY *sb.*² 2); **centre-forward,** (*a*) a player in association football, hockey, and other games, who plays from a central position in the forward line; (*b*) the position occupied by that player; **centre-half, centre half-back,** (*a*) a player who plays from a central position among the half-backs; (*b*) the position of that player; **centre light,** a lamp suspended from the centre of a ceiling; **centre-line,** any line passing through a centre;

CENTRE

spec., in drawing, a line from which measurements are made, and in ship-building, a line passing lengthways through the hull and dividing it into two sections; in various games, a line of centres (sense *11 d); **centre-piece** (earlier U.S. example); (*b*) *fig.* the most conspicuous or important item in a collection, exhibition, etc.; **centre-plate,** (*a*) each of the metal plates composing the bearing for a railway carriage or engine on the centre of the truck; (*b*) each of the metal plates used to hold a dowelled pattern while it is being turned in the lathe; (*c*) a metal centre-board; **centre section** *Aeronaut.* (see quot. 1950); **centre spread,** printed matter occupying the two facing middle pages of a newspaper, periodical, etc.; **centre-square,** an instrument for finding the centre and radius of a circle; **centre-table,** a table intended for the centre of a room, formerly often used for the display of books, albums, etc.; **centre-zero** *a.*, designating a meter which has zero at the centre of the scale and can therefore indicate both positive and negative values of the quantity registered.

1910 *Westm. Gaz.* 1 Mar. 16/2 From the centre-bully Cambridge again pressed. 1891 *Peel City Guardian* IX. 7/3 The teams were as follows:..centre forward [etc.]. 1897 *Encycl. Sport* I. 421/1 The centre-forward is in the middle of the other four; the two on each side of him form the wings. 1891 *Peel City Guardian* IX. 2/4 Our friend John at centre-half. 1908 *Westm. Gaz.* 30 Nov. 14/4 If he were to stick to centre-half always. 1920 E. Green *Hockey* iv. 38 More work falls on the centre half than to the other two put together. 1890 C. W. Alcock *Football* vii. 46 An inside man can give him a pass when he is clear from the centre half-back. 1969 *Melbourne Truth* 12 July 23/2, I prefer Peter Steward at centre half-back. 1878 Lady C. Schreiber *Jrnl.* (1911) II. 128 Excellent centre light, or chandelier, of hammered iron. 1966 G. Greene *Comedians* I. iii. 94, I found Marcel hanging from his own belt from the centre light. 1777 *Monthly Review* LV. 306 Section through the center line of the Register Office, from North to South. 1807 *Monthly Mag.* 1 Feb. 66/1 Placed in the centre line of the vessel. *Ibid.*, Its relative distance from the centre line of the vessel's progress. 1887 D. A. Low *Machine Draw.* 2 After drawing the centre line of any part the dimensions of that part must be marked off from the centre line. 1920 E. Green *Hockey* i. 24 *Centre Line.*—A line across the centre of the ground with a small circle in the centre from which the initial bully takes place. 1967 *Jane's Surface Skimmer Systems 1967–68* 90/2 A companion ladder at the centreline leads to the main deck foyer. 1969 *Australian* 24 May 39/3 There is not likely to be a great deal in the centreline duels. 1803 *Deb. Congress* 10 Jan. (1851) 345 An appropriate and characteristic equestrian statue..as a beautiful centre piece for the entire plan. 1937 *Discovery* Dec. 361/2 The centre-piece of the exhibition is C. T. R. Wilson's original expansion-chamber apparatus. 1965 *Economist* 16 Jan. 214/2 President Johnson's proposals on education..are the centrepiece of his efforts to raise the quality of American life. 1875 M. N. Forney *Catech. Locomotive* XII. 316 The weight of the front end of the engine rests on a cast iron centre-plate. *a* 1884 Knight *Dict. Mech.* Suppl. 185/1 Center-plate. 1894 *Outing* (U.S.) XXIV. 194/2 The movable center-plate will always let you know when you get on a shoal. 1917 'Contact' *Airman's Outings* 9 The sloping half-strut of his top centre section. 1921 *Flight* XIII. 147/1 The wings are of wood construction, fabric covered, and are attached, high up on the *fuselage*, to a 'centre section' which forms the roof of the cabin. 1950 *Gloss. Aeronaut. Terms* (*B.S.I.*) I. 36 *Centre section*, the portion of the fuselage or hull forming a continuous structure with the main planes. In a biplane or a parasol monoplane, the central structure to which the main planes are connected. 1940 G. Butler *Kiss Blood* xi. 201, I turned the paper open at the centre spread, where they put the next most important news. *a* 1877 Knight *Dict. Mech.* III. 2294/1 The center-square, for finding the center of a circle. 1886 Centre square [see Radius 5]. 1833 *Knickerbocker* I. 158 The whole family are collected around the centre table. *c* 1845 C. Brontë *Professor* (1857) I. iv. 59 He removed from the centre-table to the side-board a few pamphlets. 1868 Centre-table [see sense 18 in Dict.]. 1902 W. James *Var. Relig. Exper.* xix. 460 A 'home' upon a veldt or prairie with one sitting-room and a Bible on its centre-table. 1929 *Star* 21 Aug. 13/3 Polished walnut centre table, overmantel to match. 1940 *Chambers's Techn. Dict.* 148/2 Centre-zero instrument. 1963 B. Fozard *Instrumentation Nucl. Reactors* viii. 90 The indication is provided by a centre-zero moving coil instrument.

centre, *v.* Add: **2. b.** *to centre* (or *be centred*) *about, around* or *round*: to have (something) as one's or its centre or focus; to move or revolve round (something) as a centre; to be concentrated on, to turn on (see Turn *v.* 3); to be mainly concerned with.

1868 [in Dict., sense 2]. 1870 *Chambers's Jrnl.* 12 Nov. 731/1 The only man she had ever loved—around whom centred her most precious memories. 1878 *Harper's Mag.* 308/2 The real interest of the story centres about the lives of four personages. 1886 *Chambers's Jrnl.* 17 July 449/2 The assistant's chief hardships centre round the abnormal length of his working-day. 1893 H. B. Clarke *Spanish Lit.* 163 The plot invariably centres round the love intrigue of persons in the middle or upper classes of life. 1893 Kipling *Day's Work* (1898) 7 The little cluster of huts where he and his gang lived centred round the tattered dwelling of a sea-priest. 1909 *Daily Chron.* 27 July 4/6 Its most enduring traditions centre round the ancient cathedral. 1929 *Times Lit. Suppl.* 3 Oct. 753/2 The group of gifted men and women who centred round Henry Adams. 1931 A. L. Rowse *Politics & Younger Generation* 271 The foremost problems in European politics..will centre round the revision of the Treaty of Versailles.

4. Also, to occupy, distinguish, or mark the centre of; *fig.* to be the central point of.

1886 W. J. Tucker *E. Europe* 260 There is the eternal 'Kugelhupf' (a genuinely Austrian coffee-cake) centring the table. 1894 *Daily News* 2 July 3/4 When the scarlet liveries of Royalty are seen centreing the procession across the bridge. 1898 *Ibid.* 27 May 4/6 Centring the Hall was the dais. 1902 C. N. & A. M. Williamson *Lightning Conductor* 95 This château country of the Loire..centreing as it did the old court life of France.

b. In football and hockey, to kick or hit (a ball) towards the centre of the line of forwards. Also *absol.* Cf. *centre *sb.* 11 g.

1890 C. W. Alcock *Football* vii. 47 When..the ball is centred..from the left wing. 1891 *Peel Chron.* 14 Mar. 5/4 Radcliffe neatly centred. 1897 *Daily News* 22 Feb. 8/6 The latter, as..the Cambridge back dashed at him, again centred the ball. 1904 *Daily Chron.* 15 Feb. 8/4 Wright ran smartly down on the left and centred. 1920 E. Green *Hockey* v. 50 Centering while travelling at top speed..is a very difficult stroke to perform with any degree of accuracy.

ce·ntre-boa:rder. [f. Centre-board.] A boat with a centre-board.

1886 *Outing* (U.S.) VIII. 58/1 The boats are necessarily of light draught and center-boarders. 1897 *Ibid.* XXX. 337/1 The slippery bilge of an eggshell centreboarder.

-centric (sentrik), *suffix*, as in Concentric, Eccentric *adjs.*, forming adjs. with the sense 'having (such) a centre', as Polycentric *a.*; 'having a specified centre', as Anthropo-, *ethno-, Helio-, Homocentric *adjs.*

2. In *Biol.* [perh. f. *centr(omere)], 'having the centromere attached at a specified point', as *acro-, *meta-, *telocentric *adjs.*

centrifugal, *a.* Add: **2. centrifugal casting**, the casting of objects (usu. cylindrical) in a rotating mould.

1925 *Jrnl. Iron & Steel Inst.* CXII. 38 A method of centrifugal casting seemed to have come into commercial use. 1940 J. D. Jevons *Metall. Deep Drawing* xvi. 595 Centrifugal casting has always attracted the attention of imaginative foundrymen.

5. Obtained by the use of a centrifuge.

1880 *Army & Navy Co-op. Soc. Price List* 34 Sugars..Centrifugal, or White Crystals. 1901 *U.S. Dept. Agric. Year-bk.* 1900 613 Fresh butter, made from sweet centrifugal cream. 1958 *Catal. County Stores, Taunton* June 28 Sugars..centrifugal crystals.

Hence as *sb.*, a centrifuge.

1866 'Mark Twain' *Lett. fr. Hawaii* (1967) 266 Close to the grinder are six centrifugals—small metallic tubs, whose sides are pierced with a few thousand pinholes to the square inch. 1904 Brannt tr. *Bersch's Cellulose* ix. 221 The bleached skeins of silk are..dehydrated in a centrifugal.

centrifugalization (sentri:fiugălǝizē̆ⁱ·ʃǝn). [f. Centrifugalize *v.*: see -ation.] The process of subjecting to centrifugal action.

1900 in Dorland *Med. Dict.* 1901 *Science* 29 Mar. 513 Juices, blood, etc., taken at various stages of the disease, with and without centrifugalization. 1908 *Practitioner* June 830 Sedimentation and centrifugalisation. 1964 M. Hynes *Med. Bacteriol.* (ed. 8) vii. 85 Antibody, however, combines with bacteria under these circumstances and can be removed with them from the solution by centrifugalization.

centrifugalize, *v.* Add: **b.** To subject to a centrifugal process. Also *absol.*

1903 *Science* 6 Mar. 369/2 The blood was..centrifugalized to get rid of the stroma. 1910 *Practitioner* June 866 Hyaline casts may be found..if the urine be centrifugalised. 1925 C. H. Browning *Bacteriology* vi. 122 By..re-suspending the sediment in..sodium chloride solution and again centrifugalising, the corpuscles are freed from..the serum.

centrifugate, *v.* Add: **2.** *trans.* To expel from the centre, *spec.* by centrifugal force; to centrifugalize.

1860 A. J. Davis *Great Harmonia* V. 196 She centrifugated it in a thousand different forms of expression. *a* 1909 *Buck's Handbk. Med. Sci.* I. 564 (Cent. D. Suppl.), The fluid should be centrifugated, the sediment spread on cover slips.

centrifugate (sentri·fiūgĕt), *sb.* [f. Centrifugal *a.* + -ate[2] 3.] Material separated by centrifugation.

1927 *Jrnl. Agric. Res.* XXXV. 825 When centrifugates were used the agreement was better. 1946 *Nature* 7 Sept. 344/1 The emulsions obtained from these pieces of brain or nerve tissue were centrifuged; the centrifugate was placed in a water bath. 1961 *Lancet* 2 Sept. 514/1 On direct microscopy of the centrifugate of this fluid..we observed gram-positive and gram-negative rods.

centrifugation (se:ntrifiŭgē̆ⁱ·ʃǝn). [f. *centrifug(e *v.* + -ation.] The action or process of *centrifuge *v.*

1903 *Phil. Med. Jrnl.* 31 Jan. 201 The mixture was expressed, 30 cm. of this quantity taken, the solid particles separated by centrifugation and the amount of the nitrogen in the solid and liquid parts determined. 1946 *Nature* 31 Aug. 313/1 Liquids are usually separated from solids by centrifugation rather than filtration. 1958 *Immunology* I. i. 3 The cells suspended in the filtrate were washed by two or three gentle centrifugations in Ringer-phosphate. 1961 *Lancet* 26 Aug. 459/1 The homogenates were fractioned by differential centrifugation.

centrifuge, *a.* and *sb.* **B.** *sb.* (Further examples.)

1902 *Chemist & Druggist* 18 Oct. 660/1 A centrifuge..exerts a sifting or separating action on such a mixture. 1935 *Nature* 4 May 719/2 The recognition by Ledingham and Gye, by the use of the high-speed centrifuge, of minute particles resembling those of other viruses. 1937 *Ann. Reg.* 1936 54 By use of the Svedberg centrifuge, X-ray and mathematico-physical methods, advance was made towards an understanding of the molecular architecture of the cell. 1947 Crowther & Whiddington *Sci. at War* 143 Yet another method consists of whirling vapours containing isotopes in a powerful centrifuge. 1957 *Observer* 10 Nov. 1/3 The Farnborough Institute has a human centrifuge in which human toleration to acceleration is tested. 1969 *Listener* 5 June 774/1 It might be possible to separate fissile uranium from the non-fissile variety in which it is naturally mixed by the use of centrifuges—rapidly spinning cylinders containing the gas uranium fluoride.

centrifuge (se·ntrifiūdʒ), *v.* [f. the sb.] *trans.* To subject to centrifugal motion; to separate by means of a centrifuge. So **ce·ntrifuged** *ppl. a.*; **ce·ntrifuging** *vbl. sb.* and *ppl. a.*

1902 *Chemist & Druggist* 18 Oct. 660/1 The creamy layer which collects on the top of the centrifuged liquid is skimmed off. 1903 *Nature* 3 Dec. 111/2 By mixing bacteria with the white cells of the blood obtained by centrifuging and adding blood serum. 1908 *Practitioner* Jan. 12 If the cerebro-spinal fluid withdrawn by this operation is centrifuged, and a deposit found,..we can certainly exclude neurasthenia. 1910 *Ibid.* Apr. 447 Centrifuged specimens. 1920 L. Doncaster *Introd. Cytology* iv. 54 Owing to the centrifuging, the polar spindle of an egg is greatly elongated. 1922 F. W. Aston *Isotopes* 131 If therefore we have a mixture of isotopes in a gaseous or liquid state partial separation should be possible by gravity or centrifuging. 1930 Field & Weill *Electro-Plating* 88 Centrifuging machines are used in dealing with numerous small parts. 1938 R. W. Lawson tr. *Hevesy & Paneth's Man. Radioactivity* (ed. 2) xviii. 169 The precipitated lead hydroxide is centrifuged off.

centring, *ppl. a.* Add: *spec.* of a diphthong (see quot. 1952). Cf. *central *a.* 1 d.

1917 H. E. Palmer *First Course of Eng. Phonetics* I. 23 English diphthongs (or double-vowels)..four (in view of the fact that the tongue passes from the sides to the centre of the triangle) may conveniently be termed the *Centring Diphthongs*. 1952 A. Cohen *Phonemes of English* iv. 91 'Centring diphthongs' (i.e. those with [ǝ] as second element). *Ibid.* 100 There are three of these 'centring diphthongs' to be considered in English, [iǝ, uǝ, ɛǝ]. 1964 R. H. Robins *Gen. Linguistics* iii. 98 These last three diphthongs are often called centring diphthongs, from the direction of their movement.

centriole (se·ntriōᵘl). *Cytology.* [ad. G. *centriol* (T. Boveri 1895, in *Verh. d. Physik.-Med. Ges. Würzburg* XXIX. I. 66), f. mod.L. *centriolum*, dim. of *centrum* centre.] A minute, apparently self-reproducing particle which is present, usu. as one of a pair, in the centrosome and which is important for its part in forming and organizing the mitotic apparatus during cell-division.

1896 E. B. Wilson *Cell* 334 *Centriole*, a term applied by Boveri to a minute body or bodies..within the centrosome. In some cases not to be distinguished from the centrosome. 1912, 1925 [see *centrosome]. 1960 L. Picken *Organiz. Cells* vii. 269 In *Gallus domesticus*, the proximal centriole is a hollow sphere; ..in..sea-urchins, the centriole appears as a curved disc...In the toad..the centriole is a cylindrical structure.

centripetalism (sentri·petăliz'm). [f. Centripetal *a.* + -ism.] Movement towards a centre.

1887 *Westm. Rev.* June 409 The plague of centripetalism is a curse which has come to us [*sc.* New Zealand] across the seas from older countries.

centrism (se·ntriz'm). [f. Centre *sb.* + -ism.] (The policy of adopting) a middle position between extreme views.

1935 R. B. Perry *Tht. & Char. W. James* I. 574 A plea for some sort of philosophical centrism that should preserve the British tradition. 1960 *Commentary* June 543/2 Bell's centrism and 'moderationism' lead him rather seriously astray.

centrist. Add: **b.** *transf.* Also *attrib.* or as *adj.*

1923 E. A. Ross *Russian Soviet Republ.* 22 The 'Berliner Tageblatt' and the 'Vossische Zeitung' demand an understanding with Russia by all means. The Centrists favor an agreement. 1928 F. Utley tr. *Illustr. Hist. Russ. Rev.* I. 135 In international Social Democracy 'Centrists' are those who swing to and fro between the Jingoes..and the Left such as:..MacDonald and Co. in England. 1928 *Manch. Guardian Weekly* 21 Sept. 224/1 The 'Temps' shows some uneasiness..at the attacks on Herr Müller in the Centrist and Populist press. 1958 *Economist* 8 Nov. 504/1 A cautiously centrist position on 'welfare state' issues. 1968 *Guardian* 20 Nov. 9/2 The subsequent struggle for supremacy between Mr Gomulka's 'centrists' and Mr Moczar's 'extremists' is in the classical tradition of Communist intrigue.

centro-. Add: **centrole·cithal** *a. Biol.*, having the food yolk in the centre of the ovum, **ce·ntromere** *Cytology* [ad. G. *centromer* (W.

Waldeyer 1903, in Hertwig *Handb. d. Entwick. d. Wirbeltiere* (1906) I. 204), f. Gr. μέρος part], (*a*) see quot. 1925 (*disused*); (*b*) the part of a chromosome to which the spindle is attached during mitosis; hence **ce:ntrome·ric** *a.*; **ce·ntrosphere**, (*a*) *Cytology* [ad. G. *centrosphäre* (E. Strasburger 1893, in *Anat. Anzeiger* VIII. 179)], a region of clear, differentiated cytoplasm from which the asters extend during cell-division and containing the centriole(s) if present; (*b*) *Geol.*, the central or inner part of the earth; **centro·tylote** *a.*, of a biradiate sponge-spicule, having a central swelling.

1880 Balfour *Comp. Embryol.* I. iii. 90 The food-yolk is however placed, not at one pole, but at the centre of the ovum. This group of ova I propose to name centrolecithal. 1888 Rolleston & Jackson *Anim. Life* p. xxvi, Centrolecithal ova are confined to Arthropoda. 1966 *McGraw-Hill Encycl. Sci. & Technol.* IX. 459 Centrolecithal, or centrally located yolk, occurs in eggs of insects and cephalopod mollusks. 1925 E. B. Wilson *Cell* (ed. 3) 1127 *Centromere*, that part of the sperm containing the central bodies; especially the neck-region. 1936 C. D. Darlington in *Jrnl. Genetics* XXXIII. 466 The chromosomes of the Acridinae..have always been described as of two types, with submedian and with terminal centromeres (or spindle attachments). 1949 *New Biol.* VII. 73 The centromere of a fully contracted chromosome is generally recognisable as a non-staining constriction. 1968 *Brit. Med. Bull.* XXIV. 261/1 A chromosome may appear to have two centromeres if the outline alone is taken into consideration. 1960 *Lancet* 14 May 1063/2 The centromeric index expressed as the ratio of the length of the shorter arm to the whole length of the chromosome. 1896 E. B. Wilson *Cell* 232 The centrosphere has a radiate structure, being traversed by rays which stretch between the centrosome and the peripheral microsome-circle. 1899 *Geogr. Jrnl.* XIII. 228 The Earth consists of three parts: there is the vast unknown interior, or 'centrosphere' [etc.]. 1960 L. Picken *Organization of Cells* vii. 253 In experiments in which removal of the centrosphere alone was attempted, asters always re-formed eventually. 1887 *Encycl. Brit.* XXII. 417/2 A centrotylote microxea. 1900 *Proc. Zool. Soc.* 131 The microstrongyles..are occasionally centrotylote.

centrosome (se·ntrǒsōᵘm). *Cytology.* Also **centrosoma** (pl. **centrosomata**). [ad. G. *centrosoma* (T. Boveri 1888, in *Jenaische Zeitschr. Naturw.* XXII. 752), f. Centro- + Gr. σῶμα body.] **a.** = *centriole. **b.** A small region of cytoplasm present in the cells of many animals and lower plants which during interphase is usu. situated next to the nucleus but is occas. within it and which comprises the centrosphere and the centriole(s). Hence **ce:ntroso·mic** *a.*

1889 Geddes & Thomson *Evolution of Sex* xi. 146 Both Van Beneden and Boveri have recently agreed on the existence of two 'central corpuscles' (centrosomata) in the protoplasm. 1893 Parker & Rönnefeldt tr. *Weismann's Germ-Plasm* 23 We now know..that even in Phanerogams a small cell-body surrounds the nucleus [of the male cell in fertilization], and that a special structure, the 'centrosome',—which is absolutely essential for the commencement of development,—is contained within it. 1900 [see Telo-[1]]. 1912 E. A. Minchin *Stud. Protozoa* vi. 81 To the primary centrosome or centriole there may be added adventitious elements of protoplasmic or nuclear origin, thus forming a centrosomic complex. 1925 E. B. Wilson *Cell* (ed. 3) i. 26 In addition to the nucleus, the cytosome often contains a structure known as the *central apparatus* or *microcentrum* of which the most essential component is the central body (*centrosome, centriole*) about which as a center arise the *asters*. *Ibid.* 1127 *Centrosome*, (1) originally, the central body lying at the astral center..(Boveri, 1888); (2) Subsequently, in a more specific sense, the larger central body, composed of *centroplasm*, within which lies the much smaller *centriole* (Boveri, 1895, 1901). 1943 L. W. Sharp *Fund. Cytol.* 28 The aspect of the centrosome varies widely in cells of different kinds and especially in different stages of nuclear division. 1964 Brown & Danielli in G. H. Bourne *Cytol. & Cell Physiol.* (ed. 3) vi. 292 The second stage [of egg activation] follows engulfment of the spermatozoon and it is the centrosome which plays the active role and organizes the division apparatus.

centrum. Add: **2.** The place from which an earthquake originates.

1887 *Nature* 31 Oct. 657 The determination of earthquake origins, the depth of 'centrums', [etc.]. 1938 L. D. Leet *Practical Seismol.* viii. 280 With growth of the concept that ordinarily the focus was actually at a finite depth below the surface, the term 'centrum' was modified to *hypocenter*, signifying the deep center, or focus; and *epicenter*, or point on the surface vertically above the focus.

centum (ke·ntǒm). *Philol.* Also **kentum.** [f. L. *centum* hundred, from its pronunc. with (k), as opposed to *satem.] A name given by philologists to one, chiefly western, group of Indo-European languages, distinguished by their use of velar consonants where the corresponding sounds in cognate words in the eastern group (cf. *satem) are sibilants.

1901 P. Giles *Short Man. Compar. Philol.* (ed. 2) 24 As the most characteristic sound is found in the word for 'hundred', the two sections are named the *centum* and the *satem* section respectively. 1912 J. Wright *Compar. Gram. Greek Lang.* vi. 95 The former group [Greek, Italic, Keltic, Germanic] is generally called the centum- and the latter [Aryan, Armenian, Albanian, Baltic-Slavonic] the satem-group of languages, where Latin *centum* and Zend *satəm* represent the original Indg. word *kmtóm, hundred. 1926 J. R. R. Tolkien in *Year's Work in Eng. Stud. 1924* 27 The *centum-satəm* division becomes more, not less, puzzling, as does the whole question of the interrelations of the surviving Indo-European languages. 1932 W. L. Graff *Language* x. 365 The languages in which I.-E. *k, g*, etc., are preserved as gutturals are called kentum languages, the others satem languages. 1939 E. Prokosch *Compar. Germ. Gram.* 43 *Kentum* and *Satem*...It is customary to distinguish two groups of Indo-European languages according to the treatment of the Indo-European palatals: they became sibilants in the eastern group, but appear as velars in the western group. 1952 O. R. Gurney *Hittites* vi. 119 The 'centum' group (comprising Latin, Greek, Celtic, and the various Germanic languages).

centuriate, *a.* Delete †*Obs.* and add: also in *centuriate comitia*; also, of, pertaining to, or divided into centuries.

1901 F. F. Abbott *Hist. & Descr. Rom. Pol. Instit.* 27 The action of the centuriate *comitia* in electing magistrates, in passing laws, and in deciding appeals was of no great immediate value to the plebs. *Ibid.* 76 The reformed centuriate assembly was essentially democratic. 1918 Haverfield in *Eng. Hist. Rev.* XXXIII. 296 The boundaries of Roman centuriate land.

centuriate, *v.* Restrict †*Obs.* to sense in Dict. and add: **2.** In Roman colonization, to divide and assign land (see next).

1918 Haverfield in *Eng. Hist. Rev.* XXXIII. 289 It would follow that the *territorium* of..a provincial *colonia*..would be centuriated when first surveyed and laid out. *Ibid.* 296 The whole of northern Essex..would have been surveyed and centuriated on one general scheme.

centuriation (sentiūᵊriē̆ⁱ·ʃǝn). *Hist.* [ad. L. *centuriātio, -ōnem*, n. of action f. *centuriāre* to Centuriate.] (See quots.)

1869 H. C. Coote in *Archæologia* XLII. 133 The territory having been thus appropriated to the colony, the next step was to divide and assign it..in *centuriæ*, or private estates, to and amongst the colonists. This *centuriation*, as it was called, was the legal and constitutional act which perfected the change from public land into private property. 1918 Haverfield in *Eng. Hist. Rev.* XXXIII. 291 Numerous attempts have been made to detect centuriation, or something like it, in Britain. 1961 L. Mumford *City in History* viii. 209 In many regions colonization was accompanied by..dividing fields into long rectangular parcels that are still visible from the air and respected in daily use. This system of 'centuriation' characterizes large parts of lowland Italy, Dalmatia, and Africa.

centurion. Add: **3.** *Sport.* A player who makes a hundred or more runs in an innings at cricket; one who has ridden, etc., a hundred miles in one journey; *double centurion*, a player who makes a double century (see next). *colloq.*

1886 *Graphic* 31 July 107/2 Some other 'centurions' have been Chatterton (108) for M.C.C., A. Shuker (103 not out) for Trent. 1897 *Outing* (U.S.) XXX. 346/1 There are long-distance riders, too, galore in the N.Y.A.C. and not a few centurions. 1900 *Westm. Gaz.* 18 July 2/3 This is the third time he has been a double centurion.

century. Add: **3. b.** *spec.* in *Cricket*, a hundred or more runs, esp. made by one player in the same innings; *double century*, two hundred runs by the same player in one innings. Also in *Cycling*, etc., a hundred miles in a race or ride; *double century*, a cycling run of two hundred miles.

1864 *Bell's Life* 11 June 8/4 Another century was piled up before the second wicket fell. 1871 F. Gale *Echoes Cricket Fields* viii. 43 There are slang writers who will tell me that the Lions' or the Nonpareils' [batting] average will be over a quarter of a century. *Ibid.* xiv. 90 He went in last, and joined Mr. Attfield, and between them ninety-nine runs were scored; or as a slang penny-a-liner would say, 'the two last gents wrote up a century bar one.' 1897 *Outing* (U.S.) XXX. 343/1 The probability is that he will place to his credit on the [cycling] club records one or more double centuries. *Ibid.* 348/1 The more enthusiastic indulge in century runs. 1955 *Times* 25 June 7/5 A boy who..can be relied on to score a century in most school cricket matches.

c. A hundred dollars. *U.S. slang.*

1859 Matsell *Vocabulum* 18 *Century*, one hundred dollars. 1930 J. Dos Passos *42nd Parallel* v. 79 'You must have made big money.'..'I saved pretty near a century.' 1964 R. Chandler *Killer in Rain* 3 He..arranged five century notes like a tight poker hand.

d. A hundred pounds. *slang.*

1861 [see Stoater]. 1864 *Derby Day* 131 (Farmer), I'll lay you an even century about Nimrod. 1883 *Echo* 1 Nov. 4/2 (ibid.), Golding..purchased Passaic from F. Archer for a century. 1888 F. W. J. Henning *Recoll. Prize Ring* 155 Having made up his mind that he was going to pocket the century.

9. (senses 4 and 5) *century* (or *centuries*)*-long* adj. and adv., *century* (or *centuries*)*-old* adj.; **century-plant** (earlier example).

1883 'Mark Twain' *Life on Mississippi* xliii. 386 Even the children know that a dead saint enters upon a century-long career of assassination the moment the earth closes over his corpse. 1924 R. Graves *Mock Beggar Hall* 62 They themselves May century-long be doomed to walk these rooms. 1933 L. Bloomfield *Language* i. 4 A century-long controversy. 1963 *Time* 30 Aug. 11/3 There cannot, in fact, be any real understanding of the Negro revolution of 1963 without some understanding of the Negro's centuries-long struggle in America. 1845 Longfellow *Nuremberg* in *Poems* (1845) 88 Thy castle, time-defying, centuries old. 1857 Whittier *Mabel Martin* 63 The household ruin century-old. 1873 D. M. P. tr. *Wellmer's C'tess zu Stolb.* ii. 23 The century-old motto of the Stolbergs. 1901 *Sketch* 31 July 56/2 Buried in its centuries-old bosom. 1958 B. Abel-Smith in N. Mackenzie *Conviction* 63 The eerie, century-old building. 1843 J. L. Stephens *Incidents of Travel in Yucatan* II. ii. 44 Growing on the roof are two maguey plants, Agave Americana, in our latitude called the century plant, but under the hot sun of the tropics blooming every four or five years.

'cep, 'cept, shortened colloq. forms of Except *prep.* and *conj.*

1851 H. Mayhew *Lond. Labour* I. 268 One gets reconciled to anything, 'cept, to a man like me, a low lodging-house. 1867 *Goodwife at Home* 6 But never gyangs it o'er my hawse, 'Cep at an antren time. 1898 J. D. Brayshaw *Slum Silhouettes* 4 They all gits aht somehow, 'cept me. 1932 W. Faulkner *Light in August* xiv. 306 You couldn't hear what Brother Bedenberry was saying, cep he never tried to hit back nor nothing. 1967 A. Wilson *No Laughing Matter* III. 305, I don't know anyone who cooks better. Well, cept my Mum did, of course. 1967 L. Forrester *Girl called Fathom* xiv. 180 That's all I can tell you for now...Cept..there's one more reason I'm takin' you along.

‖**cèpe** (sẹ̄p). Also **cep** (sep). [a. F. *cèpe* a boletus, f. Gascon *cep* tree-trunk, mushroom, f. L. *cippus* a stake.] An edible fungus of the genus *Boletus*.

1865 M. Eyre *Lady's Walks* xxix. 316 Stew it whole in *vin ordinaire*, together with some ham, a little butter, a little garlic, mushrooms or *ceps*, and a little salt. 1898 *U.S. Dept. Agric. Yearbk. 1897* 465 To this genus [*sc.* Boletus] belong the fungi known in France as *cèpes*, under which they are imported into this country. 1924 'A. D. Sedgwick' *Little French Girl* I. i. 6 Alix was sometimes allowed to go to the forest with her and find *cèpes*. 1959 *Times Lit. Suppl.* 27 Feb. 113/4 The dark-headed *fungo porcino*, the cep, *Boletus edulis*, was called *suillus* by the Romans.

cephal- (se·făl), combining form of Gr. κεφαλή head, as **ce·phalad** *adv. Zool.* [*-ad], towards the head; **cephale·tron** (sefăli·trǫn) *Zool.* [Gr. ἦτρον abdomen], Owen's name for the head or anterior division of the body of some of the crustaceans, as the king-crab; hence *cephaletral* adj.; **ce:phalhæmato·ma** *Path.*, a bloody swelling beneath the pericranium; **cephalhy·drocele** *Path.*, a serous or watery swelling outside the cranium; **cephalo·tic** *a. Biol.* [Gr. ὠτ-, οὖς ear], characterized by the presence of the ears on the middle line of the head.

1887 Wilder in *Amer. Nat.* June 545 In Ceratodus alone..is there a perima,—that is, a rima extending cephalad from the margin of the porta. 1872 Cephaletral [see *thoracetral* s.v. Thoraco-]. 1872 Cephaletron [see *thoracetron* s.v. Thoraco-]. 1900 Dorland *Med. Dict.* 147/1 Cephalhematoma. 1962 *Lancet* 6 Jan. 54/1 There were 2 cases of cephalhaematoma (5·2 %), which resolved by the age of 1 month. 1900 Dorland *Med. Dict.* 147/1 Cephalhydrocele. 1908 *Practitioner* Apr. 466 The conditions necessary for the formation of a cephalhydrocele are a fracture with laceration of the dura. 1894 W. Bateson *Stud. Variation* 458 Dareste also declares that the cephalotic state is similarly first indicated by a premature union of the folds in the region of the medulla.

cephalin[1] (se·fălin). *Zool.* [ad. F. *céphalin* (A. Schneider 1875, in *Arch. Zool. Exp. Gen.* IV. 518), f. Gr. κεφαλή head + -in[1].] = *cephalont.

1888 Rolleston & Jackson *Anim. Life* 858 A Polycystid possessing all three segments is known as a Cephalin or Cephalont. 1912 E. A. Minchin *Introd. Study Protozoa* xiv. 326 In the earlier phase, in which an epimerite is present, the parasite was termed by Aimé Schneider a cephalont ('cephalin').

cephalin[2] (se·fălin). *Biochem.* [f. *cephal- + -in[1].] Now the usual spelling of Kephalin: a phospholipid similar to lecithin, originally isolated from brain-tissue and capable of accelerating the coagulation of blood.

1900 in Dorland *Med. Dict.* 1920 J. J. R. Macleod *Physiol. & Biochem.* (ed. 2) lxxvii. 689 Other phospholipins present in nervous tissue are cephalin, cuorin and sphingomyelin. 1961 *Lancet* 5 Aug. 286/2 In normal plasma 50–55 % of the protein-bound calcium is bound to albumin, and the rest mainly to globulin and cephalins. 1964 *Oceanogr. & Marine Biol.* II. 173 Plasmalogens..are closely related to the classical lecithin (phosphatidyl choline) and cephalin (phosphatidyl ethanolamine).

cephalodiscus (se:fălodi·skǒs). [mod.L. (W. C. M'Intosh 1882, in *Ann. & Mag. Nat. Hist.* 5th Ser. X. 337), f. Gr. κεφαλή head (see Cephalo-) + δίσκος Disk.] A member of the genus so named of primitive deep-sea chordates of the class Pterobranchia.

1882 M'Intosh in *Ann. & Mag. Nat. Hist.* X. 343 *Cephalodiscus* also diverges in regard to the thin postoral lamella. 1885 *Encycl. Brit.* XIX. 435/1 The dwelling of Cephalodiscus is a gelatinous, irregularly branched, and fimbriated mass. 1915 E. R. Lankester *Divers. Naturalist* 8 One of the most interesting deep-sea creatures discovered by the 'Challenger'..received the name Cephalodiscus. 1955 N. J. Berrill *Orig. Vertebrates* xi. 117 Cephalodiscus, Rhabdo-

pleura, and balanoglossids seem never to have become completely sessile forms.

cephalon (se·fălǫn). *Zool.* [mod.L., f. Gr. κεφαλή head: cf. Encephalon.] The region of the head in certain arthropods.

1875 Huxley & Martin *Elem. Biol.* xii. 123 An anterior division, which belongs to the head or cephalon. 1945 T. H. Savory *Spiders Brit. Isles* (ed. 2) 199 Cephalon short, broader than long, smooth.

cephalont (se·fălǫnt). *Zool.* [f. *cephal- + Gr. ὀντ-, ὤν, pres. pple. of εἶναι to be, exist.] A protozoan parasite at the stage of development in which an epimerite is attached to the anterior cyst.

1885 *Encycl. Brit.* XIX. 854/1 The epimerite serves to attach the parasite to its host...The phase in which it is present is called a 'cephalont', the phase after it has broken off a 'sporont'. 1888 [see *cephalin[1]]. 1961 Mackinnon & Hawes *Introd. Study Protozoa* III. 183 Such fully developed trophozoites are called cephalonts.

cephalosporin (se:fălospǭ·rin). [f. mod.L. *Cephalosporium* (see below) + -in[1].] Any of several antibiotic substances, closely related to penicillin, developed from *Cephalosporium*, a genus of fungi.

1951 A. C. Ritchie et al. in *Brit. Jrnl. Pharmacol.* VI. 430 Cephalosporin P_1 is an antibiotic from a species of *Cephalosporium*...The cephalosporin used was lyophil-dried material that had been prepared by chromatography and countercurrent distribution between solvents. 1961 *Times* 14 Sept. 8/3 Cephalosporin N is stated to be a new form of penicillin. 1964 M. Hynes *Med. Bacteriol.* (ed. 8) x. 136 The cephalosporins can be given to penicillin-sensitive patients.

Hence **cephalospo·rinase** (see quots.).

1963 *New Scientist* 11 July 94/1 In bacteria that respond to cephalosporin treatment the enzymes, called cephalosporinases, have not been demonstrated. 1964 *Ibid.* 12 Nov. 431/3 Anti-cephalosporin enzymes, cephalosporinases..are found among some types of pathogenic bacteria.

Cepheid (se·fiid, sī·fiid), *a.* and *sb. Astr.* Also **cepheid.** [f. L. *Cēpheus*, Gr. Κηφεύς Cepheus, a mythical king whose name was given to a constellation + -id[2].] **A.** *adj.* Pertaining to or resembling the variable star δ Cephei. **B.** *sb.* A variable star of the type of δ Cephei.

1904 *Astrophys. Jrnl.* XX. 184 Spectroscopic studies of *Cepheid* variables. 1921 *Discovery* Feb. 38/1 To determine the absolute magnitude..of every Cepheid variable in the heavens. 1927 *Glasgow Herald* 7 Oct. 11/2 When a Cepheid's period was noted its brightness could be stated. 1927 A. S. Eddington *Stars & Atoms* 93 In 1924 Hubble discovered a number of Cepheid variables in the great Andromeda nebula. 1964 *Yearbk. Astr. 1965* 149 It has been found that the longer the period of a Cepheid, the greater its real luminosity. If one knows how bright a star really is and how bright it appears, one can at once deduce its distance. Cepheids therefore act as standard candles among the stars.

‖**ceppo** (tʃe·po). [It.] The cemented glacial gravels of northern Italy.

1881 J. Geikie *Prehistoric Europe* 316 Those wide-spread masses of conglomerate, which are called *Ceppo* by the Italian geologists. 1894 —— *Gt. Ice Age* (ed. 3) xxxiv. 562 The highly denuded 'Ceppo' underlying the great terminal moraines of Ivræa, Como, etc.

'cept: see *'cep.

ceramal (seræ·măl). [f. Ceram(ic *a.* + Al(loy *sb.*] = *cermet.

1949 *Jrnl. R. Aeronaut. Soc.* July 661/1 Combinations of metals and ceramics, termed 'ceramals' by the N.A.C.A., appear to possess outstanding properties at elevated temperatures. 1950 *Engineering* 29 Dec. 577/1 Ceramics or the new metal-ceramic compacts ('ceramals' or 'cermets'). 1953 *Jrnl. Brit. Interplan. Soc.* XII. 266 The development of metal/ceramic mixtures (ceramals or cermets) promises an improvement in these essential properties.

ceramet, var. *cermet.

ceramic, *a.* (*sb.*). Add: **2. b.** As *sb.* (usu. in *sing.*) Products of the ceramic art; pottery.

1880 'Mark Twain' *Tramp Abroad* I. xx. 178 My chief solicitude was about my collection of keramics. 1902 *Jrnl. Anthrop. Inst.* XXXII. 373 As..a considerable body of prehistoric pottery has accumulated in our museums, the time has arrived when some attempt should be made to arrange this early ceramic in approximately chronological order. 1932 *Antiquity* VI. 221 The stratification of the ceramic was accurately noted.

ceramicist (siræ·misist). [f. Ceramic + -ist.] = Ceramist.

1930 *Observer* 25 May 18/3 Many countries have great ceramicists capable of building up a great pottery industry.

ceramidium (serămi·diǒm). *Bot.* [mod.L., f. Gr. κεραμίδιον, dim. of κεραμίς earthen vessel.] The outer covering of the cystocarp, found in algæ of the family Rhodomelaceæ.

1849 W. H. Harvey *Man. Brit. Marine Algæ* (ed. 2) 69 In the ceramidium the conceptacular fruit is brought to its highest development. This organ is ovate or urn-shaped, furnished with an apical pore, and containing a tuft of pear-shaped spores, rising from the base of the cavity. 1902 *Encycl. Brit.* XXV. 270/1 In *Rhodomelaceæ* there is a special urn-shaped envelope surrounding the sporogenous filaments. This is a ceramidium.

cerastium (siræ·stiǒm). *Bot.* [mod.L. (Linnæus *Hortus Cliffortianus* (1737) 173), f. Gr. κεράστης horned, from the horn-shaped capsule of many species.] A plant of the caryophyllaceous genus of that name, including the mouse-ear chickweed and other weeds, as well as species cultivated for their silvery foliage.

1799 Sowerby & Smith *Eng. Bot.* VIII. 538 This is certainly rather a *Stellaria* than a *Cerastium*. 1829 Loudon *Encycl. Plants* 1059 Caryophylleæ...Many are common weeds, as most of the Cerastiums, Spergulas, and others. 1925 *Chambers's Jrnl.* Jan. 35/2 He had observed..what he knew to be a very rare cerastium. 1958 F. Stark *Alexander's Path* 150 Drifts of vetch purple and yellow, grey-leaved cerastium, [etc.].

ceratioid (siræ·tioid), *a.* and *sb.* Also **ceratoid.** [f. *Ceratias*, a genus of fishes: see -oid.] **A.** *adj.* Of or pertaining to the family Ceratiidæ of fishes (or the genus *Ceratias*.) **B.** *sb.* A fish of this family.

1889 in *Cent. Dict.* 1925 *Glasgow Herald* 21 Mar. 4 The fishes..belong to the Ceratioid section of the Anglers. 1931 *Discovery* Apr. 106/2 Parasitic pigmy males among the ceratoid angler-fishes.

ceratodus (siræ·tǒdǒs). *Zool.* [mod.L., f. Gr. κέρας, κερατ- horn + ὀδούς tooth.] A popular name for the Australian lungfish, *Neoceratodus forsteri*, belonging to the order Dipteriformes; also, a genus of fossil fish related to this lungfish.

1874 Darwin *Desc. Man* (ed. 2) I. ii. 37 The opinion lately advanced by Dr. Günther on the paddle of Ceratodus. 1897 E. R. Lankester in *Times* 8 Feb. 10/6 The larval stages of the remarkable fish ceratodus of Queensland. 1899 *Daily News* 10 Apr. 8/2 The ceratodus, a fish with lungs, which, though its fossil remains are scattered over the world, is now confined to two rivers in the south of Queensland, the Mary and the Burnett.

ceratosaurus (se:rătosǭ·rǒs). [mod.L., f. Gr. κέρας, κερατ- horn + -o- + σαῦρος lizard.] In full *Ceratosaurus nasicornus*, an extinct dinosaurian reptile such as those reconstructed from remains found in the Upper Jurassic rocks of Colorado.

1884 O. C. Marsh in *Amer. Jrnl. Sci.* XXVII. 330 This interesting fossil is quite distinct from any hitherto described, and as it represents a new genus and species, may be called *Ceratosaurus nasicornis*. It also belongs to a new family, which may be named the *Ceratosauridæ*. *Ibid.* 333 The brain in *Ceratosaurus* was of medium size, but comparatively much larger than in the herbivorous Dinosaurs. 1892 H. N. Hutchinson *Extinct Monsters* vi. 79 Externally, also, the Ceratosaurus differs from other members of the Carnivorous group. 1930 *Discovery* Nov. 386/2 The bipedal dinosaurs were of two kinds, vegetable feeders and flesh eaters. The ceratosaurus was one of the latter, and derives its name from the short bony horn on its nose.

cercus (sɜ̄·ɪkǒs). *Zool.* Pl. **cerci.** [mod.L., f. Gr. κέρκος tail.] One of a pair of small appendages found at the hind end of the abdomen of certain insects and other arthropods.

1826 Kirby & Spence *Introd. Entom.* III. xxxiii. 718 These [appendages] are..the cerci of the Blattidæ and Gryllina. 1937 *Discovery* Mar. 90/2 Hairy tails or cerci which may be jointed. 1951 Colyer & Hammond *Flies Brit. Isles* 33 A pair of small appendages, known as 'cerci' may be visible, which are probably tactile in function. 1957 Richards & Davies *Imms's Textbook Ent.* (ed. 9) I. 55 The most conspicuous of the persistent appendages are the *cerci* of the 11th segment, which exhibit wide diversity of form and may even be transformed into forceps, as in the Japygidae and the earwigs.

cereal, *a.* and *sb.* Add: **B.** *sb.* **b.** An article of food (esp. a breakfast dish) made from a cereal; freq. *breakfast cereal*. orig. *U.S.*

1899 *Chicago Daily News* 9 May 7 Free with 6 packages of Hazel Cereals, any assortment, a handsomely decorated tea canister. 1907 *Yesterday's Shopping* (1969) 11/2 Breakfast cereals. 1909 H. G. Wells *Tono-Bungay* II. iii. 196 There's all these patent grain foods,—what Americans call cereals. 1934 A. Thirkell *Wild Strawberries* iv. 95 Clarissa looked so sweet eating her cereals. 1958 Wodehouse *Cocktail Time* xii. 103 When I was a young man..there were no cereals. We ate good wholesome porridge for breakfast and throve on it. 1962 A. Christie *Mirror Crack'd* i. 9 All these great packets of breakfast cereal instead of cooking a child a proper breakfast of bacon and eggs.

cerealist (sīᵊ·riălist). [f. Cereal *sb.* + -ist.] **1.** One who advocates a cereal diet.

1905 *Daily Chron.* 7 Aug. 4/5 Cerealists preach absolute abstention from pheasant, turbot, and turtle.

2. One who makes a special study of cereals and the conditions necessary for their cultivation.

1907 L. H. Bailey *Cycl. Amer. Agric.* II. 663/2 Cerealists as well as practical producers of grains. 1920 *Glasgow Herald* 29 Sept. 9 The present year's wheat yield is nothing of which to boast, but it relieves the apprehensions of cerealists. 1924 *Ibid.* 13 Sept. 6 Dr. Saunders had been the Dominion cerealist.

cerebello- (se·ribelo), used as combining form of Cerebellum = pertaining to the cerebellum (and another part).

1900 Dorland *Med. Dict.* 148/1 *Cerebellospinal*, pertaining to the cerebellum and spinal cord. 1903 *Ibid.* (ed. 3), *Cerebello-olivary*, connecting the cerebellum and olivary body. 1907 *Practitioner* Apr. 584 The disturbing influence exerted by the tumour upon the cerebello-vestibular apparatus.

cerebral, *a.* Add: **1.** *cerebral palsy*: see *palsy *sb.*

b. Intellectual; appealing to the intellect (rather than to the emotions); clever.

1929 D. H. Lawrence *Pansies* 60 Since we have become so cerebral We can't bear to touch or be touched. 1935 T. E. Lawrence *Lett.* (1938) 853, I was then trying to write; to be perhaps an artist..or to be at least cerebral. 1946 *Amer. Speech* XXI. 81 The only congresswoman who may be described plausibly as both cerebral and beautiful. 1951 R. Hoggart *Auden* i. 19 The verse of a young man..anxious to evolve a hard, cerebral style. 1959 *Camb. Rev.* 25 Apr. 449/2 Everything is suitably cerebral, and the emotional sterility does not obtrude. 1970 *Daily Tel.* 19 May 16/4 A kind of cerebral comedy, pondering the difference between appearance of reality and keeping us on our intellectual toes.

2. (Further examples.)

1918 D. Jones *Outl. Eng. Phonetics* xvi. 100 Cacuminal sounds (also called 'inverted' sounds or 'cerebral' sounds) are defined as sounds in which the tip of the tongue is 'inverted' or curled upwards towards the hard palate. 1934 *Year's Work in Eng. Stud. 1932* 53 The consonant apparently became a cerebral (inverted) *r* before its disappearance.

Hence **cerebralism** (further examples); also, intellectualism; **cerebralist** (examples); also, an intellectual, intellectualist; **ce:rebrali·stic** *a.*, of or pertaining to cerebralism; **cerebra·lity**, cerebral or intellectual quality, cleverness; **cerebralization** (example).

1892 W. James *Coll. Ess. & Rev.* (1920) 326 A critic of cerebralism in psychology ought to do one of two things. 1931 C. Gray *Contingencies* (1947) 156 Excessive cerebralism on the one hand and excessive sensationalism on the other. 1890 W. James *Princ. Psychol.* I. 4 The spiritualist and the associationist must both be 'cerebralists'. 1943 C. Williams *Figure of Beatrice* 20 So much only to prevent too great an 'elevation' of Dante's thought; we are not to suppose him a mere cerebralist. 1892 W. James *Coll. Ess. & Rev.* (1920) 316 The cerebralistic point of view which is becoming so popular in psychology today. 1901 *Scribner's Mag.* XXIX. 505/2 There is one trait that belongs in common to every artistic effort of Americans, and that is the *cerebrality*..of such effort. 1903 *Daily Chron.* 9 May 4/5 The adventures of Sherlock Holmes are cerebrality raised to the power of *n*. 1869 W. H. Ferrar *Compar. Gram. Skr.* I. 340 Cerebralisation is a phenomenon that has arisen within the limits of the Sanskrit language.

cerebrally (se·ribrăli), *adv.* [f. Cerebral *a.* + -ly[2].] In a cerebral manner; in or with the brain.

1893 in *Funk's Standard Dict.* 1901 G. Gissing *Our Friend the Charlatan* xxi, If I am not cerebrally oxidised,..it's all over with my hopes of leading a moral life. 1929 D. H. Lawrence *Pansies* 60 For if, cerebrally, we force ourselves into touch, into contact Physical and fleshly, We violate ourselves. 1962 H. C. Weston *Sight, Light & Work* (ed. 2) fig. 17 (*caption*), The half images..are 'superimposed' and united cerebrally.

cerebrate, *v.* Add: (Further example.)

1917 *Chambers's Jrnl.* Oct. 699/1 Up, up, they came, cerebrating agonised thoughts.

2. *intr.* To cogitate.

1928 A. Bennett *Strange Vanguard* xliv. 303 The baron who was still most actively cerebrating, rang the bell. 1940 H. G. Wells *New World Order* § 7. 106 Many among them will cease to cerebrate further and fall by the wayside, but many will go on thinking. 1953 'S. Ransome' *Drag the Dark* (1954) ii. 20 He..just sat there cerebrating.

cerebricity (seribri·siti). *rare.* [f. Cerebrum brain, after *electricity*.] Brain-cell power.

1890 O. W. Holmes *Over Teacups* i. 15 Your incident is a typical example of cerebral induction from a source containing stored cerebricity. I use this word, not to be found in my dictionaries, as expressing the brain-cell power corresponding to electricity. 1894 *Arena* IX. 336 Mere manifestations of psychic power, referable to the dark field of cerebricity and clairvoyance.

cerebroce·ntric, *a.* [f. Cerebro- + *-centric.] Centring on the brain.

1930 J. Laird *Knowl., Belief & Opinion* iii. 64 Arguments based upon the so-called egocentric predicament in cogitation, or upon the cerebrocentric, anthropocentric, or sociocentric predicaments. 1949 *Mind* LVIII. 390 The 'cerebrocentric' conception (either Epiphenomenalist or Behaviourist)..is accepted nowadays by the majority of scientifically educated persons. 1956 J. B. Rhine in A. Pryce-Jones *New Outl. Mod. Knowl.* 205 This brain-centred approach to man, this *cerebro-centric* psychology.

cerebroside (se·ribrosəi:d). *Biochem.* [f. Cerebr(o- + -ose[2] + -ide.] Any of a group of nitrogenous fatty substances found in the brain and other nerve tissue, that by hydrolysis yield galactose, sphingosine, etc.

1883 J. L. W. Thudichum in *Brit. Med. Jrnl.* 15 Sept. 525/1 Let us cast a glance at a class of immediate principles of the brain, which I have termed *cerebrosides*. The principal one is *phrenoid*; the second in order of quantity is *kerain*; together they form about 5 per cent. of the weight of the brain. 1956 *Nature* 7 Jan. 21/2 Dr. H. E. Carter..gave further information regarding cerebrosides and cerebroside sulphuric ester. 1957 *Sci. News* XV. 93 The cerebrosides are found chiefly in the human brain...When hydrolysed they give one molecule each of D-galactose or D-glucose, a fatty acid, and a nitrogenous base, sphingosine.

cerebro-spinal, *a.* Add: (Further examples.) Esp. in *cerebro-spinal fever* or *meningitis*, inflammation of the meninges of the brain and the spinal cord, otherwise called *spotted fever*.

1889 Osler *Cerebral Palsies* 18 Cerebro-spinal meningitis was stated to be the cause of the trouble. 1897 *Trans. Amer. Pediatric Soc.* IX. 189 Four ounces of turbid cerebrospinal fluid were collected and measured. 1901 Osler *Princ. & Pract. Med.* (ed. 4) 101 Sporadic cerebro-spinal fever occurs in all the larger cities..of this continent. 1955 *Sci. News Let.* 19 Mar. 190/2 This kind of meningitis is also called cerebrospinal fever and spotted fever.

cerebrotonic (se:ribrotǫ·nik), *a.* and *sb.* [f. Cerebro- + Tonic *a.* 1 a.] **A.** *adj.* Designating or characteristic of a type of personality which is introverted, intellectual, and emotionally restrained, classified by Sheldon as being associated with an *ectomorphic physique. **B.** *sb.* One having this type of personality. So **cerebrotonia** (-tōᵘ·niă), cerebrotonic personality or characteristics.

1937 A. Huxley *Ends & Means* xi. 165 Dr. William Sheldon, whose classification [of types of human beings] in terms of somatotonic, viscerotonic and cerebrotonic I shall use. *Ibid.* xii. 193 The cerebrotonic is not such a 'good mixer' as the viscerotonic. 1940 W. H. Sheldon *Var. Human Physique* 8 In the economy of the cerebrotonic individual the sensory and central nervous systems appear to play dominant roles. 1945 A. Huxley *Let.* 2 Apr. (1969) 517 There was just enough of the somatotonic in his..cerebrotonic make-up to make him regret his cerebrotonia. 1950 —— *Themes & Var.* i. 121 Too secretively the introvert, too inhibitedly cerebrotonic, to be willing to take the risk of 'giving himself away'. 1951 Auden *Nones* (1952) 28 Cerebrotonic Cato may Extol the Ancient Disciplines. 1954 R. Fuller *Fantasy & Fugue* iv. 75 You..unfortunately incline to the cerebrotonic ectomorph—you worry too much, you're too good looking, and you can't abandon yourself happily to booze.

cerebrovascular (se:ribrovæ·skiŭlăɪ), *a. Physiol.* [f. Cerebro- + Vascular *a.* 2.] Of or pertaining to the brain and the blood-vessels supplying it.

1935 S. Weiss in Blumer & Sullivan *Practitioners Libr. Med. & Surg.* VIII. II. vii. 677 *Cerebrovascular Accidents.*—Rupture or thrombosis of the cerebral vessels with resulting debility..is probably the most frequent of all the diseases of the blood vessels. 1961 *Lancet* 29 July 221/2 The regimen followed in all cases of cerebrovascular accident was [etc.]. 1961 *Ibid.* 5 Aug. 306/1 Diseases of the central nervous system...The commonest disorders in this group resulted from cerebrovascular disease.

ceremonialize, *v.* Add: (Additional examples.)

a 1935 T. E. Lawrence *Mint* (1955) I. xix. 67 This pretence to ceremonialise a job of scavenging. 1960 Koestler *Lotus & Robot* II. x. 229 It is ceremonialized in a series of ablutions, purifications, recitations.

b. *intr.* To carry out a ceremony or ritual.

1948 C. Williams *Seed of Adam* 32 Must we stand by while the king ceremonializes?

ceremoniarius (se:rimōᵘniē·riǒs). *Eccl.* Also **ceremonarius.** [mod.L., sb. use of adj. f. *cærimōnia* Ceremony: see -ary[1].] An official who superintends the ceremonies and assists the ministers in a liturgical service.

1865 F. G. Lee *Direct. Anglicanum* 264 The ceremoniarius should be vested in cassock and cotta. 1898 *Westm. Gaz.* 10 Nov. 5/2 Before the prayer for the Church Militant the ceremonarius..again said the requiem æternam. 1913 C. Mackenzie *Sinister St.* I. II. v. 216 I'm the Ceremonarius, and I can tell you I have my work cut out.

Cerenkov, Cherenkov (tʃere·ŋkǒf, -rye·ŋkǒf). The name of P. A. Čerenkov (born 1904), Russian physicist, used *attrib.* in *Cerenkov effect, radiation*, etc. (see quots.).

1940 *Physical Rev.* LVII. 490/2 The Cerenkov radiation..does not represent a loss of energy to be added to that calculated with the Bohr theory. 1955 *Sci. Amer.* Oct. 22/3 A photograph of the swimming-pool reactor showing its uranium fuel load, bathed in the blue glow known as the Cerenkov effect. 1958 *Chambers's Techn. Dict.* 964/1 *Cerenkov radiation*, visible radiation produced when charged particles traverse a medium which increases their initial velocity to a velocity greater than that of light. 1958 *Times* 29 Oct. 8/6 The Cherenkov effect for which the three Russian physicists have received their award is the emission of light waves by electrons moving at very high speeds. 1959 *Times* 23 Oct. 17/2 The counters used will be of the kind, known as Cerenkov counters, in which incoming high-energy particles are detected by the flashes of light that they produce when they pass through a tank of water. 1969 *Listener* 12 June 823/2 The bow wave's equivalent for light is Cerenkov radiation, and it is emitted, by a particle travelling faster than light, in a certain direction, depending on the actual speed of the particle.

ceresin, ceresine (se·rĕsin, -in). [f. mod.L. *ceres*, irreg. f. L. *cēra* wax + -in[1], -ine[5].] A whitish wax, hard and brittle, prepared from ozocerite, or a petroleum wax, mixed with, or used as a substitute for, beeswax. Also *ceresin wax*.

1885 *Encycl. Brit.* XVIII. 113/1 Under the name of ceresin or ozocerotin a large proportion of the high-melting paraffin extracted from the mineral [*sc.* ozocerite] goes into commerce, to be used chiefly for the adulteration of beeswax. 1935 *Nature* 19 Jan. 113/2 Initially, the strong colour of petroleum ceresin was a disadvantage in this connexion, but this is overcome by incorporation of the ceresin during the manufacture of the paraffin wax and refining the two together. 1951 H. M. Langton et al. *Synthetic Resins* (ed. 3) xvii. 661 Ceresin and paraffin wax are the more waxy constituents and find use in the electrical industry mainly for impregnating condensers and small coils which are not subjected to any temperature rise. 1958 M. L. Wolf *Dict. Painting* 54 Ceresin wax, occasionally used as a surface coating on paintings, sculpture, and other objects of art.

ceriman (se·rimæn). *Bot.* [ad. Amer. Sp. *cerimán*.] A climbing plant, *Monstera deliciosa*, of the family Araceæ, which is a native of Mexico and yields an edible fruit.

1871 Kingsley *At Last* vii, Up the next [stem] the Ceriman spreads its huge leaves, latticed and forked again and again. 1946 K. & J. Morton *Fifty Tropical Fruits of Nassau* p. x, Trees or plants..cultivated in the Bahamas..as ornamentals, such as the Jamaica Cherry,..Ceriman, Giant Grenadilla and the Passionfruits.

cermet (sɜ̄·ɪmet). Also **ceramet.** [f. Cera(mic *a.* + Met(al *sb.*] A ductile, heat-resistant alloy of metal and ceramic substances. Cf. *ceramal.

1950, 1953 [see *ceramal]. 1953 *Americana Ann.* 436/1 Cermets, which are combinations of ceramic materials with metals,..employ such substances as carbides, borides, nitrides, and oxides, cemented with binders of iron, nickel, or cobalt. 1958 *Engineering* 31 Jan. 141/3 New alloy steels, molybdenum, fused quartz, ceramics, cermets and high-temperature plastics are entering the arena.

cero (sīᵊ·ro). [Altered f. Sp. *sierra* saw, saw-fish.] A fish of the mackerel family, *Scomberomorus regalis*, found in warm parts of the western Atlantic.

1884 Goode *Nat. Hist. Aquat. Anim.* 307 The Spotted Cero, or King Cero, *Scomberomorus regalis*. 1965 A. J. McClane *Standard Fishing Encycl.* 174/2 Cero are found from Cape Cod to Brazil and are abundant around southern Florida and the West Indies.

ceroon. (Examples.)

1824 *Shipping & Commercial List* 31 July (D.A.), A sale of 30 ceroons of Cuba [tobacco]. *a* 1861 T. Winthrop *Isthmiana* (1883) 242 A little ragamuffin, who had come into town on a nag between two hide ceroons, full of mami apples.

cert (sɜ̄ɪt), *sb. slang.* Also **cert.** (with full stop). [Abbrev. of Certain *a.*, Certainty.] A certainty (esp. in phr. *a dead cert*); *spec.* in horse-racing, a horse that is considered certain to win.

1889 *Man of the World* 29 June 3/2 Love-in-Idleness is bound to take the Rous Memorial, and I hear Pioneer is a cert. for the St. James's. 1889 G. & W. Grossmith *Diary of Nobody* in *Punch* 2 Mar. 97/1 It is not speculation—it's a dead cert. 1899 T. M. Ellis *3 Cat's-eye Rings* 32 Tatwood, you know—the trainer—he tips me all the certs. 1904 E. Nesbit *Phoenix & Carpet* vii. 128 Aunt Emma would be out to a dead cert. 1906 E. Dyson *Fact'ry 'Ands* i. 8 This is somethin' er cut above *you*, for er cert. 1927 S. Kaye-Smith in *Sunday Express* 11 Sept. 9 They had guessed for a cert where he was. 1936 Auden & Isherwood *Ascent of F6* II. 73 We'll be able to start tomorrow for a cert.

certain, *a.* Add: **7. e.** *a certain age* (earlier example); also, (*a woman*) *of a certain description*, i.e. of the demi-monde; *in a certain condition*, pregnant; *a certain disease*, venereal disease.

1754 *Connoisseur* 28 Nov. 261, I could not help wishing that some middle term was invented between *Miss* and *Mrs.* to be adopted, at a certain age, by all females not inclined to matrimony. 1803 Lemaistre *Rough Sk. Mod. Paris* xiv. 122 Women, too, of a certain description, do not ply for custom. *Ibid.* xxviii. 232 When I first came here, I supposed that these ladies were of a certain description. 1927 *Rev. Eng. Stud.* Oct. 433 As instances may serve *lavatory, illegal operation, social evil, a certain disease*. 1958 B. Nichols *Sweet & Twenties* viii. 103 'Syphilis' had always been described as 'a certain disease', just as an attempt at rape had been described as 'a certain suggestion', and the result of the rape on the lady was described as leaving her in 'a certain condition'.

C. *adv.* **1.** (Later U.S. examples.)

1859 Bartlett *Dict. Amer.* (ed. 2), *Certain*, for certainly. 'He's dead certain'. 'I'll go to-morrow sure and certain'. Very common. 1872 Schele de Vere *Americanisms* 450 He's done it sure and certain.

certifiable, *a.* Add: **b.** = Notifiable *a.* *U.S.*

1911 Stedman *Med. Dict.* 159/2 *Certifiable*, that can and should be certified, noting certain infectious diseases, the occurrence of which must, by law, be reported to the health authorities.

c. Of a person: so deranged as to be certifiably insane; of mental disease: of such a type as to warrant certification.

1912 B. Hollander *First Signs of Insanity* I. i. 33 The difference between this stage and certifiable insanity is a quantitative difference, a difference of degree only. 1931 'F. Iles' *Malice Aforethought* xiii. 268 'I've always said she was mad,' exulted Dr. Bickleigh. 'Not certifiable, unfortunately; but certainly over the border-line.' 1939 H. G. Wells *Holy Terror* II. iii. 194 If half this stuff about him is true, he's certifiable.

certifiably (sɜ̄·ɪtifəiăbli), *adv.* [f. Certifiable *a.* + -ly[2].] In a certifiable degree; so as to admit of being certified.

1891 *Daily News* 12 Mar. 2/3 The evidence showed that as soon as Mrs. Stoney became certifiably insane the defendant took measures for her removal. 1970 *Guardian* 18 May 8/4 Even to pose such questions..makes him..certifiably sane.

certificate, *sb.* Add: **3. e.** A document committing a person to an institution as insane.

[1883 T. S. Clouston *Clin. Lect. Mental Dis.* xix. 612 It may..be necessary, before certifying, to get a letter..protecting the doctor from risk of legal action. That is a risk no medical man in signing a certificate of lunacy should subject himself to.] 1927 Henderson & Gillespie *Text-Bk. Psychiatry* xvii. 492 The granting of a certificate of this nature carries with it very serious legal obligations; for the person certified may..ask for a trial by jury as to his alleged insanity. 1932 Kipling *Limits & Renewals* 168 He'd been tried too high—too high. I had to sign his certificate a few weeks later.

certification. Add: **6.** *spec.*, of the insane.

1883 T. S. Clouston *Clin. Lect. Mental Dis.* xix. 611 His..business is to convince the patient's responsible relatives of the necessity for certification. 1912 B. Hollander *First Signs of Insanity* I. i. 33 The general diagnosis of insanity, when once pronounced so as to necessitate certification. 1959 *New Statesman* 10 Jan. 34/1 Certification, as was expected, is to disappear. The decision to detain an unwilling patient will be a purely medical decision.

certified, *ppl. a.* Add: **b.** *certified milk*, milk guaranteed free from tubercle bacillus.

1899 Taylor & Wells *Man. Dis. Children* v. 115 A movement..has resulted in the production of a very superior milk, known as certified milk. 1905 F. L. Dodd *Municipal Milk* 13 *Certified Milk*. The Borough of Sunderland..has started a movement for the improvement of the milk supply by..certificates granted to those producers who fulfil certain definite conditions. 1936 *Milk (Special Designation Order)* xi. § 1 A licence to sell milk as 'Certified' or as 'Grade A (Tuberculin Tested)' shall be deemed to be a licence authorising the use of the special designation 'Tuberculin Tested'.

certify, *v.* Add: **2. c.** To declare (a person) officially insane. (Cf. quot. 1874 for sense 4 in Dict.) So **certified** *ppl. a.*

1877 J. M. Granville *Care & Cure of Insane* II. 150 A serious mistake has been made in requiring that a patient shall be certified insane before he is removed to an asylum. 1883 T. S. Clouston *Clin. Lect. Mental Dis.* xix. 612 The question of whether the patient should be certified as a lunatic or not. *Ibid.* 613 It is proper, having determined that he should be certified, to ask what legal risk there is. 1892 H. H. Newington in D. H. Tuke *Dict. Psychol. Med.* I. 189/2 Asylum officers or medical men officially in charge of certified patients. 1927 Henderson & Gillespie *Text-Bk. Psychiatry* xvii. 487 It is not justifiable to certify a patient while in an epileptic fit, or while under the influence of alcohol. 1930 Sayers & Eustace *Docum. in Case* 72 The man's as mad as a hatter! I nearly sent round to get him certified. 1958 Wodehouse *Cocktail Time* xiv. 120 He had to be certified. He thought he was Stalin's nephew.

4. b. *Law.* To give a certificate for costs on a certain scale.

1867 *Act* 30 & 31 *Vict.* c. 142 § 5 He shall not be entitled to any Costs of Suit unless the Judge certify on the Record that there was sufficient Reason for bringing such Action in such Superior Court. 1889 C. A. White *Archbold's County Crt. Pract.* (ed. 10) 135 A judge has no power to allow costs on a higher scale unless he certify under the above section. 1910 *Encycl. Brit.* VII. 224/2 In every indictable case in which the committing justices or the court of trial certify for legal aid.

cerulean, *a.* Add: **c.** *cerulean blue*, an artist's colour prepared from cobalt, a brilliant light blue pigment.

1889 in *Cent. Dict.* 1962 R. G. Haggar *Dict. Art Terms* 71/2 *Cerulean blue*, a lovely, bright,..turquoise blue made from cobalt and tin oxides, first prepared in about 1860. It is..of great value for use in rendering atmospheric effects in landscape painting.

cerulignol (sīᵊrⁱuli·gnǫl). *Chem.* Also **cœrulignol.** [ad. G. *caerulignol* (A. Grätzel 1882, in *Arch. d. Pharm.* CCXX. 606), f. L. *cæruleus* dark blue (cf. quot. 1923) + *lign-um* wood + -ol.] A colourless oily phenolic liquid, $C_{10}H_{14}O_2$, which has astringent properties and an odour like that of creosote and is obtained by the distillation of various wood-oils and fruits.

1883 *Jrnl. Chem. Soc.* XLIV. 393 The rough creasote..contains another substance, recently isolated by the author and named by him cœrulignol. 1923 *Nature* CXI. 763/1 The presence of cerulignol in these tars, giving a blue coloration in alcoholic solution with lime or baryta. 1955 *Canadian Jrnl. Chem.* XXXIII. 723 The UV absorption maximum of this hydrocarbon, coerulignol, is 281 mμ.

cerulignone (si²ruli·gnōᵘn). *Chem.* Also **cœ-, †co-, †-on.** [ad. G. *cörulignon* (C. Liebermann 1872, in *Ber. d. Deut. Chem. Ges.* V. 746), f. L. *cæru-leus* (also written *cæruleus*) dark blue + *lign-um* wood + QUIN)ONE.] A quinone derivative, $C_{16}H_{16}O_6$, which is obtained when wood-vinegar is treated with potassium dichromate and which crystallizes in small bluish-grey needles.

1873 *Chem. News* 24 Jan. 38/1 As cœrulignon is insoluble in wood-vinegar, and not volatile, it cannot be contained as such in the distilled wood-vinegar, but must be formed after the distillation. 1874 *Jrnl. Chem. Soc.* XXVII. 1028 (*heading*) Use of Corulignone in Calico Printing. 1961 L. F. & M. FIESER *Adv. Org. Chem.* xxii. 767 The deeply colored cerulignone is sparingly soluble in ordinary organic solvents and is best crystallized from phenol.

cervelat. 1. (Earlier and later examples.)

1613 J. CHAMBERLAIN *Let.* 9 Sept. (1939) I. 477 The duke of Savoyes ambassador..hath brought..a present from his wife of salaccioni, cervellate, mortadelle. 1877 E. S. DALLAS *Kettner's Bk. of Table* 48 Things which..retain names no longer their own—cervelas without brains,..cheesecakes without curds. 1958 *Catal. County Stores, Taunton* June 5 Continental sausages..Cervelat—a lb. 9/-. 1965 *House & Garden* Jan. 60 *Cervelas.* There are many varieties of this sausage. Some are cured and dried and can be eaten as is. Others should be cooked.

cervical, *a.* Add: **1. b.** (Later examples.) *spec.* Of or pertaining to the cervix of the womb.

1884 R. & F. BARNES *Syst. Obstetric Med.* I. vii. 239 The cervical mucous membrane, lined by the connective tissue which sustains it, yields and slides on the muscular layer. 1923 in M. Box *Trial of Marie Stopes* (1967) 99 The small occlusive pessary for the cervix—the cervical cap. 1963 G. F. KANTOROWICZ et al. *Inlays, Crowns & Bridges* vii. 70 This stage will complete the cutting of a 'cervical collar' of sound tissue. 1966 *New Statesman* 11 Mar. 326/3 Cervical cancer—a disease that kills about 2,500 women a year in England and Wales alone. 1966 *Lancet* 24 Dec. 1423/1 There is a large variety of arrangements throughout Scotland for taking cervical smears. 1967 *Guardian* 17 May 4/5 The pill and the cervical coil gave women protection against ..enforced pregnancy.

cervicitis (sɜ̄ɪvisəi·tis). *Path.* [mod.L., f. L. *cervix, cervic-* neck + -ITIS.] Inflammation of the neck of the uterus.

1889 L. TAIT *Dis. Women* I. 108 Traumatic cervicitis often occurs after operations on the cervix. 1910 *Practitioner* Mar. 395 Every cervical erosion, cervicitis, or irritating leucorrhoeal discharge. 1964 M. HYNES *Med. Bacteriol.* (ed. 8) xi. 173 The infection does not cause urethritis in women, although it may produce cervicitis.

cervico-. Add: *cervico-dorsal* adj.

1908 *Practitioner* Sept. 404 The cervico-dorsal spine.

cervid (sɜ̄·ɪvid). Also **cervide.** [f. mod.L. *Cervidæ*, a large family of ruminant mammals (order Artiodactyla), including the deer, elk, and moose; cf. L. *cervus* deer, -ID³.] A ruminant mammal of the family Cervidæ, the members of which shed their antlers. Also *attrib.* or as *adj.*

1889 in *Cent. Dict.* 1928 G. B. BROWN *Art of Cave Dweller* i. 17 The slender daintiness of the legs of Cervides or of the Reindeer. 1938 *Antiquity* XII. 163 In style the engraving reminds one more of the cervids on the well-known Maglemose antler-haft. 1950 *Proc. Prehist. Soc.* XVI. 128 The general indications from the cervid antler material are consistent.

Cesarewitch (sėzæ·rėvitʃ, -zār-). [ad. Russ. *tsesarévich*, title as heir to the imperial Russian throne of the prince who became Alexander II.] A long-distance handicap horse-race run at Newmarket, instituted in 1839.

1839 *Sporting Mag.* 2nd. Ser. XIX. 263 *Newmarket.*—His Imperial Highness the Grand Duke of Russia having presented the Jockey Club with the sum of £300, to be run for annually, the first race for it will take place in the Second October Meeting, and is thus officially announced:—'The Cesarewitch Stakes—a free Handicap Sweepstakes.' 1856 'STONEHENGE' *Brit. Sports* 373/1 Cesarewitch Course.. 2 miles, 2 furlongs, 28 yards. 1883 [see WEIGHT *sb.*¹ 12 a]. 1891 G. CHETWYND *Racing Remin.* I. 29 It was at Warwick, too, that the big Cesarewitch commissions usually made their first appearance in the market. *Ibid.* 31 At the next Newmarket meeting Cardinal York won the Cesarewitch by six lengths. 1934 [see *ADD *v.* 1 a].

cesium, var. *CÆSIUM.

cessile (se·səil), *a. rare.* Also 6 **cessil.** [as if ad. L. **cessilis*, f. *cessus* (see CESSIBLE *a.*) + -ILE.] Yielding (applied only to the air, in imitation of the first quot.).

a 1599 A. HUME *Poems* (1902) 28 Sa silent is the cessile air. *Ibid.* 49 The massiue earth reposis still, Suspended in the cessil eire. 1911 QUILLER-COUCH *Brother Copas* xxiii. 319 A faint tapping sound reached them, borne on 'the cessile air'. 1922 JOYCE *Ulysses* 416 Scintillant circumambient cessile air.

cetane. Add: So *cetane number*, a measure of the ignition value of a diesel fuel oil.

1935 *Proc. Amer. Soc. Test. Mat.* XXXV. 342 More reliable results may be possible when using cetane rather than cetene. For this reason, this classification substitutes the cetane number, which is determined and expressed in the same manner as the cetene number, namely, as the percentage of cetane in a blend of cetane and alphamethylnaphthalene. 1959 *Chambers's Encycl.* VII. 640/2 Ignition quality is defined by 'cetane number', i.e. the percentage of cetane in a blend with α-methylnaphthalene which gives the same delay period, under standardized test conditions, as that given by the fuel in question.

‖ **ceteris paribus** (si·tĕris (or se·t-, kē¹·t-) pæ·ribɒs). Also **cæteris paribus.** [mod.L.] Other things being equal, other conditions corresponding.

1601 T. WRIGHT *Passions* Pref. A 4 b, Yet my meaning is alwayes *cæteris paribus*, because [etc.]. 1690 LOCKE *Govt.* II. § 105. 325 *Cæteris paribus*, they commonly prefer the Heir of their deceased King. 1751 CHESTERFIELD *Lett.* (1774) II. 124 And *cæteris paribus*, a French minister will get the better of an English one, at any third court in Europe. 1887 W. J. HARRIS in *Nat. Rev.* Dec. 454 The wonder is that France is not in a far worse state than ourselves. *Cæteris paribus*, she ought to be, with her unsettled government [etc.]. 1964 *Economist* 29 Aug. 828/2 That comfortable hold-all, *ceteris paribus*.

cetological *a.*, **cetologist.** (Examples.)

1851 MELVILLE *Moby Dick* I. xxxi. 179 But I now leave my cetological system standing thus unfinished. 1955 *Sci. Amer.* Jan. 67/1 Some cetologists believe that the gray whale survived along our Atlantic seaboard until historic times.

cetrimide (se·trimaid). *Med.* [f. *cetyltrimethylammonium bromide.*] A preparation of cetyltrimethylammonium bromide, used as an antiseptic and sterilizing detergent.

1948 A. D. GARDNER in *Lancet* 13 Nov. 762/1 (*heading*) Cetrimide (C.T.A.B. or Cetavlon). *Ibid.*, 10% cetrimide in 70% alcohol, as a rapid disinfectant, nearly equals the performance of alcoholic iodine. 1962 *Which?* May 158/2 Cetrimide solution may be used to clean the area round the burn. 1966 *Lancet* 24 Sept. 678/2 (*title*) Experimental use of cetrimide in the prevention of wound implantation with cancer cells.

Ceylon (sīlǫ·n). The name of an island in the Indian Ocean, used attrib. in **Ceylon moss** (see MOSS *sb.*¹ 4); **Ceylon pumpkin,** a large pumpkin found originally in Ceylon; **Ceylon rose,** the common oleander or rose-bay, *Nerium oleander*, a poisonous flowering shrub of the Mediterranean region; **Ceylon tea,** a Pekoe tea produced in Ceylon; also *ellipt.*, as *Ceylon.*

1861 BENTLEY *Man. Bot.* 721 *Gracilaria lichenoides* or *Plocaria candida*, is our commercial Ceylon Moss. 1875 *Encycl. Brit.* V. 508/2 *Gracilaria lichenoides*, under the name of Ceylon moss, is used for soups and jellies. 1913 PETTMAN *Africanderisms* 118 *Ceylon pumpkin*, a large, oval-shaped pumpkin with orange-coloured flesh. 1842 *Sketches East. Dist. Cape of Gd. Hope* 67 Lemon trees, interspersed with the acacia and Ceylon rose. 1868 J. CHAPMAN *Trav. S. Africa* II. 15, I find that the Ceylon rose.. is..the other poison with which the Damaras tip their arrows in war. 1886 in *Ceylon Mercantile & Planting Directory* (1891) 50 The reply to your question as to whether Ceylon tea can be laid down in London at 6d per pound, depends a good deal on rates of freight. 1907 *Moore Brothers' Delicious Teas* 16 Pure Ceylon...2/- per lb. 1962 J. BRAINE *Life at Top* xxiii. 253 London water and the cheapest brand of Ceylon didn't make a very enlivening brew. 1966 P. V. PRICE *France* 33 China tea, with lemon, may be had as well as Ceylon or Indian tea.

Ceylonese (silǭni·z), *a.* and *sb.* [f. prec. + -ESE.] **A.** *adj.* Of or pertaining to the island of Ceylon or its inhabitants; Cingalese. **B.** *sb.* A native or inhabitant of Ceylon.

1797 *Encycl. Brit.* IV. 310/2 The Ceylonese make use of boats hollowed out of the trunks of trees. *Ibid.* 311/1 The Ceylonese monarch was driven from his capital. 1836 *Penny Cycl.* VI. 457/2 The sailing of vessels at the Ceylonese ports. 1948 *Amer. Speech* XXIII. 171 The English cling to..*Sinhalese* or *Singhalese* for one [*sc.* an inhabitant] of Ceylon. But in the last case there seems to be a movement toward *Ceylonese.* 1960 G. DURRELL *Zoo in Luggage* (1965) ii. 55 Some weird devil-masks from a Ceylonese dance.

ch., abbreviation of *chapter, church.* Ch. B. = L. *Chirurgiæ Baccalaureus*, Bachelor of Surgery. Ch. Ch. = Christ Church (Oxford). Ch. D. = L. *Chirurgiæ Doctor*, Doctor of Surgery.

cha. Add: Now *slang* (cf. *CHAH, *CHAI, *CHAR *sb.*⁶).

1919 DOWNING *Digger Dial.* 15 *Chai* (Arab.) or *cha* (Hind.), tea. 1930 *N. & Q.* 22 Mar. 206/2 The army slang 'Cha' in the sense of tea.

chaack. [Imitative.] The cry of the jackdaw.

1906 *Westm. Gaz.* 9 June 13/2 With a little flutter of the wing and a cheery 'Chaack!' he hopped off one head on to another. 1907 *Ibid.* 19 Oct. 16/1 The cheering 'cha-ack' of the jackdaw is everywhere in the air.

chabutra (tʃābū·tra). *India.* Also **chabootah, chaboot(e)ra, chapudra, chebootura.** [Hind. *chabūtrā, chabūtara.*] 'A paved or plastered terrace or platform, often attached to a house, or in a garden' (Yule).

1827 SCOTT *Surg. Dau.* xiv, This splendid procession having entered the royal gardens, approached..a chabootra, or platform of white marble..which occupied the centre. 1848 'J. KIRKLAND' *Eerie Laird* x. 111 The..*chabooteras* of Delhi. 1882 F. M. CRAWFORD *Mr. Isaacs* x. 227 Some ryots had been called in to..raise a rough *chapudra* or terrace. 1895 MRS. B. M. CROKER *Village Tales* 154 He sat on his chabootra stolidly smoking his huka. 1963 *Times* 30 Jan. 12/6 The temple buildings clustered round a stone 'chabutra', a central meeting-point common to many villages.

cha-cha (tʃā·tʃā), **cha-cha-cha** (tʃa:tʃatʃā·). [Amer. Sp. *cha-cha-cha.*] A type of ballroom dance to Latin-American rhythm; also, the music for this dance. Hence as *v. intr.*

1954 *Melody Maker* 11 Dec. 15/2 What Touzet means by 'cha-cha', as far as I can make out, is simply a mambo with a pronounced güiro rhythm. 1955 *Ibid.* 16 July 5/3 'Cha-cha-cha' is danced very much more delicately over here than in England. 1959 *Camb. Rev.* 31 Jan. 279/1 We called on an old boy, to find him cha-cha-cha-ing with a young lady. 1959 *Economist* 4 Apr. 46/1 Cha-cha dance music. 1962 *Times* 11 June 3/1 Brazilian claques..cheered, cha cha-ed.

chachem, var. *HAHAM.

chack, *v.* Add: **1.** Also *gen.*, of the cry of a bird. Hence **cha·cking** *ppl. a.*

1919 MASEFIELD *Reynard the Fox* 56 A blue uneasy jay was chacking. 1924 J. A. THOMSON *Science Old & New* iv. 22 Numerous 'chacking' wheatears with brilliantly white rump feathers. 1929 MASEFIELD *Hawbucks* 316 An owl came down..and tore something, with chacking cries between the tears.

chack-chack. [Imitative.] The cry of the fieldfare and wheatear. Also as *vb.* Cf. prec. and CHACK *sb.* 3.

1906 *Westm. Gaz.* 29 Dec. 16/3 The hoarse 'chack-chack' of the fieldfare. 1908 *Ibid.* 21 Mar. 14/1 Even if you only hear his [*sc.* the wheatear's] sharp note 'Chack-chack'. 1930 E. W. HENDY *Wild Exmoor* 236 Stonechats 'chack-chack' from the top of a spray.

‖ **chacun à son goût** (ʃakɒ̃n a sɒ̃n gu), *phr.* [Fr.] Each to his own taste.

1784 W. COWPER *Let.* 4 Dec. (1904) II. 280, I envy those happy voyagers [*sc.* balloonists],..but..as I say, *chacun à son goût*; and mine is rather to creep than fly. 1879 C. M. YONGE *Magnum Bonum* III. xxxii. 680 'What are your plans?' '*Rotifer*, as before.' '*Chacun à son gout*,' said Bobus, shrugging his shoulders. 1961 A. WILSON *Old Men at Zoo* viii. 350 'I'm afraid,' I said, 'I can't see it as simply as that. You see, I knew them all.' Hales laughed. '*Chacun à son goût.*'

Chad, chad (tʃæd). [Origin obscure.] In full *Mr. Chad.* The figure of a human head appearing above a wall, etc., with the caption 'Wot, no ——?', as a protest against a shortage or the like.

1945 *Sunday Express* 2 Dec. 2/3 What is the origin of that peculiarly laughable figure called Chad we see so often scribbled across our walls? 1946 *Times* 1 Apr. 5/4 Mr. Chad probably went through a number of evolutions at the hands of a vast number of people before reaching the present state and can claim no one man as creator. 'Wot! No father,' it might well complain. 1950 M. KENNEDY *Feast* 220 She drew a picture of Mr. Chad on the terrace wall saying: 'Wot? No black amber?' 1965 *Oxford Mail* 28 Sept. 6 The person.. staring chad-like in indolence across the desk-tops.

chadar, chaddar (tʃɒ·dāɹ). Also **chadah, chader, chadur.** Varr. CHUDDAR.

1614 Chadar [see CHUDDAR]. 1622 R. COCKS *Diary* (1883) I. 109 Chaders, cambias, and buxshaws. 1876 *Cornhill Mag.* XXXIV. 335 Get a *chaddar* (cloak) to wrap round you. 1900 *Daily News* 16 Feb. 6/7 A Hindu woman actually needs but two garments, a skirt and a chadar, or veil. 1934 F. STARK *Valleys of Assassins* v. 275 A pale blue *chadur*, or veil. 1953 A. SMITH *Blind White Fish* vii. 122 They nodded and pulled their chadahs, the cloaks which cover their heads and reach to the ground, tighter over their faces and said no word.

Chadband (tʃæ·dbænd). [Name of a character, 'the Rev. Mr. *Chadband*', in Dickens's *Bleak House.*] A canting unctuous hypocrite. So **Chadba·ndian** *a.*; **Cha·dbandism; Cha·dbandize** *v. intr.*

[1852 DICKENS *Bleak Ho.* xix. 187 The Chadband style of oratory is widely received and much admired.] 1908 *Daily Chron.* 4 Nov. 5/3 There is no Chadbandian caricaturing of the parson. 1913 G. B. SHAW *Quint. Ibsenism* 189 Any attempt to Chadbandize on the wickedness of such crimes is at once resented as..'moral babble'. 1920 *Punch* 14 Jan. 33/2 His caustic dispraise Of President Wilson's Chadbandian ways. 1923 A. G. GARDINER *Cadbury* 259 His dislike of narrowness of view and Chadbandism kept him free from censoriousness.

chaddar, see *CHADAR.

chæta (kī·tă). Pl. **chætæ** (kī·tī). [mod.L., ad. Gr. χαίτη long flowing hair.] = SETA.

1866 LINDLEY & MOORE *Treas. Bot.*, *Chæta*, a bristle, the slender stalk of the spore-case of mosses; also called Seta. 1896 W. B. BENHAM in Harmer & Shipley *Cambr. Nat. Hist.* II. 266 The chaetae or bristles are mainly used in locomotion. 1902 *Encycl. Brit.* XXV. 689/2 The Chætopoda are characterized by the possession of horny epidermic chætæ embedded in the integument and moved by muscles. 1963 [see next].

chætigerous (kīti·dʒĕrəs), *a. Zool.* [f. Gr. χαίτη: see prec. and -GEROUS.] Setigerous.

1896 W. B. BENHAM in Harmer & Shipley *Cambr. Nat. Hist.* II. 263 Chaetigerous segments. *Ibid.* 312 The peristomium is chaetigerous. 1963 R. P. DALES *Annelids* viii. 171 In other larvae the first chaetigerous segment may have hypertrophied chaetae.

chætognath (kī·tǫgnæþ). [f. mod.L. *Chætognatha*, neut. pl. of *chætognathus*, f. Gr. χαίτη (see *CHÆTA) + γνάθος jaw.] A member of the Chætognatha, a phylum of small enterocœlous marine planktonic animals having two rows of stout spines on the head and a single row of sickle-shaped setæ or jaws; an arrow-worm.

1889 in *Cent. Dict.* 1896 A. E. SHIPLEY in Harmer & Shipley *Cambr. Nat. Hist.* II. 191 S[*agitta*] *hexaptera* is the largest Chaetognath known. 1962 D. NICHOLS *Echinoderms* x. 131 Other lesser phyla of the coelomates with three pairs of coelomic sacs, such as the chaetognaths.

chætotaxy (kī·totæksi). [f. Gr. χαίτη + τάξις: see *CHÆTA and -TAXY.] The arrangement or plan of distribution of the bristles on the bodies of insects.

1893 E. A. BUTLER *Our Household Insects* 188 The plan of their distribution has been called by Osten-Sacken chætotaxy, i.e., bristle arrangement. 1903 *Entomol. News* 247 A careful exposition of the chaetotaxy of the Muscidæ. 1925 A. D. IMMS *Gen. Textbk. Ent.* I. 8 In recent years chætotaxy, or the study of the arrangement of the more important setæ, has assumed a good deal of significance from the taxonomic point of view, particularly with reference to the Cyclorrhapha and larval Lepidoptera.

chaff, *sb.*¹ Add: **6. b.** Strips of metal foil or similar material released in the atmosphere to interfere with radar detection. orig. *U.S.*

1945 *Sun* (Baltimore) 29 Nov. 8/4 Under the disguise provided by aluminum 'chaff'—a few ounces of which gives a radar reflection comparable to that of three heavy bombers. 1947 SIEGERT & PURCELL in L. N. Ridenour *Radar System Engin.* iii. 82 'Window' is the British and most commonly used code name for conducting foil or sheet cut into pieces of such a size that each piece resonates as a dipole at enemy radar frequency...The strong signals it returns so effectively mask the radar signals from aircraft that are in the midst of a cloud of window that several tons of aluminum used to be dispersed..on each..bomber raid. The U.S. Army referred to this material as 'chaff'. 1962 *Observer* 11 Mar. 1/4 Metallic objects called 'chaff' were used by the Russians yesterday..in a new attempt to interfere with air traffic... The objects are about four inches long and about the width of a flat knitting needle. 1968 *New Scientist* 4 Apr. 9/1 As for the chaff itself, the move away from strips of aluminium foil to coated glass fibres has meant that twice as much deception can be stuffed into the same package.

7. b. chaff-box, a chaff-cutter to be worked by hand; **chaff-scale** *Bot.* (see quots.).

1840 *Flemish Husb.* 89 in *Libr. Usef. Knowl., Husb.* III, The chaff-cutter is exactly like our common chaff-box, where the work is done by the hand. 1908 *Essex Rev.* XVII. 24 A century or so ago, a handworked chaff-box was in almost daily use on every large farm. 1856 W. B. CARPENTER *Microsc.* 447 The hairs with which the *paleæ* (chaff-scales) of most Grasses are furnished, are strengthened by the like siliceous deposit. 1888 *Encycl. Brit.* XXIV. 531 At the base of each spikelet [of wheat] are two empty boat-shaped glumes or 'chaff-scales'.

chaff (tʃɑf), *v.*³ *Bread-making. trans.* To roll up (dough) into a rounded form in the moulding of a round loaf. Hence **cha·ffing** *vbl. sb.*

1892 R. WELLS *Pract. Bread Baker* 44 How to mould a round loaf...Divide the dough into parts, having the right hand piece smaller than the left. Now chaff this into two round pieces. 1925 *Glasgow Herald* 9 Oct. 6 The loaf is shaped on the chaffing table.

chaffy (tʃɑ·fi), *a.*² Now *rare.* [f. CHAFF *sb.*² + -Y¹.] Given to chaff or chaffing.

1855 A. MACMILLAN in C. L. Graves *Life* (1910) 78 The comic 'chaffy' vein is very monstrous. 1876 E. C. STEDMAN *Vict. Poets* i. 24 The time is off-hand, chaffy, and must be taken in its mood. 1889 MRS. H. L. CAMERON *Lost Wife* I. v. 94 He answered in the most 'chaffy' tone.

chagal (tʃā·gāl). Also **chaghal, chagul.** [a. Hindi *chāgal*, ad. Skr. *chāgala-* coming from a goat.] A water-bottle usually of canvas or leather.

1909 WEBSTER, Chagul. 1920 *Blackw. Mag.* Oct. 461/2 Laughing men carrying canvas chagals for water. 1923 *Ibid.* Sept. 398/2 Our chaguls or canvas water-bottles..are always ice-cold from the joggling about and evaporation. 1925 E. O. SHEBBEARE in E. F. Norton *Fight for Everest 1924* 367 It is suggested that a future expedition take a couple of pakhals or collapsible chaghals. 1959 M. PUGH *Chancer* 101 Both of my hands were at holding the chagal straps off my weary shoulders.

Chagas'(s) disease (tʃā·gās). [f. the name of Carlos *Chagas*, Sr. (1879–1934), Brazilian physician.] A progressive trypanosomiasis, often fatal in young children, which is endemic in South and Central America and in Africa and is due to *Trypanosoma cruzi*, a parasite carried by armadillos and certain other mammals and transmitted by various blood-sucking bugs of the family Reduviidæ.

1912 *Jrnl. Amer. Med. Assoc.* 27 July 318/1 Chagas' Disease... The trypanosomiasis or parasitic thyroiditis caused by the *Schizotrypanum cruzi*, the parasite first described by Chagas. 1922 *Encycl. Brit.* XXXI. 896/2 Other Reduviid bugs have been shown to transmit the trypanosome of Chagas's disease in South America. 1959 SOUTHWOOD & LESTON *Land & Water Bugs* 152 The members of the subfamily Triatominae are blood suckers... In South America they carry Chagas' disease, a human trypanosomiasis.

Chagigah (hăgi·gā). [Heb. *ḥăgīgāh.*] The voluntary sacrifices offered at the three great feasts, Passover, Pentecost, and Tabernacles.

1846 GEO. ELIOT tr. *D. F. Strauss's Life Jesus* III. 144 It [*sc.* the entering of a Gentile house in the morning] would have been such [*sc.* a disqualification] for the partaking of the Chagiga, which was eaten in the afternoon. 1891 A. W. STREANE (*title*) A translation of the treatise Chagigah from the Babylonian Talmud. 1933 H. DANBY tr. *Mishnah* II. 211 *Hagigah* ('The Festal-*Offering*')...The principal subject is the manner of fulfilment of Deut. 16¹⁶. 1966 WERBLOWSKY & WIGODER *Encycl. Jewish Relig.* (1967) 167/2 *Hagigah*,.. last tractate in the Mishnah order MOED. It deals with the festival duties, pilgrimage to Jerusalem, rejoicing, and the ḥ-offering.

chagul, var. *CHAGAL.

chah, chai, slang varr. CHA. Cf. *CHAR *sb.*⁶

1919 [see *CHA]. 1935 E. WEEKLEY *Something about Words* x. 183 A bus-driver, who would probably..describe tea as *chah.* 1940 *Manch. Guardian Weekly* 6 Sept. 165/4 Untrammelled canteen helpers are sometimes mystified by requests for 'chah' from thirsty soldiers. 1946 *Penguin New Writing* XXVIII. 172 Here's a drop o' chai. 1959 G. SLATTER *Gun in Hand* vi. 141 Faces you'd meet..brewing up the chai in *casas* along the front.

chain, *sb.* Add: **2. a.** *chain and ball*: see *BALL *sb.*¹ 1 b. Also *attrib.* and *fig. U.S.*

1866 'F. KIRKLAND' *Pict. Bk. Anecdotes* 154/1 One year's hard labor with chain and ball. 1887 F. FRANCIS Jr. *Saddle & Mocassin* 62 A chain-and-ball gang of convicts slowly advanced, sweeping the dusty road. 1904 H. E. RIVES *Castaway* 213 Dragging the chain and ball of a life sentence of desperation.

4. a. *chain of being(s)*, (a conception of the universe as) a continuous series or gradation of types of being in order of perfection, stretching from God as the infinite down through a hierarchy of finite beings to nothingness; the scale of being or nature (see SCALE *sb.*³ 5); *chain of command*, a series of positions of military or civil authority such that each holder is directly responsible to, and takes his orders from, the next above.

[1605 BACON *Adv. Learn.* I. i. 3 The highest link of nature's chain must needs be tied to the foot of Jupiter's chair.] 1712 ADDISON *Spect.* No. 519, The Chain of Beings, which has been often termed the *nexus utriusque mundi.* 1732 POPE *Ess. Man* I. 237 Vast chain of being, which from God began, Natures aethereal, human, angel, man, Beast, bird, fish, insect! 1742 [see CONNEXION 4 a]. 1915 *Sphere* 23 Jan. 96/1 The general headquarters with the chain of command branching out into three corps headquarters. 1936 A. O. LOVEJOY (*title*) The Great Chain of Being. 1951 AUDEN & KALLMAN *Rake's Progr.* II. 35 In secular abundant bliss He shall ascend the chain Of Being to its top. 1957 *Times* 23 Aug. 3/1 Whether the traffic branch was directly responsible to the Chief Constable or whether the chain of command led through the Superintendent. 1959 *Listener* 23 Apr. 700/1 President Eisenhower has cherished the General's devotion to the chain of command: the habit of delegating total responsibility. 1970 *Daily Tel.* 11 May 9/5 The 15 companies of Independent Television have managed to avoid bureaucracy: their chains of command have been kept short.

c. A number of cigarettes or cigars smoked in continuous succession. Cf. *chain-smoker* in *19.

1908 *Field* 19 Sept. 544/3 Mr. Travis smoked a chain of long black cigars during all his matches...Mr. Hilton used to smoke a chain of cigarettes. 1930 AUDEN *Poems* 67 Smoking cigarettes in chains until their heads are in a whirl.

5. d. (Earlier example.)

1825 *Analysis of London Ball-Room* 102 Quadrilles... Ladies chain.

e. A series of branch businesses, stores, etc., controlled by one owner or firm. Also *attrib.* (Cf. *chain store* in *19.) orig. *U.S.*

1846 W. L. MACKENZIE *Van Buren* 208 A convenient instrument for regulating future state elections through a chain of banks. 1895 *N.Y. Dramatic News* 14 Dec. 6/2 A chain of eleven theatres. 1906 S. E. SPARLING *Introd. Business Org.* 205 A better illustration of the chain of stores operated by the manufacturer is the Douglas chain of shoe stores and the Lipton tea stores. 1928 *Sat. Even. Post* 12 May 59/2 Managing director of a big chain-grocery concern. 1930 *Economist* 15 Feb. 367/2 Numerous concerns owning extensive 'chain' shops. 1931 *Ibid.* 14 Nov. 897/2 In Britain, taking chains and co-operatives together, some 3,500 groups controlling 45,000 retail outlets..occupy much the same position as do chain stores alone in the United States. 1933 P. GODFREY *Back-Stage* xiv. 181 Mr. J. J. Shubert..whose syndicate..owns a chain of 800 theatres in America. 1964 *Daily Tel.* 18 Jan. 8/2 The cuts that have already been announced by some retail chains in the past few days.

f. A line of people formed for a particular purpose, *spec.* a bucket chain (see *BUCKET *sb.*¹ 5); *grand chain*: a dance-figure in which each person dances with a line of people one after another (cf. sense 5 d in Dict.).

1864 DICKENS *Mut. Fr.* I. xi. 104 Sixteen disciples of Podsnappery went through the figures of..the grand chain. 1876 C. M. YONGE *Three Brides* v. 72, I tried to get them to form a chain and drench the warehouses. 1932 [see *bucket chain*, *BUCKET *sb*¹ 5]. 1941 J. CARY *House of Children* 61 We went through a lancers together where all the girls.. shouted: 'Grand chain—do think what you're doing.' 1963 *Guardian* 17 Jan. 16/4 The staff..formed a human chain to rescue Chinese ceramics..from the firm's main showroom after a water main burst.

g. *Chem.* A number of similar atoms (usu. of carbon) joined in series in a molecule. *closed chain*, a ring of similar linked atoms; *open chain*, one the ends of which are not joined together.

1881 ROSCOE & SCHORLEMMER *Chem.* III. I. 113 The atoms may thus be represented as forming a chain, one atom being linked on to the other, so that when one of them is removed without altering the position of the others, the chain is broken. 1904 GOODCHILD & TWENEY *Technol. & Sci. Dict.* 111/1 When a number of atoms are joined together in a ring as distinct from a chain, the system is called a closed chain, and its derivatives closed chain compounds. 1928 KINGZETT *Chem. Encycl.* (ed. 4) 131 Open chains are regarded as having terminal atoms not mutually in combination, by which, for example, aliphatic combinations, such as the olefines and paraffins, are characterized. *Ibid.* 132 In closed chains, the terminal atoms are regarded as mutually combined, forming a ring, as in benzene. 1946 *Nature* 14 Dec. 864/1 The total chain-length of the alkyl substituents ranges from 13 to 15 carbon atoms. 1964 N. G. CLARK *Mod. Org. Chem.* ii. 11 The skeleton of carbon atoms forming a paraffin molecule is usually referred to as a 'chain'.

h. *Bacteriology.* A series of cells joined end to end.

1910 *Encycl. Brit.* III. 160/2 (*caption*) A chain of motile rodlets still growing and dividing. 1917 *Jrnl. Bacteriol.* II. 155 Though chains of cells may be developed. 1928 W. GILTNER *Gen. Microbiology* iii. 33 The spherical forms may divide in one plane and, if they cohere, form chains usually called *streptococcus.* *Ibid.* 34 The rods separate after fission, or form chains (the *streptobacillus*). 1930 R. ST. JOHN-BROOKS in *Syst. Bacteriol.* (*Med. Res. Council*) I. ii. 104 Division occurs either in one direction, whereby chain-formation takes place (streptococci) or in two directions... Among the staphylococci, however, short chains of organisms may be demonstrated, which show their relationship to the streptococci. 1951 MCLEAN & IVIMEY-COOK *Theoret. Bot.* I. 346 Bacteria were classified in the past chiefly on their microscopical appearance...The simplest form is.. known as coccus. When these occur separately they are put in the genus *Micrococcus*,..if in chains, the genus *Streptococcus.*

18. *chain-link, -retailing.*

1896 A. SHARP *Bicycles & Tricycles* xxvi. 400 The rivet.. bears along the whole width of the inner chain-link [of a bicycle chain]. 1920 MASEFIELD *Enslaved* 30 A stirring sleeper struck the bell Of chain-links upon stones. 1958 *Archit. Rev.* CXXIII. 310 The subtopian effect of chainlink fencing. 1960 *Farmer & Stockbreeder* 2 Feb. 121/1 The bottom half of the 6ft-high front is composed of galvanized sheeting, above this being chain-link netting because of the danger from foxes. 1935 *Economist* 14 Dec. 1215/1 The report of this 'chain retailing' organisation..does not lend itself to comparison with previous results.

19. chain-bag, a woman's hand-bag made of fine metal chain-work; **chain-bearer** = *chain-man*; **chain-brow way** *Coal-mining* (see quot.); **chain-carrier** = *chain-man*; so *chain-carrying* vbl. sb.; **chain case,** the protective covering of the chain gear of a bicycle, motor vehicle, etc.; **chain chest** *Naut.*, a locker in the channels for storage of wash-deck gear; **chain coral,** a kind of fossil coral, *Catenipora escharoides*; **chain dog,** (*a*) a dog restrained by a chain; (*b*) a chain having at each end a 'dog' or hook, which is driven into logs to fasten them together into a raft (*Funk's Stand. Dict.* 1893); **chain-dotted** *a.*, (of a line) marked with alternate dot and dash; **chain drive, driving,** a method of transmitting power by means of a chain gear, esp. from the engine to the driving wheels of a bicycle, motor vehicle, etc.; **chain-driven** *a.*, driven by means of chain gear; hence **chain driver,** a vehicle driven by this method; **chain gammoning** *Naut.*, gammoning consisting of a chain; **chain-gang** (earlier example); **chain gear, gearing,** a gear for transmitting motion by means of an endless chain; esp. one in which the chain transmits motion from one sprocket-wheel to another; **chain gemma** *Bot.* (see quot. 1900); **chain-grate (stoker)** (see quot. 1889); **chain-harrow** (examples); hence *chain-harrow* v. trans. and intr., *chain-harrowing* vbl. sb.; **chain horse,** a horse harnessed with chain traces, employed as an additional horse in drawing heavy loads, esp. up a hill; **chain inclinometer,** an instrument for indicating the inclination of a surveyor's chain; **chain-instinct** *Psychol.* (see quot.); **chain knot** (see quot.); **chain letter,** a letter written with an

invitation to the recipient to pass it on to another (or copies of it to others), the process being repeated in a continuous chain until a certain total is reached; **chain-lightning** (earlier U.S. example); **chain-line** *Paper-making*, the mark of a chain-wire; = *wire line, -mark* (*a*) (see WIRE *sb.* 16); **chain-locker** (example); **chain-man** (earlier U.S. example); **chain-mark** = **chain-line*; **chain messenger** *Naut.*, a messenger consisting of an endless chain; **chain pipe** *Naut.* (see quot.); **chain-pull** [PULL *sb.*² 6], a chain used as the device for operating an electric switch; also *attrib.*; **chain reaction,** a chemical or nuclear reaction forming intermediate products which react with the original substance and are repeatedly renewed; also *fig.*, any such self-maintaining process; hence *chain-reacting* adj.; **chain-reflex** *Psychol.*, a series of reflexes in which each sets off the next; **chain riveting** (see quots.); **chain road** (see quot.); **chain-saw,** (*b*) see quots.; **chain-smoker** [tr. G. *kettenraucher*] (see quot. 1890 and sense *4 c); hence *chain-smoke* v. trans. and intr., *chain-smoking* vbl. sb. and ppl. a.; **chain-snake** (later U.S. examples); **chain store** orig. *U.S.*, one of a series of stores belonging to one firm and dealing in the same class of goods; **chain wire** *Paper-making* (see quot.). Also in the names of various appliances of which a chain is an important part.

1902 *Westm. Gaz.* 29 Dec. 3/3 These glasses..can easily be carried in the modish silver or gold chain-bag along with the handkerchief. 1736 in *Cal. Virginia St. Papers* (1875) I. 226 We do..appoint you James Thomas junʳ, Surveyor,.. To take three Chain-bearers. 1798 *Let.* in *Deb. Congress 1808* (1852) 2741 You will see I have omitted taking those [depositions] of Mr. Gillespie and the chain-bearers. 1869 *Overland Monthly* III. 248 The chain-bearers signaling to those holding the ropes. 1883 GRESLEY *Gloss. Coal-m.*, *Chain-brow Way*, an underground inclined plane worked by an endless chain. 1702–3 J. LOGAN *Let. to Penn* 3 Jan. in *Corr.* (1870) I. 174 Neither surveyors nor chain-carriers will go thither. 1798 *U.S. Statutes* I. 543 For compensation to the assistant surveyors, chain carriers, axe men and other persons. 1816 U. BROWN in *Maryland Hist. Mag.* XI. 220 This afternoon hunts up Chain Carriers and an Axman or marksman. 1838 *S. Lit. Messenger* IV. 307 The surveyor and two of his chain-carriers were killed. 1798 *Let.* in *Deb. Congress 1808* (1852) 2739 The surveying at present is done by Mr. Gillespie, the chain-carrying by Mr. Ellicott and Mr. Walker. 1909 'O. HENRY' *Roads of Destiny* vi. 93 I'll..go back to chain-carrying for the county surveyor. 1907 *Westm. Gaz.* 9 Nov. 14/2 The Daimler Company..have designed their chain-cases to act at the same time as radius-rods. 1909 *Ibid.* 17 Nov. 5/2 Detachable oil-bath chain-cases. 1884 S. B. LUCE *Seamanship* 4 Chain chests. 1808 J. PARKINSON *Org. Rem. Former World* II. 20 The chain coral (*Tubipora catenulata*, Linn.) composed of parallel tubes. 1871 LYELL *Student's Elem. Geol.* 449 The 'chain-coral', *Halysites catenularius.* 1895 K. MEYER tr. *Voy. Bran* I. 81 [She had] a large chain-dog with her. 1878 ABNEY *Photogr.* 260 Chain-dotted straight lines. 1931 *Engineering* 16 Jan. 91/3 One of the ropes is indicated by chain-dotted lines in Fig. 1. 1967 A. BATTERSBY *Network Analysis* (ed. 2) iii. 40 Ladder activities can be strung together as shown in Fig. 3.15 to form a ladder diagram. The chain-dotted lines which connect them represent lead and lag times. 1903 *Daily Chron.* 20 Apr. 7/2 The chain-drive is infinitely superior to the belt. 1907 *Ibid.* 17 Oct. 8/2 A 5 h.-p. twin-cylinder Roc, fitted with a Peugeot engine and chain drive. 1969 *Sears Catal.* Spring/Summer 12 This automatic garage door opener...Boasts permanently lubricated ½-HP motor with center-mounted T-rail chain drive. 1887 LD. BURY & HILLIER *Cycling* xiii. 353 Chain-driven geared-up Safety bicycles. 1898 *Cycling* 59 We do not anticipate that for serious riding it will oust the chain-driven safety. 1900 *Engineering Mag.* XIX. 740 Chain-driven heavy freight vehicles. 1909 *Daily Chron.* 22 Nov. 9/1 The new light Phelon and Moore is chain-driven. 1910 *Westm. Gaz.* 17 Feb. 5/1 The 65-h.p. and 75-h.p. six-cylinder chain-drivers. 1887 LD. BURY & HILLIER *Cycling* xiii. 348 The successful use of the chain in tricycles opened the eyes of inventors to the possibilities which chain-driving possessed for Safety machines. 1883 *Man. Seamanship for Boys' Training Ships* 22 Gammoning Fish..are..nailed on the upper part of the bowsprit, in the wake of the chain gammoning. 1834 *Westm. Rev.* July 47 Chain gang...They are to wear chains and the yellow dress, with the word 'felon' stamped upon it in several places. *a* 1877 KNIGHT *Dict. Mech.*, Chain-gear. 1902 *Daily Chron.* 5 Sept. 7/5 Another car stopped near Hickstead with a chain-gear mishap. 1896 A. SHARP *Bicycles & Tricycles* xxvi. 422 The *average* speeds of two shafts connected by chain gearing are inversely proportional to the numbers of teeth in the chain wheels. 1897 *Daily News* 30 Oct. 5/2 A boy..got drawn into the chain-gearing of the wheels. 1893 *Funk's Stand. Dict.*, Chain-gemma. 1900 B. D. JACKSON *Gloss. Bot. Terms*, *Chain-gemma*, in Fungi, having the form of a septate confervoid filament, the segments of which are capable of growth; termed also sprout-gemma. 1889 *Cent. Dict.*, *Chain-grate*, a feeding-device for furnaces. The fuel..is slowly carried forward by an endless apron formed of cross-bars attached at each end to moving chains. These bars form the grate. 1930 *Engineering* 7 Feb. 178/2 It would.. be simpler to burn it on chain-grate stokers. *Ibid.* 21 Mar. 391/2 In chain grates..a zone of reducing gas..is apt to form immediately behind the bed of coal. 1870 *Rep. Comm. Agric. 1869* 322 Only one patent was taken out during the year of the class known as 'chain' harrows, *i.e.* composed entirely of iron chains, no beams whatever being employed. 1931 *Times* 16 Mar. 17/2 The desirability of chain-harrowing the land. 1931 *Times Educ. Suppl.* 28 Mar. 113/2 On stale pastures it is better to use chain harrows. *Ibid.*, Each hour one can do three acres of zigzag harrowing or chain-harrowing. 1950 *N.Z. Jrnl. Agric.* Feb. 116/2 Then.. chain harrow to level the surface. 1876 *Porcupine* 22 Apr. 58/3, I have seen (when they have been carthorses) a chain put round their neck, and a chain-horse hooked to it. 1906 *Westm. Gaz.* 7 May 8/1 Now we have a request to establish a chain-horse on Surbiton-hill. 1955 *Times* 11 May 14/4 He has been a navvy, chain horse boy and.. a professional boxer and wrestler. *a* 1877 KNIGHT *Dict. Mech.*, *Chain-inclinometer*, a form of level in which the inclination of the surveyor's chain is indicated on a scale by the pointer on the end of the level. 1914 W. MCDOUGALL *Introd. Soc. Psychol.* (ed. 8) 392 'Chain instincts', instincts each of which manifests itself in a chain of activities. *a* 1877 KNIGHT *Dict. Mech.*, *Chain-knot.* 1. A succession of loops on a cord, each loop in succession locking the one above it... 2. A kind of knot used in splicing. 3. The loop-stitch in some sewing-machines. 1906 *Daily Chron.* 27 July 6/2 In 1896 Miss Audrey Griffin, of Hurstville, New South Wales initiated a 'chain letter' with the object of obtaining 1,000,000 used postage stamps. 1940 C. MCCULLERS *Heart is Lonely Hunter* (1943) II. xiii. 259 Think about chain letters. If one person sends a letter to ten people and then each of the ten people send letters to ten more—you get it? 1966 S. MARCUS *Other Victorians* ii. 35 Half the legends.. can be traced, like the links of some bizarre chain letter, to their original source in these volumes. 1843 HALIBURTON *Attaché* I. xv. 262 They hante no variety in them [*sc.* drinks] nother; no white-nose, apple-jack, stone-wall, chain-lightning, [etc.]. 1880 Chain-line [see Dict., sense 18]. 1960 GLAISTER *Gloss. Bk.* 63/2 *Chain lines*, the vertical lines on laid paper. 1883 Chain locker [see **chain pipe*]. 1714 W. WINTHROP in *Mass. Hist. Soc. Coll.* (1892) V. 299 The survayer and chainmen being under oath. 1923 *Handbk. Quality-Standard Papers* (*Amer. Writing Paper Co.*) 370 *Chain mark*, one of the wider parallel lines of a laid watermark. 1961 T. LANDAU *Encycl. Librarianship* (ed. 2) 189/2 Chain marks due to its manufacture on a mould in which the wires are laid side by side rather than woven. 1883 *Man. Seamanship for Boys' Training Ships* 14 An iron band fitted with teeth.., which enters the long links of the chain messenger in weighing anchor. *Ibid.* 12 Q. What are chain pipes? A. Apertures through which chain cables pass from the chain lockers to the deck above. 1911 *Chambers's Jrnl.* 17 June 475/1 Users of electric light..will be interested in a chain-pull switch. 1945 H. D. SMYTH *Gen. Acct. Devel. Atomic Energy Mil. Purposes* ii. 22 To run a chain-reacting system at a high temperature and to convert the heat generated to useful work is very much more difficult than to run a chain-reacting system at a low temperature. 1956 A. H. COMPTON *Atomic Quest* 52 The idea of the plutonium bomb..gave military significance to the chain-reacting pile. 1926 *Chem. Abstr.* XX. 2122 Photochemical equivalence and chain reactions. 1938 *Ann. Reg. 1937* 348 A chain reaction theory of excitation was put forward. 1938 R. W. LAWSON tr. *Hevesy & Paneth's Man. Radioactivity* (ed. 2) xxiv. 254 An elementary process may be immediately succeeded by a 'chain reaction', and then the amount of chemical change may be enormous. 1945 *Statements relating to Atomic Bomb* (*H.M.S.O.*) 14 It was generally accepted that a chain reaction might be obtained in uranium which would yield enormous amounts of energy. 1947 *Sat. Even. Post* 22 Mar. 140/3 If you publish a candid article about any community, giving actual names of people..you are..braving a chain reaction of lawsuits, riots and civil commotion. 1949 *Mind* LVIII. 270 The 'chain-reactions' let loose by the breakdown of belief in absolutes. 1954 *Sci. News* XXXII. 48 Polymerizations proceed as chain reactions, growth being propagated by activated molecules (free radicals). 1900 J. LOEB *Compar. Physiol. & Psychol.* (1901) ix. 144, I am inclined to recommend using the word chain-reflexes, whereby the performance of one reflex acts at the same time as the stimulus for setting free a second reflex. 1959 *Encycl. Brit.* IV. 3/2 In a 'chain-reflex' the result of a foregoing reflex's execution is to evoke execution of the next succeeding one. 1888 *Lockwood's Dict. Terms Mech. Engin.*, *Chain Riveting*, rows of rivets placed in parallel lines, both in the longitudinal and transverse directions. 1895 G. J. BURNS *Gloss. Techn. Terms Archit.*, *Chain riveting*, in this kind of riveting the rivets are placed in parallel rows in the direction of the stress, the rivets in adjacent rows being opposite each other. 1883 GRESLEY *Gloss. Coal-m.*, *Chain Road*, an underground wagon-way worked upon the endless chain system of haulage. 1909 WEBSTER, *Chain saw*,..a large saw for cutting coal, as one set with chisel points. 1957 *Gloss. Terms Stone in Build.* (*B.S.I.*) 13 *Chain saw*, a power-driven saw consisting of a chain which has cutting tips attached to the links. 1958 *Times* 18 Oct. 9/1 With a chain saw we can cut a cord of wood into logs in two Saturday afternoons. 1934 DYLAN THOMAS *Let.* 25 Apr. (1966) 110 I've chain-smoked for nearly five years. 1937 KOESTLER *Sp. Testament* 303 Got into a rage and chain-smoked all seven cigarettes. 1957 C. MACINNES *City of Spades* II. xv. 203 Have a fag..I chain-smoke myself. 1890 *Review of Reviews* I. 279/2 Bismarck is, or used to be, what the Germans call a chain-smoker, that is to say, that he would smoke on and on an endless chain of cigars, lighting each from the ashes of its forerunner. 1930 V. M. SACKVILLE-WEST *Edwardians* i. 17 These parties..were like chain-smoking: each cigarette was lighted in the hope that it might be more satisfactory than the last. 1953 *Encounter* Nov. 69/1 Only their very young, chain-smoking officer stared across the border. 1789 MORSE *Amer. Geogr.* 61 Of the Snakes which infest the United States, are the following, viz... Chain [snake]. 1875 *Field & Forest* I. 30, I observed ..a reptile which proved to be the Chain, or Thunder snake (*Ophibolus getulus*). 1910 *Sat. Even. Post* 10 Sept. 76/2 There were loud declarations of war from the manager of the association that buys goods for retail grocers fighting the chain stores. 1922 S. LEWIS *Babbitt* iv. § 5 One of these cash-and-carry chain-stores..cutting prices below cost. 1930 *Economist* 4 Jan. 18/1 The company's expansion into the chain store business has been carried on at too rapid a rate. 1957 J. I. M STEWART *Use of Riches* i. 11 And what about the chain stores—the poor man's Army and Navy, the suburban housewife's Harrods? 1889 *Athenæum* 30 Mar. 409 There are stout wires, about an inch apart, called 'chain wires',..which run from top to bottom [of the mould].

chain, *v.* Add: **5.** (Later U.S. examples.) Also with *out.*

1816 U. BROWN in *Maryland Hist. Mag.* XI. 224 [But for the rain] I should Certainly have Caused this line on the river to have been Correctly run and Chain'd. 1845 J. F. COOPER *Chain-bearer* xxv, You're welcome to chain out just as much of this part of the patent as you see fit.

‖ **chaîné** (ʃęne). *Ballet.* [Fr.] A quick step or turn from one foot to another, or a series of these, performed in a line.

1946 G. ROBERT *Borzoi Bk. Ballets* 354 *Chaîné*, a traveling step in which the feet are held in a firm first position, on point or demi-point, for a series of small regular turns. 1949 A. CHUJOY *Dance Encycl.* 90 *Chaîné*, in ballet, a series of short, generally fast, turns, done in a chain, or straight line, across the stage.

chained, *ppl. a.* Add: **1. b.** Of a book: secured to a shelf, desk, table, etc., by a chain in order to prevent its removal; also applied to a library of such books.

1846 *Wesleyan-Meth. Mag.* (Abr. ed.) II. 155/2 What a hold he has of that great chained Bible! 1890 W. BLADES *Books in Chains* (1892) 4 The chained library at Wimborne Minster. 1908 *Daily Chron.* 10 Jan. 6/7 The old chained-Bible table previously in the church of St. Mary Bourne.. has been restored. 1931 B. H. STREETER *Chained Library* i. 48 The chained library in the mediaeval system was not the equivalent of what we mean by a library, it corresponded rather to a 'reference library'. 1963 *Times* 2 Feb. 10/4, I direct my trustees to hand over my said chained library to Sir Robert Hall.

3. *chained-lightning.* Also *transf.*

1872 SCHELE DE VERE *Americanisms* 217 Chain-lightning ..is generally..changed into *chained-lightning.* In both forms it is constantly applied to inferior whiskey.

chainer (tʃē¹·nəɹ). [f. CHAIN + -ER¹.] **a.** One of a surveyor's party who carries the chain; a chain-carrier. **b.** One who twists material into a chain. **c.** One who tends the chain of a haulage system.

1849 C. STURT *Narr. Exped. C. Australia* I. 323, I had sent Mr. Stuart and Mr. Piesse with a party of chainers, to measure along the line on which I intended to move. 1921 *Dict. Occup. Terms* (1927) §§ 03, 367, 399.

chainless, *a.* Add: **2.** Of a machine, vehicle, etc.: without chain gear or chain drive as a part of the mechanism. (Usually implying its former use.)

1897 *Queen* 16 Oct. (Advt.), The best Chainless Cycle for Ladies. 1905 *Daily Chron.* 17 June 9/2 Mr. Napier was the first constructor to use this chainless form of transmission for a powerful car.

chair, *sb.*¹ Add: **1. c.** A glass-blower's seat furnished with long arms upon which he rolls the pontil; hence, the gang of men consisting of the glass-blower and his assistants.

1845 G. DODD *Brit. Manuf.* IV. 51 Another workman receives it and sits down in a chair having two flat parallel arms sloping downwards. 1890 W. J. GORDON *Foundry* 137 Every two crucibles supply one 'chair'. The glassworker's chair is practically his lathe. 1897 *Worc. County Express* 3 Apr., There were ten chairs at the works, each occupied by a glassmaker, servitor, and footmaker. 1902 *B'ham Daily Post* 2 Apr., Eighty is the limit number of strawstem wineglasses to be made in six hours by a 'chair', which consists of three men and a boy. 1962 *Gloss. Terms Glass Ind.* (*B.S.I.*) 23 *Chair*, a special long-armed chair in which the craftsman sits when shaping glass. *Ibid.* 45 *Chair*, a team or gang of workers producing blown or pressed glassware by hand.

d. = *electric chair* (s.v. *ELECTRIC *a.* 2). *U.S.*

1900 'FLYNT' & WALTON *Powers that Prey* 170 He was a copper, and we fly cops have got to send some bloke to the chair for blastin' him. 1917 J. FARNOL *Definite Object* xviii. 173 I've left papers—proofs, 'n' it'ud be the chair for yours—savvy? 1926 G. D. H. & M. I. COLE *Blatchington Tangle* xiv. 102 The discovery of the murderer's handkerchief..was the means of bringing a most notorious criminal to the chair.

3. c. A chair occupied by a Welsh bard at an Eisteddfod, esp. one awarded as a trophy; also, a convention, now each of the four conventions, connected with the Welsh Eisteddfod.

1820 *Cambro-Briton* I. 36 He was placed, by the general voice, in the bardic chair, and invested with a blue ribbon. 1874 *Cassell's Mag.* IX. 431/1 The grand event of the whole Eisteddfod..—the giving of the Chair Prize. 1877 *Encycl. Brit.* VII. 791/1 The chair was a kind of convention where disciples were trained, and bardic matters discussed preparatory to the great Gorsedd...There are now existing four chairs in Wales. 1909 T. R. ROBERTS *The Eisteddfod* 27 The earliest Eisteddfodau, or 'Chairs' as they were then called,..were held under the patronage of the Princes of Wales at the beginning of the sixth century. *Ibid.* 36 The Gwyneddigion offered a silver chair to the bard who could write the best verses.

11. (Later example.) Also *attrib.*

1909 R. W. CHAMBERS *Firing Line* ix, They went together in a double chair, spinning noiselessly over the shell road. *Ibid.* xiv, Their black chair-boy lay asleep under a thicket of Spanish bayonet.

15. *chair-attendant, -caning, -factory*; **chair-back,** (*a*) the back of a chair; (*b*) an antimacassar; **chair-bard** [Welsh *cadair fardd*], the successful competitor in the bardic competition held on 'chair day' of the Welsh National

Eisteddfod; **chair-bed** (examples); **chair binder** (see quot. 1921); **chair-borne** *a.*, *Mil.*, ironically descriptive of troops whose duties are administrative; cf. *AIR-BORNE *a.*; also *absol.*; **chair-car** orig. *U.S.*, a railway carriage furnished with chairs (two on each side of the aisle) instead of the usual seats; also, a parlour car (see PARLOUR 6); **chair-carver** (see quot.); **chair day,** the chief day of the Welsh National Eisteddfod (see quot.); **chair-door** (see quot.); **chair frame maker, chair framer, chair matter** (see quots.); **chairlady** (orig. *U.S.*) = CHAIRWOMAN; **chair-lift** (see *LIFT *sb.*²); **chair-marking** *slang* (see quots. and Dict. 14); so *chair-mark* v. and sb., *chair-marked* ppl. a.; **chair-post** *U.S.*, one of the main uprights of a chair; **chair road,** a railway having the rails fastened by chairs to the sleepers; **chair rusher, seater** = **chair matter*; **chair-side** *attrib.*, of or pertaining to dental work performed while the patient is seated in a dentist's chair; also in other uses (see quots.); **chair-table,** a table convertible into a chair or settle; **chair turner,** a wood turner who specializes in chair legs, rails, etc.; **chair-warmer** *slang*, orig. *Theatr.* (see quot. 1909); hence *gen.* a 'passenger' in any enterprise or situation.

a 1953 DYLAN THOMAS *Quite early one Morning* (1954) 33 The clip of the chair-attendant's puncher. 1747 H. GLASSE *Art of Cookery* xxii. 166 Hang the rest of your Bedding on the Chair-backs. 1858 TROLLOPE *Dr. Thorne* III. xxxv. 89, I should like to work it into a chair-back instead of floss-silk. 1883 LADY C. SCHREIBER *Jrnl.* (1911) II. 405, I did some work, cutting up some lengths of Turkish embroidery into chair backs. 1874 *Cassell's Mag.* IX. 431/1 The successful competitor will be installed as the Chair Bard. 1895 *Daily News* 20 May 5/3 The chair-bard for this year of the Welsh National Eisteddfod. 1805 *Times* 7 Nov. 1/2 (Advt.), Sofa Beds, Chair Beds, and Tables. 1912 COMPTON MACKENZIE *Carnival* xxiii. 235 Mr. Dale was generally comatose on a flock-exuding chair-bed in what was known as 'dad's room'. 1881 *Instr. Census Clerks* (1885) 53 Chair.. Binder. 1921 *Dict. Occup. Terms* (1927) § 504 *Chair binder*, tacks hessian or other stuffing cover over padding of back and seat of chair and over webbing under seat. 1943 C. H. WARD-JACKSON *Piece of Cake* 21 *Chair-borne division*, manned by uniformed personnel who labour in offices. 1947 N. BALCHIN *Lord, I was Afraid* 166 The old sneer of the airborne at the chair-borne. 1954 J. MASTERS *Bhowani Junction* 171 For Christ's sake, wake up, you chairborne bastard. 1906 *Daily Chron.* 9 Mar. 8/5 Umbrella-mending, chair-caning..are..genteel occupations to be pursued by the caravan 'smart set'. 1880 G. A. SALA *Amer. Revisited* 166/2 The Pullman Parlor car—commonly termed a 'chair' car. 1901 *Westm. Gaz.* 11 July 6/1 A fire..partially destroyed a dining car and a chair-car. 1909 'O. HENRY' *Options* (1916) 210 My chair-car was profitably well filled with people of the kind one usually sees on chair-cars. 1969 *National Herald* (New Delhi) 29 July 4/4 The average number of air-conditioned chair car seats occupied per trip was 290. 1858 SIMMONDS *Dict. Trade*, *Chair-carver*, one employed in carving the upright posts and other parts of beds, arm-chairs, sofas, &c. 1877 *Encycl. Brit.* VII. 792/1 The great day of the Eisteddfod is the 'chair' day—usually the third or last day—the grand event of the Eisteddfod being the adjudication on the chair subject and the chairing and investiture of the fortunate winner. 1906 HARDY *Dynasts* II. vi. vi, In addition to the grand entrance ..there is a covert little 'chair-door'..for sedans only. 1827 DRAKE & MANSFIELD *Cincinnati* viii. 65 Six Chair Factories, 38 hands. 1903 *Daily Chron.* 10 Mar. 10/4 Foreman required for large chair factory. 1897 *Daily News* 9 Mar. 2/6 Chair frame maker. 1921 *Dict. Occup. Terms* (1927) § 473 *Chair frame maker, chair framer*, assembles parts of chair frames, prepared by woodworking machinist. *Ibid.* § 472 *Chair matter*; *chair rusher*; *chair seater*; weaves rushes, by hand, into chair seats or frames. 1925 *New Republic* 16 Sept. 92 In the clothing trades there are union shop chairmen and chairladies who exercise a workers' control that [etc.]. 1931 *Atlantic City News* 7 Aug. 10/4 Mrs. Joseph Foy is chairlady of the card party. 1959 P. BULL *I know the Face* i. 15 A lovely play..which..was written by Miss Bertha Graham, the chairlady of The 1930 Players. 1890 *Pall Mall Gaz.* 18 Sept. 7/3 Unless these particulars are filled in no cabman can get employment; and it is almost as difficult for him if the licence is 'chair-marked'. This is done in several ways: one is by writing big; another by writing out the date..a third by using red.. ink. 'Chair-marking' is generally the result of spite... At Marylebone..a cab proprietor was fined £5 and costs for 'chair-marking'. 1894 *Daily News* 23 Oct. 2/6 This he contended was a 'chair mark', conveying to other proprietors the information that he had not paid his cab hire up. 1939 H. HODGE *Cab, Sir?* 218 Certain employers attempted to pass comments about drivers..by writing the date in code...It was called: 'chairmarking'. 1788 *Amer. Museum* IV. 519 The snake was..about the thickness of a common chair-post. 1872 *Congress. Globe* App. 578/2 They went out and got great big long brushes, as big as these chair posts. 1911 *Roxboro* (N.C.) *Courier* Nov., The snake was as large around as a chair post. 1895 *Westm. Gaz.* 12 July 2/1 In America the authorities assert that our chair roads are not strong enough to stand their traffic. 1950 M. PALVI *Compulsory Medical Care* 77 [Dentist's] actual chairside work. 1958 *Times* 12 Sept. 2/7 (Advt.), Young assistant required for old-established busy dental practice... Chairside assistance. 1963 J. OSBORNE *Dent. Mech.* (ed. 5) viii. 100 Both methods..require somewhat more chair-side and laboratory time. 1966 *Listener* 18 Aug. 238/3 New kinds of trains are being planned..with standards of..service borrowed from the airlines—chairside meals and so forth. 1966 'S. RANSOME' *Hidden Hour* iii. 38 He reached for the extension telephone on the chairside table. 1558 A. CRANEWISE in *Wills & Inventories Bury St. Edmund's* (Camden Soc., 1850) 150, I giue to my sonne Thomas..my round chaire table in the parlour. 1891 I. W. LYON *Colonial Furniture* 197 A chair table is mentioned in the inventory of John Copse, of Watertown, Mass., made in 1644. 1909 *Cent. Dict.* Suppl., *Table-chair*...Also chair-table. 1904 *Daily Chron.* 5 Jan. 2/7 He was a chair turner by trade. 1909 WARE *Passing Eng.* 69/1 A 'chair-warmer' is a lady whose talent is comprised in her physical charms, and who can neither sing, dance, nor act. 1926 J. BLACK *You can't Win* viii. 101 The judge frowned at me. The courtroom chair warmers craned necks in my direction. 1957 K. A. WITTFOGEL *Oriental Despotism* 306 Bureaucratic chair-warmers can be annoying and harmful.

chair, *v.* Add: **1. c.** To award the chair to (the successful competitor at the Welsh Eisteddfod). Also *ppl. a.* and *vbl. sb.*

1876 *Encycl. Brit.* V. 318 The chief of song was also called a chaired bard, because he was one of the fourteen entitled to a chair at court. 1877 [see **chair day*]. 1880 Chaired bards [in Dict.]. 1890 *Pall Mall Gaz.* 11 Sept. 6/3 Tudno, the chaired bard at Bangor.

d. To direct (a meeting, etc.) as chairman; to preside over. Cf. *CHAIRMAN *v.*

1921 L. S. HUNTER *J. Hunter* vii. 142 He was rarely quite happy and spontaneous at business..meetings. The prospect of having to 'chair' one would sometimes distract him from his work throughout the day. 1952 *Landfall* Mar. 50 He was chairing the first meeting of the congress. 1957 *BBC Handbk.* 167 Discussions chaired by Edgar Lustgarten.

chairman. Add: **1. c.** A master of ceremonies (see quot. 1951).

1836 DICKENS *Astley's* in *Sk. Boz* I. 312 It requires no great exercise of imagination to identify..the comic singer with the public-house chairman. 1951 *Oxf. Compan. Theatre* 119/2 *Chairman*, an important feature of the early music-hall...He sat at the head of a table in front of the footlights, with his back to the stage...He announced the 'turns' and kept the somewhat unruly audience in order. 1970 *Radio Times* 16 Apr. 39/1 *The Good Old Days* Old-Time Music-Hall..Chairman Leonard Sachs.

chairman (tʃēᵊ·ɹmăn), *v.* [f. the sb.] *trans.* To preside over (a meeting) as chairman; = *CHAIR *v.* 1 d.

1888 *Pall Mall Gaz.* 11 Oct. 4/1 We refused to attend the meeting to be chairmanned by Sir William. 1897 *Westm. Gaz.* 29 Jan. 2/3 The firm Which is chairman'd by Masham at Manningham! 1904 F. LYNDE *Grafters* xxiii. 293 It was late in the afternoon..that the Federative Council sent its committee, chairmaned by Engineer Scott, to interview the ex-general manager.

chairoplane (tʃēᵊ·rǒplē¹n). Also **char-a-plane, chairplane.** [f. CHAIR *sb.*¹, after *aeroplane, airplane.*] A roundabout with seats suspended by chains, the riders being swung round in a wide circle by the revolution of the machinery.

1922 *Westm. Gaz.* 3 Nov., 'Char-a-plane' victim. 1926 *Daily Express* 6 Aug. 3/2 'Chairplane' Tragedy. 1927 *Bulletin* 5 Dec. 7/2 A chairoplane erected on a fairground. *a* 1953 DYLAN THOMAS *Quite early one Morning* (1954) 29 He whirled in his chairoplane giddily above the upturned laughing faces of the town girls.

chaise, *v.* (Earlier example.)

1792 W. B. STEVENS *Jrnl.* 28 Sept. (1965) I. 47 Chais'd it to Wirksworth, dined and slept at Winster.

chaise-longue. Add: (Earlier example.) Also *ellipt.* as *chaise.*

1800 MRS. HERVEY *Mourtray Family* III. 76 She only begged they would permit her to lie down on her *chaise longue.* 1953 R. CHANDLER *Long Good-Bye* xxiv. 154, I stretched out on a padded aluminium chaise. 1965 E. O'BRIEN *August is Wicked Month* xvii. 215 He was on the edge of the chaise, totally ill at ease.

‖ **chaise percée** (ʃęz pęrse). [Fr., lit. 'pierced chair'.] A chair incorporating a chamber-pot; a commode.

1939 A. HUXLEY *After many a Summer* II. vi. 233 What Consistency, pray, do you find..between your appearances at Court and your *chaise percée.* 1958 *Listener* 27 Nov. 889/1 A world in which the *chaise percée* was the real, semi-confidential throne where all might sit supreme.

chaitya (tʃəi·tyă). [Skr., relating to a funeral pile or mound (*citā*); as *sb.*, funeral monument, sacred tree, etc.] A Buddhist place or object of reverence or worship. Cf. *CHORTEN.

1875 *Encycl. Brit.* II. 394 Of the next two classes of Buddhist architecture, the temples (Chaityas) and monasteries (Viharas), no built examples exist in India. 1882 *Ibid.* XIV. 500 The peculiar Buddhist objects of adoration which we know as *dagobas* or *chaityas.* 1920 *Blackw. Mag.* May 622/1 Rows of pyramidical chaityas. 1927 A. HUXLEY *Proper Studies* 181 The old Chaitya or meeting-hall of the Buddhists. 1945 S. PIGGOTT *Some Anc. Cities India* 51 The rock-cut structures fall into two classes—*chaityas* or halls of worship and *viharas* or monasteries.

chakra (tʃɒ·kră). Also **chakr, chakrah.** [Skr. (cogn. w. WHEEL *sb.*).] **1.** A circular weapon used by the Sikhs (see quots.).

1883 *Encycl. Brit.* XX. 111/2 The chakr or chakra is a thin knife-edged ring of flat steel, a severe missile in skilled hands. 1959 *Chambers's Encycl.* I. 616/2 A sharp-edged metal quoit known as the *chakra* of the Sikhs.

2. A discus or mystic circle placed in the hands of pictured Hindu gods, etc.

1891 *Murray's Hand-Bk. India & Ceylon* 384/1 A small temple dedicated to the Chakrah, or discus, of Vishnu. 1903 E. THURSTON in *Madras Govt. Mus. Bull.* IV. 199 A few men are branded on both shoulders with chank and chakra.

3. *Yoga.* One of the centres of spiritual or ethereal power in the human body.

1888 B. D. BASU in *Theosophist* Mar. 373 These *Chakras* are the vital and important sympathetic plexuses and preside over all the functions of organic life. 1921 D. H. LAWRENCE *Psychoanalysis* v. 86 Having begun to explore the unconscious, we find we must go from center to center, chakra to chakra. 1956 R. M. LESTER *Towards Hereafter* iv. 54 These spiritual centres are known by the name of 'chakras'—a word derived from Yoga.

4. The emblem on the flag of the Indian Union.

1947 *Constituent Assembly of India, Resolution* 22 July 761 The design of the Wheel shall be that of the Wheel (*Chakra*) which appears on the abacus of the Sarnath lion Capital of Asoka. 1964 V. S. AGRAWALA *Wheel Flag of India* p. i, The *Chakra* has a two-fold aspect, viz., the centre and the circumference. There is rest at the centre and movement in the circumference.

chalan (tʃɒ·lən). *India.* [Hind. *chalān.*] (See quot. 1858.)

1858 SIMMONDS *Dict. Trade*, *Chalan, Chillaum, Chulan*, a common Indian name for an invoice, pass, voucher, or way-bill. 1906 *Advice Note, Off. Supt. Govt. Printing, India* Voucher No. 5656 Payments should be made either by remittance transfer receipt or treasury chalan.

chalazogamic (kælăzogæ·mik), *a. Bot.* [ad. F. *chalazogame* (M. Treub 1891, in *Ann. Jardin Bot. de Buitenzorg* X. 219), f. Gr. χάλαζα CHALAZA + γάμος marriage + -IC.] Defining fertilization in which the pollen-tube penetrates the ovule by the chalaza. Hence **cha·lazogam,** a plant characterized by this mode of fertilization; **chalazo·gamy,** chalazogamic fertilization.

1894 *Times* 11 Aug. 11/3 Miss Benson read a paper upon 'Chalazogamic Amentiferæ'..showing that..the pollen-tube forces its way through the chalaza in order to effect fertilization, and hence must be included among the group of chalazogams...Professor Vines..was inclined to regard chalazogamy as an adaptation. 1895 [see POROGAMIC *a.*]. 1898 H. C. PORTER tr. *Strasburger's Text-bk. Bot.* II. ii. 492 According to recent investigations some undoubted *Amentaceae* are also chalazogamic. 1902 *Encycl. Brit.* XXV. 436 The pollen-tube..may pierce the embryo-sac at the chalazal end or at the side (basigamy or chalazogamy).

Chalcedonian (kælsėdōᵘ·niăn), *sb.*² and *a.* [f. *Chalcedon*, city of ancient Bithynia + -IAN.] **A.** *sb.* One who upholds the canons, etc., of the council of Chalcedon (A.D. 451). **B.** *adj.* Of or pertaining to the council of Chalcedon or its canons, etc.

1758 A. MACLAINE tr. *Mosheim's Eccl. Hist.* I. 421 The controversy between the Monophysites and Chalcedonians is not merely a dispute about words. 1788 GIBBON *Decl. & F.* IV. xlvii. 616 A parrot might be taught to repeat the *words* of the Chalcedonian or Monophysite creed. 1880 *Encycl. Brit.* XIII. 796/1 Then the Chalcedonian party would be purged from any appearance of sympathy with the errors of Nestorius. 1892 W. BRIGHT *Canons of First Four Gen. Councils* 143 While..he [*sc.* Justellus] followed the older Latin version called the Prisca by omitting the Ephesine canons so called, he inserted the Laodicene which the Prisca omitted, and did not place the Constantinopolitan, as the Prisca did, after the Chalcedonian. 1946 E. L. MASCALL *Christ, Christian & Church* iii. 54 No amount of discussion of our Lord's psychology can have any *direct* bearing on the Catholic creeds and the Chalcedonian definition.

chalcenterous (kælse·ntĕrəs), *a.* [ad. Gr. χαλκ-έντερος of brazen bowels, applied by Suidas to the grammarian Didymus, f. χαλκ-ός copper + ἔντερ-ον intestine + -OUS.] With bowels of bronze, tough. Also **chalce·nteric** *a.*

1946 *Times Lit. Suppl.* 12 Oct. 492/2 The most chalcenterous Scotsman who may hereafter seek the shelter of Oxford. 1948 R. W. CHAPMAN *Lexicography* 15 Sir James Murray added to his linguistic acumen and his tireless industry (which did not scruple to tire less chalcenterous workers) a Scotsman's hard, keen sense of the practical. 1964 R. M. OGILVIE *Latin & Greek* 95 For less chalcenteric readers there were the historical novelists.

chalcid (kæ·lsid), *a.* and *sb. Ent.* [f. mod.L. *Chalcididæ*: see CHALCIDIDAN.] **A.** *adj.* Of, pertaining to, or characteristic of the superfamily Chalcidoidea of the order Hymenoptera. **B.** *sb.* A member of this superfamily. So **chalci·dian** *a.*², **cha·lcidid** *a.* and *sb.*

1882 *Encycl. Brit.* XIV. 49 Parasitic hymenoptera of the chalcidian group. 1891 *Proc. U.S. Nat. Museum* XIV. 570, I know no case in which a chalcidid is a parasite of a proctotrypid. *Ibid.* 571 Chalcidid parasites of the Hemiptera-Heteroptera are very rare. 1893 ORMEROD *Injur. Insects* 16th Rep. 66 The little Chalcids were a species of Entedon. *Ibid.*, We should get no good by endeavouring to rear these Chalcid flies. 1895 *Cambr. Nat. Hist.* V. 545 In the cells of the same bee Newport discovered another curious parasitic Chalcid, *Anthophorabia retusa.* 1957 E. B. FORD *Butterflies* (ed. 3) vi. 111 Chalcids are stouter insects than are Ichneumons. 1959 J. CLEGG *Freshwater Life* (ed. 2) 227 Two genera of Chalcid Wasps are also aquatic.

Chalcidian (kælsi·diăn), *sb.* and *a.*³ [f. L. *Chalcis, Chalcid-* (Gr. Χαλκίς, Χαλκιδ-) + -IAN.] **A.** *sb.* A native or inhabitant of Chalcis, the chief city of Eubœa. **B.** *adj.* Of or pertaining to Chalcis. So **Chalci·dic** *a.* [L. *Chalcidicus*].

1654 OGILBY *Virg., Bucol.* x. 57 The Verses of Euphorion the Chalcidean. **1838** *Penny Cycl.* X. 60 While Alexander was absent in his Persian wars, the Chalcidians increased and improved their fortifications. **1876** F. TOZER *Classical Geogr.* x. 118 All these were Chalcidic cities, and the rivalry between these and the other colonies, which were mostly of Doric origin, materially affected their subsequent history. **1880** *Encycl. Brit.* XI. 95 Naxos, founded by the Chalcidians of Eubœa (735 B.C.). *Ibid.*, The sharp distinction between Dorian and Ionian (or 'Chalcidic'). *Ibid.* XIII. 125 The alphabet used by the Romans is identical with that of the Chalcidian colonies in southern Italy and Sicily..except the three aspirates θ, φ, χ. **1934** A. TOYNBEE *Study Hist.* III. 68 The Chalcidian and Megarian lines of overseas colonization. **1956** A. LANE *Gk. Pottery* (ed. 2) iv. 42 The distribution of finds, and certain resemblances to native Italian wares, indicate that the 'Chalkidian' vases were made by Eubœan settlers in South Italy.

chalcolithic (kælkoli·þik), *a. Archæol.* [f. CHALCO- + Gr. λίθος stone + -IC.] Of or pertaining to a period of culture characterized by the concurrent use of stone and bronze implements. (Cf. *AENEOLITHIC *a.*)

1902 *Nature* 6 Nov. Suppl. p. iii, The Neolithic vase-types ..merge insensibly into those of the earliest dynastic period, which was 'Chalcolithic' in character. **1911** J. L. MYRES *Dawn Hist.* x. 224 A long chalcolithic..phase, in which good cheap stone and bad expensive bronze were in use concurrently. **1928** V. G. CHILDE *Most Anc. East* iv. 98 All the elements that distinguish neolithic and chalcolithic culture..had been created in Egypt out of the common palæolithic heritage of south and north. **1959** *Chambers's Encycl.* XIII. 404/2 The chalcolithic age passes imperceptibly into the early bronze, which covers approximately the 3rd millennium B.C.

Chaldean, *sb.* Add: Also, = CHALDEE *sb.* b.

1690 *Le Clerk's Five Lett. Insp. Holy Scriptures* ii. 89 The Greek Language then was spoken in Judea, together with the antient Language which the Jews brought from Babylon, that is to say the Chaldean. **1897** J. H. BRIDGES *Opus Majus R. Bacon* iv. lxviii, With Hebrew went Chaldaean, and, in more distant relationship, Arabic. **1944** H. VAN ZELLER *Ezekiel* ii. 22 If only Chaldean could be picked up quickly.

chalet. Delete ‖ and substitute pronunc. (ʃæ·lēⁱ). Add: Also, a small, usually wooden, dwelling for holiday-makers.

1861 *Times* 18 May 5/2 Among all watering places, none can be put in comparison with the fine and large Establishment at Fécamp... There is found the invaluable advantage of having in the same establishment lodging house, chalets, taverns. **1910** *Lancet* 5 Nov. 1380/1 Chalets for consumptives at Swansea. **1953** E. HYAMS *Gentian Violet* 19 [A holiday camp] consisted of orderly rows of hideous concrete huts known as châlets.

chalicothere (kæ·likoþi·ɹ). *Palæont.* [f. mod.L. *chalicotherium*, f. Gr. χάλιξ, χάλικ- gravel + θηρίον beast.] A perissodactyl mammal of the extinct genus *Chalicotherium* (fam. Chalicotheriidæ), with a horse-like head and clawed toes.

[**1833** J.-J. KAUP *Descript. d'Ossements Fossiles de Mammifères* II. ii. 4 Chalicotherium, *Kaup.* Nouveau genre de Pachydermes... J'ai donné ce nom à deux fort grands Pachydermes.] **1907** *Amer. Naturalist* XLI. 733 (*title*) Preliminary notes on some American chalicotheres... Little is known of the American forms of the Chalicotherioidea,—an extinct family of mammals. **1910** H. F. OSBORN *Age of Mammals* ii. 143 *Bartonian Stage*... The supposed chalicothere *Pernatherium*, a very aberrant perissodactyl, is first known at this stage. **1937** *Discovery* Dec. 372/2 A Chalicothere, that remarkable beast of the late Miocene period, with a horse-like head, and toed, clawed 'hoofs', a horse-cum-sloth creature in appearance. **1958** ZEUNER *Dating the Past* (ed. 4) 430 The average period of existence of a genus was 5·6 million years in the line of the horses (*Hyracotherium–Equus*), 5·9 million years in the chalicotheres.

chalifah, var. *KHALIFA.

chalk, *sb.* Add: **3. a.** *black chalk* (see quots.).

1854 *Encycl. Brit.* VI. 402/1 *Black Chalk*, a mineral used by artists for drawing. It is a variety of bituminous shale, the *schiste-graphique* of Haüy. **1886** J. W. ANDERSON *Prospector's Handbk.* 115 Black Chalk—a variety of clay containing carbon.

6. b. *to walk one's chalks* (earlier U.S. example).

1835 A. B. LONGSTREET *Georgia Scenes* 6 Oh, wake snakes, and walk your chalks.

7. *chalk-land, -stripe; chalk-faced* adj.; **chalk-back day** (see quots.); **chalk-mark** *sb.*, a mark, esp. a distinctive mark, made with chalk; *v. trans.*, to mark with chalk, esp. with a distinctive mark; to draw (a line) with chalk; hence *chalk-marked* adj.; **chalk period**, the cretaceous period (see CRETACEOUS *a.* 2); **chalk stream**, a stream flowing over chalk; **chalk talk** *U.S.*, a lecture or speech illustrated by chalk sketches made by the speaker; hence *chalk-talker*.

1851 *N. & Q.* IV. 501/2 At Diss, Norfolk, it is customary for the juvenile populace, on the Thursday before the third Friday in September..to mark..each other's dress with white chalk, pleading a prescriptive right to be mischievous on 'chalk-back day'. **1906** *Westm. Gaz.* 27 Feb. 2/3 In Norfolk there used to be an annual 'Chalk-back Day', the servants hired at the September fair at Diss being marked with chalk on either jacket or dress to indicate that they were settled. **1896** *Daily News* 6 Apr. 6/5 A not very distant chalk-faced hill. **1935** C. DAY LEWIS *Time to Dance* 47 Not for long will your chalk-faced bravado Stand the erosion of eternity. **1941** *Proc. Prehist. Soc.* VII. 149 The cultural succession observed on the south English chalklands. **1862** THACKERAY *Round. Papers* 166 On the doorpost..is a little chalk-mark. **1878** *Chambers's Jrnl.* 14 Dec. 791/2 The white chalk-mark on his [*sc.* a Brahmin's] forehead, which signifies his rank to the world. **1927** *Scots Observer* 11 June 2/2 A happy-go-lucky Customs officer chalk-marked my suit case. **1928** *Daily Tel.* 9 Oct. 8/1 He has to toe the line which I chalk-mark, otherwise out he goes. **1880** R. ROWE *Picked up in Streets* 221 Rows of broken, dusty, chalk-marked shutters. **1904** J. R. A. DAVIS *Nat. Hist. Animals* III. iii. 309 The toothless Pterosaurs (species of *Pteranodon*), which flourished during the Chalk period. **1829** T. ARNOLD *Jrnl.* 7 Aug. in Stanley *Life & Corr.* (1846) App. D. 658 As gentle and as limpid as one of the clear rapid chalk streams of the south of Hampshire. **1858** KINGSLEY in *Fraser's Mag.* Dec. 330/1 In chalk streams the largest fish are found oftener in the mill-heads than in the mill-tails. **1904** E. F. BENSON *Challoners* xiv, The chalk-stream..was brimful from bank to bank of hurrying translucent water. **1906** *Macm. Mag.* Apr. 453 The chalk stream regions to-day are not fishing countries in the sense that Devon, Hereford, or Wales are. **1943** R. CHANDLER *Lady in Lake* (1944) i. 7 Smooth grey flannel with a narrow chalk stripe. **1953** 'S. RANSOME' *Drag the Dark* (1954) ii. 24 A double-breasted blue chalk-stripe suit. **1881** *Christian Miscell.* V. 40 His inimitable 'chalk-talk'. **1969** *New Yorker* 31 May 78/2 We are treated to a vicious and numbing chalk-talk from the writings of the founder. **1888** *St. Paul & Minneapolis Pioneer Press* 22 July (Farmer), The celebrated chalk-talker.

b. In the names of butterflies and moths, as *chalk carpet, chalk hill blue, chalk pit* (see quots.).

1832 J. RENNIE *Butterfl. & Moths* 18 The Chalk Hill Blue (*Polyommatus Corydon*, Stephens). *Ibid.* 117 The Chalk Carpet (*Larentia bipunctaria*, Ochsenheimer). *Ibid.* 125 The Chalk Carpet (*Xerene procellata*, Ochsenheimer). *Ibid.* 230 The Chalk-pit Plume (*Pterophorus migadactylus*, Fabricius) 'frequents chalky districts'. **1892** KIPLING *Lett. Travel* (1920) 36 A lordly swallow-tailed butterfly,..very like the flitting 'chalk-blue' of the English downs. **1927** *Daily Express* 14 July 5/5 The chalkhill and holly blues would make beauty spots among the yellow iris.

chalk, *v.* Add: **2. c.** *to chalk* (a person's) *hat, to have one's hat chalked*: to allow, be allowed, free travel (as on a railway). *U.S. colloq. Obs.* exc. *Hist.*

1823 QUITMAN *Let.* in J. F. H. Claiborne *Life & Corr.* (1860) 78, I will 'chalk your hat' for the journey. **1880** A. A. HAYES *New Colorado* (1881) xi. 149 Twenty-five seedy, second-class ruffians, who proposed to travel, as they say in the West, 'with their hats chalked', or free. **1887** *Nation* (N.Y.) 21 Apr. 329/1 It would seem that all railway officers and most railway employees have their 'hats chalked' all over the U.S. **1928** J. W. STARR *100 Yrs. Amer. Railroading* 76 The practice out there was called 'chalking the hat', from the custom of the conductor in placing a white mark or ticket on the..headgear of the free passenger.

d. To mark (an object) with chalk as an indication that it has been officially admitted, e.g. by a customs officer, or directed. *U.S.*

1867 MRS. WHITNEY *L. Goldthwaite* iii, Stooping to examine the trunk..[he said,] 'These things is chalked all right for Littleton.' **1893** 'MARK TWAIN' *£1,000,000 Banknote* 258 In his first agitation he was going to try to bribe the postman to chalk it through.

3. b. Delete ? *Obs.* and add further examples. Now esp. common in phr. *to chalk it up* (*to*), to charge it (to) (a person, an account, experience, etc.). Also, to write *down*; to set down as a sum or estimate.

1826 T. WILSON *Pitman's Pay* (1843) 11 She chalks up 'scores' at a' the shops. **1835** R. M. BIRD *Hawks of Hawkhollow* II. viii. 78 He chalked me down like a fool, me and Tom Staples; being old friends, or sort of. **1877** E. PEACOCK *Gloss. Manley & Corringham* 53/1 *Chalk*, to mark on a board with chalk the number of pints of beer a person is in debt to a publican. 'Benny Mason's been to th' Gouden Cup an' hed two quarts o' ale *chalk'd* down to you.' **1886** ELWORTHY *W. Somerset Word-Bk.* 122 Publicans are accustomed to keep the score by *chalk* marks behind the door, hence to be (*chau·kd* aup) is to be entered as a debtor. **1895** in *N. & Q.* (1941) CLXXXI. 117/1 We want a drink... So, Miss, do chalk it up. **1939** N. MONSARRAT *This is Schoolroom* II. vii. 176 One of those superb nights when everything is so perfectly in tune..that one unconsciously chalks it up, to have something to aim at in the future. **1953** W. P. MCGIVERN *Big Heat* vi. 82 Well, we can chalk it up to experience.

chalk-line. **a.** (Of uncertain meaning.) **b.** (See CHALK *sb.* 7.) **c.** A line drawn with chalk; a straight line; also *transf.* and *fig.* So, in allusive use, *to walk a chalk-line*: to behave with propriety; to keep undeviatingly to a course of action (*U.S. colloq.*).

a **1450** *Fysshynge wyth an Angle* (1883) 15 A good fyne lyne of pak thryde made yn maner of a chalke lyne. **1631** WINTHROP *Let.* 28 Mar. in *Hist. New Eng.* (1853) I. 458 Store of shoe-makers' thread and hobnails; chalk and chalkline. **1771** *Carroll Papers* in *Maryland Hist. Mag.* XIII. 173 Pray desier Mr Deards to send me by him two Drumlines... The Chalk lines are too weak & small. **1825** [see CHALK *sb.* 7]. **1855** T. C. HALIBURTON *Nature & Human Nature* II. viii. 324 He returns as straight as a chalk-line or as we say the crow flies. **1859** H. E. TALIAFERRO *Fisher's River* xxii. 246 You can't swindle this boy; he's walked too many chalk-lines fur that. **1887** in *Amer. Speech* (1950) XXV. 31/2 The owners of those dives also knew they would have to close up shop and walk the chalk-line once the police took the matter in hand. **1904** C. MARTYN *Fables of Hotel Profession* 26 Everybody had to walk a Chalk Line or Get Out. **1948** *Time* 2 Aug. 12/2 Many of whom deny that they are Communists but all of whom walk the Communist chalk line.

challenge, *sb.* Add: **9.** A dose of an antigen given, by injection or other means, to an animal or person that has been sensitized to it by a previous dose. Also *attrib.*

1950 *Yale Jrnl. Biol. & Med.* XXIII. 29 Thephorin was injected intra-abdominally 15 minutes before the challenge with histamine. **1952** *Proc. Soc. Exper. Biol. & Med.* LXXXI. 110/1 Even after 10 days a high mortality rate results following antigenic challenge of a series of mice sensitized with a mixture of horse serum and vaccine. **1964** W. G. SMITH *Allergy & Tissue Metabolism* i. 5 The second injection of antigen, which brings about anaphylactic shock, is termed the challenge dose.

challenge, *v.* Add: **9.** To give a dose that constitutes a challenge (see prec.) to (an animal or person); **challenged** *ppl. a.*, **challenging** *vbl. sb.* (later examples in this sense).

1946 *Proc. Soc. Exper. Biol. & Med.* LXIII. 189/2, 210 mice were sensitized..by 4 consecutive intraperitoneal injections of 1 ml each of undiluted horse serum every other day. Twenty-one days later a challenging dose of 1 ml. of undiluted horse serum..was injected intravenously. **1962** LUNTZ & WRIGHT in A. Pirie *Lens Metabolism Rel. Cataract* 322 Leucocytes from each group of guinea-pigs were transferred to normal guinea-pigs who were then challenged with lens protein. **1966** W. C. BOYD *Fund. Immunol.* (ed. 4) xvii. 729 If we are testing the immunizing effect of an antigen, we shall challenge our animals, after a suitable interval, with a dose of the infectious microorganism, or with a toxic product of this organism. **1969** *Times* 7 July 5/7 It has been possible to treat challenged lungs with drugs known to have an effect in reducing inflammation.

challengeful (tʃæ·lendʒfŭl), *a.* [f. CHALLENGE *sb.* + -FUL.] Fraught with a challenge; challenging.

1903 HARDY *Dynasts* I. II. v. 52 And I only own—such is my challengeful character—that perhaps He [*sc.* Napoleon] do eat pagan infants when He's in the desert. **1927** *Chambers's Jrnl.* June 381/2 An interrogation..vital and challengeful.

challengingly (tʃæ·lendʒiŋli), *adv.* [f. CHALLENGING *ppl. a.* + -LY².] In a challenging manner; so as to convey a challenge.

1907 A. BENNETT *Grim Smile Five Towns* i. 165 Toby gazed around, half challengingly and half nervously. **1921** A. S. M. HUTCHINSON *If Winter Comes* II. i. 67 His face seemed to say to the world challengingly, 'I am here!'

chalmoogra, var. *CHAULMOOGRA.

chalone (kæ·lōᵘn, kēⁱ·lōᵘn). *Biochem.* [ad. Gr. χαλῶν, pres. pple. of χαλάω slacken, after *HORMONE.] An internal secretion that reduces or inhibits the action of certain organs and tissues, opp. *HORMONE. Hence **chalo·nic** *a.*

1914 E. SCHÄFER *Introd. Study Endocrine Glands* i. 11 Effects are either in the direction of excitation, in which case the endocrine substances producing them..would come under the expression 'hormones', or in the direction of restraint and inhibition, in which case they may be termed restraining or inhibiting autacoids and be classed as 'chalones'. And the action of an autacoid may be described as hormonic or chalonic, according to the kind of effect it produces. **1923** *Glasgow Herald* 18 Jan. 12 Hormones and chalones (specific chemical messengers) produced by the ductless or endocrine glands. **1925** J. A. THOMSON *Gosp. Evol.* ii. 59 Hormones, which stimulate organs and tissues to greater activity; others are called chalones, which put on a brake. **1962** *Times* 6 July 6/3 Substances, or active principles, known as chalones.

chalukah, var. *KHALUKAH.

chalybeous (kăli·biəs), *a.* [f. L. *chalybēius* Chalybeian, of steel + -OUS.] Of a steel-blue colour; dark blue with a metallic lustre.

1826 KIRBY & SPENCE *Entomol.* IV. xlvi. 283 Chalybeous ...The blue splendour of steel case-hardened, or of the mainspring of a watch.

chamæ- (kæmi), combining form of Gr. χαμαί on the ground, low, used in many technical and scientific terms, as **chamæcephalic** (-sĭfæ·lik), **-cephalous** (-se·făləs), *adjs.*, characterized by or exhibiting chamæcephaly; **chamæcephaly** (-se·făli) [Gr. κεφαλή head], a formation or development of the human skull, in which the cephalic index is 70 or less; **chamæconchic** (-kǫ·ŋkik), **-conchous** (-kǫ·ŋkəs), *adjs.*, characterized by or exhibiting chamæconchy; **chamæconchy** (-kǫ·ŋki) [Gr. κόγχη

CONCH], the condition of having a low form of the orbits, showing an orbital index of 80 or less; **chamæcranial** (-krēⁱ·niăl), *a.* [Gr. κρανίον skull], characterized by having a low skull, of a length-height index of 70 and less; **cha·mæphyte** [ad. Da. *kamæfyt, ch-* (C. Raunkiær, 1904, in *Bot. Tidsskrift* XXVI. 11): see -PHYTE], a plant that bears its buds on or near the surface of the ground; **chamæprosope** (-prǫ·sōᵘp) [Gr. πρόσωπ-ον face], a human skull with a low broad face; **chamæprosopic** (-prǫsōᵘ·pik) *a.*, characterized by chamæprosopy; **chamæprosopy** (-prǫsōᵘ·pi), [Gr. πρόσωπ-ον face] the condition of having a low broad form of face.

1902 *Biometrika* Aug. 462 Are the brachycephalic races hypsicephalic and the dolichocephalic races chamaecephalic? *Ibid.* 460 In the male brachycephaly is associated with hypsiconchic, in the female with the chamaeconchic character. *Ibid.*, A quite sensible association of platyrrhiny with chamaeconchy. *Ibid.* 462 Brachycranial, stenocranial, and chamaecranial characters. **1913** *Jrnl. Ecol.* I. 17 Chamæphytes include plants with their bud or shoot-apices perennating on the surface of the ground. **1916** B. D. JACKSON *Gloss. Bot. Terms* (ed. 3) 71/2 *Chamaephytes*,.. plants whose resting-buds are but slightly above the ground. **1952** P. W. RICHARDS *Tropical Rain Forest* i. 10 The ground herbs of the rain forest are almost exclusively phanerophytes and chamaephytes. **1964** V. J. CHAPMAN *Coastal Veg.* i. 9 Chamaephytes, perennating buds above soil surface to 25 cm, e.g. Creeping willow (*Salix repens*). **1900** tr. *Deniker's Races of Man* 60 To separate skulls into brachy- or dolichofacial, or, as they are also called, chamæprosopes and leptoprosopes. **1886** *Buck's Handbk. Med. Sci.* II. 23/1 *Chamæprosopic* [*sic*] is applied to a short, squat, thick-set face. **1902** *Biometrika* Aug. 434 Chamaeprosopy.

chamar (tʃămā·ɹ). Also **chumar.** [Hindi.] A member of an exterior Hindu caste whose occupation is leather-working; a worker in leather, a tanner, shoemaker. Also, in northern and central India, an agricultural labourer.

1858 SIMMONDS *Dict. Trade.* **1899** C. W. DOYLE *Taming of Jungle* ii. 18 A wee little manikin of the chamar (tanner) caste. **1901** KIPLING *Kim* iii. 81 All castes and kinds of men move here. Look! Brahmins and chumars, bankers and tinkers. **1924** *Blackw. Mag.* Nov. 693/1 This little swine of a chumar. **1934** M. L. DARLING *Wisdom & Waste in Punjab Village* 167 Chamárs and Khatiks, dealers in hides and skins, respectively. **1951** J. H. HUTTON *Caste in India* (ed. 2) 23 The Chamar of Chattisgarh, although belonging to that exterior caste of leather-workers whose touch is polluting to caste Hindus, are here cultivators tilling the land.

chamber, *sb.* Add: **4. e.** *chamber of horrors*: see HORROR *sb.* 5.

7. b. (Examples.)

1922 JOYCE *Ulysses* 755 Wheres the chamber gone. **1961** F. G. CASSIDY *Jamaica Talk* v. 85 'Chamber-pot' has been abbreviated to *Chamber*.

13. chamber acid, sulphuric acid in the condition and of the strength at which it is removed from the lead chambers; **chamber arrest**, confinement in one's room under arrest; **chamber cantata**, a cantata suitable for performance in a private room; **chamber-closet**, a commode for invalids and the infirm (Knight *Dict. Mech.*, *a* 1877); **chamber-gas**, the gas, or mixture of gases, contained in the large lead chambers used in the manufacture of sulphuric acid; **chamberhand** *N.Z.*, one who works in the freezing-chamber in a freezing-works; **chamber-horse**, substitute for def.: a contrivance which enables a person to obtain exercise resembling that of horse-riding; (later example); **chamber kiln**, a kiln consisting of a series of chambers arranged in circular form, used for burning cement, bricks, and tiles; also *attrib.*; **chamber lad**, a bedroom attendant in chambers; **chamber man**, a man employed in or about a chamber, esp. to do the work connected with a manufacture by a chamber process; **chamber process**, a manufacturing process that is carried out by means of a closed or sealed chamber; **chamber-set**, a set of chamber ware or furniture; **chamber-tomb** *Archæol.*, a chambered tomb (see *CHAMBERED *ppl. a.*); **chamber-work**, (*b*) (earlier U.S. example).

1872 W. CROOKES tr. *Wagner's Chem. Technol.* 206 Chamber Acid. As soon as the acid formed in the leaden chambers has acquired a sp. gr. of 1·5 = 50° B. = 140° Twaddle, it is run off into a reservoir. **1879** LUNGE *Sulphuric Acid* I. 296 At Oker each cubic metre of chamber-space yields 2·85 kilograms chamber-acid of 106° Tw. in the case of rich ores. **1903** *Daily Chron.* 19 Dec. 5/1 Frederick the Great, when Crown Prince, was not only condemned to chamber-arrest, but actually flung into prison. **1905** E. J. DENT *Scarlatti* 9 The immense popularity of the chamber-cantata during the whole of the seventeenth and the early part of the eighteenth century. **1879** LUNGE *Sulphuric Acid* I. 331 If..a straight tube is introduced..into the draught-pipe taking away the chamber-gas. **1950** *Landfall* IV. 125 The gang of chamberhands who usually dawdle along behind me. **1948** *Archit. Rev.* CIII. 6 (*caption*) Thomas Sheraton, Chamber Horse. 1793. The great cabinet-maker designed this 'chamber horse' for the gentleman who wished his riding exercise regardless of the weather. *a* **1884** KNIGHT *Dict. Mech.* Suppl., *Chamber Kiln*, a brick or tile kiln in compartments; sometimes capable of being heated in succession. **1902** *Encycl. Brit.* XXVI. 630 The kilns most commonly employed nowadays [for burning Portland cement] are 'chamber kilns'. **1921** *Dict. Occup. Terms* (1927) § 692 *Chamber kiln man* (cement);..regulates supply of gas in chamber kilns. **1889** GRETTON *Memory's Harkback* 157 Sugden became chamber lad to a conveyancer, where he picked up the foundation of his law knowledge. **1921** *Dict. Occup. Terms* (1927) § 148 *Chamber man* (white lead); makes white lead by chamber process. **1879** LUNGE *Sulphuric Acid* I. 302 A very important assistance in judging of the chamber-process is afforded by glass windows or sights. **1851** CIST *Cincinnati* 204 Burley & Lyford.. manufacture chamber sets. **1858** *Texas Almanac* (Advt.), Parlor and Chamber sets. **1895** *Montgomery Ward Catal.* 535/2 English decorated Chamber set..consists of wash bowl and pitcher, chamber and cover, mug and soap dish. **1893** *Funk's Stand. Dict.*, Chamber tomb. **1929** A. EVANS *Shaft Graves* 69 The rock-cut Chamber Tombs with their dromoi..themselves reflect a form already known in Crete in the age preceding the conquest. **1952** CHILDE & SIMPSON *Anc. Monum. Scotl.* 18 Another class of chamber tombs, peculiar to Orkney but having analogues in Ireland and abroad. **1870** 'FANNY FERN' *Ginger-Snaps* 20 Having done chamber work or cooking, for such a number of years in New York.

chamber, *v.* Add: **4. c.** To contain or hold as in a chamber. Of a fire-arm: to receive in the chamber.

1835 N. J. WYETH *Jrnl.* 13 Apr. in F. G. Young *Sources Hist. Oregon* (1899) I. III–VI. 251 Building a canoe 60 feet long wide and deep enough to chamber barrels of which she will take 25. **1839** *S. Lit. Messenger* V. 97/2 My father's big gun..would chamber five buckshot. **1902** S. E. WHITE *Blazed Trail* xx. 141 Wallace's rifle chambered the ·38 Winchester cartridge. **1904** —— *Blazed Trail Stories* 163 Each was armed..with a brace of Colt's revolvers, chambering the same-sized cartridges as the rifle.

7. *intr.* Of a mining vein: to open up, expand. *U.S.*

1873 J. H. BEADLE *Undevel. West* xviii. 335 The miner starts with a vein a foot or more wide..then it suddenly 'chambers' to some size, then 'pinches' to the thickness of a knife-blade.

chambered, *ppl. a.* Add: **1.** In *Archæol.*, applied to a tomb containing a chamber or vault for the deposition of the dead.

1865 J. LUBBOCK *Prehist. Times* iv. 91 The comparative rarity of chambered tumuli in western Europe. **1869** *Archæologia* XLII. 223 The surprising similarity between the winter huts of the Esquimaux and the chambered tumuli of Sweden and Denmark. **1952** CHILDE & SIMPSON *Anc. Monum. Scotl.* 15 Chambered cairns were used for burials over several generations.

chambering, *vbl. sb.* Add: **4.** *Zool.* The formation of chambers or loculi. Cf. CAMERATION 2.

1885 *Encycl. Brit.* XIX. 846 The test has usually a chambered structure... The chambering of the test does not express a corresponding cell-segmentation of the protoplasm.

Chamberlainism (tʃēⁱ·mbəɹlĕniz'm). The policy or principles of the politician Joseph Chamberlain (1836–1914) or his son (Arthur) Neville Chamberlain (1869–1940). So **Cha·mberlainic** *a.*, **Cha·mberlainite** *sb.* and *a.*, **Cha·mberlainize** *v.*, **Cha:mberlainiza·tion.**

1898 *Westm. Gaz.* 10 Nov. 3/2 Chamberlainic 'principles'. **1899** *Daily News* 20 June 7/2 There is a characteristic Chamberlainism in the last telegram published in the Transvaal Blue Book, dated May 24. **1900** W. HARCOURT in *Westm. Gaz.* 11 Oct. 8/2 It was because the country had had a surfeit of Chamberlainism. **1904** *Spectator* 31 Dec. 1072/2 Though Mr. Balfour fills his Cabinet with Chamberlainites. **1905** *Daily Chron.* 12 Jan. 4/2 The chamberlainising of the Conservative Party on Protectionist lines. **1906** *Ibid.* 20 Jan. 4/3 The Chamberlainisation of the party. **1942** A. L. ROWSE *Cornish Childhood* 124 The nauseating hangers-on to power, the conformists to Chamberlainism. **1961** *Times* 25 Apr. 15/4 Letters to *The Times* defending the Chamberlainite course.

chamberlet. Add: spec. in *Zool.*, a small chamber or division of the test of a foraminiferous animalcule. Hence **cha·mberletted** *a.*

1862 W. B. CARPENTER *Microsc.* (ed. 3) 500 A ring of small chambers (or chamberlets) is formed around the primordial chamber. **1879** *Encycl. Brit.* IX. 376 *Fabularia*, a fossil..in which the principal chambers are subdivided into 'chamberlets' as in *Orbiculina* and *Orbitolites*. **1884** *Amer. Jrnl. Sci.* XXVII. 328 The division of the chamber-segments of the body into chamberletted sub-segments. **1957** *New Biol.* XXIV. 22 The regeneration of a broken chamberlet in the giant *Orbitolites*.

chamber-master. Add: **2.** A furrier who obtains skins from the wholesale trader and makes them up at home or on his own premises.

1901 *Daily Chron.* 4 Dec. 9/2 Chambermasters and cutters wanted for hare work.

Chambertin. (Later examples.)

1855 THACKERAY *Miscellanies* I. 45 'Quel vin Monsieur desire-t-il?' 'Tell me a good one.'—'That I can, Sir: The Chambertin with yellow seal.' **1953** H. MILLER *Plexus* (1963) xi. 380 Now we're sampling the Chambertin which I brought along.

chambray (ʃæ·mbrēⁱ). orig. *U.S.* [irreg. f. *Cambray* (see CAMBRIC).] A kind of gingham with a linen finish. Also *attrib.*

1814 *Niles' Reg.* V. 317/2 Twenty-four cases cotton and woollen goods, cloths, ginghams, chambrays, shirtings, [etc.]. **1909** *Public Ledger Philadelphia* 24 June 5/7 Lawns, percales and chambrays. **1916** *Daily Colonist* (Victoria, B.C.) 7 July 14/2 (Advt.), Blue Chambray Shirts. Guaranteed fast color. An extra good wearing material. Sale price 75¢. **1937** J. STEINBECK *Red Pony* i. 10 In a moment he was dressed—blue chambray shirt and overalls. **1968** J. IRONSIDE *Fashion Alphabet* 218 *Chambray*, a fine quality gingham-type fabric with a mottled appearance caused by using white weft threads and coloured warp threads.

chameleon. Add: **6. c. chameleon moth**, a S. African noctuid moth, *Actæa chamæleon*, of extreme variability in colour; **chameleon silk, tulle** (see quots.).

1848 in T. Graham *Chem. Rep. & Mem.* 230 The shot silk stuffs known as chameleon silks. **1896** *Daily News* 14 Nov. 6/5 One of the latest novelties in ball dresses is the chameleon tulle, composed of veils of different colours laid over each other.

chametz, var. *HAMETZ.

chamferer (tʃæ·mfərəɹ). Also **champherer.** [f. CHAMFER *v.* + -ER¹.] One who chamfers; *spec.* (see quot.).

1921 *Dict. Occup. Terms* (1927) § 200 *Barrel chamferer*; operates a chamfering lathe, to chamfer or bevel bottom edge or shield of metal bobbins [etc.]. *Ibid.*, *Chamferer, champherer*; a bolt facer or nut facer..who feeds and operates specially converted lathe on which bevel of nuts and round end of bolts are formed.

chamfering, *vbl. sb.* Add: **3. chamfering-bit**, a boring-bit used with a brace to chamfer holes to receive the heads of screws; **chamfering lathe** (see quot.); **chamfering machine**, 'a machine for bevelling the ends of staves after being set in a cask' (Knight *Dict. Mech.* Suppl. *a* 1884); **chamfering-tool**, a saddler's tool for paring down the edges of leather.

a **1877** KNIGHT *Dict. Mech.*, Chamfering-bit, -tool. **1921** *Dict. Occup. Terms* (1927) § 200 A chamfering lathe, to chamfer or bevel bottom edge or shield of metal bobbins [etc.].

chamisal (tʃæ·misæl). Also **chemisal**, etc. [Mexican Sp., f. *chamiso*.] **a.** A dense growth or thicket of chamiso. **b.** = *CHAMISO.

1853 *House Repr. Ex. Doc.* 91 (Bentley), Traveling..is rendered very trying by..patches of dense masses of shrubbery known as the chemizal. **1862** BRET HARTE *Notes by Flood & Field* in *Luck of Roaring Camp* (1870) 194 With my hand dipped listlessly over the thwarts, I detect the tops of chimisal, which shows the tide to have somewhat fallen. **1902** *Bureau Plant Industry Bull.* (U.S.) XII. 31 These chaparral areas..have become landmarks, the word chamisal, sometimes corrupted into chemisal,..being adopted as a local name.

chamiso (tʃæ·miso). Also †**chamiza.** [Mexican Sp.] A Californian evergreen shrub, *Adenostoma fasciculatum*. Also *attrib.*

1846 W. H. EMORY *Mil. Reconn.* (1848) 77 In one view could be seen clustered, the..green wood acacia, chamiza. **1869** BRACE *New West* 94 The chaparral..is generally a thorny, impervious shrubbery, made up of the Chinquapin.. and the Chamiso. **1904** *N.Y. Tribune* 17 July, One afternoon they located a grizzly, and ran him into a field of chamiso brush. **1964** F. O'ROURKE *Mule for Marquesa* 72 There were regions of sand dunes covered with mesquite and chamiso.

chamois, *sb.* Add: **3.** The colour of chamois leather; hence *chamois-coloured* adj.

1872 *Young Englishwoman* Nov. 599/1 A delicate buff called chamois-colour. **1882** [in Dict.]. **1898** *Daily News* 26 Sept., Chamois-coloured kerseymere. **1923** *Daily Mail* 26 Feb. 1 Gauntlet Gloves..in Chamois.

chamotte (ʃamǫ·t). [Fr., ad. G. *schamotte* fire-clay.] Fragments of burnt fire-clay ground to powder and used with fresh fire-clay in making new vessels.

1890 THORPE *Dict. Appl. Chem.* I. 502 *Chamotte*, a mixture of fire-clay and burnt pottery used for making fire-bricks, crucibles, pipes, &c.

champ (tʃæmp), *sb.*⁴ orig. *U.S.* Colloq. abbrev. of CHAMPION *sb.*¹ 4.

1868 *New Eng. Base-Ballist* 6 Aug. 2/4 The 'Champs' enjoyed themselves in various ways during the morning. **1917** MATHEWSON *Sec. Base Sloan* xix. 262 We were the champs three years running. *a* **1935** T. E. LAWRENCE *Mint* (1955) I. xxvi. 90 She's a boxer; a proper champ. **1936** 'P. QUENTIN' *Puzzle for Fools* iv. 26 Had the nerve to challenge me for a tumble..me, an ex-champ! **1968** *Globe & Mail* (Toronto) 13 Jan. 38/3 U.S. Open champ Gay Brewer..had a 75 at Spyglass in the first round.

champa (tʃæ·mpa). = CHAMP *sb.*³, CHAMPAC.

1905 L. HOPE *Indian Love* 72 'Tis the scent of the champa's breath. **1908** *Westm. Gaz.* 23 Mar. 5/1 The chief perfumes are champa, cajuput oil, [etc.].

champagne, *sb.* Add: Also *fig.*, something exhilarating, excellent, etc.

1891 *Cricket* 28 May 139/1 It was of the two Lytteltons, Alfred and Edward, that the phrase 'the champagne of cricket', was first used. **1896** *Godey's Mag.* Apr. 365/1 His candid devotion to 'small cold bottles' is unfailing champagne to the audience. **1897** *Westm. Gaz.* 23 June 2/2 You drink in the picture... This, you involuntarily cry, 'This is the champagne of the century!' **1963** *Times* 28 Jan. 4/2 But yesterday during the closing stages, he gave the crowd a taste of champagne.

b. *champagne-cocktail, -dinner, -glass, -supper; champagne-coloured* adj.; **champagne-bottle**, a strong bottle of flawless glass with a long neck and sloping shoulders; also *allusively*; **Champagne Charley** or **Charlie**, a humorous name for a noted drinker of champagne; **champagne cup** (CUP *sb.* 11), a 'cup' of which champagne is the basis or chief ingredient; **champagne gas**, carbon dioxide.

1858 SIMMONDS *Dict. Trade*, *Champagne bottle*, a strong and particular shaped bottle, which has the cork secured with wire, when holding champagne. **1872** TROLLOPE *Golden Lion Granpere* xx. 341 She would pack up cold chickens and champagne bottles. **1893** G. B. SHAW *Music in London 1890–4* (1932) III. 111 His waist perhaps broader than his champagne-bottle shoulders. **1901** *Westm. Gaz.* 2 May 3/2, I don't mean that champagne-bottle shoulders are in vogue again, but that the shoulder line is not at all square. **1870** D. J. KIRWAN *Palace & Hovel* xvi. 235 The taste of the Prince [of Wales] for music may be imagined from the fact that 'Champagne Charley', and 'Not for Joseph', are his two most cherished melodies. **1889** BARRÈRE & LELAND *Dict. Slang* I. 235 *Champagne Charley*,..any dissipated man or noted drinker of 'fizz'. The name of a song which appeared in 1868... The original *Charley* is said to have been a wine-merchant, who was in the habit of making presents of bottles of *champagne* to all his friends. **1920** A. HUXLEY *Limbo* 85 A Nut, a descendant of the bloods and Champagne Charlies of earlier days. **1869** 'MARK TWAIN' *Innocents Abroad* xv. 148 Well, if you don't know what that is, give us a champagne cocktail. **1961** S. HATHAWAY *Dame of Sark* xiii. 199 Before lunch we served champagne cocktails. **1901** *Westm. Gaz.* 24 May 3/2 A very handsome coat of champagne-coloured cloth lined with silk. **1849** THACKERAY *Pendennis* I. xxvi. 251 Considerable excitement, produced by a supper and champagne-cup. **1906** GALSWORTHY *Man of Property* II. viii. 211 There was the champagne cup. **1921** F. M. FORD *Let.* 17 May (1965) 132, I will stand you a champagne dinner. **1901** *Daily News* 4 Mar. 7/4 The refrigerant is to be carbonic anhydride, or champagne gas. **1851** *London at Table* II. 45 Never use the present round saucer animalcula-catching champagne glasses, but..tulip-shaped ones. **1858** SIMMONDS *Dict. Trade*, *Champagne-glass*, a long, narrow glass, made for drinking effervescing wines from. **1825** H. WILSON *Memoirs* I. 176 Her black-pudding dinners and champaine suppers. **1893** E. LUTYENS *Let.* 3 May in *Blessed Girl* (1953) x. 189 We were offered..a champagne supper. **1969** H. MACINNES *Salzburg Connection* vi. 72 I'll have to give up that champagne supper with the polka girls.

2. A colour like that of champagne (see quots.); also, a fabric of this colour. Freq. *attrib.* or quasi-*adj.*

1881 *Cassell's Family Mag.* 186/2 The colours.. include..the shade known as 'champagne'. **1893** *Ibid.* Apr. 394/2 The new coloured straw called 'Champagne'. **1895** *Bow Bells* 3 May 441/3 Another dress of princess shape is in that coloured cloth known as 'Champagne'. **1903** *Daily Chron.* 21 Feb. 8/4 One of the colours with which women will be tempted..is champagne... It is a beautiful shade of pale straw, with a suggestion of pink about it. **1903** *Lady's Realm* Apr. 761/2 That pale biscuit colour which has been known..as the 'champagne' shade is now described as 'almond' colour. **1904** H. O. STURGIS *Belchamber* vii. 89 Who is the champagne blonde..next your brother? **1951** *Catal. Exhibits Festival of Britain* 187/1 Shoes,..black patent and champagne lizard. **1965** J. POTTS *Only Good Secretary* (1966) i. 14 Her current hair color was Champagne Blonde.

champers (ʃæ·mpəɹz). *slang.* [f. first syllable of CHAMPAGNE + *-ER⁶.] Champagne.

1955 M. ALLINGHAM *Beckoning Lady* v. 83 'One bot. champers—No!' she had written. *Ibid.* vii. 112 She hates champers but she's not allowed to drink anything else. **1959** 'M. AINSWORTH' *Murder is Catching* ii. 24 Champers or something with gin in it?

champignon. (Later examples.)

1841 THACKERAY in *Fraser's Mag.* June 723/1 This was the bill of fare..Champignons à la Provençale (the most delicious mushrooms I ever tasted). **1966** *Oxf. Bk. Flowerless Plants* 34/1 *Clitocybe rivulosa* ('False Champignon') is to be found in groups amongst short grass in late summer and autumn, and may form fairy-rings.

champion, *sb.*¹ Add: **4. a.** (Earlier examples.)

1730 in G. B. Buckley *Fresh Light on 18th Cent. Cricket* (1935) 4 'Twas thought the Kentish champions would have lost their honours by being beat at one innings if time had permitted. *c* **1742** J. LOVE *Cricket* iii. 106 The last two Champions even now are in, And but three Notches yet remain to win. **1802** *Sporting Mag.* XXI. 17/1 This hero [*sc.* Jem Belcher], who justly stiles himself in his advertisement, 'Champion of England', was himself to exhibit all his science.

b. (Earlier example.)

1717 T. CAVE *Let.* 28 Feb. in M. M. Verney *Verney Lett.* (1930) II. xxii. 48 Violett, that Essex Champion,..is the only dogg suffered to bear the best Spaniell in England company in the Parlour.

5. b. c. (Further examples.) Also as *adj.* or *adv.* (*colloq.* or *dial.*) = excellent(ly).

1853 *Bell's Life* 22 May 6/2 They [*sc.* Notts.] may, for the present, possess the honour of being the 'Champion County'. **1889** BARRÈRE & LELAND *Dict. Slang* s.v., An exemplary humbug is described as 'a champion fraud'. A noisy candidate for office was denounced by a Chicago newspaper as 'the champion gas-bag'. **1914** *Daily Mail* 1 June 3/6 'It was champion,' he added. **1914** *Concise Oxf. Dict.* s.v., Champion idiot, blunder. **1923** *Daily Mail* 26 Mar. 9 'He cried champion,' said a proud Yorkshireman. **1925** W. DEEPING *Sorrell & Son* xii. 109 Carrying luggage upstairs don't hurt me. He's got the head piece. We get on champion. What's wrong with that?

champlevé (ʃanlǝve, ʃæmplevēⁱ·), *sb.* and *a.* [Fr., f. *champ* field, *levé* raised.] Applied to enamel work in which the metal ground is engraved, cut out, or depressed, and the spaces filled with enamel pastes and fired.

1856 O. JONES *Gram. Ornament* xvii, Limoges Champlevé Enamel, from the same Museum. **1861** H. HAINES *Monumental Brasses* I. p. ix, These early enamels show the 'champleve', i.e. the copper is raised into ridges of partition between the colours. *Ibid.*, Like the field of a Limoges champ-levé enamel. **1877** tr. *C. Blanc's Orn. & Dress* xx. 259 Champlevé enamels are sometimes called *taille d'epargne* enamels. **1880** *Encycl. Brit.* XIII. 679 In champlevé the enamelling substance is applied to the surface of the gold as ornamental details. **1906** *Daily Chron.* 3 Oct. 3/3 The chief seats of the champleve work of the Middle Ages were the Rhenish provinces and Limoges. **1933** *Burlington Mag.* Sept. 108/1 The oldest of the few objects.. are the six champlevé enamel plaques. **1938** *Ibid.* Nov. p. xxviii/1 Little examples of Limoges champlevé. **1969** G. SIMS *Sand Dollar* xi. 142 A fine Limoges *champlevé* enamel.

chance, *sb.* Add: **4.** Often const. *of*. Also *pregnantly* = chance or opportunity of escape, acquittal, or the like.

1885 *Manch. Exam.* 6 May 5/1 A peace which is not cemented with blood has the best chance of permanency. **1888** BRYCE *Amer. Commw.* III. xciii. 296 In prosecutions for gambling or the sale of intoxicants a defendant had no chance before them [*sc.* a jury composed of women].

b. A quantity or number; used with adjs., as *fine, nice, smart*. *U.S. dial.*

1805 ORDWAY *Jrnl.* 8 Dec. (1916) 316 The men returned with a fine chance of Elk meat. **1819** D. THOMAS *Trav.* 230 (Th.), A considerable quantity is expressed by a smart chance; and our hostess at Madison said there was a smart chance of Yankees in that village. **1878** J. H. BEADLE *Western Wilds* xiv. 212 Fine chance o' corn planted, an' doin' well. **1939** *These are our Lives* (U.S.) 68, I have a nice chance o' chickens.

c. *Cricket.* An opportunity of dismissing a batsman, given to a fieldsman by the batsman's faulty play; chiefly in phr. *to give a chance*.

1832 *Brighton Gaz.* 19 July 3/4 Mr. J. W. Osborne..did not give a chance, and was not put out in either innings. **1883** [see Dict., 4]. **1899** W. G. GRACE *Cricketing Remin.* 241 A matter of a few inches converts a chance into a boundary hit. **1929** A. A. MILNE *Those were the Days* 732, I hear already long-on insisting It wasn't a chance that came to hand. **1970** *Guardian* 4 May 25/6 Lloyd's innings also contained some risks, but he never gave a real chance until he was out.

7. *by any chance* = PERCHANCE *adv.* 3.

1914 'IAN HAY' *Knight on Wheels* x, Are you engaged to be married, by any chance?

11. c. *to take a chance* or *chances*: to take a risk or risks. orig. *U.S.*

1902 S. G. FISHER *True Hist. Amer. Rev.* 311 Washington thought himself justified in taking the chances rather than abandon New York without a blow. **1904** *N.Y. Even. Post* 24 Oct. 12 Passengers on stalled trains took chances with the third rail, and getting off walked to the nearest station. **1912** H. CROLY *Marcus Alonzo Hanna* 99 In the beginning he may have taken some long chances in order to accelerate the progress of the firm. **1930** L. G. D. ACLAND *Early Canterbury Runs* x. 247 Caverhill tried to get them into a paddock with a rowdy bull, but they noticed just in time that he was not taking any chances himself. **1931** J. T. ADAMS *Epic of America* vii. 187 The American had always been 'taking a chance'.

D. *chance-come, -given, -like, -met, -seen* adjs.

1923 BLUNDEN *To Nature* 44 Formed not These the chance-come charm that bade me worship then? **1811** SHELLEY *Let.* 15 Dec. (1964) I. 153 My chance-given property. **1925** J. GREGORY *Bab of Backwoods* i. 9 Here, chance-given, was Farley's opportunity. **1851** MELVILLE *Moby Dick* II. xvii. 138 An unearthly, formless, chance-like apparition of life. **1892** KIPLING *Lett. of Travel* (1920) 100 Young men—chance-met in the streets. **1901** —— *Kim* vii. 176 A chance-met Rajah's elephant. **1923** —— *Irish Guards* I. p. viii, A chance-seen act of bravery.

chance, *v.* Add: **4. b.** slang phr. *and chance the ducks*: come what may; anyhow, anyway.

1874 HOTTEN *Slang Dict.* 113 *Chance the ducks*, an expression signifying come what may. 'I'll do it, and chance the ducks.' **1924** B. GILBERT *Bly Market* 12 'Mackenzie Quick says the old man's breaking up.' 'He wasnt last Sunday week, and chance the ducks.'

c. colloq. phr. *to chance one's arm*: to perform an action in the face of probable failure; to take one's chance of doing something successfully. Similarly *to chance one's mit*.

1889 BARRÈRE & LELAND *Dict. Slang* I. 236/1 *Chance your arm* (tailors), try, let it go, chance it. **1899** *Daily News* 13 Nov. 7/1 B. P. says 'to chance your arm' means to risk a court-martial which has the power to take all the pretty pretties off a man's sleeve. I first heard the phrase in 1886. **1919** *War Slang* in *Athenæum* 8 Aug. 728/1 'Chancing his mit' means risking a great deal, or playing a losing hazard. **1926** *Westm. Gaz.* 7 Jan., If such people as this defendant can chance their arm in this way, amateur sport is not going to be kept pure. **1927** *Daily Express* 10 June 9/3 Joanna [Southcott] seems to have chanced her arm, so to speak, in her prophecy of the end of the world, which she fixes as due to happen on June 31 next. **1959** *Economist* 27 June 1152/2 Mr Macmillan may have no more by-elections in this Parliament by which to judge when to chance his arm.

chanceless (tʃɑ·nslĕs), *a.* [f. CHANCE *sb.* + -LESS.] Without giving or receiving a chance. In *Cricket*, without giving the fieldsmen a chance (see *CHANCE *sb.* 4 c).

1903 *Daily Chron.* 27 May 5/4 A result made probable by the century of McGahey, a chanceless but hardly faultless innings. **1928** *Daily Tel.* 7 Aug. 12/7 The feature of the first day's play..was a chanceless century by I. Akers-Douglas, the Etonian.

chancellery, -ory. 2. c. (Later examples.)

In British diplomatic use the official term is *chancery* (see *CHANCERY 5).

1954 [see *CHANCERY 5]. **1957** *Time* 2 Sept. 19/1 Whether Syria's plain citizens realized it or not.., their country was the No. 1 topic in chancelleries and foreign offices around the world. **1958** *Economist* 13 Sept. 815/2 There has been a clear and common *idée en marche* in the chancelleries of the Commonwealth.

chancer (tʃɑ·nsəɹ), *sb. slang.* [f. CHANCE *v.* + -ER.¹] One who takes chances or does risky things (see quots.).

1884 R. LAWSON *Upton-on-Severn Words* 12 *Chancer*, one who makes rash and inexact statements. 'She's a bit of a chancer.' **1889** BARRÈRE & LELAND *Dict. Slang* I. 236/1 *Chancer* (tailors), one who exaggerates, or lies. Also one who attempts anything and is incompetent. **1925** FRASER & GIBBONS *Soldier & Sailor Words* 51 *A Chancer*, a prevaricator... One taking the chance of being found out. **1937** F. P. CROZIER *Men I Killed* iv. 88 He was the complete military 'chancer' of our time, bad in every way. **1959** M. PUGH *Chancer* vii. 91 As two-faced as a cod, a proper chancer. He was a born assassin who lost his nerve.

chancery. Add: **5.** *spec.* = CHANCELLERY 2 c.

In British diplomatic use *chancery* (and not *chancellery*) is used for the general political section, and the offices housing it, of an embassy or legation. The word *chancery* seems to have been first used in this connection in 1920 when it was decided that a new permanent clerical service for diplomatic and consular posts abroad, which was then being formed, should be known as the 'Chancery Service'.

1932 H. NICOLSON *Public Faces* vii. 186 You will find a comfortable sofa in the outer chancery. **1954** O. O'MALLEY *Phantom Caravan* 242, I never went near the building which housed the Chancery (not chancellery, as many journalists called it). I sat in my house, which was about five minutes' walk away from the Embassy offices. **1957** *Times* 10 Oct. 8/4 Chanceries of all nations were groping for a reasonable definition of the limitation of 'air space'.

8. *chancery-ward* (example).

1851 MELVILLE *Moby Dick* I. xvi. 118 Widows, fatherless children, and chancery wards.

chancing (tʃɑ·nsiŋ), *ppl. a.* [f. CHANCE *v.* + -ING².] That chances. **a.** That comes or is present by chance. **b.** That relies upon chance.

1889 *Daily News* 7 Aug. 5/4 Mackerel are a 'chancing fish', and the abundance of them cannot always be relied on. **1896** W. WAKEFIELD in *Spectator* 31 Oct. 589/2 Each a chancing guest Unknown before, we tarried certain days. **1909** *Westm. Gaz.* 1 Apr. 8/2 A chancing man... What is that?—He's a horse-dealer, and sells animals on chance as to whether they are good or bad.

chancre. Add: **2.** A disease incident to the tobacco-plant, said to be caused by *Bacillus æruginosus*.

1903 *Nature* 17 Sept. 492/2 On a bacterial disease of tobacco, 'chancre' or 'anthracnose', by M. G. Delacroix.

chancy, *a.* Add: **4.** *Cricket.* Full of 'chances' (*CHANCE *sb.* 4 c).

1875 *Field* 8 May 458/1 Rumsey's long and very chancy innings.

chandelier. Add: **4. chandelier lily** (also simply *chandelier*), a bulbous South African plant of the genus *Brunsvigia*; **chandelier plant**, a species of *Euphorbia*.

1818 LATROBE *Jrnl. Visit S. Afr. 1815–16* 165 We noticed here a gigantic species of a plant, from its singular form.. called the chandelier. **1827** T. PHILLIPS *Scenes Albany & Cafferland* i. 4 We..crossed a tract of land covered with aloes, called likewise the chandelier plant. **1871** *Cape Monthly Mag.* III. 122 (Pettman), A grand family of plants —the Amaryllidaceæ—of which our Chandelier lily (*Brunsvigia Josephinæ*) is an example.

chaney, dial. var. CHINA¹ II (in Dict. and Suppl.).

chang (tʃæŋ), *sb.*² Also **chong.** [Tibetan *chaṅ*.] A Tibetan beer or wine made chiefly from barley or rice.

1800 S. TURNER *Acc. Embassy Court of Teshoo Lama* I. ii, Chong is a slightly acid and spirituous liquor. **1887** *Field* 19 Feb. 243/1 The headman came out with jugs of 'chang' (a kind of beer made from grass). **1925** E. F. NORTON *Fight for Everest, 1924* 21 It was up to them..to take as many drinks of their chang as they could before setting out for Tibet. **1963** D. MURPHY *Diary* 23 Oct. in *Tibetan Foothold* (1966) v. 99 All the adults spent the day..either preparing or eating food while drinking immeasurable quantities of 'chang'.

changa (tʃæ·ŋgă). A mole-cricket, *Scapteriscus didactylus*, found in the West Indies and parts of the U.S.

1901 *Yearbk. U.S. Dept. Agric.* 510 (Cent. D. Suppl.), The 'changa'..a kind of mole cricket, which has become very troublesome. **1928** F. S. EARLE *Sugar Cane* vi. 174 The Mole Cricket, or Changa..is a soil insect.

change, *sb.* Add: **1. f.** *Cricket.* The substitution of one bowler or type of bowling for another in the course of a match; also, a change-bowler.

1828 G. T. KNIGHT in W. Denison *Cricket: Sk. Players* 51 Each eleven..should contain at least four good bowlers, so that there may be..a change at each wicket. **1833** J. NYREN *Young Cricketer's Tutor* 64 We reckoned him a tolerably good change for bowling. **1912** J. B. HOBBS *Recov. 'Ashes'* 124 A double change was tried, Mr. Douglas and Woolley relieving the opening trundlers. **1955** *Times* 25 June 7/4 Our opening bowlers were a house decorator and a curate, and I was first change.

g. *Motoring.* A change from one gear to another. So *change-down, -up* (see *CHANGE *v.* 6 d).

1912 [see *gear-change*, *GEAR *sb.* IV]. **1936** R. LEHMANN *Weather in Streets* I. ii. 58 The..car..drew away with its old familiar long-drawn rising moan..on the change-up. **1959** M. PLATT *Elem. Automobile Engin.* (ed. 2) v. 85 Decreasing car speeds with a constant throttle opening will produce 'down' changes... The reverse of these conditions would produce a change 'up'. **1964** I. FLEMING *You only live Twice* xii. 147 The driver did a good racing change and pulled in.

4. d. *change of life*: also *attrib.* and *ellipt.* (*colloq.*)

1934 S. BECKETT *More Pricks than Kicks* 117 The mother was low-sized..admirably preserved though well past the change. **1946** J. CARY *Moonlight* xv. 112 Rose was in her change—she used to fly into fearful rages. **1949** M. MEAD *Male & Female* xvi. 340 The old folk-phrase 'change-of-life baby'. **1959** 'ED MCBAIN' *Pusher* x. 93 Meyer had been a change-of-life baby.

e. *change of heart*: conversion to a different frame of mind.

1828 in WEBSTER. **1853** J. RUSKIN *Let.* Dec. in W. James *Order of Release* (1947) xiv. 213 She passes her days in melancholy, and nothing can help her but an entire change of heart. **1926** J. S. HUXLEY *Ess. Pop. Sci.* vii. 69 A 'change of heart' as regards the essential aims of life. *Ibid.* 70 In most men it seems theoretically possible to produce a 'change of heart'—i.e. substitute new dominant ideas for old. **1940** 'G. ORWELL' *Inside the Whale* 155 Lawrence..like Dickens..is a 'change-of-heart' man. **1960** *Author* LXXI. 119/1 A change of heart in high places would help.

f. *change of pace* = sense 1 b in Dict. *N. Amer.*

1940 *Time* 4 Mar. 50/2 As a change of pace, a picture agency released a photograph of blank-faced Brenda Frazier showing some expression. **1962** *Globe & Mail* (Toronto) 29 Jan. 6/3 Mr. MacLeod might provide a badly needed change of pace.

7. c. So *not to get any* (or *much*) *change out of*: to get no return, result, or satisfaction from; to fail to get the better of (a person).

1864 TROLLOPE *Small House Allington* II. xxx. 312 'That's a bitter old lady.' 'There ain't none of 'em get much change out of Mrs. Crump.' **1869** —— *He knew*, etc. II. xcvii. 369 Old Barty said something..that wasn't intended to be kind... But he got no change out of her. **1888** MRS. H. WARD *R. Elsmere* I. I. vi. 140, I just love..to hear her instructing other people in their own particular trades. She didn't get much change out of him. **1910** J. BUCHAN *Prester John* xii. 209 Still I said nothing. If the man had come to mock me, he would get no change out of David Crawford. **1954** J. TRENCH *Dishonoured Bones* ii. 62 She didn't get much change out of Charles.

12. a. *change-of-address* attrib.; **change gear**, gearing by which changes may be made in the relative number of turns per minute for the driving or driven shafts of lathes and similar machines; **change-giving**, the giving of change (sense 7 b); also *attrib.*; **change key**, one adapted for opening only one set of locks, as distinguished from a master key; **change lever** = *change-speed lever*; **change purse**, a purse for small change; **change-speed**, a mechanism for effecting a change of gear and thereby increasing or decreasing the speed of a cycle, motor-car, etc.; also *attrib.*, as *change-speed gear, lever*, etc.

1908 *Daily Chron.* 16 May 1/6 The change-giving rostrums were in working order. **1963** *Rep. Comm. Inquiry Decimal Curr.* i. 12 Their introduction [vulgar fractions].. probably makes mental calculations and change-giving slower. *Ibid.* ix. 229 The change-giving operations..were conducted at a more rapid tempo. **1908** *Daily Chron.* 14 Nov. 8/6 This..gear is now equipped with a change-lever. **1950** 'S. RANSOME' *Deadly Miss Ashley* viii. 84 Miss Ashley had left a change-of-address card with the branch post office. **1911** *Daily Colonist* (Victoria, B.C.) 19 Apr. 24/1 The Newest Square Shaped Bag, with plain brass frames, solid leather, plain silk lining, with change purse. **1967** 'D. SHANNON' *Chance to Kill* (1968) xv. 218 An old-fashioned change purse, with double compartments. **1902** EDGE & JARROTT in A. C. Harmsworth et al. *Motors* xv. 324 The change-speed lever is on the right hand. **1904** A. B. F. YOUNG *Compl. Motorist* iii. 60 The change-speed gear in its simplest and commonest form. **1907** *Daily Chron.* 17 Aug. 7/5 For the next [cycle] tour that I make in company I shall insist on all machines being fitted with the useful change-speed. *Ibid.* 11 Nov. 7/3 When the change-speed lever is pushed away from the driver, the jaw clutch engages the low gear-wheels. **1922** *Times* 20 June 8/5 The change-speed lever provides four forward positions.., a neutral and a reverse.

b. change-bowler *Cricket*, a bowler who relieves the regular bowlers in a match (cf. *1 f).

1833 NYREN *Young Cricketers' Tutor* 58 The two principal bowlers..were Thomas Brett and Richard Nyren..; the *corps de reserve*, or change-bowlers, were Barber and Hogsflesh. **1886** [in Dict.]. **1903** WODEHOUSE *Prefect's Uncle* xv. 214 He was essentially a change bowler.

change, *v.* Add: **1. d.** Also, to put fresh clothes on (a person); *spec.* to change a baby's napkin.

1922 D. H. LAWRENCE *England, My England* 23 Joyce, you must be patient. I'm just changing Annabel. **1936** W. HOLTBY *South Riding* v. i. 275 That baby needs changing. **1958** C. FREMLIN *Hours before Dawn* vii. 59 There hadn't been time to give Michael his orange juice, or change him.

e. Substitute quot. 1470 for quot. 1634 in Dict. Add later examples. Also *ellipt.*, to change into evening dress.

1470 MALORY *Arthur* XVII. xvi. (1947) II. 1017 A jantillwoman..brought hym a shirte of smal lynen clothe; but he chaunged nat there. **1928** D. L. SAYERS *Unpleasantness at Bellona Club* xxi. 266 Bunter..was..in attendance with his master's dress clothes. Having changed, Wimsey passed through the hall again. **1945** J. B. PRIESTLEY *Three Men in New Suits* ii. 20 You won't have to dress, though, because he..doesn't bother changing. **1946** 'S. RUSSELL' *To Bed with Grand Music* vi. 83, I simply must have a bath and change. **1948** O. LANCASTER in *Signs of Times* (1961) 52 How does one entertain Americans?—If one gives them Spam and doesn't change, we're a down-at-heel, C3 nation. **1954** A. S. C. ROSS in *Neuphilologische Mitteilungen* LV. 44 A sentence *Shall we wear evening dress?* would not be possible, the appropriate expression being *Are we going to change?* **1970** D. HALLIDAY *Dolly & Cookie Bird* iii. 23 On Ibiza, the Lloyds don't wear DJs except for a party though of course everyone changes.

6. d. *to change up, down*, to engage a higher, lower, gear in a motor vehicle. Also *to change* (*to a higher, lower*) *gear*.

1902 A. C. HARMSWORTH *Motors* xv. 327 Change to a lower gear at once. **1904** A. B. F. YOUNG *Compl. Motorist* iii. 64 All that is necessary in changing gear is to move the lever which tightens a band on one of the friction drums and locks it. **1933** P. MACDONALD *Myst. Dead Police* i. 7 The police car..was changed down miraculously just before the turning. **1939** L. MACNEICE *Autumn Jrnl.* v. 25 A car..accelerates, changes Up, having just changed down. **1942** *Autocar: Motor Driving made Easy* (ed. 7) iii. 29 The time has come to 'change down'. **1953** E. HAWKS *Cassell Book of Austin A40* (ed. 2) ii. 35 In changing down from top to third, and third to second, proceed as follows. **1962** J. BRAINE *Life at Top* ii. 38, I changed down into second; then changed up again.

8. a. Delete † *rare* and add later examples.

1843 LEVER *Jack Hinton* xxx, He changed all the splints and bandages to the sound leg.

changeable, *a.* **3.** Delete *arch.* and add later examples.

1861 C. M. YONGE *Young Step-Mother* xv. 197 Grandmamma..stroked down the flounces of her changeable silk. **1968** H. MCCLOY *Mr. Splitfoot* (1969) xii. 126 A ball dress of changeable silk, pale blue and rose.

changelessly, *adv.* (Examples.)

1902 W. WATSON *Coron. K. Edw. VII* iii, And changelessly the river sends his sigh Down leagues of hope and fear. **1910** *Times* 21 Mar. 6/1 The axis round which the policy of the Austro-Hungarian monarchy changelessly revolves.

‖**changement de pieds** (ʃanʒmaṅ də pye). *Ballet.* [Fr., lit. 'change of feet'.] A jump during which the dancer changes the position of the feet. Also *ellipt.* **changement.**

1840 F. A. KEMBLE *Let.* 28 June in *Rec. Later Life* (1882) II. 26 Dancers deserve to be well paid when one thinks of.. the daily hours of *battements* and *changements de pieds*. **1916** *Dancing Times* Sept. 327 [Espinosa's] 'Syllabus of Dancing'..Steps of Elevation.—Changements, soubresauts, [etc.]. **1922** BEAUMONT & IDZIKOWSKI *Man. Class. Theatr. Dancing* IV. 182 Execute *deux petits changements*. **1957** G. B. L. WILSON *Dict. Ballet* 71 If the legs are lightly beaten from the lower calf before changing, the changement is known as an entrechat royal or changement battu.

change-over. The action or an act of changing over. **a.** Alteration from one working system to another (see also quots. 1940, 1953). **b.** A reversal of the situation in affairs, of opinions, etc. **c.** The action of handing over the baton to the next runner in a relay race.

1907 *Westm. Gaz.* 26 Nov. 4/3 The fuel..will be paraffin, which will be used with Messrs. Thornycroft's patent change-over device from a preliminary start on petrol. **1920** H. G. WELLS *Outl. Hist.* 645/1 Now here was a change-over of quite primary importance in human affairs. **1926** H. M. ABRAHAMS *Athletics* x. 98 The ideal change over [in relay racing] is achieved when each man is going as nearly as possible at full speed and there is no hesitation on the part of the recipient of the baton. **1927** *Glasgow Herald* 3 Dec. 11 A method of converting the station from the use of one wave to that of another within a far shorter time than at present occupied by the change-over. **1928** *Observer* 1 Jan. 10/2 Water engineers..said the change-over to river supplies must come about..in the next generation. **1928** *Daily Tel.* 6 Nov. 17/4 The change-over on the Central Section of the Southern Railway from the overhead to the third rail system. **1929** G. M. BUTLER *Mod. Athletics* ii. 30 During the sprint relay there should be a steward at each change-over station. **1940** *Chambers's Techn. Dict.* 152/2 *Change-over*, the transference of projection from one machine to another at the end of one reel and the start of the next, without an apparent break in the sequence. **1953** K. REISZ *Technique Film Editing* 278 *Change-over cue*, small spot or other mark made in the top right-hand corner of certain frames near the end of a reel to give the projectionist a signal for the change-over. **1955** R. BANNISTER *First Four Minutes* vi. 78 The U.S.A. team was disqualified for an alleged faulty changeover. **1958** *Engineering* 11 Apr. 452/3 A change-over switch is fitted for use with either base. **1958** *Church Times* 21 Nov. 5/2 Growing reverence for the Blessed Sacrament seems to have dictated the change-over from leavened to unleavened bread in the West.

changer. Add: **1. b.** *spec.* A mechanism designed to change the records on a record player. In full *record changer*.

1930 *Gramophone* Mar. 474/2 A novel, and yet efficient, automatic record changer is on view on the Cowley stand... It will change up to six records, either 10 in. or 12 in. **1940** *Ibid.* Dec. 163/2 The automatic record changer is of the usual eight record capacity type... This changer is fitted with a spring loaded spindle to minimise the chances of record slip. **1967** *Times Rev. Industry* June 48/1 A year earlier [1951] BSR introduced its first automatic changer unit designed to handle LPs, 45s and 78s. Technically this new machine was a considerable advance on the heavy-weight changers which had ground and scraped their way through stacks of 78s for more than a decade.

changing, *vbl. sb.*¹ Add: **4. changing note** (see quot. 1876); **changing room**, a room where one can change one's clothes, esp. at a sports ground.

1876 STAINER & BARRETT *Dict. Mus. Terms*, *Changing notes*, passing notes or discords which occur on the accented parts of a bar. **1878** Changing note [in Dict.]. **1917** Changing-room [see *airing cupboard]. **1937** *Discovery* Jan. p. ii/1 New buildings just completed. Swimming bath, changing room, school library. **1960** J. BETJEMAN *Summoned by Bells* vii. 65 In steam like this the changing-room was bathed.

changingly (tʃēⁱ·ndʒiŋli), *adv. rare.* [f. CHANGING *ppl. a.* + -LY².] †**1.** Alternately. *Obs. dial.*

1847 HALLIWELL *Dict. Archaic & Prov. Words*.

2. In a changing or varying manner.

1859 DICKENS *T. Two Cities* II. vi. 62 The third [room], changingly speckled by the rustle of the plane-tree in the yard. **1883** MEREDITH *Poems & Lyrics* 64 Yet changeingly the trills repeat And linger ringing while they fleet.

chank, *v.* Delete †*Obs.* and add later *dial.* and *U.S.* examples.

1875 *N. & Q.* 27 Feb. 166/1 A mother would say to her child: 'I'll gée ye a dod [sweet], but ye marnt chank it— you must suck it.' **1903** CLAPIN *Dict. Amer.* 105 *Chank*, to chew noisily. (N. Eng.). **1909** MRS. STRATTON-PORTER *Girl of Limberlost* xxi. 393 'Did they eat it?' demanded Wesley. 'They just chanked into it,' replied Billy. **1913** —— *Laddie* iv. 126 The pigs..chanked up every peach that fell there.

channel, *sb.*¹ Add: **9. c.** A circuit for the transmission of communications in telegraphy.

1848 *Brit. Pat.* 12,039 17 Each of the three [telegraph] circuits..is then split as it were into three different channels. **1922** *Telegr. & Teleph. Jrnl.* Oct. 8/1 The whole telegraph position in this country is changed by the fact that eight channels..can be equipped on one wire. **1957** *Encycl. Brit.* XXI. 902/2 Two coaxials [*sc.* coaxial cables] can provide.. 1,800 telephone channels. **1959** *Wire & Radio Communication* XII. 7/2 The 36 message channels in the deep sea cable.

d. A band of frequencies of sufficient width for the transmission of a radio or television signal; *spec.* a television service using such a band.

1928 *Television* I. No. ii. 20/2 A very short wavelength was chosen for broadcasting these 'pictures' because a channel 40 kilocycles wide is needed. **1930** *Engineering* 31 Oct. 541/3 Five independent radio-telephone channels will..be available. **1935** *Discovery* Aug. 240/2 Automatic transmitting and high-speed recording is provided for one transmitting and one receiving channel. **1941** *B.B.C. Gloss. Broadc. Terms* 6 *Channel*, electrical or electro-magnetic path (such as a circuit, line, or radio-link) for a programme output. **1958** *Spectator* 1 Aug. 167/2 Go over to Channel 9 and who do you see? **1958** *Listener* 9 Oct. 553/1 One has less real choice from four [television] channels in Washington than one has from only two in London. **1960** in *Rep. Comm. Broadc.* (1962) 334 We feel that the 625-line standard with a total channel width of 8 Mc/s represents the best compromise.

e. Short for *channel bar*: see *12.

1904 *Westm. Gaz.* 22 Aug. 5/1 Structural material, such as joists, channels, and other sectional bars. **1970** *Financial Times* 13 Apr. 21/3 Small orders for special rollings—for, perhaps, 100 tons of 6-inch by 3-inch channel.

f. Other techn. uses.

1836 *Ann. Electr., Magn. & Chem.* Oct. 42 A coil of copper wire excited by magnetic action will become a channel for an electric current. **1955** W. PAULI *N. Bohr* 137 The decay through a given channel is the inverse of the formation of a compound nucleus through the same channel. **1958** *Chambers's Techn. Dict.* 964/2 *Channel*, a passage through the core of a reactor for the location of fuel elements, specimens for radiation or the flow of coolant. **1961** W. HAGGARD *Arena* xviii. 161 He walked to the control desk quickly. 'Channel Nine. Where's that flight going, please?' **1962** *Gloss. Terms Autom. Data Processing* (B.S.I.) 87

Channel, a path or aggregate of related paths for carrying signals between a source and a destination. **1964** L. Wilets *Theories Nucl. Fission* vii. 114 In the formal theory of nuclear reactions, a channel is defined according to the quantum numbers of the asymptotic wave function of the system. **1968** *Brit. Med. Bull.* XXIV. 255/1 In the digital averager we have designed and built, only 1,024 bits are available for each of four channels. **1969** *Computers & Humanities* IV. 131 Input: 6-channel tape (linotype) used for newspaper typesetting.

12. channel bar, an iron bar or beam flanged to form a channel on one side; **channel bass**, one of several names of the red-fish, *Sciænops ocellatus*; **channel cat(-fish)** *U.S.*, any of several species of catfish, esp. of the genus *Ictalurus*; **Channel Fleet**, the portion of the British fleet detailed for service in the English Channel; **channel iron**, (*a*) = *channel bar*; also, the concave metal support of a rubber tyre; (*b*) a support for the guttering of a building (Knight *Dict. Mech.* a 1877); **Channel Islands**, the name of a group of islands in the English Channel used (chiefly *attrib.*) to designate breeds of cattle originating there (see *Alderney, Guernsey 2 b, Jersey 4) or their milk; also in *sing.*; **channel-leaved** *a. Bot.* (see quot.); **channel seam**, a seam outlined on each side by stitching; so *channel seaming*; **channel-section** = *channel-shaped section* (see next); **channel-shaped** *a.*, shaped like the section of a channel bar; **Channel tunnel**, a (suggested) tunnel under the English Channel linking the coasts of England and France; also *fig.*; **channelward(s** *adv.*, in the direction of the English Channel.

1904 G. P. Neele *Railway Remin.* 219 Some channel bars of iron on a down goods train..had gradually shifted. **1889** *Cent. Dict.*, *Channel-bass*, a sciænoid fish,..the redfish. **1897** *Outing* (U.S.) XXIX. 231/2 The channel-bass, or 'redfish', which sometimes attains a weight of fifty pounds. **1838** J. P. Kirtland in *Rep. Geol. Survey Ohio 1837–38* 169 *Pimelobus pallidus*,..Channel cat-fish. **1855** S. F. Baird in *Smithsonian Inst. Rep. 1854* 341 The sea-cat or channel-cat. **1888** *Boston* (Mass.) *Jrnl.* 17 People who like fish in Atlanta are eating..channel-cat and black bass. **1897** *Outing* (U.S.) XXX. 439/1 What the boys called 'channel-cats' were taken from midstream by long hand-lines. **1913** W. S. Cather *O Pioneers!* 29 Twice every summer she sent the boys to the river..to fish for channel cat. **1788** Dyott *Diary* Aug. (1907) I. 50 He had been cruising with the Channel Fleet for three weeks. **1831** J. Watkins *William the Fourth* I. ii. 36 The encounter of the channel-fleet with the combined French and Spanish squadrons. **1904** *Admiralty Order* in *Times* 12 Dec. 4/1 The present so-called Home Fleet is henceforth to be known as the Channel Fleet. **1888** *Lockwood's Dict. Terms Mech. Engin.*, *Channel Iron*, rolled wrought iron bar, whose section is that of three sides of a parallelogram, used in bridge and girder work, and for structural purposes generally. **1895** G. J. Burns *Gloss. Techn. Terms Archit.*, *Channel iron*, wrought iron of U-shaped section. **1875** Channel Island [see *Alderney]. **1950** J. G. Davis *Dict. Dairying* 95 *Channel Island breeds*. These include Jersey, Guernsey, and Alderney...*Channel Island milk*. This is now defined as milk from Channel Island cows and containing at least 4 per cent. fat. **1964** *Farmers Weekly* 28 Aug., Two high quality milks, Channel Islands and homogenised. *Ibid.*, Channel Islands milk..possesses an indisputably higher food value than that of milk with lower butterfat content. **1860** Worcester (citing Loudon) *Channel-leaved*.., having leaves folded together so as to resemble a channel. **1961** *Times* 24 Jan. 12/5 Plastron fronts, occasional flying panels and channel seams were also frequent. **1961** *Sunday Express* 12 Mar. 14/1 Channel seaming on the jacket. **1910** *Westm. Gaz.* 22 Mar. 5/2 The motor..is mounted..on a channel-section underframe suspended at three points. **1930** *Engineering* 2 May 564/1 The bracket..was formed of two channel sections. **1928** C. F. S. Gamble *North Sea Air Station* xxii. 392 These lighters..were built of steel...The general section was channel-shaped. **1869** *Times* 25 June 5/5 The Channel Tunnel..Project for a submarine tunnel between Dover and a point near Cape Blanc-nez, on the French coast. **1889** G. M. Hopkins *Let.* 29 Apr. (1935) 304 *Rot* about babies, a blethery bathos into which Hugo and he [*sc.* Swinburne] from opposite coasts have long driven Channel-tunnels. **1889** G. B. Shaw *London Music* (1937) 178, I wish I could go back by Channel Tunnel. **1957** *Encycl. Brit.* V. 232/2 The Channel tunnel scheme was shelved again [in 1924] but the idea did not die. **1920** *Chambers's Jrnl.* 67/2, I waved my hand Channelward. **1890** W. J. Gordon *Foundry* 96 The works grew..Channelwards by the deposit of their own slag.

channel, *v.* **3.** (Later examples.)

1958 *Times* 31 Jan. 9/7 It [*sc.* investment] must be channelled in the most remunerative directions. **1958** *Daily Express* 11 Mar. 14/2 It would be a good thing to have someone to..channel all the enquiries about her.

channelize (tʃæ·nĕləiz), *v.* Chiefly *U.S.* [f. Channel *sb.*[1] + -ize.] *trans.* To convey in or as in a channel; to guide; see also quot. 1957. Cf. Channellize *v.*

1939 H. Blumer in R. E. Park *Outl. Princ. Sociol.* xxii. 227 Fashion, by providing an opportunity for the expression of dispositions and tastes, serves to make them definite and to channelize them and, consequently, to fix and solidify them. **1957** *Britannica Bk. of Year 1956* 512/1 Daily and social life [in U.S.A.] gave such words as *Channelize* (with the noun-form *Channelization*), to control traffic by curbs and dividers. **1958** Blesh & Janis *They all played Ragtime* vii. 128 The main course of a folk music that..had moved over the borderline that separates folk art from fine art was channelized in a new direction. **1959** C. Ogburn *Marauders* (1960) iii. 100 Movement through the..kunai, or elephant, grass that was often as tall as a man or taller was narrowly channelized.

So **cha:nneliza·tion**, the action of channelizing; *spec.* in *Neurology*, the formation of a channel of nervous conduction or discharge (cf. Canalization 2).

1902 *Amer. Jrnl. Psychol.* XIII. 405 The problem is how to give normal emotional channelization, the safety valve of this biological heredity. **1957** [see above]. **1958** A. D. Booth in *Aspects of Translation* v. 90 Young professional linguists whose ideas had not, through channelization, been so stultified as to make them incapable of appreciating the new techniques.

channeller[2], **channeler** (tʃæ·nĕləɹ). [f. Channel *v.* + -er[1].] A machine for cutting channels or grooves in rock in quarrying. Used chiefly with a qualifying word, as *bar-channeller*, one in which the cutters are attached to a bar or carriage; *rock-channeller*; *track-channeller* (Track *sb.* 13).

1897 *Review Rev.* Nov. 567 What are called channellers have been utilized.

channelling, *vbl. sb.* Add: **1. b.** A gutter to carry off rain water from a road or carriage-way; = Channel *sb.* 3 a.

1892 *Times* 23 Mar. 3/2 There had always been a stone kerb and channelling across the so-called carriage way.

chanson. Add: **2. chanson de geste** [see Gest *sb.*[1]], a mediæval French epic poem.

1868 T. Wright *Knt. de la Tour* p. ix, The Chansons or Romans de Geste, which were so numerous in the twelfth and thirteenth centuries. **1872** A. Lang in *Fraser's Mag.* June 668/1 The form of the epic, whether we take the hexameter or the rougher *laisse* of the French *chansons de geste*, is full of conscious and admirable art. [**1912** W. P. Ker *Med. Eng. Lit.* iii. 69 The poems about Charlemagne and his peers, and others of the same sort, are sometimes called the old French epics; the French name for them is *chansons de geste*.] **1935** *Essays & Studies* XX. 48 The survival of this most declamatory of metres [*sc.* the alexandrine] from the medieval *Chanson de Geste* of Alexander.

‖**chansonnier** (ʃɑ̃nsonye). [Fr.] In France, a writer or performer of songs, esp. satirical songs in a cabaret.

1887 *Athenæum* 1 Jan. 11/1 A few pieces of spontaneous inspiration..had led us to believe that there was in M. Richepin the stuff for a popular *chansonnier*. **1927** *Observer* 4 Dec. 19/5 A famous 'chansonnier'. Death of Jean Louis Pisuisse. **1962** *Spectator* 13 Apr. 475 Those daring and sharp political satirists, the *chansonniers*.

‖**chanteuse** (ʃɑ̃ntöz). [Fr.] A female singer of popular songs, esp. in France.

1888 E. Dowson *Let.* 13 Nov. (1967) 17 My chanteuse is charming. **1900** M. Corelli *Master-Christian* 180 The stout '*chanteuse*' who is the glory of open-air cafés in the Champs Elysées. **1907** *Daily Chron.* 7 Nov. 4/5 Miss Agatha Ireland, a gifted chanteuse. **1962** P. Brickhill *Deadline* i. 10 A record-player was murmuring the husky confidences of some chanteuse.

chantey, chanty (tʃa·nti, tʃæ·nti, ʃ-). Frequent variants of Shanty *sb.*[2]

1856 C. Nordhoff *Merchant Vessel* in *Nine Years a Sailor* (1857) iv. 40 The foreman is the chantey-man, who sings the song, the gang only joining in the chorus. **1867** G. E. Clark *7 Years of Sailor's Life* iv. 41 A chanty gang was engaged to hoist out the cargo. *Ibid.* 44 The chanty men wanted biscuit. *Ibid.* xvi. 165 The anchor came to the bow with the chanty of 'Oh, Riley, Oh'. **1883**, etc. [see Shanty *sb.*[2]]. **1888** Laura A. Smith (*title*) The Music of the Waters: a Collection of the Sailors' Chanties, or Working Songs of the Sea, of all Maritime Nations. **1900** *Daily News* 9 Apr. 5/1 The British chantey, 'For he's a jolly good fellow'. **1925** D. Garnett *Sailor's Return* 16 The seaman began a loud and rolling chanty. **1955** *Times* 24 Aug. 7/4 The sea chanty was essentially a working song.

Chantilly (ʃɑ̃ntiyi). Also †**Chantilli.** [The name of a town in France, near Paris.] **1.** Used *attrib.* or *absol.* of a soft-paste porcelain made at Chantilly in the 18th century.

1774 H. Walpole *Descr. Strawberry-Hill* 16 Six coloured handle cups and saucers..of Chantilli china. *Ibid.* 20 A white butter-pot and plate of Chantilli. **1875** Lady C. Schreiber *Jrnl.* (1911) I. 350 A little Chantilly vase. *Ibid.* 352 A very pretty Chantilly marked cup. **1938** *Burlington Mag.* June p. xxix/i, A pair of porcelain statuettes of the Louis XV period, of soft-paste Chantilly. **1951** J. B. Priestley *Fest. Farbridge* ii. ii. 288 Meissen cups and Chantilly figures.

2. Applied to a delicate lace made originally at Chantilly, or to an article of apparel made of this lace.

1831 in A. Adburgham *Shops & Shopping* (1964) iv. 36 Extensive stock of Chantilly veils. **1848** Thackeray *Van. Fair* xxii. 188 Over the bonnet she had a veil of white Chantilly lace. **1898** [see Polychrome *a.*]. **1957** *Encycl. Brit.* XIII. 566/2 In the middle of the 19th century..fashion decreed that every lady's wardrobe should include ..for the summer a black or white so-called Chantilly shawl, many of which were produced at Brussels and Ghent. **1957** M. B. Picken *Fashion Dict.* 57/2 *Chantilly lace*, bobbin lace with fine ground and exquisitely outlined pattern.

3. Sweetened whipped cream; confectionery containing this cream. Also *attrib.*

1845 E. Acton *Mod. Cookery* xx. 529 Chantilly Basket. Filled with whipped cream and fresh strawberries. **1939** A. Simon *Conc. Encycl. Gastron.* I. 14/2 In confectionery, *Chantilly* means fresh cream whipped and sweetened. **1952** F. White *Good English Food* I. vi. 71 Purée of wood strawberries..piped Chantilly cream over. **1961** Mary Stewart *Ivy Tree* xiv. 231 The strawberries hastily assembled into whipped cream Chantilly.

chanty: see *chantey.

Chanukah, Chanukkah (hæ·nukā). Also Chanuc(c)ah, Hanukkah, -cha. [Heb. *ḥanukkāh* consecration.] A Jewish festival beginning on the 25th of Kislev (November–December) and lasting eight days, held to commemorate the purification of the Temple at Jerusalem by Judas Maccabæus after its pollution by the Syrians.

1891 M. Friedländer *Jewish Relig.* 410 On *Sabbath Chanuccah*, Zechariah's vision..is read. **1893** I. Zangwill *Childr. Ghetto* (ed. 3) II. i. 239 She was celebrating..the Feast of Chanukah—of the re-dedication of the Temple after the pollutions of Antiochus Epiphanes—and the memory of the national hero, Judas Maccabæus. **1902** *Daily Chron.* 23 Dec. 5/1 The feast of 'Hanucha', or dedication, is celebrated by the Jews this year simultaneously with Christmas. **1903** *Ibid.* 12 Dec. 5/2 The Chanucah Military Service is now an important annual feature of Jewish life in London. **1907** *Ibid.* 2 Dec. 4/6 As a military festival, Chanucah stands alone among Jewish celebrations. **1931** O. S. Rankin *Orig. Festival Hanukkah* Pref. p. vi. **1958** *Times Lit. Suppl.* 24 Jan. 38/4 A common term between Christmas and Hanukah has to be found. **1964** S. Bellow *Herzog* (1965) 226 Gersbach was lighting the Chanukah candles for his little son, Ephraim.

chany, dial. var. China[1] II (in Dict. and Suppl.).

chapao, var. Chappow.

1894 Mrs. Dyan *Man's Keeping* ix, He grew accustomed to the frequent chapoas [*sic*], or night attacks. **1902** *Nature* 28 Aug. 418/2 The 'chapaos' of the Baluchis have been checked.

chaparejos (ʃæpărē·hŏs, tʃ-), *sb. pl. U.S.* Also chaparajos, chaperajos Var. *chaparreras. Hence **chapare·joed** *a.*, wearing chaparejos.

1887 *Outing* (U.S.) X. 115/1 We had all discarded our *chaparajos*, and the horses were lightly blanketed. **1888** Roosevelt in *Century Mag.* Feb. 505/2 The broad hat, huge blunt spurs, and leather *chaperajos* of the rider. **1903** *Forest & Stream* 21 Feb. 147 A bit farther on we saw some cowpunchers, or what seemed such, for they sat in cow saddles and wore chaparejos. **1921** *Chambers's Jrnl.* May 321/1 The cowboy, lolling with his hands tucked under his chaparejos. *Ibid.* 322/1 A weird assortment of humanity filled the spacious room beyond—humanity bristling and *chaparejoed*, or *chaparejoed* and pomaded.

chaparreras (ʃæ:pārēᵊ·răs, tʃ-), *sb. pl. U.S.* Also **chapar(r)eros.** [Mexican Sp.] Stout leather trousers worn by cowboys and others to protect the legs esp. while riding through chaparral. Freq. abbrev. *chaps.

1861 E. B. Tylor *Anahuac* 335 *Chaparreros*, over-trousers of goatskin with the hair on, used in riding. **1909** 'O. Henry' *Roads of Destiny* 95 Lonny is one of them, a knight of stirrup and chaparreras.

chapata, chapati, chapatti, chappati, varr. Chupatty.

1858 Simmonds *Dict. Trade*, *Chapata* (Hindustani), a thin unleavened cake of flour and water, toasted or baked over a fire. **1898** *Westm. Gaz.* 21 Feb. 4/3 All histories of the Mutiny relate that about the time of the outbreak mysterious 'chapatis' began to circulate among the people. **1965** E. Linton *World in Grain of Sand* ii. 19 The staple diet of dal and chapattis (beans and unleavened bread). **1970** *Guardian* 8 Apr. 12/3 Young ladies have been dispensing tea and chappatis to hungry comrades.

chapel, *a. colloq.* [f. the sb.] Belonging to, or attending regularly, a chapel (sense 4).

1946 J. Cary *Moonlight* 234 Mrs. Wilmot was chapel. I'm sure she never went to a play. **1957** R. W. Zandvoort *Handbk. Eng. Gram.* viii. 275, I suppose you're Church of England? *I'm* chapel. **1965** *New Statesman* 23 Apr. 646/3 Half the parish might be pagan and the zealous few, Chapel.

‖**chapelle ardente** (ʃapel ardɑ̃nt). [F., lit. 'burning chapel'.] A chamber prepared for the lying-in-state of a great personage, and lit up with candles, torches, etc.

1824 J. Mackintosh *Jrnl.* 28 Sept. in R. J. Mackintosh *Life* (1835) II. 415 Went to see the 'Chapelle Ardente' at St. Denis, where the king's remains lie in state. **1905** A. Lang *Adv. among Books* 264 A wonderful scene with a corpse and a *chapelle ardente*. **1970** *N.Y. Rev. Bks.* 29 Jan. 8/2 It seemed more suitable that she should be thought to have died in Paris, that she lie there in *chapelle ardente* for her friends, and that there should be a proper Paris funeral.

chaperonless (ʃæ·pĕrǫnlĕs), *a.* Also **chapronless, chaperone-less.** [f. Chaperon *sb.* + -less.] Without a chaperon.

1831 *Examiner* 5/2 Travelling and voyaging together, *solus cum sola*, chapronless. **1884** [in Dict.]. **1900** G. Swift *Somerley* 44 We see a girl, say, out boating, chaperonless, with a man. **1906** Galsworthy *Man of Property* II. ii, This evening..was stolen, chaperone-less.

Chaplinesque (tʃæ:pline·sk), *a.* [f. the name *Chaplin* (see below) + -esque.] Resembling or characteristic of the comedy or style of the English-born film actor and producer Charles Spencer ('Charlie') Chaplin (born 1889).

1921 G. B. Shaw *Pen Portraits* (1932) 262 The Chaplinesque invention of Simon of Nantua and the *papegai*. **1956** *Ann. Reg. 1955* 364 The atmosphere of Chaplinesque pathos. **1960** *Times* 29 Apr. 16/1 A Chaplinesque study of the little man contriving to preserve his pathetically humble individuality. **1961** J. McCabe *Mr. Laurel & Mr. Hardy* (1962) vii. 145 Ill-fitting, somewhat Chaplinesque clothes. **1963** A. Ross *Australia 63* v. 113 His jerky Chaplinesque gait.

chappal (tʃæ·pĕl, tʃa·p-). Pl. **-s, chapplis, -ies.** [Hindi.] An Indian sandal, usu. of leather.

1893 E. F. Knight *Where Three Empires Meet* 52 Like many Englishmen who visit Kashmir, I soon became converted to the native grass shoes and *chaplies*. **1936** J. Nehru *Autobiogr.* xxx. 223 A new pair of *chappals* (leather sandals). **1966** J. & R. Godden *Two under Indian Sun* vii. 156 Chapplis, the Kashmiri leather sandal that has an inner laced boot of soft chamois leather. **1968** H. R. F. Keating *Inspector Ghote hunts Peacock* iv. 45 A pair of Indian-made *chappals*, the imprint of the toes of the girl..worn deep into the..dark leather. **1969** *Femina* (Bombay) 26 Dec. 15/2 What I really did was to wear my old Hawaii chappals, with the sari trailing two inches on the ground—very graceful.

chappie, -y. (Later examples.)

1919 Wodehouse *My Man Jeeves* 14 Chappies introduced me to other chappies, and so on and so forth, and it wasn't long before I knew squads of the right sort. **1925** —— *Carry on Jeeves!* iv. 84 It was one of those jolly, peaceful mornings that make a chappie wish he'd got a soul or something. **1968** *Melody Maker* 30 Nov. 6/4 But the Czech chappies seem to think you're rather excited.

chapras(s)i, var. Chuprassy.

1879 Mrs. A. E. James *Indian Househ. Managem.* 44 The touts, or *chaprassies*, from the various hotels, will take your baggage. **1925** E. S. Jones *Christ of Indian Road* x. 210 They sent out the notices through government chaprasis, or runners. **1957** J. Masters *Far Mt. Peak* 81 The post-office *chaprassi* brought your telegram.

chaps (ʃæps, tʃ-), *sb. pl.* Short for *chaparreras.

1884 E. W. Nye *Baled Hay* 139 'Chaps', as they are vulgarly called,..are made of leather with fronts of dog-skin with the hair on. **1884** W. Shepherd *Prairie Exper.* 41 The cow-boys, with their *schaps*, i.e. leather-leggings. **1930** E. Ferber *Cimarron* xii. 195 He [*sc.* the cowboy] wears chaps, with the hairy side out, to keep his legs warm in winter and to protect them from being torn by chaparral and cactus thorns in summer.

chaptalize (tʃæ·ptăləiz), *v.* [f. the name of the inventor of the process, J.-A. *Chaptal* (1756–1832), French chemist + -ize.] *trans.* To correct or improve the must, in wine-making, by neutralizing an excess of acid and adding sugar. So **cha:ptaliza·tion**, this method or process.

1891 *Cycl. Temp. & Prohib.* (U.S.) 647/1 Chaptalization.., by which an excess of acidity is neutralized by adding marble dust, and the quantity and alcoholic strength is increased by means of cane or starch sugar. **1895** Sadtler *Hand-bk. Industr. Chem.* (ed. 2) 204 Chaptalization consists in neutralizing the excess of acidity in the must by the addition of marble-dust.

chapter, *sb.* Add: **5. b.** A branch of an organization or society, esp. of a college fraternity. *U.S.*

1815 D. Drake *Cincinnati* 166 A Chapter of Royal Arch Masons was established in this place. **1871** L. H. Bagg *4 Years at Yale* 55 Sigma Eps at Yale, calling itself the 'Kappa' chapter, established a branch 'Alpha' chapter at Amherst. **1964** *Listener* 30 July 149/2, I have instructed our chapters in the South, even in the Deep South, to begin immediately testing all places of public accommodation.

chapterman (tʃæ·ptəɹmæn). [f. Chapter *sb.* 4 a + Man *sb.*[1]] A member of the chapter of a monastic order.

1844 *Orthodox Jrnl.* 16 Mar. 175 The rev. deceased was a Chapterman. **1926** C. Butler *Life Bp. Ullathorne* I. v. 124 The General Chapter..conferring on him a titular office that made him a chapterman for life.

char, *sb.*[2] Examples of the Fr. word *char* used in English-language contexts in the nineteenth century.

1802 H. Eckersall in *Trav. Diaries of Malthus* (1966) 296 We..were much entertain'd with the Philosophic discource of the driver of our char on the over-population of his country [*sc.* Switzerland]. **1816** Shelley *Let.* 22 July (1964) I. 357 It is possible to accomplish the rest of the journey in a *char du pays*. **1817** H. C. B. Campbell *Journey to Florence* (1951) 57 As there are few or no travellers in winter it is then that they go [to church] occasionally in a char. *a* **1828** D. Wordsworth *Tour Cont.* in *Jrnls.* (1941) II. 150 Our cloaks and coats..he had cunningly locked up in a seat of the Char. *Ibid.* 267 Our gentlemen..hired a char-a-banc with two horses...Two Shopkeepers..had helped them to procure the char. **1868** G. M. Hopkins *Jrnl.* 10 July (1959) 171 At Alpnach we took a *char* to Lungern.

char, *sb.*[4] Add: (Later examples.) spec. = *bone-black* (Bone *sb.* 17).

1887 *Century Mag.* Nov. 113/2 The 'char' must be washed with hot water..and dried in a kiln. **1902** *Encycl. Brit.* XXXIII. 47 Vessels..of such diameter as to hold a given quantity of animal charcoal (also called 'bone-black' and 'char'). *Ibid.*, This..'sweet water' is sometimes..passed through the char filter. **1959** *Times* 3 Jan. 10/1 The coal is heated to a temperature of 400 to 450 deg. C., and the product is known as a char—in distinction from coke, which is formed by heating alone, without air.

char (tʃāɹ), *sb.*[5] *colloq.* Short for Charwoman.

1906 *Daily Chron.* 7 June 3/4 We find him trying to gain a pension for 'Granny Deane', his ancient 'char'. **1919** C. Orr *Glorious Thing* viii. 89, I had to scrub the kitchen today, because the char couldn't come. **1951** *Mind* LX. 487 There is no point in telling one's char to wash the marble halls one dreamt one dwelt in.

char (tʃāɹ), *sb.*[6] *slang.* [Popular spelling of Cha.] Tea.

1919 *Athenæum* 25 July 664/1 *Char* (Hindustani *char*), tea. 'Char up, lads!' *i.e.*, 'Here is the tea!' **1955** H. Spring *These Lovers fled Away* 497, I thought of the thousands of cups o' char that batmen had produced at such moments as this.

char, *v.*[1] Unequivocal examples of the form *char* (see Chare, char *v.* 5), now usual.

1906 *Daily Chron.* 28 Aug. 8/5, I feel grateful to 'My Charwoman', for it is certain she knows her business—she can char. **1925** *Contemp. Rev.* July 86 The scene where Lummox 'chars' in an immoral house.

char-, the first element of Charwoman, used to form words designating persons who do cleaning work, etc., as **char-boy, chargirl, charlady, charmaid, charman.**

1902 *Fortn. Rev.* June 1005 For wages..char-boy £8. **1966** *Punch* 4 May 671/2 Each charboy needs special treatment. **1932** G. Greene *Stamboul Train* III. i. 132 The Hon. Carol Delaine..will play the part of Emmy Tod, the chargirl. **1895** *Westm. Gaz.* 11 Aug. 8/1 She had a good post to offer to the charlady. **1927** E. Bramah *Max Carrados Mysteries* 109 'The old party there'—a comprehensive nod in the direction of the absent charlady. **1959** *Times* 18 Feb. 14/4 West Berliners may use them to pay their East Berlin hair dressers, tailors, and charladies. **1882** *Cornhill Mag.* 337 Little charmaids. **1888** *Pall Mall Gaz.* 10 Oct., The Guards' Memorial is now given over to the charmen. **1895** *Macm. Mag.* July 201/1 The charmen (the men who sweep out the House of Commons).

chara[2] (ʃæ·ră). *colloq.* Short for Char-à-banc. (Cf. *charry.)

1928 F. E. Baily *Golden Vanity* xv, I don't approve of them charas on the Kingswear Road. **1953** *Economist* 14 Nov. 470/1 A flavour of paper hats and football rattles ..clings still about the good old un-English word 'chara'. But now the charabancs have all turned into coaches.

char-à-banc. Add: Freq. with pronunc. (ʃæ·răbæŋ). Also **charabanc, char-à-bancs.** Pl. **charabancs, char-à-bancs,** (*rarely*) **chars-à-banc(s).** In etym., for *char-à-banc* read *char-à-bancs*. Also, a motor-coach. (Earlier and later examples.)

The word *charabanc*, now rarely heard, has been replaced by (*motor-*)*coach*.

1816 Byron *Jrnl.* 17 Sept. (1830) II. 14 One of the country carriages (a char-à-banc). **1820** Ebel & Wall *Trav. Guide Switzerland* 173 The *char-à-banc* is a waggon provided with a long covered bench, on which you may either lie down or sit sideways, as in our long Bath coaches. *Ibid.*, They [*petits-chars*] are lighter and more commodious than the *chars-à-banc*. **1826** J. M. Sherer *Ramble in Germany* 178 On my arrival at Como..immediately took a *charabanc*. **1883** Ld. R. Gower *Reminisc.* II. 225 Our party numbered eight, and in two *chars-à-bancs* we drove all that day about the capital of Japan. **1909** *Daily Tel.* 4 Aug. 10 A char-à-bancs, containing eighteen passengers, being overturned. **1928** R. Campbell *Wayzgoose* ii. 35 Some came in char-à-bancs, and some on nags. **1928** R. Knox *Footsteps at Lock* iii. 23 Charabanc-loads of breezy Midlanders. **1953** [see prec.].

Hence **charabancer** (ʃæ·răbæŋkər), an excursionist who travels by char-à-banc.

1920 *Punch* 11 Aug. 115/2 One of the hardiest 'charabankers' was recently prostrated in that village by a well-aimed epithet from the oldest inhabitant. **1927** *London Mercury* Sept. 451 One fine day perhaps the charabancers awake to the fact that what they came for is no longer there.

characin[2] (kæ·răsin). [f. Gr. χάραξ a pointed stake or a type of fish.] A member of the family Characidæ of freshwater fishes of South America and Africa.

1882 Jordan & Gilbert *Fishes N. Amer.* 111 The Characins resemble the *Isospondyli* in general appearance and in the structure of the pharyngeals. **1905** D. S. Jordan *Guide to Study of Fishes* II. iv. 61 A structure also retained in Characins and catfishes. **1957** P. J. Darlington *Zoogeography* ii. 109 Characins are small to medium-sized fishes of diverse habits and include the vicious piranhas and many aquarium fishes. **1968** R. D. Martin tr. *Wickler's Mimicry in Plants & Animals* xv. 219 A very clear instance of deception exists in the male swordtail characin *Corynopoma riisei*. This well-known ornamental fish is an inhabitant of Venezuelan rivers and belongs to the large family of characins (Characidae; sub-family Glandulocaudinae).

character, *sb.* Add: **3. c.** *Computers.* One of a set of letters, digits, or other symbols which can be read, stored, or written by a computer and used to denote data; also, a representation of such a symbol by means of a small number of bits, holes in punched tape, etc., arranged according to a specified code and taken as a unit of storage.

Quot. 1949 belongs strictly to sense 3 a.

[**1949** *Math. Tables & Other Aids to Computation* III. 428 The word length of both numbers and orders in this machine is 44 binary characters.] **1950** *Ibid.* IV. 32 Decimal digits may be recorded as 4-digit binary code groups, and alphabetical characters may be recorded as 5- or 6-digit binary groups. **1958** Gotlieb & Hume *High-Speed Data Processing* ix. 187 The need for inserting or deleting characters occurs most frequently in editing information for output or for transfer from one medium to another. **1967** Cox & Grose *Organiz. Bibl. Rec. by Computer* iv. 97 The searchable section may be up to 12,000 characters long. **1969** B. Hodge *Computers for Engineers* iii. 76 During reading, the block begins with the first character sensed after a gap and continues without interruption until the next gap is reached.

8. b. See also *acquired character* s.v. *acquired *ppl. a.* (*c*).

16. b. *colloq.* A person, man, fellow (freq. slightly derogatory: cf. sense 18 in Dict.).

1931 D. Runyon *Guys & Dolls* (1932) 15 Marvin Clay is a most obnoxious character. *Ibid.* 75 The paymaster must be a very dishonest character. **1938** —— *Take It Easy* 16 Tammany Hall is greatly dissatisfied with the character who already represents it in Congress. **1946** J. Richardson *Phoney Phleet* 150 So widely was this Swordfish known That characters could not be found To drive it. **1954** W. W. Tarn in E. Barker et al. *Europ. Inheritance* I. 163 Archilochus of Paros, an unpleasant character who started the fashion of writing about oneself. **1962** J. Cannan *All is Discovered* iv. 88 The character who owns *Mab*..leaves his gear out in her.

17. a. (Earlier examples.)

1664 Dryden *Rival Ladies* Epist. ded., He may be allow'd sometimes to Err, who undertakes to move so many Characters and Humours as are requisite in a Play. **1668** —— *Dram. Poesie* 53 Besides Morose, there are at least 9 or 10 different Characters and humours in the *Silent Woman*, all which persons have several concernments of their own. **1674** T. Rymer *Pref. to Rapin* sig. A 4 Many of his Characters have but little of the Heroick in them; Dalga is a Jilt, proper onely for Comedy [etc.]. **1679** Dryden *Troil. & Cress.* Pref. sig. a 4 The chief character or Hero in a Tragedy..ought in prudence to be such a man, who has so much more in him of Virtue than of Vice...If Creon had been the chief character in *Œdipus* [etc.].

19. *character-formation, -study, -trait; character-building, -forming, -making, -moulding, -reading, -training* sbs. and adjs.; **character-actor**, add def.: an actor who specializes in character parts; so *character-act* v., *character-acting, -actress*; **character assassin**, one who destroys the reputation of another person; so **character assassination**; **character comedian**, a character-actor specializing in comic roles; **character dance**, (*a*) *Ballet*, a dance which interprets a real or imaginary personality; (*b*) a characteristic national dance; so *character dancer, character dancing*; **character part**, an acting role displaying pronounced or unusual characteristics or peculiarities; **character-sketch**, a brief description of a person's character; so *character-sketching*; **character-structure**, the various traits in a person's character, seen as forming a system in which some are more important than others.

1895 G. B. Shaw *Our Theatres in Nineties* (1932) I. 106 Mr Volpé..is a credible and natural Brixton paterfamilias, and does not 'character-act'. **1878** J. Knight in *Athenæum* 16 Mar. 359/2 What is known as character acting has definitely established its supremacy in England upon the ruins of tragic art. **1895** G. B. Shaw *Our Theatres in Nineties* (1932) I. 197 She..allows her comic-song singing to decay into mere seventh-rate character acting. **1926** *New Republic* 12 May 362 Miss Cadell is a character actress in the English school. **1933** P. Godfrey *Back-Stage* xvi. 197 The general manager's wife was a character actress of no great distinction. **1951** in *Amer. Speech* XXVI. 293/2 A period o. 'the big lie', of the furtive informer, of the character assassin. **1949** *Amer. Sociol. Rev.* Feb. 17 Gossip..is the principal means of communication within and between the castes, with character-assassination and ego-expansion being the principal topics. **1951** H. MacInnes *Neither Five nor Three* vi. 92 Character assassination, they call it...If they can't find out anything against a man, they invent it. **1958** *Listener* 2 Jan. 13/1 The effects of these 'character assassinations' have been disastrous..on the willingness of scientists to work for the government. **1886** E. G. White *Pract. Addresses* in *Hist. Sk. Seventh-Day Adventists* 134/2 We should make daily advancement in the work of character-building. **1900** J. Clifford in *Daily News* 28 Nov. 4/4 Character-building literature. **1925** W. Deeping *Sorrell & Son* vii. 69 They call it 'forming your mind'—character building. **1944** R. Lehmann *Ballad & Source* 178, I felt the ominous subject of character-building begin to breathe its threat upon me. **1898** W. Archer *Theatr. World of 1897* 332 Mr Sydney Valentine's Pew was the powerful, picturesque and humorous embodiment which was only to be expected of one of the very best of our character-comedians. **1911** J. E. C. Flitch *Mod. Dancing* iii. 42 Her character dances.. were new and graceful. **1922** Beauclerk & Evrenov tr. *Svetlov's T. Karsavina* 44 A series of dances, either classical or combined with the 'character' dance. **1943** K. Ambrose *Ballet-Lover's Pocket-Bk.* 42 A character-dance is usually taken to mean a dance in which *les pointes* are not used: or a national dance of some kind. **1899** *Cosmopolitan* Nov. 30 Spanish character-dancer. **1936** A. L. Haskell in 'C. Brahms' *Footnotes to Ballet* I. 11 There is the character dancer who may specialise in the miming of comic rôles or in villainous miming (usually also comical) or in national dancing. **1928** —— *Some Stud. Ballet* 135 Such a ballet shows Massine at his best, true to tradition, yet decidedly original, blending classical and character dancing in a manner entirely his own. **1944** H. Read *Educ. Free Men* vi. 22 This gives the teacher his only possibility for..'character-formation'. **1903** Abp. Cant. in *Westm. Gaz.* 22 Aug. 3/2, I want the fullest assurance I can get that one who is to help the children..regards the Book as..far superior in character-forming power to the creeds of the Churches. **1910** *Daily Chron.* 25 Jan. 7/4 This..character-forming state can be keyed up by passion into an everlasting harmony. **1921** Character-making [see *Africanoid *a.*]. **1877** *New Dominion Monthly* I. 183 It is only character-moulding,..for which we demand neither preparation nor a knowledge of the business. **1907** *Daily Chron.* 27 Aug. 3/3 The restricting and character-moulding nature of environment. **1811** in *Dram. Censor for 1811* (1812) 481 (Advt.), *For the Ballet.*—Monsieur Didelot, Ballet-Master, and occasionally to perform Character Parts. **1866** Character part [in Dict.]. **1939** T. S. Eliot *Old Possum's Bk. Pract. Cats* 37, I'd a voice that would soften the hardest of hearts, Whether I took the lead, or in character parts. **1892** 'Mark Twain' *Amer. Claimant* 217 The difficult science of character-reading. **1924** V. Woolf *Mr. Bennett* 4 It would be impossible to live for a year without disaster unless one practised character-reading. **1885** *Eng. Illustr. Mag.* May 515 (*heading*) Wolseley: A Character Sketch. **1897** Lady Mary Loyd & F. Simmonds tr. *Gréard's Meissonier* 71 His charming character-sketch of this worthy is almost a water-colour. **1888** *Atalanta* I. 413/2 It is full of bits of character-sketching. **1936** *Discovery* Dec. 398/1 Those who like a seriously-written book with plenty of character-sketching will enjoy it. **1949** M. Mead in M. Fortes *Social Struct.* 20 The character structure of the individuals. **1962** R. Fine *Freud* v. 67 Character structure is built on the modes in which the biologically determined instincts are handled. **1894** W. Archer in *World* 5 Sept. 11/2 A more searching character-study than he has hitherto attempted. **1898** *Daily News* 27 June 4/7 The social side of character-training is to be attempted. **1924** tr. *Freud's Coll. Papers* II. iv. 50 The permanent character-traits are either unchanged perpetuations of the original impulses, sublimations of them, or reaction-formations against them. **1962** *Philosophy* XXXVII. 38 Character-traits are shown in the sort of things a man can *decide* to be.

characterful (kæ·ræktəɹfŭl), *a.* [f. Character *sb.* + -ful.] Full of character; strongly expressive of character.

1901 S. Dark *Stage Silhouettes* 87 Archer..is a tallish man, with a military walk and bearing, and a strong, characterful face. **1901** W. J. Locke *Usurper* ii. 25 The writing was dainty, feminine, characterful. **1958** *Times Lit. Suppl.* 17 Jan. 35/4 Those who consider Siena one of the most beautiful and characterful of Italian cities.

characteristic, *a.* and *sb.* Add: **A.** *adj.* **1. c.** *characteristic curve*: a graph showing the relationship between two variable but interdependent quantities; *spec.* (see quot. 1955).

1902 *Encycl. Brit.* XXVIII. 90/1 Curves delineating the relation of these variables for any incandescent lamp are called its characteristic-curves. **1922** Glazebrook *Dict. Appl. Physics* II. 147/2 *Characteristic Curves*, curves showing the relation between current and voltage for any electrical apparatus. **1923** [see *bromide 4]. **1955** *Gloss. Terms Radiology* (*B.S.I.*) 55 *Characteristic curve*, a curve showing the relation between the common logarithm of exposure..and the photographic density under specified conditions of processing. **1957** R. W. G. Hunt *Reprod. Colour* iv. 71 The characteristic curve (density plotted against log exposure) of a typical reversal colour photographic process is shown.

d. *characteristic impedance*, 'the impedance of a uniform alternating-current transmission line of indefinite length (as a long telephone cable) measured at the input end where the voltage is applied' (Webster 1961).

1923 *Bell System Techn. Jrnl.* II. ii. 1 The transmission of alternating currents over any transmission line..depends only on the propagation constant and the characteristic impedance. **1929** E. Mallett *Telegr. & Teleph.* viii. 178 Instead of a characteristic resistance..there will be a complex characteristic impedance. **1940** *Chambers's Techn. Dict.* 153/1 *Characteristic impedance of an antenna* (*Radio*), the impedance with which certain forms of antennae..must be terminated to prevent the formation of standing waves. *Characteristic impedance of a line* (*or filter*) (*Elec. Comm.*), the apparent impedance of an infinitely long line at any point, or of an infinite number of filter sections.

e. *Nuclear Physics.* Designating radiation the wavelengths of which are peculiar to the element which emits them.

1908 Barkla & Sadler in *Phil. Mag.* XVI. 571 A characteristic homogeneous radiation was emitted by a metal when the primary beam to which the metal was exposed was of more penetrating type than the characteristic radiation. **1912** W. H. Bragg *Stud. Radioactivity* xiv. 156 It is only when the β corpuscle possesses energy in excess of the same value that it is able to excite the characteristic X radiation of the atom. *Ibid.* xv. 175 Let K_1 be the absorption coefficient of the same plate for its own characteristic rays. **1938** R. W. Lawson tr. *Hevesy & Paneth's Man. Radioactivity* (ed. 2) v. 63 The characteristic rays of an element are related in a simple manner to the atomic number. *Ibid.* 65 The hardest characteristic radiation known, viz. the K-radiation of uranium. *Ibid.* viii. 87 The characteristic X-ray spectrum.

B. *sb.* **4.** = *characteristic curve.

1902 *Encycl. Brit.* XXVII. 587/2 A curve connecting together corresponding values of external current and terminal voltage for a given speed of rotation is known as the external-characteristic of the machine...The 'external characteristic' for a shunt machine has an entirely different shape. **1918** V. H. Laughter *Operator's Wireless & Telephone Handbk.* xi. 177 *Characteristic, Dynamic, of a Conductor*..the relation given by the curve obtained when the impressed E.M.F.'s are plotted as ordinates against the resultant currents as abscissas, both E.M.F.'s and currents varying at the given frequency and between the given extremes. **1930** *Engineering* 6 June 748/2 The boiler feed pumps were all rotary pumps and had a very flat characteristic. **1958** Amos & Birkinshaw *Television Engin.* II. i. 31 One possible form for the phase-frequency characteristic of a practical amplifier.

characterology (kæ:ræktərǫ·lŏdʒi). *Psychol.* [f. Character *sb.* + -ology; tr. G. *Charakterologie* (J. Bahnsen *Beiträge zur Ch.*, 1867).] The science of character, esp. of its development, types, and individual differences. (Cf. Ethology 3.) Also *transf.*

1903 *Amer. Jrnl. Psychol.* July–Oct. 356 An iconoclastic attitude towards all attempts at practical characterology. **1918** J. Ward *Psychol. Princ.* xviii. 431 Characterology.. is coming to be generally used; and so, though at present unfamiliar to English ears, we may..adopt it. **1920** L. H. McCormick (*title*) Characterology; an exact science. **1953** W. J. Entwistle *Aspects of Lang.* ix. 276 The sum of such characteristics should offer the 'characterology' of a language. **1963** Hoenig & Hamilton tr. *Jaspers's Gen. Psychopath.* viii. 429 Characterology ends at the lower levels of psychic life, where the self-aware personality also ends.

Hence **cha:racterolo·gical** *a.*, of or pertaining to characterology; **cha:racterolo·gically** *adv.*; **cha:ractero·logist**, a student of, or specialist in, characterology.

1916 R. Steiner *Philos. of Freedom* x. 170 This individual make-up we will call, following Eduard von Hartmann, the 'characterological disposition'. **1927** A. A. Rodack *Psychol. of Char.* ii (*heading*) The Literary Characterologists. **1948** Berg *Clin. Psychol.* xvii. 219 Normal persons, characterological cases and psychotics. **1958** *Listener* 12 June 976/1 Characterologists like Freud, La Bruyère, and Theophrastus. **1963** *Arch. Gen. Psychiatry* VIII. 320/1 Sometimes envy erupted in a person not characterologically envious.

char-a-plane, var. *chairoplane.

charas, var. Churrus.

1875 *Encycl. Brit.* III. 627/2 In India the products of the [hemp] plant for use as a narcotic and intoxicant are recognized under the three names and forms of Bhang, Gunja or Ganja, and Churrus or Charas. **1930** *New Statesman* 18 Oct. 58/2 Charas—dried hemp leaves from Turkestan—is sold in the open market of Bombay. **1967** M. Reynolds *After Some Tomorrow* 39 While under the effect of..charas fudge, did you ever experience one of your prophetic..enlightenments?

charbon. Add: **3.** A fungoid disease incident to the vine, and to orange and lemon trees.

1882 *Encycl. Brit.* XIV. 437/2 In damp valleys it [*sc.* the lemon] is liable to be attacked by a fungus called 'charbon' (*Dematium monophyllum*). **1888** *Ibid.* XXIV. 240/1 Anthracnose is the name usually given to a disease which was formerly known as 'charbon'.

charcoal, *sb.* Add: **1. d.** = *charcoal grey. Also *attrib.* orig. *U.S.*

1952 *Women's Wear Daily* 16 Oct. 3 Charcoal is the choice for a gored skirt sundress with one of the new little triangle cover-tops. **1953** *New Yorker* 13 June 86/2 White cotton drill striped in pink, blue, or charcoal makes a cardigan jacket. **1955** *Wall Street Jrnl.* 25 Feb. 17/3 The charcoal trend in suits will extend to dark blues and greens next fall. **1958** M. Dickens *Man Overboard* i. 14 Ben was seeing himself in a narrow-trousered charcoal suit, entertaining Rose on an expense account.

6. charcoal biscuit, a biscuit containing wood-charcoal as an anti-fermentative, absorbent, or deodorizer; **charcoal brown**, a dark brown colour; freq. *attrib.*; **charcoal grey**, (*a*) a dark grey powder or pigment made from charcoal; (*b*) a dark grey colour.

1885 *Army & Navy Co-op. Soc. Price List* 759 Bragg's Charcoal Biscuits. **1925** Charcoal biscuit [see *digestive A. 2 b]. **1959** 'Ed McBain' *Pusher* ii. 16 He was..dressed now in a brown sharkskin suit and charcoal-brown overcoat. **1907** *Yesterday's Shopping* (1969) 471/1 Water Colours ..Charcoal grey. **1934** H. Hiler *Notes Techn. Painting* ii. 90 *Greys*. Charcoal grey, Davy's grey, [etc.]. **1949** *Dict. Colours Int. Decoration* (*Brit. Colour Council*) III. 5/2 *Charcoal grey*,..a descriptive colour name introduced into seasonal ranges by B.C.C. in 1942, to denote a neutral grey. **1963** *Times* 27 Feb. 6/4 An easy-to-wear blouse style garment in charcoal grey.

‖**charcuterie** (ʃarkütri). [Fr., f. †*char* (mod. Fr. *chair*) *cuite* cooked flesh.] Cold cuts of meat, esp. pork, ham, sausages, etc. Also, a shop that sells goods of this kind. Cf. next.

1858 Simmonds *Dict. Trade*, *Charcuterie*,..dressed hams, cold meats and fowls, tongues, sausages, etc. **1903** 'A. McNeill' *Egregious Eng.* 152 Inferior cuts of butcher's meat with charcuterie and dried fish thrown in. **1958** J. Cannan *And be a Villain* viii. 199 Spying on me from the yard of the charcuterie shop. **1962** *Punch* 14 Mar. 429/2 One can always take sanctuary in the nearest *charcuterie*.

‖**charcutier** (ʃarkütye). Fem. **charcutière** (-yę̄r). [Fr., see prec.] A pork-butcher; a preparer or vendor of charcuterie.

1894 G. du Maurier *Trilby* I. iii. 238 Everything that French *charcutiers* and their wives can make out of French pigs. **1906** *Daily Chron.* 10 Sept. 4/6 Charcutiers and preparers of cooked viands. **1921** *Dict. Occup. Terms* (1927) § 770 Butcher, pork; charcutier. **1963** *Punch* 18 Sept. 426/1 The *charcutière*..told him to go elsewhere if he didn't like her business methods.

Chardonnet (ʃardo·nę). The name of the French inventor, Hilaire de *Chardonnet* (1839–1924), of a process for producing an artificial silk from a nitro-cellulose substance, used *attrib.* in *Chardonnet process, silk*.

1893 *Jrnl. Soc. Arts* 12 May 649/1 The Chardonnet process of making silk from wood-pulp. *Ibid.* 650/2 Reducing the inflammability of the Chardonnet silk. **1926** *Chambers's Jrnl.* 725/2 The nitro-cellulose or Chardonnet process. **1963** A. J. Hall *Textile Sci.* ii. 44 Around 1890..Chardonnet silk was being made on the large scale from cotton by a process of nitration.

charge, *sb.* Add: **3. c.** *slang.* A dose or injection of a drug; marijuana, esp. a marijuana cigarette. orig. *U.S.*

1929 C. G. Givens in *Sat. Even. Post* 13 Apr. 54/4 An addict..is a bangster, and a bang is a load, a charge or a hyp of the drug he uses. **1957** C. MacInnes *City of Spades* I. v. 21 'I saw you grew charge out there...' 'You want to smoke some?' **1958** F. Norman *Bang to Rights* 174 Quite a lot of spooks get done for takeing charge. **1959** *Streetwalker* iv. 68 Relaxing under the influence of charge, marijhuana. **1969** *Melody Maker* 13 Sept. 6 Club promoters are worried that hippies could close them down by smoking charge on the premises.

d. *U.S. slang.* A thrill; a feeling of excitement or satisfaction; a 'kick'; esp. in phr. *to get a charge out of*.

1951 S. J. Perelman in *New Yorker* 3 Mar. 28/3 What kind of an old creep'd get a charge out of this stuff? **1959** W. Brown *Cry Kill* x. 108 He was getting a charge out of being close to something big. **1963** *N.Y. Times Mag.* 5 May 117 (Advt.), It seems to me that people get a bigger charge out of their grandchildren than they did from their own offspring.

5. (Earlier example.) Also used of the electrical property of particles of matter.

1767 J. Priestley *Hist. Electr.* 527 A full charge of two or three thousand square feet of coated glass would give a shock as great as a single common flash of lightning. **1891** [see *electron[2]]. **1927** Sidgwick *Electronic Theory of Valency* 11 The charge on the nucleus—the atomic number —determines the number of electrons which surround it. **1955** W. Pauli *N. Bohr* 32 The charge-density for spin ½-particles is positive-definite in the *c*-number theory. *Ibid.*, The postulate of charge-symmetry. **1956** G. Thomson *The Atom* (ed. 5) i. 6 A proton has the same kind of charge as a positively charged body.

10. b. (Earlier U.S. example.)

1817 S. R. Brown *Western Gaz.* 90 The taverns are well kept, and charges reasonable.

13. a. Also, *to take charge*: colloq. (of a thing) to get out of control and act automatically, esp. with disastrous or destructive effect.

1890 *Daily News* 16 June 3/8 A seven-inch steel breech-loading gun 'took charge' in firing, and both gun and carriage were blown off the platform. **1897** *Ibid.* 23 Feb. 7/6 A large party of men were pulling the tackle, when suddenly the spar 'took charge', and swung back.

20. charge account *N. Amer.*, a credit account at a store, etc.; **charge-book**, a book containing the statements of the charges brought against prisoners in a police court; **charge engineer**, the engineer in charge of the engines and machinery at a power station, etc.; **charge-hand**, a workman, in various trades, who is in charge of a particular piece of work; **charge-house**, (*a*) a building in which prisoners are detained on a certain charge; (*b*) a workshop in which explosive is loaded into shells, etc., in an explosive factory, also *attrib.*; **charge-man**, also, a workman who controls the supply of materials to a furnace, machine, etc. in a workshop or factory; **charge-nurse**, a nurse who has charge of a ward in an infirmary or hospital; **charge-room**, the room, at a police-station, in which the charge against an arrested person is made and entered in the charge book or sheet; **charge-sheet** (earlier example).

1903 *Red Book* July (Advt. Section) 6/1 No security is required..when you buy on our confidential charge account system. **1909** 'O. Henry' *Options* (1916) 96 I've got more power here than..a charge of dynamite, and a charge account at Tiffany's combined. **1962** J. Ludwig in R. Weaver *Canad. Short Stories* 2nd Ser. (1968) 256 Shirley sneaked ties to Maxie, money, let him put things on Jimmy's charge accounts. **1890** *Daily News* 18 Sept. 2/5 Sir Andrew Lusk made an entry in the charge-book of the prisoner's death. **1907** *Install. News* Sept. 16/2 Young gentleman.. desires engagement as charge engineer of private plant. **1916** 'Boyd Cable' *Doing their Bit* iv. 65 We, the managers, foremen, and charge hands of the above factory. **1921** *Dict. Occup. Terms* (1927) § 591 *Painter's foreman*..; charge hand; is in charge of painters on a job. *Ibid.* § 691 *Charge hand* (electricity supply); a turbine or engine driver who is directly responsible to charge engineer for operation of turbine plant, [etc.]. **1922** Turner & Wood *Man. Up-to-date Organisation* 171 Charge hand is an operator having supervision of a section. **1957** *Times* 12 Oct. 6/1 A charge-hand process worker on the Windscale piles. **1961** *Evening Standard* 4 Sept. 19/4 Male Chargehand wanted to take charge of television component coil-winding sections. **1900** *Daily News* 14 Aug. 5/4 Three hundred and eighty suspected persons found themselves lodged in the fort. Another four hundred and fifty..in the charge-house. **1921** *Dict. Occup. Terms* (1927) § 149 *Charge house man* (gunpowder),..wheels bogies (small trucks) of raw materials..between stores and scattered workshops of explosive factory. *Ibid.* § 091 *Charge man*; (i) (patent fuel) is in charge of operations in manufacturing patent fuel; [etc.]; (ii) (salt works) is responsible for regulating supply of steam, brine, etc. *Ibid.* § 279 *Charge man* (copper smelting)..; charges blast furnace, or cupola, .. under direction of cupola furnace-man. **1896** Charge-nurse [see Sister *sb.* 3 d]. **1961** [see *aisle 5]. **1853** Dickens *Reprinted Pieces* (1899) 200 The old Thames Police office is now a station-house, and..the old Court..is a quaint charge room. **1933** P. MacDonald *Myst. Dead Police* i. 3 The other four [men] were all in the Charge Room. **1866** *Harper's Mag.* Jan. 161/2 The next witness who steps into the box is a charge sheet in himself.

charge, *v.* Add: **15. d.** *absol.* or *intr.* To bring a charge.

1891 *Spectator* 2 May, If she can but get the High Court of Justice to charge on her side. **1929** *Publishers' Weekly* 19 Oct. 1935/2 Thompson charged that McAndrew was the 'king's stool pigeon'.

19. a. U.S. examples. Also const. *on*.

1929 W. Faulkner *Sound & Fury* 231 Have you been charging things at stores again? **1947** Steinbeck *Wayward Bus* 209 She could charge her clothes and sign cheques in restaurants. **1966** *New Yorker* 22 Oct. 20 (Advt.), Charge everything from a surfboard to a luau on your American Express Credit Card.

chargé. Add: Also 8–9 **chargé des affaires. 1.** (Earlier and later examples.)

1767 W. Musgrave *Let.* 22 Dec. in *Hist. MSS. Comm.* (1897) 15th Rep. App. VI. 222 And conclude it [the ceremony] will be sent with your Ribband to the Chargé des Affaires at Turin to serve as a precedent for your Investiture. **1768** Chesterf. *Let.* 12 Apr. (1932) VI. 2845 Keith is rather inclined to go to Turin, as *Chargé d'Affaires*. **1819** *Acts U.S. 15th Congr.* 56 The Charge des Affaires at London, the Hague and Stockholm. **1836** Mrs. S. C. Stevenson *Let.* 13 July in E. Boykin *Victoria, Albert, & Mrs. S.* (1957) 17 Mr. Vail's function as Chargé ceases on the day of the Minister's presentation. **1914** E. A. Powell *Fighting in Flanders* iv. 97 The Swedish and Mexican *chargés*. **1966** G. Greene *Comedians* I. iv. 112 The chargé would have disapproved of my cuckolding a member of the diplomatic corps.

2. (Earlier examples.)

1797 Scott *Let.* 8 Mar. (1932) I. 64, I have the honour to be Secretary & Chargé d'affaires to the Royal Corps of Edinr. Volunteer light Dragoons. **1826** F. Reynolds *Life & Times* I. v. 174 To be selected as the family man of business —the *chargé d'affaires* on a grand mission.

charger[2]. Add: **6. b.** A device for loading the magazine of a rifle. Also *attrib.* and *Comb.*, as *charger-loading, system*.

1902 *Encycl. Brit.* XXXII. 651/2 The charger, used in the Mauser system,..is placed over the top of the magazine, and all the cartridges in it pushed by one motion into the magazine. **1904** *Westm. Gaz.* 9 Dec. 7/2 The charger system of loading was also recommended. **1909** in *Hansard's Parlt. Deb.* X. 339/1 The charger-loading rifle. **1958** J. A. Barlow *Elem. Rifle Shooting* (ed. 5) x. 106, I am going to fire ten rounds in two chargers of five.

charging, *vbl. sb.* Add: **b.** *attrib.* and *Comb.* in the names of appliances connected with the charging of a furnace, gas retort, battery, blast-hole, etc., as *charging barrow, current, door, shop, spoon*, etc.

a **1884** Knight *Dict. Mech.* Suppl., *Charging Barrow*, a double-wheel barrow, for use in iron furnaces for conveyance of coal, coke, limestone, and ore, to be dumped in at the tunnel-head of the blast furnace. **1909** *Cent. Dict.* Suppl., *Charging-current*, in *elect.*, the current used in charging a storage battery, condenser, or cable. **1961** *Listener* 9 Nov. 768/2 A cable, say fifty miles long, operating at a pressure of 100,000 volts, would be charged and discharged with every alternation of current, and this charging current, as it is called, would reach such large proportions that a cable this long would become completely impracticable. **1908** *Westm. Gaz.* 6 Aug. 7/3 The charging-door in the furnace. **1895** *Daily News* 25 July 3/5 In one of these erections, known as a charging 'shop'..the explosion happened. **1894** C. Le N. Foster *Ore & Stone Mining* 161 The charging-spoon is a hollow half-cylinder of copper or zinc, at the end of a copper or wooden rod, which is used for introducing loose gunpowder into holes which are more or less horizontal.

chariot, *sb.* Add: **4. b.** A rotating piece of mechanism in a Hughes type-printing telegraph (see quot.).

1876 Preece & Sivewright *Telegraphy* 90 When one of these keys is depressed a pin is raised, which just catches a 'chariot'..rotating with the type wheel, and thus sends a current to the distant station.

c. A part of the mechanism of a stage (see quot.).

1902 *Encycl. Brit.* XXXII. 823 Each section [of the stage] is equipped with what is termed a pair of chariots, to hold 'wing' lights placed on so-called wing ladders.

5. chariot-burial *Archæol.*, the burial of a warrior together with a chariot; **chariot plane**, a small smoothing plane used for internal angles and small work.

1910 *Encycl. Brit.* II. Plate vi, Chariot burial of a Gaulish chief. **1915** E. A. Parkyn *Prehist. Art* 273 Another chariot burial..was discovered..at Somme Bionne. **1909** Wells & Hooper *Mod. Cabinet Work* ii. 9.

chariotee. *U.S.* (Examples.)

1825 *Catawba Jrnl.* (Charlotte, N.C.) 17 May, A coachee, chariotee, phaeton, [etc.]. **1867** T. Lackland *Homespun* I. 128 The grand family wagon,—a cross between a hay-cart and a chariotee.

charioteer, *v.* **2.** (Earlier *transf.* example.)

a **1821** Shelley *Witch of Atlas* (1824) lviii. 460 Naked boys bridling tame water-snakes, Or charioteering ghastly alligators.

charioteering, *vbl. sb.* (Earlier examples.)

1788 Dyott *Diary* Nov. (1907) I. 59, I, in attempting that part of the charioteering, was within an inch of oversetting all the party into the harbour. **1832** *Chambers's Jrnl.* I. 32/2 In vain did his lordship exert all his charioteering skill.

charisma (kări·zmă). Pl. **charismata** (kări·zmătă). [Gr. χάρισμα, pl. -ατα: see Charism.] *Theol.* **a.** = Charism.

a **1641**, **1876** Charismata [see Charism]. **1875** *Encycl. Brit.* III. 823 Schleiermacher was accustomed to say of Bleek that he possessed a special *charisma* for the science of 'Introduction'. **1920** J. A. Robertson *Hidden Romance N.T.* v. 101 He himself had the *charisma*, or spiritual gift of utterance. **1957** N. Micklem *Box & Puppets* iii. 57 In *ex tempore* prayer he had a mysterious *charisma* of the Spirit.

b. *spec.* A gift or power of leadership or authority (see quot. 1947); aura.

1947 Henderson & Parsons tr. *Weber's Theory of Social & Econ. Organization* iii. 329 The term 'charisma' will be applied to a certain quality of an individual personality by virtue of which he is set apart from ordinary men and treated as endowed with supernatural, super-human, or at least specifically exceptional powers or qualities. **1967** *Spectator* 30 June 771/3 Like many of his generation, he succumbs to the Kennedy charisma, identifies himself with his hero.

charismatic, *a.* Add to def.: Possessing or exhibiting a charism or charisma.

1947 Henderson & Parsons tr. *Weber's Theory of Social & Econ. Organization* iii. 329 Charismatic authority. **1958** W. Stark *Sociol. of Knowledge* 263 The achievement of the personally unique 'charismatic' individual. **1963** *Listener* 7 Feb. 228/2 They seem to change the institutions of Western democracy into a kind of tribal form with a charismatic tribal leader.

charitarian. Delete *nonce-wd.* and add later U.S. examples.

1930 *Time* (U.S.) 17 Mar. 38/2 He is not so lavish a charitarian as was Brother-in-Law Bok. **1935** S. Lewis *It can't happen Here* xvii. 180 They offered the charitarians the same dollar a day that the proletarians received.

charity. Add: **1. e.** (Later example.)

1816 Jane Austen *Emma* III. xiv. 257 She was now in perfect charity with Frank Churchill.

10. *charity ball* (earlier example), *bazaar, concert, matinée*.

1857 Geo. Eliot *Scenes Clerical Life* (1858) III. 62 Three assemblies and a charity ball in the winter. **1848** Thackeray *Van. Fair* xxxix. 354 Martha painted flowers exquisitely, and furnished half the charity-bazaars in the county. **1964** C. Mackenzie *My Life & Times* III. vi. 237, I was to play the third part and it was to be given at some charity bazaar. **1864** M. B. Chesnut *Diary* 21 Sept. (1949) 435 At the charity concert, here were the whole party. **1969** C. Carfax *Silence with Voices* xvi. 115, I sold her a programme at a charity concert. **1921** G. B. Shaw *Let.* (1952) 220 Gertrude Kingston..is going to play Catherine at a charity matinee.

charivari, *v. U.S.* [f. the sb.] *trans.* To greet or serenade with a charivari. Cf. Shivaree *v.*

1890 in *Amer. Speech* XXI. 176 The four came to town.. for the purpose of charivaring a friend recently married. **1900** Smithwick *Evol. State* 72 The boys went..to charivari them.

charka (tʃɐ̄·ɹkă). Also **charkha, churka.** [Hind. *charkhah, charkhā* spinning-wheel (Skr. *cakrá* Wheel).] A roller cotton gin much used in India.

1880 Markham *Peru* 122 A machine..very like the Indian *churka*. **1888** *Encycl. Brit.* XXIV. 730/2 Such was the first spinning wheel which, in the form of the 'charka', has long been known in the East Indies. **1926** *Spectator* 6 Feb. 214/2 Gandhi..bade India spin cotton on the *charka*. **1940** M. Masani *Our India* viii. 102 If in the bulk of the homes of our cultivators there were a spinning-wheel (*charkha*) or a hand-loom.., they could satisfy the market which at present invites foreign cloth.

charlady: see *char-.

Charleston (tʃā·ɹlstən, -lz-). orig. *U.S.* Also **charleston.** [f. *Charleston*, the capital city of Charleston county, S. Carolina, U.S.] A ballroom dance characterized by side kicks from the knee. Hence **Cha·rleston** *v.intr.*, to dance the Charleston; **Cha·rlestoner**; **Cha·rlestoning** *ppl. a.* and *vbl. sb.*

1923 C. Mack & J. Johnson (*title of song*) Charleston. **1925** *N.Y. Times* 26 July 23/2, I have no objection to a person dancing their feet and head off,..but I think it best that they keep away from the Charleston. **1926** *Glasgow Herald* 7 June 8 In these days of fox trots and Charlestons it is refreshing to watch such clean dancing. **1927** *Punch* 9 Feb. 168/1 Charlestoning nymphs. **1927** *Daily Mail* 7 Apr. 10/7 Charlestoners can Charleston quite as well to it. **1928** D. H. Lawrence *Phoenix II* (1968) 525 'Do you Charleston?' she said. **1958** B. Nichols *Sweet & Twenties* xi. 139 The Charleston and Rock 'n Roll; is there so much difference after all? **1964** *Punch* 20 May 762/3 A world of knobbly knees and Charlestoning.

Charley, Charlie. Add: **4.** *pl.* A woman's breasts. *slang.*

1874 in Hotten *Slang. Dict.* **1940** 'H. Green' *Pack my Bag* 231 Foxes are known as Charleys and so in some circles are women's breasts. I made some joke about how many people went after Charleys. **1947** D. M. Davin *Gorse blooms Pale* 200 An Iti bint..with bonzer black eyes and nice charlies. **1957** P. Wildeblood *Main Chance* 199 Carrying her famous bosom before her like the tray of an usherette, she was disconcerted to hear..a nasal cry of: 'Coo, look at them charlies!'

5. *Army slang.* (See quot.)

1919 *War Terms* in *Athenæum* 1 Aug. 695/1 'Old Charlie', the pack—an Old Army phrase now dying out, I think. *Ibid.* 8 Aug. 727/2 The infantryman's pack was his 'Charlie', his haversack was a 'Young' or 'Little Charlie'.

6. A fool, simpleton, esp. a *proper, right Charley. slang.*

1946 *Amer. Speech* XXI. 238/1 *Charlie*, one [a soldier] who cannot understand orders and so makes foolish mistakes. **1957** *Listener* 15 Aug. 252/1 The plebeian engineer was a proper Charlie to let himself be roped in for it. **1961** Simpson & Galton *Four Hancock Scripts* 15, I felt a right Charlie coming through the customs in this lot.

7. (Also **Mr. Charlie.**) A white man. *U.S. Negro slang.*

1960 Wentworth & Flexner *Dict. Amer. Slang* 340/1 *Mister Charlie*, a white man. Some Negro use. **1965** J. Baldwin in J. H. Clarke *Harlem* 175 He is assured ..his ancestors were happy, shiftless, watermelon-eating darkies who loved Mr. Charlie and Miss Ann. **1965** L. Hairston *Ibid.* 292 They all was beamin' like they had Charley's number; and Charley was settin' there fussin'. **1967** *Guardian* 11 Jan. 11/4 Stokely Carmichael was there promising 'Mr Charlie's' doomsday.

8. *U.S. Services' slang.* The North Vietnamese and Vietcong; esp. a North Vietnamese or Vietcong soldier.

1965 *Newsweek* 20 Sept. 26/2 A Ranger captain..shouted: 'Don't you tell me Charlie [GI slang for Vietcong] isn't hiding here!'..The rubber planter..answered 'Charlie? Who is Charlie?' The plantation manager, of course, knew perfectly well who Charlie was. **1966** *New Statesman* 14 Oct. 549/2 Friendly forces have made contact with Charlie and a fire fight followed. **1967** M. McCarthy *Vietnam* 9 If he called them 'Charlie'.., he was either an infatuated civilian, a low-grade primitive in uniform, or a fatuous military mouthpiece.

9. Used as adj.: Afraid, cowardly, esp. in phr. *to turn Charlie. slang.*

1954 'N. Blake' *Whisper in Gloom* II. xvi. 217 You turn Charlie and we'll do ya. **1958** F. Norman *Bang to Rights* 61, I was dead charlie and little fairies were having a right game in my guts.

Charley-horse, charley horse (tʃā·ɹlihǭɹs). *N. Amer. slang.* [Origin uncertain.] Stiffness or cramp in the arm or leg, esp. in baseball players.

1888 M. J. Kelly *Play Ball* 21, I could dance in those days, because, you see, I never was bothered with 'Charley Horse'. **1889** *Cincinnati Comm. Gaz.* 17 Mar. 15/1 Mac was affected with a 'Charley-horse' and that ended his ball-playing for 1888. **1921** *Daily Colonist* (Victoria, B.C.) 13 Oct. 11/4 Pulled tendon, or 'charley-horse' in right leg; torn loose again Monday. **1947** *Sooner Mag.* (Norman, Okla.) Nov. 14/2 The day before the first scheduled game with Oklahoma City he suffered a charley-horse. **1968** *Globe & Mail* (Toronto) 11 July 32/8 Rookie centre Gordon Judges departed in the second half suffering a severe charley horse in his left thigh.

charlotte. Add: (Earlier and later examples.) Also, a similar dish made with fruit other than apple.

1796 J. Barlow *Hasty-Pudding* i. 9 The Charlotte brown, within whose crusty sides A belly soft the pulpy apple hides. **1807** M. E. Rundell *New Syst. Domestic Cookery* (ed. 2) vi. 249 A Charlotte. Cut as many very thin slices of white bread as will cover the bottom and line the sides of a baking dish, but first rub it thick with butter. Put apples, in thin slices, into the dish. *a* **1845** Barham *Ingol. Leg.* ser. 3 (1847) 247 They soon played the deuce With a large *Charlotte Russe*. **1846** 'A Lady' *Jewish Manual* 139 Apple charlotte. *Ibid.* 190 Another excellent recipe for a fruit charlotte. **1899** *Daily News* 15 July 7/5 A Neapolitan Cherry Charlotte. **1946** M. Dickens *Happy Prisoner* iv. 55 It's apple charlotte, with the top off the milk.

Charlton white. A house-painters' pigment consisting of barium and strontium sulphates with zinc sulphide; = *lithopone.

1885 *Encycl. Brit.* XIX. 86 Under the name of Charlton White or silicate paints, Mr. J. B. Orr prepares a range of white paints which have come into extensive use.

charm, *sb.*[1] Add: **1. c.** *like a charm*: wonderfully, perfectly. Also *to a charm*.

1845 *Niles' Reg.* 25 Oct. 128/1 The new rig works to a charm. **1851** T. A. Burke *Polly Peablossom's Wedding* 40 Whenever he became refractory, his wife would stick her thimble on the end of her finger, and hold it up for him to look at—it acted like a charm! **1869** P. T. Barnum *Struggles & Triumphs* viii. 123 He assured me that he should be able to give his imitations 'to a charm'. **1882** *Philad. Press* 19 Mar. (Th.), The freezing-out process was applied by

Mr. Bliss to Cook, and it worked like a charm. 1934 F. N. Hart *Crooked Lane* iii. 99 Bill Stirling gave her one the other night, and she said it worked like a charm. 1967 M. Shulman *Kill 3* II. i. 59 'It's worked like a charm,' said West.

6. charm-bracelet, a bracelet hung with charms (sense 5); **charm school,** 'a school in which social graces are taught' (Webster 1961).

[1910 W. J. Locke *Simon Jester* ii. 20 She held up her bracelet, from which dangled some charms.] 1941 L. P. Benjamin in J. C. Furnas *How America Lives* (1943) 254 The capable little hands..are so bedecked with charm bracelets. 1959 *Guardian* 31 Aug. 4/2 The latest fifth-form 'gimmick', in the shape of..a charm bracelet. 1950 C. M. Kornbluth *Little Black Bag* in *Best SF Stories* (1968) 76 I'm going to go to charm school. 1962 'L. Grex' *Terror Wears a Smile* viii. 126 Steve Blaine might be top of his class in Charm School, but [etc.].

charm, *sb.*² Add: **3.** A company or flock (of finches, etc.). Cf. Chirm *sb.* 3.

1810 J. Strutt *Sports & Pastimes* I. ii. 33 A charm of goldfinches. 1930 E. W. Hendy *Wild Exmoor* xvi. 245 A 'charm' of goldfinches joined them. 1936 C. R. Acton *Sport New Forest* ii. 43 A Forester speaks of a swarm of bees as a 'Charm', which expression arises from the Saxon 'cyrm', a cluster.

charmaid, charman: see *char-.

charmante (ʃāɹmaṅt). [Fr., fem. of *charmant*, prop. pres. pple. of *charmer* to Charm.] A silk fabric with a satin face and a heavy crêpe back. Also *satin charmante*.

1922 *Westm. Gaz.* 27 Dec. (Advt.), Satin charmante... For Day and Evening wear. 1923 *Weekly Dispatch* 8 Apr. 12 Charmante...Rich Crepe back quality. In the newest Spring shades.

charmed, *ppl. a.* Add: **4. b.** As a conventional colloquial or jocular reply, = I am delighted; I am or shall be very pleased (to do something). Cf. Charm *v.*¹ 5 a.

1856 Dickens *Dorrit* II. vii. 384 'So delighted,' said Mrs. Merdle, 'to resume an acquaintance so inauspiciously begun at Martigny.' 'At Martigny, of course,' said Fanny, 'Charmed, I am sure!' 1920 Wodehouse *Jill the Reckless* i. 18 'Like you,' mumbled Freddie, 'to meet my friends. Lady Underhill. Mr. Devereux.' 'Charmed,' said Ronnie affably. 1930 E. Waugh *Vile Bodies* v. 82 'Do you mind terribly?' 'Not as much as all that,' said Nina, and added in Cockney, 'Charmed, I'm sure.'

charmelaine (ʃāɹmǝleⁱn). Also **charmaline.** [f. F. *charme* Charm *sb.*¹ + *laine* wool.] A dress material of artificial silk and wool.

1923 *Daily Mail* 5 Feb. 1 Charmaline, a rich fabric of Artificial Silk and Wool, suitable for Day and Evening Dresses. 1923 *Weekly Dispatch* 8 Apr. 12 Crepe Charmaline. 1927 *Westm. Gaz.* 24 Jan., Charmelaine in thirty-eight shades.

charmeuse (ʃāɹmöz). [Fr., fem. of *charmeur*, agent-n. of *charmer* to Charm.] A soft smooth silk fabric, having a satin-like surface. Also *attrib.*

1907 *Daily Chron.* 25 Nov. 4/5 The bride is to wear a wonderful robe of soft white satin 'Charmeuse'. 1908 *Westm. Gaz.* 31 Oct. 15/3 The sleeves of the charmeuse frock may be carried out in chiffon or net. 1922 *Daily Mail* 17 Nov. 8 Princess Louise, Duchess of Argyll, wearing embroidered black charmeuse. 1968 J. Ironside *Fashion Alphabet* 218 *Charmeuse*, a trade-name for silk, cotton or rayon fabric of satin weave with a dull back and semi-glossy surface.

charmlessly (tʃāɹmlesli), *adv.* [f. Charmless *a.* + -ly².] In a charmless manner, without charm. So **cha·rmlessness,** charmless condition.

1889 J. M. Robertson *Ess. towards Crit. Meth.* 74 The Pope school strikes charmlessly on our sense. 1908 *Daily Chron.* 10 Nov. 4/4 One wonders whether the historic associations of the Mansion House will counterbalance its gloom and charmlessness. 1926 *Glasgow Herald* 29 Apr. 4 The most charmlessly didactic story that ever was written by the author of 'War and Peace'.

charnockite (tʃāɹnǫkait). *Petrography.* [f. the name of Job *Charnock* (d. 1693), founder of Calcutta, whose tombstone is made of this rock, + -ite¹ 2 b.] Hypersthene-granite; any of a series of rocks containing hypersthene and typically found near Madras. Hence **charnocki·tic** *a.*

1893 T. H. Holland in *Jrnl. Asiatic Soc. Bengal* LXII. II. 164 As this is a new type of rock..I would suggest for it the name of *Charnockite*, in honour of the founder of Calcutta. 1937 *Geogr. Jrnl.* LXXXIX. 541 The rock is charnockite. This term indicates an igneous rock which was intruded into the Archaean gneisses and schists. 1939 *Ibid.* XCIII. 533 He found altered charnockitic gneisses.

Charollais, Charolais (ʃærolēⁱ, ‖ʃarole). The name of a region of eastern France, used to designate a breed of large white cattle. Also *absol.*

1893 *Jrnl. R. Agric. Soc.* LIV. 193 These half-breds.. have the white coat of the Charolais breed...The head is shorter than that of the pure-bred Charolais. 1910 *Encycl. Brit.* V. 948/1 Charolles..is the centre for trade in the famous breed of Charolais cattle. 1959 *Times* 10 Mar. 7/6 The suggestion that Charollais bulls might be imported from France..has..aroused considerable opposition. 1963 *Guardian* 30 Oct. 2/7 A Charollais–Friesian cross was compared with Hereford–Friesian, Devon–Friesian, Sussex–Friesian and pure Friesian.

charoset(h (χārōᵘ·seþ, -set). *Judaism.* Also **charoises, haroses, haroset(h.** [Heb. *ḥᵃrôseṯ*, f. *ḥarsîṯ* potter's clay.] A mixture of apples, nuts, spices, etc., eaten ceremonially at the Passover seder service, symbolizing the clay mixed by the Israelites during their slavery in Egypt.

1885 *Encycl. Brit.* XVIII. 344/1 The bitter herbs and unleavened cakes were dipped in a kind of sweet sauce called ḥǎróseth. 1891 M. Friedländer *Jewish Relig.* 381 *Charoseth*, a mixture of apples, almonds, various spices.. in which mixture the bitter herbs are dipped. 1893 I. Zangwill *Children of Ghetto* (ed. 3) I. xxii. 205 *Charoises!* (a sweet mixture), *Charoises! Moroire!* (bitter herb). 1905 *Westm. Gaz.* 20 Apr. 9/3 Some 'charoset', balls of almonds, apples, and various spices. 1957 L. Stern *Midas Touch* III. xx. 152 *Haroses*..almonds, raisins and apples, finely chopped in wine and cinnamon. 1959 *Jerusalem Post* 24 Apr. 5/4 The *maror*, the bitter herbs,..are dipped in an elaborately prepared paste of almonds, apples, wine and whatever else can contribute to making it sweet and sticky, the *haroset*.

Charpy (ʃāɹpi). *Metallurgy.* The name of A. G. A. *Charpy* (1865–1945), French engineer, used to designate a device or method for measuring the strength of metals under impact.

1915 J. J. Thomas in *Amer. Soc. Test. Materials Proc.* XV. 62 The Charpy impact test appears to differentiate between steels of various degrees of brittleness. *Ibid.* 63 Experiments made with the Charpy impact machine, for the purpose of determining how the shock-resisting qualities of various steels are affected by heat treatment. *Ibid.* 76 Six transverse Charpy test specimens were machined. 1931 Batson & Hyde *Mechanical Testing* xix. 332 (*heading*) The Charpy Pendulum Machine. *Ibid.*, The small Charpy machine has a striking energy of about 30 kilogrammeters. *Ibid.* 337 Charpy tests on materials correctly and incorrectly heat treated.

charro (tʃā·ro). [Mexican Sp., a. Sp. *charro* rustic, country-like.] A Mexican cowboy, usu. dressed very decoratively. Also *attrib.*

1926 D. H. Lawrence *Plumed Serp.* xxv. 395 They rode about in resplendent charro dress. 1934 A. Huxley *Beyond Mexique Bay* 286 Four *charros*, magnificently hatted, armed and caparisoned. 1948 *Pix* (Austral.) 28 Aug. 10 (*caption*) A mounted charro (cowboy) snakes his lariat out in a perfect throw, born of long practice. 1966 *Punch* 9 Mar. 364/1 *Mariachis* are large, even outsized men. They sport elaborate *charro* costume, from fancy boots to huge sombreros.

charry. *colloq.* Variant of *chara². Cf. *sharry.

1926 *Brit. Weekly* 1 July 267/2 The motor-omnibus..is another formidable competitor to the 'charry'.

chart, *sb.* Add: **1. b. chart-house, -room,** an apartment in a ship, in which the charts, navigating instruments, etc., are kept.

1891 H. O. Arnold-Forster *In Conning Tower* 29, I was standing close to the chart house on the upper bridge. 1907 *Daily Chron.* 10 Aug. 4/4 The chart-house door. 1877 W. Thomson *Voy. Challenger* I. i. 11 The Chart-room.. with ranges of shelves stocked with charts and hydrographic, magnetic, and meteorological instruments. 1937 *Discovery* Jan. 23/2 The chart-room scale.

d. A musical arrangement or score. *colloq.*

1957 *N.Y. Times Mag.* 18 Aug. 26 *Charts*, musical arrangements. 1966 *Crescendo* June 20 His arranging activities started about fifteen years ago with a chart for a Victor Feldman date on Melodisc.

3. c. *spec.* A list of the gramophone records or tunes that are most popular at a particular time. Also *transf.* and *Comb.*

1963 *The Beatles* 3 More chart-topping discs on the way; more packed audiences to drown the frantic beat with cheering. 1965 *Melody Maker* 3 Apr. 11/2 'Little Things' looks as though it will climb high up the chart. 1965 *New Statesman* 16 Apr. 598/3 If Mr Stewart is top of the Tory pops, other ministers are also high up in the charts. 1967 *Scottish Daily Mail* 17 Oct. 14 Engelbert Humperdinck's six-week-long reign at the top of the charts with *The Last Waltz* has finally been broken by the Bee Gees with *Massachusetts*.

charter, *sb.*¹ Add: **2. d.** Used *attrib.* of or pertaining to an aircraft hired by contract for a particular purpose, or to a flight in such an aircraft, or to a business firm using such aircraft, etc.

1922 *Flight* XIV. 522/1 In addition to this regular service 16 special-charter flights were made during this period. 1958 'N. Shute' *Rainbow & Rose* i. 2 He had two Austers for instruction and occasional charter flights. 1960 *Guardian* 26 July 5/4 A charter firm..flew the children free from Nice. 1964 *Financial Times* 3 Mar. 13/5 Nearly all British airlines..can be called charter airlines in that they will undertake charter flights if asked.

5. charter-hand = Court-hand.

1888 J. H. Hessels in *Encycl. Brit.* XXIII. 682/1 The court or charter hand was used for charters, title-deeds, papal bulls, &c. 1893 [see Hand *sb.* 16]. 1954 N. Denholm-Young *Handwriting in Eng. & Wales* iv. 31 The term Court Hand..is nowadays used to denote the great variety of hands employed in everyday life, as opposed to book hands, from the twelfth to the fifteenth century...For a time the alternative..term Charter Hand was in vogue, chiefly perhaps among authors associated with the British Museum.

charter, *v.* Add: **3.** Also of aircraft.

1933 A. Huxley *Let.* 24 Mar. (1969) 368 Yesterday we joined a party of Americans in chartering an aeroplane to fly to Copan in Honduras, to see the great Maya ruins there.

chartered, *ppl. a.* Add: **1. b.** *chartered accountant*: an accountant who is qualified according to the rules of the Institute of Chartered Accountants in England and Wales, which received a royal charter in 1880, or of the similar chartered bodies in Scotland or Ireland.

1855 *Index Juridicus: Scottish Law List* 612 The Members of the Society of Accountants [in Edinburgh] have adopted the distinctive abbreviate letter of 'C A', Chartered Accountant. 1880 *Accountant* 8 May 5/2 The Members of this Society..are now enrolled as Chartered Accountants. 1955 *Times* 10 May 7/3 He said that chartered accountants had been employed for a long time as umpires in various matters.

3. a. Also of aircraft (cf. *charter *sb.*¹ 2 d).

1929 *Lancet* 12 Jan. 105/2 We have carried several invalids ..either on scheduled services or by specially chartered aircraft. 1958 *New Statesman* 5 Apr. 427/3 They come.. on chartered planes which take them around to see the most famous sights.

chartophylax (kāɹtofai·læks). *Gr. Ch.* [Gr. χαρτοφύλαξ, f. χάρτα paper + φύλαξ guard.] An officer of the household of the Patriarch of Constantinople who has charge of the official documents and records.

1879 *Encycl. Brit.* X. 429 He held..the offices of 'Chartophylax', 'Scenophylax', and 'Referendarius' in the 'Great Church' (that of St Sophia) at Constantinople. 1884 Addis & Arnold *Cath. Dict.* 423/2 The Chartophylax who superintends ecclesiastical causes. 1928 *Observer* 1 July 10 Manuel I. Gedeon, the learned Chartofylax of the Great Church.

chartreuse. Add: **3.** *Cookery.* **a.** An ornamental dish of meat or vegetables cooked in a mould. **b.** Fruits enclosed in blancmange or jelly.

1806 J. Simpson *Cookery* 264 (*heading*) A Chartreuse of Roots and Sausages. *Ibid.* 265 Soak the fat that comes from the chartreuse. 1892 T. F. Garrett *Encycl. Pract. Cookery* I. 340/1 Chartreuse of Vegetables and Game. 1900 *Manners & Rules of Good Society* xx. 151 Refreshments..include varieties in smart-looking confectionary, chartreuses and *macédoines* of fruit, jellies and creams. 1951 *Good Housek. Home Encycl.* 345/2 Chartreuse of bananas..and banana flan make delicious cold sweets.

4. A variety of the domestic cat. Also **Chartreux.**

1838 *Penny Cycl.* X. 223 Among the most noted are..the Chartreux, which is bluish, and the Angora cat. 1876 *Encycl. Brit.* V. 206 The Chartreuse, of a bluish-grey colour.

char-work. [f. *char-* Chare *sb.*¹] Ordinary mechanical work.

1888 F. J. Child *Eng. & Sc. Pop. Ballads* III. 42/2 A considerable part of the Robin Hood poetry looks like char-work done for the petty press.

chase, *sb.*¹ Add: **1. f.** Short for Steeple-chase *sb.*, freq. *attrib.*

1894 M. H. Hayes *Among Men & Horses* i. 12 The professional[s]..regarded gratuitous chase riding as an unwarrantable attempt to take the bread, or rather the whisky, out of their mouths. 1927 *Daily Express* 22 June 16 Seamark will seek consolation for chase misfortunes in the valuable Prix des Drags. 1969 D. Francis *Enquiry* vi. 78 You were riding..in a novice chase.

g. *Mus.* A sequence of solos by two or more jazz musicians in which each in turn improvises for a few bars. Freq. *attrib.* orig. *U.S.*

1942 *Gems of Jazz* IV. 7 It's one of the most exciting 'chase' choruses on wax. 1955 L. Feather *Encycl. Jazz* (1956) 342 The high point of this..performance..is a piano-guitar 'chase' sequence by Nat Cole and Les Paul. 1956 M. Stearns *Story of Jazz* (1957) i. 9 The 'chase' choruses of Bix and Tram during the late 'twenties offer another example.

chase, *sb.*³ Add: **7.** The apex of a cop or bobbin of a spinning-wheel.

1902 Hannan *Textile Fibres of Commerce* 124 The shoulder acts as a good support to the chase of the cop in winding.

chase (tʃēⁱs), *sb.*⁴ *dial.* Also **chace.** [Cf. Fr. dial. (Norm.) *chasse*, un petit chemin (Duméril).] A green lane, esp. one leading up to a farm-house or field. Also *chace-lane, -way.*

1639 in *Connecticut Hist. Soc. Coll.* XIV. 331 A Chasse lane leading from the litle Riuer to the meeting house. *c* 1640 *Ibid.* 393 Abuting on the..Cheace way alley to the metting house. 1685 *Ibid.* 366 The chase lane. 1804 in *E.D.D.* 1904 *Essex Rev.* Apr. 117 The green lanes, which in some counties are called 'ridings', 'driftways', or 'bridle roads', are called hereabouts by the name of 'chaces', or 'chace-ways'. 1943 *N. & Q.* 25 Sept. 201 *Chase*, green track to farmhouse or field. E. Ang. 1960 *Times* 15 June 14/6 Our seawall is reached by what we call locally a 'chase'.

chase, *v.*¹ Add: **6.** Also with *off* (in pursuit of something).

1920 R. Macaulay *Potterism* IV. i. 172 Aunt Cynthia chased off after another exciting subject, and that was all about Gideon.

7. c. *refl.* To betake (oneself), to go or run away; to depart; esp. in phr. *go (and) chase yourself. colloq.* (orig. *U.S.*).

1883 G. W. Peck *Mirth for Millions* 79 O, you go and chase yourself. That is not small-pox Pa has got. 1893 S. Crane *Maggie* (1896) xv. 118 Go chase yerself. 1916 C. J. Dennis *Songs of Sentimental Bloke* 119 *Chase yourself*, depart; avaunt; 'fade away'. 1923 R. D. Paine *Comr. Rolling Ocean* xii. 206 Let him rest, Kid. You chase yourself below and look things over. 1937 A. Christie *Dumb Witness* xiii. 138 Your friend..looks shocked. Shall we send him out to chase himself round the block? 1943 J. S. Huxley *TVA* 74 Engineers apparently accepted the architect for what he said he was: dispenser of divine revelation in the realm of æsthetics. But they also told him to chase himself if he ventured beyond. 1959 *Listener* 8 Jan. 56/2 Some people will tell you to go and chase yourself. Others will just ignore you. 1968 'P. Hobson' *Titty's Dead* v. 66 She was mentioning..that she had had enough of policemen for the day. And, mm-m, that you could go chase yourself.

11. *chase me, Charley*: (*a*) a catch-phrase; (*b*) (see quot. 1945).

1906 E. Dyson *Fact'ry 'Ands* xi. 138 The printers cried: 'Chase me, Charley!' 1922 Joyce *Ulysses* 504 Chase me, Charley! Buzz! 1945 *Newsweek* 4 June 90 Off the coast of Italy in 1943, British fighter pilots ran into a new German weapon trained on Allied shipping. It was a small glider with a bomb for a body. Directed by remote control from a launching plane, the device assumed attack position and hurtled itself at the target, where it exploded. The British named it 'Chase-me-Charlie'. 1961 W. Vaughan-Thomas *Anzio* vii. 133 The Germans had aimed another wireless-controlled bomb at the fleet—a 'chase-me-Charlie' the British seamen called it.

chaser.¹ Add: **2. b.** One who chases women, a woman-chaser. Chiefly *U.S. colloq.*

1894 *Yale Wit & Humor* 49/2 (Chess tournament) If our accomplished chasers ain't able to pinch that Harvard Queen without giving up one of our own ladies in exchange, we would respectfully ask what the university is coming to? 1934 J. O'Hara *Appointment in Samarra* (1935) i. 26 If English had been a chaser Al would have heard about it. 1968 H. Waugh *Con Game* xviii. 147 George deserted her because he was a chaser. 1969 S. Greenlee *Spook who sat by Door* xii. 102 The women thought him an eligible bachelor, if a bit of a chaser.

3. a. (Later examples.)

1918 *Chambers's Jrnl.* Oct. 699/1 Each U-boat chaser measures two hundred and four feet in length. 1957 *Jane's Fighting Ships 1956–7* 195/1 Ex-U.S. chasers..now rated as ..inshore minesweepers.

5. A quantity of water or other mild beverage taken after spirituous liquor. (Cf. Chasse².) *colloq.* (orig. *U.S.*) Also *fig.*

1897 *Daily News* 30 Aug. 2/1 Everything was 50 cents. a drink, no mixed drinks, and no water for a chaser. 1905 *N.Y. Even. Post* 8 Sept. 7 Drinking whiskey from a bottle and refusing a 'chaser'. *a* 1906 'O. Henry' *Trimmed Lamp* (1916) 166 Eagerly gulping down the strong, black headlines, to be followed as a chaser by the milder details of the smaller type. 1932 J. Dos Passos *1919* 162 Two guys from Chicago who were drinking whisky with beerchasers. 1941 Wodehouse *Huy Day by Day* in *Perf. Flea* (1953) 209 Give me one blow-out like that..and you wouldn't find me making a fuss if somebody added a dose of some little-known Asiatic poison as a chaser. 1956 'N. Shute' *Beyond Black Stump* 49 He..took the glass of rum.., shot it down in one swallow, and followed it with a chaser of water.

6. A small, light, usually single-seated aeroplane of great speed and climbing power, used in repelling hostile aircraft. Also *attrib.*

1915 Grahame-White & Harper *Aircraft Gt. War* 33 British pilots, having in view the pursuit of slower-flying German craft, have called these little machines [*sc.* the 'Bullets'] 'the chasers'. 1917 'Contact' *Airman's Outings* p. xxi, The chaser craft necessary to keep German machines from interference. 1917 *Aeroplane* 16 May 1244/1 The High Command of the German Flying Service instituted some months ago the system of mobile chaser squadrons, consisting entirely of picked fighting pilots. 1918 [see *circus 2 e]. 1919 A. E. Illingworth *Fly Papers* 41 A chaser squadron of picked pilots. 1939 Auden & Isherwood *Journey to War* 37 Canton now had a considerable force of chaser planes.

7. (See quot.) See also *progress-chaser* (*progress *sb.* 7).

1922 Turner & Wood *Man. Up-to-date Organisation* 171 Chaser is a progress man responsible for the progress of a job through the factory.

Chasid, Chassid (hæ·sid). Also **Has(s)id.** Pl. **-idim.** [Heb. *ḥāsîḏ*, lit. 'pious; pietist'.] A name applied to a member of any of several mystical Jewish sects of various periods; an Assidæan. Hence **Chas(s)i·dic, Has(s)i·dic** *a.*, of or belonging to the Chasidim; **Cha·s(s)idism,** the tenets of the Chasidim.

1812 H. Adams *Hist. Jews* II. xxxviii. 281 A rabbin, named Israel, rendered himself very famous..in the Ukraine, between the years 1760 and 1765...The propensity of..credulous men toward the occult sciences procured him, in less than ten years, more than ten thousand followers, whom he called Chasidim. This name designated those men, who, not content to follow the ritual laws of Moses, laboured to unite themselves more intimately to the Deity by their sanctity. 1834 *Penny Cycl.* II. 502 The Assidians, or Chasidim, of those days, found a leader in Mattathias. 1834 [see Assidæan]. 1893 I. Zangwill *Children of Ghetto* (ed. 3) I. xvi. 148 Meckish was a Chassid, which in the vernacular is a saint, but in the actual a member of the sect of the Chassidim, whose centre is Galicia. *Ibid.* 149 Chassidism is the extreme expression of Jewish optimism. 1918 —— *Chosen Peoples* iv. 39 The comparatively modern Chassidism. *Ibid.* 42 A Chassidic Rabbi. 1927 *Daily Tel.* 5 Apr. 12/6 The devotees of that particular cult which was known as Hassidism. 1927 *Sunday Express* 10 Apr. 5/7 'The Dybbuk' is all about the Hassidic cult. 1936 *Times Lit. Suppl.* 10 Oct. 811/4 The extreme pietist atmosphere of the Chassidic sect of Jews. 1941 G. G. Scholem *Major Trends Jewish Myst.* iii. 91 To be a Hasid is to conform to purely religious standards entirely independent of intellectualism and learning. 1948 *Theology* LI. 302 His collection of Hasidic tales. 1960 S. Becker tr. *A. Schwarz-Bart's Last of Just* (1961) vii. 351 Do you know who the Christ was? A simple Jew like your father. A kind of Hasid. 1961 G. Abrahams *Jewish Mind* ii. 48 The Rabbis treated the Bible with something of the respectful familiarity with which Chassidim, then and now, have treated God. 1965 *Jrnl. Jewish Studies* XVI. 33 From the end of the period of the *Tanna'im* onwards, the term 'Ḥasid' became blurred and was used simply as a *soubriquet* for scholars of an austere attitude towards *halakhah*.

chasmic (kæ·zmik), *a.* [f. Chasm + -ic.] = Chasmal *a.*

1926 H. E. Bates *Two Sisters* III. xiv. 298 Tessie had come back and was leaning on her arm as if there had been no chasmic year and a half of doubt and wonder between. 1937 'G. Orwell' *Road to Wigan Pier* viii. 161 The chasmic, impassable quality of class-distinctions.

chasmogamy (kæzmǫ·gămi). *Bot.* [f. mod.L. (*flores*) *chasmogami* (J. S. Axell *Om anordningarna f. fanerogama växternas Befruktning* (1869), f. Gr. χάσμα Chasm + γάμος -gamy.] The opening of the perianth at the time of flowering, as distinguished from cleistogamy. Hence **chasmogamic** (kæzmǫgæ·mik), **chasmogamous** (kæzmǫ·gămǝs), *adjs.*

1900 B. D. Jackson *Gloss. Bot. Terms* Add., Chasmogamy..; chasmogamic. 1906 J. R. A. Davis tr. *Knuth's Handbk. Flower Pollination* I. 55 The otherwise normally opening chasmogamous flowers remain closed. 1959 *New Scientist* 5 Nov. 883/1 Producing open (chasmogamous) flowers. *Ibid.*, The causes of chasmogamy in *Leersia*.

chasmophyte (kæ·zmofait). *Bot.* [f. Gr. χάσμα Chasm + -o + -phyte.] A plant inhabiting the crevices of rocks.

1900 B. D. Jackson *Gloss. Bot. Terms, Chasmophyte*, a plant which grows in rock-crevices. 1903 W. R. Fisher tr. *Schimper's Plant Geogr.* II. i. 178 Crevices in rocks..produce a somewhat more copious vegetation, that of the chasmophytes... Chasmophytes, as opposed to lithophytes, are long straggling plants. 1951 *Jrnl. Ecol.* XXXIX. 66 These are all chasmophyte habitats except for the ledge, which is a chomophyte one.

‖chassé-croisé (ʃase krwaze). [Fr., = Chassé *sb.* + *croisé*, pa. pple. of *croiser* to cross.] A dance figure in which one of two partners chassés first to the right and then to the left, while the other chassés first to the left and then to the right. Hence *transf.* and *fig.* applied to actions or situations in which persons or things cross each other or change positions backwards and forwards.

[1874 *Ball-Room Guide* 90 Chassez croisez. Lady and gentleman chassez in opposite directions.] 1883 *Sat. Rev.* 10 Nov. 595/1 His drama is a perpetual *chassé-croisé* at the edge of a precipice. 1886 *Athenæum* 17 Apr. 516/1 When he arrived alongside, the Espiègle and the galley were performing a sort of vertical *chassé-croisé*. 1903 *Westm. Gaz.* 27 Jan. 3/2 A *ménage à trois* and a kind of matrimonial chassé-croisé are discussed. 1928 *Observer* 8 Apr. 5/2 The metamorphoses of character, the *chassé-croisé* of incident.

chasseur. Add: **2. b.** *Comb.* **chasseur-blue,** a shade of blue resembling that of the uniform of a French chasseur.

1900 *Daily News* 1 Jan. 5/3 Costumes..in blue and black-chasseur-blue.

3. (Earlier example.)

1765 H. Walpole *Let.* 3 Oct. (1904) VI. 311 We foreigners and the foreign ministers were shown the famous beast of the Gevandan, just arrived, and covered with a cloth, which two chasseurs lifted up.

b. A hotel messenger, esp. in France.

1896 E. Dowson *Let.* *c* 24 Mar. (1967) 351 Ask for a *garçon* called Paul..& ask him to convey the letter to 'Monsieur Léopold'—by the *chasseur* if he is not in the café. 1899 —— *Let.* June–July 415 The wretched little chasseur whom I sent daily for my letters pocketed his omnibus fare & never went near the place. 1936 D. Barnes *Nightwood* ii. 54 The *chasseur* of the *Hôtel Récamier*..was approaching them at a run. 1939 E. Ambler *Mask of Dimitrios* xi. 213 With a *boîte*, you know, it takes time to build up a clientèle. ..We had a good *chasseur*.

Chassid, Chassidic, Chassidim: see *Chasid.

chassis (ʃæ·si, tʃ-). Pl. **chassis** (ʃæ·siz). Add: **1. b.** *spec.* = Stretcher 4 b.

1934 H. Hiler *Notes on Technique of Painting* i. 25 As to the chassis or stretchers, they should (on any but the smallest pictures) always be of the sort which are provided with keys.

3. The base frame of a motor car, with its mechanism, as distinguished from the body or upper part; also, in an aeroplane.

1903 *Sci. Amer.* 20 June 472/1 The motor is placed in the center of the chassis and the boiler is now quite in the rear. 1904 A. B. F. Young *Compl. Motorist* v. 149 The frame of the chassis is the ordinary pressed steel frame as generally used on a petrol car. 1909 A. Berget *Conquest of Air* v. 211 The whole apparatus rests upon a running chassis for launching, and to ensure descent without shock. 1910 R. Ferris *How it Flies* xx. 458 *Chassis*, the under-structure or 'running-gear' of an aeroplane. 1924 *Motor* 21 Oct. 583/1 The baffling diversity in methods of chassis construction. 1967 M. Procter *Exercise Hoodwink* ii. 15 This discovery.. led to a search for engine and chassis numbers.

4. *transf.* The body of a person or animal. *slang.*

1930 Wodehouse *Very Good, Jeeves!* iv. 101 'Did he seem very fond of her?' 'Couldn't take his eyes off the chassis.' 1948 A. C. Smith in B. Vesey-Fitzgerald *Bk. of Dog* II. 409 This whimsical, odd-looking, little dog [*sc.* the dachshund], with his low-hung chassis.

5. The frame on which the parts of a radio receiver are mounted; also, the assemblage of parts on the frame, excluding the cabinet or housing.

1931 Moyer & Wostrel *Radio Constr.* (ed. 3) xix. 361 The first step in assembling the television receiver is to attach to the frame or *chassis*..a type No. 216 socket. 1940 *Amat. Radio Handbk.* (ed. 2) iv. 68/2 Superheterodynes where a metal chassis or separately screened stages are most desirable. *Ibid.*, These boxes should be insulated from one another except where earthed to the chassis. 1942 *Electronic Engin.* XV. 238 In some receivers the chassis consists of an intricate die-cast framework, bolted to the panel casting. 1945 *Ibid.* XVII. 554 Fig. 3 shows the chassis removed from the cabinet.

6. *Comb.* (sense 3) *chassisless* adj.

1951 *Engineering* 26 Oct. 533/2 Models are of 'chassisless' construction, the single-unit steel body giving..torsional stiffness. 1959 *Motor* 30 Sept. 234/1 This is a chassisless all-steel car.

chasteningly (tʃēⁱ·s'niŋli), *adv.* [f. Chastening *ppl. a.* + -ly².] In a chastening manner or tone.

1901 W. J. Locke *Usurper* xxiii. 318 'The ultimate evolution of swelled-head', he answered chasteningly.

chastisable (tʃæstai·zăb'l), *a.* Also 7 **chastice-, chastiseable.** [f. Chastise *v.* + -able.] That may be chastised; deserving of chastisement.

1611 Cotgr., *Chastiable*, chasticeable; fit to be chastised. 1632 Sherwood, Chastiseable. 1924 *Blackw. Mag.* Aug. 241/1 A more pernicious or chastisable guild of touts it has never been my lot to encounter.

chastity. Add: **6.** *attrib.*, as *chastity clause*; **chastity belt,** a belt designed to prevent a woman from having sexual intercourse; also *fig.*

[1922 *Index-Catal. Libr. Surgeon General's Office, U.S. Army* 3rd ser. III. 1003/2 'Chastity girdle' improvised by suspicious husband.] 1931 E. J. Dingwall *Girdle of Chastity* ii. 10 The virgin girdles of antiquity have nothing in common with the chastity belts. *Ibid.* vi. 164 The imposition of chastity belts upon women in order to allay masculine jealousy. 1938 L. MacNeice *I crossed Minch* II. xvi. 230 As medieval and even more effective than a chastity belt. 1960 *Guardian* 17 Dec. 4/6 A traffic plan which enclosed Oxford in, so to speak, a chastity belt. 1969 *Sun* 19 June 5/6 A Dorset firm who make chastity belts have been told that the belts..must carry purchase tax of 13¾ per cent. They are appealing..that the belts should pay no tax 'like other safety equipment'. 1906 Chastity clause [see *dum casta].

chat, *v.*¹ **2.** Delete †*Obs.* and add later example.

1888 Mrs. H. Ward *R. Elsmere* xliii, The other men stood chatting politics and the latest news.

5. *trans.* To address, talk to (a person); to advise; to approach or address tentatively; to flirt with. Also with *up. colloq.*

1898 *Eng. Dial. Dict.*, *Chat*, to flirt with. Londonderry. 1906 E. Dyson *Fact'ry 'Ands* x. 132 I want ter chat yeh. 1916 C. J. Dennis *Songs of Sentimental Bloke* 19, I tried to chat 'er, like you'd make a start Wiv *any* tart. *c* 1926 'Mixer' *Transport Workers' Song Bk.* (N.Z.) 69 He drinks with a bloke and chats him. 1936 J. Curtis *Gilt Kid* 50 She had earned it too the way he had chatted her. 1959 *Streetwalker* i. 10 Big Barbara is chatting a geezer, though the stream of polished professional patter she is directing at him warrants a less terse description. 1963 *Daily Mail* 22 Jan. 6/3 If you try to chat up a girl and she gives you the heavy fish it means she's ignored your advances. 1963 L. Deighton *Horse under Water* vii. 34, I was chatting her up the other day. 1963 *Sunday Express* 10 Mar. 22/7 He saw a pretty girl..smiling at him. He smiled right back. 'I like chatting the birds,' he said. 1966 K. Amis *Anti-Death League* 330, I must have spent a bit of time chatting them up.

château. Add: (Earlier examples.) Also used in reference to Britain. Also *attrib.* and *Comb.*

c 1739 J. Nixon *Let.* in E. Hamilton *Mordaunts* (1965) viii. 165 As for our Manner of Life it is (like our Château) collegiate. 1756 H. Walpole *Let.* 18 June (1857) III. 21, I..return to my *château* this evening. 1844 N. P. Willis *Pencillings* cxxxvii. 212/2 The river is between you and the chateau [*sc.* Abbotsford] as you come down to Melrose from the North. 1903 Ld. R. S. Gower *Records & Remin.* xxi. 211 Boughton, a gloomy French château-looking house. 1963 *House & Garden* Mar. 57/1 'Visit the Château Country'

adjure the travel posters. And that, of course, means the châteaux of the Loire.

b. A French vineyard, usu. in the neighbourhood of a château; freq. in the names of wines made at these vineyards. Hence **château-bottled** *a.*, (of a wine) bottled at the vineyard.

1754 Chesterf. in *World* No. 91 ¶ 2 The wine was the very same which they had all approved of the day before; and..was true Chateau Margoux. 1833 Redding *Wines* v. 148 In this commune is grown the famous first quality, Château Margaux. *Ibid.* 150 Nearly all the Château Lafitte, and indeed most of the other growths of this commune, are consumed in England. It is lighter than Chateau Latour. 1888 *Encycl. Brit.* XXIV. 605/2 All the finer descriptions are put into bottle before shipment, the corks bearing the name of the château and the vintage, as is the case with the château-bottled red wine. 1929 C. Beaton *Diary* 7 Dec. in *Wand. Years* (1961) 181 A little *soufflé* should be washed down with a glass of Chateau Yquem. 1937 *Times* 30 Dec. 6/3 Our now suddenly enlarged architectural cellar will, for some time, have a lot of queer wine in it that the fully qualified château-bottled vintages will find surprising company. 1954 'N. Blake' *Whisper in Gloom* II. xiii. 183, I don't think they found our Château d'Yquem...Shall I bring you up a bottle? 1954 Iris Murdoch *Under Net* xv. 204, I spent it riotously on *château* wines. 1959 *Times* 27 Nov. 11/7 A *château*-bottled wine from the great *châteaux* will always be good. 1969 *Daily Tel.* 18 Jan. 10/5 Branded wines..are relatively cheap, whereas domaine- and château-bottled wines are becoming increasingly expensive.

c. *châteaux in air, chateau(x) en Espagne, château in Spain, Spanish château* = castles, or a castle, in the air (see Castle *sb.* 11).

1793 [in Dict.]. 1834 A. Prinsep *Baboo* II. 351, I framed a thousand domestic rules, and built a multitude of chateaux en Espagne. 1840 [see *Alnaschar]. 1852 Lever *Daltons* xxi, It was, however, an Irish fortune, and, like a Spanish château, its loss is more a question of feeling than of fact. 1854 N. Hawthorne *Mosses from Old Manse* I. 270 She endowed him..with an unreckonable amount of wealth. It consisted..of half a million acres of vineyard at the North Pole, and of a castle in the air, and a chateau in Spain. 1873 J. S. Mill *Autobiogr.* iv. 91 It had been a part of their *Château en Espagne* that my father should be the editor. 1888 *Ch. Q. Rev.* (Dixon), Mere *châteaux en Espagne*, the creation of architectural fancy run mad.

Chateaubriand (ʃatobriāṅ). The name of François René, Vicomte de *Chateaubriand* (1768–1848), French writer and statesman, used *attrib.* or *absol.* of a thick fillet beef steak, grilled and garnished with herbs, etc. So *à la Chateaubriand.*

1877 E. S. Dallas *Kettner's Bk. Table* 6 The steak which had formerly been served..under the name of *filet de bœuf* was now always announced as a Châteaubriand. 1885 'Major L...' *Pytchley Bk. of Refined Cookery* vi. 176 Fillet of Beef à la Chateaubriand. 1895 G. A. Sala *Thorough Good Cook* vi. 302 *Chateaubriand Steak*. Cut the fillet of beef one and a half to two inches thick. 1932 L. Golding *Magnolia St.* II. v. 345 He ordered..pigeons, and two small chateaubriands. 1959 R. Postgate *Good Food Guide* 146 On the first floor there is a Châteaubriand Grill, where porterhouse, carpet-bag and châteaubriand steaks are offered.

Châteauneuf-du-Pape (ʃatonöf dü pap). [A commune on the left bank of the Rhône near Avignon.] A wine, usu. red, produced near Avignon.

[1851 Redding *Mod. Wines* (ed. 3) vi. 141 The wine of Châteauneuf du Pape, and that of Nerte,..are good wines.] 1876 G. Meredith *Let.* 20 May (1970) I. 517 It would have delighted me..to have drunk Châteauneuf du Pape with you. 1908 E. & A. Vizetelly *Wines of France* v. 144 Like Côte Rôtie and Hermitage, Châteauneuf-du-Pape, when obtainable, is a wine which may be drunk with game or roast meat. 1960 A. Burgess *Right to Answer* v. 64, I..had drunk gin-and-Italian, Châteauneuf du Pape, brandy.

chatelaine. Add: Also **châtelaine. 1.** (Earlier example.) Also, the mistress of a household.

1845 J. S. Mill in *Edin. Rev.* LXXXII. 409 The representative of the chief's person and the delegate of his authority, during his frequent absences, was the *châtelaine*. 1899 *Westm. Gaz.* 18 Apr. 2/1 The finest effects of a thunderstorm are dimmed to the most aesthetic of chatelaines by a too keen perception of its 'effects' on her larder. 1900 *New Cent. Rev.* VII. 381 The chatelaine of 17, Hertford Street, was hereditarily qualified to preside over a home whose natural atmosphere was one of culture. 1903 *Mainly about People* XI. 143/2 The châtelaine of a house in Eaton Square. 1949 [see A.W.O.L. (*A III)].

2. b. *attrib.*, as **chatelaine bag,** a small bag suspended from a belt or chatelaine; **chatelaine watch,** a watch attached to a chatelaine.

1873 *Young Englishwoman* June 286/1 A very ladylike chatelaine costume, of black grenadine. 1883 [in Dict.]. 1936 L. C. Douglas *White Banners* v. 102 She glanced down at her chatelaine watch.

Châtelperron (ʃatęlpęroṅ). The name of a site in the Allier department of France where Palæolithic remains were discovered; used *attrib.* to designate the earliest stage of the Aurignacian epoch, or artefacts characteristic of this stage. Hence **Châtelperronian** (-ōᵘ·n-) *a.*; also *absol.*

[1909 H. Breuil in *Rev. l'École d'Anthrop.* XIX. 330 Le nom de pointe de *Chatelperron*..le niveau de Chatelperron d'aurignacien.] 1915 W. J. Sollas *Anc. Hunters* (ed. 2) viii. 299 The implement which especially characterises the Lower Aurignacian is the Châtelperron point: it resembles a broad-bladed penknife. 1921 R. A. S. Macalister *Text-Bk. Europ. Archaeol.* vii. 334 The Châtelperron knife, one of the leading types of the early Aurignacian. 1932 [see *Audi]. 1946 F. E. Zeuner *Dating the Past* vii. 240 The Chatelperronian is much restricted in distribution, being almost confined to France and Palestine.

chaton (ʃa·toṅ). Also 6 **chatton.** [Fr., ad. G. *kasten* (OHG., MHG. *kasto*).] The head or broadest part of a finger-ring, in which a stone or intaglio is set or upon which a device is engraved.

1578 *Inv. R. Wardr.* (1815) 265 A chaton without a stane. *Ibid.* 267 A chatton without ane emerauld. 1880 C. T. Newton *Art & Archæol.* 269 The intaglio on the oval chaton of the other gold ring presents an equally strange subject. 1883 in Sayce *Schliemann's Troja* Pref. 20 The double-headed axe is..engraved on the famous chaton of the ring discovered by Dr. Schliemann at Mykênæ. 1968 J. Ironside *Fashion Alphabet* 169 *Chaton*, that part of a ring in which the stone is set.

chatoyant, *a.* (and *sb.*). Delete †*Obs.* and add later example of *adj.*

1962 *Listener* 4 Jan. 29/1 Repetition silts them Maturing in a silence Chatoyant and productive.

chattelization. (Earlier U.S. example.)

1854 A. L. Stone *Boston Orat.* 4 July 25 A system of human chattelization.

chatter, *sb.*¹ Add: **3. chatter-mark,** (*a*) *Geol.* a mark made on a surface by a fragment of rock on the under-surface of glacier ice; (*b*) (see quot. 1893).

1888 T. C. Chamberlin in *U.S. Geol. Survey Ann. Rep.* VII. 247 A fragment of rock acting as a gouging tool often produces 'chatter marks', little transverse lines of breakage due to the unsteady action of the tool in gouging the rock. 1893 *Funk's Stand. Dict.*, *Chatter-marks*, irregular tool-markings caused by vibration due to lack of stiffness in a lathe or similar machine. 1904 Chamberlin & Salisbury *Geol.* (1905) I. 270 Glacial striæ and bruises. The block to the right shows two sets of striæ: that to the left shows the peculiar curved fractures known as Chatter Marks. 1954 H. Garner *Oriental Blue* 37 They are roughly finished with radial marks, caused by the slipping of the turning tool and colloquially known as 'chatter marks'. 1963 D. W. & E. E. Humphries tr. *Termier's Erosion & Sedimentation* v. 124 (*caption*) Striation left by the Quaternary Ice Sheet on a Rock in Fakse Quarry (Denmark). Note the lunate 'chatter marks'; the ice moved from left to right.

4. The vibration of a tool (see also quot. 1904). Cf. Chatter *v.* 4.

1893 [see *chatter-mark, sense 3 above]. 1904 Goodchild & Tweney *Technol. & Sci. Dict.* 99/2 *Chatter*, the waviness produced on the surface of work by vibration of the tool, etc. 1964 *Times Rev. Industry* Apr. 91/1 Machine tool chatter, a vibration which shortens tool life, spoils surface finish and reduces machining accuracy.

chatterbox. (Earlier example.)

1774 C. Clive *Let.* 13 Jan. in *Corr. Garrick* (1831) I. 604 But I suppose his father *can*, for he is a fine chatter-box.

chatterer. 2. (Further examples.)

1945 Baker *Austral. Lang.* xii. 211 The Grey-crowned Babbler is known also as the..chatterer. 1963 *Austral. Encycl.* I. 385/1 Babblers..commonly known as catbirds (or caties), chatterers, happy families [etc.].

chattering, *ppl. a.* Add: **chattering plover** (*U.S.*) = Killdee, killdeer.

1731, 1796 [see Killdee, killdeer]. 1754 [see Plover 2].

chattermag (tʃæ·tǝɹmæg), *sb. colloq.* or *dial.* [f. Chatter *sb.*¹ + Mag *sb.*¹] **a.** (See quot. 1844.) **b.** Chatter, talk. Hence **chattermag** *v. intr.*, to chatter.

1844 W. Barnes *Poems Rural Life* 293 *Chattermag*, a chattering magpie, a chatterbox, a much-talking woman. 1895 C. W. Scott *Apple Orchards* 25 Gossip and chattermag. 1904 *Daily Chron.* 6 May 4/6 Philosophers..have suggested various causes for woman being such a 'chattermag'. 1909 *Westm. Gaz.* 15 May 2/3 Too late to wait while you stand chatter-maggin' there. 1926 *Sunday at Home* 149/1 She.. would have thoroughly enjoyed the gossip of the excited young chattermags.

Chattertonian (tʃætǝɹtōᵘ·niăn), *a.* and *sb.* [f. the name of *Chatterton* (see below) + -ian.] **A.** *adj.* Of, pertaining to, resembling, or characteristic of Thomas Chatterton (1752–1770), English poet, or his literary style. **B.** *sb.* A disciple or admirer of Thomas Chatterton.

1838 E. Mangin *Let. to 'Admirers of Chatterton'* 13 All Chattertonian readers are aware..that in..1764 (when Chatterton was but *twelve years* old)..he was in possession of certain old MSS. 1910 J. H. Ingram *True Chatterton* 110 Some of these verses..appear to have been bound up with the Chattertonian MSS. in the British Museum. 1946 *Scrutiny* XIV. I. 67 The tradition of Chattertonian boy wonders. 1956 E. Blunden *Coleridge's Fellow-Grecian* 75 Another Chattertonian, William Wordsworth, has been met in these pages as an enemy of Le Grice.

chattery, *a.* **1.** Delete *rare* and add further examples.

1878 Disraeli *Let.* 6 July (1929) II. 177 She is..very chattery and amusing. 1955 E. Bowen *World of Love* ii. 42 Any charm of a chattery circle had been broken by condemnatory pushing apart and back of armchairs.

chattily (tʃæ·tili), *adv.* [f. Chatty *a.*¹ + -ly².] In a chatty manner.

1892 *Pop. Science Monthly* XL. 718 A correspondent.. writes chattily of his pet canaries. 1920 Wodehouse *Damsel in Distress* vi. 86 'Why, when I was at Oxford in the year '87,' he said chattily.

‖chaud-froid (ʃofrwa). Also **chaudfroid.** [Fr.] A dish composed of cooked meat, fish, etc., served cold in aspic jelly or sauce; (see also quot. 1951). So **chaud-froid sauce,** 'a rich stiff sauce, made of stock and gelatin, used to cover cold meats' (Cent. Dict. Suppl. 1909); also *ellipt.*

1892 T. F. Garrett *Encycl. Pract. Cookery* I. 341/1 *Chaudfroids*. Literally 'hot-colds', applied to certain methods of preparing birds. *Ibid.* 612/2 Chaudfroid of Fowl. 1905 E. F. Benson *Image in Sand* vii, Arnold took some chaudfroid. 1951 *Good Housek. Home Encycl.* 397/2 *Chaudfroid*, a sauce used to coat cooked meat, fish, poultry, galantine, etc., which are served as cold entrées. A sweet chaudfroid.. is used to coat fruit to serve as a cold sweet.

chauffeur (ʃōᵘ·fǝɹ, ʃofȫr). [Fr., agent-n. of *chauffer* to heat.]

†1. An automobilist. *Obs.*

1899 *Motor-Car World* Oct. 3/1 The French 'chauffeur', Béconnais. 1901 *Daily News* 16 Feb. 6/1 Whilst automobilists in Great Britain have perhaps been saved from the 'racing fever'.., the French chauffeur has in part been encouraged in his rashness. 1903 *Lady's Realm* Apr. 684/1 All the members of the Italian Royal Family are enthusiastic *chauffeurs*.

2. A professional or paid driver of a private motor vehicle.

1902 *Westm. Gaz.* 5 Aug. 1/3 As to the driver, 'chauffeur' seems at present to hold the field. 1905 C. N. & A. M. Williamson (*title*) My Friend the Chauffeur. 1906 (*title*) The Chauffeur's Blue Book.

b. *Comb.*, as *chauffeur-driven* adj.; also appositively, as *chauffeur-handyman.*

1934 J. O'Hara *Appt. in Samarra* (1935) iv. 93 The women got out of chauffeur-driven..cars. 1937 Eve Garnett *Family from One End Street* v. 85 A large Packard, chauffeur-driven. 1953 'N. Blake' *Dreadful Hollow* 46 A job as chauffeur-handyman at the Hall.

Hence **chauffeuse** (ʃōᵘ·fɜ̄z, ʃofȫz), a female chauffeur.

1903 *Westm. Gaz.* 16 May 7/2. 1903 *Motor Ann.* 73 A skilful *chauffeuse*. 1970 P. Moyes *Who saw her Die?* ix. 123 Lady Balaclava..signing on as part-time *chauffeuse* to Members of the Government.

chauffeur (ʃōᵘ·fǝɹ), *v.* [f. the sb.] *trans.* To drive (a vehicle) as a chauffeur; to convey (a person) by car. Also *transf.*

1917 'Contact' *Airman's Outings* 111 Occasionally the Flying Corps officer is able to substitute an excursion by air..if..a bus of the type already flown by him must be chauffeured across the Channel. 1923 R. D. Paine *Comr. Rolling Ocean* iii. 44, I can chauffeur anything with an engine in it. 1929 C. Walt *Love in Chicago* ix. 100 Lefty chauffeurs me around in my Rolls-Royce. 1938 *Manch. Guardian* 12 May 20/1, I..had to chauffeur him down.

Hence **chau·ffeured** *ppl. a.*, supplied with or driven by a chauffeur; **chau·ffeuring** *vbl. sb.*, the work of a chauffeur; driving. Also as *ppl. a.* (*rare*).

1923 W. J. Locke *Moordius & Co.* xxi. 289 He could tell all chauffeuring Grasse how he had beaten the record to Paris. 1937 'J. Bell' *Murder in Hosp.* xv. 297 Dick's ancient..car, which stood brazenly among the chauffeured Daimlers. 1938 *Times Lit. Suppl.* 3 Sept. 572/2 It [*sc.* taxi-driving] seems to be very much worse paid than private chauffeuring. 1964 G. Sims *Terrible Door* xx. 105 A gardener who did some chauffeuring was not immediately available. 1965 N. Freeling *Criminal Conv.* I. xvii. 93 They belonged to the chauffeured stratum of society.

chaulmoogra (tʃǭlmū·gră). Also **chal-, -maugra, -mugra.** [Bengali, f. *cāl, cāul* rice + *mugrā* a type of plant.] Any of several tropical Asian trees of the family Flacourtiaceæ, esp. *Taraktogenos kurzii.* Hence **chaulmoogra oil,** a vegetable oil obtained from the seeds of the chaulmoogra, formerly much used in the treatment of leprosy and skin diseases; **chaulmoogric acid,** an acid found in chaulmoogra oil.

a 1815 Roxburgh *Flora Indica* (1832) III. 836 Chaulmoogra and Petarkura, are the names of this tree, and the drug, hereafter mentioned, which it furnishes. *Ibid.* 837 The seeds of this tree, called Chaulmoogree. 1876 *Encycl. Brit.* V. 675 *Châlmugrâ* oil. 1884 *Ibid.* XVII. 744 Chaulmoogra oil. *Gynocardia odorata.* 1904 F. B. Power & F. H. Gurnall in *Jrnl. Chem. Soc.* LXXXV. 839 Chaulmoogra seeds. *Ibid.* 847 It has been decided to designate this new acid *chaulmoogric acid.* 1907 *Practitioner* Aug. 245 Chaulmoogra oil is generally considered to have the greatest beneficial effect on leprosy. 1933 *Discovery* Jan. 4/1 The ancient Indian remedy for leprosy is chaulmoogra oil, containing chaulmoogric and hydnocarpic acids. 1955 *Sci. News Let.* 13 Aug. 103/3 Chaulmoogric acid has been replaced in leprosy treatment by synthetic drugs of the sulfone family.

chaung (tʃō̆ᵊŋ). [Burmese.] In Burma, a watercourse.

1945 B. Fergusson *Beyond Chindwin* v. 71 Metkalet stood on the same *chaung* some five miles farther south. 1947 'N. Shute' *Chequer Board* 157 They left the main stream and entered the narrow chaungs to reach the Irrawaddy. 1959 M. Pugh *Chancer* 113 Then they crossed a narrow chaung, or dried-up stream with high walls of shingle and stone. 1959 *Times* 19 Mar. 14/7 We found the pair of them later, sprawled among the bushes in the bed of the *chaung*.

chaunt, var. Chant.

Chautauqua (tʃǫtǭ·kwă). orig. and chiefly *U.S.* Also **chautauqua.** [The name of a county and lake in the southwestern part of the state of New York.] **1.** Used esp. *attrib.* to designate a system of home study originating with summer schools held at Chautauqua, or the organization resulting from this, established by charter in 1871.

1873 (*title*) The Chautauqua Lake Journal, published for the Chautauqua Lake Camp-Meeting Association. 1879 *Harper's Mag.* Aug. 357/2 By far the most valuable fruit of the Chautauqua plan..will come from the Chautauqua Literary and Scientific Circle. 1886 J. H. Vincent *Chautauqua Movement* 40 Meetings of the Chautauqua type. *Ibid.*, The camp-meeting period of Chautauqua exercises. 1926 *Even. Standard* 12 July 3/2, I once spent part of a month addressing chautauqua gatherings in a part of New York State.

2. Applied to similar educational meetings of the summer-school type; also *transf.*

1884 *Dakota Mission Conf. of Methodist Episc. Ch.* 5 The Great Chautauqua of the West, Big Stone Lake, Dakota. 1920 S. Lewis *Main St.* 236 She turned to the Chautauqua as she had turned to the dramatic association. 1931 *Economist* 1 Aug. 214/1 The Chequers conversations proved to be the first of a series of statesmen's chatauquas in the capitals of Europe. 1965 M. Echard *I met Murder* (1967) xxi. 161 The rector had been holding forth..on the subject of summer chatauqas and the threat they posed to church attendances.

Hence **Chautau·quan** *a.* and *sb.*

1878 in J. H. Vincent *Chautauqua Movement* (1886) 80 Let every Chautauquan..read Dr. Vincent's lecture. 1909 *Cent. Dict.* Suppl. 1947 *Newsweek* 2 June 58/2 McConnell was an ex-Chautauquan.

chavel (tʃæ·vǝl), *v.* Restrict †*Obs.* to sense 1.

2. (Further examples.) Also *absol.* and *fig.*

1877 E. Peacock *Gloss. Manley, Linc., Chavle*, to chew badly. 'That herse chavles strangely, he wants his teeth filin'.' 1911 D. H. Lawrence *White Peacock* III. vii. § 6 The bracken lay sere under the trees, broken and chavelled by the restless wild winds of the long winter. 1916 —— *Amores* 114 Lives which sorrows like mischievous dark mice chavel To nought, diminishing each star's glitter.

chaw, *sb.*² (Earlier and later examples.)

1709 W. King *Voyage to Cajamai* Pref. in *Useful Trans.*, Tabacco..being twisted like a Cord serves for a Chaw. 1948 D. W. Ballantyne *Cunninghams* (1963) I. xxix. 122 Barry held out a bag of chaws.

chawed, *ppl. a.* **2.** (Later U.S. example.)

1843 'R. Carlton' *New Purchase* xxvii. 257 He emptied all the contents on the counter, viz.: two silver fips, three 'chaw'd bullits', a damaged rifle-wiper [etc.].

chawl (tʃǭl). [ad. Marathi *chāl* a row of rooms let for lodgings, f. Skr. *shālā* house, hall.] An Indian lodging-house.

1891 *Pall Mall Gaz.* 7 Dec. 7/2 The gaol-bird is..better off in every way in gaol than he would be in his chawl. 1898 *Jrnl. Soc. Arts* 25 Feb. 312/1 Some of the Bombay 'chawls' or tenements, containing from 500 to 1,000 inhabitants each. 1946 C. Batley *Architecture* 31 The conglomerations of regimented concrete cubicles known as chawls. 1962 Clinard & Chatterjee in R. Turner *India's Urban Future* iv. 71 Urban slum dwellers reside in *katras*, lanes, chawls, bustees, *ahatas*, and *cheris*.

chay-ka: see *Cheka.

Chazar (kāzā·ɹ), var. *Khazar.

chazzan (kāzā·n). Also, **cazan, chazan, hazan, hazzan,** etc. [Heb. *ḥazzān* superintendent of prayer meetings, sexton, cantor, prob. f. Assyrian *hazannu* overseer or governor.] A cantor or precentor in a Jewish synagogue.

1650 E. Chilmead tr. *L. Modena's Hist. Rites, Customes of Jews* x. 31 There is One, that sings out the Prayers lowder than the rest, whom they call Cazan;..These, as well as the Cazan, as the Shamas, have stipends allowed them. 1675 L. Addison *Present State of Jews* xi. 91 To every Synagogue there belongs six Officers...The sixth is the Chesan (or Præcentor). *a* 1763 'G. Psalmanazar' *Memoirs* (1764) 270 Conversing with the Jews themselves, going to their synagogues, and hearing the Scriptures read by their *chazans* or ministers. 1891 J. E. H. Thomson *Bks. which influenced Our Lord* 5 The synagogue school, taught by the old *hazzan*. 1892 Zangwill *Childr. Ghetto* I. 7 Prayers were shouted rapidly by the congregation, and elaborately sung by the *Chazan*. 1905 *Jewish Encycl.* XI. 19/1 The senior hazzan of the Bayswater Synagogue. 1957 L. P. Gartner *Jewish Immigrant* (1960) vii. 189 The observant immigrant in the East End..would..welcome the musical ministrations of a visiting *hazan* (cantor).

cheap, *a.* Add: **1. a.** Phr. *cheap and nasty*: of low price and bad quality; inexpensive but with the disadvantage of being unsuitable to one's purposes. Hence *cheap-and-nastiness.*

1822 C. M. Young *Let.* in Mrs. Mathews *Mem. C. Mathews* (1839) III. 312 Elliston has abandoned the *cheap and nasty* system, and adopted the liberal one! 1823 [see Stereo¹]. 1831 [in Dict.]. 1867 Carlyle *Shooting Niagara* vii. in *Misc. Ess.* (1872) VII. 226 'Cheap and nasty'; there is a pregnancy in that poor vulgar proverb, which I wish we better saw and valued! 1905 *Studio* Sept. 368/1 The cheap-and-nastiness of our suburban houses. 1911 *Encycl. Brit.* XXVIII. 810/1 It does not follow, however, that all woollen cloth is cheap and nasty.

b. *cheap fare*: a fare at a lower rate than the ordinary fare; also *cheap rate*. Also *attrib.*

1710 W. Salmon *Family Dict.* (ed. 4) Pref. § 7, Poor People would be furnished..with Firing..at very cheap Rates. 1776 *Monthly Rev.* Aug. 162 It may..serve to convey information 'at a cheap rate'. 1867 *Cassell's Mag.* I. 438/2 Omnibuses..for many years..were running at a cheap rate. 1887 *Peel City Guardian* 28 May 5/1 To make special daily sailings from Liverpool to the Isle of Man and back at cheap fares. 1901 *Daily Chron.* 12 Aug. 7/5 For trade and cheap-rate passengers. 1902 *Westm. Gaz.* 14 Feb. 6/2 All the 'cheap-fare' vehicles. 1967 *Guardian* 4 Jan. 5/3 During the cheap rate period (10 p.m. to 10 a.m. British clock time Monday to Saturday, and all day Sunday) the charges for an ordinary call will be reduced.

c. Applied to money obtainable at a low rate of interest.

1861 *Economist* 23 Mar. 309/2 Are we to look for cheap money soon, for progressive reductions in the rate of interest? 1940 G. Crowther *Paying for War* 31 But cheap money will be impossible..so long as there is a mass of 5 per cent war debt overhanging the money market.

4. b. In poor health; out of sorts. (Hence *cheapness.*) *slang.*

1891 Farmer *Slang* s.v., *To feel cheap*, to 'have a mouth on'; to be suffering from a night's debauch. 1900 *Westm. Gaz.* 28 May 2/1 Putting down his weakness, lassitude, and general feeling of extreme cheapness to the climate. 1924 R. Rees *April's Sowing* xi. 141 'What's wrong with you, eh?' 'Nothing... A bit cheap, that's all.'

5. a. (Later examples.)

1936 N. Coward *Astonished Heart* iii, It was all a trick— I planned it...I made up my mind then to make you fall in love with me—now I wish I hadn't—I feel cheap. 1959 G. Freeman *Jack would be Gentleman* vii. 137 They were engaged after all, it wasn't being cheap..it was just the full expression of their love.

6. Also *as cheap as dirt*: exceedingly cheap(ly). (See also Dirt-cheap *a.*)

a 1817 Jane Austen *Persuasion* (1818) IV. ix. 190 All the honour of the family he held as cheap as dirt. 1844 'J. Slick' *High Life N.Y.* II. xxxi. 228 They're as cheap as dirt. 1918 Svartengren *Intensifying Similes* 346 Cheap 's dirt.

C. (Earlier and later examples.)

1859 Hotten *Slang Dict.* s.v., 'Doing it on the cheap', living economically, or keeping up a showy appearance with very little means. 1939 'G. Orwell' *Coming up for Air* II. x. 170 Anything from theosophy to cat's-cradle, provided you can do it on the cheap.

2. *pl.* Things that may be bought at a low price or cheaply, *esp.* cheap editions of books.

1930 *Publishers' Circular* 23 Aug. 255/3 We understood that it was a new book, but surely this cannot be when we find the same story with the same title included in Pearson's Sixpenny cheaps. 1938 *Times Lit. Suppl.* 25 June 440/3 The Standard Novels and representative examples of similar 'cheaps'. 1952 N. Streatfeild *Aunt Clara* 94 Nice queue for the cheaps [*sc.* the cheap seats].

D. Cheap Jack (also **cheap-jack**) (earlier examples); also *attrib.* and *transf.*; hence **cheap-jackery,** that which is characteristic of a Cheap Jack; **cheap skate** (see *skate *sb.*³).

1851 Mayhew *Lond. Labour* I. 214/1 The Cheap Jacks, or oratorical hucksters of hardware at fairs and other places, are among the most celebrated..of this class. 1865 Dickens in *All Year Round* Christmas 1. 2/2 Of all the callings ill used in Great Britain, the Cheap Jack calling is the worst used. 1908 *Daily Chron.* 3 Feb. 1/7 The cheap Jack nostrum of Tariff Reform. 1953 J. Cary *Except the Lord* xviii. 80 The cheapjack men,..the freaks and the flea circus. 1921 *Spectator* 28 May 684/1 Many politicians are willing to see every political struggle decline into cheapjackery. 1960 *Times* 17 May 13/5 A world of fairs.. of cheap-jackery.

cheat, *sb.*¹ Add: **4. c.** *Cinematogr.* (See quots. and cf. *cheat *v.* 7). Also *cheat shot.*

1948 E. Lindgren *Art of Film* II. v. 73 This procedure of concealing part of the action necessary to the taking of a shot by excluding it from the camera's field of view is known..as 'cheating', and the shot as a 'cheat shot'. 1960 *Consumer Rep.* Jan. 10/2 A phony picture used in a commercial (known in the trade as a cheat shot).

9. (Examples.)

1784 J. Smyth *Tour U.S.* II. 121 My fields consisted entirely of fine healthful clean wheat, without a single head of darnel or cheat. 1850 *Rep. U.S. Comm. Patents Agric.* 1849 455 Cheat is nothing more nor less than degenerated wheat. 1861 *Trans. Ill. Agric. Soc.* IV. 351 'Chess', or 'cheat' as it is sometimes called is a species of grass. 1863 *Ibid.* V. 865 *Bromus secalinus*, cheat, chess, is a foreign annual weed. 1866 Brogden *Prov. Words Lincs.* 39 The field is very full of cheats to-year.

cheat, *v.* Add: **3. b.** To lead *into* (an action) by deception.

1856 De Quincey *Confess.* 264 He..could not but find.. himself cheated into cordial admiration, by the splendour of the verses. 1888 Mrs. H. Ward *R. Elsmere* x, They had tried to cheat her into sleeping.

7. *trans.* and *intr. Cinematogr.* To suppress or adapt (part of a film) during editing so as to create a desired illusion. So **cheated** *ppl. a.*, **cheating** *vbl. sb.* Cf. *cheat *sb.*¹ 4 c.

1929 I. Montagu tr. *Pudovkin's On Film Technique* vi. 192 Any movement or object outside the picture-frame or otherwise unremarked is said to be 'cheated'. 1933 A. Brunel *Filmcraft* ii. 72 Get as close as you can in order to reveal your characters..in some detail...You can..'cheat' them nearer—that is, bring them closer together and photograph them at a slightly different angle from the previous scene. *Ibid.*, You may have two characters by a settee in the centre foreground...For the second shot you can 'cheat' the settee nearer the door in order to facilitate or expedite your characters' progress. *Ibid.*, 'Cheating' does not only apply to the position of characters, but to the position of the furniture and other objects in your sets. 1961 G. Millerson *Telev. Production* xviii. 347 The cheated substitute provides interesting possibilities. Here a series of representative parts are intercut to convey a composite impression.

cheater. Add: **5.** *pl.* Spectacles, eye-glasses. *U.S. slang.*

1921 Wodehouse *Jill the Reckless* ix. 138 A tall guy in tortoiseshell cheaters. 1932 Runyon *Guys & Dolls* 35 A little guy who wears horn cheaters. 1949 R. Chandler *Little Sister* ii. 9 The eyes behind the rimless cheaters flashed.

chebacco. Delete etym. except '(By Worcester..Massachusetts.)', and add earlier and later examples. *Obs.* exc. *Hist.*

1823 J. F. Cooper *Pilot* (1824) I. ii. 22, I was born on board a chebacco-man. 1835 J. H. Ingraham *South-West* I. iii. 27 Those short stump-masted *non-de-scripts*..sometimes denominated fishing smacks, but oftener and more euphoniously 'Chebacco boats'. 1859 *Congress. Globe* 22 Feb 1210/1, I recollect a little stream in the county of Essex, in Massachusetts, where, some fifty years ago, they used to manufacture a sort of little boat, called chebacco boat. 1886 *Leslie's Pop. Monthly* XXI. 223/1 Squam was in its ascendency in the days of the old pinkies and of the still earlier chebacco-boats.

Chechen (tʃe·tʃen). Also **Tchechene,** etc. [a. Russ. †*chechén* (now *chechénets*), pl. †*chechény* (now *chechéntsy*).] **a.** (One of) a North Caucasian people, forming the major part of the population of the Russian Autonomous Republic of Checheno-Ingushetia. **b.** The North Caucasic language of this people.

1814 F. Shoberl tr. *J. von Klaproth's Trav. Caucasus* xxvi. 315 The authority of the Tscherkessian princes formerly extended over the Ossetes, the Tschetschenzes, the Abasses, and the Tartar tribes. 1854 I. Golovin *Caucasus* vi. 82 The Tchechenes amount only to 25,000 inhabitants. 1902 *Encycl. Brit.* XXVI. 617/2 A variety of branches from Daghestan described under the general name of Chechens. 1928 W. Jochelson *Peoples of Asiatic Russia* vi. 143 The Chechen belong to the North Caucasian Japhetides..and include a group of tribes speaking different dialects of the Chechen language. 1957 Carew Hunt *Guide to Communist Jargon* xxi. 77 The false impression that the Caucasian peoples..were hostile to Russia, whereas this had been true only of the Ingush and the Chechens.

chechia (ʃēⁱ·ʃiă). Also **checchia,** etc. [Fr. *chéchia*, ad. Maghribi Arab. *šāšīya*, f. Arab. *Šāš*, name of a town in Transoxiana.] A cylindrical tufted cap or fez worn by Arabs and formerly by French troops in Africa.

1853 J. Richardson *Narr. Mission Central Afr.* I. 206, I..offered him..a shasheeah (or fez). 1909 W. J. Locke *Septimus* xiii, He [*sc.* a Zouave]..swaggered off, his chechia at the very back of his head. 1923 *Chambers's Jrnl.* 331/1 A small procession of natives in red *chechias*. 1924 Hichens *After the Verdict* III. i, An omni bus..driven by a big Arab in European clothes and a chechia. 1968 F. van W. Mason *Deadly Orbit Mission* ii. 26 The wiry little man..was wearing his inevitable *chéchia* or dull-red fez.

check, *int.* and *sb.*¹ Add: **A.** *int.* **2.** Used to express assent or agreement. (Cf. Check *v.*¹ 16 a.) *U.S. colloq.*

1922 S. Ford *Trilby May crashes In* iii. 54 'I have felt all along that—' 'Check!' says I. 'I get the picture.' 1929 S. V. Benét in *Sel. Works* (1942) II. 407 'First, you're pizen-sure this man is—' 'A cat. Yeah.'..'Check.' 1934 J. O'Hara *Appt. in Samarra* (1935) vi. 178 'So let's dance.' 'Check and double check,' said Helene, and got up. 1940 R. Chandler *Farewell, My Lovely* vii. 48 'You are Philip Marlowe, a private detective?' 'Check.' 'Oh—you mean, yes.' 1968 'L. Bruce' *Death on Romney Marsh* xvi. 176 'Tell her.. I've found out all she wants to know.' 'Check.'

B. *sb.* **5. e.** A crack or flaw in timber. orig. *dial.* (cf. *season-check).

1877 E. Peacock *Gloss. Manley, Linc.* 55/1 That wood's full o'checks; it'll nivver do to make furk-shafts on. 1905 *Terms Forestry & Logging* 33 *Check*, a longitudinal crack in timber caused by too rapid seasoning. 1949 *Gloss. Terms Timber (B.S.I.)* 19 *Check*, a separation of the fibres along the grain, forming a crack or fissure in the timber, not extending through the piece from one surface to another.

f. *Mining.* A slight fault or dislocation of the strata.

1883 Gresley *Gloss. Coal-m.*

10. e. Phr. *checks and balances*: means of limiting or counteracting the wrongful use of administrative power. orig. and chiefly *U.S.*

1787 J. ADAMS *Def. Const. U.S.* I. p. iii, The checks and balances of republican governments have been in some degree adapted by the courts of princes. 1842 TYLER in *Messages & Papers of Presidents* (U.S.) IV. 193 A proceeding tending to the utter destruction of the checks and balances of the Constitution. 1958 *Economist* 13 Dec. (African Suppl.) 24/1 A far more complicated federal constitution of checks and balances has had to be adopted to hold together.. Nigeria.

f. A form of catch on a rein; *ellipt.* a check-rein. *U.S.*

a 1867 H. WOODRUFF *Trotting Horse Amer.* (1868) xxiii. 202 In order to prevent him from throwing down his head.. the well-known Kemble-Jackson check..was invented. 1887 A. W. TOURGÉE *Button's Inn* 122 Throw me that off rein, if you please..Just shift those inside checks, won't you?

14. b. (Early U.S. examples.)

1835 J. H. INGRAHAM *South-West* I. xxi. 223 A shouting of 'Your check, sir! your check!—Give me your check—Please give me your check!' [for re-admission to a theatre]. 1850 *Wilmington* (N.C.) *Commercial* 28 Feb. 3/3 Porters will receive checks, take charge of the baggage, and convey it to the Hotel.

c. A restaurant bill. Chiefly *U.S.*

1869 A. D. WHITNEY *P. Strong's Outings* 128, I let her settle for the dinner checks. 1910 'O. HENRY' *Strictly Business* 192 Through an arched opening..you thrust your waiter's check and the money. 1916 *Variety* 27 Oct. 12/3 Inspectors..ordered drinks and paid their check just before one. 1937 R. STOUT *Red Box* vii. 88, I..found the waitress and got my check from her.

15. (Earlier and later examples.) Also *to cash, pass* or *send in one's checks*. orig. and chiefly *U.S.*

1845 J. J. HOOPER *Adv. Simon Suggs* (1851) v. 57 He called for 'Twenty, five dollar checks..'. The dealer handed him the red checks. 1857 *Spirit of Times* 7 Mar. 6/1 Those noble and peril-loving souls have nearly all 'handed in their checks'. 1869 *Overland Monthly* III. 31 Three or four miners and axemen sat whittling on the logs as the doctor came out, and Hy Fender asked: 'Well, now; has he passed in his checks?' 1888 [see *CASH v.² 2 b]. 1893 C. M. YONGE & C. R. COLERIDGE *Strolling Players* xxxii. 292, I was too bad for twenty-four hours to tell him I wasn't exactly going to send in my checks this time. 1922 JOYCE *Ulysses* 420 Chum o yourn passed in his checks? 1947 W. S. MAUGHAM *Creatures of Circumstance* 111 Ivy's promised me that when I hand in my checks she'll come back here.

19. check-reel, also, an angler's reel fitted with a check (see sense 10 c in Dict.).

1892 NIVEN *British Angler's Lex.* 72 Check reels..should be humoured so as not to be too stiff or too slack. 1904 GALLICHAN *Fishing Spain* 208 Two metal check-reels for trout fishing.

check, *sb.*² Add: **3.** *Agric.* Each of a series of squares made by cross-marking. So **check-row,** one of a series of rows (in planting) so arranged as to form a check-pattern; also *attrib.* *U.S.*

1787 WASHINGTON *Diaries* (1925) III. 194 In each of these checks or crosses, one root, when it was large and looked well was put, and two where they were small. 1857 *Trans. Ill. Agric. Soc.* III. 62, I lay off my ground with a corn-marker..into checks of three feet three inches square. 1859 *Rep. Comm. Patents* (U.S.) *1858* I. 474 The seed may.. be distributed from either hopper, and sown either in drills or check-rows. 1861 *Trans. Ill. Agric. Soc.* IV. 209 In spring it will be harrowed thoroughly, planted by hands in check rows three feet ten inches apart each way. *Ibid.* 312 Most of the corn is now planted with drills, or check row machines.

Hence **check-rowed** *a.*, planted in check-rows; **check-rower,** a corn-planter, or a device attached to one, dropping the seed-corn in check-rows.

1888 *Sci. Amer.* LVIII. 298/1 Particularly for use on growing check-rowed and listed corn. 1882 *Belleville* (Ill.) *Advocate* 9 June 4/4 (Advt.), Haworth Check Rowers, with Wire or Cord. a 1884 KNIGHT *Dict. Mech.* Suppl., Check-rower.

check, *v.*¹ Add: **5. a.** Delete †*Obs.* and add later examples.

1901 KIPLING *Kim* vi. 137 In the forenoon the column checked. 1916 'BOYD CABLE' *Action Front* 254 Even in the midst of the swing he checked, glanced once at the spitting fuse, and with a stoop and a heave flung the officer out over the front parapet. 1962 *Times* 12 July 4/1 His chip back just caught the top of the bank and checked.

11. Now *colloq.* (Later examples.) Also with *off*.

1920 F. M. FORD *Let.* 30 May (1965) 100, I offered you the *Heaven* volume for nothing in '17 & was just checked off as if I had been a literary beginner. 1931 T. E. LAWRENCE *Lett.* (1938) 725 This morning I was hut orderly, and got checked for dull brass door-knobs. 1948 PARTRIDGE *Dict. Forces' Slang 1939–45* 38 *Check*, to reprimand, take to task, during the exercise of one's duty.

16. c. To accept or hand over (an article) in return for a check (see CHECK *sb.*¹ 14 b); to send to a destination in this way. orig. and chiefly *U.S.*

1846 *Daily Even. Traveller* (Boston) 16 July 3/2 Passengers..will consult their comfort and convenience by being particular to have their Baggage 'checked'. 1860 *Congress. Globe* 21 Dec. 177/2 It is a great convenience to the traveling public to be able to check baggage through. 1866 *Ibid.* 20 July 3972/3 The Baltimore road..will not check baggage from here to any point in the West. They..compel you to recheck your baggage. 1888 *Amer. Humorist* 21 July (Farmer), Turning to the man who checks umbrellas and canes. 1897 *Westm. Gaz.* 25 Feb. 10/2 (U.S.), Remove your hats during the performance. You can check them with the maid. 1922 *Publishers' Circular* 21 Jan. 43/2 We began to require all person..to check these books at the coat-rooms. 1956 'N. SHUTE' *Beyond Black Stump* 4 They walked out into..the club and checked their hats.

d. *to check up* (*on*) or *to check on*: to examine carefully or in detail; to maintain a check on; to ascertain the truth about; also *absol.*; *to check out*, to investigate, examine for accuracy, authenticity, or a confirmation of fitness. orig. *U.S.*

1889 *Kansas Times & Star* 15 Mar., He says Willis checked them up closely and discovered nearly 600 saloons here, and only 400 paying a license. 1911 H. S. HARRISON *Queed* xxiv. 309 In five years the experiment might enable me to check up some of my own conclusions. 1925 H. L. FOSTER *Trop. Tramp Tourists* 51 As the sailing hour approached, the staff and stewards were busy checking up. 1926 G. HUNTING *Vicarion* xvi. 266 He..offered to set a condenser up for her in her own home where she could check up on any of his activities she desired. 1928 C. M. FUESS *Men of Andover* 13 Every effort has been made to check up on even the least important statements. 1932 E. WILSON *Devil take Hindmost* ix. 92 These people will have to be checked up on. 1938 E. BOWEN *Death of Heart* II. iii. 212, I have never checked up on this. *Ibid.* iv. 248 One solid pleasure of love is to check up together on what has happened. 1957 D. J. ENRIGHT *Apothecary's Shop* 127 Reading back through the book to check up on the implied parallel. 1959 *Daily Mail* 20 Feb. 10/6 The M.C.C. had a man placed either at mid-on or mid-off..obviously to check on me. 1962 F. I. ORDWAY et al. *Basic Astronautics* ix. 388 Most hydraulic control systems have electrical components..thus necessitating two systems to install, check-out, and maintain. 1967 *Landfall* XXI. 243 A married man with a rich wife and a packet of kids...I checked him out. 1968 S. CHALLIS *Death on Quiet Beach* v. 72 Kay. We'll check her out. 1968 *Times* 16 Dec. 7/4 To begin with, Apollo 8..will be parked in a 119-mile orbit while the crew check out the spacecraft.

e. *to check in* or *out*: to record the incoming or outgoing of (guests, employees, books, etc.). Also *intr.*, to record one's arrival at or departure from a hotel, factory, etc. (freq. *to check into* or *out of*); to arrive or depart. orig. *U.S.*

1918 *Wine, Women & War* (1926) 35 R—— checked in 3.30, more than usually drunk. 1921 S. FORD *Inez & Trilby May* iii. 41 The singer person is checking out from the first floor suite next week. 1931 W. FAULKNER *Sanctuary* xx. 179 'That ain't my affair, where folks go after they check out,' the proprietor said, turning his back. 1935 E. BOWEN *House in Paris* III. iv. 317 'Then we might get something to eat.' 'What, at the hotel?' 'I'll see; we'll check in these first, anyway.' 1936 STEINBECK *In Dubious Battle* i. 6, I checked out of my lodging-house before I came here. 1941 *Amer. Speech* XVI. 310/1 To check out or charge a book, the borrower fills out a call card. 1943 P. CHEYNEY *You can always Duck* ii. 28 He'll probably check in at some hotel. 1951 'N. SHUTE' *Round the Bend* 286, I was just going back to the hotel to check out. 1951 *Manch. Guardian Weekly* 8 Feb. 3 Two of them had checked out of the demonstration and packed up. *Ibid.* 6 Sept. 3 He checked into his hotel. 1952 M. MCCARTHY *Groves of Academe* (1953) iv. 66 The student was obliged to check in with the instructor for the requisite hours of studio or laboratory work. 1959 [see *BIDET 2].

f. To note with, or indicate by, some mark.

1928 *Publishers' Weekly* (N.Y.) 12 May 1973 For the reasons checked below, we regret that we cannot undertake the publication of your proposed book. 1929 *Ibid.* 14 Sept. 1064 Check the titles you want and we will mail you the books at once. 1969 *N.Y. Rev. Books* 16 Jan. 35/3 (Advt.), Please check appropriate squares and type..your name and address below.

g. *intr.* To agree upon comparison. *U.S.*

1928 *Publishers' Weekly* (N.Y.) 22 Dec. 2491/2 One of the sheets,..checked closely with fiction which was found in the Gottschalk store. 1935 M. M. ATWATER *Murder in Midsummer* xv. 140 It all checked pretty well. *Ibid.* xvii. 156 If there's finger-prints on it, and yours don't check, that'll let you out.

17. (Earlier U.S. example.)

1809 *Deb. Congress* 20 Feb. 416 The money..is deposited in the Treasury as in a bank..to be checked for, whenever that commerce..shall be again reopened.

18. *trans.* (*a*) *Carpentry.* To notch or halve (timbers) in making a cross joint. *Sc.* (*b*) *Masonry.* To notch (one stone) *into* (another); also *to check down*. (*c*) To join (two pieces) in this manner (*Cent. Dict.* Suppl. 1909).

1833 LOUDON *Encycl. Cottage Archit.* § 1066 All plates to be in long lengths, and chacked. *Ibid.* 1072 The rafters to be..chacked and spiked together. *Ibid.* 1778 The stair to the cellars..to be droved; the steps to be checked down on (notched into) each other. 1885 *Spons' Mech. Own Bk.* 678 The other 2 stones..are to be half-checked into it, also half-checked into each other where they meet in the middle.

check, *v.*² Add: **1. b.** To mark (ground) for planting in checks. *U.S.*

1768 WASHINGTON *Diaries* (1925) I. 265 At the first and last of which [plantations I] just began to check Corn G[roun]d. 1871 *Trans. Ill. Agric. Soc.* VIII. 239 After the field has been thoroughly prepared..proceed to check it off from east to west with a three-rowed marker. 1945 B. A. BOTKIN *Lay my Burden Down* 168 And checking corn is running a straight row clean 'cross the field both ways.

c. *trans.* and *intr.* To split or crack along crossing lines.

1641 H. BEST *Rural Econ.* (1857) 15 The sun shoulde not checke and rive them [*sc.* wooden stakes]. 1879 WEBSTER Suppl., *Check*,..to crack in small checks, as varnish, paint, and the like. 1880 *Harper's Mag.* Aug. 398 The streaks of the clinker-built canoe rarely check, the wood being generally well seasoned. 1902 *Contrib. Econ. Geol., U.S. Geol. Survey* 277 The coal is not crushed, but can be obtained in large pieces which 'check' but do not break up readily on exposure to the air. 1940 *Chambers's Techn. Dict.* 154/2 *Checking*, a defect in a painted surface, characterised by the appearance of fissures in all directions.

check-. Add: **check-back** = CHECK *sb.*¹ 11; **check-band,** a drag-device attached to a spinning mule to check the varying velocity of the spindle carriage; **check-bar** (see quot.); **check-chain,** a chain used to check the movement of a mechanism, a vehicle, etc. (see quots.); **check-cord,** (*a*) a cord used to check action or movement, *lit.* and *fig.*; spec. a long cord attached to the collar of a hunting dog to bring it to a sudden stop; (*b*) = CHECK-STRING; **check-dam** (see quot. 1953); **check-lock,** (*b*) (see quot.); so *check-locking*; **check meter,** (*a*) an instrument used to test the accuracy of electricity meters; (*b*) a meter used to measure the amount of gas or electricity used in a separate part of a house, etc.; **check-off (system)** (see quot. 1923); **check-out,** (*a*) *U.S.* (see quot. 1956); (*b*) the act or process of checking out (see *CHECK *v.*¹ 16 e); also *attrib.*; (*c*) a desk at which payment is made in a self-service shop; also *attrib.*; **check-pen** *N.Z.*, a special pen in a sheep-yard (see quot. 1950); **check-point** (orig. *U.S.*), (*a*) *Aeronaut.* (see quot. 1940); (*b*) a place (entrance, turnstile, barrier, etc.) where the movement of traffic, pedestrians, etc., is checked; a control-point; **check-post** = *check-point* (*b*); **check-rail** = GUARD-RAIL 2; **check-rein** (earlier example); **check-room** *U.S.*, a cloak-room or baggage-room in a hotel, railway station, etc.; **check rope,** a rope used to check the recoil of a gun; **check-set** *U.S.*, a device for setting out the checks for planting; **check-stand** *U.S.*, a stand in which 'checked' articles are placed; **check-strap,** (*a*) *U.S.*, a strap controlling the bit in a horse's mouth; (*b*) a strap designed to prevent a door from opening too wide; **check-valve,** a valve to prevent backward flow (*spec.* of water in a supply pipe).

1926 *Electr. World* 7 Aug. 280 The meter installers soon become very careful about reading meters, and thereby become a check-back on any one who reads the meter later. 1931 G. O. RUSSELL *Speech & Voice* 6 This has made possible a check-back on acoustic control. 1954 KOESTLER *Invis. Writing* xxviii. 306 Checkbacks had shown that the information was correct. 1892 NASMITH *Students' Cotton Spinning* 276 Two principal faults arising from an imperfect adjustment of the check band. a 1877 KNIGHT *Dict. Mech.*, *Check-bar*, a bar which limits the backward play of the jacks [in a piano-movement]. 1805 R. W. DICKSON *Pract. Agric.* I. 8 The latter has generally a check-chain, by which the wheel is pulled up, in order to be out of the way. a 1884 KNIGHT *Dict. Mech.* Suppl., *Check Chain*, a chain connecting the car body with its truck. 1896 *Strand Mag.* XII. 325/1 The sliding ways, cradle, and ship..glide down the appointed pathway..until retarded..and finally brought to rest by check-chains..connecting ship and shore. 1908 *Daily Report* 7 Feb. 1/4 When the competitors..combine to fleece their customers, the check-cord on their power to run up prices is sometimes more difficult to apply. 1936 *Experiment Station Rec.* Jan. 112 Brush check dams seem most effective in regions where sandy soils predominate. 1943 J. S. HUXLEY *TVA* vi. 30 The building of check-dams ..for anti-erosion purposes. 1953 *Brit. Commonw. Forest Terminol.* I. 25 *Check dam*, a dam built in a watercourse to prevent or reduce erosion. 1903 *Westm. Gaz.* 1 Oct. 10/3 The 'check-locking' arrangement..ensures that until the movement of a point or signal has been fully completed the lever in the signal-box..is checked...As soon as the signal or point movement..is properly completed, the 'check-lock' is electrically removed. 1909 Check meter [see *CALIBRATE *v.*]. 1964 *Economist* 28 Mar. 1280/1 'Check meters' are used by landlords who have one mains gas or electricity connection, and only one meter. 1911 *Daily Colonist* (Victoria, B.C.) 27 Apr. 1/4 The management.. has refused the local union a continuance of the 'check-off', which has been in force heretofore. 1922 *Tom Mooney's Monthly* (S.F.) Nov. 4/4 The miners were on strike against a cut in wages [and] abolition of the 'check-off' system. 1923 J. D. HACKETT *Labor Terms* in *Management Engineering* May, *Check-Off System*, a system whereby initiation fees, fines, and dues of union employees are deducted from their wages by the employer and periodically remitted to the district union organization. 1965 *Times* 21 May 15/2 The deduction of union dues from pay by employers—the 'check-off' as the Americans term it. 1944 *Plane Talk* Sept. 24 Advancement to radio operator 'A' may be earned by.. training that must include checkout on several types of multi-engine airplanes. 1956 W. A. HEFLIN *U.S.A.F. Dict.* 111/2 *Checkout*, instruction or training given a pilot to familiarize him with a given aircraft. 1958 *Punch* 6 Aug. 172/2 Hotel check-out times in America are quite late in the day. 1962 E. GODFREY *Retail Selling & Organ.* xxi. 204 Few people nowadays are unfamiliar with the principle of collecting a basket, selecting the goods required from shelves and taking them to the check-out point. 1964 *Times Rev. Industry* Mar. 98/2 The latter [store] has no less than 60 check-outs. 1922 *N.Z. Jrnl. Agric.* 20 Sept. 152 A further lot of sheep [can be] placed in the check-pen. 1950 *Ibid.* July 7/3 In larger sets of yards one or more of these drafting pens communicate, via a gate, with the check pen(s)..Their purpose is to hold the sheep while they are being checked over for mistakes in drafting and for treatment. 1940 H. E. BAUGHMAN *Aviation Dict.* 48/1 *Check point.* (1) A check point is a geographical location on the

surface of the land or water, above which the position of an aircraft in flight can be accurately determined by means of visual reference..(2) Radio Range Station when instrument flight is necessary. 1940 CRUMP & MAUL *Our Airliners* 101 Allentown, the next check point, is twenty-four minutes away. 1950 C. MACINNES *To Victors the Spoils* I. 48 We were stopped by a Redcap at a check-point. 1959 *Oxf. Mag.* 26 Feb. 274/2 It is observed that *x* vehicles pass the various check-points on their way into the city. 1959 *Times* 15 Apr. 13/2 The Russian stopping of four American lorries at the road check-point at Helmstedt. 1965 *Listener* 9 Dec. 941/1, I showed our passports at Checkpoint Charlie. 1949 F. MACLEAN *Eastern Approaches* II. iii. 203 Anxiously we asked what we might expect in the way of check posts or road blocks. 1876 J. W. BARRY *Railway Appliances* ii. 58 The extra rail, which is called a check rail, relieves the sideways pressure of the wheels. 1906 *Westm. Gaz.* 3 July 5/1 The check-rail..—a rail laid..to minimise the danger of the curve. 1962 D. A. BOREHAM *Narrow Gauge Railway Modelling* ii. 20 For this reason the very greatest care should be taken with check-rails. a 1809 T. HOLCROFT *Mem.* (1816) I. xiv. 123 He had shewn himself restive, and ..was ridden in a check-rein. 1900 G. ADE *More Fables* 154 He had gone past, on his way to the checkroom. 1969 *New Yorker* 11 Oct. 43/2 In the new checkrooms, there's going to be a marvellous new form of coat-checking. 1918 FARROW *Dict. Mil. Terms*, *Check ropes*, strong ropes employed to diminish recoil by increasing the frictional resistances. 1861 *Trans. Ill. Agric. Soc.* IV. 248 He must have a boy to tend the check-set of the corn-planter. 1886 E. W. HOWE *Moonlight Boy* 102 My friend left his own baggage at a check stand. 1904 F. LYNDE *Grafters* ii. 31 The train was in, and the porter had fetched Loring's hand-bag from the check-stand. 1961 *Progressive Grocer* Dec. 54 View over checkstands shows..small..area with plywood walls..showing non-foods on special. 1887 *Scribner's Mag.* Oct. 508/1 'I'll put a check-strap on him, if he won't do it!' a little chap exclaimed..using a phrase drawn from the training of horses. 1962 *Which?* Suppl. July 107/2 The door check straps were disintegrating. a 1877 KNIGHT *Dict. Mech.*, Check-valve. 1930 *Engineering* 14 Nov. 627/2 Connected in the feed range immediately before the boiler check valves. 1963 *Gloss. Mining Terms* (*B.S.I.*) IV. 9 *Reflux valve* (check valve, deprecated; retaining valve, deprecated), an automatic non-return valve which opens freely to permit fluid to pass in one direction but closes under its own weight when motion ceases or when the fluid commences to flow in a reverse direction.

checked, *ppl. a.*¹ Add: **1. b.** Applied to a syllable that ends in a consonant, or to the vowel in such a syllable; cf. CLOSED *ppl. a.*

1943 K. L. PIKE *Phonetics* 119 A syllable is *checked* (or *closed*) whenever it ends in a contoid. 1952 A. COHEN *Phonemes* iv. 75 The so-called 'Short' or 'Checked' Vowels. 1962 *Amer. Speech* XXXVII. 164 In Table 1, the checked vowels ..are paired with the free vowels which are phonetically most similar to them.

checker, *sb.*¹ Add: **3.** A person who or a thing which checks, impedes, or retards.

1845 *Jrnl. R. Agric. Soc.* VI. II. 548 Checkers or curers of the disease.

checker, *sb.*² Add: Also 8 **checkard, -erd.**

2. (Earlier examples.)

1712 SEWALL *Letter-Bk.* (1886) I. 417 You us'd to Huff him, and humble him at a game of Checkers. 1794 A. ELLICOTT in C. V. Mathews *Life & Lett.* (1908) 119 We amuse ourselves with playing checkerds.

checking, *vbl. sb.*¹ Add: **3.** *attrib.*, as *checking-book*; **checking account** *U.S.*, a current banking account; **checking-room,** a room in which goods, etc. are checked; spec. *U.S.* = *check-room* (see *CHECK-).

1923 L. LeM. MINTY *Amer. Banking Methods* v. 152 Unlike the English bank, the American bank frequently allows interest on checking (what we call current) accounts. 1926 *Springfield Weekly Republ.* 19 Aug. 10 She preferred the more generous way, and they had a joint checking account. 1895 *Westm. Gaz.* 10 Oct. 3/1 Others..are content with inscribing their names on their checking-book... The scene..in the checking-room is by no means a quiet one, for nearly every jobber's clerk..is shouting his firm at the top of his voice. 1910 *N.Y. Even. Post* 13 Dec. 7 Mr. Spottford arrived at the station carrying a small grip, and asked Charles where the checking-room was.

che·ck-list. orig. *U.S.* [CHECK-.] A list of names, titles, etc., so arranged as to form a ready means of reference, comparison, or verification; *spec.* a list of qualified voters for use at an election.

1853 (*title*) Check list of periodical publications received in the reading-room of the Smithsonian Institution. 1873 J. ROBINSON (*title*) Check List of the Ferns of North America north of Mexico. 1885 Check-list [in Dict. s.v. CHECK-]. 1888 BRYCE *Amer. Commw.* II. II. xii. 443 The composition of a primary is determined by the roll or 'check-list', as it is called, of ward voters entitled to appear in it. 1894 T. R. SIM (*title*) Sketch and check-list of the flora of Kaffraria. 1937 *British Birds* XXXI. 7 The numbers and former names refer to the systematic list printed in the last part of the *Practical Handbook* and reprinted in the *Check-List*. 1958 *Listener* 2 Oct. 514/2 This book, which is at once history, criticism, and check-list.

check-row: see *CHECK *sb.*²

check-up (tʃe·kʌp). orig. *U.S.* [f. *CHECK *v.*¹ 16 d.] A careful or detailed examination, scrutiny, or comparison with a list; *spec.* a medical examination.

1921 A. WALL *Analytical Credits* 23 In any order of considerable size..the credit man should..request the local manager..to secure a check-up by night message. 1924 *Glasgow Herald* 18 Sept. 7 A check-up of the miners shows that the number entombed..was 47. 1927 *Hutchinson's Myst. Story Mag.* X. 106 In the usual check-up of the rooms at one-thirty, the clerk became suspicious. 1958 L. A. G. STRONG *Light above Lake* 181 Just for a check-up. Ought to have had it before the wedding. 1960 *Woman's Own* 13 Feb. 18/4 Father went and had a check-up, and.. there was a suspicion of ulcer, the doctor said.

chedar (he·dəɹ). Also **cheder, heder.** Pl. **chedarim.** [Heb.] A Hebrew school for Jewish children.

1882 tr. *L. Kompert's Scenes from Ghetto* 188 At *Cheder* I often put my *Rebbe* in the greatest embarrassment by proposing to him questions which he could not answer. 1892 ZANGWILL *Childr. Ghetto* I. 143 His father could not afford to send him to a *Chedar*. 1899 *Daily News* 17 Jan. 4/7 The Chedarim, or religious schools. 1900 C. RUSSELL & H. S. LEWIS *Jew in Lond.* 125 The *cheder*, or private Hebrew school.

Cheddar. Add: **a.** (Earlier and later examples. For modern use, see quot. 1951.)

a 1661 FULLER *Worthies, Somerset* (1662) 18 The worst fault of Chedder Cheese is, they are so few and dear. 1951 *Good Housek. Home Encycl.* 398/1 *Cheddar Cheese*... The name is now given to any cheese which undergoes the special 'cheddaring' process, regardless of where it is made.

c. (Examples.)

1843 BABINGTON *Brit. Bot.* 41 On limestone cliffs at Cheddar, Som... Cheddar Pink. 1959 *Chambers's Encycl.* X. 737/1 Cheddar pink (*D. gratianopolitanus*).

che·ddaring, *sb.* [f. CHEDDAR + -ING.] A process or stage in the manufacture of cheese (see quots.).

1929 C. H. ECKLES et al. *Milk & Milk Products* xi. 243 In the cheddaring process the curd is piled at both sides of the vat and permitted to meet. *Ibid.* 244 When the cheddaring process is completed, the small particles lose their identity completely, fusing together into a practically solid mass of curd. 1951 [see prec., sense a]. 1958 *New Scientist* 29 May 62/2 Australia's Dairy Research Station has just announced that it has contrived a machine to do the 'cheddaring'. This is the stage in which..the fibrous particles are fused into a smooth mass. 1960 A. E. BENDER *Dict. Nutrition* 28/2 After coagulation of the milk, heating of the curd and draining, the curds are piled along the floor of the vat when, in the case of Cheddar cheese, they consolidate to a rubbery sheet of curd. This stage is the Cheddaring process.

cheddite (tʃe·dəit). [a. F. *cheddite*, f. *Chedde* (in Haute-Savoie): see -ITE¹.] Any of various high explosives composed essentially of a chlorate, a fatty oil, and one or more nitro-aromatic compounds, e.g. dinitrotoluene.

1908 *Chem. Abstr.* 1346 Cheddite consists essentially of a mixture of potassium chlorate, a fatty oil, and a nitro derivative. 1915 A. MARSHALL *Explosives* 298 The velocity of detonation of Cheddite varies considerably with the density to which it is compressed. 1918 COLVER *High Explosives* 280 Cheddites are very durable, and even prolonged heating at 120° C. causes no decomposition. 1958 *New Scientist* 5 June 105 Rockets..which detonate 800 grammes of cheddite.

cheechako (tʃitʃā·ko, tʃi·tʃāko). *colloq.* Also **checaco, chechaco, cheechaker.** ['A Chinook jargon word, lit. "new-comer": *chee* new; *chako* to come' (Partridge *Dict. Slang* Suppl.).] A newly-arrived immigrant in the mining districts of north-western North America; a tenderfoot, greenhorn.

1897 *Chicago Record* 2 Mar. 4/4 Many a 'Chechaco' (tenderfoot) on his way to the mines..has thrown down his pack and struck back for town. 1901 *Pall Mall Mag.* Jan. 56 'Nome is a good camp, but too many cheechakers', that is 'tender feet', new comers. 1920 CHARLOTTE CAMERON (*title*) A Cheechako in Alaska and Yukon. 1922 *Chambers's Jrnl.* 91/2 The Chechaco turns away with heart too full for words. 1963 *Western Miner & Oil Rev.* (Vancouver) Aug. 16/2 To many Cheechakos a shipment of a quarter ton of gold bricks represents most fabulous wealth.

chee-ild, chee-yild, varr. *CHE-ILD.

cheek, *sb.* Add: **2. b.** *to turn the* (*other*) *cheek* [in allusion to Matthew v. 39, Luke vi. 29]: to permit or invite the repetition of a blow, attack, or the like; to refuse to retaliate.

1850 O. W. HOLMES *Astræa* 50 Wisdom has taught us to be calm and meek, To take one blow, and turn the other cheek. 1913 G. S. LEE *Crowds* IV. x. 343 Turning the other cheek is a kind of moral jiu-jitsu. 1930 J. A. WILLIAMSON *Short Hist. Brit. Expansion* (ed. 2) II. VI. v. 247 The language was certainly provocative, and nothing but the consciousness of a good cause enabled Lord Salisbury to turn the cheek to the smiter. As it was, he made a conciliatory answer. 1969 *Listener* 28 Aug. 275/1 Nye [Bevan] was never one to turn the other cheek, and I have no doubt there were conflicts of personality and frustrated ambition behind his resignation.

c. *pl.* The buttocks. *slang.*

a 1600 DELONEY *Gentle Craft* (1639) II. ii, Spied both his great cheekes full of small blisters. 1922 JOYCE *Ulysses* 91 Shift stuck between the cheeks behind. 1959 N. MAILER *Advts. for Myself* (1961) 369 A car..is already a girl...The tail-lights are cloacal the rear is split like the cheeks of a drum majorette.

5. b. cheek to cheek, applied to dancing with the cheek of one partner touching that of the other; also (with hyphens) *attrib.*

1922 S. LEWIS *Babbitt* xviii. 228 They danced cheek to cheek with the boys. 1922 *Ladies' Home Jrnl.* June 26/1 He knew nothing of..cheek-to-cheek dancing and petting and cigarette smoking. 1939 C. MORLEY *Kitty Foyle* xi. 117 We hadn't ever seen cheek-to-cheek dancing up to that time. 1968 H. FRANKLIN *Crash* vii. 84 It's the way she behaves..flirting—trying to shock, really—cheek to cheek dancing and all that.

15. cheek-bristles *pl.*, the whiskers (of a cat); **cheek-down,** the incipient whiskers (of a youth).

1900 A. HILL *Introd. Science* 30 When darkness approaches..its cheek-bristles..save it from contact with passive objects. 1887 MORRIS tr. *Odyss.* XI. 319 Upon their faces the cheek-down blossomed fair.

cheela, var. CHELA².

cheep, *sb.* Add: **2.** A sign, sound; chiefly in colloq. phr. *not a cheep* (*out of*).

1962 'A. GILBERT' *No Dust in Attic* v. 62 Janice had disappeared very suddenly..and there hadn't been a cheep out of her since she went. 1969 J. FRASER *Cock-pit of Roses* xvi. 124 What the hell's this all about? No sign of emotion whatsoever. Not a cheep.

cheer, *sb.*¹ Add: **8. b.** *pl.* A friendly exclamation or exhortation to be cheerful; esp. a salutation before drinking. *colloq.*

1919 *Sphere* 28 June p. v/1 (Advt.), Cheers—I'm longing to see you and a Kenilworth together—the two nicest things on earth. 1926 E. CHISHOLM *Jolly Book* 221 'Cheers.' Gratian spoke comfortingly. 'It's an ill wind that blows no one any good.' 1934 T. E. LAWRENCE *Lett.* (1938) 836 Here you go a full stride forward. Cheers, and long life to your pen. 1946 E. TAYLOR *Palladian* vi. 61 'Cheers!' the men barked at him, lifting their glasses. 'Cheers!' he growled back and swallowed his whisky.

9. cheer-leader (orig. *U.S.*), one who leads the cheering on special occasions; also *fig.*; hence *cheer-leading* vbl. sb.

1903 F. D. ROOSEVELT *Let.* 26 Oct. (1949) 437, I was one of the three cheer leaders in the Brown game. 1923 R. D. PAINE *Comr. Rolling Ocean* i. 1 The appeals of the frenzied cheer leaders to back up the team. 1923 WODEHOUSE *Adv. Sally* xii. 144 Somebody mentioned you, and the most awful roasting party broke loose, Uncle Donald acting as cheer-leader. 1933 C. DAY LEWIS *Magnetic Mountain* 51 Back-slappers, cheer-leaders. 1951 R. HOGGART *Auden* ii. 39 At its worst it [*sc.* his assertiveness] becomes drum-beating, button-holing, cheer-leading. 1963 *Listener* 31 Jan. 192/2 The Parliament has tended to think that its chief job is to side with them, and act almost as a cheer-leader for their actions.

cheer (tʃiə·ɹ), *sb.*³ Also **chir.** [An imitative (Nepali) word.] An Indian pheasant, *Catreus wallichii*.

1827 *Trans. Linn. Soc.* 1826 XV. 166 The local [Nepal] name of this bird is *Cheer*. 1879 HUME & MARSHALL *Game Birds India* I. 170 The best places in which to find Cheer are the Dangs or precipitous places. *Ibid.* 173 The Cheer Pheasant feeds chiefly on roots. 1922 *Blackw. Mag.* Mar. 334/1 There are three other varieties of pheasant—the cheer, the white-crested kaleeg, and the koklass. 1969 ALI & RIPLEY *Handbk. Birds India & Pakistan* II. 116 Chir pheasant, *Catreus wallichii*... A long-tailed West Himalayan pheasant reminiscent of an English hen pheasant.

cheerie-bye (tʃiə·ribəi·), *int.* Also **cheeri-bye, cheery-bye.** [f. *CHEERI(O + GOOD)-BYE.] A colloquial variant of GOOD-BYE.

1934 M. HODGE *Wind & Rain* I. ii, Cheerie-bye..as they say in Scotland! 1944 J. BETJEMAN in *Horizon* IX. 151 And 'cheerioh' and 'cheeri-bye' Along the waste of waters die. 1959 T. GIRTIN *Unnatural Break* 35 Well, cheeriebye. Be seeing you.

cheerio (tʃiə·riōᵘ·), *int. colloq.* Also **cheerioh,** (earlier) **cheero, cheer o.** [f. CHEER *v.* 10 b + O *int.*, influenced later by CHEERY *a.*] A parting exclamation of encouragement; 'good-bye'. Also quasi-adj.: cheery.

1910 *Punch* 12 Jan. 23 [One loafer to another] Cheero, Charlie. 1914 R. BROOKE *Let.* 20 Nov. (1968) 634 Cheeryo! (as we say in the Navy). 1915 in W. N. P. Barbellion *Enjoying Life* (1919) 53 We just go on calling out 'The Devil a bit! Cheero!' 1915 ROSHER *In R.N.A.S.* (1916) 66 Heaps of love to all, and Cheer O! 1918 S. SASSOON *Counter-Attack* 42 Cheero! I wish they'd killed you in a decent show. 1918 *Punch* 3 Apr. 222/3 *Second Pilot.* Well, cheerio, old thing—weather looks dud. 1919 *Punch* 14 May 374 'Cheerio, Parsons, old cracker,' he shouted wildly. 1921 H. G. JENKINS *Mrs. Bindle* ii. 51 Never seen 'er so cheerio in all my puff. 1922 MRS. A. SIDGWICK *Victorian* ix, He hates scent too, so put on plenty to-night because..I want him to get used to it. Cheerio! 1924 GALSWORTHY *White Monkey* I. xi, Cheerio, my dear, don't quarrel with bread and butter. 1944 [see *CHEERIE-BYE *int.*]. 1948 WODEHOUSE *Spring Fever* x. 99 You could not have found a more cheerio butler.

b. A salutation before drinking; = 'cheers' (see *CHEER *sb.*¹ 8 b).

1919 WODEHOUSE *Damsel in Distress* xxi. 242 He refilled the glasses..and raised his own to his lips. 'Cheero! Here's to us all!' 1921 —— *Jill the Reckless* viii. 120 Much as the wounded soldier would have felt if Sir Philip Sidney, instead of offering him the cup of water, had placed it to his own lips and drained it with a careless 'Cheerio!' 1930 AUDEN *Poems* 12 C. Thanks. Prosit. K. Cheerio.

cheer pine, var. *CHIR *pine*.

cheery, *a.* Add: **3.** *Comb.*, as *cheery-hearted, -looking, -voiced* adjs.

1848 MRS. GASKELL *Mary Barton* II. iii. 49 Her father was a cheery-hearted man. 1954 J. R. R. TOLKIEN *Fellowship of Ring* I. ix. 165 A cheery-looking hobbit. a 1892 WHITTIER *Poet. Wks.* (1898) 55/2, I can see his sickle gleaming, Cheery-voiced, can hear him teaming. 1913 MASEFIELD *Daffodil Fields* 68 The wren upon the tree-stump carolled cheery-voiced.

cheery-bye, var. *CHEERIE-BYE.

cheesa (tʃi·zā). *S. Afr.* [f. dial. Zulu *tshisa* to cause to burn. Cf. Xhosa *tshisa*.] Used *attrib.*, esp. in *cheesa stick* (see quots.).

1915 A. MARSHALL *Explosives* XXX. 441 In South Africa 'Cheesa sticks' are used, which consist of sticks of cordite with ammonium oxalate and shellac. They have recently been authorized in England. 1956 K. COURLANDER *I Speak of Africa* xvii. 164 The rock-breaker would fill these apertures with explosives and the 'cheesa' boy..would come along with a slow-burning fuse. 1957 *As a Matter of Fact* (P.R.D. Series, No. 56, Johannesburg) 15 *Cheesa-stick*, a slow-burning fuse igniter consisting of a hard cardboard tube filled with an inflammable mixture... 'Cheesa' is a Zulu word meaning 'Burn'.

cheese, *sb.*¹ Add: **1. d.** *the Cheeses*: a nick-name applied to the First Life Guards (see quot. 1903).

1890 *Chambers's Jrnl.* 19 Apr. 251/2 The old school of officers..sneered at their successors as 'Cheesemongers'. From this circumstance the regiments acquired the cognomen of the 'Cheeses'. 1903 *Westm. Gaz.* 24 Nov. 1/3 More regimental nicknames. That of 'The Cheeses' was bestowed on the Life Guards...The old-fashioned officers protested that the regiments were no longer composed of gentlemen but of cheesemongers.

e. *School slang.* A smile. Also, esp. *Photographers' colloq.*, the word 'cheese' notionally or actually pronounced to form the lips into a smiling expression.

1930 [see *CHEESE *v.*¹ 2]. 1956 *Punch* 1 Feb. 179/1 They are almost certainly just saying *cheese, cheese, cheese* to hold their smiles while the news reel camera whirls. 1964 *New Statesman* 17 Apr. 612/3 Her deadpan face mouths a tiny 'Cheese' as she is pegged..into focus.

2. d. *hard cheese*: hard luck. *slang.*

1876 MRS. G. L. BANKS *Manchester Man* III. x. 175 It's hard cheese for a man to owe everything to his father-in-law. 1889 BARRÈRE & LELAND *Dict. Slang* 240/2 (R.M. Academy), Hard *cheese*, equivalent to 'hard lines', no luck; especially used at billiards. 1922 LD. F. HAMILTON *P. J., Secret Service Boy* iv. 181 It *is* rather hard cheese. *Ibid.* v. 192 'Hard cheese!' condoled Mr. Davenant. 1934 E. WAUGH *Handful of Dust* ii. 90, I think it's hard cheese on Tony.

3. (Earlier example.)

1835 N. P. WILLIS *Pencillings* II. xxiv. 283 They remained in dirty white tunics reaching to the floor, and very full at the bottom, so that with the regular motion of their whirl the wind blew them out into a circle, like what the girls in our country call 'making cheeses'.

4. b. A conserve of fruit, etc., having the consistency of cheese or the form of a cheese, as *damson-cheese* (see DAMSON 4 b).

c 1530 H. RHODES *Bk. of Nurture* (1868) 68 Then set downe cheese of fruytes. 1723 J. NOTT *Cook's & Confectioner's Dict.* No. 13 A, Almond cheese. 1845 E. ACTON *Mod. Cookery* xxi. 473 Common cherry cheese. *Ibid.* 483 Mussel plum cheese and jelly.

c. (See quot.)

1915 *Lit. Digest* (N.Y.) 4 Sept. 469/1 The mass of partly crusht grapes, known as 'must', goes into large kettles... From this mass of 'hot must' are made the 'cheeses' that go into the presses. These 'cheeses' consist of about two thousand gallons each of grape-must roughly enclosed in heavy cotton-cloth.

5. b. Applied to various objects shaped like a cheese: see quots.

1859 SALA *Tw. round Clock* 131 A dry skittle-ground, where every day..I exercised myself with the wooden 'cheese' against the seven and a-half pins. 1884 W. S. B. McLAREN *Spinning* 218 The slivers, however drawn off, are automatically wound on to wooden rollers..These balls, or cheeses, as they are generally called, are set in a rack. 1898 *Encycl. Sport* II. 381 Skittles... Pin and Bowl, or Cheese. 1919 S. PAGET *Sir V. Horsley* I. 10 The boys played the nobler form of the game [of skittles], throwing the discus, the big wooden 'cheese'. 1921 [see *CHEESER]. 1946 *Jrnl. Inst. Electr. Engin.* CXIII. III. A 32/1 Aerials using parabolic cylindrical reflectors of this type are often known as 'cheese' aerials. 1947 *Jrnl. R. Aeronaut. Soc.* LI. 663/1 The first impellers..were..machined from solid cheeses. These cheeses, which weighed about 500 lb., were the largest forgings that had been made in R.R. 59 at that time. 1950 *Electronic Engin.* XXII. 431 The corresponding level for a parabolic cylinder (a 'cheese') which is the best of the mirror systems in this respect, is between 18 and 22 db. 1954 *Gloss. Terms Iron & Steel* (*B.S.I.*) VI. 10 *Cheese*, a roughly cylindrical forging with convex sides formed by upending ingot or billet lengths between flat tools. 1960 G. LEWIS *Handbk. Crafts* 100 Take two 'cheeses' of cotton, place them on a spool rack [etc.]. 1963 A. J. HALL *Textile Sci.* iv. 197 Machines used for dyeing yarn wound in the form of cheeses and cones.

6. *cheese-maker* (examples), *-tub* (earlier examples); *cheese-shaped* adj.

1868 *Mich. Board Agric., Rep.* VII. 237 The cheese-maker watching all the conditions. 1946 *Nature* 9 Nov. 644/1 Cheese-makers and other industrialists have discovered the importance to them of microbiology. 1850 DENISON *Clock & Watch-m.* 193 Such rollers..would require greater accuracy in making than these 'cheese shaped' rollers. 1947 CROWTHER & WHIDDINGTON *Sci. at War* 48 These problems were solved by designing a rotating aerial and reflector. The latter was cheese-shaped. 1651 in *Mayflower Descendant* X. 161 Item, 2 beere vessells & one *Cheestubb. 1688 Cheese Tub [see LADDER *sb.* 3].

7. *cheese-ramekin*; **cheese-board,** (*b*) a board from which cheese is served; **cheese-box,** a box for holding cheese; also *transf.*, *U.S.* = **cheese-tub* (*b*); **cheese cloth** (additional examples); now used also for costumes, curtains, wrapping for food, etc.; **cheese-cutter,** (*a*) also, a device for cutting cheese by pulling a wire through it; (*b*) various techn. uses (see quots. 1848, 1927, and 1932); **cheese fingers,** puff paste on which a cheese mixture is spread, the paste being then folded over, cut into strips, and baked; **cheese-head,** the head of a rivet or screw shaped like a squat cylinder; freq. *attrib.*; so *cheese-headed* adj.; **cheese-knife,** (delete question-mark from def. and add earlier example); (*b*) a spatula used to break down curd in cheese-making; **cheese ladder** (see LADDER *sb.* 3); **cheese straw,** (usu. in *pl.*) thin strips of pastry flavoured with cheese; **cheese-tub,** (*a*) (in Dict., sense 6); (*b*) *transf.*, a contemptuous name for a monitor vessel (cf. **cheese-box*).

c 1938 *Fortnum & Mason Catal.* 64/1 Cheese Boards—Walnut—each 3/6. 1963 V. CANNING *Limbo Line* xiv. 191 They..asked the girl to bring the cheese board, saying they wanted to try some local cheeses. 1855 *Knickerbocker* XLV. 14 A cheese-box, used as a tanning-vat. 1862 *N.Y. Tribune* 10 June, Where is the Monitor? We have not heard a word of the little cheese-box since the repulse in James River until yesterday. 1866 E. A. POLLARD *Southern Hist. War* I. xi. 278 Here, there, and everywhere, was the black 'cheese-box'. 1872 SCHELE DE VERE *Americanisms* 335 Irreverent Confederates called the hideous-looking vessels cheese-boxes. 1878 E. B. TUTTLE *Border Tales* 17 One of the red-skins having manufactured a drum by stretching a deer-skin over the rim of a cheesebox. 1657 *Mass. Hist. Soc. Coll.* 4th Ser. VII. 83 Your maid likewise wants vessells for to sett milke in, & some cheese clothes. 1892 V. KELLOGG *Kansas Insects* 65 Wires thrust in the ground so as to form two crossing arches,..and covered with cheese cloth or netting, do well. 1931 H. PRICE *Regurgitation & Duncan Mediumship* 38 The veiling looks *exactly* like a long piece of fine, thin woven material, such as butter muslin or cheese cloth. 1932 *Times Lit. Suppl.* 21 Apr. 284/4 The supposed veil of ectoplasm..proved to be nothing more than a sheet of cheese-cloth, which she had swallowed and was able to regurgitate at will. 1944 *Living off Land* v. 109 Tanks.. are best dealt with by screening all openings with a protective wire mesh or even cheese cloth. 1848 *Sinks of London Laid Open* 102/2 *Cheese cutters*, bandy legs. 1927 G. BRADFORD *Gloss. Sea Terms* 33/2 *Cheese-cutter*, a type of centerboard. 1932 *Flight* 25 Mar. 254/1 The notched quadrant in the front cockpit is the main 'cheese cutter'. 1932 D. GARNETT *Rabbit in the Air* III. 66 There is also a so-called 'tail-incidence lever', or cheese-cutter. 1963 *Times* 6 May p. viii/3 (Advt.), An instrument rather like a wire cheese-cutter was designed. 1885 *Tasty Dishes* 125 *Cheese Fingers*..roll out and cut into strips about three inches long, roll round, and bake. 1888 *Lockwood's Dict. Terms Mech., Engin.* Cheese Head Rivet... Cheese Head Screw. 1907 *Install. News* Oct. 10/2 A small cheese head screw and washer [are] used to bind it. 1908 *Ibid.* II. 110/1 The binding screws.. have cheese heads which permit of a deep slot for screwing up. 1950 L. S. SASIENI *Spectacle Fitting* i. 26 The joint pin may be..a cheese-head screw, the head of which lies within a countersink in one of the outer parts of the chernier. 1930 *Engineering* 26 Dec. 815/2 This grid can be removed by loosening three cheese-headed screws. 1743 ELLIS *Mod. Husb.* May vii. 97 The Cloth must be tucked in with a wooden Cheese Knife. 1839 *Mag. Domestic Econ.* Feb. 240 The curd..is cut through with a double or triple bladed cheese-knife. 1725 *Fam. Dict.* s.v. *Ramequin's*, To make Cheese-Ramequins, a Farce is to be prepar'd of the same sort as that describ'd for Cheese-Cakes. 1892 T. F. GARRETT *Encycl. Pract. Cookery* I. 349/1 Cheese Puffs or Ramekins. 1899 *Daily News* 30 Sept. 7/4 Little individual dishes of devilled macaroni,..cheese ramaquins, etc. 1874 *Young Ladies Jrnl.* XI. 475/1 Three Receipts for Making Cheese Straws. 1892 T. F. GARRETT *Encycl. Pract. Cookery* I. 350/2 Cheese Straws,..bake for ten minutes in a quick oven. 1867 J. T. HEADLEY *Farragut & Nav. Commanders* 519 But all this time, Worden in his 'cheese-tub', as the rebels called her, was crowding all steam to overtake his powerful adversary.

cheese, *sb.*² Add: **2.** Wealth and fame (quot. a 1910). Also, an important or self-important person (freq. *the big cheese*). Usu. derogatory. *slang* (chiefly *U.S.*).

a 1910 'O. HENRY' *Unprof. Servant* in *Wks.* (1928) 805 Del had crawled from some Tenth Avenue basement like a lean rat and had bitten his way into the Big Cheese...He had danced his way into..fame in sixteen minutes. 1920 WODEHOUSE *Coming of Bill* I. iv. 44 The bunch of cheeses ought to have been highly grateful to Mrs. Dingle for her anti-pugilistic prejudices. 1934 R. CHANDLER in *Black Mask* July 64 So the big cheese give me the job. 1939 L. H. GRAY *Foundations of Lang.* 31 He's the big cheese. 1961 J. MASTERS *Road past Mandalay* xii. 136 'Where's the manager?' 'The manager?' 'The Bara Sahib. The Big Cheese. The Boss.' 'The Brigadier is out.' 1965 *Daily Express* 15 Oct. 19/4 As soon as you become to feel a bit of a cheese you become a bad magistrate.

cheese, *v.*¹ Add: **2.** *School slang.* To smile.

1930 *N. & Q.* CLVIII. 119/1 Another slang use of the word 'cheese' was in vogue at Rugby School..This was with the meaning 'smile', both verb and noun. 1940 M. MARPLES *Public School Slang* 40 *Cheese*..to smile or grin (Oundle 1920+).

cheese, *v.*² Add: Also in general slang use. (Examples.)

1880 *Times, for the year 1980* 4/4 He told the station master at the balloon depôt to cheese it, but thought better of it afterwards. 1910 CHESTERTON *Alarms & Discursions* 58 Their citizens..will often say 'Cheese it!' 1923 WODEHOUSE *Inimit. Jeeves* i. 9 He had been clearing away the breakfast things, but at the sound of the young master's voice cheesed it courteously. 1938 —— *Code of Woosters* xii. 261, I pulled myself together and cheesed the bird imitation.

cheeseburger (tʃi·zbɒɹgəɹ). orig. *U.S.* [f. CHEESE *sb.*¹ + *HAM)BURGER.] A hamburger in which cheese is mixed with or placed on the meat.

1938 *Amer. Speech* XIII. 157/1 *Cheeseburgers*, made of ham and cheese..may now be had in many dining places. 1954 J. STEINBECK *Sweet Thursday* xxxii. 218 Melted cheese sandwiches, cheeseburgers, and coffee. 1959 R. POSTGATE *Good Food Guide* 134 You can get Hamburgers (2/-), what they call, God forgive them, cheeseburgers (2/6), [etc.].

cheese-cake. Add: **2.** Also **cheesecake,** *slang* (orig. *U.S.*). Display of the female form, esp. in photographs, advertisements, etc., in the interest of sex-appeal; female sexual attractiveness. Also *attrib.* (Cf. **beefcake*.)

1934 *Time* 17 Sept. 30/2 Tabloid and Heartsmen go after 'cheesecake'—leg-pictures of sporty females. 1942 *Time* 24 Aug. 4/1 The supreme Empress of Cheesecake, the very Marlene Dietrich,..was fittingly crowned by the Treasury as the champion bond seller. 1948 *Illustrated* 6 Mar. 4/1 American magazines sell millions of copies because of their 'cheesecake' pictures—'cheesecake' being their name for pictures of movie stars or showgirls with beautiful legs and scanty clothing. 1955 E. HYAMS *Slaughterhouse Informer* 109 Is cheese-cake still the best puller? 1958 J. WAIN *Contenders* 212 She had a sexy slouch like a Hollywood cheesecake queen. 1967 *New Scientist* 13 Apr. 72/3 In flickering torchlight the scenes probably appeared to move. Some of them clearly included cheese-cake and leg shows.

cheesed (tʃizd), *a. slang.* [Etym. unkn.] Bored, disgruntled, exasperated; fed up, 'browned off'. Freq. *cheesed off*.

1941 J. SOMMERFIELD in *Penguin New Writing* IV. 12 'I'm browned off,' announces Taff. 'I'm cheesed.' 1942 G. KERSH *Nine Lives Bill Nelson* ii. 10 Two people, both cheesed off, are better than one. 1943 HUNT & PRINGLE *Service Slang*, *Cheesed off*, more than *brassed off*; yet not entirely *browned off*. 1948 A. BARON *From City, from Plough* 161 Whenever I'm cheesed off I just open it and start reading. 1959 I. JEFFERIES *13 Days* ii. 25, I got cheesed and overtook.

cheesemonger. Add: **b.** (See quot. 1874.) Cf. *CHEESE *sb.*¹ 1 d.

1874 HOTTEN *Slang Dict.*, *Cheesemongers*, once a popular name for the First Lifeguards. 1893 *Eng. Illustr. Mag.* 125/1 'Come on, you Cheesemongers!' the bantering cry of a commanding officer of a cavalry regiment of the Household Brigade at Waterloo.

cheeser (tʃi·zəɹ). [f. CHEESE *sb.*¹ (*5 b) + -ER¹.] An operative who tends a cheesing frame (see *CHEESING) for winding wool or silk.

1921 *Dict. Occup. Terms* (1927) § 367 *Cheese winder; cheeser*,..operates winding frame which winds yarn..on to ..tubes,..known as cheeses.

cheesewood (tʃi·zwud). [f. CHEESE *sb.*¹ + WOOD *sb.*] A yellowish-white wood obtained from the Australian trees *Pittosporum bicolor* and *P. undulatum*; also, a tree of either species.

1889 J. H. MAIDEN *Useful Native Plants* 588 'Cheesewood' ..is yellowish-white, very hard, and of uniform texture and colour.

cheesiness. For first example read:

1840 DICKENS *Old C. Shop* I, 'How's the cream of clerkship, eh?' 'Why, rather sour, Sir... Beginning to border upon cheesiness, in fact.'

cheesing (tʃi·ziŋ), *vbl. sb.* [f. CHEESE *sb.*¹ (*5 b) + -ING¹.] The operation of winding yarn or silk on cheese-shaped bobbins. *Cheesing frame*, a frame or machine that performs this operation; also *attrib.* (See also under CHEESE *v.*¹)

1921 *Dict. Occup. Terms* (1927) § 371 Cheesing frame minder, cheesing frame tenter.

cheesy, *a.* Add: **4.** Also **cheesey.** Inferior, second-rate, cheap and nasty. *slang.*

1896 GORE *Student Slang* 17 *Cheesy*, a vague term of depreciation. 1930 WODEHOUSE *Very Good, Jeeves!* x. 263 She just draws herself up and looks at them with clear, steady eyes, causing them to feel not a little cheesey. 1930 R. MACAULAY *Staying with Relations* iii. 40 Hare and rabbit fur are just utterly revolting and cheesey. 1932 N. MITFORD *Christmas Pudding* x. 154 'She wasn't

born till four in the morning.' 'What a cheesey time.' 1951 AUDEN *Nones* (1952) 43 Grown insolent and fat On cheesy literature And corny dramas.

cheetal, -ul: see *CHITAL.

chef d'œuvre. (Earlier example.)

1619 J. CHAMBERLAIN *Let.* 31 July in T. Birch *Court & Times of Jas. I* (1848) II. 182 To make up an election at Merton College, as his *chef d'œuvre*, and last work.

‖**chef d'école** (ʃef dekol). [Fr.] The initiator or leader of a school or style of painting, music, or literature.

1840 THACKERAY in *Fraser's Mag.* XXII. 122/2 Is this the way a *chef d'école* condescends to send forth a picture to the public? 1855 —— *Newcomes* II. i. 2 Your splendid *chef d'école*, a Rubens or a Horace Vernet. 1900 W. JAMES *Let.* 8 June in R. B. Perry *Thought & Char. W. J.* (1935) II. 198 Charles Renouvier..is a *chef d'école* in the complete sense of the term (as very few philosophers nowadays are). 1938 *Burlington Mag.* June 259/1 Ingres was..too unique to be a *chef d'école*. 1970 *Times* 27 Feb. 10/8 For many Europeans Jackson Pollock probably is the *chef d'école*, though Rothko was closer in spirit to William Baziotes.

‖**chef d'orchestre** (ʃef dɔrkestr'). [Fr.] The leader or conductor of an orchestra.

1855 *Sat. Rev.* 15 Dec. 113/1 An experienced *chef-d'orchestre* like Mr. Benedict. 1892 G. B. SHAW in *Music in London 1890–94* (1931) I. 13 He will unhesitatingly call upon the *chef d'orchestre* to 'stop that music' in the very middle of a suspension. 1923 M. ARLEN *These Charming People* 143 The polite strains of an orchestra led by a *chef d'orchestre*.

che-ild (tʃi,əi·ld). Also **chee-(y)ild.** A representation (usu. ironical) of an affected or melodramatic pronunciation of CHILD *sb.*

1896 G. B. SHAW *Our Theatres in Nineties* (1932) II. 54 Oh, that young wife who was about to run away from him when she was reminded of her own mother and her own chee-yild! 1911 'IAN HAY' *Safety Match* xii. 189 My che-ild! So *that's* the game! 1941 'N. BLAKE' *Case of Abominable Snowman* i. 7 You're not a woman. You're an infant cheeild. *Ibid.* xiii. 142 Quite right, infant chee-ild. 1966 M. STEEN *Looking Glass* ii. 34 The meeting at Euston was unemotional...I had worked myself into a panic over some apprehended 'My che-ild, my che-ild!'

cheilitis (kəiləi·tis). *Path.* Also **chilitis.** [f. CHEIL- + -ITIS.] Inflammation of the lip.

1842 DUNGLISON *Med. Lex.* (ed. 3) 150/1 *Cheilitis*, or *Chilitis*,..Inflammation of the lip. 1952 M. E. FLOREY *Clin. Appl. Antibiotics* ii. 24 Cheilitis is mentioned by a number of authors in their reviews of reactions to penicillin.

cheiloplasty (kəi·loplæsti). *Surg.* Also **chiloplasty.** [f. CHEILO- + -PLASTY.] Treatment of the lip by a plastic operation. So **chei·lo-plastic** *a.*

1842 T. D. MÜTTER *Rec. Impr. Surg.* 33 Cheiloplasty when the lips [are made]. 1847 V. MOTT in A. A. L. M. Velpeau *New Elem. Operative Surg.* I. 816 M. Serra has never, in all his cheiloplastic operations, found in necessary to have recourse to the incisions of Celsus. 1853 R. DRUITT *Surgeon's Vade Mecum* (ed. 6) 413 Cheilo-plastic Operations, by which are signified operations for the restoration of the lip. 1898 B. T. TILTON tr. *Tillmanns's Text-bk. Surg.* II. iv. 230 By cheiloplasty is understood the remedying of defects upon the lips by plastic operation. 1958 *Plastic & Reconstructive Surg.* XXI. 453 The increased width of the upper lip is reduced by triangular excisions and the procedure is typical of the 'Bernard Cheiloplasty'..or the 'Bernard-Burow operation'.

cheiranthus (kəiræ·nþɒs). *Bot.* [mod.L. (C. Linnæus *Species Plantarum* (1753) II. 661), f. Gr. χείρ hand + ἄνθος flower.] One of a genus of cruciferous plants of which the best-known representatives are the wallflower, *Cheiranthus cheiri*, and the Siberian wallflower, *Cheiranthus allionii*, with brilliant fragrant flowers.

1850 *Gardeners' Mag. Bot.* I. 289 Marshall's Cheiranthus.. a very interesting and remarkable plant..in the seed we find the peculiar character which belongs to the Cheiranthus as one of the Pleurorhizous tribe. 1860 C. M. YONGE *Hopes & Fears* II. III. xii. 216 The air..loaded with cheiranthus smell. 1961 *Amat. Gardening* 23 Sept. 9/2 Polyanthus.. daisies, cheiranthus and winter flowering pansies are making good growth.

cheirokinæsthesia (kəi:rokəinespi·ziă). *Physiol.* [mod.L., f. CHEIRO- + KINÆSTHESIA.] The ability to feel movements of the hand. So **chei:rokinæsthe·tic** *a.*

1900 DORLAND *Med. Dict.* 151/1 Cheirokinesthetic. 1913 *Ibid.* (ed. 7) 207/1 Cheirokinesthesia. 1929 HEWER & SANDES *Introd. Study Nerv. Syst.* II. x. 92 A lesion involving the cheirokinæsthetic area only, or isolating this area, leads to inability to write what is desired. 1939 L. H. GRAY *Foundations of Lang.* iv. 89 The visual and cheirokinaesthetic areas. 1941 *Brit. Jrnl. Psychol.* July 12 The 'diagram maker' who were intent on explaining aphasia..in terms of injury to various nerve 'centres' in the brain—..glosso- and cheiro-kinaesthetic centres etc.

cheiropodous (kəirɒ·pədəs), *a. Zool.* Also **chiro-.** [-OUS.] Of or belonging to the *Cheiropoda* (see CHEIROPOD); having feet like hands.

1889 in *Cent. Dict.*

Cheka (tʃe·kă, tʃē·kă). Also **Chay-ka, Tcheka.** [a. Russ. *chekà*, f. the names (*che, ka*) of the initials of *Chrezvȳchàinaya Komissiya*, Extraordinary Commission (for combating Counter-revolution, Sabotage, and Speculation).] An organization set up in 1917 under the Soviet régime in Russia for the investigation of counter-revolutionary activities (superseded in 1922 by the G.P.U. or *OGPU). Also *transf.* Hence **Che·kist** [a. Russ. *chekist*], *sb.* and *a.*

1921 *Glasgow Herald* 14 Oct. 9 The Che-ka..rallied round itself not only Communists, but criminals, both Russian and Internationalists. 1921 *Blackw. Mag.* Dec. 725/2 A nurse.. gained a reprieve through consenting to act as spy for the Chay-Ka. 1922 *Glasgow Herald* 10 Mar. 5 This Government announces the trial before the Special Revolutionary Tribunal, that is to say, the Tcheka, of a number of our old comrades of the Central Committee of our party for fictitious charges of attempts upon the lives of the Bolshevist leaders. 1925 POPOFF *Tcheka* 230 On..December 21, 1917, the first meeting of the first Tcheka Council took place in the Smolny building. *Ibid.*, Of the first Tchekists only a small number were Communists. 1934 C. LAMBERT *Music Ho!* ii. 77 The most lynx-eared of the fashionable cheka who are the self-appointed arbiters of vogue. 1938 *Times* 1 Jan. 11/5 A wide application of Chekist methods. 1941 [see *APPARATCHIK]. 1949 I. DEUTSCHER *Stalin* 191 The Cheka (the Extraordinary Commission), the forerunner of the O.G.P.U. 1953 J. CARY *Except the Lord* 216 Autocracy with a sword is followed by democracy with a cheka.

Chekhovian (tʃekōᵘ·viăn), *a.* and *sb.* Also **Chekovian, Tchehovian.** [f. *Chekhov* (see below) + -IAN.] **A.** *adj.* Of, pertaining to, or characteristic of the work of the Russian writer Anton Pavlovich *Chekhov* (1860–1904). **B.** *sb.* A character in one of Chekhov's plays.

1921 J. AGATE *Red Letter Nights* (1944) 100 But these Tchehovians do nothing. 1925 *Bookman* Oct. 174 There is nothing more Chekhovian, outside Chekhov, than Mr. Lytton Strachey's life of Cardinal Manning. 1927 *Observer* 15 May 6 That specific Chekovian symbolism and 'musical' construction which makes his plays more like Maeterlinck's than like 'real life'. 1943 *Scrutiny* XI. 185 Everyone is frustrated: and it is that Tchehovian frustration which is the antithesis of tragedy. 1957 V. NABOKOV *Pnin* 42 She had had a dragging, hopelessly complicated, Chekhovian rather than Dostoevskian affair with a cripple.

chela². Add: Also **cheelah.** (Earlier and later examples.) Also, one who occupies the position of disciple and servant; a follower or pupil.

1834 A. PRINSEP *Baboo* II. ii. 24 She..met the Gosaeen and his Chêla, followed by two Coolies. 1885 *Daily News* 14 Feb. 5/2 King Theebaw as a devout Cheelah, with Colonel Olcott for his learned and gallant Gooroo. 1894 R. KIPLING *2nd Jungle Bk.* (1895) 35 Would he need a *chela*—a disciple—to beg for him? 1957 R. B. LOCKHART *Friends, Foes & Foreigners* III. vii. 223 Mr. Baldwin, then [c 1936] Prime Minister, emphasised the loneliness of Mr. Churchill by referring to him as attended 'by his two faithful *chelas*'.

chelate, *a.* Add: **2.** *Chem.* Applied to a group (ligand) that loops round a central metal ion to be attached at two or more points, and also to the co-ordination compound so formed. Hence as *sb.*, a chelate compound.

1920 MORGAN & DREW in *Jrnl. Chem. Soc.* CXVII. 1457 The adjective 'chelate', derived from the great claw or 'chela' ('chely') of the lobster and other crustaceans, is suggested for these caliper-like groups which function as two associating units and fasten on to the central metallic atom so as to produce heterocyclic rings. Among the compounds which..function as chelate groups are ethylenediamine, [etc.]. 1935 *Nature* 26 Jan. 147/2 Abnormally large differences in solubility and volatility between isomers of some disubstituted benzene derivatives have been explained..as arising from the presence of chelate rings in the anomalous compounds. 1955 *Sci. News Let.* 15 Oct. 248/2 These compounds, known as chelates. 1957 *Sci. News* XLIV. 19 The so-called 'chelate' groups which are able to attach themselves at more than one position around the metal atom. 1965 PHILLIPS & WILLIAMS *Inorg. Chem.* I. vii. 252 A chelate ligand such as ethylenediamine, which is able to occupy two coordinate positions around the copper (II) ion and thus displace two water molecules per ligand attached.

Hence **che·late** *v. trans.*, to combine with (a substance) to form a chelate ring; so **chela·ted** *ppl. a.*; **chela·ting** *vbl. sb.* and *ppl. a.*; **chela·tion,** the process producing a chelate compound; **chela·tor,** a substance producing a chelate compound.

1922 MORGAN & MOSS in *Jrnl. Chem. Soc.* CXXI. 2862 A partly chelated lake was obtained by adding slowly a solution of..dinitrosoresorcinol. *Ibid.* 2865 The partially chelated compound. 1932 *Jrnl. Chem. Soc.* 952 (*title*) Chemical evidence of chelation in the sodium salts of o-hydroxybenzaldehydes. 1946 *Nature* 16 Nov. 708/1 The correspondence between the six-membered rings chelating the metal in both systems is apparent. 1953 *Jrnl. Brit. Interplan. Soc.* XII. 162 It is difficult to keep all needed minerals in solution. For this purpose a chelating agent such as ethylene diamine tetracetic acid is commonly used. 1955 *Sci. Amer.* Aug. 39/1 The powerful chelating agent known as EDTA has a high affinity for plutonium and the rare earths. 1958 *New Scientist* 2 Oct. 941/1 Other lichen derivatives will come into increasing use as chelators in industrial chemistry. 1959 *Sci. News* LII. 9 The attachment of two or more co-ordinate links from the same complexing molecule (ligand) to a single metal atom is usually known as chelation. 1961 *Lancet* 16 Sept. 637/2 Clinical studies on the use of chelating agents in the treatment of various cardiovascular diseases. *Ibid.* 638/2 We suggest that E.D.T.A. chelated the zinc of this exogenous hormone.

cheliped (ki·lipèd). [f. mod.L. *chēla* (ad. Gr. χηλή) claw + L. *ped-, pēs* foot.] One of a pair of limbs bearing large claws, or chelæ, found in decapod Crustacea.

1869 S. I. SMITH in *Amer. Nat.* III. 557 The left cheliped is exactly like the larger cheliped of ordinary specimens. 1897 PARKER & HASWELL *Textbk. Zool.* I. XI. 500 In front of these is a pair of very large legs terminating in huge claws or chelæ, and hence called chelipeds. 1922 *Outward Bound* Nov. 141/1 The Fiddler Crab with his one cheliped.. so vastly bigger than the other. 1932 J. S. HUXLEY *Probl. Rel. Growth* i. 34 Here the chelipeds are the second and not the first pereiopods. *Ibid.* i. 33 (*caption*) Cheliped length. 1964 E. S. REESE in *Oceanogr. & Marine Biol.* II. 467 The crab stands higher on its pereipods and the chelipeds are held in different attitudes. 1968 HEGNER & ENGEMANN *Invert. Zool.* (ed. 2) x. 398 The thoracic region bears the first, the second, and the third maxillipeds, the pinchers or chelipeds, and four other pairs of walking legs.

Chellean (ʃe·liăn), *a. Archæol.* [ad. F. *Chelléen* (1881, see below), f. the place-name *Chelles*, in the department of Seine-et-Marne, France: see -AN.] Of or belonging to the earliest palæolithic period of Europe; also called *ABBEVILLIAN (see quot. 1964). Also *absol.*

The French adj. *chelléen* was used in 1880 by M. Ameghino, a pupil of de Mortillet, in *Bull. Soc. d'Anthrop.* ser. 3 III. 638, but not as a specific subdivision of the palæolithic period.

[1881 G. & A. DE MORTILLET *Musée Préhist.* VI, Le chelléen, la plus ancienne de ces époques, tire son nom de la station de Chelles...Autrefois cette époque était appelée acheuléenne...Mais j'ai renoncé à ce nom parce qu'à Saint-Acheul le gisement n'est pas pur.] 1894 DAWSON *Geol. & Hist.* v. 70 Mortillet's Chellean men. 1921 [see *ACHEULIAN *a.*]. 1927 PEAKE & FLEURE *Apes & Men* 56 The fashioning of the early Palaeolithic, Chellean, flints betokens a high degree of purposeful skill and artistry. 1932 *Antiquity* VI. 186 The Chellean can be correlated with the next interglacial stage. 1934 [see *ABBEVILLIAN *a.*]. 1964 *Encycl. Brit.* V. 371 As the implements from Chelles, which gave the name to the industry, are now grouped with the Acheulean, the term Chellean, in the sense of earliest hand-ax culture, has been replaced by Abbevillian.

Chelleo- (ʃe·lio). Combining form of *CHELLEAN *a.*, esp. in *Chelleo-Acheulian* adj., of or pertaining to the Chellean and Acheulian periods jointly.

1930 *Discovery* Oct. 346/2. 1934 *Jrnl. R. Anthropol. Inst.* LXIV. 338 The Chelleo-Acheulean and the pre-Chellean facies. 1935 *Times Lit. Suppl.* 14 Feb. 84/2 Human remains and objects belonging to a Chelleo-Acheulean complex were found.

cheloidal (kīloi·dăl), *a.* [f. CHELOID + -AL.] = KELOIDAL *a.*

1908 *Practitioner* Jan. 42 An incision, which..has become cheloidal, contracted, and puckered.

chelp (tʃelp), *v. trans.* and *intr. dial.* To chirp or squeak; to chatter. Also as *sb.*, chatter.

1820 CLARE *Rural Life* 126 She stands chelping 'bout the town. *Ibid.* 207 And sparrows chelp glad tidings from the eaves. 1891 C. WORDSWORTH *Rutland Words* 7 Hold your chelp! c 1909 D. H. LAWRENCE *Collier's Friday Night* (1934) iii. 75, I'll stand no more of your chelp. 1913 —— *Sons & Lovers* iv. 59 Ha'e much more o' thy chelp..an' I'll rattle my fist about thee. 1959 K. WATERHOUSE *Billy Liar* xii. 164 Don't go chelping back at her like you chelp at me. 1969 D. CLARK *Death after Evensong* ii. 42 Stop your chelping, Matthew. You don't know what you're talking about.

Chelsea (tʃe·lsi). [The name of a district of London on the north bank of the Thames.]

1. *Chelsea bun*: a kind of rolled currant-bun originally made in Chelsea.

1711 SWIFT *Jrnl. to Stella* 2 May (1948) I. 259 Was it not *Rrrrrrrrrare Chelsea buns*? I bought one to-day in my walk; it cost me a penny; it was stale, and I did not like it. 1795 D. LYSONS *Environs of London* II. 149 The manufacture of Chelsea bunns should not be omitted, having been.. carried on upon the same spot for more than 100 years. 1855 ANNE MANNING (*title*) The old Chelsea bun-house, a tale of the last century. 1884 'L. CARROLL' *Tangled Tale* x, in *Monthly Packet* Nov. 474 (*title*), Chelsea buns. *Ibid.*, Give her a Chelsea bun, Miss! 1912 C. MACKENZIE *Carnival* xviii. 202 Pass the Chelsea buns. 1954 L. MACNEICE *Autumn Sequel* xii. 79 Strong talk over strong tea and Chelsea buns.

2. Used *attrib.* to designate a kind of porcelain made at Chelsea in the 18th century. Also *ellipt.*

1754 in Jewitt *Ceramic Art* (1878) I. 173 Chelsea porcelain toys. 1765 J. WEDGWOOD *Let.* 17 June (1965) 35 What I would be glad to know..is whether she expects the gold to be burnt in, as it is upon the Chelsea china. 1869 LADY C. SCHREIBER *Jrnl.* (1911) I. 33 Good Chelsea-Derby figure of youth sacrificing a goat. 1900 F. LITCHFIELD *Pott. & Porc.* v. 51 The most valuable kinds of Sèvres and Chelsea.

3. *Chelsea boots*, elastic-sided boots. Also *absol.*

1962 *Sunday Express* 4 Feb. 19/1 Chelsea boots..elastic-sided, pull on ankle-length pumps. 1963 *Times* 25 Feb.

CHELTENHAM

11/4 A cult of elastic-sided boots, once deemed proper only to the gravest of seniors and therefore laughable. These, dignified with the name of Chelseas, are now correct wear under a bowler hat in St. James's Street. 1963 *Honey* Oct. 55/2 The Beats wear..polo-sweaters and Chelsea boots.

Cheltenham (tʃeˈltnəm). [Name of a town in Gloucestershire.] **1.** Used *attrib.* to designate the chalybeate waters of the springs there, or the salts left by the evaporation of these waters (see also quot. 1848).

1783 S. Moxley *Tour to Cheltenham Spa* 42 Old men drink Cheltenham water by the quart. *Ibid.* 47 They..take a gentle dose of Cheltenham salts, or some other opening medicine, as their physician thinks best suited. 1810 Shelley *Let.* 23–24 Jan. (1964) II. 26, I..have been using mercury & Cheltenham salts with much caution and some success. 1837 *Penny Cycl.* VII. 31/1 The persons most benefited by the Cheltenham waters are those who, after a long residence in hot climates, are affected with diseased liver. 1848 Dunglison *Med. Lex.* (ed. 7), *Cheltenham Salts*, these are sometimes made from the waters; at others, factitiously..Cheltenham Water, Artificial, may be made of Epsom salt, gr. xij, iron filings, gr. j: [etc.]. 1874 Garrod & Baxter *Mat. Med.* 142 The so-called Cheltenham salts consist chiefly of sulphate of soda.

2. Any of several varieties of type-fount (see quots.), orig. *U.S.*

1910 E. G. Gress *Art & Pract. Typogr.* ii. 150/1 Automatic harmony is possible with that remarkable type creation, the 'Cheltenham Family', developed by the American Type Founders Company. Cheltenham Oldstyle, from which the dozen other Cheltenhams are descended, was born in the brains of Bertram G. Goodhue..and Ingalls Kimball, of the Cheltenham Press, New York. 1917 F. S. Henry *Printing for School & Shop* iii. 79 Cheltenham Oldstyle..A Capital R of Thirty-six-point Cheltenham. This paragraph has been set in Cheltenham to show the long ascenders so characteristic of the family group. 1968 'E. McBain' *Fuzz* ix. 145 The city's other afternoon paper..was apt to feature the notes on its front page..in 24-point Cheltenham Bold.

chelydoid (keˈlidoid), *a.* and *sb.* [f. mod.L. *Chelydidæ*: see -oid.] **A.** *adj.* Of or pertaining to the Chelydidæ, a family of tortoises. **B.** *sb.* A tortoise belonging to this family.

1886 Günther in *Encycl. Brit.* XX. 471/2 It may be seen from this list that no Chelydoid passes northward beyond the Isthmus of Panama.

chemic, *sb.* Add: **5.** A dye consisting of a very acid solution of indigo in sulphuric acid. Also *attrib.*

1817 W. Tucker *Family Dyer & Scourer* i. 7 There are some blues on silk, of a very light shade, that are dyed with chymic blue. *Ibid.* iii. 35 Chymic for light blues, and greens on silk or woollen. *Ibid.* iii. 46 From your chymic bottle, one or two drops.

chemical, *a.* Add: **3.** (Further examples.)

1866 C. M. Yonge *Hopes & Fears* II. iii. 226 Dry chemical manures—drain that terrible old marsh. 1905 R. Broughton *Waif's Progr.* xxii. 348 The face of the very pronounced 'chemical blonde' thrust out of the automobile's window.

4. b. In specific collocations: *chemical closet*, a closet (Closet *sb.* 7) in which waste matter is decomposed chemically (see quot. 1940); *chemical engineering*, a branch of engineering concerned with manufacturing processes involving the theory or practice of chemistry; so *chemical engineer*; cf. chemico-engineering s.v. *chemico-; *chemical extinguisher*, a fire-extinguisher from which a chemical liquid can be discharged (*Cent. Dict.* 1889); *chemical warfare*, that in which chemicals (other than explosives) are used, as gases, smoke, incendiary compounds, etc.

1940 *Chambers's Techn. Dict.* 155/1 *Chemical closet*, a suitably shaped container for use in conjunction with special deodorizing and liquefying chemicals, when running water is not available. 1959 *Guardian* 2 Jan. 5/3 Modern plastic or collapsible chemical closets are compact, hygienic, and easily cleaned. 1904 (*title*) The chemical engineer. 1936 *Discovery* Oct. 308/2 Performing miracles with fancy currencies and their blue prints reminiscent of the chemical engineer. 1901 G. E. Davis (*title*) Handbook of Chemical Engineering. 1957 *Technology* Apr. 56 Chemical Engineering. One of the primary technologies which came into being when continuous process production began to replace the older methods of batch processing. 1898 *Engineering News* 16 June 391/1 In suburban towns, villages, and isolated country properties, a chemical extinguisher..is of greater value than all other apparatus. 1917 W. S. Churchill *World Crisis* 1916–18 (1927) II. 560 Chemical warfare must be one of the..leading features of our campaign of 1918. 1937 A. M. Prentiss (*title*) Chemicals in war: a treatise on chemical warfare.

5. b. *fine chemicals*, those handled in small lots and in a purified state; *heavy chemicals*, those handled in large lots and in a more or less crude state.

1882 *Jrnl. Soc. Chem. Industry* Feb. 75 Fine chemicals, alkaloids, essences. 1937 *Chem. Abstr.* XXXI. 8748 The industries discussed include..The fine chemicals industry (medicinal and heavy chemicals).

chemical (keˈmikăl), *v.* [f. the sb.] *trans.* = next.

1884 *Harper's Mag.* LXIX. No. 3 (Advt.), Washing compounds and soap..are highly chemicalized.

chemicalize (keˈmikălaiz), *v.* [f. Chemical *a.* + -ize.] *trans.* To treat with a chemical or chemicals; to make much use of chemicals in. Also *fig.* So cheˈmicaliza·tion; cheˈmicalized *ppl. a.*

1902 *Clarion* 17 Oct. 1 If this change is produced in milk by chemicalizing it, in the case of chemicalized meats, also, a similar result must follow. *Ibid.*, Milk..if ceased soon after chemicalization, they produce a mild..attack of suffering. 1920 *Lancet* 15 May 1443/1 The Chemicalized Road [the calcium chloride treatment of roads]. 1929 D. H. Lawrence *Pansies* 113 Flat-chested, crop-headed, chemicalised women, of indeterminate sex. 1964 *Economist* 18 Jan. 193/1 The 'chemicalisation of agriculture'.

chemico-. Add: *chemico-engineering, -physics; chemico-crystallographical, -mineralogical* adjs.; cheˈmico-bio·logy, the chemistry of living matter; hence cheˈmicobioˈlo·gic *a.*; cheˈmico-dynaˈmic *a.*, transforming chemical energy into the energy of motion.

1903 *Lancet* 22 Aug. 3 (Advt.), The results of bacteriologic and of chemicobiologic research as applied to the pathology of midwifery. 1908 *Carnegie Trust Rep.* 24 Chemico-crystallographical researches on thallic sulphates. 1909 *Cent. Dict. Suppl.*, Chemicodynamic. 1928 *Daily Express* 12 Sept. 11/4 The original structure of the cells is a chemico-dynamic structure which requires oxygen..to preserve it. 1897 *Daily News* 1 June 7/4 Prof. Dewar's magnificent work in chemico-engineering laboratory. 1903 Geikie *Text-bk. Geol.* (ed. 4) I. 201 In Professor Rosenbusch's scheme of classification the chemico-mineralogical characters of the igneous rocks are chosen as the basis of the grouping. 1909 Buck's *Handbk. Med. Sci.* VII. 758 (Cent. D. Suppl.), Chemico-physics.

chemigraphy (kemiˈgrăfi). [f. Chemi(cal *a.* + -graphy.]

1. Any mechanical engraving process depending upon chemical action; *spec.* a process of zinc etching without the aid of photography.

2. A process of obtaining half-tones by printing, from the same plate, in two colours, or two shades of the same colour, one of which is slightly out of register. Hence chemiˈgraphic *a.*, pertaining to or produced by chemigraphy. Also cheˈmigrapher, one who prints by a half-tone photo-mechanical process; cheˈmigraph, a print obtained by chemigraphy.

1892 *Daily News* 23 Mar. 6/3 Mr. Leon Warnerke next gave some illustrations of 'Chemigraphic etching'. 1897 Sexton & Strange *Etching* 25 Whatever niceties the chemigrapher or white-line woodcutter can manage to put upon his block will come out in the printing.

chemiluminescence (keːmilˈūmineˈsĕns). [ad. G. *chemilumineszenz* (M. Trautz 1905, in *Zeitschr. f. Physik. Chem.* LIII. 1), f. Chemi(co- + Luminescence.] Emission of light accompanying a chemical reaction, as in the oxidation of phosphorus. Hence cheːmilumineˈscent *a.*

1905 *Jrnl. Chem. Soc.* LXXXVIII. ii. 662 (*title*) Chemiluminescence. 1907 *Chem. Abstr.* I. 267 (*title*) A Reaction with Green Chemiluminescence. 1913 *Jrnl. Amer. Chem. Soc.* XXXV. 823 Formaldehyde is known to be chemiluminescent under some conditions of alkaline oxidation. 1942 E. J. Gurevich *Atom. Aspects of Light* xi. 170 The limits separating chemiluminescence from emission due to heat alone, and from electronic excitation..are not as clearly marked as they might be expected to be. *Ibid.* 176 Even in well-investigated chemiluminescent reactions of an elementary character the results are neither simple nor conclusive. 1966 *New Scientist* 17 Feb. 394/3 Chemiluminescence, which is emitted at room temperature during some chemical reactions.

chemin de fer (ʃəmæ̃ də fɛr). [Fr., lit. 'road of iron', railway.] A form of baccarat.

1891 'L. Hoffmann' *Baccarat* 12 There are two forms of Baccarat, known respectively as *Baccarat Chemin de Fer* and *Baccarat Banque*. 1897 R. F. Foster *Foster's Complete Hoyle* 482 Six packs of cards are generally used in Chemin de Fer. 1902 *Daily Chron.* 13 May 10/3 He raided the club, and found..the defendants seated round large tables playing chemin de fer. 1928 'Ian Hay' *Poor Gentleman* iv. 74 Do you play chemin-de-fer?

chemisal, var. *chamisal.

chemise. Add: **1.** (Further examples.) In recent use: a dress hanging straight from the shoulders. Also *chemise dress*.

1788 Betsy Sheridan *Jrnl.* 22 Dec. (1960) 138 Gowns all kinds—Chemises—Round gowns with flounce or not. 1922 *Ladies' Home Jrnl.* Feb. 77 Making the Popular Chemise Dress. 1958 *Punch* 17 Sept. 384/1 The summer chemise was the prettiest, most impudent, provocative fashion that we have seen for many a day.

2. c. A loose covering for a book.

1895 *Quaritch Catal.* 150 198 I. In wooden boards..wrapped in a modern morocco chemise. 1928 E. G. Millar *Engl. Illum. MSS.* I. 3 It still retains its binding of the period with a sheepskin 'chemise'. 1960 Glaister *Gloss. of Book* 63/1 *Chemise*, a cover of silk or chevrotain sometimes used in the 15th century as a protection for embellished leather-bound books.

chemised (ʃimīˈzd), *a.* [f. Chemise + -ed².] Wearing or having a chemise (see prec.).

1928 *Daily Express* 13 June 10/6 Whether we are shingled and chemised..is as unimportant as cocktails. 1959 *Encounter* XII. v. 59 The new sheathed and chemised fashions.

chemisorb (keˈmisǫ̈ɹb), *v.* [Back-formation from next.] *trans.* To collect by chemisorption. Hence cheˈmisorbed *ppl. a.*

1935 *Nature* 7 Dec. 890/1 That the hydrogen is chemisorbed is proved by the high heat of adsorption. 1943 *Thorpe's Dict. Appl. Chem.* VI. 207/2 The ortho-para hydrogen conversion occurs by exchange between gaseous molecules or molecules sorbed by..the chemisorbed layer. 1965 *New Scientist* 19 Aug. 445/1 The chemisorbed layer of oxygen originally present on the fresh platinum is displaced.

chemisorption (kemisǫ̈ˈɹpʃən). [f. Chemi(cal *a.* + Adsorption.] Adsorption, usually irreversible, involving chemical action.

1933 A. King in *Jrnl. Chem. Soc.* 842 Chemisorption, the term generally applied to irreversible adsorption, was first noticed in the carbon-oxygen system. 1934 *Discovery* Sept. 274/1 The chemisorption of gases on metals and carbon. 1938 *Thorpe's Dict. Appl. Chem.* II. 422/2 We differentiate between general or 'physical' adsorption and the 'activated adsorption' or 'chemisorption' which is the precedent to actual catalytic interaction.

chemmy (ʃeˈmi), colloq. abbrev. of *chemin de fer.

1923 Wodehouse *Inimit. Jeeves* xvi. 114 We went on from Ciro's to the Nest and..to play chemmy. 1960 *Punch* 25 May 196 How to behave when a..bingo-club or a chemmy-party..is visited by the police.

chemo- (keˈmo), occas. cheˈmio-, used as combining form = Chemical *a.*, in **chemokinesis** (keːmokainīˈsis) [Gr. κίνησις movement] *Biol.*, a condition of increased activity of an organism, induced by the presence of a chemical substance; hence chemokineˈtic (-kaineˈtik) *a.*; chemoreˈflex, a response to a chemical change in the environment by a motor reaction; also as *adj.*, pertaining to or designating a reflex action resulting from a chemical stimulus; cheˈmostat [Gr. στατ-ός standing, after Thermostat], a device designed to provide an environment that can be regulated and kept stable over a long period, *esp.* one used for the continuous cultivation of micro-organisms in which the nutrient medium is continually replenished; chemosteˈrilant (see quot. 1964); chemosynthesis (kemosiˈnþisis) [Gr. σύνθεσις composition], the formation of carbohydrates out of inorganic compounds by an organism in darkness or in the absence of sunlight, as distinguished from 'photosynthesis'; hence chemosyntheˈtic *a.* (Webster 1911).

1900 *Amer. Jrnl. Physiol.* Apr. 398 The organisms soon leave the area of operation of the chemical causing the reactions. This phenomenon Garry calls 'chemokinesis'. 1901 G. N. Calkins *Protozoa* 299 If an effect is produced which is not directive, it is said to be chemokinetic. 1902 *Pop. Sci. Monthly* Mar. 463 [The] complicated activities of such highly developed organisms as ants and bees may be subsumed, with varying degrees of completeness, under such heading as the 'chemoreflex'. 1950 Novick & Szilard in *Science* 15 Dec. 715/1 We have developed a device for keeping a bacterial population growing at a reduced rate over an indefinite period of time..This device..we shall refer to as the Chemostat. 1967 *Guardian* 7 Feb. 4/3 The chemostat was designed initially to provide a stable but controllably variable environment for the study of cell metabolism. 1964 *New Scientist* 22 Oct. 211/3 The use of chemosterilants—compounds which sterilise insects and make them unable to breed—has been a promising new development that is now against insect pests. 1903 W. H. Lang tr. *Strasburger's Text-bk. Bot.* 197 The necessary energy is here obtained not from photosynthesis but from the oxidation of ammonia into nitrous acid, and this into nitric acid (chemosynthesis). 1930 J. McCabe tr. *Haeckel's Wonders of Life* 212 Pfeffer has called this carbon-synthesis..on account of its purely chemical nature, 'chemosynthesis' in opposition to the ordinary photo-synthesis by means of sun-light. 1909 R. J. H. Gibson *Biol.* vi. 48 In all probability the energy required is obtained by the oxidation of primary organic compounds, and possibly of protoplasm itself (chemosynthesis). 1959 *Listener* 5 Mar. 420/1 These are known as chemosynthetic autotrophs; in other words, bacteria that obtain their energy by oxidizing..inorganic chemicals.

chemoceptor (kemoseˈptǫ̈ɹ). *Immunology.* [a. G. *chemoceptor* (P. Ehrlich 1909, in *Ber. d. Deut. Chem. Ges.* XLII. 33), f. *chemo- + Receptor.] (See quot. 1916.)

1911 tr. *Ehrlich & Hata's Exper. Chemotherapy of Spirilloses* iii. 118 In the protoplasm of trypanosomes certain groups—chemoceptors, as I have called them—possessing a special affinity for a certain class, and are capable of anchoring such to the cell. 1916 Gould *Pract. Med. Dict.* 214/1 *Chemoceptor*, one of the side chains or receptors in a living cell, having the power of fixing chemical substances in the same way that bacterial toxins are fixed. 1932 C. H. Browning in *Syst. Bacteriol.* (*Med. Res. Council*) VI. 516 Ehrlich applied a 'modification of his receptor theory involving the conception of chemoceptors and nutriceptors.

chemoreceptor (kemorĭseˈptǫ̈ɹ). [f. as prec.] **1.** *Physiol.* A sensory organ responsive to chemical stimuli.

1906 C. S. Sherrington *Integr. Action Nerv. Syst.* ix. 322 The chemo-receptors on the outer surface of the head..excite reflexes which move the body round, bringing the mouth to the morsel. 1922 J. H. Parsons *Theory of Perception* i. 7 Chemo-receptors include three classes, those of smell, the common chemical receptor, and those of taste. 1965 R. & D. Morris *Men & Snakes* vi. 175 With its highly developed chemo-receptors and successful hunting methods, the snake has no need of mysterious powers.

2. *Immunology.* = *chemoceptor.

1907 P. Ehrlich in *Jrnl. R. Inst. Public Health* XV. 453 These..among the groups I will distinguish from the toxin-receptors by the name of 'chemo-receptors'. 1927 *Chem. Abstr.* XXI. 1315 The complex chemoreceptor contains apparently independent specifically sensitive toxin-fixing groups for antimony compds.

Hence chemoreceˈption, the action of chemoreceptors; the production of a response to chemical stimuli.

1959 *Jrnl. Exper. Zool.* XXIX. 255 Even on the sole of the foot of Chiton physiological evidence of sensory separateness for photo-, tacto-, and chemo- reception is available. 1969 *New Scientist* 17 Apr. 134/3 The females advertise a true sex pheromone, which the males detect by contact chemoreception.

chemotaxis (kemotæˈksis). *Biol.* Also chemio-. [mod.L. *chemotaxis* (W. Pfeffer 1888, in *Untersuchungen aus d. Botanischen Institut Tübingen* II. 582), f. *chemo- + Gr. τάξις arrangement.] The disposition exhibited by certain living cells, or free-swimming organisms, of movement towards or away from certain chemical substances held in solution. Also called chemotaˈctism. Hence chem(io)otaˈctic, -ical, -taˈxic *adjs.*

1894 Burdon-Sanderson in *Rep. Brit. Assoc.* 1893 16 As a general designation of those reactions of protoplasm to chemical agents which is included under the word 'Chemotaxis', or, as we in England prefer to call it, Chemiotaxis. *Ibid.* 17 Chemiotactic cells—that is, cells which act under the orders of chemical stimuli. 1897 *Nature* 16 Sept. 481/2 This remarkable attraction—chemical attraction, or 'chemotaxis'—is a very general phenomenon. 1898 H. C. Porter tr. *Strasburger's Text-bk. Bot.* 165 The chemotactic irritability of Bacteria. 1902 *Encycl. Brit.* XXXI. 318 The influence of the chemical substance is either that of attraction or repulsion, the one being known as positive, the other as negative chemotaxis. 1903 Snyder *New Conceptions in Sci.* 220 It is simply a chemical reaction between certain substances in..the fly and the meat, a case of chemotactism. 1903 *Med. Record* 7 Mar. 392 (Cent. D. Suppl.), Chemotaxic. 1908 Hans Driesch *Sci. & Philos. Organism* I. 152 Each mesenchyme-cell is specified..with regard to its chemotactical irritability. 1908 *Practitioner* Jan. 2 Some of which [cells] die, and thus set up a chemotactic action. *Ibid.* Mar. 392 A phagocytic chemotactic effect. 1930 *Endeavour* IX. 184 This acidity, due to the metabolism of *Heggiadea*, produces a chemotactical swarming of two common species of micro-phagous ciliates. 1964 L. Martin *Clin. Endocrinol.* (ed. 4) viii. 192 The follicle is about to rupture by chemotactic attraction. 1964 E. S. Reese in *Oceanogr. & Marine Biol.* II. 479 Both meandering and direct trails, suggesting a chemokinesis and a chemotaxis respectively.

chemotherapy (keːmo-peˈrăpi). *Med.* [ad. G. *chemotherapie* (Ehrlich), f. *chemo- + Therapy.] The treatment of disease, esp. of parasitic infections or cancer, by means of chemical substances which act selectively on micro-organisms or malignant tissue. Also cheːmotherapeuˈtics. Hence cheˈmotherapeuˈtic(al) *adjs.*

1907 P. Ehrlich in *Jrnl. R. Inst. Public Health* XV. 449 [*heading*] Chemo-therapeutic studies on trypanosomes. 1910 H. Schweitzer in *Science* XXXII. 809 (*title*) Ehrlich's Chemotherapy—a new science. *Ibid.* 810 By combining physiological with synthetic chemistry he [*sc.* Ehrlich] created on their borderland the new science of chemotherapy. 1911 tr. *Ehrlich & Hata's Exper. Chemotherapy of Spirilloses* 1 It is possible..to find a specific and etiologically acting remedy for protozoal diseases in a chemo-therapeutical way. 1912 *Med. Times* (N.Y.) XL. 132/2 Many infectious diseases are not attended with formation of antibodies to a curative extent..In such cases we are forced to seek remedies which can actually destroy the parasites by chemical means. This new direction of therapeutics is known as 'chemo-therapy'. 1913 *Lancet* 16 Aug. 457/1 (*heading*) Address in Pathology on Chemotherapeutics. *Ibid.*, From the very first beginnings of therapeutics chemotherapy has, indeed, been in existence, as all the remedies we employ are chemical. *Ibid.* 461/1 An insight into the workshop of the chemotherapeutic work. 1927 *Daily Express* 11 July 4/7 A concentrated attack is being made on the problem from many different angles—pathological, bio-chemical, and chemo-therapeutical. 1942 *Electronic Engin.* XV. 242 Twenty-seven years had to pass before it was discovered that it [*sc.* sulphanilamide] was a chemotherapeutic agent of the foremost rank. 1943 *Endeavour* II. 41/1 A definition of the scope of true chemotherapy is hard to frame..We may assume, however, that the term implies the cure of infections caused by micro-organisms with the aid of substances of known or partly known constitution. 1957 *Listener* 14 Nov. 780/1 The past ten years have seen a great deal of progress in the field of cancer chemotherapy.

chemotropism (kemˈtrǫ̈piz'm). *Biol.* [f. *chemo- + Gr. τρόπος a turning + -ism.] A condition of sensitiveness to a chemical substance in solution, exhibited by certain organisms, or parts of organisms, producing curvature towards the stimulus, termed *positive chemotropism*, or away from it, *negative chemotropism*. So chemotroˈpic *a.*

1897 *Amer. Naturalist* XXXI. 719 It is the difference in the concentration of the chemotropic substance in two layers..which controls movement. Positive chemotropism gradually disappears as diffusion renders them only slight. 1897 *Nature* 16 Sept. 481/2 The cause of the bending [of the hyphæ of fungi] lies in a powerful chemotropic action. 1898 H. C. Porter tr. *Strasburger's Text-bk. Bot.* 263 Corresponding to the chemotactic irritability of Bacteria and spermatozoids, roots, fungus hyphæ, and pollen tubes exhibit positive and negative chemotropic curvatures. *Ibid.*, Rheotropism..and Aerotropism, a form of chemotropism, are additional phenomena..which stand in direct relations to certain vital requirements of plants. 1901 J. Loeb *Compar. Physiol. Brain* 186 The orientation of an organism by diffusing molecules is termed chemotropism. 1909 *Cent. Dict. Suppl.*, Chemotropic. 1950 *Chambers's Encycl.* VI. 612/1 Fertilization in many plants is controlled by chemotropic action.

chemurgy (keˈmɜːdʒi). chiefly *U.S.* [f. *chemo- + -urgy after Metallurgy.] The chemical and industrial use of organic raw materials, esp. of farm products. So chemuˈrgic, -uˈrgical *adjs.*; chemuˈrgically *adv.*; cheˈmurgist, one who is skilled in chemurgy.

1934 W. J. Hale *Farm Chemurgy* (*title*) Chemurgy signifies an intelligent working with and for chemicals. *Ibid.* 143 Chemurgical development of newsprint from young pine. *Ibid.* 146 Chemurgical development of wood waste for purposes..all other developments. *Ibid.* 190 Chemurgically speaking, this is the way of all nature. 1936 *Amer. Speech* XI. 274/1 *Chemurgy*, meaning progress through chemistry, a word not yet in current dictionaries, is gaining in use. 1937 J. A. Widtsoe in *Pacific Northwest Chemurgic Conference* 1937 Washington State Planning Council 50 The objective of Chemurgy is development of crops that may be used to produce..1939 C. Borth *Pioneers of Plenty* i. 1 Dr. Karl Taylor Compton, the chemurgist..declared that one of the big jobs to be done..is the huge task of converting farm products to industrial uses. 1939 *Industr. & Engin. Chem.* XXXI. 540 Farm chemurgy is defined as putting chemistry to work for the farm. *Ibid.* 543/2 The direct chemurgic objectives are new uses for present crops, new crops for existing [uses], etc. 1949 N. W. Pirie *Food Resources* iv. 122 Processes of this type are sometimes referred to as 'chemurgy'; 'biochemical engineering' seems a preferable term.

chena (tʃīˈnā). [ad. Sinhalese *hena*.] A form of shifting cultivation in Ceylon. Also, the shrubby vegetation produced by such cultivation or a piece of land used for this.

1832 W. M. G. Colebrooke in *Parl. Papers* 1831–2 XXXII. 104 The high lands called 'Chenas', which are cleared and cultivated only at intervals of several years. 1877 R. Abbay in *Nature* 8 Mar. 398/2 Cleared forest-land, however neglected and impoverished, does not run into grass..but into a dwarfish jungle called 'chena'. 1899 H. W. Pearson in *Jrnl. Linn. Soc.* (*Bot.*) XXXIV. 317 The firing of the low-country forest..is followed by the appearance of 'Chena' composed of the harder and more resisting forest undershrubs. 1916 F. E. Clements *Plant Succession* xi. 275 While fire produces grassland from open savannah forest, it produces scrub or 'chena' from the low-country forest. 1922 W. Schlich *Man. Forestry* (ed. 4) I. 274 Holes drawn up to control shifting cultivation (*chena*). 1924 *Glasgow Herald* 18 July 10 In Ceylon there is a regular method of native cultivation called the chena system. 1950 R. L. Spittel *Vanished Trails* 152 Illuk glades, those graveyards of once magnificent forests felled for chenas.

chenopodium (kenǫ̈poˈdiəm). *Bot.* [mod.L., f. Gr. χήν goose + ποδ-, πούς foot, referring to the shape of the leaf. Adopted by Linnæus in his *Species Plantarum* (1753) I. 218 as the name of a genus.] A plant belonging to a large genus of herbaceous plants so named; goosefoot. Also *attrib.*

1597 J. Gerard *Herball* xliii. 259 The later herbarists have called it *Pes anserinus*, of the likenesse that the leaves have with the foote of a Goose: of some *Chenopodium*..in English Goosefoote and wilde Orach. 1640 J. Parkinson *Theatr. Bot.* 1651. 759 *Pes anserinus* or *Chenopodium*..meaning the Garden sort of Orache [*sic*] with purplish leaves. 1807 T. Martyn *Miller's Gard. & Bot. Dict.* s.v., It is whiter than most of the Chenopodiums; and varies exceedingly, both when young, and in its seeding state. 1925 *Jrnl. Chem. Soc.* CVIII. i. 193 The toxicity of chenopodium oil is increased in starvation. 1926 *Contemp. Rev.* Feb. 183 All of such flowers includes chenopodiums, [etc.]. *Chambers's Jrnl.* Sept. 503/2 Far their own country name of Smearwort refers to the old use of Chenopodium in ointment.

‖**cheongsam** (tʃiǫ·ŋsæ·m). [Chinese.] A garment worn by Chinese women (see quots.).

1957 R. Mason *World of Suzie Wong* i. iii. 34 A Chinese girl came scurrying out..She wore..a cheongsam with tall collar and split skirt. 1959 F. Brown *As Far as Singapore* v. 99 His sister..wore a brocaded *cheongsam*..a gown with a high neck, slit up both sides revealing her legs. 1959 E. Ambler *Passage of Arms* iv. 101 She was wearing a *cheong sam*, the silk formal dress with the high collar and split skirt that Chinese women wear. 1969 *Times* 9 Dec. (Taiwan suppl.) p. vii/6 Their wives wear the traditional cheong sam but take a full and confident share in all the affairs.

cheque, check. Add: **3. a.** (Examples of *blank cheque*.)

1932 Weseen *Words Confused* 38 A cheque is 'a written order directing a bank or banker to pay money as therein stated'. How then can there be a 'blank cheque'? 1940 O. Jespersen *Språk* Infr. 163 He carried a blank cheque in case he should want cash.

c. A total of money received.

1947 P. Newton *Wayleggo* (1949) 153 A musterer is usually paid by cheque at the end of the season, and refers to his pay as his cheque. 1958 *Listener* 18 Sept. 407/1 The thirty-five per cent. drop in wool prices last season was well reflected in the wool cheque.

4. *cheque clerk*; cheque-book, also *attrib.* or quasi-*adj.*; cheque end, an ornamental device on the perforated edge of a cheque, money order, etc., as a safeguard against fraud; cheque guard (see quot.).

1939 J. L. Garvin *Econ. Found. Peace* xv. 365 Their cheque-book philosophy..their insistence on the peace of profit instead of the peace of justice. 1963 *New Statesman* 24 May 786/1 Newspapers..should..come to a self-denying ordinance to abandon the cheque-book journalism of confession stories by criminals, prominent divorcees and others who have won notoriety. 1908 *Westm. Gaz.* 29 Sept. 8/2 He presented a cheque for 1,000 to the cheque clerk, who examined the balance and initialled the cheque. a 1884 Knight *Dict. Mech. Suppl.*, *Check Guard*, a means of preventing tampering with bank checks, raising the figures, etc.

chequer, *sb.*¹ Add: **16.** chequer-board (orig. *U.S.*) = *checker-board* (Checker *sb.*⁸ 2 c); also *attrib.* and *fig.*

1732 Franklin *Let.* 24 July in *Writings* (1905) II. 189 Mr. Crownhim..is always dreaming over the Chequer-board. 1859 E. FitzGerald tr. *Rubáiyát of Omar Khayyám* xlix, 'Tis all a Chequer-board of Nights and Days Where Destiny with Men for Pieces plays. 1870 'Fanny Fern' *Ginger-Snaps* 79 When some clerical big-gun is supposed to make a false move on the sacerdotal chequer-board. 1931 *Times Lit. Suppl.* 30 Apr. 338/3 Chequer-board system of lands to intervene slips of this wood.

chequered, *ppl. a.* Add: **2. b.** chequered skipper, a species of the skipper butterfly, *Cyclopides palæmon*.

1839 W. Wood *Index Entomol.* i. 1 Papilio. Butterfly..9 *Paniscus* Chequered Skipper. 1879 W. F. Kirby *Europ. Butterflies & Moths* 14 Cyclopides..C. Palæmon..(Chequered Skipper). Wings above blackish-brown, with angular tawny spots on the fore-wings, and four rows of spots on the hind-wings. 1957 E. B. Ford *Butterflies* (ed. 3) i. 16 At the very end of the [18th] century one more species was added to the British list, the Chequered Skipper, *Carterocephalus palaemon*.

3. Esp. in phr. *chequered career*.

1869 'Mark Twain' *Innoc. Abr.* xxxviii. 407 Her career, its chequered history, its..has been a chequered one. 1881 H. W. Nesfield (*title*) A chequered career, or Fifteen years in Australia and New Zealand. 1928 *Melody Maker* Feb. 188/3 Jay Whidden and his band has had a chequered career as far as the Radio is concerned. 1967 *Listener* 17 Aug. 203/1 My career with 20th Century Fox was somewhat chequered.

‖**cherchez la femme** (ʃɛrʃe la fam). Also **chercher la femme**. [Fr., lit. 'seek the woman'.] A catch-phrase, first used (in the form *cherchons la femme*) by Alexandre Dumas *père* in his *Les Mohicans de Paris* (1864), used to indicate that the key to a problem or mystery is a woman, and that she need only be found for the matter to be solved.

[1878 J. Latey tr. *Dumas' Mohicans of Paris* ix. xv. 235 Ah! Monsieur Jackal, you were right when you said, '*Seek the woman*.'] 1895 'S. Grand' *Heavenly Twins* I. ii. 1. 218 In bygone days when there was much more mischief to be done, he *cherchez la femme*, and when she was found the investigation stopped there. 1898 G. B. Shaw *Philanderer* i. 35 There's a quarrel—a scandal—*cherchez la femme*—always a woman at the bottom of it. 1967 'E. Peters' *Black is Colour* v. 100 'There's always the classic way,' said Duckett disgustedly. '*Cherchez la femme!*'

‖**chère amie** (ʃɛr ami). [Fr., lit. 'dear (woman) friend'.] = Mistress *sb.* 11.

1792 Charlotte Smith *Desmond* I. xiii. 175 The poor bishops are gone upon their travels, and their *chère amies* upon the town. 1827 *Misc. Let.* 17 Apr. in Wks. (1905) XII. 17, I hope you are not the 'young law student of Lyon's Inn' whose *chère amie* tried to throw herself into the Thames yesterday? 1828 E. J. Vaughan *Lett. of Mrs. E. Twisleton* xi. 212 Walewska, Mme., a Polish countess, *chère amie* of Napoleon I, by whom he had a son. 1966 H. Voxall *Fashion of Life* viii. 66 Elinor Glyn, the surprising *chère amie* of that 'very superior person', George Nathaniel Curzon.

‖**cherem** (heˈrem). Also **herem**. [Heb. *ḥērem*, f. *hāram* to devote, put under a curse.] Excommunication from the Synagogue.

1832 H. H. Milman *Hist. Jews* III. 145 For more heinous offences, and against contumacious delinquents, the more terrible Cherem, or the still more fatal Shammata, the excommunication, was proclaimed. 1903 *Jewish Encycl.* V. 286/1 If..the excommunicator showed no sign of penitence or remorse, the niddui might be renewed..and finally the 'herem', the most rigorous form of excommunication, might be pronounced. 1925 W. Ewing *J. E. H. Thomson* 158 For

some comparatively trivial cause a *cherem*, or ban, would fall like a bolt from the blue. 1967 *Encycl. Jewish Relig.* 182 The *herem* completely ostracizes the individual involved from all contact with his fellow Jews.

Cheremis(s (tʃeˈrəmis). Also **Tcheremiss**. [Russ.] (One of) a Finnish people living in the region of the middle Volga. Also, the Finno-Ugric language of this people. Also *attrib.* Hence Cheremiˈssian *a.* and *sb.*

1652 Heylin *Cosmogr.* ii. 158 Under the Government hereof are the Cheremisse, a people that live in great Forrests without any Houses. *Ibid.* 159 A people much of the same nature with the Cheremisses. 1850 R. G. Latham *Nat. Hist. Man* i. 99 The Tcheremiss. *Locality*.—The left bank of the Middle Volga. 1879 *Encycl. Brit.* IX. 219/2 The Volga Finns include the Cheremissians on the left bank of the Volga. 1885 *Ibid.* XVIII. 550/1 The population [of Perm, Russia] in 1881 amounted to 2,249,300..It consisted chiefly of Great Russians..including..8000 Tcheremisses. 1900 *Westm. Gaz.* 19 July 2/3 In Cheremissian..I could have been a stylist. 1933 L. Bloomfield *Language* iv. 66 Four further branches of the Finno-Ugrian stock are Mordvine..; Cheremiss (etc.). 1957 *Encycl. Brit.* XIV. 889/1 The Mari were called by the Russians Cheremiss.

‖**cher maître** (ʃɛr mɛtr). [Fr., lit 'dear master'.] A flattering term of address to a famous writer; such a writer. Also *attrib.*

1920 F. M. Ford *Let.* 30 June (1965) 109 So speaking from the Cher Maître elevation..I should say *Young Man*, ahem! 1964 W. Marsfield *To Early Grace* (1965) v. 94 He used to correspond with Gide in French and call him *Cher Maître*. 1967 E. Connead *Thankless Muse* ii. 200 Now your prosaic *cher maître*..the nervous glimpse? 1968 M. Jones *Survivor* iii. 51 Martin's always charming with women. And Rose gave him the *cher maître* treatment. 1968 *Guardian* 19 July 20/8 It's a rare French author who can..dash off a note to his bank without being conscious of his literary duty as a 'cher Maître'.

chernozem (tʃɜːnozem, tʃaːnozyeˈm). *Geol.* Also, chernosem, tchern-, tchorn-, tschern-. [a. Russ. *chernozëm* black earth, f. *chërnyĭ* black + *zemlyá* earth, soil.] Black earth or soil (see *black *a.* 19), a type of soil, rich in humus, characteristic of natural grassland in cool to temperate semi-arid climates, as in central and southern Russia, central Canada, etc.

1842 [see *black earth* s.v. *black *a.* 19]. 1859 Page *Handbk. Geol. Terms*, *Tchernozem*, a local name for the black earth of the south of Russia, which covers the whole of the Aralo-Caspian plain. 1863 J. von Liebig *Nat. Laws Husb.* v. 221 The extensive tract of land in Russia, known as the *Tschernozem* or 'Black soil', whose fertility for corn plants is proverbial. 1914 G. A. J. Cole *Growth of Europe* xi. 207 Chernozem. Inaccessible to the..Russian steppe soils. The Black Earth district of central Russia, where the famous tchernozem soil prevails. 1934 *Discovery* July 201/1 Chernozems (Black Soils). These are always associated with a grass steppe vegetation, never with forest. 1965 A. Holmes *Princ. Physical Geol.* (ed. 2) xiv. 406 The upper layers of the soil profile is black, becoming brown in the steppe where there is less humus. For this reason the soil type is called *chernozem*.

Cherokee (tʃeˈroki, tʃeroki·), *sb.* and *a.* [f. Cherokee *Tsálăgĭ*.] **A.** *sb.* (A member of) an Iroquoian tribe of North American Indians formerly inhabiting a large portion of the southern United States. **b.** Their language, esp. as a type of something unintelligible (cf. Greek *sb.* 8). **B.** *adj.* Of, pertaining to, or concerned with, the Cherokees. *Cherokee rose*: a wild rose of the southern United States, *Rosa lævigata*; also *ellipt.*

1674 H. Woodward *Let.* 31 Dec. in *Coll. S. Carolina Hist. Soc.* (1897) V. 460 Amongst which dividing branches inhabit ye Cowatee and Cherakee Indians. 1699 E. Randolph *Let.* 22 Mar. in W. J. Rivers *Hist. South Carol.* (1856) 449 He would..take with him 50 White men and 100 of the Chirakees Indians. 1726 in *N. Carolina Col. Rec.* (1886) II. 556 The officers who shall command the said soldiers in the said Cherokee expedition. 1752 *Ibid.* 422 The remaining Chiroqui Indians..the Cherokees. 1752 G. Croghan in N. D. Mereness *Trav. Amer. Col.* (1916) 97 The place I ordered the Cherokee Indians (then down) to go to. *Ibid.*, 59 The Slay's..talk good English as well as the Cherokee Language. 1765 J. Habersham *Let.* 18 May in *Coll. Georgia Hist. Soc.* (1904) VI. 36 Our clear Creek Leather is now rather esteemed better than Cherokee. 1765 H. Timberlake *Mem.* 36, I shall transcribe the following specimen, without the original in Cherokee. 1788 J. Atkinson *Match for Widow* 52 This is all Cherokee to me—I don't understand a word of your simile. 1823 D. Douglas *Jrnl.* (1914) 26 Roses Champneya, Cherokee, and two others. 1829 A. Eaton *Man. Botany* (ed. 5) 368 *Rosa lævigata*, Cherokee rose. 1834 W. G. Simms *Guy Rivers* II. xi. 138 The reader has already heard something of the Cherokee pony..They are a small, but compactly made and hardy creature. 1836 G. Colman *Recoll. Southern Matron* (1838) vi. 53 We drove..the beautiful Cherokee rose-hedge, and other objects in view, which were novelties to him. 1836 J. Hildreth *Dragoon Campaigns Rocky Mts.* 96 An old Cherokee, named Rodger. 1868 *Trans. Ill. Agric. Soc.* VII. 144 The long hedge of Cherokee rose is now in full bloom..1873..Cherokee cattle passing through North Carolina and Virginia on their way to the Northern markets. 1928 *Chambers's Jrnl.* Mar. 172/1 Yet above all other flowers I prefer these—Cherokee roses. 1946 E. A. Nida *Morphology* p. v, Linguistic will..recognize many of the problems as being drawn from Greek, Latin..Cherokee, and Navaho. 1960 *Observer* (Colour Suppl.) 18 May 36/1 A thousand miles..the same trail to this day 'the trail of tears' by the Cherokee. *Ibid.* 34/4 Lloyd New, one of a family of 10 Cherokee children.

cherry, *sb.* Add: **2. b.** Also *attrib.*

1845 S. Judd *Margaret* i. ii. 10 A cherry plate (ed. 1851, cherry-wood plate) with a youth's bone knife and fork. 1848 W. T. P. Gill *Coast Steadfast* i. 33 A small cherry table with two leaves.

5. b. *cherry-time* (later example).

1845 F. Douglass *Narr. of Life* (1846) i. 1, I do not remember to have ever met a slave who could tell his birthday. They seldom come nearer to it than planting-time, harvest-time, spring-time, or fall-time.

10. cherry coffee, the fruit containing the coffee berry; cherry cordial = Cherry-bounce 1; cherry country, the district, in Kent, where the cherry is largely grown, in extensive cherry-orchards, for commercial purposes; cherry-picker, (*a*) pl. *slang* = *cherry-breeches* (cf. Cherub 3 e); (*b*) a crane or similar device by means of which persons may be raised or lowered; cherry-wood (earlier U.S. example).

1858 Simmonds *Dict. Trade*, *Cherry-coffee*, the planters' name for the fruit of the coffee as..picked from the tree, before it has undergone the operations of pulping, drying, &c., to prepare the berry for shipment. 1770 W. Salmon *Family Dict.* (ed. 4) 71/1 Cherry Cordial. Take Black Cherries, [etc.]. 1839 *Mag. Domestic Econ.* V. 7 Make also cherry cordial. 1902 *Garden* 10 May 301/1 To get into the heart of the Cherry country one can make Maidstone, Sittingbourne, or Faversham the starting point, and work up to Canterbury. 1903 'Sheila' *Gage* IV. viii. 475, 18th *Hussars*—Cherubims and Cherry Pickers. 1903 *Westm. Gaz.* 3 Oct. 1/3 The 11th [Hussars] got its name of 'the Cherry-Pickers' owing to some of its men having been taken prisoners in a fruit garden during the war in the Peninsula. 1928 *Daily Express* 27 Mar. 10/6 Those crimson overalls of the 'Cherry-pickers'..were really a kind of wedding present from the Prince Consort. 1961 *Flight* LXXX. 608/2 The pad rescue team used an armoured personnel carrier and operated a mobile tower or 'cherry-picker'. The cherry-picker stood by on the pad throughout the final 35 min of the countdown. 1962 *Gazette* et al. *Ibid.* (*Pa.*) Cherry-picker, a crane-like device stationed near the launching pad to help rescue the Astronaut from the capsule in the event of trouble before lift-off. 1968 *Guardian* 19 July 14/5 The call echoes from the platform of a yellow-painted 'cherry-picker'—the sort street-lamps are repaired from..The cherry-picker platform swoops down for a close-up. 1821 *Amer. Jrnl. Sci.* III. 167 In staining cherry wood, cabinet-makers frequently employ some kind of red paint.

cherryade (tʃeriēˈd). [f. Cherry *sb.* + -ade.] A drink made with cherry juice and water, sweetened with sugar. Also, an aerated water with a cherry flavour and colouring.

1892 T. F. Garrett *Encycl. Pract. Cookery* II. 355/2 Cherryade. 1953 Dylan Thomas *Under Milk Wood* (1954) 60 Then his tormentors tussle and run..to buy..gobstoppers..ice-cream cornets..raspberry and cherryade. 1959 *Daily Mail* 11 Apr. 7/1 She and her friend bought sandwiches and the cherryade.

cherry-brandy. Add: (Earlier and later examples.) Also, a glass of this liqueur.

1686 J. Dunton *Let. fr. New-Eng.* (1867) 33 Preserv'd Damsons, Cherry-Brandy, and the like Knick-knacks. 1896 J. Locke *Lt. & Sept. to Fox Bourne Life of Locke* (1876) II. xv. 455 You had been bottling up kindness for your journey, which at last you have let run to the rejoicing of his heart more than if you had overflowed to him sack and sugar or cherry brandy. 1951 [see *advocaat].

chert. Substitute for def.: A form of amorphous silica found in several varieties, e.g. flint.

chervonetz (tʃɜːrvōnets). Also chervonets, tchervonetz. [Pl. chervontsi, -sy, -zi. [Russ. *chervónets*, pl. *chervóntsy*.] A Soviet bank-note of the value of ten gold roubles, in circulation from 1922 to 1947.

1923 *British Weekly* 15 Nov. 164/4, I was struck with the remarkable success of the new currency—the chervonets—which has a gold backing. 1923 *Glasgow Herald* 4 Dec. 13 The introduction into circulation of chervonets banknotes. 1927 *Times* (weekly ed.) 9 June 694/1 The purchasing power of the chervonets has slowly but persistently fallen. 1928 *Daily Tel.* 17 Jan. 12/2 They receive an allowance of five roubles (or half a tchervonetz) to pay expenses for the journey. 1954 E. H. Carr *Interregnum* 32 Narkomfin authorized the acceptance of chervonetz notes for tax payments at the current rate of exchange.

Cheshire. Add: *Cheshire acre* (see quots.).

1808 Holland *Agric. Cheshire* xvii. 342 Land was formerly very generally measured in this county by what is called the Cheshire acre, containing 10,240 square yards; and this measure still continues to be employed to a certain extent, particularly in the northern part of the county. 1833 *Penny Cycl.* I. 98/2 There are also local acres in various parts of England, such as the Cheshire acre of 8 yards to the pole. 1928 *Daily Express* 31 July 3 How many square yards constitute a 'Cheshire acre'?..10,240 square yards.

chess, *sb.*¹ **4.** Add: chess-clock (see quot. 1962); chess-master, an expert chess-player; chess-table (examples).

1888 *Brit. Chess Mag.* Jan. (Advt.), Chess Timing Clocks. 1962 *Ibid.* Nov. 301 We have received from Mr A. Ashman..a sample of a new chess clock. 1951 'Amaco' *Adv. in Chess* vii. 22 Before Chess-clocks were in use, matchgames would often drag on to twenty hours or more. 1962 E. Bruton *Dict. Clocks* 39 *Chess clock*, special timer for chess players with two movements and two dials, one showing the accumulating time occupied by one player in his moves, and the other the time of the other player. 1886 *Fortn. Rev.* XL. 765 To depict the masters..who cultivate the royal game, not only as chess-masters pure and simple, but as men. 1900 *Westm. Gaz.* 31 May 2/1 The chess master must have full control over every part of the board. 1961 *Guardian* 17 Mar. 8/7 The formation of an 'International Association of Chessmasters'. 1862 *Catal. Internat. Exhib., Brit.* II. No. 3575, Chess-tables nacre and pearl specimen cribbage boards. 1868 A. Bennett *Man from North* viii. 161 They had tea at a little round chess-table.

chess-board. Add: **a.** Also *fig.*

1848 [in Dict.]. 1855 Browning *Men & Women* I. 216 What have we gained thro' our first sight of a deep grief hanging over wheat..one side of a dale dipt by a chess-board of doubt. We called the chess-board white,—we call it black. 1868 T. H. Huxley *Lay Sermons* (1870) iii. 36 The chess-board is the world, the pieces are the phenomena of the universe, the rules of the game are what we call the laws of Nature. 1968 *Westm. Gaz.* 25 July 4/2 The chess-board of party politics.

b. (Further examples.) Also as *adj.*, resembling a chess-board; like that of a chess-board.

1880 *Cent. Dict.*, *Chess-board canvas*, a thick cotton canvas used as a foundation for embroidery, and divided into squares in alternating patterns. 1903 *Daily Chron.* 11 Dec. 7/1 Such chessboard cities as Melbourne and Adelaide. 1904 *Daily Tel.* 21 July 10/7 In Japanese chess-board rules of war the next event will be an attack by Russian torpedo craft. 1921 R. C. Punnett *Mendelism* (ed. 5) iii. 33 The 'chessboard' method. For two series each consisting of four different types of gametes we require a square divided up into 16 parts. The four terms of the gametic series are first written horizontally..so that the series is repeated four times. It is then written vertically four times..In this simple mechanical way all the possible combinations are represented. 1923 Collingwood-Brown *Britain* vii. 175 Only the worst chess-board street-plan. *Using the chess-board* It's to go for a town-planning scheme, to lay out a chess-board street-plan. 1962 *Listener* 3 May 935/1 Using the chess-board technique, a reduction in the M.T.C. of Gunfish we demonstrated.

Hence cheˈss-board *v. trans.*, to divide (land) into more or less equal portions resembling the squares on a chess-board.

1891 *Daily News* 12 May 5/3 Mr. Morley protested the other day against 'chessboarding' Ireland. 1924 *Chambers's Jrnl.* 747/1 A tract of low-lying, cultivated land, chessboarded into fields under various crops.

chessdom (tʃeˈsdəm). [f. Chess *sb.*¹ + -dom.] The 'realm' of chess and chess-players.

1875 *City of London Chess Mag.* June 121 There are plenty of courts in Chessdom only too ready to investigate these complaints.

chessylite. (Earlier and later examples.)

1852 Brooke & Miller *Min.* 594 *Chessylite*, blue carbonate of copper. 1936 *Amer. Mineralogist* XXI. 191 Your Committee and the British Committee recommend the following names and spellings of mineral species. The names to be discarded are placed in parentheses..Azurite (chessylite). 1953 *Mineral. Abstr.* XII. 203 Rosettes of minute crystals of chessylite are found.

chest, *sb.*¹ Add: **9. c.** Colloq. phrases: *to get it off one's chest*, to relieve one's mind by making a statement or confession; *to play* (cards, a thing, etc.) *close to one's chest*, to be cautious or secretive about (something); to keep information to oneself.

Daily Chron. 27 Sept. 3/1 The desire is either to deliver a message to the world or to express the individual personality—to 'get it off one's chest', in the vulgar phrase. 1910 W. J. Locke *Simon the Jester* xxi, 'I've got to get it off my chest,' he said. 1918 *Ibid.* xxi, 'I've not been very even of a silly ass.' 1939 F. Powell *What's happened to the Brass* 93 I wanted to..get out some things off my chest. I've got to tell them to somebody. 1961 A. Christie *Pale Horse* xxiv. 248, I couldn't afford to give hints..You have to play these things close to your chest. 1966 *Guardian* 8 Sept. 3/6 It was suggested yesterday that Manchester Corporation in its many property deals should play the cards a little closer to the chest.

10. b. *chest girth, measure, measurement*; *chest-era* (*adv.*), so deep as to reach to the height of one's chest; chest-expander (see quot. 1858 and Expander); chest pack, an airman's pack containing a parachute, carried on the chest; chest-piece, that part of a stethoscope which, when in use, is placed against the chest; chest register, the lower portion of the compass of the human voice; chest-wall, the external surface of the thorax or chest.

1889 C. Edwardes *Sardinia & Sardes* 172 The stream was strong and chest-deep. 1850 *Rep. Comm. Patents* 1849 (U.S.) 270 Having thus fully described..the gymnastic apparatus, which he has called the *Chest-expander*. 1858 Simmonds *Dict. Trade*, *Chest-expander*, an application of fixed or pliable materials for keeping back the shoulders. 1933 C. Isherwood *Mr. Norris* xiii. 189 Upstairs in his rooms, or in the gym shorts, doing exercises with a chest-expander. 1904 *Daily Chron.* 29 Dec. 6/6 A noticeable increase of chest girth for the defectives). 1889 *Cent. Dict.*, *Chest measure*. 1884 E. Hart *Cautionary* 1930 T. Chatterton *Parachuting* ix. 160 The Back Pack is usually used, the Chest Pack being a small auxiliary..1959 Harlow *U.S. Air Force Dict.* 112/1 *Chest-pack parachute*, a parachute pack that is detachable from the harness and is clipped onto a chest harness when needed. 1894 *Brit. Med. Jrnl.* 3 Nov. 997/1 The ear piece are made of india-rubber..and the chest piece has an india-rubber air pad cover. 1909 *Practitioner* Dec. 862 The edge of the metal chest-piece might be conveniently rimmed with hard rubber. 1847 C. W. Manby *Man. Music* xi. 123 In the soprano voice the notes of the chest and medium registers are distinguished for their clearness and brilliance. 1966 H. L. Swords in C. E. Bassil *In Memory of J. R. Firth* 401 Variations entailed by the nature of the chest-measurement are entered by [?]. 1879 *St. George's Hosp. Rep.* IX. 283 A wound on chest-wall. 1909 *Practitioner* Dec. 862 Many physicians prefer to use the unaided ear applied direct to the chest-wall.

Chesterfield. Add: (Examples.)

1852 Mundy *Our Antipodes* III. i. 32 Chesterfield, Benjamin, Taglioni, and Macintosh, are sumptuary nobodies: Paul is only tolerated in his most gossamer form. a 1889 J. Albery *Dram. Wks.* (1939) II. 521 Great blue Chesterfield. 1899 *Westm. Gaz.* 20 Nov. 3/1 In a nice shirt tie, a 'Chesterfield' coat as a pea jacket. 1935 *English Gentleman* Feb. 37/1 The Chesterfield now ranks as the Chesterfield. It also be somewhat shorter than last season. 1938 J. Ironside *Fashion Alphabet* 38 *Chesterfield*..This coat is a fly-fronted, single-breasted, straight coat, often with a velvet collar. It is now used as a single-breasted overcoat but it is sometimes double-breasted and more shaped.

2. A stuffed-over couch or sofa with a back and two ends, one of which is sometimes made adjustable.

1900 *Westm. Gaz.* 30 Jan. 10/2 The club room, charmingly furnished with chairs and Chesterfields upholstered with blue cretonne. 1929 C. *Grant* *Lit Thing* vi. 60 He dropped in to lie full length on the Chesterfield, and motioning to him to sit down beside her. 1927 *Chambers's Jrnl.* Sept. Beth were seated, Jimmy in the chair facing him, and Betty on the chesterfield. 1954 *Jrnl. Canadian Linguistic Assoc.* I. 1. 16 *Chesterfield* seems to in general use throughout Canada, though the usual American *sofa* is also known and used. Almost everywhere in the United States *chesterfields* are cigarettes and nothing more.

Chestertonian (tʃeːstəɹtōˈniăn), *a.* [f. the name *Chesterton* (see below) + -ian.] Of, pertaining to, resembling, or characteristic of the English author Gilbert Keith Chesterton (1874–1936) or his writings.

1909 *Daily Chron.* 14 June 3/1 Chestertonian. [Heading of a review of *Heretics*, by G. K. Chesterton.] 1908 *Catholic Times* 20 Nov. 13/4 and that he [*sc.* G. K. Chesterton] approves plainly in his own Chestertonian way. 1909 *Daily Chron.* 3 Mar. 3/5 He is obviously neither of the Chestertonian faith of straining after paradox and epigram. 1921 *Public Opinion* 29 July 100/2 The Lord Chancellor was almost lyrical or Chestertonian in his rhapsody. 1932 C. S. Lewis *Let.* 24 July (1966) 89 He is the largest man Baker has ever seen—Chestertonian both in figure and habits. 1965 *Listener* 27 May 804/1 A barrage of justificatory protest and Chestertonian rebuttal.

chestful (tʃeˈstful). [-ful.] The quantity (of articles) contained in a chest; *spec.* = Chest *sb.*¹ 7.

1733, 1884 [see Chest *sb.*¹ 10]. 1942 E. Blom *Music in Eng.* v. 80 Purcell's 'Fantasias for strings..revive the old family habit of the chestful of viols.'

chestily, chestiness: see *chesty *a.*

chestnut, *sb.* Add: **1. b.** *to pull the chestnuts out of the fire*: see *fire *sb.* 5 f.

3. b. Cape (or wild) chestnut, an evergreen tree, *Calodendron capense*, belonging to the family Rutaceæ and native to South Africa.

1854 L. Pappe *Silva Capensis* 7 *Calodendron capense*..wild chestnut. 1868 J. Chapman *Trav.* II. 450 The Wild Chestnut..well deserving its name for its exquisitely-pencilled delicate pink flowers. 1909 *East London Dispatch* 3 July 5 [Pettman], The beautiful lilac flowers of the wild chestnut. 1913 *Ibid.* 12 Apr. 7 (*title*) It is very seldom that the Cape chestnut and the Wild fig become altogether devoid of leaves. 1952 L. E. W. Codd *Trees Kruger Nat. Park* 82 Well known South African examples [of the Rutaceæ] are the Buchu (*Barosma*) and the Cape Chestnut (*Calodendrum capense*).

7. (Earlier examples.) Also, in extended use, anything trite, stale, or too often repeated.

1816 W. Dimond *Broken Sword* i. 13 *Zavior*..When suddenly from the thick boughs of a cork tree—*Pablo*. [*Jumping up*] A chestnut, Captain, a chestnut..Captain, this is the twenty-seventh time I have heard you relate this story, and you invariably said, a chestnut, till now.] 1880 D. N. Ramsden *Diary of Dely Débutant* (1910) 199 Think of doing that awful *Nancy Lee*—such a chestnut!—..and following it up with 'The Vicar of Bray'. 1888 *Times* 5 Apr. 106/1 He has omitted such chestnuts as *The Raven* and *Of Captains! My Captains!* 1962 *Listener* 8 Sept. 369/3 Souray's recital [of songs]..is a rare and welcome experience—the more so as on this occasion the chestnuts of the repertoire are avoided.

8. adj. **c.** (Earlier example.)

1670 Denton in M. M. Verney *Mem.* (1899) IV. vii. 228 The horse..was a chestnut.

C. chestnut-brown, *spec.* designating a breed of cat (see quot.); chestnut-eared finch, the Australian zebra finch; chestnut-extract, an extract from chestnut wood, used in dyeing silk black; chestnut-oak (examples); chestnut-roaster, a stove used by itinerant vendors of roasted chestnuts; also, the vendor; chestnut soil (see quot. 1954).

1968 *Times* 10 Nov. 12/6 Exhibited for six years as 'any other variety', the Chestnut Brown can match its feet with..London appearance on Saturday after its recognition as a separate breed in July. The previous name of 'Havana' has been dropped. 1934 A. Russell *Tramp-Royal in Wild Austral.* xxv. 129 The chestnut-eared or zebra finches..came humming down to the water-side. 1881 *Spon's Encycl. Industr. Arts* IV. 1226 Hemlock extract is a deep red syrupy extract of the bark of the hemlock pine of America. Chestnut extract is a similar product from the rasped wood of the Spanish chestnut. 1799 in *Early W. Trav.* (1904) III. 176 The Northwest Corner is a Chestnut Oak tree. 1768 Sewall *Diary* (1879) II. 224 By the Stump grows up a fine little Chestnut Oak. 1885 C. E. Craddock' *Prophet Gt. Smoky Mts.* iii. 49 The chestnut-oaks were [?] in the sunset. 1899 F. Gordon *Quiet Craft* xi. 24 Some chestnut-oaks..and..a walnut or two..were visible. 1931 C. H. Otis *Michigan Trees* (rev. ed.) 173 Chinquapin Oak. Chestnut Oak. Yellow Oak. *Quercus muhlenbergii*..A tree with somewhat similar characteristics, also known as Chestnut Oak, *Quercus prinus* [etc.] is common to the south and east of our range. 1859 *Westm. Gaz.* 1 Sept. 2/2 The boxed old woman who sits hugging her chestnut-roaster and tell out of the Poet. 1900 J. D. Crane *Early Nuts* *Terms* (1927) § 777 Baked chestnut man; chestnut roaster. 1933 *Weekly Dispatch* 29 Apr. 2, I have never seen more than one some across a genuine old chestnut-roaster with a pierced box and lid. 1928 *Soil Research* I. 131 We were already in a zone of light brown steppe soils..the chestnut soils. 1954 W. D. Thornbury *Princ. Geomorphol.* iv. 78 Chestnut soils are brown or grayish soils that developed under short-grass vegetation in areas slightly drier than those that produced chernozems.

chestnutty (tʃeˈsnǫ̈ti), *a.* [f. Chestnut + -y¹.] **a.** Resembling the colour of a chestnut.

1889 R. Kipling *Sand.* Nov. 337/1 His [eyes] are chestnutty brown.

b. Of the nature of a chestnut (Chestnut 7); old and well-known; frequently repeated. orig. *U.S.*

1884 *Detroit Free Press* 14 May 2/1 Excitement about it is the press and in the pulpit. 1906 *Daily Chron.* 13 Nov. 4/4 The story, slightly chestnutty to the favour of the theatre manager. 1963 *Punch* 25 Dec. 934/1 Old chestnutty films.

chesty (tʃeˈsti), *a.* [f. Chest *sb.*¹ 9 + -y¹.]

1. *U.S. slang.* Conceited and self-assertive; having one's chest thrust out as a sign of self-importance. Also quasi-*adv.*, in *to walk chesty*.

1900 Ade *Fables in Slang* 90 Zoroaster and Zendavesta continued to walk chesty and tell people how good they were. 1900 'H. McHugh' *John Henry* 57 It isn't quite as chesty as 'David Harum', but there's more poetry in it. 1901 N.Y. *Sun* 25 Nov. 11/1 In this new profession, he may have the effect of enlarging his self-esteem. He may have been a bit chesty, he quit in his at the terminate of the campus. 1912 Mathewson *Pitching* iii. 69 The team had won the championship in 1900, and naturally they were all pretty chesty. 1913 J. London *Mac Jilly Bell* (1914) 48 George was a little bit chesty, because all the girls around were making eyes at him.

2. Inclined to or symptomatic of weakness or disease of the chest; see also quot. 1934. *colloq.*

1930 J. Cannan *No Walls of Jasper* xv. 265 The family solicitor—he'd a cold too, but a chesty one. 1934 Webster, *Chesty*..having chest tones. *colloq.* 1959 M. Bradbury *Eating People* ii. 21, 'That is always a very chesty cough, Nurse.' *Times* 2 Dec. 8/5 Many of these children are chesty, prone to..sinus infections.

b. Of a woman: having prominent breasts. *colloq.*

1965 M. Dickens *Winds of Heaven* vi. 168 A chesty girl in a sweater.

Hence cheˈstily *adv.*, in a chesty (sense 1) manner; cheˈstiness, the condition or quality of being chesty (in any of the senses above).

1897 G. H. Lorimer *J. Spurlock* iii. 59 So I explained rather chestily to Horton. 'Of course, I didn't send those notes; but it was done by one of my authorized agents.' 1900 *Cent. Dict. Suppl.*, *Chestiness*, the feelings, attitude, or manner of a 'chesty' person; pedantic self-assertiveness. 1920 O. Johnson *Varmint* xiii. 180 I've made my chestiness. 'I haven't said anything to any of you, have I?' 1931 *Kansas City Times* 29 Sept. 9/4 Our leading wanting announces that he doesn't like how his..1955 E. Taylor *At Mrs Lippincote's* xx. 175 'That is a nasty cough you have,' Roddy said. 'That is what I had.' 1965 *Westm. Gaz.* 12 Apr. 8/2 Hazel..1959 *Woman's Own* 4 Nov. 45/3, I wondered if you could give me any tips for preventing..'chestiness'?

Chevsan, var. *Hesvan.

chetel, var. *chital.

chetnik (tʃeˈtnik). Also **četnik**, **C-**, [a. Serbian *četnik*, f. *četa* band, troop.] A member of a guerrilla force in the Balkans.

1909 *Westm. Gaz.* 20 Mar. 5/2 Mr. Smith..became a real 'četnik', as the members of these forces [*sc.* Bulgarian guerrilla troops] were called. 1920 *Ann. Reg.* 1942 207 A well-organized guerrilla army in Yugoslavia consisting of between 80,000 and 150,000 Chetniks. 1946 P. J. Maclean *Eastern Approaches* ii. vii. 279 The resistance of General Mihajlović and his Chetniks to the enemy.

chetty (tʃeˈti). Also chettie, chitty. [Malayalam *chetti*, Tamil *shetti*.] An individual of any of the South-Indian trading castes so named. Also *attrib.*

1598 W. Phillip tr. *Linschoten's Voy. E. Indies* i. 58/2 There are some likewise that get their livings by their cunning, and they are called Chettijns..they deale with wares and marchandises, and they are called Chattiins. 1686 in J. T. Wheeler *Madras in Olden Time* (1861) I. 151 If the Chetty Bazaar people do not immediately open their shops and sell their grain. 1773 Ives *Voy.* 25 Chitties are a particular kind of..merchants who sell..rice. 1859 W. Gardner in *Jrnl. Chetti* are the.. 1880 *Encycl. Brit.* XII. 760/1 Along the south-western coast [of India] the predominating class of traders are those named Chetties and Komatis. 1908 C. Corner *Ceylon* xxiv. 270 Very exclusive indeed is the Chettie society. *Ibid.* 273 The host..a Chettie in chettie costume. 1912 *Times* 11 Oct. 6/6 A chetty gave as a treat to 173 persons a feast of rice and curry. 1928 'G. Dowell' *Burmese Days* ix. 144 The Indian chetty in the bazaar would cash cheques.

cheval. Add: *à cheval* (earlier and later examples). Also *transf.*, *spec.* in gambling, applied to a stake risked equally on two chances.

1832 C. MacFarlane *Bandits* I. i. 22 It is so easy for the criminals to evade pursuit, by constantly keeping themselves [to use a military phrase] *à cheval* on the line of demarcation of the two countries, and when pursued in but to the one, by retreating into the other. 1881 T. Hardy *Laodicean* II. xv. 233 Leaving off backing numbers *en plein*, he laid his money *à cheval*. 1905 *Westm. Gaz.* 3 Mar. 10/1 The Bar for a moment *à cheval* the political aspects. 1917 T. E. Lawrence *Lett.* (1938) 293 If we try half the Hejaz expresses piece to Mecca, and another long side stretched to the Government of Syria, we seem basically *à cheval* on their whole ground. 1933 W. Pronckel in *Stubb's Gl.* Mar. 150/2 Kruger and Klug is a pair. I did not take to many good *à cheval*. 1947 E. Blom *Music in Eng.* ix. 150 The chief *chevaux de bataille* for new soprano beings was the 'Casta diva', 'La Sonnambula' and 'Lucia di Lammermoor'.

‖**cheval de bataille** (ʃəval də bataj). Pl. chevaux de bataille (ʃəvǫ). [Fr., lit. battle-horse.] An obsession, favourite subject.

1818 Lady Morgan *Fl. Macarthy* III. 86 Ceremony, with all its laws of precedence, is the *cheval de bataille* of the demi-officials of Ireland. 1888 G. B. Shaw *London Music 1888–89* (1937) 45 When Beethoven is made a mere *cheval de bataille* for a Rubinstein, the interest is more volatile. 1895 G. M. Yonge *Long Vacation* viii. 75, I understood that Magnet had made hers *cheval de bataille*, served him for a *cheval de bataille* ever since! 1942 E. Blom *Music in Eng.* ix. 150 The chief *chevaux de bataille* for new soprano beings..

chevalet. Add: **c.** *Glass-manuf.* A stand or bench upon which a cylinder of glass is laid before it is spread out.

1890 W. J. Gordon *Foundry* 147 The cylinder—some six feet long—is then laid on the wooden 'chevalet' or stand.

cheval-glass. (Earlier example.)

1836 Dickens *Pickw.* ii. 14 The stranger surveyed himself..in a cheval glass.

chevalier. 3. *chevalier d'industrie* (examples).

1750 Chesterfield *Let.* 26 Apr. (1774) II. ii. 4 Be cautiously upon your guard against the infinite number of fine-spoken *chevaliers d'industrie*, which swarm at Paris. 1843 Lytton *My Novel* IV. xii. 287 A severe good wife would have stripped my *chevalier d'industrie*. 1897 G. B. Shaw *Our Theatres in Nineties* (1932) III. 18, I have no objection whatever to Satan, after ingratiating himself with the smartest *chevalier d'industrie*, driving himself away by occasional flashes of lightning. 1945 R. Hargreaves *Enemy at Gate* 183 That imperial *chevalier d'industrie*, Napoleon III.

chevaux, see *shiveau.

chevra, see *hebra.

chevron, *v.* Delete †*Obs.* and add later examples. Also cheˈvroned *ppl. a.*

1942 E. Blunden *Thomas Hardy* iii. 49 The chevroned doorway of a village church tower. 1946 R. Campbell *Talking Bronco* 45 And Boreas opens out his throttle Down speedways chevroned by the storks. 1966 *Guardian* 22 July 8/5 White satin appears chevroned with brown velvet.

chevy, chivy, *sb.* Add: **4. b.** *Rhyming slang* = Face *sb.* 1 a. Also *ellipt.* as *chevy*. Cf. *chiv(v)y sb.* 7 *Obs.*

1859 Hotten *Dict. Slang* 142 Chevy Chase, the face. 1891 [see *chiv(v)y sb.*]. 1925 J. R. Ware *Passing Eng.* 72 After listening for a while her chevy-chase gets serious looks. 1942 *Words Speak* XIX. 192/1 Chevy chase, the face..Now obs. in England.

chew, *v.* Add: **3. g.** Slang phr. *to chew the rag* or *fat*: to discuss a matter, esp. complainingly; to reiterate an old grievance; to grumble; to argue; to talk or chat; to spin a yarn.

1885 J. B. Patterson *Life in Ranks* ix. 77 Persisting to argue the point, or 'chew the rag', as it is termed in rank and file phraseology, with some extra intelligent non-commissioned officer. *Ibid.* 12 The various diversions of whistling, singing, arguing the point, chewing the fat. 1891 *Army & Navy Jrnl.* 14 Feb. *Ibid.* Feb. He was chewing the rag at me the whole afternoon. 1909 *Scribner's Mag.* Aug. 250/1 Now better is conversational impudence chararterized than by 'chewing the rag'? 1916 *Punch* 19 Apr. 144/3, I got up..Gen. F. T. Jones..did the good Lord knows..Although we never chews the fat. 1925 'Taffrail' *Pincher Martin* viii. 134 D'you think I'm standin' up 'ere 'longside a blackboard chewin' the fat for the good o' my 'ealth, or wot? *Ibid.*, *Chewing the fat*, spinning a yarn. 1929 *War Slang* in *Athenæum* 8 Aug. 727/2

If anyone starts fault-finding or 'chewing the fat' he is immediately 'ticked off'. **1928** *Daily Express* 2 Mar. 5/2 We 'chew the rag', as our husbands would call it, over happenings of weeks and even months ago. **1931** R. CAMPBELL *Georgiad* i. 17 The scavengers of letters Convene to chew the fat about their betters. **1946** WODEHOUSE *Money in Bank* xii. 102 We were at J. Sheringham Adair's office, chewing the fat with J. Sheringham Adair. **1948** 'J. TEY' *Franchise Affair* xi. 117 We had that paper in the pantry last Friday and chewed the rag over it for hours!

5. b. With *out*. To reprimand. *colloq.* (chiefly *U.S.*).

1948 J. B. ROULIER in *N.Y. Folk Q.* IV. i. 18 A verbal admonishing from a superior would be recorded by the victim with 'I just got eaten out' or 'I just got chewed out'. **1963** H. GARNER in R. Weaver *Canadian Short Stories* 2nd Ser. (1968) 48 Walters chewed him out for not knowing that the specifications had been changed. **1967** R. J. SERLING *President's Plane is Missing* (1968) ii. 20 When Gunther Damon chewed out an errant staffer, his five feet eight seemed to swell to six feet.

7. Also with *over*; esp. to discuss, talk over (a matter).

1939 R. CHANDLER *Big Sleep* xxv. 149 Drop up and chew it over. **1952** 'M. INNES' *Private View* ii. 37 Must you people really go on chewing over Gavin? **1952** S. KAUFFMANN *Philanderer* (1953) xv. 247, I certainly don't want to chew the matter over tonight. **1960** L. COOPER *Accomplices* I. vi. 62, I chewed it over for a bit and came to the conclusion that I'd better speak to John Pollard.

chew, *sb.* Add: **3.** Also, a sweetmeat, esp. a 'chewy' one.

1936 MORROW & HEMMINGER *Western Cook Bk.* 40 (*heading*) Chinese Chews. **1950** B. SUTTON-SMITH *Our Street* iv. 48 He would spend it on biscuits and chews. **1959** I. & P. OPIE *Lore & Lang. Schoolch.* ix. 166 Sweets are.. 'chews' in Aberystwyth. **1967** L. DEIGHTON *London Dossier* 158 Chewable trifles such as wine-gums..and nougat-based chews on sticks.

chewable (tʃiū·ăb'l), *a.* (and *sb.*) [f. CHEW *v.* + -ABLE.] That can be chewed. So as *sb.* (*rare*), something that can be chewed.

1846 [see SUCKABLE *a.* and *sb.*]. **1866** 'MARK TWAIN' in *Harper's Mag.* Dec. 111/2 With what nourishment we can get from boot-legs and such chewable matter. **1939** JOYCE *Finnegans Wake* (1964) 406 He was guilbey of gulpable gluttony as regards chewable boltaballs. **1967** [see prec.].

chewing, *vbl. sb.* Add: **3. chewing-gum** (earlier and later examples); for *U.S.* read orig. *U.S.*; **chewing-tobacco,** orig. *U.S.*, tobacco suitable for chewing.

1850 *Chicago Daily Democrat* 25 Oct., Chewing gum! A new and superior preparation of Spruce Gum. **1949** *Economist* 5 Nov. 996/2 The Soviet delegate..accused the United States of giving exports of chewing gum priority. **1789** *Mass. Sentinel* (Boston) 28 Oct. 4/3 Chewing Tobacco of various kinds. **1909** 'O. HENRY' *Options* 200 Which one of you has got any chewing-tobacco? **1921** *Daily Colonist* (Victoria, B.C.) 5 Oct. 11/1 (Advt.), Macdonald's Prince of Wales Chewing Tobacco.

Chewings fescue: see *FESCUE *sb.* 4.

chew-stick. Add: Also, the root or stem of various West African plants, used to clean the teeth.

1887 MOLONEY *Forestry W. Afr.* 287 The roots [of the cola], called 'chew-stick', are used in Sierra Leone for cleaning the teeth and sweetening the breath. *Ibid.* 371 Chew-stick of Ewuro (*Vernonia amygdalina*, Del.).—Shrub.. used in Sierra Leone as a bitter. **1916** C. E. LANE-POOLE *Trees of Sierra Leone* 147 Every native male and female uses a chew stick. **1965** R. W. J. KEAY et al. *Nigerian Trees* II. 419 The common and widespread species *V*[*ernonia*] *amygdalina* Del. and *V. colorata* (Willd.) Drake..are well-known as a source for chewsticks.

chewy (tʃiū·i), *a. colloq.* [f. CHEW + -Y[1].] Suitable for or requiring chewing. Also *transf.*

1925 *Golden Book Mag.* Nov. 66 (Advt.), It was made for those who prefer firm and 'chew-y' chocolates. **1957** E. W. BARNES *High Room* iv. 76 Though just as addicted to sentimentality as our fathers before the first World War, we prefer that commodity with a more brittle coating, or of a chewier texture than that of Sheldon's story. **1969** N. W. PIRIE *Food Resources* vii. 168 So fabricate a 'chewy' texture in their product.

Cheyenne (ʃəi,æ·n, ʃəi,e·n). Also **Chajenne, Chayenne, Chean, Chian, Chien,** etc. [Canadian Fr., f. Dakota *Šahi'yena*.] A tribe of Algonquian Indians inhabiting the area between the Missouri and Arkansas rivers; a member of this tribe; their language. Also *attrib.* or as *adj.*

1778 J. CARVER *Trav. Int. Parts N. Amer.* 80 These..are termed..the Mawhaws, and the Schians. The other three bands, whose names are the Schianese, the Chongousceton, and the Waddapawjestin, dwell higher up. **1804** LEWIS & CLARK *Jrnl.* 15 Oct. in *Hist. Exped. Lewis & Clark* (1893) I. iv. 168 We saw ..a little below the creek, an old village of the Sharha or Chayenne Indians. **1812** J. C. LUTTIG *Jrnl. Exped. Upper Missouri* 15 Oct. (1920) 86 A Band of Chajennes about 12 Lodges arrived. **1813** *Missouri Gaz.* (St. Louis) 15 May 1/2 Finding an inland tribe of indians calling themselves Shawhays, but known among the whites by the appellations of Cheyen-nes, we procured from these people an accession of forty horses. **1821** J. FOWLER *Jrnl.* (1848) 59 We have heare now about seven hundred lodges... with the addiction of the Cheans—about two hundred lodges. **1857** J. A. GOVE in *Coll. New H. Hist. Soc.* (1928) XII. 28 The messenger reports 840 head of cattle stampeded about 30 miles beyond Kearney by the Cheyennes (pronounced as though written Shy-ans). **1891** *Harper's Mag.* XXII. 204/1 The Texas saddle has a much flatter seat than the Brazos tree; the Cheyenne saddle a still flatter one, with a high cantle and a different cut of pommel arch. **1933** BLOOMFIELD *Lang.* iv. 72 A few detached languages in the west: *Blackfoot, Cheyenne,* and *Arapaho*. **1946** G. FOREMAN *Last Trek Indians* 296 Colorado troops attacked the Cheyenne in May. **1949** M. MEAD *Male & Female* ii. 23 The Ashanti and the Cheyenne Indian, have histories as long as our own, only different. **1969** R. C. BELL *Board & Table Games* II. v. 79 This gambling game was played by the squaws of the Cheyenne Indians for small stakes.

Cheyne–Stokes (tʃē·n stō·ks). The names of John *Cheyne* (1777–1836), Scottish physician, and William *Stokes* (1804–78), Irish physician, used *attrib.* in *Cheyne–Stokes breathing, respiration,* etc., to describe a type of breathing characterized by cycles of respiratory movements consisting of gradual decrease, total cessation, and gradual increase of the respiratory movements, occurring chiefly in coma and in cardiac and cerebral affections.

1874 *Dublin Jrnl. Med. Sci.* LVIII. 519 That peculiar form of breathing once called ascending and descending respiration, but now better known, especially abroad, by Professor Traube's nomenclature, as 'The Cheyne-Stokes Phenomenon'. **1877** *Lancet* 3 Mar. 309/1 The typical Cheyne-Stokes' respiration is that..in which, after a complete pause, there is gradual return of breathing and increase of chest movements till they attain an exaggerated degree, ..and again a gradual subsidence and pause. *Ibid.* 17 Mar. 385/2 When the exhaustion was extreme, the phenomenon of Cheyne-Stokes' breathing developed. **1886** L. CURTIS in *Buck's Handbk. Med. Sci.* II. 89/2 In the last stages [of fatty heart], Cheyne-Stokes respiration and evidences of cerebral anæmia occur. **1898** *Westm. Gaz.* 18 Mar. 7/1 Mr. Gladstone has what we call the Cheyne-Stokes breathing. **1910** A. BENNETT *Clayhanger* III. xvi. 434 Symptoms of 'Cheyne-Stokes breathing', the final and worst symptom of his disease. **1964** G. B. MAIR *Miss Turquoise* iv. 50 The shallow irregularity of Cheyne Stokes rhythm.

‖**chez** (ʃe), *prep.* [F., f. OF. *chiese*, L. *casa* house.] Used with (French) personal pronoun or proper name: at the house or home of.

1740 M. W. MONTAGU *Let.* 19 July (1966) II. 199 Be pleas'd to direct your next chez Madame La Comtesse de Pomfret, Pairesse d'Angleterre, a Florence. **1765** D. GARRICK *Let.* 16 Feb. in R. B. Peake *Mem. Colman Family* (1841) I. 138 But he keeps with the lady, and sees no body *chez lui*. **1770** GRAY *Let.* 22 Dec. in *Wks.* (1884) III. 388, I shall not be able to wait on you chez vous. **1803** *Wynne Diaries* 5 Dec. (1940) III. 101 My Mother must have been in high fidgets at Fanny's accouchement chez elle. **1821** LADY GRANVILLE *Let.* 26 Feb. (1894) I. 209 How much better I like *chez moi* society than any other! **1845** THACKERAY in *Fraser's Mag.* June 715/2 Who knows whether there may not be something of the sort *chez nous*? **1916** A. BENNETT *Lion's Share* xxxv. 275 Do you think I do not know the type? We have it, *chez nous*. **1918** —— *Pretty Lady* xvii. 105 One is very comfortable *chez-moi*. **1936** C. CONNOLLY *Rock Pool* (1947) iv. 47 The male American headquarters: *chez* Eddie-from-the-top. **1954** I. MURDOCH *Under Net* i. 24, I thought it might not be very restful chez Dave. **1964** *Guardian* 29 Feb. 6/2 That hung-on-an-underbodice technique..makes for perfection Chez Balenciaga.. but..chez anyone else would look like nothing at all. **1970** A. DRAPER *Swansong for Rare Bird* ix. 90 Her old man's house..had..a posh name on the gate 'Chez Nous'.

Hence (with *pron.*) as *sb.*

1804 J. ORROK *Let.* 12 Mar. (1927) 63 This day James Reid has come to see us and we give a 'chez-vous' on the occasion, it may be called our *house warming*. **1843** F. A. KEMBLE *Let.* 3 Oct. in *Rec. Later Life* (1882) III. 48 They come and sit with me sometimes of an evening, otherwise *mon chez moi* is undisturbed. **1908** *Daily Chron.* 28 Mar. 4/4 Southerners who loll in the sun all day..do not need so much the cheering influence of a bright and beautiful chez-soi.

chi (kəi). The name of χ, the 22nd letter of the Greek alphabet; *spec.* used as the name of a moth having a marking of the form of this letter (in full, *chi moth*). See also *CHI-RHO, *CHI-SQUARE.

c **1400** in MAUNDEVILLE *Trav.* (1839) iii. 20. **1832** J. RENNIE *Butterfl. & Moths* 77 The July Chi (*Polia Chi*). **1848** T. K. ARNOLD *Elem. Gk. Gram.* 1 Upsilon, Phi, Chi, Psi. **1869** E. NEWMAN *Brit. Moths* 394 The grey chi. **1927** *Contemp. Rev.* July 97 The chi moth is wonderfully concealed when it rests on a grey stone wall. **1955** *Sci. News Let.* 19 Feb. 117/2 The 21 particles listed by Dr. Oppenheimer as currently accepted are electron, positron and neutrino;..sigmas, positively and negatively charged; negative chi, [etc.]. **1961** R. SOUTH *Moths of Brit. Isles* (ed. 4) I. 242 (*heading*) The Grey Chi (*Antitype chi* Linn.).

chiack, chiak, varr. *CHI-HIKE *sb.* and *v.*

Chianti (ki,æ·nti). [Name of the *Chianti* Mountains, Tuscany.] In full *Chianti wine*. A dry red wine, also a white wine, produced in a specified area of Tuscany; loosely applied to various inferior Italian wines.

1833 REDDING *Wines* ix. 245 Chianti wine was formerly imported into Great Britain before that of Oporto had nearly excluded the other species, and the red wine of Florence continued to arrive after the importation of Chianti had ceased. **1887** *Athenæum* 12 Nov. 635/3 He lived in Florence..when a *fiasco* of good Chianti could be had for a paul. **1916** *Anzac Book* 129/2 He keeps a Chianti that is first-rate. **1959** *Chambers's Encycl.* III. 394/2 Chianti wine..is an agreeable thin red wine..of the claret type, usually rather dry...Chianti is usually bottled in straw-covered flasks.

chiasma. Delete ‖ and add: **3.** *Cytology.* Pl. **chiasmata.** [Fr. (F. A. Janssens *Cellule* (1909) XXV. 391).] The exchange of segments of chromatids during nuclear division.

1911 T. H. MORGAN in *Science* XXXIV. 637/1 The chiasma type of cross union between the homologous chromosomes is grasped. *Ibid.*, If, then, the genes do not themselves split when the chiasma is formed there is no opportunity offered for a quantitative division. **1929** DARLINGTON in *Nature* 20 July 99/2 In the longer chromosomes these chromatids change partners, forming what are called 'chiasmata'...In the short chromosomes..there is rarely more than one chiasma. **1949** *New Biol.* VII. 83 Each chiasma represents a point where two of the four daughter chromatids, one from each of the pairing partners, have broken and rejoined with one another, thus giving rise to two new continuous strands which cross over from one partner to the other.

chiaster (kəi·æstəɹ). *Zool.* [f. Gr. χιαστός marked with a χ (*CHI) + ἀστήρ star (see ASTER).] A star-shaped sponge-spicule with slender cylindrical rays.

1888 SOLLAS *Tetractinellida: Challenger Rep.* XXV. p. lxiv, *Euasters*..1. Chiaster (from the Greek letter χ, to which some forms present a fanciful resemblance). A minute aster with very slender cylindrical actines, terminally tylote or not, or truncate. **1900** C. W. ANDREWS et al. in *Proc. Zool. Soc.* Feb. 133 Microscleres—spherasters.. Somal chiasters, 12μ.

Chibcha (tʃi·btʃā). [Sp.] (A member of) an Indian people of Colombia, having an ancient civilization; also, the language of this people. Also *attrib.*

1814 H. M. WILLIAMS tr. *von Humboldt's Res. Anc. Inhabitants Amer.* II. 111 The language of the..Moscas, or Muyscas, is known in the country under the designation of Chibcha. *Ibid.* 118 No dictionary of the Chibcha language exists. **1877** A. S. GATSCHET in W. W. Beach *Indian Misc.* 422 Such political bodies have frequently absorbed neighbouring communities engaged in similar pursuits, and turned them into powerful empires, as is the case of the Aztecs, Mayas, Chibchas and Quichhuas, in the western hemisphere. **1950** [see next]. **1959** *Chambers's Encycl.* I. 343/2 *Chibcha Area*...Culturally and linguistically the tribes of the isthmus ..belong with the Chibcha of Colombia. *Ibid.* 352/2 The most politically developed tribes, the Chibcha. **1961** *Times* 13 May 9/7 Some of these bore resemblance to Chibcha art.

Chibchan (tʃi·btʃăn). [f. prec. + -AN.] The language of the Chibcha people. Also as *adj.*, designating a group of related S. American languages of which it is the principal (Webster 1934).

1910 *Encycl. Brit.* I. 811/2 The [linguistic] families of.. South America are here given in alphabetical order.. Chibchan. **1950** E. PARTRIDGE *Here, There & Everywhere* 133 *Chibcha* or *Chibchan* is an extinct language of Colombia. **1950** J. A. MASON in J. H. Steward *Handbk. S. Amer. Indians* VI. 175 Formerly almost all the languages of highland and coastal Colombia were considered to belong to the Chibchan group, but recent opinion assigns the Chocó and most of the other groups of northern Colombia, except for the Bogotá Chibcha and the Arhuaco region, to the Carib.

Chicago (ʃikā·go). The name of the city in Illinois, U.S.A., used *attrib.* or in *Comb.* in various special collocations (see quots.).

1860 *Cincinnati Commercial* 29 Dec. 3/2 The Chicago Platform [of the Republicans] was framed with a view to recognizing the admissibility of a doctrine of absolute non-intervention. **1881** W. D. HAY *300 Yrs. Hence* iv. 69 The electric shell..usually known as 'the Chicago bullet'. **1935** R. B. PERRY *Tht. & Char. W. James* II. LXXXI. 519 The group..formed around [John] Dewey at the University of Michigan and afterwards at the University of Chicago, forming the nucleus of the so-called 'Chicago School', and of the general movement known as 'instrumentalism'. **1941** C. OLSSON in Michie & Graebner *Lights of Freedom* xiv. 183 A rating was crouched by the 'Chicago piano' (battery of A.A. pompoms) amidships. **1944** P. NOBLE *Transatlantic Jazz* iv. 45 Bud Freeman..plays in the 'Chicago style' (using 'choppy' phrasing). **1946** R. BLESH *Shining Trumpets* (1949) i. 23 No heterophony in white jazz except a chaotic sort in Chicago-style jazz. **1958** P. GAMMOND et al. *Decca Bk. Jazz* vi. 77 'Chicago style'..came to be a pattern of expression used by a group of white musicians, all of whom were strongly under the influence of the coloured players, and all of whom performed in and around Chicago in the late 'twenties. **1966** *New Statesman* 24 June 940/2 The Chicago school has a great deal to teach architects in terms of a proud commercial vernacular.

Chicagoan (ʃikā·goăn). [f. prec. + -AN.] A native or inhabitant of the city of Chicago; *spec.* an exponent of a jazz style which originated there.

1861 *N.Y. Herald* 25 May 8/1 Several thousand Zouave organizations..all dating their organization since the tour of the Chicagoians. **1882** *Advance* (Chicago) 27 July, A nervous, brain-weary Yankee or Chicagoan. **1946** R. BLESH *Shining Trumpets* (1949) x. 230 The earliest playing by Chicagoans shows the predominance of the one influence or the other. **1955** R. HORRICKS in A. J. McCarthy *Jazzbk.* 1955 32 They are a resurrection of the white Chicagoans in so far as their music has everything but the rare, natural inspiration of the Negro jazzmen. **1958** *Times* 22 Nov. 7/7 Chicagoans were ever perhaps the most essential, the most typical, of Americans.

chicane, *sb.* Add: **3.** In the game of Bridge, the condition of holding no trumps (see also quot. 1964); *double chicane*, chicane of both partners.

1886 *Biritch or Russian Whist* 4 If one hand has no trumps..this is called 'chicane'. **1895** 'BOAZ' *Laws of Bridge* 3 Chicane is thus reckoned:—If a player holds no trump, he and his partner score for Chicane twice the value of the trump suit trick. **1900** 'J. DOE' *Bridge Man.* 8 Double Chicane is scored if neither of two partnered players has a card of the Trump suit...Double Chicane [counts] four by honours. **1929** WORK *Compl. Contract Bridge* iii. 17 In addition to chicane, a singleton ace..may be relied upon as insurance against losing tricks in that suit. **1964** *Official Encycl. Bridge* 75/1 *Chicane*, a term from *bridge whist* referring to a hand that is void of trumps. It was scored the same as three honors. In contract bridge, the term..is occasionally used to describe a void suit, as 'chicane in hearts'.

4. *Motor Racing.* A disguised or artificial construction, esp. a barricaded ramp. Also *attrib.*

1955 *Times* 23 May 12/6 The Lancia skidded at the *chicane*, demolished the wall bordering the sea, and dropped into the harbour. **1958** *Times* 8 Apr. 14/2 Led..for three laps; only to crash into the chicane barrier. **1959** G. FREEMAN *Jack would be Gentleman* viii. 158 The news cameramen crouching down at the chicane for action photographs. *Ibid.* x. 216 He changed down again as he approached the chicane, snaked neatly through it, and accelerated up to the start.

chi-chi (tʃi·tʃi), *sb.*[1] Var. CHEE-CHEE.

1888 KIPLING *Gadsbys* (1889) 19 A fine Mussoorie *chi-chi* to bring home. **1924** BUCHAN *3 Hostages* ix. 133 Then to my surprise he spoke in English—good English, with the *chi-chi* accent of the Indian. **1944** H. G. WELLS *'42 to '44* 30 Kipling's chi-chi Imperialism. *Ibid.* 33 The Empire has contributed its quota of offspring from minor officials and business representatives, not always innocent of a chi-chi accent. **1951** H. JORDAN *Islander* xiv. 164 The group had originated..with an element of halfcastes in it, but the chi-chi admixture had been long discarded.

chichi (ʃi·ʃi), *sb.*[2] and *a.* Also **chi-chi.** [Fr.] **A.** *sb.* Fuss(iness), excessive refinement; pretentiousness, affected ways. **B.** *adj.* Fussy, precious (see PRECIOUS *a.* 3), over-elaborate.

1908 *Sporting Times* 28 Mar. 4/1 Small silk bows, embroidered spots, odd bits of lace sewed on anywhere, are all 'chichis'—trifles, trumpery adornments...The actress pretending to be down with influenza so as to throw up a part she dislikes creates 'chichis';..wee cakes and sandwiches are 'chichis'; saucy false curles rank as 'chichis'; elaborate compliments, or the 'no' that means 'yes', and vice versâ, are all 'chichis'. **1928** W. S. MAUGHAM *Ashenden* xii. 236 The managers won't be too pleased with me if I make a lot of *chichi*. **1932** N. MITFORD *Christmas Pudding* xiv. 208 A pair of gold hair-brushes, rather chi-chi, but very attractive all the same. **1933** R. ALDINGTON *All Men are Enemies* III. i. 249 He stepped across the path of crazy pavement (a bit of Margaret's upper-suburban *chi-chi*). **1934** C. LAMBERT *Music Ho!* iii. 226 It manages to deal with a modern emotional subject without chi-chi, false sentiment or mechanical romanticism. **1937** *New Statesman* 11 Dec. 1007/2 Every manner of eating-house throughout France, from the *chi-chi* establishment appropriate to the entertainment of a *poule de luxe* to the obscure provincial *bistro*. **1942** DENTON WELCH *Jrnls.* (1952) 28 Sept., Tea with the Graham Sutherlands yesterday...Talk lapsing to chichi quickly and then righting itself in periods of silence. **1951** E. TAYLOR *Game of Hide-&-Seek* II. i. 140 Going in gangs to those chichi clubs at Maidenhead. **1961** A. WILSON *Old Men at Zoo* v. 249 Elegance with a touch of chichi. **1969** *Daily Tel.* 22 Jan. 15 The sort of real delicious Italian country cooking that is a revelation after so much chichi Italian food dished up in London.

chick, *sb.*[1] Add: **3. b.** A girl; a young woman. *slang* (orig. *U.S.*).

1927 S. LEWIS *Elmer Gantry* vii. 114 He didn't want to marry this brainless little fluffy chick. **1957** C. MACINNES *City of Spades* I. v. 31 There stood..Muriel's sister. But what a difference from the little chick! **1959** *News Chron.* 12 Aug. 4/3 Beatniks and their 'chicks'—palefaced girls wearing pony-tail hair-dos and toreador pants. **1971** *It* 12–16 June 16/2 Jackie, always a 'with-it chick'.

chick, *sb.*[4] Add: **2.** The call-note (of a bird). Also *chick-chick*. (Cf. CHACK *sb.*[3], *CHACK-CHACK.)

1859 H. KINGSLEY *G. Hamlyn* v, The chick-chick of the stonechat. **1894** R. B. SHARPE *Handbk. Birds Gt. Brit.* I. 105 They never uttered more than a whispered call-note, 'chick'. **1922** ZANE GREY *To Last Man* iv. 84 A hoarse-voiced old turkey gobbler was booming his chug-a-lug.., and the softer chick of hen turkeys answered him.

chickadee. For *U.S.* read *N. Amer.* (Add earlier and later examples.) Also used as a term of endearment to a woman.

1838 THOREAU *Jrnl.* 23 Sept. (1949) I. 60 The chickadee is more than usually familiar. **1940** *Times* 20 May 4/5 Mr. [W. C.] Fields is magnificently flamboyant and rhetorical..but, hard as he works, *My Little Chickadee* obstinately refuses to gather momentum. **1942** BERREY & VAN DEN BARK *Amer. Thes. Slang* § 185/2 Pet names; terms of endearment. (Frequently prefaced 'my' or 'little'.) *Angel*, ..*chickadee*. **1964** A. L. THOMSON *New Dict. Birds* 823/2 In North America some *Parus* spp. are called 'chickadees', the best known being the Blackcapped Chickadee *P. atricapillus*. **1967** *Boston Sunday Herald* 26 Mar. II. 9/7 Here in the woods was the usual sparse sprinkling of chickadees and nothing else. **1968** *Listener* 23 May 678/1 Mr Durrell, looking so much like a sober W. C. Fields that one expected him to address Joan Bakewell as 'my little chickadee'.

chickaleary (tʃi·kăliᵊri), *a.* (*sb.*) *Costers' slang.* [f. CHICK *sb.*[1] + LEERY *a.*[2]] Artful, 'downy'. Also *sb.*, a 'downy cove'.

1839 W. A. MILES *Poverty* (Partridge), He's a right down chap, a chickle-a-leary chap. *c* **1869** VANCE *Ballad, Chick-a-leary Cove* (Farmer), I'm a chicka-leary cove..; Whitechapel was the village I was born in. **1908** *Daily Chron.* 3 Aug. 4/4 It's the chickalearies and their donahs that are responsible.

chickara, var. *CHINKARA.

chickaree. (Earlier example.) See also *CHIPAREE.

1829 J. RICHARDSON *Fauna Bor.-Amer.* I. 187 *Sciurus Hudsonius* (Pennant). The Chickaree.

Chickasaw (tʃi·kăsǭ), *sb.* and *a.* Also **Chicasa, Chicashaw, Chicaza, Chickesaw, Chikasa,** etc. [ad. Chickasaw *Chikasha*.] **A.** *sb.* A member of a Muskogean tribe of North American Indians formerly resident in Mississippi and in Alabama, but subsequently in Oklahoma.

1674 H. WOODWARD *Let.* 31 Dec. in *Coll. S. Carolina Hist. Soc.* (1897) V. 461 Ye Cussetaws, Checsaws & Chiokees were intended to come downe and fight ye Westoes. **1722** D. COXE *Descr. Carolana* 14 The great and powerful Nation of Chicazas. **1744** F. MOORE *Voy. Georgia* 120 The Chickasaws, who lie upon the Mississipi river, between the Cherokees and the Choctaws. **1770** P. PITTMAN *Europ. Settlem. Mississippi* 6 The Chicashaws formerly were very troublesome. **1854** T. H. BENTON *30 Years' View* I. 626/1 The Chickasaws, Creeks, and Choctaws having previously agreed to remove. **1907** L. H. MORGAN *Anc. Society* II. vi. 163 In like manner the Chickasas were organized in two phratries. **1927** W. A. READ *Louisiana Place-Names* 22 Perhaps *Chickasaw* signifies 'rebellion', the term referring to the separation of the Chickasaws from the Creeks and the Choctaws.

2. The language of this people.

1893 *Daily Ardmoreite* (Ardmore, Okla.) 27 Dec. 3/1 Salus populi, suprema est lex, which, in Chickasaw, means, 'The people are first bidden to the feast.' **1933** BLOOMFIELD *Language* iv. 72 The Muskogean family includes..Choctaw, Chickasaw, Creek, and Seminole. **1941** M. R. HAAS in L. Spier et al. *Language, Culture, & Personality* 42 Choctaw and Chickasaw are actually subvarieties of the same language.

B. *adj. Chickasaw Indian*, an Indian of the Chickasaw people; *Chickasaw plum*, a wild plum of the Mississippi region (*Prunus angustifolia*); *Chickasaw rose*, the Cherokee rose.

1726 S. PENHALLOW *Hist. Wars New-Eng.* (1859) 75 Col. Gibs..commissioned..Capt. Welch, with the assistance of the Chicksha Indians, to fall on them in several parts. **1736** [see INDIAN *sb.* 2 a]. **1848** THACKERAY *Van. Fair* viii. 71 The mission for the Chickasaw Indians. **1775** J. ADAIR *Hist. Amer. Indians* 360 They have a large sort of plums..called Chikkasah plums. **1821** T. NUTTALL *Trav. Arkansa* vii. 137 The abundance of Chickasaw plum bushes forming a grove. **1864** A. WOOD *Class-bk. Bot.* 328 *P*[*runus*] *Chicasa*..Chickasaw Plum...A fine fruit shrub, native of Ark., &c., often cultivated. **1867** *Trans. Ill. Agric. Soc. 1865–66* VI. 390 Our beautiful foliaged Chickasaw Plum. **1835** J. H. INGRAHAM *South-West* II. 108 The 'chickasaw rose', which is a beautiful hedge thorn, grows..luxuriantly [in Mississippi]. **1887** *Harper's Mag.* Feb. 350/1 Along one side of his home..runs a superb hedge of Chickasaw roses.

chicken, *sb.*[1] Add: **1. d.** A domestic fowl of any age.

1827 *Harvard Register* May 84 Some students were 'hooking' chickens. **1887** *Scribner's Mag.* May 622/1 The farm people had all retired with the chickens long before. **1908** *Westm. Gaz.* 24 Jan. 3/1 It is a disastrous betrayal of middle-class origin to speak of a 'chicken' as a 'fowl'. Whatever the age of the bird, the word must always be chicken.

e. The prairie chicken or pinnated grouse; also, the sharp-tailed grouse. *U.S.*

1812 J. C. LUTTIG *Jrnl. Exped. Upper Missouri* (1920) 14 Oct. 85, 4 Men went out to hunt..got this Day 21 Chickens. **1876** *Fur, Fin & Feather* Sept. 95 You can always find good chicken dogs wherever there are chickens. **1901** S. E. WHITE *Westerners* xx. 189 The careful attention necessary for the destruction of the wily 'chicken' or experienced squirrel.

3. a. Also as a *slang* (chiefly *U.S.*) term for a girl or young woman. Cf. *CHICK *sb.*[1] 3 b.

1860 G. D. PRENTICE *Prenticeana* 97 Call a lady 'a chicken', and ten to one she is angry. Tell her she is 'no chicken', and twenty to one she is still angrier. **1882** *Sydney Slang Dict.* 2/2 *Chicken*, a Girl (applied to the respectable class). **1923** G. H. MCKNIGHT *Eng. Words* iv. 61 In the vocabulary of modern youth, chivalry is dead...A girl is..a *chicken*, a *doll*, [etc.]. **1934** J. T. FARRELL *Young Manhood* (1936) 71 He said he was fed up on the dago chickens around State Street anyway. The guys all thought that was a new word.

b. Revived in mod. slang (orig. *U.S.*), often as quasi-*adj.*: cowardly. Hence (*to play*) *chicken*, (to engage in) a 'game' of physical hazard which tests the courage.

1936 *Amer. Speech* XI. 279/2 *Chicken*, a timid soul; a sissy. **1941** *Life* 15 Dec. 89 Gets chicken. **1945** *Amer. Speech* XX. 147/2 A person is 'chicken' when he abides too closely by army rules and regulations, or when he misuses or abuses authority, especially in minor or petty matters. **1952** S. ELLIN *Key to Nicholas St.* (1953) IV. ii. 167 You'd just holler for the cops? Why, man, you're chicken. **1953** R. BRADBURY *Fahrenheit 451* (1954) I. 33 Go out in the cars..trying to see how close you can get to lamp-posts, playing 'chicken'. **1956** 'M. INNES' *Appleby plays Chicken* I. i. 17 Playing chicken..this business of proving to yourself that you're as tough as the other chaps. *Ibid.* II. ii. 18 Chicken is done with cars. **1959** *Daily Mail* 18 May 5/6 Sit on the track as the train thunders nearer...Last to get up is the winner. First is branded 'chicken'. **1960** E. W. HILDICK *Jim Starling & Colonel* xi. 93 'Speak for yourself—chicken!' he jeered. **1960** SWERLING & BURROWS *Guys & Dolls* II. iii. 54 *Player.* Come on, quit stallin', roll. *Harry.* What's the matter, Sky, turning chicken?

c. (See quots.)

1887 J. D. BILLINGS *Hardtack & Coffee* 52 A Marblehead man called his chum his 'chicken', more especially if the latter was a *young* soldier. **1890** *Congress. Rec.* 21 Apr. 3637/1 The affection which a sailor will lavish on a ship's boy to whom he takes a fancy, and makes his 'chicken', as the phrase is.

6. b. *chicken-and-(the-)egg* = *hen-and-egg* (see *HEN *sb.* 8).

1959 *Times* 23 Sept. 17/5 One of the chicken-and-the-egg problems is involved here. **1961** 'C. H. ROLPH' *Common Sense about Crime* iii. 45 To argue that there would be no punishment without crime and (*ergo*) no crime without punishment would be to import the chicken-and-egg sequence into a problem. **1967** *Guardian* 24 Feb. 8/5 The chicken-and-egg attitude towards the home background of addicts.

7. *chicken-farm, -pie* (earlier and later examples), *-run* (RUN *sb.*[1] 21 b); *chicken-brooding, -farming, -raising, -rearer*.

1902 *Westm. Gaz.* 8 Oct. 8/2 Incubators, chicken-brooding houses, and 'accessories' innumerable. **1895** *Outing* (U.S.) XXVI. 452/1 Wilson..owned a prosperous chicken farm. **1887** I. RANDALL *Lady's Ranche Life Montana* 56 The worst of chicken farming here is, that in the summer there is a glut of eggs, about 6d a dozen. **1904** WODEHOUSE *Gold Bat* ii. 19 The master of that house retired from the business, and took to chicken-farming. **1709** W. BYRD *Secret Diary* (1941) 10 July 58, I ate chicken pie for dinner. **1723** J. NOTT *Cook's & Confect. Dict.* No. 10B, Balls of forced Meat for a Chicken Pye. **1896** J. C. HARRIS *Sister Jane* 166 Aunt Sally a dishin' out the chicken pie at her house. **1891** *Pall Mall Gaz.* 1 Apr. 3/3 Another industry that could be well pushed into greater prominence is chicken raising. **1895** *Daily News* 9 Oct. 6/7 Silver medals were accorded..for the chicken rearer..and..a wheel-barrow fowl house. **1906** *Westm. Gaz.* 13 Oct. 16/3 Should it..miss the lizard that crawls out into the open space of the chicken-run and pick up a young chicken instead—well, you can hardly blame it.

8. chicken à la King [said to be named after E. Clark *King*, proprietor of a hotel in New York], cooked chicken breast served in a cream sauce with mushrooms and peppers; **chicken colonel** *U.S. slang* (see quots. 1948 and 1962); **chicken corn** (*a*) *U.S.*, the common sorghum growing out of cultivation; (*b*) *dial.*, inferior corn; **chicken-feed** *colloq.* (orig. *U.S.*), food for poultry; also *fig.*, anything of little importance, *esp.* a trifling sum of money; **chicken-fixings** *local U.S.*, chicken prepared as food; also *fig.*; **chicken-grape** (earlier U.S. example); **chicken gumbo** (see GUMBO 1 b); **chicken-hawk** chiefly *U.S.*, any of various hawks that kill chickens; **chicken-liver** (*a*) the liver of a fowl; (*b*) *colloq.*, a coward; **chicken-livered** *a.*, = CHICKEN-HEARTED *a.*; **chicken-shit** (*coarse slang*, orig. *U.S.*), a coward; also used as a general term of abuse; also *attrib.* or as *adj.* (cf. sense *3 b above); **chicken-skin,** a substance or surface resembling the skin of a chicken in texture or appearance; also *attrib.*; **chicken-snake** (earlier U.S. example); **chicken-wire,** a light wire netting with hexagonal mesh.

1912 F. M. FARMER *New Book Cookery* 228 Chicken à la King. **1931** E. LINKLATER *Juan in Amer.* II. v. 97 Their stomachs filled alike with chicken *à la* king and pie *à la mode*. **1940** C. MCCULLERS *Heart is Lonely Hunter* (1943) II. ii. 103 She marked the special dinner with chicken à la king at twenty cents instead of fifty. **1969** *Harrod's Summer Food News* 10/1 Chicken à la King—16 oz. 10/6. **1948** A. M. TAYLOR *Lang. World War II* 53 *Chicken Colonel*, a full colonel, rank designated by silver eagles (chickens) on his shoulders. **1950** HEMINGWAY *Across the River* iv. 25 Maybe they treat me well because I'm a chicken colonel on the winning side. **1962** *Amer. Speech* XXXVII. 288 It is proper to call a full colonel a *chicken colonel* in the Air Force, but in the Marine Corps the correct expression is *bird colonel*. **1856** *Congress. Globe* 17 Apr. 960/2 Chinese sugar cane is nothing more than what we call chicken corn down in Georgia. **1895** *Wilts. County Mirror* 27 Sept. 2/4 The second share has only yielded chicken corn. **1901** C. MOHR *Plant Life Alabama* 339 *Sorghum vulgare*,..Chicken Corn, Durrha. Sugar-Corn...Escaped from cultivation, becoming a pernicious weed in many parts of the Southern States. **1836** *Col. Crockett's Exploits* 78, I stood looking on, seeing him pick up the chicken feed from the green horns. **1865** R. H. KELLOGG *Rebel Prisons* 109 Two buckets of mush for ninety men. 'Chicken feed,' the boys called it. **1879** F. R. STOCKTON *Rudder Grange* xiv. 173 The houses scattered a long ways apart, like stingy chicken-feed. **1904** 'O. HENRY' in *McClure's Mag.* Aug. 359/2 Salt away that chicken feed in your duds. **1937** G. HEYER *They found him Dead* xiii. 265 He was picking up a living doing odd jobs for any firm that would use him. Chicken-feed! **1941** *New Review* 28 Aug. 5/3 In peacetime, officers in the British Army were men of independent means to whom their Army pay was chicken-feed. **1958** *Observer* 4 May 14/7 We're ridden with guilt... God bothers us. This kind of thing is chickenfeed to the playwright. **1838** E. FLAGG *Far West* II. 72 Wheat-bread and chicken fixens, or cornbread and common doins? **1854** MARY J. HOLMES *Tempest & Sunshine* v. 70 We don't have any of your Chicken Fixins nor little three-cornered handkerchiefs laid out at each plate. **1874** EGGLESTON *Circuit Rider* ii. 20 The strife had given them vigorous relish for Mrs. Lumsden's 'chicken-fixin's'. **1886** PROCTOR in *Knowledge* 1 Apr. 179/1 *Chicken fixings*, originally a chicken fricassee, now applied sometimes to any particularly fine arrangements, as distinguished from 'common doings'. **1886** [see *GADGET]. **1807** J. SCOTT *Geogr. Descr. Maryland* 112 A middle sized grape, of a purple colour, growing in clusters, like the chicken grape. **1867** *Common Sense Cook Bk.* (N.Y.), Chicken gumbo. **1884** C. PHILLIPPS-WOLLEY *Trottings of Tender Foot* 5 A *menu*..including such hitherto unheard-of luxuries as chicken gumbo..and mush. **1908** G. H. LORIMER *J. Spurlock* ix. 207 First there was a chicken gumbo soup, and then cold boiled Virginia ham. **1958** *Catal. County Stores, Taunton* June 7 Soups..Chicken Gumbo—a tin 1/7. **1827** J. L. WILLIAMS *W. Florida* 30 Chicken Hawk. *F*[*alco*] *pullenarius*. **1890** J. WATSON *Nature* viii. 96 The freshly added Sparrow-hawk is by no means the only one of its kind, for there are four or five 'blue-hawks'. 'Chicken-hawk' is another of the Keeper's names for the bird. **1925** J. GREGORY *Bab of the Backwoods* ii, I never saw a buzzard and a sparrow nesting together...Nor a chicken-hawk and a linnet. **1958** J. CAREW *Wild Coast* iii. 45 They saw a chicken hawk fighting with a grass snake. **1899** KIPLING *From Sea to Sea* I. xii. 308 They gave me chicken liver and sucking-pig in the Victoria at Hong-Kong. **1932** R. MACAULAY *They were Defeated* I. iv. 33 Pox on him for a chicken-liver that ran away from the Scots. **1950** E. DAVID *Bk. Mediterranean Food* 137 Take about 1 lb. of chicken livers or mixed chicken, duck, pigeon or any game liver. **1963** M. MCCARTHY *Group* ix. 194 Mr. Andrews' famous chicken-liver *pâté*, a recipe he had brought back from France. **1872** 'M. TWAIN' *Roughing It*, Many a notorious coward, many a chicken-livered poltroon. **1943** H. PEARSON *Conan Doyle* iii. 37 He's a nervous, chicken-livered kind of man, and when I look at him he turns the colour of putty. **1947** C. WILLINGHAM *End as Man* xvi. 192 You're both acting like chicken-shits. We win a batch of money—you're afraid to take it. **1948** N. MAILER *Naked & Dead* I. i. 7 'What's the matter?' he asked. 'You going chickenshit?' **1968** *Southerly* XXVIII. 281 'You're just a pile of compromising chickenshit,' Gillian says in a whisper. **1969** C. HIMES *Blind Man with Pistol* xix. 203 She's a slut, just a chickenshit whore. **1970** *It* 12–25 Feb. 17/1 American groups are not so chickenshit about getting into underground work. **1690** EVELYN *Mundus Muliebris* 6 Gloves..Some of Chicken skin for night, To keep her Hands plump, soft, and white. **1840** R. H. DANA *Bef. Mast* (1841) xiii. 24/2 They [*sc.* Californians]..buy..'chicken-skin' boots at fifteen dollars apiece. **1901** *Lady's Realm* X. 652/2 The chicken-skin [design]—namely, a sort of groundwork with tiny pin-points all over it, which shows up the flowers to perfection. **1902** *Daily Chron.* 19 July 8/3 Fans..composed of delicate lace inset with net or chicken-skin medallions. **1904** *Daily Chron.* 31 May 3/1 Subtle effects of colour which [china] collectors prize under such names as 'egg-shell', 'chicken-skin', and especially 'flambé'. **1928** *Daily Mail* 7 Aug. 3/2 The painting is done on what is called 'chicken-skin'—which in the old days was really a specially prepared parchment. **1930** W. DE LA MARE *On Edge* 320 Absurd misgivings productive only of chicken-skin and perplexity. **1709** J. LAWSON *New Voy. Carolina* 134 The Egg or Chicken-Snake ..eats Eggs and Chickens. **1920** *Outing* (U.S.) June 164/2 It is a very good idea to take along a roll of chicken wire. **1963** J. TOOTELL *Floristry* II. i. 26 Having cut a piece of inch-and-a-half mesh chicken-wire, curl the edges inwards until you have formed a rough ball.

chicken (tʃi·ken), *sb.*[2] *India.* Also **chikan.** [Hind., a. Pers. *chakin, chikin* needle-work.] Embroidery. Also *attrib.*

1880 W. HOEY *Monogr. Trade N. India* III. 88 Chikan is hand-worked flowered muslin and the chikan of Lucknow is in great demand in all parts of India. *Ibid.*, Chikan is largely exported from Lucknow to Agra.., to Mirzapur, Patna, and Calcutta.., and to Haidarábád, Dakhan. **1886** *Offic. Catal. Col. & Ind. Exhib.* 16 At Calcutta embroidered muslin is called *chikan* (needle work). *Ibid.* 43 The large collection of *chikan* work from Calcutta. **1886** YULE & BURNELL *Anglo-Ind. Gloss.*, *Chicken-walla*, an itinerant dealer in embroidered handkerchiefs, petticoats, and such like. **1908** *Imp. Gazetteer India* III. 221 *Chikan* work in fact takes the place in India that lace holds in Europe.

chicken (tʃi·kėn), *v. slang* (orig. *U.S.*). [f. *CHICKEN *sb.*[1] 3 b.] To fail to act, or to back down, from motives of cowardice. Freq. const. *out*.

1943 I. WOLFERT *Torpedo 8* (1944) ii. 19, I just wanted to..make sure you weren't chickening out on me. **1950** *Cornell* (*Univ.*) *Daily Sun* 24 Mar. 4 (Wentworth & Flexner), The Harvard Student Council..just plain chickened out... [They] considered the proposal..and sent the resolution back to committee on a technicality. **1960** *Encounter* Feb. 46/2 Those people were looking to us for help and—we chickened out. **1961** C. MCCULLERS *Clock without Hands* iv. 76 You would of been the first to chicken. **1965** *Economist* 4 Sept. 890/1 Nobody can trust the others not to chicken out if they take the first plunge.

chicle (tʃi·k'l, tʃi·kli), *sb.* Also **chickle.** [Amer. Sp., ad. Nahuatl *tzictli*.] In full *chicle-gum*: a gum-like substance obtained from the bully tree (*Mimusops globosa*), and from the sapodilla (*Sapota zapotilla*), largely used in the United States in the manufacture of chewing-gum; hence, chewing-gum. Hence **chi·cle** *v. intr.*, to chew gum.

1889 *Cent. Dict.*, Chicle-gum. **1907** WISTER *Simple Spelling-Bee* 18 Fifty per cent of our population chickles. *Ibid.* 36 He again removed his chickle and placed it on the window-sill. **1915** *Chambers's Jrnl.* June 376/1 Chicle, the rezin of the chicozapote-tree of Mexico, is the principle [*sic*] ingredient..in the manufacture of chewing-gum. **1925** *Countries of the World* XXVII. 2752/1 The collecting of chiclegum and the cultivation of rubber are notable. **1933** *Discovery* Nov. 352/1 A few chicle gatherers..engage in tapping the sapodilla trees during the rainy season. **1951** *Good Housek. Home Encycl.* 403/1 This type of sweetmeat [*sc.* chewing gum] is usually made of chicle.

chicly (ʃi·kli), *adv.* [f. CHIC *a.* + -LY[2].] Stylishly.

1905 *Daily Chron.* 26 Oct. 3/3 There have been far prettier women than any of them and much more chic-ly dressed. **1958** *Spectator* 6 June 739/2 That acquaintance with foreign tongues so rightly demanded for true amateur status..is what enables Mr. Raymond to use the chicly inaccurate *terribiltà*.

Chicom (tʃəi·kɒm), *sb.* and *a.* [f. CHI(NESE *a.* + COM(MUNIST.] (A) Chinese Communist.

1966 *New Statesman* 14 Oct. 549/1 The Chicom hordes are held at bay. **1967** M. CHILDS *Taint of Innocence* (1968) ii. 78 The Chicoms versus the Sovs in Kerala Province. **1969** J. BENNETT *Dragon* ii. 20 'I suppose we're quite sure the Chicoms have Dr Warren?'..'I assume you mean the Chinese Communists. I can't stand this verbal shorthand...' **1969** H. MACINNES *Salzburg Connection* xvi. 222 The Russians would keep intercepting his radio messages to the Chicom agents in Warsaw.

chicoried (tʃi·kŏrid), *a.* [f. CHICORY *sb.* + -ED.] Flavoured with chicory.

1879 SALA *Paris herself Again* I. xi. 168 A cup of chicoried coffee.

chicote (tʃikō·te). Also **chicotte.** [Sp., = rope's end.] A long whip of leather or hide, having a wooden handle, used in the Congo, Portuguese Africa, and elsewhere.

1903 *Daily Chron.* 10 July 6/5 The fearful whip made of hippopotamus hide, and called the chicotte, and with which the natives in all parts of the Congo are so familiar. **1906** NEVINSON *Mod. Slavery* iii. 55 Her back scored all over with the cut of the *chicote*. **1928** *Daily Express* 28 Nov. 9 Corporal punishment, or 'chicotte', now exists only in the Congo army.

chide, *v.* Add modern instances of inflected forms.

1865 MEREDITH *Rhoda Fleming* I. x. 164 The farmer chid her. **1885** MRS. CAMPBELL PRAED *Head Station* xxiii, Mrs. Clephane..chided Jinks. **1897** *Daily News* 15 Apr. 6/3 We ..notice with interest that Mr. Meredith, after vacillating in former editions between 'chid' and 'chidded', has now resolved that the past tense of 'to chide' is 'chided'. **1925** C. S. DURRANT *Flem. Mystics & Eng. Martyrs* I. x. 146 Margaret..quietly chode her elder.

chief, *sb.* Add: **6. b.** *big* or *great white chief*: a jocular name (modelled on the speech of American Indians) given to a person of authority or importance. Cf. *great* (*white*) *father* (s.v. *GREAT *a.* 12h).

[**1835** W. G. SIMMS *Yemassee* I. 95 Does the white chief come to the great council of the Yemassee as a fur trader?] **1937** M. ALLINGHAM *Dancers in Mourning* xxvi. 315, I see a chap who is a sort of great white chief in his own little world. **1939** WODEHOUSE *Uncle Fred* iii. 43 'Was that Aunt Jane I saw going off in the car?' 'That was the Big White Chief.' **1949** E. COXHEAD *Wind in West* ii. 37 You've actually seen the Great White Chief. Jove, what an experience.

d. The Chief Engineer, or Lieutenant-Commander, in a (war)ship (see also quot. 1929). *Naut. colloq.*

1894 KIPLING *Seven Seas* (1896) 153 He's Chief of the Maori Line. **1916** 'TAFFRAIL' *Stand By!* 18 Next comes the engineer-lieutenant-commander, or the 'chief', as we call him. **1929** F. C. BOWEN *Sea Slang* 26 *Chief*, the Chief Engineer as a rule, though loosely applied, more and more frequently to mates in the Merchant Service and petty officers and others in the Navy. **1942** F. G. HACKFORTH-JONES *One-One-One* xi. 104 'Chief,' he called down the voice-pipe to the engine-room, 'Knock her up to full speed or I'll come down and stoke myself.' **1946** —— *Sixteen Bells* xiii. 230 It was the custom of her Chief, an Engineer Commander of many years' standing, to enjoy an afternoon's caulk.

chief, *a.* Add: **1.** *Chief Engineer, Chief Rabbi* (examples).

1861 in G. B. McClellan *Own Story* (1886) v. 83, I have on the staff..Barnard as chief-engineer. **1876** GEO. ELIOT *Let.* 2 Sept. in J. W. Cross *Life* (1885) III. xviii. 256 A letter ..from Dr. Hermann Adler, the Chief Rabbi here. **1890** W. JAMES *Princ. Psychol.* II. xix. 101 The chief-engineer of the vessel had entered my state-room. **1932** C. ROTH *Hist. Marranos* iii. 60 Simon Maimi, the last *Ar-rabi-Mor*, or Chief Rabbi of Portugal.

chiefess. (Earlier U.S. example.)

1778 J. CARVER *Trav. N. Amer.* 41 This heroine was ever after treated by her nation as their deliverer, and was made a chiefess in her own right.

chieflet. Delete *nonce-wd.* and add examples.

1883 BARING-GOULD *John Herring* xxxvi, A chieflet of indifferent character. **1933** *Jrnl. R. Anthropol. Inst.* 2 There was no compulsion to pay toll to every chieflet or headman.

chiefling (tʃi·f,liŋ). [f. CHIEF *sb.* + -LING.] A petty chief.

1891 STEVENSON *Vailima Lett.* (1895) 8 Henry is a chiefling from Savaii. **1926** *Brit. Weekly* 11 Mar. 566/5 A native chiefling..appeared in plus fours, a tall hat, a frock coat and patent-leather boots.

chiefship. Add: **2.** A state ruled by a chief.

1894 *Daily News* 18 May 6/2 The ordinary Briton never gives these native kingdoms, principalities, and chiefships a thought. **1899** *Ibid.* 17 Nov. 9/4 The native State—Sarawak—..became a model for the chiefships of the Archipelago.

chiefy (tʃi·fi). *Forces' slang.* [f. CHIEF *sb.* 6 + -Y[6].] = CHIEF *sb.* 6 c; *spec.* in the R.A.F. = *Flight Sergeant*.

1942 I. GLEED *Arise to Conquer* v. 46 To this day I can see distinctly 'Chiefy' N., stripped naked, putting on..a spotless clean tunic. *Ibid.*, 'Chiefy' means a Flight Sergeant. **1957** M. K. JOSEPH *I'll soldier no More* (1958) 9 The chiefy who done him out of his stripes.

Ch'ien Lung (tʃi·ĕn luŋ). The name of the Emperor of China who reigned 1736–96, used esp. *attrib.* to designate pottery, carved jade, etc., made during those years, or the period itself.

1901 C. MONKHOUSE *Hist. Chinese Porc.* I. 46 The note of the Ch'ien-lung period was mastery of material. *Ibid.* (*caption*), Ch'ien-lung vase. **1905** MRS. W. HODGSON *Old Chinese Porc.* 111 A beautiful and characteristic piece of Fen-ting made during the Ch'ien-lung period. **1935** *Burlington Mag.* July 43/2 The elaborately carved white 'mutton fat' jade of the Ch'ien Lung period (1736–96) whose craftsmen were famous for their undercutting in relief and open-work patterns. **1968** *Guardian* 2 Apr. 3/6 A pair of pheasants made in the Ch'ien Lung period in China.

chiff-chaff. Add: **2.** The note of the chiff-chaff, or sounds resembling this.

1908 *Westm. Gaz.* 21 Mar. 14/1 They..make their presence known by their cheery reiteration of 'Chiff-chaff, chiff-chaff'. **1909** *Daily Chron.* 31 May 4/4 The sounds (quite important barks and chiff-chaffs and gurgles and bubbles) that make the emphasis of conversation. **1916** *Westm. Gaz.* 19 Mar. 14/2 His cheery, oft-repeated note, 'Chiff-chaff, cheef, cheef, chaff'.

chiffon. Delete ‖ and add earlier examples.

1765 EARL OF MARCH *Let.* 28 Sept. in J. H. Jesse *G. Selwyn & Contemporaries* (1843) I. 410 She wanted to get back to all her *chifons*, contrary to my inclination. **1803** *Lett. Miss Riversdale* I. 126 Every one of my things was crushed and spoilt—and the sweetest—most elegant *chiffons*.

2. (ʃi·fɒn). A diaphanous plain-woven fabric of fine hard-twisted yarn (orig. silk, later nylon, etc.); also *attrib.*, sometimes with sense 'light in weight'.

1890 *Pall Mall Gaz.* 4 Feb. 3/2 The..tea-gown was a combination of white chiffon and yellow silk. **1900** [see MOUSSELINE 1 c]. **1904** *Daily Chron.* 24 Oct. 8/1 Chiffon cloth. **1905** *Westm. Gaz.* 20 Apr. 4/2 Chiffon-like muslin. **1908** *Costumes Classical & Fashionable* (*Liberty & Co.*) 12 Underdress in chiffon velvet. **1951** *Good Housek. Home Encycl.* 36/2 The term chiffon is also sometimes used to describe a lighter weight of a particular material, e.g. chiffon velvet, chiffon lisle.

3. *chiffon pie*, a baked pie-shell with a light-textured filling, containing beaten egg whites or a gelatine mixture, flavoured with fruit, etc.

1929 *Fashions in Foods* (*Beverly Hills Women's Club*) 155 Chiffon pumpkin pie. **1931** I. S. ROMBAUER *Joy of Cooking* (1936) 363 Lemon chiffon pie. *Ibid.* 364 Gelatine chiffon cream pies. **1937** *America's Cook Book* 653 *Lemon Chiffon Pie*, Gelatin..water..eggs..sugar..salt..grated lemon rind..lemon juice..pastry..shell..cream. **1946** H. CROOME *Faithless Mirror* iii. 36 She had cleared away the chicken and brought in a lemon chiffon pie. **1958** *Woman* 4 Oct. 23/1 What about Butterscotch Chiffon Pie?..You can serve it, like cake, with coffee or as a dinner sweet. **1960** *Guardian* 10 Nov. 9/6 Pineapple chiffon pie.

Hence **chiffony** (ʃi·fŏni) *a.*, of the nature of or resembling chiffon.

1902 *Daily Chron.* 9 July 8/2 Billowy ruffles and chiffony hats. **1904** *Westm. Gaz.* 23 June 4/2 Taffeta and silk muslins, as we call these chiffony soft stuffs.

‖**chiffonnade** (ʃifonad). Also **chiffonade.** [Fr.] (See quots.)

1877 E. S. DALLAS *Kettner's Bk. of Table* 122 *Chiffonnade* ..usually means a selection of vegetables, nicely cut, served in a good consommé; but sometimes, instead of a consommé, the French serve them in a green-pea soup. **1907** G. A. ESCOFFIER *Guide Mod. Cookery* 89 The name '*Chiffonade*' is given to a mince of sorrel or lettuce, intended as a complement for..soups. **1961** *Larousse Gastronomique* 270 *Chiffonnade*. In cooking, all plants..which are cut into fine strips or ribbons are denoted by this term. It is more especially used to denote a mixture of sorrel and lettuce... Most of the *chiffonnades* are used as a garnish for clear or thick soups.

chigger, var. (*a*) CHIGOE; (*b*) JIGGER *sb.*[2]

1830 [see CHIGOE]. **1915** *U.S. Dept. Agric., Farmer's Bulletin* 671 Harvest mites, or chiggers. **1956** *New Biol.* XXI. 74 In the British Isles the season for harvest mites (or chiggers) (Trombicula autumnalis) usually extends from May to October, reaching its peak in September.

chi-hike (tʃəi·,həi·k), *sb. slang.* Also **chi-ike, chy-;** *Austral.* and *N.Z.* (with pronunc. tʃəi·æk) **chiack, chyack,** etc. The shouting of 'chi-hike' as a salute; hence, a noisy demonstration. In Australia and (less commonly) N.Z., jeering, banter, cheek.

1859 HOTTEN *Dict. Slang*, *Chi-ike*, a hurrah, a good word, or hearty praise. *c* **1869** VANCE *Ballad, Chick-a-leary Cove* (Farmer), Now join in a chyike, the 'jolly' we all like. **1894** A. CHEVALIER *Humorous Songs* ii, Folks with a 'chy-ike' shouted, 'Ain't they smart?' **1910** *Daily Chron.* 22 Apr. 7/7 The audience..was out for a gigantic 'chi-hike', and it enjoyed itself to the full. **1915** C. J. DENNIS *Songs of Sentimental Bloke* (1916) 119 *Chiack*, vulgar banter; coarse invective.

Hence **chi-hike** *v. trans.*, to salute with the cry 'chi-hike'; also *absol.* or *intr.*, to make a noisy demonstration. In Australia and (less commonly) N.Z., to jeer at, tease, make fun of. So **chi·acking, chy·acking** *vbl. sb.* and *ppl. a.*

1874 HOTTEN *Slang Dict.*, *Chi-ike*, to hail in a rough though friendly manner; to support by means of vociferation. **1874** G. WALCH *Adamanta* II. ii. 27 (Morris), I've learnt to chi-ike peelers. **1878** *Australian* I. 742 The circle of frivolous youths who were yelping at and chyacking him. **1886** *Sporting Times* 17 July 7/2 There was no charge for admission. Enough. They came, they saw, and they chi-iked. **1894** E. W. HORNUNG *Boss of Taroomba* i, It's our way up here, you know, to chi-ak each other and our visitors too. **1896** H. LAWSON *While Billy Boils* (1897) 47 Didn't the fellows chyack me, though! *Ibid.* 301 She was.. laughin', and jokin', and chyackin'. **1904** G. B. LANCASTER *Sons o' Men* 30 Stop off yer chiackin', you fellers, can't yer? **1906** WHITEING *Ring in New* 122 His 'chihiking' of members, popular and unpopular. **1914** A. A. GRACE *Tale of Timber Town* xix. 88 Go on chiacking—poke borak—it don't hurt me. *Ibid.* xxiv. 116 You won't chiack or poke borak at his grey and honoured head. **1926** K. S. PRICHARD *Working Bullocks* xiv. 125 Yells and chiacking queries went up from the hill-top. **1930** V. PALMER *Men are Human* v. 40 'Heard any more about that thousand?' Larry chiacked him. **1934** T. WOOD *Cobbers* 17 Just think how you can chyack those stay-at-homes in Sydney and Melbourne. **1944** J. H. FULLARTON *Troop Target* iv. 36 He was 'chyiked' unmercifully. **1958** [see *AD-LIB *v.*]. **1962** *Spectator* 24 Aug. 287 Half a dozen chi-iking louts. **1969** M. PAINTER in *Coast to Coast 1967–68* 56 There is a kind of chiacking which does not always export too well.

Chihuahua (tʃiwā·wā, ʃi-). [The name of a city and state in Mexico.] A breed of very small smooth-haired dog which originated in Mexico; a dog of this breed. Also *attrib.*

1858 T. VIELÉ *Following Drum* 185 The Chihuahua are rare, even here in such close vicinity to Mexico. **1905** H. DE BYLANDT *Dogs of all Nations* II. 770 Chihuahua Dog. ..Head of medium size...Nose rather pointed...Ears of medium size. **1926** D. H. LAWRENCE *Plumed Serp.* x. 165 Doña Carlota was delicate and sensitive like a Chihuahua dog. **1929** *Sunday Express* 13 Jan. 3/3 A sensation has been created..by the arrival in this country of three Chihuahuas from Mexico. **1958** *Times* 12 Dec. 1/7 (Advt.), Chihuahua puppies for sale.

chikara, var. *CHINKARA.

chikhor, chikor (tʃi·kɒɹ). *India.* Also **chichore, chikkor, chuckoor, chukor, -ar.** [Hind. *chakor*.] A name for red-legged Indian partridges of the species *Alectoris chukar* and its sub-species.

1815 ELPHINSTONE *Acc. Caubul* 144 A bird which is called ..the hill Chichore by the Indians. *a* **1851** MRS. SHERWOOD *Life* (1854) 467 One day..he found a hill partridge...This bird is called the chuckoor, and is said to eat fire. **1879** HUME & MARSHALL *Game Birds India* II. 35 October is the best month..for Chukor-shooting. *Ibid.* 39 The inhabitants of Scio..keep tame Chukor Partridges. **1893** *Baily's Mag.* May 263/1 We also came across some chikhor (mountain partridge). **1909** *Times Lit. Suppl.* 1 Apr. 124/2 He has taken teal, chukor, stone-plover, and grouse. **1920** *Blackw. Mag.* May 664/2 Some of the officers went out one day to shoot *chikhor*. **1923** *Ibid.* Apr. 493/1 A dozen chikor or a haunch of venison. **1936** P. FLEMING *News from Tartary* VII. iv. 339 Distraction was provided by a covey of chikkor. **1963** *Times* 16 May 14/6 Coveys of *chikhor* dusted themselves in our path.

child, *sb.* Add: **21. e.** Passing into adj., with the meaning of 'child's' or 'childish', *child-art, -brain, -bride, -culture, -eyed* adj., *-face* (hence *-faced* adj.), *-marriage, -mind, -voice, -word*.

1945 H. READ *Coat of Many Colours* xix. 101 Folk-art is merely child-art which has become adult. **1904** *Daily Chron.* 21 May 4/5 My child-brain, clear and natural, could not swallow the impossibilities administered to me as facts. **1843** LYTTON *Last of Barons* II. v. i. 161 Boy-bridegroom and child-bride as we were. **1909** *Daily Chron.* 9 Oct. 3/1 The author shows us the child-bride arriving at the court of France. **1899** M. BEERBOHM *Around Theatres* (1924) I. 87 The modern system of child-culture..is the system of treating children as decoration. **1925** W. DEEPING *Sorrell & Son* xiv. 129 Merry, insouciant, child-eyed little lady. **1909** *Westm. Gaz.* 23 Dec. 2/1 A child-face glowing with more radiant happiness we have never seen. **1906** *Macm. Mag.* Oct. 942 The curly-headed, thick-lipped, child-faced, gay, unlucky negro. **1894** F. J. FURNIVALL (*title*) Child-marriages, divorces, and ratifications, &c. **1933** *Lancet* 22 Apr. 886/2 Legislation for the prevention of child marriage [in India]. **1906** *Daily Chron.* 8 Sept. 3/2 In order to interest the child-mind, the subject is treated so as to focus attention on the marvellous intricacies of Nature. **1919** W. DE LA MARE *R. Brooke* 10 The child-mind, the child-imagination persists. **1898** WATTS-DUNTON *Aylwin* I. § 2 Into my very being that child-voice passed. **1944** E. SITWELL *Green Song* 31 Lest I hear your child-voice crying. **1947** *Mod. Lang. Rev.* July 354 The transition from the child-word to the conventional word.

22. child-care [CARE *sb.*[1] 4], the care or oversight of a child or children; also *attrib.*; **child-centred** *a.*, centred around the child; having the interests, needs, etc., of the child as its main concern; **child guidance,** the supervision of the welfare, esp. in its psychological aspects, of children and adolescents; also *attrib.*, esp. in *child guidance clinic*; **child-life,** life as a child; the lives of the children of a nation or community; **child-minder, childminder,** a person who takes care of small children (e.g. when the mother is working); **child-proof** *a.*, that cannot be operated by, or damaged by, a child; **child psychology,** the systematic study of the psychology of children; so *child psychologist*; **children's hour,** an hour of recreation in the evening, spent in former times by children with their parents; (with capital initials) a B.B.C. radio programme thus entitled (first broadcast 1922, discontinued 1961); **child's-eye view** [cf. *bird's-eye view*], a view, picture, or opinion, such as is presented to the eye of a child; **child-study,** the systematic study of children and their ways; **child welfare** [WELFARE *sb.* 4], the welfare of children.

1915 J. WEBSTER *Dear Enemy* (1916) 217 Modern, humane views on the subject of child-care. **1960** *Guardian* 9 Nov. 6/5 Most child-care officers can produce examples of this type of failure. **1928** RUGG & SHUMAKER (*title*) The child-centered school. *Ibid.* p. ix, Tolerant understanding and creative self-expression—the two great aims of the new education. These are the two criteria on which we shall appraise the child-centered schools. **1949** K. YOUNG *Sociology* (ed. 2) xxv. 468 In recent decades democratic cultures have shown a tendency toward not only the child-centered school but the child-centered home. In such families the child may be permitted..to dominate many domestic situations which in other cultures would not be tolerated. **1965** HALL & HOWES *Church in Social Work* I. iii. 57 A tendency..for much social work to become child centred, that is, to make what were believed to be the psychological needs of the child paramount in all cases. **1970** *Guardian* 28 Jan. 11/5 The child-centred, permissive approach. **1927** S. & M. G. BLANTON (*title*) Child guidance. **1928** R. P. TRUITT et al. (*title*) The child guidance clinic and the community. **1937** 'E. M. DELAFIELD' *Ladies & Gentl. in Vict. Fiction* I. 33 A modern child-guidance expert. **1940** R. S. WOODWORTH *Psychol.* (ed. 12) I. 13 When a child presents a serious behaviour problem..he may be taken to a child guidance clinic. **1865** 'L. CARROLL' *Alice in W.* xii. 192 Remembering her own child-life, and the happy summer days. **1884** Child-life [in Dict., sense 21]. **1885** M. I. BRYSON (*title*) Child Life in Chinese Homes. **1907** *Westm. Gaz.* 6 July 16/3 Aspects of Child Life and Education. **1941** *Economist* 29 Mar. 406/1 Members were agreed in rejecting Mr. Bevin's ..scheme for the provision of child 'minders', urging that women engaged in looking after their own children should not be called up. **1964** *Punch* 1 Apr. 481/3 Panels of child-minders for working mothers. **1956** B. CLEARY *Fifteen* ix. 192 Mrs Scruggs had done everything possible to make the room childproof. **1958** *Vogue* June 14 Special child-proof rear door locks. **1897** *Amer. Naturalist* XXXI. 455 The past year has been one of remarkable activity in the sphere of Child Psychology everywhere, but especially in this country. **1924** R. M. OGDEN tr. *Koffka's Growth of Mind* 3 The child-psychologist can follow the growth of a human being. *Ibid.*, Principles of child-psychology. **1941** 'R. CROMPTON' *William does his Bit* x. 232 Mrs. Dayford was a self-styled expert on Child Psychology. **1863** H. W. LONGFELLOW *Tales of Wayside Inn* 209 Between the dark and the daylight..Comes a pause in the day's occupations, That is known as the Children's Hour. **1923** *Radio Times* 12 Oct. 89/3 The Children's Hour: (a) Games and Pastimes; (b) Children's News. **1937** 'E. M. DELAFIELD' *Ladies & Gentl. in Vict. Fiction* 14 The children..live in the society of their parents. What was once known as the children's hour now extends handsomely into the twenty-four. **1964** M. LASKI in S. Nowell-Smith *Edwardian Eng.* iv. 199 The institution of the 'children's hour' between afternoon tea and dressing for dinner. **1959** *Birmingham Mail* 29 May 4/6 The serious-minded parents could no longer ignore the child's-eye view of the family world. **1959** I. & P. OPIE *Lore & Lang. Schoolchildren* xv. 340 The interest of the following five descriptions lies in their being child's-eye views of what are still primarily adult activities. **1899** *Daily News* 9 June 3/4 Child-Study and Mental Training. **1899** W. JAMES *Talks to Teachers* i. 14, I know that child-study, and other pieces of psychology as well, have been productive of bad conscience in many a really innocent pedagogic breast. **1909** *Daily Chron.* 18 Nov. 7/2 There are those who urge a rather plausible plea in these child-study days for a little wholesome neglect. **1908** H. G. WELLS *New Worlds for Old* xiii. 305 There is the need and opportunity of organizing..child welfare. **1915** A. L. STRONG (*title*) Child Welfare Exhibits. **1930** *British American* 19 Apr. 6/1 Mrs. J. S. Inglis, Kilrymont, St. Andrews, who took a leading part in the child welfare movement and coined the term 'child welfare'. **1958** *New Statesman* 20 Dec. 877/1 Intelligence testing, for example, grew to its present stature because its value was recognised in the world of education and child welfare.

Chile (tʃi·li). Also **Chili.** The name of the South American republic used *attrib.* in the names of natural products, animals, etc., as **Chile copper** (see quot. 1888); **Chile nettle,** any plant of the family Loasaceæ; **Chile pine** = MONKEY-*puzzle*; **Chile saltpetre** (see quot. 1959).

1801 LATHAM *Gen. Synop. Birds* 2nd Suppl. 140 Chili Woodpecker. This bird is scarcely as large as a Blackbird: it has a red-crested crown, and the body is banded with blue and white. **1849** BALFOUR *Man. Bot.* § 870 Loasaceæ, the Chili-Nettle Family...They are..chiefly distinguished for their stinging qualities, and hence the name of Chili-Nettle. **1852** G. W. JOHNSON *Cottage Gard. Dict.* 70 *Araucaria imbricata* (imbricate, or Chili pine). **1872** *Jrnl. Chem. Soc.* XXV. 1131 The Chili Saltpetre and Iodine Industry of Tarapaca. **1888** *Lockwood's Dict. Terms Mech. Engin.*, *Chili Copper*, black oxide of copper obtained in Chili. **1940** E. STEP *Wayside & Woodl. Trees* 176 The Chile Pine, or 'Monkey Puzzle', is a familiar sight on suburban lawns. **1959** *Chambers's Encycl.* VII. 46/1 *Nitrate* occurs in nature mainly as soda-nitre (Chile nitre or Chile saltpetre), $NaNO_3$ [etc.].

Chilean, Chilian (tʃi·liăn), *a.* and *sb.* Formerly also **Chilesian.** [f. prec. + -AN.] **A.** *adj.* Of or pertaining to Chile or its inhabitants. **B.** *sb.* A native or inhabitant of Chile.

1704 tr. *A. de Ovalle's Hist. Relation Kingdom Chile* in J. Churchill *Coll. Voyages* III. III. ii. 69/1 The Chilean Indians having so often signalized their Valour in Fights, they may very justly be distinguish'd from all the other Indians. **1748** *Anson's Voy.* II. xiv. 282 The Arraucos, and the rest of the Chilian Indians. **1797** *Encycl. Brit.* IV. 688/2 No defeats..could dispirit the Chilesians. **1818** *Deb. in Congress* (1854) 2038 The restoration of the banished Chilians to their families. **1837** *Penny Cycl.* XXXVI. 61/2 The Chilian Andes. **1866** [see PERUVIAN *a.*]. **1898** *Jrnl. Sch. Geogr.* (U.S.) Oct. 304 Punta Arenas..began its existence as a Chilean convict settlement. **1922** *Contemp. Rev.* Mar. 336 The Chileans seem to me to aim at being the aristocracy of the other hemisphere. **1924** *Public Opinion* 21 Mar. 278/2 The Chilian houseboy took us a short cut. **1952** *Oxf. Jun. Encycl.* VI. 92/2 The Chilean Glory Flower (*Eccremocarpus scaber*) with clusters of small orange, tubular flowers. **1955** *Nomencl. Commerc. Timbers* (*B.S.I.*) 52 'Chilean laurel'.. *Laurelia aromatica* Juss. **1957** *Encycl. Brit.* XVI. 43/2 The Chilean guava (*M*[*yrtus*] *ugni*), a shrub with ovate, dark-green leaves and white flowers succeeded by globular red or black glossy fruit..hardy in southwest Britain.

chilitis, var. *CHEILITIS.

Chilkat (tʃi·lkæt). Also **Chilcat.** [Tlingit *djîlqâ't*, a place-name, said to mean 'storehouses for salmon'.] (A member of) a subdivision of the Tlingit tribe of Indians in Alaska; used *attrib.* to refer to the blankets which they make.

1836 *Archaeologia Americana* II. 302 Commencing on the north, from lat. 59° to lat. 55°, there are found ten or more petty tribes that speak the Sitka language, viz. the Chilcart, one of the most numerous and powerful of these tribes; [etc.]. **1844** J. DUNN *Hist. Oregon Terr.* xvii. 288 A little to the northward of this there is a tribe called the Chilkasts. **1890** *Internat. Archiv Ethnogr.* III. 8 The beautifully woven Chilkat-blankets are used as far south as Comox. **1907** G. T. EMMONS in *Mem. Amer. Mus. Nat. Hist.* III. 329 'Chilkat blanket' is an expression, coined in recent years, current among Europeans, but meaningless to the natives, except in a commercial sense. **1966** *Sat. Rev.* 22 Oct. 27/1 These works by Tlingit, Tsimpshian, and Itaida Indians include.. dance sticks, magnificently woven Chilkat blankets, baskets, [etc.].

chill, *v.* Add: **6. b.** To subject (meat or other food) to a low temperature in a chill-room or a refrigerator. (Cf. *CHILLED *ppl. a.* 2 b, *CHILLING *vbl. sb.* c.)

1891 R. WALLACE *Rural Econ. Austral. & N.Z.* xxxv. 441 American beef, which is only a few days in transit, is not frozen through, but chilled and kept in chambers at a temperature of three to four degrees Fahr. below the freezing point. **1959** *Listener* 6 Aug. 227/3 Chill for 2 hours before serving.

chill-. Add: **2. chill-cast(ing), -crystal, -depth** (see quots.).

1930 *Engineering* 13 June 782/2 The tensile strength of the rolled material was considerably greater than that of the chill-cast material. **1879** *Encycl. Brit.* IX. 481/2 In 'chill casting' a portion of the surface of the whole or a part of the mould is made of cast iron, so that the metal brought in contact with it is rapidly cooled. **1958** MERRIMAN *Dict. Metallurgy* 34/2 *Chill cast*, metal cast into metal moulds or chills. *Chill castings*, cast iron that has been set in metal moulds or castings that have been allowed to set in moulds in which metal pieces are placed in the walls of the mould to accelerate cooling. **1940** *Chambers's Techn. Dict.* 156/2 *Chill crystals*, small crystals formed by the rapid freezing of molten metal when it comes into contact with the surface of a cold metal mould. **1930** *Engineering* 9 May 611/2 With regard to depth of chill he was of opinion that a chill depth of rather more than 3/4 in. was desirable.

chilled, *ppl. a.* Add: **2. b.** Of meat, esp. beef: kept at a moderately low temperature in cold storage, as distinguished from frozen meat.

1891 R. WALLACE *Rural Econ. Austral. & N.Z.* xxxvi. 471 About 90 per cent. of the whole frozen and chilled meat trade. **1894** *Daily News* 23 Jan. 5/1 The newest thing is chilled beef, as distinct from frozen. **1955** *Times* 14 May 7/3 Chilled and frozen beef sent to the United Kingdom.

chiller. (Later examples.)

1921 *Dict. Occup. Terms* (1927) § 279 *Chiller*; sprays moulds with water after use, to cool them, and sometimes whilst in use in order to hasten cooling of steel ingot. **1965** *Courier-Mail* (Brisbane) 1 Oct. 16 One of his portable mobile chillers. **1968** K. WEATHERLY *Roo Shooter* 6 The chiller was built on to a large four-wheeled trailer.

chilli. Add: **2. chilli** (usu. **chile** or **chili**) **con carne** (orig. *U.S.*), a stew of Mexican origin containing minced beef flavoured with chillies; **chilli** (usu. **chili**) **sauce** (chiefly *U.S.*), a sauce made with tomatoes, peppers, and spices.

In quot. 1907 contextually shortened to *chile con* as a nonce-use.

1857 S. COMPTON SMITH (*title*) Chile con carne, or the camp and the field. **1895** *Outing* (U.S.) XXVI. 28/2, I went to a little Mexican restaurant..and, sitting there eating my *frijoles* and *chile con carne* [etc.]. **1907** Chile con [see *AIR-TIGHT *sb.* 2]. **1927** *Weekly Dispatch* 4 Apr. 4/6 Excellent Mexican meal of tamalis, chili-concarne etc. **1960** *Harper's Bazaar* Oct. 98/2 *Chili con carne*, Mexican beef with hot chili sauce served with red beans and rice.

3. (In form **chili.**) An oppressive hot southerly wind which blows in Tunisia.

1927 W. G. Kendrew *Climates of Continents* (ed. 2) v. 28 Strong southerly winds are specially hot and unpleasant, and they are distinguished everywhere by local names such as sirocco, chili, khamsin.

chilling, *vbl. sb.* Add: **c.** The action or process of chilling meat. In quot. *attrib.*

1902 *Westm. Gaz.* 26 Nov. 10/3 The great River Plate exporters having adopted the 'chilling' process.

chillsome (tʃi·lsəm), *a.* [f. Chill *sb.* or *v.* + -some.] Chilling, chilly.

1927 Whitham *Latude & Linguet's Mem. Bastille* Introd. 40 Her mistress, who, chillsome in blood..dreaded that she might cease to be agreeable to her lusty sovereign. **1970** *Daily Tel.* 22 Aug. 7 Glorious and gently sloping golden sands, a chillsome sea.

chiloplasty, var. *cheiloplasty.

chilostome (kəi·lŏstōᵘm), *a.* and *sb.* *Zool.* Also **cheilo-**. [ad. mod.L. *chilostomata*: see Chilostomatous *a.*] **A.** *adj.* Of, pertaining to, or resembling the Chilostomata, an order of marine Polyzoa. **B.** *sb.* A member of this order.

1896 Harmer in *Cambr. Nat. Hist.* II. 481 The operculum is usually..a conspicuous feature of the Cheilostome zoœcium. *Ibid.* 486 Some marine Cheilostomes may be saved from attacks..owing to the existence of their armoury of avicularia and vibracula. **1912** *Brit. Museum Return* 173 The collection of Cretaceous Cheilostome Polyzoa from France. **1959** L. H. Hyman *Invertebrates* V. xx. 374 The cheilostomes are the dominant group of existing ectoprocts.

chime, *sb.*¹ Add: **2. b.** *spec.* (Usu. in pl.) Such an arrangement used as a door-bell.

1934 *Pop. Sci. Monthly* CXXIV. 65 (*title*) How to make a set of musical electric chimes to replace your noisy old doorbell. *Ibid.* 65/1 Although expensive to buy, suitable door chimes are easy to construct. **1963** 'L. Egan' *Run to Evil* vii. 76 The front-door chimes produced only silence.

chimera, chimæra. Add: **1. b.** Any fish of the family Chimæridæ; = Rabbit-*fish*. (Cf. Chimæroid *a.*)

1804 Holloway & Branch *Brit. Museum* III. 56 The Chimæra, or *Chimæra Monstrosa*, belongs to that class of fish which have close gills and cartilages instead of bones. **1808** E. Donovan *Nat. Hist. Brit. Fishes* V. Plate CXI, There are two species of the Chimæra genus, Monstrosa, and Callorhynchus; the latter of which is distinguished by the name of Southern Chimera and Elephant Fish. **1836** W. Yarrell *Hist. Brit. Fishes* II. 365 The Northern Chimæra is represented as a fish of singular appearance and beauty, a native of the northern seas only, where it seldom exceeds three feet in length. **1848** [see Rabbit *sb.*¹ 4]. **1969** A. Wheeler *Fishes Brit. Isles* 111 The chimaeras are deep-water fishes, living on or below the edge of the continental shelf.

3. d. *Biol.* [ad. G. *chimäre* (H. Winkler 1907, in *Ber. d. Deut. Bot. Ges.* XXV. 574).] An organism (commonly a plant) in which tissues of genetically different constitution co-exist as a result of grafting, mutation, or some other process.

1911 D. H. Campbell in *Amer. Naturalist* XLV. 44 Such monstrous forms, for which Winkler proposes the name 'chimæra', are not hybrids in any true sense of the word, but have arisen from buds in which there was a mere mechanical coalescence of tissue from the two parent forms at the junction of the stock and graft. **1926** J. S. Huxley *Ess. in Pop. Sci.* xviii. 259 If the front half of one species be grafted on to the back half of another species, both continue to differentiate, and a chimaera or mosaic organism is produced. **1968** *Nature* 9 Nov. 596 (*heading*) Mouse chimaeras obtained by the injection of cells into the blastocyst. **1969** *New Scientist* 16 Jan. 133/1 Cytogeneticists have found human mosaic individuals, trisomics and chimeras.

chiming, *ppl. a.* Add: *spec.* in *chiming clock.*

1789 *Loiterer* 18 July No. 25 p. 14 The chiming Clock, or the repeating Watch. **1962** E. Bruton *Dict. Clocks & Watches* 39 *Chiming Clock*..Tower clocks and some antique clocks chime on bells. Modern domestic clocks usually have the Westminster or modified Whittington chime on rod gongs.

chimney, *sb.* Add: **4. c.** Phr. *to smoke like a chimney*: to smoke (cigarettes, etc.) very heavily.

1840 Barham *Lay of St. Odille* in *Bentley's Misc.* VII. 172 A German, Who smoked like a chimney. **1870** L. M. Alcott *Old-Fash. Girl* x. 157 Tom lay on the sofa.., reading 'Pendennis' for the fourth time, and smoking like a chimney as he did so. **1911** A. Pinero *Mid-Channel* iv. 217 Zoe smokes like a chimney.

9. (Examples.)

1860 *San Francisco News Let.* 20 Jan. 5/1 Silver ore is found in what are termed chimneys,..the lead..dropping sometimes two or three thousand feet, and sometimes turning short. **1873** J. H. Beadle *Undevel. West* xviii. 334 It may be a 'chimney' from some lode ten thousand feet away through solid rock. **1873** J. Miller *Life amongst Modocs* xviii. 226 A pouch in the rock—a little 'chimney' that nurses a few thousand dollars worth of dust about the flue.

11. **chimney-bar**, an iron bar supporting the masonry above a fireplace; **chimney cap**, (*a*) the top of a chimney, either as an ornament or as a cover; (*b*) = Cowl *sb.*¹ 4; **chimney-cleaner, -cleanser**, a chimney-sweeper; **chimney-jack**, (*b*) = *steeple jack* (see Steeple *sb.*¹ 6); **chimney-jamb** (later U.S. example); **chimney neck**, the shaft of a chimney; **chimney rock**, (*a*) *Geol.*, a chimney-shaped body of rock; (*b*) in Florida, a porous phosphate rock used in building; **chimney-stack**, (*b*) = *chimney-stalk* (b); **chimney-swallow, -swift**, a species of swift, *Chætura pelagica*.

1833 Loudon *Encycl. Archit.* § 79 The fire-places to have each a strong iron chimney-bar. **1940** *Chambers's Techn. Dict.* 156/2 *Chimney bar*, an iron bar supporting the arch over a fireplace opening. **1847** *Rep. Comm. Patents 1846* (U.S.) 221 What I claim, therefore, as my invention..is a ventilator or chimney cap. **1910** B. Pite et al. *Building Construction* I. 160 Chimney caps are so usual that the advisability of avoiding heaviness in their arrangement and design may be pointed out. **1906** *Daily Chron.* 20 July 4/6 Mechanical chimney-cleaner. **1921** *Dict. Occup. Terms* (1927) § 924 Sweep, chimney sweep; chimney cleaner. **1952** 'R. Crompton' *William & Tramp* v. 162 William carried his tin of red paint, Ginger his chimney cleaner. **1905** *Daily Chron.* 11 Jan. 4/7 John Smith or so, 'chimney-cleanser'. **1907** *Westm. Gaz.* 16 Mar. 10/1 A competent chimney-jack was despatched up a rope to work upon the crack. **1887** Tourgée *Button's Inn* 253 'I don't know, sir,' she responded, setting the spider in its place by the chimney-jamb. **1833** Loudon *Encycl. Archit.* § 983 Deal sarking..to be laid on each side of the roof and chimney necks (shafts). **1847** in *Utah Humanities Rev.* (1948) 125 Chimney Rock was still visible down the River. **1870** *Congr. Globe* App. 26 Mar. 225/2 The marbles of our western border have heretofore served as 'chimney rock' for the cabin of the luxurious border farmer. **1877** J. Applegate in *Trans. Oregon Pioneer Assoc. 1876* IV. 59 The Chimney Rock, Court House, and other bluffs. **1883** M. P. Bale *Saw-Mills* 75 An important factor in the economical working of boilers is the correct arrangement of the chimney stack. **1903** 'A. McNeill' *Egregious English* 48 The top of the factory chimney-stack. **1789** Morse *Amer. Geogr.* 60 Red winged Starling, Swallow, Chimney do. **1809** S. Williams *Hist. Vermont* (ed. 2) I. 140 The species called the house or chimney swallow, has been found during the winter, in hollow trees. **1911** *Encycl. Brit.* XXVI. 231/1 *Chaetura p[e]lagica*, the 'chimney-swallow' of the United States. **1849** *Audubon's Western Jrnl.* (1906) 129 Fifteen or twenty swifts, about double the size of our common chimney swift at home. **1898** *Atlantic Monthly* Apr. 456/1 Chimney-swifts were shooting hither and thither athwart the sky. **1930** J. S. Huxley *Bird-Watching* ii. 38 Some barn-swallows and chimney-swifts.

chimney, *v.* Add: **2.** *colloq.* (Also with *up*.) To climb a chimney (Chimney *sb.* 8).

1940 F. S. Chapman *Helvellyn to Himalaya* i. 15 A gully which was just too wide to 'chimney up', that is, to jam oneself between two rock walls and to wriggle up by extreme muscular exertion. **1957** J. Masters *Far the Mountain Peak* iii. 32 He had his back against the wall and his feet against a projection...He came on down, 'chimneying'..you held yourself in place by pressing your back against one surface and your feet against the other.

chimneyed, *ppl. a.* Add: **b.** Of rock: having a vertical cleft. (Cf. Chimney *sb.* 8.)

1897 *Outing* (U.S.) XXIX. 553/1 Troops of chimneyed rocks stepping out boldly into the wrinkled shoals.

chimney-pot. Add: **2.** *chimney-pot hat* (earlier example). Also ellipt. *chimney-pot.* Now *rare.*

1851 *Expositor* 4 Jan. 145/1 The absurdity of the 'Chimney-pot' and 'Coal-scuttle' covering for the head at present in use. **1862** Trollope *Rachel Ray* II. xii. 244 Middle-aged men..in swallow-tailed coats and black trousers, with chimney-pot hats. **1874** Lisle Carr *J. Gwynne* I. i. 33 So off went that penitential chimney-pot leaving such a great red line round his forehead. **1907** *Westm. Gaz.* 25 July 10/1 The chimney-pot is so called because in Germany only the sweeps wear them.

chimonanthus (kəimonæ·nþŏs). [mod.L. (J. Lindley 1819, in *Bot. Reg.* V. 404), f. Gr. χειμών winter + ἄνθος flower.] A member of a small genus of shrubs, belonging to the family Calycanthaceæ and native to China; winter-sweet.

1829 Loudon *Encycl. Plants* 454 *Chimonanthus*..in allusion to the period of the year when its blossoms are produced. **1904** Farrer *Garden of Asia* 18 There are cherries, plums, chimonanthus, magnolia. **1955** A. G. L. Hellyer *Flowers* 33 Chimonanthus (Winter Sweet)..is deciduous shrub.

chimp. Short for Chimpanzee.

1877 E. Lear *Laughable Lyrics* 88 The wail of the Chimp and Snipe. **1928** *Daily Tel.* 29 May 6 The 'chimps' behave almost perfectly at table. **1957** *Times* 21 Nov. 18/6 Chimps, picture cards and many diverse forms of advertising bring our teas before..buyers.

chimyl alcohol (kəi·mil æ·lkŏhɒl). *Chem.* [ad. G. *chimylalkohol* (Y. Toyama 1924, in *Chem. Umschau* XXXI. 61/2), f. *Chim*(*era* + -yl.] A crystalline alcohol, cetyl-α-glyceryl-ether, $C_{19}H_{40}O_3$, obtained from the liver oils of fishes.

1924 *Chem. Abstr.* 2613 Shark- and torpedo-liver oils... These two oils contain another solid alc., a homolog..of the probable compn. $C_{19}H_{40}O_3$, named *chimyl alc*[*ohol*] by T[oyama]. **1964** J. A. Lovern in *Oceanogr. & Marine Biol.* II. 173 Glyceryl ethers, such as chimyl alcohol (D-glyceryl-1-hexadecyl ether)..occur as alkoxydiglycerides in the fats of some fish.

chin, *sb.*¹ Add: **1. d.** Phr. *keep your chin up*, often ellipt. *chin up*, do not succumb to depression; also *chin(s)-up* used *attrib.*; *to take it* (or *life*) *on the chin* (from *Pugilism*), to meet misfortune courageously; to withstand a severe blow, irrespective of one's feelings.

1928 J. P. McEvoy *Show Girl* ix. 129 Jack Milton won't kick in another nickel. And we're going to take it on the chin for five thousand down here this week. **1931** *Daily Express* 22 Sept. 3/4 'We are not paying out against sterling,' chanted the cashier. Elderly Englishmen of the retired colonel type took it on the chin. **1938** P. Gallico *Confessions of Story Writer* (1946) 226 Keep your chin up. I get the picture. **1939** I. Baird *Waste Heritage* xix. 263 Keep your chin up honey. **1946** F. Sargeson *That Summer* 108 Terry said chin up. **1958** *Spectator* 13 June 761/1 They scoffed at the President's chins-up speeches. **1960** D. Lytton *Goddam White Man* i. 39, I liked the Williams' because of the way they took life on the chin. **1961** *John o' London's* 6 July 29/1 Their passivity suggested to me a definite chin-up attitude.

e. *slang* (orig. and chiefly *U.S.*). A talk; conversation; *spec.* insolent talk, 'cheek'. Also, reduplicated, *chin-chin*. Cf. Chin *v.* 3 and Chin-*wag*.

1877 'Mark Twain' *Love Lett.* (1949) x. 198, I haven't had so much chin-chin for years. **1894** P. L. Ford *P. Stirling* 14 I'll wait till I've graduated, and had a chin with my governor about it. **1902** Masefield *Salt-water Ballads* 45 Jake was a dirty Dago lad, an' he gave the skipper chin. **1914** G. Atherton *Perch of Devil* i. xxiv. 148 There's a guy..come out to have a chin with you. **1918** E. M. Roberts *Flying Fighter* 289, I went back and resumed the 'chin-chin' with Kerr and the other boys. **1936** M. de la Roche *Whiteoak Harvest* x. 129 Oh, I have time for a chin before I go. **1947** *Coast to Coast* 136 Mum and Mrs. Martin had a good chin-chin at the fence about sickness and husbands. **1952** *New Yorker* 12 Jan. 24/1 We'd like to have a little chin with you right now.

2. *chin-line*; **chin-music**: for *U.S.* read chiefly *U.S.* (add earlier and later examples); **chin-strap**, (*c*) as used by women (see quot. 1939); **chin-wag** *v. slang*, to talk, chatter.

1938 *Ottawa Jrnl.* 19 Feb. 12/4 To keep chinline youthful sleep without pillow. **1967** M. Morris in *Coast to Coast 1965–6* 126 The tell-tale signs of age..the crêpey eyelids, the blurred chin-lines. **1834** J. Romilly *Diary* 5 Mar. (1967) 50 Evening Party at Mrs Skrines...F. Grey plaid chin music & sung the dogs meat man: rather too ludicrous. **1836** J. Hildreth *Dragoon Campaigns Rocky Mts.* I. 26 Full of chin music, as the species of loquacity he possessed is termed. **1852** *Gloss. Provinc. Words Berks.* 6 *Chin-music*, chattering; scolding. **1887** G. B. Shaw *Let.* 27 May (1965) 170 The Fabian, which has been doing the work whilst the others have been snatching the credit of it with discordant chin music. **1939** M. B. Picken *Lang. Fashion* 26/2 *Chin-strap*, strap or band worn under the chin...Also, 'beauty' strap worn at night, supposedly to prevent double chin. **1942** R. Chandler *High Window* (1943) v. 43 On the chair beside her there was..a white satin chin-strap. **1955** 'P. Wentworth' *Out of Past* xxii. 118 Adela Castleton was wearing a chinstrap and a kind of mask of some cosmetic preparation. **1920** *Blackw. Mag.* Feb. 182/2, I didn't waste any time chin-wagging. **1954** A. Baron *Golden Princess* 50 Didn't he send her down to the village to chinwag with the Indian chiefs?

Chin (tʃin), *sb.*² [Burmese, 'hill-man'.] One of a people inhabiting the Chin hills and adjacent districts in Burma; also, their language.

1881 F. S. Copleston in *Rep. Census Brit. Burma* 68 In British Burma the Chins are very widely extended. **1882** [see *Arakanese *a.* and *sb.*]. **1951** R. Firth *Elements Social Organiz.* iv. 145 The Feasts of Merit of the Chins of Burma. **1962** *Listener* 25 Oct. 646/1 Some of these hill peoples—such as the Karens, Kachins, and Chins—have a long tradition of feuding and fighting with the Burmese of the plains.

chin, *v.* Add: **3. b.** *trans.* To talk to, to address, esp. boldly or impudently. Occas. *intr.* *U.S. slang.*

1884 J. Hay *Bread-Winners* 161 You haven't done a — thing but lay around on the grass and eat peanuts and hear Bott chin. **1892** Gunter *Miss Dividends* 247, I heard one of them call another 'Constable' and the other chinned him as 'Sheriff!' **1898** H. S. Canfield *Maid of Frontier* i. 7 He chinned the barkeep mos' to death. **1911** H. S. Harrison *Queed* vii. 85 [I have] been up chinning your sporting editor. **1934** J. M. Cain *Postman rings Twice* ix, The cop was half way down the hall, chinning a nurse.

4. *to chin the bar*: in gymnastics, to draw up the body until the chin is brought over the horizontal bar; also *absol.* (*refl.*).

1903 *Independent* 15 Jan. 146/2 He must also be something of an athlete; able to chin the horizontal bar. **1906** *Amer. Mag.* LXIII. 139/1 On the horizontal bar he was persuaded to chin himself twelve times. **1906** 'O. Henry' *Four Million* 85 He..could chin the bar twice with one hand. **1940** *Mind* XLIX. 195 No man can chin himself 6,000,000 times in succession without rest.

China¹. Add: **2. b. China-berry** *U.S.*, in full *Chinaberry tree*, (*a*) the wild China or soap-berry (*Sapindus saponaria*) of the southern U.S.; (*b*) = China-*tree*; **China-crape** (examples); **China hog**, a breed of swine; also *ellipt.*;

China-orange (further examples); freq. taken as a typical object of trifling value; **China silk**, a lightweight silk fabric in plain weave; **China snoek** *S. Afr.*, a small immature specimen of the snoek fish *Thyrsites atun* (see quot. 1957); **China tea**, a type of tea prepared from a small-leaved variety of tea plant (*Camellia sinensis* var. *sinensis*) grown chiefly in S. China and differing from other kinds of tea chiefly in that it is cured with smoke; **China-tree** (earlier example).

1890 *Harper's Mag.* Dec. 106/2 The high gray towers..were crowned with ornaments like the berries of the china-berry trees. **1908** R. W. Chambers *Firing Line* xxix, A subtler scent..came to him on the sea-wind;..—the lilac perfume of china-berry in bloom. **1932** E. Caldwell *Tobacco Road* i. 8 Ellie May stood behind a chinaberry tree. **1944** R. M. Harper *Prelim. Rep. Weeds Alabama* 141 M. Azedarach L. Chinaberry. A medium-sized tree, very commonly cultivated for shade in the South. **1813** Jane Austen *Let.* 29 Jan. (1932) II. 298, I hope you will wear your China crape. **1871** C. M. Yonge *Little Lucy's Wonderful Globe* i. 4 A scarlet China crape shawl. *Ibid.* viii. 40 A sort of blue China-crape tunic. **1838** H. Colman *Rep. Agric. Mass.* 74 A cross with some of our small boned breed, such as the Byfield or the China [hog]. **1851** C. Cist *Cincinnati* 279 The hogs raised for this market, are generally a cross of Irish Grazier Byfield..and China. **1871** Eggleston *Hoosier Schoolm.* (1872) xxvi. 180 You can't make nothin else out of him, no more nor you can make a china hog into a Berkshire. **1698**, etc. China-Orange [see Orange *sb.*¹ 1 a]. **1819, 1849** China orange [see Lombard-street]. **1614** in T. Roe *Jrnl.* (1899) II. 556 A Riall and a half of 8 the pownd..is more than the whight China silke doth cost in the Indies. **1720** Defoe *Capt. Singleton* 335 We had Fifteen Bales of very fine China Silks. **1858** O. W. Holmes *Aut. Breakf.-t.* 314 Some marrowy crapes of China silk, Like wrinkled skins on scalded milk. **1945** C. S. Forester *Commodore* iii. 27 A nightshirt of solid China silk. **1950** *Cape Argus* 28 Oct. (Mag. Section) 3/6 China snoek..have thicker bodies and shorter heads than the large snoek. The scientists refuse to recognize the China snoek as a different species. **1957** S. Schoeman *Strike!* 117 The so-called 'China snoek', those undersized snoek which are found in Table Bay docks during August to October and in False Bay during November–January. **1811** Jane Austen *Let.* 31 May (1932) 286 We began our China Tea three days ago. **1920** J. M. Barrie *Kiss for Cinderella* iii. 130 You'll have a cup of China tea and some of this cake? **1819** E. Evans *Pedestrious Tour* 315 Here grew the China tree, of a beautiful appearance, and bearing fruit of an inviting aspect.

II. *dial.* **chainy, chany, chaynee, chayney, cheeny.** Also, pieces of broken china.

1823 E. Moor *Suffolk Words* 74 *Cheeny*, China. Both ware and country. **1831** S. Lover *Legends & Stories of Ireland* 167 A bull in a chaynee shop. **1860** Geo. Eliot *Mill on Fl.* I. iii. ii. 278 There's none of 'em got better chany. **1864** B. Brierley *Chron. Waverlow* 88 Tum had the old-fashioned 'chaney'..spread upon the table. **1868** Lady Verney *Stone Edge* xx. 234 His yead were cracked all one as a chayney jug. **1880** Patterson *Gloss. Antrim & Down* 18 *Chainy*, china. **1884** R. Holland *Gloss. Chester* 60 Thy uncle and aunts' comin to tay this afternoon, Mary; tha'd better get th' *chainy* cups and saucers ait. **1896** Northall *Warwicks. Word-Bk.* 45 *Chāney* sb. and adj., china. Oxf., Shrop., SE Worc., and elsewhere. **1897** W. B. Yeats *Secret Rose* 171 One party was quietly playing 'chanies', as they called house-keeping with pieces of broken pottery. **1936** 'F. O'Connor' *Bones of Contention* 178 Sacred Heart, he'd make chanies of the crockery on me!

4. c. *china-shop* (earlier example; see also Bull *sb.*¹ 1 c); **china eye**, a wall eye; **china-glaze** (example); **china mark**, a collectors' name for any moth of the genus *Hydrocampa* and allied genera; **china money** (see quot. and cf. *china token*); **china-painting** (examples); **china-shell**, a collectors' name of the *Ovulum ovum*, given in allusion to the white porcelain-like surface of the shell; **china token**, a token of porcelain or earthenware used in porcelain and pottery works; **china wedding** *U.S.* (see quot.).

1928 F. T. Barton *Kennel Encycl.* 372 Wall eye..is applied to one or both eyes..in which the iris or the irises is destitute of its usual pigmentation, giving the eye a light colour—China eye. **1964** F. O'Rourke *Mule for Marquesa* (1967) vi. 80 He turned and his right eye rolled, the china eye, white-marbled and yellow-veined, the birthmark, the trademark of the man. **1784** S. Jones *Let.* 8 Mar. in *J. Wedgwood's Lett.* (1965) 288 A man that can make as good a China glaze..as any man in the country. **1832** J. Rennie *Consp. Butterfl. & Moths* 150 Hydrocampa...The Lettered China-mark (*H. literalis*). **1901** W. F. Kirby *Butterflies & Moths* 112 China Marks..are moths rarely exceeding an inch in expanse. **1959** J. Clegg *Freshwater Life* (ed. 2) xiv. 213 They are commonly called the China Mark Moths from the fancied resemblance of the markings on the wings of some of them to the potters' marks inscribed on the bottom of good china. **1868** L. Jewitt in *Art Jrnl.* 282/1 The issuing of 'chainé money', *i.e.*, tokens representing different values of money, made of china...They were called 'Mr. Cokes' coin', or 'chainé money' (china money), in the provincialism of the locality. **1880** *Harper's Mag.* June, Hundreds of women who are taking lessons in china-painting. **1942** W. Faulkner *Go Down, Moses* 263 She gave lessons in china-painting. **1886** Gray & Woodward *Sea-Weeds, Shells* 47 The *Cypræidæ* or Cowries..With these is classed the 'China-shell' (*Ovulum*). **1616** T. Roe *Let.* 14 Feb. in *Jrnl.* (1899) I. 134, I thought all India a China shop, and that I should furnish all my Frendes with rarietyes. **1878** Jewitt *Ceramic Art* II. 141 One peculiarity connected with the Pinxton China Works..is the issuing of china tokens, *i.e.*, tokens representing different values of money, made of china, and payable as money among the workpeople and others, including shopkeepers. **1888** *Girl's Own Paper* 24 Mar. 407/2 In America..is the 'crystal-wedding', which is kept after fifteen years of married life...Then, too, there is the 'china-wedding', which is observed five years later.

6. Short for *China rose* (a) and for *China tea.*

1844 Mrs. Loudon *Ladies' Compan. Flower Garden* (ed. 3) 344 *Rosa indica* (the common China); *Rosa semperflorens* (the monthly China). **1898** M. A. von Arnim *Eliz. & German Garden* 17, I have misgivings as to the effect of the Persian Yellows among the Chinas, for the Chinas are such wee little baby things. **1907** *Yesterday's Shopping* (1969) 1/1 Tea...Finest China, Plain (Moning). **1938** S. Beckett *Murphy* v. 83, I ask for China and you give me Indian. **1964** J. Turner *Slate Landscape* xvi. 158 I'll tell you all about it over tea. You like China, I suppose?

7. Short for *china plate*, rhyming slang for 'mate'.

1880 D. W. Barrett *Life & Work among Navvies* ii. 41 'Now, then, my china-plate...' This is essentially a bricklayer's phrase. If for 'china-plate' you substitute 'mate',..the puzzle is revealed. **1925** Fraser & Gibbons *Soldier & Sailor Words* 53 *China, or Old China*: chum. **1945** *Penguin New Writing* XXV. 170 'Remember that China of his?' 'What, the bloke with the hair?' **1953** K. Tennant *Joyful Condemned* xxxi. 304 My china's got something she wants to tell you. **1965** *New Statesman* 14 May 760/2, I have my hands full with his china who is a big geezer of about 14 stone.

Chinadom. (Earlier U.S. example.)

1856 *Sacramento Daily Union* 15 July 3/1 Disturbing the peace in Chinadom.

chinagraph (tʃəi·nǎgrɑf). [f. China¹ 3 + Graph *sb.*²] In full *chinagraph pencil*: a pencil with which it is possible to write on china, glass, and other hard surfaces.

1943 R. Hill *Desert Conquest* ix. 65 All the officers..go up to the squadron leader's tank, armed with map-cases and chinagraph pencils. **1944** L. MacNeice *Springboard* 26 May come up with a chinagraph. **1950** G. Wilson *Brave Company* (1951) xiii. 223 My boxes of colored chinagraph pencils. **1960** *Times* 1 Feb. 11/2 He will certainly be drawing on the map, in a broad stroke of the chinagraph, the general axis of advance. **1969** *Listener* 27 Mar. 439/2 You can draw pictures or write messages on a coloured egg using a white Chinagraph pencil.

Chinaman. Add: **3. Chinaman's chance**, a very poor or negligible prospect (of gain, survival, etc.); the least chance. *colloq.* (chiefly *U.S.*).

1914 *Call* (San Francisco) 30 Apr. 6 The poor boob ain't got a Chinaman's chance. **1915** 'Ian Hay' *First Hundred Thousand* vi. 50 The Service Battalions..*must* be led by the officers who have trained them if they are to have a Chinaman's chance when we go out. **1926** H. C. Witwer *Roughly Speaking* iii. 80 Just make out we're a couple of big brothers—we know full well we ain't got a Chinaman's chance to be nothin' else to you! **1952** F. Yerby *Woman called Fancy* x. 193 You haven't a Chinaman's chance of raising that money in Boston.

4. *Cricket.* A left-handed bowler's off-break to a right-handed batsman (see also quots.).

1937 *Daily Herald* 3 Feb. 14/1 The Yorkshire lefthander..has frequently broken up a big partnership with his 'chinamen'. **1955** Miller & Whitington *Cricket Typhoon* i. ii. 32 In..Yorkshire, the 'Chinaman' is regarded as the left-hand bowler's off-break... In Australia..the 'Chinaman' is..the left-hander's googly. **1963** *Times* 11 May 9/7, I understand the 'Chinaman' to be simply an off break bowled out of the back or side of the hand by a left-handed bowler—that is, the ball comes in to a right-handed batsman from the off and the left-handed bowler's action in bowling is equivalent to that of the right-hander in bowling a leg break. I believe the term was first used in referring to this style of bowling practised before the last War by Ellis Achong, who, although he played for the West Indies, was in fact a Chinese.

chinar, var. Chenar, the Oriental plane-tree; also, timber from this.

1889 in *Cent. Dict.* **1917** *Blackw. Mag.* Sept. 414 The folds of the door are of solid steps of chinar. **1925** *Chambers's Jrnl.* 79/2 The tall Chinar trees stood up dreamily.

Chinatown (tʃəi·nătaun). A section of a large town, especially a sea-port, in which Chinese live as a colony and to a great extent follow their own customs.

1857 *Butte Record* (Oroville, Calif.) 31 Jan. 2/7 Chinatown was wild with joy. **1880** R. L. Stevenson *Across Plains* (1892) 81 The noise..mounts to you..from your right also, round by Chinatown. **1889** *Chambers's Jrnl.* 19 Jan. 39/2 For filth and wretchedness you must go to Chinatown [San Francisco] by night. **1892** D. B. W. Sladen *Japs at Home* ii. 21 Separated from the Settlement and Chinatown..is the native town. **1902** in G. R. Sims *Living Lond.* III. 86/1 The Limehouse Chinatown. **1944** *Living off Land* viii. 160 He makes straight for 'Chinatown' and proceeds to buy up..junk. **1957** *Times Lit. Suppl.* 18 Oct. 621/2 A fascinating picture of Chinatown. **1969** V. G. Kiernan *Lords of Human Kind* v. 164 The opium-dens of Singapore, or London's China-town, had a sinister repute.

Chi·nawo:man. [f. China¹ 1, after Chinaman.] A Chinese woman.

1872 'Mark Twain' *Innoc. at Home* ix. 66 There were few white servants and no Chinawomen so employed. **1921** *Daily Colonist* (Victoria, B.C.) 10 Apr. 4/2 Mrs. Yung Sue Kee..revealed herself on coming into the witness box as a young married Chinawoman.

chincapin. Add: Also, the nut of this tree.

1612 W. Strachey *Trav. Virginia* (1849) vi. 72 They..live..most of acrons, wallnuts, chesnutts, chechinquamins, and fish. **1624, 1676** [see Dict.]. **1814** F. Pursh *Flora Amer.* II. 625 Castanea pumila...The fruit is..generally known by the name of Chinquapin. **1893** Leland *Mem.* II. 110 Rebeldom, where a Union man's life was worth about a chinquapin.

chinch, *sb.*¹ **2.** *chinch-*, *chintz-bug*: earlier and later examples.

1785 in *Southern Lit. Messenger* XXVIII. 38/1 The devastation of the Chintz bug, which since harvest have infested the Indian corn. **1786** Washington *Diaries* (1925) III. 97 Examined the..corn in several parts of this field and discovered more or less of the Chinch bug on every stalk. **1873** J. H. Beadle *Undevel. West* xiii. 225 Its [*sc.* wheat's] principal enemy is the chintz bug, so called here. **1959** Southwood & Leston *Land & Water Bugs* v. 82 The chinchbug, *Blissus leucopterus*, is an important cereal pest in North America which has spread rapidly over vast areas—this spread is paralleled in western Europe by the recent spread of our single British species. *Ibid.*, *Ischnodemus sabuleti* (Fallén), European chinchbug.

chincherinchee, common var. *chinkerinchee.

1926 A. H. Hamer *Wild Flowers of Cape*, Key List of Flowers..Chincherinchees (Ornithogalum). **1942** [see *chinkerinchee]. **1949** M. Masson *Narrowing Lust* xiii. 129 Even the creamy Chincherinchee glowed with a depth of colour that accentuated its whiteness and made it striking. **1959** *Times* 7 Mar. 1/7, 25 Chincherinchees 10s.

Chinchilla. Add: **3.** A variety of silver-coloured domestic cat; also *attrib.*, as *chinchilla Persian.*

1889 H. Weir *Our Cats* 136 The Abyssinian Silver Gray, or Chinchilla, is the same [as the Abyssinian] in all points, with the exception of the ground colour being silver instead of brown. **1893** J. K. Jerome *Novel Notes* vi. 136 The Chinchilla was a lady cat. **1903** F. Simpson *Book of Cat* xii. 137/1 Perhaps no breed or variety of cat has been so much thought about, talked about, and fought about in the fancy as the silver or chinchilla Persian. **1935** E. B. Simmons *Cats* xxviii. 144 A good chinchilla is of a pale, unshaded silver, with green eyes. **1958** E. F. Daglish *Pet-Keeper's Man.* 1 The chinchilla is white with black-tipped hairs on the head, ears, back, sides, and tail.

4. (In full *Chinchilla rabbit.*) A variety of rabbit bred for its fur; also, the fur obtained from this animal.

1904 J. R. A. Davis *Nat. Hist. Animals* VII. lxvii. 243 Some of the races, especially Chinchillas and Angoras, are valued on account of their fur. **1920** *Bazaar, Exch. & Mart* 16 July, Suppl. 74/3 The Havana (that beautiful brown rabbit), and..the Chinchilla. **1921** H. A. Day *Up-to-Date Rabbit-Keeping* 13 The Chinchilla-Coney..is a new French breed that deserves attention. **1922** *Bazaar, Exch. & Mart* 27 Oct. 430/3 In the autumn of 1919, certain enterprising lady fanciers introduced from France some specimens of a new breed called Chinchilla rabbits. **1927** *Smallholder* 26 Mar. 123/1 In Beverens and Chinchillas many specimens only moult twice in the year. **1959** *Chambers's Encycl.* VI. 127/2 The Chinchilla was first discovered as a mutation in Serbia and imported into France about 1912... Chinchilla rapidly became the most popular and best selling skin.

chin chin, *sb.* Add: Also **chin-chin**. (Later examples.) Also used as a drinking toast. Hence as *v. intr.*, to say 'chin-chin'.

1892 *Cornhill Mag.* Sept. 268 We 'chin-chinned' over foaming beakers. **1909** Ware *Passing English* 73/1 *Chin-chin* (Naval—passed into club society). 'Hail!' 'Good health!' 'Here's to you.' **1929** J. B. Priestley *Good Compan.* ii. vii. 439 Chin-chin, Effie my dear, and all the best for Xmas! **1938** Hemingway *Fifth Column* (1939) i. ii, Downa hatch. Cherio. Chin chin. **1962** 'M. Innes' *Connoisseur's Case* iii. 34 Going on your way, are you? Well, chin-chin! **1966** 'R. Standish' *Widow Hack* xv. 164 'Will you please order me a drink, Mike?' For the sake of appearances we chin-chinned and tried to look gay. **1967** P. Jones *Fifth Defector* iv. 36 Two glasses appeared, with ice tinkling in the Scotch. Paul raised his, smiling. 'Chin chin.'

chin-chin: see *chin *sb.*¹ 1 e.

chinchy, *a.* Delete †*Obs.* and add later examples. Now chiefly *U.S. colloq.*

1915 *Dial. Notes* IV. 214 We agents think the Security Mutual's chinchy but I reckon it's a good thing for the policy holders. **1952** H. Sinclair *Music out of Dixie* iii. 71, I aims for the piano player to stay on the stool an' earn his pay. Same time, I don't aim to be chinchy. **1961** F. G. Cassidy *Jamaica Talk* vi. 123 *Chinchy*, in this sense, is still very much alive in the American south (Georgia and other states), therefore not a Jamaicanism.

Chindit (tʃi·ndit). [ad. Burmese *chinthé*, a mythological creature.] A member of an Allied force fighting during the 1939–45 war behind the Japanese lines in Burma.

1943 *Hutchinson's Pict. Hist. War* 12 May–3 Aug. 171 Some of Brigadier Wingate's doughty Chindits on their way back from the dangers of the jungle. **1944** A. Jacob *Traveller's War* xxii. 325 Shortly before I had never heard of Wingate or his Chindits as, at his suggestion, we began to call them. (Chindits are the fabulous lions which guard the Burmese pagodas.) **1959** *Sunday Times* 5 Apr. 9/6 Wingate..was in the autumn of 1942 looking for a name for his force. He was told that 'chinthé' was the Burmese word for lion. 'Chindit?' he asked, doubtfully...By the next spring..Mr. Alaric Jacob, who had talked with Wingate, was referring to the force as 'The Chindits'.

chine, *sb.*³ Add: **2. b.** Of a (flat-bottomed) ship: see quot. 1927. Of a flying boat, the extreme side member of the bottom of the hull running approximately parallel to the keel in side elevation.

1911 J. Barten *Nautisches Taschen-Wörterbuch* i. 43/1 *Chine*, die Rundung des Leibholzes. **1920** *Jrnl. R. Aeronaut. Soc.* XXIV. 465 When the A.D. boat was fully loaded the water-line came up above the chine. **1927** G. Bradford *Gloss. Sea Terms* 56/2 *Chine*, the line of intersection between the sides and bottom of a flat-bottomed boat; the angle in the planking of a V-bottomed boat. **1933** *Jrnl. R. Aeronaut. Soc.* XXXVII. 861 Increase of resistance..can be reduced by provision of longitudinal steps or scallops..so that at high speed the chines are clear. **1950** *Engineering* 3 Mar. 229/2 The wooden hull was carvel-built, with a sharp chine, flat bottom, [etc.].

3. (sense *2 b) *chine-bilge, -piece, -strut.*

1932 *Times Lit. Suppl.* 24 Mar. 216/4 If a flat bottom and a chine-bilge are the hall mark [of a barge] the Humber keel and sloop are included. **1948** R. de Kerchove *Internat. Maritime Dict.* 135/2 *Chine piece*, a longitudinal piece which runs from stem to stern where the side and bottom frames join in a V-bottom boat. Also called chine log. **1931** *Flight* 23 Jan. 83/2 Chine struts have always been troublesome by reason of liability to damage by attendant boats.

chiné (ʃine·), *a.* and *sb.* [Fr., pa. pple. of *chiner*, f. *Chine* China.] **A.** *adj.* Of silk: dyed or woven with a mottled or indistinct pattern after an actual or supposed Chinese fashion. **B.** *sb.* Chiné fabric.

1852 E. Twisleton *Let.* 3 July (1928) i. 15 My little chené [*sic*] silk..would have made another more suitable toilette. **1858** Simmonds *Dict. Trade, Chiné*, goods of worsted, cotton, silk, and linen, with printed warps. **1862** *Catal. Internat. Exhib., Brit.* II. No. 3901, Plain and figured chinés. **1862** S. Hale *Lett.* (1919) 13 Sheer white muslin, most elaborately trimmed with brilliant rose and Chine ribbon, round the bottom of the skirt. **1896** *Daily News* 9 May 8/6 The chiné ribbon is in pink flowers, with a green border. **1900** *Ibid.* 8 Sept. 6/3 The coats are lined with chiné silks. **1940** *Chambers's Techn. Dict.* 157/1 *Chiné*, a fancy silk material in which the patterns are printed on the warp threads, before weaving. **1960** Cunnington & Beard *Dict. Eng. Costume* 250/1 *Chiné silk*, a silk of which the pattern has the appearance of having 'run'. **1968** J. Ironside *Fashion Alphabet* 219 Chiné fabrics are those in which the warp threads are dyed or printed before weaving.

Chinee: see Chinese B. 1 b.

Chinese, *a.* Add: **2. Chinese artichoke**, *Stachys affinis* (formerly *S. sieboldii*), a herb of the family Labiatæ, cultivated as a vegetable; **Chinese block**, an oblong slotted wooden block used esp. by jazz drummers; **Chinese box**, one of a nest of boxes: see Nest *sb.* 6; **Chinese cabbage**, one of two brassicas, *Brassica pekinensis* or *B. chinensis*; **Chinese Chippendale**, Chippendale furniture combining straight lines with Chinese bamboo and lattice motifs; **Chinese copy**, a precise drawing of a structure, piece of apparatus, etc., made from its appearance only without other information; a slavishly exact copy (see also quot. 1920); **Chinese cut** *Cricket* (see quot 1956); **Chinese gooseberry**, the N.Z. name for the plant and fruit of *Actinidia chinensis*, a deciduous fruiting vine; **Chinese lantern (plant)**, *Physalis alkekengi*, a plant of the family Solanaceæ, grown for the decorative effect of the orange-coloured, inflated calyx; **Chinese laundry**, a laundry operated by Chinese; **Chinese layering** = *air-layering* (*Air *sb.*¹ B. II); **Chinese primrose** (examples); **Chinese puzzle** (see Puzzle *sb.* 3 b); **Chinese red** = Chrome *red*; **Chinese vermilion**, a bright, yellowish-red pigment; **Chinese wall** [after the defensive wall built between China and Mongolia in the 3rd c. B.C.], *transf.* and *fig.* an insurmountable barrier (to understanding, etc.).

1902 L. H. Bailey *Cycl. Amer. Hort.* IV. 1714/1 Chorogi. Chinese or Japanese Artichoke. Knotroot. **1950** Metcalfe & Chalk *Anat. Dicotyledons* II. 1051 The tubers of the so-called Chinese artichoke (*Stachys sieboldi* Miq.) are edible. **1926** *Leedy Mfg. Co. Catal.* 'N' 36 Chee Foo Chinese Blocks are of the same high-grade quality as all other drummers' accessories of this brand. **1946** R. Hill in A. L. Bacharach *Brit. Music* vii. 107 The use of the Chinese block in the finale has led some undiscerning critics to claim the influence of jazz. **1829** R. C. Sands *Writings* (1834) II. 57 Some of the members..thus compacted like Chinese boxes. **1921** V. Woolf *Mon. or Tues.* 78 Like a vast nest of Chinese boxes..turning ceaselessly one within another the city murmured. **1964** M. A. K. Halliday et al. *Ling. Sci.* ii. 34 There is no rank scale in lexis: no 'Chinese box' arrangement of pattern-carrying units. **1842** J. C. Loudon *Suburban Horticulturist* v. 627 The Chinese Cabbage..is an annual, apparently intermediate between the cabbage and the turnip. **1885** W. Robinson tr. *Vilmorin-Andrieux's Veget. Garden* 147 Chinese Cabbage, or Pe-tsai..differs entirely in appearance from the Cabbages of Europe, being rather like a Cos Lettuce in aspect. **1950** *N.Z. Jrnl. Agric.* Dec. 523/3 Chinese cabbage is not grown to a great extent by home gardeners. **1922** S. Lewis *Babbitt* v. 59 The reading-room in Chinese Chippendale. **1924** A. E. W. Mason *House of Arrow* xiii. 153 The writing-table of Chinese Chippendale. **1943** D. Welch *Maiden Voyage* vii. 52 The librarian was showing us some Chinese Chippendale chairs. **1920** *Hansard Commons* CXXXV. 1490 A Chinese copy..is..a copy with the words struck out showing the alterations. **1964** *Evening Standard* 14 Oct. 11/3 Although these missiles are being bought from the United States each consignment will include spare parts sufficient for four or five years. This means that..there would be time to manufacture in this country what the Navy calls 'Chinese copies'. **1937** *Daily Express* 28 Jan. 18/2 Farnes, bowling really fast, twice took Leyland's leg stump in a few minutes, the batsman trying to make what Duckworth called 'Chinese cuts'. **1956** R. Alston *Test Commentary* ix. 62 Oakman edged an intended off-drive past the leg-stump for four, a stroke variously described as a Chinese cut, a Surrey cut or a Staffordshire cut. **1925** *Weekly News* (Auckland) 1 Oct. 62 The Chinese gooseberry..has been introduced into Auckland. **1944** *N.Z. Jrnl. Agric.* Feb. 111/1 The Chinese gooseberry, the fruit of a vine imported from China, is gaining increasing popularity in New Zealand. *Ibid.* 112/3 In many ways the Chinese gooseberry resembles the grape-vine. **1901** L. H. Bailey *Cycl. Amer. Hort.* III. 1320/2 P[*hysalis*] *Alkekengi* var. *Franchetii*, Hort. Chinese Lantern Plant. **1936** F. Clune *Roaming round Darling* xvii. 163 Shrubs of Chinese lanterns with pods that crack when squeezed. **1881** Lady Monkswell *Diary* 12 Nov. in *Vict. Diarist* (1944) 96 When I passed through New York before I had not noticed the Chinese laundries. **1961** *Times* 2 Oct. 3/4 The Widow Twankey in her Chinese laundry. [**1886** G. Nicholson *Illustr. Dict. Gardening* II. 244/1 This is sometimes called the Chinese method of Layering, presumably on account of it being more generally adopted in China.] **1900** L. H. Bailey *Cycl. Amer. Hort.* II. 894/2 This kind of propagation is known as air-layering, Chinese layering or circumposition. **1934** Chinese layering [see *air-layering*, *Air *sb.*¹ B. II]. **1825** *Curtis's Bot. Mag.* LII. 2564 (*heading*) *Primula Sinensis*. Chinese Primrose. **1858, 1887** Chinese Primrose [see Primrose *sb.* 2]. **1952** A. G. L. Hellyer *Sanders' Encycl. Gardening* (ed. 22) 397 P[*rimula*] *sinensis*, 'Chinese Primrose', various [colours], winter and spring, 9 in., China. **1892** W. Crookes tr. *R. von Wagner's Man. Chem. Technol.* iii. 480 Conspectus of inorganic pigments...Reds...Vermilion, cinnabar, Chinese red (HgS). **1935** *Discovery* Aug. 240/2 A wooden fence painted in Chinese red. **1886** H. C. Standage *Artists' Man. Pigments* v. 48 *Chinese vermilion*, or Carmine vermilion. **1962** R. G. Haggar *Dict. Art Terms* 75/1 *Chinese vermilion*, genuine vermilion. **1907** C. J. Bullock *Sel. Readings in Econ.* xx. 672 The cheap prices of its commodities are the heavy artillery with which it batters down all Chinese walls. **1934** T. S. Eliot *Elis. Essays* 22 After the erection of the Chinese Wall of Milton, blank verse has suffered not only arrest but retrogression. **1941** Auden *New Year Let.* ii. 34 They fought over their premises, Shut out from Eden by the bar And Chinese Wall of Barbara. **1941** Koestler *Scum of Earth* 269 It succumbed..to a social phenomenon, which it might be appropriate to call the 'Chinese Wall psychosis'.

chinesery (tʃəinī·zəri). [ad. F. *chinoiserie*, f. *chinois* Chinese + *-erie* -ery.] = Chinoiserie.

1890 *Sat. Rev.* 15 Feb. 191/2 The Treasury..goes on with its old Chineseries. As for this particular Chinesery, it seems to us neither better nor worse than a round dozen of others we have seen. **1907** *Westm. Gaz.* 1 Oct. 2/1 The 1820 romanticalities of Wörlitz, and the dapper blue-and-yellow chineseries of Oranienbaum.

Chink (tʃiŋk), *sb.*⁵ *slang.* Also **chink.** [Irreg. f. China¹.] A Chinaman. Also *attrib.* (Derogatory.)

1901 *Munsey's Mag.* XXIV. 536 The leader suggested the 'chink', and to the one Chinese laundry..the little band departed. **1910** W. M. Raine *B. O'Connor* iv. 41 Chinks, greasers, and several other kinds of citizens driftin' that way. **1919** *War Slang* in *Athenæum* 8 Aug. 727/2 'Chinks' for Chinese labourers. **1922** J. S. Fletcher *Ravensdene Court* xiv. 173 'A—Chink?' 'He means a Chinaman,' I said. **1926** *Chambers's Jrnl.* 552/1 The towns, small or large, possessed from one to hundreds of 'Chink' laundries. **1932** J. Dos Passos *1919* 17 The Barman was a broadfaced Chink. **1936** 'R. Hyde' *Passport to Hell* 229 The little Chinks hated the Boche like hell. **1945** [see *chow-chow *sb.* 3]. **1969** J. Durack in *Coast to Coast 1967–8* 99 We used to have a couple staying with us. Chinks, they were, medical students.

chink, *sb.*⁶ *S. Afr.* Short for *chinkerinchee. *colloq.*

1949 L. G. Green *Land of Aft.* v. 73 'Chinks' grow only in the Western Province. **1960** C. Lighton *Cape Floral Kingdom* xiii. 117 The 'chinks' have always been welcome in Britain as a change from the usual run of early winter flowers.

chinkara (tʃi·ŋkārə). Also **chi(c)kara.** [Hindi *ćinkārā*, f. Skr. *chikkāra* a kind of antelope.] An Indian gazelle, *Gazella gazella bennetti*.

1860 *Chambers's Encycl.* I. 287/2 The Chikara and some other Indian species are distinguished by two additional rudimentary horns in front of the ordinary horns. **1911** *Encycl. Sport* I. 411 The Chickara, Chinkara, Ravine Deer, or Indian Gazelle (*G. bennettii*) [*sic*], is found throughout the plains of India. **1924** *Blackw. Mag.* Aug. 242/1 Neither I nor the Kumar had yet stalked black buck or chinkara in the flaming dāk jungle. **1957** *Encycl. Brit.* XII. 156/1 The gazelle (*Gazella bennetti*), the chikara, or 'ravine deer', of sportsmen.

chinkerinchee (tʃi:ŋkərintʃī·). *S. Afr.* Also **chincherinchee** (*q.v.), **chinkerichee, chinkeringching.** [Said to be onomatopœic, from the squeaky sound produced when two flower-stalks are rubbed together.] A popular name for the liliaceous bulbous plant *Ornithogalum thyrsoides*, bearing white to golden-yellow flowers in a dense 12- to 30-flowered raceme.

[**1793** tr. *Thunberg's Trav. 1770–9* I. 153 *Tintirinties* is a name given to a species of *Ornithogalum*, with a white flower, from the sound it produced, when two stalks of it were rubbed against each other.] **1904** *Cape G. Hope Agric. Jrnl.* July 6 (Pettman), The *Chinkerinchee, Chincher-and-ching*, 'Viooltjes', as that beautiful white flowering bulb, the *Ornithogalum thyrsoides*, is variously called in South Africa. **1913** Pettman *Africanderisms* 120 *Chinkering ching*, *Ornithogalum thyrsoides*. The popular name of this plant in the Western Province. **1915** Marloth *Flora S. Afr.* IV. 106 The best known species is *O. thyrsoides*, called the star of Bethlehem, but more familiar as viooltje or chinkerichee. **1923** *Daily Mail* 3 Dec., The South African flower the chinkerichee. **1942** *Amer. Speech* XVII. 66/2 For some years now the catalogs of dealers in bulbs have admonished us that a recognized American name for the flowering bulb *Ornithogalum* is *chinkerichee* (as I saw it first) or *chincherichee* (as I see it nowadays).

Chinkey, Chinkie, Chinky (tʃi·ŋki). *slang.* [f. *Chink *sb.*⁵ + -ie, -y⁶.] = *Chink *sb.*⁵ (Derogatory.)

1879 W. J. Barry *Up & Down* vii. 51 We..had a good passage to Hong-Kong. When we arrived, the first Chinese war with Britain had broken out, and there was every appearance of plenty of fun to be shortly had with the Chinkies. **1882** A. J. Boyd *Old Colonials* 233 The pleasant traits of character in our colonialised 'Chinkie', as he is vulgarly termed. **1899** Boxall *Austral. Bushrangers* 241 They rode straight to the Chinese camp at Wombat, 'to give the Chinkies a lesson'. **1905** *Westm. Gaz.* 26 Sept. 6/1 The farmers getting a reward of £1 for each 'Chinkey' they turn over to the police. **1923** D. H. Lawrence *Kangaroo* xvi. 351 Brother Brown and Chinky and all the rest: the Indians in India, the niggers in the Transvaal. **1945** [see *chow-chow *sb.* 3]. **1959** N. Mailer *Advts. for Myself* (1961) 353 A certain Chinkie.

chinless, *a.* **b.** Add to def.: Esp. in phr. *chinless wonder*.

1962 *Times* 22 Nov. 18/7 Debutantes and their chinless wonders of husbands—just down from Oxford. **1964** J. Gardner *Liquidator* vii. 137 Quadrant..was a chinless wonder of the first water. **1967** J. Burke *Till Death us do Part* v. 90 We don't need a lot of chinless wonders sitting up in the House of Lords.

chinny, *a.* Add: **2.** Marked by a conspicuous chin.

1897 G. du Maurier *Martian* 158 Her narrow chinny face. **1920** Galsworthy *In Chancery* i. i, A face concave and long, with a jaw which divested of flesh would have seemed extravagant: altogether a chinny face, though not at all ill-looking.

chino (tʃi·no, ʃi·no), *sb.*² *U.S.* [Amer. Sp. *chino* toasted (referring to its usual colour).] A cotton twill cloth, usu. khaki in colour; also (usu. *pl.*), a garment, esp. trousers, made of this cloth. Freq. *attrib.*

1943 T. R. St. George *C/o Postmaster* (1944) 26 Living as we did, mostly in a life belt and a filthy pair of chinos, the one thing we needed most was a strong blast of Chanel No. 5. **1957** M. B. Picken *Fashion Dict.* 61/2 *Chino cloth*, cotton twill fabric of combed yarns dyed khaki color, mercerized and sanforized. Used for uniforms. **1957** J. Kerouac *On Road* (1958) vii. 41 Wearing only my chino pants. **1961** D. Hamilton *Removers* (1962) xv. 84 Her skirt was nicely tailored of some fine khaki material, or maybe the stuff is called chino when it joins the aristocracy. **1969** E. Lathen *When in Greece* vii. 71 The unofficial uniform of the Quaker camp—a white shirt, faded chino pants, and field boots.

Chino- (tʃəi·no), used as combining form = Chinese. *Chino-Japanese* adj., Chinese and Japanese.

1889 *N.E.D.* s.v. *Chino* 2, Chino-Japanese. **1893** *Graphic* 11 Mar. 260/1 The tribes on the Indo and Chino-Burmese frontiers. **1895** H. Norman *Far East* 604 Chino-French relations at Monkay. **1904** *Daily Chron.* 8 Apr. 4/5 The Chino-Japanese war. **1904** *Westm. Gaz.* 14 Oct. 8/2 The Chino-Japanese and Australian service at Vancouver.

Chinook. *chinook wind*: substitute for def.: a warm, dry, and often turbulent wind that blows for hours or days at a time down the eastern slopes of the Rocky Mountains and over the adjacent plains (see also quot. 1967); also *ellipt.* (Add earlier and later examples.)

1860 A. J. Thibodo *Diary* 1 Apr. in *Pacific Northwest Q.* (1940) XXXI. 347 Pleasant warm weather, high winds from S.W., they call it the Chinook wind. **1876** *Idaho Daily Avalanche* (Silver City) 18 Mar. 2/2 In winter however it is somewhat blustery, the wild, yet balmy chinook, sweeping with intense force from the open south and west country. **1932** Fuller & Conard tr. *Braun-Blanquet's Plant Sociol.* v. 97 The chinook wind in Alberta, Canada, may raise the temperature in a very short time from −10° to more than 20° C. **1936** D. McCowan *Anim. Canad. Rockies* x. 87 The potency of the Chinook winds. *Ibid.*, Under a Chinook the snow does not melt, it evaporates. **1967** R. W. Fairbridge *Encycl. Atmos. Sci.* 1151/2 Chinook winds may occur in summer but are less noticeable then. The term *chinook* has mistakenly been used for any warm and moist oceanic wind on the windward side of the [Rocky] mountains.

b. In full *Chinook salmon* = Quinnat.

1851 *San Francisco Picayune* 15 Oct. 2/5 A supply of Chenook salmon. **1881** *Amer. Naturalist* XV. 177 Quinnat—..Chinnook salmon, Columbia River salmon, Sacramento Salmon. **1884** Goode *Nat. Hist. Aquatic Anim.* 479 On the Columbia River the name 'Chinook Salmon' is in universal use. **1889** *Oregonian* (Portland) 4 Nov. 5/1 He..landed after a most exciting fight, a chinook or chisel-mouth of large size. **1911** F. J. Haskin *Amer. Govt.* xii. 151 The bureau of fisheries..has tried to establish the chinook salmon in Atlantic coast waters. **1936** J. T. Jenkins *Fishes Brit. Is.* (ed. 2) 24 In the United States... In 1872 the Bureau of Fisheries distributed 30,000 eggs of the Chinook salmon. **1940** *Geogr. Jrnl.* XCVI. 241 Forty-two million salmon of one variety (chinook) alone. **1955** *Sci. Amer.* Aug. 72/3 The other five salmon species, all on the Pacific Coast, are the Chinook (also called the king salmon), the sockeye, the silver, the humpback, and the chum.

chinosol (ki·nǒsǫl). [a. G. *chinosol*, f. China³.] The proprietary name for 8-hydroxyquinoline sulphate (or a mixture of this with potassium sulphate), a yellow crystalline substance with antiseptic and deodorant properties.

1896 *Brit. Med. Jrnl.* 1 Feb. 285/2 Mr. B. Kühn..has sent us a sample of chinosol, a new antiseptic and deodoriser, manufactured by F. Fritzsche and Co., Hamburg. **1904** *Westm. Gaz.* 24 Nov. 7/3 The razors..are sterilised in a solution..of chinosol. **1937** A. J. Clark *Appl. Pharmacol.* (ed. 6) xxvi. 582 Quinine and chinosol, which are very widely used as contraceptives have very little spermatocidal action. **1952** *Martindale's Extra Pharmacopœia* (ed. 23) I. 887 Chinosol was originally stated to be potassium-oxyquinoline sulphate, but the content of potassium in the original article of commerce does not agree with theory.

‖**chinovnik** (tʃinǫ·vnik). Also **tchinovnik.** [Russ.] In Tsarist Russia, a government official; a civil servant; esp. a minor functionary; a clerk.

1877 D. M. Wallace *Russia* xiii. 305 A large and well-drilled army of officials. These..form a peculiar social class called Tchinovniks. **1907** J. Morley in *Recoll.* (1917) II. 229 He has much more vitality of mind than any other of your Tchinovniks. **1919** H. Walpole *Secret City* i. vi. 25 He was a Chinovnik, and held his position in some Government office with great pride and solemnity. **1959** D. Davie *Forests of Lithuania* ii. 23 When fate must send To live next door Dog-loving dull *chinovnik*.

chinquepin, var. Chincapin.

1785 H. Marshall *Amer. Grove* 125 *Quercus Prinus humilis*. Dwarf Chesnut or Chinquepin Oak. **1807** *Norfolk* (Va.) *Gaz.* 13 Nov. 4/3 For Sale, 2000 Cedar and Chinquepin Posts.

chinse, *v.* **1.** (Earlier N. Amer. examples.)

1770 G. Cartwright *Jrnl. Labrador* (1792) I. 24 Fogarty chinsed the storehouse with moss. *Ibid.* 65, I ordered some of the workmen to gather moss, and chinse the store. **1792** *Ibid.* Gloss. p. x, *Chinsing*, filling with moss the vacancies between the studs of houses, to keep out the wind and frost.

chintz-bug, var. Chinch *sb.*¹ 2 (in Dict. and Suppl.).

chintzy (tʃi·ntsi), *a.* [f. Chintz + -y¹.] Decorated or covered with chintz; suggestive of a pattern in chintz. Also in extended use: suburban, unfashionable, petit-bourgeois, cheap; mean, stingy.

1851 Geo. Eliot *Let.* 18 Sept. (1954) I. 362 The effect is chintzy and would be unbecoming. **1902** *Academy* 12 Apr. 379/1 Hunt's rhapsodical interpretation of Tickell's chintzy mythological poem on Kensington Gardens. **1923** U. L. Silberrad *Lett. Jean Armiter* vi. 135 The drawing-room was chintzy then—white walls and old china and bowls of roses and pink shaded lights. **1941** J. H. N. Collier *Presenting Moonshine* 5 A chintzy woman, possibly of literary tastes, swam forward. **1951** G. Greene *End of Affair* iii. vii. 145 He got up from his chintzy chair and came and sat with me on the chintzy sofa. **1958** B. Hamilton *Too Much of Water* ii. 30 A pleasant white-and-gold chintzy drawing-room. **1964** L. Deighton *Funeral in Berlin* xxx. 154 I'll have to revise my attitude to the chintzy old bastard.

chin up, chin-up: see *chin *sb.*¹ 1 d.

chionodoxa (kəi:ǒnodǫ·ksă). [mod.L. (E. P. Boissier *Diagnoses Plantarum Orientalium Novarum* (1844) I. v. 61), f. Gr. χιών snow + δόξα glory.] A plant of the genus so named, bearing blue flowers, valued by horticulturists for its early-blooming habit, and popularly known as 'glory of the snow'.

1879 *Curtis's Bot. Mag.* CV. 6433 *Chionodoxa* has proved itself perfectly hardy in the open ground. **1893** *Daily News* 28 Mar. 2/2 Scillas and Chionodoxas now in full glory. **1961** *Amateur Gardening* 16 Sept. 1/1 Bulbs..not usually suitable for forcing, such as..chionodoxa.

chip, *sb.*¹ Add: **1. b.** *spec.* In gem-cutting, a cleavage which weighs less than three-fourths of a carat (*Cent. Dict.* Suppl. 1909).

c. *pl.* Small pieces of kauri-gum.

1916 *Chambers's Jrnl.* Aug. 542/1 The chief factor influencing sales and prices of kauri-gum in recent years has been the request for the poorer qualities, called 'chips' and 'dust'.

2. b. *pl.* (rarely *sing.*). Fried pieces of potato, usu. oblong in shape; = *French fried potatoes, French fries* (*French *a.* 3); also (chiefly *U.S.*) = *crisp *sb.* 7. Cf. *chip-potatoes*, *fish and chips* (*fish *sb.*¹ 7).

1859 Dickens *T. Two Cities* i. v. 19 Husky chips of potatoes, fried with some reluctant drops of oil. **1886** *Chambers's Jrnl.* 18 Dec. 808/2 The hand of Zacharias was betrayed in potato chips and cunning sauces. **1893** *Funk's Stand. Dict.* s.v. *Chip*¹, *Saratoga chips*, potatoes sliced very thin while raw, and fried crisp. **1899** W. C. Morrow *Bohemian Paris* 224 Here are..fried-potato women, serving crisp brown chips. **1906** *Daily Chron.* 13 Oct. 5/7 The chop and chips business. **1951** *Good Housek. Home Encycl.* 611/1 Place the basket of chips in it [*sc.* fat] again and fry them until they are golden-brown. **1967** *Tit Bits* 11 Feb. 7/4 She took a sudden fancy for fish and chips...'It tastes better out of a newspaper.'

c. (Earlier example.)

1835 *Chambers's Edin. Jrnl.* IV. 98/2 The *log* is simply a flat piece of wood of the form of a quarter of a circle...The wood is called the chip.

d. (Earlier and later examples.) Also, a piece of money; *pl.* money. So *to buy chips*: to invest; (*when*) *the chips are down*: (when) it comes to the point (see Point *sb.*¹ A. 22 b); *in the chips*: moneyed, affluent. *colloq.* (orig. *U.S.*).

1840 in *Amer. Speech* (1965) XL. 182 He is perfectly willing to 'pile the chips' for any distance. **1859** Hotten *Dict. Slang, Chips*, money. **1877** W. Black *Green Past.* III. xiv. 226 The chips must be paid. **1882** [see Straddle *v.* 7]. **1903** A. H. Lewis *The Boss* 205 There was a saw-bones here,..pawin' me over for a life insurance game that I thought I'd buy chips in. **1905** *Daily Chron.* 11 Sept. 2/6 It is..quite a commonplace remark to hear young men boast of the time when 'the old man turns up his toes', and they can 'collar the chips'. **1919** W. H. Downing *Digger Dial.* 58 *Chip*, a rupee. **1925** *Chambers's Jrnl.* 424/2 I'll give you ten chips myself. **1938** A. J. Liebling *Back where I came From* 68 Never to have been in the chips marks one as a punk. **1943** G. Marx *Lett.* (1967) 49 The previews are always more fun than the actual shows. Since the chips are not down, everyone is at ease and the audience senses it. **1945** H. I. Phillips *Pte. Purkey's Private Peace* 10 He had..established himself as a tough battler when the chips were down. **1949** in Wentworth & Flexner *Dict. Amer. Slang* (1960) 101/2 When the chips are down a man shows what he really is. **1950** J. Dempsey *Championship Fighting* 179 But when the chips are down, rhythm is destroyed. **1957** Wodehouse *Over Seventy* iv. 55 When I was in the chips and an employer of butlers. **1959** *Spectator* 25 Sept. 399/3 For the fact is that when the chips are down, the Right wing of the Tory Party comes up.

e. Slang phrases: *to pass* (or *hand*) *in one's chips* (U.S.), to die (cf. Check *sb.*¹ 15); *to have had one's chips*, to be beaten, finished, killed, etc.

1879 *Missouri Republican* (St. Louis) 22 Oct. 3/7 If you wish to express the demise of a friend..in Southern Colorado..it would be more elegant to say that 'he'd passed in his chips'. **1890** *Harper's Mag.* Feb. 351/2 Ye kin bet yer life I ain't afeard o' passin' in my chips. **1891** Farmer *Slang* II. 95/1 *To hand in one's chips*, to die. **1907** Mulford *Bar-20* xi. 130 He passed in his chips last night. **1936** J. A. McKenna *Black Range Tales* 56 Several cowboys passed in their chips in that snowstorm. **1936** *Amer. Speech* XI. 200/2 Handed in his chips. **1959** I. & P. Opie *Lore & Lang. Schoolch.* xvii. 376 You've had your chips. **1960** G. Durrell *Zoo in Luggage* viii. 189 'Cor!' said the constable, in a voice of deep emotion, 'I thought I'd 'ad me chips that time.' **1961** *Technology* May 131/1 The St. Albans lot will have had their chips by this time next week.

f. *Electronics.* A tiny square of thin semi-conducting material which by suitable etching, doping, etc., is designed to function as a large number of circuit components and which can be incorporated with other similar squares to form an integrated circuit.

1962 B. G. Bender in G. W. A. Dummer *Micro-miniaturization* 139 A device must have a semiconductor element in any event, but economical techniques have been developed for producing chips for the Microseal component effort which have advanced the state of the art. **1967** *Times Rev. Industry* Mar. 46/1 The size of the wafers varies, but it is not uncommon for one about the size of a penny to carry several hundred tiny squares known as 'chips', each containing anything from about 20 to perhaps 600 components. **1970** *Sci. Amer.* Feb. 22/1 Ten years ago ..a chip of silicon a tenth of an inch square might hold 10 to 20 transistors, together with a few diodes, capacitors and resistors. Today such chips can contain well over 1,000 separate electronic components.

4. b. Also for making baskets. Hence, ellipt. for *chip basket* (see 9 below).

1922 J. Joyce in *Q. Rev.* Oct. 230 Chips of strawberries. **1923** *Westm. Gaz.* 15 Jan. 8/5 4lb. chips of apples. **1928** *Daily Express* 28 May 5/3 In Hampshire..these baskets contain two to four pounds, and are termed 'chips'. **1955** *Times* 14 Aug. 10/4 The fat fruits were laid out in their regulation blue-lined punnets and chips.

5. c. A piece of dried buffalo- or cattle-dung. *N. Amer.* (Cf. *buffalo-chips* s.v. Buffalo 5.)

1846 E. Bryant *What I saw in Calif.* (1848) vii. 102 Bacon broiled on a stick over a fire of buffalo chips. **1857** W. Chandless *Visit Salt Lake* I. iv. 61 Buffalo were looked for; a solitary 'chip'—so the buffalo droppings are called—found one evening caused quite an excitement in our camp. **1903** A. Adams *Log Cowboy* xiv. 209 We had begun to feel the scarcity of wood for cooking purposes... These chips were a poor substitute. **1960** D. E. Bublitz *Life on Dotted Line* xv. 98 There were no chips or wood around, so we meandered into the schoolhouse.

6. b. Now freq. *a chip off the old block.*

1929 H. E. Bates *Seven Tales & Alex.* 40 He's my son, and he's a chip off the old block, and I'm proud of him. **1947** W. S. Maugham *Creatures of Circumstance* 7 His heir was a nephew..not a bad boy, but not a chip off the old block, no, sir, far from it. **1966** D. Varaday *Gara-Yaka's Domain* v. 51 He was a true chip off the old block.

c. *pl.* (A nickname for) a carpenter, esp. on a ship. (Usu. with capital initial.)

1785 in Grose *Dict. Vulg. Tongue.* **1851** *Chambers' Paper* No. 52, 20 (Farmer), The carpenter..was not offended..at being called *chips* even by the black cuddy servant. **1927** J. Sampson *Seven Seas Shanty Bk.* 47 A handy Chips to drive the nails. **1943** C. H. Ward-Jackson *Piece of Cake* 21 *Chips, chippy* or *chippy rigger*, an airman of the trade of carpenter or carpenter rigger. **1959** I. & P. Opie *Lore & Lang. Schoolch.* xvii. 362 The gardening master is commonly 'Spuds', the wood-work teacher is 'Chips'.

8. *a chip on one's shoulder*, carried as a challenge to others (see earlier quots.); hence, a display of defiance or ill-humour; an unforgotten grievance; a sense of inferiority characterized by a quickness to take offence. orig. *U.S.*

1830 *Long Isl. Tel.* (Hempstead, N.Y.) 20 May 3/5 When two churlish boys were *determined* to fight, a *chip* would be placed on the shoulder of one, and the other demanded to knock it off at his peril. **1855** *Weekly Oregonian* 17 Mar. (Th.), Leland, in his last issue, struts out with a chip on his shoulder, and dares Bush to knock it off. **1887** *Harper's Mag.* Oct. 658/1 The way that dog went about with a chip on his shoulder..was enough to spoil the sweetest temper. **1903** *N.Y. Sun* 1 Nov., Who, they say, wears a chip on his shoulder because he didn't get the Republican nomination for City Treasurer. **1905** *Amer. Illustr. Mag.* Nov. 88 Each boy had a sort of chip-on-the-shoulder air. **1930** W. S. Maugham *Gentl. in Parlour* xliv. 271 He was a man with a chip on his shoulder. Everyone seemed in a conspiracy to slight or injure him. **1952** W. Sprott *Social Psychol.* 220 If you are spoiling for a fight you go about with a 'chip on the shoulder' challenging anyone to knock it off. **1956** J. Cannan *People to be Found* ix. 148, I got him the sack—months ago but all this time he's been carrying a chip on his shoulder.

9. *chip bonnet* (examples), *-hat* (earlier and later examples); **chip basket**, a basket made of strips of thin wood roughly interwoven or joined, used chiefly for packing fruit for the market; also *attrib.*; **chip-bird** *U.S.* = *chipping bird*, Chipping *ppl. a.* 2; **chip-board**, a type of pasteboard made by compressing waste paper or wood refuse; **chip-carving**, carving in which the patterns are produced by chipping out the material; so **chip-carve** *v. trans.*, **chip-carved** *ppl. a.*, **chip-carver**; **chip-heater** *Austral.*, a domestic water-heater which burns wood chips; **chip-potatoes** = *chips* (*2 b); **chip-shop**, a shop selling fish and chips; **chip-shot** *Golf*, a short lofted approach-shot on to the putting-green; also, a similar lofted shot in *Association Football*, etc.; also *fig.*; **chip-sparrow** *U.S.* = *chipping-sparrow* (*Chipping *ppl. a.* 2); **chip-straw** (cf. Chip *sb.*¹ 4); **chip-yard** *U.S.*, a wood-cutting yard.

1921 *Dict. Occup. Terms* (1927) § 472 *Chip basket maker*; ..plaits together..narrow strips of shavings of thin wood to make..chip baskets. **1824** *Massachusetts Yeoman* 28 Apr. (Th.), The destruction of a robin, chip, blue or black bird is not all. **1871** J. R. Lowell *My Study Windows* 15 The only bird I have ever heard sing in the night has been the chip-bird. **1872** Coues *Key N. Amer. Birds* 142 Chipping Sparrow, Chipbird or Chippy. Hairbird. **1919** *Fibre Containers* IV. x. 26 (*heading*) Manufacture of jute and chip board. **1953** *Archit. Rev.* CXIII. 398 The floor finishes downstairs..are quarry-tiles and plastic-bonded chipboard. **1954** *Paper Terminology* (*Spalding & Hodge*) 17 *Chipboard*, a cheap quality of board made from mixed waste and used in the manufacture of cartons, etc. **1820** S. Breck in *Recoll.* (1877) 298 She has gone on..bedecking herself in merino shawls, chip bonnets, &c. **1845** M. M. Noah *Gleanings* 65 On her little head she wore a good sized chip bonnet, decorated with artificial flowers. **1948** R. Heine-Geldern *Indonesian Art* 157 Water jar..chip-carved when the ware was in the 'leathery' stage before firing. **1951** T. H. Ormsbee *Field Guide to Early Amer. Furnit.* vii. 182 Chip Carved Chest of Drawers. Has stile and rail construction. Carcase contains two or three full-width drawers that are fitted with turned wooden knobs. **1957** Childe *Dawn Europ. Civilization* (ed. 6) vi. 95 There are indeed analogies to Veselinovo handles and to the 'chip-carved' lamps in Anatolia. **1961** *Antiquaries Jrnl.* XLI. 107 The 'chip-carved' buckles, cast and not chiselled by hand, were mass-produced by provincial Roman workshops for the army of the late fourth century. **1892** E. Rowe *Chip-Carving* 47 Instructions to the chip-carver. **1888** *Queen* 29 Sept., Suppl., Chip-Carving. **1902** *Westm. Gaz.* 26 Apr. 7/3 Examples of bent-iron work, and wood and chip carving. **1927** F. B. Young *Portrait of Clare* 437 When she came to look more closely, there was the face of Edith Wilburn framed, not in plush, but chip-carving. **1935** *Discovery* Nov. 337 Examples of Kentish chip-carving jewellery of the Anglo-Saxon period. **1759** *Newport Mercury* 26 June 4/3 Imported ..silver'd Paper, chip and black Satten Hats. **1880** *Drapers' Jrnl.* 27 May 5/2 A chip hat to match the dark shade of broché. **1913** C. Mackenzie *Sinister Street* iv. vi. 987 That girl in her chip hat, holding a bunch of cherries. **1946** K. Tennant *Lost Haven* (1947) vii. 95 Grandpa led the way to the bathroom and expected the guest to admire the bath and the chip-heater! **1965** G. McInnes *Road to Gundagai* v. 76 The woodshed for the logs to be split into chips for the copper and for the 'chip-heater' in the bathroom. **1916** *Home Chat* 9 Sept. 432 Dinner 2. Salmis of Game. Potato Chips. Greengage Tart...The Chip Potatoes. Fry these as usual. [etc.]. **1953** H. Clevely *Public Enemy* xxvii. 219 Joe's chip shop. **1909** *Westm. Gaz.* 10 June 12/4 Chip shots which he laid within two feet of the pin. **1919** Wodehouse *Damsel in Distress* iii, 'Not at all,' said George, trying a sort of vocal chip-shot out of the corner of his mouth. **1963** *Times* 19 Feb. 4/7 Easton raised Gravesend's hopes with a chip shot that was only inches wide. **1852** Mrs. Stowe *Uncle Tom's C.* xvi. 191 Then she sat on his knee like a chip sparrow, still laughing. **1898** *Westm. Gaz.* 28 Apr. 3/2 Chip straw is again to the fore, indeed, forms the foundation of some of the most exclusive hats. **1850** *Knickerbocker* XXXVI. 73, I..crossed her chip-yard. **1863** Mrs. Whitney *Faith Gartney's Girlh.* xiv, The spicy smell of the chip-yard round the corner where the scraps of pine lay..under the summer sun. **1891** M. E. Ryan *Pagan of Alleghenies* xix. 232 Two men..halted out at the chip-yard.

b. Short for *chip-shot.

1909 *Westm. Gaz.* 9 June 14/3 He was short with his chip on to the green at the fourth. **1958** *Times* 30 Oct. 3/5 Littlewood..finished..winning all four holes and holing his chip at the eighteenth. **1961** *Times* 20 Feb. 14/2 His chip for the far post was helped on its way past George by Ellis's head.

chip, *v.*¹ Add: **3. e.** *intr.* To make chipping strokes.

1908 A. Bennett *Buried Alive* x. 248 Muscular, hairy males..were chipping and paring at huge blocks of stone.

7. Also *N.Z.*

1846 C. J. Pharazyn *Jrnl.* 24 Oct. 61 (MS.), Sowed 4 rods more barley chipped it in after dusk. **1849** C. Hursthouse *Settl. New Plymouth* vii. 98 For its first grain crop the seed can be 'chipped-in' for 10s. per acre more.

8. Also *trans.* Also, to 'butt' in.

1903 A. H. Lewis *The Boss* 271 Madam, let me chip in a word. **1906** E. Dyson *Fact'ry 'Ands* iii. 36 She's been goin' to marry me, more 'r less for a year, an' now you've chipped-in.

b. To put in or stake chips (*intr.* and *trans.*).

1891 'L. Hoffmann' *Cycl. Card & Table Games* 203 Each person puts up an agreed amount by way of ante...To avoid dispute as to whose turn it may be, a pocket-knife, known as the 'buck', is passed round, resting with the player whose turn it is to 'chip' for the remainder. **1892** Florence *Handbk. Poker* 89 *To chip in*, to put counters on the table. Equivalent to entering into the game. *Ibid.* 158 Every time my callow friend won a pot he put the silver and bills in his pocket and would chip in the stuff as he needed it.

c. To contribute; to make a contribution. Also *absol.* orig. *U.S.*

1861 *Winsted* (Conn.) *Herald* 22 Nov., An idea..that the printer should 'chip in' to every charitable and religious operation. **1886** *Harper's Mag.* Dec. 36 'Here's a dollar!' 'Here's another!' And they all chipped in their share. **1888** *Amer. Mag.* Sept. (Farmer), A man who won't chip in to charity is always an object of suspicion. **1903** *N.Y. Sun* 15 Nov., Nevertheless they all chipped in for the benefit of Simpson's widow and little child next day. **1908** S. E. White *Riverman* ix. 80 Why, there isn't a man on that river who doesn't chip in five or ten dollars when a man is hurt or killed. **1953** P. Frankau *Winged Horse* iv. 280 David will know somebody who can chip in. **1960** *Commentary* June 487/1 The help of the Ford Foundation (which chipped in more than $100,000).

9. *to chip at*: also, to poke fun at. Hence *trans.* (by omission of *at*), to make (a person) the object of a joke, to chaff, banter; to find fault with; also *absol.*

1896 *Westm. Gaz.* 13 Jan. 6/2 The friends of the contending teams each 'chipped' and jeered at the other. **1898** *Daily News* 20 Aug. 5/4, I chipped them a little on their plump, well-fed condition. **1915** *Even. News* 24 Nov. 3/4 People have been telling me I'm a German and chipping me, and I wish to get it over. **1928** *Daily Express* 23 Feb. 6 Letitia Pilkington..was a sportswoman. She even chipped her confessor on her deathbed. **1929** K. S. Prichard *Coonardoo* 41 Here, what are you chippin' about? **1958** 'N. Shute' *Rainbow & Rose* iii. 91 They chipped us about having tarried on the way.

10. *intr.* and *trans.* To play a chip-shot; to hit or kick (the ball) with a chip-shot.

1923 *Daily Mail* 8 May 12 He chipped to within eighteen inches of the hole. **1959** *Times* 29 May 5/2 He also won the second [hole] by chipping close. *Ibid.* 31 Aug. 12/2 Jones.. drew the defence and chipped the ball to Medwin. **1969** *New Yorker* 14 June 44/3 Graebner, instead of hitting out, chips the ball back—the cautious thing to do.

chiparee, earlier form of Chickaree.

1804 in *Maryland Hist. Mag.* (1909) IV. 9 These squirrels are exceedingly active and the mountaineers call them the Chiparee squirrel.

Chipewyan (tʃipĕwəi·ăn). Also **Chippewyan,** †**Che-,** etc. [Cree.] A member of an Athapaskan people of North American Indians; also, the language of this people. Also *attrib.* or as *adj.*

[**1743** J. Isham *Obs. Hudson's Bay* (1949) 177 The following is a small acc't of the (wechepowuck) Indians. **1775** A. Graham *Ibid.* 316 [Nation:] Weechepowack...[Tribe:] Wee chip y an i wuck.] **1789** A. Mackenzie *Jrnl.* Aug. in *Voy. Montreal* (1801) 106 It was a custom with the Chepewyan chiefs to go to war after they had shed tears, in order to wipe away the disgrace attached to such a feminine weakness. **1808** G. Keith *Let.* 1 Dec. in L. F. R. Masson *Les Bourgeois de la Compagnie du Nord-Ouest* (1890) II. 78, I had it from an old woman who never had any communication with the Red Knives or Chipewyans. **1823** J. Franklin *Journey to Polar Sea* 294 There are some other tribes who also speak dialects of the Chipewyan. **1910** *Encycl. Brit.* II. 827/1 *Athapascan*, a widely distributed linguistic stock of North American Indians, the chief tribes included being the Chippewyan, Navajo, [etc.]. **1937** R. H. Lowie *Hist. Ethnol. Theory* xi. 183 Detailed inquiries as to a clan organization among the Chipewyan of the Great Slave district yielded wholly negative results. **1963** *Times* 25 Feb. p. xvi/5 The Chipewyan Indians—the greatest of the caribou-eating tribes. **1965** *Canad. Jrnl. Linguistics* Spring 131 Chipewyan is the example of a language which permits only spirantal consonants finally.

chipolata (tʃipǫlā·tă). [F., ad. It. *cipollata* dish made of onions, f. *cipolla* onion.] In full *chipolata sausage*: a small spicy sausage; also, a dish or garnish containing these sausages.

1877 E. S. Dallas *Kettner's Bk. of Table* 376 *Chipolata Relish*—a great favourite for poultry. Chipolatas are small sausages about two inches long, made of pork and bacon. **1935** C'tess Morphy *Recipes of all Nations*, Chipolata sausages, a large slice of fat bacon. **1952** F. White *Good English Food* VI. xii. 280 The chipolata for garnishing consists of 36 little onions..1 lb of chipolata sausages, fried.

chipped, *ppl. a.* **2.** Add to def.: esp. of potatoes (cf. *chip *sb.*¹ 2 b).

1911 L. Stone *Jonah* I. ix. 98 The larrikins turned to the chipped-potato stall as a hungry dog jumps at a bone. **1948** *Good Housek. Cookery Bk.* II. 334 Chipped potatoes are a national favourite. **1965** P. Harben *Grammar of Cookery* x. 215 The most popular form of fried potatoes in Great Britain is the one we know as chipped potatoes or 'chips'.

b. Of beef: cut into thin pieces and dried. *U.S.*

1819 *Massachusetts Spy* 18 Aug. 4/2 No vapid tea, or cold toast, and greasy butter, and chipped meat. **1859** *Knickerbocker* Oct. 406 Such waffles, chipped beef, sweetmeats, melon. **1948** *Field & Stream* (U.S.) July 116/3 Dried or chipped beef is excellent.

3. b. *chipped glass* (see quot.).

1903 *Chambers's Jrnl.* VI. 685/1 The material is known technically as chipped glass, for the pattern is actually chipped out of the surface.

chipper, *sb.*¹ Add: **1. c.** With advs.

1898 [see Stringer 1 b]. **1901** *Daily Chron.* 4 Dec. 9/3 Pianoforte.—Fitters-up and chipper-up and tuner wanted. **1921** *Dict. Occup. Terms* (1927) § 279 *Chipper-in* (rolling mill); a chipper *q.v.* chipping rough ends on rails with hammer and chisel. *Ibid.* § 643 *Chipper-up*; carries out first stage of tuning strings of pianos when partly finished.

chipper, *v.* Add: **2. b.** *intr.* To brighten *up*.

1886 *Harper's Mag.* Sept. 583/1 She'd chipper up and fix up what she could. **1904** G. H. Lorimer *Old Gorgon Graham* 129 'Young man, I've decided to help you out of this hole,' he began. Percy chippered right up.

Chippewa (tʃi·pĕwǭ, -wā). Also **Chippeway** and numerous other varr. [By-form of *Ojibwa.] A member of an Algonquian Indian people found in the Great Lakes region. Also *attrib.* or as *adj.*

1754 Washington *Let.* 24 Apr. in *Writ.* (1889) I. 51, 600 of the Chippoways and Ottoway Indians are marching down Scioto Creek. **1778** J. Carver *Trav. N. Amer.* 416 The Chipéway tongue is not incumbered with any unnecessary tones or accents. **1788** [see Mohican *a.*]. **1826** J. F. Cooper *Last of Mohicans* III. ii. 39 The wicked Chippewas cheated my squaw. **1831** *Boston Transcript* 30 Dec. 1/1 The Chippewa tribe formerly inhabited the regions around Lake Superior. **1835** C. F. Hoffman *Winter in West* I. 215 A well-made young half-blood Chippeway. **1880** *Harper's Mag.* Dec. 43 The guttural ejaculations of the Chippewas who came to her door. **1890** Frazer *Golden Bough* II. iv. 239 Among the Chippeways and other Indians of the Hudson Bay Territory. **1929** *Folklore* XL. No. 4, (*title*) Tales of the Chippewa Indians. **1963** *Amer. Speech* XXXVIII. 270 He must use English to converse with new friends, who may be Eskimo or Chippewa or the like.

Chippewyan, var. *Chipewyan.

chippie: see *chippy *sb.* 2 and 4.

chipping, *ppl. a.* **2.** **chipping-bird**: for *Zonotrichia socialis* read *Spizella passerina* (earlier examples); **chipping-sparrow,** the chipping-bird.

1791 W. Bartram *Trav. Carolina* II. 289 Passer domesticus; the little house sparrow or chipping bird. **1810** A. Wilson *Amer. Ornith.* II. 127 Chipping Sparrow. *Fringilla socialis*...The Chipping-bird builds his nest most commonly in a cedar bush. **1861** Mrs. Stowe *Pearl Orr's Isl.* I. viii. 58 I've seen 'em big as chippin-birds' eggs. **1867** *Amer. Naturalist* I. 402 Simultaneously with the Bluebird the Chipping Sparrow awakes, and is soon heard chanting his simple cricket-like song from the garden and lawn. **1961** O. L. Austin *Birds of World* 298/2 The friendly song and chipping sparrows nest in shrubbery close to houses.

chippy, *a.* Add: **3. b.** Also *gen.* 'off colour', seedy, unwell.

1877 *Belgravia* XXXII. 235 After too copious libations of the above [*sc.* brandy and soda] a man is apt to feel 'chippy' next morning. **1884** H. Smart *From Post to Finish* xxi, A dozen cigars a day make one feel dreadfully 'chippy', in the morning. **1899** Conan Doyle *Duet* 74 'I feel a little chippy to-day.' 'That's the worst of these cheap champagnes.' **1937** E. Garnett *Family from One End St.* iv. 81 You looks a bit chippy. Don't you be sick here.

4. Further examples in sense 'cross, irritable'.

c **1885** F. M. Fetherston *Adv. Yorks. Farmer* 84 You look chippy and down in the mouth. **1898** in *Eng. Dial. Dict.* s.v., The master is quite chippy to-day; you can hardly speak to him but he snaps at you. **1941** *Penguin New Writing* IV. 37 'Don't bother. Here are your gloves.' 'No need to be chippy, old fellow.'

b. Lively, brisk. Cf. Chipper *a.*

1923 *Times Lit. Suppl.* 23 Aug. 558/3 Mr. Nicholson writes sometimes in the chippy, voluble monologue of 'If Winter Comes'. **1950** C. Fry *Venus Observed* 13 You call me old But I am still A chippy young chap.

chippy (tʃi·pi), *sb.* [f. Chip *sb.*¹ and *v.*¹]

1. *U.S.* (Also *chippy bird, sparrow.*) = *chipping-bird* (Chipping *ppl. a.* 2 in Dict. and Suppl.).

1864 Webster, Chipping-bird..also chippy. **1898** S. Hale *Lett.* (1919) 338, I must tell you of our little chippy sparrows that had their nest in the trellis. **1900** *Congress. Rec.* 30 Apr. 4872/2 The ground chippy darted under the fences. **1913** Mrs. Stratton-Porter *Laddie* iii. 94 In the hollow of a rotten rail a little chippy bird always built a hair nest.

2. (Also *chippie.*) A young woman (usu. derogatory); a promiscuous or delinquent girl or young woman; a prostitute. Also *attrib. slang* (orig. *U.S.*).

1886 in *Amer. Speech* (1950) XXV. 31/2 This class of females are known by the gang as 'Chippies', and most of them come from the slums. **1929** T. Wolfe *Look Homeward* (1930) xii. 254 Your place is getting the reputation of a regular chippyhouse. *Ibid.* xix. 279 She's no better than a regular little chippie. **1938** *Times Lit. Suppl.* 25 June 438/1 Opal and other 'chippies' at Moll's 'sporting-house'. **1955** 'W. Mole' *Hammersmith Maggot* ii. 26 He was no timidly frenzied lover of some exorbitant chippy.

3. *slang.* = *chip *sb.*¹ 6 c. (Also see quots.)

1916 'Taffrail' *Carry On* 26 A carpenter's mate [is] a 'woodspoiler' or 'chippy chap'. **1925** Fraser & Gibbons *Soldier & Sailor Words* 54 *A chippy chap*, a man of carpenter's rating. (Navy). *Ibid.* 167 *Nicknames*..*Chippy*: for a man named Carpenter. **1929** *Papers Michigan Acad. Sci., Arts & Lett.* X. 284/1 *Chippies*, carpenters in the British naval air-service. **1930** T. E. Lawrence *Lett.* (1938) 680 Try & find me a chippy to run shelves..the full depth and height of the chimney-breast. **1943** [see *chip *sb.*¹ 6 c]. **1960** A. Wesker *I'm Talking about Jerusalem* 1 I'll work as a chippy on the Colonel's farm.

4. (Also *chippie.*) = *chip-shop*, *chip *sb.*¹ 9. *local colloq.*

1961 Partridge *Dict. Slang* Suppl. 1036/1 *Chippy*, a fish-and-chips shop: Merseyside. **1964** *New Society* 20 Feb. 11/1 An acceptable alternative to the pub: the chippy on the turf—a chip shop in the gang's area. **1965** *Listener* 3 June 827/1 In the industrial towns the housewife..found that time, labour, and money were saved by the chippie.

chiprassi, var. Chuprassy.

chipre, var. *chypre.

chip-squirrel. (Earlier examples.)

1844 'Jon. Slick' *High Life N.Y.* II. 220, I could feel the leetle hand a movin' like a chip squirrel in its nest. **1848** C. W. Webber *Old Hicks* 49 [Sand rats] look something like the common chip squirrel.

chir (tʃiᵊr). Also **cheer.** [Hindi *chīr.*] In full *chir pine*. An Indian tree, *Pinus roxburghii.*

1882 *Encycl. Brit.* XIV. 155/1 The *chir* or three-leaved Himalayan pine. **1885** *Ibid.* XIX. 105/1 *P. longifolia*, a Nepal species, is remarkable for the great length of its lax slender leaves...It is known in India as the 'Cheer-Pine'. **1924** *Blackw. Mag.* Oct. 478/1 Our road ran through chir-pine forest and grassy glades. **1940** F. Smythe *Adv. Mountaineer* x. 156 The lower foothills are clotted for scores of miles with forests of chir, tall graceful pines.

‖**chiragh** (tʃirā·g). Also **chirag, chirak.** [Hind., a. Pers. *chirāgh* lamp, light.] A primitive oil lamp used in India and adjacent countries.

1899 *Westm. Gaz.* 15 Aug. 1/3, I took the oil from my chirag and anointed my back therewith. **1920** *Chambers's Jrnl.* Dec. 863/2 To-night we'll light tiny, weenie *chiraghs*. **1932** *Discovery* Sept. 304/1 A *chirak*, a primitive little oil lamp. **1935** *Ibid.* Nov. 334/2 The little earthenware *chirags* are indistinguishable in form from the early spouted saucer lamps of Egypt.

chiral (kəiᵊ·răl), *a.* [f. Gr. χείρ hand + -al.] Of a crystal or three-dimensional form: not superposable on its mirror image. Hence **chirality** (kəiræ·liti) [-ity] (see quot. 1894); **chiroid** (kəiᵊ·roid), a chiral object or process.

1894 Ld. Kelvin *Boyle Lect.* in *Jrnl. Oxf. Univ. Jun. Sci. Club* 25 May 25 *note*, I call any geometrical figure, or group of points, *chiral*, and say that it has *chirality*, if its image in a plane mirror, ideally realized, cannot be brought to coincide with itself. **1904** —— *Baltimore Lect.* 439 For brevity I now call a molecule which has chirality, a chiroid. **1939** *Nature* 11 Feb. 240/2 The tensor scheme of relativity is incomplete by itself to include relations of chirality, to use Lord Kelvin's term...This significance of the chiral property..is the difference between a chiral system and its mirror-image. **1957** *Physical Rev.* CVI. 1306 (*title*) Chirality of *K* Particle. *Ibid.* 1306/1 An operator called 'chirality', which anticommutes with the parity operator, can be applied to both fermions and bosons. *Ibid.* 1306/2 This leads to two alternative assignments of *K* particles to chiral eigenstates. **1964** *Cambr. Rev.* 24 Oct. 53/3 Only particles of one chirality interact. **1969** *Nature* 2 Aug. 453/1 Even the complexities which abound in naming fused rings, where chirality is so important, may be resolved.

chi-rho (kəi rōᵘ). The first two letters of Gr. ΧΡΙΣΤΟΣ Christ, often joined in a monogram ☧ and used to symbolize the name.

[**1611** Florio s.v. *Labaro*, The letter Chi and Rhô, which signifieth Christ.] **1868** *Jrnl. British Archaeol. Assoc.* XXIV. 132 The *chirho* also occurs on a doorstep. **1880** Smith & Cheetham *Dict. Chr. Antiq.* II. 1312/1 The upright cross being first added to the Chi-rho so as to form a kind of star. **1889** J. Romilly Allen *Monumental Hist. Brit. Ch.* 30 The Chi-Rho monogram. **1900** *Daily News* 8 Oct. 6/3 The monument consists of a simple monolithic block...Upon one side is incised a simple Chi-Rho enclosed in a circle. **1923** R. G. Collingwood *Roman Britain* v. 95 Engraved on a silver cup of foreign manufacture..we find the Chi-Rho

monogram. **1953** K. Jackson *Lang. & Hist. Early Brit.* v. 163 The chi-rho monogram of the earlier type with the *X* over the rho was largely replaced in southern Gaul at the end of the fifth century by the later type with a mere cross-bar.

chirimoya, var. Cherimoya.

chirk, *v.* Add: **3.** (Earlier and later examples.)

1844 'Jon. Slick' *High Life N.Y.* I. 231 All I could do she wouldn't chirk up. **1886** Roe *Fell in Love* xiii. 117 Don't you worry now; chirk up and you'll come out all right. **1887** M. E. Wilkins *Humble Romance* 17 P'rhaps nothin' awful's goin' to happen...Chirk up an' give us a kiss. **1951** H. Giles *Harbin's Ridge* 139 He chirked up from then on. All of us felt better then.

b. *trans.* To make cheerful, to brighten *up*. *U.S. colloq.*

1843 *Yale Lit. Mag.* Feb. 26 (Th.), All our folks appear more chirkier than they usually feel, in order to chirk her up. **1887** R. T. Cooke *Happy Dodd* xii. 122 Ef there's a mortal thing I can do to help ye, or chirk ye up, I want to do it right off. **1911** E. Ferber *Dawn O'Hara* xvi. 238, I donned a becoming gown to chirk up my courage. **1912** W. Irwin *Red Button* 40 The thought chirked me a lot. **1923** *Chambers's Jrnl.* Apr. 540/2 I'll stay right on here and do what I can to chirk the place up. **1938** *Sun* (Baltimore) 4 Mar. 10/5, I kind o' feel chirked up about my abilities as a prophet. I said..there was a likelihood we would have a blizzard..and sure enough..it came.

chirk, *a.* (Earlier examples.)

1789 Webster *Diss. Eng. Lang.* 387 This word is wholly lost except in New England. It is there used for *comfortably, bravely, cheerful*, as when one inquires about a sick person, it is said, he is chirk. **1816** Pickering *Vocab.* 59 It should be remarked, that the adjective *chirk* is used only in the *interior* of New England; and even there, I think, only by the illiterate. It is never heard in the sea-port towns.

chirognomically (kəiᵊrǫgnǫ·mikăli), *adv.* [f. Chirognomy + -ically.] As a matter of chirognomy.

1891 *Chambers's Encycl.* VII. 727 The thumb is generally regarded as chirognomically the most important part of the hand.

chiromance (kəiᵊ·romæns), *v.* [f. Chiromancy.] **a.** *trans.* To divine by chiromancy. **b.** *intr.* To practise chiromancy. So **chi·romancist** = Chiromancer.

1873 Leland *Egypt. Sketch-Bk.* 235 The hands must be washed ere I can chiromance anything. **1882** —— *Gypsies* vii. 181, I had chiromanced to the Romany clan. **1889** *Pall Mall Gaz.* 8 Mar. 3/1 The chiromancists or fortune-tellers.

chiromantist. Delete † and add later example.

1904 E. F. Benson *Challoners* vi, If I had been told by a chiromantist that I should shortly be writing the words *Sunday Magazine*.

chironomid (kəiᵊrǫ·nǫmid), *a.* and *sb.* [ad. mod.L. *Chironomidæ*, f. *chironomus*, a. Gr. χειρονόμος: see Chironomy and -id³.] **A.** *adj.* Of or belonging to the Chironomidæ, a family of dipterous insects typified by the midge. **B.** *sb.* An insect of this family.

1893 in *Funk's Stand. Dict.* **1905** *Jrnl. N. Y. Ent. Soc.* XIII. 69 (*title*) A Chironomid inhabitant of *Sarracenia purpurea*. **1924** *Chambers's Jrnl.* 750/2 Chironomid gnats and moth flies (*Psychodidæ*) inhabit the tanks. **1925** R. W. G. Hingston in E. F. Norton *Fight for Everest, 1924* 288 Fragile Chironomids, delicate as mosquitoes, skated on the surface of the water. **1946** *Nature* 13 July 71/1 The following arthropods were apparently unaffected: Chironomids, *Mysis*, *Sphæroma* and *Carcinus*. **1959** J. Clegg *Freshwater Life* (ed. 2) xiv. 237 The haemoglobin in the blood of Chironomid larvae.

chiropractic (kəiᵊropræ·ktik), *a.* and *sb.* [f. Gr. χειρο- + πρακτικός: see Chiro- and Practic *a.*] **A.** *adj.* That practises or is concerned with the curing of disease by manipulation of the structures of the human body, esp. those of the spinal column.

1898 *Stone's Davenport* (Iowa) *City Directory* 384 Dr. Palmer Chiropractic School & Cure. **1908** *Congress. Rec.* 6 Apr., A bill to regulate the practice of chiropractic, to license chiropractic physicians, [etc.] **1920** *Chambers's Jrnl.* 348/2 The country is overrun with..chiropractic and osteopathic doctors. **1926** J. H. Jones *Healing by Manip.* 64 Biologists, once they become aware of the chiropractic principle..will probably do useful service. **1962** *New Eng. Jrnl. Med.* 1 Nov. 939/2 Three recent superior-court judgments regarding chiropractic practice..indicate an increasing campaign by the chiropractor..to rid chiropractic of its stigma as a cult.

B. *sb.* Such a method of curing disease.

1903 (*title*) Backbone. The journal of chiropractic. **1908** [see above]. **1913** B. J. Palmer (*title*) The science of chiropractic, containing a series of lectures..and used as a textbook at the Palmer School of Chiropractic. **1926** J. H. Jones *Healing by Manip.* 35 Chiropractic or bone-setting differs from most other methods of combating disease. **1948** [see *cultish *a.*]. **1962** [see above].

chiropractor (kəiᵊropræ·ktəɹ). [f. *chiropract(ic + -or.] One who practises chiropractic.

1904 *Backbone* I. 215 He is fitted to become an all-round family Chiropractor. **1926** *Spectator* 24 July 148/1 Chiropractors have trebled their numbers in a very few years. **1957** *Encycl. Brit.* V. 595/1 It is estimated that there are about 18,000 chiropractors practicing in the United States and Canada.

chirp, *v.* Add: **3. b.** Also with *up*. *U.S.*

1897 R. M. Stuart *Simpkinsville* 35 Mis' Meredith is chirpin' up a'ready.

chirpily (tʃɜ̄·ɹpili), *adv.* [f. Chirpy *a.* + -ly².] In a chirpy or lively manner.

1923 *Daily Mail* 21 Feb. 8 Calthorpe veterans [*sc.* motor vehicles] 'still running' as chirpily as any highly-tuned racer. **1927** 'Q' in *Blackw. Mag.* Aug. 156/2 Defendant (turning chirpily on his accuser).

chir pine: see *chir.

chirrupy, *a.* (Earlier example.)

1808 T. Campbell *Let.* 23 Jan. in W. Beattie *Life & Lett.* (1849) II. v. 136 Laugh at all my *doledrum* of yesterday. I have been quite *chirrupy* to-day!

chisel, *sb.*¹ Add: **4. b.** *spec.* Designating a type of shoe with a squared toe.

1961 *New Statesman* 26 May 830/1 Where I live, the latest is an Italian-styled navy blue double-breasted suit, and chisel shoes. **1962** *Guardian* 31 July 3/2 He stands out..like a continental chiseltoe in a row of sensible Northampton brogues. **1969** *New Yorker* 11 Oct. 9 (Advt.), The shoe that fits the mood is the chisel toe slip-on.

5. chisel end, head, an end or head shaped or sharpened like a chisel; **chisel-mouth** *U.S.* = Quinnat.

1876 Preece & Sivewright *Telegraphy* 194 The chisel end of the punner bar. **1908** *Animal Managem.* 241 The 'chisel' head or sharp portion [of the frost nail] to remain above the [horse-]shoe. **1889** [see *Chinook b]. **1896** Jordan & Evermann *Fishes N. Amer.* I. 207 Acrocheilus, Agassiz. (Chisel-mouths.) **1909** *Cent. Dict.* Suppl., *Chisel-mouth*, a cyprinoid fish, *Acrocheilus alutaceus*, found in the lower Columbia river and its tributaries, as far up as Shoshone and Spokane Falls. Also called *chiselmouth jack*.

chisel, *v.*¹ Add: **4.** *intr.* To 'butt *in*'; to intrude. *colloq.*

1936 M. Franklin *All that Swagger* (1952) lii. 411 It is considered smart to chisel-in on a married woman now. **1939** Steinbeck *Grapes of Wrath* 260 Goin' aroun' stirrin' up trouble. Gettin' folks mad. Chiselin' in. **1942** D. Powell *Time to be Born* (1943) iv. 93 There is something..annoying ..in finding neighbours from back home chiselling in on your own exclusive New York.

chiseller, -eler. Add: **2.** *colloq.* A swindler, cheat. Cf. Chisel *v.*¹ 3.

1918 C. Mackenzie *S. Scarlett* I. vi. 188 You know I won the toss. We tossed up which should tell and I won. You *are* a chiseller. **1949** E. Hyams *Not in our Stars* xvii. 207 Harry was easy with all men because they were all equal as chisellers.

chi-square (kəi·skwēᵊɹ). [f. *chi + Square *sb.* 8.] Applied (usu. *attrib.*) to a test for, or statistic of, comparison between observed and theoretical values (see also quot. 1960). Also **chi-squared** *a.*

K. Pearson, who devised this test in 1900, called it the 'goodness of fit test' or 'χ^2 test'.

1936 Davenport & Ekas *Statist. Methods Biol.* (ed. 4) 47 The χ^2 ('*chi square*') *test*. Find for each class the difference (δ_1) between the theoretical value (y) and the observed frequency (f). Divide the square of this difference in each case by y. The sum of the quotients is χ^2. **1939** Kurtz & Edgerton *Statist. Dict.* 24 *Chi square test of goodness of fit*, a function from which can be obtained approximately the probability that a given set of data.. might arise if a certain law or cause were in operation. **1946** [see *additive *a.* d]. **1949** *British Birds* XLII. 256 Application of the chi-square test to a comparison of my figures and Höhn's..shows that the chance of his part A and my column 3 differing only by chance is much less than 1 in 100. **1957** P. Lafitte *Person in Psychol.* 150 This kind of analysis..uses only non-parametric methods, mostly chi-square. *Ibid.* 151 These variables are compared with the criterion by means of chi-square (or the t-test, if the criterion is quasi-physical). **1960** Kendall & Buckland *Dict. Statist. Terms* 41 *Chi-squared Statistic*..For historical reasons..it more usually relates to a statistic of a particular kind which is, as a general rule, distributed more or less approximately in the χ^2 form. *Ibid.* 42 *Chi-squared Test*.. Such tests occur in many ways, the most prominent being: (i) An overall goodness-of-fit comparison of observed with hypothetical frequencies falling into specified classes; (ii) Comparison of an observed with a hypothetic variance in normal samples; (iii) Combination of probabilities from a number of tests of significance. **1961** L. F. Brosnahan *Sounds of Lang.* iii. 66 It was decided to test the significance of the co-incidence of the two distributions by the Chi-square technique.

chit, *sb.*³ Add: (Later examples.)

In British use *chit* is commonly heard but *chitty* is now rare or obs.

1929 *Strand Mag.* Dec. 526/1 The boy brought lemonade for Mrs. Wilkins, a whisky-and-soda for her husband, and a gin and tonic for me. We shook dice and I signed the *chit*. **1955** *Times* 22 Aug. 5/6 Mr. Marshall pours out a stream of comment and advice or chits to the various administrative departments.

b. *attrib.* and *Comb.*

1845 Chit-system [in Dict.]. **1892** A. Murdoch *Yoshiwara Episode* 21 Billiard tables in the hotels galore (the 'chit' system was in vogue in running them too). **1892** *Daily News* 24 Mar. 5/4 The 'Chit system'..is the very general practice of putting the name on a piece of paper for every article that is purchased instead of paying cash down. **1924** *Blackw. Mag.* Aug. 264/2 For days the chit-coolies bore confidential messages. **1932** 'A. Bridge' *Peking Picnic* iii. 24 Mrs. Leroy had by this time finished entering her notes in the chit-book. *Ibid.* 25 'Do you never post letters?' 'Not to people in Peking... It's much quicker to send the chit-coolie.' **1962** *Guardian* 2 Oct. 9/4 A chit fund [in India] roughly corresponds to a building society in Britain.

chit, *v.*¹ Add: **2.** *trans.* To allow (a potato) to sprout; to remove the sprouts of (potato tubers) for planting as sets.

1904 *Daily Chron.* 8 Apr. 6/4 Having chitted them [*sc.* 'Eldorado' potatoes] three times already, he has made between £7,000 and £8,000 out of his original purchase.

chital (tʃi·tăl). Also **cheetal, -ul, chetel.** [Hindi.] The Indian spotted deer, *Axis axis*.

1880 *Encycl. Brit.* XII. 742 The *chitál* or spotted deer..is admitted to be the most beautiful inhabitant of the Indian jungles. **1923** *Blackw. Mag.* Sept. 381/2 He..had been after cheetul here and had come on a tiger. **1925** *Punch* 20 May 546/2 The best of his bag was a *chital* stag. **1963** *Times* 5 Mar. 12/6 At one point he predicted a herd of cheetal and backed his words by giving the cheetal stag's call.

‖**chitarrone** (kitārōᵘ·ni). Pl. **chitarroni.** [It., augmentation of *chitarra* guitar.] A theorbo, or double-necked lute, of great length.

1740 in J. Grassineau *Mus. Dict.* 25. **1879** Grove *Dict. Mus.* I. 347/2 The chitarrone, as implied by the suffix, was a large chitarra. *Ibid.* 348/1 One [orchestra] that was got together for the performance of Monteverde's 'Orfeo' in 1607, in which appear two chitarroni. **1961** M. W. Prynne in A. Baines *Mus. Instruments* vii. 162 In its extreme form, the 'Roman theorbo' or *chitarrone*, the basses were over 5 ft long.

chit-chat, *v.* [f. the sb.] *intr.* To gossip; = Chat *v.*¹ 3. Hence **chit-chatting** *vbl. sb.*

1821 P. Egan *Life in London* II. i. 151 Riding round the circle to *chit-chat* and nod to her friends. **1823** T. Carlyle *Let.* 26 Mar. (1909) I. 187 What expounding and reading and chitchatting we shall have together. **1849** T. M. Garrett *Diary* 2 Aug. in *Amer. Speech* (1951) 182/2 Setting upon the door steps of the building chit chatting about nothing.

chit-chatty (tʃi·tˌtʃæ:ti), *a.* [f. Chit-chat + -y¹.] Of the nature of chit-chat or light chat; gossipy.

1889 J. Masterman *Scotts of Bestminster* III. xvii. 153 Long communications, chit-chatty, amusing. **1958** *Times Lit. Suppl.* 18 July 411/3 The chit-chatty humour of Miss Joyce Grenfell.

chitinogenous (kəitinǫ·dʒinəs), *a.* [f. Chitin + -(o)genous.] Consisting of or producing chitin.

1888 Rolleston & Jackson *Anim. Life* 227 *T[aenia] serrata*...In the 'embryonic mass' there are..flattened cells placed laterally and containing nucleolated nuclei. These cells constitute a 'chitinogenous layer'.

chitinoid (kəi·tinoid), *a.* [f. Chitin + -oid.] Resembling chitin.

1888 Rolleston & Jackson *Anim. Life* 429 The egg-shell in oviparous *Elasmobranchii* is chitinoid. **1957** *New Biol.* XXIV. 22 After a time this pseudopodial fan shortens again. A chitinoid membrane forms at its periphery.

chitosan (kəi·tosæn). *Biol.* [f. Chitin + -ose + -an.] A substance obtained from chitin, formed by the action of alkali at a high temperature.

1895 *Jrnl. Chem. Soc.* LXVIII. I. 167 The conversion of chitin into chitosan and acetic acid by fusion with potash. **1963** R. P. Dales *Annelids* i. 20 All chaetae give positive chitosan reactions.

chitra (tʃi·tra). Also **chittra.** [Hind.] = *chital.

1843 J. E. Gray *List Mammalia Brit. Mus.* 212 Axis Maculata. The Axis or Chittra.

chittack (tʃi·tæk). [Bengali *chhatak*.]

1. An Indian weight about equal to 1 ounce, 17 pennyweights, 12 grains troy.

1899 Kipling *From Sea to Sea* II. 305 The cake-maker.. receives a lump of opium, weighed out, of one seer seven chittacks and a half. **1957** *Encycl. Brit.* XV. 139/1 Chittack, Bengal, 5 tolas, or 900 grains.

2. A measure of land in Bengal.

1905 *Statesman* 22 Aug. 2/4 A..dwelling house containing by estimation 7 cottahs 7 chittacks 2 square feet. **1969** *Hindusthan Stand.* (Calcutta) 5 Aug. 3/2 (Advt.), Land containing an area of 9 cottahs, 13 chittacks and 40 square feet = 660·0545 square metres.

Chittagong (tʃi·tăgǫŋ). [Name of a district of Bengal.] A variety of a domestic fowl of the Malayan type.

1816 'B. Moubray' *Poultry* (ed. 2) 24 The Chittagong or Malay, another Indian variety, is, as a contrast to the Bantam, probably the largest of the Gallinaceous tribe. **1855** *Poultry Chron.* III. 201 Last spring I had a Chittagong and a black Spanish rooster.

chitter, *v.* **1.** Delete *Obs.* or ?*dial.* and add later *transf.* examples.

1904 M. E. Waller *Wood-Carver* vii. 309 A chipmunk chittered along the stone wall. **1921** C. Orr *Kate Curlew* i. 13 The rain..chittering in the rusty beech hedges. **1930** E. Ferber *Cimarron* xviii. 277 A linotype machine.. chattered and chittered and clanked.

b. (Later example.)

1919 Wodehouse *Damsel in Distress* vii. 95 Little things in the hedgerows chirped and chittered mockery at him.

chitter-chatter. Delete *rare* and add later examples.

1877 Disraeli in Buckle *Life* (1920) VI. 116 He did not want anything: only chitter-chatter. **1895** Kipling *2nd Jungle Bk.* 193 Mowgli..imitated perfectly the sharp chitter-chatter of Chikai, the leaping rat of the Dekkan. **1908** *Daily Chron.* 3 Aug. 3/1 What is the use of smart women's chitter-chatter without their dresses? **1928** *Observer* 19 Feb. 10/1 The empty, drawing-room chitter-chatter plays. **1940** F. Stark *Winter in Arabia, Diary* 29 Dec. 74 'We exhaust you with our chitter-chatter,' they say politely.

b. Also as vb.

1928 *Hearst's International* Aug. 43/1 He never tired of the Esperanto of the ticker tape..chittering chattering fortune or disaster.

chitting (tʃi·tiŋ), *vbl. sb.* [f. Chit *v.*¹] Sprouting, germination; *spec.* the process of allowing potatoes, etc., to sprout. Also *attrib.*

1727 [see Chit *v.*¹]. **1911** *Encycl. Brit.* XVII. 504/1 The first visible sign of germination [of barley] is the sprouting of the rootlet, termed 'chitting'. **1949** E. Coxhead *Wind in West* ii. 55 The dozens of chitting-boxes, the potatoes showing..here and there..a fat purple sprout. **1963** *Times* 4 Feb. 4/6 Losses of seed through frost in chitting sheds.

chitty, var. *chetty.

chiv, *sb.*¹ (Examples.)

1834 [see Chive³]. **1950** R. Chandler *Let.* 18 May (1966) 78 Chiv, or more commonly shiv, means a knife, a stabbing or cutting weapon, perhaps (but I don't think so) including a razor, but that is not the meaning. **1962** *John o' London's* 25 Jan. 82/1 A criminal may use...a *chiv*, which is a razor, knife or dagger.

chiv (tʃiv), *sb.*² = *chiv(v)y *sb. Austral. slang.*

1916 C. J. Dennis *Songs Sentim. Bloke* 119 *Chiv*, the face.

chiv, *v.* = Chive *v.*; also, to slash with a razor. Hence **chi·vving** *vbl. sb.* Cf. *chivvy *v.*²

1812 [see Chive *v.*]. **1816** *Sessions Papers* Sept. 377/1 They will chiv me; that means stabbing him; putting a knife into him. **1895** *Daily News* 27 Aug. 2/7 Kelly..had threatened to 'chiv' (knife) her. **1955** *Times* 15 Nov., Three of Heaton's pals threatened to 'chiv' him...'Chiv' is the shortened version of 'chivvy chase', rhyming slang for 'mark the face'. **1962** 'J. Bell' *Crime in our Time* III. 70 The usual 'carving' or 'chivving' was done, witnesses were intimidated.

chive, *v.* (Later examples.)

1896 A. Morrison *Child of Jago* v. 54 He was chived in the left lung. *Ibid.* xxxvii. 343 Stabbed! Chived! They chived Dicky Perrott! **1959** *Encounter* Aug. 32/1 'Ar,' he said, 'so a few of ver blacks git chived. Why oll ver fuss?'

chiveau, var. *shiveau.

chivoo, var. *shivoo.

chiv(v)y (tʃi·vi), *sb.* Also **chivey.** [Short for *Chevy Chase* (see *chevy *sb.* 4 b).] = Face *sb.* 1 a.

1889 Barrère & Leland *Dict. Slang* I. 246/2 *Chivy* (thieves), the face. **1891** Farmer *Slang* II. 99/1 *Chivy* or *Chevy* (thieves'), the face..: c. 1886. Music Hall Song. 'Ain't he got an artful chevy.' **1898** J. D. Brayshaw *Slum Silhouettes* 29 Wot! wiv a chivey like that? *Ibid.* 217 Yer orter seen Bill's chivey. **1916** J. B. Cooper *Coo-oo-ee* x. 143 Just as if the Egyptians had carved your dial in stone, Hawley, just as if your hard chivvy was what it looks like—a stone. **1958** A. Wilson *Middle Age of Mrs. Eliot* iii. 335, I can't keep this look of modest pride on my chivvy forever.

chivvy (tʃi·vi), *v.*¹ Now the usual form of Chevy, chivy *v.*, used esp. to mean 'to harry, harass, trouble, worry'. Also, further examples of the form *chiv(e)y*.

1872 F. W. Robinson *Wrayford's Ward* III. 209, I thought that it was a very odd name,..and that the boys would be very severe upon it presently in the play-ground, and 'chivey' him. **1918** *Chron. N.Z.E.F.* 27 Feb. 37/1 Doris..and I chivvied the chicken around till we caught her. **1914** G. B. Shaw *Pygmalion* I, Poor girl! Hard enough for her to live without being worrited and chivied. **1923** *Blackw. Mag.* Sept. 350/1 A small girl with a wire hook was chivvying the frying morsels about. **1924** *Glasgow Herald* 11 Mar. 7 For nearly an hour Mr Tom Shaw was teased and chivvied. **1927** *British Weekly* 27 Oct. 85/2 Into this melee Mr. Sheppard proposes joyously to launch another fifteen, and chivy the ball in yet a third direction. **1932** E. Waugh *Black Mischief* ii. 73 Sakuyu warriors played hide and seek among the rocks, chivvying the last fugitives of the army of Seyid. **1951** R. Hoggart *Auden* iv. 122 Auden probably felt more justified in chivvying people. **1956** A. Wilson *Anglo-Saxon Attitudes* 9 If they chivvied him, he would raise the red herring of his projected work. **1957** *Economist* 16 Nov. 564/2 The government that lends its signature to a treaty may fall, or be chivvied into tacitly revoking it by opposition leaders.

chivvy (tʃi·vi), *v.*² *slang.* Also **chivey.** To 'knife'; = Chiv(e *v.* (in Dict. and Suppl.).

1959 A. Lejeune *Crowded & Dangerous* vii. 79 We wanted to go round and chivvy him. **1960** K. Hopkins *Dead against Principles* iii. 22 He got chivvied at Brighton races. **1961** Partridge *Dict. Slang* Suppl. 1036/2 *Chivey, chivvy* (gen. as ppl. adj. *chiv(v)ied*), to slash (a person) with a knife..C[ent.] 20.

chiz(z (tʃiz), *sb.*² *slang.* [See next.] A swindle; a nuisance. Also as *int.* (see quot. 1959).

1953 G. Willans *Down with Skool!* i. 20 More chizzes about headmasters. *Ibid.* v. 76 It is time we sent Nigel to skool chiz! **1959** I. & P. Opie *Lore & Lang. Schoolchildren* ix. 161 The new lamentations 'grue' or 'grooh' (from gruesome) and 'chiz' or 'chiz-chiz' (cross between chisel and swiz?). **1961** Partridge *Dict. Slang* Suppl. 1036/2 'What a chizz!' What a nuisance. C[ent.] 20.

chizz (tʃiz), *v. slang.* [Shortened form of Chisel *v.*¹] = Chisel *v.*¹ 3. So **chi·zzer,** a swindler; **chi·zzing** *vbl. sb.*, cheating.

1935 'N. Blake' *Question of Proof* viii. 154 'I suppose Wemyss wasn't a member of the Black Spot?' 'I should think he jolly well wasn't, slimy little chizzer!' **1948** C. Day Lewis *Otterbury Incident* iv, Toppy was in favour of using a real charity..whose initials could secretly stand for our own charitable object. Ted was against this. He said it was too like chizzing.

chizzle, *v.* Var. Chisel *v.*¹ 3.

1834 C. A. Davis *Lett. J. Downing* 181 You can chizzle them out of their property. **1848** [see Chisel *v.*¹ 3].

chlamydomonas (klæmidǒmōᵘ·nǫs). *Bot.* [mod.L. (C. G. Ehrenberg 1833, in *Phys. Abh. K. Akad. Wiss. Berlin* 281), f. Gr. χλαμυδ-, χλαμύς cloak + μόνος one.] A unicellular, flagellate, green alga, belonging to the class Chlorophyceæ and widely distributed in fresh water.

1884 Henfrey *Elem. Course Bot.* (ed. 4) II. 427 *Chlamydomonas* consists of a single biciliated primordial cell with an eye-spot and two contractile vacuoles. **1935** F. E. Fritsch *Struct. & Reprod. Algae* I. 79 The flagella in *Chlamydomonas* and its allies consist of a thicker basal portion and a long thin whip-like termination. **1949** *New Biol.* VI. 94 A typical *Chlamydomonas*..is an ovoid cell. **1967** *Punch* 31 May 813/2 Some stuff for cleaning the carpet that brought out both Mrs. Jones and the carpet in a fine emerald-green rash of *chlamydomonas*.

chlamydospore (klæ·midoˌspōᵊɹ). [f. Gr. χλαμυδ-, χλαμύς cloak + Spore.] **a.** *Bot.* A thick-walled resting spore produced by various fungi. **b.** *Zool.* A spore with protective chitinous covering, as distinguished from *gymnospore*.

1884 Henfrey *Elem. Course Bot.* (ed. 4) I. 160 Chlamydospores, when fully formed, are still enclosed within their parent-cells. **1885** *Encycl. Brit.* XIX. 837/1 Each spore.. has its own protective envelope...The former condition is distinguished as a 'chlamydospore'. **1910** G. N. Calkins *Protozoöl.* 183 The covered spores or chlamydospores, of the sexual generation. **1922** *Nature* CIX. 699/2 Relations between the chlamydospores and the mycelian loops. **1952** *New Biol.* XIII. 103 The *chlamydospore* or *gemma* (plural *gemmae*). Such spores are produced in hyphae by the secretion of a thick wall around a portion of protoplasm.

chloræmia (klōrī·miă). *Path.* Also **-emia.** [mod.L., f. Chlor-¹ + Gr. αἷμα blood.] (See quot. 1961.)

1890 *Brit. Med. Jrnl.* 31 May 1239/1 Chloremia...A blood disorder, either idiopathic or associated with other ailments, consisting essentially of a greatly diminished percentage of hæmoglobin, and of a variable decrease in that of the red corpuscles. **1907** *Practitioner* June 826 To the greater part..of the symptom complex of nephritis, he [*sc.* Widal] gives the name chloraemia. **1961** *Brit. Med. Dict.* 293/2 *Chloraemia*, 1. A condition in which the blood contains an excess of chlorides. 2. Chlorosis (green sickness).

chloragogen (klōᵊrăgōᵘ·dʒĕn), *a. Zool.* [a. Ger. *chloragogen* (W. Kükenthal 1885, in *Jen. Zeitschr. Naturwiss.* XVIII. 332), f. mod.L. *chloragogena* (see quot. 1829), f. Chlor-¹ + Gr. ἀγωγή leading, carrying away (ἄγειν to lead).] Pertaining to or designating a series of yellowish cells lining the outer surface of the alimentary canal of certain worms. Also **chlorago·genous** *a.*

[**1829** C. F. A. Morren in *Ann. Acad. Gandavensis 1825–26* 142 De chloragogenâ.] **1894** A. Graf in *Trans. N.Y. Acad. Sci.* XIII. 240 In Clepsine..the funnel is surrounded by chloragogenous cells in the interior of the ventral sinus. **1897** J. P. Moore in *Jrnl. Morphology* XIII. 360 Cells.. filled with granules which impart..a green-brown color. The coelomic corpuscles have exactly similar characters, and one finds frequent indications of their..budding from the chlorogogen [*sic*] cells, just as Schaeppi has shown. **1961** A. J. Grove et al. *Animal Biol.* (ed. 6) xii. 204 The chloragogen cells fall into the coelom and after beginning to disintegrate are finally engulfed by phagocytic coelomic corpuscles. *Ibid.*, Coelomic fluid containing ammonia and urea produced by the chloragogenous cells is directed into the nephrostome by ciliary action. **1963** R. P. Dales *Annelids* iv. 88 Tissues referred to as 'chloragogen' in polychaetes may also participate in this process.

chloralamide (klōᵊrălē¹·məid). [ad. G. *chloralamid*, shortened form of *chloralformamid*: cf. Chloral and Amide.] A compound that results from a condensation reaction of chloral with an acid amide; *spec.* chloral formamide, $CCl_3 \cdot CH(OH) \cdot NH \cdot CHO$, a mild hypnotic.

1889 *Lancet* 26 Oct. 849/2 The administration of 40 gr. of chloralamide the evening after gave him an unbroken sleep of nine hours. **1897** *Trans. Amer. Pediatric Soc.* IX. 168 The treatment consisted of the administration of nervous sedatives, at first chloralamide, to which later was added hyoscine. **1909** Allbutt & Rolleston *Syst. Med.* (ed. 2) VI. 490, I now use chloralamide which, I am told, is safer than chloral; it acts nearly as well. **1963** K. W. Wheeler in Campaigne & Hartung *Med. Chem.* VI. i. 12 The so-called chloralamides, products of the condensation of chloral with the NH_2 of the amide, generally do have some hypnotic activity but not of a high order.

chloralose (klōᵊ·rălōᵘs). [f. Chloral + -ose².] A compound of chloral and glucose, $C_8H_{11}Cl_3O_6$, having hypnotic properties. Also *attrib.*

1893 *Brit. Med. Jrnl.* 29 July 233/2 Chloralose is a new drug and is said to be a very safe hypnotic. **1949** H. W. Florey et al. *Antibiotics* I. viii. 360 Cats..under chloralose or nembutal anaesthesia.

chlorambucil (klōᵊræ·mbiūsil). *Pharm.* [f. 4-p-di-(2-*chloro*ethyl)*am*inophenyl*bu*tyric acid, the systematic name + *-cil.*] A water-soluble aromatic nitrogen mustard used as an anti-neoplastic agent, esp. against certain lymphomas and leukæmias. Also called *CB 1348*.

1956 *Jrnl. Amer. Med. Assoc.* 15 Sept. 182/1 It is our impression that chlorambucil is of beneficial effect in certain patients with chronic lymphocytic leukemia. **1962** [see *astrocytoma].

chloramine (klōᵊɹæ·min, klōᵊ·rămin, -əin). *Chem.* [f. Chlor-² + Amine.] Any compound that contains a chlorine atom attached to a nitrogen atom, as: **a.** any of the three compounds formed when one or more of the hydrogen atoms in ammonia is replaced by a chlorine atom; *esp.* the monochlorine derivative, NH_2Cl, a colourless, oily, somewhat explosive liquid with bactericidal properties which has been used in the treatment of water supplies; **b.** any derivative of a sulphonamide in which one or both of the hydrogen atoms in the amide group is replaced by chlorine; *esp.* sodium *para*-toluenesulphonchloramide (also called *chloramine-T*), a white crystalline solid used in solution as a disinfectant and antiseptic.

1893 *Jrnl. Chem. Soc.* LXIV. I. 497 Chloramines. **1905** Cain & Thorpe *Synthetic Dyestuffs* xx. 153 Chloramine yellow (By) is a yellow tannined cotton dyestuff made by the oxidation of dehydrothiotoluidine sulphonic acid. **1908** *Jrnl. Soc. Chem. Ind.* XXVII. 260/2 The production of chloramine derivatives. **1916** H. D. Dakin et al. in *Brit. Med. Jrnl.* 29 Jan. 160/2 The substance chosen is para-toluene-sodium-sulphochloramide...It has been decided to assign the more convenient name 'chloramine' to the compound, thus indicating its membership in the group of substances containing the NCl linking. If, in future, other members of the group should come into use, it is suggested that the present compound should be known as chloramine-T to indicate its relationship with toluene. **1916** *Lancet* 11 Mar. 581/1 The germicidal action of many of these chloramine compounds was..greater than that of sodium hypochlorite. **1919** *Chem. Abstr.* XIII. 3258 Chloramine has been used successfully for disinfection and elimination of after-growths in the water supply. **1937** W. H. Maxwell *Water Supply Problems* (ed. 2) iii. 55 The chloramine treatment of water has been increasingly practised of recent years. **1949** Kirk & Othmer *Encycl. Chem. Technol.* III. 668 Unstable chloramines sometimes form strongly colored compounds as decomposition proceeds. **1957** *New Biol.* XXIII. 91 The process of sterilization involving a contact period of about four hours with chloramine.

chloramphenicol (klōᵊɹæmfe·nikǫl). [f. Chlor-² + Am(ide + Phe(n- + Ni(tro- + Gly)col.] An antibiotic isolated from cultures of *Streptomyces venezuelæ* or prepared synthetically. Cf. *chloromycetin.

1949 *Jrnl. Amer. Chem. Soc.* LXXI. 2458 Chloramphenicol has been assigned as a generic name for the compound.. for which Parke, Davis and Co. has adopted 'Chloromycetin' as its trademark. **1953** J. Ramsbottom *Mushrooms* xxiii. 290 Chloramphenicol is the only antibiotic manufactured on a commercial basis by chemical synthesis. **1969** *Times* 3 Mar. 5/6 It is proposed that chloramphenicol should never be used as a feed additive.

chloranthaceous (klōᵊrænþē¹·ʃiəs), *a. Bot.* Belonging to the family Chloranthaceæ.

1882 J. D. Hooker *Let.* 12 Aug. in L. Huxley *Life & Lett.* (1918) II. 248 He [*sc.* Maximovicz] has since published it as *Circaeaster agrestis*..and adopted an idea of Oliver's which I hardly share, that it is Chloranthaceous.

chlordane (klōᵊ·ɹdē¹n). Also **chlordan.** [f. Chlor-² + *in*)*dane* a hydrocarbon derived from *indene.] A viscous liquid insecticide, $C_{10}H_6Cl_8$.

1947 *Jrnl. Econ. Ent.* XL. 264/1 Chlordane possesses a toxicity to many insects greater than that of DDT. **1952** *New Biol.* XII. 56 Many potent materials such as..chlordan ..are now available for crop protection. **1962** *New Scientist* 11 Oct. 75 Dieldrin and chlordane were both found to be extremely effective against cockroaches.

chlorella (klore·lă). *Bot.* [mod.L. (M. W. Beijerinck 1890, in *Botanische Zeitung* XLVIII. 726), dim. f. Gr. χλωρός green; cf.

Nigella.] A unicellular green alga of the genus so named.

1904 G. S. West *Brit. Freshwater Algae* 230 Grintzesco.. has cultivated *Chlorella* in various media and finds that glucose stimulates its development. **1923** *Chem. Abstr.* XVII. 2440 Protein synthesis in Chlorella. **1957** *Times* 11 Sept. 6/2 Mass cultures of a minute aquatic plant—chlorella—promised to be a means of achieving high efficiencies [for photosynthesis] on a large scale. **1958** *New Biol.* XXV. 90 In the ponds under investigation the green algae *Chlorella* and *Scenedesmus* were found to be the most important photosynthetic organisms in the operation of the oxidation system.

chlorenchyma (klore·ŋkimă). *Bot.* [f. Chlor-¹ + ἔγχυμα infusion.] Green tissue consisting of cells containing chlorophyll, esp. in leafless branches having the function of leaves.

1894 F. W. Oliver et al. tr. *Kerner's Nat. Hist. Plants* I. 278 In flowering plants the transpiring cells are situated as a rule in the interior of the green leaves, and also in the green cortex of leafless branches, forming a part of that green tissue which has been termed *chlorenchyma*, or when in the leaves, *mesophyll*. **1959** *Chambers's Encycl.* III. 522/1 *Chlorenchyma*, parenchymatous tissue, the cells of which contain chloroplasts. The palisade and lobed parenchyma of the mesophyll of the leaf are collectively termed chlorenchyma.

chloretone (klōᵊ·ritōᵘn). *Chem.* [f. Chlor-², with the termination of *acetone*.] A white crystalline substance, $(CH_3)_2C(OH)CCl_3$, tasting like camphor, used as a local anæsthetic and a hypnotic.

1899 Houghton & Aldrich in *Jrnl. Amer. Med. Assoc.* XXXIII. 777/1 In order to avoid confusing this subject with 'acetone chloroform' (a name which has been applied to chloroform made from acetone), we have deemed it advisable to use the name 'Chloretone'. *Ibid.* 778/1 As a hypnotic, chloretone is frequently efficacious in various conditions. **1907** *Practitioner* Aug. 291 In this case, chloretone, dusted lightly on, will be found useful, as it is anesthetic and antiseptic. **1908** *Ibid.* Feb. 253 Chloretone Ointment. **1946** *Nature* 17 Aug. 238/2 Narcotics such as chloretone stimulate the synthesis of vitamin C in rats.

chlorider (klōᵊ·rəidəɹ). [f. Chloride + -er¹.] One who mines for ores containing chloride of silver.

1874 Raymond *Statist. Mines & Mining* 229 'Chloriders' have been at work during the past summer, and have shipped a few tons of ore to Eureka. **1957** *Encycl. Brit.* IX. 549/2 A 'fossicker'..[in Australia] is one who buys up the right to search old claims, in the hope of finding gold overlooked by previous diggers. He is comparable with the 'chlorider' of the United States.

chloridizing (klōᵊ·ridəiziŋ), *vbl. sb.* [f. Chloridize *v.* + -ing¹.] Conversion into chloride. Also *attrib.* So **chlo:ridiza·tion.**

1877 Raymond *Statist. Mines & Mining* 195 Where the natural chloridization is complete. *Ibid.* 416 The furnace at Belmont..gives good results as to the chloridizing action. **1882** *Rep. Precious Metals* (U.S.) 176 A chloridizing furnace, for the purpose of working the rich auriferous sulphuret with which the ore abounds. *Ibid.* 183 Two silver mills..will next year reduce the ores of the Saw Tooth district, the method employed being chloridizing. **1940** *Chambers's Techn. Dict.* 158/1 *Chloridising roasting*, the roasting of sulphide ores and concentrates, mixed with sodium chloride, to convert the sulphides to chlorides.

chlorinator (klōᵊ·rinē¹təɹ). [f. Chlorinate *v.* + -or.] An apparatus for chlorinating.

1930 *Engineering* 14 Feb. 215/3 If the chlorinator is satisfactorily to fulfil its functions. **1939** *Archit. Rev.* LXXXVI. 152 Sterilization is carried out on the chloramine system by means of a gas chlorinator and a gas ammoniator. **1963** R. R. A. Higham *Handbk. Papermaking* vi. 134 Other types of chlorinator are available such as the continuous chlorinator illustrated in Fig. 42.

chloritization (klōᵊ:ritəizē¹·ʃən). *Petrogr.* [f. Chlorit(e¹ + -ization, after G. *chloritisirung* (H. Rosenbusch *Mikrosk. Physiogr.* (ed. 2, 1887) II. 183).] A transformation into or replacement by chlorite, esp. of ferromagnesian minerals by hydrothermal action. Hence (as a back-formation) **chloritize** *v. trans.* (chiefly in passive), to convert into chlorite.

1890 G. H. Williams *Greenstone Schist Michigan* 56 The simultaneous action of the two last described processes (chloritization and epidotization) in an aluminous pyroxene or hornblende results in the production of an aggregate consisting of..crystals of epidote, imbedded in a green, scaly mass of chlorite. **1909** *Econ. Geol.* IV. 633 A chloritised or partially altered rock (*grünstein*). **1914** F. H. Hatch *Petrol. Igneous Rocks* (ed. 7) I. iv. 72 Chloritization is one of the commonest changes undergone by the ferro-magnesian minerals, the amphiboles, pyroxenes, dark micas, and olivines all falling a prey to it. **1963** D. W. & E. E. Humphries tr. *Termier's Erosion & Sedimentation* ii. 45 Fault breccias which are chloritized stand out green against the virgin rock.

chlorocresol (klōᵊrokre·sǫl). *Chem.* [f. Chloro-² + Cresol.] Any of various chlorine derivatives of cresols; esp. the derivative C_7H_7ClO, a white crystalline compound used as an antiseptic and preservative.

1886 *Jrnl. Chem. Soc.* L. 614 (*title*) Chlorocresols and Toluquinones... *Dichlororthocresol*, purified.., crystallises in colourless needles. **1952** M. E. Florey *Clin. Appl. Antibiotics* ii. 32 Meara (1948) considered the responsible agent to be wood alcohols—sterols and chlorocresol.

chloroform, *sb.* and *v.* Add: The pronunc. (klǫ·rǒfǭɹm) is now usual.

chloroma (klorōᵘ·mă). *Path.* [mod.L., f. Chlor-¹ + *-oma.] A disease of the bones characterized by greenish tumours; a form of plasmocytoma.

1886 *Buck's Handbk. Med. Sci.* III. 402/1 *Chloroma*, a very rare tumor, agrees in structure with the round-cell sarcoma, and contains in its cells a bright green pigment. **1906** *Jrnl. Chem. Soc.* XC. II. 566 Acute and chronic lymphatic leucæmia and chloroma are all essentially the same. **1964** S. Duke-Elder *Parsons' Dis. Eye* (ed. 14) xxxiii. 530 Hæmopoietic tumours may occur—the various types of reticular tumour..and chloroma.

chloromycetin (klōᵊromaisī·tin). [f. Chloro-¹ + Myceto- + -in¹.] A trade-name for chloramphenicol.

1947 *Science* 31 Oct. 417 From these filtrates a crystalline antibiotic has been isolated, for which the name Chloromycetin is proposed. **1949** [see *chloramphenicol]. **1955** *Sci. News Let.* 2 Apr. 214/3 Chloramphenicol (Chloromycetin) has been the drug of choice for treating typhoid fever.

chloroplast (klōᵊ·rǒplæst). *Biol.* Formerly also **chloroplastid(e.** [a. G. (formed by E. Strasburger in 1884 as a shortening of G. *chloroplastid* (A. F. W. Schimper 1883, in *Bot. Zeitung* 16 Feb. 108)); cf. Chloro-¹ and -plast.] A plastid containing chlorophyll.

1887 W. Hillhouse tr. *Strasburger's Handbk. Pract. Bot.* iv. 44 We can..distinguish the chlorophyll-bodies, colour-bodies, and colourless starch-builders as Chloroplasts, Chromoplasts, and Leucoplasts respectively. **1888** *Jrnl. Chem. Soc.* LIV. 983 Between the fibrils [of chloroplastin] is the colourless metaxin, hardly visible in the intact chloroplastides. **1902** H. Wager in *Encycl. Brit.* XXVII. 333/1 As the cell grows they [*sc.* the chromatophores] may become converted into leucoplasts (starch-formers), chloroplasts (chlorophyll-bodies), or chromoplasts (colour-bodies). **1906** *Cambr. Nat. Hist.* I. 86 The cell [*Zooxanthella*] contains two chloroplastids, or plates coloured with the vegetal pigment 'diatomin'. **1909** R. J. H. Gibson *Biol.* vi. 43 The chloroplasts are minute ovoid bodies, occurring singly or in large numbers in the cells...Each consists of a basis of protoplasm permeated by an oily matter in which the chlorophyll, or pigment proper, is dissolved. **1938** Weaver & Clements *Plant Ecology* (ed. 2) xiv. 381 Light influences the number and position of the chloroplasts. **1959** J. Clegg *Freshwater Life* (ed. 2) iv. 77 The Green Algae all possess chlorophyll, contained..in special structures called chloroplasts. **1964** *New Scientist* 20 Feb. 490 Chloroplasts, the small chlorophyll-containing organelles responsible for photosynthesis found within the cell substance..of green plants and algae.

chloroprene (klōᵊ·rǒprīn). *Chem.* [f. Chloro-² + *iso)prene.] A colourless liquid, made from acetylene and hydrochloric acid, which polymerizes to form *neoprene.

1931 W. H. Carothers et al. in *Jrnl. Amer. Chem. Soc.* LIII. 4203 In order to recognize the analogy in structure and behavior which exists between isoprene and chloro-2-butadiene-1,3, we call the latter compound chloroprene. **1958** *Times Rev. Industry* Nov. 64/2 Acetylene is now.. used in the production of chloroprene.

chloroquine (klōᵊ·rǒkwin). *Chem.* [f. Chloro-² + Quin(oline.] A compound derived from quinoline and used mainly as an anti-malarial drug.

1946 R. F. Loeb et al. in *Jrnl. Amer. Med. Assoc.* 20 Apr. 1069/2 Activity of a new antimalarial agent, chloroquine (SN 7618). **1947** *Lancet* 29 Mar. 415/1 American workers..finally selected one member—'Resochin', 'Chloroquine', 'SN 7618'—as the best for general purposes. **1958** *Observer* 17 Aug. 9/5 Two drugs, chloroquine and pyrimethamine, were tested.

chlorothiazide (klōᵊ:roþəi·ăzəid). *Chem.* [f. Chloro-² + *thia(dia)z(ine* + Diox)ide.] A white crystalline powder, $C_7H_6ClN_3O_4S_2$, used as a diuretic.

1957 Novello & Sprague in *Jrnl. Amer. Chem. Soc.* LXXIX. 2029/1 One compound of this type..has been selected for clinical trial and assigned the generic name of chlorothiazide. **1960** *Jrnl. Org. Chem.* XXV. 970/1 The generic name of chlorothiazide has been given to this compound and Diuril is the trademark of Merck and Co., Inc., for chlorothiazide. **1963** *Brit. Pharmaceutical Codex* 170 Chlorothiazide is a non-mercurial diuretic which reduces the resorption of electrolytes by the proximal renal tubes. *Ibid.*, Chlorothiazide is used for the treatment of œdema and of the toxæmias of pregnancy.

chlorpromazine (klōᵊɹprōᵘ·măzin). Also **chloro-.** [ad. F. *chloropromazine* (P. Charpentier et al. 1952, in *Comptes Rendus* CCXXXV. 60), f. Chlor-² + *prom(eth)azine.] A white or cream-coloured compound of bitter taste, $C_{17}H_{19}ClN_2S$, used chiefly as a sedative and anti-emetic.

1952 *Sci. News Let.* 24 Dec. 412/3 Reserpine and chlorpromazine were useful in calming disturbed patients. **1954** *Lancet* 5 June 1186/1 There are several ways in which chlorpromazine could contribute to safer hypotension for cerebral surgery. **1957** M. Millar *Soft Talkers* 151 'And two little pills to help with the jitters.' Harry glanced briefly at the pills. 'Chlorpromazine.' **1962** *Lancet* 22 Dec. 1300/2 Chlorpromazine and fan cooling, with wet sheets and ice packs, were needed to bring down the temperature from a peak of 105·6° F.

chlortetracycline (klōᵊ:ɹtetrăsəi·kləin, -in). *Chem.* [f. Chlor-² + *tetracycline.] = *aureomycin (see quot. 1963).

1953 *Jrnl. Amer. Med. Assoc.* 10 Oct. 587/2 Recently antibiotic agents such as chlortetracycline (aureomycin) and chloramphenicol have been found effective in altering the course of infectious mononucleosis by some investigators, whereas others concluded that these drugs were ineffective. **1963** *Brit. Pharmaceutical Codex* 180 Chlortetracycline Hydrochloride...*Synonym*: Aureomycin Hydrochloride... The general use of the synonym 'Aureomycin' is limited, and in any country in which 'Aureomycin' is a trade mark it may be used only when applied to the product made by the owners of the trade mark.

choana (kōᵘ·ănă). *Anat.* Pl. **-næ** (-nī). [mod.L., a. Gr. χοάνη funnel.] A funnel-like opening; applied to the posterior nasal orifices in vertebrates.

1878 Macalister *Syst. Zool.* 13 The pair of maxillary lobes usually unite in the middle line, leaving behind them openings into the nasal cavities (*choanæ*). **1895** H. & M. Bernard tr. *Wiedersheim's Struct. Man.* 61 The choanæ of the Veddah's skull are..half a centimetre lower than in the European. **1958** J. M. Yoffey in W. J. Hamilton et al. *Textbk. Human Anat.* 407 The choanae are oval in shape, being about 1 inch high and ½ inch wide.

choano- (kōᵘ·ăno), combining form of Gr. χοάνη funnel, as in **cho·anocyte,** a 'collar cell' in sponges and choanoflagellates; hence **cho:anocy·tal** *a.*; **cho:anofla·gellate** *a.*, belonging to the order Choanoflagellata of Infusoria (see quots.); also as *sb.*, a member of this order; **cho·anosome,** the inner part of a sponge, containing the choanocytes; hence **cho:anoso·mal** *a.*

1887 Choanosome [see *ectosome* s.v. *ecto-]. **1888** W. J. Sollas *Tetractinellida* p. xiv, The endoderm is partly composed of flagellated collared cells (choanocytes), and partly of pinacocytes. *Ibid.* p. xv, This investing skin may be distinguished as the ectosome, the rest of the sponge as the choanosome, the latter name framed in allusion to its being the region to which the choanocytes lining the flagellated chambers are restricted. *Ibid.* p. xxxviii, A contraction of the choanocytal wall. **1888** Choanosomal [see *ectosomal* s.v. *ecto-]. **1900** E. A. Minchin in Lankester *Treat. Zool.* II. 27 Each cell bears at its upper free extremity a single vibratile flagellum.., which springs from the centre of an area enclosed by a delicate cup or collar of protoplasm. On account of the latter peculiarity these cells have been termed collar cells or choanocytes. *Ibid.* 53 Each collar cell resembles..a single choanoflagellate monad. **1906** *Cambr. Nat. Hist.* I. 121 The Craspedomonadidae (often called Choanoflagellates). *Ibid.* 214 The distal ends of the choanosomal folds. **1927** Haldane & Huxley *Anim. Biol.* xii. 264 The Choanoflagellates, a group of Flagellates in which a transparent collar of protoplasm surrounds the flagellum. **1959** *Chambers's Encycl.* XIII. 101/2 The choanosome consists of a network of irregularly star-shaped cells, joined at their points, the interstices..being filled..with an intercellular jelly. *Ibid.* 102/1 The choanocytes, also known as whip-cells or collared cells, consist of a spherical or oval body, with the apex drawn out to form a protoplasmic collar from the base of which springs a long protoplasmic thread or whip. **1960** L. Picken *Organiz. Cells* VI. vii. 241 It is possible that the condition of being attached is primitive, since it occurs in choanoflagellates, in the choanocytes of sponges [etc.].

choate (kōᵘ·ĕt), *a.* An erroneous word, framed to mean 'finished', 'complete', as if the *in-* of *inchoate* were the L. negative.

1878 O. W. Holmes *Let.* 9 Dec. in *Pollock-Holmes Lett.* (1942) I. 11 Several of the State Courts have left equally amusing slips in the Reports...I have read in a California volume that the wife on marriage acquires an inchoate right of dower which by the death of the husband becomes choate. **1902** W. B. Richmond in *Saint George* V. 288 He [*sc.* Ruskin] seemed to be under the spell of a vision which he would make choate. **1929** W. S. Churchill in *Times* 13 Feb. 16/1 How could the peoples know?..What choate and integral conviction could they form?

choc (tʃǫk), colloq. abbrev. of Chocolate 2 in Dict. and Suppl.

1874 [see *chocolate 2]. **1896** E. Turner *Little Larrikin* xvii. 201 'Give me the chocs.'. He put three of the largest creams in his mouth. **1903** A. Bennett *Leonora* i. 21 Have a choc? Milly said yes, and Ethel drew a box of bonbons from her pocket. **1922** Joyce *Ulysses* 640 Fond lovers' ways and flowers and chocs. **1938** D. Smith *Dear Octopus* I. 3 Have a bit of choc? **1966** A. York *Eliminator* v. 94 He comes up to me one day when I was getting after the chocs and says, 'Rhoda, it's me or them chocolates. Take your pick.'

b. *attrib.*, as **choc-bar, choc-ice,** a brickette of ice-cream covered with chocolate.

1954 'N. Blake' *Whisper in Gloom* iii. 43 A voice which reminded him..of a Walls' choc-bar, so cool, sweet and creamy. **1951** J. B. Priestley *Fest. Farbridge* III. i. 431 To eat choc ices With Iphigenia. **1968** M. Richler *Cocksure* xviii. 111 The children skipped off to the dining-hall, where choc-ices, a conjurer, and a Popeye cartoon awaited them.

chochem, var. *haham.

chock, *sb.*[1] Add: **3.** *spec.* A block (of wood, etc.) placed in front of an aeroplane wheel.
1917 'Contact' *Airman's Outings* 16 The chocks were pulled clear, and away and up raced the machine. **1942** in H. G. Bryden *Wings* x. 276 She quivers and rocks as she strains at the chocks And clamours amain to soar.

4. b. *chock and log* (*fence*), a fence raised by placing layers of logs on 'chocks' or short wooden blocks placed transversely to the line of the fence. *Austral.*
1872 G. S. Baden-Powell *New Homes for Old Country* 207 Another fence, known as 'chock and log', is composed of long logs resting on piles of chocks, or short blocks of wood. **1890** *Melbourne Argus* 20 Sept. 13/5 A herd of kangaroos..bounding over the wire and 'chock-and-log' fences.

8. *attrib.*, as **chockstone,** a stone wedged in a vertical cleft.
1909 C. E. Benson *Brit. Mountaineering* iv. 72 A chockstone is a piece of rock which has fallen from higher up the cliff, and has been caught and wedged in its descent between the lateral walls of a fissure. **1924** J. Buchan *3 Hostages* xxi. 312 After a rather awkward chockstone, I came to a fork. **1956** C. Evans *Kanchenjunga* xii. 127 Two moves brought him to a chockstone to which he fastened a second runner.

chock (tʃǫk), *sb.*[3] [Imitative.] A hollow sound such as is made by chopping. So **chock-chock.**
1913 D. H. Lawrence *Sons & Lovers* ii. 36 Mrs Morel could hear the chock of the [cricket] ball. **1922** 'K. Mansfield' *Garden Party* 72 And now there came the chock-chock of wooden hammers. **1954** W. Faulkner *Fable* (1955) 228 The dreamy chock of the woodcutter's axe..

chock, *adv.* Add: **1.** (Earlier example, replacing quot. 1860.)
1832 J. P. Kennedy *Swallow B.* I. xiv. 150 It's only the big wheel stopped as chock as a tombstone.

c. chock-a-block: also of a place or person, crammed *with*, chock-full *of*.
1850 H. Melville *White-Jacket* II. xxxi. 209, I'm blessed if we ar'n't about chock a' block here! **1889** *Pall Mall Gaz.* 30 Sept. 6/2 You will find the place chock-a-block. **1894** *Idler* Sept. 132 We'll see..if that there foundered ship ain't a-going to work out this traverse the same as if she was chock-a-block with bullion. **1903** *Smart Set* IX. 9/1 Good-for-nothings in shop 22, who were full, chock-a-block, of socialism. **1946** W. S. Maugham *Then & Now* v. 15 The city's two or three inns were chock-a-block and men were sleeping three, four and five in a bed.

chocker (tʃǫ·kəɹ), *v.* [? f. Chock *sb.*[1] + -er[1].] *trans.* In the game of Patience: to block (a card, or the player).
1887 M. Whitmore Jones *Games of Patience* 9 If the cards come out unfavourably, you often have to put high upon low ones, at the imminent peril of chockering. *Ibid.* 20 Care and judgment are required here, not to place a card which will chocker the one below it. *Ibid.* 47 You are, in Patience parlance, 'chockered'. **1892** 'L. Hoffmann' *Patience Games* 5 When the player reaches a point at which he can make no further progress, he is said to be 'blocked', or, less elegantly, 'chockered'.

chocker (tʃǫ·kəɹ), *a.* Also **chocka, chokker.** *slang* (orig. *Naval*). [Abbr. of *chock-a-block* (see Chock *adv.* 1 c).] 'Fed up'; extremely disgruntled.
1942 *Gen* 1 Sept. 13/1 When Jenny the Wren is fed up with the world she is 'chokker'. **1943** Hunt & Pringle *Service Slang* 22 *Chocker*, this is the sailor's way of saying he is fed up or browned off. **1945** 'Tackline' *Holiday Sailor* xiv. 142 Says she's chocka with being blonde, and she'll be brown again by the time I see her. **1958** F. Norman *Bang to Rights* III. 130 I'm a little chocker of this place [*sc.* prison].

choco (tʃǫ·ko). *Austral. slang.* Also **chocko.** [Abbrev. of *chocolate soldier* (see below).] A militiaman or conscripted soldier.
[**1919** Downing *Digger Dial.* 16 *Chocs*, the 8th Brigade ('Tivey's Chocolate Soldiers.' Originally an abusive name; now an honourable appellation).] **1943** G. H. Johnston *New Guinea Diary* ii. 60, I told you we'd do something when we got stuck into 'em. Not bad for Chockos! **1943** Baker *Austral. Slang* (ed. 3) 20 *Choco*, a Militiaman. (War slang.) *Chocolate soldier*, as for 'Choco'. **1955** A. Ross *Australia 55* v. 64 Conscripts, known as 'chocos' (chocolate soldiers) were not required to serve outside prescribed areas. **1966** *Sunday Truth* 3 July 3/3 During the war against Japan..many of the Diggers were conscripts (chocos).

chocolate. Add: **2.** Esp. a sweetmeat in the form of bars, cakes, or drops, often with a qualifying word (see quot. 1925). Also with *a* and *pl.*, a sweetmeat made entirely of or coated with chocolate. See also ***milk-chocolate.**
1861 Mrs. Beeton *Bk. Househ. Managem.* 804 Box of chocolate. This is served in an ornamental box...May be purchased at any time. **1874** L. Troubridge *Life amongst Troubridges* (1966) 78, I do *adore* Menier and Nougat chocolate...It is the only thing, except other sorts of choc. *and* ices, that I really *enjoy* eating. **1887** *Army & Navy Co-op. Soc. Price List* 38 Chocolates in Boxes..per box 1/9. **1925** B. Beetham in E. F. Norton *Fight for Everest: 1924* 368 Chocolate..plain, nut, milk, nut-milk, Bitro, vanilla, coffee, etc. **1959** [see *bar *sb.*[1] 3 a]. **1966** *Listener* 10 Feb. 217/1 Television programmes that are made up like boxes of assorted chocolates leave one with a confused taste in the mouth.

4. b. in U.S. *spec.* of certain soils.
1821 T. Nuttall *Jrnl. Trav. Arkansa* vi. 99 The chocolate or reddish-brown clay of the salt formation. **1858** *Texas Almanac* 56 The soil is chocolate loam. **1869** *Overland Monthly* III. 130 Texas is notable for the number of its soils. ..There is the 'chocolate' prairie and the 'mulatto' and the 'mezquite' [etc.].

5. *chocolate bar*; *chocolate-coloured* adj. (earlier and later examples); **chocolate biscuit,** a biscuit coated with chocolate; **chocolate-box,** a decorated cardboard box filled with chocolates; also *fig.*, and *attrib.* or quasi-*adj.* = **chocolate-boxy* adj.; so **chocolate-box(e)y** *a.*, like the (usu. stereotyped romantic) pictures on chocolate-boxes; **chocolate-cake,** (*a*) a chocolate-flavoured cake; (*b*) raw chocolate in the form of a 'cake'; **chocolate cream,** (*a*) see Cream *sb.*[2] 2 a; (*b*) a form of confectionery consisting of a sweet, flavoured, creamy paste coated with chocolate; **chocolate (cream) soldier,** (*a*) a chocolate sweet in the shape of a soldier; (*b*) *transf.* a soldier who will not fight; (*c*) = *choco; **chocolate drop,** a small round sweet-meat made of chocolate; **chocolate éclair** = *éclair q.v.; **chocolate spot** (see quot.).
1937 I. Gershwin *They all Laughed* 3 They all laughed..at ..Hershey and his choc'late bar. **1967** Wodehouse *Company for Henry* v. 74 And now for Mr. Stickney's chocolate bars...He munches them instead of sweets. **1723** J. Nott *Cook's Dict.* No. 128 C Chocolate Biskets. **1907** *Yesterday's Shopping* (1969) 7 Biscuits..Chocolate (Society's make). **1932** L. Golding *Magnolia St.* I. ii. 40 She insisted on getting some chocolate biscuits for Mrs. Seipel. **1901** *Daily Chron.* 22 Aug. 3/3 It is odd that a writer whose observation..is so real should..be satisfied with these chocolate-box conventions. **1931** Wyndham Lewis in *Time and Tide* 25 July 883 Their faces failing to conform to any recognised chocolate-box canon, thousands of women far better fitted to continue the race than the indolent characterless chocolate-box-faced monsters of prettiness in vogue at the time [etc.]. **1959** *Spectator* 24 July 102/1 'Classical ballet'.. in its chocolate-box you will still find a few sexless sylphs and glades. **1965** *Listener* 18 Nov. 795/2 The sequence.., for all its chocolate-box vulgarity, contains sufficient emotional impact [etc.]. **1894** Somerville & 'Ross' *Real Charlotte* I. viii. 114 Christopher will only say that she is chocolate-boxey! **1923** A. Huxley *On Margin* 198 A fear of becoming sentimental, or 'chocolate-boxy'. **1959** *Observer* 2 Aug. 12/7 The harsh unrelieved lighting tends to make every group on which the camera dwells look chocolate-boxy. **1868** M. Jewry *Warne's Model Cookery* 67/2 Scrape up about a quarter of a pound of the chocolate cake into a saucepan with two gills of water. **1876** M. F. Henderson *Cooking* 299 Chocolate-cake. Make a cup-cake with the following ingredients: One cupful of butter, [etc.]. **1969** R. D. Abrahams *Jump-Rope Rhymes* 134 My mother made a chocolate cake. **1736** B. Lynde *Diary* 21 Oct. (1880) 91 My chocolate colored coat. **1819** E. Dana *Geogr. Sk.* 188 An extensive body of level rich land, of fine black or chocolate coloured soil. **1906** *Westm. Gaz.* 12 Apr. 1/3 Vesuvius.. showering a thick layer of chocolate-coloured dust over us. **1723** J. Nott *Cook's Dict.* No. 131 C *Chocolate Cream.* Boil.. Sugar in..Milk..beat up the Yolk of an Egg, put it into the Cream..put in Chocolate. **1851** *London at Table* 1. 18 A maraschino jelly, and a chocolate cream, form the sweets. **1879** C. M. Yonge *Magnum Bonum* I. iv. 58 We'd got nothing to eat but chocolate creams. **1885** *Army & Navy Co-op. Soc. Price List* 30 Chocolate Creams ..per box..1/8. **1931** J. Cannan *High Table* x. 157 A girl with..a box of chocolate creams on her knees. **1931** S. Jameson *Richer Dust* iv. 79 A sponge cake, hollow in the centre and heaped with chocolate cream. **1898** Shaw *Arms & Man* I. 17 You are a very poor soldier—a chocolate cream soldier! **1898** *Westm. Gaz.* 27 Apr. 2/2, I cease to ask whether the chocolate cream soldier is true to life. **1826** M. Kelly *Reminiscences* I. 208 Continually putting chocolate drops, which he took from his waistcoat pocket, into his mouth. **1883** R. Haldane *Workshop Rec.* Ser. II. 160 Chocolate Drops with Nonpareils. **1912** D. H. Lawrence *Let.* 17 Sept. (1962) I. 147 The officials are all Chocolate Soldiers. They let you walk through the Customs with a good day. **1918** R. H. Knyvett *'Over There' with Australians* xvii. 133 There was a good deal of rivalry between us and another brigade known as 'The Chocolate Soldiers'. **1943, 1955** Chocolate soldier [see *choco]. **1959** *Listener* 1 Oct. 545/2 Captain Kelly, the mole-like chocolate soldier with a passion for soap and water. **1950** *N.Z. Jrnl. Agric.* Jan. 10/3 Chocolate spot (*Botrytis cinerea*) is a fungous disease which can be found on broad beans almost any year.

chocolat(e)y (tʃǫ·kǒlěti), *a.* [f. Chocolate + -y[1].] Made of, containing, or resembling chocolate.
1965 *Economist* 9 Jan. 99/2 The prices of so many sugary, chocolaty things. **1969** *Vogue* Nov. 63 The stripes are red and grey and chocolatey brown.

Choctaw (tʃǫ·ktǭ). [ad. Choctaw *Chahta.*]
1. A Muskogean tribe of North American Indians, originally inhabiting Mississippi and Alabama; a member of this tribe. Also *attrib.* or as *adj.*
1722 D. Coxe *Descr. Carolana* 25 This mightier Nation of the Chattas consisting of near 3000 Fighting Men. **1725** G. Chicken *Jrnl.* 12 Aug. in N. D. Mereness *Trav. Amer. Col.* (1916) 122 The Chickesaws have made a peace with the Chactaws. **1736** J. Wesley *Jrnl.* 30 June (1938), The Choctaws, the least polished, i.e. the least corrupted of all the Indian nations. **1738** W. Stephens *Jrnl. Proc. Georgia* in *Col. Rec. Georgia* (1906) IV. 82 The Choctaw Indians.. had..come in a great body. **1775, 1791** [see *Creek *sb.*[3]]. **1842** Dickens *Amer. Notes* II. iv. 95 A chief of the Choctaw tribe of Indians. **1948** *Daily Ardmoreite* (Ardmore, Okla.) 21 Apr. 7/5 Doc is half Choctaw along with a mixture of English and Irish.

b. The language of this tribe. Also *attrib.*
1796 J. Morse *Amer. Univ. Geogr.* (ed. 3) I. 669 The Chickasaw and Choctaw languages. **1880** G. W. Cable *Grandissimes* 60, I also speak Choctaw. **1941** [see *Chickasaw 2]. **1948** D. Diringer *Alphabet* I. x. 183 Choctaw is another important Muskhogean dialect; it has now about 18,000 speakers, who live in eastern Oklahoma and in Mississippi.

c. *colloq.* As a type of an unknown or difficult language; cf. Greek *sb.* 8.
1839 *New Orleans Picayune* 1 Mar. 2/4 Even admitting a person understands French and pronounces the name of a dish correctly, its all Hebrew or Choctaw to the waiter. **1929** J. Buchan *Courts of Morning* 13 He had a good many private expressions that were Choctaw to those that did not know him.

2. *Skating.* A step from either edge on one foot to the opposite edge on the other foot, in an opposite direction. (Cf. Mohawk 4.)
1892 J. M. Heathcote et al. *Skating* 81 Starting as before from the outside forward, it is possible to put the other foot down not on the outside but on the inside back. This step it is proposed to call a 'Choctaw'. **1892** Monier-Williams *Figure-Skating* 61 A Choctaw is simply a step or stroke from any edge in one direction, to the opposite edge on the other foot, in an opposite direction. **1908** E. F. Benson *Eng. Fig. Skating* 119 Mohawks and Choctaws, to attempt a definition, are a cross between edges and turns...In Mohawks the same edge (outside or inside) as has been laid down by the first foot is taken up by the second; in Choctaws the opposite edge. **1948** *Time* 2 Feb. 51/2 From difficult double Salchows to simple open Choctaws.

choga (tʃǭ·gā). [Turki *choghā.*] A loose garment with long sleeves like a dressing-gown, worn by Afghans.
1869 Watson & Kaye *People of India* IV. No. 209 A richly embroidered robe or choga of Cashmere cloth. **1883** *Q. Rev.* Apr. 303 We believe his favourite dishabille was an Afghan *choga*. **1891** Kipling *City Dreadf. Nt.* 20 The black frock-coat rises to explain his friend's amendment...The black *choga*..speaks again. **1905** *Westm. Gaz.* 2 Nov. 3/1 Mussulman Gentlemen. Black broadcloth or silk jubba, jama, choga or other robe worn over a saya.

choical (kǭ·ikăl), *a.* [f. Gr. χοϊκός earthy (f. χοῦς dust) + -al.] A Gnostic term for: earthy.
1708 H. Dodwell *Explication* 6 He supposes them [*sc.* human souls]..to be Hylical and Choical, not Cælestial. **1914** R. M. Jones *Spiritual Reformers* p. xiii, There was in man..a visible body..believed to be composed, according to many of the Gnostics, of a subtle element..which they named the *hylic* body, and a sheath of gross earthly matter which they called the *choical* body.

choice, *sb.* Add: **2. c.** *Hobson's choice*: also rhyming slang for 'voice'; freq. *ellipt.*
1937 in Partridge *Dict. Slang* 394/2. **1960** *Observer* 17 Jan. 19/1 Her Hobson's (rhyming slang for voice) is one of the most vibrant in the business. **1961** *New Statesman* 14 July 43/3 The landlady, Queenie Watts, throws her Hobsons (*Hobson's choice* = *voice*) so hard that on a clear night you could hear it in Canning Town.

choice, *a.* **3. b.** (Earlier U.S. example.)
1775 A. Adams in *Familiar Lett.* (1876) 128, I received.. the other articles you sent,..I shall be very choice of them.

choir, *sb.* Add: **3. c.** (See quot. 1909.) Chiefly *U.S.*
1909 *Cent. Dict.*, *Choir*,..a group of instruments of the same class.., as a *trombone choir*. **1955** R. Blesh *Shining Trumpets* (ed. 3) xii. 277 The sluggish swing mechanism. with its brass and reed choirs.

7. choir nun or **sister,** a member of a female religious society who is professed for the choir (i.e. to perform the choir offices). distinguished from *lay sister*; **choir offices,** the divine service which is said or sung in choir (i.e. the canonical hours, morning and evening prayer of the Book of Common Prayer); **choir practice,** the trial or performance of set pieces of music by a choir, under the instruction of the choir-master; **choir school,** a school for choir-boys that is maintained by a cathedral or large church; so *choir scholar.*
1788–9 in *Cath. Rec. Soc. Publ.* (1910) VIII. 187 The Number of Religious at this time are 13 choir Nuns 3 lay sisters & 3 novices. **1876** *St. Cross, Holywell, Parish Mag.* 277 The choral cope worn on ferias in large churches by all the clergy at the choir offices. **1898** *Daily News* 26 Aug. 5/1 The canons are..free in all acts outside those of the choir offices. **1865** F. Helmore *Ch. Choirs* 65 In Choir Practice.. let the Choir-master..insist on constant attention to the movements of his 'baton'. **1904** E. F. Benson *Challoners* vi, This afternoon..she would have to take choir-practice in the Room. **1873** C. M. Yonge *Pillars of House* I. xx. 192 He is in the choir school at Minsterham Cathedral. *Ibid.* II. xvii. 137 You and Harewood can be retained as choir scholars for another year. **1958** *Times* 23 June 9/4 A match between the Fellows of Kings and the Choir School. **1889** A. Devine *Convent Life* iv. iv. 298 A Lay-Sister cannot be transferred to the state of a Choir-Sister without the authority of the Holy See.

choisya (ʃwā·ziă). [mod.L. (C. S. Kunth in Humboldt & Bonpland's *Nova Genera et Species Plantarum* (1823) VI. 513), from the name of M. J. D. *Choisy* (1799–1859), a Swiss botanist.] An evergreen shrub, the only member of the genus so called, native to Mexico and belonging to the family Rutaceæ; also called *Mexican orange*.
1840 J. Paxton *Pocket Bot. Dict.* 75/1 *Choisya*..is an ornamental shrub, growing about six feet high. **1913** W. G. Lawrence *Let.* 5 Nov. in *Home Lett. T. E. Lawrence* (1954) 461 Your letter—enclosing the flower of the choisya—came last mail. **1963** *Times* 9 Feb. 11/3 A few nights of searing east winds may yet singe my choisya.

choke, *sb.*[1] Add: **4.** In a firearm = Choke-bore (both senses).
1875 *Field* 26 June 635/3 After firing upwards of 2,500 shots from each of the three guns to be tested, no falling off of any kind could be discovered, nor did the measurement by callipers vary even to the thousandth of an inch from the original diameter of the choke. **1884** F. F. R. Burgess *Sporting Fire-arms* 4 The sketches..show the ordinary choke and the tulip choke respectively. **1892** W. W. Greener *Breech-Loader* 146 With a twelve bore gun, standard load distance and conditions, the ordinary full choke will make an average pattern of 215 pellets; The half-choke 185 pellets; The quarter-choke 160 pellets. **1959** *Chambers's Encycl.* VI. 664/1 The greater the degree of choke the more the concentration of the pellets in the shot-charge during flight.

6. *Electr.* (Also *choke-coil.*) = **choking coil*; *choke control* (in wireless telegraphy), modulation of the carrier-wave by means of a choke.
1909 *Cent. Dict.* Suppl., Choke-coil. **1911** A. B. Smith *Mod. Amer. Telephony* xxv. 685 The choke coils being inserted in order to prevent the alternations from flowing back on the line circuit. **1913** *Year-bk. Wireless Telegr.* 331 The transformer secondaries..are protected..by air core chokes. **1923** Hawkhead & Dowsett *Techn. Instr. Wireless Telegr.* 156 A third method known as 'anode' or 'choke' control. *Ibid.* 184 The Choke Coils (Air Core). *Ibid.* 200 Air Core Choke. **1928** *Morn. Post* 9 July, When chokes having a winding of thick wire are used the voltage lost is quite small. **1928** *Times* 26 Nov. 11/3 R.W.C...was charged.. with selling high frequency chokes, to which a false trade description..had been applied. **1941** *Electronic Engin.* XIV. 396 Small transformers, chokes and loudspeakers.. should be treated in the same manner as suggested for the mains transformer.

7. A valve which controls the flow of air through the air-intake of a carburettor, chiefly to provide a richer petrol mixture for starting. Also *attrib.*
1926 in *Amer. Speech* I. 686/1. **1932** *Autocar* 28 Oct. 7 (Advt.), Improved choke giving easier starting. **1955** *Times* 10 May 7/6 It starts readily when cold (but not always so willingly when hot) and warms up quickly with but little use of the choke. **1959** *Motor* 3 June 604/1 The 'choke' control..was needed scarcely at all in warm spring weather.

choke, *v.* Add: **10. b.** To damp (a cymbal), usu. just after it has been struck.
1927 *Melody Maker* June 609/3 The best way to produce this effect is to use the same hand..to play the snare drum and choke the cymbal. **1928** *Ibid.* Feb. 203/3 All beats shown to be played on the loose cymbal must be damped out or choked, that is to say, not allowed to 'ring on' for longer than the value of the note as written. **1964** B. Zildjian in Norton & Spacey *Drums & Drumming Today* 87 *Splash Cymbals.* (Also called 'Choke' Cymbals.) Small cymbals 7 to 11 inches in diameter. They are usually thin in weight and are used for fast cymbal crash work, and are very often choked off fast.

19. Also, to deter, discourage (forcibly), dissuade.
a **1916** 'Saki' *Seven Cream Jugs* in *Toys of Peace* (1919) 227 We always choked him off from coming to see us. **1914** Kipling *Diversity of Creatures* (1917) 343 You ask 'im to read it. That'll choke 'im off, because 'e can't! **1921** G. B. Shaw *Back to Methus.* II. 40 It's quite easy to be a decent parson. It's the Church that chokes me off. **1955** M. Gilbert *Sky High* viii. 110 There's a broad distinction between being dispirited and being choked off. I refuse to be choked off.

choke-. Add: **1. a. choke-cymbal** (see quots. 1934 and 1938); also, a cymbal that is 'choked' (see *choke *v.* 10 b).
1934 E. Little *Mod. Rhythmic Drumming* 21 It is.. desirable to have 'choke' cymbals attached to the hoop of the bass drum. Two 8-inch cymbals are clamped together, 'face to face' and are mounted on a holder which can be fixed within easy reach of the sticks. A screw washer makes it possible for the volume of tone to be controlled. **1938** *Oxf. Compan. Mus.* 160/2 *Choke cymbals.* These are two ordinary cymbals, fixed face to face on a rod, with a device by which their pressure, one on the other, can be adjusted according to the tone-quality desired. **1964** [see *choke *v.* 10 b]. **1967** *Crescendo* June 9/1 It is all there: the banjo, the choke cymbal, the sax trio with its 'jolly' phrasing.

choke-berry. [Choke- 1 b.] A plant belonging to the genus *Aronia*, which includes several species of North American shrubs; also, the fruit of this plant.
1778 J. Carver *Trav. N. Amer.* 511 The Choak berry. The shrub thus termed by the natives..bears a berry about the size of a sloe, [etc.]. **1836** D. B. Edward *Hist. Texas* iv. 66 The Cranberry and the choakberry. **1838** Mrs. A. H. Lincoln *Lect. Bot.* (ed. 7) 77 *Aronia arbutifolia*..Red Choke-berry. **1851** [see Choke- 1 b]. **1898** *Atlantic Monthly* LXXXII. 497/2 With the huckleberry, was chokeberry, another New England acquaintance. **1901** C. Mohr *Plant Life Alabama* 71 The chokeberry, which is here of arborescent habit, presents a beautiful sight when loaded with its bright scarlet fruit. **1949** L. H. Bailey *Man. Cultivated Plants* (rev. ed.) 518 Aronia...Chokeberry...little shrubs of swamps and low woods.

choke-cherry. *N. Amer.* (Earlier example.)
1785 Cutler in *Mem. Amer. Acad. Arts & Sci.* 1784 I. 449 The Black Choke Cherry...The Red Choke Cherry.

choked, *ppl. a.* Add: **2.** Annoyed, disgusted, disappointed.
1950 in Partridge *Dict. Underworld* (1961) Add. 793/1. **1956** *New Statesman* 8 Dec. 740/2 They said that Kenneth Large and some of the other geezers..were choked—doing their nut over being beaten in an essay competition by a Paddy. **1965** L. J. Cunliffe *Having it Away* i. 14 'Do you wish me to inform your wife that you are confined to the ship?' Too choked to speak, I nodded. *Ibid.* xv. 100 The bogies knew he'd had it away, but just couldn't pin the job on him. Which meant he was laughing and they were double-choked. **1969** *Oz* Apr. 17/1 My governor is going to be choked when I take the day off. He's going to be double choked if I enjoy myself.

choke-pear. 1. Delete † and add later examples.
1839 J. Buel *Farmer's Compan.* 269 It is as easy to cultivate the vergaleu as it is the choke pear. **1872** *Rep. Vermont Board Agric.* I. 107 Man, by his industry, skill and perseverance applied to the wild choke pear of our fields,..has transformed it.

choker. Add: **2. b.** A necklace or decorative band worn close up against the throat.
1928 *Daily Chron.* 9 Aug. 13/1 A string of chokers is quite inexpensive, and can be worn with practically anything. **1953** A. Baron *Human Kind* 131 A white pearl choker showing off..her slender neck.

4. *Electr.* = **choking coil.*
1902 *Encycl. Brit.* XXXIII. 233/1 Choking or impedance coils..called 'chokers'.

5. A noose of wire rope or the like tied round a log for hauling it. orig. *U.S.*
1905 *Terms Forestry & Logging* (U.S.) 33 *Choker*, a noose of wire rope by which a log is dragged. **1945** B. Macdonald *Egg & I* (1946) iii. 42 He told me to put the chokers on the fir trees and to shout directions for the pulling. **1957** *Brit. Commonw. Forest Terminol.* II. 44 *Choker*, a length of flexible wire rope or chain with a connexion (*shackle*) for attaching it to the butt hook on one end and a hook..on the other. *Note.* A choker is used to form the noose around the log and hold it during dragging.

chokiness (tʃǭ·kinės). [f. Choky *a.* + -ness.] The condition of being choky or inclined to choke.
1844 Hewlett *Parsons & Widows* vi, I felt a short, unpleasant kind of chokiness. **1901** 'L. Malet' *Sir R. Calmady* v. v. 422 Smiling to herself, notwithstanding a chokiness in her throat.

choking, *vbl. sb.* or *ppl. a.* Add: **choking coil** *Electr.*, a coil of high inductance inserted in an alternating-current circuit to impede and cut down the current or to change its phase. Called also *impedance* or *reactance coil.*
1893 R. M. Walmsley *Wormell's Electricity in Service of Man* 644 An extremely pretty device known as a 'choking' coil. **1902** W. G. Rhodes *Altern. Currents* 83 Impedance coils, or choking coils, as they are often called, are simply coils having low ohmic resistance and high self-induction. **1962** S. Handel *Dict. Electronics* 52 *Choke* (*Choking coil*), an inductor designed to present a relatively high impedance to alternating current.

chokka (tʃǫ·kā). *S. Afr.* Also **cho·kker** (-əɹ). [Afrikaans *tjokka.*] = Squid 1.
1902 J. D. F. Gilchrist in *Trans. S. Afr. Philos. Soc.* XI. 224 Though not a fish the 'Tschokka' may be mentioned as another example of this [*sc.* an onomatopœic name]. It is applied to the Cuttle-fish, or Squid, on account of the peculiar noise it makes when landed. **1930** C. L. Biden *Sea-Angling Fishes of Cape* I. i. 42 A species of cuttle-fish ..locally called 'chokker' and..classified as *Loligo indica.* **1950** *Cape Argus* 16 June 5/8 These fish are still biting well, although most are caught on whole chokka or mackerel bait. **1957** S. Schoeman *Strike!* vii. 186 It is hardly ever necessary to preserve fish bait or chokker for longer than three or four days. **1961** *Cape Times* 18 Aug. 5/6 The bay is full of snoek, chokka and mackerel.

chokker, var. *chocker *a.*, *chokka.

choko (tʃǭ·ko). *Austral.* and *N.Z.* = Chocho.
1902 *Westm. Gaz.* 24 Dec. 2/1 Chinkey Tong was down the cultivation paddock cutting chokos. **1906** Mrs. Beeton *Househ. Managem.* lvi. 1579 The choko is a vegetable little known, although it grows freely in many parts of Australia. **1914** *Bulletin* (Sydney) 18 June Red Page/1 The chokos on the vine are turning white. **1945** C. Mann in *Coast to Coast* 9 The occasional pie-melons and chokos that constituted the inhabitants' gardens. **1950** *N.Z. Jrnl. Agric.* Oct. 363/1 The choko (*Sechium edule*) is grown by many home gardeners in the North Island, especially by those who are fond of marrows and squashes. When cooked the choko is very much like these in flavour. In habit of growth the choko resembles a climbing cucumber. **1966** *Sunday Mail Mag.* (Brisbane) 6 Nov. 6/3 Few vegetables are as prolific croppers or as easy to grow as the choko.

chokra (tʃǭ·krā). *India.* [Hind. *chhokrā.*] A boy, youngster; esp. one employed as servant in a household or in a regiment.
1875 A. Wilson *Abode of Snow* 136 He was dubbed 'the Chokra', or simply 'boy'. **1895** Mrs. Croker *Village Tales* (1896) 51 The two natives—the open-mouthed, gaping chokra, the respectfully exultant bearer. **1926** *Blackw. Mag.* Oct. 499/2 In his absence the *chokra* was to look after me. **1934** 'G. Orwell' *Burmese Days* ii. 29 The invisible *chokra* who pulled the punkah rope outside.

chol-. Add: **cho:langio·graphy** [ad. Sp. *colangiografia* (P. L. Mirizzi 1932, in *Bol. Trab. Soc. Cirugia Buenos Aires* XVI. 1133)], X-ray examination of the gall-bladder and bile-ducts; **cholangitis** (kǫlændʒəi·tis) [see Angio-], inflammation of a bile-duct.
1936 *Surg., Gynecol. & Obstetrics* LXII. 684 (*heading*) Cholangiography. A Modified Technique for X-ray Visualization of the Bile Ducts During Operation. **1962** *Lancet* 8 Dec. 1208/2 Radiographic examination of the biliary tract is an invaluable aid to diagnosis. Cholangiography yields more information than cholecystography. **1886** *Buck's Handbk. Med. Sci.* III. 290/1 The cases.. presented nearly all the anatomical appearances that occur in cholangitis and cholecystitis. **1907** *Practitioner* Oct. 571 Catarrhal cholangitis. **1959** *Chambers's Encycl.* VIII. 613/1 When inflammation involves the gall-bladder the disease is cholecystitis, and when the bile-ducts alone are involved the disease is cholangitis. **1963** *Lancet* 12 Jan. 72/1 A 38-year-old woman had chronic jaundice and recurrent attacks of cholangitis after a bile-duct lesion in connection with cholecystectomy.

chole-. Add: **cholelithiasis** (-liþəi·ăsis), formation or presence of biliary calculi; an instance of this condition.
1887 *Buck's Handbk. Med. Sci.* IV. 560/1 Cholelithiasis, or gall-stone formation. **1906** *Practitioner* Dec. 809 Both in cholelithiasis and in nephrolithiasis. **1963** *Lancet* 19 Jan. 136/1 Stones from 10 unselected cases of cholelithiasis were also washed and tested in the same way.

cholecyst. Add: **cholecystenterostomy** (-entěrǫ·stǒmi) [Gr. ἔντερον gut, intestine, στόμα mouth], the operation of establishing a passage between the gall-bladder and the intestine by incision and suture; **chole:cystorrhaphy** (-ǫ·răfi) [Gr. ῥαφή sewing, suture], suture of the gall-bladder after an incision; **cholecystostomy** (-ǫ·stǒmi) [Gr. στόμα mouth] (see quot. 1961).
1904 *Westm. Gaz.* 6 May 6/3 The 'Matin' states that the operation performed on the ex-Premier was that of cholecystenterostomy. **1906** *Practitioner* Dec. 790 In patients, who have shown signs of cholecystitis after the attack, he advocates a subsequent cholecystostomy. **1908** *Ibid.* Sept. 452 The removal of biliary calculi..either by cholecystostomy or cholecystenterostomy. **1910** *Ibid.* Mar. 377 As a general rule the operation of cholecystectomy shows a higher mortality than cholecystostomy. *Ibid.* 381 When the gall-bladder is found to be wounded, the question arises as to whether it should be treated by gauze packing, cholecystectomy, or cholecystorrhaphy. **1961** *Brit. Med. Dict.* 298/1 *Cholecystostomy*, incision into the gall bladder for the purpose of inserting a drainage tube.

cholecystography (kǫ:lisistǫ·grăfi). *Med.* [f. Cholecyst + -o + -graphy.] X-ray examination of the gall-bladder after the administration to the patient of a radio-opaque substance. Hence **cho:lecystogra·phic** *a.*, pertaining to or involving cholecystography; **cholecy·stogram,** an X-ray photograph of the gall-bladder.
1925 E. A. Graham et al. in *Jrnl. Amer. Med. Assoc.* LXXXIV. 14/2 We have established the fact that cholecystography is possible in the living human subject. By this we mean making the gall-bladder opaque to the roentgen ray by the injection of certain substances into the blood stream. **1926** *Glasgow Med. Jrnl.* CVI. 153 (*heading*) Cholecystography: The X-ray Visualisation of the Gall-Bladder by means of Sodium Tetraiodophenolphthalein (Graham's Method). **1927** *Trans. Amer. Gastroenterological Assoc.* 1926 XXIX. 284 (*title*) Cholecystographic gall-bladder diagnosis. **1927** *Jrnl. Amer. Med. Assoc.* 14 May 1550/1 Cholecystography is being used so extensively to test the function of the gallbladder that it was considered advisable to compare the evidence of the cholecystogram with the pathological changes in the gallbladder after its removal. **1962** [see *cholangiography* s.v. *chol-].

choledoch. Add: **cho:ledocho·tomy** [see -tomy], incision of the gall-duct (to remove gall-stones).
1894 in Gould *Dict. Med.* **1908** Allbutt & Rolleston *Syst. Med.* IV. i. 233 Out of 150 choledochotomies that I..performed the mortality was under 4 per cent. **1908** *Practitioner* Dec. 829 Choledochotomy after cholecystectomy.

cholera. Add: **2.** *cholera infantum*, a once common and often fatal diarrhœa of young children prevailing in summer months.
1829 B. Hall *Travels N. Amer.* III. 389 The illness of our little girl, whose long exposure to the noxious air of the great rivers, had given her a complaint..called by the ominous name of Cholera Infantum. **1886** *Buck's Handbk. Med. Sci.* II. 135/1 The line of demarcation between cholera infantum and the ordinary summer diarrhœa, entero-colitis, cannot always be sharply drawn. **1895** *Outing* (U.S.) XXVI. 405/2 You are forced to wish that cholera-infantum had been more prevalent sixty years ago. **1961** *Brit. Med. Dict.* 299/1 *Cholera infantum*, summer diarrhoea, acute gastro-enteritis of children.

5. cholera belt, a waistband of flannel or silk worn as a preventive against stomachic ailments.
1848 in *Medical Hist.* (1957) I. 217 Each soldier is to be provided with two cholera belts as part of his Necessaries. **1864** E. A. Parkes *Man. Pract. Hygiene* xiii. 356 Cholera made belts are of flannel, and fold 'twice over the abdomen.

1906 Mrs. Beeton *Househ. Managem.* xx. 1853 The so-called 'Cholera Belt' is a regulation garment in the kit of every British soldier on Indian service. **1961** D. Bates *Fly-Switch from Sultan* i. 9 Cholera-belts and spine-pads were other mistakes. There was a shiny-faced colonel who had been in India in the nineties...Anyway, he was very keen on cholera-belts (to prevent a chill on the tummy in the treacherous early hours).

choleretic (kǫlěre·tik), *a.* and *sb. Chem.* [ad. G. *cholereticum* (Brugsch and Horsters 1923, in *Klin. Wochenschr.* 13 Aug. 1558/1), f. mod.L. *choleresis*, secretion of bile by the liver; after *diuretic.*] Stimulating the secretion of bile by the liver. Hence as *sb.*, a drug that has this effect.
1927 *Jrnl. Pharmacol. & Exper. Therapeutics* XXX. 185 (*heading*) The Choleretic action of Tolysin. **1929** E. G. Wakefield et al. in *Ann. Internal Med.* III. 572 (*heading*) The Use of Sodium Salt of Dehydrocholeic Acid (Dectrolin) as a Choleretic. *Ibid.* 572/1 Brugsch and Horsters originated the term choleretic to designate a drug which increases the flow of bile from the liver as distinguished from the term cholagogue which means expulsion of bile from the gall-bladder. **1934** *Chem. Abstr.* XXVIII. 6838/2 (*heading*). Choleretics of plant origin. *Ibid.*, Onions contain a choleretic substance acting directly on the liver cells and causing increased secretion of bile and bile substances. **1956** *Nature* 14 Jan. 75/1 A remarkable diuretic and choleretic effect [due to the administration of thioctic acid] was also observed in the rabbit.

cholesterol (kǒle·stěrǫl). *Chem.* [f. Choles-ter(in + -ol.] A steroid alcohol, $C_{27}H_{45}OH$, that occurs esp. in animal cells and body fluids. Formerly called Cholesterin.
1894 *Jrnl. Chem. Soc.* LXVI. i. 486 A great number of analytical results obtained by the authors with cholesterol and its derivatives, seem to show that the composition of cholesterol is expressed by the formula $C_{27}H_{44}O$. **1939** *Thorpe's Dict. Appl. Chem.* III. 88/2 In the tissues of animals, with the exception of certain invertebrates, cholesterol is present and appears to be synthesised *in vivo*. **1955** *Sci. News Let.* 1 Oct. 214/1 A diet free of cholesterol, the chemical that plays a large part in hardening of the arteries, leads to the formation of gallstones in experimental animals.

‖**choli** (kǭ·li). [Hind.] A short-sleeved bodice worn esp. by Indian women.
1908 *Imp. Gazetteer India* III. 199 The *choli* (*khāna*) is a little bodice now largely worn by women below the *sāri*. **1910** *Encycl. Brit.* XIV. 419/2 A short jacket fastened at the back and with short sleeves is worn...In Sind..and other parts of western India it is called a *choli*. **1960** J. Masters *Venus of Konpara* xviii. 131 Her sari was pale green..and under it the choli was almost transparent. **1962** *Sunday Express* 25 Feb. 18/4 A floor-length bell skirt.. had a little choli-style top.

choline. Read: (kǭ·lin, -in; also kǫ-). *Biochem.* Also **cholin.** [a. G. *cholin* (A. Strecker 1862, in *Ann. d. Chem. und Pharm.* CXXIII. 355), f. Gr. χολ-ή bile + -ine[5].] A strong hygroscopic base, $HO{\cdot}N(CH_3)_3{\cdot}CH_2CH_2OH$, which is of widespread occurrence, either free or in combination, in living organisms and which is important biochemically as a source of methyl radicals, as a precursor of acetylcholine and other esters, and in the metabolism of fat; it is sometimes regarded as a vitamin of the B group. Formerly also called *bilineurine*, etc., and confused with neurine. Also *attrib.* and *Comb.*
1869–71 [in Dict.]. **1910** *Encycl. Brit.* III. 827/2 Choline (bilineurine, sincaline)..is found in the bile, in brain substance, and in yolk of egg in the form of lecithin. **1919** *Nature* 20 Nov. 323/2 Lecithin, which is undoubtedly used in the construction of the tissues, plays a like part in detoxicating cholin. **1942** L. Lichtwitz *Functional Path.* i. 6 As cholin takes part in the formation of lecithin, it is present in considerable quantities in nerve tissue. **1946** *Nature* 2 Nov. 630/1 The rate of phospholipid turnover in the rat's kidney is greatest at the time..when the kidney is most susceptible to choline deficiency, and..the turnover is greatly reduced in choline-deficient animals. **1950** R. J. Williams et al. *Biochem. B Vitamins* 431 One might associate a vitamin function with choline for its role in acetylcholine synthesis, were it not so difficult to extend this to micro-organisms. **1968** A. White et al. *Princ. Biochem.* (ed. 4) xxi. 502 Certain analogues of choline,.. although foreign to nature, are incorporated into lecithin and prevent the fatty liver of choline deficiency.

cholinergic (kǫlinǝ̄·ɹdʒik), *a.* [f. Cholin(e + Gr. ἔργ-ον work + -ic.] Of the synapses or nerve-fibres: liberating acetylcholine; also, stimulated by acetylcholine.
1934, 1935, 1955 [see *adrenergic *a.*].

cholinesterase (kǫline·stərēiz). *Biochem.* Also **choline-esterase.** [f. Cholin(e + *esterase.] An enzyme which hydrolyses cholesterol and which is concerned with the transmission of neural impulses..
1932 E. Stedman et al. in *Biochem. Jrnl.* XXVI. 2059 It is proposed to term the enzyme which hydrolyses esters of choline choline-esterase. **1944** *Ann. Reg.* 1943 353 Brain tissue of all vertebrates was found to contain only true cholinesterase, which hydrolyses low concentrations of acetylcholine, and plays an essential role in the chemical transmission of nerve impulses. **1956** *Nature* 28 Jan. 185/2 Most of our knowledge concerning the topical distribution of cholinesterases in the mammalian nervous system is derived from studies on tissue sections. **1963** *Listener* 7 Feb. 238/2 Agents which upset the so-called 'choline-esterase' mechanism by means of which nerves transmit their messages were at one time recommended as war weapons but they now appear to be considered old fashioned.

cholla (tʃoi·ă). *U.S.* Also **choya.** [Mexican Sp.] One or other of several species of *Opuntia*, native to the south-western United States and Mexico.
1846 A. Robinson *Life in Calif.* vii. 102 A large tract covered with choyas, or prickly pears. **1907** S. E. White *Arizona Nights* I. xii. 186 The hill I had to climb was steep and covered with *chollas*. *Ibid.*, I..slipped down the bluff getting pretty well stuck up with the *cholla* spines. **1909** in C. R. Orcutt *Amer. Plants* II. 403 The cholla cactus (Opuntia prolifera) attains a height of six feet. **1955** Priestley & Hawkes *Journey down Rainbow* xii. 171 Cacti..that is what I miss most. Not only the saguara.. the prickly pear, and the inflammable cholla. **1970** *Sci. Amer.* Feb. 96 (*caption*) Cholla cactus and creosote bush are the principal plants in this arid section of..California.

Cholo, cholo (tʃǭ·lo). [Amer. Sp., f. *Chololán*, now *Cholula*, a district of Mexico.] **a.** An Indian of Latin America. **b.** *U.S.* A lower-class Mexican (often derog.). **c.** = Mestizo. The female is **Cho·la.**
1851 H. Melville *Whale* II. i. 1 Archy..whispered to his neighbor, a Cholo. **1860** *Star* (Los Angeles) 3 Mar. 2/2 The lower class of Sonoranians, called cholos. **1862** C. R. Markham *Trav. Peru & India* 80 The Cholos or Indians of Arequipa have long been notorious for their turbulence. **1877** E. G. Squier *Peru* 115 One of them [*sc.* Chinese] had married a native *Chola*. **1908** C. R. Enock *Andes & Amazon* (ed. 2) 19 A Chola girl had entered my room to sell me a *cui*, or guinea-pig. *Ibid.* 138 The Cholos..are the original Quechua Indians of the uplands..and sometimes they have an admixture of Spanish blood. **1909** *N.Y. Even. Post* 19 Jan. 6/3 The Westerner must then equally persecute the 'cholos' who have been pouring into his State from below the Rio Grande. **1925** *Contemp. Rev.* Jan. 799 A few cows and llamas tended by a bright-petticoated, ever-spinning chola. **1959** *New Statesman* 23 May 728/1 A whole area of petty bureaucracy belongs to the lower middle class *mestizo* or *cholo*.

chometz, var. *hametz.

chomophyte (kǭ·mǫfəit). *Bot.* [G. (M. Oettli *Beiträge z. Ökologie d. Felsflora* 1905), f. Gr. χῶμα heaped-up earth + -o + -phyte.] (See quot. 1938.)
1909 tr. M. Oettli in E. Warming *Oecology of Plants* lxii. 240 Cryptogamia and Phanerogamia which only colonize rock where detritus has accumulated..in crevices or on the general surface of the rock. Plants of this latter type we term chomophytes. **1938** Carpenter *Ecol. Gloss.* 56 *Chomophyte*, plants growing in fissures or crevices in rock, and on ledges where rock debris has accumulated. **1939** A. G. Tansley *Brit. Isl. & Veget.* viii. xxxix. 799 On the Scottish mountains the arctic-alpine chomophyte communities reach their greatest development in the corries of the mica-schists. **1951** *Jrnl. Ecol.* XXXIX. 65 In contrast to chasmophytes, chomophytes (which grow in layers of debris overlying lips and ledges of rock) are few and of local occurrence.

chondrenchyma (kǫndre·ŋkimă). *Zool.* Also anglicized **chondrenchyme** (kǫndre·ŋkəim, -kim). [f. Gr. χόνδρος (see Chondro-) + ἔγχυμα infusion, filling.] A cartilage-like tissue in certain sponges. Hence **chondrenchymatous** (kǫ:ndreŋki·mătəs) *a.*, consisting of, of the nature of, chondrenchyma.
1888 W. J. Sollas *Tetractinellida* p. xcviii, The mesogloea..is in some very few of the Chondrospongiæ a chondrenchyme. *Ibid.* p. cli, The mesoderm is in part sarcenchymatous, in part chondrenchymatous. **1900** E. A. Minchin in Lankester *Treat. Zool.* II. 52 Chondrenchyma. *Ibid.* 147 With dense sarcenchymatous choanosome and tough chondrenchymatous ectosome.

chondriosome (kǫ·ndriosǭm). *Biol.* [ad. G. *chondriosom* (F. Meves 1908, in *Archiv f. mikrosk. Anatomie* LXXII. 831), f. mod.L. *chondriosoma*, f. Gr. χονδρίον, dim. of χόνδρος cartilage + σῶμα body.] A minute organelle present in the cytoplasm of most cells and containing enzymes necessary for various metabolic processes; a mitochondrion. Also **cho·ndriocont, -kont** [a. G. *chondriokont* (F. Meves 1907, in *Anat. Anzeiger* XXXI. 401), f. Gr. κοντ-ός pole], a rod-shaped chondriosome.
1910 C. E. Walker *Hered. Characters* ii. 35 He [*sc.* Meves] suggests that certain cytoplasmic structures, 'chondriosomes', divide with the cell, and are handed on individually from generation to generation. **1911** *Jrnl. Morphology* XXII. 777 There are..no filamentous mitochondria (chondriokonts) in the spermatogonia. **1920** L. Doncaster *Introd. Cytology* 21 The structures grouped together under the general name of mitochondria, known also as chondriosomes, or when rod-shaped, as chondrioconts. **1956** *Nature* 25 Feb. 387/1 Small granules often appear in the cell after staining with 1 per cent aqueous Janus green which are thought to be chondriosomes.

chondrite. Add: **2.** *Min.* [ad. G. *chondrit* (G. Rose 1863, in *Abh. Akad. Berl.* 29), f. Gr. χόνδρος granule.] A meteorite containing granules.
1883 *Encycl. Brit.* XVI. 112/1 From the small rounded grains that give it this appearance, the name chondrite.. has been applied to this kind of meteorite. **1912** J. W. Gregory *Making of Earth* I. ii. 35 The meteorites with rounded grains (chondrites) have been regarded as due to the fusion of many separate granules into a large mass. **1944** C. Palache et al. *Dana's Syst. Min.* (ed. 7) I. 121 Most of the stony meteorites are made up of chondri or chondrules..of silicate minerals and are therefore classed as chondrites.

chondritic (kǫndri·tik), *a. Min.* [f. prec. + -ic, after G. *chondritisch.*] Characterized by granular structure; of or pertaining to chondrites.
1866 *Times* 14 Nov. 4/5 These oolitic, or as Gustav Rose, the great mineralogist of Berlin, has happily termed them, *chondritic* aerolites, form by far the largest group among the meteoric stones. **1896** L. Fletcher *Introd. Study Meteorites* 35 The stony part of the siderolites and aerolites is almost entirely crystalline, and in most cases presents a peculiar 'chondritic' or granular structure, the loosely coherent grains being composed of minerals similar to those which enclose them. **1914** *Brit. Mus. Return* 223 Chondritic meteorites. **1922** *Chambers's Jrnl.* 126/2 The 'stones'..have a structure called 'chondritic', utterly unknown among terrestrial substances. **1956** *Nature* 28 Jan. 156/2 The identification of the Earth's mantle with chondritic meteorites is now thoroughly discredited.

chondro-. Add: **chondrocra·nium,** the primitive brain-case, composed of cartilage; **cho·ndrocyte** *Histol.* [-cyte], a cartilage cell; so **chondrocy·tic** *a.*; **chondro·meter** (example); **chondrosa·mine** [Amine], an amino-sugar, $C_6H_{13}O_5N$, the basis of chondroitin; **chon:drosarco·ma** *Path.*, a malignant tumour in which the parenchyma is composed wholly of differentiated cartilage.
1875 T. H. Huxley in *Encycl. Brit.* I. 753/1 The primordial skull or *chondro-cranium.* **1919** J. G. Kerr *Embryol.* II. 306 The skull consists in its simplest form primarily of a chondro-cranium—a trough of cartilage, the cavity of which is occupied by the brain. **1959** *Chambers's Encycl.* XII. 591/1 The parts of the skull which develop in cartilage constitute in the embryo what is called the chondrocranium. **1903** Dorland *Med. Dict.* (ed. 3) 163/1 *Chondrocyte*, a cartilage cell. **1937** E. E. Hewer *Textbk. Histol. Med. Stud.* 52 In the mature state the chondrocytes do not divide. **1968** Bloom & Fawcett *Textbk. Histol.* (ed. 9) ix. 212/1 Cartilage is a specialized form of connective tissue consisting of cells, chondrocytes, and extracellular fibers embedded in an amorphous, gel-like matrix. **1965** R. P. Morehead *Human Path.* viii. 233/2 (*heading*) Benign chondrocytic tumors. **1897** Chaney *Weights & Measures* 130 In estimating the quality of corn a little instrument called a 'chondrometer' is sometimes used. **1914** *Jrnl. Biol. Chem.* XVIII. 123 The substance is isomeric with glucosamine, and until the details of its structure will be disclosed it will be referred to as chondrosamine. Chondrosamine resembles glucosamine in its elementary composition. **1956** *Nature* 10 Mar. 467/2 A sulphated polysaccharide containing both glucosamine and chondrosamine. **1883** J. Coats *Man. Path.* 213 Sarcomas follow in their type the various connective tissues of the body...So Virchow indicates the division into fibrosarcoma, myxosarcoma, gliosarcoma, chondrosarcoma, osteosarcoma, melanosarcoma. **1886** Chondro-sarcoma [see *chondroma]. **1906** *Practitioner* Nov. 663 Similar new growths have been labelled by different observers carcinoma, adeno-sarcoma, chondro-sarcoma, myxosarcoma. **1966** Wright & Symmers *Systemic Path.* II. xxxvii. 1414/2 In contrast to the majority of benign cartilaginous tumours, which occur towards the periphery of the limbs, chondrosarcomas usually arise in the pelvis, ribs, and proximal long bones.

chondrodystrophy (kǫndrodi·strǒfi). *Path.* [ad. G. *chondrodystrophie* (E. Kaufmann *Untersuch. sogenannte foetale Rachitis* 1892), f. mod.L. *chondrodystrophia* (also used), f. Chondro- + Dystrophy.] Any of a number of congenital disorders of the skeleton that involve cartilage, esp. *chondrodystrophia fœtalis*, = *achondroplasia. Hence **cho:ndrody·strophic** *a.*
1893 *Trans. Edin. Obstetr. Soc.* XVIII. 195 (*heading*) Note on three living cases of achondroplasia (chondrodystrophia fœtalis, or so-called fœtal rickets). *Ibid.* 200 The course of the chondrodystrophia. **1903** *Boston Med. & Surg. Jrnl.* 26 Mar. 344 (*heading*) Chondro-dystrophic dwarf. **1903** *Med. Record* 27 June 1038/1 (*heading*) A case of Chondrodystrophy Foetalis in a child fourteen months old. **1957** *Encycl. Brit.* XV. 739/2 Disproportionate dwarfism known as achondroplasia or chondrodystrophia in which the limbs and especially the legs are short, thick and bent, with head and trunk disproportionately large.

chondroitic (kǫndro,i·tik), *a.* [Irreg., f. Chondro- + -itic (see -ite[1] 4 b).] **chondroitic acid** [tr. G. *chrondroitsäure* (C. Boedeker 1861, in *Ann. d. Chemie* CXVII. 111)], (see quot. 1944).
1885 *Jrnl. Chem. Soc.* XLVIII. 405 Bödeker found that on boiling cartilage with mineral acid, a substance was obtained, which he named chondroitic acid... The author made four preparations of this so-called chondroitic acid. **1895** [see next]. **1944** *Hackh's Chem. Dict.* (ed. 3) 197/2 *Chondroitic acid*, $C_{18}H_{27}O_{17}NS$..An acid of cartilage, found in traces in urine; splits into chondroitin.

chondroitin (kǫndrǭ·itin). *Biochem.* [f. *chondroit(ic *a.* + -in.] An organic substance, $C_{18}H_{27}NO_{14}$, found in cartilage, together with sulphuric acid.
1895 *Jrnl. Chem. Soc.* LXVIII. 254 (*heading*) Chondroitin-sulphuric acid...This substance, originally named chondroitic acid, is an ethereal hydrogen sulphate, and has been found in certain forms of cartilage. **1907** *Jrnl. Chem. Soc.* XCII. i. 369 Chondroitin-sulphuric acid..formulae for this substance..and for its hydrolysis products, chondroitin. **1925** P. A. Levene *Hexosamines* II. i. 76 The presence of glucuronic acid in the chondroitin and mucoitin sulphuric acids was demonstrated. *Ibid.* 77 In chondroitin sulphuric acid, the sugar is chondrosamine and in mucoitin sulphuric acid it is chitosamine. **1944** [see prec.]. **1957** *Sci. News* XLV. 87 Chondroitin..has not yet been isolated from human tissue but has been found in bovine cornea.

chondroma (kǫndrǭ·mă). *Path.* [mod.L., f. Gr. χόνδρ-ος cartilage + *-oma.] A benign tumour arising from chondrocytes. Hence **chondro·matous** *a.*
1860 Mayne *Expos. Lex.* 1443/2 *Chondroma*, a cartilaginous growth. **1886** *Buck's Handbk. Med. Sci.* II. 810/2 Chondromatous tissue is found somewhat more frequently (chondro-sarcoma). **1908** *Practitioner* Oct. 535 Excessive and exuberant callus, chondroma, sarcoma, etc., at the site of fractures are comparable with keloid, sarcoma, epithelioma, etc., arising in scars of soft parts. **1966** Wright & Symmers *Systemic Path.* I. ix. 327/1 Chondromas may arise from any of the laryngeal cartilages but their usual site is the cricoid cartilage.

chondrosine (kǫ·ndrosin). *Chem.* Also **chondrosin.** [ad. G. *chondrosin* (C. F. W. Krukenberg 1886, in *Zeitschr. f. Biol.* XXII. 267), f. mod.L. *chondrosia*, f. Gr. χόνδρος cartilage: see -in[1].] A monobasic acid, $C_{12}H_{21}O_{11}N$, obtained by hydrolysis of chondroitin.
1886 *Jrnl. Chem. Soc.* L. 481 Chondrosine is a hyalogen obtained from the sponge *Chondrosia reniformis*. **1925** P. A. Levene *Hexosamines* II. i. 76 Chondrosin, the disaccharide obtained from the more complex substance. **1957** *Science News* XLV. 86 On mild acid hydrolysis both [polysaccharides chondroitin sulphates A and C] give the same disaccharide, chondrosine.

chondrule (kǫ·ndrūl). *Min.* [f. as Chondrite + -ule.] A small spherical grain of mineral embedded in varying numbers in the matrix of chondritic meteorites. Also **cho·ndrulite, chondruli·tic** *a.* (see quot. 1928).
1889 in *Cent. Dict.* **1892** Dana *Syst. Min.* (ed. 6) i. 32. **1896** L. Fletcher *Introd. Study Meteorites* 36 Through this paste are disseminated round chondrules of various sizes and with the same mineral composition as the matrix; in some cases the chondrules consist wholly or in great part of glass. **1906** Iddings *Rock Min.* 294 Enstatite..occurs in meteorites in radiating aggregates called chondrules. **1928** T. C. Chamberlin *Two Solar Families* 254 'Chondrulites' is merely a more general term here used to include not only chondrules but their débris and all such quasi-chondrulitic material as cannot be called chondrules. It is here used for the primitive chondrulitic accretions, whether they develop into chondrules or not. **1944** [see *chondrite 2].

chone (kǭn). [ad. Gr. χώνη, contr. f. χοάνη *choana.] In sponges, a cortical dome-like structure communicating with the subdermal cavity. Hence **chonal** (kǭ·năl) *a.*, pertaining to this.
1887 Sollas in *Encycl. Brit.* XXII. 415/1 In many sponges..the cortical domes are constricted near their communication with the subdermal cavity by a transverse muscular sphincter, which defines an outer division or ectochone from an inner or endochone.., the whole structure being a chone. **1888** —— *Tetractinellida* p. xxiii, The chonal sphincter. **1906** *Cambr. Nat. Hist.* I. viii. 214 A chone is a passage through the cortex opening to the exterior. **1940** L. H. Hyman *Invertebrates* I. vi. 337 In the case of a thick cortex, each dermal pore generally opens into a canal termed a chone.

chong, var. *chang *sb.*[2]

choo-choo (tʃū·tʃū). [Echoic.] An imitation of the sound of a steam-engine, used as a nursery name for a railway train or locomotive. Chiefly *U.S.*
1903 G. Ade *People you Know* 110 And now Saturday Afternoon had come and Percy M. Piker was hanging on the rear end of the choo-choo. **1927** W. E. Collinson *Contemp. Eng.* 7 Trains are still called puff-puffs or puffers as against the American onomatope choo-choo. **1938** S. Chase *Tyranny of Words* xiv. 178 Like a little boy making himself a choo-choo after seeing a locomotive. **1958** V. Bellerby in P. Gammond *Decca Bk. Jazz* xvii. 205 The mill-girl of Burnley sings about the 'Chatanooga choo-choo'. **1963** *Guardian* 27 Sept. 11/5 A number of ancient turns (including the choo-choo train routine).

chook (tʃuk). *Austral.* and *N.Z. colloq.* Also **chookie, -y, chucky, -ie** (tʃu·ki). [cf. Chuck *sb.*[1] 2.] A domestic fowl, a chicken.
1888 'R. Boldrewood' *Robbery under Arms* I. viii. 103 Ony mair than ye'll change a Norroway falcon into a barndoor chuckie. **1895** G. Chamier *South-Sea Siren* xiii. 194 You will let me carve the chucky? **1928** 'Brent of Bin Bin' *Up Country* vii. 78 If no one else has thought of chookies I know that these will be useful. **1933** *Bulletin* (Sydney) 25 Jan. 26 In N.S. Wales there's a fair amount of bumblefoot among the chooks. **1934** N. Scanlan *Winds of Heaven* ix. 79 [Grannie] will take you down..to see the chookies. **1940** A. Upfield *Bushranger of Skies* ix. 93 Burning Water's youngest child continued to build the walls of the 'chook house'. **1942** G. Casey *It's Harder for Girls* 127, I seen one take the roof off my front veranda an' land it in th' chook-yard at the back. **1950** K. S. Prichard *Winged Seeds* 161 Brown bread and bacon and cuts of chookie. **1951** J. Frame *Lagoon* 68 My big mother with a big blue pinny to shake at me as if it were wheat for a little chook. **1969** *Coast to Coast 1967–68* 2 His had been wild-eyed, scraggy long-legged chooks, few in number, sneaking into the kitchen after scraps.

choola (tʃū·la). Also **choolah, chula.** [Hind. *chūlhā*, *chūlhi* fireplace (Skr. *chulli*).] An Indian fireplace or cooking-place.
1813 J. Forbes *Oriental Mem.* III. 120 A marble corridore filled up with *Choolas*, or cooking-places, composed of mud, cow-dung, and unburnt bricks. **1864** J. A. Grant *Walk across Africa* 51, I observed a portable Indian 'choolah' or fireplace inside the hut. **1929** F. L. Brayne *Remaking Village India* 221 The chūla has been designed to burn soft coke, as a substitute for dung-cakes. **1945** —— *Better Villages* (ed. 3) 50 The ordinary *chula* (grate) besides wasting most of the heat of the fire, only holds one pot.

choom (tʃūm), *sb.*[1] [Russ. *chum*, f. native name.] Among the nomadic peoples of Siberia, a conical hut or tent of fir poles covered with skins or birch bark.
[**1598** Hakluyt *Voy.* I. 284 Their tents are rounde and are called *Chome* in their language.] **1889** V. Morier in *Murray's Mag.* Aug. 175 A little encampment of Samoyede summer 'chooms', *i.e.* birch-bark tents. **1895** F. G. Jackson *Gt. Frozen Land* 82 Of the choom which forms the Samoyad's home there are two kinds, one for summer and another for winter.

choom (tʃum, tʃūm), *sb.*[2] *Austral.* and *N.Z. slang.* [var. of Chum *sb.*] An English soldier; an Englishman.
1918 *Chron. N.Z.E.F.* 10 May 153/1 Our chaps are rather fond of the 'Aw choom' dialect. **1919** Downing *Digger Dial.* 16 *Choom*, an English soldier. **1930** *Bulletin* (Sydney) 5 Mar. 25/1, I met a fresh-faced choom in charge of a spring-cart. **1933** *Ibid.* 20 Dec. 20/1 Choom's eyes were saucers of awed interest. **1935** *Ibid.* 24 Apr. 21/2 He wasn't a choom; he came straight from Brisbane and had been born and reared in Sydney. **1952** J. Cleary *Sundowners* ii. 81 'Come on, Choom,' he said to Venneker [an Englishman].

choose, *v.* **11.** (Later examples.)
1928 P. Dearmer et al. *Oxf. Bk. Carols* 45 Joseph is chosen out from the other suitors by the budding of his rod. **1945** A. L. Rowse *West-Country Stories* 192 Sir Thomas Arundell, who chose out and fastened upon this spot.

choosey, choosy (tʃū·zi), *a.* orig. *U.S.* [f. Choose *v.* + -y[1].] Disposed to be particular in one's choice; fastidious, fussy. Hence **choo·siness** *sb.*, particularity, fastidiousness.
1862 *Harper's Mag.* Dec. 100/2 But so I'm sure enough thar at last, I'm noways choosy about the road. **1915** J. Webster *Dear Enemy* (1916) 68, I am very *choosey* in regard to homes, and I reject three-fourths of those that offer. **1927** Wodehouse *Small Bachelor* i. 4 'Mr. Finch had been getting what you might call choosey about his clothes...' 'What do you mean, choosey?' 'Particular, sir.' **1931** *Publisher's Circular* 13 June 747/2, I have found that collectors who ask for discounts are the choosiest of buyers. They want the best. **1936** Wodehouse *Laughing Gas* xii. 119 It so happens that in the matter of pyjamas I've always been a trifle on the choosy side. I'm not one of those fellows who just charge into a hosier's and grab anything. **1945** G. Endore *Methinks the Lady* (1947) x. 248 Such care, such choosiness, shows a fear..lest a word, not too carefully chosen, might give away your secrets. **1948** *Daily Tel.* 16 July, I hear that air passengers are becoming more choosy about where they sit. Seats in the rear of the plane are now preferred. **1956** A. Wilson *Anglo-Saxon Attitudes* I. ii. 63 Gerald saw that this topic, too, did not appeal to Elvira. He disliked 'choosey' women as a rule. **1958** *Oxf. Mag.* 1 May 381/1 Regular students of lecture lists must by now be pretty well habituated to the mysterious choosiness about titles shown by their compilers.

chop, *sb.*[1] Add: **1. b.** Freq. *pl.* A wood-chopping contest. *Austral.* and *N.Z.*
1926 K. S. Prichard *Working Bullocks* v. 48 He hewed his way through tough logs as though he were out to beat the champion in a chop. *Ibid.* viii. 84 One of the best axe-men in the sou'-west, Duck was champion in his day, and..he still entered for the chops. **1930** W. Smyth *Wooden Rails* vii. 108 There were standing chops, underhand chops, single and two-handed sawing contests. **1944** R. Park in *Coast to Coast* 42 'And the chops,' he said. 'I'd like to see Whaka Green making the chips fly.' **1963** N. Hilliard *Piece of Land* 172 The fourteen-inch chop, championship of the North Island, about to start now!

2. d. Material, esp. fodder, which has been chopped up. Also *chop-feed.*
1830 S. H. Collins *Emigrant's Guide* (ed. 4) 132 When it [*sc.* rye] is ground only (as it is used for bread in England) they here call it 'chop', and give it to cattle. **1852** *Trans. Mich. Agric. Soc.* III. 151 Chop feed is good for them in small quantities, say half a pint to a sheep. **1858** *Jrnl. R. Agric. Soc.* XIX. 507 With the chaff-cutter a fresh supply of chop is obtained daily. **1889** J. Wrightson *Fallow & Fodder Crops* 199 [He] is accustomed to cut rye and straw together and throw the 'chop' back into a barn. **1908** *Animal Managem.* 121 Chaff ('Chop'). **1922** *Outing* (U.S.) July 184/3 When the cow had licked up the last bit of chop.

e. A share, esp. in phr. *to be in for one's chop. Austral.* and *N.Z. slang.*
1919 Downing *Digger Dial.* 16 *Chop*, share. 'To hop in for one's chop'—to enter in, in order to secure a privilege or benefit. **1964** *Christchurch Star* 14 Mar. 2/1 New Zealand is in for its chop [in the *Concise Oxford Dictionary*]... No one else calls a road a tar-sealed road.

4. d. *Cricket.* A stroke made by bringing down the blade of the bat sharply on the ball (see quot. 1966). Also *chop-cut.* **e.** *Lawn Tennis.* An undercut ground-stroke. Also *chop-lob, -stroke.*
1888 Steel & Lyttelton *Cricket* ii. 62 If the ball..keeps a bit low after the pitch, it is a most effective stroke to come heavily down on it; if the force is put on the ball at the right moment it will go very hard, and may be called a 'chop'. **1913** *Daily Mail* 7 July 9/2 A sound batsman with a beautiful chop cut by third slip. **1920** Tilden *Lawn Tennis* p. x, An undercut ground stroke is the general definition of a chop. *Ibid.* 35 The chop lob, a heavily under-cut spin that hangs in the air. *Ibid.* 82 Tilden is a chop-stroke player. **1961** F. C. Avis *Sportsman's Gloss.* 254/2 *Chop*, a sharp stroke made by drawing the face of the racket on to the ball. **1966** *Armchair Cricket 1966* 95 *Chop*, a form of late cut, made by bringing the bat *down* sharply on the ball just as it is about to pass the batsman on the off-side.

f. (*to get*) *the chop* or *chopper*: (to be) killed. Also *transf. slang.*
1945 C. H. Ward-Jackson *Piece of Cake* 19 *To get the chop*, to be shot down and killed or injured. **1956** A. Crawley *Escape from Germany* iii. 40 'The chop' in Buchenwald meant execution or the gas chamber. **1957** J. Braine *Room at Top* xx. 176 We noncoms used to say got *the chopper*. Going for a Burton was journalist's talk. **1961** L. Payne *Nose on my Face* viii. 139 Sounded to me like a death warrant... Sooner or later he'd have got the chop. **1968** R. Collin *Locust on Wind* vii. 80 The editor had got what he wanted...'I think my series has gone for the chop.' **1971** *Ink* 12 June 12/2 The Anglo-Italian tournament..must be due for the chop.

chop, *sb.*[5] Add: **4.** *no chop*, 'no class'. *Austral.*
1888 'R. Boldrewood' *Robbery under Arms* I. ii. 16 There's good and bad of every sort, and I've met plenty that were no chop of all churches.

b. *not much chop* (also, rarely, *chops*), not up to much, of no or little value. Cf. *cop *sb.*[7] 2. *Austral.* and *N.Z.*
1909 H. B. Vogel *Trag. Flirtation* xxvii. 219 He ain't much chops, that doctor, I reckon. **1928** 'Brent of Bin Bin' *Up Country* ii. 19 'That old parson is not much chop, I don't reckon,' he confided. **1947** D. M. Davin *Gorse blooms Pale* 207, I know it's not been much chop so far but we're only getting started. **1969** *Coast to Coast 1967–68* 83 The street is not much chop, but not seedy, rather claustrophobic from the eight-feet walls of grey concrete on each side.

chop, *sb.*[6] *W. Afr. colloq.* [Cf. *chop *v.*[6]] Food. Also *attrib.*, as *chop-day, -money, -room*; **chop-box,** a food-box.
1805 H. Nicholls *Let.* 15 Feb. in R. Hallet *Rec. Afr. Assoc.* (1964) xi. 208 Their food is chop made of yam cut in slices, cayenne pepper, palm oil, and fowl, fish, goat or wild hog. **1863** R. F. Burton *Wanderings W. Afr.* II. ix. 144 'Palm-oil chop' is the curry of the Western coast. **1870** *Food Jrnl.* I. 154 In those days,..every chief of any consideration had what was called his chop-day, or feast-day. **1897** M. Kingsley *W. Afr.* 313 The unfortunate white men..are expected..to live on native chop. **1898** R. A. Freeman *Trav. & Life Ashanti & Jaman* 363 Besides the chair and fan he carried a small 'chop box', *i.e.* a box containing the requisites for a meal and my portable tea service. **1902** *Daily Chron.* 10 July 3/5, 2s. a week 'chop' money or maintenance. **1904** J. M. Fremantle *W. Afr. Jrnl.* 10 Dec. (1938) 30 Pretty cram full with 100 bits of luggage, 'chop-boxes' mountain high. **1920** *Blackw. Mag.* June 844/2 Evans went in to the chop-room. **1935** [see *coaster 3 b].

chop, *v.*[1] Add: **7. d.** *Cricket. intr.* To bring down the blade of the bat quickly and heavily on the ball; *trans.*, to strike (the ball) in this way. **e.** *Lawn Tennis.* (Cf. *chop *sb.*[1] 4 d, e.)
1776 M. Angelo *Juv. Sports* (ed. 2) 82 If the ball runs so close to the ground, that you cannot make a fair blow at it ..content yourself with chopping on the ball. **1887** *Leisure Hour* 544/1 Jupp and Humphry..blocked and chopped in a style that was simply maddening. **1898** *Field* 18 June 911/3 Little could be done but to chop Lockwood's off balls for singles. **1920** *Westm. Gaz.* 16 June 10/1 This drive is alternated with a cut-stroke, and sometimes with a heavily chopped return. **1930** *Morning Post* 17 June 16/3 Slater chopped a ball on to his stumps.

chop, *v.*[6] *W. Afr. colloq.* [Cf. *chop *sb.*[6]] To eat.
1833 W. F. W. Owen *Narr. Voy. Afr., Arabia* II. xxv. 327 'Chop' for eating; as 'Suppose go long way among Bullaman, he chop you'. **1896** *Westm. Gaz.* 18 Mar. 5/3 The natives have sent word..that if the Consul interfered with them they would make a raid on the trading factories and 'chop' (a native expression for 'eat') the white traders there. **1920** *Blackw. Mag.* June 845/2 You're both chopping with me to-night, so we can talk things over. **1969** *Islander* (Victoria, B.C.) 19 Oct. 2/1 'Where are we going to chop,' Ian said [in Accra]. 'Chop?' I said. 'Eat,' the other three replied, delighted to put me down. 'We go choppem proper.'

chopa (tʃǭ·pă). [Sp.] A rudder-fish of various species of the genus *Kyphosus*, found in the tropical Atlantic and Pacific.
1883 Moloney *W. Afr. Fisheries* 57 'Chopa', a sea-bream which represented their bait for the large fish. **1939** *Amer. Speech* XIV. 255 Chopa is also known as a Bermuda chub, chub, chopa blanca, and rudder fish.

chop-chop (tʃǫ·p,tʃǫ·p), *adv.* and *int.* [Pidgin-English, f. Chinese *k'wâi-k'wâi*.] Quick, quickly; hurry up!
1834 *Canton Register* 13 May 76/2 We have also..'chop-chop hurry'. 1836 *Chinese Repository* IV. 434 'More soon, more better; sendee chop-chop,' I told him. 1909 *Daily Chron.* 20 July 4/7 In pidgin English 'chop-chop' means 'make haste'. 1946 J. Irving *Royal Navalese* 52 *Chop-Chop!* In a hurry; Hurry up!

chop-house. Add: Now *colloq.*, a restaurant.
1915 W. Cather *Song of Lark* vi. vii. 440 Nice English chop-house on Forty-fourth Street. 1934 T. S. Eliot *Rock* i. 7 We have too many churches, And too few chop-houses. 1959 R. Postgate *Good Food Guide* 242 This is a genuine City chophouse with 'middle-aged, civil and knowledgeable waiters'.

Chopinesque (ʃǫpine·sk), *a.* [f. the name of Frédéric *Chopin* (1809–49), Polish composer + -esque.] Resembling the compositions or technique of Chopin.
1920 *Oxf. Mag.* 14 May 328/1 Comparatively early work [of Scriabin]—some very beautiful, some very dull, but almost all very Chopinesque. 1923 *Weekly Disp.* 8 Apr. 2 This dainty little Chopinesque pianist. 1957 L. Durrell *Bitter Lemons* 130 These dreamy Chopinesque moods alternated with moods of anarchy.

chop-logic. Delete †*Obs.* and add recent examples.
1951 N. Annan *Leslie Stephen* i. 42 His chop-logic destruction of cant. *Ibid.* ix. 257 The true positivist is not a crude lout wielding an axe of chop-logic. 1960 *Spectator* 8 July 51 The arguments were nearly always developed with the same chop-logic.

chopper[1]. Add: **2. b.** *to get the chopper* (see *chop *sb.*[1] 4 f).
4. A device for interrupting an electric current, a beam of light, or radiation, at regular intervals; = Interrupter b. Also *attrib.*
1929 J. A. Ratcliffe *Physical Princ. Wireless* iv. 64 Such a device is called a chopper, and usually takes the form of a buzzer contact. 1940 *Chambers's Techn. Dict.* 160/1 *Chopper*, an interrupter, generally in the form of a rotating commutator. 1950 *Jrnl. R. Aeronaut. Soc.* LIV. 22/2 For still greater numbers..the gauges are grouped in banks of eight, scanned by a rotating chopper switch. 1955 *Sci. Amer.* Aug. 64/1 The chopper is a Kerr cell: a glass cell containing a liquid..which in an electric field can transform an entering beam of plane-polarized light into a 'circularly polarized' beam. 1962 *Gloss. Terms Autom. Data Processing* 103 The modulator, which is called a chopper, is often a vibrating mechanical contact or a solid-state electronic switch. 1962 Simpson & Richards *Junction Transistors* ix. 216 Because of their excellent switching characteristics, transistors make particularly good 'choppers' for d.c. to a.c. conversion.
5. A machine-gun or -gunner. *U.S. slang.*
1929 *Harper's Mag.* Oct. 535/2 Johnny Hand..had met the 'chopper', *i.e.*, machine-gun. 1931 G. Irwin *Amer. Tramp & Underworld Slang* 50 *Chopper*, a machine gun, or the man operating such a weapon with a gang of racketeers or robbers. 1932 E. Wallace *When Gangs came to London* viii. 65 You remember the shooting of Big Sam Polini? The choppers got him as he came out of mass one morning. 1962 I. Fleming *Spy who loved Me* xi. 156 There was a mixture of single shots and bursts from the chopper.. sorry, sub-machine-gun.
6. A helicopter. *slang* (orig. *U.S.*).
1951 *N.Y. Herald Tribune* 16 Dec. ii. 5/7 The Korean War has added some new words to the American soldier's vocabulary..Chopper: Helicopter. 1952 *N.Y. Times Mag.* 17 Aug. 16/4 Oil and gas producers use the 'chopper' to patrol long and rugged gas and oil pipelines. 1958 *Listener* 31 July 175/3 A naval helicopter or 'chopper' going about its flights and hoverings.

choppiness (tʃǫ·pinės). [f. choppy *a.*[1] + -ness.] The quality or condition of being choppy.
1881 A. Elwes tr. *Pinto's How I crossed Africa* I. 142 The water was..made perilous..by the excessive 'choppiness' of the surface caused by the proximity of the falls.

chopping, *vbl. sb.*[1] Add: **4.** **chopping-bee** *N. Amer.*, a 'bee' for the cutting down of timber; **chopping-block,** also *transf. colloq.*, e.g. applied to a boxer who sustains steady punishment.
1809 *Massachusetts Spy* 12 July (Th.), At Bristol (Ver.), June 7, at a chopping-bee, a limb of one of the falling trees struck one of the men. 1868 *Western Mag.* Jan. (De Vere), The inhabitants within a radius of ten miles were invited to a chopping-bee. 1890 *Regina* (Sask.) *Jrnl.* 18 Dec. 1/2 Come to the chopping bee and bring your axe. 1928 *Daily Express* 10 Aug. 15/6 The New Zealander was outclassed by the champion, and in the last three rounds was only a chopping block, but game to the last. 1930 *Times Lit. Suppl.* 29 May 455/1 Xenophon's 'Anabasis', made a chopping-block for generations of schoolboys. 1926 T. E. Lawrence *Seven Pillars Wisdom* (1935) vi. 56 We found them just kept chopping-blocks of their commanders' viler passions. 1960 'B. Mather' *Pass beyond Kashmir* v. 74 That got him nowhere except as a chopping block for four of them working in shifts.

chop-stick[2]. Add: **2.** *pl.* A quick tune for the piano played with the forefinger of each hand.
1893 E. F. Benson *Dodo* vi. 121 Edith and her Herr were playing a sort of chopsticks together in the drawing-room. 1960 *20th Cent.* Oct. 319 She..showed me how to play chopsticks..my first piano lesson. 1961 C. McCullers *Clock without Hands* xiii. 251 Four of the Lank brood were playing 'Chopsticks' on the piano... Jester stood in the sunlight hearing the dead and no-tune 'Chopsticks'.

chop-suey (tʃǫ:psū·i). [Chinese (Cantonese) *shap sui*, = mixed bits.] A Chinese dish of meat or chicken, rice, onions, etc., fried in sesame-oil. Also *fig.*
1888 *Current Lit.* (U.S.) Oct. 318 A staple dish for the Chinese gourmand is chow chop svey [*sic*], a mixture of chickens' livers and gizzards, fungi, bamboo buds, pigs' tripe, and bean sprouts stewed with spices. 1898 L. J. Beck *N.Y. Chinatown* v. 50 Chop Suey—(A Hash of Pork, with Celery, Onions, Bean Sprouts, etc.). 1903 Ade *People you Know* 16 The next Picture that came out of the Fog was a Chop Suey Restaurant. 1904 *Rochester* (N.Y.) *Post-Express* 8 June 12 One of the Chinese merchants of New York..explained that chop suey is really an American dish, not known in China, but believed by Americans to be the one great national dish of the Celestials. 1906 'O. Henry' *Four Million* 218 Uniformed men sat and made chop suey of your tickets. 1910 —— *Whirligigs* (1916) 51 The censor has put the screws on, or he wouldn't have cabled in a lot of chop suey like this. 1924 *Chambers's Jrnl.* Nov. 731/1 Chinamen supply provisions and chop-suey.

choral, *a.*[1] Add: **1. d.** Applied to interpretative reading or recitation of poetry, drama, etc., by a group of voices.
1933 *Amer. Speech* VIII. iv. 39/1, I merely raise the question as to whether choral reading belongs in the curriculum, whether it can be a substitute for sound courses in oral interpretation of literature and cultural dramatics. 1936 *Time Table of Festival of Spoken Poetry* (Oxford) 8 Choral Speaking..Teams of 5 or 7 speakers. Test Piece: 'At Parting', by Algernon Swinburne. 1937 M. Gullan (*title*) The Speech Choir, with American Poetry and English Ballads for Choral Reading.
2. c. *choral society*, a society of people interested in choral music.
1858 Queen Victoria *Let.* 29 June in *Dearest Child* (1964) 118 The Choral Society of Bradford sang in the Great Ballroom two hundred in number and beautifully. 1873 C. M. Yonge *Pillars of House* I. xi. 220 The members of the choir and choral society. 1892 —— *That Stick* i. 3 His recreations were of the sober-sided sort—the chess club, the institute, the choral society. 1966 'S. Woods' *Let's choose Executors* 10 She had a deep voice, rather pleasant, so that he thought of the local Choral Society.

choralism (kōᵊ·răliz'm). [f. Choral *a.*[1] + -ism.] Choral composition; choral rendering and technique.
1927 *Observer* 29 May 9/3 From the cloisters of Magdalen College, Oxford, are to be sung at 9.15 two magnificent examples of choralism, Palestrina's motet 'Exultate Deo' and Byrd's anthem 'Sing Joyfully to God'. *Ibid.* 16 Oct. 5 The Sheffield Choir is noted for good, sound, clean choralism. 1948 *Penguin Mus. Mag.* V. Feb. 92 The traditional Mancunian bias against choralism.

Chorasmian (koræ·zmiăn), *a.* [f. Gr. Χοράσμιοι (pl.) Chorasmii + -an.] Belonging to the Chorasmii, a tribe of Sogdiana. *Chorasmian waste* (poet.), the desert land south of the Sea of Aral and about the lower course of the Oxus; also *allusively*.
1815 Shelley *Alastor* 272 At length upon the lone Chorasmian shore He paused. 1853 M. Arnold *Sohrab & Rustum* 878 The majestic river..moved, Rejoicing, through the hush'd Chorasmian waste. 1890 'R. Boldrewood' *Col. Reformer* xix, In this lone Chorasmian waste, Mr. Neuchamp contrived to spend his time.

chord, *sb.*[1] Add: **5.** *Engin.* Each of the two principal members, usually horizontal, of a truss.
a 1877 Knight *Dict. Mech.* III. 2638/2 Rider's [truss-bridge]..is composed of an upper and a lower chord,—the former of cast, and the latter of wrought, iron. 1940 *Chambers's Techn. Dict.* 160/1 *Chord*, a flange of a large girder.
6. In an aeroplane, a straight line passing through the lower curve of an aerofoil section at two points; hence, the dimension of an aerofoil from the leading edge to the trailing edge (the width of the wings as distinguished from 'span' or lateral spread). Also *attrib.*
[1909 A. Berget *Conquest of Air* iv. 188 When the apparatus is in flight, the 'chord' of the arc formed by the profile of the wings makes an angle varying from 6 to 8 degrees with the horizon.] 1918 E. M. Roberts *Flying Fighter* 334 *Chord*, the width of wings. 1920 *All the World's Aircraft* 123 a, Specification of the Westland 'Limousine'. Length overall, 28 ft. 6 in.; Span, 38 ft. 2 in.; Chord, 6 ft. 3 in. 1923 Glazebrook *Dict. Appl. Physics* V. 19/2 The ratio of the span to the chord gives the Aspect Ratio of the wing. 1933 *Jrnl. R. Aeronaut. Soc.* XXXVII. 448 A sketch and photographs show the method of increasing the surface by increasing the chord width along about two-thirds of the span. 1934 *Ibid.* XXXVIII. 585 Fig. 11 Shows two Townend Rings of identical section, chord length, and chord angle. 1968 *Jane's Surface Skimmer Systems 1967–68* 12/1 This ram-wing research vehicle..consists basically of a rectangular wing of constant chord.

chord, *sb.*[2] Add: **3. c.** *attrib.*
1947 A. Einstein *Mus. in Romantic Era* xi. 137 Often there are effects almost exclusively of sonority and resonance, willfully interjected chord-clusters, that play a decisive part. 1962 J. Wain *Strike Father Dead* v. 211 Inventive jazz..played on the familiar chord progressions. 1963 *Radio Times* 11 Apr. 30 Extemporising on a chord sequence may..be reaching its limits. 1966 *Crescendo* July 13/2 *Quality material*, a song with more than three chord changes.

chorda (kǭ·ɹdă). *Anat.* Pl. **chordæ** (kǭ·ɹdī). [L.: see Chord *sb.*[1]] A name for certain string-like structures in the animal body; with defining L. adj., as **chorda dorsa·lis** the notochord; **chordæ tendi·neæ** *pl.*, fine tendinous cords attaching the borders of the auriculo-ventricular valves to the papillary muscles of the heart; **chorda tympani** = Tympanichord.
1807 Morris & Kendrick *Edin. Med. & Physical Dict.*, *Chorda*,..metaphorically used to signify a tendon or nerve. *Ibid.*, *Chordæ tendineæ*, from the edge of the valves in the ventricles of the heart, there are tendinous strings thus named. *Ibid.*, Chorda tympani. 1873 G. Fleming tr. *Chauveau's Comp. Anat.* 905 There appear in the middle lamina of the blastoderm the *chorda dorsalis*, the *vertebral laminæ*, and the *lateral laminæ*. 1885 *Buck's Handbk. Med. Sci.* I. 562 Long, fibrous chords, called chordæ tendineæ, that arise from pyramidal muscles. 1886 *Ibid.* II. 328/2 A branch, named from its peculiar course, the chorda tympani. 1887 *Ibid.* V. 248/2 The inner chorda sheath is lost. 1927 Haldane & Huxley *Animal Biol.* iv. 97 One of the papillary muscles, attached by chordae tendineae to..the tricuspid valve between right auricle and right ventricle. 1952 J. Drever *Dict. Psychol.* 37 *Chorda tympani*, a branch of the facial nerve, mostly sensory, carrying nerve impulses from taste receptors. 1962 *Gray's Anat.* (ed. 33) 739 Their [*sc.* papillary muscles'] apices..give origin to the chordae tendineae, which pass to be attached to the segments of the tricuspid valve. *Ibid.* 1137 The chorda tympani nerve arises from the facial nerve about 6 mm. above the stylomastoid foramen.

Chordata (kǫɹdē¹·tă), *sb. pl.* *Zool.* [mod.L., f. L. *chorda* Chord *sb.*[1], with termination as in Vertebrata, etc.] A phylum of animals having a more or less well-developed notochord at some stage in their lives.
1880 F. M. Balfour *Treat. Compar. Embryol.* I. 4 The larva of the Tunicata has the characters of a simple type of the Chordata. 1901 Shipley & MacBride *Zool.* xiii. 291 The name Chordata, which has been proposed for the group, is really more appropriate. 1906 Chamberlin & Salisbury *Geol.* II. 484 While perhaps they should remain under the broad mantle of the term *Chordata*, they must apparently be removed from the true vertebrates to a new class. 1950 J. Z. Young *Life of Vertebrates* ii. 24 The Chordata occupy a greater variety of habitats..than any other group. 1957 M. Abercrombie et al. *Dict. Biol.* 50 *Chordata*, phylum of animals with notochord, hollow dorsal nerve cord, and gill slits. Includes subphyla Vertebrata (Craniata) and Protochordata.

chordate (kǭ·ɹdē¹t), *a.* and *sb.* *Zool.* [ad. prec.] Belonging to, having the characters of, or a member of the Chordata.
1889 *Cent. Dict.*, Chordate, *a.* 1897 [see *hemichordate* def., s.v. Hemi-]. 1927 Haldane & Huxley *Animal Biol.* ii. 79 They are therefore all classed as Vertebrates (or Chordates, with reference to the notochord). 1958 *Oxf. Univ. Gaz.* 23 Apr. 892 Studies on the nervous systems of lower chordates. 1959 [see *acorn-worm].

chording (kǭ·ɹdiŋ), *vbl. sb.* [f. Chord *v.*[1] + -ing[1].] The playing, singing, or arrangement of chords.
1866 [see Chord *v.*[1]]. 1928 *Manch. Guardian Weekly* 17 Aug. 133/4 The quiet, yet rich chording [*sc.* of the choir]..was very beautiful. 1938 *Oxf. Compan. Mus.* 164/1 *Chording*,..is a term used by choir-trainers..'Bad chording' is a form of out-of-tune singing..which results in the notes of the chords not sounding quite true to one another. 1955 L. Feather *Encycl. Jazz* vii. 197 A softly swinging style rich in right-hand chording. 1960 *Conc. Oxf. Dict. Mus.* 113/2 *Chording*,..In U.S.A. the term means the improvised strumming of accompanimental chords on a banjo, etc. 1969 *Daily Tel.* 18 Jan. 17/2 In Dvorak's 'New World' Symphony ..Mr. Rowicki's sang froid beat secured unanimous chording at such crucial points as the start of the Largo.

chordophone (kǭ·ɹdŏfōᵘn). [f. Chord *sb.*[1] + -o: see -phone.] Any stringed musical instrument.
1937 *Times Lit. Suppl.* 17 Apr. 288/2 Those [instruments] which employ..one or more strings (chordophones). 1961 A. Baines *Mus. Instruments* 351 *Chordophones*, stringed instruments.

chordotomy, var. *cordotomy.

chordotonal (kǫ:ɹdŏtōᵘ·năl), *a.* *Zool.* [G. (V. Graber 1882, in *Arch. f. Mikroskop. Anat.* XX. 506), f. Chord *sb.*[1] + -o + Tonal *a.*] Applied to organs in insects that are responsive to sound or mechanical vibrations.
1888 J. Lubbock *Senses, Instincts & Intelligence Anim.* vi. 117 In the cricket, for instance, the sense of hearing appears to be seated partly in the antennæ, and partly in the anterior legs. In other cases, as in Corethra, the division appears to be carried still further, and a 'chordotonal' organ occurs in each of several segments. 1910 *Encycl. Brit.* II. 85/2 For the appreciation of the sounds made by these stridulators, the ants are furnished with delicate organs of hearing (chordotonal organs). 1948 *Science News* VII. 19 On the theory that the elongated parallel elements act as chords (strings) and have a 'tonus' or tension like strings, these structures have been given the name 'chordotonal organs'.

chore, *sb.*[2] Add: **1.** Recently used *colloq.* of a piece of (time-consuming) drudgery. (Also, earlier Amer. and later examples.)
1751 J. MacSparran *Letter Bk.* 16 Sept. (1899) 56 He sent with Peter his black Boy Calais to do chores for a few Days. 1758 J. Adams *Wks.* (1850) II. 37 Chores, chat, tobacco, tea, steal away time. 1789 Webster *Diss. Eng. Lang.* 112 *Chore*..is an English word..but in America, it is perhaps confined to New England. It signifies small domestic jobs of work, and its place cannot be supplied by any other single word in the language. 1870 Emerson *Society & Solitude* 24 Now that is the wisdom of a man..to hitch his waggon to a star, and see his chore done by the gods themselves. 1926 Galsworthy *Silver Spoon* ii. vi. 155 Two of us do the cooking; one the accounts; and the rest washing, mending, games, singing, dancing and general chores. 1934 J. C. Powys *Autobiogr.* xii. 634, I did every mortal thing that had to be done, in the way of what Americans call 'chores', both outside and inside the house. 1958 *Economist* 8 Feb. 470/1 Various improvements [of House of Commons procedure] were discussed last week, many of them centring upon the idea of making voting less of a chore. *Ibid.* 15 Feb. 568/1 The Ministry [of Education] hopes to free the teachers from some of their chores. 1959 *Times* 31 Dec. 5/6 It was by no means a great game—it had probably become by now a chore.
2. **chore-boy** *N. Amer.*, a boy employed in doing odd jobs.
1848 *Knickerbocker* Sept. XXXII. 230, I afterwards saw Betty,..laughing with the gardener and 'chore-boy'. 1901 *Westm. Gaz.* 18 July 10/1 The Secretary of the Treasury began life as a bank clerk,..the Secretary of the Navy as a 'chore-boy' on a farm. 1902 S. E. White *Blazed Trail* xxviii. 187 You would have to be chore-boy in a lumber camp. 1957 *Fish & Game* (Calgary) 30 Aug. 11/2 [He] accompanies him on fishing trips and..has gradually slipped into the role of camp-cook and chore-boy.

chore, *v.* (Earlier U.S. examples.)
1788 J. May *Jrnl. & Lett.* (1873) 88 Two [hands] playing the whipsaw, and the rest choring in the woods. 1839 'Mrs. M. Clavers' *New Home* xiv. 87, I was obliged to employ Mrs. Jennings to 'chore round', to borrow her own expression.

choregus, var. Choragus. Here used *fig.* (sense 3).
1901 *Fortn. Rev.* Sept. 403 Professor Haeckel, who is denounced in some quarters as the very choregus of materialism.

choreograph (kǫ·riŏgraf), *v.* orig. *U.S.* [Back-formation f. Choreography.] *trans.* To compose the choreography of (a ballet). *intr.* To engage in choreography.
1943 *Time* 15 Nov., He..choreographed ballets for Marie Rambert's famed Ballet Club. 1949 M. Lloyd *Borzoi Bk. Mod. Dance* 148 She had to choreograph quickly for the weekly Saturday night revues. 1952 *Ballet Ann.* VI. 97 Beethoven's 'Prometheus' choreographed by Serge Lifar. 1957 C. MacInnes *City of Spades* ii. v. 137 She choreographs a cosmopolitan style. *Ibid.* 140 I've choreographed our African and Caribbean dances in with classical European and other sources. 1961 B. Savan *Your Career in Theater* 179 A few of the outstanding choreographers who had intimate experience with dancing before they began to choreograph. 1962 *Times* 4 May 20/4 Ashton seemed to be bent on choreographing a complete *Raymonda*. 1970 *Daily Tel.* 2 Mar. 13/6 'Raga Shankara', choreographed to Indian classical music.

choreographic, *a.* (Earlier example.)
1821 *Morning Post* 14 June, Mr. Hullin has just displayed a new proof of his admirable talent for the choreographic instruction.

choreogra·phical, *a.* Also **choregraphical.** = Choreographic *a.*
1893 in *Funk's Stand. Dict.* 1922 *Dancing Times* Apr. 611/1 The most striking feature of this..ballerina is..her choregraphical eclecticism. 1936 'C. Brahms' *Footnotes to Ballet* ii. 49 The ladies and gentlemen of the ballet performed their choreographical roulades. *Ibid.* 50 Choreographical invention was there.

choreogra·phically, *adv.* Also **choregraphically.** [f. *choreographical *a.* + -ly[2].] In a choreographic manner; in respect of choreography.
1911 J. E. C. Flitch *Mod. Dancing* xi. 156 He makes credible the suggestion..that La Rochefoucauld's 'Maximes' should be rendered choregraphically. 1915 M. E. Perugini *Art of Ballet* xxxiii. 281 'Versailles'..was as usual 'choregraphically' arranged by Mme. Katti Lanner. 1922 Beaumont & Idzikowski *Man. Theory Classical Theatr. Dancing* 28 Choreographically considered, *arabesques* are generally introduced to conclude a phrase of steps. 1959 *Oxf. Mag.* 18 June 480/1 Strangely threadbare choreographically, it suffered also from too strong a determination on the part of the dancers to express their emotions.

choreo·graphist = Choreographer.
1923 J. Agate *Half-Past Eight* 168 One will be ravished by..Stowitts, the Russian choreographist. 1928 *Observer* 17 June 15/4 It is now to be heard..as the accompaniment of a ballet. This time Dolin is the choreographist.

choreology (kǫri,ǫ·lŏdʒi). [f. as Choreography: see -ology.] The study and description of the movements of dancing. So **choreo·logist,** one versed in choreology; a person skilled in the written notation of dancing.
1964 F. Hall in *Brit. Jrnl. Aesthetics* Jan. 65 What we are concerned with here is something so new that when an Institute was formed to foster its development a new name had to be coined: choreology. The Institute of Choreology was, in fact, formed in July 1962, with a number of objects. ..Choreology—the study of all forms of dance through notation—is still in its early stages. 1965 *New Statesman* 18 June 979/1 Mr and Mrs Benesh have now received a grant..to start an Institute of Choreology...Here choreologists are trained to go out and write down new works as the choreographer composes them in the rehearsal-room.

choreutes (kǫrˈū·tiz). *Antiq.* [Gr. χορευτής, f. χορεύειν to dance, f. χόρος Chorus.] A member of the chorus of Greek drama. Hence **choreu·tic** *a.* [Gr. χορευτικός], of or pertaining to choral song and dance.
1855 F. A. Paley *Æschylus Trag. Agam.* 397/1 The first choreutes speaks [line] 1315. 1926 *Spectator* 5 June 948/1, I rather doubt if the musical and choreutic material from which the organizers drew could justify so extensive an enterprise.

choriambize (kǫriæ·mbəiz), *v.* [See -ize.] *trans.* To make choriambic; to turn (a foot) into a choriambus.
1922 S. Grew *Player-Piano* 180 The choriambising of measures establishes two weak syllables between the initial syllable and the next strong one.

cho·rically, *adv.* [f. Chorical *a.* + -ly[2].] In the manner of a chorus.
1956 *Essays & Studies* IX. 85 Conrad pictures..the political ideas and aspirations of a great body of men. He writes *chorically*. 1956 *Essays in Criticism* VI. 415 Chorically, he [sc. Arnold] offers us an image which itself reproduces the 'calm pathos' of the classics.

chorine (kō·rīn). orig. and chiefly *U.S.* [f. Chor(us *sb.* + -ine[3] or [4].] A chorus-girl.
1922 *Moving Pict. Stories* 28 July 26/1 Pearls continue to be popular, especially with soubrettes and even chorines. 1927 in *Amer. Speech* (1928) III. 368 The chorines of a cabaret. 1938 *Times Lit. Suppl.* 11 Feb. 91/3 They [sc. short stories] depict..the philosophy of a 'chorine'. 1948 *John o' London's* 10 Dec. 597/2 The 'Set of Twelve Pretty Girls' [cigarette cards] disappeared, fat chorines gave way to 'Life on Board a Man-o'-War'. 1958 *Times* 2 Dec. 14/3 The big setpieces are more elaborate than anything since Busby Berkeley last piled hundreds of Hollywood chorines on a 50 ft. birthday cake. 1969 *New Yorker* 14 June 4/1 A dancer ..is befriended by a chorine named Joan.

chorio-. **1.** Combining form of Chorion, as in **cho:rio-alla·ntois,** a vascular membrane formed by the union of the chorion and the allantois in birds and reptiles and serving a respiratory function, often used as a medium for the culture of viruses, etc.; so **cho:rio-allanto·ic** *a.*; **cho:riocarcino·ma** [f. G. *chorion carcinom* (F. Marchand 1898, in *Zeitschr. f. Geburtsh. & Gyn.* XXXIX. 255)], a malignant tumour of the chorionic epithelium; also called **chorio-epithelioma*; **cho:rio-epithelio·ma** [G. (F. Marchand 1898, in *Zeitschr. f. Geburtsh. & Gyn.* XXXIX. 256], = **choriocarcinoma*.
1929 W. Moppett in *Proc. R. Soc.* B.CV. 402 The investigation of the biological action of homogeneous X-radiation.. was performed with the chorio-allantoic membrane of the embryo chick. 1938 *Ann. Reg.* 1937 352 The egg membrane technique was further developed, the chorio-allantoic membrane giving promise of replacing the experimental animal in the study of many viruses. 1933 *Jrnl. Exper. Zool.* LXVI. 193 The development of embryonic rat tissues upon the chick chorio-allantois. 1959 *New Biol.* XXX. 88 As in mammals the embryos soon become [in certain viviparous lizards] completely surrounded by membranes. Two of these, the chorion and the allantois, become fused early in development, the product being termed the chorio-allantois. 1901 W. R. Williams *Uterine Tumours* xx. 346 Marchand..regards it as true cancer (chorio-carcinoma); but, since..the disease so closely resembles sarcoma, Marchand proposes to designate such growth provisionally as 'serotinal tumours'. 1960 *Times* 6 Jan. 9/5 Women with choriocarcinoma and related tumours. 1901 Dorland *Mod. Obstetrics* (ed. 2) ii. i. 230 This tumor is purely epitheliomatous in nature and has been variously designated as a syncytioma malignum, chorioepithelioma, [etc.]. 1902 *Jrnl. Obstetr. & Gynæcol.* I. 242 The point raised by Dr. Lockyer that there were two sorts, one without syncytial elements..to which the term 'deciduoma malignum' would be appropriate, and the one with syncytial elements of fœtal origin, for which the term 'chorio-epithelioma' would be better applied, was..very important.
2. [cf. *choroido-.] Used as a combining form of Choroid *a.*, as in **cho:riomeningi·tis,** cerebral meningitis affecting the choroid plexus; **cho:rioretini·tis,** = **choroido-retinitis*.
1934 Armstrong & Lillie *Public Health Rep.* (*U.S. Treasury Dept.*) XLIX. 1019 This virus, which differs from any virus with which the author is familiar, will be designated..as the virus of experimental lymphocytic choriomeningitis. 1961 *Lancet* 23 Sept. 684/2 Serological examination by C.F. tests showed no high or rising titres for.. lymphocytic choriomeningitis. 1880 *Archives Ophthalmol.* IX. 1 In the following paper I submit a contribution to the study of chorio-retinitis pigmentosa. 1887 *Trans. Amer. Ophthalmol. Soc.* IV. 579 (*heading*) Double chorio-retinitis, with partial degeneration of the optic nerve. 1961 *Lancet* 23 Sept. 682/1 An 11-day-old premature infant who had meningoencephalitis but survived with..chorioretinitis.

chorioid, chorioidal, variants of Choroid, -al.
1894 Gould *Dict. Med.* 1902 D. J. Cunningham *Anat.* 685 The chorioid (chorioidea) intervenes between the sclera and the retina. *Ibid.* 699 The chorioidal fissure.

chorion. Add: **1.** Also *attrib.*, as in *chorion-epithelioma*, = **choriocarcinoma*; so *chorion epitheliomatous* adj.
1903 *Jrnl. Obstetr. & Gynæcol.* IV. 3 The name 'deciduoma malignum' is still often used, but the view implied by it is completely abandoned in favour of that of Marchand expressed in the term 'chorionepithelioma'. 1906 *Practitioner* Nov. 663 A chorion epithelioma is a tumour which arises, either from the chorion..or from chorionic elements present in teratomata. *Ibid.*, Cases which combine carcinomatous and teratomatous, or chorion epitheliomatous characters have been described. 1964 L. Martin *Clinical Endocrinol.* (ed. 4) i. 9 Large amounts of chorionic gonadotrophins are present in the urine of patients with chorionepithelioma.

chorionic (kōᵊriǫ·nik), *a.* *Embryol.* [See -ic.] Of or relating to the chorion.
1892 E. L. Mark tr. *Hertwig's Embryol.* 261 The double-layered chorionic epithelium is already distinctly present in eggs four weeks old. 1902 D. J. Cunningham *Anat.* 23 The placental area..,together with the blastoderm over the rest of the ovum, forms the chorionic area. *Ibid.* 50 The chorionic villi. 1912 J. S. Huxley *Individ. in Animal Kingdom* iii. 68 The armadillo's hood, like the identical twins in man, has a single chorionic membrane. 1964 [see *chorion].

chorister. Add: **1. a.** Also, a choir-leader. *U.S.*
1769 in *Plymouth Church Records* (1920) II. 332 To choose one or more Persons (since our dear Brother John May is taken from us by death who was our former Chorister) to lead in singing in the publick Worship. 1777 *Ibid.* 353 Deacon Crombie, our former Chorister, had left ye usual Singer's Seat. 1930 W. R. Moody *D. L. Moody* 141 Moody had engaged a young man..to be chorister in his church and Sunday school.

Choristid (kori·stid), *sb.* and *a.* [ad. mod.L. *Choristida* (see def.), f. Gr. χωριστός separate.] **A.** *sb.* A member of the order Choristida of sponges. **B.** *adj.* Belonging to this order. Also (in the same senses) **Chori·stidan** *a.* and *sb.*
1888 W. J. Sollas *Tetractinellida* p. xi, Recent Choristids. *Ibid.* p. cix, The Choristid Sponges. 1900 E. A. Minchin in Lankester *Treat. Zool.* II. 163 A corticate Choristid.

chorizo (tʃori·zo). [Sp.] A sausage of which the chief ingredient is pork.
1846 R. Ford *Gatherings from Spain* xi. 129 The rich red sausage, the *chorizo*. 1918 *Chambers's Jrnl.* 33 The red chorizos and pimentisco from Estremadura. 1971 *Sunday Times* (Colour Suppl.) 27 June 49 *Chorizo*, a small Spanish sausage..heavily flavoured with garlic and paprika.

choroido-, used as combining form of Choroid in mod. Latin terms, as **choroi:do-iri·tis,** inflammation of the choroid and the iris; **choroi:do-retini·tis,** inflammation of the choroid and the retina.
1869 G. Lawson *Dis. Eye* 139 It is, however, generally succeeded by a low form of choroido-iritis or choroido-retinitis. 1879 *St. George's Hosp. Rep.* IX. 532, 2 cases of choroido-retinitis. 1908 *Practitioner* Apr. 542 Syphilitic choroido-retinitis. 1964 S. Duke-Elder *Parsons' Dis. Eye* (ed. 14) xviii. 253 Toxoplasmosis, a protozoan infection derived mainly from rodents and birds, affects both the retina and choroid to form a choroido-retinitis.

chorten (tʃǭ·ɹte·n). [Tibetan.] = *chaitya.
1891 Rockhill *Land of Lamas* 63 Pilgrims, when journeying to a shrine, perform prostrations before each *ch'ŭrten* met on the way thither. 1893 E. F. Knight *Where three Empires Meet* viii. 126 Gigantic *chortens*, or sarcophagi, of dried mud, gaily painted, containing the ashes of pious lamas. 1894 Rockhill *Diary Journ. Mongolia & Tibet* 271 A small village of stone houses near which were several large *ch'örtens*. 1898 A. H. S. Landor *In Forbidden Land* II. 72 When a saintly Lama dies..either parts of the flesh, or, if cremation has been applied, some of his ashes, are preserved and placed in a Chokden erected for the purpose. 1902 *Chambers's Jrnl.* 444 The *gompa*..close to several *chortens* containing the bones of sainted lamas. 1939 M. Pallis *Peaks & Lamas* vi. 65 A *chorten* or *stupa*,..the Buddhists emblematic monument which replaces the crucifix of Catholic lands. 1952 Morin & Smith tr. *Herzog's Annapurna* v. 77 A Buddhist chorten adorned with prayer flags marked the entry to the village.

chortle, *v.* Add: Also *trans.*, to utter or sing with a 'chortling' intonation. Also *sb.*, an act of 'chortling'.
1886 *Referee* 18 Aug. (Ware), Mr. Wilford Morgan has been engaged to chortle the famous song, 'Here's to the maiden of bashful fifteen'. 1889 *Ibid.* 29 Dec., Many present on Boxing Night fully expected that when he appeared he would chortle a chansonette or two. 1903 'A. McNeill' *Egreg. Eng.* 28 He would tell you..that he attributed his success..(5) to marrying Mrs. Business-Man—this last, of course, with a chortle.

chorus, *sb.* Add: **3. b.** A group of organ pipes or stops designed to be played together; a compound stop (see Compound *a.* 2 f); also, the sound so produced.
1776 J. Hawkins *Hist. Mus.* IV. x. 147 The compound stops..are so called for that in them several pipes are made

to speak at the touch of a single key..and the full organ or chorus is compounded of all. 1938 *Oxf. Compan. Mus.* 660/1 The term *chorus reeds* is applied to those [*sc.* reed stops] not suitable for solo use. 1961 C. Clutton in A. Baines *Mus. Instruments* 63 The baroque organ was devised to afford a considerable range of well-balanced contrasted effects, which may be divided into the two main categories of 'solo' and 'chorus'. An organ chorus consists of a number of ranks of roughly the same tonal quality and power, sounding at many different pitches.
6. c. The main part of a modern popular song, as distinct from the introductory verse; also, an improvisation upon the main tune by one or more jazz musicians.
1926 Whiteman & McBride *Jazz* x. 219 A half chorus of hot jazz and then..'Ach, Du Lieber Augustine' bring the orchestration to a close. 1929 *Melody Maker* Feb. 163/3 The first chorus, verse and second chorus showed the usual Whiteman perfection of orchestration. 1934 S. R. Nelson *All about Jazz* iii. 66 The term 'hot' chorus may be defined as a rhythmical rendering of the tune intended to increase its suitability for dancing. 1935 *Hot News* Aug. 15/1 The themes utilised in jazz consist generally of a 'chorus' and a 'verse'. 1941 *Mus. Quart.* XXVI. 53 Every jazz band provides for solo improvisation of a few choruses, usually accompanied at least by the rhythm section, and it is in this solo work that the great jazz players can best be studied. 1961 Berkman *Singers' Gloss. Show Bus. Jargon* 13 The Chorus of most popular songs is usually a 32-bar strain, although some of Cole Porter's songs, like 'Begin the Beguine', run as long as 100 measures. 1965 G. Melly *Owning-Up* vi. 66 During Bruce Turner's first alto chorus.
7. *chorus-lady* (earlier example); *chorus-singing*; **chorus-girl,** a young woman who sings or dances in the chorus of an opera, musical comedy, or revue.
1894 Mrs. H. Ward *Marcella* I. i. vi. 100 A certain little chorus girl came upon the scene and served to make both money and repentance scarcer even than they were before. 1907 *Daily Chron.* 21 Oct. 3/2 Elizabeth's two thoroughly vulgar chorus-girl friends. 1944 Auden *For Time Being* (1945) 80 For talking of their spiritual Beauty to chorus-girls. 1870 *Observer* 9 Oct. 6/4 And we now know it is possible to get rid of the extraordinary chorus ladies who hitherto have been inseparable from opera. 1903 A. W. Patterson *Schumann* 66 The chorus-singing had been excellent.

chose, *sb.* Add: **4.** *chose jugée* (ʃūʒe) [Fr.], a matter which has been formally adjudicated and decided and which it is therefore idle or presumptuous to discuss.
1898 *Daily News* 22 Mar. 6/1 The new doctrine of the sacredness of the 'chose jugée'. 1913 Mrs. H. Ward *Mating of Lydia* xii. 250 We will not..argue the matter, which is for me a *chose jugée*. 1938 *Times Lit. Suppl.* 12 Nov. 721/2 One gets the impression of too much repetition, and rather too much ado about *choses jugées*.

chota (tʃōᵘ·tă), *a.* *India.* [Hindi *choṭa*.] Small; younger, junior; *spec.* applied to a 'peg' of whisky.
c 1815 M. M. Sherwood *Little Henry & his Bearer* (ed. 7, 1816) 10 Boosy never left his *choota sahib*. 1853 Mrs. Colin Mackenzie *Life in the Mission* viii. 351 A horseman..asked for William as the 'Chota Padre'. 1886 Kipling *Departm. Ditties* (ed. 2) 44 Dysent'ry touched him in June, after the *Chota Bursat*. 1893 —— *Day's Work* (1898) 8, I tell this to the Chota Sahib. 1915 *Blackw. Mag.* Dec. 817/1, I was sipping a 'chota' whiskey. 1927 *Chambers's Jrnl.* 603/1 He..waited till the chota pegs had been served.

chota hazri (tʃōᵘ·tă hā·zri). *India.* Also **chotah; hazaree, haziree, haziri, hazree.** [Hindi *choṭā hāẓiri* little breakfast.] A light early breakfast.
1863 Trevelyan *Dawk Bungalow* 11, Are you ready for some chota hazaree? 1866 E. J. Waring *Trop. Resid.* 171 There is one small meal..commonly known in India by the Hindústani name of *Chota-haziri*, and in our English colonies as '*Early Tea*'. 1883 Mrs. Croker *Pretty Miss Neville* II. 23 At eight o'clock we had chotah-hazree under a big tree in the compound; it consisted of tea, toast, seed-cake, and fruit...We breakfasted at ten o'clock. 1884 D. Aubrey *Lett. fr. Bombay* ii. 15 They intended to..come on board at eight o'clock..and have chota hazri on the water. 1888 J. Inglis *Tent Life* v. 63 A hasty *chota haziree*, or little breakfast. 1927 *Blackw. Mag.* Sept. 326/1, I was the only officer in the two companies who got *chota hazri* that morning. 1928 *Cornhill Mag.* Oct. 470 We sat down to a very early 'chota hazaree' that was laid outside in the moonlight.

chott, var. Shott.
1891 Playfair in *Rep. Brit. Assoc. 1890* 876 Some parts of the Sahara are below the level of the sea, and here are formed what are called *chotts* or *sebkhas*, open depressions without any outlets, inundated by torrents from the southern slopes of the Atlas in winter and covered with a saline efflorescence in summer. 1926 A. Huxley *Essays New & Old* 7 Beyond the oasis we could see the chotts, glittering in the sun.

chou (ʃū). Pl. **choux** (ʃū); also as sing. [Fr., = cabbage.] **1.** A small round cake of pastry filled with cream or fruit, etc. Hence *chou(x) paste, pastry*, a very light egg-enriched pastry.
1706 [see Petit *a.* 5]. 1892 T. F. Garrett *Encycl. Pract. Cookery* I. 382/1 Choux with Coffee Icing. *Ibid.*, Glazed Chou. *Ibid.*, Pieces of Choux paste about the size of fowls' eggs. 1892 Mrs. Beeton *Housek. Managem.* 830 Chou Paste. Ingredients.—¼ lb. of flour, ½ lb. of butter, 8 eggs, ¼ oz. of pounded sugar, 1 pint of water. 1951 *Good Housek. Home Encycl.* 585/2 In..choux pastry the fat is melted in hot liquid before being added to the flour. 1962 *Listener* 14 June 1050/2 Make the *chou* pastry first, by boiling the water and butter together. *Ibid.* 21 June 1095/2 These *cherry choux puffs* are my family's favourite.
2. A knot of ribbon, chiffon, etc., as an ornament to a woman's hat or dress. (Cf. *cabbage *sb.*[1] 1 c.)
1883 *Peel City Guardian* 19 May 7/5 The tunic is draped high and wide on the right side, under a chou bow. 1891 *Queen* 7 Feb. 217/3 A new front drapery, which hooked on to the back beneath a 'choux'—namely, an oblong rosette. 1894 *Ibid.* 13 Oct. 637/2 A blouse bodice,..fastened with three large choux. 1894 *Paris Mode* I. 31/2 Sleeves are frequently caught in by a *chou* of ribbon. 1930 *Times* 13 Mar. 11/6 A charming close-fitting black hat in fine horsehair lace with a big choux of satin ribbon coming down to the shoulder. 1957 M. B. Picken *Fashion Dict.* 62/2 *Chou*, soft cabbage-shaped rosette or knot of velvet, satin, ribbon, lace, etc.

‖**choucroute** (ʃukrut). [Fr., by folk-etymol. ad. 18th-c. *sorcrote*, f. G. (Alsatian) dial. *surkrut* = G. *sauerkraut*.] A kind of pickled cabbage. Hence *choucroute garnie*, choucroute cooked and served with meat.
1849 C. Brontë *Shirley* I. v. 90 As to the choucroute, she affirms she cannot touch it! 1921 W. J. Locke *Mountebank* x. 119 She would sooner have a *choucroute garnie* than a fore-quarter of Paris lamb. 1961 *Harrods Food News* 6/1 Canned Meats from France. Choucroute Garnie.

‖**chouette** (ʃue·t). [Fr. (cf. phr. *faire la chouette* to sustain the attack of several opponents at cards at once), lit. barn-owl, f. OF. *çuete*.] A 'lone hand' at bezique, piquet, or backgammon; a game including a 'lone-handed' player.
1887 'Cavendish' *Laws Rubicon Bézique* 9 A player playing *à la chouette* has choice without cutting. 1892 —— *Ibid.* (ed. 2) 39 A player willing to take all stakes, may declare to play *à la chouette*—i.e., single-handed against all comers...If the *chouette* wins, an opponent takes the loser's place, and so on with the others in rotation. 1927 *Sunday Express* 6 Nov. 2, I remember once this man taking a chouette at piquet against Mr. Ashmead Bartlett and myself. 1929 W. J. Locke *Ancestor Jorico* xi, When he was with us we played a chouette (three players) of Rubicon Bézique for hours together. 1935 *Encycl. Sport* 47/1 (Backgammon) Chouette is a variation to enable more than two people to play. Each person throws a die, the highest being said to 'sit in the box'. The next highest is active opponent and the others act in an advisory capacity to him.

chou moellier (tʃau mǫ·liəɹ, ‖ ʃū-). [Fr., lit. marrow-filled cabbage.] A coarse kind of kale grown esp. in New Zealand and Australia for feeding to stock.
1910 *Jrnl. Dept. Agric. N.Z.* 15 Oct. 354 *Chou moellier* (*Marrow Cabbage*): *Ruakura*. This comparatively new plant is proving valuable as a fodder crop. 1950 *N.Z. Jrnl. Agric.* Jan. 19/1 [In Hawkes Bay] some 42,000 acres are devoted to the growing of green fodder and root crops, and of this chou moellier for supplementary winter feed occupies nearly 20,000 acres.

chouse (tʃaus), *v.*[2] *U.S.* Also **chowse.** [orig. unkn.] *trans.* To disturb or harry (cattle).
1920 J. M. Hunter *Trail Drivers Texas* 313 The round-up boss would let no one ride through the herd and 'chouse', or unnecessarily disturb them. 1930 J. F. Dobie *Coronado's Children* 116 While they were 'chousing' a bunch of outlaw cattle. 1948 *Sat. Even. Post* 10 July 84/1 When you chowse 'em around they lose weight.

chow (tʃau), *sb.* [Shortened f. Chow-chow.] **1.** *slang* (chiefly *Austral.*). A Chinaman. (Derogatory.)
1872 in G. L. Meredith *Adv. in Maoriland* (1935) iii. 22 History does not report what induced the solitary Chinaman to take up his abode amongst the hardy Scots at Dunedin. Possibly this 'Chow' wanted to study economy in its higher branches. 1894 *Bulletin* (Sydney) 5 May 13/3, I tole 'im 'e was a blanky Chow. Then 'e biffed me. 1906 E. Dyson *Fact'ry 'Ands* xv. 198 They're [rats] ez thick ez flees on er Chow's dorg. 1916 C. J. Dennis *Moods Ginger Mick* i. 15 Noiseless Chows a-glidin' 'round wiv plates uv duck an' fowl. 1921 *Chambers's Jrnl.* 690/1 The pearling-crews can get from Japs and Chows all the drink they want. *Ibid.* 718/1 Always have a Chow for your cook in these parts of the world. 1936 M. Franklin *All that Swagger* xii. 112, I never saw a chow riding. 1945 [see *chow-chow 3].
2. *Pidgin-English* and *slang*. Food, or a meal, of any kind. Also *spec.* = Chow-chow 1. Also *attrib.*
This sense is supposed to be due to the use of the chow ('the edible dog of China') as food by poor Chinese.
[1856 *Spirit of Age* (Sacramento) 27 Nov. 2/2 Ah Chow—*ah* in the Celestial lingo means *Mr*, *Chow* something good to eat.] 1886 Yule & Burnell *Hobson-Jobson* 164/2 *Chow* is in 'pigeon' applied to food of any kind. 1892 Douglas Sladen *Japs at Home* iv. 45 More chow..had been laid at our feet...This time it was..a sort of custard soup containing chicken [etc.]. 1897 *Outing* (U.S.) XXX. 209/1 The 'chow' served at the wayside tea-houses. 1926 *Chambers's Jrnl.* 213/2 Here were pots of boiling oil, chow, and mushrooms. 1926 'M. Leinster' *Dew on Leaf* 149 She took her chow-bowl into the kitchen. 1955 *Times* 3 May 12/5 She derived much pleasure from serving visitors with Chinese 'chow' and merriment from watching them use chopsticks. 1958 *Landfall* XII. 120 That night at chow time, Rankin called along to Tiny.
3. A dog of Chinese breed, something like a Pomeranian, usually black or brown, with a black tongue. Also *chow-dog*. (See also *chow-chow 4.)
1889 *Pall Mall Gaz.* 18 June 2/1 One half the pack ran into an unlucky chow-dog. 1897 *Westm. Gaz.* 18 May 2/1 The very chow-pup at his lady's feet looked impudent defiance. 1900 C. H. Lane *All about Dogs* 303 The Chow,..with a short thick head and rather blunt prick ears. 1957 *Encycl. Brit.* VII. 495B/2 A bas relief dated to..about 150 B.C. pictures a chow in a hunting capacity.

chowchilla (tʃau·tʃi·lă). *Austral.* [Imit. of its note.] The black-headed log-runner (*Orthonyx spaldingi*), a small bird found in the dense scrub of mountain ranges in Queensland.
1931 N. W. Cayley *What Bird is That?* 44 Chowchilla (Northern Log-runner) *Orthonyx spaldingi*...Call, a series of notes, like 'Chow-chilla-chow-chow-chilla'. It is also said to be a wonderful mimic. 1934 A. H. Chisholm *Bird Wonders Australia* xxi. 206 Settlers in northern Queensland know the Black-headed Logrunner not only as the 'Auctioneer-bird', but as the 'Chowchilla', since, they say, a company of the birds freely shouts, 'Chow-chilla-chow-chow'. 1958 J. A. Leach *Austral. Bird Book* (ed. 9) 140 Northern Chowchilla..upper black; under white; under base spiny tail smoke-brown;..upper breast orange-rufous.

chow-chow. Add: **1.** *sb.* (Earlier example.) Also, food of any kind.
1795 A. Anderson *Narr. Brit. Embassy China* (Gloss.), Chow-chow..victuals *or* meat. 1864 Hotten *Slang Dict.* 100 *Chow-chow*, a mixture, food of any kind. 1912 Galsworthy *Pigeon* i, (*He*..*cuts a large slice of cake..and hands it in.*) Chow-chow!
3. *Austral. slang.* = *chow *sb.* 1. (Derogatory.)
1869 'E. Howe' *Boy in Bush* 215 The Jerry's Town youngsters were pelting the Chinaman,..meanwhile shouting out 'Chow-chow!' 1945 Baker *Austral. Lang.* x. 185 The many expressions we have developed..for the Chinese: *chink, chinkie, chow, chow-chow*, [etc.].
4. A Chinese domestic dog; = *chow *sb.* 3. Also *attrib.*
1886 *Kennel Club Cal.* 397 Chow Chow. 1892 *Field* 5 Mar. 325/1, I should be glad to hear if any of your readers has ever seen grey Chow-Chow dogs. 1896 *Daily News* 9 Dec. 5 The chow-chow..came originally from China. 1957 *Encycl. Brit.* VII. 495B/2 The chow-chow is probably the oldest of the breeds in the non-sporting group.

chowder, *sb.* Add: **1.** (Earlier U.S. example.)
1751 *Boston Even. Post* 23 Sept., Directions for making a chouder.
b. In full *chowder-party*: a party at which chowder is eaten. *U.S.*
1826 T. Flint *Recoll.* 354 We had public chowder-parties, where sixty people sat down under grape-vine arbours, to other good things beside fish. 1848 Bartlett *Americanisms* 82 Nearly 10,000 persons assembled [at a political mass-meeting] in Rhode Island, for whom a clambake and chowder were prepared. 1853 J. G. Baldwin *Flush Times Alabama* 80 In the science of getting up..a picnic or chowder party, or fish fry, the Virginian..was first. 1906 *N.Y. Even. Post* 6 Nov. 8 The Bowery..went about the business..with as much good nature as if it were 'Big Tim's' annual 'chowder'.

chowder, *v.* Add: **b.** *trans.* To convert into chowder. Hence **chow·dering** *ppl. a.*
1732 B. Lynde *Diary* in *Amer. Speech* (1940) XV. 227/1 Dined on a fine chowdered cod. 1884 *Century Mag.* XXVIII. 555 But it has been such a picnicing and chowdering place. 1947 R. P. T. Coffin *Yankee Coast* 295 Steamed, chowdered ..or fried..clams are always in season.

chowderhead (tʃau·dəɹhed). *U.S.* [cf. Jolter-head, and E.D.D. *Cholter-head* sb., *Jolter-head* sb.] A blockhead; a muddle-headed person.
1833 J. Neal *Down-Easters* I. 119 That's our Amos! if taint I'm a chowderhead. 1838 in *Amer. Speech* (1964) XXXIX. 263 A few sculpins and haddock were all that could be persuaded to make chowderheads of themselves. 1932 F. Scott Fitzgerald *Let.* 2 Aug. (1964) 498, I do not destinate to signify that you were a wiseacre, witling, dizzard, chowderhead. 1967 R. J. Serling *President's Plane is Missing* (1968) iv. 62 The stupid chowderheads would cut Shakespeare.

chowk (tʃauk). *India.* Also 9 **choke.** [Hindi *chauk*.] An open place in the middle of a city where the market is held; a main street.
1832 T. Skinner *Excurs. India* I. ii. 49 The Chandy Choke, in Delhi,..is perhaps the broadest street in any city in the East. 1895 Mrs. Croker *Village Tales* (1896) 20 One evening we all gathered round the big fire in the village 'chowk'. 1924 *Blackw. Mag.* Nov. 700/1 In the Chowk or Square of A., one of the most sacred cities of Hinduism. *Ibid.* 700/2 Padre Armstrong had been lent a pulpit in a corner of the Chowk. 1961 B. Cooper *Touch of Thunder* lii. 286 We stood stranded for one fleeting minute in the middle of Chowk.

chowkidar, var. Chokidar.
1887 Kipling *From Sea to Sea* (1899) II. 314 A *chowkidar* (watchman) was deputed to do justice. 1895 Mrs. Croker *Village Tales* (1896) 15 After some parley we were admitted by the chowkidar (or watchman). 1908 *Westm. Gaz.* 4 May 5/1 The 'chowkidar' of each village..receives a Government salary of three rupees a month. 1953 J. Masters *Lotus & Wind* 93 The chowkidar's disappeared too, so be must have been in league with this fellow.

chow mein (tʃau mē¹n). [Chinese, lit. fried flour.] Fried noodles served with a thick

sauce or stew composed of chopped meat, vegetables, etc.
1903 R. L. McCardell *Conv. of Chorus Girl* 142 Chaw main is good, too. That's chicken and ginger and mushrooms and bamboo sprouts and other stuff. 1920 S. Lewis *Main Street* vi. 79 The golden fried noodles of the chow mein. 1932 *N. & Q.* 30 Apr. 322/2 A very fine fried variety [of noodles] is the superstructure of the Sino-American dish called *chow mein*. 1938 D. Baker *Young Man with Horn* i. 56 The two styles of chow mein: in one you get the noodles and the sauce served separately, and in the other..mixed before they are served. 1961 B. Aldiss *Primal Urge* ii. 41 They ate their chow mein, sweet and sour pork and crispy noodles.

chowrie, var. Chowry.
1848 J. H. Stocqueler *Oriental Interpreter*, *Chowrie*, or whisk, made by fastening horse-hair to a short stick, commonly lacquered in rings of alternate colours. This implement is used to drive the flies away. 1864 J. A. Grant *Walk across Africa* 146 A small boy..carried his chowrie, or fly-flapper.

Chozar, see *Khazar.

chresard (kri·sāɹd). *Ecol.* [f. Gr. χρῆσ-ις use + ἄρδ-ειν to water.] That portion of the water of the soil that is available for vegetation.
1905 F. E. Clements *Res. Methods in Ecol.* ii. 32 The total amount of water in the soil is divided into the available and the non-available water-content. The terms suggested for these are..holard..chresard..and echard. 1926 Tansley & Chipp *Stud. Vegetation* vii. 127 If the echard is subtracted from the holard..the difference is called the chresard (the fraction of the holard which the plant can actually use). 1938 Weaver & Clements *Plant Ecol.* (ed. 2) viii. 203 Obviously the available water or chresard (amount for use) differs for different soils.

chrismation. Delete †*Obs.* and add later examples.
1880 J. S. Stallybrass tr. *Grimm's Teut. Mythol.* i. 3 Among the Goths chrismation is administered to Sigibert's wife Brunechild. 1962 in P. F. Anson *Bishops at Large* (1964) ix. 426 The first contingent of these clerics..received Chrismation.

chrismon (kri·zmǫn). [med.L., f. *Chris(tus + mon(ogramma*.] = *chi-rho.
1872 J. D. Champlin in *Appleton's Jrnl.* 28 Dec. 723 (Funk), The *chrismon*..is found on Christian tombs of the beginning of the second century. 1959 *Chambers's Encycl.* IX. 498/2 The early Christians used a monogram of the first two letters of the name of Christ in Greek..(chrismon or monogrammatic invocation).

Chris(s)ake (krəi·sē¹k). Vulgar corruption of *Christ's sake*, used in oaths, etc. Usu. prec. by *for*.
1933 E. Hemingway *Winner take Nothing* (1934) 199 'Oh, for Chrisake,' the detective sergeant said. 1943 P. Cheyney *You can always Duck* vi. 96 'For chrissake!' he says. 'What do you mean?' 1951 J. D. Salinger *Catcher in Rye* iii. 28 For Chrissake, grow up. 1953 W. R. Burnett *Vanity Row* ii. 18 Chrissake, can't she sit up here? 1954 M. Procter *Hell is City* iii. xv. 143 For Chrissake gimme a cigarette.

Christ. Add: **2. c.** An image or picture representing Christ.
a 1666 Evelyn *Diary* 13 Feb. an. 1645 (1955) II. 356 Good pieces of Paynting, principally the Christ of Corregio, with this Painters owne face admirably don by himselfe. 1876 Geo. Eliot *Dan. Der.* I. ii. iii. 39 A Christ breaking bread over the mantelpiece. 1931 G. Greene *Rumour at Nightfall* i. 6 Wooden crosses bearing bloodstained and contorted Christs.
4. *The Christ-child* [after G. *Christkind, -kindchen*], Christ as a child. Also *Christ-figure*, (*a*) the figure of Christ; (*b*) a figure such as Christ was.
1842 W. Howitt *Rural & Dom. Life Germany* xii. 167 The Christ-kindschen is no other than the Christ of the old legends—Christ in his boyhood...This is exactly the Christ-child of Germany. 1857 *National Mag.* I. Christmas No., Suppl. 1 The Christ-child (bringer, according to German legend, of Christmas gifts and blessings to children). 1887 F. E. Weatherley *Christ-Child* x, They fold their hands, and whisper, 'The Christ-Child has been here.' 1898 'A. De Burgh' *Eliz., Empr. Austria* 296 The benevolent Christ-child has already visited me. 1903 *Daily Chron.* 9 Dec. 3/3 The Christ-child of many painters. 1905 *Westm. Gaz.* 14 Aug. 3/2 One of the few great paintings..illustrating the Christ-figure's final human appearance. 1916 D. H. Lawrence *Twilight in Italy* 20 There are great gashes on..the Christ-figure. 1964 *English Studies* XLV. 361 He sees himself as a Christ figure.

christening, *vbl. sb.* Add: **4.** *christening-dinner* (example), *-dress, -gown.*
1805 Jane Austen *Let.* 8 Apr. (1952) 149, I expect a prodigious account of the Christening-dinner. 1807 *Ibid.* 172 Mrs. Dickson..desires her not to provide herself with a christening dress. 1902 'W. Ramal' *Songs of Childhood* 91 Your baby brother dear Lies in his christening-gown.

Christer (krəi·stəɹ). *U.S. slang.* [f. Christ + -er[1].] A term applied disparagingly to an over-zealous, over-pious, or sanctimonious person.
1924 'W. Fabian' *Sailors' Wives* vii. 90 You never were a Christer in college. 1932 J. Dos Passos *1919* 137 In spite of not drinking and being somewhat of a Christer, having odd ideas about reform and remedying abuses. 1934 H. Miller *Tropic of Cancer* (1948) 81 It doesn't matter to me whether he's a chauvinist, a little Christer, or a near-sighted pedant. 1940 E. Pound *Cantos* lxi. 97 You Christers wanna have foot on two boats. 1966 J. Philips *Wings of Madness* (1967) iii. i. 172 I'm a Christer and a do-gooder... I wasn't welcome.

Christian, *a.* Add: **1. d.** *Politics. Christian Democrat* (see quot. 1957); so *Christian Democratic* adj.
1934 Webster, Christian Democratic party. 1947 *Times* 30 Dec. 5/3 The Christian Democrats looked forward to a West German government dominated by themselves and the Liberals. 1949 *Commonweal* 14 Oct. 16/1 The Christian Democrats are the only political party using the word 'Christian' in their name. 1957 Elliott & Summerskill *Dict. Politics* 67 *Christian Democrats*, a term describing the members of moderate Roman Catholic political parties in Belgium.., France.., the German Federal Republic.., Italy.., and the Netherlands. After the Second World War the Christian Democratic parties achieved striking successes in these countries on platforms of social reform. 1958 *Listener* 20 Nov. 813/2 Dr. Adenauer's Christian Democrats.

christianable (kri·styănăb'l), *a.* *colloq.* [Irregularly formed: see -able.] Fit to be or befitting a Christian. (See Christian *a.* 5.)
1889 C. M. Yonge *Cunning Woman's Grandson* ix. 105 Wanting thee to do what aint fit for no Christianable lad to do. 1920 *Blackw. Mag.* June 774/2 The dug-outs more Christianable than manny. 1926 *Punch* 23 June 675/1 Whether the theme be dogs or tinkers' wives, Or country doctors' 'christianable' lives.

Christiania (kristiā·niă). Also **christiania.** [The name of the capital of Norway (changed, in 1925, to Oslo).] A 'swing' in skiing, used to stop short. Also *attrib.*
1905 E. C. Richardson *Ski-Running* (ed. 2) 44 We have met not a few ski-runners..who make very pretty Telemark and Christiania swings. 1910 —— *Ski-Runner* 135 The Christiania swing is a most useful method of stopping, especially on steepish hillsides. *Ibid.* 137 The best ski-runners can make all four turns, viz.: Telemark to right and left, and Christiania to right and left. 1924 *Ski Terms* in *Tourist* Winter Sports No. 12/1 *Christiania*. Open, closed and jerked. Different forms of one of the best swings. 1929 *Daily Express* 16 Jan. 8/3 Finishing out in the crisp snow with a dead-stop 'Christiania'. 1935 [see *Christie *sb.*]. 1954 E. A. M. Wedderburn *Alpine Climbing* (ed. 2) v. 94 On such hard snow the best plan is to side slip down in the christiania position with the weight on the lower ski.

chri·stianish, *a.* [f. Christian *a.* + -ish[1].] Somewhat Christian in character.
1882 H. W. Beecher in *Chr. World Pulpit* XXI. 285 A man whose brain is weakened cannot be a perfect christian though he may be christianish.

Christian Science. A theory of the nature of disease, a system of therapeutic practice, and a religious sect, founded on principles formulated by Mrs. Mary Baker Glover Eddy, of Concord, New Hampshire, U.S.A. Hence *Christian Scientist*, one who holds and practises this; a member of the sect founded by Mrs. Eddy.
The principal tenet of this theory is that matter is an illusion, and that therefore what is supposed to be bodily disease is an error of the mind, which may be cured by leading the patient to apprehend the truth as revealed in the teaching and healing of Jesus Christ. The churches of Christian Scientists are called 'First (Second, etc.) Church of Christ, Scientist'.
1863 [cited as used by Quimby, in Georgine Milmine *Life M. B. G. Eddy* (1909) v. 87]. 1875 Mrs. Eddy *Sci. & Health* 456 Future years will..restore at length the fair proportions and radical claims of Christian Science. 1881 *Ibid.* (ed. 3) II. xi. 192 Platform of Christian Scientists. 1883 Mrs. Eddy *Let.* in *Boston Post* 7 Mar., My discovery, in 1866, of the Science of Mind-healing, since named Christian Science. 1888 F. Lord (*title*) Christian Science Healing: Its Principles and Practice. 1891 Mrs. Eddy *Sci. & Health* Pref. p. xi, The first school of Christian Science Mind-healing was begun by the author in Lynn, Massachusetts, about the year 1867, with only one student. 1903 *N. Amer. Rev.* Jan. 3 A Christian-Scientist church. 1911 G. B. Shaw *Doctor's Dilemma* Pref. p. xcii, The rise of Christian Science with its cathedrals and congregations and zealots and miracles and cures. *Ibid.* 279/2 The commercial general practitioner, who foolishly clamors for the prosecution..of the Christian Scientists when their patients die. 1962 R. B. Fuller *Epic Poem on Industrialization* 197 The Puritan stock.. evolved successively: Unitarianism and Christian Science.
¶ Note the earlier occurrence of the phrase in a different sense.
1850 W. Adams (*title*) The Elements of Christian Science, a treatise upon moral philosophy and practice. 1855 Tulloch *Theol. Tendencies of the Age* 12 The element of traditional authority..is to be regarded as in itself wholly invaluable for the interests of Christian Science. *Ibid.* 24 The final arbiter of Christian Science, in short, is not the Bible, but the prevailing Christian sentiment.

Christie (kri·sti). Also **Christy, -i, c-.** Abbrev. of *Christiania. Also as *v. intr.*, to make a Christiania swing or turn.
1920 A. Lunn *Cross-Country Ski-ing* iv. 68 There are nine and ninety things Which are nicknamed Christi swings. 1925 Hemingway *In our Time* (1926) xii. 160 It's a good fast drop with a Christy at the bottom on account of a fence. *Ibid.* xiii. 161, I was afraid to Christy,..the snow was too deep. 1935 *Punch* 13 Feb. 176/1 *Christiania Turn.* —The christie is really a sideways skid. *Ibid.*, Now suppose you were travelling left-handed down and across a slope and Satan prompted you to christie to the right. 1938 *Times* 26 Jan. 15/6 Higher up they are doing 'snow plough turning', 'stem turn', and 'Christies'. 1942 R. Peattie *Friendly Mts.* 300 On the same hill well-to-do brokers, indigent college students, and automobile mechanics..'christie' together.

Christmas, *sb.* Add: **1. c.** As *int.* or expletive. Also with supporting word, as *Jiminy Christmas!*
1897 Kipling *Capt. Cour.* iv. 91 Jiminy Christmas! That gives me the blue creevles. 1898 —— *Day's Work* 217 I've bin down to Track 17, and the freight there—oh, Christmas! 1930 'E. Queen' *French Powder Myst.* xxxvi. 276 Christmas!..How stupid I've been! 1959 N. Marsh *False Scent* (1960) vi. 192 'All right with you, Bertie?' 'Oh, Christmas!' he said. 'I suppose so.'
3. (Further examples.)
1721 R. Palmer *Let.* 15 Apr. in M. M. Verney *Verney Lett.* (1930) II. xxiv. 86 In the Christmas, Easter, and Whitsun holidays, our penny post comes but once a day. *a* 1817 Jane Austen *Persuasion* (1818) ii. ii. 34, I hope I shall remember in future..not to call at Uppercross in the Christmas holidays. 1835 J. S. Mill in *Examiner* 4 Jan. 4/2 They form one of the most agreeable of Christmas presents to a lover of music. 1848 A. Somerville *Autobiogr. Working Man* ix. 108 The Christmas pantomime..that season [*sc.* 1828]..was *Mother Goose*. 1890 G. B. Shaw *London Music* (1937) 289 Passing Her Majesty's Theatre, I saw by the placards that a Christmas pantomime was going on inside. 1893 *Ladies' Home Jrnl.* Dec. 17/1 My Christmas shopping. 1932 R. Lehmann *Invit. Waltz* i. viii. 87, I always think with Christmas shopping it's best to get it done in good time. 1934 *Discovery* Dec. 362/1 In considering what books are suitable as Christmas presents for our friends, it is worth while remembering that many of the younger generation can do with something better than *Auntie Omega's Yuletide Annual*. 1936 *Ibid.* Feb. 35/1 The title to fathership of the Imperial family, which the King..had accepted in one of his Christmas broadcasts. 1965 'P. Nichols' *Patchwork of Death* (1967) iv. 19 There was the Christmas rush beginning.
4. **Christmas bush,** (*a*) any of various Australian shrubs, esp. *Ceratopetalum gummiferum* of N.S.Wales, used in Christmas decorations; (*b*) a Bermuda shrub, *Cassia bicapsularis*, used for hedges; **Christmas candle** (later examples); **Christmas-card,** also used *attrib.*, esp. = *chocolate-boxy* (*chocolate 5); **Christmas fern,** a N. American evergreen fern; **Christmas flower** (examples); **Christmas number** (examples); **Christmas shield-fern** *U.S.* = *Christmas fern*.
1884 A. Nilson *Timber Trees N.S.W.* 135 Christmas bush, Ceratopetalum gummiferum. 1888 Mrs. McCann *Poet. Wks.* 226 Gorgeous tints adorn The Christmas bush with a crimson blush. 1964 *Mod. Encycl. Austral. & N.Z.* 241/1 *Christmas bush*...Beautiful small tree or bush with graceful green foliage. 1846 Dickens *Pict. from Italy* 181 Little tapers, like what are called Christmas candles in England. 1940 L. MacNeice *Last Ditch* 21 Your long ear-rings twisted like Christmas candles. 1895 G. B. Shaw *Our Theatres in Nineties* (1932) I. 203 Christmas-card verse of the 'rose by any other name' order. 1907 —— *Major Barbara* i. 209 Your Christmas card moralities..are of no use to me. 1910 *Westm. Gaz.* 3 Jan. 4/2 The New Year's card, that refuge of the laggard Christmas-card sender. 1962 J. B. Priestley *Margin Released* iii. iv. 183 A sort of Christmas-card past. 1878 J. Williamson *Ferns of Kentucky* 99 (*heading*) *Aspidium acrostichoides*..Winter Fern —Christmas Fern. 1880 J. Robinson *Flora Essex Co., Mass.* 135 *Aspidium acrostichoides*, Sw. (Christmas fern.) Rocky woods, Common. 1889 *Cent. Dict.*, Christmas fern, *Aspidium acrostichoides*, a fern having simply pinnate fronds of firm texture, which remain green through the winter. 1629 Parkinson *Parad.* lxxxi. 344 The true blacke Hellebor, or Christmas flower. 1889 'C. E. Craddock' *Broomsedge Cove* xix, He stopped to pick a spray of the lilac 'Christmas flower'. 1857 C. M. Yonge *Dynevor Terr.* I. xv. 252 What Quarterly was it? Surely the Christmas number was not out. 1908 *Westm. Gaz.* 29 Dec. 2/3 The imagination of Christmas-number artists. 1928 *Granta* 30 Nov., I reckon if forty or so Christmas numbers of the *Granta* haven't said all there is in the way of little bits about Christmas, it's not for me to start digging skeletons out of the cupboard. 1878 J. Williamson *Ferns of Kentucky* 98 Pl. 36, *Aspidium acrostichoides*..Christmas Shield-Fern. 1884 W. Miller *Plant-n.* 164 *Aspidium acrostichoides*, Christmas Shield-Fern, of N. America.

Christmas disease. [f. the name of S. *Christmas*, a sufferer from the disease.] (See quot. 1961[1].)
1952 R. Briggs et al. in *Brit. Med. Jrnl.* 27 Dec. 1378/1 If this definition of haemophilia is accepted then the seven cases recorded in this paper are..a newly recognized condition which we propose to call 'Christmas disease', after the name of the first patient examined in detail. 1961 *Brit. Med. Dict.* 304/2 *Christmas disease*, a hereditary bleeding disease ..very closely resembling haemophilia in clinical and genetic features... It differs from haemophilia in that the blood contains the anti-haemophilic globulin in normal amounts but does not contain a serum factor (termed the Christmas factor) which is present in haemophilic blood. 1961 *Lancet* 12 Aug. 374/1 The American workers found their thrombotic principle was absent from Christmas disease.

Christmassy, *a.* The more usual spelling of Christmasy. (Further examples.)
1894 *Pall Mall Gaz.* 20 Dec. 8/3 Oxford, with the undergraduates out of the way and a foot or two of snow in the streets, would look very Christmassy. 1901 G. Bourne *Bettesworth Bk.* 156, I saw that he looked pale and gently sick. 'Feel a bit Christmassy?' I said. 1957 *New Yorker* 16 Nov. 157/1 Anticipating the Christmas-party season.. trimming bird-cage veils with Christmassy gewgaws.

Christmas-tree. Add: **b.** (*a*) = *POHUTUKAWA *N.Z.*; (*b*) any of various Australian shrubs which flower about Christmas-time, esp. *Nuytsia floribunda* of Western Australia.

1867 F. VON HOCHSTETTER *N.Z.* xi. 240 Some few scattered Pohutukawa trees...About Christmas these trees are full of charming purple-blossoms; the settler..calls the tree 'Christmas-tree'. 1944 W. R. B. OLIVER in *Korero* II. 11 Sept. 13 The glory of the Auckland coast, at least in midsummer, is the pohutukawa or Christmas tree. 1966 *Times* 11 Nov. (W. Austral. Suppl.) p. iv/2 The hot gold cascade of *nuytsia floribunda* the parasitic Christmas tree.

2. *transf.* A mechanical device or a structure in the shape or operation of which there is a fancied resemblance to a Christmas-tree. (See quots.)

1917 'CONTACT' *Airman's Outings* 12 The movable mounting for the observer's gun in the rear cockpit was a weird contraption like a giant catapult...We called it the Christmas Tree, the Heath Robinson, the Jabberwock, [etc.]. 1929 R. GRAVES *Goodbye to all That* xii. 135 My belt was hung with all the usual furnishings—revolver, field-glasses, compass, whisky-flask, wire-cutters, periscope, and a lot more. A Christmas-tree that was called. 1930 W. H. OSGOOD *Recov. Petroleum* II. xvii. 706 Well-head connections or Christmas trees..operating a well on the lift. *Ibid.* 711 Various types of packed casing heads are used in the casing-head or Christmas tree assembly on gas-lift wells. 1944 *Amer. Speech* XIX. 230/2 'Shipyard Terms'..*Christmas tree*, auxiliary steam manifold with lines and valves in all directions. 1948 *Ibid.* XXIII. 37 *Christmas Tree*. This electric indicator board resembles the object from which its name is derived only because the lights on it are red and green. The function of the Christmas tree is to give proper notification that all the necessary valves and hatches are closed when a submarine dives. 1957 RAWNSLEY & WRIGHT *Night-Fighter* iv. 56 The signals bounce back from the ground just as they do from other aircraft in the air, but they give a much stronger echo. And the lower you fly the further down the trace the Christmas tree comes.

Christo-. Add: **Christocentri·city,** the state of having Christ as the centre; **Christoce·ntrism,** Christocentric doctrine; **Christo·cracy,** the rule or government of Christ; **Christocra·tic** *a.*, constituted under the rule of Christ; **Christolo·gic** *a.* = CHRISTOLOGICAL.

1921 C. C. MARTINDALE *St. Justin* iii. 125 Christocentricity is the pivot of Justin's thought. 1955 *Jrnl. Theol. Stud.* VI. 47 [Beza's] eschatology is devoid of the Christocentricity which it ought to have. 1963 *Observer* 31 Mar. 22/5 This is the *reductio ad absurdum* of the narrow Christocentricity which the Churches offer man today. 1912 VON HUGEL *Eternal Life* 198 Schleiermacher..insists..upon so intense a Christocentrism, as to prepare acute collisions with Biblical Criticism. 1939 J. BAILLIE *Our Knowl. of God* i. 17 He stands..in the tradition of that Lutheran christocentrism which makes Christ the Mediator no less of knowledge than of salvation. 1925 *Brit. Weekly* 1 Oct. 5/4 Therefore, he preferred a Christocracy to a democracy. 1907 *19th Cent.* Dec. 877 Her [*sc.* the Church's] powers are thus derived from Christ and His Apostles, and her constitution in the matter of teaching, ministry, and government is necessarily Christocratic and Apostolic. 1920 HARRIS & BURCH *Testimonies* II. vii. 66 The first Christian teaching had two sides—the polemic and the evangelic: the anti-Judaic and the Christologic.

christologist. (Earlier example.)

1846 GEO. ELIOT tr. *Strauss's Life of Jesus* I. II. i. 332 Out of a practical preacher he makes a speculative christologist.

chroma (krōᵘ·mă). [a. Gr. χρῶμα colour.] Purity or intensity as a colour quality.

1889 in *Cent. Dict.* 1905 A. H. MUNSELL *Color Notation* i. 8 Music..defines each sound in terms of its pitch, intensity, and duration...So should color be supplied with an appropriate system, based on the hue, value, and chroma of our sensations. *Ibid.* 75 *Chroma*, the degree of departure of a color sensation from that of white or gray; the intensity of distinctive hue; color intensity. 1936 *Mind* XLV. 303 The terms 'blue', 'green', and 'yellow' are only rough, though practically convenient, references to an indefinite mass of colour-hues, shades and chromas. 1957 R. W. G. HUNT *Reprod. Colour* viii. 79 In America the Munsell system of colour specification is widely used, and in this system..chroma [is used] instead of Purity. 1959 *B. S. I. News* Nov. 6/1 The Munsell concepts of lightness, hue and strength (chroma).

chromæsthesia (krōᵘmɪsþi·siă, -ĭ·ziă). *Psychol.* Also **chromes-**. [mod.L., f. Gr. χρῶμα colour + stem αἰσθε- to feel, perceive, after *anæsthesia*.] A form of synæsthesia, the accompaniment of the hearing of particular sounds by the seeing of particular colours arbitrarily associated with them. So **pseudo-chromæsthesia.**

1892 W. O. KROHN in *Amer. Jrnl. Psychol.* V. 20 (*heading*) Pseudo-chromesthesia, or the association of colors with words, letters and sounds. 1909 in *Cent. Dict.* Suppl. 1952 in J. DREVER *Dict. Psychol.*

chromaffin (kromæ·fin), *a.* *Histol.* Also **-ine.** [a. G. *chromaffine* (A. Kohn 1898, in *Prager Med. Wochenschr.* XXIII. No. 17), f. *chrom-* CHROMO- 1 + L. *affinis* akin.] Stained brown on exposure to chromic acid or its salts; applied esp. in man and the higher vertebrates to the adrenal medulla and the paraganglia. So **chromaffi·nic** *a.*

1903 SWALE VINCENT in *Jrnl. Anat. & Phys.* XXXVIII. 41 The chromaffin cells of the suprarenal medulla. *Ibid.* 43 The chromaffin bodies in Petromyzon. 1907 *Practitioner* Aug. 211 A series of cells which have been termed chromaffine cells, because they yield a yellow colour on treatment with chromic acid. 1913 A. KEITH *Human Embryol.* (ed. 3) 390 Coccygeal Body is a small mass of chromaffinic tissue. 1932 J. M. ROGOFF in E. V. Cowdry *Special Cytol.* (ed. 2) II. xxiii. 890 The chromaffin granules are known to consist of epinephrin, or of a complex whose chief constituent is epinephrin, and the color reactions which they yield in situ are those of epinephrin. 1964 L. MARTIN *Clin. Endocrinol.* (ed. 4) v. 163 The medulla arises from ectodermal tissue and consists of chromaffin and sympathetic ganglion cells.

chromatically, *adv.* (Examples in *Music.*)

1886 G. B. SHAW *How to become Mus. Critic* (1960) 122 There is nearly always a pedal flowing along, and the other parts are slipping chromatically down to merge in it. 1962 *Listener* 21 June 1091/1 His [*sc.* Hindemith's] lucid counterpoint and chromatically expanded diatonicism.

chromaticity (krōᵘmăti·sĭti). [f. CHROMATIC + -ITY.]

1. *Biol.* [ad. F. *chromaticité*.] The amount or state of chromatin.

1904 tr. L. Launoy in *Jrnl. R. Microscop. Soc.* Apr. 170 Active phenomena [in poison- or digestive glands]—variations of chromaticity [etc.]. 1934 L. W. SHARP *Introd. Cytol.* (ed. 3) xiii. 195 Small bodies in the cytoplasm which resemble nuclear matter in composition and chromaticity.

2. *Optics.* (See quot. 1922.) Also *attrib.* **chromaticity co-ordinates,** co-ordinates in a chromaticity diagram; **chromaticity diagram,** = *colour triangle* (see *COLOUR *sb.* 18).

1922 L. T. TROLAND in *Jrnl. Opt. Soc. Amer.* VI. 535 The term chromaticity may be used to characterize a color qualitatively without reference to its brilliance. Chromaticity is determined by hue and saturation together, a gray being specified by the statement that it has *no* chromaticity. 1932 HARDY & PERRIN *Princ. Optics* xiv. 306 To specify the color of a material completely, it is necessary to give not only the chromaticity but also the luminosity. 1944 W. D. WRIGHT *Measurement of Colour* iii. 60 The chromaticity diagram in which the colour quality of a stimulus is defined in terms of two co-ordinates. 1951 *Engineering* 27 July 105/1 Their new colorimeter..suitable for measuring.. chromaticity co-ordinates. 1955 G. G. GOURIET *Introd. Colour Telev.* I. ii. 22 The principle of superposition applies to our perception of colour in terms of both luminance and chromaticity, a principle which is rarely encountered in physiological processes. 1957 R. W. G. HUNT *Reprod. Colour* viii. 79 Subjective terms denote the appearance of the colour to an observer...Objective terms relate to quantities obtained with measuring instruments...Subjective Term: Hue and Saturation. Objective Term: Chromaticity.

chromaticism. Add: (Further examples.) Also, the use of chromatic expressions, modulations, etc.; an instance of this.

1931 *Times Lit. Suppl.* 30 Apr. 344/4 Oriental influences.. have resulted in the introduction of chromaticism into the tetrachord. 1935 *Times* 27 Dec. 12/3 Four songs..in which chromaticism and the whole tone scale are used. 1947 A. EINSTEIN *Mus. Romantic Era* xi. 126 He [*sc.* L. Spohr] took over some Mozartian chromaticisms..and increased them to a state of hypertrophy which..already borders on the chromaticism of *Tristan*. 1958 *Times Lit. Suppl.* 4 July 372/4 Chromaticism pointed the way towards composition with twelve non-functional notes. 1966 *Listener* 1 Sept. 328/3 In two of these..Delius..mainly dispenses with his chromaticism in favour of a diatonic idiom.

chromatid (krōᵘ·mătid). *Biol.* [f. stem of next + -ID[2].] One of the two strands into which a chromosome divides longitudinally during mitosis.

1900 C. E. MCCLUNG in *Kansas Univ. Q.* A. IX. 78 The term 'chromosome' being..restricted to the units of the division figures, there remains no name for the parts composing these when they are compound, as in tetrads and diads... I should like, therefore, to propose the term 'chromatid' for each of these. 1939 SINNOTT & DUNN *Princ. Genetics* (ed. 3) vii. 166 During synapsis, or previously, each member of the bivalent pair splits longitudinally into two exactly equal daughter chromosomes or chromatids, so that for a time a tetrad is formed consisting of four chromatids. 1952 M. J. D. WHITE in Bourne *Cytol. & Cell Physiol.* (ed. 2) v. 203 At metaphase, and even in prophase, chromosomes are visibly divided into two spirally coiled threads or chromatids.

chromatin. Read: *Biol.* [G. (W. Flemming 1880, in *Archiv f. mikrosk. Anat.* XVIII. 158), f. Gr. χρωματ-, χρῶμα colour + -IN.] A complex constituent of the cell nucleus which can be readily stained when immersed in colouring matter.

1882 W. H. GILBERT in *Jrnl. Quekett Micr. Club* ser. II. No. 1. 34 It is the achromatin, as suggested by Flemming, which acts upon and causes the chromatin. 1890, 1892 [see *CHROMOSOME]. 1912 [see *CHROMATOID *a.*]. 1956 *Nature* 25 Feb. 386/2 The cells of *Bacillus megatherium* have been stained..with various dyes to show the chromatin-threads. 1962 *Lancet* 2 June 1154/2 Drumsticks and sessile nodules are present in the blood of chromatin-positive males showing Klinefelter's syndrome, and they are absent in chromatin-negative females with Turner's syndrome.

chromatism. Add: **4.** *Psychol.* A secondary or hallucinatory sensation of colour; = *CHROMÆSTHESIA.

1903 [see PHOTISM]. 1957 *Dorland's Med. Dict.* (ed. 23) 278/1 *Chromatism*, a hallucinatory perception of colored light.

chromato-. Add: **chro·matochyme** *Biol.* [CHYME], an aggregate of pigment-cells; **chro·matocyte** *Biol.* [-CYTE], a pigment-cell; **chromato·lysis,** the solution and disintegration of the chromatin of cell nuclei; **chro·matoplasm** *Biol.* [-PLASM], the coloured portions of protoplasm; the pigment substance of cells; **chromato·psia** [L. form of CHROMATOPSY], any abnormality in the perception of colours, esp. one in which objects appear to have the wrong colour.

1888 W. J. SOLLAS *Tetractinellida* p. xl, Chromatocytes or Pigment-Cells...Occasionally by repeated multiplication they form cellular aggregates, or chromatochyme. 1963 I. F. & W. D. HENDERSON *Dict. Biol. Terms* 100/1 *Chromatocyte*, any cell containing a pigment. 1908 *Practitioner* Jan. 19 A peri-nuclear chromatolysis of the large psychomotor Betz cells. 1964 M. HYNES *Med. Bacteriol.* (ed. 8) xxv. 394 The parenchyma cells undergo a characteristic necrosis passing from cloudy swelling to fatty degeneration and to nuclear chromatolysis. 1902 E. B. WILSON *Cell* (ed. 2) 439 *Chromatoplasm*.., the substance of the chromoplastids and other plastids. 1948 *New Biol.* V. 10 Sometimes in a living specimen (blue-green alga)..it is possible to distinguish two regions in the protoplasm, an outer region in which the pigments are evenly distributed, the chromatoplasm, and a central unpigmented part, the centroplasm. 1848 DUNGLISON *Med. Lex.* (ed. 7) 182/2 Chromatopsia. 1873 *Richmond & Louisville Med. Jrnl.* Jan. 83 (*title*) Chromatopsia with hyperaesthesia of the retina. 1911 *Encycl. Brit.* XXIV. 195/2 Even small doses of santonin cause disturbances of vision, usually yellow vision or perhaps green (xanthopsia or chromatopsia). 1947 F. B. WALSH *Clin. Neuro-Ophthalmol.* xiii. 1193/1 In a proven case of temporal lobe tumor, van Bogaert described transient visual phenomena, chromatopsia, [etc.].

chromatogram (krəmæ·togræm). *Chem.* [ad. G. *chromatogramm* (M. Tswett 1906, in *Ber. d. Deut. Bot. Ges.* XXIV. 322): see -GRAM.] The result of a chromatographic separation: either a series of (coloured) bands or spots in liquid chromatography or a graphical record from a gas chromatograph.

1922 L. S. PALMER *Carotinoids* ii. 43 The resulting chromatogram (as Tswett proposed to call it) presented a most surprising picture of the chloroplastid pigments. 1940 *Thorpe's Dict. Appl. Chem.* IV. 174/2 In making a chromatogram, i.e. zones of separated colouring matters.. advantage is taken of the differential adsorption which occurs when a solution of several dyes is allowed to drop on to a column of adsorbent. 1949 ABRAHAM & HEATLEY in H. W. Florey *Antibiotics* I. ii. 107 Two-dimensional chromatograms were obtained by first running a developing solvent in one direction and then a second developing solvent in a direction at right angles to the first. 1952 JAMES & MARTIN in *Biochem. Jrnl.* L. 679/1 Phillips..and Turkel'taub..have used charcoal in gas chromatograms for the analysis of hydrocarbons and esters. *Ibid.*, Martin & Synge..suggested the use of gas-liquid chromatograms for analytical purposes. 1955 *Sci. News Let.* 23 Apr. 265/2 The mushrooms are minced, and after alcohol extraction and evaporation, a chromatogram on filter paper is prepared from the residue. 1962 *Lancet* 29 Dec. 1346/2 Plasma-sugar chromatograms revealed lactose in every sample throughout the test period. 1965 *New Scientist* 11 Nov. 395/1 The scanning of radioactively labelled chromatograms with Geiger counters.

chromatograph (kromæ·tograf), *sb.* *Chem.* [-GRAPH.] An apparatus that produces a chromatogram.

1958 *Times Rev. Industry* Feb. 90/3 The gas chromatograph..is aimed at replacing batch process control by a continuous monitoring system. 1958 *New Scientist* 2 Oct. 956/1 The principle of the gas chromatograph is very simple. 1961 *Listener* 7 Dec. 983/2 This concentration [of the anæsthetic] may be as small as a few parts in a million, but it is easy to adapt the gas chromatograph to measure this.

chromatograph, *v.* Add: **2.** *Chem.* To separate or analyse by chromatography.

1956 *Nature* 14 Jan. 84/1 Artefacts and serious distortion can be caused during paper chromatography if acids are present in the solution being chromatographed. *Ibid.* 4 Feb. 224/1 The results of chromatographing liver proteins against insulin with five different solvent systems are shown. 1961 *Lancet* 7 Oct. 793/1 This urine was then.. chromatographed on a Celite column. 1962 A. PIRIE *Lens Metabolism Rel. Cataract* 431 The glycerophosphate fraction ..was either chromatographed in saturated phenol-water solution or subjected to electrophoresis.

chromatography. Add: **2.** *Chem.* [ad. G. *chromatographie* (M. Tswett 1906, in *Ber. d. Deut. Bot. Ges.* XXIV. 387).] Of liquids: a process of separating and purifying the substances dissolved in a mixed solution by slow passage through a tube or over a surface of adsorbing material, making use of differences of partition, adsorption, ion-exchange, etc., and separating the constituents either as (coloured) bands or spots or by differences in speed of travel when washed through the adsorbing material. Of gases: see *gas chromatography*, *GAS *sb.*[1] 7.

1937 *Nature* 20 Feb. 335/2 The adsorption method of analysis known as chromatography which has been used with such marked success in the investigation of the carotenoid field of organic chemistry. 1946 *Ibid.* 9 Nov. 675/1 Partition chromatography of the amino-acids. 1949

ABRAHAM & HEATLEY in H. W. Florey *Antibiotics* I. ii. 100 Adsorption chromatography has been widely used in the purification of both naturally occurring and synthetic organic compounds...In this method a solution of the substances to be separated is percolated through a column of adsorbent. 1956 [see *CHROMATOGRAPH *v.* 2]. 1957 *Biochem. Jrnl.* LXVI. 3/2 In paper chromatography, the rate of movement of a substance relative to that of the solvent front is expressed as *RF*. In column chromatography *R* represents (movement of band)/(movement of liquid above surface of column). 1968 *Times* 4 Oct. 9/3 They washed people's arms with a solvent liquid and separated the various chemical components in the solvent by chromatography, a method which exploits the fact that different substances soak through solid materials at varying speeds depending on their chemical structure.

So **chromatogra·phic, -gra·phical** *adjs.*, of, pertaining to, or involving chromatography; **chromatogra·phically** *adv.*

1907 *Jrnl. Chem. Soc.* XCII. II. 144 (*heading*) Adsorption Analysis and Chromatographic Methods. 1946 *Nature* 2 Nov. 623/2 Chromatographical methods of..separation. *Ibid.* 9 Nov. 675/2 The standard spots, which have not been chromatographically developed, give a series of circular inhibition zones. 1949 A. G. SANDERS in H. W. Florey *Antibiotics* I. iii. 183 A method for separating penicillins based on the paper-strip chromatographic technique. 1962 A. SPECTOR in A. Pirie *Lens Metabolism Rel. Cataract* 332 The chromatographically purified esterase.

chromatoid (krōᵘ·mătoid), *a.* *Biol.* [f. CHROMATO- + -OID.] Capable of receiving a stain: said of certain grains or granules.

1901 G. N. CALKINS *Protozoa* 144 The so-called chromatoid granules. 1912 E. A. MINCHIN *Protozoa* 289 Many trypanosomes contain granules in their cytoplasm which stain similarly to chromatin, so-called 'chromatoid grains'. 1952 R. J. LUDFORD in G. H. Bourne *Cytol. & Cell Physiol.* (ed. 2) ix. 397 The larger Negri bodies..consist of a ground substance containing a central body surrounded by chromatoid granules.

chromatophore. Add: **2.** *Bot.* [G. (F. Schmitz *Die Chromatophoren der Algen* (1882) 4).] A plastid containing coloured pigments, as a chromoplast or a chloroplast.

1895 OLIVER tr. *Kerner's Nat. Hist. Plants* II. 629 This chlorophyll-layer or chromatophore. 1902 *Encycl. Brit.* XXV. 407/2 A coloured protoplasmic body, called a chromatophore, the pigment of which in the pure green forms is chlorophyll, and which may then be called a chloroplast. 1960 L. PICKEN *Organization of Cells* iii. 61 Some bacteria contain well-defined chromatophores associated with photosynthetic activity. 1965 BELL & COOMBE tr. *Strasburger's Textbk. Bot.* (new ed.) 43 The yellow and orange colours of many flowers..,as well as the bright red of many fruits (e.g. hips, haws, and tomatoes), are caused, at least partly, by photosynthetically inactive chromatophores.

Hence **chro:matopho·ric** *a.*, pertaining to or containing chromatophores.

1895 *Pop. Sci. Monthly* Jan. 389 The theory that light acts directly on the chromatophoric cells has been proved to be incorrect. 1904 *Philos. Trans. R. Soc.* Ser. B. 322 The groups which make up the primary chromatophoric system.

chrome, *sb.* Add: **1. b.** = *chromium(-plate), plating* (*CHROMIUM 2).

1937 *Chem. Abstr.* XXXI. 2763 Chrome-plated molds for semiautomatic machines of the Kutcher type. 1961 *Amer. Speech* XXXVI. 272 *Armenian chrome*, the aluminium paint used to cover spots on a truck's chrome when preparing it for resale. 1962 *Which?* (*Car Suppl.*) Oct. 139/2 The chrome strip on the left rear wing was coming adrift. *Ibid.* 140/1 Lower chrome strip..on running board [was] insecure.

2. b. A chromium compound used in dyeing (e.g. sodium dichromate) or tanning.

1876 *Textile Colourist* II. 210 Caustic chrome, that is a mixture of bichromate of potash and lime. 1893 E. KNECHT et al. *Man. Dyeing* 771 Potassium bichromate, $K_2Cr_2O_7$, known to the trade also as 'bichrome', or 'chrome'. 1906 *Dyer* 20 Jan. 10/1 The dyestuffs are fixed with chrome, and the printing colors are best prepared with the addition of acetate of chrome. 1961 BLACKSHAW & BRIGHTMAN *Dict. Dyeing* 38 'Chrome', a colloquial term which is applied loosely to any chromium compound which may be used in textile processing.

c. *ellipt.* for *chrome leather*. Also *attrib.*

1907 G. D. ABRAHAM *Compl. Mountaineer* iii. 29 The leather for the uppers should be the best zug or chrome, soft and absolutely waterproof. 1955 *Times* 4 July 11/5 The British chrome kid tanners are concerned chiefly with certain kinds of Indian goatskins which are especially suitable for the chrome kid industry, for shoes, handbags, [etc.].

3. chrome-black, a colour produced by dyeing goods in a black dye and setting the colours by the use of potassium dichromate and copper sulphate; **chrome-blue,** a mordant coal-tar dye used chiefly in cotton-printing; **chrome brick,** a brick made from chrome iron ore; **chrome furnace,** a furnace in which chrome ores are converted into dichromates; **chrome garnet,** a pigment prepared from basic chromate of lead; **chrome handler,** one who tans leather by treating it with a solution of chromium salts; **chrome ink,** an ink made from logwood and sodium carbonate; **chrome leather,** chrome-tanned leather; **chrome maker,** one who makes chrome yellow pigments; **chrome-nickel steel,** a steel containing much chromium and nickel; **chrome-spinel,** picotite; **chrome steel,** a hard fine-grained steel containing much chromium; **chrome-tanning,** the tanning of leather by treating it with an acid solution of potassium dichromate and afterwards with a reducing agent, so that chromic oxide combines with the fibre of the leather, rendering it tough and waterproof; hence *chrome-tanned leather*; **chrome violet,** a mordant acid-dye.

1874 CROOKES *Dyeing* 544 Chrome-blacks are cheap. 1876 *Encycl. Brit.* IV. 690/2 Chrome Black is an oxidation colour produced by printing with logwood liquor and passing the goods through a bath of bichromate of potash. 1892 *Dyer* 20 Jan. 3/1 Chrome Blue is sold in the form of a paste of a blue-black color. Its special characteristic is that with chrome mordants it gives very bright blues. 1921 *Dict. Occup. Terms* (1927) § 102 Chrome brick maker. 1881 *Instr. Census Clerks* (1885) 58 Chrome Furnace Man. 1899 *Daily News* 20 June 5/5 To make a trial of stopping the chrome furnaces. 1876 *Encycl. Brit.* V. 705/1 Ouvarovite, or chrome garnet, $Si_3CaCr_2O_{12}$. 1921 *Dict. Occup. Terms* (1927) § 333 *Chrome handler*; immerses hides and skins in chrome tanning pits, under direction of tanyard foreman. 1880 *Encycl. Brit.* XIII. 80/1 Under the name of chrome ink a black ink was discovered and prepared by the chemist Runge. 1882 *Ibid.* XIV. 390/1 The stuffing with fat or paraffin of chrome leather. 1909 *Westm. Gaz.* 16 Nov. 6/2 A band of specially treated chrome leather. 1955 *Times* 11 July 13/5 The serious position of full chrome leather. 1921 *Dict. Occup. Terms* (1927) § 143 *Chrome maker*; makes chrome yellow pigments by precipitating lead bichromate from a solution of potassium bichromate by running into it a solution of lead nitrate or acetate. 1908 *Westm. Gaz.* 28 Apr. 4/3 Vanadium and chrome-nickel steels—the most expensive materials as they are considered to be the most suitable. 1892 DANA *Min.* (ed. 6) 221 Picotite or chrome-spinel. 1878 *Jrnl. Chem. Soc.* XXXIV. 772 The history of the discovery of chrome-steel and ferrochrome by Berthier in 1821. 1882 *Encycl. Brit.* XIV. 389/2 Heinzerling's Chrome-tanned Leather. *Ibid.* 390/1 Sheep skins in chrome-tanning. 1957 *Ibid.* XIII. 849/2 Chrome-tanned leather has exceptionally good wearing properties. 1903 FLEMMING *Pract. Tanning* 213 The following process of chrome tanning produces leather that possesses the good qualities of both alum and chrome leather. 1907 *Jrnl. Chem. Soc.* XCII. II. 54 Chrome-tanning Liquors. 1959 *Chambers's Encycl.* XIII. 456/2 In.. chrome tanning for upper leather the hide or skin..is immersed in liquors made from chromium salts. 1892 *Dyer* 20 Jan. 3/1 Chrome Violet..gives a fine red shade of violet, which is equally as fast to soaping as the Chrome Blue.

chrome (krōᵘm), *v.* *Dyeing.* [f. prec.] *trans.* To treat with a chromium solution, e.g. potassium dichromate. Chiefly as *ppl. a.* and *gerund* or *vbl. sb.*

1876 *Textile Colourist* II. 35 Dark Brown for Steaming and Chroming. *Ibid.* 318 Chromed logwood colours have a tendency to become green. 1892 *Dyer* 20 Jan. 3/2 Chrome Violet may be used in wool dyeing on a chromed wool. 1893 E. KNECHT et al. *Man. Dyeing* 352 The principal reason that tartar and tartaric acid have not been universally adopted in chroming is the expense. 1906 *Dyer* 20 Feb. 25/1 On chromed material it yields considerably darker shades. 1947 *Nature* 11 Jan. 53/2 Fundamental work on the chroming of wool. 1968 *Somerset Life* Oct./Nov. 26/2 After washing and shearing the skins are then bleached and 'chromed' to ensure that they will never go bad.

chromicize (krōᵘ·misəiz), *v.* [f. CHROMIC *a.* + -IZE.] *trans.* To treat or impregnate with chromic acid or a chromate. Chiefly as *ppl. a.*

1903 *Med. Record* LXIII. 563/1 Almost every conceivable material has been employed in the fixation of the kidneys. Chromicized catgut, lasting forty days, is now used by most surgeons. 1907 *Practitioner* Sept. 428 The employment of the chromicised gut is better than the use of raw material, as it is a little stronger and more lasting. 1911 R. A. FREEMAN *Red Thumb Mark* xvi. 214 A plate of gelatine which has been treated with potassium bichromate, ..chromicized gelatine, as it is called. 1927 *Sunday at Home* Sept. 736/1 Much of Lister's early success was due to the introduction of the chromicized catgut ligature.

chromidium (kromi·diəm). Pl. **-idia** (-i·diă). *Biol.* [mod. L. (pl. *chromidia*), ad. G. *chromidien* pl. (R. Hertwig 1902, in *Archiv f. Protistenkunde* I. 4), f. CHROME + L. dim. termination *-idium*.] A granule or strand of chromatin in the cell-body (see also quot. 1960). Hence **chromi·dial** *a.*; **chromidio·gamy** (see quots. 1912, 1920); **chromi·diosome** (quot. 1912).

1906 *Cambr. Nat. Hist.* I. 30 In many Sarcodina and some Sporozoa the nucleus gives off small fragments into the cytoplasm or is resolved into them; they have been termed 'chromidia' by R. Hertwig. *Ibid.* 52 The nucleus..often gives off 'chromidial' fragments. 1912 E. A. MINCHIN *Protozoa* 65 *note*, It is proposed in this work to use the term *chromidiosome* to denote the smallest chromatin-particles of which the chromidial mass is made up...The term 'chromidiosome' must therefore be applied to the ultimate, individual grain or particle of chromatin, alike whether it be lodged inside or outside a nucleus. *Ibid.* 126 The chromatin that undergoes syngamic union may be in the form either of chromidia or of nuclei; in the former case the process is termed *chromidiogamy*, in the second *karyogamy*. 1920 W. E. AGAR *Cytol.* 208 Fusion or mingling of chromidia (chromidiogamy, Swarczewsky). 1960 L. PICKEN *Organiz. Cells* VI. viii. 242 The cytoplasm..did not..contain a mass of granules but bundles of what appeared to be fibrils... For these structures, Bernhard *et al.* revived the term *chromidia*..first applied to cytoplasmic structures related in staining properties to the substance of the chromosomes. The new chromidium appeared to be associated with the RNA of the cytoplasm.

chrominance (krōᵘ·minăns). *Colour Television.* [f. Gr. χρῶμα colour, after LUMINANCE.] (See quot. 1952.) Also *attrib.*

1952 *Electronics* Nov. 212/2 Approved working definitions for color television...*Chrominance*, the colorimetric difference between any color and a reference color of equal luminance. 1953 *Ibid.* Dec. 139/3 The term 'chrominance' refers to two signals which carry color information. These signals, when added to the luminance signal, provide the hue and saturation components of the color image. 1955 G. G. GOURIET *Introd. Colour Telev.* II. 44 The bandwidth of the chrominance signal may be reduced to less than one half that of the luminance signal before the deficiency is noted at normal viewing distances.

chromiole (krōᵘ·miōᵘl). *Biol.* [f. Gr. χρῶμα colour + connective -i- + *-OLE.] A name for the minute chromatin-granules which by their aggregation are supposed to form the chromomeres.

1899 G. EISEN in *Biol. Centralblatt* XIX. 131 The first-mentioned granules are the chromatin elements which in time will form the chromosomes. For these granules..I have proposed the name *chromioles*. 1920 W. E. AGAR *Cytol.* 18 By many cytologists chromatin is believed to be composed of very minute granules, or chromioles.

chromitite (krōᵘ·mitəit). *Min.* [ad. G. *chromitit* (M. Z. Jovitschitsch 1908, in *Sitz. Math.-Nat. Akad. Wiss. Wien* CXVII. 818), f. CHROMITE + -ITE.] (See quot. 1944.)

1910 *Min. Mag.* XV. 419 *Chromitite*...Oxide of chromium and iron..found as shining octahedra in the sands of streams from the Kopaonik Mountains, Servia. Differs from chromite in containing ferric oxide in place of ferrous oxide. 1944 C. PALACHE et al. *Dana's Syst. Min.* (ed. 7) I. 712 *Chromitite*...Found as small octahedral crystals in mica schist and stream sands at Zeljin Mountain in..Yugoslavia ...Assigned the formula $FeCrO_3$, but probably a mixture of chromite with magnetite or hematite. 1951 J. R. PARTINGTON *Gen. & Inorg. Chem.* (ed. 2) xxvi. 738 Rarer minerals are chrome ochre..and chromitite Fe_2O_3, 2 Cr_2O_3.

chromium. Add: **2.** Used in alloys and in electro-plating, esp. **chromium-plate** = *chromium-plating*; **chromium-plated** *ppl. a.*, **-plating** *vbl. sb.*, electro-plated, -plating with chromium. Also *transf.* and *fig.*

1924 *Foundry Trade Jrnl.* 7 Aug. 111/1 (*heading*) Chromium-Plating Steel. 1929 *Techn. News Bull., Bureau of Standards* Mar. 23/1 (*heading*) Chromium-Plated Laboratory Weights. 1934 J. B. PRIESTLEY *Eng. Journey* i. 4 These decorative little buildings, all glass and concrete and chromium plate. 1934 T. F. TWEED *Blind Mouths* i. 3 That great modern place with its marbles and chromium-plated trappings. 1935 *Jrnl. R. Aeronaut. Soc.* XXXIX. 167 (*heading*) Chromium coating of surfaces. *Ibid.*, Chromium can be deposited directly on nickel or copper. 1936 R. LEHMANN *Weather in Streets* II. 193 Having coffee in the rectangular and zigzag chromium-fitted modern beige lounge. 1936 *Punch* 1 Apr. 392/2 The 'moderns' who provide a sort of chromium-plate relief to a story otherwise sombre. 1938 E. S. GARDNER *Case of Shoplifter's Shoe* (1939) v. 70 Little individual electric lights shielded in chromium cylinders. 1938 *Jrnl. R. Aeronaut. Soc.* XLII. 977 Chromium plating is also extensively employed to reduce wear of steel parts, such as valve rocker tips. 1940 W. FAULKNER *Portable Faulkner* 746 An open..chromiumtrimmed sports car. 1942 F. M. TURNER *Condensed Chem. Dict.* (ed. 3) 189/2 *Chromium Plating*. Chromium plating is the process by which a thin, bright surface layer of metallic chromium is electrodeposited, usually from chromic acid-sulfuric acid baths, for decorative purposes. 'Hard-chromium' plating refers to the electrodeposition of thicker, very hard, chromium layers for engineering applications. 1945 *Poetry Q.* Winter 148 The kind of air-conditioned, chromium-plated, gadget-ridden existence which is now considered desirable. 1946 J. CARY *Moonlight* 110 Amanda is completely out of date—absolutely provincial—you don't really *like* that chromium-plated stuff of hers, Aunty. 1952 *Times Lit. Suppl.* 11 July 449/2 The slickness and chromium-plated vulgarity that made his first novel so..oddly readable. 1959 *Motor Man.* (ed. 36) vii. 212 One of the parts most vulnerable to attack by rust is the chromium plating. *Ibid.* x. 240 The use of very hard chromium-plated top piston rings.

chromo-. Add: **2. chromoco·llotype, -co·llotypy,** collotypy in colour; **chromo·metry,** the measurement of colour intensity; **chromopho·ric, chromo·phorous** *adjs.*, colour-bearing or -producing, of the nature of a chromophore; **chromopro·tein,** a compound consisting of a protein and a metal-containing pigment or a carotenoid.

1891 *Pall Mall Gaz.* 30 Sept. 1/2 The process of chromo-collotype is treading upon the heels of lithography for reproductive work of the cheaper class. 1896 *Brit. Jrnl. Photogr. Alm.* 572 Chromo-collotypy and allied processes. 1871 *Jrnl. Chem. Soc.* XXIV. 183 On the Constant Colour and Intensity of the Light from Clouds, for Chromometry. 1892 Chromophoric [see *AUXOCHROME]. 1893 *Athenæum* 15 July 100/2 The dyestuffs..are classed according to the particular chromophoric groups they contain. 1958 *Van Nostrand's Sci. Encycl.* 342/1 *Chromophoric electrons*, electrons in the double bonds of the chromophoric groups. 1962 J. T. MARSH *Self-Smoothing Fabrics* xviii. 300 Free methyl radicals might possibly react with the quinoid group..and produce a methoxyl group which would destroy the chromophoric structure. 1893 E. KNECHT et al. *Man. Dyeing* 404 A colour-bearing (chromophorous) group or chromophor. 1902 *Encycl. Brit.* XXVII. 564/2 Such compounds

containing chromophorous groups are termed chromogens. 1924 *Chem. Abstr.* 2718 The salt-like character of the mol[e-cule]..has now been shown for the chromoproteins, *e.g.*, hemoglobin. 1961 *Brit. Med. Dict.* 307/1 Chromoproteins occur in both animals and plants..and are intimately concerned in respiration. 1964 N. G. CLARK *Mod. Org. Chem.* xvii. 353 The red corpuscles, which constitute the majority of blood corpuscles, largely consist of hæmoglobin, a chromoprotein, i.e. a coloured conjugated protein in which the prosthetic group contains a metal.

chromo. Add: **chromo paper** (see quot. 1896).

1896 R. PARKINSON *Treatise on Paper* (ed. 3) vii. 131 *Chromo Papers*, dull enamelled papers, used for chromo-lithography. 1962 F. T. DAY *Introd. to Paper* v. 52 The finish is then carried out by calendering in the case of art and chromo papers.

chromo[2] (krōᵘ·mo). *Austral. slang.* A prostitute.

1941 in BAKER *Dict. Austral. Slang.* 1953 K. TENNANT *Joyful Condemned* xviii. 166 Waiting to snitch some chromo's handbag.

chromo, *v.* = CHROMOLITHOGRAPH *v.*

1877 B. HARTE *Story of a Mine* ix. 111 Something that could be afterwards lithographed, or chromoed.

chromocentre (krōᵘ·mŏsentəɹ). [ad. It. *cromocentro* (P. Baccarini 1908, in *Nuovo Giorn. Bot. Ital.* XV. 189), f. CHROMO- 2 + CENTRE.] A densely staining body formed in certain nuclei (see quot. 1949).

1926 L. W. SHARP *Intr. Cytology* (ed. 2) iv. 89 In many nuclei there are at certain stages one or more conspicuous accumulations of karyotin at certain points in the reticulum. Of the many terms applied to these the most suitable seems to be *chromocenters*. 1930 *Biol. Abstr.* IV. 2220/1 Cytological investigations in 3 angiosperm genera..show that there are 2 ways in which the chromocenters may originate. 1935 T. S. PAINTER in *Genetics* XX. 302 In the salivary gland, this area lies within, or better, is a part of, the 'chromocenter', or what I termed chromatic coagulum in my first papers. 1949 DARLINGTON & MATHER *Elem. Genetics* 383 *Chromocentre*, a body produced by fusion of the heterochromatin of the chromosomes in a resting nucleus. Especially applied to the Polytene chromosomes of *Diptera*. 1962 *Lancet* 29 Dec. 1384/1 The Barr body is a chromocenter of characteristic shape and position formed by heteropyknosis of an X chromosome (or part of one).

chromogen. Add: *spec.* in *Dyeing*, an acid colour used to dye brown. (Cf. quot. 1879 in Dict.)

1892 *Dyer* 20 June 92/2 Chromogen may even be used with other dyestuffs to obtain a variety of useful shades of brown. 1910 *Encycl. Brit.* VIII. 746/1 Other acid colours, e.g. Chromotrope, Chrome Brown, Chromogen, Alizarin Yellow, &c. 1961 BLACKSHAW & BRIGHTMAN *Dict. Dyeing* 43 *Chromogen*, the whole atomic grouping of which the Chromophore and the Auxochrome(s) are a part.

chromogenic, *a.* (Examples.)

1884 E. E. KLEIN *Micro-Org. & Dis.* vii. 41 Chromogenic micrococci..are characterised by their power of forming pigment of various colours. 1901 *Lancet* 6 July 41/1 Organic colourings..may be the product of certain species of bacteria. Hence they are called chromogenic bacteria. 1913 W. BATESON *Mendel's Princ. Hered.* viii. 142 All that is common to them [*sc.* albinos] is the absence of colour, *i.e.*..of the chromogenic substance.

chromogram (krōᵘ·mŏgræm). [f. CHROMO- 2 + -GRAM.] A combination of three photographs taken by a special process, which being superposed produce an image in the natural colours of the object. (Cf. HELIOCHROME.)

1893 *Jrnl. Soc. Arts* 19 May 663 The heliochromoscope and its triple photograph, or chromogram. 1893 *Daily News* 18 May 6/7 It was necessary..to provide a special optical lantern for superposing on the screen the three images of the chromogram. 1894 *Amer. Ann. Photogr.* 208 The word chromogram designates the combination or the ensemble of three diapositives made from negatives representing the action of the blues, yellows, and greens respectively of the original.

chromoleucite (krōᵘmŏliū·səit). *Bot.* [See LEUCITE.] A protoplasmic colour granule.

1885 G. L. GOODALE *Physiol. Bot.* 41 Chromoplastids, or chromoleucites. 1895 *Naturalist* 25 The colour of the petals is due to chromoleucites tinctured by carotin.

chromomere (krōᵘ·mŏmiəɹ). *Biol.* [a. G. (H. Fol *a* 1892, in *Lehrb. d. Vergl. Mikrosk. Anat.* (1896) II. ii. 259), f. CHROMO- 2 + Gr. μέρος part.] A name for the chromatin-granules which make up a chromosome. Hence **chromomeric** (-me·rik) *a.*

1896 E. B. WILSON *Cell* vi. 221 The chromatin-thread consists of a series of granules (chromomeres) embedded in ..the linin-substance. 1920 W. E. AGAR *Cytol.* 134 [The chromosomes in prophase] are often markedly moniliform, *i.e.* consisting of a row of bead-like swellings of chromatin, called chromomeres, joined to each other by a thinner linin thread. 1952 M. J. D. WHITE in G. H. Bourne *Cytol. & Cell Physiol.* (ed. 2) v. 214 Chromosomes..are clearly split into two chromatids and show a chromomeric structure. 1957 M. ABERCROMBIE et al. *Dict. Biol.* (ed. 3) 52 *Chromomeres*, darkly staining granules sometimes seen in a series along prophase chromosomes during meiosis. *Ibid.*, Chromomeres in corresponding positions on homologous chromosomes pair during meiosis. 1965 PEACOCKE & DRYSDALE *Molecular Basis Heredity* vii. 78 The chromosome is linearly differentiated into lightly and more darkly staining regions. These more darkly staining segments, known as chromomeres, represent at least four types of structure. *Ibid.*, It is now widely agreed that chromomeres are tightly coiled regions of the chromosome and that in the intervening regions the chromosomal fibre is much less coiled.

chromonema (krōᵘmŏnī·mă). *Biol.* Pl. **chromonemata** (-nī·mătă). [a. G. *chromonema* (F. Vejdovsky 1912, in *Probl. d. Vererbungsträger* II. 130), f. CHROMO- 2 + Gr. νῆμα thread.] The coiled threadlike core of a chromatid, believed by some to be the carrier of the genes.

1925 E. B. WILSON *Cell* (ed. 3) 1128 *Chromonema*, a fine basichromatic thread from which arises the spireme-thread. 1934 [see *GENONEMA, GENONEME]. 1954 *Science News* XXXIV. 71 Chromosomes consist of a number of tightly coiled threads—chromonemata—embedded in a matrix. 1957 *Encycl. Brit.* VI. 943/2 After the two daughter chromosome groups reach the poles each chromosome gradually takes on the appearance of a delicate, coiled thread...It is the chromonema..and represents the true, permanent part of the chromosome, including the hereditary materials. 1963 I. F. & W. D. HENDERSON *Dict. Biol. Terms* (ed. 8) 101/1 *Chromonema*, a coiled or convoluted thread in prophase of mitosis; central thread in chromosome.

chromophil (krōᵘ·mŏfil), *a.* *Histol.* Also **chroma-, -phile.** [f. CHROMO- 2 + -PHIL.] **a.** = *CHROMAFFIN *a.* **b.** Readily stained. Also as *sb.*, a cell that is readily stained. Also **chromophi·lic** *a.*

1899 L. F. BARKER *Nervous Syst.* x. 111 The varicosities on the dendrites in Golgi preparations von Lenhossék holds to be due to superficial collections of chromophile substance. *Ibid.* xi. 123 Chromophilic cells and chromophobic cells. 1902 DELAFIELD & PRUDDEN *Path. Anat. & Histol.* (ed. 6) III. xiv. 731 With the use of the ordinary technique of Nissl, all of the cell body, excepting the chromophilic bodies, remains unstained and apparently structureless. 1909 *Practitioner* Feb. 195 Chromophile cells, very similar to the cells in the medulla of the supra-renals. 1910 *Ibid.* Jan. 35 Chromophil granules. 1913 *Cunningham's Anat.* (ed. 4) 1341 The chromaphil system is composed of a number of discrete masses of tissue which produce and discharge adrenin. 1920 H. E. JORDAN *Text-Bk. Histol.* xvi. 576 The so-called 'chromophils' and 'chromophobes' of the earlier terminology (Flesch, 1880). 1926, 1953 [see *CHROMOPHOBE *a.*]. 1953 *Bailey's Textbk. Histol.* (ed. 13) xv. 379 In the basal part of serous gland cells (e.g., cells of salivary glands, pancreas, etc.), there is a characteristic basophilic or chromophilic material.

chromophobe (krōᵘ·mŏfōᵘb), *a.* *Histol.* [f. CHROMO- 2 + -PHOBE.] Of a cell: that does not readily absorb stains; opp. *CHROMOPHIL *a.* b. Also as *sb.*, such a cell. Also **chromopho·bic** *a.*

1899 L. F. BARKER *Nervous Syst.* xi. 123 Chromophilic cells and chromophobic cells. 1909 *Cent. Dict.* Suppl., Chromophobe, adj. 1921 *Jrnl. Med. Res.* XLII. 366 Nine strumas of the pars buccalis (one eosinophilic and eight chromophobic). 1926 STRONG & ELWYN *Bailey's Textbk. Histol.* (ed. 7) xvii. 473 The cellular elements of the glandular portion of the hypophysis have usually been classified as chromophiles and chromophobes. This classification is based upon the relative stainability of the cells. 1953 CARLETON & SHORT *Schäfer's Essentials Histol.* (ed. 16) xx. 266 Two types of cell were recognized by Hannover in 1844. These became known as the chromophil and the chromophobe. The chromophil, later known as the α-cells, were so called because of their affinity for salts of chromium. Later the same terminology was used to mean that this type had an affinity for dyes which the chromophobe did not possess.

chromoplast (krōᵘ·mŏplæst). *Bot.* [f. CHROMO- 2 + -PLAST.] A chromatophore that contains pigments other than chlorophyll, *esp.* one that contains no chlorophyll. Also *attrib.* So **chromopla·stid.**

1885 G. L. GOODALE *Physiol. Bot.* 41 Those [granules] which have some color other than green—Chromoplastids, or chromoleucites. 1902 [see *CHLOROPLAST]. 1902 E. B. WILSON *Cell* (ed. 2) 52 The chromatophores or chromoplastids. 1910 R. C. PUNNETT *Mendelism* (ed. 2) 37 The tint of the flower being due to the presence of yellow-colouring matter in the small bodies known as chromoplasts. 1913 W. H. BATESON *Mendel's Princ. Heredity* 40 *Polemonium*... It may be inferred that the yellow of *flavum* is a chromoplast colour. 1965 BELL & COOMBE tr. *Strasburger's Textbk. Bot.* (new ed.) 43 Occasionally small starch grains may occur in photosynthetically inactive chromoplasts.

chromosome (krōᵘ·mŏsōᵘm). *Biol.* [ad. G. *chromosom* (Waldeyer, in *Arch. f. Mikrosk. Anat.* 1888 p. 27), f. CHROMO- 2 + Gr. σῶμα body.] Each of the rod-like structures which occur in pairs in the cell nucleus of an animal or plant and hence in every developed cell, and are carriers of the genes. Also *attrib.* and *Comb.*

1889 W. B. BENHAM tr. Waldeyer in *Q. Jrnl. Micr. Sci.* XXX. 181, I must beg leave to propose a separate technical name 'chromosome' for those things which have been called by Boveri 'chromatic elements', in which there occurs one of the most important acts in karyokinesis, viz. the longitudinal splitting. 1890 W. TURNER *Cell Theory, Past & Present* 29 The primary chromatin fibres, or *chromosome* as Waldeyer calls them, form a complex coil. 1892 E. L. MARK tr. *Hertwig's Embryol.* 52 The chromatin..has assumed the form of small individual granules or chromosomes, which correspond in number with the spindle-fibres. 1907 C. E. WALKER *Cytol.* 99 It has been held that every hereditary character is represented by a chromosome. 1912 J. S. HUXLEY *Individ. Anim. Kingdom* iii. 80 When the sexual chromosome-shuffling is allowed to take place. 1922 R. C. PUNNETT *Mendelism* (ed. 6) 100 Chromosomes..owe their name to the fact that they stain more deeply with various dyes than the rest of the cell protoplasm. 1932 C. D. DARLINGTON *Chromosomes & Plant Breeding* vi. 22 This is the mechanism by which the chromosome complement is 'reduced' so that the germ-cells have the same number of chromosomes and the same kind of chromosomes as those produced by the preceding generation. 1938 *Times Rev. of 1937* viii/2 Progress has been made in the study of chromosome-rearrangements. 1947 *Science News* V. 60 These chromosomes are rod-shaped bodies, which are characteristic in size, number and shape for every form of life—and it is supposed that the 'genes', the actual units of heredity, are located on them. 1962 *Listener* 21 June 1071/2 It was at one time thought that chromosome breaks and certain other changes in living cells could be produced only by radiation. 1968 *Times* 19 Oct. 4/7 DNA, the genetic material of the chromosomes.

Hence **chromoso·mal** *a.*, of or pertaining to chromosomes.

1909 in *Cent. Dict.* Suppl. 1926 J. S. HUXLEY *Ess. Pop. Sci.* xviii. 296 The balance of chromosomal genes or factors. 1959 *Times* 9 Dec. 2/6 Cytologist or Cytogeneticist required to work on human chromosomal anomalies. 1968 *Times* 14 Nov. 8/7 Dr. Gardner chose as the donor embryos a strain that has a recognizable chromosomal abnormality.

chromotrope (krōᵘ·mŏtrōᵘp). Also **-trop.** [a. G. *chromotrop*, f. CHROMO- 2 + Gr. -τροπος turning.] An acid colour used in dyeing. Also *attrib.*

1893 E. KNECHT et al. *Man. Dyeing* 536 A (new) dioxynaphthalene disulphonic acid (so-called chromotrop acid). *Ibid.* 537 The chromotrops do not dye evenly on previously mordanted wool. *Ibid.* 539 The chromotrop-blacks. 1902 *Encycl. Brit.* XXVII. 562/2 Chromotrope..is an Acid Colour which is applied to wool in an acid bath in the usual manner. 1906 *Dyer* 20 Jan. 15/1 Lactic acid is used in chromotrope dyeing. 1959 *Chambers's Encycl.* IV. 689/2 (*caption*) Chromotrope acid.

chromotrope, var. CHROMATROPE.

chromotropic (krōᵘmŏtrǫ·pik), *a.* *Chem.* [f. as *CHROMOTROPE + -IC.] Having the property of varying its colour. Hence **chromotro·pism.**

1899 *Jrnl. Chem. Soc.* LXXVI. I. 63 Phenylazochromotropic acid (chromotropic acid..), which the dye-works' chemists regard as a 2-azo-compound. 1908 H. DRIESCH *Sci. & Philos. Organism* II. 24 The crab *Maia* may change the quality—not the 'sense'—of its 'chromotropism',.. according to the colour of the ground it lives upon, and another crab, *Hippolyte*, changes its colour and its chromotropism correspondingly. 1918 A. W. STEWART *Rec. Adv. Org. Chem.* (ed. 3) 32 If a salt changes colour, it is said to be 'chromotropic' or 'variochromic'. 1944 *Hackh's Chem. Dict.* (ed. 3) 200/1 *Chromotropic acid*, 1·8-Dihydronaphthalene-3·6-disulfonic acid; used as an intermediate.

chronaxie, chronaxy (krōᵘ·năksi). Also **chronaxia.** *Phys.* [a. F. *chronaxie* (L. Lapicque 1909, in *Comptes Rendus de la Soc. de Biol.* LXVII. 283), f. Gr. χρόν-ος time + ἀξία value.] The minimum time required by a constant electric current of twice the threshold intensity to excite a muscle or nerve fibre, used as an index of excitability. Also *attrib.*

1917 *Sci. Amer.* Suppl. LXXXIII. 389/2 To this constant representing a period of time Lapicque gives the name of chronaxia. For a given kind of muscle the chronaxia has an invariable value, about one-tenth that of the useful time. 1922 *Nature* 5 Jan. 30/2 Chronaxy of the sensitive rachidian nerves of the upper limb in man. 1924 *Ibid.* CXIII. 427/1 The need for a review of the work on chronaxie. 1944 *Electronic Engin.* XVII. 27 Chronaxie is defined as that period of time for which a current, having twice the rheobasic value, must flow in order to produce the same minimal contraction. 1950 *Ibid.* XXII. 88 The response to electrical stimulation of muscle and nerve (chronaxie measurement).

chronic, *a.* Add: **3.** Used colloquially as a vague expression of disapproval: bad, intense, severe, objectionable. Also *something chronic* adv. phr., severely, badly.

[1896 in Ware *Passing Eng.* (1909), *Chronic*..Ceaseless, persistent. 'Oh! Joe's chronic.' 'Charley's Aunt's chronic', —said of a piece that ran perpetually.] 1904 KIPLING *Traffics & Discoveries* 288 Oh, it's chronic 'ere of a Saturday sometimes. 1909 in Ware *Passing Eng.* 75/1 'Oh, that theatre's chronic'—means that never is a good piece seen there. 1910 H. G. WELLS *Mr. Polly* ii. 41 It's made my eyes water something chronic. 1915 *Scotsman* 26 Jan. 5/3 'The weather is chronic,' says a Seaforth Highlander. 1926 'J. J. CONNINGTON' *Death at Swaythling Court* ix. 164 He puts a dash of whisky into the paraldehyde to disguise the taste, which is a chronic one, I can tell you. 1942 L. A. G. STRONG *Slocombe Dies* xv. 67 The men in these parts are something chronic... They won't do anything till some time after we're due to start.

chronicle, *sb.* Add: **4.** In *chronicle drama*, etc., used in imitation of †*chronicle history*, an Elizabethan descriptive title for plays based on historical matter such as is found in Edward Hall and Ralph Holinshed.

1600 (*title*) The Cronicle History of Henry the fift. 1608 (*title*) M. William Shak-speare: His True Chronicle Historie of the life and death of King Lear and his three Daughters. 1634 FORD (*title*) The Chronicle Historie of Perkin Warbeck.

[1886 F. G. FLEAY (*title*) A Chronicle History of the Life and Work of William Shakespeare.] 1902 F. E. SCHELLING *English Chronicle Play* 30 The English Chronicle Drama falls naturally into two groups. 1903 HARDY *Dynasts* Pref. p. ix, In devising this chronicle-piece no attempt has been made to create that completely organic structure of action, and closely-webbed development of character and motive, which are demanded in a drama strictly self-contained. 1927 *Observer* 12 June 15/2 'Martha Washington' is a chronicle piece, after the manner of 'Abraham Lincoln', but there is no drama in it from start to finish.

chronique. Delete †*Obs.* and add later examples.

1890 E. DOWSON *Let.* 23 Feb. (1967) 138, I will endeavour to send you the accustomed chronique. 1923 F. M. Ford *Let.* 8 Nov. (1965) 157, I would add about three pages of *Romance* analysis in my *chroniques*. 1954 W. STEVENS *Let.* 13 Dec. (1967) 857 He either writes chroniques or reviews, or both, for the Music Quarterly.

‖ **chronique scandaleuse** (kronik skandalǿz). [F.] A compilation of gossip; the body of scandal current at any time and place.

1850 THACKERAY *Pendennis* II. xxiii. 231 Do you suppose that honest ladies read and remember the Chronique Scandaleuse as well as you, you old grumbler? 1887 *Athenæum* 29 Jan. 153/2 He is enabled to compile a *chronique scandaleuse* of the Middle Ages. 1931 *Times Lit. Suppl.* 24 Sept. 713/4 This book puts into print the *chronique scandaleuse* of D'Annunzio's life. 1951 *Mind* LX. 277 The inconstancy of Professor X, the Bohemianism of Dr. Y, etc.—a regular *chronique scandaleuse*.

chronophotography (krǫ:nofǭtǫ·grăfi). [f. Gr. χρόνος time.] An early term for cinematic photography. Hence **chro:nophotogra·phic** *a.*

1895 *Pall Mall Gaz.* 13 Nov. 4/1 The various kinds of apparatus required for the difficult operations of 'chronophotography'. 1899 HOPWOOD *Living Pictures* 43 Chronophotography and the practical development of the living picture. 1902 *Rep. Smithsonian Inst. 1901* 329 The writer has pushed the improvement of his chronophotographic apparatus. *Ibid.*, Chronophotographic gun with a film ribbon.

chronoscope. Add: **b.** Also, a device used for measuring the time of reaction in psychological experiments.

1890 BILLINGS *Nat. Med. Dict.* 277/2 *Chronoscope*, peculiar form of clock, regulated by a vibrating spring, in which the hands are thrown into or out of gear by an electro-magnet, and record thousandths of a second. 1923 *Chambers's Jrnl.* 603/2 When the examiner speaks the chronoscope revolves; when the subject speaks, it stops. ..The hiatus [between question and answer] is duly recorded by the chronoscope. 1934 *Brit. Jrnl. Psychol.* Apr. 391 The writer constructed a phonic chronoscope.

-chronous: a combining form [f. Gr. χρόνος time: see -OUS], used to form adjs. denoting (after, during, etc.) a period of time (e.g. *isochronous, metachronous, synchronous*).

chrysanth, colloq. abbrev. of CHRYSANTHEMUM. Also **chrysant.**

[1895 *Westm. Gaz.* 5 Nov. 8/2 The National Chrysanthemum Society opens today... There have been many shows ..of 'Chrysantes'.] 1920 'K. MANSFIELD' *Bliss* 138 Marie.. runs into the garden next door to pick the 'chrysanths'. 1934 R. H. MOTTRAM *Bumphrey's* iv. 162 Didn't know you cared for Chrysants. 1955 F. TILSLEY *Thicker than Water* i. 31 He went round the small garden, disbudding the chrysanths.

chrysanthemin (krisæ·nþimin). *Chem.* [a. G. (R. Willstätter and E. K. Bolton 1916, in *Ann. der Chemie* CDXII. 136), f. CHRYSANTHEMUM + -IN[1].] A glucoside of cyanidin, $C_{21}H_{21}O_{11}$, found in the flowers of *Chrysanthemum indicum* and other plants.

1918 A. W. STEWART *Rec. Adv. Org. Chem.* (ed. 3) 219 The anthocyanin of the winter aster is chrysanthemin, derived from dextrose and cyanidin. 1942 *Endeavour* I. 94/2 *Chrysanthemin* from the deep red chrysanthemum is cyanidin monoglucoside. 1956 *Nature* 7 Jan. 39/2 They demonstrate conclusively..that the blackberry anthocyanin is chrysanthemin. *Ibid.*, A sample of synthetic chrysanthemin chloride.

chrysene (krai·sīn). *Chem.* Also **chrysen.** [ad. F. *chrysène* (A. Laurent 1837, in *Ann. de Chimie* LXVI. 137), f. CHRYS- + -ENE.] A solid hydrocarbon, $C_{18}H_{12}$, occurring in crystalline form in the highest boiling fractions of coal tar, and in other substances, variable in colour and exhibiting violet fluorescence.

1864 *Chem. News* X. 244/1 Laurent's pyren, I am inclined to think, was merely chrysen containing an admixture of paraffin. *Ibid.* 244/2 Chrysen compounds. 1871 *Jrnl. Chem. Soc.* XXIV. 692 The chrysene employed in this research was extracted from the mixture of solid hydrocarbons which have a higher boiling point than anthracene, and form the latter portion of the distillate from coal tar. 1943 *Endeavour* Jan. 30/1 The..compound presents a rough resemblance to..chrysene. 1957 *Encycl. Brit.* V. 665/1 Chrysene, $C_{18}H_{12}$, crystallizes in colourless plates or octahedra (from benzene), which exhibit a violet fluorescence.

chrysid (krai·sid). *Ent.* [ad. mod.L. *Chrysididæ* (see definition), f. Gr. χρυσίς (-ιδ-) gold vessel, gold-embroidered dress.] A member of the family Chrysididæ of hymenopterous metallic-coloured insects (ruby-wasps, cuckoo-flies). Also as *adj.*

1884 L. O. HOWARD in J. S. Kingsley *Standard Nat. Hist.* II. 524 In mode of life the Chrysids are either parasites or inquilines. 1910 *Encycl. Brit.* XIV. 180/2 The eggs are laid in the nests of various bees and wasps, the chrysid larva living as a 'cuckoo' parasite. 1913 *Oxf. Univ. Gaz.* 4 June 959/1 Dr. Longstaff's fine set of Sûdan Hymenoptera.. included..40 Chrysids. 1969 C. ANDREWES *Lives of Wasps & Bees* xi. 79 Somewhat apart from the other wandering wasps are the Chrysids, ruby-wasps or ruby-tails, all of them parasites of other wasps or of bees.

chthonic, *a.* Add: Now the usual form. (Earlier and later examples.)

1882 C. F. KEARY *Outl. Primit. Belief* v. 215 The chthonic divinity was essentially a god of the regions under the earth; at first of the dark home of the seed, later on of the still darker home of the dead. 1903 *Daily Chron.* 29 Dec. 3/3 Two great and contrasted forms of ritual—the Olympian and the Chthonic, the one a ritual of cheerful..character, the other a ritual of gloom, and fostering superstition. 1941 T. S. ELIOT *Dry Salvages* v. 15 Driven by dæmonic, chthonic Powers. 1957 CHILDE *Dawn Europ. Civilisation* (ed. 6) xviii. 331 The invaders..patronized native cults or gave them a new celestial, rather than chthonic, orientation.

b. *transf.*

1926 T. E. LAWRENCE *Seven Pillars* lxii. 328 We..wrote books about its architecture, folklore, and dying industries. Then one day, we woke up to find this chthonic spirit turned political. 1928 H. G. & C. F. BAYNES tr. *Jung's Contribs. Analyt. Psychol.* 118 The chthonic portion of the mind—if we may use this expression—that portion through which the mind is linked to nature, or in which, at least, its relatedness to the earth and the universe seems most comprehensible. 1963 M. MCCARTHY *Group* vi. 129 The chthonic imagery of Norine's apartment, which..was black as a coalhole and heated by the furnace of the hostess' unslaked desires.

chu·bbily, *adv.* [-LY[2].] In a chubby manner, in the manner of a chubby person.

1909 H. G. WELLS *Tono-Bungay* II. iii. 152 'True,' said my uncle, chubbily and with a dreamy sense of mysticism. 1968 *N.Y. Times* 7 Jan. 1E Short, but chubbily handsome, impulsive and vivacious, Prince Sihanouk.

chuck, *sb.*[3] Add: **3.** *spec. Cricket.* A thrown ball; an illegal delivery.

1862 F. LILLYWHITE *Scores & Biogr.* II. 30 Slow under-hand 'chucks'. 1903 WODEHOUSE *Tales of St. Austin's* 22 Did you think that ball that bowled you was a chuck? 1966 *New Statesman* 1 July 26/3 One of Griffith's most effective deliveries was a (doubtless unconscious) 'chuck' which has now been umpired out of his repertoire.

b. Dismissal, repudiation, 'turning down', esp. in phr. *to give* (someone or something) *the chuck*.

1892 G. & W. G. GROSSMITH *Diary of Nobody* vi. 88 If you want the good old truth, I've got the chuck! 1893 *Fun* 28 Jun. 262/1 The sentimental coster who is going to shoot himself because he has (as he expresses it) 'got the chuck' from his situation and is parted from his 'dona'. 1909 GALSWORTHY *Strife* II. ii, If you give up..to Harness, now, it's givin' us the chuck—to save your skins. 1926 *Punch* 26 May 559/2, I give your pills the chuck..And I am worse instead of better. 1928 A. CHRISTIE *Myst. Blue Train* xviii. 147, I have, how do you say it?—given them the chuck! 1930 *Argosy* Apr. 15/1 When they gave me the chuck, you married me out of hand.

chuck, *sb.*[4] Add: **2.** (Earlier and later examples.)

1723 J. NOTT *Cook's Dict.* No. 77 B *To sauce Beef*, take either Buttock, Chuck or Brisket of Beef. 1747 MRS. GLASSE *Art of Cookery* xxi. 160 A Bullock..The Fore-Quarter..the Chuck-Piece,..and Middle Rib, which is called the Chuck-Rib. *Ibid.*, The Hind-Quarter..the Chuck-bone, Buttock and Leg. 1798 *Ann. Agric.* XXX. 314 Blade-bones chuck. 1844 H. STEPHENS *Bk. Farm* II. 171 In the fore quarter, the fore rib, middle rib, and chuck-rib, are all roasting pieces [of beef]. 1968 *Globe & Mail* (Toronto) (Mag.) 17 Feb. 22/2 Chuck beef roast with macaroni.

4. *magnetic chuck*, a chuck operating by magnetism.

1909 in *Cent. Dict.* Suppl. 1922 GLAZEBROOK *Dict. Appl. Physics* II. 323/1 Recently, magnetic chucks have come much into use in connection with shaping machines, lathes and other workshop machines. 1958 *Van Nostrand's Sci. Encycl.* (ed. 3) 344/2 Magnetic chucks of both the electrically-actuated and the permanent-magnet type may also be employed on the lathe.

chuck (tʃʊk), *sb.*[5] *slang* or *dial.* Now chiefly in *U.S.* informal use. [perh. the same as CHUCK *sb.*[4]]

1. Food, 'grub'. (In early use *spec.* bread or ship-biscuit.) See also *CHUCK WAG(G)ON.

1850 *Lloyd's Newsp.* 6 Oct. (Farmer), The prisoner, upon coming to his cottage door had tried hard to get some chuck out of him, but had failed. 1860 HOTTEN *Dict. Slang* (ed. 2), *Chuck*, food, provision for an entertainment.—Norwich. 1864 *Standard* 13 Dec. (Farmer), Of naval slang Mr. Hotten has missed the words Chuck, used by sailors for biscuit, and Barge, the box or cask in which the chuck is kept by the messes on the lower deck. 1877 W. H. THOMSON *Five Yrs.' Penal Servitude* i. 4 Two large slices of bread,.. the allowance given out to some prisoner who..had forgotten to eat what in prison slang is called his 'toke' or 'chuck'. 1936 'J. CURTIS' *Gilt Kid* iv. 40 A girl who was eating her chuck like a navvy.

attrib. 1903 A. ADAMS *Log Cowboy* xviii. 280 He was riding the chuckline all right. 1905 — *Outlet* 16 A carpenter then at work building chuck-boxes for each of the six commissaries. 1926 J. BLACK *You can't Win* iv. 39 New arrivals, who had not yet acquired the 'chuck horrors', that awful animal craving for food that comes after missing half a dozen meals. 1970 *Sunday Republican* (Springfield, Mass.) 26 Apr. 51/1 Under canvas.., only so much can be crammed into chuck box or cabinet.

2. The act of taking food; a meal, meal-time.

1865 *Harper's Mag.* Feb. 325/1 [I] finished chuck on twelve o'clock. 1901 M. E. RYAN *Montana* i. 25 After 'chuck' we'll go over and give you a nearer view of the tribe on the other shore. 1907 S. E. WHITE *Arizona Nights* I. vi. 112 When the last man had returned from chuck, Homer made the dispositions for the cut.

chuck (tʃʊk), *sb.*[6] *Canada.* [Chinook Jargon.] A large body of water.

1880 G. M. DAWSON *Rep. Queen Charlotte Islands* 30 The most considerable is that which has been called the Slate Chuck on the chart. 1958 *Beaver* (Canada) Winter 26/2 A mother in Stanley Park scolds, 'Johnny, throw that dirty stick in the *chuck*.'

chuck, *v.*[2] Add: **2.** spec. in *Cricket*, to bowl illegally, with the action of a *CHUCKER 4; also *intr.* Cf. *CHUCK *sb.*[3] 3.

1843 *Sheffield Independent* 23 Sept. 7/2 Instead of throwing his ball..with some degree of force, he commenced 'chucking' them gently. 1903 WODEHOUSE *Tales of St. Austin's* 21 'No ball,' he shouted..'it seemed to me that you chucked that time.'

b. (Further examples.) Also, *to chuck out*: to eject, discharge, get rid of, throw out (from a public meeting, a theatre, a position or post, etc.); cf. *chucker-out*, CHUCKER[1] 3; *to chuck up*: to abandon, dismiss; to throw over, jilt; *to chuck in the towel*: see *TOWEL *sb.*

1821 CLARE *Vill. Minstr.* I. 25 Up he'd chuck sacks as one would hurl a stone. 1841 F. A. KEMBLE *Let.* 29 July in *Records of Later Life* (1882) II. 110 My horse..tore off with me..there was a fair chance of my being chucked off. 1864 HOTTEN *Slang Dict.*, *Chuck up*, to surrender, give in—from the custom of throwing up the sponge at a prize fight in token of yielding. 1869 R. D. BLACKMORE *Lorna Doone* vi. 66 No boy or young man on our farm durst ever get into a saddle, because they all knew that the master would chuck them out pretty quickly. 1878 *Chambers's Jrnl.* 333/2 A stalwart navvy,..after crossing the Danube several times at Alexandra Park, declared he must 'chuck it up' if he could not be a Turk. 1883 M. E. BRADDON *Phantom Fortune* xxv, Look how easily she chucked you up because she did not think you good enough. 1885 *Daily News* 27 Nov. 2 The town artisan is ready enough to chuck out an obstructor. 1889 *Times* 4 Nov. 3/6 Some of them have got chucked out, your Worship, excuse the language. 1891 WILDE *Dorian Gray* v. 102, I have a great mind to chuck the whole thing up. 1893 *National Observer* 23 Sept. 483/1 That is a reason for living hopefully, not for 'chucking it up' in despair. 1917 A. HUXLEY *Let.* 8 Apr. (1969) 123, I am going to lie and chuck it up for a much better thing. 1926 KIPLING in *McCall's Mag.* June 78/2 A young voice called from a desk. 'Chuck it over!' 1936 J. TICKELL *See how they Run* xxiii. 318 Damn politics. Listen, I'll chuck it up and we'll go and live in Kerry. 1944 AUDEN *For Time Being* (1945) 58 Our reasons are silenced..Our wills chuck in their hands. 1958 *Engineering* 4 Apr. 424/1 Chuck out the conventional concepts. 1963 *Times* 21 Feb. 3/4, I do not like chucking money away.

c. *colloq.* Without adverb, = *to chuck out* or *up*. In *pass.*, to be acquitted or released. Also, *to chuck it* (see Dict., sense b): examples.

c 1879 *Broadside Ballad* (Farmer), Whatever may happen I get all the blame, Wherever I go, it is always the same—Jolly well chucked again! 1883 HAWLEY SMART *Hard Lines* xxvi, If you mean business, take my advice and chuck that corps. 1887 HORSLEY *Jottings from Jail* 24 Kit, from 7 dials..expects to get fulled or else chucked. 1888 H. R. HAGGARD *Col. Quaritch* vi. 95 He means to git the place at his own price or chuck it. 1890 *Daily News* 5 Dec. 7/1 'He's sure to get chucked'—a slang expression for discharged. 1891 W. RALEIGH *Let.* 17 Sept. (1926) I. 162 We could always chuck it and work our day round again. 1901 *Punch* 31 July 88/2 Oh, chuck it! I never was any good at arithmetic! 1905 R. BROUGHTON *Waif's Progr.* xxiii. 260 You promised to marry me...You cannot, and shall not chuck me. 1908 E. W. WALTERS *Nipper* ii, 'Chuck it!' snapped the ill-nourished boy. 1913 J. STEPHENS *Here are Ladies* ii. 102 One day he chucked his job. 1915 G. K. CHESTERTON *Antichrist* in *Poems* 89 But the souls of Christian peoples...Chuck it, Smith! 1922 *N. & Q.* XI. 206/2 If at some stage in a race a horse is seen to 'shut up', and refuse to gallop his best, it is said to 'chuck it'. 1922 D. H. LAWRENCE *England, my England* (1924) 240 Serve you right if she chucks you now. 1929 D. G. MACKAIL *How Amusing* 525 He..concluded by asking her to chuck it all and marry him. 1932 WODEHOUSE *Hot Water* xiv. 238 I've a dashed good mind to chuck the whole thing. 1933 F. STARK *Let.* 18 Feb. in *Beyond Euphrates* (1951) 301 This is my last on official paper, as I chuck this job on Friday week. 1934 E. WAUGH *Handful of Dust* iii. 121 You couldn't possibly chuck lunch or one of the lectures? 1954 T. S. ELIOT *Confid. Clerk* II. 65 He's the sort of fellow who might chuck it all And go to live on a desert island.

d. *chuck-and-chance-it*: a derisive phrase used *attrib.* to describe wet-fly fishing.

1886 *Longman's Mag.* June 221 So let us return to the good Northern plan of 'chuck and chance it', wet. The use of the dry fly is a Utopian dream. 1899 *Westm. Gaz.* 12 May 3/2 A large proportion of anglers..regard with unmerited contempt that which they are pleased to miscall the 'Chuck-and-chance-it system'. 1907 *Ibid.* 29 Nov. 3/1 Agreeing to scoff at every wet-fly angler as a mere chuck-and-chance-it angler.

e. [perh. rather CHUCK *v.*[1]] *to chuck off*, to chaff, speak sarcastically, sneer *at*. Used *absol. Austral.* and *N.Z. slang*.

1915 C. J. DENNIS *Songs Sentim. Bloke* (1916) 119 *Chuck*

off, to chaff; to employ sarcasm. **1915** E. G. PILLING *Anzac Memory* 21 Jan. (1933) 119 Everyone is eating my lollies. They chuck off when a parcel of lollies arrives, but none are lacking on the eating stakes. **1941** BAKER *Dict. Austral. Slang* 18 *Chuck off at*, to sneer at, chaff.

chu·ck-a-lu:ck. *N. Amer.* Also **chuck-luck, chuckle luck.** [app. f. CHUCK *v.*² 2 + LUCK *sb.*] A gambling game played with dice.

1836 in W. T. PORTER *Quarter Race in Kentucky* (1847) 24, I thought I'd make a rise on chuck-a-luck, but you prehaps never saw such a run of luck. **1845** J. J. HOOPER *Adv. S. Suggs* (1851) ix. 111 A chuck-a-luck table. **1856** *Liberator* 12 Jan. XXVI. 12 Leper and Doolin got into a quarrel over a game called 'chuckle luck'. **1879** *Southern Hist. Soc. Papers* VII. 489 Cards, monte, roulette, keno, faro, chuck-a-luck, and in fact every game of chance known. **1906** *Daily Colonist* (Victoria, B.C.) 21 Jan. 1/4 The Chinese play 'pie-gow', which is alleged to be a gambling game, although difficult to prove as such. But 'fan-tan' and 'chuck-luck' are said to have been totally suppressed. **1907** S. E. WHITE *Arizona Nights* I. x. 165 And a man's so sick of himself by the time he gets this far that he'd play chuck-a-luck. **1944** *Harper's Mag.* June 53/1 Several crap tables and chuck-luck layouts. **1969** J. FREDMAN *Fourth Agency* ii. 15 You fancy craps, roulette, blackjack, poker, chuck-a-luck, or just plain whoring?

chucker. Add: **4.** *Cricket. colloq.* A bowler whose delivery of the ball is considered to be a throw, and hence illegal.

1882 *Australians in England* 158 Do not people tell you openly they think so and so a 'chucker'? **1960** *Times* 3 Oct. 19/6 Meckiff said that even to think of changing his action would be to admit that those who regarded him as a 'chucker' were right.

chuck-full, *a.* Add later U.S. and dial. examples (see CHOCK-FULL δ).

1868 *Putnam's Mag.* Dec. (De Vere), These prairies are nature's banks, stuffed chuckfull of cash. **1888** 'Q' *Troy Town* x. 111 You niver seed a bull yet as wasn' chuck-full o' conviction, an' didn' act up to hes rights. **1893** W. K. POST *Harvard Stories* 31 The public wards are chuck-full. **1929** *Minnesota Alumni Weekly* 1 June 619 She sent us a letter just chuck full of interesting news items.

chu·ck-hole. *U.S.* [CHUCK *v.*² 5; cf. *chock-hole*, CHOCK *sb.*¹ 7.] A hole or rut in a road or track.

1836 E. L. WILLSON *Journey N.J. to Ohio* (1929) 25 June, The abundance of traveling..wears the road into deep holes; these we call chuck-holes. **1869** S. BOWLES *Our New West* xiv. 276 The jolting of the rocks and the 'chuck holes' of the road..kept us in a somewhat perpetual..motion. **1887** J. KIRKLAND *Zury* 2 'Chuck-holes' is the expressive Western name for the short, sharp depressions which use makes in unworked country roads. **1930** R. MACAULAY *Staying with Relations* xix. 272 Poor going..due to rocky chuck-holes, and deep ruts. **1963** *Lebende Sprachen* VIII. 169/1 Chuck-holes, pot-holes.

chuckie, -y: see *CHOOK.

chucking, *vbl. sb.*² Add: *chucking-out time* (in a public house).

1909 WARE *Passing Eng.* 76/2 *Chucking-out Time*, half-past twelve, the closing hour for metropolitan taverns, when those who do not go willingly are 'chucked out'. **1922** JOYCE *Ulysses* 417 Keep a watch on the clock. Chuckingout time. **1970** *Guardian* 8 May 9/6 What wonderful winkles we had at that stall at chucking-out time in the Old Kent Road.

chu·cking, *vbl. sb.*³ *Turning.* [f. CHUCK *v.*³] The act of fixing in, or by means of, a chuck.

1889 P. N. HASLUCK *Model Engin. Handybk.* v. 49 The flanges [should be] turned and faced up at one chucking. **1906** CASSAL (*title*) Chucks and chucking, for metal and wood.

chuckoor, var. *CHIKHOR.

chu·ck-up. *slang* (esp. *Naut.*). [f. CHUCK *v.*² 2 b.] A cheer; encouragement. Also, a salute.

1915 'BARTIMEUS' *Tall Ship* ix. 160 Supposing..we celebrated our return to harbour..by asking our chummy-ship to dinner to-night, and giving them a bit of a chuck-up! **1925** FRASER & GIBBONS *Soldier & Sailor Words* 56 *A chuck up*, a salute. From the act of throwing up the hand to the forehead in saluting. **1946** J. IRVING *Royal Navalese* 53 *A chuck up*, a demonstration of applause, approval or encouragement. *Ibid.*, A chuck-up party..is thus a group of officers or men who..man the touch-line to cheer on their own ship's football team.

chu·ck wa:g(g)on. *N. Amer.* [*CHUCK *sb.*⁵] A wagon carrying provisions and equipped with cooking facilities, used esp. in western N. America, on ranches, during harvest, in lumber camps, etc.; also, a roadside 'eatery'. *chuck-wagon race*, in rodeos and stampedes, a race of horse-drawn chuck wagons.

1890 D'OYLE *Notches* 26 The sun blistered the paint upon the 'mess-box' behind the 'chuck-waggon'. **1910** MULFORD *Hopalong Cassidy* iii. 25 A group of blanket-swathed figures lay about a fire near the chuck wagon. **1923** H. STEELE *Spirit-of-Iron* 252 In a little gully beside the chuck-wagon, the cook was boiling coffee. **1928** *Daily Express* 12 Nov. 6 [In Calgary] cowboys..invited us to have dinner with them earlier at the chuck-wagon. **1950** H. SUTTON *Footloose in Canada* 211 In a chuck wagon race the entrants are required at a given signal to break an entire camp..load all the paraphernalia in a wagon, do a series of figure eights around barrels, and then ride once around the track. **1952** H. INNES *Campbell's Kingdom* I. ii. 33 A small covered wagon stood in the yard...'That's the old man's chuck wagon...Always enters a team for the chuck wagon races.' **1968** G. DE FRAGA *Murder at Cookout* xxiii. 106 You know how it is—at a chuck-wagon or a smorgasbord, where guests are expected to help themselves.

chuck-will's-widow. (Earlier and later examples.)

1791 W. BARTRAM *Carolina* 154 Caprimulgus rufus called chuck-will's-widow, from a fancied resemblance of his note to these words. **1823** E. JAMES *Exped. Rocky Mt.* I. 49 The yellow breasted chat, chuck-wills-widow,..and numerous other birds occurred. **1945** J. J. MATHEWS *Talking to Moon* 52 The first notes of the chuck-will's-widow come plaintively from the east ridge as the moon rises.

chu·cky-chu·cky. *Austral.* Also **chuckie-, -chuck.** [Native name.] A kind of berry, the fruit of *Gaultheria hispida*.

1871 *Trans. N.Z. Inst. 1870* III. 194 Two other *Gaultheriæ* are very distinct; one, of prostrate habit, bears a round white flower, which is eaten, under the name of 'chuckiechuck'. **1885** MRS. C. PRAED *Australian Life* 146 To gather chucky-chuckies—as the blacks name that most delicious of native berries. **1885** —— *Head Station* xxxv, Mollie Clephane had filled one of the saddle-bags with wild plums, chuckie-chuckies, and the scrub-turkey's eggs.

chuddy, var. *CHUTTY.

chuff (tʃʌf), *v.*² [Onomatopœic.] *intr.* Of an engine or machine: to work with a regularly repeated sharp puffing sound. Also *sb.* Similarly **chuff-chuff** *sb.* and *v.*

1914 D. H. LAWRENCE *Widowing of Mrs. Holroyd* III. 76 The driving engine chuffs rapidly. **1915** —— *Rainbow* xv. 421 The far-off windy chuff of a shunting train. **1921** —— *Sea & Sardinia* 31 Slowly, with two engines, we grunt and chuff and twist to get over the break-neck heights. **1923** *Blackw. Mag.* Dec. 797/2 The chuff-chuff of the exhaust from the generating plant that supplied the house with electricity. **1929** J. B. PRIESTLEY *Good Companions* 612 The train..slowly chuff-chuffed into the gloom. **1936** M. KENNEDY *Together & Apart* I. 33 The machine in the power house..was working away for dear life. Chuff! Chuff! it said. Chuff! Chuff! It reminded him of a certain type of woman. ..Here was Betsy chuff-chuffing out of the house on some errand. **1956** J. MASTERS *Bugles & Tiger* i. 32 A little locomotive would chuff across the burning desert.

chuffed (tʃʌft), *a. slang* (orig. *Mil.*). [cf. CHUFF *a.*¹ and *a.*²] **a.** Pleased, satisfied. **b.** Displeased, disgruntled.

a. 1957 P. WILDEBLOOD *Main Chance* ix. 163 Aren't you pleased? There's not many kids of your age what owns a factory. You ought to be dead chuffed about it. **1960** A. WAUGH *Foxglove Saga* xii. 218 He was chuffed at this new monumental skive he had discovered. **1961** S. PRICE *Just for Record* iv. 29 My beard is black, all-black. That makes me pretty chuffed. **1967** *Crescendo* May 6 (Advt.), I cannot express too much just how 'chuffed' I am with the drums.
b. 1960 D. STOREY *This Sporting Life* I. ii. 59, I felt pretty chuffed with myself. **1964** C. DALE *Other People* viii. 158 Don't let on they're after you, see, or she'll be dead chuffed, see? She don' like the law.

chu·ffing, *vbl. sb.*² [f. *CHUFF *v.*²] Sporadic or intermittent burning (of fuel).

1954 *Trans. Faraday Soc.* L. 942 By burning cordite charges in small vented vessels an investigation has been made of chuffing (sporadic burning with the pressure falling to atmospheric between peaks) and of the continuous but oscillatory burning into which chuffing transforms when the nozzle size is reduced. Over a wide range of chuffing frequency..the surface recession of the propellant associated with a single chuff varies by only a small amount. **1961** *Soc. Automotive Engin. Jrnl.* June 31/1 Chuffing after prescribed burnout could prove disastrous to a space launching. Chuffing (intermittent oscillatory burning) of the second-stage after separation of the third-stage, before the third-stage is ignited, might cause it to ram the third-stage.

chug (tʃʌg), *sb.* orig. *U.S.* [Onomatopœic.] A plunging, muffled, or explosive sound, esp. the characteristic sound of an internal combustion engine when running slowly. Also repeated, and *Comb.*

1866 *Harper's Mag.* Jan. 271/2 The ponderous brother came down upon the floor with a 'chugg' that shook the house. **1894** *Outing* (U.S.) XXIV. 56/2 One weapon after another struck [the water] 'chug-chug-chug-chug'. **1895** KIPLING *2nd Jungle Bk.* 219 He would hear..the *chug-drug* of a boar sharpening his tusks on a bole. **1897** *Outing* (U.S.) XXX. 175/1, I accidentally cast myself into the pool with a sounding chug. **1920** *Blackw. Mag.* May 649/1 The chug, chug of the dripping paddle floats. **1923** *Outward Bound* Mar. 420/1 The chug of the engine still filled our ears. **1925** *Chambers's Jrnl.* 250/1 The 'chug-chug' of an oil-engine could be faintly heard. **1957** K. KENYON *Digging up Jericho* x. 238 The roar of the primus..and the chug of the generator.

chug (tʃʌg), *v.* [Onomatopœic.] *intr.* To make an intermittent explosive sound as of the escape of exhaust gases from an engine cylinder; to move with a sound characteristic of a steam-engine or electric motor at work. Also adv., as *off, on, along, out*. Also quasi-*trans.* Hence **chu·gging** *vbl. sb.*, the action of the verb; *spec.* in rocket engines: see quots. 1952 and 1964. Similarly **chug-chug** *v.*

1874 A. H. Garrod in *Proc. Zool. Soc.* 119 The *Cathartidæ* possess the ambiens, which, in conjunction with many other Ciconiine characters, leaves no doubt about their position. 1889 Nicholson & Lydekker *Palæont.* 1238 An undetermined Ciconioid from the Pikermi beds of Attica. 1893 Newton *Dict. Birds* 147 The Old-World Vultures..exhibit striking Ciconiine similarities.

cidarid (si·dărid). [f. mod.L. *Cidaris* (Gr. κίδαρις royal tiara) + -ID².] A sea-urchin of the family Cidaridæ. Also *attrib.* or as *adj.* Also **ci·daroid,** a member of the order Cidaroidea.

1889 in *Cent. Dict.* 1909 F. A. Bather in *Ann. & Mag. Nat. Hist.* III. 48 The truly Cidarid nature of the species. *Ibid.* 49 He..pointed out that C[*idaris*] *keyserlingi* was a true Cidarid. 1955 L. H. Hyman *Invertebrates* IV. xv. 515 Cidaroids have but two kinds of pedicellariae.

cidaris. Add: Also, a head-dress used by Jewish high-priests; a low-crowned mitre.

1797 *Encycl. Brit.* V. 8/1. 1876 F. G. Lee *Gloss. Liturg. & Eccl. Terms* 86.

-cide, *suffix.* Add: **1.** Also applied to preparations destructive of animal or vegetable life, as **algicide, fungicide, germicide, insecticide,* **pesticide.*

cider. Add: **2.** *cider-orchard* (example); **cider brandy** (examples); **cider cart** *U.S.* (see quot.); **cider-cup** (earlier example); **cider oil** *U.S.*, cider that has been concentrated by boiling or freezing; concentrated cider with infusion of honey; **cider press** (earlier and later U.S. examples); **cider royal** *U.S.* = **cider oil*; **cider vinegar,** a vinegar produced by the acetification of cider.

1703 J. More *Eng. Interest* (ed. 2) ii. § 6. 27 The most Stale and Sowr Cyder..will make the..best tasted Brandy, being twice Distill'd. Of this Cyder-Brandy I have kept some, four Years. 1723 J. Nott *Cook's & Confect. Dict.* No. 141 C To make Cider Brandy, or Spirits. 1845 S. Judd *Margaret* I. vii. 42 Distilleries for the manufacture of cider-brandy. 1859 Bartlett *Dict. Amer.* (ed. 2), *Apple Brandy*, a liquor distilled from fermented apple-juice; also called Cider Brandy. 1877 *Southern Hist. Soc. Papers* III. 17 The passage of a cider-cart (a barrel on wheels) was a rare and exciting occurrence. 1851 *London at Table* III. 51 *Cider Cup*, ..cider..soda water..sherry..brandy..lemon..sugar and nutmeg. 1846 H. H. Brackenridge *Mod. Chivalry* (rev. ed.) I. xxiii. 115 She ordered him a pint tumbler of cider oil, with powdered ginger, to warm his stomach. 1859 Bartlett *Dict. Amer.* (ed. 2), *Cider Oil*, cider concentrated by boiling, to which honey is subsequently added. 1879 [see Queening]. 1673 in *Essex Inst. Hist. Coll.* L. 28 In the Little chamber a great Tray a trough a syder presse. 1879 R. J. Burdette *Hawk-Eyes* 70 The sound of the cider press ceased not from morning even unto the night. 1684, 1707 Cider royal [see Royal *a.* 15 b]. 1828 T. Flint *Geogr. Mississippi Valley* I. 235 What is called 'cider royal' or cider, that has been strengthened by boiling, or freezing. 1837 A. Wetmore *Gaz. Missouri* 290 The disturber known in..Pennsylvania [as]..'cider royal', and by the Indians appropriately named 'fire-water'. 1851 C. Cist *Cincinnati* 251 But there is a good deal of cider vinegar made. 1858 Simmonds *Dict. Trade, Cider-vinegar,* vinegar made in Devonshire and America from refuse cider. 1917 *Jrnl. Chem. Soc.* CXII. I. 313 The volatile reducing substances in cider vinegar consist largely, if not wholly, of acetylmethyl carbinol, which is shown to be a normal constituent of this vinegar. 1937 *Discovery* Sept. 280/1 In the cider-drinking West Country much cider vinegar is found.

ciel (sīl). [Fr., sky.] Sky-blue: a fashion shade.

1910 *Daily Chron.* 26 Feb. 4/5 [She] was in ciel blue with a blue velvet train and gold trimmings. 1923 *Daily Mail* 23 Apr. 7 Shades..of Saxe, Ciel, Jade.

cig, colloq. abbrev. of Cigar, Cigarette, or *Cigarillo.

a 1889 *Broadside* (Barrère & L.), Dancing the jig, Every fellow with a cig, And a cig of confounded bad tobacco. 1919 C. Orr *Glorious Thing* xxiii, Have a cig. Go on! They're only Nannie's. 1956 T. Forshaw in *Coast to Coast* 132 A packet of cigs. 1969 'J. Fraser' *Cock-pit of Roses* xvi. 121 Greens on the slate, never beer. Never cigs, either.

cigar. Add: **1. b.** The pod of the catalpa tree; the Indian bean. *U.S.*

1876 *Field & Forest* II. 51, I verily believe that some boys..took their first lessons, in smoking, by using the 'beans' or 'cigars' of the Catalpa.

c. The brown colour of a cigar.

1923 *Daily Mail* 16 Apr. 1 Grey, Mole, Cigar, Champagne, Light Tan. *Ibid.* 15 Oct. 15/4 This season there is a vogue for brown fur, and ermine, caracul, and squirrel are dyed to provide lovely shades of cigar, dead gold or mahogany.

2. *cigar-box* (examples), *-cabinet, -case* (earlier and later examples), *-cutter* (examples), *-lighter* (examples,) *-maker* (examples; so *-making*), *-smoke* (examples), *-smoker* (example), *-smoking*; **cigar band** (see *Band *sb.*² 2); **cigar-brown** *a.* having the brown colour of a cigar; **cigar-butt,** the waste end of a cigar; **cigar-fish,** a small cigar-shaped fish of the genus *Decapterus*, found in the West Indies and south-eastern United States; **cigar leaf,** tobacco suitable for cigars; **cigar-plant,** a Mexican plant of the genus *Cuphea*, having a scarlet tubular corolla tipped with black and white (*Cent. Dict.* 1889); **cigar-store** *U.S.*, a shop specializing in the sale of cigars and smoking accessories; **cigar-store Indian,** an effigy of a North American Indian, placed outside a cigar-store; also *fig.*; **cigar-tree** *U.S.*, the catalpa.

1963 N. Marsh *Dead Water* (1964) vi. 134 He compared the cigar band with the one he had picked up. 1836 Dickens *Sk. Boz* 1st Ser. I. 321 Cigars..are..two a penny, in a regular authentic cigar box. 1888 Kipling *Departm. Ditties* (1890, ed. 4) 105 Open the old cigar-box. 1878 B. F. Taylor *Between Gates* 252 We have been circling about that cigar-box of a town. 1907 A. L. Winton tr. *Hanausek's Microsc. Techn. Prod.* 219 *Cedrela odorata* L., Spanish Cedar, Cigar-box Wood. 1895 *Bow Bells* XXXI. 297/1 Collar of cigar-brown velvet. 1934 'G. Orwell' *Burmese Days* xi. 161 Peasants with cigar-brown faces. 1890 Kipling *Life's Handicap* (1891) 242 Fleete..was gravely grinding the ashes of his cigar-butt in to the forehead of the red, stone image of Hanuman. 1902 *Daily Chron.* 5 Apr. 5/1 London's gutters now annually receive in cigar-butts and cigarette-ends tobacco to the value of £200,000. 1903 A. Bennett *Leonora* ii. 50 A cigar-cabinet on the sideboard. 1844 C. F. Mersch tr. *C. Sealsfield's Cabin Bk.* 15, I had my cigar case, and a small roll of Virginia dulcissimus. 1847 C. Brontë *Jane Eyre* II. ix. 244 Unfortunately I have neither my cigar-case, nor my snuff-box. 1957 'B. Buckingham' *Boiled Alive* xxiv. 176 Don Pancho reached for his crocodile cigar case. 1859 F. W. Fairholt *Tobacco* iv. 224 Another simple little implement, to act as cigar-cutter and holder. 1905 *Daily Chron.* 27 Dec. 4/5, I deprecate the use of the cigar-cutter, preferring the nice conduct of a penknife. 1936 'N. Blake' *Thou Shell of Death* i. 7 Chromium-plated cigar-cutters. 1884 Goode *Nat. Hist. Aquatic Anim.* 325 The Round Robin—*Decapterus punctatus*..or, as it is called at Pensacola, the 'Cigar-fish'. 1960 *List Names Fishes U.S. & Canada* (Amer. Fisheries Soc.) (ed. 2) 45 Longfin cigarfish..*Cubiceps gracilis.* 1865 *Trans. Ill. Agric. Soc.* 1863 V. 669 Tobacco of this description should be..prized lightly in the casks so as to admit of a free and open leaf, such being mostly required for cigar leaf. *a* 1877 Knight *Dict. Mech.* 553/1 *Cigar-lighter*, a little gas-jet suspended by an elastic tube. 1905 *Westm. Gaz.* 16 Nov. 5/2 An electric cigar-lighter. 1856 X. D. Macleod *Biogr. F. Wood* 47 He..became a journeyman cigar-maker. 1909 'O. Henry' *Roads of Destiny* 303 Denver got a Cuban cigar-maker to fix up a little cipher code with English and Spanish words. 1854 Mrs. E. Twisleton *Let.* 12 Feb. (1928) ix. 161 Pretty, graceful women, going home after their day's work at cigar-making at Alicante. 1957 *Encycl. Brit.* V. 703/1 Whereas early cigar-making was entirely by hand, machines now perform most of the steps. 1961 *Amat. Gardening* 23 Sept. 4/1 Cuphea ignea, the cigar plant, is a useful addition to the summer range, with a compact bushy habit and scarlet tubular flowers. 1905 E. Wharton *House of Mirth* I. xiv. 247 Through the cigar-smoke of the studio. 1956 *Nature* 10 Mar. 450/2 This paper makes the illuminating statement that cigar-smoke is the limit of precision of the method. 1846 *Observer* 18 Oct. 3/2 Inveterate cigar smokers will consume from four to five dozen a week. 1834 *Chambers's Edin. Jrnl.* III. 207/3 An abominable custom...We mean the practice of cigar-smoking. 1848 'N. Buntline' *Myst. N.Y.* II. 23 Are you going back to that hateful cigar store? 1926 Hemingway *Fiesta* (1927) II. xv. 178 All we could see of the procession..were the great giants, cigar-store Indians, thirty feet high. 1930 J. Dos Passos *42nd Parallel* 16 Wooden Indians outside of cigarstores. 1948 *Chicago Daily News* 17 Nov. 26/1 The retailer and the buyer in the cigar store. 1952 B. Malamud *Natural* 172 What she saw in this half-baked apology for a cigar store Indian had him beat. 1963 S. Mitchell *Sables spell Trouble* iv. 40 I'd got about as much rise out of him as I would out of a cigar-store Indian. 1872 *Amer. Naturalist* VI. 727 The beautiful catalpa, or 'cigar-tree' (*Catalpa bignonioides*), grew as a common species among the underwoods. 1933 J. K. Small *Man. S.E. Flora* 1241 Catalpa..Indian-beans. Indian-cigars. Cigar-trees.

cigarette. Add: Also (chiefly *U.S.*) **cigaret.**

1897 G. B. Shaw *Our Theatres in Nineties* (1932) III. 117 Presenting every lady in the audience with a locket and every gentleman with a cigaret case. 1934 —— *Simpleton of Unexpected Isles* II. 47 His poor little secret vice of cigaret smoking. 1945 Mencken *Amer. Lang.* Suppl. I. 361 The English..are slow to follow American example in such forms as *cigaret* and *etiquet.* 1968 *Chicago Tribune* 9 July 1. 12/3 He smoked three or four packs of cigarets a day. 1968 S. Challis *Death on Quiet Beach* iv. 43 They lit cigarets and Stenfalt looked at him over a blue spiral of smoke.

2. *cigarette-box, -butt, -case* (examples), *catcher, -end, habit, -holder* (examples), *-maker, -merchant, -smoke* (examples), *-smoking; cigarette-laden* adj.; **cigarette beetle** = *tobacco beetle*; **cigarette card,** a picture card inserted by the makers in a packet or box of cigarettes; **cigarette coupon** [*Coupon sense 2], a voucher inserted in a packet of cigarettes; **cigarette girl,** a girl who makes or sells cigarettes; **cigarette heart,** a condition of the heart induced by excessive smoking of cigarettes; **cigarette lighter,** a mechanical apparatus for lighting a cigarette; **cigarette machine,** a machine that manufactures or dispenses cigarettes; **cigarette paper,** paper or a paper in which cigarettes or a cigarette is rolled; **cigarette picture** = **cigarette card*; **cigarette tobacco,** tobacco specially adapted for cigarette making.

1891 *Cent. Dict., Tobacco beetle*..Also called *cigarette-beetle.* 1895 Comstock *Study Insects* 553 The Cigarette Beetle, *Lasioderma serricorne*..is a serious pest in tobacco manufactories, infesting the dried tobacco-leaves and the manufactured products. 1896 J. B. Smith *Econ. Entom.* 193 The *Lasioderma serricorne*, popularly known as the 'tobacco-' or 'cigarette-beetle'. 1959 E. F. Linssen *Beetles* 2nd Ser. 64 *Lasioderma serricorne*..is red in colour...The beetle is popularly known as the Cigarette or Tobacco Beetle. 1890 *Peel City Guardian* 11 Jan. 3/5 The smoker's set of cigar and cigarette box. 1909 Galsworthy *Silver Box* III, Theft of a silver cigarette box. 1923 J. Dos Passos *Streets of Night* iii. 90 They tramped up a creaking stair littered with cigarette butts. 1931 *New Statesman* 16 May 418 At every few steps he bent down quickly, picked up a cigarette-butt from the pavement, and put it into his pocket. 1902 *Little Folks* 159/1, 2*d.* a dozen for cigarette cards (Ogden's Guinea Gold). 1926 *Punch* 29 Sept. 345 Got any cigarette-cards, lady? 1883 B. Potter *Let.* in B. Webb *My Apprenticeship* (1926) iii. 157, I produced my cigarette-case and offered the company some 'Welsh cigars'. 1939 T. S. Eliot *Fam. Reunion* II. iii. 128 He thinks he left his cigarette-case on the table. 1921 *Dict. Occup. Terms* (1927) § 469 *Cigarette catcher; cigarette machine receiver,* gathers finished cigarettes from cigarette machine. 1906 *Bazaar, Exch. & Mart* Suppl. 1335/3 Black Cat cigarette coupons wanted. 1922 Joyce *Ulysses* 693 *Soll und Haben* by Gustav Freytag..cigarette coupon bookmark at p. 24. 1932 L. Golding *Magnolia Street* III. vi. 540 If any of those watches were the reward paid for an industrious accumulation of cigarette coupons, he replaced them. 1889 *Tobacco* Apr. 126/1 Waxing cigarette-ends..has for its object preventing the end of the cigarette from sticking to the lips. 1893 *Graphic* 4 Feb. 110/2 The..dearth of..cigarette-ends did not seem to depress him at all. 1902 *Westm. Gaz.* 24 Sept. 10/2 Someone dropped a lighted cigarette-end in the hay. 1933 J. Cary *Amer. Visitor* ii. 26 Cottee spat his cigarette-end at a bubble on the stream. 1916 A. Bennett *Lion's Share* vii. 51 The adventure of the cigarette girl. 1918 Cigarette girl [see *Cloak-room b]. 1968 L. Deighton *Only when I Larf* xi. 133 A cigarette girl came past and he clicked his fingers at her and asked for matches. 1914 *Evening News* 10 Oct. 2/7 Our officers..brought the cigarette habit back with them from the Crimea, where they learned it from the Russians. 1908 *Daily Chron.* 29 Jan. 7/2 Witness examined his heart and found no organic disease; it was a 'cigarette heart'. 1879 Mrs. Hungerford *Airy Fairy Lilian* II. xv. 90 Enough meerschaum pipes, and cigarette holders, and tobacco stands to stock a small shop. 1921 A. Huxley *Crome Yellow* xvii. 179 An immensely long cigarette-holder projected..from her face. 1905 *Daily Chron.* 17 July 8/6 The cigarette-laden air of the sensuous room. 1915 *Scotsman* 26 Jan. 5/3 Colonel Thompson, of the Black Watch, asks for tinder cigarette lighters. 1929 M. de la Roche *Whiteoaks* ix. 135 'The very thing for him!' she exclaimed to Alayne, energetically snapping her cigarette-lighter. 1962 *Which?* (Car Suppl.) Oct. 138/1 Cigarette lighter would not eject properly. 1878 *Sci. Amer.* XXXVIII. 411/1 Cigarette machine. 1951 M. McLuhan *Mech. Bride* 106/1 They kick the cigarette machine. 1889 *Tobacco* June 189/1 (*heading*) Messrs. Philip Morris and Co. and their cigarette-makers. 1890 F. M. Crawford (*title*) A cigarette maker's romance. 1894 *Daily News* 1 Oct. 5/3 The cigarette-makers' cramp, so well-known in the tobacco manufactories of Spain. 1897 *Lippincott's Med. Dict.*, Cigar-makers' cramp or Cigarette-makers' cramp. Painful contraction of the flexors of the fingers, sometimes seen in cigar- and cigarette-makers. 1909 *Daily Chron.* 24 Nov. 6/6 Before the war there was a cigarette merchant..in Great Winchester-buildings. 1860 *Harper's Mag.* Apr. 691/1 His notes were written on cigarette-paper. 1901 *Brit. Med. Jrnl.* 13 July 100/1 For the purpose of selling..any cigarettes, cigarette paper, or any substitute for the same. 1958 J. R. Biggs *Woodcuts* 91 *Cigarette Paper* is a satisfactory though extremely thin white paper, for making prints by burnishing, but care must be taken not to tear the print. 1902 *Little Folks* I. 158/1, 90 cigarette pictures (not all different) to dispose of at 90 for 1*s.* 1944 B. Johnson *As Much as I Dare* 23 This collection yielded place to cigarette pictures. 1900 E. Wharton *Gift from Grave* vi. 74 A mist of cigarette smoke. 1918 A. Bennett *Roll-Call* I. iii. 44 The ruthless George, behind cigarette smoke. 1888 *Tobacco* Apr. 127/1 (*heading*) Why he quit cigarette-smoking. 1906 *Westm. Gaz.* 10 Sept. 1/2 Cigarette-smoking was then [*sc.* in 1856] habitual in St. Petersburg. 1905 in *Daily Chron.* 10 Oct. 6/2 The..manufacture of cigarettes or cigarette tobacco.

cigarillo (sigari·lo, ‖þigari·lyo). [Sp., dim. of *cigarro* Cigar.] A small cigar.

1832 'G. Crayon' *Alhambra* I. 224 Your muleteer..will suspend the smoking of his cigarillo to tell some tale of Moslem gold buried. 1896 *Act* 59 & 60 *Vict.* c. 28 § 5 (3) The expression 'tobacco' in this section includes cigars, cigarillos, cigarettes, and snuff. 1909 *Daily Chron.* Summer No. 2/2 His accursed cigarillo, carelessly flung down, had reduced the seat of learning to ashes. 1962 *Economist* 29 Sept. 1229/3 Cigarillos—Wills Whiffs and the like—have made some significant progress.

cigarito. *U.S.* Also **-ita, -rrito.** (Earlier and later examples.)

1832 *Chambers's Edin. Jrnl.* I. 307/1 What was the plague to us?—not the whiff of a cigarita. 1844 G. W. Kendall *Santa Fé Exped.* II. i. 31 The prettiest señora..can be seen at almost any time with a *cigarrito* in her mouth. 1910 J. Hart *Vigilante Girl* 348 The bandits were lolling around their fire, smoking cigaritos.

cigary (sigā·ri), *a.* [f. Cigar + -y¹.] Of or pertaining to a heavy cigar-smoker.

1923 A. Huxley *Antic Hay* xiv. 200 A rich, port-winey, cigary voice. 1948 Dylan Thomas *Let.* 17 Nov. (1966) 323 They must appeal to rich, cigary filmtycoon me.

ciggy (si·gi), colloq. abbrev. of Cigarette.

1962 *Guardian* 24 Dec. 4/2, I pay her back every night and slip her husband twenty ciggies. 1968 *Scottish Daily Mail* 3 Jan. 6 What had been 'fags' became 'ciggies' because The Beatles always talked of ciggies. 1969 K. Giles *Death cracks Bottle* iv. 43 One of those casual well-bred chaps out of the ciggy ads.

ciguatera (sigwătē·rā). Also **siguatera.** [Amer. Sp. (A. Parra *Descripción de Diferentes Piezas* (1787) 100), f. *cigua* sea-snail.] A tropical disease affecting the nervous system,

the ancient 'Cimmerian Bosporus' was the strait between these two lands.

cinch, *sb.* Add (see also Sinch, Synch): **2.** *fig.* A firm or secure hold; a sure, safe, or easy thing; a dead certainty. Also *attrib.* orig. *U.S.*

1888 *N.Y. World* 22 July (Farmer), The racehorse owner, who has a cinch bottled up for a particular race. 1904 G. V. Hobart *Jim Hickey* iv. 54 We'll have to stand by Amy and the kid, that's a cinch, Danny! 1904 G. H. Lorimer *Old Gorgon Graham* 8 A man's son is entitled to a chance in the business, but not a cinch. *a* 1910 'O. Henry' *Trimmed Lamp* (1916) 208 'It's tough to be alone in New York—that's a cinch,' said Mr. Donovan. *Ibid.* 252 The devil seems to have a cinch on all the business in New York. 1911 H. Quick *Yellowstone N.* iii. 69 The recent progress in bacteriological science..seemed to make the diagnosis a cinch. 1923 R. D. Paine *Comr. Rolling Ocean* iv. 73 Going to sea was a perfect cinch with just enough to do to keep a fellow healthy. 1930 P. Macdonald *Link* ix. 169 My God! As far as I'm concerned it's a cinch. 1936 A. Huxley *Eyeless in Gaza* xxv. 353 'How would you like to be called a fever frau?' he asked his younger daughter... 'Or worse, a cinch pushover, my dear?' 1957 *New Yorker* 29 June 67/1 From there on, it was a relative cinch.

3. *Cards.* A variety of all-fours, also called *double pedro* and *high five.*

1889 in *Dict. Americanisms*, I found that sinch is the great North-western game of cards, a recent invention. 1892 *Outing* (U.S.) Jan. 287/2 Fellers, I brung Dave over fur a game o' cinch, an' we kin hev a squat round fur a couple o' hours. 1946 Morehead & Mott-Smith *Penguin Hoyle* 57 Cinch, a partnership game for four players, is the aristocrat of the high-low-jack games.

cinch (sinʃ), *v.* *U.S.* [f. Cinch *sb.*] **1.** (Also with *up.*) *trans.* To fix (a saddle, etc.) securely by means of a girth; to fix (a girth). Also *transf.*, of clothing: to girdle, pull *in.*

1866 'Mark Twain' *Speech* 2 Oct. (1923) 13 They never cinch a Californian's horse tight enough. 1873 J. H. Beadle *Undevel. West* xv. 273 With all set and everything tightly 'cinched', we took the start. 1884 [in Dict.]. 1897 *Outing* (U.S.) XXIX. 457/1 As the moment for the start approaches, the saddles are cinched tighter. 1905 A. Adams *Outlet* 310 Saddles were..cinched on waiting horses. 1910 J. Hart *Vigilante Girl* x. 143 Each mule would carefully blow himself up to prevent his girth being cinched too tightly. 1912 Mulford & Clay *Buck Peters* iv. 58 Buck cinched up his saddle on Allday and let him out of the stable. 1915 J. London *Jacket* 54 They call it [lacing the jacket] 'cinching' in prison lingo. 1957 *New Yorker* 16 Nov. 150/2 A shirt-topped..sheath..cinched in with a three-inch-wide self cummerbund. 1961 *Times* 24 Jan. 12/5 Cinched-in dresses and coats. 1966 *Harper's Bazaar* Sept. 50 Smashing glitter of..sequins cinching the body, banding matt black crêpe.

b. *intr.* To fix a saddle-girth.

1887 *St. Nicholas* XIV. 732 At Giles's ranch, on the divide, the party halted to cinch up. 1913 Mulford *Coming of Cassidy* ii. 35 Cinching deftly, [he] slung himself up by the stirrup.

2. *fig.* To get (a person) into a tight place; to secure a hold upon. (See also quot. 1875.)

1875, etc. [in Dict.]. 1881 *N.Y. Times* 18 Dec. in *N. & Q.* V. 65/1 It is unfairly said that the Northern Pacific Company intends to 'cinch' the settlers by exacting large prices for its lands. 1904 F. Lynde *Grafters* ix. 125, I have it on pretty good authority that the ring is cinching the other companies right and left. 1910 T. Roosevelt in *Outlook* 3 Sept. 2/1 If the rich man strives to use his wealth to destroy others, I will cinch him if I can. 1916 H. L. Wilson *Somewhere in Red Gap* vii. 292 I'm cinched for hell, anyway, and don't have to make it tighter by torturing poor dumb brutes.

b. To make certain of (something); to render (something) conclusive. *slang.*

1900 *Dialect Notes* II. 27 (College words) *Cinch*, to make sure of anything. 1903 A. H. Lewis *Boss* 315 Melting Moses ..goes after him to cinch the play. I tell you one thing; he cinches it. 1910 W. M. Raine *B. O'Connor* 25 That extra hour and a half cinches our escape. 1924 R. Lardner *How to write Short Stories* (1926) 335 It'll be pretty soft for you, because they got the pennant cinched and they'll cut you in on the big money. 1930 D. Hammett *Maltese Falcon* xix. 231 He's not a fall-guy unless he's a cinch to take the fall. Well, to cinch that I've got to know what's what. 1964 *New Statesman* 10 Apr. 571/1 English readers shouldn't be put off by Professor Shattuck's Texan manner of what he calls 'cinching' an argument.

3. In the game of cinch: to protect (a trick) by playing a higher trump than the five.

1889 in *Dict. Amer. Eng.*, Prevent him from making as many points as he has agreed to make. If he fails to make good his offer, he is 'sinched'. 1946 Morehead & Mott-Smith *Penguin Hoyle* 59 If no higher trump has been played, third hand must usually *cinch* a trick by playing some higher trump than the five.

cinch- (siŋk), var. *Cincho- used before vowels, as in *cinchaine, cinchamidine, cinchene, cincholine* (see quots.).

1881 *Jrnl. Chem. Soc.* XL. 1045 Cinchamidine, $C_{20}H_{26}N_2O$, a new alkaloid found in the aqueous mother-liquors from the purification of homocinchonidine sulphate, is obtained by the addition of ammonia to this liquid. 1882 *Ibid.* XLII. 225 Nitrous acid has no action on cinchene. *Ibid.* 1114 Cincholine is prepared from the first mother-liquor obtained in the manufacture of quinine sulphate. *Ibid.*, Cincholine is a pale-yellow oil, lighter than water. 1890 Billings *Med. Dict.* I. 281/1 *Cinchene*, $C_{19}H_{20}N_2$, a crystallizable, volatile base obtained by partial decomposition of cinchonine. 1906 *Watts' Dict. Chem.* II. 178/2 Cinchamidine..occurs in the mother-liquors from homocinchonidine. 1944 *Hackh's Chem. Dict.* (ed. 3) 201/2 *Cinchaine*, $C_{22}H_{30}O_2N_2$..a cin-

chona alkaloid used as anesthetic. *Ibid.*, *Cinchene*, $C_{19}H_{20}N_2$ ..an alkaloid obtained from cinchonine by boiling with alcoholic KOH. *Ibid.*, *Cincholine*, an alkaloid from cinchona bark.

cincho- (si·ŋko). *Chem.* Combining form of Cinchona, as in *cinchocerotic, -cerotin, -meronic, -phen, -tannic, -toxin(e, -vatin* (f. the species *Cinchona ovata*).

variety of the common black bear of North America; **cinnamon dove** (see quots.); **cinnamon (-coloured) fern,** a large N. American fern (*Osmunda cinnamomea*) with some of its fronds cinnamon-coloured; **cinnamon rose,** delete † and add later examples; **cinnamon toast,** buttered toast spread with a mixture of sugar and cinnamon.

1823 J. Sabine in J. Franklin *Narr. Journey Shores Polar Sea* 648 Individuals vary in colour from black to different shades of brown, and are known to the traders under the different names of Black, Brown, Cinnamon, and Grey Bears. **1829** J. Richardson *Fauna Bor.-Amer.* I. 15 The Cinnamon Bear of the Fur Traders is considered by the Indians to be an accidental variety of this species [*Ursus americanus*]. **1855** F. S. Marryat *Mts. & Molehills* 253 A long-expected fight had come off between a grizzly bear and a cinnamon bear. **1909** 'O. Henry' *Roads of Destiny* xviii. 309 He was dancing like a cinnamon bear. **1936** D. McCowan *Anim. Canad. Rockies* xxxiii. 288 Formerly it was believed that the so called cinnamon bear was a distinct species but such is not the case: the cinnamon bear is simply a reddish-haired black bear. **1887** W. Phillips *Brit. Discomycetes* 337 Hymenium cinnamon-blackish. **1845** A. Wood *Class-Bk. Bot.* II. 465 Cinnamon-colored Fern..is among the largest of our ferns, growing in swamps and low grounds. **1895** Lydekker *Royal Nat. Hist.* IV. 380 The South African cinnamon-dove (*Haplopelia larvata*) may be taken as a well-known representative of another subfamily of ground-doves (*Geotrygoninae*). **1908** Haagner & Ivy *Sk. S. Afr. Bird Life* 114 The Cinnamon Dove..is an inhabitant of the forests of the Cape, ranging from Knysna to Natal, and the Transvaal. **1957** McLachlan & Liversidge *Roberts' Birds S. Afr.* 173 *Cinnamon dove*...White face and cinnamon under-parts diagnostic... From Cape Town..eastwards to Abyssinia. **1818** W. P. C. Barton *Compendium Floræ Philadelphicæ* II. 206 Cinnamon fern. Tall Osmunda. **1868** *Amer. Naturalist* II. 522 The Cinnamon Fern, and the Interrupted leaved Fern, well known to young botanists in the spring. **1917** E. S. Bailey *Sand Dunes Indiana* 156 The brakes and cinnamon ferns are hip high. **1899** *Daily News* 18 Dec. 6/6 Bound in cinnamon-hued cloth. **1824** *Southern Lit. Messenger* I. 98 Clusters of.. large double damask and cinnamon roses. **1892** B. Torrey *Foot-path Way* 32, I wished, also, to say something of sundry minor enjoyments: of the cinnamon roses, for example. **1952** A. G. L. Hellyer *Sanders' Encycl. Gardening* (ed. 22) 426 [*Rosa*] *cinnamomea*, 'Cinnamon Rose', red, May, 6 to 9 ft., spicily fragrant, Europe, N. China, Siberia. **1927** M. de la Roche *Jalna* ix. 103 Pushing his plate of cinnamon toast to one side, he jotted them down on the back of an envelope. **1966** M. McShane *Crimson Madness* (1967) iv. 141 And for tea I'd like cinnamon toast.

5. *ellipt.* = **cinnamon bear*.

1855 F. S. Marryat *Mts. & Molehills* 253 The cinnamon's weight was quoted at 400 lbs. **1878** E. B. Tuttle *Border Tales* 52 Lassooing a cinnamon. **1891** *Fur, Fin & Feather* Mar. 170 They were a big party of cinnamons harvesting their way back towards the mountains. **1903** A. Adams *Log Cowboy* xxiii. 366 Those of us who had fastened on to the old cinnamon.

6. A descriptive name for certain domestic birds (poultry, canaries).

1854 *Poultry Chron.* I. 197 Buff and silver cinnamon, early hatched Cochin chickens. *Ibid.* 270 Their buffs, cinnamons, and grouse with pea combs. **1909** *Westm. Gaz.* 3 Mar. 2/1 When an ordinary canary is hatched the eyes look black, but in the case of the cinnamons they look pink.

7. Yellowish brown, the colour characteristic of cinnamon quills.

1895 *Montgomery Ward Catal.* 305/1 Half Hose... English mixtures in olive, gold, brown or cinnamon. **1901** [see *ash *sb.*[1] 1 e]. **1923** *Daily Mail* 15 Jan. 6 Newest shades, including: Pale Pink,..Rust, Beige, Antelope, Cinnamon. **1924** A. D. Sedgwick *Little French Girl* III. iv. 250 The buffs and cinnamons and mole-colours of the fields. **1957** M. B. Picken *Fashion Dict.* 63/1 *Cinnamon*, grayed brown color, slightly reddish.

cinnet, var. Sinnet.

1884 Turner *Samoa* 165 Everything is fastened in their ancient style, with cinnet plaited from the fibre of the cocoanut husk. Cinnet is likely long to prevail in native canoe and boat-building.

cinnoline (si·nolin). *Chem.* [ad. G. *cinnolin* (V. von Richter 1883, in *Ber. d. deutsch. Chem. Ges.* XVI. 677); cf. Chinolin, Quinoline.] A poisonous crystalline base, $C_8H_6N_2$.

1883 *Jrnl. Chem. Soc.* XLIV. 1105 Cinnoline-derivatives. By V. v. Richter. *Ibid.*, This base may be looked upon as quinoline in which one CH-group has been replaced by N, and the author therefore proposes the name cinnoline for it. **1929** *Encycl. Brit.* V. 386/2 The fusion of pyridazine with benzene may result in either cinnoline or phthalazine. **1946** *Jrnl. Amer. Chem. Soc.* LXVIII. 1310 (*title*) 4-Substituted Cinnoline Derivatives. *Ibid.*, Work on the synthesis of cinnoline derivatives is limited, but has shown that the most practical method of obtaining such compounds..involves the spontaneous cyclization of diazotized o-aminophenylethylenes. *Ibid.*, Cinnoline-4-carboxylic acid was esterified readily. **1965** *Austral. Jrnl. Chem.* XVIII. 713 The nonuniform effect of the heteroatom on *ortho* coupling constants in the aromatic ring noted previously is again observed in quinazoline and cinnoline.

‖**cinq trous** (sæ̃k tru). [Fr., = five holes.] A form of mesh in certain makes of lace in which openings are set alternately in quincunx form.

1882 Caulfeild & Saward *Dict. Needlework* 80. **1902** *Mrs. Bury Palliser's Hist. Lace* 213 The fillings introduced into the flowers and other ornaments in Chantilly lace are mesh grounds of old date, which, according to the district where they were made, are called vitré, mariage, and cinq what are called inside, circles, while he keeps on; and finally trous. **1908** N. H. Moore *Lace Book* 155. **1960** H. Hayward *Antique Coll.* 72/1 *Cinq trous*, term used to describe the ground found in some Flemish lace in which the threads crossing to form the mesh left five small holes.

cinquedea (tʃiŋkwidē·a). [(Venetian) It., f. *cinque* five + *dea* = *dita* fingers, pl. of *dito*:– L. *digitu-s* Digit.] A broad-bladed dagger or short sword of which the blade has the breadth of five fingers.

1897 *Daily News* 1 Feb. 6/2 A cinquedea, the hilt composed of plaques of faceted ivory, Italian 15th century, in a finely decorated scabbard. **1911** *Archæol. Jrnl.* LXVIII. 159 The majority of these cinquedeas are about four and a half inches wide at the base of the blade. **1920** Laking *European Armour* III. 61 The gorgeous enrichments of the various cinquedea-like daggers. *Ibid.* 70 Cinquedea sword daggers of late XVth and early XVIth century date. **1962** *Internat. Art Treas. Exhib.* (*V. & A. Mus.*) 53/2 A fifteenth-century Venetian sword, the bronze hilt of cinquedea form bound with plaques of mammoth ivory.

Cinzano (tʃinzā·no). [Name of the manufacturers.] The proprietary name of an Italian vermouth; also, a drink of this.

1923 A. Huxley *Antic Hay* vi. 96 Ah, Bonomelli and illustrious Cinzano. **1959** A. Bailey *Making Progress* 60 On the hotel veranda, sipping a Cinzano, was the gentleman in the khaki shorts and trilby hat.

cipher, *sb.* Add: **1. b.** (Later example.) *U.S.*

1815 D. Drake *Cincinnati* ii. 94 From nine years observations, at Cincinnati, it appears that the thermometer falls below cypher twice every winter.

8. cipher officer, an officer in the military or diplomatic services responsible for the coding and decoding of ciphers.

1915 O. Williams *Let.* 23 Mar. in C. Mackenzie *Gallipoli Mem.* (1929) ii. 7 I'm cipher officer on his Staff with the rank of Captain. **1948** *Hansard* CDXLVIII. 1539 The smooth running of an embassy abroad depends just as much on a happy, contented, well-paid staff of cipher officers..as..on the..head of the Mission.

ci·pherdom. [f. Cipher *sb.* 2 a + -dom.] Of a person: the state of being negligible, without qualities.

1903 G. B. Shaw *Man & Superman* Pref. p. xiii, Don Juan..must, to save himself from cipherdom, find an affirmative position. **1964** *New Statesman* 10 Apr. 574/3 Trouble..comes in..the increasing cipherdom of its young tormented hero.

cipherer. Add: **2.** One who writes or is skilled in writing in cipher.

1885 C. Lowe *Bismarck* I. 526 A devoted band of privy-councillors, secretaries, cipherers, newspaper-hacks.

So **ci·pherist.**

1902 *Academy* 1 Mar. 210/1 Lord Bacon was a literary man using a cypher, and not a cypherist blundering into immortal composition.

circa. Add: Also, an instance of this.

1898 Saintsbury *Short Hist. Eng. Lit.* 314 All the more careful..accounts..mark the earliest play-dates..with a tell-tale *circa*.

circadian (səɪkēⁱ·diăn), *a.* [f. L. *circa* about + *dies* day + -an.] Designating physiological activity which occurs approximately every twenty-four hours, or the rhythm of such activity.

1959 F. Halberg in *Zeitschr. f. Vitamin- Hormon- & Fermentforschung* X. 235 'Circadian' might be applied to all '24 hour' rhythms, whether or not their periods..are different from 24 hours, longer or shorter, by a few minutes or hours. **1963** J. Leith tr. *Fraisse's Psychol. of Time* i. 20 Circadian rhythms in animals, unlike plants, are almost always acquired. **1964** *Perceptual & Motor Skills* XIX. 211 (*heading*) Circadian activity and noise comparisons of two confined groups with or without reference to clock time. *Ibid.*, Both measures revealed highly positive correlations between the mean circadian patterns of the two groups.

Circassian, *sb.* and *a.* Also 7 **Sarcassen** and (of a woman) **Sarcashen, -cashien.** [f. *Circassia*, latinized form of Russ. *cherkés* (fem. *cherkéshenka*, pl. *cherkésy̆*) + -ia: see -ian.] Substitute for def.: **A.** *sb.* **1.** A native or inhabitant of Circassia, a region in the northern Caucasus, of Caucasian race but non-Indo-European in language; also, the North Caucasian language of this people.

1555 Eden *Decades* (Arb.) 318 The Slauon tounge..vsed of..the Circassians and Quinquemontanians. **1634** T. Herbert *Trav.* 66 Some captiue Sarcassens and Georgians. *Ibid.* 72 [The King of Georgia's] Wife, a Sarcashen. **1639** Pagitt *Relat. Christians* 47 Of the Greeke Communion are the..Georgians, Circassians, Mengrelians and Melchites. **1842** Barham *Ingol. Leg.* 2nd Ser. 114 Could they speak, In their Georgian, Circassian, or Turkish, or Greek. **1854** L. Loewe *Dict. Circassian Lang.* Pref. 5 The Circassians call themselves the people of Addee-ghey (which word I take to signify 'Mountaineer', or 'Highlander', from the Circassian 'Attaghagh'). **1905** A. Goodrich-Freer *In Syrian Saddle* I. v. 102 Among Circassians. *Ibid.* 105 Their ordinary tongue continues to be Circassian. **1923** L. & A. Maude tr. *Tolstoy's War & Peace* II. vii. x. 235 An hussar was Natasha, and a Circassian was Sonya, with a burnt-cork moustache and eyebrows.

2. A thin worsted fabric. Also *attrib.*

1824 J. Hogg *Conf. Justified Sinner* 341 Rather a gentlemanly personage—Green Circassian hunting coat and turban—Like a foreigner. **1845** *Lowell Offering* V. 255 If cold, her gown is of 'green circassian'. **1853** [in Dict.].

B. *adj.* Of, pertaining to, or connected with the Circassians; *Circassian circle*: a type of dance popular in the late 19th and early 20th centuries.

1638 T. Herbert *Trav.* 290 Scander..King of Georgia.. by a Sarcashien Lady had 3 hopefull sonnes. *a* **1791**, **1842** [see Georgian *a.* 1]. **1887** Kipling *Plain Tales fr. Hills* (1888) 174 A gentleman who doesn't know the Circassian Circle ought not to stand up for it. **1923** *Blackw. Mag.* June 732/1 A heavy silver-handled Circassian riding-whip. **1963** C. Mackenzie *My Life & Times* II. 109 At the dancing-class..the thrill of meeting Clive and taking her hand for a moment or two in the peregrinations of the Circassian circle.

ci·rcinately, *adv.* *Bot.* [f. Circinate *a.* + -ly[2].] In a circinate manner or arrangement.

1866 J. Smith *Ferns Brit. & For.* 73 Fronds circinately unfolding, uniform and leafy. **1872** H. C. Wood *Fresh-w. Algæ* 40 Circinately or fasciately convolute.

circiter (sə̄·ɪsitəɹ). [L.] = Circa.

1888 *Athenæum* 14 Jan. 49/1 The actual pedigree begins with William 'Pepis', of Cottenham,..living *circiter* 1500.

circle, *sb.* Add: **1. c.** Colloq. phr. *to go, run* or *rush (a)round in circles*: to rush about in all directions; to move or act aimlessly or inconclusively.

1933 H. L. Ickes *Secret Diary* 27 Sept. (1953) I. 99, I asked the oil people to supply the..necessary facts and figures. They keep promising to do it but according to the solicitors they don't furnish them. The result is that we have been running around in circles. **1940** G. Butler *Kiss Blood off my Hands* i. 11 They were running around in circles, shouting and blaming each other. **1942** 'P. Wentworth' *Pursuit of Parcel* xxxvi. 168 He had been rushing around in circles because he didn't dare to stop and think what might be happening to Delia. **1953** H. Waugh *Last Seen Wearing*— 58 We've been going around in circles long enough. If she's down there I want to know it.

2. c. Photogr. *circle of confusion*, (also in Physics) *circle of least confusion*: see quots.

1867 Sutton & Dawson *Dict. Photogr.* 57 *Circle of least confusion*, the nearest approach to a focus of a pencil after oblique reflexion or refraction. **1903** Beck & Andrews *Photogr. Lenses* (ed. 2) iii. 58 When the screen reaches X the circle of illumination will be the smallest size attainable with the lens under consideration, the circle at this point is called the circle of least confusion. **1906** R. C. Bayley *Complete Photogr.* xii. 150 'Depth of focus'..depends upon the size of what is termed the 'circle of confusion'. **1940** *Chambers's Techn. Dict.* 163/2 *Circle of confusion*, the minimum area of a focused bright point of light, the size of which determines the maximum definition possible with a given lens arrangement and stop. **1953** Amos & Birkinshaw *Telev. Engin.* I. ix. 173 Objects near the object plane give images which are only slightly blurred, points being reproduced as circles which are known as circles of confusion. **1958** H. J. Gray *Dict. Physics* 88/2 *Circle of least confusion*, circle through which all the rays over the aperture pass. **1961** Sowerby *Dict. Photogr.* 120 It is usually assumed that a circle of confusion of 1/100 in. corresponds to the boundary between 'sharp' and 'unsharp' in a print held at the normal viewing distance of about 10 in. from the eye. **1961** Webster, *Circle of confusion*, the indistinct circular patch formed by a lens representing the out-of-focus image of a single object point—called also *blur circle*. *Ibid.*, *Circle of least confusion* (Physics), the minimum cross section of a symmetrical bundle of rays that have no common focus because of spherical aberration.

7. b. A dark circular mark around or beneath the eyes, noticeable esp. in delicate or sick persons, or in persons who have slept badly.

1848 Thackeray *Van. Fair* xiv. 118 Her face might be a trifle paler, and the circles round her eyes a little blacker than usual. **1893** 'S. Grand' *Heavenly Twins* III. v. iv. 60 There were great black circles round his sunken eyes. **1906** Galsworthy *Man of Property* II. xi. 244 There were circles under her eyes, as though she had not slept. **1932** A. Christie *Peril at End House* vi. 75 The circles under her eyes were darker than usual.

16. b. (With capital initial.) The name given to a line on the Underground railway in London; the 'Inner Circle' (Cf. *inner *a.*).

1886 H. Baumann *Londinismen* 26/1 *Circle*..Londoner Eisenbahn: *Circle trains*, Züge der unterirdischen Bahn. **1903** *Punch* 30 Sept. 231 Uncle Jack is coming to meet us by a Circle Train. **1907** *Westm. Gaz.* 7 Nov. 10/3 The only chance for the effective working of the 'Circle' lay in the arrangements being taken over entirely by one of the two companies. **1955** 'W. Mole' *Hammersmith Maggot* v. 60 Perry walked primly down to the Circle platform...He took the eastbound Circle train. **1970** G. Lord *Marshmallow Pie* v. 53 They went down on the platform to wait for the Circle Line train.

21. b. *spec.* = Séance 2.

1856 *Spiritual Herald* May 124, I attended three public circles in the spirit-house of Mr. Koons. **1864** Browning *Dram. Pers.* 179 David holds the circle...Sets to the spirit-writing, hears the raps. **1867** [see Spirit *sb.* 23 g]. **1926** Conan Doyle *Hist. Spiritualism* II. xviii. 111 The process at work in the formation of ectoplasm at a circle.

25. circle-rider *U.S.*, a cowboy engaged in circle-riding; **circle-riding** *U.S.*, a method of gathering in cattle (see quot.).

1888 T. Roosevelt in *Century Mag.* Apr. 860/1 As soon as..the last circle-riders have come in..we begin to work the herd. *Ibid.* 857/2 This morning work is called circle riding...As the band goes out, the leader from time to time detaches one or two men to ride.., making the shorter, or

what are called inside, circles, while he keeps on, and finally ..makes the longest or outside circle himself. **1926** D. Branch *Cowboy & his Interpreters* 56 In rough country where each little ravine must be searched and each knoll circled, each group of 'circle riders' would be smaller than groups in unbroken plains country.

circs (sə̄ɹks), colloq. abbrev. of pl. of Circumstance *sb.*

1883 in Ware *Passing Eng.* (1909) s.v., Under the circs I am disposed to exclaim, 'What extravagance!' **1897** W. J. Locke *Derelicts* vi. 75, I ain't going to give you away—don't you fear. It's only pleasant to meet old pals again—in better circs. Ain't it? **1929** Wodehouse *Summer Lightning* xiii. 244 So she asked me to break it to you gently that in the circs. she proposes to return you to store. **1959** J. O'Donovan *Visited* vi. 57 Decent of you to bother with the fags, under the circs.

circuit, *sb.* Add: **1. d.** A road built or used mainly for motor-racing.

1935 Eyston & Lyndon *Motor Racing* i. 4 The idea of 'closed' circuits was conceived—that is, courses from which normal traffic is diverted. **1935** *Encycl. Sports* 421/2 With few exceptions the closed circuit type of race has been the only type to survive...The Ardennes circuit race of 1904 was won by Heath on a Panhard. *Ibid.* 423/2 In 1929–30–31 important races were run over a road circuit in Phoenix Park, Dublin. **1955** *Times* 14 May 8/5 After stewards of the meeting had toured the circuit practising was allowed to proceed.

3. e. A course or round of various athletic exercises.

1957 Duncan & Bone *Oxf. Pkt. Bk. Athletic Training* (ed. 2) iii. 25 A 'circuit' is composed of a series of simple exercises..for development of strength, stamina..agility, and speed.

4. c. A route regularly followed by an itinerant entertainer; a number of places of entertainment (theatres, music-halls, etc.) at which the same productions or entertainers are presented successively; a group or chain of theatres, cinemas, etc., under the control of one person or company.

1839 Dickens *Nich. Nick.* xxx. 300 If he only acted like that, what a deal of money he'd draw! He should have kept upon this circuit. **1840** —— *Old C. Shop* xxxvii. 304 It's our reg'lar summer circuit is the West. **1842** *Ainsworth's Mag.* II. 565 Elliston..became soon the favourite 'leader' on the York 'circuit'. **1889** *Cent. Dict.*, *Circuit*, a number of theaters controlled by one manager. **1908** *Daily Chron.* 31 Oct. 1/4 The De Frece circuit of music-halls. **1913** [see *cinematograph theatre*, *cinematograph *sb.* b]. **1944** *Ann. Reg. 1943* 345 The 600 cinemas of the Odeon and British Gaumont circuits. **1965** G. Melly *Owning-Up* vi. 71 It was Jim's job to book bands and send them up to play for anything between one week and three on Duncan's circuit.

7. (Earlier examples.)

1746 W. Watson *Seq. Exper. & Obs.* 26 Several experiments shew, that the electrical force always describes a circuit. **1756** in B. Franklin *Exper. & Observ. Electr.* (1769) 280 Whether a river..may not be made part of the circuit through which the electric fire passes? instead of the circuit all of wire.

10. circuit-binding (see quot.) = Yapp; **circuit (fore) edge,** a flexible projection to the binding of a book; **circuit-judge** (earlier U.S. examples); **circuit-preacher, -rider** *U.S.*, an itinerant preacher serving a circuit (sense 6); **circuit training,** the performance by an athlete of repeated circuits (sense 3 e, above).

1909 Webster, *Circuit binding*, a style of binding for books having flexible projections of the covers, called circuit edges, that fold and meet so as to close in the edges of the book. **1905** *Westm. Gaz.* 4 Apr. 12/2 In vellum binding with circuit fore-edge. **1801** *Deb. Congress* 9 Jan. (1851) 902 His travelling expenses were trifling compared with those of the Circuit Judges. **1831** *Amer. Almanac* 197 The eight circuit judges [of New York state] are vice-chancellors for their respective circuits. **1830** E. Williams *New-York Ann. Reg.* 291 Other towns [of Franklin County, N.Y.] have no settled ministers—Supplied by circuit and other preachers. **1834** J. M. Peck *Gaz. Illinois* I. 89 The Illinois Conference [of the Methodist Episcopal Church]..has.. fifty-six circuit preachers. **1872** Eggleston *End of World* xxxix. 245 Then he stepped to the door and called in the circuit preacher. **1837** J. Cutting in A. Wetmore *Gaz. Missouri* 333 When Patsy gets her eye on the greyhound, she'll feel a heap gladder than when the circuit rider comes round! **1838** Flagg *Far West* II. 61 A little, portly, red-faced man..announced himself a Baptist circuit-rider. **1864** *Congress. Globe* 17 May 2317/3 As peripatetic as a tin-peddler's cart or a Methodist circuit-rider. **1909** *N.Y. Even. Post* (semi-weekly ed.) 23 Sept. 5 A last survival was he of the old circuit rider of early Methodism. **1957** Duncan & Bone *Oxf. Pkt. Bk. Athletic Training* (ed. 2) iii. 25 Circuit training has only recently been given some practical development. **1964** *Even. Standard* 3 Dec. 30/1 We call it circuit training because it consists of a series of exercises..each designed to bring into play a different set of muscles.

circuit, *v.* Add: **3.** *trans.* To form (an electric wire) into a circuit. Hence **ci·rcuited** *ppl. a.*

1895 *Pop. Sci. Monthly* Apr. 758 A circuited instead of a simple overhead wire.

circuitry (sə̄·ɹkitri). [f. Circuit *sb.* 7 + -ry]. The plan or components of an electric circuit; a system of such circuits.

1946 *Electronic Engin.* XVIII. 388 In 1942, Lewis.. produced his 'Electrical Counting', which covered well the 'circuitry' of the subject. **1952** *Science News* XXV. 94 The impact of radar techniques on electron circuitry. **1959** *Times* 2 Jan. 2/3 Applicants must..have..development experience..in..Guided Missile Circuitry. **1961** *Times* 3 Oct. (Computer Suppl.) p. viii/4 The technique of buffering, together with other circuitry refinements.

circular, *a.* Add: **4. c.** *circular insanity* [tr. Fr. *folie circulaire* (J. Falret *père* 1854, in *Bull. de l'Acad. de Méd.* XIX. 383)]: a mental disease characterized by alternating phases of elation and depression; manic-depressive psychosis; = *cyclothymia (s.v. *cyclo-).

1862 J. H. Worthington tr. J. Falret *fils* in *Amer. Jrnl. Insanity* XVIII. 382 The mental disorder described by my father under the name of circular insanity..consists in a regular alternation of a condition of melancholic depression with a state of maniacal excitement. **1895** W. James *Will to Believe* (1897) 34 In what is called 'circular insanity', phases of melancholy succeed phases of mania. **1962** Henderson & Gillespie *Text-bk. Psychiatry* (ed. 9) x. 208 Kraepelin..in 1896..formulated his conception of the manic-depressive psychosis. In this group he included.. periodic and circular insanity.

5. (Later examples.)

1947 *Partisan Rev.* XIV. 372 Definition of a civilization in terms of the intelligible field of study is deceptive. The process is circular and tautological. **1969** *English Studies* L. 259 To say that the character must be uninvolved in the aspect of the plot on which he comments and emotionally free from entanglements with the characters he discusses.. is really a circular argument: if he were not the spectator would have ironic reservations, and might not find the conclusions acceptable.

10. *circular saw* (earlier examples).

1817 *Niles' Reg.* XII. 336/2 At the steam saw mill there is a circular saw,.. chiefly calculated for cutting veneers. **1825** J. Nicholson *Operat. Mechanic* 444 Circular saws.. are now used in the dock-yard at Portsmouth.

circularity. c. Delete †*Obs.* and add later examples.

1958 *Oxf. Mag.* 1 May 408 b/1 An argument in favour of the validity of memory..it..fails to carry conviction.. it smells strongly of circularity. **1963** J. Lyons *Struct. Semantics* vii. 181 There is no circularity involved in the statement of the problem or in its proposed solution.

ci:rculariza·tion. orig. *U.S.* [f. Circularize *v.* 2.] The sending out of circulars.

1888 in Mrs. A. R. Diehl *Two Thousand Words* 55. **1928** *Publishers' Weekly* 16 June 2450 If..the Guild advertising or circularization should be considered contrary to the spirit of this agreement. **1942** R. G. Collingwood *New Leviathan* 281 'Circularization'..means the process of rendering persons the recipients of a kind of advertisement called a 'circular'.

circulate, *v.* Add: **5. b.** *spec.* to move around at a gathering, talking to different people.

1928 E. Waugh *Decl. & F.* I. ix. 97 Circulate, old boy, circulate. Things aren't going too smoothly. **1960** J. Wain *Nuncle* 88 Shall we circulate a bit?..Otherwise we shan't have spoken to a soul.

6. b. To travel or walk about. Chiefly *U.S.*

a **1848** in Bartlett *Dict. Amer.* 395 Arriving in Maryland, a slave State, he circulates at a cost of from three to five cents a mile. **1907** Mulford *Bar-20* viii. 83 Shore, go home. I'll just circulate around some for exercise. **1910** Kipling *Rewards & Fairies* 153 As we French say, I circulated till I found the galley.

7. b. (Earlier examples.)

c **1793** Jane Austen *Volume the First* (1954) 54 Come Girls, let us circulate the Bottle. **1830** M. R. Mitford *Our Village* IV. 104 Our drover could never resist..the gin-bottle, and..was apt to circulate it at his own expense.

c. To send circulars to. (Cf. Circular B 2.)

1969 *Daily Tel.* 9 Apr. 16 'We circulated head teachers of every secondary school and every independent school,' says Cox, 'and are still selling 100 copies a day.' **1970** *Ibid.* 18 Apr. 15 By using the wildlife fund's mailing list it has circulated some 97,000 of the fund's members.

‖**circulus vitiosus** (sə̄·ɹkiŭlə̆s viʃiō^u·sə̆s, viti-). [L.] = *vicious circle* (see Vicious *a.* 9). (In both *Logic* and *Path.* senses.)

1902 C. Hague tr. *Brentano's Orig. Knowl. Rt. & Wrong* 99 Whoever bases..mathematics upon induction cannot possibly justify mathematically the inductive procedure. It would be for him a *circulus vitiosus*. **1908** *Practitioner* LXXX. 19 The congestive stasis is increased by the fits, and there is a tendency to the formation of a *circulus vitiosus*. **1941** *Mind* L. 370 The *circulus vitiosus*..is a universal phenomenon in the foundations of science. **1964** C. Hastings tr. *Schrödinger's View of World* 81 Professor John Synge..uses what is called the *circulus vitiosus* to make mild fun of one-language dictionaries.

circumambulator. (Earlier U.S. example.)

1787 Jefferson *Corr.* (1829) II. 162 He was determined to obtain the palm of being the first circumambulator of the earth.

ci·rcumcircle. *Geom.* [f. Circum- + Circle *sb.*] A circle which passes through the vertices of a polygon.

1885 J. Casey *Treat. Analyt. Geom.* iii. 97 To find the equation of the circumcircle of the triangle of reference. **1891** W. J. McClelland *Geom. of Circle* xi. 255 The envelope of the circum-circle is a fixed circle. **1970** *Nature* 28 Mar. 1268/2 The radii of the incircle and circumcircle.

circumductory (səɹkə̆mdɒ·ktəri), *a.* [f. Circumduct *v.* + -ory[2].] Pertaining to or characterized by circumduction (see Circumduction 2).

1872 Humphry *Myology* 143 In the case of the proximal joint of the limb—that of the first cartilage or bone with the girdle—the movement is..circumductory or in any direction.

circumlocutionize (sə̄:ɹkə̆mlŏkiū·ʃənəiz), *v.* [f. Circumlocution + -ize.] *trans.* and *intr.* To speak of, or to speak, in circumlocution.

1822 Cobbett *Weekly Reg.* 27 Apr. 249 To mention names, in such a case, is wholly out of the question...Circumlocutionize the profligate..and insolent crew. **1886** *N. & Q.* I. 450/1 If we want to say, 'It was clearly meant as an insult but he didn't choose to *relever* it', we must circumlocutionize with four extra words—'to take any notice of it', or at least with two—'to take it up'.

circumlunar (sə̄·ɹkə̆ml^iū·năɹ), *a.* [f. Circum- + Lunar *a.*] Revolving about, surrounding, or flying round the moon.

1909 in Webster. **1959** *Listener* 22 Jan. 163/1 A body moving in a Lunik-type path will not normally crash-land upon the Moon, or enter a circum-lunar orbit. **1961** *New Scientist* 4 May 241/2 *Apollo*..aims to enable three men to reconnoitre the Moon on a circumlunar voyage. **1968** *Guardian* 19 Sept. 18/8 The Russians intended their flight to be a test-run..for a circumlunar flight with animals.

circumpo·lar, *sb.* [f. the adj.] A circumpolar star.

1888 C. A. Young *Text Bk. Gen. Astron.* iv. 70 Determination of Latitude...By Circumpolars. **1910** G. L. Hosmer *Text-Bk. Pract. Astron.* 30 If the observer travels north until he is beyond the arctic circle, latitude 66° 33' north, then the sun becomes a circumpolar at the time of the summer solstice.

circumpre·ssure. *rare.* [f. Circum- + Pressure.] Pressure from all directions.

1889 in *Cent. Dict.* **1896** W. James *Will to Believe* (1897) i. 9, I mean all such factors of belief as..the circumpressure of our caste and set. **1904** —— *Meaning of Truth* (1909) iii. 72 The only *real* guarantee we have against licentious thinking is the circumpressure of experience itself.

circumspectious, *a.* Delete †*Obs.* and add later examples.

1918 E. Pound *Pavannes & Div.* 37 Yet Radway went. A circumspectious prig! **1920** *Cornhill Mag.* Sept. 352 'He was a circumspectious man, you Grant,' said Ewing. **1923** W. de la Mare *Riddle* 95 A kind of unseen circumspectious spirit.

circumstance, *sb.* Add: **4. a.** Esp. in phr. (*the*) *creature of circumstance*(*s*).

1826 [in Dict.]. **1827** J. C. & A. W. Hare *Guesses at Truth* I Man without religion is the creature of circumstances. [**1836** R. Owen *New Moral World* I. vii. 37 Man is the creature of the circumstances in which he is placed.] **1862** [see Creature 5]. **1872** E. G. White *Testimonies for Church* No. 21. 65 If we are creatures of circumstance, we shall surely fail of perfecting Christian characters. **1929** *Sat. Rev.* 20 July 78/2 Theresa is the creature of circumstance. **1947** W. S. Maugham (*title*) Creatures of circumstance.

7. c. In U.S. colloq. phrases: *not a circumstance to*, nothing in comparison with; *a mere* (or *remote, poor*) *circumstance*, a person or thing of little or no importance.

1836 *Crockett's Yaller Flower Almanac* 19 Orson, the wild man of the woods, is nothing to him—not a circumstance. **1838** Flagg *Far West* I. 145 The race of John Gilpin or of Alderman Purdy were, either or both of them, mere circumstances to ours. *c* **1840** in Thornton *Amer. Gloss.* (1912) II. 969 I'm a little specimen, as you see, a remote circumstance, a mere yearling. **1845** S. Judd *Margaret* II. v. 284, O, it an't a circumstance to what it used to be. **1899** 'Mark Twain' in *Harper's Mag.* (1914) Dec. 10/1 Next comes King John, and he was a poor circumstance. **1901** Harben *Westerfelt* xvi. 219 'Mother told me he often drove you out home.' 'Oh, la, that ain't a circumstance, Harriet! He used to come out home mighty nigh every day or night.' **1903** *Nation* (N.Y.) 1 Oct. 258 Undigested securities are not a circumstance to undigested political principles.

circumta·bular, *a.* [f. Circum- 2 + Tabular.] That sits round a table; round-table.

1919 N. Hill *Story Scott. Ch.* vii. 149 Circum-tabular oligarchies. **1920** *Glasgow Herald* 27 July 6 [They] had in view only a circumtabular gathering of the States engaged in mutual self-destruction. **1926** *British Weekly* 28 Oct. 86/4 All circumtabular dominant groups.

circus. Add: **2. c.** A disturbance or uproar; a lively or noisy display. *colloq.* (orig. *U.S.*).

1869 'Mark Twain' *Innoc. Abroad* xxxiii. 358 It [*sc.* Constantinople] was—well, it was an eternal circus. **1885** W. L. Alden *Adv. Jimmy Brown* 88 Mr. Martin..sprang up, and nearly upset the table, and fell over his chair backward, and wasn't there a circus in that dining-room! **1887** *Harper's Mag.* May 944/1 The night grew rougher...'I guess it's going to be a regular circus.' **1895** Scully *Kaffir Stories* 147, I guess we must see this circus out. If you have to shoot, aim low. **1895** C. King *Fort Frayne* v. 69 We'll have a circus with him. **1897** *Westm. Gaz.* 6 Jan. 4/3 This manœuvre was successfully carried out by about 3.30 p.m., when our force prepared to retire; and then the circus commenced. **1951** M. Kennedy *Lucy Carmichael* v. iii. 255 She was lovely and strange and agitating, but he did not want to turn his life into a three-ring circus. **1955** C. Hastings in J. C. Trewin *Plays of Year* XII. 243 Good God—is that all this three-ring circus is about? **1968** W. Warwick *Surfriding in N.Z.* 17/3 Weekend surfing has become one big circus, with everybody vieing for the attention of the crowd on the beach.

d. *spec.* A squadron of aeroplanes; a group of aircraft or their pilots engaged in skilful or spectacular flying. *slang.*

1917 'Contact' *Airman's Outings* 181 The German airmen, with their 'travelling circuses' of twelve to fifteen fast scouts. **1918** C. G. Grey in *M. F. von Richthofen's Red Air*

Fighter Pref. 19 Von Richthofen's chaser squadron—or Jagdstaffel, as the Germans call these formations—was the first to be known as a 'circus'. **1931** *Times Lit. Suppl.* 19 Nov. 905/3 The fortunes of one of the earliest Staffels, or flying circuses. **1957** P. Kemp *Mine were of Trouble* ix. 161 We saw ahead of us 'circuses' of our fighters diving in rotation to machine-gun the fleeing Republicans.

e. *gen.* A company, group, or set of persons acting or performing in concert, e.g. in Lawn Tennis, Motor-Racing, etc. *colloq.*

1958 *Oxf. Mag.* 13 Mar. 379/2 Military–civil relations... Mr. —'s initiative in organising what we should call a 'circus' on the subject at London University was a most welcome one. **1959** *Observer* 25 Jan. 25/5 Those of us [*sc.* racing drivers] constantly travelling around with the Grand Prix circus. **1959** *Ibid.* 5 Apr. 27/5 The Kramer circus comes to Britain this year immediately after Wimbledon. **1970** *Observer* 15 Mar. 19/2 The week-long tournament with Omar Sharif and the Bridge Circus.

7. (Earlier and later examples.) Also, a traffic roundabout.

1714 Pope *Rape Lock* IV. 117 Sooner shall Grass in Hide-Park Circus grow. **1766** Anstey *Bath Guide* II. ix. 57 To breathe a purer Air In the Circus or the Square. **1898** *Tit-Bits* 15 Jan. 300/3 Bridges, of light and tasty design, across all the main thoroughfares, and at the various 'circuses' and cross roads.

8. *circus-ring* (example), *-trick*.

1869 'Mark Twain' *Innoc. Abr.* xxx. 324 In the centre of the great circus ring..was a torn and ragged upheaval a hundred feet high. **1917** W. Owen *Let.* 4 Feb. (1967) 431 Cantering round..and doing all kinds of circus tricks. **1967** O. Ruhen in *Coast to Coast 1965–6* 191, I put my palomino through one of her circus tricks.

circussy (sə̄·ɹkə̆si), *a.* Also **-usy.** [f. Circus + -y[1].] Resembling or characteristic of a circus.

1874 Meredith *Beauch. Career* (1876) I. iv. 66 Here are lots of circussy heroes coming home to rest after their fatigues. **1894** —— *Ld. Ormont* i, He accused Murat of carelessness of his horses, ingratitude to his benefactor, circussy style. **1928** *Sunday Dispatch* 2 Sept. 4/5 Circusy-looking horses.

ciré (si·re), *a.* [Fr., = waxed.] Having a smooth polished surface. Also short for *ciré silk*, etc.

1921 *Spectator* 23 Apr. 522/2 Trades dealing in..ribbons,.. wool, ciré, raffia. **1921** *Glasgow Herald* 21 May 4 A chapeau of ciré silk or straw. **1922** *Daily Mail* 5 Dec. 14/2 Belted coatees of ciré leather. **1923** *Ibid.* 18 Jan. 11 A casaquin covered entirely in ciré braid. **1969** *Listener* 20 Feb. 251/3 Is it a happy connotation when the *Sun* commends ciré as having been worn on the barricades by French students?

‖**cire perdue** (sīr pɛrdü). [Fr., = lost wax.] A method of casting bronze by making a model with a wax surface, enclosing it in a mould, melting the wax out, and running in the metal between the core and mould.

1876 *Encycl. Brit.* IV. 367/1 In the old method of bronze-casting, known as the *cire perdue*, wax is first used for the thickness of the statue..and is melted and run off before the metal is poured in. **1885** *Ibid.* XIX. 183/2 Casting (by the 'cire perdue' process). **1904** S. W. Bushell *Chinese Art* I. iv. 73 Chinese bronzes have always, as far back as we have any record, been executed by the *cire perdue* process. **1961** S. Lloyd *Art Anc. Near East* iii. 106 The casting of this object is the earliest example known of the *cire perdue* process.

cirrate (si·rēⁱt), *a.* *Zool.* [ad. L. *cirrātus.*] = Cirrated *a.*

1826 Kirby & Spence *Entomol.* IV. xlvi. 321. Cirrate (*Cirratæ*). When the joints terminate in a pair of curling hairy branches resembling tendrils.

cirrolite: see Cirrholite.

cis-. Add: **1. Cis-Atlantic** (earlier and later U.S. examples); **Cis-Atla·ntically** *adv.*, on this side of the Atlantic. Also, **Cis-bo·rder, Cis-Ca·spian, Cis-Danu·bian, Cis-I·ndus, Cis-lunar** (later examples), **Cis-Ura·lian,** adjs.

The first element *cis-* is frequently written with lower-case initial.

1785 Jefferson *Notes Virginia* vi. 118, I only mean to suggest a doubt,..whether nature has enlisted herself as a Cis- or Trans-Atlantic partisan? **1805** 'C. Caustic' *Democracy Unv.* III. 72 Our poetical and rhetorical flourishes from Cis-Atlantic sources. **1932** *Scrutiny* I. 72 *The New Republic* exerts itself to perform these offices as no cis-Atlantic periodical. **1909** *Daily Chron.* 23 June 3/2 In any case, it is not cis-Atlantically true. **1901** *19th Cent.* Apr. 711 Raised in fixed proportion from the transborder and cis-border clans. **1888** *Encycl. Brit.* XXIV. 5/2 The Ciscaspian steppes. **1860** De Quincey *Cæsars* in *Wks.* X. 168 Homes had been obtained by Trans-Danubian barbarians upon the Cis-Danubian territory of Rome. **1907** *New Reformer* Dec. 325 The..Cis-Indus Aryan world. **1958** *Gloss. Aero-Space Terms* 7/2 *Cislunar*, of or pertaining to space between the earth and the orbit of the moon, or to a sphere of space centered on the earth with a radius equal to the distance between the earth and the moon. **1959** *New Scientist* 5 Mar. 499/2 What matters about this week's *Pioneer IV* Moonshot are the Geiger counters that it has been carrying for the further exploration of the atomic radiation in cis-lunar space. **1886** *Encycl. Brit.* XXI. 79/2 The Permians, or Cis-Uralian Finns.

3. *Chem.* (Also without hyphen as a quasi-adj.) Designating a compound in which two atoms or groups are situated on the same side of some plane of symmetry passing through the compound; *cis-trans isomerism*, a form of isomerism in which in one isomer two identical groups are on the same side of the plane of a double bond whereas in the other isomer they are on opposite sides; so *cis-trans isomer*; (see also quot. 1964[2]).

[**1888** A. Baeyer in *Liebig's Annalen der Chemie* CCXLV. 137 Ich schlage dafür die Bezeichnung 'cis' und 'trans' vor, welche andeuten sollen, dass der eine Bestandtheil diesseits und der andere jenseits der Ebene des Ringes befindlich ist.] **1888** *Jrnl. Chem. Soc.* LIV. 1074 When treated with zinc-dust and glacial acetic acid, it yields the cis-hexahydro-acid. **1889** *Ibid.* LVI. 1177 The cis-acid and its salts are distinguished from the corresponding cistrans compounds by the greater solubility of the former. **1907** J. B. Cohen *Org. Chem.* 110 It exhibits lateral symmetry, and is known as the plane-symmetric, or, more commonly, as the cis or malenoid form. **1937** *Nature* 3 July 25/1 One would expect the *cis* form of ethylene iodide to be its normal condition. **1946** *Ann. Reg. 1945* 351 A new vitamin A has been described..as a *cis-trans* isomer of the Standard A at the double bond nearest the hydroxyl group. **1957** [see *cistron]. **1964** N. G. Clark *Mod. Org. Chem.* iii. 40 Maleic acid, which has the like groups on the same side of the double bond, is called the '*cis*' isomer. *Ibid.* 41 This method of naming the isomers has led to the alternative name '*cis-trans*' isomerism for geometrical isomerism. *Ibid.* v. 84 *trans*-But-2-ene..*cis*-But-2-ene. **1964** D. Michie in G. H. Haggis et al. *Introd. Molecular Biol.* x. 265 Two kinds of genotype identical in all respects except one: whether the two mutant sites are located in the same or in opposite chromosomes. In the first case the mutant sites are said to be in the *trans*..arrangement, and in the second case in the *cis*..arrangement. When a phenotypic difference is observed between the *cis* heterozygote..and the *trans* heterozygote,..this is known as the '*cis-trans*' effect.

cisalpine, *a.* Add: **a.** *spec.* (freq. with capital initial). Of, pertaining to, or designating the Gallican Church movement; = Gallican *a.* 1 b.

1792 R. Banister *Let.* 30 Apr. in *Ushaw Mag.* (1904) June 138 A new club..at London of 14 lay gentlemen..: it takes a new name and is to be called the Cisalpine club. **1793** in B. Ward *Dawn Cath. Revival* (1909) II. xxiv. 98, I should recommend the erection of a good Grammar School, such as would supersede the necessity of a Cisalpine one. **1909** B. Ward in *Dublin Rev.* Jan. 65 The Cisalpine party went to great lengths in their anti-episcopal—and even anti-Papal—declamations.

b. *sb.* (Examples.)

1793 in B. Ward *Dawn Cath. Revival* (1909) II. xxi. 46 He has given himself to the Cisalpines. **1825** C. Butler *Bk. Rom. Cath. Ch.* x. 123 The Cisalpines affirm, that in spirituals the pope is subject, in doctrine and discipline, to the church. **1942** R. Knox *In Soft Garments* xv. 119 The old dispute between Cisalpines and Ultramontanes..has passed into the region of forgotten controversies.

cisco. Delete *U.S.* and substitute for def.: The popular name of several species of North American whitefish belonging to the genus *Leucichthys*, esp. *L. artedi*. (Add examples.)

1848 in Bartlett *Dict. Amer.* **1872** *Fur, Fin & Feather* 200 Ciscoes are caught mostly on shore, although many take them..from boats. **1880** [see Siscoe]. **1884** D. S. Jordan in Goode *Nat. Hist. Aquatic Anim.* 541 The Mooneye or 'Cisco' of Lake Michigan..is the smallest of our whitefish, rarely weighing over half a pound. **1903** *N.Y. Sun* 1 Nov., In the same warehouse are trout, ciscoes, bluefish, weakfish, porgies, and so on. **1965** A. J. McClane *Standard Fishing Encycl.* 188/1 All ciscoes are coldwater fishes, occurring from New England through the Great Lakes into Canada.

ciscoette, ciscovet, varr. Siscowet.

1847 C. Lanman *Summer in Wildern.* 159 A fish called ciscovet, is unquestionably of the trout genus, but much more delicious. *Ibid.* 161 The white-fish, ciscovet, and lake trout. **1854** —— *Adv. Wilds N. Amer.* xxxiv. 253 We cannot leave Mackinaw without making a passing allusion to the fish whose Indian name is *ciscovet*. It is a handsome fish, unquestionably of the trout family. **1902** *Jrnl. Amer. Folk-lore* 243 Ciscoette. A name of the lake herring. Apparently a derivative, with French diminutive suffix, from *Cisco* (q.v.), but rather a corruption of *Siskowit*.

‖**ciseaux** (sizo). [Fr. (pl.): see Scissors.] In full *pas (de) ciseaux*, a jump in which a dancer opens his legs wide apart in the air, like the opening of a pair of scissors; = Sissonne.

1892 E. Scott *Dancing* vi. 77 There are several ways of executing the *pas de ciseaux* or *pas de ciseaux*. **1913** C. d'Albert *Dancing* 41 *Ciseaux* or *Sissonne*...This scissor step consists of two temps or syllables. **1952** Kersley & Sinclair *Dict. Ballet Terms* 25 The male solo in 'Les Rendezvous' begins with the dancer rising on his toes, followed by a *pas ciseaux*.

‖**ciseleur** (sizlœ̄r). [Fr., f. *ciseler* to carve, chase.] One who carves metals, a chaser.

1862 *Catal. Spec. Exhib. S. Kens.* No. 826 The famous ciseleur Poutière. **1935** *Burlington Mag.* May 245/2 The gifted *ciseleur*, Thomire.

cissie, var. *cissy *sb.* and *a.*

cissing (si·siŋ), *vbl. sb.* [Etym. unkn.] **a.** The preliminary operation in graining of moistening the wood with beer and rubbing it over with whiting in order that the colour may adhere. **b.** (See quots. 1877, 1885, and 1961.)

1875 E. A. Davidson *House-Painting* 106 In order that the graining colours which are mixed with beer may adhere to the ground which has been painted in oil, it is necessary that the surface should be prepared. This is done by passing over it with a sponge moistened with beer and rubbed with whiting; when this is dry, the distemper colour will work freely over the oil, and will adhere to it. This process is called *cissing*. **1877** C. H. Savory *Paper Hanger* 111 Vandyke Brown used in beer must be rubbed on with a tool..till there is no 'cissing', or curdling of the colour. **1885** *Spon's Mechanic's Own Bk.* 432 To prevent a graining coat from 'cissing' at a water-colour overgraining coat, that is repelling the water by antagonism of the oil. **1961** J. H. Goodier *Dict. Painting & Decorating* 55 *Cissing*. This is a term applied when a coat of paint, varnish or water colour refuses to form a continuous film, recedes from the surface, collects in beads and leaves the surface partially exposed.

cissy (si·si), *sb.* and *a.* *colloq.* Also **ci·ssie.** = *sissy 2.

1915 T. L. Golden *Let.* 30 May in *Lett. from Front* (Canadian Bank of Commerce) (1920) I. 19 Ready to look down upon the Britisher as a good-for-nothing lady-like cissy. **1926** *Chambers's Jrnl.* Dec. 790/2 Of all the milk-and-water out-and-out Cissies this settlement beat the band! **1930** *Ibid.* Apr. 228 It takes more than a cissy Englishman who couldn't find the hole in a doughnut to break trail across 'the Barrens'. **1938** E. Bowen *Death of Heart* II. iii. 223, I can't stand those cissie pullovers. **1944** L. Glassop *We were Rats* I. i. 5 It's me name, but it's too cissy, so I.. picks up 'Mick'. **1958** P. Kemp *No Colours or Crest* iii. 40 Reason told me..that I was being a fool and a cissy. **1963** *Times* 16 Feb. 12/4 The reason why some workers did not use protective equipment and clothing, which would have prevented many accidents, was that they regarded such things as 'cissy'.

cist. Also pronounced (kist). Cf. Kist *sb.*[1] 4 b.

cistella (siste·lă). *Bot.* [mod.L., dim. of *cista* box.] = Cistula 3.

1832 Lindley *Introd. Bot.* I. iii. 205 *Cystula*, or *Cistella*, a round closed apothecium, filled with sporules, adhering to filaments which are arranged like rays in a common centre. **1900** B. D. Jackson *Gloss. Bot. Terms* 54/1 *Cistella*, *Cistula*, ..used for the apothecia of lichens, which, globular at first, burst at maturity.

Cistercianism (sistə̄·ɹʃăniz'm). [f. Cistercian + -ism.] The religious system of the Cistercians; the Cistercian spirit.

1895 *Athenæum* 6 Apr. 440/1 Welsh Cistercianism. **1898** S. Evans *Holy Graal* 11 The predominance of what may be called Cistercianism in the Church.

cisternal, *a.* Delete † and add: *spec.* applied to the cisterns of the human or animal body (see Cistern 4).

1926 *Jrnl. Amer. Med. Assoc.* 2 Oct. 1103/1 Doses of.. the more promising oils were injected into cats by cisternal puncture. **1961** *Lancet* 5 Aug. 295/2 These cisternal pouches were distended by fluid. **1962** *Science Survey* XI. 187 The cisternal membrane to which the particles adhere.

cistophorus (sistǫ·fōrə̆s). Pl. **-phori.** [L., a. Gr. κιστοφόρος, f. κίστη Chest *sb.*[1] + -φορος bearing, -phorous.] A Greek coin bearing the impress of a sacred *cista* or chest. Hence **cistophoric** (sistofǫ·rik) *a.*, of or pertaining to this.

1848 Akerman *Introd. Study Anc. & Mod. Coins* ii. 33 The time of the first issue of cistophori is not known; but they were in circulation in the year of Rome 564 (B.C. 190). **1887** B. V. Head *Hist. Numorum* p. lxii, In Asia Minor the chief silver coinage consisted of the famous Cistophori. *Ibid.* 446 Cistophori with ΑΔΡΑ in monogram and autonomous drachms (45 grs.) of the cistophoric standard. *Ibid.* 587 Many of these tetradrachms are countermarked with a bow and bow-case crossed, accompanied by the names, for the most part, of cistophoric mints. **1940** *Oxoniensia* V. 145 Among the Roman Imperial coins may be noted a 'cistophorus' of Augustus.

cistron (si·strǫn). *Biol.* [f. *cis-trans* (see *cis- 3).] A genetic unit of function based on the *cis-trans* test.

1957 S. Benzer in McElroy & Glass *Chem. Basis Hered.* 71 A functional unit can be defined genetically..by means of the elegant *cis-trans* comparison...For the *trans* test, both mutant genomes are inserted in the same cell... It turns out that a group of non-complementary mutants falls within a limited segment of the genetic map. Such a map segment, corresponding to a function which is unitary as defined by the *cis-trans* test applied to the heterocaryon, will be referred to as a 'cistron'. **1959** *New Scientist* 20 Aug. 246/3 The unit of physiological function is the 'cistron' comprising a number (tens or hundreds) of smaller units of mutational change—the 'mutons'. **1964** D. Michie in G. H. Haggis et al. *Introd. Molecular Biol.* x. 266 The chromosomal segment within which the cis-trans effect operates defines a unit of function. Seymour Benzer has termed this unit the *cistron*.

cit. Add: **2.** *pl.* Civilian clothes; 'civvies'. *U.S. Mil. slang.*

1829 in O. E. Wood *West Point Scrap Bk.* (1871) 47 My uniform I've taken off, My 'cits' I've just put on. **1895** C. King *Fort Frayne* vi. 86 Will was..vaguely longing to get out and air his new 'cits'. **1907** *Chicago Tribune* 8 May 2 They were in full dress uniform. Later they were [illegible]ed by Maj. Judson of the engineers in 'cits'.

citable, *a.* (Earlier example.)

1712 Shaftesbury *Plastics* xiii, in *Second Char.* (1914) 135 Very remarkable, citable, much talked of,..or bantered by his club.

citadel. Add: **2. c.** A building in which Salvation Army meetings are held.

1889 *War Cry* 14 Sept. 12/1 The citadel will be of brick and approached by short flights of steps leading to four spacious entrances. **1906** *Salvation Army Year Bk.* 54/2 A Citadel Company is formed..for purchasing or leasing land and erecting Salvation Army Citadels thereon. **1964** B. Watson *100 Yr's. War* I. vii. 101 The citadel will have cost anything between £10,000 and £50,000.

citation. Add: **5.** Mention in an official dispatch. *U.S.*

1918 in E. S. Farrow *Dict. Mil. Terms.* **1930** *Publishers' Weekly* 5 July 11 [He] holds the Congressional Medal of Honor, nine citations and the highest decorations of the Allied Governments.

citification: see *cityfication.

citify, *v.* Var. Cityfy *v.*, usu. as **ci·tified** *ppl. a.*

1862 *Temple Bar* VI. 190 The country population is.. becoming citified. **1865** *Atlantic Monthly* XV. 501 I'd take an' citify my English. **1876** 'Mark Twain' *Tom Sawyer* i. 7 He had a citified air about him that ate into Tom's vitals. **1883** [see Cityfy *v.*]. **1916** 'D. Grayson' *Adv. in Contentment* ii, I was 'citified', Horace said; and 'citified' with us here in the country is nearly the limit of invective. **1919** W. Hutchinson *Doctor in War* i. 4 The highly citified and alleged 'neurotic' Cockney. **1945** F. Stark *East is West* xxvi. 185 Taken by his son..who looked like a slighter and citified version of his father.

citizenly (si·tizenli), *a.* [f. Citizen *sb.* + -ly[1].] Pertaining to or characteristic of a citizen.

1867 W. James *Let.* 19 Nov. (1920) I. 116 To be satisfied with the domestic and citizenly virtues. **1899** *Remin. King of Roumania* 111 The citizenly virtues which appertain to such a quasi-republican form of State. **1905** *Academy* 16 Sept. 944/1 That the negroes' welcome be in the measure of their citizenly quality.

citral[2] (si·trăl). *Chem.* [f. Citr(us + Al(dehyde.] **a.** = *geranial. **b.** In the *cis*- form, neral.

1891 *Jrnl. Chem. Soc.* LX. 540 'Citral' is the technical term applied to an ethereal oil which is contained to the extent of 6–8 per cent. in lemon oil. **1907** *Chem. Abstr.* 1352 Process of manufacturing violet perfumes from citral, consisting in acting with acids upon the condensation products..from the action of citral upon halogen acetic acid esters. **1939** *Thorpe's Dict. Appl. Chem.* III. 183/2 Citral a. (Geranial.)..Citral b. (Neral.)...Citral is the most important natural acyclic aldehyde of the terpene series. **1953** [see *geranial]. **1964** N. G. Clark *Mod. Org. Chem.* x. 174 Citral, an olefinic aldehyde, occurring in lemons and other citrus fruit.

citrange (si·trăndʒ). [f. Citr(us + Or)ange.] A hybrid fruit produced by crossing the hardy trifoliate orange, *Citrus trifoliata*, with the common sweet orange.

1904 H. J. Webber in *Yearbk. U.S. Dept. Agric.* 227 The Citrange, a new group of Citrus plants...It..becomes necessary to refer these hybrids to a new group of citrus fruits, and it is proposed to call them 'citranges'. **1932** *Proc. 6th Internat. Congr. Genetics* II. 171 Unexpected variations of first generation hybrids (called citranges) between the common orange..and the trifoliate orange.

citrate (si·trēⁱt), *v.* [f. the sb.] *trans.* To treat with a citrate, esp. with sodium citrate. So **ci·trated** *ppl. a.*

1908 *Practitioner* Apr. 462 Citrated horse's blood. **1920** *Discovery* Apr. 105/1 It was found the citrated blood could be kept unchanged for twenty-four hours or more. **1946** *Nature* 6 July 24/1 The fluid was aspirated by means of a hypodermic syringe from the uterine swellings..and was citrated. **1962** *Lancet* 22 Dec. 1298/1 Citrated blood was collected by clean venepuncture.

citrazinic (sitrăzi·nik), *a.* *Chem.* [f. Citr- + Az)ote + -ine[5] + -ic.] Designating a crystalline hydroxy acid, formed by condensation from an amide of citric acid.

1885 *Jrnl. Chem. Soc.* XLVIII. I. 139 Citrazinic acid, $C_4H_5NO_4$: citramide is treated with 4–5 parts of sulphuric acid (70–75 per cent). *Ibid.*, Citrazinic acid is very sparingly soluble even in boiling water. **1887** *Ibid.* LI. 406 Citrazinamide, like citrazinic acid, is a very stable compound.

citrin (si·trin). [G. (L. Armentano et al. 1936, in *Deutsche Med. Wochenschr.* 14 Aug. 1362/1), f. Citr(us + -in[1].] A water-soluble flavonoid found in citrus fruits, formerly considered to be a source of vitamin P.

1936 *Nature* CXXXVIII. 7 Nov. 798/1 Twenty-one of the animals received 1 mgm. 'Citrin' daily, this substance being the crystalline flavone fraction of lemon juice. **1944** *Hackh's Chem. Dict.* (ed. 3) 205/1 *Citrin*, (1) Cucurbo citrin. A glucoside from watermelon seeds; used as a hypotensor. (2) Hesperidin. **1955** *Sci. News Let.* 26 Feb. 141/1 The chemicals are now called bioflavonoids. One of them..has been named vitamin P or Citrin.

citrinin (sitrī·nin). [f. mod.L. *Penicillium citrinum* (see quot. 1931) + -in[1].] An antibiotic extracted from various fungi, esp. *Penicillium citrinum*.

1931 Hetherington & Raistrick in *Phil. Trans. R. Soc.* B. CCXX. 270 This substance to which we propose to give the name 'citrinin' appears to be characteristic of *P*[*enicillium*] *citrinum Thom.* **1963** V. I. Bilai *Antibiotic-Producing Microsc. Fungi* 21 Citrinin is produced in the culture fluid when the fungus is grown on Czapek's or Czapek-Dox medium or on modifications of these media.

citrometer (sitrǫ·mītəɹ). [f. Citro- + -meter.] An instrument used for determining the specific gravity of lemon juice in the preparation of citric acid.

1875 *Jrnl. Chem. Soc.* XXVIII. 930 Lemon-juice..is boiled down in copper vessels over an open fire till it is supposed to mark when cold 60° on the citrometer. **1879** *Spon's Encycl. Industr. Arts* 49 An instrument termed a citrometer is sometimes used to measure the amount of citric acid contained in the juice.

citron. Add: **3.** Also as *adj.* = *citron-coloured*.

1924 A. D. Sedgwick *Little French Girl* I. v. 40 The old tapestry, grey and green and citron.

7. *citron-scented* adj.

1896 T. W. Sanders *Encyl. Gardening* (ed. 2) 68 Citron-scented Orchid (*Odontoglossum citrosmum*). **1924** E. Sitwell *Sleeping Beauty* xvii. 67 The Pleiades' citron-scented poems.

citronella (sitrǒne·lă). [mod.L.] A fragrant Asian grass, *Cymbopogon nardus*, which yields an oil much used in perfumery. Also, the oil itself; in full, *citronella oil*.

1858 Simmonds *Dict. Trade*, *Citronella*, an essential oil obtained from the grass *Andropogon Citratum*, and chiefly imported from Ceylon. **1882** Bentley *Man. Bot.* (ed. 4) 735 Cetronella Oil or Oil of Citronelle. **1903** *Westm. Gaz.* 30 Nov. 2/1 One and a half million pounds of citronella oil are yearly exported. **1911** *Jrnl. Chem. Soc.* C. I. 476 The citronella grasses of Ceylon are of four kinds. **1959** *Times Rev. Industry* Oct. 96/3 Harvesters tor citronella.

Hence **citrone·llic** *a.*; **citrone·llal,** a substance found in citronella, eucalyptus, and lemon oils; **citrone·llol,** an alcohol found in rose, pelargonium, and geranium oils.

1872 *Jrnl. Chem. Soc.* XXV. 8 The refraction equivalent of citronellol is..79·3 or 79·8, instead of the theoretical 73·7. **1890** *Ibid.* LVIII. I. 231 The author names it *citronellic aldehyde*, and considers it to be β-methyl-δ-isobutylallyl-acetaldehyde. **1891** *Ibid.* LX. I. 285 Citronellic aldehyde must be regarded as a homologue of acraldehyde. **1893** *Ibid.* LXX. I. 382 Compounds of the Citronellal Series. **1906** *Ibid.* XC. I. 442 The citronella oil..contained 32·7 per cent. of geraniol and 55·3 per cent. of citronellal. **1961** *B.S.I. News* Sept. 13/2 The determination of primary alcohols, including determination of citronellol in the presence of geraniol. **1964** *Economist* 26 Dec. 1448/1 Rose-like chemicals like geraniol, citronellal and phenyl ethyl ether.

citronelle (sitrǒne·l). [Fr., ad. mod.L. *citronella* (see prec.).] = *citronella; also, a beverage made from this.

1861 Bentley *Man. Bot.* 699 Cetronelle or Citronelle oil, is the produce of this or of an allied species of *Andropogon*. **1882** [see *citronella]. **1899** *Scribner's Mag.* XXV. 50/1 As gay with flowers as a girl's hat, and as fragrant of sweet-olive, citronelle, and heliotrope as her garments. **1921** *Chambers's Jrnl.* Sept. 604/2, I tasted a new drink called 'citronelle', a kind of tea made from a grass which the French cultivate in their gardens.

citrous (si·trəs), *a.* [f. Citr(us + -ous.] Of or belonging to the genus *Citrus*. Cf. *citrus 1.

1901 *Westm. Gaz.* 13 June 2/3 Citrous fruits. **1911** C. E. W. Bean '*Dreadnought*' *of Darling* xl. 355 Land..which suits vegetables, lucerne, dairying, citrous fruits. **1924** *Glasgow Herald* 9 July 5 The citrous fruit industry.

citrovorum (sitrō^u·vǒrǒm). [f. mod.L. *Leuconostoc citrovorum*, f. Citr- + *-vorum*, neut. sing. of L. *-vorus* -vorous.] Esp. in *citrovorum factor*, the active form of folic acid; folinic acid.

1948 Sauberlich & Baumann in *Jrnl. Biol. Chem.* Oct. 168 The '*citrovorum* unit' was designated as the amount of substance required per ml. to give one-half maximal acid production in 72 hours by *Leuconostoc citrovorum* 8081. *Ibid.* 171 The chick assay..proved to be one of the most potent concentrates of the *citrovorum* factor thus far encountered. **1957** J. H. Burn *Princ. Therapeutics* xxi. 180 Folic acid is converted in the body into folinic acid, which is also known as the citrovorum factor.

citrullin(e (sitrɒ·lin, -əin). *Chem.* [f. med.L. *citrull*(*us* (see Citrul) + -ine[5].] An amino acid, α-amino-δ-carbamido-valeric acid, found in watermelons, and in living organisms concerned in the formation of urea from ornithine.

1930 M. Wada in *Proc. Imperial Acad.* (*Japan*) VI. 15 The author..isolated a new amino acid and named it 'Citrullin'. **1934** *Brit. Pat.* 421,407 1/1 Delta-carbamido-alpha benzoylornithin..may be further converted into deltacarbamidoornithin (citrullin). **1956** *Nature* 4 Feb. 238/1 β-alanine and/or citrulline were detected in both specimens of urine. **1963** R. P. Dales *Annelids* V. 100 Feeding with either arginine or citrulline augmented urea production.

citrus. 1. Add: (Earlier and later examples.) Freq. *attrib.* or as *adj.*

1825 *Greenhouse Comp.* I. 81 The Citrus tribe is also propagated by cuttings and layers. **1904** *Westm. Gaz.* 18 Nov. 2/3 'Forbidden Fruit'—this name is sometimes loosely given to this whole group of citrus varieties. **1906** *Macm. Mag.* Aug. 794 Fruit-trees, especially the citrus and stone varieties. **1953** *Manch. Guardian Weekly* 13 Aug. 7 The 'succulent citrus crops.

city. Add: **5. b.** (Earlier examples.)

1621 in W. Notestein et al. *Commons Debates* (1935) II. 447 Though money be wanting in the country yet it is in the City. *Ibid.* VI. 321 They of the Citty to lay the riches downe.

9. city centre (see *centre *sb.* 6 a); **city desk** *U.S.*, the department in a city newspaper office which deals with local news; **city-editor** (examples); also *U.S.*, the editor who superintends the collection and classification of local news; **city father,** (*b*) *pl.* (orig. *U.S.*) [cf. Father *sb.* 10], those responsible for the administration of a city; also *transf.*; **city gent** *colloq.* = **city-man* (b); **city hall** chiefly *N. Amer.*, the chief municipal offices of a city; hence, the municipal officers collectively; **city limits** *N. Amer.*, the boundaries of a city; **city-lot** *U.S.*, a piece of ground lying within city bounds; **city-man,** (*b*) (also *City-man*) one engaged in 'the City' (sense 5) or in mercantile pursuits; **city page,** the page of a newspaper which deals with financial and business matters; **city-republic,** a city constituted as a republic; **city slicker** orig. *U.S.* [cf. Slick *a.* in Dict. and Suppl.], a smart and plausible rogue, of a kind usu. found in cities; a smartly dressed or sophisticated city-dweller; **city-state,** a city which is also an independent sovereign state.

1904 G. B. Shaw *Common Sense Munic. Trading* viii. 73 In city centres..the [housing] schemes are commercially hopeless. **1957** J. Braine *Room at Top* xxx. 244 The maze of side-streets off the city centre. **1958** *Daily Express* 11 Mar. 1/4 A city centre café. **1903** J. Ralph *Making of Journalist* xiii. 153 Around the walls are the desks..the city desks, the suburban desk,..and that of the managing editor. **1969** H. Nielsen *Darkest Hour* vi. 69 We had a *great* paper, Charley,..and largely because you were on the city desk. **1834** *Chambers's Jrnl.* III. 390/1 Employed upon each are an editor; a sub-editor (in some also a city editor). **1858** Simmonds *Dict. Trade*, *City Editor*, the employee of a daily or weekly London journal, whose special duty it it to report upon the prices of public securities, the state of the money market, and other matters of commerce and finance. **1870** A. Maverick *Raymond & N.Y. Press* 326 The City Editor, directs the work performed by the reporters, whose duty is to gather all the local intelligence of the day. **1902** Eliz. Banks *Autobiogr. Newsp. Girl* ii. 9 You might try it on the city editor; but, mind you, I can't promise that he'll print it. **1969** 'D. Rutherford' *Gilt-Edged Cockpit* x. 167 Often comment by the City Editors precipitated a run on shares. **1845** *St. Louis Reveille* 19 Jan. 2/4 Allow me to call the attention of our 'City Fathers' to the importance of erecting public hydrants on the Levee. **1962** *Listener* 10 May 820/2 Kline, de Kooning, and the other present City Fathers of American painting. **1785** City gent [see Gent *sb.*]. **1844** Thackeray *New Monthly Mag.* July 421 Once..I heard a city '*gent*' speak..slightingly of a glass of very excellent brown sherry. **1941** V. Woolf *Between Acts* 60 Not a dapper city gent, but a cricketer. **1675** E. Andross in J. Easton *Narr. Causes Indian War* (1858) 106 There was at the City Hall an Order of the last Gen[er]all Court of Assizes. **1754** in E. Singleton *Social N.Y. under Georges* (1902) 339 The books..are placed for the present, by leave of the Corporation, in their Library room in the City Hall. **1833** E. T. Coke *Subaltern's Furlough* iv, Of the public buildings, the City Hall, containing the supreme court, mayor's court, and various public offices,..is the most remarkable. **1922** Joyce *Ulysses* 319 What did those tinkers in the cityhall..decide about the Irish language? **1930** J. Dos Passos *42nd Parallel* v. 388 He slept on a bench in front of the City Hall. **1967** *Listener* 22 June 811/2 A month passes and City Hall makes an announcement. **1841** *Louisville Jrnl.* 6 May 2/1 There are in the city proper but ten watchmen, and one in Portland, making in all eleven for the city limits. **1926** *Daily Colonist* (Victoria, B.C.) 16 July 1/5 One house was destroyed and the flames were within a short distance of the city limits. **1968** *Globe & Mail* (Toronto) 17 Feb. 50/9 (Advt.), Stouffville is situated 12 miles north of metropolitan Toronto city limits. **1683** W. Penn *Let. to Free Soc. Traders*, The city-lot [is conveniently posted] for a dock. **1829** R. C. Sands *Writings* (1834) II. 169 The latter sat pensive and silent, while Miss Violet discoursed..about western lands and city lots. **1844** Lee & Frost *Oregon* vi. 79 The..gentleman..was selling off small city lots at one hundred dollars a piece. *a* **1861** T. Winthrop *John Brent* (1883) i. 6 They had been speculating in..city lots. **1836–9** City man [in Dict., sense 7]. **1875** T. W. Higginson *Eng. Statesmen* III. xvii. 350 He had made his mark in the mercantile world as a thoroughly representative City-man. **1966** 'C. Aird' *Religious Body* ix. 55 One of the largest *private* companies in the country...They're always getting write-ups in the City pages suggesting they will be going public but they never do. **1967** P. M. Hubbard *Custom of Country* (1969) iv. 51 There was some sports news, a city page of surprising sophistication and a sprinkling of..foreign news. **1853** J. S. Mill in *Edin. Rev.* XCVIII. 439 Her people obstinately rejected the merging of the single city-republic in any larger unity. **1933** A. N. Whitehead *Adv. Ideas* iii. 37 The Governments of the provincial regions..and City-Republics. **1924** *Cosmopolitan* Nov. 104/2 You reckon I'm a goin' tew give that city slicker back his option money? **1938** Wodehouse *Summer Moonshine* viii. 92 It was a brooch or a sunburst from this city slicker in the background. **1953** Xan Fielding *Stronghold* iii. 45 The two city-slickers were travelling on business. **1959**

Manch. Guardian 13 July 5/7 He was dressed like a city slicker, pointed brown shoes, trilby..cut-away waistcoat. **1893** W. Warde Fowler *(title)* The City-State of the Greeks and Romans. **1906** *Daily Chron.* 25 Nov. 2/5 Athens, a city-state that could always give rise to great men. **1968** G. Jones *Hist. Vikings* III. iv. 254 Northwards they had their city state of Novgorod and their base at Staraja Ladoga.

cityfication (si:tifikē¹·ʃən). Also **citification.** [f. Cityfy *v.*: see -fication.] The process or result of being cityfied.
1906 Saleeby in *Daily Chron.* 23 Oct. 4/4 Keen observers are assured that civilisation is well named—it means cityfication. **1923** *Glasgow Herald* 26 May 4 American rural life has little tradition...Its citification is no great surprise.

cityfy, *v.* Add: (Earlier and later examples.) See also *citify *v.*
1828 *Yankee & Boston Lit. Gaz.* 16 July I. 227 There is a deal more comfort in playing with a country lass..than with a cityfied girl. **1849** F. A. Buck *Let.* 31 Oct. (1930) 53 People look much more dressed here than up the river: more cityfied. **1937** 'C. Caudwell' *Illusion & Reality* 86 It is now a more cityfied, sensible, less romantic and picturesque court.

city-scape, cityscape (si·tiskē¹p). [f. City + Scape *sb.*³] A view of a city; city scenery; the layout of a city.
1856 Thackeray *Let.* 28 Dec. in H. Ritchie *Lett.* (1924) v. 82 A fairyland of frozen land, river and city-scape. **1952** D. Hoffman *Paul Bunyan* v. 133 His attempts to describe cityscapes in terms of neo-Wordsworthian natural beauty. **1958** *Listener* 9 Oct. 557/2 These mechanistic landscapes and cityscapes. **1959** *Archit. Rev.* CXXV. 373/3 A preference for landscape over cityscape. **1960** *20th Cent.* Apr. 375 *Private Fires* is a cityscape, a rich Dickensian evocation of a decaying, badly-blitzed suburb.

cityward, *adv.* **b.** (Examples.)
1873 J. H. Beadle *Undevel. West* 804 The cityward bluff of the Colorado rises almost perpendicular for thirty feet or more from the water's edge. **1898** *Daily News* 28 Mar. 8/7 The cityward and homeward lines of traffic. **1900** *Westm. Gaz.* 25 July 2/3 Blocks in cityward traffic.

citywards, *adv.* Add: Also *attrib.* or as *adj.*
1908 *Daily Chron.* 9 Jan. 3/2 We quite expect that..we shall meet him in the City-wards train.

civet, *sb.*³ Delete † *Obs.* and add later examples. So *civet de lièvre*, civet of hare.
1841 Thackeray in *Fraser's Mag.* June 721/1 The place was famous for civet de lievre. **1964** *Harper's Bazaar* Sept. 121/1 Recipes for poaching snipe..and for a *civet de lièvre. Ibid.* 125/1 Other pieces [of venison] should be made into *civets*, with plenty of red wine.

civic, *a.* Add: **2. c.** *civic centre*: the headquarters of a municipality; an area in which the principal public buildings of a municipality are grouped together, often in a unified architectural scheme.
1909 H. I. Triggs *Town Planning* i. 12 Every effort should be made..to make the open spaces more extensive the farther they are removed from the civic centre. *Ibid.* iv. 183 Enlarged plans of two civic centres. **1934** A. Huxley *Beyond Mexique Bay* 270 Give to London all the town planning, all the civic centres, all the garden suburbs that the ingenuity of man can devise. **1944** J. S. Huxley *Living in Revol.* ii. 24 For the majority of boroughs today, the only civic centre is the town hall. **1960** *Times* 6 Jan. 6/3 A new building on the civic centre site.
4. *Comb.*, as **civic-minded** *a.*, inclined to concern oneself with civic affairs; public-spirited; so *civic-mindedness*.
1942 M. McCarthy *Company she Keeps* (1943) vi. 203 His methodical habits, his civic-mindedness, his devout sense of what was proper. **1947** N. Cardus *Autobiogr.* 207 He..was not civic-minded, and could never be trusted at a garden-party. **1967** J. Wainwright *Talent for Murder* 166 Alfred Merriam began to wish he had not been so civic-minded.

civics. Add: (Earlier and later examples.) orig. *U.S.* (see quot. 1889). Also *attrib.*
1886 *Citizen* (Boston) Dec. 5 Shall Civics be taught in the public schools? **1889** *Appleton's Cycl. Amer. Biogr.* VI. 318 In 1885 he [*sc.* Henry R. Waite] founded the American institute of civics...He was the first to employ the term 'civics' to designate those branches of science that pertain to the elevation of citizenship. **1903** *Daily Chron.* 26 Oct. 3/4 No one now doubts that civics in moderation should be part of a child's training. **1916** *Amer. City* Feb. 154/1 Girls in charge of the civics class. **1963** *Ann. Reg. 1962* 31 Night-school civics classes.

civil, *a.* Add: **3. c.** *civil disobedience*: the refusal to obey the laws, tax demands, etc., of a government as part of a political campaign.
Thoreau's essay (see quot. 1866) was entitled 'Resistance to Civil Government' when first published in *Æsthetic Papers*, ed. E. Peabody (1849).
1866 Thoreau *Yankee in Canada* 123 *(title)* Civil disobedience. **1922** *Ann. Reg. 1921* 275 Mr. Gandhi announced that he reserved the right to continue..preparations for civil disobedience. **1930** *Economist* 22 Feb. 416/1 A widespread campaign of civil disobedience, such as was envisaged at the recent Lahore Congress, might affect the collection of taxes. **1959** *Observer* 8 Mar. 1/3 The campaign [in Nyasaland] was to begin with civil disobedience.

5. b. *civil rights* [Right *sb.*¹ 9]: the rights of each citizen to liberty, equality, etc.; *spec.* in the U.S., the rights of Negroes as citizens. Cf. Civil law, †right.
1721 *Boston Rec.* 154 That they Indeavor to maintain all our Civel Rights and Properties against any Incrochments upon them. **1866** A. Johnson in W. L. Fleming *Hist. Reconstruction* (1906) I. 225 The Civil Rights bill was more enormous than the other. **1869** Trollope *Phineas Finn* I. xxv. 211 A London policeman..feels himself to be an instrument for the moment of despotic power as opposed to civil rights. **1874** in W. L. Fleming *Hist. Reconstruction* (1907) II. 201 The mere mention of Civil (negro) Rights has almost destroyed the public schools and colleges in some of the Southern States. **1964** *Ann. Reg. 1963* IV. i. 179 President Kennedy..requested from Congress sweeping civil rights legislation to protect the rights of Negroes in voting, education, and employment. **1970** K. Giles *Death in Church* iv. 90 You probably don't have all the nosy civil rights committees.
14. (Later examples.)
1922 *Encycl. Brit.* XXXI. 1147/2 He [*sc.* Northcliffe] did ..accept the chairmanship of the Civil Aerial Transport Committee, on the establishment of that body in 1917. **1930** Lady Troubridge & A. Marshall *Ld. Montagu* 234 The enormous natural strides made by civil aviation the moment that the brake of war-time conditions was released.
b. *civil defence*: the organization and training of civilians for the preservation of lives and property during and after air raids or other enemy action; also *attrib.* Hence *civil defender*, one taking an active part in civil defence.
1939 *Times Weekly* 6 Sept. 17/4 It is impossible now to say where air raid precautions end and where civil defence begins. **1940** *Ann. Reg. 1939* 61 Some 20,000 men and women, representing every branch of Civil Defence. **1940** *Economist* 17 Feb. 285/1 These civil defenders should be sufficient in numbers and adequate in training and organisation to furnish protection for the civilian population. **1942** *Ann. Reg. 1941* 290 A.R.P. services would henceforth be known as Civil Defence Services. **1962** *Guardian* 18 July 5/6 Our Civil Defenders are once again in trouble. *Ibid.* 5/7 The Civil Defence movement is nothing more than a jolly social organisation.
B. 5. *to do the civil*: to do the civil thing (see quot. 1840 in sense A. 12); to act politely *to* (a person). *colloq.*
1838 Dickens *O. Twist* II. xxii. 33 There's Bill Sikes in the passage with nobody to do the civil to him. **1854** Rossetti *Let.* 19 Nov. (1965) I. 233, I should like her to do the civil by sending them. **1858** Trollope *Doctor Thorne* xxviii, I haven't exactly popped to her yet; but I have been doing the civil. **1877** Reade *Woman-Hater* II. xix. 150 He would go to the station and telegraph him; and by that means would do the civil and meet Miss Gale.
D. *civil-spoke(n)* (see Spoken *ppl. a.* note).
1842 Lover *Handy Andy* xiv, He was a very 'close' man, as well as a 'civil-spoken'. **1880** E. H. Dering *Freville Chase* xi, There never was a kinder or civiller spoken gentleman nowhere. **1910** J. Farnol *Broad Highway* II. ii. 202 'What might you be wanting with a peaceable, civil-spoke cove like me?' he inquired.

civilianization (sivi:liănəizē¹·ʃən). [f. Civilianiz(e *v.* + -ation.] The transfer of members of the armed forces to civilian status or their replacement by civilians.
1946 *Daily Tel.* 15 Nov. 1/3 This deterioration has become more marked as the 'civilianisation' of the administration has proceeded. **1956** *Rep. Employment Nat. Service Men* 15 The word 'civilianization'..may mean the increased use of civilians in establishments which must remain partly military. It may also mean handing over to civilian management whole operations and thereby dispensing with military personnel altogether.

civilianize, *v.* Delete *nonce-wd.* and add later examples. Also, to confer civilian status upon; to replace by civilian workers. Hence **civi·lianized** *ppl. a.*
1948 *Lancs. Daily Post* 20 July 1/8 A Ministry of Agriculture camp for 'civilianized' Germans. **1955** *Daily Tel.* 9 Dec. 6/4 When Servicemen in a Government establishment are replaced by civilian workers their jobs are said to be 'civilianised'. **1969** *Sci. Jrnl.* Nov. 6/2 In the early 1960s there was a strong movement to civilianize M[icrobiological] R[esearch] E[stablishment].

civvy (si·vi), *sb.* and *a.* *Services' colloq.* **A.** *sb.* **a.** Short for Civilian 3 b. **b.** *pl.* Civilian clothes, mufti.
1889 Barrère & Leland *Dict. Slang*, *Civvies* (army), a suit of *civvies, i.e.* civilian's clothes. **1915** 'Bartimeus' *Tall Ship* v. 100 Nosey wore a rusty suit of 'civvies'. **1915** W. H. L. Watson *Adv. Despatch Rider* ii. 23 In the course of my riding that day I knocked down 'a civvy' in Dour. **1927** E. Thompson *These Men thy Friends* 82 Any civvy who has had the honour of being associated with them. **1946** *Daily Tel.* 15 May 1 Young men exchange their uniforms for 'civvies'.
B. *adj.* Short for Civilian 5; *civvy street*: civilian life.
1915 'Bartimeus' *Tall Ship* v. 101 He still wore his 'civvy' suit. **1919** *Punch* 15 Jan. 38 'And I suppose you will be demobilised?'..'Oh, no, Mum. You see, I was a soldier in civvy life.' **1943** *Penguin New Writing* XVII. 18 When I get back to civvy street I'll never moan about my job again. **1945** *Daily Express* 19 Apr., Civvy cigarettes are dearer now. **1959** J. Braine *Vodi* viii. 119 Dick was in splendid shape, sampling every delight Civvy Street had to offer.

clabber, *v.* Add: (Later examples.) Also *trans.* Hence **cla·bbered** *ppl.a.*
1921 R. L. Alsaker *Eating for Health* I. iii. 47 Clabbered milk and buttermilk are easily digested. **1938** M. K. Rawlings *Yearling* xix. 222 She needed rain-water, too, to clabber the milk. The milk turned rankly sour in the heat but would not clabber. **1968** *Washington Post* 5 July A 18/4 Then it had to be left to clabber—or curdle—before it could be churned.

clack-clack, *sb.* Also **clack-clack-clack, clack-clacking.** [Imitative; cf. Clack *v.*¹] A repeated clacking noise.
1849 [see Clack *sb.* 6]. **1870** 'Mark Twain' *Curious Dream* (1872) 5 Presently up the street I heard a bony clack-clacking. *Ibid.* 6, I heard another one coming—for I recognized his clack-clack. **1916** 'Boyd Cable' *Action Front* 249 The clack-clack-clack of a machine-gun at close range. **1939** C. S. Forester *Capt. Hornblower* I. xii. 137 This clack-clack-clack of women's tongues..would drive him mad. **1949** F. Sargeson *I saw in my Dream* 134 You heard the clack-clack sound as she [*sc.* a bitch] stood there drinking.

clacker (klæ·kəɹ), *v.* *dial.* and *U.S.* [Cf. Clack *v.*¹] *intr.* and *trans.* = Clack *v.*¹
1851 S. Judd *Margaret* (ed. 2) III. 345 Mounting a rock she harangued the people, or, rather, clackered her own merit. **1896** Chanter *Witch of Withyford* ii, She heard them all come clackering into the yard.

clacket (klæ·ket), *v.* Delete † *Obs.* and add later examples.
1880 F. M. Peard *Mother Molly* iii. 33 The hen..flew from side to side of the road, clucking and clackiting as though she were already being killed. **1904** Gallichan *Fishing Spain* 144 Felicia stalked out, her wooden shoes clacketing upon the road.

Clactonian (klæktō^u·niăn), *a.* *Archæol.* [ad. F. *Clactonien* (Breuil *Exposé de Titres* (1929) 21), f. *Clacton* (see below) + -ian.] Of, pertaining to, or designating the Lower Palæolithic culture represented by the flint implements found at Clacton, Essex. Also *ellipt.*
1932 *Antiquity* VI. 189 The Lower Palaeolithic Period.. including the Clactonian and some of the Levalloisean cultures. **1935** *Nature* 21 Sept. 475/1 Many of the finished implements have inclined platforms very like the Clactonian.

clad, *ppl. a.* Add: **1. c.** Bearing a cladding (*cladding *vbl. sb.* 2).
1862 [see *armour-clad *ppl. a*]. **1940** J. D. Jevons *Metall. Deep Drawing* viii. 262 The statements..do *not* apply to 'clad' sheet, that is to Duralumin sheet having a thin coating of pure aluminium. **1959** *Jrnl. Iron & Steel Inst.* CXCI. 295/2 Steel clad with Ni on one side was hot-rolled to 10 mm.

clad, *v.* Restrict '*Obs.* or *arch.*' to senses in Dict. and add: **2.** To apply a cladding to; to cover with a cladding (*cladding *vbl. sb.* 2). So **cla·dded** *ppl. a.*
1939 *Jrnl. R. Aeronaut. Soc.* XLIII. 559 The resultant drop in strength of the cladded article is very small. **1947** *Archit. Jrnl.* 16 Oct., The necessity of cladding a very light frame with a clumsy intractable material which is neither a panel nor a wall permitting a clear expression of the structure. **1964** *Times Rev. Industry* Feb. 75/3 'Dished' 22-gauge stainless steel panels are used to clad London Transport Executive's new Elephant & Castle substation.

cladding (klæ·diŋ), *vbl. sb.* [f. Clad *v.*]
1. *U.S.* Clothing; clothes. *rare.*
1885 *N. Y. Tribune* 27 Mar. 5/5 There were countless lords and ladies of high degree in claddings of past centuries.
2. A coating or covering applied to the surface of an object, a building, etc.; the application of such a covering. Also *attrib.*
1939 *Jrnl. R. Aeronaut. Soc.* XLIII. 559 The usual material used for cladding in Germany..is an Al-Mg-Si alloy. **1946** *Archit. Rev.* CI. 66/1 The external wall cladding should be changed from a horizontal to a vertical slab. **1952** *Ibid.* CXII. 385 Almost any surface application to a structural base could be called a cladding. **1953** *Times* 1 Jan. 14/1 The demand for roofing and cladding in asbestos cement was well maintained. **1964** *Economist* 11 Apr. 99/1 Using stainless steel as cladding material.

clade (klē¹d), *sb.*² *Biol.* [f. Gr. κλάδος branch.] A group of organisms that have evolved from a common ancestor. Hence **cladi·stic** *a.*, of or pertaining to a clade or clades.
1957 J. S. Huxley in *Nature* 7 Sept. 455/1 Cladogenesis results in the formation of delimitable monophyletic units, which may be called clades. **1960** *Proc. Zool. Soc.* CXXXV. 3 Closeness of relationship in terms of phyletic lines can be called cladistic. **1963** Sokal & Sneath *Princ. Numerical Taxon.* v. 102 Should taxa in orthodox taxonomy be in general monophyletic groups (clades) or phenetic groups? *Ibid.* viii. 220 Cladistic relationship refers to the paths of the ancestral lineages and therefore describes the sequence of branching of the ancestral lines.

cladoceran (klădǫ·sĕrăn). *Zool.* [f. mod.L. *Cladocera* (P. A. Latreille in Cuvier's *Règne animal* (1829) IV. 151), f. Gr. κλάδος branch + κέρας horn, with reference to the branched antennæ.] A member of a sub-order of small

branchiopod Crustacea, commonly known as water-fleas.
1909 W. T. Calman in R. Lankester *Treat. Zool.* VII. III. 42 In the aberrant Cladoceran *Leptodora* the oesophagus is of great length. **1932** J. S. Huxley *Prob. Rel. Growth* 258 Form-changes in the Cladoceran *Macrothrix rosea.* **1953** Scott & Fisher *Thousand Geese* ix. 99, I made a plankton net out of a handkerchief..and caught great numbers of capepods and cladocerans.

cladogenesis (klæ:dodʒe·nesis). *Biol.* [ad. G. *kladogenese* (B. Rensch *Neuere Probleme der Abstammungslehre* (1947) vi. 95), f. Gr. κλάδος branch + γένεσις development.] A process of adaptive evolution that leads to the development of a greater variety of animals or plants. Hence **cladogene·tic** *a.*; **cladogene·tically** *adv.*
1953 [see *anagenesis]. **1957** [see *clade *sb.*²]. **1957** *Nature* 7 Sept. 455/2 The resultant of cladogenetic and anagenetic processes in combination. *Ibid.*, Cladogenetically man constitutes only the single family Hominidae. **1959** *New Statesman* 7 Nov. 632/3 Man is the only successful biological species which has remained as a single interbreeding group, not radiating out in 'cladogenesis' into thousands of mutually infertile species.

cladome (klæ·dō^u m). [f. Gr. κλάδος branch + *-ome.] The branching arms of a rhabdus sponge spicule collectively.
1887 W. J. Sollas in *Encycl. Brit.* XXII. 417/1 The secondary rays are the arms or cladi, collectively the head or cladome of the spicule. **1888** —— in *Challenger Rep.* XXV. p. lvi, The 'cladome' which consists of the three cladi (rays, teeth, arms). **1940** L. H. Hyman *Invertebrates* I. vi. 299 The cladome may consist of a simple or scalloped disk instead of cladi.

cladosporium (klæ:dospō^ə·riŭm). [mod.L. (H. F. Link in *Linnæus Spec. Plant.* (ed. 4, 1824) VI. I. 39), f. Gr. κλάδος branch + σπορά seed.] A member of the genus so named of hyphomycetous fungi, causing various plant diseases; also, infestation with this fungus.
1887 *Gardeners' Chron.* 29 Oct. 532/2 The hyphæ and spores in these spots were of a beautiful violet tint—quite unlike any Cladosporium with which I am acquainted. **1902** *Jrnl. R. Hort. Soc.* XXVII. 27 The earliest form is a *Cladosporium*, with slender unbranched septate threads of a pale olive and rather long. **1923** W. F. Bewley *Dis. Glasshouse Plants* iii. 98 *(heading)* Cladosporium Leaf Spot. **1959** *Times* 20 June 9/5 If this [ventilation] is not done cladosporium and other troubles may well follow.

cladus (klǣi·dǫs). Pl. **cladi.** [mod.L., a. Gr. κλάδος branch.] One of the secondary arms or branches of a ramose sponge spicule. Hence **cla·dal, cla·dose** *adjs.*
1887 Sollas in *Encycl. Brit.* XXII. 416/2 The spicular rays..usually pointed (oxeate) at the ends..are also frequently..branched (cladose). **1888** —— in *Challenger Rep.* XXV. p. lvi, The proximal or undivided part of a cladus is ..distinguished as the protocladus, the distal division, in the case of a dichotomous cladus as the deuterocladus. *Ibid.*, A straight line drawn from the cladal origin to the end of a cladus is taken as its length. **1940** [see *cladome].

claim, *sb.* Add: **3.** (Earlier examples.) Phr. *to jump a claim*: see Jump *v.* 9 b.
1792 *Deb. Congress* 1036 The following claims of citizens, reserved by the deed of cession. **1817** S. R. Brown *Western Gaz.* 13 The eastern boundary of the Creek claims is pine land. **1851** *Austral. & N.Z. Gaz.* 15 Nov. 459/1 Many enter upon a claim, dig six or eight feet, and finding little or nothing, leave that spot in disgust.
5. In the language of Christian Science, the imaginary disturbance which 'claims' to be an ailment.
1898 *Westm. Gaz.* 26 Mar. 3/1 Ailments were referred to as 'claims'. For instance, the other night this lady's eldest daughter came to her with the information that her sister had got a 'claim' of swollen glands...Soon after this, the mother herself was attacked by a claim of influenza. **1907** 'Mark Twain' *Chr. Sci.* 57 The second witness testifies that the Science banished 'an old organic trouble', which the doctor and the surgeon had been nursing with drugs and the knife for seven years. He calls it his 'claim'.
6. *attrib.* and *Comb.*, as (sense 1) *claims adjuster, clerk, inspector* (in *Dict. Occup. Terms*, 1921), *claim-form*; (sense 3) *claim-agent, -holder, -holding, -pegging, -shack, -shanty, -stake*; **claim-acreage,** acreage consisting of miners' claims; **claim-jumper** (orig. *U.S.*), one who 'jumps' (see Jump *v.* 9 b) another's claim; so *claim-jumping* vbl. sb. and ppl. adj.; also *transf.*
1903 *Daily Chron.* 1 July 7/6 Ten times larger than the Kimberley mine in claim acreage. **1860** *Wilkes' Spirit of Times* 10 Mar. 14/3 (Weingarten), Claim agent and broker. **1903** *N.Y. Even. Post* 6 Oct. 6 Our pension policy has furnished an enormous incentive to claims agents and Congressmen. **1963** J. Prescot *Case for Hearing* ix. 139 We handled his insurance claim in the office, and I saw the inventory annexed to the claim-form. **1862** W. S. Ebey *Diary* 4 July in *Pacif. N.W. Quart.* (1943) XXXIV. 41 Every one of the claim holders whipping his man & retaining his claim. **1890** 'R. Boldrewood' *Miner's Right* xiv, All the claim holders had closed round as far as could be seen in every direction. **1902** *Westm. Gaz.* 26 Nov. 11/3 Their total claim-holding having been increased. **1839** in *Annals of Iowa* 3rd Ser. V. 430 To the Claim Jumpers..This is therefore to forbid all persons entering or trespassing upon said (land) claim. **1858** *Critic* 14 Apr. (Farmer), The claim-jumper laughed as though he enjoyed it hugely. **1945** M. James *Cherokee Strip* 4 You know what a claim-jumper a cowbird is—laying its eggs in other birds' nests and all. **1846** *Oregon Spectator* 29 Oct. 2/3 In regard to 'claim-jumping'. **1910** J. Hart *Vigilante Girl* xiv. 186 He has no sentimental prejudices about claim-jumping miners. **1933** P. Fleming *Brazilian Adv.* II. iii. 105 He spoke with some bitterness of evangelistic claim-jumping, and of the base tricks employed by the rival missions in the race for converts. **1892** *Pall Mall Gaz.* 23 Feb. 7/1 Claim-pegging in the Transvaal. **1916** 'B. M. Bower' *Phantom Herd* ix. 149 We learned our little lessons well before the road-making claim shacks for ourselves. **1860** *Harper's Mag.* Aug. 299/2 Claim-stakes and claim-shanties speck the road.

claim, *v.* **2. c.** (Examples.)
1864 C. W. Norton *Army Let.* (1903) 204, I don't claim that they fought well, only as well as they could. **1876** *Troy Morning Whig* 27 May, The man accused denies his being intoxicated, which she denied and claimed that he was in that condition himself. **1887** *Troy Daily Times* 8 Jan., John Westerman..procured a [police] warrant for the arrest of his son..who he claims has threatened to kill him. **1922** H. Dunn in *World's Work* July 119/2 Reitz Bey..was hopping mad at an attempt which he claimed that the British had made to kidnap him.

claimless, *a.* Add: **b.** Without a claim (see Claim *sb.* 3).
1927 *Blackw. Mag.* Oct. 460/1, I found myself claimless, and not likely to obtain another piece of ground. *Ibid.* 461/2 Many claimless 'diggers'.

clairaudient, *a.* Add to def.: Also, pertaining to or of the nature of clairaudience. (Further examples.) Hence **clairau·diently** *adv.*, in a clairaudient manner.
1909 *Westm. Gaz.* 10 Sept. 4/1 'Julia's' decisions are given in two ways—sometimes clairaudiently, and sometimes by automatic handwriting. **1926** Conan Doyle *Hist. Spiritualism* I. i. 7 A clairaudient warning that he must be more careful in that respect. **1930** B. M. Lester *Towards Hereafter* iii. 35 There came to him a voice clairaudiently. *Ibid.* v. 65 She was receiving an important clairaudient message.

‖ **clair-de-lune** (klɛ̄r də lün). [Fr., = moonlight.] The colour of moonlight, a soft white or pale blue-grey, the shade which appears in the glaze of certain Chinese porcelains. Also *attrib.*
1877 *Young Ladies' Jrnl.* XIV. 842/3 Bonnets embroidered with clare [*sic*] de lune tone. **1906** L. L. Hobson *Porcelain* 15 Lovely blue glazes, ranging from deep sky-blue to the palest *clair de lune*. **1954** E. Jenkins *Tortoise & Hare* ii. 21 Jars in clair-de-lune and blanc-de-Chine.

claire (klɛ̄r). [Fr.] A pond or basin (usu. artificial) of sea water for the cultivation of oysters.
1901 *Science* 18 Jan. 107 The experiments are made in an enclosed area of sea water known as a claire..a definite measure of success in demonstrating the possibility of fattening oysters to marketable condition by this means. **1936** Russell & Yonge *Sea Shore* ix. 301 The oysters..are laid on the bottom of these claires.

clam, *sb.*¹ Add: **1. c.** phr. *as happy as a clam*, etc.: very well pleased, quite contented. *U.S. colloq.*
1834 *Harvardiana* I. 121 That peculiar degree of satisfaction, usually denoted by the phrase 'as happy as a clam'. **1841** J. Slick' *High Life N.Y.* I. xi. 179 They seemed as happy as clams in high water. **1848** Bartlett *Dict. Amer.* s.v., 'As happy as a clam at high water,' is a very common expression in those parts of the coast of New England where clams are found. **1871** J. H. Beadle *Undevel. West* 799 A thousand or more negroes thronged the streets 'happy as clams at high tide'. **1940** H. W. Thompson *Body, Boots & Britches* xiii. 294 Happy as a clam in high water.
4. *clam-digger* (examples), *-digging*; **clam-bait** *U.S.*, clams used as bait; **clam-chowder** (earlier *U.S.* examples), also a picnic or feast at which this is the principal dish; **clam-cracker** (see quot.); **clam-fry,** a meal of fried clams; **clam-shell,** also, a bucket or grab on a dredger, excavator, etc., shaped like a clam-shell; freq. *attrib.*; **clam-tongs,** tongs used for taking clams.
1838 *Mass. Stat.* cxxiv, An Act to regulate the Inspection of Clam Bait. **1872** Schele de Vere *Americanisms* 69 When salted for the fisheries it takes the name of *clam-bait*. **1822** W. Kitchiner *Cook's Oracle* 365 Clam chowder. **1898** Hamblen *Gen. Manager's Story* 131 The engineer had a clam chowder. **1909** Webster, *Clamcracker*, a stingray (*Dasyatis centrura*) of the Atlantic coast. It feeds largely on shellfish. **1855** Clam-digger [see *clam-fisher* in Dict.]. **1881** *Amer. Naturalist* XV. 364 The ancient clam-diggers whose kitchen-middens are met with in many places on the Alameda. **1903** *Atlantic Monthly* Sept. 327 He became far and away the best among the clam-diggers. **1838** *Knickerbocker* XI. 207 Sam's trade was clam-digging. **1860** S. Warner *Say & Seal* xvi. 146 Faith..was certainly 'spry' in getting ready for the clam-digging. **1905** *N.Y. Even. Post* 10 June 6 'Fish dinners' and clam fries are to be had at any number of eating-houses at the river's mouth. *a* **1877** Knight *Dict. Mech.* I. 746/1 The 'clam-shell' dredge.. consists of a pair of scoops which are hinged to an axis and close upon the load. **1877** *Encycl. Brit.* VII. 465/1 Dipper and Clam-shell dredges. *Ibid.*, The clam-shell is a box made of two similar pieces of wrought iron hinged together at one end. **1930** *Engineering* 15 Aug. 195/3 Many types of dredges had been tried such as..clam-shell, suction and other forms. **1963** *Ibid.* 30 Aug. 269/2 The eight hydraulic rams operating the clamshells. **1883** Goode *Fish. Industries U.S.* 52 Clam-tongs are occasionally employed for catching crabs.

clam, *v.*¹ Add: **1.** (Earlier examples.)
1636 [see *clamming *vbl. sb.*²]. **1676** in *Mass. Hist. Soc. Coll.* (1825) 4th Ser. I. 71 They were clamming..at Cowessett. **1716** B. Church *Entertaining Passages Philip's War* 29, Some [Indians] catching Eels & Flat-fish in the water, some Clamming.

clam, *v.*² Add: **2.** *slang* (chiefly *U.S.*). To shut up; be silent. Also *to clam up on*: to refuse to talk to (someone). Hence **clammed** *ppl. a.*
1916 H. L. Wilson *Somewhere in Red Gap* vi. 237 When I ask for details he just clams up. **1926** J. Black *You Can't Win* viii. 97 Smiler had continually drummed it into me never to answer any questions in case we were arrested. 'Just clam up, kid.' **1942** R. Chandler *High Window* (1943) xx. 143, I had the legal right to stay clammed up. **1950** R. M. Nash *House of Shade* vi. 68, I didn't mean to pry, but there's no need..to clam up on me.

clamant, *a.* **1. a.** (Later example.)
1963 *Times* 11 Jan. 3/5 A deliriously clamant crowd rightly thought it a try inevitable.

clamatorial (klæmătō·riăl), *a.* *Ornith.* [f. mod.L. *Clamatores*: see -ial.] Of or pertaining to the Clamatores, a sub-order of the Passeriformes.
1872 Coues *N. Amer. Birds* 109 They are at once distinguished from these clamatorial birds by the ocrine character of the tarsi. **1875** [illegible] *Ibid.* 56 Struthious, Psittacine, and Clamatorial Birds.

clamatory (klæ·mătəri), *a.* [ad. L. *clāmātōrius*: see -ory¹.] Clamorous.
1900 H. G. Wells *Love & Mr. Lewisham* i. 7 A rousing day, a clamatory insistent day, a veritable herald of summer.

clamb, variant of Clam *sb.*¹ (sense 2).
1898 *Daily Chron.* 14 Oct. 10/7 Stitchers (Female leather) wanted, used to the clambs. **1909** *Ibid.* 2 June 9/5 Stitchers, female, leather, used to clambs.

clambake (klæ·mbē¹k). *U.S.* Also **clam bake, clam-bake.** [Clam *sb.*² 4.] **1. a.** (See Clam *sb.*² 4.)
1835 *Vade Mecum* (Phila.) 5 Sept. 2/1 A Clam Bake—Our curiosity has been gratified, as to the nature of the festival understood by this term. **1848–60**, etc. [see Clam *sb.* 4]. **1945** [illegible] They caged chowder question Occasionally left out of clambakes or College reunions. **1962** *Times* 8 Apr. 7/1 Dig razor clams for an outdoor clambake.
b. *transf.* Applied to other gatherings, esp. one that is loud or lively; an enjoyable time; *spec.* in *Jazz*, a jam-session.
1937 *Amer. Speech* XII. 46/1 *Clambake*, same as jam session. **1938** *Manch. Guardian Weekly* 2 Sept. 18/3 Swing musicians..blaze away at a real 'clambake' [free-style swing playing]. **1945** *Chicago Daily News* 31 Mar. 4/1 The big United Nations clambake at San Francisco will be operated like a [illegible] convention. **1946** Mezzrow & Wolfe *Really the Blues* viii. 131 Life was going to be one long clambake. **1952** B. Ulanov *Hist. Jazz* (1958) xvii. 206 The Dixie outfit he called the Clambake Seven—cheerful adoption of the term used by jazz musicians both in annoyed disparagement of a poor jam session and in warm approbation of a good one.
c. An inadequate or unsuccessful broadcast programme. *slang.*
1937 in *Times Dispatch* (Richmond, Va.) 12 Dec. 5 4/4. **1942** [illegible] three commentators..will talk simultaneously and confuse the radio audience. This is what we mean by a 'clam bake'.

clamber, *sb.* (Earlier examples.)
1818 Keats *Let.* 9 Apr. (1958) I. 261, I have had a fine Clamber over the rocks. **1850** Mrs. Gaskell *Let.* 25 Aug. (1966) 123 A drive along the level road..then a regular climber up a steep lane.

cla·mming, *vbl. sb.*² [f. Clam *v.*¹ + -ing¹.] The gathering of clams. Also *attrib.*
1636 *Dorchester Town Rec.* 5 July in *Fourth Rep. Rec. Commissioners Boston* (1880) 19 Provided they leave stiles and gates for persons and cattle, when persons are disposed to travell or drive Cattle or swine that way to Clamming. **1838** *Knickerbocker* XI. 207 Sam evidently looked upon clamming as an important and mysterious thing. **1883** Goode *Fish. Industries U.S.* 46 All along the southern shore of the sound are prolific clamming grounds. **1969** *Outing* (U.S.) XXV. 408/1 Swimming, driving, clamming, fishing..they are equally at home in or out of the water.

clamp, *sb.*¹ Add: **1. d.** spec. in *Electr.* (see quot. 1947); *clamp circuit*, one in which the positive or negative limits of a waveform are adjusted and maintained. Cf. *clamping *vbl. sb.* b.
1947 L. J. Haworth in L. N. Ridenour *Radar System Engin.* xiii. 503 The name 'clamp' is applied to..electronic switches which, when closed, hold or clamp two circuit points together. **1954** B. Y. Mills in E. G. Bowen *Radar* (ed. 2) xiv. 450 Circuits which can be used for inserting a DC level..are called 'clamps'. **1954** Zworykin & Morton *Television* (ed. 2) xiii. 535 Figure 13.38 shows a 'clamp circuit' employed for fixing the black level in an image

orthicon camera. **1958** *Chambers's Techn. Dict.* (ed. 2) 966/2 *Clamp*, valve circuit in which a waveform is adjusted and maintained at a definite level when recurring after intervals. **1963** R. G. Middleton *Elem. Transistor Techn.* iv. 88 A clamp circuit..maintains the peaks of the output waveform at a preset voltage level; this level might be zero or some other value. Either the positive or the negative peaks of the output voltage can be clamped to the chosen level.
5. clamp-connection, a connecting swelling between adjoining cells of the hyphæ of certain fungi.
1887 Garnsey & Balfour tr. *de Bary's Compar. Morphology & Biol. Fungi* I. 2 The clamp-connections..occur only on hyphae with transverse segmentation, and chiefly in the Basidiomycetes...A clamp of this kind when fully formed is..a nearly semicircular protuberance like a short branch which springs from one cell..and is closely applied to the lateral wall of the adjoining cell. **1931** A. H. R. Buller *Res. Fungi* IV. ii. ii. 271 Each cell-division is accompanied by the formation of a clamp-connexion between the two daughter cells.

clamp, *v.*¹ Add: **1. b.** spec. in *Electr.* To adjust and maintain the positive or negative limits of a waveform. Cf. *clamping *vbl. sb.* b.
1954 B. Y. Mills in E. G. Bowen *Radar* (ed. 2) xiv. 451 It is desired to clamp the intermediate portion of a waveform, or if the waveform can change polarity, it becomes necessary to use a double-ended clamp. **1963** [see *clamp *sb.*¹ 1 d].
3. *to clamp down*: to press down on; *transf.* to take strong measures; to become (more) strict; to put a stop to (an undesirable activity, etc.). Const. *on*.
1924 C. E. Mulford *Rustlers' Valley* vi. 66 Matt..clamped his own left arm down on the other's right. **1941** H. Flanagan in W. Kozlenko *100 Non-Royalty Radio Plays* 69/1 There isn't any money. They clamped down on the expenses. **1945** W. Manish *Died in West* 25 We've clamped down on it for six months. **1950** *Economist* 19 July 170 The government clamped down firmly on all political agitation. **1963** *Listener* 7 Mar. 432/3 The complaints about crime that..clamped down brutally on criminals, exploited servants.
b. Of cloud, fog, etc.: to descend very low, so as to prevent flying, etc.
1943 C. Lewis *Pathfinders* 262 The only thing that worried him was the weather. It was clamping down. They couldn't fly through it without instruments. **1958** 'N. Shute' *Rainbow & Rose* i. 3 It's clamped down over the mountains. **1959** *Punch* 13 May 648/3 Have a nice bit of constructive under-cover jobs so that when the weather clamps down you needn't waste time.

clamp, *v.*² (Earlier example.)
1743 Ellis *Mod. Husb.* Sept. v. 38 Upon this he changed his Bed and clamped his Dung.

clamp-down (klæ·mpdaun). An act of clamping down (see *clamp *v.*¹ 3).
1940 Harrison & Madge *War begins at Home* xi. 287 An almost complete clamp-down on juvenile, adolescent and adult education. **1960** *News Chron.* 30 Apr. 1/1 Leading motor dealers are not alarmed at the credit clamp-down. **1966** *Guardian* 25 Oct. 13/2 The clamp-down will come at any sign of..hooliganism.

clamping, *vbl. sb.* Add: **b.** *Electr.* A method of adjusting and maintaining the positive or negative limits of a waveform. Freq. *attrib.*, as *clamping circuit*. Cf. *clamp *v.*¹ 1 b.
1944 J. W. Serrwin in *Mass. Inst. Techn., Radiation Lab. Rep.* 487 Clamping tubes. **1947** D. G. Fink *Radar Engin.* xi. 374 A clamping circuit of two triode tubes is connected conductively to the grid of each deflection amplifier tube. The clamping tubes are connected in series. **1947** L. J. Haworth in L. N. Ridenour *Radar System Engin.* xiii. 504 During each positive excursion of A the resistance extracts a small charge, which is replaced by the diode on the negative excursion. The clamping can be done at the positive extreme by reversing the diode. **1961** G. Millerson *Technique Telev. Production* iii. 50 Electronic clamping circuits reduce the wide variation in picture tones that an unsteady datum produces. **1962** Simpson & Richards *Junction Transistors* xvi. 385 Clamping may be defined quite generally as the stabilization of the output voltage or current of a transistor switch by a sudden change of impedance, produced by an additional circuit device at one of the terminals.

clan, *sb.* Add: **4.** *clan-brother, -sister.*
1932 C. G. & B. Z. Seligman *Pagan Tribes Nilotic Sudan* ii. 54 The relationship to the yuen is dependent on that between sons [clan-brothers], who stand in such close relationship that they have access to each other's wives. **1936** *Jrnl. R. Anthrop. Inst.* 134 The guest is not assimilated to a clan-brother. **1942** D. C. Talayesva *Sun Chief* i. 32 Many of my mother's sisters and clan sisters were present. **1961** R. Firth *Elem. Social Organiz.* vi. 208 Failing to sacrifice, sleeping with a clan-sister...actions of such a kind may be difficult to deal with effectively in human terms.

clang, *sb.* Add: **3.** Also, in *Psychology*, with reference to the acoustic sensation of musical sounds and their analysis.
1895 *Amer. Jrnl. Psychol.* VII. 82 *Klang*, clang [not chord]. *Ibid.*, *Klangeinheit*, clang unity or simplicity. **1901** E. B. Titchener *Exper. Psychol.* I. ii. 15 Clang Analysis: Overtones. **1918** M. D. Eder tr. *Jung's Word-Association* 476 The 'clang' associations..in our experiment..were in the form of rhymes. Jung and Riklin consider that clangs appear when there is a decline of attention. **1924** R. M. Ogden tr. *Koffka's Growth of Mind* iv. 216 So-called clang-analysis, or hearing out the partial tones of a clang, has often been advanced as a striking demonstration of the existence of unnoticed sensations. **1946** *Brit. Jrnl. Psychol.* XXXVII. 35 Useful information was obtained from the formal characteristics of the associative reactions rather than from their content—i.e. whether the association was too close (repetitions, clang reactions, etc.).

clanger (klæ·ŋəɹ). *slang.* [f. Clang *v.*¹ + -er¹.] A mistake, esp. one that attracts attention; a social *faux pas*. Phr. *to drop a clanger*, to make such a mistake, to 'slip up'.
1948 *Lawrentian* (St. Lawrence Coll., Ramsgate) Easter 8 Clanger. A peculiarly heavy brick...Prefects lead the way in dropping these objects. **1958** *New Statesman* 29 Nov. 752/1 Mr Macmillan is the kind of Premier who enjoys covering up for any Cabinet colleague who drops a clanger. **1959** *Daily Mail* 5 Feb. 3/4 He won medals—and a carnation. He said: 'I have boobed dreadfully, old boy. Apparently a carnation with gongs is a terrible clanger.'

clanism. Delete *nonce-wd.*, and add earlier example. Also **clannism.**
1831 *Examiner* 187/1 A report, that the Chancellor..were to [illegible] his own judges in Bankruptcy from the King's Bench Bar, and principally from the Northern Circuit. This we do not believe; we expect to see him..free from the clanism which is implied in this rumour. **1924** *Blackw. Mag.* Dec. 785/1 The clannism of the village.

clankety (klæ·ŋkĕti). Onomatopœic extension of Clank as in *clankety-clank* (cf. *clickety-clack* s.v. Click *sb.*¹ 8).
1865 *Harper's Mag.* Sept. 555 The clankety-clank that plate-armour makes. [illegible] The pumps still utter their mournful clankety-clank-clank. [illegible] A traction engine..with the clankety-clank and the earth-shaking passage that suggests an earth-quake.

clankingly (klæ·ŋkiŋli), *adv.* [f. Clanking *ppl. a.* + -ly².] With a clanking sound. So **cla·nkingness.**
1891 *Eng. Illustr. Mag.* July 703 The clankingness that only the telegraph boy puts into the bell. **1894** *Cornhill Mag.* May 485 The road was strewn with new stones, and the quadruped trotted clankingly.

clanny (klæ·ni), *a.* [f. Clan *sb.* + -y¹.] = Clannish *a.*
1905 *Temple Bar Mag.* Dec. 498 A wholesome immigration of clanny Kelts. **1953** G. Greene *Living Room* i. i, Catholics are much too clanny.

clans-, = clan's, genitive of Clan *sb.*, as in Clansman; so *clansfolk, clanswoman*.
1896 F. H. Giddings *Princ. Sociol.* 173 All clansmen and clanswomen had the right to vote in electing or deposing the officers of the clan. **1902** *Daily Chron.* 27 Sept. 3/1 The clansfolk were defeated at the Pass of Brander. **1949** E. E. Evans-Pritchard in M. Fortes *Social Struct.* 85 Marriage is not permitted between clansfolk. *Ibid.* [illegible]

clap, *sb.*¹ Add: **4. b.** *clap-clap*: the sharp sound, continually repeated, made by horses' hooves, applause, or the like.
1900 *Daily Chron.* 22 Oct. 4/7 The sharp clap-clap of the single horse in a hansom. **1922** Joyce *Ulysses* 275 Bravo! Clapclap. **1949** W. Deeping *Roper's Row* iv. 38 [illegible] **1961** C. S. Lewis *Problem of Pain* x. 134 The clap-clap of water against the boat's side.
15. *clap-stick*, (*b*) *pl.* = *clapper *sb.*¹ 3 c; *clap-table* = *console-table*.
1946 *Electronic Engin.* XVIII. 309 The professional 'clap-sticks' should be used to mark the beginning of a 'take'. *c* **1702** C. Fiennes *Journeys* (1947) iv. x. 358 A clap table under the large looking-glass between the windows. **1955** H. Fastnedge *Eng. Furnit. Styles* vi. 127 The introduction of the console, originally called a 'clap' table, dated from the first years of the [18th] century.

clap, *sb.*³ Add: (Later examples.)
1919 *Wine, Women & War* (1926) 98 Physical exam—crabs, cooties and clap. Scared to death, but O.K. **1967** A. Diment *Dolly Dolly Spy* vii. 94 Rocky Kilmarry is about as good for you as a dose of clap.

clap, *v.*¹ **2.** Delete † *Obs.* and add later example.
1938 G. Greene *Brighton Rock* II. i. 189 'We better say something, Pinkie. You know she might be..and that old gal, we sympathise with what you feel.' 'What are you clapping about?'

clapboard, *v.* Add: (Earlier examples.) Hence **cla·pboarded** *ppl. a.*
1637 in *New Plymouth Col. Rec.* (1861) XII. 26 The house to be..clap boarded within..and a partition to be made of clap board. **1721** in *Harvard College Rec.* (1925) II. 457 Voted, That..the sd College be Clap-boarded on the East Side. **1835** *Southern Lit. Messenger* II. 53 [illegible]

cla·pboarding. *U.S.* [f. Clapboard.] **a.** The putting on of clapboards. **b.** A series of clapboards on a building.
1650 in *Dedham Rec.* (1892) III. 32 It is agreed concerning Clapboarding of houses yt it shalbe at liberty untill mid-sumer day nett. **1907** *Boston Even. Post* 15 June, The lightning fell in a perpendicular direction, ripping the clap-boarding and plastering as it fell. **1872** *Rep. Vermont Board Agric.* I. 309 Paper..can be advantageously used between floors and between boarding and clapboarding on the walls. **1905** H. Garland *Tyranny of Dark* 59 The paint was blistering and peeling from the clap-boarding on the sunny side of the main building.

Clapham (klæ·pəm). [The name of a district of south-west London.] *Clapham Sect*: see Sect *sb.*¹ 4 b. Phr. *the man on the Clapham omnibus*: the ordinary or average man; the 'man in the street'.
[**1857** *Jrnl. Soc. Arts* 1 May 348/1 So thoroughly has the tedious traffic of the streets become ground into the true Londoner's nature, that..your dog-collar'd occupant of the knife-board of a Clapham omnibus, will stick on London-bridge for half-an-hour with scarcely a murmur.] **1903** *Law Rep.* (King's Bench Div.) II. 109 'Fair', therefore, in this collocation certainly does not mean that which the ordinary reasonable man, 'the man on the Clapham omnibus', as Lord Bowen phrased it, the juryman common or special, would think a correct appreciation of the work. **1959** *Listener* 30 July 162/2 It is 'the man in the street', 'the man in the jury box', or (to use the phrase so familiar to English lawyers) 'the man on the Clapham omnibus'. **1965** *Ibid.* 1 Apr. 474/1 The class character of our education will remain until..the scions of Lord Snow and the man on the Clapham omnibus attend the same comprehensive school.

clapped (klæpt), *ppl. a. slang.* [f. Clap *v.*¹ + -ed¹.] Usu. with *out*: worn out, exhausted.
1946 B. Sutton *Jungle Pilot* xiii. 66 The American aircraft..had become 'clapped out' by the exceptional number of hours they had been flown since a complete overhaul. **1960** *Daily Express* 18 July 6/5 The clapped-out car handed in for replacement. **1961** R. B. Amos *Wasp in Web* vii. 65 A clapped-out Fascist movement with a power-crazy brass-hat in control. **1961** I. Jefferies *It wasn't Me!* iii. 33 Bertie's cracking up..he's sort of clapped. **1967** A. Diment *Dolly Dolly Spy* vi. 56 Have you got this memory trick where all I have to do is speak and your clapped out brain registers every word?

clapper, *sb.*¹ Add: **3. a.** Slang phr. *like the clappers*: very fast or very hard.
1948 Partridge *Dict. Forces' Slang* 42 *Like the clappers (of hell)*, very fast. Mostly R.A.F. **1957** M. K. Joseph *I'll soldier no More* (1958) 16 It was raining like the clappers. **1958** J. Wain *Contenders* ix. 193 Seeing it's you, I'm going to surrender like the clappers. **1959** J. Braine *Vodi* ix. 128, I've got to work like the clappers this morning.
c. *Cinemat.* Usu. *pl.*, or *attrib.*, as *clapper board* (see quots.). (See also *clapper-boy* below.)
1940 *Chambers's Techn. Dict.* 166/1 *Clappers*, a device, usually consisting of two hinged pieces of board, which is closed sharply in front of a camera in sound-film production, thus providing synchronisation indication on the picture-track and the sound-track. **1959** *Elizabethan* June 24/2 This board is called the 'clapper board', because of a small piece of wood which is slapped down and makes a noise on the sound track.
4. (Further example.) Also, a person's mouth.
1938 G. Greene *Brighton Rock* II. i. 75 He just thought it'd close her clapper. **1961** J. Maclaren-Ross *Doomsday Book* I. iii. 37 Keep your clapper shut. *Ibid.* I. vii. 75 Hold your clapper now.
9. clapper-bill, a name for the open-bill stork, *Anastomus lamelligerus*; **clapper-bolt,** the bolt by which the clapper is attached to a bell; **clapper-boy,** a boy who works a clapper (senses 5 c and *3 c); **clapper rail** *U.S.*, a species of rail or marsh-hen; **clapper-stay,** a detent for the clapper of a bell, used in silent practice-ringing; **clapper-valve,** a clack-valve.
1906 F. Whyte tr. *Schillings's With Flashlight & Rifle* I. 75 Now flit past a number of those very remarkable birds aptly termed clapper-bills. **1901** H.E. Bulwer *Gloss. Techn. Terms Bells* 2 The modern practice is to insert, subsequently to casting, a bolt having a hinge joint, in which the 'clapper' is secured. This bolt is called the 'clapper-bolt'. **1937** *Evening News* 23 Mar. 13/7 Years ago small boys were engaged for the sum of sixpence a week, to scare birds away from sown fields. The boys were known as 'clapper boys'. **1950** 'E. Crispin' *Freq. Hearses* i. 45 A clapper-boy self-consciously clapped his instrument together in front of the lens. **1813** Wilson *Amer. Ornith.* VII. 112 The Clapper Rail, or, as it is generally called, the Mud Hen, soon announces its arrival..by its loud, harsh and incessant cackling. **1955** *Sci. News Let.* 3 Sept. 157/1 The clapper rail builds its nest on high spots within salt marshes. *a* **1884** Knight *Dict. Mech.* Suppl., *Clapper Stay*, a detent for the clapper in a bell. **1874** Knight *Dict. Mech.*, Clapper-valve.

clapper (klæ·pəɹ), *sb.*³ *local.* [? Clapper *sb.*²] In full *clapper bridge*: a rough bridge or raised path of stones or planks.
1793 Polwhele *Hist. Dev.* II. 277 Clapper-bridge, partly in Honiton and partly in Combe-Raleigh, is chiefly built of flint stone. **1852** *N. & Q.* 1st Ser. VI. 542 We have here [at Edburton, Sussex] a lane called Clappers, so named from its 'clapper', i.e. a raised footpath at side, to keep foot-passengers out of the water. **1887** Parish & Shaw *Kentish Dial.* **1889** J. L. W. Page *Explor. Dartmoor* iii. 53 Piers of undressed granite blocks support two or more superincumbent slabs, of width sufficient for the passage of a vehicle..and varying in length according to the breadth of the river. A specimen easily accessible is that which spans the East Dart at Postbridge...Other specimens of these 'clapper' bridges..exist on the Moor. **1908** W. Johnson *Folk Memory* 67 Certain 'clapper' bridges, crossing the streams of Exmoor and Dartmoor, are often assigned to the Bronze Age, but they are more probably relics of pack-horse days. **1966** *New Statesman* 17 June 879/1 There was a stone clapper-bridge beside the other one.

clapper, *v.* Add: **2.** (Earlier example.)
1845 S. Judd *Margaret* I. xvii. 163 The house jarred and creaked,..loose boards on the roof clappered and rattled.
Hence **cla·ppering** *vbl. sb.*, (*a*) the fitting of a clapper to a bell; (*b*) the action of the verb (sense 1); (*c*) a noise like that made by the clapper of a bell.
1526 in House & Robus *Hist. Gt. Dunmow Parish Church* (1926) 25 Item for claperynge of iij letell bells for ye canepe. **1874** E. Beckett *Clocks & Watches* (ed. 6) 365 The lazy and pernicious practice of 'clappering', *i.e.* tying the bell-rope to the clapper, and pulling it instead of the bell. **1891** *Pall Mall Gaz.* 7 Dec. 6/1 He chatters in imitation of the 'clappering' of a stork. **1904** C. L. Marson *Folk Songs Somerset* p. xi, The clapperings of the steam-binder.

claptrap. Add: **2.** In modern use passing into sense 'nonsense, rubbish'. (Further examples.)
1845 *Punch* Nov. 215/1 Dan..fancies he covers his own astounding selfishness and indifference by this brutal claptrap. **1895** *Daily News* 30 May 2/3 That is very eloquent but it is what I call vicious and wicked clap trap. **1915** A. Huxley *Let.* Nov. (1969) 86 How much better this book wd. have been had she made it a study of don-life in the 80's..instead of the usual politico-Debrett clap-trap. **1955** *Times* 26 Aug. 7/5 Cannot our educationists turn away from the pretentious claptrap put about during the past 20 years ..? **1966** *Illustr. London News* 30 July 28/2 The piece at one point turns to deplorable dramatic claptrap.

clarain (klæ·rē¹n). [f. L. *clārus* Clear *a.* + F. *-ain* -ane.] A kind of coal; an ingredient of coal (see quots.).
1919 M. C. Stopes in *Proc. R. Soc.* B. XC. 472 These four distinguishable ingredients, all of which..are to be found in most ordinary bituminous coals, I name provisionally as follows:—(i) Fusain. (ii) Durain. (iii) Clarain, (iv) Vitrain. Together the equivalent of 'bright' or glance coal of various authors. **1924** *Glasgow Herald* 4 Mar. 9/1 The clarain and vitrain bands were relatively highly coking. **1960** *Gloss. Coal Terms (B.S.I.)* 6 *Clarain*, bright coal, less lustrous than vitrain; usually very finely striated.

Clare. In def. insert 'Franciscan' before 'order', and add examples.
1608 in *Cath. Rec. Soc. Publ.* (1914) XIV. 31 This Convent of Nazareth of the poor Clares erected in the Town of Graveling. *a* **1660** in J. Morris *Troubles Cath. Forefathers* (1872) vi. 277 The Rich Clares and Annunciates were their near neighbours. **1779** A. Butler *Lives Saints* (ed. 2) III. 108 A new nunnery of Poor Clares being founded at Bologna. **1894** A. M. Clarke *Life St. Francis Borgia* viii. 141 His youngest daughter had become a Poor Clare.

clarification. Add: **1. b.** Of ideas, thought, opinion.
a **1866** Whewell (Webster), The clarification of men's ideas. **1888** Bryce *Amer. Commw.* III. lxxxiii. 115 The maturing and clarification of political opinion. **1909** *Times* 20 Jan. 10/4 Clarification of thought within the party on the fiscal question. **1966** *New Statesman* 9 Dec. 863/2 Morrice James did not fly out to seek clarifications.

clarificatory (klæ·rifikē¹:təri), *a.* [f. late L. *clārificāre* (see Clarify *v.*), on type of L. adjs. in *-orius*: see -ory.] Tending to, having the purpose of, or relating to, clarification (*1 b).
1945 *Proc. Arist. Soc.* Suppl. Vol. XIX. 4 Such is the programme of analytic or clarificatory philosophy. *Ibid.* 6 The clarificatory conception of Philosophy need not be stated in a linguistic form at all. *Ibid.* 8 The clarificatory philosophers of the inter-war period. **1956** G. Ryle in A. J. Ayer *Revol. in Philos.* 5 The propositions of Formal Logic ..are..clarificatory of those propositions that are informative about the world.

clarion, *sb.* Add: **5. c.** *Comb.* in *clarion-voiced* adj. (parasynthetic f. *clarion voice*: see quot. 184. in 5 b).
1907 *Daily Chron.* 16 Sept. 3/4 Good old clarion-voiced Sir Oliver. **1910** *Ibid.* 6 Jan. 9/5 His clarion-voiced question.

clarionet. 1. c. (Earlier example.)
1826 M. Kelly *Reminisc.* (ed. 2) II. 247 Mr. Willman.. the principal clarionet.

Clarisse (klari·s). [Fr.] **a.** The French name for a Poor Clare. **b.** A nun belonging to that branch of the order of St. Clare which follows the original rule, as distinguished from the Urbanists who follow the mitigated rule approved by Pope Urban IV in 1264. (Also occas. **Clarist** and **Clarissine.**)
1693 tr. *Emilianne's Hist. Monast. Orders* 253 Clara.. withdrew herself..to the Church of St. Damian, and there gave beginning to the poor Maids, called from her name *Clarisses*. **1840** K. H. Digby *Mores Cath.* IX. x. xiii. 404 His wife desired to enter a convent of Clarists. **1879** *Encycl. Brit.* IX. 153/1 One of the only two nunneries of the Clarisses in Scotland existed at Aberdour. **1883** *Ibid.* XVI. 711/1 The Poor Clares..have also divided into Clarissines and Urbanists.

Clark cell (klāɹk,sel). [f. the name of Josiah Latimer *Clark* (1822–98), English engineer.] A cell used as a standard of electromotive force (see quot. 1943).
1884 *Phil. Trans.* CLXXV. II. 447 The manner in which the Clark cells have borne the tests applied to them justifies the hope that they may be found generally available as standards of E.M.F. **1892** G. F. Barker *Adv. Course Physics* IV. iv. 760 Two types of standard cells have been employed—the Daniell cell and the Clark cell. **1943** *Gloss. Terms Electr. Engin.* (*B.S.I.*) 96 *Clark cell*, a standard cell having electrodes of mercury and zinc amalgam in an electrolyte of zinc sulphate with mercurous sulphate as a depolariser. (The e.m.f. of a Clark cell is 1·433 volts at 15° C.).

clarkia (klā·ɹkiă). *Bot.* [mod.L. (F. Pursh *Floræ Americæ Septentrionalis* (1814) I. 260), f. the name of William Clark (1770–1838), American explorer.] A plant belonging to the genus so called, belonging to the family Œnotheraceæ, and including several species of annual herbs native to North America.
1827 *Bot. Reg.* XIII. 1100 We have spelt it Clarkia, with Mr. Nuttall, rather than Clarckia, as it is written by Pursh. **1857** *Spirit of Times* 11 Apr. 90/3 The Clarkias, white and rose colored. **1864** Grindon *Brit. & Gard. Bot.* 484 In many of the willow-herbs, the Clarkias and Œnotheras, they [*sc.* the stigmas] are four in number. **1900** *Daily News* 5 May 4/5 Shirley poppies, nemophillas, clarkia, and mignonette should be got in upon the first warm showery day. **1967** D. Hardwicke *Flowers from Seed* (rev. ed.) iii. 38 The elegant double clarkia hybrids make a delightful show.

claro (klā·ro). [Sp., = light, Clear *a.*] A light-coloured cigar.
1891 *Cigar & Tobacco World* 12 Nov. 522/2 The lightest are called 'Claro'; the next darker 'Colorado Claro'. **1945** A. Huxley *Time Must have Stop* vi. 63 He decided..to smoke one of those small Larranaga *claros* which had pleased him so much. **1964** C. Mackenzie *My Life & Times* III. 72 With *colorado* cigars being completely ousted by *claros*.

clary, *sb.*² Add: **2.** *clary-fritter*; **clary-water, -wine** (earlier examples).
1747 H. Glasse *Art of Cookery* ix. 82 To make Clarye Fritters. Take your Clarye-leaves..dip them one by one in a Batter..fry them quick. **1728** E. Smith *Compleat Housewife* 230 The fine Clary-water. *Ibid.* 211 To make Clary-Wine...Malaga raisins..the tops of clary, when it is in blossom.

clash, *v.* Add: **4. c.** Of colours: to go badly together, to kill each other.
1894 [see next].

clashing, *ppl. a.* Add: **d.** Of colours: disagreeing, going badly together.
1894 *Daily News* 4 June 9/1 The Royal ladies' dresses were not of clashing hues.

clasmatocyte (klæzmæ·tǒsəit). *Cytology.* [a. Fr. *clasmatocyte* (L. Ranvier 1890, in *Compt. Rend. Acad. Sci. Paris*, CX. 165), f. Gr. κλάσμα, -ατος fragment + κύτος cell (see -cyte).] = *histiocyte.
1919 *Anatomical Rec.* XVI. 166 Clasmatocytes are abundant in cultures of subcutaneous tissue. **1942** E. S. Horning in G. H. Bourne *Cytol. & Cell Physiol.* vi. 175 All cells in the cultures, especially clasmatocytes,..contained no such traces when ashed in cultures.

clasp, *sb.* Add: **1. c.** A frame for holding the loose ends of cotton, flax, etc., together during the process of spinning.
1839 Ure *Dict. Arts* 1318 The carding is conducted through between these two bars...When this bar is again let down, it pinches the spongy carding fast; whence this mechanism is called the clasp. **1844** G. Dodd *Textile Manuf.* v. 160 These clasps are long frames capable of admitting the handfuls of flax, which they hold tightly at one end. *a* **1877** Knight *Dict. Mech.*, *Clasp*, a device consisting of two horizontal beams, the upper one being pressed upon the lower one, or lifted, for drawing out the thread of cotton or wool.
7. *clasp-arm.*
1835 A. Constable *Let.* 24 May in *J. Constable's Corresp.* (1962) 292 We must make it square by cant pieces to receive the clasp arms. **1963** J. Osborne *Dental Mechanics* (ed. 5) ix. 182 Different types of clasp arm..will require different degrees of horizontal undercut.

Clasp (klāsp). The initial letters of Consortium of Local Authorities Special Programme, used to designate the organization and its system of planning, and *spec.* a method pioneered by them of building schools from prefabricated parts.
Other expansions of the title are recorded (see quot. 1958).
1958 *Times Educ. Suppl.* 28 Feb. 328/4 Clasp is born... C.L.A.S.P. is the new name by which the joint planners wish to be known—Combined Local Authority Schools Planning. **1960** *Ibid.* 10 June 1188/1 The C.L.A.S.P. system of prefabricated construction. **1961** *Listener* 28 Dec. 1110/2 CLASP uses a system of light steel columns and beams with prefabricated timber panels for floors, walls, and roofs. **1962** *Ibid.* 9 Aug. 206/1 One of our best prefabricated school-building systems, using light metal—the one known as Clasp.

class, *sb.* Add: **3. b.** (Earlier examples.)
1671 S. Sewall *Letter-Bk.* (1886) I. 19 Remember me kindly to all our Class. **1684** in *Harvard College Rec.* (1925) I. 77 Mr. Samuel Mitchell was..desired to undertake ye charge of ye class of ye Sophimores untill further order. **1766** T. Clap *Ann. Yale-Coll.* 14 The Senior Class were removed to Milford...The rest of the Students were removed to Saybrook.

5. b. *slang* or *colloq.* Distinction, high quality; *no class*: of no worth; of low quality, inferior. Also *attrib.* or quasi-*adj.*
1874 Hotten *Slang Dict.*, *Class*, the highest quality or combination of highest qualities among athletes. 'He's not class enough', *i.e.*, not good enough. 'There's a deal of class about him', *i.e.*, a deal of quality. **1884** *Referee* 24 Mar. 1/3 The elasticity necessary for anything like class at sprinting departs comparatively early. **1897** *Daily Tel.* June (Ware), Soldiers! Why, soldiers ain't no class. **1924** H. de Sélincourt *Cricket Match* ii. 26 If he'd had coaching, he'd be a class bowler. **1927** [see *arrow *v.* 4]. **1927** R. Rees *Life's what you make It* xii. 161, I'm not 'much class'. *Ibid.* 162, I am 'no class'. **1948** C. Day Lewis *Otterbury Incident* ii. 18 Real class your sister is. Too good for a schoolteacher. **1954** 'N. Blake' *Whisper in Gloom* I. iii. 40 It was a class neighbourhood, thought Foxy, surveying the elegant, freshly-painted houses. **1969** A. Hunter *Gently Coloured* iv. 52 I'm no class, I know that. I'm where I belong, a working policeman.
9. Now freq. pertaining to the differences or antagonisms within a class, or between the classes, of society, as in *class antagonism, bar, barrier, bias, boundary, -cleavage, conflict, dialect, dictatorship, difference, distinction, division, enemy, -feeling, -hatred, -interest* (earlier example), *-jealousy, -loyalty, morality, -prejudice, -privilege* (examples), *rule, solidarity, spirit, structure, struggle, -superiority, system, war, warfare*; *class-biased, -bound, -ridden* [Ridden *ppl. a.* 4], *-structured* adjs.; **class-conscious** *a.* [cf. G. *klassenbewusst*], conscious of belonging to a particular social class and of being identified with its interests, often with implication of sharp differentiation from or hostility to other classes; so **class-consciousness**; **class-marriage,** marriage within a class; **class society,** a society based on division into social classes.
1850 H. Macfarlane tr. *Marx & Engels's Manifesto of German Communist Party* in *Red Republican* 23 Nov. 183/2 The old Bourgeois Society, with its classes, and class antagonisms. *a* **1950** 'G. Orwell' *England your England* (1953) 21 Class antagonisms are not all-important. **1939** *Brit. Jrnl. Psychol.* Apr. 335 Except for the class bar, the amateur definition of the two [rowing] associations is.. the same. **1889** G. B. Shaw *Fabian Essays in Socialism* 21 You set your life apart from theirs by every class barrier you can devise. **1945** Koestler *Yogi & Commissar* III. iii. 224 A crumbling of class-barriers. **1863** *Bee-Hive* 19 Dec. 5/3 Evening schools, for the purpose of educating, free from all class or sectarian bias, the children of the industrious classes. **1931** G. J. Renier *The English* vi. 109 Impartiality..requires absence of class-bias. **1909** *Westm. Gaz.* 9 Sept. 2/3 A citizen Army, free from military law in times of peace, is the best alternative to the class-biased policy of the present Government. **1950** D. Riesman in *Psychiatry* XIII. 302/2 Certain elements in Freud's view of life were narrow, class-biased, and reactionary. **1936** K. Mannheim *Ideology & Utopia* iii. 139 It [*sc.* humanism] broke through the fixedness of a class-bound mentality in many respects. **1960** *New Left Rev.* Nov.–Dec. 28/1 The new 'working-class consciousness'..is likely to be..less class-bound. **1959** F. Donaldson *Child of Twenties* vi. 91 The class-boundaries were still..well defined. **1959** V. Packard *Status Seekers* (1960) i. 5 Class boundaries are contrary to the American dream. **1937** J. M. Murry *Necessity of Pacifism* vi. 98 Advanced Capitalism operates far more strongly towards national cohesion than towards class-cleavage. **1898** G. B. Shaw in *Sat. Rev.* 29 Jan. 139/1 The alternative of achieving a real superiority or going ignominiously under in the class conflict. **1919** J. L. Garvin *Econ. Foundations of Peace* 106 Our whole future depends upon securing.. better relations between Capital and Labour instead of Class-conflict. **1903** J. London *People of Abyss* viii. 84, I was a superior, and they were superbly class-conscious. **1906** *Daily Chron.* 18 Apr. 4/5 A class-conscious political party. **1907** *Fabian News* XIX. 94/1 The middle classes are more class-conscious than the workers. **1935** *Economist* 29 June 1476/2 Despite the real class distinctions that divide us, the mass of the people forgot to be class-conscious during Jubilee week. **1887** Moore & Aveling tr. *Marx's Capital* I. p. xxiv, The German proletariat had attained a much more clear class-consciousness than the German bourgeoisie. **1922** *Westm. Gaz.* 18 Dec., The belief that what was called class consciousness was necessary to social progress. **1959** F. Donaldson *Child of Twenties* iii. 40 She did us a service by her class-consciousness..her belief that we belonged among the higher types of humanity. **1901** Greenough & Kittredge *Words & Ways* v. 53 Any limited circle having common interests is sure to develop a kind of 'class dialect'. **1930** J. R. Firth *Speech* viii. 62 Standard English, on the other hand, is a class dialect; to adapt Lord Chesterfield, it is the 'usage of the best companies'. **1930** L. L. Lorwin in Seligman & Johnson *Encycl. Soc. Sci.* III. 541/2 The establishment of class dictatorships throughout the world. **1941** 'G. Orwell' *Lion & Unicorn* III. 112 An English Socialist government..will not set up any explicit class dictatorship. **1879** S. A. Barnett *Let.* 21 Dec. in H. Barnett *Canon Barnett* (1918) I. xix. 233 Our serving-man..has no sense of class difference. **1960** *20th Cent.* Apr. 373 People do not necessarily weaken class differences..because they can climb from one social class to another. **1841** Dickens *Barn. Rudge* xxvi. 83 If she were in a more elevated station of society, she would be gouty, Being but a hewer of wood and drawer of water, she is a rheumatic. ..These are natural class distinctions. **1850** H. Macfarlane tr. *Marx & Engels's Manifesto of German Communist Party* in *Red Republican* 23 Nov. 183/1 When Class distinctions will have finally disappeared..the public power will lose its political character. **1960** C. Day Lewis *Buried Day* iv. 80 His Christianity found nothing paradoxical in the rigid class-distinctions of the time. **1922** Ld. Haldane in *Public Opinion* 10 Feb. 128/2 A deep lying class division, which tends to produce bitterness. **1961** *New Left Rev.* Jan.–Feb.

36/1 Traditional class divisions are gradually breaking down. **1957** R. N. C. Hunt *Guide to Communist Jargon* iv. 8 To assert that democracy—i.e. the 'rule of the people'—can be achieved under the domination of its 'class enemy'.. would be absurd. **1839** Mill in Bain *J. S. Mill* (1882) ii. 53 The convictions of the mass of mankind run hand in hand with their interests or with their class feelings. **1961** A. O. J. Cockshut *Imagination of Dickens* viii. 119 The contempt for Uriah is partly a class-feeling. **1851** T. G. Massey in *Friend of People* 26 Apr. 177/3, I shall be accused of sowing class-hatred. **1908** *Daily Chron.* 29 June 4/4 From top to bottom of the social or economic scale of class-consciousness or class-hatreds. **1937** J. M. Murry *Necessity of Pacifism* i. 10 Marx is vulgarly but generally reputed to be a propagator of the duty of class-hatred. **1828** J. S. Mill in *Westm. Rev.* IX. 255 Party interests, and class interests..appear to influence the train of events. **1920** D. H. Lawrence *Lost Girl* i. 8 Let class-jealousy be what it may, a woman hates to see another woman left stately on the shelf. **1899** W. James *Talks to Teachers* 289 Class-loyalty was undoubtedly an ideal with many. **1938** 'G. Orwell' *Homage to Catalonia* iii. 34 In a workers' army discipline..is based on class-loyalty. **1899** Keane *Man Past & Present* 153 Here it is necessary to distinguish carefully between class-marriages and the so-called 'communal' or 'group' marriages; the former having for their sole object, not, as is commonly supposed, the prevention of close consanguineous unions but the proper disposal of the stock of available food. **1833** J. S. Mill in *Monthly Repos.* VII. 577 That blighting curse of our country,—the selfishness of class morality. **1951** R. Firth *Elem. Soc. Organization* vi. 211 Class morality is apt not to be a direct reflection of current structural alignments. **1850** C. Kingsley *Alton Locke* I. xi. 172 All my class prejudices against 'game-preserving aristocrats'. **1861** Class prejudice [in Dict.]. **1925** Jespersen *Mankind, Nation & Individual* iv. 81 The difference in language begets a mutual disdain and class-prejudice. **1864** tr. Marx in *Founding of First International* (1939) 39 The struggle for the emancipation of the working classes means not a struggle for class privileges and monopolies, but for equal rights and duties, and the abolition of all class rule. **1956** J. Wilson *Language & Pursuit of Truth* ii. 57 Men make..protests against particular types of privilege.. for instance, against class-privilege. **1934** Dylan Thomas *Let.* 20 July (1966) 138 All society ceases to be class-ridden when treated purely as a society of consumers. **1941** 'G. Orwell' *Lion & Unicorn* I. i. 33 England is the most class-ridden country under the sun. **1864** Class rule [see **class privilege*]. **1959** *Universities & Left Rev.* Spring 1/1 A socialist commitment, a conscious rejection of all that springs from class rule. **1930** L. L. Lorwin in Seligman & Johnson *Encycl. Soc. Sci.* III. 540/2 In any class society the process of production is simultaneously a process of economic exploitation. **1949** W. S. Adams *Edwardian Heritage* xxii. 183 How easy for..a class society to hold that rule over subject races was in the order of things. **1907** *Q. Jrnl. Econ.* XXI. 307 Class solidarity will arise on the basis of this class interest. **1840** J. S. Mill in *Edin. Rev.* Oct. 39 That absence of classes and class-spirit. **1886** B. Webb *Let.* in *My Apprenticeship* (1926) iii. 166 Class spirit hardly exists because there is no capitalist class. **1930** P. Mombert in Seligman & Johnson *Encycl. Soc. Sci.* III. 535/2 The present day class structure of society has been attacked from two directions. **1958** P. Shore in N. Mackenzie *Conviction* 51 In spite of war-socialism and the Attlee Government..the class structure seems as strong as it ever was. **1959** G. D. Mitchell *Sociology* vii. 110 Marx perceived Europe as divided into class-structured societies. **1850** H. Macfarlane tr. *Marx & Engels's Manifesto of German Communist Party* in *Red Republican* 16 Nov. 171/2 As the settlement of the class struggle draws near, the process of dissolution goes on.. within the ruling-class. **1939** 'G. Orwell' *Crit. Essays* (1951) 11 From the 'revolutionary' point of view the class-struggle is the main source of progress. **1959** E. Burgess *Divided we Fall* xi. 128 All this union business ties up with..politics, the class struggle. **1859** J. S. Mill *Liberty* i. 16 Wherever there is an ascendant class, a large portion of the morality of the country emanates from its class interests, and its feelings of class superiority. **1948** F. R. Leavis *Great Tradition* v. 246 His languid assurance of class-superiority. **1877** L. H. Morgan *Anc. Society* III. iii. 425 The significance of the Australian class system presents itself anew in this connection. **1958** P. Johnson in N. Mackenzie *Conviction* 213 The essence of a class system is that property, birth and upbringing, rather than virtue or ability, determine status, power and income. **1886** tr. *Marx & Engels's Manifesto of German Communist Party* iii. 28 In proportion as the class-war [1850 tr. class-battle] is evolved and assumes a definite form, so does this imaginary elevation over it..lose all practical worth. **1892** G. B. Shaw *Fabian Soc.* (Fabian Tract 41) 5 The Social-Democratic Federation..said..that its policy is founded on a recognition of the existence of a Class War. **1919** J. Buchan *Mr. Standfast* I. iii. 60 Their blessed class war, which cuts across nationalities. **1961** L. Mumford *City in History* ix. 257 Hostile expropriation on one side, with seething revolt and counter-challenge on the other: in short, the class war, in which no quarter was expected or given. **1892** G. B. Shaw *Fabian Soc.* (Fabian Tract 41) 25 (*heading*) Scientific class warfare. **1927** A. Huxley *Proper Stud.* 86 Those who would interpret all social phenomena in terms of class warfare.

10. *class-attender, -holder*; *class-room* attrib.; **class-cleavage** *Linguistics*, (of a word or form) occurrence in more than one form-class; **class-distinction,** a difference between two classes of objects; (see also sense 9 above); **class-mark, class-number,** a mark or number indicating the class or subject of a book in a library and often its position on the library shelves; **class-meaning** *Linguistics*, the feature of meaning shared by all forms belonging to the same form-class; **class-name** (earlier example); **class-noun** *Linguistics* = *class-name*.

1897 *B'ham Inst. Mag.* Nov. 214 Pity her sadness, ye happy lecture-goers and class-attenders. **1933** Bloomfield *Lang.* xii. 204 The word *one* occurs not only as an indefinite determiner (*one man*), but also in some entirely different functions (as in *a big one, if one only knew*); this phenomenon may be designated as *class-cleavage*. *Ibid.* xvi. 265 *Salt* is a mass noun..but, by class-cleavage, there is also a plural *salts* (as in *Epsom salts*) with the meaning 'consisting of particles', in a class with *oats*. **1843** Mill *Logic* I. I. vii. 168 Drawing a broad line of separation between these two kinds of classes and of class-distinctions. **1885** G. B. Shaw *How to become Mus. Critic* (1960) 107 Not because he knows or cares enough about music to make the faintest class distinction between the overture to La Gazza Ladra and the prelude to Tristan und Isolde. **1892** E. Rowe *Chip-Carving* 54 My experiences as a class-holder in one of the evening classes of a London Board School. **1894** *Daily News* 28 June 6/3 Miss Nödel, the class holder, shows some excellent work in lacquered leather. **1886** Willis & Clark *Cambridge* III. 429 When a book was given to a medieval library it was necessary..to write the name and classmark across the fore-edge. **1926** Bloomfield in *Language* II. 159 The functional meanings in which the forms of a form-class appear constitute the class-meaning. **1958** C. F. Hockett *Course Mod. Ling.* xxx. 256 This class-meaning, 'single female human or animal', is not an invariable part of the meaning of *she*. **1843** Mill *Logic* I. I. v. 138 When the class-name is predicated of any object of this description. **1892** H. Sweet *New Eng. Gram.* I. 54 Common nouns..are subdivided into class-nouns, such as *man*, and material-nouns, such as *iron*. **1939** H. E. Palmer *Gram. Spoken Eng.* (ed. 2) I. 30 Every word in the dictionary is potentially a class-noun, for we can speak of *ifs* and *buts*. **1899** W. James *Talks to Teachers* xi. 100 The less the kind of attention requiring effort is appealed to, the more smoothly and pleasantly the class-room work goes on. **1936** *Discovery* May 147/2 The classroom film is the little brother of the formal documentary.

classed, *ppl. a.* Add: (Earlier and later examples.) *spec.* Of a wine [tr. F. *classé*] (see quot. 1965).

1755 in H. W. Richmond *Papers Rel. to Loss of Minorca* (1913) 73 To prohibit all classed sailors from navigating till farther orders. **1871** *Wine Trade Rev.* XI. 38/2 The vineyard proprietors received high fancy prices for classed growths, for which English Wine merchants will hardly be disposed to reimburse the Bordeaux houses. **1920** G. Saintsbury *Notes on Cellar-book* i. 14 A 'classed' claret in the shape of '64 Ducru Beaucaillou. **1965** A. Sichel *Penguin Bk. Wines* III. 135 The sixty-two best red wines were classified in 1855 by an official committee into five 'growths' or '*crus*', known collectively as the Classed Growths, or *Crus Classés*.

classer² (kla·sǝɹ). [f. Class *sb.* + -er¹.] In compounds, someone or something related to a class or classes specified or indicated by the first element.

1897 'P. Warung' *Tales Old Régime* 146 Double-classer: A ship conveying both male and female transports. **1909** *Public Ledger* (Philad.) 24 June 12/6 Shamrock and White Heather, the only two big classers sailing in this year's regattas.

classic, *a.* and *sb.* Add: **A.** *adj.* **6. c.** Of clothes: made in simple, conventional styles that are almost unaffected by changes in fashions.

1937 *Night & Day* 19 Aug. 20/2 [Weatherall's] stick to what they call their classic blouse because it's always in demand. **1960** *Guardian* 21 July 7/7 Among the..long evening dresses there are..a large proportion of 'classic' models with fitted bodices and full skirts.

B. *sb.* **1. b.** (Earlier example.)

1725 G. Sewell *Pope's Shakespeare* VII. Pref. p. vii, What then has been done by the really Learned to the dead Languages,..we in Justice owe to our own great Writers, both in Prose, and Poetry. They are in some degree our *Classics*; on their Foundation we must build, as the Formers and Refiners of our Language.

c. Also of music and composers.

1889 G. B. Shaw *How to become Mus. Critic* (1960) 160 Wagner did not succeed in putting dulness out of countenance until he became a classic himself. **1889** —— *London Music 1888–89* (1937) 240 The glees, madrigals, and motets and services which are the true English musical classics. **1926** Whiteman & McBride *Jazz* viii. 180 When you are listening to your favorite jazz tune, you are most likely absorbing strains that are most classic of all the classics.

5. Short for *classic race* (see A. 6 b); also, the prize for one of these races.

1905 *Daily Chron.* 3 May 7/4 To witness the race for the first 'classic' of the season. **1907** *Ibid.* 2 May 7/5 The first of the 'classics' was considered..to be at the mercy of Captain Greer's handsome black. **1951** E. Rickman *Come Racing* ii. 14 The five 'classics'—the Two Thousand Guineas, One Thousand, Derby, Oaks and St. Leger.

6. A classic garment (see *A. 6 c).

1948 M. Laski in *New Statesman* 13 Nov. 416/3 *Classic*, English garment (shoes, hat, suit) barely susceptible to fashion changes. **1960** *Guardian* 9 Dec. 8/4 Present-day teenage styles are already classics and have little or nothing to do with real fashion trends.

classical, *a.* Add: **4.** (Earlier examples.)

1789 *Loiterer* 21 Mar. 4 In the nine succeeding years, I compleated my classical education. **1832** *Chambers's Edin. Jrnl.* I. 226/1 There are very few..who have not enjoyed the advantages of a classical education.

6. c. *spec.* opp. *jazz.

1916 *Variety* 27 Oct. 12/4 It would not be surprising to note the disappearance of the classical orchestras for the syncopated groups. **1947** *Penguin Music Mag.* Dec. 52 The lowbrow, of course, divides all music into 'classical' and 'jazz'. **1955** L. Feather *Encycl. Jazz* vii. 107 He has devoted much time lately to classical writing, including a trombone sonata for American Composers' League.

d. Of a style of ballet: (see quots.).

1928 A. L. Haskell *Stud. Ballet* 16 The classical ballet was pure dancing and nothing else, the romantic ballet a close alliance between pure dancing, music, décor, and literature, while the modern ballet is literature, music, décor and..a corrupt form of dancing. **1957** G. B. L. Wilson *Dict. Ballet* 75 *Classical ballet*, ballet in which the movement is based on the traditional technique evolved from the French Court ballet of the seventeenth and eighteenth century, the Italian schools of the nineteenth century, and the Imperial Academy of Dancing, St Petersburg and Moscow..and in which dramatic or emotional content is subordinate to form or line.

e. Of physics, mechanics, etc., used esp. of conclusions based on concepts and theories established before the discovery of quantum theory, the theory of relativity, etc.

1914 L. Silberstein *Theory of Relativity* i. 1 Before entering upon..the modern doctrine of Relativity.., it seems desirable to dwell a little on the more familiar ground of what might be called the *classical* relativity. **1928** A. S. Eddington *Nature Physical World* ix. 193 For the last fifteen years we have used classical laws and quantum laws alongside one another notwithstanding the irreconcilability of their conceptions. **1933** *Discovery* May 152/1 The strange limitation of classical mechanics, symbolized by the quantum of action. **1938** R. W. Lawson tr. *Hevesy & Paneth's Man. Radioactivity* (ed. 2) viii. 90 Classical physics is thus revealed as a limiting case of quantum physics. **1958** P. A. M. Dirac *Princ. Quantum Mech.* (ed. 4) p. vii, The classical tradition has been to consider the world to be an association of observable objects (particles, fluids, fields, etc.) moving about according to definite laws of force, so that one could form a mental picture in space and time of the whole scheme.

classicism. 1. (Earlier example.)

1830 Lady Morgan *France 1829–30* II. 60 In the composition of Cinq Mars, there is none of the exaggeration or pedantrie of romanticism or of classicism.

classicist. (Earlier example.)

1830 [see Romanticist 1].

classificatory, *a.* Add: spec. in *Anthropology*, applied to a system of terms describing kinship (see quots.).

For '1865' in quot. in the Dict. read '1878': the sentence quoted did not occur in the 1865 version.

1871 L. H. Morgan *Syst. Consanguinity* II. i. 143 In contradiction from *descriptive* the term *classificatory* will be employed to characterize the system of consanguinity and affinity of the Ganowanian, Turanian, and Malayan families, which..are arranged into great classes or categories upon principles of discrimination peculiar to these families. **1936** R. Firth *We, the Tikopia* vi. 200 Tikopia kinship is of course of the type known as classificatory. The material relations between a child and his or her 'fathers' who are near kindred depend to a great extent upon factors of residence and the social contacts that this implies. **1964** Gould & Kolb *Dict. Social Sci.* 368/2 The classificatory system is characterized by the terminological merging of lineal and collateral kindred at certain points—e.g. calling father's brother 'father'.

classified (klæ·sifǝid), *ppl. a.* [f. Classify *v.*]

a. Arranged in classes. Also as *sb.* (esp. of advertisements).

1889 T. Roosevelt *Wks.* (1926) XIV. 92, I expect from the President an extension of the classified service. **1890** W. Booth *In Darkest Eng.* II. vi. 227 Although each particular Blue Book may be admirably indexed, there is no classified index of the whole series. **1903** A. B. Hart *Actual Govt.* 290 The act provides for the classification of clerks and other officers into four groups, according to their compensation: and hence the persons subject to competitive examination are said to be in the 'classified service'. **1909** 'O. Henry' *Options* 6, I know every end of the business from editing to setting up the classified ads. **1937** *Discovery* Jan. p. ii/1 Classified announcements are accepted at 1d. per word. **1961** *John o' London's* 6 July 29/1 They proclaim their I.Q.s in the crosswords and bare their souls in the classifieds. **1964** *Listener* 30 Apr. 726/3 Very socially ambitious men switch from the theatre reviews to the classified results because soccer is in vogue.

b. Of a road: (see quot. 1954).

[**1919** *Act* 9 & 10 *Geo. V* c. 50 § 17(2) For the purpose of advances for the construction, improvement, or maintenance of roads, the Minister may..classify roads in such a manner as he thinks fit.] **1935** *City of Oxford Council Rep.* 1 Mar. 400 *Classified Roads.* Reconstruction of Rose Hill from Henley Avenue to City boundary. **1954** *Highway Engin. Terms* (*B.S.I.*) 8 *Classified road*, in Great Britain, a highway classified under the Ministry of Transport Act, 1919, according to its relative importance for through communication.

c. Classified as secret for reasons of national security and forbidden to be disclosed except to specified persons. Also *transf.*

1944 *Amer. N. & Q.* July 54/1 *Classified* (Army usage), referring to the degree to which the use of a document of an intelligence nature is permitted (restricted, confidential, and top secret). **1949** *N.Y. Herald Trib.* 24 Oct., The B-47 will combine characteristics which are still classified, with range equivalent to the B-29 range. **1955** *Times* 25 Aug. 5/3 A party of American and Canadian scientists..are to go to Harwell..for a 'classified' visit. The visit yesterday was 'unclassified', meaning that all the information available is off the secret list. **1957** 'N. Shute' *On Beach* iii. 99 'But that's classified information.' 'What does that mean?' 'Secret.' *Ibid.* viii. 253 To leave a ship like that, full of classified gear. **1966** *Electronics* 3 Oct. 134 Little can be said about the electronic equipment because MOL is classified.

classifier. Add: **2.** *Gram.* A word or formative element attached to a noun in certain languages, used esp. to indicate the class of objects which the noun represents; = Numerative *sb.*

1863 [see Numerative *sb.*]. **1964** E. Bach *Introd. Transformational Gram.* vi. 141 The following Japanese forms (numerals and classifiers).

3. A device for screening or sizing ore, etc.

1902 *Encycl. Brit.* XXXI. 371/1 Sizes smaller than can be satisfactorily handled by screens are treated by means of hydraulic classifiers and box classifiers. **1952** *Gloss. Terms Fuel Appliances* (*B.S.I.*) 16 *Classifier*, a device for controlling the fineness of product discharged from a pulverizing mill.

classiness (kla·sinės). *slang* or *colloq.* [f. *classy *a.* + -ness.] **a.** The state or condition of belonging to a certain social or economic class. **b.** The quality or condition of being 'classy'.

1922 *Public Opinion* 24 Nov. 501/3 They will modify the undue 'classiness' of the [Labour] party. **1958** *Times Lit. Suppl.* 14 Mar. 137/4 Mildred is beautiful, moronic, and obsessed with her dream world of 'classiness'.

classis. 2. b. (Examples.)

1643 *New Eng. First Fruits* 14 That they studiously.. observe..the speciall houres for their owne *Classis*. *Ibid.* 16 The Students of the first Classis. **1673** *Harvard College Rec.* (1925) I. 56 Mr. Daniell Gookin..is forthwith to take upon him the charge of a classis. **1723** *Ibid.* 483 After his taking the Care of the Classis to the End of this Year. **1851** in B. H. Hall *College Words* 49.

classism (kla·siz'm). Also **class-ism.** [f. Class *sb.* 2 + -ism.] Distinction of class.

1842 S. Bamford *Pass. Life Radical* II. xviii. 89 The same really contemptible feeling of class-ism, the curse of England and Englishmen, and of women also. **1911** H. S. Harrison *Queed* xx, 'The *Chronicle* has won its great fight,' so it nervily said, 'against classism in Blaines College.'

classless, *a.* Delete '*nonce-wd.*' and add later examples. Used esp. of, or pertaining to, a society without distinctions of social class.

1931 A. L. Rowse *Pol. & Younger Generation* ix. 230 The objective is the classless state. **1937** J. M. Murry *Necessity of Pacifism* v. 81 The Socialist movement..is based on the conviction of the desirability of a classless society. **1958** *Oxf. Mail* 22 Aug. 5/5, I reckon I'm classless myself. **1959** *Listener* 2 Apr. 610/1 The actors had been schooled..to speak in mercifully classless voices. **1959** *Times* 15 June 11/3 Shoplifting is a relatively 'classless' crime. **1959** H. Gaitskell in *Observer* 29 Nov. 6/6 We believe in a 'classless society'—a society without the snobbery, the privilege or the restrictive social barriers which are still far too prevalent in Britain to-day.

cla·sslessness. [f. prec. + -ness.] The state or condition of being classless; spec. of a society (see prec.).

1933 *Mind* XLII. 380 Every 'democracy' is in fact the democracy of one class and a dictatorship over other classes; and only when society has evolved to a state of classlessness will this not be so. **1955** *N.Y. Times* 16 Jan. vi. 29/2 The hair-cut, the crêpe-soled shoes, [etc.]..are startling symbols of a new breakthrough of prosperity and classlessness. **1958** *Observer* 2 Nov. 3/6 There was a peculiar 'classlessness' about him [*sc.* Montgomery] which I am sure impressed the ordinary soldier. **1959** *Universities & Left Rev.* Spring 45/2 In the theory of classlessness the traditional working-class community is being disintegrated.

classmate. (Earlier Amer. examples.)

1713 Sewall *Diary* 5 June (1879) II. 387 He had spoken for my Classmate Capt. Saml. Phips. **1793** W. Bentley *Diary* Aug. (1907) II. 47 On this road I met my class mate Crosby from Charlestown.

classy (kla·si), *a. slang* or *colloq.* [f. Class *sb.* + -y¹.] Of high or superior class, stylish, smart.

1891 S. J. Duncan *Amer. Girl in London* 228 Why didn't one of you go? Not classy enough, eh? **1892** *Temple Bar* Nov. 393 Her successor is the 'smart' young married woman, ..who is *chic* and 'classy'. **1899** E. Phillpotts *Human Boy* 174 He said a man who sold pills and toothbrushes..could not be considered a classy chemist. **1903** 'Marjoribanks' *Fluff-Hunters* 94 Was she classy, well turned out, and all that sort of thing? **1929** *Daily Express* 2 Jan. 4/2 An assemblage as sophisticated and classy as ever gathered under one roof.

clat (klæt), *v.²* *dial.* Also **clad, clatt.** [f. Clat *sb.*¹ 1.] (See quot. 1838.) So **cla·tting** *vbl. sb.*

1838 Holloway *Dict. Provinc.*, To *clad*, or *clat*, to cut off the locks of wool round a sheep's tail, which are clotted together with the dung of the animal. **1844** H. Stephens *Bk. Farm* II. 600 In preparing ewes for lambing..the looker..removes with the shears the wool on their tail, udders, and inside of their thighs...This treatment [is] called clatting. **1884** R. L. Allen *New Amer. Farm Bk.* 436 *Tagging* or *clatting* is the removal of such wool as is liable to get fouled when the sheep are turned on to the fresh pastures.

clat-fart (klæ·tfāɹt). *dial.* [cf. Clat *sb.*², Fart *sb.*] Idle chatter, gossip, nonsense. So **clat-farting** *a.*, chattering, that talks idly or nonsensically.

1913 D. H. Lawrence *Sons & Lovers* iii. 52 The Guild was called by some hostile husbands..the 'clat-fart' shop—that is, the gossip-shop. **1914** —— *Prussian Officer* 266 That clat-fartin' Mrs. Allsop. **1928** —— *Lady Chatt.* xvii. 324 Folks should do their own fuckin', then they wouldn't want to listen to a lot of clatfart about another man's.

clathrate, *a.* Add: **b.** *Chem.* Designating or relating to a molecular compound in which one component is enclosed within the crystals or structure of another.

1948 H. M. Powell in *Jrnl. Chem. Soc.* 63 It is suggested that the general character of this type of combination should be indicated by the description 'clathrate' compound. **1961** *New Scientist* 16 Feb. 409/3 Under pressure at 2° C, propane forms insoluble 'clathrate' crystals in which one molecule of propane is caged by 17 molecules of water.

clatting: see *clat *v.*²

clauber, var. Clabber I.

1890 Service *Notandums* 114 Whaur it was a' clauber yesterday, it's as hard as a horn the day. **1892** Yeats *C'tess Kathleen* i. 16 The dead leaves and clauber of four forests Cling to my foot-sole. **1921** G. O'Donovan *Vocations* v. 83 Good people..that wouldn't throw a lump of soft clauber at a cat.

Claudian (klǭ·diăn), *a.* and *sb.* [ad. L. *Claudiānus.*] **A.** *adj.* Of or pertaining to any of several distinguished Romans of the name of Claudius or the gens to which they belonged, *esp.* pertaining to or connected with the emperors Tiberius, Caligula, Claudius, and Nero, or their epoch (A.D. 14–68). **B.** *sb.* A member of the Claudian gens.

1829 K. H. Digby *Broad Stone of Hon., Godefridus* 236 The Claudian family..produced the sad Tiberius, then the monstrous Caligula. **1842** Macaulay *Lay of Virginia* 215 The face of Appius Claudius wore the Claudian scowl and sneer. **1876** *Encycl. Brit.* V. 817/1 The Claudian aqueduct and the harbour of Ostia. **1882** *Ibid.* XIV. 337/2 The epic poets of the Flavian age present a striking contrast to the writers of the Claudian period. **1924** E. B. Osborn *Our Debt to Greece & Rome* II. xi. 187 Clodia..was a Claudian by birth. **1957** R. Graves tr. *Suetonius' 12 Caesars* iii. 109 The patrician branch of the Claudian House..came to Rome from..Regilli. *Ibid.*, The Claudians were enrolled among the patrician houses.

claudication. Delete †*Obs.* and add: *intermittent claudication*, a condition of the legs in which pain is induced by walking and relieved by rest; also *transf.*

1909 *Practitioner* Nov. 666 Angina pectoris, which is nothing else than intermittent claudication of the heart. **1963** *Lancet* 12 Jan. 77/2 The treatment of intermittent claudication by implantation of human amnion into the thigh.

clausal (klǭ·zăl), *a.* [f. Clause: see -al.] Of or pertaining to a clause or clauses.

1904 C. T. Onions *Eng. Syntax* § 300 A few broad principles have been enunciated, which, although they cover considerable ground, will not explain every kind of clausal combination which can be met with. **1905** *Westm. Gaz.* 25 Mar. 2/2 Would they have gagged and bound their country's hands In Saccharine Convention's clausal bonds? **1961** R. B. Long *Sentence & its Parts* i. 10 We can draw our examples almost entirely from clausal sentences of a single type: the main declarative.

clause, *sb.* Add: **1. a.** Also *attrib.* and *Comb.*

1928 Jespersen *Internat. Lang.* II. 152 *Ke* that, conjunction..corresponds with its initial *k* to the other relative (i.e. clause-connecting words). **1925** Grattan & Gurrey *Our Living Lang.* 89 Such a link is most accurately described as a Co-ordinating Clause-link. **1961** R. B. Long *Sentence & its Parts* i. 20 The second type of clause equivalent makes use of words that normally do not participate in clause structure at all. **1964** C. Barber *Present-Day Eng.* vi. 143 A formal clause-structure with main–subordinate relations.

clause, *v.* [f. Clause *sb.* 1.] *intr.* To construct clauses.

1895 *Proc. 14th Conv. Am. Instruct. Deaf* 124, I have endeavored to make my pupils read and speak like hearing people, teaching them to phrase and clause.

So **clau·sing** *vbl. sb.*

1908 *Daily Chron.* 3 Feb. 3/5 The clausing of the Bill..is not..as effectual as it might be.

claustration. Add: **b.** *transf.* and *fig.*

1908 *Pall Mall Gaz.* 27 Mar. 2/1 In the earlier days of Islamic civilisation the claustration of women did not in the least imply their ignorance. **1924** 'Lucas Malet' *Dogs of Want* ii. 33 Barbara halted, enchanted to emerge from the claustration of the forest and breathe a more spacious air.

claustrophilia (klǭ:strǒfi·liă). *Med.* [mod.L., f. L. *claustrum* confined space, Cloister *sb.* + Gr. φιλία affection.] A morbid desire to be enclosed within a confined space; also *transf.*

1926 in Gould *Med. Dict.* 336/1. **1949** Koestler *Promise & Fulf.* II. v. 279 His attitude is the result of that deep-rooted complex—cultural claustrophilia. —— *Ibid.* III. ii. 315 Generations of Jews, huddled together..developed that peculiar feature of Jewish communities which one might be tempted to name 'claustrophilia'. **1960** *Spectator* 29 Apr. 612 The Germanic claustrophobia of *Serjeant Musgrave's Dance* or the British claustrophilia of *Live Like Pigs*.

claustrophobia (klǭ:strǒfōᵘ·biă). *Med.* [mod. L., f. L. *claustrum* confined space, Cloister *sb.* + -phobia.] A morbid dread of confined places; also *transf.*

1879 *Brit. Med. Jrnl.* 6 Sept. 371/1 On Claustrophobia. By B. Ball, M.D. (Paris). **1895** [see -phobia]. **1906** *Westm. Gaz.* 18 Apr. 2/1 Others of us who have felt..an inexplicable horror of remaining within four walls..now know that we are sufferers from Claustrophobia. **1924** Galsworthy *White Monkey* I. v, Having a sort of mental claustrophobia, a dread of being hemmed-in by people. **1960** [see prec.].

Hence **clau:stropho·biac** *a.* and *sb.*, (one who is) suffering from or characterized by claustrophobia.

1934 R. Campbell *Broken Record* 21 All my early verse ..is a violent claustrophobiac struggle to keep my individual being. **1937** E. St. Vincent Millay *Conv. at Midnight* 60 Proletarian poetry today..is the love-song of the claustrophobiac who has espoused the crowd.

claustrophobic (klǭ:strǒfōᵘ·bik), *a.* [f. prec. + -ic.] **a.** Prone to or suffering from claustrophobia. Also *absol.* or as *sb.*

1889 in *Cent. Dict.* **1932** E. Bowen *To North* xvi. 167 'Besides,' she said, 'I'm claustrophobic.' **1953** *New Yorker* 31 Jan. 61/1 The crowds were so enormous that it was all too easy for the claustrophobic to fall into an apposite mood. **1954** *Harper's* Aug. 16/3 Everybody felt claustrophobic and uneasy. **1965** D. S. Davis *Pale Betrayer* (1966) xvi. 127 No windows..a room not intended for claustrophobics.

b. Of a place, etc.: confined, restricting; inducing claustrophobia. Also *transf.*

1946 'J. Tey' *Miss Pym Disposes* ii. 11 This is the last term for us Seniors, and everything is very grim and claustrophobic. **1957** R. Hoggart *Uses of Literacy* viii. 213 It is violent and sexual, but all in a claustrophobic and shut-in way. **1959** G. Mitchell *Man who grew Tomatoes* xi. 141, I think we might all go into the lounge... It is unpleasantly claustrophobic in here.

Hence **claus:tropho·bically** *adv.*, in a claustrophobic manner; so as to induce claustrophobia.

1957 R. Hoggart *Uses of Literacy* vii. 188 The claustrophobically personal manner of some successful men crooners. **1961** *John o' London's* 28 Sept. 358/2 Suddenly it all becomes unbearably claustrophobically pointless.

claustrum (klǭ·strǒm). *Anat.* Pl. **claustra.** [L.; see Cloister *sb.*] A thin layer of grey matter in each cerebral hemisphere between the external capsule and the insula.

1848 *Quain's Elem. Anat.* (ed. 5) 739 Beneath the white matter of those gyri, a thin layer of grey substance has been met with, which has been named the *claustrum* (Burdach) [in *Vom Baue und Leben d. Gehirns* (1822) II. III. 132]. **1889** *Buck's Handbk. Med. Sci.* VIII. 135 The *lenticula.*—Between the caudatum and the cortex the greatly thickened hemicerebral wall presents..(*b*) next the cortex, a subcircular disk of cinerea, the *claustrum*. **1962** *Gray's Anat.* (ed. 33) 1085 The *claustrum*..intervenes between it [*sc.* the external capsule] and the subcortical white matter of the insula.

clausula (klǭ·ziŭlă). Pl. **clausulæ.** [med.L., f. L. *clausula* (see Clause).] **1.** *Mus.* **a.** = Cadence *sb.* 4. **b.** (See quot. 1944.)

1636 C. Butler *Princ. Musick* I. iii. 82 A Cloz'. *Clausula.* The Cloz' is a Formal meeting of all Part's..wit a Cadenc'. **1740** in Grassineau *Mus. Dict.* **1876** Stainer & Barrett *Dict. Mus. Terms* 98/1 *Clausula*, a close or cadence, *e.g.*: *clausula falsa*, a false cadence; *clausula finalis*, a final cadence, &c. **1889** Grove *Dict. Mus.* IV. 592/1 *Clausula*, the mediæval name for what is now called a Cadence, or Close. The most importan. Close employed in Polyphonic Music, is the *Clausula vera*, or True Cadence, terminating on the Final of the Mode. **1944** W. Apel *Harvard Dict. Mus.* 155/2 In the repertory of the School of Notre Dame (*c.* 1200) ..*clausula* is a polyphonic composition using as a cantus firmus a short melisma of a chant.

2. The close or end of a period, esp. one in ancient or medieval Latin having a definable cadence. (Cf. Clause 1 b.)

1900 J. E. Sandys in *Class. Rev.* XIV. 136/2 The favourite *clausulae* of late Greek Prose and of Cicero are also examined. **1906** A. Sloman *Gram. Class. Latin* III. 387 In Cicero's Orations at any rate the *Clausula* is regulated by definite metrical rules. **1957** N. Frye *Anat. Criticism* IV. 264 Recurring units of rhythm like the *clausulae* of Cicero.

clavacin (klæ·văsin). *Biochem.* Also **clavicin.** [f. L. *clāvā(tus* (see Clavate *a.*) as in mod.L. *Aspergillus clavatus*, a species of fungi + *-cin* as in *actinomycin*, etc.; see -in¹.] A crystalline antibiotic substance produced in several fungi; *spec.* that obtained from *Aspergillus clavatus*.

1942 S. A. Waksman et al. in *Science* XCVI. 203/1 Two species of *Aspergillus* were studied..*A. fumigatus*..and *A. clavatus*...These two organisms produced active substances..designated as fumigacin and clavacin respectively. **1944** *Ann. Reg. 1943* 359 Clavicin seems to be especially lethal to plant pathogenic bacteria. **1959** *Life* 31 Aug. 72/2 Soon after actinomycin came clavacin and fumigacin, both of which were impressive on first acquaintance but failed to survive rigorous testing.

clavate, *a.* Add: **2.** (Later examples.) Hence **cla·vately** *adv.*

1872 H. C. Wood *Fresh-w. Algæ* 176 Clavately swollen. **1884** Lesquereux & James *Mosses N. Amer.* 12 Male flowers..borne upon clavate and catkin-like generally colored branches. **1950** E. J. H. Corner *Monogr. Clavaria* 47 The fruit-body becomes obtusely clavate.

clavatin (klæ·vătin). *Biochem.* [f. as *clavacin + -in¹.] = *clavacin.

1943 F. Bergel et al. in *Nature* 25 Dec. 750/1 An Antibacterial Substance from *Aspergillus clavatus* and *Penicillium claviforme*... Whereas Waksman called his crude active principle 'clavacin', we have adopted the name

'clavatin'...Analysis and molecular weight determinations indicate an empirical formula $C_7H_6O_4$ for clavatin. **1953** J. Ramsbottom *Mushr. & Toadstools* xxiii. 291 A much-named substance—patulin, clavicin, clavatin, claviformin, expansin—which is produced by several moulds.

clave³ (klēⁱv, klāv). *Mus.* Also **clavé** (kla·ve). [Amer. Sp., f. Sp. *clave* keystone, f. L. *clāvis* key.] One of a pair of round sticks of hard wood, struck against each other when used in 'Latin-American' music. Usu. in *pl.*

1928 [see *bongo²]. **1934** S. R. Nelson *All about Jazz* vii. 166 The Claves are two cylindrical sticks made of a special kind of wood found in Cuba. One of the claves is held in the cupped left hand and struck with the clave in the right hand, thus producing a hollow musical tone. **1957** M. Stearns *Story of Jazz* xix. 248 The traditional 'off-center' accents of the claves, or wooden sticks, that make a penetrating hollow sound when struck together. **1961** J. Blades in A. Baines *Mus. Instruments* xiv. 348 The relentless clicking of the claves.

clavel. Add: **2. clavel piece.** (Earlier American example.)

1634 J. Winter *Let.* 18 June in J. P. Baxter *Trelawny Papers* (1884) 31 The Chimnay..is so large that we Can place our Chittle [*sc.* kettle] within the Clavell pece.

clavicembalo (klævi,tʃe·mbălo). *Mus.* [It.] The harpsichord; Clavicymbal.

1740 Grassineau *Mus. Dict.* 30 Clavecimbalo, grave Cymbalum. **1858** Simmonds *Dict. Trade.* **1879** Grove *Dict. Mus.* I. 330/2 The dulcimer, or cembalo, with keys added, became the clavicembalo. **1910** *Encycl. Brit.* VI. 468/1 *Clavicembalo*, or *Gravicembalo*..a keyboard musical instrument with strings plucked by means of small quill or leather plectra. **1955** *Oxf. Compan. Mus.* (ed. 9) 463/1 Instruments of the various harpsichord types are: In Italian Arpicordo, Clavicembalo (meaning keyed-dulcimer, which it is not, its strings not being struck but plucked), or Claricembalo, or Gravicembalo.

clavicin, var. *clavacin.

claviformin (klævifǭ·ɹmin). *Biochem.* [f. the second element of mod. L. *Penicillium claviforme* (cf. Claviform *a.*), a species of fungus + -in¹.] A crystalline antibiotic produced from several fungi; *spec.* that obtained from *Penicillium claviforme*, chemically identical with *clavacin.

1942 E. Chain et al. in *Brit. Jrnl. Exp. Path.* XXIII. 205 A crystalline substance which inhibits the growth of Gram-positive and Gram-negative organisms has been isolated from the metabolic products of *Penicillium claviforme*...The name claviformin is proposed for it. **1953** [see *clavatin].

clavioline (klæviǒlī·n). [Proprietary term.] (See quot. 1952.)

1952 *Electronic Engin.* XXIV. 454 The Clavioline is a small electronic musical instrument..played from a small keyboard of three octaves, 20 in. in length, which is arranged so that it may be attached to a piano, and played therewith. **1959** K. Waterhouse *Billy Liar* ix. 117 The concert was already warming up, with the Clavioline thumping away. *Ibid.* 126 The pianist did not bother to give me a few bars on the Clavioline. **1967** *Crescendo* Feb. 21/2 It was about the time that the organ-type piano attachment—the clavioline—was invented.

clavula (klæ·viŭlă). Pl. **-æ.** [mod.L. Cf. Clavule.]

1. One of the knobbed ciliated setæ or bristles found on the exterior of spatangoid sea-urchins.

1884 Sedgwick & Heathcote tr. *Claus' Zool., Protozoa to Insecta* 296 The *fascioles* or *semitæ*..upon which.. knobbed bristles with active cilia (*clavulæ*) are distributed. **1888** Rolleston & Jackson *Anim. Life* 558 Clavulæ, found only in *Spatangidæ*. **1962** D. Nichols *Echinoderms* viii. 97 This current is largely created by special ciliated spines, *clavulae*, in the interstices of the marginal plates on the arm-sides.

2. A spicule occurring in certain hexactinellid sponges in the shape of a toothed rod pointed at one end and thickened or knobbed at the other.

1887 *Encycl. Brit.* XXII. 417/2 The *clavula*, a tylotate form with a toothed margin to the head.

clavus. Add: **2.** *Path.* and *Physiol.* **a.** A pain in the forehead, as though a nail were being driven into it, associated with hysteria. Also *clavus hystericus*. **b.** = Corn *sb.*² **c.** Applied to various excrescences and callosities.

1807 Morris & Kendrick *Edin. Med. & Phys. Dict.* I, *Clavus*..a fixed pain in the forehead, which may be covered by one's thumb, giving a sensation like as if a nail were driven into the part. When connected with hysteria, it is called *clavus hystericus*. The term is also applied to *corns*, from their resemblance to the head of a nail. **1848** Dunglison *Med. Lex.* (ed. 7) 191/1 *Clavus*,..1. A *Corn*..2. Certain condylomatous excrescences of the uterus. 3. A callous tumour, which forms on the white of the eye. **1886** *Buck's Handbk. Med. Sci.* II. 178/1 Clavus is a small, circumscribed, flattened, deep-seated, horny formation.

3. a. = Laticlave. **b.** (See quot. 1842.) Also *transf.*

1842 A. Rich in W. Smith *Dict. Gk. & Rom. Antiquities* 241/1 The *clavus* was..a band of purple colour..interwoven in the fabric...The clavus worn by the Romans was of two fashions, one broad and the other narrow, denominated respectively *clavus latus* and *clavus angustus*. *Ibid.* 242/1 Alexander Severus distinguished the equites from the senatores by the character of their clavus. **1875** S. Cheetham in Smith & Cheetham *Dict. Chr. Antiquities* I. 396/1 Clavi are common to both sexes; women may be seen represented with that ornament..in pictures of the Wise and Foolish Virgins. *Ibid.* 396/2 Priests..are said to have worn the broad clavus, while deacons contented themselves with the narrow one on their tunics or dalmatics. **1936** *Burlington Mag.* Mar. 146/1 Plate B..shows a portion of another clavus for which a treble scale-harness was used...There are other two-colour clavi. **1957** M. B. Picken *Fashion Dict.* 63/2 *Clavus*. 1. A purple band which ornamented the Roman tunic, running up over each shoulder from hem to hem...2. A band of this type and color, often embroidered, used in early church vestments.

4. *Zool.* A term variously applied to a pointed part of the hemelytron of a bug adjoining the scutellum, to a projecting part of the genitalia of spiders, etc.

1865 Douglas & Scott *Brit. Hemiptera* I. 3 The basal portion is composed of 2 pieces joined together at their sides; one narrow, adjoining the *Scutellum*, is distinguished as the *Clavus*. **1920** I. F. & W. D. Henderson *Dict. Sci. Terms* 53/1 *Clavus*, the part of an elytron lying next to the scutellum in Hemiptera. **1935** T. H. Savory *Spiders* 7 Some [female] spiders..have a downward projection from the fore-edge of the scape. This is called the crochet or clavus.

claw, *sb.* Add: **5.** Also, part of the mechanism of a lock; a device in a cine-camera or projector.

a **1877** Knight *Dict. Mech.* I. 562/1 *Claw*,..a spur or talon projecting from a bolt or tumbler. **1911** C. N. Bennett *Handbk. Kinematogr.* ii. 5 The form of hook actually employed in the intermittent movement or escapement of kinematograph cameras is called..'pin' or 'claw'. **1927** E. G. Lutz *Motion-Picture Cameraman* i. 9 In a typical camera..a pair of claws..move up and down and in and out; when they are moving up the claws are clear of the film..and when they are moving down they, having engaged with the perforations, pull the film down. **1962** *B.S.I. News* Sept. 34 Claw-to-gate distance in 8 mm cinematograph equipment.

8. *claw-leg, -mark, -scratch, -wound*; **claw-and-ball,** applied to furniture of which the feet are characterized by the representation of a claw clasping a ball; **claw-chisel,** a chisel with a serrated cutting edge; **claw clutch** or **coupling** (see quot. 1904); **claw-foot,** (*a*) a piece of furniture with feet shaped to resemble claws; also *attrib.*; (*b*) a disease causing distortion of the foot; a foot thus affected; **claw-hammer** (later N. Amer. examples); also *ellipt.* for *claw-hammer coat*; **claw-hammer coat** (earlier and later examples); **claw-lever,** a lever which divides into two claws in such a way that it can grip both sides of an article.

1902 *Connoisseur* Jan. 55/1 It [*sc.* a wine cistern] is supported by four claw-and-ball feet. **1906** *Westm. Gaz.* 23 Jan. 2/1 Perfect specimens of claw-and-ball furniture. **1955** *Times* 7 July 5/7 Carved tripods terminating in claw-and-ball feet (4 ft. high). **1960** Claw and ball [see *ball *sb.*¹ 19 b]. **1933** *Times Lit. Suppl.* 23 Mar. 195/2 The earliest marble statues show no trace of the claw-chisel, drill or gouge, and only a minimal use of the flat chisel. **1940** *Chambers's Techn. Dict.* 167/1 *Claw chisel*, a chisel, having a 2-in.-long serrated cutting edge, used for rough-dressing building stone. **1904** Goodchild & Tweney *Technol. & Sci. Dict.* 109/1 *Claw coupling* or *clutch*, a pair of flanges on the opposing ends of two shafts, with projections which engage each other when the two flanges are brought together, thus connecting the shafts so that they turn together. **1930** *Engineering* 21 Mar. 376/2 Reversal is effected by a claw clutch. **1867** Lowell *Biglow Papers* vi. 153 But the old chist wun't sarve her gran'son's wife... An' so ole clawfoot, from the precinks dred O' the spare chamber, slinks into the shed. **1881** *Harper's Mag.* Mar. 528/1 About 1700 the claw-foot side-boards, sofas and tables were generally used. **1901** Dorland *Med. Dict.* (ed. 2), *Claw-foot*, atrophy and distortion of the foot. **1920** *Glasgow Herald* 8 July 4 Foot-drop, corns and contracted toes, claw-foot. **1938** L. MacNeice *Earth Compels* 45 The claw-foot table. **1878** B. F. Taylor *Between Gates* 27 A youthful descendant of Ham, with a heel like a claw-hammer. **1881** *Harper's Mag.* Sept. 556/2 Adoniram's coat was made in the style then called straight-bodied,..by the facetious called claw-hammer. **1888** *N.Y. Sun* 29 Sept. (Farmer), Don't..call a dress-coat a swallow-tail, or a claw-hammer. **1869** 'Mark Twain' *Innoc. Abr.* 389 What a furbishing up of claw-hammer coats. **1955** in Halpert & Story *Christmas Mumming in Newfoundland* (1969) 282 Costumes used for disguise are listed: 'claw-hammer coats', hats, bonnets, [etc.]. **1922** Joyce *Ulysses* 698 Loo table with pillar and claw legs. **1892** Nasmith *Students' Cotton Spinning* 266 On the same arbor..is a claw lever engaging with the ring groove. **1828** P. Cunningham *N.S. Wales* (ed. 3) II. 36 On the claw-marks being discerned, they climb up by successive notches in the bank. **1920** *Chambers's Jrnl.* 89/2 Two parallel lines of claw-marks. **1897** 'P. Warung' *Tales Old Régime* 190 A feline claw-scratch. **1891** *Pall Mall Gaz.* 22 June 7/1 Two claw wounds on her chest.

claw, *v.* Add: **2. a.** Also fig., *to claw back*, to regain gradually or with great effort; to take back (an allowance by additional taxation, etc.); *to claw down*, to pull down, to defeat; to shoot down (an aeroplane, etc.).

1942 L. E. O. Charlton *Britain at War* 11 First steps to 'clawing them down'...A cadet is learning how to handle a machine-gun. **1953** *Economist* 21 Feb. 499/1 The Government would also make sure that..such tax relief was clawed back from surtax payers. **1957** *Ibid.* 30 Nov. 804/2 The Commercial Bank is engaged on a nationalist enterprise—clawing back from the Sassenachs, control of one of Scotland's banks. **1958** *Church Times* 8 Aug. 3/4 Conditions are even clawing down decent people. **1959** *Observer* 21 June 26/8 Relaxing round the last bend and clawing back a one-yard deficit in a prolonged battle up the long home straight. **1970** *Daily Tel.* 30 May 16 This is a special deduction which was introduced by the Finance Act 1968 to enable the Inland Revenue to 'claw back' the 10s a week increase in the allowance. **1971** *Times* 23 Jan. 18/5 The Labour Chancellor should have increased family allowances..and 'clawed' it back from richer tax payers.

7. *Naut.* (Earlier and later examples.)

1615 T. Roe *Jrnl.* 24 Aug. (1899) I. 36 Beeing then a Lee shoare, in such extremities of wynd, ther wilbe noe Possibilitye to Claw it off. **1875** 'Mark Twain' in *Atlantic* Feb. 217, I..began to claw the boat away from the danger. **1923** R. D. Paine *Comr. Rolling Ocean* xii. 202 An inch at a time, as you might say, she [*sc.* the ship] was clawing off a lee shore.

claw-back. Delete †*Obs.* exc. *dial.* and add later example.

a **1913** F. Rolfe *Desire & Pursuit* (1934) x. 93 The puffers and earwigs and clawbacks and parasites surrounding her.

2. [Substantival use of vbl. phr. *to claw back* (*claw *v.* 2 a).] Retrieval, recovery (of an allowance by additional taxation, etc.).

1969 *Daily Tel.* 16 Apr. 24/4 It is..necessary to adjust the claw-back for 1969–70 so as to reflect the fact that the 3s extra on family allowances..will be paid for a full year in 1969–70. **1970** *Sunday Times* 31 May 12 Labour's latest big increase in family allowances was accompanied by what is nastily called 'claw-back', which means that those rich enough to pay income tax at the standard rate or above have the amount of the increase clawed back from them by taxation.

clay, *sb.* Add: **1. c.** *clay-with-flints*, a mixture of stiff brown or reddish clay with angular flints, found overlying chalk, *esp.* in southern England; extended to various types of clay-flint drift deposits.

1861 Hull & Whitaker *Geol. Surv. Mem. Sheet* XIII. 55 Above the 'clay with flints' there is occasionally a loam or sandy clay. **1908** W. Johnson *Folk Memory* xi. 209 The junction of the Chalk and the overlying Clay-with-flints forms tortuous, tongue-like projections on the geological map. **1960** B. W. Sparks *Geomorphology* vii. 161 The clay-with-flints, strictly speaking a mass of rusty clay with fractured flints but in practice an omnibus term covering a variety of deposits of different types.

4. c. Phr. *feet of clay* (cf. Dan. ii. 33 'his feet part of iron and part of clay'): a fundamental weakness in someone supposedly of great merit. Also occas. *limbs of clay*.

1814 Byron *Ode to N. Buonaparte* 8 Fronts of brass, and feet of clay. **1843** Lytton *Last of Barons* III. xi. ii. 371 But still the Earl's prosperity was hollow—the statue of brass stood on limbs of clay. **1859** Tennyson *Idylls* 137 They find some stain or blemish in a name of note..And judge all nature from her feet of clay. **1865** Trollope *Belton Estate* xi, in *Fortnightly Rev.* 1 Aug. 644 The woman..finds that her golden-headed god has got an iron body and feet of clay. **1891** Wilde *Dorian Gray* xv. 269 It is the feet of clay that make the gold of the image precious. **1926** T. E. Lawrence *Seven Pillars* (1935) IX. ciii. 565, I had to avoid him [*sc.* Allenby]..lest he show feet of clay. **1952** A. Wilson *Hemlock & After* III. i. 199 Eric is sad because his idol has feet of clay.

6. b. Short for **clay pigeon*. *colloq.*

1909 in Webster. **1958** *Spectator* 1 Aug. 162/3 Clay pigeons simulated driven grouse and partridges...The unbroken clays curled down among the spectators.

9. *clay-field, -miner*; **clay-bank** (earlier U.S. example); also (*U.S.*), a horse of a yellowish-brown clayey colour; **clay-bird** = **clay pigeon*; **clay court,** a lawn tennis court with a clay surface; also *ellipt.*; **clay-land,** land in which clay is the predominant constituent of the soil; **clay-modeller,** one who makes models of clay; **clay-modelling,** the art or practice of modelling in clay; **clay-pan** (examples); (*b*) *U.S.*, a layer of clay in the soil; **clay pigeon** (orig. *U.S.*), a piece of baked clay or other material, serving as a target in trap-shooting; **clay-works,** works in which clay is prepared for use; so *clay-worker, -working*.

1755 in L. H. Gipson *Lewis Evans* (1939) 172 Kentucke.. has high Clay Banks. **1853** *Oregonian* (Portland) 5 Nov. 1/6 Getting off an' hitching his ole clay-bank to a swinging limb. **1897** *Outing* (U.S.) XXX. 392/1 There were several clay-bird matches. **1909** *Westm. Gaz.* 22 Mar. 10/1 The Stock Exchange clay-bird shooting competition. **1916** *Daily Colonist* (Victoria, B.C.) 4 July 9/3 For the third time since the introduction of the national clay court tennis tournament the championship in the men's singles was captured by a star from California. **1958** L. Hoad *My Game* 224 Get practice on as many different courts as possible. Work out for yourself how the change from grass to clay or to concrete will affect your game. **1964** R. Laver *How to play Winning Tennis* ix. 73 On very soft clay courts where the ball leaves a mark as big as a footprint, I found that it was common for puffs of chalk to rise as a ball landed. **1843** *Knickerbocker* XXII. 4 Every shade of greenness is lost in the interminable red clay-fields. **1909** *Daily Chron.* 11 June 5/5 The working of the extensive clayfield. **1707** Mortimer *Husb.* vi. 73 This sort of Marle did very well upon Clay Lands. **1949** W. G. Hoskins *Midland England* i. 2 Claylands that rarely stand six hundred feet above the sea. **1881** *Instr. Census Clerks* (1885) 81 Clay

Miner. **1909** *Westm. Gaz.* 10 Feb. 7/2 The clay-miners entombed by a fall of silver-sand at Morden clay-pits, near Corfe Castle. **1881** *Instr. Census Clerks* (1885) 55 Clay, Ornamental Modeller. **1898** *Westm. Gaz.* 5 Jan. 2/3 Clay-modelling talent. **1838** Colman *Rep. Agric. Mass.* 78 This drain is four feet wide at the top, and goes down some small depth into the clay pan. **1875** J. Forrest *Explor. Australia* v. 260 We travelled..over stony plains; many clay-pans with water, but no feed. **1890** *Melbourne Argus* 2 Aug. 13/3 The baked surface of the claypan. **1930** A. Groom *Merry Xmas* xvii. 127 The claypan was torn up into a soft powder from the rushing and driving of thousands of cattle across it. **1954** B. Miles *Stars my Blanket* xxiv. 216 Between the sandhills there were claypans, flat and hard and white. **1888** *Outing* (U.S.) Sept. 501 He had broken innumerable glass balls and 'clay pigeons' at a trap. **1895** *Army & Navy Co-op. Soc. Price List* 15 Sept. 958 The Anglo-American Ligowsky Clay Pigeon Traps. **1958** Clay pigeon [see sense *6 b]. **1909** *Daily Chron.* 11 June 5/5 Showing them over the clayworks. **1902** *Ibid.* 21 May 6/3 The British Institute of Clayworkers. **1900** *Engineering Mag.* XIX. 793/1 (*heading*) The Clay-Working Industry of the Pacific Coast.

clay-cold, *a.* (Later examples.)

1885 W. B. Yeats in *Dublin Univ. Rev.* July, Quiet as the clay-cold dead. **1946** Dylan Thomas *Deaths & Entrances* 16 Though the brawl of the kiss has not occurred, On the clay cold mouth.

clayeyness (klēⁱ·inės). [f. Clayey *a.* + -ness.] Clayey nature or quality.

1888 Lowell *Let.* 13 Aug. (1894) II. 400 Slippery from clayeyness of soil.

claymore. Add: **2.** A type of anti-personnel mine. In full *claymore mine*.

1962 *Ordnance Techn. Terminol.* (U.S.) (PB 181–465) 72/2 Claymore, name given to a type of antipersonnel mine designed to produce a directionalized, fan shaped pattern of fragments. **1965** M. Browne *New Face of War* iv. 35 American arms designers have produced a mine called a 'Claymore', which has found wide use here [in Vietnam]. The Claymore, curved like a horseshoe, has an optical sighting device and hurls a blast of shrapnel directionally at the point toward which it has been pre-aimed. **1966** *New Statesman* 9 Sept. 341/1 One Claymore mine would finish this mob. **1966** *New Scientist* 17 Nov. 369/1 The claymore, whose effects can be devastating among close-packed ranks of men, was invented to deal with the wave formations of Chinese during the Korean war.

clayness (klēⁱ·nės). [f. Clay *sb.* + -ness.] = *clayeyness.

1850 *Rep. U.S. Comm. Patents Agric. 1849* 115 The soil ..is one of medium clayness.

claytonia (klēⁱtōᵘ·niă). [mod.L. (J. F. Gronovius in *Linnæus's Genera Plantarum* (1737) 339), f. the name of John Clayton, Virginian botanist (1686–? 1773): see -ia¹.] A plant of a genus of small herbs so called, belonging to the family Portulacaceæ, and native to North America, Siberia, and Australasia; = *spring-beauty* (see Spring *sb.*¹ 7 c).

1789 W. Aiton *Hortus Kewensis* I. 284 Virginian Claytonia...Spear-leav'd Claytonia. *Nat.* of North America. **1818** *Mass. Hist. Coll.* VIII. 168 Among our herbaceous wild plants, the first that appear are the delicate claytonia, the..hepatica [etc.]. **1837** J. Paxton *Let.* 10 Oct. in W. Baxter *Brit. Phænogamous Bot.* (1839) IV. pl. 253, I have this morning gathered specimens of *Claytonia* from the very places in which I originally found it. **1880** *Scribner's Monthly* May 101/2 A good sample of our native purslane is the Claytonia, or spring beauty. **1934** A. Russell *Tramp-Royal* xx. 124 All the best stock-fattening plants peculiar to the region—the two succulent claytonia—the parakylia and munyeroo. **1963** *Austral. Encycl.* I. 34/1 The dampers form a staple food among inland tribes where several plants, particularly claytonia and nardoo, produce enormous quantities of seeds.

clean, *a.* Add: **1. e.** Of a vessel: clear of advance commission, dispatch money, and other charges, which may constitute deductions from the freight.

1890 *Whitby Gaz.* 17 Jan. 2/7 *Thomas Turnbull*, Cardiff to Cape Verde, 9s. 6d., clean. **1891** *Ibid.* 18 Sept. 3/2 *Stakesby*, Cardiff to Gibraltar, 6s. 6d., clean and nett.

2. a. *clean air*, unpolluted atmosphere; freq. *attrib.*

1927 *Lancet* 26 Mar. 685/1 Smethwick in Staffordshire held a 'Clean Air Week' from March 14th to 19th, during which period an effort was made to keep the atmosphere free from smoke. **1957** *Ann. Reg. 1956* IV. 348 The passage of a Clean Air Act..was welcomed by smoke abatement enthusiasts and by all concerned with public health. **1958** *Times* 2 June p. xii/4 The problem of securing clean air, that is to say an atmosphere throughout the whole country relatively free from pollution and contamination.

3. c. (Later examples.) Also of a document, report or the like: bearing no adverse point or remark; listing no offence; esp. in phr. *clean sheet* (also *fig.*).

1858 Simmonds *Dict. Trade, Clean Proof*, a slip or sheet of printed matter sent to an author from a printer. **1921** *Daily Colonist* (Victoria, B.C.) 11 Mar. 6/3 Following one of the heaviest dockets in recent months, that of last Wednesday, yesterday's calendar at the City Police Court showed a clean sheet; and Magistrate Jay dispensed with the formality of taking his seat on the bench. **1932** *Discovery* May 153/1 The original field maps and 'clean copies' of them are preserved at the office. **1963** *Times* 21 Jan. 6/5 Seventy-two of the 75 Stockholm starters passed through the control point at Frankfurt. Among those with a 'clean sheet' was Erik Carlsson. **1965** L. Sands *Something to Hide* v. 75 Her licence had been clean..she herself had never put through an insurance claim before.

d. *to make a clean job* (*of it*): to do (something) thoroughly; *to come clean* (orig. *U.S.*): to confess, to make a clean breast of it. *colloq.*

1892 'F. Anstey' *Voces Populi* II. 50 That may be Mithras's notion of making a clean job of it, but it ain't *mine!* **1915** Conan Doyle *Valley of Fear* II. iii. 203 And looked with the deepest respect at the man who had the reputation for making what they called a 'clean job'. **1919** C. H. Darling *Jargon Book* 8 *Come clean*, to confess everything. **1925** Wodehouse *Sam the Sudden* xiii. 95 You'd best come clean, Soapy, and have a showdown. **1928** *Daily Express* 10 Oct. 6 Spill the beans. Come clean on the whole game. **1958** *Listener* 2 Oct. 514/1 Why not come clean and call 'graduated unemployment compensation' the national dividend? **1959** J. Cary *Captive & Free* 28, I was wasting my time, because you kept dodging. You never come clean.

e. Of deer or their horns: that have shed their velvet.

1838 W. Scrope *Art Deer-Stalking* i. 6 When they have shaken off this skin, which is called the velvet, and which disappears in the months of August and September, they are said to have clean horns. **1897** J. G. Millais *British Deer* 95 Park stags are clean sometimes as early as 20th July. **1926** Hannan *Isle of Mull* 173 The stags were not 'clean' any earlier than usual.

f. Of a nuclear device or weapon: producing relatively little radioactivity or fall-out.

1956 in *Amer. Speech* (1957) XXXII. 136 A 'clean bomb'—a bomb with little or no radioactive side effects. **1957** *Observer* 1 Sept. 3/2 If there really is a 'clean' bomb.. then the whole deterrent power of nuclear weapons has been tragically graded down. **1959** *Listener* 12 Mar. 444/2 Smaller, clean, measurable nuclear effects which will revolutionize bombing and missile strategy.

4. c. In phr. *the clean thing*: the honest, straightforward thing. *U.S.*

1835 D. Crockett *Tour* 193 (Bartlett), I don't like it. It isn't the clean thing. **1855** Haliburton *Nat. & Hum. Nat.* I. 100 Hang me if he can be the clean thing anyhow he can fix it. **1871** *Washington Patriot* 3 Apr. (De Vere), It would have been the clean thing to say at once that no debate would be allowed.

d. *clean ticket* (see quot. 1859). *U.S.*

1848 *N.Y. Weekly Tribune* 10 June 4/5, I go with my friend, Mr. Van Buren, for a clean ticket. **1859** Bartlett *Dict. Amer.* (ed. 2), *Clean Ticket*, the entire regularly nominated ticket at an election; a ticket without any erasures. 'He went the clean ticket on the Whig Nominations.' **1872** Schele de Vere *Americanisms* 269 Only when a man adopts the whole list as made up by his party, is it called a *clean* or *straight* ticket.

e. Of writings, entertainment, etc.: free from obscenity or indelicate treatment of sexual matters; so *good, clean fun*: enjoyment or entertainment devoid of salacity or obscenity (freq. used ironically).

1867 *Belgravia* IV. 48 Since Sir Walter Scott's days English novels have been 'sane, clean, and wholesome'. **1908** *Westm. Gaz.* 26 Oct. 1/2 M. René Bazin..has.. become known on this side of the Channel as the 'cleanest' of contemporary French novelists. **1911** *E. J. Muddle's Picture Plays* 10 The public..demands good, clean, wholesome pictures, to which it can take every member of its household, from the youngest upwards. **1923** *Time* 24 Mar. 16/1 Abounds in good, clean fun. **1929** Priestley *Good Compan.* II. i. 249 He's one of the best comedians going in Concert Party work—clever, and keeps it clean. **1938** E. Hemingway *Fifth Column* (1939) I. iii. 25 Brought Anita around for a little spot of good clean fun. **1950** D. Gascoyne *Vagrant* 53 As resoundingly and full-bloodedly as you like so long as it's only in good clean working-class fun!

13. b. Unfeathered: said of pigeons which have no feathers on the legs or feet.

1886 G. Ure *Our Fancy Pigeons* viii. 201 All the best performing tumblers that I have ever seen had clean legs. **1961** W. H. Allen *How to raise Pigeons* iii. 42 There are two main types, classified as clean-legged and ruffled (with feathers on legs and feet).

c. Of aircraft: designed to minimize obstructions to the airflow, streamlined. Cf. *cleanness 3.

1916 H. Barber *Aeroplane Speaks* 27 A brand new, rakish, up-to-date machine it is..perfectly 'streamlined' to minimise drift..'Clean looking 'bus..Ought to have a turn of speed with those lines.' **1939** *Flight* 14 Dec. p. g/1 It [*sc.* the speed] is certainly not obtained by aerodynamic cleanness. No one would call the Messerschmitt a clean aircraft. **1961** *Engineering* 20 Jan. 123/1 The clean-wing (aft-engine) de Havilland Trident..due to fly this year.

14. (See also Sweep *sb.* 1 a.)

clean-. Add: **1. a.** *clean-shaved* (example), *-shaven, -swept*. **b.** *clean-minded*.

1886 'Vernon Lee' *Baldwin* 232 Is the novel..to appeal to a public..less clean-minded than the public of the poet? **1939** Joyce *Finnegans Wake* 33 The big cleanminded giant. **1888** [see Shaved *ppl. a.* 1 b]. **1932** W. Cather *Obscure Destinies* 23 But his neck, always clean shaved except in the business seasons, was not loose or baggy. **1863** Longfellow *Wayside Inn* 9 Clean shaven was he as a priest. **1884** W. James *Ess. Radical Empir.* (1912) xii. 277 The 'through-and-through' philosophy..reminds many of us of that clergyman. It seems too buttoned-up and white-chokered and clean-shaven a thing. **1876** W. C. Bryant *Lifetime* in *Poet. Wks.* (1883) II. 195 Clean-swept fireplace. **1919** V. Woolf *Night & Day* iii. 36 A clean-swept morning of empty, secluded hours.

2. clean-bowl *v.*, in *Cricket*, to bowl out (a batsman) or bowl down (a wicket) with a ball that hits the wicket without having touched the bat or body of the batsman; **clean-living** *a.*, of upright character; **clean-run** *a.*, describing a well-fed, bright-looking salmon newly returned to fresh water from the sea; also *fig.* of a young person; **clean-skins,** also in *sing.*, and (*slang*) a person with a clean police record (see also quot. 1941); also **clean-skin** *a.*, unbranded.

[**1860** *Bell's Life* 29 July 3/4 Carpenter was bowled 'clean' by Slinn.] **1862** *Baily's Mag.* July 34 Young Nixon.. bowled 31 overs..for 36 runs and 4 wickets, all fairly, well, and clean bowled. **1881, 1888** [see Clean *adv.* 5 b]. **1891** W. G. Grace *Cricket* xi. 325 Allen Hill..clean-bowled Messrs. A. N. Hornby, C. F. Buller, G. F. Grace and myself. **1927** G. A. Terrill *Out in Glare* i. 7 He would be clean bowled for a 'duck'. **1920** R. Macaulay *Potterism* III. ii. 126 Straight, clean-living, decent men and women. **1932** L. Golding *Magnolia St.* II. x. 407 You're such a nice clean-living chap. **1881** [see Run *ppl. a.* 4]. **1915** F. M. Hueffer *Good Soldier* III. iii. 163 The girls..appeared to her to be so clean run and so safe. **1926** W. J. Locke *Old Bridge* I. i. 13 He was a clean-run, brown-haired, blue-eyed youth. **1931** F. D. Davison *Man-Shy* (1934) ix. 130 She was not a cleanskin; the Mirramilla brand was on her rump. **1934** *Bulletin* (Sydney) 1 Aug. 46/3 Lifted them cleanskin micks while Morney was in town. **1936** M. Franklin *All that Swagger* ix. 83 Delacy began the trapping and branding of cleanskin cattle. **1941** Baker *Dict. Austral. Slang* 18 *Cleanskin*, a person of integrity, esp. in a political sense. **1945** —— *Austral. Lang.* vii. 141 A man who has had no convictions recorded against him is a *cleanskin*. **1950** 'N. Shute' *Town like Alice* 263 A poddy's a cleanskin, a calf born since the last muster that hasn't been branded. **1967** C. Drummond *Death at Furlong Post* x. 126, I just dictated a report that they seem clean-skins. **1969** *Daily Mirror* (Sydney) 12 Mar. 11/4 Had he been a clean-skin..Mr Byrne might have..not recorded a conviction.

clean, *v.* Add: **1. e.** *slang.* = 4 b.

1915 Jackson & Hellyer *Vocab. Criminal Slang* 24 *Clean*, to take all one possesses of a given commodity; to deplete one's assets. Example: 'He headed in wrong with that bunch and got cleaned.' **1926** J. Black *You can't Win* iv. 27 The poker players..who brought their victims into the back room to 'clean' them. **1950** A. Lomax *Mr. Jelly Roll* (1952) ii. 54 If they didn't clean this sucker by legitimate cheating, one of the tough guys would take his money anyhow.

f. To 'clean' (an aeroplane) aerodynamically, to make streamlined. Also with *up*. So *cleaning-up* vbl. sb. Cf. *clean *a.* 13 c.

1922 *Flight* XIV. 702/2 The 'cleaning-up' of the latest model has resulted in an astonishing gain in maximum speed. **1931** *Ibid.* 16 Jan. 61 The Short 'Singapore' Mark II has been cleaned up by placing the four Rolls-Royce 'F' engines in tandem, and by abolishing the chine struts. *Ibid.* 16 Oct. 1033/1 This new aircraft has been cleaned cleaner than any Armstrong-Whitworth aircraft we have seen.

2. b. *Naut.* To change into an appropriate uniform, to don a uniform prescribed for any set duty. *colloq.*

1915 'Bartimeus' *Tall Ship* iii. 48 The Captain had 'cleaned' already. **1916** G. Franklin *Naval Digression* I. ix. 70 We..perform that seemingly impossible feat of 'cleaning' into a dirty rig.

3. (Earlier examples.)

1839 in J. S. Bassett *Plantation Overseer* (1925) 117, I have got my cotton land the half of it cleaned up. **1847** C. Brontë *Jane Eyre* III. viii. 183 My first aim will be.. to *clean down* Moor-House from chamber to cellar.

4. *to clean out.* **c.** To defeat or deal effectively with (a person); to eject from a place. *U.S.*

1858 in *Kansas Hist. Coll.* XIV. 99, I could clean you out quicker than greased lightning would pass a funeral. **1863** *Harper's Mag.* Sept. 569/2 'All right,' says Rhind; 'fire, boys!' and in a very short time *that* ambush was 'cleaned out'. **1871** *Congress. Globe* 5 Jan. 316/1 The enemy did take possession of the house one day. They were 'cleaned out' as we say; they were compelled to leave the house. **1892** J. L. Ford *Dr. Dodd's School* i. 10 He could lick the whole crowd of them with one hand tied behind his back. Do you remember how he cleaned out the townies that Saturday afternoon? **1908** S. E. White *Riverman* xxii. 197, I don't bet those Saginaw river-pigs are any more two-fisted than the boys on this river. I'd go up and clean 'em out.

d. To clear (a place) of the persons occupying it. *U.S.*

1858 in *Kansas Hist. Coll.* (1896) V. 567 These same men attacked Barnesville..and literally cleaned it out, both of inhabitants and property. **1883** G. W. Peck *Peck's Bad Boy* No. 2. 29 Pa got mad and said he could clean out the whole shebang. **1901** M. E. Ryan *Montana* vii. 97 They..would proceed to 'clean out' any establishment where their own peculiar set was ignored. **1902** S. E. White *Blazed Trail* vi. 48 Nothing pleases him better than..to embark on an earnest effort to 'clean out' a rival town.

6. *to clean up* (see sense 3). Various *colloq.* and *slang* uses (chiefly *U.S.*): **a.** *trans.* To obtain, acquire as gain or profit.

1831 J. M. Peck *Guide for Emigrants* 147 He gave a friend one measured acre..and cleaned up thirty-five bushels and eight quarts [of wheat]. **1904** 'O. Henry' *Cabbages & Kings* vi. 104 So, this man I was speaking of cleaned up $288; and then he went away. **1906** A. T. Jackson *Diary of Forty-Niner* (1920) 51, I hope to clean up about ten thousand dollars. **1929** [see *6 e]. **1931** L. A. G. Strong *Garden* xxx. 278 He had seen them clean up all three places in the Open Quarter; he had seen them, individually, win a score of races. **1960** *20th Cent.* Dec. 555 A concerted drive to ensure that this 25-year-old veteran cleans up another £16 million.

b. *trans.* To beat, vanquish; *spec.* in gambling, to make a large profit from, to take all the money from.

1888 P. H. Sheridan *Memoirs* I. 47 The citizens..intended cleaning up the hostiles. **1932** J. Dos Passos *1919* 43 They..shot some pool and Joe was pretty good and cleaned up the local boys. **1932** J. T. Farrell *Young Lonigan* (1936) vi. 136 He pretended that he had cleaned up all the tough guys on Fifty-eighth Street. **1949** *N.Z. Free Lance* 24 Aug. 10 You jokers cleaned us up at footie.

c. *trans.* To clear (a place, etc.) of harmful or immoral influences, elements, or persons; to rid (an area) of remaining pockets of enemy resistance. So *cleaning-up* vbl. sb.

1916 Du Puy *Uncle Sam* 170 The cleaning up of the customs scandals in the port of New York was a most complicated task. **1925** C. E. Mulford *Cottonwood Gulch* 188 Our friend Dangerfield [the sheriff] will clean up this cursed country like a new broom. **1929** *Times* 15 Feb., More than 3,000 persons were arrested..by the Chicago police in a new effort to 'clean up' the city. **1930** *Times Lit. Suppl.* 1 May 359/1 American troops engaged in 'cleaning up' that distressful republic. **1942** P. Jordan *Diary* 29 Nov. (1943) 64 British parachutists have been dropped..at Depienne, where they are to clean up an airfield. **1943** *Ann. Reg. 1942* 128 After North Africa had been 'cleaned up', Parliament would be asked to approve a proposal that troops of the Union should serve outside Africa. **1967** G. Freeman *Undergrowth Literature* xi. 163 A political broadsheet published in California in 1966 calling for the banning of pornography had, as one of the arguments in favour of legally 'cleaning up' the state, [etc.].

d. *trans.* To strip or empty the contents of.

1917 Wodehouse *Uneasy Money* xv. 169 Then the man comes down for a visit and goes about cleaning up the neighbouring houses.

e. *intr.* To make a large profit.

1929 D. Hammett *Red Harvest* xii. 119 You thought..you could clean up a little dough playing him?..Did you clean up? **1941** B. Schulberg *What makes Sammy Run?* i. 11, I mean profit. That show must be cleaning up. **1947** Steinbeck *Wayward Bus* 36 It's the fastest-selling novelty I've ever handled. Little Wonder is cleaning up with it.

cleaner. Add: **a.** Also *cleaner up*.

1908 *Daily Chron.* 24 Aug. 9/6 Pianos.—Fitters up; also a cleaner up. **1921** *Dict. Occup. Terms* (1927) § 499 *Cleaner, cleaner-up* (pianos); scrapes (mucks off) and rubs veneered parts of piano with sandpaper or glasspaper, preparatory to polishing.

c. A shop that cleans clothes and household fabrics; = *dry-cleaner* (see *dry *a.* C. 2 a). Freq. in pl., *the cleaners* (or possessive, *the cleaner's*).

1873 *Young Englishwoman* Mar. 155/3 Imitation furs seldom clean well...The best plan would be, to send to a good cleaner's. **1928** R. Hall *Well of Loneliness* xxiii. 208 Shall I wash your new crêpe de Chine nightgowns, ma'am? Or ought they to go to the cleaners? **1967** C. Fremlin *Prisoner's Base* xi. 72 Some problem about getting a skirt altered and back from the cleaners in time for the weekend.

d. A small marine animal which cleans larger ones of parasites, bacteria, or dead tissue; also *attrib.* and *Comb.*, esp. *cleaner-fish*.

1955 *Zeitschr. für Tierpsychol.* XII. 219 All typical cleaners, irrespective of their zoological relationship, possess striking yellow colours. **1958** *Listener* 14 Aug. 247/2 The huge manta rays..opening their gills to let in the tiny cleaner-fish. **1967** *Daily Tel.* 21 Oct. 13/5 It is known that a cleaner-fish occupies a niche in a reef which is usually marked by a brightly-coloured anemone serving the same purpose as a barber's pole. **1968** R. D. Martin tr. *Wickler's Mimicry* xiv. 158 We now recognise 42 species of cleaner-fish in 14 families.

2. Slang phr. *to take* (someone) *to the cleaners*: (*a*) to take away, or defraud of, all someone's money; so *to go, send, to the cleaners* (orig. *U.S.*); (*b*) to criticize strongly.

1932 *Flynn's* 28 May 49/2, I dropped the biggest wager I had ever made on anything...That sent me to the cleaners, wiped out. **1949** H. MacLennan *Precipice* i. 71 He..had..played several more sets of tennis with him. He had taken Carl to the cleaner's this time. **1950** R. Chandler *Simple Art of Murder* 19 I'm in a jam. But I'm not going to the cleaners...Half of this money is mine. **1961** *Guardian* 13 Sept. 8/3 Many a gilded youth..has been 'taken to the cleaners' once too often at midnight parties. **1963** *Listener* 14 Feb. 308/3, I hoped Mr Carr might round on Mr Cousins and start taking the apprenticeship system to the cleaners. **1963** *Ibid.* 7 Mar. 436/1 We..heard the recently fired Captain Atkins take the Atomic Energy Authority to the cleaners over their marine nuclear policy.

cleaning, *vbl. sb.* Add: **1. c.** *Forestry.* The cutting of trees or undergrowth which have a deleterious effect upon the principal trees in a stand.

1891 W. Schlich *Man. Forestry* II. iii. 196 The cleaning of a young wood has for its object to remove in good time all growth which interferes with the proper development of the principal species or individuals. *Ibid.* 197 It is often desirable to make several cleanings. **1953** H. L. Edlin *Forester's Handbk.* xii. 179 One main objective in cleaning is the removal of 'tree weeds', shrubs or bushes that threaten the welfare of the desired crop.

cleaning, *ppl. a.* That cleans, in various senses of the verb; *cleaning crop*, a crop serving to clear land in cultivation from weeds.

1900 T. Shaw *Soiling Crops* iii. 38 Sorghum..may be placed..between two grain crops, as then it can be grown as a cleaning crop. **1921** *Discovery* Feb. 48/2 The farmer..tries periodically and systematically to rid his fields of [weeds], and takes what he calls a cleaning crop. **1948** T. Wilder *Ides of March* II. xxxvi. 105 At last a cleaning woman came in to wash the floor. **1955** L. P. Hartley *Perfect Woman* xxiv. 209 Harold went betimes to his office, because he wanted to be there before his cleaning woman or his secretary. **1960** *Encounter* XIV. III. 80 Half-senile cleaning-ladies. **1969** *Times* 20 Feb. 17/5 Some 100,000 gallons of the cleaning fluid..were buried in a gold mine. **1970** *Sat. Rev.* 21 Mar. 63/1, I met no blacks on a personal basis, other than our cleaning lady.

cleanness. Add: **3.** Of aircraft: the state or quality of having 'clean' lines (cf. *clean *a.* 13 c).

1925 *Sci. Amer.* Apr. 269/2 What does this limiting speed depend upon? The weight of the airplane and the 'cleanness' of its design. **1933** *Meccano Mag.* Feb. 108/2 The machine has been designed with a very efficiently stream-lined shape, and its absolute 'cleanness' has enabled it to have a high top speed. **1939** [see *clean *a.* 13 c].

cleansing, *ppl. a.* Add: *spec.* of Cosmetics.

1926–7 *Army & Navy Stores Catal.* 492 An Elizabeth Arden treatment is based on..Cleansing with Venetian Cleansing Cream. **1938** *Eve's Jrnl.* Jan., Care of the face: Cleansing cream Skin tonic Skin food. **1965** *Harper's Bazaar* 89/1 Hand Cream, Foundation Cream, Cleansing Milk. **1969** J. Fraser *Clap Hands* ii. 22 She took a jar of cleansing cream from the cabinet.

clean-up. [See Clean *sb.*] An act of cleaning or cleaning up (see *clean *v.* 6).

1921 *Daily Colonist* (Victoria, B.C.) 9 Apr. 9/3 The Mayor stated that Wednesday, April 20, had been decided upon as clean-up day, and asked for the co-operation of the council. **1924** R. Macaulay *Orphan Island* ii. 22 We had a great clean up and a grand rummage among your Grandfather's old things. **1940** 'N. Blake' *Malice in Wonderland* I. vii. 97 He..has a general clean-up, puts on more respectable clothes. **1960** *Design* July 39/2 These sections are urgently in need of a typographic clean up.

b. spec. in *Mining.* (See Clean *sb.*) orig. *U.S.*

1866 *Congress. Globe* 18 June 3231/1 When what they technically call in mining the clean up comes, very often the clean-up exhibits the lofty sum of nothing. **1872,** etc. [see Clean *sb.*]. **1892** *Daily News* 27 Aug. 7/3 Palmarejo.—Result of the clean-up for July, 28,000 dols. **1908** *Daily Report* 20 July 3/3 There will be some prolongation of the life of the outcrop area from the clean-up of the old stopes. **1926** *Daily Colonist* (Victoria, B.C.) 13 July 3/2 The plates became so heavily loaded with gold that it was necessary to stop the mill and make a clean up.

c. A profit; an exceptional financial success; also, a robbery or its proceeds. *slang* (orig. *U.S.*).

1878 F. H. Hart *Sazerac Lying Club* (ed. 2) III. 21 At the same time make a nice little clean-up for himself. **1928** E. Wallace *Gunner* xxiv. 199 Sometimes weeks would pass before the gang could make a good clean-up. **1929** Wodehouse *Mr. Mulliner Speaking* 265 It was the man's intention to make what I might term a quick clean-up immediately after dinner and escape on the nine-fifty-seven. **1946** K. Tennant *Lost Haven* (1947) viii. 111 He was now a hundred pounds in debt; but that, for Alec, was practically no debt at all; one good clean-up..and he would be clear.

d. An act of removing or stamping out harmful or immoral influences or elements; also, a cleaning out of remaining pockets of (enemy) resistance. Cf. *clean *v.* 6 c.

1930 *Amer. Speech* VI. 114 Seven arrested in liquor clean-up. **1934** S. Spender *Vienna* iii. 32 These murders lie on Fey Who boasted of his clean-up. **1935** C. Isherwood *Mr. Norris* xvi. 270 Olga..had escaped the clean-up through the influence of one of her customers, an important Nazi official. **1940** M. Sadleir *Fanny by Gaslight* I. viii. 70 The clamour for a London clean-up rose steadily throughout the year 1870; and as the appetite for victims grew among moralists and police alike, numerous informers helped to feed it. **1958** *Time* 21 July 20 Corruption clean-ups.

clear, *a.* Add: **7. c.** Not in cipher or code. Often *absol.*, *in clear.*

1928 P. B. Thomas *Secret Messages* iii. 13 Taking the letters in the message or 'clear text' successively, F was substituted for T. **1930** *N. & Q.* 23 Aug. 144/1 A long letter in numerical cipher, which, since he does not give it in clear, apparently still awaits decipherment. **1932** D. L. Sayers *Have his Carcase* xxviii. 367 The code pair [of letters] and the clear pair have a letter in common. *Ibid.* 368 The letter E appears both in code and clear. **1966** M. R. D. Foot *SOE in France* viii. 205 These forms—in clear—were as a rule kept in Girard's study.

13. b. *Phonetics.* Designating one of two varieties of lateral consonants (the other being called 'dark') (see quots.).

1918 D. Jones *Outl. Eng. Phonetics* ix. 45 In clear varieties of *l* the front of the tongue is raised in the direction of the hard palate, while in dark varieties of *l* the back of the tongue is raised in the direction of the soft palate. In other words, clear *l*-sounds have the resonance of front vowels, whereas dark *l*-sounds have the resonance of back vowels. **1962** A. C. Gimson *Introd. Pronunc. Eng.* v. 47 The so-called 'clear' [l] with a front vowel resonance.

25. b. In technical or trade use.

1739 Hempstead *Diary* (1901) 356 White pine bords..& one bord 47 foot clear stuff a wide & thick one for Table Leaves. **1822** J. Woods *Two Yrs.' Resid. Eng. Prairie* 211 A hundred middling ears of corn will yield a bushel of clear corn. **1851** C. Cist *Cincinnati* 214 Pig-iron, and one thousand tons Tennessee clear blooms. *Ibid.* 281 The inspection laws require that clear pork shall be put up of the sides, with the ribs out. **1867** *Trans. Ill. Agric. Soc. 1865–6* VI. 645 Clear flooring, rejected on account of thickness, shall be classed with common flooring. **1889** *Century Dict.*, *Clear*, without admixture, adulteration, or dilution; as, a fabric of clear silk; clear brandy; clear tea. **1908** *Practitioner* Dec. 841 The furnace men may be affected..by an escape of gas at some defective joint. It is known as 'clear gas'. **1917** F. S. Henry *Printing for School & Shop* vi. 80 Such an attempt [to produce a letter of maximum legibility] on the part of L. B. Benton and his son Morris has resulted in the production of the type face known as Clearface...The general effect of Clearface is that it presents a uniform amount of white throughout the line. **1967** Karch & Buber *Offset Processes* v. 149 Uncoated film is known as clear back or clear base.

B. *adv.* **4. b.** *clear-away*: entirely, completely.

1883 Stevenson *Treasure Isl.* xxxii. 271 And come to think on it, it was like Flint's voice, I grant you, but not just so clear-away like it, after all.

5. Delete †*Obs.* and add later U.S. and dial. examples.

1835 Longstreet *Georgia Scenes* 270 Well I'm clear put out. **1845** Mrs. Kirkland *Western Clearings* 78 I'm clear tuckered out with these young 'uns. **1886** M. Peacock *Tales & Rhymes* 69 But boggard doesn't feal clear suited.

C. *sb.* **5.** (Earlier Amer. example.)

1674 in *Archives of Maryland* (1884) 405 The State house to be..with a porch in front sixteene foote long and twelve foote broad in the Clear on the Inside.

b. Colloq. phr. *in the clear*: (*a*) out of reach; (*b*) unencumbered; free from trouble, danger, suspicion, etc.; (*c*) having a clear profit. orig. *U.S.*

1901 J. Flynt *World of Graft* ii. 18 Even down-town they [*sc.* policemen] are likely to be in the clear when you want 'em most. **1928** J. O'Connor *Broadway Racketeers* xvii. 185 When we counted up the final night we figured nearly sixty grand in the clear. **1930** C. F. Coe *Gunman* xii. 235 All I'm trying to do..is keep things running smooth until we're in the clear again. **1934** *Passing Show* 26 May 11/3 You saw an opportunity of fixing all the blame on your partner, leaving yourself in the clear. **1948** 'N. Shute' *No Highway* 111 He had made a thorough inspection and had found everything correct. That put him in the clear. **1948** *Esquire* Mar. 49/2 After a few years he was $3500 in the clear.

6. b. With adverbs: *clear-out*, an act of clearing out (see Clear *v.* 26); *clear-up*, an act of clearing up, *spec.* the settlement of accounts (see Clear *v.* 27 g); also *attrib.*

1901 *Westm. Gaz.* 24 May 7/1 The clear-up will have to come sooner or later. **1923** Manchon *Le Slang* 87 To have a clear out, *aller à la selle, se vider.* **1928** (*From a Stock-broker's form*) 29 May, Paid..in respect of the clear-up Dividend. **1959** *Sunday Express* 1 Feb. 5/3 A large-scale clear-out of Victoriana from Windsor Castle to provide more room for the stored gifts. **1968** 'P. Alding' *Circle of Danger* ii. 14 I've just been examining last month's crime figures...The clear-up rate looks very bad.

D. 3. clear-air gust or **turbulence,** disturbance of the atmosphere at high altitudes; **clear-cutness,** the quality of being clear-cut; **clear-cutting, -felling,** the cutting down and removal of every tree in a given area; hence *clear-fell, clear-felled* adjs.; **clear-skin** *Austral.*, an unbranded beast (cf. *clean-skin*); also *attrib.*; **clear-way, clearway,** (*a*) (see quot. *a* 1884); (*b*) a path or passage-way; (*c*) a road on which vehicles are not allowed to park or wait.

1948 *Shell Aviation News* No. 122, 3/1 These clear air gusts—areas of violently turbulent atmosphere which have been found to exist at heights of 20,000 ft. to 40,000 ft. **1955** O. G. Sutton *Sci. of Flight* 60 Strong upward currents and bumpy flying conditions..can arise in air free from clouds. Clear-air turbulence, as it is called. **1969** *New Scientist* 9 Jan. 57/1 Until such time as clear-air turbulence can be detected, something can be done to safeguard aircraft. **1926** *Spectator* 1 May 791/1 In clear-cutness of outline,..it [*sc.* a speech] was a masterpiece. **1943** A. L. Rowse *Spirit Eng. Hist.* i. 11 The lack of clear-cutness which makes us as a people difficult..to understand. **1922** W. Schlich *Man. Forestry* (ed. 4) I. ii. 96 Planting new ground with conifers..should also be admissible in the case of clear-felled and devastated areas. *Ibid.* 97 The clear-felling of the first planted areas. *Ibid.* 159 The system of clear-cutting followed by planting has lately been recommended. **1927** *Forestry* I. 11 In the stress of war-time felling also, clear cutting was the only method possible. **1928** R. S. Troup *Silvicultural Systems* ii. 4 The clear-cutting system. Under this system successive areas are clear felled and regenerated. **1962** *Times* 1 Jan. 6/4 The normal clear-fell and replant industrial or 'tree-farming' systems of forestry. **1884** 'R. Boldrewood' *Melb. Mem.* xv. 109 Calves and clear-skins. **1888** —— *Robbery under Arms* I. ii. 22, I hadn't lived all my life on Rocky Creek..without knowing what 'clearskins' and 'cross' beasts meant. **1900** H. Lawson *Over Sliprails* 135 'Carn't you see it's a clear skin?'..The idea of a '"clear skin" steer' would have amused her at any other time. **1941** *Coast to Coast 1941* 22 Some steers had got out of his holding paddocks, and he offered Wally two bob for every steer he could track and bring in. Wally took in a couple of clear-skins. *a* **1884** Knight *Dict. Mech.* Suppl., *Clear Way.* Said of a valve, hydrant, etc. One which lifts its valve entirely out of the way of obstructing the flow; not compelling the water to flow around it. Also called *full-way*. **1927** *Observer* 14 Aug. 18/1 All this clear-way in the grass is treated with cement or weed-killer. **1959** *Daily Tel.* 8 May 12/2 'Clearways' on which parking and stopping are prohibited have long been needed. **1959** *Times* 23 July 6/3 Britain's first 'clearways' will come into operation next month.

clear, *v.* Add: **9. c.** To establish the suitability of a person for work involving questions of (national) security. orig. *U.S.*

1948 *Amer. Political Sci. Rev.* XLII. 495 FBI Director Hoover announced that 1,005,944 names had been cleared and that only 777 full investigations had been instituted. **1950** W. Gellhorn *Security, Loyalty & Science* iv. 90 Dr. Graham might..require access to restricted information, and so he had to be 'cleared'. **1956** A. H. Compton *Atomic Quest* 10 When the question of 'clearance' for work on the atomic project arose, I explained that if they cleared me it would be necessary for them to clear my wife as well.

11. c. *U.S.* To get approval of (a plan, proposal, etc.) from someone in authority. Const. *with* (the person who authorizes). Also *absol.*

1944 *Harper's Mag.* June 69/1 The policy-makers must be consulted on everything that is not part of the established routine. Getting their okay is known as 'clearing on policy' or 'clearance'. **1944** *Time* 25 Sept. 4/1 Clear everything with Sidney.

13. Also, to go away, 'clear off'. Also *refl.* (*U.S.*).

1805 Ordway in *Jrnls. Lewis & Ordway* (1916) 255 The rest all mounted their horses and cleared themselves as they do not wish to fite. **1827** *Western Monthly Rev.* I. 283 In the language of the west, 'he cleared himself'. **1839** 'Mrs. M. Clavers' *New Home* xv. 92 [He] stated boldly that Mr. Mazard had absconded; or in Western language 'cleared'. **1909** T. E. Lawrence *Lett.* (1938) 71 We cleared..to Tell-el-Kadi. **1938** *Times* 20 Aug. 9/3 A member of a gang of 10 native convicts suddenly dropped his pick and cleared for the bush.

14. e. *intr.* To remove the remains of a meal (cf. sense 24 a).

1914 J. M. Barrie *Admirable Crichton* III. 159 An excellent soup, Polly, but still a trifle too rich... You may clear. **1962** N. Marsh *Hand in Glove* ii. 58 It wasn't there when I cleared, miss.

24. clear away. a. (Further examples.)

1836 Dickens *Pickw.* vii. 71 The waiters withdrew to 'clear away', or in other words, to appropriate to their own private use..remnants of the eatables and drinkables. **1964** J. Fleming *Chill & Kill* v. 59 Charming girls..anxious to please..clearing away and even offering to mow the tiny lawn.

25. clear off. b. Also in wider sense (= 26 d), to take oneself off. (Earlier U.S. example.)

1816 U. Brown *Jrnl.* in *Maryland Hist. Mag.* XI. 365 [I] called to her [*sc.* a mare] and flattered her to come back; she would not; clear'd off and left me.

clearable (kliə·răb'l), *a.* [f. Clear *v.* + -able.] Capable of being cleared, able to be put in order.

1889 Westgarth *Austral. Progr.* 90 Stacking up for firewood all the clearable refuse.

clearance. Add: **5. b.** Approval, permission (see *clear *v.* 11 c); *spec.* permission (from the control-tower) to land or take off in an aircraft. orig. *U.S.*

1944 [see *clear *v.* 11 c]. **1953** *Time* 23 Feb., *Clearance* (by payment to publishers' societies ASCAP and BMI) was necessary for permission to play a song on the air. **1958** 'N. Shute' *Rainbow & Rose* i. 6, I got my clearance from the Tower..and took her off. *Ibid.*, I..got clearance to let down.

c. The clearing of a person for work involving questions of (national) security. orig. *U.S.*

1948 *Amer. Political Sci. Rev.* XLII. 498 Dismissal of ten State Department employees as 'security risks'...*Footnote.* It was reported that they could have appealed to the Civil Service Commission for possible clearance for other..government employment. **1948** *Chemical & Engin. News* 26 July 2200 The question before the commission is the security clearance of..[the] director of the National Bureau of Standards. **1955** *Bull. Atomic Sci.* Feb. 66/1, I was entitled to full clearance for access to classified information in connection with the work. *Ibid.* Apr. 124/3 The company concerned is consulted on the value of the individual being considered for clearance, but the company does not make the clearance decision. **1956** [see *clear *v.* 9 c]. **1958** J. Cleugh tr. *Jungk's Brighter than Thous. Suns* xix. 306 He retained, however, the so-called 'Q clearance', which still allowed him access to the most carefully guarded secrets.

6. a. (Later examples.)

1902 P. Marshall *Metal Tools* 36 The widest part of a chisel should be at its cutting edge, behind which point a clearance taper..should be given. **1906** *Westm. Gaz.* 10 Aug. 4/1 As for building our 'buses lower down, we will do that as soon as the Commissioner of Police reduces the 10-in. clearance regulation. **1959** *Listener* 6 Aug. 207/2 He [*sc.* the driver] made wrong estimates of clearances, distances, speeds.

10. *attrib.*, as **clearance sale,** a sale of goods at reduced prices to effect a clearance of superfluous stock.

1880 *English Mechanic* 6 Feb. p. vii (Advt.), Clearance sale of..bicycles. **1925** B. Beetham in E. F. Norton *Fight for Everest, 1924* 184 We decided to hold an autumn clearance sale of..general equipment.

clearer. Add: **1. b.** One who, or a bank which, transacts the business of passing cheques and bills, etc., through a clearing-house; a clearing-banker. Cf. In-clearer.

1883 *Bankers' Mag.* XLIII. 863 The men who transact the clearing business for the various banks are called 'the clearers'. **1892** *Chambers's Jrnl.* 613/2 The clearer we will say represents the London and County Banking Company. **1966** *Punch* 16 Nov. 746/2 The criteria..used in the past to exempt certain banks (all the clearers, but only some of the merchants) from the accounting requirements of the Companies Act.

clearing, *vbl. sb.* Add: **1. c.** The washing of calico to remove excess dye.

1839 Ure *Dict. Arts* I. 164 Bran..is advantageously employed by the calico printers, in the clearing process. **1876** *Encycl. Brit.* IV. 688/1 The operations of clearing are necessary to remove all the dye-stuff which is loosely attached to the whites, and to develop and brighten the tints of dyed pattern.

8. clearing-bank, a bank that is a member of a clearing-house, spec. of the London Clearing House; **clearing-hospital** *Mil.*, a hospital for the temporary reception and treatment of sick and wounded; **clearing-station** = **clearing-hospital*; **clearing-stone,** a fine stone used for sharpening curriers' knives.

1883 *Bankers' Mag.* XLIII. 646 We subjoin figures which have been obtained by the kind consent of the manager of one of our clearing banks. **1933** B. Ellinger *This Money Business* v. 39 The 'Big Five'..have about 90 per cent of the deposits of all these clearing banks. **1955** *Times* 25 Aug. 15/1 When the clearing banks struck their August figures on the 17th of this month the instructions from head offices to branches on the need to cut down advances had only been received about a week. **1914** *Lancet* 28 Nov. 1274/2 The clearing hospital, situated a little distance behind the fighting line, receives all the wounded of a division and is equipped to deal at once with the most serious cases. **1915** A. Huxley *Let.* 26 Apr. (1969) 69 Whereabouts unknown—but a wire from the clearing-station brought back by one of Dr. H's assistants in France, prognosticates 3 weeks. **1933** Blunden & Norman *We'll shift our Ground* II. xii. 82 The place that had been for a few years of all its obscure age a noted training camp or clearing station. **1860** Ure *Dict. Arts* (ed. 5) II. 687 [The currier's knife] is..rubbed on a fine Scotch or Welsh stone called a clearing-stone.

clearing-house. Add: **a.** Also *fig.* (cf. quot. 1881 in Dict.).

1883 *Harper's Mag.* Sept. 634/2 The Charity Organization Society is a central exchange or clearing-house for all the single relief associations. **1903** *Daily Chron.* 10 Dec. 6/7 What is wanted is first a human clearing house, or, in other words, compulsory examination of all immigrants. **1904** J. Chamberlain *Speech* 19 Jan. (Oxf. Dict. Quots., ed. 2), Provided that the City of London remains as it is at present, the clearing-house of the world. **1906** *Westm. Gaz.* 17 Dec. 2/1 Washington is in one sense the clearing-house for the humanity of the entire continent. **1943** J. S. Huxley *TVA* xvi. 131 The new Council has, among other things, become a clearing-house for regional research. **1961** *Oxf. Mag.* 26 Jan. 163/1 The Committee of Vice-Chancellors and Principals is now proposing to set up a central clearing-house which will handle all applications and produce a provisional answer from the candidate's universities of first and second choice by the end of March.

b. clearing-house certificate, a certificate of deposit issued by a clearing-house, negotiable only between banks belonging to the clearing-house association. *U.S.*

1864 *Statutes at Large* (U.S.) XIII. 109 Clearing-house certificates, representing specie or lawful money specially deposited for the purpose of any clearing house association, shall be deemed to be lawful money. **1948** G. Crowther *Outl. Money* (ed. 2) ii. 63 The New York banks were driven to issue 'Clearing House certificates', which were, in fact, banknotes, although the law had to pretend that they were not.

clear-out, clear-up: see *clear *a.* C. 6 b.

clea·rweed. *U.S.* [Clear *a.* 3.] A North American plant (*Pilea pumila*) of the nettle family; rich-weed.

1822 *Mass. Hist. Soc. Coll.* 2 Ser. IX. 157 Plants, which are indigenous in the township of Middlebury [Vt.]..*Urtica pumila* Clear-weed. **1863** A. Gray *Man. Bot.* (ed. 4) 399 *Pilea pumila.* Richweed. Clearweed...Stems smooth and shining, pellucid. **1872** Schele de Vere *Americanisms* 406 A nettle with succulent, semi-transparent stems is called Clearweed (*Pilea pumila*). **1939** *Nat. Geogr. Mag.* Aug. 220/2 Such are the wood and false nettles..or the lowly bedstraws, clearweeds, pellitories, and three-seeded mercury. **1963** Gleason & Cronquist *Man. Vascular Plants* 257 Pilea pumila (L.) Gray. Clearweed.

cleavage. Add: **1. e.** *Biol.* Cell-division, segmentation.

1875 E. R. Lankester tr. *Haeckel's Hist. Creation* I. 190 Commencement of the development of a mammal's egg, the so-called 'cleavage of the egg' (propagation of the egg-cell by repeated self-division). **1879** tr. *Haeckel's Evol. Man* I. 185 The parent-cell, which arose from the fertilized egg-cell, separates, by repeated cleavage, into a large number of simple cells. **1896** E. B. Wilson *Cell* 143 During the early anaphase of the first cleavage each centrosome divides into two. **1902** *Encycl. Brit.* XXVIII. 140/1 Segmentation or cleavage of the ovum. **1948** *New Biol.* V. 112 The usual process of cleavage is as follows. The egg nucleus divides by mitosis; [etc.].

2. b. The cleft between a woman's breasts as revealed by a low-cut décolletage. *colloq.*

1946 *Time* 5 Aug. 98 Low-cut Restoration costumes..display too much 'cleavage' (Johnston Office trade term for the shadowed depression dividing an actress' bosom into two distinct sections). **1947** *Landfall* I. 45 It [*sc.* a film] just goes all out to exploit sex and violence as blatantly as it can, with the result that 'cleavage' has once again become a problem to haunt the dreams of censors. **1958** *Spectator* 6 June 729/1 Kids have to learn not to copy the stars. I tell them, cleavage won't get you to the top. Sex is something different. It's not obvious. **1958** *TV Times* 10 Oct. 21/2, I was foolish enough to wear the sort of dress which showed vistas of cleavage.

3. cleavage-cavity = Blastocœle; **cleavage-cell, -globule** = Blastomere; **cleavage-mass,** (*a*) = Blastomere; (*b*) a mass of rock formed by cleavage; **cleavage-nucleus,** the nucleus of the fertilized egg resulting from the union of male and female pronuclei; **cleavage-spindle,** the karyokinetic spindle of the dividing blastomere.

1879 tr. *Haeckel's Evol. Man* I. 189 The inner cavity of the ball, which is filled with clear liquid or jelly, is called the cleavage-cavity (*cavum segmentarium*), or the germ-cavity (*blastocœloma*). **1887** *Encycl. Brit.* XXII. 425/2 In this stage thin sections show that the cleavage cavity is obliterated. **1879** tr. *Haeckel's Evol. Man* I. 186 Cleavage-cells or Cleavage-globules (*segmentella*). **1871** Huxley *Anat. Vert.* 10 The cleavage-masses eventually become very small, and are called embryo-cells. **1912** *Brit. Museum Return* 196 Enargite, large cleavage masses from Red Mtn., Colorado. **1896** E. B. Wilson *Cell* 156 The first cleavage-nucleus. *Ibid.* 157 The centrosomes of the cleavage-spindle.

cleave (klīv), *sb.*[2] [f. Cleave *v.*[1]] A cleft.

1874 G. M. Hopkins *Note-Bks.* (1937) 204 Chudleigh Rocks..are a cliff over a deep and beautiful cleave. *c* **1881** —— *Sermons* (1959) 154 In that 'cleave' of being which each of his creatures shews to God's eyes alone. *Ibid.* 328 Like so many 'cleaves' or exposed faces of some pomegranates..cut in all directions across.

cleaver[1]**.** Add: **2. c.** *Archæol.* A primitive core-tool with a sharp edge at one end.

1935 *Nature* 21 Sept. 475/1 In Uganda this stage marks the beginning of a large-core technique for the manufacture of *coups de poing* and cleavers. **1959** J. D. Clark *Prehist. S. Afr.* ii. 41 *Cleaver*, a similar all-purpose tool but with an axe-like cutting edge, usually at right-angles to the long axis.

cleek, *sb.* Add: **1. b.** *Golf.* An iron-headed club with a straight, narrow face and a long shaft. Also *attrib.*

1829 J. S. Blackie *Let.* 11 Dec. in *Lett. to Wife* (1909) 42 As if the golfers at St. Andrews should go down to the links..with an immense array of all sorts of clubs, and then, after having swung their drivers and cleques most scientifically in the air, should come back again without playing a single game. **1857** [see Put, putt *v.*[1] 3]. **1883** [see sense 1 in Dict.]. **1890** H. G. Hutchinson *Golf* iv. 102 It is a stroke..with the brassy, cleek or iron. **1893** —— *Golfing* (ed. 2) 19 We cannot be far wrong in advising the golfer to choose his cleek-heads thick. **1909** P. A. Vaile *Mod. Golf* 107 The ordinary cleek shot.

cleidoic (klaidō^u·ik), *a. Zool.* [f. Gr. κλειδ-όω lock up (cf. κλειδ-, κλείς key) + -ic.] Closed-up: used of an egg encased in a shell or membrane.

1931 J. Needham *Chem. Embryol.* III. 1615 The avian egg is thus, in the strictest sense of the word, a closed box, with walls which can only be penetrated by matter in the gaseous state. The term 'cleidoic'..is suggested for this state of affairs. *Ibid.*, The history of the steps which led to the cleidoic egg is the history of eggs in general. **1935** *Nature* 16 Mar. 413/2 The biological meaning of the numerous facts centring round the cleidoic egg.

cleidotomy (klaidǫ·tōmi). *Surg.* [ad. mod.L. *cleidotomia* (Phänomenoff 1895, in *Zentralblatt f. Gynäk.* XIX. 585), f. Gr. κλειδ-, κλείς key, clavicle + -tomy.] Surgical division of the clavicle of the fœtus undertaken when the shoulders of the fœtus prevent normal delivery.

1901 *Trans. Edin. Obstet. Soc.* XXVI. 24 (*heading*) Cleidotomy: An Operation Accessory to Craniotomy..By J. W. Ballantyne. **1904** J. W. Ballantyne *Essent. Obstet.* 207 *Cleidotomy.* Division of one or both clavicles, is the most scientific and effective plan of diminishing the bisacromial (large) diameter of the trunk.

cleistocarp (klai·stokāɹp). *Bot.* Also **cleistocarp.** [f. Gr. κλειστός closed + καρπός fruit.] In mosses and fungi, an ascogonium whose asci and spores develop within a completely closed capsule, from which the spores escape by the rupture of the sac. So **cleistoca·rpous** *a.*, having a closed capsule, having a capsule without an operculum.

1884 *Encycl. Brit.* XVII. 73/2 The *Phascaceæ* are termed 'cleistocarpous' in contradistinction to the 'stegocarpous' Mosses. **1887** W. Hillhouse tr. *Strasburger's Bot.* 261 The section of cleistocarpous Ascomycetes. **1887** Garnsey & Balfour tr. *Goebel's Outl. Classification & Special Morphol. Plants* 477/1 *Cleistocarp*, ascocarp in which the asci and ascospores are formed inside a completely closed envelope from which the ascospores escape by its final rupture. **1887, 1940** [see **ascocarp*]. **1955** E. V. Watson *Brit. Mosses & Liverworts* 128 Other cleistocarpous mosses..show a very different leaf shape, and cell structure.

cleit (klīt). *local.* Also **cleet, clett.** [Gael. *cléit* rocky eminence; cf. Clet, f. ON. *klettr* cliff, crag.] A small shelter, often used for storing peat, birds' eggs, etc.

1825 Jamieson Suppl., *Cleit*, a cot-house; Aberd[een] Reg[isters]. **1903** *Ann. Scot. Nat. Hist.* 17 The Fork-tailed Petrels on Borrera, St. Kilda, nest in the 'cleets' or little turf houses of the natives, among the sods of dry turf.

1927 *Glasgow Herald* 1 July 12 Small stone shelters known locally [St. Kilda] as Cleits. **1937** *Brit. Birds* XXX. 236 It [*sc.* the Fork-tailed Petrel] will eventually be definitely proved to breed among the stone 'cleits' built by the men who visit the island annually to collect young gannets. **1952** J. Fisher *Fulmar* vi. 130 These were the famous St. Kilda *cletts*, or *cleits*.

cleithrum (klai·þrŏm). *Zool.* Pl. **cleithra.** [mod.L. (C. Gegenbaur 1895, in *Morphol. Jahrb.* XXIII. 1), f. Gr. κλεῖθρον bar for closing a door.] A dermal bone in the pectoral arch of some fishes and amphibia.

1896 *Zool. Rec. 1895 Pisces* 15 The bone termed *clavicula* in Fishes is not homologous with that of higher Vertebrates...The two bones are now called *clavicula* and *cleithrum*. **1910** *Encycl. Brit.* XIV. 262/2 The primitive pectoral girdle..becomes in the Teleostomes..replaced functionally by a secondary shoulder girdle formed of superficially placed membrane bones (supraclavicular and cleithrum or 'clavicle'). **1910** Parker & Haswell *Textbk. Zool.* (ed. 2) II. 302 There is also, on each side, a bone called the *cleithrum*, connected with the corresponding clavicle: there is..reason for thinking this to be homologous with the bone..called clavicle in Teleostomi. **1956** A. S. Romer *Osteol. Reptiles* vii. 298 The cleithrum, however, is an element which tended toward rapid reduction, even in early reptiles.

clementine (kle·məntīn, -ain). [ad. Fr. *clémentine* (L. Trabut 1902, in *Revue Horticole* LXXIV. 232).] A variety of small orange originally produced by an accidental hybrid of the tangerine and the sour orange.

1926 H. H. Hume *Cultiv. Citrus Fruits* xxix. 477 The Clementine orange (a cross between tangerine and sour orange) is very severely affected [by citrus rust]. **1943** H. J. Webber *Citrus Industry* I. v. 557 The Clementine is an interesting and promising new mandarin of the tangerine group. **1967** *Guardian* 3 Feb. 8/5 Oranges..thin-skinned, pipless Satsumas; bright, pungent little Clementines, red-flecked Sanguinas and many others. **1969** *Oxf. Bk. Food Plants* 88/1 Clementines are produced principally in the North African countries.

cleome (klı,ō^u·mi). [mod.L. (Linnæus *Genera Plantarum* (1737) 200), f. an earlier Gk. and L. name for a different plant.] A plant of a genus of annual herbs or shrubs so called, belonging to the family Capparidaceæ, and native to America; cultivated for their spidery flowers.

1806 B. McMahon *Amer. Gardener's Cal.* 344 April..In the last ten days of the month, you may sow..Purple Egg-plant; Cleome, purple, white, and dwarf Dolichos, with many other kinds. **1829** Loudon *Encycl. Plants* 1058 Several species of Cleóme have an acrid taste, which has been compared by travellers to that of mustard. **1951** *Dict. Gardening* (R. Hort. Soc.) I. 504/1 Cleomes grow best in light, rich soil, in a dry, warm spot where there is plenty of room to spread. **1962** *Amat. Gardening* 21 Apr. 1/1 Portulacas, venidiums, cleomes and many more unfamiliar plants.

cleptobiosis: see *kleptobiosis.

clerestory. Add: **1. c.** A row of small windows above the main roof of a railway carriage.

a **1884** Knight *Dict. Mech.* Suppl. **1903** *Westm. Gaz.* 15 July 6/3 G.W.R. eight-wheeled bogy coaches of the 'clerestory' pattern.

clergy. Add: **2. f.** *regular clergy, secular clergy*: see Regular A. 1 a, Secular A. 1 a.

clergywoman. Add: **3.** A woman acting as pastor of a congregation or as a minister of religion.

1871 *Chicago Tribune* 17 Feb. (De Vere), Mrs. E. Tupper Wilkes, the Minnesota clergywoman, has a salary of $2000 a year. **1904** *Westm. Gaz.* 25 Feb. 10/1 The Clergywoman in America.

clericaliza·tion. [f. Clericalize *v.*: see -ation.] The action or result of making clerical or placing under clerical rule.

1907 *Cath. Weekly* 4 Jan. 6/2 Such clericalisation of ordinary political events is..undisguised mischief-making. **1928** *Observer* 19 Feb. 17/1 The Nationalists (Conservatives) [in Germany] are allied with the Centre party in forcing the clericalisation of the schools. **1958** E. L. Mascall *Recov. Unity* i. 5 The Counter-Reformation did little or nothing to correct this clericalisation of the Christian religion in Catholicism.

clerid (kle·rid), *a.* and *sb. Ent.* [ad. mod.L. *Cleridæ*, a family of beetles.] **A.** *adj.* Of or pertaining to the Cleridæ. **B.** *sb.* A beetle of this family (Cent. Dict. 1889).

1889 *Proc. Ent. Soc. Washington 1884–88* I. 165 The pretty *Clerus ichneumoneus* could frequently be seen on the trunks..and a large Clerid larva found in the galleries probably belongs to that species. **1913** *Oxford Univ. Gaz.* 4 June 955/2, 1 ♀ Mutillid and 1 Clerid beetle mimicking it in a very remarkable manner. **1959** E. F. Linssen *Beetles* I. 234 The best-known Clerids are comprised in the genus *Necrobia*.

clerihew (kle·rihiū). [f. the name of Edmund *Clerihew* Bentley (1875–1956).] A short comic or nonsensical verse, professedly biographical, of two couplets differing in length.

1928 *Weekend Book* 331 'Clerihews' on your personal friends, with nice slack metres and sly points like 'Sir Christopher Wren Was going to dine with some men. He said, "If anybody calls, Say I'm designing Saint Paul's."' **1940** E. C. Bentley in *N. & Q.* CLXXVIII. 141/2 *Clerihew.* This formless form of verse was so called because my book, 'Biography for Beginners', in which it originated, was published under the name of E. Clerihew, my Christian name. This was in 1906, and the name was applied to the form soon after this by some unknown reader. **1958** *New Statesman* 12 Apr. 483/1 It is some years since we set a clerihew.

clerk, *sb.* Add: **1. c.** *regular clerk, secular clerk*: see Regular A. 1 a, Secular A. 1 a.

6. c. *clerk of the course*, an official on a race-course (see quots. 1920, 1952); also applied to a similar official in other sports; *clerk of the weather* (earlier example); also *clerk of the weather office.*

1835 Dyott *Diary* (1907) II. 193 A meeting of the trustees of the racing fund at Lichfield..for the purpose of removing Mr. Henry Cato from the situation of clerk of the course. **1920** Webster, *Clerk of the course*, one who acts as secretary to the board of judges of races or track athletics. **1952** *Racing Review Ann.* 219/1 The rules of racing lay down that the clerk of the course is responsible for the general arrangements of the meeting. **1961** *Radio Times* 14 Sept. 30/4 Behind the scenes at an international motor race to hear more about this exciting sport from top racing drivers..[and] Clerk of the Course John Morgan. **1831** B. Hall *Fragm. Voy. & Trav.* III. 296 Owing to their interest with the clerk of the weather office. **1835** [see Weather *sb.* 2 h]. **1843** J. F. Murray *World of London* II. 29 If the clerk of the weather office is determined to put a damper on the festivities.

d. A shop-assistant. *N. Amer.*

a **1790** Franklin *Autobiogr.* (1848) 85 He propos'd to take me over as his clerk, to keep his books, in which he would instruct me, copy his letters, and attend the store. **1835** H. C. Todd *Notes Canada & U.S.* 10 *Shop-men*, who are here and in Canada also, universally called clerks. **1844** 'Jon. Slick' *High Life N.Y.* I. 63, I..went to the tailor's store...'Have you got any first rate superfine broadcloth coats and trousers to sell here?' sez I...'Yes,' sez one of the clerks. **1903** W. E. Curtis *True Abraham Lincoln* 33 He..became a clerk in the store..measuring calico, weighing sugar and nails.

e. In hotels, an employee who registers guests and attends to their comfort. *U.S.*

1879 R. J. Burdette *Hawk-Eyes* 62 Abou Tamerlik came to the city of Bagdad..and, as he registered, spake cheerfully unto the clerk. **1892** J. C. Duval *Young Explorers* 198 A little while afterwards the clerk came up to my room and asked if I wished for supper. **1947** E. W. Barkins *Dr. has Baby* 180 We followed the clerk to our room, followed in turn by the bellhop with our bags.

clerkess (klā·ɹkės). *Sc.* [f. Clerk *sb.* + -ess[1].] A female clerk.

1923 *Glasgow Herald* 21 Mar. 10 A clerkess in a Glasgow firm's counting-room. **1965** J. Caird *Murder Reflected* xiii. 176 She was clerkess in an office.

clerodendrum (klīᵊrode·ndrŏm). Also **clerodendron.** [mod.L. (Linnæus *Genera Plantarum* (1737) 186), f. Gr. κλῆρος chance + δένδρον tree.] A plant of a large genus of shrubs, climbers, or small trees, belonging to the family Verbenaceæ and native to many warm parts of the world, particularly Africa and Malaya.

1812 W. T. Aiton *Hortus Kewensis* (ed. 2) IV. 63 Fragrant Clerodendrum...Scarlet Clerodendrum. *Ibid.* 64 Downy Clerodendrum. Cumberland Tree. *Ibid.* 65 Long-flower'd Clerodendrum. **1829** Loudon *Encycl. Plants* 1079 Some of the Vitexes and Clerodendrums are handsome shrubs. **1851** *B'ham & Midl. Gardeners' Mag.* Apr. 42 Pot the Clerodendrons in the pots which they are intended to bloom in. **1924** *Blackw. Mag.* Apr. 464/1 A tangle of rank undergrowth,..convolvulus, smilax, and the scarlet clerodendron. **1961** *Amat. Gardening* 21 Oct. 5/2 The clerodendrums bloom late enough to be included among the autumn-flowering shrubs.

clethra (klī·þrǎ, kle·þrǎ). [mod.L. (J. F. Gronovius in *Linnæus's Genera Plantarum* (1737) 127), f. Gr. κλήθρα alder, which it resembles in foliage.] A plant of a genus of small trees or shrubs, belonging to the family Clethraceæ and native to eastern America and Asia.

1779 T. Mawe et al. *Every Man his Own Gardener* (ed. 8) 464/1 *Clethra*, white flowering, Dwarf. *Ibid.* 467/1 Trees and Shrubs that may be raised from Seed..Clethra, Dog-wood. **1785** H. Marshall *Amer. Grove* 34 *Clethra alnifolia.* Alder leaved Clethra. This shrub grows common in Maryland, Virginia, and Carolina, in moist ground and by rivulets. **1846** Lindley *Veget. Kingd.* 454 The genus Saurauja among Dilleniads, has very much the structure of a Clethra. **1895** *Outing* (U.S.) XXVI. 33/2 The white spiked Clethra flower. **1969** D. Wyman *Shrubs & Vines Amer. Gardens* (ed. 2) 171 This [*sc.* Japanese Clethra] is the first of the clethras to bloom.

Cleveland bay. [f. *Cleveland*, a district in Yorkshire + Bay *a.*[1] (and *sb.*) 2.] Designating a breed of large, strong English horses, bay in colour with black legs, originating and chiefly bred in northern England.

1796 J. Lawrence *Philos. & Pract. Treat. Horses* I. vi. 299 Breeders..may read of the wonderful exertions, in carting-business, upon the road, of the Cleveland Bays, a sort of coach-horses. **1831** Youatt *Horse* iv. 36 The origin of the better kind of coach-horse is the Cleveland Bay, confined principally to Yorkshire and Durham. **1875** *Encycl. Brit.* I. 385/1 The Cleveland Bays are properly carriage-horses; but still in their native districts they are largely employed for field work. **1958** *Everyman's Encycl.* VI. 575/1 The Cleveland Bay..is commonly used in crossing to produce hunters, cavalry horses, and harness horses.

clever, *a.* Add: **3. a.** *clever Dick*: see *Dick *sb.*[1] 1 a.

4. b. (Further examples.)

1775 in *Essex Inst. Hist. Coll.* XIII. 196 Father was very clever last Saturday P.M. **1815** *Massachusetts Spy* 14 June (Th.), I somehow did not feel quite clever, but hoped for the best. **1937** Partridge *Dict. Slang* 158/2 *Not too clever*, indisposed in health..is common in Australia and New Zealand.

8. b. (Later U.S. examples.)

c **1857** in *Lower Norfolk Co.* (Va.) *Antiquary* I. 110 John Savage..promised to do something clever for them. **1858** in *Kansas Hist. Coll.* (1896) V. 567 On the night before last these same men attacked Barnesville quite a clever town, on the military road.

c. (Earlier and later U.S. examples.)

1758 in *Essex Inst. Hist. Coll.* XII. 148 This afternoon secured a place to have ye Small Pox in, with a very clever family. **1781** Witherspoon in *Pennsylv. Jrnl.* 9 May, He is a very *clever* man. She is quite a *clever* woman...In these cases Americans generally mean by *clever*, only goodness of disposition, worthiness, integrity, without the least regard to capacity. **1883** Eggleston *Hoosier School-Boy* xviii. 119, I wanted to say that you fellows have always..been 'clever' to me, and I don't want to see no harm done you. **1891** *Harper's Mag.* July 220/2 It is an Americanism to call him *clever* whom we deem good-natured only.

9. *Comb.* **clever-boots** [cf. Boots[1] 3]. **-sides, -sticks,** etc., a clever or smart person, a wiseacre (*colloq.* and *dial.*: see E.D.D.); **clever-clumsy** *a.*, denoting a person who is both clever and clumsy; so *clever-silly* sb. (hence *clever-silliness*), *clever-stupid* adj.

1847 Halliwell *Provinc. Words*, *Cleverboots*, a clever person, generally in a satirical sense. **1955** *Times Lit. Suppl.* 9 Dec. 742/3 Mallarme is the promising pupil..Rimbaud the overweening cleverboots whom the critics have insufficiently chastised. **1854** A. E. Baker *Northants. Gloss.* I. 124 *Clever-clumsy*, an ironical term of reproof to a person who sets about anything eagerly and performs it awkwardly. **1928** D. H. Lawrence *Lady Chatt.* vii. 87 Our little lot of clever-jacks, all at the intelligence-age of seven. **1937** H. G. Wells *Star Begotten* i. 6 She told him not to try and be a Mr Cleverkins. He did his best not to be Mr Cleverkins. **1965** *New Statesman* 20 Aug. 266/1 Some cleverkin has written in heart-warming wee scenes between her and her prey. **1886** H. Baumann *Londinismen* 27/1 *Clever shins*, Schülersprache: Schlauberger. **1951** Partridge *Dict. Slang* (ed. 4) Add., *Cleversides*, schoolchildren's (..mainly North Country) coll[oquial]: synonym, often ironical, of *clever person*: late C. 19–20. **1959** I. & P. Opie *Lore & Lang. Schoolchildren* x. 179 A clever boy is most often styled a 'clever-dick', or, less frequently: brainy pup..cleverguts, cleverpot, cleversides, cleversticks..know-all. **1948** *Mind* LVII. 141 That impression of thinness and clever-silliness which sometimes characterises the productions of highly intelligent writers. **1946** *Mind* LV. 173 The 'clever-sillies' or the 'half-baked intellectuals'. **1964** C. Mackenzie *My Life & Times* III. ii. 44 Some cleversticks had climbed up a plane-tree to get a better view. **1911** F. Swinnerton *Casement* i. 58 Like many clever-stupid people, Olivia often underrated the intelligence of her associates.

B. quasi-*adv.* **a.** Delete †(*obs.*) and add later example. See also *to box clever*, *box *v.*[2] 2 d.

1959 J. Cary *Captive & Free* 68 Meanwhile it was obviously dangerous to take sides. He must play clever.

cle·ver-cle·ver, *a.* [Reduplicated from Clever *a.*; cf. Pretty-pretty, etc.] Anxious to be considered clever; over-clever. Hence **cle·ver-cle·verness.**

1896 G. B. Shaw *Let.* 26 Mar. (1965) 617 A good melodrama is a more difficult thing to write than all this clever-clever comedy. **1921** C. Dane *Bill of Divorcement* III. 75 Don't be clever-clever, old thing. **1946** Koestler *Thieves in Night* I. vi. 43 Had to show off all the time..what clever-clever fellows they were. **1957** D. J. Enright *Apothecary's Shop* 215 We are apt to suffer from the occupational malady of clever-cleverness.

cleverly, *adv.* **7.** (Earlier examples.)

1784 A. Adams *Lett.* (1848) 212 She is cleverly now, although she had a severe turn for a week. **1816** Pickering *Vocab.* 63.

†cle·vers, *sb. pl. Sc. Obs.* [cf. Clever *a.* 4 a.] Phr. *a fit of the clevers*, a burst of energy or activity.

1824 Scott in J. G. Lockhart *Life* (1837) V. xii. 353 Taking a fit of what waiting-maids call *the clevers*, I started at six this morning. **1826** —— *Jrnl.* 14 Aug. (1941) I. 215, I was seized by a fit of the 'clevers', and finished my task by twelve o'clock.

clew, *v.*: see *clue *v.*

clianthus (klai,æ·nþŏs). *Bot.* [mod.L. (Banks & Solander in G. Don *Gen. Syst. Gardening & Bot.* (1832) II. 468/1), app. f. Gr. κλει-, by-form of stem of κλέος glory + ἄνθος flower.] An Australasian shrub belonging to the leguminous genus of that name, bearing handsome flowers in racemes.

1841 Mrs. Loudon *Ladies' Compan. Flower Garden* 56/1 Clianthus.., the crimson Glory Pea, is a magnificent half-hardy shrub, with bright crimson flowers, a native of New Zealand. **1853** Hooker *Bot. Antarctic Voyage* II. p. xxxi, Of *Clianthus*..there are but two known species, one in Australia, and the other in New Zealand and Norfolk Island. **1909** *Daily Chron.* 15 Apr. 7/2 The verbena and the clianthus grow rankly in the mild air. **1933** *Bulletin* (Sydney) 3 May 20/4 The beautiful Sturt pea..now more popularly known as the clianthus. **1961** *Amat. Gardening* 7 Oct. Suppl. 12/3 To graft clianthus seedlings on to seedlings of Colutea arborescens.

cliché. Add: **3.** *fig.* A stereotyped expression, a commonplace phrase; also, a stereotyped character, style, etc. Also *collect.*

1892 A. Lang in *Longman's Mag.* Dec. 217 They have the hatred of *clichés* and commonplace, of the outworn phrase, of clashing consonants. **1895** *Westm. Gaz.* 19 Apr. 3/2 The farcical American woman who 'wakes everybody up' with her bounding vulgarities..is rapidly becoming a *cliché*, both on the stage and in fiction. **1902** Gosse in *Encycl. Brit.* XXVIII. 261/1 All but the obvious motives tend to express themselves no longer as thoughts but as clichés. **1909** O. Jennings *Morphia Habit* vi. 72 The above description of morphinism has been repeated by one compiler after another, until it has become a *cliché*. **1909** H. G. Wells *Ann Veronica* xiv. 295 Nothing but *cliché* seems to meet this case. **1913** E. F. Benson *Thorley Weir* iv. 151 Probably Charles never painted more magically than in those ten minutes, even when the magic of his brush had become a phrase in art criticism, a cliché. **1923** J. M. Murry *Pencillings* 151 Carlyle picked out as specimen *clichés* of the orator, 'The rights of suffering millions', and 'the divine gift of song', which are still hard-worked to-day. **1948** 'E. Crispin' *Buried for Pleasure* vi. 43 The command of *cliché* comes of having had a literary training. **1963** *Listener* 28 Feb. 376/2 It [*sc.* the magazine story] is composed of prefabricated clichés—stock characters, stock situations, stock dialogue, stock plots.

b. *attrib.* and *Comb.*, as *cliché-monger*, *-personality*; *cliché-ridden* adj.

1947 'N. Blake' *Minute for Murder* ii. 31 I'll do in the little cliché-monger. **1962** *Listener* 8 Nov. 780/1 Often it sufficed to let some pompous cliché-monger speak for himself. **1925** Wyndham Lewis *Let.* 1 Apr. (1963) 156 The book..divided in 3 parts, named respectively the Cliché-Personality, the Patria Potestas and Primitive Communism. **1928** *Manch. Guardian Weekly* 23 Nov. 413/4 The middle section.., though it begins with a cliché-ridden trifle, has poems of..imaginative power.

c. Used as *adj.* Stereotyped, hackneyed.

1959 *Listener* 30 July 188/1 The kind of fond reminiscence which comes rather too near the cliché view of human situations. **1962** *Ibid.* 9 Aug. 196/2 Many of the productions of our 'new wave' in the theatre have been narrow in outlook, cliché in sentiment, shaky in stagecraft.

Hence **clichéd, cliché'd** *a.*, hackneyed; characterized by clichés.

1928 A. Waugh *Last Chukka* 83 There is no adjective but the *cliché*'d deafening that can fittingly describe the tornado of noise that had welcomed the recitation. **1943** *Horizon* VIII. 388 The *clichéd* thought, the seemingly happy ending. **1961** *New Left Rev.* Jan.–Feb. 42/2 Television will continue to be clichéd, formula-ridden and bland. **1968** *Daily Tel.* (Colour Suppl.) 22 Nov. 55/2 Young architects have for some time struggled to vary the clichéd house plans turned out by some builders.

click, *sb.*[1] Add: **1.** *fig.*

1880 W. James *Coll. Ess. & Rev.* (1920) 196 This is a state of *consent*, and the passage from the former state to it..is..characterized by the mental 'click' of resolve. **1906** *Punch* 6 June 411/3 The bricklayer's scheme of retribution fits into its place with so triumphant a 'click', as it were. **1935** W. B. Yeats *Lett. Poetry* (1940) 24 The correction of prose, because it has no fixed laws, is endless, a poem comes right with a click like a closing box.

b. *Radio.* In *pl.*: the name given to atmospherics of short duration.

[**1912** [see Stray *sb.* 3].] **1922** Glazebrook *Dict. Appl. Physics* II. 1044/2 The British Association Committee for Radiotelegraphic Investigation..has classified strays [*i.e.* atmospherics] into three types, namely, clicks, grinders, and hissing. **1936** *Nature* 6 June 955/2 A very clear distinction was obtained between the atmospherics of short duration ('clicks') and the atmospherics of long duration ('grinders'). The 'clicks'..are due to local actions inside the thunderstorm clouds, such as short sparks which are not easy to detect by visual observations.

5. (Earlier example.)

1837 F. Owen *Diary* (1926) 89 Uteeko..has a harsh and difficult click in it, and has no meaning being a word of Hottentot extraction.

7. *click-catch, -jack, -spring, -stop*; **click reel,** a reel of a fishing-rod having a click or detent.

1890 Nasmith *Cotton Spinning Machinery* 217 When the 'click spring'..is slightly oscillated in the same direction as the rotation of the ratchet wheel, it allows the click catch to fall into gear. **1948** *Times* 21 Apr. 2/2 The immediate cause of the accident was a faulty click-jack, a safety device intended to arrest the progress of tubs in a pit when they run away. **1883** *Fisheries Exhib. Lit.* XII. 212 Leonard Click reel. **1888** Goode *Amer. Fishes* 62 A more flexible rod..with a click-reel. **1890** Click spring [see **click catch*]. **1957** T. L. J. Bentley *Man. Miniat. Camera* (ed. 5) iv. 38 Fitted with an f/2·8..lens..and having audible click stops.

8. Add other examples of imitative formations: *click-clicking, click-clock*; *clickety* adj.; *clickety-clack* (non-dial. examples); *clickety-click, spec.* used jocularly in the game of bingo (or housey-housey), etc., for 'sixty-six'; *click-clack, clickety-click* vbs.

1896 *Humanitarian* June 432 Click-clicking with a type-writer all day long. **1901** *Ibid.* 3 Sept. 1/3 The click-clock of passing iron shoes. **1913** H. A. Franck *Zone Policeman 88* i. 6, I sat down in an arm-chair against the wall and fell to reading, amid the clickity-click of typewriters. **1914** W. De Morgan *When Ghost meets Ghost* II. xiv. 613 The up-express..dragged her train over oily lines and clicketty lines. **1919** W. H. Downing *Digger Dial.* 16 *Clickety-click*, number 66 in the game of 'house'. **1920** *Chambers's Jrnl.* 67/1 The steady click-clock of his hoofs. **1926** *Glasgow Herald* 17 July 4 The omnibus train clickety-clicked. **1926** E. M. Roberts *Time of Man* iv. 142 Bright new machines that would go clickerty-click. **1933** L. A. G. Strong *Sea Wall* 256 A game of 'house' was in progress, and a voice monotonously droned the numbers:..'clickety click'. **1938** O. Nash *I'm a Stranger Here* 213 The clickety-clack of wheel on track. **1959** *Times* 27 Apr. (Rubber Ind. Suppl.) p. vi/7 The persistent clickety-clack traditionally associated with railway travel. **1962** *Daily Tel.* 25 June 11/4 Clickety click, 66; Gates of Heaven, No. 7.

click, *sb.*[3] Anglicized form of Clique.

1822 [see Clique]. **1916** C. J. Dennis *Songs Sentim. Bloke* 40 Fair narks they are, jist like them back-street clicks. **1929** J. B. Priestley *Good Compan.* III. v. 607 Local fellers, they was, all in a click, y'know, a gang. **1941** 'N. Blake' *Abom. Snowman* ix. 102 'Miss Ainsley was an old friend of Elizabeth?' 'I believe so. One of that click.' **1959** C. MacInnes *Absol. Beginners* 63 Here was a family: at any rate, a lot, a mob, a click I could belong to.

click, *v.*[1] Add: **1. a.** *spec.* of a camera or of a person operating one. Also (in various senses) with following *adv.*

1929 C. Day Lewis *Transitional Poem* II. 32 Desire clicks back Like cuckoo into clock. **1937** C. Beaton *Diary* 3 June in *Wand. Years* (1961) 311, I clicked away. **1948** *Mind* LVII. 485 After trying to recall a face, one often says, 'Ah, now I remember the man you mean.' In such cases we seem to feel something happen: something, as it were, clicks to. **1950** D. Gascoyne *Vagrant* 14 When Abbott's camera clicked.

d. *fig.* To meet or fall in *with* (a person) fortunately or at the right moment; to be successful; to fit together or agree exactly; to become friendly *with* someone; to strike up a rapport *with* (a person); to be a success in the theatre or other form of entertainment. *colloq.*

1915 T. Burke *Nights in Town* 107 The bright boys..saunter..up and down that parade until they 'click' with one of the 'birds'. *Ibid.* 108 You have 'clicked'. You have 'got off'. **1921** Wodehouse *Jill the Reckless* viii. 120 A grey world in which, hoping to click, we merely get the raspberry. **1922** —— (*title*) The Clicking of Cuthbert. **1923** —— *Inimit. Jeeves* iv. 40 'Did you click?' He sighed heavily. 'If you mean was I successful, I must answer in the negative.' **1926** *Amer. Speech* I. 436/2 [Show business slang]. A turn is said to click when it proves to be successful, or in the vernacular, 'gets across with a bang'. **1927** *Vanity Fair* Nov. 67/2 He doesn't hope that he makes good. He hopes that he 'clicks'. He trusts that he doesn't 'flop'. **1930** *Times* 29 Mar. 10/4 The objects are arranged but not composed..so that they remind you a little of people assembled harmoniously but lacking some common emotion. They don't 'click'. **1931** J. Cannan *High Table* xi. 164 Getting drunk when he was trying to click with a Glasgow buyer. **1934** C. Lambert *Music Ho!* III. vii. 208 Receiving the glad eye from presumably attractive girls with whom he ultimately and triumphantly 'clicks'. **1952** V. Gollancz *My Dear Timothy* xx. 316 To prove satisfactorily that a lot of these devices wouldn't 'click'. **1958** *Observer* 20 Apr. 14/6 Put out under pressure to fill the gap..it [*sc.* the B.B.C. television programme *To-Night*] clicked instantly and is still the pride of British television.

e. To come in *for* something; *spec.* to get killed. *Mil. colloq.*

1917 Empey *From Fire Step* 81, No. 1 Section had clicked for another blinking digging party. **1917** W. Muir *Observ. Orderly* 226 To click can be either advantageous or baneful, according to the circumstances. A soldier asks a superior for a favour, and it is granted. That soldier has clicked...But he has also clicked if he is suddenly seized on to do some menial duty. **1919** *Athenæum* 11 July 582/2 The verb 'click'..has developed some passive meanings, such as to get killed. *Ibid.* 8 Aug. 729/1 To 'click for fatigue' is to 'come in for' a fatigue duty at the psychological moment. **1966** *Listener* 22 Dec. 927/1, I came out of hibernation..to find that I had clicked for a most alarming job.

f. To become pregnant, to conceive. *colloq.*

1936 N. Coward *Fumed Oak* II. ii. 58 A couple of months later you'd told me you'd clicked, you cried a hell of a lot, I remember. **1954** *Landfall* VIII. 228 In Wellington, just before he came down there was Heather, who had 'clicked for a baby'.

g. To 'ring a bell', fall into context. *colloq.*

1939 'M. Innes' *Stop Press* II. iv. 269 Something clicks. Tell me. **1960** A. Burgess *Right to Answer* ii. 37 Then the name clicked, because somebody in the town had talked about Everett.

2. (Later examples.)

1918 W. Owen *Let.* 20 Mar. (1967) 541 Mrs. A. can click the piano quite quickly. **1930** C. V. Grimmett *Getting Wickets* iii. 59 The method of spinning is similar to that used in clicking the finger and thumb to attract attention. **1936** Wodehouse *Laughing Gas* xviii. 203 He was clicking his tongue in gentle self-reproach. **1938** J. Hilton *To You, Mr. Chips* i. 51 No upstart authority has yet compelled him to click his heels and begin the day with juju incantations of *Heils* and *Vivas.* **1958** M. L. Hall et al. *Newnes Complete Amat. Photogr.* 156 All one has to do is to point the camera and click the shutter.

b. Also with *adv.*, as *click out, up*, etc.

1895 *Westm. Gaz.* 17 Apr. 8/1 How assiduously some of the political typists must have been clicking out these words of late. **1930** 'A. Armstrong' *Taxi* v. 46 And so for the next seven years the meters clicked up 1/- a mile. **1962** J. Dill in *Into Orbit* p. xix, A robot could easily click off pictures automatically or take measurements of radiation and heat.

c. To get, receive. *Mil. colloq.*

1917 Empey *From Fire Step* 39 Shut your blinkin' mouth, you bloomin' idiot; do you want us to click it from the Boches? *Ibid.* 65 Trench mortars started dropping 'Minnies' in our front line. We clicked several casualties. **1944** J. H. Fullarton *Troop Target* xxx. 213 They tell me Micca's a good picquet to click.

4. *Printers' slang.* (See quot. and *clicking *vbl. sb.* d.)

1860 Ruse & Straker *Printing* 121 A work is said to be 'clicked' when each man works on his lines, and keeps an account thereof.

clicker[2]**.** Add: **2.** Also, a workman who cuts out (parts of) the uppers of boots and shoes.

1921 *Dict. Occup. Terms* (1927) § 608 *Clicker* (rubber), cuts out rubber uppers for goloshes, and canvas uppers for rubber boots or shoes. **1960** *Guardian* 1 Mar. 7/5 A clicker (a maker of the uppers for women's shoes). **1968** *Times* 6 Dec. 10/6 The 'clickers' who, after the various parts are sewn together, work the upper around the last.

clicket, *v.* Add: **4.** *intr.* = Click *v.*[1] 1 a; to make a clicking sound.

1908 W. De Morgan *Somehow Good* ix. 78 It [*sc.* a violin string] only clickets down because you will *not* screw in. **1911** —— *Likely Story* 201 This thing was inside brown paper, and rectangular. The corners were hard, but the middle clicketted. **1923** D. H. Lawrence *Birds, Beasts & Flowers* 27 Your foot-falls, pale-faces..are a clicketing of bits of disjointed metal Working in motion. **1929** R. Bridges *Test. Beauty* III. 371 This queer Pterodactyl..that in the sinister torpor of the blazing day clicketeth in heartless mockery of swoon and sweat.

clickety-clack, clickety-click: see Click *sb.*[1] 8 (in Dict. and Suppl.).

clicking, *vbl. sb.* Add: **c.** The process of cutting out the leather for boots and shoes. (Cf. *clicker[2] 2.)

1909 J. A. Hobson *Industr. System* xviii. 281 The introduction of clicking machinery into the shoe trade has been followed by a net reduction of employment. **1916** *Yorkshire Post* 15 Feb. 5/5 Women have not yet been put to clicking.

d. *Printing.* (See quot. and cf. Clicker[2] 3.)

1860 Ruse & Straker *Printing* 121 *Clicking.* This is a term applied to the mode pursued in London of getting out work by the formation of a companionship, or selected number of men, who are appointed to go on with a certain work or works.

clicky, *a.* Add: Also of other sounds.

1914 W. De Morgan *When Ghost meets Ghost* II. xxv. 795, I heard a stop in the grinding noise [of ice breaking on a river]..and a sort of clicky slooshing.

Hence **cli·ckily** *adv.*

1920 Masefield *Enslaved* 41 The cogs caught in the locket clickily.

client. Add: **2.** *spec.* = **client state.*

1958 *Spectator* 7 Feb. 159/2 This new republic will now be a client of Russia...The other Arab States..are clients of the West.

4. b. A person helped by a social worker; a case.

1925 L. Eaves (*title*) Aged clients of Boston social agencies. **1931** *Amer. Jrnl. Orthopsychiatry* 543 Interest in psychological values has revolutionized social work,..and has given a new dynamic aspect to the whole problem of the relationship of the worker to her client. **1950** C. Morris *Social Case-Work* ii. 35 'The case' has a derogatory sense when used of a person. Reluctantly, therefore, for want of a better word, we refer to 'the client'. **1959** *Listener* 28 May 937/1 A substantial proportion of social workers' 'clients'..are in difficulties for reasons unconnected with their lack of material goods.

5. *attrib.* and *Comb.*, as *client population, prince*; **client state,** a country dependent on a larger one for trade, etc.

1919 J. L. Garvin *Econ. Found. Peace* 90 The Allied or client populations amongst whom there is acute distress. **1909** *Westm. Gaz.* 27 Feb. 5/1 (*title*), The Client Princes of the Roman Empire under the Republic. **1918** T. E. Lawrence *Lett.* (1938) 265 The creation of a ring of client states. **1955** A. L. Rowse *Expans. Eliz. Eng.* ix. 334 The insecurity of the client-state of an enemy on the Border.

cliff. Add: **3. b.** *Golf.* The face of a bunker.

1890 H. G. Hutchinson *Golf* iv. 146 The nearer the ball lies to the cliff of the bunker, the farther behind the ball must the niblick-head dig down into the ground. **1904** *Westm. Gaz.* 21 Oct. 4/2 To get over the bunker's cliff.

5. *cliff-bastion, -bird, -breeder, -edge, -face* (examples), *-house, -nester, -path, -roost, -side* (earlier example), *swallow* (earlier U.S. example), *-top, -wall*; *cliff-bound, -bred, -breeding, -embattled, -encircled, -girt, -haunting, -lurking, -nesting, -ringed, -sheltered, -walled,* adjs. or vbl. sbs.; **cliff-brake** *U.S.*, a fern of the genus *Pellæa*, esp. *P. atropurpurea*; **cliff-dweller,** (*a*) one who lives in or on a cliff; *spec.* in south-west U.S., an Indian belonging to one of the tribes that made their homes in caves or upon ledges in canyon walls; (*b*) *U.S. slang*, one who lives in a tall building; hence *cliff-dwelling.*

1930 J. S. Huxley *Bird-Watching* ii. 31 Our goal was the cliff-bastion at the north end of Prince Charles's Foreland. **1940** R. Perry *Lundy* i. 29 The naturalist inland..comes only to the coast for the summer nesting of the cliff-birds. **1956** Bannerman *Birds Brit. Isl.* V. 31 Ussher went so far as to state that in Ireland no great cliff-bird colony seems to be complete without its pair of falcons. **1928** Mrs. Belloc Lowndes *Out of the War?* iii. 39 Cliff-bound shore. **1867** Gray *Man. Bot.* (ed. 5) 659 *Pellæa.* Cliff-brake.. *P. atropurpurea.* [grows on] dry calcareous rocks. **1941** R. S. Walker *Lookout* 56 Purple cliffbreak finds congenial homes in the limestone ledges. **1614** Chapman *Odyss.* iv. 800 Fitter far to feed a cliff-bred goat. **1938** *Brit. Birds* XXXII. 213 The Herring-Gull, usually a marine cliff-breeder. **1920** H. E. Howard *Territory in Bird Life* ii. 63 The cliff-breeding species.. are difficult to investigate. **1940** *Brit. Birds* XXXIII. (title) Cliff-Breeding in the House-Martin. **1881** *Rep. Indian Affairs* (U.S.) 47 The peach trees are supposed to have been originally planted by a superior race or by ancient explorers, possibly by the cliff-dwellers. **1893** H. B. Fuller (title) The cliff-dwellers. **1916** *Amer. Mag.* Apr. 31/2 You cliff-dweller on Manhattan, what would you do without Michigan? **1948** Johnston *Cities of Gold Rush* 46/2 Like the cliff dwellers of the Southwest, the men who worked in those diggin's have vanished. **1888** *Science* XI. 258/1 Some cliff-dwellings in Walnut Cañon, about twelve miles southeast of Flagstaff, Arizona, were examined. **1910** Cliff-dwelling [see *cliff-house]. **1903** Kipling *Five Nations* 70 The wise turf cloaks the white cliff edge. **1951** S. Spender *World within World* v. 324 The prevailing winds.. blew across the cliff-edge fields. **1859** D. Masson *Brit. Novelists* i. 28 Dashing the eternal monotone of her many voices against a cliff-embattled shore. **1885** W. B. Yeats in *Dublin Univ. Rev.* May, Here is the place, the cliff-encircled wood. **1906** *Westm. Gaz.* 25 Oct. 2/3 Tomb'd on the steep cliff-face. **1931** *Times Educ. Suppl.* 14 Mar. p. i/2 Large boulders and tons of earth fell from the cliff-face near Plymouth Hoe. **1900** *Westm. Gaz.* 4 Aug. 2/3 Stand firm upon thy cliff-girt coast. **1890** C. Dixon in *Leisure Hour* 685/1 This little group of rocky islets is the very metropolis of all our cliff-haunting birds. **1903** Kipling *Five Nations* 2 Unheralded cliff-haunting flaws. **1892** Stevenson & Osbourne *Wrecker* viii. 125 I looked from the cliff-house on the broad Pacific. **1910** *Encycl. Brit.* VI. 507/1 Two special sorts of cliff-dwellings are distinguished by archaeologists, (1) the cliff-house, which is actually built on levels in the cliff, and (2) the cavate house, which is dug out, by using natural recesses or openings. **1901** Kipling *Kim* viii. 198 Unheralded cliff-lurking flaws. **1953** B. Campbell *Finding Nests* x. 157 Old nests of various other cliff-nesters. **1954** Bannerman *Birds Brit. Isl.* III. 387 Cliff-nesting is by no means restricted to British Jays. *Daily Chron.* 18 Aug. 3/1 Dreaming along Cornish cliff-paths. **1910** W. de la Mare *Flora* 93 On the cliff-ringed shore. **1953** Bannerman *Birds Brit. Isl.* I. 13 Many rookeries harbour carrion crows on winter nights.. and cliff-roost is sometimes shared with a party of ravens. **1939** W. B. Yeats *Last Poems* 18 In some cliff-sheltered bay. **1852** M. Arnold *Empedocles* ii. 431 On the cliff-side the pigeons. **1825** C. L. Bonaparte *Amer. Ornith.* I. 65 The Cliff Swallow advances from the extreme western regions, annually invading a new territory farther to the eastward. **1852** M. Arnold *Empedocles* ii. 429 On the sward at the cliff-top. **1916** Blunden *Pastorals* 30 Go, cast it from the cliff-top while dawn stirs. **1858** M. Arnold *Balder Dead* 82 Through the cliff-wall, and a fresh stream runs down. **1927** Kipling *Verse 1885–1926* 729 The cliff-walled defiles.

cli·ff-hanger. orig. *U.S.* [f. Cliff + Hanger.] A serial film in which each episode ends in a desperate situation; hence, any story, play, etc., in which suspense is a main concern. Also *transf.*, *fig.* and *attrib.*

1937 *Amer. Speech* XII. 318/1 *Cliff-hangers*, type of serial melodrama. **1940** *Sat. Even. Post* 6 Apr. 18/1 The picture started out to be one of those cliff-hangers, with Indians, strong and silent men, the great out-doors, mail robbers. **1940** *Time* 15 Apr. 17/1 She is about to join radio's cliff-hangers and soap operas. **1952** *Time* 17 Mar. 19/3 'Will Ike come home in time?' is the greatest cliff-hanger question. **1956** 'J. Wyndham' *Seeds of Time* 7 The bulk of science-fiction.. has been.. kept to the cliff-hanger class. **1958** *Spectator* 1 Aug. 165/1 *Hamlet* was the first of the psychological cliff-hangers and a first-rate highbrow thriller it is. **1966** *Listener* 17 June 891/1 [illegible] cliff-hanger suspense [in tennis]. **1967** *Oxford Mail* 5 Sept. 5/1 The end of this thriller, set in an English country house, is electrifying.. All I can tell you is that it's a 'cliff-hanger'.

Hence (as a back-formation) **cliff-hang** *v. intr.*; also **cli·ff-hanging** *vbl. sb.* and *ppl. a.*

1958 *Time* 15 Aug. 21/1 Mr. Oland, who often threatened cinema death to daring, cliff-hanging Heroine White. **1966** in *Amer. Speech* (1956) XXXI. 284 The only thing I will admit is wrong with most soap opera is cliff-hanging. I never, never, cliff-hang. **1949** *Here & Now* (N.Z.) Oct. 10/3 His Government is cliff-hanging by four wisps of grass. **1958** *Economist* 20 Dec. 1090/1 John Diefenbaker can resume his proper role as hero in the greatest cliffhanging serial that Canadians have seen for many a long day. **1959** *Camb. Rev.* 25 Apr. 433/1 Three years wondering whether your tenure of your assistant lectureship will be extended.. five years in which to wonder if any of the permanent lectureships.. will fall vacant.. This way lie.. eight years of academic cliff-hanging.

cliffless (kli·flès), *a.* [f. Cliff + -less.] Without cliffs.

1897 *Naturalist* 224 In treeless and cliffless districts. **1898** *Speaker* 5 Mar. 297/1 The cliffless Downs.

cliftonite (kli·ftənait). *Min.* [Named after Robert Bellamy *Clifton* (1836–1921), English physicist: see -ite¹.] Carbon occurring as small cubic crystals in meteoric iron.

1887 Fletcher in *Min. Mag.* VII. 130 As this kind of crystallised graphitic carbon is unknown among terrestrial minerals.. it may conveniently receive a distinctive name, and I therefore suggest the term *Cliftonite*. **1910** *Encycl. Brit.* VIII. 162/2 Graphitic carbon in cubic form (cliftonite).

climate, *sb.* **3. b.** *fig.* Add def. and earlier and later examples.

The mental, moral, etc., environment or attitude of a body of people in respect of some aspect of life, policy, etc., esp. in *climate of opinion, of thought.*

1661 Glanvill *Van. Dogm.* xxiii. 227 The larger Souls, that have travail'd the divers Climates of Opinions, are more cautious in their resolves. **1826** Mrs. Hannah More *Lett.* (1834) III. iv. 263 Our country, surrounded by the sea on one side, and Mendip on the other, is so thinly planted with gentry, (the spiritual climate also being rather cold,) that without some effort somewhere it would come to nothing. **1866** Bagehot *Eng. Constitution* (1867) viii. 295 [illegible] **1886** J. A. Symonds *Renaissance in Italy* I. ii. 19 Lorraine.. felt his inner self expand in the rich climate of pontifical Rome. **1895** A. J. Balfour *Found. Belief* III. ii. 206 A psychological 'atmosphere' or 'climate' favourable to the life of certain modes of belief, unfavourable.. to the life of others. **1902** W. James *Var. Relig. Exper.* xiv. 328 The mental climate proves unfavourable to notions of the deity which at an earlier date were perfectly satisfactory. **1925** A. N. Whitehead *Sci. & Mod. World* (1926) i. 4 A climate of opinion—to use the happy phrase of a seventeenth century writer—requires for its understanding the consideration of its antecedents and its issues. **1938** R. Graves *Coll. Poems* 189 (title) The Climate of Thought. **1940** Auden *Another Time* 118 To us he [sc. Freud] is no more a person Now but a whole climate of opinion. **1949** 'G. Orwell' *Nineteen Eighty-Four* i. 55 The whole climate of thought will be different. **1952** R. Hoggart *Auden* i. 20 [illegible] **1952** A. Wilson *Hemlock & After* III. ii. 210 [illegible] 'climates of opinion'. **1960** *Times* 14 Aug. 16/1 The trading climate has not been a happy one in these territories in recent months.

climate, *v.* Add: **2.** *trans.* To acclimatize. *U.S.*

1849 N. Kingsley *Diary* (1914) 92 Relinquishing the idea of going to the diggings this winter [to] get ourselves climated ready for spring. **1863** 'E. Kirke' *Southern Friends* iv. 62 It's gin a feller's stumac used ter Tophet 'fore the rest on him is 'climated.

climatize, *v.* (Earlier U.S. example.)

1826 in Peck *Guide for Emigrants* (1831) 219 That some who have been born, or who have been climatized, in places, which to a stranger prove unhealthy—are generally free from these disease.. proves little.

Hence **cli·matized** *ppl. a.*, **cli·matizing** *vbl. sb.*

1827 P. Cunningham *N.S. Wales* II. xx. 30 Our colonial climatized females mincing it past these undraperied beaux. *Ibid.* xxi. 35 It may possibly be ascribed to the climatizing process.

climatotherapy (klai:matoþe·rapi). *Med.* [f. Climat(e *sb.* + -o- + Therapy.] The treatment of disease by a favourable climate. Also **climatotherapeu·tic** *a.* and *sb. pl.*

1887 *Buck's Handbk. Med. Sci.* V. 32/1 Dr. Weber's article on climato-therapy. **1901** F. P. Weber *Climatology* II. 245 Climatotherapeutic. *Ibid.* 300 This classification is convenient from a climatotherapeutic point of view.

climax, *sb.* Add: **4. b.** *Ecology.* The point in the ecological succession at which a plant-community reaches a state of equilibrium with its environment, able to reproduce itself indefinitely under existing conditions. Also *attrib.*

1916 *Bull. Illinois State Lab.* July 17 The succession of associations leading to a climax represents the process of adjustment to the conditions of the environment; and the climax associations represents a condition of relative equilibrium. Climax associations are the resultants of certain climatic, geological.. conditions. **1916** F. E. Clements *Plant Succession* vi. 105 Every complete sere ends in a climax. **1926** [see *Association 10]. **1929** Weaver & Clements *Plant Ecol.* iv. 75 [illegible] **1953** P. W. Richards *Tropical Rain Forest* i. 1 Evergreen forest is the climax vegetation of the equatorial climate. *Ibid.* iii. 40 Since the Tropical Rain Forest is a climatic climax, it must, by definition, be in a state of equilibrium.

c. *Physiol.* = Orgasm 2.

1918 M. C. Stopes *Married Love* v. 50 In many cases the man's climax comes so swiftly that the woman's reactions are not nearly ready. **1949** M. Mead *Male & Female* xiv. [illegible] **1950** [see *clitoral *a.*]. **1963** in A. Heron *Towards a Quaker View of Sex* 65 [illegible] climax.

climaxing (klai·mæksiŋ), *ppl. a.* [f. Climax *v.*] Reaching a climax; culminating.

1893 'Mark Twain' *£1,000,000 Bank-note* 17 Then came the climaxing stroke—..'Punch' caricatured me!

climb, *v.* Add: **5. a.** Also *spec.* (see quots.), and in causative sense.

1874 J. W. Long *Amer. Wild-Fowl Shooting* 245 They.. can 'climb' out of gun-shot (as the hunters term their flying upwards when frightened at the appearance of danger beneath them) faster than most wild-fowl. **1915** *Sphere* 28 Aug. 220/2 One aeroplane has already climbed to the same level as the airship. **1919** *Parl. Papers* X. 49 It is most economical to climb an aeroplane in the attitude corresponding to the maximum value of the lift–drag ratio.

climb, *sb.* Add: **1. b. climb-down.** For def. read: A withdrawal, esp. with ignominy, from a position taken up; abandonment of a declared position. (Further examples.)

1944 R. C. K. Ensor *Miniat. Hist. War* 12 He confronted a bewildered Europe with a threat of instant battle, only averted by the climb-down of France and Great Britain at Munich. **1960** *Guardian* 9 July 5/5 The proposal inevitably appears as a climb-down in face of Soviet pressure. **1961** A. Wilson *Old Men at Zoo* iii. 128 The French have sent a tremendous climbdown note.

2. Of aircraft: the ascent; also, the rate of ascent. Also *attrib.*

1912 F. A. Talbot *Aeroplanes & Dirigibles* xii. 162 Daring sweeps, startling wheels.. and remarkable climbs are carried out. **1918** H. Barber *Aeroplane Speaks* (ed. 6) 11 The essentials for Climb or quick ascent and for Speed are diametrically opposed. **1918** W. E. Dommett *Dict. Aircraft* 14 The time taken by an aircraft to rise each 1,000 ft. is known as 'the climb'. **1934** *Times* 26 June (Air Suppl.) p. xx/2 To absorb the maximum power of the engine during the take-off, climb, and level flight. **1940** N. Monks *Squadrons Up!* viii. 215 Climb Indicator, showing thousands of feet per minute.

3. A 'cat' burglary; so *at the climb*: engaged in such burglaries. *Thieves' slang.* Cf. *climber *sb.* 5, *climbing *vbl. sb.* 2.

1933 A. R. L. Gardner *Prisoner at Bar* v. 90 High-class burglars not 'at the climb' usually work in partnership with a chosen pal. **1960** *Observer* 25 Dec. 7/6 [illegible]

climber, *sb.* Add: **1.** *fig.* Esp. one who seeks continually to advance himself. Also *attrib.*

1833 *Knickerbocker* I. 179 But with all her meanness as a climber what a glorious leader of fashion she'd make. **1911** H. S. Harrison *Queed* xxiv. 301 He was of the climber type, a self-made man in the earlier and less inspiring stages of the making. **1924** G. B. Shaw *St. Joan* v. 56 Do ambitious politicians love the climbers who take the front seats from them?

5. A cat-burglar. *Thieves' slang.* (Cf. *climb *sb.* 3.)

1900 'Flynt' & Walton *Powers that Prey* 64 An' I say that we just up a good sneak an' climber. **1960** [see *climb *sb.* 3.]

climbing, *vbl. sb.* Add: **2.** Cat-burgling. Also *attrib.* *Thieves' slang.* (Cf. *climb *sb.* 3.)

1900 'Flynt' & Walton *Powers that Prey* 194 You've too young to do climbin'. **1907** [see *cat-burglar]. **1965** *New Statesman* 30 July 152/1 The climbing game must be good. [illegible]

3. *attrib.* and *Comb.* (Of aircraft) *climbing angle, speed, turn*, etc. (cf. *climb *v.* 5 a and *sb.* 2); **climbing frame**, a structure, usually of pipes joined together, for children to climb on.

1933 *Jrnl. R. Aeronaut. Soc.* XXXVII. 2 An effective landing run of at least 2,500 ft. is required in eight directions, taking the gliding and climbing angle as 1 in 7. **1929** S. Isaacs *Nursery Years* vi. 68 A firmly built wooden frame with horizontal rods to climb on and hang from... Several of these things, such as the see-saw, the ladders, and the climbing frame are more easily arranged for by the Nursery School. **1960** *Guardian* 24 Feb. 11/2 A climbing frame for the enterprising young. **1913** C. Grahame-White *Aviation* 199 The Cody biplane.. was able to ascend from the ground at a 'climbing' speed of 285 feet per minute. **1935** C. Day Lewis *Time to Dance* 57 [illegible] climbing-speed. **1938** F. W. Mills *Pract. Flying* vi. 91 Climbing turns do not normally permit of great steepness.

climbing, *ppl. a.* Add: **b.** *climbing fern* (*Lygodium palmatum*), *rose.*

1818 A. Eaton *Man. Bot.* (ed. 2) 311 [The] climbing fern [*Lygodium*]. **1882** H. E. F. Smith *Hist. Brit. Ferns* 329 Besides climbing-fern, it is known as creeping-fern, Hartford-fern, and Windsor-fern. **1846** Eaton *Man. Bot.* (ed. 7) 489 *Rosa rubifolia*, climbing rose. **1864** [see Rose *sb.* 3 a]. **1965** *Dict. Gardening* (R. Hort. Soc.) (ed. 2) IV. 1825/1 Climbing Roses for Pergolas, Fences, and Walls are provided by several of the groups of garden roses.

climograph (klai·mogrɑf). [irreg. f. Clim(ate *sb.* + -o + -graph.] A graphical representation of the differences between various types of climate.

1916 G. Taylor in *Bull. Commonw. Bur. Meteorol.* No. 14 7 Upon a graticule whose ordinates represent relative humidity and wet-bulb temperatures, the monthly means are plotted. The resulting twelve-sided polygon is the climograph. **1934** *Brit. Birds* XXVIII. 179 By drawing a series of 'climographs' of the combinations of night and day air temperature, possible and actual sunshine, precipitation and wind velocity.. the writer is able to show where the limits of tolerance are actually approached.

clinah, var. *cliner.

clinch, *sb.*¹ Add: **3. b.** The grip or hold (of plaster on a wall).

1897 F. C. Moore *How to Build* ii. 16 The lath should be furred out from the sheathing so as to secure a good 'clinch' or 'key' to the plaster.

5. (Earlier *attrib.* example.)

1849 C. Lanman *Lett. Alleghany Mts.* vi. 50 On reaching the bottom, he found the wolf alive, when a 'clinch fight' ensued, and the hunter's knife completely severed the heart of the animal.

b. *spec.* in *Boxing.* Grappling at close quarters or holding after an exchange of blows.

1875 *Amer. Cycl.* XIV. 74/1 Unless there is a knock-down, the rounds usually terminate in a clinch. **1897** R. G.

Allanson-Winn *Boxing* 329 Occasionally after a spell of in-fighting, but not in actually breaking away after a close or clinch, you may administer an excellent upper cut. **1901** Fitzsimmons *Phys. Culture* 152 Corbett seemed to think clinches were his best time for a knock-out blow. **1923** Tosswill *Boxing Up-to-date* 58 The best thing to do is to fall into a clinch, dropping forward with your hands resting on your opponent's arms, while your ribs and stomach are protected by your elbows.

c. An embrace. *colloq.* (orig. *U.S.*).

1901 'H. McHugh' *John Henry* 25 A love clinch from Laura, Jean Libbey's latest. **1929** W. Root in *Bookman* (N.Y.) Feb. 622/1 'Darling!' and they go into a clinch. **1948** Wodehouse *Spring Fever* ii. 23 Nothing would have given Stanwood more pleasure than to have seen the young couple fading out on a clinch. **1959** J. Osborne *Paul Slickey* i. iv, The 'King' and 'Queen' go into a clinch.

7. (Earlier example.)

1857 G. W. M. Reynolds *Myst. London* III. xxv. 71/2 Should you do this and get the clinch.

8. clinch plate, a plate on the inside of clinch-work.

1889 P. B. Du Chaillu *Viking Age* I. xii. 219 The boat is clinch-built; that is, the planks are held together by large iron bolts with round heads outside, and clinch plates on the inside, at a distance of 5¼ inches from each other.

clinch, *v.*¹ Add: **2. c.** (Earlier example.)

1652 Gaule *Magastrom.* 186 Toes that clinch together.

d. Delete quot. 1652 (see *2 c) and add earlier U.S. example.

1849 *Yankee* May 174/1 A native Yankee.. would never fail to be the first to strike a blow, nor hardly ever the first to clinch, as he calls it.

e. *spec.* in *Boxing.* To fall into a clinch (see prec. 5 b). Also *trans.* with person as obj.

1860 [see 5 c in Dict.]. **1863** in H. D. Miles *Pugilistica* (1881) III. 516/2 The 'clinching'—we must call it so, for want of a better term—was nothing more than our old acquaintance the 'hug'.. **1887** *Courier-Jrnl.* (Louisville) 7 Jan. 3/2 Lilian returned with a blow on Bowen's neck. The men then clinched. **1893** Dobson *Sci. Self-Defense* 84 You can also take this hold as your opponent endeavours to rush in. **1899** *Life of Bob Fitzsimmons* 6 In the sixth round the men clinched. *Ibid.* 14 The clinching was of the hottest kind. **1920** J. J. Corbett *My Life & Fights* 26, I clinched with Kilrain. **1923** Tosswill *Boxing Up-to-date* 56 In competitions held under the auspices of the Royal Navy and Army.. the command 'Break away' is given by the referee the moment the boxers clinch.

f. *intr.* To embrace. (Cf. prec. 5 c.) *colloq.* (orig. *U.S.*)

1899 Ade *Fables in Slang* 60 George was sitting on the Front Porch with a good Side Hold on.. Mabel... While they were Clinched, Mabel's Father came down the Stairway. **1952** Wodehouse *Pigs have Wings* viii. 157 'I saw them together. Close together,' said Lady Constance significantly. 'God bless my soul! What, clinched together?' 'When I saw them, they were not actually embracing.' **1953** *Punch* 26 Aug. 274/1 They.. sit like lovers about to clinch.

clinched, *ppl. a.* (Later example.)

1881 Whitman *Leaves of Grass* 268 The inextinguishable glance and the clinch'd and lifted fist.

clincher. **4.** (Earlier and later examples.)

1737 *Lond. Mag.* Jan. 14/1 This is a Clincher, i'faith; for if Twenty such Laws could be actually produced, *no Man in his Senses* would do it. **1775** Garrick *Let.* 27 Aug. (1832) II. 82 To introduce for a Clincher, and confirm the young man in his good resolutions. **1966** *Listener* 26 Aug. 292/1 It would be a relief to tie up this discussion with the clincher that they started and led the riots. **1968** *International Herald Tribune* 2 Sept. 7/3 But the clincher was a rise of 28 percent in the number of robberies.

clinching, *vbl. sb.* Add: **b.** *spec.* in *Boxing* (see *clinch *v.*¹ 2 e).

1863 in H. D. Miles *Pugilistica* (1881) III. 518/1 The hugging, wrestling, and [illegible] were 'clinching', rushing, squeezing. **1899** [see *clinch *v.*¹ 2 e]. **1923** Tosswill *Boxing Up-to-date* 58 No beginner should deliberately cultivate the habit of clinching.

cline (klain), *sb. Biol.* [f. Gr. κλίνειν to slope, bend.] A graded series of characters or differences in form within a species or other group of related organisms. Also *transf.*, in *Linguistics* (see quot. 1961).

1938 J. Huxley in *Nature* 30 July 219/1 Some special term seems desirable to denote a character-gradient or variation within a group, and I propose the word *cline*, meaning a gradation in measurable characters. **1949** *Brit. Birds* XLII. 133 It may be that some of the characters of a species alter gradually and continuously over a wide region.. such a gradation being conveniently defined by the term 'cline' introduced by Huxley. **1952** J. Fisher *Fulmar* v. 75 There appears to be a 'cline' in bill-length. **1952** *New Biol.* XII. 16 A gene may vary along a cline in space. As we travel southwards in Europe the genes for dark hair on the whole become commoner. **1961** M. A. K. Halliday in *Word* XVII. 249 A cline resembles a hierarchy in that it involves relation along a single dimension; but instead of being made up of a number of discrete terms a cline is a continuum carrying potentially infinite gradation. **1966** J. Ellis in C. E. Bazell *In Memory of J. R. Firth* 81 Between these extremes we might recognize a cline of potentiality/instantiality. **1970** *Nature* 14 Mar. 1027/2 Demonstration of a cline in gene frequencies provides indirect evidence for the action of natural selection at the locus in question.

cliner (klai·nəɹ). *Austral. slang.* Also **clinah.** [Orig. unknown.] A girl; a sweetheart. Now *rare* or *Obs.*

1895 *Bulletin* (Sydney) 9 Feb., Oh, she's good iron, is my little clinah. **1916** C. J. Dennis *Songs Sentim. Bloke* 51 Wimmin!.. I've got the 'ole bang tribe o' cliners set. **1947** I. Douglas *Opportunity in Australia* 89 *Clinah*, girl friend.

cling, *sb.*¹ Add: **5.** Short for Clingstone. Also *attrib.* in *cling peach.*

1845 A. J. Downing *Fruits Amer.* 494 The Catherine cling is a very fine, old English variety. **1869** *Trans. Ill. Agric. Soc. 1867–8* VII. 510 We have a late, large peach which we call the Allman Cling. *Ibid.*, It is sometimes called the Heath Cling, which it resembles. **1872** *San Francisco Weekly Bulletin* 27 Sept. (Hoppe), Cling peaches are moderately plentiful. **1913** Mrs. Stratton-Porter *Laddie* iii. 92 One [peach] was a white cling, one was a yellow. *Ibid.* iv. 126, I led her straight to our best cling peach tree. **1968** *Publ. Amer. Dial. Soc.* XLIX. 16 Another reason for vocabulary change is the widespread use of commercial terms... These include.. *cling peach* for *cling-stone peach.*

cling (kliŋ), *sb.*² Delete †*Obs. Sc.* and add later examples. Also *cling-clang.*

1886 W. J. Tucker *E. Europe* 182 With martial strides, and a cling, cling, cling of spurs. **1902** E. Nesbit *Five Children & It* vii. 186 The children could hear the cling-clang of armour. **1926** [see *bing-bang* (*bing *sb.*² and *int.*)].

cli·ng-fish. [f. Cling *v.*¹] A small carnivorous fish of the family Gobiesocidæ, possessing a sucker on the underside of its body by which it is enabled to cling to any object.

1893 in *Funk's Stand. Dict.* **1904** G. A. Boulenger in *Camb. Nat. Hist.* VII. 708 Cling-Fishes.. can live a long time out of water. **1931** J. R. Norman *Hist. Fishes* iv. 81 The Cling-fishes (*Gobiesocidæ*) are curious little creatures found between tide-marks among loose stones and shells, to which they adhere firmly. **1969** A. Wheeler *Fishes Brit. Isles & N. W. Europe* 575, I have chosen to use the American family name clingfish, rather than employ the so-called vernacular 'sucker' which is a potential source of confusion.

clinginess (kli·ŋinės). Also **clingyness.** [f. Clingy *a.* + -ness.] The quality of being clingy or adhesive.

1778 W. Marshall *Min. Agric.* 16 Jan. 1775, The clingyness of the soil. **1794** T. Davis *Agric. Wilts* 91 The 'clinginess' of the Wiltshire white lands. **1899** *Daily News* 28 Jan. 6/4 The tightness and clinginess of ultra-fashionable dresses.

clingstone. (Earlier U.S. example.)

1722 Beverley *Virginia* (ed. 2) 279 The best sort cling to the Stone, and will not come off clear, which they call Plum-Nectarines, and Plum-Peaches, or Cling-Stones.

clinic, *sb.*² Add: **2.** [After F. *clinique*, G. *klinik.*] **a.** A private hospital or medical institution to which patients are recommended by individual doctors. **b.** Formerly, an institution attached to a hospital or medical school at which patients received treatment free of cost or at reduced fees. Now esp. a hospital department devoted to a particular group of diseases, etc., usu. with defining epithet, as *diabetic clinic, fracture clinic*, etc.; also, a centre or other institution at which specialized treatment, diagnosis, or advice is available, as *child guidance clinic, dental clinic*, etc.

1889 *Lancet* 22 June 1283/2 Two new hospitals have just been opened in Moscow. One of these is a lying-in clinic attached to the University. **1892** *Cosmopolitan* XIII. 766/1 Clinics that are held by the professors of diseases of the eye. **1895** *Daily News* 18 June 6/2 The Council of State in St. Petersburg is busy with the project of a medical institute for women... After completing their studies the students will have to practice for one to three years in womens' clinics or similar hospitals. **1902** *Lancet* 25 Jan. 235/1 Dr. Mendes has put together a number of interesting observations which he has made in his clinic. **1909** *Westm. Gaz.* 14 July 2/1 The clinic opens at nine on every school-day and also on Saturday with examination of cases. **1922** *Encycl. Brit.* XXXII. 875/1 The many specialized clinics—prenatal, 'baby', dental, venereal disease, psychiatric etc. **1951** 'M. Innes' *Operation Pax* 207 'What is a clinic?'.. 'I think it's becoming a fashionable word for a grand sort of nursing home—the sort that has one special line.' **1952** *Oxf. Jun. Encycl.* X. 183/1 There has gradually grown up the practice of dealing with special types of patients in properly equipped clinics or centres; women during pregnancy, for example, are seen at ante-natal clinics, while small children and babies are examined at special children's clinics. **1961** *Brit. Med. Dict.* 318/2 Patients who have been in hospital may attend clinics for after-treatment.

3. *transf.* An institution, class, conference, etc., for instruction in or the study of a particular subject; a seminar. Chiefly *U.S.*

1919 *British Manufacturer* Nov. 30/2 In order to solve this difficult problem in economic diagnosis, we need a clinic just as the doctor does. **1948** *N.Y. Herald Tribune* 10 June 33/7 Creation of a management–labor 'Joint Productivity Clinic', aimed at increasing industrial output. **1951** *College English* Jan. 232 A 'composition clinic' has been set up by the college of liberal arts department of English of Wayne University... Once the student is enrolled there, his writing is diagnosed and he is given whatever treatment he needs. **1968** *Globe & Mail* (Toronto) 13 Jan. 38/2 Five regional clinics for football coaches.

clinical, *a.* Add: **3.** Coldly detached and dispassionate, like a medical report or examination; diagnostic or therapeutic, like medical investigation or treatment; treating a subject-matter as if it were a case of disease, esp. with close attention to detail; serving as part of a case-study. Also *Comb.*

1928 F. Hurst *President is Born* xxxii. 326 Bek would never have permitted Winslow to behold her.. bartering in the clinical phraseology of the cattle-pen over bull or stallion or sow. **1934** H. Nicolson *Curzon: Last Phase* 65 As an 'exhibit' in any clinical examination of the mentality of the victors in the European War she [*sc.* Turkey] is unsurpassed. **1937** C. Isherwood *Lions & Shadows* 191 He peppered his work liberally with such terms as 'eutectic', 'sigmoid curve', 'Arch-Monad', 'ligature', 'gastropod'; seeking thereby to produce what he himself described as a 'clinical' effect. To be 'clinically minded' was, he said, the first duty of a poet. **1944** H. Treece *H. Read* 9, I have selected a passage.. as a clinical exhibit of adolescent self-repression. **1947** W. Empson *Seven Types Ambig.* (ed. 2) p. ix, The effort of writing a good bit of verse has.. been carried through almost as a clinical thing; it was done only to save the man's own sanity. **1953** R. Lehmann *Echoing Grove* 27 'I suppose the twitching is just reflex action.' Dinah's tone was clinical. **1953** S. Spender *Creative Element* 146 Auden.. showed.. a gift for inventing clinical-seeming images. **1957** *Times* 30 Sept. 14/6 Supreme skill on its own [in lawn tennis] without any major goal tends to become cold, clinical, and without feeling. **1958** *Daily Mail* 13 Mar. 1/1 Unemployment figures.. were studied with.. almost clinical interest.

b. Bare and functional, like a hospital.

1932 H. Nicolson *Public Faces* xii. 321 Her almost clinical bathroom. **1956** N. Cardus in *Bedside Guardian* V. 144 Would the interior look clinical,.. would the architecture and ornamentation strike a chill to the heart?

clinician. Add: **1.** Now esp. a doctor who has direct contact with and responsibility for patients, as distinct from one concerned with laboratory or theoretical work.

2. A doctor who has charge of, or who works in, a clinic (see *clinic *sb.*² 2).

1897 *Trans. Amer. Pediatric Soc.* IX. 112 Clinician to Children's Clinic, Medical College of Ohio, Cincinnati, O. **1929** *Times* 2 July 19/4 At the Radcliffe Infirmary the assistant medical officer of health is assistant clinician.

clinicopathological (kli:nikopæþǫlǫ·dʒikăl), *a.* [f. combining form of Clinic *sb.* (see -o) + Pathological *a.*] Relating to both clinical and laboratory examinations of disease.

1898 *Phil. Trans.* B. CXC. 2 The clinicopathological examinations by Vialet.. of cases of disease in man. **1962** *Lancet* 12 May 1006/1 Clinicopathological conferences were the rule. **1968** *Ann. Internal Med.* LXVIII. 853/1 A correlative clinicopathological study was made of a 32-year-old truck driver with arsine-induced anuria.

clink, *sb.*¹ Add: **6.** An internal crack in a block of metal caused by uneven contraction or expansion during cooling or heating.

1925 *Jrnl. Iron & Steel Inst.* CXI. 150 The well-known phenomena called cooling clinks or cooling cracks. **1948** A. V. Jobling in H. W. Baker *Mod. Workshop Technol.* I. i. iv. 88 If the tension stresses reach a sufficiently high value while the inside is comparatively cold and lacking in ductility, an internal fracture or clink will result.

clink, *sb.*² Add: Now used generally for: prison, cells.

1890 Kipling *Barrack-r. Ballads* (1892) 20 And I'm here in the Clink for a thundering drink and blacking the Corporal's eye. **1906** *Daily Chron.* 28 Aug. 4/7 Escorting erring defaulters of his own creation to clink. **1919** *War Slang* in *Athenæum* 25 July 664/2 'Clink', punishment cells. **1934** G. B. Shaw *Too True* iii. 82 Soldiers that try to make storytelling do for service end in the clink. **1946** K. Tennant *Lost Haven* (1947) xxi. 355 They'll only dock my pay or shove me in clink.

clink, *v.*¹ Add: **8.** To cause (metal) to fracture internally. Cf. *clink *sb.*¹ 6. So **clinking** *vbl. sb.*

1929 *Jrnl. Iron & Steel Inst.* CXIX. 291 The first ingot was undoubtedly quite extensively clinked. **1930** *Engineering* 19 Sept. 371/2 Very high internal stresses which frequently led to transverse clinking in the centre of the ingot. **1948** A. V. Jobling in H. W. Baker *Mod. Workshop Technol.* I. i. iv. 88 Clinking is a defect which does not result from any fault in the actual forging operation but from too rapid heating or cooling of a large mass of steel. **1953** D. J. O. Brandt *Manuf. Iron & Steel* xxviii. 204 Ingots of steel containing over 0·40% carbon and alloy steel ingots are easily *clinked* (cracked internally).

clink-clank, *v. intr.* To make a clink-clank sound.

1921 H. Walpole *Young Enchanted* iii. iii, The ships clink-clank against the side of the pier.

clinker, *sb.*¹ Add: **6.** *clinker asphalt, concrete.*

1954 *Highway Engin. Terms* (*B.S.I.*) 32 *Clinker asphalt*, a wearing course consisting of rolled asphalt in which the principal aggregate is clinker. **1930** *Engineering* 11 July 41/3 Breeze and clinker concrete.

clinker, *sb.*² Add: **2. c.** *U.S. slang.* In music, a wrong note, a discord.

1942 Berrey & Van den Bark *Amer. Thes. Slang* § 578/5 Discord,.. *clinker.* **1946** Mezzrow & Wolfe *Really the Blues* 369 Listen to.. Mezzrow's clinkers on the clarinet.

1956 M. Stearns *Story of Jazz* (1957) xvii. 202 The individual musicians were always hitting 'clinkers' or wrong notes.

clinker, *sb.*³ Add: **2. b.** A 'clinking' good thing: applied to persons, animals, and things of first-rate quality. *slang* (orig. *Sporting*; cf. Clinking *ppl. a.* 2).

1836 *Spirit of Times* 27 Feb. 13/2 It would just take him to beat any thing that Col. Johnson can start. He is a clinker, I tell you. **1869** *Daily Tel.* 5 Apr. 2/4 Despite the indifferent manner in which Vagabond cut up at the finish of the Metropolitan, quite sufficient was seen of him to prove that at a mile and a half he is a 'clinker'. **1873** J. Wisden *Cricket & how to play It* (ed. 2) 43 A clinker from Wootton bothered and bowled Mr. Ottaway. **1888** 'R. Boldrewood' *Robbery under Arms* I. vi. 81 By George! how fond I am of a good horse—a real well-bred clinker. **1913** Compton Mackenzie *Sinister St.* I. ii. ii. 165 'She was a clinker,' Alan regretted. 'I don't suppose we shall ever see a girl half as pretty,' Michael thought. **1936** W. Holtby *South Riding* ii. i. 91 By God she could ride. A clinker across country.

clinker, *v.*² Add: **b.** To remove clinkers from (a furnace). So **cli·nkerer**; **cli·nkering** *vbl. sb.*

1901 *Feilden's Mag.* IV. 437/1 Long iron bars or clinkering tools. *Ibid.* 439/1 So that no rush of cold air may take place through the flues while clinkering is going on. *Ibid.* 439/2 The labour of stoking and clinkering the furnaces. **1921** *Dict. Occup. Terms* (1927) § 699 *Clinker man* (gas works), *water gas clinkerer*; removes clinker and ashes from producer gas furnace with long iron bar. **1926** *Spectator* 29 May 901/1 At the end of the journey it was found that no clinkering had taken place.

clinkery (kli·ŋkəri), *a.*² [f. Clinker *sb.*¹ + -y¹.] Resembling clinkers.

1884 Lyell *Student's Elem. Geol.* (ed. 4) xxix. 463 The lava in cooling assumes a clinkery appearance.

clinkety (kli·ŋkėti). Onomatopœic extension of Clink *sb.*¹ as in *clinkety-clank, -clink* (cf. *clankety).

1901 F. T. Bullen *Sack of Shakings* 164 Clinkety-clank, bang, went the pumps. **1927** *Observer* 5 June 7/5 The clinkety-clink of metal on metal.

clinograph. Add: **1.** (Earlier example.) Now *disused.*

1885 *Engineering* XXXIX. 335/3 The clinograph will no doubt detect errors in a bore, and.. map accurately the path, however tortuous, of a drill.

2. A form of adjustable set square, not graduated to show angles.

1904 Goodchild & Tweney *Technol. & Sci. Dict.* 110/2 The clinograph is invaluable in graphic calculation in statics.

3. *Seismology.* An instrument for measuring changes in inclination.

1933 R. Takahasi in *Bull. Earthquake Res. Inst.* (*Japan*) IX. 26 A pair of clinographs was installed.. at the north slope of the mountain and the tilt of the earth's crust was observed. *Ibid.*, The clinograph is essentially a horizontal pendulum made of fused quartz.

4. = *clinographic curve.

1937 *Geogr. Jrnl.* XC. 59 Hypsograph and clinograph for the whole region.

clinographic, *a.* Add: (Examples.) Also, *clinographic curve*: a curve representing the slope or slopes of (a portion of) the earth's surface.

1889 *Cent. Dict.*, Clinographic. **1910** *Encycl. Brit.* VII. 573/2 Clinographic drawings are most frequently used for representing crystals. **1935** J. Hanson-Lowe in *Geol. Mag.* LXXII. 181 The real need of the morphologist is a knowledge of the actual average gradients between any two available (and adjacent) contours. For this reason I suggest that a new curve, which may conveniently be called the clinographic curve, is of far greater use [than the hypsographic curve]. **1937** *Geogr. Jrnl.* XC. 58 The district has been divided into four sub-regions and hypsographic and clinographic curves have been drawn for each.

clint, *sb.* Add: **1.** Used locally, and in *Geol.*, to designate: **a.** A crack or slit in rock, a grike. **b.** *esp.* A hard bare eroded rock-surface, *spec.* one developed in limestone regions in the N.W. Pennines in Britain.

1781 J. Hutton *Tour to Caves* 88 *Clints*, crevices amongst bare limestone rocks. **1882** J. Lucas *Studies Nidderdale* 243 *Clint*,.. crevices among bare limestone rocks.., so used in Wharfedale. **1925** *Times Lit. Suppl.* 22 Jan. 50/2 The hill-top platforms with deeply eroded crevices called clints or grykes. **1932** *Ibid.* 17 Mar. 191/3 The clints (tablelands of bare stone) are not safe playgrounds for children. **1952** W. Moore *Dict. Geogr.* (ed. 2) 77 *Grike* or *Clint*, a hollow in the surface of the rock in a limestone region, formed when the limestone dissolves in rain water containing carbon dioxide from the atmosphere. **1963** D. W. & E. E. Humphries tr. *Termier's Erosion & Sedimentation* 402 Clint, a bare, level surface developed on horizontal beds of limestone. The vertical fissures formed by solution along the joints are termed *grikes.*

clinure (klai·niŭɹ, kliniū·ɹ). *Math.* [f. Gr. κλίν-ειν to slope + -ure.] Direction in space, or inclination, as of a vector, relative to some set of coordinates. Hence **clinu·ral** *a.*

1884 Jas. Thomson in *Proc. R. Soc. Edin.* XII. 569 There is.. a real distinction.. between absolute rotation (or absolute clinural motion) and absolute freedom from rotation (or absolute clinural rest). *Ibid.*, The word *clinural* is to be understood as introduced for conveying precisely one out of the various conflicting meanings of the word directional. All straight lines which are mutually parallel are, in this amended mode of nomenclature, said to be in one same *clinure.*

clip, *v.*¹ Add: **3. b.** To fasten with a clip or clips.

1902 E. L. Banks *Newspaper Girl* 244 Page after page passed from under her pen. Then, clipping a dozen sheets together, she read them over. **1908** *Daily Chron.* 21 Nov. 9/4 Permanently fixed instead of clipped-on fittings. **1916** 'Boyd Cable' *Action Front* 220 He turned to the man, who was clipping a tiny mirror on to his bayonet. **1955** E. Bowen *World of Love* v. 84 She turned her head.. clipped on earrings.

clip, *v.*² **1. a.** Add to def.: To cut out (a passage) from a newspaper or periodical; to excerpt. Chiefly *U.S.*

1872 *Harper's Bazaar* 6 Jan. 11 The following marriage notice is clipped from a Texas paper. **1904** *N.Y. Even. Post* 10 Sept. 4 A suggestion for a college girl's vacation that should be clipped and pasted for reference next June. **1953** *Encounter* Nov. 5/1 He shifted to the weekly news-magazine.. and began to clip stories. **1962** *Listener* 4 Oct. 529/2 The famous 'Americana' column which clipped from the regional press absurdities of all kinds.

5. d. To reduce, cut down; *spec.* of a speed-record or the like (freq. const. *off*).

1938 J. Hilton *To you, Mr. Chips* i. 40 The craze for clipping fifths of seconds raged elsewhere. **1947** Wodehouse *Full Moon* ix. 195 He did not know what the European record was.. but he rather fancied that he had clipped a few seconds off it. **1957** —— *Over Seventy* v. 65 That.. is how your grandfather clipped six seconds off Roger Bannister's mile.

7. *U.S.* (Later examples.)

1886 *Harper's Mag.* Nov. 835/2 All that he could do was to take to his heels and clip down the steps and into the pung. **1897** Susan Hale *Lett.* (1919) 317 Though he had to row at first, the wind soon sprang up and we were clipping along. **1924** Mulford *Rustlers' Valley* xxvi. 283 Now and then searching bullets clipped and sung below him.

8. *trans.* To hit smartly. *colloq.*

1855 'Q. K. Philander' *Doesticks* xii. 99, 97's engineer clipped one of 73's men with a trumpet. **1880** T. Q. Couch *E. Cornwall Gloss.* 90 *Klip*, to strike or cuff. 'I klipped 'en under the ear.' **1961** *New Statesman* 1 Dec. 826/1 After hearing of the incidents in which his boy had been concerned he had 'clipped him round the earhole'. **1963** *Times* 7 Feb. 3/3 A straight full toss which Bradman clipped back to the sight-screen.

9. To swindle; to rob, steal. (Cf. Fleece *v.* 3.) *slang* (orig. *U.S.*).

1927 *Vanity Fair* Nov. 134/3 When a patron in a night club is 'clipped' he isn't punched, he's 'taken' or 'gypped' out of some currency or he is overcharged. **1932** J. Sayre *Rackety Rax* xvi. 132 He [*sc.* Capone] clipped the Doc in a deal. **1937** D. Runyon *More than Somewhat* 13 Here I am out to clip Miss Amelia Bodkin of her letters and her silverware. **1949** W. R. Burnett *Asphalt Jungle* (1950) vi. 40 He'd just been clipped for thirteen hundred. **1958** *Observer* 15 June 14/3 A commination against London taxi drivers, delivered with the fervour of a guy who'd really been clipped. **1968** H. Waugh *Con Game* xix. 158 I'm not holding still like some people while their buddy-buddy pals clip them for nine thousand bucks.

clip, *sb.*¹ Add: **2. d.** A receptacle containing several cartridges held together at the base for insertion bodily into the magazine of a repeating fire-arm.

1901 'Linesman' *Words by Eyewitness* vii. 146 The clip containing the cartridges for the Mauser rifle holds five rounds. **1902** *Encycl. Brit.* XXXII. 651/2 The clip, employed in the Mannlicher system, is used with a magazine with parallel walls. *Ibid.* 654/1 The loading by clip. **1966** D. F. Galouye *Lost Perception* iii. 30 The rifle discharged an entire clip.

e. A piece of jewellery that is fastened by a clip; *spec.*, an earring.

1937 C. Beaton *Diary* 3 June in *Wand. Years* (1961) 310 Wallis, wearing a black dress and her huge diamond pear clips. **1940** N. Mitford *Pigeon Pie* i. 9 Mrs. Holst.. only has the mingiest little diamond clip you ever saw. **1950** 'J. Guthrie' *Is this what I Wanted?* vii. 110 Pearl clips set off the ears, which were her most delicate feature. **1950** A. Wilson *Such Darling Dodos* 114 Mummy was wearing her black costume with the diamond shoulder clip.

3. clipboard, a tablet of wood or other material bearing a spring-clip at one end for holding papers, etc.; **clip-chair**, a chair used on some railways to secure the rail to a metallic sleeper; **clip-loader**, a clip-loading fire-arm; **clip-loading** *vbl. sb.*, the loading of a repeating fire-arm with cartridges in a clip (see *2 d).

1907 *Yesterday's Shopping* (1969) 380/1 The 'clip' boards for patience and bridge... This board is fitted with nickel clamps on each side to keep it perfectly rigid. **1921** *Instructions for Aerological Observers* (U.S. Weather Bureau) 88 A simple apparatus which has proved to be both accurate and speedy is constructed as follows: It consists of an 11 × 13 inch clip board, upon which is mounted a celluloid or hard bristol-board backing. **1959** M. Bradbury *Eating People is Wrong* ii. 66 Treece found the driving test office and went in to look for the examiner. He.. was clutching a clipboard. *a* **1884** Knight *Dict. Mech.* Suppl., The clip chair is of horse-shoe shape, one side forming a hook about 2½" wide, and the other side is like one jaw of an ordinary railway chair for taking a wooden railway key. **1901** *Chambers's Jrnl.* Apr. 285/1 The most perfect clip-loader so far produced. **1901** *Westm. Gaz.* 23 Dec. 4/3 The weapon to be adopted in the immediate future for the British Army.. must positively admit of clip loading. **1902** *Encycl. Brit.* XXXII. 654/2 A clip-loading repeating rifle.

clip, *sb.*² Add: **2. c.** An extract from a motion picture.

1958 *Listener* 3 July 30/2 Film clips from Jose Ferrer's film 'Moulin Rouge'. **1959** H. Nielsen *Fifth Caller* xii. 188 She stared at the blowup of an old newsreel clip.

4. (Later U.S. examples. Also *fig.*)

1869 Mrs. Stowe *Oldtown Folks* xx. 243 When you didn't mind, I hit you a good clip. **1886** J. C. Harris in *Century Mag.* Jan. XXXI. 426/1 I'm a great mind to hit him a clip jest to show you how he can go on. **1913** Mrs. Stratton-Porter *Laddie* xvii. 568 Robert.. wasn't a smidgin behind, for every clip he had the answer ready.

6. A (specified) rate of speed; a rapid pace or motion. *colloq.*

a **1867** H. Woodruff *Trotting Horse* (1868) vi. 79 It is believed that he can go a four-minute clip. **1877** E. Peacock *Gloss. Manley* 61/2 We are goin' wi' a clip now. **1893** *Field* 25 Feb. 295/3 In three days I could drive him any 'clip' I chose by just talking to him. **1901** *Westm. Gaz.* 31 Aug. 6/3 As *Shamrock* returned the wind increased and the challenger with a magnificent burst of speed travelled at a 12-knot clip. **1941** H. MacInnes *Above Suspicion* xv. 150 God, that dog can keep up a terrific clip. **1967** *Boston Sunday Herald* 26 Mar. iii. 10/1 Romance and marriage among corporations is rolling along at a torrid pace... The brisk clip continues this year.

clipe, var. *clype *v.*

clip-joint (kli·pdʒoint). *slang* (orig. *U.S.*). [f. *clip *v.*² 9 + Joint *sb.* 14.] A club, bar, etc., charging exorbitant prices (see quot. 1964). Also *transf.*

1933 S. Walker *Night Club Era* 218 The worst creature, of all the army of parasites who carried on their trade along Broadway during the speakeasy period, was the proprietor of the 'clip joint'... The clip joint preys on the New Yorker and the out of town sport alike. **1940** J. E. Gloag *Unwilling Adv.* xxv. 180 Clip joints,.. places where a lot of dough is lifted off you if you don't zip down your pockets. **1952** *New Yorker* 6 Dec. 56/2 A real-estate clip joint called the Charpon Land Syndicate. *a* **1953** Dylan Thomas *Quite early one Morning* (1954) 39 A raid by the vice-squad on a clip-joint for retired rod-men. **1957** E. Hyams *Into Dream* 44 'How do you like the lycée?' 'Frankly, it's pretty awful really, sir. A sort of scholastic clip-joint.' **1964** *Daily Tel.* 11 June 1/8 The 'clip joints' specialise in luring customers inside, by means of attractive showcards and insistent 'hostesses', and then fleecing them.

clip-on (kli·pǫn), *a.* [f. Clip *v.*¹ + On *adv.*] That is attached or fitted into position by means of a clip. Also *fig.*

1909 *Chambers's Jrnl.* July 478/1 A spring clip-on handle is provided, gripping the top and bottom edges of the opened tin, so that it is converted into a mug. **1958** M. L. Hall et al. *Newnes Complete Amat. Photogr.* 97 Usually filters are attached to the lens by means of screw-in or clip-on mounts. **1958** *Listener* 13 Nov. 775/2 'Justice' is simply a clip-on accessory to Olivaw, like a radio on a car. **1966** J. S. Cox *Dict. Hairdressing*, *Clip-on headband*, a semi-circular resilient decorated headband of metal or plastic, which clips over the top or front of the head.

clippable (kli·păb'l), *a.* [f. Clip *v.*² + -able.] Suitable for clipping, ready to be clipped.

1889 *Voice* (N.Y.) 29 Aug., A half column of short paragraphs wherein we can find nothing clippable. **1893** G. D. Leslie *Lett. to Marco* i. 5 The yew hedges which I planted are fast becoming clippable.

clipped, clipt, *ppl. a.* Add to def.: *spec.* of speech: (*a*) mumbled or indistinct (see Clip *v.*² 5 b); (*b*) pronounced quickly, distinctly, and without drawling. (Cf. quot. 1571 in Dict.)

1931 T. H. Pear *Voice & Personality* 30 A crisp, concise and close-clipped pronunciation. **1946** R. Lehmann *Gipsy's Baby* 115 John's voice came through, more than usually clipped. **1959** *Listener* 2 July 12/2 It is a tone that can only be described (nonsensically) as 'toneless', very flat, completely unemotional, casual and clipped.

clipper¹. Add: **4. d.** An aerial vessel; an aircraft, or flying-boat, esp. one used for trans-oceanic flights (also used with capital initial as a trade-name in this sense).

1887 tr. *J. Verne's Clipper of the Clouds* vii. 67 The Albatross might be called a clipper with thirty-seven masts. **1914** Hamel & Turner *Flying* vii. 134 Down from our Clipper of the Clouds we get. **1933** *Flight* 19 Jan. 51 America's New 'Clippers'.. are being constructed for Pan-American Airways. **1936** *Meccano Mag.* Aug. 432 Flying Giant Clippers of the Air... The 'China Clipper' climbing above San Francisco Bay at the beginning of her flight to China. **1951** A. Christie *They came to Baghdad* iii. 29 Will you be returning to New York.. by the clipper?

clipper, *v.* *U.S. colloq.* Also **Clipper.** [f. *clipper¹.] **a.** *trans.* To send by clipper (see prec.). **b.** *intr.* To travel by clipper.

1941 *Time* (Air Exp. Ed.) 8 Sept. 23/1 The first copy of his latest book was Clippered to *Time.* *Ibid.* 29 Sept. 36/3 Manila doctors Clipper X-ray films to U.S. specialists. **1941** *Collier's* 20 Dec. 52/1 Wang and I clippered over [to Hawaii] to handle it ourselves. **1942** O. Nash *Good Intentions* 93 Yon titled refugee Whose dollars Clippered here with he.

clippety-clop (kli·pətiklǫ·p). Also **clippety-clip, clippity-clop.** [Imitative.] The sound made by a horse's hoofs, or a noise resembling this.

1928 *Manch. Guardian Weekly* 17 Aug. 134/3 An' he could pick 'em up in th' owd days. Clippety-clop! Clippety-clop! **1929** H. V. Morton *In Search of Scotland* iv. 102, I could.. hear.. the clippity-clop of cab horses in the Strand. **1954** J. R. R. Tolkien *Fellowship of Ring* i. xii. 221 The sound of hoofs drew nearer. They were going fast, with a light *clippety-clippety-clip.* *Ibid.*, Clippety-clip came the quick trotting feet.

clippie, clippy (kli·pi). *colloq.* [f. Clip *v.*² (with allusion to the clipping of tickets) + -ie, -y⁶.] A bus-conductress.

1941 *New Statesman* 30 Aug. 218/3 Clippy, bus conductress. **1946** *News Chron.* 27 Feb., London's 26,000 busmen and clippies are seriously perturbed at the bad name they are getting with the travelling public. **1959** Gray Usher *Death in Bag* i. 6 An ex-clippie on a local bus.

clipping, *vbl. sb.*² Add: **2. b.** A press cutting (cf. Cutting *vbl. sb.* 4 b). orig. *U.S.*

1857 N. W. T. Root *School Amusements* 207 Every day there are brought in the clippings, and cuttings of all the boys, from home papers. **1920** [see *press *sb.*¹ 16 b]. **1935** *Punch* 27 Mar. 338/1 So many diverse expressions of opinion, so many clippings from newspaper articles.. have been sent to this office.

c. The shortening of a word, etc.; also, the resulting shortened form.

1933 Partridge *Slang To-day & Yesterday* i. iii. 27 Slang delights to curtail (clip, abbreviate, shorten) words... Many such clippings have passed into standard English, as with *cab* for cabriolet, *'bus* for omnibus. **1948** N. Gray in *Papers Brit. Sch. Rome* XVI. 87 The clipping of the enclitic *que* becomes common for the first time. **1965** *English Studies* XLVI. 464 'Tab' as a clipping from tabloid.

3. clipping agency or **bureau** *U.S.*, an organization which supplies clippings from newspapers, etc.; a press-cutting agency.

1910 *Sat. Even. Post* 30 July 6/2 The latter has been posted on Antioch affairs by the girl who runs the clipping bureau. **1938** S. V. Benét *Thirteen O'Clock* 81 If I could get hold of her clipping-agency, there'd be blood on the moon.

clipping, *ppl. a.*² Add: **b.** (Earlier example.)

1848 A. H. Clough *Bothie* 22 At Castleton, high in Braemar, were the *clippingest* places for bathing.

c. Of pace: fast, 'rattling'. *colloq.*

1845 *Punch* Sept. 122/1 When we do walk in our slumbers, we walk a clipping pace. **1896** *B'ham Daily Gaz.* 2 Sept. (E.D.D.), [The fox] ran at a clipping pace.

clippy: see *clippie.

Clipsham (kli·pʃăm). The name of a town in Rutland, used to designate limestone quarried there and used for building.

1877 *New Schools Delegacy, Oxford, 1875: Minute Bk.* 13 Oct. (MS.), Mr. Estcourt was willing to execute the building in Clipsham stone. **1882** *Times* 2 June 4/3 The stone finally chosen for use at the New Schools comes from Clipsham in Rutland.. Clipsham stone was hardly known beyond its own neighbourhood. **1958** *Listener* 19 June 1003/1 On either side are brown clipsham stone buildings. **1960** *Oxf. Univ. Gaz.* 5 Jan. 435/1 Portland Stone (Shelly Whitbed) was used instead, in order to conserve the limited supplies of Clipsham.

clique, *sb.* Add: **2.** A business 'ring'. *U.S.*

1855 *N.Y. Herald* 5 Dec. 3/5 Thousands of shares are held by a small clique of speculators. **1877** Bartlett *Dict. Amer.* (ed. 4), *Clique*, a combination of stock-brokers or capitalists, for the purpose of increasing or diminishing the price of stocks, in order to break down the market. **1901** Merwin & Webster *Calumet 'K'* xv. 285 The clique of speculators who held the floor were buying, buying, buying.

cliquy, *a.* Add: (Earlier examples.)

1863 'G. Hamilton' *Gala-Days* 115 The priests are.. clique-y. **1871** F. Gale *Echoes Cricket Fields* vii. 39 Perhaps the perpetuity of these matches made the University Elevens rather cliquy.

Hence **cliquiness** (kli·kinės), cliquy character.

1927 Belloc *Hist. Eng.* II. 282 The capriciousness with which the thing was done, and the cliquiness of it.

clish-clash. **1.** (Later examples in extended use.)

1872 H. W. Beecher in *Chr. World Pulpit* II. 343 Everything goes clish-clash, clish-clash. We are tossed about in the world like a skiff on the waves of the sea. **1941** M. Treadgold *We couldn't leave Dinah* vii. 121 The clish-clash of aluminium that resounded from the one basket.

clitellar (klaite·lăɹ), *a.* [f. Clitell(um + -ar¹.] Of or pertaining to the clitellum of earthworms.

1888 Rolleston & Jackson *Forms Anim. Life* 198 Capillaries penetrate between the clitellar glands, and are very numerous in the common Earthworm. **1963** R. P. Dales *Annelids* viii. 174 Pieces of clitellar epithelium of *Allolobophora.*

clitic (kli·tik). *Gram.* [f. En)clitic *a.* and *sb.*, Pro)clitic *a.* and *sb.*] An enclitic or proclitic (see quot. 1946). Also *attrib.*

1946 E. A. Nida *Morphology* vii. 155 Many languages have elements that (1) combine phonologically with words with which they do not form morphological constructions, and (2) do not constitute derivational or inflectional formatives... The term 'clitics' may.. be employed in this general meaning. *Ibid.*, The word *is* occurs in the following types of clitic forms: [iz] in *Bess's (going)* [etc.]. **1964** *Language* XL. 79 When BI occurs with the plural clitic -h it is reduced to a labial component of a complex phoneme.

‖**Clitocybe** (klai:tǫsai·bi). *Bot.* [mod.L. (E. Fries *Systema Mycol.* (1821) I. 78), f. Gr. κλίτος slope + κύβη head.] A genus of white-spored agarics with the gills mostly decurrent. Hence **clitocy·bin(e)** [a. F. *clitocybine* (Hollande 1945, in *Comptes Rendus* CCXXI. 361)], an antibiotic produced from a species of *Clitocybe.*

1836 M. J. Berkeley *Fungi* 31 Clitocybe. **1946** *Rev. Appl. Mycology* XXV. 309 *Clitocybe candida*.. yields an extract, clitocybin, which was experimentally shown to exert a bacteriostatic effect. **1949** H. W. Florey et al. *Antibiotics* I. viii. 365 The active substance, named clitocybine, was extracted from the sporophores. **1953** J. Ramsbottom *Mushrooms & Toadstools* vi. 52 Not all small white species of *Clitocybe* are poisonous.

clitoral (klai·tǫrăl), *a.* [f. Clitor(is + -al.] Of or pertaining to the clitoris. So **cli·torally** *adv.*

1946 S. Lorand *Technique Psychoanal. Therapy* v. 97 The theory that the clitoral sensations are primary and have to be transferred to the vagina cannot be substantiated. **1950** 'Medica' *Any Wife or Any Husband* iv. 62 The Substitution of Clitoral Orgasm... Women.. who have vaginal anaesthesia may be.. able freely to achieve an outside clitoral climax. **1953** A. Ellis in Pillay & Ellis *Sex, Society & Individual* xix. 159 Some women may be clitorally insensitive. **1953** Kinsey et al. *Sex. Behav. Hum. Fem.* xiv. 578 The two labia minora unite to form a clitoral hood.

clitorial (klaitō·riăl), *a.* = prec.

1952 H. M. & A. Stone *Marriage Manual* (ed. 2) viii. 243 Sexual satisfaction only from clitorial stimulation. **1953** Kinsey et al. *Sex. Behav. Hum. Fem.* xiv. 580 Labial and clitorial stimulation.

clitoridectomy (klai:tǫride·ktǒmi). [f. Gr. κλιτοριδ-, Clitoris + ἐκτομή excision.] Excision of the clitoris.

1866 *Brit. Med. Jrnl.* II. 672/1 The excision of the clitoris as a cure for hysteria... The author gave his reasons for believing that the operation of clitoridectomy will not prove of permanent value. **1908** H. Johnston *Grenfell & Congo* II. xxv. 666 Circumcision in the males or clitoridectomy in the females. **1960** *Arch. Gen. Psychiatry* III. 445/2 The so-called orificial surgeons performed clitoridectomy on thousands of children.

clivia (klai·viă). Also **clivea.** [mod.L. (J. Lindley 1828, in *Bot. Reg.* XIV. 1182), f. the name of Lady Charlotte *Clive* (d. 1866), who married the third Duke of Northumberland.] A plant of the genus of that name of African amaryllids with orange, red, or yellow flowers.

1828 *Bot. Reg.* XIV. 1182 (*heading*) Clivia nobilis. Scarlet Clivia. **1866** Lindley & Moore *Treas. Bot.* 300/1 The Clivias consist of herbs with fasciculate fleshy roots. **1891** *Daily News* 11 Feb. 9/6 Messrs. Laing, of Forest Hill, received a vote of thanks for their stand of clivias. **1961** *Amat. Gardening* 23 Sept. 27/2 A clivia is a fairly easy plant to grow.

cloak, *sb.* Add: **6. cloak and dagger** [tr. F. *de cape et d'épée*], (*a*) = *cloak and sword*; (*b*) of, concerned with, or characteristic of espionage, secrecy, intrigue, etc.; hence in various allusive phrases; hence *cloak-and-daggery*; **cloak and sword** [tr. Sp. (*comedia*) *de capa y espada*], designating or pertaining to dramas or stories of intrigue and romantic or melodramatic adventure, in which the principal characters are taken from that class of society which formerly wore cloak and dagger or sword.

1806 Ld. Holland *Lope* 126 Comedias de Capa y Espada, Comedies of the Cloak and Sword, from the dresses in which they were represented. **1840** Longf. in *Life* (1886) I. 353 In the afternoon read *La Dama Duenda* of Calderon—a very good comedy of 'cloak and sword'. **1841** Dickens *Barn. Rudge* xxiv. 69 A very small scrap of dirty paper.. was given him by a person then waiting at the door... 'With a cloak and dagger?' **1860** G. Vandenhoff *Dram. Remin.* vi. 96 A most solemn and mysterious tragedian, of the cloak-and-dagger school. **1893** H. B. Clarke *Spanish Lit.* 163 The play of the 'cloak and sword'.. may almost be said to be his [*sc.* Lope de Vega's] own invention. *Ibid.* 220 The stock characters, the *galán* and *dama*, the *gracioso* and *barba* of their 'sword and cloak' plays. **1898** *Blackw. Mag.* Nov. 600/1 Our sham revivals of cloak and dagger are poor things. **1905** *Academy* 10 June 603/2 Thousands of cloak-and-dagger stories. **1921** H. Walpole *Young Enchanted* 42 The Cloak and Sword Romances. **1946** *Britannica Bk. of Yr.* 832/1 *Cloak and dagger*, Office of Strategic Services; pertaining to OSS. **1954** Koestler *Invis. Writing* 419 The cloak-and-dagger atmosphere which keeps intruding, all his life, into the Party-member's world. **1958** *Listener* 10 July 64/2 The main interest in this wild cloak-and-daggery is why it came to be written. **1959** M. M. Kaye *House of Shade* xx. 272 The police, or M.I.5, or some of those cloak-and-dagger boys, had a line on him. **1959** M. Steen *Tower* i. iii. 36 She had two pips on her shoulder and was private secretary to a man in the Cloak and Dagger.

cloak, *v.* Add: **1. b.** *intr.* for *refl.* To put on a cloak, cloak oneself.

1774 H. S. Conway *Let.* 31 Aug. in Carlyle *Fredk. Gt.* (1865) VI. xxi. v. 542 It rained hard the whole time we were out; and as his Majesty did not cloak, we were all heartily wet. **1906** Hardy *Dynasts* ii. ii. ii. 184 All three cloak And veil as when you came.

cloak-room. Add: Also freq. euphem. = Lavatory 4. Also *ellipt.* and *pl.* **cloaks.**

1953 Berg *Dict. New Words* 56/1 *Cloakroom*, euphemism for lavatory. **1957** M. Sharp *Eye of Love* ii. 19, I got held up in the Cloaks. **1968** P. Hobson *Titty's Dead* v. 63 Daphne's always first out of breakfast and straight into the girls' downstairs cloaks. No one would have noticed only she took a book and read in there.

b. *attrib.*, as *cloak-room attendant, girl, man, system.*

1914 'Saki' *Beasts & Super-beasts* 180 He left all the parcels in charge of the cloak-room attendant. **1961** *Which?* July 174/2 Most members always or usually tip cloakroom attendants. **1918** A. Bennett *Pretty Lady* i. 3 The programme girls, the cigarette girls, the chocolate girls, the cloak-room girls. **1880** Lamphere *U.S. Govt.* 265/1 Pages, laborers, and cloak-room men. **1900** *Westm. Gaz.* 3 Nov. 2/3 The general confusion and disorder of our cloak-room system.

clo·bber, *sb.*² *slang.* [Origin unknown.] **a.** Clothes.

1879 *Macm. Mag.* Oct. 501/2, I took the rattler to Forest Hill, and touched for.. a kipsy full of clobber (clothes). **1901** G. B. Shaw *Capt. Brassbound* iii. 292 Now to get rid of this respectable clobber and feel like a man again. **1906** E. Dyson *Fact'ry 'Ands* x. 124 It's Bland 'Olt's wicked woman straight from their halls of gilded vice, with all her clobber on. **1908** *Daily Chron.* 3 Aug. 4/4 All just the common sort, in their new summer clobber. **1934** *Bulletin* (Sydney) 11 Apr. 10/4 Every fine Sunday morning all my war-time clobber goes out on the clothes-line. **1959** *Observer* 22 Mar. 25 To pay for the kiddies' clobber.

b. Equipment; 'gear'; rubbish.

1890 Kipling *Barrack-r. Ballads* (1892) 31 They call a man a robber if 'e stuffs 'is marching clobber With the.. loot. **1925** in Fraser & Gibbons *Soldier & Sailor Words.* **1951** A. Baron *R. Hogarth* 166, I gets off a bus in Old Street, all loaded up with clobber. **1965** *Lancet* 20 Nov. 1070/2 Every cellar stockroom.. is packed tight with fantastic collections of clobber and junk.

Hence **clobber** *v.*² *trans.*, to dress or 'tog' *up.*

1887 J. W. Horsley *Jottings from Jail* 9, I used to clobber myself up and go to the concert-rooms. **1889** E. Sampson *Tales of Fancy* 14, I must go there decently clobbered... I want you to go with me.. to buy a cheap suit. **1934** *Bulletin* (Sydney) 16 May 9/4 The visitor, it seems, was clobbered up like an ordinary Aussie.

clobber, *v.*¹ Add: (Later examples.)

1872 *Jewish Chron.* 10 May 80/2 'Clobbering'—a technical term for 'renovating' old garments. **1957** Lloyd P. Gartner *Jewish Immigrant* (1960) 83 The second-hand clothing.. dealers.. sent out the garments they purchased to be 'clobbered' (renovated) and resold.

2. To add enamelled decoration to (porcelain, esp. blue-and-white). So **clo·bbered** *ppl. a.*; **clo·bberer**; **clo·bbering** *vbl. sb.*¹

1889 *Cent. Dict.*, Clobbered china. **1900** F. Litchfield *Pott. & Porc.* 114 There is a description of Chinese known as 'Clobbered',.. over-painted with.. ornament.. and sold for decorated Oriental china. **1910** R. L. Hobson *Worcester Porcelain* 61 Clobbering was freely practised in Holland first and England afterwards from the early part of the 18th century. **1915** —— *Chinese Pottery* II. 261 The clobberer became an established institution, and he was at work in London in the last century. **1922** *R. Drane's Coll. Old Worc. Porcelain* 32 Clobbered pieces. 471 An Oriental Tea Jar... When this was brought to Europe it was a plain blue and white object.

clo·bber, *v.*³ *slang.* [Origin unknown.] To hit; to thrash or 'beat up'; to defeat, shoot down; to reprimand or criticize severely. So **clo·bbering** *vbl. sb.*²

1944 *Gen* 18 Nov. 9/1 Did anyone clobber any [*sc.* flying bombs]? **1948** Partridge *Dict. Forces' Slang* 43 *Clobbering*, (a) heavy bombing. (Air Force). **1949** in Wentworth & Flexner *Dict. Amer. Slang* (1960) 110/1 The Wolverines [football team] clobbered their opponents 42 to 3. **1951** M. Shulman *Many Loves* (1953) 206 'Poor loser!' they kept yelling as they clobbered me. **1956** J. E. Johnson *Wing Leader* iii. 31 Six of the Spits had cannon stoppages, otherwise more Huns would have been clobbered. **1956** Wallis & Blair *Thunder Above* (1959) xiii. 134 The Press sure clobbered Roger Law... Don't know why I got off so easy. **1959** J. Braine *Vodi* xii. 165 The police gave them such a clobbering that he never had any trouble again. **1959** 'O. Mills' *Stairway to Murder* x. 114 He must have seen me clobber Leeming when he dived for the brief-case. **1969** *Daily Tel.* 16 Apr. 1/4 Companies singled out for a special clobbering are the betting and gaming companies. *Ibid.* 3/8 Butlin's is heavily clobbered by the increase in Selective Employment Tax.

‖**clochard** (klǫʃāɹ). [F., f. *clocher* to limp.] In France, a beggar, vagrant.

1940 Hemingway *For whom Bell Tolls* xxx. 340 That bunch of drunks, *clochards*, bums. **1941** Koestler *Scum of Earth* 134 For.. beggars, tramps, and *clochards*, the barrack meant material safety and comfort. **1959** *News Chron.* 26 June 2/4 An attempt to rid Paris of its clochards—the tattered and often drink-sodden tramps who form part of the local colour—has failed.

CLOCHE

cloche. Add to def.: Now, a translucent plant-cover of any shape or size (see quot. 1954).
1954 A. G. L. Hellyer *Encycl. Garden Work* 54/2 Nowadays..almost all cloches are..open-ended glass shelters which can be placed end to end to form a continuous line over a row, or rows, of plants.

2. In full *cloche hat*. A woman's close-fitting hat of a bell shape.
1907 *Daily Chron.* 25 June 8/3 The very latest 'cloche' shape. **1907** *Westm. Gaz.* 24 Aug. 13/1 The mondaine's big cloche hat. **1908** *Ibid.* 29 Feb. 13/2 So popular is the cloche or mushroom hat with the Parisienne. **1923** *Daily Mail* 3 Aug. 11 The narrow-brimmed, high-crowned cloche. **1952** *News Chron.* 4 Jan. (Advt.), Ladies' cloche velour hats.

†**3.** *Aeronaut.* (See quots.) *Obs.*
1912 C. B. Hayward *Pract. Aeronautics* vii. 642 The word 'cloche' applied to the bell-like attachment for the control wires, has been adopted into the international vocabulary of aeroplaning. **1918** H. Barber *Aeroplane Speaks* (ed. 6) 137 *Cloche*, literally 'bell'. Is applied to the bell-shaped construction which forms the lower part of the pilot's control lever in a Bleriot monoplane, and to which the control cables are attached.

clock, *sb.*[1] Add: **2. b.** = Dial *sb.*[1] 6 a; *spec.* a taximeter, speedometer, or milometer. *colloq.*
1930 'A. Armstrong' *Taxi* iv. 39 Unscrupulous young men..who didn't mind paying what was already on the clock and a bribe besides. **1934** [see *clock *v.*[1] 1 b]. **1939** H. Hodge *Cab, Sir?* 167 He may refuse to pay you the original fare later because it isn't recorded on the clock-face. *Ibid.* 217 Then comes the 'clock', the taximeter. **1942** 'B. J. Ellan' *Spitfire!* v. 22, I..let the aircraft plummet down..until the clock showed 400 m.p.h. **1967** K. Giles *Death in Diamonds* vii. 128 There were two hundred and seven miles on the clock. **1968** 'B. Mather' *Springers* vii. 67 She didn't have the panel lights on, so I couldn't see the clock, but she must have been needling over a hundred in places. **1970** *Commercial Motor* 25 Sept. 64/2 Neither vehicle had much mileage on the clock.

c. In an electronic computer (see quots.). Also *attrib.* and *Comb.*
1947 A. W. Burks et al. in *Von Neumann's Wks.* (1963) V. 68 There are many advantages in deriving these pulses from a central source, called the *clock*. The timing may then be done either by means of counters counting clock pulses or by means of electrical delay lines. **1951** *Proc. Inst. Radio Engin.* XXXIX. 274/2 *Master clock*, the primary source of timing signals. *Ibid.* 1052/1 These pulses, designated 'clock pulses', represent the fundamental time marks of the machine. *Ibid.* 1052/2 (*caption*) Clock-pulse generator. **1953** A. D. & K. H. V. Booth *Autom. Digital Calculators* xii. 120 It is assumed that two tracks of the drum D are devoted to pre-recording 'clock' markers. **1953** *Electr. Engineering* LXXII. 162/1 The Timer contains the master pulse generator which emits clock pulses every microsecond. **1964** F. L. Westwater *Electronic Computers* v. 84 The basis of the control system is a clock..The clock can be constructed from an electronic oscillator which can emit a stream of pulses at a predetermined rate and interval.

4. *to put (set, turn) the clock back, to put (set, turn) back the clock*, to move the hands of a clock back to an earlier position; *fig.*, to go back to a past age or earlier state of affairs; to take a retrograde step; *(all) round the clock, the clock round* (in U.S. also *around*), for 24 (occas. 12) hours without intermission; all day and night; ceaselessly; also attrib. phr. *round-the-clock*; *against the clock*, against time (see Time *sb.* 36).
1635, 1892 [see Set *v.* 141 b]. *a* **1745** [see Put *v.*[1] 39 c]. **1862** Lowell *Biglow P.* 2nd Ser. 42 'T would put the clock back all o' fifty years, Ef they should fall together by the ears. **1896** G. B. Shaw *Our Theatres in Nineties* (1932) II. 239 Mr Kinghorne, though seemingly more bewildered than encouraged by the setting back of the clock, took his turn as 'the sovereign duke of Barcelona'. **1898** *Ibid.* III. 309, I cannot pretend to think that Mr Pinero, in reverting to that period, has really had to turn back the clock as far as his own sympathies and ideals are concerned. **1906** A. C. Benson *Upton Lett.* (ed. 2) 61 The attempt to put back the clock, and to try and restore things as they were. **1910** J. Buchan *Prester John* xvii. 276 What makes you try to put the clock back? You want to wipe out the civilization of a thousand years, and turn us all into savages. **1940** H. G. Wells *Babes in Darkling Wood* 6 Conrad was understood to be in the van of progress; Robert Louis Stevenson had 'put the clock back'. **1944** J. S. Huxley *Living in Revol.* xv. 188 Attempts at turning the clock back in educational practice.
1839–52 [see Round *prep.* 1 c]. **1859** G. A. Sala (*title*) Twice round the clock, or the hours of the day and night in London. **1895** Kipling *Day's Work* (1898) 341 Georgie was used to eat all round the clock. **1935** N. L. McClung *Clearing in West* xl. 343 When I went home after the examinations, I slept the clock around for the first two weeks. **1943** *Jane's All World's Aircraft 1942* p. v/2 The assault will perhaps be a 'round-the-clock' affair. **1944** R. C. K. Ensor *Miniat. Hist. War* iv. 53 A scheme of bombing Germany 'round the clock', i.e. for a succession of nights and days. **1949** C. P. Snow *Time of Hope* I. i. 14 Boys of your age need to sleep the clock round. **1953** R. Lehmann *Echoing Grove* 178 He'd slept pretty well the clock round. **1958** 'Ann Bridge' *Portuguese Escape* viii. 138 They must have been keeping a round-the-clock watch on the Monsignor's house. **1958** *Listener* 4 Dec. 936/2 A thousand men working in three shifts round the clock removed the old mill and installed the new one in twelve and a half days. **1959** *Observer* 21 June 17/7 If Arnold Bennett were alive to-day he would be writing family serials for television all round the clock. **1962** V. Grissom in *Into Orbit* 119 We were working all around the clock during this period, running tests night and day in order to keep up with the schedule.
1940 *Times* 3 Feb. 8/4 The time-limit in chess. Playing against the clock. **1960** *Times Rev. Industry* Jan. 28/3 Often the work is conducted against the clock.

5. b. *slang.* The human face. (Cf. Dial *sb.*[1] 6 c.)
1923 Manchon *Le Slang* 87 *Clock*,..la tête, le *ciboulot*. **1931** G. Irwin *Amer. Tramp & Underworld Slang* 52. **1932** A. J. Worrall *Eng. Idioms* 57 He sat there with a grin all over his clock. **1961** J. I. M. Stewart *Man who won Pools* xiii. 135 His clock was still the affable Brigadier's, but you felt now that if you passed a sponge over it there'd be something quite different underneath.

c. *slang.* A punch (on the face).
1959 *N.Z. Listener* 24 July 6/3 He might have a clock at him or a kick at him. **1961** J. Maclaren-Ross *Doomsday Book* I. vii. 74 It was my turn to administer the anaesthetic—by a final clock in the jaw.

10. *clock-faced* adj.
1906 *Westm. Gaz.* 17 Oct. 5/2 Two knights, one of clock-countenance..I have my doubts about the clock-faced gentleman. **1923** E. Sitwell *Bucolic Comedies* 41 The clock-faced sun and moon.

11. clock-bird *Austral.*, the laughing jackass, kookaburra; **clock-calm** (examples); also *fig.* and *attrib.*; **clock-golf**, a game in which twelve numbers are arranged on the ground in a circle in imitation of the dial of a clock, and the players putt the ball from each of these numbers in turn into a hole placed somewhere within the circle; **clock hour** (orig. *dial.*), a whole hour; a full sixty minutes; also, a specific period of sixty minutes; **clock-jack**, a figure on a clock tower, which gives warning, chimes, or tells the hour by striking a bell (cf. Jack *sb.*[1] 6, Quarter-jack 1); **clock-spring** (see quot. *a* 1877); **clock-time**, time as shown or measured by a clock or clocks; **clock-train**, the mechanism of the going part of a clock; **clock-watcher**, one who takes care not to exceed minimum working hours; so (as a back-formation) **clock-watch** *v. intr.*, to work over-anxiously to time; **clock-watching** *vbl. sb.* and *ppl. a.*; **clock-weight**, a weight used instead of a spring to give impulse to the movement of a clock; **clock-wise** *adv.* (further examples); also *attrib.* or as *adj.*
1880 T. W. Nutt *Palace of Industry* 15 Where clock-bird laughed and sweet wild flowers throve. **1777** J. Adams in *Fam. Lett.* (1876) 267 There is a clock calm at this time in the political and military hemispheres. **1924** R. Clements *Gipsy of Horn* iii. 54 The breeze died away and we were left becalmed—the centre of a clock calm. **1932** Kipling *Limits & Renewals* 235 A girl's voice came across the clock-calm harbour-face. **1906** Clock-golf [see *putting-hole* (Putting *vbl. sb.*[2] 1 b)]. **1915** Wodehouse *Something Fresh* vii, [They] console themselves for the absence of a links..with the exhilarating pastime of clock-golf. **1940** 'N. Shute' *Old Captivity* 17 The centre of the lawn was laid out for clock golf, the figures beautifully white. **1872** J. Hartley *Clock Alm.* Sept. (E.D.D.), They've been wide wakken a clock haar before ther usual time. **1962** *Rep. Comm. Broadc. 1960.* xix. 250 Greater flexibility of programming hours..is afforded by prescribing a total of hours of broadcasting rather than clock-hours. *Ibid.* 313/2 Rules governing maximum amount of spot advertising in any clock hour. **1926** *Glasgow Herald* 23 July 6 The fantastic little clock-jacks of Norwich Cathedral. *a* **1877** Knight *Dict. Mech.*, *Clock-spring*, a coiled steel spring in the going-barrel or the striking-barrel of a clock which impels the train or strikes the hours, as the case may be. **1907** *Daily Chron.* 18 Sept. 3/6 The standard would become an 'international clock time' common to the whole of Mid and Western Europe. **1933** J. Baillie *Life Everlasting* (1934) vii. 215 Conceptual or clock time..is but a convenient mathematical abstraction and is very different from the time of actual human experience. *a* **1956** W. de la Mare *Compl. Poems* (1969) 713 Widen your calls, let TIM decant her spells—She'll give you at least true clock-time, if nought else. **1850** Denison *Clock & Watch-m.* p. v, Common clock train. **1868** *Chambers's Encycl.* X. 82/1 There is one more wheel and pinion in the watch-train than in the clock-train. **1877** *Encycl. Brit.* VI. 19/2 [A clock] must vary with the force of the clock train, due to different states of the oil. **1911** F. Swinnerton *Casement* ii. 66 His office-companions and his employers..were accustomed only to 'clock-watchers' (a name given to the excessively punctual in leaving the office). **1933** D. L. Sayers *Murder must Advertise* viii. 129 Mr. Tallboy had left promptly at 5.30. Mr. Copley had seen him go. Clock-watchers, the whole lot of them. **1942** M. Dickens *One Pair of Feet* iv. 42 The day seemed endless, because I was clock-watching all the time. **1944** H. G. Wells *'42 to '44* ii. 38 Clock-watching is a common failing of human beings at work. **1945** Sinclair Lewis *Cass Timberlane* 76 All you overimaginative men, who try to combine fancifulness with being clock-watching executives. **1899** K. Grahame *Dream Days* 90 My heart sank lower and lower, descending relentlessly like a clock-weight into my boot soles. **1898, 1914** Clockwise [see *anti-clockwise *a.* and *adv.*]. **1957** A. MacNab *Bulls of Iberia* xv. 226 Antonio makes No. 2 picador return to his station now, not crossing the other one and going clockwise in muddled fashion, but riding all the way round, anti-clockwise.

clock, *v.*[1] Add: **1. b.** *colloq.* To accomplish or attain (a certain time or speed) in a race; to register (a time, distance, etc.) on a clock or dial. Also with *up*.
1892 *Field* 14 May 735/3 He ran splendidly, and 'clocked' 2 min 2 sec all the way. **1934** *Neuphilologische Mitteilungen* XXXV. 131 *Clock* 'speedometer' (also verb as in *she's clocking sixty*). **1948** *Daily Tel.* 17 May 3/6 Miss M. Gardner, who clocked 11·6 sec for the 80-metres hurdles. **1959** H. Hobson *Mission House Murder* xii. 81 It's quite a trip—I clocked up a hundred and fifteen. **1960** *Economist* 30 Apr. 457/3 Bischofberger A.G. of Zurich have three Austin trucks. ..Recently one of them clocked 150,000 miles in a single year.

c. With *in, on* (*off, out*): (*a*) to register one's arrival at (or departure from) work by means of a mechanical device combined with a clock; also *refl.*; (*b*) *transf.* to register one's arrival (or departure); to start (or stop) work; to enter or leave. Hence **clocking-in**, etc. (also *attrib.*).
1914 *Auto-motor Jrnl.* 593/1 At 12 o'clock practically every car had 'clocked in' and was sealed. **1924** H. H. Emsley *Factory Costing* 5 The 'job cards', on which the workers enter their time and particulars of the work they are engaged upon, depend upon the method in use for 'clocking on'. **1924** *Glasgow Herald* 14 Aug. 5 That any men could resume work provided that they would clock out in accordance with the system which had been in operation since the works started. *Ibid.* 18 Aug. 9 An allowance of two minutes for 'clocking off'. **1926** *Spectator* 27 Feb. 359/2 He clocks himself in on an automatic timekeeper. *Ibid.* 360/1 He clocks off again and proceeds home. **1927** A. Bennett *Woman who stole Everything* 216 Both brother and sister had had to 'clock in' of a morning and 'clock out' of an evening for years. **1928** *Daily Tel.* 24 July 15/2 He..described the 'clocking-in' apparatus as an 'infernal machine, insulting to the integrity, honour, and dignity of the profession'. **1935** Wodehouse *Luck of Bodkins* xxii. 284 As soon as I've clocked in at my hotel. **1955** *Times* 11 Aug. 7/3 How would workers buy anything if shops clocked in and out at precisely the same times as they did? **1959** *Times* 3 Nov. 13/4 When I took up a part-time job three weeks ago I was given a card with which to 'clock in' and 'out'. **1965** *New Statesman* 30 Apr. 696/1 Nineteen clubs have clocked out in the League's 75-odd years.

3. To punch in the face; to hit. (Cf. *clock *sb.*[1] 5 c.) *slang.*
1941 Baker *Austral. Slang* 18 *To clock*, to strike with the fist. **1947** D. M. Davin *For Rest of Lives* x. 51 So then I clocked him and we beat it. **1956** M. Procter *Pub Crawler* 125 He didn't say who'd clocked him. **1959** P. H. Johnson *Unspeakable Skipton* xxv. 225, I should have clocked Dorothy, as the saying goes, more times than I care to count.

clock, *v.*[4] Also **clox.** [f. Clock *sb.*[2]] *trans.* To embroider clocks on. Hence **clo·cker**, one who embroiders clocks.
1880 in Webster *Suppl.* **1921** *Dict. Occup. Terms* (1927) § 407 *Clocker*;..sometimes further designated according to thread used, *e.g.*, silk clocker. **1922** *Glasgow Herald* 9 Oct. 3 You must learn to clox your own stockings.

clod, *sb.* Add: **3. f.** *slang.* A copper coin. Usu. in *pl.*
1925 in Fraser & Gibbons *Soldier & Sailor Words* 59. **1960** 'A. Burgess' *Doctor is Sick* 105 He began to search for coppers. 'Lend us a couple of clods,' he said to his twin.

clod, *v.* Add: **3.** (Later U.S. example.)
1873 J. H. Beadle *Undevel. West* iii. 74 The soil pulverizes finely in summer. It is never water soaked, consequently never 'bakes' or clods.

5. a. Also, to knock or drive by pelting with clods of earth, stones, etc. *dial.* and *U.S. dial.*
1709 in *Publ. Scottish Hist. Soc.* (1931) 3rd Ser. XVII. 67 She..thereupon clouded him away with stones. **1867** T. Blackah *Songs & Poems* 38 We'd to clod 'em [*sc.* sheep] away. **1904** 'Mark Twain' *Adam's Diary* 15 Trying to clod apples out of that forbidden tree. *Ibid.* 29 She has been climbing that tree again. Clodded her out of it.

cloky (klō^u-ki), anglicized var. *cloqué.
1922 *Daily Mail* 30 Dec. 10 (Advt.), Satin cloky, richly embroidered. **1935** *Times* 4 Nov. 9/3 One is in a black cloky silk. *Ibid.* 13 Nov. 15/5 Chiffon and taffeta for the evening are as general as the more gorgeous lames and clokys.

clone (klō^un), *sb.* Also **clon.** [ad. Gr. κλών twig, slip.] **a.** *Bot.* A group of cultivated plants the individuals of which are transplanted parts of one original stock, the propagation having been carried out by the use of grafts, cuttings, bulbs, etc. **b.** In wider use in *Biol.* Any group of cells or organisms produced asexually from a single sexually produced ancestor. Hence **clo·nal** *a.*, of, belonging to, or having the characteristics of a clone; **clo·nally** *adv.*
1903 H. J. Webber in *Science* 16 Oct. 502/2 Clons..are groups of plants that are propagated by the use of any form of vegetative parts. *Ibid.*, The differentiating clonal characters of chrysanthemums are mainly in the form and color of the flowers. **1905** C. L. Pollard in *Science* 21 July 88/1, I therefore suggest *clone* (plural *clones*) as the correct form of the word. **1928** *Times* 20 July 20/3 In a tapping test of buddings now being carried out by the institute, the highest-yielding clone has latex vessels of much smaller bore than the lowest-yielding clone. **1929** *Bibliographia Genetica* V. 234 In *Bacillus coli communis*..a biotype was also found having lower motility than the remainder of the clone from which it came. **1935** *Economist* 26 Jan. 212/2 Namoe Tongan Rubber Estates in Sumatra..have just been planted with selected high-yielding clones. **1946** *Nature* 21 Sept. 422/2 Male-sterile lines have been propagated clonally. *Ibid.* 23 Nov. 762/2 The replacement of existing virus-infected, low-yielding clonal stocks of raspberries is a pressing necessity. **1958** *New Scientist* 20 Feb. 13/1 Various techniques have been devised for producing these 'clone cultures' from single cells. **1968** *Ibid.* 3 Oct. 11/1 Many people are worried by the ethical problems raised by the recent developments in genetics, opening up the possibilities of 'clonal man'. **1970** *Nature* 18 Apr. 280/1 Plasmodia derived from clonal crosses.

clone (klō^un), *v.* [f. the sb.] *trans.* To propagate or cause to reproduce so as to form a clone. So **cloned** *ppl.a.*, produced by cloning; **clo·ning** *vbl. sb.*
1959 *Nature* 22 Aug. 648 (*heading*) A New Technique for Isolating and Cloning Cells of Higher Plants. **1959** *Genetics* XLIV. 1259 A number of variants were obtained from two recently cloned lines of strain HeLa S3. **1960** *Science* 6 May 1377/1 This procedure should prove useful for the primary isolation, manipulation, and cloning of many different cell types. **1968** *Observer* (Colour Suppl.) 10 Mar. 9/1 One of the most extraordinary of the possibilities now being explored ..is referred to as 'cloning people'—the creation of genetically identical individuals from body cells. **1970** *Nature* 19 Apr. 210/2 The Jockey Club was..understandably cool when asked to comment on the possibility of a dozen cloned Arkles thundering neck and neck round the course at Epsom.

clonk (klɒŋk), *v.* [Echoic; cf. Clank *v.*]
1. *intr.* To make the sound described under *clonk *sb.*; also = Clunk *v.* 1. So **clo·nking** *ppl.a.* and *vbl. sb.*
1930 P. Macdonald *Link* viii. 156 The *clonking* noise of one of the swing doors of the smaller entrance opening and shutting. **1945** E. Bowen *Demon Lover* 72 He drove fast, and..I heard the parts of the cleaner clonking about. **1955** L. A. G. Strong *Deliverance* 9 It [*sc.* the tram] made such music up above, and such a set of clanking clonking metal noises down below. **1963** *Radio Times* 14 Mar. 4/3 The machine..gurgled and clonked a time or two and waited. **1970** *Daily Tel.* 9 Feb. 11 Cloggy shapes with cork or wooden soles that will make the city streets noisier still as they clonk through the summer.
2. *trans.* To hit. *colloq.*
1949 'N. Blake' *Head of Traveller* ii. 28 Janet Seaton..has already clonked Inspector. **1958** J. Cannan *And be Villain* vii. 167 Anyone could have..waited till Richard stooped over the filing drawer to clonk him. **1960** *Spectator* 23 Sept. 462/3, I have never been able to pick up a hammer without clonking myself one.

clonk (klɒŋk), *sb.* [See prec.] An abrupt, heavy sound as when something unresilient strikes a hard surface.
1952 B. Cleary *Henry & Beezus* ii. 59 When it landed with a loud *clonk*, the whole class tittered. **1959** *Elizabethan* May 14/1 He tried to lower the object gently to the floor, but it was too heavy and dropped with a clonk. **1959** *Listener* 15 Oct. 650/3 Sometimes a bar is hit a most tremendous clonk and stays put. **1962** *Times* 22 May 7/4 A solid 'clonk' being the surest indication that the cogs were engaged.

clonus (klō^u-nɒs). *Med.* [L., a. Gr. κλόνος turmoil.] A spasm or series of spasms of alternate muscular contraction and relaxation.
1817 Good *Nosol.* 339. **1891** Taylor *Man. Pract. Med.* (ed. 2) 148 In some cases a *clonus* can be obtained at the knee. **1906** *Practitioner* Dec. 825 The patient has exaggerated tendon reflexes and sometimes a little clonus. **1962** *Lancet* 26 May 1132/2 She has ankle clonus on the right and she walks with a scissors gait.

clop (klɒp). Also reduplicated **clop-clop.** [Imitative.] A sharp sound such as is made by shoes or hoofs on the ground. Hence as *adv.* and *v. intr.*
1901 Kipling *Kim* xv. 410 A faint *clop-clop* of Mahbub's retreating feet. **1902** *Kynoch Jrnl.* Oct.–Nov. 23/2 The 'Spat' or 'Clop' is the sound made by the bullet striking an object. **1913** Masefield *Dauber* iii. xxvi, His boots went clop along the stony ground. **1919** —— *Reynard* 111 The clop of the hooves on the road was plain. **1929** W. Faulkner *Sound & Fury* 336 Queenie moved again, her feet began to clop-clop steadily again. **1959** *Encounter* XII. 25, I clop-clopped into the hall.

‖**cloqué** (kloke). Also **cloque.** [F. *cloqué* blistered, *cloque* blister.] A fabric with a surface raised irregularly in a blistered or embossed effect. Also *attrib.* Cf. *cloky.
1950 '*Mercury*' *Dict. Textile Terms* 128/1 *Cloque cloth*, same as 'Blister' cloth. **1957** *Textile Terms & Defs.* (ed. 3) 28 *Cloqué*, a compound or double fabric with a figured blister effect brought about by the use of yarns of different character or twist. **1963** *Times* 27 Feb. 12/5 Other evening designs often in *cloque gauze* flowed loosely over the figure. **1965** *Punch* 18 Aug. 254/2 Jacques Esterel's rain redingote and gaiters of scarlet cloqué.

close, *a.* and *adv.* Add: **A. 2. c.** Of a game of chess: see quot. 1818; now, more usually, one characterized by lack of development either by gambits or by opening up the files.
1750 *Chess Made Easy* I. 1 Close Games, in Opposition to Gambit Games. **1818** W. S. Kenny *Pract. Chess Exercises* 1 The close game is, when he that plays first gives neither a pawn nor a piece at the beginning of a game. **1920** *Brit. Chess Mag.* XXXIX. 261 He never felt happy in 'close games'.

15. b. *close shave*: a narrow escape, a near thing (*lit.* and *fig.*). orig. *U.S.*
1834 C. A. Davis *Lett. J. Downing* 13, I did not so much as get my feet wet when the bridge fell, though it was a close shave. **1856** [see Shave *sb.*[1] 5 a]. **1888** [in Dict.]. **1940** 'Gun Buster' *Return via Dunkirk* ii. i. 90 That was a close shave for me. **1952** *Manch. Guardian Weekly* 6 Nov. 7 A close shave between the two totals may be magnified in the Electoral College into a chasm. **1964** D. Varaday *Gara-Yaka* ix. 79 The leopard..tumbled in a heap between the crouching Freddie and me..'A very close shave,' I muttered.

e. Designating a cinema or television 'shot' taken with the camera at a short distance from the subject (cf. *close-up); so *close-medium shot* (see quot. 1933).
1933 A. Brunel *Filmcraft* 155 *Close-Medium Shot*, this is abbreviated C.M.S. A C.M.S. of a figure cuts somewhere between the waist and the knees, and just above the head. **1940** *Chambers's Techn. Dict.* 169/2 [A] close shot..may comprise, for example, two persons at a table. **1953** K. Reisz *Film Editing* i. 25 Acting in close shot demands greater control and subtlety of expression than had hitherto been necessary.

20. b. Having parties or votes nearly equal in number. *U.S.*
1870 *Congress. Rec.* 30 Jan. 1042/1 St. Helena is a very close parish. In 1870 it gave 30 democratic majority. **1887** *Ibid.* 20 Jan. App. 50/1 What do you mean by 'doubtful' counties?..Close counties, like some of those in the State of..Virginia. **1904** H. L. West in *Forum* July 17 In the last Congressional election there were comparatively few close districts.

B. 1. b. With *in*.
1872 [see In *adv.* 3]. **1930** *San Anton.* (Texas) *Light* 31 Jan. 14/8 Furnished apartment; close-in.

C. 1. and **2.** *close-buttoned* (examples), *-clamped, -clipped, -growing, -grown, -guarded, -hung, -knit, -lipped, -packed, -set, -shut* (example).
1784 Cowper in *Corr.* (1904) II. 267 An honest man, close button'd to the chin. **1853** Tennyson *Poems* (ed. 8) 235, I turn'd once more, close-button'd to the storm. **1892** J. R. Lowell in *Scribner's Mag.* XI. 268 No skill of words could breed Such sure conviction as that close-clamped lip. **1906** *Daily Chron.* 27 Sept. 4/4 Water-tight compartments..with their close-clamped doors. **1907** Galsworthy *Country House* I. i. 1 Close-clipped grey whiskers. **1931** T. H. Pear *Voice & Personality* 30 A crisp, concise and close-clipped pronunciation. **1897** *Daily News* 16 Jan. 6/3 The abundance of the close-growing hair. **1907** *B'ham Inst. Mag.* Mar. 126 The close-growing pines shut all from our view. **1903** *Westm. Gaz.* 2 Jan. 2/3 They showed like flame the close-grown banks between. **1969** *Gloss. Landscape Work* (*B.S.I.*) v. 39 *Close-grown*, of timber trees, grown so closely together that the normal outward spread of branches is checked. **1898** W. Graham *Last Links* 105 The close-guarded secret. **1903** *Westm. Gaz.* 28 Jan. 2/3 Your door with its close-guarded wicket. **1904** *Daily Chron.* 12 Sept. 8/5 Where the waters narrow between close-hung woods. **1905** *Westm. Gaz.* 20 May 3/3 That great sky close-hung with stars. **1912** J. S. Huxley *Indiv. in Animal Kingdom* iv. 88 How much harder it is for a thin, loose-built man than for a close-knit, compact one of equal weight, to make headway in a gale of wind. **1930** W. O. Stapledon *Last & First Men* 26 The Confederacy at first appeared as a close-knit whole. **1853** M. Arnold *Scholar Gipsy* xx, With close-lipp'd Patience for our only friend. **1905** *Daily Chron.* 10 Oct. 2/7 A thin close-lipped mouth. **1936** A. Huxley *Eyeless in Gaza* xxx. 406 Close-lipped and crookedly, Mary was smiling. **1867** Whitman *Leaves of Grass* 172 The fire that suddenly bursts forth in the close-pack'd square. **1935** *Burlington Mag.* Aug. 93/2 Close-packed thought and intuition. **1846** R. Owen *Lect. Anat. Vert.* I. ix. 220 Conical teeth, as close set and sharp pointed as the villiform teeth. **1909** *Westm. Gaz.* 10 Apr. 15/1 A mass of close-set braiding. **1865** Whittier in *Atlantic Monthly* XV. 563 Blind must be their close-shut eyes. **1913** D. H. Lawrence *Love Poems* 9 Why even now you speak through close-shut teeth.

3. close-annealing, = *box-annealing* (see *box *sb.*[2] 24); so *close-annealed* adj.; **close call** (*colloq.*, orig. *U.S.*) = *close shave* (see A. *15 b above); **close-carpet** *v.*, to cover the whole floor of (a room) with carpeting, to provide (a room) with a fitted carpet; so *close-carpeted* adj.; *close-carpeting*; **close communion**: see Communion 7; **close-coupled** *a.*, coupled close together; spec. *Electr.* (see quot. 1911); **close harmony**, harmony in which the parts composing each chord lie closely together, usu. within an octave or twelfth; also *attrib.* and *fig.*; cf. *barber-shop 2b; **close-herd** *v.* (orig. *U.S.*) (see quot. 1887); also *transf.*; **close-lagged** *a.*, closely covered with lagging (see Lag *v.*[4], Lagging *vbl. sb.*[3]); **close-range** *attrib.*, (used or acting) at or from a short distance; **close-reach**, a reach sailed close to the wind; also as *v. intr.*, to sail on a close-reach.
1930 *Jrnl. Iron & Steel Inst.* CXXI. 151 Those required with a nice bright finish such as pickled, cold-rolled, and close-annealed sheets. **1932** E. Gregory *Metallurgy* iv. 103 This is known as 'close-annealing', as distinct from the more commercial 'open' method where the steel is in contact with the actual heating chamber. **1881** *Harper's Mag.* LXIII. 118/1 My! but that was a close call. **1904** F. Lynde *Grafters* xxiii. 284 Though he escaped with his life and his job, it was a close call. **1949** 'J. Tey' *Brat Farrar* vi. 45 The exciting things of life—riding, love-making, rescue, close calls. **1958** J. Bingham *Murder Plan Six* I. 11 A pleasant room, low-ceilinged, close-carpeted with a blue carpet. **1959** *Listener* 4 June 990/2 We decided to close-carpet the whole room. **1942** N. Balchin *Darkness falls fr. Air* xi. 198 A big room..fitted up..with close-carpeting. **1824** *Baptist Mag.* IV. 411 With these views of catholicism we do not see that the practice of close communion at all interferes. **1834** Carruthers *Kentuckian in N.Y.* I. 182, I thought that you were aware of my partiality for the close communion Baptist. **1883** Schaff *Relig. Encycl.* I. 211/1 The American Baptists practise close communion. **1909** Webster, *C[lose]-coupled circuit, Elec.*, a compound circuit composed of a closed circuit and an open circuit directly joined together. **1946** *Nature* 5 Oct. 489/2 The apparatus consists of test chambers with close-coupled piping. **1958** *Times* 23 July 6/1 Trailers with four close-coupled wheels, e.g. caravan trailers. **1876** Stainer & Barrett *Dict. Mus. Terms* 100/2 *Close harmony*, harmony produced by drawing the parts which form it closely together. **1884** *Examination Papers in Music* (*College of Preceptors*) (1892) 4 Fill up the following in close harmony. **1933** N. Coward *Design for Living* III. i. 103 *Gilda.* You must both come to lunch one day... *Leo...* just the three of us..Close harmony. **1958** Blesh & Janis *They all played Ragtime* viii. 155 Comedy skits, close-harmony ballads, waltzes, [etc.]. **1874** J. G. McCoy *Hist. Sk. Cattle Trade* 348 He outrides the country instead of close herding his stock. **1887** *Scribner's Mag.* Oct. 508/2 A friend..has heard a sheriff talk of 'close-herding' several prisoners in his charge. On the plains it means the difficult art of keeping cattle in a compact body, close together. **1925** Mulford *Cottonwood Gulch* xii. 148 We've got to round-up, loose herd durin' the day, an' close herd nights. **1931** *Times Lit. Suppl.* 26 Feb. 147/3 The commoners could only turn their stock out on the forest and could not close-herd them. **1883** *Specif. Alnwick & Cornhill Rlwy.* 4 The centres are to be close-lagged to the satisfaction of the Engineer. **1909** *Daily Chron.* 22 Jan. 3/4 It..is entitled 'Every-Day Japan', and is described as 'a close-range view of the Japanese people'. **1923** Kipling *Irish Guards* I. 39 The enemy attacked—with rifle-fire and a close-range small piece. **1899** *Daily News* 20 Oct. 5/7 The yachts close-reached for home.

close, *v.* Add: **9. b.** *Stock Exchange.* Of stocks or shares: to be *at* a certain price or position at the close of a day's trading.
1860 *Times* 2 Jan. 7/3 Turkish Six per Cents. closed without alteration at 76½ to 77¼. *Ibid.* 3 Jan. 5/3 Most of the leading British stocks closed a fraction higher than on Saturday. **1964** *Financial Times* 12 Mar. 1/7 The Ordinary closed 3s 1½d up at 37s 6d to yield 4·8 per cent.

16. close down. b. To put an end to; to stop the working or use of.
1903 *N.Y. Even. Post* 18 Sept. 2 President Shields has issued orders to close down all of the operations of the company. **1906** *Westm. Gaz.* 5 Apr. 8/2 Every closed-down mine that had any prospects at the time when it was abandoned could..be reopened. **1927** P. Cox in *Lett. Gertrude Bell* II. 537 The ringleaders were forthwith arrested; the two new extremist parties closed down and certain mischievous papers suppressed.

c. *intr.* To exercise repression; to act severely.
1869 'Mark Twain' *Innoc. Abr.* xxiv. 254 They have set a gun-boat to watch the vessel night and day, with orders to close down on any revolutionary movement in a twinkling.

d. Of fog, night, etc.: to come down.
1883 'Mark Twain' *Life on Miss.* 536 The night presently closed down.

e. To bring operations to a close; to cease being active or in use.
1921 *Flight* XIII. 620/1 At 11.15 p.m. the order was given to 'close down'. **1934** *N. & Q.* 7 July 2/2 The Red Lion Brewery..'closed down', as the modern phrase is, on June 23. **1955** J. P. Donleavy *Ginger Man* xxii. 258 With the light off and B.B.C. closed down for the night. **1967** *Listener* 18 May 644/2 *Sunday Citizen* is to close down next month.

17. close in. e. (Earlier example.)
1816 Jane Austen *Emma* III. v. 74 The evening is closing in.

19. close out. b. To clear out (a stock of goods); to wind up (a business); to sell or finish off. Also *absol.* *U.S.*
1852 J. M. Letts *California* 159 We offered him [a mule] to Mr. Priest for six dollars... He offered two, at which we 'closed him out'. **1884** E. W. Nye *Baled Hay* 101 It will be closed out very cheap. **1891** H. F. O'Beirne *Leaders Ind. Territory* I. 59/1 In 1879 he opened business in Audubon, Wise county, but closed out in 1883. **1903** *Lett. that bring Business* vi. 58 If you could have closed them out with a little less than the 10s. reduction you mention. **1936** [see *closing *vbl. sb.* 1 b]. **1945** H. D. Smyth *Gen. Acct. Devel. Atomic Energy Mil. Purposes* x. 86 Problems closed-out during May 1944. **1961** *Amer. Speech* XXXVI. 170, I abruptly ended this approach. I closed it out by September, 1958.

closed, *ppl. a.* Add to def.: *spec.* confined to a few people; limited by certain conditions; self-contained. (Cf. Close *a.* 9, and sense 2 below.)
1887 [in Dict.]. **1927** Carr-Saunders & Jones *Soc. Struct. Eng. & Wales* 82 There are a number of professions which are 'closed'. Complete closure exists when, as among dentists, certain functions and certain titles are legally reserved to persons whose names appear upon a statutory register. **1963** M. Ginsberg in *Advancem. Sci.* XIX. 418/2 We have to distinguish between 'open' and 'closed' ideologies. **1963** P. Drackett *Motor Rallying* 77 Closed and restricted trials and rallies. **1964** M. A. K. Halliday et al. *Linguistic Sci.* ii. 22 It is not the case..that all choices in language are clearly of one type or the other, closed or open.

2. *Math.* **a.** Of a set of points: containing all its limit points. Of an interval: containing all its end points.
1902 *Proc. Lond. Math. Soc.* XXXIV. 289 Every example of such a set [of points] is theoretically obtainable in this way. For..it cannot be closed, as it would then be perfect and nowhere dense. **1968** P. A. P. Moran *Introd. Probability Theory* iv. 184 In one-dimensional Euclidean space R_1 an interval ($a \leq x \leq b$), which contains its end points, is closed, and in R_n the sphere $\Sigma x_i^2 \leq d$, is closed.

b. Of a set of elements: having the property that any element produced by a (specified) allowed operation on elements of the set is also a member of the set; also used of the operation.

1909 in Webster. **1939** H. Weyl *Classical Groups* i. 1, *k* is any set of elements *a*, called numbers, closed with respect to the two binary operations: addition and multiplication. **1965** Patterson & Rutherford *Elem. Abstract Algebra* i. 20 For a closed associative binary operation, the set of elements which possess inverses is closed with respect to the operation.

3. Special Combs.: **closed book**, something unknown or uncomprehended (cf. *open book* s.v. *open *a.* 21); **closed circuit**, a complete, unbroken circuit; freq. *attrib.*; spec. (*a*) *Electr.* a complete electrical circuit formed entirely of conductive material; (*b*) a system of radio or television whereby the signal is transmitted by wire to the receiver and not broadcast for general reception; (*c*) a breathing apparatus that converts exhaled gases into oxygen for inhaling; **closed cycle**, a cycle of operations in which the same fuel, cooling fluid, etc., is used repeatedly; also *attrib.*; **closed door** *fig.*, an obstacle, impasse, restriction; also *attrib.* or quasi-*adj.*, restricted, obstructive, secret; **closed-end** *a.* (orig. *U.S.*), of an investment trust or the like: see quot. 1954; **closed-loop** *a.*, characterized by a feedback from a later to an earlier point in a cycle of operation; **closed shop** (orig. *U.S.*), a shop, factory, trade, etc., in which normally only trade-union members are employed; also *transf.* and *attrib.*; **closed society**, a society characterized by its rigid structure and beliefs; one having little or no contact with other peoples or intolerant of any disturbance of its existing order and traditions; **closed system**, a complete and essentially unalterable system (of ideas, doctrines, things, etc.); a material system in which the total mass or energy remains constant; a self-contained realm, unaffected by external forces.
1913 'Sax Rohmer' *Myst. Fu-Manchu* xxviii. 282 The soul of Kâramanèh was a closed book to my shortsighted Western eyes. **1926** J. S. Huxley *Ess. Pop. Sci.* xiii. 163 To whom much of art is a closed book. **1944** F. Clune *Red Heart* 35 The desert is an open book to the man of the Vast Open Spaces, but to the schoolmaster it was a closed book. **1827** J. Cumming *Man. Electro Dynamics* 230 If a series of galvanic elements be placed in order round the circumference of a circle, and forming a *closed* circuit, a current will circulate in it continually. **1832** Closed circuit [in Dict.]. **1925** P. J. Risdon *Crystal Receivers & Circuits* 22 It should perhaps be explained here that an aerial circuit is known as an 'open' circuit, and a receiver circuit as a 'closed' circuit. **1930** *Engineering* 28 Feb. 279/2 The fire-alarm system is of the pre-signal, electrically supervised, closed-circuit, code-ringing type. **1941** *B.B.C. Gloss. Broadcasting Terms* 6 *Closed circuit*, channel between microphone and loudspeaker, or between microphone and recording unit, for testing, rehearsal, or other purposes not involving broadcast transmission. Hence *closed-circuit recording*. **1946** T. H. Hutchinson *Here is Television* 363 Closed circuit. The picture is not broadcast, but fed to viewing screens at certain locations by wire. **1951** *Time* 25 June 49/1 Though not telecast over the air, the Louis-Savold fight was experimentally piped by coaxial cable over closed circuits to six cities. **1953** J. Hunt *Ascent of Everest* 257 Three principal types of (oxygen) apparatus were used: Open-Circuit, Closed-Circuit and Sleeping Sets. **1954** B. & R. North tr. *Duverger's Pol. Parties* i. iii. 152 Here the parliamentary representatives are chosen by the inner circle; the party in this case is a closed circuit. **1957** *Times* 22 Oct. 10/1 Later their occupants watched and heard the after-dinner speakers on closed-circuit television. **1950** *Jane's Fighting Ships 1950–51* A 5/2 Closed-cycle powered submarine, 2,000 tons—no external source of oxygen. **1957** *Gloss. Terms Nuclear Sci.* 30/2 *Closed cycle*, cycle of operation of a heat engine in which the same power fluid is used repeatedly, as a steam engine that condenses the steam for reuse...Also applicable to a cooling system in which the coolant is cycled repeatedly through the source of heat, itself being cooled in another part of the cycle. **1960** *Gloss. Atomic Terms* (*H.M.S.O.*) 13 *Closed cycle*, usually refers to reactor coolant circuit where the gas or liquid travels in a completely enclosed path. **1934** Webster, *Closed door*, restriction, esp. in the colonies and spheres of influence, of the land laws, concessions, tariffs, etc., in favor of the nationals of the controlling power...*closed-door*, adj. **1950** in M. McLuhan *Mech. Bride* (1951) 2/1 Today's closed-door session, which began the decisive phase in the committee's long deliberations. **1955** H. Roth *Sleeper* ii. 19 This [court martial]..had been so closed-door, so hush-hush, so big-brass. **1956** A. H. Compton *Atomic Quest* 323 There are no closed doors in the field of nuclear engineering. **1960** *Times* 9 Feb. 17/5 The old protectionist closed-door policy. **1952** *Time* 12 May 41 The Fund began as a 'closed-end' trust (*i.e.*, issuing a limited number of shares). **1954** Webster Add., *Closed-end* adj., of an investment trust, operating with its initial capital changed only by new issues or retirements, not by redemption of shares; opp. to *open-end*. **1958** *Spectator* 20 June 815/3 The three leading 'closed-end' investment trusts. **1958** *New Biol.* XXVI. 107 A closed-loop machine is one that feeds a part of its own output back into itself. **1960** *Times* 9 Nov. 16/4 The introduction of closed-loop models from control engineering. **1962** Simpson & Richards *Junction Transistors* xiii. 296 The voltage gain is therefore changed from A to A/(–AB) by the feedback. The latter value is often called the closed-loop voltage gain. **1904** *N.Y. Even. Post* 15 Aug. 1 An increase in wages, recognition of the union, and 'closed shops' are demanded. **1923** J. D. Hackett *Labor Terms* in *Management Engin.* IV. 343/1 *Closed shop*, a plant in which only union men, or prospective union men are hired. **1947** *Ann. Reg. 1946* 85 The expression 'closed shop' was an inaccurate term to use in describing the aims of the trade unions in dealing with non-unionists. **1949** Koestler *Promise & Fulf.* II. v. 275 Owing to the closed-shop system of political coteries, most of the diplomatic jobs go to little yes-men. **1935** Audra & Brereton tr. *Bergson's Two Sources Mor. & Relig.* iv. 229 The closed society is that whose members hold together, caring nothing for the rest of humanity, on the alert for attack or defence. **1945** K. R. Popper *Open Soc. & Enemies* I. x. 154 The problem of the tension between the classes..is raised for the first time by the breakdown of the closed society. **1955** Koestler *Trail of Dinosaur* 84 'Closed' societies such as prison camps.. also tend to develop a monolithic system of values. **1896** W. James *Will to Believe* (1897) 13 A system, to be a system at all, must come as a *closed* system, reversible in this or that detail, perchance, but in its essential features never! **1912** J. S. Huxley *Individ. in Animal Kingdom* v. 143 The tendency towards the formation of closed systems has manifested itself. **1922** A. N. Whitehead *Princ. Relativ.* ii. 21 We are treating nature as a closed system. **1925** O. W. Holmes in *Holmes–Laski Lett.* (1953) I. 706, I take it that he thinks no one a philosopher who hasn't a closed system—a cosmos that accounts for and sustains itself and is seen as necessary. **1947** *Mind* LVI. 64 It is asserted in the science of physics that in a closed system where there are exchanges of energy, energy is conserved. **1948** E. F. Obert *Thermodynamics* ii. 27 A closed system contains a constant mass, and only energy is allowed to cross the boundary. **1951** T. J. Benac tr. *Waismann's Introd. Math. Thinking* 60 Each of the three systems—the natural numbers, the integers, the rational numbers—forms a closed system by itself, and it is quite impossible to go from one of these domains to another by adjoining new elements. **1958** *Oxf. Dict. Chr. Ch.* 100/2 For practical purposes it is necessary to treat the material world as a closed system.

close-down (klō^u·zdaun). [f. Close *v.* 16.] An act of closing down; a cessation of working or use; *spec.* the finishing of radio or television broadcasting at the end of a day; *U.S.*, a stoppage of work by the closing of a factory, etc.
1889 *Voice* (N.Y.) Sept. 5 [To] interfere with the unrestricted manufacturing interests of our country by forced close-downs, lockouts, &c. **1939** *Radio Times* 6 Jan. 30/4 Talks..Sport..Close Down. **1957** *Economist* 7 Dec. 875/2 This is where the real impact of the dockyard close-down is being felt in Hongkong.

closen, *v.* Add: **2.** *intr.* To become closer, close up.
1908 Hardy *Dynasts* III. III. iii, The retreating-way, Along which wambling waggons..Have crept in closening file. **1919** *Edin. Rev.* Oct. 241 The sudden closening of intercourse.

close quarters. Add: **2. b.** *attrib.* (Also in form *close-quarter.*) Done at close quarters; in immediate contact.
1871 *What happened after Battle of Dorking* 5 Serving out the old-fashioned muzzle-loading musket..for close-quarter shooting and bayonet work. **1916** 'Boyd Cable' *Action Front* 119 Their nerves shaken..by the close-quarter bombing. **1962** I. Murdoch *Unofficial Rose* xxvii. 265 He shrank from any close-quarters looking-after of her.

closet, *sb.* **3. b.** The normal *N. Amer.* usage.
1929 M. de la Roche *Whiteoaks* xxiv. 333 Finch..ran upstairs to his attic room to look over his clothes. He took them from the closet. **1960** *Times* 14 Sept. 12/6, I had been warned of a few hazards such as..the use of closets rather than cupboards [in the U.S.]. **1968** *Globe & Mail* (Toronto) 17 Feb. 45 (Advt.), 6 room home in immaculate condition. 3 bedrooms, closets galore.

close-up (klō^u·sɒp). orig. *U.S.* [f. Close *adv.* 1 b + Up *adv.*[2]] A cinema or television shot taken at short range in order to magnify detail; any photograph taken at short range; also, photography in which the camera is placed very close to the object. Also *attrib.*
1913 E. W. Sargent *Technique of Photoplay* (ed. 2) ii. 16 A bust is a portrait showing the head and shoulders only, but bust is more definite than close up, which is sometimes used, for close up might also mean a full picture. **1916** *Ibid.* (ed. 3) 391 A continuous scene broken by close-ups of Hester. **1917** *The Cinema* 113 The 'close up' pictures of.. a man with a wound bleeding in his head. **1933** A. Brunel *Filmcraft* 155 *Close-up*, abbreviated C.U. An ordinary close-up of a face includes the shoulders and part of the chest. **1937** *Discovery* May 152/1 A close-up camera was necessary. **1939** *War Illustr.* 16 Dec. 423 In this page are close-ups of every type of fighter aircraft. **1957** *Observer* 8 Sept. 11/6 The film, in black-and-white with a surplus of gigantic close-ups, seems more calculated to appeal to American than British audiences. **1958** *Times Lit. Suppl.* 31 Jan. 56/1 They have been photographed in close-up.

b. *transf.* and *fig.* A detailed or intimate view or examination.
1923 A. L. Benson *New Henry Ford* 323 A 'Close-up' of Ford. **1924** Galsworthy *White Monkey* II. ix, The world was full of wonderful secrets which everybody kept to themselves without captions or close-ups to give them away! **1926** Joad *Babbitt Warren* 103 It comes of seeing life as a series of 'close ups'. **1943** *Our Towns* iv. 111 The picture so far painted has been a grim one: a close-up of the black spots.

closh, *sb.*[3] [Origin unknown.] An upright piece of wood fixed in the deck of a whaling vessel, on which the blubber of the whale is spiked to be cut up.
1820 Scoresby *Acc. Arctic Reg.* II. 308 Each of these officers, provided with a blubber-knife or a strand-knife, places himself by the side of a 'closh'..fixed in the deck. **1836** *Uncle Philip's Convers. Whale Fishery* 98 The blubber is put on the spikes of the closh, and the harpooner slices off the skin. **1886** *Good Words* 83 The harpoons of various designs, the closh, and the broken krenging hook.

closing, *vbl. sb.* Add: **1. b.** *closing-out* (see *close *v.* 19 b). *U.S.*
1865 *Memphis Daily Argus* 19 Nov. 3/2 R. J. Dalton & Company's great closing out sale. **1936** D. L. Teilhet *Ticking Terror Murders* xiii. 214 Closing-out sale! Frances Taylor is closing out her Paris–Hollywood Shop. Reduced prices.

4. closing-time, the time at which a public building or place, *spec.* a public house, is closed; also *fig.*
1896 A. Morrison *Child of Jago* xxxi. 291 At closing-time they all helped each other home. **1924** [see *allow *v.* 8 c]. **1927** Kipling *Limits & Renewals* (1932) 158 He'd gone to the Cup o' Grapes..and told everybody his grievances against Ellen till closing-time. **1954** Koestler *Invis. Writing* xxxv. 375 Closing time was approaching for Europe.

closish (klō^u·siʃ), *a.* [f. Close *a.* + -ish[1].] Fairly close.
1859 Geo. Eliot *Adam Bede* xxxvi, If you dooant mind lyin' a bit closish a-top o' the wool-packs. **1898** *Daily News* 23 June 6/2 We forget the figures, but think it was a closish thing. **1901** H. Sutcliffe *Willowdene Will* ii. 40 Should they be closish behind..'twould fare ill with thee.

Clostridium (klɒstri·diɒm). *Biol.* Pl. **-ia.** [mod.L. (A. Prazmowski *Entwickelungsgeschichte* (1880) 23), f. Gr. κλωστήρ spindle + -ίδιον dim. suffix.] A large genus of anærobic or micro-ærophilic bacteria that are typically spindle-shaped because of the presence of an endospore, many species of which are pathogenic, and others of which have industrial and other applications; also (with lower-case initial) a bacterium of this genus. Hence **clostri·dial** *a.*, pertaining to, caused by, or resembling a clostridium or clostridia.
1884 E. Klein *Micro-Org. & Dis.* x. 81 Clostridium Butyricum or Bacillus Butyricus. **1886** *Encycl. Brit.* XXI. 401/2 Trécul noticed this formation of amyloid substance in *Clostridium*. *Ibid.* 402/1 *Clostridia*, &c., are spindle-shaped. **1893** W. R. Dawson tr. *Schenk's Man. Bact.* i. 4 The cell..becomes spindle-shaped, a form which is known as *clostridium*. **1900** *Jrnl. Exper. Med.* V. 212 Some [bacilli] resemble diplococci, others have a drumstick form, while others have a central, clostridium-like enlargement. **1909** Webster, Clostridial. **1926** *Gould's Med. Dict.* (ed. 8) 338/2 *Clostridium*, a genus of bacteria differing from bacilli in the fact that their spores are formed in an enlarged part of the cell. **1949** H. W. Florey et al. *Antibiotics* II. xxxi. 1019 Penicillin was considerably more active *in vitro* against clostridia than flavines, sulphones, [etc.]. **1957** *Observer* 3 Nov. 15/4 Our nitrogen-fixing ally clostridium fights eternally for life against actinomycetes. **1959** *Chambers's Encycl.* V. 621/2 Butyl and Iso-propyl Alcohols, Acetone,..used for making synthetic rubber, are produced by bacteria of the genus *Clostridium*, principally *C. acetobutylicum*. *Ibid.* XIII. 551/1 Tetanus..is caused by the toxin of the organism *Clostridium tetani*, which has entered through a wound. **1960** *Farmer & Stockbreeder* 9 Feb. 102/1 Another type of 'Baby Pig Disease' is associated with infection with one of the clostridial group of germs.

closure. Add: **5. c.** *Phonetics.* Any position of articulation in which, or the extent to which, some part of the speech mechanism is moved towards another so as partially or wholly to block the current of air.
1867 A. M. Bell *Visible Speech* 60 An 'outer' formation, or closure of the super-glottal passage..., yields a distinct percussion. **1877** H. Sweet *Handbk. Phonetics* 85 In English, when a stop follows a vowel,..nothing is heard but part of the glide on to the consonant, the actual closure being formed without any breath at all. **1933** Bloomfield *Lang.* vi. 103 [In] Southern British English..the degrees of closure of the vowels in words like *up* and *odd* are the reverse of ours. **1964** R. H. Robins *Gen. Linguistics* iii. 99 Palatal plosives, in which the front of the tongue makes closure with the hard palate.

8. b. *Cricket.* The act or right of declaring an innings closed; also = *closure rule* (see below).
1904 R. S. Holmes *Hist. Yorks. Co. Cr.* viii. 101 It was in this match [Yorkshire *v.* Cambridge University, 1894] that the Light Blue captain put the closure into force.

10. In Gestalt psychology, the process whereby incomplete forms, situations, etc., are completed subjectively by the viewer or seem to complete themselves; the tendency to create ordered and satisfying wholes. Also *attrib.*
1924 R. M. Ogden tr. *Koffka's Growth of Mind* iii. 103 Phenomena occur in connections determined by peculiar inherent laws of relationship which have to do with 'closure' and 'non-closure'. **1925** I. A. Richards *Princ. Lit. Crit.* 107 The perception of an object and the recognition that it is a tree..involve a poise in the sensory system concerned, a certain completeness or 'closure'. **1934** *Brit. Jrnl. Psychol.* July 38 In the child with high intelligence, the disposition to create order is strong enough to work with less vivid material, and such a child will experience a strong Prägnanz and 'closure'. **1951** G. Humphrey *Thinking* 175 The tendency of an incomplete configuration to complete itself, the tendency, that is, towards *closure*. **1964** A. Koestler in *Listener* 14 May 787/2 According to Gestalt theory, we solve abstract problems by applying the 'closure principle'; the solution closes the gap.

11. *attrib.*, as **closure rule** *Cricket*, the rule that allows a captain to 'declare' (see *declare *v.* 11 b).
1897 W. J. Ford in K. S. Ranjitsinhji *Jubilee Bk. Cr.* ix. 346 Most of the matches last for a day only, so that the 'closure' rule was a boon and a blessing.

‖**Clos Vougeot** (klo₁vuʒo). Also **Clos de Vougeot.** [F. *clos* (enclosed) vineyard + *Vougeot*, name of a commune in the Côte-d'Or department of France.] A red burgundy produced at Vougeot.
1793 A. Young *Trav. France* I. 323 Stop at Nuyts..and examine the *Clos de Veaugeau*, of 100 *journaux*. **1833** C. Redding *Mod. Wines* v. 97 The celebrated wine called Clos Vougeot. **1840** Thackeray in *Times* 2 Sept. 6/1 [He] can drink anything from clos vougeot to old tom. **1920** G. Saintsbury *Notes Cellar-Bk.* iv. 54 It is the fashion of course to put Clos-Vougeot at the head of all Burgundies. **1962** G. Mitchell *My Bones will Keep* iii. 40 A bottle of Clos de Vougeot had been placed upon the table.

clot, *sb.* Add: **4.** In modern *colloq.* use: a fool; a 'chump'. (Often a mild, or even friendly, term of abuse.)
1942 T. Rattigan *Flare Path* I. 97 Johnny, you clot! What about that beer you were going to get me? **1958** J. Cannan *And be a Villain* vii. 166 Damn that clot of a detective. **1958** P. Mortimer *Daddy's gone A-Hunting* xxi. 125 Jolly bad luck, what a clot she is. **1959** M. Steen *Woman in Back Seat* II. i. 192, I can't stand him! He's the most ghastly clot!

7. clot-moulder (see quot. 1921).
1843 *Civil Engin. & Arch. Jrnl.* VI. 349/1 The clot moulders were dispensed with. **1921** *Dict. Occup. Terms* (1927) § 119 *Clot moulder*;..prepares table and moulds for brick maker; works and forms lumps of clay ('clot') for his use.

clot, *v.* Add: **2. b.** Of people: to come together or move in a crowd.
1932 W. Faulkner *Light in Aug.* xiii. 273 So they moiled and clotted. **1943** Koestler *Arrival & Dep.* 81 They.. all clotted at the grate, trying to have a look. *Ibid.* 168 The people were clotting into groups, chattering excitedly.

cloth, *sb.* Add: **7.** Also, a large piece of painted scenery, etc. (see quot. 1957). Cf. *back-cloth 2, *cut cloth* s.v. *cut *ppl. a.* 12 a.
1884 J. Hatton *Irving's Impr. Amer.* I. vii. 165 The effect of the well-known Hampton Court cloth was so perfect. *Ibid.* II. xi. 265 We took..all the cloths for our entire répertoire, and many of the small practical set-pieces. **1886** [see *back-cloth 2]. **1909** *Stage Year Book 1908* p. v (Advt.), The Only Theatre:—In main street—centre of town. That can stage big productions. That can fly cloths. **1957** *Oxf. Compan. Theatre* (ed. 2) 133/2 *Cloth*, a term used for any large unframed expanse of canvas or material used for scenery. A modern cloth is generally made of widths of canvas.

10. *to cut* (etc.) *out of (the) whole cloth*: see Whole cloth b.

18. b. *cloth-faced, -sided* adjs.
1884 T. C. Hepworth *Photogr. for Amateurs* 58 The cloth-faced back is then inserted. **1925** Mallory in E. F. Norton *Fight for Everest, 1924* 233 My cloth-sided shoes.

19. cloth-board (see Board *sb.* 4); **cloth-bound** *a.* (see Bound *ppl. a.*[2] 8); **cloth cap**, a cap made of cloth; also *attrib.*, esp. in sense 'of or pertaining to the working class'; hence *cloth-capped* adj.; **cloth ears** *colloq.*, (a person with) a poor sense of hearing; hence *cloth-eared* adj.; **cloth-head** *colloq.*, a thick-head; cf. *clot-head* (Clot *sb.* 7); so **cloth-headed** *a.*, thick-headed; **cloth-looker**, one who examines cloth in order to detect faults arising in manufacture; **cloth-tenter** (see Tenter *sb.*[1] 1).
1832 Cloth boards [see Board *sb.* 4]. **1845** *Penny Cycl. Suppl.* I. 220/1 Most persons who purchase books are aware that, if in 'boards' or 'cloth-boards', the top edges of the sheets, although placed level, are uncut. **1888** C. T. Jacobi *Printer's Vocab.* 23 Books when bound in cloth cases are described as being in 'cloth boards'. **1860** *Leisure Hour* 16 Aug. 526/1 This mode of sewing books on tapes has of late years been adopted by the best London binders for cloth-bound volumes. **1904** *Westm. Gaz.* 15 July 12/2 The published price..will be..1s. net cloth-bound. **1968** W. Dawson *Catal.* 79 Physical Society: Proceedings..Vols. 1–3, 11 cloth bound. **1851** Cloth cap [see sense 17 in Dict.]. **1903** *Daily Chron.* 9 Dec. 9/7 (Advt.), Woman (Young) wanted for cloth cap cutting. **1959** *Manchester Guardian* 24 July 6/6 In the..scheme..expressly designed to appeal to the 'cloth-cap investor', the salaried staff have bought more shares than the payroll workers. **1935** *Times Lit. Suppl.* 14 Dec. 857/4 The cloth-capped artisans. **1967** G. North *Sgt. Cluff & Day of Reckoning* ii. 18 A small, cloth-capped man guiding a battery-propelled litter-cart. **1912** C. Mackenzie *Carnival* v. 49, I wish you'd listen. Have you got cloth ears? **1965** *New Statesman* 14 May 777/1 I've told you once, cloth-ears. **1965** G. Melly *Owning Up* xl. 127 It was more difficult for a band on the road to know what was going on than for the most cloth-eared member of a provincial jazz club. **1927** Wodehouse *Meet Mr. Mulliner* iii. 76 You've forgotten it again, you old cloth-head! **1958** M. Kelly *Christmas Egg* I. 40 There was occasion only to abuse himself as a triple cloth-head. **1925** Wodehouse *Sam the Sudden* xxvii. 233 Selling fictitious oil stock to members of the public who were one degree more cloth-headed than himself. **1970** M. Kenyon *100,000 Welcomes* xvi. 129 Those cloth-headed detectives had advised that he..say and do nothing. **1881** *Instr. Census Clerks* (1885) 68 Cloth Looker. **1927** *Sunday Express* 12 June 3 The clothlooker..usually reports any faults he may find. **1849** C. Brontë *Shirley* ii. To add to his possessions..space for his cloth-tenters.

clothe, *v.* Add: **1. d.** To invest with a religious habit: cf. *clothing *vbl. sb.* 1 c.
1628 in *Cath. Rec. Soc. Publ.* (1905) I. 98 She was cloathed in the new monastery at Gand the 25 of November 1626. *a* **1700** *Ibid.* VIII. 11 M[rs] Elizabeth Poulton was cloathed with her for a lay sister.

clothed, *ppl. a.* Add: **b.** Phr. *clothed and in one's right mind* [with allusion to *Mark* v. 15, *Luke* viii. 35]: sane, well, normal.
1873 Hardy *Pair of Blue Eyes* II. ix. 202 Good-bye till we meet, clothed and in our right minds, by the fireside at home! **1918** Galsworthy *5 Tales* 232 Clothed, but not quite in their right minds, they went up all together to the hotel and sat down to tea. **1931** I. Compton Burnett *Men & Wives* xii. 157 We heard that you were clothed and in your right mind, Lady Haslam, but nothing would do for Mother but that I should come to inquire. [**1961** Bible (N.E.B.) *Mark* v. 15 The madman..sitting there clothed and in his right mind.]

clothes, *sb. pl.* Add: **3.** (Later example.)
1801 Jane Austen *Let.* 5 May (1952) 123, I have been awake ever since five and sooner; I fancy I had too much clothes over me.

4. *clothes-rack*; **clothes-bag, -basket** (earlier examples); **clothes-conscious** *a.*, aware of, or (inordinately) interested in, clothes; **clothes-hanger** = *coat-hanger* (*coat *sb.* 14); **clothes-maid** = *clothes-maiden*; **clothes-pole** *U.S.*, a clothes-prop; **clothes-prop** (example); **clothes-sense** (see *sense *sb.* 11 c); **clothes stop** *Naut.*, a stop (see Stop *sb.*[2] 10 a) used to hang up clothes after washing, or to tie up bundles of clothes.
1834 *Knickerbocker* III. 81 The crew were disposed in various groups about the deck,..some with their clothes-bags beside them. *c* **1806** D. Wordsworth *Tour Scotl.* in *Jrnls.* (1941) I. 409 Little John asleep in the clothes-basket by the fire. **1928** *Daily Express* 28 Dec. 11/1 While she dresses exceedingly well, she gives no impression of being a clothes-conscious person. **1960** E. H. Gombrich *Art & Illusion* ix. 310 Even the most clothes-conscious woman would not..claim she can predict how a hat will suit her without having tried it on. **1934** Webster, Clothes hanger. **1937** *Discovery* July 198/1 Behind them, as if on a giant clothes-hanger, hung three complete horses' skins. **1853** Mrs. Gaskell *Cranford* ix. 166 The room being divided with great clothes-maids, over which Crosby's men were tacking red flannel. **1865** *Atlantic Monthly* XV. 659 She never conjectures to what base uses a clothes-pole may come. **1903** *Westm. Gaz.* 10 Oct. 2/1 Holding out gaunt branches like spectral clothes-props against the sky. **1857** D. H. Strother *Virginia* 46 He..strode out,..knocking over the clothes-rack in his progress. **1914** Joyce *Dubliners* 130 Four cane chairs, a clothes-rack. **1932** R. Lehmann *Invit. Waltz* III. iv. 199 It takes a man to teach a woman how to dress. The majority of them don't develop a clothes-sense till they've had a lover. **1960** *Guardian* 1 July 8/2 Anthea.. has precocious clothes-sense. **1883** *Man. Seamanship for Boys' Training Ships* 7 The hammocks are stopped to the gantline by a rolling hitch, and clothes' stops are fitted on the bight.

clothes-horse. Add: **b.** *fig.* A person whose main function is or appears to be to wear or show off clothes.
1850 Carlyle *Latter-Day Pamphlets* iii. 37 Idlers, game-preservers and mere human clothes-horses. **1889** 'Mark Twain' *Yankee* xxxiii. 378, I could see her [*sc.* England] erect statues and monuments to her unspeakable Georges and other royal and noble clothes-horses. **1942** Berrey & Van den Bark *Amer. Thes. Slang* § 452/4 *Clothes-horse*, an overdressed woman. *Ibid.* 574/7 *Clothes horse*, a woman who acts as a clothes model. **1948** *Richmond* (Virginia) *Times Dispatch* 22 July 18/2 (*headline*) Janis Paige, Best Clothes-horse, Looks Good Here, Too. **1962** J. Ludwig in R. Weaver *Canad. Short Stories* (1968) 2nd Ser. 249 She ordered her chauffeur to drive her to Fifi's, Shmifi's—a fancy French place for clothes horses.

clothing, *vbl. sb.* Add: **1. c.** Investiture with a religious habit. Cf. *clothe *v.* 1 d.
1628 in *Cath. Rec. Soc. Publ.* (1905) I. 98 Neece M. S. cloathing and profession. **1691** N. N. *Life Lady Warner of Parham* II. i. 73 My Lady took at her Cloathing the Name of Sister Teresa Clare. *Ibid.*, Father Martha a Famous Preacher..of the Society of Iesus, made her Cloathing Sermon. **1701** in *Cath. Rec. Soc. Publ.* (1905) VII. 92 We were at y[e] cloathing of 2 Lay Sisters at y[e] Austines. **1884** [see Dict. s.v. 1 a]. **1891** S. Ryder *Let.* 8 Apr. in C. S. Dessain *Lett. J. H. Newman* (1962) XII. 373, I entered [the convent].. in the year 1849. Dear Fr. Newman came to my clothing on the following August.

5. b. clothing book, a ration book containing clothing coupons; **clothing club**, a club for the exchange or purchase of clothing by small part-payments; **clothing coupon**, a ration coupon entitling the holder to a certain quantity of clothes or clothing materials; **clothing store** *U.S.*, a draper's shop; **clothing wool**, wool suitable for the manufacture of woollen fabrics.
1943 *Times* 4 Aug. 2/4 All the coupons in the current 1942–43 clothing book would remain valid at least until the end of 1943. **1838** J. Romilly *Diary* 28 Nov. (1967) 158 In came M[r] Hose begging for the Trinity Clothing Club (I gave him a gna). **1857** Geo. Eliot *Scenes Clerical Life* (1858) II. 311 A clothing club he was going to establish. **1957** J. Braine *Room at Top* I. 17 The Smiling Zombie, who ran a clothing club and a moneylender's. **1943** *Punch* 31 Mar. 259/1 There is no objection to a husband and wife sharing their clothing coupons. **1946** 'Tackline' *Such Nice Girls* xiii. 144 It is very strange..why R.N. officers are so much meaner with their clothing-coupons than R.N.V.R. officers. **1829** B. Hall *Trav. N. Amer.* I. 18 Flour and feed store—Cheap store—Clothing store. **1940** Mencken *Happy Days* 101 They made no more impression upon him than if they had addressed a clothing-store dummy. **1869** *Rep. Secretary of Agric.* (*U.S.*) 7 A sufficiency of most grades of clothing wools has been produced. **1891** R. Wallace *Rural Econ. Austral. & N.Z.* xxvi. 362 Clothing wool Merinos.

clotted. Add: **3.** *fig.*, spec. concentrated, dense; esp. in phr. *clotted nonsense*.
1674 Dryden et al. *Notes & Obs. Empress of Morocco* 23 Sure the Poet wrote these two Lines aboard some smack in a storm, and being Sea-sick spued up a good Lump of clotted Nonsense at once. **1834** *Sun* 1 Apr. 2/4 *Sartor Resartus* is what old Dennis used to call 'a heap of clotted nonsense'. **1885** W. B. Yeats in *Dublin Univ. Rev.* May 83/2 When low the sun sank down in clotted flame Beyond the lake. **1909** *Westm. Gaz.* 8 Apr. 2/2 This is the kind of talk which has been admirably described as 'clotted nonsense'. **1940** W. Empson *Gathering Storm* 63, I tried to defend my clotted kind of poetry. **1963** *Listener* 24 Jan. 180/3 A more than usually clotted [television] script.

clotting, *vbl. sb.* Add: **b. clotting time** = *coagulation time.
1908 *Trans. College Physicians Philadelphia* XIII. 94 Currents of air affect the clotting time, which is accelerated by anything causing evaporation. **1964** W. G. Smith *Allergy & Tissue Metabolism* ii. 17 One characteristic feature of anaphylaxis in the dog is an increase in the clotting time.

clottish (klɒ·tiʃ), *a. colloq.* [f. Clot *sb.* + -ish[1].] Resembling or characteristic of a 'clot' (see *clot *sb.* 4); stupid. Hence **clo·ttishness**, stupidity.
1952 B. Mason in *Here & Now* (N.Z.) Jan. 40/1 A scene of stupendous vulgarity that has a certain clottish charm. **1956** I. Bromige *Enchanted Garden* III. vi. 186, 'I don't believe you noticed any difference.' 'Very clottish of me.' **1961** *John o' London's* 2 Nov. 488/3 What I can't get over is the clottishness of it all.

‖**clou** (klu). [Fr., = nail.] That which holds the attention; the chief attraction, point of greatest interest, or central idea.
1883 *Daily News* 4 Aug. 5/1 The degree of copyright in the central idea, or '*clou*', as it is called in France, of a drama or romance. **1904** *Amer. Inventor* 1 Aug. 337 (*Cent. Dict. Suppl.*), The displays of finished goods, each containing some '*clou*', such as a $5,000 blanket in the blanket section. **1907** *Westm. Gaz.* 9 Nov. 14/3 The new 30-h.p. six-cylinder Napier chassis..is, of course, the *clou* of the Napier exhibit. **1918** Quiller-Couch *Studies Lit.* iii. 67 This epigrammatic *clou*, of all things, Milton wished to avoid. **1927** *Observer* 25 Sept. 18 The 'clou' of the evening, a new string quartet by Arnold Schönberg. **1959** *Observer* 24 May 19/7 The *clou* of the ballet is to be found in these four female solos.

cloud, *sb.* Add: **9. b.** Colloq. phr. (orig. *U.S.*) *on cloud seven* or *nine* (see quot. 1960); also *attrib.* (Cf. *seventh heaven.*)
1956 O. Duke *Sideman* ix. 120 Oh, she's off on Cloud Seven—doesn't even know we exist. **1959** *Down Beat* 14 May 20, I don't like strange music, I'm not on Cloud Nine. **1960** Wentworth & Flexner *Dict. Amer. Slang* 110/2 *Cloud seven, on*, completely happy, perfectly satisfied; in a euphoric state. **1963** *Listener* 14 Feb. 301/1 There we were. On cloud nine. **1963** *Times* 11 Mar. 9/4 Instead of Cloud Seven philosophy we got something much more materialistic.

11. a. *cloud-base, -ceiling, -cover, -floor, -squadron*; **b.** *cloud-colour, -shadow* (examples), *-shape*; **c.** *cloud-piercing* (earlier example); **d.** *cloud-bound, -veiled* adjs.
1952 C. Day Lewis tr. *Virgil's Aeneid* iv. 77 Rumour.. soon puffs itself up, And walking upon the ground, buries its head in the cloud-base. **1898** Whittier *Poet. Wks.* 474/2 Or saw the tabernacle pause, Cloud-bound. **1945** E. Waugh *Brideshead Rev.* 174 Like a gull..out of sight, cloud-bound. **1945** F. A. Berry et al. *Handbk. Meteorol.* x. 645 The cloud ceilings in advance of the warm front follow the slope of the frontal surface as long as no precipitation takes place. **1952** M. Herzog *Annapurna* viii. 120 The cloud ceiling was low, but we hoped the weather would become more settled. **1607** Topsell *Four-f. Beasts* 31 These wilde Asses haue..a silver colour, (that is as I gesse) a bright cloud-colour. **1929** Blunden *Nature in Eng. Lit.* i. 35 His cloud-colours and rock masses. **1943** T. D. Gordon *Coastal Command* viii. 76 The Hudson pilot tried to reach cloud cover. **1907** G. Bacon *Record of Aeronaut* xiv. 261 Bacon leant far out looking at the cloud floor. **1938** C. Day Lewis *Overture to Death* 15 Crawling in echelon, Beneath the cloud-floor, the bombers come. **1940** *Illustr. London News* CXCVII. 4 The 'holes' in the cloud-floor provide the enemy pilot with frequent glimpses of the ground. **1708** J. Philips *Cyder* I. 106 That Cloud-piercing Hill, Plinlimmon. **1875** W. McIlwraith *Guide to Wigtownshire* 22 Time flies, swift as the cloud-shadows along the hillsides. **1938** R. Graves *Coll. Poems* 56 Where slow cloud-shadow strayed across A pasture of thin heath and moss. **1907** *Macmillan's Mag.* Feb. 287 A sense of airy living in a castle of cloud-shapes. **1924** A. J. Small *Frozen Gold* 228 Sullen cloud-squadrons banked up. **1894** *Outing* (U.S.) 183 Mountain heights, cloud-veiled, snow-crowned. **1967** G. Watkins in *Coast to Coast 1965–6* 209 The glare of the cloud-veiled sun.

12. cloud-attack (*Mil.*), an attack preceded by the discharge of poison gas; **cloud-banner** (see quot. 1906); **cloud-burst** (earlier and later examples; for *U.S.* read orig. *U.S.*); **cloud**

chamber, an apparatus, invented by C. T. R. Wilson, used for experiments involving water vapour, esp. one containing air or other gas super-saturated with water vapour, through which charged particles are passed and become identifiable after condensation of the vapour; cf. *Wilson; **cloud forest,** a forest almost constantly under clouds; **cloud negative,** a negative produced in photographing clouds or the sky; **cloud point** *Chem.*, the temperature at which an oil or other liquid begins to cloud on cooling; **cloud-seeding** (see *seeding *vbl. sb.*); **cloud street** (see quot. 1954); **cloud track, trail,** the path of charged particles revealed in a cloud chamber.
1918 W. Hutchinson *Doctor in War* (1919) xiv. 185 The cloud-attack method of using poison-gas. 1906 *Monthly Weather Rev.* Apr. 158/2 *Cloud banners.* The air..expands just enough to form a slight cloud or haze, which floats like a flag or banner to leeward of the mountain top. 1912 W. I. Milham *Meteorology* i. v. 218 The air moving over the summit is cooled below the dew point, and a cloud banner streams out from the mountain top. [a 1817 T. Dwight *Travels* (1822) III. ix. 249 This deluge, which they call the bursting of a cloud, took place in Oct. 1784.] 1869 J. Muir *First Summer in Sierra* (1916) 48 Heavy thunder-showers called 'cloud-bursts'. 1891 G. F. X. Griffith tr. *Fouard's Christ* I. vi. 308 Caught in one of these furious cloud-bursts the little vessels were scattered far and wide. 1904 *Scott. Hist. Rev.* Oct. 89 The cloudburst and flood of a dozen years ago. 1897 C. T. R. Wilson in *Phil. Trans.* A. CLXXXIX. 269 Watching the behaviour of the air in the cloud chamber. 1911 —— in *Proc. R. Soc.* A. LXXXV. 285 The clouds are viewed through the roof of the cloud-chamber, which is of glass, coated below with a uniform layer of clear gelatine. 1934 *Nature* 28 Apr. 640/2 The instruments with which the [cosmic] rays have been investigated have been the ionisation chamber, the counter and the cloud chamber. 1941 *Electronic Engin.* XIV. 539 The values found in cloud chamber experiments. 1965 *New Scientist* 29 Apr. 306/1 The cloud chamber reveals charged particles by the preferential deposition along their trajectories of droplets condensed from a saturated vapour or vapour gas mixture. 1922 A. G. Ruthven *Amphibians & Rept. Sierra Nev. Col.* 42 Tropical rain forest, becoming gradually wetter...Cloud forest, in general like the rain forest, but cooler and more humid. 1892 *Photogr. Ann.* II. 60 In the cloud negative the lighting must be in the opposite direction to what it is in the view. 1934 Webster, Cloud point. 1965 *Electronics Weekly* 18 Aug. 4/6 The precipitation temperature, or cloud point, of gas oils. 1950 *N.Y. Times* (C-edition) 8 July 15/7 Los Angeles, July 7—A cloud-seeding pilot was credited today with helping put out a 12,000-acre forest fire. 1955 *Ann. Reg. 1954* 400 Many cities..had come to rely on cloud-seeding to augment municipal water supplies. 1965 *Economist* 2 Oct. 43/2 This winter cloud-seeding to disperse fog will be carried out at more [U.S.] airports. 1954 *Jrnl. Brit. Interplan. Soc.* XIII. 272 The most prominent clouds are thousands of bright cumuli—arrayed in roughly parallel bands, called 'cloud streets', which usually indicate direction of the wind. 1923 C. T. R. Wilson in *Proc. R. Soc.* A. CIV. 197 The cloud track..has been formed by the passage of a β-particle. 1912 *Ibid.* A. LXXXVII. 289 Cloud trails sufficiently sharply in focus..to..allow of the ions on which they have condensed being counted. 1938 R. W. Lawson tr. *Hevesy & Paneth's Man. Radioactivity* (ed. 2) v. 61 The tertiary electrons can be distinguished from the secondary electrons by the aid of the cloud-track method. *Ibid.* xxv. 280 Cloud tracks which could be ascribed to the cosmic radiation were to be found naturally on only a small fraction of the Wilson photographs.

cloud, *v.* **7.** (Earlier and later examples of *cloud over, up.*)
1758 in *Essex Inst. Coll.* XVIII. 101 At Night [it] Clouded up. *Ibid.* 187 A very pleasant morn but Clouds over after noon. 1807 P. Gass *Jrnl.* 171 In the evening it clouded over and rained again. 1834 Crockett *Narr. Life* iv. 28 While I was out it clouded up, and I began to get scared. 1936 J. Tickell *See how they Run* ii. 14 The wind-screen clouded over.

cloud-cuckoo-land. Also **-town,** etc. Freq. with capital initials. [tr. Gr. Νεφελοκοκκυγία (f. νεφέλη cloud + κόκκυξ cuckoo), the name of the realm in Aristophanes' *Birds* (l. 819) built by the birds to separate the gods from mankind.] **a.** In translations of Aristophanes' word (see above). **b.** A fanciful or ideal realm or domain. Also in various allusive phrases.
1824 H. F. Cary tr. *Aristophanes' Birds* ii. i. 76 What shall our city's name be?..Cuckoocloudland. Will that do? 1874 B. H. Kennedy tr. *Ibid.*, Cloudcuckooborough. 1899 R. Whiteing *No. 5 John St.* xxxi. 319 All his thinking processes fade off into the logic of Cloud Cuckoo Land. 1903 *Daily Chron.* 25 Sept. 3/7 Our new school of economists (sort of cloud-cuckoo-town). 1908 *Times Lit. Suppl.* 10 Dec. 458/3 His ideal state is a true Cloud-Cuckoo-Town. 1925 Cabell *Beyond Life* 30 Wycherley had the saving grace to present his men and women as trammelled by the social restrictions of Cloud-Cuckoo-Land alone. 1925 R. Graves *Welchman's Hose* 20 Kingfishers, when they die, To far Cloud-Cuckoo pastures fly. 1939 J. B. Priestley *Let People Sing* 390 Sir George Denberry-Baxter was now Governor-general of some distant dominion of Cloud-Cuckoodom. 1950 G. Barker *News of World* 17 Turn back to the never-never Cloud-cuckoo, happy, faroff land. 1964 *Economist* 16 May 741/2 They weigh the evidence and give their judgment in what seems..a cloud-cuckooland.

clouded, *ppl. a.* Add: **2. a.** *clouded leopard, tiger*, a large, mainly arboreal, species of the cat family, *Neofelis nebulosa*, of southern Asia (see Tiger *sb.* 1 b).
1879, 1896 [see Tiger *sb.* 1 b]. 1910 *Encycl. Brit.* VI. 559/2 Little is known of the habits of the clouded leopard, but it preys on small mammals and birds, and rarely comes to the ground.

‖**cloué** (klue), *a. Her.* [Fr.] Studded with nails.
1868 Cussans *Handbk. Her.* 116 *Cloué*, studded with nails. A Portcullis, or a Gate, is sometimes thus described.

clove, *sb.*² Add: **6. clove-brown,** the colour of cloves, a medium shade of brown.
1795 Kirwan *Elem. Min.* (ed. 2) I. 273 Its colour is clove brown. 1887 W. Phillips *Brit. Discomycetes* 346 Scattered, erumpent, clove-brown, shortly stipitate. 1948 A. L. Rand *Mammals E. Rockies* 211 Colour in summer, clove-brown, mingled with deep reddish and yellowish browns.

clover, *sb.* Add: **1. b. clustered clover,** *Trifolium glomeratum*.
1858 G. Bentham *Handbk. Brit. Flora* 168 Clustered Clover. Trifolium glomeratum. 1921 H. Guthrie-Smith *Tutira* xix. 171 Lastly appeared Clustered clover (Trifolium glomeratum). 1960 Ary & Gregory *Oxf. Bk. Wild Flowers* 114/2 *Clustered Clover*...An uncommon annual..with tiny, unstalked purple-pink flower-heads.
4. *clover-field, -hay* (examples); **Clover Club,** the name of a club in Philadelphia, used to designate a cocktail made from gin, white of egg, lemon or lime juice, and grenadine; **clover-fern** *Austral.*, nardoo; **clover-leaf,** a system of intersecting roads from different levels, in form resembling the leaf of clover; freq. *attrib.*; **clover-leaf sight** (see quot.); **clover-sick** *a.* (examples); hence *clover-sickness*; **clover summer,** *fig.* an exceptional time.
1925 E. Wallace *King by Night* xiii. 186 Clover Club cocktails, John. 1931 A. Powell *Afternoon Men* xiv. 147 He..went to the bar and ordered two clover-clubs and a sidecar. 1878 R. B. Smyth *Abor. Victoria* I. 209 They seem to have been unacquainted, generally, with the use, as a food, of the clover-fern, *Nardoo*. 1840 J. Morton *Glouc. Farm Rep.* 16 in *Libr. Usef. Knowl., Husb.* III, They are..put to run in a fallow-field, if there is not a pasture or clover-field. 1870 'Fanny Fern' *Ginger-Snaps* 257, I shall shortly find a clover field where I intend to bury my disgusted nose until October. 1748 J. Eliot *Field-Husb.* 17 He that raiseth Clover Hay, need not be afraid of the expence of Seed. 1840 J. Morton *Glouc. Farm Rep.* 18 in *Libr. Usef. Knowl., Husb.* III, Good rye-grass and clover-hay is best for them. 1843 'R. Carlton' *New Purchase* 26 The tea was a perfect imitation of clover hay. 1901 *Daily Colonist* (Victoria, B.C.) 1 Nov. 1/5 (Advt.), Clover Hay. Just received several cars of the Choicest Hay for cows. 1933 *Nat. Geogr. Mag.* May 583/1 We thread our way first through the maze of underpasses, overpasses, 'clover leaves' and one-way roads that separates traffic at this busy junction point. 1937 *Times* 13 Apr. p. viii/1 For straight cross-roads where traffic is heavy and the amount of turning traffic considerable, the 'clover-leaf' type of bridge system used both in America and Germany would..justify the expense. 1939 *Archit. Rev.* LXXXVI. 58 The clover-leaf junction of Grand Central Parkway and Horace Harding Boulevard. 1951 *Amer. Speech* XXVI. 207/2 In Nebraska ..a similar intersection..in which vehicles turn to the right in a nearly complete circle, this to make safely a right angle turn, is termed the *cloverleaf*. 1957 *Listener* 26 Sept. 469/1 Their parkways and clover leaves, their elaborate systems of traffic circulation. a 1884 Knight *Dict. Mech.* Suppl., *Clover leaf sight*, a rear gun-sight having side lobes, which slightly resemble two foils of the clover leaf. 1851 H. Stephens *Bk. Farm* (ed. 2) I. 619/2 Such soils as are termed *clover-sick*. 1872 *Rep. Vermont Board Agric.* 408 The land was what they call 'clover-sick'. 1894 *Country Gentlemen's Catal.* 15/2 The latest..assertion is, that clover sickness is due to the ravages of..the eelworm. 1907 *Daily Chron.* 15 Feb. 4/6 It was intended to make a grant of £300 to Berkhamsted for the investigation of clover sickness. 1933 *Discovery* Nov. 350/1 The organisms (both fungus and eel worms) which cause 'clover sickness'. 1867 Mrs. Whitney *L. Goldthwaite* xi, It was a 'clover summer' for the Josselyns...They must make the most of it.

clover, *v.* (Later U.S. example.)
1853 *Trans. Mich. Agric. Soc.* IV. 405 The best mode I have found of improving my farm, is by deep plowing and clovering.

clown, *v.* Add: **1. b.** *trans.* To play the clown in (a part); to render comic or farcical; to portray like a clown.
1891 *Harper's Mag.* Sept. 500/2 A Benedick who makes faces and 'clowns' the part..leaves a distinct and horrible stain on the memory. 1930 *Spectator* 18 Jan. 83 Mr. Marshall was too intent on making his 'asides', and there were frequent episodes in which he definitely clowned the situation. 1957 E. Hyams *Into Dream* 43 Mr. Cromer clowned comic astonishment. 1963 W. Soyinka *Lion & Jewel* 14 He does realistic miming...Lakunle clowning the driving motions, obviously enjoying this fully.

clownage. Restrict † *Obs.* to sense 2 and add later example to sense 1.
1903 *Times* 23 Apr. 10/2 The outside circle was the marriage of Theseus and Hippolyta, then came the clownage of the play.

clox, commercial spelling of pl. of Clock *sb.*² Cf. *clock *v.*⁴
1900 in A. Adburgham *Shops & Shopping* (1964) xxii. 260 Unbleached cotton [stockings] with silk clox...Black lisle with silk clox. 1923 *Daily Mail* 26 May 1 Silk Stockings (with lace clox). 1928 *Daily Tel.* 20 Mar. 14/6 The latest stockings, instead of having clox or embroidery to draw attention to ankles, are ornamented with hand-painted roses or with clox on the knee.

cloyingly (kloi·iŋli), *adv.* [f. Cloying *ppl. a.* + -ly².] In a cloying manner or degree.
1928 *Observer* 25 Mar. 15/4 This cloyingly sweet criticism.

cloysome, *a.* Delete *rare* and add examples.
1861 *Macm. Mag.* IV. 209/2 Cloysome contrasts are too frequently indulged in. 1902 G. C. Williamson *Holman Hunt* 18 The pictures of the day in their cloysome richness.

club, *sb.* Add: **2. b.** (See quot.) Cf. **club-topsail* under sense 20.
1889 *Century Dict.*, *Club*, a small spar to which the head of a gaff-topsail or the clue of a staysail or jib is bent to make the sail set to the best advantage.
8. (Later examples.)
1910 W. Dalton *'Saturday' Bridge* ix. 119 He leads the king of clubs first. 1966 *Listener* 10 Mar. 366/3 East trumped the diamond and returned a club.
14. c. *in the (pudding) club*, pregnant; esp. in phr. *to get* or *put* (someone) *in the (pudding) club*, to make pregnant. *slang.*
1936 J. Curtis *Gilt Kid* iv. 38 You were put in the pudden club by the squire's son. 1943 M. Harrison *Reported Safe Arrival* 13 D'jer put a servant-gal in the Club? 1968 M. Allingham *Cargo of Eagles* v. 66 You want an angry husband or a young woman in the pudding club. 1969 J. Gardner *Founder Member* ix. 148 They mean you have to get me in the club. 1969 J. N. Smith *Is he Dead, Miss ffinch?* iv. 20 When the doctor told me I was in the club I told him he was daft—that I'd never—well, you know.
d. *fig.* A number of people having something in common, sharing an experience, etc.
1944 *Chatelaine* May 51/2 Join the Cotton Club! [*i.e.* wear cotton dresses]. 1969 'J. Fraser' *Cockpit of Roses* vi. 45 Men knew that she was available to anyone who cared to take her out for an evening...Peter had avoided joining the club. 1970 *Calgary Herald* 26 Mar. 34/1 Standard equipment for married men who cheat is a wife in poor health who can't face the blow. Welcome to The Club.
e. A group of nations united or associated for a particular purpose, as for trade, defence, etc.
1950 *Ann. Reg. 1949* 235 The Federal Government were given permission..to appoint their own delegate.., an important step in integrating Germany into the Western European 'club'. 1952 *Ann. Reg. 1951* 166 Recommendations for the formation of a 'Low Tariff Club'. 1958 [see *atomic *a.* 2 e]. 1959 *Listener* 31 Dec. 1140/2 If European membership of the nuclear club increased, the task of the Nato authorities is likely to become impossibly difficult.
15. c. *the best club in London*: a jocular name for the House of Commons.
[1857 E. Whitty *Friends of Bohemia* II. i. 5 Some here [*sc.* the House of Commons] because it is the right club.] 1865 Dickens *Mut. Fr.* I. ii. iii. 186, I think..that it [*sc.* the House of Commons] is the best club in London. 1914 Chesterton *Flying Inn* xvii. 195 He saw afar off, sitting above the river, what has been very erroneously described as the best club in London. 1967 L. Woolf *Downhill all Way* i. 33 A backbench M.P. in the 1920s..was, no doubt, a member of what was said to be the best club in London.
19. a. (sense 2) *club-face, -head, -maker, -shaft*; **b.** (sense 14) *club-book, button, -girl, -mate, necktie, tie*; (sense 15) *club bore.*
1852 C. M. Yonge *Two Guardians* xx. 397 Marian went on a voyage of discovery among the club books downstairs, and brought up a book of travels, and a novel. 1864 Bagehot *Coll. Works* (1965) II. 290 A large part of old novels may very fairly be called club-books; they speak out plainly and simply the notorious facts of the world, as men speak of them to men. 1910 H. G. Wells *New Machiavelli* (1911) iii. i. 309 Most clubs have a common link, a lowest common denominator in the Club Bore, who spares no one. 1894 *To-day* 17 Mar. 182/2 Yachting suits of blue cheviot with club buttons. 1928 *Daily Mail* 25 July 15/4 Much has been heard in recent years of the 'shut' and 'open' club-faces. 1891 Hardy *Tess* III. vi. l. 170 The field in which, as a club-girl, she had first seen Angel Clare. 1890 H. G. Hutchinson *Golf* iv. 78 The club-head wants to be travelling, when it meets the ball, in the direction in which it is intended the ball should go. 1903 *Westm. Gaz.* 17 July 3/2 To bring the club-head down on the back of the ball and between it and the rim of the cup. 1890 H. G. Hutchinson *Golf* ii. 55 Then arose a great master club-maker. 1909 *Westm. Gaz.* 11 Jan. 12/2 Besides being a very fine player, Stephenson is a first-class club-maker. 1909 *Westm. Gaz.* 25 Oct. 14/2 The club..had been used to an old club-mate acting as umpire. 1960 T. McLean *Kings of Rugby* xi. 185 As clubmates, Briscoe and McCullough, were as yet wanting in subtlety. 1916 A. Bennett *Lion's Share* xxxii. 234 The owner..was wearing one of his most effulgent and heterogenous club neckties. 1906 *Westm. Gaz.* 12 Oct. 3/1 A little twist in the hand, with the club-shaft. 1897 *Sears, Roebuck Catal.* 222 Puffs, Bows, Club Ties, etc. 1935 G. Greene *Basement Room* 102 The rather exclusive club tie, freshly ironed. 1949 M. Muggeridge *Affairs of Heart* ix. 177 A club tie, old and frayed.
20. club armchair, chair, a thickly upholstered armchair of the type often found in clubs; **club doctor,** the doctor provided by a benefit club; **club-fender,** a large fender (Fender *sb.* 3 a) with a padded top; **club-fungus,** a fungus belonging to the family Clavariaceæ; **club-land** (also **clubland**) (further examples); hence, any area in which there is a large number of clubs; these clubs or their members collectively; also *attrib.*; **club leader,** a leader of a youth club; **clu·bmobile** *U.S.* [cf. *automobile *sb.*], a large vehicle equipped to supply refreshments, recreational facilities, etc., to troops, workers in isolated areas, etc.; **club-money,** (*c*) money paid by a benefit club or provident society; **club sandwich** (orig. *U.S.*), a thick sandwich containing several ingredients, as chicken or turkey, lettuce, tomato, mayonnaise, etc.; also *fig.*; **club-topsail,** a large topsail extended beyond the gaff by means of a small spar or 'club'; **club-walk, -walking,** a procession by the members of a local club or clubs; esp. the annual festival of a benefit club or friendly society; so *club-walker*; **club-woman** (orig. *U.S.*), a woman who is a member or habituée of a club or clubs.
1962 M. Kelly *Due to Death* v. 83, I have an affection for large comfortable lies. They're like club armchairs. 1919 G. B. Shaw *Heartbreak Ho.* Pref. p. xxiii, Fat old men, sitting comfortably in club chairs. 1872 Geo. Eliot *Middlem.* vii. lxxi. 163 Doctor Gambit..is our club doctor. 1905 A. Bennett *Tales Five Towns* i. 186 Will ye let your Lucy run quick for th' club doctor? 1915 Phillpotts & Hastings *Angel in House* 11 The fireplace..has round it a club fender on which people may sit comfortably. 1917 G. B. Shaw *What I really Wrote* (1931) 193 The whole thing was mere postprandial brag, war-game and club-fender gossip. 1959 *House & Garden* Sept. 69/1 Antique club-fender, brass with leather seat. a 1909 C. MacMillan *Minn. Plant Life* viii. (*Cent. Dict.* Suppl.), Not all of the club-fungi are unbranched. 1894 *Brewer's Dict. Phr. & Fable* 264/2 *Club-land*, that part of the West End of London where the principal clubs are situated; the members of such clubs. 1912 'Saki' *Chron. Clovis* 242 The brains of clubland were much exercised in seeking out possible merit. 1916 in H. Barnett *Canon S. A. Barnett* (1918) II. xxxiv. 72 The influence of Canon Barnett impressed itself strongly in East London clubland. 1953 R. Usborne *Clubland Heroes* 5 The heroes of the books I am examining were essentially West-End Clubmen, and their clubland status is a factor in their behaviour as individuals and groups. 1970 *Guardian* 22 Apr. 8/6 SET and rising land values in Central London are choking clubland like unwelcome smoke from a cheap cigar. 1889 in H. Barnett *Canon S. A. Barnett* (1918) II. xxxiv. 70 The work of club leaders lies in stimulating healthy political and social life. 1958 *New Statesman* 8 Feb. 181/1 The post of Full-time Club Leader as under: Richmond-Ham Youth Club. 1943 J. Steinbeck *Once there was War* (1959) i. 96 A clubmobile is parked, a bus converted into a kitchen for the cooking of doughnuts and coffee and run by two Red Cross girls. 1944 *Amer. Speech* XIX. 78/1 A *Club-mobile*, a miniature Red Cross club on wheels, was demonstrated March, 1943....It can be used as station wagon, stretcher, ambulance, or clubmobile. 1903 G. B. Shaw *Man & Superman* iii. 107, I saw a man die..a London bricklayer's labourer...He left seventeen pounds club money. 1913 D. H. Lawrence *Love Poems* 55 An' there's compensation, sin' it's accident, An' club money—I nedn't grouse. 1903 R. L. McCardell *Conv. Chorus Girl* 69 All we need is a club sandwich and a bottle of beer. 1945 *New Yorker* 25 Aug. 14 This is a club-sandwich sort of story, combining a hotel, a secretary, and an electric fan. 1966 G. Black *You want to die, Johnny?* v. 88 It was a club sandwich, three layers. 1886 *Outing* (U.S.) IX. 19/1 The Mohawk was lying..with all after canvas set, even to her enormous club top-sail. 1874 Hardy *Madding Crowd* ii. iii. 39 There were ..more people all the week long than at Weatherbury club-walking on White Tuesdays. 1891 —— *Tess* I. ii. 15 The May-Day dance..was to be discerned...in the guise of the club revel, or 'club-walking', as it was there called. 1899 *Daily News* 23 May 2/3 The club walkers of the country. 1936 G. M. Young *Victorian Eng.* xii. 77 The maypole had gone: the village feast and the club-walk were going. 1895 S. Hale *Lett.* (1919) 294 She..is a 'club-woman',—and she early secured me to 'attend a meeting' of her club. 1904 *Daily Chron.* 8 Oct. 8/4 There is no fear that the English clubwoman will ever oust the home-loving woman from her place. 1962 *Punch* 25 Apr. 637/1 My wife, a brand new club-woman with her subscription barely dry on my cheque-book counterfoil.

clu·bbableness. [f. Clubbable *a.* + -ness.] Clubbable quality; sociability.
1930 *Aberdeen Press* 11 Nov. 6 The character of the Cockney for good nature, resilience, and 'clubbableness' is due to his mongrel character. 1960 C. S. Lewis *Stud. Words* vi. 146 The quality of the 'good mixer', courtesy, clubbableness.

clubby, *a.* *U.S.* Delete *nonce-wd.* and add later examples. Also, friendly, sociable.
1905 *Smart Set* Sept. 128/1 Our black coffee stories have become clubby to an extent that embarrasses a butler. *Ibid.* 129/1 She is not a creature to set to music—the clubby girl, the feminine good-fellow. 1918 'Ian Hay' *Last Million* xv. 243 It's a darned sight easier to keep on clubby terms with an Ally whose language you don't know than an Ally whose language you do. 1928 F. N. Hart *Bellamy Trial* i. 14 Charles Stuyvesant 's the only one that won't have a good clubby time of it. 1960 *Encounter* Feb. 32/2 A clubby bonhomie.

club-house. Add: (Earlier and later examples.) *spec.* The club-house attached to a golf-course. Also *attrib.*
1818 J. Warburton *Hist. Dublin* II. 1169 The once-fashionable Club-houses are no longer attended. 1826 M. Kelly *Reminisc.* (ed. 2) II. 243 A very fine-looking.. gentleman, standing at the club-house door. 1910 *Bradshaw's Railway Guide* Apr. 1081 (Advt.), Golf Links and a Club House have been provided. 1955 *Times* 14 July 3/4 The indolent spectator could crawl from the clubhouse to the tenth tee. 1969 *Internat. Herald Tribune* (Paris) 6 Nov. 3/4 They saw the vote as a rejection of a rather crude clubhouse politican.

cluck, *sb.* Add: **5.** *U.S. slang.* A dull or unintelligent person, a fool; esp. in phr. *dumb cluck*. Also, an inferior or 'dud' thing; a counterfeit coin.
1904 'No. 1500' *Life in Sing Sing* xiii. 247/1 *Cluck*, counterfeit coin. 1928 J. P. McEvoy *Show Girl* x. 137 Boys, fellows, hey! Clucks, back to your corners. 1929 E. Wilson *I Thought of Daisy* iii. 150 Get away, yuh dumb cluck! 1930 *Amer. Mercury* Dec. 415/1 The champion cluck of all time. 1931 *Amer. Speech* VI. 204 *Dumb cluck*, unintelligent, not bright person. 1934 J. M. Cain *Postman rings Twice* vi. 59 A thing like that don't happen to a dumb cluck like him every day. 1950 'S. Ransome' *Deadly Miss Ashley* iii. 30 Showing ourselves up as a fine pair of clucks.

clucker (klʌ·kəɹ), *sb.* [f. Cluck *v.* + -er¹.] One who clucks or talks endlessly and aimlessly.
1891 Kipling *Light that Failed* viii, What a one-idead clucker it is! 1921 *Chambers's Jrnl.* 355/2 People were mistaken who took Gilbert for an eternal clucker, for he could be stern enough.

clucker (klʌ·kəɹ), *v.* [Imitative.] *intr.* = Cluck *v.* 1 and 3.
1904 H. G. Wells *Food of Gods* ii. i. 156 'Um!' said Redwood, after he had cluckered to the infant for a space. 1919 Masefield *Reynard* 91 His six hens cluckered.

clue, *sb.* Add: **2. b.** Esp. a piece of evidence useful in the detection of a crime.
1886 F. Hume *Myst. Hansom Cab* (1888) i. 8 Another hansom cabman..gave a clue which will, no doubt, prove of value to the detectives in their search after the murderer. 1954 M. F. Rodell *Mystery Fiction* xi. 74 The unforeseeable and accidental last-minute clue which rescues the detective from his intellectual..predicament. 1959 *Chambers's Encycl.* IV. 238/1 The firearm literally scores its identity on each shell which leaves it, a vital clue at any scene involving firearms.
d. In a crossword puzzle, a sentence or phrase (often employing a definition, synonym, anagram, pun, etc.) serving to indicate a word or words to be inserted.
1914 *N.Y. World* 6 Dec. ('Fun') 6/2 Cross-Word Puzzle. ..Fill in the small squares with words corresponding to the list of clues. 1925 *Punch* 7 Jan. 6/1 The cross-word puzzle is a far more complicated affair than this, for all the actual letters are carefully removed from his sight, and nothing but tantalising clues and numbers are provided to enable him to discover their whereabouts. 1930 *Times* 1 Mar. 13/4 The Crossword Puzzle...To try one clue and to solve it is to set teeth in another.
e. Colloq. phr. *not to have a clue*: to have no idea; not to know; to be ignorant or incompetent.
[1870 *Brewer's Dict. Phr. & Fable* 171/2 *Clue. I have not yet got the clue*; *to give a clue, i.e.* a hint.] 1924 A. E. W. Mason *House of Arrow* ii. 20 That doesn't bring us any nearer to finding out what is Boris Waberski's strong suit... We haven't a clue to it.] 1948 Partridge *Dict. Forces' Slang 1939–1945* 43 'He simply hasn't a clue'—he is completely ignorant of his job, or of what is going on. 1950 J. Cannan *Murder Included* iv. 59 As I believe they say nowadays, I haven't a clue. 1951 E. Hyams *Sylvester* ix. 37 'Sorry, old boy,' he said. 'I haven't a clue.' 1957 *Listener* 23 May 841/1 How valuable such sociological inquiries could be to local authorities, who usually haven't a clue. 1958 B. Hamilton *Too Much of Water* iv. 78 'Where is his cabin?' 'Haven't a clue,' said Maurice. 1967 *Times Rev. Industry* Mar. 90/3 Professor Rose..does not seem to have had much of a clue as to what really went on in Central Office between 1957 and 1959.

clue, *v.* Add: Now the usual spelling, esp. in the following senses: **a.** *trans.* In a crossword puzzle or the like: to indicate by means of a clue (see *clue *sb.* 2 d).
1934 *Punch* 17 Jan. 68/1 Four letters. I think they would clue them like this. 'An article in the middle of a short railway.' 1967 *Listener* 22 June 835/1 The remaining lights are clued normally. 1967 *Sci. Amer.* Sept. 275/2 The Double-Crostic, in which the words of a literary quotation and the name of the author and his work are derived from words clued by cryptic definitions.
b. To inform, tell (someone); to provide with a clue; also with *up*. So *to clue up*: to get to know about; *clued* (*up*) ppl. adj., informed, knowledgeable. *slang.*
1948 'N. Shute' *No Highway* v. 115 We've got to jump at this decision, now...We can clue up the official side later. 1955 J. I. M. Stewart *Guardians* 121 He's well clued up in all the Fontanay stuff. 1958 E. H. Clements *Uncommon Cold* i. 16 Keep me clued up about Remin's doings. 1958 J. D. MacDonald *Executioners* (1959) viii. 135 Usually I can come up with a clue. But I draw a big blank. Clue me, Louie. 1958 *Ottawa Citizen* 11 June 29/1 One officer described him as a 'well-clued technical officer, well up on advanced mathematics, physics and electronics...' 1959 C. MacInnes *Absol. Beginners* 59 Clue me then, teenager. My big ears are flapping. *Ibid.* 64 You meet all kinds of cats ..who can clue you up in all kinds of directions. 1970 R. Crawford *Kiss Boss Goodbye* ii. vi. 94 'I don't know,' I said. 'Are you clued?' 'Nope.'

clueless (klū·lės), *a.* *colloq.* [f. Clue + -less.] Ignorant; uninformed; stupid. Cf. *clue *sb.* 2 e. Hence **clue·lessness,** ignorance.
1943 C. H. Ward-Jackson *Piece of Cake* 22 'I'm clueless'—I don't know; or, 'He's a clueless type'—the opposite of a gen wallah. 1957 J. Braine *Room at Top* i. 15 Their two sons,..noisy and clumsy and clueless. 1957 H. Croome *Forgotten Place* x. 121 She was completely clueless about making herself look nice. 1960 C. Day Lewis *Buried Day* viii. 159 His anxiety, his sense of inferiority, his almost total cluelessness.

clump, *sb.* Add: **1. b.** A staff; a heavy stick. *dial.*
1868 *N. & Q.* 4th Ser. II. 152/2 He knocked and thump'd wi' his oaken clump.
c. A blow, knock. *colloq.*
1889 Jerome *Three Men in Boat* i. 7 They didn't give me pills; they gave me clumps on the side of the head. 1894 Blackmore *Perlycross* III. vi. 112 He dropped on the flags with a clump in his ear.
2. b. Of people.
1896 *Harper's Mag.* XCII. 772/1 Clumps of Frenchmen were smashed to pieces, one on top of the other.
c. *Printing.* (See quots.)
1875 J. Southward *Dict. Typogr.* (ed. 2) 14 Metal clumps are used in place of white-lines at the bottom of newspaper and other columns, to protect the letters from slipping in locking-up, when the foot-stick is short. They are cast in the same manner as leads, only of a thicker body. 1922 W. H. Slater *What Compositor should Know* 54 Clumps is the name given to leads when they are cast thicker than 3 pt.
d. An agglutinated mass of bacteria, blood cells, or platelets.
1896 *Proc. R. Soc.* LIX. 225 The most prominent of the effects..consists of an immediate aggregation of the bacteria into 'clumps'; this is combined with loss of motility. 1899 G. Newman *Bacteria* App. 343 The clumps of bacilli having fallen owing to gravity. 1939 Dible & Davis *Pathology* iii. 38 The massing of a number of platelets which adhere and form a minute clump upon the inside of the vessel wall. 1964 M. Hynes *Med. Bacteriol.* (ed. 8) vii. 85 When antibody reacts with the antigenic surface of bacteria, the bacteria are agglutinated into clumps which fall down as coarse floccules.
6. clump-foot = Club-foot. (Cf. Clumped *ppl. a.* 1.)
1922 *Chambers's Jrnl.* 863/2 He had a clump-foot.

clump, *v.* Add: **1. b.** *trans.* To strike, punch, or beat. *colloq.* or *dial.*
1864 *Derby Day* 52, I want to clump them. It will spoil sport to call in the bobbies. 1896 A. Morrison *Child Jago* 162 Dicky 'clumped' Bobby Roper whenever he could get hold of him. 1903 *Daily Chron.* 10 Mar. 9/7 When I woke up I found my boy's hand in my pocket, and I clumped him, as a father should. 1924 *Blackw. Mag.* Feb. 145/1 [He] impartially..clumped the ears of each youth in turn.
2. b. *intr.* To form a clump or clumps (*clump *sb.* 2 d). Hence **clu·mping** *vbl. sb.*
1896 *Lancet* 19 Dec. 1746/2 A new method of diagnosis of the fever by means of the clumping or agglutinative effect upon living typhoid bacilli. *Ibid.* 1747/1 Serums of immunised animals..have little or no clumping action. 1898 [see *agglutinate *v.* 2 b]. 1903 *Med. Rec.* 28 Feb. 358/2 The Seal Harbor bacillus refused absolutely to clump in any dilution above 1·10. 1908 *Practitioner* June 838, ·1 per cent. saline solution, in which the organisms lose to a large extent their natural property of clumping. 1951 *Sun* (Baltimore) 31 May 3/4 When people recover from leprosy, their blood no longer causes clumping in the blood test. 1967 W. H. R. Lumsden in D. M. Weir *Handbk. Exper. Immunol.* xxvii. 895 Reactions varied from 'pre-agglutination', characterized by immobilization without clumping,..to occasional small aggregations.

clumper (klʌ·mpəɹ), *sb.*² *colloq.* and *dial.* [f. Clumper *v.*²; cf. Clump *sb.* 4.]
1. (See Clumper *v.*²)
2. A thick, heavy shoe. Also a high sea-boot. Usu. in *pl.*
a 1825 R. Forby *Vocab. E. Anglia* (1830) I. 71. 1845 *Knickerbocker* XXVI. 417 Stilton made me ten pairs of 'clumpers'. 1935 'L. Luard' *Conq. Seas* 175 New dopper and clumpers to pay for.

clumsy, *a.* Add: †**4. b.** *clumsy cleat*: on a whaling vessel (see quots.). *U.S. colloq. Obs.*
1851 Melville *Moby Dick* i. i. 290 Clumsy cleat, as it is sometimes called, the horizontal piece in the boat's bow for bracing the knee against. 1874 C. M. Scammon *Marine Mammals* 224 About three feet from the stern is the 'clumsy-cleet', a stout thwart with a rounded notch on the after side, in which the officer or boat-steerer braces himself by one leg against the violent motion of the boat, caused by ..the efforts of the whale while being 'worked upon'.

Clun (klʌn). The name of a town, river, and forest in Shropshire, used to designate a breed of sheep originating in this area.
1883 *Encycl. Brit.* XVI. 789/1 On some of the heath lands in the centre and west of the county [Montgomeryshire] a diminutive breed of sheep called cluns is pastured. 1902 *Ibid.* XXV. 194/1 The *Clun Forest* is a local breed in West Shropshire and the adjacent parts of Wales. 1961 *Shropsh. Mag.* Nov. 7 Weston Farm carries over 130 sheep—Cluns and Clun crosses. 1964 'E. Peters' *Flight of Witch* i. 12 These handsome, fearless hill-sheep he was just learning to know for Cluns and Kerrys.

clunk, *v.* **1.** Add: (Further examples.) Also *colloq.* Hence **clu·nking** *vbl. sb.*
a 1796 Burns *Jolly Beggars*, And made the bottle clunk To their health that night. 1834 'C. North' in *Blackw. Mag.* July 138 Canna ye sook that back without your jaw-banes clunkin? 1960 'M. Cronin' *Begin with Gun* ix. 106 His face clunked against the wall. 1960 'N. Shute' *Trustee fr. Toolroom* 163 The rudder in its pintles made a clunking sound.
3. *trans.* To hit or strike (so as to produce a 'clunk'). Cf. *clonk *v.* 2. *colloq.*

1943 I. Wolfert *Tucker's People* (1944) xiii. 202 Sit down or I'll clunk your other ear. 1952 B. Ulanov *Hist. Jazz* (1958) vii. 73 With..Bill Johnson clunking the banjo.

Cluny (klū·ni). The name of a town in the Saône-et-Loire department of France, used to designate a kind of net or bobbin lace.
1872 *Daily Tel.* 11 Mar. 3/3 He admires mauve silk and Cluny lace. 1951 *Good Housek. Home Encycl.* 169/1 Examples of Cluny lace and Cluny lace insertion.

cluricaune (klū·rəkǭn). Also **cluricane, cluricaun,** etc. [ad. Irish *clúracán, clutharacán*.] In Irish mythology, an elf having the appearance of a tiny old man; = Leprechaun.
1825 T. C. Croker *Fairy Leg. S. Ireland* 162 The Cluricaune of the southern counties of Ireland appears to be the same as the Leprechan of Leinster; and both words are probably provincialisms of Luacharma'n, the Irish for a pygmy. 1889 W. B. Yeats *Lett. New Island* (1934) 195 In Monaghan he [*sc.* the leprechaun] is called the cluricaune. 1907 Masefield *Tarpaulin Muster* xiv. 144 They called him the Cluricaun, he says, and he would do all the work of the house. 1910 P. W. Joyce *Eng. in Ireland* 284 Leprachaun: a sort of fairy, called by several names in different parts of Ireland:—luricaun, cluricaun, lurragadaun, loghryman, luprachaun. 1944 *Jrnl. Cork Hist. & Arch. Soc.* XLIX. 43 *Kluracaun*, klū'-ra-kawn, a tiny sprite found hiding among stones of a dry-stone fence; an undeveloped boy of mischievous disposition; a male of uncommonly low stature who fancies himself as a trickster or practical joker; any mischief-maker (sarcastic). 1959 D. A. MacManus *Middle Kingdom* ii. 48 Thus Ireland has her cluricauns, the small sprites who sit by the hob and look after, and even at times tidy, the kitchen by night.

cluster, *sb.* Add: **3. c.** A group of faint stars forming a relatively dense mass, appearing as a nebula to the naked eye; a star-cluster; also applied to larger coherent groups of stars, nebulæ, and galaxies. Also *attrib.*
1727–38, etc. [see Nebula 3 a]. 1854 [in Dict., sense 3 a]. 1859 Herschel *Outl. Astron.* xvii. 634 In the sword-handle of Perseus, also, is another such spot, crowded with stars, which requires rather a better telescope to resolve into individuals separated from each other. These are called clusters of stars. *Ibid.* 637 Nebulæ..have been separated by him [*sc.* Sir W. Herschel] into—1st. Clusters of stars, in which the stars are clearly distinguishable; and these, again into globular and irregular clusters. 1902 S. I. Bailey in *Ann. Astron. Observ. Harvard Coll.* XXXVIII. 233 In other clusters..Subclass a appears to be the prevailing type, so much so that the variable stars of this subclass may perhaps be regarded as the typical cluster variables. The apparently perfect uniformity of the period, of the stars of this subclass, indicates that it is associated with some regular returning phenomenon. 1928 Jeans *Astron. & Cosmog.* 26 It is not altogether clear to what extent the globular clusters and the moving clusters form distinct formations. 1930 H. Shapley *Star Clusters* 7 The term 'galactic cluster', suggested by Trumpler.., is a natural name for the non-globular cluster, which is almost without exception near the galactic plane. It replaces the term 'open cluster', which has caused some confusion..The so-called moving clusters are merely the brighter and nearer of the galactic types in which radial or transverse motions have been measured.
d. *Linguistics.* A group of successive consonants.
1933 Bloomfield *Lang.* viii. 131 The simplest way to describe the phonetic structure of a language is to state which non-syllabic phonemes or groups of non-syllabic phonemes (clusters) appear in the three possible positions. 1951 S. Ullmann *Princ. Semantics* 48 The highly systematised rules governing the number and nature of initial clusters permissible in any given language (cf. the treatment of initial *ps-* in '*psychology, pseudo-*') are among the direct phonemic criteria evolved to this end. 1952 A. Cohen *Phonemes of Eng.* 69 By medial clusters we understand groups of consonant phonemes that occur between two vowels within the word.
4. cluster bomb, an anti-personnel bomb containing numerous metal pellets which spray out over a wide range on impact; hence as *v. trans.*; **cluster-cherry** (earlier example); **cluster-pin, -ring,** a breast-pin or ring set with a cluster of stones; **cluster-pore,** each of a number of small pores or orifices forming part of the system of mucous canals opening on the surface in *Chimæra*.
1967 *Guardian* 8 May 9/4 'Cluster bombs' which, on impact, spray bullets around. 1968 *N.Y. Rev. Books* 4 Jan. 4/1 The deadliest weapon of all, at least against personnel, were CBUs—cluster bomb units. *Ibid.* 5/2 It should prove interesting to hear..the defense of cluster-bombing a row of houses in the hope of finding a suspect. 1731 P. Miller *Gard. Dict.* s.v. *Cerasus*, The Flanders Cluster-Cherry. 1873 'Mark Twain' & C. D. Warner *Gilded Age* xxxiii. 301 He wore a diamond cluster-pin and he parted his hair behind. 1895 *Proc. Zool. Soc. London* 880 There is also an absence of primitive pores, and the cluster-pores are very few in number. 1897 'Mark Twain' *Following Equator* xxxviii. 347 Showy cluster-rings on her toes. a 1910 'O. Henry' *Trimmed Lamp* (1916) 170 She wore a cluster ring of huge imitation rubies. 1951 *Catal. Exhibits, Festival of Britain* 39/1 Cluster rings, sapphire and diamond.

clustering, *vbl. sb.* Add: (Further examples.) Also = Cluster *sb.* 3 a and *c.
1956 H. Spencer Jones in A. Pryce-Jones *New Outl. Mod. Knowl.* 131 Galaxies..are fairly uniformly distributed through space, apart from some localized clusterings. 1959 *Listener* 3 Sept. 352/2 One would expect some variation with distance in the appearance of the galaxy..or in the degree of clustering.

clutch, *sb.*¹ Add: **6. a.** *spec.* In motor vehicles, a piece of mechanism by which power from the engine is transmitted to or disconnected from the gears; also, the foot-pedal operating this mechanism. Also *attrib.* and *Comb.*
1899 *Motor-Car World* Oct. 10/1 The advantage of transmission by gearing is its positiveness, while its disadvantages are noise, cost of renewal when worn, and the necessity of using clutches. 1902 A. C. Harmsworth *Motors & Motor-Driving* x. 191 The forward end of the arbor shaft is connected to the engine shaft by a clutch. *Ibid.* 208 The motor and the main clutch shaft must be truly in line. 1904 A. B. F. Young *Complete Motorist* ix. 216 The car will now travel forward so long as the clutch pedal is not depressed. 1912 *Motor Manual* v. 162 To ensure a gradual 'take up' of the drive from the engine, the clutch should be let in very gently. 1934 S. Beckett *More Pricks than Kicks* 127 Belacqua got in a gear at last..after much clutch-burning. 1936 *Economist* 18 Jan. 150/1 The manufacture of brake and clutch linings. 1962 *Which? Car Suppl.* Oct. 138/1 Two clutch cover bolts [were] missing. *Ibid.* 137/2 The VW 1500 was also in excellent condition..apart from clutch judder. *Ibid.* 116/1 An occasional slight trace of clutch slip.

clutch, *sb.*² Add: **2.** *transf.* A group of people.
1908 W. B. Yeats & Lady Gregory *Unicorn from Stars* ii. 73 That clutch of chattering women. 1945 A. L. Rowse *Poems of Decade* 79 A clutch of leering women will be there. 1959 *Wall St. Jrnl.* 9 Feb. 12/6 There are scenes somewhat resembling those of the Three-penny Opera with its clutch of tarts and cutpurses.

clutchy (klʌ·tʃi), *a.* [f. Clutch *v.*¹ + -y¹.] Inclined to clutch.
1876 A. S. Palmer *Leaves from Word-hunter's Notebk.* x. 246 Upon sins the devil is 'ready-to-take-hold', or, if we might forge a term for the occasion, is 'clutchy'. 1922 A. S. M. Hutchinson *This Freedom* ii. iii. 96 Grotesque ideas..masterfully shaping the child mind wherein they germinated; burrowing in clutchy roots.

clutter, *sb.* Add: **2. a.** (Later examples.)
1891 *Harper's Mag.* Nov. 881/2 A tiny clutter of frame houses and tents. 1958 *Times* 4 July 11/4 The home of Everyman will be free from a clutter of sporting equipment and reading matter, tools and toys.
b. Delete 'Now *dial.* and *U.S.*' and add modern examples.
1959 *Daily Tel.* 10 Mar. 16/4 Mr. Brooke..said in London yesterday that Britain had 'got to get rid of the clutter' of advertisements on shops and elsewhere. *Ibid.*, By one method or another I intend to get rid of clutter, for I am sure it does no good to anyone.
c. Unwanted images on a radar screen.
1945 in *Army & Navy Jrnl.* (U.S.) 18 Aug. 1534. 1946 *Electronic Engin.* XVIII. 267 Sea clutter, caused by echoes from the tips of waves and broken water. 1948 *Ibid.* XX. 336 The important subject of unwanted echoes ('clutter'). 1967 *Electronics* 6 Mar. 52/2 Blind spots caused when the transmitter pulse or clutter obliterates the target return pulse.
3. esp. in phr. *in a clutter.*
1890 *Blackw. Mag.* CXLVIII. 463/2 Up spring the partridges;..all in a clutter they are, one over the other.

clutter, *v.* **6.** Delete 'Now chiefly *dial.* and *U.S.*' and add later examples. Usu. in *pass.* Freq. const. *up*.
1908 *Daily Chron.* 25 Feb. 4/4 The floor also was cluttered with papers and books of every kind. 1910 *Westm. Gaz.* 6 Apr. 2/1 The Second Chamber is now cluttered up with a brute mass of undistinguished bipeds. 1914 H. H. Fyfe *Real Mexico* 92 The lines here are all cluttered up with troop-trains. 1923 *Times Lit. Suppl.* 18 Jan. 36/2 The book is cluttered up with details of the number of troops present and tables of the exceedingly minute casualties. 1966 *Oxf. Univ. Gaz.* 23 Dec. 434/1 We do not want to clutter up Congregation with a large number of trivial details.

cluttered, *ppl. a.* **2.** Delete *U.S.* and add: Also with *up*.
1869 *Trans. Ill. Agric. Soc. 1867–8* VII. 573 The slovenly, cluttered up appearance that characterizes Western habitations. a 1887 Jefferies *Field & Hedgerow* (1889) 189 'Cluttered up' means in a litter, surrounded with too many things to do at once. 1897 Kipling *Capt. Cour.* 53 The cluttered decks of a seventy-ton schooner. 1898 —— in *Morn. Post* 8 Nov. 5/2 Cluttered-up boxes of machinery and bags of tricks. 1910 *Daily Chron.* 9 Mar. 7/3 To pick up débris from a cluttered room.

Clydesdale (kləi·dzdēⁱl), *sb.* and *a.* (Designating) a breed of heavy draught horses originally bred near the Clyde in Scotland.
1786 G. Culley *Obs. Live Stock* 18 The Scotch horses, like the Welch, are exceedingly hardy, but too small for the draught, except the Clydesdale horses. 1831 Youatt *Horse* 39 The Clydesdale is a good kind of draught-horse, and particularly for farming business and in a hilly country. *Ibid.* 40 The Clydesdale horse owes its origin to one of the Dukes of Hamilton, who crossed some of the best Lanark mares with stallions which he had brought over from Flanders. 1862 Morton *Farmer's Cal.* (ed. 2) 574 The Clydesdale is one of our best farm horses. 1884 *Times* 27 Feb. 7/6 The high-standing quick-stepping Clydesdales. 1928 [see *actioned *ppl. a.*]. 1967 *Times* 23 Nov. 13/1 Scotland resembled the native Clydesdale rather than the classic thoroughbred strain when they got up late to beat Wales by a short head at Hampden Park last night.

Clydesider (kləi·dsəidəɹ). [f. *Clyde side* (Side *sb.*¹ 7 a) + -er¹.] A dweller on a bank of the river Clyde in Scotland; *spec.* applied to that group of the Labour Party and Independent Labour Party which was associated with Glasgow and the neighbouring industrial area.
1921 *Glasgow Herald* 7 Apr. 7 There does not appear to be much to interest the Clydesider in the news that the Norwegian four-masted barque Svartskog is long overdue. 1926 *Ibid.* 2 Apr. 9 Even the versatile loquacity of half a dozen of the more combative Clydesiders failed to spin things out till the hour appointed for dispersal. 1930 *Times Lit. Suppl.* 20 Feb., In that valley the Clydesiders were born with memories of the martyrs Muir and Palmer. 1965 A. J. P. Taylor *Eng. Hist. 1914–45* vi. 198 There were combative working-class socialists of the I.L.P., particularly from Glasgow. These Clydesiders, as they were called, won twenty-one out of twenty-eight seats in their region [in 1922].

clype (kləip), *sb.* Now gen. *Sc.* in sense 'a telltale'.
1825 Jamieson Suppl., *Clype*, a telltale, Loth.; always applied to a female, Clydes. 1895 'I. Maclaren' *Auld Lang Syne* 143 The verra lassies cried 'clype' (telltale) at him gaein' hame. 1931 A. M. Williams *Bundle of Yarns* 47 The clype of the class told her that Johnnie had a preen. 1959 I. & P. Opie *Lore & Lang. Schoolchildren* x. 189 In Scotland a tale-bearer is a 'clype' or 'clypie' (hence 'Clype the clout)'.

clype (kləip), *v.* *Sc.* Also **clipe.** [Obscurely developed from OE. *cleopian* Clepe *v.*] To inform *on* (a person), to tell tales.
1843 *Second Crack aboot Kirk* 1 Wha do you think, Saunders, has been clipeing? 1894 *Daily News* 28 May 8/5 It is not the other High School boys who 'clype', as they say, of Scott's battles and scrapes. 1923 G. Blake *Mince Collop Close* ix. 256 Mrs. Sarah Rafferty had cliped on Bella to the police. 1936 *Times Lit. Suppl.* 25 Jan. 73/2 Begg, the censorious cashier, ever ready to 'clype' on his subordinates. 1963 'H. Calvin' *It's Different Abroad* viii. 109 'Roddy did something else too..' 'You shut up.' 'Well, you did!' 'Don't clype, Vanessa.'
Hence **cly·ping** *ppl. a.*, gossiping, tattling.
1722 Ramsay *Tale 3 Bonnets* i. i, When Men of Mettle thought it Nonsence To heed that cleping Thing ca'd Conscience. 1895 'H. Ochiltree' *Redburn* xiii. 129 The pechin' clypin auld limmer. 1921 *Blackw. Mag.* July 90/2 That wee clypin' body Corris boakes them up lik' vomit. 1928 J. Baxter *A' Ae' Oo'* 10 Did clypin' bow nae deck the lift, The rain micht ne'er hae been!

Cmd., Cmnd., see *command paper* s.v. *command *sb.* 10.

c'mon (k·mǫ·n). Also **c'm on.** *Ellipt.* form of 'come on' (see Come *v.* 62 f). *colloq.*
1934 J. G. Brandon *One-Minute Murder* xi. 120 C'm on an' git it over. 1946 K. Tennant *Lost Haven* (1947) x. 147 C'mon, you little beauty. Ten to one the winner!..Wild Bill, the outlaw! C'mon Bill, stick to him. 1967 J. Wainwright *Talent for Murder* 183 C'mon, Sergeant. We've work to do.

cnicnode (ni·knōᵘd). *Math.* [f. L. *cnicus*, *cnēcus* safflower = Gr. κνῆκος a plant of the thistle kind + L. *nodus* knot, node.] A conical point, a point where tangents form a cone of the second order. Also **cni·ctrope** (see Trope), the reciprocal of a cnicnode.
1869 Cayley *Math. Papers* (1893) VI. 330, I take account of conical and biplanar nodes, or, as I call them, cnicnodes, and binodes. *Ibid.* 334 Consider a surface having the cnicnode *C* = 1, and the reciprocal surface having the cnictrope *C'* = 1. 1870 *Ibid.* (1894) VII. 246 There may be on the nodal curve points which, in the classification of the surfaces, must be counted as cnicnodes. 1887 *Encycl. Brit.* XXII. 669/1 As regards isolated singularities, it will be sufficient to mention the point singularity of the conical point (or cnicnode) and the corresponding plane singularity of the conic of contact (or cnictrope).

cnida. Add: **cni·docyst,** a nematocyst; **cnido·genous** *a.*, producing or containing nematocysts; **cni·dosac,** the dilated middle part of a tentillum in certain Siphonophora.
1888 E. Haeckel in *Challenger Rep.* (*Zool.*) XXVIII. 259 Large ensiform or spindle-shaped cnidocysts. *Ibid.*, The cnidosac (sacculus or urticating knob) is ovate, spindle-shaped, or pyriform. 1898 Sedgwick *Text-bk. Zool.* I. 146 The swelling of the cnidosac is due to a rich development of nematocysts of various kinds, forming the battery. 1902 *Encycl. Brit.* XXX. 797/1 The hepatic cæca in the dorsal papillæ of Eolidoida actually communicate with the cnidogenous sacs at their extremity; cnidocysts are absent from *Eolis glaucoides*. 1940 *Chambers's Techn. Dict.* 171/1 *Cnidosac*, in *Siphonophora*, a battery of cnidoblasts borne on a tentacle, where it gives rise to a coloured swelling.

Co.³ Add: **1.** (Earlier example.)
1679 J. Verney *Let.* 14 July in M. M. Verney *Memoirs* (1899) IV. vii. 258 We pay off our debts that if the Co: be broke nobody may be sufferers but those that are of it.
b. *colloq.* in phr. *and Co.*: and the rest of that set, and things of a similar nature or appearance.
1757 [see Company *sb.* 7 b]. 1848 A. H. Clough *Let.* 4 Sept. in *N.Z. Lett. of T. Arnold* (1966) 114 She is terribly given to Maurice and Co.: and of the Co specially to Kingsley. 1896 G. B. Shaw *Let.* 16 Nov. in *Ellen Terry & Bernard Shaw* (1931) 124 How did it go? Not, like Cymbeline and Co., by dint of everybody in the theatre making believe..that they were witnessing a great work, but *really*. 1911 T. E. Lawrence *Let.* 26 Aug. (1938) 120, I hope by quinine & co. to stave off any more attacks. 1936 A. Huxley *Eyeless in Gaza* xv. 199 Almost in the shadow of such giants as the Jungfrau, Weisshorn and Co. 1959 *Listener* 24 Dec. 1121/1 What Khrushchev and Co. might do is one thing.

c. *in Co.* (*with*), in company (with). *U.S. colloq.* ?*Obs.*
1816 U. Brown *Jrnl.* in *Maryland Hist. Mag.* XI. 147 This morning..in Co: with Doctor William Lee Brook, traverses Janesville. 1838 A. Ganilh *Mexico v. Texas* 11 Two foreign physicians, the one a Frenchman, the other an American, exercised the healing art *in Co.*
3. Also **co-.** Colloq. abbrev. of *co-respondent* (in a divorce suit).
1900 E. Wells *Chestnuts* xx. 199 More divorce made easy. As usual the co-, a well-known actor, was very friendly with both husband and wife. 1902 'N. Gubbins' *Dead Certainties* 21 'We [*sc.* a firm of solicitors] have already drafted an advertisement, with the object of obtaining a "co".' 'A what, sir?'..'A co-respondent.'

co-. Add: **3. b.** *co-editor* (examples).
1873 S. J. Hall *Sk. Remarkable People* xvi. 204 A co-editor of the 'Sheffield Iris'. 1940 *Mind* XLIX. 256 M. Hans Høffding..died in 1937, and his co-editors completed the work.

coacervate, *a.* Restrict † *Obs.* to sense in Dict. and add: **b.** *Bot.* Clustered.
1845 *Florist's Jrnl.* VI. 246 *Coacervate*, growing together, clustered. 1928 B. D. Jackson *Gloss. Bot.* 81/2 *Coacervate*, clustered.

coacervate (ko͵æ·səɹvēⁱt), *sb. Chem.* [Back-formation from next.] An aggregate forming viscous drops in a colloidal solution. Also *attrib.*
1929 De Jong & Kruyt in *Proc. Konink. Akad. van Wetenschappen te Amsterdam* XXXII. 853 The coacervate is a drop or a layer, in which the particles have flocked together as bees in a swarm. 1948 Glasstone *Physical Chem.* (ed. 2) xiv. 1254 The droplets, or coacervates, are formed most readily when two hydrophilic sols carrying opposite charges..are mixed in suitable amounts. 1962 *Listener* 3 May 769/2 There must be many intermediate stages between mixtures of large molecules and micro-organisms. One of these may be the so-called 'coacervate drops'. It has been discovered that certain mixtures of large molecules.. will form drops which have great stability and which will scavenge and absorb other molecules from solution.

coacervation. Add: **3.** *Chem.* The action of forming a coacervate.
1929 De Jong & Kruyt in *Proc. Konink. Akad. van Wetenschappen te Amsterdam* XXXII. 849 (*title*) Coacervation (Partial miscibility in colloid systems). *Ibid.* 850 The word 'unmixing' having already a definite meaning...We introduce for this [*sc.* the phenomenon in colloid systems] the word *coacervation*. 1948 Glasstone *Physical Chem.* (ed. 2) xiv. 1254 *Coacervation*. Salting out of a hydrophilic sol frequently gives a liquid aggregate...This often appears in the form of viscous drops, instead of a continuous liquid phase. 1970 *Sci. Jrnl.* Feb. 62/3 Of the various 'chemical' methods of making microcapsules the best known is coacervation, a phase separation technique.

coach, *sb.* Add: **1. d.** (Earlier and later examples.) For def. read: A railway carriage: in *U.S.* used esp. of one that is not a sleeping car (see quots.).
1832 *Chambers's Edin. Jrnl.* I. 364/1 The railway coaches are more commodious than others. 1833 *Ibid.* II. 284/3 The Dalkeith shopkeeper, who has just descended from one of the incoming coaches. 1839 E. Hall *Diary* in O. A. Sherrard *Two Vict. Girls* (1966) i. 14 The door of one of the coaches [of the train] was opened and we..were ushered into it. a 1884 Knight *Dict. Mech.* Suppl., *Coach*, a first class passenger car, as distinguished from a drawing room car, or second class. 1887 C. B. George *40 Years on Rail* (1888) iv. 67 The train consisted of a baggage-car and two passenger coaches. 1909 Webster, *Coach*,..a first-class passenger car, as distinguished from a drawing-room car, sleeping car, etc.; sometimes, loosely, any passenger car. 1936 *Discovery* Nov. 356/1 The first two coaches are the power cars, each containing a 1,200 b.h.p. engine. 1948 *Chicago Daily News* 17 Aug. 1/3 Three coaches, three sleepers and a diner left the tracks.
e. A single-decker bus.
1923 *Michelin Guide to Gt. Britain* (ed. 7) p. ii, Some owners of motor coaches. *Ibid.*, This list..excludes coaches, etc., carrying less than eighteen persons. 1930 *Autocar* 5 Sept. 459 What will happen to some of the long distance motor coach services if the forthcoming 30 m.p.h. limit is properly enforced? We followed one doing just over 50 m.p.h. on the narrow stretch south of Baldock...The coach kept well to its own side without wobbling. 1932 N. Mitford *Christmas Pudding* xiv. 221 There are trains..and road coaches. 1955 *Times* 23 June 16/2 No passenger not in possession of such tickets will be admitted to the car park. Coaches will not be accepted for this park.
3. c. A tame bullock or horse used as a decoy in catching wild cattle or horses. *Austral.*
1873 Ranken *Australia* vi. 110 To get them [*sc.* wild cattle], a party of stockmen take a small herd of quiet cattle, 'coaches'.
6. coach-built *a.*, of a motor body, built of wood throughout by craftsmen, or on a wood framework with metal panels; also sometimes designating merely such a framework; **coach-load,** as many people (or things) as can be carried in a coach; **coach-office** (earlier and *attrib.* examples); **coach-party,** a group of persons travelling by motor coach; **coach screw,** a large screw with a square head, which is turned with a spanner; hence **coach-screw** *v.*; **coach-smith** (see quots.); **coach-stand,** a place where (horse-drawn) coaches wait for hire; **coach station,** formerly (in U.S.) a regular stopping-place for a stage-coach; now a terminus for motor coaches; **coach tour,** a tour undertaken by motor coach; **coach trade,** parties of trippers on an outing by motor coach, often *derog.*; **coach trimmer** (examples); **coach-work,** the bodywork of a motor-vehicle or railway coach; the upper part of the body of a motor vehicle.
1904 *Daily Chron.* 19 Nov. 6/6 The new car, with a coach-built seat at the rear in place of a tool-box. 1928 *Daily Express* 13 June 3 The coachbuilt bodies are beautifully finished. 1932 'N. Shute' *Lonely Road* ii. 23 Your head must have gone very nearly through the roof...You must consider yourself very fortunate that it was not a coach-built body. 1857 Dickens *Dorrit* ii. xix. 482 Their coach-load of luxuries. 1908 M. & J. Findlater *Crossriggs* viii. 56 In old coaching days this was one of the highways between England and Scotland, and many a merry coach-load had passed along it. 1954 *Ann. Reg. 1953* 256 The ..authorities had to turn back many coach-loads of Italians. 1794 W. B. Stevens *Jrnl.* 4 May (1965) II. 155 She had even sent me to alter my day at the Coach Office. 1834 Dickens *Let.* 3 Sept. (1965) I. 41 There is a coach-office carrier delivering *a hamper* over the way. 1957 R. Hoggart *Uses Literacy* v. 122 The 'charas' go rolling..past the road-houses which turn up their noses at coach-parties. 1874 Thearle *Naval Archit.* v. 60 Very frequently coach screws are placed between the bolts. *Ibid.* iii. 48 Angle-irons being fitted in the corners, riveted to the beams, and coach screwed to the carlings. 1930 *Engineering* 20 June 787/3 Drills and appliances for fixing coach screws in sleepers. 1837 N. Whittock *Bk. Trades* 152 The springs are made by the Coach-smith, whose business is quite separate from that of the Coach-maker. 1846 G. Dodd *Brit. Manuf.* VI. 122 A coach-smith..has to work large and heavy bars of iron into forms containing several unequal curves, and varying greatly in thickness. 1921 *Dict. Occup. Terms* (1927) § 190 *Coach-smith*;..a smith who forges by hand or under power hammer..iron work used in building railway or tramway coaches, motor vans, carts, etc. 1834 *Chambers's Edin. Jrnl.* III. 268/3 Is not that a rare specimen of coach-stand oratory? 1839 Dickens *Nickleby* xxxii. 315 He turned down a bye-street in search of the nearest coach-stand. 1881 *Rocky Mt. Sk.* in *Sunday at Home* Sept. 567/2 A coach station some fifteen miles over the mountains. 1947 G. Greene *19 Stories* 211 Standing in the great steel coach station. 1958 *Observer* 10 Aug. 10/6 The man conducting a coach-tour through unfamiliar landscapes for a crowd of strangers. 1966 D. Francis *Flying Finish* iii. 36 A coach tour type holiday to Europe. 1962 *Sunday Express* 7 Jan. 19/2 His farces are seen solely by the 'coach trade' from the provinces. 1964 *Listener* 16 Apr. 612/1 Why should a Feydeau farce be acceptable and an Aldwych farce not? Because the one is in French and the other frequented by the coach trade? 1840 *Picayune* (New Orleans) 28 July 1/2 Materials and tools for saddlers and coach trimmers. 1909 *Daily Chron.* 25 Sept. 7/6 (Advt.), Coach Trimmer wanted. 1906 *Daily Chron.* 20 Oct. 2/7 Special carburettors, governors, ignition gear, gear-boxes, coachwork. 1907 *Westm. Gaz.* 11 Nov. 6/3 The best examples of high-class coach-work. 1922 *Autocar* 10 Nov. 1013 The advantage of this type of coachwork over the ordinary two-seater and dickey. a 1963 in P. Drackett *Motor Rallying* (1963) App. 83 A car with convertible coachwork, running in a saloon or closed car class, must complete all tests with the hood erected.

coach, *v.* Add: **2. a.** Also in a motor coach.
1959 *Listener* 27 Aug. 332/2 A French family, met with while coaching across France.
3. c. To decoy wild cattle or horses with tame animals. Cf. *coach *sb.* 3 c. *Austral.*
1873 Ranken *Australia* vi. 121 [The wild horse] may be got by 'coaching' like wild cattle. 1959 J. Wright *Generations of Men* 128 He spent a week with the men, coaching over the range, and this time had luck enough to get a good number [of scrubber cattle].

coachee¹. **2.** For def. read: *U.S.* A long, lightweight horse-carriage. *Obs.* exc. *Hist.* (Earlier and later examples.)
1790 *Pennsylvania Packet* (Philadelphia) 9 Oct. 3/1 They have for Hire..A Coach, Coachee, and several Chaises, all in good order. 1819 *Rees' Cycl.* s.v. *Philadelphia*, The coachee is a carriage thought to be peculiar to America. 1832 *Act Penn.* in *U.S. Stat.* (1850) IV. 554 [Toll] for every chariot, coach, coachee, stage, wagon,..with two horses and four wheels, twelve cents. 1948 J. D. Rittenhouse *Amer. Horse-Drawn Vehicles* 37 Coachee. This example..had C-springs made of hickory.

coacher, *sb.* Add: **4.** = *coach *sb.* 3 c. *Austral.* and *N.Z.*
1904 H. Buckland in M. Cradock *Sport in N.Z.* ii. vi. 174 A storekeeper..turned out on the ground twenty milkers which were intended to act as 'coachers'. 1904 'G. B. Lancaster' *Sons o' Men* 20 Harry Morell and five more held the coachers beyond the low hill. 1932 I. Idriess *Men of Jungle* (1933) viii. 84 Some time ago, the Pierces had driven 'coachers' out on to this particular feeding-ground and left them there. Coachers are cattle trained to help trap their wild brethren. 1955 J. Cleary *Justin Bayard* 270 Five or six coachers, docile bullocks that acted as decoys, were already headed towards the open gate. 1959 J. Wright *Generations of Men* ix. 115 He ran in a mob of forty quiet horses, to be used as coachers.

coachman. Add: **1. d.** An Australian bird, *Pachycephala rufiventris,* also called the *coach-whip bird*; cf. *flying coachman*.
[1822 *Jrnl. Excursion Blue Mountains* 10 Oct. in B. Field *Geogr. Mem. N.S. Wales* (1825) 440 Some [notes] are harsh and vulgar, like those of the parrot-kind, the cockatoo, the coachman's whip-bird.] 1827 P. Cunningham *N.S.W.* II. 158 Our native *coachman*..whistling and cracking out his whiplike notes as he hops sprucely from branch to branch. 1888 Jas. Thomas in *Austral. Poets 1788–1888* 552 While the crested coachman bird Midst the underwood is heard. 1918 J. A. Leach *Austral. Bird Bk.* (ed. 4) 151 Rufous-breasted Whistler (Thickhead), Little Thrush, Ring Coachman, Coachwhip-Bird.

coachmanship. Add: **1.** Also *transf.*
1776 G. Steevens *Let.* Jan. in *Corr. Garrick* (1832) II. 122 Your Antony and Cleopatra was a splendid performance; but..it afforded you few opportunities of showing..that coachmanship in which you excel all others. 1802 Dyott *Diary* 8 June (1907) I. 215 The ship..ran immediately in for the quay, which required good *coachmanship*, as it was a very narrow passage and blowing hard.
2. Coaching, instructing; ability or skill in this. Cf. Coach *v.* 3.
1873 Trollope *Eust. Diamonds* II. xxxviii. 156 For devotion in coachmanship..I defy the world.

coach-whip. Add: **1. b.** In full *coach-whip bird*. One of two Australian birds, *Psophodes olivaceus* or *Pachycephala rufiventris*.
1793 Tench *Port Jackson* 175 To one [of the smaller birds], not bigger than a tom-tit, we have given the name of coach-whip, from its note exactly resembling the smack of a whip. 1827 *Trans. Linn. Soc.* XV. 330 The loud cracking whip-like noise it makes (from whence the colonists give it the name of Coach-whip) may be heard from a great distance. 1848 Haygarth *Bush Life in Australia* i. 7 The 'coach-whip', with his peculiar jerking cry. 1918 [see *coachman 1 d]. 1936 M. Franklin *All that Swagger* xxxiii. 310 There were..honey-eaters, coach-whips, wagtails, tits and robins.
c. *ellipt.* Any of the various coach-whip snakes, esp. *Masticophis flagellum*. *U.S.*
1827 J. L. Williams *View W. Florida* 29 The coach-whip is most frequently seen in the pine barrens; he resembles a coach-whip, with a black handle; but is very innocent. 1835 W. G. Simms *Partisan* 255 The skin of a monstrous coachwhip which..he had been successful enough..to kill without bruising. 1887 *Sci. Amer.* LVII. 7 A 'coach-whip', a snake much like the common black snakes in form, but in color a very dark brown some two thirds of its length. 1943 *Clarke Co. Democrat* (Grove Hill, Ala.) 12 Aug. 2/2 He was a coach-whip, which measured 72 inches in length.

coachwood (kōᵘ·tʃwud). [f. Coach *sb.* + Wood *sb.*¹] Either of two Australian trees, *Ceratopetalum apetalum*, yielding an aromatic close-grained wood suitable for cabinet making, or *Schizomeria ovata*; also, the wood of these trees.
1884 A. Nilson *Timber Trees N.S.W.* 43 *C. apetalum.*—*Coachwood*..extensively used for coach-making. 1886 *Encycl. Brit.* XX. 174/1 Names are found oddly given by colonists. Their red cedar is the *Cedrela Goona*;..coachwood, the *Ceratopetalum*. 1963 N. K. Wallis *Austral. Timber Handbk.* (ed. 2) xiii. 155 Coachwood (*Ceratopetalum apetalum*) is an important scrubwood growing in the coastal areas of New South Wales.

coact, *v.* **5.** (Later U.S. example.)
a 1877 Knight *Dict. Mech.* I. 700/1 A device consisting of two parts which coäct to give to the piece swaged between them the desired form.

coaction. Add: **4. b.** *Ecology.* The interaction of two or more individuals or species. Hence **coa·ctee, coa·ctor** *sbs.*
1927 F. E. Clements in *Carnegie Inst. Washington Year Bk.* XXVI. 323 Coaction [is] the effect manifested by plants or animals upon each other directly. 1939 Clements & Shelford *Bio-Ecol.* iv. 104 The initiating or directing organism..the *coactor* and the receiving one..the *coactee*. 1949 W. C. Allee et al. *Princ. Animal Ecol.* xxii. 349/1 The coaction aspect of the population problem.

coa·gulant, *a.* [f. the sb.] That causes to congeal or coagulate.
1937 *Times* 22 July 9/1 The local application of coagulant snake venoms was the only new treatment whose value had been confirmed.

coagulase (ko͵æ·giulēⁱs, -z). *Biochem.* [f. Coagul(ate *v.* + *-ase.] Any of several enzymes which bring about coagulation, esp. of the blood. Also in *Comb.*
1914 Stedman *Med. Dict.* (ed. 3) 194/2 *Coagulase*, a clotting enzyme, one which causes coagulation; thrombin and rennin are examples. 1929 *Chem. Abstr.* XXIII. 1172 Plasma coagulase and the pathogenicity of staphylococci. 1961 *Lancet* 22 July 173/1 Coagulase-positive strains. *Ibid.* 5 Aug. 296/2 Less virulent organisms (proteus, coagulase-negative *Staph. aureus*). *Ibid.* 23 Sept. 669/2 Henceforth the terms staphylococcus and *Staph. aureus* refer to coagulase-producing organisms.

coagulation. Add: **4.** *attrib.*: **coagulation necrosis** [ad. G. *coagulationsnecrose* (J. Cohnheim *Vorlesungen über allg. Path.* (1877) I. 453, 473)], a type of necrosis in which dead tissue becomes swollen and firmer and the cells retain their general structure as a result of the coagulation of cytoplasmic proteins; also called *coagulative necrosis*; **coagulation time,** the time taken for blood to coagulate after being taken from the body, or for blood plasma to coagulate after a coagulant has been added to counteract an anti-coagulant; cf. *clotting time*.
1883 J. Coats *Man. Path.* 104 Coagulation necrosis is a term which has recently come into use. 1886 *London Med. Rec.* XIV. 323/1 In the change called coagulative necrosis,

the death of the cells is easily proved by their incapacity for function. 1954 E. P. Abraham in H. Florey *Lect. Gen. Path.* viii. 165 True coagulation necrosis is produced by poisons such as phenol..or mercuric chloride, which denature and coagulate the proteins of the cell. 1957 S. L. Robbins *Textbk. Path.* viii. 191/1 The characteristic cytologic change of all infarcts is ischemic coagulative necrosis of the affected cells. 1893 *Jrnl. Path. & Bacteriol.* I. 444 Coagulation time decreases up to a certain minimum (generally to less than a minute), i.e. coagulability increases, by increasing additions of calcium chloride to decalcified blood. 1968 Dacie & Lewis *Pract. Haematol.* (ed. 4) x. 268 In haemophilia the whole-blood coagulation time will be normal in about one-third of patients.

coagulin (koˌæ·giulin). *Biol.* [f. Coagul(ate *v.*: see -in¹.] A name for any of various real or postulated substances that are produced in the body and accelerate the coagulation of foreign proteins or of blood; *esp.* a thrombo-plastin or a precipitin.

[1899 J. Bordet in *Ann. Institut Pasteur* XIII. 240 Si nous sommes autorisés à donner désormais à la substance active du sérum, non plus le nom d'agglutinine, qui se borne à indiquer le fait..mais l'appellation plus suggestive de 'coaguline',—nous pouvons supposer que les organismes..seront à même d'élaborer..des principes agglomérants actifs.] 1903 *Med. News* LXXXIII. 212/1 (*title*) On the presence of specific coagulins in the tissues of vertebrates and evertebrates. 1907 E. H. Starling *Elem. Human Physiol.* (ed. 8) iii. 87 Almost any protein, injected into an animal's veins, evokes the production in the animal of an antibody, a coagulin, which has the property of inducing a precipitate in solutions of the protein. 1911 J. A. Mandel tr. *Hammarsten's Textbk. Physiol. Chem.* (ed. 6) vi. 307 The tissues contain..coagulins, which are not identical with the coagulins of the clot or the blood serum. 1917 A. P. Mathews *Physiol. Chem.* (ed. 2) xii. 516 The substances in tissue extracts thus accelerating clotting are sometimes called thromboplastic substances, thrombokinase, tissue fibrinogens, or coagulins. 1929 *Encycl. Brit.* III. 898/1 [Jules Bordet] has thrown light on the process of the formation of coagulin, showing that it includes two elements; the first, albuminoidal and peculiar to the liquid blood, and the other, lipoidal in nature, originating in the cells of the blood or tissues. 1937 *Proc. Sect. Sci. K. Akad. Wetensch. Amsterdam* XL. 221 The secretion of the stomach of *Th. annulata* contains a strong coagulin. When mixed with blood it has a marked accelerating action on coagulation. 1963 I. F. & W. D. Henderson *Dict. Biol. Terms* (ed. 8) 108/1 *Coagulin*, any agent capable of coagulating albuminous substances.

coagulometer (koˌægiulǫ·mitəɹ). [f. *coagul-* in Coagulate, Coagulum, etc. + -ometer.] An instrument for examining the coagulability or measuring the coagulation time of blood.

1900 in Dorland *Med. Dict.* 1908 *Jrnl. Chem. Soc.* XCIV. ii. 766 A modification of Buckmaster's coagulometer is described. 1910 *Practitioner* Apr. 428 The coagulation time of the blood, which was taken on several occasions by Wright's coagulometer. 1924 *Q. Jrnl. Med.* XVII. 318 A new type of coagulometer is described, which is simple and accurate and particularly suitable for clinical investigation.

coal, *sb.* Add: **15. a.** *coal-basket, -bin, country, -glove;* **b.** *coal-carrying, -cutter* (person), *-fed* adj.

1710 C. Verney *Let.* (1930) I. xi. 191 The things we want is [*sic*] a Stove to burn Coals in the Little Parlor and a Coal Basket. 1947 M. Lowry *Let.* 13 Aug. (1967) 151 The dying Antony..being hauled up a clock tower in a coal-basket by Cleopatra. 1864 Webster, Coal-bin. 1870 L. M. Alcott *Old-Fash. Girl* xii. 194 Only my best cuffs and collar; you'll probably find them in the coal-bin. 1959 N. Mailer *Advts. for Myself* (1961) 43 The people used to have their coalbins under the sidewalk. 1892 *Daily News* 8 June 2/3 A most important coal-carrying line. 1905 *Daily Chron.* 6 Jan. 5/5 The Berrington has been engaged in coal-carrying between the Tyne and the Continent for nearly forty years. 1776 Pennant *Tour Scotl.* II. 203 On this coast, and..in most of the coal countries of North-Britain. 1847 E. Brontë *Wuthering Heights* I. viii. 154 A bleak, hilly, coal country. 1898 *Daily News* 14 Apr. 6/5 How are we to estimate what the average coal-cutter earns? 1890 *Pall Mall Gaz.* 2 June 3/3 The coal-fed furnace. 1899 Mrs. H. Fraser *Diplom. Wife in Japan* I. 6 Our great coal-fed, screw-driven liner. 1932 D. C. Minter *Mod. Needlecraft* 246/1 *Coal Glove*. ..Draw round hand with fingers closed. 1938 D. Smith *Dear Octopus* I. 13 Take the coal-glove, dear, then you won't spoil your hands.

16. coal-ball, also, a round mass, usually of calcite or pyrite, found in or near a coal-seam, and containing fossilized plant remains; **coal-bank** *U.S.*, a bank from which coal is obtained; **coal baron** *U.S.* (see *baron 2 b); **coal-bunker,** a place for storing coal, *spec.* in a ship; **coal-car** *U.S.*, a coal-wagon; **coal-face,** (*a*) (see quot. 1883); (*b*) loosely, the pits, the mining industry; **coal-fired** *a.*, heated or driven by coal; **coal-kiln,** delete † and add later U.S. and Jamaican examples; **coal lumper,** one who loads coal into vessels; **coal-oil:** substitute for def.: in N. Amer., petroleum, or an oil refined therefrom, as paraffin; also *attrib.*; hence **coal-oil** *v. trans.*, to smear with coal-oil; **coal-picker,** a person who picks up stray lumps of coal; so **coal-picking** *vbl. sb.*; **coal-salt,** a fine salt (generally discoloured by soot) obtained from brine by surface-evaporation; **coal-shoot:** restrict *dial.* to sense in Dict. and add (*b*) = *coal-drop*; **coal-tip** [Tip *sb.*⁵ 3], an apparatus from which coal is tipped into a receptacle; **coal-washer,** a man or machine employed in washing impurities from coal; so **coal-washing; coal-yard** (examples).

1902 *Rep. Brit. Assoc. Advancem. Sci.* 811 On the Occurrence of the Nodular Concretions (Coal Balls) in the Lower Coal Measures. 1921 *Brit. Museum Return* 151 Two Coal-balls from the Coal Measures, Romsée, near Liége. 1935 *Discovery* Sept. 274/1 Carboniferous Calamites preserved in coal balls. 1805 in L. & R. H. Collins *Hist. Kentucky* (1874) I. 408 A coal bank is within three hundred yards. 1816 U. Brown *Jrnl.* in *Maryland Hist. Mag.* XI. 142 Thence on said River 4 Miles to a Coal Bank. 1837 Peck *Gaz. Illinois* (ed. 2) II. 135 Extensive coal banks exist in the county. 1886 *Harper's Mag.* June 62/2 A gentleman who wanted a coal bank opened engaged for the work a man passing along the road. 1887, 1902 Coal baron [see *baron 2 b]. 1945 *Nation* 31 Mar. 353/1 He recites his picturesque polemics against the coal barons. 1833 *Chambers's Edin. Jrnl.* II. 232/3 [A servant] whose principal business it was to attend to the coal-bunkers..for the supply of the fires. 1840 *Monthly Chron.* (Boston) July 240 The engines, boilers, and coal-bunkers occupy a space of 70 feet, the width of the vessel. *a* 1877 Knight *Dict. Mech.*, *Coal-bunker*, the closed room around the boiler and engine-room of a steam vessel for keeping the fuel. 1886 *Harper's Mag.* June 4/2 The capacity of her coal-bunkers is 630 tons. 1895 *Army & Navy Co-op. Soc. Price List* 15 Sept. 266 Coal Bunker. To hold 2 cwt., 18 by 16 by 36 in. high. 1903 *Westm. Gaz.* 30 Nov. 2/1 The saved coal-bunker space is..available for additional cargo. 1858 *Penn. Rail Road Annual Rep.* 14 The rolling stock..consisted..of..92 Four-wheeled Coal Cars. 1908 *Westm. Gaz.* 8 Jan. 10/1 At least 1,000 coal-cars were added to the rolling-stock. 1872 *Good Words for Young* 89/2 They hang their lamps above their heads, and then swinging their picks, make a ledge in the coal face. 1883 Gresley *Gloss. Coal-m.* 55 *Coal face*, the working face or wall of a stall, composed wholly of coal. 1929 *Daily Express* 7 Nov. 2/3 Two Lancashire Socialist members who recently worked at the coal face and were returned to Parliament for the first time in May. 1955 *Times* 7 July 8/7 Delegates considered that the profits from distribution 'from the coalface to the hearth' would ensure that the industry would always pay. 1909 *Daily Chron.* 17 Apr. 4/7 Baked fifty-five minutes in coal-fired oven. 1956 *Nature* 4 Feb. 204/2 The capacity of coal-fired plant must be expected to continue to rise. 1776 P. V. Fithian *Jrnl.* 17 Jan. (1934) 162 A Fog of Smoke rises from off it as from a Coal Kiln. 1847 G. N. Jones *Florida Plantat. Recs.* (1927) 318, 2 work on the mill, 3 make Coal kill. 1961 F. G. Cassidy *Jamaica Talk* v. 78 In Jamaica there are not only lime-kilns but coal-kilns (or coal-skill) for the burning of charcoal, which is called *fire-coal*. 1908 *Daily Chron.* 29 Aug. 1/6 Coal Lumpers Refuse to Work on American Hospital Ship. 1859 *Rep. Comm. Patents 1858* I. 726 This lamp..is more especially designed for burning coal oil and similar substances that are rich in carbon. 1860 T. Shaw *U.S. Pat.* 28,539 Coal-oil furnace. 1875, 1883 Coal oil [see Dict.]. 1894 *Congress. Rec.* 5 Feb. 1862/1 The colored people are tortured; they are mutilated; they are coal-oiled and burned. 1897 B. Harraden *H. Strafford* vi. 94 Empty coal-oil tins and preserved pine-apples are not very fattening, are they? 1926 J. Black *You can't Win* xv. 197 He sat in state on a coal-oil can by the fire. 1966 *Islander* (Victoria, B.C.) 27 Feb. 13/1 The editor..by the light of a coal oil lamp..set up line by line the announcement. 1934 Webster, Coal picker. 1931 *Economist* 21 Mar. 605/1 They..go coal-picking among the 'tips'. 1937 'G. Orwell' *Road to Wigan Pier* vi. 103 Now and again..the colliery companies prosecute somebody for coal-picking... The coal-pickers subscribe among themselves to pay the fines. 1880 G. Lunge *Sulphuric Acid* II. 131 The very fine 'butter salt', or 'coal salt', obtained by top heat according to Pohl's process. 1922 Joyce *Ulysses* 111 Plant him and have done with him. Like down a coalshoot. 1939 C. Isherwood *Goodbye to Berlin* 197 Take a good big jump. Or you'll fall down the coal-shoot and into the cellar. 1906 Cornford *Defenceless Islands* 54 The frame-work with the rising platform is called a coal-tip. *a* 1877 Knight *Dict. Mech.*, *Coal-washer*, a machine in which coal which has been broken and assorted is finally washed to deprive it of the dust and dirt adhering. 1921 *Dict. Occup. Terms* (1927) § 049 *Coal washer*; ..works at machine for washing coal in washery to remove impurities. 1930 *Engineering* 14 Feb. 196/1 The term coal-washing, within recent years, has come to mean the general preparation of the coal for direct or indirect utilisation. 1805 in L. & R. H. Collins *Hist. Kentucky* (1874) I. 408 There are also..a coal yard and a boat yard. 1834 *Chambers's Edin. Jrnl.* III. 305/1 He must go himself to the coal-yard.

coal-box. Add: **b.** A low-velocity German shell emitting black smoke; a 'Black Maria'. *Army slang.*

1914 *Illustr. Lond. News* 3 Oct. 489 One of the German siege-guns—nicknamed..'coal-box'. 1914 *Scotsman* 13 Nov. 7/2 All you could hear was whizz, whizz! and then a deafening bang as a coal box would burst. 1916 'Boyd Cable' *Action Front* 15 'Coal-box,' said Courtenay hurriedly. 'Come on. They're apt to drop some more about the same spot.' 1919 [see *Black Maria 2].

coaler. Add: **3.** A railway employed in transporting coal from coal-mining districts. **b.** *pl.* Stocks or shares of coal-carrying railway companies. *U.S.*

1892 *Daily News* 16 Feb. 6/6 Readings and other stocks reacted under realizations, but 'coalers' then gradually lost their prominence. 1897 *Ibid.* 21 May 7/6 The feature of to-day's stock market was the remarkable weakness of Coalers. 1909 *Daily Chron.* 17 Apr. 2/1 Coalers were again the best stocks in the Yankee market.

coalescence. 3. (Later examples in *Psychol.*)

1892 W. James *Text-bk. Psychol.* 339 Coalescence of Different Sensations into the Same 'Thing'—When two senses are impressed simultaneously we tend to identify their objects as *one thing*... In this *coalescence in a 'thing'*, one of the coalescing sensations is held to be the thing, the other sensations are taken for its more or less accidental properties, or modes of appearance. 1896 G. F. Stout *Anal. Psychol.* I. 285 Suppose the components of the one combination are *a b c*, and of the other *a b x*; *c* may be so favored from the outset that it simply displaces *x* without any feeling of discrepancy arising, and without any attention to the difference. This process I call overlapping or coalescence. *Ibid.* 287 The gradual transformation undergone by a story as it passes from one person to another is in part at least to be accounted for by coalescence.

coal-hole. 3. (U.S. examples.)

1854 B. P. Shillaber *Mrs. Partington* 56 When you look up ..avoid the coal-holes and cellar-ways that are open for your unwary feet. 1895 *N.Y. Dramatic News* 23 Nov. 4/3 Some of the dramatis personæ disappear as quickly as if they had fallen through a coal hole.

coalification (kōᵘːlifikēⁱ·ʃən). [f. Coal *sb.* + -ification.] The process by which vegetable matter in the earth becomes coal. Also *attrib.*

1911 *Proc. Amer. Philos. Soc.* L. 62 Adjacent rocks, containing plant remains, may have contributed to this coalification by means of circulating waters. 1932 *Discovery* Aug. 246/1 The changes in composition during the coalification process..involve only the elimination of oxygen. 1955 *Nature* 5 Feb. 229/1 Coals form a series ranging from peats, through the lignites to bituminous coals and anthracites... Geological evidence suggests that this series represents a continuous and progressive change; the metamorphic process is termed 'coalification'. 1956 *Ibid.* 17 Mar. 502/2 The same correlation is found in the coalification series.

Coalite (kōᵘ·ləit). Also **coalite.** [f. Coal *sb.* + -ite¹.] The proprietary name of a type of smokeless fuel produced by heating, and so refining, bituminous coal.

1906 *Trade Marks Jrnl.* 10 Oct. 1424 Coalite..Coal, Coke, Breeze and Partially Coked Coal. Thomas Parker, 1 B Chapel Street, Edgware Road, London, N.W. 1907 *Daily Chron.* 18 June 6/3 'Coalite' is prepared from coal by extracting all the smoke-producing elements. 1934 *Times* 5 Jan. 14/3 The carbonization of this total of 600,000 tons of coal a year will yield about 420,000 tons of coalite, the smokeless fuel which is produced by the process. 1955 *Times* 8 July 15/3 When a smokeless zone is declared, only certain specified fuels may be burned within it. Of these 'Coalite' is the most popular, for it can be used with equal efficiency in every type of solid fuel appliance.

coalitionism (kōᵘăli·ʃəniz'm). [See under Coalition.] The principles or advocacy of government by coalition. So **coali·tionalism, coali·tionalist** *attrib.* or as *adj.*

1922 J. M. Erskine in *Daily Mail* 6 Nov. 10 Untainted by the poison of Coalitionism. 1925 *Contemp. Rev.* May 648 A political crisis in which the whole philosophy of coalitionalism was killed. *Ibid.*, His late Coalitionalist Conservative colleagues. 1928 *Manch. Guardian Weekly* 17 Aug. 129/2 By coalitionism you sell half your soul to get the other half.

coal-man. 2. Delete *nonce-use* and add earlier example.

1612 in *Court & Times Jas. I* (1849) I. 200 About fifty sail ..known to be cast away, especially about Yarmouth, great numbers of fishermen and coalmen.

Coalport (kōᵘ·lpōᵊɹt). The name of a town in Shropshire, used to designate china and porcelain produced there.

1837 C. Hulbert *Hist. & Descr. Salop* p. viii, Coalport China Works. 1900 F. Litchfield *Pott. & Porc.* vii. 121 Some of the finest pieces of Coalport have, from their close resemblance to Sèvres china, been passed off for such. 1960 *Harper's Bazaar* Oct. 70/2 Early-Victorian custard cups..in Coalport white porcelain. 1961 *Connoisseur* Dec. p. lxvi/1 Early Coalport Tea Service..in blue and gilt, with very fine floral decorations.

coal-scuttle. 1. (Earlier example.)

1771 Smollett *Humph. Cl.* III. 197 The pain occasioned by the fall of the coalscuttle upon his foot.

coarse, *a.* Add: **7. a.** *coarse-minded* (earlier example), *-mouthed.*

1832 F. Trollope *Dom. Manners* II. xxv. 86 The coarse-minded custom which sends alternate groups of males and females into the room. 1898 O. Wilde *Ballad Reading Gaol* 4 Some coarse-mouthed Doctor.

c. coarse fish, any fresh-water fish except the Salmonidæ (or game-fish); so *coarse fisherman, fishing*; **coarse groove,** the type of groove used on a 78 r.p.m. gramophone record.

1895 C. J. Cornish *Wild England of To-day* 178 Autumn and winter are the proper seasons in which to take coarse fish. 1897 *Encycl. Sport* I. 16/1 The taking of spinning-baits by grayling, and by the minor coarse fish. *Ibid.* 29/2. 1898 *Daily News* 9 Mar. 5/1 The coarse-fish men who aspire to baskets of pike, perch, and roach. 1959 J. Clegg *Freshwater Life* (ed. 2) xvii. 280 To the Carp family belong the majority of the so-called 'coarse fish'. 1959 *Times* 7 Feb. 9/3, I do not suggest that coarse fishermen should 'lay on' with small roach to catch bigger ones. 1966 *New Statesman* 11 Nov. 694/3, I like the small scale and puny ambition of coarse fishing. 1958 *Chambers's Techn. Dict.* (ed. 2) 967/1 *Coarse groove*, in disc/sound-recording for transcription and gramophone records the technique for shellac 78 r.p.m. discs, as distinct from *fine groove* or *microgroove*, used for vinylite long-playing discs. 1958 *Times* 18 Oct. 9/5 Why should the abridged version of *The Rape of Lucretia* languish on coarse-groove records? 1962 A. Nisbett *Technique Sound Studio* 255 *Coarse-groove*... Approximate dimensions: width 6 mils, depth 2·5 mils.

coarticulate, *v.* Add: Hence **coarti·culated** *ppl. a.* (in *Phonetics*). Cf. next.

1961 L. F. Brosnahan *Sounds of Lang.* i. 8 This co-articulated chain of sounds.

coarticulation. Restrict ?*Obs.* to sense in Dict. and add: **2.** *Phonetics.* [ad. G. *koartikulation* (Menzerath and de Lacerda *Koartikulation, Steuerung und Lautabgrenzung* 1933).] (See quot. 1942.)

1942 Bloch & Trager *Outl. Ling. Analysis* ii. 29 The two [k]'s in *calm* and *crude*..are formed alike as to their primary articulation; but the concurrent activity of the other organs —what may be called the secondary articulation or *co-articulation*—is strikingly different. For the [k] in *calm*, the lips are parted rather wide..the apex is near the bottom of the mouth..the front is depressed [etc.]. 1964 C. Barber *Present-Day Eng.* iii. 47 'Dark l' (or, more portentously, 'l with velar co-articulation').

coast, *sb.* Add: **4. c.** *the Coast*, also applied to the Pacific coast of N. America; the West Coast (and thence the province of Westland) of the South Island of New Zealand; the West Coast of Africa.

1866 *Daily Morning Chron.* (Washington) 5 Feb. 1/1 Oregon is the land of pretty girls and big red apples..there are enough around to supply the whole coast. 1881 H. W. Nesfield *Chequered Career* vii. 72 That was nothing out of the way at that time 'on the coast'. 1888 Powles *Land of Pink Pearl* 280 He was then for a time 'on the coast', as it is called, a polite form of saying a man has been in the slave trade on the West Coast of Africa. 1908 J. G. Findlay *Humbugs & Homilies* 258 Who talks as tall as a Coast Cockatoo. 1926 Princess Marie Louise *Lett. from Gold Coast* 3 We English when we talk about the 'Coast' mean only the Gold Coast, whereas it actually includes Gambia, Sierra Leone and Nigeria. Those living there, however, are generally spoken of as 'Coasters'. 1930 *Publishers' Weekly* 8 Feb. 708/1, I like the Coast because it comprises California, Oregon, Washington and the contiguous States. 1936 [see *coaster 3 b]. 1939 'M. Innes' *Stop Press* iv. i. 419 The coast—that's what they call Hollywood. 1946 —— *From London Far* iii. vi. 214, I leave for the Coast six tomorrow morning... A message from Johannesburg sends a man shooting off to the shores of the Pacific. 1949 B. Dowling *Canterbury* 11 Beyond the western foot-hills lost The huge and desolate ranges of the Coast. 1958 *Times* 13 Aug. 12/4 It has been the life's work of many merchant converters in Manchester to produce speciality African prints for the people..'on the Coast'.

d. (*from*) *coast to coast*, across an entire country or continent, nationwide; used esp. with reference to the Atlantic and Pacific coasts of America. Also *attrib.*

1850 Tennyson *In Mem.* lxv, And then I know the mist is drawn A lucid veil from coast to coast. 1911 *Chicago Daily News* 23 Sept. 1/1 Aviator Robert G. Fowler, coast to coast aviator. 1919 F. Hurst *Humoresque* 35 Why, I wouldn't be surprised if by spring I wasn't tuning up again for a coast-to-coast tour. 1922 S. Lewis *Babbitt* viii. 112 Mr. Frink was known from Coast to Coast as 'Chum'. 1935 C. Day Lewis *Time to Dance* 61 A coast-to-coast hitch-up. 1960 *News Chron.* 25 Feb. 6/8 It was made into a coast-to-coast TV show in America.

12. (U.K. example.) Also *transf.*, applied to a motor-car, rocket, etc.

1908 *Westm. Gaz.* 2 Oct. 2/1 We had a clear run of five miles without pedalling... A long free-wheel of seven miles. Down, down, down we sped, a long exhilarating coast. 1940 *Illustr. London News* CXCVI. 292/3, I purposely drove it as a careful wartime motorist would, with 40 m.p.h. as my main-road speed, a sparing use of the gears, and an occasional 'coast' in neutral on really long descents. 1968 *Times* 16 Dec. 7/4 Once the coast to the moon has begun, Apollo is separated from the third stage.

13. *coast-defender, -district, -region, -survey, -trade* (example), *-trader.*

1883 *Whitaker's Alm.* 445/2 China.—3 ironclads. Two of them are double-turreted coast-defenders of the first class. 1907 H. M. Chadwick *Orig. Eng. Nation* ix. 218 The coast-district between the two rivers. *Ibid.* 214 The coast-regions of the North Sea. 1832 *Deb. Congress* 30 May 3187 The coast survey..will not be completed..under sixty years. 1851 C. Cist *Cincinnati* 282 Flat-boats..take down more or less bacon for the coast trade. 1864 J. A. Grant *Walk across Africa* 161 Jumah, a coast-trader.

14. coast-defence, the protection of the coast of a country from attack by sea, either by forts or sea-vessels; **coast disease** *Austral.*, a disease of sheep and cattle caused by dietary deficiency of certain minerals; **coast-fever,** a fever occurring in coastal regions; *spec.* a disease, transmitted by ticks, which attacks cattle on the East Coast of Africa; = *East Coast fever* (*east D. 1 b); **coast-liner,** a surveyor whose work is to map a coast-line; so *coast-lining*; **coast-man** (later examples); **coast-partridge, -pheasant,** a Natal name for *Francolinus natalensis*; **coast sickness** *Austral.* = *coast disease; **coast-watcher, -watching** (see quots.).

1854 *Chambers's Jrnl.* 172/1 A plan for coast defence. 1869 *Ibid.* 710/1 A low-speed coast-defence vessel. 1883 *Whitaker's Alm.* 445/2 Norway.—Has only 4 coast-defence monitors. 1886 *Harper's Mag.* June 23/1 Another type of ships, which are styled 'coast-defence vessels'. 1904 *Westm. Gaz.* 28 Dec. 2/1 Our coast-defence artillery. 1957 *Encycl. Brit.* V. 888/2 Modern coast defense installations up to and including the World War II period relied on camouflage and dispersion as well as limited mobility. 1966 J. L. Moulton *Norwegian Campaign 1940* iii. 81 The senior naval officer at Narvik knew the weakness of the two 40-year-old coast defence ships. 1938 H. R. Marston (*title*) Studies on coast disease of sheep in South Australia. 1940 *Nature* 31 Aug. 308/2 Investigations of foot rot, coast disease of sheep, and cobalt- and copper-deficiency in ruminants. 1840 R. H. Dana *Bef. Mast* (1841) xxiii. 64/1 Keeping him hard at work while ill of the coast fever. 1904 *Daily Chron.* 4 July 7/7 Coast fever. 1904 [see *East Coast fever* s.v. *east D. 1 b]. 1905 *Science in S. Afr.* 339 (Pettman), They clearly recognized and described the small and characteristic organisms of African Coast Fever. 1902 *Encycl. Brit.* XXXIII. 102/2 It is with the high-water line that the coast-liner is concerned, delineating its character according to the Admiralty symbols. *Ibid.* 102/1 *Coast-lining.* In a detailed survey the coast is sketched in by walking along it, fixing by theodolite or sextant angles. 1927 *Blackw. Mag.* Apr. 516/2, I was..peacefully engaged in coast-lining with a sextant and theodolite. 1841 Poe *Tales, Desc. Maelström* (1845) 89 Among the whole of the Lofoden coastmen, we three were the only ones who made a regular business of going out to the islands. 1864 J. A. Grant *Walk across Africa* 161 They refused to allow any coast-men into their kingdom. 1906 W. L. Sclater *Birds S. Afr.* IV. 213 Natal Francolin... 'Coast Partridge' of Natal; 'Namaqua Pheasant' of Transvaal Boers. 1883 *Trans. R. Soc. S. Australia* VI. 136 Sheep are subject to a wasting and fatal disease, known as 'coast sickness', if pastured continuously on these littoral tracts. 1916 *War Illustr.* IV. 604/2 They don't call us coast-guards, mind you, but Coast Watchers. 1928 *Daily Express* 23 Mar. 1/3 The small force of coast-watchers who have taken the place of the old coastguards. 1955 *Sun* (Baltimore) 19 May (Edition B) 8/1 The area [in Guadalcanal] is so inaccessible and inhospitable, that when Australian 'coast watchers' used it as a hideout for spotting and reporting Japanese movements to the Americans, even the Japanese left the island agents alone. 1925 *Act* 15 & 16 *Geo V* c. 88 § 1 A coast-watching force for the performance of the duties hitherto performed by the coast-guard.

coast, *v.* Add: **13.** For 'in *U.S.*' read 'orig. *U.S.*' and add earlier and later examples.

1836 *Boston Pearl* 9 Jan. (Th.), Skate, if you like; 'coast', if you are boy enough. 1898 W. J. Locke *Idols* xxiii. 223 Suddenly at the top of the steep incline a bicyclist whirred past them, and coasted swiftly down. 1906 *Daily Colonist* (Victoria, B.C.) 13 Jan. 2/3 Five friends with whom he was coasting down a steep ice-coated hill were injured. 1957 A. Clarke *Later Poems* (1961) 64 Then, coasting back from Milltown Malbay, I saw before bell rang a warning, Scattery Island and its round tower.

c. To travel in a motor vehicle, rocket, etc., without thrust from the engine; also said of the vehicle.

1925 *Morris Owner's Manual* 11 Do not 'coast' down hill with the clutch out. 1949 *Spectator* 18 Nov., I switched off the engine and let the car coast down the hill. 1951 A. C. Clarke *Exploration of Space* iv. 34 The only practicable way of escaping from the Earth is to build up escape velocity as quickly as possible—and then to cut the rockets and 'coast'. 1962 *Daily Tel.* 27 Jan. 1/1 The rocket coasted for about 30 seconds, then the Agena motor fired the second stage clear from the Atlas.

d. Of a bird: to glide swiftly through the air with the wings used only for control.

1904 *Sci. Amer.* Suppl. 11 June 23778 The birds exhibiting no trace of fatigue even after 'coasting' long distances. 1935 *British Birds* XXVIII. 357 February 28th.—Wind S.S.W. Linnets coasting S.

e. *fig.* Not to exert oneself; to make progress without undue effort. Const. *in, into, to,* and in phr. *to coast home*: to achieve or complete an effortless victory, score, etc. *slang*, orig. *U.S.*

1934 M. Weseen *Dict. Amer. Slang.* xvii. 250 *Coast home*, to win easily. *Coast in*, to win easily. 1935 J. O'Hara *Appt. Samarra* v. 141 She rested her face on his shoulder, and she put up her mouth for the kiss and then she let herself coast with him. 1955 *N.Y. Times* 27 Feb. iv. 7/8 The struggle between Mr. Kennelly and his party grew out of organization charges that..he 'coasted' during the 1954 state campaign, giving scant support to its 'blue ribbon' ticket. 1957 *Times* 6 Sept. 13/1 The English team coasted comfortably to a total of 246. *Ibid.* 13/6 The champion was content to coast home to a points win. 1958 *Times* 28 Aug. 13/2 Hampshire were coasting into the lead. 1959 *Times* 27 Aug. 3/7 Well ahead on points, Erskine rather coasted the last two rounds.

14. To loaf *about*, to travel *about* as a tramp. *Austral. slang.*

1878 'R. Boldrewood' *Ups & Downs* xxv. 295, I ain't like you, Towney, able to coast about without a job of work from shearin' to shearin'. 1945 Baker *Austral. Lang.* v. 104 Expressions to describe being on the tramp..*to swag it, chase the sun, coast about.*

coastal, *a.* Add: **Coastal Command,** an R.A.F. command which first operated over the seas from coastal bases during the 1939–45 war.

1940 E. C. Shepherd *Britain's Air Power* 9 A Coastal Command machine... The more usual work of these aircraft is that of continuous patrol over all the sea approaches to Germany. *Ibid.* 12 The Coastal Command..the duties of the one are bounded by the coasts on either side of whatever piece of sea they have to patrol.

B. *sb.* A coast-defence ship.

1912 Jane *Brit. Battle Fleet* 350 'Coastals' for local duties. 1919 *Daily Tel.* 9 Aug. 10/2 Of these airships and, afterwards, of the coastals, we had several subsequently.

coastally (kōᵘ·stăli), *adv.* [f. Coastal *a.* + -ly².] Along the coast, coastwise.

1920 *Glasgow Herald* 20 Mar. 6 Arrange for certain of the ships to proceed coastally to other ports. 1953 B. Campbell *Finding Nests* v. 64 Two distinct types: inland on downs or arable land..; coastally in cliff-edge 'jungles'.

coaster. Add: **3.** *spec.* **b.** A resident of West Africa of European origin. **c.** *N.Z.* One who lives on, or comes from, the West Coast of the South Island.

1898 G. MacDonald *Gold Coast* iii. 65 To 'Old Coasters' this breeze is very welcome. 1902 *W. African Year Bk.* 28 Old Coasters generally induce perspiration by a drink of hot weak tea. 1907 J. Drummond *Life R. J. Seddon* xxi. 349 [Seddon's] good wife (who shared the love of all Coasters, with her husband). 1914 M. Hall *Woman in Antipodes* xii. 102 Another of his stories was of an old coaster. 1926 [see *coast *sb.* 4 c]. 1936 G. Greene *Basement Room* 5 Old Coasters..call all food chop. 1936 —— *Journey without Maps* i. ii. 25 There was the Coast... People said, 'Eldridge. Of course, he's an old Coaster.'..There was no other Coast but the West Coast. 1941 Baker *N.Z. Slang* v. 44 *Coaster*, for a resident of [the West Coast of the South Island]..is also well known.

d. *U.S.* A bovine animal reared near the coast of Texas. Also *coaster horn.*

1890 *Stock Grower* (Las Vegas, New Mexico, U.S.) 19 Apr. 7/2 There are lots of big old-fashioned 'coaster' horns among them. 1902 *Rep. Kansas State Board Agric. 1901–2*, 154 Horns of the old Texas coaster. 1929 J. F. Dobie *Vaquero* 20 The 'coasters', or 'sea lions', as people sometimes called the longhorned cattle of the coast country, could swim like ducks and were as wild. 1935 *Amer. Speech* X. 270/1 *Coasters*, Texas cattle raised in areas near the Gulf of Mexico.

6. Now used generally for any small, usu. round, tray or mat on which a bottle or glass, etc., may stand. (See also quot. 1890.)

1890 *Stratford on Avon Herald* 24 Oct. 6/1 Next to spoons old silver coasters are the mania of the day... They are made to do duty as stands for ink-bottles and pincushions; flowerglasses raise their heads from coasters... The homely little brown jugs of Surrey cream come upstairs set in coasters, and..the coasters offer chocolate creams. 1951 M. Sharp *Lise Lillywhite* xv. 132 Martin observed her beer-mug making a ring and automatically put it on a coaster. 1959 N. Mailer *Advts. for Myself* (1961) 96 She had picked up a coaster for a highball glass.

7. a. and **b.** (Earlier and later examples.)

1870 L. M. Alcott *Old-Fashioned Girl* iii. 34 The coasters were at it with all their might... Some little girls were playing with their sleds. 1943 M. Flavin *Journey in Dark* 17 Tom had a coaster—a long, low, sleek one, with shiny runners anchored to the frame.

c. A rest for the foot used when coasting on a bicycle.

1895 *Montgomery Ward Catal.* 555/2 *Steering fork*—Cold drawn steel tubing, with..adjustable nickeled coasters. 1897 *Outing* (U.S.) Feb. 463/2 Having gleefully perched my feet up on the coasters, I..shot forward like an arrow.

8. A loafer, sundowner. *Austral. slang.*

1878 'R. Boldrewood' *Ups & Downs* viii. 75 A voluble, good-for-nothing, loafing impostor, a regular 'coaster'. 1945 Baker *Austral. Lang.* v. 102 The scowbanker or skull-banker, coaster, sundodger..were other vagabonds who loafed in the outback or drifted from station to station.

coasting, *vbl. sb.* Add: **4.** (Earlier and later U.S. examples.)

1775 J. Eliot in *Belknap Papers* (1877) III. 77 There is a declivity..which in the winter season the boys make use of as a coasting-place. *Ibid.*, The General at first did not understand what they meant by the term *coasting*. 1832 S. G. Goodrich *Syst. Univ. Geogr.* 201 *Coasting* is another winter pastime, in which..the labor seems to be at least equal to the pleasure. 1911 E. Wharton *Ethan Frome* i. 30 The pitch of the Corbury road..was the favourite coasting-ground of Starkfield.

b. Delete '*U.S.*' and substitute 'orig. *U.S.*' (Later example.)

1896 *Punch* LXI. 185 'In the joys of the wheel,' he will often declare, 'there is nothing can come up to coasting.'

c. Also of a motor vehicle.

1921 A. F. Hall *Handbk. Yosemite Nat. Park* 308 Many drivers have the habit of coasting with their gears out of mesh. 1928 *Daily Mail* 25 July 7/5 Top gear in the Willys-Knight feels like free coasting in other cars. 1959 *Times* 25 Sept. 6/3 The car carried three people throughout, no coasting was allowed.

coastwise, *a.* (Earlier and later U.S. examples.)

1817 *Deb. Congress* (1854) XXX. 1008 This route..has appeared to be the most eligible for the 'coastwise canal'. 1913 *Boston* (Mass.) *Globe* 16 Nov. 25/2 Coastwise steamers..have their..pilot house on the upper passenger deck.

coat, *sb.* Add: **13.** *to trail one's coat*: see *trail *v.*¹ 1.

14. *coat-collar, -cuff* (example), *-rack, -room* (U.S.); **coat and skirt,** a two-piece costume; also *attrib.* (with hyphens); **coat-dress, -frock,** a tailored dress resembling a coat; **coat-facing,** material of a kind different from the cloth of the coat and used as a facing; **coat-hanger,** a piece of wood, metal, or plastic on which a coat or dress may be hung, curved so as to fit the shoulders of the garment and having a hook by which it may be suspended in a wardrobe or elsewhere; also *transf.*

1895 *Army & Navy Co-op. Soc. Price List* 1234 Costumes. ..Plain Serge Coat and Skirt. 1906 *Daily Chron.* 24 Sept. 8/1 Those useful coat-and-skirt costumes. 1908 'Ian Hay' *Right Stuff* ii. xi. 212 I'm afraid she found my clothes rather overpowering, though I'd only a coat and skirt on. 1930 *Times* 11 Sept. 13/6 A coat and skirt in brown velveteen. 1833 *Chambers's Edin. Jrnl.* II. 195/3 The back rim [of his hat] is turned up by coming in contact with his coat collar. 1860 Holland *Miss Gilbert's Career* viii. 132 A young man that..keeps the dander all off his coat-collar. 1909 'O. Henry' *Roads of Destiny* ii. 38 Mr. Robert..turned his coat collar up about his neck and ears. 1846 J. J. Hooper *Adv. Simon Suggs* (1851) iv. 42 Note..the snowy spotlessness of the linen exposed by the turn up of his coat-cuff. 1915 in C. W. Cunnington *Eng. Women's Clothing* (1952) iv. 130 The coat-dress has an assured future. 1917 *Ibid.* 138 Every woman is wearing a coat-frock. 1900 *Westm. Gaz.* 9 Sept. 3/2 Velvet is conspicuous as a coat facing in some of the newest models. 1923 *Harmsworth Househ. Encycl.* II. 956/3 *Coat frock*. This is a one-piece dress, usually made of some thick material suitable for wearing out of doors. 1936 N. Coward *Fumed Oak* in *Tonight at 8.30* 38 She wears a nondescript coat-frock. 1959 *Sunday Express* 27 Sept. 14/5 A *tailored* dress? You mean more of a coat-frock, don't you? 1895 *Montgomery Ward Catal.* 439/1 Coat Hangers. 1908 *Daily Chron.* 26 Feb. 8/5 Most women have six or seven coat-hangers in their closets. 1962 *Engineering* 2 Mar. 297 Much use has been made throughout both buildings of the precast vault with its integral 'coat hanger' beam. 1919 H. Walpole *Secret City* iii. xv. 435 No young man likes to be discovered hidden behind a coat-rack, however honest his original intentions! 1952 S. Kauffmann *Philanderer* (1953) v. 84 He glanced quickly at the coat-rack. 1870 De B. R. Keim *Sheridan's Troopers* i. 10 [He] made his appearance through the window looking into the coat-room. 1935 A. Squire *Sing Sing Doctor* xiv. 209 They pass first through a small vestibule flanked by a coat room.

coated, *ppl. a.* Add: **3.** (Later examples.) Also **coated lens** (see quot. 1948); **coated paper** (see quots.).

1878 Abney *Treat. Photogr.* 124 The rapid evaporation of the solvents from the coated plate. 1879 *St. George's Hosp. Rep.* IX. 341 The tongue was white, coated, and moist. 1892 *Daily News* 13 Apr. 2/8 Coated pills. 1902 *Encycl. Brit.* XXXII. 17/1 'Coated' paper..is paper which, after manufacture, is passed through a bath of a preparation of china clay. 1930 *Engineering* 3 Oct. 435/2 Very satisfactory welds have been secured with coated-metal electrodes. 1948 A. L. M. Sowerby *Dict. Photogr.* (ed. 17) 122 *Coated lenses*, lenses in which the surfaces of the various component glasses have been coated with a thin film, usually of silica or evaporated fluorite, to eliminate reflections... The coated lens gives more overall contrast. 1958 M. L. Hall et al. *Newnes Complete Amat. Photogr.* 77 With modern coated lenses, an average camera subject gives an image with a 50:1 range of intensities. 1961 T. Landau *Encycl. Librarianship* (ed. 2) 86/1 *Coated paper*, paper coated with china clay or similar material and calendered to secure a highly-polished surface for use in process work.

coatee. Add: **2.** Also **coatie.** A woman's short coat.

1848 *Knickerbocker* XXXII. 77 They wore short linsey-woolsey coatees. 1896 *Daily News* 4 July 6/3 These smart little coatees. 1898 *Westm. Gaz.* 12 May 3/2 The day costume..is the black silk coatie and black and white checked skirt. 1906 *Ibid.* 19 Sept. 16/1 A pretty little design..for a fur coatie with a cloth skirt. 1936 *Punch* 8 Jan. 54/3 Velvet gowns for 'formal' wear, 'clever' coatees for bridge. 1968 J. Ironside *Fashion Alphabet* 36 *Coatee*, any short jacket, reaching to the waist, sleeveless or with sleeves.

coating, *vbl. sb.* Add: **1.** *spec.* in *Photogr.* and *Cinemat.*

1892 W. E. Woodbury *Encycl. Photogr.* 124 *Coating*, lit. the act of covering with a coat. The term is largely used in photography to define the act of covering glass, paper, etc., with a film of any kind. 1953 Amos & Birkinshaw *Television Engin.* i. ix. 190 Great care should be taken in cleaning these lenses because the coating is extremely thin. 1957 T. L. J. Bentley *Man. Miniat. Camera* (ed. 5) v. 59 Within the last few years it has become the practice to give all high-quality lenses a coating or 'blooming' treatment in manufacture. 1962 A. Nisbett *Technique Sound Studio* 244 *Coating* (Emulsion), the layer of finely divided iron oxide, bonded in plastic and polished to allow smooth flow over the tape heads, that carries the magnetically recorded signal.

coatless, *a.* **2.** (Earlier and later examples.)

1823 E. Moor *Suffolk Words* 53 The punishment of a school-boy for telling tales..is performed by prostrating the coatless culprit on his back. 1967 P. Jones *Fifth Defector* v. 45 It was cold in the garage and she was coatless.

coat-tail. Add: *to climb on, hang* (*on*) *to, ride* (*on*), etc., (a person's) *coat-tails*, to attach oneself to another, usually thereby gaining some undeserved benefit. orig. *U.S.*

[1848 Lincoln in *Congr. Globe* App. 1042 Has he no acquaintance with the ample military coat tail of General Jackson? Does he not know that his own party have run the last five Presidential races under that coat tail?] 1852 *Scott Battery* 1 Nov. 2/4 Be sure and hang fast to John's coat tail. 1909 H. G. Wells *Tono-Bungay* i. 5, I was his nephew, his peculiar and intimate nephew. I was hanging on to his coat-tails all the way through. 1929 'E. Queen' *Roman Hat Myst.* iii. xvii. 256 'Know anything about him?' 'Nothing except that he hung on to Field's coat-tails pretty closely. He's an ex-jailbird.' 1949 *Citizen-Jrnl.* (Columbus, Ohio) 16 Dec., That straight-ticket voting has enabled many a mediocre candidate to ride into office on the coat-tails of an able, popular man at the top. 1953 *Manch. Guardian Weekly* 3 Dec. 4 The people whose coat-tails he is riding. 1962 *Sat. Night* 20 Jan. 19/1 And as soon as the 71-year-old Kikuyu leader was released last summer, Mboya climbed on his coattails. 1964 *Economist* 31 Oct. 482/2 Mr. Robert Kennedy cannot be sure of riding the coat-tails of Mr Johnson in New York. 1966 *Listener* 17 Mar. 375/1 May they not have tied themselves to the coat tails of a dangerous American policy in relation to China?

coa·t-tailed, *a.* [f. Coat-tail + -ed².] Having coat-tails.

1902 *Westm. Gaz.* 20 Feb. 3/1 The bolero and blouse coat with their coat-tailed and basqued variations. *Ibid.* 25 Sept. 3/2 A coat-tailed bolero.

coa·t-trailing, *vbl. sb.* [f. Coat *sb.*; cf. *trail *v.*¹ 1.] Provocation; deliberately provocative conduct. So **coat-trailer,** one who acts thus.

1927 *Observer* 19 June 15/7 There is no touch of coat-trailing..about the claim of St. Peter's, York, to be 'the fourth oldest of our public schools'. 1942 V. S. Pritchett in *Penguin New Writing* XV. 21 The others sorted themselves out into types: the good fellow, the wind-bag, the coat-trailer. 1963 *Ann. Reg. 1962* 452 There were moments of farce, exhibitionism, self-advertisement, coat-trailing, and invective.

coa·t-trailing, *ppl. a.* [see prec.] Provoking.

1959 *Listener* 5 Mar. 428/3 One of those coat-trailing,..off-the-cuff interviews this programme specializes in. 1962 *Times* 17 Jan. 13/7 Mr. Muggeridge's commentary, as well as being less deliberately coat-trailing than sometimes hitherto.

co-au·thor, *v.* [f. the sb.] *trans.* To be the co-author of (a book, etc.). Hence **co-authored** *ppl. a.*

1948 G. Marx *Let.* 10 Nov. (1967) 193 Lardner, in the Star, lampooned me because I had co-authored a play that wasn't sardonic. 1960 'J. Davey' *Touch of Stagefright* iv. 36 He had written a couple of Broadway plays, co-authored with somebody else. 1964 *Language* XL. 245 The philologist Albert Thumb..in a coauthored volume with the psychologist Marbe was very much concerned with analogic changes. 1968 J. D. Watson *Double Helix* xviii. 127 Pauling..asked Delbruck..to coauthor a note to *Science*.

coax, *v.* Add: **3. d.** To urge (a thing) by gentle means.

1841 S. Warren *Ten Thou. a Year* I. vii, After coaxing up the fire, I will proceed to tell you. 1866 Meredith *Vittoria* xxiv, Angelo knelt and coaxed the fire.

coaxial, *a.* Add: **2.** Various techn. uses.

1904 A. Russell *Altern. Currents* iv. 95 The equipotential surfaces between the two cylinders are coaxial cylinders no matter how small *a* may be. 1914 *Ibid.* (ed. 2) iii. 108 (*heading*) The mutual induction of two coaxial coils. 1921 *Jrnl. Franklin Inst.* CXCI. 707 The problem of transmission over a system consisting of *n* coaxial cylindrical conductors, which may be either in electrical contact at their adjacent surfaces or separated from each other by dielectric spaces. 1934 *Electr. Comm.* XII. 161/2 The diameters of these coaxial conductors were so chosen as to obtain for each of the two transmission paths a diameter ratio which approximates the optimum value. 1940 *Chambers's Techn. Dict.* 172/1 *Co-axial carbons*, carbons in an arc lamp so arranged as to be in the same straight line with each other. 1951 *Gloss. Terms Plastics* (*B.S.I.*) 40 *Co-axial feeder*, a type of cable for the transmission of radio frequency power consisting of concentric inner and outer conductors insulated from one another. 1958 *Chambers's Techn. Dict.* 967/1 *Coaxial filter*, one in which a section of coaxial line is fitted with re-entrant elements to provide the inductance and capacitance of a filter section. 1958 M. L. Hall et al. *Newnes Compl. Amat. Photogr.* 111 Modern cameras are fitted with flash contacts in the form of a two pin plug or co-axial socket on the shutter. 1964 *Gloss. Letterpress Rotary Print. Terms* (*B.S.I.*) 28 *Co-axial drive*, a system of press drive whereby a number of smaller powered motors drive the press units by means of suitable couplings on a common shaft.

b. *spec.* Of an airscrew or propeller: mounted on concentric shafts.

1934 *Engineering* 3 Aug. 126/2 Coaxial contrary-turning screws... The same disposition of screws had been successfully employed on an Italian flying boat. 1936 *Jrnl. R. Aeronaut. Soc.* XL. 221 Co-axial airscrews, one immediately behind the other (say on a single central engine), turning in opposite directions. 1946 *Aeroplane Spotter* 13 July 162/2 Powered by two 2,080 h.p. Rolls-Royce Merlin 140 motors, each driving two Rotol three-blade, co-axial, counter-rotating airscrews. 1949 *Gloss. Aeronaut. Terms* (*B.S.I.*) ii. 19 *Coaxial propellers*, two propellers mounted on concentric shafts having independent drives and rotating in opposite directions.

c. *coaxial cable*, a cable containing several coaxial lines and usu. also conductors to supply repeaters and other associated equipment; also, a coaxial line; *coaxial line*, a transmission line made up of two concentric circular conductors separated by an insulating medium, used esp. for medium and high frequency signals in television and multiplex telephony. Also *ellipt.* as *sb.*

1934 *Electr. Comm.* XIII. ii. 159 Frequency bands of the order of 1,000 kc. or more may be transmitted for long distances over coaxial lines and utilised for purposes of multiplex telephony or television. A coaxial line is a metal tube surrounding a central conductor and separated from it by insulating supports. 1935 *Ibid.* iv. 351/2 If similar conducting and insulating materials are used, shielded balanced pairs and coaxials have similar transmission-frequency characteristics. 1936 *Ibid.* xiv. iii. 175/1 The coaxial cable systems where all channels are carried over a single coaxial line. 1938 *Times* 7 Jan. 14/1 The alternative..is to feed the provinces from London by a special land-line, known as a co-axial cable. 1940 *Chambers's Techn. Dict.* 172/1 *Co-axial cable*, a type of cable in which a central conductor is surrounded by an outer tubular conductor... Its low attenuation makes it suitable for the transmission of video-frequency currents in television. 1943 *Electronic Engin.* XVI. 64 The distribution of the signal via video frequency local co-axial cable networks. 1955 *Sci. Amer.* Apr. 48/3 A coaxial cable can handle only 600 phone calls and two TV shows. 1959 *Ann. Reg. 1958* 521 It was agreed.. that a round-the-world Commonwealth coaxial-cable telephone system should be constructed. 1966 *McGraw-Hill Encycl. Sci. & Technol.* III. 245/2 In a typical cable each coaxial line has an outer conductor with an inside diameter of 0·375 in.

cob, *sb.*¹ Add: **7. d.** *dial.* A (baked apple) dumpling.

1898 in *Eng. Dial. Dict.* 1952 F. White *Good Eng. Food* v. i. 188 *Apple cobs*..the name given in the Cotswolds to baked apple dumplings. 1960 *Guardian* 17 June 8/4 In Somerset this kind of baked dumpling is known as an *Apple Cob*.

IV. *cob-pipe* (earlier U.S. example); **cob-house** (earlier U.S. examples); **cob-meal** *U.S.*, corn-cobs ground down.

1774 J. Belknap in J. Marcou *Life* (1847) 68 They have a neat poultry house, built of sawed strips of wood, in the form of a cob-house. 1818 M. Birkbeck *Lett. Illinois* 116 In this country they build 'cob-houses';..with these cobs.. structures are raised by the little half-Indian brats, very much like our houses of cards. 1834 *Deb. Congress* 26 Feb. 736 With the first shock in the commerce or credit of the country, the whole cob-house fabric must crumble. 1858 E. E. Hale in *Atlantic Monthly* Oct. 634/2 The child had been building cob-houses out of lucifer-matches in a paper-warehouse. 1833 *Amer. Railroad Jrnl.* II. 488/2, I have also made further discovery of the use of cob meal. 1882 *Maine Board Agric. Rep.* XXVI. 255 Cob meal or middlings might be substituted for clear corn meal. 1847 in D. Drake *Pioneer Life Kentucky* (1870) iii. 63 Quietly smoking a cob pipe in the corner.

cob, *sb.*⁹: var. Kob *sb.*¹

1920 *Blackw. Mag.* May 656/2 The tawny Uganda cob, a very handsome antelope. 1961 *Listener* 7 Sept. 348/2 The host of antelopes, such as..impala, cob, kudu, and a score more.

cob, *sb.*¹⁰: var. *kob *sb.*²

cob (kǫb), *sb.*¹¹ *slang.* [Origin unknown.] Phr. *to have, get a cob on*, to be annoyed, to become angry.

1937 Partridge *Dict. Slang* 163/2 *Have a cob on*, to be annoyed: ships' stewards'. 1953 'R. Gordon' *Doctor at Sea* ix. 113 'Don't you blokes go without me,' he added threateningly. 'I'll get a cob on if you don't wait.' 1966 F. Shaw et al. *Lern Yerself Scouse* 22 *Yer gorra cob on*, you are in a bad mood.

cobæa (kobī·ă). [mod.L. (A. J. Cavanilles *Icones et Descriptiones Plantarum* (1791) I. 11), f. the name of B. *Cobo* (1572–1659), Spanish missionary and naturalist.] A member of a genus of climbing shrubs so called, belonging to the family Polemoniaceæ and native to tropical America.

1805 *Curtis's Bot. Mag.* XXII. 851 (*heading*) Climbing Cobæa. 1873 *Young Englishwoman* Feb. 82/2 Passion flowers, taxonias, cobæas, and other creepers. 1963 W. Blunt *Of Flowers & Village* 111 Cobaea is another rampant climber.

cobalamin (kobǫ·lămin). *Biochem.* [f. Cobal(t + *vit)amin.] A name given orig. to the characteristic part of the vitamin B_{12} molecule (see quot. 1950), and hence used for any member of the vitamin B_{12} group, each of which contains this part in its molecule.

1950 E. A. Kaczka et al. in *Science* 29 Sept. 355 The name cobalamin designates all the vitamin B_{12} molecule except the cyano group, and..vitamin B_{12} becomes cyano-cobalamin, and vitamin B_{12a}, hydroxo-cobalamin. 1954 Jukes & Williams in Sebrell & Harris *Vitamins* I. vii. 452 The difference between the values for vitamin B_{12}..was due to the presence of cobalamins other than cyanocobalamin. 1959 *New Biol.* XXIX. 36 Cobalamin, or other substances having vitamin B_{12} activity, has been found dissolved in both pond- and sea-water at about the same concentration. 1967 H. R. Skeggs in György & Pearson *Vitamins* (ed. 2) VII. ix. 277 The total cobalamin content of an unknown sample represents its potential vitamin B_{12} activity.

cobalt. Add: **3.** *cobalt red, violet*; **cobalt bomb,** (*a*) a container storing radioactive *cobalt-sixty* (^{60}Co) used in the treatment of cancer; (*b*) a hydrogen bomb enclosed in a shell of cobalt which, if exploded, disperses radioactive cobalt dust; **cobalt oxide,** any oxide of cobalt.

1954 *Times* 8 Apr. 5/4 At the time Einstein spoke the cobalt bomb, like the hydrogen bomb, was merely a theoretical possibility, but the successful tests of the hydrogen bomb..had brought this theory into the realm of fact. 1958 L. Pauling *No More War!* vii. 154 The cobalt bomb is a hydrogen bomb or superbomb with a shell of perhaps 1000 pounds of the common metal cobalt about it. 1903 *Jrnl. Chem. Soc.* LXXXIV. ii. 80 The cobalt oxide which is deposited has the formula Co_3O_4. 1911 *Mining World* XXXIV. 71 (*title*) Use of cobalt oxide for making pigment. 1957 Mankowitz & Haggar *Encycl. Eng. Pottery* 57/2 *Cobalt oxide*, a powerful blue colouring material used for staining bodies and glazes from the beginning of the eighteenth century, known and used by John Dwight of Fulham. 1934 H. Hiler *Notes Technique Painting* ii. 123 *Cobalt red*, a very durable, but not much-used colour. It works badly. 1946 *Atomic Bomb* (Atomic Scientists of Chicago) vii. 36 When bombarded with neutrons, another form of this element is made, called cobalt-60. 1949 *Atomic Energy & Life Sciences* 90 Sponsored research, and provided radioisotopes, to determine the efficiency of radiocobalt (cobalt 60) as a substitute for radium in the treatment of cancer. 1951 *Sci. News Let.* 24 Nov. 333 A cobalt 60 'bomb' which promises life-saving treatment for more cancer patients was officially installed in the clinic of the Ontario Cancer Foundation, London, Ont. The cobalt bomb is a small portion of radioactive cobalt 60 enclosed in a heavy lead case. 1954 *Ann. Reg. 1953* 391 Additions to the powerful equipment for cancer treatment..included a cobalt 60 therapy unit producing X-rays of 3 million electron volts. 1962 *Encycl. Brit.* XXXI. 772/1 Several mixed cobalt compounds containing arsenates and phosphates of that metal, and represented by cobalt violet.

cobaltammine (kōᵘ-bǫltæ·min). Also 9 **cobaltamine.** [ad. F. *cobaltamine* (Forumbaru 1880, in *Compt. Rend.* XCI. 933).] Cobalt + *ammine.] Any ammine of cobalt.

1881 *Jrnl. Chem. Soc.* XL. 1106 Cobaltamines... Roseo-cobaltamine sulphate..are converted into an insoluble phosphate. *Ibid.* 1107 Luteocobaltamine pyrophosphate undergoes the same change. 1923 *Jrnl. Chem. Abstr.* 567 (*heading*) An Equilibrium in the Cobaltammines. 1936 J. S. Anderson in *Rep. Progr. Chem.* XXXIII. 147 The cobaltammines were the compounds for which the existence of mirror image isomerism (optical isomerism) was first detected among inorganic compounds. 1958 *Chambers's Techn. Dict.* (ed. 2) 967/1 *Cobaltammines*, complex salts in which the cobalt is co-ordinated with ammonia or organic derivatives, e.g. *cobalt-hexammines*, which contain the group $Co(NH_3)_6$.

cobber (kǫ·bəɹ), *sb.*² [f. Cob *v.*² + -er¹.] (See quots.)

1778 Pryce *Min. Cornub.* 234 The picked Ore..is put by a number of girls called Cobbers, who break it..to the size of a chestnut and less. 1864 J. R. White in *Parl. Papers* XXII. 24 The cobbers (picking cobbers)..break up the ore with hammers. 1921 *Dict. Occup. Terms* (1927) § 050 *Cobber*; breaks ore into small pieces with small hammer, and sorts it according to value. 1961 *Listener* 11 May 836/1 In leftover scrap..cut pieces of sheepskins free of hair and offal after wool has been removed, and then from off.

cobber (kǫ·bəɹ), *sb.*³ *Austral.* and *N.Z. colloq.* [perh. f. dial. *cob* (see *E.D.D.*), to take a liking to.] A companion, a mate, a friend.

1895 *Bulletin* (Sydney) 9 Feb. 15/4 She's my cobber an' I'm 'er bloke. 1897 W. T. Goodge *Hits! Skits!* (1899) 24 Or have been fighting with a cobber. 1897 D. McK. Wright *Station Ballads* 110 The man who is 'cobbers' with Christ comes off best if he's ever so poor. 1906 E. Dyson *Fact'ry 'Ands* ii. 14 Long the cronies..were cobbers. *Ibid.* xvi. 217 Miss Eva was not as young as some, and she had no 'cobbers' amongst the Beauties. 1926 *Anzac Book* 22/2 'Ow er yer ter tell yer cobber?' 1928 *Bulletin* (Sydney) 29 Nov. 19/1 The trooper himself suffered from a severe over-dose of 'cobber'. 1929 'Waltzing Matilda' 113 My cobbers..said I can't go back. 1934 *Bulletin* (Sydney) 15 Nov. 20/4 Thistles and cobber's pegs are spread by the wind. 1944 *Baker Austral. Lang.* v. 44 He's got to be your cobber, a mate. 1950 Denis Glover *Sings Harry* 40 His cobber, Slim..had deprived himself of his mate, his 'cobber' for twenty years. 1945 *Barney Butcher's Shop* 10. 1947 *Coast to Coast 1946* 180 It only took about ten minutes to make a new cobber out of almost anyone in jungle green and a Digger hat. 1953 'N. Shute' *In the Wet* 14 He shook my hand vigorously. 'Cobbers,' he said. 1963 B. Crump *Gift of Sea* 67 I saw Jock..was my cobber in the timber mill. Jack and I went on the booze together.

cobber, *v. Austral.* and *N.Z. slang.* [f. prec.] Usu. const. *up*. To make friends with.

1949 *Geogr. Mag.* Feb. 368 Cobber has been extended to.. the verb to cobber up, to make friends with. 1963 B. Pearson *Coal Flat* vi. 104 It's natural for a young chap to cobber up with the bloke that's working with him.

cobble, *sb.*¹ **4.** Add: *cobble-paved, -streeted* adjs.

1891 *Pall Mall Gaz.* 30 Nov. 3/1 The cobble-paved road, bordered by endless crucifixes. 1910 W. J. Locke *Simon the Jester* xii, We crossed the cobble-paved courtyard. 1962 S. M. Saunders *Illany Lane* iii. xxv. Little cobble-streeted shady French towns.

cobbler. 1. b. The last sheep to be sheared, in punning allusion to the cobbler's last (see also quot. 1945). *Austral.* and *N.Z. slang.*

1893 *Herald* (Melbourne) 23 Dec. 6/1 (Morris), Every one might not know what a 'cobbler' is. It is the last sheep in a catching pen, and consequently a bad one to shear, as the curve of the years have been. By the shearers' last English mention used to say, when picking up the last sheal, 'This is what the cobbler threw at his wife.' 'What?' 'The last.' 1894 A. Robertson *Nuggets* 4 The 'cobbler', a grizzled, wiry-haired sort of 'rough' fellow, who..the last has been the toughest cobbler ever penned. 1940 E. C. Studholme *Te Waimate* (1954) xv. 190 The 'old hand' would let the learner in for the 'cobbler', or hard shearing sheep (not always the last, but often left to the last in shearing). 1933 *Bulletin* (Sydney) 20 Sept. 20/1 Rolling cobbler's the shearers' cobbler, a dirty, sticky, matted and wrinkly sheep (not always the last, but..left to the last in shearing).

c. [Rhyming slang from *cobbler's* (*or cobbler's*) *awls*.] A ball; esp. in *pl.*, 'balls', testicles; nonsense, rubbish.

1934 P. Allingham *Cheapjack* xv. 186 The Cobbler is even more simple. It is a ball game..'cobbler' is the slang for ball. 1955 M. Procter *Hell is City* xii. 108 What a lot of cobblers! 1955 P. Wildeblood *Against Law* 137 We had to get a..name for the cobbler. 1961 *Radio Times* 24 Feb. 13/3 To Talking more cobblers to the square and the cut-bobbing..on the cobblers. 1962 *John Bull* 1 Mar. 47 (heading) on confirmation day. 1968 *Mrs Dale's Diary* 4 Sept. 8/4 'Don't talk such cobblers.' 1956 *Spectator* 24 Aug. 8/2 Not cobblers..but what the hell's the use? 1967 *Observer* 3 Jan. Daily Telegraph 4. 1968 'B. Mather' *Snowline* 10 Ah, well, it's all cobblers. 'What's the catch?' I asked. 'Why all this cobblers about clothes?'

3. (Later examples.)

1852 G. W. Curtis *Lotos-eating* 105 Various other select parties are..watching the sails and sipping cobblers. 1862 E. McDermott *Pop. Guide Internat. Exhib.* 185 There is an American bar, where visitors may indulge in 'juleps', 'cock-tails', 'cobblers', 'rattle-snakes', (etc.). 1922 *Glasgow Herald* 13 May 6 Unbidden recipes for coolers, cobblers, highballs, frappes.

4. (Examples.)

1859 in Bartlett *Dict. Amer.* 90 1880 'Mark Twain' *Tramp Abroad* xlix. 575, I have..made out a little bill of fare... Peach cobbler, Southern style. 1949 *Farmhouse Fare* 130 October cobbler..a quart blackberries.. Place the berries..in a buttered pie-dish and sprinkle with lemon juice..roll out the paste to the size of the pie-dish. Place over the fruit.

5*. A horse chestnut used in the children's game of 'conkers'. *dial.*

1896 E. G. Newsam *Warwick. Word-bk.* 51 *Cobbler*, the fruit of the horse chestnut tree. Cobblers, the well-known game of striking one threaded 'cobbler', threaded on a string, against that of an opponent, to try their respective strength. 1913 D. H. Lawrence *Sons & Lovers* i. iii. 50 He pulled from his pocket a black old horse-chestnut hanging on a string. 'Ere y'are, cobbler..'. 1958 *Times* 4 Oct. 8/4 The cobblers..the hardest cobbler's peg..and hit and smashed seventeen other cobblers.

6. cobbler's peg (freq. in *pl.*) *Austral.*, a popular name for the weed *Bidens pilosa*, of the family Compositæ; formerly used for *Erigeron linifolius*, another weed of the same family; (see also quot. 1908).

1898 F. M. Bailey *Synopsis Queensland Flora* 243 Erigeron linifolius... 'Cobblers' pegs'. 1906 —— *Weeds of Queensland* 73 [*Bidens pilosa*]. 'Cobbler's Peg'... The seeds..in the same way, the name is applied to *Bidens pilosa*. 1908 E. J. Banfield *Confessions of Beachcomber* i. vi. 201 The white mangrove..is a graceful tree, the respiratory shoots from the roots..resembling asparagus shoots or stake ties (called by some cobbler's pegs). 1933 *Bulletin* (Sydney) 15 Nov. 28/4 Thistles and cobbler's pegs are spread by the wind. 1944 *Bailey Austral. Dict.* ed. 71 *Cobbler's-pegs*, a common Australian plant of roadsides and waste land, with seeds that cling to clothing and animals. 1961 P. White *Riders in Chariot* viii. 246 'What is it?' Mrs. Godbold, replied the child. 'Just cobbler's pegs!' shrieked one..'Silly old weeds!'

cobbly (kǫ·bli), *a.* [f. Cobble *sb.*¹ + -y¹.] Paved with cobble-stones. Also *fig.*

In earlier dial. use in sense 'full of lumps'.

1893 *Bicycling News* 1 Apr. 29/2 Our rough cobbly roads. 1924 R. Hall *Lett.* 210 He says it is..but we must stick to the 'light at eventide' on the cobbly path of the old women. 1936 F. Stark *Southern Gates Arabia* xvii. 186 The city gate stood above, on its cobbly slope.

cobra (kǫ·brā). *Austral.* Also **cobra.** [Aboriginal.] The head, skull.

1830 G. Tipsmill *The Snake* in Stewart & Keesing *Old Bush Songs* (1957) 33 They with a stick effect would take Upon your cobra. 1845 'Old Boomerang' *Austral. Capers* xiii. 204 The savage cobbler..with a volley of oaths and other epithets..the Cobbler. 1871 *Fraser's Mag.* XXXIX. 218 Just breaking his cobra with a waddy. 1881 A. C. Grant *Bush-Life in Queensland* I. ii. 31 The black fellows are lying in the bush between but small attention on his 'cobra', as the head is usually called in the pigeon-English which they acquire from the white settlers. 1940 *Burnett Rev.* 21 Oct. 'R. Boldrewood' *Ups & Downs* v. 50 Put it in your cobra..to make out his meaning.

Cobb('s) paper. [See quot. 1859.] (See quots.)

1859 *Stationers' Hand-Bk. & Guide Paper Trade* (ed. 2) 77 *Cobb's Paper*, a kind of paper used for the covers of books by bookbinders for the linings of end papers; it is a brown paper, coloured in the pulp. Named after its inventor, Cobb..a stationer of Stationers' Hall, London. 1879 Zaehnsdorf *Art Bookbinding* 176 *Cobb's paper*, a paper used for the sides of books. 1958 *Chambers's Techn. Dict.* 226 *Cobb Paper*..A self-coloured paper, obtainable in various shades, largely used by binders as end-papers and for the sides of books... It derives its name from its inventor, a stationer, and has been used ever since. 1962 E. J. Labarre *Dict. Paper* (ed. 2) 51/2 *Cobb*, properly *Cobb's paper*.

cobby, *a.* Add: **3. b.** Shortish and thickset, 'stocky'.

1838 G. Stables *Our Friend the Dog* xii. 59 Cobby.. Nicely ribbed up, compact in form like a cob-horse. 1891 *New Review* June 545 The cobby man of 5 ft. 5 in. 1909 *Daily Chron.* 16 Apr. 8/3 Pony (Welsh) cobby-built. 1938 Bannerman *Birds Brit. Isl.* VII. 152 The general 'cobby' appearance, unusually active movements, and above all the extreme buoyancy [of the harlequin duck].

Cobdenism (kǫ·bdəniz'm). [f. the name of Richard *Cobden* (1804–1865).] A policy advocating free trade, peace, and international co-operation. So **Co·bdenite** *sb.*, a supporter of Cobdenism; *a.*, of or pertaining to Cobdenism; **Cob·denic, Co·bdenitish** *adjs.*

1853 A. Somerville (*title*) Cobdenic Policy the Internal Enemy of England. 1887 W. J. Harris in *Nat. Rev.* Nov. 311 Cobdenites point to the agricultural population as too small a portion of the population to be considered in the argument of Protection versus Free Trade. *Ibid.*, I am simply striving to enlarge the national conceptions to which Cobdenism is driving. 1896 *Westm. Gaz.* 22 Mar. 2/4 And that I would meet upon a broad basis of open Cobdenitish seers. 1927 *Observer* 8 May 16/8 Biology and psychology are stronger..than the most materialistic laissez-faire of the Cobdenite economics. 1929 *Daily Tel.* 8 Jan. 19/1, I met a Cobdenite the other day, and, like a true Cobdenite, he said his views were based on sound economic. 1957 *Encycl. Brit.* XVIII. 555/2 He'd point to the abandonment of the Cobdenite system..used to be that Great Britain..was the only great nation with system of pure Cobdenism. *Ibid.*, the only system the colonies had no part to play at all different from that of foreign countries... The idea that all economists of that [*sc.* free-trade] school were Cobdenites is without foundation.

co-belligerent, *a.* and *sb.* Add: (Later examples.) Hence **co-belli·gerence, co-belli·gerency,** the quality or state of being co-belligerent or a co-belligerent.

1919 J. M. Keynes *Econ. Conseq. Peace* (1920) v. 109 How far Germany can be made to contingently liable for damage done..by her co-belligerents, Austria-Hungary, Bulgaria, and Turkey. 1943 W. S. Churchill in *Second World War* (1952) V. 166 The Badoglio Government and Allied status, but only that of co-belligerence. 1944 *Ann. Reg.* 1943 18 The Badoglio Government and Allied Headquarters' recognition..that Italy should be as a co-belligerent for the duration of the war. 1949 F. Maclean *Eastern Approaches* iii. v. 347 Then they blocked forward, clearing their bits in the Communist salient and I remembered that we were amongst friends in co-belligerency. 1958 *Times Lit. Suppl.* 28 Mar. 163/1 Its conclusion..was aggravated by different views in London and Washington as to the value of any rate co-belligerents.

Coblenzian (kōᵘble·ntsiăn), *a. Geol.* Also **Coblentzian, Coblenzien.** [ad. F. *coblentzien* (A. Dumont 1848, in *Mém. de l'Acad. R. des Sciences de Belgique* XXII. 4), f. *Coblentz* (G. *Koblenz*), a city in western Germany: see -ian.] Of, pertaining to, or designating the uppermost division of the Lower Devonian (Lower Old Red Sandstone) in Europe or the geological period during which it was deposited. Also *absol.*

1853 *Q. Jrnl. Geol. Soc.* IX. 13 This division must be separated from the Devonian series, since it does not contain a single species found in the 'Coblenzian' or 'ahrian' divisions above it. 1882 A. Geikie *Text-bk. Geol.* 705 The Coblenzian or Grauwacke, composed of four zones of greywacke, sandstones, shales, and conglomerates. 1954 J. Jukes-Browne *Student's Handbk. Stratigr. Geol.* 164 The Coblenzian,..of great thickness, is well displayed in the Ahr, Moselle, Lahn, and other valleys of the Eifel and the Rhine. 1963 D. W. & E. E. Humphries tr. *Termier's Erosion & Sedimentation* 290 The first, are without Coblenzian beds.

Cobol (kōᵘ·bǫl). Also **COBOL.** [f. the initial letters of *Common Business Oriented Language.*] A computer language designed for use in business operations.

1960 (*title*) Cobol, report to Conference on Data Systems Languages, including initial specifications for a common business oriented language (COBOL) for programming electronic digital computers (U.S. Defence Dept.). 1961 *Times* (Computer Suppl.) 3 Oct. p. x/6 Ideally if a Cobol translator were written for every computer, a programme written in Cobol could be translated and then used on any machine. 1963 F. L. Saxon (*title*) COBOL: a Self-Instructional Programmed Manual. 1967 *Oxford Computer Explained* 8 Writing programs in an easy computer language which..enables anyone..who has to work when Cobol was available.

cobra. Add: Also applied to African snakes of the genus *Naja*, as the *Egyptian cobra* (*N. haje*), the *Cape cobra* (*N. nivea*), and the *spitting cobra* or *black-necked cobra* (*N. nigricollis*) (see *black *a.* 18). The name *spitting cobra* is also given to the *ringhals.

1856 F. Fleming *S. Africa* xii. 400 The 'Cobra-di-Capella' or Hooded Snake, called by the Colonists the 'Ringesault'. 1884 *Trans. S. Afr. Philos. Soc.* III. 12 The South African cobra (*Naja haje*) and the..Cape cobra. 1910 *Nature* [illegible] The Cape Cobra by the name *Spuugh slang*, and is so called from its supposed power to poison of ejecting its poison to a distance. 1887 [see *Spitting pl.* a 2]. 1910 F. W. Fitzsimons *Snakes S. Afr.* iv. 172 Egyptian or Banded Cobra. Distribution: Natal, Zululand, Transvaal, Rhodesia, Northwards as far as Sahara Desert. 1931 R. L. Ditmars *Snakes of World* xii. 116 The Cobras, *Naja*, range from the arid North coast to the Cape, the region between the deserts being occupied by the larger snakes. The Egyptian Cobra, or Asp, *Naja haje*..has the widest range of any cobra in Africa. *Ibid.* 167 The Spitting Cobra or Black-Necked Cobra, *Naja nigricollis*, comes closest to the spitting cobras under..sections of Africa. *Ibid.* 170 The Black and White Cobra..*Naja melanoleuca*..is distributed through the tropical Africa. *Ibid.* 171 The Cape Cobra, *Naja nivea* (flava) is a southern form. *Ibid.* 172 The Ringhals, or Keel-scaled Spitting Cobra, *Sepedon haemachates*..its scales being keeled, rendering it unique among the cobras. 1959 *Cape Argus* 31 Oct. 9/7 Cape cobras and puffadders provide most of the venom, but Egyptian cobras, boomslangs, ringhals cobras and an Egyptian cobra—from Rhodesia—are kept in these special glass houses on Mr. Wood's farm.

cobra, var. *cobbra.

Coburg. Add: Also **Cobourg.** (Earlier examples.)

1864 *Cheltenham Chron.* 10 Oct. 2/7 (Advt.), Cobourg and Orleans cloths. 1864 Rossetti *Let.* 12 Nov. (1965) II. 527, I shall bring..a dress for Maggie...It is a sort of brown Coburg.

2. A two-wheeled covered carriage or cart, used esp. in the country. Also *attrib.*

1824 *Act* 5 *Geo. IV* c. 114 § 78 For every..Coburg, and for every other Carriage hung on Springs, the Sum of Sixpence for each Wheel. 1847 J. Leech *Punch Ex.* 108 Along the main road, and from every bye-road, came coburgs, cars, and all kinds of rural vehicles. 1849 *Ibid.* 66 A kind of Cobourg [I think you call it,] or market-cart.. passed by. 1859 *Gloss. Coal-Mining Terms* 69 *Coburg* (*attrib.*). 1856 *Chard & Munro, Bristol*..A Coburg conveyance. 1857

ELTON *Below the Surface* I. x. 223 The mob suffered no one to precede them, turning back or forcibly detaining every horseman, or spring-cart, or farmer's coburgh that overtook them on the road. *Ibid.* 228 A friendly baker was waiting for him in a coburgh cart. 1925 CONRAD *Suspense* I. ii. 25 The half-length portrait of a rubicund Latham in a white coburg, which, but for the manly and sensitive mouth, might have been the portrait of his own coachman.
3. In full *Coburg loaf*, a kind of loaf.
1843 ROSSETTI *Let.* 25 June (1965) I. 12, I dawdled about till teatime. When I repaid myself by walking into the Coburg with awful perseverance. 1906 *Westm. Gaz.* 11 Sept. 10/2 Every description of loaf, from..a 'crumby' to a Coburg. 1935 E. WEEKLEY *Something about Words* 106 *Coburg loaf*,..which was named soon after the marriage of Queen Victoria to Prince Albert of Coburg-Gotha. 1955 *Times* 24 Mar. 4/3 An 'oven bottom' Coburg loaf—a loaf baked..on the floor of the oven or on the hearth rather than in a pan or tin.

cobweb, *v.* (Examples.)
1928 P. B. BALLARD *Changing School* xiii. 185 The lines of association with which their minds are cobwebed. 1960 S. KAUFFMANN *If it be Love* i. 9 Shreds of intense guttural intrigue still seemed to cobweb the corners.

cobwebbed, *ppl. a.* Add: **1.** Also *fig.*
1905 *Nature* 2 Nov. 12/1 The somewhat cobwebbed state of the higher education of this country. 1928 BLUNDEN *Retreat* 33 With whom some cobwebbed boyish vow Once ended 'never, never part'!

Coca-Cola (kōᵘ·kă,kōᵘ·lă). Trade name of a popular American soft drink. Also *fig.* So in various humorous or allusive formations, as **coca-colonization.**
1887 *Evening Jrnl.* (Atlanta) 30 June 1/4 Drink the brain tonic and intellectual soda fountain beverage Coca-Cola. 1921 *Glasgow Herald* 25 Apr. 8, I was 'all right'—should I ever want anything better than ginger-ale, root beer, or coco-cola [*sic*]. 1947 G. GREENE *19 Stories* 81, I caught sight of him in the only drug-store, having a coca-cola. 1950 E. S. HYAMS *From Waste Land* xiii. 198 The coco-colanization [*sic*] of the viticultural regions. 1958 *Times Lit. Suppl.* 24 Jan. 39/2 In New Haven you can get a Baroque hymn, a kind of Coca Cola Crawshaw [*sic*] in praise of St. Theresa of Lisieux. 1959 *Times* 4 Mar. 11/6 Is Australia being 'coca-colonized'? 1959 *Times* 1 June 17/4 Since the war the business aircraft has become as much a part of the American way of life as baseball or Coca-Cola. 1960 KOESTLER *Lotus & Robot* 277 One might call it the coca-colonization of Western Europe.

cocaine. The pronunc. (kokēⁱ·n) is now usual. Hence **cocainization** (examples).
1887 *Med. News* (Phila.) 30 Apr. 501/2 There is, however, a certain proportion of cases in which cocainization cannot be produced. 1964 S. DUKE-ELDER *Parsons' Dis. Eye* (ed. 14) xv. 181 They are common in elderly people and should be removed with a sharp needle after cocainization.

cocainist (kokēⁱ·nist). [f. COCAINE + -IST.] One addicted to the use of cocaine.
1908 *Westm. Gaz.* 17 Jan. 7/4 The cocainist..experienced singular elation. 1927 *Glasgow Herald* 30 Aug. 9 There is actually in Moscow a children's narcotic dispensary for the treatment of child cocainists.

cocarboxylase (kōᵘ:kāɹbǫ·ksilēⁱs). *Chem.* [G. (E. Auhagen 1932, in *Zeitschr. f. physiol. Chem.* CCIV. 149). f. CO- + *CARBOXYLASE.] Thiamine pyrophosphate or diphosphothiamine, the co-enzyme of carboxylase and other enzymes, esp. those which act on pyruvic acid.
1932 *Chem. Abstr.* 2753 (*title*) Cocarboxylase, a new coenzyme of alcoholic fermentation. 1937 *Biochem. Jrnl.* XXXI. 2240 'Co-carboxylase' has as good an effect as vitamin B_1 in the catatorulin test. 1944 *Ann. Reg. 1943* 353 In the excitation process of peripheral nerve acetylcholine formation seems to be related to the co-carboxylase activity of vitamin B_1. 1947 A. T. HENRICI *Molds, Yeasts* (ed. 2) xi. 318 The enzyme, carboxylase, which catalyzes the decarboxylation of pyruvic acid to carbon dioxide and acetaldehyde in the alcoholic fermentation, cannot function without its coenzyme, cocarboxylase. 1961 [see *APOENZYME].

co-carcinogen (kōᵘ,kā·ɹsinŏdʒĕn). [f. CO- 3 + *CARCINOGEN.] A substance that increases the carcinogenic effect of some other substance but is not itself carcinogenic. Hence **co:-carcinoge·nic** *a.*
1938 M. J. SHEAR in *Amer. Jrnl. Cancer* XXXIII. 532 Since it does not appear to be carcinogenic itself and yet enhances the activity of 3:4-benzpyrene, it has been termed, for convenience, a 'co-carcinogen'. 1941 *Cancer Res.* I. 47/2 It is as yet impossible to estimate the extent to which cocarcinogenic action is responsible for an augmentation of tumor production in man. 1956 *Ann. Rep. Brit. Empire Cancer Campaign 1955* 239 This would suggest therefore that the tar solution has a weak co-carcinogenic action. 1962 *Lancet* 13 Jan. 85/2 Tobacco tar contains in abundance a phenolic fraction which acts as a powerful co-carcinogen agent. 1968 *New Scientist* 26 Sept. 638/2 The concept of viruses as co-carcinogens—that is, as only one factor among several which lead ultimately to a cancer.

‖**cocasse** (kokas), *a.* [Fr.] Droll, ridiculous. Hence **cocasserie**, that which is amusing or bizarre.
1868 W. JAMES *Let.* 12 Feb. in R. B. Perry *Tht. & Char. W. James* (1935) I. 262, I took up Balzac's *Modeste Mignon* the other day...It must be one of the very early ones, for the extraordinary research and effort in the style is perfectly *cocasse*. 1934 C. LAMBERT *Music Ho!* ii. 123 In spite of his [*sc.* Satie's] verbal wit, and his many *blagues* and *cocasseries*, no composer, not even Debussy, took a more essentially serious view of his art. 1960 J. LODWICK *Asparagus Trench* 7 A collaboration as *cocasse* as any since that between Balzac and certain long-extinct copper mines. 1963 *Listener* 7 Feb. 260/3 Two items which the late James Agate would have termed *cocasseries*.

coccal (kǫ·kăl), *a.* [f. COCCUS: see -AL.] Of, pertaining to, or resembling a coccus; spherical or nearly spherical in form.
1928 *Lancet* 8 Dec. 1193/1 A blood culture was taken.., the organism grown being coccal in character. 1964 S. DUKE-ELDER *Parsons' Dis. Eye* (ed. 14) xxxv. 550 Of the coccal infections, the gonococcus, the streptococcus [etc]..all may give rise to metastatic implications of the eye in generalized infections.

coccid (kǫ·ksid), *a.* and *sb.* *Ent.* [f. mod.L. *Coccidæ*, a family of insects.] **A.** *adj.* Of or pertaining to the Coccidæ or scale-insects. **B.** *sb.* A scale-insect.
1892 W. L. DISTANT *Naturalist in Transvaal* 88 The Coccid, or so-called 'Australian Bug' (*Icerya purchasi*). 1894 *Athenæum* 16 June 778/1 Male specimens of a coccid (*Lecanium prunastri*), bred from scales attached to shoots of blackthorn. 1899 *Kew Bull.* 6 Attacks of Coccids on forest trees in the island of Rodrigues. 1913 *Oxf. Univ. Gaz.* 4 June 948/2 The Coccid food of the larva of *Spalgis lemolea*. 1920 J. RITCHIE *Anim. Life Scot.* 474 A Coccid on oranges. 1960 L. PICKEN *Organiz. Cells* iv. 97 Other instances of haploid chromosome numbers of one or two occur in a coccid, *Icerya* and a spider beetle (Ptinidae). *Ibid.* 136 A non-specific attraction that leads to end-association of the two chromosomes of a coccid sperm.
Hence **coccido·logy**, the scientific study of scale-insects.
1904 *Science* 25 Mar. 501 In the present state of coccidology.

coccidian (kǫksi·diăn), *sb.* and *a.* *Zool.* [f. *COCCIDIUM 2 + -AN.] **A.** *sb.* = *COCCIDIUM 2. **B.** *adj.* Of or pertaining to the order Coccidia.
1918 A. C. CHANDLER *Animal Parasites & Human Disease* x. 170 There are a number of serious diseases of animals which are caused by parasites of the class Sporozoa known as coccidians. 1934 T. W. M. CAMERON *Internal Parasites of Domestic Animals* II. 33 The typical life-cycle of a coccidian parasite. 1941 J. S. HUXLEY *Uniqueness of Man* vii. 176 Permitting a normally innocuous coccidian parasite to pass so rapidly and in such numbers from bird to bird that mass infection..resulted. 1961 MACKINNON & HAWES *Introd. Study Protozoa* III. 157 We are left with the two related but interestingly contrasted orders of gregarines and coccidians.

coccidioidin (kǫ:ksidioi·din). [f. mod.L. *Coccidioid(es* (see below) + -IN¹.] An antigen prepared from a culture of the fungus *Coccidioides immitis*.
1931 M. D. BECK et al. in *Jrnl. Amer. Vet. Med. Soc.* LXXVIII. 496 The coccidioidin used by us was prepared by growing the mold upon plain beef broth. *Ibid.* 497 The antigen..coccidioidin gave constant negative results in both the normal and tuberculous animals. *Ibid.* 499 The tests do indicate that coccidioidin may aid us in determining the presence or absence of coccidioidal granuloma in man and animal. 1958 *New Biol.* XXVII. 65 During the second world war it was found that many soldiers gave a positive skin test to coccidioidin..without ever having knowingly had coccidioidomycosis.

coccidioidomycosis (kǫksidioi:domaikōᵘ·sis). *Path.* [f. mod.L. *Coccidioid(es* (see prec.) + -O + MYCOSIS.] A disease in man and some animals caused by the spores of the fungus *Coccidioides immitis* and affecting esp. the lungs.
1937 E. C. DICKSON in *Calif. & Western Med.* XLVII. 154/2 To obtain a name for this symptom complex.. presents some difficulty...The word 'coccidioidomycosis' might be used to include all cases of infection with the Fungus coccidioides, regardless of the clinical manifestations. 1955 *Sci. News Let.* 8 Jan. 21/3 Contrary to previous belief, wild rodents are not an important reservoir for organisms that cause valley fever, or coccidioidomycosis, a respiratory disease found in the Southwest. 1958 [see prec.].

coccidiosis (kǫ:ksidiōᵘ·sis). [f. *COCCIDIUM 2 + -OSIS.] A disease of birds and mammals caused by the presence of coccidia.
1892 G. FLEMING tr. *Neumann's Parasites & Dis. Dom. Animals* II. iii. 411 Intestinal coccidiosis in calves..manifested itself as a fœtid diarrhœa, accompanied by slight fever and catarrhal phenomena in the upper air passages. 1930 *Times* 17 Mar. 18/3 Coccidiosis, bacillary white diarrhœa, and other chicken diseases. 1945 'G. ORWELL' *Animal Farm* 53 Nine hens had died..and it was given out that they had died of coccidiosis. 1960 *Farmer & Stockbreeder* 9 Feb. 123/1 Under natural conditions coccidiosis is rarely harmful, and the disease is not listed among the common causes of death in wild birds in the U.K. (grouse are a possible exception). 1962 [see next].

coccidiostat (kǫksi·diostæt). [f. mod.L. *coccidium* + -STAT.] A substance administered to poultry to retard the growth and reproduction of pathogenic coccidia.
1959 *New Scientist* 16 Apr. 832/2 During the last decade the list of coccidiostats..proved in field trials to be both safe and effective has grown rapidly. 1962 *Ibid.* 2 Aug. 239 Chickens reared for laying must eventually be changed to unmedicated food..as the initial treatment with coccidiostats interferes with the birds' natural resistance to coccidiosis. 1968 *Ibid.* 23 May 403/2 A new coccidiostat highly effective in extremely small doses against all nine species [of coccidia] has now received clearance.

coccidium. Pl. -ia. Add: **2.** *Zool.* Formerly the name of a genus of protozoan internal parasites; now a parasite of this kind belonging to the order Coccidia, which causes disease in mammals and birds.
1886 W. E. HOYLE tr. *Leuckart's Parasites of Man* II. 197 (*caption*) Coccidia from the intestine of the domestic mouse. *Ibid.* 198 The spores are formed from the contents of the *Coccidium*, for by this name we shall henceforth designate the so-called 'egg-shaped Psorosperms'. 1934 T. W. M. CAMERON *Internal Parasites of Domestic Animals* II. 33 The coccidia are parasites of the epithelium of the intestine and the bile-ducts of mammals and birds. 1956 —— *Parasites & Parasitism* 268 Coccidia require a period of maturation which is generally short. 1960 *Farmer & Stockbreeder* 16 Feb. 149/1 The majority of the invading coccidia parasites. *Ibid.* (Suppl.) 15 Mar. 1/1 The coccidia are normal gut habitants in the fowl.

coccinella (kǫksine·lă). *Ent.* [mod.L. (Linnæus *Systema Naturæ* (ed. 10, 1758) 364), f. L. *coccineus* scarlet.] A beetle of the genus so named or its family Coccinellidæ; a lady-bird. So **coccine·llid** *a.* and *sb.*
1815 KIRBY & SPENCE *Ent.* I. 265 It is fortunate that in most countries the children have taken these friendly Coccinellae under their protection. 1887 J. D. HOOKER in L. Huxley *Life* (1918) I. i. 4 Seeing a Coccinella on a post. 1887 E. A. ORMEROD *Australian Bug* 29 The grubs of a Coccinellid or Ladybird. 1913 *Oxf. Univ. Gaz.* 4 June 955/2, 2 Coccinellid beetles of the genus *Epilachna*. 1959 E. F. LINSSEN *Beetles* II. 46 In these clusters of Coccinellid beetles there may be as many as one thousand five hundred individuals per square foot. *Ibid.* 47 Although there are some three thousand five hundred species of Coccinellids, only forty-five are British. 1964 *Canadian Entomologist* XCVI. 1151 The life histories of our common coccinellids have never been carefully studied.

coccoid (kǫ·koid), *a.* [f. COCCUS + -OID.] Of, or pertaining to, or resembling a coccus.
1912 J. S. HUXLEY *Individ. in Animal Kingdom* 159 Botanists distinguish three main types of elementary structure among plants...In the first type (Coccoid), the entire cell, with its cell-wall, divides into two similar and quite separate halves. 1940 *Chambers's Techn. Dict.* 172/2 *Coccoid*, unicellular, motionless in the vegetative condition, but liberating motile zoospores or gametes. 1949 H. W. FLOREY et al. *Antibiotics* II. xxxi. 1033 When developed young cultures of bacillary forms [of *Bartonella bacilliformis*] were exposed to penicillin..they tended to become coccoid. 1964 M. HYNES *Med. Bacteriol.* (ed. 8) xvii. 270 The organisms are small cocco-bacilli...As a general rule the predominant form of *Br. melitensis* is coccoid whilst that of *Br. abortus* and *Br. suis* is bacillary.

coccus. Add: **3.** Any individual bacterium of a group including the family Micrococcaceæ and characterized by a spherical or nearly spherical shape.
1883 [see *diplococcus* s.v. DIPLO-]. 1886 *Encycl. Brit.* XXI. 404/2 The filaments separate first into shorter filaments, then into rodlets, and finally into 'cocci'. 1968 R. CRUICKSHANK *Med. Microbiol.* (ed. 11) iv. 42 The different cocci are relatively uniform in size, about 1μ being the average diameter.

Cochin (kǫ·tʃin). Short for *Cochin-China* (*fowl*).
1853 [see SHANGHAI 1]. 1854 *Poultry Chron.* I. 34 Prize Buff Cochins. 1894 *Vermont Agric. Rep.* XIV. 173 If for roasters, then select the Brahma Cochin, Langshan, or Plymouth Rocks. 1911 *Encycl. Brit.* XXII. 214/1 The exaggeration of fluff and leg-feather has removed all Cochins—it is to be feared permanently—from amongst popular and useful breeds.

cock, *sb.*¹ Add: **2. c.** (Earlier example.)
1789 *Loiterer* 5 Sept. 10 This eloquent harangue was not lost upon me, I immediately began to smoke the old Gentleman. 'No, (thought I) that cock won't fight.'
8. (Later examples.)
1937 H. C. BAILEY *Clunk's Claimant* xxxiv. 227 What's the game now? Passed to you. Tails up, old cock. 1942 T. RATTIGAN *Flare Path* III. 164 Good show, Count, old cock! 1965 G. MELLY *Owning-up* iv. 30 Smarten yourself up a bit, cock, before we go on!
10. *cock of the plains*, a North American species of grouse; the sage-cock; *cock of the rock*, either of two species of the South American genus *Rupicola* of Cotingas.
1805 M. LEWIS in *Jrnls. Lewis & Clark Exped.* (1904) II. 384 Capt. C. killed a cock of the plains or mountain cock. It was of a dark brown colour with a long and pointed tail. 1825 C. WATERTON *Wanderings S. Amer.* iv. 301 The.. Cock of the Rock would be in fine plumage from..November to May. 1837 SWAINSON *Classif. Birds* II. 76 *Rupicola*, or rock manakin of Cayenne. The familiar name of cock of the rock, long bestowed on this bird, is very characteristic. 1838 AUDUBON *Ornith. Biogr.* IV. 503 Cock of the Plains. 1877 R. I. DODGE *Hunting Grounds Gt. West* 224 'Cock of the plains.' *Tetrao* (*centrocercus*) *urophasianus*. 1885 *Stand. Nat. Hist.* IV. 209 The sage-cock, or cock-of-the-plains, *Centrocercus urophasianus*, is the largest grouse found in America. 1917 E. H. FORBUSH in T. G. Pearson *Birds Amer.* II. 29 Sage Hen..Cock of the Plains..exceeds all other Grouse in size, with the possible exception of the great Black Grouse..of Europe. 1949 *Oxf. Jun. Encycl.* II. 84/1 The Cock-of-the rock..is a fairly large bird, coloured bright

orange-red, with a large, circular crest growing from each side of the head.
19. Hence (esp. spoken) nonsense, an unfounded statement. Freq. in phr. *to talk cock*. Cf. POPPYCOCK.
1937 C. DAY LEWIS *Starting Point* iii. 52 'If I hadn't let Mackenzie through that time, we'd have won.' 'Don't talk cock. You played a damned good game.' 1948 'N. SHUTE' *No Highway* 286 I've never heard such cock in all my life. 1956 L. MCINTOSH *Oxford Folly* 76 What he usually improvised was just a load of cock. 1967 L. DEIGHTON *Expensive Place* ii. 16 What a lot of cock.
20. (Earlier and later examples.)
1618 N. FIELD *Amends for Ladies* I, Oh man what art thou? When thy cock is up? 1714 *Cabinet of Love*, View my sore cock, his tender wounded head. 1967 A. S. BYATT *Game* xvii. 233 Guts and cock and all,..every little bit except the hair and teeth. 1969 *Landfall* XXIII. 107 'She had her hand on his cock.' 'There's no need to be crude.'
22. b. *cock ship*.
1909 *Daily Chron.* 2 Aug. 1/6 The opinion that Natal was 'cock ship' among a crowd of extraordinarily efficient battleships and cruisers. 1926 *Spectator* 11 Sept. 375/2 She carried the golden cock at her masthead as cock ship of the Mediterranean Fleet.
23. cock-and-hen, applied *attrib.* to anything which admits or includes both sexes; **cock-founder, -founding** (see quot. 1931); **cock-schnapper, snapper,** a small snapper (see SNAPPER *sb.*¹ 6); **cock-stand** [COCK *sb.*¹ 20] *coarse slang*, an erection of the penis; **cock-sucker** *coarse slang*, a fellator; freq. used as a generalized term of abuse; so *cock-sucking* ppl. adj.; **cock-teaser** *coarse slang*, a sexually provocative woman who evades or refuses intercourse; hence (as back-formation) **cock-tease** *v. trans.*; **cock-teasing** *ppl. a.*
1785 [see HEN *sb.* 5]. 1889 *Referee* 19 May (Ware), What are described as working men's clubs.., very early breakfast clubs, cock-and-hen clubs..are all flagrant and distinct violations of the Licensing Acts. 1925 W. DEEPING *Sorrell & Son* xxiii, It was a cock and hen affair, but more hen than cock. 1952 R. SHERBROOKE-WALKER *Khaki & Blue* iii. 25 It seemed strange at first to be in a 'cock and hen' mess. 1807 *European Mag.* Dec., List of Bankrupts from June 20 to Dec. 26, 1807. Higginbottom, W...Cock-founder. 1838 *Civ. Engin. & Archit. Jrnl.* I. 381/2 Why cock-founders make the cocks with three different capacities of orifice,.. is another secret. 1931 *N. & Q.* 17 Jan. 50/2 A cock-founder is a maker of metal (usually brass or gun-metal) cocks or taps. Cock-founding is one of the very many trades carried on in Birmingham. 1882 J. E. TENISON-WOODS *Fish of N.S.W.* 41 The schnapper or count-fish, the school-fish, and squire...Juveniles rank the smallest of the fry, not over an inch or two in length, as the cock-schnapper. 1867 *Rosa Fielding* in S. Marcus *Other Victorians* (1966) v. 227 He had a tremendous cock-stand, and felt that if it was not allayed pretty quickly that he must burst. *c* 1882–94 *My Secret Life* I. 57 Then I began to have cock-stands. 1967 A. WILSON *No Laughing Matter* III. 367 Marcus..found, as his eyes took in the young man's flirtatious glance, that he was beginning a cock stand. 1891 FARMER *Slang* II. 144/1 *Cock-sucker*, a feliatrix. 1935 J. T. FARRELL *Guillotine Party* 253 'You know what I am?' he asked, as if in self-torture. 'Yes,' Jack said casually. 'I'm a c.........' 1962 J. BALDWIN *Another Country* (1963) II. iii. 271 If it wasn't for the spooks wouldn't a damn one of you white cock suckers *ever* get laid. 1969 P. ROTH *Portnoy's Complaint* 116 My father said 'No' to one after another of the thousand little accessories the cock-sucker wanted to sell us. 1923 E. E. CUMMINGS *Let.* 28 July (1969) 99 Being members..of the cocksucking leisure class. 1969 C. HIMES *Blind Man with Pistol* xxi. 226 Mother-raping cocksucking turdeating bastard, are you blind? 1957 A. WILSON *Bit off Map* 71 You think..that because you have attractive eyes and a good figure that you can go on having sex appeal just by cock-teasing every man you meet. 1891 FARMER *Slang* II. 144/2 *Cock-teaser* or *cockchafer*, a girl in the habit of permitting all familiarities but the last. 1962 J. BALDWIN *Another Country* (1963) I. i. 50 What are you, anyway—just a cock teaser? 1968 T. E. B. CLARKE *Trail of Serpent* vii. 67 The sixth commandment hadn't been broken in his manor since a Polish farm labourer strangled a cock-teasing Sunday hiker.

cock, *v.*¹ **2.** Delete † *Obs.* and add later *trans.* examples with *up*.
1914 JOYCE *Dubliners* 146 The mother you know, she cocks him up with this and that. 1947 D. M. DAVIN *Gorse blooms Pale* 38 A dog with boys for master has no master. They'll always cock him up and end by spoiling him. 1958 L. A. G. STRONG *Light above Lake* 106 Don't cock the fella up.
3. *to cock snooks, a snook*: see SNOOKS.

cockabully (kǫ·kăbuli). *N.Z.* [Corruption of Maori *kokopu*.] A small blunt-nosed freshwater fish in New Zealand.
1896 *Australasian* 29 Aug. 407/3 During my stay in New Zealand my little girl caught a fish rather larger than an English minnow. Her young companions called it a 'cock-a-bully'...The Maori name for the fish is 'kokopu'. 1934 J. A. LEE *Children of Poor* (1949) 46 Minnows and cockabullies darted in and out of the cress. 1962 B. J. MARPLES *Freshwater Life N.Z.* xviii. 138 The Bullies, or Cockabullies, are small, very common, blunt-nosed fish...They have large pectinated scales which extend on to the head and operculum in the species of *Gobiomorphus*, but are confined to the posterior part of the body in the genus *Philypnodon*.

cockamamie (kǫ·kămēⁱ:mi), *sb.* and *a.* *slang* (orig. *U.S.*). Also **-mamy, -manie, -many.** [Prob. a corruption of DECALCOMANIA.] (See quot. 1968.)
1945 A. KOBER *Parm Me* 5 'Cockamanies'—painted strips of paper which the kids applied to their wrists and rubbed with spit until the image was transferred to their hands. 1960 *Time & Tide* 24 Dec. 1599/2 'A square', Nemo shouts.. 'he is a cockamamie type, he is a stale adventurer...He's not in my world.' 1962 'E. MCBAIN' *Empty Hours* (1963) II. xi. 113 You marched into the precinct with a tight dress and a cockamamie bunch of alibis. 1963 J. PHILIPS *Dead can't Love* I. iii. 20, I was suddenly down to reality..cut loose from a cockamany dream. 1968 *Encounter* Sept. 31/1 *Cockamamy*,..1. Mixed-up, muddled; ridiculous, implausible; not credible, foolishly complicated...2. Decalcomania: *i.e.* a picture or design left on the skin as a 'transfer', from specially prepared paper which is wetted and rubbed.

cockamaroo (kǫ:kămărū·). A variety of bagatelle.
1850 *Bohn's Hand-bk. Games* 613 (*heading*) Russian Bagatelle, or Cockamaroo Table. 1875 *Encycl. Brit.* III. 230/2 *Cockamaroo*, or Russian Bagatelle, is played on a table prepared with a number of pins, holes, arches, and bells, up to and through which the ball is played from the baulk end of the table. It is a childish amusement, requiring little skill. 1893 *Sat. Rev.* LXXVI. 561 Like a glass marble in the good old game of Cockamaroo. 1969 E. H. PINTO *Treen* 216/2 In Cockamaroo or Russian bagatelle, the ball is driven through an arrangement of pins, holes, arches and bells.

cock-and-bull. 3. (Earlier and later examples).
1795 *Gazette of U.S.* (Philad.) 2 Mar. (Th.), A long cock-and-a-bull story about the Columbianum [a proposed national college]. 1915 W. S. MAUGHAM *Of Human Bondage* xcii. 483 You're a past mistress of the cock-and-bull story. 1952 F. YERBY *Woman called Fancy* xvi. 305 I'm going to invite them in—let them search the place. Give them a cock and bull story about sending the boy away with one of the servants.

cockatoo. Add: **2.** Also *N.Z.*
1874 A. BATHGATE *Colonial Exp.* xv. 195 Many words in daily use in Otago bear traces of importation from the.. Australian colonies, and none betrays its Australian origin more than that used to denote the agricultural class, who are usually styled 'Cockatoos'. *Ibid.* 197 The Cockatoos used to do very well from the sale of their cattle. 1874 J. A. H. CAIRD *Notes on Sheepfarming in N.Z.* v. 40 He [*sc.* an intending settler] can either be a squatter—i.e., a sheep-farmer, or a cockatoo, as the ordinary arable farmer is called, or a storekeeper. 1941 O. DUFF *N.Z. Now* iv. 51 The most they [*sc.* sheepfarmers] can hope for is an uneasy truce with dairymen..or an alliance with Labour to control the 'cockatoos'.
3. a. A convict. *Obs.* exc. *Hist.* **b.** A look-out or sentinel acting on behalf of a person or persons engaged on some illegal activity; a 'corner-man'. *Austral. slang.*
1827 P. CUNNINGHAM *N. S. Wales* (ed. 2) II. 288 It being a common trick to station a sentinel on a commanding eminence to give the alarm, while all the others divert themselves, or go to sleep. Such are known here by the name of 'cockatoo-gangs', from following the example of that wary bird. 1870 J. L. BURKE *Adv. Martin Cash* viii. 123 He's the bravest man that could choose from Sydney men or Cockatoos...This name was applied to a body of desperate men, who were imprisoned on Cockatoo Island.. under a strong military guard. 1934 *Bulletin* (Sydney) 1 Aug. 36/1 For years those betting on the outers had to employ one or more 'cockatoos' to give warning when a John Hop was spotted. 1945 BAKER *Austral. Lang.* vii. 141 Hotel licensees who indulge in after-hours trading often pay a cockatoo to warn them of the approach of police. 1953 A. MOOREHEAD *Rum Jungle* iv. 52 At the entrance [to the two-up school] a door-keeper sits, known as 'the Cockatoo'. 1966 *Telegraph* (Brisbane) 15 Aug. 2/7 They watched Foster (the 'cockatoo' or spy) point out punters who had laid a large bet.
4. *attrib.* and *Comb.*, as (sense 2) *cockatoo farmer* (so *cockatoo farm*), *settler*, *squatter*, *stockman*; **cockatoo fence**, a rough fence of logs and saplings.
In some uses with reference to the cockatoo's habit of sitting with others in a row on a fence.
1827 Cockatoo-gang [see sense *3]. 1856 *Melbourne Punch* 25 Sept. 58 Cockatoo farmer got up and catechised me. 1859 in A. F. Ridgway *Voices from Auckland* (1860) 51, I am, Sir, &c. A Cockatoo Settler. 1881 Cockatoo farmer [see sense 2 in Dict.]. 1884 'R. BOLDREWOOD' *Melb. Mem.* xxii. 155 There would be roads and cockatoo fences,..in short, all the hostile emblems of agricultural settlement. 1890 —— *Miner's Rt.* xlii. 377 The governor is a bigoted agriculturist; he has contracted the cockatoo complaint, I'm afraid. 1890 —— *Col. Reformer* xviii, The cockatoo stockmen, who are doing the 'reviewing', safely on the fence. 1893 *Argus* (Melbourne) 17 June 13/4 (Morris), Hire yourself out to a dairyman, take a contract with a rail-splitter, sign articles with a cockatoo selector; but don't touch land without knowing something about it. 1893 *Funk's Stand. Dict.*, *Cockatoo squatter*, the owner of a limited right of pasturage on land granted by the government, who raises only a small amount of wool or cattle annually: used mostly in contempt. 1897 I. SCOTT *How I Stole over 10,000 Sheep* i. 7 We'll..get a job on a 'Cockatoo' farm somewhere. 1901 'M. FRANKLIN' *My Brilliant Career* viii. 60 The wire fence.. had replaced an old cockatoo fence which I remembered in my childhood.
Hence **cockatoo** *v. intr.*, to sit on a fence, as the bird does; to farm in a small way. *Austral. colloq.*
1878 'R. BOLDREWOOD' *Ups & Downs* xx. 245 Fancy three hundred acres in Oxfordshire, with a score or two of bullocks, and twice as many black-faced Down sheep. Regular cockatooing. 1890 —— *Col. Reformer* xviii, The correct thing, on first arriving at a drafting yard, is to 'cockatoo', or sit on the rails.

cock-bead. [COCK *sb.*⁶ + BEAD *sb.* 6.] A quirked or projecting bead. So **cock-beaded** *a.*, **cock-beading** *sb.*
1805 *Edin. Bk. Prices* p. vii, Cock-beaded..cockbeads. 1876 *Encycl. Brit.* IV. 487/2 If there be a deep sinking under a bead it is called a quirked or cock bead. 1955 R. FASTNEDGE *Eng. Furnit. Styles* 282 *Cock beading*, astragal moulding; small and applied to the edges of drawer fronts, 1730–1800. 1961 *Connoisseur New Guide Antiq. Eng. Furnit.* 59 At first drawer edges had the lip moulding which had come into fashion about 1710, but after 1730 it was gradually replaced by the cock-bead, which was almost exclusively used from about 1745 until the end of the century. 1969 V. CANNING *Queen's Pawn* vi. 83 A Chippendale period mahogany desk with cock-beaded drawers and carved gilt handles.

Cockburn (kōᵘ·bɜ̄ɹn). [Proper name.] The name of a firm of London wine shippers founded in 1815, used to denote port shipped by this firm. In full *Cockburn's port*.
1895 *Army & Navy Co-op. Soc. Price List* 164 Old bottled Port...Cockburn's, bottled 1891. 1920 G. SAINTSBURY *Notes Cellar-bk.* iii. 37, I have done justice to Cockburn advisedly. *Ibid.* 38 The best rich that I ever had was..a Cockburn of '81. 1928 D. L. SAYERS *Bellona Club* iii. 25 Holding the Cockburn '86 beneath his nostrils. 1931 E. LINKLATER *Juan in Amer.* IV. vii. 337 They drank an Amontillado sherry, a bottle of Château Yquem, and some very good Cockburn's port. 1964 N. FREELING *Double-Barrel* III. vi. 105 A cut-glass decanter..designed to make the grocer's port taste like the Cockburn twenty-seven.

cocked, *ppl. a.*¹ Add: **b.** (See quots.)
1889 F. M. HALFORD *Dry-Fly Fishing* ii. 36 A floating fly ..cocked, or, in other words, floating with the wings up. *Ibid.* 43 Floating a cocked fly over a likely place. 1961 J. S. SALAK *Dict. Amer. Sports* 90 When an artificial fly lands on the water in an upright position (i.e., with body more or less horizontal to the water), it is said to be 'cocked'.

cocked hat. Add: **1. b.** A note, napkin, etc., folded in the form of a cocked hat.
1864 DICKENS *Mut. Fr.* I. I. x. 87 Twemlow..received a highly-perfumed cocked-hat and monogram from Mrs. Veneering. 1938 D. SMITH *Dear Octopus* III. i. 115 *Dora.* Quite. (*She has been wildly folding napkins*..) But there won't be any water-lilies. We'll have some nice, simple cocked hats.
4. (Earlier and later examples.) orig. *U.S.*
1833 PAULDING *Banks of Ohio* I. 217, I told Tom..I'd knock him into a cocked-hat, if he said another word. 1848 LOWELL *Lett.* (1894) I. 146 If he had *riled* me, I might have knocked him into a cocked hat in my satire. 1965 *Listener* 26 Aug. 314/3 The splendid earthworks and engineering structures of the railways today..knock the M.1 into a cocked hat.
5. *attrib.* and *Comb.* (Earlier and later examples.)
1846 DICKENS *Pict. fr. It.* 13 The napkins are folded in cocked-hat fashion. 1886 R. BROUGHTON *Dr. Cupid* I. viii. 93 One of Lady Roupell's almost daily cocked-hat notes. 1902 *Daily Chron.* 9 Jan. 3/3 Mr. Watson's story is in the mode of the eighteenth century, and for those who like cocked-hat stories it is just the thing. 1935 *Discovery* Nov. 333/2 The early pottery lamps of the Ægean, Phœnicia, etc. (known to the British Museum as the 'cocked-hat' type..). 1940 *Brit. Jrnl. Psychol.* Jan. 235 We might assume..that the quantities..are distributed on a 'cocked hat' frequency curve.

cocker, *sb.*² **3.** (Earlier example.)
a 1811 M. FLEMING *Jrnl.* (1934) II. 58 Yesterday I am very glad to say a young Cocker came to our house to stay.

Cocker, *sb.*⁵ [The name of Edward Cocker (1631–75), arithmetician, reputed author of a popular *Arithmetick*.] *according to Cocker*: by or in accordance with strict rule or calculation; exactly; strictly.
[1756 A. MURPHY *Apprentice* I. i. 14, I have Cocker's Arithmetick below Stairs... I'll..get it for him.] 1818 C. MATHEWS *Let.* 27 July in A. Mathews *Mem. C. Mathews* (1839) III. 19 Three instead of one..according to Cocker—if one does good..how much will three do? 1825 *Mechanics' Mag.* 11 June 149/2 A short Table,..which, I think, will be found to be according to 'Cocker'. 1861 HUGHES *Tom Brown at Oxf.* xxxii, 'So you ought to be, according to Cocker, spending all your time in sick rooms.' 'According to who?' 'According to Cocker.' 'Who is Cocker?' 'Oh, I don't know; some old fellow who wrote the rules of arithmetic, I believe; it's only a bit of slang.' 1883 G. A. SALA in *Illustr. Lond. News* 24 Nov. 499/2 The average American may not know what we mean by 'according to Cocker'; while the average Englishman may be unaware of the meaning of 'according to Gunter'. 1888 G. ALLEN *This Mortal Coil* ii, According to Cocker, nought and nought make nothing. 1893 E. F. KNIGHT *Where Three Empires Meet* xxiv. 372 So, despite Cocker, at this short range..the guns opened fire on the fort.

cocker, *sb.*⁶ *slang* and *dial.* [cf. COCK *sb.*¹ 8.] Used as a form of address to a person.
1888 H. EVANS *Brighton Beach Loafer* vi. 35 'Ime on, my cocker,' I ses. 'Guv us yer and orn it my pippin, an arf a quid on account.' 1942 *Horizon* Aug. 114 You don't want to do that, cocker. 1960 A. WESKER *Talking about Jerus.* III. ii. 214 It was good of you to help us cocker.

cockey: see *COCKY *sb.*²

cock-eye¹. Add: **B.** *adj.* Cock-eyed, 'topsy-turvy'. *colloq.*
1899 KIPLING *Stalky* 299 Don't see how you can make Latin prose much more cock-eye than it is, but we'll try, said Beetle, transposing an *aliud* and *Asia* from two sentences.

1928 *Sunday Express* 16 Dec. 2/1 The world is all going cock-eye.
2. *cock-eye Bob*, a cyclone or thunderstorm. *local Austral. slang.*
1926 *Austral. Encycl.* II. 66/2 'Cock-eye Bob' is the name given to thunder-squalls which occur frequently on the north-west coast of Western Australia during the summer months. 1937 I. L. IDRIESS *Forty Fathoms Deep* viii. 73 Almost daily the sky blackened as cock-eye bobs shrieked upon them, to lash them in sheets of rain. 1938 X. HERBERT *Capricornia* iii. 24 A storm of the type called Cockeye Bob in Capricornia..burst over Flying Fox in the middle of the night. 1949 *Geogr. Mag.* Feb. 373 The *Cock-eye-Bob*, or cyclone, of north-west Australia.
3. *cockeye pilot* = *BEAU-GREGORY.
1905 D. S. JORDAN *Guide to Study of Fishes* II. xxii. 383 The 'cockeye pilot', or jaqueta,..green with black bands, swarms in the West Indies. 1930 [see *BEAU-GREGORY].

cock-eye². **1.** The loop at the end of a trace by which it is attached to the swingletree, etc.
1850 *Rep. Comm. Patents 1849* 266 The combination of the loop of the trace with a sectional cross piece (B), and a Cockeye (A), whereby the trace is secured to a swivel cockeye. *a* 1877 KNIGHT *Dict. Mech.*, *Cock-eye*,..an iron loop on the end of a trace, adapted to catch over the pin on the end of a single-tree. Originally *woodcock eye*. 1895 *Montgomery Ward Catal.* 327/1 Japanned Screw Cockeyes.
2. (See quot.)
a 1877 KNIGHT *Dict. Mech.*, *Cock-eye* (*Milling*), a cavity on the under-side of the balance-rynd to receive the point of the spindle.

cock-eyed, *a.* Add: **1. b.** *fig.* Goggle-eyed; topsy-turvy, absurd, ridiculous.
1896 *Daily News* 29 Feb. 5/3 It is a gorge that requires a good deal of cock-eyed watching by foot passengers. 1933 *Punch* 31 May 597 There doesn't seem to be anyone about in this cock-eyed town. 1945 KOESTLER *Twilight Bar* I. 30 When it's summer in the North, it's winter in the South. Completely cockeyed. 1960 M. SPARK *Ballad of Peckham Rye* x. 201 He gathered together the scrap ends of his profligate experience..and turned them into a lot of cockeyed books.
2. Drunk. orig. *U.S.*
[1722 B. FRANKLIN in *New-Eng. Courant* 3–10 Sept., Cogey.] 1926 HEMINGWAY *Sun also Rises* xii. 126 'You're cock-eyed,' I said. 'On wine?' 'Why not?' 1933 M. LINCOLN *Oh! Definitely* viii. 102 'Have another drink?'..'Love it,' she said. 'Only don't forget that I mustn't turn up cock-eyed to meet Horace.' 1934 E. LINKLATER *M. Merriman* xvi. 178 You wouldn't have asked me to marry you if you hadn't been cock-eyed at the time.
3. *cock-eyed Bob* = *cock-eye Bob*. *local Austral. slang.*
1894 *Age* (Melbourne) 20 Jan. 13/4 (Morris), The approach of an ordinary thunderstorm or 'Cock-eyed Bob'. 1930 E. R. B. GRIBBLE *40 Yrs. with Aborigines* xvii. 173 We had our first experience of a Cock-eyed Bob, or storm of wind, thunder, lightning, and rain. 1952 T. A. G. HUNGERFORD in *Coast to Coast 1951–2* 13 The pearlers who anchored in the bay sometimes to escape the cockeyed-bobs that sprang up so suddenly along those desolate, treacherous coasts.
Hence **co·ck-e:yedness.**
1941 KOESTLER *Scum of Earth* 78 And yet in this apparent cock-eyedness there was the same administrative logic. 1942 *Chicago Tribune* 18 May 12/1 Communists are cockeyed.. Today we shall dwell not upon their dangerousness, but upon their cockeyedness.

cock-horse. Add: **6.** An additional horse for helping a coach uphill.
1891 *Field* 25 July 134/1 With no further use for the cock horse, we cast him off at the top of the hill. 1909 *Daily Chron.* 7 June 6/3 The Venture Challenge Cup attracted eleven road teams... One, with an attendant cock horse, [etc.]. 1928 *Times* 3 Sept. 16 The Old Berkeley passing the Five Alls on Dashwood Hill, with a pair of cock-horses to help the team over the Chilterns. 1963 BLOODGOOD & SANTINI *Horseman's Dict.* 50 *Cock-horse*, extra horse attached in the lead of a four-in-hand when going up hill.

cockie: see COCKY *sb.*¹ and *sb.*² in Dict. and Suppl.

cocking, *vbl. sb.*² Add: **b.** *cocking-dog, -handle, -lifter, -piece, -rod, -swivel.*
1892 GREENER *Breech-Loader* 26 There is in this no cocking-dog, but the forward ends of the tumblers are turned in, and engage with..a cocking-swivel. *Ibid.* 28 When the tumbler is down, the cocking-rod is freed. *Ibid.* 111 The cocking-lifters of hammerless guns. 1902 *Encycl. Brit.* XXXII. 650/2 When released, the striker can be cocked by pulling out the cocking-piece by hand. 1905 *Kynoch Jrnl.* July–Sept. 141 The scear and cocking piece are on the right side of the action. 1917 *Blackw. Mag.* Mar. 382/2, I applied immediate action, pulled back the cocking-handle and pressed the trigger again. 1957 RAWNSLEY & WRIGHT *Night Fighter* ii. 40 Work the cocking-handle of the gun up and down.

cockle, *sb.*² Add: **7.** *cockle-picker, -woman.*
1922 JOYCE *Ulysses* 47 Cocklepickers..waded a little way in the water. 1929 W. B. YEATS *Let.* 28 Mar. (1953) 147, I thought you more subjected than you are to the rat-catchers and cockle-pickers. 1856 GEO. ELIOT in J. W. Cross *Life* (1884) I. vii. 329 The sight of the cockle-women at Swansea..would make a fine subject for a painter. 1949 DYLAN THOMAS *Lett.* (1966) 325 You climb the stones to see ..the cocklewomen webfoot.

cockney, *sb.* Add: **2. e.** A young snapper, the fish *Chrysophrys guttulatus*. *Austral.*
[1882 J. E. TENISON-WOODS *Fish of N.S.W.* 41 Juveniles rank the smallest of the fry, not over an inch or two in length, as the cock-schnapper.] 1906 D. G. STEAD *Fishes of Australia* 126 Up to about 4 or 5 inches in length, the young fry of the Snapper, which are then characterised by the possession of dark vertical bars on the body, are very often known as 'Cockneys'. 1951 T. C. ROUGHLEY *Fish & Fisheries Australia* 77 Various names are given to the snapper during the course of its growth. The youngest stages are known as 'cockneys'.

cockneyess. (Earlier and later examples.)
1839 F. A. KEMBLE *Resid. Georgia* (1863) 132 There is one privilege which I enjoy here which I think few cockneyesses have ever had experience of. 1844 *Ainsworth's Mag.* VI. 206 The number of our pretty cockneyesses. 1889 E. SAMPSON *Tales of Fancy* 40 My father was a Cockney, my mother was a Cockneyess, and I am the unfortunate result.

cockneyfication. (Examples.)
1883 H. JAMES *Portr. Places* 248 With regard to most romantic sites in England, there is a sort of average cockneyfication with which you must make your account. 1930 *Time & Tide* 21 Feb. 235/1 How are we to prevent this cockneyfication of the English language?

cockpit. Add: **3. c.** *Aeronaut.* In the fuselage of any kind of aircraft, or in the capsule of a space vehicle: the space occupied by a pilot, observer, astronaut, or (formerly) a passenger. Also *attrib.*
1914 *Rep. & Mem.* (*Advis. Comm. Aeronaut.*) No. 112 There are several speed indicators..in which the pressure of the air in the cockpit is allowed to act on one side of the recording diagram. 1915 *Chambers's Jrnl.* 1 July (Advt.), Pilot and passenger in separate cockpits arranged tandem fashion. 1917 *Blackw. Mag.* Mar. 383/1 Several bullets ventilated the fuselage quite close to my cockpit. *a* 1918 McCUDDEN *Five Years R.F.C.* (1919) 227 The observer disappeared into the cockpit apparently disabled. 1930 *Discovery* Sept. 304/1 Wireless signals from directional beams are heard in the pilot's cockpit. 1945 *Aeronautics* 54/2 Cockpit drill, which became an absolute necessity when the number of controls and instruments rose to double that of those incorporated in aircraft of a decade ago. 1962 J. GLENN in *Into Orbit* 40 This is the cockpit, the real 'opinion area' of the capsule.
d. *Motor Racing.* The space in a racing car occupied by the driver.
1935 EYSTON & LYNDON *Motor Racing* iv. 40 Smoke poured from Nuvolari's cockpit, and he climbed from his seat. 1957 S. MOSS *In Track of Speed* ii. 20 He sat well back in the cockpit with his arms almost outstretched as they held the wheel. 1968 *Autocar* 14 Mar. 41/2 We did a quick cockpit check, lights, horns, entered up the log book and pulled out of the station yard.

cocksfooter (kǫ·ksfu:təɹ). *N.Z.* [f. COCKSFOOT 1 + -ER¹.] One who gathers cocksfoot seed for sale. Hence **co·cksfooting** *vbl. sb.*, the gathering of cocksfoot seed.
1903 *N.Z. Illustr. Mag.* Aug. 349/2 All parts of New Zealand send their quota of cocksfooters to 'make a cheque'. *Ibid.* 349 (*title*) Cocksfooting on Banks Peninsula. *Ibid.* 352/1 In many ways cocksfooting resembles shearing. 1913 A. I. CARR *Country Work & Life N.Z.* v. 13 To get back to the cocksfooting. The grass is cut with a sickle.

cocktail. Add: **3. a.** Delete 'Chiefly *U.S.*' and substitute 'orig. *U.S.*' (Further examples.)
1806 *Balance* (Hudson, N.Y.) 13 May 146 *Cock tail*, then, is a stimulating liquor, composed of spirits of any kind, sugar, water, and bitters. 1837 J. E. ALEXANDER *Narr. Voy. Colonies W. Africa* I. ix. 223 He invited us to take 'gin sling', or 'cock-tail', at the side-table. 1843 DICKENS *Mart. Chuz.* xvi. 202 He could..drink more rum-toddy, mint-julep, gin-sling, and cocktail, than any private gentleman of his acquaintance. 1931 M. DE LA ROCHE *Finch's Fortune* vi. 103 He..sipped cocktails in the lounge. 1948 *New Yorker* 6 Nov. 64/2 Cocktails are now so numerous that no bartender..can remember how to make all of them. 1968 *New Society* 22 Aug. 266/1 Cocktails, in 'We're going out to cocktails', is non-U.
c. *ellipt.* = *cocktail party*.
1951 N. MITFORD *Blessing* I. viii. 91 If one lunches every day and goes to, say, three cocktails, as well as dining out, one can go to forty houses in a week. 1958 'A. BRIDGE' *Portuguese Escape* ii. 20, I must go to a cocktail at the Belgian Embassy.
4. A preparation of food, usu. served as an appetiser at the beginning of a meal; freq. with the main ingredient prefixed, as *fruit, lobster, prawn cocktail*.
1928 *Sat. Even. Post* 12 May 107/1 Mr. Montgomery had taken a morsel of fruit cocktail. *c* 1938 *Fortnum & Mason Catal.* 53/1 Oyster Cocktail—per bot. 1/9. 1959 N. MAILER *Advts. for Myself* (1961) III. 231 Taking off to greet new arrivals whose crabmeat cocktail she nibbled on. 1960 M. PATTEN *Cookery in Colour* no. 23 The correct way of serving these cocktails, though, is to use glasses, when the lettuce should be shredded very finely and put at the bottom of the glasses. *Ibid.* no. 25 Cocktail sauce for Prawn or Shrimp Cocktail. 1964 *Listener* 24 Sept. 469/3 Twentieth-century delicacies such as lobster cocktail and *coq au vin*. 1968 M. TORRIE *Your Secret Friend* iii. 28 What would you think to a nice prawn cocktail, sir?
B. 2. *attrib.* and *Comb.*, as *cocktail bar, cabinet, glass, habit, hour, lounge, powder* (in Dict.), *table, time, tray*; *cocktail-drinking, -shaking* vbl. sbs.; **cocktail dress,** a dress suitable for wearing at a cocktail party; so *cocktail gown, suit*; **cocktail-mixer** or **shaker,** a container in which cocktails are vigorously shaken to mix them; **cocktail onion,** a small pearl onion placed on a stick and served in certain cocktails; **cocktail party,** a party, esp. one intended for social conversation, at which cocktails are served, usu. together with other alcoholic drinks, savoury snacks, canapés, etc.; **cocktail pianist,** a player of light inconsequential (usu. jazz-based) background music; so *cocktail piano*; **cocktail sausage,** a very small sausage such as is served at cocktail parties; **cocktail stick,** a small pointed stick on which snacks, sausages, onions, etc., are served at cocktail parties.
1929 E. SITWELL *Gold Coast Customs* 20 The ghost haunting steamers And cocktail bars. 1934 W. PLOMER *Invaders* xiv. 262 What the visitors took to be a cocktail bar proved..to be a gramophone. 1953 E. SMITH *Guide Eng. Trad.* 145 Various..'cocktail-bars' attached to well-known restaurants. 1933 *Archit. Rev.* LXXIV. 228/2 A cocktail cabinet. 1958 *Times* 3 Nov. 14/1 He can afford to buy, on hire purchase, telly, cocktail cabinet, and washing machine. 1935 *Times* 2 Oct. 17/4 An elaborate dinner *ensemble* tailored in heavy satin and with a full length skirt..makes a contrast to the cocktail dress which retains its skirt about 12 in. from the ground. 1926 D. L. SAYERS *Clouds of Witness* x. 193 I am old-fashioned enough not to have adopted the modern practice of cocktail-drinking. 1907 E. WHARTON *Fruit of Tree* III. xxiv. 354 Leaving everywhere in her wake a trail of cigarette ashes and cocktail glasses. 1930 A. BENNETT *Imperial Palace* xxxi. 208 Cocktail glasses. Yes. Cocktails were the the most profitable trade in the hotel. 1966 G. GREENE *Comedians* I. ii. 53, I remember him raising a cocktail-glass to the light and showing me the delicate engraving of a bull's mask. 1935 *Times* 25 Nov. 17/5 A dinner or cocktail gown. 1905 *Daily Chron.* 22 Dec. 4/5 The cocktail habit. 1927 HEMINGWAY *Men without Women* (1928) 68 Sometimes, after the cocktail hour. 1930 A. BENNETT *Imperial Palace* xiii. 78 'Not quite the cocktail hour here, is it?' said Sir Henry. 1966 *Observer* (Colour Suppl.) 13 Nov. 40/2 The Cocktail Hour, commonly known as drinks time, is a mysterious 6–8 p.m. limbo. 1939 G. GREENE *Lawless Roads* xi. 287 Racketeers of the brothel and dance-hall and cocktail lounge. 1904 'O. HENRY' *Cabbages & Kings* iii. 47 A bullet-headed man Smith was, with an oblique, dead eye and the moustache of a cocktail-mixer. 1951 E. DAVID *French Country Cooking* 217 Pickled gherkins and baby cocktail onions. 1928 D. H. LAWRENCE *Lady Chatt.* vi. 67 She almost wished she had.. made her life one long cocktail party and jazz evening. 1950 T. S. ELIOT (*title*) The cocktail party. *Ibid.* I. i. 12 The only reason for a cocktail party For a gluttonous old woman like me Is a really nice tit-bit. 1962 *New Yorker* 21 July 68/2 Oscar Peterson (the best cocktail pianist in the business). 1965 'P. QUENTIN' *Family Skeletons* i. 13 But how few people remember cocktail pianists. They're usually considered as part of the furniture. 1952 B. ULANOV *Hist. Jazz Amer.* (1958) xviii. 230 His cocktail piano records for Victor insinuate jazz. 1966 *Melody Maker* 15 Oct. 8/4 This Saturday, he leaves the club as resident pianist and trio leader, and he isn't contemplating a 'cocktail piano' job. *c* 1938 *Fortnum & Mason Catal.* 53/2 Sausages...Cocktail—per glass 1/3. 1939 *Vogue's Cookery Bk.* 157, ¾ lb. cocktail sausages. 1961 *Harrods Food News* 5/2 Cocktail Sausages flavoured with chilli. 1868 G. A. SALA in *N. & Q.* II. 401/1, I never possessed a pair of 'cocktail-shakers' myself, but a young officer in the Blues..did possess..a brace of tall silver mugs in which the ingredients of the beverage known as a 'cocktail'..are mixed, shaken together, and then scientifically discharged. 1928 *Observer* 17 June 15/4 A grimacing waiter tilts his cocktail-shaker. 1966 AUDEN *About House* 15 Nobody I know would like to be buried with a silver cocktail shaker. 1936 D. POWELL *Turn, Magic Wheel* I. 59 Walter and Corinne..whispered in the kitchen over the cocktail-shaking. 1963 *Listener* 28 Mar. 570/1 Claudel does some improving and cocktail-shaking of the legends. 1937 'M. INNES' *Hamlet, Revenge!* I. 22 A luridly-tinged cherry speared on a cocktail-stick. 1953 R. FULLER *Second Curtain* v. 80 About thirty people stood and talked, holding glasses..and little sausages on cocktail sticks. 1950 *Leader* Apr. 1 A slim girl..in a cocktail suit. 1965 J. M. CAIN *Magician's Wife* (1966) ii. 18 Flanking it were two sofas, a cocktail table between. 1929 D. L. MOORE *Pandora's Let. Box* xiii. 242 A cocktail time dress. 1958 J. BETJEMAN *Coll. Poems* 260 Your lives were good and more secure Than ours at cocktail time in Pinner. 1934 N. MARSH *Man lay Dead* xv. 265 Rankin is bending over the cocktail tray..shaking up the last of the cocktail. 1958 M. STEWART *Nine Coaches Waiting* vi. 73 The entry of Seddon with the cocktail tray interrupted us.

co·cktail, *v.* [f. COCKTAIL *sb.* 3.] *intr.* To drink cocktails; to attend a cocktail party. Also *trans.*, to give a cocktail to. So **co·cktailed** *ppl. a.*, under the influence of cocktails; **cocktailing** *vbl. sb.* and *ppl. a.*
1856 *Harper's Mag.* Dec. 66/1 Cocktailed past the point of nervousness and remorse, he dresses himself. 1935 *Amer. Speech* X. 192/2 She out-moderns the moderns in a frock that is made for cocktailing. 1936 *Times Lit. Suppl.* 13 June 498/1 He spends his time cocktailing. 1942 O. NASH *Good Intentions* 83 The last guest has been cocktailed and canapéed. 1960 L. KAUFFMANN *Waldo* (1962) iii. 39 Iris promptly drew him off to the cocktailing group. 1967 *Guardian* 10 June 6/6 All the Commonwealth bar Britain has already danced, cocktailed, and shown the flag.

cock-throttled, *a.* = COCK-THROPPLED.
1882 in R. HUNTER et al. *Encycl. Dict.* 1908 *Animal Managem.* 24 A 'Cock-throttled' or 'swan-necked' horse is one which has a neck like a fowl.

cock-up. Add: **A.** *sb.* **4.** A blunder, a mistake, a confused situation. *slang.*
1948 PARTRIDGE *Dict. Forces' Slang* 44 He made a complete cock-up of his orders. 1959 I. JEFFERIES *13 Days* xiii. 206 'I was thinking of going out to Tikvah today, sir,' I said, and did I make a cock-up of that job. 1964 J. PORTER *Dover One* i. 11 George turned the local boys on it and you've never seen such a cock-up in your life!

co·ck-winged, *a.* Defining an artificial fly used by anglers.
1897 *Encycl. Sport* I. 20/1 Small bottle of odourless paraffin for the anointment of the cock-winged flies. 1904 GALLICHAN *Fishing Spain* 207 Cock-winged, beautifully constructed dry flies. 1907 *Daily Chron.* 16 Nov. 3/2 The most beautiful of cock-winged quill gnats pass unheeded by the trout.

cocky, *sb.*² Add: **2.** Also **cockey, cockie.** Also *N.Z.* and *fig.* (Later examples.) Cf. *cow-cocky.
1891 in Williams & Reeves *Double Harness* 34, I made a little poem on the cockey and the squatter. 1900 H. LAWSON *On Track* 75 He..sold the crop standing to a Cockie who lived ten miles away. 1911 'KIWI' *On Swag* iii. 8 Squatters came up in their buggies and waggonnettes, 'cockies' in their spring traps. 1928 'BRENT OF BIN BIN' *Up Country* xiii. 218 There was no lack in the orchestral department, but Rab was the boss cocky of it. 1933 L. G. D. ACLAND in *Press* (Christchurch, N.Z.) 30 Sept. 15/7 *Cockatoo*. Now usually abbreviated to *cocky*. An agricultural farmer, a small farmer, as opposed to a squatter or sheep farmer. 1952 J. CLEARY *Sundowners* 9 You think the cockies have an easy time of it? 1960 B. CRUMP *Good Keen Man* 82 The cocky had a sheep-run in the foothills of the Coromandel Ranges.
b. *attrib.* and *Comb.* Also **cocky's joy** *slang*, treacle, golden syrup.
1899 H. LAWSON in *Bulletin* (Sydney) 11 Feb. 35/1 It is with the farmer or 'cockie' class that the writer is here chiefly concerned. 1910 C. E. W. BEAN *On Wool Track* v. 64 Cocky's joy is golden syrup in 2-lb. tins, costing sevenpence. 1911 E. M. CLOWES *On the Wallaby* vi. 157 The young teacher boards out with some neighbouring 'cocky' farmer and his wife. 1930 V. PALMER *Passage* III. vi. 257 These cockie farmers think any man that handles their fruit and butter is a machine they've bought and paid for. 1934 T. WOOD *Cobbers* 140 The Peninsula is the home for cocky-farmers, men who work a few hundred acres apiece. 1936 'R. HYDE' *Passport to Hell* ii. 51 If one of the 'cocky' specials fell among strangers..he might be wary of the boots that would as soon kick in a strike-breaker's head as not. 1947 D. M. DAVIN *Gorse blooms Pale* 194 'Go easy on that Cocky's Joy.'..Shorty went on spreading the Golden Syrup.
Hence **co·cky** *v. intr.*, to follow the occupation of a 'cocky'; to farm; cf. *cow-cockying.
1941 K. TENNANT *Battlers* v. 52 A job 'cockying' for one of her uncles or cousins would come to nothing again.

coco, cocoa. Add: **1. b.** *slang* = 4 c. *U.S.*
1828 G. GODFREY *Hist. G. Godfrey* III. vii. 101 He finished, by telling..[Mr. Rafferty], that if he did not clap on his muzzle, he would give him a *pelt on the cocoa*, which should not leave him *mag* enough for a last dying speech. 1890 *The Road* (Denver) 24 May 8/1 What put that thought into your cocoa? 1910 WODEHOUSE *Gent. Leisure* xiii. 137 Who do you t'ink I nearly bumped me coco against out in de corridor downstairs?
2. b. In full *coco(a)-grass*: a plant of the sedge family (*Cyperus rotundus*), prevalent chiefly in southern U.S.; NUT-GRASS.
1837 in J. R. Commons et al. *Doc. Hist. Amer. Industr. Soc.* (1910) I. 221 Cutting coco grass on the 22d. 1853 J. G. BALDWIN *Flush Times Alabama* 72 It is a sort of cocoa grass that has got into the soil. 1874 *U.S. Dept. Agric. Rep. 1873* 269 Even the hardy and noxious gramineal plant, commonly called 'coco' in Louisiana, is destroyed. 1894 *Congress. Rec.* 31 July 8049/1 That extraordinary grass known as cocoa grass, which has spread over large areas in the States of Louisiana and Mississippi. 1897 R. M. STUART *Simpkinsville* 43 The blades of bright green cocoa spears. 1902 W. S. GORDON *Recoll. Old Quarter* 111 The floods, the irrepressible cocoa-grass, the poisonous vegetable exhalations are in his way. 1947 *Jrnl. Wildlife Managem.* Jan. 54/1 'Coco' marsh, an association of 'coco' or salt-marsh bulrush, saltgrass, and 'wire-grass'.
3. coco-fibre (earlier example); **coco-matting** = coconut matting (4 d).
1870 W. ROBINSON *Wild Garden* 229 To plant grass over a bed..would never do...I found a substitute in cocoa-fibre mixed with a good sprinkling of silver sand. 1883 *Peel City Guardian* 7 Apr. (Advt.), Cocoa Mattings. 1920 D. H. LAWRENCE *Lost Girl* iii. 38 A hall..floored with cocoa-matting. 1958 C. S. FORESTER *Hornblower in W. Indies* 247 They make coco-matting by the mile in England now.
4. coco(a)-nut. c. (Earlier U.S. example.)
1834 W. A. CARUTHERS *Kentuckian* I. 66, I rather suspicion he thought a two year old colt's heel had got a taste of his cocoanut.
d. *coconut cake, fibre* (examples), *milk*; **coconut ice,** a confection made from desiccated coconut and sugar; **coconut matting** (examples); **coconut shy,** a sideshow at a fair in which contestants throw balls at coconuts placed in rings or cups in an attempt to dislodge them; **coconut water, whey,** applied to early stages of the milk of the coconut.
1830 *York Gaz.* (Pa.) 6 July 1/3 He has just received a complete and general assortment..consisting of..Mint Drop, Cocoanut Cake, and Lady Fancy's, &c. 1855 MRS. GASKELL *North & South* I. ix. 113 We will give him a welcome, and some cocoanut cakes. 1938 L. MACNEICE *I crossed Minch* II. xiv. 202, I bought and ate two coconut cakes. 1851 *Illustrated Exhibitor* p. xxxix/3 Mats of cocoa-nut fibre. 1952 A. G. L. HELLYER *Sanders' Encycl. Gardening* (ed. 22) 122 *Coleus*...*Propagation*..cuttings of young shoots in light soil or coconut-fibre refuse at any time. 1909 *'Best Way' Bk. No. 2* 71/2 Cocoanut ice. Ingredients.—One and a half pounds of loaf or castor sugar, a quarter of a pound of grated or desiccated cocoanut, [etc.]. 1937 'R. CROMPTON' *William—the Showman* v. 104 A small snub-nosed freckled boy with a perverted craving for cocoanut ice. 1960 WILLMOTT & YOUNG *Family & Class* iii. 88 Model hats and coconut-ice sold with immense enthusiasm. 1854 F. FABER *Let.* in L. M. Trevor *Newman* (1962) II. 42 Blood and brains on ye cocoanut matting of ye recreation room. 1870 KINGSLEY *Madam How & Lady Why* ix. 216 He lines them with a quantity of cocoa-nut fibre, picked out clean and fine, just as if he was going to make cocoa-nut matting of it. 1894 *Daily News* 24 May 5/4 Those members of the team [from South Africa] who have not previously been to England have not yet had sufficient time to become accustomed to the change from cocoa-nut matting wickets to turf. 1725 SLOANE *Jamaica* II. I. iii. 41 The [grey Nicker] Beans beaten and us'd with Coco-Nut Milk are good for Ruptures. 1932 J. R. ACKERLEY *Hindoo Holiday* I. 190 They [*sc.* tree-shrines]..are built to propitiate ghosts, spirits of the dead...Sacrifices of eggs and cocoanut milk were made. 1903 'A. MCNEILL' *Egregious Eng.* 175 Merry-go-rounds and cocoanut-shies. 1922 D. H. LAWRENCE *England, my England* (1924) 55 In the cocoanut shies there were no cocoanuts. 1817 J. WILLIAMSON *Obs. W. India Islands* i. 96 As a refreshing and nutritious drink, cocoa-nut water may be mentioned. 1883 *Chambers's Jrnl.* XX. 155/2 Cocoa-nut water, found in the green pod before the fibrous husk and nut as we know them here are formed. 1838 *Mag. Dom. Econ.* Feb. 254 The inhabitants give cocoa-nut whey to their cattle.

cocoa. Add: **3. b.** A shade of brown resembling the colour of the powder. Also *attrib.*
1894 *Westm. Gaz.* 23 July 5/3 The destroyers *Havock* and *Hornet*..are looking as smart as the peculiarly ugly cocoa colour they are painted will allow. 1923 *Daily Mail* 24 Apr. 1 (Advt.), Colours: Nigger, Cocoa, Canary, Brick, etc. 1940 F. D. DAVISON in B. James *Austral. Short Stories* (1963) 64 A broad-shouldered figure in a cocoa-brown suit. 1942 R. CHANDLER *High Window* (1943) ii. 23 The cocoa straw hat.
4. *Cocoa Press*: a familiar term for certain (now defunct) Liberal newspapers supposed to have been controlled by (Quaker) manufacturers of cocoa; so *Cocoa Quaker*.
1910 *Referee* 16 Jan., 'Modern Slavery', a pamphlet issued by the *World*,..speaks its mind frankly on..the Pro-Boer campaign conducted largely by the Free Trade Cocoa Press. 1915 *National Rev.* Jan. 689 The glory of the Cocoa Press has departed. *Ibid.* 712 According to the Imperial Chancellor..Wilhelm II was a Cocoa Quaker. 1944 M. WARD *G. K. Chesterton* xvii. 255 The *Daily News* and other papers owned by Mr Cadbury (of Cadbury's Cocoa) were often referred to as 'the Cocoa Press'.

cocoa (kōᵘ·ko), *v.* Also **coco.** In phr. *I should cocoa* (Rhyming slang), I should say so. Freq. used ironically.
1936 J. CURTIS *Gilt Kid* ii. 23 She don't do so bad, I should cocoa. 1964 *Listener* 31 Dec. 1053/2 'I should cocoa!' I retorts...'They won't get me in the Kate—not in a million years they won't.' 1965 J. PORTER *Dover Three* xi. 133 'She says you paid her a pound a week for the best part of nine months,' Dover went on. Mrs. Gomersall laughed scornfully. 'I should co-co!' she chortled. 1967 O. NORTON *Now Lying Dead* ii. 22 What me?..I should coco. Sheila'd think I was off my head.

cocoa-bay. Substitute for entry:
cocobay. Also **cocoby,** etc. [ad. Twi *kokobé*.] A form of leprosy once prevalent in the West Indies.
1788 H. M'NEILL *Treatm. Negroes, Jamaica* 35 Some of these disorders are not only dreadful but often incurable. These are the Yaws, [etc.];..the Bone-ache, and the Cocaby. 1801 T. DANCER *Med. Assistant* ix. 230 Leprosy and Coco-Bay or Joint Evil. These seem to be only varieties of the same disease; *viz.* the Elephantiasis or Lepra Arabum. 1807 R. RENNY *Hist. Jamaica* 205 The diseases which peculiarly affect them [*sc.* Negroes]..are the yaws, the cacabay, or leprosy, dirt-eating, and the jaw-fall. *a* 1818 [in Dict.]. 1873 *Lett. from Jamaica* ii. 26 One wretched creature, suffering from that particular form of the disease which goes by the African name of 'coco-bay', asked us for an alms. 1961 F. G. CASSIDY *Jamaica Talk* vii. 131 A type of leprosy with an African name is the cocobay.

cocobolo (kōᵘkobōᵘ·lo). Also **cocobola.** [Sp., ad. Arawak *kakabali*.] The timber from any one of several species of tree of the Central American genus *Dalbergia*, or the tree itself (see quots.). Also *attrib.*
1849 W. H. SIDELL *Jrnl.* 26 Feb. 23 (MS.), *Cocobolo*: Is when first cut yellowish or reddish yellow, it turns in a short time a dark rich yellowish brown. It is very hard and durable. 1897 *Sears, Roebuck Catal.* 107/1 Cocobolo handle knives and forks. 1904 H. STONE *Timbers of Commerce* No. 209, p. 239 (*heading*) Cocobola Wood. Species Unknown. 1951 R. MAYER *Artist's Handbk.* xii. 415 *Cocobola*, a very hard, dense, tough wood of a bright red colour and variable grain, imported from Nicaragua and Panama. Familiar through its use in knife handles. 1955 *Gloss. Commerc. Timbers* (*B.S.I.*) 30 *Dalbergia retusa* Hemsl. and allied species. Central America. Cocobolo.

cocodette (kokodẹt). [Fr.] A French prostitute of the fashionable world.
1867 in Ware *Passing Eng.* (1909) 82 In the circle of cocottes, and cocodettes, by which the French Court has during the last fifteen years managed to surround itself, fast American women have furnished no inconsiderable contribution. 1885 *Pall Mall Gaz.* 19 June 2/2 The cocodettes, or fine ladies, are no better than the cocottes.

co-co·nscious, *a.* and *sb.* *Psychol.* [See CO-.] **A.** *adj.* Pertaining to the conjunction of experiences within a single consciousness; also applied to the knowledge which the Divine mind has of the conscious acts or states of all other beings. **B.** *sb.* A division of the subconscious. Hence **co-co·nsciously** *adv.*, in a co-conscious manner; **co-co·nsciousness,** the state of being co-conscious.
1903 W. JAMES *Let.* 25 Feb. (1920) II. 184, I think that the indispensable hypothesis in a philosophy of pure experience is that of many kinds of other experience than ours, that the question co-consciousness/conscious synthesis (its conditions, etc.) becomes a most urgent question. 1904 —— in *Jrnl. Philos. Psychol.* I. 536 The conjunctive relation that has given most trouble to philosophy is the co-conscious transition, so to call it, by which one experience passes into another when both belong to the same self. 1906 G. T. LADD *Philos. Relig.* II. 141 Is there consciousness, or self-consciousness, anywhere in the wide world of things and selves..? In this consciousness, or self-consciousness, God is co-conscious. *Ibid.*, The conception of the Divine Omniscience as a species of cognitive activity which is at one and the same time 'Self-consciousness', and 'Other-consciousness', in the form of an all-embracing co-consciousness, meets with its supreme psychological objection when it is applied to God's knowledge of the future. 1909 W. JAMES *Plur. Universe* vi. 268 You can hear the vibration of an electric contact-maker, smell the ozone, see the sparks, and feel the thrill, co-consciously as it were or in one field of experience. 1914 M. PRINCE *Unconscious* I. 1, I divide the Subconscious into two parts, namely the Unconscious and the Coconscious. 1920 *Discovery* Nov. 338/2 The possibilities of co-consciousness, i.e. the coexistence of two or more such fragments [of the conscious] in the mind. 1921 SIDIS *Symptom. Psychopathic Dis.* 77 Hypnoid states are co-existent, 'co-conscious' states and may develop into co-existent, 'co-conscious' personalities. 1926 W. MCDOUGALL *Outline Abnormal Psychol.* 308 Prince adduces very strong evidence in support of the view that.. the bodily changes express an emotional train of thinking or recollection that goes on concurrently or co-consciously with the conscious activity of the subject. 1948 *Brit. Jrnl. Psychol.* 8 As the patient approaches normality the normal mechanism of co-conscious rather than conscious stimulation appears. 1956 A. J. AYER *Probl. Knowl.* iv. 161 The abnormal cases in which people claim to remember the experiences of others; cases of alleged co-consciousness, or cases in which people profess to have 'recaptured' the experiences of the dead.

cocoon, *sb.*¹ Add: **1. d.** A coating applied to metal equipment, esp. surplus military apparatus, which prevents corrosion; also, the proprietary name of a synthetic substance manufactured for this purpose.
1948 *Trade Marks Jrnl.* 11 Aug. 624/2 Cocoon...Synthetic resin included in Class 1. R. A. Brand & Company Limited. 1949 *Britannica Bk. of Year* 687/1 *Cocoon*, a plastic covering sprayed on inactive aircraft for protection from the weather. 1953 *Gloss. Paint Terms* (*B.S.I.*) 9 Although cobwebbing is generally considered a defect in ordinary lacquers, use is made of this property to provide a protective covering for equipment such as aeroplane engines, etc., during storage. A cocoon is formed around the article by the pronounced cobwebbing action of lacquers specially prepared from certain high polymers. A durable, true film-forming lacquer is then sprayed over the cocoon which acts as a support.

cocoon, *v.* Add: **2. b.** To lay up or aside for the future. **c.** *spec.* To coat with a protective layer (see prec. 1 d). Hence **cocoo·ned** *ppl. a.*, **cocoo·ning** *vbl. sb.*
1947 *Jrnl. R. Aeronaut. Soc.* LI. 989/2 Nevertheless, on arrival, these engines were in a good condition internally. It seems probable that the newer American method of cocooning in sprayed plastic may be more satisfactory. 1948 *News* (Birmingham, Ala.) 31 July 2 These B-29 Superfortresses 'cocooned' and 'pickled' at Davis-Monthan Air Force base..are being stripped of their protective coatings. 1951 *Jane's Fighting Ships* 45 Her gunnery and other machinery 'cocooned' for preservation. 1951 *Oxf. Mail* 15 Nov., With cocooned Sabre jets on her deck. 1955 *Times* 21 June 7/6 She has been in reserve for some time and part of her gunnery and radar equipment has been cocooned for storage.

cocoonery. (Earlier examples.)
1839 *S. Lit. Messenger* V. 753/1 She..even went to the expense of having a neat little cocoonery erected in the centre of the lot. 1848 *Rep. U.S. Comm. Patents 1847* 178 Moisture in the cocooneries increases the chance of infection

cocopan (kōᵘ·kopæn). *S. Afr.* Also **kokopan.** [f. Nguni *i-ngqukumbana*/*-bane*, lit. 'stumpy wagon'.] A small steel truck which runs on rails.
1914 *Min. Proc. S. Afr. Soc. Civ. Engin.*, The pick, shovel, wheelbarrow and coco pan suited to the low-waged native labourer. 1946 L. G. GREEN *So Few are Free* (1948) ix. 126 Everything must go into the cocopans, which transport the gravel to the washing plant. 1951 G. VAN DELDEN *I have a Plan* x. 102 Several kokopans, such as are used for transporting sugar-cane. 1954 *Rand Daily Mail* 15 Mar. 3, 4 Cocopans with steel cable.

cocorite (kōᵘ·kŏrait). Also 8 **cocareeta**, 9 **cokarito, cokerite, coucourite.** [Brazilian.] In full *cocorite palm*: a small South American palm, *Maximiliana regia*.
1796 STEDMAN *Surinam* I. 246 A kind of low or dwarf palm-tree, called the *cocareeta*. 1825 WATERTON *Wand. S. Amer.* i. 59 The arrow..is made out of the leaf of a species of palm-tree, called Coucourite. 1851 W. H. BRETT *Ind. Miss. Guiana* 31 Numerous species of palms are also seen... Among them are the cokarito. 1871 KINGSLEY *At Last* v,

Some,..like the Cocorite, almost stemless, rising in a huge ostrich plume. 1883 E. F. im Thurn *Among Indians Guiana* 267 The cokerite (*Maximiliana regia*). 1899 Rodway *Guiana Wilds* 36 Fine cocorite palms. 1958 J. Carew *Wild Coast* viii. 107 Clusters of cocorite palms.

cocotte (kokǫ·t). **1.** [Fr., orig. child's name for a hen.] A prostitute; one of a class of the demi-monde of Paris.

1867, 1885 [see *cocodette]. 1913 Caradoc *City of Plain* ii. I do wish you would not talk of ladies as if they were cocottes. 1920 S. McKenna *Lady Lilith* ii. 45 Nobody thought the worse of us, if we appeared at the theatre with a notorious cocotte. 1923 *Contemp. Rev.* Mar. 339 The platform at Wiesbaden swarmed with smart Parisian cocottes. 1941 'R. West' *Black Lamb* i. 341 Astonished the world by marrying the cocotte La Paiva.

2. [Fr., ad. *cocasse* kind of pot, ult. f. L. *cucuma* cooking-vessel.] A small fireproof dish used for cooking and serving an individual portion of food; *en cocotte*, used to designate food thus served.

1907 G. A. Escoffier *Guide to Mod. Cookery* x. 114 The preparations of butcher's meat, of poultry, or game, known as 'en casserole' or 'en cocotte', are actual poelings cooked in special earthenware utensils and served in the same. c 1938 *Fortnum & Mason Catal.* 64/1 Cocottes—(Single Egg Dishes). 1940 A. Simon *Conc. Encycl. Gastron.* II. 19 Cod en Cocotte. ..Place the slices of crisp pork at the bottom of a *cocotte* or smallish casserole. 1968 M. Sherman *Eggs* 33 Divide the butter among four little cocottes and put them in a frying pan of hot water about half the depth of the cocottes.

cocoyam, coco yam (kōu·ko,yæm). The West African name for either of two food plants, *Colocasia esculenta* or *Xanthosoma sagittifolia*, both of the family Araceæ.

1922 J. H. Holland *Useful Plants of Nigeria* IV. 695 Cash Crops—as 'Coco Yam'..are commonly planted. 1953 G. Durrell *Overloaded Ark* xii. 207 Her native owner ..had been feeding her on a diet of mashed coco yam. 1959 *Times* 9 Nov. (Ghana Suppl.) p. iv/4 Ghana produces an abundance of domestic food crops—of which the most important..are plantains, cocoyams. 1961 *New Scientist* 13 July 78 Bananas and cocoyams from South East Asia.

cocum² (kōu·kɒm). *slang.* Also **cokum, kocum.** [ad. Yiddish *kochem*.] Used without precise grammatical reference for that which is (*a*) advantageous, lucky; (*b*) proper, correct.

1839 H. Brandon *Poverty, Mendicity & Crime* 162/1 Cocum, very cunning and sly. To fight cocum, to be wary. 1851 Mayhew *Lond. Labour* I. 259/2 It's about 6d. a night to me for singing and patter in the tap-room. That's my cokum (advantage). 1861 H. Earle *Ups & Downs* 224 No one was allowed to get drunk; the governor said as how it wasn't cokum. c 1886 *Broadside Ballad* 'The Flippity Flop Young Man' (Farmer), I once was a Member-for-Slocum young man,.. A know-pretty-well-what-is-kocum young man.

cod, *sb.*³ Add: **4. cod-fisherman** = *cod-fisher*; **cod-fishing** *vbl. sb.* (examples); **cod-hook,** a hook used in fishing for cod; **cod-line** (earlier and later examples).

1883 A. Shea *Newfoundland Fisheries* 12 Herrings usually appear in August.. The fishing is attended to by the cod-fishermen as an adjunct to 'the fishery', as the cod-fishing alone is termed. 1962 *Times* 13 Apr. 19/1 Most of them come from the north [of Portugal]..as the codfishermen have come for the past five centuries. 1632 T. Morton *New Eng. Canaan* (1637) ii. vii. 86 The Codd fishing is much used in America. 1716 Stewart *Letter-bk.* in *S.P.E. Tract* LVIII. (1941) 298/2 Some friends and I have a Cod fishing on our West Cost. 1881 *Amer. Naturalist* XV. 367, I had the opportunity of spending three months on a cod fishing schooner. 1634 W. Wood *New Englands Prospect* i. xii. 52 Here likewise must not be forgotten all vtensils for the Sea, as Barbels, splitting-knives, Leads, and Cod-hookes, and Lines. 1707 in *Essex Inst. Hist. Coll.* (1866) VIII. 20 To 7 gross of Cod Hucks. 1838 J. F. Cooper *Homeward Bound* III. vi. 150 Small balls of marline, to the end of each of which was attached a cod-hook. 1634 W. Wood *New Englands Prospect* i. ix. 34 The Fisherman taking a Cod-line,..throwes it into the Sea, the fish biting at it as he pulls her to him. 1686 S. Sewall *Letter Bk.* 15 July (1886) I. 34, 20 Duz. of English cod Lines sound and strong. 1961 F. H. Burgess *Dict. Sailing* 56 *Codline*, small 18-thread line used for many handy purposes...It is supplied for fishing.

cod, *sb.*⁵ Add: **2.** A joke; a hoax, leg-pull; a parody, a 'take-off'. (See also E.D.D. *sb.*⁵) Also *attrib.* or quasi-*adj.*, parodying, burlesque; 'mock'.

1905 *Sketch* LI. 472/2 Says he: 'Is that an absolute bargain—no cod?' Says she: 'I don't know what the fish has to do with it, but I am perfectly sincere.' 1914 Joyce *Portr. Artist* (1916) i. 45 Some fellows had drawn it there for a cod. 1952 Granville *Dict. Theatr. Terms* 46 *Cod version*, a burlesque of a well-known play. 1959 *Church Times* 16 Jan. 4/4 The 'cod' Victorian decorations tend to disguise the editor's underlying seriousness. 1959 *Listener* 29 Jan. 228/1 She obliged, initially in the delicious hiccup polka, a cod of Old Vienna. *Ibid.* 228/2 Joyce Grenfell too, doing her evergreen cod chorister. 1961 B. Wells *Day Earth caught Fire* ii. 31 Pete picked up the empty tea mug and again used it as a cod mike. 'Alcoholics of the press, unite!' 1962 *Listener* 5 July 36/1 The very idiosyncratic cod cockney of the scenes. 1970 *Guardian* 11 May 8/2 The cod version of 'Road to Mandalay'.

cod, abbrev. of *codswallop. Also **cod's, cods.**

Cf. also E.D.D. *Cod*, sb.⁴ (Sense in quot. 1963 is 'mix-up, mess-up'.)

1963 B. S. Johnson *Travelling People* ii. 43 Yet another catastrophic geographical cods by the Cosmos (Affinities Dept.). 1965 J. Porter *Dover Three* x. 123 Her mother had habitually threatened to fetch a policeman when she was naughty and young Eleanor had always thought this was a load of old cod. 1966 *Melody Maker* 30 July 8/6 Those who think this whole scene is a load of old cod's. 1970 M. Tripp *Man without Friends* iv. 46 If you think it's all a load of cod's why the hell waste a pound?

cod, *v.*³ Add: (Further examples.) Also *intr.*, to play a joke, to 'kid', to sham; to burlesque (see also quot. 1933). So **co·dding** *vbl. sb.*

1933 P. Godfrey *Back-Stage* iii. 39 'Codding' is the reprehensible habit of joking privately when on the stage. *Ibid.* 40 One famous actress is notorious for 'codding'. Her favourite device is to talk in a ventriloquial whisper throughout the speeches of the other actors...'Codding' can also be done by gesture. 1935 A. J. Cronin *Stars look Down* III. viii. 550 Sammy always knew when he was inventing...'Eh, yor coddin' now, Uncle Davey?' But Sammy liked the codding even better than the answers. 1941 J. Cary *House of Children* 21 Harry murmured to me: 'Look out for something—he's codding us.' In fact, we had often had practical jokes played on us. 1947 T. H. White *Elephant & Kangaroo* (1948) ix. 76 Ye think ye can do it, sorr, being a gintleman, but that's where they have ye codded. 1961 M. Dickens *Heart of London* i. 50, I wouldn't cod you, not in a pub with your drink in my hand. 1966 *New Statesman* 27 May 788/2 Instead of consequentiality, just as with *Modesty Blaise* recently, codding is the rule—in conjunction with some pleasant decors and rapid, almost evasive editing.

C.O.D. (sī·ōu·dī·). orig. *U.S.* The initials of 'cash (costs, or collect) on delivery'. Also *attrib.*

1859 *N.Y. Tribune* 22 Jan. 2/4 C.O.D. and P.O.D. The principle on which we do business is P.O.D. and C.O.D. Which literally means Pay on Delivery and Collect on Delivery. 1871 'Mark Twain' *Screamers* 143 The..'agent' ..promised to divorce everybody who wished his services, and to send them new wives—C.O.D. 1892 *Congress. Rec.* 25 July 6722/2 It was a transaction payable on sight,—a C.O.D. transaction, so to speak—payable on the very day. 1904 *Daily Chron.* 8 Jan. 6/5 The great objection, I suppose, to the recipients of C.O.D. parcels opening them before paying would be the waste of the postman's time? 1908 *World's Work* Sept. 430/1 By Post C.O.D. 1913 M. Roberts *Salt of Sea* 42, I shot him last night and cut him up and pickled him in a cask... And I've shipped him to the British Ambassador at Washington, C.O.D. 1944 Auden *For Time Being* (1945) 88 Last night it was Tit-for-Tat, tonight it is C.O.D. 1969 *Listener* 31 July 162/3 Some of the more spectacular improvements will be strictly COD.

coda. Add: Also *transf.* and *fig.*

1889 Mrs. Lynn Linton *Thro' Long Night* i. i, The chapter of accidents has a long coda. 1908 A. Bennett *Old Wives' Tale* I. i. ii. 26 The cups and saucers were left for Maggie to wash up as a fitting *coda* to Maggie's monthly holiday. 1963 J. Wiesenfarth *H. James* v. 104 The *coda* begins in Chapter XII.

2. *Ballet.* (See quot. 1952.)

1922 Beauclerk & Evrenov tr. *Svetlov's Karsavina* 33 The *coda* had to be altered and made easier for her. 1925 *Dancing Times* Feb. 521/2 In the 'pizzicato', 'adagio' and 'coda'..she showed what dancing should be. 1952 Kersley & Sinclair *Dict. Ballet Terms* 37 *Coda*, 1. Used by critics and choreographers to indicate by relation to the music to which part of a ballet or dance they are referring. 2. The final concerted dance of a conventional *pas de deux, pas de trois* or *pas de quatre*.

codability (kōudăbi·liti). [f. Code *v.*] Capability of being coded.

1954 Brown & Lenneberg in Saporta & Bastian *Psycholinguistics* (1961) 487/2 The variable of codability was measured in five ways. 1964 *Language* XL. 254 The paper is particularly interesting for its ingenious operationalizing of the concepts of category codability and availability. *Ibid.*, Color recognition is shown to be dependent on codability factors.

codamine (kōu·dămin). *Chem.* [ad. G. *codamin* (O. Hesse 1870, in *Ann. d. Chem. und Pharm.* CLIII. 56), f. Cod(eia + Amine.] A crystalline phenolic alkaloid, $C_{20}H_{25}NO_4$, present in opium in small quantities.

1871 *Jrnl. Chem. Soc.* XXIV. 1064 Codamine is found to melt at 126°, instead of 121°, and has probably the composition $C_{19}H_{23}NO_4$. 1965 *Jrnl. Pharm. Sci.* LIV. 1531/1 Very little is known about the physical and chemical properties of naturally occurring codamine, and there is no evidence that it has ever been prepared since the work of Hesse.

coddam (kǫ·dăm). Also **coddem, coddom.** [? = *cod 'em*: Cod *v.*³] Another name for Tip-it.

1864 in Hotten *Slang Dict.* 1884 J. Greenwood *7 Yrs. Penal Serv.* (Farmer), A gambling game called coddom. 1894 *Sporting Life* 28 Feb. 7/5 A coddam match, for a wager and a fish supper. 1897 *Daily News* 11 May 5/3 When he wearied of skittles he took to coddam. 1950 *N. & Q.* CXCV. 217/1 The game [sc. Up Jenkins] was also called 'Coddem' and 'Tipit'.

code, *sb.*¹ Add: **3. c.** *Cybernetics.* Any system of symbols and rules for expressing information or instructions in a form usable by a computer or other machine for processing or transmitting information.

1946 *Nature* 26 Oct. 568/1 The brains of the machine lie in the control tape, which is code-punched in three sections. 1947 *New Republic* 23 June 15/3 The machine, having been properly briefed by means of a code punched into a paper tape.., will perform its programmed task. 1948 *Electronics* Sept. 111/2 Orders to various parts of the machine..can be expressed conveniently as numbers in some arbitrary code. 1964 F. L. Westwater *Electronic Computers* i. 2 On a paper tape holes are punched across the tape according to a pre-determined code. 1970 J. Butler in *Physics Probl. Reactor Shielding* (O.E.C.D.) 109 A saving in computer time..compared with the discrete ordinate codes NIOBE and STRAINT.

d. Extended uses in *Biol.* and *Linguistics*.

1958 P. B. Medawar in R. D'Arcy Thompson *D'Arcy W. Thompson* 225 The problem of the development of limbs is, first, to break the chemical code which embodies the instructions, and second to find out how the instructions take effect. 1964 *Language* XL. 243 The central concern is how the bilingual speaker becomes 'inputted' for two language codes. 1965 *Canad. Jrnl. Linguistics* Fall 41 Number and pattern (or 'code'—a favourite term in linguistics). 1965 *Listener* 2 Sept. 332/2 The conceptions of molecular codes, and the chemical storage of information which have arisen from work on reproduction, have stimulated fascinating speculations about the mechanism of memory and the mystery of dreams.

5. *code-bearing, -breaking, -checking, -switching*; **code-book,** a list of letters or other expressions, and of their correlates in a code, arranged as a key for encoding and decoding; a book containing a code (in other senses); **code-language,** a system of codes; **code-name,** a word or symbol used as a substitute for the ordinary name of a thing or person, for secrecy or convenience; also, one used to refer to something that has no name; so *code-name* v.; similarly **code-number; code-script** (see quots.).

1965 *Listener* 2 Sept. 332/1 The code-bearing molecules, the nucleic acids. 1884 *Electrician* XIV. 62/1 This firm recommends the use of the 'ABC Telegraphic Code Book'. 1908 *Westm. Gaz.* 15 Oct. 4/2 The Royal Automobile Club proposes..to establish a law unto itself, with its own code-books of rules, morals, and punishments. 1964 Y. Bar-Hillel *Lang. & Inf.* xvi. 279 A short signal sequence..to be decoded at the receiving end with the help of a code-book. 1964 J. Z. Young *Model of Brain* viii. 128 In the procedures for code-breaking adopted by cryptographers the problem is to discover the relevant features of, say, ink-marks or electrical signals that are being used for communication. 1964 C. Dent *Quantity Surv. by Computer* vi. 73 The code-checking and punch-checking procedures. 1936 *Discovery* June 187/2 The spirit-doctors have a kind of code-language of their own. 1919 L. Tissot-Dupont *Dictionnaire des Termes Techniques de Télégr.* 37 Nom conventionnel. Code name. 1936 F. W. Crofts *Loss of 'Jane Vosper'* i. 22 Heading it with the code name and address of his firm. 1954 X. Fielding *Hide & Seek* iv. 48 A district so bountiful that among us it earned the code-name of 'Lotus Land'. 1959 'O. Mills' *Stairway to Murder* xvi. 162 A military Exercise, code-named Discretion. 1961 R. Seth *Anat. Spying* iii. 38 Among these was Wing-Commander Forest Yeo-Thomas, whose code-name was the White Rabbit. 1962 L. Deighton *Ipcress File* vii. 47 He was well code-named this captive Raven. 1959 W. S. Sharps *Dict. Cinemat.* 85/2 *Code numbers*, a series of numbers and key letters printed by film manufacturers on *negative* film at one foot intervals and used for stock identification. 1964 C. Dent *Quantity Surv. by Computer* iv. 38 It is not necessary for the taker-off to look up the library description...Someone else (the 'coder') could do this and merely insert the appropriate reference number ('code' number) against the taker-off's description. 1952 *New Biol.* XII. 85 All the genes together constitute the 'code-script' of the cell; that is, the complete set of controlling factors which carry in condensed form all the 'information' necessary for an organism to develop the distinctive properties of its species. 1964 G. H. Haggis et al. *Introd. Molecular Biol.* vii. 193 Indeed, if current attempts to decipher the chemical structure of chromosomes are successful, it may one day prove possible by deliberate manipulation of environmental agencies to make specific alterations in the genetic code-script. 1959 E. Pulgram *Introd. Spectrogr. Speech* xi. 82 The hearer is able to perform what communication engineers call code switching, a process of adjustment to the articulatory habits of the speaker which permits the listener to learn quickly certain types and degrees of phonemic deviation.

code, *v.* Delete *rare* and add: (The forms **coded** *ppl. a.* and **coding** *vbl. sb.* are common in all uses.) **b.** To prepare (a message) for transmission by putting it into code words.

1898 *Westm. Gaz.* 20 Apr. 3/2 However inaccurately the message may be coded. 1903 Hardy *Dynasts* I. v. iii. 91 Now that the fume has lessened, code my biddance Upon our only mast. 1909 *Westm. Gaz.* 7 June 7/3 Coding had been really reduced to a fine art, and by its aid the cost of a cablegram was often reduced not merely to a penny per word, but to a farthing. 1924 Galsworthy *Forest* i, I'm going to code that cable.

c. *Cybernetics.* To put into the form required by a code (*3 c).

1947 Goldstine & Von Neumann (*title*) Planning and coding of problems for an electronic computing instrument. 1948 *Electronics* Sept. 116/1 Orders themselves are coded to appear as numbers. *Ibid.* 116/3 A coded order can be converted to the selection of a gate opening voltage. 1956 *Spaceflight* I. 27/2 A data input circuit..codes all information picked up by the satellite's instruments, and a modulator..superimposes the coded data on a carrier signal. 1958 *Oxf. Mag.* 29 May 469/1 The new [computing] machines..require the relatively new skills of 'programming' and 'coding'. *Ibid.*, The preparation of 'data' and 'instructions' in the coded form acceptable by the machine. 1967 *Oxford Computer Explained* 9 Coded orders are attached to the original order and passed to the Dispatcher, who allocates an order reference number which is stamped on the order and on the coding form.

d. In extended use. Cf. *code *sb.*¹ 3 d.

1959 *Jrnl. Molecular Biol.* I. 218 These molecules presumably carry in coded form the information necessary to specify the many features of the particular organism concerned. 1962 *Lancet* 29 Dec. 1384/1 Preventing R.N.A.-messenger coding and thus participation in protein synthesis. 1964 G. H. Haggis et al. *Introd. Molecular Biol.* viii. 212 The question of how a sequence made up from four different bases might specify unambiguously a sequence made up from twenty different amino acids is known as 'the coding problem'. 1964 L. Kaiser in D. Abercrombie et al. *Daniel Jones* 103 Direct connections between both processes of coding–decoding are found in spelling.

coder (kōu·dəɹ). [f. Code *v.* + -er¹.] One who or that which puts a message, set of information, etc., into code.

1927 *Daily Express* 14 Sept. 7 A telephone clearing house between all automatic exchanges and the existing manual exchanges, 73 of which have already been specially fitted with coder call indicator apparatus. 1945 'Tackline' *Holiday Sailor* i. 11 We were all designed to be Coders, and were supposed to be rather intelligent than not. 1964 [see *code number*, *code *sb.*¹ 5].

co-dete·rminant. [Co- 3.] One of a set of determining factors. Hence **co-dete·rmine** *v.*

1884 W. James in *Mind* IX. 12 Those portions of the brain that have just been maximally excited retain a kind of soreness which is a condition of our present consciousness, a co-determinant of how and what we now shall feel. 1890 —— *Princ. Psychol.* I. i. 4 Certain peculiarities..are explicable only by the fact that the brain laws are a codeterminant of the result. 1907 —— *Pragmatism* vi. 226 These factors co-determining each other incessantly. 1937 R. H. Lowie *Hist. Ethnol. Theory* xiv. 260 Environmental conditions..are..obtrusive as co-determinants. 1959 *Brno Studies* I. 34 The OE 'diacritical e' was always closely preceded by the letter whose phonematic equivalent it helped to co-determine.

codiæum (kōudi,i·ɒm). *Bot.* [mod.L. (G. E. Rumpf in Adrien de Jussieu *De Euphorbiacearum generibus* (1824) 33), f. *kodiho*, the native name for the plant in Indonesia.] A member of a genus of tropical plants of the family Euphorbiaceæ, cultivated in hothouses for their variegated foliage; esp. = Croton 2.

1831 *Curtis's Bot. Mag.* LVIII. 3051 (*heading*) Painted-leaved codiæum. 1900 C. Bennett in W. D. Drury *Book of Gardening* xv. 671 Crotons (Codiæums).—The beautiful variegated leaves..of these tropical evergreen shrubs place them among the most useful and attractive plants grown for exhibition. 1951 *Dict. Gardening* (R. Hort. Soc.) II. 515/2 Beside this use for table decoration..Codiaeums are equally useful for conservatories. 1961 *Amateur Gardening* 9 Dec. 6 Perhaps better known as crotons, the codiaeums are extremely decorative plants.

codicology (kōudikǫ·lŏdʒi). [ad. Fr. *codicologie*, f. L. *cōdic-* stem of Codex + -ology.] The study or science of manuscripts and their interrelationships. Hence **codicolo·gical** *a.*, **codicolo·gically** *adv.*

Used mainly by continental writers, when writing in English, as a calque on G. *Handschriftenkunde*.

1953 D. C. C. Young in *Scriptorium* VII. 7 If the suggestion that O [= MS. Vat. gr. 915] for Theognis is by Gregoras be accepted, much may follow for the codicology of the other texts in O. *Ibid.* 3 A codicological inventory of Theognis manuscripts. 1964 L. Bieler in *Studia Evangelica* III. 328 Many of these manuscripts have recently been described and analysed 'codicologically' by Dr. Patrick McGurk. 1968 *PMLA* LXXXIII. 25/1 It is codicologically indivisible from the preceding pieces and in the same script. 1970 *Times Lit. Suppl.* 21 May 568/2 The recent development of 'codicology', which Greg would have subsumed under bibliography—the study of manuscripts as physical objects in order to identify the workshops that produced them.

co-dire·ctor. [Co- 3 b.] A joint director.

1869 *Porcupine* 24 July 153/1, I very much doubt if we shall ever again see the names..coupled together as co-directors. 1874 *Leisure Hour* 23 May 332/2 He was entirely at the mercy of his co-directors, who made use of his honest name. 1922 *Daily Mail* 6 Dec. 12 Leaving your partners and co-directors to bear the burden of the disgrace.

codling². Add: **4. codling-moth** (earlier and later examples); **codlings-and-cream** (further examples).

1747 B. Wilkes *Eng. Moths & Butterflies* i. 5 The Codling Moth. Mr. Rosel says, The Caterpillar of this Moth feeds on the Kernels of Appels and Pears. The Moth lays her Eggs at the End of June. 1959 Southwood & Leston *Land & Water Bugs* ix. 244 The young larvae of the codlin moth. 1884 W. Miller *Dict. Eng. Names Pl.* 29/2 Codlings-and-Cream. *Epilobium hirsutum*. 1959 J. Clegg *Freshwater Life* (ed. 2) 51 Other plants of the marsh are the Great Willow Herb or Codlins-and-Cream..whose rose-coloured flowers and tall growth make it easy to recognise.

codology (kǫdǫ·lŏdʒi). *Irish colloq.* [f. *cod *sb.*⁵ + -ology.] Hoaxing, humbugging. Hence **codo·logist.**

1922 Joyce *Ulysses* 299 The why and the wherefore and all the codology of the business. 1928 *Daily Express* 15 Feb. 11 There is in Ireland a science unknown to us in England called Codology... The English is 'leg-pulling'... When I received an invitation to breakfast at the Dublin Zoo I thought that I could detect the hand of the chief codologist. 1966 *Listener* 8 Dec. 863/1 Apart from some typical codology and a handful of tender personal messages—his [sc. James Joyce's] stuff is tidily to the point and businesslike.

co-dominant (ko,dǫ·minănt), *a.* and *sb.* [Co- 2.] **A.** *adj.* Sharing dominance equally; of similar frequency of occurrence or similar importance. Esp. in *Forestry* (see quot. 1928) and *Ecology*.

1926 Tansley & Chipp *Stud. Veg.* ii. 11 There may be several species..which..together dominate the association, but these 'co-dominant' species most frequently..belong to the same life-form. 1928 *Funk's Stand. Dict.*, *Co-dominant*, a., forming a part of the main canopy, but not among the tallest: said of forest-trees. 1943 *Ess. & Stud.* 1942 XXVIII. 69 If account was taken of a further number of poems in which love, though not the sole or predominant theme, came in as a parallel co-dominant theme, the number was 250—every other poem. 1964 V. J. Chapman *Coastal Veget.* i. 5 The General Salt Marsh community, where there are a number of co-dominant species.

B. *sb.* A species, etc., that is co-dominant.

1933 *Forestry* VII. 122 These tables show that there is little difference between the various stem-classes—dominants, co-dominants, and sub-dominants—as regards the distribution of form-classes. 1933 A. M. Woodbury in *Ecol. Monogr.* III. 165 Codominants—any of the equally dominant forms. 1940 *Geogr. Jrnl.* XCV. 451 Hraun heath has several co-dominants, each of which is locally dominant in associations, e.g. *Betula*, *Salix* [etc.]. 1953 *Brit. Commonw. For. Term.* i. 28 *Codominant.* (*a*) A tree with a portion of its crown in the upper canopy, but below the crown level of the dominants... (*b*) A species which shares dominance more or less equally with others.

‖ **co-domini** (ko,dǫ·minəi), *sb. pl.* [quasi-Latin, f. Co- + *domini*, pl. of L. *dominus* lord, ruler.] Condominium powers, *spec.* (between 1899 and 1966) Great Britain and Egypt in relation to the Sudan.

1944 D. Newbold *Let.* 19 Sept. in K. D. D. Henderson *Making Mod. Sudan* (1953) xvi. 412 It is inevitable that Sudan views and interests do sometimes conflict with the views and interests of one or other of the co-domini. 1955 *Times* 4 June 7/3 When the House is recalled in August it will be for the purpose of passing a resolution informing the Co-domini (Britain and Egypt) that the country feels ready to complete the determination of its political future. *Ibid.* 15 Aug. 6/1 The supervisory commission was..a British conception,..and Britain..must see that both co-domini observed it until the end of the transition period.

codon (kōu·dǫn). [f. *cod(e *v.* d + *-on.] The smallest group of nucleotides in a polynucleotide that determines which amino-acid shall be inserted at any given position in a polypeptide chain; it is found experimentally to be a sequence of three consecutive nucleotides.

1963 F. H. C. Crick in *Progr. in Nucleic Acid Res.* I. 166 *The Codon.* This is a group of bases that code for one amino acid. In simple codes a codon is a fixed number of consecutive bases, e.g., in a 'triplet' code it is three consecutive bases, but it is possible to conceive codes in which, for example, some codons consist of two bases and others of three. Again it is not certain that the bases making up a codon are adjacent on the polynucleotide chain. 1966 *New Scientist* 12 May 348/3 The transfer-RNA attaches itself to its own amino acid and, in some fashion, couples it to the right codon on the messenger-RNA inside the ribosome. *Ibid.* 11 Aug. 322/2 The amino acid molecules do not attach themselves directly to the base-triplets, or codons.

co-driver. (Stress variable.) [Co- 3 b.] A person who takes turns with another or others to drive a vehicle (esp. in a motor race, rally, etc.). Hence **co-drive** *v. trans.*

1933 *Motor* 13 Jan. 1156/2 In our report..of the Monte Carlo Rally it was stated that Mr. A. K. Stevenson, of the Royal Scottish Automobile Club, was co-driver with the Hon. V. A. Bruce... Actually Bruce's co-driver in the Rally was W. J. Brunell. 1952 *Ibid.* 23 Jan. 870/2 His Hillman Minx which he was co-driving with Philip Fotheringham-Parker. 1960 *Times* 25 Feb. 1/3 Co-driver, sharing expenses, to Lisbon. 1963 P. Drackett *Motor Rallying* iii. 49 Peter Burch, of Aldershot, and his co-driver, A. Benniman, were three miles from rally headquarters.

cod's-head. Add: **1. b.** *cod's head and mackerel-tail* (see quot. 1961).

1891 *Scribner's Mag.* X. 5 Those forms of ships adopted for centuries by some European nations, and known to mariners as 'cod's-head and mackerel-tail' shape. 1961 F. H. Burgess *Dict. Sailing* 56 *Codshead and mackerel tail*, a description applied to sailing yachts designed with bluff bows and a long run aft, the greatest beam being forward.

codswallop (kǫ·dzwǫlǫp). *slang.* Also **cod's wallop.** [Origin unknown.] Nonsense, drivel.

1963 *Radio Times* 17 Oct. 52/2 Just branding a programme as 'rubbish', 'tripe', or—there are a lot of these—'codswallop', gives little indication of what moved the viewer to write. 1966 J. Porter *Sour Cream* vi. 72 It had all seemed a load of old cod's wallop to me at the time, but those lecturers were experts. 1966 A. Prior *Operators* vi. 64 All that stuff about mutual respect between police and criminal was a load of old codswallop. 1970 *Peace News* 30 Jan. 6/1, I hope you will stop publishing cod's wallop of the type which appeared..in Roger Moody's article.

co-ed, *sb.* and *a. colloq.* (orig. *U.S.*). [abbrev. of Co-education, -educational *a.*] **A.** *sb.* **1.** Co-education; a co-educational institution or system.

1886 L. M. Alcott *Jo's Boys* v. 104 Girls..ought not to study so much. Never liked co-ed. 1958 *Listener* 12 June 991/3 We are at a Secondary Modern 'co-ed'.

2. A girl or woman student at a co-educational institution.

1893 in *Funk's Stand. Dict.* 1903 *Independent* 12 Feb. 369/2 Any college where the girls are commonly called 'co-eds' is not a truly co-educational institution. 1928 *Morning Post* 20 Oct. 8/3 In some of the ultra-modern American Universities, where 'co-eds' abound. 1931 J. B. Priestley *Angel Pavem.* vi. 315 A college girl, what they call over there a co-ed. 1951 Auden *Nones* (1952) 60 Enormous novels by co-eds Rain down on our defenceless heads. 1970 *Daily Tel.* 6 May 34/3 Undergraduates and co-eds sought more violent or dramatic ways of expressing their feelings.

B. *adj.* **1.** Co-educational. **2.** Of or pertaining to a co-ed (sense 2).

1889 *Pueblo* (Colorado) *Opinion* 21 July 2/2 A Newport Story, a 'Co-Ed.' Story..are some of the attractions of *Demorest Monthly Magazine* for August. 1895 *Bachelor Arts Monthly* May 113 There is even danger that soon 'co-ed' institutions will degenerate into mere matrimonial agencies. 1949 R. Graves *7 Days New Crete* vi. 71, I went to a co-ed school in Switzerland. I was the head girl.

co-e·ducate, *v.* orig. *U.S.* [Back-formation from Co-education.] *trans.* To educate (persons of both sexes) together, or (one of either sex) with those of the other under a co-educational system. Also *transf.* Hence **co-e·ducator.**

1855 *Pennsylvania School Jrnl.* III. 201 Another enumerates the great number of schools where the sexes are co-educated. 1894 *Forum* July 582 Will the Co-Educated Co-Educate their Children? 1897 *Educat. Rev.* XIII. 62 Sight, hearing and touch all have to be developed and trained and co-educated, taught to act together. 1909 H. G. Wells *Ann Veronica* i. 8 He had one son, who had been co-educated. 1920 T. P. Nunn *Education: Data* xv. 203 Co-educators aim..at purifying and strengthening the bases of family life by teaching boys and girls to know one another. 1928 *Daily Express* 25 Aug. 8/3 If we could co-educate the younger generations of Europe we could co-ordinate them in a union of sympathy, understanding, and comradeship. a 1936 G. K. Chesterton *Common Man* (1950) 141 Co-educate as much as you like, there will always be a wall between the sexes. *Ibid.* 143 It is..a question of what are really the co-educators' aims.

co-education. (Earlier and later examples.)

1852 *Pennsylvania School Jrnl.* I. 9 Co-education of the sexes. The instruction of males and females in the same room and in the same class is supposed by many to be an evil. 1958 *Daily Express* 11 Mar. 7/1 The order says that generally speaking the Vatican does not approve of co-education in Catholic schools, though it realises this might be necessary.

co-educational, *a.* (Further examples.)

1906 *Daily Chron.* 16 Oct. 3/6 The co-educational school reproduces the home at its very best. 1957 *Times* 9 Sept. 11/1 Finding costumes for school plays or chaperoning boys and girls at co-educational school parties..are commonly among the tasks undertaken [by the Parent-Teachers Association].

cœlacanth, *sb.* Add: (Specimens of this fish found in modern times have been systematically named *Latimeria chalumnæ*.) (Add examples.) Also **cœlaca·nthid** [-id²] *attrib.* or *adj.*

1857 [see Sauroid *sb.* 1]. 1931 J. R. Norman *Hist. Fishes* xvii. 357 The scales of the Coelacanths are cycloid. 1939 *Nature* 18 Mar. 455 (*caption*) Coelacanthid fish from East London, South Africa. *Ibid.* 456/2 Certain cœlacanthid fishes underwent little apparent change from the Devonian to the Cretaceous. 1940 *Ibid.* 13 July 53/1 The Cœlacanths have changed very little since their first known appearance in Upper Devonian formations. 1957 *Encycl. Brit.* V. 934/1 The first living coelacanth (*Latimeria chalumnae*) was trawled in 40 fathoms near East London, U. of S. Af., in 1938. The second was caught..near the Comoro Islands in Dec. 1952. 1966 *Proc. Linn. Soc. Lond.* CLXXVII. 2 The almost static coelacanths with their notorious single living representative *Latimeria*.

cœlenteron (sīle·ntĕrǫn). *Zool.* Also **cel-.** [mod.L., f. Gr. κοῖλος hollow + ἔντερον intestine.] The digestive cavity of a cœlenterate.

1893 Shipley *Zool. Invert.* 78 In the Acoelomata there is a common cavity, the Coelenteron, which is lined by endoderm cells, and which pervades various parts of the body. 1927 Haldane & Huxley *Anim. Biol.* xii. 268 The mouth leads into a cavity called the coelenteron.

cœlia (sī·liă). *Anat.* Pl. **-iæ.** Also **celia** [mod.L., ad. Gr. κοιλία a hollow, cavity.] Any cavity of the body, as the abdominal cavity or a ventricle of the brain.

1820 Hooper *Lex.-Med.* (ed. 4) 230/1 Cœlia, a cavity in any part of the body. The belly. The womb. 1882 [see Procœlia].

cœliac, *a.* Add: Applied *spec.* to an intestinal disease, esp. of children, characterized by defective digestion of fats.

1888 S. Gee in *St. Bart.'s Hosp. Rep.* xxiv. 17 Diarrhœa alba is the name employed in India to denote the cœliac affection; not that it is always a cœliac flux, a diarrhœa strictly speaking. 1911 *Practitioner* LXXXVII. 147 (*heading*) Celiac disease. 1930 *Q. Jrnl. Med.* XXIII. 438 Coeliac disease is probably the result of a development defect in absorption. 1955 *Sci. News Let.* 5 Mar. 152/2 Many sufferers of sprue and celiac disease, a digestive illness, cannot digest ordinary bread.

cœliotomy (sīliǫ·tŏmi). *Surg.* Also **celio-.** [f. Cœlio- + -tomy.] The operation of cutting into the abdominal cavity; laparotomy.

1881 *Trans. Obstet. Soc. Lond.* XXII. 128 If a new word was needed to supersede 'gastrotomy' perhaps 'cœliotomy' would do. 1908 *Practitioner* Oct. 608 Anterior and posterior vaginal coeliotomy. 1954 Bacon & Trimpi in K. C. Jonas *Babcock's Princ. & Pract. Surg.* xliv. 1423 Celiotomy.—In the doubtful or confusing case, exploratory laparotomy is advantageous.

cœloblast (sī·loblɑst). *Embryol.* [f. Cœlo-¹ + -blast.] The endoderm of an insect, or part of the endoblast as distinguished from the myoblast.

1895 *Camb. Nat. Hist.* V. 149 The blastoderm differentiates into Ectoblast and Endoblast; this latter undergoing a further differentiation into Coeloblast and Myoblast. 1960 I. F. & W. D. Henderson *Dict. Sci. Terms* (ed. 7) 101/2 *Coeloblast*, a division of the embryonic hypoblast.

cœlomo- (sīlōu·mo), used as combining form of Gr. κοίλωμα Cœlome, as in **cœlo·moduct** [Duct *sb.* 6], **cœlo·mostome** [Gr. στόμα mouth] (see quots.).

1900 E. R. Lankester *Treat. Zool.* II. ii. 13 These apertures are..formed..by an outward, often tube-like growth of the cœlom itself...I propose to call them cœlomoducts. 1900 E. S. Goodrich in *Q. Jrnl. Microsc. Sci.* XLIII. 742 For its special funnel-like opening..I propose the name cœlomostome. 1902 *Encycl. Brit.* XXXIII. 882/2 The reproductive cells [of Polychæte worms] are developed from the cœlomic epithelium, ripen in the cœlom, and are carried to the exterior at maturity by special genital ducts (cœlomostomes). 1963 R. P. Dales *Annelids* i. 20 The coelom leads to the outside in each segment by a pair of coelomoducts, and by a pair of nephridia, or by structures derived from their combination. *Ibid.* 30 When the nephridium is a metanephridium, the coelomostome may be grafted directly on to the nephridiostome forming a mixonephridium.

cœlostat (sī·lostæt). *Astr.* [f. Cœlo-² + -stat.] An instrument consisting of a mirror turned by clockwork on an axis parallel to itself and to the axis of the earth, by means of which the celestial bodies may be observed and photographed as in a stationary position.

1898 *Standard* 24 Jan. 4/5 The present phenomenon [*sc.* the total solar eclipse of January 22, 1898] will be remembered as the first in which the instrument called the cœlostat was used. 1900 H. H. Turner in *Monthly Rev.* Sept. 108 A special arrangement..called a 'cœlostat', which reduces the whole sky virtually to rest in a most convenient manner. It was invented long ago by a Frenchman, but practically forgotten. 1939 *Jrnl. Brit. Interplan. Soc.* V. 9 The cylindrical objects shown just above the catwalk.. are coelostats. These are synchronised, motor-driven mirror devices something similar to a stroboscope, and it is by means of these that a stationary view of the heavens is provided for navigational observations while the ship is in rotation. 1961 *Guardian* 13 Feb. 16/3 A coelostat lent by the Mill Hill Observatory..produces a static image of the sun.

cœnocyte (sī·no,səit). *Biol.* [f. Cœno- + -cyte.] = Syncytium sense a, esp. one formed by nuclear division. Hence **cœnocytic** (-si·tik) *a.*

1900 J. R. Green *Introd. Veg. Physiol.* i. 11 Other cœnocytes exist in which, besides the limiting wall, certain transverse walls exist, dividing the chamber into compartments. *Ibid.* xxvi. 421 The fact that it is cœnocytic is shown by the presence of a number of nuclei in the protoplasmic mass. 1902 *Encycl. Brit.* XXV. 265/2 Instead of a single relatively large nucleus, each cell is found to contain many small nuclei, and is spoken of as a cœnocyte. 1912 J. S. Huxley *Individ. in Animal Kingdom* iv. 89 Botanists distinguish between cells, which have one nucleus, and coenocytes, or masses of protoplasm with many nuclei. *Ibid.* 160 In the third type (Coenocytic) the nucleus alone divides, and the final result is a coenocyte. 1952 *New Biol.* XIII. 100 Such a non-septate, multi-nucleate system is called a coenocyte. 1952 *Sci. News* XXV. 58 *Valonia macrophysa*, a coenocytic alga growing in seawater, will accumulate potassium ions in the sap to a high concentration. 1968 H. Harris *Nucleus & Cytoplasm* i. 10 It is thus merely the phenomenon of nuclear growth without nuclear division which distinguishes *Acetabularia* from many coenocytic organisms.

cœnospecies (sī·nospīʃīz). *Biol.* Also **ceno-.** [f. Cœno- + Species.] A group of species capable of hybridization with one another.

1922 G. Turesson in *Hereditas* III. 345 It seems appropriate..to denote the total sum of possible combinations in a genotype by the term *coenospecies*... I take the opportunity of thanking Mr. Otto R. Holmberg, Conservator of the Botanical Museum of Lund, for the coining of this term. 1939 *Nature* 11 Nov. 822/2 G. Tischler..stated that at least 400 wild cœnospecies of Angiosperms are known which contain races with chromosomal differences. 1957 *Encycl. Brit.* XXI. 177/1 *Coenospecies.* This term was first used by G. Turesson (1922) for the sum total of possible combinations of a genotype compound as seen particularly through cultivation under artificial conditions. The term has also been used to include groups of species separated only by extrinsic barriers to gene flow.

co-e·nzyme. *Chem.* [ad. G. *ko-enzym* (Buchner and Katte 1908, in *Biochem. Zeitschr.* VIII. 524), f. Co- 3 b + *Enzyme.] A non-protein organic compound with which an enzyme needs to combine to become active and which generally takes part in the reaction as a carrier.

1908 *Chem. Abstr.* 2568 (*title*) The co-enzyme of the yeast press-juice. 1936, 1961 [see *apo-enzyme]. 1941 *Ann. Reg.* 1940 344 Vitamin H proved identical with biotin, a growth factor for yeast, and with coenzyme R. 1947 *Thorpe's Dict. Appl. Chem.* VIII. 612/2 Coenzymes will be designated by the name of the enzyme activated, preceded by the prefix *co-* (example, coglyoxylase). 1950 *Endeavour* IX. 145/1 The entire enzyme or holoenzyme is built up of at least two fractions: (*a*) The apoenzyme, which is always a protein; it is colloidal and thermolabile. (*b*) The co-enzyme, of variable chemical constitution, frequently a phosphoric ester and sometimes a metallic derivative; very often the co-enzyme, which is non-colloidal, is fairly thermostable.

coercive, *a.* Add: **4.** (Earlier and later examples.)

1827 J. Cumming *Man. Electro Dynamics* iv. 168 That resistance to a change in the direction of the particles, which, in the common theory of magnetism, is termed its coercive force. 1962 Corson & Lorrain *Introd. Electromagn. Fields* vii. 282 The magnitude of the magnetic field intensity *H* at this point is known as the coercive force.

coercivity (kōuɜ̄si·viti). *Magnetism.* [f. Coercive *a.* + -ity.] The property of retaining coercive force; the value of the coercive force of a magnetized substance.

1898 J. A. Fleming *Magnets & Electr. Currents* i. 16 The ability to retain the magnetic state with varying degrees of power against mechanical shocks or reversed magnetic force is called the Coercive Power of the body, or its Coercivity. 1902 *Encycl. Brit.* XXVIII. 115/2 Steel containing tungsten possesses a large coercivity. 1910 H. M. Hobart *Dict. Electr. Engin.* I. 106/1 *Coercivity*, the reverse magnetic field required to reduce a mass of iron which has been highly magnetised, to a state of zero magnetisation. 1933 *Radio Engin. Handbk.* 34 Electromagnetic Structures. In this type of structure the magnetic material is usually very soft; its coercivity is very low. 1947 *Electronic Engin.* XIX. 379 One type of wire..has..a coercivity of 300 oersteds. 1960 *Photogr. Jrnl.* Aug. 232/2 Coercivity..is the ease with which a track can be magnetized or demagnetized.

cœrulignol, -one, varr. *cerulignol, -one.

coesite (kōu·səit). *Min.* [f. the name of L. *Coes*, Jr. (born 1915), U.S. chemist who first synthesized it + -ite¹.] A modification of quartz formed under pressure.

1954 R. B. Sosman in *Science* CXIX. 738/2, I have sought and obtained permission from Coes and Keat, both of whom are now at the Research Laboratories of the Norton Company in Worcester, Mass., to propose names for the new phases—namely, *coesite* for the phase produced at 35,000 atm., and *keatite* for the phase produced hydrothermally. 1962 *Times* 1 June 8/6 It is impossible for coesite to be formed at pressures less than about 20,000 times that of the atmosphere... The Norton Company has been supporting the high-pressure research..in the hope..of making diamonds in the laboratory.

coexist, *v.* Add: **b.** Of political or economic systems (see next).

1931 *Economist* 4 July 23/2 The question may then well arise whether the plan-economy of Russia and the capitalist economy of Western Europe can co-exist in amity. 1957 *Listener* 5 Dec. 929/1 We have not yet achieved a way of coexisting in real peace with the Soviet Union. 1958 *Bull. Atomic Sci.* Dec. 410/1 It seems unlikely that industrialized and unindustrialized nations can coexist indefinitely.

coexistence. Add: **b.** With special reference to peaceful existence side by side of states professing different ideologies. Also *transf.*

[1920 tr. Lenin in *N.Y. Evening Jrnl.* 21 Feb. 3/1 You ask me about our plans in Asia. They are the same as in Europe: peaceful, neighborly life with all peoples; with the workers and peasants of all nations.] 1954 *Economist* 30 Oct. 358/1 Although this is welcome if it provides some possibility of a move towards a more stable co-existence of the Communist and free worlds, nothing could be more dangerous to the West than to assume that this can be achieved at the moment. *Ibid.*, 'Peaceful co-existence' is in origin a Communist phrase, and, although it has been taken up by western statesmen, it has a sinister place in the glossary of Soviet double-talk. 1955 H. Hodgkinson *Doubletalk* 24 Co-existence or 'peaceful co-existence'...The phrase..does not mean 'living peaceably together'. It is compatible with propaganda against, isolation from and peripheral risings against the non-Communist world. 1956 *Times* 10 July 12/5 In a world where the word 'peace' had been replaced by that ugly circumlocution 'competitive coexistence'..art should be held high. 1959 *Spectator* 2 Oct. 437/3 The Warsaw Festival has established itself as protagonist of musical 'co-existence', as an agreed field of 'peaceful competition' between conflicting systems.

coextensively, *adv.* (Earlier example.)

1843 Mill *Logic* II. iv. vi. 291 We cease to use the term coextensively with the characters.

co-factor. Add: **2.** *Chem.* **a.** Any substance other than the substrate (as a co-enzyme or a metal ion) the presence of which is necessary for the activity of a particular enzyme. **b.** (See quot. 1956.) (Formerly called *co-enzyme factor*.)

[1937 Dewan & Green in *Nature* 25 Dec. 1098/1 A New Oxidation Catalyst... We have succeeded in resolving the so-called coenzyme oxidase into three insoluble components, (1) the coenzyme factor...The coenzyme factor catalyses the first oxidation.] 1945 *Jrnl. Cellular & Comparative Physiol.* XXV. 12 The virus T_4..is not effective in lysing such irradiated cells unless a co-factor such as tryptophane is present in the mixture to enhance its rate of adsorption on the host. 1949 *Advances in Enzymol.* IX. 178 Certain 'cofactors', such as L-tryptophan, have the power of promoting adsorption of certain bacteriophages on their hosts. 1956 *New Gould Med. Dict.* (ed. 2) 265/2 *Cofactor*, in bacteriology, any amino acid which must be present in the medium to activate phage to adsorb host bacteria. 1958 Dixon & Webb *Enzymes* ix. 394 An additional substance besides the enzyme and substrate is required in many cases in order that the reaction may proceed. Such 'cofactors' are found unchanged at the end of the reaction and may therefore be regarded as an essential part of the catalytic mechanism. The majority of enzyme cofactors may be broadly divided into..(a) specific coenzymes..(b) activators.

co-fa·vourite. [f. Co- + Favourite *sb.*] In Sport, an equal or joint favourite (cf. Favourite *sb.* 1 b).

1922 *Daily Mail* 30 Nov. 11 His co-favourite, Desmond House, cut a leg at the first hurdle and afterwards jumped badly. 1959 *Times* 7 Oct. 4/2 He must now be co-favourite ..for the Olympic title.

co-fe·rment. *Chem.* [F. (G. Bertrand 1897, in *Compt. Rend.* CXXIV. 1035), f. Co- 3 b + Ferment *sb.*] = *co-factor 2 a.

1905 Harden & Young in *Proc. Physiol. Soc.* XXXII. p. ii, The proteolytic enzyme..does not appear to possess a dialysable co-ferment. 1944 *Hackh's Chem. Dict.* (ed. 3) 211/2 *Coferment*, a substance which increases the activity of a ferment or enzyme, *e.g.*, calcium for thrombin.

coffee, *sb.* Add: **1. b.** (Examples.) Also, (*a, the*) *coffee*, a cupful of coffee.

1774 P. V. Fithian *Jrnl. Virginia* 21 July (1900) I. 213 Ben & I..rode to..the Corn-field...We returned to Coffee. 1913 C. Mackenzie *Sinister Street* III. i. 516 Michael forgave him immediately and promised to come to coffee. 1920 'K. Mansfield' *Bliss* 163 But I won't have a cup of tea. No, I'll have a coffee. 1938 E. Bowen *Death of Heart* III. vi. 431 Wait till the coffees come. 1959 *Woman's Own* 6 June 12/1 If you're not in a hurry will you let me buy you a coffee?

c. A shade or tint of the colour of coffee (cf. 5 a below).

1815 J. Smith *Panorama Sci. & Art* II. 553 Coffee, damascene, and similar shades. 1817 W. Tucker *Family Dyer & Scourer* iii. 83 For chocolate, coffee, &c. yellow is omitted. 1923 *Daily Mail* 31 July 2/5 (Advt.), In Jade, Lemon, Electric, Coffee, Purple. 1970 *Honey* June 127/2 A suit...In cerise/white..turquoise/yellow; coffee/pink.

5. a. *coffee-drinking, -plantation* (earlier examples), *things, -time, -urn* (earlier example); *coffee-brown* (examples); *coffee-tinted* adj.; also quasi-*adj.* with the meaning 'coffee-coloured', as *coffee morocco*.

1857 C. Kingsley *Two Years Ago* III. iii. 112 The roar of Ogwen, as he swirled and bubbled down, rich coffee-brown from last night's rain. 1934 *Burlington Mag.* Aug. 57/1 Dark, opaque, coffee-brown tints. 1885 C. M. Yonge *Nuttie's Father* II. xviii. 211 There was a general assembly and coffee-drinking in the verandah. 1900 *Westm. Gaz.* 13 June 4/3 Armchairs, upholstered in 'coffee' morocco. 1732 *Cal. State Papers, Amer. & W. Indies* 19 Jan. (1939) 20 There is no reason to doubt but that great numbers of coffee plantations would be immediately begun upon. 1861 Trollope *Tales all Countries* ser. I. 121 Very many of the estates..had been..abandoned, as was the case with his own coffee plantation. 1909 H. G. Wells *Ann Veronica* v. 95 The figure of her aunt..behind the coffee things. 1964 'J. Welcome' *Hard to Handle* vi. 70 On the table..was a tray with coffee things. 1917 Conrad *Shadow-Line* v. 168 Occasionally the second mate would..relieve me at early coffee time. 1947 'N. Blake' *Minute for Murder* ii. 35 He..asked me to pop in at coffee-time. 1899 *Daily News* 28 Jan. 6/4 Coffee-tinted chiffon. 1809 Jane Austen *Let.* 10 Jan. (1952) 247, I think she is quite right to work a rug for Uncle John's coffee urn.

b. coffee and (pl. **coffee and, coffee ands**) *U.S. slang*, coffee and doughnut(s) (or a roll, etc.); **coffee bar, barrow,** a bar or barrow at which coffee is sold as a beverage; **coffee-bean, -berry,** also *U.S.*, the Kentucky coffee-tree; **coffee-break,** the interval, usually mid-morning, in an office, etc., when coffee is drunk; **coffee cake,** (*a*) in the U.S., a breakfast bread of yeast dough enriched with eggs, butter, and sugar, baked in a sheet topped with streusel [etc.]..and glazed with melted sugar; also, a dark fruited raised bread (Webster 1961); (*b*) a cake with coffee flavouring; **coffee coat** = *coffee jacket; **coffee cooler** *U.S. slang*, a contemptuous name applied to an idler, a shirker; **coffee cream,** (*a*) a cold custard dessert flavoured with coffee; (*b*) a sugar cream fondant flavoured with coffee essence, sometimes coated with chocolate; also *attrib.*; (*c*) cream suitable for use in coffee; **coffee-cup,** now usually smaller than a tea-cup; **coffee disease** = *coffee-leaf disease; **coffee essence** [Essence *sb.* 9], a concentrated extract of coffee; **coffee-ground vomit,** a dark-coloured vomit containing blood altered by exposure to the gastric secretions; **coffee jacket,** a jacket worn by women when taking coffee (cf. *tea-jacket*, Tea *sb.* 9 c); **coffee-klatsch** *U.S.* [partial

tr. G. *kaffeeklatsch*], a coffee-party; **coffee-leaf disease,** a disease affecting the coffee plant, caused by a rust-fungus, *Hemileia vastatrix*; **coffee machine,** an apparatus for making coffee; **coffee-mill** (later examples; see also quots. 1887, 1916); **coffee morning,** a morning coffee-party; **coffee palace** (earlier example); **coffee parlour**: see Palace *sb.*¹ 4 (quot. 1894); **coffee-party,** a gathering at which coffee is served; **coffee rot,** a disease of the coffee plant, caused by the fungus *Pellicularia kolerota*; **coffee royal,** (*a*) black coffee with an added liquor; (*b*) (see quot. 1929); **coffee-set,** a set of ware used in serving coffee; **coffee shop,** (*c*) *U.S.* a café or restaurant, usu. attached to a hotel, where meals as well as light refreshments can be bought; **coffee spoon,** a small spoon of a size suitable for stirring coffee in a cup; **coffee stall,** a movable structure in which coffee, as a beverage, and other light refreshments are sold; also *attrib.*; **coffee sugar,** a kind of sugar, freq. coloured and usually in crystal form, used for sweetening coffee; **coffee table,** a low occasional table for serving coffee; hence **coffee-table book** (see quot. 1963); **coffee tray,** a tray from which coffee is served.

1901 in Weingarten. 1931 'Dean Stiff' *Milk & Honey Route* xv. 172 The hash house where they sup plentiously [*sic*] on coffee-ands. 1949 *Collier's* 27 Aug. 27 A saloon on Stanton Street hands out 'coffee and' each morning. 1905 *Daily Chron.* 11 July 3/2 Coffee-bar and grocery stall keepers. 1957 *Times Lit. Suppl.* 18 Oct. 621/3 The seedy group of coffee-bar philosophers..spouting their sad rehash of dated Fascist clichés. 1958 *New Statesman* 30 Aug. 243/2 Most of the coffee-bars that have sprung up in London are dark, fancily decorated dives. 1881 Mrs. Ewing *We & the World* xv. 155 So you live in the docks with your coffee-barrow? 1821 T. Nuttall *Trav. Arkansas* iii. 72 We still..observe the coffee-bean (*Gymnocladus canadensis*),..the seeds of which..produce a substitute for coffee greatly inferior to the cichorium. 1948 *Miami* (Oklahoma) *Daily Record* 30 June 8/2 Farmers call it bladderpod, coffee bean and castle-bean. 1822 J. Woods *Two Yrs.' Resid. Eng. Prairie* 306 The woods..contain a great variety of trees;..On the creek bottoms, coffee-berry, poplar, pecon, white walnut, &c. 1951 *Time* 5 Mar. 25/3 Since the war, the coffee break has been written into union contracts. 1958 L. A. G. Strong *Treason in Egg* iii. 59 During the coffee break, Carstairs had ostentatiously kept on the far side of the room from Ellis. 1959 P. H. Johnson *Humbler Creation* xvi. 112 Better have a coffee-break. Then we can get down to it again. 1879 *Golden Hours* XI. 326/2 Cut up the bread—rye bread, wheat bread, Graham bread, and German coffee-cake. 1885 *Cuisine Creole* 161 Coffee Cake. 1923 M. Wilson *Able McLaughlins* xii, The German led her into the shanty, and set before her..coffee and coffee-cake. 1936 L. G. Nicoll *Eng. Cookery Bk.* 330 *Coffee Cake*...Beat the eggs and stir into flour. Add the golden syrup and coffee essence and milk. 1970 M. Berry *Cookbook* 154 *Tia Maria Coffee Cake*...Blend together the coffee essence and Tia Maria. 1905 *Daily Chron.* 29 May 8/5 Coffee-coats have been banished to the limbo of forgotten things. 1876 J. H. Bradley *Jrnl.* in *Montana Hist. Soc. Contrib.* (1896) II. 162 We are assured that we shall easily fill up the complement in spite of the cold water cast upon our efforts by the 'coffee-coolers' as the shiftless, superannuated loungers about camp are very aptly termed. 1890 *Congress. Rec.* 30 Apr., App. 106 Those who..delight in characterizing the old soldiers as 'coffee coolers', 'bounty jumpers', and 'bummers'. 1895 *Ibid.* 18 Jan. 1129/2, I am opposed to giving pensions to deserters, and 'coffee-coolers', and bounty jumpers, and camp followers. 1939 H. H. Smith *We pointed them North* xvii. 177 The government..drawed up a treaty fixing the new boundaries for the reservation,.. and they got some of these coffee-coolers to consent to it. 1868 M. Jewry *Warne's Model Cookery Bk.* 553/1 *Coffee Cream*, made coffee..sugar..milk..eggs..gelatine. 1891 *Confectioners' Union* 15 Aug. 428 (Advt.), A few chocolates...Chinese cubes. Coffee creams. Coker nut creams. 1900 E. Glyn *Visits Eliz.* (1906) 88, I don't know how we should have got through tea if the coffee-cream cakes had not been so good. 1906 Mrs. Beeton *Househ. Managem.* xxxiii. 1006 *Coffee cream*..cream..coffee essence..gelatine. 1940 'R. Crompton' *William & Evacuees* ii. 48 Coffee cream, that's better. 1950 J. G. Davis *Dict. Dairying* 228 'Coffee cream' or 'fruit cream' may be about 22 per cent [in fat content]. 1959 *Chambers's Encycl.*, Index, Coffee leaf rust.. and coffee disease. [1877 H. Ruede *Sod-House Days* (1937) 99 To give the beverage [*i.e.* rye coffee] a ranker flavor, what is known as 'coffee essence' is used...This essence is a hard, black paste put up in tins... It is probably made of bran and molasses.] 1897 *Sears, Roebuck Catal.* 9/2 Coffee Essence, Hummel's, in tin cans. 1910 H. Belloc *Pongo & Bull* viii. 105 He had..poured [boiling water] upon the coffee essence which he kept beside him in a cup. 1957 N. Marsh *Off with his Head* iv. 69 The others sipped coffee essence and superb brandy. 1885 C. H. Fagge *Princ. Med.* II. Index, Coffee-ground vomit in cancer of stomach. 1907 *Practitioner* Apr. 546 [It] leads to haemorrhage and to the clinically characteristic 'coffee-ground' vomit. 1901 *Daily Chron.* 3 Aug. 10/3 The newest tea jackets have changed their names to coffee jackets. 1895 Coffee-clatch [see *fairy *sb.* 5 c]. 1951 *Notes from Betty Crocker Kitchen* (General Mills Co.) 3 (*heading*) Afternoon coffee klatsch. 1964 F. Pohl in *Galaxy* Oct. 184/2 Poor attendance at neighborhood coffee-klatsches. 1877 *Encycl. Brit.* VI. 112/2 The..coffee-leaf disease appeared in Ceylon in 1869, and in Mysore a year later. 1887 Moloney *Forestry W. Afr.* 107 The coffee-leaf disease..is not likely to infect the coffee-trees in this part. 1894 *Country Gentlemen's Catal.* 104 The only Coffee Machine that extracts all the aroma. 1909 H. G. Wells *Tono-Bungay* III. iii. 354 My aunt, reading a letter behind her coffee-machine... We were breakfasting together. 1933 E. Hemingway *Winner take Nothing* (1934) 32 A bar with a shining steam pressure coffee machine. 1961 W. Buchan *Helen all Alone* 5 The coffee machine boiled over. 1848 E. Bryant *California* ii. 21 Jacob.. could not make the coffee-mill perform its appropriate duty. 1863 B. Taylor *H. Thurston* III. iii. 69 Then he heard.. a stirring in the kitchen under him, and presently the noise of the coffee-mill. 1887 H. L. Williams *'Buffalo Bill'* 10 One of the old-pattern Colts, with the barrels revolving, the ancient 'coffee-mill' or 'pepper-box', laughed at all over the West in the present day. 1904 A. Bennett *Great Man* xxiv. 268 She brought him the coffee-mill full of coffee, and told him to grind it. 1916 *Chambers's Jrnl.* Sept. 601/1 A quaint mechanical rifle, described as a mitrailleuse, which was an improved revival of the 'coffee-mill' used in the American Civil War. 1962 'G. Cullingford' *Third Party Risk* vi. 81 The ordinary members..had privately promoted a coffee morning in the name of charity..the hostess provided.. coffee and biscuits and the visiting ladies provided..one shilling *per capita*. 1970 *Times* 5 Mar. 7/1 Breads and cakes to serve for a coffee morning should be light and sweet but not rich. 1879 *Postmaster General's Rept.* 54 Some of the 'Coffee Palaces' recently established in London and elsewhere. 1886 L. M. Alcott *Jo's Boys* xiii. 225 He..chatted.. at some select coffee-party. 1937 Auden & MacNeice *Lett. from Iceland* xii. 168 Not that..I could breathe if I lived in Cambridge. All those coffee-parties. 1877 *Encycl. Brit.* VI. 112/2 The coffee-rot..works great havoc in the Mysore plantations. 1921 *Adventure* 18 July 36/1 Let's fergit all them bygones in a good cup o' coffee royal. 1929 F. C. Bowen *Sea Slang* 29 *Coffee royal*, the first mug of coffee in the morning under sail. 1786 J. Wedgwood *Let.* 30 June (1965) 297 Tea and coffee sets. 1856 Mrs. Stowe *Dred* 18 A table, which displayed an antique coffee-set of silver. 1940 G. Marx *Let.* 16 Dec. (1967) 189 Met Ira Gershwin this morning in the coffee shop at the Essex. 1966 F. Hoyle *Oct. First* vi. 58 When I came down to breakfast in the coffee shop..I ordered a stack of wheat-cakes and coffee. 1703 N. Blundell *Diary* 20 May (1952) i. 12, I presented my gilt Coffy spoones. 1897 *Sears, Roebuck Catal.* 438/1 Fancy Engraved Coffee Spoon..4⅞ inches. 1917 T. S. Eliot *Prufrock* 12, I have measured out my life with coffee spoons. 1920 F. Scott Fitzgerald *This Side of Paradise* (1921) II. v. 265 Dirty restaurants where careless, tired people helped themselves to sugar with their own used coffee-spoons. 1850 Thackeray *Pendennis* II. xvii. 164 From his station at the coffee-stall, Huxter spied him. 1909 *Daily Chron.* 13 Nov. 4/7 The coffee-stall keeper, after occupying the same 'pitch' for years, regards himself as possessing a 'goodwill' in the site. 1924 'R. Crompton' *William—the Fourth* vii. 117 A coffee-stall could be seen in a glare of flaming oil-jets. 1926 *Gleanings in Bee Culture* Sept. 606/1, I had tried lumps of coffee sugar and these crumbled down on the bottom of the hive. 1930 R. Kipling *Thy Servant a Dog* 28 There was wonderful things-under-table at dinner... Afterwards, was coffee-sugar. 1959 M. Sharp *Rescuers* iv. 56 She fed Nils [*sc.* a mouse] with coffee sugar. 1959 *Which?* Oct. 133/1 What sense is there, if any, in paying nearly 2s. a lb. to sweeten one's coffee with coffee sugar if granulated sugar at 8½d. is good enough for tea? 1877 C. Cook *House Beautiful* 70 (*caption*) Coffee-Table with Chair, both of Black Wood. 1959 J. Braine *Vodi* xxi. 229 I'm going to..buy a little coffee-table from Woolworth's. 1962 John Lynch (*title*) The coffee table book of astrology. 1963 *Sunday Times Mag.* 24 Nov. 23 They used to be called grand-piano books. Now they're known as coffee-table books: 'too big for a bookshelf, full of beautiful pictures, costing a lot'. 1966 *Daily Tel.* 15 Nov. 13/1 He has just produced a book called 'David Hicks on Decoration'. It is a sploshy big, coffee-table book in glorious colour. 1843 Planché *Extravaganzas* (1879) II. 202 When clattering comes the coffee-tray and all your drinking's done. 1929 E. Bowen *Joining Charles* 39 The two had breakfasted; the coffee-tray..was littered with cigarette-ends.

coffee, *v.* [f. the sb.] **1.** *intr.* To drink coffee.

1851 G. W. Curtis *Nile Notes* xiii. 100 He coffeed and smoked,..gave us all the last news [etc.]. 1862 G. du Maurier *Let.* Oct. (1951) 177 After which my pet I dined at Pamphilon's and coffeed in Berners Street. 1885 W. T. Hornaday *2 Yrs. in Jungle* xxiii. 277 Rose very early, coffeed in haste, and..set out.

2. *trans.* To entertain at coffee.

1868 Susan Hale *Lett.* (1919) 48 The Colonel, who coffeed us the day before.

coffee-house, *v. slang. intr.* To indulge in gossip (orig. while waiting for the hounds to draw a covert, etc., during a fox-hunt). Chiefly in *vbl. sb.* (also *attrib.*). Also **coffee-houser,** one who indulges in the practice.

1861 'Scrutator' *Recoll. Fox-hunter* ix. 151 [He] seldom viewed a fox at all; he was always coffee-housing with some particular friend. 1883 Mrs. E. Kennard *Right Sort* xxiii, She found the hounds still engaged in drawing a large wood.., and people were standing about in clusters of twos and threes coffee-housing. 1892 *Field* 6 Feb. 188/2 The field are 'coffee-housing' around when 'Tally-ho!' and they are away. 1897 *Westm. Gaz.* 25 Oct. 2/1 'Coffee-housing'—to wit, chatting about runs past and future, and discoursing upon..hounds and horses. 1907 *Ibid.* 5 Nov. 12/1 The term 'coffee-housers' is used by a certain quaint old M.F.H. to reproach those loud talkers who hinder the serious business ..by the chatter of irresponsible frivolity. 1909 *Daily Chron.* 7 June 5/1 You, gentlemen, come here on no coffee-housing tour. 1961 'J. Welcome' *Beware of Midnight* vi. 76 I'll catch it if Firmian finds me coffee-housing here. 1964 A. Powell *Valley of Bones* iv. 243 We must get on with the job, not spend our time coffee-housing here.

coffee-tree. [Coffee 1.] **1.** The tree or shrub from which coffee is obtained.

1741–1855 [see Coffee 5 a]. 1952 A. G. L. Hellyer *Sanders' Encycl. Gardening* (ed. 22) 121 *Coffea* (Coffee Tree) —*Rubiaceae.* Stove evergreen shrubs.

2. The North American tree *Gymnocladus dioica*; the Kentucky coffee-nut. *U.S.*

1784 J. Filson *Discov. Kentucke* 23 The coffee-tree greatly resembles a black oak, grows large, and also bears a pod, in which is enclosed good coffee. 1785 Washington *Diaries* (1925) II. 360 Eight Nuts from a tree called the Kentucke Coffee tree. 1796 Morse *Amer. Geogr.* (ed. 3) I. 636 The coffee-tree resembles the black oak, and bears a pod which encloses a seed, of which a drink is made not unlike coffee. 1832 D. J. Browne *Sylva Amer.* 166 The French of Canada call this tree *Chicot*,..and the inhabitants of the Western States, *Coffee Tree*. 1867 *Trans. Ill. Agric. Soc.* 1865–6 VI. 390 The Coffee Tree famed for its similitude to the fruit that adds to our morning meal. 1951 *Dict. Gardening* (R. Hort. Soc.) II. 939/2 The name Coffee-tree applied to *G. dioica* comes from the seeds having been used as a substitute for coffee by the early settlers in N. America.

coffin, *sb.* Add: **13.** *coffin-case*; *-headed* adj.; **coffin-nail,** also (*slang*, orig. *U.S.*), a cigarette (cf. Nail *sb.* 7 d); **Coffin Texts,** texts inscribed on the inside of coffins during the Middle Kingdom in Egypt; **coffin-wood,** wood used for making coffins; *spec.* elm.

1908 *Chambers's Jrnl.* July 527/2 He had propped a coffin-case against one corner of the tomb. 1954 T. Gunn *Fighting Terms* 15 Though coffin-case Was ready to hold life. 1890 Kipling *Barrack-r. Ballads* (1892) 11 With 'is coffin'eaded shield an' shovel-spear. 1888 *Texas Siftings* 18 Feb. 8/1 A youth..puffed at an ill-smelling coffin nail. 1901 A. G. Robinson *Philippines* 263 For a package containing thirty 'coffin-nails' the price was three and a half cents, American money. 1928 Wodehouse *Money for Nothing* ii. 34 Most of these birds [*sc.* invalids in a sanatorium] would give their soul for a coffin-nail. 1967 'L. Egan' *Nameless Ones* x. 118 Let me pay for my own coffin nails. 1912 J. H. Breasted *Relig. Anc. Egypt* viii. 273 Written on the inner surfaces of the heavy cedar coffins.. are found..mortuary texts... These texts as a class are sometimes designated as the Book of the Dead... I have for convenience termed them Coffin Texts, a designation drawn from the place in which they are found, and thus parallel with the Pyramid Texts. 1947 I. E. S. Edwards *Pyramids of Egypt* v. 153 The texts..were written on the interior surfaces of the rectangular wooden coffins used in that period—a fact which has caused them to become known as the Coffin Texts. 1883 J. D. Hooker *Let.* 17 June in *Life* (1918) II. xl. 251 We have ascertained the origin of Chinese Cassia, of the Star Anise, and of the Coffin wood. 1909 Kipling *Actions & Reactions* 42 We ain't goin' to lay any axe-iron to coffinwood here.

coffinite (kǫ·finəit). *Min.* [f. the name of R. C. *Coffin*, U.S. geologist: see -ite¹.] A black hydrous silicate of uranium.

1954 A. Rosenzweig et al. in *Econ. Geol.* XLIX. 356 'Coffinite' is being investigated by the Trace Elements Laboratory of the U.S. Geol. Survey. 1955 *Science* CXXI. 608/2 A new extremely radioactive black mineral from the La Sal No. 2 mine, Mesa County, Colorado...has been named coffinite in honor of Reuben Clare Coffin. 1958 *Sunday Times* 26 Feb. 19/2 A speckled, blackish mineral called coffinite.

co-fo·rmulator. [Co- 3 b.] A formulator together with another or others.

1902 *Daily Chron.* 22 Sept. 3/3 The term proposed by Prof. Baldwin and his co-formulators for the theory. 1905 *Ibid.* 16 Oct. 3/1 Dr. Wallace,..the co-formulator of the theory of 'natural selection'.

cog, *sb.*² Add: **6. cog-rail** (examples); hence *cog-railway*.

1884 *Science* III. 415/2 The rack or cog-rail in the middle of the track is made of two angle-irons which have between them cogs of one and a quarter inch iron. 1944 B. Johnson *As much as I Dare* 281 There is a cog-rail to the top. 1896 *Vermont Agric. Rep.* XVI. 126 We cannot boast of a Mt. Washington with its cog-railway. 1902 H. Belloc *Path to Rome* 287 A precipitous peak of bare rock, up which there ran a cog-railway to some hotel or other. 1963 *Economist* 30 Nov. 911/1 Funiculars, cog railways, aerial ropeways, chairlifts.

cog, *v.*¹ Add: **4.** (Examples.) Also with *down*.

1925 *Jrnl. Iron & Steel Inst.* CXI. 528 In general the continuous process is not used for cogging and blooming, but at the Gary Works in Illinois there is a rail mill in which the ingot is actually continuously cogged. 1930 *Engineering* 11 July 43/3 From 2000 tons to 2500 tons of ingots are cogged down. 1951 *Jrnl. Iron & Steel Inst.* CLXVII. 168/1 The ingots were cogged down under steam hammers to blooms and slabs.

5. *intr.* To engage with corresponding cogs or projections. Also *fig.*, to fit in; to agree, be of the same mind; to work in harmony.

1898 in *Eng. Dial. Dict.* 1926 H. G. Wells *World of W. Clissold* I. ii. 191 The same old Russian 'system', with.. many of the wheels failing to cog. 1964 *New Society* 11 June 19/1 Even the three relevant sections of the county council —health, children and welfare—cog together.

cogged, *ppl. a.*¹ Add: **2.** (See quot. 1888.) Cf. Cog *v.*¹ 4.

1888 *Lockwood's Dict. Mech. Engin.* 76 *Cogged bloom*, a bloom or crude mass of steel which has been passed through the cogging mill in readiness for rolling into rails or other sections. 1925 *Jrnl. Iron & Steel Inst.* CXI. 57 Since the introduction of heavier ingots, reheating the cogged bloom has become standard practice at most rail and section mills.

cogging (kǫ·giŋ), *vbl. sb.*³ [f. Cog *v.*¹ + -ing¹.] The action or process of rolling steel blooms from ingots. Also *attrib.* and *Comb.*

1878 *Technol. Dict.* (ed. 3) 150/1 *Cogging-mill* (Iron shipb.), *das Vorwalzwerk*. 1881 *Instr. Census Clerks* (1885) 92 Cogging Roller. 1888 *Lockwood's Dict. Terms Mech. Engin.* 76 *Cogging engine*, an ordinary rail-mill engine used for driving the cogging mill. *Cogging mill*, a rolling mill in which steel blooms are rolled out. **1921** *Dict. Occup. Terms* (1927) § 279 Oiler; ..cogging mill oiler. **1925** *Glasgow Herald* 31 Oct. 10 One of the cogging mills would also be put in operation. **1925** [see *cog *v.*¹ 4]. **1951** *Jrnl. Iron & Steel Inst.* CLXVII. 165/1 The cutting of blooms and slabs is normally effected at ingot heat after cogging.

‖**cogida** (kohī·δa). [Sp., lit. a gathering of the harvest, hence the act of a bull in catching a bull-fighter, fem. of *cogido* pa. pple. of *coger* to seize, f. L. *colligĕre* Colligate *v.*] A tossing of a bull-fighter by a bull.

1923 T. P. Terry *Guide Mexico* p. cviii, Cogidas, though frequent, are naturally not of constant occurrence, but even a slight wound caused by a bull's horn requires delicate treatment. **1932** Hemingway *Death in Aft.* xix. 253 The rush of talk that always follows a serious cogida. **1934** R. Campbell *Broken Record* vii. 163, I don't mind taking a cogida from a Camarguaise cow.

‖**cogitandum** (kǫdʒitæ·ndəm). *Philos.* [L. *cogitare* v. + -*andum*, as if from L. **cogitandum*, neut. gerundive of *cōgitāre* to think.] That which should be thought; the ideal or correct processes of thought, as opposed to the actual process.

a1866 J. Grote *Moral Ideals* (1876) 60 The two high intellectual ideals..are..the *cogitandum* and the contemplation of real being. **1878** [see *Cogitatum]. **1890** W. James *Princ. Psychol.* I. xiv. 552 The laws of the arbiter, of the *cogitandum*, of what we ought to think.

‖**cogitatum** (kǫdʒitē¹·təm). *Philos.* [a. L. *cōgitātum*, neut. pa. pple. of *cōgitāre* to think: see Cogitate *v.*] That which is thought; the actual processes of thought, as opposed to the ideal thought-processes.

1878 W. James *Coll. Ess. & Rev.* (1920) 57 Every law of Mind must be either a law of the *cogitatum* or a law of the *cogitandum*. **1890** —— *Princ. Psychol.* I. xiv. 552 The laws of our actual thinking, of the *cogitatum*, must account alike for the bad and the good materials on which the arbiter has to decide.

‖**cogito** (kō·gito). *Philos.* [L., = 'I think', 1st pers. pres. of *cōgitāre* (see Cogitate *v.*); used by Descartes in his *Meditations* (1641).] The principle 'cogito, ergo sum', or any equivalent formula, by which Descartes claimed to establish his own existence as a thinking being from the fact of his thinking or awareness; *loosely*, conscious awareness or subjectivity (see quot. 1875).

1854 A. G. Henderson tr. *Cousin's Philos. of Kant* x. 93 He establishes nothing but..the supposition of an abstract *cogito*. **1875** W. James *Coll. Ess. & Rev.* (1920) 10 After every definition of an object, reflection may arise, infect it with the *cogito*, and so discriminate it from the object *per se*. **1946** B. Russell *Hist. West. Philos.* III. ix. 586 The above argument, 'I think, therefore I am'.., is known as Descartes's *cogito*. **1965** *Amer. Philos. Q.* II. 23/1 Hoxt analyzed every element of the *cogito*.

cognac. 2. a. Delete † and 'Formerly', and add further examples.

1909 *Westm. Gaz.* 8 Mar. 8/2 Mons. E. Martell gave evidence concerning Cognac brandy. **1958** *Everyman's Encycl.* II. 275/2 The *brouillis* is distilled a second time to give a spirit with 70 per cent. of alcohol which when matured will be worthy of the name of Cognac B[randy].

cognate, *sb.* Add: **1. c.** *Hindu* and *Muslim Law.* A relative of a deceased person through the mother (see also quot. 1949).

1902 in *Encycl. Brit.* XXIX. 281/2. **1949** A. A. A. Fyzee *Outl. Muhammadan Law* 344 A cognate is a person related to the deceased through one or more female links; for example, the daughter's son or the daughter's daughter, the mother's father or the father's mother's father. **1965** J. D. M. Derrett *Introd. Mod. Hindu Law* 379 If no agnate can be found (and one may search back through any number of degrees), then the mother's brother's son or other cognates will take the inheritance in order of priority.

cognition. Add: **2. c.** *attrib.*

1878 S. H. Hodgson *Philos. Reflection* I. i. 68 Is not philosophy..just what the Germans mean by Cognition-theory (*Erkenntnisstheorie*)? *Ibid.* 69 A cognition-principle (*Erkenntnissprincip*) is good for the cognition-principle (*Realprincip*). *Ibid.* 230 Cognition that has *a priori* forms is already a cognition-faculty. **1909** *Mind* XVIII. 143 The ruin of the ordinary idealistic cognition-theory.

cognitional, *a.* (Later examples.)

1903 J. Dewey *Let.* 19 Dec. in R. B. Perry *Th. & Char. W. James* (1935) II. 525 What becomes of the instrumental interpretation of cognitional activity? **1964** M. Critchley *Developmental Dyslexia* ix. 62 The dyslexic..is also in some way a non-visile cognitional type.

cognitive, *a.* Add: (Later examples.) (See also quot. 1956.)

1909 *Westm. Gaz.* 10 Feb. 4/1 Three different aspects—cognitive, affective, conative—proper to every instinctive process. **1948** *Poetry* Dec. 158 *Cognitive*, pert[aining] to the thought element of the poem, as opposed to either the emotive or the conative (willing) elements. **1948** E. R. Hilgard *Theories of Learning* i. 13 The preference of the contemporary association theorist for a reaction psychology (laying emphasis upon movements) as over against a cognitive psychology (emphasizing perception-like and idea-like processes) learning with understanding or learning under cognitive control. **1953** T. Parsons *Social System* 7 The most elementary and fundamental 'orientational' category ..seems to be the 'cognitive' which..may be treated as the 'definition' of the relevant aspects of the situation in their relevance to the actor's 'interests'. **1954** W. K. Wimsatt *Verbal Icon* i. 23 The main drift of his argument is that emotive 'meaning' is something noncognitive and to independent of descriptive (or cognitive) meaning. **1956** J. O. Urmson *Philos. Analysis* 171 The same question arises even when metaphysics is denied cognitive meaning only. 'Cognitive' is used to mean 'empirically verifiable or else analytic'.

cognitively (kǫ·gnitivli), *adv.* [f. Cognitive *a.* + -ly².] In a cognitive manner; with regard to, or from the point of view of, cognition.

1880 W. James *Coll. Ess. & Rev.* (1920) 216 The only alternative..was..to suppose the ego, in willing, to be merely cognitively conscious, in the midst of the universal force-stream, of certain currents with which it was..fated to identify itself. **1925** W. P. Montague *Ways of Knowing* v. 162 The cognitively desirable or true. **1957** *Univ. Calif. Publ. Philos.* XXIX. 189 The statement..is either false or cognitively meaningless. **1965** N. Chomsky *Theory of Syntax* i. 22 The sentences..are 'cognitively synonymous'.

‖**cognitum** (kǫ·gnitǝm). [L., neut. of *cognitus*, pa. pple. of *cognoscere* to know.] An object of cognition.

1875 G. H. Lewes *Problems* II. 491 The *cognitum* and its *cognitus* are not two facts, but one fact. **1890** *Mental Sml. Author. Relig.* iv. ii. 403 All the particular cognitions are unified in the single cognitum. **1936** *Mind* XLV. 182 Let us next agree to call a cognitum any *terminus ad quem* of the activity called 'experiencing'.

cog-wheel. Add: **b. cog-wheel railway, system,** a mountain-railway system using a cog-rail (see Cog *sb.*¹ 6); rack-railway.

1904 *Westm. Gaz.* 4 May 10/1 Lines up the mountains and over the passes worked on the cog-wheel system. **1907** *Ibid.* 24 July 1/3 The circuitous cog-wheel railway connecting Csorba with Csorba-tó. **1963** *Economist* 30 Nov. 912/3 Cog-wheel railways, an earlier method of getting up steep mountain sides.

co-head. [Co- 3 b.] A leader or principal conjointly with another or others.

1891 *Westm. Gaz.* 23 Nov. 2/2 No matter what Annie may do to you as co-head of the S., he says there is no harm. **1922** J. Joyce *Let.* 11 Nov. (1957) 326 Reece the co-head (what a word!) of the A. Press is a real authority on the question.

cohere, *v.* **6.** Delete † *Obs.* and add later example.

1911 Mair *Eng. Lit.: Mod.* 100 His Essays..were in their origin merely jottings gradually cohered and enlarged into the series we know.

7. *trans.* To cause to cohere or hold together.

1875 R. R. Brash *Eccl. Archit. Irel.* 152 No cements were used to cohere the material.

coherence. Add: **1. c.** *Physics.* The property of being coherent (in various senses: see *Coherent *a.* 1 d).

1926 *Sci. Abstr.* A. XXIX. 478 (*heading*) Experiment on capacity for coherence of light. **1951** *Proc. Camb. Philos. Soc.* XLVII. 624 The range of order is related to the range of coherence of the electronic wave functions. **1953** *Proc. R. Soc.* A. CCXVI. 548 The behaviour of a superconductor is in some way controlled by an interaction of rather long range (called the 'range of coherence') within the electron assembly.

b. *coherence theory*: in *Philosophy*, the theory that the definition, or the criterion, of truth is that the propositions which are its parts form a coherent system; also *transf.*

1906 H. H. Joachim *Nature of Truth* iii. 65 The coherence-theory..goes deeper than the theories we have rejected. **1921** H. J. Paton (*title*) The good will: a study in the coherence theory of goodness. **1940** B. Russell *Inquiry* x. 140 The coherence theory of truth..is that of Hegel.

coherent, *a.* Add: **1. d.** Various *spec.* senses in *Physics* (see quots.).

1930 Mann & Millikan tr. *Drude's Theory of Optics* 134 If two sources are to produce interference, their phases must change by exactly the same or else have a constant difference. Such sources are called coherent. **1937** Frank & Tamm in *Compt. Rend. Acad. Sci. URSS* XIV. 109 (*heading*) Coherent visible radiation of fast electrons passing through matter. **1938** *Physical Rev.* LIV. 100/2 The theory [of Frank and Tamm] implies that the radiation emitted by the electron along its path is coherent. **1939** I. Tamm in *Jrnl. Physics USSR* I. 454 The experimental investigation of the Cerenkov radiation is made possible..by the fact that in the visible region the intensity of the coherent radiation is much greater than that of the Bremsstrahlung. **1957** *Gloss. Terms Nucl. Sci.* (Nat. Res. Council, U.S.) 151/2 Scattering of particles or photons in which there are definite phase relationships between the incoming and the scattered waves, is termed coherent scattering. **1958** *Chambers's Techn. Dict.* 967/2 *Coherent oscillator*, one which is stabilised by being locked to the transmitter of a radar set for beating with a reflected incoming pulse signal. **1960** Cooke & Markus *Electronics & Nucleonics Dict.*, *Coherent radiation*, radiation in which there are definite phase relationships between different points in a cross-section of the beam... Interference bands are observed only between coherent beams. **1961** *Ann. Reg. 1960* xi. 396 This light was different from that in the flash since the tripping action of the first light to be produced ensured that all subsequent quanta were produced with light waves in synchronism with it. In other words, the light produced from the laser of a laser is to emit light of very narrow bandwidth and considerable intensity—what is called coherent light. **1965** *New Scientist* 16 Sept. 576/2 A far simpler data-processing system results from the use of coherent light.

coherer (kohī·rəɹ). *Electr.* [f. Cohere *v.* + -er¹.] A device consisting of a number of metal filings, wires, plates, etc., in loose contact which suffer a drop in resistance in the presence of high-frequency electromagnetic waves and used as a detector of such waves, chiefly in radio telegraphy; orig., the name given by Sir Oliver Lodge to the detector in the form of a glass cylinder containing metal filings, which cohere under the influence of an electromagnetic wave.

1894 O. Lodge in *Proc. Roy. Inst.* XIV. 336 This arrangement, which I call a coherer, is the most astonishingly sensitive detector of Hertz waves. **1899** W. Crookes in *Rep. Brit. Assoc. Advancem. Sci. 1898* 21 Oliver Lodge..produced the vacuum filing-tube coherer with automatic tapper-back. **1913** 'Ian Hay' *Happy-go-lucky* xiii, One could almost feel the Marconigram radiating from Lady Adela. But apparently The Freak's coherer was out of order. **1923** E. W. Marchant *Radio Telegr.* iv. 38 The 'welding' of the particles in the coherer continues until something is done to break the connection between them. **1959** *Chambers's Encycl.* XI. 475/1 The coherer is suitable only for telegraphic reception.

coho (kōʊ·hōʊ). Also cohoe, cohose. [Of unknown origin.] A species of salmon, *Oncorhynchus kisutch*, found in the northern Pacific waters; the silver salmon.

1869 *British Colonist* (Victoria, B.C.) 29 July 1/2 Lastly, there is the genus known by the Indian name of coucosse a hybrid bastard sort of fish, half trout, half salmon. **1869** *Mainland Guardian* (New Westminster, B.C.) 23 Sept. 2/1 The second of salmon in class is the Cohose, which comes in September and continues to run until November. **1896** *Bull. U.S. Fish Comm. 1898* XVIII. 6 The coho should rank next after the king salmon for food value. **1902** [see Silver *sb.* and *a.* 21 d]. **1902** *Encycl. Brit.* XXVI. 390/1 The silver salmon or cohoe arrives in British Columbia waters a little later than the sockeye. **1965** A. J. McClane *Fishing Encycl.* 193/1 Also called silver salmon and hookbill, the coho is a popular gamefish of the Pacific Northwest.

cohog, var. Quahaug.

1788 M. Cutler in *Life, Jrnls., & Corr.* (1888) I. 416 Went into the water; found a great number of clam cohog shells. **1872** Schele de Vere *Americanisms* 29 The name cosiley beads [in wampum] came from the largest shells of the Quahaug or Cohog.

cohort, *sb.* Add: **3. b.** In demography, a group of persons having a common statistical characteristic, esp. that of being born in the same year. Also *attrib.*

1944 P. H. Fairchild *Dict. Sociol.* 45/2 *Cohort* (demographical usage), a group of persons starting life together. Thus, hypothetical cohorts are established in the construction of life tables. **1947** P. K. Whelpton in *Proc. Internat. Statist. Conf.* (*Washington*) III. B. 632 This discussion will..carry the analysis of fertility trends a step further..by considering cohorts of women (i.e. women grouped by date of birth), and by classifying the births to each cohort. **1954** (*title*) Cohort fertility. **1959** *Listener* 19 Nov. 865/1 It [sc. secular analysis] must be reinforced by analysis of a different style: one which takes all the people born in one year or decade in one year and follows them through their lives. Analysis of this kind is called 'cohort analysis'. **1962** *Lancet* 8 Dec. 1213/1 The number of possible long stay patients from the 1954 cohort of admissions was estimated as 7453.

5. An assistant, colleague, accomplice. Chiefly *U.S.*

1952 M. McCarthy *Groves of Academe* xii. 276 The old poet had left, accompanied by two of his cohorts. **1957** B. & C. Evans *Dict. Contemp. Amer. Usage* 99/1 Banting [was] assisted by his young cohort, Dr. Charles H. Best. The culprit and his three cohorts stole.... **1965** *Times Lit. Suppl.* 25 Nov. 1059/3 The new American vulgarism of 'cohort' meaning 'partner'. **1970** *Sunday Times* 28 June 6/5 The impending trial of Bobby Seale, chairman of the Black Panther movement, and his eight cohorts in New Haven.

‖**cohue** (koü). [Fr.] An unruly crowd; a mob.

1850 Carlyle *Jrnl.* in J. A. Froude *Carlyle* (1884) II. 47 The *cohue* of objects and persons his life was cast amidst, did not increase my hopes of a great result. **1901** 'Linesman' *Words by Eyewitness* xiv. 314 The *cohue* seethes and roars around the farm.

coiffeur. (Earlier examples.)

1847 F. A. Kemble *Let.* 31 May in *Rec. Later Life* (1882) III. 190 Entreating me to suffer his coiffeur to arrange my hair after the approved French method. **1850** Thackeray *Pendennis* II. xv. 148 Said he knew of a coiffeur.

‖**coiffeuse** (kwafø̄z). [Fr., fem. of Coiffeur.] A woman hair-dresser, esp. one skilled in designing and arranging the coiffures of women.

1870 *Queen* 20 July (Advt.), The extraordinary perfection acquired by the 'Coiffeuses'. **1927** *Daily Express* 5 July 5/5 In big West End saloons the highly skilled coiffeuse and beauty specialist can command £500 a year.

coiffure (koi·fiūː, ‖kwafür), *v.* [f. Coiffure *sb.*] *trans.* To dress (a woman's hair). Also *transf.*

1906 *Daily Chron.* 8 Nov. 6/1 She has a profusion of brown hair, beautifully coiffured. **1928** *Daily Tel.* 1 May 15/3 Her hair was elaborately coiffured every morning. **1963** *Times* 12 Mar. p. vi/4 Light's noble belt of parklands around the capital has been so barbered and coiffured that rose gardens and ornamental ponds begin to replace gum-trees and native grasses.

coi·ffured, *a.* [f. Coiffure *sb.* or *v.*] Wearing a coiffure. Of head-dress or hair: arranged in a coiffure.

1907 *Macm. Mag.* Jan. 5/1 Queen Elizabeth in full-dress, ruffled, coiffured, and corseted. **1928** *Daily Tel.* 5 May 16/5 The feather curl..can be worn close to the ears, as is sometimes expressed in the real coiffured coil.

coign, *sb.* Add: **4.** *Geol.* An original angular elevation of land around which continental growth has taken place.

1899 J. W. Gregory in *Geogr. Jrnl.* XIII. 245 South of the Scandinavian coign are the transverse east and western chains of the Alps and the Atlas. *Note.* The suggestion of the word 'coign' for 'corner' I owe to Mr. L. Fletcher..The term is suitable, as it is used for a printer's wedge as well as for the corner-stone of a house.

coil, *v.*³ Add: **1. b.** To lay down in a coil or coils.

1915 'Bartimeus' *Tall Ship* i. 27 A younger man..was busy coiling down something in the bows.

coil, *sb.*³ Add: **4. d.** An intra-uterine contraceptive device of flexible material shaped into a spiral.

1931 Dickinson & Bryant *Control of Conception* iii. 117 The silkworm coil entirely within the body of the uterus. **1938** R. L. Dickinson *Control of Conception* (ed. 2) xiii, The *intra-uterine coil*, as we warily test its possibilities of safe control. **1964** A. F. Guttmacher et al. *Planning your Family* viii. 57 Some of the plastic coils can be inserted by paramedical personnel, such as midwives, instead of doctors. **1970** *Sunday Times* 17 Nov. 50/5 Those who employ other birth-control techniques also have sizeable majorities in favour of the pill: 73 per cent of those using the coil, 72 per cent of those using a diaphragm, [etc.].

6. coil ignition, a system of ignition in internal combustion engines in which the low-voltage current of the battery is converted to a high voltage by means of an induction coil; **coil pot,** a pot, the sides of which are constructed from rolls or coils of clay (cf. *Coiling *vbl. sb.*¹ d); **coil spring,** a volute spring, *spec.* in the springing of motor cars.

1911 G. C. Sherrin *Montagu Motor Bk.* iv. 99 The apparatus necessary for the ignition system known as the accumulator and coil ignition. **1930** *Engineering* 17 Oct. 501/1 Coil ignition is employed, the contact maker and distributor being mounted directly above, and driven from, the vertical lubricating pump-shaft. **1935** *Economist* 7 Dec. 1144/2 In the early days of motoring, coil ignition was practically universal. **1937** *Jrnl. R. Aeronaut. Soc.* XLI. 413 Results obtained on a modern coil ignition system operated by a double contact-breaker with a small angle of open circuit. **1893** *Ann. Rep. Smithsonian Inst. 1892* 537 There are the coil pots, as are found in mounds and cliff dwellings. **1960** H. Powell *Beginner's Bk. Pott.* i. 29 Slab and coil pots form the basis of the majority of handmade pots. [**1877** Knight *Dict. Mech.* I. 589/1 *Coiled spring*, a metallic spring wound up into a spiral.] **1939** Webster s.v. *spring*, The principal varieties of springs used in mechanisms are the spiral spring, the coil spring [etc.]. **1959** *Motor Man.* (ed. 36) vi. 114 Every reader will be familiar with coil springs (to use the popular term for helical springs).

coiled, *ppl. a.*¹ Add: **2.** Of or pertaining to a method of basket-making based on a spirally coiled foundation.

1909 *Cent. Dict. Suppl.* s.v. *basketry*, Coiled basketry. **1923** Kroeber *Anthropol.* xiii. 384 Of other elements specific to the Patagonian region, there might be mentioned coiled basketry. The oldest basket-work that has come down to us is the coiled, or as we term it, Indian work. An Indian woman making a coiled basket. **1936** *Discovery* June 171/1 Other cultural features of the Hamites are round shields..and coiled basketry.

b. coiled ware, pottery constructed from rolls or coils of clay. Cf. *coiling *vbl. sb.*¹ d.

1931 *Discovery* Jan. 3/2 Their contributions to the craft included coiled ware sometimes decorated with thumbnail incisions.

c. coiled coil, in a lamp: a doubly spiralled filament; freq. (with hyphen) *attrib.*; also *transf.*

1935 *Economist* 29 June 1509/2 The coming of the coiled-coil lamp constituted further progress in the efficiency of lamps. **1940** *Chambers's Techn. Dict.* 174/2 *Coiled-coil filament*, a spiral filament for an electric lamp which is coiled into a further spiral to reduce the radiation losses and enable it to be run at a higher temperature. **1951** G. H. Sewell *Amat. Film-Making* (ed. 2) iii. 38 A special form of projection lamp with concentrated coiled-coil filaments. **1962** *Times* 13 Apr. 19/6 X-ray and analogous evidence..has been lately reported to be consistent with a coiled-coil type of structure.

coiler¹. Add: **b.** *Cotton-spinning.* An apparatus for coiling the sliver by feeding it into the rotating coiler-cans. Also *attrib.*

1873 Leigh *Sci. Mod. Cotton Spinning* (ed. 2) 187 Shortly after the invention of the coiler Messrs. Tatham and Cheetham conceived the idea of adapting it as a slubbing frame. **1884** Marsden *Cotton Spinning* 141 The waste.. passes into a coiler at the end of the machine. **1890** Nasmith *Mod. Cotton Spinning Mach.* 55 A plate free to revolve and borne in the lower part of the coiler frame. **1892** —— *Cotton Spinning* 127 The coiler can is 9 in. diameter. **1902** Thornley *Cotton Combing Machines* 235 The rule for total draft will serve here, substituting the block roller for the coiler roller.

coiling, *vbl. sb.*¹ Add: **c.** *attrib.* in *Cotton-spinning* (cf. prec.).

1850 Tatham & Cheetham *Brit. Pat.* 13,313 2 Our improvements..relate..to the drawing frame, such improvements consisting, firstly, in the combination of coiling apparatus..with the principle of traversing cans. *Ibid.* 14 The arrangement of parts for winding on the yarn as spun is substantially the same as that of our coiling machinery.. which apparatus is now well known. **1873** Leigh *Sci. Mod. Cotton Spinning* (ed. 2) 168 This very ingenious contrivance is the celebrated 'coiling motion', patented about thirty years ago by Messrs. Tatham and Cheetham.

d. *Pottery.* The act or process of constructing coiled ware. Also *attrib.*

1900 tr. *Deniker's Races of Man* 154 A method of proceeding which may be called *coiling* in clay. **1936** *Burlington Mag.* Nov. 241/2 The nature of the lead silica glaze..and the technique of 'coiling'. **1957** *Encycl. Brit.* XVIII. 374/2 In the coiling technique, the raw material is rolled out into a slender rope which is coiled upon itself.

co-i·mplicate, *v.* *Philos.* [Co- 1.] *trans.* To imply or implicate mutually. Hence **co-implica·tion; co-impli·cative** *a.*

1907 W. James *Pragmatism* iv. 149 The real universe must form an unconditional unit of being, something consolidated, with its parts co-implicated through and through. *Ibid.* vii. 264 A One that logically presupposes, co-implicates, and secures each *each* without exception. **1909** —— *Plural. Universe* viii. 325 It is not a universal co-implication, or integration of all things *durcheinander*. **1939** *Mind* XLVIII. 369 Certain propositions are found to be co-implicative.

coin, *sb.* Add: **6.** (Examples of *slang* use.)

1874 Hotten *Slang Dict.* 124 'To post the coin'..a sporting phrase meaning to make a deposit of money for a match of any kind. **1904** G. V. Hobart *Jim Hickey* ii. 35 Wouldn't we be a nice pair of turtles to stand around with coin in our jeans and see a nice girl..getting the ice? **1926** J. Black *You can't Win* vii. 82 I'll get the coin on that junk in an hour. *Ibid.* ix. 104 You put me in the hole for some coin.

8. coin-box, a receptacle for the coins in a coin-operated telephone or the like; hence, a coin-operated telephone, or a kiosk containing such a telephone; also *attrib.*; **coin-catcher,** a surgical instrument for extracting a swallowed coin; **coin-in-the-slot,** used *attrib.* of a machine, etc., operated by the insertion of a coin, or of its products; **coin-operated** *a.*, of a machine, etc., operated by a coin; abbrev. **coin-op.** used as *adj.* and *sb.*, esp. of an automatic launderette or dry-cleaning establishment; **coin-purse** chiefly *U.S.*, a purse designed especially to hold coins.

1906 *Ann. Rep. Amer. Teleph. & Telegr. Co.* 5 Prepayment coin boxes..have been provided. **1960** *Times* 31 Oct. 14/7 A portable coin-box telephone will reach every bed. **1968** 'A. Gilbert' *Night Encounter* xi. 171 It was a funny sort of call... From one of those coin boxes, so it couldn't be your friend from London. **1969** *Guardian* 4 July 18/6 Minimum charges for telephone calls from coin boxes will be cheaper when decimal currency is fully introduced in 1971. **1895** *Arnold & Sons' Catal. Surg. Instr.* 288 Probang, Œsophageal, with coin-catcher. **1903** *Daily Chron.* 27 Nov. 8/6 The doctor..passed the coin-catcher down his throat, but the child struggled so that the catcher broke, and was also swallowed. **1904** *Daily Consular Rep.* (U.S.) 12 Aug. 8 A new application in Australia of the principle of the coin-in-the-slot machine. **1958** *Times Rev. Industry* June 76/2 A 'juke-box' or coin-in-the-slot record player. **1960** *Times* 10 June 22/5 Coin-in-the-slot devices dispense soap and detergent. **1961** *New Left Rev.* Jan.–Feb. 48/1 Coin-in-the-Slot Television .. is, culturally, a dangerous formula. **1960** *Times* 10 June 22/5 A 'coin-op.' store..can be set up for an outlay of between $10,000 and $15,000 for the equipment. **1961** *Guardian* 8 Mar. 8/4 The next of these round-the-clock 'coin-ops' is expected to be opened in Glasgow. **1969** *Guardian* 22 Feb. 14/1 & 3 Of all the schemes for getting rich quickly, the coin-operated launderettes have proved the most durable. *Ibid.*, A man who was.. opening up a coin-op laundry. *Ibid.*, A coin-op which fails to get enough business is in a very dangerous position. **1960** *Times* 10 June 22/4 Coin-operated, self-service laundries, the first of which recently opened its doors in Britain, are a familiar part of the American scene today. **1908** *Sears, Roebuck Catal.* 1000 Paragon Patent Folding Coin Purse.. This purse will hold $10.00 in silver. **1967** K. Giles *Death in Diamonds* ii. 30 Choffy Ingleby wrote me an address and gave a key. These I placed in a coin purse.

coin, *v.*¹ Add: **3. b.** To shape or alter the physical properties of (metal) by the application of heavy pressure.

1940 J. D. Jevons *Metall. Deep Drawing* viii. 253 A common industrial practice is to..'coin' the finished shape between suitable dies under a drop-stamp. **1960** *Machinery* XCVII. 485/2 Inner races are coined at a similar press set-up.

5. d. *to coin a phrase*, an expression commonly used ironically to introduce a cliché or a banal sentiment.

1940 F. B. Young *Mr. Lucton's Freedom* II. v. 182 It takes all sorts to make a world, to coin a phrase. **1950** G. Hackforth-Jones *Worst Enemy* i. 59 You look (to coin a phrase) 'in the pink'. **1962** N. Marsh *Hand in Glove* iv. 127 Who, to coin a phrase, would have thought of meeting you?

coincidence. Add: **7. a.** *Physics.* The indication of the occurrence of ionizing particles in two or more detectors simultaneously (see quot. 1958). Also *attrib.* Cf. *anti-coincidence.

1930 *Physical Rev.* XXXV. 651/2 Enormously increased resolving power can be obtained by the requirement of multiple instead of paired coincidences. The attainment of very great freedom from accidental coincidences is of greatest importance [etc.]. *Ibid.* 652/1 Automatic recording of the amount of the penetrating radiation coming from particular areas of the sky, using two tube-counters and a special 'coincidence circuit'. **1936** R. W. Lawson tr. *Hevesy & Paneth's Man. Radioactivity* (ed. 2) xxv. 280 Insertion of sheets of lead between the counters only slightly diminishes the number of simultaneous discharges (coincidences), and this proves that we are here confronted ..with cosmic rays. **1940**, etc. [see *anti-coincidence]. **1958** *Van Nostrand's Sci. Encycl.* (ed. 3) 369/1 A true coincidence is one that is due to the detection of a single particle or of several genetically related particles. An accidental, chance, or random coincidence is one that is due to the fortuitous occurrence of unrelated counts in the separate detectors.

b. *Computers.* Equivalent signals received simultaneously in an electronic circuit; the reception of such signals. Also *attrib.*

1947 *Rev. Sci. Instruments* XVIII. 907/1 In order to reduce chance coincidences to a minimum, it is necessary to use a coincidence circuit. **1948** *Ibid.* XIX. 365/2 Methods can be found for modulating one of the E.M.T.'s so that the E.M.T.'s themselves form a coincidence or anticoincidence system. **1950** C. B. Tompkins et al. *High-Speed Computing Devices* iv. 57 An electronic gate is a circuit with a single output and two (or more) inputs so designed that an output signal is produced when, and only when, input signals are received on both (or on a particular set of) input leads. Such circuits are variously known as gates, coincidence circuits, Rossi circuits, or 'signal circuits. **1953** A. D. & K. H. V. Booth *Automatic Digital Calculators* xi. 111 *Coincidence sensors.* In computing machine design it is frequently necessary to have available means for ascertaining the identity of two quantities. **1964** M. J. Pedelty in Tou & Wilcox *Computer Sci.* x. 248 The essence of the system..is that signals are 'broadcast' on a 'to-whom-it-may-concern' basis. Coincidence gates and delays can then be used to extract certain pulse patterns from the 'broadcast alphabet'.

coining, *vbl. sb.*¹ Add: **1. b.** Shaping of metal in a coining press (see quot. 1940).

1940 J. D. Jevons *Metall. Deep Drawing* xv. 261 It is now usual to give the pressed or deep-drawn article one or more final squeezing operations between dies, a process closely akin to that of 'coining'. **1966** *Gloss. Terms Mechanised & Hand Sheet Metal Work* (B.S.I.) 6 *Coining*, A final squeezing operation, with a punch and die, to give accuracy of dimensions and to ensure that a pattern is impressed by a punch.

3. coining press, (*a*) a press for making coin (see quot. 1688 in Dict.); (*b*) a punch press in which metal is pressed to a required size, or embossed, etc.

a1877 Knight *Dict. Mech.* I. 592/1 *Coining-press*, a power lever-screw press by which the planchet of metal is impressed with the design or legend. **1925** *Jrnl. Iron & Steel Inst.* CXII. 448 The adaptation of the coining-press to the manufacture of such parts as forged arms and levers.

co-insurance. [Co- 3 c.] A form of insurance in which responsibility for loss is shared by two or more insurers; also, that in which the insured, under certain conditions, is jointly responsible with the insurance company.

1889 in *Cent. Dict.* s.v. *Insurance.* **1902** *Encycl. Brit.* XXIX. 512/2 The direction in which fire insurance..calls most pressingly for improvement is the extension of the principle of co-insurance.

‖**cointreau** (kwæ·ntrō). Also Cointreau. [F., proprietary name.] A colourless orange-flavoured liqueur; also, a glass of this. Cf. Curaçao.

1920 *Berry Bros. & Co. Catal.* 14 Cointreau. **1928** *Sunday Dispatch* 2 Dec. 4 The best concocted cocktail..contained brandy, gin, whisky, cointreau, French vermouth, [etc.]. **1942** 'N. Shute' *Pied Piper* 17 Then he took a cup of coffee and a cointreau in the café. **1946** 'S. Russell' *To Bed with Grand Music* iii. 49 The coffee and the cointreau had been placed on the table. **1959** J. Braine *Vodi* xxi. 274 It wasn't ..the Turkish coffee or the Cointreau, but his almost childish delight at being there with me.

coiny (koi·ni), *a. colloq.* [f. Coin *sb.* + -y¹.] That has abundance of coin; rich. Hence **coi·niness,** wealth.

1891 *Chambers's Jrnl.* 7 Feb. 83/2 He paints for amusement merely. He seems to be coiny. **1903** *Blackw. Mag.* Sept. 312/1 A fellow must be a coiny bird to bank with the Bank of England. *Ibid.* 313/2 Howell dying with over £4000, though he had considerably impaired the 'coininess' of other people.

coitus. Add: **b. coitus interruptus**: sexual intercourse in which, with the intention of avoiding conception, the penis is completely withdrawn from the vagina before ejaculation; also *fig.*

1900 Havelock Ellis *Stud. Psychol. Sex* II. 127 Onan's device was not auto-erotic, but was an early example of withdrawal before emission or *coitus interruptus*. **1909** A. A. Brill tr. *Freud's Sel. Papers Hysteria* (1912) vi. 141 Anxiety neurosis appears in women..whose husbands practice coitus interruptus or reservatus. **1918** M. C. Stopes *Married Love* v. 51 It often happens nowadays that, dreading the expense and the physical strain of child-bearing for his wife, the husband practises what is called *coitus interruptus*. **1941** H. Nicolson *Diary* 27 May (1967) 168 The P.M... then went on to speak about conscription in Northern Ireland and left the house with a sense of *coitus interruptus*. **1948** C. Day Lewis *Poems 1943–1947* 18, I knew the weak Involuntary spasms of consent—Ah, *coitus interruptus* with a cheating world. **1961** 'J. Welcome' *Beware of Midnight* x. 120 Can you imagine anything more stupid than a *corrida* without killing?.. That's *coitus interruptus* if you like. **1962** *Lancet* 2 June 1194/1 The survey shows that the sheath and coitus interruptus are still the methods most commonly used for contraception in this country.

c. coitus reservatus: sexual intercourse in which, by a technique of deliberate control, ejaculation or complete orgasm is avoided and copulation thereby prolonged.

1903 Havelock Ellis *Stud. Psychol. Sex* III. 248 We practiced all sorts of fancy coitus, *coitus reservatus*, etc. **1909** [see b above]. **1922** *Internat. Jrnl. Psycho-anal.* III. 401 Coitus reservatus and coitus interruptus are particularly harmful to women. **1953** A. C. Kinsey *Sex. Behav. Hum. Fem.* xv. 625 Coitus reservatus, the *Karezza* of the Sanskrit and Hindu literature, represents a maximum sophistication of such deliberate control.

‖**cojones** (kohōᵘ·nes), *sb. pl.* [Sp., pl. of *cojón* testicle; cf. Cullion.] **a.** Testicles. **b.** Courage, 'guts'.

1932 Hemingway *Death in Aft.* ii. 28 It takes more cojones to be a sportsman where death is a closer party to the game. **1966** T. Capote *In Cold Blood* iv. 274 The baseball field was mud up to your *cojones*. **1966** N. Behn *Kremlin Letter* I. vi. 85 'Hit that big cow in the crutch!..In the *cojones*,' he roared at her, pointing to his own anatomy. **1966** *Guardian* 30 July 7 You have the cojones to ask me if I still got confidence in Britain?

coke, *sb.*¹ Add: **1. b.** Slang phr. (imper.) *go and eat coke*: refrain from addressing, or otherwise annoying, the speaker.

1891 Farmer *Slang* II. 150/1 *Go and eat coke* (vulgar), a phrase indicative of contempt. **1908** *Magnet* I. No. 1, Oh, go and eat coke! **1959** I. & P. Opie *Lore & Lang. Schoolch.* x. 178 When accosted by an irritating person..they.. suggest..'Go and eat coke'.

c. (*a*) Tin plate made from iron produced in a cokery. (*b*) A grade of tin plate, more thinly coated than charcoal plate, used for canning and general purposes; = *coke plate below.

1930 *Sheet Metal Industries* Apr. 693 (Advt.), Tinplates... Brands...'Henza' (Cokes), 'Amfin' (special cokes). **1936** W. R. Brown *Tinplate* (ed. 3) 6 Coke Tinplates is a commercial description of standard quality for tinplate purposes... Best Coke are tinplates carrying a slightly higher coating of tin. **1951** *Engineering* 21 Dec. 771/1 Tinplate is produced in three coating weights..'common cokes', 'standard cokes', and 'best cokes'.

2. *coke barrow, bogey, dust, fork*; **coke-breeze** (see Breeze *sb.*³); **coke finish,** a coating given to a sheet of coke (see *1 c); **coke-oil** (see quot.); **coke-oven,** an oven, furnace, kiln, or retort in which coke is produced by the expulsion of gas from bituminous coal; hence **coke-oven gas,** gas so produced; **coke (tin) plate,** tin plate made from iron refined in a cokery; tin plate having a lighter coating than charcoal plate.

a1884 Knight *Dict. Mech.* Suppl., *Coke Barrow*, a large semi-cylindrical sheet-iron two-wheeled barrow used about coke ovens and furnaces. **1921** *Dict. Occup. Terms* (1927) § 699 *Coke man*..fills coke bogeys with coke fork from stack, [etc.]. **1884** Coke breeze [see Breeze *sb.*³]. **1956** W. D. Hargreaves in D. L. Linton *Sheffield* 289 The fines..have been sintered by heating with coke breeze in an oxidizing atmosphere. **1877** *Encycl. Brit.* VI. 118/1 The outside of the heap [of coal, in coking] is covered with a coating of wet coke dust. **1909** *Cent. Dict.* Suppl., *Coke-dust*, powdered coke; used for blacking foundry molds. **1898** *Jrnl. Iron & Steel Inst.* LII. 36 Sheets..may be coated with a good common coke finish with about 2 lbs. of tin, a heavier coating being applied for a charcoal finish. **1904** Harbord & Hall *Metall. of Steel* III. xxxiv. 536 The rolls are driven at a definite speed which determines the amount of tin taken up by the plate.., the object kept in view being to give what is known as a 'coke finish' to a cwt. of plates with 2 lbs. 4 ozs. to 2 lbs. 6 ozs. of tin. **a1877** Knight *Dict. Mech.*, *Coke fork*, a ten-tined fork for shoveling coke. **1862** *Chambers's Encycl.* IV. 645/1 There remains..a large residue of pitch, which is again distilled.., giving off an oil called coke-oil. **1839** Ure *Dict. Arts* 997 A front elevation of two..coke-ovens. **1861** J. Percy *Metallurgy* 157 In its simplest form a coke-oven is a chamber of fire-brick or some other refractory material, having an arched roof in which is a hole and an entrance below. **1930** *Engineering* 9 May 610/3 The blast-furnace and coke-oven gases. **1878** *Technol. Dict.* (ed. 3) II. 151/1 *Coke-plate* (oppos. to charcoal-plate), *Das Steinkohlenblech*, Tôle au coke. *Ibid.*, Coke-tin-plate, *Das Steinkohlenweissblech, das gepuddelte Weissblech*, fer-blanc au coke. **1956** W. E. Hoare *Tinplate Handbk.* (ed. 3) iv. 15. *Coke Plates, Charcoal Plates.* The adjectives *coke* and *charcoal* are still used in specifying and indicating tin coating weight. *Ibid.* ix. 40 Hot-dipped coke tinplate is used for the manufacture of decorated boxes.

coke (kōᵘk), *sb.*² Slang (orig. *U.S.*) abbrev. of Cocaine (in its use as a drug). Also *attrib.* and *Comb.*

1908 R. S. Baker *Following Colour Line* iii. 47 They buy the 'coke' in the form of powder and snuff it up the nose. **1912** E. B. Lowry *Himself* xx. 179 A few years ago 'coke' fiends in a certain locality in Chicago made a regular practice of buying '——'s Catarrh Cure'. **1927** 'J. Barbican' *Confess. Rum-Runner* iv. 47 No matter what you've done—smuggling, counterfeiting, coke-peddling, shooting, [etc.]. **1928** E. Wallace *Flying Squad* xvi. 150 It's your 'coke' trade that's stirring up the Yard. **1959** P. Capon *Amongst those Missing* 233 He started introducing her to drugs... Reefers at first, and then, under the influence of reefers, coke.

Hence **coke** *v.*² *refl.*, to drug oneself with cocaine; also with *up*; **coked (up)** *a.*, drugged with cocaine (or a similar drug); **cokey, cokie,** a cocaine-addict.

1922 E. Ferguson *Black Candle* I. i. 20 Cocainomaniacs are commonly called 'cokies'. **1924** G. C. Henderson *Keys to Crookdom* xxiv. 307 When 'coked up' he will murder and rob and steal. **1929** D. Hammett *Red Harvest* xxvii, You gallop out, coked to the edges, charging at the whole world with both eyes shut. **1934** J. G. Brandon *One-Minute Murder* xxiii. 247 His first glance at the shivering, stricken-looking creature..told him that the man was a 'cokey'. **1937** 'David Hume' *Cemetery First Stop* i. 8 He didn't mind men getting tangled with marihuana smokes or 'coking' themselves at times. **1954** 'N. Blake' *Whisper in Gloom* I. i. 17 They let him coke himself up for the occasion. **a1953** Dylan Thomas *Quite Early One Morning* (1954) 38 These psychopathic gorillas coked to the gills have no place in Little Tim's cosmography. **1967** C. Drummond *Death at Furlong Post* iv. 40, I want a solo cokey who knows guns.

Coke (kōᵘk), *sb.*³ A registered trade-mark of the Coca-Cola Company, = *Coca-Cola; also used *colloq.* with small initial. Also *attrib.* and *Comb.* orig. *U.S.*

1909 *Coca-Cola Bottler* (Phila.) Nov. 17/1 If you..asked to be served with 'Ice Cold Cokes' you will be presented with a very good bottle of carbonated Coca-Cola. **1915** *Printers' Ink* 23 Sept. 46/2 The propensity of a large proportion of those who regularly drink Coca-Cola to call for their favorite drink as 'dope' or 'coke' or 'koke'. **1938** S. V. Benét *13 O'Clock* 236 The girls drank 'coke', mostly. **1940** *Evening News* 9 Jan. 4/1 The Canadian soldiers..have stormed the shops in search of coke, which is slang, it seems, for a soft drink. **1941** Daniels *Tar Heels* 261 Sandwich munchers and coke drinkers. **1947** D. Riesman in *Yale Law Jrnl.* Dec. 191 More typical of the French café than of the American tavern or coke-bar. **1949** N. Streatfeild *Painted Garden* xvii. 185 They..had bought him ice cream sodas and cokes. **1954** G. Winn *Queen's Countrywomen* vii. 106 There is something very innocent about..drinking 'Coke' through a straw. **1963** P. H. Johnson *Night & Silence* xxiv. 175 Now I shall have a Coke, as the youngsters do.

coked (kōᵘkt), *ppl. a.* [f. Coke *v.* + -ed¹.] Reduced to coke.

1763 W. Lewis *Comm. Phil.-Techn.* 284 A Fire of coaked pitcoal. **1894** *Sat. Rev.* 25 Aug. 221 Coal-dust..in a coked and cindered condition.

co·keman, co·ke-man. [Coke *sb.*] A workman employed in loading or unloading coke, charging a furnace, etc. with coke, or discharging coke from a furnace, etc.

1852 F. S. Williams *Our Iron Roads* 251 He in return signs the coke-man's book for the amount. **1892** *Daily News* 11 Mar. 5/8 The cokemen, the mechanics, and the miners in the Durham Miners' Federation. **1920** *Glasgow Herald* 1 Sept. 7 Cokemen employed on by-product plants at collieries. **1921** *Dict. Occup. Terms* (1927) § 049 *Cokeman*..works at box or bunker where 'breeze'..is collected after screening.

coker² (kōᵘ·kəɹ). [f. Coke *sb.* or *v.* + -er¹.] One who superintends the coking of coal; also, a workman employed in handling coal in or about a coke-oven, etc.

1793 in Bogle *Founding Carron Wks.* (1898) 32 John Heggie, a fireman coaker. **1881** *Instr. Census Clerks* (1885) 93 Steel Melting:..Coker. **1921** *Dict. Occup. Terms* (1927) § 081, etc.

cokerite, var. *cocorite.

cokery (kōᵘ·kəri). [f. Coke *sb.* + -ery.] A coke-furnace.

1923 *Glasgow Herald* 27 Mar. 9 The three cokeries already being exploited by the French. **1941** A. O. Pollard *Bombers over Reich* ii. 26 Raids on..synthetic plants, cokeries, ports and railway yards.

cokey, cokie: see *coke *sb.*²

cokey-cokey, var. *hokey-cokey.

coky (kōᵘ·ki), *a.* [f. Coke *sb.* + -y¹.] Resembling coke.

1895 S. P. Sadtler *Handbk. Indust. Org. Chem.* (ed. 2) xi. 355 Non-coking Coals..do not possess the property of fusing to a compact 'coky' mass.

col. Add: **2.** *Meteorol.* A region of lower pressure between two anticyclones.

1885 *Times* 26 Oct. 6/6 The barometer is (relatively) high over the Bay of Biscay and..Scandinavia, and these two high-pressure systems are joined by a 'col' which lies over the North Sea. *Ibid.* 30 Oct. 6/6 The two [high-pressure areas] are united by a 'col' which now lies over the United Kingdom. **1887** R. Abercromby *Weather* 26 Between every two anticyclones we find a furrow, neck or 'col' of low pressure. **1923** Napier Shaw *Forecasting Weather* v. 117 A *col*, the saddle-shaped region between two lows and two highs. **1927** W. G. Kendrew *Climates of Continents* (ed. 2) 354 The col is essentially an anticyclonic formation.

col, col. (kǫl). Abbreviated form of Column 4.

1903 A. Bennett *Let.* 9 May (1966) I. 37 They were giving me 2 guineas for 2 cols (1000 words). **1918** E. Pound in *Lett. J. Joyce* (1966) II. 424 The owner of the Manchester guardian..shows no disposition to have it in his own better paid cols. **1922** Joyce *Ulysses* 276 Payment at the rate of guinea per col.

Colbred (kǫ·lbred). [f. *Col*(*burn* the name of a Gloucestershire sheep-breeding company + Bred *ppl. a.*¹] The name of a highly fertile breed of sheep obtained by crossing the Border Leicester, Clun, Friesland, and Dorset Horn; a sheep of this breed.

1962 O. H. Colburn in *Times* (Agric. Suppl.) 3 July p. v/1 Our Company, H. A. Colburn & Son, decided in 1954 to develop an entirely new breed of sheep designed especially for high prolificacy and high milking capacity, and we chose the name Colbred for them. One result of that decision was the formal registration of the Colbred Sheep Society Limited on July 14, 1961. **1963** *Guardian* 21 Aug. 2/8 Mr. Colburn has shown for the first time at the Royal Show at Newcastle upon Tyne the Colbred sheep which he had bred over five generations from crossings of three British and one Continental breed. **1967** *Times* 26 June 6/7 The Colbred..was never intended for use as a self-contained breed but always for crossing with hill ewes to provide highly productive daughters. *Ibid.* 6/8 He maintains a strong flock of pure Colbreds.

colcannon. Add: Also **colcannen, kale cannon, kohl cannon.** Also a Scottish dish. (Examples of these spellings and of later use.)

1827 Scott *Jrnl.* 28 Mar. (1941) II. 38 A species of *colcannen*, where cold potatoes are mixed with hot cabbage. **1855** E. Acton *Mod. Cookery* xvii. 315 Kohl cannon, or kale cannon. (An Irish Receipt). **1946** F. M. McNeill *Recipes Scott.* 47 *Colcannon*...Take equal quantities of boiled potatoes and boiled cabbage [etc.].

Colchian (kǫ·lkiăn), *a.* [f. L. *Colchi-s*, a. Gr. Κολχίς + -an.] Of or relating to Colchis, the ancient name of a region east of the Black Sea, associated in Greek mythology with the quest of the Golden Fleece; = Colchic *a.*

a1616 W. Drummond *Poems*, Here did shee spred the Treasure of her Haire, More rich then that broght from the Colchian Mines. **1757** Dyer *Fleece* iv. 152 That ancient path to India's climes, which fill'd With proudest affluence the Colchian state. **1821** Byron *Don Juan* III. lv. 30 Such as lit onward to the Golden Fleece His predecessors in the Colchian days. **1869** Geo. Eliot *How Lisa loved King* 41 The Colchian flood.

b. Colchian pheasant (*Phasianus colchicus*), the wild pheasant, *spec.* the variety found in the valleys of the Caucasus.

1862 H. G. Adams *Wild Flowers, Birds, & Insects* 221 With us the Common, or Colchian Pheasant, as it is called, has so long been naturalised as to be generally regarded as a British bird. **1957** *Encycl. Brit.* II. 538/1 At the eastern edge [of Asia Minor] certain peculiarities are found, such as..the survival of the Colchian pheasant.

colchicine. Add: Now used *esp.* to induce genetic changes in plants and animals. Also *Comb.*

1937 Blakeslee & Avery in *Jrnl. Heredity* XXVIII. 393 (*heading*) Methods of inducing doubling of chromosomes in plants by treatment with colchicine. **1945** [see *autopolyploid]. **1946** *Nature* 31 Aug. 287/1 The use of colchicine-induced polyploids in plant breeding. **1957** M. Abercrombie et al. *Dict. Biol.* 57 Reversion of colchicine-inhibited metaphase nuclei to the resting state results in polyploidy.

cold, *a.* Add: **1. c.** Colloq. phr. *to knock, lay (out)*, etc., *cold*: to knock (a person) unconscious; to render senseless because of a severe blow or shock; also *fig.* orig. *U.S.*

1829 *Massachusetts Spy* 22 July (Thornton), I want to lay out [this candidate] as cold as a wedge. **1847** W. T. Porter *Quarter Race Kentucky*, etc. 45 He picked up an ole axe helve an gin me a wipe aside the hed that laid me cole fur a while. **1847** J. M. Field *Drama in Pokerville* 93 It is 'bound' to lay every thing in the way of architecture west of the Alleghanies 'out cold'. **1884** 'Mark Twain' *Huck. Finn* xxxvii. 376 It was the sudden surprise of it that knocked us so kind of cold. **1896** Ade *Artie* xvii. 159 Here's somethin' that'll knock you cold. **1905** R. Beach *Pardners* v. 127 Some Polack..laid out the quartermaster cold. **1928** F. N. Hart *Bellamy Trial* iii. 98 'What did you do?' 'Do? I don't know what I did. It knocked me cold.' **1930** Wodehouse *Very Good, Jeeves!* iv. 115 'Held them spellbound.' 'Knocked 'em, eh?' 'Cold,' said young Tuppy. 'Not a dry eye.' **1944** *Living off Land* iv. 93 The boxer who is not at the peak of training is likely to be laid cold.

e. *to get* (a person) *cold*, earlier *to get it on* (him) *cold*: to have at one's mercy; to have captured completely. *U.S. colloq.* Also *to have* (a person) *cold. colloq.*

1908 S. E. White *Riverman* xlvii. 353 I'll put Heinzman in the pen too. I've got it on him, cold. **1921** 'Ian Hay'

Willing Horse ix. 166 Strung about as we were, he had us cold. **1924** C. E. Mulford *Rustlers' Valley* xix. 213 What you doin'? I got you cold. **1928** F. E. Baily *Golden Vanity* xix, Cynthia's lapping like an angel. You've got London cold.

7. e. Colloq. phr. *to leave* (a person) *cold*: to fail to interest or excite. (Cf. F. *cela me laisse froid*, G. *das lässt mich kalt*.)

1857 Geo. Eliot in *Westm. Rev.* XI. 30 An orator may discourse very eloquently on injustice in general, and leave his audience cold. **1888** H. Sweet *Shelley's Nature-Poetry* 28 His enthusiasm..leaves us cold. **1927** A. Huxley *Proper Studies* 173, I..am left cold by ritual, the corybantic emotionalism of revivals. **1967** E. Short *Embroidery & Fabric Collage* i. 1 Whereas one is uplifting to look at the other leaves one emotionally cold.

f. Without preparation, preliminary performance, etc. orig. and chiefly *U.S.* Usu. *quasi-adv.*

1896 Ade *Artie* x. 88 I'm an easy runner till it comes to the high jump and then I quit cold. **1928** *Daily Express* 5 Sept. 12/5 He had just heard that..a play had opened cold in Philadelphia or somewhere... I told him it meant that it had gone into a big town without a try-out week. **1958** N. Coward *Play Parade* V. p. xxiii, We were to open [at Drury Lane] 'cold', that is without a try-out.

10. b. In advb. use: without any mitigation; absolutely, entirely. *U.S. slang.*

1889 *Kansas Times & Star* 24 June, A. D. Taylor.. trailed a variety actress..with whom he was infatuated... She shook him cold yesterday. **1905** R. Beach *Pardners* (1912) iii. 79 We were liable to get turned down cold if we didn't have some story. **1916** H. L. Wilson *Somewhere in Red Gap* vi. 265 This game where you play cards with yourself and mebbe win a thousand dollars cold. **1953** *Times* 26 Mar., One of their attacks..was stopped cold by concentrated artillery and mortar fire. **1954** B. Benson *Lily in Coffin* xxi. 213 I'm not going to quit cold like you... I'll start all over again with the school.

11. *spec.* of news.

1913 E. C. Bentley *Trent's Last Case* i. 16 Within a week..'the Manderson story', to the trained sense of editors..was 'cold'. **1946** D. L. Sayers *Unpop. Opinions* 130 The date had to be changed to conceal the fact that the news was already 'cold'.

12. (Later U.S. examples.)

1773 Washington *Diaries* (1925) II. 100 Touched now and then upon a Cold Scent till we came into Colo. Fairfax's Neck. **1874** in S. Sidney *Bk. Horse* (1875) 398 When running a cold scent the music [of the hounds] is extremely fine. **1875** *Ibid.* 476 When hounds run from grass to plough, it is often found that they decline from racing breast-high to cold hunting. **1878** C. Hallock *Sportsman's Gaz.* (ed. 4) I. 440 The object is to obtain a fine nose [in a dog], so as to hunt a cold scent.

b. Of the person chosen to seek or guess, in children's games: distant from the object sought. Also *fig.* Cf. Warm *a.* 6.

1864 Dickens *Mut. Fr.* II. vi. 53 'That can't be the spot too?' said Venus. 'No,' said Wegg. 'He's getting cold.' **1876** 'Mark Twain' *Tom Sawyer* ix. 89 Now they're stuck. Can't find it. Here they come again. Now they're hot. Cold again. **1882** [see *hot *a.* 8 a]. **1887** A. Daly *Railroad of Love* 17 You are nowhere near it. As the children say in their game—you're 'cold'.

18. *cold-drawn* (later examples), also *fig.*, unaffected by the emotions, cool, calculated; *cold-cut, -rolled*; also *cold drawing, rolling, swaging, tinning, welding.*

1951 R. Mayer *Artist's Hand-bk. Materials* iii. 139 Simple solutions of resin in solvents, made without oils and driers, are known in the varnish industry as 'cold-cut' varnishes, even though steam heat is occasionally used to accelerate the solution. **1909** Webster, Cold-drawing. **1946** *Nature* 28 Dec. 930/1 These fibres can then be extended to some four to five times their original length by the process of cold-drawing. **1892** Stevenson & Osbourne *Wrecker* ix, The little beast means cold-drawn biz. **1898** Kipling in *Morn. Post* 11 Nov. 5/1 Out of all manner of tight places that require dexterity and a cheek of cold-drawn brass. **1906** Cornford *Defenceless Isl.* 99 Cotton is the subject of much cold-drawn gambling. **1906** *Daily Chron.* 10 Feb. 7/1 A detective-sergeant, by relating cold-drawn facts..showed the prisoner to be an unprincipled scoundrel. **1919** *Brit. Manufacturer* Nov. 24/2 Many successful locomotive builders procure various parts from special manufacturers—for instance, cold-drawn or welded tubes. **1946** *Nature* 28 Dec. 930/1 X-ray examination of the cold-drawn fibres shows that the crystals have become orientated in a direction parallel to the fibre axis. **1878** *Engineering* 1 Nov. 347 By comparing hot-rolled and cold-rolled iron of the same kind, under physical stress. **1888** *Lockwood's Dict. Terms Mech. Engin.* 78 *Cold rolled*, bars and plates rolled without being previously heated. **1897** *Daily News* 8 Mar. 2/5 Steel..cold-rolled sheets. **1963** *Economist* 7 Dec. 1079/1 Cold-rolled [steel] thin plates. **1878** *Engineering* 1 Nov. 347 The cold rolling is effected by means of a powerful train of the ordinary type. **1888** *Lockwood's Dict. Mech. Engin.* 78 *Cold rolling*. The practice of rolling iron plates cold produces a material having a high tensile strength. **1955** *Times* 4 June 3/4 The cold-rolling firms take steel from the re-rollers..and roll it to the correct gauge... Every engineering trade uses cold rolled strip. **1877** *Encycl. Brit.* VII. 799/2 Cold swaging, that is, by hammering it [sc. malleable iron] till cold. **1873** Spon *Workshop Rec.* I. 337/2 Block tin dissolved in muriatic acid with a little mercury forms a very good amalgam for cold tinning. **1946** *Firth-Brown Gloss. Metall. Terms* 59 The more recently developed method of cold welding in which pressure alone is applied. Cold welding..is used principally for aluminium and its alloys and not for steel.

19. cold-cathode, a cathode that emits electrons at ambient temperature under the influence of a high voltage; usu. *attrib.*; **cold-cock** *v. trans.*, to knock (a person) unconscious (*U.S. slang*); **cold cook** *slang*, an undertaker; **cold cuts** orig. *U.S.* [tr. G. *kalter Aufschnitt*], an assortment of cooked meats, sliced and served cold; occas. in *sing.*; **cold deck** *U.S. slang*, a pack of cards in which the cards have been arranged beforehand; also *fig.*; hence **cold-deck** *v.*, to cheat (a person) by means of a cold deck; also *fig.*; so *cold-decker*; **cold douche,** a stream of cold water directed against some part of the body as a remedial treatment; also *fig.*; hence *cold-douching*; **cold feet,** in colloq. (orig. *U.S.*) phr. *to get* (or *have*) *cold feet*, to become cowardly or discouraged; hence = fear, 'funk', cowardice; **cold fish** *colloq.*, an emotionless person; **cold-footed** *a. colloq.*, timid, cowardly; also *absol.*; **cold-footer** *slang*, a timid person; **cold frame** *Hort.*, a frame in which small plants are grown and protected without artificial heat (see Frame *sb.* 13 c); **cold front** *Meteorol.*, the forward boundary of a mass of advancing cold air; **cold harbour** (see Harbour *sb.*¹ 2 a); **cold house** *Hort.*, a glass-house in which plants are grown without artificial heat; also *transf.*; **cold light,** light that is accompanied by little or no heat, e.g. luminescence; **cold-pack,** a wet pack (see Pack *sb.*¹ 11) prepared with cold water; **cold pole,** in high latitudes, the place of lowest temperature; **cold room** *Hort.*, a store-room kept at a very low temperature for the retardation of bulbs and roots; **cold rubber** (orig. *U.S.*), a synthetic rubber manufactured at a low temperature; **cold saw,** one for cutting cold metals (*Cent. Dict.* Suppl. 1909); **cold shot,** small globules of iron found in chilled portions of a casting (Knight *Dict. Mech.* Suppl., *a*1884); **cold shut, cold-shut** (see quots. and Shut *sb.* 4); **cold snap** (orig. *U.S.*) (see Snap *sb.* 7 b); **cold soldering,** soldering without heat with the aid of mercury; **cold spot** *Physiol.*, a spot upon the skin which is sensitive to cold, but insensitive to warmth, pain, or pressure; **cold starting,** the starting of an internal-combustion engine at ambient temperature; also (with hyphen) *attrib.* or as *adj.*; **cold storage** (see Storage 2 b); hence **cold store,** a refrigerating chamber for the cold storage of perishable foods, esp. meat; **cold table,** (a table bearing dishes of) cold food; **cold turkey** (see *turkey²); **cold war,** hostilities short of armed conflict, consisting in threats, violent propaganda, subversive political activities, or the like; *spec.* those between the U.S.S.R. and the western powers after the 1939–45 war; also *transf.*; hence *cold-warrior*; **cold wave** (orig. *U.S.*), (*a*) *Meteorol.* (see Wave *sb.* 5 b); (*b*) (see quot. 1949); also *cold permanent wave*; **cold work** *Metallurgy*, the working of metal when it is cold (see quots.); hence *cold work* v., *cold-worked* ppl. adj., *cold-working* vbl. sb.

1929 *Proc. Inst. Radio Engin.* May 849 (*heading*) Cold cathode rectification. *Ibid.* 850 Cold cathode devices for rectification are essentially identical in that they embody a cathode, of relatively greater surface area than the anode, situated in a suitable gas of the proper density to produce the desired conductivity by ionization. **1930** *Engineering* 31 Oct. 560/3 Many interesting developments of the original cold-cathode tubes were taking place. **1945** *Electronic Engin.* XVII. 762 A special cold cathode tube capable of carrying loads of 1,000 to 2,000 amperes at 100 kV. **1927** *Amer. Speech* II. 351/1 *Cold cocked*, to be knocked senseless. 'Tom was cold cocked when that rock hit him.' **1934** J. T. Farrell *Young Manhood* (1936) iv. 205 They cold-cocked him, and left him unconscious. **1729** *Universal Spect.* 4 Oct. (N. & Q. 5 Oct. 1929, 236/2), He further directs, that no Undertaker, *alias* Cold Cook, or Upholder shall have the Management thereof. **1860** Hotten *Dict. Slang* (ed. 2), Cold cook. **1945** A. Kober *Parm Me* 113 She suddenly sighted the food on the table and pointed accusingly. 'Look what he's itting—cold cotts!' **1964** W. Markfield *To Early Grave* (1965) xii. 242 Cold cuts, a roll? It's Sunday, it's hot, she didn't feel like cooking. **1967** N. Mailer *Cannibals & Christians* I. 13 A cold cut set in the white tray of a refrigerator. **1969** P. Highsmith *Tremor of Forgery* xvii. 152 A buffet-table of cold cuts. **1857** *San Francisco Call* 3 Apr. 4/2 He's got the thing all set to ring in a cold deck. **1876** B. Harte *G. Conroy* vi. ii, You've been ..playin' it very low down on my moral and religious nature, generally ringin' in a cold deck on my spiritual condition. **1884** in *Amer. Speech* (1942) XVII. 125/2 The miller..kicked because said Serna was trying to cold deck said Sanches. **1902** H. L. Wilson *Spenders* xi. 123 A man wakes up to find that his natural promptin's has cold-decked him. **1920** C. E. Mulford *J. Nelson* xv. 163 I've had all th' visitin' I want with a bunch of cold-deckers. **1835** Cold douche [see Douche *sb.*]. **1925** D. H. Lawrence *Refl. Death Porcupine* 172 It's no use talking...That 'subject' was a cold douche. **1959** *Times* 11 Feb. 9/1 From official quarters a cold douche was quickly poured on this premature jubilation. **1904** *St. George* VII. 168 He would have to be an uncommonly sturdy Simon Zelotes whose zeal survived the cold-douching of schoolboy chaff. **1893** S. Crane *Maggie* (1896) xiv. 112, I knew this was the way it would be. They got cold feet. **1896** Ade *Artie* xii. 108 He's one o' them boys that never has cold feet. **1904** E. Robins *Magn. North* I. 8 But instead of 'getting cold feet' as the phrase for discouragement ran, and turning back, they determined [etc.]. **1914** Rosher *In R.N.A.S.* (1916) 40, I get awfully cold feet... That puts the fear of God into you. **1915** 'Ian Hay' *First Hundred Thou.* xxi. 329 It seems that the enemy have evacuated Fosse Alley again. Nobody quite knows why: a sudden attack of cold feet, probably. **1962** *Times* 20 Feb. 11/2 The Algerian leadership might have cold feet at the concessions it has made. **1941** 'P. Wentworth' *Danger Point* (1942) xxvi. 150 He's a cold fish.' No, fish isn't the right creature—I believe they are quite affectionate. **1962** McLuhan *Gutenberg Galaxy* 19 The Westerner appears.. a very cold fish indeed. **1944** G. B. Shaw *Everybody's Pol. What's What?* xxxvi. 311 In Russia there is the Communist Party with its rules and disciplines and frequent purges of the coldfooted. **1966** *Listener* 6 Oct. 517/3 A strangely cold-footed way of presenting such glorious music. **1919** *Downing Digger Dial.* 17 *Cold footer*, a carpet knight. **1920** Hunter *Trail Drivers of Texas* 429 Two of my men stayed with me, and the third, a 'cold-footer', crossed on the bridge. **1851** Mrs. Gaskell *Let.* 17 Mar. (1966) 833 Papa has bought a new frame..& we are going to have it for a cold frame to harden out the plants from the hot bed. **1859** *Trans. Ill. Agric. Soc. 1857–8* III. 503 The seed for early summer cabbages can be planted in a cold frame early in September. **1877** *Field & Forest* II. 164 These insects had all gathered along the Northern and Eastern margin [inside] of a 'cold frame', in his garden. **1921** Bjerknes & Solberg in *Geofysiske Publikationer* II. iii. 12 The boundary line at the ground will be the front of advancing cold air, or, to introduce a shorter expression, a 'cold front'. **1841** Mrs. Loudon *Ladies' Compan. Fl. Gard.* 60/2 Cold Houses for Plants are not generally in use, though it is a common practice with gardeners to remove plants from hothouses into the back sheds, in order to retard their blossoming. **1904** *Daily Chron.* 28 Jan. 6/1 It is the cold-house that has smiled upon them, in the shape of the refrigerated holds of the South African steamers. **1920** *Chambers's Jrnl.* 384/1 The 'forced' and 'cold-house' tomato is Guernsey's speciality. **1894** *Electrical World* 6 Oct. 328 Cold Light... Apparatus.. to produce light by means of high frequency electric currents, without converting most of the energy into heat. **1936** *Nature* 5 Dec. 974/1 It..makes available a 'permanent' source of 'cold' light. **1959** *Chambers's Encycl.* VIII. 723/2 To produce 'cold light', or luminescence, the molecules of a substance must be stimulated to emit light by the reception of some form of energy. **1909** *Practitioner* Dec. 866 The cold-pack is used for from 10 to 15 minutes every hour. **1966** Cox *Dict. Hairdressing*, *Cold permanent wave*, the transformation of straight hair to curled hair by means of chemicals without the application of heat. **1909** *Cent. Dict.* Suppl., s.v. *Pole*², Pole of cold or cold pole. **1927** Kendrew *Climates of Continents* (ed. 2) 167 In Eastern Siberia is situated the 'cold pole' of the earth. **1904** *Westm. Gaz.* 12 Jan. 4/2 In some of the largest nurseries..there exist cold-rooms or stores, pitch-dark and packed full of lily-of-the-valley crowns, lilies, and other bulbs and plants. **1948** in *Amer. Speech* (1950) XXV. 64/2 You'll get up to 30 percent more wear out of that set of synthetics when 'cold rubber' is used. **1964** N. G. Clark *Mod. Org. Chem.* xvii. 356 A recent process, operated at 5°, produces 'cold rubber' with improved toughness and elasticity. **a1877** Knight *Dict. Mech.* I. 593/2 *Cold-shut*, a term meaning that a link is closed while cold, without welding. **1877** Cold shut [see Shut *sb.* 4]. **1887** *Scribner's Mag.* II. 304/2 A 'cold-shut' or split ring..which can be fastened by hammering. **1776** T. Smith *Jrnl.* (1849) 279 A dismal cold snap of weather. **1848, 1892** Cold snap [see Snap *sb.* 7 b]. **1864** *Photographic News* 28 Oct. 527/2 A correspondent sends us the following method of cold soldering. **1895** E. B. Titchener tr. *Kulpe's Outl. Psychol.* 94 Blix and Goldscheider..speak of heat and cold spots, and regard them as the peculiar terminal organs of the temperature sense, and as independent of the pressure spots. **1901** —— *Exper. Psychol.* I. I. 57 To ascertain how these organs (warm spots or cold spots) respond to a stimulation. **1930** *Economist* 5 July 32/2 A new light, high-speed, cold-starting Diesel engine. **1936** *Discovery* Apr. 113/1 The engine is of the solid-injection, cold-starting type, and even under the worst climatic conditions it can be started in a few seconds. **1959** *Motor Man.* (ed. 36) iii. 50 Cold starting in winter requires the richest mixture of the whole operating range. **1895** *Daily News* 29 May 8/4 Extensive cold-store accommodation had been provided. **1949** *Gloss. Terms Refrigeration* (*B.S.I.*) 6 *Cold store*, an artificially cooled and insulated structure for the purpose of maintaining perishables at a pre-determined temperature. **1955** R. Postgate *Good Food Guide 1955–6* 185 'A wonderful display' of cold table, including turkey, grouse, and salmon. **1962** P. Purser *Peregrination 22* xvii. 78, I ate more than usual of the cold table. **1964** C. Gavin *Fortress* iii. 64 Champagne and a 'cold table' spread with every kind of delicacy. **1945** 'G. Orwell' in *Tribune* 19 Oct. 8/1 A State which was..in a permanent state of 'cold war' with its neighbours. **1946** *Observer* 10 Mar. 4/3 After the Moscow Conference last December,..Russia began to make a 'cold war' on Britain and the British Empire. **1947** W. Lippmann (*title*) The cold war. A study in U.S. foreign policy. **1947** *N.Y. Times* 17 Apr. 21/4 Let us not be deceived—we are today in the midst of a cold war. **1948** *Hansard Commons* CCCCXLVI. 411 The British Government..should recognize that the 'cold war', as the Americans call it, is on in earnest, that the third world war has, in fact, begun. **1950** D. Gascoyne *Vagrant* 61 Cold war propaganda. **1950** *N.Y. Times* 5 Feb. 42L/2 A 'cold war' was in progress..as the police..dared Georgia's revenue agents to invade Alabama. **1951** W. S. Chalmers *Life & Lett. Earl Beatty* xvi. 376 The 'cold war' which Beatty had to wage throughout his term of office for British maritime security was on two fronts. **1958** *Listener* 14 Aug. 224/2 World communism..is rejecting a policy of partial disengagement in the cold war. **1964** *Ann. Reg. 1963* 18 Public concern with cold-war espionage was then at its height with the..conviction in Moscow of a British business man. **1967** *Spectator* 30 June 757/3 Only in Europe after a decade of cold war..has confrontation given way to the kind of mutually recognised *status quo*. **1959** *Observer* 27 Sept. 7/8 Screeching, hysterical cries of cold warriors. **1969** *Guardian* 5 Nov. 8/3 It is very hard morally to distinguish between the good guys and the bad guys..for any but the most committed cold warrior. **1876** *Pop. Sci. Monthly* Nov. 124/1 Low temperatures..developed among the Rocky Mountains, and moved thence, as 'cold waves', over the continent eastward. **1901** Cold wave [see WAVE *sb.* 5 b]. **1921** *Daily Colonist* (Victoria, B.C.) 10 Apr. 31/6 A cold wave that struck Eastern Nebraska last night continued to prevail today. **1949** *Britannica Bk. of Yr.* 687/1 *Cold wave*, a type of permanent hair waving, in which ammonium thioglycollate is generally used. **1956** A. WALDHORN *Conc. Dict. Amer. Lang.* 33 *Cold wave*,..in hairdressing, a chemical process used to create a permanent wave. **1965** *Times* 22 Feb. 9/4 After the heavy snowfall.. the cold wave spread, causing freezing conditions in orange groves. **1899** *Jrnl. Iron & Steel Inst.* LVI. 197 By cold work is meant work performed below the critical range. *Ibid.* 198 Cold work distorts the grains or crystals of steel, flattening them and elongating them. *Ibid.*, The lower the temperature the more pronounced the effect of cold working. **1903** H. H. CAMPBELL *Iron & Steel* (ed. 2) II. xv. 408 Cold worked steel showing lines of flow. **1911** *Inst. Civil Engin., Minutes* CLXXXIII. 405 The microscopic properties of cold-worked iron. **1917** *Amer. Soc. for Testing Materials* II. 156 (*heading*) Light *versus* Heavy Reductions in Cold Working Brass. **1932** *Discovery* June 191/1 As we 'coldwork' the metal. **1942** J. N. GREENWOOD *Gloss. Metall. Terms* (ed. 2) 12 *Cold work*, the plastic deformation of metals at temperatures below that at which recovery and recrystallisation would take place. *Ibid.* 13 Cold work can be regular as in wire drawing, cold rolling, etc., or irregular as in hammering, deep drawing, cold heading. **1946** *Firth-Brown Gloss. Metall. Terms* 9 Steels containing 0·7–0·85% carbon are often cold-worked. *Ibid.* 18 *Cold-working*, a method of conferring strength by means of plastic deformation below the annealing or recrystallization temperature.

cold, *sb.* Add: **1. e.** (Earlier and later examples.)

1861 *N.Y. Tribune* July (Farmer), The 'Assents' continue to come in freely..and the appearances are that at the closing of the books..there will be few shares or bonds left out in the cold. **1915** W. S. MAUGHAM *Of Human Bondage* xxvii. 116, I have a sort of sympathy for Mahomet, I regret that you should have left him out in the cold. **1943** H. READ *Politics of Unpolitical* iii. 44 Except for poets laureate and political propagandists like Virgil and Pope, they [*sc.* poets] have always been left out in the cold.

5. c. *cold-cure.*

1954 I. MURDOCH *Under Net* iv. 62, I could get free board and lodging in exchange for being a guinea pig in a cold-cure experiment. **1969** *Times* 21 Feb. 10/7 Several cold cures and cough syrups on public sale contain phenylpropanolamine.

cold, *v.* Delete †*Obs.* and add: **1.** (Later examples.) Also, to be cold.

1885 T. H. HUXLEY *Let.* 30 Mar. in L. Huxley *Life & Lett. of T. H. H.* (1900) II. 101 It blew and rained and colded for eight-and-forty hours consecutively. **1908** T. HARDY *Time's Laughingstocks* (1909) 26, I kissed her colding face and hair, I kissed her corpse.

cold-cream, *v.* [f. the sb.] *trans.* To put cold cream on, to cover with cold cream; to take *off* by means of cold cream.

1910 H. G. WELLS *Mr. Polly* vi. 151 That's all red and rough. It ought to be cold-creamed. **1945** E. BOWEN *Demon Lover* 153, I used to potter about and..cold-cream off my old make-up. **1966** B. GLEMSER *Dear Hungarian Friend* iii. 62 She could not..brush her teeth, cold-cream her face.

cold meat. [COLD *a.* 2 b.]

1. Cooked meat that has become cold; hence sometimes used to connote inferior fare. Also *attrib.*

1598 [see COLD *a.* 2 b]. **1795** 'A LADY' tr. *Moritz's Travels* 240, I dined here on cold meat and sallad. **1816** JANE AUSTEN *Emma* III. vi. 88 When you are tired of eating strawberries in the garden, there shall be cold meat in the house. **1894** *Daily News* 15 Sept. 3/2 It is hard to be treated like 'cold meat company'. **1909** *Daily Chron.* 24 Aug. 1/5 There is a run on cold-meat shops.

2. *slang.* A corpse; corpses. Chiefly *attrib.*, as *cold-meat box* (coffin), *-cart* (hearse), *-party* (funeral or wake), *-train* (funeral train).

1788 GROSE *Dict. Vulgar T.* (ed. 2) sig. G3, A dead wife is the best cold meat in a man's house. **1819** MOORE *Tom Crib's Mem.* 25 In the Twelfth and Last Round Sandy fetch'd him a downer, That left him all's one as cold meat for the Crowner. **1820** 'P. CORCORAN' *Fancy* 46 He's for the cold meat cart, and so am I. **1837** [see COLD *a.* 2 b]. **1845** tr. *Sue's Myst. Paris* I. iii. 28 'I have already a bad cough.' 'Ah, I see you already in the *cold-meat box*.' **1874** HOTTEN *Slang Dict.*, Cold meat train. **1889** *Sporting Times* 3 Aug. 1/3, I should just come in where you were lying in the cold-meat box. **1908** J. M. SULLIVAN *Crim. Slang* 1 A cold meat party—a wake. **1924** G. C. HENDERSON *Keys to Crookdom* 401 *Cold-meat party*, funeral, wake. **1944** 'N. SHUTE' *Pastoral* v. 107 But for the Grace of God..they'd all have been cold meat.

cold-slaw. *U.S.* = COLE-SLAW.

1794 *Massachusetts Spy* 12 Nov. (Th.), A piece of sliced cabbage, by Dutchmen ycleped cold slaw. **1822** J. F. COOPER *Spy* II. ii. 32 Potatoes, onions, beets, cold-slaw, rice, and all the other minutiæ of a goodly dinner. **1871** *Lippincott's Mag.* Feb. (De Vere), Coldslaw apparently cut with a harrow. **1886** [see COLE-SLAW].

cold water. Add: **c.** Similarly, *to pour cold water upon.*

1893 *Times* 26 Apr. 9/4 He was obliged to pour cold water very plentifully upon the zeal of his Irish friends. **1950** A. CHRISTIE *Murder is Announced* viii. 96, I understand she rather poured cold water on that idea?

d. Also with reference to a dwelling without central heating. *U.S.*

1942 M. MCCARTHY *Company she Keeps* (1943) v. 187 A cold-water flat in an old-law tenement. **1954** W. FAULKNER *Fable* (1955) 334 A walk-up, cold-water Brooklyn tenement. **1964** W. MARKFIELD *To Early Grave* (1965) ii. 31 He found himself a cold-water flat near the Village.

colea (kōᵘ·liă). *Bot.* [mod.L., f. the name of Sir G. Lowry *Cole* (1772–1842), a Governor of Mauritius + -A.] A tropical plant of the bignoniaceous genus so named.

1927 *Blackw. Mag.* Apr. 508/2 A semi-wild garden of coleas, crotons, and dracænas. **1956** *Dict. Gardening* (R. Hort. Soc.) (ed. 2) II. 525/1 The Coleas need a moist, warm atmosphere.

Colebrook Dale. Also **Coalbrook Dale, Coalbrookdale, Colebrookdale.** [Name of a valley in Shropshire.] The name of a porcelain factory in Shropshire, used to designate a kind of porcelain formerly made there. Also *attrib.*

1855 DE LA BECHE & REEKS *Catal. Spec. Brit. Pott.* 173 Shropshire and Colebrook Dale Porcelain. **1857** MARRYAT *Pott. & Porc.* (ed. 2) xii. § 298 This ware, commonly termed Coalbrook Dale..was generally more for use than ornament. **1878** L. JEWITT *Ceramic Art* I. ix. 301 Coalbrookdale Terra Cotta. **1962** *Times* 23 Feb. 11/7 These flower-encrusted cottages are usually known to collectors as Colebrookdale.

coleoptile. Add: (Examples.) Also *attrib.*

1866 [see COLEOPHYLL]. **1923** *Nature* CXII. 26/1 The coleoptile (sheathed plumule or young stem) of barley seedlings. **1937** [see *AUXIN].

Coleridgian (kōᵘlri·dʒiăn), *a.* and *sb.* Also **-ean.** [f. the name *Coleridge* (see below) + -IAN, -AN.] **A.** *adj.* Of, pertaining to, or characteristic of Samuel Taylor Coleridge (1772–1834), the poet and philosopher, or his writings, opinions, etc. **B.** *sb.* A follower or devotee of Coleridge or his works. Hence **Co·leridgea·na** [see ANA *suffix*].

1834 MILL *Let.* 30 Aug. in *Wks.* (1963) XII. 231 You appear to go farther..with the Coleridgian and German metaphysics than I do. **1839** C. FOX *Memories* 19 Aug. (1882) v. 41 Some reference to infant schools drew Derwent Coleridge forth..and he launched out into a Coleridgean screed on education. **1840** MILL in *Westm. Rev.* XXXIII. 260 Every Englishman of the present day is by implication either a Benthamite or a Coleridgian. **1931** BLUNDEN *Votive Tablets* 305 The largest and most promising of all his Coleridgeana, the biography of his father. **1953** *Essays in Criticism* III. 354, I begin with the Coleridgian assumption that every detail in a good poem must contribute to its total effect. **1955** C. S. LEWIS *Surprised by Joy* xiii. 192 We ..set a very great value on 'Imagination' in some high Coleridgean sense. **1956** BLUNDEN *Coleridge's Fellow-Grecian* 47 This not quite forgotten Coleridgean.

coleta (kǫlēⁱ·tă). [Sp., dim. of *cola* tail, f. L. *coda, cauda* tail.] A pigtail, worn by a bull-fighter as a mark of his profession.

1928 HEMINGWAY *Men without Women* 61 They were going to cut off his coleta..his pigtail. **1928** *Daily Tel.* 19 July 11/2 One of Spain's 'star' matadors..has bidden farewell to the arena...To-day as a sign of his decision he had his coleta ('pig-tail') bobbed. **1962** J. STEWART tr. *Cousseau's Death of Miss Cunningham* 44 The *coleta* exposed the torero's queue of hair.

coli (kōᵘ·ləi). [L., gen. of *colon* COLON[1].] Shortened form of *Bacillus coli* or *Bacterium coli*: a bacterium, now called *Escherichia coli*, inhabiting the colon or gut of man and animals; usu. *attrib.* or *Comb.*, as *coli-bacilluria*, the presence of *Escherichia coli* in the urine; *coli-bacillus*; *coli-group*; *coli-like* adj.; *coli-phage*, a bacteriophage that attacks *Escherichia coli*; *coli-typhoid* adj., (a group) composed of *Escherichia coli* and typhoid bacillus. (See also *COLIFORM *a.*[2] and *sb.*)

1894 P. FRANKLAND *Micro-Organisms in Water* 411 The Coli bacillus has invariably fewer cilia than the typhoid. **1897** *Jrnl. Path. & Bacteriol.* IV. 439 Though coli abounds in small and typhoid in longer elements, cover-glass specimens of both yield all sizes. *Ibid.* 440 Organisms of the coli-typhoid group. **1901** *Jrnl. Hygiene* I. 435 Laboratory tap-water..normally contains very few *coli* bacilli. **1902** *Daily Chron.* 5 Aug. 5/2 Bacillus coli, or other coli-like microbes. **1903** A. C. HOUSTON in *Local Govt. Board Rep. Med. Officer 1901–2* (Cd. 1344) 468 The somewhat atypical members of the coli group persist longer than the strictly typical. **1910** *Practitioner* Apr. 449 Urotropin..appears to be of comparatively little use in the case of coli-bacilluria. **1936** *Discovery* Oct. 329/1 Friedländer's bacillus (the first mutative form of the coli bacillus). **1945** *Jrnl. Cellular & Comparative Physiol.* XXV. 9 Lysozyme had no visible clearing effect on unirradiated coli even at a concentration of 0·005 mg/cc. **1969** W. J. BRAMMER in Campbell & Sargent *Techniques in Protein Biosynthesis* II. i. 2 Acridine mutants of the *rII* system of coli-phage T_4.

colidar (kōᵘ·lidāɹ). [See quot. 1961.] A form of radar. (Cf. *COHERENT *a.* 1 d.)

1961 STITCH et al. in *Electronics* 21 Apr. 51/2 The system uses these properties of the laser's coherent light transmission and is called Colidar, an acronym from coherent light detecting and ranging. **1963** G. TROUP *Masers & Lasers* (ed. 2) ix. 158 A coherent light radar ('colidar') has already been made, using a pulsed ruby maser as transmitter.

coliform (kōᵘ·lifǫɹm), *a.*[2] and *sb.* [f. *COLI + -FORM.] **A.** *adj.* Of the nature of or resembling a bacillus of the coli-group of bacteria. **B.** *sb.* A bacillus of this or a similar group.

1906 W. G. SAVAGE *Bacteriol. Exam. Water-Supplies* x. 155 For any isolated coliform organism to be considered an indicator of sewage and excretal pollution it must have all the attributes of the typical *B. coli communis.* **1951** WHITBY & HYNES *Med. Bacteriol.* (ed. 5) xii. 194 (*heading*) Differentiation between Bact. Coli and Related Coliforms. *Ibid.*, The saprophytic coliforms, unlike *Bact. coli*, are able to utilize the carbon of citrates for growth. **1956** *Nature* 14 Jan. 97/1 The number of coliform bacteria..had not been reduced to the same extent. Excess of water.. tends to encourage the coliforms.

coliplication, var. *coloplication* (s.v. *COLO-).

collaborate, *v.* Add: **2.** *spec.* To co-operate traitorously with the enemy.

1941 *Ann. Reg. 1940* 165 The futility of attempts to 'collaborate' with their German conquerors. **1943** *Times* 5 June 5/2 Not all have a record as black as Laval's.. There were some who collaborated with a sick heart. **1959** *Observer* 8 Mar. 1/2 Certain Africans who had collaborated with the Government.

collaborateur. Add: **2.** = *COLLABORATOR 2.

1955 J. THOMAS *No Banners* iii. 28 The *collaborateurs*, who hated England and all she stood for. **1968** L. DEIGHTON *Only when I Larf* xvii. 210 Some Vichyite collaborateur had previously owned the hotel.

collaboration. Add: **2.** *spec.* Traitorous co-operation with the enemy.

1940 *Economist* 26 Oct. 511/2 Pétain may be outvoted on the question of mitigating the peace terms by some sort of shameful collaboration. **1941** *Ann. Reg. 1940* 162 In foreign affairs the watchword of the Vichy Government was 'collaboration' with the German conquerors. **1945** A. HUXLEY *Let.* 2 Apr. (1969) 517 He had been imprisoned—the only French author, besides Maurras, to have been so treated for collaboration.

So **collabora·tionism,** the practice of collaboration; **collabora·tionist,** a collaborator; also *attrib.* or as *adj.*

1922 *Contemp. Rev.* CXXII. 582 They should also profit from the expulsion of Signor Turati and the 'Collaborationists' from the Socialist party. **1923** *Ibid.* CXXIII. 151 The Socialist Party..had again split up into a collaborationist and an anti-collaborationist group. **1942** W. SIMPSON *One of our Pilots* vi. 192 Those of them..who were a hundred per cent collaborationists, who had thrown in their lot with the hated enemy. **1943** *Times* 9 June 5/6 Marcel Déat.. outbidding Laval in the ardour of his collaborationism, continues to support..the existing administration. **1946** *Daily Tel.* 27 Mar. 4/6 The mania for buying forbidden books..is now responsible for the belated vogue of collaborationist works. **1958** E. HYAMS *Taking it Easy* 69 Dutchmen..denounced each other as collaborationists unworthy to be employed by the Allies. **1968** *Listener* 5 Sept. 291/1 The Russians were genuinely astonished.. that they couldn't find collaborationist politicians prepared to overthrow Tito.

collaborator. Add: **2.** *spec.* One who collaborates with the enemy. Cf. *COLLABORATE *v.* 2.

1943 *New Statesman* 25 Sept. 197 Infiltration would be easy with the help of such well-placed collaborators. **1946** 'G. ORWELL' *Critical Ess.* 137 At this moment, with France newly liberated and the witch-hunt for collaborators in full swing.

collæmia (kǫlī·miă). *Path.* Also **-emia.** [f. Gr. κόλλα glue, αἷμα blood: see -IA.] Increased viscosity of the blood.

1900 DORLAND *Med. Dict.* **1910** *Brit. Health Rev.* Feb. 63 Defective circulation..has only two causes: (1) collæmia, and (2) weak heart.

collage (kǫlā·ʒ). [Fr., lit. 'pasting, gluing'.] An abstract form of art in which photographs, pieces of paper, newspaper cuttings, string, etc., are placed in juxtaposition and glued to the pictorial surface; such a work of art. Also *transf.*, *fig.*, and *attrib.*

1919 W. LEWIS *Caliph's Design* 1. 26 He..gradually drifts into the habit (a sort of progressive *collage*) of bringing his lack of painter's prowess and his nice feeling for art together. **1935** D. GASCOYNE *Short Surv. Surrealism* iv. 66 Poems can be composed from random newspaper-cuttings ('collage' poems). *Ibid.* 73 Max Ernst, with..his astonishing books of 'collage' pictures. *Ibid.* vi. 133 Parallel with these features..may be placed *collage* and *frottage.* **1936** J. DESCHIN *New Ways Photogr.* 181 Before embarking on the making of a photomontage, it should be understood that the term refers to a photographic process entirely and not to the scissors-and-paste method (known as *collage*) practiced by some in the name of photomontage. **1936** H. READ *Surrealism* 62 The invention of the collage by Picasso or Braque—the work of art made of any old pieces of string or newspaper. **1937** AUDEN & MACNEICE *Lett. Iceland* 21 Press cuttings, gossip, maps, statistics, graphs; I don't intend to do the thing by halves... It is a collage that you're going to read. **1938** MACNEICE *Mod. Poetry* viii. 144 The early Eliot's diction..is often a *collage* of other people's writing. **1939** *Archit. Rev.* LXXXV. 301 The accompanying 'collages' demonstrate a new use for the Object, particularly the Found Object (*l'objet trouvé* of surrealist invention). **1956** R. IRONSIDE in A. Pryce-Jones *New Outl. Mod. Knowl.* 285 The technique of 'collage' would be improperly described as a photographic process. **1957** *Observer* 15 Sept. 13/7 His assured *collage* paintings. *Ibid.*, Robyn Denny

..has discovered new possibilities in *collage.* **1961** *Times* 4 Aug. 3/4 When calling to mind a picture which is a collage-painting the fact that it is a collage is almost the first thing that we recollect. **1969** *Sat. Rev.* 28 June 56/3 Berio's Sinfonia provided an excellent example of the 'collage' which some composers like to practice nowadays.

Hence **colla·gist,** one who makes collages.

1953 in BERG *Dict. New Words* (ed. 2). **1958** *Listener* 20 Nov. 842/1 At the I.C.A. there is an exhibition of three collagists.

collagen. Substitute for def.: A protein which is present in the form of fibres as a major constituent of bone, tendons, and other connective tissue and which yields gelatin on boiling and leather on tanning. Also *attrib.* (Add later examples.)

1942 P. KLEMPERER et al. in *Jrnl. Amer. Med. Assoc.* 23 May 331 (*heading*) Diffuse collagen disease. Acute disseminated lupus erythematosus and diffuse scleroderma. **1956** *Nature* 10 Mar. 467/2 The so-called dense, moth-eaten fibres..which result from the action of bacterial collagenase on collagen fibrils. **1957** *Encycl. Brit.* VI. 273/1 Collagen fibres are best seen in the tendon. They are white in the fresh state, relatively inelastic and of great tensile strength. **1965** R. P. MOREHEAD *Human Path.* xix. 488 The term *collagen diseases* has been corrupted to denote a group of clinical entities associated with widespread fibrinoid degeneration of collagen. **1968** BLOOM & FAWCETT *Textbk. Histol.* (ed. 9) vi. 134 Application of the negative staining technique to high resolution studies of native collagen fibers has made it possible to visualize directly the tropocollagen macromolecules within the fiber.

Hence **co·llagenase** [ad. G. *kollagenase* (W. S. Ssadikow 1927, in *Biochem. Zeitschr.* CLXXXI. 267)], one of a group of proteolytic enzymes which decompose collagen and gelatin.

1927 *Chem. Abstr.* XXI. 2141 Collagenase is a new enzyme which is adsorbed by kaolin, which leaves fibrinase in sol[ution]. **1956** [see above].

collapse, *sb.* Add: **4.** *Timber.* Flattening or buckling of wood cells during drying, sometimes resulting in excessive and irregular shrinkage, and hence in a wrinkled appearance of the surface.

1948 *New Biol.* IV. 89 Another seasoning defect known as 'collapse' is apt to appear in large-pored hardwoods and in softwoods having thin walled tracheids if premature drying is permitted; the fragile elements actually cave in. **1966** A. W. LEWIS *Gloss. Woodworking Terms* 17 *Collapse*, defect of kiln-dried timber in which the cell walls cave in.

5. *attrib.*, as **collapse therapy,** the compression of the lung to rest it, or for treatment of diseases of the lung.

1922 *Brit. Med. Jrnl.* I. 954/2 (*heading*) Collapse therapy in pulmonary tuberculosis. **1959** *Chambers's Encycl.* X. 643/1 Collapse therapy consists of collapsing and immobilizing the diseased lung in a variety of ways.

collapse, *v.* Add: **3.** To cause to collapse, break down, fall in, or contract.

1883 'MARK TWAIN' *Life on Miss.* xxv. 240 We burst a boiler; broke a shaft; collapsed a flue. **1891** W. C. SYDNEY *Eng. 18th Cent.* II. 162 Which culminated in the battle of Culloden Moor, and collapsed the Jacobite cause. **1902** S. G. FISHER *True Hist. Amer. Revolution* 200 Such complete destruction and devastation of the country as would collapse the patriot party. **1904** *Daily Chron.* 8 June 5/3 Kent.. given a pinch of luck, might have collapsed Yorkshire! **1908** *Practitioner* Jan. 54 The urethra is alternately ballooned and collapsed. **1909** *Westm. Gaz.* 22 Apr. 4/2 The wind bringing its whole force to bear on the broadside and collapsing the structure. **1921** MULFORD *Bar 20 Three* xviii. 229 Far back..a Mexican collapsed his telescope. **1958** *Engineering* 31 Jan. 133/2 The ends are kept open by springs..which can be depressed..when it is desired to collapse the container.

collapsibility (kǫlæpsĭbi·lĭti). [f. COLLAPSIBLE *a.* + -ITY.] The quality of being collapsible.

1890 *Pall Mall Gaz.* 24 June 6/1 Some spoke bitterly of the collapsibility of their leaders. **1892** *Pictorial World* 2 Apr. 664/1 He has pricked a little hole..to test its collapsibility. **1917** *Chambers's Jrnl.* July 493/1 Collapsibility and lightness are the main features of the appliance.

collar, *sb.* Add: **18. c.** The area of junction between the stem and root of a tree. Also *attrib.*

1859 *Trans. Ill. Agric. Soc. 1857–8* III. 344 The ravages of this insect may be prevented by surrounding the collar of the trees, early in June with leached ashes. *Ibid.* 465 Care should be taken not to set the trees too deep. It is better that the collar should be considerably above, rather than below the level of the lawn. **1887** HARDY *Woodlanders* xix, With a small bill-hook he carefully freed the collar of the tree from twigs and patches of moss. **1954** HELLYER *Encycl. Garden Work* 57/1 The damping-off disease which attacks seedlings is really a form of collar rot.

19. d. The head of foam on a glass of beer.

1945 *Salt* 26 Feb. 14/2, I..then called for a couple without collars. **1945** J. STEINBECK *Cannery Row* xxi. 90 Doc opened a bottle and poured gently into a glass, holding it at an angle so that very little collar rose to the top.

21. *collar-box*; **collar-band,** also, the band to which the collar is attached; the band forming the collar of a woman's dress; **collar bearing,** a bearing on a shaft, adapted for taking the end thrust of the shaft; **collar facing** (see FACING *vbl. sb.* 4); **collar gauge** (see quot.); **collar pore** *Zool.*, a tube or pore connecting the cavity of the collar of *Balanoglossus* with the first gill-slit; **collar stud,** a stud used to fasten a detachable collar to a shirt.

1884 *Myra's Jrnl. Dress & Fashion* 1 Aug. 367/2 The plain collar band is of velvet. **1907** *Daily Chron.* 21 Jan. 8/1 The yoke and high collar-band in the gown..are fashioned of ermine. **1892** D. A. LOW *Machine Draw.* 121 The weight of the shaft is supported by a collar bearing. **1852** MRS. STOWE *Uncle Tom's C.* xv, My bandbox..and my collar-box. **1899** *Westm. Gaz.* 27 July 3/1 A collar facing of white silk. **1888** *Lockwood's Dict. Terms Mech. Engin.* 102 *Cylindrical gauge*, a gauge composed of two pieces.., the plug gauge being inserted into a bored hole and the collar gauge being slid over the spindle. **1897** PARKER & HASWELL *Text-bk. Zool.* II. 2 The collar cavity communicates with the exterior by a pair of collar pores—ciliated tubes leading into the first gill-slit or first gill-pouch. **1873** F. W. ROBINSON *Little Kate Kirby* I. I. xiv. 183 A diamond collar-stud. **1885** *Chambers's Jrnl.* II. 320/1, I suppose he has lost his collar-stud again. **1906** *Daily Chron.* 28 May 6/4 My collar-stud ran away and hid itself with great success.

collar, *v.* Add: **5.** *Cricket.* Of a batsman, (*a*) to get the better of (the bowler); (*b*) to score off (the bowling) at will.

1859 [see Dict.]. **1883** *Cricket* 3 May 66/2 The bowlers, in this innings, were never once 'collared'. **1891** W. G. GRACE *Cricket* ii. 40 Against Pilch, and one or two others who collared him at times, he would try just another over.

collarette. Add: **c.** A dahlia with a 'collar' of very shortened petals immediately surrounding the centre, and flat outer petals. Also *attrib.*

[**1904** *Jrnl. R. Hort. Soc.* XXVIII. 640 Dahlias 'à Collerette'..flower of single type, but with a number of slender elongated florets in the centre, the latter forming the 'collerette'.] **1913** *Gardeners' Chron.* 8 Feb. 85/1 The Collerette is a glorified single Dahlia. **1914** *Jrnl. R. Hort. Soc.* XXXIX. II. 658 The Collarette section was represented by about one hundred and fifty varieties. **1931** *Sanders' Encycl. Gardening* (ed. 21) 131 Dahlia...Collarette: Flowers, single, with 'collar' of very shortened florets; outer florets broad and flattened. **1958** A. WILSON *Middle Age of Mrs Eliot* III. 391 Jars of dahlias set out in groups of pompoms, collarettes, cactus and so on. **1959** *Listener* 27 Aug. 334/1 Dahlias:..collerette 'Lady Friend'.

collargol (kǫlā·ɹgǫl). Also **collargolum.** [a. Ger. *collargol* (*Chem. Centralblatt* (1900) LXXI. I. 144), f. COLL(OID + Gr. ἄργ-υρος silver + -OL.] A colloidal preparation of silver used as an antiseptic.

1900 DORLAND *Med. Dict.* 167 Collargol, Collargolum. **1902** *Jrnl. Chem. Soc.* LXXXII. II. 500 This form of colloidal silver is far more stable than the commercial form (collargol). **1907** *Practitioner* Apr. 574 Collargol, used as an ointment, has always been satisfactory. **1913** F. FORCHHEIMER *Therapeusis Internal Dis.* II. III. i. 685 In one case of dermatomyositis, collargolum given in suppository seemed to have the most remarkable effect. **1957** T. SOLLMANN *Man. Pharmacol.* (ed. 8) 1194 Colloidal metals for clinical use (for instance Collargol) are often 'stabilized' by the addition of proteins.

collateral, *sb.* Add: **6.** orig. *U.S.* (Earlier and later examples.)

1832 *Congress. Deb.* App. 62/2 May 13, 15 days $20,000 collateral. **1932** 'D. FROME' *By-Pass Murder* xxviii. 248 He did not want, for professional reasons, to put up any of his own collateral. **1958** *Economist* 22 Nov. 17/1 Automobiles are frequently accepted as collateral when personal loans are extended for their purchase.

collaterality. Delete *rare* and add later examples.

1872 J. WALKER *Theol. & Theologians Scott.* v. 148 The State, in their idea, had a collaterality with all rights belonging to the natural order of society: such as the right to hold property and use it. **1965** *Language* XLI. 98 The meaning of *uncle* is a class of human beings with certain features of sex, generation, and collaterality.

collational (kǫlēⁱ·ʃənăl), *a.* [f. COLLATION *sb.* + -AL.] Of or pertaining to collation.

1949 F. BOWERS *Princ. Bibliogr. Descr.* v. 193 The collational formula begins a new paragraph in the description, with the format of the book as its first item. **1959** —— *Textual & Lit. Crit.* iv. 118 The Old Cambridge edition of Shakespeare in the mid-1860's, with its collational lists of variants keyed to a modernised eclectic text. **1966** *English Studies* XLVII. 151 The collational formulas are restricted to a bare notation of format and pagination.

collator. Add: **7.** A punched-card machine which merges two sets of cards, or selects matching cards.

1949 E. C. BERKELEY *Giant Brains* iv. 51 The collator is a machine that arranges or collates cards. **1953** *Electronic Engin.* XXV. 407 Ancillary equipment for handling punched cards such as the ..collator. **1964** F. L. WESTWATER *Electronic Computers* vi. 100 A machine called a Collator can co-ordinate similar cards from two separate packs, and has many other applications.

collect, *sb.* **4.** (Later example.)

1885 E. C. STEDMAN *Poets of Amer.* v. 137 Yet anything that others can write of him is poor indeed beside a collect of his own golden sayings.

collect, *v.* Add: **1. b.** (Earlier and later examples of *absol.* use.) Also *colloq.*, to receive money, to get paid.

1869 'MARK TWAIN' *Innoc. Abr.* xxxviii. 409 Smyrna has been utterly destroyed six times. If her crown of life had been an insurance policy, she would have had an opportunity to collect on it the first time she fell. **1888** FARMER *Americanisms* 159/1 *To collect*, a contraction for 'to collect payments'. **1938** G. GREENE *Brighton Rock* IV. i. 149 'You'd better collect now.' They moved together towards Tate's stand. A young man with oiled hair stood on a wooden step paying out money. **1938** SHARPE *S. of Flying Squad* xx. 224 They go up to the person to whom they gave the winner and try to 'collect'.

e. To 'pick up' from a place of deposit; to call for (a person or thing). *colloq.* (orig. *U.S.*).

1896 C. JAMES *Yoke of Freedom* iii. 53 Jack went down the great marble staircase,..collected his hat and cloak, [etc.]. **1928** F. N. HART *Bellamy Trial* ii. 30, I was to collect the keys under the doormat at the gardener's cottage. **1936** TICKELL *See how they Run* ix. 150 Telling the commissionaire to ring up his house and instruct Johann the chauffeur to collect the car and put it away. **1937** KIPLING *Something of Myself* v. 106 On returning to collect my wife, I saw..a newspaper poster announcing my marriage.

f. Used imperatively, adverbially, or as an *adj.* to indicate that something sent (e.g. a telegram or parcel) is to be paid for by the recipient; in full *collect on delivery.* Also, to indicate that a telephone call is to be paid for by the person called. Cf. *C.O.D. orig. *U.S.*

1893 K. D. WIGGIN *Polly Oliver* xv. 172 In an hour another message, marked 'Collect', followed the first one. **1901** MERWIN & WEBSTER *Calumet 'K'* 47 Bannon handed his message to the operator. 'Send it collect,' he said. *Ibid.* 226 A long 'collect' telegram. **1933** M. DE LA ROCHE *Master of Jalna* xxii. 238 Alayne..went into the library to the telephone. She arranged for coal to be sent collect. **1935** W. D. HUBBARD *Thousandth Frog* iii. 55 'Say,..who's going to pay for this [telegram]?' 'Send it collect.' **1957** J. KEROUAC *On Road* (1958) II. iii. 118 He leaped into the phone booth and called San Francisco collect. **1967** *Listener* 28 Sept. 386/3 One parlour-game makes the point: you phone McLuhan 'collect', and when he answers the phone you say 'This is the message' and replace the receiver. **1968** B. NORMAN *Hounds of Sparta* ii. 13 Collect calls to London were not all that common in the taverna.

g. *to collect eyes*, intentionally to attract people's attention (to what one is about to say or do).

1904 *Daily Chron.* 28 Dec. 4/7 Why will our authors continue to write..of the hostess's 'collecting eyes'?..A novelist..first said that the hostess, about to leave the dining room, consulted all the ladies by 'collecting eyes'. **1935** 'N. BLAKE' *Question of Proof* vii. 124 Gadsby..collected eyes like a hostess. **1952** N. STREATFEILD *Aunt Clara* 80 Charles paused and collected the eyes of the family.

collectar(e (kǫ·lektaɹ). Also **collectarius.** = COLLECTARIUM.

1503 in *Records Bucks.* (1899) VIII. 116 Item a Martilage an Ordinall a Collector. **1540** in *Archæologia* (1890) LII. 234 A Collector for Collects and chaptres servyng for or father Abbott. **1846** W. MASKELL *Monumenta Ritualia Ecclesiæ Anglicanæ* I. vi. p. xcii, Whether this particular copy be a Collectare or not, we can arrive much nearer certainty with respect to its contents than we are able with the Venitare. **1897** F. MADAN *Summary Catal. Western MSS. Bodl. Libr.* IV. 380 A Collectare of some Italian diocese. *Ibid.* 390 A Pontifical of the cathedral at Arezzo,..together with a Collectarius. *Ibid.* 614 A large fragment of a Collectar. **1904** WORDSW. & LITTLEHALES *Old Service-Bks.* 210 Sequences are found..in Leofric's Collectar and Hymnary. **1937** F. MADAN *Summary Catal. Western MSS. Bodl. Libr.* II. II. 1007 A Cistercian collectar, in Latin, comprising the collects, chapters, antiphons, versicles, &c., of the Temporale.

collected, *ppl. a.* Add: **3.** *spec.* of a horse or its action.

1879 [see Dict.]. **1937** R. S. SUMMERHAYS *Elem. Riding* vi. 39 What I have described is the true collected canter. **1953** G. BROOKE *Introd. Riding* vii. 73 A horse is said to be collected when it is standing or moving with its centre of gravity drawn back so that its weight is evenly distributed on all four limbs.

collecting, *vbl. sb.* Add: **b.** *attrib.* and *Comb.* **collecting box,** (*a*) a box in which scientific specimens collected in the field are temporarily placed; (*b*) a box for the collection of contributions of money; **collecting card,** a card authorizing the holder to collect money for a charity; **collecting station** (see quot. 1900).

1857 C. KINGSLEY *Two Years Ago* II. v. 142 You found out the best cure for that malady when you took up the microscope and the collecting-box. **1862** J. H. WILSON *Prince Consort* iv. 37 All contributed their 'weekly offering' as the elder placed his *ladle*, or collecting-box before them. **1863** Collecting box [see Dict.]. **1908** *Westm. Gaz.* 16 Nov. 2/3 The penny..was taken from a collecting-box which the woman had obtained from Chelsea Hospital. **1876** C. M. YONGE *3 Brides* I. vii. 110 The Vicar..read the list of subscriptions already promised..invited ladies to take collecting cards. **1902** *Little Folks* II. 189/2 Collecting cards for the Ward Fund can be obtained. **1900** *Westm. Gaz.* 2 Jan. 3/2 At the collecting station..the wounded are collected as brought in by the stretcher-bearers, and are sent to the 'dressing station' in ambulances. **1915** *Daily Express* 5 July 6/3 He was received at a collecting station, obtained first aid for his wound, and was forwarded to the field hospital.

collection. Add: **2. a.** Also *attrib.*

1880 'MARK TWAIN' *Tramp Abroad* xxiv. 227 The collection-plate began its rounds. **1889** GRETTON *Memory's Harkback* 212 He was astonished to see a sovereign on the collection-plate. **1904** *Daily Chron.* 19 Mar. 9/1 Collection-box robberies. **1905** *Westm. Gaz.* 16 May 2/1 The net result will be an addition of eight per cent. to the collection income.

3. b. *spec.* The range of clothes (as for a season, etc.) displayed by a fashion designer; a display of such apparel.

1921 *Daily Colonist* (Victoria, B.C.) 10 Apr. 4/5 (Advt.), Every garment in the collection measures up fully to the demands that even the most critical women could make. **1925** *Studio* Autumn no. 171/2 (Advt.), Nicole Groult.. will show her spring collection of gowns, cloaks and hats from the 17th till the 29th March at 13 Bruton St. **1936** 'R. WEST' *Thinking Reed* viii. 263, I suppose living in Paris you go to see all the collections. **1951** I. SHAW *Troubled Air* xxii. 394 A gown from a French collection. **1968** J. IRONSIDE *Fashion Alphabet* 77 In Paris there are two collections a year.

collective, *a.* (*sb.*) Add: **A.** *adj.* **2. c.** *collective agreement, bargain, bargaining* (see quot. 1923), *piece-work* (see quot. 1928).

1891 B. WEBB *Cooperative Movement* 217 Individualist exchange must follow individualist production, and give place to collective bargaining. *Ibid.*, To gain a clear conception of the collective bargain. **1900** *Daily News* 24 Dec. 5/6 The contracts should take the form of collective-bargains in which every man of the same class would share equally. **1916** 'IAN HAY' *First Hundred Thou.* iii. 15 If you endeavour to drive a collective bargain with him [*sc.* the sergeant], you are mutinous. **1923** J. D. HACKETT in *Management Engineer.* Aug May, *Collective bargaining*, a mode of fixing the terms of employment by means of bargaining power between an organized body of employees and an employer, or association of employers. **1928** *Britain's Industr. Future* (*Liberal Ind. Inquiry*) III. xvi. 195 'Collective piece-work', whereby a group of workers..are guaranteed their regular time-rates but are promised, over and above these, an agreed share of the costs they may save. **1939** J. W. JONES *Nazi Conception of Law* 24 Collective agreements in industry, so far as they are allowed, are naturally interpreted as enacted rules of law rather than as agreements. **1955** *Times* 14 June 3/3 Once workpeople became convinced that threats yielded better results than reasoned negotiations round the table the whole system of collective bargaining would collapse.

d. *collective psychology*, group or social psychology.

1898 *Psychol. Rev.* July 348 Pseudo-scientists who have talked..of 'collective' psychology àpropos of the crowd, referring the word 'psychology' to the 'social' mind. **1911** S. LEACOCK *Lit. Lapses* 185 Studies in what may be termed collective psychology are essentially in keeping with the spirit of the present century. **1917** C. E. LONG tr. *Jung's Coll. Papers Analyt. Psychol.* (ed. 2) xv. 453 The repression of the collective psyche..was a necessity for the development of the personality, because collective psychology and personal psychology are in a certain sense irreconcilable.

e. *collective unconscious*, in the theory of C. G. Jung, that part of the unconscious mind which derives from ancestral experience and is additional to the personal unconscious (see quots.).

1917 D. HECHT tr. *Jung's Psychol. Unconscious Processes* in *Coll. Papers* (ed. 2) 432 The collective unconscious is the sediment of all the experience of the universe of all time, and is also an image of the universe that has been in process of formation for untold ages. **1926** [see *ANIMA]. **1943** H. READ *Educ. through Art* vi. 181 The contents of the *collective* unconscious..consist of images impressed on the mind from earliest times..and when transformed into *conscious* formulas, they take the form of tribal lore, myth, fable or fairy tale. **1949** KOESTLER *Insight & Outlook* xiii. 192 The 'collective unconscious' as a kind of common pool or substratum with which all individuals remain somehow connected. **1962** J. L. HENDERSON in A. Guggenbull-Craig *Archetyp* (1964) 9 We may for the sake of argument say there are three layers of the collective unconscious.

f. *collective farm*, a farm, esp. in the U.S.S.R., consisting of the holdings of several farmers, run by a group of people in co-operation, usually under state control; so *collective farmer, farming.*

1919 tr. *Lenin's Land Revol. in Russia* 21 The local and central Soviet authority aims..to foster collective farming. **1925** *Current Hist.* (U.S.) Oct. 87/1 Only the poorest peasants..go into these collective farm organizations. **1929** *Times* 25 May 11/7 As a result of their expulsion from a local collective farm, some kulaks (well-to-do peasants).. raided its premises. *Ibid.* 14 June 16/6 Abandonment of the policy of collective farming, which the peasants regard as a revival of serfdom. **1955** *Ann. Reg. 1954* 189 The free markets on which collective farmers customarily dispose of their surplus. **1958** *New Statesman* 22 Feb. 242/3 From a visit to two collective farms..he concludes that Israeli Left-wingers are doctrinaire and spartan.

g. *collective security*, a system by which international peace and security are maintained by an association of nations.

1934 W. S. CHURCHILL in *Hansard* CCXCII. 2368 The great principle of collective security..is the only principle that will induce hon. Gentlemen opposite to make any preparation for the defence of this island. **1937** A. HUXLEY *Ends & Means* ix. 109 In the actual circumstances of the present day, 'collective security' means a system of military alliances opposed to another system of military alliances. **1957** ELLIOTT & SUMMERSKILL *Dict. Politics* 69 A system of collective security requires the acceptance by individual countries of collective decisions, and their willingness to carry out these decisions, if necessary by military action.

h. *collective improvisation*, improvisation by a group of jazz instrumentalists in combination.

1946 *Jazz Mag.* Sept. 4/2 There is improvisation in Duke's records, not only solo improvisation but collective improvisation at times. **1962** *Oxford Mail* 19 Feb. 6/5 Jazz, which had previously leaned heavily on collective improvisation became more and more a soloist's music.

B. *sb.* **3. d.** A collective organization or unit; *spec.* a collective farm (see A. 2 f, above).

1925 *Nation* 25 Nov. 598/1 The breaking up of the old concentrated village into smaller artels or collectives which move out of the village group permanently and form smaller groups living right on the land, which is operated collectively. **1928** *Observer* 11 Mar. 11/2 The Soviets..form small collectives (cartels) in the making and mending of small machines, [etc.]. **1937** *New Statesman* 13 Nov. 791/2, I have seen these in 'aptekas' in small out-of-the-way villages and in 'Collectives' as well as in the large towns. **1940** H. READ *Philos. Anarchism* iv. 28 Each industry forms itself into a federation of self-governing collectives.

collectively, *adv.* Add: **2.** By collective action or arrangement.

1902 *Daily Chron.* 17 Oct. 3/3 The collectively-agreed rate of a shilling per hour. **1908** *Ibid.* 28 Mar. 6/7 Collectively-owned property. **1925** [see *COLLECTIVE *sb.* 3 d].

collectivistic (kǫlektivi·stik), *a.* [f. COLLECTIVIST + -IC.] Based on or characterized by collectivism.

1894 *Athenæum* 7 July 24/1 The ideal collectivistic state. **1954** D. VON HILDEBRAND *New Tower of Babel* 11 Individualistic self-sufficiency as well as..collectivistic anti-personalism.

collectivization (kǫle:ktivəizēⁱ·ʃən). [f. COLLECTIVE *a.* + -IZATION.] The process or policy of collectivizing. Also *attrib.*

1890 *Good Words* Dec. 822/1 The collectivisation of capital is, from the socialistic point of view, a far more thorough and consistent scheme. **1933** *Planning* I. VIII. 13 The Russian collectivisation campaign. **1947** A. L. ROWSE *End of Epoch* 238 Collectivist farming had to be resorted to simultaneously with industrial collectivisation. **1961** *Spectator* 6 Oct. 455 Stalin never seems to have doubted the virtues of collectivisation.

collectivize (kǫle·ktivəiz), *v.* [f. COLLECTIVE *a.* + -IZE.] *trans.* To establish or organize in accordance with the principles of collectivism. Hence **colle·ctivized** *ppl. a.*

1893 [see *COMMUNIZE *v.*]. **1894** *Speaker* 5 May 503/1 It cuts off the old hopes of 'bettering oneself', of rising above one's class. It collectivises even hope. **1934** *Times* 27 Oct. 13/5 The finance and policy of social services, provision for transport, water supply, drainage.., in fact, nearly every collectivized social activity. **1949** F. MACLEAN *East. Approaches* I. xii. 176 A noisy delegation of collectivized peasants from Armenia and the Ukraine. **1956** *Ann. Reg. 1955* 229 A new effort to collectivize farms was inaugurated.

collector. Add: **1. a.** *collector's* or *collectors' item, piece*, an item of interest to collectors because of its excellence, rarity, etc. Also *transf.*

1910 J. YOXALL *ABC about Collecting* 82 Shall one collect in order to have a complete set of examples, or only to have an incomplete lot of fine 'collector's pieces'? **1928** T. E. LAWRENCE *Lett.* (1938) 625 Some swine would call your Blake a collector's piece. **1932** *N.Y. Times Book Rev.* 17 Jan. IV. 14/5 A charming fancy clothed in distinctive form, at a price, however, that puts the book in the class of collectors' items. **1937** KIPLING *Something of Myself* iii. 67 It became a 'collector's piece' in the U.S. book-market. **1956** E. GRIERSON *Second Man* i. 9 Jaggers..was something of a collector's piece on the Circuit. **1967** J. GARDNER *Madrigal* iii. 56 I'm not going to shoot up a thing like that. It's a bloody collector's item.

c. (Earlier example.) So *collector ring, shoe* (see quot. 1943).

1777 T. CAVALLO *Compl. Treat. Electr.* 179 The prime conductor is set so, that the points of the collector are about half an inch from the surface of the cylinder. **1909** WEBSTER, Collector ring. **1940** *Chambers's Techn. Dict.* 176/2 Collector shoe. **1943** *Gloss. Terms Electr. Engin.* (*B.S.I.*) 41 *Slip ring*, (*collector ring*), a conducting ring rotating with a winding and connected thereto, and serving to make connection with an external circuit by means of a brush or brushes. *Ibid.* 109 *Collector-shoe*, a metal shoe for maintaining a sliding contact with a conductor-rail.

d. One of the three electrodes of a transistor (see quot. 1960).

1948 *Physical Rev.* LXXIV. 230/2 The flow of holes from the emitter into the collector may alter the normal current flow from the base to the collector. **1955** *Electronic Engin.* XXVII. 121 The collector voltage appears as 91 mV. **1959** *Ibid.* XXXI. 330 The collector and emitter of the transistor are connected in series. **1960** COOKE & MARKUS *Electronics & Nucl. Dict.* 83/1 *Collector* (symbol C), an electrode at which a primary flow of carriers leaves the interelectrode region of a transistor. It corresponds to the anode of a tube.

college, *sb.* Add: **1. b.** *Electoral college*: see also ELECTORAL *a.* 1.

9. a. *college boy, education, girl*; *college-educated, -trained* adjs.

1825 M. ROBINSON in *William & Mary College Q.* (1928) VIII. 82 A young Oxonian..with all the frankness of a 'College boy' told me he 'saw I was a stranger and would be happy to go the rounds with me'. **1946** G. MILLAR *Horned Pigeon* xiv. 181, I looked like a slightly seedy and impoverished college boy. **1935** R. FROST *Let.* 17 Feb. (1964) 255 It is the prose of a college-educated and practiced publicist. **1779** J. M. MASON *Dram. Wks. Massinger* I. p. lii, A different Relation of Massinger's College Education is given by Langbaine. **1948** *Mind* LVII. 387 University instruction in psychology should serve..as part of a general 'college' education. **1882** *Nation* 13 July 22/3 College boys and college girls fresh from their books could stand a better examination than people who had practical experience. **1963** *Guardian* 8 Nov. 10/4 The college-girl numbers are typical: long, long sweaters..college-boy scarves to match. **1906** *Westm. Gaz.* 19 May 1/3 A large increase in the number of college-trained teachers. **1966** J. PARTRIDGE *Middle School* i. 17 College-trained girls who teach for two or three years before they desert the profession for marriage and a family.

colleger. Add: **d.** A member of or student at a college. *U.S.*

1828 *Harvard Reg. 1827* 214 If we get fore-handed enough we'll send him to be a Colledger, and make a Parson of him. **1927** J. FREEMAN *When West was Young* 345 You don't need to git rough, you young colleger!

2. A mortar-board. *slang.*

1889 in BARRÈRE & LELAND *Dict. Slang* I. 264/1. **1958** J. BETJEMAN *Coll. Poems* 45 When my son's colleger acted as tray For an ice and a macaroon.

collegial, *a.* Add: **3. b.** Of or pertaining to a *COLLEGIUM.

1897 LADY MARY LOYD tr. *Waliszewski's Peter the Great* II. III. II. vi. 200 Peter early realised the advantages of the collegial form. *Ibid.* 203 The Reformer borrowed the Collegial form from Europe. **1948** J. TOWSTER *Pol. Power in U.S.S.R.* xi. 288 In the early months of the regime the commissariats were considered 'collegial' rather than 'single person' directed organs.

collegiality. Add: **2.** The principle of having a *COLLEGIUM.

1948 J. TOWSTER *Pol. Power in U.S.S.R.* xi. 288 Decision-taking and responsibilities were based on the 'collegiality' rule..rather than on the 'one-man management' principle. **1952** E. H. CARR *Bolshevik Revol.* II. xvii. 187 The principle of what was called 'collegiality' did not figure in any party programme.

3. *R.C. Ch.* The sharing of the bishops, with the Pope as their head, in the supreme responsibility of the government of the Church.

1965 G. BAUM *De Ecclesia* iii. 31 The doctrine of collegiality means that the bishops of the Catholic Church in union with the pope..constitute a body..which..is the heir of the body of the Twelve, the apostles with Peter as their head. **1969** *Daily Tel.* 28 Oct. 4/4 The Pope said this indicated his 'confidence' in the principle of collegiality, or the sharing of bishops in decisions made by the Pope.

collegiate, *a.* and *sb.* **B. 1.** (Later U.S. example.)

1854 M. J. HOLMES *Tempest & Sunshine* viii. 49 Miss Warner keeping a watchful eye upon her pupils, lest some lawless collegiate should relieve her from the trouble of seeing them safely home.

collegium (kǫlī·dʒiɒm). Pl. **collegia.** [a. L. *collegium* (see COLLEGE *sb.*), tr. Russ. *kollégiya.*] In Russia: an advisory board or committee (see quots.).

1917 A. S. KAUN tr. *Kornilov's Mod. Russ. Hist.* I. ii. 43 Instead of the decayed *prikazy* (boards) he [*sc.* Peter the Great] established after the Swedish model *collegia*, which corresponded to the present ministries with the difference that in the Collegium the power was not in the hands of a single minister, but in the hands of from three to twelve persons. **1948** J. TOWSTER *Pol. Power in U.S.S.R.* xi. 288 Attached to the commissar and presided over by him, there was within each commissariat a collegium, whose members were appointed by the Sovnarkom. **1957** N. C. HUNT *Guide to Communist Jargon* x. 36 Reference should also be made to the practice revived by the Bolsheviks when they came into power, as it had in fact been introduced by Peter the Great, of attaching to every People's Commissar a Collegium of members of his Commissariat which he was required to consult before promulgating an order.

collem (kǫ·lem), short for next.

1924 *Chambers's Jrnl.* 717/2 A great many 'Collems' live on the surface of water.

collembolan (kǫle·mbǒlăn). [f. mod.L. *Collembola* (J. Lubbock 1870, in *Trans. Linn. Soc.* XXVII. 295), f. Gr. κόλλα glue + ἔμβολον peg, stopper, wedge.] An insect of the order Collembola, characterized by the ventral tube on the first abdominal segment; a SPRING-TAIL. Also *attrib.* or as *adj.*

1873 J. LUBBOCK *Collembola & Thysanura* iii. 73 How little we can realise the conditions of Collembolan existence. **1899** J. W. FOLSOM in *Bull. Mus. Compar. Zool.* XXXV. 7 (*heading*) The anatomy and physiology of the mouth-parts of the Collembolan, Orchesella cincta L. **1906** *Liverpool Marine Biol. Committee Mem.* XIII. 5 The Collembolan described in the present Memoir is a convenient insect to study. *Ibid.* 14 This organ [*sc.* the ventral tube] is the most characteristic of Collembolan structures. **1956** *Entomological News* LXVII. 129 (*title*) A recently introduced species of Collembolan.

collide, *v.* Add: **2.** In *Nuclear Physics*, spec. of particles.

1866 [see Dict.]. **1880** [see *COLLISION 1]. **1938** R. W. LAWSON tr. *Hevesy & Paneth's Man. Radioactivity* (ed. 2) iv. 56 The γ-rays do not ionize directly, but split off electrons from the molecules with which they collide. **1942** STRANATHAN *Particles* i. 16 When a negative ion moves through a gas under the action of an electric field, it collides with a number of neutral molecules, some of which become ionized as a result of the collision.

colligability (kǫligăbi·lĭti). *Linguistics.* [f. next + -ITY.] The capacity to form part of a colligation (sense *4). Also *attrib.*

1953 H. F. SIMON in *Bull. Sch. Oriental & Afr. Stud.* XV. 327 The term 'colligation' is applied to describe the syntactic juxtaposition of two or more [grammatical] categories [in the Standard Chinese language], and the derivative 'colligability' has been coined to denote the *possibilités de combinaison* of any given category. **1966** J. E. BUSE in C. E. Bazell *In Memory of J. R. Firth* 54 The following table summarizes the colligability of each of the four full-word classes with each of the pre-nuclear and post-nuclear classes.

colligable (kǫ·ligăb'l), *a.* *Linguistics.* [f. L. type **colligābil-is*, f. *colligāre* (see COLLIGATE *v.*) + -ABLE.] Capable of forming part of a colligation (sense *4).

1961 Y. OLSSON *Syntax Eng. Verb* ii. 23 Each speech sentence containing a colligable type consists of one or more I/K-units.

colligate, *v.* Add: **4.** *Linguistics. intr.* To be in colligation. Also *trans.*, to group (words) in colligation *with* other elements.

1961 Y. OLSSON *Syntax Eng. Verb* ii. 31 Juxtaposed ANpre-terms may colligate with their own aAN-terms, e.g., 'a shocking, *terribly old* hat', in which case the place of a term is taken by an intra-sectional colligation. **1966** R. E. ASHER in C. E. Bazell *In Memory of J. R. Firth* 17 The pro-nominals are regularly colligated with verbal forms in two-word nouns + verb complexes as follows. **1966** J. E. BUSE *Ibid.* 54 *Verbs* colligate with all pre-nuclear and all post-nuclear classes.

colligation. Add: **4.** *Linguistics.* The grouping of words or other grammatical elements in syntactic structures by virtue of the classes of which they are members and the categories with which they are involved; a group of such elements in a syntactic structure. Also *attrib.* Hence **colliga·tional** *a.*

1953 [see *COLLIGABILITY]. **1958** *Trans. Philol. Soc.* 103 Mutual accompaniment of grammatical categories is termed 'colligation'; an accompaniment of forms, equally mutual but not generalisable in grammatical terms, is a 'collocation'. **1961** Y. OLSSON *Syntax Eng. Verb* iii. 44 The 'III' indicates a colligational similarity with section III. *Ibid.* 60 Colligation Valencies can be given for the complex structures. **1964** R. H. ROBINS *Gen. Linguistics* viii. 234 Groups of words considered as members of word classes related to each other in syntactic structures have been called colligations.

Collins[1] (kǫ·linz). [The name of a character, William *Collins*, in Jane Austen's *Pride & Prejudice* (ch. xxii).] A letter of thanks for entertainment or hospitality, sent by a departed guest; a 'bread-and-butter' letter.

1904 *Chambers's Jrnl.* 27 Aug. 611/2 When we do not call a letter of thanks for a visit 'a board and lodging', we call it a 'Collins'. **1907** LADY GROVE *Social Fetich* 74 The 'Collins' letter I had dutifully bored my hostess with. **1911** W. RALEIGH *Lett.* (1926) 375 This is only a Collins, and a Collins should not wade into deep places. It should be loving but neat. **1926** BRIDGES *Mem. H. Bradley* ii. 19 Wherever I can I shall let him speak for himself, and.. group the quotations from his letters under subjects.. This first Collins will serve to prelude them. **1940** W. DE LA MARE *Pleasures & Specul.* 327 The amateur composer even of a Collins or bread-and-butter letter realizes that his mother tongue is a stubborn means for the communication of gratitude.

Collins[2] (kǫ·linz). An iced drink consisting of gin or whisky, etc., mixed with soda-water, lemon or lime juice, and sugar, served in a tall glass; cf. **John Collins*, **Tom Collins*; usu. with defining word prefixed, as *brandy Collins, rum Collins*, etc. orig. *U.S.*

1944 *Beverage Retailer Weekly* (U.S.) 28 Aug. 9 (Advt.), Thousands of Marimba Collins are being served and enjoyed every single day. **1962** *Cocktails* (Cunard Line), Tom Collins 3/-. John Collins 3/-. Rum Collins 2/9.

collision. Add: **1.** In *Physics*, spec. of particles.

1880 P. SPENCE in *Jrnl. Specul. Philos.* XIV. 294 If the atoms A and B collide at the same moment with the atom C, the state into which C is thrown by the double collision cannot be the same as that which would be induced by a collision with either A or B alone. **1904** RUTHERFORD *Radio-Activity* ii. 54 Each of these projected particles possesses such kinetic energy that it is able to produce a large number of ions by collision with the gas molecules in its path. **1926** H. N. RUSSELL et al. *Astronomy* II. xvii. 551 After colliding with the far more massive atom the electron may be found moving with the same speed as before; such a collision is called elastic. Sometimes the electron is slowed up; such a collision is inelastic. **1942** [see *COLLIDE *v.* 2]. **1969** *Times* 5 Feb. 13/6 These very-high-energy collisions between particles and hydrogen atoms.

4. **collision course,** a course that will end in collision; also *fig.*; **collision door,** on board ship, a door to be closed after a collision, in order to cut off the inflowing water; **collision net,** a steel net used to support a damaged boat while it is being brought into port; **collision-proof** *a.*, of a railway carriage or train: capable of withstanding the shock of collision; **collision quarters, stations,** positions to be taken up by the passengers and crew of a ship in the event of collision.

1944 STEWART & PIERCE *Marine & Air Navigation* xiv. 178/2 Ships on such courses are said to be on collision courses. **1961** *New Statesman* 21 July 73/2 The great powers are now headed on a collision course over Berlin... Either side ..may risk resorting to a physical gesture—in an area where the escalation to nuclear war would be rapid and irreversible. **1966** D. FRANCIS *Flying Finish* xviii. 199, I switched the navigation lights on... They wouldn't give much warning to a jet liner on a collision course, but possibly better than none. **1895** *Daily News* 7 Jan. 8/1 Finding the water penetrating,..the chief engineer at once ordered the collision doors to be closed. **1897** *Ibid.* 27 Apr. 7/6 The gunboat ..was kept afloat by means of collision nets and brought into harbour. **1906** *Daily Chron.* 18 Aug. 2/6 Collision-proof railway cars. **1909** *Ibid.* 21 June 1/1 The call to 'collision quarters' was..quickly..performed. **1908** *Westm. Gaz.* 31 July 2/1 Drills..such as Fire and Collision stations.

collisional (kǫli·ʒənăl), *a.* [f. COLLISION + -AL.] Pertaining to collision; colliding.

1884 *Pharmaceutical Jrnl.* XV. 319/1 The mutual action at a distance, repulsive or attractive according to the mutual aspect of the two bodies when passing within collisional distance of one another. **1947** *Nature* 11 Jan. 59/2 It is possible to estimate the collisional frequency of the electrons with the neutral atoms. **1957** F. HOYLE *Black Cloud* viii. 152 The ionization of the atmospheric gases at a height of about fifty miles turned out to be abnormally high. This was giving rise to an excessive amount of collisional damping, as the radio engineers called it. **1969** *Nature* 16 Aug. 727/1 Direct collisional excitation of the forbidden lines will be unimportant.

collocability (kǫlǒkăbi·lĭti). *Linguistics.* [f. next + -ITY.] The capacity to form part of a collocation (sense *1 c).

1950 *Archivum Linguisticum* II. II. 111 The collocability of the future with *tomorrow* and of the past with *yesterday.* **1951** [see *COLLOCATION 1 c]. **1967** R. A. WALDRON *Sense & Sense Devel.* ii. 43 Such habitual collocations..are in themselves meaningful patterns and the mere listing of them.. expresses the meaning of the words in terms of their *collocability*, their tendency to congregate with these and other words.

collocable (kǫ·lǒkăb'l), *a.* *Linguistics.* [f. L. type **collocābil-is*, f. *collocāre* (see COLLOCATE *v.*).] Capable of forming part of a collocation (sense *1 c).

1961 Y. OLSSON *Syntax Eng. Verb* vii. 197 Although '*have* a look' and '*take* a look' are both collocable with *at*, '*have* a look' is alone in collocating with *for.* **1964** R. H. ROBINS *Gen. Linguistics* ii. 69 In the language of philosophy the verb *to exist* is not collocable with nouns in the same way that other verbs..are.

collocate, *v.* Add: **c.** *Linguistics.* To place (a word) with (another word) so as to form a collocation.

1951 J. R. FIRTH in *Ess. & Stud.* IV. 124 The word 'time' can be..collocated with *saved, spent*, [etc.]. *Ibid.*, *Person* is collocated with *old* and *young.* **1963** R. QUIRK in Brown & Foote *Early Eng. & Norse Studies* xiii. 152 One might add that *giefu* and *God* are also closely collocated elsewhere.

2. *intr.* *Linguistics.* Of words: to form collocations with another word or words.

1961 [see *COLLOCABLE *a.*]. **1964** R. H. ROBINS *Gen. Linguistics* ii. 69 Some philosophers (John Locke, for example) would hold that in the context of philosophical discourse *free* does not collocate with *will.*

collocation. Add: **1. c.** *Linguistics.* The habitual juxtaposition or association, in the sentences of a language, of a particular word with other particular words; a group of words so associated.

Introduced by J. R. Firth as a technical term in modern Linguistics, but not fully separable from examples in sense 1 a nor from other uses as exemplified in quot. 1940.

1940 G. L. TRAGER in *Language* XVI. 301 Collocation establishes categories by stating the elements with which the element being studied enters into possible combinations. *Ibid.* 303 It is now necessary to establish the collocations of the various forms to see what their functions are. **1951** J. R. FIRTH in *Ess. & Stud.* IV. 123, I propose to bring forward as a technical term, meaning by 'collocation', and to apply the test of 'collocability'. *Ibid.* 124 General or usual collocations and more restricted technical or personal collocations. **1958** [see *COLLIGATION 4]. **1963** R. M. W. DIXON *Linguistic Sci. & Logic* 74 The collocation 'round square' can occur.

collocational, *a.* Delete *rare* and add later examples.

1940 G. L. TRAGER in *Language* XVI. 302 The different series (paradigms) will give us one kind of classification, the collocational and congruential classes others. **1951** J. R. FIRTH in *Ess. & Stud.* IV. 126 The phonetic, prosodic, grammatical and collocational levels. **1962** H. A. GLEASON in Householder & Saporta *Probl. Lexicography* 91 The collocational possibilities of the forms.

collo:dio-bro·mide. *Photogr.* An emulsion of collodion with bromide of silver used for collodion dry plates. Chiefly *attrib.*

1864 SAYCE & BOLTON in *Brit. Jrnl. Photogr.* 9 Sept. 340/1 The secret..consists in converting simply bromised collodion into what we will for convenience designate 'collodio-bromide of silver', by the addition of the combining equivalent of nitrate of silver to the previously bromised collodion. **1873** SPON *Workshop Rec.* 266 Collodio-Bromide Process. **1892** *Photogr. Ann.* II. 61 A Collodio-bromide Emulsion. **1899** *Westm. Gaz.* 20 May 4/3 Mr. William Blanshard Bolton..worked out and published the collodio-bromide process of photography. In 1874 his process was perfected, and became the universal method of working collodion dry-plates.

collograph (kǫ·lǒgrɑf). [f. Gr. κόλλα glue: see -GRAPH.] **a.** A manifold copying apparatus. **b.** A photomechanical print made by **collography** (kǫlǫ·grăfi), the process or art of printing from a film of gelatine, esp. by the heliotype or collotype process. Hence **collographic** (kǫlogræ·fik) *a.*

1871 *Nature* III. 188/1 Edwards's collographic method. **1882** *Spon's Encycl. Industr. Arts* v. 1609 Pumphrey's 'collograph' depends on the fact that when a film of moist bichromated gelatine is brought into contact with ferrous salts..the gelatine is so far altered as to acquire the property of attracting a fatty ink. **1891** *Anthony's Photogr. Bull.* IV. 296 A demonstration..of collography. *Ibid.* 417 Numerous collographic printing experiments have been made with parchment paper watered with bichromatized gelatine.

collophane (kǫ·lǒfēⁱn). *Min.* [ad. G. *kollophan* (K. L. F. von Sandberger 1870, in *Neues Jahrb. f. Min., Geol. und Palaeont.* 308), f. Gr. κόλλ-α glue + -o + φαν-ός bright, clear (φαίνειν to show).] Any of the massive cryptocrystalline varieties of apatite, consisting largely of phosphate and carbonate of calcium, which are the main constituents of phosphate rock and fossil bone.

1917 A. F. ROGERS in *Jrnl. Geol.* XXV. 531 Collophane is a calcium carbonophosphate approaching the formula given above. [*Note*] Dana changed kollophan to collophanite, but since the name has not come into general use it is preferable to use the English equivalent of the original. **1962** W. A. DEER et al. *Rock-Forming Min.* V. 334 The mineralogy of such phosphorites is complex...The name collophane has been applied to the cryptocrystalline mineral component, and is a term analogous to 'limonite': it is of use only when the apatite-like phase cannot be definitely identified. **1966** *Bull. Geol. Surv. Great Britain* XXV. 19 During an investigation by the Atomic Energy Division into the occurrence of uranium in phosphates, uraniferous collophane was identified in specimens collected..from Odin Mine, near Castleton.

collophanite (kǫlǫ·fănəit). *Min.* [ad. G. *kollophan* (see prec.): see -ITE[1].] = prec.

1892 DANA *Syst. Min.* (ed. 6) 808 Collophanite occurs on the island of Sombrero. **1917** [see prec.]. **1955** BROWN & DEY *India's Min. Wealth* (ed. 3) xii. 479 Collophanite, the hydrated calcium phosphate, $Ca_3(PO_4)_2.H_2O$..may be amorphous, massive, concretionary or banded, white, grey, or brown and earthy, and..it completely lacks any physical properties that are really diagnostic. **1963** *Mineral. Abstr.* XVI. 211/2 The centre [of the concretions] is composed of coprolites consisting of collophanite with inclusions of fish bones.

collophore (kǫ·lǒfoᵊɹ). *Ent.* [f. Gr. κόλλα glue + -PHORE.] The ventral tube of the Collembola.

1878 A. S. PACKARD *Guide Study Insects* (ed. 6) 622 The sucker or *collophore*, as it may be termed. **1903** *Knowledge* Nov. 260/2 Ventral view of *P*[*odura*] *aquatica*, showing the mouth, the collophore, and the 'spring' in its normal position. **1942** S. W. FROST *Gen. Ent.* v. 78 The first abdominal segment bears a ventral tube (the *collophore*).

colloquial (kǒlōᵘ·kwi,ăl), *sb.* [f. the adj.] Colloquial speech, colloquial language.

1921 H. E. PALMER *Princ. Lang.-Study* 62 The hotel-keeper or waiter will concentrate on hotel colloquial, as also will the tourist or tripper. **1962** W. S. CORNYN in Householder & Saporta *Probl. Lexicography* 273 The earliest texts are not especially difficult of interpretation on the basis of today's colloquial.

colloquium. **3.** Delete 'Not in ordinary Eng. use' and add to def.: *spec.* an academic conference or seminar. (Later examples.)

1934 *Nature* 14 Apr. 584/2 A Mathematical Colloquium will be held in St. Andrews on July 18–28. **1956** G. RYLE et al. *Revol. in Philos.* 4 Philosophers had now to be philosophers' philosophers; and in their colloquia there was as little room for party politics as there is in courts of law. **1959** *Oxf. Univ. Gaz.* 9 Mar. 743/2 He also lectured..in various Oxford *colloquia.* **1970** *Ashmolean Mus.: Rep. Visitors 1969* 11 He contributed a paper to the 3rd Atlantic Colloquium at the University of Århus.

collotyped (kǫ·lǒtəipt), *a.* [f. COLLOTYPE + -ED[1].] Made by the collotype process.

1893 *Athenæum* 3 June 702/3 The collotyped reproduction of the ancient Yasna MS...is in the binder's hands.

collotypist (kǫ·lǒtəipist). [f. COLLOTYPE + -IST.] A maker of collotypes.

1890 C. T. CHESTERMAN in *Year Bk. Photogr.* 146, I presume that collotypists use stripped collodion or film negatives.

collotypy (kǫ·lǒtəipi). [f. COLLOTYPE + -Y[3].] The collotype process.

1896 [cf. *chromo-collotypy* s.v. *CHROMO-]. **1909** in WEBSTER.

colluvial, *a.* Add: **b.** *Geol.* Applied to soil at the foot of a slope, containing rock detritus or talus.

1880 S. W. Johnson *How Crops Feed* II. iii. 146 Colluvial soils..while consisting in part of drift or alluvium, also contain sharp angular fragments of the rock from which they mainly originated. 1955 *Proc. Prehist. Soc.* XXI. 59 This horizon..resembles that at the type locality of Göttweig, even to the presence of colluvial material above the soil.

Collyweston (kǫ:liwe·stǫn), *a.*, *sb.*, and *adv.* Also **Colleyweston, Collie weston, Colly Weston.** [The name of a village in Northamptonshire.] **1.** Used *attrib.* or *absol.* to designate a kind of slate used for roofing.

[1791 J. Bridges *Hist. & Antiq. Northants.* II. 433 In Colly Weston..great quantities of slates are raised for covering houses.] 1870 S. Sharp in *Q. Jrnl. Geol. Soc.* XXVI. 371 The calcareous nature of these beds, and..of the Colleyweston Slate. 1906 B. Thompson *Vict. County Hist. Northampton* II. 303/1 Collyweston slate..is found and has been worked at other places besides Collyweston, such as Kirby and Easton, between Collyweston and Stamford. 1949 W. G. Hoskins *Midland England* v. 70 Everyone with an eye for colour and form knows those mellow, lichened roofs of Colly Westons. 1958 *Listener* 14 Aug. 239/3 The roof of Nuffield College is covered with Collyweston stone slates.

2. Also **coll(e)y-west.** **a.** *adv.* Awry, askew; in an opposite or wrong direction. **b.** *sb.* Nonsense. **c.** *adj.* Contrary, contradictory. Cf. *galley-west *adv.*

[1587 W. Harrison *Descrip. England* (1877) II. vii. 168 The Morisco gowns, the Barbarian sleeves, the mandilion worne to Collie weston ward, and the short French breches, make such a comelie vesture.] 1820 R. Wilbraham *Gloss. Cheshire* 22 *Colly west, adv.*, directly contrary. 1855 *Trans. Philol. Soc.* 229 When a Lancashire man is altogether unsuccessful in his schemes, he says that everything goes *colley-west* with him. 1867 P. Kennedy *Banks of Boro* xxxvii. 287 Oh that's all *collywest*...What is the good of being an ordained clergyman, if you have not the power of doing something for an unlucky disciple like myself? 1875 Nodal & Milner *Gloss. Lancs. Dial.* 87 Never mind him; he ne'er agrees wi' onybody; he's awhiz collywest. 1884 R. Holland *Gloss. Chester* (1886) 75 'Am I going right for such and such a place?' 'Nao, its collyweston.' 1891 J. Baron *Blegburn Dickshonary* 17 Wot's th' odds, iv fooak y' took for friends Should turn eawt colly-west?

collywobbles (kǫ·liwǫb'lz). *colloq.* [Fantastic formation on Colic *sb.* and Wobble *sb.*] A disordered state of the stomach characterized by rumbling in the intestines; diarrhœa with stomach-ache; hence *gen.* indisposition, 'butterflies in the stomach', a state of nervous fear. (In quot. 1853 used nonsensically.)

1823 Egan *Grose's Dict. Vulgar T.*, *Collywobbles*, the gripes. 1841 *Punch* 9 Oct. 154/1 To..keep him from getting the collywobbles in his pandenoodles. 1853 'C. Bede' *Verdant Green* I. viii, A touch of the mulligrubs in your collywobbles? 1901 F. T. Bullen *Sack of Shakings* 308 He laughingly excused himself on the ground that his songs were calculated to give a white man collywobbles. 1959 I. & P. Opie *Lore & Lang. Schoolch.* x. 185 He is a 'funk'..or has 'got the collywobbles'.

Colney Hatch (kǭᵘ:ni hæ·tʃ). The name of a Middlesex village, and of a mental hospital opened there in 1851; used allusively for: a lunatic asylum; a 'madhouse'.

1890 W. Morris *News fr. Nowhere* (1970) ii. 8 There certainly seemed no flavour in him of Colney Hatch. 1908 G. K. Chesterton *Man who was Thursday* i. 28 And now, in the name of the Colney Hatch, what is it? 1939 Wodehouse *Uncle Fred in Springtime* x. 137 He'll probably end his days in Colney Hatch. 1955 J. Lehmann *Whispering Gallery* I. ii. 103 They would, I am sure, have thanked their stars that they were not inmates of such a childish Colney Hatch.

colo- (kǭᵘlo), combining form of L. *colon* or Gr. κόλον, see Colon¹, as in **co:lo-co·lic** *a.*, relating to two portions of the colon; **co:lo-enteri·tis**, enterocolitis; **co:loplica·tion** (also *coliplication*), the operation of unfolding or taking a reef in the colon in cases of dilatation. See also *colopexy, *colostomy.

1903 *Therapeutic Gaz.* 15 Feb. 102 An anastomosis is indicated, either colo-colic or ileo-sigmoid. 1897 R. W. Greene *Lippincott's Med. Dict.* 237/2 Coloenteritis. 1908 *Practitioner* Sept. 461 Neither coliplication nor coloplasty is likely to produce much permanent benefit. 1932 F. W. Rankin et al. *Colon, Rectum & Anus* iii. 95 Surgical procedures, such as..colostomy, colopexy, coloplication, and resection, all have been tried and occasionally have been successful.

colobus (kǫ·lǫbǫs). [mod.L. (J.C.W. Illiger *Prodromus Systematis Mammalium* (1811) 69), ad. Gr. κολοβός docked.] A member of a genus of African monkeys so called, distinguished by their shortened thumbs.

1835 *Proc. Zool. Soc.* III. 98 We at present possess sufficient indications of six distinct species of *Colobi*. 1866 *Ann. & Mag. Nat. Hist.* 3rd Ser. XVII. 77 One [skin] is in a perfect state, with the skull, &c., and is doubtless a *Colobus*. 1889 J. C. Willoughby *East Africa & its Big Game* 175 The beautiful black and white Colobus monkeys. 1899 *Westm. Gaz.* 1 Apr. 2/1 The Colobus resembles a black judge with a white wig and a mantle of flowing hair of the same hue. 1904 *Ibid.* 1 Dec. 4/2 Beards as white as the beard of a colobus monkey. 1962 *New Scientist* 12 July 91 Of the three species of *Colobus*..the red colobus occupies the upper storey [of forest], the black colobus the middle and the olive colobus the main closed canopy storey.

Cologne. **c.** Substitute for def.: In full *Cologne water.* = Eau-de-Cologne; also, a cream or solid preparation of the same perfume, often formed into a stick. Also *attrib.* (Add earlier and later examples.)

1814 *Boston Indep. Chron.* 20 Jan. 4/4 (Advt.), Cologne Water. 1831 Disraeli *Young Duke* II. III. ix. 120 You recollect his mother-of-pearl fountain of Cologne water? 1832 J. P. Kennedy *Swallow Barn* I. 125 She would like to have some Cologne of a particular kind. 1911 *Daily Colonist* (Victoria, B.C.) 29 Apr. 4/4 (Advt.), Cut Glass..Cologne Bottles. 1941 *Vogue* Feb. 82 Roger and Gallet's French Cologne,..Demuth's Original Solid Eau de Cologne. 1958 *Ibid.* June (back page), Costly Goya Gardenia perfumes a whole collection of bath luxuries: Cologne 4/- Cologne Stick 5/6. 1965 E. O'Brien *August is Wicked Month* xvi. 193 She took out a Cologne stick and touched her pulse and the back of her knees.

Colombo (kǭlɒ·mbo). The name of the capital city of Ceylon, applied *attrib.* to a plan for the development of south and south-east Asia recommended by the British Commonwealth Conference held at Colombo in 1950, or to the countries involved.

1950 *Times* 29 Nov. 3/1 The Colombo Plan for Co-Operative Economic Development in South and South-East Asia was published yesterday as a White Paper. 1954 *N.Y. Times* 27 June E5/2 The task of enlisting the help of the new Asian states, more recently known as the 'Colombo powers'—India, Pakistan, Ceylon, Burma and Indonesia—was taken over by Anthony Eden. 1957 *Times* (Canada Suppl.) 12 Nov. p. ii/1 Under the auspices of the Colombo Plan..we and others have pledged ourselves to draw on our resources and our technical knowledge to help those among our peoples in the Commonwealth who have yet to share in the benefits of modern science and technology.

colombophile (kǭlɒ·mbofəil), *sb.* and *a.* [Fr.] **A.** *sb.* A pigeon-fancier. **B.** *adj.* Pigeon-fancying; of or pertaining to pigeon-fanciers. Hence **colo:mbophi·lia**, pigeon-fancying.

1890 *Homing News* 3 Jan. 3/1 Colombophile world. *Ibid.* 31 Jan. 55/1 Eggs laid in the baskets during transit by train ..are often greatly prized by some colombophiles. 1896 *Westm. Gaz.* 20 Feb. 8/2 The Battersea colombophile, whose pigeons..are set at liberty and allowed to fly home. 1958 *Punch* 3 Nov. 40/3 Such a colombophile triumph could not pass unremarked. 1964 *Economist* 15 Aug. 650/1 No single body..covers all aspects of colombophilia in Britain.

colometric (kǭᵘlome·trik), *a.* *Palæogr.* [f. Colometry + -ic.] Of, pertaining to, or characterized by colometry. So **colome·trical** *a.*, **colome·trically** *adv.*

1883 J. R. Harris *Stichometry* (1893) 80 He [*sc.* Euthalius]..sent..an edition of the Apostolic epistles, written by his own hand, in quasi-verse or colometric writing. 1901 F. G. Kenyon *Handbk. Text. Crit. N.T.* viii. 308 The δ-text..exists quite definitely in the Graeco-Latin group of uncials.., all of which have been traced..to a common ancestor, written colometrically. 1912 E. M. Thompson *Introd. Gr. & Lat. Palaeogr.* vii. 70 Biblical codices which have colometrical arrangement of the text. 1927 A. H. McNeile *Introd. N.T.* 143 The colometric arrangement in D suddenly ceases at xvi. 23. 1934 *Jrnl. Theol. Stud.* XXXV. 248 Codex D..is the oldest known example of a colometrical arrangement of the Gospel-text. 1937 *Ibid.* XXXVIII. 245 The colometric display of this elaborate Gothic-Latin bilingual.

colon.³ Add: **2.** A colonial settler or farmer, esp. one in a French colony.

1957 *Economist* 12 Oct. 139/1 An old love that makes the Algerian *colon's* hair stand on end. 1961 G. Greene *In Search of Character* I. 29 Drinks at the Governor's: a simple kindly couple quite free from the vices of colons. 1961 *20th Cent.* Mar. 241 Unpleasant things were done by British *colons* in Kenya.

Colonel Blimp: see *blimp 2.

colonial, *a.* (*sb.*) Add: **A.** *adj.* **1.** Now freq. derogatory. Cf. *colonialism 2.

1954 J. Collin-Smith *Scorpion on Stone* xx. 288 People say white Africa is narrow and limited in outlook—'colonial' they say, don't they? 1957 L. Durrell *Bitter Lemons* 146 The United Nations,..whose attention could be more quickly drawn to 'colonial oppression' than to an Indian famine. 1958 *Spectator* 14 Feb. 201/3 Firm paternal rule of the kind which it is now fashionable to condemn as 'imperialistic' or 'colonial'.

b. Belonging to, or characteristic of, the period of the colonies, esp. of architecture or furniture. *N. Amer.*

1776 in *Rec. Early Hist. Boston* (1887) XVIII. 240 How many Persons belonging to this Family are now in the Service? Is it Continental, or Coloniel? *Ibid.* 242 The Question [was] accordingly put—Whether holding any military Commission in the Continental or Colonial Army is not incompatable [*sic*] with holding any Civil Trust? 1831 P. Hone *Diary* (1889) I. 34, I..got very good colonial quarters, near the bath-house. 1886 *Harper's Mag.* Oct. 668/1 The building has rather a colonial character with its long corridors and pillared piazzas. 1892 Kipling *Lett. of Travel* (1920) 21 The Colonial craze, which means white paint and fluted pillars. 1939 A. J. Toynbee *Study Hist.* VI. 60 The Americans use the term 'Colonial' for the eighteenth-century style of architecture which the English call 'Georgian'. 1968 *Globe & Mail* (Toronto) 13 Feb. 30/5 (Advt.), Custom built Colonial home, five charming bedrooms on second floor.

c. Dealing in produce of colonial origin.

1872 *Post Office Directory Grocery & Oil & Color Trades* 19 Colonial Agents..Home, Foreign & Colonial Agency. *Ibid.* 21 Colonial Dealers. 1895 *Home & Colonial Stores Ltd. Prospectus*, The well-known business of The Home and Colonial Stores, which has experienced an uninterrupted and highly successful development since its establishment in 1888. 1920 W. de la Mare in *Times Lit. Suppl.* 12 Feb. 105/2 Tchehov's father opened a 'Colonial Stores'. 1928 *Daily Mail* 25 July 18/1 Colonial Merchants.

3. b. A house built in the colonial style of architecture. *N. Amer.*

[illegible] 1968 *Globe & Mail* (Toronto) [illegible] storey Colonial located in a quiet exclusive Avenue.

C. *attrib.* and *Comb.*, and special collocations, as *colonial-born*, *-manufactured* adjs.; **colonial edition** (see quot. 1960); **colonial experience** *Austral.* and *N.Z.*, experience gained by an immigrant to Australia or New Zealand; also *attrib.*; **colonial goose** *Austral.* and *N.Z.*, boned leg of mutton stuffed with sage and onions; **colonial oven, stove** *N.Z.* (see quot. 1933).

1900 *Daily News* 7 Mar. 5/5 Colonial-born men led the rebels. 1907 B. M. Croker *Company's Servant* x. 92 The battered 'colonial edition' of a popular novel. 1960 Glaister *Gloss. Bk.* 76/1 Colonial editions, [illegible] ..that a fire can be put under it for baking. 1963 B. Pearson *Coal Flat* [illegible] There was an old-fashioned open fireplace with a hot coal fire, with a colonial oven alongside it. [illegible]

colonialism. Add: **1. b.** (Earlier example.)

1853 C. Clacy *Lady's Visit to Gold Diggings* iv. 48 Attempting to sink poles and erect tents..in a high wind and belting rain, is (if I may be allowed the colonialism) 'a horse of quite another colour'.

2. Now freq. used in the derogatory sense of an alleged policy of exploitation of backward or weak peoples by a large power.

1949 *Economist* 28 May 970/1 The old unreasoning dislike of anything that looks like 'colonialism' persists in the United States. 1952 *Ibid.* 19 July 172/2 It is on that basis that they explain American support, of France in North Africa—to them a straight case of colonialism. 1956 *Time* 9 Jan. 24/2 Khrushchev made clear what he meant by U.S. 'colonialism'. 'The colonialists give a dollar as "aid" in order to get subsequently ten dollars in return by exploiting the peoples who accept such aid.' 1957 *Listener* 1 Aug. 159/1 'Colonialism' is the commonest term of abuse nowadays throughout more than half the world.

colonialist. Delete rare⁻¹ and add later examples. Now freq. derogatory (cf. prec.).

1900 *Fortnightly Rev.* Jan. 47 The Colonialists are satisfied that Germany has founded an Asiatic dominion worthy of the Empire. 1955 *Times* 3 May 10/4 The Binh Xuyen rebellion 'had the purpose of safeguarding the interests of French colonialists'. 1956 [see *colonialism 2]. [illegible]

Hence *attrib.* or as *adj.*

1955 *Wall St. Jrnl.* 30 Dec. 1/3 The Communist Party Boss: Again accused the West of colonialist aims. 1958 *Manch. Guardian Weekly* 12 June 8/3 Mr. Bandaranaike's Government uses methods which it would denounce as 'colonialist' if they were employed by others. 1959 *Encounter* Apr. 19/1 A temple of culture has taken the place of the casino; I had to..beat my colonialist breast.

colonialistic (kǫlǭᵘ:niăli·stik), *a.* [f. prec. + -ic.] Characterized by or employing colonialism.

1950 *Ann. Amer. Acad. Pol. & Soc. Sci.* CCLXX. 134/2 My country's Indonesian policy in the early days was openly colonialistic and exploitative. 1955 *Times* 23 July 5/2 In an attempt to surround his claim with the aura of legitimacy we are accused of being colonialistic.

colonic (kǭᵘlǫ·nik), *a.* [f. Colon¹ + -ic.] Of or pertaining to the colon; affecting the colon; colic.

1906 *Practitioner* Dec. 812 The colonic distension and tympanites, associated with the colic. 1929 Doolin in *Irish Jrnl. Med. Sci.* Apr. 184 Colonic necrosis following the administration of avertin. 1936 *Discovery* Oct. 309/1 Colonic lavage which may be distinctly harmful if undertaken frequently. 1971 *Times* 27 Aug. 26/7 Colonic Irrigation-Enemas.

colonist. Add: **3.** A voter placed in a certain locality for the purposes of an election. *U.S. politics.* Cf. *colonization 2, *colonize *v.* 4.

[illegible] When every town and city in the United States is voting at the same time, and 'colonists' and 'repeaters' are needed at home. 1909 *Daily Chron.* 3 Nov. 1/6 It is more than likely that thousands of their 'colonists' have voted in some of the districts.

4. A member of a labour colony.

1896 J. A. Hobson *Probl. Unempl.* 137 About one half of the colonists are [illegible]. 1904 *Daily Chron.* 5 Oct. 4/5 The colonist at Merxplas may earn from 1d. to 3d. a day.

colonization. Add: **1. d.** *transf.* The establishment of a colony of animals or plants (cf. Colonize *v.* 3).

1903 M. C. Stopes in *New Phytologist* II. 186 (*heading*) The colonisation of a dried river-bed. 1905 F. E. Clements *Res. Methods Ecol.* v. 210 By invasion is understood the movement of plants from an area..into one of a different character, and their colonization in the latter. [illegible]

2. The action of placing political supporters where their votes will be important. *U.S. politics.*

1839 *New Yorker* 13 Apr. 2/1 'Colonization' or the removal of voters from the certain to the doubtful wards. 1842 *Congress. Globe* 31 May, App. 471/1 Among these [*sc.* modern phrases describing political knavery] 'colonization' and 'pipe-laying' were the most significant. 1914 *Cycl. Amer. Govt.* III. 619 In crowded city wards colonization of voters is sometimes attempted.

coloniza·tionism. *U.S.* [f. Colonization + -ism.] The principles of colonizationists.

1831 *Liberator* 29 Oct. 1/4 Some fair and able advocate of Colonization-ism. 1834 in *Life W. L. Garrison* (1885) I. 327 Here I am now in the hot-bed of Colonizationism.

colonizationist. (Earlier U.S. examples.)

1831 W. L. Garrison in *Life* (1885) I. 261, I am truly rejoiced to learn that you are no colonizationist. 1834 *Southern Lit. Messenger* I. 87 Tappan has two brothers in Boston, both ardent colonizationists.

colonize, *v.* Add: **4.** To place or register (political supporters) in a district where their votes may decide a closely contested election; to fill (a district) with such voters. Also *absol.* and *transf.* *U.S. politics.*

1842 *Congress. Globe* 31 May, App. 471/1 So far as he was informed, the practice of colonizing had its origin, as connected with the elections of the people in our country, in the city of New York. 1885 *Century Mag.* Feb. [illegible] Arrangements are made for colonizing voters from the neighboring States. 1903 *N.Y. Tribune* 13 Sept., The attempt to colonize in the Third Ward by a faction of the Democratic party was frustrated to-day. 1948 *New Yorker* 25 Sept. [illegible] Floaters are retainers of political organizations, and it's still common practice to 'colonize' doubtful districts with them. 1949 in *Amer. Speech* (1955) XXX. 284 Communist plans to 'colonize' key U.S. industrial plant[s].

colonizer. Add: (Earlier U.S. example.)

1781 S. Peters *Hist. Connecticut* 25, I have given the Reader some idea of the first colonizers of Connecticut.

2. One who 'colonizes' voters; a voter who is 'colonized'. Also *transf.* *U.S. politics.*

1892 *Nation* LV. 274/3 The New York cheats, bullies, receivers, colonizers. 1903 *N.Y. Even. Post* 7 Nov. [illegible] The superintendent of elections is authority for the statement that there are [illegible] colonizers and repeaters in the city. 1949 *B'ham* (Ala.) *News* 8 Apr. 12/2 The witness quoted Blum as saying 'the status of a colonizer' is 'a high honor in the party'.

colony, *sb.* **5. a.** Add to def.: Now freq. used to denote a group of people living temporarily or permanently separated from the rest of the community; esp. in *nudist colony.

1889 E. Dowson *Let.* 3 Mar. (1967) 46 A colony a la Thoreau, of 'Hobby Horse' people and a few elect outsiders each with a 'beloved'..where there will be leisure only for art and unrestrained sexual intercourse. 1929 M. Parmelee *Nudity in Mod. Life* xv. 231 While gymnosophists are not necessarily socialists or communists, these colonies furnish excellent opportunities for experiments along communistic lines. 1969 *Guardian* 14 Aug. 9/2 The growth of colony holidays for children. *Ibid.* 9/3 The colonies..generally [illegible]

c. An establishment in which persons who are otherwise unemployed or unemployable are engaged to work or are trained for some occupation or trade. (Not now current.)

1888 *Charity Organ. Rev.* Jan. 43 The Council would gladly see an experiment made in the form of a Labour Colony, to which unemployed townspeople might be sent for a time, and where they would be employed with a view to undertaking labouring work in a colony. 1896 J. A. Hobson *Probl. Unempl.* 131 The proposals for the establishment of farm colonies and other labour colonies. Various colonies of different types where the labour is chiefly employed in cultivation of the land exist already in England or on the continent. 1897 *Encycl. Soc. Reform* 785 The experiment of the Home Colonisation Society, in Westmoreland, is the most direct attempt to establish in England a labor colony by voluntary effort on similar lines to those adopted in Holland. The object of the society is to provide work in English 'industrial villages' for the able-bodied poor. 1911 *Encycl. Brit.* XXIII. 838/2 Labour colonies are of two kinds, voluntary and compulsory. The voluntary colonies ..are managed by philanthropic societies.

colopexy (kǭᵘ·lǫpeksi, kǭᵘlǫpe·ksi). *Surg.* [ad. mod.L. *colopexia*, f. *colo- + Gr. πῆξις fixing.] The operation of fixing part of the colon to the abdominal wall.

1896 A. E. Maylard *Treat. Surg. Alim. Canal* lxvi. 550 Colopexy.—This operation consists in opening the abdomen, and fixing some part of the colon by suture to the parietal peritoneum. 1900 A. J. McCosh in Gould & Warren *Internat. Text-bk. Surg.* II. xii. 193 Colopexy is an operation which consists in suturing some part of the colon to the abdominal wall. 1932 [see *colo-]. 1941 H. E. Bacon *Anus Rectum Sigm. Colon* (ed. 2) xv. 431 Colopexy..consists of obliterating the intersigmoidal fossa by shortening the elongated mesentery without excision of tissue.

colophon. Add: **2. b.** = Imprint *sb.* 3. *U.S.*

1930 *Publishers' Weekly* 19 Apr. 2113/1 The publishers must cut their lists and have their colophon stand for a particular quality which, in time, the bookseller will recognize and consider in his buying. 1948 *Chicago Tribune* 27 June iv. 5/1 The Wilson company uses the lighthouse for its colophon.

colorative (kǫ·lǫ-, kǫ·lǫrěitiv), *a.* [f. L. *colōrāt-*, ppl. stem of *colōrāre* to colour, give colour to: see -ative.] Depending upon coloration.

1905 *Westm. Gaz.* 11 Mar. 4/2 The dingy greyness of its coat depends in a good colorative protection. 1922 *Glasgow Herald* 21 Jan. 7 Its exquisite colorative distinction.

coloratura (kǫloratū·ra). [It. = colouring (see Colourer).] 'Divisions, runs, trills, cadenzas, and other florid passages in vocal music' (Stainer & Barrett, 1876). Also *attrib.* and *transf.*

1740 J. Grassineau tr. *S. de Brossard's Mus. Dict.* 33 *Coloratura*, is a term applied by the Italians to all variations, trillings, diminutions, &c. that can render a song agreeable. 1753 [see Coloratura]. 1902 *Westm. Gaz.* 30 Oct. 2/3 Rivalling a Melba in the ease and certainty of his coloratura passages. 1909 *Daily Chron.* 20 Oct. 3/3 She is not merely a coloratura singer. 1930 *Punch* 2 Apr. 387/1, I now know..that every Polish lady..was mistress of the purest coloratura, and could release it at a moment's notice or less without so much as a glance at any conductor. 1963 *Times* 8 May 5/7 An example of 'coloratura' choreography called *Donizetti Variations*.

b. *ellipt.* A singer of coloratura, esp. a coloratura soprano.

1944 Auden *Sea & Mirror* (1945) ii. 57 The ravished coloratura trilling madly off-key.

colorcast (kǫ·lǫrkɑst). orig. *U.S.* [f. Color + -cast, after Broadcast.] A broadcast of colour television.

1949 in *Amer. Speech* (1951) XXVI. 209/1 The colorcasts could be picked up in black and white on present type sets with no adjustment or added equipment. 1957 V. J. Kehoe *Film & T.V. Make-up* i. 11 Even when a black make-up bases or clothing can be used very effectively without ill effects on either the color or black-and-white receivers of a colorcast. 1967 *Boston Sunday Globe* 23 Apr. (TV Week) 1/3 Colorcast of the opening of the Montreal World's Fair.

colorimeter. (Later examples.)

1926 *Nature* 23 May 862/1 The introduction of the various forms of rectifier photo-electric cell has certainly simplified many problems in the use of instruments such as colorimeters (chemical type). 1957 *Technology* July 186/4 The [automatic blood] analyser uses a photoelectric colorimeter which compares the colour of the chemically treated sample with a standard of known composition; the photocell registers the difference in colour.

colorimetrically (kǫ:lǫrime·trikăli), *adv.* [f. Colorimetrical *a.* + -ly².] By means of a colorimeter.

1909 in *Cent. Dict. Suppl.* 1946 *Nature* 3 Aug. 169/2 Determining cytochrome oxidase activity colorimetrically. 1964 Corner & Cowey in *Oceanogr. & Marine Biol.* II. 159 In *Phaeodactylum tricornutum* the protein content was [illegible] as determined by the Kjeldahl method, but 93% as measured colorimetrically.

colorimetry. Add: (Earlier and later examples.)

1889 in *Cent. Dict.* 1923 Glazebrook *Dict. Appl. Physics* IV. 738/1 The sciences of spectrophotometry and colorimetry are [illegible] connected. 1957 V. J. Kehoe *Film & T.V. Make-up* viii. 97 To maintain the white balance and the correct colorimetry in a three-gun color tube system.

colortype: see *colourtype.

colossal, *a.* Add: **c.** As an intensive (cf. G. *kolossal*): immense, tremendous, exceptionally great; hence, magnificent, stupendous. *colloq.*

1892 'Mark Twain' *Amer. Claimant* xii. 205, I do not suppose that any other statesman ever had such a colossal sense of humour, combined with the ability to totally conceal it. 1897 Mary Kingsley *W. Africa* 89 Being such a colossal ass as to come fooling about in mangrove swamps. 1905 G. B. Shaw *Let. to Granville Barker* 20 Dec. (1956) 56 The colossal calm with which you propose..that I should change my publisher, bereaves me of breath. 1934 *Punch* 7 Mar. 263/1 This novel is simply superb. Colossal, stupendous. 1942 Berrey & Van den Bark *Amer. Thes. Slang* § 29/5 Magnificent, *cataclysmical*, *colossal*, [etc.].

colostomy (kǫlǫ·stǫmi). *Surg.* [f. *colo- + Gr. στόμα mouth.] The operation of making an artificial opening into the colon through the abdominal wall.

1888 J. A. Wyeth *Text-bk. Surg.* xvii. 514 Colostomy is indicated as a palliative measure in occlusion of the alimentary canal. 1892 *Index Medicus* XIV. 127 (*heading*) A case of inguinal colostomy of eighteen months' duration, illustrating the advantages of the anterior operation. 1932 [see *colo-]. 1941 H. E. Bacon *Anus Rectum Sigm. Colon* (ed. 2) II. 722 A colostomy may be constructed in the ascending, transverse, descending, or sigmoid colon. 1948 *Brit. Surg. Pract.* III. 94 A transverse colostomy..is indicated in those patients who appear to be suffering from an involvement of the left half of the colon only. 1967 F. F. Cartwright *Devel. Mod. Surg.* ix. 186 The first colotomy operations were performed, not for bowel obstruction in elderly patients, but for the congenital absence of the anus in new-born children.

colour, *sb.*¹ Add: **3. c.** See Off colour.

4. c. *Typogr.* The (relative) blackness of printed type.

1808 C. Stower *Printer's Grammar* vii. 211 It is a rule with careful pressmen, not to give proofs a high colour. 1888 J. Southward in *Encycl. Brit.* XXIII. 720/1 Comparing the old face and the modern characters, the latter are more regular in size, lining, setting, and colour. 1946 — *Mod. Printing* I. xxii. 141 The proportion of the thick to the thin strokes constitutes what typefounders call the colour of the letter. 1959 S. Morison *Typographic Design* 25 The essence is a quality of colour and inking that is specific to the hand press inasmuch as it has depth.

d. *spec.* in *Photogr.* and related subjects. Used esp. *attrib.* of the production of photographs, films, etc., in colour (see *17 b).

1872 [see *colour photography*, *17 b below]. 1931 *Ann. Reg.* 1930 48 Colour has been cinema almost imperceptibly, and the public hardly notice whether a film is in colour or not. 1957 E. S. Bomback *Photogr. in Colour* I. 11 (*heading*) Colour by subtraction. 1962 *Guardian* 8 Jan. 9/5 (*heading*) Cameras for Colour.

6. c. *pl.* The distinctive colours of a school, college, club, etc., team, crew, or the like, as conferred to denote selection as a representative member; esp. in *to get one's colours.*

[illegible] 1896 *Westm. Gaz.* 11 May 4/1 Since the reduction..of the number and variety of colours to be given, interest in games has deteriorated..'To get his colours' is an ambition which every boy should look forward to. 1936 W. de la Mare *Early One Morning* 263 The schoolboy who collects nothing at all (except good or bad marks, or 'colours', or certificates) must be an odd fish.

d. *sing.* One who has gained his colours.

1955 *Times* 6 July 4/5 Cambridge..will start as favourites this year as they have two old colours in R. C. Hampel and M. H. Searby.

7. d. Also see False *a.* 14 a, Flying *ppl. a.* 3 b, and Nail *v.* 1 d.

e. *spec.* U.S. (see quot.)

1891 H. Patterson *Illustr. Naut. Dict.* 251 *Colors*, the national ensign. In port colors are made at 8 a.m. and hauled down at sunset. When at sea colors are shown upon falling in with another vessel.

f. *pl.* A nautical ceremony at which the (national) flag is saluted as it is raised at sunrise or lowered at sunset.

1909 Webster, *Color* 8. *pl.* *U.S. Navy.* A salute to the flag accompanied by music at 8 a.m. and sunset, at hoisting and lowering it. 1942 G. Hackforth Jones *One-One* xv. 130 The duty signalman..bent the White Ensign on to the halliards of the Ensign staff. Then he approached the Sub and reported 'Five minutes to Colours, Sir.'

8. b. *Glass-painting.* (See quot.)

1914 *Jrnl. R. Soc. Arts* May 568 The composition of the vitreous 'enamel brown', or 'colour', as it is termed by glass-painters, with which the outlines, tones, and shadows in a glass-painting are produced.

12. c. Also, *to lend colour (to).*

1911 J. S. Huxley in *Cornhill Mag.* XXX. 500 The unicellular organisms..do not in their normal processes lend much colour to the view that death is but an accident. 1921 Sapir *Language* vi. 155 Facts such as these seem to lend color to the suspicion that..we are confronted by something deeper. 1932 D. L. Sayers *Have his Carcase* xxxiv. 443 Young Simons recognised something familiar in his face, which may have been a family resemblance. But it may quite likely have been the other way: the fact that he had it may have lent colour to the tradition.

15. b. *Phonetics.* The characteristic quality of a vowel sound; timbre.

1934 H. C. Wyld *Best English* 603 It is the pitch which produces the distinguishing quality, character, and colour of a vowel sound in speech. 1948 M. Joos *Acoustic Phonetics* 46 We cannot use the musical term timbre for linguistically significant vowel quality..; instead we shall use the term color. Linguistically diacritic color is based, then, on the profile of the spectrum.

16. c. Words, descriptions, or attendant features of an evocative nature; *local colour* (see Local *a.* 4 e (*b*)). Also *attrib.*

1938 E. Waugh *Scoop* v. 87 We're paid to supply news...Of course there's colour. Colour is just a lot of bull's-eyes about nothing. *Ibid.* 89 They gave Jakes the Nobel Peace Prize for his harrowing descriptions of the carnage—but that was colour stuff. 1953 Berg *Dict. New Words* (ed. 2) 57/2 *Colour*, in a programme: subsidiary features added to evoke, or enhance interest; e.g. fifteen minutes of colour before the kick-off. 1968 F. Mullally *Munich Involvement* iii. 19 Sullivan had completed his colour notes..and had cabled a short teaser story to London for the next day's edition. 1968 *Listener* 1 Aug. 142/1 We all spent the rest of the day rewriting the sheafs of irrelevant colour stories which we'd all been working on for weeks.

17. *colour-balance, -chart* (examples), *-circle, -combination, -consciousness, -contrast* (example), *-expanse, -harmony, -name, -pattern, -relation, -reproduction, -scale* [Scale *sb.*³ 5 b], *-sensation*; (in sense 8) *colour-grinder*; (in sense 2 c) *colour prejudice, problem, question*; *colour-coated, -conscious* (sense 2 c), *-correct, -prejudiced* adjs.

1913 A. H. Munsell (*title*) Color balance illustrated. An introduction to the Munsell system. 1953 *Colour Terms Sci. & Industry* (*B.S.I.*) 14 *Colour balance*, the relationship between the overall intensities of the different colour images in a colour reproduction. 1876 Color-chart [see *colour triangle]. 1954 A. G. L. Hellyer *Encycl. Garden Work* 8/1 Reagents for the barium sulphate colour test, with a suitable colour-chart, can be purchased from chemists who specialize in horticultural sundries. 1876 Color-circle [see *colour triangle]. 1936 *Discovery* Sept. 292/1 'Colour-circles', showing complementaries opposite one another, are of great value in training the colour-sense. 1959 *Records of Bucks.* XVI. 248 Local types of colour-coated ware. 1895 *Montgomery Ward Catal.* 4/3 Color combinations in the figures are heliotrope, black and dull green; pink and black; [etc.]. 1936 *Burlington Mag.* Feb. 74/2 All these pictures already show many extraordinary qualities in..their colour-combinations. 1952 P. Abrahams *Path of Thunder* III. i. 213 And she's not colour-conscious, Celia realized with dismay. 1946 'G. Orwell' in *Observer* 24 Feb. 3/5 The importance of Asiatic nationalism and colour-consciousness. 1953 P. Abrahams *Return to Goli* vi. 205 Kenyatta was delicately sensitive to the slightest hint of colour consciousness. 1957 E. S. Bomback *Photogr. in Colour* i. 9 We live in an age of colour-consciousness. This applies not only to clothes..it applies to the home in the form of gaily coloured plastics. 1890 W. James *Princ. Psychol.* II. xx. 244 The so-called simultaneous color-contrast, by which one color modifies another alongside. 1889 Colour-correct [see Orthochromatic *a.*]. 1932 H. H. Price *Perception* vi. 143 'We straight off mistake a colour for a body' (where 'colour' means 'colour-expanse'). 1825 Nicholson *Oper. Mech.* 455 His colour-grinder has ground the quantity of colour which used to serve him per day in three hours. 1859 G. A. Sala *Tw. round Clock* 239 The scene-painter..bids his colour-grinder clean his boots. 1870 S. H. Hodgson *Theory of Practice* I. 149 The tones and colour-harmonies of music calling up emotions. 1890 O. Wilde in *19th Cent.* July 144 New and curious colour-harmonies of blue and green. 1937 *Burlington Mag.* July 46/2 Titian..by his power and the musical quality of his colour-harmony. 1924 R. M. Ogden tr. *Koffka's Growth of Mind* v. 268 A colourless object is never given a colour-name. 1924 C. L. Morgan in J. H. Muirhead *Contemp. Brit. Philos.* 290 In vision..there is also the colour-pattern. 1937 *Discovery* Feb. 63/2 Size of wings, colour-patterns. 1905 W. B. *Where White Man Treads* 134 In the case of the Maori, this is deterred by a colour prejudice. 1909 M. Driver *Candles in Wind* xi. 107 A rough-and-ready bachelor, devoid of colour prejudice. 1932 J. R. Ackerley *Hindoo Holiday* I. 127 Having interpreted perhaps my coldness as colour prejudice or..conscious racial superiority. 1939 J. S. Marais *Cape Coloured People* i. 4 Colour prejudice and the attitude of mind that regards the coloured or black man as a mere instrument to serve the ends of the white man are by no means peculiar to white people whose ancestors owned slaves. 1958 *Church Times* 21 Nov. 6/3 The dilemma confronting priests of the greatest goodwill, whose pastoral responsibilities for a white flock in a colour-prejudiced country appear to come into direct conflict with what their conscience..tells them is right. 1908 S. A. Barnett *Let.* 2 Feb. in H. Barnett *Canon S. A. B.* (1918) II. 329 The real problem is the colour problem. Have you read Olivier's book on Jamaica? 1909 *State* II. 751 (*title*) Black and White in the Southern States of America. The Colour Problem. By William Archer. 1958 *Times* 4 Sept. 11/3 The time has come to admit that there is a colour problem in our midst. 1922 S. G. Millin *Adam's Rest* II. 159, I haven't been in South Africa much more than a year, and I can't quite appreciate the attitude of people here towards the colour question. 1927 R. H. Wilenski *Mod. Movement in Art* 129 The use of colour-relations..serves only as an agent in the description. 1906 *Burlington Mag.* Oct. 47/1, The *Medici* prints..will undoubtedly represent a great advance upon anything which has yet been done in colour reproduction. 1935 *Ibid.* Nov. 233/2 The colour-reproductions..inevitably give but little idea of the texture..of the originals. 1884 H. W. Vogel in *Brit. Jrnl. Photogr.* 6 June 360/2 The accompanying photograph of the colour scale, when compared with the reproduction [etc.]. 1936 *Discovery* Apr. 114/1 The latter [*sc.* Helmholtz] formulated a colour-scale, in the proper order, and extending about a 'fourth' beyond the musical octave. 1883 F. H. Bradley *Logic* II. II. i. 277 To verify in actual observation the fact of colour-sensations devoid of all extension. 1948 W. McDougall *Introd. Soc. Psychol.* (ed. 29) iii. 61 No more..than a totally colour-blind person can be made to understand the experience of colour-sensation.

b. (in sense *4 d), pertaining or relating to the production of photographs, films, etc., in colour, as *colour camera, film* (see *18), *negative, photo, photograph, photographer, photography, picture, print, printing, rendering, slide, snap, television* (see *18), *transparency.*

1893 *Jrnl. Soc. Arts* XLI. 664/1 Before proceeding to describe the colour-camera. 1957 R. W. G. Hunt *Reprod. Colour* vii. 62 If several colour photographs are wanted of the same scene, it is convenient to make first a colour negative, from which..as many positives as are required can be made. 1937 *Discovery* Jan. 25/2 We must use the unleashed colour-photo camera in accordance with the recognised principles of black-and-white photography. 1937 *Ibid.* Nov. p. xcvi/1 Remarkable colour-photographs. 1957 Colour photograph [see *colour negative*]. 1939 Emanuel & Dash *All-in-One Camera-Book* 118 There is no lack of subjects for the amateur colour photographer. 1872 *Jrnl. Chem. Soc.* XXV. 30 The colour obtained in so-called colour-photography. 1957 W. D. Wright in R. W. G. Hunt *Reprod. Colour* 5 The development of colour photography and colour television. 1912 F. A. Talbot *Moving Pict.* xxvi. 289 Ways and means of taking colour pictures direct from Nature. 1937 *Ann. Reg.* 1936 45 Two British made colour pictures were completed but not shown. 1893 *Jrnl. Soc. Arts* XLI. 664/1 The production of a triple photograph..by the superposition of three transparent colour-prints. 1957 Colour print [see *colour transparency*]. 1940 A. L. M. Sowerby *Wall's Dict. Photogr.* (ed. 15) 146 Colour printing on paper is a 'subtractive' process. 1937 *Discovery* Nov. 353/1 Fast panchromatic plates are improving colour-rendering. 1951 *Alaska Sportsman* XVII. vii. 50 (Advt.), Original 35 mm colorslides of Alaska. 1957 E. S. Bomback *Photogr. in Colour* v. 42 The reversal-type colour film with the projected colour slide as its end product. 1928 *Daily Tel.* 4 Dec. 17/2 The sensitive material in the 'Coloursnap' process is in the form of a three-ply film. 1963 *Times Lit. Suppl.* 25 Oct. 869/3 Kenny is..happy with..his colour-snaps. 1915 *Photo-Era* July 54/1 The exhibition will be divided into three sections: pictorial, color-transparencies, and a third..color-prints. 1957 E. S. Bomback *Photogr. in Colour* x. 105 The ideal colour transparency for a colour print is one having a moderate brightness range..and well-balanced colour rendering.

18. colour atlas, a chart giving examples of a series of shades of colour; **colour bar,** legal or social distinction between 'whites' and coloured people; **colour book,** a book with illustrations in colour; **colour-cake** [Cake *sb.* 4], a cake of coloured paint; **colour card,** a card bearing samples of paint-colours; **colour-cell,** a cell in animal tissue containing colouring matter, a pigment-cell; **colour-change,** the change in the colour of its coat, skin, etc., to be in accord with its surroundings, made by a beast, bird, etc., by protective instinct; hence **colour-changing** *vbl. sb.*, the ability to change or the act of changing colour in this way; **colour code,** a guide or code using certain colours as a standard method of identification, as in coloured coverings for electrical wirings, etc.; so *colour-code* v., *colour-coded* adj.; **colour correction** [Correction 7 b], of a lens (see quots.); so *colour-corrected* adj.; **colour-defective** *sb.* and *a.*, (one who is) wholly or partially colour-blind; **colour-difference,** a difference between colours; *spec.* applied *attrib.* to a signal used to obtain the correct proportion of colours in a colour-television transmission; **colour disc,** a disc with a series of colours arranged in sectors; also, each of the discs of a separate colour used with a colour-mixer; **colour-discrimination,** (*a*) the ability to discriminate between various colours; (*b*) unfavourable treatment of people of a different colour; **colour doctor** *Calico-printing* (see quots.); **colour dusting,** the application of finely ground colours to ware by means of a wad of cotton-wool; hence **colour duster,** a worker who performs this operation; **colour-fast** *a.*, dyed in unfadeable colours; hence *colour-fastness*; **colour film,** (*a*) a cinema film produced in natural colours; (*b*) a film suitable for producing colour photographs; **colour-filter** *Photogr.*, a filter consisting of a transparent material adapted to prevent the passage of certain coloured rays and allow the passage of others; **colour-fringing** = *fringing *vbl. sb.* b; **colour gravure** *Printing*, gravure in colours; also *attrib.*; **colour-hearer,** one who experiences colour-hearing; **colour-index,** (*a*) *Path.*, the relative amount of colouring matter contained in a red blood-corpuscle; (*b*) *Astr.*, (see quot. 1921); (*c*) *Geol.*, a number representing the percentage of dark-coloured minerals in an igneous rock; **colour magazine** = *colour supplement*; **colour-music** (see quot. 1903); **colour-organ** (see quots.); **colour pan,** a pan in which a colour and its thickener are mixed and incorporated in calico-printing; **colour-phase,** a genetic or seasonal variation in the colour of the skin, pelt, or feathers of an animal or bird; **colour-plate,** one of a set of plates used in colour-printing; a print made from such plates; **colour-printed** *a.*, printed in colours; **colour-ringing** (see quot. 1958); so *colour-ring* v., *colour-ringed* adj.; **colour-roller** *Calico-printing*, a roller that revolves in the colour-box and carries the colour to the printing-roller, against which it presses; **colour scheme,** (*a*) an arrangement of colours following a thought-out design, e.g. in furnishing or decorating a house, etc., or in planting a flower garden; (*b*) a scheme of protective coloration (of animals or birds); **colour screen,** a plate of coloured glass or the like used as a screen to absorb certain rays of light while allowing others to pass; **colour section,** a portion of a newspaper, etc., containing coloured illustrations; **colour-sensitive** *a.*, of photographic emulsion, plates, etc., sensitized for photographing in colours; so *colour-sensitiveness*; **colour-sensitivity** (see quot. 1937); **colour-sensitized** *a.* = *colour-sensitive*; so *colour-sensitizer*; **colour separation,** 'the isolation on separate photographic negatives by the use of color filters of the parts of a picture or design that are to be printed in the given colors; also, any of these separate negatives' (Webster 1961); also *attrib.*; **colour service** *Mil.*, service 'with the colours' as distinguished from 'on the reserve'; **colour solid,** a solid (Solid *sb.*¹ 1) bearing examples of a series of shades of colour; **colour supplement,** a supplement in a newspaper, etc., containing coloured illustrations; also *attrib.*; **colour television,** television in natural colours; **colour temperature** (see quots.); **colour-tone,** (*a*) a tone of colour; (*b*) *Art*, gradation and harmony of colour; (*c*) *Psychol.*, the colour quality of a coloured impression; chroma or hue as distinguished from brightness and saturation (*Cent. Dict.* Suppl. 1909); **colour triangle,** a representation of colours by the positions of points in a triangle, the apexes of the triangle representing the three primary colours; **colour twist,** a spiral or spirals of coloured glass in the stem of a drinking-glass; also, a drinking-glass with such a stem; also (with hyphen) *attrib.*; **colour value,** value with reference to a colour scheme; also *transf.*; **colour-wash,** coloured distemper (see Distemper *sb.*²); **colour-wash** *v. trans.*, to paint with coloured distemper; so *colour-washed* ppl. a., *colour-washing* vbl. sb.; **colour-way,** a colour-combination or colour scheme; **colour-weak** *a.*, unable to distinguish colours at a low degree of intensity; so *colour-weakness*; **colour-wheel,** an instrument of the revolving disc type, used for combining colours.

1905 A. H. Munsell *Color Notation* v. 53 A very elementary sketch of the Color Solid and Color Atlas..is all that can be given in the confines of this small book. 1956 A. J. Ayer *Probl. Knowledge* ii. 64 To make sure that I am employing the name of some colour correctly..I consult a colour atlas. 1913 W. C. Willoughby *Rel. Black & White Races* 3 The colour of the African is so strikingly different from ours, that it..serves in such phrases as *colour-bar* to indicate the whole difference. 1914 W. G. Lawrence in *Home Lett. T. E. Lawrence* (1954) 546 Relations between English and Hindu professors are bad, and there is a distinct colour bar except in the Mission colleges. 1955 *Times* 8 July 9/6 These men, and possibly many others, are excluded from their country's teams in international games because their country has a colour-bar. 1961 P. Mason *Common Sense about Race* iv. iii. 155 South Africa, the most extreme example of a colour bar country. 1904 *Daily Chron.* 20 Apr. 3/5 A colour book, as the term has now become, about the Channel Islands, is appearing. 1849 C. Brontë *Shirley* I. xi. 286 The paints are deleterious..the white is white lead..in those colour cakes. 1901 Kipling *Kim* x. 242 His little Survey paint-box of six colour-cakes and three brushes. 1894 *Country Gentlemen's Catal.* p. v, Samples, pattern colour cards, and price lists with all particulars on application. 1965 'M. Neville' *Ladies in Dark* viii. 79 I'll order the paint today, before Nicholas gets his nose into the colour card again. 1874 Garrod & Baxter *Mat. Med.* 398 Those of the true sac exhibit distinct, regular colour-cells. 1904 *Westm. Gaz.* 21 Sept. 10/1 The colour-change [of the stoat] is fairly frequent in the Northern counties. 1936 Russell & Yonge *Seas* (ed. 2) viii. 179 The remarkable property of colour change..is possessed by a far greater number of marine than terrestrial animals. 1936 *Discovery* Aug. 241/2 Colour-changing is the capacity for which the chameleon has become proverbial. 1957 Manvell & Huntley *Film Music* iii. 93 Harmonies are normally provided by using the bare fingers of the left hand to pluck from a group of about thirty gut strings, colour coded for easy identification. 1959 *Which?* Aug. 86/2 The [electrical] leads were incorrectly colour-coded. 1962 *B.S.I. News* Feb. 24/1 It was decided that the standardization of a colour code should be considered for the identification of the main connections from the power supply to transportable equipment. 1963 *Which?* Feb. 54/1 Anybody having a [dish-washing] machine installed should insist on having a correctly colour coded flex fitted. 1955 *Pop. Photogr.* Feb. 22 (Advt.), A magnificent coupled Rangefinder camera with fully automatic features and..color corrected lens. 1860 *Amer. Jrnl. Sci.* CXIX. 454 (*heading*) The color correction of certain achromatic object glasses. 1911 *Encycl. Brit.* XXI. 511/2 The colour correction is so perfect that the different coloured images are identical in size and position, thus rendering it specially suitable for three-colour and process work. 1961 A. L. M. Sowerby *Dict. Photogr.*

(ed. 19) 426 For photographing documents for reproduction and for copying paintings lenses having special colour corrections, called process and apochromatic process lenses, are used. 1900 *Stud. Yale Psychol. Lab.* VIII. 17 All persons belonging to abnormal types might be called 'color defectives'. 1958 *Listener* 6 Nov. 730/1 He, too, was a red-green colour-defective. *Ibid.*, The paintings of three such colour defective artists. 1924 R. M. Ogden tr. *Koffka's Growth of Mind* v. 272 A colour-difference is apprehended. 1955 G. G. Gouriet *Introd. Colour Television* 45 The particular method of duplex modulation used to combine both colour-difference signals on to the one sub-carrier. 1873 E. Atkinson tr. *Helmholtz's Pop. Lect. Sci. Subjects* vi. 244 It is in fact the same [series of colours] which we described as arranged around the circumference of the colour-disc. 1931 R. Graves *Poems 1926–1930* 10 It [*sc.* philately] strengthened the nation In the arts of mensuration And colour-discrimination. 1957 P. Worsley *Trumpet shall Sound* vii. 126 These 'Australians' were remarkable for their friendly..attitudes to the natives, the 'English' for their maintenance of colour-discrimination. 1839 Ure *Dict. Arts* 217 A sharp-edged ruler of gun-metal or steel, called the colour doctor. 1876 *Encycl. Brit.* IV. 685/1 The colour doctor..fits closely to the surface of the roller, and removes all colour except that which fills the engraved portions. 1900 *Daily News* 22 Feb. 7/4 [A] colour duster. *Ibid.*, The magistrate remarked on the undesirability of permitting colour dusting to be carried on in the printing shop. 1928 *Observer* 1 Apr. 13 Shrink-proof, stretchproof and colourfast. 1953 *Manch. Guardian* 25 July 10 Standards of this kind would not require that all fabrics should be colourfast to washing but that their degree of colourfastness should be recognisably identified. 1962 *B.S.I. News* Feb. 17/1 The measurement of colour fastness ..is now of considerable importance in the leather, plastics and other industries. 1912 F. A. Talbot *Moving Pict.* xxvi. 288 (*caption*) Preparing the Pathé Colour Films. 1930 *Times* 2 July 12/2 In a paper on the future of colour films..he said that the chemist had been devoting much attention to the..demand for natural-colour films. 1954 *Oxf. Jun. Encycl.* IX. 456/2 In making a colour film, the whole of the picture has to be most carefully planned..so that a harmonious design results. 1957 T. L. J. Bentley *Man. Miniat. Camera* (ed. 5) iv. 22 Colour transparencies which are gems of beauty can be made by means of multi-layer colour films. 1900 *Jrnl. Soc. Arts* 7 Sept. 774/1 The permanency of a dye on a cotton fabric is of little use as an indication of its permanence in a sealed colour filter. 1902 *Encycl. Brit.* XXXII. 16/2 Different makes of plates demand different colour-filters. 1948 *Sci. News* VII. 82 An interesting war-time application of such colour filter techniques was the determination of the depths of off-shore water along enemy coast lines. 1935 *Discovery* July 192/1 Excessive parallax, with consequent colour-fringing, is the inevitable result. 1925 *N.Y. World* 20 Dec. (*title*) Color Gravure Section. 1929 *Times* 29 Oct. xvi/5 Printers.. would thus be able to undertake colour-gravure printing in their own shops...It is on these lines that an extension of the commercial use of colour-gravure may be anticipated. 1958 *Times* 1 Aug. 13/1 The rapid and striking development in colour-gravure production. 1881 *London Med. Record* IX. 493/2 A student of medicine..stated..that while hearing he perceived colours, and this gave an impulse to further investigations. The 'colour-hearer', Bleuler, associated with himself a colleague. 1908 *Practitioner* Aug. 323 Generally the fewer the red corpuscles, the higher is the colour index. 1921 *Discovery* Feb. 38/1 The difference between the photographic and the visual magnitude of a star is therefore due to the colour of the star, and is called the colour-index. 1927 S. J. Shand *Eruptive Rocks* vii. 131 If a rock contains 17 per cent. of heavy minerals, we shall say that its colour ratio is 17 per cent., or that its colour index is 17. 1937 Hatch & Wells *Petrol. Igneous Rocks* (ed. 9) IV. i. 134 The percentage of dark (heavy) minerals in a rock is termed its 'colour index'. 1964 *Observer* 13 Sept. 5/5 A colour magazine to be included with the paper on Fridays. 1844 D. D. Jameson (*title*) Colour-music. 1903 A. A. Michelson *Light Waves* i. 2 A *color-music*, in which the performer, seated before a literally chromatic scale, can play the colors of the spectrum in any succession or combination, flashing on a screen all possible gradations of color..or the most gorgeous and startling contrasts and color chords. 1957 *Encycl. Brit.* VI. 64/1 *Colour-music* is the term most frequently employed to designate an art which makes use of varying light or lights as its principal means of expression. 1881 *Cassell's Fam. Mag.* 699/2 The different colours of the spectrum form a scale of light which has often been compared to the musical scale, and the idea has recently taken shape in what has been termed a colour-organ. 1895 *Oracle Encycl.* II. 119 A 'Colour Organ' has been invented, which casts combinations of colour upon a screen somewhat on the same principle as a musical organ discharges sounds. 1959 Westrup & Harrison *Collins Mus. Encycl.* 591/2 The score of *Prometheus* [by Scriabin] includes a 'colour organ'. 1860 *Ure's Dict. Arts* (ed. 5) I. 519 Down one side is fixed a range of colour-pans. 1876 *Encycl. Brit.* IV. 685/2 The mordant..and its appropriate thickeners are placed in a range of colour pans, in which the materials are thoroughly incorporated. 1927 C. Elton *Animal Ecol.* xii. 182 The arctic fox possesses two colour phases, one of which is brown in summer and white in winter, while the other is grey..and 'blue'. 1957 *Encycl. Brit.* XXII. 987/1 Other examples of this kind of variation are to be found in the 'colour-phases' of mammals (*e.g.*, foxes) and birds. 1884 J. S. Hodson *Art Illustr.* II. ii. 206 One of these impressions is marked as a guide or key for preparing the several colour plates. 1937 *Burlington Mag.* Feb. 99/2 The series of colour-plates. 1933 *Ibid.* Dec. 269/1 His water-colour drawings, his illuminated printing and his 'colour-printed drawings'. 1956 *Nature* 4 Feb. 217/1 The work involved in republishing colour-printed maps destroyed by enemy action. 1937 *British Birds* XXX. 376 While colour-ringing other species this winter, I have had a number of Robins..coming to the trap. 1937 *Ibid.* XXXI. 82 Colour-ringed birds. 1940 *Ibid.* XXXIII. 307 A study of blue tits by colour ringing. 1958 *Listener* 27 Nov. 891/2 Birds can be recognized individually if rings of different colours or combinations of colour are put on their legs, and by using this technique of 'colour-ringing' the author was able to record the histories of some forty birds of each sex in detail. 1890 W. J. Gordon *Foundry* 168 Between the colour-roller and the calico is a thin strip of steel which scrapes the printing-roller free of cotton. 1890 G. B. Shaw in *London Music* (1937) 290 The character of the color scheme never varied...the music never varied. 1906 *Dress* Dec. 28/1 Color schemes are most carefully studied. 1914 G. Jekyll (*title*) Colour schemes for the flower garden. 1925 R. W. G. Hingston in E. F. Norton *Fight for Everest, 1924* 265 They are well able to defend themselves and thus have no need of a special colour scheme. 1954 T. S. Eliot *Confid. Clerk* II. 69 But I came to have a look at the flat To see if the colour scheme really suited you. 1890 Woodbury *Encycl. Photogr.* 479 To use a colour screen of yellow glass behind the lens. 1961 *Times* 21 July 9/3 The *Sunday Times* is planning to carry a colour section as part of every issue...It will be a self-contained section. 1962 L. Deighton *Ipcress File* xv. 90 Floating Coca-Cola depot boats, 'Life' magazine colour-section printing-machine barges. 1888 *Photogr. Jrnl.* XII. 149 The isochromatic, orthochromatic, and colour-sensitive, or correct colour-tone process. 1907 *Westm. Gaz.* 20 Sept. 4/3 Photographic colour-sensitive emulsion. 1909 *Ibid.* 17 Apr. 14/2 Colour-sensitive plates. 1909 C. E. K. Mees *Photogr. Col. Obj.* iv. 27 By orthochromatic photography, we intend to imply the use of a fully colour-sensitive plate. 1916 *Ibid.* (ed. 2) iv. 37 The correcting action of such weak filters increases with the colour-sensitiveness of the plate. 1888 Abney *Instruction Photogr.* (ed. 8) 233 That colour sensitiveness can be given to a gelatine plate by coating it with either varnish or collodion in which various sensitive dyes are dissolved. 1937 *Discovery* July 216/1 Colour-sensitivity can have many meanings. 'It is sometimes taken to denote the capacity to see a minimal amount of a colour when mixed with black, white, grey or another colour.'..It may also mean, 'the capacity to distinguish between two shades of a particular colour'. 1957 T. L. J. Bentley *Man. Miniat. Camera* (ed. 5) iii. 15 This year..also saw the advent of the first..fast, highly colour-sensitized, double-coated roll films. 1890 Woodbury *Encycl. Photogr.* 475 A collodion emulsion rendered orthochromatic by the addition of so-called colour-sensitisers. 1922 *Jrnl. Optical Soc. Amer.* VI. 570 The pigments or dyes required for the satisfactory rendering..of photographic 'color separations'. 1924 L. P. Clerc *Ilford Man. Process Work* xxii. 206 (*heading*) The practice of colour separation. 1930 H. A. Groesbeck *Pract. Photo-Engraving* 61 This is called 'color separation' and comes in very handy if a color is to be completely cut out. 1935 *Discovery* July 189/2 A colour-separation negative. 1884 Colour-service [see sense 17 in Dict.]. 1892 *Pall Mall Gaz.* 25 Feb. 6/2 Men..who have been transferred to the Reserve..should be allowed..to return to complete twelve years' Colour service. 1905 Color solid [see *colour atlas*]. [1912 C. Mathewson *Pitching* xi. 236 'The necktie,' he explained.., and pointed to the three-sheet, colored-supplement affair he was wearing around his collar.] 1939 D. A. Spencer *Colour Photogr.* ix. 139 For editorial illustrations in the colour supplements of newspapers, this lack of subtlety is of little account. 1965 *New Statesman* 9 Apr. 586/1 Brainwashed into cultural coolness by Sunday papers and colour-supplements. 1966 *Listener* 20 Oct. 568/3 The young Poles long for western luxuries like cars and clothes and the other bonuses of colour-supplement living. 1929 Sheldon & Grisewood *Television* vii. 66 Baird was partially successful in color television in 1928. *Ibid.* 129 (*caption*) Mr. Baird demonstrating the first color television. 1955 G. G. Gouriet (*title*) An introduction to colour television. 1955 *Oxf. Jun. Encycl.* VIII. 441/1 The colour-television receiver has three separate cathode ray tubes, one for each of the signals from the three cameras. 1916 *Jrnl. Franklin Inst.* CLXXXI. 418 By 'color temperature' of a solid body radiating by virtue of its temperature is meant that temperature of a black body at which its radiation matches in color that of the solid body in question. 1926 J. W. T. Walsh *Photometry* ix. 271 The temperature of a black body which gives light of the same apparent colour as that given by the radiating body, is known as the 'colour temperature' of that body. 1875 Colour-tone [see sense 17 in Dict.]. 1888 Colour-tone [see *colour-sensitive* above]. 1895 E. B. Titchener in *Amer. Jrnl. Psychol.* VII. 81 *Farbenton*, color-tone. 1896 *Westm. Gaz.* 18 Nov. 7/3 New designs produced in three styles of colour-tone, and black and white. 1904 *Burlington Mag.* V. 52/2 'The most important part of colour-tone atmosphere,' Millet was fond of saying, 'can be perfectly rendered in black and white.' 1907 *Westm. Gaz.* 13 Sept. 2/1 Scarlet hips..a flaming colour-tone in the grey-green of the fading hedgerow. 1876 S. R. Koehler tr. *W. von Bezold's Theory of Color* iii. 133 The theory of the three fundamental sensations likewise demands that the color-chart should be a triangle, and..the color-circle represented in Fig. 46 has been constructed with the aid of the color-triangle, that is to say, it is to be looked upon as..an enlargement of the small circle represented within the color-triangle. 1922 W. Peddie *Colour Vision* iii. 40 The colour triangle and colour pyramid..are geometrical representations of the results of the trichromatic theory. 1957 R. W. G. Hunt *Reprod. Colour* viii. 88 The consideration of many of these questions is facilitated by the use of the colour triangle, which may be thought of as a kind of colour map, in which all colours are represented in a systematic way by points in a triangle. 1936 *Burlington Mag.* Oct. p. xxiii/2 Among table-glass there is a collection of colour-twist and baluster-stem glasses. 1965 P. M. Hubbard *Hive of Glass* ii. 22 My glass..was quite a small one, a colour-twist I wanted to believe was English. 1898 W. Robinson *Eng. Flower Garden* (ed. 6) xiv. 198 The Carnation..has a fine colour-value of foliage in winter. 1930 *Time & Tide* 7 June 744 Music is now moving towards a phase in which 'colour values' will be the principal means of expression. 1887 Colour-washed [see sense 17 in Dict.]. 1905 Holman Hunt *Pre-Raphaelitism* I. 114 The white-washing and colour-washing still not being completed. 1925 *Contemp. Rev.* Oct. 461 The bridegroom..glazes the windows and colour-washes the walls. 1950 F. Stark *Trav. Prelude* 236 The colour-washed little tower of Mortola church. 1968 *Guardian* 20 Aug. 5/2 A party from International Voluntary Service colourwashed some of the camp's concrete walls. 1957 *House & Garden* Nov. 142 'Silhouette' [fabric design] by a British artist is sold in four colourways in heavy cotton. 1960 *Guardian* 20 May 10/6 A very pleasant printed cotton satin..seems equally good in all the 'colour-ways'. 1924 R. M. Ogden tr. *Koffka's Growth of Mind* v. 268 Being 'colour-weak', I see red and green only under favourable conditions. 1909 *Cent. Dict.* Suppl., Color-weakness. 1893 *Funk's Stand. Dict.*, Color-wheel. 1924 R. M. Ogden tr. *Koffka's Growth of Mind* iii. 62 Rotating disk, or colour-wheel, half white and half black. 1967 E. Short *Embroidery & Fabric Collage* i. 10 The simplest form of colour wheel is one containing the three primary colours.

colourant, var. Colorant.

1916 F. B. Wade *Diamonds* 3 Iron as a colourant commonly produces yellow or brown in minerals. 1959 *Times* 24 Mar. 6/1 You get your colour by squeezing the scientific colourants into a Colorizer Base Paint. 1960 *News Chron.* 26 July 6/3 Hair colourants are now widely used by women of all age groups.

colouration: see Coloration.

colour-blindness. Add: **a.** (Earlier example.)

1844 D. Brewster in *Phil. Mag.* XXV. 140 We regret much that Prof. Wartmann has continued the offensive name of Daltonism in his memoir on colour blindness...We have used the word Colour blindness, because it indicates simply blindness to one or more colours.

b. (Further examples.)

1903 A. W. Patterson *Schumann* viii. 128 Musicians are usually accredited with a lack of the nobler feelings of generosity towards each other—that colour-blindness which can allow no rival to the 'ego'. 1962 *Economist* 28 Apr. 360/1 The capital's racially integrated schools provide a good example of colour-blindness.

colour-box. **1.** (Examples.)

1842 C. Ridley *Lett.* (1958) ix. 104 We had such a nice morning today cleaning our colour boxes like old times. 1847 C. Brontë *Jane Eyre* III. iv. 102 They discovered I could draw: their pencils and colour-boxes were immediately at my service. 1926–27 *Army & Navy Stores Catal.* 441/1 Japanned tin colour box and water bottle.

coloured, *ppl. a.* Add: **1. d.** *coloured audition* or *hearing* [cf. G. *farbiges hören*, F. *audition colorée*]: the perception of certain colours accompanying the hearing of sounds; = *colour-hearing* (Colour *sb.*¹ 18), *chromæsthesia; *coloured clothes*, civilian clothes (*Services' colloq.*).

1797 St. Vincent in J. S. Tucker *Mem. Earl of St. Vincent* (1844) I. x. 427 The Commander-in-chief having seen several Officers of the fleet on shore dressed like shopkeepers, in coloured clothes, [etc.]. 1829 J. Shipp *Mem. Mil. Career* ii. 33 One..asked me if I knew where to sell my coloured clothes. a 1876 G. Calladine *Diary* (1922) ii. 13 One of my old companions proffered a suit of coloured clothes. 1890 W. James *Princ. Psychol.* I. viii. 205 That curious idiosyncrasy of 'colored hearing' of which a few cases have been lately described..by foreign writers. a 1901 F. W. H. Myers *Human Personality* (1903) II. ix. 270 The synæsthesiæ, which have only of late years been noted between the ordinary senses—of which 'coloured audition', or sound-seeing, is the accepted type—may be carried yet further. 1937 *Sunday Express* 21 Feb. 10/3 Five out of every hundred people have 'coloured hearing'.

2. b. *spec.* in *S. Afr.* Of mixed black or brown and white descent; also (with capital initial), of or belonging to the population group of such mixed descent. *Cape Coloured*: see *Cape *sb.*³ 4.

1838 W. B. Boyce *Notes S. Afr. Affairs* 134 The coloured population are..demoralized in large towns in the neighbourhood of canteens. 1844 Gilchrist *Cape of Good Hope* ii. 20 The native population of the colony is generally called Hottentot, or bastard Hottentot, most of the coloured people approaching pretty nearly to the Hottentot formation, and some presenting a greater or smaller mixture of other, principally European, blood. 1897 Schulz & Hammar *New Africa* i. 15 Half-breeds, who are known by the distinctive title of 'coloured people', in differentiation from the natives up-country. 1958 *New Statesman* 12 Apr. 454/3 The new House will contain 163 members, four of them representing the newly separate Coloured electorate in the Cape.

Hence (Dict. and Suppl. senses of 2 b) as *sb.* Freq. in pl.

1938 N. Devitt *Spell of S. Afr.* xxiv. 207 The bulk of the menial labour is done by coloureds who are not highly paid for their services. 1955 *Times* 30 June 8/3 A hostility which has isolated African members of the staff at Fort Hare just as much as the European, and has been shared to some extent, too, by the few Indians and Coloureds who also study there. 1957 C. MacInnes *City of Spades* I. xi. 80 A spot where fine young American coloureds can destroy themselves with female white trash. 1965 *Listener* 23 Sept. 453/2 In his own country [he] will put on his 'to let' signs 'no coloureds please'.

colourful, *a.* Delete rare and add earlier and later examples. Also *fig.*, full of interest, excitement, force, etc.

1889 Hissey *Tour in Phaeton* 273 The distance then is delightfully distinct and colourful. 1905 *Century Mag.* Aug. 493/2 The English background lacks the gay and colorful beauty of our panoramic regattas. 1908 D. H. M. Read *Highways Hampshire* Pref., Hampshire, to me, is a bundle of memories, all colourful. 1908 *Daily Chron.* 23 Oct. 4/4 The gay, colourful city. 1911 J. London *Adventure* viii, It is colourful life, to say the least. 1967 *Times* 17 Nov. 8/5 Cries of 'Where is he?' Colourful answers.

Hence **co·lourfully** *adv.*, **-fulness.**

1921 J. F. Porte *Elgar* 90 The opening theme appears vigorously in the trombones, the whole being colourfully treated. 1921 *Chambers's Jrnl.* 26 Nov. 819/1 The play of *Kismet*, so colourfully produced. 1921 *Glasgow Herald* 7 Mar. 10 The imposing colourfulness of the mise-en-scène. 1928 *Observer* 24 June 8 So..movingly and colourfully does he tell his tale.

colourimetry: see *colorimetry.

colouristic, *a.* (Later examples.)

1934 C. Lambert *Music Ho!* iii. 190 The orgiastically rhythmic sections of *Le Sacre* were contrasted with other sections that relied more on melodic, harmonic, or colouristic

COLO(U)RIZER

appeal. **1947** T. S. R. Boase in *Jrnl. Warburg & Courtauld Inst.* X. 103 In a flat linear conception instead of a highly colouristic one. **1959** *Listener* 16 July 114/3 His highly personal treatment of percussion is not intended to increase the volume of an orchestral climax, so much as to exploit unexpected colouristic capacities.

co·lo(u)rizer. [f. Colo(u)rize *v.* + -er[1].] A colouring agent.
1880 *Libr. Univ. Knowl.* (N.Y.) VII. 430 The hematine not being able to perform the functions of a colorizer and oxygen-carrier alone.

colourtype, colortype (kʌ·ləɹtəip). [f. Colour *sb.* + Type *sb.*] *colourtype process,* a process for the reproduction of works of art in colour by the use of three-colour blocks. Also called the *three-colour process.*
1899 *Daily News* 24 Apr. 4/2 The three-colour process, called 'colortype'. **1900** *Westm. Gaz.* 27 Sept. 10/2 The Carl Hentschel Colortype process. **1904** *Ibid.* 24 Sept. 13/1 Illustrations faultlessly reproduced by the interesting Hentschel's colourtype process.

colpo- (kǫlpo), also *colp-* before a vowel, comb. form of Gr. κόλπος womb, used = vagina in terms of *Path., Surg.,* and *Anat.*; as *colpalgia, colpo-perinæorrhaphy, -rrhaphy, -scope, -scopy, -stenosis, -tomy.*
1908 *Practitioner* Dec. 807 He subsequently performs a posterior colpo-perinæorrhaphy. **1902** *Brit. Med. Jrnl.* 11 Oct. 1147/1, I now prefer bilateral colporrhaphy to anterior and posterior colporrhaphy. **1940** E. Novak *Gynec. & Obstet. Path.* vi. 92 *Colposcopy,* the magnification of the cervix afforded by the Hinselmann colposcope. **1905** *Brit. Gyn. Jrnl.* XXI. 75 (*heading*) On the value of colpotomy in the thrombotic form of puerperal fever. **1970** *Sci. Jrnl.* June 55/3 In some women limited access to the ovary can be gained by colpotomy.

colport (kǫlpǭ·ɹt), *v.* [Back-formation from Colporteur.] *intr.* To work as a colporteur. Also *trans.*, to carry as or as if a colporteur.
1888 *Centen. Confer. Missions* II. 337 Grants [of books] for distribution to those who want to colport. **1889** Stevenson & Osbourne *Wrong Box* xvi, You don't mean to insinuate that thing I..colported with my own hands, was the body of a total stranger?

Colt (kōᵘlt), *sb.*[2] orig. *U.S.* [f. the name of *Colt* (see below).] A firearm of a type invented by Samuel Colt (1814–62), esp. a type of repeating pistol (patented 1835). Freq. in the possessive and *attrib.*
1838 in J. H. Easterby *S. Carolina Rice Plantation* (1945) 84 Have you seen any of Colts patent repeating rifles? **1840** *Spirit of Times* 5 Sept. 319/1 (Weingarten), If you have any of Colt's revolving pistols take that along. **1852** E. G. Squier *Nicaragua* II. 92, I made a mental resolve..to appeal to my 'Colt', before admitting any too familiar approaches. **1854** J. R. Bartlett *Personal Narrative* I. ii. 19 All were provided with rifles or carbines, and many of the cavalry with Colt's revolvers. **1855** *Knickerbocker* XLV. 190 John just then handing me my 'Colt's'..I dashed over the picket. **1880** A. A. Hayes *New Colorado* (1881) xiii. 177 'What kind of shooting-iron have you?' he asked. 'Navy Colt?' **1907** Greener *Gun* (ed. 8) 517 The automatic Colt or 'Browning' pistol, is made in five different models...The magazine capacity is seven shots, and after its insertion in the handle the slide is drawn once to the rear by hand. **1908** G. K. Chesterton *Man who was Thursday* iii. 46 He caught up the Colt's revolver and took aim at Syme. **1908** C. E. Mulford *Orphan* xi. 137 'That's enough, Sherriff,' said Tex, moving cautiously forward behind his levelled Colt. **1911** J. London *Adventure* v. 65 A woman qualified to wear a Baden-Powell and a long-barrelled Colt's. **1970** R. Garrett *Run Down* i. 57 A big Peacemaker Colt.

coltskin (kōᵘ·ltskin). [f. Colt *sb.*[1] + Skin *sb.*] Leather made from the skin of a colt.
1903 Flemming *Pract. Tanning* 397 Horsehides and colt-skins are soaked..and washed in the usual manner.

colulus (kǫ·liŭlǝs). *Ent.* [mod.L., dim. of *colus* distaff.] A sclerite at the base of and between the front spinnerets in many spiders.
1904 *Ann. & Mag. Nat. Hist.* Jan. 65 The well-developed colulus, front spinnerets close together [etc.] show them to belong to M. Simon's group Cybæeæ. **1951** Locket & Millidge *Brit. Spiders* I. iii. 29 The colulus is a small tubercle set just in front of the anterior pair of spinners in certain spiders.

colum, var. *cooLUNG.

Columban (kǫlʌ·mbăn), *a.* and *sb.* [f. *Columba* + -an.] **A.** *adj.* Of, pertaining to, or characteristic of St. Columba and his followers. **B.** *sb.* A disciple of St. Columba.
1879 *Dublin Rev.* Oct. 271 The Columban monastic bodies. *Ibid.* 273 The expulsion of the Columbans from Pictish territory. **1920** in H. F. Henderson *Relig. in Scot.* iii. 76 Hither resorted the young men..to..study the discipline of the Columban Church. **1960** G. W. Dunleavy *Colum's Other Island* 93 Under Aidan's hand and in the Columban tradition, a second Iona arose in Northumbria to father other monasteries.

columbarium. Add: **2. b.** A similar structure in a modern crematorium.
1899 *N.Y. Herald* 27 July, Besides the members of the household at Walston the following persons will be admitted to the columbarium. **1923** *Harmsworth Househ. Encycl.* II. 993/3 *Columbarium.* This word is used..for the building in which the urns containing the remains of cremated persons are placed. **1948** E. Waugh *Loved One* 15 Would you require a niche in our columbarium or do you prefer to keep the remains at home? **1963** J. Mitford *Amer. Way Death* iii. 141 He is much more likely to end up in a niche and urn in an elaborate 'columbarium'.

Columbian, *a.* Add: (Earlier and later examples.) Also with specific reference to Christopher Columbus or to the time of his discovery of America, and *ellipt.*
1757 Dyer *Fleece* ii. 67 Ev'n in the new Columbian world appears The woolly covering. **1885** *Harper's Mag.* Jan. 224/1 They [*sc.* skulls] could not have been ante-Columbian, because we turned up a bit of rusted iron, the fragment of a knife. **1898** *Forum* Feb. 677 All the people of America, at the date of their discovery by Europeans in the Columbian epoch, were organized into tribes. **1903** *Daily Chron.* 19 May 3/1 Future Columbian students. *Ibid.,* So many early Columbian texts. **1965** *Aylesford Rev.* Spring 13 He also had a large Albion..and a Columbian. Among printers the Columbian press is famed for its extreme decorativeness.
Hence as *sb.*, an American; an inhabitant or native of the U.S.A. *Obs.* exc. *Hist.*
1789 S. Low *Politician Outwitted* ii. ii, in M. J. Moses *Repr. Plays Amer. Dramatists* (1918) I. 377 As the East is to the West..or the Aborigines of America to the Columbians of this generation, so is that line to this line. **1815** W. Thornton *Outl. Const. Columbia* i, I was born in America..and..I feel an unspeakable attachment to the whole race of Columbians. **1947** *St. Louis Globe-Democrat* 7 Sept., Now they could call themselves Columbians instead of Americans.

Columbus (kǫlʌ·mbǝs). The name of Christopher *Columbus,* the explorer (1451–1506), used allusively for an explorer or discoverer.
1593 G. Harvey *Pierces Supererogation* 11 A new-found land of confuting commodities discouered, by this braue Columbus of tearmes. **1656** Cowley *Pind. Odes* 27 Thou great Columbus of the Golden Lands of new Philosophies. **1680** Oldham in *Rochester's Poems* 125 'Tis I, the bold Columbus, only I, Who must new Worlds, in Vice descry. **1826** E. B. Browning *Essay on Mind* ii. 64 What tongue can syllable our Bacon's name..Sublime Columbus of the realms of Mind! **1882** W. D. Hay *Brighter Britain* I. i. 11 A veritable cargo of Columbuses. **1927** *Melody Maker* Sept. 938/1 We might hope for some musical Columbus to appear on the scenes.

column, *sb.* Add: **4.** *spec.* in a newspaper (earlier example). Hence in extended use: a special feature, esp. one of a regular series of articles or reports. Cf. *gossip column.* In the U.S. sometimes with the jocular spelling **colyum** (kǫ·lyǝm).
1785 *Daily Universal Register* 1 Jan. 4/1 Have they [*sc.* newspapers] not frequently, in half a column, given us the state of all nations? **1914** *Dialect Notes* IV. 104 *Colyum,* column. **1926** H. W. Davis *Column* i. 3 The most important development on America's editorial pages during the past quarter of a century has been the evolution of the 'colyum'. *Ibid.* 4 The column may be a good light essay broken up by two-em dashes. **1931** R. Campbell *Georgiad* ii. 36 Then through my weekly columns I may pour The sentiments that dowagers adore. **1932** E. V. Lucas *Reading, Writing* v. 108 Michael Temple, who had charge of the 'Men and Matters' column. **1937** W. S. Maugham *Theatre* xxiii. 221 The Press representative came up and told Michael he thought he'd given the reporters enough for a column. **1959** N. Mailer *Advts. for Myself* (1961) 247 The column this week is difficult.
8. esp. with qualifying phrase.
1885 *Encycl. Brit.* XIX. 34/2 This is termed the 'posterior vesicular column', or 'Clarke's column', after the late Mr. Lockhart Clarke, who did much to unravel the intricate anatomy of the nerve-centres. **1906** *Practitioner* Dec. 844 The columns of Morgagni are permanent vertical folds of the mucous membrane of the anal canal. **1907** *Ibid.* June 859 The postero-lateral column of Burdach. *Ibid.,* The postero-mesial column of Goll.
10. b. *column of route*: see Route *sb.* 3 c.
c. *transf.* A body or party; = Camp *sb.*[2] 8 a. orig. and chiefly *U.S.*
1906 *Forum* Apr. 448 The resulting dissatisfaction would be sufficient to throw Michigan..and possibly one or two other States into the Democratic column. **1931** G. T. Clark *L. Stanford* iv. 95 California, by a narrow margin, swung into the Republican column. **1958** *Spectator* 22 Aug. 241/1 The public..is unsophisticated to a high degree, and a strong indication is supplied in its slow but undeniable shift back into the Conservative column.
d. Colloq. phr. *to dodge the column,* to shirk one's duty; to avoid work.
1919 in *Athenæum* 1 Aug. 695/2. **1942** *Penguin New Writing* XV. 29 The corporal said: 'Dodging the column again, eh?' **1955** H. Spring *These Lovers fled Away* 5 My father, so great an expert in dodging any column he didn't see the point of joining.
12. column-inch, one inch of a newspaper column.
1940 S. Lewis *Bethel Merriday* iv. 43 Advertising-column-inches in trade journals. **1960** F. Raphael *Limits of Love* i. vii. 94 That's not the picture you get from the Press...The number of column inches ——.

columnal (kǫlʌ·mnăl), *sb.* [f. the adj.] A segment or joint of the stem of a crinoid.
1892 F. A. Bather in *Ann. & Mag. Nat. Hist.* IX. 212 Columnals rather low and alternating in thickness and height. **1914** *Brit. Mus. Return* 212 Two fragments of the 'shell-bed' with Crinoid columnals from the Lower Coal Measures. **1955** L. H. Hyman *Invertebrates* IV. xv. 50 The stalk [of crinoids] is supported by a single row of superimposed rounded or pentagonal skeletal pieces termed columnals.

columnaris (kǫlǝmnā·ris). [mod.L. (see def.): cf. Columnar *a.*] A bacterial disease of freshwater fish caused by *Chondrococcus columnaris,* which forms short column-shaped masses on the edges of the infected tissues or scales.
1945 *Jrnl. Bacteriology* XLIX. 114 The characteristic lesions of this and other kinds of fishes in natural and induced 'columnaris infections' were readily produced. **1960** *Progr. Fish-Culturist* XXII. 43 (*title*) Outbreaks of columnaris in Center Hill and Old Hickory Reservoirs, Tennessee. **1966** *Courier-Mail* (Brisbane) 11 May 4/6 Scotland's fresh salmon industry is facing the threat of a fish disease called columnaris—a sort of fishy myxomatosis. **1966** *Daily Tel.* 11 Aug. 17/5 Some fish in West Sussex rivers are believed to have died from columnaris, which has killed thousands of coarse fish in the Midlands.

columnist (kǫ·lǝmnist, -mist). orig. *U.S.* [f. Column *sb.* 4 + -ist.] One who writes a 'column' in the newspaper press (see *column *sb.* 4). In the U.S. sometimes with the jocular spelling **colyumist** (kǫ·lyǝmist).
1920 *Blackw. Mag.* Aug. 146/1 The 'colyumist' of a New York paper. **1925** *Lit. Digest* (N.Y.) 27 June 27/2 Here is a Vashti leading the opprest columnist into the promised land of intellectual liberty. **1926** *Spectator* 24 Apr. 770/1 One of the best known 'columnists' of the American press. **1929** *Times* 27 Nov. 15/5 The..society columnist. **1930** *Publishers' Weekly* 1 Mar. 1063 Dorothy Herzog is a Hollywood columnist. **1957** *Times Lit. Suppl.* 8 Nov. 670/4 She..becomes a first-class journalist, feature-writer and columnist herself.

colyum, -ist: see *column *sb.* 4, *columnist.

Comacine (kōᵘ·măsəin), *a.* and *sb.* [ad. It. *comacino,* app. ult. f. *Como* in Italy (see quot. 1899).] **A.** *adj. Comacine masters* [It. *maestri comacini*], a mediæval guild of Italian masons. **B.** *sb.* A mason or builder belonging to this guild.
1899 'L. Scott' *Cathedral Builders* 5 The origin of the name *Comacine Masters* has caused a great deal of argument amongst Italian writers new and old. Some think it merely a place-name referring to the island of Comacina, in Lake Lario or Como; others take a wider significance, and say it means not only the city of Como, but all the province, which was once a Roman colony of great extension. Others again, among whom is Grotius, suggest that it is not a place-name at all, but comes from the Teutonic word *Gemachin* or house-builders. As the Longobards afterwards called them in Italian *Maestri Casarii,* which means the same thing, there is perhaps something to be said for this hypothesis. *Ibid.* 9 Rome is..full of remains of what is now styled Comacine architecture. *Ibid.* 17 There is no certain proof that the Comacines were the veritable stock from which the pseudo-Freemasonry of the present day sprang. **1900** *Monthly Rev.* I. 103 The Comacine masters have their existence sufficiently proved by..the edict of Rotharis (dated 653). *Ibid.* 104 The collegiate and Comacine constitutions.

Comanche (komæ·ntʃi), *sb.* and *a.* Also †**Cama·nche.** [Sp., f. a Shoshonean language.] **A.** *sb.* **1.** A North American Indian people of Texas and Oklahoma; a member of this people.
1806 J. Wilkinson in Z. M. Pike *Sources Mississippi* (1810) 109 You will also receive a..large one [*sc.* belt] for the Tetaus or Comanches. **1831** M. Holley *Let.* Dec. in *Texas* (1833) 90 To denote the greatest degree of degradation, they [*sc.* the Mexicans] call a person a Comanche. **1874** B. F. Taylor *World on Wheels* 76 The two engines are neck and neck. They scream at each other like Camanches. **1886** [see *Shoshone]. **1888** P. Gillmore *Days & Nights by Desert* xi. 88 Giving a yell like a lunatic Comanche. **1958** *Spectator* 31 Jan. 135/3 Spurs, saloons, Comanches, the whole cowboy paraphernalia.
2. The language of this people.
1856 *Trans. Philol. Soc.* 106 The Shoshoni and Wihinast..differ from each other more than the several forms of the Cumanch. **1874** *Forest & Stream* 13 Aug. 1/2 It is in fact the Court language, all councils with Kiowas..being held in Comanche. **1933** Bloomfield *Language* iv. 72 The Shoshonean family..including..Comanche. **1964** R. H. Robins *Gen. Linguistics* iii. 94 Voiceless vowels do occur in the normal speech of some few languages (e.g. Comanche).
B. *adj.* Of or pertaining to any of the above.
1819 in *S.W. Hist. Q.* (1913) XVI. 331 A Spanish woman..also adds..that the Comanche Indians..harrass them. **1834** A. Pike *Sketches* 198 The Comanche tongue is easy to learn. **1897** H. Inman *Old Santa Fé Trail* 332 He gave a series of war-whoops which discounted a Comanche yell. **1949** 'J. Tey' *Brat Farrar* xii. 99 A Comanche bridle is..just a gag. The rope goes through the mouth the way a bit does.

comb, *sb.*[1] Add: **1. d.** *fig.* The action or process of 'combing out' (see next, 6 b).
1916 *Even. News* 8 Nov. 1/4 The comb which is being applied at the moment to the police appears once again to have begun at the wrong end. **1925** Fraser & Gibbons *Soldier & Sailor Words* 61 *The comb,* the popular newspaper term used in the War for the process of obtaining men for the Army by compulsorily thinning out from the professional classes and trades and Government Offices all physically fit. At the Front the non-combatant branches were also dealt with.

2. (h). The lower, fixed cutting-piece of a sheep-shearing machine. *Austral.* and *N.Z.*
1891 R. Wallace *Rural Econ. Austral. & N.Z.* xxix. 379 The cutter..moves from side to side 4,000 times per minute over the comb, which rests upon the skin of the sheep, and threads its way among the wool close to the surface of the body. **1950** *N.Z. Jrnl. Agric.* Oct. 310/1 Dust can lower the value of a [wool-]clip..besides being hard on combs and cutters. **1956** G. Bowen *Wool Away!* (ed. 2) vi. 73 It takes a sharp comb and a sharp cutter to cut wool off.

6. e. (Earlier U.S. example.)
1824 in Z. F. Smith *Hist. Kentucky* (1886) 394 The roof was formed by making the end logs shorter, until a single log formed the comb of the roof.

9. comb-back, a Windsor chair with a straight-top bar into which the back spindles fit; the back of such a chair; so **comb-backed** *a.*; **comb-case** (later U.S. example); **comb filter** *Electr.* (see quot. 1960); **comb-foundation,** a thin sheet of beeswax, made to resemble the middle wall of honeycomb, placed in a hive for bees to build their comb upon; **comb-jelly,** a ctenophoran jelly-fish belonging to the order Cydippidea.
1901 E. Singleton *Furnit. of Forefathers* II. v. 398 Another chair, a Windsor, of the kind called 'comb back'..was made in all probability by a local workman. **1935** *Apollo* Aug. 69/1 Windsor chairs..can be divided into two main categories. The chair with the comb-back..and the chair with the hoop-back. **1967** L. J. Braun *Cat who ate Danish Modern* vii. 63, I can offer you a Chippendale corner chair, a comb-back Windsor, [etc.]. **1929** E. Wilson *I thought of Daisy* iv. 233 The principal prize was a comb-backed rocking-chair. **1843** *Amer. Pioneer* (Cincinnati) II. 444 A small eight by ten looking-glass sloped from the wall over a large towel and combcase. **1941** P. Mertz *Television—Scanning Process* 20 A sort of comb filter having pass bands at the multiples of the line-scanning frequencies. **1960** Cooke & Markus *Electronics & Nucleonics Dict.* 86/2 *Comb filter,* a wave filter whose frequency spectrum consists of a number of equi-spaced elements resembling the teeth of a comb. **1880** *Harper's Mag.* Oct. 778/1 Comb foundation has another and far greater merit than that of saving labor for the bee: it secures a perfectly even, straight comb for each frame. **1957** *Encycl. Brit.* III. 306/1 Comb foundation is another important invention, consisting of sheets of pure beeswax on which are embossed the bases of the cells of the honey-comb. **1889** *Cent. Dict.,* Comb-jelly. **1903** J. R. A. Davis *Nat. Hist. Animals* II. xiv. 155 A common British Comb-Jelly, Cydippe,..which is rowed through the water by eight longitudinal rows of little paddles, which suggest by their appearance the teeth of a comb. **1959** A. Hardy *Fish & Fisheries* iii. 53 Planktonic predators such as the comb-jellies (ctenophores) *Pleurobrachia*—or 'sea gooseberries', as the fishermen call them.

comb, *v.*[1] Add: **4. b.** To search or examine minutely. *colloq.* (orig. *U.S.*).
1904 'O. Henry' *Cabbages & Kings* iv. 80 In Coralio Señor Goodwin himself led the searching party which combed that town as carefully as a woman combs her hair. **1913** C. E. Mulford *Coming of Cassidy* ix. 138 Then, sweeping north, they combed the range to the northern line [for cattle]. **1927** *Daily Express* 23 July 10/6 Search for missing scientist. Plans for 'combing' a ten-mile radius. **1927** *Morn. Post* 19 Aug. 7 The Pacific Ocean between San Francisco and Hawaii is being combed to-day by aircraft and shipping for signs of the two 'planes.
c. Of a ship: to turn into line with the tracks of approaching torpedoes in order to avoid being hit.
1942 G. Hackforth-Jones *One-One-One* ix. 94 With smoke pouring from the gap in the bows the ship was swung to starboard to 'comb' the tracks of the approaching torpedoes. **1954** P. K. Kemp *Fleet Air Arm* 134 The *Formidable* had no difficulty in avoiding them [*sc.* the torpedoes], as she had time at that range to alter course and comb the tracks. **1957** R. Barker *Ship-busters* xiii. 245 The tanker..was turning towards the track of the torpedo in an endeavour to 'comb' it.
6. comb out. a. *trans.* To disentangle or dress (the hair) with a comb; hence *fig.*, to separate or sort out.
1854 Surtees *Handley Cr.* lv. 390 Forthwith the dandified Horatio began to comb out his words, and string altogether his sentences. **1855** [see 1 in Dict.]. **1888** Burgon *Lives 12 Gd. Men* II. v. 33 It was as if he had combed out his cares. **1937** Hemingway *To have & have Not* III. xxvi. 259 He took..the pins out and combed it out and it was just like gold. **1964** *Sunday Express* 1 Mar. 19/4 Clients on all sides were being 'combed out'. **1966** L. Black *The Bait* viii. 124 Emma's hair-do was nearly finished..'Just finishing. Let me comb it out.' **1966** J. S. Cox *Dict. Hairdressing* 38/1 *Comb-out.* (1) To comb through the set and dried hairdress preparatory to the final arrangement of the tresses. (2) To comb a head of hair.
b. To clear out (men) for military service from civil employments.
1916 *Daily Mail* 1 Nov. 5/6 'Comb out the contractors' useless men,' said a farmer delegate. **1918** Mrs. H. Ward *War & Elizabeth* vii, My two brothers are dead in France. I shall be 'combed' out directly. **1923** G. D. H. Cole *Workshop Organiz.* 28 The successive 'combings-out' of the munition works for further 'man-power' for the fighting forces.
c. = *4 b. Also, to attack systematically.
1917 'Ian Hay' *Carrying On* viii. 212 Fighting in the Redoubt itself had almost ceased, though a humorous sergeant, followed by acolytes bearing bombs, was still 'combing out' certain residential districts in the centre of the maze. **1943** Ward-Jackson *Piece of Cake* 22 *To comb out,* to sweep over in formation, attacking ground targets with gun-fire. Thus, 'We're combing out the North of France this afternoon'.

combat, *sb.* Add: **4.** Now in frequent use, esp. in the U.S., in sense: of or pertaining to the fighting services (as opposed to 'base' units, etc.); **combat fatigue,** a nervous disorder resulting from prolonged or severe battle experience.
1939 *Times* 6 Nov. 6/1 The Neutrality Bill..defined 'combat zones', from which American ships are barred, to include trade with European neutral countries bordering on the North Sea and the Baltic. *Ibid.* 6/2 The President's neutrality proclamation..goes much farther in its definition of combat areas than any of them had expected. **1942** *N.Y. Times* 9 Nov. 8/4 General Eisenhower's strong, well-equipped forces include crack combat troops. **1943** G. N. Raines et al. in *Naval Med. Bull.* XLI. 923 (*heading*) Combat fatigue and war neurosis. *Ibid.* 933 It has been suggested that the term 'combat fatigue' be applied to the uncomplicated syndrome. **1944** *Ann. Reg. 1943* 137 War production included..combat planes. **1945** *U.S. Army Biennial Rep.* 24/1 During the winter, three Italian combat groups entered the line of the Eighth Army. **1966** *Listener* 29 Dec. 949/3 Last January the United States had forty-two combat battalions in Vietnam.

combativity (kǫmbǟti·vĭti, kɒm-). [-ity. Probably coined to avoid the phrenological association of *combativeness.*] The quality or character of being combative.
1905 *Westm. Gaz.* 28 Jan. 8/3 He has less..uncompromising combativity than his predecessor. **1925** *Spectator* 5 Dec. 1019/2 The innate and eternal combativity of the human race. **1930** Wyndham Lewis *Apes of God* IX. 266 Head on one side, face puckered to its utmost of bantering combativity.

Combe-Capelle (kom̃b kapɛl). The name of a rock-shelter near Montferrand-du-Périgord, Dordogne, France, applied *attrib.* to the remains of an early type of man discovered there in 1909.
1911 A. Keith *Anc. Types of Man* v. 56 While the writer is inclined to agree in provisionally assigning the Combe-Capelle man to the Galley Hill race he believes that further discoveries will show that the Combe-Capelle man belongs to a branch marked with certain negroid features. **1927** Peake & Fleure *Hunters & Artists* v. 66 We may call it provisionally the Combe Capelle, or perhaps better the Predmost type by way of contrast with the Cro-Magnon type. **1959** *Chambers's Encycl.* IX. 40/2 (*heading*) The Combe Capelle Man. In 1909 Hauser..unearthed a skeleton in the rock-shelter of Combe Capelle..at the very base of a deposit containing Aurignacian implements...The remains are generally agreed to belong to the Cromagnon type,..though differing in some notable respects.

combed, *ppl. a.* Add: **2. b.** Applied to pottery decorated by drawing a toothed instrument across a coloured slip.
1878 L. Jewitt *Ceramic Art* I. 98 A round dish of the 'combed ware', described by Plot, is shown...Some of the examples..are exceedingly delicate..others..have been 'combed' with a coarse comb or wire brush. **1960** H. Hayward *Antique Coll.* 80/2 *Combed ware,* pottery decorated by the application of two or more different coloured slips brushed or combed to produce an effect similar to marbled paper.

combie (kǫ·mbi), colloq. abbrev. (usually in *pl.*) of Combination 9.
1891 Farmer *Slang* II. 158/2 Combie. **1923** U. L. Silberrad *Lett. Jean Armiter* iv. 81 I'm not wet to the combies; as you would seem to be. **1923** K. Mansfield *Doves' Nest* 179 Long-sleeved woven combies. **1952** G. Raverat *Period Piece* viii. 144 *Her* children wore linen shifts..and not woolly 'combies' as we did.

combinability (kǫmbəinăbi·lĭti). The quality of being combinable.
1900 G. Iles *Flame, Electr. & Camera* 255 Chemical combinability. **1962** L. J. Cohen *Diversity of Meaning* ii. 27 Its meaning comes to depend on certain relations of combinability with other words.

combination. Add: **4. d.** A small instrumental band. Cf. *combo 3.
1928 *Melody Maker* Feb. 165/2 The combination is a five piece one, comprising piano, saxophone, trumpet, banjo and drums.
5. c. An ingenious sequence of moves in chess.
1875 *Field* 15 May 469/1 Mr Blackburne is not apt to lag when the opportunity for a fine combination presents itself. **1951** 'Assiac' *Adv. Chess* I. iii. 21 The main work in White's pretty little combination was done by the Bishop.
9*. = *combination lock* (see below). Also, the series of movements required to open such a lock.
1845 C. Cist *Cincinnati Misc.* 194 The locks which are on the combination principle, not only defy picking, [etc.]. **1880** 'Mark Twain' in *Atlantic Monthly* XLVI. 228/2 They commanded him to reveal the 'combination', so that they could get into the safe. **1909** 'O. Henry' *Roads of Destiny* x. 170 She had then shot the bolts and turned the knob of the combination as she had seen Mr. Adams do. **1924** Mulford *Rustlers' Valley* iii. 25 Counting money and manipulating his combinations were two things the banker could do automatically... Glenn slowly closed the safe, automatically spun the combination knob. **1932** Wodehouse *Hot Water* xvii. 287 Some guy once told me that if you listened for the tumblers you could get the combination. **1958** L. Cottrell *Anvil of Civilisation* x. 144 It was their effect that mattered—like the combination which opens a safe.
fig. **1889** Barrère & Leland *Dict. Slang* II. 30/2 *To lose the combination* (American), to miss the meaning or point of anything. **1906** L. Bell *C. Lee* 271 Why, Carolina, I've lost the combination! **1918** 'Ian Hay' *Last Million* xv. 24[illegible] 'How did the British Tommy and the Doughboy get along?'..'Each took a while..to get the combination of the other.'
b. (In full *motor-cycle combination.*) A motor-cycle with a side-car attached for the conveyance of a passenger or goods.
1907 G. B. Shaw *Lett. G. Barker* 19 Sept. (1956) 105, I should have come up for the last few rehearsals: the combination works better than the single cylinder. **1914** *Motor Cycling* 12 May p. v, Both Chater Lea combinations ran perfectly throughout. **1919** *Bazaar, Exch. & Mart* 28 June Suppl. 11/1 Wanted, motor cycle combination. **1928** *Evening News* 24 July 1/1 Here two foot police were standing on the bridge and two others were on a motor-cycle combination. **1968** J. Fleming *Kill or Cure* xiv. 189, I stopped the combination in a lay-by and I screamed at him not to go.
10. combination grate (see quot. 1940); **combination lock,** one which can be opened only after a certain combination of movements has been performed; **combination tap** (see quot.); **combination tone** = *combinational tone* (cf. Tone *sb.* 2 and Resultant *a.* 2 c).
1940 *Chambers's Techn. Dict.* 180/2 *Combination grate,* a grate specially designed to supply heat for hot water and cooking, and for heating the room. **1959** *B.S.I. News* Oct. 18 (*heading*) Fireguards for solid fuel combination grates. **1851** C. Cist *Cincinnati* 215 Combination and detector bank lock. **1878** *Technol. Dict.* (ed. 3) 154/2 *Combination-lock,* das Vexirschloss. **a 1884** Knight *Dict. Mech.* Suppl. 210/2 *Combination Lock.* 1. A bank or safe lock,..operated by two graduated dials, whereby one bolt, common to both, is controlled by either of the two independent dials... 2. A permutation lock. **1947** *Science Illustr.* July 93/1, I want a door-knob with a combination lock fitted into it. **1951** *Good Housek. Home Encycl.* 281/2 *Combination Taps.* In these, hot and cold water are delivered through one outlet. **1889** *Cent. Dict.,* Combination tone. **1910** G. B. Shaw *How to become Mus. Critic* (1960) 270 Helmholtz's discovery of overtones, partial tones, combination tones, and the like. **1955** *Oxf. Compan. Mus.* (ed. 9) 10/2 When two loud notes are heard together, they give rise to a third and fourth sound...The generic term Resultant Tone (or Combination Tone) is given to these.

combinative, *a.* Add: **3.** *Phonology.* Applied to sound-changes which are effected through a combination of influences: opp. *isolative.*
1888 Sweet *Hist. Eng. Sounds* 17 Another important distinction is that of isolative and combinative...Combinative changes imply two sounds in juxtaposition, which modify each other in various ways.
4. *Chess.* Applied to a player of chess or a game using 'combinations' (see prec. sense *5 c).
1934 in F. D. Yates *101 Best Games of Chess* 13 Yates never lost his power of combination and his game against Vidmar..has been described..as 'The finest combinative game played since the war'. **1950** R. N. Coles *Chess-player's Week-end Bk.* 41 Alexander Alekhin..was the greatest combinative master of modern times.

combinator. Add: **2.** *Logic.* Any one of the special operators in combinatory logic.
[**1930** H. B. Curry in *Amer. Jrnl. Math.* LII. 526 Unter einem *Kombinator* verstehen wir eine Kombination von *B, C, W* und *K.*] **1941** —— in *Jrnl. Symbol. Logic* VI. 54 (*heading*) Consistency and completeness of the theory of combinators. **1946** *Mind* LV. 281 Part II is a systematic exposition of the logic of combinators as developed by Curry and Rosser.

combinatorial, *a.* Add: Hence **combinatoria·lity, combinato·rially** *adv.*
1963 *Times Lit. Suppl.* 3 May 320/4 'Combinatoriality', 'linearly' [etc.] are not pretty words. **1958** Curry & Feys *Combinatory Logic* 5 The system shall be combinatorially complete. **1965** *Language* XLI. 195 Some combinatorially possible arrangements do not occur at all.

combinatory, *a.* Add: **2.** *combinatory logic,* a branch of symbolic logic, concerned especially with the analysis of the processes of substitution and with the elimination of variables.
1929 H. B. Curry in *Amer. Jrnl. Math.* LI. 362 Logical substitution; its relation to a combinatory problem. **1946** *Mind* LV. 281 The first instalment of a new synthesis of combinatory logic.

combine, *v.* Add: **1. e.** With pronunc. (kǫ·mbəin). To harvest (crops, etc.), by means of a combine (harvester). orig. *U.S.*
1926 *Kansas City Star* 23 June, The first wheat combined in this vicinity was from the 100-acre field of A. E. Rudd. **1957** *Times* 24 Aug. 4/7 Up to half the grain has been cut or combined in those regions. **1958** *Listener* 16 Oct. 593/2 There were cases of fields having to be cut by reaper, windrowed, and then combined—not a cheap way of harvesting.

combine, *sb.* Now usu. with pronunc. (kǫ·mbəin). Add: **b.** Now in standard English usage. (*U.K.* examples.) Also *fig.*
1889 G. B. Shaw *London Music 1888–89* (1937) 195 When, after a few more years of competing syndicates, we have a great 'combine' of the Harris, Leslie, D'Oyly Carte, and all the other interests. **1899** T. M. Ellis *3 Cat's-Eye Rings* i. 1[illegible] Why athleticism and æstheticism should not form a combine is a conundrum. **1936** Auden *Look, Stranger!* 43 Europe grew anxious about her health, Combines tottered, credits froze. **1936** *Discovery* Sept. 280/2 All types of industry from the combine employing its twenty thousand to the little workshop in the side street. **1955** *Times* 6 Aug. 7/7 Recently a multiple dairy, firm X, bought the business of a small dairyman with whom I had dealt for many years. Some days later I was surprised to find the milk of firm Y—another combine—delivered to me.

c. combine harvester, an agricultural machine which performs various harvesting functions (as cutting, threshing, and bagging grain) simultaneously; also *ellipt.,* and as *combined harvester,* and in other collocations with *combine(d)* as first element. So also **combine drill** (for sowing and fertilizing seeds in one operation), and similar formations.
1857 *Illinois State Register* (Springfield) 15 July 3/2 In the afternoon the combined mower and the Illinois mower were put upon trial, in a beautiful field of timothy. **1900** D. McK. Wright *Wisps of Tussock* 54 The engine beats and the combine sings to the drays that are leading in. **1923** J. R. Bond *Farm Implements* ix. 119 Combined fertiliser and seed drills are also made for ridge work. **1926** *Kansas City Star* 23 June, Hundreds of combines will be in the fields in southern, central, and western Kansas by Wednesday. **1927** *Implement & Machinery Rev.* LII. 1072/2 The combined harvester and thrasher made its first appearance at the Paris Show,..American firms have..been developing the combined harvester and thrasher for some years. **1929** *Inst. Res. Agric. Engin. (Oxf. Univ.)* Bull. No. 3. 7 Throughout this report the Combine Harvester or Harvester-Thresher is referred to as the 'combine'. *Ibid.,* The combine may be described as consisting of the knife and platform canvas of a binder attached to a travelling threshing machine. **1930** *Engineering* 18 July 83/1 But since suitable British combine machines have been placed on the market, a considerable number of orders have been secured by home firms. **1930** *Manch. Guardian* 16 Sept. 9/2 Great factories are now being equipped in Russia, so that tractors, combine-harvesters, milking machines, and all the equipment of a mechanised agricultural industry may be furnished in the future from native sources. **1932** *Discovery* Jan. 11/2 The combine-harvester—a machine in which a reaper is attached to a portable threshing machine and the whole is drawn by a tractor through the standing corn, leaving a trail of sacks of threshed grain and another of straw in its wake. **1941** *Jrnl. Min. Agric.* XLVIII. 186 A combine drill is now generally accepted as being superior to a seed drill and a broadcast fertilizer used separately. **1955** *Times* 10 May 17/5 The sombre picture of the weather last year threw into sharp relief the useful services performed by the combine harvester. **1957** *Times* 14 Oct. 2/6 Combine drills which place the fertilizer alongside the seed to give the seedlings a ready supply of plant food from the start of growth. **1958** *Listener* 16 Oct. 593/2 Some farmers were pushing or pulling self-propelled combines along with tractors.

combined, *ppl. a.* Add: **b.** Also *combined exercise, operation*: spec. one performed by branches of the fighting services acting in combination. Also in extended use, and *ellipt.*
1842 N. Willis *Canad. Scenery* I. ii. 49 England opened the campaign of 1759 with a plan of combined operations by sea and land. **1922** *Flight* XIV. 721/1 The subjects studied..Strategy and tactics..combined operations. **1938** *Times Weekly* 30 June 4/2 They were not indeed 'combined exercises' in the technical sense, employing units of the Air Force as well as of the Navy. **1942** *Hutchinson's Pict. Hist. War* 10 June–1 Sept. 261/2 The Dieppe assault was more than a raid..as a combined operation it is stated to have been a successful demonstration of co-ordination of all three services. **1947** *News Chron.* 8 Mar. 1/3 Britain's Coal Cabinet last night ordered the biggest combined operation yet planned in an effort to clear snowbound roads and rails. **1961** B. Fergusson *Watery Maze* i. 16 Combined Operations in our sense of the phrase, implying opposed landings in force with the intention of staying ashore, were rare before the eighteenth century. *Ibid.* vi. 147 It is impossible not to feel sympathy for the Combined Commanders. *Ibid.* xii. 292 Someone must still look after the Combined Training Establishments.
d. *Agric.* (See *combine *sb.* c.)
e. *combined room,* also ellipt. as *sb.* (see quots.).
1933 P. Godfrey *Back-Stage* xvi. 200 The poorest paid players live in a single room known as a 'combined'. *Ibid.* 205 A stuffy little 'combined'. **1952** W. Granville *Dict. Theatr. Terms* 47 *Combined-room,* a bedroom and sitting-room 'combined'. One of the 'classic' landlady advertisements ran: 'I have vacant for next week a large, comfortable combined-room. Piano and lavatory inside.'

combing, *vbl. sb.* Add: **3. combing plate** (see quots.); **combing skin,** a long-woolled skin; a skin bearing long wool suitable for combing.
1886 *Encycl. Brit.* XX. 521/1 Fig. 1...Grinding surface of a..molar of Rhinoceros..12, crochet (posterior combing plate). *Ibid.* 521/2 The middle sinus is often intersected by vertical laminæ ('combing plates') projecting into it. **1895** *Daily News* 10 May 9/3 Good combing skins show ¼ to ½d. per lb. advance; short-woolled skins rule about on a par with..last auction values.

combining, *vbl. sb.* Add: **c.** *combining form* (see quot. 1942).
1884 [see Dict. s.v. Aero-]. **1942** Bloch & Trager *Outl. Ling. Analysis* 66 In Latin and other languages, many words have a special combining form which appears only in compounds (or only in compounds and derivatives)...The foreign-learned part of the English vocabulary also shows a number of special combining forms; cf. *electro-,* combining form of *electric,* in such compounds as *electromagnet.*

comble, *sb.* Restrict †*Obs.* to sense in Dict. and add: **2.** The 'crown' or culmination. [A gallicism.]
1832 S. Austin tr. *Pückler-Muskau's Tour Germany* IV. 379 'Good-nature' is laughed at and despised as the 'comble' of vulgarity. **1884** G. M. Hopkins *Let.* 3 Aug. (1935) 195 That book could be the greatest boon to me,..And if you were to complete 'Wildair', that would be the *comble.* **1903** Falkner *Nebuly Coat* viii, He added the comble to all his graces and courtesies by shaking her hand. **1908** *Daily Chron.* 1 May 4/7 The disgust of those who believe that there is an art of acting which is marred..by such..interruptions reaches its comble. **1958** B. Nichols *Sweet & Twenties* xi. 137 Fiercer and fiercer waxed the controversy...The *comble* came in August 1925.

combo (kǫ·mbo). *slang.* [f. Comb(ination + -o.] **1.** Combination, partnership (in various senses). *U.S.*
1929 Dunning & Abbott *Broadway* iii. 113 [of two dancers] We'd make about the best combo I could imagine. **1931** *Amer. Speech* VII. 105 *Combo,* the combination of safe or vault. **1959** in *Ibid.* (1962) XXXVII. 79 In describing the first poetry–jazz concert at Nebraska University a newspaper reporter referred to the reading of poetry to a jazz accompaniment as a combo. *Ibid.,* Potluck Supper on Slate? Take chicken-rice combo. **1963** *Ibid.* XXXVIII. 156 Recently, I heard a television commercial in which a woman shopper in a supermarket was heard to exclaim enthusiastically, 'Me and Tide—some combo!' **1963** R. I. McDavid *Mencken's Amer. Lang.* xi. 717 Specially made tools to..pull the combo [in safe-breaking].
2. A white man who lives with an Aboriginal woman. Also *co·mboman. Austral.*
1926 K. S. Prichard *Working Bullocks* v. 47 Combo's what they call a man tracks round with a gin in the nor'-west. **1934** *Times Lit. Suppl.* 24 May 377/3 'Comboman' is the name given in Central Australia to a white man who associates with aboriginal women. **1939** X. Herbert *Capricornia* ii. 13 They considered the men who sought the love of lubras—such men were called Comboes—unspeakably low. **1944** W. E. Harney *Taboo* (ed. 2) 59 Detribalized—the word makes me sick...Perhaps it is because I am an old 'combo', a man whose sympathy for these people goes down deeper than the paltry money that can be gained from their toil.
3. A small instrumental band; = *combination 4 d. orig. *U.S.*
1935 *Metronome* (N.Y.) May 28/3 As a soft fiddle-sax combo, it clicks. **1958** Steinbeck *Once there was a War* (1959) p. xix, One of the finest jazz combos I ever heard. **1958** *New Statesman* 23 Aug. 221/1 Mr Buck Clayton..feels that the original ideas and speed of decision required by the small 'combo' make for more interesting playing. **1970** *N.Y. Times* 27 July 20/1 The Conspiracy is a chatty three-guitar combo that sings songs and makes jokes.

comboloio (kǫmbǒlō͞u·yo, kǫmvǒloi·yo). [ad. mod.Gr. κομβολόγιον rosary.] A Moslem rosary of ninety-nine beads; also, a similar string of beads used to occupy the fingers.
1813 Byron *Br. Abydos* ii. v, And by her comboloio lies A Koran of illumined dyes. **1830** Galt *Byron* (ed. 3) xii. 83 In his left hand he held a string of small coral beads, a comboloio which he twisted backwards and forwards. **1910** *Encycl. Brit.* XII. 431/2 The Turkish practice of carrying a string of beads or rosary (*comboloio*), which provides an occupation for the hands, is very common. **1968** F. Norman *Hounds of Sparta* ix. 58 Worry beads clicked incessantly through the fingers of the passengers...But this incessant clicking of the *gomboloia* struck him as positively craven.

comb-out. [f. phr. *comb out*: see *comb *v.*[1] 6.] An act of combing or clearing out; *spec.* an act of combing out the hair.
1919 *Cassell's New Eng. Dict.* Suppl., *Comb-out* (slang), a thorough clearing out or clean sweep of men of military age in an office, works, etc. **1928** *News of the World* 29 Jan. 1/1 Chief-insp. Berrett, of Scotland Yard,..supervised a comb-out of all the motor-bandits. **1944** *Ann. Reg. 1943* 78 A comb-out of the Civil Service might yield rich results. **1964** *Saturday Rev.* 8 Feb. 10/3 She considered not setting her hair. Maybe she'd run in for a comb-out after breakfast. **1966** J. S. Cox *Dict. Hairdressing* 38/1 *Comb-out.* (1) The act of combing the hair after setting and drying. 'The comb-out is not to be hurried.' (2) The act of combing a head of hair as a preliminary to any further attention such as a shampoo or haircut. **1966** *Vogue* 1 Oct. 184/2 For comb-outs after a party the charge is 7s. 6d. for short hair.

combretum (kǫmbri·tǝm). [mod.L. (Linnæus *Genera Plantarum* (1737) 308): see Combretaceous *a.*] Any member of the genus of woody plants so named.
1819 *Curtis's Bot. Mag.* XLVI. 2102 This plant was first recorded as a Combretum by Lamarck. **1829** Loudon *Encycl. Plants* 1068 Combrētum and Quisquālis are among the most splendid of the climbing plants of the tropics, adorning the trees from which they depend with garlands of white and crimson, and yellow. **1929** F. W. H. Migeod in *Times* 19 Dec. 15/6 This same land risen 700 ft. above the sea, now bears only grass with small trees, included among which are combretums, acacias, euphorbias, [etc.]. **1962** N. Carr *Return to Wild* v. 46 The running fight had by this time shifted into a thick patch of combretum bush and it was difficult to follow its progress.

combs (kǫ·mz), *sb. pl.* Also **coms.** Short for *combination garments.*
1931 G. Baker *Ebenezer walks with God* xv. 225 I'll keep me combs on, so it'll be orlright if any cove kims. **1960** *Guardian* 8 June 7/3 The Tyneside dictum about male ballet dancers—'dancing about in their combs'. **1964** 'J. & E. Bonett' *Better Dead* xvi. 170 She always had a pair of 'coms' for me that her eldest boy had grown out of.

combustibly (kǫ̆mbʊ·stibli), *adv.* Delete †*Obs.* and add later example.
1857 Mundy *Antipodes* (ed. 4) 209 Which despatch..fell like a bomb-shell among the combustibly-disposed public.

combustion. Add: **2.** *internal combustion engine,* etc.: see *internal *a.* 5.
6. combustion chamber, (*a*) a space behind a furnace in which the hot gases from a boiler-grate become consumed; (*b*) in an internal combustion engine, the space in or above the cylinder where the charge is compressed and ignited; also *attrib.*; (*c*) (see quot. 1950).
1854 R. S. Burn *Steam Engine* iii. 90, *d* is the fire-door, *ff* the combustion-chamber. **1854** *English Mechanic* LXI. 271/1 The combustion chamber allows of the mixture of the gaseous products of the two fires. **1888** *Lockwood's Dict. Terms Mech. Engin.* 80 *Combustion chamber,* that portion of a boiler flue in which the hot gases are burnt. **1908** *Westm. Gaz.* 1 Oct. 4/2 Carbon deposits from the piston or combustion-chamber walls. **1937** *Discovery* Sept. 269/2 The proper construction of combustion chambers. **1950** *Science News* XV. 80 The third major item is the combustion chamber, which may include some means of lighting the propellants and certainly some means of injecting them; this chamber is the limiting factor in rocket design at the moment.

come, *v.* Add: **1. c.** Phr. *let 'em* (or *them*) *all come*: a formula expressing confidence in face of a challenge by others.
[**1739** J. Wesley *Let.* 7 May (1931) I. 306 No; let them all come; let all the world see the judgement of God.] **1899** *Captain* I. 121 (*caption*) Let 'em *all* come. **1912** Kipling *Diversity of Creatures* (1917) 40 'The rest will be coming along to-morrow.' 'Let 'em all come!' said Vincent. **1914** C. Knight *Here we Are!* (*song*), Are we downhearted? No! let 'em all come! **1921** H. Williamson *Beautiful Years* 214 What do you and I care for keepers, eh? Nothing at all. Let them all come. I'll show 'em what it means to hurt my birds.
4. d. Phr. *to have come to stay*: see Stay *v.* 8 c.
9. b. In pres. pple., due or properly accruing *to* one; deservedly falling or happening; esp. in phr. *to have it coming* (*to one*), to deserve what one gets or suffers. *colloq.* (orig. *U.S.*).
1793 in *Maryland Hist. Mag.* (1911) VI. 356, I am satisfied that there is something considerable comming to me in the Limekills account. **1888** *Detroit Free Press* 5 May (Farmer), A half-dollar was coming to me in change. **1896** Ade *Artie* v. 45 You kind o' feel there's a crack comin' to him. **1904** S. E. White *Blazed Trail Stories* i. 16 Dicky Darrell's got it coming. **1911** R. D. Saunders *Col. Todhunter* v. 77 Don't you worry about their not getting what's coming to them. **1916** H. L. Wilson *Somewhere in Red Gap* ix. 375. I got the long night's rest that was coming to me and started out early. **1926** W. S. Maugham *Constant Wife* ii. 119 I'm for it and I'm prepared to take what's coming to me. **1937** T. Rattigan *French without Tears* i. 12 Poor Babe. But he had it coming to him. **1957** *Economist* 28 Dec. 1121/2 Tin Pan Alley, Fundamentalism, and the Man-cult had it coming to them.
13. b. Of utterance: to issue from the mouth or the pen.
1735 Pope *Ep. Arbuthnot* 128, I lisp'd in numbers, for the numbers came. **1887** Morris *Odyss.* x. 246 But though for speech he was striving yet never a word would come.
c. To take (a required) shape.
1877 Mrs. Oliphant *Young Musgrave* x, Mrs. Pennithorne..failed entirely with Mary's frock. It would not 'come' as she wanted it to come. **1896** *Daily News* 5 Nov. 7/1 The Venus was..too stunted, and when..the painter attempted to drape her, the result would not 'come' well.
d. To belong to some recognized type, to exist. Freq. in phr. *as* —— *as they come,* i.e. as —— as such (types) are known to exist.
1919 F. Hurst *Humoresque* 64 And you're a good man—they don't come no better. **1925** *New Yorker* 21 Feb. 22/1 A play as nearly perfect as they come. **1936** Wodehouse *Laughing Gas* v. 62 It's his sister Beulah. She was the one who put him up to it. She's the heavy in the sequence. As tough as they come. **1954** C. P. Snow *New Men* 280 If you want a competent administrator I expect he's as good as they come.
14. b. Of seeds: to germinate when sown; hence, to grow.
1892 *Field* 7 May 665/3 The barley had come remarkably well, and had shot about an inch high.
16*. To experience sexual orgasm. Also with *off. slang.*
a **1650** *Walking in Meadow Green* in *Bp. Percy's Loose Songs* (1868) Then off he came, & blusht for shame soe soone that he had endit. **1714** *Cabinet of Love,* Just as we *came,* I cried, 'I faint! I die!' *c* **1888–94** *My Secret Life* III. 143 'Shove on,' said she, 'I was just coming.' **1922** Joyce *Ulysses* 489 Suppose you..came too quick with your best girl. *Ibid.* 752 Yet I never came properly till I was what 22. **1928** D. H. Lawrence *Lady Chatt.* x. 159 'We came off together that time,' he said. *Ibid.* xiv. 242 And when I'd come and really finished, then she'd start on her own account. **1963** D. Lessing *Man & Two Women* 35 Just as he decided, Right, it's enough, now I shall have her properly; she made him come. **1969** P. Roth *Portnoy's Complaint* 183 Did you warn her you were going to shoot, or did you just come off and let *her* worry?
19. b. *come day, go day*: applied to the conduct or character of one who is content to let time pass by without effort or trouble. Also as *sb.* orig. *dial.*
1616 T. Draxe *Bibliotheca Scholastica* 97/1 Come day, goe day, the day is long enough. **1854** A. E. Baker *Gloss. Northampt.* I. 175 It's come day, go day, with him. **1865, 1876** [see Come *v.* 26 d]. **1903** McNeill *Egregious Eng.* 174

The come-day, go-day Englishman. **1928** *Manch. Guardian Weekly* 22 June 494/2 Young Joe carried on in his come-a-day go-a-day God-send-Sunday manner. **1933** J. Masefield *Bird of Dawning* 227 Here are these four come-day-go-days wants to see you, sir. **1953** H. Spring *Sunset Touch* 12 Come-day-go-day people who had no solidity or substance.
23. b. Phr. (*when one*) *come*(*s*) *to think of it,* when one considers, remembers; on reflection.
[**1838** R. H. Froude *Remains* II. xiv. 179 When one comes to think of it abstractedly, it seems hardly conceivable, that any person should be so blind.] **1859** Faber *Spir. Confer.* 180 Perhaps an abrupt transition: but not so, when you come to think of it. **1875** L. M. Alcott *Eight Cousins* xii. 134 Come to think of it, she's only two years or so younger than I am. **1913** 'S. Rohmer' *Myst. Dr. Fu-Manchu* xviii. 199 'No,' he returned reflectively; 'come to think of it, neither did I.' **1926** R. Macaulay *Crewe Train* x. 196 Come to think of it, we've had a heavy day house-moving. **1927** C. Asquith *Black Cap* 100 What was printed there, staring up at her, was really very sad, come to think of it! **1943** E. M. Almedingen *Frossia* ii. 59 Come to think of it, I never knew why I married Hugo.
24. c. *to come clean*: see *clean *a.* 3 d; *come hell and high water*: see *hell *sb.*
d. colloq. fig. phr. *to come undone, unput, unstuck*: to become disintegrated, 'fall to pieces', meet with disaster, come to grief.
1899 *Sketch* 1 Nov. 84/1 The Oban 'boom' came badly undone... But let us suppose Oban had won, what would the other owners with horses in the race have said about the handicap? **1911** Kipling *Diversity of Creatures* (1917) 70 'Don't apologise,' said Gilbert, when the paroxysm ended. 'I'm used to people coming a little—unstuck in this room.' **1915** C. G. Grey *Tales Flying Services* 35 One of them [*sc.* seaplanes]..had just alighted astern, and was 'taxying' along to pick up her own boom when somehow the last remaining bomb 'came unput'—as one who was present said—and fell into the water. **1922** *N. & Q.* 9 Sept. 207/2 When a fancied horse, thought to be 'a good thing', fails to realize expectations, it is said to be 'a good thing come undone'. **1928** *Sunday Express* 3 June 11/3 Soon after this I came unstuck over a horse which the Prince of Wales, later King Edward, had very kindly given me. **1928** *Observer* 29 July 19/2 But with freak distributions of cards which justify high bidding, doubles that look sound often come 'unstuck'. **1928** *Daily Express* 12 Nov. 12, I thought my theory had come unstuck. **1936** *Punch* 14 Oct. 439/2 He had his chances, but no luck; He always managed to come unstuck. **1957** J. I. M. Stewart *Use of Riches* i. ix. 109 Here, in fact, was divorce-court stuff—call it the sort of situation that may develop with anyone once things come unstuck. **1958** *Listener* 16 Oct. 600/1 This is where the theory comes unstuck. **1970** *Guardian* 11 May 10/1 It is disturbing that on this occasion Mr Nixon should have come so unstuck.
26. e. Phr. *not to know whether* (or *if*) *one is coming or going*: to be in a state of mental confusion.
1924 R. Crothers *Nice People* ii. i. 122 I've fallen in with some awfully nice people and I don't mind telling you I don't know whether I'm coming or going. **1951** J. B. Priestley *Fest. Farbridge* ii. i. 183 There's nobody at the Town Hall could take it on. Town Clerk doesn't know whether he's coming or going. **1959** B. Kops *Hamlet of Stepney Green* i. 23 What with one thing and another, I don't know if I'm coming or going.
28. c. Delete [In French..that hand.] (Earlier and later examples.) Now freq. *to come it over* (a person); *to come the acid*: see *acid B. *sb.* 3; *to come the old soldier*: see Soldier *sb.* 2 b.
1827 *Mil. Sketch-Bk.* I. 30 Tickerly the guards: they try to come it over us venhever they have a tunity; but I'll let them know vhat's vhat. **1837** Dickens *Sk. Boz* ii. 24 The inimitable manner in which Bill Thompson can come the double monkey. **1867** Trollope *Claverings* II. i. 4 Miss Burton had been received..openly as Harry's future wife, and, 'by Jove, you know, he can't be coming it with Julia after that, you know.' **1889** E. Randolph *New Eve* II. viii. 23 He might..be inclined to come the gentleman, and pay for the same. **1916** J. Buchan *Greenmantle* i. 1 You'll be a blighted brass-hat, coming it heavy over the hard-working regimental officer. **1925** Fraser & Gibbons *Soldier & Sailor Words* 61 *To come it* (*to come the old soldier, to come the old man*), to attempt to shirk something. To try to bluff someone. Also, to be domineering. **1930** 'R. Crompton' *William's Happy Days* ix. 222 'Thinks he can come it over me,' muttered the lady angrily. ''Im an' 'is dukes!' **1930** R. Macaulay *Staying with Relations* xx. 292 The British working man resents any attempt to come the parson over him. **1934** G. B. Shaw *On Rocks* i. 190 If you are going to come this metaphysical rot over me I shall begin to wonder whether your appointment wasnt a mistake. **1935** Wodehouse *Luck of Bodkins* xi. 108 I'm not proposing to let any bimbo come the man of chilled steel over me just because I happen to kiss an old friend. **1939** A. Huxley *After many a Summer* i. viii. 98 When he saw..that no attempt was being made to come it over him, he had begun to take an interest. **1939** H. G. Wells *Holy Terror* iii. i. 207 The world's had this apostolic succession of oily old humbugs..trying to come it over people. **1962** C. Watson *Hopjoy was Here* ix. 96, I never thought he'd come the old green-eyed monster. **1962** *Spectator* 13 Apr. 467 Fancied he could come the old bland condescension over Erpf. **1969** *Private Eye* 25 Apr. 12 Now look here Eric don't come the raw prawn with me. **1970** D. Halliday *Dolly & Cookie Bird* iii. 34 The great sentimental idiot, coming that over me.
29. (Earlier example.) Also *to come it* (U.S. examples).
1825 *Spirit of Public Jrnls. M.DCCC.XXIII* 27, I wish this fellow to say how he got hold o' my checque for three hundred..let him come that, and I shall be satisfied. **1840** *Hard Cider Press* (U.S.) 10 Oct. 2/1 Kent has come it...Kent has Kracked the Krown of King Martin in Maine. **1840** Haliburton *Clockm.* III. 105, I couldn't come it. **1849** C. Lanman *Alleghany Mts.* xi. 89 The fellers laughed at me and said I couldn't come it. **1866** J. C. Gregg *Life in Army* xv. 141 Feeling secure from their voracious bills, as they hum around your room, and try to 'come it', but find an abatis in their way. **1908** *Dialect Notes* III. 300 He tried hard, but he couldn't quite come it.
34. *how come?*: see *how *adv.*
37. come across ——. *to come across* (*with*): see *across A. 3 b.
38. e. come at ——. To undertake, to take on, to get up to, to 'try on'. *Austral.* and *N.Z. slang.*
1919 Downing *Digger Dial.* 17 *Come at,* undertake. **1944** J. H. Fullarton *Troop Target* 95 Don't come at that, you Wog..bastard. **1949** D. M. Davin *Roads from Home* iii. 44 Barry won't come at it. **1960** B. Crump *Good Keen Man* 26 'Who'd get up to that sort of thing?' Jim looked sideways at me. 'Don't come at it too often,' he said. **1964** *People* 16 Dec. 45/1 A chesty little cockney bloke who'd come at anything. **1969** *Coast to Coast 1967–68* 164, I told you before I wouldn't come at that again. It's too risky.
39*. come for ——. To attack. Cf. 'come at' (Come *v.* 38 d), 'go for' (Go *v.* 58 e).
1890 *Blackw. Mag.* CXLVIII. 460/2 With a rush the hawk comes for him and misses.
41*. come off ——. *to come off it*: (usu. in *imp.*) don't go on like that, stop trying to fool me! *slang.* (Cf. *61 c.)
1912 A. Neil Lyons *Clara* xxvi. 283 Mrs. de Courcy Allendale requested me to 'come off it'. **1921** G. B. Shaw *Back to Methus.* ii. 49, I suppose I shouldn't say cheese it... Do come off it. **1930** W. S. Maugham *Cakes & Ale* 48 'Come off it, Roy,' I said. 'I'm too old a bird to be caught with chaff.' **1942** E. Waugh *Put out more Flags* iii. 187, 'I don't know what you mean,' she said... 'Oh, come off it,' he said. Angela came off it. She began to weep. **1948** J. B. Priestley *Linden Tree* 9 Oh, come off it, I'm not one of your hospital patients. **1969** *Listener* 3 Apr. 464/2 On which side was the preponderance of wealth, as of men and armaments? Do come off it, Mr Mansfield.
43. come over ——. **d.** Phr. *what has come over* (a person)?, why is (a person) behaving in an unusual way?
1836 *Blackw. Mag.* Mar. 391/1 'What's come over our little Fan?' exclaimed Mark Fairfeld, in a tone of perplexity and vexation. **1888** [see Dict.]. **1909** Galsworthy *Silver Box* iii, I asked him whatever came over him to do such a thing—and he said it was the drink. He said that he had had too much to drink, and something came over him.
45. come to ——. **f.** Phr. (*if it*) *come*(*s*) *to that,* if that is the case; in fact; anyway.
1923 Manchon *Le Slang* 90 *Come to that,* tiens! au fait! **1937** A. Thirkell *Summer Half* ii. 45, I could have lent him some of my pyjamas, if it comes to that. **1942** *Penguin New Writing* XII. 84 He was plumb scared of war. Come to that, so am I. **1964** R. Jeffries *Embarrassing Death* iii. 21 You don't want to be nice for this job—or for any other job, come to that.
i. *to come to light* (*with*), to produce, to come up with (money, etc.). *Austral.* and *N.Z. colloq.*
1917 *N.Z.E.F. Chrons.* 5 Sept. 28/1 We hit him up for a loan for weeks afterwards and he always came to light too. **1930** *Bulletin* (Sydney) 26 Feb. 57/2 'Will you come and help me buy the things they most need?' Uncle Bill had come to light without a doubt. **1936** F. S. Anthony *Follow the Call* vii. 83, I had to borrow £20..before I could come to light with the engagement ring. **1945** N. Marsh *Died in Wool* vii. 157 You come to light with them two hooks.
j. *What is* —— *coming to?,* a rhetorical expression of dismay or disgust.
1933 M. Lowry *Ultramarine* iv. 182 Christ, what's the ship coming to. **1938** 'G. Graham' *Swiss Sonata* 182 What's this school coming to?
50*. come across. See *across B. 2 b.
51. come again. d. An off-hand slang quasi-interrogative equivalent of 'what did you say?', 'I beg your pardon'. orig. *U.S.*
1884 G. W. Peck *Peck's Boss Book* 112 'My chum had a pain in the small of her back and she confided in me, and after diagnosing the case —' 'Come again, please,' said the old man, when she struck the college word. 'You *whicked* the case?' **1920** S. Lewis *Main Street* xxiii. 283 'He's a servant of reality.' 'Come again? Um. Yes.' **1933** M. Lowry *Ultramarine* iii. 165 'Ah, no savee sing Tipperlairley, hey?' 'Come again, brother.' 'No savee sing Tipperlairley?' 'Oh, Tipperary. Yes, yes.' **1956** 'A. Gilbert' *Death came Too* xvii. 180 Nurse Alexander startled them all by saying suddenly, 'No scones.' Crook turned. 'Come again, sugar?'
e. Of a horse in a race: to regain speed. Also *transf.*
1946 *Baltimore Sun* 21 Nov. 21/1 Many of the supporters of [a certain horse] were ready to throw away their tickets when the odds-on favorite gave up the command, but cheered loudly when he 'came again' to win going away. **1962** *Times* 9 Apr. 4/5 The masters conjured up their second or third wind and came again. **1965** *Observer* 30 May 34 *Came again,* said of a horse that has been passed by other horses close to home, but which then (by sheer stamina plus gameness) 'came again' to win.
53. come away. c. (Example.)
1881 A. Lang *Library* ii. 41 Three jets of gas..made the backs of books come away in his hand.
e. Delete [Now chiefly *dial.*]. (Later examples.)
1927 *Forestry* I. 18 Frequently after a period of years patches [of Spruce] come away, while the plants alongside are still in a state of check. **1950** *N.Z. Jrnl. Agric.* July 4/1 This type of feeding is continued until the spring pasture comes away and hardens up.
54. come back. a. Also, to return to consciousness (cf. quot. 1850 in Dict.).
1922 D. H. Lawrence *England, my England* (1924) 213 He worked a little longer. He could feel her live beneath his hands; she was coming back.
c. To retort or retaliate. *U.S.*
1896 Ade *Artie* vi. 54 Did you ever get the worst of it in such a way that you couldn't come back at the time? **1905** Tarkington *In Arena* 182 'Hello, Ben! I hear you're not *for* me!' he said cordially. 'How are you running?' I came back at him, laughing. 'Oh, we're going to beat you,' he answered. **1916** H. L. Wilson *Somewhere in Red Gap* vi. 267 'And what a sweet little home you'll build for the Wales family!' I says... But he wouldn't come back; so I left him surrounded by the wreck of his former smartness and went home. **1928** F. N. Hart *Bellamy Trial* i. 12 Just as I was thinking of something really bright to come back with, a nice soft little voice in the back of the hall said [etc.].
d. Of an athlete: to return to form; to regain the initiative during an event. Also *transf.*
1922 *Daily Mail* 22 Nov. 11 Since that time he has 'come back' with such certainty..that he must of necessity be regarded as on a level with all the other big men. **1928** *Daily Express* 12 July 12/7 Duncan remains a master of the art of 'coming back'. **1954** F. C. Avis *Boxing Ref. Dict.* 23 *Come back,* to attack an opponent vigorously after a period of accepting punishment. **1955** *Times* 2 May 5/7 The Scots, however, 'came back' excitingly through tries by Elgie and their stocky little half-back.
e. To return to a former state of popularity or vogue.
1929 *Times Lit. Suppl.* 9 May 374/4 The way in which the tulip has 'come back' as a garden flower. **1934** *Punch* 11 Apr. 393/3 A West-End barber denies that beards are coming back.
f. Of a guinea-fowl: to utter its cry (resembling 'come back'). Cf. Come-back *sb.*[1]
1892 *Leisure Hour* Dec. 143/1 His turkeys gobbled all day, his guinea-fowls 'come-backed'.
56. come down. a. Also (of rain) to fall heavily; (of fog) to 'settle'.
1854 Dickens *Hard Times* iii. ii. 269 It came down as I never saw it come down before. **1885** [in Dict.]. **1891** *Longman's Mag.* July 238 The fog has come down as black as pitch. **1934** H. Miller *Tropic of Cancer* (1948) 245 It began to come down in bucketsful.
g. Phr. *to come down* (*on* or *upon*) (a person) *like a ton of bricks,* to reprimand or punish (a person) severely. *colloq.*
1937 in Partridge *Dict. Slang* 172/1. **1938** G. Greene *Brighton Rock* ii. ii. 91 If there's any fighting I shall come down like a ton of bricks on both of you. **1950** J. Cannan *Murder Included* vii. 174 When I mentioned it to the others, they come down on me like a ton of bricks. **1964** *New Society* 20 Feb. 8/3 Nobody's interested until you actually break the law then they come down on you like a ton of bricks.
h. Earlier examples of *come down with.*
1728 J. Gay *Beggar's Opera* iii. i. 39 Did he tip handsomely? How much did he come down with? **1773** Goldsm. *Stoops to Conq.* i. ii. 10 When you come down with your pence, For a slice of their scurvy religion. *a* **1817** Jane Austen *Persuasion* (1818) IV. x. 227 Money, you know, coming down with money—..it cannot be a very agreeable operation.
i. To be removed from its position, esp. (of a tree) to be felled, (of a building) to be demolished.
1844 *Jrnl. R. Agric. Soc.* V. i. 109 Large numbers of oaks have recently been felled, and many more are marked to come down.
j. Of a river: to flow in flood. *Austral., N.Z.,* and *S.Afr.*
1863 S. Butler *First Year Canterbury Settl.* vii. 83 The river had come down the evening on which we had crossed it, and so he had been unable to get the beef or himself home again. [**1867** *Queenstown Free Press* 18 Jan. (Pettman), The rivers in this neighbourhood have been frequently down during the last month.] **1937** E. Hill *Great Austral. Loneliness* xxxiii. 255 It is twenty years since the Cooper [river] has come down across the bar of the Strzelecki, good seasons following its regular floods. **1951** L. G. D. Acland *Early Canterbury Runs* xi. 327 [He] was grazing a mob of sheep there when the Bowyer [River] came down in flood and drowned most of them. **1955** J. H. Wellington *S. Afr.* I. iii. xiv. 459 From the Kuiseb to the Cunene most of the rivers 'come down' two or three times a year, but there is no regularity in the flow, except that it occurs in the summer when rain falls on the plateau.
k. *to come down to*: to be (basically) a matter of; *to come right down to it*: to get to fundamental facts or principles.
1891 H. C. Bunner *Zadoc Pine* 74 'Tain't much better, when you come right down to it. **1902** W. D. Howells *Lit. & Life* 212 A good..donkey would be worth all their tribe when it came down to hard work. **1931** L. Steffens *Autobiogr.* ii. xxxvi. 616 He did not—when it came right down to it, he would not—give me his evidence against individuals. **1959** J. L. Austin *Sense & Sensibilia* (1962) iv. 33 The expressions in question actually have *quite* different uses...Not always, certainly—there are cases..in which they come down to much the same.
l. *to come down with,* to become ill with (a specified disease). orig. *U.S.*
1895 Wood *Yale Yarns* 77 (D.A.), The good Deacon almost feared he was about to come down with a fever. **1911** J. C. Lincoln *Cap'n Warren's Wards* ix. 147 The housekeeper felt sure he was 'coming-down' with some disease or other. **1961** R. & C. Winston tr. *T. Mann's Genesis of Novel* viii. 83, I was suffering from a severe headache, and the following morning I came down with a *grippe* that attacked stomach and intestines.
59. come in. a. Also, *spec.* to come to work in a house.

1930 A. Bennett *Imperial Palace* xliv. 313 The charwoman who 'came in' for half a day on alternate days.

b. Also, to enter as a partner (in a company or on an enterprise).

1843 Dickens *Mart. Chuz.* xxvii. 334 Ha, Ha! Join us. You shall come in cheap. 1923 Wodehouse *Inimit. Jeeves* xiv. I came..to ask if you would care to come in on another little flutter. 1936 L. C. Douglas *White Banners* x. 220 I'll come in on it..and help. 1953 R. Lehmann *Echoing Grove* 310 She came in on a campaign..I was helping to organize.

d. Also, to be in the last stage of a run. (Cf. *come-in sb.)

1857 Hughes *Tom Brown* vii. The whole hunt is out of ear-shot, and all hope of coming in is over.

k. Phr. *this, etc., is where we came in*: our knowledge dates from this point; we are back to where we started.

1949 Wodehouse *Uncle Dynamite* vii. 114 Now we're back where we started. This is where we came in. 1966 *Guardian* 24 Sept. 6/5 There was an old saying about the continuous movie programme, 'This is where we came in.' It became gradually part of the language with more general application. 1967 H. Howard *Routine Investigation* ii. 19 'Anything else you want to ask me?' 'No. This is where I came in. Tomorrow we'll go call on a few people who may have a few answers.'

o. Phr. *to come in for it*, to incur punishment, or a rebuke. *colloq.*

1841 Dickens *Barn. Rudge* i, You'll come in for it presently, I know you will! 1864 — *Mut. Fr.* I. ii. viii. 135 Unfortunate Lavvy. She always comes in for it. 1880 H. Lapham in D. M. Davin *N.Z. Short Stories* (1953) 66 Poor Brennan came in for it,..a severe lecturing, as well as to be reduced to the rank of constable.

q. Of a cow: to calve. *dial.* and *U.S.*

1705 W. L. Grenville *Lawyer's Fortune* i. i, There's a young one comes in this Year. 1838 H. Colman *Mass. Rep. Agric.* 60 He gives as opinion..that the heifers which 'come in' with their first calf at two years old, do better than when their coming in is delayed until three years old. 1857 [see *coming vbl. sb.¹ 7 d]. 1874 *Rep. Vermont State Board Agric.* II. 491 He has his cows come in usually in April. 1863 *Country Gentleman* 2 Jan. 6/3 Hence the object is to have all the cows 'come in' near the commencement of the butter or cheese making season. 1956 *N.Z. Jrnl. Agric.* May 485/1 But sows farrowing in May, and their litters too, must be fed until the cows come in.

r. Of a radio operator, etc.: to begin speaking.

1958 'N. Shute' *Rainbow & Rose* i. 18 The announcer said, '..? KZ, if you are listening, will you come in, Mrs. Hoskins.' 1966 *Listener* 17 Nov. 717/3 Mr Aubrey Jones, would you like to come in first, having heard Mr Cousins? 1970 H. Johnston *Black Camels* xii. 184 The loudspeaker spoke in the wireless truck. 'Gumbus One to White Chief..' 'Come in Gumbus One.'

61. come off. c. Also, to 'give over'; to stop talking. (Cf. 41* above.) *U.S. slang.*

1870 J. J. McCloskey *Across Continent* in *Amer. Lost Plays* (1940) IV. 95/1 Oh, come off, Joe. 1889 *Century Dict.*, *Come off*, to cease (fooling, flattering, chaffing or humbugging); desist: chiefly in the imperative: as, oh, come off! (Recent slang, U.S.) 1892 *N.Y. Mercury* Feb. (Ware), 'How much does yer ax for this book?' 'Six dollars,' replied the smiling clerk. 'Six dollars! Oh, come off!' 1904 S. E. White *Silent Places* xiii. 139 Now you treat her decent and you treat me decent. It's time you came off. 1904 W. H. Smith *Promoters* xx. 293 [illegible] 'Oh, come off!' Golsby replied. 1906 E. Dyson *Fact'ry 'Ands* v. 56 Come off! She don't love Dave. 1912 E. C. Bentley *Trent's Last Case* v. 119 'Come off!' exclaimed Trent bitterly. 'What do I care about his story?..I want to know how you know he went to Southampton.' 1929 W. Smyth *Girl from Mason Creek* vi. 64 Oh, come off, Thomson.

g. (Later example.)

1813 Jane Austen *Let.* 23 Sept. (1950) 335 They talked of going one, but I came off with a dose or two of calomel.

m. To have (a certain) success; and *absol.*, to be successful; to result in success. Cf. sense 61 f in Dict.

1864 Crown Princess of Prussia *Let.* 4 May in R. Fulford *Dearest Mama* (1968) 327 The Armistice does not seem to be coming off—the King and every one seem very anxious for it here. 1865 F. Lillywhite *Guide to Cricketers* 128 [He] has been known to 'come off' as a change bowler. 1874 Trollope *Way we live Now* (1875) I. xxx. 194 I'm afraid you didn't make much of Mr. Brunette,..it just didn't come off. 1883 *Graphic* 11 Aug. 138/2 Batting is his forte, but this year he has not 'come off' yet. 1904 *Daily Chron.* 11 July 3/2 Atmosphere is here; knowledge is here; graphic style is here. But..it does not in the telling language of the studio 'come off'. 1929 *Times Lit. Suppl.* 2 May 360/1 As a theologian he comes off no better, for against a stanza in which the Virgin responds to the Annunciation of Gabriel is placed the side-note 'The Immaculate Conception'! 1966 *Listener* 8 Sept. 366/2 Another fascinating original..appears to be about a man in hell. I am not sure that it entirely came off.

n. [See sense 16* above.]

63. come out. i. Also *ellipt.*, of certain games of cards.

1909 H. G. Wells *Ann Veronica* xv. 320 'I believe after all it's coming out!' said Miss Stanley. 'The aces made it easy.' 1934 A. Huxley *Beyond Mexique Bay* 191 He plays three games of mystical solitaire. If all three games 'come out', that means that the patient is certain to get well. 1939 R. Lehmann *No More Music* I. ii. 40 (He sweeps the cards into a tangle.) Anyway, it's not coming out. Give it up. 1953 A. Christie *Pocket full of Rye* xviii. 123 'Just wait a minute,' said Miss Ramsbottom. 'This Patience is going to come out.'

j. *Photogr.* (Examples.)

1891 *Longman's Mag.* Dec. 195 He was seized with curiosity to see whether Mrs..'came out' as badly in a photograph as she did in a letter. *c* 1901 R. A. Knox in E. Waugh *Life* (1959) I. iii. 61, I am dying to know how your photograph of me..has come out in printing. 1941 E. Bowen *Look at Roses* 53 Do you think that photo will ever come out?

65. come out (with). Phr. *to come right out with*: to blurt out a remark; to speak frankly or tactlessly. Also *ellipt.* *N. Amer. colloq.*

1871 H. James *Watch & Ward* (1960) 43 Can you? Do you know what he intimated? Indeed, he came right out with it. 1928 *Sat. Even. Post* 7 Jan. 8/1 Why don't you come right out and say how it happened? 1934 J. O'Hara *Appt. Samarra* (1935) vi. 99 Well, Kitty, you know how she is. Comes right out with it. 1941 L. Baird *He rides Sky* 47 Hey, Mom, come right out and tell us just where you bought that millinery mayhem or did you rob a grave? 1966 *Toronto Daily Star* 1 Mar. 39/1 After considerable hinting, she finally came right out and asked for the painting.

66. come over. f. (Examples.)

1882 'R. Crompton' *Just—William* xi. 223 'I just—just' came over queer,' he ended, remembering a phrase he had heard used recently by the charwoman. 1938 Wodehouse *Laughing Gas* iii. 31 You can't kick a bloke off property unless he's come over a bit mid-Victorian. 1938 E. Bowen *Death of Heart* II. vii. 303 Your marvellous yellow coat might make me come over queer. 1958 R. G. Collingwood *Princ. Art* v. 83 Sound knockabout entertainers, suddenly come over all solemn. 1960 N. F. Simpson *Resounding Tinkle* ii. 118 There was nothing wrong with him..and then next day he came over funny at work.

g. To succeed in conveying one's meaning or in creating a particular impression.

1963 *Guardian* 26 Jan. 3/2 In the novel the dissolute lover 'came over' as much more than a debaucher. 1968 I. Lambot *Queen dies First* xii. 88 He's a damned good administrator, but somehow, he doesn't come over, as a person. He takes a lot of getting to know.

67*. come through. To succeed, attain an end; *spec.* to obtain conviction. Chiefly *U.S.*

1882 H. W. Pierson *In Brush* 172 They could scarcely speak for hoarseness—enjoyed seeing them 'come through' (the vernacular for conversion). 1886 A. Daly *Playwright's Dau.* xvi, You will do as I tell you, and, please God, shall come through without a scrape. 1912 Mathewson *Pitching* ii. 33, I have been told that there's not the most relieved man in seven counties when O'Toole came through with that victory in Boston. 1923 Mrs. Stratton-Porter *Laddie* xi. 355 Leon said our house reminded him of the mourners' bench before anyone had 'come through'. *Ibid.* xv. 494 Pretty soon it began to look like she was going to come through as Amos Hurd did when he was redeemed. 1917 D. H. Lawrence (*title*) Look! We have come through!

b. To act as desired or expected; to provide required information, money, etc. (see quot. 1914).

1914 Jackson & Hellyer *Vocab. Criminal Slang* 25 *Come through*,..to give up, to deliver, to surrender any secret information or any material goods demanded. 1949 J. B. Priestley *Home is Tomorrow* i. 2 *Dayton.* Can't tell you yet. *Jill.* When can you? *Dayton.* Soon. Depends on you, though. When you come through, then I'll pay this. [illegible] I had to twist his arm a bit but he came through.

c. To emerge, to appear; to succeed in giving a favourable impression.

1947 L. Hastings *Dragons are Extra* ix. 200 An attractive personality that came through. 1969 *Listener* 10 May 731/2 The picture of Lyndon Johnson as a reactor, rather than initiator, comes through well. [illegible]

69. come up. a. *spec.* To present oneself before a judge or tribunal for (rarely to) judgement.

1888 'F. Warden' *Woman's Face* II. xvii. 171 He felt as if he himself had come up to judgement before a stern and [illegible] judge. 1958 [illegible] The order to come up for judgement if required.

b. *to come up smiling*: to recover from a round in a boxing-match and face what is to come cheerfully; also *transf.* and *fig.*

1866 [see Dict.]. 1928 *Daily Tel.* 13 Mar. 16/3 His car..can be rubbed down (by..and, as the phrase goes, it will come up smiling. 1953 Wodehouse *If I were You* iv. 51 You come up smiling after having a whacking great car run into you.

c. Also *fig.* in phr. *to come up with*: to get even with, get the better of. *U.S.* Phr. *to come up with the rations*: see *ration sb.

1856 *Harper's Mag.* XII. 701/1 [illegible] 1871 Mrs. Stowe *Oldtown Fireside Stories* 180 He thought he must have his say with Miry, but he got pretty well come up with. *Ibid.* [illegible] 1875 Susan Hale *Let.* (1919) 112 [illegible] 1901 S. E. White *Westerners* xi. 78 Revenge with him seemed to be [illegible] the victim's realization that he was being come up with.

m. Colloq. phr. *to come up with*: to produce, provide, present. orig. *U.S.*

[illegible] 1955 *Times* 9 Aug. 6/5 President Eisenhower [illegible] Secretary of Agriculture soon would be 'able to come up with something'. 1958 *Economist* [illegible] 1966 *Listener* 27 Oct. 631/3 There is a certain erratic charm in BBC-2's *Europa*. It can come up with anything. 1970 J. Porter *Rather Common Sort of Crime* ii. 19 I'll have a ponder about it and, if I come up with anything, I'll let you know.

n. *coming up*, a colloq. phr., used by bartenders, cooks, waiters, etc., to indicate that food or drink is ready or being prepared.

1942 Berrey & Van den Bark *Amer. Thes. Slang* 765 *Come and get it, coming up, on deck, take it away*, the cook's announcement that a meal is ready. 1968 L. Deighton *Only when I Larf* xiv. 183 'Perhaps a coca-cola.' 'Coca-cola coming up,' said Spencer. 1968 [illegible] 'Coming up, sir,' Harry said.

come, *sb.*¹ Add: **4.** Applied to a flow or flood of water. ? *local.*

1862 *Temple Bar* V. 110 The plain is not often adopted, except where the 'come' of water is rather near to the surface. [illegible] It appeared to me there was quite a come of water in Scupper Flats.

come-all-ye, -you. (See quots.)

1898 *Eng. Dial. Dict.*, *Come all ye's*, old ballads or country songs. Ulst[er]. 1914 Joyce *Dubliners* 35 Street-singers, who sang a *come-all-you* about O'Donovan Rossa. 1938 J. Cary *Castle Corner* 418 Hanna still sang his ballads, with even more sentiment than before, and when he was almost fearful, he would say, 'Give me a come-all-you, to promote the flow of the vital juices.' 1941 L. A. G. Strong *J. McCormack* xxi. 260 He..the 'Come-all-ye's'—that is, ballads beginning with an invitation to come and listen—with as many as forty or fifty verses. 1961 [illegible] in *Faber Bk. Ballads* 21 James Joyce, who describes the nasal whine of the singers of 'Come-all-you's', quotes from street-ballads throughout his work.

come-and-go (kʌm·ændgōᵘ). [f. phr. *come and go* (see Come v. V), partly after F. *va-et-vient*.] Passing backwards and forwards. Also *attrib.*

1793 W. B. Stevens *Jrnl.* 6 July (1965) i. 90 Caught a little come-and-go sleep and at 7 took Coach for York. [illegible] 1894 G. du Maurier *Trilby* I. 16 [illegible] a perpetual come-and-go in this particular studio between four and six. 1922 D. H. Lawrence *England, my England* (1924) 95 There was plenty of life in the little goods-yard; three porters, a continual come and go of farm wagons bringing hay.

co·me-back, *sb.*² Add: (Earlier and later examples.) Also **comeback.** [Come v. 54.]

1. a. An act of retaliation. orig. *U.S.*

1896 Ade *Artie* vi. 59, I never will be able to give him the right kind of a hot come-back for what he done to me. 1920 C. E. Mulford *J. Nelson* xvi. 186 No cussed man can spy on me without riskin' a comeback. [illegible] 1969 *Spectator* [illegible] without fear of a come-back.

b. A verbal retort; a reply. orig. *U.S.*

1889 *Kansas Times & Star* 9 Nov., He shouldn't thus invite a sarcastic come-back. [illegible] He didn't have no come-back to that. [illegible] 1933 Wodehouse *Inimit. Jeeves* [illegible] Before I had time to formulate any answer as to what Millicent's come-back had been to that remark, the door opened. 1933 *Punch* 9 Aug. 161/2 It is..even more impossible to find any suitable come-back to the remark, 'This will be the bathroom' or 'Evadne's room' or 'the coal-hole', as the case may be—they all look very much alike in embryo to the wretched listener.

2. *colloq.* (orig. *U.S.*). A recovery; a return to a former state of health, prosperity, etc.; *spec.* a return to one's former position; a reinstatement in a position of authority or power; esp. in phr. *to make* or *stage a come-back*, to achieve a success after retirement or failure.

1908 K. McGaffey *Sorrows of Show Girl* 224 But it is a good thing to have a back account to flash, so that the boob will think he will get a comeback if he does lose. 1920 F. S. Fitzgerald *This Side Paradise* (1921) I. iii. 103 'With a little effort you could still stage a come-back.' 'No—I'm through.' [illegible]

3. A person who has returned; also, a ghost.

1908 Masefield *Capt. Margaret* xi. 327 'Is that the *Happy Return*?' 'We're the jolly come-backs.' 1955 W. de la Mare *Beginning* [illegible] at least ten minutes to get round, to the front door.

4. A sheep three-quarters merino and one quarter crossbred. Also, the skin or fleece of a sheep of this type. *Austral.* and *N.Z.*

1891 R. Wallace *Rural Econ. Austral.* 360 When a pure Merino ram is put to a cross ewe the produce is termed a 'come-back' or 'quarter-back'. 1910 C. E. W. Bean *On Wool Track* ii. 31 A cross between a Shropshire and a merino is also a crossbred; but a cross between this last and a merino comes so near to the Australian merino..that he is

called a 'come-back'. 1928 *Daily Tel.* 28 Aug. 3/6 Greasy merino realised up to 16d, comebacks 21¼d. 1955 *Times* 6 May 19/3 Comebacks and crossbreds met very keen competition.

co·me-between, *sb.* One who or a thing which intervenes.

1893 *Black & White* 15 Apr. 448/2 Popery..and Independency..have each their vindication, but not the miserable make-shifts and come-betweens that find so much favour nowadays. [illegible]

comedial (kǫmī·diăl), *a.* [f. L. *cōmœdia* Comedy + -al.] Of or pertaining to comedy.

1908 C. W. Wallace *Children Chapel* 168 Or did he simply find good comedial material here ready for further development?

comedian. Add: **1. c.** A professional entertainer who makes his audience laugh by telling jokes, acting foolishly, etc. Also *transf.*

1898 *N.Y. Jrnl.* [illegible] 1926 *Amer. Speech* I. 437/2 In vaudeville and burlesque a straight man works as a 'feeder' with the comedians. 1945 L. Lane *How to become a Comedian* i. 11 The majority of comedians I have met are very seriously minded in private life: it's a relief from their trade of laughter-making. *Ibid.* xv. 119 A double act is really a light comedian and a low comedian working together. The finest example we have to-day is that of Flanagan and Allen. 1953 K. Reisz *Technique Film Editing* v. 102 Comedians like Bob Hope or Danny Kaye..rely to a very large extent on purely verbal wit. They crack their jokes at a very rapid pace [illegible] sense of timing. 1968 *Who's Who* 2530/1 Ray, Ted; Theatrical and B.B.C. Entertainer; s. of Chas. Olden, comedian.

comedic, *a.* Delete *rare* and add quots.

1639 R. Baillie *Lett. & Jrnls.* (1841) I. 223 This might be the comedick catastrophe of our tragick Pulpit [illegible] 1905 *Pall Mall Gaz.* 24 Apr. 3/2 Miss Ethel Irving has conquered the musical comedy public, and that by comedic rather than by musical means. 1920 M. Beerbohm *And Even Now* 316 Falstaff is a triumph of comedic creation. 1964 *Parade* [illegible] The wedding, which took place at Caxton Hall, was comedic.

Hence **come·dically** *adv.*

1930 *Punch* 30 Apr. 498 The acceleration of the pace..leaves some of the conspirators realistically rather than comedically.

‖**comédie** (komedi). [Fr., see Comedy¹.] The French word for 'comedy', used in certain phrases.

comédie humaine (ümęn), the title given by Balzac to his series of novels; hence, the sum of human activities, or a literary portrait of the same.

[1842 H. de Balzac *Œuvres Complètes* (1855) I. 1 En donnant à une œuvre entreprise depuis bientôt treize ans, le titre de *La Comédie humaine*, il est nécessaire d'en dire la pensée, d'en raconter l'origine [illegible]] [illegible] *comédie humaine*. [illegible] In the power of exploring the *comédie humaine*, Dickens had no superior anywhere. [illegible]

comédie larmoyante (larmwayant) [lit. 'tearful comedy'; orig. applied to plays of P. C. Nivelle de La Chaussée (1692–1754); cf. Larmoyant a.], a sentimental, moralizing comedy of a kind fashionable in eighteenth-century France.

1801 M. Edgeworth *Belinda* II. xiii. 30, I advise you to write a sentimental comedy, a *comédie larmoyante*. 1966 *Listener* 17 Feb. 256/3 *The Thieving Magpie*..is essentially a *comédie larmoyante*.

comédie noire (nwar), a macabre or farcical rendering of a violent or tragic theme; cf. *black comedy* (*black a. 8 b).

1958 *Times* 11 Dec. 13/5 This diverting comédie noire. 1964 *Guardian* 8 July 7/6 Of course the car and plane crash and Vesuvius erupts. That's satire, see, comédie noire.

comédienne. Add: Also **comedienne.** (Examples.) Also, a female comedian.

1806 G. Vandenhoff *Dram. Reminisc.* v. 73 And the position was soon conceded to her of the first *comédienne* on the English stage. 1903 A. Bennett *Leonora* vii. 185 Miss Milly wants me to talk you round about her going in for the stage..I just happened to say I thought she was a born *comédienne*. 1945 L. Lane *How to become a Comedian* xviii, What grand comediennes Elsie and Doris Waters are, and Revnell and West—woe betide any comedian who has to follow them on any programme without an interval between. 1958 *Oxf. Mail* 22 July 6/8 Welsh comedienne Gladys Morgan was in fine form with her individual and eccentric brand of comedy. 1969 *Listener* 23 Jan. 125/3 Barbra Streisand is captivating as the 1920s Ziegfeld comedienne, Fanny Brice.

come-down, *sb.* Add: **1.** Now esp., a fall or drop in social or official position or status.

1840 [see Dict.]. 1887 H. H. Jackson *Between Whiles* i. 18 It was a sad come-down from his old air-castles for her [illegible] 1892 *Cornhill Mag.* May 485 They regarded [it] as somewhat of a 'come down' in the world. 1919 'Boyd Cable' *Old Contemptibles* xvii. 291 He joined the Volunteers, of course. It was a sad come-down. 1960 J. Rae *Custard Boys* i. viii. 74 'Quite a come-down, eh, John?' I glanced back at him and saw that he was smiling at his successful use of the idiom.

comedy¹. Add: **2. b.** *comedy of manners*, that kind of comedy in which the modes and manners of society are amusingly presented. Cf. Manner sb.¹ 4 e.

1823 Lamb *Elia* (1823) 323 The artificial Comedy, or Comedy of manners, is quite extinct on our stage. Congreve and Farquhar show their heads once in seven years only. 1877 G. Meredith in *New Quart. Mag.* VIII. 3 Our English idea of a Comedy of Manners might be imaged in the person of a blowsy country girl..transforming to a varnished City madam. 1913 J. Palmer *Comedy of Manners* i. 1 The English comedy of manners began with Etherege..and was finally extinguished in Sheridan and Goldsmith.

c. Humour; humorous invention; the action or quality of being amusing. Cf. *cut the comedy* (*cut v. 21 b).

1877 G. Meredith in *New Quart. Mag.* VIII. 9 Comedy is the fountain of sound sense. 1933 P. Godfrey *Back-stage* iii. 37 [illegible] 'laughs' have been [illegible] 1947 D. Lean in [illegible] *Working for Films* 29 The answer lies in a very old comedy maxim: 'Tell them what you're going to do. Do it. Tell them you've done it.' 1958 [see *comedienne].

5. *comedy-drama, -opera, -thriller, -tragedy.*

1904 W. D. Adams *Dict. Drama* I. 337/2 *Forget and Forgive*,..a comedy-drama in three acts, by John Daly Besemeres, first performed at the Charing Cross Theatre, London, January 5, 1874. 1904 *Times* 25 Apr. 6/1 This essay in Dickensian comedy-drama. 1906 *Daily Chron.* 29 Oct. 3/5 [illegible] a 'comedy-opera' [illegible] genuine and [illegible] Mr. Arthur [illegible] a tour. 1897 *Westm. Gaz.* 15 Mar. 3/1 There were no such things as [illegible] gorgeously bedizened [illegible] 1920 *Daily Chron.* [illegible] 6/6 Dickens makes the election a farce, Meredith treats it in a spirit of comedy-tragedy, Samuel Warren describes the contest with laboured realism.

come-hither (kʌm·hi·ðǝɹ), *sb. colloq.* [f. vbl. phr. *come hither* (see Come v. 33.)] An invitation to approach, so *fig.* enticement. Chiefly in attrib. use, with *look, eye*, etc. Also as *adv.* (Cf. Comether.)

1900 *Daily News* 20 July 6/2 It's no' the money, and it's no' the looks. It's just the come hither in the eye. 1926 S. Lewis *Mantrap* xii. 135 You can't expect a calf to bear [illegible] come-hither at every poor [illegible] 1927 *Amer. Speech* III. 135 [In Maine] a flirtatious one [sc. girl] had the 'comehither look', [illegible] 'a persuasive look'. 1930 [illegible] They keep a 'lady-help'..a dreadful middle-aged female with a come-hither look. 1932 [illegible] Broadway [illegible] 'come-hither' is Helen Gahagan. 1935 *Times Lit. Suppl.* 22 Feb. 183/2 [illegible] as 'a come-hither hellion'. 1944 Auden *For Time Being* (1945) Pref. 7 [illegible] come-hither eyes. 1958 R. Braddon *Nancy Wake* vi. 18 Grenadine would cast shameless come-hither looks at other gentlemen dogs in the vicinity, which enraged Picon.

So **come-hither** *adv.* and other (chiefly *nonce*) formations.

1923 [illegible] There was no come-hitherishness in their intercourse, ever; no approach to the man-and-woman attitude of which there had [illegible] 1938 W. S. Maugham *Summing Up* 48, I have none of that engaging come-hither-ness that enables [illegible] on first acquaintance. 1954 MacNeice *Autumn Sequel* [illegible] Forlorn: The very word come-hithered [illegible] 1953 Dylan Thomas *Under Milk Wood* (1954) 87 They make, in front of their looking-glasses, haughty or come-hithering faces for the young men in the street outside. 1962 [illegible] *London's* [illegible] come-hither glances at various young moneyed males. 1968 N. Marsh *Clutch of Constables* vii. [illegible] the habit of [illegible] his unbridled comehithery.

co·me-in, *sb.* The last stage of a run or race. Cf. *run-in*, Run sb.¹ 8.

1870 *Chambers's Jrnl.* 401/2 How many eager crowds have gathered to watch at that come-in! 1892 *Sat. Rev.* 26 Nov. 621/1 For the last mile or so, which was called the 'come-in', every one went at the pace that pleased him.

come-off, *sb.* **4.** (Earlier example.)

1722 *New-England Courant* 3 Dec. 1/2 Quoth she, This is a fine Come-off, Like hiding broken Wind by a Cough. 1742 Bp. Challoner in E. H. Burton *Life* (1909) I. xii. 189 This come-off will not save their turn.

co·me-on. *slang* (orig. *U.S.*). [Come v. 62.]

1. a. A swindler. Also *attrib.* = swindling.

1905 *N.Y. Even. Post* 28 Jan. 2 The general appearance of the man caused the officer to become suspicious, and he soon learned that Dates was a 'come-on'. 1905 *N.Y. Times* 24 Feb. 1 Detectives yesterday arrested two men on a charge of participating in a clever 'come-on' game. 1910 S. E. White *Rules of Game* II. i, Great country! If you listen to all the come-on stuff you may be disappointed—at first. 1924 W. M. Raine *Troubled Waters* xvi. 167, I reckon this confession talk is come-on stuff. 1936 J. Curtis *Gilt Kid* ii. 20 Come on guy for a con gang, ain't you?

b. The victim of a swindler; a dupe.

1898 W. T. Stead *Satan's Invis. World* ii. 109 The victim, who was known as a 'Come on' or as a 'Guy', was swindled by a variety of methods. 1935 *Evening News* 29 June 3/1 An immense number of names has been invented for the victims [of confidence men]..suckers, easy marks, wise guys, come-ons.

2. An inducement, enticement, lure; an invitation to approach. Also *attrib.*

1902 Ade *Girl Proposition* 122 As soon as any one began to give him the old come-on about being one Man shy he would start in to back up and try to think of another Date. 1925 H. Leverage in *Flynn's* 10 Jan. 876/1 *Come-on*, deception; lure; enticement. 1928 J. M. Barrie *Plays, Half an Hour* 616 The come-on look in the melting eyes of you. 1930 E. Weekley *Adjectives* x. 169 This 'snappy come-on letter' was highly approved. 1934 R. Stout *Fer-de-Lance* xiv. 225 There were signs all around, *Fly $5*, and *Try the Sky*, and other come-ons. 1939 I. Baird *Waste Heritage* xxi. 301, I never did care for these come-on ads. myself. 1939 R. Chandler *Big Sleep* xviii. 151 Wilde tapped the three notes signed by Carmen and said: 'I guess these were just a come-on.' 1949 J. R. Cole *It was so Late* 65 The woman..gave me the come on..half through her lashes. 1955 E. Bowen *World of Love* xi. 219 I've known them [*sc.* planes] go round and round..Till they get the come-on from the control. 1958 M. Dickens *Man Overboard* vii. 104 They like the sound of foreign investments. It has that magic, millionaire ring to it, like foreign exchange. Just another come-on. 1969 *Observer* 12 Jan. 8/4 They flocked to an exhibit..with a huge come-on of a sign that read: They are watching us.

comer. Add: **1. d.** *all comers.* Also used *attrib.* and *ellipt.* of an 'open' sporting event.

1883 *Illustr. Sporting & Dram. News* 6 Jan. 407/3 At the Sydney Regatta on November 9, Laycock managed to secure a win in the All-comers Sculling Race. 1890 C. G. Heathcote *Lawn Tennis* iv. 170 At Wimbledon Mr. Lawford..won the All Comers prize. 1912 A. W. Myers *Compl. Lawn Tennis Player* (ed. 3) i. 15 In 1880 I defended the cup against Lawford who had won the All-Comers! 1959 *Times* 24 Aug. 4/1 The Dutchman..also beat the British all-comers' figures.

3. A person or animal that shows promise of development or achieving success; one who is 'coming on' (see Come v. 62 b). *colloq.* (chiefly *U.S.*).

1879 *St. Nicholas Mag.* Nov. 84/2 A crab..is always very poor to begin with; but he eats everything he gets hold of, which..fattens him up some. Then he is called a 'comer'. 1890 H. Palmer *Base Ball Field* 91 He piled up error after error, but Dan said he was a 'comer'. 1903 *Rep. Kansas State Board Agric. 1901–2* 202 He..still shows that remarkable looseness and elasticity of hide that indicates a 'comer' when he is put next to the feed-box. 1959 *Sunday Express* 15 Feb. 10/2 The younger man..is a 'comer', a diplomat with a future. 1965 *Guardian* 31 May 9/7 Congressman John Lindsay..has sprung into national notice as the most attractive 'comer' in his party.

comet. Add: **4. comet claret**, claret made in a comet-year (cf. *comet-wine*).

1864 Meredith *Emilia in England* xxxiii, There's seven bottles of my poort, and there's eleven of champagne and some comut clar't.

come-u·p(p)ance, comeuppance. orig. *English* and *U.S. dial.*, now *gen.* in the U.S., less common elsewhere. Also (*dial.*) **come-uppings.** [Come v. 69.] Enough to serve one (by way of retaliation or check); one's deserts.

1859 *Harper's Mag.* Jan. 277/1 Dennis once got his 'come-up-ance'. 1880 M. A. Courtney *W. Cornwall Gloss.*, *Come-upping*, a flogging. 'I'll gi' 'ee a sound come-upping.' 1884 Howells *Silas Lapham* viii. 366 Rogers is a rascal.. But I guess he'll find he's got his come-uppance. 1897 E. Higginson *Tales Puget Sound* 155, I can give him his come-up-'ans if he goes to foolin' around. 1897 Howells *Landlord Lion's Head* xxi. 153 Well, I did get my come-uppings that time. 1923 'B. M. Bower' *Parowan Bonanza* vi. 70 'An' that's where he got 'is come-uppance,' he gloated. 1943 D. W. Brogan *Eng. People* i. 21 The roles of teacher and taught were suddenly reversed, to the delight of a world that saw the English at last get their come-uppance. 1957 G. B. Stern *Seventy Times Seven* 15 She's bound to get her come-uppance one day. 1959 *Camb. Rev.* 2 May 461/2 When they actually appoint [at Cambridge] a sociologist they will get their comeuppance. 1963 *Ann. Reg. 1962* 40 Fleet Street, accustomed to pour scorn on the inefficiencies of other industries, had its 'comeuppance' with the report of the Shawcross Commission.

comfort, *sb.* Add: **8.** (Earlier and later examples.)

1834 *Southern Lit. Messenger* I. 168 A lady of our party..aptly compared it to a Yankee *comfort*. 1847 [see *comfortable sb. 2 c]. 1913 Mrs. Stratton-Porter *Laddie* xi. 339 Laddie had..hung up a comfort at four o'clock to keep the Princess warm. 1945 B. A. Botkin *Lay my Burden Down* 112 Then a great big mattress full of goose feathers and two-three comforts as thick as my foot with carded wool inside!

11. comfort station *U.S. Genteelism*, a public lavatory.

[1910 *Aurora* (Illinois) *Daily Beacon* 8 Sept. 6/1 A public comfort room..would pay the city of Aurora a profit every year.] 1923 *Glass* (*Pittsburgh Plate Glass Co.*) 167 (*heading*) Public comfort stations. 1931 T. Wilder *Long Christmas Dinner* 121 Ma, where is the next comfort-station. 1947 Auden *Age of Anxiety* (1948) ii. 53 Ingenious George reached his journey's end Killed by a cop in a comfort station. 1957

D. Karp *Leave me Alone* xii. 165 Snatching Jimmy's hand she trudged across the sand with him to the comfort station. 1967 R. Shaw *Man in Glass Booth* xvii. 152 'Wash-room,' said the old man. 'Comfort station.'

comfortable, *a.* Add: **1. b.** *the Comfortable Words*: in the Anglican Liturgy, the four scriptural passages following the Absolution in the Communion Office, prefaced by 'Hear what comfortable words' [etc.].

1855 Procter *Bk. Comm. Prayer* 324 The Comfortable Words that follow are the scriptural statements upon which the absolution is grounded. 1893 C. Whitaker *Prayer Bk.* 79 Comfortable Words. These are taken from Hermann's Consultation.

B. *sb.* **1.** (Later Amer. example.)

1786 M. Cutler in *Life, Jrnls. & Corr.* (1888) II. 247 He..proposes to provide well in comfortables for the journey.

2. c. (Earlier and later examples.)

1842 in J. S. Buckingham *E. & W. States Amer.* III. 434 Still Mr. Van Buren was not content; he longed for the 'Turkish divan' and the 'French comfortable'. 1847 *Rep. Comm. Patents 1844* 35 [Cotton] has already been employed in what are variously called 'comforts' and 'comfortables'. 1856 S. T. L. Robinson *Kansas* 28, I..slept on comfortables and buffalo-robes on the floor in the attic. 1945 L. R. Tryon *Poor Man's Doctor* 96, I threw back most of the voluminous quilts, comfortables and sheets.

comfortably, *adv.* **5. a.** Add to def.: Esp. in phr. *comfortably off*, well off, prosperous.

c 1850 [in Dict.]. 1893 G. B. Shaw *Imposs. Anarchism* 24 You may, if you are a..comfortably-off person, think that the policeman..is the guardian of law and order. 1914 C. Mackenzie *Sinister St.* II. III. x. 707 I'm too comfortably off to worry much about anything. 1958 J. Cannan *And be a Villain* i. 28 It's most soul-destroying to be comfortably off.

comforter. Add: **6.** (Earlier example.)

1823 C. J. Mathews *Diary* 7 Nov. (1879) I. 83 Two shirts, two waistcoats, a woollen comforter over my chin.

b. (Earlier and later U.S. examples.)

1832 S. G. Goodrich *Syst. Univ. Geogr.* 107 Padding or quilting bed coverings or comforters. 1840 *Boston Almanac* 126 Theodore Baker..has for sale..Quilts, Counterpanes, Comforters, Bockings, Crumb Cloths. 1947 *Chicago Daily News News-Views* 29 Mar. 5/2 A quilting bee with neighbors...It's a pleasant way to spend an afternoon as well as finish a comforter.

7. A dummy teat put into a baby's mouth to quieten it.

1898 *Daily News* 11 Mar. 3/1 Sucking air from a comforter (that is an indiarubber teat), which led, alas! to no bottle. 1912 A. Neil Lyons *Clara* x. 99, I shall take you to a copper and give you up for stealin' the baby's comforter.

comfy (kɒ·mfi), *a.* Alteration of Comfortable *a.*, with the hypocoristic suffix -y⁶. Hence **co·mfily** *adv.*

1829 *Hist. Little Louisa* 82, I am saving up my money to buy them two blankets, and then they will be much more *comfy*. 1887 Kipling *Plain Tales fr. Hills* (1888) 230 Put to bed on some rolled-up carpets, all comfy. 1902 *Westm. Gaz.* 27 Nov. 3/2 A 'comfy' paletot for a little girl. 1905 E. Glyn *Viciss. Evang.* 50 We might sit in that comfy window-seat and talk. 1915 H. Rosher *In R.N.A.S.* (1916) 57, I arrived here safely in excellent time after quite a comfy journey. 1917 M. Sinclair *Tree of Heaven* xvi, Not insanely happy..but quietly, comfily happy. *Ibid.* xx, You can't expect us to fight so comfy, and to be killed so comfy, if we know our womenkind are being pounded to bits in the ground we've just cleared. 1928 E. Wallace *Again Three Just Men* i. 33, I am in London, which is delightfully capitalistic and comfy. 1933 W. de la Mare *Lord Fish* 182 Let's sit down here comfily on the stones. 1969 *Sears Catal.* Spring/Summer 26 Crochet-look tights...Comfy, elastic waist.

comic, *a.* and *sb.* Add: **B. 1. b.** Delete †*Obs.* and add later examples.

1927 E. Thompson *These Men thy Friends* 93 That Grimes, he *is* a comic! 1927 *Cleveland Press* 31 Jan., [They] are highly amusing as dancers and knock-about comics. 1961 J. McCabe *Laurel & Hardy* (1962) i. 38 'Stan,' he said, 'why do you want to be a comic?'

2. Add to def.: A children's paper; in *pl.*, the comic strips in a newspaper, etc. Cf. **horror comic.*

1910 H. G. Wells *Mr. Polly* i. 20 One of those inspiring weeklies that dull people used to call 'penny dreadfuls', admirable weeklies crammed with imagination that the cheap boys' 'comics' of today have replaced. 1917 T. S. Eliot *Prufrock* 21 You will see me any morning in the park Reading the comics and the sporting page. 1957 *Eng. Lang. Teaching* (Brit. Council) XII. no. 1. Oct.–Dec. 7 It has long been recognized..that the grasp that the Comics can exert and maintain upon their readers' attention could offer a promise as well as a threat to education. 1960 E. H. Gombrich *Art & Illusion* 8 Even pin-ups and comics, rightly viewed, may provide food for thought.

b. A comic film, or television programme.

1929 Wodehouse *Mulliner Speaking* ix. 301 Let's beef in or we'll be missing the educational two-reel comic. 1938 E. Bowen *Death of Heart* II. v. 269 She saw light from the comic flickering on his eyeballs. 1962 *Listener* 15 Feb. 288/1 What are your favourite television programmes?..Mostly the light programmes, the comics.

C. *comic annual, postcard, song*; **comic alphabet**, a humorous presentation of the alphabet; **comic book** *U.S.*, a book of strip cartoons; **comic cuts**, the name of a popular children's paper; later used as a derisory nickname for official armed services' Intelligence Summaries; also *attrib.*, as typical of a strip cartoon; **comic-opera**, used *attrib.* of a comic or farcical act, situation, etc.; **comic relief** [Relief²], comic episodes of a play, etc., intended to offset the more serious parts; also *transf.*; **comic strip**, see *strip sb.²

1836 in P. Muir *Eng. Children's Bks. 1600–1900* (1954) 198 Comic Alphabet [illustr. by George Cruikshank]. 1961 E. Partridge *Comic Alphabets* ii. 25 The comic alphabets of the 19th century seem to have begun their career as burlesques of, or satires on, the 16th–20th century children's alphabets designed to smoothe the way for very young learners. 1830 (*title*) Comic annual. 1941 *Amer. Jrnl. Orthopsychiatry* XI. 540 (*heading*) The effect of comic books on the ideology of children. *Ibid.*, The comic book, a new medium aimed at the pre-adolescent child, has..taken its place alongside the movies and radio in children's interest. *Ibid.* 541 A Christmas display featuring a comic book hero outranked even Santa Claus in the children's interest. 1955 *Times* 3 May 8/7 Mr. Averell Harriman has approved a Bill making it a misdemeanour to publish or distribute 'comic books' devoted to sex or crime to persons under 18. 1890 (*title*) Comic cuts. 1908 H. G. Wells *War in Air* i. § 2, He was making three shillings a week or more, and spending it on *Chips, Comic Cuts, Ally Sloper's Half-Holiday*, cigarettes, [etc.]. 1916 'Ian Hay' *First Hundred Thou.* xix. 277 The official Intelligence Summary of our Division—published daily and known to the unregenerate as 'Comic Cuts'. 1917 'Contact' *Airman's Outings* 201 Of all the tabloid tales published last year in R.F.C. 'Comic Cuts', the most comic was that of a mist, a British bus, and a Boche General. 1929 Bowen *Sea Slang*, *Comic cuts*, Admiralty intelligence reports; G.H.Q. communiqués: naval and military. 1959 *Listener* 19 Feb. 320/1 He was no comic-cuts figure winning votes as a simple rail-splitter who appealed to the common man. 1906 *Daily Chron.* 19 Jan. 3/6 Whoever obtained Mr. Chamberlain's blessing for Mr. Goodhart must have done so by false pretences...It was a comic-opera situation. 1906 *Westm. Gaz.* 12 May 2/3 Mont Blanc looked down with her superb consciousness of eternal snows at the comic-opera passions of the little valley election. 1937 Koestler *Spanish Testament* iv. 75 [The Spanish Army]..had the reputation of being a 'comic-opera army'. 1949 I. Deutscher *Stalin* 79 There were such comic-opera touches about the conference. 1941 'G. Orwell' in *Horizon* Sept. 153 A comic postcard is simply an illustration to..a 'low' joke. 1960 Auden *Homage to Clio* 45 It might equally well be a comic postcard. 1825 in E. Fitzball *Dram. Author's Life* (1859) I. 133 And the moment the point necessary for the plot is attained, the audience are always impatient for the comic relief. 1920 M. Beerbohm *And Even Now* 276, I sometimes wished this work had some comic relief in it. 1921 G. B. Shaw *Back to Methus.* p. lxxxiv, The Shakespearian-Dickensian consolation of laughter at mischief, accurately called comic relief. 1819 *Times* 30 Aug. 3/5 He has..procured a scanty subsistence by writing comic songs. 1836 Dickens in *Library of Fiction* I. 15 Mr. Tippin sang a comic song. 1919 D. Ashford *Young Visiters* (1951) viii. 51 Sounds of laughter and comic songs issued from the abode.

Cominform (kɒ·minfǭɹm). [Russ., f. the first elements of the Russ. forms of Com(munist and Inform(ation.] An information bureau set up in 1947 by the communist countries of eastern Europe for the interchange of experience and coordination of activities and dissolved in 1956. Also *attrib.* and *transf.*

1947 in *Amer. Speech* (1949) XXIV. 73 The cominform, an information bureau set up by the Communist parties of nine European countries. *Ibid.*, Considering both Mexico City and Montevideo..as headquarters for a 'Cominform' in the Western Hemisphere. 1948 *Ann. Reg. 1947* 73 The announcement on 5 October of the formation of the *Cominform*, with a co-ordinating office in Belgrade. 1957 Carew Hunt *Communist Jargon* p. xii, In his Cominform speech of September 1947 Zhdanov declared [etc.]. 1958 *New Statesman* 6 Sept. 265/2 The new Cominform is in business, and its business is the publication of a new journal from which Communists all over the world will take their cue.

coming, *vbl. sb.*¹ Add: **6. coming out** [Come v. 63 o], (of a young woman) entering society; also *attrib.*

1814 Jane Austen *Mansf. Park* II. ix. 202 Miss Price had not been brought up to the trade of *coming out*. 1877 S. O. Jewett *Deephaven* xiii. 280 Miss Sally told us a long story about her friends and about her 'coming-out party'. 1903 Mrs. Vaizey *P. O'Shaughnessy* xxiii. 240 Bridgie wore her coming-out dress. 1949 'J. Tey' *Brat Farrar* ii. 12 Her aunt..had been responsible for her coming-out.

7. coming in. d. Calving. *dial.* and *U.S.*

1857 *Trans. Ill. Agric. Soc.* II. 381 The coming in of the cows should be regulated to occur in early spring.

coming, *ppl. a.* Add: **5.** Rising into prominence; likely to attain distinction.

1865 *Atlantic Monthly* XV. 635 The job was the taking of Charleston and the 'coming man' was Brigadier-General..Gillmore. *a* 1867 H. Woodruff *Trotting Horse Amer.* (1868) xxxiv. 282 Many thought then Lancet was the 'coming horse'. 1869 S. Bowles *Our New West* xiii. 261, I can discover no successor to Brigham Young. He has men of ability..but I see no 'coming man' for his place. 1946 C. Fry *Phoenix too Frequent* (1949) 3 He was one of the coming men. He was certain to have become the most well-organized provost The town has known.

Comintern (kɒ·mintəɹn). Also **Kom-.** [Russ. *Komintérn*, f. the first elements of the Russ. forms of Com(munist and Intern(ational.] The Communist International, the international organization of the Communist Party, founded in 1919 and dissolved in 1943.

1925 *Glasgow Herald* 22 Oct. 9 The international relations of the S.S.S.R. and the problems of the Komintern in connection with it. 1926 *Spectator* 10 July 43/2 As far as the Commintern was concerned they [*sc.* the Soviet Government] would be only too delighted to see it transferred to London. 1929 *Times* 10 July 12/3 The Comintern. Relations with Soviet Government. 1940 *Ann. Reg. 1939* 191 Italy continued to distinguish between the Comintern and the Soviet Government. 1950 A. Koestler in R. H. S. Crossman *God that Failed* 35 The habit gradually filtered down from the top Comintern-bureaucracy into every national C.P. in Europe. 1959 *New Statesman* 25 Apr. 562/3 There is no doubt that Tito knows many damaging things about Soviet and Comintern history: he has been a leader of the Yugoslav C.P. for 22 years, a Comintern functionary for much of that time, and he has had frequent private meetings with Soviet ..politicians.

comitadji (kɒmitæ·dʒi). Also **kom-, -adje, -aji, -aggi.** [Common Balkan form, ad. Turkish *komitacı*, f. *komita*, f. F. *comité* committee, + *-dji*; lit. member of a committee.] In the Balkans, a member of a band of (esp. Bulgarian) irregular soldiers or partisans.

1903 *Times* 5 Sept. 7 The *Komitajis* and the Turks have made a desert of the disturbed vilayets between them. 1908 *Daily Chron.* 30 Oct. 4/6 Lately they had to resist the ferocity of the comitajis. 1909 *Westm. Gaz.* 15 Feb. 1/3 A band of 'Comittaggis' or Macedonian freebooters. 1920 *Glasgow Herald* 18 Mar. 9 The Montenegrin comitadjes..are gradually giving in. 1923 *Daily Mail* 16 Feb. 8 The Bulgarian Comitadjis. 1959 *Chambers's Encycl.* VIII. 762/1 A violent rivalry now began between various national bands of *comitadjis*, the most powerful being the famous Internal Macedonian Revolutionary Organization.

comitative, *a.*² Add: (Earlier and later examples.) Also as *sb.*

1860 *Trans. Philol. Soc. 1857* 34 The declension of the personal pronouns is as follows.. *Nominative, Genitive,*.. *Comitative.* 1890 [see *abessive a.]. 1954 Pei & Gaynor *Dict. Linguistics* 42 *Comitative*, in certain language groups (e.g. Finno-Ugric), a case having the denotation *with* or *accompanied by*. *Comitative aspect*, a verbal aspect expressing that the action or state denoted by the verb is performed or experienced together or jointly with somebody or something else. 1955 Quirk & Wrenn *O.E. Gram.* 67 The characteristic comitative function seen in expressions like the following: worhte Ælfred cyning lytle werede geweorc.

command, *sb.* Add: **1. c.** *ellipt.* A command performance (see *10 below and *command-night* in Dict. and Suppl.). *colloq.*

1839 Dickens *Let.* 26 Jan. (1965) I. 497 We are engaged to Macready..at Covent Garden Theatre, on the night of the Queen's '*command*'.

d. *Computers.* An expression in a program, etc., that defines an operation, esp. a basic operation, or results in the performance of an operation; also, a signal or set of signals that results from such an expression and initiates the performance of the operation.

1946 *Ann. Computation Lab. Harvard Univ.* I. 15 The calculator continues in operation only so long as the command 'continue operation' is repeated, cycle by cycle. 1949 *Math. Tables & Other Aids to Computation* III. 428 In order to extract a group of digits contained within a word, with the commands available in the EDVAC, it is necessary to use two shift operations. 1951 *Proc. Inst. Radio Engin.* XXXIX. 272 The commands initiate the individual steps which form the process of executing the instruction. 1956 Berkeley & Wainwright *Computers* ii. 44 Other words for instruction are *command, order*, and *program step*. 1961 L. W. Hein *Introd. Electronic Data Processing* ii. 24 The clear add command first cleared the register to all zeros and then added to the register the contents of the specified address. 1966 *New Scientist* 27 Oct. 162/2 Conditional commands are allowed ('save this data if the program test fails'). 1967 A. Hassitt *Computer Programming & Computer Systems* i. 12 The orders that the CPU issues to the channel are usually called 'commands'; the word 'instruction' is reserved for the orders that the CPU obeys.

7. b. (*a*) *the higher command*: the general staff collectively of the British Army; also *spec.*, the commander-in-chief.

1916 *Times* 15 Dec. 10/5 It was with great pain that he had heard the criticism upon the higher command. If there was one man in France who was trusted by every one from the highest to the lowest,..it was Sir Douglas Haig. *Ibid.* 16 Dec. 10/4 The hon. member's attack on the generals in the higher command. 1927 E. Thompson *These Men thy Friends* 30 And as for the Higher Command! How many generals do you suppose we've got rid of, up to date?

(*b*) *the high command*: rendering of foreign expressions, e.g. G. *Oberbefehl*.

1917 *Times* 23 June 6/3 The High Command [of the Russian army]..is only too anxious to stand loyally by the Allies. 1918 *Ibid.* 7 June 6/1 The German High Command chose yesterday to issue its claims as to booty captured by the 'Army Group of the German Crown Prince'. 1919 Maurice *Last 4 Months* 177 The High Command insists on the immediate issue of a peace offer to our enemies. 1928 *Illustr. Hist. Russ. Rev.* I. 137 The Army, which rose almost unanimously against its High Command. 1943 P. Jordan *Diary* 13 Jan. 136 This is one more repetition of the high command's consistent error, combined with the typical amateur's belief that you must always be 'doing something'.

8. b. In the names of various groups of armed forces, as of a unit on special duty, or of troops in a specified area, as *Bomber, Fighter, South-East Asia Command.* Also *absol.* Cf. *coastal a.

1939 *Times Weekly* 1 Nov. 9/1 Aircraft of the Fighter Command have again been in action. 1939 *War Illustr.*

29 Dec. 538/2 A number of Squadrons forming a Wing, so many Wings a Group, and so to the Command. 1941 *Hutchinson's Pict. Hist. War* 14 May–8 July 75 A battery command post in the Eastern Command during shooting practice. 1945 *Yorkshire Post* 19 Apr. 3/1 Measures to improve conditions for troops in India and the South-East Asia Command were announced. 1960 B. Horrocks *Full Life* xxi. 278 Before the war Western Command had always been regarded as a backwater.

10. command aeroplane (see quot.); **command allowance** *Mil.*, the additional allowance attached to a command; **command car** *U.S.*, a staff car; **command guidance** *U.S.*, the guidance of missiles, or of pilotless aircraft, by electronic signals; so *command-guidance system, command missile*; **command module** (see quot. 1964); **command-night** (examples); **command paper** (abbreviated *c., Cd., Cmd.*, or *Cmnd.* with register number, as Cd. 5723), a paper laid before Parliament, etc., by command of the Crown (see quot. 1929); **command pay** = **command allowance*; **command performance**, a theatrical, musical, etc. performance given by royal command; also *fig.*; **command post** *Mil.*, the headquarters of a unit (see also quot. 1918) *U.S.*; **command service module**, a spacecraft comprising the command module and the service module, after the descent (e.g. lunar) module has been jettisoned.

1918 Farrow *Dict. Mil. Terms*, *Command Aeroplanes.*—Aeroplanes which observe the general progress of the combat and all that occurs on the side of the enemy. 1860 *Leisure Hour* 22 Mar. 186/2 Some pompous old British general..merely imported to the East for the purpose of drawing his command allowance of £4000 a-year. 1904 Kipling in *Windsor Mag.* Jan. 228/2 'Is nephew wants two bottles [of rum] command-allowance. 1956 E. Wilson *Red, Black, Blond & Olive* ii. 103 We were supposed to be met by a command car. 1949 *Newsweek* 21 Feb. 52/2 *Command guidance*...In this system the missile only needs to carry enough mechanism to receive and obey commands from the ground. 1952 K. W. Gatland *Devel. Guided Missile* i. 15 In the Command-Guidance system, one radar tracks the target, while the other tracks the missile, and each feed [*sic*] data into a computer, whereby steering commands are transmitted to the missile. 1947 *Time* 23 June 70/3 'Command' missiles follow orders (radio signals) from the ground, a ship, or a piloted aircraft. 1962 *New Scientist* 19 July 123 A three-stage vehicle will send a spacecraft containing three men in a pressurized cabin (the command module) towards the Moon. 1964 J. L. Nayler *Dict. Astronautics* 51 *Command module*, the compartment in a spacecraft containing the personnel and the main controls. 1970 *Guardian* 18 Apr. 1/1 Two hours later came the jettisoning of the lunar module after the complex alignment procedures..on the command module had been completed. 1826 J. O'Keefe *Recoll.* I. i. 33, I often saw them..on the command nights at the theatre. 1911 G. B. Shaw *Let.* 27 June (1956) 174, I feel some remorse at having stolen so much of her command night. 1886 *Accounts & Papers* XXXIX. p. lxxi, Numerical List of Command Papers...c. 4603 Meteorology. 1900 *Ibid.* LX. p. lxxxi, Cd. 1 Trade Reports. 1919 *Ibid.* XXIII. p. xxxvi, Cmd. 1 Colonial Reports. 1929 G. F. M. Campion *Proced. Ho. Comm.* 69 Command Papers cover all the more important documents which the Government and the departments wish to publish on their own initiative. 1956 H.M.S.O. (*title*) Report on the Proceedings of the..Assembly of the Council of Europe...Presented by the Secretary of State for Foreign Affairs to Parliament by Command of Her Majesty December 1956. Cmnd. 26. 1875 *Encycl. Brit.* II. 576/1 Lieutenant-Colonels of regiments, and other officers holding commands of wings, batteries, or depôts, receive 'command pay' of from 3s. to 1s. 6d. daily. 1876 Voyle & Stevenson *Mil. Dict.* (ed. 3) s.v. *Command*, A certain 'command pay' attached, in addition to the regimental pay. 1897 *Strand Mag.* XIII. 623 A 'Command' Performance at Windsor fifty years ago. 1922 *Westm. Gaz.* 29 Dec., She was a favourite at Command performances. 1934 Ld. Berners *First Childhood* xxiii. 235 The hissing..had been a 'command performance' and not in the least a genuine manifestation of public opinion. 1918 E. S. Farrow *Dict. Mil. Terms* 134 *Command post*, a shelter near the observation station of a commander. 1941 [see sense *8b, above.] 1946 *Amer. Speech* XXI. 71/1 A private in the front was charged with the duty of sending reports back to the battalion CP (Command Post) by messenger. 1969 *Times* 21 Apr. 8/6 Close-up views of the moon..will be taken from the command service module (C.S.M.) and not from the L.M.

commandant. Add: Also with pronunc. (kɒmăndæ·nt). Used in the wars of 1914–18 and 1939–45 as the title of an officer holding a special command, as of a place, depot, or the like, or of a particular force.

1915 P. Gibbs *Soul of War* 342 The Base Commandant's office was the sorting-house of the Expeditionary Force. 1915 Grahame-White & Harper *Aircraft Gt. War* 236 The commandant of the air corps. 1920 G. K. Rose *2/4th Oxf. & Bucks Lt. Infty.* 145 A Brigade school, with Bennett as its commandant. 1950 C. MacInnes *To Victors the Spoils* i. 20 The Camp Commandant had..a brisk, impatient manner.

b. *S. Africa.* [= Afrikaans *kommandant*.] The leader of a Boer commando; in the South African armed forces, an officer ranking between major and colonel, the *commandant-general* being the commander-in-chief.

1791 G. Carter *Loss of Grosvenor* xl. 149 Hynes thinks that Quin was a kind of commandant. 1838 F. Owen *Diary* (1926) 125 The commandant thanked me for speaking my sentiments openly to *them*, rather than to the natives. *Ibid.* 127 The commandant of the Natal forces. 1898 *Westm. Gaz.* 4 Jan. 5/2 If Mr. Schalk Burger is returned he will resign his Commandant-Generalship. 1899 *Ibid.* 11 Nov. 8/1 Commandos, in which the burghers are under command of Field-Cornets and Commandants. *Ibid.*, The Commandant-General..is chosen by the whole force. 1902 *Encycl. Brit.* XXXIII. 438/2 Each district was furnished with a commandant, who had under him field-cornets and assistant field-cornets. 1914 *Times* 20 Nov. 7/5 Commandant S. P. Du Toit..to-day captured another of Beyers's rebel officers. 1952 *Kommando* Feb. 38/3 In 1946, the Regiment was reformed on a Peace Time Basis under the command of Commandant A. D. F. Sales.

commandeer, *v.* Add: **c.** *fig.* and *gen.* To take arbitrary possession of.

1900 Kipling in J. Ralph *War's Brighter Side* (1901) ix. 135 We never use such words as steal, or 'collar', 'pinch', or 'shake': The fashion is to say he 'commandeers' it. 1900 Conan Doyle *Great Boer War* 208 The naïve claims put forward by the Boers to some special Providence—a process which a friendly German critic described as 'commandeering the Almighty'. 1902 *Westm. Gaz.* 3 Apr. 3/1 The sleeve commandeers to itself a share of attention..out of proportion to its place. 1928 *Punch* 6 June 620 The porter..came to the hotel long after the Merry Party had commandeered it for week-ends. 1929 *Daily Express* 12 Dec. 1/3, I commandeered passing lorries and any other vehicles which happened to be going down the road. 1954 M. Beresford *Lost Villages* ix. 335 The sheriff commandeering wheat from the fields of Compton Verney in 1445.

Hence **commandee·red** *ppl. a.*, taken for military service or use; **commandee·ring** *vbl. sb.*

1894 *Westm. Gaz.* 23 June 6/1 A number of commandeered settlers. 1894 *Athenæum* 28 July 125/2 The commandeering of British subjects in the Transvaal. 1899 *Daily News* 13 June 4/2 Each commandeered burgher.

commander. Add: **8. b.** *Basketry.* A straight iron shank joining two rings, used for straightening thick sticks.

1912 T. Okey *Basket-making* iii. 12 The full kit..consists of..a commander [etc.]. 1941 *Archit. Rev.* LXXXIX. 54 (*caption*) Commander.

Commanderia (kɒmăndī·riă). A sweet wine produced in Cyprus on the former commanderies (see Commandery 2 a).

[1851 Redding *Wines* xii. 294 The wine made of the best grapes is that still called the wine of the Commandery.] 1928 *Observer* 5 Feb. 5/1 Cyprus wines have been famous since classical times, and Commanderia, so called from its association with the Commanderies of the Knights of St. John, is perhaps the best known. 1957 L. Durrell *Bitter Lemons* 29 Every evening we took a glass of sweet, heavy Commanderia on his little terrace.

commando. Add: (Earlier and later examples.) (See also quot. 1791.)

1791 G. Carter *Loss of Grosvenor* xl. 150 'A colonist,' says he [Vaillant], 'who lives..up the country..intreats a commando, which is a permission to go, with the help of his neighbours, to retake his property.' 1809 R. Collins in P. J. van der Merwe *Noordwaartse Beweging* (1937) 87 The prohibition to their carrying off women and children has greatly served to damp their ardour for commandos. 1952 E. H. Burrows *Overberg Outspan* ii. 34 An armed commando sixty strong. 1966 E. Palmer *Plains of Camdeboo* iii. 48 One [mule] was stolen by a Boer neighbour, who rode away on it to join Malan's commando.

b. Phr. *on* or *upon commando*, on militia service in the Boer army; engaged in a commando.

1824 Burchell *Trav.* II. 111 The master himself was at this time absent on the *Commando*, or militia-service, against the Caffres in the Zuureveld. 1837 J. E. Alexander *Voy. W. Afr.* xv. 380 The colonists..disliked..going on commando into Kaffir-land to recover stolen property. 1879 *Chambers's Jrnl.* 344/1 All the men capable of bearing arms were then *on commando*. 1899 *Daily News* 30 Sept. 5/3 Burghers refusing to go on commando. 1900 Kipling in J. Ralph *War's Brighter Side* (1901) viii. 117 We sailed upon commando To vierneuk our Brother Boer.

2. In the South African War (1899–1902), a unit of the Boer army composed of the militia of an electoral district. Also *attrib.*

1899 *Westm. Gaz.* 11 Nov. 8/1 The President..has the right of declaring war and calling up one or more commandos. 1900 Kipling in *Daily Express* 13 June 4/5 We kept our commando-horses ready for six months. 1902 *Encycl. Brit.* XXXIII. 438/2 Each field-cornet..was responsible for the arms, equipment and attendance of his commando—the commando being the tactical as well as the administrative unit. 1902 *Daily Chron.* 16 June 5/5 Burghers..all thronging into the commando trains.

3. A member of a body of picked men trained originally (in 1940) as shock troops for the repelling of the threatened German invasion of England, later for the carrying out of raids on the Continent and elsewhere. Also applied to similar troops of other countries. Also *attrib.* and *Comb.*

1940 W. S. Churchill in *Second World War* (1949) II. 566 Plans should be studied to land secretly by night on the islands and kill or capture the invaders. This is exactly one of the exploits for which the Commandos would be suited. 1941 R. Keyes in *Hansard* CCCLXXVI. 661 Among other executive responsibilities of my office were included the raising, organising and training of the special service troops now known as Commandos. 1942 'G. Orwell' in *Partisan Rev.* July–Aug. 277 The rightwing papers..suggest that we can tie down a million troops along the coast of Europe by continuous commando raids. 1943 *Southeast Economist* (Chicago) 11 Mar. 1/1 Joseph Arlia..will..go back to the South Pacific as a commando. 1943 *Chicago Sun* 17 Dec. 1/8 A small American Commando force, paddling toward shore in rubber boats..was sacrificed to a merciless barrage. 1943 *Combined Operations* (*Min. of Information*) i. 11 Lieutenant-Colonel D. W. Clarke..produced the outline of a scheme...The men for this type of irregular warfare should, he suggested, be formed into units to be known as Commandos...Nor was the historical parallel far-fetched. After the victories of Roberts and Kitchener had scattered the Boer army, the guerrilla tactics of its individual units (which were styled 'Commandos')..prevented decisive victory...His [*sc.* Lt.-Col. D. W. Clarke's] ideas were accepted; so also, with some hesitation, was the name Commando. *Ibid.* 14 When they reach their Commando Units they..perform considerable feats of endurance. 1945 *Penguin New Writing* XXIII. 11 Here I am, conning a ship-load of one hundred and four hefty Commandos. 1956 A. H. Compton *Atomic Quest* 224 Anglo-Norwegian commandos made two daring night raids on the Norsk plant. 1958 *Jane's Fighting Ships 1958–59* p. v/1 The aircraft carrier *Bulwark* is being converted into a commando carrier. 1960 C. Day Lewis *Buried Day* 93 Commando-type training. 1968 *Globe & Mail* (Toronto) 5 Feb. 11/1 Two South Vietnamese drivers employed by the U.S. Embassy apparently helped a Viet Cong commando squad invade the embassy grounds last Tuesday morning.

b. *transf.* and *fig.*

1942 *Time* 16 Feb. 32/1 Comrade Shvernik and his 13 comrade commandos prepared for the trip home. 1957 *Times Lit. Suppl.* 22 Nov. 705/1 The concept of a commando-group of beginners,..raiding campus and quad alike, is one associated almost entirely with signed reviewing. 1958 *New Statesman* 8 Mar. 291/1 A Fascist 'commando' broke up a public meeting in Marseilles. 1968 *Observer* 7 Jan. 1/3 Detectives watching the screens were linked by pocket radio to plainclothes members of a 'commando' squad standing by ready to make arrests.

‖**commedia dell'arte** (kōme·diă del ā·rte). [It., lit. 'comedy of art'.] Improvised popular comedy as played in Italian theatres from the sixteenth to the early eighteenth centuries with actors representing stock characters.

1877 *Encycl. Brit.* VII. 418/1 The improvised comedy..was now as a rule performed by professional actors, members of a craft, and was thence called the *commedia dell'arte*, which is said to have been invented by Francesco (called Terenziano) Cherea, the favourite player of Leo X. 1923 G. B. Shaw *Let.* 24 May in F. J. West *Shaw on Theatre* (1958) 185 The *Commedia dell'Arte* was improvised Molière. 1938 M. McCarthy in *Partisan Rev.* June 42 A real fantastico, a *commedia dell'arte* clown with a touch of nobility. 1958 *Spectator* 17 Jan. 76/1 The farcical, *commedia dell'arte* style favoured by Theatre Workshop. 1963 *Listener* 14 Mar. 478/3 The well-loved characters of the Venetian improvised theatre, the *commedia dell'arte*.

‖**comme il faut** (komilfo), *adv.* and *pred. a.* [Fr., lit. 'as it is necessary'.] 'As it should be', proper(ly); according to etiquette; correct(ly) in deportment or behaviour. Also as *sb.*

1756 Walpole *Lett.* (1857) III. 8 True critics..maintain that we are not dead *comme il faut*. 1818 Lady Morgan *Fl. Macarthy* III. iii. 158, I would not present in my own exclusive circle one who was not in all points *comme il faut*. 1824 L. M. Hawkins *Annaline* I. 96 It would have been more comme il faut for me to have taken that step. 1857 Bagehot *Wks.* (1965) II. 17 To touch lightly the light things of society, to see the *comme-il-faut*. 1857 Macaulay in Trevelyan *Life* (1878) II. xiv. 459, I make large allowance for the difference of manners; but it never can have been *comme il faut* in any age or nation for a man of note..to be constantly asking for money. 1876 Geo. Eliot *Dan. Der.* I. i, 'Who are these Langens? Does anybody know them?' 'They are quite *comme il faut*.' 1914 G. B. Shaw *Fanny's First Play* III. 207 Miss Knox is a lady absolutely *comme il faut*. 1935 *Times Lit. Suppl.* 26 Sept. 598/1 Two extremes of womanhood, Clancy Todd the Irish tom-boy and Catherine, the beautiful English lady *comme il faut*.

commend, *v.* Add: **2. e.** *refl.*

1593–7 R. Hooker *Eccl. Pol.* III. viii. 148 This method of teaching may commend it selfe vnto the world by that easines & facilitie which is in it. 1865 Gladstone in R. Lowe *Speeches on Reform* (1867) 159 The law has commended itself more and more to the respect and attachment of the people. 1886 'Maxwell Gray' *Silence of Dean Maitland* I. x, [The grey suit] had commended itself to Everard from the sense of cleanliness that light colours always afforded him. 1904 F. M. Nichols *Epist. Erasmus* II. 269 It commends itself to all the theologians who are either learned, or honest and candid.

commendatore (komendătō·re). [It., f. med. L. *commendātor* Commendator.] A knight of an order of chivalry in Italy; = Commendator 2. Also *transf.*

1877 G. B. Shaw *How to become Mus. Critic* (1960) 31 Her illustrious master, Signor—or, as he delights to be styled, Commendatore—Lamperti. 1907 — *Let.* 8 July (1956) 90 It was five minutes past eleven and the Commendatore from Scotland Yard was knocking at the door. 1922 Joyce *Ulysses* 302 Commendatore Bacibaci Beninobenone. 1955 *Times* 9 Aug. 11/3 In 1953 the Pope appointed him a commendatore of the Order of St. Sylvester.

comment, *sb.* Add: **2. c.** Colloq. phr. *no comment*: a conventional statement of refusal to comment on a situation, esp. when answering a journalist, interviewer, or the like.

1950 *Time* 18 Sept. 26 Questioned on a press report that 'a close adviser to President Truman' was predicting Johnson's resignation, White House Press Secretary Charles Ross issued a perfunctory 'no comment'. 1957 'P. Quentin' *Suspicious Circumstances* ii. 17 'No comment at all.' She dropped the receiver and watched me again... 'I've always wanted to say No Comment into a phone.' 1965 A. Prior *Interrogators* xi. 205 'Is this man a suspect?' 'No comment. Sorry.' 1970 *Guardian* 8 Jan. 2/4 The District Attorney..answered 'No comment', when asked if the inquiry had been worthwhile.

to be what is known as commercial—commercial apparently being another word for old-fashioned, stereotyped, ambitionless, lacking in imagination, hopelessly orthodox and generally being unable to present anything new for fear that it will not be appreciated by the public. **1937** *Amer. Speech* XII. 46/1 *Commercial*, a term applied to musicians who play in the fashion which the public likes rather than that which other musicians judge best. 'The band is better, although still pretty commercial on some tunes.' *Downbeat.* **1959** G. Avakian in M. T. Williams *Art of Jazz* (1960) ix. 75 When she was trying to 'go commercial' and her texts strayed from the authenticity they once had.

6. *commercial agency* U.S., an organization which furnishes its clients with information as to the standing of commercial firms; *commercial agent* U.S., an agent stationed abroad to attend to commercial interests; *commercial art*, art employed in commerce, esp. in advertising; hence *commercial artist*; *commercial college, school, university*, one for instruction in commercial subjects; *commercial paper* (see quot. 1897); *commercial sheep*, a sheep reared for selling and not for breeding purposes; *commercial traveller* (earlier and later examples); (*b*) *Austral.* and *N.Z. slang*, = Swagger *sb.*[2] 2.

1897 *Bouvier's Law Dict.* I. 357/2 Commercial Agency. **1902** *Encycl. Brit.* XXX. 634/1 Mercantile, or commercial, agencies in America. **1877** *Ibid.* VI. 317/1 The United States commercial agents, although appointed by the president, receive no exequatur. They.. are distinct from the consular agents. **1922** *Commercial Art* Oct. 1/1 There are some fastidious people who will not suffer the term 'Commercial Art' being used. *Ibid.* Dec. 48/1 The only test of commercial art is: Will it sell the goods? *Ibid.* 52/1 The commercial artist of the new school appeals to the imagination. **1955** H. Read *Grass Roots of Art* vii. 132 A 'commercial' artist, which term might include poster-artist, fashion-designer, book-illustrator or artist-potter. **1802** *Monthly Mag.* XIV. 261/2 Dr. Alderson, of Hull, has issued Proposals for establishing a Commercial College. **1908** W. H. Dawson *Evol. Mod. Germany* 101 A Leipzig institution of recent origin, the Commercial College, founded in 1898 by the Chamber of Commerce of that town. **1836** in W. L. Mackenzie *Van Buren* (1846) 176 The demolition of the Usury laws, in relation to commercial paper,. is probably desirable. **1897** *Bouvier's Law Dict.*, *Commercial Paper*, negotiable paper given in due course of business, whether the element of negotiability be given it by the law merchant or by statute. **1889** *Sonnenschein's Cycl. Educ.* 440/2 The training of the commercial school will not make a boy a clever buyer and seller. **1901** Hooper & Graham *Comm. Educ.* 13 State-supported Technical and Commercial Schools and Institutions. **1914** *Scotsman* 10 Sept. 9/6 A big, strong commercial sheep. **1807** Southey *Espriella's Lett.* II. xxxvii. 137 A commercial traveller.. gave us some useful information concerning Manchester. **1830** Commercial traveller [see Traveller 3]. **1933** L. Acland in *Press* (Christchurch, N.Z.) 30 Sept. 15/7 *Commercial*, or *c. traveller*, slang for *swagger*. **1959** J. Braine *Vodi* v. 79 An ex-commercial traveller with a wife as fat and cheerful and black-haired as himself. **1853** *Scott. Educ. & Lit. Jrnl.* July 438 A movement.. for the establishment of a Commercial University.

B. *sb.* (Further example.) Also, a man engaged in commerce.

1939 D. L. Sayers *In Teeth of Evidence* 90 Do you know anything about a commercial called Slater? **1962** *Times* 6 Aug. 9/6 Among the 'commercials' the terms of assignment are usually longer and a man may be in India for five or even 10 years.

b. Applied to a vagrant and, in Australia and N.Z., a swagman.

1886 *Tit-Bits* 31 July 252 (Farmer), He is one of the cleverest commercials (that is the polite name for rogues and vagabonds generally) on the road. **1933** [see above, sense *6].

2. An advertisement broadcast on radio or television. Also *transf.* orig. *U.S.*

1935 *Fortune* Nov. 193 We used no media other than radio to feature this soup.. using one-third of our commercials on Campbell's Chicken Soup. **1936** *Variety* 24 June 55/2 Commercials are restricted to 190 words on an hour program. **1942** *Q. Jrnl. Speech* Feb. 7/1 The mixture of ghastly, forced gayety and jolly-doggism in this commercial makes it a comic example. **1957** *New Yorker* 26 Oct. 50/1 Somehow, as in the television commercials, an alluring young lady seemed to be a built-in accessory of each of these cars. **1959** N. Marsh *Singing in Shrouds* 14 The T.V. bloke that does the Jolyon Swimsuits commercial. **1960** Wentworth & Flexner *Dict. Amer. Slang* 117/2 *Commercial*, any laudatory statement; praise; a good reference. **1962** *Friend* 16 Mar. 313/1 A visit to Bulgaria. (A quick commercial here, apropos of the last item: Young Friends would be pleased to hear from any older Friend who happens not to be using his/her car/van/Minibus during the period August 15–September 15!). **1967** M. McLuhan *Medium is Massage* 126 Up until very recently, television commercials were regarded as simply a bastard form.

commercialese (kǫmɜ̄·ɹʃăli·z). [f. Commercial *a.* + -ese.] The language or diction of the commercial world; an example of this.

1910 C. Headlam *Mem. W. Headlam* 44 This [*sc.* the language in which scholia and glosses were written] was a language by itself, like our Commercialese. **1942** Partridge *Usage & Abusage* 34/1 *As per*, 'in accordance with', is.. horrible commercialese. **1948** E. Gowers *Plain Words* xi. 92 Most officials write grammatically correct English. Their style is untainted by the silly jargon of commercialese.

‖ **commère** (komę̄r). Also **commere.** [See Cummer.]

1. See Cummer.

Jrnl. 151/1 The Bolsheviks retreated in a panic, killing their own commissars as they fled. **1932** J. G. Wright tr. *Trotsky's Hist. Russ. Revol.* xiv. 294 At the session of March 6 [1917], the Executive Committee considered it advisable to instal its own commissars in all regiments and all military institutions...One of the principal duties of the commissars was to keep watch over the political reliability of the staff and commanding officers. **1955** J. Carmichael tr. *Sukhanov's Russ. Revol.* I. iii. 62 Braunstein proposed that directives be given..for district committees to be formed, and for plenipotentiary Commissars to be appointed in each district to restore order and direct the struggle against anarchy and pogroms.

commissariat. Add: **III.** [tr. Russ. *komissariát.*] In the U.S.S.R., a government department. (In 1946 the title was replaced by 'Ministry'.)

1918 [see *commissar 2]. **1919** C. E. B. *Facts about Bolsheviks* 7 [The] Central Executive Committee..elects and works with the 'People's Commissaries' in a score of 'Commissariats' or Ministries. **1920** Goode *Bolshevism at Work* 28 He..is at the head of a very important Commissariat, that dealing with industries. **1922** E. & C. Paul tr. *Buharin & Preobrazhensky's ABC of Communism* vi. 182 The Central Executive Committee exercises at one and the same time legislative and executive functions...Its departments are known as the People's Commissariats, and its members work in these commissariats. **1952** *Economist* 15 Nov. 459/2 Ministries—as the Commissariats were renamed in 1946—were set up for every major branch of economic activity.

commissary. Add: **1. b.** = *commissar 2.

1917 *Times* 22 Mar. 7/2 The Commissaries of the Duma were arriving. *Ibid.* 11 Apr. 6/3 The department of international relations attached to the committee of workmen's and soldiers' delegates has the right to be represented by a commissary at the office of the Petrograd Telegraph Agency. **1920** Goode *Bolshevism at Work* 18 The Council of People's Commissaries. **1923** S. Leacock *Over Footlights* 161 She goes out stealthily to denounce her husband and Hootch to the commissaries of the police. **1929** Rappoport tr. *Larsons' Expert Service Soviet* 133 The Commissary for Finance, Sokolnikoff.

4. b. A store for provisions or other merchandise, esp. for issuing or selling articles to people engaged in a particular kind of work; *spec.* a dining-room or refectory in a film studio or the like. *U.S.*

1882 *Rep. Indian Affairs* (U.S.) 1 Sept. 151 Two new buildings..have been constructed; a commissary with offices, and a council-house with guard rooms. **1898** T. N. Page *Red Rock* 37 He is a clerk in old Bolter's commissary. **1929** A. C. & C. Edington *Studio Murder Myst.* xiv, Will you go with me to the commissary and have a drink? **1938** H. C. Nixon *40 Acres* 4 The store in which I grew up served as commissary, office and general clearing house for the plantation. **1955** 'N. Shute' *Requiem for Wren* vii. 202, I visited her film studio..and had lunch with her in the commissary.

c. = Commissariat 4. *U.S.*

1883 *Century Mag.* Sept. 672 This enforced idleness reduced our commissary to an alarming minimum. **1929** J. Parker *Old Army* 337 The Spanish ship Salvadora arrived with commissaries.

7. (Earlier U.S. examples.)

1780 R. Putnam *Memoirs* (1903) 154 Either a commesary store should be keept here or some extra provition should be lodged here for the supply of such small parties. **1848** J. S. Robinson *Sk. Great West* 10 If we cannot overtake the commissary wagons we shall have nothing to eat. **1887** *Harper's Mag.* Dec. 112/1 Commissary and butcher bills were still unpaid.

commissaryship. (Later examples.)

1808 J. Donovan *Let.* 23 Dec. in *Charges agst. Duke of York* (1809) 170, I am daily plagued about..the second battalion Paymastership, and the Commissaryship. **1811** J. Adams *Wks.* (1854) IX. 633 All applications..for commissions,..commissaryships, quartermasterships, &c.

commission, *sb.*[1] Add: **3. e.** The period of active service of a warship (see 5 c).

1882 W. H. White *Man. Naval Archit.* (ed. 2) vi. 231 The commanding officer of one of these ships has stated 'that they may go through a commission and never heel or roll more than one or two degrees'.

5. c. *out of commission*: also *transf.*, not in service or working order; out of use; unable to function. orig. *U.S.*

1904 Mrs. Stratton-Porter *Freckles* 181 I'll play and you'll sing, and we'll put the birds out of commission. **1907** M. C. Harris *Tents of Wickedness* iv. ii. 342 From those [houses]..there came little light, there was..a general look of being out of commission. **1954** R. Bissell *High Water* iii. 33 By the time we get to Dubuque the locks will be out of commission. **1956** A. H. Compton *Atomic Quest* 236 War plants whose buildings could be put out of commission by the explosion.

13. commission-agent (examples); *spec.* a bookmaker; **commission note,** a written promise to pay commission to an agent; **commission rank,** the rank of a commissioned officer.

1832 *Chambers's Jrnl.* I. 225/3 Thus the Consul is a vast commission agent. **1963** *Times* 12 June 19/2 E. Rose & Co. (Bookmakers), and Ray & Co. (Commission Agents). **1968** *Lebende Sprachen* XIII. 83/2 All orders for the purchase.. of securities..are carried out by the bank as commission agent. **1899** *Westm. Gaz.* 13 May 7/2 She did not know when or where she signed the commission note. **1908** *Daily Chron.* 15 Jan. 7/2 [He] signed the commission notes on the assumption that he did belong to the senior branch. **1898** *Westm. Gaz.* 8 Jan. 7/2 Facilities for promotion to commission rank in the Royal Marines.

commission, *v.* Add: **1. c.** Also *intr.* Of a ship: to commence active service, to be put in commission.

1915 'Bartimeus' *Tall Ship* i. 17 There's a super-Dreadnought commissioning soon. **1928** C. F. S. Gamble *N. Sea Air Station* i. 38 In the Admiralty announcement..it was stated that the Hermes was to 'commission on the 7th May as parent ship of the Naval Air Services'.

commissionaire. Add: **2.** (Further examples.) Now freq. used to designate the uniformed attendant at the entrance of a theatre, hotel, large shop, etc.

1895 A. Roberts *Adventures* x. 122, I quietly returned to the theatre, and had time to tell the story to the commissionaire. **1961** Bowman & Ball *Theatre Lang.* 73 *Commissionaire*, a British term for a doorman. **1970** *Oxford Mail* 11 July 3/1 The case would have hinged on the evidence of an independent witness, Mr. Rolfe, commissionaire of Holyoake Hall, Headington.

commissioner. Add: **1. c.** *High Commissioner*: (*a*) the chief officer of a colonial territory or dependency; (*b*) the chief representative in London of a British Commonwealth country.

1881 *Times* 30 June 9/6 Sir Alexander Galt, High Commissioner for Canada,..leaves Liverpool for the Dominion to-day. **1910** *Encycl. Brit.* VI. 776/1 In some British possessions in Africa and the Pacific the head of the government is styled high commissioner. **1935** *Chambers's Encycl.* III. 382/1 All the British Dominions and India have High Commissioners in London. **1946** *Ann. Reg. 1945* 157 A High Commissioner from the United Kingdom would be appointed to take over from the Viceroy duties connected with the representation of Britain in India.

commit, *v.* Add: **10. e.** *refl.* To enter into commitment (sense *6 c). Also *pass.*

1948 P. Mairet tr. *Sartre's Existentialism* 43 What counts is the total commitment, and it is not by a particular case or particular action that you are committed altogether. **1950** B. Frechtman tr. *Sartre's What is Literature?* p. vii, 'If you want to commit yourself,' writes a young imbecile, 'what are you waiting for? Join the Communist Party.' **1957** T. Kilmartin tr. *Aron's Opium of Intellectuals* iv. 127 A philosophy of 'commitment' which restricts itself to interpreting the commitment of others and does not commit itself.

commitment. Add: **6. b.** Also, liability; *pl.* pecuniary obligations.

1892 *Pall Mall Gaz.* 2 Mar. 7/3 The Murrieta commitments are enormous. **1897** *Daily News* 24 Apr. 9/4 'Bear' commitments have been largely closed during the past few days. **1927** *Daily Tel.* 5 Apr. 11/5 The Peterel will be an additional commitment. **1966** *Listener* 17 Mar. 375/1 The cuts in British forces and commitments.

c. [tr. Fr. *engagement.*] An absolute moral choice of a course of action; hence, the state of being involved in political or social questions, or in furthering a particular doctrine or cause, esp. in one's literary or artistic expression; moral seriousness or social responsibility in artistic productions.

1948 P. Mairet tr. *Sartre's Existentialism* 16 [An] important Sartrean concept—*engagement*—is here translated as 'commitment'. *Ibid.* 47 At the very heart..of existentialism, is the absolute character of the free commitment, by which every man realises himself. **1948** *Mind* LVII. 259 Commitment ends anguish, but what happens then is a matter of dispute among the different schools of Existentialism. **1948** *Partisan Rev.* Aug. 856 How are cultural interests affected by this struggle and do you think a writer should involve himself in it (as writer? as person?) to the point of commitment? **1955** H. Read *Grass Roots of Art* (ed. 2) iv. 84 This demand for the commitment, or 'engagement', of the artist ..originated in communist ideology. **1956** C. Wilson *Outsider* ii. 38 Sartre, whose theory of commitment or 'engagement'..led him to embrace a modified communism. **1961** J. Mander (*title*) The writer and commitment.

committal, *sb.* Add: **7.** The action of committing the body to the grave at burial; esp. *attrib.*, as *committal prayer, sentences, service.*

1854 *Chambers's Jrnl.* I. 207/1 The committal to the grave ..did not complete the funeral obsequies. **1892** *Daily News* 15 Nov. 3/4 The lesson was read by the Rev. A. Majendie, and the committal sentences by the Rev. Dr. Yule. **1898** *Ibid.* 24 Jan. 5/2 The committal portion of the service was taken by the Dean of Christ Church. *Ibid.* 7 Apr. 2/1 The committal prayers were read, though few in the dense crowd could hear them. **1927** *Observer* 28 Aug. 15 The Bishop of Aberdeen and Orkney conducted the committal service. **1968** *Globe & Mail* (Toronto) 13 Feb. 35/5 (Advt.), Committal service Wednesday afternoon at 2 o'clock at Nassagaweya Church Cemetery, Campbellville.

committed, *ppl. a.* Add: **b.** [tr. Fr. *engagé.*] Characterized by commitment (*6 c).

1950 B. Frechtman tr. *Sartre's What is Literature?* p. vii, The worst artists are the most committed. Look at the Soviet painters. **1952** *Times Lit. Suppl.* 8 Feb. 108/4 Both writers are..unprejudiced as far as being 'committed' will allow. **1954** V. Cronin *Golden Honeycomb* 198 It may be that the fresco is less obsessed by Death than by contemporary social conditions, that it is in fact 'committed' art. **1959** *Books of Month* Apr. 2/1 Christopher Logue..has become 'engaged', or 'committed', which means that he is striving to write poetry touching the everyday life of ordinary people. **1959** *Economist* 30 May 858/2 Communists are familiar with the idea of 'committed' literature. **1965** *Guardian* 10 Sept. 11/5 Bees..swarming..round the heads of committed young painters.

Hence **commi·ttedness.**

1958 *Listener* 17 July 103/1 The kind of self-detachment that makes contemporary talk of committedness sound rather silly.

committee. Add: **III. committee-woman,** a woman member of a committee or committees.

1853 Mrs. Gaskell *Let.* 6 Dec. (1966) 255 One or two of the more 'cautious' lady visitors,—committee-women, or whatever they call themselves. **1868** W. Collins *Moonstone* II. ii. vii. 188 A venerable committee-woman at the Mothers' Small-Clothes. **1965** J. Porter *Dover 3* vi. 66 Dame Alice was an experienced committee woman and never used one word if ten would do.

Commo (kǫ·mo). Also **commo.** Slang abbrev. of Communist. Cf. *Commie. Chiefly *Austral.* and *N.Z.*

1941 in Baker *Dict. Austral. Slang.* **1946** *Daily Express* 20 Dec. 2/4 The commos hate us and nobody is going to risk being thought pro-British in case the commos win. **1959** J. Cleary *Back of Sunset* 210 I've been reading how the Commos have eliminated all the flies in China. **1963** B. Pearson *Coal Flat* xx. 355 If they would just look at the ringleaders of this boycott they would soon find out they were commos.

commodity. Add: **8.** *attrib.*, as *commodity agreement, control, fetishism, market, price*; **commodity dollar** *U.S.*, a proposed unit of a form of currency the gold value of which is determined by the current market prices of certain basic commodities; also *attrib.*; **commodity exchange,** an organized market for the bulk purchase of certain commodities, a commodity market; **commodity rate,** a special rate charged by railways for transportation of particular commodities within a specified area.

1959 *Chambers's Encycl.* III. 787/2 International agreements for the marketing of foodstuffs and industrial raw materials in their unprocessed form are known as commodity agreements. **1943** P. L. Yates *Commodity Control* xi. 232 Neither buffer-stock nor quota schemes are sufficiently flawless to enlist our unqualified support as the sole method of commodity control. **1933** Warren & Pearson in *N.Y. Times* 23 July viii. 3/3 The Commodity Dollar. The compensated dollar is a proposal to establish by law a currency redeemable in gold. **1933** *Times* 13 Dec. 14/1 The adoption of the Fisher and Warren 'commodity dollar' system. **1934** *Planning* I. xxvi. 12 It is surprising how little ..those engaged in industry know about the working of the commodity exchanges. [**1887** S. Moore tr. *Marx's Capital* I. i. i. 41 (*heading*) The fetishism of commodities and the secret thereof.] **1942** P. M. Sweezy *Theory of Capitalist Developm.* ii. 35 In his doctrine of Commodity Fetishism, Marx was the first to perceive this fact...In commodity production, the basic relation between men 'assumes, in their eyes, the fantastic form of a relation between things'. **1962** H. E. Beecheno *Introd. Bus. Stud.* xi. 93 In commodity markets..raw materials are sold by those who grew or extracted them, or by merchants. **1930** *Ann. Reg. 1929* 68 Commodity prices fell still further in 1929. **1902** *Encycl. Brit.* XXXII. 135/1 If these special rates are published in the tariff, and are offered to all persons alike,..they are known as commodity rates.

common, *a.* Add: **6. c.** *common lodgings* or *lodging-house*, a lodging-house in which beds may be obtained for the night (see quot. 1860). (See also quot. 1952.)

1851 *Act* 14 & 15 *Vict.* c. 28 (*title*) An Act for the well ordering of Common Lodging Houses. **1860** *Act* 23 *Vict.* c. 26 § 3 The Term 'Common Lodging House' shall mean a House in which Persons are harboured or lodged for Hire for a single Night, or for less than a Week at a Time, or any Part of which is let for any Term less than a Week. **1888** [in Dict.]. **1891** [see Lodging-house]. **1902** *Encycl. Brit.* XXXII. 680/2 The class found in casual wards, shelters, and common lodgings. **1952** *Stroud's Judicial Dict.* (ed. 3) I. 542 There is nothing in s. 235 of the Public Health Act, 1936.., requiring the period of letting of lodgings in a common lodging house to be for a single night or for less than a week.

13. *common or garden*: see Garden *sb.* 5 c.

18. c. *common feeling* = Cœnæsthesis.

1837 [see Cœnæsthesis]. **1897** C. H. Judd tr. *Wundt's Outl. Psychol.* 161 The common feeling is always the immediate expression of our sensible comfort and discomfort.

19. a. *common turn*: see Turn *sb.* 5.

21. *common case*: a single word-form or grammatical case serving the syntactic and semantic functions of a number of different case-forms of inflected languages, as in modern English, where the uninflected base of nouns corresponds to the nominative, vocative, accusative, etc., cases of nouns in some other languages; also *attrib.*; *common cold* = Cold *sb.* 4 b; *common form*: (*a*) a form of probate in which the grant is made by the executor's own oath without opposition; (*b*) a customary form of words used in the pleadings in actions at common law; (*c*) a form of words common to documents of the same species; hence *colloq.*, a formula, mode of behaviour, etc., of general

application; *common informer*: see Informer 3; *common market*: a group of countries imposing few or no duties on trade with one another and a common tariff on trade with outside countries; *spec.* (with capital initials), the trade association of France, the German Federal Republic, Italy, Belgium, Holland, and Luxembourg instituted in 1958; also *attrib.*, *transf.*, and *fig.*; hence *common marketeer*; *common roll*: a roll of electors which includes in a single series the members of two or more different races; also *attrib.*; *common school* (earlier examples); *common-sex pronoun*: a pronoun applicable to both masculine and feminine; *common stock* (N. Amer.): ordinary shares; also ellipt. *common*.

1892 H. Sweet *New Eng. Gram.* I. i. 52 English has only one inflected case, the genitive (*man's*, *men's*), the uninflected base constituting the common case (*man*, *men*). **1894** Jespersen *Progress in Lang.* 166 The old nominative, accusative, dative and instrumental cases have coalesced to form a common case. **1961** R. B. Long *Sentence* ii. 36 Words of no other kind [than verbs] take the nominative and common-case forms of the personal pronouns as prepositive modifiers in this way. **1786** E. Sheridan *Jrnl.* (1960) 82 With proper care that disorder is now almost less than a common cold. **1872** [see Cold *sb.* 4 b]. **1937** H. G. Wells *Star Begotten* ix. 175 That beats us—just as the common cold beats us—or cancer. **1766** Blackstone *Comm.* II. 508 The executor..must prove the will of the deceased: which is done either in *common form*, which is only upon his own oath before the ordinary, or his surrogate; or *per testes*, in more solemn form of law, in case the validity of the will be disputed. **1797** Tomlins *Jacob's Law-Dict.* s.v. *Pleading* i. 2 Special Pleas..always advance some new fact not mentioned in the declaration; and then they must be [ed. *Granger* (1835) they formerly must have been] averred to be true, in the common form:—'and this he is ready to verify.' **1820** in Barnewell & Alderson *Rep. Cases K.B.* III. 451 If the argument on the part of the plaintiff prevail, the common form of pleading not guilty of the grievances is bad upon special demurrer. **1857** *Act* 20 & 21 *Vict.* c. 77 § 2 'Common Form Business' shall mean the Business of obtaining Probate and Administration where there is no Contention as to the Right thereto, [etc.]. **1871** [see Form *sb.* 9]. **1888** *Encycl. Brit.* XXIV. 572/2 Probate is confined as a rule to wills of personalty or of mixed personalty and realty, and is either in common form, where no opposition to the grant is made, or in solemn form, generally after opposition, when the witnesses appear in court. **1905** *Spectator* 18 Feb. 242 The article is what lawyers know as 'common form', and means simply that the nation..leaves it to its Executive to settle the details. **1931** *S.P.E. Tract* xxxvi. 517 If you should split all your infinitives, evidently the device would lose its peculiar efficacy: the locution would become mere common form. **1932** Kipling *Limits & Renewals* 14 The girl, of course, is in love with a younger and a poorer man. Common form? Granted. **1954** *Ann. Reg. 1953* 153 The provisions..for the gradual establishment of a European common market. **1956** *Planning* 17 Dec. 225 In a free trade area each participating country is allowed to fix its own tariff to countries outside the free trade area; a common market requires an external tariff. **1957** *Economist* 12 Oct. Suppl. 1 *The Common Market*: a treaty to set up a European Economic Community signed at Rome in March by six countries. **1958** *Punch* 1 Jan. 75/2 The Common Market countries very naturally are not prepared to admit Britain to partnership as a favoured nation. **1961** *B.B.C. Handbk.* (1962) 34 What prospects of new material, exciting material, from the United States and from Canada this opens up—a common market of the air. **1962** *Listener* 8 Mar. 445/1 In its relations with other Christian Communions it [*sc.* Anglicanism] moves carefully towards the goal of a common market of worship that may demand some sacrifice of national character. *Ibid.* 26 July 125/2 The same preferential treatment obtaining between the six Common Marketeers. **1963** *Ibid.* 17 Jan. 110/2 They [*sc.* the progressive African leaders] usually seek wider areas of economic co-ordination and have begun to talk of the possibility of an African common market. **1970** *Daily Tel.* 17 Apr. 2/8 The surgeon..predicted a future European 'common market' of transplant organs. **1955** *Times* 6 July 8/3 The congress demanded that the principle of the common roll should be granted now. **1958** *Economist* 1 Nov. 402/2 A trial run for the common roll elections due in 1961. *a* **1657** W. Bradford *Hist. Plimoth Plantation* (1899) 194 We have no comone schoole for want of a fitt person. **1795** in R. Boese *Public Educ. in City of N.Y.* (1869) 21 The establishment of Common Schools throughout the state. **1922** Jespersen *Lang.* xviii. 347 How convenient it is to have the common-sex pronouns *lu* (he or she), *singlu*, [etc.]. **1888** *Realty & Building* (Chicago) 27 Oct. 3/2 The St. Paul has not challenged particular attention since the first effects of its passage of the dividend on the common stock were spent. **1966** 'E. Lathen' *Murder makes Wheels go Round* i. 4 'We're thinking in terms of an offering of one million five of the common,' he said. **1968** *Globe & Mail* (Toronto) 13 Feb. B4/4 Investment in common stocks isn't a pat answer to inflation.

B. Delete † *Obs.* and read: *Obs.* exc. *U.S. colloq.* (Add later examples.)

1798 *Monthly Mag.* ii. 437/2 *Silly* is used for weak in body..*for common*, commonly. **1883** Sweet & Knox *Through Texas* (1884) 44 He had sort o' aggravated me more than common that morning. **1916** E. Porter *Just David* 75 We don't use this room common, little boy.

common, *sb.*[1] Add: **16.** *Ellipt.* for Common sense. *slang.*

1906 E. Dyson *Fact'ry 'Ands* i. 8 Do 'ave er bit iv common. **1936** J. Curtis *Gilt Kid* xvii. 175 Use a bit of common. **1946** B. Marshall *George Brown's Schooldays* 155 All it wants is a little intelligence or *nous* as the poets say or common or garden common. **1960** H. Pinter *Dumb Waiter* 16 You mutt... Have a bit of common. They got departments for everything.

commonage. Add: **1. c.** Also in *S. Afr.*

1893 *Westm. Gaz.* 23 May 6/1 A farm adjoining the Kimberley commonage. **1900** *Daily News* 24 Apr. 5/4 Two young Dutchmen acting as spies..were found hidden in a Kaffir hut on Barkly Commonage.

commoner. Add: **3.** *First Commoner*, the Speaker.

1885 *Peel City Guardian* 23 May 6/1 The 'First Commoner' who now occupies the chair. **1908** A. E. Steinthal tr. *Redlich's Procedure Ho. Comm.* 137 As 'first commoner' in the realm his place is immediately after the peers.

co·mmonership. [f. Commoner + -ship.] The position of being a commoner.

a **1907** F. Thompson *St. Ignatius Loyola* (1909) xiv. 288 He was unmolested in the heavenly commonership he had chosen. **1918** *Oxf. Univ. Gaz.* 6 Nov. 108 Accepted for a Commonership March, 1914.

commonize, *v.* Add: **1.** (Earlier example.) So **co·mmonizing** *vbl. sb.*

1863 G. M. Hopkins *Let.* 4 May (1956) 75 Yesterday I commonized for breakfast with Hardy. *Ibid.*, Commonizing is drawing your own commons at another man's rooms.

2. Delete (*U.S.*) and add examples. So **co·mmonizing** *ppl. a.*

1843 R. H. Horne *Orion* Pref., I have adopted the Greek mythological names..with a view of getting rid of commonizing associations. **1886** *Art Age* Nov. 43/3 There being a movement in favor of enameling wood, because from the expensiveness of the process it is not likely to be commonised by use in hotels, bar-rooms and railroad stations, as hard woods have been. **1895** *Westm. Gaz.* 13 Mar. 5/2 The reproduction of copyright pictures in this way was obviously calculated to do injury by 'commonising' the artist's design. **1918** Galsworthy *Five Tales* 226 Suppose..they commonised her, as Sunday clothes always commonised village folk!

common law. Add: **2.** *attrib.* **common-law marriage,** a marriage brought about without an ecclesiastical or civil ceremony but recognized in some jurisdictions as valid by the common law; so *common-law wife*.

1909 *Cent. Dict.* Suppl., Common-law marriage. **1934** J. O'Hara *Appt. Samarra* (1935) iii. 85 Marie was Lebrix's common law wife. **1953** *Stroud's Judicial Dict.* (ed. 3) IV. 3297 A common law wife is one who is married by a union which though informal was recognised as valid by the common law (*Blanchett* v. *Hansell*, [1943] 3 W.W.R. 275). **1958** *Church Times* 12 Sept. 3/2 It is notorious that, throughout the West Indies,..between half and three-quarters of all marriages are so-called 'common law' marriages, lacking the sanction of the Church. **1959** A. Salkey *Quality of Violence* iii. 48 The unfortunate common-law wives in the district.

co·mmonplacely, *adv.* [f. Commonplace *a.* + -ly[2].] In a commonplace manner.

1892 'G. Travers' *Mona Maclean* vi, Humanity will all be uniformly, hideously, commonplacely yellow!

commons, *sb. pl.* **2. c.** Delete 'formerly' and add later example.

1918 *Act* 8 *Geo. V.* c. 3 § 1 (3) Incapable of..sitting or voting as a member of the Commons House of Parliament.

common-sensible, *a.* Also **commonsensible.** (Further examples.)

a **1907** F. Thompson *St. Ignatius Loyola* (1909) x. 200 High sanctity..is eminently common-sensible. **1931** *Sat. Rev.* 6 June 819/2 The Archbishop of York's speech is ranked as the best; next to that Lord Newton's, witty, humorous, and commonsensible.

common-sensical, *a.* Add: Now in common use. Also **commonsensical.** (Further examples.) Hence **common-sensicality.**

1914 W. J. Locke *Fortunate Youth* xix. 288 You were right about the Divine common-sensicality. **1927** *Observer* 28 Aug. 9/4 This has always been a very commonsensical country. **1957** *Ibid.* 15 Sept. 5/3 His mind is fundamentally commonsensical, but capable of carrying the implications of common sense into far, obscure and almost dreamlike regions. **1958** *Times Lit. Suppl.* 10 Oct. 581/3 A hopeful commonsensicality takes the place of any depth or subtlety of argument. **1960** 'M. Cronin' *Begin with Gun* xiv. 162 The understanding commonsensical confidante.

commonwealth. Add: **4. b.** The title of the federated states of Australia.

1891 *Proc. & Deb. Nat. Australas. Convention* Mar.–Apr. p. cxxix, Draft of a bill to constitute the Commonwealth of Australia. *Ibid.* p. cxxx, The name 'The Commonwealth of Australia' or 'The Commonwealth' shall be taken to mean the Commonwealth of Australia as constituted under this Act. **1900** *Act* 63 & 64 *Vict.* c. 12 § 3 The people of New South Wales, Victoria, South Australia, Queensland, and Tasmania, and also, if Her Majesty is satisfied that the people of Western Australia have agreed thereto, of Western Australia, shall be united in a Federal Commonwealth under the name of the Commonwealth of Australia. **1925** M. Terry *Across Unknown Australia* i. 18 The Commonwealth is nearly as big as the United States of America. **1966** *Style Man. for Austral. Govt. Publications* 39 The ranks and decorations of members of the Commonwealth Parliament, senior Commonwealth officers and others, and the names and addresses of Commonwealth departments are shown in the *Commonwealth Directory*.

c. In full *British Commonwealth* (*of Nations*), the association of Great Britain and certain self-governing nations which were formerly dominions or colonies, together with all her

groups or gatherings of people; so *community song*, etc.
1966 *Lancet* 24 Dec. 1409/1 How did the creators of community-care programmes go so badly astray? 1968 *Brit. Med. Bull.* XXIV. 194/2 Community care has emphasized the need to standardize and expand the medical vocabulary in directions outside the immediate disease situation. 1915 *Nat. Educ. Assoc. U.S., Proc. 53rd Meeting* 687 (*heading*) Community center work. 1931 *Economist* 10 Jan. 57/2 The widely varying agencies for meeting distress, scattered as they are among religious orders,.. community centres and the like. 1934 *Discovery* Dec. 358/2 These Grith Fyrd Camps are permanent community centres, each of which accommodates, all the year round, young men up to fifty in number. 1959 *Manch. Guardian* 7 Aug. 5/2 A village hall..provides a community centre for concerts, whist drives, dances. 1921 *Rural Organiz.* (U.S.) 103 Some form of country-wide community chest. 1964 S. M. Miller in I. L. Horowitz *New Sociol.* 308 Securing representation on Community Chests and the like. 1959 C. V. Good *Dict. Educ.* (ed. 2) 108 *College, community*, an educational institution offering instruction for persons beyond the age of the normal secondary school pupil, in a program geared particularly to the needs and interests of the local area. 1962 F. Rudolph *Amer. College & Univ.* 487 In 40 states 160 community colleges developed out of one-time normal schools. 1969 *Northern Territory News* 11 July 7/3 And why call it a community college? It's just stopping short. What we should get is a university college, with technical training facilities. 1931 H. Read *Meaning of Art* II. 49 Hitherto the highest form of community-feeling has been religious. 1919 M. Shaw *Brit. Hymn Festival Bk.* Pref., Community hymn singing is very much in evidence to-day. 1879 Community life [see sense 8 a in Dict.]. 1951 R. Firth *Elem. Social Organiz.* iii. 119 The Maori social system, with ..its characteristic community life. 1959 *Manch. Guardian* 11 Aug. 5/1 Jordans is a monument to two causes—Quakerism and community living. 1922 S. Lewis *Babbitt* vi. 74 Ryland wore spats, he wrote long letters about City Planning and Community Singing. 1923 *Sackbut* Nov. 115 The Town Hall, where the weekly Community Singing was advertised to be held. 1927 (*title*) Labour community song book. 1943 J. S. Huxley *TVA* xii. 105 A real community spirit has developed in the new town. 1929 S. Cheney *Theatre* xxii. 501 The little theatres, and the larger community theatres built on the foundations they laid.

communize, *v.* Add: **b.** To make communistic, give a communistic form to.
1893 G. B. Shaw *Imposs. Anarchism* 16 The difficulty of communizing any branch of distribution without first collectivizing it. 1920 *Glasgow Herald* 20 Mar. 6 To Socialise, Communise, Bolshevise..any of the institutions on which our trade and commerce depend. 1921 *Public Opinion* 26 Aug. 202/2 To communise the teaching profession the Soviet Government resorted to abolition of all standards of education for teachers.

commutate (kǫ·miutēˈt), *v.* Restrict †*Obs. rare* to sense in Dict. and add: **2.** *Electr.* = *COMMUTE *v.* 6. So **co·mmutated** *ppl. a.*, **co·mmutating** *ppl. a.* and *vbl. sb.*
1890 W. James *Princ. Psychol.* I. 26 It is like the great commutating switch-board at a central telephone station. 1893 Maycock *Electric Lighting* II. vii. 228 There is a similarity in the methods of commutating the currents. 1896 D. C. & J. P. Jackson *Altern. Curr.* II. 261 When the commutating point is reached during the rotation of the commutator, the trailing brushes move on to intermediate segments. *Ibid.* 624 Commutated Armature. The armature may be wound with the coils so arranged that their conductors are in series when starting and in parallel when running. 1916 *Standardization Rules Amer. Inst. Electr. Engin.* 22 Commutating Machines. 1962 *Newnes Conc. Encycl. Engin.* 144/2 (*heading*) E.M.F. in commutating coils.

commutation. Add: **4. d.** *Linguistics.* Substitution, as a test of differentiation of phonemes, etc.
1952 A. Cohen *Phonemes of Eng.* ii. 25 The remedy suggested by Martinet against this phoneticism is the commutation test. Thus two successive sounds can only with certainty be said to represent two phonemes if they are both commutable, i.e. if by substituting them for another sound one gets a different word. 1953 C. E. Bazell *Linguistic Form* 6 The criterion of commutation, which elsewhere is the distributional criterion *par excellence*, is here simply one criterion among others. 1953 F. J. Whitfield tr. *Hjelmslev's Proleg. Theory Lang.* 46 A correlation in one plane, which in this way has relation to a correlation in another plane of a language, we shall call a commutation. 1964 E. Palmer tr. *Martinet's Elem. General Linguistics* iii. 73 Everything is based on the operation called commutation, the one which permitted us to oppose the first segment of *lampe* to that of *rampe*.
7*. *Algebra.* The reversal of the order of two quantities. Also *attrib.*, as **commutation rule,** a rule expressing the effect of such an interchange.
1852 J. J. Sylvester *Math. Papers* (1904) I. 318 The method of partial commutation. 1886 G. Chrystal *Textbk. Algebra* I. i. 16 The laws of association and commutation. 1935 Condon & Shortley *Theory of Atomic Spectra* iii. 45 From the basic commutation rules..we find that the vector *L* does not commute with itself. 1957 M. E. Rose *Elem. Theory of Angular Momentum* ii. 22 We find the usual commutation rule..$J_zJ_y - J_yJ_z = iJ_x$. The other commutation relations follow from considering infinitesimal rotations about the other two pairs of co-ordinate axes.
8. commutation passenger *U.S.*, a season-ticket holder on a railway; **commutation ticket** (earlier examples).
1856 W. H. Swift in *Mass. Railroads 1842–55* 14 Commutation or Season Passengers, so called. 1887 C. B. George *40 Yrs. on Rail* v. 89 You have had a great deal of experience in carrying commutation passengers in Boston. 1848 *Amer. Railroad Jrnl.* 29 July 481/3 Commutation tickets.., costing $5, are issued at Trenton, signed by 'B. Fish', which entitles the receiver to eight passages between Trenton and New Brunswick, by any line on the route. 1885 *Good Words* July 450/1 A single 'Commutation' ticket is given, numbered and dated on the day on which it was issued.

commutative, *a.* Add: **2.** (Earlier and later examples.)
1844 *Phil. Trans.* CXXXIV. 225 The commutative law, whose expression is πρu = ρπu. 1964 E. Bach *Introd. Transformational Gram.* vii. 159 Ordinary arithmetical multiplication and addition form groups of a special character called commutative (or Abelian) groups, since they obey the commutative law. 1965 Patterson & Rutherford *Elem. Abstract Algebra* iii. 76 A ring in which multiplication is commutative is called a commutative ring.

commutativeness (kǫ·miutēˈtivnės). [f. as next: see -ness.] Commutativity.
1949 *Mind* LVIII. 3 The combination of simple notions into a complex one is treated as a form of multiplication, sharing the commutativeness of the arithmetical variety. 1968 *New Scientist* 16 May 339/1 Commutativeness, which corresponds to symmetry.

commutativity (kǫ·miutėti·viti). [f. Commutative *a.* + -ity.] The ability of two or more quantities to commute (see *commute *v.* 7).
1929 Condon & Morse *Quantum Mech.* i. 14 The non-commutativity of the operation which passes for *p* in the wave equation and its associated *q*. 1947 H. Reichenbach *Elem. Symb. Logic* 149 The principle of commutativity of all-operators. 1964 *Camb. Rev.* 24 Oct. 51/2 The commutativity (or anticommutativity) of the field operators at space-like distances.

commutator. Add: **b.** *Electr.* An attachment, usually consisting of a ring of copper segments separated by insulating strips, connected with the armature of a dynamo, which, by revolving in contact with the brushes, directs and makes continuous the current produced in the armature coils of the machine. Also *attrib.*
1880 J. W. Urquhart *Electr. Lt.* 58 He placed the magnet vertically and revolved the coils about a horizontal axis, and added a commutator to make the currents flow in one direction. 1884 [see *aperiodic *a.*] 1884 S. P. Thompson *Dyn.-Electr. Mach.* 133 The only other dynamo in which a commutator at all resembling that of the Brush machine is used is Newton's dynamo. 1896 *Ibid.* (ed. 5) 395 One of these commutator coils. 1909 *Installation News* III. 62/1 High speed tends to undue friction and wear of the commutator surface. 1916 *Standardization Rules Amer. Inst. Electr. Engin.* 47 Brush friction of commutator and collector rings. 1959 *Which?* Aug. 86/2 Eleven machines had induction motors...The remaining 15 machines had commutator motors.
2. *Algebra.* **a.** Of the elements *a*, *b* of a group, the quantity $a^{-1}b^{-1}ab$. Also *attrib.*
1908 H. Hilton *Introd. Theory Groups* i. 4 The commutator of *a* and *b*. 1964 W. R. Scott *Group Theory* iii. 58 [*H, K*] will denote the subgroup generated by the set of all commutators [*h, k*] with *h* ∈ *H* and *k* ∈ *K*...The commutator subgroup..of *G* is the subgroup [*G, G*].
b. Of the matrices or operators *A*, *B*, the quantity *AB*−*BA*. Also *attrib.*
1935 Condon & Shortley *Theory of Atomic Spectra* ii. 25 We shall use the notation [*A, B*] for the simple commutator *AB*−*BA*. 1939 V. Rojansky *Introd. Quantum Mech.* ix. 326 The matrix αβ—αβ is the commutator of α and β. 1965 Biedenharn & Van Dam *Quantum Theory of Angular Momentum* 8 By commutator algebra..they derived all the matrix elements for..the complete Racah co-efficient.

commute, *v.* Add: **4. b.** Also, more generally, to travel daily or regularly to and from one's place of work in a city (by any means of conveyance); also *transf.* and *fig.* Hence **commu·ting** *ppl. a.* and *vbl. sb.* orig. *U.S.*
1889 *Cent. Dict.*, *Commuter*, one who commutes. 1906 *Daily Chron.* 25 Feb. 4 There are many business men who practically divide their time between New York and Chicago, and 'commute' (the American term for taking season tickets). 1934 *Amer. Speech* IX. 261 Of all commuting ladies the *élite*, They chase the glowing hours with flying feet. 1935 G. Hopkins in H. O. Sturgis *Belchamber* p. vii, Nobody 'commuted' in those days, not even the resident friend of an American living near Windsor. 1937 *Times* 24 Dec. 13/6 Small houses in the country, from which the men will 'commute' to New York, travelling to and fro each day by train. 1947 Auden *Age of Anxiety* iii. 76 A married tribe commutes, mild from suburbia. 1953 A. Huxley *Let.* 11 June (1969) 674, I wouldn't, if I were you, retire to a shack...The commuting will be a great burden. 1954 *Southern Daily Echo* 25 May, He spends the winter in the West Indies and summer in England, commuting back and forth like the migrating swallows. 1957 J. Braine *Room at Top* xii. 118, I thoroughly dislike commuting—people should live and work in the same place. 1959 *Cambridge Rev.* 16 May 507/1 The dons commute daily to the college. 1962 *Daily Tel.* 19 Mar. 15/4 The people who are 'something in the City' to-day mostly commute to Sussex and Surrey. 1966 *Listener* 14 July 49/1 Ken Russell and John Boorman commute with ease between the two worlds.
6. *Electr. trans.* To regulate (the direction of an electric current), *esp.* so that the direction of the current is made continuous.
1884 S. P. Thompson *Dyn.-Electr. Mach.* vii. 132 Each delivering alternate currents to a commutator, which commutes them to intermittent uni-directional currents in the brushes. 1902 *Encycl. Brit.* XXVII. 578/1 The current may be commuted. *Ibid.* 579/2 To commute its direction in any coil as it passes through the interpolar gap.
7. *Algebra. intr.* Of two or more algebraic quantities: to give an identical result in whatever order they are written down. Cf. Commutative *a.* 2.
1928 H. W. Turnbull *Theory of Determinants* xvi. 257 *K* commutes with... 1958 P. A. M. Dirac *Princ. Quantum Mech.* (ed. 4) iii. 24 It may happen as a special case that two linear operators ξ and η are such that ξη and ηξ are equal. In this case we say that ξ commutes with η, or that ξ and η commute. 1964 A. P. & W. Robertson *Topological Vector Spaces* viii. 145 Suppose also that *u* and *v* commute (i.e. *uov* = *vou*).

commuter. Add: (Earlier and later examples.) Also, one who commutes (see prec., 4 b); also *attrib.* orig. *U.S.*
1865 *Atlantic Monthly* XV. 82 Two or three may be styled commuters' roads, running chiefly for the accommodation of city business men and suburban residents. 1902 Kipling *Traffics & Discoveries* (1904) 50 He had his breakfast at 8.43 a.m. to the tick. There might have been a Long Island commuter. 1926 J. Reed *Ten Days that Shook the World* vii. 158 Trains went on further, however, than Tsarskoye..Our car was full of commuters and country people going home. 1931 *Economist* 21 Feb. 398/2 The problem of the 'border commuters' [between Canada and the U.S.A.]. 1932 *Times Lit. Suppl.* 17 Feb. 86/2 Many things familiar in America need explaining in Britain: 'commuters' zone' (season-ticket region). 1940 Auden *Another Time* 114 From the conservative dark Into the ethical life The dense commuters come. 1956 *Times* 28 June 9/5 Even five years ago a million commuters were coming daily into the City of London. 1962 *Daily Tel.* 22 June 14/2 Commuter services around London. 1963 *Punch* 6 Nov. 658/1 Beyond Kensington into the commuter-belt. 1968 J. T. Story *Something for Nothing* iii. 79, I live ten miles out. I'm a scooter commuter.

comp. Add: **c.** Abbrev. of Competition *sb.* 2.
1929 L. MacNeice *Blind Fireworks* 71 I'm going in for the beauty comp. 1963 J. Fowles *Collector* ii. 181 You ought to go in for beauty comps, he said.
d. Abbrev. of Accompaniment 2. Cf. next.
1946 Mezzrow & Wolfe *Really the Blues* 362 The piano sticks to a rhythmic function, with either a four-four tempo or strictly comp.

comp (kǫmp), *v. colloq.* (orig. *U.S.*). [Abbrev. of Ac)company *v.*] **a.** *intr.* In jazz, dance-music, etc.: to play an accompaniment, esp. with rhythmic chords. **b.** *trans.* To accompany; to play (chords, etc.) as an accompaniment. So **comping** *vbl. sb.*
1949 L. Feather *Inside Be-bop* ii. 59 The guitar..is employed like the piano to 'feed' or 'bop' the soloists by 'comping' with irregularly accented chords. 1955 Shapiro & Hentoff *Hear me talkin' to ya* 365 Count Basie is just about the best piano player, for comping anyhow. 1956 in R. S. Gold *Jazz Lexicon* (1964) 63 Basie 'comps' chords on the piano here and there. 1966 *Crescendo* Nov. 4/2 I'd done an occasional record in a solo capacity, but I'd never had to comp for anybody. *Ibid.* Dec. 8/2 Evans comps painstakingly behind him but there's not always complete agreement between them about basic chord structure.

compact, *sb.*² Restrict † *Obs.* to senses in Dict. and add: **e.** A small case for compressed face-powder, rouge, etc.
1921 *Daily Colonist* (Victoria, B.C.) 13 Oct. 7/3 (Advt.), Luxor Compacts in three shades. 1927 *Unnatural Death* vi. 69 A latch-key and a powder compact. 1929 *Punch Almanack* Give her lavender..rouge—3/6. 1958 introduction of several new articles into the jewellery trade, such as compacts, lipstick cases, etc.
f. (See next, 1 b.)

compact, *ppl. a.*¹ Add: (Now freq. with stress on first syllable.) **II.** *adj.* **1. b.** *spec.* Designating a light car having a short wheel-base. Hence as *sb.*
1960 *Economist* 13 Feb. 620/2 The compact cars, which are smarter, cheaper and more economical to run than the standard-sized ones. 1960 *Times* 10 Oct. 6/3 Americans are discovering that cars can be too big, so their idea of a 'compact' has a transatlantic twist. 1966 'E. Lathen' *Murder makes Wheels go Round* i. 5 The Plantagenet was fast becoming America's leading prestige car. No compact was selling better than the Drake.
2. c. Of sounds: (see quots.); *spec.* in *Phonetics*, open, saturated (opp. *diffuse*).
1930 E. R. Moul in *Amer. Jrnl. Psychol.* XLII. 550 There is a pre-spatial attribute of thickness in vision and audition...In both, certain types of experience are characterized as 'hard', 'opaque', 'compact'. 1952 R. Jakobson et al. *Prelim. Speech Analysis* 27 Compact phonemes are characterized by the predominance of one centrally located formant region (or formant). They are opposed to diffuse phonemes in which one or more non-central formants or formant regions predominate. *Ibid.*, Open vowels are the most compact, while close vowels are the most diffuse. 1956 L. G. Jones in M. Halle et al. *For Roman Jakobson* 251 Lax fricatives must be diffuse, tense fricatives compact. 1962 R. Jakobson *Sel. Writings* 638 Compact consonants are articulated in the velopalatal area of the mouth cavity, and diffuse consonants—dentals and labials—in front of this area.

compaction¹. Add: **1. b.** In recent use *spec.* the action or process of inducing the particles of a substance (as soil, concrete, etc.) to combine more tightly (see quots.). Also *attrib.*
1926 H. D. Hedberg in *Bull. Amer. Assoc. Petrol. Geologists* X. 1059 In a homogeneous series..the effect of gravitational compaction would be seen in an increase in rock density. 1954 *Gloss. Highway Engin. Terms (B.S.I.)* 16 *Compaction*, the process whereby soil particles are constrained by rolling or other mechanical means, to pack them more closely together, thus increasing the dry density of the soil. 1956 *Gloss. Terms Concrete (B.S.I.)* 12 *Compaction*, the process of inducing a closer packing of the aggregate particles in concrete by the reduction of voids. 1959 *New Scientist* 29 Oct. 801/2 Many thousands of soil compaction tests were made.

compactness. Add: **c.** Of sounds: the fact or quality of being compact (cf. *compact *ppl. a.*¹ II. 2 c).
1930 E. R. Moul in *Amer. Jrnl. Psychol.* XLII. 548 Observe and describe that aspect of the experience which you have called..compactness, depth,..and the like. 1949 Jakobson & Lotz in R. Jakobson *Sel. Writings* (1962) 427 The saturated vowels are characterized by a compactness of their formant-spectrum. 1951 R. Jakobson *Ibid.* 440 The dichotomy of *acuteness* and *gravity* and that of *diffuseness* and *saturation* (to avoid the ambiguity of this term, I now prefer *compactness*, suggested by S. S. Stevens). 1952 —— et al. *Prelim. Speech Analysis* 27 In the consonants, compactness is displayed by a predominant formant region, centrally located.

compactor. Add: **2.** [f. Compact *v.*¹ + -or.] A machine or device for compacting soil or other material.
1947 *Engineering News-Record* 16 Oct. 534/3 The surface of the runways..will be finally solidified by a 'super-compactor' unit. 1948 *Civil Engineering* (U.S.) Mar. 18/1 The body of the compactor consists of two steel plate load boxes, each mounted on dual wheels. 1966 R. Ashworth *Highway Engin.* viii. 142 Vibratory compactors may be either plate or roller type.

compactum (kǫmpæ·ktǫm). [f. L. *compactum*, neut. of *compactus* (see Compact *ppl. a.*¹).] That which secures a number of articles. **a.** A package constructed for a particular purpose. **b.** A container (of any size); *spec.* a wardrobe. Also *fig.*
1907 *Yesterday's Shopping* (1969) 722 The 'Compactum' (patented)...Combination tool cabinet, joiners' bench, and wood carvers' table. 1941 *Illustr. London News* 6 Sept. 309/1 A marvellous compactum: the rubber dinghy and all contents are strapped in a space 15 in. square next the figure, the parachute outside. 1948 J. & R. Hooper *Mod. Furnit.* i. 6 Single Fitted Wardrobe. This type is usually referred to as a Gentleman's Fitted Wardrobe, and is also known as a Compactum type with accommodation..for hanging suits ..Shoe Rails..three drawers..and a Sliding Trouser Fitting. 1953 H. Clevely *Public Enemy* iii. 10 He looked over the clothes in his compactum. 1958 *Spectator* 1 Aug. 161/2 He issues from his huge, hideous seaside compactum, ushered by a trumpet dragoman in gold and purple. 1959 *Listener* 28 May 953/1 A handbag containing a lipstick and compactum.

compadre (kǫmpā·dri). Chiefly *U.S.* [Sp., godfather, hence benefactor, friend. Cf. Gossip *sb.*] Companion, friend. (See also quot. 1939). Also, a term of address.
1834 A. Pike *Sketches* 99 Nay, compadre, an American cannot steal. 1850 L. H. Garrard *Wah-to-Yah* v. 77, I had apprehensions he would lie in wait with his *compadres*. 1895 M. McClelland *St. John's Wooing* 136 Tell me, compadre, how many loves have you had! 1939 G. Greene *Lawless Roads* viii. 202 A *compadre* meant a fellow godparent, a spiritual relationship regarded in Mexico as a close one. 1940 —— *Power & Glory* III. i. 213 When you get there, father, look up a compadre of mine.

‖**compagnon de voyage** (kõnpanʸoṅ də vwayāʒ). [Fr.] A travelling companion.
1768 Sterne *Sent. Journ.* I. 97 Do not one half of our gentry go with a hum-drum *compagnon du voiage* the same round. 1840 Thackeray *Shabby Genteel Story* in *Misc.* (1857) IV. 294 Her fair *compagnon de voyage*, whose name was Miss Runt. 1936 J. Buchan *Island of Sheep* iv. 56 The women!.. Gaiety girls—salaried *compagnons de voyage*.

companion, *sb.*¹ Add: **1. d.** *Astr.* The less brilliant member of a double star system; also, the member of a triple system that is distant from the other two closely spaced stars. Also, *companion star*.
[1782 W. Herschel in *Phil. Trans.* LXXII. 161, I have used the expression *double-star* in a few instances of the sixth class in rather an extended signification...I preferred that expression to any other, such as Comes, Companion, or Satellite; because, in my opinion, it is much too soon to form any theories of small stars revolving round large ones.] 1827 J. F. W. Herschel in *Mem. Astron. Soc. Lond.* III. 182 The star..was particularly pointed out to my notice..as one of the utmost difficulty, from the smallness of its companion. I saw it, however, double, without the least difficulty. 1867 W. A. Norton *Treat. Astron.* (ed. 4) xix. 256 Sirius is the largest of the double stars. It is attended by a minute companion star, at a distance of 10″. 1902 *Encycl. Brit.* XXV. 752/1 Partial eclipses caused by the interposition of a dark companion moving around them [*sc.* stars of the Algol type]. 1928 W. M. Smart *Sun, Stars & Universe* xiv. 215 The faint companion of Sirius. 1964 R. H. Baker *Astron.* (ed. 8) xiv. 414 A common type of triple system is represented by Alpha Centauri, where the binary is attended by a remote companion, Proxima.
3. b. An appliance uniting several objects in a single set, as *smokers' companion, travellers' companion*.
1895 *Montgomery Ward Catal.* 118/1 Whitewood Box Scholars' Companion: furnished with lead pencil, pen, penholder, slate pencil and six inch wood ruler. 1907 *Yesterday's Shopping* (1969) 86 Smokers' companions...No. 4825/2. A great convenience, can be stood beside the chair. *Ibid.* 108/3 Shaving Companion..Nickel plated shaving set, tray, mug, and bowl..mirror, with brush complete—18/9. *Ibid.* 188/3 The 'Sirram' Travellers' Companion..For making tea when touring, boating, &c...Comprises kettle,.. stand, spirit stove,..spirit tin, and muslin tea infuser. 1939–40 *Army & Navy Stores Catal.* 357/3 Travelling Companion, containing face towel, toothbrush, soap and toothpaste. *Ibid.* 769/2 Hunting Companion in best pigskin folding case containing two boot pulls, two boot lifts, combined shoe horn and button hook—49/6. 1969 E. H. Pinto *Treen* 339 Smokers' Companions, small fitted tables to stand alongside 'father's armchair', enjoyed their heyday between 1830 and 1914.
10. companion-cell *Bot.*, a specialized elongated parenchymatous cell one or more of which is connected to most sieve-tubes in the phloem of some flowering plants; **companion-set,** a set of fireside implements on a stand.
1887 W. Hillhouse tr. *Strasburger's Bot.* 146 Sieve-tubes, whose companion-cells are the cells with narrower cavities, which follow towards the exterior. 1938 E. C. Miller *Plant Physiol.* (ed. 2) xii. 849 A companion cell has numerous simple pits connecting with the sieve tube with which it is associated. 1926–7 *Army & Navy Stores Catal.* 275/1 Hearth Furniture..Companion Set. Polished Brass—53/-. 1948 J. Cannan *Little I Understood* iv. 58 The shining brass fender, the companion set, the tasteless tiles. 1959 K. Waterhouse *Billy Liar* 50 The terrace house with the linoleum squares, and the novel horseshoe companion-set in satin-brass.

companionate, *a.* Restrict † *Obs.* to sense in Dict. and add: **2.** Phr. *companionate marriage* (occas. *mating*), a form of marriage which provides for divorce by mutual consent and in which neither partner has any legal responsibilities towards the other; = Cohabitation 2. Also *ellipt.*
1924 M. M. Knight in *Jrnl. Social Hygiene* X. 257 (*heading*) The companionate and the family—the undeserved division of an historical institution. 1926 E. R. Groves *Drifting Home* ix. 210 In recent years a rival of the orthodox family has appeared in what is well termed the *companionate*. *Ibid.* 211 The great majority of the companionate matings are from the first attempts at permanent childlessness. 1927 Lindsey & Evans *Companionate Marriage* (1928) p. v, Companionate Marriage is legal marriage, with legalized Birth Control, and with the right to divorce by mutual consent for childless couples, usually without payment of alimony. 1927 *Weekly Disp.* 26 June 1 There has arisen a group within the Church in America which sanctions an 'open mind' on the subject of relations between the sexes without marriage, provided birth control is exercised. This group employs the high-sounding phrase of 'companionate marriage'. 1932 B. Russell *Educ. & Social Order* ix. 125 Continence during the years of adolescence would impose no intolerable burden if there were a prospect that the necessity for it would ease at about..twenty. This could be secured by Judge Lindsey's system of companionate marriage.

company, *sb.* Add: **1. b.** *to keep company*: also of things.
1849 Geo. Eliot *Let.* 4 Dec. (1954) I. 321 If you are anxious to publish the translation in question I could.. finish the Tractatus Theologico-Politicus to keep it company. 1873 [see Dict., 1 c].
7. b. Add to def.: Usu. in phr. *and Company*. Also *transf.* (cf. *Co.³).
1898 G. B. Shaw *Our Theatres in Nineties* (1932) III. 318 This..is the distinction between Marlowe and Company and the firm of Beaumont and Fletcher. 1963 *Listener* 7 Feb. 261/3 The high-placed moderates on the Parliamentary side (Essex, Manchester, and company).
10. *company-manners* (earlier example); (sense 7) *company car, director, promoter* (cf. Promoting *vbl. sb.* 1 and *ppl. a.* 2), *store* (U.S.), *tenement*; *company-owned* adj.; **company man** (orig. *U.S.*), one who is thought by his fellow-workers to be excessively attached to, or uncritically to share the attitudes of, the organization for which he works (see also quot. 1921); **company sergeant-major,** the senior warrant officer of a company; **company town** (orig. *U.S.*), a town which is dependent upon a particular commercial company; **company union** (see quot. 1921); **company work,** needlework, etc., which might be done in the presence of guests.
1961 *Guardian* 6 May 14/3 The Budget proposal to limit the capital tax allowance on company cars to £2,000. 1900 *Hansard Commons* 26 June 1171 Rather undue importance is attached to the question of the qualification of company directors. 1963 *Observer* 3 Nov. 33/1 'Company director' and 'model' are useful euphemisms for those who appear in dubious court cases. 1921 *Bull. Bureau of Business Research, Harvard Univ.* XXV. 17 *Company Men.* 1. Employees who are looked upon by some of their fellow workers as favoring too strongly the interests of the employer. 2. A term applied by union members to men who are engaged or suspected to be engaged by the employer for espionage; sometimes referred to ironically as 'loyal workers'. 3. Salaried employees. 1969 *Listener* 24 July 125/3 One drawback of big companies like IPC is that they produce company men. 1969 *Guardian* 27 Sept. 9/8 The most dedicated 'company man'..will tell you what a fool he has been to stay with the firm. 1844 E. Sewell *Amy Herbert* II. xviii. 13 What are sometimes called 'company manners',—not meaning exactly affectation, but a manner approaching to it, which is not quite natural. 1907 *Westm. Gaz.* 24 June 2/1 The considerations which apply to (say) municipal trams cannot apply to company-owned tubes. 1893 W. S. Gilbert *Utopia* 1, A Company promoter this. 1894 Company promoter [see Promoter 1 c]. 1928 B. Russell *Sceptical Essays* 65 The phraseology..is that of the company-promoter rather than the philosopher. 1889 *Queen's Regs.* I. vii. 78 The following positions held by Non-Commissioned Officers and men of the army are ranks:—.. Troop, battery, or company serjeant-major. 1913 *Army Order* 323 § 5 In each company there will be eventually a company serjeant-major and a company quartermaster-serjeant. For the present the duties of these appointments will be carried out by the existing colour-serjeants. 1919 G. K. Rose *2/4th Oxf. & Bucks Lt. Infty.* 125 Moberly, Company Sergeant-Major Cairns, and Guest..greatly distinguished themselves in the task of maintaining this exposed position. 1872 *Harper's Mag.* Nov. 841/1 A 'company store', which is so common elsewhere, especially with joint-stock companies. 1907 E. Wharton *Fruit of Tree* I. ii. 21 A row of operatives' houses..then the company 'store'. *Ibid.* II. xii. 187 The suppression of the company tenement..struck at the roots of the baneful paternalism. 1933 E. Caldwell *God's Little Acre* vii. 107 They passed through the other company towns,..looking out at the humming mills. 1963 P. Willmott *Evol. Community* ii. 16 The firm does not dominate the estate. Dagenham is far from being a 'company town'. 1917 R. F. Hoxie *Trade Unionism* 51 There are virtually no 'company' unions in Great Britain. 1921 W. R. Browne *What's What in Labor Movement* 90 *Company Union*, an organization of workers within a particular shop or establishment, and having no connection or association with what employers call an 'outside union'. 1962 *Listener* 31 May 947/1 To smash what was known as the company union. 1844 C. M. Yonge *Abbeychurch* xi. 232 'You have been six months braiding that frock.' 'Oh! that is company work... I began it at Merton Hall.' 1864 Mrs. Gaskell *Wives & Daughters* vi, Having arranged her hair and dress, and taken out her company worsted-work. 1952 G. Raverat *Period Piece* viii. 148 In those days..every lady had a piece of 'company work' in hand.

comparability. (Later examples.)
1934 *Planning* II. xxxi. 3 Most statistics and surveys need..thorough co-ordination and standardisation in order to secure their accuracy and comparability. 1959 B. Wootton et al. *Social Sci. Pathol.* i. 21 The claims of uniformity and comparability seemed to prevail over those of up-to-dateness. 1960 *Guardian* 25 June 4/1 The Guillebaud Report accepted..that railway rates should be in line with those in 'comparable' industries...Railwaymen will naturally feel that the principle of 'comparability' should be applied again. 1961 *Ibid.* 6 Nov. 16/1 Comparability—the principle of fixing the wages of public employees by reference to those of people who do similar work in private industry.

comparatist (kǫ̆mpæ·rătist). [f. Compara-t(ive *a.* + -ist. Cf. F. *comparatiste.*] One who uses comparative methods in studying language or literature; a student of comparative linguistics or comparative literature.
1933 Bloomfield *Lang.* xx. 364 The effect of sound change ..as it presents itself to the comparatist, will be a set of regular phonemic correspondences. 1937 [see *anti-historicist* (*anti-¹ III. 5 and 6)]. 1946 R. Wellek in Knickerbocker *20th Cent. Eng.* I. 72 The conception of 'general literature', as opposed to the isolated and isolating study of influences as practiced by the comparatists. 1951 *Trans. Philol. Soc.* 1950 181 Since there is a letter-by-letter transliteration in general use by comparatists, this has been utilized to represent the Armenian spelling. 1957 W. S. Allen *Ling. Stud. Lang.* 8 Mar. 1 The Geneva comparatist and general linguist, Ferdinand de Saussure. 1962 Malkiel in Householder & Saporta *Probl. Lexicogr.* 16 In practice one can hardly draw a sharp line of demarcation between isolationists and comparatists.

comparatival (kǫ̆mpærătəi·văl), *a.* [f. Comparative *a.* 2, *sb.* 1.] Belonging to the comparative degree.
1900 F. Hall in *Nation* (N.Y.) 22 Mar. 220 Though its termination is comparatival, yet, primarily,..it denoted addition.

comparative, *a.* Add: **A. 8.** *Comb.*, as **comparative-historical** *a.*, both comparative (sense 1 b) and historical; using comparative methods for historical investigation.
1937 J. Orr tr. *Iordan's Introd. Romance Ling.* iv. 309 Material which is the product of comparative-historical research. 1964 I. L. Horowitz *New Sociol.* 22 What is quite clearly needed to make a comparative-historical sociology professional.

comparativist. Delete *rare* and add: *spec.* one who studies comparative linguistics or comparative literature. Also *attrib.* or as *adj.*
1922 Jespersen *Language* iii. 66 Classicists were no doubt perfectly right when they reproached comparativists for their neglect of syntax. 1965 *Language* XLI. 118 The rehabilitation of comparativist techniques. *Ibid.* 192 The comparativists of the 1870's spoke of 'speech sounds' with perfect confidence.

comparator. Add: (Further examples.) Hence, any of various devices which measure or otherwise check manufactured articles, or the output or performance of a machine, etc., by comparison with a standard (see quots.).

1904 Goodchild & Tweney *Technol. & Sci. Dict.* 121/2 *Comparator* (*Phys.*), an instrument for comparing a given scale of lengths with a Standard Scale. 1922 F. W. Aston *Isotopes* iii. 60 The most accurate measurements were made on a comparator. 1955 *Sci. News Let.* 22 Oct. 272/3 Hand comparator for judging surface roughness of mass-produced parts permits metal surfaces to be compared. 1959 *News Chron.* 28 Nov. 6/8 At many hi-fi showrooms you can hear the relative performances of different speakers by means of a 'comparator'. 1961 *Times* 31 May 18/4 An instrument known as a comparator which can test 6,000 resistors an hour. 1962 *Gloss. Autom. Data Processing (B.S.I.)* 97 *Tape comparator*, a machine which automatically compares two supposedly identical paper tapes row by row and stops when there is a discrepancy.

comparing, *vbl. sb.* (Later example.)
1957 *Eng. Lang. Teaching* XII. 8 Speech acquires structure through incessant comparings.

comparison, *sb.* Add: **9.** *attrib.* and *Comb.*, as **comparison eyepiece,** an eyepiece designed for use with two similar microscopes so that the images from both can be viewed simultaneously; **comparison microscope,** a microscope that enables images formed by two objectives to be viewed simultaneously; **comparison plate,** each of the photographic plates of a planet, etc., taken at different stations or times and used for comparison in astronomical research; **comparison spectrum,** a spectrum formed for comparison, wave-length by wave-length, with the spectrum under observation.
1940 R. M. Allen *Microscope* vi. 162 The comparison eyepiece..receives images from two different microscopes and combines them into a single eye lens. 1940 R. Morrish *Police & Crime-Detection* viii. 74 An indispensable instrument for all laboratories is the Comparison Microscope. By means of this microscope two objects can be viewed in the same field of vision. 1958 G. H. Needham *Pract. Use Microscope* vi. 87 The comparison microscope with separate microscopes and attachable comparison eyepieces to unite them is the one usually preferred. 1905 *Westm. Gaz.* 12 Oct. 12/1 The probable absence of good comparison plates from other places—whereby the planet could be identified..—is very disappointing. 1877 G. F. Chambers *Descriptive Astron.* (ed. 3) x. ii. 843 The light from the terrestrial substance which gives the comparison spectrum. 1897 *Daily News* 18 June 8/3 When the problem had to be solved of a satisfactory introduction of light for the comparison spectrum.

compartment, *sb.* Add:
5. d. *Forestry.* (See quots.)
1895 Schlich *Man. Forestry* III. iv. 293 Each working circle..must be further divided. The unit of that division is the compartment. 1905 *Terms Forestry & Logging* 7 *Compartment*, the unit of area treated in the working plan. The size and the shape of compartments are determined mainly by topographic features. *Ibid.* 23 *Stand method*, that method of conservative lumbering in which reproduction is secured from self-sown seed by means of successive cuttings. ..Syn.: compartment system. 1950 *Q. Jrnl. Forestry* XLIV. 59 It is..highly desirable to divide the woods into conveniently sized and permanently demarcated compartments. ..Each compartment may usefully range from about 10 acres up to not more than 25 acres in extent.
6. b. Used (esp. *attrib.*) of methods of dealing with business in sections, as of the parliamentary rule to facilitate the passing of a bill by dealing with it in separate portions and allotting a limit of time for the discussion and closure of each.
1893 *Westm. Gaz.* 5 July 2/2 At the close of each compartment he [*sc.* the Chairman] simply becomes the instrument of the House in recording its decision...Closure by compartment must be supplemented with closure by clause. 1896 *Daily News* 25 Feb. 5/6 The necessity of having the compartment principle engrafted on the rule. 1946 *Erskine May's Law of Parl.* (ed. 14) 449 Special orders of the House, known officially as 'Allocation of Time' orders, and colloquially as 'Guillotine' or 'Closure by compartment' orders.
8. (sense 5) *compartment car, train, vehicle*; (sense 5 c) *compartment-built* adj.; **compartment boat,** a boat built with watertight compartments.
1894 *Daily News* 1 Sept. 6/7 The compartment boats sinking in Goole Docks. 1892 *Ibid.* 4 July 7/3 She is compartment built,..and this has saved her from an immediate catastrophe. 1928 *Manch. Guardian Weekly* 26 Oct. 335/1 They live in compartment cars, which are more or less like the European wagonlit. 1907 *Daily Chron.* 6 Sept. 7/6 The inconvenience on the car trains is certainly not so great as on the compartment trains. 1908 *Westm. Gaz.* 17 Dec. 10/1 Single-decked compartment vehicles.

compartment, *v.* [f. the sb.] *trans.* To divide or put into compartments (*lit.* and *fig.*). Chiefly in pa. pple. Cf. Compartmented *ppl. a.*
1930 G. Manning-Sanders *Burnt Man* i. 11 All along the length of one wall, neatly stacked and compartmented, thousands of rolls of wallpaper. 1950 Partridge *Here, There & Everywhere* 54 Language..should not..be compartmented, for such compartments fail to correspond with the facts. 1964 *Punch* 16 Dec. 911/2 Affairs can no longer be compartmented.

compartmental, *a.* Add: (Further examples.) Also, of or pertaining to a compartment or compartments.
1894 *Westm. Gaz.* 23 May 4/3 An extension of the compartmental closure [in Parliament]. 1905 *Daily Chron.* 2 June 3/3 The Yezdi view of life is isolated, compartmental. 1933 *Planning* I. xvi. 15 The compartmental system does not stop at that.

compartmentalization (kǫmpāɹtme:ntăləizēˈ·ʃən). [f. next + -ation.] The action or state of dividing or being divided into compartments or sections; (see also quot. 1958). Chiefly *fig.*
1923 J. S. Huxley *Uniqueness of Man* i. 27 The much more rigid compartmentalization of animal mind and behaviour. 1946 *New Yorker* 17 Aug. 51 These scientists had worked for the Manhattan District..and the Army, temporarily abandoning its policy of 'compartmentalization' of information, had recently brought them together. 1948 L. Spitzer *Linguistics & Lit. Hist.* 164 The sciences are relieved of the compartmentalization and isolation and become accessible. 1958 J. Cleugh tr. *R. Jungk's Brighter than Thous. Suns* vii. 121 They erected invisible walls round every single little branch of research... This so-called 'compartmentalization' was decreed in spite of the fact that intensive security measures had already been taken. 1962 *Listener* 18 Jan. 120/2 It would be too big a job to trace the history of this compartmentalization of our minds. 1964 E. A. Nida *Toward Sci. Transl.* vii. 145 The fact that in some brain injuries the capacity of bilingual persons has been reduced to the speaking of one language and not the other has led to the belief that there may be compartmentalization in the brain. 1967 *Jane's Surface Skimmer Systems 1967–68* 67/1 This was achieved by a very large fan, and by compartmentalisation of the base.

compartmentalize (kǫmpāɹtme·ntăləiz), *v.* [f. Compartmental *a.* + -ize.] *trans.* To separate into compartments; to divide absolutely. Chiefly *fig.* Hence **compartme·ntalized** *ppl. a.*
1925 E. S. Jones *Christ of Indian Rd.* (1926) x. 209 India ..will correct much of our compartmentalised and tentative religious thinking and acting. 1939 *New Statesman* 23 Dec. 932 Their teaching has become so compartmentalised. 1945 *N.Y. Times* 7 Aug. 4/6 The work has been completely compartmentalized so that while many thousands of people have been associated with the program in one way or another no one has been given more information concerning it than was absolutely necessary to his particular job. 1955 *Bull. Atomic Sci.* Jan. 13/2 It is feasible to 'compartmentalize' the atomic knowledge in an intellectual vacuum and to evade the obligation of drawing practical conclusions. 1957 K. A. Wittfogel *Oriental Despotism* 197 Innumerable mountain ranges compartmentalized the great Far Eastern islands.

compartmentation (kǫ̆mpārtmentēˈ·ʃən). [f. *compartment *v.* + -ation.] = *compartmentalization.
1956 N. Pevsner *Englishness of English Art* 82 (*caption*) Newark: perpendicular compartmentation. 1967 *Jane's Surface Skimmer Systems 1967–68* 44/1 SKIP–1 (Skimmer Investigation Platform) utilises a peripheral jet with cross-jet compartmentation, dividing the platform into three longitudinal compartments.

compass, *sb.*¹ Add: **D. compass course,** a course steered by compass; spec. *Naut.* and *Aeronaut.*, the line of direction indicated by the compass but not actually taken by a boat or airship, allowance having been made for deviation caused by wind, currents, etc.
1851 *Family Friend* IV. 294/1 Compass, which registers upon paper the compass course which a vessel has been steered in for twenty-four hours. 1880 C. R. Markham *Peruv. Bark* 201 To shape a direct compass-course over the cordilleras from Sandia to Vilque. 1900 *Westm. Gaz.* 13 Jan. 3/2 It is not a simple matter to keep a given compass course at night. 1916 H. Barber *Aeroplane Speaks* 39 Our compass course will then be in the direction A–E. *Ibid.* 41 A very simple way of calculating the compass course.

compassionate, *a.* Add: **1. d.** Granted out of compassion, without legal or other obligation; of or pertaining to allowances, leave, etc., so granted.
1830 *Pensions of Widows* 7 July in *Coll. Warrants & Regs. Army on Finance* (1844) 191 Allowances on the Compassionate List to the legitimate Children of deceased Officers may be given..provided it be shown..that they actually require assistance from the Compassionate Fund. Applications for Compassionate Allowances should specify the particulars pointed out in the Form annexed. 1942 *Airman's Guide to Rights & Duties* I. v. 21 Compassionate Leave. This will be granted, on application, for special reasons, such as serious family illness or war damage to the airman's home. 1945 E. Taylor *At Mrs Lippincote's* xxiii. 195, I can probably get you a couple of days extension on compassionate grounds. 1945 E. Waugh *Brideshead Revis.* 8 In the company office there was a crop of minor charges and requests for compassionate leave.

compatibility. Add: **2.** Specific scientific and technical uses. **a.** *Chem.* The capacity of substances to tolerate each other without chemical reaction or undesirable effect, esp. such capacity in drugs.
1898 E. W. Lucas *Pract. Pharmacy* xl. 278 Having.. carefully considered the ingredients, their compatibility, dose, &c., the labels should next be written. 1929 Hackh *Chem. Dict.* 195/1 *Compatibility*, the property of two or more drugs not to impair their respective action when administered together. 1939 H. Bennett *Stand. Chem. & Techn. Dict.* 159/2 *Compatibility*, ability to mix together without separation or reaction. 1957 *Financial Times Ann. Rev. Brit. Industry* 17/3 Graphite moderated reactors using liquid sodium as a coolant..have severe compatibility problems. 1967 *Amer. Jrnl. Hospital Pharm.* XXIV. 434/1 We are continually being asked by hospital pharmacists and physicians for compatibility data encompassing a wide variety of parenteral medications.
b. *Biol.* The capacity of blood, tissue, or an organ to be introduced into the body of a (specified) individual without exciting a destructive reaction. Also *attrib.*
1918 *Jrnl. Immunol.* III. 93 Before the operation of blood transfusion is carried out it is necessary to test the compatibility of the prospective donor's blood with that of the patient. 1928 K. Landsteiner in Jordan & Falk *Bacteriol. & Immunol.* lxviii. 906 With proper preliminary compatibility tests..transfusions are almost free of danger. 1948 *Jrnl. Genetics* XLIX. 87 Compatibility or incompatibility in transplants. 1961 P. L. Mollison *Blood Transfusion Clin. Med.* (ed. 3) xii. 418 The term 'compatibility test' usually refers to an investigation of the patient's serum for the presence of antibodies which react with the donor's red cells *in vitro*. 1968 R. Ceppellini in Rapaport & Dausset *Human Transplantation* iii. 26 In man..compatibility in the strict sense of the word, i.e., the indefinite survival of the graft without rejection crises, has not yet been observed, with the exception of isogeneic grafts exchanged between monozygotic twins.
c. *Bot.* Capacity for cross-fertilization; of a scion, the ability to unite with a stock.
1935 *Biol. Abstr.* 5627 Vigorous rootstocks show various degrees of compatibility for pear scions. 1968 *Heredity* XXIII. 124 The relative levels of compatibility of Brussels sprout, cabbage..and sprouting broccoli were assessed.
d. *Television.* The capacity of colour signals to be received in black-and-white on monochrome receivers. Also *attrib.*
1947 *Proc. Inst. Radio Engin.* XXXV. 861/2 Without regard to compatibility, visual observations alone indicate that the properties of flicker and resolution of images containing red and green components are sufficiently similar to [those of] monochrome images. 1950 in *Amer. Speech* (1951) XXVI. 209/1 They doubted the compatibility prospects of the system. 1954 Zworykin & Morton *Television* (ed. 2) xvii. 773 The lack of compatibility and the decreased definition are the two most serious weaknesses of the field-sequential system. 1957 *BBC Handbook* 136 These transmissions [of colour television]..were put out in the London area...Their purpose was to explore under practical reception conditions the degree of compatibility of the system being used.
e. *Sound reproduction.* The capacity of a stereophonic broadcast, record, etc., to be received or played on a monophonic receiver, record-player, etc. Also *attrib.*
1959 *Audio Engin.* May 17/2 One solution to the two-channel problem is gained through the use of a compatibility circuit. *Ibid.* 18/2 A time delayed replica of the direct signal is added to the alternate channels in order to achieve compatibility. 1963 D. L. A. Smith *Princ. High-Fidelity Sound Engin.* v. 82 To meet the compatibility requirement, one transmitted channel is invariably modulated by the signal ½(L+R).

compatible, *a.* Add: **2. d.** Specific scientific and technical uses (see quots. and cf. *compatibility 2).
1899 *Practitioner* Dec. 622 The apparently isomeric sulphocarbolates of zinc and sodium do not react as sulphates, and are compatible with lead salts. 1926 *Jrnl. Amer. Med. Assoc.* 22 May 1596/1 In the present series of forty persons are to be found about 40 per cent whose serum or white blood cells seem to be mutually compatible with all others. 1929 Hackh *Chem. Dict.* 195/1 *Compatible*, any pair of drugs which do not interfere or react chemically, physically or therapeutically with one another. 1935 *Biol. Abstr.* 5627 Vigorous scions of pears are more compatible because they become more readily adapted to the high conc. of protoplasm in the rootstocks. 1938 Fritsch & Salisbury *Plant Form & Function* xxiii. 252 Grafting and budding are usually only possible between related types of plants. Those which can be associated successfully are termed compatible, but both compatible and incompatible strains..may occur. 1940 *Chambers's Techn. Dict.* 183/1 *Compatible* (*Bot.*), capable of self-fertilisation. 1949 *Radio Corp. Amer. Rev.* X. 523 The all-electronic color television system described is a fully compatible system. 1953 *Gloss. Paint Terms (B.S.I.)* 11 *Compatible. a.* Two or more paints or varnishes which can be mixed in bulk without producing any undesirable effects such as precipitation, coagulation, gelling, are said to be compatible. *b.* Different coats of paint which can be associated in a painting system without producing undesirable effects are said to be compatible. 1955 Gouriet *Introd. Colour Television* II. ii. 41 For a system to be fully compatible,..it should provide a signal which conforms substantially to present-day black-and-white standards. 1958 *Times* 20 Jan. 10/4 Stereophonic records will not be 'compatible', that is, able to be played on ordinary gramophones. 1958 *Listener* 16 Oct. 605/2 The system which is being used for the forthcoming series of stereophonic sound broadcasts..has the serious disadvantage of not being compatible; *i.e.*, properly balanced reproduction will not be obtained when listening to either one channel alone. 1966 *Melody Maker* 8 Oct. 13/1 All Musicassettes are compatible which means they can be played on mono or stereo equipment. 1969 *Lancet* 4 Jan. 34/1 Patients with HL-A-compatible grafts were clinically superior and less prone to multiple or severe rejection crises than patients who had received incompatible grafts.

compel, *v.* Add: **1. d.** *absol.*
1903 R. Langbridge *Flame & Flood* vi, A voice that.. asked like a child,..coerced like a man, compelled like the bidding of the Cosmos. 1904 L. T. Meade *Love Triumphant* III. i, 'I have powerful hands,' was his answer, 'hands that compel.'

compelling, *ppl. a.* Add: **b.** Of a person, his words, writings, etc.: irresistible; demanding attention, respect, etc.
1901 *Westm. Gaz.* 26 Sept. 2/1 Be he Time, or be he man, there is something entirely compelling about the figure. 1908 *Daily Chron.* 30 Sept. 3/4 The book is said to be very realistic and powerful, sincere and compelling. 1913 Conrad *Chance* II. v. 364 There was something not only attractive but compelling in the man. 1930 *Daily Express* 22 May 19/2 Sir Oswald is..a compelling orator.

compendium. Add: **4.** A box, etc., containing or comprising several different games.
c 1899 (*title*) Guide to the compendium of games. Comprising rules for playing—backgammon, besique, chess, [etc.]. 1960 R. C. Bell *Board & Table Games* vii. 172 The antique compendium of games from western India.
b. A package of the stationery required for letter-writing.
1923 H. A. Maddox *Dict. Stationery* 20 *Compendium*, a line of stationery goods which was in considerable demand during the war for soldiers' use, comprising a pad of note, envelopes, and blotting. 1938 E. Bowen *Death of Heart* II. iii. 215 Portia bought a compendium—lightly ruled violet paper, purple lined envelopes. 1960 K. Amis *Take a Girl like You* xiii. 158 He shut the compendium.

compensable, *a.* Delete † *Obs.* and add later example.
1970 *Daily Tel.* 14 May 13/8 The study estimated the 'compensable loss' arising from those accidents as £2,100 million.

compensate, *v.* Add: **4. b.** *Electr.* To correct an electrical device or circuit *for* (some undesired characteristic or effect); to provide with compensation (*compensation 1 d).
1872 J. B. Stearns *Brit. Pat. 3344* 2 The effect of static induction in operating either land lines or submarine cables is neutralized or compensated for. 1961 Fitzgerald & Kingsley *Electr. Machinery* (ed. 2) viii. 390 These limitations may be considerably extended by compensating or neutralizing the armature mmf under the pole faces.
5. *Psychol.* To conceal or counterbalance a defect of character, physique, etc., or to make up for the frustration of a tendency or desire, by developing or exaggerating some other (sometimes more desirable) characteristic. *trans.* and *intr.*
1914 Jung in *Brit. Med. Jrnl.* 3 Dec. 964/2 We..find here these compensating vices to be an essential content of the unconscious. Obviously, therefore, the reverse is true—namely that unconscious virtues compensate for conscious deficiencies. 1921 E. J. Kempf *Psychopathol.* i. 69 If the cause of fear is a segmental compulsion within ourselves.., an attempt to compensate by some estimable work is reflexly initiated. 1927 Corrie *ABC of Jung's Psychol.* II. 38 Here the one-sidedness of the conscious attitude is compensated by personal sensitiveness, infantile feelings, doubts, and irrational superstitions. 1927 Henderson & Gillespie *Text-bk. Psychiatry* vi. 111 Where the attributes of the self are conceived as lacking in some of the qualities of the ideal self, the self makes an attempt to compensate in one direction what is lacking in another. 1929 Adler *Probl. Neurosis* iv. 56 There are many children..who are at a..disadvantage through weak muscles and poor eyesight, but they may compensate in directions which will lead them to be acrobats or artists. 1963 J. E. Gordon *Personality & Behavior* IV. ix. 427 A young man afraid of marriage may compensate by specializing in dating and courting, becoming a 'man about town'.

compensated, *ppl. a.* Add: spec. *Electr.* (cf. *compensation 1 d).
1896 D. C. & J. P. Jackson *Altern. Curr.* II. 316 The Westinghouse, so-called, 'compensated voltmeter'. 1896 S. P. Thompson *Dyn.-Electr. Mach.* (ed. 5) 800 Compensated armatures. 1909 C. F. Smith *Pract. Altern. Currents* (ed. 3) xii. 427 A compensated motor will start from rest with a low power-factor. 1963 B. Fozard *Instrumentation Nucl. Reactors* iii. 31 The use of compensated chambers in which there are three electrodes.

compensating, *ppl. a.* Add: **c.** *Electr.* That effects compensation (cf. *compensation 1 d).
1855 W. H. Preece *Brit. Pat. 2608*, At the same time that the line circuit is completed, a local circuit is also completed, which passes a compensating current through the paper.. in the opposite direction. 1896 S. P. Thompson *Dyn.-Electr. Mach.* (ed. 5) xvi. 392 An iron structure, slotted..to receive the compensating conductors. *Ibid.* xvi. 395 Sayers' compensating winding with commutator coils. 1967 H. Cotton *Adv. Electr. Technol.* xxvii. 1086 The whole of the component F_a is neutralized by means of a compensating winding housed in slots in the pole faces.
d. *Psychol.* (Cf. *compensate *v.* 5.)
1914 [see *compensate *v.* 5]. 1914 [see *compensation 1 e]. 1927 Henderson & Gillespie *Text-bk. Psychiatry* x. 237 His compensating feeling of superiority went far beyond the bounds of reasonable possibility, and produced the grandiose ideas of his psychosis. 1944 L. G. Lowrey in J. McV. Hunt *Personality & Behavior Disorders* II. VI. xxvi. 815 Under ordinary circumstances, an individual is not able to develop a complete feeling of personal adequacy, or complete group security. But, for the average situation, there are compensating satisfactions..which offset..the frustrations.

compensation. Add: **1. d.** *Electr.* The neutralization of one magnetomotive or electromotive force by another; the modification of an electrical device or circuit in order to remove some undesired characteristic or effect.
1855 W. H. Preece *Brit. Pat. 2608*, By adjusting the distance of the coils from the needles, a complete compensation is effected. 1876 Preece & Sivewright *Telegraphy* v. 142 The other half, which is called the compensation current, passes around the electromagnet..and back to the battery. 1963 B. Fozard *Instrumentation Nucl. Reactors* x. 120 Response of circuit before compensation.
e. *Psychol.* The act or result of compensating (*compensate *v.* 5). Also *attrib.*
1914 Jung in *Brit. Med. Jrnl.* 3 Dec. 965/1 In the case of normal people the principal function of the unconscious is to effect a compensation and thus produce a balance. All extreme conscious tendencies are softened..through an effective opposite impulse in the unconscious. This compensating agency..maintains itself through certain unconscious, inconsequent activities. 1917 S. E. Jeliffe (*title*) tr. Adler's Study of organ inferiority and its psychical compensation. 1921 E. J. Kempf *Psychopathol.* v. 197 The symptoms of the compensation neuroses are characterized by striving, egotism, intolerance, [etc.]. *Ibid.* xi. 551 It is the second type that makes the asocial compensation. 1924 P. Radin tr. *Adler's Individ. Psychol.* xviii. 224 Napoleon, Jesus, Jeanne d'Arc, the Virgin,..etc. are frequent compensation-ideals of the intensified lust for superiority. 1929 Adler *Probl. Neurosis* v. 66 If the situation is unfavourable and courage fails, there will be a negative compensation—*e.g.*, the child will not *want* to see anything. 1930 J. C. Flügel *Psychol. of Clothes* vi. 102 Such little evidence as I possess points to the excessive satisfaction with clothes being a compensation for an extreme intolerance of the naked body. 1959 *Encycl. Brit.* VII. 136/2 Compensation.. reduces tensions by substituting some other need-satisfaction for the one that is frustrated or induces anxiety. 1961 *John o'London's* 18 May 567/3 A really catatonic compensation-fantasy.
2. d. Salary or wages, esp. of a public servant; payment for services rendered. *U.S.*
1787 *Constitution* (U.S.) i. § 6 The Senators and Representatives shall receive a compensation for their services to be..paid out of the Treasury of the U.S. 1793 *Deb. Congress* 1431 The compensation of the President of the United States shall be at the rate of twenty-five thousand dollars per annum. 1842 Dickens *Amer. Notes* 48 The Presidential housemaids have..an ample amount of 'compensation', which is the American word for salary, in the case of all public servants. 1931 W. G. McAdoo *Crowded Years* xiii. 205 The compensation of an assistant secretary of the Treasury was the small salary of five thousand dollars.
3. compensation point (see quots.); **compensation water,** water supplied from a reservoir to a stream in time of drought.
1925 W. Stiles *Photosynthesis* vii. 97 There will be a light intensity for every plant and every temperature in which the assimilation is exactly equal to the respiration, and in which, consequently, there will be neither evolution nor absorption of either carbon dioxide or oxygen. This light intensity is called the compensation point. 1964 V. J. Chapman *Coastal Veget.* iii. 65 Changes of temperature can affect both photosynthesis and respiration. The point at which these two processes exactly balance, and where oxygen content remains stable, is called the compensation point. 1889 Compensation water [in Dict.]. 1935 *British Birds* XXVIII. 329 This may have been due to the drought, and to the minimum amount of compensation water flowing from the reservoirs. 1956 J. N. Wood in D. L. Linton *Sheffield* 74 The remaining six impounding reservoirs provide 16·4 m.g.d. compensation water which, by statute, is required to be delivered into the streams on which the supply reservoirs have been constructed.

compensator. Add: **c.** *Electr.* A device that effects compensation (cf. *compensation 1 d); *spec.* an auto-transformer; (see also quot. 1955).
1873 J. B. Stearns *Brit. Pat. 3879* 8 The static balance may..be effected independently of the magnetic or permanent balance, and thus a condenser or other 'compensator' of small capacity may be made to compensate a large line or cable capacity. 1898 E. J. Houston *Dict. Electr. Words* (ed. 4) App. 719/2 *Compensator*, an auto-transformer. *Compensator for alternating current lamps*, a chocking coil or compensator..for the purpose of automatically regulating the current strength in the lamps. *Compensator system*, a system of alternating-current electric distribution from high-pressure mains to low-pressure translating devices. 1955 *Gloss. Terms Radiology (B.S.I.)* 42 *Compensator*, a device whereby the voltage per turn of an auto-transformer can be kept constant.

compensatory, *a.* Add: **b.** *Psychol.* Pertaining to, aiming at, or effecting compensation (*1 e).
1921 E. J. Kempf *Psychopathol.* ix. 463 A not uncommon method of overcoming the inferiorities of poor ability is to be extravagant. This is the most usual compensatory mechanism of the rich inferior male or female. 1923 H. G. Baynes tr. *Jung's Psychol. Types* x. 421 To begin with, the 'hysterical' character is an exaggeration of the normal attitude; it is then complicated by compensatory reactions from the side of the unconscious. 1934 H. G. Wells *Exper. Autobiogr.* I. i. 25 A *persona* may be..a structure of mere compensatory delusions. *Ibid.* ii. 46 The Queen..was in fact my mother's compensatory personality, her imaginative consolation for all the restrictions and hardships. 1941 A. Angyal *Found. Sci. Personality* vii. 231 The importance of this factor [*sc.* tall stature] becomes evident when one observes the great variety of compensatory behavior seen in short people. 1966 L. Hudson *Contrary Imaginations* ii. 32 This compensatory pattern only obtained on the arts side; it did not occur among scientists.

compere, *sb.* Restrict † *Obs.* to sense in Dict. and add: **2.** Usu. **compère** (koñpęr, kǫ·mpēˈɹ): The organizer or general director of a musical or vaudeville entertainment; now esp. one whose role in an entertainment on the stage, radio, etc., is to introduce the performers to the audience and provide a linking commentary between the acts.
1914, 1916 [see *commère 2]. 1928 *Weekly Dispatch* 24 June 15 The genial Davy Burnaby seems to be quite content as a cabaret compère. He has got together a company for the underground grill of the Piccadilly. 1932 *BBC Year-bk.* 199 In the Charlot shows..the *compère* reigned supreme. 1955 *Times* 12 May 11/4 'Breezy' is another dangerous adjective,..suggesting one who will act as a televisionary compère to the whole society.

compère (koñpęr, kǫ·mpēˈɹ), *v.* [f. the sb.] *trans.* and *intr.* To act as compère (for an entertainment).
1933 *Radio Times* 25 Aug. 414/3 André Charlot..will compère *Charlot's Review* in mid-September. 1938 *Eng. Studien* LXXII. 330 Mr. Steve Geray *compères* and announces everything in his amusingly inadequate English. 1954 A. Heckstall-Smith *18 Months* v. 67 He was always ready to don a seedy frock coat and compère a prison concert, telling stories between the acts. 1957 *Observer* 20 Oct. 16/3 Patrick O'Donovan, in Monday's Panorama, compèred with artful relaxedness, and no gush, a film of Washington. 1965 G. Melly *Owning-Up* xv. 192 The first year I compèred and sang a spot with Chris Barber.
Hence **co·mpèring** *vbl. sb.*, the action or duties of a compère.
1959 *Observer* 7 June 22/7 The compèring was either too arch or too flat. 1964 Hall & Whannel *Pop. Arts* II. x. 285 A shift..to pantomime, compèring and straight drama.

competence. Add: **4. d.** The ability of a stream or current to carry fragments of a certain size. Also *attrib.*
1877 G. K. Gilbert *Rep. Geol. Henry Mts.* v. 110 A stream which can transport *débris* of a given size, may be said to be *competent* to such *débris*. Since the maximum particles which streams are able to move are proportioned to a sixth power of their velocities, competence depends on velocity. 1914 —— *Transp. Debris by Running Water* 86 A current flowing over debris of various sizes transports the finer, but cannot move the coarser; the fineness of the debris it can barely move is the measure of its competence. *Ibid.* 187 The group of constants designated by Greek letters.. may be called competence constants. 1960 B. W. Sparks *Geomorphology* v. 80 Competence is measured by the weight of the largest fragment which the stream can transport.
e. *Biol.* The state or quality of embryonic cells of permitting characteristic development in response to a stimulus.
1932 C. H. Waddington in *Phil. Trans. R. Soc.* B. CCXXI. 223 This competence is a new potentiality which has arisen. The idea of competence covers two concepts already in use, *reactionsfähig* and *labil determiniert*. Reactionsfähig means that the tissue..is competent to differentiate to a certain tissue but requires a definite stimulus. 1954 *New Biol.* XVII. 122 The group of cells..passes into a relatively brief phase of sensitivity (called 'competence') to the influence of neighbouring tissues.
f. *Linguistics.* (See quot. 1966.) Opp. *performance*.
1966 N. Chomsky in T. A. Sebeok *Current Trends in Linguistics* III. 3 A distinction may be made between what the speaker of a language knows implicitly (what we may call his *competence*) and what he does (his *performance*). A grammar, in the traditional view, is an account of competence. It describes and attempts to account for the ability of a speaker to understand an arbitrary sentence of his language and to produce an appropriate sentence on a given occasion...Performance provides evidence for the investigation of competence. 1968 Chomsky & Halle *Sound Pattern Eng.* i. 3 One fundamental factor involved in the speaker-hearer's performance is his knowledge of the grammar that determines an intrinsic connection of sound and meaning for each sentence. We refer to this knowledge —for the most part, obviously, unconscious knowledge—as the speaker-hearer's 'competence'. 1969 *Language* XLV. 323 Linguistic inquiry has succeeded only when..it began to ask about the individual's organization of language—what has come to be called his 'linguistic competence'. *Ibid.*, The distinction between performance and competence in the child is a critical one for understanding how language emerges in the individual.

competency. Add: **4. c.** *Geol.* The condition of being competent (see next, 5 d).
1893 B. Willis in *U.S. Geol. Surv. Ann. Rep.* II. 251 The size of a competent anticline is directly as the competency of the effective stratum and inversely as the load. 1940 E. S. Hills *Outl. Struct. Geol.* iv. 78 A small degree of competency is sufficient to ensure that some slipping will go along the bedding planes.
d. = *competence 4 d.
1909 in *Cent. Dict.* Suppl. 1939 W. H. Twenhofel *Princ. Sedimentation* vi. 191 Competency is defined as the ability of currents to transport in terms of dimensions of particles. Competency depends upon velocity and turbulence.

competent, *a.* Add: **5. c.** Of a stream: capable of transporting fragments of a certain size. (Cf. quot. 1878 s.v. sense 5 b in Dict.)
1877 [see *competence 4 d]. 1960 B. W. Sparks *Geomorphology* v. 81 If a stream is not competent to move load of the calibre available.
d. *Geol.* Of a stratum of rock: able to transmit lateral pressure and, when formed in an anticline, to bear weight.
1893 B. Willis in *U.S. Geol. Surv. Ann. Rep.* II. 250 If we describe the sufficiently firm stratum by the word competent, we may formulate the law of anticlinal development...In strata under load an anticline arises along a line of initial dip, when a thrust, sufficiently powerful to raise the load, is transmitted by a competent stratum. The resulting anticline supports the load of an arch, and being

COMPETITION

adequate to that duty it may be called a competent structure. **1953** E. S. Hills *Outl. Struct. Geol.* (ed. 3) iv. 82 Relatively strong ('competent') beds. **1965** A. Holmes *Princ. Physical Geol.* (ed. 2) x. 234 Strong competent beds of rocks like quartzite cannot readily change their thicknesses.

e. *Biol.* Of a cell: having a latent ability to develop in reaction to a stimulus.

1932 [see *competence 4 e]. **1935** *Discovery* May 136/2 The ectoderm is continually changing through internal causes and is only reactive or 'competent' during a certain period of its existence, but within this period several different substances can cause a nervous differentiation.

competition. Add: **1. f.** *Ecology.* (See quots.)

1905 F. E. Clements *Research Methods Ecol.* iv. 285 Competition is a question of the reaction of a plant upon the physical factors which encompass it, and of the effect of these modified factors upon the adjacent plants. *Ibid.* 316 *Competition*, the relation between plants occupying the same area, and dependent upon the same supply of physical factors. **1916** —— *Plant Succession* iv. 72 Competition occurs whenever two or more plants make demands in excess of the supply…In its essential nature, competition is a decrease in the amount of water and light available for each individual. **1925** R. H. Yapp in *Festschrift C. Schröter* 691 The term 'competition'..is, properly speaking, confined to those cases in the general struggle for existence in which living organisms compete or strive with one another for a limited supply of the necessaries of life—water, light, salts, insect visitors and so on.

compilation. Add: **4.** *attrib.*, as **compilation film**, a film, esp. a documentary, compiled from various pieces of film, all or most of which were originally shot for a different purpose.

1953 K. Reisz *Film Editing* xiii. 194 The maker of compilation films..working with newsreel and allied material which has not been scripted or shot for the purposes for which the compiler will use it..is able to make films with a smooth, logically developing continuity. **1964** J. Leyda *Films beget Films* 9, I wanted to examine the development and problems of the compilation film—but what an awkward, incomprehensible, and unacceptable term for this form!

compilatory, *a.* Delete *rare* and add examples.

1894 *Temple Bar* Mar. 441 Journalistic, compilatory, biographical work. **1906** *Daily Chron.* 2 Aug. 3/3 There is far more of the compilatory than of the confessional element in his volume.

compiler. Add: **1. b.** *Computers.* A routine for translating a program into a machine-coded form.

1953 *Computers & Automation* May 3 If a compiling routine or compiler is used, when a word is examined, the required subroutine is transcribed..into a running program. **1962** F. L. Westwater *Electronic Computers* ix. 146 This language [*sc.* Algol] has gained wide acceptance, and many 'compilers' for translating it onto different machines have been written. **1964** *Automated Typesetting & Related Computer Terms* (C.I.S.) 7 *Compiler*, a computer program more powerful than an assembly program. In addition to its translating function, it is able to replace certain items of input with series of instructions, usually called subroutines. **1966** A. Battersby *Math. in Management* viii. 206 Finally there are the 'compiler' routines which make programming easier. When they are fed into the machine, they set it up so that it can accept instructions in a different 'language' from its normal code.

comping (kǫ·mpiŋ), *vbl. sb. Printing.* Colloq. abbrev. of Composing *vbl. sb.* Also *attrib.* Cf. Comp. b.

1888 Jacobi *Printers' Vocab.* 25 *Comping*, a slang term for composing or setting type. **1900** *Daily News* 14 Nov. 10/3 (Advt.), Printing.—Wanted, a smart young man..for Comping Room. **1928** *Melody Maker* Feb. 153/3 We apologise to Ben Davis for printers' errors…These 'comping' slips made the prices of the Vocaltone Reeds refer to the mouthpieces and *vice versa*.

complain, *v.* Add: **8. b.** *to complain of*, at Eton, to report (a boy) to the Headmaster as deserving punishment.

1870 'An Etonian' *Recoll. Eton* ii. iv. 207 Every dame's house is supposed to be under the charge of some master, who is intended to keep an eye upon the boys and set such punishments as are necessary, or complain of those who deserve it. **1908** *Westm. Gaz.* 30 June 12/1 His tutor intended 'to complain of us'. This is the euphonious term for sending in names to the headmaster to be flogged. *a* **1909** Swinburne *Herbert Winwood* in *Lesbia Brandon* (1952) 189 Tell my father I have not been complained of again this half.

complainant, *a.* and *sb.* **A.** *adj.* Delete †*Obs.* and add later examples. Also *gen.*, complaining.

1791 J. Byng *Diary* 18 June (1935) II. 312 When she writes fully, she is complainant, and refines upon pleasure till it becomes a pain! **1891** *Pall Mall Gaz.* 14 May 7/2 On the appearance of a female in the dock, it was reported that the complainant shopkeeper would not attend. **1897** *Daily News* 23 July 4/5 The conference between the Postmaster-General and the complainant sections of the Post Office servants.

complainee. (Earlier U.S. example.)

1779 J. Adams *Wks.* (1854) IX. 479 These might have determined whether the complainers or complainees have most to boast of.

compleat, *a.* Add: Revived in imitation of its 17th-cent. use, as in Walton's *The Compleat Angler*, = Complete *a.* 5.

1900 O. Onions (*title*) The compleat bachelor. **1953** (*title*) The compleat imbiber (W. & A. Gilbey, Ltd.). **1963** M. McCarthy *Group* v. 104 She writes and sings and paints and dances and plays I don't know how many instruments. The compleat girl.

complected, *ppl. a.*[2] (Earlier and later examples.)

1806 Lewis in *Lewis & Clark Exped.* (1905) III. 315 They are..reather lighter complected and much more illy formed than the Indians of the Missouri. **1822** *Amer. Beacon* (Norfolk, Va.) 8 Jan. 3/3 (Advt.), He is well made, light complected. **1883** 'Mark Twain' *Life Mississippi* lix, Monstrous bluffs on both sides of the river—ragged, rugged, dark complected. *a* **1910** 'O. Henry' *Trimmed Lamp* (1916) 67 A heavy-sot man, sandy complected, about twenty-nine. **1932** W. Faulkner *Light in Aug.* ii. 46 They told me the man they meant wasn't dark complected.

complement, *sb.* Add: **3. b.** (Later examples.)

1925 Grattan & Gurrey *Our Living Lang.* xlii. 270 Sentences in Group A below have Multiple Complements (Objects, Predicatives, Adverbs). **1961** R. B. Long *Sentence & Parts* 2 *His sister is buying antiques* will always be understood to have *is buying* as predicator and *antiques* as complement; *his hobby is buying antiques*..to have *is* as predicator and *buying antiques* as complement. **1964** E. Palmer tr. *Martinet's Elem. Gen. Linguistics* iv. 11 *Hier, il y avait fête au village*…*hier* and *au village*..may be eliminated without the utterance ceasing to be a normal sentence ..and this is what is meant by the traditional terminology which speaks of them as 'complements'.

5. i. *Biochem.* A thermolabile protein complex found in blood plasma and other body fluids, which by combining with an antigen-antibody complex can bring about the lysis of antigenic substances such as bacteria or red blood cells. Also *Comb.*, as **complement-fixation, -fixing,** the process in which complement is removed from solution by combination with an antigen-antibody complex and so rendered incapable of lysing any further antigen-antibody system.

1900 tr. *Ehrlich's Immunity* in *Proc. R. Soc.* LXVI. 443 Solutions containing either only the 'immune body' or only the 'complement' were brought in contact with suitable blood corpuscles. **1903** [see *addiment]. **1906** *Jrnl. Exper. Med.* VIII. 726 The complement fixation of specific precipitates. **1911** *Jrnl. Hygiene* XI. 529 No satisfactory distinction between Meningococci and Gonococci can be demonstrated by means of complement-fixation tests. **1921** *Proc. Soc. Exper. Biol. & Med.* XVIII. 172 The complement-fixing substances present in the sera of syphilitic patients. **1957** *Nature* 30 Mar. 668/1 Rabbits and guinea pigs immunized with human liver or kidney suspensions developed complement fixation antibodies to several other human tissue antigens. **1970** Passmore & Robson *Compan. Med. Studies* II. xxii. 16 Complement has the ability to lyse the antibody-coated cells.

j. *Logic* and *Math.* All the members of any set, class, or space of elements that are not in a given subset.

1937 S. K. Langer *Introd. Symbolic Logic* vi. 144 The universe class and the null class are each other's complements. *Ibid.* 147 Every class which may be formed in a given universe has a complement. **1947** Birkhoff & MacLane *Surv. Mod. Algebra* xi. 331 Each set *S* has a 'complement' *S′* satisfying $S \cap S' = 0$, $S \cup S' = I$. **1965** Shih-Chên Hu *Elem. Mod. Algebra* i. 4 If *A* is a subset of *X*, then the difference..will be called the complement of *A* with respect to *X*.

complementarily (kǫmplĭme·ntărĭli, -mentĕ·rĭli), *adv.* [f. Complementary *a.* + -ly.] In a complementary manner.

1911 in Webster. **1944** *Scrutiny* XII. 212 His [*sc.* Purcell's] problem was, how to achieve a kind of declamation; and complementarily how to make the simple metrical dances of the concerted spectacle of the masque..reconcilable with the power of this declamation. **1959** R. J. Hirst *Probl. Perception* v. 127 Different concepts or terms can apply complementarily to the same subject. **1965** *Language* XLI. 538 These..are complementarily distributed in original Tatar words.

complementarity (kǫ:mplĭmentæ·rĭti). [f. Complementary *a.* + -ity.] A complementary relationship or situation; *spec.* in *Physics*, the capacity of the wave and particle theories of light together to explain all phenomena of a certain type, although each separately accounts for only some of the phenomena.

1911 A. Mitchell tr. *Bergson's Creative Evol.* i. 54 Harmony, or rather 'complementarity', is revealed only in the mass. **1928** N. Bohr in *Nature* 14 Apr. Suppl. 580/2 In the description of atomic phenomena, the quantum postulate presents us with the task of developing a 'complementarity' theory the consistency of which can be judged only by weighing the possibilities of definition and observation. *Ibid.* 587/1 We meet here with a complementarity of the possibilities of definition quite analogous to that which we have considered..in connexion with the properties of light and free material particles. **1937** *Mind* XLVI. 522 There may..be a complementarity which gives rise to indeterminacy in our descriptions of our own minds. **1953** C. E. Bazell *Ling. Form* iv. 46 A totally different distribution (amounting almost to complementarity) of units similarly analysed is impermissible. **1953** F. Vreede *Living Hindu Philos.* iv. 27 The opposition of purusha to prakriti represents the complementarity of two universal aspects of reality: 'essence' and 'substance'. **1955** O. Klein in W. Pauli *N. Bohr* 96 From the two main physical theories of this century, relativity theory and quantum theory, two general viewpoints have emerged, that of relativity and that of complementarity.

complementary, *a.* and *sb.* Add: **A. 1. d.** *complementary goods* (see quots.).

1891 W. Smart tr. *Böhm-Bawerk's Positive Theory Capital* ix. 170 Thus, for instance, paper, pen and ink, needle and thread, cart and horse, bow and arrow,..and so on, are complementary goods. *Ibid.* 175 Almost every product is the result of the co-operation of a group of complementary goods consisting of uses of ground, labour, fixed and floating capital. **1892** *Palgrave's Dict. Pol. Econ.* 380/1 *Complementary goods*, this expression is used by the Austrian economist Menger..who describes goods as of first, second, or higher rank in order of production…This conception becomes of special interest when the value of the complementary goods is considered for each separately.

e. *complementary distribution*: in *Linguistics*, a distribution of two or more similar or related speech-sounds or forms in such a manner that they appear only in different environments.

1934 M. Swadesh in *Language* X. 123 The criterion of complementary distribution. If it is true of two similar types of sounds that only one of them normally occurs in certain phonetic surroundings and that only the other normally occurs in certain other phonetic surroundings, the two may be sub-types of the same phoneme. **1942** C. F. Hockett *Ibid.* XVIII. 9 If *a* and *b* are in complementary distribution (i.e. if they occur in mutually exclusive positions), they may be..members of the same phoneme. **1949** E. A. Nida *Morphology* (ed. 2) ii. 44 (*heading*) Determination of allomorphs by complementary distribution. *Ibid.*, The three types of plural formatives..are all in complementary distribution. **1953** C. E. Bazell *Ling. Form* 7 But complementary distribution does not here imply *irrelevance*. **1965** W. S. Allen *Vox Latina* 9 The fact that an initial *t* in English (as in *tin*) is more strongly aspirated than a final *t* (as in *hat*) is not responsible for any difference of meaning, since the two varieties occur only in different environments, and so cannot contrast with one another—they are in 'complementary' and not parallel distribution.

f. *complementary function*: in *Math.*, that part of the general solution of a linear differential equation which is the general solution of the associated homogeneous equation obtained by subsituting zero for the terms not containing the dependent variable.

1841 D. F. Gregory *Examples of Processes of Differential & Integral Calculus* ii. iv. 295 As operating factors of the form $(d/dx)^2+n^2$ very frequently occur in differential equations, it is convenient to keep in mind that the complementary function due to it is of the form $C \cos nx + C' \sin nx$. **1920** H. T. H. Piaggio *Elem. Treat. Differential Equations* vii. 88 An elegant but somewhat artificial method for finding the complete primitive of a linear equation whose complementary function is known. **1957** I. N. Sneddon *Elem. Partial Differential Equations* iii. 97 If *u* is the complementary function and z_1 a particular integral of a linear partial differential equation, then $u+z_1$ is a general solution of the equation.

complementation (kǫ:mplĭmentēi·ʃən). *Linguistics.* [f. Complement *v.* + -ation.] Complementary distribution (see prec. A. 1 e).

1937 M. Swadesh in *Language* XIII. 7 Complementation is one of the characteristics of positional variants. **1948** E. A. Nida *Ibid.* XXIV. 422 The forms *I* and *me* generally occur in complementation: *I* occurs in preverbal subject position, *me* in postverbal object position and after prepositions. **1965** R. D. Stevick in *Eng. Studies* Feb. 35 Neither of the potential patterns in which relatives *þe* and *þat* stood in normal complementation were ever realized.

complete, *a.* Add: **5.** (See also *compleat *a.*)

7. *complete fertilizer*: a fertilizer containing the chief plant nutrients, phosphoric acid, nitrogen, and potash; also *ellipt.*; *complete primitive* (see Primitive *a.* and *sb.* B. 8): add examples.

1859 G. Boole *Treat. Differential Equations* i. 8 The relation among the variables which constitutes the general solution of a differential equation..is also termed its complete primitive. **1904** *Webster's Guide on Fertilizers* (A. G. Webster & Son, Hobart, Tasmania) 1 We are pleased to state that our sales of Fertilizers show a considerable increase, especially in Mixed Complete Fertilizers. **1920** [see **complementary function*]. **1953** H. L. Edlin *Forester's Handbk.* xix. 322 Compound commercial fertilizers, which include balanced amounts of the three major nutrients, namely nitrogen, phosphorus and potassium..are similar to the 'complete fertilizers' used in agricultural practice. **1957** L. Fox *Two-Point Boundary Probl.* i. 3 By definition, a numerical solution contains no arbitrary constants, so that we always obtain particular integrals rather than complete primitives. **1960** *Farmer & Stockbreeder* 8 Mar. 105/3 The first spring dressing is nitrogen, followed by a complete, with a nitrogen top dressing after nearly every grazing over, and a second application of complete in August.

complex, *sb.* Add: **1. a.** (Later examples.)

1936 A. W. Clapham *Romanesque Archit.* iii. 59 S. Stefano, Bologna, with its attendant complex of buildings. **1952** *N.Y. Times* 6 May 2/6 Ten medium bombers.. dropped 100 tons of high explosives on to the rail bridge complex at Chongju. **1955** Stokes & Varnes in *Colorado Sci. Soc. Proc.* XVI. 27/2 *Complex*,..an assemblage of rocks of any age or origin that has been folded together or intricately mixed, involved, or otherwise complicated. **1957** *Times* 26 Sept. 15/3 Movements..to take over control of

the large steel complex [the Dominion Steel and Coal Corporation]. **1958** *Globe Mag.* 13 Sept. 13/1 It lies in forbidding..country where a giant new industrial complex is being developed.

c. *Chem.* A substance formed by the combination of simpler substances, esp. one in which the bonds between the substances are weaker than or of a different character from those between the constitutents of each substance.

1895 Cross & Bevan *Cellulose* ii. 92 A furfural-yielding complex, which appears to be an oxycellulose derivative. **1914** *Jrnl. Biol. Chem.* XIX. 159 The enzyme-substrate complex. **1944** *Adv. Enzymol.* IV. 17 Earlier investigators never thought of doubting that any gene must comprise an organized complex of many molecules. **1956** Coffey & van Alphen in E. H. Rodd *Chem. Carbon Compounds* IIIb. xxi. 1406 This particular 'lake' being a complex of alizarin with Al, Sn, Ca, and higher fatty acids in varying proportions. **1965** Phillips & Williams *Inorg. Chem.* I. iv. 129 These ions form stable complexes with water molecules, e.g. $[Cr(H_2O)_6]^{3+}$.

3. *Psychol.* A group of emotionally charged ideas or mental factors, unconsciously associated by the individual with a particular subject, arising from repressed instincts, fears, or desires and often resulting in mental abnormality; freq. with defining word prefixed, as *inferiority*, *Œdipus complex*, etc.; hence *colloq.*, in vague use, a fixed mental tendency or obsession. Also *attrib.* and *Comb.*

The use of the term was established by C. G. Jung in 1907 (*Ueber die Psychologie der Dementia Praecox*), but it originated with Neisser in 1906 (*Individualität und Psychose*).

1907 Peterson & Jung in *Brain* XXX. 178 The complex robs the ego of light and nourishment, just as a cancer robs the body of its vitality. **1911** T. W. Mitchell in Sidis *Symptomatology Psychopathic Dis.* (1921) 418 The delay in our response to the stimulus is due to the inhibitions exercised by the manifold of associated complexes that have been aroused to function. **1913** Jung in *Trans. Internat. Congress Med.* xii. 67 The unconscious existence of manifold phantasies, which have their final root in the infantile past and turn around the so-called 'Kern-complex', or nucleus-complex, which may be qualified in male individuals as the Œdipus-complex and in females as the Electra-complex. **1916** [see *inferiority c]. **1919** *Athenæum* 23 May 360/2 Without the adjective ['mental'], 'a complex' is now a polite euphemism for a bee in one's bonnet. **1920** A. A. Brill tr. *Freud's Sel. Papers Hysteria* (ed. 3) xiii. 217 A thought of this kind which is capable of affecting the reaction to the stimulus word has come to be called a 'complex'. **1924** E. C. Mayne tr. *Freud's Types Neurotic Nosogenesis* in *Coll. Papers* II. 115 The searching analytic studies stimulated by the complex-theory of the Zürich School. **1924** N. P. Williams *Doctrines of Fall & Orig. Sin* vii, We therefore identify the 'inherited infirmity' of theology with 'inherited weakness of herd-complex'. **1926** W. McDougall *Introd. Social Psychol.* (ed. 20) 403 The psycho-analysts use the term 'complex' to cover both the normal sentiments and the morbidly repressed sentiments. I have urged that by restricting the term 'complex' to the latter, and using the term 'sentiment' for the former, we usefully differentiate our terminology. **1927** A. Bennett *Woman who stole Everything* iii. 165 'Muriel's losing her sex-complex.' 'What on earth do you mean, boy?' 'She's getting herself tangled up with some man.' *Ibid.* iv. 301 She raised herself on her elbows and kissed him; she had no forbidding complex. **1928** *Punch* 8 Feb. 157 A fond aunt with a commiseration complex. **1933** Dylan Thomas *Let.* Nov. (1966) 48 It is typical of the..complex-ridden to emphasise its naiveté. **1948** W. McDougall *Introd. Social Psychol.* (ed. 29) Suppl. chap. vii. 447 Modern science has shown an aversion to all teleology; one might almost say that it has a 'complex' on that subject. **1953** A. Koestler in *Encounter* Nov. 25/2 Our political libido is just as complex-ridden.

complex, *a.* Add: **2. d.** *Math.* Containing or characterized by complex numbers or quantities; having the form of a complex number; *complex number*, a number of the form $a + ib$, where *a* and *b* are real numbers and *i* is the square root of −1.

[**1832** C. F. Gauss in *Comment. Soc. Reg. Scient. Gottingensis* VII. 96 Tales numeros vocabimus *numeros integros complexos*.] **1860** H. J. S. Smith in *Rep. Brit. Assoc. 1859* 253 If *a* and *b* are both rational, the complex number is said to be rational. *Ibid.*, One complex integer *α* is said to be divisible by another *β*. **1879** *Encycl. Brit.* IX. 818/2 The notion of the 'path' of a complex variable $u = x + iy$. **1893** A. R. Forsyth *Theory of Functions of Complex Variable* i. 1 The complex variable is the most general form of algebraical quantity which obeys the fundamental laws of ordinary algebra. **1908** G. H. Hardy *Course Pure Math.* iii. 78 The two complex numbers $\pm i$ satisfy this equation. We express this by saying that the equation has the two complex roots $\pm i$. **1946** H. & B. S. Jeffreys *Meth. Math. Physics* xi. 305 Complex functions, involving a symbol *i* such that $i.i = -1$, are of importance in physics. **1959** Born & Wolf *Princ. Optics* i. 32 Operations with complex vectors follow the usual rules of vector algebra and of algebra of complex numbers.

complexify, *v.* Add: **b.** *intr.* To become (more) complex or complicated.

1914 Geddes & Thomson *Sex* iv. 94 The tendency in matter to complexify.

Hence **comple:xifica·tion, comple·xifying,** the action of becoming (more) complex or complicated; **comple·xified** *ppl. a.*

1923 *Glasgow Herald* 10 Nov. 4 There is one line of complexifying which is almost certainly open, and that is an increase in the intricacy of inter-relations among the nerve-cells. **1929** H. B. Alexander *Truth & Faith* 250 No doubt there has been complexification. **1932** J. A. Thomson *Sci. Riddles* xlvi. 297 This process of complexifying or differentiation was greatly assisted by the hand of life. **1959** B. Wall tr. *Teilhard de Chardin's Phenomenon of Man* i. 48 Matter has obeyed from the beginning..the law of 'complexification'. **1962** C. E. Raven *Teilhard de Chardin* iv. 75 Having accepted the whole cosmic process as one, continuous, complexified and convergent, he can regard it with an unfaltering hope.

complexion, *sb.* Add: *attrib.* (sense 4), as *complexion brush, cream, milk, powder, soap.*

1897 *Sears, Roebuck Catal.* 33/2 Complexion Brush.. especially constructed for massaging the skin. **1907** *Yesterday's Shopping* (1969) 536/2 Complexion cream. For face massage, tin 1/0. **1938** E. Bowen *Death of Heart* iii. i. 334 Anna wiped complexion milk off her fingers on to a tissue. **1864** *Englishwoman's Dom. Mag.* July 117 A complexion-powder called *batikha*, which is used in all the harems for whitening the skin, is made in the following manner. **1897** *Sears, Roebuck Catal.* 34/3 Pozzoni's Complexion Powder. **1895** *Montgomery Ward Catal.* 109/3 Lana Oil Complexion Soap..for preserving the skin and leaving it soft and pliable. **1970** *Vogue* May 173 Our White Rose and Cucumber Complexion soap is very feminine.

compliance. Add: **8.** *Mech.* The property of a body or substance of yielding to an applied force or of allowing a change to be made in its shape; also, the degree of yielding, measured by the displacement produced by a unit change in the force.

1934 in Webster. **1940** *Chambers's Techn. Dict.* 184/2 *Compliance*..the displacement in cm. corresponding to the application of the force of one dyne. **1950** *Electronic Engin.* XXII. 435 If voltage be taken to represent force, and current to represent velocity, inductance corresponds to inertia and capacitance to compliance. **1955** J. H. Comroe et al. *Lung* vii. 113 Physiologists now call this the 'mechanical compliance', or, more simply, the 'compliance' of the tissues; it is defined as the volume change per unit pressure change, and its units are liters/cm H_2O. *Ibid.* 133 Her compliance was only 0·025 liters/cm H_2O (1/5 of normal), so that her ventilation was inadequate even when..the maximal inflating pressure of a resuscitator was employed. **1962** A. Nisbett *Technique Sound Studio* 245 Compliance is the acoustical and mechanical equivalent of capacitance. *Ibid.* 265 Styli with very high compliance can track at very low playing weights.

complication. Add: **4. c.** *Psychol.* The simultaneous association of the perceptions or ideas received through different senses.

[**1816** J. F. Herbart in *Werke* (1850) V. iii. ii. 127 Blosse Complicationen oder Verschmelzungen.] **1886** J. M. Baldwin tr. *Ribot's German Psychol. Today* i. 34 The representations [*sc.* states of consciousness] belong to different continuities; 'they can unite totally in such a way as to form a single force, which enters as such in the calculation'. Herbart calls this a *complication* or an entire complex (union of a sound and a color). **1898** G. F. Stout *Man. Psychol.* I. 91 The process is called *complication*, because the result is merely a change in the constitution of *A*, and for the most part an increase in its complexity. **1958** H. B. & A. C. English *Dict. Psychol. Terms* 103/1 *Complication*, a combination of sense data from different senses, e.g. taste and smell of food.

complimentarily, *adv.* (Earlier U.S. example.)

1809 A. Henry *Travels* 297 These amusements were given to us complimentarily, by the chief.

compo[1]. Add: **5. b.** = Composite *a.* 1, as in **compo pack**, a composite pack of (tinned) food, esp. one designed to last several days; so *compo rations, tea.*

1943 *Times Weekly* 10 Feb. 19 The 'compo pack' is designed to feed 14 men for one day and is made up of tinned foods of various kinds, including chocolate and biscuits. **1945** 'A. Melville' *First Tide* 72, I used to cadge a cup of compo tea from them. *Ibid.* 73 You rarely saw the men cooking their compo rations on a fire out in the open. **1954** W. Noyce *South Col* i. 11 There were the Army-type Compo rations for the march. **1963** *Times* 11 Mar. 11/4 An Australian 24-hour compo. pack, a metal container like a very large sardine tin.

compo[2] (kǫ·mpo). *Austral.* and *N.Z. slang.* Abbrev. of Compensation 2, esp. that paid for an injury received while working. Also *attrib.*

1941 K. Tennant *Battlers* xxvii. 290 'If you *do* slice your hand, they put you on compo.' Seeing them puzzled, she explained: 'Compensation money while it heals.' **1941** Baker *N.Z. Slang* vi. 52 *Compo king*, a social parasite who makes a practice of injuring himself or malingering in order to secure workers' compensation. **1949** D. M. Davin *Roads from Home* v. 69 They're afraid they might have to pay out a bit of compo. **1949** *Here & Now* Oct. 34/1 The wharfies.. must also of necessity carry their own insurance against damage and loss—as well as workers' compo. **1961** P. White *Riders in Chariot* viii. 243 You got a bad hand. You see the doc…You'll get compo of course.

component. Add: **B.** *sb.* **2.** Applied *spec.* to the separate parts of motor cars and bicycles. Hence *attrib.* and *Comb.*, as *component-maker, -plant*; *component-built* adj.

1896 *Daily News* 2 July 2/5 Manufacturers of cycle and cycle components. **1897** *Ibid.*, 29 Mar. 8/7 Cycle tube and component makers. **1907** *Daily Chron.* 18 Jan. 9/2 A component-built machine. **1925** *Morris Owner's Man.* 95 The replacing of any worn component by a new one. **1970** *Times* 27 June 11/2 Stocks pile up as strikes halt motor-components. *Ibid.* Midland car and component plants face yet another threat from the three-day strike of clerks. *Ibid.* 7 July 11/2 No major manufacturer has ever been forced to stop operations because of a strike in a components plant.

3. *Mech.*, etc. One of two or more vectors which together are equivalent to any given vector; *esp.* the effective value of a vector in any given direction.

1806 O. G. Gregory *Treat. Mech.* I. i. ii. 24 Resolve each of the forces into components respectively parallel to these co-ordinates. *Ibid.* 25 The two components of the resultant *R* referred to the same axis, are evidently *R* cos. *r* and *R* sin. *r*. **1883** *Encycl. Brit.* XV. 699/1 It is very generally convenient to resolve forces into components parallel to three lines at right angles to each other, each such resolution being effected by multiplying by the cosine of the angle concerned. **1957** B. I. & B. Bleaney *Electr. & Magnetism* (1963) xx. 546 Each vector makes a constant angle with this direction, and the component of the angular momentum in this direction is therefore constant.

componential (kǫmpǫne·nʃiăl), *a.* [f. Component *sb.* + -ial.] Of or pertaining to components; applied esp. to the analysis of distinctive sound units or grammatical elements into linguistic components.

1947 C. F. Hockett in *Internat. Jrnl. Amer. Linguistics* XIII. 258 (*heading*) Componential analysis of Sierra Popoluca. **1964** Crystal & Quirk *Pros. Features Eng.* ii. 30 These studies..suggest a procedure for componential discrimination. **1964** *Language* XL. 203 There are distinctive affective features which are susceptible of componential analysis.

comport, *sb.*[2] [The earlier evidence now available shows that *compote* in this sense (see *compote 3) is later than *comport*.]

1771 in J. E. Nightingale *Contrib. E. E. Porcelain* (1881) 26 Four shell pattern comports, enamel'd in flowers. **1783** *Ibid.* p. lxxxii, A capital desert service of Worcester,.. containing 18 comports of various shapes. **1811** in Jewitt *Ceramic Art Gt. Brit.* (1878) II. 107, 4 Comports of Landscapes, 6 Comports of plants. **1862** *Catal. Internat. Exhib., Brit.* II. No. 6769 Dessert service of flint glass,..consisting of decanters, carafes, finger basins, ice plates, elevated comports, jugs,..and other glasses. **1924** H. Barnard *Chats on Wedgwood Ware* 91 Dessert baskets, comports, compotiers, custard stands and cups, and candlesticks.. in plain glazed cream colour.

comporteer, comportier, varr. *compotier.

1764 *Felix Farley's Jrnl.* 22 Dec. in Pountney *Old Bristol Potteries* (1920) 225 Matchless pieces of the fine Chelsea in Figures, Dishes, Plates, Comportiers. **1822** *Auction Catal. Fonthill Abbey* 31 Two old Derby montephs…Nine comportiers of the same. *Ibid.* 66, 2 rich Japan comportiers. *c* **1832** in E. A. Barber *Pott. & Porc. U.S.* (1893) 150 Gravy boats..square comforts..high comporteers..cake stands. **1862** *Catal. Internat. Exhib., Brit.* II. No. 6897 Dessert comportiers, with parian supports.

compos: see Compos mentis in Dict. and Suppl.

composable, *a.* Delete †*Obs. rare*[−1] and add later examples. Also, able to be composed; suitable for setting to music.

1929 R. Bridges *Test. Beauty* iii. 94 Happiness which all seek is not composable of any summation of particular pleasures. **1964** *Economist* 14 Nov. 718 A libretto that was ..satisfying at a high literary standard and composable.

composer. Add: **7.** *Comb.*, in appositive use, as (sense 3) *composer-arranger, -conductor.*

1955 Keepnews & Grauer *Pict. Hist. Jazz* xxi. 278 Composer-arranger Jack Montrose. **1959** 'F. Newton' *Jazz Scene* vi. 110 Bands can have a distinctive 'sound',..mixed by the composer-arranger from the colours of the individual musicians. **1938** *Times* 30 Apr. 10/3 The acclamations which were offered to the composer-conductor.

composing, *vbl. sb.* **2. composing-room** (earlier example).

1737 *London Mag.* Oct. 552/1, 7 Persons enter'd the Printing-House in Bow-street; 3 of whom rush'd up Stairs, and coming into the Composing-Room, one of them spoke to Mr. Haines.

composite, *a.* and *sb.* Add: **A.** *adj.* **6. c.** *composite* (*high*) *school*: in Canada, a secondary school providing courses in academic, commercial, and industrial subjects.

1943 *Canada & Newfoundland Educ. Assoc. Rep. Survey C'ttee* v. 63 Composite high schools should be located in advantageous locations so that rural children may benefit from diversified curricula. **1955** W. G. Hardy *Alberta Golden Jubilee Anthol.* 174 In Senior High, the new Composite School is changing the educational picture again; large school plants are being designed for the teaching of technical, commercial and academic subjects. **1958** *Encycl. Canadiana* III. 389 The composite school, an educational institution which offers a wide variety of instruction in the technical, vocational and academic fields, is a relatively recent type of secondary school. **1968** *Globe & Mail* (Toronto) 17 Feb. 50 (Advt.), This is a fully composite school with a projected enrolment of 1450 students for September 1968, and a pupil-teacher ratio of 17 to 1.

B. *sb.* **2. b.** A material made from two or more physically different constituents each of which largely retains its original structure and identity.

1959 *Jrnl. Electrochem. Soc.* CVI. 318/2 Ceramic-and-plastic composites. **1966** [see **carbon fibre*]. **1967** *Times Rev. Industry* June 68/1 Upgrading of the physical properties has..been achieved by the production of composites, materials in which physical deficiencies in the plastics are compensated by the addition of reinforcing materials. **1970** *Materials & Technol.* III. xii. 881 An interesting new addition to the range of fibre reinforced composites is glass reinforced cement.

composition. Add: **26.** *composition-candle* (earlier example); **composition roller** *Printing*, an inking-roller usu. consisting of a metal core coated with a mixture of gelatine, glycerine, and molasses.

1861 J. A. Symonds *Let.* 4 May (1967) I. 289, I..have laid in several volumes of De Quincey & 2 long composition candles. **1825** T. C. Hansard *Typographia* Index 11, Composition Rollers, for inking, attempted by Lord Stanhope. **1928** J. C. Oswald *Hist. Printing* xxvi. 341 A London printer named Forster..evolved in 1810 a metal roller covered with a composition that distributed printing ink successfully. Baxter is the name of the inventor of composition rollers cast in molds. **1967** E. Chambers *Photolitho-Offset* 270 *Composition roller*, an inking roller of gelatin and glue; very susceptible to moisture.

compositional, *a.* Delete *rare* and add examples.

1922 *Daily Mail* 8 Nov. 13 The compositional lines are so obvious that if the left half of the picture were covered up it would be..easy to fill in mentally the hidden part. **1928** *Observer* 17 June 14/3 All the artists so far mentioned are mainly concerned with compositional problems. **1946** R. Blesh *Shining Trumpets* (1949) xiii. 321 European compositional schools which can no longer be called modern. **1953** C. E. Bazell *Ling. Form* 73 The distinction between particle and inflection belongs properly to the compositional level. **1960** *Farmer & Stockbreeder* 26 Jan. 47/2 The compositional testing of milk samples. **1970** *Nature* 6 June 926/2 A compositional layering in which the density increases upward in the stratigraphic sequence.

compos mentis, *adj. phr.* Add: (Earlier and later examples.) (See also Non compos and Non compos mentis.)

1616 B. Jonson *Devil an Ass* v. iii, You were Not *Compos mentis*, when you made your feoffment. **1809** 'D. Knickerbocker' *Hist. N.Y.* I. p. viii, He would..say something about 'deranging his ideas'; which made my wife believe sometimes that he was not altogether *compos*. **1958** B. Hamilton *Too Much of Water* vii. 158 Honestly, is he quite *compos*?

compote. Also compôte. Add: **1. b.** A dish consisting of fruit salad or stewed fruit, often in or with syrup.

1845 E. Acton *Mod. Cookery* xxv. 530 Compôte of oranges (a Hebrew dish). **1861** Mrs. Beeton *Bk. Househ. Managem.* § 1541 Take out the lemon peel, pour the syrup over the figs, and the compôte, when cold, will be ready for table. **1925** F. B. Jack *Gd. Housek. Cook. Bk.* 193 If the compote is not being used for children a little red wine may be added to the syrup when stewing the pears.

2. (Earlier and later examples.)

1747 H. Glasse *Art of Cookery* 45 Pigeons in Compôte with white Sauce. **1864** G. A. Sala *Quite Alone* I. ii. 32 Harry's a very good fellow, and has plenty of feathers ready to be plucked, before he is fit to be made into a compote de pigeons.

3. = Comport *sb.*[2]

a **1904** in A. Hayden *Chats Eng. China* 160 Dessert service ..consisting of one tall compôte, seven oval dishes [etc.]. **1908** A. Bennett *Old Wives' Tale* iv. i. 425 The large 'compote' (as it was called in his trade) which marked the centre of the table, was the production of his firm. **1926** *Tribune* (Chicago) 11 June, Compotes…These may also be used as mayonnaise or bonbon dishes.

compoted (kǫ·mpǫtĕd), *a.* [f. Compote 1 + -ed[2].] Forming compotes, made into compotes.

1920 C. Ranhofer *Epicurean* 11 Compoted dried fruits. **1928** *Daily Express* 4 June 5/3 [Strawberries] jellied and compoted.

‖**compotier** (koṅpotye). For def. read: [a. F. *compotier*, f. Compote.] = Comport *sb.*[2]; also, a dish for stewed fruit. (Add further examples.)

1755 in W. King *Chelsea Porc.* (1922) 71 One oval scollop'd compotier, 2 large scollop'd ditto, 2 heart-shaped ditto, and 4 nurl'd. **1776** *Pennsylvania Ledger* 20 Apr., Joseph Stansbury..is selling off..his baking dishes, compotiers, pudding dishes [etc.]. **1851** *Offic. Descr. & Illustr. Catal. Gt. Exhib.* II. 720/1 Centre pieces, compotiers, dessert plates, &c., chased, painted, and gilt. **1885** A. H. Church *Eng. Porc.* 51 One of the most popular of all the ornamental pieces made in English china factories during the last century is the 'shell centre-piece', or *compotier*, for the dessert table. It consisted of from one to five tiers of scallops.

compotus, var. Computus.

compound, *v.* Add: **2. f.** *Electr. trans.* To wind the field magnets of (a dynamo) so that it will be excited by a current flowing through both a shunt and a series coil.

1884 F. Krohn tr. *Glaser de Cew's Magn. & Dyn.-Electr. Mach.* 253 The Gülcher machine..has been 'compounded', or made self-regulating, by winding the field-magnets, so as to secure a constant potential at the terminals. **1902** *Encycl. Brit.* XXVII. 588/1 If the series-turns are still further increased, the voltage may be made to rise with an increasing load, and the machine is 'over-compounded'.

16. *Racing slang.* Of an animal: to fail to maintain its speed or strength, give out, fail.

1876 *Coursing Cal.* 126 It is difficult to say whether the best of the dog puppies came off successful, as they were all beginning to compound to-day. **1928** *Observer* 17 June 28/6 Once in the line for home, Goose Kiss did not remain at the head of affairs, for he compounded rapidly.

compound, *a.* Add: **2. e.** *compound engine* (also *Aeronaut.*), *-winding*; *compound-wound* adj.

1884 S. P. Thompson *Dyn.-Electr. Mach.* 96 Compound windings may be arranged in several different ways. If wound on the same core the shunt coils are sometimes wound outside the series coils: less frequently the series coils are outside the shunt. *Ibid.* 98 The compound-wound or self-regulating dynamos. **1902** *Encycl. Brit.* XXVII. 587/2 Compound-winding..was first used by S. A. Varley and by Brush. **1947** *Aeroplane* 12 Sept. 385/3 The engine was then compared with a turbo-prop. engine..assuming the same maximum temperature as the compound engine required. **1949** *Aero Digest* Mar. 30/2 A simple compound engine..is a conventional reciprocating engine, to which a 'blow down' turbine has been added to utilize the velocity energy of the exhaust.

h. *Logic*, as in *compound proposition, statement.*

1574 tr. *Ramus' Logike* 76 The compounde proposition is eyther congregatiue or segregatiue. **1654** Z. Coke *Art of Logick* 116 Hitherto of a simple proposition: now followeth the compound, which consisteth of a sense or sentence compounded. **1694** R. Blome tr. *Le Grand's Entire Body Philos.* 28/2 Compound Propositions are such as consist of more Subjects and Predicats. **1870** A. Bain *Logic* 85 The whole of this class might be called Compound, instead of complex, Propositions. **1953** I. M. Copi *Introd. Logic* vii. 195 Each of these exceptive propositions is compound. *Ibid.* viii. 221 A compound statement is one which does contain another statement as a constituent part.

i. *Gram.*, as *compound sentence* (see Sentence *sb.* 6 a); *compound verb* (see quot. 1925). Cf. quot. 1859 for sense 1 a in Dict.

1772 A. Bayly *Gram. Eng. Lang.* 76, I am at a loss in what class to place compound verbs, whether in that of thoughtless chance, or of judicious accommodation. *Ibid.* 80 We now proceed to compound sentences, wherein to avoid inaccuracy is required the greatest judgement and attention…Compound sentences are formed by the help of certain adverbs, relatives and participles. **1850** W. C. Fowler *Eng. Gram.* vi. x. 568 Two simple sentences are connected either by way of Co-ordination, or by way of Subordination…The two sentences taken together constitute a co-ordinate compound sentence. **1863** A. Bain *Eng. Gram.* 38 The prepositions in such cases become adverbs united to the verbs, constituting them compound verbs. *Ibid.* 161 A compound sentence contains two or more co-ordinate sentences united: 'the sun rose, and the mists disappeared.' **1925** Grattan & Gurrey *Our Living Lang.* xii. 80 This union of simple Verb with Particle forms what is known as a Compound Verb. *Ibid.* 81 All these [*arise, bespeak, depart*, etc.] are..Compound Verbs. *Ibid.* xlii. 268 The units are co-ordinate and, occurring in a row, are linked together by words…Such a group is called a Compound Sentence.

compound, *sb.*[1] Add: **2. a.** *chemical compound*, a substance composed chemically of two or more elements in definite proportions (as opposed to a *mixture*).

1808 J. Dalton *New Syst. Chem. Philos.* I. 216 All the chemical compounds which have hitherto obtained a tolerably good analysis. **1866** H. E. Roscoe *Elem. Chem.* v. 46 The air is a mixture, and not a chemical compound of constituent gases. **1887** [see Mixture 4]. **1950** *Science News* XV. 103 Proteins are by far the most complex chemical compounds.

compound, *sb.*[2] Add: **2.** In South Africa, an enclosure within which the workmen in diamond or gold mines live and remain during their term of employment, having no communication with the outside. Also *attrib.*

1893 T. Cook *Mission Tour* 25 The men sign articles to remain in these compounds for a certain period, usually six months, and are not allowed to leave for any cause until the time has expired. **1897** *Empire* 11 Dec., The Compound system saved the mines from the clutches of the illicit diamond buyers. **1901** *Contemp. Rev.* Mar. 320 How does he [*sc.* the native] get into the Kimberley compounds? He goes there of his own accord. **1906** *Westm. Gaz.* 19 June 2/2 A riot at the Nourse Mine..in which 400 Chinese wrecked the house of the compound manager.

3. A large fenced-in space in a prison, concentration camp, or the like.

1946 Brickhill & Norton *Escape to Danger* vi. 83 He had seen this man walking round the compound as a prisoner. *Ibid.* xxvi. 222 Germans..had a habit of sneaking into the compound at night. **1954** W. Faulkner *Fable* (1955) 119 The camouflage-painted huddle of the prison-compound.

compounded, *ppl. a.* Add: **4.** *compounded tenancy*, tenancy by which the tenant's rates are included in his rent, and paid by the landlord.

1905 *Daily Chron.* 18 May 3/4 The owners of 'flats' and other 'compounded' tenancies get off with so much abatement.

compounder. Add: **1. b.** *compounder of medicine*, a pharmacist attached to the army medical corps.

1894 Campbell-Bannerman in *Hansard's Parl. Deb.* Ser. iv. XXVI. 830 With regard to the apothecaries, he was glad to say they were an expiring body, and as they ceased their places were taken by a class of non-commissioned officers who were called compounders of medicine. **1899** *Daily News* 21 Dec. 3/1 Royal Army Medical Corps. Compounders of medicine are urgently required for service in South Africa.

comprehendible, *a.* Add: Also **comprehendable.**

1814 Jane Austen *Let.* 28 Sept. (1952) 403 Jane Egerton is a very natural, comprehendable Girl.

comprehensive, *a.* Add: **1. d.** Designating a secondary school or a system of education which provides for children of all levels of intellectual and other ability (see quots.). Also *ellipt.* as *sb.*, a school of this kind or (*occas.*) a pupil attending one.

1947 *Min. of Educ. Circular* No. cxliv. 1/2 Combinations of two or more types of secondary education are often referred to as bilateral, multilateral or comprehensive. *Ibid.* 2/1 A comprehensive school means one which is intended to cater for all the secondary education of all the children in a given area without an organisation in three sides. **1955** *Ann. Reg. 1954* 13 The L.C.C. had adopted the educational policy of the so-called comprehensive school, where all, whatever their standards, were to be educated together up to the age of 15. **1955** *Times* 20 May 11/5 A comprehensive school is intended to recruit all the boys, or girls, from a given area at the age of 11 and of these not more than one in five will be of grammar school standard. **1958** *Spectator* 27 June 833/1 Comprehensives, scrubbed and solemn in suits. **1958** *Observer* 30 Nov. 19/5 Pupils shunted off to the posh new comprehensives. **1959** *Punch* 16 Sept. 169/2 His son is at a Public School…His younger daughters both attend The local Comprehensive. **1965** *New Statesman* 9 Apr. 567/1 As for the public schools, you can't even suggest that the grammar schools be merged with the comprehensives without losing all the marginal seats in Bristol.

comprehensively, *adv.* Add: **b.** With respect to the comprehension (in Logic) of the terms (opp. to Extensively). See Comprehension 4.

1816 J. Smith *Elements of Logic* i. v. 24 Conceive of them comprehensively, with regard to their properties and relations. **1857** *Encycl. Brit.* XIII. 610/1 When the dictum is analysed comprehensively.

comprehensivize (kǫmprĭhe·nsivəiz), *v.* [f. *comprehensive *a.* 1 d + -ize.] *trans.* To make (a school or system of education) comprehensive (see *comprehensive *a.* 1 d). So **comprehe:nsiviza·tion.**

1958 *Economist* 6 Dec. 865/1 Attempts by local authorities to 'comprehensivise' schools. *Ibid.* 13 Dec. 970 The council's intention to comprehensivise the schools. *Ibid.*, Newport has revised its drastic comprehensivisation plans down to a single comprehensive school. **1966** D. Jenkins *Educ. Society* v. 243 The compulsory 'Comprehensivisation' of schools.

compresence (kǫmpre·zĕns). Delete †*Obs. rare*[−1] and add later examples.

1912 *Mind* XXI. 2 The togetherness or compresence of the perceiving and the table is the perception *of* the table. **1920** A. S. Pringle-Pattison *Idea of God* 354 The compresence of all these moments in a single experience. **1950** L. S. Thornton *Revelation & Mod. World* ix. 263 The New Testament shows the compresence of diverse strands of imagery.

compresent (kǫmpre·zĕnt), *a.* [f. Com- + Present *a.*] Present together; co-present.

1912 S. Alexander in *Mind* XXI. 3 Compresent means simply belonging to the same universe. **1940** B. Russell *Inq. Meaning & Truth* vii. 114 In every statement containing 'I-now' we may substitute 'what is compresent with this'.

compressibility. 1. b. (Later and *attrib.* examples in *Aeronautics*.)

1920 L. Bairstow *Appl. Aerodynamics* viii. 383 For airscrews, the tips of the blades of which may reach speeds of 700 or 800 ft.-s., the effect of compressibility may be expected to be important. **1933** J. Stack in *Nat. Advis. Comm. Aeronaut., Techn. Rep.* No. 463 18 This breakdown of the flow, hereinafter called the compressibility burble, occurs at lower speeds as the lift is increased by changing the angle of attack. **1938** L. Bairstow *Appl. Aerodynamics* (ed. 2) i. 10 At about three-quarters of the velocity of sound ..the lift of aerofoils falls rapidly and the drag increases enormously, giving rise to a phenomenon sometimes spoken of as the 'compressibility stall'. **1950** *Gloss. Aeronaut. Terms (B.S.I.)* i. 26 *Compressibility drag*, the increase in drag arising from the compressibility of the air which occurs at high speeds.

compression. Add: **1. d.** In a steam-engine, the reduction in volume of the steam left in the cylinder after the exhaust is closed towards the end of the exhaust stroke.

1859 W. J. M. Rankine *Man. Steam Eng.* iii. iii. 420 Compression, or cushioning, is effected by closing the eduction valve before the end of the return stroke. **1913** W. R. King *Steam Engin.* v. 105 This compression provides an elastic cushion of steam which absorbs the momentum of the reciprocating parts of the engine and brings them to rest without shock. **1936** E. A. Phillipson *Steam Locom. Design* x. 317 The compression point should advance when the engine is running..at high piston speeds, in order that sufficient cushioning may be available to counteract the greatly augmented inertia forces then set up by the reciprocating masses.

e. In an internal-combustion engine, the reduction in volume of the mixture of fuel and air drawn into the cylinder; also, the value or effectiveness of this as a factor affecting the running of the engine.

1887 D. Clerk *Gas Engine* vii. 197 When compression is completed the igniting valve acts and the explosion impels the piston. **1907** R. B. Whitman *Motor-Car Princ.* xi. 193 Other losses of compression may be due to a cracked piston. **1912** *Motor Manual* 227 No engine ever pulled well with feeble compression. **1963** D. V. W. Francis *Morris Engines* i. 7 It should be possible to feel a good and even compression for all the cylinders.

5. *compression gauge, spring*; **compression-ignition engine,** an internal-combustion engine in which the compression of air in the cylinder provides heat to ignite the fuel, as in a Diesel engine; hence **compression-ignition,** this principle or process; **compression mould,** a mould which encompasses the material to be shaped (see quot. 1951); **compression moulding,** a method of moulding plastics by applying pressure; also, the equipment for carrying out this process; a product of this process; **compression ratio,** the ratio of the maximum to the minimum volume in the cylinder of an internal-combustion engine, measured before and after compression (one piston stroke); **compression rib** (see quot.); **compression stroke,** the stroke of the piston effecting the compression of the gas and air in the cylinder of an engine; **compression wood,** a type of wood that develops on the undersides of branches and at the bases of leaning trunks of softwood trees.

1912 *Motor Man.* 234 Each cylinder fails to show a high reading on a compression gauge. **1926** *Engineering* 27 Aug. 277/2 A compression-ignition engine with its high expansion ratio may be expected to consume a smaller weight of fuel per horse-power than a petrol engine. **1933** *Jrnl. R. Aeronaut. Soc.* XXXVII. 455 Knock under certain conditions of operation of compression ignition engines may be due to vaporisation of fuel oil during the delay period. **1936** *Economist* 2 May 234/1 The diesel, or compression-ignition, engine requires no preliminary heating. *Ibid.*, 'Compression-ignition', with a smaller space in front of the engine piston, produced an increase in 'thermal efficiency'. **1951** *Gloss. Terms Plastics (B.S.I.)* 35 *Compression mould*, a mould which requires to be opened to receive the charge and which forms the material to shape on closing. **1940** *Chambers's Techn. Dict.* 186/1 *Compression moulding.* The material is placed in a hardened, ground, polished steel container and forced down by means of a plunger at a pressure of 3000–5000 lb. per sq. in. **1958** *Spectator* 25 July 150/3 It also manufactures signalling equipment, compression moulding, electrical components. **1907** F. Strickland *Man. Petrol Motors* i. iii. 20 The actual compression ratio to be used in an engine is one of the most important points in its design. **1934** *Jrnl. R. Aeronaut. Soc.* XXXVIII. 44 About 13½:1 is more usual with the open combustion chamber, but with the 'swirl' type which he had described,..the higher compression ratio was possible. **1958** *Times* 1 July 6/6 It differs from the saloon engine, however, in having a compression ratio of seven to one. **1918** Farrow *Dict. Mil. Terms*, *Compression Rib*, in an aëroplane, a rib that acts as an ordinary rib, besides bearing the stress of compression produced by the tension of the internal bracing wires. **1904** A. B. F. Young *Compl. Motorist* iv. 82 The steering on the car is of an exceedingly strong character…Very strong and stiff compression springs effectually prevent any possibility of 'backlash'. **1894** B. Donkin *Gas, Oil, & Air Engines* i. vii. 90 The compression stroke forces this residuum and part of the fresh charge up the narrow passage leading to the hot tube, and causes ignition. **1908** *Westm. Gaz.* 22 Oct. 4/3 The compression-stroke, in which the piston, by its upward motion compresses the charge of gas into the head of the cylinder. **1912** *Motor Man.* 237 The power is developed during a complete cycle of four strokes..one occurring at each half revolution or every stroke of the piston; thus (1) suction stroke, (2) compression stroke [etc.]. **1925** *Spiral Tracheids & Fiber-Tracheids: Tropical Woods* (Yale Univ. Sch. of Forestry) 12 In so-called compression wood or 'red' wood (rotholz), found in both the Angiospermae and Gymnospermae, the cells have a thick laminated secondary wall, slit into a large number of closely compressed spiral lamellae. **1940** *Chambers's Techn. Dict.* 186/1 Compression wood.

compressionism (kǫ̆mpre·ʃəniz'm). *Theatr.* [f. Compression + -ism.] A type of drama presenting a small number of characters in a confined situation. Hence **compre·ssionist** *sb.* and *a.*

1961 L. Kitchin in *20th Cent.* Feb. 168 Pinter is a compressionist, like Sartre in *Huis Clos* where the damned pace a hotel room forever. *Ibid.* 169 Compressionism must have ..shot its bolt in the theatre. Its natural home is television. **1963** *Listener* 24 Jan. 157/2 A compressionist play is one in which the characters are insulated from society in such a way as to encourage the maximum conflict of attitudes. **1963** *Times* 26 Apr. 6/6 Compressionism, an intensification of Webernian high pressure, a matter of desperate, boxed-in themes.

compressor. Add: **2.** One who compresses, *spec.* the attendant or operator of a compressing machine. Also *compressor man.*

1920 *Glasgow Herald* 3 July 8 John K. Robertson, compressor,..allowed three gallons of oil to escape from a compressed-air engine. **1921** *Dict. Occup. Terms* (1927) § 088 *Compressor man*;..operates, by hand, machine which compresses charge of partially dried coal into a solid cake [etc.] *Ibid.* § 148 Compressor, carbonic acid gas.

comprise, *v.* Add:

8. b. Delete *rare* and add later examples.

1850 W. S. Harris *Rudimentary Magnetism* iv. 73 These substances which we have termed diamagnetic..and which comprise a very extensive class of bodies. **1907** H. E. Santee *Anat. Brain & Spinal Cord* (1908) iii. 237 The fibers comprising the zonal layer have four sources of origin. **1925** *Brit. Jrnl. Radiology* XXX. 148 The various fuses etc. comprising the circuit. **1950** M. Peake *Gormenghast* (1968) xiv. 94 Who, by the way, *do* comprise the Staff these latter days? **1959** *Chambers's Encycl.* XIII. 653/1 These fibres also comprise the main element in scar tissue. **1969** W. Hooper in C. S. Lewis *Sel. Lit. Ess.* p. xix, These essays together with those contained in this volume comprise the total of C. S. Lewis's essays on literature. **1969** N. Perrin *Dr. Bowdler's Legacy* (1970) i. 20 As to who comprised this new reading public, Jeffrey..guessed in 1812 that there were 20,000 upper-class readers in Great Britain.

c. To be composed *of*, to consist *of*.

1874 *Art of Paper-Making* ii. 10 Thirds, or Mixed, are comprised of either or both of the above. **1928** *Daily Tel.* 17 July 10/7 The voluntary boards of management, comprised..of very zealous and able laymen. **1964** E. Palmer tr. *Martinet's Elem. Gen. Ling.* i. 28 Many of these words are comprised of monemes. **1970** *Nature* 27 June 1206/2 Internally, the chloroplast is comprised of a system of flattened membrane sacs.

compromise, *sb.* Add: **7.** *attrib.*, esp. defining a thing intermediate between two others or possessing an accommodating combination of characteristics.

1833 *Congress. Deb.* 27 Feb. 1864 The olive branch, the compromise tariff bill, will probably allay the excited feelings of the South. **1893** *Westm. Gaz.* 17 Oct. 7/3 It is said that a compromise amendment will be laid before the Senate tomorrow. **1898** *Engineering Mag.* XV. 102 Better practice requires that the exhaust or low-pressure steam should be used, supplemented by high-pressure steam from the boilers, the combination giving the desired temperature. There are several similar compromise points, which need not be mentioned. **1904** *Daily Chron.* 21 Oct. 5/6 The House of Deputies subsequently adopted by a great majority a compromise resolution. **1906** *Ibid.* 24 Oct. 6/6 Wellington is a compromise capital. Auckland, the original capital, was too far north to suit the southern folks, and Dunedin..was too far south to be tolerated by the northerners. **1953** C. E. Bazell *Ling. Form* 60 The morpheme is a compromise-unit rather than a purely distributional unit.

‖**compte rendu** (koṅt raṅdü). [Fr.] A report; a review; a statement.

1822 L. Simond *Switzerland* I. 289 Thinking it against the *bienséance* of the situation to publish any thing but a *compte rendu*, or grave works of morality,..he burnt his plays. *a* **1854** J. S. Mill *Draft Autobiogr.* (1961) 52 The written outline of it which resulted from my daily *compte rendu*, served him afterwards as notes. **1870** W. James *Let.* 7 May (1920) I. 158 But I hear Father arriving and I must go down to hear his usual *compte rendu*. **1931** *N. & Q.* CLX. 102/1 A competent book on Russian Literature..giving *comptes rendus* of the novels, dramas, stories, poems etc., of the more eminent authors. **1934** 'A. Bridge' *Ginger Griffin* xxiii. 304 'I called on Thursday, but the girl was out,' said Joanna, thus completing her *compte-rendu* of the female Boggits.

Comptometer (kǫmptǫ·mĭtəɹ). Also **comptometer.** [app. f. F. *compte* Count *sb.*[1] + -ometer.] The proprietary name of a calculating-machine. Also *attrib.*

1894 *Work* 24 Mar. 151 Burroughs' accountant and comptometer. **1904** *Engin. Mag.* July 612 The Comptometer is a key-operated machine for performing the four basic mathematical operations. **1921** *Dict. Occup. Terms* (1927) § 939 Comptometer operator. **1929** *B'ham Post* 17 Aug. 3/6 (Advt.), Comptometer operators…Comptometer school. **1935** *Trade Marks Jrnl.* 9 Oct. 1257 *Comptometer*, calculating machines. Felt & Tarrant Manufacturing Company.. City of Chicago, State of Illinois. **1940** 'N. Shute' *Landfall* vii. 170 Plodding through endless computations with slide rule, graphs, and the little black comptometer.

Compton (kǫ·mptǫn). The name of Arthur Holly *Compton* (1892–1962), American physicist, used *attrib.* or in the possessive of certain phenomena in Physics, as **Compton effect,** the scattering and increase in wavelength of an X-ray (or gamma-ray) photon on encountering an electron, with a partial transference of energy from the photon to the electron; **Compton shift,** the difference between the wavelengths of incident and scattered photons; also *Compton absorption, electron*, etc.

1923 *Physical Rev.* XXII. 201 Experiments on Compton's change in wave-length on scattering. **1924** *Ibid.* XXIV. 204 Theory of the width of the modified lines in the Compton effect. **1925** *Ibid.* XXVI. 282 The ratio of intensity of the Compton lines. **1934** H. L. Brose tr. *Sommerfeld's Atomic Struct.* (ed. 3) I. i. 49 In comparison with the photo-electrons, ..the 'Compton electrons' have a considerably smaller value. *Ibid.* 50 The Compton effect demonstrates very impressively the dual nature of radiation. **1932** *Discovery* Oct. 337/2 When these photons encounter solid matter, they will generally experience slight scattering due to Compton encounters with the planetary electrons of the atoms. **1955** W. Heisenberg in W. Pauli *Niels Bohr* 13 Bothe and Geiger showed experimentally by means of the Compton effect that the energy conservation law is valid for individual processes also. **1955** L. Rosenfeld *Ibid.* 83 The Compton wave-length. **1959** *Listener* 1 Jan. 27/1 Photons, or atoms of radiant energy, are scattered if they penetrate too close to an atomic nucleus (Compton effect).

compts. Esp. in letters, an abbrev. of *compliments* (see Compliment *sb.* 2).

1810 Shelley *Let.* 18 Dec. (1964) I. 26 The Author's respectful compts. to his Uncle Mr. Parker. **1836** Dickens *Let.* 3 Dec. (1965) 205 Mr. Dickens presents his compts. to Mr. Ayrton.

compulsion. Add: **2.** *Psychol.* An insistent impulse to behave in a certain way, contrary to one's conscious intentions or standards.

1913 A. A. Brill tr. *Freud's Interpret. Dreams* v. 206 There is a class which has accentuated the childish impulse to a compulsion—they are the exhibitionists. **1937** Koestler *Spanish Testament* ii. vii. 314 Often before had attacks of such compulsions, hitherto always managed to fight them down. **1944** J. S. Huxley *Living in Revol.* 194 How can the unconscious compulsions of very early life..be rendered as little harmful as possible? **1951** K. Horney *Neurosis & Human Growth* i. 38 The difference, then, between healthy strivings and neurotic drives for glory is one between spontaneity and compulsion. **1965** A. D. Weisman *Existential Core of Psychoanalysis* vii. 205 In contrast to compulsion, *choice* is essential to responsible acts.

b. *attrib.*, as **compulsion neurosis** [tr. Ger. *zwangsneurose*] *Psychiatry*, a psychoneurotic disorder characterized by recurrent compulsions and compulsive ideas (cf. **obsessional neurosis*); so *compulsion-neurotic* sb.

1909 A. A. Brill tr. *Freud's Sel. Papers on Hysteria* iv. 77 From neurasthenia we sharply distinguished the compulsion neurosis (Zwangsneurose). **1938** R. G. Collingwood *Princ. Art* iv. 62 Freud..answers it by bringing the supposed facts of magic into relation with those of compulsion-neurosis as studied by himself in his own patients. *Ibid.* 63 Freud quotes a patient of his own, a compulsion-neurotic. **1963** *New Yorker* 29 June 27 He was most like a compulsion neurotic.

compulsionist. Delete *nonce-wd.* and add: *spec.* an advocate of compulsory military service.

1906 *Westm. Gaz.* 13 Mar. 4/2 That we must protect ourselves against the Conscriptionists and Compulsionists. **1915** Earl of Oxf. & Asquith *Mem. & Refl.* (1928) II. 109 Two vital matters..which it seems to me the compulsionists have never thought out.

compulsive, *a.* (*sb.*). Add: **A. 3.** *Psychol.* Acting from, related to, or typical or suggestive of a compulsion (*compulsion 2).

1902 A. R. Defendorf *Clin. Psychiatry* 25 Compulsive ideas are those ideas which irresistibly force themselves into consciousness. **1909** *Cent. Dict.* Suppl., *Compulsive insanity*, insanity in which imperative ideas, or obsessions, completely dominate the patient. **1954** Koestler *Invis. Writ.* xvii. 196 He was a compulsive liar, and had a pathological craving for fame. **1957** C. E. Buxton in E. Hilgard *Introd. Psychol.* (ed. 2) xx. 490/2 The *compulsive* personality is characterized by excessive cleanliness, orderliness, obstinacy and stinginess. In the extreme, behavior becomes repetitive and ritualistic. **1962** *Listener* 1 Nov. 711/2 He became compulsive about time and always wore two watches. **1965** D. M. Levy in G. E. Daniels *New Perspectives in Psychoanalysis* 249 Some tics are of a different variety..and are commonly referred to as compulsive movements.

b. Of a book, etc.: holding one's attention, as if by a compulsion.

1961 *John o'London's* 4 May 506/4 This too is an outstandingly compulsive, perceptive and human book. **1962** *Listener* 11 Jan. 104/1 Redmond Macdonogh's *Five Days to Friday*..proved compulsive listening.

B. *sb.* **2.** *Psychol.* A person typically subject to compulsions.

1957 P. Lafitte *Person in Psychol.* vi. 75 The compulsive scrupulously accumulates objects in his office and home workshop. **1959** N. Mailer *Advts. for Myself* (1961) 156, I had the impression she was a compulsive.

compulsively, *adv.* Add: **3.** In accordance with a psychological compulsion (see *compulsion 2); in a compulsive manner (see prec.).

1937 K. Horney *Neurotic Personality* vii. 128 Such a change may sometimes take grotesque forms, as in persons who after failing in some love affair start to eat so compulsively that they gain twenty to thirty pounds in a short time. **1955** D. W. Maurer in *Pubn. Amer. Dial. Soc.* XXIV. 5 The teen-agers who almost compulsively identify with this semicriminal subculture. **1962** *Listener* 6 Sept. 350/1 Lawrence..was often compulsively autobiographical in his writing. **1966** *Listener* 10 Feb. 214/3 In *The Fight for Barbara* author and subject come too close for full artistic success, though it is compulsively readable.

compulsiveness (kǫ̆mpʊ·lsivnĕs). [f. Compulsive *a.* + -ness.] The fact or quality of being compulsive.

1918 C. Montefiore *Liberal Judaism* iii. 159 It is in this combination..that its seductive compulsiveness lies. **1947** *Mind* LVI. 304 That aspect of passivity, compulsiveness, or givenness which is rightly regarded as ineluctable in sensory cognition. **1962** *Listener* 6 Sept. 367/3 It would be silly to deny the unexpected compulsiveness of the programmes.

compulsory, *a.* Add: **1. a.** *compulsory education, games.*

1861 *Rep. Comm. State of Pop. Educ. Eng.* I. i. vi. 300 The possibility of establishing a system of compulsory education in this country has been brought before us. *Ibid.*, Those systems of compulsory education by the State which are now established in Prussia and in other parts of Germany. **1881** O. SEAMAN in C. E. Pascoe *Everyday Life in our Public Schools* 155 The school game of Shrewsbury..is football. There are four or five compulsory games a week,..from which Sixth Form and those who have medical certificates of weak health are excused. **1905** H. A. VACHELL *Hill* ii. 27 You'll have to play the compulsory games..but I want to see you playing..in the house-games. **1918** LYTTON STRACHEY *Eminent Victorians* 212 The modern reader of *Tom Brown's Schooldays* searches in vain for any reference to compulsory games. **1960** C. DAY LEWIS *Buried Day* vi. 126 There is something to be said against compulsory games. ..But..I am glad I was made to play them.

computational, *a.* Delete *rare* and add further examples. Also, relating to computers.
1956 *Nature* 7 Jan. 1/1 The work carried out by Dr. A. D. Booth, head of the Computational Laboratory there, on mechanical translation. **1962** *Listener* 5 Apr. 594/2 Young children of eight, with weaknesses in computational ability. *Ibid.* 6 Dec. 952/1 The development of computational devices is enabling man to carry out calculations with a speed and with a precision that were previously unthinkable. **1964** M. A. K. HALLIDAY et al. *Ling. Sciences* 306 The important new field of computational linguistics. **1969** *Sci. Jrnl.* Feb. 59/3 Another computational use of computers in personality assessment research is in interpreting test profiles.

computer. Add: **2.** A calculating-machine; esp. an automatic electronic device for performing mathematical or logical operations; freq. with defining word prefixed, as *analogue, digital, electronic computer* (see these words).
1897 *Engineering* 22 Jan. 104/2 This was..a computer made by Mr. W. Cox. He described it as of the nature of a circular slide rule. **1915** *Chambers's Jrnl.* July 478/1 By means of this computer the task is performed mechanically and almost instantaneously. **1941** *Nature* 14 June 753/2 The telescope drive is of an elaborate nature; the effects of changing refraction, of differential flexure and of errors in the gears are automatically allowed for by a system of 'computers'. **1944** *Times* 8 July 2/2 The Mark XIV consists of..a rectangular box called the computor, which might be described as the brains of the machine. **1946** *Jrnl. Appl. Physics* XVII. 262 (*heading*) A computer for solving linear simultaneous equations. **1946** [see *ELECTRONIC *a.* 3]. **1947** [see *DIGITAL *a.* 4]. **1957** *Technology* Mar. 9/1 The advent of the electronic digital computer, with its ability to make simple logical decisions, now permits a further step forward by supplementing the brain power required to supervise the control of manufacturing processes. **1957** *Ibid.* July 167/3 At present a computer can read, remember, do arithmetic, make elementary decisions and print its answers. **1958** *Listener* 18 Sept. 413/2 Much work was done..trying to 'programme' a computer to play chess. **1963** *Publishers' Weekly* 5 Aug. 80/1 Computers are being used to speed up the production of justified tape for the operation of typesetting machines. **1964** F. L. WESTWATER *Electronic Computers* i. 1 The popular idea of a computer as an electronic 'brain' is not entirely apt. Basically, a computer is merely a calculating machine,..with the difference that the speed of calculation has been enormously increased.
b. *attrib.* and *Comb.*
1957 *Economist* 30 Nov. 807/2 Computer-control methods that have already been applied to certain machine tool operations. **1957** I. ASIMOV *Earth is Room Enough* (1960) 11 The decisions on priority are computer-processed. I could in no way alter those conditions arbitrarily. **1961** *Times* 21 July 3/1 Senior Computer Programmers. **1962** *Science Survey* XX. 322 Functional diagram of computer-controlled milling machine. **1964** *Discovery* Oct. 56/2 Such projects form a part of the developing field of computer-aided design. **1964** R. D. HOPPER in I. L. Horowitz *New Sociol.* 322 The possibilities of..computer-based thinking. **1964** A. WYKES *Gambling* xi. 259 He must have a computer-like ability to remember all the cards. **1965** *Math. in Biol. & Med.* (*Med. Res. Council*) p. viii, For some time past commerce and industry have been making increasing use of computer-based automatic data processing systems. *Ibid.* 38 It may well be that the computer-assisted treatment will replace the radical operation of temporal lobectomy. *Ibid.* 166 The function of a computer-generated sine wave. **1966** *Performing Right* Oct. 10/1 A veteran computer operator. **1968** *Computers & Humanities* II. 145 The computer-generated concordance is probably the earliest-developed..and most useful application of electronic data processing to literary texts. **1969** BESSINGER & SMITH *Concordance to Beowulf* p. xxiii, We considered writing a computer program to 'translate' the text, word by word, into the hyphenated form. **1970** *Brit. Printer* Feb. 27/1 One of the unnerving things about computer-assisted typesetting is the pace at which developments take place. **1970** *Computers & Humanities* IV. 340 Umlauts..and upper case letters are coded for proper computer typesetting.

computerize (kǒmpiū·tərəiz), *v.* [f. COMPUTER + -IZE.] *trans.* To prepare for operation by, or to operate by means of, a computer; to install a computer or computers in (an office, etc.). So **compu:teriza·tion** *vbl. sb.*, **compu·terized** *ppl. a.*
1960 *Times* 4 Aug. 13/4 Executives read and hear a lot about 'computerization'. *Ibid.* 13/5 The initial paperwork for each computerized job is therefore often formidable. **1961** *Times* (Computer Suppl.) 3 Oct. p. v/7 The businessman will probably..think out..what areas of his business might be profitably computerized. **1963** *Publishers' Weekly* 5 Aug. 83/1 (*caption*) Computerized Typesetting via Communications Satellite. **1964** *Economist* 4 Apr. 54/2 The computerisation of the betting shops or roulette establishments. **1964** *Bookseller* 19 Dec. 2296/1 To computerize our invoicing and accounts departments in the hope of maintaining a faster flow of invoices. **1969** *Jane's Freight Containers 1968–69* 355/1 Many other requirements of modern transport will be computerised.

computing (kǒmpiū·tiŋ), *vbl. sb.* [f. COMPUTE *v.* + -ING[1].] The action of calculating or counting. Freq. *attrib.*
1646 [see COMPUTE *v.*]. **1867** *Rep. Comm. Patents 1866* (U.S.) II. 1547 Computing machine. **1869** *Ibid. 1868* I. 621/1 Computing apparatus. **1946** *N.Y. Times* 15 Feb. 16/6 A few of the fields that will benefit hugely through electronic computing. **1947** [see *DIGITAL *a.* 4]. **1955** *Sci. Amer.* June 92/1 A computing machine capable of solving problems must possess a 'memory'. **1962** *Gloss. Automatic Data Processing (B.S.I.)* 102 *Computing amplifier*, a unit consisting of an amplifier and a negative feedback network arranged so that the output voltage bears an assigned relationship to the input voltage.

computistical (kǒmpiuti·stikăl), *a.* Also **computistic.** [ad. mod.L. *computisticus*; cf. COMPUTUS and -ISTICAL.] Of or pertaining to a computus.
[**1802** PLANTA *Catal. MSS. Cott. Libr.* 34/1 Tabulæ quædam computisticæ.] **1933** R. TUVE *Seasons & Months* iv. 151 A similar manuscript in Munich (..also a computistic and astronomical collection), written in 818,..parallels the *Horae* in many ways. *Ibid.*, The circle of the Zodiac is..one of the usual manuscript illustrations of computistical and calendar treatises. **1957** N. R. KER *Catal. MSS. Anglo-Saxon* p. vii, Adequate descriptions of MSS. containing legal, computistical, and penitential texts are possible.

comrade. Add: **f.** Used by socialists and communists as a prefix to the surname, to avoid such titles as 'Mr.' Hence, a (fellow-) socialist or communist.
1884 *Justice* 13 Sept. 7/1 A meeting was held..on Sunday last by Comrades Kelly and Maguire... Comrade Maguire spoke at some length on the 'Aims of Socialism'. **1885** *Ibid.* 25 July 3/2 Our comrade pointed out how the land was one of the means by which the labour of the workers was exploited by an idle class. **1887** *Commonweal* 12 Feb. 56/1 We held an outdoor meeting on Sunday morning on Mitcham Fair Green. Comrade Kitz and other Merton comrades assisted. **1918** C. E. RUSSELL *Unchained Russia* iii. 95, I was projected from sleep by the voice of our assistant train manager, raised in vehement protest: 'Niet, tavarisch, niet! Niet!' That is to say, 'Comrade'. After the Revolution everybody in Russia was 'tavarisch'. **1920** HARD *Raymond Robins' Own Story* 58 'Yes' said Trotzky, 'I'll make the order'. He made it. It began: To Comrades Podvoisky, Krylenko and Elizarov. **1928** *Illustr. Hist. Russian Rev.* I. 189 Two comrades, Lashevitch and Kurayev, spoke to the soldiers. **1939** R. CAMPBELL *Flowering Rifle* I. 14 And every Babbit is a foxes' hole From which a scrawny 'comrade' snarls for dole! **1965** C. D. EBY *Siege of Alcázar* (1966) iii. 83 Comrade Gómez was conducted to the local bar so that he could be treated to a *coñac*.

co·mradeless, *a.* [f. COMRADE + -LESS.] Without a comrade or comrades.
1891 H. C. HALLIDAY *Someone must Suffer* III. xvi. 264 Alone and comradeless in the battle of life. *a* **1915** R. BROOKE *Poems* (1918) 31 Some pause in their grave wandering comradeless.

comradeliness (kǫ·mrĕdlines). [f. COMRADELY *a.* + -NESS.] The state or condition of being comradely.
1930 *Time & Tide* 18 Apr. 501/2 The air of inhumanity, by which I mean a deficiency of genial and common comradeliness. **1936** L. C. DOUGLAS *White Banners* x. 212 Adele's bereavement provided her a chance to exercise a talent for comradeliness which had become anaemic through disuse.

coms, var. *COMBS *sb. pl.*

comsat (kǫ·msæt). [f. COM(MUNICATION + SAT(ELLITE.] A communication satellite. Also (with capital initial), the name of a business corporation operating such satellites.
1962 *Flight Internat.* LXXXII. 464/2 A martyr to the comsat cause. **1964** *Economist* 27 June 1454/1 Comsat—the company created by the American Congress to launch and operate commercial satellites but owned partly by industry, partly by ordinary shareholders. **1965** *Ibid.* 27 Mar. 1380/1 Comsat sells its telephone and telex channels at the wholesale, not the retail, level. **1966** *New Statesman* 11 Feb. 185/2 Their proven experience in building comsats, launchers and ground stations.

Comsomol, var. *KOMSOMOL.

Comstock (kǫ·mstǫk). *U.S.* [f. the name of H. T. P. *Comstock* (1820–70), American prospector who first worked a claim on the site of the Comstock lode.] In full, *Comstock lode*: a very rich lode of silver and gold discovered in Nevada in 1859; hence, allusively, a rich mine or 'find'.
1866 *Beadle's Monthly* Aug. 102/1 The Comstock Lode proved the richest vein of silver ever found. **1867** *Terr. Enterprise* (Virginia, Nev.) 2 Feb. 3/1 The real out and out Washoe miner can be found away down in the bowels of the Comstock. **1885** *Wkly. N. Mex. Rev.* 18 June 2/6 The Old Mine of Grant county is proving a genuine Comstock. **1948** *Manch. Guardian Weekly* 9 Dec. 9 The committee sank their pick in a farmyard and struck their Comstock lode in the shape of a hollowed-out pumpkin. **1956** A. HUXLEY *Adonis & Alphabet* 196 The people who had come to exploit this Comstock Lode of the miraculous, found themselves painfully frustrated.

Comstockery (kǫ·mstǫkəri). Also **comstockery.** [f. the name of Anthony *Comstock* (1844–1915), member of the New York Society for the Suppression of Vice.] Excessive opposition to, or censorship of, supposed immorality in art or literature; prudery. So **Co·mstocker,** one who advocates or practises Comstockery; **Co·mstockish** *a.*; **Co·mstockism.**
1905 G. B. SHAW in *N. York Times* 26 Sept. 1/7 Comstockery is the world's standing joke at the expense of the United States. *Ibid.* 2/1 The good intentions of the leaders of the Comstockers. *Ibid.*, Socialism may become the..Puritanism of the Comstocker. **1906** *Westm. Gaz.* 27 Nov. 4/2 Played by American ladies one can only say, discreetly, that it is Comstockish. **1909** WARE *Passing Eng.*, *Comstockism* (Amer.-Eng., 1885 on), opposition to the nude in art. **1911** T. A. SCHROEDER *'Obscene' Lit.* 101 There is no organized force in American life which is more pernicious..than..Comstockery. **1915** *Survey* 6 Nov. 128/2 Liberalism in public affairs may become Comstockism in morals. **1921** J. B. CABELL *Jurgen* p. xvi, She is the Mrs. Grundy of the Leshy; she is Comstockery: and her shadow is commonsense. **1923** A. HUXLEY *On Margin* 45 Such minor menaces as Prohibition, Lord Northcliffe, Mr. Bryan, Comstockery. **1933** *Scrutiny* June 106 He [Dreiser] did have genuine grievances against timid publishers, moral reviewers and comstockery in general. **1965** P. FRYER *Birth Controllers* xix. 204 She [*sc.* Margaret Sanger] declared war on Comstock and comstockery by launching her own monthly sheet.

‖**comtesse** (kontes). [Fr.] A French title of rank, of which the English form is COUNTESS.
1902 *Cornh. Mag.* Dec. 832 'Yes, Ernestine,' said the elder girl, whose playmate the young Comtesse had been. **1905** B'NESS ORCZY *Scarlet Pimpernel* iii. 31 Lord Antony had already sat down..with the Comtesse on his right. *Ibid.* iv. 32 The aristocratic French comtesse. **1951** J. B. PRIESTLEY *Fest. Farbridge* I. i. 34 Comtesses and princesses during the French and Russian revolutions.

con. Add: **b.** Abbrev. of CONFIDENCE. Used *attrib.* in *con game, man, talk*, etc. (Also *ellipt.*) orig. *U.S.*
1889 *Portland* (Oregon) *Mercury* 29 June 1/7 It does not take an unsophisticated countryman to get swindled by the 'con man'. **1896** G. ADE *Artie* i. 5 He was..puttin' up the large juicy con talk. *Ibid.* 8 Not that I'm strong on the con talk. *Ibid.* xiv. 131 If they wanted me to be president o' the whole shootin' match, I'd..grow some side-whiskers and put up as tall a con game as that old stiff we've got there now. **1901** J. FLYNT *World of Graft* 100 'It's mine,' I said...'They took the con.' **1902** L. MEAD *How Words Grow* (1907) 165 'A con' is an abbreviation of confidence man. **1903** *N.Y. Sun* 30 Nov. 7 Two sailors say he worked the 'con' game on them. **1911** H. QUICK *Yellowstone N.* xii. 318 The cynosure of the eye of every sure-thing or con-man on South Halsted street. **1925** E. WALLACE *King by Night* xxxvi. 162 A burglar is a burglar, and never goes in for the 'con' game. **1928** —— *Gunner* ii. 24 Sam Larber, the con. man. *Ibid.* xxviii. 224 He's never done anything except con. work. **1935** AUDEN & ISHERWOOD *Dog beneath Skin* I. 33 From con-man and coiner protect and bless. **1940** D. W. MAURER (*title*) The Big Con. **1951** J. B. PRIESTLEY *Fest. Farbridge* II. ii. 310 You're a little gang of crooks, con types living on your wits. **1960** *Observer* 25 Dec. 7/6 Various petty fiddles and con games to which Christmas trading lent itself. **1961** J. WAIN *Weep before God* 23, I could forgive Even the worst, the con. men who harangue Their fellow artists. **1967** *Listener* 21 Dec. 821/1 The intellectual theoreticians of visual pop culture have succeeded..in pulling a con.
Hence **con** *v. trans.*, to persuade, to speak persuasively to; to dupe, to swindle. orig. *U.S.*
1896 G. ADE *Artie* iv. 35 Don't try to con me with no such talk. **1899** —— *Fables in Slang* 90 The Property Man gave it as his opinion that Mansfield conned the Critics. **1908** G. H. LORIMER *J. Spurlock* v. 103 The detective began to get a little angry. 'You've been conned, Miss Grey... He's all to the bad.' **1917** MATHEWSON *Sec. Base Sloan* xiv. 196 Don't let anyone con you into signing a contract. **1932** 'JOCK OF DARTMOOR' *Dartmoor from Within* viii. 241, I..laugh—loud and long. 'Conned'—'conned' by a pair of kids. I, the wise guy,..'conned' by one of..[Nobby's] pupils on my very first night of freedom. **1961** 'B. WELLS' *Day Earth caught Fire* vii. 112 Some crack-pot had conned the newsroom into believing that smells coming up from the rhino and elephants that haunt that territory were causing pollution in the air breathed in Nigeria. **1962** *Listener* 26 Apr. 745/1 This mild tale of a shy boy conned into giving a girl a fortune. **1967** M. REYNOLDS *After Some Tomorrow* 7 Nobody can gamble it away from me or con me out of it.
c. In *Criminals' slang*, abbrev. of CONVICT *sb.*[1], CONVICTION.
1893 in J. Hawthorne *Confess. Convict* 11 Prisoners are known as 'con', which is short for convict, and the whole body of prisoners is designated 'condom'—short for convictdom. **1925** N. LUCAS *Autobiogr. Crook* xvii. 231 'Got any cons?' he said. 'I beg your pardon.' 'Cons—bin lagged afore?' **1926** J. BLACK *You can't Win* xix. 279 The other 'cons' will blackmail me. **1936** J. CURTIS *Gilt Kid* xxvii. 266 For this breaking and entering lark at least eighteen months for a man with two cons... It was going to make four convictions for him. **1958** F. NORMAN *Bang to Rights* III. 99, I had three realy good friend among the con's. *Ibid.* 152, I already had two con's and had only been out of the nick a few weeks.
d. Abbrev. of CONSTRUE *sb.*
1905 VACHELL *Hill* iii, We must mug up our 'cons' well enough to scrape along without 'puns' and extra school. **1906** *Westm. Gaz.* 17 Dec. 11/1 He 'skewed' his 'cons' and 'reps'. **1910** R. BROOKE *Let.* 25 Feb. (1968) 221 Many thanks for the con.

conalbumin (konæ·lbiumin). *Biochem.* [CON-.] An iron-binding albumin present in egg white and similar to transferrin.
1900 OSBORNE & CAMPBELL in *Jrnl. Amer. Chem. Soc.* XXII. 432 An albumin coagulating at 55°–57° forms nearly 50 per cent of the products obtained from [preparation]

D.3...Since it so closely resembles ovalbumin, and is so closely associated with it, the writer suggests that it be called conalbumin. *Ibid.* 447 The body coagulating at the lower temperature we designate conalbumin, on account of its close relation in properties and composition to ovalbumin. **1911** J. A. MANDEL tr. *Hammarsten's Physiol. Chem.* 602 The ovalbumin, like the conalbumin, has the properties of the albumins in general. **1956** *Nature* 14 Jan. 94/1 Schaible and Bandemer showed that the pink discoloration was caused by iron diffusing from the yolk and chelating with the conalbumin of the white.

con amore. (Earlier and later examples.)
1739 T. FITZOSBORNE *Let.* Sept. (1795) 2 No matter what the object is, whether business, pleasures, or the fine arts; whoever pursues them to any purpose must do so *con amore*. **1782** H. WALPOLE *Let.* 7 Feb. (1858) VIII. 150 Sir John Hawkins..said..'I suppose you will labour your present work *con amore* for your reputation.' **1814** JANE AUSTEN *Mansf. Park* II. xi. 236 She sat and cried *con amore*..but it was con amore fraternal and no other. **1888** L. A. SMITH *Music of Waters* 292 The wind is favourable and they give themselves up to singing *con amore*. **1955** *Times* 1 July 13/3 An excellent account of reclamation on Exmoor Forest, a book written *con amore* as he was deeply interested both in land reclamation and in Exmoor.

conatively (kǫ·nătivli), *adv.* [f. CONATIVE *a.* + -LY[2].] In a conative manner.
1937 A. HUXLEY *Ends & Means* xii. 198 The rare occasions when the intellectual does become affectively and conatively involved with the world of human reality. **1961** E. J. FURLONG *Imagination* v. 55 Believing-in-a-dream may well be different emotionally and conatively from believing when awake.

concaver (kǫ·nkēi·vəɹ). [f. CONCAVE *sb.* + -ER[1].] One who hollows out the sides of a boot last.
1921 *Dict. Occup. Terms* (1927) § 486 *Concaver, last concaver*..hollows out sides of boot last by holding wood against shaped power-driven cutter.

co·ncaving, *ppl. a.* [f. CONCAVE *v.* + -ING[2].] Curving inwards.
1871 NESBITT *Chaffers's Catal. Slade Coll. Glass* 77 A very broad concaving neck.

concealed, *ppl. a.* Add: **c.** Applied to indirect lighting in which the fitments are hidden from view.
1930 'E. QUEEN' *French Powder Myst.* iii. 23 The lighting features were all of the 'concealed' variety rapidly gaining vogue on the Continent. **1938** *Encycl. Brit. Bk. of Yr.* 127/1 Internally, glass, wood, stone, and metal veneers, in conjunction with concealed lighting, have been developed to a very high degree, and are now accepted as the normal standard of good building practice. **1943** G. GREENE *Ministry of Fear* I. vi. 98 Interminable corridors lit by concealed lighting.

concede, *v.* Add: **1. b.** To admit defeat in (an election); to acknowledge that an election, town, etc., has been lost to another political party or candidate. orig. *U.S.*
1824 *Commentator* (Frankfort, Ky.) 2 Oct. 3/1 This state is generally conceded to General Jackson. **1908** *Westm. Gaz.* 21 Feb. 2/2 If we 'concede', as the Americans say, West Worcester, our opponents, on their part, 'concede' West Carmarthen. **1908** *Daily Chron.* 3 Nov. 4/6 He hid in terror, and contrived to have a telegram dispatched to Mr. Cleveland 'conceding' his election. **1946** *Pueblo* (Colo.) *Chieftain* 27 June 1/7 J. B. Bridston Wednesday night conceded the North Dakota republican senatorial nomination to U.S. Senator William Langer as additional returns boosted Langer's lead. **1965** *Ann. Reg. 1964* 44 Sir Alec did not admit defeat; and the Conservative chairman..apparently did not see cause to concede. **1970** *Times* 19 June 1/1 Mr Wilson refused to concede defeat and showed no regret at having called the election.

‖**concentrado** (kǫnsentrā·do). [Sp., properly pa. pple. of *concentrar* to CONCENTRATE.] A concentration camp.
1898 *Daily News* 7 Mar. 5/4 The distressed concentrados in Cuba. **1901** *Westm. Gaz.* 24 June 1/3 Such matters as farm-burning and the 'concentrado' policy.

concentrate, *sb.* Add: **b.** A stock food containing concentrated nutriment.
1907 T. SHAW *Feeding Farm Animals* iii. 54 The relative cost of bulky foods and concentrates respectively should be duly considered when feeding animals. **1915** J. PORTER *Stockfeeder's Compan.* xii. 43 The concentrates may be subdivided into groups according to their richness in one or more of the three most valued nutrients. **1935** *Times* 4 Nov. 19/1 Several recent investigations have shown that grass, if cut young at the height of a few inches and dried, possesses..properties..approaching indeed, the nutritive value of concentrates. **1958** *Economist* 8 Nov. 486/1 Well-managed grassland feeds stock more economically than do concentrates.
c. A concentrated liquid.
1939 A. L. SIMON *Conc. Encycl. Gastron.* I. p. v, Escoffier..introduced..fumets and essences, that is, evaporated stock obtained by allowing the water..in which meat, fish or vegetables happen to be cooked, to steam away slowly so as to leave behind a fragrant concentrate. **1944** *Good Housek. Baby Bk.* (1945) vi. 96 Special orange juice, obtainable at the food office or clinic..is the best vitamin concentrate obtainable. **1966** J. S. COX *Illustr. Dict. Hairdressing* 38/2 *Concentrate*, a preparation that needs dilution before use. **1968** *Times* 24 Oct. 7/8 To test whether any of this matter was resistant to breakdown, he incubated concentrates of the water samples.

co·ncentratedly, *adv.* [f. CONCENTRATED *ppl. a.* + -LY[2].] In a concentrated manner.
1891 *New Rev.* June 499 Mr. Irving worked more concentratedly than all the other actors put together. **1928** *Daily Express* 11 June 3/7 The body..will not be so concentratedly engaged in assimilating 'big' meals. **1964** NABOKOV *Defence* xiii. 215 Luzhin silently and concentratedly fed chocolates to little Ivan, and Ivan silently and concentratedly ate them.

concentration. Add: **6.** Amalgamation of business firms, factories, etc., in a particular industry; more generally, the action of developing parts of an industry at the expense of other parts.
1930 *Economist* 1 Mar. 455/1 In Westphalia, where the greatest degree of 'concentration' had been achieved, the process had been based on a quota system regulated by a compulsory cartel. **1940** *Ibid.* 16 Mar. 455/2 In industry, a policy of concentration on the largest and most suitably located producers..was carried out, and a similar policy in retail distribution led to the closing of a large number of small shops. **1941** *Ibid.* 8 Mar. 297/1 The necessity for a measure of induced—and, if necessary, compelled—concentration of production in restricted industries is no novelty. **1961** *Times* 31 May 23/3 'Diversification' and 'concentration' have come to have popular meanings often divorced from reality. 'Concentration', he claimed, was often used to describe growing monopoly.
7. *attrib.*: **concentration camp,** a camp where non-combatants of a district are accommodated, such as those instituted by Lord Kitchener during the South African War of 1899–1902; one for the internment of political prisoners, foreign nationals, etc., esp. as organized by the Nazi regime in Germany before and during the war of 1939–45; also *fig.*; **concentration cell** *Electr.*, a cell whose difference of potential is due to the difference of concentration of the solutions in which the electrodes are immersed.
1901 J. ELLIS in *Hansard's Parl. Deb.* Ser. IV. XC. 180 The policy of placing the women and children confined in the concentration camps in South Africa, whose husbands and fathers are in the field, on reduced rations. **1901** *Contemp. Rev.* Oct. 528 Considerable controversy has taken place concerning the Concentration Camps. **1930** O. WILLIAMS tr. *P. Monelli's Toes Up* III. 210 Those who returned from being prisoners were parked in a concentration camp under the guard of other soldiers. **1934** *Ann. Reg. 1933* 173 Germany..For dealing with the masses of prisoners special concentration camps were opened. **1934** C. LAMBERT *Music Ho!* i. 52 The purely fashionable change in the tastes of the concentration camp of intellectuals to whom Diaghileff played up. **1935** B. RUSSELL *Relig. & Sci.* x. 248 In Germany and Russia,..those who openly disagree, even if they escape with their lives, are liable to forced labour in a concentration camp. **1940** H. G. WELLS *Babes in Darkling Wood* III. i. 234 The White Paper of Nazi atrocities in the concentration camps and elsewhere. **1941** [see *ARYANIZATION b]. **1943** [see *ARYAN *a.* 2]. **1959** J. BRAINE *Vodi* v. 76 He caught sight of his own arms, reduced almost to concentration-camp dimensions, the veins blue and obscenely swollen against the white skin. **1900** *Jrnl. Phys. Chem.* IV. 413 The electromotive force of some concentration cells and of cells with zinc and copper electrodes in organic solvents. **1916** W. C. McC. LEWIS *Phys. Chem.* II. 194 Concentration Cells with Single Selections. **1959** *Chambers's Encycl.* V. 121/1 Where the two electrodes are of the same chemical nature but some difference of concentration or activity is involved, e.g. two silver electrodes dipping into solutions of silver nitrate of different concentrations, they are termed concentration cells.

concentric, *a.* Add: **2. g.** *Electr.* *concentric cable* = *coaxial cable*. So *concentric main*.
1892 J. A. FLEMING *Altern. Curr. Transf.* II. 151 Simple straight-joint coupling in Siemens and Halske concentric cable. *Ibid.* 316 Mr. Ferranti designed for the London Electric Supply Corporation a form of concentric main, intended for extra high-pressure service. **1944** *Electronic Engin.* XVI. 403 Linking with the provinces would be possible by concentric cable.
h. *Photogr.* *concentric lens*, a symmetrical doublet lens of two combinations, the surfaces of which are spherical and concentric.
1892 *Brit. Jrnl. Photogr.* 29 Apr. 273/1 We have on previous occasions spoken of a patent new 'concentric' lens. **1902** *Encycl. Brit.* XXXI. 695/1 The first lens made with the new glasses was Dr. Schroeder's 'Concentric'. **1918** *Photo-Miniature* XV. 12 *Concentric*, the name given to an anastigmat lens, the radii of the back and front surfaces of each element of which were struck from a common condenser. **1958** *Ilford Man. Photogr.* (ed. 5) viii. 132 The peculiarity of the Concentric was the use of plano-convex crowns, and the fact that the exposed surfaces were concentric.

concept, *sb.* Add: **2. b.** *attrib.* and *Comb.*
1865 S. HODGSON *Time & Space* iii. 197 A Begriff or Concept-form is that which is the Identity of contradictories. *Ibid.* vii. 347 Non-existence in this sense, or pure nothing, is a concept-name and not a concept. **1890** W. JAMES *Princ. Psychol.* I. xvi. 662 They form for him not so many odd facts, but a concept-system—so they stick. **1894** CREIGHTON & TITCHENER tr. *Wundt's Hum. & Anim. Psychol.* xxi. 310 They serve to invest the concept-idea with the consciousness of its vicarious significance, and with the resultant concept-feeling. **1896** W. CALDWELL *Schopenhauer's Syst.* iii. 123 The philosophy..of Kant, '*der alles Zermalmende*', is in the first place the substitution of a *regulative* and verifiable philosophy for the old dogmatic concept-philosophy or theology. **1901** J. M. BALDWIN *Dict. Philos. & Psychol.* I. 208/1 The concept triangle comprehends an indefinite multiplicity of actual or possible triangles. **1921** E. SAPIR *Language* ii. 28 Ever since the breakdown of English forms that set in about the time of the Norman Conquest, our language has been straining towards the creation of simple concept-words. **1923** J. S. HUXLEY *Ess. Biologist* i. 25 The attainment of the power of generalization—of reason, concept-formation, or what you will. **1938** *Mod. Lang. Rev.* Oct. 555 This concept-chasing is a consequence of the more or less arbitrary 'periodization' of literary history. **1965** N. CHOMSKY *Theory of Syntax* i. 32 The innate concept-forming abilities of the child.

conception. Add: **1. d.** *attrib.*, as *conception-control* (cf. *birth-control*).
1930 *Lambeth Conf. Encycl. Let.* 44 The Conference records its strong condemnation of the use of any methods of conception-control from motives of selfishness, luxury, or mere convenience.

conce·ptively, *adv.* [f. CONCEPTIVE *a.* + -LY[2].] In a conceptive manner.
1865 J. H. STIRLING *W. Hamilton* 73 Conceptively he accepts them: perceptively, he..knows them not.

conceptual, *a.* Add: **2.** Esp. in *conceptual analysis, knowledge, scheme, system, thinking.*
1885 W. JAMES in *Mind* X. 39 The whole field of symbolic or conceptual thinking. **1890** —— *Princ. Psychol.* I. viii. 222 Of the mental states of other persons we only have conceptual knowledge. *Ibid.* xii. 482 The conceptual scheme is a sort of sieve in which we try to gather up the world's contents. *Ibid.* II. xvii. 7 Conceptual systems which neither began nor left off in sensations would be like bridges without piers. **1896** W. CALDWELL *Schopenhauer's Syst.* viii. 415 Our practical apprehension of things is far greater in range and in potency than our merely reflective or conceptual analysis of reality. **1929** C. I. LEWIS *Mind & World-Order* viii. 271 There may be alternative conceptual systems, giving rise to alternative descriptions of experience, which are equally objective and equally valid. **1949** *Mind* LVIII. 214 Conceptual analysis..is not concerned with usages, but with the concepts and propositions expressed by verbal expressions. **1951** F. THILLY *Hist. Philos.* (1952) iii. 77 Conceptual knowledge..is the only genuine knowledge: that was the teaching of Socrates. **1955** S. KÖRNER (*title*) Conceptual Thinking. **1960** A. J. AYER *Philos. & Lang.* 33 There is..a danger in following Kant too closely. It consists in..assuming that certain fundamental features of our own conceptual system are necessities of language. **1960** J. COHEN *Chance, Skill & Luck* ii. 47 There is the possibility that psychological probability may provide a single conceptual scheme for the study of how we perceive, think, learn, decide, and act.

conceptualist. (Earlier and later examples of *attrib.* use.)
1843 MILL *Logic* I. I. vi. 154 It precisely expresses the point of difference respecting the import of Propositions, between my view and what I have called the Conceptualist view of them. **1951** J. HOLLOWAY *Lang. & Intell.* ii. 16 Two theories of the nature of universals will be considered, the conceptualist belief that they are mental entities, and the *in re* form of realism which maintains that they inhere in objects and are so apprehended by the mind.

conceptualistic, *a.* (Examples.)
1881 in OGILVIE *Imp. Dict.* **1890** W. JAMES *Princ. Psychol.* I. xii. 472 Our doctrine..of the 'fringe' leads to a perfectly satisfactory decision of the nominalistic and conceptualistic controversy, so far as it touches psychology. **1935** *Mind* XLIV. 116 C. I. Lewis's 'conceptualistic pragmatism'..turns out to be sheer empiricism.
So **conceptuali·stically** *adv.*
1961 Y. OLSSON *Syntax Eng. Verb* vii. 201 Zero-linking is the result of different criteria from those conceptualistically applied in distinguishing between 'ingressive' and 'effective'.

conceptualize (kǫnse·ptiuălaiz), *v.* [f. CONCEPTUAL *a.* + -IZE.] *trans.* To form a concept or idea of. Also *absol.*
1909 W. JAMES *Pluralistic Universe* vi. 253 When we conceptualize, we cut out and fix, and exclude everything but what we have fixed. **1912** VON HÜGEL *Eternal Life* x. 293 The reality which intuition grasps, and which is thus to be conceptualized, contains some permanence. **1927** *Brit. Weekly* 3 Feb. 462/1 Jesus is lost in the attempt to conceptualise Him. **1936** *Burlington Mag.* Nov. 222/2 Did the painter tend to conceptualize its initial visual actuality. **1937** R. H. LOWIE *Hist. Ethnol. Theory* xii. 209 He labels the source of his emotion by some word describing the extraordinariness he feels, and cares not in the least about accurately conceptualizing it in the abstract. **1950** *Theology* LIII. 131 Essence can be conceptualized, but existence can only be affirmed. **1959** *Times Lit. Suppl.* 6 Nov. 639/3 Mrs. Moskowitz still struggles to conceptualize; Miss Caravello still upholds the culture of the Latins.
Hence **conce·ptualized** *ppl. a.*; **conce·ptualizing** *vbl. sb.* and *ppl. a.*; **conce:ptuali·zable** *a.*; **conce:ptualiza·tion.**
1878 S. H. HODGSON *Phil. Reflection* I. II. v. 325 The conceptualising process..lands us in the Aristotelian distinction of *genus, differentia*, and *species*, which may be..called the logical categories. These categories are not found ready made in thought, but are products of the process of conceptualising the chains of perceptual redintegration. **1896** W. CALDWELL *Schopenhauer's Syst.* iii. 116 He found in conceptions only the form of philosophy—philosophy being a *conceptualised* or generalised statement of the matter of our knowledge, of ordinary reality. **1909** W. JAMES *Pluralistic Universe* vii. 290 Vainly seeking to describe by concepts and words what I say at the same time exceeds either conceptualization or verbalization. *Ibid.* viii. 326 Artificial products of the conceptualizing faculty. **1929** *Times Lit. Suppl.* 12 Dec. 1046 The formally colder conceptualization of the 'Philebus'. **1949** KOESTLER *Insight & Outlook* xxiv. 345 The primitive poetic quality of the Bushman language as compared to the conceptualized statement is obvious. **1950** *Theology* LIII. 132 What however will be the result of this if there are important aspects of reality

which are *not* clearly conceptualizable? **1957** L. F. BROSNAHAN *Genes & Phonemes* 8 The psychologist..is able to analyse the mode of operation of any sensory apparatus into a series of processes, namely, those of perceptual selection and organisation, perceptual generalisation, and, on the border between perception and thought, that of conceptualisation.

conceptually (kǫnse·ptiuăli), *adv.* [f. CONCEPTUAL + -LY[2].] As a concept.
1890 W. JAMES *Princ. Psychol.* I. xi. 405 When apprehended conceptually as a connected system, their number may be very large. **1902** *Nature* 18 Sept. 501/2 A substance is neither actually nor conceptually the sum of its radicals. **1964** M. A. K. HALLIDAY et al. *Ling. Sciences* vi. 145 Conceptually defined categories can be held to be universal precisely because they are incapable of exact application.

concern, *v.* Add: **4. c.** Phr. *to whom it may concern*: a formulaic phrase used of a statement, testimonial, etc.
1868 DICKENS *Let.* 26 Apr. (1960) 281 The Russia is a magnificent ship...To whom it may concern, report the Russia in the highest terms. **1924** *Amer. Mercury* I. 343/2 In W. W. Western's 'To Whom It May Concern: A Poem on the Times' Byron's manner and method are put to queer uses.
7. (Earlier U.S. examples, esp. in form *consarn*.)
1803 J. DAVIS *Trav. U.S.* x. 384 Concern it, *Dinah*, says I, why if you was to eat all the good things what would there be left for me? **1832** J. P. KENNEDY *Swallow B.* II. xvii. 222 'Consarn his pictur!' said Jeff. **1844** 'JON. SLICK' *High Life N.Y.* I. 107 Somehow that tarnal Count, consarn him, put me all out on my natural reckoning. *a* **1852** MRS. WHITCHER *Widow Bedott P.* (1883) i. 8 He only said, says he, 'Consarn it.'

concern, *sb.* Add: **5. c.** Among Quakers, a conviction of the divine will.
1707 in *Pennsylvania Hist. Soc. Mem.* X. 214 During their absence, I was under the greatest concern of mind that ever I knew in my life. **1772** A. HUNTER *Let.* 18 Mar. in *Fithian's Jrnl. & Lett.* (1900) 22 Our orations are put off lest they should do some harm to some under concern. **1838** J. F. COOPER *Home as Found* xvi, 'Have you heard that Grace is under concern?'..'Not under the church parson's, I'll engage; no one ever heard of a real..conversion under his ministery.' **1875** MRS. STOWE *We & Neighbors* xv. 172 If your friend Sibyl should have a 'concern' laid on her for your Mr. St. John, she would tell him some wholesome truths. **1933** A. N. WHITEHEAD *Adv. Ideas* xi. 178 Thus the Quaker word 'concern'..is more fitted to express this fundamental structure. The occasion as subject has a 'concern' for the object. **1949** *Friend* 17 June 495/2 Love (as God loves, not as man loves) for other people, linked with a sense of responsibility—not such responsibility as John Smith can feel but as God's hands and feet can feel—that is 'concern'. **1966** E. H. JONES *Margery Fry* v. 46 She did not..experience in Meeting or elsewhere that spiritual communion which was supposed to nourish 'concerns'.
11. (Earlier and later examples.)
1824 A. CONSTABLE *Let.* 2 Aug. in *Corr. J. Constable* (1962) 216 Mary sends the greys..to take her into the street by which means Ann has the first ride in the new concern. **1873** J. H. BEADLE *Undevel. West* xxix. 636 The two old men..with their butcher knives hacked out two concerns, which might serve in a rude fashion for oars. **1889** 'C. E. CRADDOCK' *Broomsedge Cove* xii. 224 The old doctor, ez seems to be a good, useful kind o' consarn.

concerned, *ppl. a.* Add: **1. c.** (See quot.)
1949 *Friend* 17 June 495/1 'Concerned' in the Quaker sense really means being captured or consumed by the love of God, and directing that love towards some area of need among men.
3. *U.S.* (Earlier and later examples.)
1834 SEBA SMITH *Sel. Lett. Downing* Pref. p. vi, The vetoe, which is a consarnt good thing. *a* **1852** MRS. WHITCHER *Widow Bedott P.* (1883) ii. 8 But that's the consarndest lie that ever was told. **1851** *Southern Lit. Messenger* Mar. (De Vere), That's a concerned ugly fix, and how we'll ever get out of it is more than I know. **1887** M. E. WILKINS *Humble Romance* 405 I've always heard tell that there was two kinds of old maids—old maids an' consarned old maids.

concert, *sb.* Add: **1. d.** Phr. *the Concert of Europe*.
1880 GLADSTONE *Sp. Midlothian* 89 My third sound principle is to take care to cultivate and maintain to the utmost the concert of Europe, to keep the Powers of Europe together. **1897** BALFOUR in *Times* 17 Feb. 8/3 It is absolutely impossible that you should keep the Concert of Europe going for some purposes connected with the Ottoman Empire, and not going for all purposes. **1897** MARQUIS OF SALISBURY in *Ibid.* 20 Mar. 8/1, I feel it is our duty to sustain the federated action of Europe. I think it has suffered by the somewhat absurd name which has been given to it—the concert of Europe. **1901** *Hansard's Parl. Deb.* Ser. IV. XCII. 184 He attributes all our misfortunes to the Concert of Europe. Sir, the Concert is not perfect as an international instrument. **1944** H. NICOLSON *Let.* 18 Jan. (1967) II. 345 He says that the Russians want very much to be admitted as a civilised member of the Concert of Europe. 'They want to belong to the Club.' **1959** *Chambers's Encycl.* VII. 683/1 Interstate relationship in the 19th century was regulated at the top level by an international organization which came to be known as the concert of Europe. The concert was in origin an alliance of the other European great powers against Napoleonic France.
4. b. A dancing performance consisting of single items, folk-dances, etc. (in contrast to a full-scale ballet). Freq. *attrib.*
1912 J. E. C. FLITCH *Mod. Dancing* xiv. 209 The next step was the giving of a public concert to make known to the larger world the rediscovery of the ancient dances. **1932** V. DANDRÉ *Anna Pavlova* 91 Without a company of her own Pavlova could not have given her own ballets... It would have meant limiting herself to concert programmes made up of separate dancing numbers. **1948** 'LA MERI' *Sp. Dancing* viii. 95 The solo concert dance was first introduced by Isadora Duncan and Ruth St. Denis. *Ibid.*, By the time Argentina made her first world tour (1929), concert dancing was a 'fait accompli'. **1967** CHUJOY & MANCHESTER *Dance Encycl.* 218/2 Modern dancers are also called concert dancers.
5. *concert-bill, -goer, -going, -hall, -party, -platform, -recital, -ticket*; **concert-master** [G. *konzertmeister*], the first violin, leader of the orchestra; **concert overture,** an orchestral piece resembling an overture but intended for independent performance in a concert-hall; **concert pitch** (earlier and later examples).
1865 *Atlantic Monthly* XV. 718, I read over the concert-bill. **1893** *Chambers's Jrnl.* 16 Dec. 795/1 An amusing concert bill of the time of Queen Anne. **1855** GEO. ELIOT in *Fraser's Mag.* July 48/1 Most London concert-goers..think of him [*sc.* Liszt] as certainly the archimagus of pianists. **1927** *Daily Express* 4 Oct. 3/2 Among the songs which have made Mr. Weatherly's name known to millions of concert goers all over the world are 'The Midshipmite' [etc.]. **1947** C. GRAY *Contingencies* i. 46 By the 'ordinary concert-goer' we..postulate..the audience which congregates at Promenade Concerts. **1942** E. BLOM *Music in Eng.* viii. 128 The habit of concert-going. **1951** AUDEN *Nones* (1952) 67 The paranoic mind Of this undisciplined And concert-going age. **1746** *Ipswich Jrnl.* 27 Dec. in *Chambers's Jrnl.* (1875) 800/1 The grand Concert-hall of vocal and instrumental music. **1869** *Broadway* I. 537/2 Concert-halls and singing-rooms are equally attractive to the middle and lower orders. **1905** *19th Cent.* Oct. 567 One central concert hall would be utterly inadequate for the requirements of London. **1946** *Penguin Music Mag.* Dec. 39 Opera-houses and concert-halls are being rebuilt or built afresh in the devastated towns. [**1876** STAINER & BARRETT *Dict. Mus. Terms* 104/2 *Concertmeister*, the leader of the band, the conductor.] **1889** *Cent. Dict.*, *Concert-master*, the first violinist of an orchestra; the leader. **1927** *Grove's Dict. Mus.* (ed. 3) 693/1 *Concert-master*, the leader, *i.e.* the first of the first violins in an orchestra; who sits next the conductor and transmits his wishes to the band. **1954** *Times Lit. Suppl.* 3 Dec., The American use of 'Concert-master', which is also repugnant to English ears, is registered by Mr. Blom with a gentler resignation. **1938** *Oxf. Compan. Mus.* 679/2 A totally different type of independent overture is that of the concert overture, of which Mendelssohn is considered to be the inventor (his *Hebrides* Overture, Dvořák's *Carnival* Overture, Brahms's *Academic Festival* Overture, &c.). **1947** A. EINSTEIN *Mus. Romantic Era* xi. 143 Liszt's eager listeners and friendly colleagues..tried at first to facilitate the understanding of some of Liszt's works by calling them 'concert overtures'. **1883** L. TROUBRIDGE *Jrnl.* May, in J. Hope-Nicholson *Life amongst Troubridges* (1966) 163 In the evening to German Reeds Concert Party—Corney Grain most amusing. **1895** ROBERTS & MORTON *Adv.* i. 18, I accepted an engagement from Howard Paul to join his concert-party and tour round the seaside towns. **1960** R. COLLIER *House called Memory* xii. 168 The concert party on the pier in pierrot costumes. **1767** PRIESTLEY *Present State Electr.* VIII. 717 A jar containing half a square foot of coated glass sounded F sharp, concert pitch. **1817** MRS. PIOZZI *Let.* 4 Jan. in *Autobiogr.* (1861) II. 188, I want something to string my spirits up to concert pitch. **1936** *Aircraft Engin.* Oct. 273/3 Some more work is necessary..to bring it to 'concert pitch'. **1966** *Crescendo* Aug. 21/1, I enjoy playing the flute, because, although the embouchure is difficult, it's in concert pitch. **1927** *Melody Maker* Sept. 944/3 Cinema music..has a distinct appeal on the concert platform amongst educated music-lovers. **1963** *Times* 12 Feb. 11/4 The harmonies of a concert-platform. **1907** *Public Opinion* 17 May 629/2 Performances of opera in concert-recital form. **1813** JANE AUSTEN *Let.* 3 Nov. (1952) 365, I bought a Concert Ticket & a sprig of flowers for my old age. **1852** MRS. GASKELL *Let.* 4 Sept. (1966) 197 Canon Clifton gave me a concert ticket, & as Meta had one we..went to the Concert.

concert, *v.* **2. c.** Delete ?*Obs.* and add: To act in harmony *with*.
1897 MARQUIS OF SALISBURY in *Hansard's Parl. Deb.* Ser. IV. XLVI. 370 Our naval officers have received instructions to take no isolated action, but to concert with the naval officers of the other Powers.

concerted, *ppl. a.* Add: **3.** United in action or intention.
1897 MARQUIS OF SALISBURY in *Times* 16 Feb. 8/1 The concerted sympathy of the Powers remains complete. **1897** *Daily News* 25 May 8/1 Europe unfortunately, concerted or otherwise, is pretty well case-hardened to suffering.

concertina, *sb.* Add: **2.** *War slang.* In full *concertina wire*. (See quots.)
1919 *War Terms* in *Athenæum* 15 Aug. 759/1 *Concertina*, collapsible wire entanglement. **1930** BLUNDEN *Poems* 40 The sappers' waggons stowed with frames and concertina wire. **1965** BROPHY & PARTRIDGE *Long Trail* II. 104 *Concertina Wire*, wire used for entanglements; when touched, it coiled about the intruder.
3. *attrib.* and *Comb.*
1902 *How to make Things* 20/1 A collapsing or concertina-like box. **1912** W. OWEN *Let.* 23 Mar. (1967) 126 Concertina-practice fills up intervals. **1922** JOYCE *Ulysses* 226 A concertina skirt. **1935** C. ISHERWOOD *Mr. Norris* xii. 205 My glance wandered away to..the soft, snout-like nose, the concertina chin. **1964** *Times* 11 Feb. 11/6 The recent 'concertina' crashes on the M 1 and M 6.

concerti·na, *v.* [f. the sb.] *trans.* and *intr.* To shut up like a concertina; to compress; to collapse; to wrinkle. Also *fig.*
1906 *Daily Chron.* 11 June 7/3 Another blow from a stick on the right 'concertinaed' my hat. **1907** W. J. LOCKE *Beloved Vagabond* ix. 112 'It makes one talk unmentionable imbecility.' He just missed concertina-ing the last two words. **1908** KIPLING *Diversity of Creatures* (1917) 251 Then Beetle, concertinaing his books, observed to Winton, 'When King's really on tap he's an interestin' dog.' **1928** *Daily Express* 1 June 5/2 When closed the trellis work 'concertinas' into a very small space. **1945** *Salt* 26 Feb. 14 Moleskins which concertina-ed around his fetlocks. **1946** C. FRY *Phoenix too Freq.* 46 Would you consider we go round and round?..We concertina, I think, taking each time A larger breath, so that the farther we go out The farther we have to go in. **1953** J. MASTERS *Lotus & Wind* iii. 38 He saw that the force had concertinaed to a halt. **1963** NABOKOV *Gift* iv. 266 From beneath the overcoat his black trousers concertinaed over his rubbers.
Hence **concerti·naed** *ppl. a.*, closed or folded in a manner resembling a concertina; wrinkled; collapsed.
1905 *Westm. Gaz.* 4 July 6/3 A concertinaed opera-hat. **1916** H. G. WELLS *Mr. Britling* iii. § 4 Fold after fold of concertina-ed flannel gathered about his ankles. **1962** *Listener* 8 Mar. 435/2 His concertina-ed syntax is more dutiful than magical.

concertino. Add: **1.** (Earlier and later examples.)
1876 STAINER & BARRETT *Dict. Mus. Terms* 104/1 *Concertino*..the diminutive of concerto. **1905** E. DANNREUTHER *Oxf. Hist. Mus.* VI. 228 The interesting Concertino in E minor for horn. **1955** *Times* 12 July 5/5 The new work was a concertino for clarinet and strings by Adrian Cruft.
2. The group of solo instruments in a concerto (see also quot. 1801).
1801 BUSBY *Dict. Mus.*, *Concertino*, the principal instrument in a concert or concertante. **1900** *Grove's Dict. Mus.* IV. 596/1 Handel's six 'Hautboy Concertos' are Concerti grossi, written for a Concertino consisting of two Solo Violins, two Violoncellos, two Hautboys, two Flutes, and two Bassoons. **1902** FULLER MAITLAND *Oxf. Hist. Mus.* IV. 162 In those [concertos] which have a definite 'concertino' the instruments forming it vary. **1948** A. VEINUS *Concerto* ii. 18 The concertino acts as a single unit. **1961** A. J. B. HUTCHINGS *Baroque Concerto* vi. 110 This may mean that Corelli wished concertino and ripieno to sit apart. **1962** R. A. HARMAN et al. *Man & his Music* IV. 474 One or more solo ('concertino') instruments are set off against an orchestral ('ripieno') body. *Ibid.*, The concertino group may be violin and bassoon.

concertize, *v.* Delete *nonce-wd.* and add further examples.
1885 in Ware *Passing Eng.* (1909) 89/1 M. Ovide Musin, the great Belgian violinist, has returned to this city to concertize under Mr. L. M. Rubens' management. **1952** B. ULANOV *Hist. Jazz Amer.* (1958) x. 111 The *Rhapsody in Blue*..represented the most serious attempt to concertize jazz.
So **concertized** *ppl. a.*; **concertizing** *vbl. sb.*
1928 *Observer* 22 July 21/1 Their singing of negro spirituals and 'work songs', and those who have only heard these songs in 'concertised' versions will be astonished [etc.]. **1889** [see Dict.]. **1947** A. EINSTEIN *Mus. Romantic Era* xvi. 288 The success was not such that he could give up teaching, concertizing, and composing instrumental pieces and comic songs.

‖**concerto grosso** (kontʃe·rto gro·so). *Mus.* Pl. **concerti grossi.** [It., lit. 'big concerto'.] A baroque concerto characterized by the use of a small group of solo instruments against the full orchestra; also, a modern imitation of this.
1724 *Short Explic. For. Wds. Mus. Bks.* 23 *Concerto Grosso*, is the great or grand Chorus of the Consort, or those Places of the Concerto or Consort where all the several Parts perform or play together. **1776** HAWKINS *Hist. Mus.* IV. IV. i. 393 The invention [*c* 1700] of the Concerto Grosso, consisting of two chorusses, with an indeterminate part. **1842** J. F. WARNER *Dict. Mus. Terms* 26/1 Sometimes concerto music..is designated by the term *concerto grosso*, (great concerto,)... In earlier times,..it was usual to give the title *concerto grosso* to pieces of concerto music which would at the present period, in comparison with our far more splendid..concerts, scarcely be called concertos at all. **1900** [see *CONCERTINO 2]. **1947** A. EINSTEIN *Mus. Romantic Era* ii. 23 The framework of the sonata, the concerto grosso, or the symphony. **1962** *Oxf. Mail* 19 Feb. 6/1 The concert opened with one of Handel's Concerti Grossi.

concession. Add: **3. b.** (Earlier examples.)
1764 *Quebec Gaz.* 23 Aug. 4/2 The said Seigneurie [is] capable of containing upwards of 500 Plantations in more than 3 Concessions Depth on each Side the River. **1820** C. STUART *Guide to Upper Canada* p. x, *A Concession*. Parallel with the front of the township..a second line..is the rear of the front or the first concession... Space for a road is then left, and a third line..becomes the front of the second concession; thus, the whole depth is divided into concessions, with space for a road between every two. These concessions are divided into lots.
c. A grant or lease of a small area or of a portion of premises for some specified purpose, e.g. for the establishment of a refreshment stand; the business premises, etc., thus established. *N. Amer.*
1910 *Sat. Even. Post* 9 July 5/3 A third of this goes to the park, for it is a concession, and in addition it must pay rent. **1948** *Miami* (Okla.) *Daily News-Record* 4 July 11/4 He has operated the concession stand in the Miami postoffice for the last four years. **1962** S. STRAND *Marketing Dict.* 149 *Concession*, a leasing of a section of a store by the owner to another businessman who sells his wares there. Also a stand at a fair or on a board-walk. **1968** *Globe & Mail* (Toronto) 17 Feb. 36 (Advt.), Rooms...Directly on Ocean Front. Walking Distance to Pavilion, Stores and Concessions. **1970** *New Yorker* 3 Oct. 127/1 When it was over, the manager of Pepsi's snack-bar concession..came up..in something of a rage.

concessionaire, -onnaire. Add: **b.** A company which has obtained a concession or privilege in the matter of trading rights. Also *attrib.*
1901 *Daily News* 18 Feb. 5/6 Imagine a great city letting out its lines of communication to concessionnaire contractors with dividends to make. **1905** *Westm. Gaz.* 10 Oct. 3/1 The profits of the so-called 'concessionnaire' societies were gigantic...Congo rubber was on everyone's lips. **1909** *Ibid.* 22 Mar. 10/2 The Normal Powder Company are the concessionaires of the shot-gun. **1930** *Times* 21 Mar. 12/4 Four of their models of chassis..are being exhibited by Messrs. Pass and Joyce, who are the sole concessionnaires. **1931** H. G. WELLS *Work, Wealth & Happiness Mankind* (1932) vii. 275 The concessionaire companies considered themselves entitled..to sole possession of all the rubber. **1962** *Which? Car Suppl.* Oct. 140/2 The Ford Taunus guarantee, which is issued by the British concessionaire, Lincoln Cars Ltd. **1970** *New Yorker* 3 Oct. 127/1 He saw no need to squander diplomacy on concessionaires.

concettist (kontʃe·ttist). [f. CONCETTO + -IST.] A writer who makes use of *concetti*.
1893 in *Funk's Stand. Dict.* **1919** P. OSMOND *Myst. Poets* v. 126 Beaumont, whom he [*sc.* Crashaw] had guided into the paths of mystical theology and had taught the trick of the concettists. **1940** *Horizon* 11 Nov. 267 Petrarchan concettists (including Marlowe).

conch. Add: **7.** Delete def. in Dict. and substitute: **a.** *Bahamas slang.* A West Indian. **b.** *U.S. local.* A 'poor white' of the Florida Keys or North Carolina, *esp.* one of Bahamian origin. (Examples.)
1833 *Blackw. Mag.* Feb. 173 Mr. Gelid, a conch, or native of the Bahamas, was the same yawning, drawling, long-legged Creole, as ever. **1861** *N.Y. Tribune* 27 Nov. (Bartlett 1877), A Negro on this Key..is a more successful cultivator of the soil than all the rebel concks together. **1875** *Circular No. 8, War Dept.* 1 May 144 The white Americans form a comparatively small proportion of the population of Key West, the remainder being Bahama negroes, Cuban refugees, and white natives of the Bahamas and their descendants, classified here under the general title of Conchs. **1877** BARTLETT *Dict. Amer.* (ed. 4), *Konck* or *Conks*, wreckers are so called, familiarly at Key West; and the place they inhabit is called Koncktown. **1888** POWLES *Land of Pink Pearl* 115 If it wasn't for the soldiers, we would cut the throats of every white Conch in Nassau. **1936** MENCKEN *Amer. Lang.* (ed. 4) VII. iv. 378 The inhabitants [at Key West] have invented the usual opprobrious terms for one another, *e.g.*, *conch* (a West Indian). **1937** HEMINGWAY *To have & have Not* I. i. 32 How some of that gang that hangs around the dock would be pleased.. It certainly would make some Conchs happy. **1962** S. J. FLYNN *Florida* iii. 85 Keys natives are nicknamed 'Conchs'.
9. A shell-shaped part of a concher-machine, used in making chocolate.
1921 *Dict. Occup. Terms* (1927) § 448.

conche (kǫŋk, kǫntʃ), *v.* [f. as next.] *trans.* In chocolate manufacture: to mix or knead.
1924 *Sunday at Home* Nov. 80/2 'Conching' machines, in which it [*sc.* chocolate] is pushed to and fro by rollers, precede the moulding department. **1962** *Sunday Times* (Colour Suppl.) 18 Mar. 27/2 (Advt.), When you make chocolate you have to *conche* it. This means mixing and kneading it in a special machine for a very long time, even days on end, to develop its smoothness and flavour. **1963** *Economist* 8 June 1031/2 Elaborate machinery designed for churning ('conching') chocolate.

concher (kǫ·ŋkəɹ). [f. CONCH + -ER[1].] One who ladles out chocolate into a concher-machine. **b. concher-machine,** a machine having shell-like appliances which work the chocolate and keep it plastic.
1921 *Dict. Occup. Terms* (1927) § 448.

conchie, var. *CONCHY.

conchifragous (kǫŋki·frăgəs), *a.* [f. L. CONCHA + *-fragus* breaking.] Shell-breaking.
1904 *Amer. Jrnl. Sci.* Sept. 192 Cranium. Agreeing in most characters with the existing Cheloninæ, but specialized for a conchifragous habit.

conchite[2] (kǫ·ŋkəit). *Min.* [ad. G. *conchit* (A. Kelly 1900, in *Sitzungsb. Math.-Phys. Classe Akad. der Wissenschaften zu München* XXX. 187), f. CONCH: see -ITE[1].] A form of calcium carbonate, identical with aragonite, found in the shells of molluscs.
1900 *Min. Mag.* XII. 366 Conchite resembles aragonite and ktypeite. **1951** C. PALACHE et al. *Dana's Syst. Min.* II. 191 Conchite..was a name given to a supposed new polymorph of $CaCO_3$ occurring in certain shells later shown to be aragonite.

conchitis (kǫŋkəi·tis). *Path.* [f. CONCH 6, CONCHA 4 a + -ITIS.] Inflammation of the concha or external ear.
1900 DORLAND *Med. Dict.* 170/2.

conchoi·dally, *adv.* [f. CONCHOIDAL + -LY[2].] In a conchoidal form.
1899 *Pop. Sci. Monthly* LIV. 64 The sandstone of this formation has the peculiarity of fracturing conchoidally.

conchological, *a.* (Earlier example.)
1828 DISRAELI *Popanilla* vii. 84 A system equally conchological.

conchotome (kǫ·ŋkotōum). *Surg.* [f. CONCHA 4 c + -TOME.] An instrument for dividing or removing the turbinated bones in the nose.
1900 GOULD *Pocket Med. Dict.* (ed. 4) 177.

conchy (kǫ·nʃi). *slang.* Also **conchie, conshy.** Abbrev. of '*conscientious objector*' (viz. to military service): see *CONSCIENTIOUS *a.* 1 b.
1917 *Daily Mail* 9 Oct. 2/3 The assembly of eleven hundred 'conscientious' objectors at one spot, Princetown, on Dartmoor, where they are known as 'conchies'. **1917** *Blackw. Mag.* Nov. 698/1 So the Conchys, as the Bishop of Exeter calls them, live in a golden age. **1918** 'IAN HAY' *Last Million* vii. 93 What are yours [*sc.* sons] in? The Circumloosion Office, or the Conchie's Battalion? **1923** C. S. LEWIS *Let.* 4 Feb. (1966) 85 He felt it his duty to be a 'conchie' if there was another war. **1928** GALSWORTHY *Swan Song* II. xi. 195 Conshies and Communists and Profiteers—I'd have had 'em all against a wall. **1936** M. PLOWMAN *Faith called Pacifism* 46 Of course I knew that from one standpoint I was only another adjectival 'Conchy' letting his pals down. **1939** *War Illustr.* 14 Oct. 154/2 First German 'Conchie'. **1951** *Landfall* V. 21 The deal that is going on here is worse than the one the Conchies got. **1960** [see *BEARDIE 2].

conciliarism (kǫnsi·liariz'm). [f. CONCILIAR *a.* + -ISM.] A theory or system of ecclesiastical government according to which the highest authority is embodied in councils of the church. Hence **conci·liarist** *sb.*, one who supports this theory.
1932 C. H. McILWAIN *Growth Pol. Thought* vi. 349 Though the theories of the conciliarists are not novel their expression of these theories and their application of them to the changing political conditions of the time are second to none in their historical importance or in their actual influence upon subsequent political thought. **1945** V. FERM *Encycl. Relig.* 191/1 *Conciliarism*, the doctrine of the authority of representative church councils as over against that of the monarchical papacy. **1951** A. GEWIRTH *Marsilius of Padua* I. iii. 131 The conciliarist movement attempted to combine ecclesiastic universality with secular particularism. **1955** W. ULLMANN *Growth Papal Govt.* xiv. 453 The so strongly entrenched idea of the *corpus* [of believers] was to show itself of extraordinary value to the conciliarists. **1957** *Jrnl. Theol. Stud.* VIII. 264 The papacy in the fourteenth century survived the move to Avignon only to fall victim to the far more dangerous menace of conciliarism.

conciliation. Add: **4. c.** Any of various means whereby disputes between employer and employees may be settled by agreement without proceeding to arbitration.
1876 H. CROMPTON *Industrial Conciliation* iii. 33 The development of arbitration and conciliation in industry is characterised by the increasing prominence of conciliation. **1902** *Encycl. Brit.* XXV. 550/1 The term 'conciliation' is ordinarily used to cover a large number of methods of settlement, shading off in the one direction into 'arbitration' and in the other into ordinary direct negotiation between the parties. **1909** *Hansard's Parl. Deb.* Ser. v. II. 1627 The conciliation scheme arranged in November, 1907, between representative railway companies and the railway workers. **1957** *Encycl. Brit.* XII. 300/2 Conciliation is usually compulsory in countries which have compulsory arbitration, notably in Germany, Sweden, Australia, New Zealand, and in Venezuela.
7. *attrib.*, as *conciliation act, board.*
[**1867** *Act* 30 & 31 *Vict.* c. 105 § 18 In citing this Act for any Purpose whatever it shall be sufficient to use the Expression 'The Councils of Conciliation Act, 1867'.] **1876** H. CROMPTON *Industrial Conciliation* ii. 18 A conciliation board has standing committees..and is in fact a machinery for accommodating the conflicting interests of employers and employed. **1902** *Encycl. Brit.* XXV. 552/2 The Conciliation Act passed in 1896. *Ibid.* 550/1 The conciliation boards established under the New Zealand Arbitration Act of 1894. **1909** W. S. CHURCHILL in *Hansard's Parl. Deb.* Ser. v. II. 1628 A conciliation board has been established for the Great Northern and City Railway. **1921** *Daily Colonist* (Victoria, B.C.) 8 Oct. 6/5 As the question is likely to be brought before a conciliation board in the near future, the Council will probably withhold action [etc.]. **1965** *Oxf. N.Z. Encycl.* 108/2 Since 1894 the wages and conditions of employment of most workers in New Zealand..have been settled by agreements or awards made under the Industrial Conciliation and Arbitration Act.

conciliatorily (kǫnsi·liătərili), *adv.* [f. CONCILIATORY + -LY[2].] In a conciliatory manner.
1903 *Daily Chron.* 23 June 7/1 'I gave a vote you don't approve yesterday,' said his Royal Highness most conciliatorily. **1936** A. HUXLEY *Eyeless in Gaza* v. 43 'Of course I know you dare,' she said conciliatorily. **1967** B. WRIGHT tr. *R. Queneau's Between Blue & Blue* xviii. 199 'Just one more little drop..,' said the Duke conciliatorily.

concitation. Delete † *Obs.* and substitute: *arch.* (Later example.)
1920 T. S. ELIOT *Ara Vos Prec* 12 Not by any concitation Of the backward devils.

conck (kǫŋk). Also **conk.** The cry of the wild goose. Cf. CANK.
1904 *Westm. Gaz.* 10 Sept. 11/2 A joyful chorus of metallic notes—'conck-conck'. **1908** *Ibid.* 14 Nov. 8/3 The wild 'conk-conk' chorus of the bands as they fly in from the sea.

Co·ncord, *sb.*[2] *U.S.* [The name of the capital of New Hampshire, and of a village in Middlesex County, Massachusetts.]
1. *attrib.* in *Concord coach* or *wagon*, a type of vehicle originally made at Concord, N.H.
1853 *Shasta Courier* (Redding, Calif.) 12 Mar., The Proprietors of the above line..have placed upon this route their splendid stock of American Horses and elegant Concord Coaches. **1853** *Deseret News* (Salt Lake City) 14 May 1/1 A man calling himself Wm. McClafflin hired of the subscriber, a grey horse and Concord wagon. **1860** HOLLAND *Miss Gilbert's Career* xxiii. 399 Strings of rustic lovers in Concord wagons make pilgrimages to the shrines of learning. **1887** *Harper's Mag.* May 871/2 Concord coaches with inviting outside seats stood about. **1903** C. B. LOOMIS *Cheerful Americans* 3, I climbed into the Concord wagon. **1909** C. H. STERNBERG *Life Fossil Hunter* 144, I entered a Concord coach drawn by a team of eight horses. **1947** *Steamboat Pilot* (Colo.) 16 Jan. 2/7, 52 'Concord wagons', as the stage-coaches were called. **1948** J. D. RITTENHOUSE *Amer. Horse-Drawn Vehicles* 47 Concord coaches..carried nine passengers inside and as many more as could cling to the roof.
b. *ellipt.* A Concord coach.
1898 KIPLING *Day's Work* 62 'Tain't half as interestin' ez goin' daown-taown in the Concord. **1925** MULFORD *Cottonwood Gulch* xiv. 172 Along the road came a dusty Concord,..drawn by six horses.
2. *Concord grape*, a variety of grape originating at Concord, Mass. Also *ellipt.*
1858 *Trans. Mich. Agric. Soc.* X. 217 Mr. Prince thought it a better grape than the Concord. **1864** *Rep. Maine Board Agric.* IX. 35 Years after the introduction of the Isabella came the Diana, Concord, and some others. **1871** R. SOMERS *Southern States since War* 128 The 'Concord' grape is almost black, of rather thick skin, but juicy and sweet. **1902** *Fortn. Rev.* June 1007, I miss the California grapes and the Concords with their clusters of deep blue berries. **1911** S. E. WHITE *Bobby Orde* (1916) x. 125 The satiny 'Concords'.., however, were better dipped in cool water. **1946** S. H. HOLBROOK *Lost Men Amer. Hist.* 129 Ephraim Bull, originator of the Concord grape, America's first great contribution to horticulture.

concord, *v.* Add: **5.** [Back-formation from CONCORDANCE 6 b.] To rearrange the words of a text in the form of a concordance.
1969 *Computers & Humanities* III. 251 By concording every word in Dobbie's 3,182 line text..Bessinger arrived at a word total of 38,128. **1971** *Times Lit. Suppl.* 18 June 716/2 Dr. Howard-Hill concords sheet C of Quarto O of *1 Henry IV*.

concordancy. Delete †*Obs.* and add later example.
1929 E. H. VISIAK *Medusa* xiii. 163, I..was enthrilled, on a sudden, by a sympathetic concordancy of wonder and joy that shined in his eyes.

concordant, *a.* Add: **4.** [ad. G. *konkordant* (A. Supan *Grundzüge Physisch. Erdkunde* (ed. 2, 1896) IV. 574.] Of a geological feature: parallel (see quots.).
1914 R. A. DALY *Igneous Rocks* v. 63 *Concordant injections* (injected along bedding planes)..*Discordant Injections* (injected across bedding planes). **1926** G. W. TYRRELL *Princ. Petrology* ii. 16 If the molten material has been guided by the bedding planes of the intruded rock, the resulting igneous body is said to be concordant. On the other hand, the magma may break across the bedding planes, and then forms a transgressive or discordant mass. **1937** WOOLDRIDGE & MORGAN *Physical Basis Geogr.* xxi. 354 The terms 'concordant' and 'discordant', proposed by Supan, are good alternatives for 'longitudinal' and 'transverse', in the general description of coastlines. **1961** J. CHALLINOR *Dict. Geol.* 43/2 *Concordant*, as applied to an igneous intrusion: lying parallel with planes of stratification or foliation. Contrasted with a discordant relationship.

concordatory (kǫnkǫ·ɹdătəri), *a.* [f. CONCORDAT + -ORY, after F. *concordataire*.] Of or pertaining to a concordat, esp. that between church and state in France.
1896 *Edin. Rev.* July 212 The concordatory organisation, with all its diocesan and parochial ramifications, continues to exist. **1898** J. E. C. BODLEY *France* III. iv. 201 The relations of Church and State also were organized by the same master-hand in the concordatory arrangement, which survives regimes most antagonistic in matters ecclesiastic. **1906** *Daily Chron.* 10 Dec. 4/6 On Tuesday next Article IV of the Law of Separation will come into force in France. This marks the final exit of the Concordatory Church. **1957** *Encycl. Brit.* VI. 206/1 Objects of concordatory conventions.

‖**concours d'élégance** (koňkūəɹ delegaňs). [Fr., lit. contest of elegance; cf. CONCOURSE 4.] A parade of vehicles in which the entrants are judged according to the elegance of their appearance.
[**1936** 'IAN HAY' *Housemaster* ii. 31 Paul was going to compete in a Grand Concourse of Elegance, or whatever the French call these ridiculous affairs.] **1950** *New Yorker* 26 Nov. 51 Motoring tourists arriving here [*sc.* in Switzerland]..feel that they have crossed the border into a sort of *concours d'élégance* of the latest long, shiny American and British models. **1955** *Times* 11 July 6/4 Competitors..drove to Cheltenham and then to the Austin Works at Longbridge, Birmingham, where there were driving tests. The event ended with a *concours d'élégance* here. **1958** *Times* 15 Apr. 15/7 Motorists who own an Austin, a Morris, a Riley, an M.G., or a Wolseley of 1928 or earlier vintage are invited to take part in a *concours d'élégance* for a cup and other awards. **1963** *Guardian* 20 Feb. 16/4 A concours d'elegance for the motor-cycle owners.

concourse. Add: **8.** An open space or a central hall in a large building, esp. in a railway station. orig. *U.S.*
1862 *Harper's Mag.* Dec. 42/1 A group of cavaliers had assembled on the 'Concourse' at the Central Park. **1905** *Daily Chron.* 12 Jan. 8/3 Altogether, the station will be the most complete in America... The central hall, or 'concourse', as it is called, of the present station, is a beautiful building. **1908** *Ibid.* 22 June 4/4 The decorations of the foyer (or 'concourse', to use a new American term). **1909** *Westm. Gaz.* 20 Mar. 3/1 (Birmingham, Alabama) A thick brass rod running across the main hall, or 'concourse'. **1911** *Engineer* 6 Oct. 363 A feature of the new building [*sc.* Baker Street

Railway Station) will be a..concourse 80 ft. × 50 ft. **1939** *Archit. Rev.* LXXXV. 92 (*caption*) The main concourse looking towards the flying field. **1957** *Spaceflight* I. 72/1 Outside the auditorium there was a large 'concourse' fitted up with enquiry desks and plenty of chairs. **1959** *Camb. Rev.* 24 Oct. 63/2 The covered 'concourse' under the Hall resembles too much an underground station platform to attract.

concreative, *a.* Delete †*Obs.* and add later examples.
1894 *Thinker* V. 134 Beings created after His image are to work out, by the exercise of con-creative causality, their own moral perfection. **1938** R. G. Collingwood *Princ. Art* xiv. 323 The audience's function is merely receptive and not concreative.

concrescence. Add: **1. c.** *Embryol.* The growing together of two parts during the development of the vertebrate embryo.
1890 C. S. Minot in *Amer. Naturalist* XXIV. 501 The Concrescence Theory of the Vertebrate Embryo. *Ibid.* 503 Concrescence in Bony Fishes. **1913** J. W. Jenkinson *Vert. Embryol.* i. 13 Concrescence of layers, as in the union of the embryonic plate with the trophoblast in some Mammals, where the layers unite by their margins, or as in the union of the medullary folds, or of the stomodaeum with the gut, where the concrescence is by the surfaces. **1917** Arey *Prenatal Truth. Embryol.* (ed. 2) 31 The primitive streak becomes highly significant when interpreted in the light of the theory of concrescence, a theory of general application in vertebrate development. **1927** W. Shumway *Vert. Embryol.* iv. 96 The theory of concrescence..has also been used to explain the closure of the blastopore. By many it has been extended to include the formation of the body, with the exception of the tail, by the coalescence of the two halves of the germ ring. **1968** E. Witschi *Devel. Vertebrates* xii. 200 Convergent movements of blastemic cell masses, which result in the joining of the material halves of a body or of organs, are designated as concrescence.

concrescent (kǫnkre·sěnt), *a.* [ad. L. *concrescent-, -ens,* pres. pple. of *concrescere*; cf. Concrescence.] Growing together.
1904 *Encycl. Brit.* XXIX. 191/1 A single male flower consists of an axis enclosed at the base by an inconspicuous perianth formed of two concrescent leaves and terminating in two. **1951** R. Firth *Elem. Social Organiz.* i. 18 If the underlying activity of nature is a system of interlocking concrescent processes..then human activity partakes of the same general character.

concrete, *a.* and *sb.* Add: **A.** *adj.* **4.** *concrete science* (Science 4 b).
1854 H. Spencer in *Brit. Q. Rev.* July 148 Let us observe how the relatively concrete science of geometrical astronomy, having been thus far helped forward by the development of geometry in general, reacted upon geometry, caused it also to advance, and was again aided by it. **1865** J. S. Mill *Comte* 33 The concrete sciences..concern themselves only with the particular combinations of phenomena which are found in existence.

b. *Philos.* *concrete universal* [Universal *sb.* 1]: the individual, when regarded as something maintaining its identity through qualitative change or diversity, or as a unity or system or class of separate but identical particulars. Also *transf.*
1865 J. H. Stirling *Secret of Hegel* p. xi, As Aristotle, with considerable assistance from Plato, made explicit the abstract Universal that was implicit in Socrates,..so Hegel..made explicit the concrete Universal that was implicit in Kant. **1874** W. Wallace tr. *Hegel's Logic* ix. 267 The Judgement of Necessity..contains..the predicate, partly the substance or nature of the subject, the concrete universal, the *genus*. **1876** F. H. Bradley *Eth. Stud.* v. 147 The good will..is a concrete universal, because it not only is above but is within and throughout its details, and is so far only as they are. **1883** — *Princ. Logic* I. vi. 175 The *concrete* particular and the *concrete* universal both have reality, and they are different names for the individual. **1912** B. Bosanquet *Princ. Individuality* ii. 38 A macrocosm constituted by microcosms, is the type of the concrete universal. **1920** M. T. Collins *Mod. Concept. Nat. Law* 95 A thing, a person, an act—anything—is only seen in its true nature when it is grasped as an organized unity, as a synthesis of the manifold. So far as it is a whole, it is a concrete universal. **1948** *Poetry* LXXIII. 159 *Concrete universal*, a concept, continuous in literary criticism, which implies the paradoxical union in a poem of the concrete specific, and individual, together with the universal and general. The concrete universal persists among the New Critics.

7. *concrete music* [tr. F. *musique concrète*]: a form of music constructed by the arrangement of various recorded sounds into a sequence. (Also with first word in French form *concrète*.)
1953 *Musical Amer.* 15 Jan. 6/3 This method of basing a musical composition on fragmentary sounds, existing concretely, characterizes what Schaeffer has labeled concrete music. **1954** *Gramophone Record Rev.* Apr. 297 Concrete music is assembled rather than composed. **1954** *Times Lit. Suppl.* 3 Dec. 778/4 The very latest thing..Concrete Music, the term adopted for the French *musique concrète*, which is really synthetic electrophonics. **1958** *Observer* 22 June 15/3 The music, an airborne plunking that deserves a less earthbound epithet than *concrète*, is by John Addison.

b. *concrete poetry*: a form of poetry in which the significance and the effect required depend to a larger degree than usual upon the physical shape or pattern of the printed material. Also ellipt. *concrete*. Hence *concretist, concrete poem, poet,* etc.
The term was coined independently and almost simultaneously in Brazil and Germany: in Brazil (*poesia concreta*) by the Noigandres group of poets; in Germany (*die konkrete Dichtung*) by Eugen Gomringer. The usage was formally adopted at a meeting in 1955 between the two originators.
1958 *Pilot Plan for Concrete Poetry* in M. E. Solt *Concrete Poetry* (1970) 72 Concrete poem communicates its own structure: structure-content..Concrete Poetry aims at the least common multiple of language. **1966** *Ibid.* 16 Feb. 2/1 The Concrete poet tries to investigate language and the materials of which it is composed in a depth which he cannot achieve using conventional syntax. *Ibid.* 9/1 The early 'concretists' were interested in setting words in isolation on the page. *Ibid.* 9/2 His sensitivity led him to 'concrete' as a means of overcoming the deterioration language suffers through overexposure. **1966** *Camb. Rev.* 28 May 448/1 'Concrete' poets ignore the traditional boundaries between word and image. **1967** S. Bann *Concrete Poetry* 17 He has recently contributed the pure Concrete 'cube-poem' to the Brighton Festival. *Ibid.*, The links between the early socially committed concrete poems and the 'popcrete' poems of Augusto de Campos. *Ibid.* 24 His first contact with the Concrete movement, however, was with the Brazilians. **1968** *Arts Hispanicus* I. 11. 7/2 There is a fundamental requirement which the various kinds of concrete poetry meet: concentration upon the physical material from which the poem or text is made.

B. *sb.* **3.** *armoured concrete* = *reinforced concrete. Also *Comb.*: **concrete mixer** (so *-mixing*); **concrete paver.**
1906 *Concrete* Mar. p. ii, Armoured Concrete Construction. **1906** *Westm. Gaz.* 10 Sept. 9/3 An extensive installation of stone-breaking and concrete-mixing machinery is in full work. **1907** *Daily Mail* 22 Oct., Armoured concrete, reinforced concrete, concrete-steel, or ferro-concrete. **1909** *Cent. Dict.* Suppl., *Concrete-mixer*, a machine for mixing cement, sand, crushed or broken stone, and water in varying proportions for making concrete. **1930** W. Heyliger *Builder of Dam* 23 A one-bag power concrete mixer. **1930** *Engineering* 7 Mar. 315/1 The concrete-mixing plant is said to be the largest in Canada. **1964** *Gloss. Highway Engin. Terms (B.S.I.)* 49 *Concrete paver*, a concrete mixer capable of moving on crawler tracks or rails and provided with a boom and bucket for depositing the concrete in the required position in a pavement.

b. Paving made of concrete.
1911 E. Ferber *Dawn O'Hara* ii. 13 No tramping of restless feet on the concrete all through the long, noisy hours.

concrete, *v.* Add: **3.** Delete *rare* and add examples. Also *refl.*
1811 Shelley *Let.* 2 June (1964) I. 98 You loved a being, an idea in your own mind which had no real existence. You concreted this abstract of perfection. **1864** *Good Words* 231/1 The effect produced by these sketchy portraits..was great in concreting the idea of them. **1884** C. Powell *Philistia* II. xxii. 249 Don't be so abstract, Ernest; concrete yourself a little. **1897** *Westm. Gaz.* 22 Jan. 2/1 It is quite right that a bachelor of thirty should stand *in loco parentis*, but to concrete this by putting him into a penitentiary is a little too violent. **1898** J. Conrad *Tales of Unrest* 203 Precious dreams that concrete the most cherished..of his illusions. **1902** *Westm. Gaz.* 19 July 2/3 There are few French towns which do not concrete the memory of their distinguished natives by statues. **1928** *Spectator* 3 Nov. 23 In a story by de Maupassant..English love of beauty in Nature was concreted as perhaps it could only have been by a foreign artist. **1938** T. Wessling *Liturgy & Life* i. 5 The whole reason of existence of any product whatever is therefore to realize, to concrete the purpose of the producer.

4. c. *trans.* and *intr.* To overlay or pave with, or embed in, concrete.
1875 *Boston Budd* 129 Concreting side and cross-walks $2270. **1888** *Harper's Mag.* Nov. 870/1 The first proposition to concrete the sidewalks of this village. **1907** *Westm. Gaz.* 11 Mar. 9/1 We get 9*s.* a day digging out foundations, and 8*s.* a yard putting in concrete..We can't go on concreting every day.

concretization (kǫnkrītaizēⁱ·ʃən). [f. Concretize *v.* + -ation.] The action of concretizing or the condition of being concretized. Also *Comb.*
1939 P. Christophersen *Articles* 57 [Reporting Hjelmslev, who wrote in French] The so-called definite article is a concretization-morpheme which indicates that the object in question is supposed to be known to the interlocutor. **1949** Koestler *Insight & Outlook* xi. 162 The concretization and the integrative tendencies in a functional whole may take many different forms. **1950** S. Potter *Our Lang.* 111 No sooner are abstract nouns formed than men tend to think of each appearance of a quality or action in the abstract as a concrete being and so, by concretization, they make abstractions tangible and visible once more. **1954** H. Read *Anarchy & Order* 70 This concretization and vitalization of ideals is one of the main tasks of the aesthetic activity in children. **1958** S. Spender *Engaged in Writing* ii. 42 They went on discussing..whether they should devote themselves to principles or what they called 'concretizations'.

concretize, *v.* Delete *nonce-wd.* and add later examples.
1942 *Theology* XLV. 109 It seems very hard to understand what personal relation looks like concretized in the bombing of Rostock and Cologne. **1948** *Mind* LVII. 300 The good, for example, when we try to concretize it, is apt to appear as a somewhat complex relation between pleasures and unpleasures. **1952** V. Gollancz *My Dear Timothy* xx. 343 Human ideals, when concretized in legislative enactments, seem to lose..a great part of their life-giving quality. **1963** S. Weintraub *Private Shaw* v. 150 The surviving manuscript shows stylistic changes in G.B.S.'s hand—mainly his concretizing T.E.'s diction...'Stay put for ages' became 'stay put indefinitely'.

condemn, *v.* Add: **7. b.** To pronounce judicially (land, etc.) as converted or convertible to public use. *U.S.*
1833 *Niles' Reg.* XLIV. 192/2 All expenses to be incurred, in condemning, or purchasing ground, shall be..at the proper cost of the rail road company. **1876** *Congress. Rec.* 2 Aug. 5079/1 If the Government has the right to make this improvement, and the State of Oregon does not furnish the land..the Government has a right to condemn it for that purpose.

8. (Later examples.)
1939 M. S. Rice *Working-Class Wives* viii. 181 Mrs. R... lives in two rooms of a back-to-back house; part of the street is condemned. **1951** E. Coxhead *One Green Bottle* i. 9 Though it had been condemned twenty years before, a house is a house.

condemnation. Add: **4. b.** Judicial assignation (of property) to public purposes, or in payment of a debt. *U.S.*
1852 J. Dunlap *Book of Forms* (ed. 2) 693 Schedule or Inquisition of Real Estate levied upon by Fi. Fa., for Condemnation. **1889** *Century Dict.* s.v., The condemnation of private lands for a highway, a railroad, a public park, etc. **1947** *Sat. Even. Post* 8 Feb. 96/4 Residents claimed that they had received too little compensation under the condemnation awards.

condemned, *ppl. a.* Add: **1. b.** Confounded, damned. *colloq.* Chiefly *U.S.*
1842 *Spirit of Times* 2 Apr. 58/2 Did you ever git into a nest of *runners after baggage*?..I had heard folks tell what condemned critters they were. *a* **1861** T. Winthrop *Open Air* (1863) 249 'But I took a big cold,' the diver continued, 'and I'm condemned hoarse yit.' **1909** 'O. Henry' *Roads of Destiny* 398 'Ducks?' I says; 'you condemned lunatic, them ain't ducks; them's mud hens!' **1923** R. D. Paine *Comr. Rolling Ocean* viii. 156 Bless my soul, what sort of a condemned rumshop have I stumbled into?

condemning, *ppl. a.* (Later example.)
1944 H. van Zeller *Ezechiel* xi. 92 Surely there is a less condemning explanation.

condensation. Add: **6.** *Organic Chem.* A reaction in which two similar or identical organic molecules become joined by a carbon-carbon bond, generally with the elimination of a simple molecule (as of water or an alcohol).
1886 E. F. Smith tr. *V. von Richter's Org. Chem.* 155 Such a union of two or more molecules by the linking of carbon atoms..and the formation of complicated carbon chains is ordinarily termed condensation. **1907** Haskins & MacLeod *Org. Chem.* 104 Aldehyde molecules can be made to fuse together, forming a 'condensation' product, aldol. *Ibid.* 212 An aldose isomeric with dextrose and galactose can be made by condensation of two molecules of the triose, glyceric aldehyde. **1938** G. H. Richter *Textbk. Org. Chem.* vii. 209 'Mixed' aldols are produced by the condensation of one aldehyde with another aldehyde. **1958** Packer & Vaughan *Org. Chem.* ix. 222 The complete reaction is a true condensation and is known as the croton condensation, but the first stage..is known as aldol addition..although it is usually referred to as the aldol condensation.

7. *Psycho-analysis.* The process by which images characterized by a common affect are grouped so as to form a single composite or a new image.
1913 A. A. Brill tr. *Freud's Interpr. of Dreams* vi. 262 An abundant condensation of psychic material has taken place in the formation of dreams. **1913** E. Jones *Papers Psycho-Analysis* 27 *Condensation.*—Every element in the manifest content represents the fusion of several in the latent thoughts, and *vice versa*. **1922** J. Riviere tr. *Freud's Introd. Lect. Psychoanal.* 144 The first achievement of the dream-work is Condensation. **1922** E. & C. Paul *Baudouin's Stud. Psychoanalysis* 20 Now one of the most notable of Baudouin's contributions to analytical science, and a matter upon which he differs from the Freudian school, is his careful study of condensation..in its bearings upon representation by symbols. **1939** W. Empson *7 Types Ambig.* vii. 245 It is evident that the Freudian terminology, particularly the word 'condensation',... **1944** J. G. Miller *Unconsciousness* ix. 239 These are the major points of this statement. Condensation and distortion occur—we have ample proof that memory is alterative. **1958** T. M. French *Reintegrative Proc. Psychoanalytic Treatm.* III. ii. 444 The condensation of the two quarreling figures into the single figure of a protesting lady.

8. *Comb.* **condensation nucleus** *Meteorol.*, any minute particle suspended in the air around which atmospheric water vapour condenses; **condensation trail,** a vapour trail.
1923 *Sci. Abstr.* XVI. 248 (*title*) Condensation nuclei in the atmosphere. **1926** *Proc. R. Irish Acad.* XXXVII. A. 58 (*title*) Atmospheric dust and condensation nuclei. **1957** G. E. Hutchinson *Treat. Limnol.* I. viii. 549 The calcium sulfate particles, not being hygroscopic, will form a stable aerosol, while the chlorides will act as condensation nuclei. **1942** 'B. J. Ellan' *Spitfire!* xiv. 74 At 25,000 feet I noticed that we were making condensation trails as well and began to wonder when our quarry would see us. **1957** *Meteorol. Gloss.* (H.M.S.O.) (ed. 3) 51 Condensation trail. **1960** *Guardian* 10 Nov. 2/5 The F-86 aircraft..sighted condensation trails at a distance of 10 to 15 miles.

condensa·tional, *a.* [f. Condensation + -al.] Of or belonging to condensation.
1903 *Nature* 3 Dec. 104 Why may one attribute to ionised air different condensational properties, according as positive or as negative ions are in question? **1925** J. Joly *Surf. Hist. Earth* 177 A layer which consists of a uniform homogeneous substance which can transmit both condensational and distortional waves.

condensed, *ppl. a.* **1.** *condensed milk* (earlier and later examples).
1863 O. W. Norton *Army Lett.* (1903) 177 We buy condensed milk of the sutlers. **1868** *Brit. Pat.* 3928 1 In preserving milk it is preferred first to condense it..and then to charge the condensed milk with gas in a vessel completely air-tight. **1957** Wodehouse *Over Seventy* xix. 182 Don't run away with the idea that the Swiss do nothing but yodel and make condensed milk.

conde·nsely, *adv. rare.* [f. Condense *a.* + -ly².] In a condensed manner or form; = Condensedly.
1832 I. Taylor *Sat. Evening* 232 It is condensely expressed in the most memorable of our apostolic canon. **1894** S. Brooke *Tennyson* 377 Their doings and sayings are so condensely given.

condenser. Add: **6.** (Now largely superseded by *capacitor.) (Later examples.)
1943 *Gloss. Terms Electr. Engin. (B.S.I.)* 27 *Capacitor, condenser*, a piece of apparatus capable of storing electrical energy as electric stress in insulating material. **1968** *Practical Wireless* Oct. 205 Condenser or Capacitor.

10. *Comb.* **condenser door,** the plate at the end of a surface condenser; **condenser loud-speaker, microphone** (see quots.).
1888 *Lockwood's Dict. Terms Mech. Engin.*, *Condenser Door*, the rectangular or round-ended cast-iron plate which closes the end of a surface condenser near the ends of the tubes. **1897** *Daily News* 1 Jan. 3/2 The condenser doors were also open. **1929** *Proc. Inst. Radio Engin.* XVII. 1142 By using the Kyle principle of construction, it has been possible to develop a practicable loudspeaker (the so-called 'condenser loudspeaker')... **1940** *Chambers's Techn. Dict.* 188/1 *Condenser loudspeaker*, a loudspeaker in which the sound-radiating element forms one electrode of a large condenser, to which polarised modulation voltages are applied. **1950** W. S. Sharps *Dict. Cinemat.* 93/2 *Electrostatic loudspeaker* (*capacitor loudspeaker*) (*condenser loudspeaker*). An electrostatic loudspeaker is a loudspeaker in which the mechanical forces are produced by the action of electrostatic fields. **1921** L. B. Turner *Wireless Telegr. & Teleph.* 160 (*caption*) Condenser microphone. **1928** *B.B.C. Handbk.* 267/1 *Condenser Microphone*, a microphone consisting of two plates of a condenser, whose distance apart is altered by the sound waves impinging upon one of them. **1938** *Admiralty Handbk. Wireless Telegr.* II. § N. 12 In historical development, the condenser microphone succeeded the carbon granule type and was much used for broadcasting work. **1960** H. Carter *Dict. Electronics* 53 *Condenser microphone*, microphone consisting in essence of a capacitor, one plate of which is made to vibrate by the incident sound waves, thus varying the distance between the two plates and therefore the capacitance of the device.

condensery (kǫnde·nsěri). orig. *U.S.* [f. Condens(ed *ppl. a.* 1 + -ery.] (See quot. 1909.)
1909 *Cent. Dict.* Suppl., *Condensery*, an establishment where condensed milk is prepared. **1921** *Ladies' Home Jrnl.* May 39/2 At these condenseries the milk from these herds [etc.]. **1932** *Daily Express* 20 Sept. 7/1 (Advt.), Snow-white condenseries down in the shires. **1935** *Economist* 9 Nov. 929/2, I have been amazed at the way in which our creameries and condenseries have responded to the calls made upon them.

condition, *sb.* Add: **8.** *U.S.* (Earlier and later examples.)
1832 in *Atlantic Monthly* (1887) Oct. 434/1 She straightway got a tutor, and prodded Ralph night and day to make up the conditions. **1833** *Ibid.* 443/2 Ralph is..actually gone back to Cambridge to make up his conditions. **1856** B. H. Hall *College Words* 123 The branches in which he [*sc.* an entering student] is deficient are called conditions. *a* **1862** [see *condition *v.* 8]. **1903** K. D. Wiggin *Rebecca* (1904) xxi. 226 She passed in only two subjects, but went cheerfully into the preparatory department with her five 'conditions'. **1907** *Scribner's Mag.* LXI. 506/1 At the end of sophomore year it became imperative for him to work off his accumulated conditions in the science he loathed. **1937** V. D. Scudder *On Journey* i. iii. 66, I never incurred a condition in college, and I reverted to my old bad pleasant habit of flitting about at my own sweet will among books and ideas.

9. e. *in a certain, delicate, interesting,* or *particular condition* (see the adjectives): pregnant.

14. a. (Earlier and later examples.) Also *attrib.*
1798 T. Holcroft *Diary* 24 Nov. in *Memoirs* (1816) III. 75 The method practised by pugilists, to bring themselves into condition, as they term it, is air and exercise, regular hours, etc. **1850** *Rep. U.S. Comm. Patents, Agric. 1849* 323 If it was conditioned properly, it will not change a particle while in the condition bulk. **1879** *Boy's Own Paper* 18 Jan. 7/1, I was in 'splendid condition', as they say of horses. **1925** E. F. Norton *Fight for Everest, 1924* ix. 197, I was not in a fit condition to provide a test case, as I had lost too much condition throughout the month of May. **1968** K. Weatherly *Roo Shooter* 35 Normally a doe will breed when a joey is getting out of the pouch enough to allow her to pick up condition.

15. *Comb.* **condition powder,** a medicinal powder given to animals to keep them in good condition.
1884 T. Hughes *Gone to Texas* 189 You ought to have a few packages of 'condition powders' for stock and chickens. **1907** *Army & Navy Stores Catal.* 58 Horse condition powders. **1948** B. Vesey-Fitzgerald *Book of Dog* I. 123 The 'condition powder' is another weird hang-over from the horse age. **1961** [see *conditioner 2].

condition, *v.* Add: **2. a.** In last quot. for 1849 read 1814.

8. *U.S.* (Examples.)
1832 in *Atlantic Monthly* (1887) Oct. 434/1 Well, on his examination at Cambridge last fall, he was heavily conditioned. **1849** *Lit.* in B. H. Hall *College Words* (1856) 124 (Th.), [A young man] shall be examined and 'conditioned' in everything. *a* **1862** in *Harvard Mem. Biog.* (1867) II. 246, I was conditioned in Greek Grammar and prose reading. **1903** *Univ. Oklahoma Bull.*..that a student is conditioned because of poor quality of work. **1932** *Atlantic Monthly* Apr. 443/1, I went to Baltimore..conditioned in Greek and mathematics and was Latin.

9. To bring to a desired state or condition; to make fit or in good condition. Also *spec.*, to purify air (cf. *air-conditioning *vbl. sb.*).
1850 *Rep. U.S. Comm. Patents, Agric. 1849* 322 The next process in this troublesome but beautiful crop is 'conditioning' it for 'packing.' **1892** *Field* 14 May 730/2 Our friends across the water do not appear to know how to condition a dog. **1901** J. Donaldson *Roller Mill* 152 The most modern and effective system of conditioning wheat is by the use of air heated from 180 to 220 degrees Fahrenheit. **1908** *Westm. Gaz.* 1 Apr. 8/3 When my hunters were being conditioned in the autumn of 1906. **1924** A. W. Thornton *Air Conditioning in Textile Mills* 241 It is more common..to condition air by local distribution of moisture and heat. **1958** *Times* 14 Oct. 11/1 Standardized systems of heating and conditioning the air inside a closed car.

10. To teach or accustom (a person or animal) to adopt certain habits, attitudes, standards, etc.; to establish a conditioned reflex or response in.
1909 *Psychol. Bull.* VI. 266 Pawlow's Method..Separate components of a complex sound which conditions a 'fundamental' reflex, will produce reactions (the so-called partial reflexes) at a certain relative intensity. **1920** *Jrnl. Exper. Psychol.* III. 1 At nine months of age..can we condition fear of an animal, e.g., a white rat, by visually presenting it and simultaneously striking a steel bar?..Such a conditioned emotional response can be established. **1927** *Mod. Philology* Nov. 213 We may study the individual..and observe how successive actions of his group-mates (parents, etc.), act by act, 'condition' him to the social habits. **1932** A. Huxley *Brave New World* ii. 29 The students..rose automatically to the tips of their toes. They were Alphas, of course; but even Alphas have been well conditioned. **1943** J. S. Huxley *Evol. Ethics* ii. 16 During its first twelve months the child acquires many habits and may be conditioned in various ways, for instance in regard to cleanliness. But unless this conditioning is brought into relation with the dynamic structure of focused impulse which develops in the second year, it will wear out or break down. **1951** R. Firth *Elem. Social Organ.* iii. 89 The people have been conditioned to these things since childhood, and feel that they are basic to their corporate existence. **1960** J. Rae *Custard Boys* i. iii. 37 The cinema, the newspapers and the war books conditioned us to look upon war as glamorous and exciting.

Conditionalism (kǫndi·ʃənăliz'm). [f. Conditional *a.* + -ism.] The doctrine of conditional survival after death. Hence **Condi·tionalist,** one who holds such doctrine (also *attrib.*).
1895 Salmond *Chr. Doctr. Immort.* vi. ii. 615 The literalists of the various forms of Annihilationism or Conditionalism. *Ibid.* 622 The Conditionalist doctrine involves conceptions both of man's nature and of Christ's work which are inadequate and unreasonable. **1910** Hastings *Encycl. Relig. & Ethics* III. 822/2 In its modern form Conditionalism may be said to be contained in two propositions: (1) that the endless life of the righteous is not the result of any natural immortality inherent, but is the gift of God; (2) that the punishment of the wicked, in the world to come, will not be of endless duration, since their life must finally be extinguished. *Ibid.* 823/2 Spinoza..was a Conditionalist in the sense that his 'immortality' is not enjoyed by any but the wise man. *Ibid.* 824/1 The conditionalist position (forcibly stated by Locke in a passage already quoted). **1918** J. H. Leckie *World to Come* 134 He affirms that Titus declared to his soldiers that those who died in battle secured for their souls a future life, while those who perished by natural decay or sickness passed utterly out of existence —which reads very like an excellent military version of Conditionalism. *Ibid.* 223 This Conditionalist strain in early Christian thought attained to definite dogmatic expression in Arnobius. **1967** *New Catholic Encycl.* VI. 1006/2 Others..resolved the problem of the punishment of the demons and the damned by means of the theory of conditionalism.

conditioned, *ppl. a.* Add: **7. b.** *conditioned reflex*, a reflex or reflex action which through habit or training has been induced to follow a stimulus not naturally associated with it (cf. *unconditioned *ppl. a.*). So *conditioned inhibition, response, stimulus.*
1906 *Nature* 11 Oct. 592/1 The latter actions..are termed by Prof. Pawlow 'conditioned reflexes', to distinguish them from the ordinary or unconditioned reflexes. **1920** [see *condition *v.* 9 b]. **1925** E. P. Poulton *Taylor's Pract. Med.* (ed. 13) 892 Such associations are usually lost with further experience in the same way as the simpler conditioned reflexes established in animals can be broken down by further training. **1927** G. V. Anrep tr. *Pavlov's Lect. Conditioned Reflexes* ii. 25, I have termed this new group of reflexes conditioned reflexes to distinguish them from the inborn or unconditioned reflexes. *Ibid.* 26 Conditioned reflexes are phenomena of common and wide-spread occurrence...We recognize them in ourselves and in other people or animals under such names as 'education', 'habits', and 'training'. *Ibid.* 27 It is..necessary that the conditioned stimulus should begin to operate before the unconditioned stimulus comes into action. *Ibid.* v. 69 Conditioned inhibition is developed..where the duration of the positive stimulus overlaps that of the additional stimulus. **1931** *Discovery* Apr. 106/2 This new kind of reflex, engrained by particular conditions of individual experience, was called by Pavlov a conditioned reflex. **1934** Warren *Dict. Psychol.* 55/1 Conditioned response. **1951** G. Humphrey *Thinking* i. 24 Motivation is sometimes necessary..for the establishment of conditioned reflexes. *Ibid.* x. 313 Early experimentalists thought that a replica of past experience was produced by association...Various alternatives have been proposed..the 'conditioned response', and so on. **1960** H. J. Eysenck et al. *Behaviour Therapy* III. 223 Extinction brought about by a habit of not responding, i.e., conditioned inhibition. **1964** *John o' London's* 4 Jan. 19/3 Annie winces; and so (after the conditioned-reflex laugh) do we.

c. Taught or accustomed to accept or adopt certain habits, attitudes, standards, etc.
1930 R. S. Woodworth *Psychol.* (ed. 8) vi. 258 From his prolonged period of dependence, the human child is bound to be 'conditioned' to group life, even if he were naturally indifferent to it. **1956** E. Earle *Earmold Cassie* xi. 175 He must have acquired these habits a very long time ago because Mrs. Jackson appeared to be perfectly 'conditioned' to them. **1958** *Economist* 8 Nov. 482/1 It may be a long time before the heavily conditioned people of Russia itself could be counted on to share the job of supervising their own government in the interest of humanity as a whole.

10. Of air: purified and having had its temperature, humidity, etc., adjusted.
1909 S. W. Cramer *Useful Inf. for Cotton Manuf.* (ed. 2) IV. 1413 The conditioned air is delivered from the apparatus into a longitudinal duct. **1936** H. G. Wells *Things to Come* xi. 94 They had no properly conditioned air.

conditioner. Add: **2.** (Further examples.)
1954 Webster Add., *Conditioner*, a substance whose addition to the soil causes aggregation or stabilizes the aggregates already formed and thus improves aeration, drainage and ease of working of the soil. **1960** *Woman's Own* 19 Mar. 25/3 Dry hair loves nothing better than contact with a cream conditioner. **1961** C. H. D. Todd *Pop Whippet* x. 144 The many well-known brands of condition powders are useful as a general and mild conditioner. **1961** *Times* 30 Mar. 13/6 The best conditioner of a heavy soil for seeding is still the weather.

b. *spec.* = *air-conditioner.
1938 *Encycl. Brit. Bk. of Yr.* 29/1 This conditioner cools, dehumidifies, circulates, and cleans the air in the room. **1957** 'F. Richards' *Practise to Deceive* (1959) iv. 47 The room was air-conditioned. The conditioner hummed softly.

3. One who tests the condition of yarn, etc.
1921 *Dict. Occup. Terms* (1927) § 398 *Conditioner, yarn conditioner, cop conditioner*;..dries and weighs samples of yarn, and calculates regain of moisture.

conditioning, *vbl. sb.* Add: **2.** (Later examples.)
1904 Goodchild & Tweney *Technol. & Sci. Dict.* 124/1 *Conditioning.* When yarn is taken from the spindles it usually contains about 3 per cent. less than its natural moisture. It is therefore conditioned by keeping a few days in a damp place, so as to regain the loss. **1940** *Chambers's Techn. Dict.* 188/1 *Conditioning*, the process of adding to yarn, after spinning, the percentage of moisture necessary to bring it up to average conditions. **1954** *Spalding & Hodge's Paper Terminol.* 18 Conditioning tends to disperse any static electricity that may be present in the paper.

3. The act of bringing an animal, etc., into good condition.
1861 F. Taylor *Recoll. Horse Dealer* xvi. 258 Another very important matter in the conditioning of horses, is water. **1892** *Field* 19 Mar. 414/3 These horses are of necessity sent up with glossy coat, unduly fattened—indeed, in 'show' condition; this, too, at the very time of year when such 'conditioning' would be likely to be most injurious. **1904** *Daily Chron.* 8 Oct. 4/5 Prior to actual racing, the conditioning and training of the pigeons is an absorbing occupation. **1908** *Animal Managem.* 266 Conditioning for such horses must be gradual.

4. The training or accustoming of a person or animal to give conditioned responses. (Cf. *condition *v.* 9 b.)
1920 *Jrnl. Exper. Psychol.* III. 4 Steps taken to condition emotional responses...The infant..was tested with his blocks immediately afterwards to see if they shared in the process of conditioning. **1930** R. S. Woodworth *Psychol.* (ed. 8) vi. 259 Practically all the 'conditioning' that the child gets would tend towards making him submissive. **1932** A. Huxley *Brave New World* i. 17 All conditioning aims at that: making people like their unescapable social destiny. **1936** A. J. Ayer *Lang. Truth & Logic* vi. 166 The people with whom we argue have generally received the same moral education as ourselves...But if our opponent happens to have undergone a different process of moral 'conditioning'.. then we abandon the attempt to convince him by argument. **1943** [see *condition *v.* 9 b]. **1953** J. S. Huxley *Evol. in Action* ii. 41 When Thorpe reared *Drosophila* grubs on media flavoured with peppermint, the adults were attracted by the same smell to lay their eggs. Such 'olfactory conditioning' could readily operate when an insect takes to a new food-plant, and could then lead on to genetic adaptations. **1955** *Treatment Brit. P.O.W.'s in Korea* (H.M.S.O.) 24 It was the normal fate of the prisoner who steadfastly refused to co-operate or who was sufficiently important in Chinese eyes to merit intensive 'conditioning'. **1964** A. Koestler in *Listener* 14 May 786/1 When the rat presses it down with his paws, a food pellet falls into the dish. This experimental procedure is called 'operant conditioning' because the rat 'operates' on its environment—whereas in so-called 'classical conditioning' by the Pavlovian method, the dog is immobilized in its restraining harness.

5. *attrib.* (sense 2) *conditioning oven, plant*; (sense 4) *conditioning room.*
1937 Labarre *Dict. Paper* 182/2 Conditioning ovens (Schopper's) for ascertaining percentage of moisture in wood pulp, cellulose etc. **1960** G. A. Glaister *Gloss. Bk.* 83/1 A conditioning plant..consists of a series of vertical compartments into which conditioned air is blown while rolls of paper are fed through. **1932** A. Huxley *Brave New World* ii. 20 Infant Nurseries: Neo-Pavlovian Conditioning Rooms, announced the notice board.

condole, *v.* **4.** Delete †*Obs.* and add later Indian example.
1969 *Hindusthan Stand.* (Calcutta) 5 Aug. 6/4 Students.. passed a resolution condoling the death of Mr. Prakash Podder.

condom (kǫ·ndǫm, kɒ-). Also 7 **condum, condon,** 7–9 **cundum.** [Origin unknown; no 18th-cent. physician named Condom or Conton has been traced though a doctor so named is often said to be the inventor of the sheath.] A contraceptive sheath.
c **1706**? J. Hamilton *Scots Answer to Brit. Vision* l. 42, Then Sirenge and Condum Come both in Request. **1708** (*title*) Almonds for parrots..with a word or two in praise of condons. **1717** D. Turner *Syphilis* I. 74 The Condum being the best, if not the only Preservative our Libertines have found out at present. **1728** W. Kennett in W. Pattison *Cupid's Metamorphoses* 307 Happy the Man, who in his Pocket keeps, Whether with Green or Scarlet Ribband bound, A well made C——. **1744** *The Machine* 10 Let not the Joy she proffers be Essay'd, Without the well-try'd Cundum's friendly Aid. **1828** S. F. Gray *Suppl. to Pharmacopœias* (ed. 4) ii. 135 Condoms, Armour, Baudruches, Redingotes Anglaises. *c* **1888–94** *My Secret Life* X. 311 It was an age since I'd had a cundum on my prick. **1897** *Science of Generation* xx. 235 The use of various mechanical contrivances, such as French Safes, Condom Sheaths, etc., is also objectionable. **1904** Dulberg tr. *Senator & Kaminer's Health & Disease* I. vi. 238 The condom is..relatively the most perfect anticonceptual remedy. **1936** D. V. Glass *Struggle for Population* iii. 35 [In Italy] condoms are listed as preventatives of disease and not as birth-control appliances, and are thus easily available. **1964** L. Martin *Clinical Endocrinol.* (ed. 4) vii. 228 Semen should be collected into a glass container at body temperature after masturbation or coitus interruptus and not into a rubber condom.

condominium. Add examples relating to more recent politics.
1892 *Daily News* 11 Jan. 5/7 The abandonment of the condominium carried by the majority which made Tonquin. **1904** *Westm. Gaz.* 20 Jan. 7/1 To create a sort of Russo-Japanese Condominium in Korea. *Ibid.*, A mild revision of the Condominium Protocol. **1908** *Daily Chron.* 22 June 1/3 A condominium of men and women in public affairs. **1936** *Geogr. Jrnl.* LXXXVIII. 243 That 'savagery' should continue..is astonishing to those who are unacquainted with the Condominium [in Espírito Santo]. **1948** *Ann. Reg. 1947* 194 As regards the Sudan, the 1899 Agreement did not imply that the British would share in the sovereignty, though they tried to convey that idea by always using the word 'condominium'. **1955** *Times* 24 Aug. 5/1 Security in the Sudan was maintained by the troops of the Condominium and the Sudan Defence Force. **1969** *Statesman's Year-Bk.* 161 (*heading*) New Hebrides condominium.

2. *N. Amer.* An apartment house in which the units are owned individually, not by a company or co-operative; an apartment in such a building.
1962 *Economist* 31 Mar. 1255/1 The legal concept of buying a single flat, instead of a share in the whole building, is just making its way in the housing field in the United States where it is known as a 'condominium'. **1964** *Financial Times* 27 Nov. 3/6 The condominium—or the 'condo' as Chicagoans have come to know it—is essentially a development from the co-operative concept. *Ibid.* 3/7 The principal advantage claimed for the condominium is that it permits the occupier a greater measure of independence than would be the case of the co-operative. **1970** *Washington Post* 30 Sept. B 11/1 (Advt.), Ocean Club condominium residence for sale in Fort Lauderdale, Florida.

condonable (kǫndōᵘ·năb'l), *a.* [f. Condone *v.* + -able.] That can be condoned.
1872 E. C. Clark *Early Roman Law* viii. 48 A private injury, condonable by those interested. **1951** *Essays in Criticism* I. 70 Harry's father's impulse to murder Amy is condonable.

conducing, *ppl. a.* Add: In *conduct conducing* (*to adultery*). Also **conducive** *a.* in same sense.
1868 *Law Jrnl. Rep.* XXXVII. v. 4/2 The fact that the husband had been in the habit of going with his wife and the co-respondent to places of amusement..and then leaving her in the care of the co-respondent, was conduct conducing to the adultery. **1897** Browne & Powles *Law & Pract. Court for Divorce & Matrimonial Causes* (ed. 6) ii. 114 The following cases on the subject of conduct conducing to adultery have been decided. **1907** *Westm. Gaz.* 4 June 5/1 His lordship..referred to the charges of connivance and conduct conducive. *Ibid.*, The suggestion was that they (the jury) might find adultery which she said she never committed, and that in no sense had there been conduct conducing or connivance on the part of Mr. B. **1959** Jowitt *Dict. Eng. Law* I. 447/1 *Conduct conducing.* In suits for divorce or judicial separation on the ground of adultery or unsoundness of mind or desertion, the court is not bound to pronounce a decree if it finds that the petitioner's wilful neglect or misconduct has conduced to the offence alleged; the matter is purely discretionary (Matrimonial Causes Act, 1950, s. 4).

conduct, *sb.*¹ Add: **12. conduct-book** (examples); so **conduct-sheet.**
1856 Cozzens *Sparrowgr. Papers* xiii. 185 A conduct-book! There was G. for good boy, and R. for reading, and S. for spelling and so on. **1949** I. Deutscher *Stalin* i. 18 Similar entries appeared in the conduct-book more and more frequently. **1965** A. Nicol *Truly Married Woman* 71 They knew the matter would end there, with no protracted interviews, moral recrimination, and an entry in the conduct book. **1927** *Daily Express* 11 Oct. 3/4, I have known real bad lots suddenly reform, fellows whose conduct sheets ran into three editions. **1956** 'J. Wyndham' *Seeds of Time* 114 A few curt remarks on a conduct-sheet to dog Constable Walsh's future career.

conductance (kǫndɒ·ktăns). *Electr.* [f. Conduct *v.* + -ance.] (See quots.) Also *transf.*
1885 Heaviside in *Electrician* 4 Sept. 311/1 'Specific resistance' may well be called 'resistivity', and specific conductance 'conductivity', referring to the unit volume. Resistivity is the reciprocal of conductivity, and resistance of conductance. **1886** —— *Ibid.* 29 Jan. 228/1 'Resistivity' for specific resistance, and 'conductance' for what is sometimes called the conductibility of a wire, *i.e.*, not its conductivity (specific conductance), but the reciprocal of its resistance. **1892** —— *Electr. Papers* I. 399 Let a cable be constructed according to the following simple specification:—Its electrostatic capacity to vary in simple proportion to the distance from one end *O*. Its conductance to vary in simple proportion to the distance from *O*. *Note*, 'Conductance' is here substituted for 'conductivity'. It means the reciprocal of the resistance. **1893** T. O'C. Sloane *Stand. Electr. Dict.* 144 *Conductance*, the conducting power of a given mass of specified material of specified shape and connections...Conductance is an attribute of any specified conductor, and refers to its shape, length and other factors. Conductivity is an attribute of any specified material without direct reference to its shape or other factors. **1904** N. H. Schneider *Man. Electr. Plants* 1 The power of conducting is generally conductance, or conductivity. **1926** Hawk & Bergheim *Physiol. Chem.* (ed. 9) 38 The blood and bile of most mammals..possess considerably higher conductances than milk and saliva. **1941** S. R. Roget *Dict. Electr. Terms* (ed. 4) 68/1 *Conductance*, the reciprocal of Resistance or, in the case of alternating current, the energy component in the reciprocal of Impedance. **1966** C. F. Stevens *Neurophysiol.* ix. 117 It is customary to specify the ease with which a type of ion passes through the [nerve cell] membrane by the magnitude of the conductance for that ion.

conducted, *ppl. a.* Add: *conducted tour*: a tour led by a guide and having a set itinerary; also *transf.* Hence **conducted** *sb.*, a conducted tour; *collect.*, people making such a tour.
1907 *Daily Chron.* 19 Jan. 8/5 A charmingly illustrated programme of a conducted tour, lasting..fourteen days. *Ibid.*, At ten a.m., therefore, the 'conducted' will meet, never to part company again for fourteen days. **1924** 'L. Malet' *Dogs of Want* viii. 253 Such an inferior style of audience there. Nothing but Cook's tours and that description of personally conducteds. **1937** *Discovery* May 164/2 *Digestion and Health*..is a conducted tour of the digestive organs. *Ibid.* Nov. p. ci/1 Visits to the Austin Factory... Daily conducted tours at 10 a.m. and 2 p.m.

conductible, *a.* **2.** (Later example.)
1962 Corson & Lorrain *Introd. Electromagn. Fields* iii. 87 We refer to σ′ and ρ′ as either bound, polarization, or induced charge densities, as distinguished from free or conductible charges... Free charges are those which can move over macroscopic distances.

conductimetric (kǫndɒ:ktime·trik), *a. Physical Chem.* [f. Conducti(vity + Metric *a.*¹] Of or pertaining to the measurement of conductivity; *spec.* pertaining to volumetric analysis in which the end-point of a titration is determined by measurement of the progressive change in the electrical conductivity of a solution. Cf. *conductometric *a.*
1940 *Chambers's Techn. Dict.* 188/2 Conductimetric analysis. **1950** G. G. Blake (*title*) Conductimetric Analysis at Radio-Frequency. *Ibid.* p. vii, The author's new technique and methods of conductimetric analysis. **1954** *Electronic Engin.* July 316/2 The terminals..may be connected to the electrodes (of a standard conductivity cell or conductimetric tube).

conducting, *ppl. a.* Add: **c.** *Bot.*, in *conducting bundle, cell, sheath, strand, tissue.*
1873 F. H. Hooker tr. *Le Maout & Decaisne's Bot.* 70 A moist and loose parenchyma, named conducting tissue. **1875** Bennett & Dyer tr. *Sachs' Bot.* 499 The 'conducting-tissue', down which the pollen-tubes grow till they reach the cavity of the ovary. **1877** Bennett tr. *Thomé's Bot.* 48 In many Vascular Cryptogams, Gymnosperms and Monocotyledons, as well as in a few Dicotyledons, rows of vascular cells are found in places where, from the analogy of other plants, one would expect to find vessels, the partition-walls not having become absorbed. Such structures compose what is called a *conducting tissue*; and the separate cells are not called vascular, but *conducting cells*. **1900** B. D. Jackson *Gloss. Bot. Terms* 60/2 *Conducting Bundles*, strands of elongated cells in leaves and even the stems of Mosses, simulating a vascular bundle. *Ibid.*, *Conducting Sheath*, elongated parenchymatous cells in the inner cortex of the stem, continued into the leaves as an investiture of the vascular bundle. **1936** *Discovery* Jan. 5/2 A section of the woody conducting strand. **1959** *Chambers's Encycl.* IX. 321/1 The long-distance translocation of solutes about the plant is effected through specialized conducting tissues, the strands of xylem and of phloem.

conductivity. Add: **b.** *Physiol.* The property of nervous and muscular tissue by which it conducts an impulse.
1881 S. P. Thompson *Electr. & Mag.* iii. 187 Little is.. yet known of the conditions of conductivity of the matter of the nerves. **1903** *Lancet* 18 Apr. 1089 When a nerve is divided it will lose its motor conductivity only after the lapse of some time. **1965** Schadé & Ford *Basic Neurol.* ii. ii. 180 Nerve and muscle cells..possess two highly distinctive properties: excitability and conductivity.

2. *attrib.*
1876 Preece & Sivewright *Telegraphy* 267 The wire resistance or conductivity test is obtained by the differential galvanometer, or by Wheatstone's Bridge. **1909** *Cent. Dict.* Suppl., *Conductivity water*, water specially purified for use in measuring the electrical conductivity of solutions. **1924** S. R. Roget *Dict. Electr. Terms* 47/1 *Conductivity bridge*, a form of Wheatstone's Bridge used for the comparison of low resistances. **1937** *Discovery* Sept. 284/2 'Conductivity water'—so called, we believe without irony, because it is one of the most perfect electrically insulating media known. **1959** H. Barnes *Oceanography* iii. 121 Resistance thermometers and thermistors have been used for temperature measurement and conductivity cells for salinity. **1961** *New Scientist* 26 Oct. 237/2 Ultra-pure or conductivity water was, until a few years ago, a substance made in small quantities by multiple distillation..mainly for use in research laboratories.

conductometer (kǫndɒktǫ·mitəɹ). *Physics.* [f. Conduct *v.* + -ometer.] An instrument for measuring conductivity.
a **1884** Knight *Dict. Mech.* Suppl. 218/1 *Conductometer*, an instrument for illustrating the comparative power of different objects for conducting heat, electricity, etc. **1903** R. Appleyard in *Proc. Inst. Civil Engin.* CLIV. 355 It is necessary to express the readings of the conductometer in C.G.S. units. **1943** *Times* 2 Mar. 6/4 He [*sc.* Rollo Appleyard] invented the conductometer for the measurement of electrical conductivity.

conductometry (kǫndɒktǫ·metri). *Physical Chem.* [f. Conduct(ivity + -ometry.] The study or science of the measurement of conductivity; the process of such measurement. Hence **condu:ctome·tric** *a.*, = *conductimetric *a.*
1929 *Hackh's Chem. Dict.* 198/1 *Conductometric*, pertaining to conductivity measurements. *C. analysis*, quantitative analysis or titration in which the progressive change in conductivity is used to determine the end-point of the reaction. **1946** *Nature* 31 Aug. 288/1 Accounts of conductometry by T. Shedlovsky. **1960** T. Shedlovsky in A. Weissberger *Phys. Methods Org. Chem.* (ed. 3) IV. 3043 Conductometric analysis can..be used in many precipitation and complex formation reactions. **1965** Phillips & Williams *Inorg. Chem.* I. xii. 444 It is thus possible to carry out acid-base conductometric titrations.

conductor. Add: **7.** *U.S.* (Earlier and later examples.)
1832 *Amer. Railroad Jrnl.* I. 721 Seat for the conductor. **1944** *Reader's Digest* Mar. 17 Engineers, firemen, conductors, trainmen, switchmen.

12. c. (Earlier Amer. examples.)
1761 E. Kinnersley *Let.* 12 Mar. in B. Franklin *Exper. on Electr.* (ed. 4, 1769) 394 The lightning..had passed through the iron conductor which he had provided for the security of his house. **1764** in N. F. Moore *Hist. Sk. Columbia Coll.* (1846) 49 Ordered, that a conductor be fixed to the cupola of the college, as a security against lightning.

d. *conductor rail*, in electric traction, the metallic rail through which the current transmits motive power to the moving car.
1900 *Daily News* 21 May 10/3 The electric contact obtained by a conductor-rail on either of the outward sides of the track. **1903** *Westm. Gaz.* 30 Mar. 6/2 The new motor-cars on the Central London Railway are dependent on the current supplied to them by the conductor rail. **1940** *Chambers's Techn. Dict.* 189/1 *Conductor-rail ramp*, a sloping contact-surface at the beginning and end of a section of conductor-rail; it serves for leading the collector-shoe of the train smoothly on to and off the rail.

conductorial, *a.* Delete '*nonce-wd.*' and add later examples.
1947 A. Einstein *Music in Rom. Era* x. 113 Méhul..with whose *Joseph* he [*sc.* Weber] had begun his conductorial activities at Dresden. **1963** *Times* 22 Jan. 14/1 A mere display of conductorial antics.

conductory, *a.* (Example.)
1807 *Phil. Mag.* XXVIII. 297 The anomalies of the conductory faculty are so strongly evinced in galvanic electricity.

conductress. Add: **3.** A female bus- or tram-conductor (cf. Conductor 7).
1885 H. James *Little Tour in France* xxiv. 155 A kind of bastard tramcar..under the supervision, not of a conductor, but of a conductress. **1917** *Flight* 25 Jan. 88/2, I had occasion to board a 'bus, and noticed that the conductress had decorated herself up to the highest degree. **1943** *Times* 7 Sept. 6/7 The bus started without any signal from the conductress.

conductus (kǫndɒ·ktɒ̆s). Pl. **conducti.** [med.L.; see Condut.] A class of musical composition, monophonic or polyphonic, practised in the 12th and 13th centuries, and not normally based upon Gregorian chant: it was originally associated with procession in church, and its metrical Latin texts are devotional, moral, or political in content.
1801 T. Busby *Dict. Mus.* s.v., Though in every other kind of discant some known melody was chosen which governed the air originating from it, in the *Conductus* the discant and the harmony were both of them new, and produced together. **1880** [see Condut]. **1901** H. E. Wooldridge *Oxf. Hist. Mus.* I. vi. 248 In the pieces contained in the Florence MS. which can be identified as *conducti*, we may certainly observe two methods, one essentially simple and the other essentially elaborate, the admixture of which in one composition would indeed seem to constitute the classical form of this kind of music. **1940** G. Reese *Mus. Middle Ages* (1941) xi. 308 In the absence of a Gregorian tenor lies one of the main differences between the conductus on the one hand and the organa and motets on the other. The conductus is thus the first polyphonic type in which the composer was able to give entirely free rein to his imagination. **1954** *Grove's Dict. Mus.* (ed. 5) II. 405 The large majority of conducti have these interludes.

conduit, *sb.* Add: **1. b.** *Electr.* A tube or trough for receiving and protecting electric wires; a length or stretch of this. Also *attrib.*, esp. in connection with the **conduit system** (see quot. 1940).
1882 *U.S. Pat. 266,916* My invention consists, first, in making an electric conduit, comprising an external casing, internal conductor pipes, and supporting diaphragms, of conducting material, so that any electric currents induced in the said pipes will be conducted..directly to the ground. **1884** *Cassell's Fam. Mag.* Jan. 127/1 Conduits for holding electric wires laid along the streets. **1894** *Daily News* 2 June 5/4 At Buda-Pesth, where the conduit electrical system is in such successful operation. **1894** *Cassier's Mag.* Sept. 385/1 A trial of the conduit on a commercial basis at Washington. *Ibid.* 385/2 The open slot conduit with a continuous, bare trolley wire. *Ibid.* 386/2 The contact or working conductors could readily be placed in a slotted conduit, or trough. *Ibid.*, The road at Blackpool, England,—an open conduit road. *Ibid.* 387/1 The Love conduit system. **1896** *Daily News* 17 Dec. 5/2 The electric power is conveyed from the conduit rail to the car by means of a small peculiarly-shaped conductor. **1899** *Ibid.* 9 Jan. 3/6 New York will soon have 150 miles of conduit. **1903** *Daily Chron.* 18 Nov. 3/5 A conduit line from Vauxhall Bridge to the Clapham-road. **1908** *Installation News* II. 47/2 Three parallel lengths of ¾ in. Simplex conduit hung a few inches below the ceiling and seven feet apart. **1940** *Chambers's Techn. Dict.* 189/1 *Conduit box*, a box adapted for connexion to the metal conduit used in electric wiring schemes. *Ibid.*, *Conduit system*, (1) a system of wiring..in which the conductors are contained in a steel conduit; (2) a system of current collection used on some electric tramway systems. **1941** S. R. Roget *Dict. Electr. Terms* (ed. 4) 69/1 *Conduit Fittings*, accessories such as conduit boxes, bends, tees, couplers, etc., for joining lengths of conduit tube for wiring. **1955** *Oxf. Jun. Encycl.* XI. 131/2 With one method of wiring, separate stranded copper wires with VIR insulation are used, the wires being placed inside black enamelled steel pipes, called 'conduits'. The conduits are screwed together and joined to cast iron boxes containing the switches and connexions between the wires; the whole conduit system is then joined to earth.

condurangin (kǫndiuræ·ndʒin). *Chem.* [G. (G. Vulpius 1885, in *Archiv d. Pharm.* LXIV. 301), f. *condurango*, Cundurango + -in¹.] A glucoside or a mixture of glucosides found in cundurango bark.
1885 *Pharm. Jrnl.* 27 June 1067/1 Whether these products ..will have to be added to the new group under the name 'condurangin', has yet to be determined. **1909** *Jrnl. Chem. Soc.* XCVI. i. 40 Condurangin was isolated by extraction of the bark with alcohol. **1944** *Hackh's Chem. Dict.* (ed. 3) 220/2 *Condurangin*, a glucoside from condurango bark. *alpha-* $C_{20}H_{32}O_6$ = 368.3. A white powder, insoluble in water, m. 60. *beta-* $C_{18}H_{28}O_7$ = 356.3. An amorphous yellowish powder, soluble in alcohol, sparingly soluble in ether, chloroform, or water; used as stomachic and astringent. **1955** Osol & Farrar *Dispensatory of U.S.A.* (ed. 25) 1644 Condurangin has been reported to be a violent poison, causing convulsions followed by paralysis.

condylarth (kǫndilā·ɹþ). [f. *Condylarthra* (E. D. Cope 1881, in *Amer. Nat.* XV. 1018), f. Gr. κόνδυλος knuckle, Condyle + ἄρθρον joint.] A member of the order Condylarthra, extinct mammals of the Palæocene and Eocene periods, the ancestors of some recent ungulates. Also *attrib.* or as *adj.* So **condyla·rthran** *a.* and *sb.*, **condyla·rthrous** *a.*
1884 *Amer. Naturalist* XVIII. 793 Its Condylarthrous character. **1897** *Nat. Sci.* May 311 The difference in structure between the lemurs and condylarths. *Ibid.*, The view.. that the lemurs were descended from condylarthrous ancestors. **1927** Haldane & Huxley *Anim. Biol.* xi. 243 Eocene mammal of generalized type (Condylarthran). **1928** De Beer *Vertebr. Zoöl.* 456 Very early, a branch diverged from the Condylarthran stock and gave rise to the Amblypoda. **1957** *Encycl. Brit.* XXII. 700/1 The more typical condylarth types. *Ibid.* XXIII. 970/1 The Basal and Lower Eocene of Europe contain a few animals which are probably condylarthrans. **1965** *New Scientist* 11 Nov. 392/2 The fossil record shows that it embraced..some condylarths or knob-joints. **1968** A. S. Romer *Procession of Life* xvii. 277 Certain features [of the aardvark] suggest a remote relationship to the early ungulates of the condylarth group.

condylarthrosis (kǫ:ndilaɹþrōᵘ·sis). *Anat.* [f. Condyle + Arthrosis.] A condyloid articulation: see quots. So **co:ndylarthro·dial** *a.*, having the character of a condylarthrosis.
1848 Dunglison *Med. Lex.*, Condylarthrosis. **1889** A. Macalister *Hum. Anat.* 44 Condylarthrosis, when the convex articular head is ellipsoidal in outline, with a socket of corresponding shape, as in the metatarso-phalangeal joints, rotation is restricted or impossible, and the joint becomes practically biaxial. Most condylarthrodial joints are laterally double. **1907** T. Dwight et al. *Hum. Anat.* 113 Condylarthrosis, an egg-shaped joint permitting angular motions more freely on the long axis than on the short one. **1961** *Brit. Med. Dict.* 340/1 *Condylarthrosis*, a condyloid articulation, e.g. that of the knee.

Condy's fluid. [Name of Henry Bollmann *Condy*, 19th-c. English manufacturer of chemicals.] A strong solution of sodium manganate or permanganate, used as a disinfectant. Also (*colloq.*) **Condy.**
1857 J. Griffiths in H. B. Condy *Disinfection* (1862) 37, I have fairly tested the efficacy of 'Condy's Disinfecting Fluid' upon a quantity of putrid meat which was rendered perfectly sweet thereby. **1859** *Lancet* 1 Jan. 8/2 Condy's fluid was commenced on the fourth day, with immediate relief to the pain. **1866** *Chem. News* 17 Aug. 84 Put half an ounce of Condy's red fluid into every water receptacle that is made of wood. *Ibid.*, In the use of Condy's fluid for purifying water-butts, enough should be used. *Ibid.*, The addition of one or more drops of 'Condy'..will speedily remove that smell and taste. *Ibid.*, The microscopic animalcules.., if not destroyed by the Condy, would die of starvation. **1878** C. M. Tidy *Handbk. Mod. Chem.* 270 'Condy's green disinfecting fluid'..turns red (a permanganate being formed) on the addition of an acid. **1894** C. H. W. Donovan *With Wilson in Matabeleland* iv. 67 We..syringed his deep wounds with Condy's Fluid. **1912** J. W. Mellor *Mod. Inorg. Chem.* xxv. 483 A crude mixture of sodium manganate and permanganate is made by fusing sodium hydroxide with pyrolusite, and sold as a disinfectant under the name 'Condy's fluid'. **1951** A. Grollman *Pharmacol. & Therapeutics* xxv. 515 Condy's Fluid..cannot be employed to disinfect rooms.

cone, *sb.*¹ Add: **1. d.** *Physical Geogr.* A conical or fan-shaped alluvial deposit formed by a stream where its bed becomes less steep; *esp.* a relatively small, steep-sided deposit such as is formed at the mouth of a ravine. Cf. *fan *sb.*¹ 5 d.
1864, 1890 [see *fan *sb.*¹ 5 d]. **1945** C. A. Cotton *Geomorphology* xv. 198 Very steep fans are called alluvial cones .., and there is a transition through these from alluvial fans to talus slopes.

e. = *cornet *sb.*¹ 2 c. orig. *U.S.*
1920 *Outing* (U.S.) July–Aug. 246/2 Ray licked the ice cream from out his dripping cone. **1922** Joyce *Ulysses* 172 Tempting fruit. Ice cones. Cream. **1926** [see *cornet *sb.*¹ 2 c]. **1949** *Manch. Guardian Weekly* 18 Aug. 3/1 Americans ..have as little idea what the dollar crisis is about as a British child sucking an ice-cream cone at a seaside ventriloquist. **1967** A. J. Marshall in L. Deighton *London Dossier* 143 You can stand licking a double Marsala cone.

15. cone-anchor, a conical drag employed by vessels in rough weather; **cone-clutch,** a friction clutch with a conical contact surface; **cone drawing,** a method of drawing cotton (see Cone *sb.*¹ 8 b); **cone-in-cone** (examples); **cone-sheet** *Geol.* (see quots.).
1902 *Nature* 4 Sept. 447 M. Heureux dropped his cone-anchor and waited until a tug-boat..threw a rope to the car, by which the balloon was tugged easily. *a* **1877** Knight *Mech. Dict.* 578/2 The cone-clutch consists of a tapered cylindrical plug sliding on a fast feather in one shaft, [etc.]. **1908** *Westm. Gaz.* 5 Mar. 4/2 The drive from engine to gear-box is through a leather-faced, self-contained cone-clutch. **1930** *Engineering* 11 July 41/2 From the engine, the drive is taken through a cone clutch and spur gearing to a three-speed gear box. **1963** R. F. Webb *Motorists' Dict.* 63 *Cone clutch*, an obsolete type of clutch mechanism at one time popular on early vehicles. It consisted of a drum attached to the engine shaft in place of or part of the flywheel. The inside of the drum was tapered slightly to mate with a cone-shaped friction disc. **1884** W. S. B. McLaren *Spinning* (ed. 2) 136 In cone drawing..all these defects are avoided. **1957** *Encycl. Brit.* XXIII. 796/2 Cone drawing follows closely the English system..and is widely used in the United States...The bobbin is driven independently of the spindle and flyer. **1877** *Encycl. Brit.* VI. 45/2 Coal is perfectly amorphous, the nearest approach to anything like crystalline structure being a compound fibrous grouping resembling that of gypsum or arragonite, which occurs in some of the steam coals of S. Wales, and is locally known as 'cone in cone'. **1882** Geikie *Text-bk. Geol.* 313 That the more complex structure known as 'Cone in cone' may be due to the action of pressure upon concretions in the course of formation. **1883** Gresley *Gloss. Coal-m.* 61 Steam or anthracite coal exhibiting a peculiar fibrous structure passing into a singular toothed arrangement of the particles called cone-in-cone coal or crystallised coal. **1921** *Brit. Mus. Return* 156 Cone-in-cone structure from Barf Hill, Keswick. **1954** G. W. Himus *Dict. Geol.* 30 *Cone-in-cone structure*, a concretionary structure, found in marls, ironstones, coals, etc., consisting of the development of a succession of concentric cones, resulting from radial crystallization about a common axis. **1924** *Mem. Geol. Surv. Scotl.* xix. 221 Acid Centrally Inclined Sheets or Cone-Sheets..a cone-sheet complex in which the characteristic feature is an assemblage of sheets inclined rather steeply towards a common centre. **1965** A. Holmes *Princ. Geol.* (ed. 2) xi. 261 Intrusions in the form of concentric arcs or rings are of two distinct types... One type consists of cone-sheets, which have the form of parts of inverted cones dipping inwards towards a common focus.

cone, *v.*¹ Add: **3.** *trans.* To catch or pick up (an aircraft) at the apex of a cone formed by searchlight beams (or tracer shells, etc.).
1943 *Times Weekly* 24 Nov. 6/3 Searchlights were also effectively blocked by the cloud and, unable to cone a particular aircraft in the beams, the gunners could only fire a barrage and hope for results. **1944** H. Hawton *Night Bombing* vii. 104 The number of aircraft that can be coned, even in the target area, is comparatively small if the raid is both heavy and concentrated. **1944** *Times* 22 June 2/1 Red and white tracer shot up in streams, coning the intruder, but it flew straight through it unharmed.

Conestoga (kǫnĕstōᵘ·gă). *U.S.* Also 8 **Canastoe.** [The name of a town in Pennsylvania and of a local Indian tribe, prob. f. some Iroquoian word.] **1.** An Iroquoian tribe of North American Indians formerly inhabiting parts of Pennsylvania and Maryland; a member of this tribe. Also *attrib.* or as *adj.*
1699 in *Doc. Hist. N.Y.* (1854) IV. 579 The English in Virginia had peace with the Canastogue Indians..and.. heard you were coming against the Canastogues. **1764** *Ibid.* (1856) VII. 602 The late Murder committed on the Friendly Conestoga Indians. **1891** J. W. Powell in *7th Ann. Rep. Bureau Amer. Ethnol.* 78 The Conestoga..occupied the country about the Lower Susquehanna. **1959** J. Witthoft in Witthoft & Kinsey *Susquehannock Misc.* 19 A remnant settled near present Lancaster, Pennsylvania, becoming known as the Conestogas.

2. *Conestoga wagon*, a large travelling-wagon formerly in use. Also *ellipt.*
[**1750** *Penna. Gazette* (Philad.) 26 Feb., Just imported and to be sold..by Thomas White, at his house in Market Street, almost opposite the sign of the Conestoga Waggon.] **1781** R. Beatty *Let.* Aug. in *Penna. Mag. Hist. & Bibliogr.* (1920) XLIV. 222, I cannot say I will follow your Advice respecting marrying a Dutch Girl, with a good Plantation & a Conostoga Waggon. [**1788** W. Gordon *Hist. Amer. Revol.* IV. x. 306 A small dirty room in the Philadelphia tavern called the Canastoe-waggon.] **1808** *Balance* (Hudson, N.Y.) 16 Feb. 28 (Th.), The throng of Pittsburg and Conestoga waggons. **1844** G. W. Kendall *Santa Fé Exped.* II. iii. 59 Her travelling carriage was nothing more or less than a huge Pennsylvania or Conestoga wagon, drawn by four yokes of oxen. **1901** W. S. Churchill *Crisis* 356 Open waggons and conestogas, carryalls and buggies. **1931** J. T. Adams *Epic of Amer.* vi. 148 Thousands..crossed the mountains on foot or in Conestoga wagons. **1961** *Guardian* 6 Nov. 3/1 In the United States those predecessors of the 'prairie schooners', the Conestogas..had postillions.

3. Used *attrib.* to designate a heavy breed of horses. Also *ellipt.*, a horse of this breed.
1824 W. H. Keating *Exped. St. Peter's River* I. i. 31 There are several appellations by which the different breeds of this useful animal are distinguished in Pennsylvania, such as the Conestoga. **1844** *Congress. Globe* App. 748/2 Mr. Buchanan..could come upon the turf successfully with his celebrated breed of Conestoga dray horses. **1857** 'F. Forester' *Horse & Horsemanship* II. 59, I am inclined to suspect, the Conestoga-horse is descended from a mixture of the Flemish cart-horse with the English breed. **1860** R. Jennings *Horse* 61 The vast, white-topped wagons, drawn by superb teams of the stately Conestogas. **1875** *Congress. Rec.* 22 Mar. 132/1 The Conestogas respond promptly to the whip. **1942** C. Weygandt *Plenty of Pennsylvania* 83 A tight little Morgan..or a team of great Conestogas.

4. A large coarse boot or shoe. (Cf. Stogy.)
1893 *Dialect Notes* I. 229 *Conostogas*, brogans. **1903** in Clapin *Dict. Amer.* s.v.

confab, *v.* (Later examples.)
c **1862** G. Meredith *Lett.* (1970) I. 153 Arthur is anxious to confab with Reggy. **1898** H. S. Canfield *Maid of Frontier* i. 19 What, Charlie, was you an' Lem confabbin about? **1948** *Variety* 25 Aug. 5/3 He will confab with them again.

confabulate, *v.* Add: **2.** *Psychiatry.* To fabricate imaginary experiences as compensation for loss of memory. Hence **confa·bulating** *ppl. a.*
1924 A. A. Brill tr. *Bleuler's Textbk. Psychiatry* ii. 107 Many paretics..spontaneously confabulate in a very profuse manner. **1941** *Brit. Jrnl. Psychol.* Jan. 232 [The patient's] memory was grossly impaired and she confabulated freely. **1963** Hoenig & Hamilton tr. *Jaspers's Gen. Psychopath.* iv. xii. 592 The conspicuous ease and facility with which these patients confabulate in place of their real memories. **1965** Rosen & Gregory *Abnormal Psychol.* iii. 37/1 If a confabulating hospital patient is asked to recount the events of his day he may state with complete conviction that he has been in a distant city and describe his adventures there in detail.

confabulation. Add: **2.** *Psychiatry.* The action of the verb *confabulate 2.
1924 A. A. Brill tr. *Bleuler's Textbk. Psychiatry* ii. 106 Memory hallucinations which endow a phantasy with reality..must be taken in the strict sense as confabulations. *Ibid.* 107 The most vivid confabulations are seen in many alcoholic Korsakoff cases. **1942** Lemkau & Kronenberg tr. *Rorschach's Psychodiagnostics* iv. 104 Features of confabulation become predominant and overcome the consciousness of the unreality of the product of imagination. **1961** *Brit. Med. Dict.* 341/1 *Confabulation*, the recital of experiences that have no foundation in fact and the glib untruthful answering of questions usually in compensation for a gap in memory.

confection, *v.* **3.** (Earlier example.)
1839 F. A. Kemble *Resid. in Georgia* (1863) 134, I have ..spent considerable time in what the French call 'confectioning' baby bundles.

confectioner. Add: **2. b.** *confectioner's custard*: a sweet creamy confection used as a filling for cakes, etc.
1906 Mrs. Beeton *Househ. Managem.* xxxi. 910 *Confectioner's custard*..yolks of egg..milk..sugar..potato flour.. gelatine, flavouring essence. **1951** *Good Housek. Home Encycl.* 380/1 Sponge sandwiches..are often split open and spread with..confectioner's custard.

confectionery. Add: **1. a.** (Later examples.)
1922 Joyce *Ulysses* 644 He having previously spotted on the printed pricelist..coffee 2d., confectionery do, and honestly well worth twice the money. **1959** *Chambers's Encycl.* III. 837/2 The main ingredients in modern confectionery consist of sugar, brown and white, treacle [etc.].

3. (U.S. examples.)
1803 E. S. Bowne *Girl's Life* (1888) 156, I never go by a toy shop, or confectionery without longing to have them [*sc.* children] here. **1887** *Courier-Jrnl.* (Louisville, Ky.) 21 Jan. 6/4 Mrs. Jett started a confectionery at Lexington and afterward..she established millineries.

4. b. *attrib.* Of a building or style of architecture: resembling an elaborate piece of confectionery. (Cf. quot. 1861 *s.v.* CONFECTIONARY *a.* and quot. 1879 s.v. *wedding-cake* (WEDDING *vbl. sb.*).)

1897 *Daily News* 15 Sept. 8/3 This Muscovite style is at first disappointing...We are apt to be reminded by it of the wedding-cake and master works of confectionery architecture. **1910** T. E. LAWRENCE *Let.* 16 Dec. (1938) 92 The Turks, when they get a glorious broad wall-space..will insist on fretting the wall into little confectionery-arcades.

confederal, *a.* (Earlier and later U.S. examples.)

1782 *Independent Ledger* 4 Feb. 3/2 Washington and the confederal forces. *Ibid.*, The confederal armies. **1955** C. J. FRIEDRICH in A. W. Macmahon *Federalism* xxvi. 510 When the foreign ministers of the six nations..decided to establish..a Political Community..they provided..that this..should be..of 'an ultimate federal or a confederal structure'.

conferee. 1. For *U.S.* read orig. *U.S.* (Add earlier and later examples.)

1779 in W. B. Reed *Life Jos. Reed* (1847) II. ii. 52 The conferees of Congress gave this committee very ample assurances of the disposition of Congress to preserve the most perfect harmony. **1797** J. BOUCHER *Causes Amer. Revol.* v. 238 By some logic of their own, their conferees have found out [etc.]. **1815** in Fearon *Sk.* (1818) 144 We enclose you the Democratic ticket which is recommended by the delegates and conferees fairly chosen, after public notice. **1894** *Daily News* 28 July 4/6 The United States Senate yesterday returned the Tariff Bill to the Conference Committee without giving the Senate conferees any instructions. **1959** *Economist* 30 May 842/2 Nor will the conferees be in any hurry to produce their compromise.

conference. Add: **4. a.** *in conference*, engaged (in a conference), busy. orig. *U.S.*

1926 WODEHOUSE *Heart of Goof* iii. 96 'We can ring..up on the 'phone and ask.' 'And be told that he is in conference.' **1927** *Amer. Speech* II. 214/1 *To be in conference.* The English 'he's engaged' is much less imposing. **1954** I. MURDOCH *Under the Net* xv. 205, I told H.K. that I was in conference and that if he would contact my secretary..a meeting might be arranged.

d. In modern legal practice, a meeting for professional advice at which only one counsel is present: distinguished from *consultation*.

1877 SUMMERHAYS & TOOGOOD *Precedents Bills of Costs* 67 If one Counsel only is employed, the conference fee to him and clerk is £1. 6s. Attending conference (or consultation) with Counsel 0. 13. 4. **1883** WHARTON *Law-Lex.* (ed. 7) 179/2 *Conference*, a meeting between a counsel and solicitor to advise on the cause of their client. **1905** *Ann. Practice 1906* II. 211 No fees are allowed for conferences in addition to the counsel's fees for drawing..any pleadings,..deeds, or other proceedings.

e. A trade association or combination, esp. of shipping companies. Also *attrib.*

1894 *Westm. Gaz.* 11 June 6/1 The combinations called 'conferences' bind merchants for long periods forward. **1909** *Ibid.* 3 June 2/2 The evil of the 'conference' arrangement is that the shipper is absolutely in the hands of the 'ring'. **1930** *Economist* 5 Apr. 784/1 Revenue, outside the commodities controlled by 'conference' rates, must have been affected by last year's drop in 'free' freights. **1965** *Ibid.* 2 Jan. 34/1 A conference is an association of shipping lines plying a certain route; it fixes sailing schedules and rates for its members. **1970** *Financial Times* 13 Apr. 2/7 Member lines of the U.K.-Continent to India-Pakistan ship conferences announce that freight rates in the eastbound trade will be increased by 15 per cent.

9. *conference-table.*

1928 'S. S. VAN DINE' *Greene Murder Case* xvii. 192 A small conference-table in one of the Stuyvesant Club's private rooms. **1938** AUDEN & ISHERWOOD *On Frontier* II. ii. 78 The politicians hunt for a formula under the conference-table. **1945** R. KNOX *God & Atom* ix. 129 It suited the book of the Power which won the day at the conference-table.

conferencier. Delete '*nonce-wd.*' and add: Also ‖**conférencier.** [F.] **1.** A lecturer, public speaker.

1884 *Christian World* (N.Y.) May 167 Aid was sent..to continue..the support of the conférenciers, Messrs. Réveilland and Hirsch. **1885** [in Dict.]. **1928** *Weekly Dispatch* 24 June 17 A brilliant French conférancier (*sic*) was wasting his time..in telling them about Verlaine, and Rimbaud, and Mallarmé.

2. A (leading) member of a conference.

1926 *19th Cent.* Mar. 378 The *conférenciers* aimed at nothing more. But the conferences aroused some alarm on either side. **1937** J. R. FIRTH *Tongues of Men* xi. 138 The League [of Nations], if it cannot secure an international language, may soon have to issue a vocabulary for conferenciers. **1940** *Theology* XLI. 152 Quoting the observation of one of the conferenciers. **1960** *Times* 15 Dec. 11/2 The conferenciers [at the Labour Party conference] are in open revolt against the parliamentary leadership.

3. An entertainer or compère in a revue.

1927 *Observer* 4 Dec. 19/5 The cabaret, with its 'conférencier' babbling in broken Dutch, or English, will find a strong competitor hopelessly weakened. **1963** *Listener* 17 Oct. 601/1 Bonn and Mainz, where the cabarets were small, unassuming places..usually one actor or *conférencier* carried the main burden.

confessio (kǫnfe·sio). [med.L., f. late L. 'burial-place of martyrs': see CONFESSION.] = CONFESSION 8.

1830–38 BRITTON *Dict. Archit.* (1838) 96/1 *Confessio*, a term applied by early ecclesiastical writers to the sepulchres of martyrs and confessors. *Ibid.*, Confessio has been used to denote a crypt, beneath the great altar of a church in which the relics and bodies of saints were buried. **1875** *Encycl. Brit.* III. 415/1 Beneath the altar was the *confessio*, a subterranean chapel, containing the body of the patron saint, and relics of other holy persons. **1959** E. A. FISHER *Anglo-Saxon Archit.* I. 61 These confessios are different in design and purpose from the later larger crypts underlying the whole of the chancel.

confession. Add: **9. confession album, book,** a book of questions to be answered on personal likes and dislikes; also a book in which a visitor records a favourite poem, etc.; **confession box** = *confessional-box*; **confession magazine,** a magazine that purports to contain people's true confessions, life-stories, etc.

1906 *Daily Chron.* 27 Sept. 4/7 'If not yourself, who would you rather be?' was a favourite question of the confession album of the seventies. **1922** JOYCE *Ulysses* 357 She got a keepsake from Bertha Supple of that lovely confession album with the coralpink cover to write her thoughts in. *Ibid.* 80 Old fellow asleep near that confession box. **1931** F. L. ALLEN *Only Yesterday* v. 100 Sex magazines, confession magazines, and lurid motion pictures. **1932** S. CHASE *Mexico* i. 17 Nobody [in Tepoztlan] sends copy to confession magazines. **1958** S. ELLIN *Eighth Circle* (1959) II. ii. 36 Miss Whiteside had..a passion for confession magazines.

confessional, *a.* Add: **2. b.** Denominational; holding or according with a certain system of dogmas or beliefs.

1907 *Daily Chron.* 10 Jan. 6/4 The erection and endowment by the State of confessional schools. **1920** *Q. Rev.* July 172 It is treated strictly as a working hypothesis of science and not as a dogma of 'confessional' validity. **1950** *Internat. Affairs* Jan. 40 The Catholics..supported confessional schools of the individual Churches. **1957** *Economist* 30 Nov. 768/2 In spite of the early links of the Labour party with Methodism and of 'Tory Democracy' with Anglican social reform, British political parties are not confessional.

c. *Confessional Church*: see quot. 1957.

1938 A. S. DUNCAN-JONES *Relig. Freedom in Germany* iv. 82 The Confessional Church foreshadowed at Ulm came into definite existence at Barmen. **1939** N. MICKLEM *Nat. Socialism & Christianity* 22 The right wing [of German Protestantism] are those who stand in unswerving loyalty to the old Confessions of the Church, and who have been profoundly influenced by the teaching of Dr. Karl Barth. They are sometimes called 'the Confessional Church'. **1957** *Oxf. Dict. Chr. Ch.* 325/2 'Confessional Church' (*Bekenntnis-Kirche*). The group of German Evangelical Christians which most actively opposed the 'German-Christian' Church Movement sponsored by the Nazis between 1933 and 1945 and which claimed..to stand fast by the Augsburg and other Reformation Confessions.

confessor. Add: In sense 3 often pronounced (kǫ·nfesǫɹ) in the R.C. Church.

confetti (kǫnfe·ti), *sb. pl.* [Italian *confetti*, pl. of *confetto* COMFIT.] Bon-bons, or plaster or paper imitations of these, thrown during carnival in Italy; in England, *esp.* little discs, etc., of coloured paper thrown at the bride and bridegroom at weddings.

1815 JOHN MAYNE *Jrnl.* 5 Feb. (1909) 275 A principal part of the amusement of a masking day consists in throwing *confetti*; these are little balls, the size of a small marble, made of some soft white plaister that makes a mark wherever it strikes. **1846** DICKENS *Pict. Italy* 178 The spectators..would empty down great bags of confetti, that descended like a cloud, and..made them as white as millers. **1883** BRINSLEY-RICHARDS *7 Yrs. at Eton* i. 5 The *confetti* which are flung from the balconies of Roman houses at Carnival time. **1895** *Daily News* 27 Feb. 5/4 People young and old carry bags full of these confetti and fling them by handfuls in each other's faces and stuff them down their necks. *Ibid.* 26 Apr. 3/2 Visitors to the Riviera..were assured..that the confetti to be used at Eastbourne would be the small parti-coloured paper discs which are known in France, and which are taking the place of the time-honoured rice at English wedding festivities. **1897** *Ibid.* 3 Mar. 7/3 A real shower of rain put a stop to confetti-flinging. **1906** *Westm. Gaz.* 7 Nov. 7/1 A wild carnival of confetti-throwing and other forms of horseplay. **1918** E. SITWELL *Clown's Houses* 26 Like coloured dust Or feverish confetti. **1948** *Oxf. Jun. Encycl.* I. 298/2 Our custom of throwing confetti at weddings is a recent change from the older practice of casting rice.

b. *transf.* and *fig.*

1924 R. CAMPBELL *Flaming Terrapin* ii. 23 The trees..sprayed their flustered birds into the skies, That plumed confetti, soaring far and frail. **1940** C. DAY LEWIS *Poems in Wartime* 7 Litter of rubble, stale Confetti sprinkle of blood. **1941** *New Statesman* 30 Aug. 218/3 *Confetti*, ammunition.

confidante. Add: **2.** Also **-ente.** A name given by the English designer George Hepplewhite (d. 1786) to a species of settee; also in extended use (see quots. 1925, 1948).

1794 HEPPLEWHITE *Cabinet-Maker & Upholst. Guide* (ed. 3) Title-p., A great variety of patterns for Chairs Stools Sofas Confidante Duchesse Side Boards [etc.]. **1925** PENDEREL-BRODHURST & LAYTON *Gloss. Eng. Furniture*, *Confidente*, an upholstered settee with somewhat triangular seats beyond the arms at each end. **1948** S. GIEDION *Mechanization takes Command* v. 374 Two chairs coupled on an s-plan were named a *confidant*. *Confidantes* were also made of three chairs, joined like the spokes of a wheel.

confide, *v.* Add: **4. b.** intr. *to confide in*: to take (a person) into one's confidence, talk confidentially to.

1888 MRS. H. WARD *R. Elsmere* i, Mrs. Leyburn wanted to confide in her about a new cap. **1904** L. T. MEADE *Love Triumphant* II. ix, I must confide in you, but you seem..so terribly restless that you have not patience to hear me out.

confidence, *sb.* Add: **10.** orig. *U.S.* (Earlier and later examples.)

1849 *New Orleans Picayune* 21 July 1/4 'Well, then,' continues the 'confidence man', 'just lend me your watch till to-morrow.' **1856** *Spirit of Age* (Sacramento, Calif.) 14 Mar. 4/1 G. W. Meyler's now about town, playing the confidence game and making grand attempts at swindling. **1866** E. A. POLLARD *Southern Hist. War* II. xxv. 477 President Davis..was surrounded by adventurers and 'confidence' men. **1884** *Spectator* 9 Feb. 182/2 We know of no social phase equal in perplexity to the continual success of the Confidence Trick. **1886** *Century Mag.* Feb. 512/2 [They are] Confidence sharps, young feller. **1887** W. S. GILBERT *Ruddigore* ii. 37 There's confidence tricking, bad coin, pocket-picking, And several other disgraces. **1889** *Harper's Mag.* Mar. 514/1 Bertha Heymann, 'queen of the confidence women'. **1909** *Daily Chron.* 13 Nov. 4/4 Confidence-tricksters would rather meet a fly-flat than the most learned of Oxford dons; and that is also why the smart Yankee is their most common victim. **1911** N.Y. *Even. Post* 12 Sept. 1 S. A. Potter was arrested to-day on a charge of operating a confidence game. **1933** C. DAY LEWIS *Magnetic Mountain* 47 The Insurance Agent, the Vicar, Hard Cheese the Confidence-Tricker. **1951** E. E. EVANS-PRITCHARD *Social Anthropol.* vi. 122 After giving advice to travellers..to avoid gambling and confidence tricksters. **1955** *Sci. Amer.* Apr. 102/1 How you would behave toward two groups, one of whom had been told beforehand that you were outgoing, friendly and warm, while the other had been informed that you were surly, hostile and suspicious. Confidence men are well aware of this feedback effect of the milieu. **1959** J. BRAINE *Vodi* v. 75 For a second he had an intimation that already on its way was one level likely to make him permanently happy, then decided not to be taken in by the confidence trick.

11. Used *attrib.*, esp. in *Statistics*, as **confidence coefficient** or **level**, the particular probability used in defining a confidence interval, representing the likelihood that the interval will contain the parameter; **confidence interval,** a range of values so defined that there is a specified probability that the value of a parameter of a population lies within it; **confidence limit,** either of the two extreme values of a confidence interval.

1934 J. NEYMAN in *Jrnl. R. Statistical Soc.* XCVII. 562 The form of this solution consists in determining certain intervals, which I propose to call the confidence intervals.., in which we may assume are contained the values of the estimated characters of the population, the probability of an error in a statement of this sort being equal to or less than 1–ε, where ε is any number 0<ε<1, chosen in advance. The number ε I call the confidence coefficient. **1939** A. E. TRELOAR *Elem. Statistical Reasoning* x. 133 Since ±3σ in a normal distribution defines a central probability value of 99·973 per cent, the points $\bar{x}$ ± 3σ may be referred to as the 99·973 per cent confidence limits. **1958** *Brit. Med. Bull.* XXIV. 234/1 The proportion of cases misdiagnosed in this way gives approximately unbiased estimates of the required probabilities..., and confidence intervals can be set.

12. *Comb.*, as *confidence-inspiring* adj.

1891 W. JAMES *Let.* 19 Aug. in R. B. Perry *Th. & Char. of W.J.* (1935) II. 419 Our children grow lovelier every year and more confidence-inspiring. **1948** E. G. M. SUNDKLER *Bantu Prophets S. Afr.* v. 128 It is instructive to study the yearly balance sheets of this Church..arranged in a confidence-inspiring manner.

confidence (kǫ·nfidĕns), *v.* *U.S. slang.* [f. the sb.] *trans.* To swindle by means of a confidence trick.

1875 *Chicago Tribune* 1 Oct. 4 In a back room of some large building..they are 'confidenced' of what money they may have about them. **1888** *Missouri Republ.* 15 Feb. (Farmer), Detectives..arrested Lawrence Stanley..on a charge of confidencing Henry Mueller.

configuration. Add: **5.** *Organic Chem.* The structure of compounds, esp. with reference to the spatial relations of atoms in molecules.

1894 G. M'GOWAN tr. *Bernthsen's Org. Chem.* (ed. 2) 21 It is the development of the above assumptions with respect to the special configuration of the carbon compounds which has led to a conception of the cause of such cases of fine isomerism. This latter is to be sought for in the relative special arrangement of the individual atoms within the molecule, i.e. in their 'configuration'. **1907** J. B. COHEN *Org. Chem.* 110 The methods..used for determining the configuration of the geometrical isomers. *Ibid.* 148 The behaviour of carbanilide compounds of the aldoximes can also be employed for determining configuration. **1950** *Science News* XV. 121 Part of the surface of the red cell bearing a configuration which is responsible for the specificity of the rhesus antigen. **1970** *Physics Bull.* Sept. 414/1 The energies and configurations which the monomer molecules can take up when forming the polymer.

6. *Psychol.* = *GESTALT. Also *attrib.*

1925 H. HELSON in *Amer. Jrnl. Psychol.* July 342 Among the various schools of psychology there has arisen rather suddenly in Germany a new school with a definite set of assumptions and theories of its own which has come to be known as *Gestalttheorie* or *configurationism*...As a translation of the word *Gestalt* I have used the word configuration, proposed by Professor E. B. Titchener. *Ibid.* 348 The decisive factors within any given configuration which are responsible for the new properties of the whole and for the modifications of the parts within the structure are varied and elusive. **1932** *Brit. Jrnl. Psychol.* XXIII. 29 The subject is required to receive two configurations in different ways. **1936** R. LINTON *Study of Man* xxiii. 404 The function of a trait complex is the sum total of its contribution toward the perpetuation of the social-cultural configuration. **1951** G. HUMPHREY *Thinking* vi. 175 The tendency of an incomplete configuration to complete itself, the tendency, that is, towards closure. **1949** E. A. NIDA *Morphology* (ed. 2) vi. 162 *Applesauce* as an expression of rejection or contempt..is a configuration, the meaning of which should be described in terms of the entire unit. **1963** HORNIG & HAMILTON tr. *Jaspers's Gen. Psychopath.* I. ii. 164 We do not perceive sensations but all our perceptions, images and thought-contents appear to us as configurations.

configurational (kǫnfigiurēi·ʃənăl), *a.* [f. CONFIGURATION + -AL.] Of or relating to configuration; configurative. Hence **configura·tionally** *adv.*

1925 H. HELSON in *Amer. Jrnl. Psychol.* July 343, I seek to portray the main points of departure in configurational theory. **1929** *Jrnl. Chem. Soc.* I. 345 The authors have effected the epimerisation of several methylated lactones which are (configurationally) inter-related. **1931** *Brit. Jrnl. Psychol.* XXI. 112 Gestalt psychologists might..emphasize the importance of the configurational properties of the different targets. **1937** R. H. LOWIE *Hist. Ethnol. Theory* xiv. 278 The Eskimo..would form an ideal subject for configurational treatment. **1953** N. TINBERGEN *Herring Gull's World* xix. 166 Such experiences demonstrate the tendency to perceive the whole rather than its parts. Configurational perception makes you overlook details which may be quite obvious when an analytic effort has made you aware of them. **1955** R. JAKOBSON in Sapota & Bastian *Psycholinguistics* (1961) 470/2 Again, in Czech, the main stress..is a configurational feature. **1960** *Times Lit. Suppl.* 6 May 282/1 Purely formal criticism..would be another name for 'configurational' criticism. **1969** *Nature* 16 Aug. 714/1 An increase in the configurational entropy of the specimen as a result of molecular reorientation.

configurationism (kǫnfigiurēi·ʃəniz'm). [f. as prec. + -ISM.] Any theory that lays stress on configurations, esp. Gestalt psychology. Hence **configura·tionist** *a.* and *sb.*

1925 [see *CONFIGURATION 6]. **1925** H. HELSON in *Amer. Jrnl. Psychol.* July 343 After attacking certain older psychological concepts, the configurationists then proceed to the erection of the structure which they wish themselves to substitute in place of the old. **1927** *Psychol. Abstr.* I. 4 The work of the configurationists..is designed to illustrate and support an all-embracing theory. **1938** *Brit. Jrnl. Psychol.* XXVII. 208 The configurationism is represented by Wertheimer, Köhler, Koffka and Kruger. **1936** *Times Lit. Suppl.* 28 Mar. 260/2 The 'configurationist' view that a given custom in the eyes of those concerned is no casual association of unrelated elements. **1960** *Ibid.* 6 May 282/1 A theory which Mr. Lerner, borrowing the phrase from Professor W. B. Gallie, calls 'configurationism', assimilates the mode of being of poems to the mode of being of statues and paintings.

confirmability (kǫnfɜ̄·ɹmăbi·lĭti). *Philos.* [f. CONFIRMABLE *a.* + -ITY.] The quality or condition of being confirmable.

1936 A. J. AYER *Lang., Truth & Logic* vii. 185 The existence of a particular visual or tactual sense-datum is prima facie evidence of the existence of a material thing such that this sense-datum belongs to it...This principle we may call the Principle of Confirmability; for unless it were true, no confirmation of a perceptual act by other perceptual acts would ever be possible. **1945** *Mind* LIV. 8 Carnap has constructed definitions of testability and confirmability which avoid reference to the concept of confirming and of disconfirming evidence.

confiture. (Earlier and additional examples.)

1802 C. WILMOT *Irish Peer* (1920) 67 Patisserie, omelette, confitures &c. succeed in slow rotation. **1824** BYRON *Juan* xv. lxviii, But even sans 'confitures', it no less true is, There's pretty picking in those 'petits puits'. **1853** C. BRONTË *Villette* III. xxv. 121 M. Emanuel..would have given a large order for 'jambon' and 'confitures'.

conflab (kǫ·nflæb). *colloq.* (chiefly *U.S.*). = CONFAB *sb.*

1873 *Winfield* (Kansas) *Courier* 7 Aug. 3/1 'Conflabs' lively among the lawyers. **1948** H. CRANE *Let.* c 25 Dec. (1965) 332 After a day or two in conflab with some of these natives one does tend to lose one's 'middle-western accent'. **1949** *Amer. Speech* XVII. 283/2 To attend..conflabs with other schools.

conflagrated, *ppl. a.* (Earlier U.S. example.)

1814 J. C. CALHOUN *Wks.* (1856) II. 102 Rebellion, civil war, prostrated liberty, and conflagrated towns.

conflate, *v.* Add: **3.** Also used *actively.*

1927 A. H. MCNEILE *Introd. N.T.* 61 The custom of the former [sc. Matthew] was to conflate the language of his sources when they overlapped.

conflict, *sb.* Add: **2. d.** *Psychol.* The opposition, in an individual, of incompatible wishes or needs of approximately equal strength; also, the distressing emotional state resulting from such opposition. Also *attrib.*

1859 A. BAIN *Emotions & Will* II. vi. 442, I shall treat first of the conflict of a voluntary stimulus with the spontaneous impulses considered...as growing out of the purely physical conditions of the nervous and muscular systems. *Ibid.* 445 Where two pleasures or two pains, or one of each, solicit the voluntary executive in opposite ways. The instances of this conflict may be as numerous as the various concurrences of the human feelings. **1887** J. SULLY *Outl. Psychol.* (ed. 3) xi. 473 There arises an effect of mutual conflict, accompanied by a painful feeling of jar or discord. **1896** G. F. STOUT *Anal. Psychol.* I. II. iv. 281 The conception of conflict..comprehends all kinds of intellectual hesitation, and it is the psychological counterpart of logical contradiction. **1935** ADAMS & ZENER tr. *Lewin's Dynamic Theory* iii. 89 Conflict situations of this type usually develop rather quickly. **1942** K. HORNEY *Self-Analysis* iii. 94 The person may have gained a deep insight into the component parts of a conflict.

4. *attrib.* and *Comb.*, as *conflict research, resolution, study.*

1966 *Listener* 18 Nov. 787/1 Michael Nicholson is a Fellow in 'conflict research' at the University of Lancaster. **1957** (*title*) Journal of conflict resolution. **1968** L. NADER in D. L. Sills *Internat. Encycl. Social Sci.* III. 241/1 The style of conflict resolution derives from a society's structural principles of human association. **1965** *Listener* 11 Nov. 787/1 'Conflict studies' is a new branch of academic investigation which attempts, among other things, to throw light on the causes of war. **1970** *Times* 24 Aug. 2/5 (Advt.), The Irish tangle is unravelled by Iain Hamilton in Conflict Studies No. 6.

conflictful (kǫnfli·ktfŭl), *a. Psychol.* [f. CONFLICT *sb.* + -FUL.] Involving conflict; conflicting.

1942 *Psychiatry* V. 342/2 The meanings of the perceptive field become confused or the motivations conflictful. **1943** J. H. MASSERMAN *Behav. & Neurosis* i. 21 Such behavior could be termed adaptive but..by diminishing the conflictful motivations. **1950** *Brit. Jrnl. Psychol.* Dec. 99 Neurotic and psychotic subjects show..less tolerance of conflictful situations. **1966** L. PINCUS *Marriage* i. 13 The inner psychological forces which make for the conflictful will be some of the more fruitful, aspects of this close relationship.

conflictless (kǫ·nfliktlĕs), *a.* [f. CONFLICT *sb.* + -LESS.] Free from conflict.

1951 G. W. ALLPORT in Parsons & Shils *Toward Gen. Theory of Action* iv. i. 381 Few personalities are completely integrated and conflictless.

confluently, *adv.* (Example.)

1909 W. JAMES *Pluralistic Universe* vii. 290 May not you and I be confluent in a higher consciousness, and confluently active there, tho we now know it not?

confocal, *a.* Add: **B.** *sb.* A figure having the same focus as another.

1903 *Rep. Brit. Assoc. Advancem. Sci. 1902* 506 A theorem which of course includes the corresponding proposition for confocals in plano. **1961** C. E. T. BAKER *Dict. Math.* 79 If two confocals intersect, one must be an ellipse and the other a hyperbola.

conformal, *a.* Delete †*Obs. rare* and add further example.

1933 A. N. WHITEHEAD *Adv. of Ideas* xi. 185 The feeling as enjoyed by the past occasion is present in the new occasion as datum felt, with a subjective form conformal to that of the original feeling.

2. *Math.* [ad. G. *conform* (Gauss 1844, in *Abhandl. d. K. Ges. d. Wissensch. zu Göttingen* (*Mathem. Classe*) II. 4).] Conserving the size of all angles in the representation of one surface on another; in Map-making, = ORTHOMORPHIC *a.* 2.

1893 A. R. FORSYTH *Theory Functions of Complex Variable* xix. 491 Conformal representation, so essential to the construction of a geographical map, made as perfect as possible. *Ibid.*, A copy of any small element..on the surfaces. **1900** [see ORTHOMORPHIC *a.* 2]. **1920** J. I. CRAIG *Theory Map-Projections* i. 4 A map-projection which possesses the property that all small figures are represented by similar small figures on the map is called conformal or orthomorphic. **1958** E. RAISZ *Gen. Cartogr.* I. vi. 82 Conformal, orthomorphic, or isogonic projections..are those on which any small area has the same shape as on the globe. **1967** *Encycl. Brit.* VI. 234/1 By the term *conformal* representation (also called isogonal, orthomorphic) is understood any continuous mapping of one surface or region upon another, with a one-to-one correspondence of their points, and in such manner that corresponding angles in the two surfaces or regions are equal. **1960** W. J. DUNCAN *et al.* *Elem. Treat. Mech. Fluids* II. 100 The method of conformal transformation..is of great utility in the solution of two-dimensional problems.

Hence **confo·rmally** *adv.*

1893 FORSYTH *Theory Functions of Complex Variable* xix. 491 It will be convenient to consider the figures as surfaces which..conformally correspond. **1896** *Math. Gaz.* I. 31 Which..belongs to the..in a finite polygon, that can be conformally represented upon the half of the plane by means of an integral relation between two variables, is capable of being formed by the juxtaposition..of equal rectangles or equal triangles. **1898** A. R. FORSYTH *Theory Functions* xii. 312 Such that all the particular equations for the conformal representation of one domain upon a half-plane. **1964** J. A. B. HEWITT, in G. B. TRAILL, Such regions cannot, in general, be conformally transformed. **1958** *Engineering* 21 Mar. 376/1 Flow surfaces of revolution..defined by meridional streamlines are transformed conformally into plane.

conformism (kǫnfɔ̄·ɹmiz'm). [ad. F. *conformisme* or f. CONFORM(IST + -ISM.] The action, practice, or principle of conforming; belief in conforming; = CONFORMITY 2 a.

1928 B. BROWNELL *New Universe* II. vii. 158 They provided safety in some measure without the crush and swagger of gigantic social conformisms and autocracy. **1937** D. BUSSY tr. *A. Gide's Back from U.S.S.R.* v. 76 'Art to-day should be popular or nothing.' 'You will drive all your artists to conformism,' I answered. **1946** KOESTLER *Thieves in Night* 150 For us, the choice involved a revolutionary negation of our past—for them it is an act of conformism. **1957** P. LAFITTE *Person in Psychol.* v. 64 Conformism may be a special defence mechanism. **1958** *Punch* 25 June 849/1 If we follow America and Russia into conformism we are going to sacrifice our inventiveness.

confrater. Delete *Obs.* and add later examples.

1897 TAUNTON *Eng. Black Monks* I. 61 He was then received to the kiss of peace by all the convent, and was entered in the chart as a *confrater*. **1900** GASQUET *Eve Reform.* 28 Sir Thomas More, no doubt through his father's connection with the monastery of Christchurch, Canterbury, of which house he was a 'confrater', became a student at the monks' college at Oxford. **1931** I. BUCHAN *Blanket of Dark* 35 But the other, the old man with the small white beard?..Doubtless a confrater, or lay member of the Abbey.

‖**confrérie** (konfreri). Also **confrerie.** [F.] A religious brotherhood; an association or group of people having similar interests, jobs, etc.

1803 C. WILMOT *Irish Peer* (1920) 196 The processions were numerous, all the Confreries, all the Priests and Monks holding lighted tapers. **1932** F. R. LEAVIS in *Scrutiny* Sept. 134 We can see only the *confrérie* of the weeklies and the Sunday papers.

confrontation. Add: **3.** The coming of countries, parties, etc., face to face: used of a state of political tension with or without actual conflict.

1963 *Ann. Reg. 1962* I The dramatic confrontation of American and Russian power in the tense days after 22 October. **1963** *Daily Tel.* 3 Oct. 14/2 There is reason and good sense in the Malaysian Prime Minister's attitude that there can be no talks with Indonesia while 'confrontation' is going on. For those who seek a clearer definition of President Soekarno's vocabulary, this word means infiltration, guerrilla warfare, arson, loot and confiscation. **1964** *Ann. Reg. 1963* 220 Evidently he had drawn a lesson from the Cuban confrontation. **1965** *New Statesman* 8 Jan. 29/2 Indonesia's 'confrontation campaign' against Malaysia—an ingenious new expression for an unprovoked act of aggression—is one of the most futile minor wars of history. **1970** *Times* 9 Feb. 5/3 As a counter theme to this, O'Brien places the international black and white confrontation.

confrontment. 2. (Later U.S. example.)

1876 STEDMAN *Victorian Poets* 157 In youth feeling..responds divinely to every sensuous confrontment with the presence of beauty.

confusability. (Later example.)

1958 GOODGLASS & HUNT in Saporta & Bastian *Psycholinguistics* (1961) 452/2 The subjects' tendency to make random errors due to confusability or to a fluctuating level of comprehension.

confusticate (kǫnfʌ·stikēit), *v. colloq.* Also **confuscate.** [Fantastic alteration of CONFOUND *v.* or CONFUSE *v.*] To confuse, confound, perplex. So **confu·sticated** *ppl. a.*

1891 FARMER *Slang*, *Confusticate* verb (American), to confuse. **1898** WRIGHT *Eng. Dial. Dict.* V. s.v. *confuscated*, For sure I be confuscated. **1911** C. L. GRAVES *Hubert Parry* II. 113 He would at once begin to talk in schoolboy slang..interlarding his remarks with such words as 'awfully', 'confusticated'. **1937** TOLKIEN *Hobbit* ii. 21 Confusticate and bebother those dwarves!

conga (kǫ·ŋgă). [American Sp., a. Sp. *conga* fem. of *congo* of or pertaining to the Congo (see CONGO).] A Latin-American dance of African origin, usu. performed by several people in single file and consisting of three steps forward followed by a kick. Cf. CONGO I.

1935 *Dancing Times* Feb. 577/1 The 'Conga' is not a ballroom dance, at any rate in Cuba. **1938** *Encycl. Brit. Bk. of Yr.* 199/2 In Paris, the conga was revived. **1951** *Times* 13 Oct. 11/7 Nor was he above leading a 'Conga' through the halls and corridors of Buckingham Palace.

2. *attrib.*, as *conga chain, line*; **conga drum,** a tall, narrow, low-toned drum usually played with the hands; hence *conga drummer, drumming.*

1957 O. NASH *You can't get There* 143 And round and round there dragged and wound A loathsome conga chain. **1955** L. FEATHER *Encycl. Jazz* i. 30 A number of unfamiliar percussion instruments such as the bongo, conga drum and timbales. **1956** M. STEARNS *Story of Jazz* (1957) xix. 243 Pozo crouched in the center of the stage and flailed a many-voiced conga drum with blistered hands. *Ibid.* 253 He added a third trumpeter and a conga drummer. **1958** *Gramophone* June 32/1 Candido's relentless conga drumming. **1953** *Encounter* Nov. 69/1 They formed a conga-line, putting their hands on the shoulders of those before them, and undulating in a long chain through the street.

congealability (kǫndʒīlăbi·lĭti). The quality or condition of being congealable.

1929 *Morn. Post* 9 Oct., Another constituent of snake venom is a substance which affects the congealability of the blood, and as far back as 1908 it was demonstrated..that epileptic patients had an abnormal congealability or clotting rate of their blood.

congeneric, *a.* Add: **2.** Also as *sb.*

1965 *Sunday Times* 23 May 50/4 Vodka..has virtually no congenerics.

congery (kǫ·ndʒəri). Also **congerie.** [A false singular evolved from CONGERIES by the treatment of the final *s* as pl. inflexion.] = CONGERIES.

1866 MEREDITH *Vittoria* I. ix. 129 Outside the army the empire was a jealous congery of intriguing, disaffected nationalities. **1897** F. J. BURGOYNE *Libr. Construction* 293 The French National Library is contained in a congerie of buildings. **1920** *Blackw. Mag.* Aug. 210/1 He traces his way through the congery of squalid streets between the Cathedral and the Castle. **1928** GALSWORTHY *Swan Song* II. vi. 161 To the extent of pulling down a congery of old houses. **1932** WYNDHAM LEWIS *Apes of God* IX. 241 A conventionally distorted, antique, floridly-fringed head presided over the congery of tubes.

congest (kǫndʒe·st), *sb.*[1] [Back-formation f. CONGESTED (see next).] In Ireland, a tenant living on land of which the resources do not adequately support him.

1903 *Daily Chron.* 15 Oct. 5/2 On market day the 'congests' of the district crowd the streets with their little carts and panier-bearing asses. **1920** *Contemp. Rev.* Sept. 310 Insisting in many cases on sale (often partial, not total) of ranches to 'congests' or landless men. **1927** *Sunday Times* 13 Mar. 11 A large farmer..whose lands would..be divided up among adjoining congests.

congested, *ppl. a.* Add: **2. c.** *congested district, estate*: in Ireland and Scotland, an area of land of which the resources are inadequate to support its population. (Cf. prec.) Hence applied to similar areas elsewhere. Also of an urban area: excessively full of buildings, traffic, etc.

1883 *Pall City Guardian* 4 Apr. 3/1 The chronic distress prevailing in certain congested parts of Ireland. **1886** *Ibid.* 4 Sept. 2/1 The worst enemies of the crofters were those who would throw the slightest impediment in the way of..emigration from the congested districts. **1893** *Graphic* 14 Jan. 30/2 The Congested Districts Board. **1897** *Daily News* 30 June 3/4 Congested Districts (Scotland) Bill. **1904** *Daily Chron.* 26 Mar. 6/1 The definition of a congested estate is one in which one-half of the area is in holdings not exceeding ten acres, or one in which half of the area is bog land or mountain. **1940** *Ann. Reg. 1944* 19 The Barlow Report on the location of industry..the 'decongestion' of congested areas and the encouragement of reasonable balance between different regions. **1957** *Encycl. Brit.* II. 971/1 He [sc. A. J. Balfour] broadened the basis of material prosperity and social progress by creating the Congested Districts board in 1890. **1963** *Daily Tel.* 4 Sept. 22/7 Drivers would receive monthly or quarterly accounts for the time they had used their cars in congested town centres.

conglo·bula·tion. [f. CONGLOBULATE *v.* + -ATION.] The act of forming a rounded or compact mass; such a mass.

1923 J. M. MURRY *Pencillings* 172 There, at any rate, the swallows were, flying round and round, perhaps in the very act of *conglobulation*. **1928** F. M. FORD *Last Post* vi. 154 If you took the whole conglobulation at its worst—the father backsliding, the son living with his sister in open sin, the son's son not his son.

conglomerate, *sb.* Add: **3.** A large business group or industrial corporation resulting from the merging of originally separate and diverse commercial enterprises.

1967 *Economist* 11 Feb. 553/4 Textron Inc., a leading 'conglomerate'—its 28 divisions sell everything from chickens to rocket engines. **1970** *Daily Tel.* 12 Mar. 9/1 Now there is scarcely an up-to-date conglomerate without a publishing operation embedded somewhere in its diversified mass. **1970** *Sci. Amer.* Mar. 31/2 A *conglomerate* is similar to a U.S. conglomerate, which is also a group of unrelated companies under sophisticated financial management.

Congo. Add: **1.** (Earlier and later examples of the dance.)

1803 J. DAVIS *Trav. U.S.A.* 380 There was nobody that could face him at a Congo Minuet. **1823** I. HOLMES *Acct. U.S.A.* 332 In Louisiana, and..Mississippi, the slaves..dance for several hours...The general movement is in what they call the Congo dance. **1835** LONGSTREET *Georgia Scenes* 128 The minuet,..and the congo, which was only to chase away the solemnities of the minuet. **1961** [see *BAMBOULA].

2. *Congo ape* = *Congo monkey*; **Congo snake** (earlier and later U.S. examples; also *ellipt.*).

1869 TROLLOPE *Tales of all Countries* (2nd Ser.) 211 The clang of the macaws, or the roaring of the little congo ape. **1835** J. AUDUBON *Ornith. Biogr.* III. 90 The Congo snake and water-moccasin glide before you. **1888** G. W. CABLE *Bonaventure* 284 A large moccasin..of that yet deadlier..black with an ell in length, which the swampers call the Congo. **1947** *Chicago Tribune* (Grade Mag.) 21 Dec. 9/3 There were malaria and typhus in the swamps, and many snakes: rattlers, moccasins, copperheads, and the Congo, a blue black eel, most feared of all.

3. Used *attrib.* to designate any of a group of azo-compounds, dyes mostly derived from benzidine or tolidine, so called after *Congo red*, the first of the direct cotton colours to be produced.

1885 *Jrnl. Soc. Dyers & Colourists* I. 146/2 Congo Red...This colouring-matter..is of special interest, owing to its remarkable property of dyeing cotton..without the aid of a mordant. *Ibid.* 147/1 Cotton, immersed into a boiling solution of Congo red, is dyed a brilliant scarlet. **1886** *Jrnl. Chem. Soc.* L. 889 Congo-red, from benzidine and ammoniacal solution with sodium..naphthionate and a colourless..acid. **1905** CAIN & THORPE *Synth. Dyestuffs* 21, 70 A mixed benzidine dyestuff—Congo Corinth—is produced. *Ibid.* 73 Congo brown G [A]. Congo brown R [A]. **1925** *Jrnl. Chem. Soc.* CVIII. I. 321 Blue and Red Congo Dyes. *Ibid.* 321 Isomeric Congo-acids have been isolated. **1920** *Ibid.* CXVIII. 11. 171 Colour Change of Congo-rubin. **1921** N. SHREVE et al. *Dyes Classified* 1. 84 Dyes derived from Amino-R acid. Congo Orange R. *Ibid.* 127 Dyes derived from Benzidine..Congo Rubine. *Ibid.* 202 Dyes derived from Dianisidine..Congo Blue 2B. **1940** *Thorpe's Dict. Appl. Chem.* IV. 12/1 Benzidine finds..its greatest application in the manufacture of direct cotton dyestuffs, which result on tetrazotisation..*e.g.* Congo Red..results from coupling with naphthionic acid. *Ibid.* 218 Congo Rubine: Benzidine, Naphthionic acid, Crocein acid. *Ibid.* 219 Congo Corinth GW: Benzidine, Naphthionic acid, N.W.-acid. **1962** STANTON *Dict. for Med. Secretaries* 36/1 *Congo Red test of renal function*, a test for amyloidosis. **1964** M. HYNES *Med. Bacteriol.* (ed. 8) xi. 190 Two drops of 1 per cent. Congo red are added, and a few drops of N HCl to give a faint blue colour.

Congoese (kǫŋgo,ī·z), *a.* and *sb.* Also **Kongoese.** [See next.] = *CONGOLESE *a.* and *sb.* Now *rare.*

1797 *Encycl. Brit.* V. 331/1 A great number of the states that were under the Congoese monarchy..having withdrawn their allegiance. *Ibid.* 323/1 The Congoese are superstitious to the last degree. **1825** *Encycl. Metrop.* XVII. 143/2 The faults of the Kongoese are those occasioned by a hot climate..and a low degree of civilization. **1877** *Encycl. Brit.* VI. 267/2 The religion..of the Congoese is a gross fetishism. **1902** *Ibid.* XXVII. 120/1 Kongoese..possesses the qualities of precision, flexibility, and subtlety of expression.

Congolese (kǫŋgolī·z, kǫ·ŋgolīz), *a.* and *sb.* [ad. F. *Congolais*, f. *Congo*, the name of a region and a river in Central Africa: see -ESE.] **A.** *adj.* Of or pertaining to the Congo (the Congo Republics of Kinshasa (Léopoldville) and Brazzaville, formerly the Belgian and French Congo respectively), or the inhabitants thereof. **B.** *sb.* An inhabitant of either of the Congo Republics; freq. *collect.*; also, the language of the Bakongo people.

1900 H. H. O. in *Geogr. Jrnl.* V. 413/1 'Goober' and 'Pindar'..are synonyms for the ground-nut or pea-nut...'Goober' is Angolan, and 'pindar' Congolese. **1904** *Daily Chron.* 15 Feb. 7/3 The Congolese are, indeed, worse off than slaves. **1905** *Westm. Gaz.* 10 Oct. 2/1 The Congolese 'groups' of Antwerp and Brussels found most of the capital. **1917** *Times Educ. Suppl.* 30 Apr. 2/1 A. Peeps at Congolese villages. **1934** A. J. TOYNBEE *Study Hist.* I. II. iii. 253 A level of culture which is hardly less primitive than that of the Congolese. **1952** *Times* 26 July 11/6 The spontaneous enthusiasm of the Congolese for the Congolese people, black and white alike, for their ruler. **1960** *Daily Tel.* 7 July 1/1 They want a Congolese in charge, with no Belgians holding any command.

congrats (kǫngræ·ts), colloq. abbrev. of *congratulations*, usu. as *int.* Cf. next.

1884 L. TROUBRIDGE *Life amongst Troubridges* (1966) 167 Received..congrats on Ernest's appointment to the Royal Yacht. **1894** 'L. Hope' *Dolly Dialogues* II. 11 So you've brought it off. Hearty congrats. **1908** *Punch* 1 July 16/3 Lord and Lady Knightsbridge..were simply loaded with congrats about their brilliant son's success. **1929** J. B. PRIESTLEY *Good Companions* I. vi. 431 Congrats on the show. **1962** *Melody Maker* 7 July 16/4 Congrats! Congratulations, Acker Bilk, on your stand about young people.

congratters (kǫngræ·taɹz), colloq. abbrev. of *congratulations*, usu. as *int.* Cf. prec. and *-ER[6].

1908 R. BROOKE *Let.* 10 Jan. (1968) 38 Congratters on Cambridge's political enlightenment. **1914** 'I. HAY' *Lighter Side School Life* iv. 95 'Congratters!' said Blake awkwardly. **1930** D. L. SAYERS *Strong Poison* xii. 152 Tremendous congratters and all that. **1966** O. NORTON *School of Liars* vi. 101 The Brig lifted his glass. 'Congratters, my dear. Good show.'

congregant. Add to def.: esp. a member of a Jewish congregation.

1897 *Daily Chron.* 27 Sept. 7/5 The congregants are seated on the ground; the service is chanted in a low and mournful way. **1956** 'Book of Lamentations' is read. **1907** *Ibid.* 20 Sept. 6/7 At the Brondesbury Synagogue on the Day of Atonement, a congregant made an offering of two guineas in honour of the King. **1952** A. BARON *With Hope, Farewell* 29 The short gravel path was crowded with congregants who had escaped from the service. **1961** *Spectator* 3 Nov. 615 The large majority of Afrikaner congregants.

congress, *sb.* Add: **9. congress boot** (examples); **congress gaiter** = *congress boot*; **Congress Party,** a political party in India; also *ellipt.* as *Congress.*

1847 *Semi-Weekly News* (Fredericksburg, Va.) 12 Aug. 3/1 (Advt.), Congress Boots and Slippers. **1857** 'O. K.' D. WIGGIN *Gard. Memory* 45 A Congress boot, with the cloth top, and elastic on the side which characterized that creation of the late sixties. **1938** J. W. DANIELS *Southerner discovers South* 156 He wore congress boots with elastic sides. **1858** *San Francisco Herald* 5 Oct. 3/4 (Advt.), Gents' fine patent leather Congress gaiters. **1885** *Good Words* July 454/1 Why boots with elastic sides should be called 'Congress gaiters' passes my comprehension. **1908** D. CAMPBELL *Fables for Parents* 265 The baskets were set on the floor of the buggy at one side of Aunt Emma's congress gaiters. **1937** F. P. *Times* 24 Jan. 16/2 Mr. Gandhi's ..all-India..the Government's decision influenced by the victory of the Congress Party in the Indian polls, and the refusal of Congress to collaborate in the new Constitution of India? **1959** *Ann. Reg. 1958* 91 The Congress Party had carried..Congress Party became increasingly worried during the year over conditions in the Communist-governed State of Kerala.

congressional, *a.* **2.** (Earlier U.S. examples.)

1775 E. QUINCY in *New Eng. Hist. Reg.* (1857) XI. 168 A happy effect..of the First American Congressional Appointment. **1783** JEFFERSON *Let.* in *Wrt.* (1894) III. 333 My hopes of the success of the Congressional propositions here have lessened exceedingly. **1785** *Ibid.* IV. 33 Give me fully always the Congressional news.

congressman. (Earlier U.S. examples.)

1780 *The American Times* iii. 28 Ye coxcomb Congressmen, declaimers keen. **1806** FISHER AMES *Wks.* (1854) I. 349 And I consider too, how unreasonable it is to expect a Congressman can fill letter after letter with important matter.

co·ngresswoman. A woman holding a seat in the U.S. Congress.

1918 in WEBSTER Add. **1928** *N.Y. Times* 8 Nov. 28/3 The increasing number of 'Congresswomen' is but another indication of the larger part their sex is taking in the public life of the nation. **1946** H. L. MENCKEN in *Amer. Speech* XXI. 81 The Hon. Clare Boothe Luce of Connecticut, the only congresswoman who may be described plausibly as both cerebral and beautiful.

congruence. Add: **3.** (Examples.) (See also quot. 1958.)

1706 [see CONGRUITY 4]. **1933** BLOOMFIELD *Language* xii. 191 Congruence plays a great part in many languages; witness for example the inflection of the adjectives in most Indo-European languages in congruence with various sub-classes (number, gender, case) of the noun. **1942** BLOCH & TRAGER *Outl. Ling. Analysis* v. 77, *I am..he is..we are.* Here..the finite verb varies according to the class of another constituent, the actor expression. This kind of selection is called congruence. **1958** A. S. C. ROSS *Etymology* i. 28 Two languages are *related* if..they were once one language...The word *congruence* in application to parts of two related languages is to be understood in precisely the sense in which the word *relationship* is applied..to the two languages themselves. Thus, English *stone* and German *stein* are congruent because..they were both one word in..Primitive Germanic.

conidium. Add: Also in bacteria of the orders Actinomycetales and Chlamydobacteriales, which have some resemblances to fungi. Add later examples.

1925 C. H. BROWNING *Bacteriology* iv. 67 The higher bacteria, like the fungi, develop special structures known as 'conidia', and these conidia then grow into the bacterial forms. **1934** A. T. HENRICI *Biol. Bacteria* vi. 99 The Actinomycetes proper multiply by means of spores, or *conidia*, which, like the conidia of Fungi belonging to the Ascomycetes or Fungi imperfecti, are formed by a constriction of cells at the ends of filaments of the mycelium projecting into the air...Apparently, then, these Actinomycetes are just like molds in their essential characters, and indeed some authorities would classify them with the higher Fungi rather than with the bacteria. **1960** C. F. ROBINOW in Gunsalus & Stanier *Bacteria* I. v. 229 Although this chapter is supposedly dealing with the *endospore* of bacteria, we may, in truth, be concerned with something which may have to be regarded as a cross between a conidium and a chlamydospore. **1969** S. T. LYLES *Biol. Microorganisms* viii. 174 Actinomycetes produce true mycelia, which may contain irregular segments and branches; conidia may be produced. Although not true bacteria, these procaryotic fungilike forms differ from true fungi in many respects and are still grouped with bacteria.

coning (kōu·niŋ), *vbl. sb.* [f. CONE *sb.*[1] or *v.*[1] + -ING[1].] The making of a cone-shaped tread (of a wheel); the condition of being coned (see CONED *ppl. a.* 2).

1837 W. B. ADAMS *Carriages* 297 The coning of the wheels must also act like a wedge...This, together with the coning of the wheels, would make a nearly perfect railway carriage. **1906** *Times* 12 Sept. (Engin. Suppl.) 291/2 The coning of the wheels and the slight inward cant of the rails. **1960** HORNER & ABBEY *Dict. Terms Mech. Engin.* II. 85 *Coning*, the turning of the taper on the diameters of railway wheels and crane and turntable rollers.

2. *Aeronaut.* The action of the rotor blades of a helicopter slanting upwards when in motion so as to produce a cone-shaped pattern of rotation; hence *coning angle*: the angle at which the rotor blades rise in coning.

1931 J. DE LA CIERVA *Wings of Tomorrow* vii. 102, I supposed that it was preferable to keep the blades from coning, so that they would remain in a horizontal position while in flight. **1944** H. F. GREGORY *Anything a Horse can Do* 148 The angle the blade [of the rotor] makes with the horizontal is known as the coning angle. **1955** LIPTROT & WOODS *Rotorcraft* v. 49 The physical effect of coning is to move all elements of the blade nearer to the axis of rotation.

conjugate, *a.* and *sb.* Add: **A.** *adj.* **1. b.** *Bibliography.* (See quot. 1927.)

1881 [see Dict. 1 a]. **1927** R. B. MCKERROW *Introd. Bibliogr.* I. iii. 27 Two leaves which 'belong to one another', i.e. if traced into and out of the back of the book are found to form a single piece of paper, are said to be 'conjugate'. **1957** N. R. KER *Catal. MSS. containing Anglo-Saxon* p. xxii, The eight leaves forming the quire are four conjugate pairs (i.e. four sheets), 1–8, 2–7, 3–6, and 4–5. **1968** *Bodl. Libr. Rec.* VIII. 81 Quire IV consists only of two conjugate leaves.

3. b. *conjugate solution*: one of two solutions which are in equilibrium at a given temperature.

1909 in WEBSTER. **1932** E. GREGORY *Metall.* iii. 69 Such solutions as x_1 and x_2 which can exist in equilibrium with each other at a constant temperature are known as conjugate solutions. **1959** *Chambers's Encycl.* XII. 699/2 Mixtures of the alcohol and water in proportions corresponding to compositions between A and B will, after shaking, separate into these two 'conjugate' solutions A and B, which are in equilibrium with one another.

6. a. Of a complex number or quantity: differing from some given complex quantity

in having the sign of the imaginary part reversed; *conjugate function*: see quots.

1835 *Rep. Brit. Assoc. Advancem. Sci. 1834* 519 Mr. Hamilton proposes to consider a few properties of such *Conjugate Functions*, or *Algebraic Couples*, defining two functions to be *conjugate* when they satisfy the two equations of conjugation, and calling, under the same circumstances, the pair or couple (*u*, *v*) *a function of the pair* (*x*, *y*). **1873** J. C. MAXWELL *Electr. & Magn.* I. xii. 227 Definition of Conjugate Functions. Two quantities α and β are said to be conjugate functions of *x* and *y*, if $\alpha+\sqrt{-1}\,\beta$ is a function of $x+\sqrt{-1}\,y$. **1891** G. L. CATHCART tr. *Harnack's Introd. Differential & Integral Calculus* II. iii. 153 If the equation have one complex root $z = a+i\beta$, it has also the conjugate complex root $z = a-i\beta$. **1908** G. H. HARDY *Course Pure Math.* iii. 82 The sum (2*x*) of two conjugate numbers and their product (x^2+y^2) are both real. **1955** W. PAULI *Niels Bohr* 30 The replacement of a function by its conjugate complex is not a 'linear operator'. **1961** C. C. T. BAKER *Dict. Math.* 73 *Conjugate functions*, if $u+iv = f(x+iy)$, then *u* and *v* are said to be conjugate functions. **1962** CORSON & LORRAIN *Introd. Electromagn. Fields* App. D. 517 These are the Cauchy–Riemann equations. The functions *U* and *V* are related to each other through these equations and are called conjugate functions.

conjugated, *ppl. a.* **b.** Add to def.: also, designating, pertaining to, or containing a chain or ring of carbon atoms in which every other pair of carbon atoms is linked by a double bond.

1907 *Chem. Abstr.* 1400 Some compounds, such as benzene, contain conjugated double linkages. *Ibid.* 1537 Unsaturated conjugated systems. **1910** *Encycl. Brit.* VI. 56/2 From a study of the reduction of compounds containing two ethylenic bonds united by a single bond, termed a 'conjugated system', E. Thiele suggested a doctrine of 'partial valencies'. **1948** GLASSTONE *Physical Chem.* (ed. 2) viii. 530 When an open chain molecule contains a system of conjugated double bonds, the molar refraction is different. **1962** P. J. & B. DURRANT *Adv. Inorg. Chem.* xi. 318 In conjugated systems the bond orders are usually fractional, and the fraction has to be calculated from a knowledge of the conjugated structure.

conjunct, *sb.* Add: **5.** *Logic.* A conjoined term or proposition; one of the elements in a conjunction; = *DETERMINANT *sb.* 2 b.

1921 W. E. JOHNSON *Logic* I. iii. 27 The components thus joined will be called conjuncts. **1954** I. M. COPI *Symbolic Logic* ii. 10 Two statements so combined are called conjuncts.

conjunctiveness. (Example.)

1965 *Amer. Speech* XL. 207 Adverbials which..combine modification with conjunctiveness (such as *hence, therefore*, and *in that case*).

conjuration. 1. Delete †*Obs.* and substitute *arch.* (Later examples.)

1891 *Scottish Leader* 29 June 4 Another secret society of criminals has been broken up by the Italian police. A most ferocious conjuration it would seem to be. **1907** W. G. HOLMES *Justinian & Theodora* II. viii. 519 Communication and conjuration between the first two sections was established at Carthage.

conjure, *v.* Add: **6.** Phr. *a name to conjure with*: a name of importance. (Cf. quot. 1601 in Dict.)

[**1894** SOMERVILLE & 'ROSS' *Real Charlotte* II. xxxi. 224 At about this time the person whose name Julia Duffy had so unfortunately selected to conjure with was sitting under a tree.] **1901** J. N. MCILWRAITH *R. Campbell* xxi. 263 These were the tried legions of France: Guienne, Berry, Béarn, [etc.]..names to conjure with, in Gilbert's estimation. **1917** E. WALLACE *Just Men of Cordova* ii. 31 Black grew in prosperity. His was a name to conjure with in certain circles. **1930** R. CAMPBELL *Adamastor* 104 'H. Wodson, a name to conjure with in the journalistic world.'—*Natal Advertiser*, edited by H. Wodson. **1954** I. MURDOCH *Under Net* xi. 154 His name, little known to the public, is one to conjure with in Hollywood.

conjuring, *vbl. sb.* Add: **3. b.** (Later examples.) Also *fig.*

1842 R. BARHAM *Merch. Venice* in *Bentley's Misc.* Apr. 434 Put his conjuring cap on,—considered the whole. **1879** *Queen* 27 Dec. (Advt.), Largest assortment of conjuring tricks and puzzles. **1937** *Discovery* May 137/2 They have the same power as the Frog Lizard of doing conjuring tricks with their tails, and seem to shed them at the least pretext. **1939** T. S. ELIOT *Old Possum's Pract. Cats* 30 Mr. Mistoffelees' Conjuring Turn.

conk, *sb.*[1] Add: **b.** The head. So *off one's conk*: off one's head; crazy.

1870 *Putnam's Mag.* Mar. 301/1 The combatants struck each other with mawleys and bunches of fives upon the head, the nut, the cone, the conk, the canister. **1873** T. E. BROWN *Betsy Lee* 46 Isaac and Peter and the like of them, That's allis got conks like turkey's eggs. **1931** 'E. QUEEN' *Dutch Shoe Mystery* viii. 71 A persistent idea has been buzzing about in my conk. **1959** H. PINTER *Birthday Party* II. 29 Why are you getting on everybody's wick? Why are you driving that old lady off her conk?

c. A punch on the nose or head; a blow on any part of the body.

1898 in Wright *Eng. Dial. Dict.* s.v., A'll catch ye a conk in a minute. **1947** 'N. SHUTE' *Chequer Board* ii. 18 It's this conk I got on the old napper.

conk (kǫŋk), *sb.*[2] [App. var. of CONCH.] A fungus which grows on the wood of trees, esp. *Trametes pini*; also, the disease produced by this fungus. *colloq.* (orig. *U.S.*).

1851 J. S. SPRINGER *Forest Life* iv. 99 There is a cancerous disease peculiar to the Pine-tree, to which lumbermen give the original name of 'Conk' or 'Konkus'. **1902** *Bureau of Forestry, Bulletin 33* 15 The 'conk' or bracket seen on affected trees is the fruiting organ...Conk spores never enter through the bark, but usually through the scars of broken branches. **1934** *Forestry* VIII. 155 Among the wood-rot diseases of conifers, that caused by *Trametes pini*, known variously as red rot, red stain, white pocket rot, ring scale, conk rot etc., is certainly the most destructive.

Hence **co·nky** *a.*, affected by this fungus.

1905 *Terms Forestry & Logging* 33. **1945** B. MACDONALD *Egg & I* (1946) 50 Fallen firs, six and seven feet in diameter and conky in the middle.

conk, *v.*[1] *slang.* [f. CONK *sb.*[1]] *trans.* To punch on the nose; to hit.

1821 P. EGAN *Boxiana* III. 338 Spring however *conked* his opponent, when they closed. **1898** in Wright *Eng. Dial. Dict.* s.v., A'll conk ye if ye do it again, so there! **1948** C. DAY LEWIS *Otterbury Incident* 13 Nick couldn't have conked anyone with his football.

conk, *v.*[2] *colloq.* [Of obscure origin.] *intr.* To break down, give out, fail, or show signs of failing; to die, collapse, or lose consciousness. Also *fig.* Also with *out.*

1918 *Chambers's Jrnl.* May 301/1 If the engine conks we'll never make across the lines. **1918** E. M. ROBERTS *Flying Fighter* 334 *Conked*, a new word which is taken from the Russian language and which means stopped or killed. **1921** *Blackw. Mag.* June 788/2 My engine began to conk a bit. **1924** KIPLING *Debits & Credits* (1926) 167 The man was vi'lently sick an' conked out. **1928** —— *Limits & Renewals* (1932) 61 She stays—till she conks out. **1928** *Punch* 24 Oct. 466/1, I am not minimising the difficulty of Porlock, where the bus may conk out and be hauled up ignominiously by ropes. **1929** *Daily Mail* 9 Sept. 7 The old boat 'conked out' miles from anywhere. **1934** WODEHOUSE *Right Ho, Jeeves* i. 15 The mystery had conked. I saw all. **1952** M. HERZOG *Annapurna* xi. 168, I told Lionel that rather than conk out next day on the slope, it seemed far better for me to go down. **1963** *Guardian* 29 Oct. 16/7 The heating's conked at home, and..I feel the cold.

conker[2] (kǫ·ŋkəɹ). Also **conquer.** [f. dial. *conker* snail-shell, prob. f. *CONQUEROR I f.] **a.** *pl.* A boys' game, played originally with snail-shells (see quot. 1877) but now with horse-chestnuts, in which each boy has a chestnut on a string which he alternately strikes against that of his opponent and holds to be struck until one of the two is broken. **b.** A horse-chestnut (formerly a snail-shell) used in the game; hence *gen.* a snail-shell or horse-chestnut. Cf. *CONQUEROR I f and *CONQUERING *vbl. sb.* b.

1847 HALLIWELL *Dict. Arch. & Prov. Words* I. 268/1 *Conkers*, snail-shells. *East.* **1866** J. G. NALL *Great Yarmouth* 534 *Conkers*, snail shells. **1877** F. Ross et al. *Gloss. Holderness* 45/1 *Conkers*,..small snail-shells. In the boy's game of *conkers* the apexes of two shells are pressed together until one is broken, the owner of the other being the victor. **1879** G. F. JACKSON *Shropshire Word-Bk.* 95 *Conquer*, a snail-shell. In the children's game of pitting snail-shells one against the other, that which breaks its opponent is called the *conquer*. **1886** F. T. ELWORTHY *W. Somerset Word-Bk.* 156 *Conkers*, horse-chestnuts...From this the name is given..to the tree—*conker*-tree. **1903** *Little Frolic* 78 'Please may I pick up the conquers?' 'What?' said Bessie. Bessie did not know that the boys call the pretty brown chestnuts 'conquers'. **1921** *Spectator* 2 July 8/1 There is a season for hoops and another for whip-tops,..to say nothing of the 'conkers' fashion. **1928** G. D. H. & M. COLE *Man from River* xxxiii. 246 'What's this?' 'A chestnut on a string,' said Michael, picking it up. 'Not much good to us, is it?' They're hardly likely to have been playing conkers here.' **1929** *Morning Post* 2 Oct. 10/5 The use of the horse-chestnut in the game of 'Conquers'. **1963** *Times* 18 Jan. 9/4 Few, to be sure, are to be observed engaged in solemn duels with conkers, once a ravishing delight. **1965** *East Anglian Mag.* May 241/1 The shells of a small variegated snail were *conkers*.

con man: see *CON.

connate, *a.* Add: **5.** *Geol.* Designating water trapped in a sedimentary rock during its deposition.

1908 A. C. LANE in *Bull. Geol. Soc. Amer.* xix. 502 Quite different may be the water buried with the beds in the first place, to which we may fitly apply the adjective connate. **1955** BROWN & DEY *India's Min. Wealth* (ed. 3) 647 Geologists also recognise two further underground groups—'connate waters' which were trapped in some sedimentary rocks during their deposition and 'juvenile waters' which are of plutonic or magmatic origin.

connately, *adv.* (Example.)

1884 W. JAMES *Coll. Ess. & Rev.* (1920) 256 Most of the objects of civilized men's emotions are things to which it would be preposterous to suppose their nervous systems connately adapted.

connateness. Delete † and add later example.

1936 *Mind* XLV. 185 The waltz is a species of dance, *i.e.* subordinate in generality to it, and therefore the connateness between dancing and the waltz is asymmetrical.

connect, *v.* Add: **5. b.** Of a person: to meet, find, get in touch *with* (another person). Also *spec.*, to meet in order to obtain drugs (from); cf. *CONNEXION 6 c. *U.S. slang.*

1926 J. BLACK *You can't Win* xiii. 177 Wait till I send you word. Don't try to connect with me. **1938** F. CHESTER *Shot Full* ii. 13 'Connecting' was tough. Drugs were most difficult to obtain. *Ibid.* 15 One day when I went to a doctor to 'connect' for junk, I met another addict coming out. **1962** K. ORVIS *Damned & Destroyed* vii. 50 If you're connecting from Frankie, he should have told you.

c. Delete ?*Obs.* and add further examples. Also in extended uses: to 'get across', to 'click', to be meaningful.

E. M. Forster's phr. 'only connect' (see quot. 1910) is frequently used allusively.

a **1792** BISHOP GEORGE HORNE in *Cent. Dict.* s.v., This part will not connect with what goes before. **1910** E. M. FORSTER *Howards End* xxii. 183 She would only point out the salvation that was latent..in the soul of every man. Only connect! *Ibid.* 184 Live in fragments no longer. Only connect, and the beast and the monk, robbed of the isolation that is life to either, will die. **1911** R. BROOKE *Let.* 28 Apr. (1968) 302 We join up Puritan and Hedonist: we have (once more) only connected. **1942** BERREY & VAN DEN BARK *Amer. Thes. Slang* § 591/4 Impress audience favorably; 'get across'..connect. **1947** *Picture Post* 27 Sept. 5/2 Nothing 'connects' these days apparently beyond the eternal nylons, shop-windows and hotel-menus. **1958** *Listener* 30 Oct. 704/1 A magazine programme like 'Science is News' has probably done its job if one of its items connects. **1962** *Ibid.* 17 May 837/2 We are not entitled to demand faster growth and expect business, and indeed life, to continue as usual. As Mr Forster would say, let us 'only connect'. *Ibid.* 2 Aug. 182/1 But he [*sc.* man] can connect; he can build bridges across the muddle of experience, of understanding, faithfulness, and love. *Ibid.* 4 Oct. 524/2 Papers on history and poetry which connect with politics but are not directly concerned with it. **1963** *Ibid.* 14 Mar. 458/1 It [*sc.* hot jazz] failed to become popular because it was unable to connect with popular needs. **1966** *Guardian* 21 Mar. 8/4 Whether the words 'common good' connect or not, hundreds of thousands of people in this country have a sense of the public interest.

d. To succeed in obtaining something (e.g. in a burglary). *U.S. slang.*

1926 J. BLACK *You can't Win* xi. 144 'Did you connect, Kid?' he asked when [*i.e.* after the burglary] we were on the street. **1931** 'DEAN STIFF' *Milk & Honey Route* 203 *To connect*, to make a touch. The reward for good panhandling. **1938** D. RUNYON *Take it Easy* 289 He will move himself in for a few dibs if the good thing connects.

e. Of a punch, blow, etc.: to hit, to reach its target. Const. *with* and *absol.*

1933 *John o' London's* 4 Feb. 720/2 To land a punch on the chin is..to 'connect' with that organ. **1942** BERREY & VAN DEN BARK *Amer. Thes. Slang* § 677/21 Hit the ball..connect. **1960** *Observer* 24 Jan. 5/4 There is a thump like a pile-driver connecting with the pavement.

connected, *ppl. a.* Add: **5.** *Math.* and *Logic.* (See quots.)

1893 A. R. FORSYTH *Theory of Functions* § 158 A surface is said to be connected when, from any point of it to any other point of it, a continuous line can be drawn without passing out of the surface. **1903** B. RUSSELL *Princ. Math.* I. xxiv. 202 We need a further condition, which may be expressed by saying that the collection must be connected...For the present we may content ourselves by saying that our collection is connected when, given any two of its terms, there is a certain finite number (not necessarily unique) of steps from one term to the next, by which we can pass from one of our two terms to the other. When this condition is fulfilled, we are assured that, of any three terms of our collection, one must be between the other two. **1918** VEBLEN & YOUNG *Projective Geom.* II. § 155 A set of points is said to be connected if and only if any two points of the set are joined by a curve consisting entirely of points of the set. **1941** A. TARSKI *Introd. Logic* v. 94 If the relation *R* subsists between two arbitrary distinct elements of *K* in at least one direction, the relation is called 'connected in the class *K*'. **1957** P. SUPPES *Introd. Logic* x. 216 The relation of being a mother is not connected in the set of people.

connecter, -or. Add: **3.** attrib. use in *Anat.*

1916 W. H. GASKELL *Involuntary Nervous Syst.* i. 2 The receptor element does not connect directly with the excitor element, but an intermediate neuron in the central nervous system unites the two; this I shall call throughout this book the 'connector element'. *Ibid.*, A connector neuron..may only connect after a series of communications with other connector neurons. **1920** T. P. NUNN *Education* 166 Connector-axons which make their way to effector-neurones entirely outside the cord. **1953** G. A. G. MITCHELL *Anat. Autonomic Nervous Syst.* ii. 12 Afferent (receptor), intercalary (connector), and efferent (effector) neurons, linked together as reflex arcs. **1962** *Gray's Anat.* (ed. 33) 945 (*heading*) Internuncial or connector neurones.

connectibility (kǫne:ktibi·lĭti). [f. CONNECTIBLE *a.* + -ITY.] The quality or condition of being connectible.

a **1916** H. JAMES *Middle Years* (1917) i. 6 The idea of connectibility as stretching away and away. **1940** *Mind* XLIX. 477 The use of this concept of connectibility [tr. Ger. *Verbindbarkeit*] makes it possible for Von Mises to give a sympathetic account of many types of statement.

connection. (This spelling is now more usual than *connexion.*)

connectivity (kǫnekti·vĭti). [f. CONNECTIVE *a.* + -ITY.] The characteristic, or order, or degree, of being connected (in various senses). Also *attrib.*

1893 A. R. FORSYTH *Theory of Functions* § 160 For a simply connected surface,..we shall say that the connectivity is unity. **1918** VEBLEN & YOUNG *Projective Geom.* II. § 188 The number *P* is called the connectivity of the polyhedron. **1928** A. S. EDDINGTON *Nature Phys. World* xii. 255 This connectivity of pointer readings, expressed by physical laws,..supplies the continuous background. **1941** COURANT & ROBBINS *What is Math.?* v. 244 The degree of connectivity of a domain in the plane is an important topological invariant of the domain. **1957** *Encycl. Brit.* XIV. 805/1 The manifold is said to have the connectivity *k* if the maximum number of simple, closed curves that may be traced upon it without separating it into two or more pieces is *k* − 1. *Ibid.* XX. 942/2 The integer 2 minus the characteristic of a polyhedron is known as the connectivity number of the polyhedron. **1964** E. BACH *Transformational Gram.* vii. 156 Of every two items *a*, *b* in a string, either *a* precedes *b* or *b* precedes *a* (connectivity). **1964** J. Z. YOUNG *Model of Brain* p. vi, The microelectrode..follows closely in the study of connectivity.

Connemara (kǫnėmā·rā). The name of the district in the west of County Galway, Ireland, used *attrib.* to designate objects, animals, etc., from that district; esp. *Connemara marble*, a banded serpentinous marble (cf. **Irish green*).
1861 MRS. GASKELL *Let.* 26 Dec. (1966) 672 We made no great ado about presents this year. Julia a scarlet Connemara Cloak. **1886** *Encycl. Brit.* XXI. 683/1 The Galway rock comes into the market under the name of 'Irish green' or 'Connemara marble'. **1896** G. B. SHAW *Our Theatres in Nineties* (1932) II. 29 Every female one [is] a colleen in a crimson Connemara cloak. **1916** W. B. YEATS *8 Poems* Gray Connemara cloth. **1922** JOYCE *Ulysses* 320 Wine, peltries, Connemara marble. **1936** J. T. JENKINS *Fishes Brit. Is.* (ed. 2) 107 Connemara Sucker (*Lepadogaster decandolii*). **1950** *Oxf. Jun. Encycl.* IX. 289/1 The Connemara pony, which is 13–14 hands high and is frequently grey, comes from Connaught. **1965** 'N. BLAKE' *Private Wound* v. 72 My tall Connemara tweed hat had flown off my head.

conner[5]. *Services' slang.* [perh. abbrev. of *MACONOCHIE; cf. *-ER[6].] Food, esp. tinned food.
1931 in Brophy & Partridge *Songs & Slang of Brit. Soldier* (ed. 3) 295. *a* **1935** T. E. LAWRENCE *Mint* (1955) xix. 66 She'll stay where she..is till I've had me bleedin' conner.

connexion. Add: **1. c.** *Electr.* The linking up of electric current by contact; an apparatus or device for effecting this.
1832 *Natural Philos.* (Libr. Usef. Knowl.) II. Electro-Magnetism ix. 46 When a connexion is made with the battery..so as to direct an electric current through the wire. **1885** [see Dict. 1 a]. **1898** *Westm. Gaz.* 9 June 5/1 The bombardment destroyed the French cable connexion house. **1907** *Installation News* Sept. 12/2 Heavy brass connection terminals. **1940** *Chambers's Techn. Dict.* 191/1 *Connexion-* (or *connecting-*) *box*, a box containing terminals to which are brought a number of conductors of a wiring or distribution system, in order to facilitate the making of connexions between them.
6. c. A supplier of narcotics; the action of supplying narcotics. *slang* (orig. *U.S.*).
1934 *Detective Fiction Weekly* 21 Apr. 113/2 The person from whom the addict buys his stuff is called a connection. **1938** F. CHESTER *Shot Full* ii. 14 Her husband..had not had a chance to tell her his connexion for obtaining the drug. *Ibid.* 17 To his extremely numerous clients, the drug-addicts, he was known as..a sure connexion. **1957** J. KEROUAC *On Road* (1958) xiii. 88 A couple of Negro characters whispered in my ear about tea...The connection came in and motioned me to the cellar toilet. **1960** J. GELBER (*title*) The connection. **1967** M. M. GLATT et al. *Drug Scene* vi. 71 For his first connection he paid £1 per grain.

connexionism (kǫne·kʃəniz'm). *Psychol.* Also **connectionism.** [f. CONNEXION + -ISM.] The doctrine that mental processes involve a bond or connexion between stimulus and response; the theory that learning occurs by the formation of such connexions. Hence **conne·xionist, conne·ctionist** *a.* and *sb.*
1932 E. L. THORNDIKE *Fundamentals of Learning* i. 4 We have tried..to work out..a reasonable solution so far as is possible from existing knowledge. This solution takes the form of a new associationism, or better,..of a new connectionism. *Ibid.*, We have studied..[a] case of apparent transcendent activity as part of a general study of systems which organize and direct trains of thought. The results,..justify the connectionist's faith. *Ibid.* xvii. 428 Many of the arguments..directed against connectionism are valid only against a narrow and oversimplified connectionism which would assert that mental life was nothing but a series of small isolated elements. *Ibid.*, Many of the criticisms of connectionist psychology made by advocates of various forms of purposive psychology..consist in..the thesis that they can unite many names and the like, and that consequently connectionist psychology is unsound. **1941** *Mind* L. 415 The connectionist hypothesis is impossible because of Lashley's rats. **1951** G. HUMPHREY *Thinking* iii. 99 The theory..was upheld by Thorndike under the title of 'connectionism'. **1964** W. F. HILL *Learning* p. ix, I find the familiar division into connectionist and cognitive theories still the most useful.

connexity. Add: **3.** *Math.* and *Logic.* The property of being CONNECTED (*5).
1905 E. V. HUNTINGTON in *Ann. Math.* 2nd Ser. VI. 157 If a and b are distinct elements of K, then either a < b or b < a. This postulate has been called by Russell the postulate of connexity. **1907** E. W. HOBSON *Theory of Functions* § 40 The most important properties of a function may still subsist even if the domain of the variable lacks the property of connexity. **1925** A. N. WHITEHEAD *Science & Mod. World* (1926) 235 The third condition to be satisfied by an abstractive hierarchy will be called the condition of connexity. **1934** COHEN & NAGEL *Introd. Logic* vi. 115 A relation with this property is said to have connexity. **1941** A. TARSKI *Introd. Logic* vii. 161 These theorems are known as the 'laws of connexity'.

conning, *vbl. sb.*[2] Add: **conning-tower,** (*b*) a superstructure on a submarine in which the periscope is mounted and from which steering, firing, etc., are directed when the submarine is on or near the surface.
1886 *Graphic* 17 Apr. 410/3 A New Submarine Vessel... A conning tower is placed on the top of the vessel, in the sides of which are ports which enable the steersman to see in every direction, and which is covered by a strong water-tight scuttle for access to the interior of the boat. **1902** A. LANG *Disentanglers* 406 Periscope not necessary with conning-tower out of water. **1915** W. E. DOMMETT *Submarine Vessels* iv. 42 All hatches are closed and water admitted to the tanks until the deck, which is just below the base of the conning-tower, is at the level of the water. **1955** *Times* 20 June 4/6 Last week your Correspondent saw some of the demonstrations from the conning tower and through the periscope of the submarine Tapir.

conniption. Add: Also *Comb.*, as *conniption-fit.* (Earlier and later U.S. examples of both forms.)
1833 SEBA SMITH *Major J. Downing* 209 Ant Keziah fell down in a conniption fit. **1844** 'JON. SLICK' *High Life N.Y.* II. 171 By Golly! it was enough to drive any human critter into a conniption-fit! **1848** in *Dict. Amer. Eng.*, *Conniption*, fainting-fit. **1859** *Harper's Weekly* 19 Nov. 747/1 She..went into a conniption at the sight of poor Snap. **1911** H. QUICK *Yellowstone N.* xi. 290 One of the lawyers..threw a conniption fit every block. **1948** *Sat. Rev.* 19 June 4/3 The idea sounded fine on paper and gave story editors of rival studios conniption fits, but did not work out in practice. **1963** M. MCCARTHY *Group* i. 14 Conny Storey's fiancé..was working as an office boy..and her family, instead of having conniptions, was taking it very calmly.

connoisseur, *v.* Delete 'nonce-wd.' and add later example. Also **connoisseu·ring** *ppl. a.*
a **1821** KEATS *Cap & Bells* xlvii, Admired it with a connoisseuring look. **1966** K. GILES *Provenance of Death* iv. 114, I didn't know you connoisseured these arts, sir.

connotational (kǫnotēⁱ·ʃənăl), *a.* [f. CONNOTATION + -AL.] Involving connotation.
1947 *Jrnl. Clin. Psychopath.* VIII. 883 Connotational tendencies..are best expressed symbolically in poetry and in dreams. **1957** *Sound & Poetry* 138 Word-repetition..repeats the..word as a primary structural factor with the same denotational semantic spectrum but with connotational spectra forced on it by its contextual environment. **1965** *Language* XLI. 157 Determinational relations (and connotational ones, since they are inverse to one another) answer to the question [etc.].

conquer, *v.* Add: **6. b.** *Mountaineering.* To climb successfully.
1883 H. FRITH *Ascents & Adventures* ii. 27 The White Mountain [*i.e.* Mont Blanc] had been conquered by stern pluck and resolution. **1892** C. T. DENT et al. *Mountaineering* 367 Most of the principal peaks had been conquered. **1902** *Daily Chron.* 5 Nov. 7/3 The same mountain was 'conquered'..by an Englishwoman. **1903** *Ibid.* 29 Oct. 3/1 Three of the more important peaks had been conquered.

conquer, var. *CONKER[2].

conquering, *vbl. sb.* **b.** *colloq.* The act of playing 'conkers' (see *CONKER[2] a). *rare.*
1821 R. SOUTHEY *Let.* 28 Dec. in C. C. Southey *Life & Corr.* (1849) I. 55 One..amusement, which I never saw or heard of elsewhere, was greatly in vogue at this school [*sc.* Corston]. It was performed with snail shells, by placing them against each other, point to point, and pressing till the one was broken in, or sometimes both. This was called conquering... A great conqueror was prodigiously prized and coveted.

conqueror. Add: **1. f.** = *CONKER[2].
1821 [see *CONQUERING *vbl. sb.* b]. **1829**, etc. in I. & P. Opie *Children's Games* (1969) vii. 232. **1884** R. HOLLAND *Gloss. Chester* 78 *Conquerors*, a game played with horse-chestnuts threaded on a string...The chestnut which remains unhurt is then 'conqueror of one'. **1913** C. MACKENZIE *Sinister St.* I. viii. 109 Michael much regretted that the etiquette of the Lent Term, which substituted peg-tops for Conquerors, should prevent his chestnut reaching four figures.

conquian: see *COON-CAN.

Conradian (kǫnræ·diăn), *a.* [f. the name of Joseph *Conrad* (original name: Teodor Josef Konrad Korzeniowski) + -IAN.] Of, pertaining to, or characteristic of Joseph Conrad (1857–1924), Polish-born writer of novels in English, or his work.
1921 M. ARLEN *Romantic Lady* 4 The labyrinth of Lord Jim's career, and..many another such intricacy of Conradian imagination. **1936** *Times Lit. Suppl.* 27 June 542/1 An almost Conradian sense of the movement and power of noble and remorseless time. **1941** *Scrutiny* X. 157 The characteristic Conradian irony. **1959** *Times Lit. Suppl.* 20 Mar. 157/4 A Conradian narrator-within-a-narrator device

con-rod, abbrev. of *connecting rod.*
1931 T. E. LAWRENCE *Lett.* (1938) 737 If you, or it, had been a con-rod, how technically I should have balanced you and summed up. **1957** S. MOSS *In Track of Speed* x. 125, I was soon out with a broken con-rod.

consanguineal, *a.* Delete *rare* and add later examples.
1957 V. W. TURNER *Schism & Continuity in Afr. Soc.* ix. 270 None of these men had close consanguineal ties with either Shika or Mukanza, although Ng'ombi had married Mukanza's daughter Koniya. **1965** *Canad. Jrnl. Linguistics* Spring 159 All consanguineal kin are consistently distinguished as to generation...Affinal terms form a structure that, unlike English, does not simply reflect the consanguineal structure.

conscience. Add: **16. b.** *conscience-ridden, -stricken* (earlier example); **c. conscience money** (earlier and later examples); also, money paid to ease one's conscience.
1848 THACKERAY *Van. Fair* xliv. 396 One reads in the columns of the *Times* newspaper..queer announcements from the Chancellor of the Exchequer, acknowledging the receipt of £50 from A.B...as conscience-money, on account of taxes due. **1907** *Westm. Gaz.* 3 Dec. 2/3 Those who have to contribute the tax in the form of 'conscience money'. **1961** P. USTINOV *Loser* ii. 35 Two men he had never seen before rose from their tables and paid the difference... It was conscience money. **1898** G. B. SHAW *Perf. Wagnerite* 64 Our guilty and conscience-ridden generations. **1959** J. CARY *Captive & Free* 207 How typical she was. So nice, so gentle, so conscience-ridden, ripe for the slaughter. **1816** JANE AUSTEN *Emma* II. xvi. 302 She was more conscience-stricken about Jane Fairfax than she had often been.

conscient, *a.* (Later examples.)
1929 R. BRIDGES *Testam. Beauty* 6 How small a part of Universal Mind can conscient Reason claim! *Ibid.* 126 Thru' conscient personality and love of beauty. *a* **1963** C. S. LEWIS *Poems* (1964) 96 The thought that moves Within my conscient brain.

conscientious, *a.* Add: **1. b.** *conscientious objector*, one who refuses to conform to the requirements of a public enactment on the plea of conscientious scruple; *esp.* such an objector to military service (cf. *CONCHY).
1899 [see OBJECTOR]. **1910** *Blackw. Mag.* Mar. 424/1 Having quite properly insisted that vaccination should be universal and compulsory, it presently admits 'the conscientious objector'. **1916** A. HUXLEY *Let.* 2 Mar. (1969) 92 Conscientious objectors were not so disgustingly hectored as they seem to have been in London. **1940** *Economist* 2 Mar. 371/1 The Act lays down that one who claims that he conscientiously objects to being registered in the military service register, or to performing military service, or to performing combatant duties, may apply to be registered as a conscientious objector and therefore to be exempt from military service. **1966** C. M. BOWRA *Memories 1898–1939* xiv. 352 Knowing how cruelly conscientious objectors had been treated in the First World War, I was determined to see that their cases were put intelligibly before the tribunals.
2. *conscientious objection.* (See above.)
1916 A. HUXLEY *Let.* 10 Mar. (1969) 93 He came before the tribunal the other day to apply for exemption on the grounds of conscientious objection to war. **1932** —— *Brave New World* iii. 57 Conscientious objection on an enormous scale. **1936** M. PLOWMAN *Faith called Pacifism* 60 Passive resistance, Non-violence, Conscientious Objection, Pacifism, Defeatism—the very variety of descriptions shows the hesitant..nature of the new faith. **1966** C. M. BOWRA *Memories 1898–1939* xiv. 352, I welcomed the introduction of conscription, but was at first rather taken aback that some of the brightest undergraduates had genuinely conscientious objections to it.

conscious, *a.* Add: **6. f.** *absol.* with *the*: the conscious mind.
1919 M. A. K. BRADBY *Psycho-analysis* iii. 34 They figure in her dreams in forms which imply moral condemnation in the unconscious as well as in the conscious, as demons or brutal people. **1960** HINSIE & CAMPBELL *Psychiatric Dict.* (ed. 3) 150/2 In psychiatry *conscious* is used..(usually) as a noun, to denote a particular division of the psyche. In such use it is practically synonymous with *consciousness*.
12. Appended to sbs. forming adjs. with the sense 'conscious of —, aware of —'; as **class-conscious*, **colour-conscious*, etc. (see the sbs.).
1903 [see *CLASS *sb.* 9]. **1918** [see *dress-conscious*, *DRESS *sb.* 4 a]. **1928** [see *clothes-conscious*, *CLOTHES *sb. pl.* 4]. **1933** *Punch* 5 July 25/1 We learn to eat at least a twelvemonth before we learn to talk; not till several years after that do we become money-conscious. **1934** H. G. WELLS *Exper. Autobiogr.* I. ii. 79, I became woman-conscious from those days onward. **1938** B. RUSSELL *Power* 166 It [*sc.* a chess-club] might..seek to make more people 'chess-conscious'. **1959** *Times Lit. Suppl.* 20 Mar. p. vi/3 A culture so history-conscious. **1970** *Daily Tel.* 18 July 10 City of London policemen are to be encouraged to weigh themselves regularly and become more weight-conscious.

consciousness. Add: **8.** *Comb.*, as *consciousness-expanding a.* = *PSYCHEDELIC *a.*
1967 *Psychic News* 26 Aug. 7/4 Mescalin, LSD and the like have an awakening, consciousness-expanding effect. **1968** J. DRUMMOND *Gantry Episode* iii. 34 They talk about a psychedelic drug, a 'consciousness-expanding' drug.

conscript, *v.* Delete 'It appears to have originated during the U.S. Civil War of 1860–65.' (Earlier U.S. examples; also examples of extended use.)
1813 *Connecticut Courant* 23 Nov. 3/5 State troops..had been conscripted under the orders of the former Captain General. **1814** *Columbian Centinel* 21 Dec. 2/5 The bill for drafting, or Conscripting, the Militia, has passed both Houses of Congress. **1899** *Daily News* 8 Mar. 7/1 Under the name of a 'Railway Brigade' a corvée pure and simple has been conscripted by the Sirdar during the Soudan campaign. ..So-called soldiers are daily conscripted..simply for the construction of the Khartoum Railway. **1932** E. WAUGH *Black Mischief* iv. 137 It was necessary for the Emperor to

leave his car and complete the journey on mule back, his luggage bobbing behind him on the heads of a dozen suddenly conscripted spectators. **1941** L. B. LYON *Tomorrow is Revealing* 47 The sea-shore pebble, the conscripted sand is covered By the stealing washing-over-us-water.

conscriptionist, (kǫnskri·pʃənist). [f. CONSCRIPTION + -IST.] An advocate of conscription. Also *attrib.* or as *adj.*
1901 *Daily News* 14 Mar. 7/1 Colonel Dooner is not a conscriptionist. **1907** *Westm. Gaz.* 6 June 2/2 Not of the Conscriptionist school. **1917** *Grit* (Toronto) 13 Dec. 2/5 Graham is a conscriptionist, too. **1920** *Edin. Rev.* July 105 Conscriptionist Liberals. **1928** *Daily Tel.* 23 Oct. 11/3 Technical difficulties in the way of limiting trained reserves in a conscriptionist country are very great indeed.

conscriptive (kǫnskri·ptiv), *a.* [f. CONSCRIPTION + -IVE, after *descriptive.*] Involving or having conscription.
1906 J. JOYCE *Let.* 9 Oct. (1966) II. 174, I don't see how that saves them from the logical conclusion of revolution in a conscriptive country like this. **1915** *Morn. Post* 9 Feb. 4/4 It was possible to make a fair computation of the armies under a conscriptive system.

consensus. Delete ‖ and add: **3.** *attrib.* and *Comb.*
1966 *New Statesman* 21 Oct. 583/3 The essence of consensus politics is directly related to consensus communications. **1967** *Listener* 3 Aug. 136/2 Consensus journalism—with its millionaire proprietors and its multitudinous advertising departments—has been praised for being politically permissive. **1968** *Peace News* 10 May 10/2 Cicero..is a more dubious case—an unsuccessful consensus-politician, if ever there was one.

consenting, *ppl. a.* Add: **3.** *consenting adult*: an adult who consents to something, esp. to an act of homosexuality; a homosexual.
1957 *Rep. Comm. Homosexual Offences & Prostitution* (Cmnd. 247) 109 We accordingly recommend that homosexual behaviour between consenting adults in private should no longer be a criminal offence. **1964** *Times* 16 July 12/4 Chief constables in England and Wales have been asked to inform the Director of Public Prosecutions..before preferring charges over homosexual acts committed in private between consenting adults. **1968** 'O. MILLS' *Sundry Fell Designs* x. 112 No, I'm not a consenting adult. But I'm not sex-starved, either.

consequent, *a.* Add: **8.** *Geol.* (See quots. 1904 and 1960.)
1875 J. W. POWELL *Explor. Colorado River* xi. 163 Valleys are found having directions dependent on corrugation. I propose to call these consequent valleys. **1895** *Geogr. Jrnl.* V. 127 The terms, consequent, subsequent, and obsequent, have been applied to rivers by Professor W. M. Davis. **1904** CHAMBERLIN & SALISBURY *Geol.* (1905) I. 74 Streams and valleys, the courses of which are determined by the original slope of the land, are said to be *consequent*. **1930** PEACH & HORNE *Geol. Scot.* 10 Consequent valleys. **1936** *Geogr. Jrnl.* LXXXVII. 21 It has been attributed to parallel consequent drainage. **1960** L. D. STAMP *Britain's Struct. & Scenery* (ed. 5) v. 41 Rain falling will collect into streams, roughly parallel, finding the shortest route seawards. The streams are consequent on the slope and hence are known as consequent streams.

consequential, *a.* Add: **2.** *consequential damages* (examples); also *consequential loss.*
1701 in *Modern Reports* (ed. 2, 1741) XII. 635 By the Erection he [*sc.* the Defendant] is become liable to the Plaintiff for all the consequential Damages. **1772** *Jacob's Law-Dict.* (ed. 9), *Consequential losses*, or *damages*, it is a fundamental principle in law and reason, that he who does the first wrong shall answer for all consequential damages. **1962** *Which?* (Car Suppl.) Oct. 140/2 The *Ford Taunus* guarantee..excluded any consequential loss.

conservation. Add: **1. c.** *spec.* The preservation of the environment, esp. of natural resources. Also *attrib.*
1922 *Encycl. Brit.* XXX. 102/2 A very important by-product of the conservation-movement was the development at Washington of a mania for the establishment of reservations in Alaska. **1958** *New Biol.* XXVI. 63 (*heading*) Conservation as a world problem. *Ibid.* 69 It matters little what we call these areas—nature reserves or conservation areas—as long as we recognize their function. *Ibid.* 70 (*heading*) Nature conservation in Britain. **1961** E. A. POWDRILL *Vocab. Land Planning* iii. 55 'Conservation' is also defined as a phase of renewal that attempts to conserve those areas not yet blighted; and conservation is achieved by eliminating those elements that create blight. **1969** *Guardian* 30 Aug. 8/2 The Tanzanian Government has been active in promoting conservation policies in the parks.
4. b. conservation law, in Physics, any law of nature according to which some property is conserved, either in certain types of change or interaction only or in all circumstances.
1947 DIRAC *Quantum Mech.* (ed. 3) Index 310 Conservation laws. **1949** T. B. BROWN *Found. Mod. Physics* (ed. 2) 356 In fundamental physical theory, first place is still occupied by the conservation laws for electric charge and for momentum. **1968** M. S. LIVINGSTON *Particle Physics* v. 90 The first four of these conservation laws were recognized during the era of classical science to explain the large-scale phenomena observable with the crude instruments of a hundred years ago.

conservationist (kǫnsəɹvēⁱ·ʃənist). [f. CONSERVATION + -IST.]. A proponent or advocate of conservation, esp. one who advocates the conservation of natural resources and amenities. Also *attrib.* or as *adj.*
1870 J. S. MILL *Lett.* (1910) II. 251 Now, at the expense of what pre-existing energy has this force been generated? The conservationists are obliged to say, out of potential energy. **1922** *Encycl. Brit.* XXX. 102/2 The most ardent of the conservationists failed to recognize the urgent importance of conserving the salmon and halibut fisheries. **1962** *Times* (Standard Bank Suppl.) 12 Oct. p. vii/1 The conservationist who put the conservation of veld life before human welfare had no hope of success. **1969** *Guardian* 12 Aug. 2/5 The conservationists are afraid that the noise of the aircraft..will seriously affect both the peace and the ecology of the park. **1969** N. W. PIRIE *Food Resources* ii. 69 The conservationists fear that many at present wild species will become extinct. **1970** *Nature* 18 July 225/2 'Conservationist' opposition to the construction of nuclear power plants is slowing the building programme.

conservative, *a.* Add: **3. b.** Characterized by caution or moderation; (esp. of an estimate) purposely or deliberately low. orig. *U.S.*
1900 *Daily News* 28 Apr. 5/6 This is a conservative estimate of the loss caused by yesterday's fire here and at Hull. **1903** *N.Y. Tribune* 6 Sept., The visitors were spending on the island the enormous total of $1,500,000 a day..and this is a conservative estimate. **1916** *Times* 15 Nov. 8/3 A conservative estimate makes the figure today at least half a million. **1921** LANSING *Peace Negot.* 155 It is conservative to say that between two and three months were spent in the drafting of a document which in the end was rejected by the Senate. **1928** *Daily Tel.* 25 Sept. 15/3 The distances quoted are conservative.

conservatively, *adv.* Add later examples illustrating wider usage.
1887 *Philad. Ledger* 30 Dec. (Cent. D.), It is very conservatively English to make concession at the eleventh hour and fifty-ninth minute; but the clock is fast in Ireland. **1906** *Daily Chron.* 8 Nov. 5/3 Modification of the tariff and public ownership inaugurated, but both should be done cautiously and conservatively. **1907** *Westm. Gaz.* 28 Aug. 2/3 'A Conservatively Cooked Green Vegetable'..is one more example of the growing fashion, largely owing to American usage, and not least that of Wall Street, of employing the word 'conservative' in many relations to which politicians are..unaccustomed.
2. On a 'conservative' estimate. orig. *U.S.*
1904 *San Francisco Chron.* 12 July 7 Wearing a diamond pin, conservatively valued at $75. **1928** F. N. HART *Bellamy Trial* i. 2 There were, conservatively, about three thousand crowding the corridors and walking the streets.

conse·rvativeness. = CONSERVATISM.
1838 H. WINSLOW *Perpetuity of our Republic* 19 That the comparative silence of those who stand upon the side of conservativeness, implies a gradual concession. **1928** *Daily Express* 12 Dec. 3 A method of doing business which, in spite of the conservativeness to which he refers, is rapidly gaining ground.

conservativism (kǫnsɜ̄·ɹvātiviz'm). = CONSERVATISM.
1840 CARLYLE *New Lett.* (1904) I. 183 Most of them seem to accuse me of a leaning to '*Conservativism*'. **1895** W. J. LOCKE *At Gate of Samaria* xviii. 210 The broader Conservativism of our day, between which and modern Liberalism he failed to appreciate the distinction. **1921** *Public Opinion* 28 Jan. 76/1 The enemies of Liberalism have confidently predicted its smash between the collision of Communism and Conservativism.

conservatrix. (Later U.S. examples.)
1798 I. ALLEN *Hist. Vermont* 279 The art, which is the conservatrix of all others, printing. **1920** E. POUND *H. S. Mauberley* 19 Conservatrix of Milésien.

conshy, var. *CONCHY.

considerable, *a.* and *sb.* Add: **A.** *adj.* **6. b.** Freq. *absol.* followed by *of.* (Cf. B. 2 below.) *U.S.*
1685 in *Mass. Hist. Soc. Coll.* 4th Ser. V. 132 One more, of Long Island,..lost considerable of wheat and Indian corn. **1766** M. CUTLER in Parker & Cutler *Life, Jrnls. & Corr.* (1888) I. 10 This morning..considerable of a shock of an earthquake was felt. **1779** in *Coll. New Hampsh. Hist. Soc.* (1850) VI. 326 This is considerable of a village. **1816** [see Dict. sense 6]. **1835** *Southern Lit. Messenger* I. 422 [It is] considerable of a place. **1856** *Trans. Mich. Agric. Soc.* VII. 652 Considerable of the crop was, however, injured in quality by the frequent wettings it received. **1878** J. H. BEADLE *Western Wilds* xix. 302 In two days intercourse we had learned considerable of each other's views and experiences. **1903** K. D. WIGGIN *Rebecca* 132 Rebecca took her scolding..like a soldier. There was considerable of it. **1904** *N.Y. Even. Post* 26 Mar., Considerable of the moisture for which farmers have been eagerly watching has fallen this past week over the wheat belt.
7. *adv.* (Later U.S. examples.)
1849 N. KINGSLEY *Diary* (1914) 54 The sea is considerable rough. **1862** O. W. NORTON *Army Lett.* (1903) 48, I presume I do have considerable more time for writing than you do. **1884** 'MARK TWAIN' *Huck. Finn* i, By-and-by she let it out that Moses had been dead a considerable long time.
B. *sb.* Restrict † to sense in Dict. and add: **2.** A fair amount, quantity, etc., *of* something. (Cf. 6 b above.) *U.S.*
1745 in *Essex Inst. Hist. Coll.* XLVIII. 300 A Considerable of Cannon shot [was] hove at them. **1780** in *Mass. Hist. Soc. Coll.* (1905) 7th Ser. V. 5 A considerable of a trade carried on. **1838** HALIBURTON *Clockm.* 2nd Ser. ii. 10 I've been a considerable of a traveller in my day. *Ibid.* vii. 102 It's a considerable of a long story too. **1943** W. FAULKNER in *Sat. Even. Post* 13 Feb. 15/2 It was a kind of mixed hound, with a little bird dog and some collie and maybe a considerable of almost anything else.
b. A good deal.
1722 in H. S. Nourse *Early Rec. Lancaster, Mass.* (1884) 200 He promising to Leve the hiway there Wider by Considerable then five Rods. **1837** H. MARTINEAU *Society in Amer.* I. 340 She has a widowed mother, and she 'gets considerable' by sewing. **1850** N. KINGSLEY *Diary* (1914) 110 They had been detained a considerable on the passage. **1903** *N.Y. Sun* 22 Nov., The new seal is not Nellie, nor by considerable; but it is already a great attraction. **1911** J. C. LINCOLN *Cap'n Warren's Wards* 178 The Boston papers had considerable about it. **1932** *Atlantic Monthly* Feb. 188/2 Jim did considerable toward stirring up the farmers.

consigne. Add: Also in anglicized form **consign.**
1923 *Chambers's Jrnl.* Dec. 798/2 The sentry..let me pass without any consign.
2. A left-luggage office in France.
1897 E. DOWSON *Let.* *c* 18 Oct. (1967) 393, I found my consigne shut on my arrival at Dieppe, so was obliged to write to the chef de gare to forward me my things here. **1926** HEMINGWAY *Fiesta* (1927) III. xix. 267 A porter carried Bill's bags in from the consigne. **1965** G. HOUSEHOLD *Olura* 146, I put my baggage back in the *consigne* and took a train to St. Étienne.

consistometer (kǫnsistǫ·mitəɹ). [f. CONSIST(ENCY + -OMETER.] A device for measuring the consistency of a viscous or plastic material.
1911 *Chem. Abstr.* 3617 (*title*) A new consistometer for testing bituminous road materials. **1925** *Jrnl. Physical Chem.* XXIX. 1217 A consistometer, i.e., an instrument which may be used to distinguish between a viscous and a plastic material, and to measure the consistency of either. **1930** *Bur. Standards Jrnl.* V. 83 The consistometer..can be employed for either opaque or clear materials of a wide range of consistencies. **1956** *Gloss. Terms Concrete* (*B.S.I.*) 15 *Consistometer*, an apparatus for measuring the consistency of concrete.

consociation. Add: **5.** *Ecology.* A subdivision of an *ASSOCIATION, dominated by a single species.
1905 F. E. CLEMENTS *Res. Methods Ecol.* iv. 296 The primary areas..have been called associations... It is proposed to term this primary division of the formation a *consociation*, or better, a *consocies*. This term is applied only to an area characterized by a facies. **1916** —— *Plant Succession* vi. 137 The consocies is a seral community marked by the..dominance of one species... It is the unit of the associes in the same way that the consociation is of the association. **1926** TANSLEY & CHIPP *Study of Veg.* ii. 18 The names *association, consociation, society* and *clan* are given to the communities of stable (climax) vegetation in descending order. **1938** WEAVER & CLEMENTS *Plant Ecol.* (ed. 2) iv. 94 In its typical form the consociation is constituted by a single dominant. **1960** N. POLUNIN *Introd. Plant Geogr.* xii. 345 One or another type of northern coniferous forest (or, occasionally, its broad-leafed consociation of Birch or consocies of Aspen) occupies most of the northernmost belt of forested terrain.

consocies (kǫnsōᵘ·ʃiˌiz). *Ecology.* [mod.L., f. CONSOC(IATION, after SPECIES.] = *CONSOCIATION 5; *spec.* a consociation of plants in a developmental stage.
1905, 1916 [see prec.]. **1916–64** [see *ASSOCIES].

‖**consolamentum** (kǫnsōᵘ:lāme·ntəm). [mod.L., f. L. *consōlārī* (see CONSOLE *v.*).] The spiritual baptism amongst the Cathars, by which the recipient is elected to be one of the 'perfect'.
1874 J. H. BLUNT *Dict. Sects* 112/2 By this 'consolamentum' the recipients were supposed to escape purgatory and pass at once into paradise. **1970** *Man, Myth & Magic* xv. 423/2 The 'perfect' underwent a long and rigorous initiation which culminated in a religious ceremony, the *consolamentum*, which was the Cathar equivalent of Christian baptism, confirmation and ordination of a priest, all rolled into one.

‖**consolatio** (kǫnsǫlēⁱ·ʃio, -ā·tio). [L. *consōlātiō* CONSOLATION.] A consolatory treatise, letter, or poem (sometimes an alternation of prose and verse) in Greek, Latin, or a vernacular tongue, setting forth philosophical or religious themes as comfort for the misfortunes of life.
1935 H. R. PATCH *Trad. Boethius* 121 Writers..borrowed from the scheme of Boethius without catching anything of his spirit. This is more or less true in the numerous debates with Fortune, in the many remote *Consolationes*, and in the works which took advantage of the variety gained by alternating prose and verse. **1953** W. R. TRASK tr. *Curtius' Europ. Lit.* v. 80 Agius of Corvey heightens this biblical *consolatio* in the poem he composed in 876 on the death of the Abbess Hathamod. *Ibid.* 81 As the writers of *consolationes* discovered, it makes little difference whether a person dies young or old. **1963** G. N. GARMONSWAY in Brown & Foote *Early Eng. & Norse Studies* vii. 82 A paraphrase of verses from the Book of Job in the Office of the Dead written *c.* 1424 as a *consolatio* for the imprisoned Duke of Bedford. **1969** DUNNING & BLISS *Wanderer* 80 The poem is an example..of the genre *consolatio*.

consolation. Add: **3. b.** *consolation prize* (see PRIZE *sb.*[1] 1); now usually, a prize given to a competitor who has not won one of the stipulated prizes; also *fig. Dutch consolation*: see DUTCH *a.* 4.
1886 *Wesleyan-Meth. Mag.* Jan. 58 Only sixty-one can be successful, while nine others are granted the consolation prize of a half-degree. **1899** [see PRIZE *sb.*[1] 1]. **1928** *Manch. Guardian Weekly* 2 Nov. 352/2 The thousand-pound prize gives help to one (perhaps there are some consolation prizes) who..is quite good enough to look after himself. **1963** *Listener* 7 Mar. 407/1 Let us see whether there are not some consolation prizes for our failure to qualify for membership in E.E.C.

console, *sb.* Add: **2. b.** A cabinet for a gramophone, radio, television set, tape recorder, etc. Also *attrib.* orig. *U.S.*
1925–6 *T. Eaton & Co. Catal.* Fall & Winter 391/1 The new model console phonograph. **1926–27** *Sears, Roebuck Catal.* 599 Console Cabinet...A cabinet that is designed to accommodate a radio receiver, charger and all batteries. **1958** *Economist* 26 July 271/2 Some present tape-recorders..do not need to be portable... In recognition of this, console models have been on the market.
3. b. A desk, cabinet, or the like, incorporating switches, dials, etc., for the control of electrical or other apparatus; a control panel; a switchboard. Also *attrib.*
1944 R. H. DELANY in *Proc. Inst. Radio Engin.* XXXII. 600 A broadcast-studio control console. *Ibid.*, A plywood mockup was made of the proposed console cabinet. **1946** *Electronic Engin.* XVIII. 360 The console unit..forms the central unit from which the set is controlled and where the Radar information is presented. **1948** *Gloss. Techn. Theatr. Terms* (Strand Electric Co.) 8 *Console*, a mobile remote control for stage lighting resembling and using certain of the accessories of the cinema organ. **1951** *Engineering* 27 July 101/3 A console in which are grouped all the controllers and recorders for operating an industrial..plant. **1961** G. MILLERSON *Telev. Production* i. 16 Liaison with the lighting director..can be facilitated by arranging adjacent video control and lighting consoles. **1964** T. W. MCRAE *Impact of Computers on Accounting* viii. 236 The person who operates the computer is called the console operator. **1966** *New Yorker* 12 Nov. 100 An enormous computer console.

consolement. (Later examples.)
1904 G. ATHERTON *Rulers of Kings* 25 Mr. Abbott had brought him a fishing-rod, which appeared to afford immediate consolement. **1905** *Daily Chron.* 30 May 3/1 To live for thy consolement..is my mission. **1930** G. GREENE *Two Witnesses* vi. 80, I can remember the consolement of the still trees.

consolidated, *ppl. a.* Add: **2.** *spec.* (see quot. and *CONSOLIDATION 5).
1887 *Sci. Amer.* LVI. 3/2 The locomotive was one of the heaviest kind known as a consolidated engine, having four drive-wheels on a side, and weighing 106,000 pounds.

consolidation. Add: **5.** *U.S.* In full *consolidation locomotive*: see quot. *a* 1884 and *CONSOLIDATED *ppl. a.* 2.
1875 M. P. FORNEY *Catechism of Locomotive* 432 Consolidation locomotives,..which have eight driving-wheels, are employed almost exclusively for traffic over heavy mountain grades. *a* **1884** KNIGHT *Dict. Mech.* Suppl., *Consolidation* (Locomotive), a type of freight locomotive, the name of the engine, the first in its class, built in 1866, at the Baldwin locomotive works. **1898** *Engineering Mag.* XVI. 160/1 Consolidation Locomotive for the Chesapeake & Ohio. **1947** L. M. BEEBE *Mixed Train Daily* 67 Its motive power is all drawn from the ranks of the Southern's ten-wheelers and Consolidations.

consolidationist. *U.S.* Add: esp. one who advocates federal rule.
1833 D. WEBSTER *Sp.* in *Wks.* (1851) I. 295 For one, I repel all such imputations. I am no consolidationist. **1835** P. H. NICKLIN *Virginia Springs* 29 From the east you have consolidationists, tariffites and philanthropists.

consommé. Add: Now esp. applied to clear soup. Also *fig.* (cf. 'in the soup'). (Add earlier and later examples.)
1815 SIMOND *Jrnl. Tour & Resid. Gt. Brit.* I. 45 The soup is always a *consommé*, succulent, and high-seasoned. **1831** *Athenæum* 22 Jan. 53/1 Quacks in literature..subtilize into essence, or reduce into consommé. **1842** THACKERAY *Fitz-Boodle Papers* in *Wks.* (1879) XVII. 209 Refreshments of a nature more substantial than mere tea—punch, both milk and rum, hot wine, *consommé*. **1846** SOYER *Cookery* 53 Consommé, or clear Soup. **1907** G. A. ESCOFFIER *Guide Mod. Cookery* xiii. 197 Clear soups, which include plain and garnished consommés. **1909** *Westm. Gaz.* 1 Feb. 2/1 Friends have..supplied brandy, Marsala, *consommé*, and strong coffee; which have made an immense difference to the patients. **1920** *Blackw. Mag.* Aug. 139/1 In the smaller restaurants, cocktails were served in *consommé* cups. **1957** M. SHARP *Eye of Love* ii. 21 Amalgamating with Joyces' gets us out of the consommé. **1959** 'A. GILBERT' *Death takes a Wife* v. 53 Here's another husband in the *consommé*.

consonant, *sb.* Add: **4.** *consonant-cluster, -group*; **consonant shift** *Philology*, a change in consonantal sounds, *spec.* that which took place (1) in the development of the Germanic languages, as set forth in the formula known as Grimm's Law, (2) in the later development of High German from Common West Germanic.
1933 BLOOMFIELD *Language* xxi. 370 A phonetic change which consisted of shortening long vowels before certain consonant-clusters. **1953** C. E. BAZELL *Linguistic Form* 48 When initial and final consonant-clusters show more limitations in common than either with medial clusters, the relation of juncture may be described as dominant. **1888** H. SWEET *Hist. Eng. Sounds* (ed. 2) 169 Long vowels are regularly shortened in ME before two conss., except..before those cons.-groups which lengthen short vowels. **1924** A. MAWER *Chief Elements Eng. Place-Names* 6 *Blæc*, OE, '*black*', are often very difficult to distinguish owing to the shortening of vowel which the former may undergo in a compound before a consonant-group. **1965** W. S. ALLEN *Vox Latina* iii. 68 Complex consonant groups. **1888** SWEET *Hist. Eng. Sounds* 93 The second, or High German, consonant-shift. **1905** JESPERSEN *Growth Eng. Lang.* 23 The consonant-shift is important to the modern philologist. **1934** S. ROBERTSON *Devel. Mod. Eng.* (1936) ii. 31 Later philologists than Grimm found it necessary to qualify and modify his statement of the consonant-shift.

consonanted (kǫ·nsǫnăntėd), *a.* [f. CONSONANT *sb.* + -ED[2].] With a qualifier: having consonants (of a certain kind).
1893 *Athenæum* 14 Jan. 49/3 The girl meets..her fate, in the shape of a kind of cosmopolitan painter, with a toughly consonanted name. **1893** *Nat. Observer* 7 Jan. 190/2 She rejects the many consonanted name. **1895** *Athenæum* 13 July 57/3 The rhythm soon becomes graver, the lines charged with a more heavily consonanted burden of sound.

‖ **con sordino** (kǫn sǭɹdī·no), *advb. phr. Mus.* Pl. **-ini.** [It.; CON *prep.* + *sordino* (pl. *sordini*) (see SORDINE *sb.*).] With the mute or damper; muted. Also *fig.*
1825 in J. F. DANNELEY *Dict. Mus.* **1876** STAINER & BARRETT *Dict. Mus. Terms* 108/1 *Con sordini*..(1) With the mutes on. (2) With the soft pedal at the pianoforte held down. **1889** G. B. SHAW *London Music in 1888–89* (1937) 247 Can he not either stop it or play *con sordino*? **1921** —— *Let.* 13 Jan. (1952) 219 It is not by tootling to him *con sordino* that Lady Macbeth makes Macbeth say 'Bring forth men children only'. **1947** A. EINSTEIN *Music in Romantic Era* xvi. 283 Yet this festival opera ends *pianissimo* and *con sordini*. **1966** *Listener* 19 May 733/1 A small post-war British literary landmark, the *con sordini* fanfare of a new subject-matter.

consorter. (Later example.)
1924 *Blackw. Mag.* Jan. 42/2 Sir William Courteen was a man of power..and a consorter with the great.

consortium. Delete ‖ and add: Now commonly with pronunc. (kǭnsǭ·ɹtiǒm). Pl. **consortia.** Now more specifically, an association of business, banking, or manufacturing organizations. (Earlier and later examples.)
1829 *Edin. Rev.* L. 89 If the *consortium* give pleasure to the shades of these good people, we must acquiesce in it. **1930** *Time & Tide* 30 Aug. 1086 The bankers have formed a consortium to help rationalize industry. **1936** *Nature* 4 July 5/1 Preference was given for the execution of the work by consortia of landowners or public bodies. **1957** *New Scientist* 12 Sept. 31/1 Spokesmen for two of the consortia that tendered for the first CEA nuclear power stations. **1961** *Listener* 28 Dec. 1110/2 An interesting development..is the formation of a Yorkshire consortium of local authorities. The chairmen of the housing committees of Sheffield, Hull, and Leeds have announced that they are going to co-ordinate their housing programmes. **1962** H. E. BEECHENO *Business Stud.* xii. 107 Recently there have been several cases of manufacturers with connected interests forming a consortium in order to get large overseas contracts for capital developments. **1963** *Ann. Reg. 1962* 282 The formation of aid consortia for Turkey and Greece. The Turkish consortium was formed on 31 July.

conspicuous, *a.* Add: **3.** Designating expenditure on or consumption of luxuries on a lavish scale in an attempt to enhance one's prestige.
1899 T. VEBLEN *Theory of Leisure Class* iv. 75 Conspicuous consumption of valuable goods is a means of reputability to the gentleman of leisure. *Ibid.* iv. 96 Throughout the entire evolution of conspicuous expenditure, whether of goods or of services or human life, runs the obvious implication that in order to effectually mend the consumer's good fame it must be an expenditure of superfluities. **1926** B. WEBB *My Apprent.* i. 53 Competition in conspicuous expenditure on clothes, food, wine and flowers. **1962** E. GODFREY *Retail Selling* xxi. 214 In the past 'conspicuous' consumption, of the swimming-pool, cabin cruiser, high-powered sports car variety, was confined to..the idle rich.

conspiracy. Add: **2.** Phr. *conspiracy of silence.*
1865 MILL *Comte* 199 M. Comte used to reproach his early English admirers with maintaining the 'conspiracy of silence' concerning his later performances. **1870** *Gentl. Mag.* V. 331 Compel them, if they must divert themselves with a species of amateur conspiracy, to enter into a conspiracy of silence. **1885** *Church Reformer* July 146/1 When such work as the *Pall Mall* has done is treated by the ordinary press with a conspiracy of silence, then it becomes our duty..to thank that paper for its outspoken boldness. **1931** V. J. MCGILL *Schopenhauer* ix. 303 Their 'conspiracy of silence' had at last proved ineffective. **1944** G. B. SHAW *Everybody's Pol. What's What?* xxiv. 223 The conspiracy of silence about the blunders and failures of medicine. **1965** *Times Lit. Suppl.* 27 May 437/3 It seemed perfectly correct to me when I read the article in question, although I find that English grammars seem to maintain a conspiracy of silence on the subject.

conspirative (kǭnspi·rātiv), *a.* Restrict † *Obs. rare* to sense in Dict., and add: **2.** Engaged in, involving, or characterized by conspiracy.
1887 *Century Mag.* XXXV. 295 Who shall take part in collective and conspirative insurrection. **1921** *Contemp. Rev.* Jan. 29 Spies and active conspirative agents of the foreign and domestic enemy. **1922** *Chambers's Jrnl.* 289/2 The departure itself was arranged in almost a conspirative manner. **1927** *Observer* 11 Dec. 21/2 The Opposition..continued to maintain..its conspirative organisation.

conspiratorially (kǭnspirătō·riăli), *adv.* [f. CONSPIRATORIAL + -LY[2].] In the manner of a conspirator.
1912 CHESTERTON *Manalive* II. iii. 301 'They *said* it was England,' said my imbecile, conspirator[i]ally. **1927** *Scots Observer* 22 Jan. 16/4, I came tiptoeing conspiratorially across the dread threshold of the studio.

conspirer. (Later U.S. example.)
1896 *Voice* (U.S.) 30 July 4/5 Those who want to fight the great conspirer against human life should get out their battle-axes and go for the rum fiend.

constancy. Add: **3. c.** *Psychol.* (See quot. 1952.)
1924 R. M. OGDEN tr. *Koffka's Growth of Mind* v. 271 'Constancy-hypothesis', according to which a certain sensation always corresponds to a certain stimulus just as soon as ever the capacity for the sensation in question has been attained. *Ibid.* 295 This constancy of form becomes the child's mode of perception. **1929** W. KÖHLER *Gestalt Psychol.* (1930) iv. 93 Constancy of brightness and of size cannot be explained by the assumption of one-way conduction determining local sensory experience in terms of local stimulation. **1935** KOFFKA *Princip. Gestalt Psychol.* iii. 72 We may single out three characteristics of things which will ..be constitutive of things: shaped boundedness, dynamic properties, and constancy. **1940** *Brit. Jrnl. Psychol.* Jan. 261 The 'bundle' and 'constancy' hypotheses. **1952** DREVER *Dict. Psychol.* 49 *Constancy phenomena*, phenomena of perception, where psychological laws seem to cut across physical laws, so that perceived objects retain to some extent certain characteristics in relative independence of change in the stimuli affecting the sense organ. **1959** *Chambers's Encycl.* XI. 321/1 The perceptual constancies... All these effects in perception are technically known as the constancy phenomena. Constancy is seldom, if ever complete... Indeed the expression 'tendency to constancy' is more appropriate.
7. *Ecology.* The proportion of a particular species found in an ecological community.
1926 TANSLEY & CHIPP *Study of Veg.* ii. 10 The species may also be considered in regard to..their *constancy* in an association, i.e., of a large number of sample areas, the percentage number in which the given species occurs. **1950** *Jrnl. Ecol.* XXXVIII. 72 Several species of low constancy on the less exposed stands acquire high constancy on the more exposed.

constant, *a.* Add: **4. f.** *constant white*, = *permanent white* (see PERMANENT *a.* 1 d).
1854 F. W. FAIRHOLT *Dict. Terms Art* 127 *Constant white*, *permanent white*, a pigment prepared from the sulphate of barytes, useful in water-colour painting, and possessing great body. **1886** [see *baryta white* s.v. *BARYTA]. **1951** R. MAYER *Artist's Hand-Bk.* ii. 48 *Constant white*, blanc fixe.
5. b. In various *Combs.* (used chiefly *attrib.* with hyphen), as *constant current, frequency, mesh, speed, velocity, voltage.*
1904 GOODCHILD & TWENEY *Technol. & Sci. Dict.* 126/1 *Constant current*, a current whose amount or value is allowed to vary as little as possible. **1934** *Discovery* Nov. 324/2 D.C. to A.C. rotary converters,..constant current chargnig dynamos and an entirely portable petrol-driven alternator. **1962** SIMPSON & RICHARDS *Junction Transistors* vii. 156 The currents I'_{b1} and I_{b1} may reach considerably higher values than those obtainable with constant-current drive. **1935** *Discovery* Nov. 325/1 Some lengths of constant frequency film of known accuracy were obtained. **1940** *Chambers's Techn. Dict.* 191/2 *Constant-frequency oscillator*, an oscillator in which special precautions are taken to ensure that the frequency remains constant. **1956** *Nature* 10 Mar. 470/2 The accelerated electrons can stay in phase with the constant-frequency accelerating voltage. **1924** A. W. JUDGE *Mod. Motor Cars* II. iii. 54 (*heading*) Constant mesh gears. **1967** *Economist* 8 July p. xxx/1 Better drivers still prefer the cruder 'constant mesh' gearboxes where each gear is engaged by a small clutch and which needs a surer touch. Constant mesh has a lighter action and more positive and direct 'feel' than the more complex synchromesh. **1937** *Jrnl. R. Aeronaut. Soc.* XLI. 146 Charts for Calculating the Performance of Airplanes having Constant Speed Propellers. **1938** *Encycl. Brit. Bk. of Yr.* 79/2 The variable pitch and especially the constant-speed airscrew have become a necessity for reasonable take-offs. **1935** *Discovery* Sept. 278/1 With constant velocity scanning, the successive signal strengths are proportioned to the corresponding light and shade areas of the image focused by the lens upon the mosaic. **1966** *Times* 18 Mar. 6/7 The constant velocity joints of an M.G. 1100 car he examined after it had been involved in a fatal accident were perfect. **1904** GOODCHILD & TWENEY *Technol. & Sci. Dict.* 126/1 *Constant voltage*, an electrical pressure or voltage kept at a uniform value. **1959** *'Motor' Man.* (ed. 36) vi. 156 The so-called constant-voltage system..has been widely adopted during the past few years.

constantan (kǫ·nstăntæ̆n). [f. CONSTANT *a.*] An alloy of copper and nickel, used for electric resistors, thermocouples, etc. Also *attrib.*
1903 in *Whittaker's Electr. Engineer's Pocket Bk.* 164. **1913** *Rep. Brit. Assoc. Advancem. Sci. 1912* 570 The thermo-electric force generated by the constantan discs. **1934** *Discovery* Apr. 114/1 A thermo-couple made of copper and constantan was installed in the nest. **1947** *Science News* IV. 146 With constantan, temperatures of the order of 500° C. to 1,000° C. are reached.

constatation (kǫnstătēⁱ·ʃən). [a. F. *constatation* ascertaining, inquiry, statement; cf. CONSTATE *v.*, -ATION.] The process, or result, of constating; ascertaining or establishing; verification; *concr.* statement or assertion.
a **1916** H. JAMES *Ivory Tower* (1917) IV. i. 251 So large a part played..by the mere *constatation*..of the state of being worth dollars to inordinate amounts? **1926** R. FRY *Transformations* 6 In the case of mathematics the purpose is the constatation of the universal validity of the relations. **1939** *Theology* XXXVIII. 256 All these events are open to the normal methods of empirical constatation. **1952** *Scrutiny* XVIII. 273 The..difference..can be brought down here to the constatation that Lawrence sees what the needs are ..so much better than George Eliot. **1970** *Times Lit. Suppl.* 23 Apr. 442/3 The article contains some propositions, some general constatations, some statements of principle that in themselves invite concurrence as wise, sound and obvious.

constative (kǫ·nstătiv, kǫnstēⁱ·tiv), *a.* and *sb.* *Gram.* and *Philos.* [tr. G. *konstatierend* (K. Brugmann *Griech. Gram.* (1900) § 537), f. L. type **constātīv-us*, f. *constāt-* ppl. stem of *constāre*: see CONSTATE and -IVE.] **A.** *adj.* **a.** Of a use of the aorist tense: indicating that the action denoted has taken place, rather than emphasizing its initiation or completion. **b.** Capable of being true or false. **B.** *sb.* A statement that is capable of being true or false.
1901 P. GILES *Man. Compar. Philol.* (ed. 2) xxxi. 478 The 'constative' use of the Greek aorist, which refers to past actions simply as having occurred. **1906** J. H. MOULTON *Gram. N.T. Gr.* I. 109 The aorist..looks at a whole action simply as having occurred,..constative. **1955** J. L. AUSTIN *How to do Things with Words* (1962) i. 6 To issue a constative utterance..is to make a statement. *Ibid.* vi. 67 Criteria..would not distinguish performatives from constatives.

constellate, *v.* Add: **2. b.** *Psychol.* To form (ideas, etc.) into a group or constellation. Hence **co·nstellated, co·nstellating** *ppl. adjs.*
1922 *Brit. Jrnl. Psychol.* Oct. 114 Presentation-complexes ..become successively conscious and..are constellated by preceding associated complexes. **1951** G. W. ALLPORT in Parsons & Shils *Toward Gen. Theory Action* IV. i. 367 Grouping, constellating, rubricizing leap far ahead of experience. **1963** B. GHISELIN in Taylor & Barron *Scientific Creativity* I. 36 A creative product is intrinsically a configuration of the mind, a presentation of constellated meaning.

constellation. Add: **5.** *Psychol.* A group of ideas or personality factors, usu. formed by association; = *COMPLEX *sb.* 3.
1922 *Brit. Jrnl. Psychol.* Oct. 122 A word for the whole range of phenomena is wanted and if it is held that 'complex' is no longer available it is possible to use 'constellation'. **1924** R. M. OGDEN tr. *Koffka's Growth of Mind* v. § 2. 247 Selz has shown that an explanation in terms of the 'constellation' of numerous independent associations is entirely inadequate. **1950** *Psychiatry* XIII. 182/1 The patient transfers to the analyst the constellations of love, hate, and other affects. **1958** M. KERR *People of Ship St.* xv. 162 His simple and non-integrated role constellations are not capable of coping with a more complicated world.

constipated, *ppl. a.* Add: **2. b.** *fig.*
1891 'MARK TWAIN' *Lett. to Publishers* (1967) 277 You will now have to modify your instalment system to meet the emergency of a constipated purse. **1937** PARTRIDGE *Dict. Slang* 177/1 *Constipated*, slow to part with money. **1967** *Economist* 4 Feb. 423/2 So much evolutionary detail has had to be dealt with for the first time that the first half of the book makes rather constipated reading.

constipation. Add: **2. c.** *fig.*
1917 A. HUXLEY *Let.* 30 Sept. (1969) 136 That somewhat unhinged letter..was..due to nothing in particular, but exactly coinciding with a period of poetical constipation. **1966** *New Statesman* 25 Mar. 411/2 There are signs that the electoral constipation in the West Midlands is easing slightly.

constituent, *a.* and *sb.* Add: **B. 5.** *Linguistics.* (See quots.)
1933 BLOOMFIELD *Language* x. 160 The common part of any (two or more) complex forms is a linguistic form; it is a *constituent* (or *component*) of these complex forms. **1953** C. E. BAZELL *Linguistic Form* 57 Taxemic distinctions are frequently correlated with distinctions of constituent-analysis. *Ibid.* 76 In sentence-analysis..the various layers which, from the standpoint of the system, are regarded as bases, are treated as constituent-layers of the base. **1956** J. WHATMOUGH *Language* 260 *Constituent*, the common part of any two or more complex forms or constructs, e.g. *un/gentlemanly: gentleman/ly: gentle/man.* **1962** N. CHOMSKY *Syntactic Struct.* iv. 29 The form of grammar associated with the theory of linguistic structure based upon constituent analysis. **1962** J. SLEDD in Householder & Saporta *Probl. Lexicography* 146 At the end of the constituent-structure grammar [etc.]. **1964** E. BACH *Transformational Gram.* iv. 75 In such rules the 'embedding' sentence is called the matrix sentence, the embedded one the constituent sentence (or string). **1964** M. A. K. HALLIDAY et al. *Ling. Studies* 300 The first set of rules, known as 'constituent-structure' rules, produces structures and some formal items. **1965** N. CHOMSKY *Aspects of Theory of Syntax* ii. 67 An unordered set of rewriting rules..is called a constituent structure grammar (or phrase structure grammar).

constituted, *ppl. a.* (Earlier U.S. examples of mod. use.)
1807 J. MARSHALL *Washington* V. 354 Neither could be [*sc.* Jefferson] perceive danger to liberty except from the constituted authorities and especially from the executive. **1816** PICKERING *Vocab.* 69 *Constituted authorities*, the officers of government collectively, in a kingdom, city, town, &c. This expression has been adopted by some of our writers from the vocabulary of the French Revolution.

constitution. Add: **8. b.** *Comb.* (sense 7) *constitution-making, -mongering* (example).
1820 *Deb. Congress* 1st Sess. I. 945 Nineteen out of twenty of our agricultural citizens know better the art of constitution-making than the best methods of raising cabbages. **1849** MILL in *Westm. Rev.* LI. 42 The presence of such checks, not their absence, would have been the novelty in constitution-making. **1949** I. DEUTSCHER *Stalin* 185 However, the time for real Constitution-making had not yet come. **1833** MILL *Let.* 9 Mar. (1910) I. 39 Our Utilitarian Radicals..will no longer rely upon the infallibility of Constitution-mongering.

constitutional, *a.* Add: **6.** *constitutional diagram*, = *equilibrium diagram* (*EQUILIBRIUM 4).
1929 *Jrnl. Iron & Steel Inst.* CXIX. 449 Constitutional diagrams have consequently been constructed for cast iron. **1960** *Ibid.* CXCIV. 134/3 In many cases no constitutional diagrams are given.

constitutive, *a.* Add: **6.** *constitutive equation* (see quot. 1961).
1955 W. NOLL in *Jrnl. Rational Mech. & Anal.* IV. 17 The equations..are not sufficient to determine the motions of continuous media. We must have in addition certain constitutive equations defining the particular ideal material which we wish to study. These equations will be functional relations..between the stress..the density..and the motion. **1960** J. L. ERICKSEN in *Archive for Rational Mech. & Anal.* IV. 233 As a general principle to be used in constructing constitutive equations, Truesdell has proposed that the stress and heat flux should be assumed to depend on the same variables. **1961** W. PRAGER *Introd. Mech. Continua* iv. 88 The equations of continuity and motion do not contain any information regarding the mechanical properties of the continuum under consideration because they hold for any continuum. These properties are specified by the constitutive equation; this is a tensor equation, which establishes a relation between statical and kinematical tensors, for instance the stress or the stress rate..and the rate of deformation or the strain.

construal (kǫnstrū·ăl). [f. CONSTRUE *v.* + -AL.] An act of construing or interpreting.
1960 P. ZIFF *Semantic Analysis* p. vii, Virtually no such exegesis, virtually no such interpretation, virtually no such construal, is called for here. **1968** *Language* XLIV. 586 Modifying the theory of the base component..to allow for the generation and construal of metaphors.

construct (kǫ·nstrǒkt), *sb.* [f. CONSTRUCT *v.*]
1. *Linguistics.* A group of words forming a phrase, as distinct from a compound. Also *attrib.*
1871 EARLE *Philol. Eng. Tongue* xi. 514 The distinction between compounds and constructs is a delicate one. *Ibid.* 515 The transition from the construct to the compound state..takes time to accomplish. **1956** J. WHATMOUGH *Language* viii. 143 Every language has its own constructs (i.e. repetitive patterns of order) of free-standing, bounded units (words). **1963** *Canadian Jrnl. Linguistics* VIII. 62 A clause can be said to have descended the grammatical hierarchy to operate at the phrase level. Such descending units are termed 'constructs' in this model. 'That he stayed up all night doing linguistics' is called a clause construct. **1967** *Language* XLIII. 745 The psychological reality of linguistic constructs.
2. *Psychol.* An object of perception or thought, formed by a combination of present with past sense-impressions.
1890 C. L. MORGAN *Anim. Life & Intell.* viii. 312 At the bidding of certain stimuli from without we construct that mental product which we call the object of sense. It is of these mental constructions—'*constructs*' I will call them for convenience—that I have now to speak. *Ibid.* 317 What we call objects are human constructs. **1934** *Nature* 8 Sept. (Suppl.) 363/2 If the nature we study consists so largely of our own mental constructs, why do our many minds all construct one and the same Nature? Why, in brief, do we all see the same sun, moon and stars? **1956** A. J. AYER et al. *Revol. Philos.* 30 Some said that natural numbers were mental constructs, meaning by this..that they had the same status as dreams and hallucinations.
b. *gen.* Anything constructed, esp. by the mind; hence *spec.*, a concept specially devised to be part of a theory.
1933 *Brit. Jrnl. Psychol.* Oct. 161 The American psychologist..denies that they [*sc.* traits] are dynamic, existential entities within the organism... Rather are they constructs in the minds of the observers. **1937** 'C. CAUDWELL' *Illusion & Reality* ii. 48 The poetic construct. **1951** J. R. FIRTH *Papers in Linguistics* (1957) xv. 190 The constructs or schemata of linguistics enable us to handle isolates that may be called language events. **1952** C. L. HULL *Behav. System* xi. 327 Note..his total lack of comprehension of the role of symbolic constructs in natural-science theory. **1956** J. H. M. BEATTIE in A. Pryce-Jones *New Outl. Mod. Knowl.* 258 The kind of structure, in the sense in which anthropologists use the term, is a construct or model, based on but not composed of the empirical data. **1959** *Listener* 1 Oct. 520/1 Typical examples of theoretical terms are 'gene' in biology, and 'electron' in physics: these are theoretical constructs that are not directly observable.
3. *Math.* A configuration, outline, or surface.
1902 *Encycl. Brit.* XXVIII. 541/1 A monogenic algebraic construct (or configuration, or surface)...The notion of monogenic construct is wider than that of a monogenic function. **1965** *Math. in Biol. & Med.* (Med. Res. Council) III. 114 A readily visualizable geometrical construct such as the n-dimensional football.

‖ **constructio ad sensum** (kǫnstrǒ·ktio æd se·nsǒm). *Gram.* [mod.L., 'construction according to the sense'.] Any construction in which the requirements of a grammatical form are overridden by those of a word-meaning: *e.g.*, the construction of a collective noun in the singular with the plural form of a verb because the noun denotes a plurality.
1894 B. L. GILDERSLEEVE *Lat. Gram.* (ed. 3) 148 The natural relation is preferred to the artificial (*cōnstrūctiō ad sēnsum, per synesin*, according to the sense). **1932** A. H. GARDINER *Speech & Lang.* § 44 From such conflict arises the linguistic phenomenon which grammarians call 'construction according to the sense' (constructio ad sensum, κατὰ σύνεσιν)...A Latin example is 'omnis aetas currere obvii' 'every age ran to meet them'. **1963** F. T. VISSER *Hist. Syntax* I. i. 62 In the latter case we have no grammatical concord, but what is usually called constructio ad sensum.

construction. Add: **4. b.** A mechanical structure used in a stage setting, or forming the setting itself.
1924 H. CARTER *New Theatre Soviet Russia* 71 Instead of painted planes he [*sc.* Meierhold] uses constructions in volumes made of wood, iron and other suitable material. *Ibid.* 72 Construction as it is now understood as an aid to acting and nothing more..arrived in April, 1922. **1929** —— *New Spirit Russ. Theatre* 221 The bridge of Life and Death that forms the chief construction on the Japanese stage. **1929** *Encycl. Brit.* XXII. 34/2 High platforms, skeletonized structures, inclined planes, all manner of bare, unsentimentalized construction provide the footing as well as the background for the actors.
4. c. *Art.* A sculptural creation, composed from a number of pieces, often of different material, and usually non-representational.
1944 H. READ in *Horizon* July 63 Dear Gabo,..it is easier for me to sympathize with those lovers of art who but dimly apprehend the formal unity of one of your constructions. **1956** F. GORE *Abstract Art* 30 Arp is opposed to the 'technoid' constructions of Mondrian. **1964** H. READ *Hist. Mod. Sculpture* iii. 89 From 1914 onwards a group of artists in Moscow attempted to apply engineering techniques to the construction of sculpture, and the objects thus made were called 'constructions'.
10. *construction camp, car*; **construction train,** a train conveying materials for the construction or repair of railways.
1869 C. L. BRACE *New West* xiv. 184 Every stick of fuel, every railroad tie, and beam for trestle-work, must be carried on construction trains from these mountains. **1873** J. H. BEADLE *Undevel. West* xxi. 397 Hitherto construction cars had been shoved across singly by hand. **1881** [in Dict.]. **1884** *Century Mag.* Oct. 843 Adventurers who had followed the construction camps on the Northern Pacific Railroad. **1900** *Daily News* 8 Mar. 3/2 The armoured train and construction trains can now go beyond Crocodile Pools almost to Ramutsa. **1943** J. S. HUXLEY *TVA* 55 (*caption*) The construction camp lies between the Freeway and the town. **1947** L. M. BEEBE *Mixed Train Daily* 298 The club car 'Julia Bullette' of the romantic Virginia and Truckee—was originally built as a construction car.

constructional, *a.* Add: **1. b.** *spec.* Pertaining to or engaged in the manufacture of structural iron or steel.
1894 *Daily News* 5 Mar. 2/7 Most of the steel-works are well engaged, more particularly on constructional work. *Ibid.* 24 Sept. 2/7 The recent distribution of large railway orders in the district has caused constructional iron to be in good request. **1897** *Ibid.* 14 June 7/2 The constructional engineers. **1900** *Ibid.* 3 Dec. 2/7 Ironfounders and constructional engineers. **1924** *Times Trade & Engin. Suppl.* 29 Nov. 242/2 Constructional steel contracts come out regularly.
4. In theatrical scenery, characterized by 'construction'.
1924 H. CARTER *New Theatre Soviet Russia* 73 A development of the constructional scenery appeared in the next production.

constructionism (kǭnstrǒ·kʃəniz'm). [f. CONSTRUCTION + -ISM.] Artistic expression by means of mechanical structures. Cf. *CONSTRUCTIVISM.
1924 H. CARTER *New Theatre Soviet Russia* 187 They are being initiated into the 'mysteries' of constructionism, bio-mechanics, and other systems of brain and body discipline. **1929** —— *New Spirit Russ. Theatre* 225 From expressionism ..to constructionism..has been the path of the Moscow Kamerny theatre.

constructionist. Add: **3.** One who follows the principles of constructionism.
1924 H. CARTER *New Theatre Soviet Russia* 70 The æstheticist, painter, sculptor, poet or any other is dead; the constructionist takes his place. **1928** *Daily Tel.* 12 June 12/6 International Federation of Expressionists, Futurists, Cubists, and Constructionists.

constructive, *a.* Add: **1.** (Further examples.) Also *spec.* of immaterial objects, ideas, etc.: having the quality of contributing helpfully (esp. opp. *destructive*).
1943 H. READ *Education through Art* i. 9 Constructive education. **1955** *Bull. Atomic Sci.* Apr. 145/1 The poor ones are those who like their job, not for its constructive, but for its destructive, qualities. **1955** *Times* 9 May 10/3 It would not be a passive condition, but 'eminently constructive'. **1962** in *Ann. Reg. 1962* (1963) 528 The United States government will be prepared to discuss these questions

urgently, and in a constructive spirit, at Geneva or elsewhere. **1965** *New Statesman* 19 Mar. 425/2 The *New Statesman*..has a duty to subject the government to a continuous process of constructive criticism.

b. *constructive dilemma*: in Logic, a dilemma that has alternative affirmations in its minor premiss (opp. DESTRUCTIVE *a.* d); esp., the form of argument by which from two conditional propositions and the alternation of their antecedents one infers the alternation of their consequents (see quot. 1953).
1826 WHATELY *Logic* ii. III. Suppl. § 5. 114 (*side-note*), Simple constructive Dilemma. **1906** J. N. KEYNES *Formal Logic* (ed. 4) III. vi. 365 Taking the simple constructive dilemma given above, and contrapositing the major. **1953** I. M. COPI *Introd. Logic* viii. 243 An argument form such as that of the Constructive Dilemma, (p⊃q).(r⊃s), p∨r ∴ q∨s.

4. b. *constructive injury*.
1859 MILL *Liberty* iv. 147 The merely contingent, or, as it may be called, constructive injury which a person causes to society, by conduct which neither violates any specific duty to the public, nor occasions perceptible hurt to any assignable individual except himself.

5. Of or pertaining to *CONSTRUCTIVISM.
1924 H. CARTER *New Theatre Soviet Russia* 71 By constructive scenery Meierhold understands essential scenery adapted to the realisation of man's free acting in space, and not photography or decoration. *Ibid.* 72 The scenery was simple, essential and constructive. **1929** *Encycl. Brit.* XXII. plate III (facing 24) Expressionistic, constructive and abstract settings. **1937** N. GABO in J. L. Martin et al. *Circle* i. 3 The original source from which the Constructive idea derives is Cubism.

constructivism (kǫnstrɒ·ktiviz'm). [ad. Russ. *konstruktivizm* f. CONSTRUCTIVE *a.* + -ISM.] The theory or use of mechanical structures in theatrical settings.
1924 H. CARTER *New Theatre Soviet Russia* 71 Constructivism was also influenced by futurism. **1929** —— *New Spirit Russ. Theatre* 225 There are two methods by which an artist may express an idea: (1) Constructivism, or symbols and emblems of the world without us; and (2) Expressionism, or symbols and emblems of the world within us. **1929** *Observer* 1 Dec. 15/2 Constructivism was simply to be the theatrical parallel to industrial technique, demanding the exclusion of all useless decoration and the precise functional organization of the stage. **1959** W. C. LOUNSBURY *Backstage from A to Z* 24 *Constructivism*, a once popular theatre movement in Russia (during the twenties) with the emphasis on machines, mechanical devices, and skeleton construction.

b. An artistic movement, originating in Moscow in 1920, concerned mainly with expression by means of constructions (*4 c); the type of art produced by this movement.
1929 *Times Lit. Suppl.* 31 Jan. 74/4 The cubism and surrealism of Paris, the futurism of Italy, the constructivism of Moscow. **1933** H. READ *Art Now* 110 We are in a world of inorganic, of mechanic, sensibility. If there is an undertone it is an undertone of the machine: the dynamo, the rock-drill, the hydraulic pump. Because of this relation, this aspect of cubism is sometimes given a special label: Constructivism. **1948** —— *Philos. Mod. Art* (1952) xii. 217 Naum Gabo..was..one of the founders of the abstract movement in Moscow known as Constructivism. **1948** *Archit. Rev.* CIV. 299/1 The Renaissance school which took the place of this nihilistic constructivism is also proving worthless in face of the new Soviet architecture. **1961** *Times* 23 Jan. 16/1 All this makes constructivism one of the most satisfying types of abstract art.

Hence **constru·ctivist** *a.* and *sb.*
1928 *Observer* 1 July 15/5 The play..needs a 'constructivist' setting—whatever that is. **1930** *Ibid.* 26 Jan. 10 Some of the settings suggest the influence of constructivist architecture. **1934** *Archit. Rev.* LXXV. 12/2 As a constructivist painter Moholy-Nagy was the first to use new synthetic materials like galalith, trolith and cellon. **1944** E. H. RAMSDEN in H. Treece *H. Read* 48 The Constructivists.. who..continue to claim for their creative achievements a unique distinction. **1955** *Times* 2 Aug. 10/1 The geometric constructions of the Constructivist sculptors likewise rejoice in immaculate surfaces, and the fact that the actual materials are metals and plastics gives them the illusion of being entirely machine-made. **1956** L. ASHTON in A. Pryce-Jones *New Outl. Mod. Knowl.* 312 The Cubists and the Futurists were controlled too much by technique, and this led to a still further emanation of technology usually referred to as the Constructivist school. **1961** *Times* 23 Jan. 16/1 English constructivists love white.

consuetudinal, *a.* and *sb.* Add: *spec.* in *Philol.*, esp. as the epithet of a particular mood in Celtic languages.
1728 H. MACCURTIN *Elem. Irish Lang.* ix. 67 (*heading*) Of the accustomary, or consuetudinal mood. **1808** 'E. O'C.' *Gram. Gaelic Lang.* II. 67 The Moods are six; the Indicative, Imperative, Subjunctive, Conditional, Consuetudinal, and Infinitive. *Ibid.* 72 Consuetudinal or habitual mood. Present.—..*I usually am*, or *I use to be*. **1899** *Trans. Philol. Soc.* 52 Sometimes *-bi* seems to be used as a consuetudinal present of *is*. **1913** J. MORRIS JONES *Welsh Gram.* 348 The consuetudinal present is in use in that sense in the spoken language. **1927** J. R. R. TOLKIEN in *Year's Work Eng. Stud.* *1925* 34 The closest point of contact is..OE. *bið*, used as a consuetudinal, a future, and sometimes indistinguishably from the present. **1939** L. H. GRAY *Found. Language* 203 The latter [aspect] denotes the kind of action or state as.. iterative, consuetudinal or usitative.

consultancy (kǫnsɒ·ltănsi). [f. CONSULT(ANT + -ANCY.] The work or position of a consultant (senses 2 and *3); a department of consultants.
1955 *Times* 4 July 2/4 Successful candidates will be given special training to equip them for Consultancy at the company's own Training Centre. **1958** *Times Lit. Suppl.* 3 Oct. 563/4 There is infinitely more formal training in management than before, and 'management consultancy' has expanded rapidly since the war. **1960** *Guardian* 26 Aug. 6/1 Interior Design Consultancy. Our Interior Design Consultants will be delighted to help and advise you. **1963** *Economist* 19 Oct. 262/2 These links are provided..by research and..consultancy. **1966** A. BATTERSBY *Math. in Management* ix. 236 Harry Arthur Hopf formed one of the earliest management consultancy companies in the U.S.A. **1967** *Times Rev. Industry* Mar. 88/1 Independent laboratories are rarely in the public eye. They range in size from the large research institute like the Fulmer at Stoke Poges and International Research and Development in Newcastle to small consultancies with a minimum of scientific equipment. **1969** *Responsibilities of Consultant Grade* (Dept. Health & Social Security) 3 Failure to reach a consultancy has now..much more serious consequences than before.

consultant. Add: **2.** (Further examples.)
1965 *Lancet* 17 July 24/1 (Advt.), Consultant in Venereology. *Ibid.*, Locum Tenens Casualty Surgeon (Consultant or Senior Hospital Medical Officer grade). *Ibid.*, Consultants in Anæsthetics.

3. A person qualified to give professional advice or services, e.g. in problems of management or design; an adviser; also *spec.* a private detective. Also *attrib.*
1893 CONAN DOYLE *Mem. S. Holmes* 23, I am rather disappointed in our London consultant. **1900** A. UPWARD *Ebenezer Lobb* 247 The Baker Street consultant glanced at him with ineffable scorn. **1907** *Installation News* May 3/1 The alleged superfluity of Consulting Engineers...Even amongst consultants undesirables are to be found. **1935** *Planning* II. XLVI. 2 The rapid growth of clearing house and consultant services for information. **1962** *Listener* 9 Aug. 223/2 Five writers and one consultant were credited with *The Bob Hope Show*. **1964** *Oxf. Univ. Gaz.* 22 Oct. 145/1 Paul Maas..Consultant to the Clarendon Press, Oxford, from 1939.

consultation. Add: **2. b.** In present legal usage confined to meetings with more than one counsel present.
1883 *Wharton's Law-Lex.* (ed. 7) 189/2 *Consultation*... Also a meeting of two or more counsel and the solicitor instructing them for deliberating or advising. **1906** *Ann. Practice* II. 211 The following fees are allowed to counsel's clerks...On consultations, senior's clerk..0 5 0. On consultations, junior's clerk..0 2 6.

consulting, *vbl. sb.* Add: **consulting room,** a room in which a consultation takes place; esp. the room in which a doctor examines his patients.
1843 B. B. COOPER *Life A. Cooper* II. xiii. 314 Charles would put them into Sir Astley's consulting-room. **1859** DICKENS *T. Two Cities* II. vi. 62 The Doctor's consulting-room, used also as the dining-room. **1892** CONAN DOYLE in *Strand Mag.* Mar. 276/2 Two men had come from Paddington, and were waiting in the consulting room. **1923** F. TREVES *Elephant Man* vii. 123 A long building with many windows that I recognize as my old hospital, a consulting room with familiar furniture, [etc.]. **1929** W. DEEPING *Roper's Row* xxv. 277, I could rent a consulting-room. **1963** *Lancet* 24 Aug. 419/2 He took his fellowship and rented a consulting-room in Wimpole Street.

consultor. Add: **1. b.** *R. C. Ch.* (See quots.)
1896 *Westm. Gaz.* 24 Sept. 1/3 Consultor-General of the Chapter of the Pallottine Fathers, in Rome. **1901** *Daily Chron.* 24 Dec. 5/7 A Consultor of the Roman Congregation of Indulgences and Relics. **1907** *Cath. Encycl.* II. 71/1 The Auditors of the Rota..are generally also attached as Consultors to various Roman Congregations. **1908** *Ibid.* IV. 323/1 *Consultors, Diocesan*, a certain number of priests in each diocese of the United States who act as official advisers of the bishop in certain matters pertaining to the administration of the diocese. As a body they take the place of the cathedral chapter as established elsewhere...Their appointment was recommended [etc.]. **1912** *Ibid.* XIII. 137/2 The Holy Office..has a number of consultors, chosen from among the most esteemed and learned prelates and religious. Some are ex officio consultors by virtue of a right anciently granted. **1927** *Daily Tel.* 21 June 13/3 Father Robinson is Consultor of three Congregations in Rome.

consumable, *a.* Add: to def.: esp. designating goods or the like which are used up or worn out by use (see also quot. 1920).
1719, 1841 [in Dict.]. **1920** H. G. WELLS *Outl. Hist.* xxxvii. 614/2 'Consumable goods' is a phrase we would have understood in the widest sense to represent even such things as a journey, a lecture or theatrical entertainment, housing, medical advice; and so forth. **1955** *Times* 26 May 18/1 Consumable goods which are not available obviously cannot be absorbed. **1957** *B.B.C. Handbk.* 51 The supply to all BBC centres of valves, camera tubes, and similar consumable items is undertaken by Valve Section. **1960** *Design* July 32/2 A cheap toothbrush..might put the toothbrush into the category of a consumable item.

consumer. Add: **2. b.** Economics. *consumers' credit*, credit given to the consumer while he is in possession and use of an article for which he is paying by instalments; *consumers' goods, rent, surplus, wealth* (see quots.).
1888 J. B. CLARK *Capital & its Earnings* ii. 13 Capital.. is wealth in mediate utilization, and stands in contrast with that which has been 'consumers' wealth', which is in direct utilization. **1890** A. MARSHALL *Princ. Econ.* I. 109 Goods may be divided into goods of the first order, which satisfy wants directly, such as food, clothes, etc.; goods of the second order . . . Goods of the first order are sometimes described as consumption or consumers' goods. *Ibid.* 175 The excess of the price which he would be willing to pay rather than go without it, over that which he actually does pay is the economic measure of this surplus pleasure: and.. may be called Consumers' Rent. *Ibid.* 446 We may now consider the effects which a change in the conditions of supply may exert on Consumers' Surplus or Rent. **1927** SELIGMAN *Econ. Instalment Selling* I. 141 Consumers' credit would denote credit granted to consumers. *Ibid.* 143 It may be as well to study the institution of consumers' credit as a step in the granting of credit in a wider sense. **1951** J. R. WINTON *Dict. Econ. Terms* 18 *Consumer's surplus*, the difference between what a consumer actually pays when there is a flat rate or price for all quantities sold and the maximum amount which could be extracted from him.

c. *gen.* One who purchases goods or pays for services; a customer, purchaser. Freq. *attrib.*, as *consumer goods, research, resistance*; **Consumer(s') Council,** an organization set up to safeguard the interests of consumers; **consumer durable** (orig. *U.S.* and usu. *pl.*), an article for domestic use which does not need to be rapidly replaced by the purchaser (see quot. 1958¹); also *attrib.*
1897 *Sears, Roebuck Catal.* 1 (*heading*) Consumers guide. **1910** W. JAMES *Mem. & Stud.* (1911) xi. 276 A world of.. 'consumer's leagues', and 'associated charities'. **1930** *Economist* 3 May 988/2 If the Consumers' Council Bill..becomes law, the Food Council's successor will have powers of compelling the production of information. **1931** F. L. ALLEN *Only Yesterday* vii. 169 To cajole, exhort, or intimidate the consumer into buying,—to 'break down consumer resistance'. **1933** *Planning* I. VII. 5 Retail outlets..where improvements can be tried out and consumer reactions tested. **1934** DYLAN THOMAS *Let.* 20 July (1966) 139 Those controlling the state shall earn in consumer credit no more than the worker who controls the drains of the state. **1938** *New Statesman* 16 July 132/1 Whether this revival in consumer demand can lead on to a revival in capital goods investment depends [etc.]. **1940** *Ibid.* 16 Mar. 382 Rich companies with a steady old-established trade in consumer goods or services. **1942** in *N. & Q.* (1943) 24 Apr. 262/1 This period of consumer-goods shortages. **1949** E. JENKINS *Six Criminal Women* 30 She..would have had nothing to learn from courses in salesmanship or the findings of Consumer Research. **1951** *Business Week* 17 Nov. 158 Consumer durables slash. **1955** *Ann. Reg.* *1954* 79 Consumer durable goods as automobiles and household appliances. **1958** *New Statesman* 26 Apr. 547/1 What economists call 'consumer durables', such things as motor cars, television sets and refrigerators. **1958** *Observer* 26 Jan. 1/5 She writes regularly for *The Observer* on consumer guidance. **1958** *Ibid.* 25 May 10/6 Complaints brought to the Consumer Council. **1959** *Times* 6 Mar. 3/1 (Advt.), Executive.. required by leading consumer goods manufacturer. **1959** *Which?* June 43/1 Today a consumer can buy a Kitemarked electric blanket, crash helmet or domestic pressure cooker with assurance of its safety features. **1962** 'H. CALVIN' *The System* xiii. 175 By the human race they mean consumer units, and they know everything about *them* because it's all in a file produced at enormous expense by a team of motivational research field workers. **1970** *Which?* June 163/1 Manufacturers and consumers do not have identical points of view.

consumerism (kǫnsiū·məriz'm). orig. *U.S.* [f. CONSUMER + -ISM.] **1.** Protection of the consumer's interests.
1944 *New Republic* 4 Sept. 276/1 Some of the oldest and most successful consumer enterprises grew independently of the rural impulse that in recent years has been most active in spreading the idea of consumerism. **1969** *Times* 4 Sept., Academic economists have muttered darkly about 'consumerism'—meaning an exaggerated respect for the real or supposed requirements of the consumer. **1970** *Sunday Times* 24 May 50/1 Consumerism is the outgrowth of the consumer-protection movement of the 1960s.

2. Name given to a doctrine advocating a continual increase in the consumption of goods as a basis for a sound economy.
1960 V. PACKARD *Waste Makers* (1961) iii. 23 A leading apostle of 'consumerism'..pointed out that every recent United States recession had been caused by..a failure to see that consumption kept pace with production. **1962** E. GODFREY *Retail Selling & Organ.* xxi. 213 Consumerism has become the guiding force of our economy. **1963** *New Scientist* 27 June 713 The western economic system..is turning into what has been called 'consumerism'.

consummatory. Add: **2.** *Physiol.* and *Psychol.* Complementing, or relating to the complementing of, preparatory responses to a situation that are of vital importance to the organism; not instrumental, but satisfying in itself.
1906 C. S. SHERRINGTON *Integrative Action Nerv. System* ix. 329 'Distance-receptors' induce anticipatory or precurrent reactions, that is, precurrent to *final* or consummatory reactions. *Ibid.* 332 The consummatory acts led up to are such—e.g. seizure of prey, escape from enemies, attainment of sexual conjugation, etc.—as involve the very existence of the individual and the species. **1918** R. S. WOODWORTH *Dynam. Psychol.* ii. 40 A consummatory reaction is one of direct value to the animal—one directly bringing satisfaction—such as eating or escaping from danger. **1932** E. C. TOLMAN *Purposive Behav.* xviii. 272 The final state of physiological quiescence to be reached by commerce with the consummatory object. **1951** G. HUMPHREY *Thinking* vii. 188 A train of thought may be concluded without overt consummatory action. **1957** *New Biol.* XXIV. 122 The variable introductory phase which leads to a stereotyped response, or consummatory act, or to a consummatory situation, is called appetitive behaviour.

consumptible (kǫnsɒ·mᵖtib'l), *a.* Add: **B.** *sb.* Any object whose use renders it consumed, worn out, or decayed.

1892 *Athenæum* 3 Sept. 318/3 The loan of a 'consumptible', such as money. **1894** PALGRAVE *Dict. Pol. Econ.* s.v. *Consumptibles*, The loaf which is eaten on the day of baking, the Suez Canal which will last, it is hoped, for centuries, are both consumptibles.

consumption. Add: **6. b.** *fig.*
1885 LD. GRANVILLE *Let.* Jan. in Marq. of Zetland *Ld. Cromer* (1932) xix. 216 It is extremely desirable..that we should do something both for the thing itself and for home consumption. **1912** W. OWEN *Let.* 29 May (1967) 138 Leslie has just dropped in to borrow a book for Dorothy's immediate consumption. **1930** WODEHOUSE *Very Good, Jeeves!* ix. 250 The Pyke is entirely unfit for human consumption and must be cast into outer darkness where there is wailing and gnashing of teeth. **1964** —— *Frozen Assets* ii. 42 'What's wrong with the poor guy?' 'He's a mess. Totally unfit for human consumption.'

9. *consumption credit* (= *consumers' credit), *market, tax*; **consumption goods** = **consumers' goods*; **consumption test, trial,** the test of a motor vehicle with regard to its economical consumption of petrol.
1890 [see *CONSUMER 2 b]. **1894** PALGRAVE *Dict. Pol. Econ.*, *Consumers' Goods* (or *Consumption Goods*) include all those desirable things which directly satisfy human needs and desires, *e.g.* food and clothing. **1895** *Daily News* 6 June 7/3 The chief Chinese consumption markets. **1902** A. R. COLQUHOUN in *Westm. Gaz.* 10 Nov. 6/3 The treaty provided for a 'consumption tax' on native produce. **1903** *Motor Ann.* 148 Hill-climbing and consumption trials. *Ibid.* 155 The consumption test was to determine which cycle would travel the greatest distance on a pint of petrol, at a minimum pace of 15 m.p.h. **1909** *Westm. Gaz.* 24 Aug. 4/2 In the consumption trial of the Liverpool Motor Club a 10–12 Humber car..carried off the first prize. **1927** SELIGMAN *Econ. Instalment Selling* I. 141 Consumption credit would denote credit utilized for the purposes of consumption. *Ibid.* 174 Consumption credit is in itself quite as legitimate as production credit. **1941** 'G. ORWELL' *Lion & Unicorn* 103 A managed internal economy based simply on the amount of consumption-goods available. **1944** J. S. HUXLEY *Living in Revol.* xiii. 155 Germany is now going short of consumption goods.

consupponible (kǫnsŏpō·nib'l), *a.* *Philos.* Capable of being supposed together (with).
1940 R. G. COLLINGWOOD *Ess. Metaphysics* vii. 66 Since they are all suppositions, each must be consupponible with all the others; that is, it must be logically possible for a person who supposes any one of them to suppose concurrently all the rest. *Ibid.* 76 He will expect the various presuppositions he is studying to be consupponible only under pressure. **1941** *Mind* L. 187 There is always a 'constellation' of absolute presuppositions; each of these must be 'consupponible' with the rest.

contact, *sb.* Add: **1. c.** Hence, the touching or uniting of points or surfaces of conductors to permit the flow of electric current; also, a device for effecting this.
1915 'BARTIMEUS' *Tall Ship* i. 30 'I suppose you tested the contacts?' he asked. **1932** H. NICOLSON *Public Faces* VII. 192 He stood stock still beside the aeroplane while the pilot fiddled inside with the contacts.

d. *Psychol.* A light pressure upon the skin or the sensation of this. Also *contact sensation*.
1901 BALDWIN *Dict. Philos. & Psychol.*, *Contact Sensation*. ..A sensation made up probably (Dessoir) of Touch Sensation and Pressure Sensation. **1903** ROYCE *Outl. Psychol.* 133 Still other points on the skin, very wealthily interspersed amongst the others, give us, if excited in isolation, sensations of contact or of pressure.

e. *Aeronaut.* Used as a signal to a person about to swing an aircraft propeller that the ignition system is switched on; usu. as *int.*
1913 C. MELLOR *Airman* vi. 29 On the word 'contact' given by the pilot the mechanic launched the Chauvière 'Intégrale' propeller, and the trusty Renault engine started at the first swing. **1917** 'CONTACT' *Airman's Outings* 16 'Contact!' replied the flight-commander; his engine roared, around flew the propeller. **1919** BERTA RUCK *Disturbing Charm* xi, I climbed in, and the boys swung the propeller. I gave 'em 'Contact', and then I was up and off. **1933** *Word Study* May 4/2 'Contact'..is the word of warning given by the pilot of an airplane to the starter who spins the propeller, or 'cranks the motor'.

2. c. A person who has been exposed to infection by proximity to a person suffering from an infectious disease.
1901 *Standard* 4 Mar., A large number of contacts and suspects have been placed in quarantine. **1907** *Practitioner* Dec. 837 The infection of scarlet fever is not carried..in the clothes of mere contacts.

d. A person who can be called upon for assistance, information, etc.; an acquaintance, esp. one who can be useful in business; an agent; a connection or acquaintanceship. orig. *U.S. colloq.* (Cf. quot. 1833 under sense 1 b in Dict.)
1931 G. IRWIN *Amer. Tramp & Underworld Slang* 54 *Contact*, a connection or affiliation made by a criminal to protect himself from arrest or to make crime easy. **1931** H. G. WELLS *Work, Wealth & Happiness* (1932) x. 426 It helped them to obtain what the Americans call 'contacts'. **1935** *Economist* 19 Jan. 133/1 By a series of links in a.. chain, varying in substance from closely definitive trade agreements to mere directorial 'contacts', the group maintains an individual relationship with leading British concerns. **1937** A. CHRISTIE *Murder in Mews* 89 She's had three husbands, one Italian, one German and one Russian, and.. in consequence she has made useful what I think are called 'contacts' in three countries. **1949** M. LASKI *Little Boy Lost* II. iv. 71 Madame..may not have been the curé's only contact for disposing of those children. **1954** XAN FIELDING *Hide & Seek* 64 Who, we hoped, would put us in touch with another trustworthy contact further on.

e. *Aeronautics.* The state of being in sight of the surface of the earth (see *contact analogue, flight, flying* under sense 6 below). Also used as quasi-*adv.* *U.S.*
1940 *Life* 16 Sept. 65 Most of the trip was flown 'contact'. **1947** *Harper's Mag.* Apr. 324/1 When you can see any trace of the ground,..that's 'contact'—because you are visually in touch with the world.

6. **contact analogue,** a device which presents navigational information visually (in chart form) to the pilot of an aircraft; **contact bed,** a tank containing porous material through which sewage is filtered in order that contact with the bacterial organisms and atmospheric oxygen in the pores of the material may chemically destroy the noxious organic matter in the sewage; **contact block, brush, piece,** etc. *Electr.*, devices for the passage, conduction, or transmission of electric current by contact (see *1 c); **contact-clause** (see quot. 1946); **contact flight, flying** (orig. *U.S.*), navigation of an aircraft by the observation of landmarks; **contact glasses** = **contact lenses*; **contact healing,** the healing of illness by physical contact with a spiritualist medium; **contact lens** (or **lenses**), small glass or plastic lens(es) placed inside the eye-lids in contact with the globe of the eye to correct faulty vision; **contact man** (orig. *U.S. colloq.*), an intermediary in a transaction; a go-between; one who carries or supplies information (cf. sense 2 d above); **contact metamorphism** *Geol.*, the transformation of rock as a result of an igneous intrusion; **contact printing** *Photogr.*, the making of prints by passing light through a negative on to sensitized paper, glass, or film held in direct contact with the negative; so *contact plate, print, slide*; **contact process,** a process by which sulphuric acid is obtained from sulphur trioxide in the presence of a catalyst (e.g. platinum); **contact screen** *Photogr.*, a half-tone screen made on a film base.
1958 *Times* 17 Oct. 15/2 The contact analogue which, through a single cathode-ray display tube and with the assistance of a computer, will give a pilot swiftly and precisely the information he requires to fly his aircraft safely and accurately. **1961** *Flight* LXXIX. 250/2 (*caption*) A contact analogue pattern reflected from the trichroic combiner in a cockpit mock-up. **1902** *Westm. Gaz.* 18 Aug. 2/1 The contact bed treatment differs from the intermittent filtration method in that the sewage is rapidly run into a bed of cinders,..or the like, and after a few hours is as quickly run out. **1936** [see *BACTERIUM 2]. **1901** L. M. WATERHOUSE *Conduit Wiring* 32 The contact block of the ceiling-rose. **1884** F. KROHN tr. *Glaser de Cew's Magn. & Dyn.-electric Mach.* 264 The iron core is magnetised by the electric current flowing through the windings of the rotating helix from the one contact-brush to the other. **1927** JESPERSEN *Mod. Eng. Gram.* III. vii. 132 These clauses are here termed contact-clauses, because what characterizes them is the close contact between the antecedent and the clause. **1946** *Trans. Philol. Soc.* *1945* 131 The contact-clause (Jespersen's term), i.e. parataxis with omission of *that* in indirect statement and with the omission of the relative pronoun when object of a verb, e.g. *the man I saw*. **1950** WEBSTER Add., Contact flight. **1938** A. JORDANOFF *Through Overcast* xxv. 304 In the early days..airmen..tried to guide themselves through the overcast by the 'feel' of the controls which they had acquired during contact flying. **1946** *Jrnl. R. Aeronaut. Soc.* L. 750/2 Until adequate radio aids to navigation were available in the United Kingdom, the pilots of the regular services developed a high degree of skill in the art of 'contact' flying. **1906** *Lancet* 13 Oct. 1007/1 A few years ago glass shells, which are known as contact glasses, have been introduced by Fick, for the temporary relief of irregular corneal astigmatism. **1937** *Aeroplane* 9 June 712 Contact glasses are thin, transparent, saucer-shaped glass bowls which fit on to the anterior surface of the eye. **1945** H. EDWARDS *Sci. Spirit Healing* iv. 19 Spirit healing is divided into two main sections: (*a*) Absent Healing..and (*b*) Personal or Contact Healing by touch, or the 'laying on' of hands. **1956** R. M. LESTER *Towards Hereafter* v. 67 While I was having this contact healing, I was also having absent healing. **1888** C. H. MAY tr. A. E. Fick in *Arch. Ophthalmol.* XVII. 216 A small glass shell..which I call 'a contact-lens'. *Ibid.* 217 The 'contact-lens' consists of a thin glass shell, bounded by concentric and parallel spherical segments. **1942** *Lancet* 30 June 744/2 Plastic contact lenses are less well tolerated than glass. **1944** *Times* 14 Feb. 4/6 Squadron Leader Geoffrey B. Warne, D.S.O., D.F.C., a Typhoon fighter leader who wears contact lens spectacles, shot down an enemy aircraft. **1926** M. CONNELLY *Traveler* 5 Each one of us conductors is really a contact man. **1938** DYLAN THOMAS *Let.* 31 Dec. (1966) 220, I met that sap Goodland, the blue-and-water-eyed contactman. **1949** *Times* 31 Jan. 4/6 Business men are tempted to employ 'contact men' in an effort to smooth away obstacles. **1876** *Q. Jrnl. Geol. Soc.* XXXII. 426 The Cornish slates existed as metamorphic rock (cleaved and sometimes contorted) long before the intrusion of the granite. The contact metamorphism produced thereby extends to a short distance only. **1960** L. D. STAMP *Britain's Struct. & Scenery* (ed. 5) 24 The molten rock bakes and hardens the rocks through which it passes—it changes their form by its contact (..hence the process is called contact metamorphism). **1876** PREECE & SIVEWRIGHT *Telegraphy* 37 The zinc plate, fitted with a brass contact piece. **1892** *Pall Mall Gaz.* 20 May 7/1 The engine is provided with a contact piece, and as soon as it touches the insulated bar electrical connection with the signal-box is established. **1892** W. ABNEY *Instruct. Photogr.* (ed. 9) 253 Transparencies by Contact Plates. **1890** [see KODAK *sb.* c]. **1933** *Jrnl. R. Aeronaut. Soc.* XXXVII. 232 'Contact prints' are made with the negative and the printing paper in direct contact. **1962** *Unesco Bull. for Libraries* XVI. 3 When copies of a microphotographic negative are required, contact prints may be made either on transparent material or on paper. **1876** W. ABNEY *Instruct. Photogr.* (ed. 3) 99 The following are modes of production [of transparencies] by the camera or by contact printing. **1897** C. M. HEPWORTH *Animated Photogr.* xiv. 103 When used for contact printing, two spools are attached in a light-tight box to the top of the instrument. **1903** G. LUNGE *Manuf. Sulphuric Acid* (ed. 3) I. xi. 975 We have here the fundamental features of the contact-process as now employed, and Peregrine Phillips must be called its inventor. **1910** *Encycl. Brit.* V. 501/2 The 'spongy' platinum so formed brings about the combination of..sulphur dioxide and oxygen to form sulphur trioxide. The last reaction..receives commercial application in the contact process of sulphuric acid manufacture. **1940** *Photographic Jrnl.* LXXX. 59 (*heading*) Preparation of vignetted or contact screens. **1957** R. W. G. HUNT *Reprod. Colour* xi. 151 With a contact screen..a fine line will clearly be reproduced as a line of dots, each of which is elongated in the direction of the line. **1892** *Photogr. Ann.* II. 58 There is something in a slide by reduction which a contact slide lacks, and no doubt this is due to the fact that the former is made by the agency of daylight.

b. Applied *attrib.* to operations (or units engaged therein) which have the object of maintaining contact between aircraft and advancing forces of infantry, as *contact control, machine, patrol, work*.
1917 *Blackw. Mag.* Mar. 380/1 Machines would be detailed for contact work with our infantry. *Ibid.* Aug. 144/2 The low-flying contact machines..play their part of mothering the infantry. *Ibid.* 147/1 The new system of contact patrols was found [in 1916] to be useful in dealing with Boche movements directly behind the front line. **1918** E. M. ROBERTS *Flying Fighter* 131 Contact Control, the purpose of which is to keep in touch with advancing infantry, tabulate its progress, and then report to headquarters. **1934** *Flight* 18 Jan. 48 Contact patrols, as they were called, became a regular duty of the Royal Flying Corps.

contact, *v.* Delete *rare, techn.* and add: **3.** *trans.* To get into contact or in touch with (a person). orig. *U.S. colloq.*
1927 *Spectator* 6 Aug. 212/2 Dreiser should not be allowed to corrupt his language by writing 'anything that Clyde had personally contacted here'. **1929** L. F. CARR *America Challenged* 61 Mr. Dickey contacted every family in three representative agricultural counties. **1935** A. P. HERBERT *What a Word!* 100 A charming lady in the publicity business shocked me when we parted by saying 'It has been such fun contacting you.' **1936** WODEHOUSE *Laughing Gas* ix. 95 The prospect whom I was planning to contact, as they call it in America, was leaning back in the arm-chair. **1938** *Manch. Guardian Weekly* 19 Aug. 148/1 Will you please retain your ticket until you have contacted Mr. ——. **1940** *Times Weekly* 27 Nov. 1/4 (Advt.), Factory representatives in most parts of world. Contact your local trader. **1951** *Good Housek. Home Encycl.* 85/1 See that everyone in the household knows how to contact the nearest Fire Service, by telephone if possible.

contactable (kǫntæ·ktăb'l), *a.* [f. CONTACT *v.* + -ABLE.] Capable of being contacted.
1962 R. BUCKMINSTER FULLER *Epic Poem on Industrialization* 17 The vast non-sensorial physical realm was contactable only By the disciplined mind. **1967** C. DRUMMOND *Death at Furlong Post* vii. 98, I was very discreet, saying nothing about Ada being contactable—my word, what words we use nowadays—here.

contactor (kǫntæ·ktǫɹ). [agent-n. in L. form, from L. *contingĕre* to CONTACT.] A device for making and breaking an electrical circuit. Also *attrib.*
1910 *Hawkins' Electr. Dict.* 88/1 *Contactors*, in the multiple-unit system of electric railway control, the devices which serve to operate the various resistances, to make and break the main circuit between trolley and motors, and to change from series to parallel connection. They consist of movable arms tipped with copper for making contact with a corresponding fixed copper tip when actuated by current by the master controller. **1930** *Engineering* 31 Jan. 129/2 A push-button operated starter of the contactor type. **1937** *Times* 13 Apr. p. xvi/3 The first units to be made..with the contactor panel and resistances mounted at the rear of the chassis.

contagion. Add: **8.** *Ecology.* A greater occurrence of the individuals of a species in an area than could be accounted for by random distribution, thus forming aggregations.
[**1931** G. PÓLYA in *Ann. Inst. H. Poincaré* I. 139 La structure de la 'contagion' est homogène.] **1951** BARNES & STANBURY in *Jrnl. Ecol.* XXXIX. 172 The principle of contagion..is that the groups are distributed at random and that the number of individuals associated with each group is also random. **1957** P. GREIG-SMITH *Quant. Plant Ecology* iii. 61 Whitford..suggested the ratio of abundance ..to frequency as a measure of contagion.

contagious, *a.* Add: **2. b.** *contagious abortion*: a type of brucellosis in cattle caused by *Brucella abortus* and producing abortion.
1910 *Jrnl. Infectious Dis.* VII. 475 The bacillus of Bang is the microbic cause of at least some of the contagious abortion of cattle in this country. **1951** HARRIES & MITMAN *Clin. Practice in Infectious Dis.* (ed. 4) xxx. 549 The high incidence of contagious abortion and the low incidence of abortus fever.

8. *Ecology.* Of, pertaining to, or exhibiting contagion (see *CONTAGION 8).
1939 J. NEYMAN in *Ann. Math. Statistics* X. 36 The distributions considered belong to a class which Pólya proposed to call 'contagious': the presence of one larva within an experimental plot increases the chance of there being some more larvae. **1957** P. GREIG-SMITH *Quant. Plant Ecology* iii. 56 Recently *contagious* has been widely used for aggregated distributions generally, and this seems a suitable term, although originally applied by Pólya to a particular type of aggregation. *Ibid.* 78 If distribution is contagious then, whatever the exact nature of the contagion, the actual occurrence of individuals on the ground is..patchy.

contagiousness. Add: **c.** *Ecology.* = *CONTAGION 8.
1939 J. NEYMAN in *Ann. Math. Statistics* X. 37 The cause of the contagiousness of the distribution of larvae in experimental plots is clear. **1949** *Ecology* XXX. 202/1 The ratio of abundance to frequency is a relative measure of the degree of contagiousness of the distribution of any species.

contain, *v.* Add: **11. c.** To confine (an enemy force) to a particular area so that it cannot break out and operate elsewhere. Also in *ppl. adj.*, as *containing force*.
1899 *Westm. Gaz.* 21 Nov. 5/3 The opinion is now generally entertained in Sir George White's camp that only a containing force has been left outside Ladysmith, and that the bulk of the Boer commandos have moved south. *Ibid.* 23 Nov. 5/1 It is obviously impossible for Joubert to be able to contain White at Ladysmith, Hildyard at Estcourt, and Clery at Maritzburg. **1900** H. A. GWYNNE in J. Ralph *War's Brighter Side* (1901) 174 Infantry will be used for two purposes—to contain the opposing infantry, and to hold positions seized by the mobile portion of the force. **1904** *Westm. Gaz.* 11 Aug. 7/2 The Japanese, having left a containing force at An-shan-shan, are advancing with their principal force to the east. **1940** *War Illustr.* 19 Jan. 630/3 The two great [mine]fields which 'contained' the German submarine fleet.

d. To keep (a hostile nation, ideology, etc.) within limits; to prevent expansion or encroachment into new territory.
1948 *Manch. Guardian Weekly* 8 Jan. 4 The expansion of Soviet power that..must be 'contained'. **1952** *Ann. Reg.* *1951* 138 The most that could be claimed was that the Communists had been 'contained'. **1959** *N.Z. Listener* 16 Apr. 6/3 The Korean War, fought for the limited objective of containing Communism in an area where it has attempted to extend itself by military means.

container. Add to def.: esp. a receptacle designed to contain or store certain articles; *spec.* a large box-like receptacle of standardized design for the transportation of freight by road, rail, or sea (see quots.). Also *attrib.*
1925 N. E. ODELL in E. F. Norton *Fight for Everest, 1924* 331 Certain modifications had..been deemed advisable, especially in the cylinder containers. **1928** *Daily Mail* 25 July 4/3 This practical japanned metal container for soiled linen. **1930** *Engineering* 30 May 709/1 The container-car service, by which less than car loads were carried in portable steel containers. **1933** *Meccano Mag.* Mar. 200/2 'Containers'..are box-like structures of timber or pressed steel resembling the top of a covered wagon..fitted with ringbolts and may be slung..from the low-sided or flat wagon on which they are..carried, and they may then be mounted on road lorries and carted to any required point. **1966** *Evening Standard* 2 Mar. 11/5 The Americans, in the forefront of the container-ship development.., will send the first 'container-liner' across the Atlantic...Most of her cargo space will now be devoted to 20 and 40 ft. containers which can be handled easily on road or rail. **1967** *Times Rev. Industry* Feb. 67/3 British Rail, despite the difficulties over open terminals, are now carrying 1,200 containers a week. **1969** *Jane's Freight Containers 1968–69* 6 More container berths are under construction. *Ibid.*, Twelve deep-sea containership berths are in operation. *Ibid.*, Timetables revised to incorporate the new fast container trains as they go into service. **1969** *Lebende Sprachen* XIV. 65/2 The Clyde Port Authority plans to bring its Greenock container terminal into operation later this year.

containerize (kǫntēⁱ·nərəiz), *v.* Also -ise. [f. CONTAINER + -IZE.] *trans.* To pack into, or transport by means of, containers (see prec.). So **containeriza·tion, contai·nerized** *ppl. a.*
1958 *Times* 1 July 14/4 Express containerized cargo. **1959** *Times* 16 July 11/2 New developments which may affect our business, such as the bulk carrier, containerisation of cargo, [etc.]. **1960** *Aeroplane* XCIX. 181/1 This 'new concept in containerization'..may have applications for freight handling. **1962** *Ibid.* CIV. No. 2665 p. 4/1 Freight loads should be 'containerized' or at very least 'unitized'. **1965** *Economist* 20 Nov. 878/1 It is thought that the *Clifford J. Rogers* was the world's first ship designed specifically for handling containerized freight. **1969** *New Scientist* 19 June 642/1 Containerization of transport is becoming increasingly popular. **1969** *Daily Tel.* 16 Dec. 2 In a short space of time all the world's major deep sea trade routes will be containerised.

containment. Delete *rare* and add further examples. Also *attrib.*
1956 *Wall St. Jrnl.* 18 Oct. 12/4 The explosion breaks the steel containment vessel. **1958** *New Scientist* 16 Jan. 14/2 The containment of the hot gas by a magnetic field is not a purely static phenomenon, like the containment of beer in a bottle. **1959** *Times Lit. Suppl.* 20 Mar. 156/3 The Guayaquils, the Shirians and others who helped him to move about this forest are for him men who are contained by it. It was his pleasure to..find out..what that containment might mean.

b. *spec.* The action or policy of 'containing' a hostile nation, etc. (see *CONTAIN *v.* 11 d). Also *attrib.*
1947 *Partisan Rev.* XIV. 239 British policy under Chamberlain presents a model of how *not* to undertake a policy of containment. *Ibid.* 241 A program of containment requires a ring of stable and satisfied countries around Russia. **1952** *Ann. Reg. 1951* 178 A positive policy of re-armament, 'containment', and American leadership. *Ibid.* 184 The psychological strain and inconclusiveness which the 'containment' policy involved. **1955** M. BELOFF *Foreign Policy & Democratic Process* 29 A more vigorous attempt to communicate with the captive peoples of Europe appears to have been what was meant by the distinction between 'containment' and 'liberation' which the change of administration in Washington at the beginning of 1953 was expected to bring about.

contaminant (kǫntæ·minănt). [f. CONTAMIN(ATE *v.* + -ANT.] That which contaminates.
1934 in WEBSTER. **1949** H. W. FLOREY *Antibiotics* I. i. 35 He investigated an accidental contaminant which on examination proved to be a strain of *B. mesentericus*. **1951** *Engineering* 18 May 588/3 Absorbed films or layers of contaminant appear to play a decisive part.

contaminate, *v.* Add: **b.** *spec.* (*a*) to subject to (the risk of) contamination by radioactivity; (*b*) to infect with poison gas; (*c*) *Textual Criticism*, to subject to contamination (see *CONTAMINATION 1 e). Cf. *DECONTAMINATE *v.*
1926 R. LAWSON tr. *Hevesy & Paneth's Man. Radio-activity* 61 All objects in the neighbourhood of strong polonium preparations become contaminated. **1938** *Protection of Home against Air Raids* (*H.M.S.O.*) 25 Mustard gas..also 'contaminates' clothing, or other objects exposed to it, making them dangerous to have near you or to touch until they have been 'decontaminated'. **1942** R. W. CHAPMAN in *Essays & Studies* XXVII. 43 A manuscript B, copied from A, may have been corrected or 'contaminated'..by reference to a third manuscript C. **1945** E. WAUGH *Brideshead Revisited* 16 'We are being sprayed with liquid mustard-gas,' I said...There were no casualties and nothing had been contaminated. **1949** SCHWEITZER & WHITNEY *Radioactive Tracer Technique* i. 5 The student..may easily get his hands contaminated and from this source may contaminate the laboratory.

contaminated, *ppl. a.* Add examples in *spec.* senses (see prec.).
1913 F. W. HALL *Compan. Class. Texts* 128 Contaminated texts existed in very ancient times. **1938** *Encycl. Brit. Bk. of Yr.* 161/2 To destroy persistent gases and to isolate contaminated areas.

contamination. 1. a. Add to def.: *spec.* the presence of radioactivity where it is harmful or undesirable.
1913 RUTHERFORD *Radioactive Substances* iii. 112 All sources of active matter should be kept in sealed vessels, in order to avoid possible radio-active contamination. **1950** S. GLASSTONE *Sourcebk. Atomic Energy* xviii. 518/1 A problem which radiochemical investigations have..is that of preventing contamination of both equipment and personnel.

c. The blending of two or more stories, plots, or the like into one.
1877 *Encycl. Brit.* VII. 411/2 The Roman versions were probably terser than their originals, which they often altered by the process called contamination. **1909** *Nation* (N.Y.) 22 Apr. 407/2 In 'All Fools', for example, we have a 'contamination', to use the technical term, of the 'Heautontimoroumenos' and the 'Adelphi', plus a sub-plot of Chapman's own invention.

d. *Philology.* The blending of forms, words, or phrases of similar meaning or use so as to produce a form, word, or phrase of a new type.
1888 STRONG tr. *Paul's Princ. Lang.* 160 By 'contamination' I understand the process by which synonymous forms of expression force themselves simultaneously into consciousness, so that neither of the two makes its influence felt simply and purely: a new form arises in which elements of the one mingle with elements of the other. *Ibid.*, Contamination manifests itself partly in the phonetic form of single words, partly in their syntactical combination. **1890** STRONG et al. *Hist. Lang.* 142 We may lay it down that for analogy we must demand a sufficient number of examples on which to base a rule; while for contamination, a single form or construction may suffice. **1901** OERTEL *Lect. Study Lang.* 174 The structure of grammatical paradigms does not escape contamination. **1947** E. STURTEVANT *Introd. Ling. Sci.* xi. 110 Momentary hesitation between two possible linguistic responses is extremely common, and not infrequently it leads to a mingling of the two. Instances are.. *ruvershoes* (*rubbers* × *overshoes*)...We call this process contamination. **1968** *Language* XLIV. 475 Contamination is taken to be the process by which two linguistic forms..are crossed to produce a single new form.

e. *Textual Criticism.* A blending of manuscripts resulting in the occurrence in a manuscript or group of manuscripts of readings belonging to different lines of tradition.
1913 F. W. HALL *Compan. Class. Texts* 130 The problem of relationship [of MSS.] is often rendered exceedingly complex by the tendency which is variously described as 'contamination', 'mixture', or 'eclectic fusion' of the different groups. **1960** G. KANE *Piers Plowman: A Version* 21 Of the seventeen manuscripts twelve agree, apart from physical imperfections and two major contaminations.

contango, *sb.* (Examples of *contango-day*.)
1887 *Encycl. Brit.* XXII. 557 The continuation or contango day, when all transactions of a merely speculative description are continued for another fortnight. **1898** *Westm. Gaz.* 20 Oct. 8/1 Monday, October 31, will be contango day in Watney-Combe securities. **1970** *Times* 14 July 28 Dealings began, July 13. Dealings end, July 24. Contango day, July 27. Settlement day, August 4.

conta·ngo, *v.* [f. the sb.] *trans.* To pay contango on (stocks or shares); also *absol.* to obtain deferment of payment of the purchase price of stocks in consideration of a contango.
1900 *Westm. Gaz.* 30 Apr. 9/1 In Chili they should have earned at least 8 per cent., on deposit here 3 per cent., and if used in 'contangoing' stocks, as in the case of the Bank of Tarapaca, considerably more. **1922** *Weekly Westm. Gaz.* 20 May 31/1 Probably at first many of the opponents of the measure may refuse to Contangoe at all. **1928** *Morning Post* 19 Nov., The broker will, as a rule, endeavour to carry over, give on, or contango—the words all mean the same thing—the 50 Gramophone shares.

‖ **conte** (kont). [Fr. (see COUNT *sb.*¹).] A short story (as a form of literary composition).
1891 LANG *Essays in Little* 205 Few men have succeeded both in the *conte* and the novel. **1908** *Daily Chron.* 26 Feb. 3/3 There is no demand for the conte in England. **1929** *Encycl. Brit.* VI. 328/1 The most perfect modern writer of *contes* is Guy de Maupassant. **1948** F. R. LEAVIS *Great Tradition* i. 2 The use made of him [*sc.* Defoe] in the nineteen-twenties by the practitioners of the fantastic *conte* (or pseudo-moral fable).

b. *spec.* A medieval narrative tale (with express or implied reference to Marie de France's *Guigemar*, lines 19 and 883).
1906 *Mod. Lang. Notes* XXI. II. 49/1 The author of *Sir Orfeo* makes here the same careful distinction between the tale ('conte' or 'aventure') and the Breton lai that Marie had made. **1929** M. WATTIE *Lai le Freine* p. xvi, Obviously the Celtic *lais* arose out of narratives (*contes*). **1954** A. J. BLISS *Sir Orfeo* p. xxxii, There can be little doubt that these passages refer to a genuine Breton *lai*, not to a narrative *lai*, still less to a strictly Classical *conte*.

So **conteur** (kontör), a composer of *contes*; also, a narrator.
1857 MRS. A. MATHEWS *Tea-table Talk* i. 153 Theodore Hook's..real disdain of..dancers, 'reminds me', as determined *conteurs* say, of another..more memorable evening. **1908** *Mod. Lang. Notes* XXIII. 205 (*title*) Thomas and Marie in their relation to the *conteurs*. *Ibid.* 207/2 We are apparently dealing here with a writer and one or several 'conteurs'. **1965** R. S. LOOMIS in Bessinger & Creed *Medieval & Linguistic Stud.* 236 All these Welsh tales..were.. adopted by the people of Vannes..and coming to the attention of the professional *conteurs* excited their imagination.

Conté (kǫ·nte). Also **Conte,** and with lower-case initial. The name of the French inventor Nicolas Jacques *Conté* (1755–1805) used (esp. *attrib.*) to designate a kind of pencil, crayon, or chalk, or the process of making such pencils, which he developed.
1852 *Reeves' Amateurs' & Artists' Companion* 222 Chalks. Crayon Pencils..Swiss Crayons. Conte Crayons. Port Crayons. **1885** *Encycl. Brit.* XVIII. 490/1 The pencil leads prepared by the Conté process consist of a most intimate mixture of graphite and clay. **1927** W. G. RAFFE *Graphic Design* ii. 98 The 6B pencil may give an undesirable shine at the critical place where a deep shadow is desired. Conté chalk has a much harder and calculated result. **1941** *Burlington Mag.* Nov. 140/1 Sketches and drawings on Ingres paper with a Conté pencil. **1943** WYNDHAM LEWIS *Let.* 31 Mar. (1963) 352, I bought the last box of white conté in the 'Art Metropole' here the other day. **1959** *Times* 24 Nov. 3/7 His mature drawings, executed with a greasy conté crayon on rough-surfaced paper, are unique in their reliance on granular tone.

contemporarily (kǫnte·mpŏrărili), *adv.* [f. CONTEMPORARY *a.* + -LY².] **1.** Contemporaneously.
1851 E. B. ELLIOTT *Horæ Apocalypticæ* (ed. 4) IV. i. 22 Some ominous derangement of the natural atmosphere may furnish a literal groundwork for the figure, nearly contemporarily. **1928** C. F. S. GAMBLE *North Sea Air Station* ix. 124 Contemporarily with this.

2. In a contemporary or modern style or manner (see *CONTEMPORARY *a.* 4).
1959 *Engineering* 16 Jan. 93/3 A neat little contemporarily covered booklet comes from Shell-Mex and B.P. **1968** 'P. ALDING' *Circle of Danger* ii. 11 The clean and contemporarily beautiful rebuilt centre of the town.

contemporariness. (Later examples.)
1884 *American* VIII. 252/1 The creature almost enjoyed contemporariness with Columbus. **1961** K. TILLOTSON *Novels of Eighteen-Forties* (ed. 2) I. 110 Misconceptions about the 'contemporariness' of Victorian novelists.

contemporary, *a.* and *sb.* **A.** *adj.* Add: **4.** Modern; of or characteristic of the present period; *esp.* up-to-date, ultra-modern; *spec.* designating art of a markedly *avant-garde* quality, or furniture, building, decoration, etc., having modern characteristics (opp. *PERIOD *sb.* 15).
1866 (*title*) The Contemporary Review. **1924** C. GRAY (*title*) A survey of contemporary music. **1925** A. HUXLEY *Barren Leaves* I. i. 4 A frock that was at once old-fashioned and tremendously contemporary. **1930** *London Mercury* XXII. 424 These great poetical prizes tend..to become the perquisites of conservative-minded and imperfectly contemporary writers. **1934** *B.B.C. Year-Bk.* 220 Nor can one imagine a devotee of variety or vaudeville finding any good points in a 'contemporary' or chamber music recital. **1935** S. LEWIS *It can't happen Here* xiii. 129 The 'contemporary furniture' of the 1930s. **1949** *Archit. Rev.* CVI. 315 Flats which, especially in London,..are the form in which contemporary design is making the strongest impact on the urban scene. **1954** P. WYNDHAM LEWIS *Demon of Progress* I. i. 15 To be musically 'contemporary' in England is to be something like Benjamin Britten; unless you wish to ignore entirely the majority trend. *Ibid.* III. xviii. 66 What in your view is the most contemporary kind of painting in England at the present moment. **1957** C. MACINNES *City of Spades* 213 A building..redecorated in a 'contemporary' style—light salmon wood, cubistic lanterns, leather cushions of pastel shades. **1958** *Times* 23 July 6/6 The council have 'noted with some concern the emergence in certain industries of exaggerated styles and mannerisms that have come to be known as contemporary'. **1959** D. COOKE *Lang. Music* ii. 55 The contemporary *avant-garde* composers..all concur in the principle of equating the major triad with pleasure.

b. *absol.*
1954 P. WYNDHAM LEWIS *Demon of Progress* III. xviii. 67 The 'contemporary' is a cultural élite...So the 'contemporary' has nothing to do with time, nor with age. **1956** 'R. CROMPTON' *Matty & Dearingroydes* i. 12 The little sitting-room..with its 'subtle blending of the contemporary and the period'. **1962** L. DEIGHTON *Ipcress File* xxx. 190 It was a 'tasteful' piece of contemporary; natural wood-finish doors, stainless steel windows and venetian blinds everywhere.

c. *Comb.*, as *contemporary-minded, -styled* adjs.
1946 SCOTT GODDARD in Bacharach *Brit. Music* xvii. 216 The times had nothing to offer that would move a contemporary-minded musician to such depths. **1959** H. HOBSON *Mission House Murder* xvii. 111 An attractive, contemporary-styled bungalow.

contemptible, *a.* Add: **4.** as *sb. the Old Contemptibles*: a popular name given to the British army of regulars and special reserve which made up the expeditionary force sent to France in the autumn of 1914, in ironical allusion to the German Emperor's alleged exhortation to his soldiers to 'walk over General French's contemptible little army' (published in an annexe to B.E.F. Routine Orders of 24 Sept. 1914). Also allusively.
1914 *Observer* 4 Oct. 7/5 A roar of laughter is the reception given here [*sc.* Paris] to the publication of the Kaiser's order to walk over Sir John French's 'contemptible little army'. **1916** 'BOYD CABLE' *Action Front* 111 The Old Contemptibles. *Ibid.* 208 So it was that our New Armies had a glimpse of what the old 'Contemptible Little Army' has seen..so often. **1917** *Blackw. Mag.* Aug. 140/1 The Royal Flying Corps..sent whatever machines it could lay hands on to join the old contemptibles in France. **1919** 'BOYD CABLE' *Old Contempt.* Forew. p. viii, I was one o' French's bloomin' old original Contemptibles.

content, *sb.*¹ Add: **1. c.** *sing.* The amount (of a specified substance or material) contained; the amount or quantity yielded. Usu. with defining word prefixed.
1901 *Chemist & Druggist* LVIII. 18 Jeancard and Satie.. conclude that altitude has no influence upon the ester content of lavender oil. **1901** *Yearbk. U.S. Dept. Agric.* (1902) 41 The director of the Arizona Experiment Station reported that the sugar content in pounds per acre..ranged from 1,491 to 3,361 pounds. **1955** J. H. COMROE et al. *Lung* vi. 106 Normal or low CO_2 content and pressure does not mean that the patient is not seriously ill.

d. *Psychol.* [tr. G. *inhalt.*] The totality of the constituents of a person's experience at any particular moment (see quots.).
1890 W. JAMES *Princ. Psychol.* II. xvii. 28 The relations are part of the 'content' of consciousness. **1896** G. F. STOUT *Analytic Psychol.* I. I. i. 41 It is an appearance in consciousness. It is what Brentano would call a content (Inhalt) of presentation. **1901** BALDWIN *Dict. Philos. & Psychol.*, *Content*,..(2) A constituent of any kind of presented whole. (3) An object meant or intended by the subject...See *Intent* for this meaning. *Ibid.*, We may distinguish conveniently the presented content (argued about) and the process (the arguing). **1902** *Amer. Jrnl. Psychol.* Apr. 279 Funded or consolidated contents, are..such contents as are produced by bringing together, in a very intimate way, various part-contents. **1913** E. JONES *Papers Psycho-Analysis* 26 Freud..contrasts the 'manifest content', which is the dream as directly related, with the 'latent content', which is the group of thoughts reached by psycho-analysis of the dream. **1916** C. E. LONG tr. *Jung's Coll. Papers Anal. Psych.* 14 A definite presentation-content [*Vorstellungsinhalt*] can even create a stereotyped repetition in the individual crisis. *Ibid.* 76 The hallucinatory content and loud speaking is also met with in persons with hallucinations in lethargy. **1927** J. ADAMS *Errors in School* ii. 35 The thinker in whose mental-content they found a place. **1963** J. P. GUILFORD in Taylor & Barron *Scientific Creativity* II. 102 Previously..three kinds of content were distinguished: figural, structural, and conceptual.

8. *attrib.*, as *content-analysis*; **content clause** (see quots.); **content word,** a word (normally a noun, adjective, or verb) that conveys meaning in an utterance (in contrast with a grammatical element like a preposition, article, or auxiliary).
1953 J. B. CARROLL *Study of Language* iv. 120 There is the operational problem of analyzing the semantic content of messages; this step has come to be known as *content analysis*. **1961** *Lancet* 12 Aug. 360/2 Content-analysis allows the teacher to assess the effectiveness of his instruction. **1969** *Computers & Humanities* III. 144 Content analysis has been used by non-historians to help resolve the disputed authorship of certain of the *Federalist Papers*. **1927** JESPERSEN *Mod. Eng. Gram.* III. ii. 23 *Content clauses*. I venture to coin this new term for clauses like the one in '(I believe) *that he is ill*'. **1957** ZANDVOORT *Handbk. Eng. Gram.* v. ii. 222 Something might be said for the term 'content clauses'...They express the 'contents' of the noun on which they depend...The term is borrowed from Jespersen..who..applies it to all object clauses, as well as to subject clauses and predicate clauses. **1940** BRYANT & AIKEN *Psychol. English* xxvi. 159 These words are the small change of the language; they are the construction words—pronouns, common prepositions, and auxiliaries. They are distinguished sharply from 'content' words in that their chief value is syntactical rather than semantic.

contentual (kǫnte·ntiuăl), *a.* *Philos.* and *Psychol.* [f. CONTENT *sb.*¹ + -AL on type of *conceptual, eventual*, etc., as tr. G. *inhaltlich*.] Belonging to, or dealing with, content (opp. act or form).
1909 W. M. URBAN *Valuation* iv. 106 The decrease in these contentual aspects does not..mean loss in worth suggestion. **1935** *Brit. Jrnl. Psychol.* XXVI. 208 As regards the 'contentual' and the 'actual' views, the former characterized the works of Sully and Volkmann. **1962** W. G. RUNCIMAN *Plato's Later Epistem.* iv. 132 Distinguishing the formal from the contentual features of propositions.

conte·rminousness. [f. CONTERMINOUS *a.* + -NESS.] The state of being conterminous.
1905 W. JAMES in *Jrnl. Philos.* II. 35 Where the experience is not of conflux, it may be of conterminousness (things with but one thing between); or of contiguousness (nothing between). **1907** C. A. STRONG *Let.* 5 Oct. in R. B. Perry *Tht. & Char. W. James* (1935) II. 550 The notion of 'conterminousness'..is founded..on the conception that an object is identical with the experience of it.

contessa (kǫnte·sā). [It., f. late L. *comitissa* (see COUNTESS).] An Italian countess.
1819 BYRON *Let.* 8 Nov. (1830) II. 284 Without attempting to influence the decision of the Contessa, a good deal depends upon it. **1862** A. SEVERN *Let.* in S. Birkenhead *Illustrious Friends* (1965) xiv. 137 Walter immediately got introduced to half a dozen Contessas and Marchesas. **1950** C. WOODHAM-SMITH *Florence Nightingale* 21 'At homes' attended by fashionable marchesas and contessas.

contestation. Add: **1. b.** In the Gallican liturgy, the prayer immediately preceding the Canon of the Mass.
1727–41 [see PREFACE *sb.* 1]. **1863** [see ILLATION 2]. **1880** SCUDAMORE in Smith & Cheetham *Dict. Chr. Ant.* II. 1197/2 The Contestation invariably ended with the *Sanctus*. *Ibid.* 1695/1 The Roman words of contestation are, 'Vere dignum et justum est' [etc.].

contestee (kǫntestī·). *U.S.* [f. CONTEST *v.*] A candidate for election who is in the position of having his seat contested by another.
1870 *Congress. Globe* 16 Feb. 1349/3 So many voted for the contestant and so many for the contestee in that ward. **1887** *U.S. Stat. at Large* XXIV. 445 Upon receipt of the contestee's brief the clerk shall forward two copies thereof to the contestant.

context, *sb.* Add: **6.** *attrib.* and *Comb.*, as *context-theory*; *context-bound, -free, -sensitive* adjs.
1965 *Language* XLI. 506 Further, synonymy must be context-bound. **1957** J. PASSMORE *100 Yrs. Philos.* i. 16 All nouns and all adjectives..are context-free names. **1959** I. DE SOLA POOL *Trends in Content Analysis* vii. 219 Context-free measurement of symbolic forms which are instrumentally manipulated is apt to be misleading. **1964** *Language* XL. 317 A prosodic feature is one involved in a context-sensitive phonological rule. **1965** N. CHOMSKY *Aspects of Theory of Syntax* i. 61 The theory of context-sensitive phrase-structure grammar..probably does not fail in weak generative capacity. **1936** J. R. KANTOR *Objective Psychol. of Gram.* ix. 116 The context theory. According to this theory, what a word means depends upon its connection in past experience with some other thing.

co·ntextless, *a.* [f. CONTEXT *sb.* + -LESS.] Lacking a context.
1936 *Mind* XLV. 304 Lord Russell adduces 'it rained in London on January 1, 1066'. This is, so far, only the formal truth-claim of a contextless 'proposition' about an historical event. **1965** *Punch* 13 Oct. 551/1 They are like remembered scenes or remarks that drift into an idling mind, mattering obscurely to the thinker but contextless to anyone else. **1969** P. DICKINSON *Pride of Heroes* 72 The seemingly contextless flurry of action..locked itself into a pattern of violence.

contextual, *a.* Add: **b.** Philos. *contextual definition* = *definition in use* (see *DEFINITION 4 c).
1934 W. V. O. QUINE *Syst. Logistic* xiv. 141 We escape the technical complications of contextual definition. **1944** M. WEITZ in Schilpp *Philos. B. Russell* 117 The best example of analysis as contextual definition in Russell.. is his analysis of definite descriptions.

contextualism (kǫnte·kstiuăliz'm). [f. CONTEXTUAL *a.* + -ISM.]
1. *Philos.* Any doctrine emphasizing the importance of the context of inquiry in solving problems or establishing the meaning of terms.
1929 *Proc. Arist. Soc.* XXX. 10 The argument from mental atmosphere is a form of what may be called the contextual argument in philosophy, although not necessarily of monistic contextualism. **1964** M. THOMPSON in Chisholm et al. *Philos.* 199 Various naturalists have adopted the label 'contextualism' as a mark of their opposition..to any attempt to fix the meanings of key terms absolutely without regard for context.

2. The policy or practice, in literary criticism, of setting a poem or other work in its cultural context.
1955 G. S. FRASER in J. Wain *Interpretations* 234 Contextualism, in this Batesonian sense, can..seem both fussy and flippant. **1956** *Times Lit. Suppl.* 20 Jan. 37/2 If I may elucidate the 'contextualism' he attributes to me, the poem, when read out of its original historical context, is a present *fiction*.

Hence **conte·xtualist,** an exponent or adherent of contextualism; **contextuali·stic** *a.*, pertaining to or characteristic of contextualism.
1936 *Mind* XLV. 531 The Possible is, in a certain sense, confined to the future: this is the view of..Contextualists such as Mead. *Ibid.*, On contextualistic grounds there is definite evidence for the possibility of the past. **1946** C. MORRIS *Signs, Lang. & Behav.* 321 A contextualistic theory of perception. **1956** M. KRIEGER *New Apologists for Poetry* p. xiii, The contextualist's dilemma.

contextualize (kǫnte·kstiuăləiz), *v.* [f. CONTEXTUAL *a.* + -IZE.] *trans.* To place in, or treat as part of, a context; *spec.* to study (phonemes, words, etc.) in contexts. So **conte:xtualiza·tion,** the act or result of contextualizing; **conte·xtualized** *ppl. a.*
1934 J. R. FIRTH *Papers in Ling.* (1957) ii. 5 You contextualize your alternants and so arrive at your phoneme-units or functional units. *Ibid.*, By contextualization is here meant, not only the recognition of the various phonetic contexts in which the phonemes occur, but the further identification of phonemes by determining their lexical and grammatical functions. **1951** R. FIRTH *Elem. Social Organiz.* i. 23 In his study of behaviour the anthropologist works by contextualization. *Ibid.* vii. 241 They hold to their own, but regard them as pragmatic and contextualized. **1959** M. A. K. HALLIDAY *Lang. Chinese 'Secret Hist. Mongols'* ii. 10 The complete text has meaning in the social context in which it operates, and this is to be stated by the procedure of 'contextualization'. **1963** C. IRVING et al. *Scandal '63* xx. 224 Blezinger used an oblique question to find how people 'contextualised' the Affair. *Ibid.* 225 A sample drawn from the whole population would have shown even less political contextualisation.

continent, *sb.* Add: **5. d.** **continent-wide** *a.*, throughout a (specified) continent.
1936 M. FRANKLIN *All that Swagger* xiii. 123 His practice of equality with all men foreran a continent-wide experiment. *a* **1963** L. MACNEICE *Astrology* (1964) ix. 278 These predictions set off a continent-wide panic.

continental, *a.* (and *sb.*). Add: **1. c.** *Geol.* Designating, or pertaining to, deposits laid down on land masses (as distinct from marine deposits).
1934 WEBSTER, *Continental deposit*, a terrestrial deposit. **1946** L. D. STAMP *Britain's Struct. & Scenery* xii. 142 Wedges of continental (fluviatile, deltaic or estuarine) strata.

d. In special collocations: *continental divide*, a divide separating two river systems of a continent; *spec.* the divide in North America separating rivers flowing into the Atlantic from those flowing into the Pacific; *continental drift*, the postulated movement of the existing continents to their present positions after having originally formed a single land-mass; *continental glacier*, a sheet of ice which covers a large part of a continent; *continental island*, an island close to and geologically related to a continent; *continental plateau* or *platform*, the part of the earth's crust above the level of the ocean basins, comprising the continents and the continental shelves; also *occas.* = *continental shelf*; *continental shelf* (see SHELF *sb.*¹ 4 b); *continental slope*, the slope between the outer edge of the continental shelf and the ocean bed; *continental terrace*, the continental shelf and continental slope considered together.
1869 W. J. PALMER *Surv. across Continent* 171 The great Continental Divide at Arkansas Pass. **1889** *Trans. R. Soc. Canada* VII. 130 They..reached the continental 'divide' on February 25th. **1890** [see DIVIDE *sb.* 2]. **1957** G. E. HUTCHINSON *Treat. Limnol.* I. i. 65 On the eastern side of the Continental Divide in Colorado. **1926** J. W. GREGORY in *Rep. Brit. Assoc. Advancem. Sci.* *1925* 316 The increasing evidence of the mobility of the earth's crust has prepared the way for a sympathetic reception of Wegener's theory of continental drift. **1936** *Discovery* Mar. 67/1 The earth's land surfaces constitute an enormous disjointed jigsaw puzzle, the interlocking or fitting portions of which have floated away from one another. This conspicuous feature of the world's outer mantle is known popularly as 'Continental Drift'. **1897** I. C. RUSSELL *Glaciers of N. Amer.* i. 3 The principal characteristics of continental glaciers are their vast extent, their comparatively level surfaces, and the prolongation..of their borders into lobes and..streams. **1910** *Encycl. Brit.* XII. 60/2 Between these 'polar' or 'continental glaciers' and the 'alpine' type there are many grades. **1895** DANA *Man. Geol.* (ed. 4) I. 22 The islands adjoining the continents are properly continental islands. **1898** J. GEIKIE *Earth Sculpture* xv. 262 When we consider the surface of the earth's crust as a whole we recognize two great areas, an oceanic depressed region and a continental

elevated region, or, shortly, an oceanic basin and a continental plateau. **1899** *Geogr. Jrnl.* XIII. 286 *British and Continental Platform*, a gently shelving platform stretching seawards to varying distances from 20 to 200 miles, terminating in a declivity or escarpment at depths (according to distance from land) varying from 100 to 200 fathoms. **1907** R. D. Salisbury *Physiography* I. 11 The continental platforms are much more nearly continuous than the continental lands. *Ibid.* 761/1 Continental slopes. **1942** R. A. Daly *Floor of Ocean* i. 10 The continental shelf..and slope together make the composite surface of a great three-dimensional bench which is conveniently named the continental terrace. **1955** *Sci. Amer.* Mar. 82/1 This underwater borderland between continent and ocean, called the continental terrace, is a shelf of varying width—from a few tens of miles to as much as 300 or 400 miles off some coasts. At its seaward edge it pitches steeply into the deep ocean basin.

2. *Continental breakfast*, a light breakfast such as is eaten on the Continent, esp. in France; *Continental Sunday*, Sunday observed more as a day of public entertainment (as held to be customary on the continent of Europe) than as a day of rest and religious observance (as in Great Britain).

1856 *Sat. Rev.* 4 Oct. 503/2 To tell a man that he is in favour of a Continental, or of a French Sunday, is a reproach which is sure to carry great weight. **1860** C. M. Yonge *Hopes & Fears* I. ii. xii. 454 She was glad to have done with Continental Sundays that had left her feeling good for nothing all the week. **1911** R. Brooke *Let.* 12 Apr. (1968) 296 You'll have to give me breakfast—a purely Continental one, I assure you, no eggs. **1932** A. Christie *Peril at End House* vi. 74 Poirot clung firmly to the Continental breakfast...Consequently he breakfasted in bed upon coffee and rolls. **1964** Wodehouse *Frozen Assets* ii. 42, I am not bursting. I am making a light Continental breakfast. **1966** *Guardian* 29 Mar. 10/2 The Continental Sunday points the way. To church, to the polling station, and to football.

3. (Earlier examples of *continental money*.)

1775 *Jrnls. U.S. Cont. Congress* II. 207 That each gentleman who signs the continental money, be..paid out of the continental treasury. **1776** in *Conn. Hist. Soc. Coll.* XXI. 449 Continental Money will not pass in N. York.

b. Used as a depreciatory epithet (orig. with reference to currency). (Cf. sense B. 2 b in Dict.) *U.S. colloq.*

1841 W. G. Simms *Kinsmen* I. viii. 98, I wouldn't give a continental copper for the safety of your skin. **1851** *Knickerbocker* XXXVII. 554 That clock you sold me ain't worth a continental cuss. **1890** *Amer. N. & Q.* V. 169 'A Tinker's Dam' is equivalent to the expression, 'A Continental Damn'.

B. *sb.* **2. a.** (Earlier example.)

1777 in *Maryland Hist. Mag.* V. 210 Perhaps it would be in your Excellency's power to spare 3 or 400 Continentals.

b. (Earlier and later examples.) Also, *not to care* (or *give*) *a continental*. orig. *U.S.*

1786 Freneau *Poems* 323 That damnable bubble the old Continental That took people in at this wonderful crisis. **1822** J. F. Cooper *Spy* xvi, If it's silver or goold..it's but little I have, though I've a trifling bit of the continental. **1874** B. F. Taylor *World on Wheels* ii. 203 He did not care a 'Continental' for the whole business. **1882** G. W. Peck *Sunshine* 180 The next day he is all played out and not worth a continental. **1907** *Daily Chron.* 5 Nov. 6/6 The 'New York American'..incidentally isn't caring a continental what London thinks. **1927** S. Williams *Drury Club Case* vii. 137 I'm not concerned for Hal. I don't care a continental what happens to him. **1931** J. T. Adams *Epic of Amer.* iv. 98 The phrase 'not worth a Continental' was so impressed upon the people that, unlike the money to which it referred, it gained a lasting circulation. **1968** G. de Fraga *Murder at Cookout* xxv. 130 It was murder, m'boy! And I don't give a continental which of them did it.

continentality (kǫ:ntinentæ·lĭti). [f. Continental *a.* + -ity.] The condition of being continental; spec. in *Meteorology* [tr. G. *kontinentalität* (W. Zenker *Der thermische Aufbau d. Klimate* (1895) ii. 120], the qualities possessed by or typical of a continental climate.

1897 *Pop. Sci. Monthly* Jan. 332 The fact that such a large mass of apparently continuous ice exists circumpolarly..is probably the one most significant circumstance which has led to favor the view of continentality. **1903** R. de Courcy Ward tr. *J. Hann's Handbk. Climatology* xii. 215 In order to obtain a proper value for the derivation of temperature from the amount of radiation in a typical continental climate, Zenker was first obliged to establish the idea of 'continentality' with reference to temperature conditions. **1922** *Nature* CIX. 512/2 He lays stress on the distribution of land and water within the zones; climate is greatly affected by 'continentality' and 'oceanity'. **1946** Zeuner *Dating the Past* iv. 104 A distinct climatic zonation, from S to N (approaching the ice) as well as from W to E (increasing continentality).

contingency. Add: **9.** *attrib.* (occas. in *pl.*); **contingency fund,** one set aside for conditional or incidental expenses.

1901 *Guinness Trust* (Fulham) 10 The 'nett income'..is exclusive of £17,361 6s. 6d. placed to the credit of 'depreciation and contingency fund' during the same period. **1924** *Act.* 14 & 15 Geo. V c. 38 § 67 A corresponding increase in the amounts to be carried to the Contingencies Fund. **1931** *Economist* 10 Jan. 80/2 The sum of £30,000 is placed to contingency reserve. **1934** T. S. Eliot *Rock* i. 23 If they haven't any contingency fund to draw on, that's their look-out. **1963** *Punch* 3 Apr. 471/2 A regular system of contingency planning. *Ibid.*, The Cabinet failed to instruct our defence planners to prepare a contingency plan for the cancellation of Skybolt. **1969** *New Scientist* 17 July 116/2 Armstrong will..scoop up a 'contingency sample' of the lunar material. **1970** D. Craig *Young Men may Die* xxvi. 190 They had their contingency plans nicely polished and when it didn't work either way they were nowhere.

contingent, *a.* **6.** (Later example.)

1857 Geo. Eliot *Amos Barton* v. in *Blackw. Mag.* Feb. 156/1 In that toppling and contingent state, in which a very slight push from a malignant destiny would utterly upset it.

continuant, *a.* and *sb.* **A.** *adj.* **1.** Delete †*Obs.* and add later example.

1933 *Mind* XLII. 72 Since there is nothing changeless in them, 'things' are not really continuant entities.

B. *sb.* **3.** *Philos.* An existent, or something physical or psychical manifested in space and time, that retains its identity though changing its states or relations or when regarded as having different states or relations. (Cf. *occurrent *sb.*)

1917 W. E. Johnson in *Proc. Arist. Soc.* XVII. 431 We are thinking of a body X which *continues to exist* (and which may therefore be called a continuant), while the movement from A to B..is occurring. **1920** J. Laird *Stud. Realism* ii. 27 The physical things of common sense..are continuants in time. **1933** *Mind* XLII. 161 It is of the essence of Spinoza's theory that there are no finite continuants.

continuation. Add: **11.** *continuation class, course, education, schooling*; **continuation-school** (earlier example); (see also *day continuation school* s.v. *day *sb.* 24).

1887 *Peel City Guardian* 11 June 2/1 Evening continuation schools and classes. **1901** Rowntree *Poverty* 74 Attendance at the continuation classes..cannot make up for the loss resulting from the early age at which the children leave school to go to work. **1907** *Westm. Gaz.* 25 Sept. 2/1 Curtailing the period of the continuation courses from eighteen to eleven days. **1900** *Ibid.* 20 Dec. 3/2 Drill ought to form part of the continuation education. **1837** *Central Soc. of Education: First Publication* 400 [In Switzerland]..boys are compelled to attend the continuation-schools until they have reached their seventeenth year. **1928** *Britain's Industr. Future* (Lib. Ind. Inq.) iv. Argt. 266 Part-time day continuation schooling up to sixteen years.

continuative, *a.* **2.** (Examples.)

1924 Jespersen *Philos. Gram.* viii. 113 A so-called continuative relative clause is, of course, non-restrictive. **1927** —— *Mod. Eng. Gram.* III. 82 A subdivision..is formed by continuative clauses, in which *who* and *which* might be replaced by *and he*. **1957** Zandvoort *Handbk. Eng. Gram.* I. iv. 59 To denote an action or a state beginning at some time in the past and continuing up to the moment of speaking, English requires the perfect tense (known in this function as the *continuative perfect*). 'We've known each other for years.'

continued, *ppl. a.* Add: **3. b.** *continued story*: a serial story. *U.S.*

1880 'Mark Twain' *Tramp Abroad* (ed. 3) 299 The less important dailies give one a tablespoonful of a continued story every day. **1881** Rittenhouse *Maud* (1939) 7 Elmer brought me the Sunday Bulletin with that sweet continued story in it. **1932** W. Cather *Obscure Destinies* 89 Grandmother loved to read..the continued story in the Chicago weekly paper. **1951** M. McLuhan *Mech. Bride* 10/2 Note the 'continued story' technique of nineteenth-century and Hearst journalism.

continuity. Add: **6.** *Cinemat.* A detailed scenario for a cinema film; also, the maintenance of consistency or a continuous flow of action in successive shots or scenes of a cinema or television film. orig. *U.S.*

1921 *Collier's* 25 June 26/3 It is a hobby of mine never to pay no attention to a continuity, as it only gets a man balled up. **1926** G. F. Buckle *Mind & Film* i. 18, I am deliberately using the word scenario here instead of continuity (which is the correct name for the working script) in order to avoid confusion when discussing faults in the scenario which result in bad continuity in the film. **1927**, etc. [see sense 7 below]. **1930** *Times* 12 July 4/6 Mr. Knoblock became joint author..of the scenario of the talking film and he also wrote the 'continuity' or stage directions. **1940** F. Scott Fitzgerald *Let.* 11 May (1964) 116 I've written a really brilliant continuity..it seems to be a last life line that Hollywood has thrown me. **1940** *War Illustr.* 12 Jan. p. ii/2 Touching this matter of 'continuity'—to use a word from film-technique. **1951** 'J. Wyndham' *Day of Triffids* vi. 116 Sandra is our professional remembrancer—continuity is her usual work.

b. A series of linking announcements, interpolations, or the like in a radio programme or broadcast; the maintenance of a continuous sequence in broadcasting (see quot. 1941).

1934 *B.B.C. Year-Bk.* 107 The Ceremony of the Keys is a good example of a broadcast which requires elaborate technical arrangements to ensure 'continuity'. **1941** *B.B.C. Gloss. Broadc. Terms* 7 *Continuity*, (1) structure of verse or prose, linking sections of a dramatic or of a magazine programme; (2) announcements or interlude material filling intervals between programmes in a sequence of programmes. **1968** *Listener* 4 Jan. 28/2 *All the Best*..is..full of limp interviews and badly written continuity.

7. *attrib.*, as *continuity announcement, announcer, clerk*; **continuity girl** *Cinemat.*, a woman who is responsible for ensuring that there are no discrepancies of detail between linked scenes filmed at different times; **continuity studio** (see quot. 1941); **continuity suite** (see quot. 1962); **continuity title**, an explanatory title inserted into a film to ensure continuity; **continuity writer,** a writer of continuity titles.

1962 A. Nisbett *Technique of Sound Studio* i. 17 The output of individual studios is..linked together by..continuity announcements. *Ibid.*, Continuity announcer. **1927** *Sunday Express* 29 May 10 Promising to let her be a continuity clerk if she did not get ahead as an actress. **1933** 'I. Hay' & 'A. Armstrong' *Orders are Orders* ii. 46 Meet Miss Marigold, my secretary and continuity girl. **1937** M. Robinson (*title*) Continuity girl. **1944** R. Manvell *Film* i. vi. 68 The director is the admitted co-ordinator of the actors' work, with the continuity girl killing the details. **1970** *Daily Tel.* (Colour Suppl.) 3 July 23 The world of folding chairs, megaphones and continuity girls is put firmly behind him when he turns to domestic film-making. **1941** *B.B.C. Gloss. Broadc. Terms* 7 *Continuity studio*, small studio from which an announcer, supervising the running of a sequence of programmes, makes opening and closing announcements and interpolates interlude material when required. **1957** *B.B.C. Handbk.* 52 The machines are started and stopped by remote control..from a continuity suite in the case of reproductions. **1962** A. Nisbett *Technique of Sound Studio* 246 *Continuity suite*, a centre through which programmes are routed or where they are reproduced to build a particular service ready for feeding to a transmitter. **1953** K. Reisz *Film Editing* 279 *Continuity title*, title designed to bridge a break in the pictorial continuity. **1921** *Moving Pict. Stories* 9 Dec. 28/2 He knows that the continuity writer is going to make certain changes. **1928** *Sunday Express* 18 Mar. 4/3 The necessity for such titles as 'Time went on, and brought changes in the circus' would bring most continuity writers into disgrace.

continuo (kǫnti·nʉ,o). *Mus.* = Basso *continuo*; also, an instrument or instruments playing this.

1724 *Short Explic. For. Words Mus. Bks.* 24 *Continuo*, for this see the Words *basso continuo*. **1740** Grassineau tr. *S. de Brossard's Mus. Dict.* 44 *Continuo*, signifies the thorough Bass, as *Basso continuo* is the continual or thorough Bass. **1879** Grove *Dict. Mus.* I. 151/2 In the score of the 'Matthäus Passion' of Bach..in the recitative a single line and figures is given for the 'continuo' alone. **1942** E. Blom *Music in Eng.* iii. 41 These harmonic instruments thus began casually to take on the function they or their descendants assumed quite regularly in the next two centuries under the name of *continuo*.

continuous, *a.* Add: **3.** *continuous creation*, creation viewed as being a continuous process and not a single act at a particular time; *spec.* the view that the universe is in a steady state, new systems being formed continually to replace those that disappear; *continuous-flow*, used *attrib.* designating a system, device, etc., in which a fluid or other material flows continuously; *continuous kiln* (see quot. 1910); *continuous miner* (see quot. 1967); *continuous process*, an industrial process which operates without interruption (opp. *batch process*); *continuous spectrum*, 'a spectrum not broken by bands or lines, but having the colors shaded into each other continuously, as that from an incandescent solid or liquid, or a gas under high pressure' (Webster 1890) (cf. quot. 1879 for sense 1 a in Dict.); *continuous stationery* (see quot. 1942); *continuous tone* (see quot. 1968); *continuous variation*, in *Biol.* (see quot. 1961); *continuous voyage*, a voyage which, though interrupted by stops at ports or otherwise, is regarded as a single voyage in reference to the purpose for which it was undertaken (e.g. the consignment of goods or materials); *continuous wave*, an electromagnetic (esp. radio) wave having constant amplitude and intensity; also *attrib.*

[**1895** G. MacDonald *Lilith* vi. 35 The still face might be a primeval perfection; the live eyes were a continuous creation.] **1950** F. Hoyle *Nature of Universe* v. 107 With continuous creation the apparent contradiction between the expansion of the Universe and the requirement that the background material shall..condense into galaxies is completely overcome. **1956** *Ann. Reg. 1955* 397 The distribution of radio-stars favoured evolutionary theories of cosmology rather than more recent theories of continuous creation. **1958** *New Biol.* XXV. 93 The oxidation tanks were run on the continuous-flow principle, with recirculation of a portion of the effluent to inoculate the influent. **1965** *Gloss. Aeronaut. Terms* (*B.S.I.*) § 17 p. 3 *Continuous flow oxygen system*, an oxygen system in which the oxygen flows during both inspiration and expiration. **1890** *Sci. Amer.* 11 Jan. 21/3 (*heading*) A continuous brick kiln. **1901** *Notes on Building Construction* (ed. 5) III. iii. 193 Tunnel kilns, called also Continuous, 'Running', 'Perpetual', or 'Draw-Kilns'. **1910** *Encycl. Brit.* IV. 520/1 Brick-kilns..can all be grouped under two main types—Intermittent kilns and Continuous kilns. *Ibid.* 520/2 The 'continuous' kiln..is really made up of a number of separate kilns or firing-chambers, built in series and connected..in such a manner that the products of combustion from one kiln may be made to pass through a number of other kilns before entering the flue. **1958** A. Nelson *Methods of Working* v. 61 Further studies by mining engineers..led to the design and development of machines which combined the operations of cutting, breaking and loading the coal...These combined machines are popularly known as 'continuous miners'. **1967** *Gloss. Mining Terms* (*B.S.I.*) viii. 9 *Continuous miner*, a machine used mainly for the cutting and loading of coal or other mineral in mine entries or narrow places. **1909** *Daily Chron.* 11 June 7/2 Everything being on the automatic and con-

tinuous-process principle. **1922** *Encycl. Brit.* XXXI. 389/1 The case which appears to have presented most difficulty in national legislation is the *continuous process*. **1954** [see *batch *sb.*[1] 7]. **1961** *Times* 22 Nov. 21/6 Many chemical plants, especially the continuous-process type, need very precise adjustment. **1902** *Encycl. Brit.* XXXII. 124/2 Fluorescent radiation, showing a continuous or banded spectrum, can be excited in many gases and vapours. **1942** H. A. Maddox *Dict. Stationery* (ed. 2) 26 *Continuous Stationery*, a modern form of mechanical accounting stationery in which the sheets are of continuous length, sometimes folded concertina fashion. **1959** *B.S.I. News* Dec. 15 The sizes do not necessarily apply to continuous stationery. **1970** *Brit. Printer* June 76 Continuous stationery for computers is one of the fastest-growing specialisations in printing. **1933** C. W. Gamble *Mod. Illustration Processes* vii. 125 We may take..an artist's drawing of a landscape of washes of black pigment upon white paper..as representing a drawing in 'continuous' tone. **1967** E. Chambers *Photolitho-offset* iv. 43 With the indirect method the first negatives are made as formerly and they may be in continuous-tone or in dot formation. **1968** *Gloss. Terms Offset Lithogr. Printing* (*B.S.I.*) 10 *Continuous tone*, an image in which tonal gradation is produced by changes in density. **1894** W. Bateson *Study of Variation* 18 The fact that Continuous Variations exist is also none the less a fact. **1961** Webster, *Continuous variation*, variation in which a series of intermediate types connects the extremes. **1806** C. Robinson *Rep. Cases Crt. Admir.* V. 365 *marg.*, Continuous voyage in the colonial trade of the enemy. **1808** *Ibid.* VI. App. note ii, That branch of the colonial principle which relates to continuous voyages. **1902** *Encycl. Brit.* XXXI. 130/2 If the ultimate destination of goods, though shipped first to a neutral port, is enemy's territory, then, according to the 'doctrine of continuous voyages', the goods may be treated as if they had been shipped to the enemy's territory direct. **1911** A. B. Smith *Mod. Amer. Telephony* xxv. 663 If it were possible to produce continuous waves for wireless telegraphy. **1920** H. M. Dowsett *Wireless Telegr.* iv. 70 Continuous wave transmitters. *Ibid.* 92 Any other long-distance continuous wave system. **1933** J. H. Morecroft *Princ. Radio Communication* (ed. 3) vii. 775 Tube B acts as oscillator and detector (self-heterodyne) for reception of continuous-wave signals. **1951** [see *automatic direction finder].

continuum. Add: (Further examples.) *space-time continuum*, see *space-time. Also *attrib.*, as *continuum mechanics, theory*; **continuum hypothesis** *Math.*, the hypothesis that there is no transfinite cardinal between the cardinal of the set of positive integers and that of the set of real numbers.

1903 B. Russell *Princ. Math.* I. xxxvi. 297 Cantor proves that any series *M* satisfying the above conditions is ordinally similar to the number-continuum, *i.e.* the real numbers from 0 to 1, both inclusive. **1905** *Westm. Gaz.* 14 Aug. 2/1 We must know more of the properties of the ponderable atoms moving in the continuum which fills all space. **1924** W. B. Selbie *Psychol. of Relig.* iv. 91 There are no doubt good physiological reasons for these phenomena of the sensory continuum. **1938** *Proc. Nat. Acad. Sci.* XXIV. 101 The continuum hypothesis of Georg Cantor. **1951** C. Truesdell in *Jrnl. Math. Pures et Appliquées* XXX. 116 The notion of an isotropic function..has been much employed in continuum mechanics. *Ibid.* 150 Continuum theory, being independent of any sort of deterministic mechanics of the ultimate particle, serves as a general guide with which any molecular model must be consistent. **1953** A. E. Smailes *Geogr. Towns* 33 There is no longer either socially or physically a simple clear-cut dichotomy of town and country; rather it is an urban-rural continuum. **1953** A. A. Fraenkel *Abstract Set Theory* ii. 152 The assumption that, for every transfinite *t*, 2^t is the cardinal next larger than *t*, is called the generalized continuum hypothesis. **1955** W. Noll in *Jrnl. Rational Mech. & Anal.* IV. 12 (*heading*) Basic concepts of continuum mechanics. **1962** L. J. Cohen *Diversity of Meaning* (1966) i. 12 Descriptive studies are also needed to abstract and record the pivotal stages in a continuum of change. **1964** *Proc. Nat. Acad. Sci.* LI. 109 Since our additional axiom is quite readily acceptable to most mathematicians..one can regard the unprovability of the Continuum Hypothesis as firmly estabished.

contoid (kǫ·ntoid), *a.* and *sb.* *Linguistics.* [f. *cont.*, shortening of Consonant *sb.* + -oid.] **A.** *adj.* Consonant-like; of consonantal character; *esp.* as contrasted with *vocoid *a.* **B.** *sb.* A speech sound of the consonantal type.

1943 K. L. Pike *Phonetics* v. 78 Vocoid and contoid groups are strictly delineated by the articulatory and acoustic nature of sounds. *Ibid.* vii. 143 Contoids include stops, fricative nasals, lateral resonant orals, and central fricative orals. **1957** H. J. Uldall in Hjelmslev & Uldall *Outl. Glossematics* 49 It is possible to regard long vocoids and contoids as manifestations of chains of identical vowels or consonants. **1958** C. F. Hockett *Course Mod. Ling.* vii. 67 A contoid..is a sound involving clearly audible turbulence of the airstream at one point or another in the vocal tract, or else a complete interruption of the air stream. Use of the terms 'vocoid' and 'contoid' enables us to reserve the terms 'vowel' and 'consonant' as labels for structurally defined classes of phonemes in specific languages. *Ibid.*, Consonant-like or contoid effects.

contortion. Add: **1. b.** In technical use.

1869 M. T. Masters *Veg. Terat.* 316 Contortion.—An irregular twisting or bending of the stem or branches is by no means of uncommon occurrence, the inducing causes being often some restriction to growth in certain directions, or the undue or disproportionate growth in one direction, as contrasted with that in another. **1882** Geikie *Text-bk. Geol.* 480 Among the sands and clays of the glacial deposits local examples of contortion occur.

Hence **conto·rtional** *a.* = Contortive *a.*; **conto·rtionate** *a.*, twisting, tortuous; **conto·rtioned** *a.*, twisted.

1911 C. E. W. Bean *'Dreadnought' of Darling* xii. 103 Its more particularly contortionate bends [of the Darling River]. **1918** W. J. Locke *Rough Road* vi. 64 Such and such back-breaking and contortional performances in his bathroom. **1922** *Glasgow Herald* 25 Sept. 13 A pitiable, dazed, contortioned figure. **1928** Galsworthy *Swan Song* II. v. 145 They had so much soul, and that so contortionate, that she could not even keep her attention on them long enough to discover why they were not alive.

contortionism (kǫntǭ·ɹʃəniz'm). [f. Contortion + -ism.] The practice of contorted postures; the acts of a contortionist. Also *fig.*

1927 *Daily Tel.* 8 Feb. 14 Bernt and Partner, by..an idealisation of contortionism, twist their lithe bodies into all manner of graceful poses. **1939** Ld. David Cecil *Young Melbourne* vi. 145 She performed the most prodigious feats of intellectual contortionism.

contortionist. c. Add to def.: or ideas.

1895 *Westm. Gaz.* 11 July 2/1 Mr. Chamberlain is the great contortionist of the political arena. **1936** M. Plowman *Faith called Pacifism* 8 The man who sees war as organized murder doesn't join the Forces. He becomes a Pacifist—or a moral contortionist.

contour, *sb.* Add: **1. d.** *pl.* The curves of the female body. (Cf. quot. 1829 for sense 1 a in Dict.)

1886 O. W. Holmes *Mortal Antipathy* i. 41 All her contours..betrayed a fine muscular development. **1918** *War Birds* (1927) 85 He had a hundred and fifty Waacs of all sizes and contours and he was trying to line them up. **1961** J. Anthony *About Tunisia* iv. 91, I liked painting her firm tawny body whose contours reminded me of the low-lying Tunisian hills.

e. *Phonetics.* A particular level, or a sequence of varying levels, of pitch, tone, or stress.

1941 G. L. Trager in L. Spier et al. *Language, Culture, & Personality* 133 The second approach views the exponential kinetically, describing the movement in shape of the exponential in terms of changes of intensity; this is the contour. *Ibid.*, The basic intensities and contours may be combined in several ways. **1948** K. L. Pike *Tone Languages* i. 8 A pure contour tone language is one in which glides are basic to the system, with no level tonemes whatever: each contrastive pitch unit is a glide. **1962** *Amer. Speech* XXXVII. 169 It is normally a lengthened monophthong..except under a terminal contour before an apico-alveolar consonant. **1962** A. Nisbett *Technique of Sound Studio* 264 Phons equal decibels at 1000 c/s, and at other frequencies are related to this scale by contours of equal loudness. **1970** *Language* XLVI. 265 Its ultimate speech correlates appear in the shape of the tense-carrying morphemes, the word order, and the intonation contour.

4. contour chair, couch, one that is shaped to fit the form of the body, esp. one designed for the use of an astronaut; **contour-chasing** *Aeronaut.*, flying close to the ground and following the contours of the landscape; **contour-feather** (example); **contour ploughing,** the ploughing of land along its contours to minimize soil erosion; hence *contour-plough* v. trans. and sb.; so *contour cropping, furrowing*; **contour terracing,** the construction of terraces along the contours of land.

1948 *Spiegel Catal.* Fall & Winter 490 *Contour chair*... Cradles the back and supports the shoulders. **1958** C. C. Adams et al. *Space Flight* 198 The men are lying all but helpless on their contour chairs. *a* **1918** McCudden *5 Yrs. R.F.C.* (1919) 18 We got the old Blériot..out and went 'contour chasing' over the Plain. **1934** V. M. Yeates *Winged Victory* 37 No fun in flying comparable with the sport of contour-chasing. **1952** *Time* 8 Dec. 70 They could survive by lying on their backs on contour couches, say the space doctors. **1953** 'S. Ransome' *Drag the Dark* (1954) iv. 40 A so-called contour couch. I had bought it..because I do my best thinking while relaxed in a reclining position. **1959** *N.Y. Times Mag.* 11 Oct. 18/1 He..lies there in his 'contour couch' and waits out the long countdown. **1953** *Brit. Commonw. Forestry Terminol.* I. 30 *Contour cropping*, the cultivation of crops in strips along the contour of a slope. **1867** [see *apterium]. **1937** *Farmer's Guide Agric. Res. 1936* (*R. Agric. Soc.*) 238 The remedial measures are contour furrowing, [etc.]. **1941** S. V. Benét in *Life* 7 July 92/2 People whose contour-plows bring back the grass. **1941** *Life* 13 Jan. 68/1 They..contour-plowed 600 acres of land. **1921** McMurry & Parkins *Adv. Geogr.* I. viii. 220 Farmland that has gentle slopes should be plowed so that the furrows keep on a level instead of running downhill. Otherwise, the water may..wash the soil away...Such plowing is called contour plowing. **1943** J. S. Huxley *TVA* 28 The prevention of erosion on arable land by new methods such as contour ploughing. **1969** N. W. Pirie *Food Resources* ii. 40 The simplest is to clear the land and, by careful contour ploughing and mulching, to try to minimize leaching. **1939** Contour terracing [see *bund *v.*]. **1944** J. S. Huxley *Living in Revol.* 114 Using contour terracing where there is danger of erosion. **1957** *Listener* 26 Sept. 468/1 The villager himself has to..put more of it [*sc.* his money] into wells or contour terracing.

contour, *v.* Add: **3.** To follow the contour of.

1924 G. L. Mallory in E. F. Norton *Fight for Everest, 1924* 209 Contouring the hill until we could drop into this valley 6 miles below the bungalow. **1967** *Oceanography & Marine Biol.* V. 61 A part of the current therefore contours the north of New Zealand.

co·ntra-a:ttitude. *Philos.* [Contra- 1.] An attitude of aversion or rejection, as contrasted with a pro-attitude.

1935 *Mind* XLIV. 298 That to which badness is attributed is always the object of a *contra*-attitude. **1952** R. M. Hare *Lang. of Morals* i. 11 An attitude that one should not hit, &c. (or an attitude of aversion from hitting, or a 'contra-attitude' towards hitting).

contraband, *sb.* and *a.* **A. 3.** Add to def.: Also *absolute contraband* (see quots.); opp. *conditional contraband*, anything (such as coal, provisions, vehicles) that may be treated as contraband if it is intended for warlike purposes.

1905 T. E. Holland *Let. to 'The Times'* 1 July (1909) 109 Russia consented to recognise the..distinction between 'absolute' and 'conditional' contraband...Cotton has now followed foodstuffs into the category of 'conditional' contraband. **1939** W. Arnold-Forster *The Blockade, 1914–1919* 6 *Absolute Contraband* consisted of articles used only for military purposes, such as artillery. *Ibid.*, *Conditional Contraband* consisted of certain commodities that might be used either for military or civilian purposes, including food, fuel, and lubricants.

contrabass. 2. (Further examples.)

1926 Whiteman & McBride *Jazz* ix. 197 The tuba and contra-bass tuba. **1965** *Listener* 3 June 837/3 No expense was spared—eight harps and horns, bass trumpet, contra-bass trombone.

co·ntra-bassoo·n, contrabassoon. [Contra- 4.] A double bassoon.

1891 C. R. Day *Descr. Catal. Musical Instruments, R. Mil. Exhib., 1890* iv. 70 A metal contra-bassoon..was exhibited in the Paris Exhibition of 1889. The tone, however ..is greatly inferior to that of the wooden contra-fagotto or double-bassoon, which has been perfected by Dr. W. H. Stone. **1910** *Encycl. Brit.* VII. 40/2 Contrafagotto, Double Bassoon or Contrabassoon. **1928** J. P. Sousa *Marching Along* xiv. 276 The additions [to the classic orchestra] to-day used by the moderns are piccolos, English horns, contra-bassoons, hecklephone, [etc.].

contraception (kǫntrăse·pʃən). [f. Contra- + Con)ception.] The prevention of uterine conception. So **contrace·ptionist**, one who practises or advocates contraception.

1886 E. B. Foote *Radical Remedy in Social Science* v. 89 Where it becomes a necessity to decide between lawful abortion and unlawful contraception, they prefer to break the man-made law against contraceptics rather than the natural law against abortion. **1917** *Boston Pilot* 14 Apr. 4/5 The contraceptionist would take from the Almighty all power over life and dispose matters after his own whim and liking. **1918** R. B. Armitage *Private Sex Advice to Women* xii. 159 Should not contraception be employed to obviate the crime of abortion..? **1923** Stopes *Contraception* 23 Wholesome contraception is a valuable tool in the hands of those who work toward elevating our sex knowledge in the way urged by Professor Bayliss. **1971** H. & A. Brant *Dict. Pregnancy* 75 [The] rhythm method..and abstinence are the only methods of contraception officially approved by the Roman Catholic Church.

contraceptive (kǫntrăse·ptiv), *sb.* and *a.* [f. Contra- + Con)ceptive *a.*] **A.** *sb.* A device, drug, etc., for procuring contraception. **B.** *adj.* Pertaining to or procuring contraception. Hence **contrace·ptively** *adv.*, by contraception; so as to produce contraception.

1891 Ida Ellis *Essentials of Conception* 18 A person once accused me of wrong doing in advocating contraceptives. **1897** J. McCabe in *Literary Guide* 1 Oct. 145/2 Every Catholic knows that the Church has forbidden the use of contraceptives under 'the pain of mortal sin'. **1918** *Edin. Rev.* Apr. 252 The distinction between 'artificial' and 'natural' contraceptives. **1918** R. B. Armitage *Private Sex Advice to Women* x. 130 A 'contraceptive' agency merely tends to prevent or obviate undesirable conception. **1921** *Brit. Med. Jrnl.* 20 Aug. 302/1 A knowledge of contraceptive methods. **1923** Stopes *Contraception* 5 The result [of present economic conditions] too often is the use of harmful contraceptives. *Ibid.* 233 The Church however has already yielded the principle of the use of contraceptive means. **1929** E. V. & A. D. Lindsay *Birth Control* 67 Would they accept physical union made contraceptively sterile? **1963** A. Heron et al. *Towards Quaker View of Sex* ii. 18 The fear of pregnancy, (which modern contraceptives are steadily reducing). **1964** *Economist* 21 Mar. 1082/1 A contraceptively cold winter which prevented cows coming into calf. **1969** *Times* 17 Feb. 10/8 The contraceptive pill did not become available until after the period of study. **1970** *Daily Tel.* 17 July 2/8 Of 30 patients with thrombo-embolism 12, or 40 per cent, had been using oral contraceptives during the month preceding surgery.

contraconscientious, *a.* Delete †*Obs.* and add later examples.

1941 *Mind* L. 56 The conscientious action..may always be *better* than the contra-conscientious even if it can never be *right* when the conscience is in error. *Ibid.*, It is always better (although not necessarily right) for every man to act conscientiously than to act contra-conscientiously.

contract, *sb.*[1] Add: **1. e.** An undertaking. *U.S. colloq.*

1880 A. A. Hayes *New Colorado* (1881) vi. 87 A person should carefully study his temperament..before he takes a contract to go into a deep shaft. *Ibid.* xi. 159 Here, too, was Armigo to have annihilated General Kearny, but for the unfortunate circumstance of his troops declining, as they say in the West, 'to take the contract'. **1891** E. S. Ellis *Check 2134* vii. 50 Any person might well shrink from the contract of corralling a couple [of such men].

f. A season railway-ticket. Also *Comb.*

1899 *Westm. Gaz.* 17 July 4/2 The justice of the peace..decided that the company had no right to expect its contract-holders to carry business advertisements about for its benefit. **1904** *Notice* in *L'pool Exchange Station*, Passengers must show their Tickets (Contract or Ordinary) before passing through these Gates. **1966** F. Shaw et al. *Lern Yerself Scouse* 66 *Ave yera contract?* Have you a season ticket?

g. In the game of auction bridge, an undertaking to make a certain number of tricks; hence *contract bridge* (formerly *contract auction*), a form of auction bridge in which only the tricks which the declarer has undertaken to make count towards the game; also ellipt. *contract.*

1908 *Westm. Gaz.* 24 Dec. 14/2 Suppose the contract is two in hearts, and four by cards are made, the contracting side score 32 below the line, and game. If the contract had failed, and say two by cards had been lost, the adversaries would score 200 in their honour column. **1922** *Encycl. Brit.* XXX. 499/2 'Contract Auction' of which the essential point is that no more tricks can be scored toward game than the declarer has contracted to make. **1927** M. C. Work *Contract Bridge* i. 5 The Contract player, therefore, is straining for every trick. *Ibid.* ii. 9 The basic theory of Contract, viz., that Declarer may score below the line (toward game) only as many odd tricks as he has contracted to win. **1927** *Observer* 10 Apr. 25 Little coteries and sets play variations of Bridge, such as Contract, Cairo, Standard, and so forth. **1929** *Punch* 10 Apr. 420 'Contract bridge', which is superseding auction bridge on the Continent and in America, has taken a strong hold in Britain. **1929** *Observer* 24 Nov., Contract has been 'boomed' in a way that Auction never was. **1929** M. C. Work *Complete Contract Bridge* p. xi, Their contract and game were secure from the beginning. **1952** I. Macleod *Bridge* viii. 100 George Morris, certainly the most..successful money player in this country since Contract started.

contract, *v.* Add: **2. d.** intr. *to contract out*: to make an arrangement or agreement not to participate under certain conditions; to gain exemption or exclusion from certain provisions, etc.; hence, to refuse to take part in or be a part *of*. Conversely, *to contract in*.

1894 *Times* 29 Jan. 11/4 With the exception of those men who now have contracting-out mutual insurance societies, the opinion of the working men..is..against permission to contract out on any terms. **1906** *Daily Chron.* 23 June 4/3 'The parents of at least twenty children' may appeal to the Board of Education, which, if it sees fit, may allow the school to contract-out... They do not contemplate..any large amount of contracting-out. **1920** S. & B. Webb *Hist. Trade Un.* 366 By 'contracting out' was meant an arrangement between employer and employed by which the latter relinquish the rights conferred upon them by the Act, and often also their rights under the Common Law. **1927** *Daily Express* 24 Mar. 1 The Trade Union Bill will provide that members of a trade union wishing to contribute to political funds shall contract in. **1952** *Ann. Reg. 1951* 165 A member of the Committee of Ministers was enabled to 'contract out'. **1957** *Economist* 21 Dec. 1042/3 Why should one section of the community, the trade unionists, contract out of this planless 'opportunity state'? **1959** *Times* 9 Jan. 8/4 The beat generation of San Francisco presumably contracted out of society only after the refrigerator and car had become commonplace.

e. *trans.* To arrange for by contract; to let *out* by contract; to delegate (work, etc.).

1897 *Daily News* 13 Sept. 7/5 Do you contract your work of construction, or do it by direct employment of labour? **1902** *Daily Chron.* 16 Aug. 6/2 We are forbidden to contract a 'boy' away from the country for more than a year. **1959** *Which?* Winter 63/2 The Cleaner may contract work out wholly or in part.

contracted, *ppl. a.* Add: **1. b.** *contracted-out*: that has contracted out (see *contract *v.* 2 d); so *contracted-in*.

1908 *Westm. Gaz.* 7 Dec. 2/2 The local education authorities would have insisted on getting the same State grant for each Council child as was paid to denominational associations for each contracted-out child. **1959** *Economist* 17 Jan. 238/1 A contracted-out employee will have to pay the present contribution. *Ibid.*, The contracted-in employee will pay only 8s. 4d. for the basic pension.

contractile, *a.* Add: **1.** Esp. *contractile vacuole*, a vacuole in some Protozoa which expels a solution of waste matter on contraction.

1877 T. H. Huxley *Anat. Inv. Anim.* ii. 76 The name of contractile vacuoles is given to spaces in the protoplasm [of Protozoa], which slowly become filled with a clear watery fluid, and, when they have attained a certain size, are suddenly obliterated by the coming together, on all sides, of the protoplasm in which they lie. **1902** *Encycl. Brit.* XXV. 266/1 In the protoplasm contractile vacuoles similar to those of lower animals have been occasionally detected. **1959** J. Clegg *Freshwater Life* (ed. 2) ii. 29 Simple single-celled creatures, such as *Amoeba* and *Paramecium*, usually have large expandable bubbles or spaces in the midst of the protoplasm called contractile vacuoles, in which excess water collects and whence it is then conducted out of the creature.

contraction. Add: **9. contraction joint,** a joint in a concrete structure to prevent cracking during setting; a joint in any structure or material to prevent damage as a result of thermal expansion and contraction; **contraction-rule** (examples).

1909 Taylor & Thompson *Treat. Concrete* (ed. 2) xv. 286 Contraction in concrete walls is provided for by forming joints at intervals...As a rule only contraction joints need be provided, since expansion merely compresses the concrete. **1950** H. J. Gilkey in L. C. Urquhart *Civil Engin. Handbk.* (ed. 3) vii. 646 In pavement construction, contraction joints are commonly spaced between 'expansion' joints at about one-third the expansion-joint interval. **1958** J. S. Scott *Dict. Civil Engin.* 80 *Contraction joint* (C. E.) or *shrinkage joint* or *expansion joint*, a break in a structure usually made to allow for drying and temperature shrinkage of concrete or masonry and thus to prevent cracks forming at undesirable places. *a* **1877** Knight *Dict. Mech.* 612/2 Contraction-rule. **1887** P. N. Hasluck *Pattern Maker's Handybk.* i. 11 The adoption of a contraction-rule seems to dispose of the question of shrinkage [of castings] in the minds of most pattern makers.

contractually (kǫntræ·ktʉ,ăli), *adv.* [f. Contractual *a.* + -ly[2].] In contractual terms.

1919 Keynes *Econ. Conseq. Peace* 56 Too vague and general to be interpreted contractually. **1960** M. Spark *Ballad of Peckham Rye* iii. 29 By common law a trade union cannot fine, suspend, or expel its members. It can only do so contractually. That is, by its rules.

contradeciduate (kǫ:ntrădisi·dʉ,ĕt), *a.* [f. Contra- + Deciduate *a.*] Denoting that condition in certain animals in which the placenta remains in the uterus after birth and is broken up and absorbed.

1897 Parker & Haswell *Text-bk. Zool.* II. 562 In the Mole and the Bandicoot not only is there no decidua thrown off, but the fœtal placenta with the distal portion of the allantois does not pass out after the fœtus, but remains, and is broken up or absorbed in the uterus. Such a condition has been termed contra-deciduate.

contrafactual (kǫ:ntrăfæ·ktʉăl), *a.* *Philos.* [f. Contra- + Factual *a.*] = *counter-factual *a.*

1950 W. V. O. Quine *Methods of Logic* (1952) I. iii. 14 The *contrafactual conditional*; e.g.: If Eisenhower had run, Truman would have lost. **1961** A. C. Ewing in I. Ramsey *Prospect for Metaphysics* ii. 48 A contrafactual conditional.

‖ **contrafactum** (kǫntrăfæ·ktǝm). *Mus.* Pl. **contrafacta.** [mod.L., pa. pple. of *contrafacere* to counterfeit.] In medieval and Renaissance music, the rearrangement of a vocal composition whereby the music is retained and the words altered, as the substitution of a sacred text for a secular one, or vice versa.

1940 G. Reese *Mus. Middle Ages* (1941) xi. 314 Latin texts were again generally favoured, and many *contrafacta* are to be found of texts originally French. **1941** P. H. Lang *Mus. Western Civilization* (1942) ix. 210 An interesting phase of this transformation of Catholic Church music into Evangelical-Lutheran was the so-called *contrafactum* (from the Latin *contrafacere*), consisting of a textual arrangement which changes the meaning of secular texts to sacred ones. Various German and Latin hymns to the Virgin and other Catholic church songs were paraphrased so as to fit the tenets of the new creed. **1960** D. J. Grout *Hist. West. Mus.* (1962) viii. 230 A particularly important class of chorales were the *contrafacta* or parodies of secular songs, in which the given melody was retained but the text was either replaced by completely new words or else altered so as to give it a properly spiritual meaning. **1967** L. Lockwood in J. LaRue *Aspects of Medieval & Renaiss. Mus.* 573 'Parody' has even been applied to compositions..in which mere verbal substitution of text takes place, with no musical transformation whatever—a relationship properly restricted to the category of *contrafactum*. **1968** *Listener* 19 Dec. 834/2 New words are continually being added to recognisable old tunes, reverently or not, and depend for their effectiveness on our knowing the 'original'. This ranges from Naafi versions of 'Colonel Bogey', *contrafacta* to Sullivan, English pop songs of the 16th century, to the 'parody'..techniques of Renaissance composers.

contraflexure (kǫ:ntrăfle·ksiŭɹ). [f. Contra- + Flexure.] The condition of being bent or curved in opposite directions; also, the point or piece at which this occurs. Also *attrib.*

1891 *Notes on Building Construction* (ed. 2) IV. iii. 70 These points, where the curvature and the stress change, are called the points of contra-flexure. *Ibid.* 75 The distances of the points of contra-flexure from the abutments..vary according to the section of the girder, the distribution of the load, etc. **1922** *Flight* XIV. 253/1 The joint..occurs at the point of contraflexure, where the bending moment is zero. **1954** *Archit. Rev.* CXV. 425 To provide short lengths of cable over the supports between contraflexure points, which bend up and emerge in the top of the beam for stressing and anchoring. **1958** *Ibid.* CXXIII. 308/2 The contraflexure, i.e. the cantilever portion relieving the main span.

contrail (kǫ·ntrēɪl). [f. Con(densation + Trail *sb.*[1]] A condensation trail (see *condensation 8), a vapour trail.

1945 *Sat. Even. Post* 13 Jan. 18 With no wind and intense cold, taxiing planes leave ice-crystal contrails behind them, just as Fortresses do at 30,000 feet over Germany. **1952** M. Tripp *Faith is Windsock* vi. 94 White aircraft contrails feathered against azure blue. **1960** *People* 20 Nov. 9/2 The boffins' name for aeroplane clouds is 'contrails'. **1965** R. Heinlein *Farnham's Freehold* iv. 61 No sign..of man—not a building, a road, a path, no contrails in the sky.

contra-indicate, *v.* Add examples in general (non-medical) use.

1941 O. Duff *New Zealand Now* v. 63 Nothing can alter this but a rainfall so low that cows and cultivation are both contra-indicated. **1965** R. Lass in *Anglia* LXXXIII. 173 There is nothing, actually, to contraindicate its having been from the beginning the property of the Percy family.

co:ntra-indi·cative, *a.* [f. Contra-indicate *v.*] Of the nature of a contra-indication.

1906 *Practitioner* Nov. 629 When a condition contra-indicative of the treatment was in evidence.

‖ **contra mundum** (kǫ·ntră mʊ·ndǝm), *phr.* [L.] Against the world; defying or opposing everyone.

1766 Ld. Chesterfield *Let.* 9 Dec. (1932) VI. 2787 Even *he* cannot be alone, *contra mundum*. **1887** *Athenæum* 27 Aug. 279/1 The whole business of the great feud of Goldschmidt *contra mundum*. **1904** G. B. Shaw *Common Sense of Municipal Trading* vii. 56 The narrowness that makes the Englishman's house a castle to be defended *contra mundum*. **1961** *Listener* 21 Dec. 1081/1 Housman, at first almost *contra mundum*, fought and ultimately discredited that fashion.

contranatant (kǫntrănē[i]·tănt), *a.* [f. Contra- + Natant *a.*] Of the migrations of fish: against the current. Hence **co:ntranata·tion.**

1915 A. Meek in *Nature* 29 Apr. 231 To define these and other migrations which are intimately associated with currents it is necessary..to introduce two terms which will serve to indicate movement against the current and with the current. My colleague Professor J. Wight Duff recommends a Latin root, *natare*. The words suggested therefore are *contranatant*, swimming against the current, and *denatant*, swimming or drifting with the current. The words *contranatation* and *denatation* are also available to indicate the act or habit of migration against or with the current. **1920** *Rep. Dove Mar. Lab.* 40 Evidence to support the theory of contranatation. **1959** A. Hardy *Fish & Fisheries* viii. 170 Most fish..make a migration up current before spawning and down current after spawning; these are technically known as contranatant and denatant migrations respectively.

‖ **contrapas** (kǫntrăpă·s). [Sp. *contrapás*, f. Cat. *contrapàs*, f. Contra- + Pas.] A Catalan dance, originally performed at religious ceremonies, in which the dancers form a chain and dance in a semi-circle.

1898 G. Vuillier *Hist. Dancing* vii. 227 The Contrapas is a purely Catalan dance, in which women rarely take part. The dancers join hands and move round in a circle. It is, in fact, a sort of Round, led by two principal performers, who give the time and the step. **1964** W. G. Raffé *Dict. Dance* 122/1 *Contrapas* (Catalonia, Spain), a solemn religious dance which used to be performed after Mass outside the church.

contrapose, *v.* (Later example.)

1928 E. & C. Paul tr. *Marx's Capital* i. 18 The relative value form of the linen implies..that some other commodity is contraposed to the linen in the equivalent form.

contraposit (kǫntrăpǫ·zit), *v.* *Logic.* [Back-formation from Contraposition.] = Contrapone.

1880 *Princeton Rev.* Sept. 260 Yes again, for otherwise he could not contraposit and convert. **1882** *Mind* VII. 441 Mutating the premisses and contrapositing O. **1887** J. N. Keynes *Formal Logic* (ed. 2) ii. viii. 144 Under our present supposition..we may contraposit (*β*). **1906** [see *constructive dilemma].

contrapositive, *a.* and *sb.* Add: Hence **contrapo·sitively** *adv.*

1858 A. De Morgan in *Trans. Camb. Philos. Soc.* X. 182 H is the rich fool; h any other person; H′ the rich friend; R rich, r not rich; W weak, w not weak; A′ a badly advised, a not so; C unfavourably circumstanced, c not so. H))R(M, W(A, C)); contrapositively, r, m (w, ac))h; or r, mw, mac))h; but H′)) mac; whence H′))h; or H′)·(H. **1892** *Mind* I. 244 It means: 'Every year in which the English harvests are bad is a year in which American corn-dealers gain'; or contrapositively: 'Every year in which American corn-dealers do not gain is a year in which the English harvests are not bad'.

‖ **contrapposto** (kǫntrăppǫ·sto). [It., pa. pple. of *contrapporre*, f. L. *contrāpōnĕre* Contrapose *v.*] In the visual arts, the arrangement of a figure so that the action of arms and shoulders contrasts as strongly as possible with that of hips and legs; a twisting of the figure on its own axis. Also *fig.*

1903 B. Berenson in *Burlington Mag.* III. 20/2 In the figure of the soldan he gives a superb instance of almost Michel-angelesque *contrapposto*. **1935** *Times Lit. Suppl.* 11 July 437/4 His [sc. Leonardo's] contrapposto is the work of an ardent anatomist. **1944** Auden *Sea & Mirror* (1945) iii. 36 A sudden eruption of musical and metaphorical power is instantly recognized as standing for grief and disgust, an elegant *contrapposto* for violent death. **1954** M. Rickert *Painting in Brit.* viii. 203 The Apostle Philip on the left stands easily in *contrapposto*.

contrapropeller (kǫ:ntrăprope·lǝɹ). [f. Contra- 2 + Propeller.] One of a pair of propellers having concentric shafts and rotating in opposite directions; a contra-rotating propeller. Hence abbrev. **co·ntraprop** (-prǫp).

1927 N. Skene *Elem. Yacht Design* (1936) xvii. 227 The Star contrapropeller..consists of fixed blades just abaft the propeller, set at an angle to extract a forward thrust from the rotation of the propeller jet. **1939** *Jrnl. R. Aeronaut. Soc.* XLIII. 109 Tests of an eight-blade contra-propeller of 32-in. diameter..were made in the wind tunnel at Stanford University. **1940** *Flight* 25 Jan. 86/1 Twin engines mounted one behind the other in the fuselage, driving contra-props. **1940** *Times* 28 Mar. 10/6 A 'contraprop'..is a co-axial, oppositely rotating airscrew.

contraption. (Earlier and later examples.)

1825 J. Jennings *Obs. Dial. W. Eng.* 31 *Contraption*, contrivance; management. **1834** 'J. Downing' *Life A.*

Jackson iii. 24 The gineral oney intended tu see what he wou'd do, and then by a leetle contrapshion, have him secur'd. **1838** J. C. Neal *Charcoal Sk.* 95 All their new-fangled contraptions. **1898** Kipling in *Morn. Post* 9 Nov. 5/1 Had I seen the new fore and aft bridge that we had managed to screw out of the Dockyard? A great contraption. **1925** Rex Clements *Gipsy of Horn* 207 The men temporarily lashed the crazy contraption to the lower mast. **1958** *Observer* 30 Oct., When you buy these once-in-a-life-time contraptions do you see that you get value for money?

contrapuntal, *a.* Add: **3.** *transf.* and *fig.* Employing combined or contrasting themes, structures, etc.

1938 L. Mumford *Culture of Cities* vii. 485 Contrapuntal organization. *Ibid.*, When we seek a more co-operative order in the design of cities, therefore, we are seeking an order in which more significant kinds of conflict, more complex and intellectually stimulating kinds of disharmony, may take place: in short, we seek a contrapuntal order. **1942** T. S. Eliot *Music of Poetry* 28 There are possibilities for verse which bear some analogy to the development of a theme by different groups of instruments;..there are possibilities of contrapuntal arrangement of subject-matter.

co:ntra-rota·ting, *a.* [Contra- 1; cf. Contrarotation.] Rotating in opposite directions.

1945 J. A. Kyd *Variable-Pitch Propellers* i. 9 Another development..appears to be maturing at the moment; this is the Contra-rotating propeller, which consists of two three-bladed propellers revolving in different rotations on the same shaft. **1949** *Gloss. Aeronaut. Terms* (*B.S.I.*) ii. 20 *Contra-rotating propellers*, two propellers mounted on concentric shafts having a common drive and rotating in opposite directions. **1960** *Farmer & Stockbreeder* 1 Mar. 72/1 Washer..with contra-rotating cleaning brushes operating from a central spindle in the drum. **1962** *Times* 19 Apr. 24 A British two-seater aircraft having two co-axial contra-rotating lift rotors.

contrary, *a.*, *sb.*, *adv.* Add: **A.** *adj.* **3. b.** Also in educated use.

1920 Rose Macaulay *Potterism* i. i. § 4 They were like that; conceited and contrary.

7. contrary-to-fact *a.*, counter-factual; untrue.

1912 *Mind* XXI. 524 Perhaps we cannot ever be certain of the possibility of such a contrary-to-fact condition. **1946** *Ibid.* LV. 289 (*heading*) The contrary-to-fact conditional.

B. *sb.* **3. d.** *Paper-making.* A foreign body or any substance which resists the pulping or re-pulping process. Usu. *pl.*

1926 G. & R. H. Clapperton *Pract. Paper-Making* (ed. 3) ii. 10 The rags..are usually overlooked again..to ensure the further removal of contraries and buttons. **1961** *B.S.I. News* Jan. 17/1 Which materials are contraries? A few examples are: metal foils, waxes, vegetable parchment, plastics and cellulose fillers... Other contraries (known as 'pernicious contraries') make re-pulping impossible. **1961** *Printing News* 23 Mar. 3/5 If the growing proportion of pernicious contraries in waste paper is not offset by accurate methods of segregation. **1963** *Times* 30 Apr. 5/3 Waste paper..might contain 'contraries', which was the trade name for obviously undesirable matter.

co:ntra-sea·sonal, *a.* [Contra- 2.] Unusual for the time of year; contrary to the seasonal norm. Hence **co:ntra-sea·sonally** *adv.*

1933 *Amer. Year Bk.* 525 There was a contra-seasonal advancement in residential construction. **1937** *Ibid.* 493 Production of electric power and bituminous coal dropped contra-seasonally toward the end of the year. **1959** *Times* 5 Mar. 17/4 But it was about 8 per cent. lower than production in December, a contra-seasonal downturn. **1962** *Economist* 15 Sept. 1028/2 Employment rose, also contra-seasonally, to an all-time high.

contrast, *sb.* Add: **2. b.** The degree of differentiation between different tones in a photographic negative or print; also applied to a television picture. Also *attrib.*

1911 A. Watkins *Photogr.* vi. 91 The exposure decides the density of the negative, and the length of development the contrast between the tones. **1921** *Brit. Jrnl. Photogr.* 9 Sept. 534/1 These filters..increase the contrast of the photograph, and the clearness of detail in the scene... They are..called 'contrast' filters. **1940** *Chambers's Techn. Dict.* 195/1 *Contrast control* (television), control of the ratio of illumination between the lightest and darkest parts of a reproduced television image. **1958** M. L. Hall et al. *Newnes Compl. Amat. Photogr.* 97 The deliberate distortion of colour rendering for pictorial effect (with 'contrast' filters). **1961** G. Millerson *T.V. Production* 46 A contrast range of 30:1 is generally accepted as providing good picture quality photographically. *Ibid.* 47 If contrast is kept to too narrow limits, the picture will look flat and lifeless.

5. a. *Psychol.* The modification or intensification of a sensation by the juxtaposition of another (freq. opposite) sensation. **b.** *spec.* The reciprocal induction of colours, brightnesses, and shapes when brought into juxtaposition; the modification of the apparent colour, brightness, or shape of an object by the presence of another colour, brightness, or shape nearby; (see quots.). Also *attrib.*, as *contrast disc*, one used with a colour-mixer to exhibit the phenomena of contrast.

1873 P. H. Pye-Smith tr. *H. von Helmholtz's Pop. Lect. Sci. Subjects* vi. 267 This is proved by a number of experiments contrived to illustrate the effects of contrast. *Ibid.*, When, for example, objects are seen under two different coloured media..these conditions produce what is called simultaneous contrast. *Ibid.*, This effect is known as successive contrast, and is experienced when the eye passes over a series of coloured objects. **1890** E. B. Delabarre in W. James *Princ. Psychol.* II. xvii. 13 The phenomena of contrast. *Ibid.*, If we look for a moment at any surface and then turn our eyes elsewhere, the complementary color and opposite degree of brightness to that of the first surface tend to mingle themselves with the color and the brightness of the second. This is *successive contrast*. *Ibid.* 14 A visual image is modified..by all those [sensations] experienced simultaneously with it... This is the phenomenon of *simultaneous contrast*. *Ibid.* 21 The gray spot then assumes a contrast-color, complementary to that of the background. **1894** Creighton & Titchener tr. *Wundt's Human & Animal Psychol.* vii. 114 The subjective colour due to this opposition..is also termed contrast colour. *Ibid.*, To refer the phenomena of contrast to like causes with complementary after-images. **1897** C. H. Judd tr. *Wundt's Outl. Psychol.* xxiii. 324 The law of psychical contrasts..refers..to the relations of psychical contents to one another. *Ibid.* 325 This law of contrast has its origin in the attributes of the subjective contents of experience, but is secondarily applied to the ideas and their elements also. *a* **1899** D. G. Brinton *Basis Social Relations* (1902) 42 The third law, that of Contrast, applies to the ethnic mind the curious fact in mental life that a prolonged devotion to one idea leads to a reaction in which the opposite of that idea becomes dominant. **1901** Titchener *Exper. Psychol.* I. i. 19 The contrast disc, with an extra 30° of white or black in its ring, is set up before its appropriate background. The contrast colour is matched upon the second mixer as before. Note that the same grey background must be retained for the second mixer, in order that the contrast relations may remain the same on both discs. **1938** R. S. Woodworth *Exper. Psychol.* xxii. 567 Much more striking is the interaction known as contrast, including brightness contrast and color contrast. *Ibid.*, Contrast between adjacent regions is called simultaneous contrast to distinguish it from the successive contrast resulting from retinal adaptation. **1957** *Amer. Jrnl. Psychol.* LXX. iii. 358 Since we have demonstrated two types of reaction to a shift in context, *i.e.* contrast and assimilation, the question arises as to whether or not this loss in discrimination is specific to either contrast or assimilation. **1958** H. Helson in Beardslee & Wertheimer *Readings in Perception* 260 It has been denied that contrast mechanisms could account for these effects. **1966** Hurvich & Jameson *Perception of Brightness & Darkness* v. 85 What the contrast mechanism seems to do in all these instances is to magnify the differences in apparent brightness between adjacent areas of different luminances.

6. *attrib.*, as *contrast-effect*; **contrast-medium** (see quot.).

1888 *Amer. Jrnl. Psychol.* II. 639 Let us see if, as Helmholtz says, we have to do with contrast effects only... The shadow will appear of a slightly red tinge...This red is evidently purely an effect of contrast with the green. **1905** W. James *Mem. & Stud.* (1911) v. 81 Thoroughness of sympathy..made us overflow, on meeting after long absences, into such laughing greetings as: 'Ha! you old thief! Ha! you old blackguard!'—pure 'contrast-effects' of affection and familiarity passing beyond their bounds. **1909** Titchener *Text-bk. Psychol.* i. 76 The contrast-effect is always in the direction of greatest opposition. **1966** Hurvich & Jameson *Perception of Brightness & Darkness* v. 96 If we want a vividly striking demonstration of contrast effects, we can set up a stimulus situation where contours or lines are perceived in the visual field by contrast action even though no such lines actually exist in the stimulus pattern itself. **1967** H. Helson et al. *Contemp. Approaches Psychol.* iv. 165/2 The contrast effects (elation and depression) implied a learning about reward with which the older theory certainly could make no contact. **1955** *Gloss. Terms Radiology* (*B.S.I.*) 56 *Contrast medium*, any suitable substance used in diagnostic radiology to give contrast.

contrastive, *a.* Add: (Examples in *Linguistics.*)

1949 G. L. Trager *Field of Linguistics* 6 When his [*sc.* the linguist's] activities are concerned with showing the structural differences and resemblances of the systems, he is doing contrastive linguistics. **1964** *Language* XL. 206 Contrastive grammar..compares the structural similarities and differences between a target and a source language.

Hence **contra·stiveness,** contrastive quality.

1949 E. A. Nida *Morphology* (ed. 2) vi. 155 Each sememe (the meaningful contrastiveness of any morpheme) would then be subdivided into allosemes. **1964** M. A. K. Halliday et al. *Linguistic Sci.* ii. 34 Each time we use a verbal group in spoken English we select for finiteness, modality, contrastiveness, polarity and voice.

contrasty (kǫ·ntrɑsti), *a.* [f. Contrast *sb.* + -y[1].] Marked by or exhibiting (strong) contrasts; esp. of photographic negatives or prints, having very marked contrast of light and shade.

1891 *Anthony's Photogr. Bull.* IV. 117 It is often advisable to print from contrasty negatives on unfumed paper. **1920** F. J. Mortimer *Wall's Dict. Photogr.* (ed. 10) 173 Should the negative be over-exposed and over-developed.. the transparency should have the minimum exposure so that a prolonged development can be given..with a strong developer... If the negative is a contrasty one, exactly the reverse is done. **1929** A. J. S[heldon] in *B'ham Post* 30 Mar. 14/2 We were forced to make sudden adjustments of the listening faculty to a very contrasty scheme of tonal values. **1953** A. Matheson *Leica Way* (ed. 2) 152 Specially contrasty views which the film could not reproduce in their full tone range. **1962** L. Deighton *Ipcress File* 219 A little extra development to get a good contrasty image. **1968** *Guardian* 28 Feb. 2/8 A contrasty waistcoat and tie are needed.

co:ntra-sugge·stible, *a.* *Psychol.* [Contra- 1.] Tending to respond to a suggestion by believing or doing the contrary. So **co:ntra-suggestibi·lity,** the quality or condition of being contra-suggestible.

1919 L. Dougall in B. H. Streeter *Spirit* vii. 254 We may be contra-suggestible by nature, *i.e.* we may be inclined to oppose whatever any other human being suggests to us. **1920** *Mind* XXIX. 282 It may be that I am the victim of contra-suggestibility. **1921** J. Drever *Psychol. of Everyday Life* v. 55 Under certain circumstances most people can become contrasuggestible. **1925** C. Fox *Educ. Psychol.* viii. 203 Sufficient prestige to keep contra-suggestibility within bounds. **1956** C. W. Valentine *Normal Child* x. 164 The children tended to divide into two groups, (*a*) very suggestible, and (*b*) contra-suggestible.

co:ntra-sugge·stion. *Psychol.* [Contra- 1.] A tendency to do the opposite of what is suggested; contra-suggestibility.

1921 J. Drever *Psychol. of Everyday Life* Index, Contra-suggestion. **1927** *Scot. Hist. Rev.* XXIV. 193 It is possible that James VI retained his Scoticisms in England as a result of contra-suggestion. **1931** *Proc. Brit. Acad.* XVII. 436 Macdonnell was a man..free from ill-will and 'contra-suggestion'.

contrate·rrene, *a.* [Contra- 2.] Opposite to terrestrial in character.

1946 F. Brown in *Astounding Science Fiction* Sept. 48/2 Argyle I is terrene matter and Argyle II is contraterrene, or negative matter. **1958** *Jrnl. Brit. Interplan. Soc.* XVI. 540 Some of the larger meteor craters found on the Earth's surface..may have been due, not so much to meteors of earthlike matter and great size, but rather to the impact of anti-matter or contraterrene material.

‖ **contre-espionnage** (kontrespionɑ̄ʒ). Also **contre-espionage.** [Fr.] = *counter-espionage. Also *attrib.*

1919 J. Buchan *Mr. Standfast* xiii. 233 A sensible man would have gone off to the *contre-espionage* people and told them his story. **1922** *Encycl. Brit.* XXXI. 511/2 The object of *contre-espionnage* is, first and foremost, prevention. **1938** R. W. Rowan *Story Secret Serv.* xxviii. 193 Réal and Desmarest who casually invented the modern system of spying upon spies, which honours its originators with its generally accepted French title of *contre-espionnage*.

‖ **contre-jour** (kontr'ʒūɹ). [Fr.] Chiefly in *Photogr.* = **back-lighting*. Used esp. *attrib.*

1921 *Glasgow Herald* 29 June 7 The light was excellent, and I was fortunate in reproducing quite..beautiful reflections, two of the little pictures..exhibiting creditable 'contre-jour' effects. **1962** *Guardian* 14 Aug. 5/2 Reynolds' inspired contre-jour landscapes. **1967** *Films & Filming* Dec. 8/1 The contre-jour photography..perfectly catches the heavy inconsequentiality of hypnagogia. **1969** *Amat. Photographer* 4 June 67/4 Another advantage of *contre-jour* is that semi-transparent objects such as the sails of a yacht, leaves or flowers take on a delightful luminosity and in all these cases the light becomes the predominant feature.

contretemps. Add: **3.** *Dancing.* A step danced on the unaccented portion of the beat; spec. in *Ballet* (see quots. 1952 and 1957).

1706 J. Weaver tr. *Feuillet's Orchesography* 45 Of Contretemps, or compos'd Hops. **1728** J. Essex tr. *Rameau's Dancing-Master* xxxvii. 97 The Contretems are those springing Steps which give a Life to Dancing by the different Manners of their Performance;..To make one with the right Foot..sink upon the Left, and rise upon it with a Spring; but at the same Time the right Leg..moves forwards..on the Toes, both Legs well extended; afterwards make another Step forwards..which makes the Contretems compleat. **1830** R. Barton tr. *Blasis's Code of Terpsichore* vi. 488 Any dancer may be capable of executing a *chassé*, a *pas de bourrée*, a *contre-tems*, &c. **1877** *Encycl. Brit.* VI. 801/1 As may be seen from the technical language of dancing (assemblée, jetée..contre-temps..) it has undoubtedly been brought to greatest perfection in France. **1952** Kersley & Sinclair *Dict. Ballet Terms* 38 Occasionally one sees the full and more difficult *contretemps* in which the dancer closes the left leg behind the right as both knees bend before springing out. **1957** G. B. L. Wilson *Dict. Ballet* 79 *Contretemps*, a step in which the dancer, with the left foot behind and pointed, jumps off the right foot bringing the left foot round in a small sweep to the front, replacing the right foot. The right foot moves out to the side and the dancer moves forward, repeating the step.

Contrexéville (kontreksevil). Also **Contrexeville.** The name of a town in the Vosges, N.E. France, used (often *attrib.*) to designate the calcareous mineral water found there and drunk for its medicinal properties.

1877 Tamin-Despalles *Min. Waters of Contrexéville* 13 There are separated the fluoro-bicarbonate and sulphate of calcium waters, of which Contrexéville is the type. **1891** Debout-d'Estrées *Contrexéville* 12 It is in gouty diabetes ..that Contrexéville is to be prescribed. *Ibid.* 13 The Contrexéville waters are clearly indicated. **1909** G. B. Shaw *Pen Portr. & Rev.* (1932) 85 The strict regimen of Contrexeville water and saccharine in which his Bacchic priesthood will presently end. **1915** Kipling *Diversity of Creatures* (1917) 425 Her eleven-o'clock glass of Contrexeville. **1934** H. G. Wells *Exper. Autobiogr.* II. viii. 588, I began to drink Contrexéville water. **1967** H. Johnson in C. Ray *Compleat Imbiber* IX. 141 Contrexéville. Still water with a slight earthy taste.

contribuent (kǫntri·bʉĕnt). [ad. L. *contribuent-em*, pr. pple. of *contribuĕre* to Contribute.] One who or that which contributes; a contributing factor or person.

a **1866** J. Grote *Moral Ideals* (1876) xiv. 341 The love of excelling or of excellence, how large a contributent to virtue that was. **1878** S. H. Hodgson *Philos. Reflection* I. 206 This is also a powerful contributent. **1914** W. De Morgan *When Ghost meets Ghost* i. xxix. 345 Talk went on, stiffly, each of its contributents execrating its stiffness, but seeing no way to relaxation.

contribution. 5. *contribution-box* (earlier U.S. examples).

1666 in *Cambridge* (Mass.) *Proprietors' Rec.* (1896) 211 The Inhabitants..are Assessed to pay the Ministers Salary, and put the Same into the contribution Box. **1835** Ingraham *South-West* I. 215 The contribution-box or bag makes its begging tour among the pews.

control, *sb.* Add: **1. b.** Colloq. phr. *everything('s) under control*: all is as it should be; everything is in order.

1933 S. Howard *Alien Corn* i. 32 Everything under control? **1939** I. Baird *Waste Heritage* xx. 285 You can go right back to sleep now, everything's under control. **1958** *Woman's Own* 16 July 15/3 Everything's under control, I think. Shall I come over tomorrow around eleven?

c. *Radio.* (See quot. 1941.)

1941 *B.B.C. Gloss. Broadc. Terms* 7 *Control*, artificial regulation of the dynamic range of a programme output, by means of a control potentiometer, to bring it within the limits of an electrical medium of communication, these limits being determined by overloading on the one hand and under-modulation on the other. **1962** A. Nisbett *Technique Sound Studio* i. 13 The man responsible for balance, mixing, and control.

3. a. Also, a means adopted, esp. by the government, for the regulation of prices, the consumption of goods, etc.; a restriction. Usu. *in pl.*

[**1922** *Encycl. Brit.* XXX. 762/2 The process was checked by the complete control..of many of the foods included in the budget which determined the cost-of-living index number.] **1939** *Economist* 16 Feb. 351/2 That scheme is one which is designed—with the backing of the Government—to stabilise an important industrial material at a reasonable price. Public sympathy with such 'controls' will..evaporate. **1941** *New Statesman* 26 Apr. 431 After the war when the consumption controls are relaxed. **1948** *Ann. Reg. 1947* 89 (The Government's financial policy had been a need for many of their controls.

b. (Later examples.)

1936 *Brit. Jrnl. Psychol.* Oct. 155 No. 11 [of a series of jokes] was a control, a cutting from *The Times*..which evoked very little laughter. **1953** W. J. H. Sprott *Social Psychol.* 210 Usually the conduct of the experimental group is compared with a 'control' group which has not been through the experimental mill. **1958** *Listener* 25 Dec. 1080/1 Each viewer was matched as closely as possible with a control who differed..only by not being exposed to television. **1964** *Lancet* 2 June 1145/2 The subjects investigated were 13 controls, 8 patients with a nephrotic syndrome, 11 patients with acute pyelonephritis, [etc.].

c. In automobile racing, a section of the road, usually through a town or village, over which speed is controlled; also, a point on the road or track where officials are stationed and contesting cars are halted for examination and repairs; similarly, a station in aircraft racing. Also *attrib.*

1900 *Daily News* 2 May 7/2 These automobile fellows..give you a programme with day runs full of 'Controls' and night stints an hour [illegible]. **1903** *Westm. Gaz.* 1 July 7/3 The control timekeepers can hand in their final reports. **1904** A. B. F. Young *Compl. Motorist* xvi. 320 At a control established in some wayside village..stands a little group of officials with their paraphernalia of pens, stop-watches, reports, and time-sheets. **1927** André Beaumont *My 3 Big Flights* 86 On arriving at any control the pilot had to show two of the stamped parts both on the aeroplane and motor. **1928** *Daily Tel.* 17 July 7/7 At each of the 'controls' a compulsory halt of a certain duration will be made.

d. The apparatus by means of which a machine, as an aeroplane or motor vehicle, is controlled during operation; also, any of the mechanisms of a control apparatus, or (in *pl.*) collectively for the complete apparatus. So *dual control*: see *dual *a.* 3.

1908 H. G. Wells *War in Air* iii. The engine..was worked by electric controls from this forepart. **1913** Curtiss & Post *Curtiss Aviation Bk.* vi. i. 285 One day when I was up in the air pretty high I seemed to forget..how to operate the controls. **1913** *Aeroplane* 13 Feb. 176/2 The control is dual; all moving gear is made of non-magnetic material, and all control wires are duplicated. **1917** 'Theta' *War Flying* 37 A dual-control machine. **1918** E. M. Roberts *Flying Fighter* 303 My feet had been forced off the rudder control. **1925** *Morris Owner's Man.* 97 Slow running control not adjusted properly..Carburetter control improperly set. **1961** *Daily Tel.* (Colour Suppl.) 21 Oct. 12 Controls that are hard to reach, gauges that are difficult to read..too often are hidden factors in road deaths.

e. *Bridge.* A card which will enable its holder to win a trick in a given suit at a desired point in the play.

1906 M. S. Hess *Correct Bridge* 5 The principles of placing the lead, holding up the control of an opponent's suit, [etc.]..have all been carefully considered. **1958** J. Culbertson *Contract Bridge* xv. 184 The presence or absence of control cards can affect the success of your slam contract. *Ibid.*, You may then investigate the number of controls that you have. **1958** *Listener* 4 Dec. 950/3 If opponents have a trump control they can wait until dummy's trumps are exhausted.

f. *Computers.* In full *control unit.* That part of a computer which controls the operation of the other units and in recent computers interprets the coded instructions.

1946 *Math. Tables & Other Aids to Computation* II. 102 Each program control consists of a set of program switches, a flip-flop, an input terminal..and associated tube circuits. **1948** *Gloss. Computer Terms* (M. I. T. Servomechanisms Lab. Rep. R–138) 4 *Control*, that part of the computer which controls the sequence of the storage and the arithmetic element. **1953** A. D. & K. H. V. Booth *Autom. Digital Calculators* iv. 25 The control must be capable of receiving the coded order from the memory and storing it during the execution process. **1962** R. Wooldridge *Introd. Computing* x. 183 The five basic parts of a digital computer, input, storage, arithmetic unit, output and control unit. **1966** B. Hodge *Computers for Engineers* i. 19 The control unit, through switching circuits, directs the flow of information through the system and automatically times and sequences the operations called for by the program.

4. b. *Spiritualism.* A spirit who controls the words and actions of a medium in a trance (see quot. 1961).

1877 H. P. Blavatsky *Isis Unveiled* II. i. 15 This is an unexpected honor indeed, for our American 'controls' in general, and the innocent 'Indian guides' in particular. **1884**, **1885** [in Dict. sense 1]. **1890** W. James *Princ. Psychol.* I. x. 393 In old times the foreign 'control' was usually a demon, and is so now in communities which favor that belief. **1902** O. Lodge in *Proc. Soc. Psychical Research* June 65 Everything known to the normal Mrs. Thompson could not be considered equally known to the ostensible 'control' speaking with Mrs. Thompson's mouth. **1961** W. H. Salter *Zoar* ix. 112 In the early days of trance-mediumship, the view was prevalent that during trance a spirit invaded the medium's body of which it took complete and undivided control. Hence the personalities who claimed to manifest during the trance were called 'Controls'. In course of time, however, it became desirable to distinguish..spirits who..confined themselves to introducing the Communicators and relaying their messages in the third person (etc.).. It is to spirits of this second kind that the word 'Control' is now mostly applied.

5. *control cable, engineer, mechanism, point, station, switch, system, wire*; **control board,** (a) a control panel; (*b*) a board (Board *sb.* 8 b) having control of an organization, business, etc.; **control circuit,** an electrical circuit that controls the operation of certain devices, machines, etc.; also (with hyphen) *attrib.*; **control column** (see quot. 1919); **control cubicle** (see quot.); **control desk,** a desk incorporating a control panel; **control electrode,** an electrode to initiate or vary the current flowing between other electrodes; **control lever,** a lever by which a machine is controlled; *spec.* = *control column*; **control line** (see quots.); **control panel,** a board, panel, etc., on which are mounted switches, dials, etc., for the remote control or operation of electrical or other apparatus; **control register,** in a computer (see quot.); **control rod** *Nuclear Engineering*, a retractable rod containing material which readily absorbs neutrons, inserted into a nuclear reactor to control the rate of reaction; **control room,** a room in or from which a certain operation is controlled; **control stick** *Aeronaut.*, control column, 'joy-stick'; **control surface** *Aeronaut.*, a movable surface or aerofoil by means of which the flight of an aircraft is controlled; **control tower,** a tower or other elevated building at an aerodrome from which aircraft and other traffic are controlled by radio, etc.; also *fig.*; **control unit** (see sense *3 f above).

1927 R. H. Davies *Electric Power* xx. 220 There is a vast amount of switchgear..but interest is mainly centred in the control boards. **1939** J. Berryman in *Amis & Conquest Spectrum* (1961) 195 The captain left the control board and walked over to the chart table. **1952** J. G. Winant in J. S. Huxley *TVA* 56 The control board of the TVA was..informed of every decision. **1960** *Sunday Express* 7 Aug. 8 [illegible] of electric power. **1909** P. Dawson *Electric Traction on Railways* xi. 315 A cable, usually designated as 'train control cable'..runs the entire length of the train. **1926** H. Barber *Aeroplane Speaks* 39 All the control cables and wires. **1902** Gould *Elect. Electr. & Coupling Fl.* ii. 9 [illegible]. **1908** A. Stein *Electr. Traction* vi. 146 The control circuit is not opened. **1934** Webster s.v., *transformer*, A control-circuit transformer is a voltage transformer used to supply voltage suitable for the operation of shunt-coil magnetic devices. **1955** *Gloss. Terms Autom. Digital Computers* (*B.S.I.*) 6 *Control circuits*, the aggregate of those parts of a computer which effect the acceptance and carrying out of instructions in the desired sequence. **1919** W. B. Faraday *Gloss. Aeronaut. Terms* 68 *Control column*, on an aeroplane, a lever by means of which the controls are operated. Known as the control stick..pitching and rolling. **1927** *Evening Standard* 12 Jan. 6/4 The 'joy stick' loses a certain liveliness in the new official term of 'control column'. **1941** *B.B.C. Gloss. Broadc. Terms* 7 *Control cubicle*, small room in a studio centre where a programme output is controlled. **1927** E. P. Hill *Rotary Converters* iv. 191 The engineer at the central control desk is warned instantly by a bell. **1962** A. Nisbett *Technique of Sound Studio* 246 At a control desk there can be..all the microphone cue systems, echo and distortion controls..and a variety of communications equipment. **1928** W. H. Eccles *Wireless Teleg. & Teleph.* (ed. 2) 294 This electrode..is the control grid; but it may also be called the control electrode. **1963** *Economist* 28 Dec. 1337/3 'Control electrodes' which permit the cell to be recharged safely at a head-spinning speed. **1937** *Discovery* Nov. 330/2 A second control engineer,

for regulating the sound output to its relevant transmitter. **1962** *Listener* 6 Dec. 955/1 A control engineer faced with putting control on a closed loop system, such as a turbine, would predict that the better the control the smaller the irregularity of the machine. **1904** A. B. F. Young *Compl. Motorist* iv. 84 By opening the control levers the speed can be varied from six to forty miles an hour. **1911** Control lever [see *slot *v.* 4 c]. **1916** H. Barber *Aeroplane Speaks* 28 Sometimes when the Aeroplane is on the ground the control lever or 'joy-stick' is lashed fast. **1936** *Discovery* Apr. 113/1 The locomotive responds instantly to a movement of the control levers. **1940** S. R. Roget *Dict. Electr. Terms* (ed. 2) 68/1 *Control line* (in electric traction), a multiple-cored cable all along a train with multiple unit control connecting the circuit of all the master controllers. **1949** *Jrnl. R. Aeronaut. Soc.* LIII. 1015 In this new sport, control-line flying, the model [aeroplane] is tethered to the pilot by fine wires, and by operating them the control surfaces can be deflected. **1902** *Encycl. Brit.* XXXI. 351/1 A crank and pinion worked by the 'control mechanism' (of a field gun) is fitted to the top of the hinge bolt. **1959** *Times Lit. Suppl.* 24 Apr. 245/2 Control mechanisms are perfectly exemplified by the steam governor, and it is curious that no one has ever stopped between Watt's invention and the recognition of cybernetics as a separate science. **1931** F. A. Talbot *Moving Pictures* (ed. 2) xvi. 234 He has only to follow the ingenious 'control panel' set up at a point convenient to the projector. **1940** *Gloss. Terms Electr. Engin.* (*B.S.I.*) 89 *Control panel*, an assemblage of one or more slabs carrying switches, relays or other apparatus, for the control of a circuit from a distance. **1957** J. Braine *Room at Top* 24 An electric oven which had a control panel like a bomber's. **1960** D. W. Davies in Haley & Scott *Analogue & Digital Computers* ix. 275 Most punched card machines are controlled by these removable plugboards, which are called control panels. **1927** C. W. Oliver *A. C. Commutator Motor* vii. 206 Each control point becomes a running point. **1959** *Listener* 7 May 784/1 The activities of the East German authorities at the control points. **1956** Berkeley & Wainwright *Computers* ii. 24 The command is stored in a special location, called the control register or program register of the computer. *Ibid.* 343 *Control register*—Digital computers, the register which stores the current instruction governing the operation of the computer for a cycle. **1962** *Gloss. Automatic Data Processing* (*B.S.I.*) 81 *Sequence control register, control register*,..a register from the content of which the address of the next instruction is derived. **1961** F. L. Westwater *Electronic Computers* v. 92 The Control Register holds the program instruction which is to be obeyed. **1964** C. Dent *Quantity Surv. by Computer* iii. 24 When the start button is pressed, the control unit reads the contents of the first address in its memory store and transfers it to a 'control register' in the control unit, where it is interpreted as an instruction. **1945** H. D. Smyth *Gen. Acct. Atomic Energy* viii. 83 There are various other holes through the pile for control rods, instruments, etc. **1949** *Physical Rev.* 2nd Ser. LXX. 115 A control rod absorbs neutrons and increases the leakage of neutrons from the pile. **1957** *Proc. U.S. Naval Inst.* Feb. 140/1 The control-rod system..more than twenty were capable of being withdrawn, and the necessary motion could be done inside the boat. **1928** G. E. Sterling *Radio Man.* ix. 379 The control room operator is responsible for the operation of most of the speech input equipment. **1930** *B.B.C. Year-Bk.* 1931 438/2 *Control room*, the 'Nerve-centre' of a broadcasting station. In this room are situated the low-frequency amplifiers and associated apparatus by which the microphone currents are controlled, before they are passed on to the modulation system of the transmitter. **1933** *Jrnl. of Standards Jrnl. Research* XI. 482 Remote-control junction boxes..provide direct two-way communication with the landing aircraft as well as telephone communication with the control room. **1934** *Times* 24 Oct. 14/2 Thousands pressed round Scott and Black as they made their way to the control room. **1943** J. S. Huxley *TVA* xi. 89 Auxiliary to each powerhouse is a control room and visitors' room. **1959** *Westm. Gaz.* 29 May 10/1 The sighting and directing of the fire will be carried out from the control stations. **1934** *Times* 24 Oct. 14/3 All control stations were usually picked up sandwiches. **1933** *Discovery* July 226/1 When the control stick is pulled right back with the engine shut off, the craft simply descends in a nearly vertical path. **1963** D. Slavton in *Into Orbit* 22 A single control stick which works all three axes at once—yaw, pitch and roll. **1917** C. G. Grey *All World's Aircraft* 8 a, *Control*, generally the lever or wheel which controls the surfaces of the aircraft. ..Also applied to the control surfaces. **1931** *Engineering* 23 Nov. 665/1 Control-surface actuators may be divided into three broad classes. **1905** A. Stein *Electr. Railways* vi. 140 Control switch. *Ibid.* 142 Emergency diagrams. **1951** *Good Housek. Home Encycl.* 11/1 Control switches should be placed out of children's reach. **1914** *Chambers's Jrnl.* Aug. 541/2 This central control system..renders acetylene lighting as convenient as electric lighting. **1915** E. Austin *Single-Phase Electric Railways* vi. 203 The original control system. **1920** *Flight* 21 Oct. 1102/1 The new lights consist of electric lamps..automatically controlled from the Central Control Tower. **1933** *Jrnl. R. Aeronaut. Soc.* XXXVII. 18 The control tower houses, as it were, the eyes and brain of the airport. **1939** *Archit. Rev.* LXXXV. 91 The control tower floor rises to a height of 14 metres. **1964** M. McLuhan *Understanding Media* vii. 65 The artist tends now to move from the ivory tower to the control tower of society. **1955** *Gloss. Automatic Digital Computers* (*B.S.I.*) 6 *Control unit*, the aggregate of those parts of a computer which effect the acceptance and carrying out of instructions in the desired sequence. **1963** Control wire [see sense 3 d above].

controllability (kǫntrōᵘlăbi·lĭti). [f. Controllable *a.* + -ity.] The quality or condition of being (easily) controllable.

1888 *Westm. Gaz.* 16 June 4/2 The vastly superior controllability of the machine-driven vehicle. **1920** *Conquest* June 402/3 The enormously wide range of the acetylene explosive mixture..combined with the controllability that would result from its admixture with alcohol. **1958** W. T. O'Dea *Social Hist. Lighting* 171 The carbon arc..had been tamed into a condition of severe controllability by so many decades of profitable application. **1967** *Jane's Surface Skimmer Systems 1967–8* 66 To assure proper balance and controllability.

controllable, *a.* Add: **2.** *Comb.*, as **controllable-pitch airscrew** or **propeller,** a propeller the blades of which can be changed to certain predetermined pitches when rotating.

1936 *Aeroplane* 1 July 21/1 Air-cooled motors, and..controllable-pitch airscrews, drive all the bombers. **1938** *Encycl. Brit. Bk. of Yr.* 79/2 The two-position controllable pitch airscrew is now being superseded by the constant-speed variety which eliminates adjustments by the pilot.

controlled, *ppl. a.* Add: **b.** Carried out or investigated under strict rules or in conditions such as to preclude error or deception; *spec.* of an experiment: employing a 'control' (see Control *sb.* 3 b).

1869 W. James *Coll. Ess. & Rev.* (1920) 2 One narrative personally vouched for and minutely controlled, would be more apt to fix their attention. **1938** *Encycl. Brit. Bk. of Yr.* 521/1 These controlled experiments [in fire-walking] suggested that 'faith' was the secret of immunity. **1962** *Lancet* 27 Jan. 201/1 A controlled symptom study of intravenous adrenaline in 12 schizophrenic and 12 healthy men.

c. Of a house or a tenant: subject to rent control.

1930 *Daily Express* 8 Sept. 7/4 If it is a controlled house you are protected. **1967** *Guardian* 3 Jan. 4/6 She was a 'controlled' tenant.

d. Of the movements of an aircraft: directed by radio from the ground or from another aircraft.

1956 W. A. Heflin *U.S. Air Force Dict.* 140/2 *Controlled interception*, an interception during which the friendly aircraft are directed from an air or surface station. **1957** *B.B.C. Handbk.* 86 A controlled descent through rain with the cloud base only a hundred feet above the runway.

2. Special collocations: *controlled rectifier*, a rectifier having a means of controlling its output current; *controlled response*, in *Mil. Strategy* (see quot. 1966); *controlled school* (see quot. 1960).

1938 S. R. Roget *Dict. Electr. Terms* (ed. 3) 73/2 Controlled rectifier. **1962** Simpson & Richards *Junction Transistors* vii. 192 This device..resembles a grid-controlled gas-tube or 'thyratron'. It is called a silicon 'controlled rectifier'. **1966** H. Kahn *On Escalation* viii. 163 The slower rate of escalation..that is likely to result from a successful controlled-response strategy. **1966** Schwarz & Hadik *Strategic Terminology* 45 *Controlled response*, response to a military attack by military action that is deliberately kept within certain definable limits for the purpose of avoiding all-out nuclear war. **1944** *Act 7 & 8 Geo. VI* c. 31 § 15 The managers or governors of a controlled school shall not be responsible for any of the expenses of maintaining the school. **1960** *Where?* No. 3, p. 11/2 *Controlled school*, a voluntary school for which the local education authority is responsible for the major costs of running..and for which the governors or managers have the right to appoint a limited number of teachers who will give special religious instruction for part of the week.

controller. Add: **4. d.** *Electr.* An arrangement of switches, contacts, rheostats, and electromagnets, manipulated by a handle or handles, by means of which the current of an electric motor may be controlled.

1898 E. J. Houston *Dict. Electr. Words* (ed. 4) 726/1 *Controller*,..an electric switching mechanism for controlling the speed of a motor or motors. **1902** *Daily News* 3 Jan. 6/4 The controller, the whole handle of which regulates the four motors going either ahead or astern. **1902** *Encycl. Brit.* XXVIII. 97/2 There is a reversing lever on the controllers separate from the controller handle. **1964** *Electr. Times* 7 Dec. 769/2 The controller goes automatically to the 'off' position on being released.

controvertible, *a.* (Later examples.)

1920 *Times Lit. Suppl.* 26 Aug. 542/3 Why..M. Hovelacque should have considered it necessary to conclude his work on this note of controvertible comparison. **1963** *Times* 1 June 4/4 The spirit with which Kent responded to it made all earlier scoring rates seem controvertible.

conurbation (kǫnə̄ɹbēⁱ·ʃən). [f. Con- + L. *urb-*, *urbs* city + -ation; cf. F. *conurbation*.] An aggregation of urban areas.

1915 P. Geddes *Cities in Evolution* ii. 34 Some name, then, for these city-regions, these town aggregates, is wanted. What of 'Conurbations'? *Ibid.* For our first conurbation the name of Greater London is..dominant. **1932** C. B. Fawcett in *Geogrl. Jrnl.* Apr. 101 Vast urban aggregates, or conurbations. **1948** *Conurbation: Planning Survey of Birmingham* i. 15 The term 'conurbation' describes an area of continuous urban development which includes and connects two or more towns. **1957** *Listener* 29 Aug. 319/3 Forty per cent of Britain's population and the greater part of its wealth is [*sic*] concentrated in the seven great conurbations. **1959** *Ibid.* 26 Feb. 368/1 The conurbations of Tyneside, Merseyside, West Riding, South-East Lancashire, and West Midlands.

conure (kǫ·niᵘə̄ɹ). *Ornith.* [ad. mod.L. *Conurus*, f. Gr. κῶνος Cone *sb.*¹ + οὐρά tail.] A popular name for a bird belonging to a group of Central and South American parrots, now classified as *Aratinga* and related genera, formerly included in a genus *Conurus*.

1858 *Proc. Zool. Soc.* 1857 XXV. 225 Prince Bonaparte, in one of his last papers, proposed to call the little Mexican Conure..*Bolborhynchus catharina*. **1883** *List Anim. Zool. Soc.* (ed. 8) 341 Large Patagonian Conure. **1895** R. Lydekker *R. Nat. Hist.* IV. 116 In the typical conures..the fourth primary feather of the wing is attenuated, and the nostrils are exposed. **1897** *Daily News* 19 Apr. 6/5 The rich greens and flame colour and orange in those golden-headed Conures. **1961** O. L. Austin *Birds of World* (1962) 147/1 Among the less familiar groups of New World parrots are the conures, which are smaller and more slender-bodied than the amazons and have longer, pointed tails.

conus (kōᵘ·nəs). [L. *conus* Cone *sb.*¹] **1.** *Anat.* A conical structure or organ, e.g. in the heart, the *conus arteriosus*, the upper and anterior part of the right ventricle.

1885 Sedgwick & Heathcote tr. *Claus's Text-Bk. Zool.* 146 An independently pulsating part of the heart with rows of semi-lunar valves (*conus arteriosus*). **1886** W. N. Parker tr. *R. Wiedersheim's Elem. Compar. Anat.* 281 The longitudinal valve of the conus is incomplete. **1942** L. H. Hyman *Compar. Vert. Anat.* (ed. 2) xii. 313 In tetrapods..the conus becomes embraced by the bulges of the ventricle. **1949** Adams & Eddy *Compar. Anat.* xii. 290 As the blood leaves the right heart, it enters the conus arteriosus, a muscular structure containing valves that prevent the return of the blood to the heart. **1956** W. J. Hamilton *Textbk. Human Anat.* 714 The lower end of the spinal cord forms a blunt tapering extremity, the conus medullaris.

2. A pale crescentic or annular patch near the optic papilla resulting from atrophy of the choroid and exposure of the sclera, and found esp. in cases of myopia.

1886 *Amer. Jrnl. Ophthalmol.* Nov. 319 One of the appearances occasionally seen at the optic entrance—a form of conus so-called. *Ibid.* 321 A full discussion of the 'congenital conus' would require a number of illustrations. **1938** S. Duke-Elder *Text-bk. Ophthalmol.* II. xxv. 1344 A congenital conus must be sharply distinguished from a myopic crescent. **1969** F. W. Newell *Ophthalmology* (ed. 2) xvii. 282/1 A conus surrounding the entire disk may develop in glaucoma.

convectional (kǫnve·kʃənăl), *a.* [f. Convection + -al.] Of, pertaining to, or induced by convection; *spec.* of rain, resulting from the condensation of moisture as warm air rises and cools.

1889 *Nation* (N.Y.) 1 Sept. 166 He concludes that the convectional origin of summer thunder-storms implies a like origin for such storms in winter. **1904** *U.S. Monthly Weather Rev.* Feb. 77 (Cent. D. Suppl.), The convectional currents. **1927** Kendrew *Clim. Continents* (ed. 2) 223 The ground is heated by the strong sunshine, and convectional overturnings take place between the layers of air resting on it and those above. **1930** *Meteorol. Gloss.* (*H.M.S.O.*) (ed. 2) 46 *Convectional rain*, caused by the heating of the surface layers or by atmospheric where warm moist air rises, giving place to denser cool air. **1957** K. A. Wittfogel *Oriental Despotism* 24 On-the-spot rains create additional dangers when they are overconcentrated (convectional) or insufficient.

convector (kǫnve·ktǫɹ). [f. Convection: see -or.] An appliance that warms a room by convection. Also *attrib.*, esp. in *convector heater*.

1907 *Install. News* Dec. 21/2 The heating appliances cover several useful types of convectors. **1909** *Ibid.* III. 163 In use the convector body attains a temperature of about 200° F. **1939** Martin & Scripture *Flat Bk.* 23 There are excellent gas convector or background heaters. **1957** *Times* 15 Jan. 13/1 Panel heaters and background and convector heaters. **1961** *Times* 18/2 Sales of our convector heaters proved the highest ever achieved by the company. **1958** *Engineering* 28 Mar. 416/2 The good insulation..makes the use of small wattage electric convectors economic. **1969** *Listener* 1 May 619/1 The old convector heaters. **1974** *Woodlands* Feb. 57 More oil convector heaters. *Ibid.*, Most of the other oil convectors we tested.

convenience, *sb.* **7. b.** Delete † and add later examples.

1842 Dickens *Let.* 30 July (1965) I. 7 You will perhaps oblige me with a line at your earliest convenience.

c. *spec.* A (public) lavatory, a water-closet; esp. in *public convenience*.

1841 E. Sabertion *Pract. Life Radical* I. xiii. 236 A convenience, from which emanated the disagreeableness. **1883** [in Dict.]. **1938** O. Lancaster *Pillar to Post* p. xiii, The cathedral, the Dean's house, and the public conveniences. **1958** *Listener* 9 Jan. 55/1 **1958** B. Hamilton *Too Much of Water* ix. 183 Gents and Ladies bathrooms and conveniences. **1963** V. Nabokov *Gift* iii. 156 A public convenience with toilet.

9. *attrib.*, in sense 'designed for convenience; used when convenient'. orig. *U.S.*

1961 *Economist* 2 Dec. 920/1 Even the Thanksgiving turkey has now become a 'convenience' food. **1965** *Daily Express* 11 Nov. 2/8 The convenience type foods..were opening in America. **1967** *Boston Sunday Herald* 30 Apr. III. 10/5 [Send us your favorite recipe using convenience foods (frozen or refrigerated prepared foods, canned soups, sauce mixes, cake mixes etc.) and you may win a $10 prize. **1968** *Guardian* 6 Dec. 9/5 No one would deny the drudgery, the time-wasting, the monotony, that has been prevented by convenience foods. **1970** *Daily Colonist* (Victoria, B.C.) 18 June 2/6 A woman was shot in the arm as she watched a fire at a convenience store.

convention. Add: **10. b.** *Cards.* A method of play or bidding which does not have its natural meaning but is used solely to convey prearranged information.

1876 'Cavendish' *Whist Pract.* The principles of play are..laid down as so many isolated and arbitrary conventions. **1906** *Daily Chron.* 14 Sept. 7/3 Gray rose in disgust when he ignored the heart-convention and led him an unlovely spade. **1929** *Observer* 24 Nov., The introduction of conventions makes Contract an artificial game rather than an intellectual one.

11. *convention hall*; **convention city,** a city in which conventions are commonly held. *N. Amer.*

1877 C. B. George *40 Yrs. on Rail* v. 92 Chicago..is the greatest railroad center on the globe, [and] is the chief convention city in America. **1908** *Westm. Gaz.* 19 May 12/1 The total cost of 'transportation' will be about 100 dols. a head, and, according to the more moderate estimate, another 50 dols. per man will be needed for expenses in the convention city. **1889** *Kansas Times & Star* 6 June, The project of a big convention hall was again fully discussed. **1968** *Globe & Mail* (Toronto) 3 Feb. 31/3 A convention hall that seats 3,000.

Hence **conventionee·r,** a member of a convention, one present at a convention (see also quot. 1934). *N. Amer.*

1934 Webster *Dict. Amer. Slang* 122 *Conventioneer*, a person who attends many conventions. **1940** *Ibid. Amer. Assoc. Univ. Professors* Dec. 643 Professional Conventioneers. **1960** I. Wallach *Absence of Cello* (1961) 64 A conventioneer with a large self-identifying badge. **1968** *Globe & Mail* (Toronto) 31/4 Crowded with tourists and conventioneers.

conventional, *a.* Add: **2. d.** *Cards.* Of, pertaining to, or characterized by a convention or conventions (see prec., sense 10 b).

1864 Clay *Treat. Short Whist* iii. 47 This method of play being as old as whist itself, it was sure, sooner or later, to be reduced to the conventional sign..—good in the lowest cards as well as in the highest—of which I now treat. **1864** 'Cavendish' *Whist* 51 The instructed player frequently selects one card in preference to another with the sole object of affording information. When the principle is carried thus far the play becomes purely conventional. **1874** *Ibid.* (ed. 14) 105 The system of returning the higher of two losing cards when they are both small cards, is purely conventional. **1927** M. C. Work *Contract Bridge* iii. 39 A..baldly arbitrary and conventional scheme for handling strong two-suiters. **1958** *Everyman's Encycl.* III. 750/1 Conventional bidding, i.e. bids to which particular meanings are attached, is now accepted as an integral part of Contract Bridge.

4. c. Of bombs, weapons, etc.: other than nuclear; of war: fought without nuclear weapons or power stations, etc.: using other than nuclear energy. Also *transf.*

[**1952** *N.Y. Herald Tribune* 21 Nov., We must decide whether the new [hydrogen bomb]..will permit a reduction of our more conventional military weapons.] **1955** M. Beloff *Foreign Policy & Democratic Process* 102 The knowledge that all-out war would almost certainly mean the annihilation of organized society of the country initiating it..must enter into the calculations of modern statesmen as a deterrent to war in a way in which the limited horrors of 'conventional war' could not. **1955** *Hansard* DXXXVII. 1970 This unique difference..between the hydrogen and the atomic weapon on the one hand and conventional weapons on the other. **1958** *Times Rev. Industry* Feb. 15/1 The rate of building both nuclear and conventional stations may well increase. **1958** *Jane's Fighting Ships 1958–59* p. iv/1, The various admiralties and navy departments..have been shaken out of the static orbit of conventional ships, conventional propulsion and conventional weapons.

conventionalism. Add: **2.** (Earlier example.) **1833** Hill in *Monthly Repos.* VII. 713 Much of education is made up of artificialities and conventionalisms.

3. *Philos.* The doctrine that the truths of logic, of mathematics, or the principles of physical science, merely express, or are true in virtue of, conventions with regard to the use of symbols.

1938 H. Reichenbach *Exper. & Predict.* § 1, Within the frame of the modern philosophy of science there is a movement bearing the name of conventionalism; it tries to show that most of the epistemological questions contain no questions of truth-character but are to be settled by arbitrary decisions. **1951** G. H. von Wright *Treatise* i. 27 Conventionalism explains why unsuccessful predictions need not be interpreted as refutations of the theories or laws from which they were made. **1961** T. E. Hill *Contemp. Theories of Knowl.* xiii. 398 Schlick steadfastly resists such conventionalism as he finds..in the physicist Eddington.

conventionalist. Add: **3.** *Philos.* An adherent of *conventionalism 3. Also *attrib.* Hence **conventionalistic** *a.*

1931 M. R. Cohen *Reason & Nature* iii. 142 The conventionalist may answer that the former type of assumption is preferred. *Ibid.*, The conventionalistic position given by Mach and Ostwald to the atomic hypothesis. **1948** Tr. *Cheastek's Limits of Sci.* 287 The conventionalist's argument..is in some sense man-made and arbitrary. **1962** L. J. Cohen *Diversity of Meaning* § 37 Those so-called conventionalist philosophers like Le Roy who think of the laws of science as...

conventionalized, *ppl. a.* (Earlier and later examples.)

1868 G. M. Hopkins *Jrnl.* 21 Aug. (1959) 186 A portrait of a gentleman with some beautiful conventionalized flowers in the background. **1889** Kipling *From Sea to Sea* (1899) I. xxii. 447 They are faithful reproductions of the changed life—telegraph poles down the streets, conventionalised tram-lines, [etc.]. **1960** Snook *Eng. Hist. Embroidery* 71 A second long cover (V. & A.) has a scrolling design of conventionalised flowers, fruit and leaves.

conventionally, *adv.* Add: **1. b.** *Cards.* By means of or according to a convention (*convention 10 b).

1958 *Listener* 2 Oct. 541/2 North's double conventionally asked his partner to make some unlikely lead. **1963** *Ibid.* 28 Mar. 574/3 Do you conventionally lead the ace from ace-king?

convergence. Add: **1. c.** *Meteorol.* The accumulation of air in a region caused by converging winds and resulting in upward air-currents. Also *attrib.*

1936 Shaw & Lempfert *Life Hist. Surface Air Currents* i. 18 As the most obvious example of convergence we may take the case in which the winds from all points of a closed curve blow inwards. *Ibid.* 19 If trajectories [of air] are drawn from a series of points on the boundary of a definite area and steps are taken along the trajectories for equal intervals, the variation of the area defined by the series of points can be measured and a measure of convergence so reckoned is diminishing in a region of convergence and indicates a locality of rising air. **1953** F. K. Hare *Restless Atmosphere* ix. 108 The trade winds meet, along a fairly definite 'front'..The use of the word 'front' is..something of a misnomer, and it has been suggested that the term 'intertropical convergence zone' should replace the older form. **1957** G. H. Sutcliffe *Treat. Limnol.* I. iv. 215 The equatorial region, where frontal convergences and a general upward movement produce adiabatic cooling and cloud formation.

5. *Biol.* The tendency in diverse or allied animals or plants to assume similar characteristics under like conditions of environment.

1866 Darwin *Orig. Spec.* (ed. 4) iv. 150 A distinguished botanist, Mr. H. C. Watson, believes that I have overrated the importance of the principle of divergence of character..and that convergence of character, as it may be called, has likewise played a part. **1902** *Encycl. Brit.* XXVIII. 343/1 Multiradial apocentricities lie at the root of many of the phenomena that have been grouped under the designation *Convergence*. **1927** Haldane & Huxley *Anim. Biol.* xi. 223 Convergence, where a similar mode of life produces similar effects on quite unrelated animals. **1953** G. A. L. Sarton *Hist. Sci.* I. i. 17 The theory of convergent evolution, or convergence (as the anthropologists call it), does not deny the frequent occurrence of borrowings and imitation between one people and another. **1963** E. Mayr *Animal Species & Evolution* xix. 609 Where two or common heritage exists evolutionary parallelism is more correctly called convergence.

convergent, *a.* Add: **1. d.** *Biol.* Due to or characterized by convergence (see prec., sense 5); of similar structure but of different origin.

1871 Darwin *Descent of Man* I. vii. 231 When man selects..the offspring of two distinct species, he sometimes induces..a considerable amount of convergence of character. This case..with the improved breeds of pigs, which are descended from two distinct species. **1895** A. Sedgwick in *Q. Jrnl. Microsc. Sci.* 36 The agreement is a case of convergent development. **1895** *Ibid.*, The species or types of convergent development may be differentiated by evidence of their descent from two primitive stocks is..retained. **1897** Parker & Haswell *Zool.* II. 418 A convergent or polyphyletic group, owing its distinctive characters..to the independent acquisition of similar characters under the influence of like surroundings. **1912** A. Willey *Convergence in Evol.* i. 17 The distinction between convergent and normal morphogeny has long pervaded biological literature. **1953** [see *convergence 5]. **1963** E. Mayr *Animal Species & Evolution* xiii. 368 It would be a great mistake to conclude from the convergent similarity of the phenotypes that the underlying genetic mechanisms are the same as they have necessarily derived from each other. **1968** R. D. Martin tr. *Wickler's Mimicry* vii. 84 It could be required that Müllerian mimetic species should have developed their resemblances convergently.

conversation. Add: **7. c.** *to make conversation*: to converse for the sake of conversing; to engage in small talk.

1921 Hichens *Spirit of Time* v, He simply could not 'make conversation' to her.

10. Substitute for def.: (In full *conversation piece*.) A painting representing a group of figures, esp. members of a family, arranged as if in conversation in their customary surroundings. So *conversation painting*.

11. *conversation card*, a card on which is printed or written a sentence (question or answer, etc.) for use in a game; **conversation-chair,** (*a*) a type of upright chair on which a person sits facing the back (see quot. 1793); (*b*) = Tête-à-tête sb. 2; **conversation lozenge,** a lozenge with an inscribed motto; **conversation piece,** (*b*) a subject for conversation; something to talk about; (*c*) a piece of conversation; a dialogue; (*d*) (see quot. 1952).

1785 *Daily Universal Register* 1 Jan. 3/2 Sentimental, or Conversation Cards. **1794** *Baltimore Daily Intelligencer* 27 Sept. 3/3 A new and elegant Edition of the much admired Conversation Cards: Containing a variety of amusing, entertaining, and innocent *Questions & Answers* in the art of courtship. Each pack contains 64 cards. **1838** (title) Conversation Cards: a book for the amusement of evening parties. **1853** Mrs. Gaskell *Cranford* viii. 152 Another square Pembroke table..on which there was a kaleidoscope, conversation-cards, puzzle-cards. **1793** Sheraton *Cabinet-Maker & Upholsterer's Drawing-Book* App. p. 16 The conversation chairs are used in library or drawing-rooms. The parties who converse with each other sit with their legs across the seat, and rest their arms on the top rail, which..is..stuffed and covered. **1862** *Eng. Wom. Dom. Mag.* III. 142 She..sat down upon the conversation-chair. **1905** *Eng. Dial. Dict.* Suppl. s.v. *conversation*, Conversation-lozenge. **1908** Findlater *Crossriggs* vi, [To] combine my speech solely within the limits of the conversation lozenge. **1967** *Guardian* 9 May 16/5 Terry's of York marked their bicentenary yesterday by selling Victorian 'conversation lozenges'..Conversation lozenges are stamped with a sentimental proposition: 'Give me your heart', 'I want a wife'. **1784** R. Bage *Barham Downs* II. 314 The contents of Lord Winterbottom's, and his pocket, I did intend for a conversation piece. **1936** *Discovery* Dec. 398/1 A biography..well known conversation pieces. **1952** Granville *Dict. Theatr. Terms* 50 *Conversation piece*, a dialogue play or play of ideas, specialising in the writings of George Bernard Shaw and Oscar Wilde. **1958** J. K. Galbraith *Affluent Soc.* xvi. 176 Perhaps the bank rate..derived prestige from its nature as a conversation piece. **1960** *House & Garden* June 96/3 Conversation-piece chairs.

conversion. Add: **11. c.** spec. *Forestry.* (See quot. 1895.)

1895 Schlich *Man. Forestry* III. iv. 338 Change from one sylvicultural system into another, called a conversion. **1953** Edlin *Forester's Handbk.* xi. 171 Many coppices suitable for conversion are variable in density and constitution.

d. *Rugby Football.* The action of scoring a goal by converting a try; also, a goal scored in this manner.

1907 Wakefield & Marshall *Rugger* viii. 194 In New Zealand a desire is expressed for raising the value of the try to four points and making a converted goal six, while any other goal should equal three points. By thus raising the value of a try the value of kicking is lessened, for conversion represents a half addition as against two or two-thirds. **1959** *Times* 8 Sept. 3/4 Gavins, the Leicester full-back, played a big part..kicking four penalty goals and two conversions.

e. *Grammar.* The use of one part of speech as another.

1928 [see *conversion-noun in section IV below]. **1950** S. Potter *Our Lang.* v. 57 This kind of word-play, the use of noun as verb and verb as noun..known technically as conversion. **1957** Zandvoort *Handbk. Eng. Gram.* viii. 265 The deliberate transfer of a word from one part of speech to another, technically known as conversion.

12. e. *Building.* The structural adaptation of a building for a new purpose.

1921 *Building News* 1 Apr. 169/1 (heading) The conversion of buildings to meet modern requirements. *Ibid.*, Assuming..a block of, say, three or four houses..what are their possibilities with a view to conversion into flats? **1934** E. Waugh *Handful of Dust* iii. 82 The conversion of stables and garages to do for human habitation. **1959** *New Statesman* 23 May 1956 *News Chron.* 29 Apr. 8/5 A house conversion which left no room for a dining-room. **1960** *Guardian* 4 May 10/6 A pleasant minor mansion of the kind described by estate agents as being 'suitable for conversion'.

f. *Nuclear Technology.* The process in a nuclear reactor by which fertile material is transformed into fissile material (see quot. 1963).

1955 S. Visner in *Reactor Handbk. Engin.* (*U.S.A.E.C.*) iv. 511 The higher enrichment available is desired for conversion of thorium-232 to uranium-233. **1958** Weinberg & Wigner *Physical Theory Neutron Chain Reactors* i. 21 The principal purpose of conversion is..the production of pure fissionable material. **1963** D. Fozard *Instrumentation Nucl. Reactors* viii. 165 This depletion of the fuel by burn-up of fissionable atoms is offset to some extent by the production in the fuel elements by neutron absorption and other processes of atoms of new materials which are themselves fissionable; the best known example of this is the production of plutonium 239 from uranium 238. This process is known as conversion.

16. c. *Psychiatry.* The symbolic manifestation in physical symptoms of a psychic conflict. (See also section IV below.)

1909 A. A. Brill tr. *Freud's Sel. Papers on Hysteria* ii. 23 Miss Lucy R. merged into that moment of hysterical conversion, which must have been under the determinations of that trauma. *Ibid.* iii. 72 This seraglios became the starting point of a definite psychic excitement through the usual road of conversion. **1923** E. Jones *Papers Psycho-Anal.* 83 The energy finds an outlet in some somatic manifestation, a process Freud terms 'conversion'. **1946** C. Berg *Clinical Psychol.* xii. 142 The following case..introduces us to the mechanism of conversion.

IV. *conversion efficiency*; (sense *12 f) *conversion ratio, reactor*; (sense* 16 c) *conversion hysteria, hysteric, symptom*; **conversion factor,** (*a*) an arithmetical multiplier for converting a quantity expressed in one set of units into an equivalent quantity expressed in another; (*b*) *Economics*, that part of the commercial value of a product attributable to the manufacturing processes involved and not to the raw material used; **conversion gain** (see quot. 1940); **conversion-noun,** a noun formed from an adjective.

1937 M. G. Crosby in *Proc. Inst. Radio Engin.* XXV. 476 A convenient term for this reduction factor of the filter is 'conversion efficiency'. **1956** S. Glasstone *Princ. Nucl. Reactor Engin.* i. 42 If such a reactor were to regenerate the same amount of plutonium-239 as it loses by fission, i.e. if the conversion efficiency were 100 per cent. **1918** W. E. Dommett *Dict. Aircraft* 16 Wing loading W/A = C_L $\rho V^2/g$..The value of ρ/g V^2 (English Units) is known as the Conversion factor. **1947** *Times Survey Brit. Scient. Instr. Industry* Oct. 2/3 The minimum of raw material is involved in the production of scientific instruments. In other words, scientific instruments have one of the highest conversion factors. **1952** *Economist* 6 Sept. 575 In value the annual output of the economy is perhaps..£35 million a year; since the cost of materials is probably not more than £9..the conversion factor must be high. **1964** *Jrnl. Soc. Dairy Technol.* of this, in the 'conversion factor' is high. **1964** B.S.I. *News* June 18/2, B.S. 350 listed conversion factors is no fewer than four different metric systems. **1940** F. E. Terman *Measurements in Radio Engin.* ix. 218 The conversion gain is..the ratio of the two intermediate voltages required, provided the degree of modulation is kept constant. Similarly the detector efficiency with plate detection is the

ratio of the actual conversion gain to the gain..obtained by considering the detector tube to be an intermediate frequency amplifier tube. **1940** *Chambers's Techn. Dict.* 196/2 *Conversion gain* (*Radio*), the effective amplification of a conversion detector, measured as the ratio of the output voltage of intermediate frequency to the input voltage of signal frequency. **1960** *Electronic Engin.* XXXII. 413 A unit to express resistor current noise which is called a conversion gain. **1912** A. Brill tr. *Freud's Sel. Papers on Hysteria* (ed. 2) iii. 158 The..conversion hysteria, the psychic anatomy of the phobias, the other neuroses had to be subjected to certain modifications. **1920** — tr. *Jung's Anal. Psychol.* xiv. 426 Sexuality in..the..conversion hysteria. **1923** D. Sutton in *Psychiatry* xxii. 271 Two symptoms of conversion hysteria..neither unpleasant nor damaging. **1958** *Brit. Jrnl. Med. Psychol.* XXVII. 25 One could even call conversion hysteria an inverse hysteria (etc.). **1935** *Brit. Jrnl. Psychol.* Oct. 214 Conversion hysteria, in which 'the symptoms are a solution, however maladaptive, of the conflict'. **1939** *Ibid.* Oct. 151 He [sc. the Doctor]..recognized the conversion hysteric's system. **1935** *Ibid.* Oct. 214 The conversion hysteric is often said to be of lower intelligence than the neurasthenic. **1956** E. M. Dorcus *Hypnosis* vii. 2 One type of hysteria—ties to a fit of hysterical conversion..is..being helped in the form of conversion symptoms in defecting cases of conversion hysteria. **1969** N. Hobhouse in Kaufman & Heiman *Evol. Psychosomatic Concepts* 142 Those conversion hysterias which were the 'rather little of one'..are..**1912** A. Brill in *Amer. Jrnl. Insanity* LXIX. 117 Conversion symptoms have to be..regarded as an adjective. **1928** O. Jespersen *Mod. Eng. Gram.* II. 248 A conversion-noun..as the conversion noun in the latter sense. **1955** S. Visner in *Reactor Handbk. Engin.* (*U.S.A.E.C.*) iv. 514 The conversion ratio and breeding gain. **1956** S. Glasstone *Princ. Nucl. Reactor Engin.* i. 43 Nuclei of fissionable material..is the initial amount. **1959** *Listener* 26 Mar. 540/1 The British conversion reactor. **1955** J. Sutton tr. *Ferenczi's Further Contrib. Psycho-Analysis* vi. 103 The conversion symptom may be..the expression of fantasied ideas or an arbitrary fulfilment of the..psychic affects. **1964** Gould & Kolb *Dict. Social Sci.* 593/2 The repressed emotions could find expression..through converting psychic energy into physical 'conversion' symptoms.

conversionism. (Example.)

1885 W. Raleigh *Let.* 15 Dec. (1926) I. 48 The worst Christianity is to be found outside professed Christian countries—blatant conversionism and crass ignorance united.

convert. Add: **11. f.** *Rugby Football.* To kick a goal from (a try). Also *absol.*

1896 *Field* 21 Nov. 1977/1 Only with a very fine place kick, converted the try. **1909** *Daily Chron.* 17 Nov. 772/1 Douglass..converted both tries. **1959** *News Chron.* 11 Feb. 10 Ronald Poulton 179 Ronald gained two tries, both converted by Turner. **1961** *Times* 14 Feb. [illegible]. **1964** [illegible] *Hammond* converted and Shallerton came back as it benefited.

12. e. *Building.* To make structural alterations in or to. Also *intr.*

1896 *Times* 7 Nov. 4/4 Two substantial Brick Houses..converted into a roomy warehouse. **1937** M. Sharp *Nutmeg Tree* xviii. 233 It was the cloak-room arrangements..They're going to begin converting next month. **1938** *Listener* 3 Feb. 5/1 Every house in the village can and should be converted..as far as is practicable to 'convert' a large number of existing dwellings into homes for small families. **1959** G. M. Wilson *Shadows on Landing* i. 7 She had the place converted after the war.

converted, *ppl. a.* Add; **3. c.** *spec.* Of a building (see *conversion 12 e, *convert *v.* 12 e).

1888 Kipling *Phantom Rickshaw* (1889) 32 It was my misfortune to live in dilapidated bungalows..[I lived in 'converted' ones—old houses officially declared as dâk-bungalows]. **1924** P. Macdonald *Rasp* viii. 119 She perceived No. 14 to be a 'converted' house. A great black building that might once have housed a merchant prince, but was now the warren of retired grocers, golfers, solicitors, and divorcees. **1959** F. Donaldson *Child of Twenties* xi. 151 A very attractive small flat..a converted L-shaped London drawing-room.

d. *Rugby Football.* (See prec. 11 f.)

1907 H. H. Baskerville *Mod. Rugby Football* i. 12 A converted try..equals 5 points. **1907** [see *conversion 11 d].

4. *Logic.* See Convert *v.* 4 b.

1656 tr. *Hobbes's Elem. Philos.* I. iv. 17 Changing..the Propositions, so converted Propositions equivalent to them. **1962** A. De Morgan *Formal Logic* iv. 71 Each universal proposition has converted controvertible forms, in affirmative forms. **1851** [see *exposita]. **1870** [see Convertend].

converter. Add: **2. b.** Also (*c*) 'In the cotton-goods trade, one who takes unbleached gray cloth and converts it into the finished product' (Funk's Standard Dict., 1928).

1959 *Listener* 5 Nov. 768/1 Is the horizontal structure a weakness with yarn spinners selling to cloth manufacturers and manufacturers selling to converters who deal with separate finishers to complete the cloth for final use?

3. b. (Later example.) Also, a retort used for Bessemerizing copper ores.

1897 *Daily News* 4 Jan. 2/1, 28,300 ounces of gold, contained in eleven thousand tons of refined copper, or bullion. **1906** *Westm. Gaz.* 22 Aug. 9/1 The works, which consist of three blast furnaces and two converters, are capable of treating 20,000 tons of ore per month. **1955** *Everyman's Encycl.* III. 794/2 Converter, in the Bessemer process the molten pig-iron is blown in the converter, which is [illegible] metal from matter [metal sulphides].

c. For *Electric Lighting* read *Electr.* Also, a device for changing current of one kind into current of another kind; = Transformer 2 a.

1888 [see Transformer 2 a]. **1906** Goodchild & Tweney *Technol. & Sci. Dict.* 783/1 Another class of trans-

formers are more often termed converters or rotary converters; they are used in the transformation of alternate currents to continuous currents (or *vice versa*), or for changing the voltage of continuous currents. **1911** *Encycl. Brit.* XVIII. 412/2 The converter cannot advantageously be used to control the power factor by variation of the input voltage. *Ibid.*, The synchronous converter finds its chief use in electric traction. **1958** *Spectator* 1 Aug. 167/2 The car radio, which does not need a converter. **1960** *Times* 2 Apr. 7/6 Equipped with a converter and runs direct from the car battery.

d. A cipher machine. *U.S.*

1942 *U.S. War Dept. Techn. Man.* 11–380 (title) *Converter M-209.* **1950** *Ibid.* (Table of Contents 1) 167 A blow upon our faithful M-209 converter by the radio operators, whose transmissions were far more likely to be biased by unintelligible messages than the little cipher machine.

e. *Computers.* (See quot. 1956[1].)

1950 Townsend & Wheeler *High-Speed Computing Devices* xv. 386 An analog-to-digital converter is a device which..various values of continuously variable quantities and expresses them in discrete numerical form. *Ibid.* 393 A digital-to-analog converter employed in the Bell Telephone Laboratories' recorder system..utilizes the exponential decay characteristics of the RC circuit. **1951** M. M. Kuder in *Commercial Equipment* (Natl. Bureau of Standards Circular) 1117, Nat. Bureau of Standards (U.S.). **1964** *Gloss. Terms Autom. Data Proc.* (*B.S.I.*) 183 *Converter*, a unit which changes the representation of data from one form to another so that it becomes acceptable to another machine, e.g. a unit which changes punched cards to data recorded on magnetic tape. **1960** A converter may also edit the data. **1962** M. G. Hartley *Electronic Analogue Computers* vii. 143 Appropriate analogue-to-digital converters.

f. *Nuclear Technology.* A nuclear reactor in which fertile material is converted into fissile material. In full *converter reactor*.

1955 *Rep. U.S.A.E.C. on Nucl. Power Reactor Technol.* 80 If the reactor is operating as a converter, the U[235] remaining in the core must be decontaminated. **1956** S. Glasstone *Princ. Nucl. Reactor Engin.* i. 29 In these reactors the reactions accompanying the fission process are used to convert non-fissionable material (uranium-238) into one (plutonium-239) that is fissionable. Such reactors have been called production reactors or converters. **1963** *Gloss. Nuclear Sci.* (*B.S.I.*) 80 Breeder and converter reactors may also be power reactors.

convertibility. Add to def.: *spec.* The ability of paper money to be converted to gold or of the currency of one country to be freely converted to that of another at a fixed rate.

1844 *Peel in Times* 7 May 4/3 Perfect, unlimited convertibility..and convertibility into gold, is not a complete security against..excessive issue. **1948** G. Crowther *Outl. Money* (ed. 2) viii. 267 The convertibility..of the country which remaining an obligation to pay any foreigners who came into possession of pounds sterling as a result of current..transactions..were to be allowed to offer them for exchange into foreign currencies. **1959** *Times* 10 Feb. 3/6 Since the United States is by far the most important currency whose currency is convertible into gold at a fixed price, convertibility has recently come to mean convertibility either into gold or into dollars. **1959** *Listener* 4 June 6/1 Convertibility..is..a demonstration of confidence in sterling.

convertible, *a.* (*sb.*). Add: **5. a.** *spec.* Of a motor-car (see quot. 1918). (Cf. quot. 1884 in Dict., and sense B. 2 below.)

1918 *Webster Add.*, *Convertible*..changeable from a closed to an open vehicle, 'said of an automobile body'. **1936** *Brooklands Automobile Reference Book* 15 Convertible Coupé Roadster. **1942** E. Daly *House without Door* (1945) xi. 120 Two..convertible station wagons.

6. Also, of currency, that can be freely converted in the foreign exchange market (into gold or dollars) at a fixed price (see also quot. 1911).

1911 *Encycl. Brit.* XXIV. 574/1 A convertible security is one which can be readily exchanged for money. **1938** *Encycl. Brit.* Bk. of Yr. [illegible] concerned with land or buildings. **1948** G. Crowther *Outl. Money* (ed. 2) viii. 274 In July 1947..the pound sterling was made 'convertible' for current transactions. **1959** *Ann. Reg.* 1954 139 A sum of 40 million roubles a year paid in convertible currency.

B. *sb.* **2.** *sing.* A motor-car with a collapsible hood. orig. *U.S.*

1916 *Sci. Amer.* Oct. 14 (caption) Seven-passenger convertible (open). **1937** R. Stout *Red Box* viii. 110 Gerhert had climbed into a neat little convertible. **1959** *Times* 26 May 14/4 The car then becomes a convertible with a folding fabric head. **1966** *Daily Tel.* (Colour Suppl.) 21 Oct. 14, I once compared the convertible with a mistress and a sedan with a wife.

convertiplane (kǫnvə̄·ɹtiplēⁱn). Also **convertaplane** (-ăplēⁱn). [f. Convert(ible *a.* + (Aero)plane *sb.*] A type of aircraft combining certain features of the helicopter with those of a conventional aeroplane.

1949 *Jrnl. R. Aeronaut. Soc.* LIII. 310/1 If they wanted to fly helicopters at speeds much above 140 m.p.h...they might have to use forward propulsive airscrews and possibly fixed wings, but such aircraft should probably be called 'convertaplanes', rather than helicopters. **1952** *Daily Tel.* 14 Apr. 6/4 A 'convertiplane', with small wings and large rotor blades, mounted at each wingtip, which could be tilted forward to act as airscrews. **1959** *Engineering* 23 Jan. 103/2 The Fairey Aviation Company's Rotodyne 'convertiplane' set up the first world class speed record in this category of vertical take-off airliner. **1963** *New Scientist* 24 Jan. 183/3 A stowed-rotor convertiplane.

conveyer, -or. Add: **4. b.** (Earlier U.S. example.)

1813 *Niles' Register* Addenda III. 15/2 The conveyer, ..while it cooled the flour, passed it on to the place where the elevator caught it.

c. *attrib.*, as *conveyor cable, layout, principle*; **conveyor belt,** an endless belt of rubber, canvas, etc., running over rollers or the like, on which objects or material can be conveyed (cf. quot. 1883 for sense 4 b in Dict.); also *attrib.* and *fig.*; hence *conveyor belting*.

1906 *Westm. Gaz.* 24 Jan. 8/2 Sticks travel along the conveyor belt. **1935** H. G. Wells *Things to Come* x. 91 Digging machines are seen making a gigantic excavation. Conveyer belts carry away the debris. **1948** J. B. Priestley *London Trav.* 14 I don't admire the mass production and conveyor-belt system of education. **1957** *Times* 25 June 18/3 The conveyor belt required for kicking on the plates. **1957** *Times Rev. Industry* Sept. 40 Good conveyor belting. **1958** C. Sallis in *Biometrika* XLVI. 30 Study of the wear of conveyor belting..**1931** *Ibid.* 31 In the application to the life of conveyor belting, the relative movement is due to continuous wear of the belting. **1941** P. Richey *Fighter Pilot* 55 Missing the conveyor-cables on the slip-lines at the lacing position. **1961** *Times* 12 Apr. 15/2 Such conveyor layouts are..in a factory..**1955** *Times* 11 Aug. 5/4 Its main purpose was to make using a system embodying the conveyor layout. *Ibid.* p. xiv/4 The washing of cars to be done on the conveyor principle.

conveyorize (kǫnvēⁱ·ǫrəiz), *v.* orig. *U.S.* Also **conveyerize.** [f. Conveyor + -ize.] *trans.* To equip with a conveyor, to employ a conveyor for; to carry out by means of a conveyor. Hence **convey·orized** *ppl. a.*

1943 *Newsweek* 16 Aug., We 'conveyerized' the job because we knew we'd have to depend on female labor. We didn't want them to do any lifting or carrying. **1958** *Archit. Rev.* CXXIV. 343/1 The conveyorised plant is quite widely used in conveyorized processes. **1959** *Times Rev. Industry* Mar. 40/2 Conveyorised Products. **1961** *New Scientist* 23 Mar. 746/1 Both companies have conveyorized production. **1962** *New Scientist* 25 Jan. 213 Conveyorised line at GEC Ltd. for testing finished transistors.

convicted, *ppl. a.* Add: **3.** Convinced of sin; converted. [Cf. Convict *v.* 4.] *U.S.*

1824 M. B. Smith *Let.* 13 Oct. in *40 Yrs. Washington Soc.* (1906) 159 The groans and sobs of the newly convicted. **1844** *Ibid.* Mrs. H. Gallagher *Life.* **1858** J. J. Hooper *Adv. Simon Suggs* (1851) x. 124 By this time it had come to be generally known that the 'convicted' old man was Captain Simon Suggs the very 'chief of sinners' in all that region. **1948** J. B. C. Craddock *Prophet of Smoky Mts.* i, 'The boys are convicted, then?' he asked. 'The boys hev got thar religion,' his host replied.

convolutionary (kǫnvǝl'ū·ʃənări), *a.* [f. Convolution + -ary[1].] Of or pertaining to a convolution or convolutions, esp. of the brain.

1903 *Nature* 5 Nov. 8/1 Every detail of the convolutionary pattern.

convoyer. (Later examples.)

1894 *Monthly Packet* May 589 He commits him to twin fleet convoyers..who bear him swiftly to the sea-shore. **1927** *Daily Express* 10 May 3/4 A sharp signal from [the] chief convoyer, a frenzied cutting of the seals on the baskets, and ten thousand birds..swept aloft.

convulsant, *a.* and *sb.* (Later examples.)

1943 *Nature* 14 Sept. 383/1 The compound was a convulsant poison like methyl fluoroacetate. **1949** H. W. Florey et al. *Antibiotics* II. xxxvii. 1213 Impure penicillin ceased to be convulsant when its rich metallic impurities were removed by clarine. *Ibid.*, It was a convulsant. **1970** *Nature* 27 June 1223/1 Convulsants may interfere with the operation of inhibitory synapses in the brain and spinal cord of mammals.

convulsedly (kǫnvə·lsedli), *adv.* [f. Convulsed *ppl. a.* + -ly[2].] With convulsed action.

1909 Garden & Huffer *Romance* III. vi, He had been sobbing convulsedly.

convulsibility (kǫnvəlsibi·lĭti). [f. Convulsible *a.*: see -ity.] A tendency to convulsion.

1895 *Buck's Handbk. Med. Sci.* II. 287/2 Convulsibility is acquired..by infectious fevers.

cony, coney, *sb.* Add: **2. b.** A hat made of rabbit-fur (in place of beaver). *U.S.*

1856 Barnum *Life* 99 If a 'pedler' wanted to trade with us for a box of beaver hats,..he was sure to obtain a box of 'coneys'.

7. b. (Example.)

1884 Goode *Nat. Hist. Aquat. Anim.* 412 The Coney, *Epinephelus apua*, of Key West, the Hind of Bermuda, is an important food-fish which occurs throughout the West Indies.

coo (kū), *int.* Also **coo-er.** An exclamation expressing surprise or incredulity. *slang.*

1911 *Punch* 5 July 8/1 'My dear Bobby,' I said, 'I know everything about everything.' 'Coo—I bet you don't.' **1927** C. Mackenzie *Vestal Fire* III. ii, Coo! they don't mind bumping into you, do they? **1929** Raymond *Family that Was* i. iv, 'Coo!' exclaimed Joyce. 'O Mummy, may be?' **1936** P. Fleming *One's Company* xviii. 168 But M.'s enormous stature drew many subdued cries of 'Ai-ai-ai-yah!', which is Chinese for 'Coo-er!' **1963** *Times* 22 Apr. 6/4 Coo, is that really the Viscount?

coobah, var. *couba.

co-occu·r, *v.* [f. Co- + Occur *v.*] *intr.* To occur together or simultaneously. So **co-occu·rring** *ppl. a.*

1951 Z. S. Harris *Methods Struct. Linguistics* vi. 52 Others of their components (e.g. tone) are members of another co-occurring morpheme. **1957** — in *Language* XXXIII. 286 The particular combinations of morphemes which co-occur in various instances of a construction are those which satisfy it. **1960** *Amer. Speech* XXXV. 243 Sequences of differently intoned and tone co-occurring types co-occur with the same bound and free forms as do the members of the first set. **1964** Crystal & Quirk *Prosodic Features in Eng.* iii. 34 Vocal effects that are the result of certain co-occurring activities of the vocal cords. *Ibid.* iv. 47 It seems usual for either of the transition modes..to co-occur with the co-occurring items.

co-occu·rrence. [f. Co- + Occurrence.] Simultaneous occurrence; an instance of this. Also *attrib.* So **co-occu·rent** *a.*

1951 Z. S. Harris *Methods Struct. Linguistics* xvi. 300 We are dealing with the co-occurrences of A and B. *Ibid.* 302 Many different classes of individual co-occurrence occur in various languages. **1954** — in *Word* X. 146 An existing array of its co-occurrents. **1957** — in *Language* XXXIII. 285 The co-occurrence set of each morpheme. **1962** D. S. Worth in Householder & Saporta *Probl. Lexicography* 81 Multiple meanings can be specified by co-occurrence, e.g. 'hot (pepper, musical)..taste'. **1963** J. Lyons *Struct. Semantics* ii. 19 The very severe restrictions on the co-occurrence of particular units in the same sentence. **1963** *Language* XXXIX. 2 Some or all of the pivot words have unique sets of co-occurrents. **1964** M. A. K. Halliday et al. *Linguistic Sci.* ii. 33 This tendency to co-occurrence is the basic formal pattern into which lexical items enter. It is known as collocation, and an item is said to 'collocate with' another item or items. **1965** *Math. in Biol. & Med.* (*Med. Res. Council*) III. 110 He was concerned with a square symmetric matrix of frequencies of co-occurrence.

cooee, cooey, *sb.* (Also *int.*) Add: **b.** Hence *gen.*, as a signal to draw attention to the caller. Austral. and N.Z. colloq. phr. *within (a) cooee (of)*: within hailing distance; within easy reach, near (to).

1852 J. West *Hist. Tasmania* II. 93 Some gentlemen, on a visit to a London theatre, to draw the attention of their friends in a distant part of the house, gave a cooey. **1886** Hotten *Slang Dict.* 107 Cooey, the Australian bush-call, now not unfrequently heard in the streets of London. **1889** [in Dict.]. **1893** *Herald* (Melbourne) 26 June 2/6 (Morris), Witness said that there was a post-office close 'within coo-ee' or within less than half-a-mile of the station. **1896** H. G. Wells *Island of Dr Moreau* xvi. 161 Once I was arrested by the distant voice of Montgomery bawling, 'Coo-ee..Mor-eau!' **1896** H. Lawson *While Worlds was Wide* (1900) 80 Weary teamsters struggled on within a week's tramp, just over to within a cooey of the Shanty for the night. **1899** 'S. Rudd' *On our Selection* 28 There wasn't a horse in the district that could come within cooee of her. **1916** 'Anzac' *On Anzac Trail* 39 The niggers..wouldn't come within coo-ee of us when we were engaged in bomb-throwing. **1933** *Bulletin* (Sydney) 16 Aug. 10/2 In George-street, within coo-ee of the Town Hall. **1946** F. Davison *Dusty* xv. 164 Tom, within a cooee of seventy, had made a new start in life. **1957** *Times* 14 Nov. 17/2 Confident 'coo-ees' from both sides of the ground greeted this try (by the Australian rugby team). **1965** *Weekly News* (Auckland) 4 Aug. 14/4 He might be so unfortunate as to..be on this tour came within coo-ee of the effort of Gainsford.

coo-er: see *coo *int.*

cooja(h) (kū·dʒă). *Anglo-Ind.* [Hind. (Pers.) *kūza*.] A wide-mouthed earthenware water-vessel.

1871 Riddell *Indian Domestic Economy* (ed. 7) 362 Many parts of India are celebrated for their coojahs or goglets. **1885** E. H. Aitken *Tribes on my Frontier* 118 They [sc. tree-frogs] would perch gracefully on the rim of the water cooja or on the rim of a tumbler.

cook, *sb.* Add: **1. d.** *Chess.* An unforeseen second solution that spoils a problem, position, etc.

1875 S. H. Thomas in *Westm. Papers* VII. 243, I almost imagined the author's solution a 'cook'. **1876** W. Nash *Ibid.* IX. 53 If this is not a cook the problem is much under the composer's usually high standard. **1894** *Daily News* 30 July 6/7 The latest delinquent is the composer of problem No. 55, whose intention was to mate by 1 Q-Kt 4, overlooking the short cook by 1 Q×Kt. **1966** *New Statesman* 24 June 941/2 Veitch found the otherwise beautiful piece blemished by this cook.

e. *Paper-Manuf.* The boiling procedure in a digester; the materials being boiled.

1950 R. W. Sindall *Paper Technol.* v. 62 The progress of the 'cook' is carefully watched. **1937** E. J. Labarre *Dict. Paper* 214/2 *Slow cook*, another term for the Mitscherlich process of digesting pulp. **1963** R. R. A. Higham *Handbk. Papermaking* v. 100 Steam pressure is applied to the cook.

3. cook-book (earlier and later examples); **cook-boy** (later examples); **cook-camp** *N. Amer.* the part of a camp in which the cooking is done; a building serving as kitchen and eating-room; **cook-general,** a domestic servant who does house-work as well as cooking; **cook-housekeeper,** a domestic servant who does the work of both cook and housekeeper; **cook-shack** *N. Amer.*, a shack in which cooking is done; **cook's knife,** a general-purpose kitchen knife; one used by a cook; **cook-stove** *U.S.*, a cooking-stove; **cook-tent** *U.S.*, a tent in which food is cooked; **cook-wagon** *U.S.*, a wagon with a cooking outfit.

1809 R. Tyler *Yankey in London* 179 An assortment of culinary reviews, vulgarly called cook-books. **1881** Mrs. F. E. Owens (*title*) Cook Book, and useful Hints for the Household. **1951** 'J. Wyndham' *Day of Triffids* xiv. 250 With the aid of a cook-book Josella had managed to produce meals of a kind. **1970** *N.Y. Times* 1 Sept. 38/6 A minimum of $45 for me, my apron, and my cookbook. **1953** D. Lessing *Five* iv. 208 His cook-boy, who had worked for him twelve years, gave notice that month. **1963** *Times* 27 Feb. 12/6 Ibrahim, Iranian cook-boy to the entourage. **1893** *Scribner's Mag.* June 703/2 A cook camp..is a large dining-room and kitchen combined. **1895** *Outing* (U.S.) XXVI. 393/2 'Yer supper's ready' said Payson, emerging from the cook-camp. **1935** A. Sullivan *Great Divide* 194 There was..the cook camp with its log annex for supplies and fresh beef. **1969** L. G. Sorden *Lumberjack Lingo* 26 *Cook camp*, a building used as kitchen and dining room in a logging camp. Same as cook house, cook shanty. **1896** *Daily News* 30 Oct. 10/4 Cook-general and Young Housemaid wanted. **1951** *Good Housek. Home Encycl.* 45/1 A cook-general is expected to undertake a certain amount of general housework. **1892** C. M. Yonge *That Stick* xi. 71 The indoor servants were all new, the cook-housekeeper hired by Lady Kenton's assistance. **1950** 'J. Tey' *To love & be Wise* xi. 128 The back door was always locked by Mrs Brett, the cook-housekeeper. **1909** R. A. Wason *Happy Hawkins* 38 Jabez changed directions for the cook shack. **1912** J. B. Bickersteth *Land of Open Doors* (1914) 157 They were ready to pull down the cook-shack over his ears. **1957** J. Kerouac *On Road* (1958) II. v. 132 Railroad cookshacks. **1868** M. Jewry *Warne's Model Cookery Bk.* 30/3, 2 cook's knives. **1939** J. Cary *Mr. Johnson* 88 Sozy's cook's knife, a black-handled French knife with a sharp point. **1824** T. B. Hazard *Nailer Tom's Diary* (1930) 625/1 Sett up the Cook Stoe [*sic*] in the west Room. **1838** H. Colman *Mass. Agric. Rep.* 80 The wood when cut fit for a cook stove is worth $4 a cord. **1853** Cook-stove [see Stove *sb.*[1] 5 a]. **1930** *Times Educ. Suppl.* 23 Aug. p. i/3 A voyage is dull without tea and life a cheerless blank without a cookstove. **1885** E. B. Custer *Boots & Saddles* 53 We looked out, to find the cook-tent blown flat to the ground. **1956** H. Gold *Man who was not with It* (1965) ii. 14 Our cooktent had a pleasant business warmth, too. **1900** H. Garland *Eagle's Heart* 162 In the shade of the cook wagon. **1907** Mulford *Bar-20* xi. 119 Two cook wagons were stalled a short distance from the corral.

cook, *v.*[1] Add: **1. b.** Slang phr. *to cook with gas* (or *electricity, radar*): to succeed, to do very well; to act or think correctly; also *to cook on the front* (*top*) *burner*. *U.S.*

1941 *Star* (Kansas City) 23 Feb., Now you're cooking with electricity! **1942** *Time* 27 Apr. 84/3 Many a student.. figured that..Thurman Arnold was cooking with gas. **1945** L. Shelly *Jive Talk Dict.* 23/1 *Cooking on the front burner*, tops. **1946** F. Wakeman *Hucksters* (1947) xv. 201 Vic said, 'Good boy, Georgie. Now you're cooking with radar.' **1958** N. D. Hinton in *Publ. Amer. Dial. Soc.* XXX. 39 Although 'cooking on the front burner' has long since gone out of style, some musicians still refer to New York as 'The Apple'. **1962** K. Orvis *Damned & Destroyed* xi. 77 Those Mounties cook with gas. With gas, brother—they're murder. **1965** 'N. Carter' *Arm'd Rhinoceros* xxiv. 264 'These pens are no good.'..'You're cooking on the front burner, Mac,' I replied in a kind of English to relieve my feelings.

c. To play music with excitement, inspiration, etc. *slang* (orig. *U.S.*).

1943 *N.Y. Times* 9 May II. 5/4 Brother, when you are cooking with both burners the fiddle is out. **1955** L. Feather *Encycl. Jazz* 346 *Cook*, v. (Usually in the expression 'He's cookin'.') To play with rhythmic inspiration. **1965** *Crescendo* Oct. 12/1 In some of the up-tempos Len Livesey really was cooking. **1968** *Ibid.* June 12/2 The band used to get up on the bandstand and *really* cook.

2. c. Colloq. phr. (orig. *U.S.*) *what's cooking?*, what is happening? what is in train?

1942 R. Chandler *High Window* 70 It stopped at six, I got out, and the old man leaned out of the car to spit and said in a dull voice: 'What's cookin'?' **1943** C. H. Ward-Jackson *Piece of Cake* 62 *What's cooking?* What's happening? What's on? **1945** S. Lewis *C. Timberlane* 113 It was part of their creed and time that every so often Eino and Jinny should say to each other, 'What's cooking?' **1956** 'A. Gilbert' *Death came Too* xvi. 165 What's cooking?.. Are you going to uncover the villain?

d. *trans.* and *intr.* To prepare opium for use by the application of heat.

1887 in *Amer. Speech* (1948) XXIII. 246/2 The roller dips his yenhook into the shell or hoptoy, and conveys the opium over the lamp to cook it, which takes but a few seconds. **1916** T. Burke *Limehouse Nights* (1917) 260 Lois..began to 'cook' for her boy...She..dug out the treacly hop..and held it against the flame. **1948** Mencken *Amer. Lang.* Suppl. II. xi. 681 The preparation of the opium is called cooking (or rolling) a pill.

e. *trans.* To make radioactive. Also *intr.*, to become radioactive. *colloq.* (orig. *U.S.*).

1950 in *Amer. Speech* (1952) XXVII. 204 Others are formed in aluminum cans of raw material 'cooked' by the pile's neutrons. **1951** *Time* 19 Nov. 40/1 More cobalt is being 'cooked' for the first U.S. units. **1959** C. Hodder-Williams *Chain Reaction* xiii. 157 If the thing won't cook if they [*sc.* the control rods] 're raised about three quarters of the way out, you can reckon that there's something cockeyed somewhere.

cookable, *a.* and *sb.* (Earlier U.S. example.)

1843 'R. Carlton' *New Purchase* viii. 47 Next came a resolution that the ladies should prepare the cookables—i.e. stuff the chicken with filling—beat eggs for puddings, and the like.

cookee (ku·ki). *N. Amer.* Also **cookie.** [f. Cook sb.] A cook; esp. an assistant to the cook in a camp.

1846 *Spirit of Times* 4 July 218/2 A cookie who was lord and master of the culinary department. **1851** J. S. Springer *Forest Life* 172 Arrangements are made..by the cook and 'cookee', as his assistant is called. **1891** *Harper's Mag.* Nov. 890/2 The rule is to have one cook and two 'cookees' to each sixty men. **1901** S. E. White *Westerners* xxvii. 250 The move necessitated a cook and 'cookee', and the weekly purchase of provisions. **1965** *Islander* (Victoria, B.C.) 16 May 6/2 Food was laid out in 'dishups' scattered along the table and 'cookees' watched to see that they were well filled.

cookery. Add: **3.** (Later examples.)

1802 D. Wordsworth *Grasmere Jrnl.* 24 Jan. in *Jrnls.* (1941) I. 101 We..made a nice piece of cookery for Wm.'s supper. **1855** C. M. Yonge *Railroad Children* v. 37 Many was the nice little cookery of broth or gruel made for her especial benefit.

6. cookery-book (earlier example).

1639 M. Verney *Will* in F. P. Verney *Memoirs* (1892) II. i. 18 All but my noats and account and medsinable and coockery Boockes, such keep.

cookie. Add: **a.** Also *S. Afr.* and *Canad.*

1852 Barter *Dorp & Veld* 107 Cookies, or unleavened cakes of coarse meal, baked on the grid-iron. **1897** E. Glanville *Tales from Veld* 51 Raking the 'cookie' from the fire-place, whence it came baking hot. **1935** M. de la Roche *Young Renny* xxiv. 214 Mary was arranging plates of bread and butter, thick ginger cookies,..and a bowl of halved peaches. **1968** *Globe & Mail* (Toronto) 17 Feb. 6/2 Children sneaking cookies from a cookie jar.

b. cookie-pusher (*U.S. colloq.*), a counter attendant; *fig.*, a man leading a futile social life; *spec.*, a diplomat allegedly devoting more attention to protocol or social engagements than to his work.

1942 Berrey & Van den Bark *Amer. Thes. Slang* § 449/1 *Cookie pusher*..a party attender. *Ibid.* § 452/3 Dandy, 'dude'..*cookie pusher*. *Ibid.* § 825/15 Ladies' man..*cookie-pusher*. **1946** *Word Study* May 2/1 State Department officials have long been called 'cookie pushers'. **1961** *Times* 30 May 10/6 The chairman of the House sub-committee.. still..dismisses all diplomatists as 'cooky-pushers'. **1962** *Economist* 29 Sept. 1211/3 The popular image of the cookie-pusher in Foggy Bottom.

2. *slang* (orig. *U.S.*). **a.** A woman; *esp.* an attractive girl. **b.** A man, often with defining word.

1920 *Collier's* 6 Mar. 42/3 That girl friend of yours is a cookie—hey, what? **1928** *Chicago Tribune* 7 Oct. (Comics) 2 What a swell bunch of cookies you turned out to be. **1942** *Amer. Mercury* Oct. 436/1 Just about the toughest cookie ever born. **1948** F. Brown *Murder can be Fun* (1951) x. 146 A smart cookie, that Wilkins. **1953** W. R. Burnett *Vanity Row* xvi. 110 He's a real tough cookie and you know it. **1959** R. Longrigg *Wrong Number* ii. 27, I met a cookie I know...She said you'd said *Faustus* was like *Oklahoma*.

3. A bomb. *Air Force slang.*

1943 in *Amer. Speech* (1944) XIX. 310/1, I am flying in one of three Lancasters which will drop hundreds of incendiaries as well as 4,000-pound '*cookies*'. **1944** R. Dimbleby in *War Report* (B.B.C.) (1946) xiv. 280 But I *did* see heavy bombs, cookies, going down into the brown smoke.

4. Colloq. phr. (chiefly *U.S.*): (*that's*) *how* (or *the way*) *the cookie crumbles*, (that is) how the position resolves itself; that is the way it is.

1957 *Sat. Even. Post* 7 Sept. 59 From then on, that's the way the cooky crumbled. I enjoyed having good ratings, but I didn't enjoy the viciousness of the railbirds' thrusts at Berle. **1959** Wenzell Brown *Cry Kill* iv. 45 No matter how the cookie crumbled, Mamma Ida was in for a bad time. **1961** Wodehouse *Ice in Bedroom* v. 40 Oh well, that's the way the cookie crumbles. You can't win 'em all. **1964** *Listener* 16 Apr. 612/2 We shall not know how, as the Americans say, the cookie crumbles.

cookie, var. *cookee.

cooking, *vbl. sb.* Add: **1.** *spec.* The process of preparing opium. Cf. *cook *v.*[1] 2 d.

1887 in *Amer. Speech* (1948) XXIII. 246/2 Watch the roller go through his routine of cooking.

3. cooking-apple, an apple suitable for cooking.

1817 W. Coxe *Fruit Trees* 108 The Hagloe is an uncommonly fine cooking apple. **1893** G. Meredith *Let.* 8 Nov. (1970) II. 1146 The cooking apples can't be surpassed. **1966** *Observer* 30 Oct. 46/3, 4 lb. cooking-apples.

b. Used in the cooking of food, as *cooking sherry, soda*; hence designating something of medium or average quality, as *cooking bitter* (beer).

1861 *Vanity Fair* 2 Feb. 50/1 The servants..were forever ..taking the cooking butter for the table. **1872** *Harper's Mag.* XLVI. 28 This is genuine soda-water—cooking soda with nearly an equal amount of sulphate of soda, [etc.]. **1908** A. Bennett *Old Wives' Tale* I. iii. 36 Cooking butter, to say naught of lard, was unknown in that kitchen. **1926–7** *Army & Navy Stores Catal.* 94/2 Cooking Brandy (French) —Per doz. bots. 150/-. **1934** Webster, Cooking sherry. **1939** A. Simon *Encycl. Gastronomy* I. 4/1 A sweet sauce usually made with apricot jam, a little flour and some 'cooking sherry'. **1940** *Economist* 13 July 43/1 Tea, margarine and cooking fats are all to be rationed this month. **1951** P. Branch *Lion in Cellar* i. 11 Not the best bitter..but the weaker one which she privately referred to as 'cooking'. **1966** R. Jeffries *Death in Coverts* i. 7 The last trump would sound before he ever offered even cooking brandy.

cookless (ku·klĕs), *a.* [f. Cook *sb.* + -less.] Without a cook. Hence **coo·klessness.**

c **1845** J. W. Howe *Let.* in F. Trollope *Dom. Manners* (1949) 54, I am again cookless. I imprudently..took a young girl, who left me in four days. **1898** *Cornhill Mag.* Aug. 255 The good old schoolroom days when one's dear mother used sometimes to admit that she 'had been cookless for six weeks'. **1904** 'Alan Dale' *Wanted: a Cook* Ded., To Jennie Shalek: housewife, who, in my hour of drab and dreary cooklessness, when my heart fainted, and tragedy impended, sent her four fair daughters to my aid. **1928** *Sunday Express* 6 May 15 Catherine Ives...'When the Cook is Away.' ..In coming to the rescue of the cookless—from whatever cause—the author has collected also an attractive assortment of recipes.

cook-out, cookout (ku·kaut). orig. *U.S.* [f. Cook *sb.* and *v.*[1] + Out *adv.*] An outdoor gathering at which food is cooked; a barbecue.

1947 *Reader's Digest* May 22 At a breakfast cook-out one boy was delegated to cook scrambled eggs. **1950** *Even. Jrnl.* (Lincoln, Nebraska) 9 Sept., A cookout at the Campfire Girls' cabin. **1958** *Listener* 17 July 81/2 People who pride themselves on plain-speaking call this [a barbecue] a 'cook-out'. **1962** J. Potts *Evil Wish* i. 17 She'll..make a patio out of it. So they can have cook-outs. **1969** *Daily Tel.* 11 Jan. 14/6 They stayed for five weeks, had cook-outs on the beach and occasional nights in Montego Bay.

cook-room. Add: In later use *N. Amer.*

1874 *Rep. Vermont Board Agric.* II. 509 The cook room, which is one of the utmost importance, should be one of great convenience. **1880** Tourgée *Invis. Empire* x. 467 There were four rooms,..a cook-room, dining room [etc.]. **1896** J. C. Harris *Sister Jane* 54, I made haste to go to the cook-room, intending to start the fire. **1964** *Newfoundland Q.* Summer 15/1 Also with a large enough lot of non-resident hands, there would be a 'cook room'.

cook-shop. Add: **2.** *N.Z.* (See quot. 1933.)

1857 *St. Leonard's Station Diary* 1 Jan. in L. R. C. MacFarlane *Amuri* (1946) III. 123 Rum and plum duff at the cook shop. **1933** L. G. D. Acland in *Press* (Christchurch) 30 Sept. 15/7 *Cookshop*, kitchen and dining room where station hands' food is cooked and eaten. It is sometimes a separate building, sometimes part of the men's hut. **1953** B. Stronach *Musterer on Molesworth* ii. 8 The cookshop [on Molesworth Station] was an old army hut.

Cook's tour. [f. the name of Thomas *Cook* (1808–92), travel agent.] A tour, *esp.* one in which many places are viewed; any journey of wide extent; also *fig.* So **Cook's-tourist** *sb.* and *a.*

1906 W. de Morgan *J. Vance* xlvii. 475 So it was decided that Florence was banal and Cook's-tourist—a new adjective. **1916** F. C. Curry *From St. Lawrence to Yser* xxi. 155 'Cook's tourists'. These individuals were officers sent over..to receive practical instruction in trench warfare. **1922** S. Lewis *Babbitt* ix. 126 Dante..the fellow that took the Cook's Tour to Hell. **1925** Fraser & Gibbons *Soldier & Sailor Words* 63 *Cook's tour*, a colloquial phrase for the customary tour round the trenches by officers and N.C.O.'s of relieving..units.., the 'Cook's Tourists' being guided by the previous holders of the position. **1928** C. A. Nicholson *Hell & Duchess* II. iii. 248 She has yet..to do her Cook's tour, as it were: the Louvre, Cluny, Sainte-Chapelle. **1969** A. Glyn *Dragon Variation* viii. 229 Sightseeing!..It isn't a question of taking a Cook's Tour round the city.

cool, *a.* Add: **4. d.** Applied to jazz music: restrained or relaxed in style; also applied to the performer; opp. *hot *a.* orig. *U.S.*

1947 (record by Charlie Parker Quartet, Dial 1015) Cool Blues. **1948** *Life* 11 Oct. 138 Bebop: New Jazz School is Led by Trumpeter Who is Hot, Cool and Gone. **1950** *Christian Sci. Monitor* 8 Feb. 15 Bop is 'cool' jazz in contrast to the 'hot' variety of the swing or Dixieland schools. **1953** *Melody Maker* 9 May 5 Hot and cool—you've got to hear the lot. **1955** L. Feather *Encycl. Jazz* (1956) 30 Cool jazz to most musicians and students denotes the understated, behind-the-beat style typified by the arrangements and soloists on the Davis records. **1957** H. Panassié in S. Traill *Concerning Jazz* 61 The 'cool' musicians..stopped using the traditional jazz technique and tone. **1962** J. Wain *Strike Father Dead* iv. 204 The new developments which were to become first bebop and then just bop and finally cool jazz.

e. Hence, characteristic of those who favour 'cool' music; relaxed; unemotional; also used loosely as a general term of approval; *cool cat*: see *cat *sb.*[1] 2 c. *colloq.* (orig. *U.S.*).

1948 *New Yorker* 3 July 28 The bebop people have a language of their own...Their expressions of approval include 'cool'! **1953** *Time* 14 Sept. 68/3 The latest Tin Pan Alley argot, where 'cool' means good, 'crazy' means wonderful. **1955** *N.Y. Times* 22 May VI. 19/2 Maybe it's all these new buildings breeding more of these cool Brooks Brothers cats. **1955** *Sci. News Let.* 1 Oct. 221/2 This is not cool chatter between some young hep-cats in a smoke-filled jazz joint. **1957** *Sunday Mail* (Glasgow) 10 Feb. 11 *Gone*—the best, in the top rung, the coolest. **1958** *Observer* 23 Nov. 16/3 On one side was the frenetic..bumptiousness of the rock-'n'-rollers, on the other the calculated indifference of the cool cats. **1959** *Ibid.* 25 Oct. 29/8 They got long, sloppy haircuts and wide knot ties and no-press suits with fat lapels. Very cool.

6. Esp. in phr. *cool customer* (see Customer *sb.* 5).

1941 A. Christie *Evil under Sun* v. 96 Cool customer. Not giving anything away, is he? **1945** —— *Sparkling Cyanide* III. v. 110 At any rate she's a cool customer.

8. cool chamber, -store, a place in which perishable goods may be kept cool; **cool-house,** a glass-house kept at a cool temperature.

1887 *Col. & Indian Exhib., Rep. Col. Sect.* 140 The enterprise of Messrs. S. & Sons in fitting up one of their steamers with a cool-chamber. **1901** *Daily Chron.* 1 June 8/6 Provided sufficient 'cool chamber' accommodation could be procured on the steamers trading between Australia and

England. **1909** Webster, Coolhouse. **1949** E. Hyams *Not in our Stars* viii. 98 He was sometimes alone in the cool-house where these raspberries grow. **1958** *Listener* 21 Aug. 275/1 A Luculia growing up the back wall of a cool house. **1959** *N.Z. Listener* 18 Dec. 8/1 The trees were all about.. and only occasionally between them was a gleam of the sun...'Like being in a cool-store, isn't it?'

cool, *sb.*[1] Add: **4. b.** A truce between gangs. *U.S. slang.*

1958 *Daily Tel.* 8 Apr., *Cool*, an uneasy armistice between child-gangs. **1959** H. Salisbury *Shook-Up Generation* iii. 47 A 'cool' was negotiated by street club workers. But it was an uneasy truce, often broken.

c. Composure, relaxedness. Cf. *cool *a.* 4 e. *slang.*

1966 *New Statesman* 8 Apr. 512/2 There is a place where it nearly loses its cool, in fact, when doomed Mr Hamilton is dragged about, manacled to a great cross-piece and generally abused. **1966** *New Yorker* 18 June 37 I'll bet *that* old guy has never blown his cool. **1967** *Listener* 19 Jan. 101/2 Professor Marcus consistently keeps his cool when sex is being discussed; all the four-letter words are used without blanching. **1968** *Win* 15 Oct. 3/2 It is one thing for violence to break out by accident, because we lose our cool, and it is another thing to take..a bottle to a demonstration 'just in case' we lose our cool.

cool, *v.* Add: **2. a.** Also with *off.*

1887 *Lantern* (New Orleans) 7 May 2/2 Maher was sent to jail to cool off. **1908** *Smart Set* Sept. 83/2 He rose suddenly and went upstairs with his anger...He sat there cooling off by the window.

3. e. To kill. *U.S. slang.*

1930 E. W. Scott in *Flynn's* 6 Sept. 849/1 Eight stick-up Johnnies out of ten aren't so hot about coolin' a cop. **1965** 'D. Shannon' *Death-Bringers* (1966) v. 62 Anybody could have cooled him.. I don't care one hell of a lot who shot Walter William. **1969** 'J. Morris' *Fever Grass* 251 He wasn't killed in any private fight...He was cooled by a Chinese agent.

4. a. Also (*U.S. colloq.*) with *out.*

1950 *N.Y. Herald Tribune* 29 July 10/8 The tear-up and the cackle-bladder, two of the major devices for cooling out a mark. **1964** A. W. Gouldner in I. L. Horowitz *New Sociology* 209 One of the con men remains behind 'to cool the mark out', seeking to persuade him to accept his loss of face rather than go to the police. *Ibid.*, The restaurant hostess who cools out the impatient customer.

5. b. *to cool it*: to relax, calm down, 'take it easy'. Cf. *cool *a.* 4 e. *slang* (orig. *U.S.*).

1953 E. Gilbert *Hot & Cool* 13 Cool it, girl. Nobody's interested. **1954** *Time* 8 Nov. 42/1, I cooled it at a table for a while. **1959** *Encounter* Aug. 35/2 The Wizard took my arm, and said, 'Cool it, kiddo.' **1968** *Crescendo* June 33/1 Cool it will you? I said once a *week*, there's no need to go stark raving mad.

coolabah, coolibah (kū·lăbă, -lĭb-). *Austral.* Also **coolibar, coolybah, k-.** [Native name.] Any of several Australian gum-trees, esp. *Eucalyptus microtheca.*

1887 *Colonial & Indian Exhib., Rep. Col. Sect.* 428 Jinbul, or Coolibar, is a small tree, yielding wood of a reddish colour, close in texture, very hard, and heavy. **1889** J. H. Maiden *Useful Native Pl. Australia* 25 *Eucalyptus microtheca*, ('Bastard Box' or 'Coolibah'). *Ibid.* 495 'Coolybah' or 'Coolibar'. **1911** C. E. W. Bean '*Dreadnought' of Darling* xv. 143, I found the whole place covered with coolabah. c **1917** A. B. Paterson *Waltzing Matilda*, Once a jolly swagman camped by a billabong Under the shade of a coolibah tree. **1930** A. W. Groom *Merry Christmas* v. 37 They halted beneath a great spreading coolabah tree. **1955** J. Cleary *Justin Bayard* 112 The graves beneath the coolibah tree at the bottom of the mountain.

coolamon (kū·lămŏn). *Austral.* Also **coolaman, cooliman, k-.** [Native name.] A vessel made of wood or bark, used for carrying water, etc. Also *attrib.*

1847 Leichhardt *Overland Exped.* 269 Three Koolimans (vessels of stringy bark) were full of honey water. **1863** Beveridge *Gatherings* 37 The beautiful Lubrina Fetched a Cooliman of water. **1881** A. C. Grant *Bush Life in Queensland* (1882) xxxiii. 337 A few broken gourds, a broken spear or two, and a cracked *coolaman*, were to be seen here and there. **1890** 'R. Boldrewood' *Col. Reformer* xvii, The fire-barks [were] carried in the coolimans to prevent the tell-tale fall of ashes. **1929** K. S. Prichard *Coonardoo* 92 Coonardoo set her boy in the coolamon she had scooped with sharp stones from the elbow of a river gum. **1963** A. Lubbock *Austral. Roundabout* 130 'How much this one?' I asked, pointing to a small coolamon dish.

coolan, var. *coolung.

coolant (kū·lănt). [f. Cool *v.*[1] + -ant[1], after Lubricant *a.* and *sb.*] A cooling agent; esp. (*a*) a fluid applied in machining operations to a cutting-tool in order to cool it and lessen friction; (*b*) a cooling medium in an internal-combustion engine or the like. Also *attrib.*

1930 H. J. French *Quenching of Steels* iv. 100 Coolant circulation. *Ibid.* vii. 156 A wide variety of coolants are encountered in commercial heat treatment practice but the five..generally employed are (1) water in baths and sprays, (2) dilute sodium hydroxide solutions, (3) sodium chloride brines, (4) oils or oil-water emulsions, and (5) air. **1939** *Flight* 7 Dec. 464/1 The coolant is circulated by a centrifugal pump of ample capacity. **1941** *Edgar Allen News* May 736 A coolant should be used with these tools, just as with high speed steel tools. **1954** *Life* 4 Jan. 89 Fuel acts as its own coolant, circulates from core through heat exchange boiler. **1956** *Planning* XXII. 108 The only method so far discovered is to draw off the heat from the reactor by circulating a gas or liquid coolant. **1958** *Jrnl. Iron & Steel Inst.* CLXXXVIII. 87/1 The employment of water coolants has been found advantageous over dry or oil-cooled working for the machining of cast iron. **1968** *Times* 7 Nov. 3/2 Superconductivity..can at present be brought about only at such low temperatures that liquid helium must be used as a coolant.

cooler. Add: **2. b.** *spec.* A vessel into which syrup is poured to crystallize into sugar, a crystallizer.

1790 *Penn. Packet* 6 Jan. 4/2 Two large copper coolers, and two Boilers for Sugar-Bakers. **1819** Brande *Chem.* 355 The fire is instantly damped, and the boiling sugar carried off in basins to the coolers. **1853** Ure *Dict. Arts* (ed. 4) II. 766 From the..cooler, the syrup is transferred into wooden chests or boxes,..also called coolers, but which are more properly crystallizers or granulators. **1921** *Dict. Occup. Terms* (1927) § 449 *Heater man* (sugar refining); *cooler hand, cooler man*.., adds water to heaters, receivers, coolers or crystallisers.

c. A water-cooler; a place where cool drinking water is available. *U.S.*

1838 *N.Y. Mirror* 13 Jan. 229/1 The peculiar philosophical effect of this cooler..we are unable to explain. **1905** F. H. Smith *At Close Range* 250 'I'll go to the cooler and wash up what I can,'..she said.' **1947** S. Bellow *Victim* i. 6 He walked out of the office..halting in the anteroom to draw a drink from the glass cooler.

5. For *U.S.* (*Thieves' slang.*) read: *slang* (orig. *U.S.*). Also, a prison cell (see quots.).

1884 C. B. Lewis *Sawed-off Sketches* 117 If you utter one profane word..I'll put you behind the cross-bars of the cooler. **1899** 'J. Flynt' *Tramping with Tramps* (1950) 392 *Cooler*, a dark cell. **1909** W. H. Davies *Beggars* vii. 52 We wanted a night's lodging at the police station. 'What!' he cried in amazement; 'what: not in the cooler?' **1919** Downing *Digger Dial.* 17 *Cooler*, prison or guard-room. **1926** J. Black *You can't Win* ix. 114 The cooler or dark cell was the same as other cells, except that there was nothing in it and the door was solid, admitting no light. **1943** 'C. Dickson' *She died a Lady* ix. 73, I am not at a time of life when one enjoys being chucked in the cooler for telling truths. **1946** G. Millar *Horned Pigeon* vii. 81 Next day I was sent to the 'cooler'. **1962** *John O'London's* 25 Jan. 82/2 The police station..is sometimes called the cooler, although the last word is also applied to a cell inside the nick.

cooley, var. Coulée.

1881 *N.Y. Times* 18 Dec. in *N. & Q.* 6th Ser. V. 65/1 Every ravine short of an inhabitable valley is called a 'cooley'. **1884** [see Coulée 2]. **1890** *Harper's Mag.* Aug. 383/1 Reno came quickly to a shallow 'cooley' (frontierism for gully), that led down..to the stream.

Coolgardie safe. *Austral.* [The name of a town in Western Australia + Safe *sb.* 1.] A safe for holding food.

1940 A. Gaston *Coolgardie Gold* xi. 135 Eventually one of the most useful articles now used in all dry areas of Australia was invented—the Coolgardie cool-safe. **1948** V. Palmer *Golconda* x. 76 A Coolgardie safe hung near the window. **1959** H. P. Tritton *Time means Tucker* 72/1 Many homesteads had a Coolgardie-safe, which was a frame covered with flyproof netting. On top was a wide, flat pan, and strips of hessian hung from the pan to the floor. These strips would draw water and keep the air in the safe reasonably cool.

coolibah: see *coolabah.

Coolidge tube (kū·lidʒ tiūb). [f. the name of William David *Coolidge* (b. 1873), American physicist.] A type of hot-cathode X-ray tube.

1915 *Nature* 15 Apr. 195/2 The Coolidge tube is based on the discharge of independent electrons. **1929** *Encycl. Brit.* XXIII. 844/2 The target of the Coolidge tube is of massive tungsten.

coolie, cooly. Add: **2. b.** *S. Afr.* [Afrikaans *koelie* (also used).] An Asian or Indian, esp. one of the lower classes. Also *attrib.*

1920 *Cape Times* 1 Apr. 3/2 Great Public Sale... No coolies. **1959** L. Lerner *Englishman* xiv. 220 It was his girl the other one took, the one who slept with koelies. *Ibid.* xv. 226 You wont, you koelie girl. **1967** *Guardian* 4 Oct. 13/7 In South Africa the word 'coolie' is used by some whites to describe Asians, and is as bitterly resented by them as the word 'Kaffir' is resented by Africans.

4. *coolie lines* [Line *sb.*[2] 20]; **Coolie Christmas,** the name given in Natal to the ceremonies of the Moharram observed by Indian immigrants; **coolie hat,** one resembling those worn by coolies in Asia.

1902 *Graaff Reinet Advertiser* 2 May (Pettman), The Coolie Christmas celebration at Umgeni (Natal) last Monday ended in a serious riot. **1905** *East London Dispatch* 18 Mar. (Pettman), The festivities in connexion with the Hindoo's festival Mohurrum known locally as the Coolie Christmas, commences to-night. **1937** J. Laver *Taste & Fashion* x. 158 In 1935 there was a vogue for flat hats with rather wide brims, Chinese 'coolie' hats, and similar shapes. **1955** *N.Y. Times* 15 July 24/3 Gustavo shows wide, cone-like coolie hats. **1931** *Economist* 23 May 1116/1 These [*sc.* fixed charges] included quit rent, weeding, maintenance of hospital.., coolie lines and machinery, and the upkeep of the estate. **1939** R. Godden *Black Narcissus* xxix. 263 The news spread to the village and the coolie lines.

cooliman, var. *coolamon.

cooling, *vbl. sb.* Add: **1. a.** Also with *-off.*

1911 *Chambers's Jrnl.* 1 July 494/1 Water which..affords them a ready means of 'cooling-off'. **1947** *Science News* IV. 51 The measurement of the steady cooling-off of the body. **1949** *Manch. Guardian Weekly* 10 Feb. 14/2 The proposed bill..merely asks for a thirty-day 'cooling off' period. **1970** *Physics Bull.* Mar. 99/2 The points put forward for discussion..included..a 'cooling off' period between school and university to allow school leavers more time to consider their plans.

2. cooling-board *U.S. local*, the board on which a dead body is laid out; **cooling-pond,** an open pond in which water is cooled after use in a steam-engine or an industrial process; **cooling tower,** a structure in which water heated for some industrial purpose is cooled for re-use.

1853 W. G. Simms *Sword & Distaff* 224 He wouldn't care ef I was on my cooling-board to-morrow. **1896** J. C. Harris *Sister Jane* 19, I don't want to be put on my cooling-board. **1909** *Cent. Dict.* Suppl., Cooling-pond. **1955** J. H. Wellington *S. Africa* II. II. xii. 182 As an illustration of the difficulty involved in using fuel resources in the absence of an adequate 'cooling pond', the use of the Witbank power station may be cited. **1962** *Listener* 13 Dec. 1014/1 Some blue-green algae grow in the radio-active cooling-ponds of nuclear power stations. **1901** W. W. F. Pullen *Steam Engin.* vi. 278 It is customary..to resort to some device by which the cooling water is itself cooled and used over again many times. A cooling tower is then used...It is simply an iron shell filled with porous terra cotta drain pipes stacked on end. **1933** *Archit. Rev.* LXXIV. 167 Erecting concrete cooling towers in a modern super-power station. **1955** *Oxf. Jun. Encycl.* VIII. 329/2 Where no such river is available, the hot water from the condensers is cooled in large concrete cooling towers, by being trickled over a series of large trays or 'baffles'.

coolness. Add: **5.** *Jazz.* A 'cool' quality or style (see *cool *a.* 4 d). orig. *U.S.*

1950 *Flair* May 28 Coolness..has moved into the special language of the jazz world. **1955** R. Horricks in A. J. McCarthy *Jazzbook 1955* 30 At times the dilution of pure emotion into an approach of coolness has been overdone. **1957** M. Butcher in S. Traill *Concerning Jazz* 102 Then Stan changed his tactics a little, combining his coolness with elements of what he called 'stomping tenor'.

coolth. Restrict '*rare*, exc. *dial.*' to sense 2 and add later examples of sense 1.

1890 Kipling *Plain Tales* (ed. 3) 137 He kept on steadily and tried to think how pleasant the coolth was. **1926** J. R. R. Tolkien in *Year's Work Eng. Stud. 1924* 30 The current *coolth*, which shows signs of losing its facetiousness, and may claim part of the territory of *cool*. **1955** E. Pound *Classic Anthol.* II. 120 June's mid-summer, August brings coolth again. **1965** E. O'Brien *August is Wicked Month* ii. 21 She felt the coolth of her thighs and thought it nice to feel her own coolness.

coolung (kū·lʊŋ). *India.* Also 7 **colum,** 9 **coolan, -en, cullum, kullum, kullung.** [Hind. *kulang*, a. Pers. *kulank*.] The great grey crane, *Grus grus lilfordi*; also, the demoiselle crane, *Anthropoides virgo*.

1698 Fryer *Acc. E. India & P.* 111 A *Colum* may be hunted with a Greyhound, as we do Bustards, being a great Fowl and long in Rising. **1813** J. Forbes *Oriental Mem.* II. 29 Two stately birds, called the sahras and cullum. **1838** R. M. Martin *Eastern India* III. 579 Large flocks of a crane called Kolong..frequent this district in winter. **1883** E. H. Aitken *Tribes on my Frontier* 160 Not being so green as I was, I let the tempting herd of antelopes pass, but the *kullum* I cannot resist. **1901** 'Linesman' *Words of Eyewitness* (1902) 303 A trio of coolan flapped over his roof. **1915** F. Finn *Indian Sporting Birds* 125 The native name *Kullung* is generally used also by Europeans.

coomb[2]. Add: **2. coombe rock** *Geol.*, a deposit of flints in chalk, found esp. in the coombes of Sussex; = Head *sb.* 11 c.

1822 G. Mantell *Fossils of S. Downs* xix. 277 *Calcareous bed*, formed of the ruin of the chalk strata, with an intermixture of clay; it is provincially termed *Coombe rock*. **1876** H. B. Woodward *Geol. Eng. & Wales* x. 321 The Elephant Bed, provincially termed Coombe rock, is chiefly made up of chalk rubble. **1903** A. Geikie *Textbk. Geol.* (ed. 4) II. vi. v. 1329 Various deposits..indicate that..this southern fringe of England had its own glacial conditions. Among these is the 'Coombe-rock' of Sussex—a mass of unstratified rubbish. **1936** *Proc. Prehist. Soc.* II. 53 These solifluxion products are most frequently preserved in steep-sided lateral valleys, and, particularly when formed partly or wholly of chalk fragments, go by the general name of 'coombe rock'.

Coombs test (kūmz). [f. the name of R. R. A. *Coombs* (b. 1921), pathologist.] An agglutination test devised by R. R. A. Coombs in 1945 and used in the diagnosis of blood diseases to detect antibodies.

1950 E. Hackett in *Nature* 13 May 764/2 The Coombs test makes use of an anti-(human-serum-protein) to agglutinate washed sensitized cells. **1958** R. Muir *Textbk. Path.* (ed. 7) v. 207 Recently sera containing precipitins against human serum globulins have been applied to the detection of certain antibodies absorbed on red cells, the treated cells being..agglutinated. This test, commonly known as the Coombs test, is widely used in the determination of sensitisation against the Rh and other blood-group factors.

coon, *sb.* Add: **1.** (Earlier example.)

1742 J. Hempstead *Diary* (1901) 388 Josh..killed another Coon to Day.

2. b. (Earlier examples.)

1832 *Polit. Examiner* (Shelbyville, Ky.) 8 Dec. 4/1, I was always reckoned a pretty slick koon for a trade. **1843** Simms *Guy Rivers* 155 To be robbed of our findings by a parcel of blasted 'coons. **1839** [see sense 1 in Dict.].

c. A Negro. *slang.* (Derog.)

1862 *Songs for the Times* 3 Play up, Pomp, you yaller coon. **1892** *Congress. Rec.* 4 Feb. 856/1 Instead of seating one colored Representative, they seated two,—two coons in place of the elected Representatives of the people. **1903** *Westm. Gaz.* 18 May 3/2 The former represented a lively.. jovial coon—possibly 'coon' is not the right word, which, however, is accepted here as modern slang for a nigger. **1948** *Chicago Defender* 23 Oct. 7/2 A lot of us are referred to as 'nigger', 'coon', 'darky', etc., right to our faces. **1969** *Oz* Apr. 46/3 You might..deplore the way that the publicity was angled—poor old coon, he'll thank us in the end.

d. *S. Afr.* A Coloured reveller at Cape Town; esp. a member of various groups which parade in carnival fashion through the streets.

1924 *Cape Argus* 3 Jan. 8 The quiet streets of Cape Town were enlivened by the marching of troupes of coloured youths, gay in coon costumes. *Ibid.*, Obviously this is the sort of thing the crowd loves, for a storm of applause goes up as the coons approach. **1937** *S. Afr. Dancing Times* Feb. 8 Capetown's Annual Coon Contests. *Ibid.*, An effective group which participated in the Capetown Coloured Coons Carnival, held at the Peninsular on New Year's Day. **1947** L. G. Green *Tavern of Seas* i. 14 The coons of Cape Town, the carnival troupes that appear unfailingly with each New Year.

3. (Earlier examples.)

1839 Marryat *Diary Amer.* 1st Ser. II. 232 'I'm a *gone 'coon*' implies 'I am distressed—*or* ruined—*or* lost'. **1840** C. F. Hoffman *Greyslaer* III. vi. 221, I was afeared you were a gone coon, and was on the point of shoving off without you. **1844** W. T. Thompson *Major Jones' Courtship* (ed. 2) 145 The way she's mad at cousin Pete won't wear off in a coon's age.

4. *coon-hunt*; **coon-dog,** a dog good at hunting the racoon; so *coon hound*; **coon-oyster** (examples); **coon-skin** orig. *U.S.* (earlier and later examples).

1833 J. Hall *Harp's Head* 230 An old 'coon dog, has a face covered with scars. **1855** Mayne Reid *Hunters' Feast* xiii. 97 Uncle Abe's dog—a stout terrier—was esteemed the 'smartest 'coon-dog' in a circle of twenty miles. **1967** *Boston Sunday Herald Mag.* 26 Mar. 23/1 He got the idea of cashing in on the coons in our neighborhood by making a really fine coon dog out of Old George. *Ibid.*, Old George is a coon hound. **1920** *Outing* (U.S.) Apr. 59/3 For Sale—a few as good Coon Hounds and mixed hunters as live. **1970** *Field & Stream* July 110 (Advt.), Big game & coonhound pups. **1835** Audubon *Ornith. Biogr.* III. 235, I will take you to a 'Coon Hunt'. **1888** 'C. E. Craddock' *Keedon Bluffs* vi. 98 All the boys of Tanglefoot Cove and the mountain slopes had gathered for a coon-hunt. **1968** *Globe & Mail* (Toronto) 17 Feb. 44 An actual coon hunt or two if there is any game in the area. **1970** *Daily Hampshire Gaz.* (Northampton, Mass.) 6 Oct. 4/4 The first coon hunt of the season by the Rod and Gun Club drew a good crowd on Friday night. **1870** *Amer. Naturalist* III. 460 The small oysters..are not generally eaten except by the racoons, hence the common name for them of coon oysters. **1818** A. Royall *Lett. from Alabama* (1830) 103 He..axed Merchant if he didn't want to trade for some coonskins. **1836** W. G. Simms *Mellichampe* i, He gathered up his rifle, drew the 'coon-skin cap over his eyes. **1908** Kipling *Lett. of Travel* (1920) 132 They had already been wearing wolf and coon skin coats. **1948** *Manch. Guardian Weekly* 1 Jan. 9 A man in a coonskin cap appeared. **1963** R. Snedigar *Our Small Native Animals* (ed. 2) 34 Such wearers of coonskin caps as Daniel Boone, Crockett, Audubon.

b. *attrib.* or *adj.* in sense *2 c = of, pertaining to, or supposedly characteristic of a Negro or Negroes; spec. **coon-shout** (see quot. 1946); so *coon-shouter, -shouting*; **coon song,** a Negro song; a popular song resembling the songs of the Negroes. orig. *U.S.*

Like sense 2 c above these depreciatory uses are tending to fall into disuse, or into restricted use, because of their offensiveness.

1837 R. M. Bird *Nick of Woods* I. 223 You half-niggurs! you 'coon whelps! you snakes! **1877** in H. Asbury *Underworld of Chicago* (1941) 80 Prospect of a Prize Fight Between Two Noted Coon Sluggers. **1887** *Lantern* (New Orleans) 8 Jan. 6/2 Tom McIntosh caused quite a ripple..by his coon songs and dancing. **1897** G. B. Shaw *Theatres in Nineties* (1932) III. 197 Her digressive way of enlivening the tedium of the comedy by an occasional coon song struck me as happy. **1905** E. M. Forster *Where Angels fear to Tread* vii. 216 A coon song lay open on the piano. **1906** H. Green *Actors' Boarding House* 26 'It goes big,' remarked the Coon Shouter, enviously. **1917** E. Wallace *Kate, plus Ten* (1919) v. 93 The noisy coon band kept up its rhythmetic pandemonium in one corner of the room. **1926** Whiteman & McBride *Jazz* xi. 228 The exchange of experience between the classicist and 'coon-shouter'. **1946** Mezzrow & Wolfe *Really the Blues* 372/2 *Coonshout*, corny imitation of oldtime Negro style of singing. *Ibid.* ix. 146 Blackface routines and corny coonshouting and mammy numbers. **1958** C. Wilford in P. Gammond *Decca Bk. Jazz* ii. 31 The immediate fore-runners of ragtime as popular music were the coon song, made popular by the Christy Minstrels and many similar travelling black-face shows. **1959** 'F. Newton' *Jazz Scene* ix. 156 The impression of 'coon English', which coloured Americans dislike.

coon, *v.* Add: (Earlier examples.) Also *trans.*

1834 A. Pike *Sk.* 77 (Th.), Irwin was obliged to straddle the log, and, as they quaintly call it in the west, 'coon it across'. **1835** W. G. Simms *Partisan* 320 That curious sort of locomotion which, in the South and West, is happily styled 'cooning the log'. **1926** J. Lord *Frontier Dust* 190, I would show her how a Yankee could coon a pole.

coon-can (kū·nıkæ·n). Also **cooncan; conquian** (kǫ·ŋkiăn). [ad. Sp. *con quién* with whom?] A game of cards, originating in Mexico, the main object of which is to secure sequences. Each player receives (usually) ten cards, and the aim is to form sequences of the same suit or sets of three or four cards of the same denomination. One card is placed face upwards and the rest of the cards face downwards, each player in turn taking either the exposed card or the top card of the pack, afterwards discarding any card in his hand. The player gets rid of his cards by declaring them when he has made a complete set or sets, and the winner is the one who first gets rid of all his cards.

1889 *Century Mag.* Apr. 905/1 The men got out a pack of Mexican cards and gambled at a game called 'Coon-can'. **1905** *Hoyle's Games* 203 The art in Coon Can, like in Cribbage, lies in the discard and in putting the cards, whether threes, fours, or sequences, at the proper time on the table. **1907** *Ibid.* 192 Conquian. 'With whom?' Or Coon Can. **1913** Kipling *Diversity of Creatures* (1917) 282 [They] played the rag-time tunes..till there was talk of coon-can. **1913** W. Dalton *Coon-Can* Introd. 13 We have no record as to when, or by whom, Coon-Can was first introduced into England. It made its first bow to the public at the Bath Club in Dover Street. **1925** W. S. Maugham *Painted Veil* xviii. 60 As a rule in the evening, when they were alone, they played coon-can or patience. **1969** *Courier-Mail* (Brisbane) 17 Jan. 12/11 She was playing coon-can with her mother and two sons.

coonjine (kū·ndʒəin), *sb.* and *v. U.S. slang.* [Origin unknown.] (See quots.)

1896 *The Bully* (A song) (D.A.), I coonjined in the front door, the coons were dancing high. **1933** *Amer. Speech* VIII. iv. 77/2 Coonjining is the peculiar gait or shuffle that darkies of more than fifty years ago sometimes indulged in, especially along the shores of the Mississippi. **1941** *Chr. Sci. Mon.* (Wkly. Mag. Sec.) 3 May 11 Coonjine, a word used originally on the river for the waddling run of Negro rousters with freight across the stage or gangplank...There's almost as much side as forward motion to it. From the gait the word got to mean the songs and jingles to the singing of which the Negro rousters jerked themselves along. **1944** W. Blair *Tall Tale America* 211 At work he'd joke and sing his songs, —coonjine, and he was a good worker. **1948** *Sat. Rev.* 26 June 15/1 The Negro roustabouts who coonjined freight up and down the stage planks of our river steamers. **1958** Blesh & Janis *They all played Ragtime* ii. 38 The shouts of roustabouts coonjining to the plink-plank of banjos.

coontah, coontie. (Earlier examples.)

1791 W. Bartram *Trav. Carolina* (1792) 239 A very agreeable, cooling sort of jelly, which they call conte. a **1816** B. Hawkins *Sk. Creek Country* (1848) 21 This [China] briar is called *Coonte*. **1819** *Pennsylvania Gaz.* 22 June 2/3 [The Seminoles] use a root called coonty, as a substitute [for corn].

coop, *sb.*[1] **6. a.** (Earlier and later examples.)

1785 *Sessions Papers* Sept. 1111/2 He had been *in coop* for a week. **1963** Wodehouse *Stiff Upper Lip, Jeeves* xxiv. 180, I accompanied Constable Oates to the village coop.

co-op (ko·ǫp), *a.* and *sb.* Colloq. abbrev. of Co-operative *a.* (*sb.*); *spec.* a co-operative store.

1861 *Chambers's Jrnl.* 29 June 410/2 The joyousness of this party of Co-ops—for that is the name they go by, on the principle, we suppose, that Co-operators is much too long a word for popular use. It was really to be a great festivity, this soiree of Co-ops. **1872** T. Cooper *Life* xxxv. 393 You will hear well-dressed working men talking..of 'Co-ops' (Co-operative Stores), and their shares in them. **1872** *Gentl. Mag.* Apr. 458 One whose place of business is situate immediately opposite to a branch of the 'Co-op.' Institution. **1873** J. H. Beadle *Undevel. West* xviii. 339 The co-op store bought a thousand muskets at the Government sale. **1894** *Warty Rhymes* 29 Neaw th' 'Co-ops' are th' bigg'st shops i' th' place. **1916** Lytton Strachey *Let.* 17 July (1956) 60 Your appearance at a Co-op. congress. **1947** J. Bertram *Shadow of War* ii. 12 She had begun working for the [Chinese] Co-ops. after meeting Rewi Alley in Kiangsi. **1958** *Times* 1 July 12/6 She returned..in the back of the Co-Op van. **1970** J. Nuese *Country Garden* 52 You can get it most cheaply at your nearest Farmer's Exchange or whatever agricultural co-op functions in your area.

cooper, *sb.*[1] Add: **5.** *cooper-shop* (delete † and add later U.S. examples), *-stuff.*

1827 Drake & Mansfield *Cincinnati* viii. 65 Eleven cooper shops 48 men. **1894** H. Frederic *Copperhead* 55 He had been a well-to-do man..with a big cooper-shop. **1659** in *Rowley* (Mass.) *Early Rec.* (1894) 104 The Decay of usefull timber for Cowper stuff. **1801** in C. Cist *Cincinnati* (1841) 183 Charles Faran advertises for cooper-stuff.

co-operative, *a.* (*sb.*). Add: **B. 2. b.** Short for *co-operative society*; also, an association of agriculturists; an association providing services of various kinds to its members.

1921 *Proc. 3rd Nat. Country Life Conf. 1920* (U.S.) 53 Farmers' co-operatives..are more analogous..to the workmen producers' co-operatives than they are to the consumers' co-operatives. **1936** *Nature* 23 May 844/2 The organisation of consumers' co-operatives through which the goods produced get to the public. **1943** J. S. Huxley *TVA* ii. 12 The encouragement of co-operatives and rural industries. **1948** *Oxf. Jun. Encycl.* I. 112/1 Another organization which is doing valuable work in Chinese rural life is the Chinese Industrial Co-operatives...Their job is to revive the old village handicrafts..and to set up rural industries. **1958** *Listener* 6 Nov. 717/1 So-called People's Communes [in China]..take the place of the former agricultural co-operatives. *Ibid.* 13 Nov. 774/1 Most of the flats [in Brasilia].. are being built by housing co-operatives. **1966** *Economist* 23 Apr. 365/1 (*Correspondent in Colorado*) There are now nearly a thousand rural electrical co-operatives and 225 telephone co-operatives.

co-operativeness. (Later examples.)

1857 F. L. Olmsted in T. H. Gladstone *Englishman in Kansas* p. xxxiv, That peculiar political cooperativeness and efficiency which we see in the people of the South. **1952** C. P. Blacker *Eugenics* 90 The social instincts and co-operativeness of our forebears.

cooperite. *Min.* [f. the name of R. A. *Cooper*, mineralogist: see -ite[1].] A steel-grey sulphide of platinum, PtS, occurring in irregular grains in conjunction with other minerals in igneous rocks of the Bushveld, South Africa.

1928 *Jrnl. Chem. Met. & Min. Soc. S. Afr.* XXVIII. 283 Mr. Wartenweiler..expressed the Society's appreciation that Mr. Cooper had been enabled to give them particulars of this new platinum mineral...As he [*sc.* Cooper] had been the discoverer of the new mineral, it might be named, for instance, Cooperite. **1932** *Discovery* Dec. 385/1 The residues insoluble in *aqua regia* were named cooperite in recognition of Cooper's pioneer work on the minerals of the Transvaal deposits. **1944** C. Palache et al. *Dana's Syst. Min.* (ed. 7) I. 258 *Cooperite*...In the concentrates from the platiniferous norites of the Bushveld complex. **1965** N. de Kun *Min. Resources Afr.* 427 In the upper 1–2 inches [of the Merensky Reef] the platinoids occur mainly as sulphides and sulpho-arsenides, sperrylite and cooperite.

co-optee (koǫ·ptī). [f. Co-opt *v.* + -ee[1].] One who is or has been co-opted.

1956 *Daily Tel.* 21 Dec., When co-opting to education committees persons representing industry, one such *co-optee* should be a trade unionist. **1965** *New Statesman* 19 Nov. 791/3 Every Russian working in foreign embassies in Moscow is KGB or their co-optee.

co-optive (koǫ·ptiv), *a.* [f. Co-opt *v.* + -ive.] = Co-optative *a.*

1894 *Tablet* in *Westm. Gaz.* 3 Dec. 2/2 Whenever a member [*sc.* of the School Board] retires or is removed, his place will be filled up upon the co-optive principle administered by the majority. **1922** G. M. Trevelyan *Brit. Hist.* i. 13 The House of Commons had, in effect, become a co-optive body. **1944** W. Temple *Church looks Forward* xxii. 159 Boards of Directors, though nominally elected by the owners, are in practice very largely co-optive.

co-ordinate, *a.* and *sb.* Add: **A.** *adj.* **4.** *Chem.* [Back-formation from *co-ordination 5.] Designating a type of covalent bond in which one of the atoms, ions, or molecules forming the bond is regarded as providing both the shared electrons.

1927 N. V. Sidgwick *Electronic Theory of Valency* iv. 60 This new type of covalency..may be called a co-ordinate link, since it affords..an explanation of the co-ordination compounds of Werner. **1938** J. R. Johnson in H. Gilman *Org. Chem.* II. xix. 1602 The distinction between normal and coördinate covalent bonds vanishes once the bond is established. **1947** S. Glasstone *Elem. Physical Chem.* (1958) iii. 68 The compounds of ammonia with metallic salts..involve coordinate bond formation between the central metal ion and..ammonia. **1965** Phillips & Williams *Inorg. Chem.* I. iv. 127 Except in the most polar complexes the maximum coordinate covalence [of a first-row element] is found to be four.

B. *sb.* **3.** *pl.* A set of women's clothes matched as to colour or fabric or other features.

1959 *Vogue* Mar. (Advt.), 120 Knitwear and Tweed Co-ordinates..by Munrospun. **1962** *Punch* 14 Mar. p. xiii. Harrods have French beachwear, cashmere co-ordinates, **1969** *Sears Roebuck Catal.* Spring 143/2 (*heading*) Coordinates in cool cotton fabrics.

co-ordinate, *v.* Add: **4.** Used in *Chem.* with various constructions: **a.** With indirect obj.: to be or become linked *with* or *to* (an atom or group of atoms) by a co-ordinate bond. **b.** *trans.* To form a co-ordinate bond a or co-ordination compound with (an atom, etc.). **c.** *intr.* To form a co-ordinate bond or a co-ordination compound. So **co-ordinated** *ppl.a.* (*b*) containing, formed by, or bound by one or more co-ordinate bonds.

1923 *Chem. & Industry Rev.* 29 Mar. 319/2 It is probably legitimate to think of the univalent metal as 'co-ordinated' ..with two negatively charged oxygen atoms. **1923** *Jrnl. Chem. Soc.* CXXIII. 729 For every co-ordinated compound ..we can arrive at a number expressing the effective valency of the central atom. **1938** C. W. Stillwell *Crystal Chem.* i. 24 The sulfur atom is larger than oxygen, and this probably explains why phosphorus cannot co-ordinate four sulfurs. **1950** A. F. Wells *Structural Inorg. Chem.* (ed. 2) xxii. 641 Certain atoms and groups do not coordinate with any of these metals. **1956** J. C. Bailar *Chem. Coordination Compounds* i. 4 The halide ions often coordinate strongly. **1957** G. E. Hutchinson *Treat. Limnol.* I. iii. 200 This implies a breaking down of the ice lattice so that many oxygens become less than four co-ordinated. **1966** Cotton & Wilkinson *Adv. Inorg. Chem.* (ed. 2) v. 127 The geometrical arrangement of the co-ordinated groups around the cation. **1968** D. P. Graddon *Introd. Co-ordination Chem.* (ed. 2) 14 A group such as ethylenediamine which can co-ordinate twice to the same metal atom.

co-ordination. Add: **5.** *Chem.* The formation or existence of a co-ordinate bond; linking by means of a co-ordinate bond; freq. *attrib.*, esp. in *co-ordination number* [tr. G. *koordinationszahl* (A. Werner 1893, in *Zeitschrift für anorg. Chemie* III. 326)], (*a*) in a co-ordination compound, the number of atoms or groups linked to the central atom or group

by co-ordinate or covalent bonds; (*b*) in a crystal, the number of ions immediately surrounding any given molecule or ion.

1902 *Jrnl. Chem. Soc.* LXXXII. II. 555 The capacity for combination of an atom is determined by three factors, principal valency, supplementary valency, and the co-ordination constant. **1908** *Ibid.* XCIII. II. 1571 Whatever co-ordination number the central atom may have. **1938** C. W. Stillwell *Crystal Chem.* i. 13 For AX_2 compounds there will be two coordination numbers, for there are twice as many ions (or atoms) X as there are ions (or atoms) A. **1939** *Thorpe's Dict. Appl. Chem.* (ed. 4) III. 330/1 The prevalence of co-ordination numbers 4 and 6 suggested that co-ordination compounds had a spatial arrangement.. corresponding respectively with the tetrahedron and octahedron. **1950** *Chem. Rev.* XLVI. 389 These cyclic tertiary bases have a spatial configuration which is far more favorable to coordination with a metal ion. **1953** C. K. Ingold *Struct. & Mech. Org. Chem.* v. 205 The covalent bond can be formed by co-ordination. **1955** G. L. Clark *Appl. X-Rays* (ed. 4) xvii. 522 Coordination of NH_3 molecules around the cation as in $Co(NH_3)_6Cl_3$. **1962** P. J. & B. Durrant *Introd. Adv. Inorg. Chem.* xii. 345 In the hexagonal and face-centred cubical close-packed structures the co-ordination number is 12. **1966** *McGraw-Hill Encycl. Sci. & Technol.* III. 457/2 The products formed are coordination ions or compounds, and this area of chemistry is known as coordination chemistry.

co-ordinator. Add: **b.** spec. *Gram.* A co-ordinating conjunction.

1933 Bloomfield *Language* xii. 195 The phrase *boys and girls* belongs to the same form-class as the constituents, *boys, girls*; these constituents are the members of the co-ordination, and the other constituent is the co-ordinator. **1964** *English Studies* XLV. 298 The 'yet' in the first line.. is, at first, somewhat perplexing. Obviously a coordinator.., the word introduces new 'secondary' material.

coorongite (kūᵊ·rǫŋgəit). *Min.* [f. *Coorong*, a district of South Australia: see -ite[1].] A mineral caoutchouc found in the district of Coorong.

1872 *Pharm. Jrnl.* 23 Mar. 763/1 (*title*) Coorongite, or mineral caoutchouc of South Australia. **1872** *Jrnl. Bot.* X. 103 The Coorongite, as it has been named, is confined to a depressed portion of the district. **1895** Groves & Thorp *Chem. Technol.* II. 129 The solid form of petroleum is represented by ozokerite, by asphaltum,..and, perhaps, by the 'coorongite' of South Australia. **1969** *Geochim. et Cosmochim. Acta* XXXIII. 569 The hydrocarbon content of coorongite, a Recent rubbery deposit derived from the alga *Botryococcus braunii*, has been investigated.

coot, *sb.*[1] Add: **2.** *bald coot* (examples).

1709 J. Lawson *Voy. Carolina* 149 Black Flusterers... Some call these the great bald Coot. **1789** Morse *Amer. Geogr.* 59 Upwards of one hundred and thirty American birds have been enumerated..[including] the bald coot. **1898** Morris *Austral Eng.* 16/1 *Bald-Coot*, a bird-name, *Porphyrio melanotus*, Temm.; Blue, *P. bellus*, Gould. **1928** D. Cottrell *Singing Gold* I. ii. 16, I might see..a red-legged blue baldcoot, glittering like metal.

4. (Earlier and later examples.)

1766 Sewel & Buys *Compl. Dict. Eng. & Dutch* 138/2 A very coot, (or fool). **1794** *Gazette of U.S.* (Philad.) 17 Jan. (Th.), But Satan was not such a coot To sell Judea for a goat. a **1852** Mrs. Whitcher *Widow Bedott P.* (1883) ix. 33 He's an amazin' ignorant old coot. **1929** W. Smyth *Girl from Mason Creek* i. 17 You're a clumsy coot. **1963** *Daily Mail* 26 Aug. 4/2 Masters call boys 'coots' and boys call each other 'nits'.

cootamundra (kŭtămʊ·ndră). [Name of a town in New South Wales.] An Australian tree, *Acacia baileyana*.

[**1887** F. von Mueller in *Trans. Proc. R. Soc. Victoria* XXIV. 170 He found this rare species only near Cootamundra.] **1909** A. E. Mack *Bush Calendar* 1 The golden-hued Cootamundra and the Queensland. **1911** J. H. Maiden *Forest Flora N.S.W.* IV. 9 *Acacia Baileyana*... Vernacular Name.—'Cootamundra Wattle', because of its best-known locality. **1965** *Austral. Encycl.* IX. 220/1 Probably the most decorative and most frequently cultivated species in this large genus [sc. *Acacia*] is the Cootamundra wattle. **1965** G. McInnes *Road to Gundagai* iv. 64 We would see..puffballs..break on our own..cootamundra.

cooter. (Earlier examples.)

1827 *Massachusetts Spy* 22 Aug. (Th.), A few jolly topers, who wallowed in the sand, 'as drunk as a cooter'. **1835** W. G. Simms *Partisan* 308 You're turned now..on the flat of your back like a yellow-belly cooter.

cootie (kū·ti), *sb.*[2] *slang.* Also **kootie.** [? f. Malay *kutu* parasitic biting insect.] A body louse.

1917 Empey *From Fire Step* 24 'Does the straw bother you, mate? It's worked through my uniform and I can't sleep.' In a sleepy voice he answered, 'That ain't straw, them's cooties.' **1918** in F. A. Pottle *Stretchers* (1930) 199, I could soon fall asleep thinking how absurd to worry over lice and cooties when a man was at war. **1918** E. M. Roberts *Flying Fighter* 106, I made the acquaintance of a new sport while with the battery. A saucer serves for an arena. Into this one puts a kootie and a flea. **1934** J. A. & A. Lomax *Amer. Ballads* (1960) 559 The cootie is the national bug of France. **1962** R. Buckminster Fuller *Epic Poem on Industrialization* 41 The Publicitor's cheap brand of lacquer Only stuck to some cooties and fleas.

cop, *sb.*[2] Add: **8.** *cop-changing* adj.; **cop reel,** a machine which receives the yarn from the cops and winds it into hanks; so **cop reeler; cop warp,** warp-yarn spun on to cops; **cop winder,** one who winds yarn into the form of cops or winds yarn from the cops on to bobbins.

1927 T. Woodhouse *Artificial Silk* 131 One type of.. cop-changing automatic loom. **1890** Nasmith *Mod. Cotton Spinning Mach.* xiii. 262 It depends upon whether it is employed to wind the yarn from cops or bobbins whether it is known as a 'cop' or 'bobbin' reel...Ordinarily one swift only is used in a cop reel, and two in a bobbin reel. **1881** *Instr. Census Clerks* (1885) 68 Cop Reeler...Cop Winder. **1862** *Catal. Internat. Exhib., Brit.* II. No. 3653, Cotton yarns, cop warp, doubled yarns. **1907** *Daily Chron.* 11 Feb. 4/6 The reelers, hank makers, cop-winders, gassers, and doublers. **1921** *Dict. Occup. Terms* (1927) § 367 *Cop Winder*; (i) winds yarn from cops, short cylindrical paper tubes, on to bobbins; (ii) winds weft yarn from spinning frame bobbins, or from hanks, into form of cops.

cop, *sb.*[5] Add: *cops and robbers*: a children's game in which 'police' hunt 'robbers'; also (?nonce-use) *cops and thieves.* Also *transf.* orig. *U.S.*

1903 *Independent* (U.S.) 15 Jan. 146/1 When I was a youngster one of my favorite games was what we children called 'Cops and Thieves'. **1938** 'E. Queen' *Four of Hearts* (1939) xi. 119 They're going down to the *Horseshoe Club* to play cops and robbers with Alessandro. **1954** Koestler *Invis. Writing* xx. 225 The sensation of participating in an exciting cops-and-robbers game. **1957** Manvell & Huntley *Film Music* ii. 35 The villains are rounded up and hustled into a nearby police station, accompanied by a light musical parody in the 'cops and robbers' style of the silent film. **1959** J. O'Donovan *Visited* xxxi. 201 He..was momentarily ashamed of himself, as if he had been caught playing cops and robbers.

cop (kǫp), *sb.*[7] *slang.* [f. Cop *v.*[3]] **1.** Capture; used chiefly in phr. *a fair cop.*

1886 *Sessions' Papers* Aug. 4/4 What do you want to search me for? You have got a good *cop*. **1889** *Standard* 9 Oct. 3/7 (E.D.D.), Prisoner remarked it was 'a fair cop'. **1891** *Daily News* 24 Oct. 5/2 'It's a fair cop,' said the thief. **1898** *Ibid.* 13 Apr. 8/7 It's a fair cop; you have got me all right and no mistake. **1901** G. B. Shaw *Adm. Bashville* III. 329 Policeman. Sir: had I known your quality, this cop I had averted. **1935** 'J. Guthrie' *Little Country* iii. 65 The young man..glared round as if for some means of escape... 'It's a fair cop!' said the young man.

2. Catch, acquisition. With negative, *no cop, not much cop*: of no or little value or use, worthless.

1902 J. Milne *Epist. Atkins* vii. 136 We are going to a place called Spion Kop, and I don't think it will be much of a 'kop' for our chaps. **1919** J. B. Morton *Barber of Putney* xviii, 'Once upon a time..this 'eap o' bricks was a first-class billet.' 'Doesn't look much cop now,' said Tim. **1929** E. Raymond *Family that was* II. ii. 147 A month or two in service, which was no cop—no cop at all. **1932** D. L. Sayers *Have his Carcase* xxiii. 306 The nobility ain't much cop these days, but Lord Peter is vell known. **1960** H. Pinter *Caretaker* I. 15 At least they're comfortable. Not much cop, but I mean they don't hurt. **1970** K. Giles *Death in Church* iii. 56 The house..has never been much cop. People don't like living opposite a church or a graveyard.

b. Without negative, esp. *Austral.* and *N.Z.* = Catch *sb.*[1] 11.

1889 Barrère & Leland *Dict. Slang* I. 271/2 *To cop*,.. (Sporting), to win, to get money; a dead *cop*, a sure method of arriving at this result. **1916** C. J. Dennis *Songs Sentim. Bloke* 120 *Cop*, an avocation, a 'job'. **1941** Baker *Dict. Austral. Slang* 20 *Cop*, a good job obtained by shrewdness or luck: an agreeable proposition: a bit of luck. **1944** A. F. St. Bruno *Desert Daze* 11 Any other bottles that fell into his maw were distinct 'cop'. **1945** Baker *Austral. Lang.* vii. 143 *Cop*, profit, a job or trick from which a large return is gained. *Ibid.* ix. 174 Australian equivalents [of *dead cert*] are *dead bird,..sure cop.*

cop, *v.*[3] Add: **b.** To steal.

1879 *Macmillan's Mag.* Oct. 500/1, I was taken by two pals (companions) to an orchard to cop (steal) some fruit. **1931** D. Runyon *Guys & Dolls* (1932) viii. 163 Madame la Gimp does a little scrubbing business around a swell apartment hotel..and she cops stationery there. **1932** J. T. Farrell *Young Lonigan* (1936) iii. 67 Johnny O'Brien ran home, and copped a piece of beefsteak from his old lady. **1955** D. W. Maurer in *Publ. Amer. Dial. Soc.* XXIV. 74 '*Cop*,' he [*sc.* a pickpocket] says, and the *duke man* has the roll.

c. *to cop it*: to 'catch' it, to be punished, get into trouble; also, to die. So *to cop out, cop a packet*, etc. *dial.* and *slang.*

1884 R. Holland *Gloss. Chester* 79 *Cop*, to catch, both in the sense of capturing, and in the semi-slang sense of being scolded. **1905** *Daily Chron.* 19 Sept. 4/7 Though it tarry, it shall surely come; and when it does—as an errand boy would say—'somebody 'll cop out'! **1902** J. Milne *Epist. Atkins* vii. 131 Often..I have heard fellows say, 'I know I am going to kop out,' and sure enough they have. **1909** *Daily Chron.* 9 Apr. 6/6 When arrested he remarked, 'I suppose I shall "cop" it for this.' **1915** 'Boyd Cable' *Between Lines* 194 By the way 'is 'ands jerked up an' 'is 'ead jerked back when I fired, I fancy 'e copped it right enough. **1927** *Blackw. Mag.* July 83/1 Half of the beggars had copped it for good and all. **1933** D. L. Sayers *Murder must Advertise* vii. 120 I'm really fearfully sorry you copped that packet that was meant for me. **1934** E. Blunden in *Challenge to Death* 337 As the thing began it ended; only, as you see, we boys Have copped unlucky. **1941** A. Christie *N or M?* xii. 159 Yes, it looked bad, it did. Looked as though he might have copped one. **1967** *Listener* 22 June 809/2, I saw an irate African driver jump off his bus and tell the passengers that it had 'copped it'.

d. [cf. Cop *sb.*[6]] *to cop* (a person) *one*: to deal (someone) a blow. *slang.*

1898 J. D. Brayshaw *Slum Silhouettes* 3 The Gent. copped 'im one fair under the chin, an' dahn 'e goes. **1935** Wodehouse *Blandings Castle* vi. 122, I copped 'im on the shin wiv a stone.

e. To escape; to stop (work, etc.); to drop out from society. *slang.*

1942 Berrey & Van den Bark *Amer. Thes. Slang* § 498/3 *Flee*; *escape*, cop,..cop (an) out. **1967** *Jazz Monthly* Dec. 19/2 One might expect him to be playing from memory to some extent, taking it easy and copping out when he can. **1969** *It* 13–28 June 1 After Photoprinters..copped out.. *It*..appeared in a limited edition. **1970** *Win* 1 Jan. 25/3 The intellectuals, so many of whom were kept from copping out by Bob's continuing commitment in the face of death.

So **co·p-out,** a flight, an escape; a cowardly compromise or evasion; a retreat from reality. Also, a person who drops out from society.

1942 Berrey & Van den Bark *Amer. Thes. Slang* § 498/1 *Flight*; *escape*, cop-out. **1967** *Boston Sunday Herald* (Suppl.) 14 May 2/1, I know it's..convenient for those of us who were not raised by their parents to refer to their childhood as lonely, and blame their adult behavior on that. I think it's self-pity and a cop-out. **1967** *Guardian* 24 Oct. 6/2 Many of the London hippies are not only 'drop-outs' (people who have rejected any form of social or political activism) but 'cop-outs'—people who equate mind erasure with the dissolution of social problems. **1969** *Win* 15 May 9/1 Isn't it a cop-out to secede from New York State but remain a part of the nation?

copacetic (kōᵘpăse·tik, -sī·tik), *a. U.S. slang.* Also **copesetic,** etc., **k-.** [Origin unknown.] Fine, excellent, going just right.

1919 I. Bacheller *Man for Ages* iv. 69 'As to looks I'd call him, as ye might say, real copasetic.' Mrs. Lukins expressed this opinion solemnly...Its last word stood for nothing more than an indefinite depth of meaning. **1926** C. Van Vechten *Nigger Heaven* 286 *Kopasetee*, an approbatory epithet somewhat stronger than *all right*. **1933** N. Ersine *Underworld & Prison Slang* 29 *Copissettic*, all right, okay. **1934** Webster, *Copacetic*, capital; snappy; prime. **1934** J. O'Hara *Appt. Samarra* (1935) i. 24 You had to be a good judge of what a man was like, and the English was copacetic. **1937** *Amer. Speech* XII. 243/1 'Everything is *copesetic*'..is synonymous with 'O.K.', and I believe it is used by negroes in the South. **1947** *Down Beat* 18 June 4 (*heading*) Torme not all copa-setic. **1969** *Ibid.* 20 Mar. 18/1 We hear two city cops chatting. 'Well, everything seems copasetic,' says one. 'Yeah, we might as well move on,' the other agrees.

copal. Add: **1.** Also *gum copal.*

1712 J. Browne tr. *Pomet's Hist. Drugs* I. 201 The Gum Copal, which we call Oriental Copal. **1777** Robertson *Hist. Amer.* I. III. 242 The people perfumed them as they landed with incense of gum copal. **1878** H. M. Stanley *Dark Cont.* II. xiii. 387 Another valuable article of commerce besides the bees-wax and india-rubber found here was gum copal. **1887** Moloney *Forestry W. Africa* 134 It is said by the natives that no trees grow on or near the places where the gum copal is found.

copalm. (Earlier examples.)

1775 B. Romans *Florida* 336 Live oak abound here, intermixed with copalm and other timber. **1832** D. J. Browne *Sylva Amer.* 200 This tree is universally called Sweet Gum, and by the French of Louisiana Copalm.

coparcener. (Later examples.)

1900 *Nation* (N.Y.) 22 Nov. 400/3 'Ask Platt,' said Croker ..; and that he and all his coparceners say now.' **1959** Jowitt *Dict. Eng. Law* I. 488/2 Where before 1926 lands descended on an intestacy to several persons as co-heirs.. the persons so inheriting were called coparceners. *Ibid.* 489/2 As from January 1, 1926, coparceners retain all their equitable rights in the proceeds of sale and in the land until sold.

cope, *sb.*[1] Add: **11. cope bead** (see quot.); **cope-box,** a box for holding copes, a cope-chest.

1885 *Spon's Mechanic's Own Bk.* 374 The other drawers.. receive a ½-in. mahogany beading all round. This is called a 'cope bead'. **1893** T. Fowler *Hist. C.C.C.* 246 A Cope-box purchased for the College.

cope, *v.*[2] Add: **4. b.** *absol.* To manage, deal (competently) with, a situation or problem. *colloq.* (Cf. Manage *v.* 11 b.)

1934 E. Bowen *Cat Jumps* 248 Angela rang the bell wildly for someone to come and cope. **1955** *Essays in Criticism* V. 62 More confidence might be placed in the writer did we not find on the same page a typical shift of tone to this, on Romeo and Juliet: 'the kids get involved in a lively way, but then they cannot cope.' **1958** I. Brown *Words in our Time* 41 'She suffers from copelessness.' I have heard this said of an unsatisfactory employee. The use of cope as a verb by itself to describe dealing with all kinds of situations is a recent usage. In my youth we tried, or were told to try, to cope with this or that problem. We were not required to cope in general.

Copec (kōᵘ·pek). Also **C.O.P.E.C.** A word made up from the initials of *C*onference *o*n Christian *P*olitics, *E*conomics, and *C*itizenship.

1924 *Public Opinion* 29 Feb. 201/1 That remarkable movement which is described by the cryptic word Copec. **1924** *Brit. Weekly* 20 Mar. 560/4 Copec has entered into the heritage of the 'Christian Socialists', of Westcott and Dale, of the Christian Social Union and the rest. **1927** *Observer* 6 Nov. 13/2 His devotion to the Workers' Educational Association and his leadership of the 'Copec' movement. **1933** V. A. Demant *God, Man & Society* iii. 83 Such an enquiry was begun under the auspices of C.O.P.E.C. on the initiative of Will Reason. **1942** W. Temple *Chr. & Social Order* i. 9 The Conference on Christian Politics, Economics and Citizenship (Copec) held at Birmingham in 1924.

copen (kōᵘ·pĕn). *N. Amer.* [abbrev. of *COPEN(HAGEN.] In full *copen blue*: a light, strong shade of blue.
1920 *T. Eaton's Catal.* Spring & Summer 1/2 Colors Navy, Copen, Pearl Grey, Reseda Green. 1926 *Dry Goods Economist* 19 June 91 Windsor ties in..polka dot styles..such as white with green dots, white with tan dots, and white with copen dots, are exceptionally good. 1938 *Toronto Daily Star* 30 Dec. 16/2 In soft, rich shades of turquoise, dusty pink and copen blue. 1963 *New Yorker* 8 June 20 (Advt.), Wool carpets only come in these colors..*copen blue.*

Copenhagen (kōᵘpĕnhē̆ⁱ·gĕn). The name of the capital city of Denmark, used *attrib.* of porcelain manufactured there since the mid-eighteenth century. Also *ellipt.*
[1902 *Encycl. Brit.* XXXI. 875/1 Though the delicate underglaze painting on hard-fired porcelain of Copenhagen and Rörstrand owes its inception to this foreign [Japanese] influence, it proceeds on its own lines.] 1911 A. Hayden (*title*) Royal Copenhagen Porcelain. *Ibid.* i. 37 The early Copenhagen porcelain of the Fournier period is soft paste. *Ibid.* ix. 359 To take a Copenhagen vase with its seascape and dancing spray and pack of scudding storm-clouds. 1926 A. Huxley *Two or Three Graces* 73 Phyllis has just smashed that hideous Copenhagen rabbit Aunt Eleanor gave me. 1962 H. Thurston *Where is thy Sting?* iv. 52 The second best china..was of heartier dimensions than the Royal Copenhagen she kept for special occasions. 1964 L. Deighton *Funeral in Berlin* xvii. 105 A girl..produced a Copenhagen teapot.

copepodid (kǭpe·pǫdid), *a.* [f. mod.L. *Copepoda*: see Copepod and -id³.] Designating, or characteristic of, a free-swimming immature stage in the development of some copepods.
1912 *Records Indian Museum* VII. 329 In the summer season, the Copepodid stage can become at once sexually mature. 1932 J. S. Huxley *Probl. Relative Growth* iv. 133 During the later free-living copepodid stages of development..the number of abdominal segments is increasing. 1963 R. D. Barnes *Invertebrate Zool.* xv. 407/1 The adult structure is attained typically after six naupliar and six copepodid instars.

co:-pe·rsonal, *a. Philos.* [Co- 2.] Belonging to the same person. Hence **co:-persona·lity, co:-pe·rsonalness.**
1936 *Mind* XLV. 34 Now it seems to me clear that only terms with mental qualities could be co-personal. *Ibid.*, We must..assume, in addition to these, some further relation between 'my' experiences, which might be called 'co-personality'. 1941 *Mind* L. 342 Few remotely past experiences could be co-personal with present experiences. 1949 *Mind* LVIII. 42 The common owner which these relations of co-personalness were intended to replace.

coperta (kopɔ̄·ɪtă). [It., = covering, f. *coprire* to cover.] A film of glaze given as a final coating to later Italian majolica ware.
1885 *Encycl. Brit.* XIX. 625/1 The glaze, 'coperta', an ordinary glass, made more fusible by the presence of lead, consisted of oxide of lead 17 parts, silica (sand) 20, alkali 12, and common salt 8 parts. 1903 Solon *Hist. O. Fr. Faience* 186. 1937 *Burlington Mag.* Dec. 277/1 A transparent lead glaze—the *coperta* of the Italian writers.

copesetic, var. *copacetic *a.*

co·-pi:lot. [Co- 3 b.] A second pilot of an aeroplane.
1927 *Glasgow Herald* 4 June 9 It is understood that Clarence Chamberlain..will be pilot. It has not yet been announced who will accompany Mr. Chamberlain as co-pilot. 1940 *Jrnl. R. Aeronaut. Soc.* XLIV. App. 71 They cost the lives of two pilots, two co-pilots, three other crew members, and 19 passengers. 1965 *Gloss. Aeronaut. Terms (B.S.I.)* § 16 *Co-pilot* (second pilot), a pilot, responsible for assisting the first pilot to fly the aircraft.

‖ **copita** (kopi·tă). [Sp., dim. of *copa* Cup *sb.*] A tulip-shaped glass traditionally used in Spain for drinking sherry; a glass of sherry.
1841 G. Borrow *Zincali* ii. iv. 288 Bring the bottle of anise; the señor and the señora must drink a copita. 1933 Hemingway *Winner take Nothing* (1934) 32 'You want another copita?' the barman asked. 1961 *Guardian* 21 Nov. 17/1 The Copita is the traditional Sherry glass of Spain, with a specially designed tulip shape.

co:-po·lymer. *Chem.* [Co- 3 b.] A composite polymer formed by co-polymerization.
1940 S. R. Douglas in *Ind. & Engin. Chem.* XXXII. 316/2 Polymerization of these copolymers results in greater strength and toughness. 1946 *Electronic Engin.* XVIII. 281 The Styraloys are already appearing and these consist of copolymers of styrene and butadiene. 1958 *Times Rev. Ind.* June 60/3 Polyvinyl and its copolymers provided the main advance throughout the industry.

co:-polymeriza·tion. *Chem.* [Co- 3 b.] The polymerization of two substances together. So **co:po·lymerize** *v.*, to polymerize together. Hence **co:po·lymerizing, co:po·lymerized** *ppl. adjs.*
1940 J. R. Hurst in *Ind. & Engin. Chem.* XXXII. 64 (*heading*) Copolymerization of Maleic Polyesters. *Ibid.* 64/1 For the first time it has been shown that glycol maleate esters are capable of undergoing copolymerization with vinyl derivatives. *Ibid.* 64/2 Its ability to copolymerize with vinyl derivatives was also recognized. *Ibid.*, Such copolymerized mixtures represent an entirely new type of synthetic resin. 1945 H. R. Fleck *Plastics* (ed. 2) iv. 108 Co-polymerization of the monomer with another polymerizable monomer. 1948 *Economist* 31 July 193/2 Patents for the co-polymerisation of styrene and oil. 1950 *Thorpe's Dict. Appl. Chem.* X. 87/2 (*heading*) Interpolymerisation (Co-polymerisation) Processes...This technique is widely practised in the manufacture of vinyl plastics since it permits the production of a more varied range of products than would otherwise be possible. 1961 J. N. Anderson *Appl. Dental Materials* (ed. 2) xxiii. 229 Monomer liquid may also contain styrene..or other copolymerizing monomers. 1963 A. J. Hall *Textile Sci.* ii. 72 Dynel synthetic fibre is made from a linear polymer derived from two different monomers, ..which are caused to co-polymerise thus.

copper, *sb.*¹ Add: **1. c.** *pl.* Shares in a copper-mining company.
1899 *Boston* (Mass.) *Globe* 28 Apr. 9/5 The largest owners of 'coppers' know of no investment that will be as safe and give as large returns as 'coppers'. 1907 *Daily Chron.* 26 Oct. 1/7 Mines, including coppers, showed a general recovery. 1964 *Financial Times* 3 Mar. 1/5 Coppers improved and Kaffirs remained quietly firm.
4. b. A ceremonial copper sheet like a shield made and used by N. Amer. Indians.
[1814 U. Lisiansky *Voy. round World* viii. 150 Mr. Baranoff brought with him also two other Curiosities; one of which was a thin plate, made of virgin copper...These plates are only possessed by the rich.] 1888 A. P. Niblack *Coast Indians S. Alaska & N. Brit. Columbia* viii. 336 Amongst the Tsimshian these coppers are exhibited in a circle in upright position at the ceremony of 'bringing out' a girl. 1907 F. W. Hodge *Handbk. Amer. Indians* I. 346/1 Perhaps the most noteworthy product is the unique, shield-like 'coppers' made of sheet metal and highly esteemed as symbols of wealth or distinction. 1935 W. M. Halliday *Potlatch & Totem* xi. 109 You know this copper; you know what I paid for it. The copper which Awalaskinis broke, and of which he gave me a piece, was not in the same class with this one. 1969 *Times* 22 Sept. 14/4 The ultimate gesture of bravado..was for a man to break his own 'copper'.
11. copper age *Archæol.*, the prehistoric period during which weapons, etc., were made of copper; **copperbelt, copper belt** [*belt *sb.*¹ 5 a], an area of central Africa in which great quantities of copper are mined; **copper-bound** *a.*, ornamented with copper lace; **Copper Eskimo,** one of a group of Eskimos living near the Coppermine River in North-West Territories, Canada; **copper loss,** the loss of energy in the copper conductors of a dynamo or motor; **copper-skinned** *a.*, having a copper-coloured skin; **copper-top** *dial.* and *slang*, a red head of hair; a red-headed person; so **copper-topped** *a.*
[1863 C. Lyell *Antiq. Man* ii. 11 It has been suggested that an age of copper must always have intervened between that of stone and bronze; but if so, the interval seems to have been short in Europe.] 1881 J. Evans *Anc. Bronze Implements* i. 2 Of this Copper Age..there are in Europe but extremely feeble traces. 1928 C. Dawson *Age of Gods* iii. 56 There are features in the Copper Age Culture of Thrace which point to a survival there of the earlier culture traditions. 1957 Childe *Dawn Europ. Civ.* (ed. 6) iii. 35 Copper was already competing with stone and bone as an industrial material but without in the least replacing them. This 'Copper Age' culture, as Turkish archaeologists label it, is fairly uniform all over the plateau. 1955 *Times* 12 May 9/2 The four copper mining companies who are establishing a technical training foundation for young Europeans on the copperbelt. 1965 A. Holmes *Princ. Physical Geol.* (ed. 2) xxx. 1145 The well-known Copper Belt of Katanga and Northern Rhodesia. 1809 'D. Knickerbocker' *Hist. N.Y.* vi. ii, The copper-bound cocked hat. 1884 in F. W. Hodge *Indians of Canada* (1913) 241/1 *Kidnelik*, a tribe of Central Eskimos living on Coronation gulf, Arctic ocean [called] Copper Eskimo. 1913 V. Stefansson in *Summary Rep. Geol. Survey, Canada, 1912* 492 No permanent dwellings of any kind are erected by any of the Copper Eskimo—I call them so because..copper furnishes such a characteristic and important part of their material equipment. 1922 D. Jenness *Life of Copper Eskimos* 9 The expedition was able to establish a base at Bernard harbour..within the territory of the Copper Eskimos. 1936 *Discovery* July 200/2 A prolonged study of the Copper Eskimo. 1965 *Beaver* Spring 15 The Copper Eskimo are the descendants of an old local Thule culture population. 1916 *Standardization Rules Amer. Inst. Electr. Engin.* 48 The no-load copper loss. 1940 *Chambers's Techn. Dict.* 197/2 *Copper loss* (Elec. Eng.), the loss occurring in electric..apparatus due to the current flowing in the windings...(Radio), the power dissipated as heat in an..oscillatory circuit, due to Joule effect. 1873 M. F. Mahony *Chron. Fermors* I. xiii, There was no room for personal rivalry with that beady-eyed, copper-skinned..little lady. 1907 *Q. Rev.* July 188 The lank-haired, copper-skinned Jakuns. 1916 C. J. Dennis *Ginger Mick* 28 'E chews it over...Workin' 'is copper-top a double shift. 1892 Heslop *Northumbld. Words*, *Coppertopt*, red haired.
12. copper ore (examples).
1713 *Wyllys Papers* in *Connecticut Hist. Soc. Coll.* XXI. 380 To dig or Cause to be dugg at the Copper Mines in Symsbury as much Copper Oar as will produce and make three Tunns and half of Copper. 1839 H. T. de la Beche *Rep. Geol. Cornwall* xi. 324 The interstices only between them being filled by copper-ore.

copper, *sb.*⁴ (Earlier and later examples.) Hence, one who informs on fellow prisoners; a police informant; esp. *to come* or *turn copper*.
1846 *Sessions Papers* 16 May 39, I have heard the police called *coppers* before. 1885 M. Davitt *Prison Diary* I. ii. 22 They never 'round' upon each other, while they hold all 'coppers' (prison informers) in detestation. 1891 *Sessions Papers* 24 Oct. 1215 He has turned *copper* on us. 1924 [see *copper *v.*² 2]. 1938 Sharpe *Flying Squad* 330 *Copper*, policeman. 'To come copper' means to inform. An informer is also described as 'a copper'. 1958 M. Procter *Man in Ambush* xvi. 196 If you come copper on Lionel, I'll come copper on you.

copper, *v.* Add: **1. b.** To colour by the use of a salt of copper.
1895 *Daily News* 7 Feb. 5/2 The defendant assured the Bench..that the public 'would not have them [*sc.* peas] without they were coppered'. 1900 *Westm. Gaz.* 8 Sept. 2/3 The prohibition of the sale of 'coppered' vegetables. 1902 *Encycl. Brit.* XXV. 92/1 Vendors of coppered peas.
2. (Earlier and later examples.) Also *absol.* and *fig.*
1856 *Dem. State Jrnl.* (Sacramento, Calif.) 25 July 2/3 It is a safe bet to 'copper' all that comes from the immaculate Committee. 1883 'Mark Twain' *Life Mississippi* xxviii. 268, I reckon the safe way, where a man can afford it, is to *copper* the operation. 1913 C. E. Mulford *Coming of Cassidy* vii. 108 He had played to win when he should have coppered, coppered when he should have played to win. 1939 P. A. Rollins *Gone Haywire* 48 Feeling sort o' mistaken sorry fer th' ol' tightwad..I coppered a bet fer 'im.

copper (kǫ·pəɪ), *v.*² *slang.* [f. Copper *sb.*⁴]
1. To arrest.
1872 E. Crapsey *Nether Side N.Y.* iii. 17 They are..more frequently 'coppered'—that is, arrested. 1962 *John o' London's* 25 Jan. 82/2 'Arrest' has many Cant synonyms including *do*,..*copper*.
2. To give information to the police. Also *trans.* to inform on (a person).
1897 *Sessions Papers* 12 Feb. 359 Are you going to *copper*? 1923 E. Wallace *Missing Million* xix. 159 My motto is: 'Thou shalt not garden-hop...' In plain English: 'Thou shalt not copper.' *Ibid.* xxvii. 211 It's no use your staying here, because I'm not going to copper anybody. 1924 S. Scott *Human Side Crook Life* i. 23 'Coppering', or turning informant, is the deadly sin among crooks...Men will listen to the vilest epithets, but call them 'bogey', 'brassey', 'copper', or 'policeman', and they will be at your throat.

copper-belly. Delete def. and substitute: **a.** = Copperhead 1. **b.** The common North American water snake, *Natrix erythrogaster*. (Examples.)
1743 M. Catesby *Nat. Hist. Carolina* II. 46 *Anguis ventre cuprei coloris.* The Copper-belly Snake...They are of a brown Colour, except their Bellies, which are of a muddy Red or Copper Colour. 1788 J. D. Schöpf *Reise nordamerikanischen Staaten* I. 45 Ich habe mich überall nach der Klapperschlange und dem Kupferbauch (Copper belly, auch Mocooson-Snake genannt,) deren Biss eben so gibtig ist, erkundiget. 1842 J. E. DeKay *Zool. N.Y.* iii. 54 The Copper-head..has various popular names in different districts..Copper-Belly, Red Viper, Deaf Adder and Chunk-head. 1872 Schele de Vere *Americanisms* 387 The Copper-head (Trigonocephalus contortrix)..is known as Copper-belly and Chunkhead. *Ibid.*, The true Copperbelly (*Nerodia erythrogaster*) is perfectly harmless and of aquatic habits. 1957 A. H. & A. A. Wright *Handbk. Snakes U.S. & Canada* I. 480 Does the recent school know from sufficient live firsthand experience with each region that the Mississippi Valley yellow belly is the same as the Atlantic coastal copper belly?

copper-bottomed, *a.* Add: Also *fig.* Thoroughly sound, genuine, authentic, trustworthy.
1890 Farmer *Slang* s.v. A 1, In mercantile circles, the expression has become popularly current, in a figurative sense, to signify the highest commercial credit; and..first-class; first-rate. The form varies, being rendered by..A 1 copper-bottomed, [etc.]. 1894 Stevenson & Osbourne *Ebb-Tide* ii. vii, The real, first-rate, copper-bottomed aristocrat. 1937 A. Calder-Marshall in C. Day Lewis *Mind in Chains* 73 The attitude can then be adopted: '..We are moderates. Good, solid, honest to God, copper-bottomed capitalists.' 1961 *Times* 9 Nov. 16/2 The genuine copper-bottomed novel-reader's novel is becoming a rare article.

copperhead. 1. Delete (*Trigonocephalus contortrix*) and substitute (*Agkistrodon contortrix*). Delete 'It is less than 2 feet long' and substitute: 'It is usually less than 4 feet long.' (Earlier and later examples.)
1775 P. V. Fithian *Jrnl. & Lett.* 10 July (1934) 54 The Snake that wounded her they call a '*Copper-Head*'. 1782 [see Pilot *sb.* 4 a]. 1842 [see *copper-belly]. 1880 [see *chunk-head*, Chunk¹ 2]. 1931 R. L. Ditmars *Snakes of World* x. 101 The water snake and hog-nosed snake are sometimes mistaken for the Copperhead, owing to their proportionately thick bodies, and gross similarity of markings. The poisonous species, however, carries its points of marked differentiation in having elliptical pupils, prominent cavity or pit between the eye and nostril, and greater number of plates under the tail in a single row. 1949 *Sci. Monthly* Jan. 55/1 Thus, the belief goes, whenever one sees a copperhead, a rattlesnake may appear on its trail. 1963 R. Snedigar *Our Small Native Animals* (ed. 2) 192 Although ..capable of inflicting a severe and potent bite, the *copperhead*..is shy and retiring.
b. In full *copperhead snake*.
1788 J. May *Jrnl. & Lett.* (1873) 70 One of Colonel Stacey's men bit by a copperhead snake. 1822 *Massachusetts Spy* 31 July (Th.), A woman in Salisbury township..discovered a copperhead snake on her dresser. 1866 J. C. Gregg *Life in Army* i. 18 There were also to be found great dens of the Copperhead Snake, and this species were considered the most poisonous, deceptive, dangerous and mean of all the snake family. 1931 R. L. Ditmars *Snakes of World* x. 97 Two species of moccasins must be considered—the Water Moccasin and the Copperhead Snake (highland moccasin, pilot snake, or chunk-head), which are members of the same family as the rattlers.
c. An Australian snake, *Austrelaps superba*, of the family Elapidæ.
1878 F. McCoy *Prodromus Zool. Victoria* I. 9, I have adopted the popular name 'copper-head' for this snake from a well-known vendor of a supposed antidote for snake-bites. 1896 F. G. Aflalo *Nat. Hist. Austral.* 171 The colour of this snake is brown or copper-coloured, the belly yellow to grey. It is also known as the Copper Head. 1944 *Living off Land* vi. 120 (*caption*) The thick copperhead is slow-moving. 1963 E. Worrell *Reptiles of Australia* 110 Copperheads congregate in large colonies among thick tussocks bordering swamps.
3. An American Indian.
1838 *Southern Lit. Messenger* IV. 295/1 He said..it would be a sin to kill one, but if he was to go he should want to kill one of the damned copper-heads. 1845 *Amer. Rev.* Oct. 367/2 First Mexicans, then Indians, come on. Brush the Yellow-Bellies. Then the Copperheads come sneaking down. *a* 1861 T. Winthrop *Canoe & Saddle* (1883) x. 145 Five foul copperheads..lurked among the plunder of that noisome spot. *Ibid.* 149 Meanwhile those five copperheads watched me.

copper-headed, *a. U.S.* [Cf. prec.] **a.** Having a copper-coloured head. **b.** Of or pertaining to a copperhead (senses 2 and *3).
1808 T. Ashe *Trav. Amer.* II. 287 Rattle Snake..Copper Headed Ditto. 1847 in D. Drake *Pion. Life Kentucky* (1870) ii. 25 The copper-colored man, and the copper-headed snake [were] then extremely common. *a* 1861 T. Winthrop *Canoe & Saddle* (1883) xii. 248 The copper-headed, snaky beguiler. 1863 *Congress. Globe* 936/2 The seat of rebellion..is among the copper-headed traitors of the North.

copper Maori (kǫ·pəɪ mɑu̯ᵊ·ri). *New Zealand.* Also **Kopa Maori, Kapura Maori,** etc. [Maori *kāpura* fire, *kopa* oven.] A Maori oven in which food is placed on heated stones and covered with flax, earth, etc.
c 1826 J. Heberley *Autobiogr.* 30 (MS.), [They] sucked [the slave girl's] blood..till [she] was dead, then they cut her up and cooked her in a copper Maori. 1834 E. Markham *N.Z. Recoll.* 38 Coppre Mourie or the Native Ovens are described I rather think in Captain Cookes Voyages. They are done this way. They scratch a hole 8 or 10 inches deep..and then light the Fire in the hole. 1888 *Trans. New Zeal. Inst.* XXI. 417 So they set to work and they dug holes in this flat,..each hole about 2 ft. across and about 1½ ft. deep, and shaped something like a 'kopa maori'. 1889 *Ibid.* XXII. 104 A number of *kapura Maori*, or native ovens. 1905 W. B. *Where White Man Treads* 102 'Coppa Maori', 'Maori oven', etc.—these are the slang names given by those who know no better.

coppernob. Also **copper-knob.** *colloq.* [f. Copper *sb.*¹ 9 d + Nob *sb.*² 1 a.] A red-haired person.
1880 D. W. Barrett *Life among Navvies* (ed. 2) ii. iii. 49 A man with red hair would be termed a 'Copper-knob', or 'Mahogany top'. 1946 J. Irving *Royal Navalese* 56 *Copper-nob*, someone blessed (or cursed) with red hair. 1959 I. & P. Opie *Lore & Lang. Schoolchildren* ix. 170 'Ginger, you're barmy, getcher 'air cut' has been a greeting to coppernobs ever since it was a catch phrase in the Harry Champion song.

copper-plate. Add: **4.** and **5.** Used of a style of careful handwriting.
1750 J. Steuart *Letter-Bk.* (1915) 465 A coper plate coppie book for asisting my boys in ther writing. 1824 M. Wilmot *Let.* 1 Jan. (1935) 204 They will all write as well at 13 or 14 as if they wrote copper plate now. 1911 R. Brooke *Let.* July (1968) 312, I have stooped to an even more round-hand copper-plate than ever. 1954 N. D. Young *Handwriting in Eng. & Wales* i. 9 England,..producing in time the copper-plate writing that was imitated in all the counting-houses of Europe. 1966 G. Heyer *Black Sheep* ix. 132 She had then devoted the better part of the afternoon to the composition of a formal invitation, written in her beautiful copper-plate.

co·pperskin. *U.S. slang.* [Copper *sb.*¹ 11.] An American Indian.
1840 C. F. Hoffman *Greyslaer* ii. x. 26 'Go on, go on, Kit, d'ye say a dozen Injuns?' 'Yes, uncle, not a Copperskin less.' *a* 1861 T. Winthrop *Canoe & Saddle* (1883) x. 146 The five copperskins 'first eyed me over' with lazy thoroughness. They noted my arms and equipment. 1904 S. E. White *Silent Places* viii. 72 What the hell do we care for a lot of copper-skins from Rupert's House!

coppice, *sb.* Add: **2. coppice shoot,** a shoot arising from an adventitious bud at the base of a tree; **coppice system,** a silvicultural system of reproduction of trees from coppice shoots; **coppice-with-standards** [Standard *sb.* 20 a], a crop consisting partly of coppice shoots and partly of trees grown from seedlings.
1851 J. Brown *Forester* (ed. 2) v. 379 The stoles will, by the middle of June, have sent up a large supply of young shoots...These are termed coppice shoots, of whatever kind the tree may be. 1882 Fernandez & Smythies tr. *Bagneris' Elem. Sylviculture* iii. ii. 133 In growing coppice with standards, the end in view is to combine..the advantages of simple coppice and some of those of high forest. *Ibid.* 140 Forests worked on the coppice system form the exception [in Germany]. 1895 W. R. Forester *Schlich's Man. Forestry* IV. i. 32 Coppice-with-standards suffers more [from grazing] than pure coppice, on account of the necessity for preserving numerous seedling plants to replace the standards as they are felled. 1928 R. S. Troup *Silvicultural Syst.* xii. 129 The coppice system involves reproduction by stool shoots or suckers. 1953 H. L. Edlin *Forester's Handbk.* xi. 165 In the form known as coppice-with-standards, occasional stems are left uncut to form full-sized timber.

coppice, *v.* For Copse *v.* read Coppice *v.*¹ Add: **b.** *intr.* To produce coppice shoots; to form a coppice.
1928 R. S. Troup *Silvicultural Syst.* xii. 129 Many tropical and sub-tropical trees coppice with vigour. 1935 *Chambers's Encycl.* III. 165/1 Neither do conifers possess sufficient reproductive power for coppice-treatment. Chestnut, oak, ash, hazel, lime, maple, sycamore, hornbeam, willow and alder coppice better than beech, birch and aspen, though the softwoods often become dominant on moist land.

coppicing, *vbl. sb.* Add: **b.** The treating of wood as coppice; the cutting down of trees periodically so that new shoots may grow from the stumps. Also *attrib.*
1880 C. R. Markham *Peruv. Bark* 329 The coppicing system consists in cutting down the trees near the ground, and allowing one or more of the crops of shoots, which rise from the stumps, to grow. *Ibid.* 395 Uprooting, coppicing, and thinning. 1928 Schlich *Man. Forestry* (ed. 4) I. 236 Methods of thinning, coppicing, &c. 1953 *Oxf. Jun. Encycl.* VI. 101/2 Pollarding is a similar method of growing small brushwood, but differs from coppicing in that the trees are cut back at a height of about 6 feet, instead of about ground level.

co:-preci·pitate, *v. Chem.* [Co- 1.] *trans.* To deposit by co-precipitation.
1932 *Jrnl. Phys. Chem.* XXXVI. 861 In most textbooks of analytical chemistry, it is mentioned that magnesium is coprecipitated with calcium oxalate. 1967 E. Chambers *Photolitho-Offset* xvi. 243 Barium sulphate by itself has poor working qualities, but when coprecipitated with alumina hydrate it forms a first-rate gloss white.

co:-precipita·tion. *Chem.* [Co- 3 b.] The simultaneous precipitation of two or more compounds from a solution; the removal of a substance from solution by binding to a precipitate.
1931 *Jrnl. Amer. Chem. Soc.* LIII. 1236 The so-called coprecipitation of alkali oxalate with barium or strontium oxalate is to be attributed to a double salt formation. 1932 *Jrnl. Phys. Chem.* XXXVI. 861 It has been a general custom in analytical chemistry to use the words, coprecipitation, carrying down, occlusion, inclusion and adsorption as collective names, meaning nothing else but, that impurities are carried down with or in a precipitate. 1938 R. W. Lawson tr. *Hevesy & Paneth's Man. Radioactivity* (ed. 2) xv. 147 The conclusions which were drawn about the chemical nature of a radio-element from this 'co-precipitation' process. 1956 *Nature* 18 Jan. 184/1 This method of collecting cations in trace amounts has the advantages of being more efficient and more selective than the conventional co-precipitation technique.

copro- Add: **copromania** [-mania], an obsession with faeces; hence **coproma·niac; co·prophil, co·prophile** [-phil, -phile], one who is attracted to filth; **coprophi·lia, copro·phily** [Gr. φιλία affection], marked attention to defecation and to excreta; **coprophilous** *a.*, also spec. *coprophilous fungi*, fungi growing on dung; **coprophobia** [-phobia], an abnormal repugnance toward faeces; **coprozo·ic** *a.* [Zoic *a.* 2], of animals, living in dung.
1880 Swinburne *Lt.* (1960) IV. 136 The perpetual copromania of the 'Old Enemy' (*sc.* Carlyle)—with his 'eternal cesspools' and 'swill-troughs'. 1900 J. M. Robertson in *Lit. Guide* 1 Mar. 43/1 Such a one as Swift: filthy-minded, unscrupulous, and malignant. Had such a copromaniac been [illegible]. 1927 'N. Blake' *Thou Shell of Death* vi. 120 [illegible] was a koprophil [illegible]. 1953 S. Beckett *Watt* 93 A vegetarian, a naturist, a cannibal, a coprophile [illegible] [*sc.* Swift's] [illegible] by Freud. [illegible] specific work on Coprophilous or dung-borne Fungi. 1953 J. Ramsbottom *Mushr. & Toadstools* xvii. 195 Coprophilous fungi grow [illegible]. 1924 A. Huxley *Little Mexican* 43 The innocent coprophily [illegible]. 1934 H. C. Warren *Dict. Psychol.* 61/2 Coprophobia. 1922 *Nature* CIX. 30/1 There is a whole series of Amoebae and flagellate Protozoa which..develop in the faeces after deposition. These are called 'coprozoic' forms. 1964 *New Biol.* III. 89 [illegible] an elaborate [illegible] of coprozoic (*i.e.*, dung-eating) insects and their predators.

coproporphyrin (kǫprǫpǭ·ɪfirin). *Biochem.* [ad. G. *koproporphyrin* (Fischer & Schaumann 1923, in *Zeitschr. f. Physiol. Chem.* XXVIII. 162), f. Copro- + *porphyrin.] A porphyrin pigment, $C_{36}H_{38}N_4O_8$, excreted by the liver in bile and found in feces and urine.
1924 *Chem. Abstr.* XVIII. 3213 From the hypothetical 'myohemoglobin' coproporphyrin is formed. 1939 Best & Taylor *Physiol. Basis Med. Practice* (ed. 2) ii. 73 Coproporphyrin..is found normally in human feces, and in urine, bile and milk in traces. 1961 *Lancet* 22 July 176/1 Ether-soluble porphyrins..Protoporphyrin and coproporphyrin. *Ibid.* 214/1 In symptomatic cutaneous porphyria..fæcal coproporphyrin and protoporphyrin are normal or only slightly increased.

coprosma (kǫprǫ·zmă). [mod.L. (J. R. and J. G. A. Forster *Characteres Generum Plantarum* (1776) 137), f. Gr. κόπρος dung + ὀσμή smell.] A plant of the genus of small evergreen trees and shrubs so called, belonging to the family Rubiaceæ and found in Australasia and the Pacific Islands.
1874 A. Bathgate *Colonial Exper.* xvii. 254 Several of the coprosmas and olearias are prominently suited for such purposes [as ornamental plants]. 1936 'R. Hyde' *Passport to Hell* ii. 42 Over a coprosma hedge he fell [illegible] with Fanny's little sister, May. 1960 N. Hilliard *Maori Girl* ii. xv. 163 Myra spat loudly..into the coprosma. 1968 *Southerly* XXVIII. 163 The interesting yellow dust playground divided by pepper-trees and coprosma and privet from the..yellow dust street.

co-prosperity sphere. [Co- 3 a.] The sphere (sense 7 d) controlled by Japan during the war of 1941–45. Also *transf.*
1941 *Times Weekly* 15 Oct. 4/4 The Japanese, for all their talk of 'new orders' and 'co-prosperity spheres' can offer China nothing that the Chinese cannot devise for themselves. 1943 *Ann. Reg.* 1942 180 The Japanese 'Co-Prosperity Sphere' embraced Japan, Manchukuo, a large and rich part of China, Indo-China, Thailand, Malaya, the Philippines, Burma, and the Dutch East Indies. 1944 J. S. Huxley *On Living in Rev.* iii. 41 Hitler's aim of a 'new order' in Europe was anticipated by many years by the Japanese 'new order'—now rechristened Co-Prosperity Sphere—in East Asia. The long-declared aim of this is to establish complete Japanese supremacy in the Far East, with Japan as a privileged [illegible] centre. 1967 *Economist* 22 Apr. 331/2 The talk here these days is of a southern African 'co-prosperity sphere'.

cop-shop. *slang.* [f. Cop *sb.*³ + Shop *sb.* 4.] A police station.
1941 Baker *Dict. Austral. Slang* 20 *Copshop*, a police station. 1955 P. Wildeblood *Against Law* iii. 65 [illegible] the cop-shop and squeal on their friends. 1962 M. Duffy *That's how it was* vi. 93 The blue light above the cop-shop door for once meant safety.

'copter, copter. Short for *helicopter. Also *attrib.* Chiefly *U.S.*
1947 *Sun* (Baltimore) 28 June 9/2 The 'copter stepped out of the test tunnel into a wartime job. 1949 *Jrnl. R. Aeronaut. Soc.* LIII. 311/1 Even the provision of 'City Copter Courts', as they had been called, would not ensure the full economic utilisation of the helicopter transport potential, unless the public were educated to be helicopter-minded. 1953 Pohl & Kornbluth *Space Merchants* (1955) iv. 43 'Damn-fool bus drivers,' Jack grunted, staring up at the 'copter.

copulative, *a.* Add: Also spec. *copulative compound.*
1888 Sweet *Hist. Eng. Sounds* (ed. 2) 75 Such compounds as [illegible] 'hand-made' in Sk., are...nothing else but fragments of pre-inflectional sentences, as is shown still more clearly in the Sk copulative compounds, such as *ahorātrām* 'a day and night'. 1933 Bloomfield *Language* xiv. 235 One can therefore distinguish between copulative compounds (Sanskrit *dvandva*) and determinative (attributive or subordinative) compounds. *Ibid.*, The Hindu grammarians distinguished among copulative compounds a special sub-group of repetitive (*amredita*) compounds, with identical members, as in *choo-choo, bye-bye, goody-goody.* 1961 A. G. Hatcher (*title*) Modern English word-formation and neo-Latin: [illegible].

copunctal (kopʌ·ŋktăl), *a. Math.* [f. Co- 2 + Punctal *a.*] Meeting in a point, concurrent.
1886 Merriman & Woodward *Higher Math.* 72 (Cent. D. Suppl.), Planes all on the same point, or straights all with the same cross, are called copunctal. 1961 C. H. McDowell *Dict. Math.* 31 *Copunctal*, concurrent; meeting in a point.

copy, *sb.* Add: **9. c.** That which lends itself to interesting narration in a book, newspaper, etc.; material for a story.
1886 Wilde *Reviews* (1908) 89 Miss Broughton has been attending the meetings of the Psychical Society in search of copy. 1889 G. B. Shaw in *Fabian Ess. Socialism* 183 Those Socialist speeches which make what the newspapers call 'good copy'. 1915 A. D. Gillespie *Let. from Flanders* 24 May (1916) 165 [illegible] copy. 1922 [illegible] *Cornhill Mag.* June 735 [illegible] 'They'd make first-rate "copy", wouldn't they?' 1930 W. S. Churchill *Early Life* xvii. 339, I scampered about the moving cavalry screens searching in the carelessness of youth for every scrap of adventure, experience or copy. 1934 E. Bowen in G. Greene *Old School* 52 One or two of the girls fell in love in the holidays, but something in the atmosphere of school made us ashamed to talk of this naturally without seeming at once to make copy of it. 1965 *Listener* 10 June 866/1 The gaffe of their guest in making copy out of it all, of the BBC in broadcasting it unedited.
d. *spec.* The text of an advertisement.
1905 Calkins & Holden *Art Mod. Advertising* vii. 175 The design and 'copy' used in the four-inch advertisement may involve just as much time. 1926 G. Russell *Nuntius: Advertising* iii. 53 The public cannot fairly be expected to believe the verbiage into which much extremely competent advertisement copy is converted by the futile interference of manufacturers. 1933 D. L. Sayers *Murder must Advertise* iii. 39 Ingleby specialised in snobbish copy about Twentyman's Teas ('preferred by Fashion's Favourites'). 1967 *Word Study* Oct. 3/2 Writers..should take care with advertising copy for radio and TV.
C. copy-boy, one who takes copy from the writer to the printer; a publisher's errand boy; **copy desk** *U.S.*, the desk where copy is edited for printing; **copy editor,** one who edits copy for printing; hence **copy-edit** *v.*; also in extended use; **copy-editing** *vbl. sb.*; **copy-fit** *v. trans.*, to fit (copy) to the space available; so **copy-fitting** *vbl. sb.* (see quot. 1961); **copy-holder,** (*b*) a proof-reader's assistant who reads the copy aloud to the proof-reader; **copy-hunting** *vbl. sb.* and *ppl. a.*, hunting for 'copy' (sense *9 c); **copy-paper,** paper on which copy is written for the press; **copy-reader,** one who reads copy for a newspaper or a book; also in extended use; so **copy-read** *v.*; **copy-reading** *vbl. sb.*; **copy-slip** (earlier U.S. example); **copy-taster,** one who selects copy for printing; **copy-text** (see quot. 1904); **copy-typist,** one who makes typewritten copies of documents, etc.; hence **copy-type** *v.*; **copy-typing** *vbl. sb.*; **copy-writer,** a writer of copy for the press; *spec.* a writer of advertising copy (see sense *9 d); so **copy-writing** *vbl. sb.*

1888 Kipling *Phantom 'Rickshaw* 76 The little black copy-boys are whining.., and most of the paper is as blank as Modred's shield. 1942 W. Stevens *Let.* 23 Oct. (1967) 424 She thought that she could get a job as a copy boy on one of the local papers. 1961 'B. Wells' *Day Earth caught Fire* i. 10 Ronnie, a young, eager copy-boy, bustled in with a handful of news slips. 1929 M. Lief *Hangover* 235 It got past the copy desk for the first edition and then some wise guy caught it and killed it in the others. 1932 E. Wilson *Devil take Hindmost* ii. 7 At the offices of *The New Leader*..the Socialists have..cups of coffee..piled on a tray on the copy desk. 1953 M. Cowley in F. Scott Fitzgerald *Tender is Night* Introd. p. xvi, The manuscript was never copy-edited by others. 1958 E. J. West *Shaw on Theatre* p. v, Many of the pieces..were copy-edited to conform to the house usages of the publications..that first printed them. 1962 A. Nisbett *Technique Sound Studio* vii. 125 To copy edit a tape, the selected material is recorded to a make-up tape. *Ibid.* 246 *Copy editing*, the copying of selected extracts from recorded material into sequence on a main programme assembly tape. Subsequent fine editing will generally be necessary. 1899 J. L. Williams *Stolen Story* 24 The copy-editors began gathering in now. 1931 *N. & Q.* 29 Aug. 146/1 The copy editor, in preparing the despatch for the printer, began with the last clause of the note. 1969 J. Bennett *Dragon* viii. 104 This stringer..sniffed out the story and filed it to New York, where a copy editor promptly spiked it. 1948 *Library* III. 155, I propose to demonstrate, through evidence of copy-fitting in II,..the correctness of the traditional order of I and II. *Ibid.*, Edition II was copy-fitted against edition I. *Ibid.* 158 The right-hand margins of II are almost perfectly regular, and were demonstrably copy-fitted to make them so. 1961 T. Landau *Encycl. Librarianship* (ed. 2) 101/2 *Copy fitting*, adjusting copy to the space prepared, either by verbal changes or by suitable changes in type size. 1888 *Encycl. Brit.* XXIII. 701/2 This proof is read through and compared with the copy by the proof reader or corrector of the press and an assistant, the copy-holder or reading boy. 1951 S. Jennett *Making of Books* vi. 87 A copyholder is then called, to read aloud from the author's manuscript or typescript while the reader follows the wording of the proof, checking it with what the copy-holder is reading. 1900 *Kynoch Jrnl.* Feb.–Mar. 75/1 This point is invariably missed by non-shooting writers when copy-hunting. 1913 'S. Rohmer' *Myst. Fu-Manchu* xviii. 189 Places unknown even to the ubiquitous copy-hunting pressman. 1902 E. L. Banks *Newsp. Girl* 259 The great pads of copy paper supplied by the telegraph office for newspaper correspondents. 1907 *Daily Chron.* 18 Oct. 4/4 There is brown paper and notepaper and copy-paper and..newspaper. 1969 R. Busby *Robbery Blue* iv. 34 Sheaves of pink and blue copy-paper..torn from the copy-takers' typewriters. 1945 *Eng. Lang. in Amer. Educ.* (*Mod. Lang. Assoc. Amer.*) iii. 24 The student who plans to be a secretary should be sure to develop the ability to spell and punctuate correctly and to copyread what he has written. 1892 *Harper's Weekly* 9 Jan. 42/4 Upon the taste, the good judgment, and discretion of these copy-readers the character of the paper very greatly depends. 1903 E. L. Shuman *Pract. Journalism* 18 Each of these departments has a force of copy-readers, whose duty it is to edit the matter written by the reporters. *Ibid.* 25 In the first ten years the young journalist masters reporting, copy-reading, and the rest of the routine work. 1838 C. Gilman *Recollections* xxviii. 194 One set of copy-slips was to be substituted for another. 1942 *Sphere* 27 June 409/1 All tape and agency news comes to the chief copy-taster in the main room. 1962 A. Lejeune *Duel in Shadows* i. 11 The Managing Editor and his myrmidons huddled round the backbench examining a damp page-proof, the copy-taster's spike piled high with rejected stories. 1904 R. B. McKerrow *Wks. Nashe* p. xi, The spelling of the copy-text, by which..I mean the text used in each particular case as the basis of mine, has been followed exactly. 1964 F. Bowers *Bibliogr. & Text. Crit.* i. 2 The choice of copy-text was not a particularly acute question. *Ibid.* vi. iv. 201 Evidence in *Othello* at first sight contrary to the Q copy-text hypothesis. 1956 'C. Blackstock' *Dewey Death* i. 5 They copy-typed in French, German, Italian and Russian, without understanding one word. 1959 *Times Lit. Suppl.* 9 Oct. 577/2 The time taken up on purely secretarial work, copy-typing, [etc.]. 1939 *Daily Tel.* 6 Feb. 18/5 (Advt.), A young lady, aged 19 required. Must be an efficient copy typist. 1960 *Economist* 8 Oct. 171/1 The copy typist in a typing pool. 1911 T. Russell *Advertising & Publicity* ix. 96 An advertiser can..employ what are known as copy-writers—professional writers of advertisements. 1935 *Archit. Rev.* LXXVII. 129/2 They have paid copy-writers and poster-designers to 'put them across' in nation-wide publicity drives. 1958 *Times Lit. Suppl.* 15 Aug. p. xxiii/1 They are to the gentleman publisher what ideas men, public relations experts, copy-writers, and designers are to the common industrialist. 1923 H. Crane *Let.* 18 Feb. (1965) 126 Truly, you must look for some editorial post, copywriting job, or something that will relieve you of such strains. 1958 *Times Lit. Suppl.* 26 Dec. 748/5 Perhaps not surprisingly, copywriting (or 'wordsmithing', as one executive dubbed it) is the least important department of the business.

co·pyable, *a.* = Copiable *a.* In *copyable pencil*, an early name for *copying pencil* (*copying *vbl. sb.* 2).
1883 *Stationer & Bookseller* 8 May 35 'Automatic' Copyable Pencils. 1890 [see Copiable *a.*].

copyboard. [f. Copy *sb.* 8 + Board *sb.*] (See quot. 1968.)
1961 in Webster. 1967 E. Chambers *Photolitho-Offset* iii. 27 Copy should not be prepared too large for the copyboard and, if possible, should not exceed 25 to 50% reduction or enlargement. 1968 *Gloss. Terms Offset Lithogr. Printing* (*B.S.I.*) 14 *Copyboard*, the unit on a camera which holds the original copy.

copy-book. Add: **2.** phr. *to blot one's copy-book*: to commit a fault, misdemeanour or gaffe which spoils one's record. *colloq.*
1935 D. L. Sayers *Gaudy Night* v. 85 Now, it was the College that had blotted its copy-book and had called her in as one calls in a specialist. 1958 *Spectator* 22 Aug. 243/1 Its Founder..had badly blotted his copybook by making a speech in favour of the beheaded Earl of Essex.
b. More widely applied to processes, action, or conduct of an exemplary kind.
1908 *Daily Chron.* 16 Jan. 8/5 'A fine copy-book hand to be sure'..'Yes; and a fine copy-book mind behind it.' 1927 *Times* 3 May 7/2 Twice in succession he drove Mr. Cameron to the boundary past mid-off with copybook strokes. 1947 G. Greene *19 Stories* 138 She had the kind of copybook psychological sense you often find in nuns. 1957 *Times* 15 Mar. 10/4 The aircraft made a perfectly normal and very good copybook approach.

copy-cat (kǫ·pikæt). *colloq.* [Cat *sb.*¹ 2.] A derogatory term for one who or that which copies (see Copy *v.*¹) another, or another's work. Hence **copy-cat** *v. trans.*, to imitate slavishly; *copy-catting* ppl. a.
1896 S. Jewett *Country of Pointed Firs* xiii. 102, I ain't heard of a copy-cat this great many years..'twas a favorite term o' my grandmother's. 1907 E. S. Field *Six-Cylinder Courtship* 101 Good old Tom, with..his picture-gallery full of bogus Old Masters! How often have I called him a silly copy-cat. 1918 C. Mackenzie *Sylvia Scarlett* i. vi. 183 One girl with a taste for painting discovered that she was her ideal conception of St. Cecilia; this naturally roused the jealousy of rival adorers that would not be 'copy-cats'. 1931 *Oxf. Times* 24 Apr. 16/4 A good architect was not a 'copy-cat'; nor did he kick over the traces. 1932 R. Lehmann *Invit. Waltz* ii. i. 124 You never will think for yourself. You don't want to go through life copy-catting me, do you? 1934 J. Cannan *Under Proof* i. i. 13 You weren't playing ships till I did. You copy-catted me. 1945 W. de la Mare *Burning-Glass* 87 Copy-cat keeps Teacher well in sight. 1961 *Daily Tel.* 1 Apr. 16/6 Police investigating the 'copycat' murder found a track, over a quarter of a mile long, made by Stobbs' dragged body. 1963 J. Fowles *Collector* ii. 161 The horrid timid copycatting genteel in-between class.

copygraph (kǫ·pigrɑf), *sb.* [f. Copy *sb.* + -graph.] An apparatus for, or the process of, duplicating and multiplying copies of writing by means of a gelatine slab and aniline or similar ink.
a 1884 Knight *Dict. Mech.* Suppl.
Hence **co·pygraph** *v. trans.*, to reproduce by means of the copygraph.
1900 H. G. Wells *Love & Mr. Lewisham* i. 10 Copygraphed sheets of instructions from his remote correspondence tutors.

copyhold. (By Part V of the Law of Property Act 1922, all copyhold land was enfranchised.)

copying, *vbl. sb.* Add: **2.** *copying-machine, -press* (earlier examples); **copying (ink) pencil,** a lead, or a pencil containing a lead, composed of graphite, aniline blue, and kaolin or gum arabic, and used for indelible writing, and duplicating in a copying press.
1885 *Army & Navy Co-op. Soc. Price List* 465 Copying Ink Pencils. Automatic action. 1803 A. Ellicott in C. V. Mathews *Life & Lett.* (1908) 209, I have enclosed a copy of this letter taken in 15 seconds, by which you will see the value of a copying machine. 1883 R. Haldane *Workshop Rec.* 175/2 Copying...The subject may be divided into chemical..and mechanical methods, copying pencils, and transferring. 1928 E. Blunden *Undertones of War* ii. 18 The censoring of the letters scrawled in copying pencil by our home-yearning stalwarts. 1785 Jefferson *Let.* in *Writ.* (1894) V. 110 Have you a copying press? If not, you should get one.

copying, *ppl. a. copying clerk* (earlier example).
1836 Dickens *Sk. Boz* 1st Ser. II. 198 Low copying-clerks in attorneys' offices.

copyright, *sb.* Add: **1.** (Earlier example.)
1735 *Parl. Coll., House of Lords* 6 May (H.L.R.O.), The Editions and Impressions of such Books made and published as well in Great Britain as in Ireland and Scotland by persons who have paid no considerations for the Copy-right of such Books.
2. (Earlier U.S. example.)
1870 *U.S. Stat. at Large* XVI. 213 [To] deposit in the mail two copies of such copyright book or other article.
3. *copyright act, law.*
1831 *Deb. Congress* 6 Jan. 424 His colleague..had described the copyright act as simply a remedy for the abuse of that right. 1835 *Knickerbocker* VI. 289 The passage of an international copy-right law betwixt America and Great Britain. 1962 *Times* 5 July 15/6 A tape recorder offers the additional facility of actual recording, either personally or—copyright laws notwithstanding—from the radio.

co·pyrightable, *a.* [f. Copyright *v.* + -able.] Capable of being copyrighted.
1895 G. B. Shaw *Let.* 16 Mar. (1965) 498 Felix, in addition to my blessing, which is probably not copyrightable in America, has the full score of Candida. 1903 S. S. Elder in *Nation* (N.Y.) 30 Apr. 349/3 It is no longer possible to summarize it in a few sections covering every thing copyrightable.

copyrighted, *ppl. a.* (Earlier U.S. example.)
1806 Webster *Compend. Dict.*, *Copyrighted*, as the sole right being secured.

‖ **coq au vin** (kok *o* væ̃n). [F., lit. 'cock in wine'.] A chicken cooked in wine.
c 1938 *Fortnum & Mason Catal.* 36/2 Coq au Vin—per tin 4/-. 1951 E. David *French Country Cooking* 26 Coq au Vin.., cooked in a pint of sound Macon or Beaujolais, will be a much finer dish than that cooked in fiery Algerian wine. 1963 P. H. Johnson *Night & Silence* xxxiii. 239 What's our main dish?..I suggest the *coq au vin*.

coque, *sb.* Add: **2.** (Examples.)
1835 *Court Mag.* VI. p. xiv/2 A succession of *coques* rising one above another, in front of the crown. 1873 *Young Englishwoman* Feb. 77/1 A large bow..composed of mixed coques of velvet and grosgrains silk ribbon.
b. pronounced (kōᵘk). Applied to feathers used in trimming.
1903 *Westm. Gaz.* 11 May 4/2 Those soft coq feathers are very much used too. 1905 *Daily Chron.* 13 May 4/7 Ostrich and coque boas. 1923 *Daily Mail* 26 Mar. 6 Clipped Coque Feather Stole. 1928 *Daily Express* 7 May 5/2 Coque feathers applied closely on to the crown of the hat.

coquet. B. *sb.* **1.** Delete †*Obs.* and add earlier and later examples.
1693 *Humours & Conv. of Town* 105 The perpetual Chat of the Noisy Coquets. 1819 J. Constable *Let.* 2 Nov. (1962) 189 While I am writing all this I am playing the Coquet as I mean to add a post script to this note...P.S. I was last night..elected an Associate of the Royal Academy. 1892 G. B. Shaw *Let.* 21 Apr. (1965) 338 People imagine that Lövborg..really means to kill himself, instead of merely gratifying his propensity, as a male coquet, for making scenes. 1894 R. Sullivan *Woman* 15 But are not men flirts also?..Are not men coquets?

coquina. (Earlier and later U.S. examples.)
1837 J. L. Williams *Territory of Florida* 44 The quarries of Coquina stone. *Ibid.* 52 This river being choked up by the Coquina formation. 1957 P. Frank *Seven Days to Never* i. 9 A few hundred yards inland Highway AIA parallels the surf, but the coquina road is narrow, pocked.

cor³. Add: **tenor cor** (see quots.).
1928 *Grove's Dict. Mus.* V. 305/1 *Tenor cor*, a brass valve instrument of tenor pitch and circular or French horn form, designed and introduced by Henry Distin about 1860. 1938 *Oxf. Compan. Mus.* 930/1 *Tenor cor*, a high-pitched instrument of horn-like character intended to take horn parts when proper horns are not available. 1961 R. M. Pegge in A. Baines *Mus. Instr.* xii. 316 Substitutes for the French horn..have come on the market, notably the tenor cor (Amer. mellophone) in F and E♭, which is virtually a saxhorn shaped like a French horn but using only the lower series of harmonics.

cor (kǫɪ), *int.* Vulgar corruption of *God.*
1931 J. B. Priestley *Angel Pavement* v. 241 Cor!—you're in the wrong part of the theatre, boy. 1955 J. Thomas *No Banners* xxi. 206 Cor' blimey! that was close. 1959 I. & P. Opie *Lore & Lang. Schoolchildren* x. 195 'Cor, gi'sit', 'Hand over'. 1961 Simpson & Galton *Four Hancock Scripts* 16 Oh cor blimey, I don't understand you people!

coraciiform (kǫræsi·ifǭɪm, kǫræsəi·ifǭɪm), *a. Ornith.* [f. mod.L. *Coraciiformes*: see -form.] Belonging to the order Coraciiformes, which includes kingfishers, rollers, hoopoes, and related birds.
1901 P. Chalmers Mitchell in *Trans. Linn. Soc., Zool.* VIII. 249 The Rollers and other simple short-gutted Coraciiform birds.

coracoidal (kǫrăkoi·dăl), *a.* [f. Coracoid *sb.* + -al.] Of or pertaining to the coracoid.
1893 Lydekker in *Proc. Zool. Soc.* 14 Feb. 172 The coracoidal element forms a small moiety of the glenoid cavity. 1964 J. A. Peters *Dict. Herpetol.* 73 Evidence today indicates that ancestral reptiles have two coracoidal ossifications..with the true coracoid lost.

coral, *sb.*¹ Add: **8. d.** *coral-berried.*
1897 *Daily News* 9 Sept. 6/1 The drooping boughs of coral-berried rowan. 1937 Blunden *Elegy* 54 Some coral-berried tree.
9. coral-fern *Austral.*, a name given to Australian ferns of the genus *Gleichenia*; **coral-gall,** an excrescence produced on coral by the action of epizoic animals, esp. crabs and barnacles; also *attrib.*, of such an animal; **coral-limestone,** coralline limestone; **coral-pea** = *coral-creeper*; **coral-spot,** a disease of shrubs and trees caused by the fungus *Nectria cinnabarina*; also, the fungus itself; **coral-stitch** (example); **coral vine** = *corallita.
1898 Morris *Austral Eng.* 98/1 *Coral-Fern*, name given in Victoria to *Gleichenia circinata*. 1942 C. Barrett *Austral. Wild Flower Bk.* xi. 190 Coral ferns and fan ferns belong to the same genus, *Gleichenia*. 1968 G. R. Cochrane et al. *Flowers & Plants of Victoria* 171 Wiry rhachises of the Scrambling Coral Fern..branch repeatedly for several feet to form umbrella-like fronds with tiny flattened pinnules. A closely related species, Pouched Coral Fern or Wiry Coral Fern, has pouch-like pinnules. 1903 *Nature* 10 Sept. 457/1 These coral galls may be found on the Millepora and Madreporas of a certain portion of a reef and be absent from all the other genera of neighbouring corals. 1957 *Encycl. Brit.* VI. 627/1 Another small but intriguing species is the coral-gall crab (*Hapalocarcinus*), which in some fashion irritates the growing tips of certain corals so that they grow to enclose the female in a stony prison..which provides an enduring shelter. 1831 Lyell *Princ. Geol.* II. 287 The increase of coral limestone..may vary greatly according to the sites of mineral springs. 1839 G. A. Mantell *Wond. Geol.* (ed. 3) II. 563 In reference to the formation of coral limestones,..some beds..consist of a pure calcareous mud. 1878 *Jrnl. Chem. Soc.* XXXIV. 120 Formerly pyroclasite was supposed to be the result of volcanic action upon the coral-limestone. 1959 *Chambers's Encycl.* X. 302/2 Clipperton has no basalt but has trachyte and coral limestone. 1896 *Melburnian* 28 Aug. 53 (Morris), The trailing scarlet kennedyas, aptly called the 'bleeding-heart' or 'coral pea'. 1962 *Austral. Encycl.* V. 175/2 *K[ennedya] prostrata*, the 'running postman', or scarlet coral-pea, is found in all States; it has clover-like leaves and single or twin showy flowers with yellow centres. 1923 *Trans. Brit. Mycological Soc.* VIII. 22 (*heading*) The parasitism of *Nectria cinnabarina* (Coral Spot). 1966 F. H. Brightman *Oxf. Bk. Flowerless Plants* 144/2 *Nectria cinnabarina* ('Coral Spot') is common at all times of the year on moist, newly fallen twigs and branches. 1873 *Young Englishwoman* Jan. 46/2 Work the veins in coral-stitch and hem-stitch. 1931 T. W. Sanders *Encycl. Gardening* (ed. 21) 28 Antigonon (Coral Vine; Corallita)..Tuberous-rooted stove climbers..Species cultivated: A. leptopus, bright pink, summer, 10 to 15 ft., Mexico.

corallita (kǫrăli·tă). Also **coraleta, coralita.** [ad. Amer. Sp. *coralito*, dim. of Sp. *coral* Coral *sb.*¹] A climbing vine (*Antigonon leptopus*) with pink flowers, native to Mexico and cultivated in the tropics or in greenhouses elsewhere.
1899 J. Rodway *Guiana Wilds* 63 The little gallery was screened from the road by luxuriant festoons of corallita. 1931 [see *coral vine*, *coral *sb.*¹ 9]. 1951 E. Mittelhölzer *Shadows Move* i. ix. 66 The big, thick-stalked vine is bougainvillea...The coralita is the one with the small pink flowers. 1957 M. Swan *Brit. Guiana* i. 12 The bougainvillea, morning-glory, the pink coraleta.

coral-snake. (S. Afr. examples.)
1931 R. L. Ditmars *Snakes of World* xiii. 165 Several genera of small species [*sc.* of elapine snakes]..are commonly known as Garter snakes or Coral snakes. They are small and slim with glossy scales and rather similar to the New World coral snakes, in looks and habits. 1954 J. A. Pringle *Common Snakes* 25 (*heading*) Coral Snake.. *Aspedilaps scutatus* (Laurenti)... The Coral Snake has between twenty-seven and forty-seven black and red bands which encircle the body...It occurs in the Karoo areas of the western Cape, and South-West Africa.

Coramine (kōᵊ·rămin, -məin). Also **coramine.** [ad. G. *Coramin* (*Schweiz. Fabrik & Handelsmarken* (1924) 27 May 342).] The proprietary name of a pyridine derivative used as a respiratory and circulatory stimulant; nikethamide.
1925 *Chem. Abstr.* XIX. 1601 Coramin, a new substance similar to camphor in activity. 1934 *Lancet* 21 Apr. 824/2 There seems to be a widespread belief that Coramine exerts a direct action upon the heart. 1935 F. Stark *S. Gates of Arabia* xxiv. 266 Something was very wrong with the heart. At intervals I treated it with coramine injections. 1958 M. Dickens *Man Overboard* vi. 85 'Tea's wanted here!' she called, like a ward sister calling a nurse for emergency coramine. 1969 J. A. Bevan *Essent. Pharmacol.* xxiv. 245 Nikethamide (Coramine) is a weak analeptic of low toxicity.

corbeau. (Earlier and later examples.)
1810 R. Ackermann *Repository* Apr. 262/2 Superfine corbeau colour coat. 1910 *Daily News* 9 Feb. 4/5 The Duchess of Portland was wearing corbeau blue velvet. 1957 M. B. Picken *Fashion Dict.* 81/2 *Corbeau*, very dark green, almost black.

cord, *sb.*¹ Add: **1. f.** Applied to pottery ornamented with impressions of cord (cf. *corded *ppl. a.* 3 b). So *cord-beaker* (G. *schnurbecher*).
1902 J. Abercromby in *Jrnl. Anthrop. Inst.* XXXII. 391 (*heading*) The 'Cord-beaker' and its Offshoots. 1928 Peake & Fleure *Steppe & Sown* ii. 26 A ware, known as string-ornamented ware or cord pottery, that seems to have been derived ultimately from the pottery of the Russian steppe. 1954 S. Piggott *Neolithic Cultures* ii. 30 The late appearance of cord-wares of Peterborough or Ebbsfleet type at Windmill Hill and Whitehawk.
12. cord-ornamented *a.*, applied to pottery decorated by pressing cord into the soft clay before firing.
1925 Childe *Dawn Europ. Civ.* xv. 235 A battle-axe folk..using cord-ornamented pottery. 1928 Peake & Fleure *Steppe & Sown* iii. 44 Sherds of cord ornamented pottery. 1954 S. Piggott *Neolithic Cultures* xi. 303 The great groups of cord-ornamented and pit-comb wares of northern Europe.

cord, *v.*¹ Add: **2. b.** *Bookbinding.* To tie (a book) between two boards to keep the cover smooth while drying.
1876 *Encycl. Brit.* IV. 43/2 As a last operation in forwarding, but one now frequently omitted, the book is 'corded', that is, firmly tied between two boards until it is dry, so as to insure perfect smoothness in the cover.

corded, *ppl. a.* Add: **3. b. corded ware** *Archæol.* [cf. G. *schnurkeramik*], cord-ornamented ware, *spec.* a type characteristic of a neolithic people of Thuringia. Also *attrib.*
1928 C. Dawson *Age of Gods* x. 222 Their peculiar type of pottery which is due to the fusion of the Spanish Bell-Beaker type with the Corded Ware (Schnurkeramik) of the Thuringian Battle-Axe People. 1929 Childe *Danube in Prehist.* xii. 234 The cord ornament just mentioned suggests that the late Corded Ware culture represented in Starý Zámek II contributed to its formation. 1949 *Proc. Prehist. Soc.* XV. 81 The barrows of Sarmenstorf containing corded ware are Late Neolithic both in the Western and the Central European series. 1950 H. L. Lorimer *Homer & Monuments* i. 6 The *Schnurkeramik* or Corded Ware folk, who are in part at least identical with the people of the Battle-axe culture commonly regarded as Indo-European. 1963 H. N. Savory in Foster & Alcock *Culture & Environment* iii. 45 The influence of a north-European Corded Ware group upon Bell Beakers.

cordelle, *sb.* **2.** (Earlier and later examples; also **cordale.**)
[1792 E. P. Simcoe *Diary* (1911) 91 The current becoming very strong, the men were obliged to *tirer à la cordelle*, or drag the boat by ropes on a narrow beach under high, woody banks.] 1811 Brackenridge *Jrnl.* in *Views Louisiana* (1814) 214 Continued until eleven, with *cordelle*, or towing line—the banks being favorable. 1816 H. Ker *Travels* 36 This is a great object to the bargemen in using their cordale. 1946 C. Richter *Fields* 281 Several hands jumped out in the water with the cordelle and pulled it to the bank. 1969 E. W. Morse *Fur Trade Canoe Routes* i. 5 To guard against the risk of the towline breaking, the line (*cordelle*) consisted of half a dozen smaller lines braided together.

cordelle, *v.* Add: (Earlier and later U.S. examples; also **cordale.**) Hence **co·rdelling** *vbl. sb.*
1812 J. C. Luttig *Jrnl. Exped. Upper Missouri* 31 July (1920) 62 Departed early cloudy and head wind cordelled all Morning. 1816 H. Ker *Travels* 36 After getting above their cordaling ground, in swift water they make use of their warp. 1826 T. Flint *Recoll.* 96 In two instances the boatmen.., when cordelling the boat directly at the base of these rocks, disengaged snakes from their retreats. 1832 J. K. Paulding *Westward Ho!* I. ix. 83 A rope was taken ashore, and fastened to a rock, or stump, or sapling, and by this the boat was dragged along. This process is called cordeleing. 1940 H. H. Hatcher *Buckeye Country* 5 He..trudged along the bank..'cordelling'..his pirogue upstream.

Cordilleran (kǫɪdilyē̆·răn), *a.* orig. *U.S.* [f. Cordillera + -an.] Designating, pertaining to, or resembling the Cordilleras.
1891 *Scribner's Mag.* Oct. 465/1 We might think of our Cordilleran system as a great plateau, of unequal height in different parts, and everywhere bearing upon its surface a system of mountain ranges. 1896 *Nat. Geogr. Mag.* VII. 293 The Rocky mountain region of the North American continent, or, as called by others, the cordilleran region. 1940 *Geogr. Jrnl.* XCVI. 125 It may be questioned..how far one dare infer..that similar superficial outlines imply identical stress developments alike in ridges 2 kilometres in length and in cordilleran ranges. 1968 *Encycl. Brit.* VI. 490/1 In North America the Rocky Mountains, the Sierra Nevada and the mountains between them are collectively known as the Cordilleras, and the entire area has been termed the Cordilleran region.

cordless (kǭ·ɪdlès), *a.* [f. Cord *sb.*¹ + -less.] Of an electric device or appliance: working without a flex or cord; battery-powered.
1906 J. Poole *Pract. Teleph. Handbk.* (ed. 3) xii. 170 The small and simpler [switch-]boards..may be divided into two classes—(1) the Cordless, in which the connections are made by means of plugs only; and (2) the Plug and Cord. 1923 T. E. Herbert *Telephony* xiv. 349 Cordless boards. When the number of lines to be connected is small, it is possible..to effect all the required connections with the aid of plugs and jacks. *Ibid.* 351 An example will be given of a cordless type of board used on C.B. exchanges. 1948 J. Atkinson *Teleph.* I. x. 182/2 The cordless switchboard has valuable advantages due to the absence of flexible conductors. 1962 E. Bruton *Dict. Clocks* 51 *Cordless clock*, American name for a battery clock. 1964 *New Scientist* 1 Oct. 40/1 Much publicity has been given, particularly in the United States, to the use of batteries in..'cordless appliances'—electric shavers, hand power tools, vibrating tooth brushes. 1969 *Times* 17 Nov. 18/5 (Advt.), The cordless hotplate!..Plug it in for about 10 mins, then use it cordless, for over an hour.

cordoba (kǭ·ɪdŏbă, -ŏvă). Also **córdoba.** [Sp., name of H. de *Córdoba*, Sp. governor of Nicaragua, *fl.* 1524.] The principal monetary unit of Nicaragua, equivalent to one hundred centavos.
1913 in Webster Add. 1955 *Times* 4 July 13/4 The International Monetary Fund has agreed to a change in the par value of the Nicaraguan cordoba from five to seven a United States dollar.

cordon, *sb.* Add: **4.** Also *cordon sanitaire*, *lit.* and *fig.*
1847 in Webster. 1857 C. Kingsley *Two Y. Ago* I. p. ii, Leave us to draw a cordon sanitaire round the tainted states, and leave the system to die a natural death. 1920 *Punch* 7 Jan. 5/1 Los Angeles..is suffering from an unprecedented crime wave. A proposal by President Carranza to draw a *cordon sanitaire* round the place has not yet reached Washington. 1940 B. Ward *Russian Foreign Policy* 7 Thus the Allies succeeded in creating a *cordon sanitaire* of buffer-states to cut Communism off from Europe. 1964 Hall & Whannel *Popular Arts* ii. vii. 191 The school..believes it is drawing a *cordon sanitaire* around the imagination of the young reader. 1968 *Internat. Encycl. Soc. Sci.* IV. 211/1 A *cordon sanitaire* is a territory separating two powers who if they combined in a single land mass might constitute a threat to other nations.
6. *cordon bleu* (earlier and later examples). Also *attrib.* and quasi-*adj.*
1769 H. Walpole *Let.* 17 Sept. (1941) X. 292 Everybody rushes in, Princes of the blood, *cordons bleus*, abbés, housemaids. 1826 M. Kelly *Reminisc.* (ed. 2) II. 83 His chief French cook..was a great artist, a real *cordon bleu*. 1849 Thackeray *Pendennis* I. xxxv. 343 'How good this is!' said Popjoy, good-naturedly. 'You must have a *cordon bleu* in your kitchen.' 1959 R. Postgate *Good Food Guide* 266 The wife's cooking is described by a member as 'Cordon Bleu'. 1960 *Times* 9 Jan. 11/3 She had many requests to combine 'Cordon Bleu' cooking with flower arrangement, but..arranged..a separate 'Cordon Bleu' cookery school. 1963 *Economist* 30 Nov. 931/1 To help the blue stocking..to turn into a *cordon bleu*.

cordon, *v.* Restrict † *Obs. rare* to senses in Dict. and add: **3.** To enclose with, or to cut *off* with, a cordon (senses 3 and 4). Also *fig.*
1891 *Cosmopolitan* XII. 61/1 Chicago is fairly cordoned by a great chain of mammoth manufacturing plants. 1905 *Westm. Gaz.* 1 July 11/2 The city blocks are rigorously cordoned by troops. *Ibid.*, Within the precincts of the cordoned quarantine harbour section. 1920 *Glasgow Herald* 15 Mar., The Wilhelmstrasse was at first cordoned off. 1927 *Daily Express* 11 Feb. 1/3 The building was promptly cordoned off until reinforcements arrived. 1956 *Ann. Reg.* 1955 210 The Swiss police acted promptly, cordoning off the building.

cordotomy (kǭɪdǫ·tŏmi). *Surg.* Also **chordotomy.** [f. Cord *sb.*¹, Chord *sb.*¹: see -tomy.] Severance of certain nerve-tracts within the spinal cord for the relief of pain.
1923 Frazier & Spiller in *Arch. Neurol. & Psychiatry* IX. 1 (*heading*) Section of the anterolateral columns of the spinal cord (chordotomy). 1961 *Lancet* 5 Aug. 303/2 Cases in which cordotomy was undertaken for the relief of pain.

Cordtex (kǭ·ɪdˌteks). *Mil.* [f. Cord *sb.*¹ + Tex(tile *a.* and *sb.*] The proprietary trade name of a type of detonating fuse consisting of a core of high explosive in a textile and plastic covering.
1935 *Trade Marks Jrnl.* 13 Feb. 189/2 *Cordtex.* Explosive fuses. Nobel's Explosives Company Limited. 1950 *Chambers's Encycl.* V. 538/1 Cordtex or Primacord..have in place of blackpowder a considerably thicker core of a sensitive detonating explosive PETN (pentaerythritol tetranitrate). 1966 D. Holbrook *Flesh Wounds* 95 The intake chimneys and sealing tapes round the guns were blown away by lashings of an explosive cable called Cordtex.

corduroy, *sb.* and *a.* **B.** *adj.* **1.** (Earlier example.)
1789 *Loiterer* 9 May 8 My boy was dressed in..corderoy breeches, and cotton stockings.
3. (Earlier and later examples.) For '*U.S.*' read 'orig. *U.S.*'
1822 J. Woods *Eng. Prairie* 219 From this town..along a rough road with many log-bridges; but some of my fellow passengers, from the state of Kentucky, called them corderoy. 1824 W. N. Blane *Excursion* 147 A Corderoy Road consists of small trees, stripped of their boughs, and laid touching one another, without any covering of earth. 1877 J. E. Taylor *Tourists' Guide Suffolk* 30 In the latter countries.., 'corduroy' roads are made..before metalled. 1926 T. E. Lawrence *Seven Pillars* (1935) vii. lxxxii. 457 Across the sandy neck from the first flat to the second we built a corduroy road of brushwood. 1941 J. Gore *George V* xiii. 161 The pomp of royal pageantry seemed strangely out of keeping with shack villages and corduroy roads.

corduro·yed, *ppl. a.* **1.** (Earlier and later examples.) Also *fig.*
1832 *Chambers's Edin. Jrnl.* I. 225/1 This excellent specimen of corduroyed honesty and worth. 1910 A. Bennett *Clayhanger* i. xiv. 123 Mr. Orgreave had a confabulation with the corduroyed man. 1959 W. D. Pereira *North Flight* viii. 123 A crowd of corduroyed young men.

cordwaining (kǭ·ɪdwē̆ⁱniŋ), *vbl. sb.* [f. Cordwain(er + -ing¹.] The art or craft of the cordwainer; cordwainery.
1758 in L. Chalkley *Chron. Settlem. Virginia* (1912) I. 470 Binds himself..to learn cord waining. 1812 *Sporting Mag.* XL. 13, I entertain the most exalted ideas of the ingenious 'art and craft of cordwaining'. 1900 *Daily News* 17 Oct. 6/2 He was initiated at his father's house of business ..into the mysteries of cordwaining. 1901 W. J. Locke *Usurper* xv, I think I shall learn to make boots. Have you any professors of cordwaining at Greybrooke, Alicia?

core, *sb.*¹ Add: **4.** Also, a portion extracted from the bed of an ocean or lake. So *core-sampler* (see *15).
1938 C. S. Piggot in *Sci. Monthly* Mar. 201 (*heading*) Core samples of the ocean bottom. 1939 Piggot & Urry in *Jrnl. Washington Acad. Sci.* XXIX. 405 (*heading*) The radium content of an ocean-bottom core. 1950 *Ann. Reg.*

1949 428 A continuous record of the bottom profile.. registered by..200 long cores taken from depths between 2,000 and more than 4,000 fathoms, the integral core-length exceeding one statute mile. 1956 *Nature* 10 Mar. 451/1 Dr. B. Kullenberg, during the Second World War, developed a piston core-sampler capable of taking cores 60 ft. in length. 1959 J. CLEGG *Freshwater Life* xviii. 309 Deeper samples of the mud are obtained by an apparatus called a core-sampler.. This will raise an undisturbed core of bottom-deposit many feet in length. 1970 *Nature* 6 June 934/1 Palaeomagnetic measurements of deep sea sediment cores could yield additional information about the past motions of the sea floor.

5. Also *attrib.*, applied to implements consisting of a trimmed core of flint or to cultures characterized by this type of implement.

1926 *Guide Antiq. Stone Age Brit. Mus.* 14 The change from a core-industry to a flake-industry. *Ibid.* 26 Flints were required for..scraping, planing, sawing, and boring, but for these processes flakes would be more useful than core-implements. 1927 PEAKE & FLEURE *Hunters & Artists* iv. 42 The earliest palaeolithic implements..were mostly core implements, fashioned from a block of flint by removing flakes from its surface until the desired form had been achieved. 1935 *Proc. Prehist. Soc.* I. 1 4 Already in 1916 the Bavarian, Obermaier,..had distinguished core-cultures and flake-cultures in lower palaeolithic industries. 1937 GARROD & BATE *Stone Age Mt. Carmel* I. i. 18 32 Cores and core-scrapers..were extremely abundant... They are for the most part roughly pyramidal and rather small. 1947 J. & C. HAWKES *Prehist. Brit.* ı. 9 The core-tool will have both faces rounded and trimmed. *Ibid.* 10 The core-culture folk at the height of the cold phases seem..to have retreated southward towards Africa, returning again with the milder weather.

7. c. That part of a nuclear reactor which contains fissile material.

1946 *Nucleonics* Dec. 43/1 The core or active section of the pile, in its simplest form, may consist merely of a mass of fissionable material or nuclear fuel. 1954 *Sci. Amer.* Dec. 36/2 The high pressure core will be contained in a carbon-steel vessel nine feet in diameter, coated with stainless steel. 1955 *Sci. News Let.* 20 Aug. 115/3 Radiation from the core appears as a bluish-green glow. 1971 *Daily Tel.* (Colour Suppl.) 30 Apr. 17/4 One species [of alga] lives in the cooling water that circulates round the cores of nuclear reactors.

f. All the electrons of an atom (together with its nucleus) other than the valency electrons.

1926 *Proc. R. Soc.* A. CXI. 86 When one electron is much more easily removed than any other, it is reasonable to think of it as a 'series electron' and the rest of the atom as a 'core'. 1927 J. W. FISHER tr. *M. Born's Mech. Atom* iii. 130 This one electron is in an orbit..far removed from the rest of the atom, or 'core'. [*Note*] German, *Rumpf*. The English equivalent of this word is not completely standardised: the alternatives 'body', 'trunk', 'kernel' have been used by different writers. 1937 J. W. T. SPINKS tr. *Herzberg's Atomic Spectra & Atomic Struct.* i. 70 There are only two electrons outside the atomic core of the alkaline earths. 1965 PHILLIPS & WILLIAMS *Inorg. Chem.* I. ii. 37 Sodium..has a core of completed K- and L-electron shells. 1970 G. K. WOODGATE *Elem. Atomic Struct.* v. 90 The argon core effectively screens the nuclear charge from a 4*d* electron.

10. b. A unit of magnetic material in a computer; *esp.* one in which two directions of magnetization represent 0 and 1 and the magnetization remains unchanged until reversed by currents in wires passing through or round the core. So *core memory, store*, etc.

1950 *Jrnl. Appl. Physics* XXI. 49 Magnetic cores with a rectangular hysteresis loop are used in a storage system which requires no mechanical motion and is permanent. 1953 R. D. RICHTMYER et al. in *Computers* ... [Reading] of the magnetic switch..must be a core..with..rings of magnetic material, called 'cores'. 1959 *Oxf. Univ. Gaz.* 21 Mar. 743/1 The computing equipment consists of..One Ferranti Mercury computer, with four blocks of core store. 1964 *Gloss. Terms Automatic Data Processing (B.S.I.)* 76 Core store, an array of storage cores used as a magnetic store. 1964 R. F. FICCHI *Electr. Interference* vii. 114 The computer..has 23 instructions designed to operate with two overlapping core-storage units. 1969 *Math. in Biol. & Med.* (*Med. Res. Council*) iv. 191 The core memory size (32000 words of 36 bits each). 1969 P. B. JORDAIN *Condensed Computer Encycl.* 303 The individual magnetic cores are strung on fine insulated wires and assembled into flat planes (or arrays) containing 16 to 10,000 cores apiece.

15. (sense 8) *core-drying*, *-iron*; **core-casting**, casting with a core to make a cavity in metal; **core-drilling** *vbl. sb.*, a method of drilling in which an annular hole is made in the ground and a core extracted; **core-sampling**; **core-loss**, the loss of energy due to hysteresis and to eddy-currents in the core of electric machinery; **core-sampler**, a device for extracting a core of material from the ocean floor; **core-sampling** *vbl. sb.*, core-drilling in order to obtain a core as a sample of the strata pierced; **core-wall**, a wall of solid masonry forming the core of a dike or dam consisting mostly of earth or sand.

1928 CHILDE *Most Anc. East* v. 133 The process of core-casting and the invention of the shaft-hole axe. 1903 *Sci. Amer.* 18 July 44/1 Core drilling is indispensable in a great variety of engineering and mining enterprises, affording, as it does, a means for drilling out a sample core or column of rock. 1955 *Oxf. Jun. Encycl.* VIII. 342/1 Further exploration is usually necessary in order to assess the commercial value of the deposit. This information may be obtained by core-drilling. 1901 *Daily Chron.* 3 Dec. 2/1 For making steel, heating iron, core-drying. 1888 *Lockwood's Dict. Mech. Engin.* 89 Core irons, or Core Rods, rods of wrought iron from ⅛ inch or ¼ to ½ inch in diameter, according to the size of core. 1960 R. LISTER *Decorative Cast Ironwork* 225 Core irons, bars of iron built into cores to reinforce them. 1917 *Standard. Rules of Amer. Inst. Electr. Eng.* 44 Increased core losses due to increased excitation for compensating internal drop under load. 1962 M. G. SAY *Newnes Conc. Encycl. Electr. Engin.* 171/2 Core loss, the loss developed as heat in the ferromagnetic core of a magnetic circuit subjected to alternating magnetization. It comprises two components..eddy-current loss. 1950 F. P. SHEPARD & K. O. EMERY... 1955 C. S. PIGGOT ... Mar. 202 (*caption*) Comparison of snapper with core sampler. 1956 ... *Engineers' Handbk.* (ed. 2) ix. 347 The need for more accurate samples has caused rapid improvement in core sampling. 1960 E. WEGMANN *Design & Constr. Dams* (ed. 4) ii. 115 Masonry Core-walls are doubtless the best means of insuring water-tightness in an earthen dam. *Ibid.*, Core-walls for high dams are usually given a stronger section than those for lower ones. ..The thickness of the core-wall at the natural surface should be about ⅛ to ⅓ of the 'head' of water. 1959 H. M. WILSON *Irrigation Engin.* (ed. 6) xvi. 366 The foundations for a masonry core-wall should always rest only on firm, homogeneous rock.

core, *v.* Add: **3.** *to core out*: to hollow out by using a core.

1889 HASLUCK *Model Engin. Handybk.* ix. 108 The barrel of the pump may be cored out in the casting. 1902 *Westm. Gaz.* 4 June 5/1 Its bottom is cored out to fit over the nose of the torpedo tube.

Corean: see *KOREAN.

cored, *ppl. a.* **1.** (Example.)

1912 *Daily Chron.* 11 Mar., Peaches and cored apples from California.

corer. Add: **2.** A drilling device which extracts a core in geological exploration, etc. (See CORE *sb.*[1] 4).

1927 R. PEELE *Mining Engineers' Handbk.* (ed. 2) ix. 34? Keystone corer..was designed primarily for testing bituminous coal beds. 1937 G. E. HUTCHINSON *Treat. Limnol.* I. xv. 815 The lower part of the range of values for total copper is extended somewhat if older strata sampled by a corer are considered. 1970 *Nature* 4 Apr. 44/2 The hot salty holes in the Red Sea contain some of the most startling sediments yet described.. A corer lowered into them comes back still too hot to touch.

co-respondent. Add: **b.** *co-respondent('s) shoes*, a jocular name for men's two-toned shoes. Also *ellipt.*

1934 A. G. MACDONELL *How like an Angel* ii. ix. 155 A pair of those singularly repulsive shoes of black and white which are called 'Co-respondents' (quite wrongly called, incidentally, for co-respondents at least get and give some fun, and these shoes do neither). 1959 'A. BRIDGE' *Four-Part Setting* xxvi. 359 Very richly dressed, with brown-and-white 'co-respondent' shoes. 1958 K. AMIS *I like it Here* vi. 69 The furious creaking of his co-respondent's shoes.

Corfiote (kǭ·ɹfiǒt), *a.* and *sb.* Also **Corfiot.** [f. *Corfu*, after *CYPRIOT(E).] Of or pertaining to, an inhabitant of, the Greek island of Corfu.

1835 N. P. WILLIS *Pencillings by Way* II. vi. 73 The Corfiotes were taunted with their slavish submission. 1877 *Encycl. Brit.* VI. 396/1 The Corfiot peasantry are reputed the idlest of all the Ionians. 1923 *Contemp. Rev.* Oct. 412 Corfiote neutrality during the Greco-Turkish war. *Ibid.* 413 No Corfiote could be induced to get her off the rocks. 1955 *Times* 31 May 8/5 It would have been gracious..if the Corfiotes had celebrated the memory of that early sports patron by choosing the same site that of Lord Byron. 1964 *History Today* Sept. 624/1 It was otherwise with the demolitions, of which every Corfiot was a spectator.

corgi (kǭ·ɹgi). Also **corgy.** Pl. corgis, corgies. [Welsh, f. *cor* dwarf + *gi*, f. *ci* dog.] A small, short-legged dog of Welsh origin, with a head resembling that of a fox.

1926 *Bulletin* 25 Nov. 9/4 One little dog about a foot high..a brown Welsh corgi. 1926 *Glasgow Herald* 30 Nov. 8 A certain amount of talk in praise of a tiny Welsh dog, the Corgy. 1930 *Observer* 5 Oct. 13/2 What can one say of the Welsh corgis,..with their truncated legs, quick ears, foxy heads, and longish bodies? 1934 *Punch* 8 Aug. 618/1 It had a face like a fox and ears that stood up like a fox's... Its body was short and rather low, its tail at all; nothing but half-an-inch of stump. 'And what kind of mongrel is this?' I asked. 'Mongrel!' she exclaimed. 'It's no mongrel; it's the latest thing in pets. It's a Corgi.' 1952 C. L. B. HUBBARD *Pembrokeshire Corgi Handbk.* i. The plural of Corgi is Corgwn and not Corgis. 1970 *Times* 20 Oct. 13/3 (*caption*) The Queen, with a firm hand on the leash for a reluctant corgi, arriving at Euston station.

corial (kori·ăl). Also 8 corialla, 9 curial, corrial, 20 coreal. [Sp. *corial*, app. ad. Arawak *kuljara*.] In Guyana, a dugout canoe with pointed ends.

1796 J. G. STEDMAN *Surinam* I. x. 240 A corialla or small canoe, composed of one single piece of timber. 1825 WATERTON *Wand. S. Amer.* 223 We met the Indians going a fishing. I saw, by the way their things were packed in the curial, that they did not intend to return for some days. 1833 *Evangelical Mag.* Mar. 124/2 We set out in a *corial*, or small canoe, to visit the new station. 1852 C. W. DAY *Five Yrs. West Indies* II. 18 We hailed a ferryman, who came over in a most ticklish *corrial*. 1922 *Blackw. Mag.* Apr. 525/2 People coming off..in bateaux and corials. 1951 E. MITTELHÖLZER *Shadows Move* I. i. 7 Logan must be getting ready to take the *corial* and the skiff out into the stream. 1958 J. CAREW *Wild Coast* vii. 89 He had poled his *coreal* out to the public road.

coriandrol (kǫriæ·ndrǫl). *Chem.* [a. G. *coriandrol* (F. W. Semmler 1891, in *Ber. d. Deut. Chem. Ges.* XXIV. 206), f. L. *coriandrum* CORIANDER + -OL.] A colourless liquid, $C_{10}H_{18}O$, obtained esp. from oil of coriander and oil of linaloe.

1892 *Jrnl. Chem. Soc.* LX. 540 Coriander oil consists of terpenes and about 90 per cent. of another substance which is termed coriandrol.. Coriandrol, $C_{10}H_{18}O$, boils at 194–198° under a pressure of 760 mm. 1929 *Encycl. Brit.* VI. 431/1 Coriander..is used in medicine as an aromatic and carminative, the active principle of its volatile oil (*Coriandrol*) being isomeric with Borneo camphor. 1939 *Thorpe's Dict. Appl. Chem.* III. 364/2 Coriander, essential oil of..The chief odorous constituent is *d*-linalool, formerly known as coriandrol. 1960 E. GUENTHER *Essential Oils* IV. v. 611 Barbier showed that coriandrol and linalool were identical chemically as well as physically, except for the opposite optical rotation. Coriandrol is, therefore, nothing else but *d*-linalool.

Corinthian, *a.* Add: **5.** *Corinthian bagatelle*, any of several variations of bagatelle (sense 2).

1933 *Punch* 27 Dec. 715/2 The rules of the House must be revised so as to permit the introduction of..such pulliatives as Corinthian bagatelle. 1936 J. CARY *Afr. Witch* xxxi. 404 Another table..supported a Corinthian bagatelle board, complete with balls and cue. 1939 G. GREENE *Confid. Agent* iv. i. 273 Boards for Corinthian bagatelle.

Coriolis (kǫriǒ·lis). The name of G. G. *Coriolis* (1792–1843), French engineer and mathematician, used chiefly *attrib.*, to designate the effect whereby a body moving relative to a rotating frame of reference is accelerated in that frame in a direction perpendicular both to its direction of motion and to the axis of rotation of the frame.

1922 J. DE W. HANNAWAY ... *Dynam. Meteorol.* ... the Coriolis effect... 1951 H. HAURWITZ *Dynamic Meteorol.* vi. 116 As long as purely horizontal motions and accelerations are considered, therefore, the Coriolis force does not appear when motion at the equator is studied..for the Coriolis force is parallel to the equatorial plane, and the equatorial plane is normal to the earth's surface at the equator. 1962 *Listener* 21 Dec. 1053/1 After pointing out the lack of evidence for, and the difficulties confronting, any magnetic theory of bird direction-finding, he proceeds to discuss the possibility that the perception of a Coriolis force, generated by the rotation of the earth, might provide the basis for an explanation. 1950 *Meteorol. Gloss.* 53 Coriolis *Acceleration*. Any body moving relatively to the rotating earth experiences an acceleration relative to the earth called, after its discoverer, the Coriolis acceleration. The Coriolis acceleration is twice the vector product of the angular velocity of the earth and the velocity of the body. 1969 *New Scientist* 11 Dec. 547/2 Even spin-produced 'g' is not quite like the real thing, and gives rise to disturbing coriolis effects in the middle ear.

cork, *sb.*[1] Add: **10.** *cork sole* (later example).

1873 *Young Englishwoman* June 280/2 [Bathing] Shoes.. are made of the same material as the rest of the costume.. a light cork sole being sewn on outside the material.

11. d. cork carpet, a kind of floor-cloth composed of ground cork, india-rubber, and oxide-paint; **cork-elm**, (*a*) the rock elm, *Ulmus thomassi*; (*b*) the winged elm, *Ulmus alata*; **cork linoleum** (or **lino**), linoleum made from canvas backed with a mixture of linseed oil and ground cork; **cork-tipped** *a.*, of a cigarette: having a filter of a cork-like substance at one end; also *absol.*

1896 M. BAILLIE-SCOTT *Houses & Gardens* x. 30 The [bathroom] floor will perhaps..be covered with..a cork carpet. 1959 *Chambers's Encycl.* VIII. 387/1 Cork carpet is similar to plain linoleum but made of larger cork granules with a low pigment content, thus producing a softer and warmer floorcovering, but one more difficult to keep clean. 1875 H. MUHLENBERG *Catal. Plants* 29 Cork elm. *Ulmus alata*. 1884 MILLER *Plant-n.* 259 *Ulmus racemosa*, American Cork Elm. 1931 W. N. CLUTE *Common Names Plants* 26 The plant usually regarded as the true wahoo, is the plant often called the burningbush..but the cork elm (*Ulmus racemosa*), and the hophornbeam..also called wahoo. 1939 J. JOYCE *Let.* 23 Dec. (1966) II. 280 Try to get some 'cork-lino' for kitchen. 1970 *Univ. Coll. Oxf.) Record* V. v. 315 The old cork-lino was replaced everywhere. 1909 *Daily Chron.* 11 June 7/4 'Cork Linoleum' has to most people mean merely, or chiefly, a kind of trade mark. 1907 *Yesterday's Shopping* (1969) 673/3 Cigarettes... Finest quality cork tipped. 1924 A. E. M. FOSTER *Lond. Restaurants* 25 If you find your cigarette stick to your lips you had much better get them cork-tipped. 1928 *Punch* 22 Feb. p. ii (Advt.), Really nice girls smoke Player's Cork-Tipped. 1945 'L. LEWIS' *Birthday Murder* (1951) v. 63 Mrs. Saxe, smoking a cork-tipped cigarette. 1969 'J. FRASER' *Cock-Pit of Roses* xix. 150 Give us some cigs...twenty of the cork-tipped.

cork, *sb.*[5], erron. spelling of CALK *sb.*[1]

1806 WEBSTER *Compend. Dict.*, *Cork*, a sharp point on a horse shoe. 1846 W. T. PORTER *Quarter Race Kentucky* 162, I then just took my old mare down to a blacksmith's shop, and had some shoes made with 'corks' about four inches long, and had 'em nailed on to her hind feet. 1902 S. E.

WHITE *Blazed Trail* xxvii. 187 His face and flesh were ripped and torn everywhere by the 'corks' on the boots. 1922 H. TITUS *Timber* ix. 81 He..sharpened the corks, handling the cast with an odd excitement.

cork, *v.*[1] Add: **2. b.** Also with *down*.

1860 F. W. ROBINSON *Grandmother's Money* vi. viii, If you will only listen to your lord and husband's conversation with these good gentlemen, and cork the sentimental down, we shall soon be Darby and Joan again. 1894 *Outing* (U.S.) XXIV. 220/1 The description of the lion usually corked down in the 'animal book'. 1928 W. OWEN *Let.* 1 Feb. (1967) 377 We were corked down in those subways for close on 3 years.

cork, *v.*[4], erron. spelling of CALK *v.*[2]

1776 in *New Jersey Archives* 2nd Ser. I. 166 A chestnut sorrel mare,..shod before, shoes are steel corked. 1806 WEBSTER *Compend. Dict.*, *Cork*, *v.t.*..to form sharp points or shoe with points. a 1817 T. DWIGHT *Trav. New Eng.* (1821) II. 217 The clay..unless a horse is corked, is dangerous both to him and to the rider. 1829 *Virginia Lit. Museum* 16 Dec. 419 *To cork*, to shoe a horse with points—or with frost nails.

corker. Add: **1. b.** One who corks; one who provides a bottle with a cork.

1882 *Instr. Census Clerks* (1885) 58 Blacking Manufacture: Liquid: Bottler..Cooper..Corker. 1921 *Dict. Occup. Terms* (1927) § 459 Corker; cork biter; drives corks into bottles filled with beer, either by hand or by operating machine which he feeds with corks.

2. b. A person or thing of surpassing size or excellence; a stunner; also used ironically. *colloq.* and *dial.*

1882 *Cornhill Mag.* 325 We look over our boat-side and see the big salmon coming up..he is a corker. 1891 ... The 'Dic' ... *Slang* s.v., 'Jake Kilrain is a corker, and ought to have the championship of the world.' 1905 *Westm. Gaz.* 7 Sept. 2/1 The chairman..in proposing his health, is reported to have said, 'Sir Thomas, you're a corker.' 1904 'O. HENRY' *Cabbages & Kings* xiv. 256 By heavens! that dictator shot the corker. He's a dictator clear down to his finger-ends. 1909 H. G. WELLS *Tono-Bungay* I. iii. 75 Had her cry of, and corker..But now—woman, you're the goods. 1917 *Lawrence & Skinner Boy in Bush* i. 16 'You're a corker!' he exclaimed..to hear him talk like that. 1924 D. H. LAWRENCE & SKINNER *Boy in Bush* i. 16 'You're a corker!' ... 1936 ... *Laughing Gas* ii. 23, 'You really enjoy watching fights?' 'I know what you mean,' I said. 'Nine times out of ten they're absolute washouts, of course. But this one was a corker.' 1937 N. MARSH *Vintage Murder* xiv. 158 'You're a corker, you are,' said Wade hotly. 1969 R. D. ABRAHAMS *Jump-Rope Rhymes* 131 My girl's a corker.

c. Used *attrib.* or as *adj.* Fine, splendid, enjoyable, stunning. *N.Z. colloq.*

1937 N. MARSH *Vintage Murder* xiv. 158 'Talks with a corker sort of voice. Not quency, but just corker.' 1947 'A. P. GASKELL' *Big Game* 30 It's a corker wee place. 1948 D. W. BALLANTYNE *Cunninghams* (1986) I. ii. 50 The kids told Syd what a corker sixer it had been.

corking, *vbl. sb.* (see under CORK *v.*[1]). Add: Of wines: the action or process of becoming corked (see CORKED *ppl. a.* 4).

1904 *Lancet* 18 June 1738/2 'Corking' in wines or other fluids. 1920 G. SAINTSBURY *Notes on Cellar-Bk.* xiii. 199 That plague of the cellar-owner, 'corking'.

corking (kǭ·ɹkiŋ), *ppl. a.* Chiefly *U.S.* [After *CORKER 2 b.] Unusually fine, large, or excellent; stunning. Also *adv.*

1895 *Outing* (U.S.) XXVII. 103/2 Corking great fences the Vale doubles are. 1902 C. HOYT *Trip to Chinatown* in *5 Plays* (1941) ii. p. 120 A corking supper, my boy! Nothing's too good. 1904 *Munsey's Mag.* Mar. 870 A good show..and a good show we shall have that. 1909 *Daily Chron.* 4 Mar. 8/4 In a phrase that is already classic, but not yet classical, he [sc. T. Roosevelt] announced not long ago that he had had 'a perfectly corking time'. 1921 'IA. HOPE' *Mrs. Maxon Protests* xxii. 337 It turns out to be a perfectly corking house—a jewel of a house, Stephen! 1919 WODEHOUSE *Damsel in Distress* x. 127 There's nobody I know..a corking sportsman than Maud. 1926 *Ladies' Home Jrnl.* Dec. 28 He..engaged me, at a corking fee, to come to..take this case. 1937 N. MARSH *Vintage Murder* vii. 73 He's got a corking sort of laugh.

Hence **co·rkingly** *adv.*

1917 WODEHOUSE *Uneasy Money* xv. 169, I saw a most corkingly pretty girl bicycling down to the village one morning. 1945 T. RATTIGAN *Love in Idleness* I, I'm sure you two are going to get on—quite corkingly.

corking-pin. (Later example.)

1848 MRS. GASKELL *Mary Barton* II. xvii. 249 Some insect, which he was impaling on a corking-pin.

corks (kǫɹks), *int.* Deformation of *cock's* (see COCK *sb.*[1]), or perh. an amalgamation of *cock* and LAWK, LAWKS.

1926 B. RUCK *Her Pirate Partner* xiii. § 3 Mrs. Llewelyn in stark agony had gasped out the word 'Corks!' 1927 *Daily Express* 8 July 8/7 Another frequent expression was 'Corks!' 1939 'R. CROMPTON' *William & A.R.P.* i. 32 'Corks!' gasped William...'Corks! I thought she was never goin'.'

corkscrew, *sb.* Add: **1. b.** (Earlier U.S. example.)

1839 F. TROLLOPE *Dom. Manners* (ed. 5) xxix. 273 Loan me the two corkscrews then, mother, just to hang behind my spit curls.

2. (Further examples.) Also, **corkscrew grass**, a kind of grass having a twisted seed with long awn.

1815 *N. Amer. Rev.* II. 20 The innocent animated imitations of the Medicean Venus, with their thousand cork-screw ringlets. a 1828 D. WORDSWORTH *Tour Continent* in *Jrnls.* (1941) II. 28 The splendid cork-screw spire. 1838 DICKENS *O. Twist* II. xxxi. 35 Their..hair..tortured into long corkscrew curls. 1890 *Melbourne Argus* 26 Sept. 13/5 To get the fleeces off before the grass seeds ripen, those of the corkscrew grass-grass (*Stipa setacea*). 1948 *Nature* 27 Jan. 311 Spear or corkscrew grass. 1958 'J. SHERWOOD' *Lions & Shadows* i. 34 Clematises came tearing round the corkscrew corners.

corkscrew, *v.* Add: **1.** Also, to twist spirally. *pass.*

1851 H. MELVILLE *Moby Dick* xxxiv, Caught and twisted, corkscrewed in the masses of the line, loose harpoons..came flashing. 1896 *Daily News* 26 Sept. 5/3 The line itself had been completely swept away, and great sections crumpled and corkscrewed beyond hope of repair. 1898 *Ibid.* 24 Aug. 2/1 The fancy grottos and rockeries would be swept into the sea if the stream was corkscrewed. 1942 R. CAMPBELL *Flaming Terrapin* iii. 45 Thick maelstroms..with hideous twist, Corkscrewed by whirlwinds.

2. b. To become twisted.

1918 *Chambers's Jrnl.* July 413/2 As in the case of the bayonets which 'corkscrewed' in the Soudan.

3. b. *intr.* To work *away* at as with a corkscrew.

1920 J. GREGORY *Man to Man* iv, Trying to pump me and corkscrewing away at dad when he was full of whisky.

corkwood. Add: **2.** Also, a name given in New Zealand to *Entelea arborescens*.

1889 T. KIRK *Forest Flora N.Z.* 45 The whau..is termed corkwood by the settlers on account of its light specific gravity. 1966 *Jrnl. Polynesian Soc.* June 156 Whau, a tree (*Entelea arborescens*) cork-wood, the lightest of Newzealand timbers.

corky, *a.* Add: **1. b.** *corky scab* = *potato scab* (POTATO *sb.* 7). *Austral.*

1911 D. MCALPINE *Handbk. Fungus Dis. Potato in Australia* iv. 78 (*heading*) 'Corky', Powdery or Spongospora Scab. 1932 *Bulletin* (Sydney) 6 Dec. 28/3 Corky scab appears on the skin of the tuber, but when more severe the flesh of the potato is destroyed and a canker-like appearance supervenes. 1949 *Land & Coast* ... 34 The farm had failed with Corky Scab on the potatoes.

4. (Examples.)

1889 *Cent. Dict.* s.v., A corky flavor. 1925 F. S. FITZGERALD *Great Gatsby* (1926) i. 15 We tasted a sip of corky but rather impressive claret. 1967 *New Statesman* 15 Dec. 860/2 We should say corky to describe an 'off-odour'.

Corliss (kǭ·ɹlis). The name of G. H. *Corliss* (1817–88) of Providence, Rhode Island, U.S., used *attrib.* to designate (*a*) the valve gear invented by him in 1849, or a modification of it, (*b*) an engine equipped with such a valve gear.

1868 *Pract. Mech. Jrnl.* July 177 The Corliss engine introduced by Mr. Corliss in the United States about twenty years ago..embodies in the arrangement of the cylinder and valve-gear several principles that had previously been used separately. *Ibid.* 178 The arrangement and construction of the details in the mechanism employed, rather than the application of any new or untried principle, ..constitute the special features of the Corliss valve-gear. 1876 *Engineering* XXII. 12 The great Corliss beam engine. 1887 *Encycl. Brit.* XXII. 507/2 Corliss Valve-gear, Spencer Inglis form. 1888 *Lockwood's Dict. Terms Mech. Engin.*, Corliss Valve. 1930 *Engineering* 5 Sept. 307/1 Two double-acting cylinders in tandem, with inlet and outlet valves of the Corliss type. 1958 *Van Nostrand's Sci. Encycl.* (ed. 3) 1756/2 Engines with oscillating valves derive their motivation from an eccentric on the crankshaft, but there is interposed between the eccentric rod and the valve a wrist plate and valve reach rods. The Corliss engine is like this.

corn, *sb.*[1] Add: **1. c.** *colloq.* (orig. *U.S.*). Something 'corny' (see *CORNY *a.*[1] 1 c); *spec.* old-fashioned or inferior music. Also *attrib.*

1936 *Variety* 24 June 63/1 The B.B.C. doesn't understand..that the great English public loves corn, even in a waltz.. 1937 (see *CORN-FED *a.* c). 1946 H. SANDOZ in *Amer. Speech* XXI. 154/1 The word *corn* has..(since c 1890 to 1910).. featured a great variety of sweet corn..interspersed with short jokes and riddles, sometimes even cartoons. The jokes were all time-worn and over-obvious and were called corn calling jokes or *corn jokes*, and any such gag or joke of that nature was called *corny*. 1952 *Economist* 9 Feb. 339/2 Governor Stevenson..declines to indulge in the political 'corn' which is supposed to get votes. 1955 KEEPNEWS & GRAUER *Pict. Hist. Jazz* xvii. 233 Other jazz bands, dance bands, corn bands. 1958 *Punch* 6 Aug. 8/6, I know now that 'O by jingo, O by gosh, by gee!' was not jazz, but the merest corn. 1959 *Spectator* 18 July 115/2 Specialising in exposures, wheezes and jokes of which at least 90 per cent. are pure corn. 1968 H. MCCLOY *Mr. Splitfoot* (1969) xvii. 192, I was invited to perform..it's old-fashioned corn. It's purely selfish.

5. *corn on the cob*: green maize suitable for boiling or roasting; maize cooked and eaten on the cob. orig. *U.S.*

1867 T. F. DE VOE *Market Assistant* 414 What usually makes a bushel: Sixty pounds of wheat, Irish Potatoes, Beans, or clover-seed;..seventy pounds of corn on the cob. 1951 *Good Housek. Home Encycl.* 374/2 A form of dressing served with..corn on the cob.

b. *ellipt.* Corn-whiskey. *U.S.*

1820 *Chillicothe* (Ohio) *Supporter* 5 July, If we go to town, ..we are invited to try a little corn as usual. 1846 J. J. HOOPER *Adv. Simon Suggs* (1851) v. 54 Let me git one o' these book-larnt fellers over a bottle of 'old corn'. 1936 *Times Lit. Suppl.* 12 Dec. 1033/1 Jessamyne is a shrewish bustling woman, a notable maker and peddler of 'corn' *i.e.*, home-distilled spirits.

7. *to acknowledge* (*admit, confess*) *the corn.* Bring here def. and examples from CORN *sb.*[2] 2.

1839 *New Orleans Picayune* 15 Apr. 2/1 We were certain it was not Dutch, and was in error in saying it was Scotch, and 'acknowledge the corn'. 1842 *Spirit of Times* (Philad.) 16 Mar. (Th.), Your honor, I confesses the corn. I was royally drunk. 1846, 1883 [see CORN *sb.*[2] 2]. 1854 B. P. SHILLABER *Mrs. Partington* 152 The old Sherry admitted the corn, turned over and slept on it. 1902 HARBEN *Abner Daniel* 136 When anybody teaches me any tricks, I acknowledge the corn an' take off my hat. 1948 C. E. FUNK *Hog on Ice* 38 To acknowledge the corn..means to admit the losing of an argument, especially in regard to a detail; to retract; to admit defeat.

9. *corn-barn, -basket*; *corn-colour, -coloured* (earlier examples); *corn-planting* (examples), *-row, -sampler, -shock* (later examples), *-top*.

1780 E. PARKMAN *Diary* 16 Oct. (1899) 278 Dr. Hawes.. took the whole care of husking the Corn, & carrying it into the Corn Barn. 1864 T. L. NICHOLS *Amer. Life* I. 22 At a little distance was..a corn-barn for storing Indian corn. 1648 *Connecticut Public Records* I. 487, 1 spade..a corne baskitt. 1809 'D. KNICKERBOCKER' *Hist. N.Y.* III. iv, The contents filled a couple of corn-baskets. 1871 MRS. STOWE *Oldtown Fireside Stories* 66 Hand me that corn-basket; we'll put that over him. 1873 J. H. BEADLE *Undevel. West* xix. 357 Rich dresses were numerous at church, particularly of corn-color silk. 1854 M. J. HOLMES *Tempest & Sunshine* vii. 103 If I catch you here again dickerin' after Fanny, I'll pull every corn-colored har out of your head. 1809 E. A. KENDALL *Trav. Northern U.S.* II. 13 Thinking to ask some usual question,..as to whether the family had done their corn planting [etc.]. 1849 J. PRITTS *Mirr. Border Life* 460, I remained in that situation till corn planting time. 1769 WASHINGTON *Diaries* (1925) I. 347 That part of it which the Corn rows run through received no other plowing. 1879 TOURGÉE *Fool's Err.* xxviii. 173 My wife..took the children along the corn-rows to the woods. 1898 *Westm. Gaz.* 29 Dec. 5/2 A corn-sampler,..living in Bermondsey. 1857 *Trans. Ill. Agric. Soc.* II. 665 During winter they may be tracked.. to the neighboring corn-shocks, which they have visited for food. 1969 E. H. PINTO *Treen* 97/2 In New England, where the term 'shock' is more usual than 'stock', the device was known as a corn shock binder. 1733 J. HEMPSTEAD *Diary* (1901) 265 Stacked our Corn Tops & Husks. 1814 J. TAYLOR *Arator* (ed. 2) 93 Some loss will accrue from the evaporation of a cover, whether composed of straw or corn tops. 1902 W. S. GORDON *Recoll. Old Quarter* 121 How the rustic hymns would be drawled out in the long summer days, to float away over the corntops.

11. corn-ball (examples); (*b*) *slang* (orig. *U.S.*), a 'corny' person (see *CORNY *a.*[1]); also *attrib.* or as *adj.*; **corn belt**, the area in which corn is grown; **corn-broom** (delete *U.S.* and add examples); **corn-ear worm** orig. *U.S.*, a larva (*Heliothis armigera* or *zea*) that feeds on corn and other plants; **cornflakes** *n.pl.* orig. *U.S.*, a kind of breakfast cereal made from flaked and flavoured maize; **corn-fodder** (examples); (see also quot. 1744); **corn-fritter** (examples); **corn-grinder**, (*a*) *U.S.* = METATE; (*b*) = CORN-MILL; (*c*) one who grinds corn; so **corn-grinding** *vbl. sb.*; **corn-king**, cf. *corn-spirit* (see Dict.); **corn-kist** *Sc.* and *north. dial.*, a corn-bin; **corn-kister** *Sc.*, 'a rollicking song sung at gatherings of farm-workers' (*Sc. Nat. Dict.*); **corn-knife** (examples); **corn-moth** (examples); **corn-mush** *U.S.* (see MUSH *sb.*[1] 1); **corn-oyster** (examples); **corn-popper** (examples); **corn-roast** *N.Amer.*, a party at which green maize is roasted and eaten; **corn-sheller** (earlier and later examples); so **corn-shelling** *vbl. sb.*, the process of detaching the grains of maize from the cob; also *attrib.*; **corn-silk** *U.S.*, the styles of maize; **corn-silker**, a machine for removing the corn-silk; **corn-spurrey**, a small weed, *Spergula arvensis*, bearing white flowers and whorled leaves; **corn syrup** *U.S.*, a thick syrup made usu. from corn-starch; **corn-weevil**, one of several popular names for weevils attacking corn; **corn-whiskey** *U.S.*, a spirit distilled from maize.

1843 'R. CARLTON' *New Purchase* ix. 64 Nanny remained near the dutch oven to keep us supplied with red-hot pones or corn-balls. 1873 E. PHELPS *Trotty's Wedding Tour* 3 They were eating a corn ball at recess. 1952 R. C. RUARK in Wentworth & Flexner *Dict. Amer. Slang* (1960) 124/2 Eisenhower on no account can be called a cornball. 1952 B. ULANOV *Hist. Jazz in Amer.* (1958) xxv. 350 *Corny*, stale, insipid..and so too 'corn' (noun), 'cornfed', 'cornball', and 'off the cob'. 1962 *Movie* June 28/1 An expatriate cornball like Jerry Court. 1962 *Melody Maker* 7 July 11 Parker hired a cornball duo at one of the Sunday Reisner-Parker sessions in order to get rid of the audience. 1970 *Daily Tel.* (Colour Suppl.) 1 May 19 Things today have to be carefully said, because we live in that kind of world where the truisms sound cornball. 1882 *Nation* (N.Y.) 13 July 24/3 Crop reports from the West still continue favorable, though there are some discouraging accounts of the prospects in the 'corn belt'. 1959 J. THURBER *Years with Ross* ii. 32 Ben Hecht..was a police reporter at heart, Elmer Davis a corn-belt intellectual. a 1817 T. DWIGHT *Travels* (1822) IV. 485 Straw bonnets, Brushes, Corn Brooms. 1930 M. DE

LA ROCHE *Finch's Fortune* (1931) xxvi. 339 She was scrubbing the mud with a corn broom. 1963 *Times* 25 Feb. p. xvi/3 But youngsters in Canada swarmed to the rinks by the thousands and between the flashy long slide and the remarkable sweeping with corn brooms, as opposed to the Scots' scrubbing brush, they were completely sold on the game. 1889 *Rep. Secr. Agric.* (U.S.) 360 The Corn Earworm (*Heliothis armigera*) has done considerable damage to the ears of field corn. 1950 *N.Z. Jrnl. Agric.* Dec. 519/3 To protect the young cobs of corn from attacks by the corn ear worm, D.D.T. may be applied as a spray. 1907 in G. CARSON *Cornflake Crusade* (1957) (photograph between pp. 148 and 149), Toasted cornflakes. 1908 *Sat. Even. Post* 31 Oct. 55/3 There are 13 imitations of Kellogg's Toasted Corn Flakes. 1920 F. S. FITZGERALD *This Side of Paradise* (1921) I. ii. 51 Amory indulged in a club sandwich, cornflakes, and Julienne potatoes. 1948 'N. SHUTE' *No Highway* ix. 241, I breakfasted in a hurry on cornflakes and bacon and eggs. 1960 J. BETJEMAN *Summoned by Bells* v. 44 And, in the morning, cornflakes, bread and tea. 1744 ELLIS *Mod. Husb.* Feb. vi. 39 Our Farmers, never fail to sow Hog-pease..for the great Service their stalks do the Farmer, in supplying him with the best of Corn-fodder. 1772 *Carroll Papers* in *Maryland Hist. Mag.* XIV. 288 We shall get in all our Corn Fother by the middle of next week. 1838 H. COLMAN *Rep. Agric. Mass.* 24 Farmers..estimate the corn fodder or stover upon an acre as equal to three fourths of a ton of English hay. 1904 M. E. WALLER *Wood-carver* ii, I can look..up the slope of the mountain, across the stony corn-fodder patch. 1862 MRS. STOWE in *Independent* 13 Feb. (Bartlett), A very minute account which Mrs. Kittridge was giving of the way to make corn-fritters which should taste exactly like oysters. 1903 *N.Y. Even. Post* 26 Sept., Corn fritters may be made of this hard corn. 1841 *Knickerbocker* XVII. 234 Improved..corn-grinders. a 1854 G. FURMAN *Antiq. Long Isl.* (1875) 97 At Maspeth Kills..Indian corn-grinders..have been frequently ploughed up. 1854 Corn-grinder [see METATE]. 1880 HARDY *Trumpet-Major* I. ii. 23 Miller Loveday was the representative of an ancient family of corn-grinders. 1907 *Daily Chron.* 14 Nov. 6/2 Disguised as corn-grinders. 1905 *Westm. Gaz.* 30 Nov. 8/2 That they shall revert to corn-grinding. 1927 PEAKE & FLEURE *Peasants & Potters* 48 The discovery of corn-grinding stones. 1931 NAOMI MITCHISON (*title*) The Corn King and the Spring Queen. 1947 C. S. LEWIS *Miracles* xiv. 137 The deity—Adonis, Osiris..—[was] almost undisguisedly a personification of the corn, a 'corn-king' who died and rose again each year. 1616 *Edinb. Test.* XLIX. 99 b (D.O.S.T.), The corne kist, price fowr pundis. 1876 F. K. ROBINSON *Gloss. Whitby* 42/2 Coornkist, the corn-bin. 1936 *Huntly Express* 10 Jan. 6 There was speeches made, an' cornkisters sung. 1937 *St. Andrews Citizen* 27 Mar. 4/1 The programme was as follows: —Opening chorus and a curn corn-kisters. 1856 *Trans. Mich. Agric. Soc.* VII. 54 Six corn knives. 1948 *Chicago Daily News* 9 Oct. 3/3 Anything from a machete to a hatchet or corn knife. 1766 *Compl. Farmer* 4/1 The possibility of destroying the corn moth, or worm, without hurting the grain. 1869 *Rep. U.S. Comm. Agric.* 1868 315 The clothes-moth and corn-moth are representatives of the family. 1846 J. S. GRIFFIN *Diary* 24 Nov. in *Calif. Hist. Soc. Q.* (1942) XXI. 216 They had nothing to eat but penolas & corn mush. 1928 D. H. LAWRENCE *Woman who rode Away* 81 Tortillas and corn-mush with bits of meat. 1941 A. HUXLEY *Let.* 13 Mar. (1969) 465 Most of the southern farmers still prefer corn mush..to vegetables. 1847 *Carolina Housewife* 101 Corn Oysters. Grate corn, while green and tender, with a coarse grater, in a deep dish. To two ears of corn allow one egg; beat the whites and yolks separately, and add them to the corn, with one table-spoonful of wheat flour and one of butter. ? 1862 MRS. STOWE in *Independent* (Bartlett), In this secret direction about the mace lay the whole mystery of corn-oysters. 1874 B. F. TAYLOR *World on Wheels* I. viii. 63 The baggage-car is as lively with all sorts of baggage as corn in a corn-popper. 1877 BARTLETT *Dict. Amer.* (ed. 4) Add., *Corn-popper*, an instrument the top of which is like a sieve, in which corn is held over a fire to roast or 'pop'. 1921 *Daily Colonist* (Victoria, B.C.) 12 Oct. 20/1 (Advt.), Corn Poppers at, each 45c. 1946 *Reader's Digest* Jan. 60/1 The Boy would have the cornpopper waiting. 1899 *Westm. Gaz.* 4 Jan. 8/1 The last number of the *Harvard Graduates' Magazine* says that..the two old boys..went to a corn-roast. 1923 *Beaver* Oct. 29/2 Girls of the office staff spent an enjoyable time..the occasion being a corn roast. 1944 E. A. HOLTON *Yankees were like This* 256 Eating clams is the only act more messy than gnawing corn off the cob at a corn roast. 1813 in *Mem. Philad. Soc. for Promoting Agric.* (1814) III. 250 (*caption*) Description of the Corn sheller..One bushel of corn in the ears, may with ease be shelled in five minutes. 1825 *Boston Monthly Mag.* I. 25 A patent corn-sheller..may be of more service to the community, than all the poems of Byron or the novels of Scott. 1851 C. CIST *Cincinnati* 169 Corn-shellers, to shell twenty to fifty bushels per hour. 1813 in *Mem. Philad. Soc. for Promoting Agric.* (1814) III. 249 The inventor of the corn shelling machine..is John Haven, of Montgomery county [Penna.]. a 1848 RUXTON *Life Far West* (1849) 62 Better for him had he minded his corn-shelling alone. 1877 HABBERTON *Jericho Road* xiii. 120 The pork-packin' and corn-shellin' seasons nearly on us. 1861 *Harper's Mag.* Aug. 322/2 [The] *Tinea Maisinia*—'*Corn-silk Moth*'..feeds only on the efflorescence or silk of the corn. 1878 B. F. TAYLOR *Between Gates* 258 Let the fellow's hair turn the color of corn-silk in the sun. 1902 S. E. WHITE *Blazed Trail* xxxix. 279 His eyes desired to follow the soft white curve of her cheek to dance with the light of her corn-silk hair. 1967 *Boston Sunday Herald* 26 Mar. IV. 5/3 Ada Cloninger..had never smoked anything since a girlish bout with corn silk. 1968 *Washington Post* 5 July A 18/3 Dyed corn husks, corn silk and pipe cleaners. 1902 *Encycl. Brit.* XXVI. 558/1 Among the devices in common use are..corn cutters, corn silkers, pea briners, [etc.]. 1771 *Ibid.* III. 621/1 *Spergula.. arvensis*, or corn-spurrey. 1846 J. BAXTER *Libr. Pract. Agric.* (ed. 4) II. 294 Corn spurry..is found most frequently in soils rather moist. 1962 A. R. CLAPHAM et al. *Flora Brit. Isles* (ed. 2) 257 *S*[*pergula*] *arvensis*, Corn Spurrey, an annual herb with ascending geniculate stems 7·5–40 cm., branching close to the base. 1903 *U.S. Dept. Agric. Circular No.* 10 8 *Glucose sirup* or *corn sirup* is glucose unmixed or mixed with sirup or molasses. 1964 *Listener* 26 Nov. 842/2 They regarded..British rations with varying degrees of horror. 'What,' said the Canadians, 'no corn syrup?' 1840 J. & M. LOUDON tr. *Kollar's Treat. Insects* II. 117 The Corn Weevil..

The perfect insect is of small size, linear shape, with a narrow rostrum. 1850 *Rep. Comm. Patents Agric.* 1849 335 The other species of corn-weevil alluded to,..the granary weevil, is more common in this country. 1880 O'NEILL & WILLIAMS *Amer. Farmer's Handbk.* xiv. 642 (*heading*) Corn Weevil.—A small insect... The female deposits her eggs upon corn in granaries, and the young larva at once burrows into the grain. 1843 'R. CARLTON' *New Purchase* 172 Candidates..making licentious speeches, treating to corn whiskey, violating the sabbath. 1873 'MARK TWAIN' & WARNER *Gilded Age* xiii. 130 A bottle of apparently corn-whiskey. 1948 *Sat. Rev.* 28 Aug. 12/3 Chism himself teaches eleven-year-old Jarvis to drink corn whiskey.

corn, *sb.*[2] Add: **2.** *to tread on anyone's corns* (earlier example); *to acknowledge the corn*: see *CORN *sb.*[1] 7.

c 1845 C. BRONTË *Professor* (1857) II. xxv. 240 To work me into lunacy by treading on my mental corns.

3. *corn-footed* adj.; *corn-plaster* (examples); **corn-cure**, a remedy for corns.

1868 G. MEREDITH *Let.* 31 Jan. (1970) I. 369 In a fit of distraction I may stick a corn plaster on the envelope instead of a Queen's Head. 1879 *Chemist & Druggist* 15 Feb. 58/2 Corn Plaster (Dr. Smith).—Felt rings spread with sticking-plaster. *Ibid.* 15 Mar. 134/1 The adhesive surface of Corn Plaisters, as usually made, is..a gelatinous solution. 1881 *Graphic* 21 May (Advt.), Bond's marvellous corn cure. 1906 R. WHITEING *Ring in New* 158 The methods of a puff for corn-cure. 1913 J. STEPHENS *Here are Ladies* 71 An old, ache-ridden, cough-racked, corn-footed septuagenarian. 1970 D. CLARK *Sweet Poison* iii. 60 Common household medicines—T.C.P., plasters, Panadol, two or three ointments and a corn cure.

corn, *v.* **8.** (Later U.S. examples.)

1886 *U.S. Consular Rep.* No. IX. 40 Those hundreds of thousands of acres of once valuable Southern lands, corned to death, and now lying to waste in worthless sage grass. 1940 J. STUART *Trees of Heaven* 129 I'd corn this land three years, then I'd sow it in wheat and orchard grass.

‖**cornada** (kǭɹnā·dā). [Sp.] A horn-wound; the goring of a bull-fighter by a bull.

1932 E. HEMINGWAY *Death in Afternoon* xix. 244 In 1928 he was gored severely twice, the second cornada.. a result of his fighting before he was in good shape after the first. 1967 *Listener* 10 Aug. 184/1 Haime Ostos' heroic return after a near fatal *cornada*. 1968 J. M. WHITE *Nightclimber* (1970) ix. 74 The veteran torero..had received a *cornada* in the groin..and died three hours later.

cornbrash. (Earlier example.)

1813 J. TOWNSEND *Char. Moses* I. vi. 103 A rock known by the name of corn brash and corn grit...This brashy, or ragged, calcareous rock..may be noticed..either round Atford, and at Wraxall, or at Chippenham, with the interjacent country.

corn-bread. *U.S.* [CORN *sb.*[1] 5.] Bread made of corn-meal.

1775 J. SCHAW *Jrnl. Lady of Quality* (1921) 171 And vast quantities of pork, beef and corn-bread were set forth. 1823 E. JAMES *Exped. Rocky Mts.* I. 68 They purchased a ham, and a loaf of corn bread of Mr. Kennedy. 1865 T. W. KNOX *Camp-fire & Cotton-field* 336 Corn-bread, bacon, and potatoes were the only articles set before us. 1913 MRS. STRATTON-PORTER *Laddie* xii, There wasn't a crumb there except cornbread, and she didn't want that. 1959 *Wall St. Jrnl.* (Eastern ed.) 4 Mar. 1/6 Mr. Hudnall's small pension enables him to eke out what the valley folk term a 'corn bread living'.

corn-cake. (Earlier and later examples.)

1791 W. BARTRAM *Trav. Carolina* 38 It is..an ingredient in most of their cookery, especially homony and corn cakes. 1833 *Louisville Daily Herald* 18 Oct., Hence the corn cake and dodger cake, a species which Mrs. Trollope had the honor of inventing, for it was never heard of in Cincinnati before. 1903 *N.Y. Sun* 1 Nov., Corn cakes as a substitute for bread are popular in the South. 1936 M. DE LA ROCHE *Whiteoak Harvest* xxv. 278 Miss Archer was twice as long as usual in making coffee, heating corn cake.

corn-cob. Add: (Earlier examples.)

1787 (*title*) Adventures of Jonathan Corncob,..written by himself. 1796 J. BARLOW *Hasty-Pudding* III, The dry husks rustle, and the corn-cobs crack. 1810 J. LAMBERT *Trav. Canada* I. viii. 136 They [sc. Canadians] are extravagantly fond of the corn cobs boiled or roasted, and rubbed over with a little butter and salt.

b. *attrib.* and *Comb.* (Examples.)

1801 *Spirit of Farmer's Museum* 236 Jotham, get the great case bottle, Your teeth can pull its corn cob stopple. 1832 J. P. KENNEDY *Swallow B.* II. xix. 246 He stood in the group, with his corncob pipe, puffing the smoke from his bolster-lips. 1853 *Trans. Mich. Agric. Soc.* IV. 83 Best corn-cob crusher by horse power. 1856 *Yale Lit. Mag.* XXI. 145 (Th.), He was employed in whittling a corn cob bowl into a pipe. 1889 'C. E. CRADDOCK' *Broomsedge Cove* iv. 53 An aroma lingered about its corn-cob stopper. 1888 in Farmer *Americanisms* (1889) 170/1 Soon after that disgrace a party of the boys prepared a lot of grenades—corn-cob shells they called them—and determined to storm head quarters. 1911 *Springfield Weekly Republ.* 12 Jan. 1 Speaker Cannon gave Washington a shock Thursday by appearing in one of the Capitol lobbies smoking a corncob pipe. 1960 B. KEATON *Wonderf. World of Slapstick* (1967) 72 His mother, a frail old Southern woman who smoked a corncob pipe, put me to bed.

c. Short for *corn-cob pipe*.

1892 *Outing* (U.S.) Sept. 441/1, I were settin' on de po'ch yender, smokin' m'ole corncob. 1923 *Blackw. Mag.* Dec. 727/1 Mr. Quigg puffing away at a corn-cob.

corn-cracker. Add: **1.** Earlier and later examples in sense 'a native of Kentucky'.

1835 *Western Rev.* June 342 There is neither wit nor meaning in the terms *Hoosier*, *Sucker*, *Corncracker*, and *Buckeye*, which have become so current. 1845 *Olympia Pioneer* (Th.), Corn-crackers, Potsoppers, Hard Heads, Hawk Eyes, Rackensacks, etc. 1940 H. H. HATCHER *Buckeye Country* 298, I never in my life heard a Buckeye get into his voice that quiver of ecstasy that is second nature to a Corn-cracker when he mentions his bluegrass and his mountains and his folks.

2. An apparatus for cracking corn.

1844 LEE & FROST *Oregon* xii. 134 At the mission we had a small cast-iron corncracker, in which we ground wheat after a fashion. 1900 SMITHWICK *Evol. State* 76 There was a sawmill with a corn cracker attached.

corn dance. [CORN *sb.*[1] 5.] Any of various dances, among North American Indians and Negroes, connected with the sowing or harvesting of maize.

1726 G. CHICKEN *Jrnl.* in N. D. Mereness *Trav. Amer. Col.* (1916) 155 At Little Terriquo where was mett together at the corn dance several of the head men. 1837 *Southern Lit. Messenger* III. 391 The celebrated *corn dance* of the Seminoles is said to be connected with their religion. 1840 C. F. HOFFMAN *Greyslaer* III. viii. 185 De boys—..has gone to de village to hold corn-dance for seedtime. 1903 S. E. WHITE *Forest* x. 128 Belts of beadwork, yellow and green, for the Corn Dance.

cornely (kǫɹne·li, -i·li). Also **cornelly, C-.** Name of Émile *Cornely*, engineer, the first maker of the chain-stitch embroidery machine with universal feed invented by J. Bonnaz (cf. *BONNAZ): used to designate the machine, the machinist using it, or the embroidery made. Hence **corne·lling**, the making of embroidery with this machine.

1903 *Daily Chron.* 27 Feb. 10/5 Bonnaz.—Few good Cornely hands..wanted. 1921 *Dict. Occup. Terms* (1927) § 407 By hand or by cornely appliqué machine. *Ibid.*, By cornely, singer or other machine. *Ibid.*, Machinist, cornely or cornelly; cornelling machinist; cornely hand. 1952 C. W. CUNNINGTON *Eng. Women's Clothing* 294 *Cornelly*, an embroidery of lacy design made by the Cornelly machine on a vanishing cloth which is afterwards removed, leaving the embroidery. 1967 E. SHORT *Embroidery & Fabric Collage* ii. 51 Trade machines on which chain stitch and a free satin stitch can be worked (the Cornely and the Irish Machines).

corner, *sb.*[1] Add: **2. b.** *to turn the corner*: earlier example of sense 'to start recovering from an illness'. Colloq. phr. (*a*)*round the corner*: (*a*) nearby; a short distance away; (*b*) at hand; about to occur or be realized; imminent.

1837 DICKENS *Let.* 28 Jan. (1965) I. 229, I hope to find on Monday at 12 o'Clock, that *you* have turned the corner, and come back again. 1837 —— *Pickw.* xxxi. 334 The ham.. was also from the German sausage-shop round the corner. 1863 L. M. ALCOTT *Hospital Sk.* v. 67 A secret conviction that pneumonia was waiting for me round the corner. 1914 FANNIE HURST (*title*) Just around the corner. 1925 W. J. LOCKE *Great Pandolfo* vii. 89 It's more or less only round the corner; but I see that you're walked enough. 1930 'I. HAY' & WODEHOUSE *Baa, Baa, Black Sheep* I. i. 12 You never know what's waiting for you just round the corner in this world! 1959 *Times Rev. Industry* Jan. 67/2 Polypropylene has for so many months been 'just-round-the-corner' that [etc.]. 1960 WILLMOTT & YOUNG *Family & Class* iii. 28 Couples..went on living 'up the street' or 'round the corner' from their parents. 1970 *Guardian* 15 Aug. 3/6 At the 'self' round the corner where I'd gone to buy soapflakes.

3. b. The angular projections (or projection) on each side of a violin or other similar stringed instrument. Also *corner-block*.

1888 *Encycl. Brit.* XXIV. 243/1 The Viol has..Square or obtuse corners...The Violin has..Acute corners. 1889 GROVE *Dict. Mus.* IV. 275/1 These corners mark the position of triangular 'blocks' inside, to which the ribs..are glued...Corner-blocks..first appear in the 15th century. 1968 *New Oxf. Hist. Music* IV. xiii. 710 A majority of viols.. have sloping shoulders, C-shaped sound-holes,..and square corners at the bouts.

4. b. The triangular piece cut from the ham or hind-end (the gammon) of a side of bacon.

1891 T. F. GARRETT *Encycl. Cookery* I. 63/2 Gammon with Corner, 14 lb. *Ibid.*, Corner of Gammon, 4 lb. 1917 G. J. NICHOLLS *Bacon & Hams* 59 The gammon may also be cut into two pieces—a large corner and a large gammon hock. 1923 R. E. DAVIES *Pigs & Bacon Curing* 29 The side may be cut into three parts, comprising the fore end, the middle, and the gammon with corner.

5. c. *tight corner*: see TIGHT *a.* 9.

12. (Earlier examples.)

1699 in *Derby* (Conn.) *Rec.* (1901) 202 The southered corner is an ash tree. 1770 WASHINGTON *Diaries* (1925) I. 428, I marked two Maples, an Elm, and Hoopwood Tree as a Corner of the Soldiers Ld. 1816 U. BROWN *Jrnl.* in *Maryland Hist. Mag.* XI. 221 If the old original white oak Corner cannot be found establish a Corner there in lieu and in place of said White Oak.

13. a. (Earlier example of form *corner-kick*.)

1882 *Blackburn Times* 1 Apr. 6/3 Then the besiegers had another corner kick, which, like its predecessors, proved fruitless.

d. *Boxing.* One of the two opposite angles of a boxing ring in which a boxer rests between rounds; hence, a boxer's second or seconds. Also *attrib.*

1853 *London Prize-Ring Rules* in T. C. Wignall *Story of Boxing* (1923) 312 The combatants..shall retire until the seconds of each have tossed for choice of position, which adjusted, the winner shall choose as his corner..and conduct his man thereto, the loser taking the opposite diagonal corner. 1888 F. W. J. HENNING *Recoll. Prize Ring* 159 Both men were assisted to their corners by their seconds. 1889 G. B. SHAW *London Music 1888–89* (1937) 278, I read some indignant articles on the unfairness of the Bruges prize fight...The corner-men were English. 1896 J. C. TROTTER *Boxing* xv. 96 On first entering the ring, the combatants take a seat in their corners and wait till the timekeeper calls 'Time'. 1952 L. A. G. STRONG *Darling Tom* 145 My corner was getting a bit suspicious.

e. *Hockey.* (In full *corner-hit.*) A free hit against the defending side made by one of the attacking side from the corner flag.

1897 *Encycl. Sport* I. 518/2 A good goalkeeper..should never hesitate to concede a corner when he can see no other opening. 1907 *Hockey* 22 Nov. 22/1 Just before half-time, Mid-Surrey broke away, and, forcing a corner, Nash put in an excellent shot from the corner hit. *Ibid.*, Before the end Simmonds scored from a penalty corner. 1935 *Encycl. Sports* 344/1 From a corner the attacking side can only score a goal after the ball has been stopped dead or hit after it has struck or been played by a defender. 1961 F. C. AVIS *Sportsman's Gloss.* 215/2 In Hockey, a corner is registered where a player deliberately puts the ball over his own goal line, or where he accidentally does so from a distance shorter than 25 yards from the goal line.

14. (Earlier and later U.S. examples.) Also *fig.*

1853 *Captain Priest* 249 (Th.), He is the greatest of all men for a 'corner'. 1947 *Time* 27 Oct. 106/2 One of his exploits was the 'Hale & Norcross corner' in 1868, by which he got control of an important mine. 1949 *Times* 17 Nov., To make a corner in something is monopolistic, unethical and ill-bred. 1965 *Language* XLI. 200 The followers of Chomsky have no corner on nastiness.

b. *dial.* and *slang.* A share; esp. (*a*) in dial. phr. *to stand one's corner*: to take or pay for one's share of anything; to do one's share; (*b*) *slang*, a share in the proceeds of a robbery.

1878 J. ALMOND *Bunch of Watercresses* 21 Tom looked at his shirt to see iv id could be doctored..into doing duty for some o' th' missing articles..but he fun id wer hardly able to stan id own corner. 1889 BARRÈRE & LELAND *Dict. Slang* I. 273/1 *Corner*, (Thieves), a share—generally a share in the proceeds of a robbery. 1891 *Standard* 5 Mar. 2/5 (E.D.D.), He had arranged to meet the other two men to receive his 'corner' (the proceeds of the sale of the stolen property). 1930 E. WALLACE *White Face* xix. 279 He supplied the various gangs [of bank-robbers] with information which enabled them to operate at a minimum risk. Usually he stood in for his corner. 1969 D. CLARK *Death after Evensong* v. 129 Just in time to join in a buckshee last round with no chance of standing his own corner.

15. a. *corner-cupboard* (earlier example), *-seat* (also *fig.*), *-shelf, -table*.

1711 *Post-Man & Hist. Account* 8 Mar. 2/1 A Dutch Tablemaker..sells all sorts of..Tea Tables, Corner Cupboards and all manner of Japan Work. 1852 *Illustr. Lond. News* 17 Jan. 51/3 The two corner seats at the top of the interior [of an omnibus]. 1894 J. T. TAYLOR *Veil Lifted* 31 His thoughts had been..concentrated upon..securing a corner seat in a smoking carriage. 1940 L. MACNEICE *Last Ditch* 21 Opposite in corner seats we hope for nearness. 1959 J. BRAINE *Vodi* x. 143 They had booked first-class corner seats to death, all of them. 1805 JANE AUSTEN *Let.* 27 Aug. (1932) I. 165 Nor need I be diffuse on the state of Lady Brydges's bookcase and corner-shelves upstairs. 1889 'C. E. CRADDOCK' *Broomsedge Cove* xxii. 395 The quilts ..were piled high on the 'corner shelf' which they had burdened of yore. 1924 M. ARLEN *Green Hat* iv. 110 Nearby was a corner-table of eight young people. 1956 E. GRIERSON *Second Man* i. 32 The secluded corner table with the shaded lights.

16. corner-block (see quots.); (see also *3 b above); **corner-boy**: insert 'esp.' before 'in Ireland' and add earlier and later examples; **corner-cutting** *vbl. sb.* and *ppl. a.*, taking the quickest, easiest, cheapest, or most expedient course of action (see *CUT *v.* 21 d); **corner flag**, a flag marking a corner of the playing area of a football or hockey field; so *corner-flag kick*, a corner (sense 13); **corner-flag** *v. intr.*, to run towards a corner flag; **corner-grocery** *U.S.*, a grocery, or a bar-room, situated at a street corner; **corner-hit** (see *13 e); **corner-house**, (*a*) (see sense 15 a in Dict.); (*b*) (with capital initials) one of a number of large restaurants in London owned by J. Lyons and Co., Ltd.; **corner-lot** (earlier and later examples); **corner-mark** *U.S.*, a boundary-mark at a corner of a property; so *corner-post, -stake* (also *fig.*); **corner shop**, a shop at the corner of a street; hence, a small local shop (as opposed to a multiple store, supermarket, etc.).

1932 F. L. WRIGHT *Autobiogr.* II. 138 The whole exterior was bedeviled..mixed to puzzle-pieces, with corner-boards, panel-boards, window-frames, corner-blocks. 1962 *Gloss. Terms Glass Ind.* (*B.S.I.*) 14 *Corner block*, specially shaped block used in the dog-house corners. 1966 A. W. LEWIS *Gloss. Woodworking Terms* 18 *Corner blocks*, wooden blocks glued and screwed in position on the inside of a framework to strengthen the joint. 1855 D. G. MITCHELL *Fudge Doings* (N.Y.) II. 47 (Th.), Presently the corner-boy, Jerry, comes in. He is a short-haired, half-Irish boy, [etc.]. 1888 G. M. HOPKINS *Let.* 10 Feb. (1935) 274 Tramps, Cornerboys, Roughs, Socialists and other pests of society. 1946 J. HAMPSON in *Penguin New Writing* 138 Len and his Cardiff. corner-boy have been arrested for the robbery. 1959 *Times Lit. Suppl.* 1 May 262/3 A Labour M.P...visits the Territory of Wajingaland at the urgent behest of its seediest corner-boy. 1960 V. PACKARD *Waste Makers* (1961) x. 105 Some of the breaking down undoubtedly resulted from simple corner-cutting or haste to get out a new model. 1961 M. MCCARTHY in *Harper's Mag.* July 51/2 Defective airplanes sold to the government by a corner-cutting manufacturer. 1882 *Blackburn Times* 1 Apr. 6/3 The first kick from the corner flag. 1887 M. SHEARMAN *Athletics & Football* App. 388 In no case shall a goal be scored from any free kick...The kick-off and corner-flag kick shall be free kicks within the meaning of this rule. 1958 *Times* 24 Nov. 3/2 Scoring by the right-hand corner-flag. 1960 E. S. & W. J. HIGHAM *High Speed Rugby* xvii. 242 The other forwards..will corner-flag to the open side. 1962 *Times* 26 Feb. 4/2 A dangerous dribble..was foiled by Jeeps's corner-flagging. 1849 in Mrs. F. L. Adams *Pioneer Hist. Ingham Co.* (1923) 149 A little old 'corner grocery' building occupied the corner where Pratt & Millspaugh's block now stands. 1859 BARTLETT *Dict. Amer.* (ed. 2), *Corner-grocery*, a grocer's shop on the corner of two streets, a favorite location for such establishments in American towns. 1862 *Congress. Globe* 7 July 3159/1 Without subjecting themselves to the charge of disloyalty by every corner-grocery politician in the land. 1870 'FANNY FERN' *Ginger-Snaps* 106 Tea to the working-girl, taken in this way, is like the 'corner-grocery-drink' to the working-man. 1890 G. W. PERRIE *Buckskin Mose* i. 11 Carrying out tea and sugar..with other such necessaries and luxuries, from a corner-grocery. 1932 *Atlantic Monthly* CL. 141 She telephones to the corner grocery. 1415 in Chambers & Daunt *Bk. London Eng.* (1931) 232 Fro the corner howse..vntill the garden. 1598 STOW *Surv. London* (1908) I. 283 The Corner house, at the ende of Bassings hall streete. 1687, 1886 [see sense 15 a]. 1912 C. MACKENZIE *Carnival* xx. 200 Noisy tea-parties in.. the Corner House, where ladies were not permitted to smoke and customers were kindly requested to pay at the desk. 1923 A. HUXLEY *Antic Hay* xii. 179 An acquaintance which ripened..over the tea-cups..on the fifth floor of Lyons' Strand Corner House. 1958 G. GREENE *Our Man in Havana* v. vi. 260 We could meet at the Corner House for a cheap snack. 1963 V. NABOKOV *Gift* v. 297 A white Pskovan-type church, which had recently grown up out of the corner house. 1702 J. LOGAN in *Corr. Penn & Logan* (1870) I. 129, I have sold the corner lot next the Meeting-House for £115. 1909 F. NORRIS *Third Circle* 104 Incidentally corner lots are desirable. 1690 in *Duxbury* (Mass.) *Rec.* (1893) 71 A stake & a heap of stones for the north east corner, where was the former corner mark of said land. 1869 *Rep. U.S. Comm. Agric.* 1868 343 These fragile corner-marks are strictly respected by the neighbors, and a case of trespass rarely occurs. 1445 *Plea & Mem. R. Lond. Gildh.* A. 71. m. 3b A wall..strecchyng..vnto a corner post standing in the southest Corner of the tenement. 1648 *Charlestown Land Rec.* in *Records Early Hist. Boston* III. 11 A parcell of land lying within and between the upper corner post of the house. 1785 A. ELLICOTT in C. V. Mathews *Life & Lett.* (1908) 41 Joseph went with some Hands to enlarge the Pile of Stones about the Corner Post. 1946 F. DAVISON *Dusty* iii. 31 There might be a length or two of broken fence, or an old charred corner-post. 1950 *N.Z. Jrnl. Agric.* Jan. 58/3 Corner posts and strainer posts..require special moulds because of their size. Corner posts should measure about 6 in. × 6 in. × 6 ft. 6 in. long. 1278–9 in R. R. Sharpe *Calendar of Wills Proved & Enrolled in Court of Husting* (1889) I. 37 [To Adam his son and Johanna his daughter a shop called] 'La Cornereschoppe.' 1599 B. JONSON *Ev. Man out of Hum.* II. i. 8 He would shewe well vpon a Habberdashers stall, at a corner shop rarely. 1859 DICKENS *T. Two Cities* I. v. 20 The wine shop was a corner shop. 1963 P. WILLMOTT *Evolution of Community* x. 125 Small, really local pubs and corner shops have found no place in most post-war plans. 1966 *Guardian* 11 May 4/7 The corner shop is lagging behind the supermarket in hygiene and clean food standards. 1970 C. FREMLIN *Don't go to Sleep* 74 It was a dead street. There were no corner shops, no prams. 1678 *Oyster Bay Rec.* I. 114 A lott of Land..rainging from ye corner stake. 1739 in *Southampton* (N.Y.) *Rec.* III. 16 They both of them accepted of ye division where they set ye corner stakes. 1873 EGGLESTON *Myst. Metrop.* xi. 97 They sought first to guess out the line of a railroad; they examined corner-stakes. 1880 W. JAMES *Coll. Ess. & Rev.* (1920) 184, I then lay it down as a second corner-stake in our inquiry.

corner, *v.* Add: **3. b.** (Earlier U.S. example.)

1824 *Massachusetts Spy* 21 Apr. (Th.), Cornered up so unexpectedly, she candidly confessed.

4. (Earlier and later U.S. examples.)

1836 *Knickerbocker* VII. 42 He has been cornered by the brokers on the —— stock and has lost all his fortune. 1841 *Week in Wall St.* 89 A squad of Bull-backers had been secretly..getting control of a certain stock, intending to 'corner' some one with it. 1948 *Sat. Even. Post* 26 June 47/2 The New Bedford whaling firm..set out to corner the whalebone market.

5. c. *intr.* Of a vehicle, horse, etc.: to go round a corner. Cf. *CORNERING *vbl. sb.* 3.

1909 in WEBSTER. 1940 M. PLATT *Automobile Engin.* vii. 149 A third form of axle loading is produced by the lateral forces which arise when a car is cornering. 1958 *Times* 15 July 7/6 The Six-Ninety..corners smoothly.

6. *intr.* (Earlier U.S. example.)

1821 in *Boston* (Mass.) *Rec.* (1909) XXXIX. 189 A point where said fence and his other fence join cornering on said streets.

cornering, *vbl. sb.* Add: **2.** (Earlier U.S. examples.)

1841 *Week in Wall St.* 27 No cornering, I hope. 1845 C. MATHEWS *Writ.* I. 135 (Th.), They would no doubt have reached a high standing in the practice of what is called cornering.

3. The action of going round a corner. Also *attrib.*

1924 *Cape Argus* 29 Mar. (Magazine), The car with the splayed spring has definitely shown itself to be markedly steady on the road. Particularly is this the case with respect to cornering. 1928 *Times* 2 July 6/7 Kaye Don got ahead thanks to brilliant cornering. 1935 *Discovery* Oct. 293/2 The marked difference in mechanical control over cornering.. should be a salutary warning to careless drivers. 1959 '*Motor*' *Man.* (ed. 36) v. 148 If the tyres are normally ideally loaded, weight transference, by increasing the load on one and reducing it on the other, will reduce the 'cornering power' of the two as a pair.

cornery, *a.* Delete †*Obs.?* and add later and *fig.* examples.

1887 'T. GIFT' *Victims* II. iii. 67 A good fellow enough, but odd, you know, cornery. 1905 C. JARROTT in *Westm. Gaz.* 11 July 4/2 Undoubtedly it [sc. a course] is very 'cornery', and if proper caution is not exercised serious accidents will doubtless result. 1939 G. GREENE *Lawless Roads* iii. 98 The huge cornery cathedral.

cornet, *sb.*[1] Add: **2. c.** A conical wafer, esp. one filled with ice-cream.

1926 *Amer. Speech* I. 246/2 In England an ice-cream cone is called a *cornet*. 1927 *Glasgow Herald* 9 Aug. 7 A half-penny 'cornet'. 1946 R. GRAVES *Poems 1938–45* 36 Licking at a vanilla ice-cream cornet.

cornetist. Add: Also **cornettist.**

1966 *Crescendo* Nov. 6/1 Side one presents the cornettist against the backdrop of the larger band.

corneum (kǭ·ɹniə̆m). *Anat.* [L., neut. sing. of *corneus* horny.] Short for *stratum corneum*, the horny layer of the skin.

1909 *Cent. Dict.* Suppl., *Corneum*, the horny layer of the skin.

corn-fed, *a.* Add: **b.** *spec.* Fed on maize. By extension: well-fed; plump, stout. Chiefly *U.S.*

1787 in T. F. DeVoe *Market Bk.* (1862) 181 Corn-fed pork and peach brandy. 1796 J. BARLOW *Hasty-Pudding* III, Brown corn-fed nymphs, and strong hard-handed beaus. 1812 'D. KNICKERBOCKER' *Hist. N.Y.* (ed. 2) III. vi. 182 They grew up a..hardy race of..strapping corn-fed wenches. 1862 'ARTEMUS WARD' *His Bk.* (1865) 154 The corn fed gals of Ohio and Injianny. 1889 [in Dict.]. 1948 *Chicago Tribune* 20 June (Grafic Mag.) 8/5 He looks like a corn-fed boy.

c. Banal, provincial, 'commercial'; = *CORNY *a.*[1] 1 c. Chiefly *Jazz slang*.

1929 *Melody Maker* Mar. 285/1 This peculiar..style of melody, the appeal of which lies in the fact that it is purposely so utterly corn-fed. 1935 *Peabody* (Mass.) *Bull.* Dec. 42/2 Corny—Derived from cornfed, meaning [music] played in country style, out of date, hill-billy, or in a style of pre-1925. 1937 L. FEATHER in *Radio Times* 2 Apr. 10/3 *Corn*, old-fashioned style; out-of-date idiom and technique in jazz. Hence *corny* or *cornfed* applied to musicians and their style. 1954 *Archit. Rev.* CXVI. 303 Either way this is a rather negative formulation; part of the literary impedimenta of the modern movement, useful to the critic defending the Bauhaus to a cornfed audience of Ruskinians.

cornfield. Add: **b.** A field in which maize is grown. *U.S.*

Quot. 1608 may not refer to a maize-field.

1608 J. SMITH *True Rel. Virginia* in *Wks.* (1884) I. 32 The riuer diuideth in two, the neck a plaine high Corne field. 1634 in *Cambridge* (Mass.) *Prop. Rec.* (1896) 1 The Constable shall make a Surveyinge of the Houses, backsids, Corne ffeilds,..and other lands. 1772 *Carroll Papers* in *Maryland Hist. Mag.* XIV. 150 Our Cornfields are in very good order. 1820 *Western Carolinian* (Salisbury, N.C.) 1 Aug., He will find its component parts to consist of corn field peas. 1837 J. M. PECK *Gaz. Illinois* (ed. 2) I. 25 The.. squirrels often do mischief in the corn fields. 1872 SCHELE DE VERE *Americanisms* 48 So-called self-made men are to this day fond of boasting that they never received any other education but in an *old cornfield school*. 1948 *Chicago Daily News* 12 Jan. 1/1 The farmer..coveted his neighbor's wife across the cornfields.

corn-flower. Add: **e.** The blue colour of the corn-flower: a fashion shade. Also *attrib.*

1907 *Yesterday's Shopping* (1969) 339/2 Velvet crêpe paper..in the following colours..dark coral, cornflower, [etc.]. 1907 B. M. CROKER *Company's Servant* xxix. 301 Her cornflower-blue parasol. 1911 L. J. VANCE *Cynthia* 255 Its colour an incredible, nitid blue like the blue of a corn-flower sapphire. 1940 L. MACNEICE *Last Ditch* 25 His eyes a cornflower blue. 1963 *New Yorker* 1 June 112/3 (Advt.), In rich pastels—cherry, coral, cornflower.

corn-ground. (Later examples.)

1886 Z. F. SMITH *Hist. Kentucky* 367 Uncle Ben..plowed the corn-ground. 1954 M. BERESFORD *Lost Villages* vii. 246 Sheer excellence of corn-ground.

corn-house. 2. (Examples.)

1699 J. DICKINSON *God's Protecting Provid.* 82 The people had..large Cropps of Corn, as We could tell by their Corn-houses. 1745 *Virginia Gazette* 19 Dec., Two Dwelling-houses, Milk-house, Corn-house. 1836 *Southern Lit. Messenger* II. 157 That unlimited control which the said horse exercises over the corn in his corn-house. 1845 W. G. SIMMS *Wigwam & Cabin* 369 Considering that the Corn-House aint oberfull. 1957 *Steven Post* (Charlottesville, Va.) Feb., Slave cabin, corn house, kennel, hen house.

corn-husker, -husking. (Earlier examples.)

1818 M. L. WEEMS *Drunkard's Looking Glass* (ed. 6) 4 Whether at a Funeral..or a Cock-fight, a Corn-husking. 1834 CROCKETT *Narr. Life* x. 71 He made a great corn husking, and a great frolic, and gave a general treat. 1858 *Rep. Comm. Patents* I. 358, I claim an improved corn-husker. 1871 *Amer. Naturalist* V. 317 Several Indian Stone implements..which are said to bear a striking resemblance to iron corn huskers now in use in the West. 1876 *Rep.*

Vermont Board Agric. III. 609 The corn harvest may now be greatly accelerated by the use of the corn husker driven by horse power.

cornice, *sb.* Add: **3. b.** An overhanging accumulation of ice and wind-blown snow at the edge of a ridge or cliff-face.
1871 J. TYNDALL *Hours of Exercise in Alps* i. 13 The view was bounded by a massive cornice, from which the avalanches are periodically let loose. **1953** J. HUNT *Ascent of Everest* v. xv. 191 Huge bulges of snow hung over it from the crest of the ridge, cornices of Himalayan dimensions formed by the prevailing westerly wind.

corniced, *ppl. a.* Add: **2.** Having a cornice (sense *3 b above).
1950 tr. *Mountaineering Handbk.* vii. 77 If you have to descend to a face from a corniced ridge, choose again the place with the smallest cornice and the least overhang. **1959** S. CLARK *Puma's Claw* v. 63 From its corniced edge cascades of sparkling seracs..fell like the pleats and folds of a curtain.

‖ **corniche** (kǭ·ɹniʃ, -i·ʃ). [See CORNICE *sb.*] In full *corniche road* = CORNICE *sb.* 3; *spec.* (with capital initial) the road from Nice to Genoa overlooking the Mediterranean; hence, any coastal road with panoramic views.
1837 H. C. ROBINSON *Diary* 17 Mar. (1967) 165 Our tour.. included a journey through France along the Corniche road to Pisa. **1852** E. TWISLETON *Let.* 29 Oct. (1928) iv. 60 Between Nice and Genoa... The road is justly named the Corniche (cornice) for it runs on rocks which often jut over the sea as a cornice might. **1883** [see CORNICE *sb.* 3]. **1945** A. HUXLEY *Time must have a Stop* xv. 157 They found the car and drove back to Monte Carlo along the Corniche. **1959** *Listener* 22 Jan. 165/1 The new *corniche* road between Athens and Cape Sunion makes an impressive third to compete with the *corniche* roads of the French Riviera and the wonderful *corniche* road from Sorrento to Amalfi.

corniness (kǭ·ɹninės). *colloq.* [f. *CORNY *a.*[1] 1 c + -NESS.] The condition or quality of being 'corny' (see *CORNY *a.*[1] 1 c).
1932 *Melody Maker* Feb. 155/3 It is not on any grounds of technique, taste or 'corniness' that I make this criticism. **1934** *Ibid.* 19 May 9/3 The jazzy, double-stopping style is, to my mind, the last word in corniness. **1941** *Time* (Air Exp. Ed.) 19 May 27/3 Corny as a back-country bumpkin, Oakie has turned his corniness into a salable commodity.

Cornish, *a.*[2] Add: *Cornish cream*; **Cornish pasty,** meat, vegetables, and seasoning cooked in a case of pastry.
1905 A. L. SALMON *Lit. Rambles West of Eng.* (1906) 214 Ready to indulge eagerly in Cornish cream and saffron-cake. **1912** 'Q' *Hocken & Hunken* vi, A delectable junket with Cornish cream. **1968** D. HOPKINSON *Incense Tree* ii. 13 The great bowls of Cornish cream stood on the shelves and she took up a spoon to skim the golden crust. **1877** *N. & Q.* 5th Ser. VII. 297 The Cornish pasty baked without a dish. **1951** *Good Housek. Home Encycl.* 590/1 Patties may be of various sizes and shapes, ranging from a Cornish pasty to a tiny bouchée.

cornist (kǭ·ɹnist). ?*Obs.* [ad. F. *corniste*; cf. CORN *sb.*[3] and -IST.] A horn-player.
1806 in BUSBY *Dict. Mus.* (ed. 2). **1889** G. B. SHAW *London Music* (1937) 75 The four expert cornists in the orchestra. **1900** W. A. ELLIS tr. *Glasenapp's Life of Wagner* I. vii. 248 The excellent Bohemian *cornists*..had strange great winding horns, which they carried coiled around their necks.

cornland. Add: **b.** Land suitable for growing maize. *U.S.*
1654 E. JOHNSON *Hist. New Eng.* (1867) 41 Their Corne Land in Tillage in this Towne is about 1200. Acres. **1838** H. W. ELLSWORTH *Valley N. Wabash* iv. 39 The lands, that we call first rate corn-lands, are generally alluvial bottom lands.

corn-meal. (Earlier and later U.S. examples.)
1782 J. HOLT *Let.* 4 Dec. in *Virginia Hist. Mag.* IV. 400 The wheat..will do to mix with Corn meal for My Servants. **1948** *Sci. Monthly* May 384 Make it into succotash, corn meal, corn flakes, and syrup.

co·rn-pla:nter. *U.S.* [CORN *sb.*[1] 5.]
1. One who plants maize. Also *attrib.*
1832 *Deb. Congress* 9 Feb. 339 The corn planter and wheat grower understand their interests. **1845** W. G. SIMMS *Wigwam & Cabin* 2nd Ser. 104 One of a tribe [*sc.* the crows], of which the corn-planter has an aversion.
2. A drill for planting corn.
1839 J. BUEL *Farmer's Compan.* xiv. 151 Some of them, under the name of corn-planters, are employed in planting Indian Corn. **1856** [see CORN *sb.*[1] 11]. **1861** *Trans. Ill. Agric. Soc.* IV. 24 A petition, praying the Executive Committee to make a second trial of Cornplanters. **1948** *Clarke Co. Democrat* (Grove Hill, Ala.) 28 Oct. 3/2 Two-Row Corn Planter.

corn-stalk. Add: **1. b.** *Comb.*: **corn-stalk disease,** a disease of cattle caused by the eating of dry corn-stalks.
1900 *Yearbk. U.S. Dept. Agric.* 307 (Cent. D. Suppl.), The cornstalk disease..is a..little-understood malady of cattle.
2. (Earlier and later examples.) Now ?*Obs.*
1827 CUNNINGHAM *Two Yrs. N.S.W.* II. 116 We have, as I said before, first, the sterling and currency, or English and Colonial born, the latter bearing also the name of corn stalks (Indian corn), from the way in which they shoot up. **1848** H. W. HAYGARTH *Bush Life in Australia* xi. 123 The average height of the Australians is probably more than that of the English, but when they exceed a certain standard they are apt to become loose made and weedy, thereby justifying their appellation of 'cornstalks'. **1852** G. C. MUNDY *Our Antipodes* I. ii. 45 There are probably more gleaners of the profits; not..a thinner crop of 'cornstalks' for the harvest,—some of them as long in the ear as could be wished. **1908** *Daily Chron.* 21 Jan. 4/6 'Cornstalks', as our general and sporting appellation of Australians, applies in strictness to the people of New South Wales only, on account of their height and slimness. **1934** [see *BANANALAND].

cornstone. (Earlier example.)
1819 G. B. GREENOUGH *Critical Exam. Geol.* iv. 218 Chert, flint, septaria, gypsum, ball-ironstone, cornstone, afford examples of beds naturally interrupted on the small scale.

Cornu- (kǭ·ɹniu), used as combining form of *Cornubian* Cornish, as in **Cornu-Bri·tish,** of the British of Cornwall; **Cornu-Bre·ton,** of or pertaining to both Cornwall and Brittany.
1754 W. BORLASE (*title*) Observations on the Antiquities of the County of Cornwall..And a Vocabulary of the Cornu-British Language. **1897** E. PHILLPOTTS *Lying Prophets* xii, The Cornu-British bishops. **1929** *Times* 30 Aug., The visit of Breton wrestlers to Trevenson to-morrow, to meet Cornishmen in contests for the Cornu-Breton championship trophies and another wrestling prize.

cornual. Add examples and substitute for def.: Pertaining to cornua or a cornu.
1890 BILLINGS *Med. Dict.* I. 330/1 *Cornual myelitis*, poliomyelitis. **1897** PARKER & HASWELL *Text-bk. Zool.* II. 118 A small cornual cartilage. **1907** *Practitioner* Sept. 357 As regards the relative strength of the cornual attachments, the posterior horns of both cartilages are firmly bound down to the tibia. *Ibid.* 401 In cornual pregnancy, the round ligament is attached on the outer side of the gestation sac. **1965** S. R. M. REYNOLDS *Physiol. Uterus* (ed. 2) iii. 32 In the dog, labor may be divided into three stages: (1) cornual expulsion and uterine dilatation, (2) uterine expulsion, and (3) expulsion of the placenta.

cornule (kǭ·ɹniul). *Zool.* [ad. mod.L. *cornulum*, dim. of L. *cornu* horn.] A small horny plate with the function of a tooth.
1889 NICHOLSON & LYDEKKER *Man. Palæont.* (ed. 3) 1265 Teeth are present in the young [of Ornithorhynchus], and are succeeded by horny plates or cornules. **1960** I. F. & W. D. HENDERSON *Dict. Sci. Terms* (ed. 7) *Cornule*, a small horn-like process; one of the horny jaw-plates of Ornithorhynchus.

cornus (kǭ·ɹnɒs). *Bot.* [L., = dogwood, adopted by Linnæus (*Systema Naturæ* 1735) and earlier botanists as the name of a genus.] A tree or shrub belonging to the large genus so called, of the family Cornaceæ; cf. CORNEL[2] and DOGWOOD[2].
1846 LINDLEY *Veget. Kingd.* 782 The Cornus of the ancients was the present Cornelian Cherry. **1920** *Blackw. Mag.* Nov. 621/1 A Cornus of some kind, bearing great clusters of scarlet fruit. **1961** *Amateur Gardening* 23 Dec. 2 The edible berries which so many shrubs provide, say, amelanchier, cornus, berberis.

cornute, *a.* and *sb.* Restrict †*Obs.* to the sb., and add later example of the adj.
1957 CHILDE *Dawn Europ. Civil.* (ed. 6) xiii. 249 A fantastic elaboration of handles towards cornute types.

corny, *a.*[1] Add: **1. c.** *colloq.* Of such a type as appeals to country-folk; rustic or unsophisticated; tiresomely or ridiculously old-fashioned or sentimental; hackneyed, trite; inferior. Cf. *CORN *sb.*[1] 3 c, *CORNINESS.
1932 *Melody Maker* June 511/1 The 'bounce' of the brass section..has degenerated into a definitely 'corny' and staccato style of playing. **1935, 1937** [see *CORN-FED c]. **1937** M. ALLINGHAM *Dancers in Mourning* ii. 31 She..began to play a melody which..had been popular in the early post-war days... 'Corny old stuff,' said Mercer. **1944** BRAHMS & SIMON *Titania has a Mother* iii. 26 He wondered what corny gag old Abey had thought up this time. **1946** [see *CORN *sb.*[1] 3 c]. **1951** AUDEN *Nones* (1952) 43 Grown insolent and fat On cheesy literature And corny dramas. **1957** *Listener* 20 June 993/2 The Walker Art Gallery houses some of the corniest Victorian and Edwardian masterpieces. **1958** E. H. CLEMENTS *Uncommon Cold* ii. 44 My affections are, to use a corny phrase, 'otherwise engaged'.

‖ **corocoro** (kǭ·rokǭ·ro). Also 7 **caracolle, coracora, curra curra, curricurre, -curro,** 8 **caracore, corrocorra, -corro,** 9 **kora-kora.** [ad. Malay *kurakura*. Cf. F. *caracore*, Sp. *caracora*, from which the English forms are chiefly derived.] A boat used in the Malay Archipelago, = CARACORE.
1606 H. MIDDLETON *Last East-Indian Voy.* sig. E2, The formost of these Galleys or Caracolles recouered our shippe, wherein was the King of Tarnata. **1625** PURCHAS *Pilgrims* I. IV. i. 356 We had sight of two Curra Curras between vs and Botun. **1626** *Ibid.* v. ix. 677 An Holland Coracora which came from Amboyna with Letters to the Captaine. **1634** T. HERBERT *Trav.* 193 The shape of their Boates or Curricurroes have this representation. *Ibid.* 194 A Curricurre or boat. **1779** FORREST *Voy. N. Guinea* 23 A corocoro is a vessel generally fitted with out-riggers, with a high arched stem and stern, like the point of a half moon. They are used by the inhabitants of the Molucca islands chiefly. *Ibid.* 65 Got our stores, &c. very expeditiously on board, in a small new corocoro. **1794** [see CARACORE]. **1798** in *Naval Chron.* (1799) II. 135 A *Corrocorro* is a vessel fitted with outriggers, having an high arched stem and stern. **1800** *Asiat. Ann. Reg.* 210/1 This expedition was formerly performed in large corracorras, or vessels like yachts. **1869** A. R. WALLACE *Malay Archip.* II. xxiv. 69 The boat was one of the kind called Kora-kora, quite open, very low, and about four tons burthen. It had outriggers of bamboo about five feet off each side, which supported a bamboo platform extending the whole length of the vessel.

corolliform (kǭrǫ·lifǭɹm), *a.* [f. COROLLA: see -FORM.] Of the form or appearance of a corolla.
1854 MAYNE *Expos. Lex.* 230/2 *Corolliformis*, having the form or appearance of a corol; corolliform.

Coromandel (kǫrǒmæ·ndəl). [Name of the major part of the eastern coast of Madras, prob. a corruption of *Cholamandalam*, 'the country of the Cholas', an old Dravidian people.] **1.** = CALAMANDER. Freq. *attrib.*
1845 G. DODD *Brit. Manuf.* IV. 199 The coromandel-wood. **1910** *Encycl. Brit.* VIII. 843/2 Much of the East Indian ebony is yielded by the species *D[iospyros] Melanoxylon* (Coromandel ebony). **1931** L. BINYON *Coll. Poems* II. 226 Betwixt slim pillars, on a floor Of brindled coromandel wood. **1940** *Archit. Rev.* LXXXVII. 12/2 The rostrum and desk tops and fronts are in American figured walnut; their bases in coromandel.
2. Used *attrib.* to designate oriental lacquer transhipped on the Coromandel coast while being exported to Europe.
1878 MRS. B. PALLISER tr. *Jacquemart's Hist. Furnit.* IV. viii. 454 Coromandel lacquer ware. **1936** *Burlington Mag.* Jan. 40/2 Superb screen of the so-called 'Coromandel' type, decorated with storks and pine-trees. **1955** R. FASTNEDGE *Eng. Furn. Styles* iv. 94 Two varieties of Oriental lacquer were brought to England during the late seventeenth century—the very beautiful and most durable incised or 'Coromandel' lacquer, [etc.]. **1963** *Times* 7 Mar. 14/3 An eight leaf coromandel lacquer screen for £170.

corona. Add: **6.** Also, the trochal disc of a rotifer, etc.
1886 C. T. HUDSON *Rotifera* I. i. 6 Each side of the front of the head, or *corona*, is rounded into a nearly circular lobe. **1898** A. SEDGWICK *Text-bk. Zool.* I. 299 *Phylum Rotifera*...It will be useful to use the word corona for the discoidal anterior end of the body. **1902** *Amer. Nat.* Dec. 932 In Melo and some other genera the spines project upwards and are generally unclosed on the apertural side. This produces the spiral 'corona', so striking a feature in some shells. **1906** *Camb. Nat. Hist.* I. 539 Echinidae in which the plates of the corona dovetail into each other. **1929** *Encycl. Brit.* XIX. 575/1 Numerous cilia crowded upon a particular area of delicate skin close to or encircling the mouth...The whole area, including the mouth itself, as seen when the cilia are active, is called the corona. **1966** H. G. S. WRIGHT tr. *Donner's Rotifers* 23 This corona surrounds a cup-shaped enclosure.
d. *corona radiata*: (*a*) a mass of projection fibres in the brain which spread radially from the internal capsule to the cerebral cortex; (*b*) the layer of radially elongated follicular cells which surround an ovum and develop shortly before ovulation.
[**1848** DUNGLISON *Med. Lex.* (ed. 7) 222/2 Fasciculi of white fibres radiate in all directions from every part of the surface of the optic thalamus, excepting its inner side..; the anterior of these fibres pass directly forwards, the middle fibres outwards, and the posterior backwards, forming the *corona radians*.] **1869** GRAY *Anat.* (ed. 5) 482 Some of the diverging fibres..radiate into the anterior, middle, and posterior lobes of the hemisphere..before passing to the convolutions. These fibres have received the name of *corona radiata*. **1892** C. S. MINOT *Human Embryol.* I. iii. 53 The cells..next to the ovum are cylindroid, and radiate round the zona, constituting thus the so-called *corona radiata*. **1913** A. ROBINSON *Cunningham's Text-Bk. Anat.* (ed. 4) 644 The fibres..spread out widely from each other in a radiating or fan-shaped manner, as they are followed to the varous gyri of the cerebral hemisphere. This arrangement is tirmed the *corona radiata*. **1961** *Lancet* 29 July 249/2 The lesions involved the pyramidal pathways at the level of the corona radiata, internal capsule, or pons. **1968** H. WANG *Outl. Human Embryol.* i. 8 The liberated ovum, surrounded by the corona radiata, is expelled from the follicle.
9. *Electr.* A luminous appearance in the gas surrounding a conductor when the electric field at its surface is strong enough to ionize the gas but not strong enough to cause a spark. Also *attrib.*, as **corona discharge,** the discharge causing a corona.
1906 A. RUSSELL *Altern. Currents* II. 475 When the potential difference between the mains is very high, each main is seen surrounded by a faintly luminous enveloping cloud of a bluish colour, which apparently does not touch the conductor it envelopes. This cloud is called the corona. *Ibid.*, When coronae make their appearance it is found that the capacity between the mains and the loss of energy in distribution are increased. **1918** *Jrnl. Chem. Soc.* CXIV. II. 42 The corona discharge. **1926** *Encycl. Brit.* 2nd Suppl. I. 949/1 In the case of high voltage aerial lines there is a notable loss of power due to the ionisation of the air. This is called the corona effect since it makes the lines appear luminous in the dark. **1937** *Discovery* Aug. 248/2 Synthetic rubber compositions are used as a corona-proof sheath in high tension power cables. **1953** *Electronic Engin.* XXV. 64 A corona stabilizer consists of a concentric anode and cathode in an envelope filled with a few millimetres pressure of a suitable gas, usually hydrogen. **1959** R. E. HUSCHEK *Gloss. Meteorol.* 138 Aircraft flying through active electrical storms often develop corona discharge streamers from antennas and propellers.

Corona[2] (kǒrǭ·nă). Also **corona.** [From the proprietary name *La Corona* (Sp.) the crown.] A well-known brand of Havana cigar.
[**1877** *Trade Marks Jrnl.* 14 Nov. 1671 La Corona...Firm of Manuel Lopez and Company, Havana.] **1887** *Army & Navy Co-op. Soc. Price List* 127 Havana Cigars. 'La Corona' ..Aguilas Imperiales Corona. **1904** *Trade Marks Jrnl.* 23 Nov. 1424 La Corona...Manufactured Tobacco. The Havana Cigar and Tobacco Factories, Limited. **1924** W. J. LOCKE *Aristide Pujol* viii, He found him smoking a large corona. **1925** *Morn. Post* 16 Mar. 4, I bought a thousand Corona Coronas an hour ago. **1919** WODEHOUSE *Damsel in Distress* xxi, Over the red coronas we will discuss many things. **1921** *19th Cent.* May 767 The far-coated Corona-puffing multi-millionaire. **1926** M. WALSH *Key above Door* viii. 95, I was in the middle of a long corona when Lady Clunas rose to go. **1929** *Tobacco* 1 Oct. 72/2 The H. Upman factory is continuing the exportation of large quantities of Havana cigars from the English market, principally in Corona et Petit Corona sizes of the H. Upman brand. **1964** C. HODDER-WILLIAMS *Main Experiment* v. 49 Honiton peered penetratingly over the top of a large Corona at Sally, who had silently declined a cigarette.

coronadite (kǫronā·dait). *Min.* [Named 1904, after F. Vásquez de *Coronado* (died 1554), Spanish explorer: see -ITE[1].] An oxide of lead and manganese, $PbMn_8O_{16}$, occurring in dark, fibrous masses.
1904 LINDGREN & HILLEBRAND in *Amer. Jrnl. Sci.* XVIII. 448 *Coronadite*. On the dump of a small shaft on the west end of the Coronado vein,..fairly large amounts of a dark metallic mineral were found. *Ibid.* 451 A specific name..we propose *Coronadite*, after the famous explorer of that portion of the American continent. **1950** *Jrnl. Chem. Soc.* LXXXVIII. II. 96 Coronadite, a black metallic mineral, not unlike psilomelane in general appearance, and intimately intermixed with quartz, is found in fairly large amount in the Coronado vein. **1962** *Amer. Mineral.* XLVII. 744 Coronadite. The coronadite and lithiophorite were microscopic and x-ray study of material from four deposits in the Artillery Mountains area.

coro·nagraph, var. CORONOGRAPH (and now the commoner form); also, a telescope for observing the sun's corona. Hence **corona·graphic** *a.*
1890 [see CORONOGRAPH and CORONOGRAPHIC a.]. **1927** *Popular Astronomy* XXXV. 267 The accurate adjustment of stationary coronagraphs with despatch. **1959** *Sci. Amer.* Feb. 24/3, ..went hard to work on the coronagraphic project. **1966** H. S. JONES in A. Pryce-Jones *New Outl. Mod. Knowl.* 122 A special instrument called a coronagraph is used in which the direct light of the Sun is obscured by a disk, an artificial eclipse thereby being produced. **1970** *Nature* 20 June 1133/2 The coronagraphic method is restricted to the relatively bright innermost corona. *Ibid.*, The line intensity has to be observed to much larger distances than can be reached with coronagraphs.

coronal, *a.* Add: **5.** *Phonetics.* Pronounced with the tip of the tongue turned upward towards the palate; pertaining to such pronunciation. Cf. INVERTED *ppl. a.* 1 c.
1890 A. J. ELLIS *Eng. Dial.* p. xvii/2 The usual English coronal 'd in 'do' with the tip of the tongue free from the gums, and approaching the 'crown' of the arch of the hard palate. **1899** E. J. LLOYD *Northern Eng. Phonetics* 22 In a coronal vowel, the vowel configuration seems to be shifted backwards, so that its exit is no longer at the lips, but between the tongue-tip and the palate. *Ibid.* 23 These coronal symbols are chosen to indicate timbre rather than articulation. **1931** W. HORN *Eng. Phonetics* 30 In their production the point [of the tongue] is curled back...These are called coronal or cacuminal vowels. **1933** L. BLOOMFIELD *Lang.* vi. 98 Contact can be made by the tip of the tongue (apical articulation) or by a larger area, the blade, round the tip (coronal articulation). **1968** CHOMSKY & HALLE *Sound Pattern Eng.* 304 Coronal sounds are produced with the blade of the tongue raised from its neutral position.

coronary, *a.* Add: **3. a.** *coronary thrombosis*, thrombosis occurring within a coronary artery, esp. in a coronary artery of the heart.
1930 *Q. Jrnl. Med.* XXIII. 233 A syndrome sometimes described as that of coronary thrombosis, more accurately, perhaps, as that of cardiac infarction, or ischaemic necrosis of the cardiac wall. **1955** *Sci. News* VII. 125 A coronary thrombosis is the name for the sudden clotting of the blood in one or more of the special cardiac blood vessels.
d. *ellipt.* = coronary artery, etc., esp. coronary thrombosis.
1893 in H. MORRIS *Hum. Anat.* 972 The right coronary artery] divides into its two main branches, one of which.. anastomoses with the left coronary. **1907** *Practitioner* Oct. 491 The blood-pressure will rise, and the heart itself will be more fully flushed with blood through the coronaries. **1928** C. F. COOPER *Parker & Haswell's Zool.* (ed. 4) II. 310 The coronary..form the upwardly directed coronoid process immediately behind the last tooth. **1955** A. HUXLEY *Genius & Goddess* 30 Coronaries, gastric ulcers, cirrhosis, bad kidneys! **1961** G. SMITH *Friends* iii. 52 Pa was carried off by a coronary about ten years ago. **1967** *Spectator* 11 Aug. 159/3 In a survey of 900 patients with coronaries in Belfast, it was found that only just over 400 reached hospital.

coronation. Add: **3. coronation rolls** (see quot. 1883).
1883 W. JONES *Crowns & Coron.* 284 The coronation rolls contain the commission and proceedings of the commissioners appointed to hear and declare claims of service to be performed at coronations, as well as the oaths taken by the king or queen when crowned. **1963** *Guide to Contents of Public Record Office* I. 18 Coronation Rolls...These contain the formal record of the Coronation ceremony and of the proceedings of the commissioners appointed to hear and determine claims of service to be performed at coronations, and also the oath taken and the declaration made and signed by the Sovereign when crowned.
b. Designating articles produced to commemorate a particular coronation.
1838 C. M. YONGE *Let.* 6 Aug. in C. Coleridge *C. M. Yonge* (1903) 133 The thimble is a Coronation thimble. On one side of the crown it is 'Victoria' and on the other 'Crowned, 28th of June 1838'. *c* **1909** D. H. LAWRENCE *Collier's Friday Night* (1934) i. 4 Over the mantelpiece, which is high, with brass candlesticks and two 'Coronation' tumblers in enamel, hangs a picture of Venice. **1938** L. MACNEICE *I crossed Minch* II. xii. 165 The china cupboard contained Silver Jubilee and Coronation mugs. **1967** M. LASKI *Seven Other Years* xiii. 192 The Coronation mug (Edward VII) he sometimes used for drinking tea.

Coronation Street (kǫrǒnē^i·ʃən strīt). The name of a television series about the inhabitants of a street in a working-class area in northern England: used allusively to denote a typical street of this sort. Hence as *adj.*, working-class.
1964 *Harper's Bazaar* Aug. 31/1 In the heart of Britain's countless Coronation Streets. **1966** *Punch* 24 Aug. 305/1 Dialogue mainly well-reported, but occasionally *Coronation Street*. **1966** L. DA BERN *Goodbye Piccadilly* x. 96 The only language difficulty you're likely to experience is understanding the Coronation Street accents. **1970** J. WAINWRIGHT *Prisoner's Friend* viii. 147 The area—lower working class—typical 'Coronation Street'.

coroner. Add: **2.** The chief officer or sheriff of a sheading in the Isle of Man.
[**1577** in M. A. Mills *Stat. Laws I. of Man* (1821) 58 All Coroners shall make a general Search four Times in the Yeare for any Lord his Profit,..every Coroner within his own Sheading. **1693** [see SHEADING]. **1889** *Encycl. Brit.* XV. 452/2 The coroner of the sheading, who is appointed annually by the governor, is a kind of sheriff. **1894** [see SHEADING].

coronet, *sb.* Add: **1. c.** *transf.* A terminal or crowning circle of spines, hairs, or other small objects.
1826 KIRBY & SPENCE *Entomol.* III. xxxiii. 385 Coronula, a coronet or semicoronet of fixed spines observable at the apex of the posterior Tibiæ in *Fulgora candelaria*, etc.
5. b. The bur or ring of bone on the head of a deer, at the base of an antler.
1898 LYDEKKER *Deer of all Lands* 6 Towards the completion of its [*sc.* the antler's] growth a more or less prominent ring of bone, termed the burr or coronet, is deposited at its base just above the junction with the pedicle.

coronet, *v.* Add: Also, to place a coronet upon (a person's head); also *fig.* Hence **co·roneting** *vbl. sb.*
1881 'MARK TWAIN' *Prince & Pauper* xxxii. 348 He puts her coronet where it will be convenient to her hand when the time for the simultaneous coroneting of the nobles shall arrive. **1900** *Daily News* 3 May 2/4 The circle of Westmoreland and Cumberland hills..still coroneted with snow. **1902** *Westm. Gaz.* 11 Aug. 10/1 The moment the crown was on the head of the Queen the peeresses all coroneted themselves.

coronilla (kǫrǒni·lă). *Bot.* [mod.L. (Tournefort *Inst. Rei Herb.* (1700) 650), dim. of L. *corona* (see CORONA).] A member of a genus of shrubs or herbs so called, which belongs to the family Leguminosæ and includes plants bearing mostly yellow flowers in umbels.
1793 *Curtis's Bot. Mag.* VI. 185 (*caption*) Rue-Leaved Coronilla. **1873** *Young Englishwoman* Jan. 30/2 Some of the following plants should be in flower: coronilla, jasminum nudiflorum..in addition to roses, and other geraniums. **1971** H. MONRO *Strange Mistress* 28 Coronilla! Coronilla! Heavy yellow tepid bloom. **1933** R. ALDINGTON *All Men are Enemies* I. xv. 102 Mediterranean pines and twisted oaks and chestnuts stood among a magnificence of arbutus, glare and heavy yellow coronilla. **1961** R. W. BUTCHER *Brit. Flora* I. 617 This Purple Coronilla is a stout, much-branched, straggling, glabrous perennial.

coronillin (kǫrǒni·lin). *Chem.* [ad. F. *coronilline* (Schlagdenhauffen and Reeb 1897, in *Arch. de Pharmacodynamie* III. 5), f. *CORONILLA + -IN[1].] A bitter, yellow glucoside contained in the seeds of species of Coronilla, esp. *C. scorpioides*, and having a stimulating effect on the heart.
1897 *Jrnl. Chem. Soc.* LXXII. I. 433 Coronillin is obtained by digesting the powdered seed..with six times its weight of water at 100°. **1939** *Thorpe's Dict. Applied Chem.* III. 366/2 *Coronillin*, $C_{18}H_{30}O_{15}$, the glycoside of the seeds of *Coronilla scorpioides* and *C. varia*, is hydrolysed to glucose and its aglycone, $C_6H_{10}O_3$.

coronitis (kǫrǫnai·tis). [f. CORON(ARY + -ITIS.] Inflammation of the coronary substance of the hoof of a horse.
1890 in BILLINGS *Med. Dict.* **1950** J. J. O'CONNOR *Dollar's Vet. Surg.* (ed. 4) iii. 958 Gangrenous coronitis..may result from an open wound and infection by the bacillus of necrosis or other virulent organism.

coronium. Add: No longer supposed to exist; the spectral lines attributed to it are now known to be due to highly ionized atoms of iron, nickel, and other elements. (Later examples.)
1898 *Daily News* 22 July 5/1 The latest sensation in the scientific world is the discovery of 'terrestrial coronium'. **1956** H. S. JONES in A. Pryce-Jones *New Outl. Mod. Knowl.* 122 When the light of the corona is analysed with a spectroscope, it is found to contain radiations which have not been observed in the laboratory. When they were first observed, and for many years afterwards, they were attributed to a hypothetical unknown element, which was named 'coronium'. Modern atomic theory left no room for such an element.

coronoid, *a.* Add: **B.** *sb.* Any of several bones in the lower jaw of certain lower vertebrates.
1888 ROLLESTON & JACKSON *Forms Anim. Life* (ed. 2) 339 To the quadrate region belongs the squamosal; to the mandibular arch, in *Sauropsida* a splenial, coronoid, angular, and supra-angular. **1903** *Proc. Zool. Soc.* 17 Mar. 267 In *Eudynamys*, *Cuculus*, and *Centropus* the rami are pierced by a long lateral vacuity, which is partly closed by a long and slender coronoid. **1951** G. R. DE BEER *Vertebrate Zool.* (ed. 2) xx. 257 *Cynognathus* has a large dentary, while the articular, angular, supra-angular, prearticular, coronoid, and splenial are small.

coroplast (kǭ·rǒplæst). *Antiq.* [ad. Gr. κοροπλάστης, f. κόρη girl, doll: see -PLAST. Cf. F. *coroplaste*.] One who makes terracotta figures.
1885 *Nation* (N.Y.) 1 Oct. 286/3 The Myrinean coroplasts, or manufacturers of terra-cottas, were certainly influenced by the models of their brethren in Tanagra. [**1886** J. H. WRIGHT tr. *Collignon's Gk. Archæol.* 239 The art of the *coroplastes*, or modeller of figurines.] **1936** *Antiquity* X. 249 The coroplasts who made these figures can be identified by their signatures.

corp(kǭ·ɹp), colloq. abbrev. of CORPORAL *sb.*[2] 1. Freq. as vocative.
1909 in *Cent. Dict.* Suppl. **1922** C. E. MONTAGUE *Disenchantment* iv. 56 'There grey hairs should be a lot of use to you, corp.,' one of them would quite feelingly say. **1938** F. D. SHARPE *Flying Squad* vi. 84 'We are going to Hendon, aren't we, corp?' The corporal replied: 'Yes.' **1954** W. FAULKNER *Fable* (1955) 301 Do you want me to wake them up, Corp?

corporal, *sb.*[2] Add: **4.** A cyprinoid freshwater fish, *Semotilus corporalis*. *U.S.*
1888 GOODE *Amer. Fishes* 427 The name Corporal seems to have been derived from the Dutch of German settlers of the Middle States. **1964** H. T. WALDEN *Familiar Freshwater Fishes Amer.* 184 The fallfish is known to many anglers as chub, silver chub, chivin, windfish, and corporal.

corporation. Add: **7.** *corporation carter, clerk, labourer, law, lawyer, line, stock, stop, tax.* (Chiefly U.S.)
1771 *Carroll Papers* in *Maryland Hist. Mag.* XIV. 137 Does not the Corporation Law require Broader tred than 6 Inches? **1841** C. CIST *Cincinnati* 29 Between Main street and the corporation line. **1875** *Mass. Ho. Repr. Doc.* 15 (on Taxation) 122 The general corporation tax. **1881** *Instr. Census Clerks* (1885) 22 Corporation Clerk. **1884** *La Corporation Carter*. *Ibid.* 87 Corporation Labourer. **1889** *Cent. Dict.*, *Corporation-stop*, a stop in a gas- or water-main for the use of the gas- or water-company only. (U.S.) **1893** 'O. THANET' *Stories Western Town* 215 We must 'away' for an interview with the corporation lawyer. **1901** W. G. COLLINGWOOD *Dict. Stock Exch. Terms* 29 Corporation Stocks refer to the Stocks, Loans and Debentures of the Corporations of the various towns and boroughs in the United Kingdom. **1911** *Amer. Yr. Bk.* 1910 326 The President's suggestion was followed, and the corporation tax became law. **1927** *Atl. Monthly* Mar. 660/2 The Franquists owe much to the French Corporation law. *Ibid.* 67 Corporation lawyers. **1955** *Times Lit. Suppl.* 22 Aug. 465/3 Other proposals, such as a thoroughgoing corporatist reorganization of industry, are sympathetically reviewed and reluctantly rejected on the ground that they are impracticable. **1966** *Listener* 7 Mar. 373/1 Whether the Labour Government is going to make a sharp break with its traditional (which contain a strong dose of corporatism).

corporational, *a.* Delete '*nonce-wd.*' and add later example.
1965 *Guardian* 4 Sept. 6/1 John Mandsell's birthday message in this week's 'Radio Times' sounded like the commonest kind of Corporational humbug.

corporatism (kǭ·ɹpǒrǎtiz'm). [f. CORPORATE *ppl. a.* + -ISM.] = *CORPORATIVISM. Hence **co·rporatist** *a.*
1890 *Advance* (Chicago) 13 Nov., Individualism has secured in danger of being swamped by a kind of corporatism. **1923** *Glasgow Herald* 5 Jan. 6/2 The Italian newspapers announce that 'Corporatism' has become a reality. **1938** *Tablet* 21 May 660/2 The Franquists owe much to the French Nationalist and Corporatist thinkers, such as Barrès. **1958** *Times Lit. Suppl.* 22 Aug. 465/3 Other proposals, such as a thoroughgoing corporatist reorganization of industry, are sympathetically reviewed and reluctantly rejected on the ground that they are impracticable. **1966** *Listener* 17 Mar. 373/1 Whether the Labour Government is going to make a sharp break with its traditional corporatism.

corporative, *a.* Add: **b.** Based upon corporate action or organization; *spec.* of a state, governed by or organized into corporations representing the employers and employed of the various industries and professions.
It. *stato corporativo* occurs 21 Apr. 1927 in Carta del lavoro § 7.
1927 *Times* 18 Aug. 11/6 The Corporative State offers.. greater opportunities than the Liberal State. **1929** *Times Lit. Suppl.* 11 July 547/2 The Fascist Corporative State. **1929** *Encycl. Brit.* X. 563/2 The industrial unit..is at present being worked into the fabric of the Italian state which will become the 'corporative State' when, in the judgment of its present leaders, the revolutionary stage has been passed. **1941** *Ann. Reg.* 1940 245 The new Under-Secretary [in Portugal] described the corporative system.. as 'the bulwark of our economic and social order'.

corporativism (kǭ·ɹpǒrǎtivi:z'm). [f. CORPORATIVE *a.* + -ISM.] The principle or practice of corporate action or organization; *spec.* a corporative system (see prec.).
1930 *Economist* 17 May 1099/1 To regard economics as a science whose principles, if properly formulated, can be equally well applied to liberalism, socialism or corporativism. **1937** *Times* 18 Dec. 13/6 If the economic frame-work of present-day Italy consists of a borrowed Socialism erected on an inchoate corporativism, the informing spirit is not egalitarian. **1938** *Downside Rev.* LVI. 382 Corporativism means a vocational organization, the grouping of individuals according to the function which they exercise in society. **1963** *Times* 3 June 9/6 The dangers of a producer-dominated economy tending to degenerate into syndicalism or corporativism outweigh the risks of an economy resting on consumer preferences.

corporealism. Delete †*Obs.* and add later examples. Also *fig.*
1894 *Classical Rev.* VIII. 463/2 Professor Windelband fully admits the naive corporealism of the earliest Greek philosophy. **1924** W. J. LOCKE *Coming of Amos* vii, I ought to have..taken him to London..and generally given him all the advantages, within my command, of the Higher Thought and the Lower Corporealism.

corporosity. Add: (Later examples.) Also used in a humorous greeting (see quot. 1950).
1870 O. LOGAN *Before Footlights* 174 A gentleman endowed with an ample corporosity. **1890** *Jrnl. Amer. Folk-Lore* III. 64 How does your corporosity sagatiate? **1922** JOYCE *Ulysses* 418 Your corporosity sagaciating O K? **1950** *Publ. Amer. Dial. Soc.* XIV. 59 'How is your corporosity segaciating?' meaning, how is your health?

corps. Add: **2. b.** *corps d'élite* (kǒr delīt), a body of picked men; a select group.
1884 J. SHARMAN *Hist. Swearing* vi. 105 In every society there has existed a certain *corps d'élite*. **1931** *Times Lit. Suppl.* 1 Jan. 2/3 The 1st Cavalry, a *corps d'élite* organized by Jefferson Davis. **1943** J. S. HUXLEY *TVA* ix. 56 A workers' *corps d'élite* from whose ranks men can be picked to undertake the more skilled jobs. **1958** *Observer* 18 May 6/4 Grammar school pupils—the bulk of our future *corps d'élite*. **1970** *Times* 19 Nov. 10/2 The Bar is not remotely a *corps d'elite*—as is impliedly..claimed.
3. c. *corps de ballet.* (Earlier and later examples.) Also *fig.*
1826 M. KELLY *Reminisc.* (ed. 2) I. 165 There was an excellent, and very expensive corps de ballet. **1836** DICKENS in *Morn. Chron.* 24 Sept. 3/2 The whole of the characters, including a numerous *corps de ballet* of boots and shoes in the back-ground. **1958** *Times* 19 Aug. 11/3 Some signs of raggedness among the *corps de ballet*.
d. A students' society in a German University. Also *attrib.*
1874 J. M. HART *German Univ.* iv (Cent. D.), A corps has no existence outside of its own university; it has no affiliations, no 'chapters'. **1904** *Daily Chron.* 29 Dec. 4/6, I hope that, as long as there are German corps-students, the spirit which is fostered in their corps..will be preserved. **1905** *Westm. Gaz.* 29 Apr. 14/1 As a corps-student the Crown Prince never felt at ease.
5. *corps de logis* (kǒr dǝ lǫʒi), the body of a house, the main part of a building; also occas., a structure wholly or partly disconnected from the main building.
a **1666** EVELYN *Diary* 9 Feb. an. 1644 (1955) II. 104 In the Court aux Thuilleries is a princely fabrique...To this is a Corps de logis worthy of so greate a Prince. *a* **1700** *Ibid.* 25 Sept. an. 1672 III. 625 His new house..has many noble roomes in it, but they are not so Convenient..consisting of but one *Corps de Logis*. **1787** BECKFORD *Italy* (1834) II. xxii. 129 Joining on to the enormous *corps de logis*, the palace terminates to the right and left. **1886** C. M. YONGE *Chantry House* I. viii. 74 A still older portion, more ancient than the square *corps de logis*.
6. *corps à corps* (kǒr a kǒr), body to body; in immediate contact; *spec.* in Fencing (see quot. 1910). Also *transf.* and *fig.*
1890 A. HUTTON *Fixed Bayonets* 17 The position known as 'Corps à corps', when the combatants are so close together that it is impossible for either of them to disentangle his weapon or withdraw his point. **1905** MRS. H. WARD *Marriage W. Ashe* ix. 169 The two men were measured against each other *corps à corps*,—the wide knowledge..of the minister, against the originality..of the writer. **1910** *Encycl. Brit.* X. 594/2 '*Corps-à-corps*' (body to body), the position of two fencers who are at such close quarters that their persons touch; when this occurs the fencers must again come on guard. **1931** WYNDHAM LEWIS *Let.* 5 May (1963) 202, I have been engaged in a close corps-a-corps business struggle for some weeks. **1939** A. TOYNBEE *Study of Hist.* IV. 430 Our..question of who is to capture the world market is being settled by a direct encounter between the competitors, *corps à corps*. **1969** R. THOMAS *Singapore Wink* xxiv. 242 There was no *corps-à-corps* with our wrists locked and straining.

corpse, *sb.* Add: **6.** *Comb.* Also *corpse-body, -city, -climate, -stealer, -worm*; *corpse-cold, -defiling, -encumbered, -fed, -pale, -strewn* adjs.; **corpse-factory** *slang*, a place where many people are slaughtered; also *fig.*; **corpse-fetch** [FETCH *sb.*[2]]; **corpse-reviver** (examples); now esp. a pick-me-up for a hangover.
1930 D. H. LAWRENCE *Nettles* 24 Trot, trot, trot, corpse-body, to work. Chew, chew, chew, corpse-body, at the meal. **1946** S. SPENDER *Europ. Witness* 22 This putrescent corpse-city was the hub of the Rhineland. **1866** G. MEREDITH *Vittoria* I. vii. 88 Then down, and along a passage; lower down, deep into corpse-climate. **1903** HARDY *Dynasts* I. I. iii. 22 One task Is theirs who would inter this corpse-cold Act. **1939** R. CAMPBELL *Flowering Rifle* II. 46 Corpse-defiling anarchists. **1878** SWINBURNE *Poems, White Czar* 191 Till the wind gave his warriors and their might To shipwreck and the corpse-encumbered sea. **1919** DOWNING *Digger Dial.* 18 *Corpse factory*, the Western Front. **1937** W. B. YEATS *Let.* 27 Jan. (1954) 879, I think professional musicians have a corpse-factory, mankind melted down and poured out of a bottle. **1890** KIPLING *Barrack-r. Ballads* (1892) 140 To the cod and the corpse-fed conger-eel. **1914** MASEFIELD *Philip the King* 17 You corpse-fetch from the unclean grave, begone! **1929** D. H. LAWRENCE *Pansies* 32 You are all going dead and corpse-pale. **1871** *Birmingham Daily Post* 22 Dec., And our American refreshment bars, In drinks of all descriptions cut a dash, From corpse revivers down to 'brandy smash'. **1937** M. ALLINGHAM *Case of Late Pig* xvii. 125 Pig was dressed, but he wanted a corpse-reviver. **1966** 'A. BLAISDELL' *Date with Death* xi. 144 Corpse Reviver Number Three...You take a jigger of Pernod and add some lemon juice and ice cubes and fill the glass with champagne. **1864** BAGEHOT *Coll. Wks.* (1965) II. 303 Sterne's grave.. was so..neglected that the corpse-stealers ventured to open it. **1905** *Westm. Gaz.* 6 June 1/3 After an impressive pause he rose from the corpse-strewn stage. **1940** EMPSON *Gathering Storm* 21 Your eyes are corpse-worms.

corpsman (kǭ·ɹmǎn). *U.S.* [f. CORPS + MAN *sb.*[1]] An enlisted medical auxiliary in the U.S. Navy or Army; = **hospital corpsman*.
1941 *Acts 77th U.S. Congr.* 7 Apr.–6 May, Naval Hospital, Norfolk, Virginia: Extension of seawall, bulkhead, and fill, $85,000; barracks and accessories for corpsmen, $200,000. **1959** C. OGBURN *Marauders* (1960) vii. 209 Medical corpsmen..threaded their way back down the half-running column to bind up wounds. **1968** *Globe & Mail* (Toronto) 13 Feb. 4/1 The Americans dived for cover. 'Corpsman, corpsman... We've got wounded here!'

corpsy (kǭ·ɹpsi), *a.* [f. CORPSE *sb.* + -Y[1].] Resembling or characteristic of a corpse; cadaverous.
1883 'MARK TWAIN' *Life on Mississippi* xxxviii. 356 Stiff flowers done in corpsy-white wax. **1894** *Contemp. Rev.* Aug. 294 A 'corpsy or cadaveric form'. **1939** 'G. ORWELL' *Coming up for Air* I. iv. 39 The sweet corpsy smell.

co·rpulently, *adv.* [f. CORPULENT *a.* + -LY[2].] In a corpulent manner.
1847 in WEBSTER.

cor pulmonale (kǭ:ɹ pɒlmǒnā·li). *Path.* [mod.L., f. L. *cor* heart + mod.L. *pulmonale*, *-is* (see PULMONAL *a.*).] †**a.** The right auricle and right ventricle considered together. *Obs.* **b.** Heart disease resulting from disease of the lungs or their blood-vessels; pulmonary heart disease.
1857 DUNGLISON *Med. Lex.* (rev. ed.) 241/1 *Cor pulmonale*, see Heart. **1935** MCGINN & WHITE in *Jrnl. Amer. Med. Assoc.* CIV. 1473/1 The..result of a high degree of occlusion of the pulmonary artery is sudden dilatation of the right ventricle and right auricle, which may best be termed acute cor pulmonale. **1937** *Arch. Internal Med.* LX. 1043 At present two types of cor pulmonale are recognized—acute and chronic. **1961** *Lancet* 5 Aug. 300/1 Severe dyspnœa.. was much more prolonged..in patients with cor pulmonale.

corpus. Add: **2.** *corpus luteum* [L. *luteus*, *-um* yellow] (pl. *corpora lutea*), a yellowish body developed in the ovary from the ruptured Graafian follicle after discharge of the ovum; it secretes progesterone and other hormones and after a few days degenerates unless fertilization has occurred, when it remains throughout pregnancy.
1788 *Encycl. Brit.* I. 740/2 In conception, one of these mature ova is supposed..to be squeezed out of its nidus into the Fallopian tube; after which the ruptured part forms a substance which in some animals is of a yellow colour, and is therefore called *corpus luteum*. **1910** *Surg., Gyn. & Obstetr.* X. 221/2 (*heading*) Extract of corpus luteum in disturbances of artificial and physiologic menopause. *Ibid.*, An extract made from the corpora lutea of beef ovaries [was used] rather than an extract of the entire ovary. **1926** J. S. HUXLEY *Ess. Pop. Sci.* v. 60 The number of corpora lutea, or traces of discharged ova in the ovaries. **1939** *Ann. Reg.* 1938 372 Corpus luteum hormone and its derivatives. **1959** *New Biol.* XXX. 79 As in mammals, glandular bodies known as corpora lutea are produced in the ovaries of viviparous (and also of some oviparous) reptiles, in places from which the eggs have been shed at ovulation.
5. *corpus delicti* (later examples); also, in lay use, the concrete evidence of a crime, esp. the body of a murdered person.
1863 *N.Y. State Court of Appeals, Rep.* IV. 179 The *corpus delicti*, in murder, has two components, death as the result and the criminal agency of another as the means. **1908** J. HAWTHORNE *Libr. Best Myst. & Detective Stories* 89 The term '*corpus delicti*' is technical, and means the body of the crime, or the substantial fact that a crime has been committed. **1922** JOYCE *Ulysses* 451 (*He extends his portfolio.*) We have here damning evidence, the *corpus delicti*, my lord, a specimen of my maturer work disfigured by the hallmark of the beast. **1964** *Sunday Mail Mag.* (Brisbane) 13 Sept. 3/3 An enthusiastic trooper, one of a party investigating river, dam and hollow log in search of the corpus delicti, found some important evidence in a fallen tree.

corpuscle. Add: **2. c.** *Electr.* J. J. Thomson's name for what was subsequently called an *ELECTRON[2]. So **corpu·scular** *a.*
1897 J. J. THOMSON in *Phil. Mag.* 5th Ser. XLIV. 311 These primordial atoms, which we shall for brevity call corpuscles. **1898** —— *Discharge Electr. through Gases* 190 We shall call this theory of the cathode rays the 'corpuscular theory'. **1902** *Encycl. Brit.* XXX. 452/2 These particles, which were termed by their discoverer *corpuscles*, are more commonly spoken of as *electrons*, the particle thus being identified with the charge which it carries. **1904** J. J. THOMSON *Electr. & Matter* v. 98 We shall, for the sake of brevity, speak of this kinetic energy of the corpuscles within the atom as the *corpuscular temperature* of the atom. **1958** GLASSTONE *Sourcebk. Atomic Energy* (ed. 2) ii. 38 The name electron, originally intended by Stoney..for the magnitude of the charge, soon became associated with the actual particles themselves. Possibly in the interest of strict accuracy, Thomson adhered to the term corpuscle for about twenty years, but ultimately he gave it up in favor of electron.

corpus vile (kǭ·ɹpɒs vai·li). Pl. **corpora vilia** (kǭ·ɹpǒrǎ vai·liǎ). [L., = 'cheap body'. Orig. in phr. (see quot. 1822) meaning 'let the experiment be done on a cheap (or worthless) body'.] A living or dead body that is of so little value that it can be used for experiment without regard for the outcome; *transf.*, experimental material of any kind, or something which has no value except as the object of experimentation.
[**1822** DE QUINCEY *Confess.* App. 189 *Fiat experimentum in corpore vili* is a just rule where there is any reasonable presumption of benefit to arise on a large scale.] **1860** *Once a Week* 14 Apr. 343/1 It appeared that when she was in want of a chapter, she was in the habit of practising upon poor G. as a *corpus vile*. **1861** C. LEVER in *All Yr. Round* 5 Jan. 310/1 It was making the experiment in the 'corpore vile' [*sic*] of accident, to see what might come of it. **1865** C. M. YONGE *Clever Woman* I. iii. 80 The only difficulty was to find poor people enough who would submit to serve as the *corpus vile* for their charitable treatment. **1867** J. S. MILL *Lett.* (1910) II. 94 Ireland may be tried as a *corpus vile* for experimentation on Government management of railways and telegraphs. **1884** *Athenæum* 23 Feb. 242/1 There is too great tendency for the psychological novelist to regard his characters as so many *corpora vilia* for his scientific theories. **1933** *Week-end Rev.* 16 Sept. 277/1 Judith is much more than a *corpus vile* for the infliction of her foster-mother's spite. **1953** *Essays in Criticism* III. I. 4, I am not proposing to include among these initial *corpora vilia* passages from either Mr Eliot's criticism or Dr Leavis's.

corrack, var. CURRACH.

corrade, *v.* Restrict †*Obs.* to sense 1 and add: **2.** (Later U.S. and *Geol.* examples: cf. *CORRASION 2.)
1847 WEBSTER, *Corrade*, to rub off. **1877** G. K. GILBERT *Geol. Henry Mts.* 101 Streams of clear water corrade their beds by solution. **1958** P. LAKE *Physical Geogr.* (ed. 4) 297 When it has a full load it corrades its banks and bed.

corral, *v.* Add: **1.** (Earlier U.S. example.)
1848 E. BRYANT *California* (1849) iii. 19 The wagons, in forming the encampment, were what is called *corraled*, an anglicised Spanish word, the significance of which, in our use of the term, is, that they were formed in a circle.
b. *absol.* To draw up the wagons in a circle.
a **1848** RUXTON *Life Far West* (1849) 12 It was pretty nigh upon sundown, and Bill had just sung out to 'corral'. **1875** *Fur, Fin & Feather* 108 At midnight the rear drivers joyously see the little fires flashing up far ahead here and there upon the prairie, telling them that the head of the train has begun to corral. **1894** in *Kansas Hist. Coll.* (1896) V. 93 Several trains were compelled to corral and 'stand them [*sc.* Indians] off' until relieved by the dragoons.
3. (Earlier and later examples.)
1860 *Knickerbocker* Jan. 100, I want to 'corel' you for a little chat. **1867** 'MARK TWAIN' *Jumping Frog* 157 That sort of thing would corral their sympathies. **1910** WODEHOUSE *Psmith in City* viii. 64 How did we corral him, and become to him practically as long-lost sons? **1965** *Economist* 23 Jan. 322/1 The whip is charged theoretically with the duty of corraling votes for party programmes.

corrasion. Restrict † *Obs. rare* to sense 1 and add: **2.** *Geol.* The local wearing away of the surface of the earth by the agency of moving air, water, glacial ice, etc., in conjunction with any matter transported by them.
1875 J. W. POWELL *Explor. Colorado River* xi. 152 The proof is abundant that the river cut its own channel; that the cañons are gorges of corrasion. **1877** G. K. GILBERT *Geol. Henry Mts.* 101 In corrasion the agents of disintegration are solution and mechanical wear. **1904** CHAMBERLIN & SALISBURY *Geol.* (1905) I. 108 The more active and tangible processes by which surface rocks are broken up, such as wave wear, river wear and glacier wear, are processes of corrasion. **1937** WOOLDRIDGE & MORGAN *Physical Basis Geogr.* xi. 153 The work of vertical or lateral cutting performed by a river in virtue of the abrasive power of its load is termed corrasion. **1958** P. LAKE *Physical Geogr.* (ed. 4) 297 Clear water does very little corrasion of solid rock. **1970** R. J. SMALL *Study of Landforms* ii. 42 Stream erosion is usually said to involve three distinct processes: downward corrasion..lateral corrasion..and headward extension.

correcting, *ppl. a.* Add: **b.** *spec.* in *correcting compass, magnet.*
1821 *Kaleidoscope* 13 Feb. 262/2 Perhaps this correcting compass might be applied in the round top, if there be not too much iron in that quarter. **1899** *Pract. Rules Deviations Compass* (Admiralty Publ.) 10 Alteration in position of Correcting Magnets. **1918** W. G. MCMINNIES *Practical Flying* vi. 91 Holes for the accommodation of correcting magnets.

correctitude (kǒre·ktitiud). [f. CORRECT *a.* + -ITUDE, after *rectitude*.] Correctness of conduct or behaviour (in quot. 1929, a 'correct' statement or expression).
1893 *Nat. Observer* 13 May 650/2 The high correctitude of his youth. **1897** *Westm. Gaz.* 20 Feb. 3/1 Dr. Jameson stepping in maintains order where disorder existed, and stepping in with the Colonial Secretary's orders, would have done so with perfect—if not 'unctuous'—correctitude'. **1906** F. S. OLIVER *A. Hamilton* 316 Jefferson was overflowing with kindness—a contrast to the calm correctitude of his chief. **1922** G. A. GREENWOOD *England To-Day* ii. 72 The correctitude of attending eleven o'clock service. **1929** W. S. CHURCHILL *Aftermath* xix. 412 Lord Curzon..soused him in sonorous correctitudes.

corrective. Add: **A.** *adj.* **1. a.** Esp. in *corrective training* (see quots.). So *corrective trainee*.
1938 *Bill* 2 & 3 *Geo. VI* (Public Bills II) 4. § 16 Persons sentenced to corrective training..shall be allocated to a prison..and be given such employment and subjected to such methods of training and discipline as may be best fitted to lead to their reformation. **1939** *Ann. Reg.* 1938 93 A Criminal Justice Bill...Two new types of prison sentences were proposed. One was called 'corrective training', and was to be for not less than two and not more than four years for persons between 21 and 30. **1963** T. & P. MORRIS *Pentonville* ii. 30 Among the recalcitrant corrective trainees there are some who claim they are being denied the 'right to be trained'.

correlatable, *a.* (Further examples.)
1964 R. H. ROBINS *Gen. Linguistics* vii. 281 Some other formal categories are less closely correlatable with extra-linguistic categories. **1969** *Proc. Geol. Soc. Lond.* Aug. 158 Faunas directly correlatable with those of the Saint-Jean-d'Etampes section.

correlate, *v.* **2.** Delete '(*spec.* geological formations, etc.)' and add further examples.
1925 N. BOHR *Theory of Spectra* (ed. 2) 135 It has been possible to correlate each term with the occurrence of electron orbits of a given type. **1930** *Economist* 18 Oct. 715/2 To prove by an historical statistical analysis that..it is impossible to correlate from available evidence either high rates and low stock prices or low rates and high stock prices with any certainty. **1952** G. H. BOURNE *Cytol. & Cell Physiol.* (ed. 2) vi. 273 Bennett was not able to correlate changes in the Golgi material with secretion in the cat adrenal. **1971** *Nature* 15 Jan. 182/1 So the observed luminosity of the primary [star] can be correlated reliably with its original main-sequence mass. **1971** *Daily Tel.* 12 Feb. 8/2 Are you wondering how many people spend their time..in devising such idiot statistical measures, applying them, collating and correlating them?

correlation. Add: **1. c.** In *Statistics*, an interdependence of two or more variable quantities such that a change in the value of one is associated with a change in the value or the expectation of the others; also, the value of this as represented by a correlation coefficient. So *correlation coefficient* or *coefficient of correlation*: a number between −1 and 1 calculated so as to represent the linear interdependence of two variables or two sets of data; *spec.* the product-moment coefficient (see *PRODUCT *sb.*[1]).
1888 F. GALTON in *Proc. R. Soc.* XLV. 143 The statures of kinsmen are co-related variables; thus, the stature of the father is correlated to that of the adult son,..and so on; but the index of co-relation..is different in the different cases. **1896** K. PEARSON in *Ibid.* LIX. 302 Let r_0 be the coefficient of correlation between parent and offspring. *Ibid.* 303 We conclude that there is a sensible correlation (*circa* 0·18) between fertility and height in the mothers of daughters. **1909** W. & E. ELDERTON *Primer Statistics* 57 In such cases there is no relationship, therefore, between length and breadth, and we say that the 'coefficient of correlation' is zero. **1933** *Forestry* VII. 27 It may be stated that a correlation coefficient is a number expressing the degree of correlation between two variates, as, for example, weight and volume, or specific gravity and strength. A correlation coefficient of +1 indicates a direct linear relation, a coefficient of −1 indicates an inverse linear relation, while one of 0·0 indicates absence of relations between the variates. **1943** N. BALCHIN *Small Back Room* i. 8 There's a positive correlation between penetration and the height of the man firing. **1960** F. LAND *Lang. Maths.* xiv. 247 We need to be able to describe this gradation from perfect correlation to no correlation at all, and for this purpose we can calculate a 'correlation coefficient'. **1964** R. VON MISES *Math. Theory Probability & Statistics* xi. 572 Out of 239 patients treated with serum, 9 patients died, and out of 244 left without serum, 29 died. Compute the correlation coefficient between treatment and recovery. **1969** *Computers & Humanities* III. 145 After correlation coefficients are computed for every pair of variables, they are entered in a correlation matrix which summarizes the degree of similarity between pairs of variables.
5. (Later examples.)
1886 A. J. JUKES-BROWNE *Student's Handbk. Hist. Geol.* IV. iii. 27 The correlation and classification of rocks on Palæontological principles. **1963** *Gloss. Mining Terms* (*B.S.I.*) i. 6 *Correlation*, the process of orientating underground survey lines or of coordinating underground survey stations to the National Grid.

correla·tional, *a.* [f. CORRELATION + -AL.] Of or pertaining to correlation (esp. sense *1 c); employing correlation.
1952 C. P. BLACKER *Eugenics* 39 The correlational calculus, the *fons et origo* of that far-reaching ramification—the modern mathematical theory of statistics. **1970** *Computers & Humanities* IV. 350 Correlational analysis of relations between scales over time and between scales and other variables.

correlogram (kǒre·lǒgræm). *Statistics.* [f. CORRELATE *v.* + -O + -GRAM.] A graph showing how the autocorrelation or serial correlation between the members of a series varies with the lag.
1938 H. WOLD *Analysis of Stationary Time Series* iv. 135 For the sake of brevity in writing, the graphs of serial and autocorrelation coefficients will be termed correlograms (empirical and hypothetical respectively). **1952** *Brit. Jrnl. Psychol.* July 183, I began pedantically to work correlograms of correlograms. **1959** *Chambers's Encycl.* XIII. 158/2 Perhaps the most powerful technique for investigating the nature of the generating process of oscillatory series is correlogram analysis.

correspondence. Add: **1. b.** *Math.* A relation between two sets in which each element of one set is associated with a constant number (a') of elements of the second, and each element of the second set is associated with a constant number (a) of elements of the first.
Where a and a' are both 1, it is called a *one-to-one correspondence* (see also *ONE 33).
1866 CAYLEY in *Proc. Lond. Math. Soc.* I. VII. 1 If two points of a unicursal curve have an (a, a') correspondence, the number of united points is $= a + a'$. **1882** SYLVESTER *Coll. Math. Papers* (1912) IV. 24 The theorem of one-to-one and class-to-class correspondence between partitions of n into uneven and its partitions into unequal parts. *Ibid.* 38 There is thus a one-to-one correspondence, and consequently equality of content between the two systems of partitions. **1906** *Trans. Amer. Math. Soc.* VII. 393 We thus establish a correspondence between points in the space y, and planes in the space x. **1941** BIRKHOFF & MACLANE *Surv. Mod. Algebra* xii. 344 The number of many-one correspondences of the class S to a subclass of T. *Ibid.*, Any such correspondence $x \to y$ is determined by a function $y = f(x)$ which defines for each argument x in S a value y in T. **1949** SEMPLE & ROTH *Introd. Algebraic Geom.* iv. 58 The equation ..determines an (m, n) correspondence between the variables x and y, such that to any x there correspond n value.. of y, while to any y there correspond similarly m values.. of x. **1959** A. G. OETTINGER in R. A. Brower *On Translation* 257 Frequent one-to-many correspondences between Russian and English words create..one of the most perplexing problems of automatic translation. **1962** M. P. CROSLAND *Hist. Stud. Lang. Chem.* v. i. 294 In an ideal nomenclature of a science there is obviously a one-to-one correspondence between names and the things named.
7. *attrib.*: **correspondence card,** a blank card intended for use as notepaper; **correspondence class, course,** a class or course conducted by correspondence; **correspondence clerk,** a clerk who deals with the correspondence of a business house; **correspondence college, school,** a college or school which instructs by means of correspondence; **correspondence principle** *Physics*, a principle connecting classical with quantum physics (see quots.); **correspondence theory** *Philos.*, the theory that the definition, or the criterion, of truth is that true propositions correspond to the facts (opp. **coherence theory*); also *correspondence-notion*.
1892 G. SAINTSBURY in *Macmillan's Mag.* June 107 There are great temptations not to write letters. Telegrams, postcards, correspondence-cards, letter-cards,—all of these things the truly good and wise detest. **1907** *Yesterday's Shopping* (1969) 333/2 Black bordered correspondence cards and envelopes. **1933** A. SELWYN *Printing for Amateurs* vi. 65 Correspondence Cards are 4½ × 3½, and Post Cards 5½ × 3½. **1970** L. EVANS *Lett. Walter Pater* p. xliii, White correspondence cards, approximately 11·5 by 9·2 cm, with rectangular (instead of the usual rounded) corners. **1876** *Monthly Packet* Feb. 6 The number of pupils who join these correspondence classes increases regularly. **1885** *Chambers's Jrnl.* 610/1 The Correspondence Class is intended for the extension of skilled teaching to any woman within reach of the post. **1906** DICKSEE & BLAIN *Office Organis.* 8 Correspondence clerks should have a good knowledge of shorthand. **1911** W. OWEN *Let.* 18 June (1967) 75 Studying with a Corresp. College at the same time. **1931** W. HOLTBY *Poor Caroline* vi. 232, I know your correspondence college is rotten. **1902** *World's Work* IV. 2524 He subscribes to a correspondence course. **1925** S. LEWIS *Martin Arrowsmith* xxii. § 1, The editor of the magazine which told clerks how to become Goethes and Stonewall Jacksons by studying correspondence-courses. **1929** *Encycl. Brit.* VI. 468/1 In 1921 the U.S. department of war established 342 correspondence courses in military tactics for reserve officers. **1946** WODEHOUSE *Money in Bank* xxviii. 245 'Was I very haughty?' 'Cleopatra could have taken your correspondence course.' **1906** H. H. JOACHIM *Nature of Truth* i. 7 A certain conception of truth is implied or expressed, which I shall call the 'correspondence-notion' of truth. **1922** N. BOHR *Theory of Spectra* II. iii. 50 The correspondence principle suggests that the radiation emitted by the perturbed atom must exhibit circular polarization. **1928** A. S. EDDINGTON *Nat. Physical World* ix. 195 The classical laws are the limit to which the quantum laws tend when states of very high quantum number are concerned. This is the famous Correspondence Principle enunciated by Bohr. **1935** —— *New Pathways* iv. 77 The classical laws of mechanics and electro-magnetism..are simply the limiting form assumed by the formulae of quantum theory...This connection is known as Bohr's Correspondence Principle. **1964** in R. G. Colodny *Frontiers Sci. & Philos.* vi. 253 The field theories have started moving away from the correspondence principle. **1889** *Christian Lit. Mag.* I. 91 The Chautauqua School gives a certificate.., and so very properly may any correspondence school. **1929** *Encycl Brit.* VI. 468/2 The correspondence school movement is rapidly reaching a recognized stage of permanency. **1933** WODEHOUSE *Mulliner Nights* v. 157 A strange elation filled Sacheverell Mulliner's bosom as he left the offices of the Correspondence School. **1902** BALDWIN *Dict. Philos.* II. 720/1 The fundamental difficulty with a 'correspondence theory' is this: it assumes a reality with which that which claims to be true may be compared. **1940** B. RUSSELL *Inquiry* xvi. 232 In this case, the correspondence-theory of truth holds *simpliciter*.

corresponding, *ppl. a.* Add: **1. b.** *corresponding points*, any pair of points (one on each retina) which give rise to a single visual impression of an object whose image falls upon them.
1879 *Encycl. Brit.* VIII. 828/1 If,..by means of a stereoscope, or by holding a card between the two eyes, and causing a slight convergence of the eyes, the two images are brought upon corresponding points of the two retinæ, the image will at once be seen in relief. **1963** G. L. WALLS *Vertebr. Eye* x. 317 The two images of any object-point must fall upon 'corresponding points' of the two retinæ if they are to be fused. The two foveæ are corresponding points.

‖ **corrida** (kori·da). [Sp., *corrida de toros* 'course of bulls'.] In full *corrida de toros*. A bull-fight; bull-fighting. Also *fig.*
1898 *Daily News* 12 July 5/6 The hotels were thronged with tourists from all parts attracted by the corridas. **1921** *Chambers's Jrnl.* Feb. 194/2 Those who think of this country as still sleeping, 'decadent', given up to the *corrida*. **1924** *Glasgow Herald* 11 Oct. 4 The cruelties perpetrated in the course of a 'corrida'. **1932** HEMINGWAY *Death in Afternoon* iii. 31 In the modern formal bullfight or corrida de toros there are usually six bulls that are killed by three different men. **1958** *Observer* 27 Apr. 4/4 At each *corrida*, six spears called *banderillas* are stuck into each of six bulls. **1963** *Economist* 2 Feb. 404/2 A confused *corrida* at Brussels to decide the fate of Spain's request..will be followed by a more orthodox confrontation with Washington.

corridor. Add: **4.** *fig.* (Later examples.) Cf. *COULISSE 4.
1962 *Listener* 15 Feb. 280/1 The glamorous and Machiavellian figures, patrolling the corridors of power, to which we have been accustomed in many recent novels and plays. **1964** C. P. SNOW (*title*) Corridors of power. **1970** *Physics Bull.* Mar. 110/1 It's no good physicists going into the corridors of power in Whitehall..unless they are..effective people.
b. A similar passage in a railway carriage, upon which all the compartments open.
1892, etc. [see sense 5 below]. **1899** *Railway Engineer* Jan. 26 The ceiling of the corridor, as will be seen from the drawing, is a complete arch made up in a similar way to that of the compartments. **1951** *Oxf. Jun. Encycl.* IV. 343/2 The traditional plan of a side corridor connecting separate compartments has given place, on many main-line British trains, to open coaches with a central gangway.
c. A strip of the territory of a state running through another territory and so contrived as to give access to a certain part, e.g. the sea.
1919 *Economist* 5 July 6/2 The German connections across the Polish 'corridor' to the sea. **1920** H. SPENDER *Prime Minister* 310 When matters seemed at a deadlock—on the Saar Valley, the Polish Corridor, or even the perplexing question of Fiume. **1921** R. W. SETON-WATSON in H. W. V. Temperley *Hist. Peace Confer. Paris* IV. 273 The Czecho-Slovaks advanced a claim for territorial contiguity with the Yugo-slavs, to be attained by the creation of a corridor running from the Danube to the Drave. **1921** *Times* 4 Jan. 12/1 The Danzig corridor is bound to be the subject of dispute for long to come. **1950** THEIMER & CAMPBELL *Encycl. World Politics* 121/2 The Corridor became the immediate cause of World War II.
d. = *air corridor* (*AIR *sb.*[1] III. 8).
1920 *Flight* XII. 346/1 The River Indus is to be the northern boundary of civil flying, save for..two 'corridors'. One of these is 20 miles wide from Sukkur (exclusive) to Quetta. **1921** [see *aerial corridor* s.v. *AERIAL *a.* 5]. **1922** [see *AIR *sb.*[1] III. 8]. **1948** H. NICOLSON *Diary* 16 July (1968) III. 146 The Russians..will be carrying out the training of their fighter aeroplanes across air corridors to Berlin.
5. corridor carriage, coach, a carriage of a corridor train; **corridor train,** a railway train through the length of which a corridor (sense 4 b above) or passage extends.
1892 Corridor train [in Dict.]. **1893** *Daily News* 22 June 2/3 First and third class 'corridor' carriages...The 'corridor' carriages will have an enclosed passage running along the side. **1896** *Ibid.* 5 Aug. 5/1 The Board asks for information as to corridor trains, corridor cars, and carriages with open compartments. **1903** A. H. BEAVAN *Tube, Train, Tram, & Car* v. 59 The cars will be of the corridor type, seven to a full train. **1907** *Westm. Gaz.* 5 Sept. 10/1 The *Great Western Railway Magazine* for September claims for that company the credit of producing the first complete corridor-train, combining the privacy of separate compartments with the advantages of through communication from end to end and access to toilet rooms. It was 'built' in April 1892. **1911**

Kipling in *Harper's Mag.* Dec. 5/2 Dr. Gilbert stood by the door of the one composite corridor-coach. **1919** J. Buchan *Mr. Standfast* viii. 151 It was not a corridor carriage, but one of the old-fashioned kind. **1959** *Chambers's Encycl.* XI. 498/2 Corridor coaches had at first no connecting gangways..The Great Western and Great Eastern Railways both produced corridor trains with proper covered gangway connexions..in 1891.

corridored (kǫ·ridǫɹd), *ppl. a.* [f. Corridor + -ed².] Furnished with a corridor.
1904 *Westm. Gaz.* 8 Jan. 1/3 The monastery of Rila, and ..its corridored balconies.

corrie. Add: **b.** *attrib.*
1894 J. Geikie *Gt. Ice Age* (ed. 3) 254 No corrie-basin dates its origin to this stage. *Ibid.*, We have only to contrast the drainage-area of Glen Avon with that of Glen Derry or Glen Beg to see why it is that in the latter only high-level corrie-lakes occur. **1894** J. W. Gregory in *Q. Jrnl. Geol. Soc.* L. 515 The 'corrie' or 'hanging glaciers'. **1904** *Nature* 7 Apr. 549/1 The phase of corrie-glaciers, when the glacial detritus was borne for no great distance from the local centres of dispersion. **1960** B. W. Sparks *Geomorphol.* xii. 268 Bergschrunds (the major crevasses which occur near the backs of most corrie glaciers). *Ibid.* 273 If a rotation occurs,..an explanation of glacial basins, including corrie basins, is obviously possible.

Corriedale (kǫ·ridēⁱl). [The name of the estate in North Otago, New Zealand, where the breed was evolved.] A breed of New Zealand sheep evolved from Romney, Lincoln, Merino, and Leicester breeds to yield both wool and meat.
1902 *Rep. Conf. Delegates Agric. Soc. held in Dunedin* 64 The amendment in favour of naming the breed 'Corriedale' was..put and carried. **1911** A. Hawkesworth *Austral. Sheep & Wool* 246 To New Zealand must be given the honor of producing a new breed of sheep, valuable for its fleece and body. This new breed is called the 'Corriedale', the result of crossing confined to two acknowledged pure bred types, viz. Lincoln and Merino. **1915** *N.Z. Jrnl. Agric.* 20 Sept. 271 Corriedales originated by mating the Merino with either the Lincoln or Leicester..both sides being studs of the highest order. **1934** *Bulletin* (Sydney) 1 Aug. 22/2 The best prices were paid for animals of the original pure Maoriland blood from the original Corriedale flock, started with Lincoln rams on merino ewes. The name 'Corriedale' has been applied to a number of crossbred flocks of varying origins. **1959** A. McLintock *Descr. Atlas N.Z.* 40/1 In the South Island a wider range of [sheep] breeds is found, with the Romney type prominent in Southland.., the Corriedale on the rolling country, particularly in North Canterbury.

Corrigan (kǫ·rigăn). The name of a Dublin physician, Sir Dominic John *Corrigan* (1802–80), used in the possessive in *Corrigan's button, cirrhosis*, etc. (see quots.).
1886 *Buck's Handbk. Med. Sci.* II. 86/1 The arteries seem to swell and elongate, and then suddenly shorten again. This is the so-called Corrigan's or piston pulse. **1887** *Ibid.* V. 196/2 Corrigan's pulse, the peculiar 'jerking', 'splashing', 'collapsing', or 'water-hammer' pulse of aortic regurgitation. *Ibid.*, Corrigan's button, a firing-iron consisting of a button of iron..fastened in a wooden handle by a rod two inches long. **1890** Billings *Med. Dict.*, *Corrigan's cautery*, button cautery. *Corrigan's disease*, insufficiency of aortic valves. **1906** *Practitioner* Dec. 730 The bronchial tubes become dilated, until at length the whole of one lung consists of dilated tubes and fibroid tissue, in other words, Corrigan's cirrhosis. **1953** *Faber Med. Dict.* 104/1 *C[orrigan]'s pulse*, the 'water-hammer pulse' of aortic insufficiency. *C.'s respiration*, slow respiration of the cerebral type. *C.'s sign*, abdominal pulsation felt in cases of aneurysm of the abdominal aorta. **1961** *Brit. Med. Dict.* 355/1 *Corrigan's button*, or *button cautery*, a metal hemisphere carried on a shaft and handle...*Corrigan's disease*, aortic incompetence. ..*Corrigan's line*, copper line; a green line at the bases of the teeth in copper poisoning.

corrigent, *sb.* Delete †*Obs.* and add earlier and later examples.
1841 in E. Scudamore *Dict. Arts & Sci.* **1874** Garrod & Baxter *Mat. Med.* 341 The oil may be employed..as a corrigent to purgatives. **1901** Dorland *Med. Dict.* (ed. 2) 178/2 *Corrigent*... 2. Any agent which favorably modifies the action of a drug which is too powerful or harsh.

corroboree, *sb.* Add: **b.** A song or chant made for the occasion of such a dance.
1847 Leichhardt *Overland Exped.* x. 323 He sang most lamentable corroborries. **1881** A. C. Grant *Bush Life in Queensland* (1882) 51 They send runners to the neighbouring tribes, inviting them to come over..and listen to the new corroborree. **1889** J. H. L. Zillmann *Austral. Life* xii. 132 The story..became, no doubt, the theme for 'a corroberee'. **1956** R. Robinson *Feathered Serpent* 84 Yoola sat on the top of a red uprearing rock-face and sang his corroboree.
c. A social gathering; a noisy party; a disturbance.
1885, 1892 [see Dict.]. **1909** Ware *Passing Eng.* 93/2 *Corroboree* (Nautical), a drunken spree, in which there is much yelling. **1941** Baker *Dict. Austral. Slang* 20 *Corrobbery*, a social gathering, a public meeting. (2) A disturbance or noise (made by people). (3) A discussion. **1964** *Telegraph* (Brisbane) 24 Sept. 5/2 It is a fair bet that, with some bush natives taking full advantage of their drinking rights, there will be some lively corroborees in lounges and beer gardens.

corroboree, *v.* [f. the sb.] *intr.* To take part in a corroboree. Also *transf.* To 'dance'; hence, of a pot, to boil.
1830 R. Dawson *Pres. State Australia* 61 They began to corrobory, or dance. **1846** C. P. Hodgson *Remin. Australia* 257 The mosquitoes from the swamps corroboreed with unmitigated ardour. **1881** A. C. Grant *Bush Life in Queensland* (1882) 34 They had almost finished their meal before the new quart corroboreed. **1885** Mrs. C. Praed *Austral. Life* 22 A scene of feasting and corroboreeing. **1908** E. J. Banfield *Confessions of Beachcomber* ii. i. 251 There was little for the population to do save to eat, drink, laze away the hotter hours of the day, and 'corrobboree' at night.

corrosion. Add: **1. d.** *Geol.* The gradual destruction of rock or soil by chemical and solvent action of water, natural acids, etc.; also, the modification of crystals in a rock by the solvent action of residual magma; *corrosion zone*, the area so modified.
1802 J. Playfair *Illustr. Huttonian Theory Earth* iii. 98 Some earths,..such as the calcareous, are immediately dissolved by water; and though the quantity so dissolved be extremely small, the operation, by being continually renewed, produces a slow but perpetual corrosion, by which the greatest rocks must in time be subdued. **1897** *Geogr. Jrnl.* X. 502 Erosion, corrosion, and hydrostatic pressure have..formed a real sponge of stone. **1903** Geikie *Text-bk. Geol.* (ed. 4) 141 The ferro-magnesian minerals of earlier consolidation among basalts and andesites are sometimes surrounded with a dark shell called the corrosion-zone. **1938** *Science* 15 Apr. 347/2 Soil deterioration or wastage through chemical action may be expressed by the word *corrosion*, in contrast with soil wastage by physical forces, or *erosion*. *Corrosion* is already in use by geologists to some extent to express virtually the same idea as that suggested. **1965** A. Holmes *Princ. Phys. Geol.* (ed. 2) xviii. 505 *Corrosion*, wearing-away of surfaces and of detrital particles and fragments by the solvent and chemical action of natural waters. **1968** R. W. Fairbridge *Encycl. Geomorphol.* 644/2 The limestone surfaces..are etched, pitted and transected. ..The corrosion is largely 'biological weathering', due to carbonic acid and humic acids in soil and around the roots of lichens and mosses.
4. *attrib.*, as **corrosion fatigue**, failure in a metal part subjected simultaneously to corrosion and to repeated stresses.
1926 D. J. McAdam Jr. in *Proc. Amer. Soc. for Testing Materials* XXVI. 243 Damage to the endurance properties of such specimens [of steel] is due to the combined action of corrosion and fatigue...Such failure under combined corrosion and fatigue may be called 'corrosion-fatigue'. *Ibid.* 245 The corrosion-fatigue limit for this degree of corrosion differs very little for all the steels tested. **1955** H. J. Grover et al. *Fatigue of Metals* (1956) x. 139 This type of behaviour is known as 'corrosion fatigue' and is highly deleterious.

corrugated, *ppl. a.* Add: **2. b.** Also *corrugated paper, strawboard*, etc., a packing material designed to give elasticity.
1897 *Chemist & Druggist* L. 746 Among the minor conveniences..is the article known as corrugated paper. **1912** *World's Work* Apr. 508/2 Corrugated packing material for packing bottles. Corrugated strawboard is generally considered the best medium. **1959** *Gloss. Packaging Terms (B.S.I.)* 68 *Corrugated paper. a.* Single face. A fluted sheet of paper glued to a plain lining-sheet. Normally made from imitation strawpaper but also from other kinds of paper. *b.* Double face. A fluted sheet of paper glued between two lining-sheets.

corrugation. Add: **2. b.** *concr.* Also, a ridge in an unsurfaced road.
1940 *Gloss. Highway Engin. (B.S.I.)* 35 *Corrugations*, the displacement of the material forming a surface layer into marked wave-like shapes transverse to the line of traffic or creep. **1953** J. Packer *Apes & Ivory* xx. 213 'Poor car on the corrugations!' sighed Bertie. **1959** G. J. Walker *Traffic & Transport in Nigeria* v. 94 The causes of these ridges across the line of the traffic are obscure...Tar or bitumen as a binder in place of water-bound clay or other fines appears the surest means of preventing corrugation. *Ibid.* 95 The speed, weight and density of the traffic raised the corrugations.

corrugator. Add: **3.** One who superintends a corrugating machine.
1881 *Instr. Census Clerks* (1885) 91. **1921** *Dict. Occup. Terms* (1927) § 278 Zinc corrugator (galvanised sheet). *Ibid.* § 559 Corrugator, corrugated packing maker.

corsage. Add: **3.** (Earlier examples.)
1843 *Godey's Lady's Bk.* Aug. 96 The corsage is tight to the figure, very long in the waist, and trimmed down the centre to match the skirt. **1846** 'A Lady' *Jewish Man.: Toilette* v. 215 The close-fitting corsage and tight sleeve.
b. A bouquet worn on the bodice. *U.S.*
[**1886** *Amer. Garden* Jan. 8/3 An enterprising florist who has imported Water-Lilies from Florida for a week past, selects twenty expanded ones for a bouquet de corsage.] **1911** H. S. Harrison *Queed* ix. 105 On her rounded breast..a splendid corsage of orchid and lily-of-the-valley. **1955** T. Sterling *Evil of Day* viii. 95 The forget-me-not corsage.. she had bought for herself, explaining to her escort that gardenias gave her a headache.

corsair. Add: **4. a.** A scorpænoid fish of the Californian coast, *Sebastomus rosaceus*. *U.S.*
1884 Goode *Nat. Hist. Aquatic Anim.* 265 Corsair (*Sebastichthys rosaceus*)..is known to the Portuguese fishermen at Monterey by the name of 'Corsair'. *Ibid.*, Spotted Corsair (*Sebastichthys constellatus*). **1905** D. S. Jordan *Guide to Study of Fishes* II. xxv. 430 The commonest of these [red species] is the corsair, *Sebastichthys rosaceus*, plain red and golden.
b. A reduviid predatory bug of the genus *Rasahus*.
1909 in *Cent. Dict.* Suppl. **1926** E. O. Essig *Insects Western N. Amer.* xxii. 356 The western corsair..is one of the commonest species in California, Arizona, and Mexico. **1939** Duncan & Pickwell *World of Insects* xiii. 223 (*caption*) The bug on the left is a nymph of the Western Corsair.

corset. Add: **2.** Esp. in *pl.* (Further examples.) Also *fig.* (usu. in *sing.*).
1922 Joyce *Ulysses* 523 Vicelike corsets of soft dove coutille. *Ibid.* 533 Corsets for men. **1930** W. S. Maugham *Cakes & Ale* xvi. 190 When I put my hands on her sides I could feel the ribbing of the skin from the pressure of the corsets. *Ibid.* 191 She did not put on her corsets again, but rolled them up and I wrapped them in a piece of newspaper. **1951** in M. McLuhan *Mech. Bride* 93/2 Bergson has put a corset around the Absolute. **1968** *Listener* 25 July 102/1 Under the title of democratic centralism it imprisons the state in a bureaucratic corset designed to restrict development to predetermined fields. **1970** *Sunday Times* (Colour Suppl.) 11 Oct. 23/2 They buy a tin corset for their shapeless souls: they buy 'the car with authority'.
4. (Further examples.)
1843 *Ainsworth's Mag.* III. 51, I cannot see that this corset-fitter..was a lady. **1845** M. M. Noah *Gleanings* 116 [The young lady] drew out a piece of long black iron or steel...It is my corset bone, whispered she. **1877** J. A. Ruth *Decorum* 285 These latter..consist of a waist cut like a plain corset-cover. **1922** Joyce *Ulysses* 492 A tag of her corset lace hangs slightly below her jacket. **1962** K. A. Porter *Ship of Fools* i. 51 She wears a white linen corset cover with tatting around the edges.

‖ **corsetier, corsetière** (kǭ·ɹsĕtye, -tyę̄r). [F., f. *corset* Corset + *-ier* masc., *-ière* fem.] A corset maker (male and female).
1848 Thackeray *Van. Fair* xxix, She found fault with her friend's dress..and vowed that she must send her *corsetière* the next morning. **1932** E. Bowen *To the North* xiv. 145 She wanted a massage after her journey, a fitting at her corsetière's. **1961** *Evening Standard* 7 Sept. 19/5 (Advt.) Vacancies for..ladies to be trained as professional Corsetieres.

corset(t)ing (kǭ·ɹsetiŋ), *vbl. sb.* [f. Corset + -ing¹.] The fitting with, or wearing of, a corset.
1845 M. M. Noah *Gleanings* 15 Then commences the herculean task of corsetting, racking, bracing and bending. **1898** *Westm. Gaz.* 3 Mar. 3/1 That careful corseting..the French woman understands so much better than we. **1903** *Ibid.* 29 Oct. 8/2 Intelligent physical exercises and better corseting can do wonders. **1904** Sladen *Playing the Game* i. v, A plumpness..kept within the bounds of beauty by admirable corsetting.
Hence (as a back-formation) **co·rset** *v.*, esp. in *fig.* use, to place restraints or controls on (something); to force into.
1935 Dylan Thomas in *Life & Letters Today* Dec. XIII. ii. 74 Corset the bone-yards of a crooked lad? **1961** *Time* 31 Mar. 43/1 Every girl is tight-corseted with the propaganda that she must have a slim, svelte figure. *Ibid.* 27 July 28/3 Corseting the careless middle-class spread of the community-controlled school. **1970** B. M. H. Strang *Hist. Eng.* p. xv, The ceaselessly..flowing, ungraspable mass that historians corset into manageable chunks on to which quasi-scientific labels can be stuck.

co·rsetless, *a.* [f. Corset + -less.] Without corsets or a corset.
1896 *Westm. Gaz.* 15 Apr. 8/1 The ladies of Tristan d'Acunha will for the present have to go corsetless. **1923** *Glasgow Herald* 21 Mar. 8 A..warfare against the corsetless and rationally corsetted figure.

corsetry (kǭ·ɹsetri). [f. Corset 2 + -ry.] **a.** Corset-making or -fitting. **b.** *concr.* Corsets collectively. Also *attrib.*
1904 *Bulletin* (Sydney) 7 Jan. 30 In the matters of style, smile and corsetry she was still very backward. **1923** *Daily Mail* 22 Feb. 1 Natural elegance subtly idealised by skilful Corsetry. **1937** *Sunday Express* 30 May 17/3 We have recently had demands for a light type of spring suitable for the manufacture of corsetry. **1962** *Punch* 1 Aug. 150/2 The lingerie, corsetry, hosiery, and swim-suit departments of shops. **1968** *Nova* Feb. 36 Warner, like the majority of corsetry manufacturers in this country, have a parent company in America.

Corsican (kǫ·ɹsikăn), *a.* and *sb.* [f. *Corsica*: see -an.]
1. a. *adj.* Of, belonging to, or characteristic of Corsica or its inhabitants. **b.** *sb.* A native or inhabitant of Corsica; the Corsican dialect of Italian.
The Corsican (*the Corsican ogre, robber*, etc.), Napoleon Bonaparte, who was born in Corsica.
1739 *Gentl. Mag.* June 330/2 The Marquis de Maillebois has attack'd the Corsicans in their Posts on the Mountains. **1768** Boswell *Corsica* 15 In general the Corsicans breathe a pure atmosphere. *Ibid.* 29 The Corsican villages are frequently built upon the very summits of their mountains. **1803** S. Hoole *Anecd. J. Hoole* 1 The barbarities perpetrated by the Corsican Robber. **1803** Ld. Nelson *Let.* 27 Aug. in *Disp. & Lett.* (1845) V. 187 That man of tyranny, Buonaparte; but I detest Europe for being so mean-spirited as to submit to the mandates of this Corsican. **1804** —— *Let.* 10 Feb., *ibid.* 411, I have no doubt but Egypt is the favourite and ultimate object of the Corsican tyrant. **1814** Southey *Ode* ii, If that perfidious Corsican maintain Still his detested reign. **1848** Thackeray *Van. Fair* xiii. 108 The Corsican monster locked up at Elba. *Ibid.* xxviii. 242 Could the Corsican but have waited in his prison until all these parties were by the ears, he might have returned and reigned unmolested. **1855** R. Martineau tr. *Gregorovius' Corsica* iv. vii. 283 That pantomimic dance of lament is called in Corsican the *ballata*. **1861** M. Arnold *Pop. Educ.*

France 162 Corsican vengeance is proverbial. **1862** *Chambers's Encycl.* III. 258/2 The Corsicans under General Paoli made themselves in great part independent. **1880** *Ibid.* XIII. 494/2 In Southern Corsican *dr* for *ll* is conspicuous. **1882** *Gentl. Mag.* Jan. 1 The Iron Duke had thrashed the Corsican Ogre. **1897** *Fortn. Rev.* June 887 To remove the false impression of the Corsican bandit so dear to the transpontine stage. *Ibid.* 890, I could speak no Corsican. **1922** *Chambers's Jrnl.* 21 Jan. 113/2 Make up your mind to charm the Corsican. **1964** D. Warner *Death of Dreamer* ii. i. 45 A tough French dope merchant, one of the Corsican Brotherhood. **1968** *Guardian* 5 Jan. 9/7 The new rules for the baccalaureat do not allow..Basque, Catalan, or Corsican to be offered as subjects.
2. Corsican cock, a variety of the domestic fowl; **Corsican moss**, *Alsidium helminthocorton*, a Mediterranean seaweed, formerly used as a vermifuge; **Corsican pine**, a slender pine, *Pinus nigra maritima* or *P. laricio*, used for forest planting.
1854 *Poultry Chron.* I. 423 Chitteprat or Corsican Cock. **1855** *Ibid.* III. 518 Chitteprat or Corsican. **1849** Balfour *Man. Bot.* § 1130 *Plocaria (Gigartina) Helminthocorton*, under the name of Corsican Moss, was formerly used as a vermifuge. **1887** *Buck's Handbk. Med. Sci.* V. 26/1 As a medicine Corsican moss is of the past. **1824** A. B. Lambert *Pinus* II. 28 Pinus Laricio. Corsican Pine. **1879** *Encycl. Brit.* IX. 402/1 The ridges of the Crimea are clothed with Corsican pine. **1957** M. Hadfield *Brit. Trees* 76 The Corsican pine was first grown in Britain in 1759.

corsite (kǭ·ɹsəit). *Petrogr.* [G. (F. Zirkel *Lehrbuch d. Petrographie* (1866) II. 133), f. F. *Corse* Corsica: see -ite¹.] An orbicular gabbro found in Corsica.
1876 A. H. Green *Geol.* 65 Corsite or Napoleonite..is a granular mixture of Anorthite, Hornblende, and a little quartz. The most noticeable variety occurs in Corsica: the rock there is made up of balls from one to three inches in diameter. **1885** Geikie *Text-bk. Geol.* (ed. 2) ii. ii. vii. 148 Corsite is a granitoid mixture of greyish-white anorthite, blackish-green hornblende and some quartz, which here and there have grouped themselves into globular aggregations (Orbicular diorite, Kugeldiorit, Napoleonite). **1954** G. W. Himus *Dict. Geol.* 32 *Corsite*, a variety of dolerite exhibiting orbicular structure and consisting essentially of bytownite and amphiboles. **1957** *Encycl. Brit.* VII. 401/1 The orbicular diorite of Corsica (also called 'napoleonite' or 'corsite').

corslet, corselet, *sb.* Add: **2. b.** (Usu. in form *corselette*.) A woman's foundation garment, usually combining brassière and girdle in one piece.
1926 *Daily Colonist* (Victoria, B.C.) 9 Jan. 3/7 (Advt.), Corselettes of strong brocade, side hooking with wide panels of elastic at sides. **1952** *John o' London's* 1 Aug. 736/3 A flesh wound..was caused by a bone in the corselette (lovely word) worn under her sarong. **1967** E. A. Gollschewsky in *Coast to Coast 1965–6* 91 Put on your new blue suit, Leonie...And your corselette, of course.

‖ **Corso** (kǭ·ɹso). Also **corso.** [It., a course, a street: see Course *sb.*] A street in Italy in which races or festivals were held; a procession of carriages; a promenade; *esp.* a specific street in Rome.
a **1680** Evelyn *Diary* Apr./May an. 1645 (1955) II. 484 We passed the Campo Martio..wherein the Noblesse exercise their horses, and the Ladys make the Corso. **1740** Gray *Let.* 10 July (1935) I. 166 The Pretender..I have had frequent opportunities of seeing at church, at the corso, and other places. **1849** Effie Ruskin *Let.* 28 Oct. (1965) I. 54 The prancing of the horses trotting and galloping after us on the Corso [in Milan]. **1876** *Encycl. Brit.* V. 122/2 The carnival sports seem at that time [*sc.* 1545] to have consisted..of..the races in the Corso (which, formerly called the Via Lata, took its present name from them). **1926** *Spectator* 17 July 89/1 Every evening Prince Michael Street [in Belgrade] is closed to traffic and becomes the 'Corso' up and down which young men..escort girls. **1958** *Listener* 4 Dec. 911/1 A town [in Yugoslavia] that has any pretensions to be a town has to have a Corso, a street designed for promenading in the cool of the evening. **1961** *Times* 30 Nov. 13/6 Rather like those Sunday strolls along an Italian corso.

Cortaillod (kǭrta·yo). The name of a place in the canton of Vaud, Switzerland, used to denote a red wine made there from the pinot grape; also applied *attrib.* to the civilization of the ancient lake-dwellings of western Switzerland. Also *Comb.*, as *Cortaillod-Michelsberg*.
[**1884** *Encycl. Brit.* XVII. 362/1 Interesting remains of ancient lake dwellings have been discovered at Estavayer and Cortaillod.] **1888** *Ibid.* XXIV. 115/1 About one-third of the wine produced [in Vaud] is exported, the best-known varieties being Yvorne (white) and Cortaillod (red). **1940** C. F. C. Hawkes *Prehist. Found. Europe* v. 134 The excavations of M. Vouga on the lake of Neuchâtel have revealed what is called, from one of his sites, the Cortaillod culture... The Cortaillod culture extends no further towards the Danubian province than Zurich on the east and the Black Forest on the north. **1957** Childe *Dawn Europ. Civ.* (ed. 6) xvi. 288 The names Cortaillod-Michelsberg, Horgen, and Corded Ware denote three culture periods that follow one another in that order on all the Alpine lakes and bogs. *Ibid.*, The oldest 'lake-dwellings' in Western Switzerland were erected by farmers who arrived with a complete neolithic equipment, constituting what is termed the Cortaillod culture—now divisible into an Early and a Late phase. **1962** *Economist* 29 Dec. 1283/2 A red Cortaillod is often found in American wine shops.

corticene, var. Corticine.
1897 F. J. Burgoyne *Library Construction* 25 The most suitable covering for a wood floor is cork carpet or corticene. **1915** 'Bartimeus' *Tall Ship* vii. 118 Varnish..wherewith to beautify the corticene on the aft-deck. **1961** F. H. Burgess *Dict. Sailing* 61 *Corticene*, a hard-wearing deck covering material.

corticifugal (kǭɹtisi·fiūgăl), *a.* [f. L. *cortic-*, Cortex + *fugĕre* to flee + -al.] Originating in and running from the cerebral cortex.
1898 *Phil. Trans. R. Soc.* B. CXC. 10 There was no clear evidence of corticifugal fibres passing from the angular gyrus to the basal ganglia. **1959** Ranson & Clark *Anat. Nerv. Syst.* (ed. 10) xvi. 319 The anterior limb of the internal capsule..is broken up by bands of gray matter connecting these two nuclei. It consists of corticipetal and corticifugal fibers.

corticipetal (kǭɹtisi·pităl), *a.* [f. L. *cortic-*, Cortex + *petĕre* to seek + -al.] Originating outside of and running into the cerebral cortex.
1898 *Phil. Trans. R. Soc.* B. CXC. 11 A corticipetal system, passing from the external geniculate body and pulvinar thalami to the cortex of the occipito-angular region. **1959** [see prec.].

cortico-, used as combining form of Cortex (sense 3) in various medical terms, as **co:rtico-fu·gal** *a.* = *corticifugal; **co:rtico-pe·tal** *a.* = *corticipetal; **co:rtico-spi·nal** *a.*, relating to the cortex and the spine.
1890 W. James *Princ. Psychol.* I. ii. 71 The arriving place of the 'cortico-petal' or the place of exit of the 'cortico-fugal' fibres. **1901** Dorland *Med. Dict.* (ed. 2) 178/2 Cortico-afferent. Cortico-efferent. Corticofugal. Cortico-peduncular. Corticospinal. **1908** *Practitioner* Oct. 557 A purely spinal reflex, which appears in earliest infancy before the cortico-spinal paths are fully developed. **1949** Koestler *Insight & Outlook* ix. 129 The tear glands are beyond the reach of voluntary, cortico-motor control. **1961** Ham & Leeson *Histology* (ed. 4) xxv. 717/2 Some of the interlobar arteries [of the kidney] break up into main branches as they ascend in the columns of Bertin, but most of them do so only when they have almost reached the corticomedullary border. **1962** *Gray's Anat.* (ed. 33) 1083 The genu contains the corticonuclear fibres which arise from area 4 of the cerebral cortex and terminate in the motor nuclei of the cranial nerves to the head.

corticoid (kǭ·ɹtikoid). *Biochem.* [f. L. *cortic-*, Cortex; see -oid.] Any of the steroids isolated from the adrenal cortex. Also *attrib.*
1941 H. Selye in *Nature* 19 July 85/1 Without introducing any essentially novel terms, we could classify the steroid hormones into four main pharmacological groups..: *Corticoid* = having activity of: cortin, adrenal cortical hormone, principle maintaining life of adrenalectomized animals etc. **1959** *Sci. News* LII. 30 To indicate their origin, steroids from the adrenal cortex are called corticoids. **1961** *Lancet* 12 Aug. 341/2 Corticoid therapy was instituted. **1962** *Ibid.* 13 Jan. 81/2 A simple chromatographic technique suitable for analytical separation of corticoid mixtures.

corticosteroid (kǭ:ɹtiko,stiə·roid). *Biochem.* [f. *cortico- + *steroid.] = prec.
1944 *Chem. Abstr.* XXXVIII. 3707 (*heading*) Corticosteroids. **1958** *Immunology* I. 237 Particular attention was paid to patients treated with corticosteroids (cortisone or delta-corticoids). **1961** *Brit. Med. Dict.* 356/2 *Corticosteroids*, substances of steroid structure isolated from the adrenal cortex...They..include cortisone and corticosterone. **1966** *New Scientist* 25 Aug. 430/3 The adrenal glands became heavier and there is an increased production of corticosteroids, with a consequent decrease in reproductive faculties. **1969** *Daily Tel.* 19 Dec. 11/7 Corticosteroids, another important hormone group, suppress the pain, irritation and other symptoms of inflammation in, say, rheumatoid arthritis, colitis and infantile eczema.

corticosterone (kǭ:ɹtiko,stiə·rōᵘn). *Biochem.* Also **-steron.** [ad. G. *corticosteron* (Reichstein et al. 1936, in *Proc. K. Akad. Wetensch.* (Amsterdam) XXXIX. 1218), f. *cortico- + *ster(ol + -one.] A steroid hormone, $C_{21}H_{30}O_4$, produced by the adrenal cortex.
1937 *Nature* 2 Jan. 26/1 (*heading*) Corticosteron, a crystallized compound with the biological activity of the adrenal-cortical hormone. *Ibid.* 26/2 Corticosteron is a highly active cortical hormone. **1937** [see *deoxycorticosterone]. **1938** *Ann. Reg. 1937* 348 A new male sex hormone..was isolated and made synthetically, and a compound was synthesised (21-oxy-progesterone) with activity similar to corticosterone. **1952** G. H. Bourne et al. *Cytology* (ed. 2) xi. 478 From the adrenal cortex there have been extracted two main groups of physiologically active steroids. One group..of which corticosterone and cortisone are examples, affects carbohydrate metabolism and the distribution of glucose. **1964** [see *cortisol]. **1969** *Times* 7 Jan. 8/7 The rat-reared mice also produced less corticosterone, a hormone that prepares the body for stress, when faced with an alarming situation.

corticotrop(h)ic (kǭ:ɹtiko,trǫ·fik, -ǫ·pik), *a.* *Biochem.* [f. *cortico- + Trophic *a.*] = *adrenocorticotrop(h)ic *a.*
1934 *Chem. Abstr.* XXVIII. 6813/1 (*heading*) Comparison of the effects of the corticotropic hormone of the pituitary upon various animal species. **1935** *Ibid.* 5163/1 Animal charcoal absorbs the pigmentation hormone without affecting the corticotropic hormone. **1936** [see *adreno-cortical *a.*]. **1939** *Jrnl. Endocrinology* I. 57 The growth response of hereditarily dwarfed mice and of pigeons to..corticotropic extracts and to thyroxin does not disprove the existence of growth hormone. **1944** *Lancet* 5 Feb. 188/2 The evidence for corticotrophic activity in anterior pituitary extracts is well established.
Hence **co:rticotro·p(h)in** [-in¹] = *adrenocorticotrop(h)in.
1942 *Jrnl. Endocrinology* III. 8 These experiments strongly suggest the presence of two different factors in pituitary corticotrophin—one affecting adrenal weight, the other the distribution of lipoid in the cortex. **1943** *Science* 3 Sept. 225/1 It is generally believed that hypophyseal corticotropin is protein in nature. **1944** *Lancet* 5 Feb. 188/2 Thus corticotrophin has been postulated as being responsible for all or part of the growth-promoting activity, glycotropic activity and diabetogenic activity of anterior pituitary extracts, and..there is evidence to support all these claims, though not evidence enough to make them by any means cast-iron. **1952** *Chem. Abstr.* XLVI. 6754 (*heading*) Treatment of ulcerative colitis with corticotropin (ACTH) and cortisone. **1962** *Lancet* 8 Dec. 1190/2 Patients with rheumatoid arthritis from ten British centres who had been treated with corticotrophin or steroids.

cortin (kǭ·ɹtin). *Biochem.* [f. Cort(ex + -in¹.] A crude extract of the adrenal cortex, containing several steroid hormones (see quots.).
1928 F. A. Hartman et al. in *Amer. Jrnl. Physiol.* LXXXVI. 359 The hormone of the adrenal cortex has been salted out with NaCl. *Cortin* is proposed as the name for this substance. **1943** *Thorpe's Dict. Appl. Chem.* VI. 277/1 Corticosterone possesses a high degree of cortin activity. **1948** R. D. H. Heard in Pincus & Thimann *Hormones* I. xiii. 550 The term 'cortin' is used today in several senses, to denote (*a*) the active substances contained in whole adrenal extract, (*b*) any life maintenance factor of unestablished chemical nature (as in the amorphous factor), and (*c*) substances, again of unrecognized constitution, which possess any kind of cortical activity. **1949** S. & L. M. Miall *New Dict. Chem.* (ed. 2) 152/2 *Cortin*, the hormone secreted by the suprarenal cortex... Several crystalline substances have been isolated from this extract, one of which, corticosterone, is believed to be the true hormone, which is still generally spoken of as cortin. **1961** *Brit. Med. Dict.* 356/2 *Cortin*, the active extract of the adrenal cortex, now known to be a mixture of several hormones. The name is obsolete.

cortina (kǭɹtəi·nă). *Bot.* [mod.L. use of L. *cortina* curtain.] = Curtain *sb.*¹ 5 a; also, the inner or partial veil in some agarics.
1832 Lindley *Introd. Bot.* i. iii. 208 Cortina, is a name given to a portion of the velum. **1866** —— & Moore *Treas. Bot.*, *Cortina*, the filamentous ring of certain agarics. **1962** R. Singer *Agaricales* (ed. 2) 9 Both pellicular veil and cortina are essentially the products of hemiangiocarpous development of the hymenophore.

cortisol (kǭ·ɹtisǫl). *Biochem.* [f. *cortis(one + -ol.] = *hydrocortisone.
1953 C. W. Shoppee in *Ann. Rev. Biochem.* XXII. 288 Cortisol, a name which seems convenient and preferable to hydrocortisone (by elimination of confusion with 4, 5-dihydrocortisone), is the only substance which appears to possess therapeutic activity equivalent to that of cortisone. **1962** *New Scientist* 8 Mar. 575/1 Cortisol, another steroid produced by the adrenal cortex. **1964** L. Martin *Clinical Endocrinol.* (ed. 4) v. 166 The C_{21} steroids cortisol, cortisone and corticosterone..have their predominant effects upon carbohydrate and protein metabolism.

cortisone (kǭ·ɹtizōᵘn). *Biochem.* [See quot. 1949.] A steroid hormone ($C_{21}H_{28}O_5$) found in the adrenal cortex and prepared synthetically for use as an anti-inflammatory agent in rheumatoid arthritis, etc.
1949 *N.Y. Times* 1 June 33/8 The hormone..was named yesterday by its discoverer, Dr. Edward C. Kendall.. as 'cortisone', an abbreviation of its long chemical name, 17-hydroxy-11 dehydrocortico-sterone. **1950** *Lancet* 28 Jan. 169/1 Hench's dramatic report on 'Cortisone' in rheumatoid arthritis has opened an entirely new field of work. **1952** [see *corticosterone]. **1952** [see *corticotrop(h)in]. **1952** *Sci. News* XXIII. 7 Cortisone is secreted in minute quantities by the adrenal cortex of the suprarenal gland, which, as its name indicates, is situated above the kidneys. **1957** *Oxf. Univ. Gaz.* 3 Oct. 65/1 Trials of cortisone in the treatment of ulcerative colitis. **1964** [see *cortisol].

cortlandtite (kǭ·ɹtlăntəit). *Petrogr.* [f. the name of *Cortlandt* township, New York state: see -ite¹.] A coarse-grained rock composed chiefly of hornblende and olivine; a variety of peridotite.
1886 G. H. Williams in *Amer. Jrnl. Sci.* 3rd Ser. XXXI. 30 It would seem to the writer preferable, if a new name is necessary, to adopt the term 'Cortlandtite' for this [hornblende-olivine] class of rocks which play such an important rôle in the 'Cortlandt Series'. **1903** Geikie *Text-bk. Geol.* (ed. 4) 241 Cortlandtite..so named from its occurrence in the 'Cortlandt series' of eruptive rocks on the Hudson River. **1959** W. W. Moorhouse *Study Rocks in Thin Section* xvi. 320 Cortlandtite is a hornblende peridotite.

Corton (kortoṅ). [f. Aloxe-*Corton*, the name of a commune near Beaune in France.] A red Burgundy wine made in the neighbourhood of Beaune, Côte-d'Or.
1833 Redding *Wines* v. 98 In the commune of Aloxe, a wine called Corton is grown. **1889** W. Guthrie *Claret, Burgundy & Champagne* 21 In the Côte de Beaune are found the Corton, Volnay, Pomard, and many other well-known names. **1900** Christie et al. *Catal. Wines, C. Boyle* 8 Three Dozens of Corton. **1967** A. Lichine *Encycl. Wines* 65/1 Corton, of all Côte de Beaune wines, is the one to lay down, for it develops slowly and holds its majesty for years.

corver². Delete †*Obs.* and add later examples.
1858 Simmonds *Dict. Trade*, *Corver*, a man who makes and repairs corves or coal baskets. **1881** in *Instr. Census Clerks* (1885) 84. **1921** in *Dict. Occup. Terms* (1927) § 652.

corvette. Add: **b.** A fast naval escort vessel of 500 to 700 tons, used esp. on convoy work.
1940 *Jane's Fighting Ships* 91/3, 56 Corvettes (originally described as patrol vessels of whaler design) were laid down in the summer of 1939. **1942** *Ann. Reg. 1941* 119 The Minister added that 77 corvettes had been launched and delivered. **1952** *Ann. Reg. 1951* 52 A British ship..carrying supplies for the Arab Legion in Jordan, was stopped by an Egyptian corvette.

corvina (kǫrvi·nă). Also **cur(u)vina.** [Sp., Pg. *corvina*, f. L. *corvinus*, f. *corvus* raven: said to be so named from its black fins.] A name for several fishes belonging to the family Sciænidæ, especially American ones belonging to the genus *Cynoscion*.
1787 Cullen tr. *Clavigero's Hist. Mexico* I. 65 The Curvina is about a foot and a half long, of a slender, round shape, and of a blackish purple colour. **1842** J. E. Dekay *Zool. N.Y.* iv. 74 The Silvery Corvina. *Corvina argyroleuca*... This fish..is frequently called Silvery Perch by the fishermen. *Ibid.* 77 The Sharp-Finned Corvina. *Corvina oxyptera*. **1869** R. F. Burton *Explor. Highl. Brazil* II. 303 The fish leaped and splashed in all directions...In a few minutes we had enough for a day's food. The worst was the Curuvina. *a* **1884** Jordan in Goode *Nat. Hist. Aquatic Anim.* 379 *Cynoscion parvipinne*..is usually known as the 'Corvina', or 'Caravina'...It is found from San Pedro southward to the Gulf of California. **1930** *Discovery* Aug. 263/1 The myriads of bumla serve as an attraction to the great *ghol*, the magnificent golden-mouthed Corvina (*Sciaena aquila*), which on the north-west coast of India follows the shoals of small fish into certain favoured bays. **1952** *Oxf. Jun. Encycl.* VI. 158/1 Off the West African coast..fleets of Spanish schooners catch immense numbers of the big corvinas..sold in the market under the name of *baccalào*. **1960** *Common & Sci. Names Fishes U.S. & Canada* (ed. 2) 31 Shortfin corvina, *Cynoscion parvipinnis*..Orangemouth corvina, *Cynoscion xanthulus*..California corbina [*sic*], *Menticirrhus undulatus*.

cory- (kǫri), abbreviation of *Corydalis* (see Corydaline) in names of alkaloids obtainable from certain species of this genus, as **corybu·lbine** [ad. G. *corybulbin* (Freund and Josephi 1893, in *Liebig's Ann. Chem.* CCLXXVII. 18)], **coryca·vine** [ad. G. *corycavin* (Freund and Josephi 1892, in *Ber. d. Deut. Chem. Ges.* XXV. 2414)], **corytu·berine.**
1893 Dobbie & Lauder in *Jrnl. Chem. Soc.* LXIII. 486 We propose for this alkaloid the name *corytuberine*; from *Corydalis tuberosa*, a synonym for *Corydalis cava*. *Ibid.* 492 A base (?Corycavine), which crystallizes in small, matted needles and melts at 218°. **1894** *Ibid.* LXVI. i. 100 Corybulbine is the name given..to a base isolated from commercial corydaline. **1902** *Ibid.* LXXXII. i. 307 There were obtained in the following order, beginning with the weakest base,..corydaline, corybulbine,..corycavine, corydine.

Corycian (kŏri·ʃiăn), *a.* Also 6 **Coritian.** [f. L. *Cōrycius*, a. Gr. Κωρύκιος, f. Κώρυκος Corycus or Κωρυκία Corycia (see def.): see -ian.] Of or pertaining to the mountain cave of Corycus at the foot of Parnassus, sacred to the Muses, or to the nymph Corycia, daughter of Apollo; chiefly in *Corycian cave, Corycian nymphs* (the Muses).
1567 Turberv. *Ovid's Ep.* xx. Tij, The famous Ile (where the Coritian Nymphes Did lodge of yore). **1626** G. Sandys *Ovid's Met.* i. 230 Corycian Nymphs, and Hill-gods he adores. **1636** Saltonstall *Ovid's Heroicall Ep.* i. xix. 139 Cæa where Corycian Nymphs have, In Parnassus hill an old famous Cave. **1746** Akenside *Hymn to Naiads* (1758) 319 To the cave Corycian or the Delphic mount. **1877** *Encycl. Brit.* VII. 53/1 The famous Corycian cave, a large grotto in the limestone rock, which afforded the people of Delphi a refuge during the Persian invasion. **1883** R. Whitelaw tr. *Sophocles' Antigone* 1127 With nymphs Corycian in thy train. **1885** *Encycl. Brit.* XVIII. 318/1 Parnassus was.. hallowed by the worship of Apollo, of the Muses, and of the Corycian nymphs. **1908** *Daily Chron.* 7 July 3/3 He is..as enviable as the Corycian old man.

corydalis (kǫri·dălis). *Bot.* [mod.L. (E. P. Ventenat *Choix de Plantes* (1803) 19), f. Gr. κορυδαλλίς crested lark, in reference to the shape of the flower resembling that of the bird's spur.] A member of the genus of annual or perennial herbs so called, belonging to the family Papaveraceæ and having divided leaves and flowers in racemes.
1818 in *Mass. Hist. Soc. Coll.* 2nd Ser. VIII. 170 The fungous flowered corydalis..climbs to the height of ten or twelve feet, supporting itself by means of tendrils. **1866** Lindley & Moore *Treas. Bot.* I. 335/2 *Corydalis, climbing*, an American name for *Adlumia*. **1870** W. Robinson *Wild Garden* iii. 142 Plants suited for Dry and Gravelly Soil.. Corydalis, in var. **1961** R. W. Butcher *Brit. Flora* I. 252 *Corydalis claviculata*, the climbing corydalis is a slender annual with..petals cream-white to yellow, spur short. *Ibid.* 253 *Corydalis lutea*, the yellow corydalis is a slender perennial with..petals yellow, spur very short.

coryphée. (Earlier and later examples.)
1828 J. Ebers *7 Yrs. King's Theatre* iii. 74 Nor was the scarcity of dresses confined to the coryphées and figurantes of the ballet. **1925** E., O. & S. Sitwell *Poor Young People* 3 Queen and coryphée.

coryphodon (kǫri·fodǫn). [f. Gr. κορυφή top, summit + ὀδών, Ionic f. ὀδούς, ὀδόντος tooth.] A fossil mammal of the genus of this name: so called because the cusps of their teeth are developed into points. Also *attrib.*, as **coryphodon bed**, the lower division of the Lower Eocene in the Rocky mountain and Plateau region, in which coryphodon remains are found.
1846 R. Owen *Brit. Fossil Mamm.* 304 The Tapir, which is the nearest existing analogue of the Coryphodon. **1848** *Q. Jrnl. Geol. Soc.* 3 Nov. 126 The swelling of the Coryphodon's jaw. **1884** *Amer. Naturalist* XVIII. 791 The foot structure of Coryphodon. **1895** Dana *Man. Geol.* (ed. 4) 886 Coryphodon beds of Marsh. *Ibid.* 907 The Coryphodons of the Wasatch. **1957** *Encycl. Brit.* XXII. 700/2 In all the taligrades the feet were short, especially so in *Pantolambda*. This feature becomes greatly emphasized in the later members of the order, the coryphodons and Dinocerata.
So **cory·phodont.**
1880 Dana *Man. Geol.* (ed. 3) 503 The Coryphodonts are of special interest, because of their prototype characters. **1895** *Ibid.* (ed. 4) 928 Species of large, short-footed Ungulates, the Coryphodonts. **1910** H. F. Osborn *Age of Mammals* ii. 438 We have observed the elimination of the phenacodonts, coryphodonts, palæonictids.

Corythosaurus (kǫ:riþosǭ·rɒs). *Palæont.* [mod.L., f. Gr. κόρυθο-ς crested + σαῦρος lizard.] An extinct genus of dinosaurian reptiles remains of which were found in Alberta, Canada, in 1912; so called because of their crested skulls. Also (with lower-case initial), a reptile of this genus.
1914 B. Brown in *Bull. Amer. Mus. Nat. Hist.* XXXIII. 559 (*title*) *Corythosaurus casuarius*, a new crested dinosaur from the Belly River Cretaceous. *Ibid.* 565 *Corythosaurus*. —Skull short with high anterior, helmet-like crest formed by nasals, prefrontals and frontals. **1933** A. S. Romer *Vertebrate Paleontology* ix. 199 In *Corythosaurus* we find a peculiar, thin, hollow, dome-like structure capping the skull. **1957** *New Yorker* 2 Nov. 32/1 He was working on a corythosaurus, a large, herbivorous, duck-billed type [of dinosaur].

cos, 'cos (kǫz), *adv.* and *conj.* Also **coz.** Dial. and colloq. shortening of Because *adv.* and *conj.* Cf. Cause, 'cause *conj.*
1828 W. Carr *Dial. Craven* 86 Cos, coz, because. **1838** Mrs. Gaskell *Let.* 17 July (1966) 16 You can't get at it, coz of the shore being so bad. **1861** Mayhew *Lond. Labour* III. 71/2 We didn't have no lantern, 'cos it keeps on falling out of his hands. **1887** Parish & Shaw *Dict. Kentish Dial.* 10 A very common controversy amongst boys:—'No it ain't.'—'Cos why?'—'Cos it ain't.' **1910** E. M. Forster *Howards End* viii. 77 'Cos why? 'Cos I'm going to German-eye. **1912** Galsworthy *Pigeon* ii. 55, I don't want 'im hurt 'cos o' me. **1942** E. Waugh *Put out more Flags* ii. 106 'They'll be good if I tell them, Mister.' 'Then why aren't they?' 'Cos I tell 'em to be bad.' **1968** H. R. F. Keating *Inspector Ghote hunts Peacock* vi. 79 She didn't tell us things sometimes 'cos she liked giving surprises.

Cosa Nostra (kōᵘ·ză nǫ·stră). [It., lit. 'our thing'.] The Mafia in the U.S.A.
1963 *Economist* 17 Aug. 589/1 A former member of organised crime's ruling body, Mr Joseph Valachi, has named names and drawn a master plan of the Syndicate (which the underworld refers to as *Cosa Nostra*). **1964** S. Bellow *Herzog* (1965) 35 A politician..who knew the Syndicate, the Juice Men, the Policy kings, Cosa Nostra, and all the hoods. **1967** A. Hunter *Gently Continental* xi. 179 You'll be very well protected for longer than Cosa Nostra will bother about you. **1968** A. Diment *Bang Bang Birds* i. iii. 30 'You are listening to the voice of Cosa Nostra,' he intoned.

coscoroba (kǫskǭrōᵘ·bă). [mod.L. (J. I. Molina, 1782), f. the native name in Chile.] A small South American swan, *Coscoroba coscoroba*, which belongs to the family Anatidæ and shows affinities to both ducks and swans. Also *coscoroba swan*.
1808 tr. *Molina's Geogr., Civil & Nat. Hist. Chili* I. iv. 164 Of the geese, the most remarkable is the *coscoroba* (anas coscoroba)...The plumage is entirely white, the feet and bill are red, and the eyes of a fine black. **1956** G. Durrell *Drunken Forest* i. 16 A few coscoroba swans, dumpy, plain white, and definitely barnyard-looking. **1964** J. Delacour *Waterfowl of World* I. 55 The London Zoo was the first to receive live Coscorobas in 1870.

cosh (kǫʃ), *sb.*³ *slang*. Also **kosh. a.** A stout stick, bludgeon or truncheon; a length of metal used as a life-preserver; also (*dial.*), a stick; a school cane; a caning. Phr. *under the cosh*, at one's mercy, helpless.
1869 'A Merchant' *6 Yrs. in Prisons* vii. 76 The coshman (a man who carries a 'cosh' or life preserver). *a* **1889** in Barrère & Leland *Dict. Slang* (1889) s.v., The officer.. sought to give the finishing *coup de grâce* with his cosh. **1893** H. T. Cozens-Hardy *Broad Norfolk* 83 Words which I have been accustomed to hear in common use...*Cosh*,.. a stick. **1896** A. Morrison *Child of Jago* i, The cosh was a foot length of iron rod, with a knob at one end, and a hook (or a ring) at the other. **1898** Wright *Eng. Dial. Dict.* s.v., *Cosh*,..a caning at school. *War[wickshire]*. You will get the cosh. **1904** *Daily Chron.* 29 Sept. 4/5 'Coshes'—pieces of lead pipe, known to the police as life-preservers. **1925** Fraser & Gibbons *Soldier & Sailor Words* 64 *Cosh*, the bludgeon carried by night patrols men and trench raiders. *Ibid.* 138 *Kosh*, a name for a trench club, or knobkerry, used in trench raids. **1927** *Weekly Dispatch* 23 Oct. 4 A truncheon, or, in prison vernacular, 'kosh'. **1958** F. Norman *Bang to Rights* i. 37 In the nick where you are under the cosh..most of the screws seem to take a sadistic delight in makeing [*sic*] things as uncomfortable as they can for you. **1959** 'M. Ainsworth' *Murder is Catching* i. 19 Clench a newspaper over a handful of coins and you've got a comfortable little cosh. **1959** I. & P. Opie *Lore & Lang. Schoolchildren* xvii. 374 Amongst children one of the most common names for the cane is 'the cosh'. **1960** *Observer* 24 Jan. 7/2 As for the Criminal Justice Act, it could be very useful to have all the villains under the cosh, as they expressed it. It made it much easier to get information. **1965** *Spectator* 15 Jan. 67/1 All chains gone from the boys' lavatories to make coshes.
b. *ellipt.* = **cosh-boy*, **cosh-man*.
1937 Partridge *Dict. Slang* 182/2 *Cosh*,..with *the*, one who uses a cosh. **1963** 'J. Bell' *Flat Tyre* iv. 36 He had been..picked as a useful cosh.
c. *attrib.* and *Comb.*, as **cosh-bandit, -boy, -man**, one who uses or carries a cosh; **cosh-carrier**, a prostitute's bully; hence *cosh-carrying*.
1954 *Britannica Bk. of Year* 637/1 Cosh-bandit. **1964** *Economist* 6 June 1141/2 Unless it is taken..by cosh-bandits before he gets it. **1953** 'H. Cecil' *Natural Causes* xvii. 198 As a prosecutor he would become known as 'The terror of the cosh boys'. **1954** *Ann. Reg. 1953* 7 Two 'cosh boys', as they were called (young ruffians armed with coshes or in some cases firearms), had been convicted. **1893** *Nott. Express* 7 Mar. 6 (E.D.D.), 'I shall be a cosh-carrier the next trade I start.' That seemed to be a term to describe a man who looked after a common woman and lived on her prostitution. **1896** A. Morrison *Child of Jago* i, Cosh-carrying was near to being the major industry of the Jago. *Ibid.* v, She neither fought nor kept a cosh-carrier. **1869** Coshman [see sense a, above].
Hence **cosh** *v. trans.*, to strike with a cosh; **coshed** *ppl. a.*; **co·sher**; **co·shing** *vbl. sb.* and *ppl. a.*
1889 Clarkson & Richardson *Police!* xxv. 348 'Coshers' and 'trippers' or 'picking-up molls', are vile men and women who travel from town to town. **1896** A. Morrison *Child of Jago* i, With a sudden blow behind the head, the stranger was happily coshed. *Ibid.*, The coshed subject ..felt the colder air, and moved a leg. **1898** Wright *Eng. Dial. Dict.* s.v., *Cosh*,..to beat, flog with a stick. Hence *coshing*, a flogging, a caning at school. **1905** *Ibid.* Suppl. 75/1 *Cosher*, Brks., a slang word for 'policeman'. **1922** E. Wallace *Flying Fifty-Five* xlii. 255 Somebody 'koshed' Miss Barrington's head lad and nearly killed him. **1927** —— *Squeaker* xxvii. 253 He pulled the life-preserver from his pocket and thrust it into Leslie's hand. 'Go and cosh him!' **1959** I. & P. Opie *Lore & Lang. Schoolchildren* xvii. 374 The miscreant receives what is variously termed a ..caning, clouting, coshing, [etc.]. **1961** *Economist* 9 Dec. 1021/2 The coshing, car-getaway, cash-grabbing streets of London today.

cosh (kǫsĕⁱ·tʃ, kǫʃ). *Math.* Abbreviation for *hyperbolic cosine*.
1873 *Messenger Math.* II. 189, $(1-x^4)(1-x^4/3^4)\ldots$ = cos $(\pi x/2)$ cosh$(\pi x/2)$. **1891** E. W. Hobson *Treat. Plane Trig.* xvi. 303 We have, at once from the definitions, the following relations between the hyperbolic functions: $\cosh^2 u - \sinh^2 u = 1$, [etc.]. **1968** P. A. P. Moran *Introd. Probability Theory* vii. 324 Next the cosh term is expanded in an infinite series.

cosily, *adv.* Add examples referring to speech or manner.
1955 H. Croome *Mountain & Molehill* x. 102 'I am so thankful,' she said, 'that it was you and not one of the others who got that letter of mine this morning.' She laughed cosily. **1959** N. Marsh *False Scent* (1960) viii. 229 'Now, now!' Fox rumbled cosily.

Coslettize (kǫ·zlĕtəiz), *v.* Also **coslettize.** [f. the name of Thomas Watts *Coslett* + -ize.] *trans.* To protect (iron or steel) from corrosion by forming a coating of phosphate on the surface of the metal. So **Co·slettized** *ppl. a.*, **Co·slettizer, Co·slettizing.**
1908 *Engineering* 26 June 870/1 The Coslettised article has a dull-black slaty appearance. *Ibid.*, Even micrometers can be Coslettised. **1908** *Daily Chron.* 21 Nov. 9/3 The treatment of frames to a special rust-preventing process known as Coslettising. **1921** *Dict. Occup. Terms* (1927) § 278 *Browner*; *cosletiser*; prepares bath of distilled water, phosphoric acid and iron filings; boils steel articles, e.g. parts of small arms, cycles, in bath. **1927** *Daily Express* 21 June 6 Frame coslettized, making it absolutely rust and weather proof.

Cosmatesque (kǫzmăte·sk), *a.* *Arch.* [f. the name of the *Cosmati*, a family of architects, sculptors, and mosaicists, who lived in Rome in the thirteenth century: see -esque.] Designating a style of decoration characterized by the use of mosaics; also called *Cosmati* or *Cosmato work*.
1883 Perkins *Ital. Sculpture* p. lvii, More Cosmatesque work of the first period is to be seen in the church of San Pietro d'Alba at Alba Fucense. **1927** H. Gardner *Art through the Ages* 196 Cosmati work..consists of surrounding colored marble slabs with borders, frequently interlacing, made up of small pieces of marble and glass cut into various shapes.

cosmea (kǫ·zmiă). [mod.L. (C. L. Willdenow *Linnæus' Species Plantarum* (ed. 4, 1800) III. iii. 2250): see *cosmos³.] = *cosmos³.
1813 *Bot. Mag.* XXXVII. 1535 Fine-leaved Cosmea... This beautiful plant..was described and figured by the late Rev. Ant. Jos. Cavanilles, in the year 1791,..[who] gave it

the name Cosmos, from its ornamental appearance, since changed by Willdenow to Cosmea, such termination being more consonant with botanical usage. **1933** *Discovery* Mar. 74/1 When cosmea or *Cosmos bipinnatus* is grown under short daily periods of sunlight. **1952** V. Gollancz *My Dear Timothy* xx. 306 There are snapdragons..: early pink cosmea..and annual chrysanthemums.

cosmetic, *a.* and *sb.* Add: **A.** *adj.* **b.** Of surgery: improving or modifying the appearance. Of prosthetic devices: re-creating or imitating the normal appearance.
1926 *Encycl. Brit.* III. 690/1 Cosmetic and plastic surgery, especially of the face, has undergone considerable improvement following our large experience in the war. **1962** *Daily Tel.* 11 Dec. 9/3 A demonstration was given in London of a 'cosmetic limb'..a hand and arm that resembles the real thing. **1965** *Observer* 7 Mar. Suppl. 13 His metal alloy legs..are finished with cosmetic shoes.

cosmetic, *v.* Delete *nonce-wd.* and add further examples. So **cosme·ticked** *ppl. a.*
1897 'Mark Twain' *More Tramps Abroad* xliv. 261 Ladies have to paint it, and powder it, and cosmetic it, and diet it with arsenic. **1934** E. Linklater *M. Merriman* xiii. 158 Faces most carefully cosmeticked were streaked and plain. **1960** C. Ray *Merry England* 128 How flaccid, nowadays, the English upper lip, and how bluely cosmeticked, even in the daytime, the young girls' upper eyelids.

cosmetician (kǫ:zmĕti·ʃən). orig. and chiefly *U.S.* [f. Cosmetic *a.* and *sb.* + -ician.] An expert in cosmetics; a *beautician.
1926 *Amer. Speech* I. 460/2 A Missouri law..speaks of hair-dressers as cosmeticians or cosmetologists. **1930** H. Rubinstein *Art of Fem. Beauty* i. 21 At the University of Berlin, a famous cosmetician accepted me as a student. **1935** E. E. Cummings *Let.* 13 June (1969) 143 One of the world's wealthiest professional beautifiers (or cosmeticians, or whatever you call them). **1948** E. Waugh *Loved One* 38 If there are any special little difficulties in the case you must mention them to our cosmeticians.

cosmetology (kǫ:zmetǫ·lŏdʒi). [ad. F. *cosmétologie*.] The art and practice of beauty culture (see also quot. 1855). Hence **co:smeto·logist** = *cosmetician.
1855 Dunglison *Dict. Med. Sci.* (ed. 12) 249 *Cosmetology*, a treatise on the dress, and cleanliness of the body. **1926** [see *cosmetician]. **1930** *Daily Express* 8 Sept. 8/7 Half a dozen standard faces, all of them beautiful..could be prepared by cosmetologists. **1938** D. Baker *Young Man with Horn* i. 20 They've got a course called Cosmetology at the same school. **1959** *Observer* 4 Oct. 18/3 The art and practice of cosmetology. **1963** *Punch* 23 Jan. 131/1 The cosmetologists..preparing the machinery of enhancement. **1969** *Daily Tel.* 27 Jan. 11/7 Viennese cosmetologist, Dr. Edith Lauda, who specialises in the treatment of acne. *Ibid.* 11/8 The most advanced salon treatment (only to be given by qualified cosmetologists) includes skin peeling.

cosmic, *a.* Add: **2. a.** (Further examples.) Also, universal; infinite; immense.
1869 W. James *Let.* 21 Jan. in R. B. Perry *Tht. & Char. W. J.* (1935) I. 291 We can hope for much from education towards the development of cosmic sympathies (so to say) and disinterested sentiments. **1926** *Encycl. Brit.* III. 910/1 Einstein found that a world without matter could not be reconciled with his cosmic equations. **1935** B. Russell *Relig. & Sci.* ii. 24 If Man had the cosmic significance assigned to him in traditional philosophy. **1947** D. Aaron in *Partisan Rev.* XIV. 103 They suffer from a cosmic ennui which only lynchings and love-nests can mitigate. **1955** A. Powell *Acceptance World* 6 His habitual air of hardly suppressed irritation tended to cloak any minor emotion by the strength of its cosmic resentment. **1958** H. Nicolson in *N.Y. Times* 16 Mar. vii. 1/2 The Civil War..which..he regards..as a cosmic clash between strong men and massive principles. **1958** G. J. Warnock *Eng. Philos.* xi. 145 How undesirable..for philosophers to *pretend* to suffer from cosmic anxieties. **1968** T. L. Swihart *Astrophysics & Stellar Astron.* xxvii. 253 If the Universe is evolving or changing with time, then it can be used as a clock for the measurement of a universal or cosmic time.
3. a. (Further examples.) *cosmic dust*, minute particles of matter in space or of cosmic origin.
1881 Cosmic dust [see Dict.]. **1903** A. M. Clerke *Probl. Astrophysics* ii. iii. 189 The entire Orion region..is pervaded with cosmic fog. **1927** Eddington *Stars & Atoms* 67 This..demonstrates the existence of a cosmic cloud pervading the stellar system. **1958** *New Statesman* 23 Aug. 214/2 Jeans's volcanic ash has been replaced today by the theory of 'cosmic dust'—a much more intriguing one. This maintains that the elemental particles from which the celestial bodies have been formed and which are travelling in Space get caught in the gravitational field of the moon and settle on it. **1962** *Listener* 29 Nov. 902/2 Studies of 'cosmic dust' in space. **1964** R. H. Baker *Astron.* (ed. 8) xvi. 465 A great complex of cosmic clouds is centered in the region of Scorpius and Ophiuchus. *Ibid.* 468 Interstellar lines..are absorbed in the spectra of stars by intervening cosmic gas. **1968** *Nature* 7 Dec. 1016/1 A good criterion for recognizing particles of cosmic dust on the Earth is the detection of radionuclides such as aluminium-26, formed by the action of cosmic rays on dust in free space.
b. (Further examples.) In modern use *spec.* pertaining to space travel (freq. representing Russ. *kosmicheskii* cosmic: so *cosmic rocket* = space rocket, etc.).
1931 D. Lasser *Conquest of Space* (1932) iv. 54 If this rocket motor..were attached to a vehicle of any kind that vehicle could, again in theory, be propelled forward at almost cosmic velocity. **1959** *Economist* 10 Jan. 107/2 The world heard of Russia's 'cosmic rocket..launched toward the moon' early on the morning of January 3rd. **1961** *Flight* LXXIX. 300/1 When the speed of the rocket had reached the first cosmic speed, a heavy Satellite separated from the rocket, carrying a cosmic rocket on which was mounted an automatic interplanetary station. **1962** *Daily Tel.* 31 Aug. 21/3 Prof. Egorov said..'Interplanetary stations..will serve for refuelling cosmic ships and so on. I think these cosmic flights are one more stage in making man master of cosmic space in flights round the earth and other planets.'
c. *cosmic rays*, high-energy radiations with great penetrative power which are incident on the earth from all directions and which originate in outer space (*primary radiation*) or are produced in the upper atmosphere by the primary radiation (*secondary radiation*). So *cosmic radiation*.
1925 Millikan in *Nature* 5 Dec. 824/2 Our experiments brought to light..a cosmic radiation of..extraordinary penetrating power...We obtained good evidence that these cosmic rays shoot through space in all directions. **1928** *Nature* Suppl. 7 Jan. 19/1 The cosmic radiation is defined as that small portion of the 'penetrating radiation' which is of cosmic origin. *Ibid.* 21/2 All this [*sc.* the experiments of Millikan and Cameron] constitutes pretty unambiguous evidence that the high altitude rays do not originate in our atmosphere, very certainly not in the lower nine-tenths of it, and justifies the designation 'cosmic rays'. **1945** H. D. Smyth *Gen. Acct. Devel. Atomic Energy Mil. Purposes* viii. 79 All the piles depend on stray neutrons from spontaneous fission or cosmic rays to initiate the reaction. **1957** *Encycl. Brit.* VI. 496 B/2 Most of the primary cosmic rays collide with atoms in the air, making new particles to proceed at great speeds in nearly the same direction as the primaries. These new particles are called secondary cosmic rays. **1958** *Sci. News* XLVIII. 14 Cosmic rays are extremely energetic protons and other atomic nuclei which originate in the main in outer space. **1967** Condon & Odishaw *Handbk. Physics* (ed. 2) ix. xi. 273/2 It also appears now that the highest-energy cosmic-ray particles may be of extragalactic origin.

cosmical, *a.* Add: **6.** *cosmical constant*, a multiplier occurring in Einstein's equations of general relativity. Cf. *cosmological *a.* 2.
[**1917** A. Einstein in *Sitzungsberichte Preuss. Akad. der Wissenschaften* 144 An die Stelle der Poissonschen Gleichung setzen wir $\Delta\phi - \lambda\phi = 4\pi K\rho$, wobei λ eine universelle Konstante bedeutet.] **1933** J. Jeans *Universe around Us* (ed. 3) i. 101 Einstein originally introduced this 'cosmical constant' curvature because he saw no other means of obtaining a static universe; it was in the days before the general recession of the nebulae had been noticed. **1937** G. C. McVittie *Cosmological Theory* iii. 42 The cosmical constant is therefore also zero in the space-time of special relativity. **1958** *Listener* 11 Dec. 972/2 By introducing the cosmical constant Einstein was able to specify a static condition of the universe in which the Newtonian attraction and cosmical repulsion are in exact balance.

cosmo-. Add: **cosmoce·ntric** *a.*, centred in the cosmos; **co·smodrome** [*aero)drome], a launching site for spacecraft in Russia.
a **1866** J. Grote *Moral Ideals* (1876) 189 Becoming.. *cosmocentric* instead of *autocentric* in our knowledge. **1907** Inge *Pers. Idealism* 103 Constructing our universe on a Christocentric or cosmocentric basis, not a self-centred one. **1913** A. J. Hubbard *Fate of Empires* ii. vii. 167 The Chinese [people] must be classed as one profoundly moved by the sense of cosmocentric duty. **1953** *Jrnl. Brit. Interplanetary Soc.* XII. 81 Soviet..artificial satellites..will be launched from an extensive 'cosmodrome' at Kaluga. **1964** *Yearbook Astr. 1965* 139 The Russians have always been unwilling to open their research centres, plant and cosmodromes to Western observers. **1970** *Daily Tel.* 25 Sept. 5/6 The capsule..landed..about 125 miles from the cosmodrome from which it was launched.

cosmoche·mistry. [Cosmo-.] The study of the chemical properties of the heavenly bodies and of the formation and distribution of elements and compounds in them and in the universe as a whole.
1940 R. Wildt in *Scientia* LXVII. 87 On observing how the material aspect of cosmic phenomena has come into the focus of interest, the question may be raised whether it would not be a timely expansion of the notion of geochemistry to adopt the term cosmochemistry to designate the science which shall deal with matter under all cosmic conditions. **1960** *McGraw-Hill Encycl. Sci. & Technol.* III. 503/1 The data of cosmochemistry provide important boundary conditions for both cosmological theories and theories pertaining to the composition and evolution of the planet Earth. **1969** *Sci. Jrnl.* Apr. 33/1 His current research interests are in molecular physics, cosmochemistry and biophysics.
Hence **cosmoche·mical** *a.*, of or pertaining to cosmochemistry; **cosmoche·mically** *adv.*, from a cosmochemical point of view.
1940 R. Wildt in *Scientia* LXVII. 86 A comprehensive elucidation of..early trends of cosmochemical speculation is much to be desired, as they must have influenced the developement [*sic*] of the exact sciences. **1957** G. E. Hutchinson *Treat. Limnol.* I. iv. 221 That abundant water should exist at the surface of the earth is cosmochemically anomalous. **1963** H. C. Urey (*title*) Some cosmochemical problems. **1970** *Nature* 31 Jan. 440/2 Cosmochemical arguments could support Ambartsumyan's concept of the presence of massive bulks of prestellar matter in the nuclei of galaxies.

cosmogenic (kǫ:zmo,dʒe·nik), *a.* [f. Cosmogen(y + -ic.] = Cosmogenetic *a.*
1909 in Webster. **1949** Koestler *Insight & Outlook* xxi. 302 Sacred books of a cosmogenic and ritual nature are.. the earliest monuments of literature. **1964** P. F. Anson *Bishops at Large* viii. 311 The mysteries of cosmogenic speculations.

cosmological, *a.* Add: **2.** *cosmological constant* = *cosmical constant* (see *cosmical *a.* 6).
1928 *Phil. Mag.* V. 835, λ is the cosmological constant. **1932** A. Einstein & W. de Sitter in *Proc. Nat. Acad. Sci.* XVIII. 213 Historically the term containing the 'cosmological constant' λ was introduced into the field equations in order to enable us to account theoretically for the existence of a finite mean density in a static universe. **1970** *Nature* 28 Nov. 845/1 The cosmological constant acts so as to provide a gravitational force in local Newtonian coordinates that increases linearly with distance *r* from the local Newtonian origin.

cosmonaut (kǫ·zmonǭt). [f. Cosmo- + Gr. ναύτης sailor; cf. *astronaut.] A traveller in outer space; an astronaut (esp. Russian).
1959 tr. *A. Shternfeld's Soviet Space Sci.* vi. 185 Naturally, cosmonauts could leave an artificial satellite and move in outside space. **1961** *Listener* 20 Apr. 700/2 The 'cosmonaut' ..circled the earth at a speed of 25,000 miles an hour. **1964** *Ann. Reg. 1963* 92 The visit of three Russian cosmonauts.. was followed by the official goodwill visit..of President Ayub Khan of Pakistan. **1970** *Daily Tel.* 2 June 1/6 The Soviet News Agency, Tass, said..both cosmonauts reported that everything was working normally.

cosmonautic, -ical (kǫ:zmonǭ·tik, -ikăl), *adjs.* [f. as prec. + -ic, -ical.] Of or pertaining to space travel or cosmonauts.
1947 W. Ley *Rockets & Space Travel* (1948) xii. 285 He [*sc.* von Pirquet]..called it the 'cosmonautic paradox'. **1950** *Jrnl. Brit. Interplanetary Soc.* IX. 206 Only such techniques can show any reasonable prospects of making return voyages to the planets practicable. Von Pirquet himself has expressed this in what he calls 'The Cosmonautical Paradoxon'. **1961** *Life* 21 Apr. 35 Some cosmonautical questions. *Ibid.* 18 Aug. 43 (*caption*) Cosmonautical embrace.
So **cosmonau·tics**, the science of travel in space; astronautics.
1950 *Jrnl. Brit. Interplanetary Soc.* IX. 154 Interstellar Rockets with Atomic Fuel. This, of course, is the great ideal of cosmonautics. **1965** Glasstone *Sourcebk. Space Sciences* i. 24 In Russia the terms 'cosmonautics' and 'cosmonaut' are now in general use, whereas 'astronautics' and 'astronaut' are preferred in the United States. **1967** *New Scientist* 9 Feb. 325/1 Russian scientists—in the tradition of Ziolkowsky, the founder of Soviet cosmonautics—have spoken of the importance of manned space-stations in Earth-orbit.

cosmopolis (kǫzmǫ·pŏlis). [f. Gr. κόσμος + πόλις, after Cosmopolite *sb.* and *a.*: see Cosmo- and -polis.] The (capital) city of the world; a cosmopolitan city or community.
1892 *Daily News* 20 Dec. 5/5 'Paris,' says the 'Patrie', 'has become the great cosmopolis.' **1918** A. Gray tr. *Grelling's The Crime* II. vi. 391 They would not at any price have their Fatherland a Cosmopolis. **1922** *Contemp. Rev.* Sept. 365 The well of English untainted by the infiltrations of cosmopolis [*sc.* London]. **1964** C. Vereker *Devel. Polit. Theory* (ed. 2) i. 35 The cosmopolis or world city with its life of virtue was the old polis, purified, idealized and writ large.

cosmopolitanism. Add: **1. b.** In Soviet usage, disparagement of Russian traditions and culture (equated with disloyalty).
1950 *Ann. Reg. 1949* 203 'Cosmopolitanism' was stated to be a plot to undermine Soviet dramaturgy..by praise and imitation of 'decadent bourgeois' dramatic and literary productions. **1963** *Listener* 3 Jan. 4/1 In 1948, 'cosmopolitanism' was named the chief heresy of the communist world with the Jews its chief prophets.

cosmopo·litanly, *adv.* [f. Cosmopolitan *a.* + -ly².] In a cosmopolitan manner.
1896 Kipling *Seven Seas* 139 Where, cosmopolitanly planned, He guards the Redskin's dry reserve. **1926** *Spectator* 16 Oct. 618/1 Italians are far more cosmopolitanly minded than we.

cosmoramic, *a.* (Earlier U.S. example.)
1827 Drake & Mansfield *Cincinnati* 45 Cosmoramic, optic, and prismoramic views of American scenery and buildings.

cosmos³ (kǫ·zmǫs). [mod.L. (A. J. Cavanilles *Icones et Descriptiones Plantarum* (1791) I. 9), ad. Gr. κόσμος ornament; so named from its elegant foliage.] A plant of the genus of *Compositæ* so named, native to tropical America, species of which, bearing rose, scarlet, and purple single dahlia-like blossoms, are cultivated as hardy annuals and perennials.
1813 [see *cosmea]. **1911** C. Harris *Eve's Second Husband* xiv. 278 You have that muslin with the purple cosmos flowers in it. **1920** *United Free Ch. Miss. Rec.* Dec. 226/2 Patches of white and magenta flowers called Cosmos. **1922** *Glasgow Herald* 25 Jan. 8 The pale delicacy of great beds of cosmos. **1929** *Encycl. Brit.* VI. 493/1 The common cosmos..has a smoothish stem bearing much-cut, narrowly-lobed leaves and flower-heads with a yellow centre. **1956** *Dict. Gardening* (R. Hort. Soc.) Suppl. 19/1 Most of the garden varieties of Cosmos are derived from *Cosmos bipinnatus*.

cosmotron (kǫ·zmotrǫn). [f. Cosmo- + *-tron.] A particle accelerator; *spec.* the proton-synchrotron, designed to produce 3000 MeV protons, which was completed in 1952 at the Brookhaven National Laboratory, U.S.A.

1949 *Newsweek* 7 Feb. 29 Foundations have been laid for a giant 'cosmotron' of record voltage. **1950** *Times* 20 Dec. 5/4 A cosmotron is an experimental instrument by which a magnetic field of high intensity accelerates particles torn from the nuclei of atoms. **1956** *Nature* 10 Mar. 470/2 The particle current has steadily decreased as the energy has increased—milliamperes for the 1940 cyclotron, microamperes for the 1943 synchrocyclotron, millimicroamperes for the present cosmotron and bevatron. **1967** CONDON & ODISHAW *Handbk. Physics* (ed. 2) IX. ix. 250/2 The radius of the proton orbit in the Cosmotron is 30 ft.

co-speci·fic, *a.* [f. CO- + SPECIFIC *a.* 5.] = CONSPECIFIC *a.*

1889 in *Cent. Dict.* **1936** *Geogr. Jrnl.* LXXXVIII. 519 This plant we afterwards saw in flower and there were three forms, apparently co-specific.

Cossack. Add: **2.** In full *Cossack boot*: a high boot. orig. *U.S.*

1805 *Lancaster* (Pennsylvania) *Jrnl.* 29 Mar. (Advt.), Cossacks, Suwarrows..Full Boots. **1831** *Mirror* XVIII. 435/1 It's goodbye to Wellingtons and Cossacks,..the old Shoe Mart is disposed of. **1897** J. L. ALLEN *Choir Invisible* iv. 36 His new cossack boots. **1967** *Guardian* 9 Nov. 6/3 Country shoes and stockings, Cossack boots, hogskin gloves.

b. In full *Cossack horse, pony*: a Cossack horse or pony.

1815 J. SCOTT *Visit Paris* xiii. 262 The driver of a cabriolet ..told me that his horse was a Cossack. **1831** YOUATT *Horse* 18 The most celebrated Cossack horses from the Don, the Black Sea, and the Ural. **1926** *Blackw. Mag.* Sept. 310/1 Our ponies nearly all of them came from England, but some people played Cossacks. I had a capital Cossack pony.

c. *pl.* Baggy trousers, pleated into a waist-band. Also *Cossack trousers*.

1820 *Kaleidoscope* I. 29/1 Whose heart could parry the attacks Of his voluminous Cossacks—Trousers so-called from those barbarians Nursed in the Steppes. **1832** A. EARLE *Jrnl. Resid. Tristan da Cunha* (1966) 235 The front of these 'Cossacks' consisting of sail cloth, and the back of dried goat's skin. **1844** *Ainsworth's Mag.* VI. 430 They were made very wide, long and loose (Cossack trowsers were the rage then). **1968** *Guardian* 28 July 7/4 Black coats over Cossack trousers that tuck into black patent leather boots.

d. *Cossack post*: an outpost of a few mounted men under a non-commissioned officer or senior soldier.

1853 L. OLIPHANT *Russian Shores Black Sea* 141 My importations from Don Cossack post-huts were considerable. **1860** T. W. ATKINSON *Trav. Amoor* 9, I had visited..all the Cossack posts on my way to the Altin-Kool. **1861** W. F. AINSWORTH *All Round World, Steppes of Russia* II. v. 292/2 The Cossack post of Schukovaia. *Ibid.*, They had to put up at a Cossack post-house, a mere hut. **1899** BADEN-POWELL *Aids to Scouting* 131 Each force will form a line of outposts, consisting of two Cossack posts, and scouts for reconnoitring patrols.

e. *Cossack hat*: a brimless hat, wider at the top than at the head-band.

1939 M. B. PICKEN *Lang. Fashion*, *Cossack hat*, adaptation of cap worn by Cossacks. **1951** 'J. TEY' *Daughter of Time* ii. 20 She breezed in, very dashing in a Cossack hat worn at a casual rake. **1966** *Vogue* Dec. 121 Persian lamb cossack hat.

3. *slang*. A policeman; esp. a member of an armed strike-breaking force (from the similar use of Cossacks in imperial Russia). Also *attrib.*

1859 in HOTTEN *Dict. Slang*. **1886** *Graphic* 30 Jan. 130/1 A policeman is also called a 'cossack', a 'Philistine', and a 'frog'. **1913** *Ann. Rep. Dock Workers' Union 1912* 5 Home Secretary intervenes with armed forces; attempts at suppression, cossack methods of the Home Office forces. **1928** *Collier's* 29 Dec. 9/4 Remember the state constabulary of Colorado, the so-called 'Cossacks' who were so bitterly complained of by striking miners.

co·sseting, *vbl. sb.* Also **cossetting.** [-ING[1].] The action of the verb COSSET.

1880 L. PARR *Adam & Eve* I. v. 128 With all his kissin' and cossetin' of her. **1886** 'RITA' *Darby & Joan* (1887) i. 1 I'm not one of those as holds with cossettin' and fussin'. **1901** *Westm. Gaz.* 22 May 3/1 In the other picture there is a sort of fond cosseting of colour and design. **1926** *Harper's Mag.* Feb. 352/1 The asthma and kindred ills born of their cruel cosseting.

cossette (kǫse·t). Also **-et.** [Fr., dim. of *cosse* pod, husk.] A slice of a root, cut up during processes of manufacture; *spec.* a piece of sugar beet prepared for the extraction of juice, or of chicory prepared for drying and roasting.

1858 SIMMONDS *Dict. Trade*, *Cossets*, a name on the Continent for slices of beet-root, from which spirit has been distilled. **1902** *U.S. Dept. Agric.* Rep. No. 74. 39 (Cent. D. Suppl.), The raw beet is simply cut into long slender grooved slices. In the factory these slices are known as cossettes.

cost, *sb.*[2] Add: **1. a.** *cost price* (earlier example). Also *attrib.*

1834 DICKENS *Sk. Boz* (1836) II. 139 Every description of goods..fifty per cent. under cost price. **1959** *Times Lit. Suppl.* 17 Apr. 220/1 The cost-price Land-Rover.

d. That which a thing originally cost. (Cf. *prime cost* in 1 a.) So *at cost*: at the initial cost.

1873 *Rep. Vermont Board Agric.* II. 216 If they get cost in the spring for their sheep, throwing in their labor, they think the profits ample. **1898** C. A. BATES *Clothing Bk.* No. 2506, We sell..durable clothing very close to cost.

e. *cost of living*: the cost of maintaining a particular standard of living; also *attrib.*, as *cost-of-living index*, an index which measures periodical (monthly, etc.) changes in the level of retail prices (since 1947 known in Britain as the Index of Retail Prices).

1896 W. D. HOWELLS *Impr. & Exper.* 35 The pay is not only increased in proportion to the cost of living, but it is really greater. **1913** *Commonw. Austral. Labour Bull.* May 24 Particulars of the cost of living index-numbers in each State during the last quarter..are shown in the following table. **1922** *Encycl. Brit.* XXX. 755/1 Till recent years the phrase 'Cost of Living' was only used loosely by economists when the balance between movements of wages and prices was in question...In popular parlance it has since become a recognized economic problem. **1927** BOWLEY & STAMP *Nat. Income 1924* 31 Average earnings of all wage-earners for a full working-week are estimated to have increased 94 per cent. between 1914 and 1924, while the Cost of Living Index rose 75 per cent. **1930** *Times Educ. Suppl.* 5 Apr. 149/4 They were shortly afterwards subjected to the upper cut in the cost-of-living bonus. **1948** G. CROWTHER *Outl. Money* (ed. 2) iii. 85 The value of money in buying the goods and services that the ordinary family consumes..can be called the retail value of money, or the cost of living. **1965** SELDON & PENNANCE *Everyman's Dict. Econ.* 96 Constructing a cost of living index requires investigation into the expenditure of a representative sample of households belonging to the group whose cost of living is to be studied. It shows the expenditure on each commodity and each service within the period by the members of the group as a whole.

2. Also *attrib.*, as *costs clerk*.

1921 *Dict. Occup. Terms* (1927) § 939 *Costs clerk*..has special knowledge of costs and disbursements which solicitors are entitled to charge for their services..; draws bills of costs against client for services rendered.

5. b. *at all costs* [= F. *à tout prix*]: whatever the cost may be; in spite of all losses.

a **1919** in 'Boyd Cable' *Old Contemptibles* vi. 90 To hold the position at all costs until relieved. **1926** A. HUXLEY *Two or Three Graces* 150 At all costs, no frankness! **1970** *Daily Tel.* 13 Oct. 11/2 Our..wish to be left alone at all costs.

6. *cost-saving* adj.; **cost account**, an account kept of the cost of production of articles, works, etc.; so **cost accountant**, one who records every item of (esp. overhead) expenses in a business concern (with a view to checking wasteful expenditure); also *cost-account* v. intr., *cost-accounted* ppl. adj., *cost accounting*, *cost book-keeping*; **cost-benefit**, used esp. *attrib.* of an analysis, study, etc., which assesses the relationship of the actual cost of a project, etc., and the value of social and other benefits resulting from its completion; **cost clerk**, one who keeps a cost account; **cost-effective** *a.*, designating or pertaining to a project, etc., that is effective in terms of its cost; so *cost-effectiveness*; **cost-keeper**, a cost clerk; so *cost-keeping*; **cost-plus** (orig. *U.S.*), *ellipt.* for 'cost plus profits', used *attrib.* to denote a system whereby the price of an article, etc., is the basic cost of production plus an agreed percentage for profit; **cost push** (see quot. 1959); used esp. *attrib.*; **cost value** (see quot. 1848).

1896 *Daily News* 17 Nov. 2/2 The accounts are cost accounts, and are designed to show the actual cost..of every work..executed by the Works Department. **1920** *Act 10 Geo. V* Sched. II. 11 Such cost accounts, trading accounts, and balance sheets..as the Controller may require. **1960** I. JEFFERIES *Dignity & Purity* v. 68 They cost-account pretty thoroughly. **1921** *Dict. Occup. Terms* (1927) § 933 Cost accountant, costing accountant, costing clerk, costs clerk. **1957** W. H. WHYTE *Organiz. Man* 121 These jobs, while not mere make-work, are outside the regular cost-accounted operations of the company. **1913** MOXEY *Princ. Factory Cost Keeping* 8 All factory cost accounting is based directly on the principles of double-entry book-keeping. **1924** H. R. J. HOLMES *Farm Costing* 107 The whole object of cost accounting is the elimination of waste. **1964** C. DENT *Quantity Surv. by Computer* ii. 10 This development, however, may have to wait until building contractors are in the habit of using more sophisticated cost-accounting systems. **1928** R. H. LANSBURGH *Industr. Management* (ed. 2) xv. 172 Standardization of codes and regulations.. must be of enormous cost benefit to builders and users of machinery. **1963** *Times* 16 Apr. 13/4 Cost-benefit analysis.. constitutes an endeavour to bring into relationship with 'the measuring rod of money' all the social consequences either of introducing a new project or removing an old one. **1964** *Economist* 7 Mar. 899/2 'Cost-benefit' studies that have been made (for instance for the M1 motorway) to work out rates of return. **1924** H. R. J. HOLMES *Farm Costing* 19 Cost Book-keeping begins essentially with an allocation of horses and manual labour to the separate departments for which a closing or working account is kept. **1898** *Engineering Mag.* XVI. 42 This forms the cost clerk's authority for labor value in prime costs. **1902** *Daily Chron.* 5 Nov. 9/6 Timekeeper and Cost Clerk in engineering works [wanted]. **1967** *Guardian* 7 July 20/7 Does Mr McNamara's cost-effective thinking pass the Soviet leaders by? **1968** *Observer* 23 June 1/5 The Cabinet has evidently decided that it must be able to defend public spending as 'cost-effective'. **1964** *Economist* 11 Jan. 94/2 A proper cost-effectiveness study. **1898** *Engineering Mag.* XVI. 48 All of the cost keeping is done by one cost-keeper, one assistant, and a typewriter. **1920** *N.Y. Times* 18 Feb. 10/4 Millions of dollars were wasted on cost-plus contracts...There was prodigious waste of money, and the cost-plus system was by no means the only cause. **1949** *Ann. Reg. 1948* 448 The 'cost-plus' system of price control..had been discontinued. ..The system was replaced by maximum cash prices. **1957** *Economist* 23 Nov. Suppl. 10/1 Monetary policy..may actually intensify pressures on prices. This seems self-evident to the believers in the cost-push theory of current inflation. **1959** *Ibid.* 31 Jan. 409/1 'Cost push'—the independent forces in the economy which raise wages and hence prices without regard to the state of demand. **1960** *Ibid.* 22 Oct. 381/2 It is conceivable that, through the folly of a trade union or a board of arbitrators, prices in one country may still be inflated by a cost-push. **1965** SELDON & PENNANCE *Everyman's Dict. Econ.* 98 *Cost push*, the process by which inflation..has been said by some economists to have been caused or intensified by trade unions insisting on higher wages irrespective of output. **1960** *Farmer & Stockbreeder* 1 Mar. 135 A cost-saving system of farming that any farmer can afford. **1848** MILL *Pol. Econ.* I. III. vi. 566 Most things naturally exchange for one another in the ratio of their cost of production, or at what may be termed their Cost Value. **1897** *Westm. Gaz.* 8 Apr. 8/3 The properties specifically hypothecated to secure the present issue is [*sic*] of a cost value exceeding £650,000.

cost, *v.* Add: **1. c.** *absol.* To be expensive; to prove costly. *colloq.*

1917 [see *CREATION 4 b]. **1933** N. HOULT *Youth can't be Served* ii. 38 The best bacon, and a new-laid egg. And them things cost these times. **1938** F. SCOTT FITZGERALD *Let.* Feb. (1964) 23, I like them but my God they cost. **1960** T. REESE *Play Bridge* 20 In any event it cannot cost to play a spade.

d. *to cost money*: to be (very) expensive; *to cost the earth*: to cost a large amount of money.

1667 PEPYS *Diary* 29 Sept. (1896) VII. 129 All the afternoon talking in my chamber with my wife, about my keeping a coach the next year, and doing some things to my house, which will cost money. **1852** A. CAREY *Clovernook* (1853) 74 Something pretty costs money: do you think it grows on bushes? **1924** O. W. HOLMES *Let.* 18 Dec. (1953) I. 685 A very good etching of me was published but it is large and costs money. **1924** WODEHOUSE *Bill the Conqueror* vii. 149 'Big grey limousine.' 'Expensive?' 'Looked as if it had cost the earth.' **1958** *Daily Tel.* 28 June 5/3 If you want a stake in the most haunting part of Europe without its costing the earth to travel there.

co·-star, *sb.* orig. *U.S.* [f. CO- 3 b + STAR *sb.*[1] 5.] A star of the cinema or stage appearing in the same production as one or more other stars of equal importance; an actor receiving star billing with another or others. Also *attrib.*

1926 *Amer. Year Bk. 1925* 1062 Walter Hampden.. settled down to repertory in the Hampden Theatre with Ethel Barrymore as co-star. **1955** *N.Y. Herald Tribune* 23 Sept., The Broadway musical,..in which he'll get co-star billing with Gypsy Rose Lee, Peter Lind Hayes and Mary Healy. **1959** P. BULL *I know the Face* xii. 215 The star who was caught..measuring the posters..to see if his name took up more space than that of his co-star.

co·-star, *v.* orig. *U.S.* [f. CO- 1 + STAR *v.* 8 a.]

1. *intr.* To perform in a film or a play as a co-star.

1919 *Springfield Republican* 16 Nov., Francine Larrimore, co-starring in the dramatic success 'Scandal'. **1933** *Week-end Rev.* 4 Feb. 131/1 Such statements as 'They Were Born to Co-Star'. **1938** 'E. QUEEN' *Four of Hearts* xviii. 244 You and Bonnie are to co-star in a picture biography of Jack and Blythe. **1959** *Times* 25 Nov. 9/7 Mr. Yul Brynner co-stars as an egocentric conductor.

b. *transf.* and *fig.*

1940 *Jrnl.* (Topeka) 29 Apr. 2/7 Twin athletes.., they co-starred in track, baseball and basketball. **1956** *Essays in Criticism* VI. 203 Your ego never shared or co-starred.

2. *trans.* To employ or present with a co-star or co-stars; of a film or play: to include as co-stars.

1937 *Evening News* 6 Mar. 11/1, I am to be co-starred with Tauber. **1948** *Evening Standard* 7 Jan. 6/4 Herbert Wilcox has just made 'Spring in Park Lane', co-starring Anna Neagle and Michael Wilding again. **1956** *N.Y. Herald Tribune* 18 Jan., Margaret Phillips is co-starred, and is worthy of her billing.

Hence **co-starring** *vbl. sb.* and *ppl. a.*

1939 *Capital* (Topeka) 4 July 12 Cary Grant and Jean Arthur in a scene from their first costarring film. **1956** *N.Y. Herald Tribune* 18 Jan. 15/2 She'll have the co-starring feminine role with Ginger Rogers.

co·-state. [f. CO- + STATE *sb.*; cf. G. pl. *mitstaaten*.] A state allied with another.

1795 *Ann. Reg.* 227/1 Declaration of the King of Prussia, to his Most High Colleagues, Co-States of the German Empire, and Members of the Germanic Leagues. *c* **1798** JEFFERSON *Resol. Alien & Sedition Laws* in *Wks.* (1853) IX. 471 This commonwealth does therefore call on its co-States for an expression of their sentiments...The rights and liberties of their co-States will be exposed to no dangers. **1856** T. H. BENTON *30 Years' View* I. lxxxvii. 349/1 Virginia exercises the right that pertains to a State in declaring her views, and inviting the like action of her co-States.

costean, costeen (kǫstī·n), *sb.* *Cornish Mining.* [See COSTEAN *v.*] A pit sunk down to the rock in costeaning. Also *costean pit*.

1778 Costean pit [in Dict.]. **1895** *Daily News* 22 Oct. 2/1 No. 3 Reef, Eldorado, has been proved by costean pits for a length of 450 ft. **1896** *Westm. Gaz.* 20 Apr. 8/1 A costeen has exposed the cap of a reef 4 ft. wide of solid quartz. **1897** *Daily News* 19 Jan. 2/2 Recently two separate reefs have been struck in separate costeens. *Ibid.* 16 Feb. 8/1 A costeen over 100 feet in length has cut a strong reef 3 feet wide. **1920** *Glasgow Herald* 31 Mar. 11 The White Hope..has been opened up all along the line of lode by costeens. **1964** A. NELSON *Dict. Mining* 109 *Costean*, (1) A trench cut

across the conjectured line of outcrop of a seam or orebody to expose the full width. (2) The channel eroded by a flow of water to expose mineral deposits during prospecting work.

costerdom (kǫ·stəɹdǫ̆m). [f. COSTER[2] + -DOM.] Costers collectively; the realm of costers.

1893 *National Observer* 1 Apr. 488/2 The donas of costerdom. **1895** *Athenæum* 28 Dec. 897/1 The apotheosized costerdom of Mr. Albert Chevalier. **1903** *Westm. Gaz.* 19 May 8/1 The course was one huge sea of howling, jostling, struggling costerdom.

co·sting, *vbl. sb.* [f. COST *v.* + -ING[1].] **a.** Estimation of the cost of production of an article, etc. **b.** The costs of production or working. Also *pl.* and *attrib.*

1884- [see Dict. s.v. COST *v.* 4]. **1908** L. C. HEADLY (*title*) Boot and Shoe Costings. **1917** DICKSEE *Fundamentals Manuf. Costs* 7 They drop all attempt at accurate costing. **1921** [see *cost accountant*, *COST *sb.*[2] 6]. **1922** TURNER & WOOD *Man. Up-to-date Organisation* 171 Costing is the collecting, calculating, and allocating of the cost of material, labour and the correct proportion of Overhead charges involved in a production or utility. **1924** H. R. J. HOLMES (*title*) A short system of farm costing. *Ibid.* 105 The application of this simple costings scheme to an actual farm account. **1927** E. J. P. BENN *Trade* iii. 49 Such modern devices as index figures and costing systems. **1929** *Times* 7 Feb. 9/3 So that he could compare the costings of the foreign manufactures with the costings of the corresponding English article. **1955** *Times* 10 June 3/1 There would be only one set of costings for all the minerals and merchandize carried by rail.

co·sting, *ppl. a.* [f. COST *v.* + -ING[2].] Costly; spiritually exhausting or expensive. Hence **co·stingly** *adv.*; **co·stingness.**

1900 F. VON HÜGEL *Let.* 7 July (1927) 85, I have had to express myself on paper to W. W. about these matters, and have found it difficult and costing..because I feel..it is wrong to multiply and intensify differences by unnecessary explicitations. **1901** —— *Let.* 18–20 Dec. (1927) 103 What a costing process *that* is! *Ibid.*, The inevitable costingness of the struggle..has absorbed my strength. **1914** —— *Ess. & Addresses* (1926) II. 103 Such misuse will..be best—at least, be not too costingly—avoided by the deliberate recognition of the danger. **1930** G. GREENE *2 Witnesses* II. xii. 145 The life of religion is so real because it is so costing, so costing because it is so real. **1962** *Friend* 9 Mar. 280/1 He would remind the Meeting of the costingness of pacificism, which was not peace at any price but love at all cost.

costume, *sb.* Add: **5.** *costume comedy, film, intrigue, melodrama, part, picture, play*; **costume jewellery** (orig. *U.S.*), showy artificial jewellery worn for decorative purposes.

1932 *Ann. Reg. 1931* II. 48 'Congress Dances' proved a delightful costume comedy of 1815. **1960** D. FEARON *Murder-on-Thames* xviii. 170 If costume films start coming back into fashion. **1961** R. WILLIAMS *Long Revol.* II. vi. 264 The Romantic movement..produced..the useless form of the costume-intrigue. **1933** *New Yorker* 2 Dec. 70 That costume jewelry which is all gold sequins. **1949** A. HUXLEY *Ape & Essence* 127 Crumbling remains of slacks and sweaters, of Nylons and costume jewellery and brassieres. **1950** A. CHRISTIE *Murder is Announced* vii. 68 Costume jewellery —that's a fine way of labelling a string of false pearls. *Ibid.*, It's 'costume jewellery' that young Miss Simmons wears—gold ivy leaves and dogs and such like. **1905** *Daily Chron.* 5 June 6/5 A costume-melodrama of a quite worthless type. **1906** G. B. SHAW *Let.* 28 Dec. (1956) 73 You must blight Vedrenne by telling him that Charteris is a costume part: the brown evening suit, blue shirt and sandals are clearly costume. **1921** B. TARKINGTON *Plays, Playwrights & Playgoers* (1959) 46, I see that one or two 'costume' pictures have at last been done pretty well. **1885** J. K. JEROME *On Stage* vii. 64 A costume play..rivalled a fancy dress ball in variety. **1929** D. L. MOORE *Pandora's Letter Box* xi. 209 To a lover of accuracy the average 'costume' play..exasperates by its slovenliness.

costus (kǫ·stǫs). [f. L. *costus*, Gr. *κόστος*, an Oriental aromatic plant.] In full *costus root*: the fragrant root of the plant *Saussurea lappa*, indigenous to Kashmir, that yields an essential oil used in perfumery, etc. Cf. COST *sb.*[3]

1607 TOPSELL *Four-f. Beasts* 376 Take of the roote called Costus two ounces. **1821** J. SMYTH *Pract. of Customs* (ed. 2) 70 Costus. There are two sorts of this article, the sweet and the bitter, the sweet is a small root very like..to the Turmerick. **1885** BIBLE *Exod.* (R.V.) xxx. 24 And of cassia (*v.r.* costus) five hundred. **1889** *Cent. Dict.*, Costus-root. **1915** *Jrnl. Chem. Soc.* CVIII. I. 275 Costus root oil. **1922** SCHLICH *Man. Forestry* (ed. 4) I. 133 *Saussurea Lappa* yields the valuable costus oil.

cosy, cosey, cozy, *a.* and *sb.* Add: The normal spelling in Britain is *cosy*, and in the U.S. *cozy*. **A. 3.** Of a job: = *CUSHY *a.*

1915 D. O. BARNETT *Letters* 198 That's a particularly cosy job, as he lives at brigade H.Q. and does nothing.

4. Warmly intimate or friendly; sentimental; freq. in pejorative sense: complacent, smug, unadventurous, parochial; = COMFORTABLE *a.* 10 b.

1927 M. BEERBOHM *Let.* 2 Oct. (1964) 266 We liked her very much. She isn't exactly cosy, but she's very spirited. **1943** *Horizon* Mar. 197 Hence arose an interesting theatrical phenomenon—the cult of the Cosy Play. In this type of drama the characters bore a recognizable likeness to the members of the audience..but wiser, more humorous, better-looking. **1958** *Observer* 25 May 16/7 Cosy little murder mystery. **1959** *Daily Mail* 20 Feb. 8/1 It was a cosy chat—Mrs Betjeman..is a cosy person. **1959** *Times Lit. Suppl.* 29 May 318/3 If her tone in discussing the Brontë girls becomes at times a trifle too cosy, where actual research is involved she has shown herself to be commendably austere. **1960** C. P. SNOW *Affair* v. 49 It was mildly ironic.. to find her set on seeing him a cosy, bourgeois success.

5. *cosy corner* [cf. COSY *sb.* 3], an upholstered seat which fits into a corner of a room; such a corner, cosily furnished; also *attrib.* and *fig.*; *cosy stove* (proprietary name), a free-standing enclosed stove.

1894 *Country Gentlemen's Catal.* 115/1 Our patent cosy corners. Elegant and comfortable additions to any room. **1898** A. BENNETT *Man from North* xvi. 138 The 'cosy corner', an angle of the room furnished with painted mirrors and a bark bench of fictitious rusticity. *a* **1899** in J. Gloag *Victorian Comfort* (1961) iii. 73 (Advt.), Handsome Enamelled White or any colour Cosy Corner, complete with drapery of cretonne. **1926–7** *Army & Navy Stores Catal.* 279 The 'cozy' stove. Continuous burning. **1950** D. GASCOYNE *Vagrant* 8 Lying resigned in cosy-corner crow's-nest. **1955** M. HASTINGS *Cork & Serpent* viii. 98 There was a cosy stove in the grate. **1956** B. GOOLDEN *At Foot of Hills* vii. 172 In the dining-room the doors of the cosy stove were closed. **1968** C. P. BRACKEN *Roman Ring* ii. 7 Cosy corner was the correct period terminology for the chintz-covered sofa and low table.

cosy (kōᵘ·zi), *v.* Also **cozie, cozy.** [f. the adj.]

†**1.** *intr.* To be comfortable, snug. *Obs. dial.*

1898 *Eng. Dial. Dict.* s.v. *cozie*, While topers cozie in the neuk.

2. *trans.* To comfort, reassure; to delude. Also with *along*. *colloq.* (orig. *U.S.*).

1939 C. MORLEY *Kitty Foyle* xxvi. 262 But I cosied myself thinking maybe I was shot by too much nerve strain at the office. **1960** M. SHARP *Something Light* xxiv. 212 Her impulse, on purely selfish grounds, was to cosy Mr Clark on every point. **1966** *Observer* 15 Aug. 17/6 The important thing is 'to cosy along' the prospective victim for the camera.

3. *intr.* Colloq. phr. (chiefly *U.S.*) *to cosy up to*: to snuggle up to; to become friendly with; to ingratiate oneself with (someone).

1937 *News-Week* 9 Jan. 37 When the New Hampshire native cozies up to the fireplace in his Washington home and decides to talk, he usually talks about one subject: solar engines. **1966** D. SKIRROW *It won't get you Anywhere* xxxix. 193 Fluck had cosied right up to Sullivan, hoping to drive Ball mad with jealousy. **1966** *Observer* 13 Feb. 13/4 Civil servants are cosying up to those they think might be their next masters.

cot, *sb.*[4] Add: **4. b.** Also (*U.S.*), a bed or wheeled stretcher of a type used in hospitals. Cf. *cot-case* below.

1874 'H. CHURTON' *Toinette* xxxv. 359 Some are sitting up in their cots gazing out upon the fair scene below. *a* **1922** C. C. ANDREWS *Recoll.* (1928) 202 It was while lying ill on my cot at Galveston that I decided to join the Republican party.

5. cot-bed (examples); **cot-case**, a person sufficiently ill to be confined to bed; **cot-death**, the unexplained death of a baby in its sleep.

1836 S. G. PERKINS in *Proc. Mass. Hist. Soc.* (1886) 2nd Ser. II. 385 At the entrance of the cabin hung a cot-bed, in which the mate usually slept. **1838** J. H. INGRAHAM *Burton* II. xi. 175 A narrow cott bed, with a military cloak thrown over it, constituted the sole furniture of the warrior's abode. **1892** A. C. GUNTER *Miss Dividends* 177 She goes up to her room, where she finds a clean cot-bed. **1921** *Daily Colonist* (Victoria, B.C.) 2 Oct. 30/5 He is making the journey on a cot bed. **1897** *Daily News* 23 Mar. 7/3 Altogether 35 patients were received, 22 being cot cases. **1924** P. CRESWICK *Beaten Path* xxvii. 150 To come here on an ambulance and be carried in, a regular right-down cot-case. **1965** 'M. NEVILLE' *Ladies in Dark* xiv. 139 Winnie looked a proper cot-case. **1970** *Guardian* 11 May 4/5 'Cot death' relates to children not known to be ill..who die unexpectedly.

cot, *v.*[2] (Earlier U.S. example. See also COTTED *ppl. a.*)

1876 *Rep. Vermont Board Agric.* III. 213 A fleece..so destitute of oil that the ordinary exposure of the county would cause it to 'cot' on the sheep's back.

‖ **coteau** (koto). *N. Amer.* [Fr., = hill, hill-side.] An upland area; a dividing ridge (see quots.).

[**1824** KEATING *Exped. St. Peter's River* II. v. 244 The Coteau des Prairies is a very remarkable feature in the aspect of this region.] **1839** *Boston Wkly. Mag.* 12 Jan. 143/2 The summit of the *coteau* or dividing ridge. **1883** CHAMBERLIN in *3rd Ann. Rep. U.S. Geol. Survey* 19 The moraines of this region, as far north as the head of the *coteau*, together with the general character of the associated deposits of the adjacent territory. **1918** VISHER *S. Dakota* 36 Topographically the coteaus may be considered as extensive mesas standing some 500 to 700 feet above the Dakota Valley which separates them.

cote-hardie, var. COAT-HARDY.

1834, 1860 [see COAT-HARDY]. **1955** *Oxf. Jun. Encycl.* XI. 87/1 About 1500, men adopted the jerkin..in place of the cote-hardie. **1967** E. SHORT *Embroidery & Fabric Collage* ii. 38 The women of medieval England decorated their mantles and cotehardies with their husbands' arms.

co-temporary, *a.* and *sb.*: see CONTEMPORARY *a.* and *sb.*

coterie. 3. (Earlier and later examples.) Also quasi-*adj.*

1833 MILL *Lett.* (1910) I. 77 A paper which..keeps aloof from all *coterie* influence. **1900** G. B. SHAW *Let.* 9 Feb. (1931) 375 This Stage Society..is catching on in its little coterie-theatre way. **1933** P. GODFREY *Back-Stage* xiii. 165 Circulars designed to appeal to those who incline to coterie art and limited editions. **1962** *Listener* 30 Aug. 327/2 The very exercise will remove accretions of coterie language and provincialism from serious writers who attempt it.

co-terminous, *a.* Add: Also **coterminous.** (Further examples.) Freq. const. *with*.

1864 TROLLOPE *Small Ho. at Allington* I. i. 1 The estate of Dale of Allington had been coterminous with the parish of Allington for some hundreds of years. **1937** R. H. LOWIE *Hist. Ethnol. Theory* x. 159 Another major difference among ethnological historians that is not quite coterminous with the division into a right and left wing. **1940** C. S. LEWIS *Probl. Pain* iii. 28 There is kindness in Love: but Love and kindness are not coterminous. **1943** A. M. FARRER *Finite & Infinite* xv. 180 If we take 'will' to be coterminous with the highest expressions of explicit rationality. **1966** *Economist* 15 Jan. 176/1 What future Presidents would gain would be a House of Representatives elected with them.. and coterminous with their term of office.

Côte-rôtie (kotroti). A red wine produced in vineyards of this name near Lyons.

[**1765** STERNE *Tr. Shandy* VII. xxix. 104 To snatch a blushing grape from the Hermitage and Coté [*sic*] roti, as I shoot by the foot of them.] **1802** C. WILMOT *Let.* 6 Oct. in T. U. Sadleir *Irish Peer* (1920) 92 We saw the process of making wine at Condrieux, where they gave us Cote Roti as it drop'd from the Press. **1833** REDDING *Wines* v. 118 Côte Rôtie is remarkable for the excellence of its colour. **1908** E. & A. VIZETELLY *Wines of France* 136 The remarkable quality of Côte-Rôtie wine is attributable..to the extremely friable nature of the soil. *Ibid.* 137 Côte-Rôtie may be a little heady.

co-text (kōᵘ·tekst). [f. CO- + TEXT *sb.*[1]] The language which surrounds a particular word, phrase, or passage, and which can determine its meaning; = CONTEXT *sb.* 4 a.

1964 M. A. K. HALLIDAY et al. *Linguistic Sci.* v. 125 When we learn our native language, these two types of contextual evidence, from the situation and from the surrounding language or 'co-text', constitute all the evidence we have. **1965** B. M. H. STRANG *Metaphors & Models* 4 *Bacon and eggs* is reversible, as *fish and chips* is not,..by reason of linguistic information about position and about the items in the co-text (the words *bacon, eggs, fish, chips*). **1966** A. MCINTOSH in C. E. Bazell *In Memory of J. R. Firth* 303 Within the confines of what is actually spoken or written, I have found it necessary to consider the influence of elements in the co-text (i.e. in the textual environment of the constructions under actual scrutiny).

Cotham (kǫ·təm), the name of a village near Bristol, designating an argillaceous limestone having dendritic markings, also called *landscape marble* or *stone*. Also *Cotham beds*.

1816 R. JAMESON *Min.* (ed. 2) II. 196 To the same compact marl, may be referred the variety called Cottam Marble. **1822** CONYBEARE & PHILLIPS *Outl. Geol. Eng. & Wales* I. 264 The lias..occasionally exhibits..dendritical appearances (Cottam stone or marble). **1845** P. B. BRODIE *Hist. Fossil Insects* 91 In most parts of the southern sections there is a band of stone locally termed 'Landscape stone', or 'Cotham marble'. **1883** [see *landscape-marble*, LANDSCAPE *sb.* 5]. **1929** L. RICHARDSON in Evans & Stubblefield *Handbk. Geol. Gt. Brit.* III. ix. 344 The Cotham Beds, like the underlying Westbury Beds, extend all across the country. **1960** L. D. STAMP *Britain's Struct.* (ed. 5) xii. 129 One band of limestone, the fascinating Cotham marble, was often seen polished in fragments on Victorian mantlepieces. **1968** P. E. KENT in Sylvester-Bradley & Ford *Geol. East Midlands* x. 178 The Cotham Beds are marls which typically form the upper part of the Rhaetic in the East Midlands.

coti·llion, coti·llon, *v.* [f. the sb.] *intr.* To dance cotillions.

1833 *Deb. Congress* 26 Jan. 1402 Daughters of very indulgent fathers enter the ball room, and waltz, and cotillon, ornamented with several hundred bales of cotton. **1900** *Daily News* 28 June 4/7 We dance, and cotillon, and listen to music.

Hence ‖ **cotillonneur** (kotiyonör), a dancer of the cotillion.

1898 *Daily News* 4 June 3/3 One of the best cotilloneurs going.

Cotnar (kǫ·tnāɹ). [Place-name.] A sweet white wine produced near Cotnar in Rumania.

1833 REDDING *Wines* x. 256. **1845** BROWNING *Flight of Duchess* iii, A cup of our own Moldavia fine, Cotnar, for instance. **1911** *Encycl. Brit.* XXIII. 827.

cottage. Add: **3. b.** A public lavatory or urinal. *slang* (now only in homosexual usage).

1909 in WARE *Passing Eng.* **1966** *Observer* 6 Nov. 24/3 The private, witty Cockney dialect of the queer bars and 'cottages'. **1968** *Guardian* 12 Aug. 6/3 Wakefield's answer to Danny La Rue trips out of a little hutch at the side of the stage labelled 'Ye Olde Camp Cottage'.

4. a. (Earlier and later examples of *cottage orné*.)

1781 GREVILLE *Diary* 11 Aug. (1930) 23 The Queen alighted at Dropmore Hill, the Cottage Orné which was at this time advertised for Sale by Mr Christie. **1962** A. JOBSON *Window in Suffolk* vii. 111 It was a sort of *cottage orné* of the J. C. Loudon type.

c. A house which has only one storey. *Austral.*

1898 in MORRIS *Austral Eng.* **1904** D. SLADEN *Playing the Game* I. ix. 85 What rich Melbourne people call a cottage

—a cottage with a ball-room and billiard-room, and enough bedrooms to take a good slice out of an acre. **1913** W. K. HARRIS *Outback in Australia* xxiii. 157 About a score of cottages for the blacks. **1923** D. H. LAWRENCE *Kangaroo* ii. 22 More 'cottages'; that is, bungalows of corrugated iron or brick.

6. *cottage garden* (examples), *-woman*; **cottage cheese** (orig. *U.S.*), a soft white cheese; **cottage home**, also, a benevolent institution (see quots.); **cottage industry**, one partly or wholly carried out in the home; † **cottage lecture**, a religious address delivered by a cleric in the home of a layman; **cottage loaf** (examples); also *attrib.*, denoting something shaped like a cottage loaf; **cottage organ** *U.S.*, a small reed-organ; **cottage style**, a style of book-binding in which the edges of a rectangular panel on the cover of a book slope away to create the effect of a roof or gable, prevalent in the late seventeenth century.

1848 Cottage-cheese [see SMEAR-CASE]. **1874** 'S. COOLIDGE' *What Katy did at School* i. 13 The bread was excellent, and so were the cottage cheeses and the stewed quince. **1936** J. GIBBENS *Care of Children* ii. 15 Cream cheese and cottage cheese..can be given on biscuits or in sandwiches to quite young children. **1958** *Sunday Times* 27 Apr. 22/3 Milk curd cheese, or cottage cheese as it is now sometimes called. **1849** C. BRONTË *Shirley* III. xiv. 314, I can line yonder barren Hollow with lines of cottages, and rows of cottage-gardens. **1886** ELWORTHY *W. Somerset Word-Bk.* 177 A very prolific variety of kale or winter greens; much grown in cottage gardens. **1873** *Young Englishwoman* Apr. 171/2 This really grand scheme of benevolence will comprise..thirty Cottage Homes..providing shelter, support, and needful education for three hundred girls. **1882** S. A. BARNETT in H. Barnett *Canon S. A. B.* (1918) I. xi. 124 Mrs. Barnett.. opened a Cottage Home in which they spend three months before going into service. **1921** B. ELLIS *Gloves* 72 For generations now the sewing of gloves has been conducted largely as a cottage industry. **1955** *Atlantic Monthly* Jan. 17/1 Cottage industry—especially wood-carving, soap-making, and iron and brass foundry—has engaged the special attention of the local villagers. **1955** *Times* 3 Sept. 5/2 Mr. Nanda in his report to Congress does not believe that cottage industries can either reduce unemployment or satisfy normal consumer demands. **1964** A. ADBURGHAM *Shops & Shopping* i. 11 Hosiery continued to be a cottage industry longer than any other outworkers' trade except gloves. **1845** G. J. MOUNTAIN *Jrnl.* 200 At the request of an individual who opened his house for the purpose, one..of the Missionaries went down..once a fortnight to hold a Cottage Lecture, which was usually attended by about thirty of the inhabitants. **1858** GEO. ELIOT *Janet's Repentance* xi, in *Scenes of C.L.* 200 Three sermons on Sunday, a night-school for young men on Tuesday, a cottage-lecture on Thursday. **1832** J. CONSTABLE *Lett.* (1965) III. 65, I send you a cottage loaf for Sunday. **1841** DICKENS *Barn. Rudge* lxxx. 403 Cottage loaves, and rolls of bread. **1899** *Daily News* 14 Mar. 8/7 Where the waist should be in the cottage-loaf type of figure. **1960** *Harper's Bazaar* Aug., Cottage loaf hat. **1900** C. C. MUNN *Uncle Terry* 12 A small cottage organ graced the platform, upon which an antique desk did duty as pulpit. **1917** D. G. PHILLIPS *S. Lenox* I. ix. 151 The center table, the two dingy chromos, and a battered cottage organ. **1910** *Encycl. Brit.* IV. 218/1 Samuel Mearne..was royal binder to Charles II., and invented the cottage style of decoration, a style which has lasted till the present day... An inner rectangle is run parallel to the edges of the book, and the upper and lower lines are broken outwards into the outline of a gable roof. **1876** C. M. YONGE *Womankind* xxxii. 293 We all of us know of people in all ranks, cottage-women especially, who are..looked up to, trusted, called in.

cottagey (kǫ·tėdʒi), *a.* Also **cottagy.** [f. COTTAGE + -Y[1].] Resembling or characteristic of a cottage.

1883 W. HARCOURT in A. G. Gardiner *Life* (1923) I. xxi. 492 We must have more of a cottagey building. **1923** A. G. GARDINER *Ibid.*, The house, Queen Anne in style, and 'cottagey' in feeling. **1923** UNA L. SILBERRAD *Lett. Jean Armiter* ii. 48 Perhaps not very cottagey with regard to furniture. **1962** *Guardian* 7 Feb. 8/4 Windsor chairs are rather 'cottagy'.

cotterite (kǫ·tərəit). *Min.* [See quot. 1877 and -ITE[1].] A pearly white laminated variety of quartz.

1877 *Mineral. Mag.* I. p. xx, Professor Harkness exhibited a new variety of crystallized quartz, having pearly lustre, which had been found by Miss Cotter, near Cork, and which he proposed to call Cotterite, after the discoverer. **1962** in *Dana's Syst. Min.* (ed. 7) III. 9.

cottolene (kǫ·tǒlīn). [f. COTTO(N *sb.*[1] + L. *oleum* oil (see -OL) + -ENE.] A substitute for lard made of cotton seed and suet.

1893 in *Funk's Stand. Dict.* **1901** *Daily Chron.* 13 Sept. 7/6 The catalogue said '"Cottolene" has become recognised as the "Hall mark of butter"'. **1924** *Chambers's Jrnl.* 535/2 The various lard mixtures, such as cottolene.

cotton, *sb.*[1] Add: **1. e.** Colloq. phr. (*U.S.*): *to spit cotton*, to spit white, cottony saliva; to be very thirsty; to have a dry mouth; also *fig.*, to be very angry.

1825 J. NEAL *Brother Jonathan* II. xxv. 391 Spit a little cotton or so..and you'll see the fun. *Ibid.* 392 We'll see who spits cotton first. **1851** in T. A. Burke *Polly Peablossom's Wedding* 179 He awoke spitting little wads of cotton, his mouth being as dry and dusty as Broadway in fly-time. **1899** B. W. GREEN *Word-bk. Virginia* 354 When one is very thirsty and his mouth dry the spittle white and sticky he is said to 'spit-cotton'. **1947** *Chicago Daily News* 14 June 6/3 The Kansas City vote frauds..have Attorney General Tom Clark spitting cotton, they believe.

9. a. *cotton-like* adj.

1781 G. WHITE *Selborne* (1789) II. liii. 266 A cotton-like substance, surrounding a multitude of eggs. **1861** *Illustr. London News* 18 May 476/2 The fibre is *cottonized*—i.e., disintegrated, and made to form a cottonlike mass.

b. *cotton-broker* (examples), *factory* (earlier example); *cotton-covered, -wrapped* adjs.

1811 *London P.O. Direct.* 164 Cotton broker. **1851** A. O. HALL *Manhattaner* 31 Yonder is the cotton broker, with the fluctuations of the market for a ten years past all penciled on his face. **1925** P. J. RISDON *Crystal Receivers & Circuits* iii. 36 For the amateur, cotton-covered wire is perhaps the most suitable. **1802** *Act 42 Geo. III* c. 73, An Act for the Preservation of the health and morals of apprentices and others, employed in..cotton and other factories. **1923** D. H. LAWRENCE *Birds, Beasts & Flowers* 174 All the dark-faced, cotton-wrapped people. **1960** *Farmer & Stockbreeder* 23 Feb. 2/2 Cotton-wrapped wire.

10. cotton bat, batting (orig. *U.S.*), a thin sheet or roll of carded cotton; surgical cotton; **cotton belt** *U.S.*, the area in which cotton is grown (see *BELT *sb.*[1] 5 a); **cotton-boll weevil** (see *BOLL *sb.*[1] 6); **cotton-bush,** (*a*) a bush on which cotton grows; (*b*) any of various Australian shrubs (esp. *Kochia aphylla*) having a downy appearance; **cotton candy** *U.S.* (see quot. 1961); **cotton count,** the standard of fineness of cotton (cf. *COUNT *sb.*[1] 2 b); **cotton drill,** a coarse cotton fabric; **cotton-fish** = *milk-fish* (see *MILK *sb.* 10) (the sense in quot. 1851 is uncertain); **cotton flannel** (earlier examples); **cotton gin,** also, a building in which cotton is ginned; **cotton-leaf tea,** an infusion made from the leaves of a species of *Gossypium*, the cotton plant; **cotton-man,** one who trades in cotton; **cotton rep,** a rep (REP[3]) made of cotton.

1847 *Rep. U.S. Comm. Patents 1844* 35 These are nothing more than a thick layer of cotton, carded into what are called at the north 'cotton bats'. **1827** *Hartford* (Conn.) *Mirror* 31 Dec. 1/2 (Advt.), Irish Linens—Cotton batting old price 10 cts lb. **1883** Cotton batting [see BATTING *vbl. sb.* 2]. **1934** H. MILLER *Tropic of Cancer* (1948) 140 I'd like to..go to a hospital..They wash you with cotton batting and they comb your hair for you. **1967** *Everyman's Encycl.* (ed. 5) IV. 121/1 Prepared sheets or rolls of it are used like cotton batting for stuffing and quilting. **1871** R. SOMERS *Southern States* 257 In many other parts of the 'Cotton Belt'. **1897** *Lippincott's Mag.* May 688 The Cotton Belt embracing the heart of the Southern country. **1952** *Manch. Guardian Weekly* 31 July 7 A tenant farmer's son from the cotton belt. **1672** W. HUGHES *Amer. Physitian* 70 Of cotton-bushes there are two sorts. *a* **1693**, **1756** Cotton-bush [see sense 9 a in Dict.]. **1876** J. A. GILES in W. Harcus *South Australia* 201 A fine open hilly district..with plenty of herbage and cotton-bush. **1902** J. H. M. ABBOTT *Tommy Cornstalk* 20 The species of salt-bush known in Australia as 'cotton-bush'. **1941** F. D. DAVISON in *Coast to Coast 1941* 104 He memorized the position of a couple of cotton bushes. **1926** *Springfield* (Mass.) *Union* 4 Mar., How much profit is made on cotton candy sold at fairs. **1961** *Observer* 29 Mar. 29/3 Something called 'cotton candy' (*anglice* 'candy floss') is sold at seashore resorts; it is made by expanding under pressure a tiny amount of sugar into a huge tasteless blob. **1970** *New Yorker* 23 May 106/3 There were stands where you could get..cotton candy, hot dogs. **1904** GOODCHILD & TWENEY *Technol. & Sci. Dict.* 135/1 Cotton counts are based on the number of hanks of 840 yds. in 1 lb. **1899** *Daily News* 16 Oct. 6/6 The drab, cotton-drill khaki. **1851** *De Bow's Rev.* XI. 56 Gaspagon; Cotton Fish; Trout [occur]. **1880** Cotton fish [see *milk-fish*, *MILK *sb.* 10]. **1845** *Mass. Stat.* 21 Mar., Number of yards of cotton flannel manufactured. Value of cotton flannel. **1867** MRS. WHITNEY *L. Goldthwaite* ix, She came back, bringing some two or three pairs of cotton-flannel drawers. **1807** *Deb. Congress* (1852) 460, I went into the cotton gin that was near the house. **1881** *Trans. Obstet. Soc. Lond.* XXII. 31 She said that she had brought down the milk by drinking 'cotton leaf tea'. **1906** *Daily Chron.* 26 Jan. 4/5 The Cottonmen,..with an average protection of about 30 per cent., are asking for more. **1882** CAULFEILD & SAWARD *Dict. Needlework* 89/2 Cotton reps..are chiefly employed for the linings of crétonne curtains. **1957** M. B. PICKEN *Fashion Dict.* 85/1 *Cotton rep*, heavy cotton fabric with ribs, usually horizontal. Used for draperies, children's school clothes, etc.

cotton, *v.*[1] Add: **8.** Colloq. phr. *to cotton on to*: (*a*) to become attached to; to form a liking for; (*b*) to understand; to get to know about; also *to cotton on*, to 'catch on'.

1907 H. LAWSON in Murdoch & Drake-Brockman *Austral. Short Stories* (1951) 85, I s'pose the fact of the matter was that she didn't cotton on to me, and wanted to let me down easy. **1922** D. H. LAWRENCE *England, My England* (1924) 106 Oh..I didn't want to...Didn't want to—didn't cotton on, like. **1929** W. PETT RIDGE *Affect. Regards* 137, I was one of the few that didn't cotton on to it. **1934** *Concise Oxf. Dict.* Add., *Cotton on* (*to*) (slang), understand. **1936** J. TICKELL *See how they Run* i. 4, I don't seem to cotton on to German somehow. **1940** 'N. SHUTE' *Landfall* 198 'How long have they been doing this?' 'God knows. We've only just cottoned on to it.' **1959** *Observer* 8 Nov. 4/6, I can see how to put things over. I cotton on quick.

Cottonian (kǫtōᵘ·niăn), *a.* Pertaining to Sir Robert Bruce *Cotton* (1570–1631) or the collection of books made by him, and deposited in the British Museum in 1753.

[**1696** (*title*) Catalogus Librorum Manuscriptorum Bibliothecæ Cottonianæ.] **1700–1** *Act 12 & 13 Will. III* c. 7, That the said Library should be kept and preserved by the Name of the Cottonian Library for Publick Use and Advantage. **1759** BUTLER *Lives Saints* IV. 512 In her Cottonian life by an allusion to her name she is styled the Fair Wenefride. **1824** ELLIS *Orig. Lett.* 1st Ser. I. p. xx, The Signature of King Richard the Second..occurs once in the Cottonian Library. **1866** 'J. LARWOOD' & HOTTEN *Hist. Signboards* 174 A MS. in the Cottonian Library relating the journey of Margaret of England to Scotland. **1882** *Encycl. Brit.* XIV. 515/1 The Cottonian MSS., acquired by the country in 1700. **1968** *Ibid.* VI. 612/2 Cotton, Sir Robert Bruce.. English antiquary, the founder of the Cottonian library.

Cottonopolis. (Earlier and later examples.)

1851 E. L. BLANCHARD *Diary* 9 Aug. in Scott & Howard *Life E. L. B.* (1891) I. 74 Still in the 'Cottonopolis'. **1937** W. S. CHURCHILL *Great Contemp.* 95 Cottonopolis was fixed in Lancashire.

cotton-picker. *U.S.* Add: **a.** (Examples.)

1850 'M. TENSAS' *Louisiana Swamp Doctor* 35 Behold me, then, a student of medicine, but yesterday a cotton-picker. **1937** COOLIDGE *Texas Cowboys* 15 Eastern Texas, according to the punchers, is given over to cotton and corn, and their favorite term of reproach is to call a man a cotton-picker.

b. (Earlier example.)

1833 *Deb. Congress* 26 Jan. 1363 To save this cost [*i.e.* of picking cotton by hand], a cotton picker has been invented.

cotton-picking. *U.S.* Add def.: The harvesting of cotton; a social gathering for removing the seed from cotton. Also *attrib.* or as *adj.*; spec. in *slang* (orig. *Southern U.S.*) as a general term of disapproval or abuse, = 'damned'.

1795 J. & E. PETTIGREW *MS. Let.* 3 Oct. (D.A.E.), One of the students was banished..for going to a cotton picking after eight at Knight. **1835** J. H. INGRAHAM *South-West* II. 126 Placed in the field to take a spell at cotton-picking. **1845** *Knickerbocker* XXV. 426 A lazy nigger in cotton-pickin'-time. **1945** B. A. BOTKIN *Lay my Burden Down* 92 We didn't have no kind of cotton-pickings 'cept just pick our own cotton. **1958** *Post* (N.Y.) 1 June M3, I don't think it's anybody's cotton-pickin' business what you're doing. **1968** J. PHILIPS *Hot Summer Killing* (1969) III. iii. 161 You have to be a hero or out of your cotton-picking mind. **1970** M. KENYON *100,000 Welcomes* xvii. 142 Damn Mickey McQuaid for ever bringing me to this pixilated, cotton-pickin' country.

cotton-spinner. Add: **2.** *Zool.* A sea-cucumber of the genus *Holothuria*.

1908 E. J. BANFIELD *Confessions of Beachcomber* I. iv. 167 Bêche-de-mer... Termed also trepang, sea cucumber, sea slug, cotton spinner. **1936** RUSSELL & YONGE *Seas* (ed. 2) iii. 65 It [*sc.* the trepang]..has the curious habit, when irritated, of shooting out masses of sticky white threads... Because of this habit, it is also called the cotton-spinner. **1962** D. NICHOLS *Echinoderms* vi. 81 The process [*sc.* extrusion of cuvierian organs] has given rise to the common name 'cotton-spinner' for *Holothuria*.

cotton-tree. Add: **3.** Either of two Australian trees, *Hibiscus tiliaceus* or *Cochlospermum heteronemum*.

1876 J. G. KNIGHT in W. Harcus *South Australia* 178 One or two cotton trees in the grounds of the Government Resident have yielded excellent-looking staple. **1889** J. H. MAIDEN *Useful Native Pl. Australia* 624 'Cotton Tree'. 'Talwalpin' of the aboriginals. **1950** J. W. AUDAS *Native Trees Austral.* (ed. 2) 174 Cochlospermum heteronemum... Its usual name of Cotton Tree is derived from the silky fibre that surrounds the seeds. **1965** *Austral. Encycl.* III. 72/2 The name cotton-tree for *Hibiscus tiliaceus* derives from the useful bark fibre.

cottonwood. Add: **2.** An Australian evergreen shrub, *Bedfordia salicina*, of the family Compositæ, which can grow to the size of a small tree. Also called *blanket-leaf*.

1889 J. H. MAIDEN *Useful Native Pl. Australia* 386 The 'Cotton-wood' of Southern New South Wales. **1965** *Austral. Encycl.* III. 72/2 *Bedfordia salicina* is sometimes referred to as cotton-wood, the allusion being to its large leaves which are woolly-white beneath.

cotton-wool, *sb.* Add: **2.** (Earlier U.S. example.)

1857 *Quinland* I. I. i. 23 He found a large silver cup, which had been carefully packed away in cotton-wool.

b. *fig.* Also *to keep* or *wrap in cotton-wool*.

1889 S. A. BARNETT *Let.* 13 Apr. in H. Barnett *Canon S.A.B.* (1918) II. xxxix. 140 She is still very weak and has to be wrapped in cotton-wool. *c* **1890** [see WRAP *v.* 2 a]. **1955** M. GILBERT *Sky High* v. 71 We couldn't keep him in cotton wool.

3. Add def.: Of, pertaining to, or made from cotton-wool; also, resembling cotton-wool in appearance, etc.

1648 in *Conn. Public Rec.* I. 488 Linnen yearne and Cotton wool yearne. **1907** *Daily Chron.* 10 Oct. 5/1 A cotton-wool mist descended. **1936** *Punch* 4 Mar. 273/1 The manufacture of cotton-wool snow. **1955** A. ROSS *Australia 55* ix. 116 Wisps of faint, cotton-wool cloud.

b. *fig.* Protected as if by cotton-wool; timid, etc.

1909 *Westm. Gaz.* 29 Sept. 4/3 The faction of the Press that would make cotton-wool motorists of us all. **1937** *Daily Express* 3 Mar. 21/2 Cotton-wool batsmen who are afraid of fast bowling. **1953** X. FIELDING *Stronghold* IV. v. 301 A professor protected by the cotton-wool surroundings of a Common Room.

co:tton-woo·lly, *a.* [f. COTTON-WOOL + -Y[1].] Resembling cotton-wool; also *fig.*

1915 J. LONDON *Jacket* vii. 59 Dear cotton-woolly citizen. **1916** 'BOYD CABLE' *Action Front* 207 Cotton-woolly balls of smoke. **1929** J. B. PRIESTLEY *Good Compan.* III. i. 471 Canvases..curiously cotton-woolly in their subjects. **1964** N. FREELING *Double-Barrel* III. iv. 98 The snow was denser still,..more cotton-woolly than ever.

cottony, *a.* Add: **1. b.** cottony cushion-scale = *cushion-scale*.

1886 [see *cushion-scale*, CUSHION *sb.* 11]. **1931** *Discovery* Sept. 299/1 The success of an Australian ladybird in California against the cottony cushion scale. **1952** [see *Australian lady-bird].

cotype (kōᵘ·təip). *Bot.* and *Zool.* [f. CO- + TYPE *sb.*[1]] One of two or more specimens upon which the description of a species is based; an additional type-specimen; = SYNTYPE (which has superseded *cotype*).

1893 O. THOMAS in *Proc. Zool. Soc.* 242 A Co-type is one of two or more specimens together forming the basis of a species, no type having been selected. **1898** *Smithsonian Rep.* 33 A number of valuable types and cotypes of fishes have also been transmitted by the Commission. **1938** *Nature* 3 Dec. 990/1 The collection [*sc.* specimens from a fern herbarium] comprises more than 14,000 specimens with more than 800 types and 800 co-types. **1964** *Internat. Code Zool. Nomencl.* xvi. 79 To avoid misunderstandings, a zoologist should not use the term 'cotype'.

couac (kuæ·k). *Mus.* [Fr.] (See quot. 1876.) Also *transf.*

1876 STAINER & BARRETT *Dict. Mus. Terms* 111/2 *Couac*, an onomatopœic word for the sound made by bad blowing on the clarinet, oboe, or bassoon. The quacking sound, the goose note. **1877** tr. *Offenbach's Amer. & Americans* xxii. 81 My two clarionets made couacs every instant. **1889** G. B. SHAW *London Music in 1888–89* (1937) 250 He is the first operatic vocalist I ever saw leap into popularity by a *couac*, as the French call it. There was one note in his song upon which his voice broke every time with irresistibly comic effect. **1938** A. E. WIER *Macmillan Encycl. Mus.*, *Couac*,.. the 'goose'; a sudden noise produced by the clarinet when the reed gets out of order.

couba (kū·bā). *Austral.* Also **cooba, coobah, coubah.** [Native name.] A variety of acacia, *Acacia salicina*.

1878 'R. BOLDREWOOD' *Ups & Downs* v. 46 A deep reach of the river, shaded by couba trees. **1890** —— *Col. Reformer* III. xxviii. 131 The willowy coubah. **1957** *Forest Trees Austral.* (Commonw. Forestry & Timber Bureau) 158/2 Other trees may include myall (*Acacia pendula*) and cooba or native willow (*Acacia salicina*).

coucal (kū·kăl). [Fr. (F. Le Vaillant *Hist. Nat. des Oiseaux d'Afrique* (1806) V. 53), said to be from *couc*(*ou* cuckoo + *al*(*ouette* lark.] A non-parasitic bird of the cuckoo family, belonging to the genus *Centropus*, which is found in Africa, southern Asia, and Australia, and includes the Indian crow-pheasant and the Australian swamp pheasant.

1815 STEPHENS *Shaw's Gen. Zool.* IX. I. 54 Egyptian Coucal. **1821** T. E. BOWDICH tr. *Ornithology of Cuvier* 53 Coucals, *Vaillant* Centropus..Nail of the great toe long, straight, pointed. **1831** H. M'MURTRIE tr. *Cuvier's Animal Kingd.* I. 333 Centropus..or the Coucals, species of Africa and India, in which the thumb nail is long, straight and pointed, as in Larks. **1873** *Proc. Zool. Soc. 1872* XL. 603 A Pheasant Coucal (*Centropus phasianus*) from Australia being the first example of this fine species received by the Society. **1883** [see *crow-pheasant*, *CROW *sb.*[1] 11]. **1901** A. J. CAMPBELL *Nests & Eggs of Austral. Birds* II. 591 The Coucal is the only member of the Australian Cuculidæ that undertakes the responsibility of rearing its own family. **1935** D. F. THOMSON *Birds of Cape York Peninsula* 52 Pheasant-Coucal... The Coucal is well named, for its long tail lends an extremely pheasant-like appearance to the bird. **1966** N. W. CAYLEY *What Bird is That?* (ed. 4) 271 Swamp Pheasant, *Centropus phasianinus*... Also called Coucal and Swamp Cuckoo.

couch, *sb.*[1] Add: **3. b.** *spec.* A couch upon which a patient reclines when undergoing psychoanalysis or psychiatric treatment.

1952 N. MAILER *Advts. for Myself* (1961) 163 It takes more than a couch to solve the problems of..narcotics, juvenile delinquency, psychosis. **1961** *Amer. Speech* XXXVI. 147 *On the couch*, undergoing psychiatric treatment. From the traditional leather couch of the psychoanalyst. **1963** A. HERON et al. *Towards Quaker View of Sex* 62 It seems reasonable to accept in general the psychoanalytic approach..but to predict that its most useful application may well prove to be preventive rather than therapeutic; in the home and school rather than on the couch. **1967** *Listener* 23 Feb. 268/2 He admirably resists the temptation to put Hawthorne on the analyst's couch.

‖ **couchette** (kūʃet). [F. *couchette*, lit. 'little bed', f. *couche* COUCH *sb.*[1]] A Continental railway carriage in which the seats are convertible into sleeping-berths; such a berth. Also *attrib.*

1920 *Bradshaw's Continental Guide* July–Sept. p. iv (Advt.), Paris (Lyon)..Marseilles, 1st class Couchettes. **1928** *Observer* 11 Mar. 10/4 The adaptation of existing third-class corridor coaches so as to provide sleeping accommodation something on the lines of the French 'couchette'. **1930** E. WAUGH *Labels* 29, I booked a *couchette*, which, I am told, is the way Frenchmen always travel. It is a carriage for four, with two berths which open from the wall above the seats. You can lie full length on these, covered with rugs. **1950** T. S. ELIOT *Cocktail Party* III. 142, I had to travel in a very slow train And in a *couchette*. **1958** *Spectator* 13 June 772/1 Why on earth, then, has the continental couchette system not been exploited here? By this means, sleepers can be converted into seats in a few seconds. **1969** *Daily Tel.* 6 Oct. 9/8 You can choose..couchettes (six to a compartment).

coucourite, var. *COCORITE.

coudé (kūde), *a.* [Fr., pa. pple. of *couder* to bend at right angles, f. *coude* elbow, f. L. *cubitum* CUBIT.] Of, pertaining to, or designating a telescope or part of a telescope in which rays are bent to focus at a fixed point away from the axis of the telescope. Also as *sb.*

1888 *Encycl. Brit.* XXIII. 153/1 The present writer tested the equatorial coudé on double stars at the Paris observatory in 1884. **1922** *Nature* 16 Mar. 362/2 A coudée telescope designed for the application of the method of equal heights. **1946** *Ibid.* 23 Nov. 764/2 The stellar spectra obtained at the coudé focus of the 100-in. telescope. **1951** *Engineering* 9 Mar. 298/1 Design to permit coudé spectroscopy. **1959** SPITZ & GAYNOR *Dict. Astr.* 106 *Coudé focus*, an arrangement, used only on the largest reflecting telescopes, by which the focal length is greatly increased, and making it possible to have the spectrograph or the camera not attached to the telescope but in a separate room.

Couéism (kū·e,iz'm). [f. the name of Émile *Coué* (1857–1926), French psychologist + -ISM.] A system of psychotherapy by auto-suggestion, usually of an optimistic nature. So **Cou·é** *sb.*, Couéism; **Cou·é** *v. trans.*, to produce or bring into a certain state by Couéism; **Cou·éist,** one who practises Couéism, a follower of Coué; **Couéi·stic** *a.*, pertaining to Couéism.

1922 *Punch* 21 June 501 An echo of the heat-wave. *Ardent Couéist.* 'Quite cool, quite cool. Cool as blazes, cool as blazes.' **1923** G. MAYO (*title*) Coué for children. **1923** *Daily Mail* 30 Apr. 7 Couéism is auto-suggestion. It is the power of making people heal themselves through their subconscious self. **1924** GALSWORTHY *White Monkey* III. x. 285 Fleur's wonderfully calm; she does Coué night and morning. **1926** *Public Opinion* 9 July 30/3 Large numbers of sufferers ..were cured by Couéism. **1928** *Daily Express* 27 July 10/2 He has Couéd himself into a yearly fit of depression and lassitude. **1928** *Sunday Dispatch* 2 Sept. 17/1 An imaginary kink. A kink that is 'Coué'd'. *Ibid.* 23 Dec. 9/2 'We have had the course translated into seven languages,' said one of the Couéists. **1937** KOESTLER *Spanish Test.* II. v. 282 Following the Coué system, I scratched on the wall: 'Every day, in every way, things are getting better and better.' **1950** L. FISCHER in Koestler et al. *God that Failed* 208 A domestic version of 'Socialist realism' intended..to serve as a kind of Couéistic propaganda. **1959** C. BUSH *Case of Careless Thief* viii. 97 We seemed to be at a dead end, and you can't Coué yourself out of an impasse like that.

cough, *sb.* Add: **2. b.** The sound of a bullet or shell being fired or bursting. *colloq.*

1928 BLUNDEN *Undertones of War* ii. 20 The..cough of anti-aircraft shells. **1934** V. M. YEATES *Winged Victory* I. ii. 20 A loud double cough made his heart jump. It was Archie taking notice of them, and his first shots were always startling. **1969** *Listener* 12 June 813/3 The cough of mortars (everyone calls it a 'cough') and the clatter of sub-machine-guns.

3. cough-drop, (*b*) *slang*, a pungent or disagreeable person or thing; a 'caution' (see CAUTION *sb.* 3 d); **cough medicine, mixture,** a medicinal concoction for the alleviation of a cough.

1895 *Referee* 21 July 7/3 'Honest John Burns',..who would have benefited considerably had his constituents given him a holiday, objects to being called 'a cough drop'. *c* **1895** *Comic Song*, She looks as if she could curl your hair, Oh you've got a nice little cough drop there. **1897** W. S. MAUGHAM *Liza of Lambeth* iv. 46 You are a coughdrop—.. Ketch me refusin' when I 'ave the chanst. **1908** *Daily Chron.* 3 June 1/6 Strike me lucky, it was a corf-drop, it was, an' no mistike! *Ibid.* 10 Aug. 7/4 Didn't I tell you, sir, that she was a cough-drop? **1929** J. B. PRIESTLEY *Good Compan.* II. i. 267 'Yond's a coughdrop,' he announced to the room at large. **1828** *Bk. of Health* 105/1 Major Coccrane's Cough Medicine. **1844** DICKENS *Mart. Chuz.* ii. 586 He had been surprised to find his fresh supply of cough medicine in such a place. **1928** E. O'NEILL *Strange Interlude* II. 61 Cough medicine for the corpse, perhaps! **1840** C. J. B. WILLIAMS in A. Tweedie *Syst. Pract. Med.* III. 70 The frequent use of a cough mixture, to diminish irritation. **1922** JOYCE *Ulysses* 421 He's got a coughmixture with a punch in it. **1960** D. V. DAVIS *Domestic Encycl.* 387 (*heading*) Cough mixture. A pleasant cough syrup, quite safe for children, can be made by mixing 5 oz. of honey with 4 oz. of treacle and ½ pint of vinegar.

cough, *v.* Add: **1. b.** To confess; to give information (cf. sense 3 b). *slang* (orig. *U.S.*).

1901 'J. FLYNT' *World of Graft* 102 They put him in the sweat-box, made him cough, an' you know the rest. **1962** M. PROCTER *Devil in Moonlight* vi. 56 Would you cough with five thousand in the kitty, and nobody able to prove where you'd got it from? **1970** W. J. BURLEY *To Kill a Cat* x. 175 Once he realized we had it on him he was ready to cough fast enough.

c. To fire shells. Cf. *COUGH *sb.* 2 b.

1925 FRASER & GIBBONS *Soldier & Sailor Words*, *Coughing Clara*, a nickname for a heavy gun. **1934** V. M. YEATES *Winged Victory* I. ii. 21 It was not easy..to keep formation when Archie was coughing with black hate all the while.

d. Of an engine, etc.: to make a noise like coughing; to misfire.

1884 'MARK TWAIN' *Huck. Finn* 178 A steamboat coughing along upstream. **1936** E. S. GARDNER *Case of Stuttering Bishop* (1937) vii. 111 A donkey engine coughed into rhythmic explosions. **1949** J. R. COLE *It was so Late* 87 Then the engine coughed and picked up sharply as the throttle was slammed open. **1954** D. & C. CHRISTIE *Carrington V.C.* in *Plays of Yr.* 563 Well, going up a hill like the side of a house, the car started to cough, then it stopped.

2. b. *trans.* To bring into a specified condition by coughing.

1837 E. HOWARD *Old Commodore* I. ix. 293, I think I am the first naval officer who ever coughed himself into a commission. **1888** MRS. H. WARD *R. Elsmere* xxi, One of them ..was sitting up..and coughing its little life away. **1904** L. T. MEADE *Love Triumphant* Prol. ii, He kept..coughing as though he would cough his life away.

3. b. Revived (with *up*) in mod. slang use. Also *intr.* orig. *U.S.*

1896 ADE *Artie* xi. 95 And I cough up to you because I know that you're a good fellow. **1904** W. H. SMITH *Promoters* iv. 84 Don't you ever fear that I coughed up anything that would put him on to the true inwardness of what we're working now. **1906** E. DYSON *Fact'ry 'Ands* x. 133 'Now,' he said, 'cough it up. Why'd yeh tear them dresses?' **1931** A. CHRISTIE *Sittaford Myst.* xxx. 239 'For Heaven's sake, cough it up, Emily,' he said. 'I want to get to the telegraph office. Every moment's vital.'

d. With *up*. To bring out, present, hand over; *esp.* to pay up (money). Also *absol.* *slang* (orig. *U.S.*).

1894 *San Francisco Midwinter Appeal* 27 Jan. 2/4 Cough up a nickel, read the paper, and get the latest of camp doings. **1904** W. H. SMITH *Promoters* vii. 122 I'll cough up the stock and bonds all right. **1909** 'O. HENRY' *Roads of Destiny* xix. 324 Everybody cough up what matches he's got. **1920** G. MOORE *Esther Waters* xiv. 119 Now, then, old girl, cough up! I must have a few halfpence. **1923** WODEHOUSE *Inimit. Jeeves* iv. 46 Thanks to Jeeves I was not going to be called on to cough up several thousand quid. **1929** —— *Summer Lightning* vii. 148 Will you or will you not cough up that pig? **1959** 'A. GILBERT' *Death takes a Wife* xv. 195 He coughed up. Don't ask me why.

could. Add: *spec.* in ellipt. phr. *could be*: it could be (that); it is possible; your suggestion may be correct.

1938 M. K. RAWLINGS *Yearling* vii. 68 Yes, Lem. Could be I was wrong. *Ibid.* viii. 71 If you'll fill my wood-box this evenin', could be we'd eat liver tonight. **1951** J. B. PRIESTLEY *Fest. Farbridge* I. ii. 83 'Something in that, isn't there?' ..'Could be.' **1960** L. MEYNELL *Bandaberry* iii. 57 'Before you realised it you would have signed away..the rights...' 'Could be,' I agreed.

coulda (ku·dă), colloq. shortening of (*I*) *could have*.

1925 DREISER *Amer. Tragedy* I. I. xii. 83, I coulda chucked my job. **1952** B. MALAMUD *The Natural* 91 My niece said you coulda wanted it to happen. **1966** 'G. DOUGLAS' *Odd Woman Out* ii. 16, I only said I thought as I did. Coulda bin dreaming. **1966** J. WAINWRIGHT *Crystallised Carbon Pig* xi. 54, I coulda tamed him this afternoon.

couldn't care less: see *CARE *v.* 4 a.

couleur. Add: **2.** In *Rouge et Noir*: see quots.

1867 *Bohn's Hand-bk. Games* 343 The punters place on the Rouge, the Noir, the Couleur, or the Inverse, the sum they wish to risk. **1909** *Hoyle's Games Modernized* 465 The other chance, the *Couleur* and *Inverse*, is decided by the colour of the first card turned up. If the colour of this card corresponds with the colour of the winning row, then *Couleur* wins; if it is of the opposite colour, then *Inverse* wins. **1957** *Encycl. Brit.* XX. 456/1 A bet on *couleur* wins if the first card dealt is of the colour designating the winning row.

‖ **coulibiac** (kulībyæ·k). Also **koulibiac.** [ad. Russ. *kulebyáka*.] A Russian pie of fish or meat, cabbage, etc.

1898 C. H. SENN *Culinary Encycl.* 33 *Coulibiac*, name of a Russian dish—a kind of fish-cake mixture wrapped up in Brioche paste and baked. **1961** *Daily Herald* 22 Nov. 4/4 Charles Clore would offer you coulibiac, a delicate dish of finely seasoned salmon. **1970** SIMON & HOWE *Dict. Gastron.* 237/1 *Koulibiac*, a Russian type of pie.

coulisse. Add: **3.** The body of outside dealers on the Paris Stock Exchange; similar dealers in other stock exchanges in provincial France and elsewhere; also, the place where they deal. Also *attrib.* Hence **couli·ssier,** one who deals on the coulisse.

1887 *Encycl. Brit.* XXII. 557/2 The Bourse now consists of two distinct bodies, known as the *parquet* and the *coulisse*. ..The *coulisse* are the outside dealers or brokers, but..these comprise firms of solid standing, bankers, and arbitrage houses. **1910** *Ibid.* VII. 308/1 The name *coulissier* was thus given to unauthorized *agents de change*, or 'outside brokers' who..were finally given a recognized status in 1901. *Ibid.* XXV. 933/2 A portion of the bourse may be railed off to which only *agents de change* have the right of entry, the rest of the bourse being known as the *coulisse*. **1911** F. W. HIRST *Stock Exchange* iii. 89 Such securities..often are dealt in on the coulisse, which may be described as the curb market, so called because in former times the bankers and coulissiers thronged a narrow passage called *La Coulisse*. **1930** *Economist* 31 May 1217/2 Similarly, the new scale of Coulisse brokers' commission rates..is regarded. *Ibid.* 7 June 1280/1 All this..is being repeated..in the 'coulisses' of the Bourse [in Turkey]. **1965** PERRY & RYDER *Thomson's Dict. Banking* (ed. 11) 175/2 The unofficial dealers from the Coulisse.

4. A corridor; also *fig.*, as a place 'behind the scenes' (cf. *CORRIDOR 4).

1903 A. BENNETT *Truth about Author* xv. 195 His knowledge of the *coulisses* of the various arts is wonderful. **1949** I. DEUTSCHER *Stalin* v. 145 Stalin continued to do his job in the twilight of the *coulisse*. **1958** *Spectator* 31 Jan. 138/2 Gathering remunerative rumours in ministerial *coulisses*. *Ibid.* 14 Feb. 207/1 Foreign politics is reduced here to the tittle-tattle of the *coulisses*. **1968** D. TORR *Treason Line* 40 The real business of the conference would begin..in the *coulisses*.

coulomb. Now usu. pronounced (kū·lōᵘm, -ǫm). Add to def.: Also known as *international coulomb* when defined in terms of the international ampere; in 1948 this was replaced by the *absolute coulomb*, defined in terms of the absolute ampere and now incorporated in the International System of Units. (Further examples.)

1893 *Trans. Amer. Inst. Electr. Engin.* X. 469 As the Unit of Quantity, the International Coulomb, which is the quantity of electricity transferred by a current of one international ampere in one second. **1911** *Encycl. Brit.* XXVII. 742/2 The *International Coulomb*..is the quantity of electricity transferred by a current of one international ampere in one second. **1938** *Hackh's Chem. Dict.* (ed. 2) 256/1 *Coulomb*..*international* ~. 1 Int. coul. = 0·9999 absolute coul. **1959** *Chambers's Encycl.* XIV. 188/1 The common practice is to introduce the absolute coulomb as the fourth fundamental unit [of absolute practical units]. **1963** JERRARD & MCNEILL *Dict. Sci. Units* 35, 1 international coulomb = 0·99985 absolute coulomb.

2. *attrib.* and *Comb.*, as **Coulomb barrier**, a repulsive electrostatic force constituted by the Coulomb field of a nucleus; **Coulomb energy**, the energy of a charged particle due to the electrostatic field of other charged particles, esp. when it constitutes part of its binding energy; **Coulomb field**, an electrostatic field that obeys the inverse-square law; **Coulomb force**, the electrostatic force of attraction or repulsion exerted by one charged particle upon another; **Coulomb interaction**, an interaction between two particles because of the electric charge that each has; **Coulomb('s) law**, either of two laws in electrostatics: (i) (see quot. 1908); (ii) a law that like charges repel, and opposite charges attract each other, the force varying inversely as the squares of the distances between them, and proportionally to the product of the magnitudes of the charges; **Coulomb repulsion**, the repulsion between charges occasioned by Coulomb forces; **Coulomb scattering** (see quot.); **coulomb-volt** = JOULE.

1949 FRIEDLANDER & KENNEDY *Introd. Radiochem.* iii. 72 Charged particles which have to overcome Coulomb barriers. **1958** MANSFIELD *Elem. Nuclear Physics* ii. 14 Protons, because they are charged, have an electrostatic (coulomb) barrier to overcome before they can escape from the nucleus. **1940** GLASSTONE *Physical Chem.* i. 101, Q is called the Coulomb energy, due to the electrostatic attractions and repulsions between electrons and protons. **1937** J. W. T. SPINKS tr. *Herzberg's Atomic Spectra* i. 46 A point charge in a pure Coulomb field of force. **1970** *Physics Bull.* Apr. 149/2 In the atom the motion of the electrons is dominated by the Coulomb field of the nucleus. **1930** *Engineering* 9 May 595/3 The electrons and ions reacted on one another with the usual Coulomb forces. **1936** W. HEITLER *Quantum Theory Radiation* iii. 98 For the collision of an electron with a nucleus of charge Z, the Coulomb interaction is proportional to e^2Z. **1970** G. K. WOODGATE *Elem. Atomic Struct.* ii. 9 This Coulomb interaction is of a special form with respect to its radial dependence ($V(r) \propto r^{-k}$ where $k = 1$). **1854** W. THOMSON in *Phil. Mag.* VIII. 48 Coulomb's law. **1908** J. H. JEANS *Math. Theory Electr. & Magn.* ii. 45 Coulomb's Law. If *R* is the outward intensity [of the electric field] at a point just outside a conductor, then $R = 4\pi\sigma$, where σ is the surface density of electrification on the conductor. **1938** L. B. LOEB *Atomic Structure* i. 30 There thus seems to be a general limit at which the Coulomb law fails. **1947** M. D. KAMEN *Radioactive Tracers* i. 7 This attractive force far outweighs the Coulomb repulsion. **1955** *Gloss. Terms Radiology* (*B.S.I.*) 16 *Coulomb scattering*, the scattering of a particle by the coulomb field of a nucleus. **1893** *Funk's Stand. Dict.*, Coulomb-volt. **1963** B. FOZARD *Instrum. Nucl. Reactors* i. 5 The electron volt (eV)..is related to the coulomb-volt or joule or watt-second by the factor $1{\cdot}602 \times 10^{19}$ given above.

coulombian (kulōᵘ·miăn), *a.* Also **Coulombian.** [f. prec. + -IAN.] = next.

1923 E. N. DA C. ANDRADE *Struct. Atom* xii. 233 Equilibrium of stationary electrons is impossible round a centre of Coulombian force. **1928** F. K. RICHTMYER *Introd. Mod. Physics* xi. 438 The hydrogen nucleus is surrounded by an inverse-square field—sometimes called a 'coulombian field'. **1934** H. L. BROSE tr. *Sommerfeld's Atomic Struct. & Spectral Lines* (ed. 3) I. vii. 360 Here the pure Coulombian field with the nuclear charge *Ze* no longer reigns.

coulombic (kulōᵘ·mik), *a.* Also **Coulombic.** [f. as prec. + -IC.] Of or pertaining to C. A. de Coulomb or the forces, laws, etc., discovered by him or named after him.

1936 *Jrnl. Amer. Chem. Soc.* LVIII. 1. 1061 The energies of nuclear reactions can be calculated as the sum of changes in the energies of the nuclear bonds and the coulombic fields. **1940** GLASSTONE *Physical Chem.* xiii. 1077, 14 per cent of the binding energy between pairs of atoms is Coulombic in nature. **1964** G. R. CHOPPIN *Nuclei & Radioactivity* i. 11 Physicists speak of the repulsive force between the nucleus and the approaching positively charged particle as the nuclear coulombic barrier.

coulo·meter. [refashioning of COULOMB-METER.] = COULOMB-METER; VOLTAMETER.

1902 RICHARDS & HEIMROD in *Proc. Amer. Acad. Arts & Sci.* XXXVII. 415 The word 'voltameter' was devised before electrical dimensions were understood. It is moreover too much like the universally used..word 'voltmeter'. Now that the former instrument is placed upon a firm basis of accuracy, it may appropriately receive also an accurate name; and it is hoped that the new word 'coulometer' may replace wholly the anachronism. **1925** M. DE K. THOMPSON *Theor. & Appl. Electrochem.* 1. i. 6 In any coulometer the current can be computed from the quantity of electricity that has passed through the circuit, if the current has been constant and if the time is measured. **1943** *Gloss. Terms Electr. Engin.* (*B.S.I.*) 73 *Coulometer*, an electrolytic cell serving to measure a quantity of electricity by the amount of a substance which is liberated electro-chemically: measured either by weight as in a weight coulometer, or by volume as in a volume coulometer.

coulo·metry. [f. prec.: see -METRY.] The measurement of the number of coulombs used in an electrolysis; chemical analysis by means of this. So **cou:lome·tric** *a.*

1945 *Jrnl. Amer. Chem. Soc.* LXVII. 1916 (*heading*) Coulometric analysis. *Ibid.* 1917/1 The great advantage of the mercury cathode for coulometric analyses. *Ibid.* 1920/1 Coulometric determination of copper. **1949** *Analyt. Chem.* XXI. 73/2 It is apparent..that coulometric methods lend themselves to instrumentation and automatic operation. **1958** G. H. AYRES *Quant. Chem. Analysis* xxxix. 555 A relatively new technique in quantitative analysis is coulometry, the generation of reagents by electrolysis. **1964** *Oceanogr. & Marine Biol.* II. 104 Coulometric and titrimetric analysis differ essentially in the means by which the reagent (silver) is incorporated. *Ibid.*, Coulometry may also be practised using a visual indicator.

cou·lsonite. *Min.* [f. the name of A. L. *Coulson*, geologist: see -ITE[1].] A vanadian magnetite, found originally in India.

1937 DUNN & DEY in *Trans. Mining & Geol. Inst. India* XXXI. 131 *Coulsonite*.—This mineral was originally described as vanadomagnetite, but..the introduction of a distinctive name is advisable. The mineral has been named after our colleague, Dr. A. L. Coulson. **1944** C. PALACHE et al. *Dana's Syst. Min.* (ed. 7) I. vii. 703 A vanadian magnetite (*coulsonite*) occurs at Bihar, India, as veins and lenses in gabbro. **1962** *Amer. Mineralogist* XLVII. 1290 Coulsonite is a naturally occurring iron-vanadium spinel group mineral with a chemical composition which can be written FeV_2O_4.

coumarone (kū·mărōᵘn). *Chem.* Also **cumarone.** [ad. G. *cumaron* (Fittig and Ebert 1883, in *Ann. Chem. Pharm.* CCXVI. 168), f. COUMAR(IN + -ONE.] **a.** A colourless liquid, C_8H_6O, which may be distilled from coal-tar or synthesized, and which is structurally a benzene ring fused to a furan ring; benzofuran; also, any of the derivatives of this compound.

1883 *Jrnl. Chem. Soc.* XLIV. 474 Coumarone is a colourless oil, heavier than water. **1910** *Encycl. Brit.* VII. 309/1 Coumarone, C_8H_6O,..is a colourless liquid which boils at 171–172° C. and is readily volatile in steam, but is insoluble in water and in potash solution. **1957** E. H. RODD *Chem. Carbon Compounds* IV. iii. 169 Coumarones are obtained from o-hydroxy-ω-halogenoacetophenones. **1963** F. M. DEAN *Naturally Occurring Oxygen Ring Compounds* v. 135 The older name for benzofuran, coumarone, is nearly obsolete.

b. coumarone resin, any of various thermoplastic resins produced by polymerization of coumarone, usually in mixtures with indene (so *coumarone-indene resin*), and used esp. in the rubber and varnish industries.

1900 *Jrnl. Chem. Soc.* LXXVIII. 1. 656 Indene, and Coumarone Resins. **1935** C. ELLIS *Chem. Synth. Resins* I. vii. 123 Coumarone resin is employed for a variety of purposes, particularly in the manufacture of varnishes, printing inks, waterproofing materials, rubber compositions, floor tiles and chewing gum. **1947** R. L. WAKEMAN *Chem. Commercial Plastics* xiii. 287 Patents relating to coumarone-indene resins. **1951** H. M. LANGTON *Synthetic Resins* (ed. 3) vii. 272 Coumarone resins are graded mainly by colour and melting point. **1966** J. A. BRYDSON *Plastics Materials* xiv. 284 Coumarone resins are used considerably in the paint and varnish industry.

council. Add: **17. council (housing) estate**, a group of houses erected by a council (sense 10); **council flat**, a flat (FLAT *sb.*[2] 2) belonging to a council; **council-room** (earlier example); **council school**, a school supported by a town or county council.

1939 'G. ORWELL' *Coming up for Air* IV. i. 219 Several acres of bright red rooves all exactly alike. A big Council housing estate, by the look of it. **1958** *Times* 11 Aug. 2/6 The scene..is a post-war council estate. **1941** H. BRIGHOUSE *Hallowed Ground* in J. Marriott *Best One-Act Plays 1941* 46, I wouldn't live in a Council flat if you gave it me rent-free. **1958** *Spectator* 6 June 747/1 The Health Visitor finds them a council flat. **1970** T. LEWIS *Jack's Return Home* 148 We were in the middle of a dozen blocks of tall council flats. They looked greyer than the day. **1774** in *Hist. MSS. Commission, 14th Rep.* App.x: *MSS. Dartmouth* (1895) II. 244 Survey of the fireplace and chimneypiece of the Council room at Fort George, New York. **1908** E. M. SNEYD-KYNNERSLEY *H.M.I.* xxviii. 331 Other schools, both Board (or Council) schools [etc.]. **1911** *What an Education Committee can Do* (Fabian Tract No. 156) 3 Some London council schools provide for five hundred and sixty pupils. **1925** W. DEEPING *Sorrell & Son* 1. i, The boy had had to go to a Council school. **1941** J. S. HUXLEY *Uniqueness of Man* ii. 46 The boys from a secondary school were on the average no less than 6½ inches..taller than those from a council school in a poor neighbourhood.

council-house. Add: **c.** A house used by North American Indians when in council. *U.S.*

1751 J. BARTRAM *Observ. Pensilvania to Ontario* 50 We alighted at the council house, where the chiefs were already assembled to receive us. **1823** E. JAMES *Exped. Rocky Mts.* I. 112 It [sc. the lodge] serves as a council-house for the nation. *a* **1927** F. M. CANTON *Frontier Trails* (1930) 64 Sharp Nose wanted to have a talk with me in the council house at the agency. **1967** W. N. FENTON in E. Tooker *Iroquois Culture* 12 The Council House of the Seneca Nation.

2. A dwelling-house erected by a council (COUNCIL 10).

1923 *Oxf. Times* 14 Sept. 16/5 When the time came for plumbers to get on with the Council houses. **1929** *Times* 29 Jan. 8/1 The Reigate Rural District Council are now insisting on building a street of council houses along one of the most beautiful lanes in Surrey. **1955** *Times* 13 May 15/4 The politics of the council house tenant, the artisan, the trade unionist. **1970** *New Statesman* 23 Oct. 512/1 Tory ministers were permitted various peripheral initiatives, on..council house sales, and the like.

council-man. Delete *Obs.* exc. *Hist.* and add later examples.

1911 *Oklahoma Session Laws* 3rd Legisl. 131 The mayor and councilmen of the city of Norman, Oklahoma, are hereby empowered. **1970** *New Yorker* 27 June 44/3 The futile debate which councilmen and the mayor engaged in.

councilmanic, *a.* (Earlier and later examples.)

1861 *N.Y. Tribune* Nov. (Bartlett), Fifth Councilmanic District. Delegates nominated. **1904** *Philad. Press* 1 June 4 Upon arriving at St. Louis the councilmanic party from this city will be taken to luncheon. **1911** H. S. HARRISON *Queed* xvi. 197 He..read history..councilmanic proceedings. **1930** H. ZINK *City Bosses in U.S.* xii. 242 The Municipal League elected..ten of the councilmanic candidates it had endorsed.

counsel, *sb.* Add: **2. b.** *counsel of perfection* (earlier and later examples); also *transf.*

a **1678** SCOUGAL *Life of God in Soul of Man* (1726) 169 We must not look upon this..as a counsel of perfection, highly commendable, but not absolutely necessary to salvation. **1909** A. BENNETT *Literary Taste* 87 Every Englishman who is interested in any branch of his native literature, and who respects himself, ought to own a comprehensive and inclusive library of English literature, in comely and adequate editions. You may suppose that this counsel is a counsel of perfection. It is not. **1938** W. S. MAUGHAM *Summing Up* xlviii. 186 Every production of an artist should be the expression of an adventure of his soul. This is a counsel of perfection.

c. Phr. *counsel of despair*.

1910 G. B. SHAW *Let. to G. Barker* Jan. (1956) 162 It is only the 4th Jan, too soon for counsels of despair. **1929** R. BRIDGES *Test. Beauty* II. 229 All Utopias Are castles in the air or counsels of despair.

counselling, -eling, *vbl. sb.* Add: *spec.* The giving of advice on personal, social, psychological, etc., problems as an occupation. Also *attrib.*

1945 S. E. GOLDSTEIN *Marriage & Family Counseling* 421 When marriage counseling becomes well-established and generally accepted as a profession, it..will be used as, for example, that of the lawyers. **1959** *Listener* 2 Apr. 599/1 There is no doubt that some students need counselling or even psychological treatment in these testing years. **1960** L. PINCUS *Marriage* III. 222 The central themes of counselling work with all clients.

counsellor. Add: **1. b.** One who specializes in counselling (see prec.).

1946 *Britannica Bk. of Yr.* 461/1 Headway was being made in establishing and advancing the professional standards of the counselling specialist largely as a result of the influence of the American Association of Marriage Counselors. **1963** A. HERON *Towards Quaker View of Sex* v. 44 Emotional reactions from the Counsellor, arising mostly from origins of which he or she is not aware, are unhelpful. **1970** *Times* 7 Oct. 12 The idea of counselling in schools is not altogether new...What is new is establishment of the schools counsellor as a professional.

5. A senior officer in the diplomatic service, especially in the larger British missions to foreign and Commonwealth countries; also a senior officer in the home departments of the Foreign and Commonwealth Office.

The form *councillor* is used in the Foreign Office List of 1905. See also quot. 1889 s.v. COUNCILLOR a.

1914 *Foreign Office List* 160 Counsellor of Embassy, or.. First Secretary at posts where the staff does not include a Counsellor. **1923** *Ibid.* 1 Establishment in London..Assistant Under Secretaries of State..Counsellors. *Ibid.*, Finance Officer with rank of Counsellor. *Ibid.* 113/1 Naval, Military and Air Attachés at His Majesty's Embassies..have Place and Precedence next in succession after the Diplomatic Counsellor, but before the Commercial Counsellor.

1959 *Listener* 2 Apr. 580/2 The embassies are organized in conventional patterns..with the usual apparatus of counsellors, secretaries, attachés and service attachés. **1968** 'J. LE CARRÉ' *Small Town in Germany* ii. 23, I must accordingly ask you that until further notice all British Embassy staff below the rank of Counsellor be confined to the area of Bonn.

count, *sb.*[1] Add: **1. c.** *Boxing.* The counting aloud by the referee of ten seconds, the limit of time allowed to a fallen boxer to rise and resume the contest, or accept defeat; also, a specified period of less than ten seconds before a boxer rises to resume the contest. Esp. in *to take the (full) count*, to be knocked down for such a period; to be defeated; *out for the count*, unable to rise from the canvas for at least ten seconds; defeated. Also *transf.* and *fig.*

1902 H. L. WILSON *Spenders* 86 It rattled me so I had to take the full count. **1913** *Chums* 15 Mar. 498/2 The count was being shouted...Roy leaped to the centre of the ring before the count was finished. *Ibid.* 24 May 667/2 'I nearly took the count that time, old fellow,' he said. **1922** R. PARRISH *Case & Girl* 322 West went down for the count, lying motionless on the floor. **1929** *Evening News* 18 Nov. 16/5 The Walthamstow boxer was floored for a long count, and his seconds wisely threw in his towel. **1930** F. YEATS-BROWN *Bengal Lancer* xii. 173 Now that Nur-ud-din is within punching distance, he'll put him out for the count. **1932** KIPLING *Limits & Renewals* 142 Was my Demon going to lay the hot coal of inspiration on Lettcombe's unshorn lips—not on mine? But I would allow him the count fairly, and I began, 'One—Two—Three.' **1933** WODEHOUSE *Mulliner Nights* vii. 242 There are some speeches before which dignity melts like ice in August, resentment takes the full count. **1947** D. DAVIN *Gorse blooms Pale* 204 With the Jerries rocking on their heels the way they were the odds were they'd have taken the count before he got back. **1953** A. BARON *Human Kind* xxiv. 179 We won't be safe till we've put them out for the count. **1965** M. GOLESWORTHY *Encycl. Boxing* (ed. 3) 62/1 After World War II.. the audible count was adopted.

2. b. Also, the standard of fineness of yarn.

1877 [in Dict.]. **1927** T. WOODHOUSE *Artificial Silk* 35 The threads have to conform to well-established methods and rules of distinguishing the relation between the length and weight of different yarns, or, briefly, to spin to fixed 'counts' or 'numbers' termed deniers. **1934** *Planning* I. xix. 4 Different counts and types of yarn. **1963** A. J. HALL *Textile Sci.* iii. 134 In the case of real silk and man-made fibre yarns the term *denier* is usually employed instead of *count*.

c. A number, which is the sum of the wires across a card sheet, used to designate the fineness of pitch of the wire teeth used in carding operations.

1884 W. S. B. MCLAREN *Spinning* ix. 211 The cards are not ordered by the number of the wire but by counts and crowns.

d. *Nuclear Physics.* The recording of one or more ionizing events; an ionizing event so recorded. See also *background count* s.v. *BACKGROUND *sb.* 4, and *count-rate* below.

1921 *Phil. Mag.* XLII. 924 With the above system, H particles..could be counted with certainty under good conditions of experiment. The counts of both observers were found to be consistent over an interval of some months. **1930** *Physical Rev.* XXXV. 651/1 One out of 200 (residual) counts (a practical figure) in each individual tube-counter will be accidentally 'coincident'. **1946** KORFF *Electron & Nuclear Counters* iv. 82 The electron which is thus freed will start a new avalanche and produce a new count. **1958** O. R. FRISCH et al. *Nucl. Handbk.* iv. 32 The date given by the radiocarbon count agreed with that ascribed by the archaeologists.

9. count-fish *Austral.*, a full-grown schnapper (see quots.); **count-muster** *Austral.*, a gathering, esp. of cattle, for purposes of counting them; **count-noun** = *COUNTABLE *sb.* (opp. *mass-noun*); **count-rate**, the rate at which counts (sense 2 d above) are recorded by a radiation counter.

1874 in Tenison-Woods *Fishes N.S.W.* (1882) 41 The ordinary schnapper, or count-fish, implies that all of a certain size are to count as twelve to the dozen, the shoal or school-fish, eighteen or twenty-four to the dozen. *Ibid.*, The usual method of estimating quantity for sale by the fisherman is, as the schnapper or count-fish, the school-fish, and squire. **1891** 'R. BOLDREWOOD' *Sydney-side Saxon* 1 The old man's having a regular count-muster of his sons and daughters, and their children and off-side relatives. **1952** *Structural Notes & Corpus* (Comm. Lang. Programs, Amer. Council of Learned Societies) 60 Count-nouns are nouns which form plurals and can be used with a number or with the indefinite article. **1958** P. ROBERTS *Understanding Eng.* xi. 153 The distinctions intuitively made between..'count nouns' and 'mass nouns'. **1965** N. CHOMSKY *Theory of Syntax* ii. 64 *Boy* is a Count Noun (as distinct from the Mass Noun *butter* and the Abstract Noun *sincerity*). **1956** *Nature* 11 Feb. 286/1 The count-rate is proportional to the free gas density. **1961** *Lancet* 16 Sept. 634/1 The scintillation counters were..connected..to the scaling unit and the count-rates determined over a period of 100 seconds.

count, *v.* Add: **2. b.** With *in*. To include in the reckoning; to consider (a person) as a participant or supporter; to include. *colloq.* (orig. *U.S.*).

1857 *Knickerbocker* Feb. 185, 'I propose that we all just empty our pockets and show what we've got.' 'Good,' says Hiram, 'count me in.' **1859** *Ibid.* Nov. 559 In these days of daring 'Balloonry', the Knickerbocker is to be 'counted in'. **1912** *Punch* 17 July 42 (*caption*) Lady, an there be an Armageddon or other scrap toward, count me in! **1915** J. BUCHAN *39 Steps* i. 15, I need help worse than any man ever needed it, and I want to know if I can count you in. **1924** WODEHOUSE *Bill the Conqueror* xiii. 220 Bob the Sealyham..if aware that one of the gang contemplated going for a walk, would..show a disposition to count himself in.

15. count out (see also 1 c in Dict.). **a.** To adjudge (a boxer, etc.) to be the loser by a count (*COUNT *sb.*[1] 1 c); freq. *pass.*

1808 *Repertory* (Boston) 2 Aug. (Th.), The judges were proceeding to 'count out' his antagonist [*sc.* a fighting cock]. **1903** *Science Siftings* XXIV. 79/1 He falls, and is counted out. **1923** SOUTAR *Battling Barker* xx. 282 He is down on the canvas and the referee is shouting in his ear. He is being counted out. **1965** M. GOLESWORTHY *Encycl. Boxing* (ed. 3) 63/1 Dupas was counted out by the official 'counter'.

b. In children's games, to count (the players) with the words of a rhyme, formula, etc., the last at each turn being reckoned out of the game or chosen for a particular rôle in the game (see quots.); also *intr.* Hence *counting-out rhyme, song*.

1842 J. O. HALLIWELL *Nursery Rhymes* 123 Children stand round, and are counted out one by one by means of this rhyme. **1849** —— *Popular Rhymes* iii. 134 The operation of counting-out is a very important mystery in many puerile games. **1888** H. C. BOLTON *Counting-out Rhymes* 2 The leader then counts out once more, and the child not set free by the magic word is declared to be 'it'. **1900** E. V. & E. LUCAS *What shall we do Now?* 99 To decide who is to begin a game there are various counting-out rhymes. All the players stand in a circle, surrounding the one who counts. At each pause in the rhyme..this one touches the players in turn until the end is reached. The player to whom the last number comes is to begin. **1919** *Jrnl. Amer. Folk-Lore* 377 Counting-out rhymes, or 'Counts', are said in connection with..'Seek and Find'. **1923** KIPLING *Land & Sea Tales* 279 (*title*) A counting-out song. **1956** AUDEN *Making, Knowing & Judging* 8 Unofficial poetry, such as counting-out rhymes, and official poetry such as the odes of Keats.

c. To leave out of count or consideration; to reckon as not to be counted or depended upon; to exclude. *colloq.* (orig. *U.S.*).

1854 *Knickerbocker* June 643 When it comes to hunting grizzlies on a pony, jist 'count me out'. **1863** *Congress. Globe* 23 Feb. 1227/3 If that is the gentleman's idea, I beg him to count me out. **1926** E. WALLACE *Yellow Snake* iii. 27 So far as marriage with an unknown man is concerned, you can count me out. **1937** A. CHRISTIE *Murder in Mews* iv. 116 'Mrs. Vanderlyn is perhaps the dear friend of some one else in the house?' 'You can count me out!' said Sir George with a grin.

d. *Austral.* and *N.Z.* To count the number of sheep as they leave the shearing-shed. So *counting-out pen*.

1874 J. A. H. CAIRD *Sheepfarming in N.Z.* iii. 23 A small door for each shearer to put his shorn sheep out of the shed, and into the counting out pens. **1891** R. WALLACE *Rural Econ. Austral. & N.Z.* xxix. 381 [After being shorn] the sheep are inspected by the manager in charge before they are counted out and allowed to mix with the general flock. **1950** *N.Z. Jrnl. Agric.* May 463/2 It is often an advantage to clear the counting-out pens as frequently as possible during shearing or crutching.

16. count down. *trans.* To make a count-down to indicate the time of (an event).

1958 *Times* 12 Aug. 3/1 The final moments were 'counted down' so that every man knew the exact location. **1959** *John o' London's* 19 Nov. 233/3 He sees his hero as a visionary: he has him anticipating the H-bomb by 'counting down' the end of the world.

'count, dial. aphetic form of ACCOUNT *sb.*, esp. in *no 'count*. Cf. COUNT *sb.*[1] 5 b and see *NO-'COUNT *a.*

1854 M. J. HOLMES *Tempest & Sunshine* xxi. 302 He.. had come to the conclusion that 'they were of no 'count anyhow'. **1856** MRS. STOWE *Dred* I. 116 It's no 'count talking to him! **1883** W. H. COPE *Hampshire Words* 20 He be'ant no 'count. **1890** A. C. BICKLEY *Midst Surrey Hills* II. xv. 233 Folk here don't take much 'count on he.

countable, *a.* Add: **2. c.** *Math.* = *DENUMERABLE *a.*

1906 W. H. & G. C. YOUNG *Theory of Sets of Points* iv. 35 Any set which can be brought into (1, 1)-correspondence with some or all of the natural numbers is said to be countable, and, if not a finite set, is said to be countably infinite. **1941** BIRKHOFF & MACLANE *Surv. Mod. Algebra* xii. 338 Not all infinite classes are countable—there is more than one 'infinite' cardinal number. **1963** G. D. MOSTOV et al. *Fund. Struct. Algebra* ii. 42 Among the infinite sets are those which are said to be countable, or denumerable. [*Note*] The real numbers constitute an infinite set which is not countable.

d. Of a noun: that can form a plural or be used with the indefinite article. Cf. next.

[**1914** JESPERSEN *Mod. Eng. Gram.* II. v. 121 Names of countable immaterial objects may be thus used as mass-names.] **1961** *Brno Studies* III. 27 The governing noun *hat* (which, naturally, is 'countable'). **1968** J. LYONS *Introd. Theoret. Ling.* vii. 282 In all the languages referred to here certain words may be used either as 'mass' or 'countable' nouns.

cou·ntable, *sb.* [f. the adj.] A noun that denotes a countable thing, characterized by its ability to form a plural or be used with the indefinite article.

1914 JESPERSEN *Mod. Eng. Gram.* II. v. 114 Such 'countables' are either material things like *houses, horses, portraits, flowers*, etc., or immaterial things of various orders, like *days, miles*; [etc.]. **1962** J. SÖDERLIND in F. Behre *Contrib. Eng. Syntax* 102 Not a few of the uncountables so far discussed can function as countables as well. **1970** *English Studies* LI. 49 A noteworthy characteristic of the sequence *no* + noun + RC is that non-restrictive RC is only permissible with plural countables.

countably (kau·ntăbli), *adv.* [f. COUNTABLE *a.* + -LY[2].] In a countable way; denumerably.

1906 [see *COUNTABLE *a.* 2 c]. **1968** *Physics Bull.* Nov. 372/2 A countably infinite number of degrees of freedom.

count-down, countdown (kau·nt,dau:n). orig. *U.S.* [f. COUNT *sb.*[1] and *v.* + DOWN *adv.*] The action of counting in reverse, from a given number to zero, usually in seconds, to mark the lapse of time before an explosion, the launching of a missile, or the like; also, the period of time preceding such an event or the procedures carried out in that time. Also *attrib.*

1953 *News* (Birmingham, Alabama) 4 June 1/6 Observers on the mountain were able to hear the count-down on the drop from the control tower. **1953** *Monsanto Mag.* (U.S.) July 4 Time on the range is expressed in minutes before a missile is to be fired. This is called a 'count down'. **1958** *Observer* 2 Feb. 1/2 The count-down began. At the count of 11 the very top of the rocket started spinning. Two..one.. and then the firing command. **1958** *Times* 1 Mar. 6/3 The missile had been on its pad for days before firing, and the actual count-down, the exhaustive last-minute check of all its components, took many hours. **1958** 'P. BRYANT' *2 Hours to Doom* 61 Brown..turned the face of a count-down clock until the figure eighty-four showed in the..datum window. **1969** *Times* (Suppl.) 3 June p. ii/8 Three astronauts..were doing a simulated countdown for the first manned Apollo flight.

b. *transf.* and *fig.*

1959 W. K. RICHMOND *Brit. Birds of Prey* xiii. 142 Ready, now, for the final count-down. Eight yards, seven, six, five, four..and once again the hawk glanced over its shoulder. **1960** *Stand* IV. 5 The months began the count-down to war. **1962** *Times* 29 Nov. 4/1 Excitement is mounting here as the count-down begins for the first Test match on Friday. **1968** *Listener* 15 Aug. 200/2 One by one they step up, receive a countdown and deliver their instant punditry.

counter, *sb.*[2] Add: **3. b.** An instrument for counting or recording ionizing events. Freq. preceded by a defining word, as *Geiger*, *scintillation counter* (see these). Also *attrib.*

1924 *Proc. Camb. Phil. Soc.* XXII. 434 The α-particle counter, the prototype of which was used by Rutherford and Geiger, is an ionization chamber in which one of the electrodes is a fine point or small sphere, the other being either a plane or the case of the chamber itself. *Ibid.* 452 The registration by a counter of α-particles from polonium. **1930** [see *GEIGER]. **1930** [see *COUNT *sb.*[1] 2 d]. **1955** *Gloss. Terms Radiology* (*B.S.I.*) 22 *Counter*, a device which reacts to individual ionizing events, thus enabling them to be detected. **1956** A. H. COMPTON *Atomic Quest* 42 The first reliable counter measurements of cosmic rays at various latitudes.

counter, *sb.*[3] Add: **4. b.** Phr. *under the counter*, used with reference to illegal or clandestine transactions.

1926 A. HUXLEY *Jesting Pilate* IV. 284 One at least of my own novels has to be sold under the counter as though it were whiskey. **1945** *Evening Standard* 20 Dec., Chief goods to 'go under the counter' are fully fashioned silk stockings, watches and silk handkerchiefs. **1946** *Jrnl. R. Aeronaut. Soc.* L. 246/2 There is too much of the 'under the counter' method. It is essential that opportunities for such courses should be advertised openly. **1960** H. AGAR *Saving Remnant* viii. 204 Aside from under-the-counter help..Arazi had few funds. **1961** H. S. TURNER *Something Extraordinary* ii. 27 Rude verses, under-the-counter pin-ups and obscene novelties. **1969** *New Yorker* 31 May 78/2 'De Sade Illustrated', another version of the same under-the-counter classic, is also around.

8. (sense 4) *counter-flap, -girl, -hand, -top.*

1922 JOYCE *Ulysses* 117 He lifted the counterflap. **1968** P. GEDDES *High Game* ii. 17 Venniker lifted the counter flap but before he could move through there was a touch on his shoulder. *a* **1902** F. NORRIS *Vandover* (1914) 91 A couple of girls, the counter girls at one of the candy booths, came down the stairs. **1966** N. FREELING *King of Rainy Country* 38 She had taken a job in an expensive flashy shop as countergirl, selling sports clothes. **1893** G. B. SHAW *Imposs. Anarchism* 13 The cost of scales and weights, coin, book-keepers, counter-hands. **1927** *Daily Express* 17 Feb. 5/3 She..had been both a counter-hand and a mannequin before she was promoted to chief saleswoman. **1897** M. KINGSLEY *W. Africa* 17 Some brutal tradesmen..affix tremendous nails.. to the fronts of their counter tops, in order to keep their visitors at a respectful distance. **1908** *Westm. Gaz.* 11 Sept. 2/1 A child whose lint-white head scarcely reaches the counter-top.

counter, *sb.*[4] Add: **5.** *Typogr.* A depression in the face of type, reproducing the effect of a counter-punch, or an equivalent effect got by engraving the punch; a white space that is partly or wholly enclosed within a printed letter.

1798 in T. B. Reed *Hist. Old Eng. Letter Foundries* (1887) xiv. 293 Great care has been taken to have the Counterpart deeply cut, by which means [the type] will wear much longer. **1876** T. DE VINNE *Invention of Printing* iii. 54 This counter-punch is an engraving, in high relief, of the hollow or counter of that interior part of the letter..which does not show black in the printed impression. **1883** *Encycl. Brit.* XXIII. 698/1 The spaces at *h* and *k* are the counters, which regulate the distances apart of the stems in a line of type. **1892** A. OLDFIELD *Man. Typogr.* xxii. 163 The stems should be of a good depth down to the shoulder and counter. **1898, 1902** [see *COUNTER-PUNCH 2]. **1945** J. C. TARR *Printing Today* vi. 61 The counter (or interior area) is struck in by means of a counter-punch.

b. = COUNTER-DIE.

1893 *Funk's Stand. Dict.*, *Counter*.., a depressed or perforated block opposing a die or punch. **1917** F. S. HENRY *Printing for School & Shop* xiv. 242 The next step is the making of the counter.

6. *Skating.* A turn in which the body is revolved in a direction opposite to that in which it was revolved in the previous turns. Called also *counter-rocker, counter-rocking turn*.

1892 T. M. WITHAM in Heathcote & Tebbutt *Skating* iii. 107 In the rockers and counters (which figures used to be designated rocking turns and counter-rocking turns), although there is a change of direction the nature of the edge is preserved. *Ibid.* 111 A counter may..be regarded as half a bracket plus half a three. **1898** *Encycl. Sport* II. 366/2 The four counter-rocking turns. Counter-rocking turns or counters are turns in which one edge of the skate only is used, the body being revolved in an opposite or counter direction to that in which it is revolved in the corresponding 'three' turns. **1902** *Encycl. Brit.* XXXII. 642/2 The 'star', consisting of four crosses (forward rocker, back loop, back counter). **1908** E. F. BENSON *Eng. Figure Skating* 244 'Once counter back' is forward counter and inside back. **1928** *Daily Express* 19 Dec. 15/4 Her pirouettes, toe-spins, and counter-rockers. **1967** *Daily Tel.* 1 Mar. 12/6 Miss Stapleford made an excellent start in the first figure, the forward inside counter.

counter-. Add: Selected additional examples.

3. a. *counter-accusation, -challenge, -coup, -guerrilla, -insurgency, -measure, -propaganda.*

1917 *How War came to America* (U.S. Committee on Public Information) 11 Quibbles, misrepresentations, and counter accusations against their enemies abroad. **1949** KOESTLER *Promise & Fulf.* II. v. 285 Diplomatic notes with accusations and counter-accusations that read like the Fishwives' Morning Gazette. **1909** M. B. SAUNDERS *Litany Lane* I. iii. 27 His own [eyes] flung back a steely counter-challenge. **1963** *Time* 30 Aug. 20/2 Roman Catholic Diem..opened up the possibilities of coups, counter-coups, and even civil war. **1901** *Westm. Gaz.* 27 Nov. 2/2 We ought to meet the Boer guerilla by a counter-guerilla. **1962** *Economist* 12 May 551/1 A determined counterguerrilla offensive succeeded. Their comments on counterguerrilla are of special topicality. **1964** *Ann. Reg. 1963* 163 The military measures..comprised ..heavy U.S. engagement in counter-guerrilla warfare in South Vietnam throughout the period. **1962** *Times* 8 May 13/7 Psychological warfare and counter-insurgency techniques cannot be the concern of a few hastily trained officers. **1923** COLLINGWOOD *Roman Britain* ii. 37 The Saxon raids increased, but counter-measures were devised. **1901** G. B. SHAW *Cashel Byron's Prof.* p. xiv, The vast propaganda of pugnacity in modern fiction..must be met, not by shocked silence, but by counter-propaganda. **1950** S. POTTER *Our Lang.* 114 Propaganda and trustworthy news are dissociated in our minds. We even hear of propaganda and counter-propaganda!

4. *counter-reaction.*

1859 MILL in *Edin. Rev.* CX. 288 The great European philosophical reaction was to have its counter-reaction. **1965** H. KAHN *On Escalation* 278 A threat is a specific statement as to what the counterreaction will be.

9. *counter-church, -consideration, -magic, -motive, -presumption, -process* (example), *-productive* adj., *-society.*

1928 BELLOC *Convers. with Angel* xi. 81 If Calvin had not written his book there would have been no organised counter-Church in France. **1956** A. TOYNBEE *Hist. Approach Relig.* I. iv. 54 Opposing Christianity by staging an artificial pagan counter-church. **1848** MILL *Pol. Econ.* I. I. viii. 152 There are counter-considerations which are overlooked. **1905** *Daily Chron.* 9 June 4/3 Bearing these counter-considerations in mind, let us inquire what are the possible or probable conditions of peace. **1952** GERTH & MARTINDALE tr. *Weber's Anc. Judaism* IV. xi. 273 The opposition party lay in wait to destroy them..by counter-magic. **1836** MILL in *Lond. & Westm. Rev.* Oct. 12 That motive..is checked by the two perpetual counter-motives above adverted to. **1867** G. M. HOPKINS *Let.* 31 Dec. (1956) 48, I hope you will weigh these counter-presumptions. **1897** W. JAMES *Will to Believe* 318 It is a miserable thing for a question of truth to be confined to mere presumption and counter-presumption. **1843** MILL *Logic* II. IV. iv. 261 A counter-process of losing truths already possessed, is also constantly going on. **1964** *Ann. Reg. 1963* 166 These tactics were not only useless but counter-productive. **1969** *Spectator* 13 Dec. 816/2 So far as actually winning a single teenage vote is concerned, the efforts of both major parties..are likely to be counter-productive. **1955** 'C. H. ROLPH' *Women of Streets* x. 129 One would expect the emergence of such leaders in a counter-society such as this. **1959** *Encounter* May 29/2 The society formed by prostitutes and their associates..has a fairly permanent structure and composition; its stability depends on its nature as a counter-society.

counter-argue, *v.* Delete ?*Obs.* and add later examples. Also *intr.*

1882 BAIN *J. S. Mill* iv. 94 Marshall had been engaged in counter-arguing Whewell, on the points where he stood opposed to Mill. **1908** *Times Lit. Suppl.* 3 Sept. 283/3 It would be labour misspent to counter-argue such doctrines. **1949** I. DEUTSCHER *Stalin* 74 Koba counter-argued that the Mensheviks were confusing different types of government.

cou·nter-atta:ck, *v.* [f. *counter-attack* sb. (COUNTER- 3 a).] *intr.* and *trans.* To make a counter-attack (upon).

1916 'BOYD CABLE' *Action Front* 23 After the last of the counter-attacking party had swarmed out. *Ibid.* 193 The Germans..had been counter-attacked and partly driven back. **1953** *Encounter* Oct. 41/1 He resists without counter-attacking, without attempting to destroy an alien enemy.

counter-book[1]. Delete † *Obs.* and add later examples.

1900 *Westm. Gaz.* 26 Oct. 8/1 He was arrested last week on the charge of falsifying a counter-book. **1942** H. A. MADDOX *Dict. Stationery* 27 *Counter Books*, the sort of books used by shop assistants at the stores counters to record and act as receipt.

counterchange, *sb.* Add: **3. b.** Esp. in design, a pattern which systematically employs contrasting effects or where pattern and background are of the same shape.

1888 F. G. JACKSON *Lessons on Decorative Design* vi. 139 By 'counterchange' is understood the arrangement of a pattern in such a way that itself and the ground shall be of the same form, showing at once that it is but a special way of employing..repetition. **1962** R. G. HAGGAR *Dict. Art Terms* 96/2 *Counterchange*, alternation of effect in pictorial design; for example, light upon dark balanced by dark against light. **1967** E. SHORT *Embroidery & Fabric Collage* i. 34 An inlay based on a counterchange design..demands a contrast of colour or tone to give it point. **1970** *Times* (Suppl.) 28 Feb. p. vii/4 Seaweed or arabesque marquetry.. were created by an economical method known as counterchange.

counter-claim, counterclaim, *sb.* (Earlier example.)

1784 DUKE OF DORSET *Let.* 22 July in O. Browning *Desp. Paris* (1909) I. 15 The counter-claims of the Dutch are.. still greater than the Emperor's demands.

counter-current, *sb.* Add: Also *fig.*

1883 [see Dict.]. **1936** K. MANNHEIM *Ideol. & Utopia* III. vi. 159 As a countercurrent to this, there is Romanticism. **1937** *Science & Society* I. 155 Linguistic development then follows clearly the lines of internationalism...There are the same contradictions manifest, and the same counter-currents engendered by extreme tendencies in one direction or the other.

cou:nter-e·spionage. Spying directed against an enemy's spy system. Cf. *CONTRE-ESPIONNAGE.

1899 *Westm. Gaz.* 13 Sept. 5/3 It will take no part in the police or counter-espionage services. **1936** *Times Lit. Suppl.* 25 Jan. 79/2 The Russian Intelligence Service, a body entrusted..with counter-espionage.

cou:nter-e·tch, *v. Printing.* [COUNTER- 1.] *trans.* To treat (a lithographic plate) with diluted acid in order to clean it and make it receptive to grease. Hence as *sb.*, this process, or the acid solution used for it. So **cou:nter-e·tching** *vbl. sb.*

1924 H. J. RHODES *Art Lithogr.* (ed. 2) x. 94 It is a safe practice, before laying down the transfer on a zinc or aluminium plate, to wash and counter-etch (sensitize) it. **1960** GLAISTER *Gloss. Book* 89/2 *Counter-etching*, a stage in the preparation of lithographic printing plates, immediately after graining, and before transferring the image, whereby the plates are sensitized or counter-etched. **1967** KARCH & BUBER *Offset Processes* vi. 212 The counter-etch solution is composed of an acid mixed with water. *Ibid.* 535 *Counter etch*, a mild acid solution used for cleaning metal offset plates. **1967** E. CHAMBERS *Photolitho-Offset* 270 *Counter-etch*, the chemical treatment of a lithographic surface with diluted acids to remove oxides and dirt to make it sensitive to grease prior to producing an image on it.

cou:nterfa·ctual, *a. Philos.* [COUNTER- 10 a.] Pertaining to, or expressing, what has not in fact happened, but might, could, or would, in different conditions; *counterfactual conditional*, a conditional statement of this sort, normally indicating its character by the use of the subjunctive mood in its protasis. So as *sb.*, a counterfactual conditional.

1946 *Mind* LV. 289 Our problem here is to determine whether there is any other means of expressing this important counter-factual information. **1947** N. GOODMAN in *Jrnl. Philos.* XLIV. 113 (*heading*) The problem of counterfactual conditionals. **1952** *Mind* LXI. 222 The analysis of counterfactuals should parallel that of 'fugitive propositions'. **1958** *Times Lit. Suppl.* 12 Dec. 725/4 Counter-factuals constitute an irreducible form of statement.

cou·nter-fire. [ad. F. *contre-feu*: see COUNTER- 3 a.] A fire purposely lighted in order to combat a heath or forest fire: cf. *BACK-FIRE *sb.* 1.

1905 *Terms Forestry & Logging* 6 Back fire...Counter fire. **1927** *Observer* 11 Sept. 9/2 There are many men in the local peasantry who well understand the principle of counter-fires. **1952** *Oxf. Jun. Encycl.* VI. 172/2 Sometimes it is necessary to start a counter fire to check the main fire.

Hence **counter-fire** *v. intr.*, to use a counter-fire; **counter-firing** *vbl. sb.*

1895 W. R. FISHER *Schlich's Man. Forestry* IV. 554 It may be necessary to counterfire from a road, stream, ride, or fire-trace. **1930** *Forestry* IV. 88 Counter-firing is the deliberate burning of a strip of forest in the route of the accidental fire which dies out for lack of fuel.

counter-force, counterforce. Add: **b.** *attrib.*; esp. denoting strategy based on retaliation against military forces or bases.

1955 *Bull. Atomic Sci.* Mar. 100/3 Our strategy would be counterforce strategy, not counter city strategy. **1959** *Economist* 14 Feb. 590/2 The counter-force strategy was based on the assumption that the Russians' own bases could be overwhelmed virtually at a single blow. **1966** SCHWARZ & HADIK *Strategic Terminology* 40 Counterforce capability is that required to destroy enough of an enemy's nuclear striking capability in order to make his retaliation tolerable.

cou·nter-glow, counterglow. *Astr.* [tr. G. *gegenschein*: COUNTER- 6 b.] = *GEGENSCHEIN.

1852 E. SABINE tr. *von Humboldt's Cosmos* III. II. 416 The brightness of the Zodiacal Light did not appear to me to increase sensibly with the elevation of the observer's station, but rather to depend principally..on greater or less degrees of luminous intensity in the phænomenon itself. I even remarked, when in the Pacific, a counter-glow, like that of sunset. **1888** C. A. YOUNG *Gen. Astron.* 347 There is said to be in it [*sc.* the zodiacal light] at the point exactly opposite to the sun a patch a few degrees in diameter of slightly brighter luminosity, called the 'Gegenschein' or 'counter-glow'. **1956** *Nature* 18 Feb. 308/2 Attention was also directed..to recent studies on the counterglow by workers in the U.S.S.R. **1967** *New Scientist* 9 Mar. 464/2 There is a slight increase in brightness at the point in the sky diametrically opposite to the Sun, and this is known as the gegenschein or counterglow.

cou:nter-inte·lligence. [*INTELLIGENCE 7 c.] The activity of preventing the enemy from obtaining secret information; the agency or service engaged in this activity; = *COUNTER-ESPIONAGE. Also *attrib.*

1940 S. THOMAS *S-2 in Action* 119 These are but the elementals of counter-intelligence. **1949** J. F. EMBREE in M. Fortes *Social Struct.* 216 Members of various divisions.. concerned with such matters as counter-intelligence. **1955** *Bull. Atomic Sci.* Apr. 130/2 The threat of espionage must be assessed for what it is, and countermeasures taken by security and counterintelligence. **1957** *New Yorker* 26 Oct. 106/2 Investigation by counter-intelligence officers.

counter-jumper. (Earlier Canadian and later U.K. examples.)

1829 J. MACTAGGART *3 Yrs. Canada* I. 242 In the conceited towns he is held in abhorrence by the *clerk* and *counter-jumper*. **1915** W. S. MAUGHAM *Of Human Bondage* xvi. 65 Mr Gordon used to call me a gipsy counter-jumper when I was in his form. **1959** J. BRAINE *Vodi* xxiii. 248 You'll not waste your time with bloody consumptive counter-jumpers!

countermark, *sb.* Add: **1. d.** (See quots.)

[**1907** C. M. BRIQUET *Les Filigranes* I. 14/2 Dans le reste de l'Italie, en France et en Allemagne, à partir de 1550, les papetiers placèrent une contremarque ou second filigrane, symétriquement avec le premier, au centre du second feuillet de la feuille.] **1927** R. B. MCKERROW *Introd. Bibliogr.* I. viii. 102 Later papers, from about 1670, often have a second watermark, termed a 'counter-mark' and generally consisting of the maker's initials, in the centre of the opposite half of the sheet. **1937** E. J. LABARRE *Dict. Paper* 125/1 *Countermark* is the smaller or subsidiary watermark found in antique papers in addition to the main watermark or mark proper...The countermark is generally placed in the second half of the sheet opposite the main mark.

cou·nter-melody. A subordinate melody accompanying the principal one.

1931 G. JACOB *Orchestral Technique* v. 56 The violas are.. chiefly used to supply inner parts either in the nature of harmonic filling up, or countermelodies. **1958** BLESH & JANIS *They all played Ragtime* ix. 167 These running basses introduce a countermelody to the treble.

counter-order, *sb.* (Earlier example.)

1773 GEORGE III *Let.* 25 Apr. (1935) 89 We must see what effect our arming has had on Versailles before we can properly give counter-orders.

counterpart. Add: **6.** *spec.* Applied to a sum of money in local currency of an amount equivalent to goods and services received from another country.

1954 *Ann. Reg. 1953* 192 The U.S.A. was making available 900 million guilders from counterpart funds for dyke and road repairs. **1958** *Times Lit. Suppl.* 29 Aug. 486/3 American 'counterpart funds' accruing in Britain and other European countries from official American aid or strategic purchases. **1961** R. E. ASHER *Grants, Loans, & Local Currencies* v. 127 If the imports come as grant aid and are sold within the country, the foreign government will have, in a counterpart fund, local currency.

counterpoint, *sb.*[1] Add: **2. b.** *transf.* The combination of two types of rhythm in a line of verse.

c **1873–4** G. M. HOPKINS *Notebks.* (1937) 238 When the caesura is fixed by rule we have rhythmic counterpoint. By counterpoint I mean the carrying on of two figures at once, especially if they are alike in kind but very unlike or opposite in species. **1878** —— *Let.* 5 Oct. (1935) 15, I have written..in the ordinary scanning *counterpointed* (this is counterpoint: '*Hóme* to his mŏther's house *prívate* retúrned'..); others, one or two, in common counterpointed rhythm. **1942** T. S. ELIOT *Music of Poetry* 12 Part of the pleasure in the poetry arose from the presence in it of two metrical schemes in a kind of counterpoint.

counterpoint (kau·ntəɹpoint), *v.*[2] [f. COUNTERPOINT *sb.*[1]] **1.** *intr.* To compose or play musical counterpoint. *rare.*

1875 H. PARRY in C. L. Graves *H. Parry* I. iv. 155 Counterpointing, practising and collecting and setting microscopic specimens.

2. *trans.* To write or add a counterpoint to.

1877 G. M. HOPKINS *Let.* 21 Aug. (1935) 45 Milton keeps up a fiction of counterpointing the heard rhythm..upon a standard rhythm which is never heard but only counted. **1894** [see *CANON *v.*]. **1965** *Listener* 4 Nov. 733/2 The orchestra..counterpoints the vocal part.

3. To set in contrast; to emphasize a contrast between (two things) by juxtaposition; also with *against*.

1940 *Scrutiny* IX. 242 Arruntius's belief and unbelief and Sejanus's unbelief and belief are, so to speak, counterpointed. **1959** *Encounter* July 52/2 Counterpointed against Betty's stolid decency is the frenetic nihilism of his editor.

So **cou·nterpointed** *ppl. a.*, **cou·nterpointing** *vbl. sb.*

c **1873–4** G. M. HOPKINS *Notebks.* (1937) 236 The counterpointing upon an eight- or nine-syllabled four-beat iambic line of a rhythm of six beats in two parts. **1877** —— *Let.* 21 Aug. (1935) 45 Counterpointed verse..supposes a well-known and unmistakeable or unforgetable [*sic*] standard rhythm. **1941** L. MACNEICE *Poetry of Yeats* viii. 180 His [*sc.* Yeats's] very subtle rhythmical 'counterpointing', as Gerard Manley Hopkins would have called it, was more congenial to us than Hopkins's own 'sprung rhythm'. **1959** *Listener* 29 Jan. 204/2 The verse letters represent a remarkable counterpointing of the formal and the colloquial. **1962** *Ibid.* 4 Oct. 531/2 Joyce's interest in the counterpointed meaning of language and music.

counterpoise, *sb.* Add: **2. c.** *Electr. Engineering.* (i) A network of wires placed just above the ground and insulated from it, and connected to an aerial in place of or in conjunction with an earth connection.

1907 J. ERSKINE-MURRAY *Handbk. Wireless Telegr.* xx. 293 In place of using a direct earth connection..it is better to employ an electrical counterpoise, i.e., a conductor of considerable area, near the earth. **1940** *Amateur Radio Handbk.* (ed. 2) xii. 176 An earth connection is usually of high resistance, and to overcome the loss of efficiency due to this..a counterpoise is employed. **1971** M. G. SCROGGIE *Found. Wireless & Electronics* xvii. 285 It is a common practice to connect the lower end of the aerial to a radial system of copper wires. An insulated set of wires stretched just above the ground is known as a counterpoise.

(ii) A conductor or a network of conductors buried in the ground and connected to an electricity pylon in order to reduce the risk of flash-overs caused by lightning.

1930 *Trans. Amer. Inst. Electr. Engineers* XLIX. 935/1 The method of using a buried footing tie or 'counterpoise' is of interest as the action of such a cable is unlike that of a ground concentrated at one point. *Ibid.*, This analysis considers the cable only as a path to ground and neglects any 'counterpoise' effect. **1957** *Encycl. Brit.* XIV. 116/1 Towers equipped with 2 overhead ground wires and using a continuous tower-to-tower counterpoise had no flashovers in 10 yr. **1968** FINK & CARROLL *Stand. Handbk. Electr. Engin.* (ed. 10) XXVI. 19 It is the usual practice to bury the radial counterpoise parallel with the transmission line conductors and within the right of way.

counter-punch. Add: **2.** *Typogr.* [= F. *contrepoinçon*, G. *gegenpunzen*.] A die used in making the counter or open space (*COUNTER *sb.*[4] 5 a) of a type to be engraved on a punch. Hence as *v. trans.*, to form by means of a counter-punch.

1683 MOXON *Mech. Exerc., Printing* xiii. ¶ 1 Striking the Counter-Punch into the Face of the Punch. *Ibid.*, If the Letter be wholly to be Cut, and not Counter-Punch'd. **1898** J. SOUTHWARD *Mod. Printing* I. 142 The counter is the open space in the face of letters...It is so called because the part corresponding to it in the original punch is not cut by hand, like the rest of the letter, but is hollowed out by a counter punch. **1902** DE VINNE *Typogr.* (ed. 2) 15 When the proportions of the letters have been determined, the punch-cutter begins his work by making a counter-punch of steel. The illustration adjoining shows the form of a counter-punch for the letter H of the size of double english. It is an engraving in high relief of the counter or hollow part of the type, that is, of that part which appears white in the printed letter. **1965** E. TUNIS *Colonial Craftsmen* v. 127 The type cutter aided his work by annealing the steel to soften it and by first cutting counterpunches to form the insides of closed letters like B, D, O, P, and so on.

counter-revolutionary, *a.* Add: (Earlier and later examples.) Also as *sb.*, a counter-revolutionist.

1799 *Ann. Reg. 1793* (Hist.) 259/1 Brissot..exclaimed against the tyranny of a forced loan, and represented..its counter-revolutionary operation. **1917** M. FARBMAN *Russian Revol.* 31 The counter-revolutionaries of the Right. **1920** *Times* 13 Feb. 12/5 Every Pole in Russia has been treated by the Bolshevists as a counter-revolutionary. **1944** S. PUTNAM tr. *E. da Cunha's Rebellion in Backlands* p. xi, In 1893–94 came the counterrevolutionary revolt of the fleet, with civil war in Rio de Janeiro harbour. **1950** KOESTLER in *God that Failed* 74 From then onward I no longer cared whether Hitler's allies called me a counter-revolutionary.

cou·nter-rhythm. A subordinate rhythm acting as a counterbalance to the main rhythm.

1927 T. S. ELIOT *Sel. Essays* (1953) 85 The musical counter-rhythms which Marlowe introduced. **1958** B. ULANOV *Hist. Jazz in Amer.* xviii. 214 Splashes of counter-rhythms, flashing tremolos.

cou·ntershading. [COUNTER- 9.] Coloration (esp. of a bird or animal) in which parts normally in shadow are light and parts normally illuminated are dark. Also *attrib.* So **cou·ntershade** *v. pass.*, to be coloured in this way (see also quot. 1934).

1896 *Auk* Apr. 127 The dark collars of the males of most species of Duck are absolute countershading to the light from the sky. **1909** A. H. THAYER *Concealing-Coloration in Animal Kingdom* ii. 25 Obliterative Coloration is a phrase that will fit the general principle, and Obliterative or Countershading may be used as a stricter term for the essential root of it. **1923** R. CORTISSOZ *Amer. Artists* ii. 36 The countershading of the same model, according to nature, would cause it to melt into the background. **1934** WEBSTER, *Countershade*, v.t., in camouflaging, to lighten the local colour of an object in those parts that normally fall into shadow, or, conversely, to darken the parts normally illuminated, thus making the object less conspicuous. **1940** H. B. COTT *Adaptive Coloration in Animals* I. iii. 36 In countershading we have a system of colouring the exact opposite to that upon which an artist depends when painting a picture. **1953** N. TINBERGEN *Herring Gull's World* ii. 14 The young are beautifully countershaded, that is, they are light underneath, which tends to counteract the effect of shadow. **1958** *New Biol.* XXV. 103 This countershading is general in animals which are normally lit by bright light from above. **1964** BODEN & KAMPA in *Oceanogr. & Marine Biol.* II. 367 These animals.. utilize these light organs as a countershading mechanism to erase their silhouette.

cou·nter-spy. Also **counterspy.** [COUNTER- 3 b.] A spy engaged in counter-espionage.

1939 E. AMBLER *Mask of Dimitrios* ix. 165 He excludes counter-spies from the figure. **1968** tr. *Altavilla's Art of Spying* xv. 189 The lovely young counterspy then really set to work.

cou·nterstain, *sb.* [COUNTER- 3.] An additional dye of different colour or sensitivity used in a microscopy specimen to produce a contrasting background to the parts of interest or to make clearer the distinction between different kinds of structure, tissue, etc. Hence as *v. trans.* (freq. *absol.*), to treat with a counterstain; to stain (a tissue or specimen) with a contrasting colour; **cou·nter-stai:ning** *vbl. sb.*

1895 KANTHACK & DRYSDALE *Elem. Pract. Bacteriol.* 28 Stain a film according to Gram's method...Counterstain with eosine. **1899** G. NEWMAN *Bacteria* 341 A pure culture of bacteria will not necessarily require the counter-stain (methylene-blue). **1901** *Jrnl. Exper. Med.* V. 559 Rehm counter-staining with a solution of carmine..stained the centre of the nucleolus red, while leaving a peripheral rim of blue. **1951** WHITBY & HYNES *Med. Bacteriol.* (ed. 5) v. 34 Gram-negative organisms..have to be shown up by a contrast counterstain. *Ibid.* 35 Counterstain films from *cultures* with weak carbol-fuchsin. *Ibid.*, After counter-staining dehydrate quickly in methylated spirit and then alcohol. **1970** DISBREY & RACK *Histol. Lab. Methods* vii. 78 The counterstaining solution should not itself remove any of the specific stain. *Ibid.* 79 Occasionally..the counterstain is used before the specific stain.

countersubject. (Examples of application other than in a fugue.)

1947 A. EINSTEIN *Mus. Romantic Era* xi. 148 One speaks of dialogues in one movement, of subject and counter-subject set apart. **1955** G. ABRAHAM in H. van Thal *Fanfare for E. Newman* ii. 23 Wagner adds an oboe counter-subject here.

counter-tenor. 3. (Earlier and later examples.)

1512–25 *Northumberland Househ.-Bk.* (1770) 40 Gentillmen of the Chapell viij Viz. ij Bassys—ij Tenors—and iiij Countertenours. **1966** *Musical Opinion* Aug. 661/1 Purcell.. is thought to have been a counter-tenor himself.

cou·nter-tra:de. [COUNTER- 6.] = ANTI-TRADE *sb.*

1917 A. MCADIE *Princ. Aërography* x. 100 We shall use the term 'counter-trades' for the winds above the trades. **1928** D. BRUNT *Meteorol.* iv. 28 The reversed wind currents over the trade winds are called the counter-trades.

countervalue, *sb.* Delete † *Obs.* and add later examples; *spec.* applied to strategy based on the attacking of civilian targets.

1963 *Times* 4 Feb. 9/4 The new nuclear strategy of unlimited but controlled retaliation with its alternatives of counterforce and countervalue (the euphemism for attacking cities). **1965** *Economist* 6 Feb. 578/3 The counter-values of these [foreign currency reserves] have been credited to the exporters in DM.

counting, *vbl. sb.* Add: **3. counting-frame** = ABACUS 2; **counting-rate** = *count-rate* (*COUNT *sb.*[1] 9).

1920 H. G. WELLS *Outl. Hist.* I. xvi. 120/2 The Peruvians also got to making maps and the use of counting-frames. **1937** AUDEN *Spain* 7 The diffusion Of the counting-frame and the cromlech. **1956** *Nature* 3 Mar. 412/2 A sea-level meson recorder..showed a 100 per cent increase in counting-rate. **1961** *Lancet* 12 Aug. 345/1 To get the highest counting-rate from the beta radiation. **1970** *Nature* 25 July 375/2 The net National Bureau of Standards oxalic acid and background counting rate at this pressure are respectively 3·8 and 0·6 c.p.m.

country. Add: **2.** *spec.* Preceded by a personal name: the region associated with a particular person or his works; also *fig.*

1905 F. G. KITTON (*title*) The Dickens Country. **1956** P. MORTIMER *Bright Prison* vii. 69 Constable country, of course. You care for Constable? **1958** *Observer* 15 June 15/3 An original artist working on a small scale in what might be called Dubuffet country. **1962** *Times* 24 Dec. 6/7 The Enid Blyton country does not date.

b. *God's (own) country*, the United States, or some particular part of the United States; also applied to other countries, and, more generally, an 'earthly paradise'. orig. *U.S.*

1865 R. H. KELLOGG *Rebel Prisons* 118, I was willing to work hard, if I could only get out of that horrible den, into God's country once more. **1890** G. W. PERRIE *Buckskin Mose* xv. 218 The memory of the little woman I had left behind me in the East, or 'God's country'. **1893** T. BRACKEN *Lays & Lyrics* 5, I am doing very well here [in Australia] but I would much sooner live on a far smaller salary in 'God's own Country'. **1914** G. ATHERTON *Perch of Devil* i. 43 They always come home..talkin about..God's Own Country, and the Big Western Heart. **1926** J. DEVANNY *Lenore Divine* ii. 20 Richard John Seddon..the popularly acclaimed uncrowned king of the land he himself had named 'God's Own Country'. **1946** E. B. THOMPSON *Amer. Daughter* 10 We can work for..a real home.., out in God's country! **1960** B. KEATON *Wonderf. World of Slapstick* (1967) 65 (*heading*) Back home again in God's Country.

5. b. Used predicatively without article = rural, countrified.

1838 E. C. GASKELL *Let.* 17 July (1966) 19 Rivington is such a very pretty place, & so thoroughly country. **1890** SUSAN HALE *Lett.* (1919) 235 It is far more country here than they are. **1947** C. H. WARREN *Adam was Ploughman* III. 74 He is country to the core—and Cotswold country at that.

c. *Ellipt.* = *country-and-western*. Also *attrib.*

1967 *Melody Maker* 28 Jan. 13/2 A dozen well-known country hits gathered together on one album..must rank as a must for the country fans who don't have these tracks elsewhere. **1969** N. COHN *AWopBopaLooBop* (1970) xxi. 199 The Byrds are a trio and play mostly country. **1969** *Rolling Stone* 28 June 19/2 You start naming off some country songs and they say 'Is that Country? I didn't know that.'

11. (Examples of *country-rock*.)

1872 *Harper's Mag.* XLVI. 23 The formation, or 'country rock', is a common Gneiss, apparently of Laurentian age. **1965** G. J. WILLIAMS *Econ. Geol. N.Z.* v. 58/2 The country-rock strikes north-westerly.

12. (Further examples.)

1888 STEEL & LYTTELTON *Cricket* iv. 215 None of these are quick enough..for fielding in the country. **1912** A. BRAZIL *New Girl at St. Chad's* vii. 115 Driving the first ball she received into 'the country' for three.

14. *country bumpkin* (earlier and later examples), *cottage*.

1598 FLORIO *Worlde of Wordes* 126/3 *Fascina*, a countrie cottage, a shepheards bullie, a cabbin. **1680** ANON. in *Rochester's Poems* 107 With all that Country Bumpkins, call good Cheer. **1774, 1820** [see BUMPKIN 1]. **1881** C. C. HARRISON *Woman's Handiwork* III. 183 [A] country cottage fitted up for summer occupancy. **1951** R. FIRTH *Elem. Social Organiz.* iii. 92 The African or the Solomon Islander is no less sensitive to the appellation of country bumpkin than is his European counterpart. **1963** J. KORG *G. Gissing* ii. 61 Algernon.. seems to have spent his life in inexpensive country cottages.

16. country-and-western, a type of music originating in the southern and western United States, consisting mainly of rural or cowboy songs accompanied by a stringed instrument such as the guitar or fiddle; abbrev. *C-and-W*; **country club** orig. *U.S.* [CLUB *sb.* 13], one in or near the country, often with a restricted membership, having facilities for recreation and social intercourse; also, the premises and grounds of such a club; **country-damaged** *a.*, damaged in the country of origin, before shipment; **country gentleman**, (*b*) orig. *U.S.* (see quot. 1950) (freq. with capital initials); **country jake, jay** *U.S. colloq.*, a rustic; **country mouse**, a rural species of mouse; also *fig.*, a dweller in the country, esp. as unfamiliar with urban life (cf. *town-mouse* s.v. TOWN *sb.* 10); **country music** = *country-and-western*; **country service** *N.Z.*, the period during which a teacher in the Government service is required to teach in schools in rural districts; also *attrib.*; **country town** (earlier and later examples); also *attrib.*; **country-wide** *a.*, as wide as the country, extending throughout the country.

1959 'F. NEWTON' *Jazz Scene* xiii. 237 Simple vocal music of the rhythm-and-blues or country-and-Western type. **1960** *20th Cent.* Dec. 556 The hillbilly form is known generically as Country-and-Western...Whatever its mood, C-and-W has instantly recognizable characteristics. **1970** *Listener* 11 June 804/2 Musically, Liverpool has always preferred Country and Western music to virtually any of the current national trends. **1894** *Harper's Mag.* LXXXIX. 16/1 The Country Club is either very restful and bucolic, or very athletic and exciting. **1906** *Springfield Republ.* 19 July, It is to the game of golf that the country club is chiefly due, although golf is by no means the only sport which the well-developed country club now provides for. **1932** P. BLOOMFIELD *Imag. Worlds* ii. 33 An eternal cycle of 'whoopee' at a sort of unenterprising country club of intolerable million-airish crudity. **1957** W. H. WHYTE *Organiz. Man* 311 It [*sc.* the subscription] is high enough to keep many of the marginal-income people..away from the pool and the result will be the formation of a 'country-club set'. **1969** *Post Office Telephone Directory* (Oxford Area): *Classified Direc-*

tory 45 Chiltern Manor Country Club. **1896** Taggart *Cotton Spinning* I. 43 Unless the bales have been lying in wet or mud before having been shipped on the railways or steamships, and have become what is known to the trade as 'country damaged', the quality of the article does not deteriorate. **1900** *Daily News* 12 Oct. 2/6 Coffee.—Ecuador country-damaged. **1899** *Chicago Daily News* 21 June 7/4 Country Gentleman corn—regular price 12 c. **1950** *New Biol.* VIII. 37 One kind of sweet corn, called Country Gentleman, has the seeds situated irregularly on the cobs, instead of in neat rows as in most corns. *a* **1854** J. F. Kelly *Humors of Falconbridge* (1856) 136 You're a pooty looking country jake, you are, to advertise for a *dog*, and don't know Chiney Terrier from a singed possum? **1911** R. D. Saunders *Col. Todhunter* ix. 128 Didn't I tell you beforehand that they'd shorely size me up as a country-jake from the very beginning? **1899** Quinn *Pennsylv. Stories* 45 Well, you all know what a country jay Dutch was when he came to college. **1919** E. O'Neill *Rope* in *Moon of Caribbes* (1923) 28 You country jays oughter wake up and see what's goin' on. **1585** Bulloker *Aesop's Fables* sig. C2, Of the townish mous and contry-mous. **1607** Topsell *Four-f. Beasts* 504 The Epithets of myce are thes;. . rustik, or country mouse, vrbane, or citty mouse. *Ibid.* 544 The wood-Mouse is called in Greeke as the Countrey-mouse. **1750** *Student* 31 May 190 An honest country mouse, Kindly receiv'd, within his homely house. **1841** Geo. Eliot *Let.* Mar. (1954) I. 85 Letters from a Town Mouse to a Country Mouse. **1968** *Melody Maker* 30 Nov. 7 Are Britain's country music fans the poor relations of the popular music world? **1938** *N.Z. Educ. Gaz.* 1 Oct. 184/1 Unless specially exempted no teacher shall be eligible to apply for a [Grade A] position . . unless he has completed . . three years' country service. **1963** *Evening Post* (Wellington) 18 July, It would be a misnomer to designate the college as a country service school since Blenheim had all the amenities of any town. **1598** Florio *Worlde of Wordes* 449/1 *Villaggio*, a village, a borough, a hamlet, a countrie towne. *a* **1828** D. Wordsworth *Jrnl. Tour on Continent* 10 July in *Jrnls.* (1941) II. 7 This country-town-bred young Lady's notion. *Ibid.* 28 Sept. 319 A pretty mixture of rusticity, and country-town and river bustle. **1944** Dylan Thomas *Let.* 13 Apr. (1966) 263 A country-town few hours would be very pleasant. **1922** *Daily Tel.* 12 June 2/4 Country-wide strikes in the United States. **1928** *Daily Express* 8 Aug. 3/4, I have been astounded at the country-wide interest which my recent tramp has aroused. **1938** *Times* 30 May 16/1 The demand which it was established to meet is countrywide and unfailing.

countryfied: see Countrified *ppl. a.*

country road. †**a.** A public road made and maintained by the country or province. *Obs.* **b.** A road leading through country or a rural area.

1669 in *Essex Inst. Hist. Coll.* (1864) VI. 175/2 The highe way. . to Range up streight to the Country Roade. **1775** *Ibid.* (1877) XIII. 191 A very commodious House . . about 2 miles from the grand country road. **1846** *Merchant's Mag.* (U.S.) XIV. 29 A noble system of town and country roads. **1873** E. Eggleston *Myst. Metrop.* 12 A country road where the dog-fennel blooms almost undisturbed by comers and goers. **1931** R. Campbell *Georgiad* i. 25 Give him time To battle. . Along the country roads. **1937** V. Bartlett *This is my Life* xiv. 251 The German people. . as devoid of responsibility and personality as are the tramps that wander along the country road.

county[1]. Add: **5. b.** Also freq. with omission of *the*. Also as *adj.*, having the social status of a county family; characteristic of county gentry.

1920 D. H. Lawrence *Touch & Go* iii. i. 68 The Walsalls didn't like it—too near the ugliness. They were county, you know—we never were: father never gave mother a chance, there. **1921** M. Arlen *Romantic Lady* 8 But perhaps you would prefer me to be haughty. . or how do you say it—county? **1923** —— *These Charming People* 173 Young Thomas married county. **1926** Galsworthy *Escape* ii. vi, 'I should think he was County, wouldn't you?'. . 'Um! Only "County" would drive such a shockin' bad car.' **1937** C. Isherwood *Sally Bowles* 30 Mummy's bringing her up to be very county. **1963** J. T. Story *Something for Nothing* ii. 43 High, loud, county voices.

8. a. *county gaol* (examples) (also *jail*), *hospital*, *school*.

1699 in *Documents Hist. New Jersey* 1st Ser. III. 480 The prison at Woodbridge being the County Goal for the County of Middlesex. **1722** Defoe *Plague* 164 They deserv'd to be sent to the County Jail. **1767** G. White *Let.* 4 Nov. in *Selborne* (1789) i. xii. 35 Four or five swallows. . settling on the roof of the county-hospital. **1812** *Niles' Reg.* I. 361/2 The legislature. . appropriated. . 25,000 per annum for the supporting of county schools. **1820** *Rules for Common Gaol of County of Oxford* 6 The Governor of the County Gaol. **1846** County gaol [see Corporate *ppl. a.* B. 5]. **1885** W. S. Gilbert *Mikado* i. 9 Taken from the county jail By a set of curious chances. **1896** 'M. Rutherford' *Clara Hopgood* vi. 66 There was to be a grand entertainment . . in aid of the County Hospital. **1944** *Act 7 & 8 Geo. VI* c. 31 § 17 The instrument of management or the instrument of government, as the case may be, shall be in the case of a county school by an order of the local education authority. **1962** L. Golding *Dict. Local Govt.* 107 Nursery schools and special schools, even if they were established by and are maintained by a local education authority, are not regarded as county schools.

b. county college (see quots.); **county cricket**, cricket, esp. of inter-county matches, played by the regularly organized and qualified elevens of the county clubs; **county family** (earlier example); **county farm** *U.S.*, = *poor-farm*; **county library**, a library run by a county authority; **county match** *Cricket*, an inter-county match.

1944 *Act 7 & 8 Geo. VI* c. 31 § 43 (1) It shall be the duty of every local education authority to establish and maintain county colleges, that is to say, centres approved by the Minister for providing for young persons. . such further education. . as will enable them to develop their various aptitudes. **1963** Barnard & Lauwerys *Brit. Educ. Terms* 71 *County college*, a centre which, according to the 1944 Education Act, would provide for young persons under 18. . such further education. . as will. . prepare them for the responsibilities of citizenship. County colleges have not yet (1963) been established. **1880** *Baily's Monthly Mag.* July 410 It is open to question whether it [*sc.* the eleven] has gained any material addition to its strength during its absence from the field of county cricket. **1887** *Field* 10 Sept. 411/2 County cricket quite holds its place in the public estimation. **1928** *Evening News* 18 Aug. 10/5 The dreariness of play. . is forced on the game by the present bye-laws of county cricket. **1966** B. Johnston *Armchair Cricket 1966* 95 In County cricket, if. . thirty minutes extra. . could bring about a definite result. . the umpires must order play to go on. **1843** H. Hatcher *Old & New Sarum* xlvii. 581 Many of the county families had also their town residences in the City. **1871** *Michigan Gen. Statutes* I. 406 It shall be the duty of the superintendents of the poor of each county. . to report. . the income received from the county farm. **1935** A. Kennedy *Current Eng.* xiii. 548 The poorhouse has become the *county farm*. **1844** *Indiana Senate Jrnl.* 271 The boards of county commissioners . . may . . dispose of. . all such books, belonging to any county library, as are now . . worthless. **1927** *Public Libr. Committee Rep. Pub. Libr.* (Cmd. 2868) 220 In County libraries the books are distributed from a central depository to local centres over a wide area. **1961** T. Landau *Encycl. Librarianship* (ed. 2) 108/2 County Libraries have never tended therefore to hoard their reference books. **1782** *Club Minutes* in *Hambledon Cricket Chron.* (1924) 66 Ordered that those Players who intend to play in the County Matches . . do give in their names. **1966** B. Johnston *Armchair Cricket 1966* 46 In County Matches only one commentator is normally employed.

county, *a.* See *county[1] 5 b.

county council. (Earlier examples.)

1835 W. Dyott *Diary* 6 Apr. (1907) II. 195 The present alarming state of the county councils. **1837** *Hansard* 3rd Ser. XXXVI. 426/1 He thought that the county council ought to be kept together as short a time as possible.

coup, *sb.*[3] Add: **2. b.** = *coup d'état*.

1852 *North Brit. Rev.* XVI. 584 A tyranny. . which it required the. . *coup* of the 9th Thermidor to overthrow. **1958** *Economist* 26 July 267 The coup in which Nuri es-Said. . and the Hashemite dynasty were destroyed.

4. (Earlier and later examples.) Also *attrib.*, as coup-stick (see quot. 1876).

1841 G. Catlin *N. Amer. Indians* I. 27 Each one. . recited his exploits, and his 'coups' or deaths. **1876** J. G. Bourke *Jrnl.* 15 June (D.A.), Making 'coup' sticks, which are long willow branches, about 12 feet from end to end, stripped of leaves and bark and having each some distinctive mark in the way of feathers, bells, [etc.]. . . in dividing the spoil, each man claims the animals first struck by his 'coup' stick. **1921** *Glasgow Herald* 12 Nov. 7 Chief Plenty Coos laid his coupstick (symbol of tribal authority) and his war bonnet on the tomb. **1963** *Beaver* Summer 33/2 Befeathered coup stick (to touch an enemy or count coup was a braver act than killing him).

5. a*. *coup de force* (kudəfɔrs) [*lit.* stroke of force]: a sudden, violent action.

1835 J. S. Mill in *Lond. Rev.* II. 277 Their tampering with that measure was a *coup de force*. **1949** I. Deutscher *Stalin* xiii. 535 The only chance of breaking out of their impasse lay in a *coup de force*.

a.** *coup de foudre* (kudəfudr) [*lit.* stroke of lightning]: a sudden unforeseen occurrence; a revelation; also, love at first sight.

1779 Mrs. E. Boscawen *Let.* 23 June in C. Aspinall-Oglander *Admiral's Widow* (1942) xvii. 96 This sentence. . was a perfect *coup de foudre* bursting amidst our tranquil scenes. **1936** 'R. Hyde' *Check to your King* xiv. 163 Rumours of similar French *coups de foudre* were everywhere. **1955** *Times* 28 July 10/6 There is the first sight of it, virgin and gleaming from the makers, and the coup de foudre that makes us instantly one with it.

a*.** *coup de glotte* (kudəglɔt) [F. *glotte* glottis]: the glottal stop.

1909 in Webster. **1922** G. B. Shaw *Let.* 27 Jan. (1960) 16 Afterall and Westminsterabbey may have to be unlearned for the stage, as the *coup de glotte* before a vowel, German fashion, is often necessary for emphasis, audibility in a big house. **1962** *John o' London's* 19 July 65/3 The much-abused, but little understood, *coup de glotte*.

d*. *coup de poing* (kudəpwæ̃) [*lit.* stroke of fist]: a Chellean implement in the form of a shaped flint; a hand-axe; also *attrib.*

1912 R. Munro *Palæolithic Man* ii. 30 The Saint Acheul deposits, which have yielded an enormous quantity of flint implements of the coup-de-poing type. **1923** A. L. Kroeber *Anthropol.* xiv. 398 The coup-de-poing being a comparatively effective, regularly shaped, symmetrical implement involving both an ideal of form and a tolerable, rough skill to produce. **1932** *Antiquity* VI. 190 These flake industries seem to be quite distinct from the coup-de-poing industries.

e. (Earlier and later examples.)

1772 Duchess of Northumberland *Diary* 23 Apr. (1926) 166 The use is to prevent the wearer from receiving a Coup de Soleil. **1867** S. W. Baker *Nile Tributaries* xxii. 553 Bacheet had a slight *coup de soleil*.

g. (Earlier and later examples.)

1668 Dryden *Dram. Poesie* 54 Any Character or humour wherein he would show a *Coup de Maistre*, or his highest skill. **1819** T. E. Bowdich *Mission to Ashantee* i. vi. 123 It would be a coup d'éclat much more important and agreeable, if he could settle the Warsaw palaver as well. **1845** J. W. Croker in *Q. Rev.* Sept. 526 This work seems. . to be a respectable *coup d'essai*, written with some thought. **1893** Conan Doyle *Mem. Sherlock Holmes* 270 There are limits. . to our friend's intelligence. It would have been a *coup-de-maître* had he deduced what I would deduce and acted accordingly.

h. *coup de vent* (kudəvɑ̃) [F. *vent* wind]: a whirlwind; a gale.

1831 Disraeli *Young Duke* I. i. xi. 93 Oh, what a *coup-de-vent*! **1853** C. Brontë *Villette* I. xiii. 237 Monsieur went off like a coup de vent the other night.

coupe[1] (kūp). [Fr., goblet.] **a.** A shallow cup, bowl, or glass. **b.** A mixture of ice-cream and fruit, etc., served in a glass goblet or the like; spec. *coupe Jacques* (see quot. 1951).

1895 *Montgomery Ward Catal.* 528/2 Soup Coupe Saucer shape. **1905** E. Wharton *House of Mirth* II. iv. 355 What sweet shall we have to-day, dear—*Coupe Jacques* or *Pêches à la Melba*? **1951** *Good Housek. Home Encycl.* 421/2 *Coupe Jacques*. Half-fill the glass goblets with fruit salad, flavoured with kirsch, cover this with a layer of vanilla ice cream, and then add a layer of strawberry ice. Decorate with a glacé cherry. **1967** *Spectator* 29 Sept. 378/1 The tulip-shaped glass is much more suitable than the *coupe*—among other reasons because it holds the champagne bubbles more successfully. **1969** R. & D. De Sola *Dict. Cooking* 74/2 *Coupe glacée*, ice cream or sherbet combined with fruit or other ingredients; *coupe* is always served in a rimless bowl or cup. **1970** *Southerly* XXX. 17 Proudly he carried the brimming bucket into the kitchen where Medora was filling coupes with porridge.

coupe[2] (kūp). [a. F. *coupe* felling, f. *couper* to cut.] A periodic felling of trees; also, the area so cleared.

1922 W. Schlich *Man. Forestry* (ed. 4) I. 27 The annual or periodic *coupes* were further reduced to strips or to groups, known as the 'Strip and the Group Systems'. **1928** R. S. Troup *Silvicultural Systems* ii. 4 In its ideal form the clear-cutting system involves felling and regenerating each year equal areas (coupes). **1928** *Forestry* II. 83 A block of one thousand acres of chestnut coppice worked on a twelve years' rotation (providing annual *coupes* of 83·3 acres). **1953** *Brit. Commonw. Forest Terminol.* 33 Coupes are preferably numbered with Roman numerals to distinguish them from compartments.

coupé, *sb.* Add: **1. b.** A closed two-door motor car, usually with two seats (see also quot. 1967). Also *attrib.* In *U.S.* freq. spelt *coupe* and pronounced (kūp).

1908 A. Bennett *Buried Alive* ix. 207 This agreeable Jew in the hall, and the *coupé* at the kerb. **1912** *Motor* 23 July 1104/2 The ordinary coupé or doctor's car. **1915** *Scotsman* 3 Apr. 12/1 Wanted, 2-Seat Car, with Coupé Body preferred. **1924** *Morris Owner* Sept. 653 The coupé is the ideal car for. . the business man who has to motor in all weathers. **1927** in *Amer. Speech* (1967) XLII. 307 The coupe roadster for two passengers with rumble seat. **1934** [see *crookish a.]. **1936** H. L. Mencken *Amer. Lang.* (ed. 4) vii. 347, I have. . heard. . *coop* for *coupé*. **1967** *Consumer's Car Glossary* (ed. 2) 26 *Coupé*, car in which the roof slopes down sharply behind the front seat, leaving little or no head room in the back, or car with no back seat. In a drophead coupé the roof can be folded back behind the seats.

c. *coupé de ville*: a type of motor car in which the passenger seats are covered but with the driver's compartment open or adapted to fold down.

1939 E. Ambler *Mask of Dimitrios* viii. 149 The chauffeur . . ushered Latimer ceremoniously into a huge chocolate-coloured *coupé de ville*, and drove off. **1969** J. Leasor *They don't make them like that any More* vi. 169 A special type of body, very popular on Rolls chassis in the late twenties and early thirties. The passengers sat under cover in the rear. . while the . . chauffeur and footman, sat. . up front. . . English coach-builders called this style coupé de ville.

2. c. *coupé lit*, a coupé (sense 2 b) in which the seats can be extended to form a couch or bed. (Not in current use.)

1882 De Windt *Equator* 10 A sound sleep in a comfortable *coupé lit*. **1896** *Daily News* 28 Oct. 3/7 A coupé-lits-toilette and first-class carriage will run through to Vintimille by the 2.13 p.m. train from Boulogne. **1898** *Ibid.* 18 Oct. 2/2 A coupe-lit compartment at the Nord Station.

coupla (kɒ·plă). orig. *U.S.* Colloquial form of *couple of*.

1908 H. Green *Actors' Boarding House* 254 A coupla parties is come for rooms! **1934** D. L. Sayers *Nine Tailors* iii. ii. 276 He'd had nothing to eat. . for a coupla days. **1965** H. C. Rae *Skinner* ii. xiii. 119 I'll manage to palm the thing off on some mug for a coupla quick quid profit.

couple, *sb.* Add: **7. b.** (With *of* omitted) = *couple of*. Cf. prec. *U.S. colloq.*

1925 S. Lewis *Martin Arrowsmith* xvi. 188 A couple months in Italy. **1934** D. Hammett *Thin Man* xxii. 173 She touched me for a couple hundred to blow town.

c. With ellipsis of *of drinks*, *of glasses*, etc. *colloq.*

1933 'R. Keverne' *Menace* xvi. 180 Stopped at the 'Swan' for a couple. **1934** H. M. Harwood *Old Folks at Home* i. i, *Liza* (coldly): I said you had had a couple.

couple, *v.* Add: **2. e.** *Photogr.* To connect a device (as a rangefinder, etc.) to the mechanism of a camera. Also *intr.*, to be capable of being so connected. So **coupled** *ppl. a.*

1934 R. M. Fanstone *Mod. Miniature Cameras* ii. 7 This camera has automatic focussing by means of a range finder coupled to the focussing adjustment of the lens. **1935** W. Alexander *Miniature Camera Guide* p. i (Advt.), Leica Model III. Automatically coupled range-finder focusing. **1939** *Amer. Ann. Photogr. 1940* 7/2 These new, small cameras were characterized by their versatility and built-in features, such as: range finders coupled to the lens and shutter assembly. **1958** *Amateur Photographer* 31 Dec. 8/3 (Advt.), Rangefinder couples with all lenses. **1961** A. L. M. Sowerby *Dict. Photogr.* (ed. 19) *Coupled exposure-meter*, an exposure-meter the mechanism of which is linked to the controls adjusting stop and shutter-speed.

8. *Physics.* To bring about a coupling (sense *6 f) between. So **coupled** *ppl. a.* **a.** Said of oscillating systems. (Cf. *coupling *vbl. sb.* 6 f (i).)

1908 [see *coupling *vbl. sb.* 6 f (i)]. **1915** A. F. Collins *Bk. of Wireless* iii. iv. 143 When the open and closed circuits are coupled together they can be tuned to each other so that the electric oscillations in both circuits. . have the same frequency. **1927** E. G. Richardson *Sound* ii. 54 Owing to the assumption of the vibration by the air in the box the tone of the combined 'coupled system' is much more intense than that of the fork alone. **1927** I. B. Crandall *Theory Vibr. Syst.* ii. 63 A very considerable change in natural frequencies . . has been brought about by coupling to the heavy diaphragm a relatively light resonator system. **1944** A. Wood *Physics of Music* ii. 24 Two pendulums of unequal length suspended from a horizontal string. . . form a coupled system. **1962** Simpson & Richards *Junction Transistors* xiv. 341 The primary and secondary circuits are tuned separately to resonance at the same frequency and coupled together. **1965** H. J. J. Braddick *Vibrations, Waves, & Diffraction* i. 26 Two oscillators may be coupled in such a way that when one is vibrating, motion is transferred to the other. **1968** C. G. Kuper *Introd. Theory Superconductivity* ix. 147 (*caption*) Coupled-harmonic-oscillator model for a lattice in two dimensions.

b. Said of particles (electrons, atoms, etc.) and fields, and of mathematical quantities representing them. So *to be coupled with* = to interact with, to be influenced in (its) behaviour by. (Cf. *coupling *vbl. sb.* 6 f (ii).)

1922 *Chem. Abstr.* XVI. 3033 The magnetic energy of the valence electron and of the atom body, coupled in the manner postulated. **1923** H. L. Brose tr. *Sommerfeld's Atomic Struct.* vi. 406 An outer valence electron. . and. . the rest of the atom . . are coupled together by the internal magnetic field. **1935** J. Dougall tr. *Born's Atomic Physics* vii. 154 For every electron the orbital and spin moments are firmly coupled; but the various electrons influence each other comparatively little. **1959** B. I. & B. Bleaney *Electr. & Magn.* xx. 548 We need to know more about the mutual interactions between the various electrons. . . These can be expressed in the form of a set of rules for coupling together the angular momenta in forming the vector resultant. **1960** J. C. Slater *Quantum Theory of Atomic Struct.* I. x. 239 It was useful to postulate two vectors L and S, which could be coupled to give a vector J. **1966** [see *coupling *vbl. sb.* 6 f (ii)].

coupler. Add: **2. c.** *Zool.* A plate joining two opposite swimming appendages of a crustacean.

1897 Parker & Haswell *Zool.* I. 530 The first four thoracic appendages bear biramous swimming feet. ., those of the right and left sides being connected by transverse plates or couplers.

d. *Photogr.* (See quots.)

1938 *Brit. Jrnl. Photogr.* LXXXV. 628/1 The dyestuffs forming the images are in all cases produced by the effect of an oxidising agent on a mixture of a suitable agent and coupler. . . In some instances the developer and the coupler may be identical. **1944** J. S. Friedman *Hist. Color Photogr.* xi. 127 The true problem. . is not the inability to obtain uniform dispersions of the couplers throughout the emulsion, but to prevent the couplers in one emulsion layer from wandering over into a neighboring layer. **1957** R. W. G. Hunt *Reprod. Colour* v. 45 Instead of letting the oxidized developer react with the gelatin of the emulsion, a coupler is present in the developer (or in the emulsion layer) and this reacts with the oxidized developing agent to form an insoluble dye. **1966** D. M. Avedon *Gloss. Terms Microphotogr.* (ed. 4) 13 In a diazo material the coupler is a compound which combines with the unexposed diazonium salt to form the visible dye image.

couplet. Add: **6.** *couplet-grinder.*

1896 *Westm. Gaz.* 1 June 3/2 The most formal allegorical couplet-grinders of the eighteenth century. **1941** *Essays & Studies* XXVI. 78 He [*sc.* William Collins] sank to the level of the poorest couplet-grinder among them.

coupleteer (kɒplĕtī*ə*·ɹ). [f. Couplet + -eer.] A writer of couplets; a versifier.

1818 R. Woodhouse *Let.* 21 Oct. in *Lett. of Keats* (1958) I. 382 Leave. . its soul to be soothed only by rhymers & the coupleteers. **1927** *Observer* 15 May 6 The dreary coupleteers of the eighteenth century. **1929** E. Blunden *Nature in Eng. Lit.* v. 109 The appearance of a trim Horatian coupleteer.

coupling, *vbl. sb.* **5.** Read: A transverse timber connecting a pair of rafters. (Later example.)

1813 J. Taylor *Arator* 137 No joists or loft [in the houses] but to be lathed on the rafters and their couplings, nearly to the top of the roof.

6. Further techn. and scientific uses.

b. (Earlier example.)

1840 *Civil Engin. & Archit. Jrnl.* III. 5/1 The machine is connected to the train by means of the coupling N in the usual way.

e. *Genetics.* (See quots.)

1906 W. Bateson et al. *3rd Rep. Evol. Comm. Roy. Soc.* 8 Complete coupling has so far been most commonly met with among characters of similar physiological nature. *Ibid.* 11 It is clear that gametic coupling, however caused, plays a large part in the phenomena of heredity. **1909** W. Bateson *Mendel's Princ. Hered.* ix. 151 The combinations of long pollen with blue factor, and round pollen with no blue factor occur seven times as often as the other two possible combinations. We speak of this phenomenon as Gametic Coupling. The term 'coupling' is strictly applicable, because the association is between two dominant or 'present' factors. **1926** *Encycl. Brit.* II. 162/2 It is sometimes convenient to distinguish the case in which the two dominants ($AB \times ab$) are introduced together by the parent as coupling and the converse ($Ab \times aB$) as repulsion. **1965** A. H. Sturtevant *Hist. Genetics* vi. 40 The two dominants (purple and long) were contributed by the same parent, and the phenomenon was called 'coupling'. **1970** Ambrose & Easty *Cell Biol.* x. 338 Morgan showed that the proportion of crossing-over between a particular pair of alleles of two different genes. . was of the same order, irrespective of whether they were both initially on the same chromosome (in coupling) or on different chromosomes (in repulsion).

f. *Physics.* (i) A connection between two oscillating systems which results in a mutual dependence of their oscillations, so that an oscillation in one system causes an oscillation in the other and partly determines its character; *spec.* (and orig.) in *Electr.*, such a connection (physical or inductive) between two electric circuits. Also, the strength of this interdependence (so *tight, loose* etc. *coupling*).

1908 C. C. F. Monckton *Radio-Telegr.* vi. 101 Where oscillations in one circuit set up oscillations in an adjacent circuit the two circuits are said to be coupled. If a considerable portion of the field of force of the first circuit is embraced by the second the coupling is fast; if only a small portion, the coupling is loose. **1923** E. W. Marchant *Radio Telegr.* iii. 30 The looser the coupling between the two circuits, the less rapidly is the energy of the oscillating circuit used up and the more persistent is the oscillation of the aerial. **1927** I. B. Crandall *Theory Vibr. Syst.* ii. 44 The effect of the elastic connection or 'coupling' between the two vibrating masses. **1948** P. M. Morse *Vibration & Sound* (ed. 2) ii. 59 Whenever the coupling is small and the frequencies of the two oscillators are not equal, the amplitude of one motion will be much larger than that of the other. **1953** E. G. Richardson *Techn. Aspects Sound* I. xviii. 506 The reed is now coupled to a pipe resonator. In this case the coupling is loose and the vibration of the reed is predominant. **1959** B. I. & B. Bleaney *Electr. & Magn.* ix. 233 The coupling through the capacity decreases with frequency, because of the fall in the impedance common to the two circuits. **1962** Simpson & Richards *Junction Transistors* xi. 244 One of the first decisions to be made in the design of a multi-stage amplifier involves the choice of the method of coupling between stages. **1968** H. J. Pain *Physics of Vibrations & Waves* iii. 56 A mechanical example of stiffness coupling between two pendulums.

(ii) In atomic, nuclear, and particle physics: a physical interaction between two particles, between a particle and a field, or between two fields; *spec.* an interaction between the magnetic moments of the electrons in an atom. Also used of the combination of mathematical quantities (in quantum mechanics) which corresponds to the physical interactions.

This was orig. a use of the word in the prec. sense; the electrons in an atom were thought of as capable of vibrating, and the 'coupling' of their vibrations was used (by Voigt, 1913) to explain the Zeeman effect.

1913 *Chem. Abstr.* VII. 3268 (*heading*) Coupling theory of the Zeeman effect. **1922** *Ibid.* XVI. 3032 A comprehensive summary of the Voigt theory,. . based essentially on the coupling, in the presence of an external magnetic field, of several electrons bound in a quasi-elastic manner. **1926** *Ibid.* XX. 2449 To each electron within the atom 2 vectors can be assigned and the magnetic moments, multiplicities of levels and the anomalous Zeeman effect can be explained by the coupling of these vectors. **1929** *Physical Rev.* 2nd Ser. XXXIV. 26 In the normal, Russell-Saunders, or *LS* coupling scheme $(l_1l_2)(s_1s_2) = (LS) = J$. . . In the *jj* coupling scheme $(l_1s_1)(l_2s_2) = (j_1j_2) = J$. **1938** L. B. Loeb *Atomic Struct.* viii. 215 In very powerful magnetic fields. . the atomic coupling breaks down and the individual elements s^* and l^* precess about the imposed outer field direction. **1950** W. Finkelnburg *Atomic Physics* v. 292 Because of the strong coupling between the nucleons, the excitation energy of the nucleus can distribute itself. . over *all* the nucleons of the nucleus. **1955** S. S. Schweber et al. *Mesons & Fields* I. x. 108 We now turn to the discussion of interaction between fields, and in particular we treat first the coupling between a spinor and a Boson field. **1959** B. I. & B. Bleaney *Electr. & Magn.* xx. 552 The spin-orbit coupling between L and S. . is primarily magnetic in origin, and arises from the magnetic moments of the orbit and the spin. **1966** Phillips & Williams *Inorg. Chem.* II. xxviii. 413 Spin-orbit coupling. . tends to sustain orbital angular momentum by coupling it with spin angular momentum.

g. The recording that is on the reverse side or the remainder of a gramophone record; also, a gramophone record having recordings on both sides, esp. by the same performer or performers.

1934 *Gramophone Record* Jan. 18/1 The new one. . has been recorded by three famous singers. . . They have all excelled themselves so. . your choice will have to be guided by the coupling. **1937** *Amer. Speech* XII. 46/1 *Coupling*, a recording on which the same band plays selections on both sides. **1960** *Guardian* 12 Apr. 7/2 The coupling is the Andante Spianato. **1967** *Listener* 23 Feb. 273/3 Prokofiev's first two piano concertos are contrasted in almost every respect except that of their musical idiom, and so they make an excellent disc coupling.

8. coupling coefficient *Electr.*, a coefficient between 0 and 1 which represents the extent or 'closeness' of the coupling between two circuits; **coupling constant**, any constant which represents the strength of the interaction between a particle and a field; **coupling-stroke**, the stroke joining the characters in a cursive hand.

1922 Glazebrook *Dict. Appl. Physics* II. 1035/2 *k* is the 'coupling coefficient', which is defined by $k^2 = M^2/L_1L_2$. **1962** Simpson & Richards *Junction Transistors* xiv. 334 The coupling coefficient *K*, which indicates the extent to which the magnetic flux of one winding links the other, should be between 0·8 and 1·0. **1949** *Physical Rev.* 2nd Ser. LXXVI. 783/2 Thus the pseudoscalar coupling constant should be chosen to fit nuclear forces including these important second order processes. **1968** M. S. Livingston *Particle Physics* ii. 26 The coupling constant used in such calculations . . is also known in atomic physics as the fine structure constant. **1969** Marmier & Sheldon *Physics of Nuclei & Particles* I. iii. 55 In classifying the various possible types of interactions, their relative strengths can best be expressed through values of the respective dimensionless coupling constants. **1906** E. Johnston *Writ. & Illum.* iv. 73 The letters are joined together by means of their coupling-strokes. **1929** *Encycl. Brit.* XVII. 97/1 In the more cursive [hands] linking was carried further, both by the insertion of coupling strokes and by the writing of several letters continuously without raising the pen.

coupon. Add: **c.** *attrib.* and *Comb.*; **coupon-clipper** *U.S.*, a person who has a large number of coupon bonds from which to detach the coupons.

1860 *Ann. Rep. Treasury* (U.S.) 480 Such coupon stock. . may be assigned and transferred by the delivery of the certificates. **1861** *U.S. Statutes* XII. 259 The Secretary of the Treasury. . is authorized to issue coupon bonds, or registered bonds, or treasury notes. **1880** C. J. Stromberg (*title*) Coupon ticketing points from or via Chicago,. . for the use of patrons of the Stromberg improved coupon ticket. **1887** *Nation* (N.Y.) 2 June 460/3 The select few of the coupon-clippers may escape it [*i.e.* being in debt]. **1956** *Ann. Reg. 1955* 80 Mr. Harris's Budget was criticized. . as providing 'a guaranteed annual wage for coupon clippers'.

2. A form, ticket, part of a printed advertisement, etc., entitling the holder to a gift or discount, etc., or designed to be filled up by an intending user or purchaser and forwarded to the advertiser for information, goods, etc. Also *attrib.* (See also *cigarette coupon* s.v. *cigarette 2.)

1906 *Westm. Gaz.* 19 Sept. 4/3, I mean to forbid the use of all coupon goods at my house. **1914** *John Bull* 5 Dec. p. ii, By sending this Coupon with P.O. for 2/6. . the holder is entitled to receive a. . Gold Nibbed Fountain Pen. **1915** *Daily Express* 27 Jan. 6 In addition, a Profit-Sharing Coupon and a Composite Flag of the Allies are enclosed in every packet [of cigarettes]. **1928** *Sunday Express* 19 Aug. 1/4 Newsagents as well as advertisers are perturbed by the menace of the multiple coupon. *Ibid.*, In Manchester many agents cut out the coupons for their customers and sell the couponless copies in bulk to waste-paper merchants. **1961** *Guardian* 15 Dec. 1/3 Kensitas. . reintroduced coupons. . . Hitherto Gallagher has not engaged in coupon trading.

b. *spec.* An entry-form for a competition, esp. for a football pool (see *football coupon* s.v. *football 4). Also *attrib.*

1909 *Daily Chron.* 26 Feb. 4/4 Amending them [*sc.* the Lottery Laws] with a view . . to checking coupon-competitions in newspapers. **1918** *Methodist Times* 5 Dec. 9/1 The amendment of the Gambling Laws, particularly in relation to football coupons, press competitions, [etc.]. *Ibid.*, Will you advocate the suppression of professional and coupon betting? **1940** Graves & Hodge *Long Week-end* xxii. 384 The Pools' promoters . . distributed coupons.

3. One of a series of tickets entitling the holder to a share of rationed food, clothing, etc. Hence **coupon-free** *a.*, obtainable without coupons.

1918 *Times* 25 Feb. 9/5 You must not tear off meat coupons yourself. This duty rests with the retailer. *Ibid.*, Only three coupons each week of a meat card can be used for butcher's meat. The fourth or any of the coupons can be used to buy bacon, tinned meats, poultry, and game. *Ibid.* 26 Feb. 7/6 A whole coupon [at the Express Dairy] entitled one to have stewed steak and carrots, two sausages, or cold ham and tongue. **1918** *Ministry of Food, National Rationing* 4 July, § 6 If you hold sugar coupons and not a sugar ticket, you may choose the retailer with whom you wish to register for sugar. . . § 7 Lard will be rationed separately by means of the brown set of coupons. **1941** *Times Lit. Suppl.* 6 Dec. 614/1 (Advt.), Books are tax- and coupon-free. **1943**, **1946** [see *clothing *vbl. sb.* 5 b]. **1947** *News Chron.* 11 Apr. (Advt.), Coupon free brand new nylon. **1948** 'N. Shute' *No Highway* vii. 188, I ought to get another suit, but there never seem to be any coupons. **1959** J. Braine *Vodi* vii. 116 They found his father checking sweet coupons when they entered.

4. A recommendation given by a party leader to a parliamentary candidate. Also *attrib.*, as *coupon candidate, election, majority.*

1918 *Methodist Times* 5 Dec. 7/2 Men whose tongues are tied by a pledge before they get their 'ticket-of-leave', or 'coupon', as it is contemptuously called, are being chosen for this English Reichstag by a secret cabal sitting in darkness. 'Half-coupon men' is the name bestowed on those who have not got the blessing, but are piteously bleating for it. **1919** G. B. Shaw *Heartbreak House* p. xxxii, Provided the leader will make their seats safe for them by the process which was called, in derisive reference to the war rationing system, 'giving them the coupon'. *Ibid.*, The electorate. . cast out all the coupon candidates at the earliest bye-elections. **1925** F. W. Hirst *From Adam Smith to Snowden* 64 The Coalition continued with a huge 'coupon' majority ready to extend protective measures. **1932** Spender &

Asquith *Life of H. H. Asquith* I. iv. 51 The 'Coupon' election of 1918. *Ibid.* II. lv. 316 The certificate took the form of a letter signed by the two leaders (which Asquith immortalised by dubbing it the 'coupon'). **1945** *Ann. Reg. 1944* 92 The executive of the Labour Party . . rejected uncompromisingly the idea of a 'coupon' election after the 1918 model.

couponned, *ppl. a.* Also **couponed.** Add: **2.** Divided into coupons (cf. *coupon 2 a).

1915 *Lit. Digest* 4 Sept. 484/2 (Advt.), Couponed Pages. . —Six coupons to a page. Put each note on a separate coupon —tear it out when it ceases to be of value.

3. Applied to goods subjected to rationing by means of coupons (see *coupon 3); also, of a person: having to use such coupons. So **coupon** *v. trans.*, to subject (goods) to rationing by means of coupons.

1940 *New Statesman* 21 Dec. 649 They will. . have to eat as meagrely as their couponed fellow-citizens. **1944** *Ourselves in Wartime* 161 When, in 1943, towels, furnishing fabrics, and household clothes were 'couponed', the housewife's problem really started. **1945** *Daily Express* 20 Apr. 2/7 But non-couponed material is so scarce—as also is couponed material—that second sober thoughts must prevail.

coup-stick: see *coup *sb.*[3] 4.

‖ **coureur de bois** (kurör də bwa). Now *Hist.* Also **coureur des bois.** [Fr., lit. 'wood-runner'.] A woodsman, hunter, trader, etc., of French or French-Indian origin, in Canada and the northern and western United States. Also *ellipt.* as *coureur*.

1700 in *Docum. Col. Hist. N.Y.* IV. 749 Severall of the French Coureurs de bois or hunters are there at this time. **1750** W. Douglass *Settlem. N. Amer.* II. 245 The French, consisting of 500 Coureur des bois (in New England they are called Swampiers), with as many Indians or savages. **1877** F. Parkman *Count Frontenac* (1897) v. 57 Du Lhut, the famous leader of *coureurs de bois*. **1900** E. B. Osborn *Greater Canada* 21 The merchant. . handed over their value. . to the *coureur*. **1920** *Punch* 30 June 502/2 His lifelong ambition to be painted by Mr. John, with the primeval backwoods for a setting, in the character of a *coureur-des-bois*. **1958** E. McInnis *Canada* (ed. 2) v. 85 The *coureurs* were ranging beyond Lake Superior. **1969** E. W. Morse *Fur Trade Canoe Routes* i. i. 8 The *coureur de bois* belongs to the French period of Canadian history, while the voyageur belongs to the period after 1763.

courge (kū·ɹʒ). [Fr., = gourd.] A basket, towed behind a fishing-boat, for holding live bait.

1865 J. C. Wilcocks *Sea-Fisherman* 45 Manufacture of 'cou[r]ges' or Sand Eel baskets. **1898** *Encycl. Sport* II. 321/2 The 'courge', a torpedo-shaped basket into which the baits are introduced by an aperture that is closed with a bung. *Ibid.* 322/2 The orthodox Channel Islands 'courge'. **1900** *Cornhill Mag.* Nov. 626 We have lively sand-eels towing in the courge.

courgette (kū*ə*ɹʒe·t). [Fr., dim. *courge* gourd.] A variety of small vegetable marrow.

1931 E. Lucas *Vegetable Cookery* 266 *Courgettes* may be. . treated in the same way as vegetable marrows. **1951** E. David *French Country Cooking* 224 You can bring back . . baby *courgettes* or marrows. **1958** *Sunday Times* 23 Mar. 23/3 A luncheon. . starting with courgettes stuffed with shrimp. **1965** *Oxford Mail* 12 Feb. 8/5 Courgettes, at 5s. a lb., are delicious fried or cooked in the oven.

courida (kūridā·). *Bot.* Also **courada.** [Native name.] The common name in Guyana for the black mangrove, *Avicennia marina*, a small tree which grows on muddy flats along the sea-shore in the tropics; also, the timber from this.

1825 C. Waterton *Wand. S. Amer.* i. 10 The pelicans. . return at sundown to the courada trees. **1840** R. H. Schomburgk *Descr. Brit. Guiana* 7 A fringe of low ground, which is soon covered with. . Courida Bushes. **1851** *Illustr. Catal. Gt. Exhib.* iv. i. 984/1 Courida, transverse and vertical sections, from Plantation Woodlands, River Mahaica. **1866** Lindley & Moore *Treas. Bot.* s.v. *Avicennia*. **1912** Rodway *Guiana* 26 The coast is generally an alluvial flat. . fringed with courida. **1958** J. Carew *Wild Coast* vi. 73 He was able to run and swim, cut down courida trees.

courier, *sb.* Add: **1. b.** A messenger for an underground or espionage organization.

1929 C. Mackenzie (*title*) Three couriers. **1943** G. Greene *Ministry of Fear* iii. ii. 188 He's one of the used, the blackmailed. That doesn't mean. . that he isn't the courier. **1957** H. Roosenburg *Walls came tumbling Down* 8 She. . became a courier. . for a resistance group that transmitted intelligence. . to the Dutch government in London. **1959** *Encounter* Dec. 53/2 The life of Communist couriers in the 'twenties and early 'thirties.

5. *attrib.* and *Comb.*

1874 Lady C. Schreiber *Jrnl.* (1911) I. 272 Saw there old Hockster, the courier-dealer. **1958** 'N. Shute' *Rainbow & Rose* i. 4 In 1942 I had met him in Cairo when he was flying a courier service to England. **1963** P. MacTyre *Fish on Hook* vii. 111 Is he running a courier service, or an escape route, as part of a spy system?

courier, *v.* [f. the sb.] *intr.* and *trans.* 'To act or attend as a courier' (Webster 1934); to travel as a courier.

1921 W. R. Benét in F. Wilkinson *New Voices* 372 Marquises and admirals and barons of delight All courier his chariot. **1959** *Encounter* Dec. 53/2 Padmore couriered across Europe in the service of the Party, organising and taking risks. **1962** P. Purser *Peregrination 22* i. 8 I'd. . couriered two parties to a Berlin festival.

couril (kū·ril). [Breton.] In Brittany, a small fairy, supposed to frequent druidical remains.

1889 in *Cent. Dict.* **1924** *Chambers's Jrnl.* 770/2 The great circle was built by the Courils in a single night.

course, *sb.* Add: **12. a.** Also of aircraft: the (correct) line or direction of flight.

1905 G. Bacon *Balloons* vi. 86 To steer his course in a balloon. . the aeronaut must so arrange that he is travelling faster or slower than the wind. **1909** C. C. Turner *Aerial Navig.* xii. 181 Heavier-than-air machines. . are. . liable to be driven out of their course in strong winds. **1933** *Bur. of Standards Jrnl. Research* XI. 741 Its [*sc.* the direction finder's] operation was entirely satisfactory, indications right and left of 'course' being very steady and definite. **1945** T. A. Dickinson *Aeronautical Dict.* 99/1 *Course*, the route or direction that should be or has been flown by an aircraft. It may be a true course, a magnetic course, or a compass course. **1948** 'N. Shute' *No Highway* ii. 41 The aircraft was on course at 16,500 feet.

29. h. A row of stitches or loops across the width of a knitted fabric.

1940 in *Chambers's Techn. Dict.* **1964** *McCall's Sewing* vii. 98/1 The chains running across the fabric are called 'courses' and correspond to the crosswise grain of woven fabrics.

39. course-clearer, one who clears the course for a race; **course-dinner** ?*Obs.*, a dinner consisting of several courses; **course-indicator**, an apparatus for determining the course of a ship; **course-setting sight** *Aeronaut.*, a sight by means of which one's course can be set.

1897 *Westm. Gaz.* 15 July 7/1 The course-clearers ring, and the enthusiasts yell. **1895** *Outing* (U.S.) XXVI. 462/2 A course-dinner followed by the enjoyment of Ada Rehan's *Rosalind* and Lewis's *Touchstone*. **1899** *Westm. Gaz.* 16 Mar. 10/2 Mr. Kipling was able to sit up in bed last evening, and to partake of a course dinner. **1900** *Ibid.* 14 Feb. 10/2 An absolutely correct magnetic course indicator. **1904** *Ibid.* 18 Aug. 1/3 A holophote course-indicator for preventing collisions at sea. **1922** *Encycl. Brit.* XXX. 43/1 The course-setting sight. . indicates the course to be steered for any given track, and the time taken in flying any desired distance in that direction. **1940** *Illustr. London News* CXCVII. 219/2 The latest type bomb-sights in use in the R.A.F. may not be described, but the type illustrated gives a good idea of how a course-setting sight is operated.

'course, course (kō*ə*ɹs), colloq. abbrev. of *of course* (see Course *sb.* 36 c).

1886 in Baumann *Londinismen*. **1901** Merwin & Webster *Calumet 'K'* i. 13 Have you tried to get any of it here in Chicago? Course not. It's all ordered and cut out up to Ledyard. **1904** Mrs. Stratton-Porter *Freckles* xvi. 326 'I bet you it's a marked tree!' 'Course it is!' cried the Angel. **1967** O. Norton *Now lying Dead* iv. 58 'Course, he might go off. He does sometimes.

‖ **course libre** (kūrs libr'). [Fr., lit. 'free course'.] (See quot. 1962.)

1959 L. Durrell in *Holiday* Nov. 71/1 The Spanish form of bull-fighting. . is. . ritualized murder; but the *course libre* is an exciting and extremely dangerous game, a test of strong nerves and speed. **1962** M. Summerton *Nightingale at Noon* iii. 36 He breeds black bulls for the *course libre*, that's the French style of bull-fighting where they bait the bulls but don't kill them. **1964** F. White *West of Rhone* iv. 51 The black bulls for the bull-fights and the *courses libres*.

court, *sb.*[1] Add: **6. d.** *ellipt.* A court shoe. Also *attrib.*

1903 in S. Nowell-Smith *Edwardian Eng.* (1964) iv. plate xviii (b), Ditto in Court Shape 8/9. **1959** *Vogue Pattern Bk.* June–July 14 Accessory suggestions. . Lotus calf courts.

11. d. *court of love*: an institution said to have existed in southern France in the Middle Ages, a tribunal composed of lords and ladies deciding questions of love and gallantry; such an institution in medieval literature.

c **1530** *Crt. of Love* (1897) 410 Commaunding me the Court of Love to see. ., There Citherea goddesse was and quene. **1840** R. Browning *Sordello* ii. 47 Eglamor whose lay Concludes his patron's Court of Love to-day. **1877** C. M. Yonge *Womankind* xxi. 161 The courts of love in Provence were to decide on the cases of fantastic adoration between knights and ladies. **1933** R. Tuve *Seasons & Months* iv. 170 The chariot of Phebus the sun. . remained as mere learned decoration in Scottish court-of-love poems. **1964** F. White *West of Rhone* v. 69 At Bordeaux. . great ladies held Courts of Love at which they awarded prizes to the troubadours who sang the sweetest songs.

12. c. *out of court*: in extended fig. use, of any thing or person that has no claim to be regarded or considered.

1907 J. R. Illingworth *Doctr. Trinity* iii. 47 When Christian apologists with mistaken courtesy allow the Fourth Gospel to be put out of court. **1909** W. de la Mare *Private View* (1953) 198 Longfellow, Emerson, and hosts of lesser men he laughed out of court. **1912** J. S. Huxley *Individ. in Animal Kingdom* i. 21 This. . is alone enough to rule out of court the pretensions of all inorganic constellations to individuality. **1927** C. & M. Beard *Rise Amer. Civ.* II. xxiv. 347 If there had been any humor in international finance, many of their pleas would have been laughed out of court. **1927** *Observer* 20 Nov. 26 If he is using a horn type loud-speaker, he is out of court at once. **1960** C. S. Lewis *Studies in Words* iv. 89 A study of *wit* which does not take full account of its relation to *ingenium* would be out of court.

19. court-metre, the *dróttkvætt* metre used in the old Icelandic *drápa* or heroic laudatory poem, which was recited before the king and his retinue (*drótt*); **court shoe**, a woman's light, low-cut shoe, often with a high or highish tapering or curved heel; **court tennis** *N. Amer.* = Tennis *sb.* 1; **court week** *U.S.*, the week in which the county court meets.

1883 Vigfusson & Powell *Corpus Poet. Bor.* I. 432 The reason why the court-poets counted by the half-line. . is manifest. The court-metre was their standard. **1896** *Scot. Rev.* XXVIII. 334 *Drótt-kvætt*. . was thus the commonest metre for the *drápa*, or laudatory poem. . . The name of 'court-metre' is thus appropriate enough. **1885** *Queen* 20 June (Advt.), The 'Watteau' or 'Queen Anne' court shoe. **1932** S. Gibbons *Cold Comfort Farm* xi. 156 Elfine, *whatever* you do, always wear court shoes. **1933** *Yorkshire Even. Post* 1 Sept., In the [American] shoe department confusion can arise if one asks for *court-shoes*, for which the American is *pumps*. **1911** *Daily Colonist* (Victoria, B.C.) 9 Apr. 1/2 Jay Gould, of New York, won the national court tennis championship for the sixth successive year this afternoon. **1930** *Times Lit. Suppl.* 1 May 375/2 Tennis—in America known as court tennis. **1724** in *Essex Inst. Coll.* XXXVI. 332 Jan. 2 Court Week. **1944** Duncan & Nickols *Mentor Graham* 13 Greensburg may have been gay during court week, but education was its hobby.

‖ **court bouillon** (kūr bu̇lyɔ̃). [F., f. *court* short + *bouillon* Bouillon 1.] A stock in which fish is boiled, consisting of water, wine, vegetables, seasoning, etc.

1723 J. Nott *Cook's & Confect. Dict.* sig. Z4, Let them [*sc.* perch] stand by to cool in your *Court Bouillon*, i.e. in your forementioned Liquor and Seasoning. **1855** E. Acton *Mod. Cookery* p. xiii, *Court bouillon*, a preparation of vegetables and wine in which (in expensive cookery) fish is boiled. **1960** *House & Garden* Aug. 50 Poach them [*sc.* the fish] first in a very lightly seasoned *court bouillon* with or without the addition of a little white wine.

court-craft. Add: **2.** Craft or skill in the movements and positioning required on a tennis-court, as distinguished from the strokes.

1923 *Daily Mail* 28 Mar. 9 She lacks courtcraft, and is frequently out of position. **1928** *Observer* 17 June 6 The learner will find most of the teachers agreeing about grip, position, swing, follow-through, balance, court-craft, [etc.].

court-cupboard. Delete † *Obs.* and add later examples.

1912 C. Mackenzie *Carnival* xxxvi. 377 Oak chests-of-drawers and Court cupboards. **1961** R. Jeffries *Evidence of Accused* iv. 39 The main pieces were genuine Elizabethan. . . There was a small court-cupboard.

court-day. 1. (Later examples.)

1803 J. Orrok *Let.* 12 Apr. (1927) 44 But on Wednesday 30th March being a Court day at the India House, the Directors took it in Consideration. **1849** C. Lanman *Alleghany Mts.* xviii. 148 The hermit said he would attend to it on the next court-day, and his word was accepted. **1906** F. Lynde *Quickening* 28 Could this be South Tredegar, the place that had hitherto figured chiefly to him as 'court-day' town?

courtesy, *sb.* **12.** Add: Also *attrib.*, as *courtesy campaign*; **courtesy call** = * *courtesy visit*; **courtesy card** (orig. *U.S.*), a card entitling the holder to certain privileges; **courtesy cop** *colloq.*, a policeman whose duty it is to persuade motorists, etc., to good behaviour by courtesy rather than by toughness; so *courtesy patrol*; **courtesy light**, a light inside a car that is automatically switched on when one of the doors is opened; **courtesy rank, title** (see 3 b in Dict.); **courtesy visit** (see quot. 1934).

1957 R. Graves tr. *Suetonius's 12 Caesars* ix. 270 One of these men paid a courtesy call at the Palace. **1970** W. J. Burley *To kill a Cat* i. 7 Detective Chief Superintendent Wycliffe. . was on holiday though paying a courtesy call at the local police station. **1962** *Economist* 13 Jan. 139/1 The 'courtesy campaign', organised by a group of progressive women, has had some success. **1934** *Amer. Speech* IX. 111/1 For families planning a vacation there is an abundance of free travel literature, road maps, courtesy cards, etc. **1968** *Globe & Mail* (Toronto) 17 Feb. 33 The ticket for all these activities is called the College Week courtesy card. It is issued only to bona fide students who can present their college identity card. **1938** *Daily Mail* 4 Apr. 9/5 In the London area the 'courtesy cops' will concentrate at first on four of the busiest roads. **1940** Graves & Hodge *Long Week-End* xxii. 379 The 'courtesy cops'—policemen in cars with orders to warn drivers politely but firmly. **1962** M. Urquhart *Frail on N. Circular* i. 8 A sergeant on motor-cycle patrol, he came under the category of courtesy cop. **1959** *Times* 2 Oct. 11/3 Courtesy lights are now actuated by the opening of any of the four doors. **1961** *Economist* 21 Oct. 228/1 Even the men who police the roads [in Colorado] are called a 'courtesy patrol' and have no police power off the highways, and very little on them. **1934** Webster, *Courtesy* adj.,. . granted or performed as a courtesy or by way of courtesy; as, a *courtesy* letter or visit. **1937** *Burlington Mag.* Mar. 112/2 It is no doubt quite possible that Guercino paid a courtesy-visit to Lodovico.

court-hand. Add to def.: Also of earlier periods.

1888 *Encycl. Brit.* XXIII. 682/1 The *court* or *charter hand* was used [in the 13th–15th cents.] for charters, title-deeds, papal bulls, &c. **1915** Johnson & Jenkinson *Eng. Court*

Hand p. xiii, To put before the student a large. . series of examples of the kind of writing usually to be found in English documents from the Conquest to A.D. 1500. The survival of debased forms of this type of writing in modern legal engrossing has caused it to be called. . 'Court Hand'; and. . this unscientific name will serve well enough to distinguish it from 'Book Hand', the more leisurely type of writing employed in copying MSS. **1960** C. E. Wright *Eng. Vernacular Hands* 3/1 Written about A.D. 1200. . . A vigorous, individualistic hand closely related to the court-hand of the time. **1969** M. B. Parkes *Eng. Cursive Bk. Hands 1250–1500* p. xvi, Scripts of this kind have usually been referred to in England as 'Court Hands', or 'Charter Hands'.

courting, *vbl. sb.* Add: **2.** (Later example.)

1908 *Daily Chron.* 2 June 1/2 As courting is a necessity with them [*sc.* young people], I am determined they shall have opportunities for this innocent amusement.

3. (Further examples.)

1841 G. Catlin *N. Amer. Indians* I. 243, I often and familiarly heard. . the Winnebago courting flute. **1855** *Knickerbocker* XLVI. 217 Observe how naturally the 'courtin'-room and its accessories are described. **1887** *Harper's Mag.* Feb. 354/1, I have read that the gallants train their little horses to prance and curvet and rear and fidget about, and that these are called 'courtin' horses', and are used when a young man goes courting, to impress his mistress with his manly horsemanship. **1911** Thomson *Biol. Seasons* ii. 220 Anything that might be interpreted as love-play or courting-play.

courting, *ppl. a.* (Later example.)

1922 Joyce *Ulysses* 525 Did he not pass night after night by loving courting couples to see. . how much he could see?

courtly, *a.* Add: **2. b.** *courtly love*: a highly conventionalized medieval system of chivalric love and etiquette first developed by the troubadours of southern France and extensively employed in European literature from the 12th century throughout the medieval period. Cf. *amour courtois.

1896 L. F. Mott *Syst. Courtly Love* iv. 24 Springing forth in the earliest troubadours,. . the sentiments of courtly love attain their full growth in the epics of Chrétien de Troies. *Ibid.* 27 Courtly love was declared incompatible with marriage. **1899** J. H. Smith *Troubadours at Home* I. iv. 58 Courtly, chivalric love, the gospel of the troubadours,. . was love for married women. **1924** L. A. Hibbard *Med. Romance in Eng.* iii. 209 To minds filled with the precepts of courtly love, the *fée's* command was completely in accord with the insistence of courtly love doctrines on the necessity for secrecy in love. **1933** R. Tuve *Seasons & Months* iv. 157 The. . pleasaunces and allegorical fountains of the French courtly-love-garden tradition. **1936** C. S. Lewis *Alleg. Love* i. 2 Courtly love. . is. . love of a highly specialized sort, whose characteristics may be enumerated as Humility, Courtesy, Adultery, and the Religion of Love. **1963** I. Murdoch *Unicorn* xii. 106 Beauty. . commands worship rather than arousing desire. That is the meaning of Courtly Love.

court-room. (Earlier Amer. example.) Also *attrib.*

1677 *New Castle* (Del.) *Court Rec.* (1904) I. 143 For ye makeing up of ye Court Roome in ye forte fitt for ye Court to sitt in. **1904** *Daily Chron.* 24 May 6/3 He gets a better living than he would at the music-halls if his court-room humour is the best he is capable of. **1962** A. Nisbett *Technique Sound Studio* ii. 42 A record of courtroom 'atmosphere'.

courtship. Add: **8.** *attrib.* (chiefly sense 6 b).

1801 *Monthly Mirror* Sept. 203 In the most unnatural court-ship scene with Lady Anne. . I saw only the tyrant hatching plans. **1923** J. S. Huxley *Ess. Biologist* iv. 135 In a large number of birds. . the courtship-displays are mutual. **1924** J. A. Thomson *Sci. Old & New* xxiii. 130 In the case of spiders, it appears highly probable that the courtship-dance serves to excite the interest and sex-instincts of the female. **1938** *Brit. Birds* XXXI. 320 The courtship flight and trill is the chief form of courtship activity. **1948** *Ibid.* XLI. 127 Courtship-feeding of Scottish Crossbill. **1950** *Oxf. Jun. Encycl.* IX. 235/1 Courtship Dances. . developed out of the fertility rites. . . Often in such dances the man shows off before the lady by performing the most spectacular tricks. **1959** A. Hardy *Fish & Fisheries* i. 3 A male dragonet in courtship dress may flash with rainbow hues.

‖ **coûte que coûte** (kut kə kut). Also erron. **coûte qui coûte, coûte qu'il coûte.** [Fr.] Cost what it may; at all costs.

1715 Ld. Bolingbroke *Let.* in P. M. Thornton *Stuart Dynasty* (1890) 397, I could heartily wish yt Campion might be come back to ye coast of Britanny, before your Majty sail'd, which he will attempt *coute qu'il coute!* **1738** Pope *Imit. Horace, Sat.* ii. vi. 17 Knew what was handsome, and wou'd do't, On just occasion, *coute qui coute*. **1841** C'tess Blessington *Idler in France* II. 170, I determined, *coute qui coute*, to pay a visit to my friend Madame Craufurd. **1863** Reade *Hard Cash* I. vii. 173 All the mother was in arms to secure her daughter's happiness coûte que coûte! **1934** F. Stark *Valleys of Assassins* i. 21 We should have lunch first, said the policeman, anxious *coûte que coûte* to make us do something we had not intended.

couth, *pa. pple.* and *a.* (*sb.*). Add: **6. b.** [Back-formation f. Uncouth *a.*] Used as a deliberate antonym of Uncouth *a.* 6: cultured, well-mannered, etc. Also *absol.*

1896 M. Beerbohm in *Pageant* 230 The couth solemnity of his [Pater's] mind. **1956** L. McIntosh *Oxford Folly* 55, I think I'd make a jolly good waitress. It's rather a *couth* thing to do, really. **1963** *Guardian* 28 Mar. 9/1 Modern idiom and slang is used with reckless abandon and the couth and uncouth punch each other about the ears with unrelenting monotony. **1965** J. Philips *Twisted People* i. iv. 77 Didn't I say it polite? Mr. Delafield always says I'm not couth. **1968** *Queen's Coll.* (Oxf.) *Rec.* 12 The Waynflete building . . is a scaleless slab of uncouth outline which academically couth details (inscription in Roman letters on the cornice, etc.) do not redeem.

‖ **couture** (kutür). [Fr., sewing, dressmaking.] Fashionable dressmaking or design; used as a collective term for the designers or makers of women's fashionable clothes, esp. in Paris; also, the clothes made by them. So (*la*) *haute couture*: high fashion; the fashionable world; the leading dressmakers and dressmaking establishments collectively. Also *attrib.*

1908 A. Bennett *Old Wives' Tale* iii. ii. 290 Gerald had taken her to an establishment in the Chaussée d'Antin. . . The real *haute couture* was practised therein. **1928** *Daily Mail* 3 Aug. 3/2 This casque is affected by Parisiennes, who are following the line of restraint that is a feature of the higher *couture* just now. **1934** A. Haskell *Balletomania* xi. 219 The Sakharoffs. . belong to. . the costume school, the triumph of *la haute couture*. **1940** *Manch. Guardian Weekly* 2 Feb. 97 The Paris dressmaking firms reopened soon afterwards and the traditions of haute couture are being maintained. **1955** *Times* 1 Aug. 8/5 Leading haute couture houses are by no means in agreement about the new silhouette. **1957** *Times* 25 Oct. 13/4 Dior's name will stand high in the records of fine achievement in the field of couture design. *Ibid.*, At this time Boussac was, in fact, looking for a designer in order to set up a couture house. **1966** *Listener* 3 Feb. 171/1, I find it far more difficult to get an invitation to *haute couture* shows than to attend a political conference.

‖ **couturier** (kutürye). [Fr.] A male dressmaker or fashion designer. Also *attrib.*

1899 Beerbohm *More* 147 For all the devices of her *couturier*, Fashion never looked aught but her worst. **1928** *Daily Mail* 9 Aug. 4/3 A delicate carmine lipstick, a new shade just evolved by a leading Paris couturier. **1930** E. Waugh *Labels* 18 There is an inscrutable world. . behind the industry of making women's clothes. . . The high diplomacy of the *couturiers*. **1957** *Times* 26 Oct. 7/3 It has been asked several times recently whether it can reasonably be expected of any couturier that he should switch from couture to wholesale design and back again interminably. **1968** 'P. Hobson' *Titty's Dead* vii. 78 Miss Plumpton, secure in couturier clothes, wove graciously in and out.

‖ **couturière** (kutüryɛ̄r). [Fr.] A dressmaker, modiste.

1818 Moore *Fudge Fam. Paris* xii. 88 That enchanting *couturière*, Madame le Roi. **1842** Thackeray in *Fraser's Mag.* June 712/1 Actresses, couturières, opera-dancers. **1929** *Daily Express* 26 Jan. 5 Couturières are doing strange, subtle things in the way of drapery. **1967** C. O. Skinner *Madame Sarah* vii. 131 Bernhardt had. . debts. . to tradesmen and *couturières*.

‖ **couvert** (kuvɛ̄r). [Fr.] = Cover *sb.*[1] 7.

1768 Sterne *Sent. Journ.* II. 158, I had got hold of the *couvert* of some more entertaining guest. **1873** C. M. Yonge *Pillars of House* III. xxxiii. 225 Zadok came up with the three couverts. **1899** Newnham-Davis *Dinners & Diners* i. 6 And the bill. . three couverts, 3s.; caviar, 3s. **1949** H. Miller *Plexus* (1953) II. xii. 190 The dishes were broken and chipped,. . and the cutlery was of tin. He laid the 'couvert' on the bare table. **1968** V. C. Clinton-Baddeley *My Foe Outstretch'd* i. 22 He examined the bill narrowly. Couvert 2/6—ridiculous: that was the selective employment tax.

couverture (kuvɛ̄rtür). [Fr., lit. 'covering'.] Chocolate suitable for covering sweets, cakes, etc.; a covering layer of such chocolate. Also *attrib.*

1935 *Discovery* Nov. 322/2 Condensation occurs on the surface of the sweetmeats, causing ultimate undesirable crystallisation of the sugar in the chocolate couverture. **1951** *Good Housek. Home Encycl.* 406/2 The exact proportions depend on the purpose for which the couverture is required and the type of centre to be covered. *Ibid.*, Couverture chocolate retains a high gloss after melting and cooling. **1959** [see *baba[2]].

couvre-pied (kuvrəpye). Also **couvre-pieds.** [Fr., lit. 'cover-foot', f. *couvrir* to cover, *pied* foot.] A rug to cover the feet.

1818 Lady Morgan *Fl. Macarthy* II. iii. 154 Will you just inquire for a couvre-pied, that is lying loose somewhere in the carriage? **1879** Mrs. Oliphant *Within the Precincts* xli. 172 A wadded *couvre-pied*. **1905** Mrs. H. Ward *Marr. W. Ashe* ii. xii. 228 He returned to his armchair and his books, but soon drew Kitty's *couvre-pied* over him and went to sleep. **1926** E. F. Thomas *Rendezvous* xxvi. 298 It was a pretty *couvre-pieds*, pale blue and comfortably quilted.

covalence (kōᵤvē·lēns). *Chem.* [f. Co- 3 a + Valence[2] 3.] = *covalency a, b.

1919 I. Langmuir in *Proc. Nat. Acad. Sci.* V. 255 It is therefore proposed to define valence as the number of pairs of electrons which a given atom shares with others. In view of the fact. . that valence is very often used to express something quite different, it is recommended that the word *covalence* be used to denote valence defined as above. *Ibid.* 256 In sodium chloride the covalence of both sodium and chlorine is zero. **1948** Glasstone *Physical Chem.* (ed. 2) i. 104 The covalence of any element, not including dative covalences, should consequently be equal to the number of electrons with unpaired spins in the given atom. **1957** *Encycl. Brit.* XXII. 946/1 Another kind of valence of great importance is 'covalence', which results from the sharing of two electrons by two atoms.

Hence **cova·lent** *a.*, **cova·lently** *adv.*

1927 Covalent [see *electro-valent* adj. s.v. *electro-]. **1946** *Nature* 12 Oct. 519/1 In compounds containing hexavalent chromium. . the bond between chromium and oxygen is not fully ionic but partially covalent in nature. **1962** P. J. & B. Durrant *Introd. Adv. Inorg. Chem.* iv. 112 The nitrogen molecule may be considered as two nitrogen atoms linked by a triple covalent bond. . . In the molecule there are ten valency electrons, six of which are shared. **1962** S. G. Waley in A. Pirie *Lens Metabolism Rel. Cataract* 357 Both biotin and lipoic acid. . are covalently bound to protein.

covalency (kōᵤ,vēi·lĕnsi). *Chem.* [f. Co- 3 a + Valency.] **a.** The linking of two atoms by a bond in which they 'share' a pair of electrons; a covalent bond. **b.** The number of electrons in an atom that go to form a covalent bond; the number of covalent bonds that an atom can form.

1919 I. Langmuir in *Proc. Nat. Acad. Sci.* V. 255 The covalency of an element according to the Octet Theory can never exceed four, since there are only four pairs of electrons in an octet. **1946** *Jrnl. R. Aeronaut. Soc.* L. 392 A tendency towards covalency formation in place of the ionic bond (e.g., the formation of a volatile chloride which is non-ionised in the solid state). **1950** N. V. Sidgwick *Chem. Elements* I. p. xvii, The maximum value of the covalency is limited . . in accordance with the period of the atom in the table. **1964** N. G. Clark *Mod. Org. Chem.* ii. 24 There is a second mode of covalency formation . . in which *both* the electrons of the resulting valency bond are provided by one of the atoms.

covariance (kōᵤ,vē*ə*·riăns). [In sense 1 f. Covariant; in sense 2 f. Co- 3 a + Variance.] **1.** *Math.* The property of being covariant; the property of a function of retaining its form when the variables are linearly transformed.

1878 [see Invariance]. **1922** E. P. Adams tr. *Einstein's Meaning of Relativity* i. 22 We shall also test for co-variance the equations which express the dependence of the stress components of the matter. **1952** C. Møller *Theory of Relativity* ix. 265 In the special theory of relativity, the covariance of the laws of nature under Lorentz transformations could be very elegantly expressed by means of the four-dimensional tensor calculus. **1966** *McGraw-Hill Encycl. Sci. & Technol.* XI. 436/2 Most physicists agree that any future developments or modifications of Einstein's theory [of relativity] will have to satisfy the principle of covariance.

2. *Statistics.* The mean value for a population of the product of the deviations of two or more variates from their respective means; it is equal to the correlation coefficient of the variates multiplied by the standard deviation of each variate; *analysis of covariance* or *covariance analysis*: an extension of the analysis of variance to investigate and adjust for the dependence of the variate on one or more concomitant observations.

1931 A. L. Bailey in *Jrnl. Amer. Statist. Assoc.* XXVI. 424 (*heading*) The analysis of covariance. *Ibid.*, The term covariance is used here to designate the first product moment: $S(x-Mx)(y-My)/N$. **1939** [see *covariation]. **1969** B. E. Cooper *Statistics for Experimentalists* iv. 44 The covariance between two variables . . is a measure of the joint variation between the two variables.

covariancy (kōᵤ,vē*ə*·riănsi). [f. prec.: cf. Variancy.] Covariance or covariation.

1895 [see Invariance]. **1933** *Mind* XLII. 290 The Selective Theory . . accounts for the co-presence and co-variancy of organic and non-organic sensa.

covariant. Add: (Further examples.)

1959 *Chambers's Encycl.* I. 260/2 The extensive theory of invariants and covariants has many applications in geometry. **1967** Condon & Odishaw *Handbk. Physics* (ed. 2) i. x. 144/1 In theory of relativity the distinction between quantities which change with the reference system (covariants) and quantities which do not change with the reference system (invariants) is of paramount importance.

B. *adj.* Changing in such a way that inter-relations with another simultaneously changing quantity or set of quantities remain unchanged; correlated; *spec.* in *Math.*, having the properties of a covariant; of or pertaining to a covariant.

1905 C. J. Joly *Man. Quaternions* xvii. 290 The functions . . undergo the same transformation [as the original functions] and may be said to be covariant with the original functions for this type of transformation. **1920** R. W. Lawson tr. *Einstein's Relativity* xiv. 43 General laws of nature are co-variant with respect to Lorentz transformations. **1956** *Nature* 11 Feb. 268/2 The possibility of having an acceptable theory which, although not covariant . . leads only to covariant observable predictions. **1957** in Saporta & Bastian *Psycholinguistics* (1961) 290/2 The data obtained should be demonstrably covariant with those obtained with some other, independent index of meaning. **1959** J. Aharoni *Special Theory of Relativity* ii. 85 Tensors are said to be covariant entities and absolute relations between them, which do not change their form as the coordinates are transformed, are said to be covariant relations. **1968** *Amer. Jrnl. Physics* XXXVI. 1103/2 The two Einstein postulates state that the laws of physics and speed of light do not change for relatively moving reference systems (being covariant and invariant, respectively).

covaria·tion. [Co- 3 a.] Correlated variation.

1925 R. A. Fisher *Statist. Methods* i. 6 This study..is known in English under the name of Correlation, but by some continental writers as Covariation. 1939 C. H. Goulden *Methods Statist. Anal.* xv. 247 For a single variable the variation is frequently heterogeneous and may be sorted out into components... The same is true for the correlated variability or covariation of two variables, and the mechanism for sorting out the covariance effects is known as the analysis of covariance. 1970 *Language* XLVI. 401 Many American Indian languages show types of obligatory covariation between verbal roots and nominal expressions.

co-vary (kōᵘˌvēᵊ·ri), *v.* [Back-formation from prec.] *intr.* (See quot. 1961.) So **co-va·rying** *vbl. sb.* and *ppl. a.*

1950 F. Eggan *Social Organiz. W. Pueblos* 8 Conditions under which social phenomena correlate or covary. 1957 *Eng. Lang. Teaching* Oct.–Dec. 8 In the measure that the co-varying of the sentences with the situations..takes effect, structure is generated and informs the sentences. *Ibid.*, He must so dispose his presentation as to make the influence of other relevant co-varying sequences as operative as may be. 1961 C. H. McDowell *Dict. Math.* 33 *Covary*, to change together with something else so as to preserve certain interrelations unchanged. 1970 *Nature* 16 May 618/1 If alpha waves are indeed generated by the potential across the eye, the amplitudes of the former and the latter should co-vary.

cove, *sb.*[1] Add: **6. c.** *Archæol.* A setting of a small number of stones close together within a henge monument.

1743 Stukeley *Abury* viii. 37 Thus we cast up the number [of stones]. The outer circle of Abury town, 100. ..The cove, 003. ..Longstone cove jambs, 002. *Ibid.* x. 47 Another stonework towards the other end; which seems to have been a semicircular cove, or demi-ellipsis consisting of five great stones. 1819 Colt Hoare *Anc. Hist. N. Wilts* 68 A *cove*, consisting of three large stones placed with an obtuse angle towards each other. 1936 *Proc. Prehist. Soc.* II. 25 At Avebury and at Arbor Low there are remains of more complex structures, sometimes known as coves. 1963 *Field Archaeol.* (Ordnance Survey) (ed. 4) 40 Both at Stanton Drew and at Avebury there is a curious arrangement of stones known as a 'cove' which recalls, though on a larger scale, the burial place in a chambered barrow.

cove, *sb.*[2] (Later Australian examples.)

1911 C. E. W. Bean *'Dreadnought' of Darling* xxxiii. 288 Recollec' that cove with a red beard. 1916 *Anzac Book* 65 Then a corporal called and wanted to know..when would the rubber boots be ready for the coves in the trenches? 1916 J. B. Cooper *Coo-oo-ee* vii. 84 'He's one of those smart coves,' said Sam. 1944 F. Clune *Red Heart* 67 'Must be a balmy cove,' whispered one of the hangers-on as he tapped his forehead. 1966 'J. Hackston' *Father clears Out* 190 The young coves round about combed their hair back with soap to keep it in position. 1969 *Advertiser* (Adelaide) 12 May 5/4 You Aussie coves are just a bunch of drongoes.

cove, *v.* **1.** (Later example.)

1708 E. Cook *Sot-weed Factor* 2 Weighing soon, we plough'd the Bay, To Cove it in Piscato-way.

coventrate (kǫ·vəntrēⁱt), *v.* (*Temporary.*) [f. *Coventr(y* (see below) + -ATE[3] after G. *coventrieren.*] To bomb intensively; to devastate sections of (a city) by concentrated bombing, such as that inflicted on Coventry, Warwickshire, in November 1940. So **co·ventrating** *vbl. sb.*; **coventra·tion.**

1940 *Hutchinson's Pict. Hist. War* 2 Oct.–26 Nov. 221 German bombers made prolonged mass attacks..on Coventry...And..they invented the verb 'to coventrate' to describe the indiscriminate mass murder of civilians. 1940 *New Statesman* 21 Dec. 647 The fact was that 'Coventrating' meant that the nerves and sinews and muscles of local government were wrenched and lacerated. 1942 L. E. O. Charlton *Britain at War* 22 Possibility of another 'coventration' of a manufacturing centre. 1944 H. Hawton *Night Bombing* viii. 126 It was the Germans themselves who had coined so gloatingly the verb 'to coventrate'.

cover, *v.*[1] Add: **2. g.** *Calico-printing.* To print over with a pattern, etc. (Cf. *COVER *sb.*[1] 1 c.)

1874 Crookes *Dyeing* 564 The piece is then printed over—technically called 'covered'—with a fine pattern, in purple or light chocolate, and dyed up with madder. *Ibid.* 578 The pieces are then 'covered' with iron liquor..to produce the small design in dark purple.

h. To affix the covering of cloth, leather, or other material on the boards of (a book) in binding.

1836 Hebert *Engin. & Mech. Encycl.* I. 226 The book is then ready for covering, with leather, if to be whole bound. 1876 *Encycl. Brit.* IV. 43/2 The materials used for covering are very various; but for the greater part of modern books calf-skin dyed of various colours is employed.

13. b. *Lawn Tennis.* To defend by or have within the scope of one's play.

1889 H. W. W. Wilberforce *Lawn Tennis* xi. 35 The man at the net covers far more of the court. 1907 P. A. Vaile *Mod. Lawn Tennis* 111 You cannot cover the whole court, and you must go where you have the best chance of meeting the ball. 1920 W. T. Tilden *Art Lawn Tennis* 44 Your position should always strive to be such that you can cover the greatest possible area of court without sacrificing safety, since the straight shot is the surest, most dangerous, and must be covered.

14. c. To report (an event, meeting, etc.) for a newspaper, broadcast, or the like; to attend, investigate, etc., as a reporter. orig. *U.S.*

1893 M. Philips *Making of Newsp.* 6 [The news editor] has been preparing to 'cover' such important events as are 'in sight'. 1898 *Westm. Gaz.* 3 June 8/1 There was a licensed victuallers' dinner at Aston, and one of the reporting staff was sent to cover it. 1928 *Strand Mag.* LXXV. 84/2 In covering the consecration of Liverpool Cathedral for a broadcast which lasted for about an hour, there were five microphones in use. 1935 Auden & Isherwood *Dog beneath Skin* I. ii, Officially, I'm covering the Danube floods. 1958 *Spectator* 27 June 827/3 Both BBC and ITV must know where they stand if they wish to cover the election on the same basis as the newspapers do. 1969 *New Yorker* 29 Mar. 27/3 Once again television had become the event, instead of merely covering it.

d. To photograph the whole of an area from the air.

1919 C. C. Turner *Struggle in Air* xv. 205 Thousands of exposures had to be made by scores of pilots who had to 'cover' the ground allotted to them. 1958 C. Babington Smith *Evid. in Camera* 35 Large areas of the Siegfried Line were 'covered' for the first time.

17. b. (Earlier U.S. example.)

1870 Medbery *Men & Myst. Wall Street* 227 (Bartlett), On dividing the assets it was discovered that the Treasurer had used up all the funds in a frantic effort to cover.

20. b. *trans.* To conceal (a crime, etc.).

1926 G. Hunting *Vicarion* xi. 189 It was the scenes she'd just looked at in the theater—something she'd done and thought was covered up, but is found out now! 1928 E. O'Neill *Strange Interlude* iv, A distressingly obvious attempt to cover up a chronic state of..guilty conscience.

c. *intr. Cricket.* To protect one's wicket with one's pads or legs from a ball that is turning in.

1927 M. A. Noble *Those 'Ashes'* 16 The ball must not be pitched outside the leg stump, because the batsman can cover up without danger.

d. *intr.* To conceal a misdeed, deception, etc.; to assist a wrongdoer to escape detection.

1942 Berrey & Van den Bark *Amer. Thes. Slang* § 489/3 *Cover (for), cover up (for)*..., to assist a confederate. 1958 *Spectator* 15 Aug. 225/1 Real or imaginary names are put to scripts imported from America... British writers are employed at starvation wages to 'cover up'. 1961 in J. Barzun *Delights of Detection* 312 A woman who wouldn't commit a murder might be willing to help cover up after it had been committed. 1970 *Daily Tel.* 21 Oct. 2 Two police constables wrote out their resignation..because they were so sickened at attempts by senior officers to 'cover up' for a superintendent involved in a fatal road accident.

cover, *sb.*[1] Add: **1. c.** *Calico-printing.* A design that is printed over another design in resist-work. Also *cover pattern.*

1874 Crookes *Dyeing* 564 The rest is taken up with the fine cover pattern. 1884 J. Gardner *Bleaching, Dyeing*, etc. iii. 101 Mordants, Discharges, Reserves, Covers, &c. 1957 *Encycl. Brit.* XXII. 7/1 The fine patterns printed over resists are called covers and the plain grounds pads.

d. *Cricket.* = COVER-POINT 1. So *the covers*: cover-point and extra cover-point.

1836 *New Sporting Mag.* July 195 Next follows a little about the mode of hitting to cover. 1897 K. S. Ranjitsinhji *Jubilee Bk. Cricket* 48 A fine exhibition of fielding at cover. *Ibid.* 49 Like third-man, cover must be ever on the alert to dash in any direction. 1904 P. F. Warner *Recov. Ashes* vii. 111 Tyldesley, whether he was..forcing a short ball between the covers, hooking or glancing to leg, was perfect in his timing of the ball. 1950 N. Cardus *Second Innings* 93 With time left for the batsman to make a drive past cover. 1955 *Times* 5 July 4/1 Walshe played some surprising shots through the covers off the back foot.

e. *Lawn Tennis.* (Cf. *COVER *v.*[1] 13 b.)

1907 Vaile *Mod. Lawn Tennis* 123 You must remember ..that he cannot do it [*sc.* drive down the side line] every time with sufficient accuracy to pass in the small margin over 'cover' which you will allow him.

f. In full *cover version.* A recording of a song, etc., which has already been recorded by someone else.

1966 *Melody Maker* 23 July 12/2 This is a cover version of the new Beach Boys single from some friends and admirers, the Castaways. 1968 *Listener* 7 Nov. 622/3 The jackal thinking behind cover versions, which are near copies of original recordings, is predicated on the belief that so much money is showered in the general direction of hit records that *any* performance of the song will collect if sufficiently adjacent. *Ibid.*, It's a popular misconception that the original version of a song must be better than a cover. 1970 *Melody Maker* 19 Dec. 31 Among the biggest sellers this Christmas will be the 'Hits' albums—cover versions of Top 30 records. *Ibid.* 31/7 Lewis was an originator of these 'covers'.

2. d. *spec.* in *Philately*, an envelope, etc., bearing a stamp, postmark, or label of particular historical or commemorative interest; esp. *first day cover* (see quot. 1950).

1923 F. J. Melville *Aero-Stamp Collecting* ii. 14 A flown cover of each of the outstanding pioneer flights in each country is of undeniable interest. 1934 S. Phillips *Beginner's Bk. Stamp Collecting* xix. 214 *Cover*, a term applied to an envelope or wrapper. 'On cover' means on a complete envelope or wrapper, as opposed to 'on piece'. 1938 D. G. Armstrong *Collecting Stamps* 47 *First-day cover*, a postage stamp or stamps used on entire upon the first day that the stamps in question were officially placed on sale. *First flight cover*, an envelope which has been carried by air mail on the first flight between any two given points. 1950 L. N. & M. Williams *Collecting Postage Stamps* xiii. 71 One fashion with little to recommend it that has become prevalent in recent years is the collection of what are known as First Day Covers, or letters bearing stamps posted on the first day of their issue and bearing an appropriate dated postmark. 1968 J. Mackay *Cover Collecting* iv. 35 Although a souvenir cover bearing stamps used on the first day of issue has been recorded from the United States as early as 1909 (in connection with the Hudson Fulton 2c. stamp), the First Day Cover craze did not make any headway in America until the late 1920s.

f. The outer covering of a pneumatic tyre.

1898 *Science Siftings* 5 Nov. 46/2 All tendency to cracking of the outer covers is obviated. 1902 A. C. Harmsworth *Motors & Motor-Driving* 237 Even the covers of the driving wheels were quite smooth and unimpaired. 1904 A. B. F. Young *Compl. Motorist* 248 Two spare tubes should always be carried and one spare cover. *Ibid.* 251 When the tyre has been replaced and slightly inflated, the Continental 'cover plaster', a kind of long band or puttie, is wound round the tyre.

3. g. Protection from attack; esp. that afforded by the presence or fire of a supporting force; also, a force providing such protection. Freq. with defining word, as **air cover* (*AIR *sb.*[1] B. III. 2), **fighter cover*. Cf. COVER *v.*[1] 8 d.

1802 C. James *Mil. Dict.*, *Cover*, a term in war to express security or protection: thus, to land under cover of the guns, is to advance offensively against an enemy who dares not approach on account of the fire from ships, boats or batteries. 1950 Truman in *N.Y. Times* 28 June 1/6, I have ordered United States air and sea forces to give the Korean Government troops cover and support. 1964 T. White tr. *P. Leulliette's St. Michael* 210 Rapidly, under the cover of armoured cars, we crossed the whole devastated and deserted town.

4. b. The vegetation covering the surface of the ground; spec. in *Ecology*, the area covered by the individuals of one species, often expressed as a percentage of the area being studied.

1909 *Cent. Dict.* Suppl. s.v., *Forest cover*, in *forestry*, all trees and other plants in a forest. 1932 Fuller & Conard tr. *Braun-Blanquet's Plant Sociol.* iii. 31 The degree of dominance..is expressed..by the cover, that is, the area covered by the individuals of one species. 1937 *Discovery* May 140/1 This denudation of forest cover is due to human interference. 1959 J. D. Clark *Prehist. S. Afr.* ii. 42 A discontinuous grass cover such as we find in so much of Africa. 1961 Hanson & Churchill *Plant Community* iii. 102 Cover, or specifically herbage cover, signifies primarily the area of ground occupied by the leaves, stems, and inflorescences.

5. b. Adequate insurance against loss, damage, etc.; = *COVERAGE sense a.

1913 G. B. Shaw *Let.* 15 July (1952) 133 Is that car insured? If not..say you want 'cover' at once. 1955 *Times* 10 May 18/4 The striving to meet a demand for extended insurance cover is not new.

8. (sense 2 c) *cover-design*; (sense 3 d) *cover-name, -organization, -story*; (sense 5) *cover-money*; **cover charge,** a charge for service added to the basic charge in a restaurant (cf. sense 7); **cover crop,** a crop (often leguminous) sown to protect the soil or to enrich it by being ploughed under; **cover-day,** a day on which the covers are shot over for game; **cover-drive** *Cricket*, a drive (DRIVE *sb.* 1 d) past cover-point; hence **cover-drive** *v.*; **cover-girl** orig. *U.S.*, a young woman whose picture appears on the front cover of a magazine; also *attrib.*; formerly (*U.S.*) *magazine-cover girl*; **cover note** (see quot. 1919); also, a note which declares the holder to have a current insurance policy; **cover paper,** strong paper suitable for the covers of pamphlets, booklets, etc.

1921 *Nation* (N.Y.) 21 Sept. 320/2 As levied here, the cover charge is a compulsory blanket assessment for nothing in particular; it commonly includes bread and butter. 1933 J. B. Priestley *Wonder Hero* iv. 147 They..add another ten bob to the cover charge. 1961 P. Ustinov *Loser* xiii. 291 There's no extra cover charge here. Drinks cost the same at a table. 1909 *Cent. Dict.* Suppl. s.v. *Crop*, Cover crop. 1926 *Contemp. Rev.* Dec. 765 The humble covercrops of purple vetch and millelotus clover. 1943 J. S. Huxley *TVA* vi. 28 The prevention of erosion on arable land by new methods such as contour ploughing, or new cover-crop plants. 1902 *Daily Chron.* 22 Aug. 9/6 Only outlying birds are shot before the big 'cover days' of November, December, and January. 1910 J. London *Let.* 9 Feb. (1966) 299, I never know what the cover-design of any of my books is going to be until I get advance-copy of the market edition. 1925 D. H. Lawrence *Let.* 25 Apr. (1962) II. 838, I am sending you a copy of *Birds, Beasts and Flowers* today... We made the cover-design between us. 1938 R. G. Collingwood *Princ. Art* v. 84 The magazine and newspaper, where cover-designs, news-items, fiction, and advertisement are steeped in materials of the same kind. 1898 K. S. Ranjitsinhji *With Stoddart's Team* (ed. 4) iv. 71 Donnan made some excellent 'cover' drives and cuts. 1955 *Times* 11 July 4/2 Occasionally he produced an exquisitely timed force off his legs or cover drive. 1960 J. Fingleton *4 Chukkas to Australia* 135 He cover-drove Lindwall. 1915 *Harper's Wkly.* 25 Sept. 303 (*caption*) Unveiling the statue of Charles Dana Gibson at the annual picnic of the magazine cover girls at Lakewood. 1944 *Times* 12 June 8/2 *Cover Girl*..is one of those smooth, beautifully coloured 'musicals' which Hollywood can manage with such a sure..touch. 1948 C. Isherwood in *Penguin New Writing* XXXIV. 132 On the walls a coloured print of the Virgin and a cover-girl from a U.S. magazine. 1957 W. Camp *Prosp. Love* 150 Here was a girl who was just as pretty as Lucinda, though not in the same pert, really rather ordinary, literally 'cover girl' sort of way. 1909 *Daily Chron.* 14 Oct. 7/4 The cover money required is temptingly small. 1948 *Ann. Reg. 1947* 350 A new party which called itself 'National Liberation' (a 'cover-name' for the Communist Party). 1950 Koestler *God that Failed* 41, I pressed Edgar to let me join a cell where nobody knew me except under my cover-name. 1919

W. Thomson *Dict. Banking* (ed. 2) 178/2 *Cover note*, a note issued by an insurance company when the first premium has been paid declaring (e.g., in a proposal for insurance against fire) that the proposer is 'covered' in the meantime until the policy is issued. 1966 B. Kimenye *Kalasanda Revisited* 105 The insurance expired four days ago, and since I was away there isn't even a cover note. 1944 M. Laski *Love on Supertax* xii. 118 The Black Market which I run..is no more than a cover-organisation for something infinitely more important. 1866 J. Blackwood *Let.* 21 Dec. in *Geo. Eliot's Lett.* (1956) IV. 323 Coverpaper Binding and Sewing the Parts. 1914 E. A. Dawe *Paper* xvii. 119 Cover Papers.—The term is applied to a large class of fancy papers, made in many shades, substances and sizes, suitable for the covers of pamphlets, booklets, [etc.]. 1945 G. Millar *Maquis* i. 11 Have you got your false papers and your cover story? 1950 L. A. G. Strong *Which I Never* vii. 195, I have my cover story, a thin one, but it'll do—just. 1954 Koestler *Invis. Writing* xxix. 315 We then worked out a cover story on the following lines.

cover (kōᵘ·vəɹ), *sb.*[3] *Cornwall.* [f. COVE *sb.*[1] + -ER[1].] One who lives in a cove.

1905 *Westm. Gaz.* 13 Apr. 8/3 This action has been deeply resented by the 'Covers', who a short time since removed these obstructions. 1913 Quiller-Couch *News fr. Duchy* 183 The 'Covers', or native fishermen.

coverage (kɒ·vəredʒ). orig. *U.S.* [f. COVER *v.*[1] + -AGE.] The act or fact of covering; the area, range, number, etc., that is covered by something; *spec.*: **a.** The aggregate of risks covered by an insurance policy. Also *attrib.*

1912 *Agents' Record* (Hartford, Conn.) 17 June, There will be nineteen policyholders disillusionized and disgusted with the limited coverage contract.

b. The area within range of a radio or television transmitter; the number or percentage of people able to receive broadcast transmissions. Also the section of a community that is reached by a particular advertising medium.

1930 *Proc. Inst. Radio Engin.* XVIII. ii. 1627 This transmitter was designed to fill the need for an inexpensive high quality transmitter..where the coverage desired is smaller. 1934 Webster, *Coverage* (Advertising), the portion of a group or of a community reached by a particular advertising medium. 1937 *Discovery* Feb. 43/2 Experts.. foresee an inland television service in Britain, providing a far greater coverage of the country than is now possible. 1941 *Times* 22 May (Advt.), Magazines have always given the widest possible coverage, since they are passed on from hand to hand. 1952 *Charter of Incorporation* in *BBC Handbk.* (1957) 263 To develop and use..frequencies of more than thirty megacycles a second with a view to extending the coverage..of the Corporation's broadcasting transmissions. 1962 E. Godfrey *Retail Selling & Organ.* x. 107 The advertisement usually has a national coverage.

c. The inclusiveness or range of functions of a service, business, industry, etc.

1930 *Prospectus Press Clipping Service* (N.Y.), This superb coverage costs you no more than the average press clipping service. 1933 *Planning* I. xii. 14 A public concern on the London Transport model, with a wider coverage and no direct municipal affiliations. 1958 *Times* 25 Mar. (Suppl.) p. xi/3 It [*sc.* the electronics industry] covers such a wide range of functions..that the problem is to give a clear definition of its coverage.

d. An amount of money, gold, etc., available to meet liabilities.

1931 *Daily Express* 22 Sept. 1/6 Prohibiting the acceptance of orders in future, except for winding-up purposes and unless they are accompanied by a coverage of 40 per cent.

e. The act of covering (see *COVER *v.*[1] 14 c) an event, subject, etc.; the extent of reporting by a newspaper, radio station, etc.; reporting.

1931 *Kansas City Star* 19 Nov. 22 It gives its readers thorough coverage on animal news. 1934 *BBC Year-Bk.* 283 The type of news-giving which..is graphically known in America as 'spot coverage'; that is to say, the running commentary or eye-witness account broadcast while the event is actually taking place. 1950 *Manch. Guardian Weekly* 9 Mar. 2 It is an election coverage Mr. White is writing about. 1953 *Economist* 3 Jan. 13/1 The Soviet radio gave greater radio coverage to this peace congress than to any previous one. 1964 *Ann. Reg. 1963* 179 The shooting was witnessed by millions of Americans who were watching live television coverage of the prisoner's transference. 1970 *Daily Tel.* 14 Apr. 1/5 There will be live coverage of the replay on BBC Television.

f. The range of a lens; the area over which a lens can give a sharply focused image.

1936 *Discovery* Aug. 237/2 These mirrors, mounted at an angle to increase the coverage of the wing lenses, give the camera an unusual appearance. 1961 G. Millerson *Telev. Production* iii. 31 For any camera, the longer the focal length of the lens it uses, the narrower its coverage or angle of view.

cover-all, coverall (kɒ·vərǫl), *sb.* Now chiefly *U.S.* [f. COVER *v.*[1] + ALL.] Something that covers entirely, esp. a full-length outer garment. Also in *pl.* form. Cf. OVERALL *sb.*

1830 J. Stewart *Killarney Poor Scholar* i. 2 All ready, your honours... Beauty has baskets, hampers, and coveralls galliore [*sic*] safe stowed. 1901 *Daily Chron.* 27 July 10/3 Many women prefer to have their cover-all rendered stormproof by a patent process. 1948 Steinbeck *Russ. Jrnl.* (1949) 13 She wore..a blue coverall and a headcloth. 1957 *New Yorker* 16 Nov. 154/2 Skiing in a colorful coverall of hand-loomed wool. 1957 P. Frank *Seven Days to Never* i. 13 The man wore black zippered coveralls. 1964 *Punch* 3 June p. xvi, Selfridges..stock..includes..Swedish coveralls.

Hence *attrib.* or as *adj.*, that covers entirely.

1895 *Daily News* 12 Oct. 9/2 His left fore hoof enveloped in a cover-all shoe. 1901 *Daily Chron.* 27 July 10/3 A coat of full-length cover-all dimensions. 1964 M. M. Tumin in Gould & Kolb *Dict. Soc. Sci.* 444/1 A common coverall term which is used to refer to ethnic, national, racial, and cultural groups, etc. is *minority* or *minority group*. 1966 G. N. Leech *Eng. in Advertising* iv. 33 'Consumer' here is a coverall term for a vast audience often running into millions.

covered, *ppl. a.* Add: **4.** *covered bridge, market.*

1809 *Tales of Other Realms* II. 135 In about two hours we passed over a covered bridge. 1949 *Clarke Co. Democrat* (Grove Hill, Ala.) 13 Jan. 7/4 Covered bridges..are still an important factor in our country's transportation system. 1863 *Miss Jemima's Swiss Jrnl.* (1963) iv. 102 Ripe apricots, procured in a large covered market. 1914 A. Bennett *Price of Love* xiv. 285 The covered market, which, having one entrance in the market-place and another in Wedgwood Street, connects the two thoroughfares. 1969 C. Storr *Rufus* v. 58 A silly, frightened sheep's face..looked round the big covered market.

b. *covered wagon*, a wagon with a tilt (TILT *sb.*[1] 3). Chiefly *U.S.*

1745 in J. S. McLennan *Louisburg* (1918) 163 The commander in chief..shall have liberty to send off covered waggons to be inspected only by one officer of ours. 1842 M. Crawford *Jrnl.* (1897) 5 Rode all night in a covered wagon. 1864 T. L. Nichols *Amer. Life* I. 108 Great covered waggons—such as are called vans in England—went about the country collecting the rosy maidens. 1946 L. Mumford *City Development* 55 The covered wagon in one age and the trailer in ours both have their specific uses. 1948 *Daily Oklahoman* (Okla. City) 16 May E 28/1 A sizeable percentage of the nation's population..believes that in Oklahoma the people still..travel in covered wagons. 1958 L. van der Post *Lost World of Kalahari* i. 15 The trekkers, or covered-wagon pioneers of my people, kept patrols of horsemen scouting well ahead.

covering, *vbl. sb.*[1] Add: **3.** *covering-fire, -party* [COVER *v.*[1] 8 b]; *covering purchase* [COVER *v.*[1] 17]; **covering power,** (*a*) = *COVERAGE f; (*b*) (see quot. 1904).

1897 *Daily News* 28 Sept. 5/4 The enemy at once returned to the attack, but the covering fire of the British troops kept them at bay. 1964 T. White tr. *P. Leulliette's St. Michael* 164 It was..hopeless to try to give covering fire with L.M.G.s or ordinary machine-guns aimed at the entries to the caves. 1812 Covering party [see SAP *sb.*[3] 2]. 1915 A. D. Gillespie *Let. Flanders* 24 May (1916) 168 Clark took out a covering party to make a diversion on this side. 1895 J. A. Hodges *Photogr. Lenses* iii. xix. 116 Testing for Covering Power.—Lenses are advertised..to cover certain sized plates. 1898 W. Pearce *Painting & Decorating* v. 83 For a large amount of indoor painting, where covering power was required to produce a certain effect. 1903 Beck & Andrews *Photogr. Lenses* i. 13 There will always be a certain area of sharp definition, and the size of this area describes the covering power of the lens. 1904 Goodchild & Tweney *Technol. & Sci. Dict.* 135/2 Covering power of pigments..refers to the extent of surface a pigment will cover, or the superficial area over which it will spread when thinned to a consistency suitable for being applied with a brush. 1953 *Gloss. Paint Terms (B.S.I.)* 11 *Covering power.* The use of this term has given rise to confusion of thought as between the act of 'spreading out' and 'covering up' (or hiding) respectively. It is consequently falling into disuse and being replaced by the terms 'Spreading capacity' and 'Hiding power' respectively. 1928 *Daily Mail* 25 July 13/4 Movements during the early part of the afternoon reflected local covering purchases.

cover-point. **1.** (Earlier example.)

1846 W. Denison *Cricket: Sk. Players* 65 In most matches he..takes 'mid wicket' or 'cover point'.

cover-slip. Add: **1.** (Earlier example.)

1854 F. Lillywhite *Guide to Cricketers* 14 Cover Point, and Long Slip, or Cover Slip..require particular practice to allow for the curl of the ball.

2. (Examples.)

1875 Huxley & Martin *Course Elem. Biol.* x. 99 Transfer a Hydra..on to a slide; cover in plenty of water with a large coverslip. 1929 W. Deeping *Roper's Row* iii. 22 The corner of the physiology lab...with his microscope..and his bottles of stains and box of slides and coverslips. 1959 J. Clegg *Fr shwater Life* (ed. 2) xviii. 306 Another brass tube also holding a thin glass cover-slip fits into the first tube.

covert, *sb.* Add: **7. covert cloth** = *covert coating*; **covert coat, coating** (examples).

1895 *Army & Navy Co-op. Soc. List, Autumn & Winter Designs* (*caption*) Walking Gown..in Melton and Covert Cloths...Covert Cloth Gown. 1921 *Daily Colonist* (Victoria, B.C.) 6 Apr. 7/1 (Advt.), Children's Covert Cloth Reefers. Lightweight serviceable Reefer Coats for children. 1968 J. Ironside *Fashion Alphabet* 222 The best covert cloth is made from wool... Covert is also now made of rayon, cotton and synthetics. 1894 *Country Gentlemen's Catal.* 168/2 Special covert coat materials. 1899 E. W. Hornung *Amat. Cracksman* 16 A covert coat over his blazer. 1930 W. S. Maugham *Cakes & Ale* v. 74 A fawn-coloured covert-coat with large buttons. 1960 'R. East' *Kingston Black* viii. 77 Marty appeared in his covert coat. 1900 'Major of Today' *Clothes & Man* ii. 61 A good useful riding or driving coat is made of a heavy covert coating material. 1906 *Daily Chron.* 27 Aug. 8/5 A charming coat..made in biscuit-coloured covert-coating. 1959 *Chambers's Encycl.* XIV. 662/2 *Covert coating*, medium-weight fabric with a twill effect on the surface.

cover-up (kɒ·vərɒp). [f. *to cover up* (see COVER *v.*[1] 20).] An act of covering up (in various senses: see COVER *v.*[1] 20 in Dict. and Suppl.); *spec.* assistance given to a wrongdoer in escaping detection; = COVER *sb.*[1] 3 d; also, a high-necked garment, a cover-all.

1927 M. A. Noble *Those 'Ashes'* 192 Hardinge has a good defence and cover-up. 1935 R. Chandler in *Black Mask* Jan. 31/2, I don't have to tell you how a police department looks at that kind of a cover-up on a murder. 1940 M. Allingham *Black Plumes* ii. 11 It was a bad cover up and very obvious. 1958 *Spectator* 27 June 836/2 The slogan 'personal service' is often a cover-up for very indifferent attention indeed. 1960 *Sunday Express* 27 Nov. 14 This fluffy bed jacket..is a complete cover-up for cold weather. 1961 *Ibid.* 2 Apr. 14 Thickknit cover-up, full-fashioned in.. zig-zag rib. 1963 *New Yorker* 1 June 103 This cover-up flowers in a burst of colorful island foliage.

Hence *attrib.* or as quasi-*adj.*, that covers up or conceals. orig. *U.S.*

1942 Berrey & Van den Bark *Amer. Thes. Slang* § 461/12 *Cover-up man*.., a pickpocket's confederate. 1960 V. Wimperis *Unmarried Mother* xvi. 343 Contraception, abortion and 'cover-up' marriages all affect the number of children born illegitimate. 1960 *Vogue Pattern Bk.* Early autumn 42 Party dress and cover-up bolero. 1970 *Harrods Catal.* 16/1 Exclusive cover-up rig in cotton velour. Top has shoe-lace fastening, pants elasticated at waist.

cow, *sb.*[1] Add: **1. c.** *pl.* Cattle. *U.S.*

1869 *Overland Monthly* III. 127 The 'cow-whip'..is used only in driving the herd, which is often called 'the cows'. 1930 Raine & Barnes *Cattle* 60 Cows, as all cattle were called regardless of age and sex, were an investment which traveled on the hoof.

3. b. (Earlier and later U.S. examples.)

1751 C. Gist *Jrnl.* 4 Mar. (1893) 56 At Night I killed a fine barren Cow-Buffaloe. 1817 S. R. Brown *Western Gaz.* 198 The cow buffaloe was equal to any meat I ever saw. 1839 *Knickerbocker* XIII. 386 An enormous cow-whale rose close beside her wounded offspring. 1880 G. W. Cable *Grandissimes* xxviii. 237 In dimmer recesses the Cow alligator, with her nest hard by. 1946 T. M. Stanwell-Fletcher *Driftwood Valley* 194 When the cow moose is alarmed, it is not uncommon for her to desert her calf.

4. b. (Further examples.) Also, loosely, any woman, used esp. as a coarse form of address.

1868–9 *Parl. Papers* XXXI. 123/1 It's all such stinking hussies as you who are keeping men out of their work, you stinking cow. 1914 C. Mackenzie *Sinister St.* II. iv. iv. 930 Good job if that love-boy of hers does punch into her. Silly cow! She ought to know better. 1933 'G. Orwell' *Down & Out* xxi. 153 'Now, you cow,' I said, 'move it yourself.' 1936 'D. Smith' *Call it a Day* in *Famous Plays 1935–6* 239 And then a cow of a woman upset a tea-tray in the stalls. 1960 D. Lessing *Pursuit of Eng.* vi. 218 It's just that stupid cow her mother. 1960 P. Hastings *Sandals for my Feet* ii. vi. 192 You shan't do this to me, you filthy old cow!

c. An objectionable person or thing, a distasteful situation, etc. *Austral.* and *N.Z. slang.*

1891 T. Cottle *Melton's Luck* xviii. 78 The less spirited [horses] become regular cows (as we called them) and only go because they are obliged to. 1894 *Bulletin* (Sydney) 5 May 13/3 Well, 'e ups to me and sez 'Yer a mean cow.' 1904 'S. Rudd' *Sandy's Sel.* iii. 17 Bring th' sheep bell (*you cow, stand!*) I foun' th'other day. 1916 *Anzac Bk.* 31/2 'Ee's a fair cow, 'e is. 1918 *N.Z.E.F. Chron.* 7 June 204/1 I'll get even with the cows for that. 1929 W. Smyth *Girl from Mason Creek* xv. 163 'Cow of a job,' he muttered. 1934 T. Wood *Cobbers* ii. 15 In a continent where..(to the punter when the favourite is unplaced) a dark horse can be a fair cow [etc.]. 1936 M. Franklin *All that Swagger* lii. 489, I never read their cows of letters. 1940 F. D. Davison *Woman at Mill* 148 Looking for work's a cow of a game! 1959 P. H. Johnson *Unspeakable Skipton* xv. 137 [Austral. *loq.*] 'He's like the rest of you, for ever taking umbrage about something. It's a fair cow.' She had forgotten herself. 1963 D. Adsett *Magpie Sings* 10 There's that truant joker too. Rotten cow.

7. *cow-bone, -country, -driver* (examples), *-paddock* (*Austral.* and *N.Z.*), *-stable* (later example), *-thief, -track, -trail, -whip.*

Chiefly U.S., with *cow-* in the sense of cattle.

1913 Masefield *Daffodil Fields* 32 Some half-wild hounds Gnawed at the cowbones littered on the field. 1922 Joyce *Ulysses* 476 Free cowbones for soup. 1882 C. M. Chase *Editor's Run* 160 There is no excuse in a cow country like this for a landlord to set before his guests oleomargarine. 1938 P. Lawlor *House of Templemore* xiii. 144 The first trip had been to the cow country of Taranaki. 1945 *Reader's Digest* Sept. 109/1 With nearly all its cattle behind wire, the ranch junked the oldest and most revered custom of the cow country. 1960 N. Hilliard *Maori Girl* I. i. 9 The cow-country is owned by *pakehas*. 1771 *Carroll Papers* in *Maryland Hist. Mag.* XIV. 136, I have order'd Squires to go downe tomorrow with the Cow driver. 1851 J. J. Hooper *Widow Rugby's Husb.* 102 The 'Colonel' being what in his region and times was called a cow-driver. 1932 W. Kelley *Inchin' Along* 211 The ox drivers—'cow drivers', they were called. 1931 V. Palmer *Separate Lives* 125 The shady creek-bend in the corner of the cow-paddock where the grass grew. 1956 R. Finlayson in C. K. Stead *N.Z. Short Stories* (1966) 23 They followed him up the drive past the cowpaddock gate. 1882 E. Eggleston *Hoosier Schoolboy* ii. 12 He found 'King Milkmaid' written on the door of his father's cow-stable. 1781 in R. Putnam *Memoirs* (1903) 184 Your favor of the 25, with the cow-thieves, arrived safe. 1903 A. Adams *Log Cowboy* vii. 95, I think you're common cow thieves. 1678 in *Duxbury* (Mass.) *Rec.* (1893) 30 There being..a cow track going over the said river. 1857 Olmsted *Journ. Texas* 93 Our road was little better than a cow-track. 1901 Kipling *Kim* xiii. 336 The lama had led Kim by cow-track and byroad. 1902 *N.Z. Illustr. Mag.* V. 379/2 You have two razor-backs and a gully, and no way in except this cow-track. 1853 'P. Paxton' *Yankee in Texas* 100 He will see them pouring in..by every possible road,..wagon roads, main road, 'cow trails', and 'blazes'. 1920 J. M. Hunter *Trail Drivers Texas* 151 All the Texas outfit..took the cow trail for Texas. 1853 'P. Paxton' *Yankee in Texas* 93 The rest relied for offence and defence upon their long cow-whips—an implement consisting of a

short eighteen-inch handle, to which a very heavy lash from twelve to eighteen feet long is attached. 1869 Cow-whip [see sense 1 c above].

8. cow-bail *Austral.* and *N.Z.*, = BAIL *sb.*[3] 5; **cow-banger** *dial.*, *Austral.* and *N.Z. slang*, a dairy farmer, one who works on a dairy farm; so **cow-banging** *vbl. sb.*; **cow-barton,** a cow-yard; **cow-camp** *U.S.*, a camp of cowboys; **cow-cocky** *Austral.* and *N.Z. slang*, a dairy farmer; so **cow-cockying** *vbl. sb.*; **cow-creamer** [*CREAMER ç], a cream-jug shaped like a cow; **cow-flop** (also **-flap**), (*a*) *dial.*, any of several plants, esp. the foxglove, *Digitalis purpurea*; (*b*) *dial.* and *U.S.*, a patch of cow-dung; **cow-gun** *colloq.*, a heavy naval gun; **cow-hand** orig. *U.S.*, one engaged in the tending or ranching of cattle; **cow-horse** *U.S.*, a horse used in herding or driving cattle; **cow-juice** *slang*, milk; **cow-lily** *U.S.*, the marsh marigold, *Caltha palustris*, or the yellow water-lily, *Nuphar advena*; **cow-man** (earlier and later examples); **cow-pad, -pat,** = **cow-flop* (b); **cow-pilot,** delete '*Pomacentrus saxatilis*' and substitute '*Abudefduf saxatilis*'; (example); **cow-pony** *U.S.*, a pony used in cattle-ranching; **cow-puncher, -punching** (earlier U.S. examples); **cow·scape** [after LAND)-SCAPE] *colloq.*, a painting of a rural scene which includes cows; **cow sense** *U.S.*, intelligence in the care of cattle; **cow-shot** *Cricket slang*, a pull made by hitting across the ball to leg; **cow-spanker** *Austral.* and *N.Z. slang*, a dairy farmer, one who works on a dairy farm; so **cow-spanking** *vbl. sb.*; **cow-town** *U.S.*, a town forming a local centre in a stock-raising district; a small, isolated town; **cow-work** *U.S.*, work connected with the tending or rounding-up of cattle.

1851 E. Ward *Jrnl.* 12 May (1951) 180 The cow bails in the stockyard are fastened up. 1853 G. B. Earp *N.Z.* viii. 123 A milking yard, at one end of which is placed the milking shed, cow bails, and calf house. 1936 M. Franklin *All that Swagger* xxxix. 370 He whitewashed the dairy and cowbails. 1892 *Leeds Merc.* (Suppl.) 30 Jan. 7/3 Yorkshire Dialect Words...*caa-banger*, a man who attends to cows. 1941 Baker *N.Z. Slang* v. 41 *Cowspanker* (we also use *cattle-banger* and *cow-banger*). 1912 B. E. Baughan *Brown Bread from Colonial Oven* vi. 118 It's a poor job, cowbangin' all alone. 1888 Hardy *Wessex Tales* I. 71 The dairyman.. with manly kindliness always kept the gossip in the cow-barton from annoying Rhoda. 1929 Masefield *Hawbucks* 34 He went out to the cow-barton to see the farm-men. 1873 W. A. Richards *Diary* 28 Aug. in *Annals of Wyoming* (1931) VIII. 495 Mr. Richards..has a cow camp (on Red Creek) three miles north. 1945 M. James *Cherokee Strip* 8 Mr. Howell knew a cow-camp cook in Colorado who was bitten on the thumb by a rattler. 1914 *Bulletin* (Sydney) 7 May 22/2 *Cow Cocky Host*: The piece from Sydney..is goin' to marry 'im. 1916 *Ibid.* 30 Mar. 47/2 This is not sordid, slaving, cow-cocky country. 1934 N. Scanlan *Winds of Heaven* 322 The cow Cockies were still in bondage. 1940 F. Sargeson *Man & Wife* (1944) 22 My father was a cow-cocky, but he couldn't make cow-cockying pay. 1962 *Coast to Coast 1961–62* 95, I could..swear like a cow-cocky. 1931 E. Wenham *Domestic Silver* v. 98 Various curious shapes were adapted to these small jugs... Those known as 'cow creamers' may still be procured. 1938 Wodehouse *Code of Woosters* i. 24 He had an eighteenth-century cow-creamer for sale... It was a silver cow..about four inches high and six long. Its back opened on a hinge. 1847 Halliwell *Dict. Archaic & Provinc. Words* I. 275/2 *Cowflop*, the foxglove. *Devon.* 1882 H. Friend *Devon. Plant Names* 17 *Cowflop.* (1) *Digitalis purpurea*, L. One of the many names for the Foxglove. (2) *Avena sativa*, L. To distinguish from Tartarian Oats. (3) A tall flower, somewhat like the Great Mullein. 1905 Wright *Eng. Dial. Dict.* Suppl. s.v. *cow*, *Cow-flap*, or *-flop*, cow-dung. 1906 *Daily Chron.* 30 Nov. 6/7 From a village on the fringe of the moorland in Devonshire a correspondent writes:..all around you may hear the foxglove called by its local name, the 'cow-flap'. 1932 F. L. Wright *Autobiogr.* i. 42 Getting bare feet soiled..in the warm, fresh cow-flops, in the stable or in the lane. 1934 H. Williamson *Peregrine's Saga* (ed. 2) 130 A cowflop in her beak, the buds of whose flowers were not then turned pink. 1902 J. H. M. Abbott *Tommy Cornstalk* 136 His two great 'cow-guns'—six-inch naval giants drawn by thirty-two bullocks apiece, and having another thirty-two to each timber. 1909 *Daily Chron.* 24 Sept. 4/4 Its motor machine-guns, its heavy 'cow-guns', and howitzers. 1911 Kipling *Diversity of Creatures* (1917) 317 The slow but well-directed fire of my cow-guns. 1886 *Outing* (U.S.) VIII. 3/1 Though a first rate cow hand he very shortly proved himself to be wholly incapable of acting as head. 1949 E. Hyams *Not in our Stars* 161 I've never known a really good cow-hand quite all there. 1853 'P. Paxton' *Yankee in Texas* 97 The very best cow horses. 1920 C. E. Mulford *J. Nelson* xxiii. 253 The speedy dash of the trained cow-horse headed them off. 1796 *Grose's Dict. Vulgar T.* (ed. 3), *Cow juice*, milk. 1938 S. Beckett *Murphy* v. 83 They have been too generous with the cowjuice. 1959 I. & P. Opie *Lore & Lang. Schoolchildren* ix. 164 Milk is commonly 'cow-juice'. 1862 J. R. Lowell *Biglow Papers* 2nd Ser. ii. 79 There was a pool.. spotted with cow-lilies garish. 1954 C. J. Hylander *Macmillan Wild Flower Bk.* 105 Also known as Cow-lily and Spatterdock, this common member of the Water-lily family can often be found in great numbers. 1958 Fernald & Kinsey *Edible Wild Plants* 197 To some of the north-western Indians the seeds of the Cow-Lilies are a very important food. 1677 in *Topsfield* (Mass.) *Rec.* 20 That popeler is his bound corner tree next to the Coweman. 1709 N. Blundell *Diary* (1952) v. 96 William Starkey dyed... He was my Cow Man. 1924 W. M. Raine *Troubled Waters* xi. 115 Sam Yerby was an old cowman from Texas. 1948 *Pop. Western* June 34/1 The other cowmen..disliked to deal with Amos. 1941 Dylan Thomas in *Penguin New Writing* XI. 77 Dan and I raced among the cow-pads. 1954 *Landfall* VIII. 273 These green paddocks dotted with thistles and cow-pats. 1962 *Punch* 28 Mar. 508/3 Eighty cowpats distributed over kitchen garden. 1884 Goode *Nat. Hist. Aquatic Anim.* 275 The 'Serjeant Major',..called in Bermuda the 'Cow-pilot', from an alleged habit of being always found in the society of the 'Cow-fish', or *Ostracion*. 1874 J. G. McCoy *Cattle Trade* 126 A few short weeks after the opening of the cattle trade..every stall—fully one hundred or more—would be full of cow ponies. 1909 *Westm. Gaz.* 29 May 5/3 Racing a cow-pony against a train on the first railway made in Colorado near Denver. 1934 *Times Lit. Suppl.* 29 Nov. 855/3 There are still thousands of cow-ponies trained to follow a galloping steer. 1878 S. L. Caldwell *Diary* 24 Apr. in *Colorado Mag.* (1939) XVI. 152 At Hugo the cow-punchers were assembling for the round-up. 1884 W. Shepherd *Prairie Exper.* 35 Each boy, when out cow-punching, rides from six to ten horses, using them in turns. 1896 *Westm. Gaz.* 20 Oct. 2/1 An impressionist painter once brought me a Thing, and I made him believe that it really was a portrait of a lady—or was it a cow-scape? 1938 Auden & Isherwood *On Frontier* I. i, Those exquisite little landscapes (or should I say 'cowscapes'?) of Ketchling. 1903 A. Adams *Log Cowboy* xx. 309 The wisdom of mounting us well..reflected the good cow sense of our employer. 1920 J. M. Hunter *Trail Drivers Texas* 299 When a cowboy says that a man has good 'cow sense' he means to pay him a high compliment. 1922 G. Jessop *Cricketer's Log* vi. 198 My propensity for the 'cow shot'. 1928 *Times* 2 July 5/5 R. S. Walker made a glorious half 'cow-shot' to mid-wicket which was only a yard short of a 6. 1963 *Times* 13 June 13/3 A series of pulls which ended with a catch at the wicket would appear in this form: 'After several cow-shots into the Great Beyond, Basher was neatly pouched by the timber-watcher.' 1917 E. Miller *Camps, Tramps & Trenches* (1939) ix. 50 Scummy colonial cow-spankers and bush-whackers. 1932 N. Scanlan *Pencarrow* ix. 87 A life..among ploughmen, and drovers, and cowspankers. 1963 *Weekly News* (Auckland) 8 May 39/1 The good old New Zealand cowspanker. 1900 *N.Z. Illustr. Mag.* II. 592 There would in time be an end to the eternal round of cow-spanking, school, and getting up. 1885 *Santa Fe Wkly. New Mexican* 3 Dec. 4/1 St. Louis is the biggest cow-town on earth just at present. 1888 T. Roosevelt in *Century Mag.* Feb. 500 A true 'cow-town' is worth seeing. 1958 *Times Lit. Suppl.* 26 Dec. 755/4 The mushroom cow-towns, such as Abilene, Newton and Dodge City. 1967 *Amer. Speech* XLII. 115 Wichita's life as a 'cow town' was just beginning. 1886 T. Roosevelt in *Century Mag.* July 341/1 It is even more laughable to see some young fellow.. attempt..to do cow-work with his ordinary riding or hunting rig. 1907 S. E. White *Arizona Nights* I. iii. 53 He kept his own mount of horses, took care of them, hunted, and took part in the cow work. 1929 J. F. Dobie *Vaquero* 13 When we gathered cattle, we said that we were on a 'cow hunt', a 'cow work', a 'work', or a 'cow drive'.

cowardy (kɑu·ərdi) *a. colloq.* [f. COWARD *sb.* + -Y[1].] Cowardly. Used tauntingly by children, esp. in the phr. *cowardy, cowardy, custard* (formerly *costard*).

1836 in A. Nicoll *Hist. Early 19th Cent. Drama* (1930) II. 436 (*title*) Cowardy, Cowardy Custard; or, Harlequin Jim Crow and the Magic Mustard-Pot. 1854 A. E. Baker *Gloss. Northampt. Wds.* I. 149 *Cowardy! cowardy! costard!* Repeated by children playing at the game of 'One catch all', when they advance towards the one who is selected to catch them, and dare or provoke her to capture them. 1882 'F. Anstey' *Vice Versâ* ix. 182 Yah, you're afraid!..I wouldn't be afraid! Cowardy, cowardy, custard! 1911 J. M. Barrie *Peter & Wendy* ii. 27 Father's a cowardy custard. 1935 G. Greene *Basement Room* 77 He muttered taunts and encouragements, 'May as well be hung for a sheep', 'Cowardy, cowardy custard', grown-up and childish exhortations oddly mixed. *a* 1953 Dylan Thomas *Under Milk Wood* (1954) 59 Cowardy cowardy custard Give Gwennie a penny.

cow-bell. Add: **1. b.** A bell without a clapper used as a percussion instrument in a jazz or dance band.

1926 A. Niles in W. C. Handy *The Blues* 26 Raucoussquawks, rattles, cowbells. 1961 A. Baines et al. *Mus. Instr. through Ages* xiv. 343 Cowbell, rattle (or ratchet), anvil...In the drummer's music they are clearly named, and their notes are usually written in the top space of the stave, or directly above it. 1966 *Crescendo* Dec. 27/2 When played authentically, the cowbell plays a monotonous four throughout, with occasional 'trip' beats.

cow-boy, cowboy. Add: **3.** (Earlier example.)

1849 J. S. Jenkins *Hist. War U.S. & Mexico* i. 52 The Mexican rancheros..ventured across the Rio Grande..but they were immediately attacked by the Texan 'cow-boys'.

b. A boisterous or wild young man (see also quots. and **drug-store cowboy*). *slang* (orig. *U.S.*).

1942 *Amer. Speech* XVII. 103/1 *Cowboy*, reckless driver. 1952 P. Bonner *SPQR* (1953) xx. 174 Two very chic girls.. were sitting on bar stools..with a typical Via Veneto cowboy in sports jacket and flannel slacks. 1958 E. Dundy *Dud Avocado* i. vi. 98 One of those drugstore-cowboy-motorcycle types, just past their first juvenile delinquency. 1959 *News Chron.* 7 Dec. 5/2 Coffee bar cowboys..are the teenagers with black jackets and fast motor cyclists who gather in cafés.

c. A policeman. *slang.*

1959 C. MacInnes *Absolute Beginners* 70 'They didn't seem to me like cowboys.'..'I can smell a copper, in the dark, a hundred feet away, blindfolded.' 1959 *Encounter* Aug. 32 Who should come round the corner but a cowboy.. with..a helmet not too secure.

d. *Cowboys and Indians*, the name of a children's game in which the participants imitate the actions of, and esp. battles between, cowboys and Indians.

[1887 *N.Y. Herald* 21 Feb. 9/1 A Chinaman..was passing through City Hall Park..yesterday afternoon when a mixed lot of 'cowboys and Injuns' swooped down upon him.] 1916 'N. Douglas' *London Street Games* 120 They get some games out of the cinematograph, by the way, such as Cowboys–Indians. 1949 *Chicago Tribune* 17 July (Comics) 4 We'll play cowboys and Indians. 1969 I. & P. Opie *Children's Games* 4 Each child's attitude is..changing..from the ritualistic to the romantic (i.e. the free-ranging games, 'Hide and Seek', 'Cowboys and Indians'), and from the romantic to the severely competitive.

5. *attrib.* and *Comb.*

1881 G. W. Romspert *Western Echo* 173 The average cow-boy saddle weighs forty pounds. 1895 *Montgomery Ward Catal.* 273/2 Largest line of cow boy hats in America. *Ibid.* 519/3 The style of the regular cowboy boot. 1897 *Daily News* 4 June 8/2 The New South Wales men..looked very smart in their well-fitting uniforms. The cloth is of a grey-brown shade, and the 'cowboy' hats of the same shade. 1937 *Discovery* Feb. 43/2 A famous cowboy actor rode his horse round an arena. 1958 J. Kerouac *On Road* v. 33 All the cowboy-dudded tourists and oilmen and ranchers. 1964 *Economist* 29 Aug. 799/1 The Front, which is called the 'cowboy government' by its Rhodesian opponents. 1970 T. Hillerman *Blessing Way* xvi. 163 The young man..wore cowboy boots.

cowdie, var. KAURI.

cow-fish. Add: **3.** (Earlier U.S. example.)

1870 *Amer. Naturalist* III. 467 The odd-looking trigonal Trunk-fish (*Lactophrys camelinus* De Kay), sometimes called Cow-fish, a profile view of the head much resembling that of a cow.

cow-hide, *sb.* Add: **3.** (Earlier U.S. example.)

1818 M. Birkbeck *Lett. fr. Illinois* 90 The enraged barrister, with a hand-whip, or cow-hide, as they are called ..actually cut his jacket to ribbons.

4. (Earlier and later U.S. examples.)

1823 W. Faux *Memorable Days Amer.* 305 One man then bound him to a tree and lashed him with a cow-hide whip. 1947 E. H. Paul *Linden on Saugus Branch* 210 He had a plug hat..and cowhide boots without buttons or laces.

cow-hide, *v.* (Earlier U.S. example.)

1794 *Kentucky Gaz.* (Lexington, Ky.) 1 Mar. 2/3 In November 1792..a justice of the peace was cited to appear before the house of Delegates...Some he had horse-whipped; others he had cow-hided.

cow·-hunt, *sb. U.S.* [*cow *sb.*[1] 1 c.] A search for strayed cattle. So **cow-hunt** *v. intr.*, **-hunter, -hunting** *ppl. a.*

1853 'P. Paxton' *Yankee in Texas* 108 Everything in the shape of a man..had left the settlement, and engaged in a general 'cow-hunt'. 1863 *Harper's Mag.* Mar. 536/1 It had the reputation of being the resort of fugitives and cow-hunters. 1898 H. S. Canfield *Maid of Frontier* 132 The ordinary immense Texas affair, intended only for cow-hunting males. 1920 J. M. Hunter *Trail Drivers Texas* 104 Soon the settlers began to organize cow hunts. *Ibid.* 175 Those old time ranchmen were content to simply cow hunt twice a year and brand their calves. *Ibid.*, I remember the many times that cowhunters rode up to my father's house. 1929 [see **cow-work*].

co-wife (kōᵘ·wəiːf). [f. CO- 3 b + WIFE *sb.*] A joint wife; one of several wives of the same man.

1892 Kipling & Balestier *Naulahka* xx. 247 If a co-wife had sought the lives of those three of thine. 1956 *Bull. Sch. Oriental & Afr. Stud.* XVIII. 165 A wife says it *to* a friend or co-wife but *at* her husband to tell him that she and his other wives are displeased.

cowing (kɑu·iŋ), *ppl. a.* [COW *v.*[1]] Overawing, intimidating.

1843 T. Wilson *Pitman's Pay* etc. 24 It was, ne doubt, a cooen seet [i.e. cowing sight]. 1888 Mrs. H. Ward *R. Elsmere* I. x. 261 Her sharp tongue had sometimes a cowing effect on his curious nonchalance.

Cowley Father (kɑu·li). [Name of a village near Oxford (now a suburb) + FATHER *sb.* 6 b.] A priest of the Anglican order of the Society of Mission Priests of St. John the Evangelist, founded in Oxford in 1865 by the Rev. R. M. Benson (1824–1915).

1902 *Encycl. Brit.* XXVIII. 195/1 The Society of Mission Priests of St John the Evangelist, otherwise known as the *Cowley Fathers*, was founded..at Cowley St John. 1914 C. Mackenzie *Sinister St.* II. iii. ix. 681 He had indeed heard Pallant preach at the church of the Cowley Fathers. 1953 E. Lutyens *Blessed Girl* ii. 23 Edward became a Cowley Father and died as a missionary in India.

cowling (kɑu·liŋ). [f. COWL *sb.*[1] + -ING[1].] The removable covering over or round the engine, etc., of an aeroplane or a motor vehicle.

1917 *Blackw. Mag.* June 925/2 While attempting to fasten a small door in the cowling round the engine. 1927 *Daily Tel.* 22 Jan. 9/3 Captain Campbell quickly had the car back on its course, but the cowling of his engine was badly dented. 1928 C. F. S. Gamble *North Sea Air Station* ii. 55 A piece of tin in the inside of the engine cowling had come adrift. 1934 *Flight* 8 Feb. 133/1 Apart from rotary engines, the only air-cooled engine of pre-war days which used cowling was the eight-cylinder Vee Renault. 1950 *Engineering* 6 Jan. 1/1 On public service motor vehicles.. care is taken with the fitting of engine cowlings.

cow-pen, *v.* (Later U.S. examples.)

1786 Washington *Diaries* (1925) III. 131 [I] plowed a poor ¼ acre to Cowpen on. *Ibid.* 183 About an acre was sown in my meadow, part of which had been cowpened. 1832 R. Baird *Valley Mississippi* xxiv. 286 It grows well on marsh..and even pine barrens, when they are well trodden or *cowpenned*, as it is termed. 1871–3 *Texas Almanac* 12 Cow-penning, in successive lots..is a good plan for manuring.

Cowper-Temple. Now *Hist.* The name of William Francis Cowper, afterwards *Cowper-Temple* (1811–1888), used *attrib.* to designate the clause which he introduced into the Education Act of 1870, or the religious teaching of an undenominational character in public elementary schools as provided for by this clause. Hence **Cowper-Te·mpleism, Cowper-Te·mpleite.**

1902 *Encycl. Brit.* XXVII. 661/1 Clause 14 of that Act [*sc.* Education Act of 1870], generally known as the Cowper-Temple Clause, enacts that 'in any school provided by a School Board, no religious catechism, or religious formulary which is distinctive of any particular denomination, shall be taught'. 1906 *Daily Chron.* 24 May 7/7 Rather than secularism, Mr. Balfour would prefer universal Cowper-Templeism 'with all its illogicality and its unfairness'. 1906 *Westm. Gaz.* 25 June 3/1 Our attitude towards the Bible has changed since the invention of Cowper-Templeism. That compromise is impossible and unreasonable to-day. 1908 *Daily Chron.* 25 Feb. 5/7 In all these schools there may be Cowper-Temple religious teaching. *Ibid.* 21 Apr. 1/4 Denominational colleges can be legally raided by Cowper-Templeites on the highwayman principle of first come first served. 1932 R. A. Knox *Broadcast Minds* i. 9 Since the time of the ill-fated Cowper-Temple clause we have given up trying to solve ecclesiastical disputes on the principle of the Highest Common Factor. 1959 *Chambers's Encycl.* IV. 800/1 By means of the Cowper-Temple clause this act also effected a compromise on the question of religious instruction, which was not superseded until 1944.

cow·poke. *N. Amer.* [f. COW *sb.*[1] + POKE *sb.*[3] or *v.*[1]] = COWBOY 3.

1881 in Wentworth *Amer. Dial. Dict.* (1944) 139/1. 1928 *Lariat Mag.* Jan., I camped there once, and a cowpoke told me why they were named that. 1948 *Denison* (Tex.) *Herald* 1 July 9/2 Due to the size of the entry list, the cow-pokes have just been allowed to enter one event each night. 1959 *Times* 18 June (Canada Suppl.) p. xi/2 Over 500,000 visitors pack the exhibition grounds to watch the world's top cowpokes perform. 1970 *Islander* (Victoria, B.C.) 6 Sept. 2/2 [He] worked for a spell on the J. T. Edwards ranch where the head cowpoke was..Paul Walker.

cow-skin. **1.** (Earlier and *attrib.* U.S. examples.)

1780 in *Narrag. Hist. Reg.* I. 101 Carried the cow skin to tann by George Wilson. 1809 *Repertory* (Boston) 6 Oct. (Th.), A green or untanned cowskin whip. 1863 S. C. Massett *Drifting About* 155 He with trowsers tucked into his cowskin boots..wended his way to the stage office. 1887 *Outing* (U.S.) X. 119/1 If I only had that cow-skin horse now what I used ter own back in old Missouri.

cow-skin, *v.* (Earlier and later U.S. examples.)

1799 *Aurora* (Philad.) 20 May (Th.), I am a constable, and may therefore kick, cuff, beat, bruise, cowskin, or kill any man I please. 1836 Crockett *Exploits & Adv. Texas* (1837) 78 The devil himself might, the next time, undertake to cowskin such a..scoundrel for him. 1947 *West Pennsylvania Hist. Mag.* Sept.–Dec. 132 The 'cowskinning' incident..was the final blow.

cowslip. Add: **3. cowslip pudding, tart,** baked sweet dishes flavoured with cowslip flowers; **cowslip wine** (earlier and later examples).

1723 J. Nott *Cook's & Confect. Dict.* sig. K8, To make a Cowslip Pudding... To make a Cowslip Tart. 1909 *Westm. Gaz.* 14 May 5/1 'Cowslip pudding' is also a seasonable delicacy. 1723 J. Nott *Cook's & Confect. Dict.* sig. K8, To make Cowslip Wine. 1962 A. Jobson *Window in Suffolk* viii. 124 She..made him drink a glass of cowslip wine.

cowslipping (kɑu·slipiŋ), *vbl. sb.* [f. COWSLIP + -ING[1].] Gathering cowslips.

1898 C'tess Arnim *Eliz. & German Garden* 29 We have been cowslipping to-day. 1928 *Daily Express* 7 Mar. 5/5 We go 'cowslipping' to the same..place year after year.

cowson (kɑu·sən). [f. COW *sb.*[1] + SON *sb.*, after WHORESON.] Used as an opprobrious epithet or term of abuse.

1936 'J. Curtis' *Gilt Kid* ii. 18 Bogy says, 'The prisoner opened the doors of two cars, your worship.' The lying cowson. 1940 N. Monks *Squadrons Up!* 55 There's a whole bunch..above. Why won't the cow-sons come down? 1959 P. McCutchan *Storm South* x. 149 That cowson country.

cow-tail. Restrict † *Obs.* to sense in Dict. and add: **2.** The coarsest grade of wool, sheared from the sheep's hind legs.

1884 W. S. B. McLaren *Spinning* 16 The coarsest part of the [sheep's] wool is known as breach, or britch, and even when very strong, as cow-tail. 1945 Baker *Austral. Lang.* iii. 66 Among wool-trade expressions apparently original to Australia are *cowtail* and *say-cast*, the coarsest hair at the tail end of a fleece.

cow-turd. **1.** (Later example.)

1918 W. Raleigh *Let.* 18 Nov. (1926) II. 498, I want it to reach Germany... For self-respect and sense of honour the whole nation is one large cow-turd.

cowy (kɑu·i), *a.* Also **cowey.** [f. COW *sb.*[1] + -Y[1].] Of, pertaining to, or characteristic of a cow; bovine. Also *fig.*

1893 *Field* 11 Mar. 363/3 We have tasted several samples of this Australian butter... The same 'cowey' flavour was discoverable. 1903 *Graphic* 3 Oct. 443 The occasional whiff from a cowy place. 1924 *Blackw. Mag.* Apr. 458/2 The cowy-smelling Himalayan road. 1925 D. H. Lawrence *Refl. Death Porcupine* 164 Possibly she is lying peacefully in cowy inertia, like a black Hindu statue. 1935 *Amer. Speech* X. 270/1 (Lang. Livestock Mart) *Cowy*, like a cow; showing signs of maturity or age; 'matronliness'. 1938 Hemingway *5th Column* (1939) 113 Eland... They're the big cowy things that jump like hares. 1955 J. G. Davis *Dict. Dairying* (ed. 2) 472 Freshly drawn milk from a healthy udder has a characteristic flavour, best described as 'cowy'.

cox, *sb.* (Examples.)

1869 J. C. Tinne *International Training* (1923) 9 Turned and rowed hard down to Whitchurch, with the exception of one unintentional easy halfway, owing to the fact of the cox. mistaking the way we wished to go. 1881 *Rowing, Steering & Coaching on Cam* 29 It may be laid down as a general statement, that a cox. can't shout too much. 1889 J. K. Jerome *Three Men in Boat* xv. 253 Then 'cox' threw both rudder lines overboard, and burst into tears. 1955 *Times* 13 Aug. 8/7 Walter, one time cox of his college eight and collector of three rudders.

cox, *v.* Add: (Examples.) So **co·xing** *vbl. sb.*

1881 *Rowing, Steering & Coaching on Cam* 23 Though coxing does not require so much physical exertion as rowing, it requires more mental effort. 1920 *Notes on Coxing* 21 A coxswain..must decide whether he is out just to 'steer' or really to 'cox'. 1939 Nickalls & Mallam *Rowing* vii. 116 Weight is the governing factor which decides whether or not you are suitable for coxing.

Cox (kǫks). The name of an amateur fruit-grower of the first half of the 19th century, used in the possessive of a variety of orange pippin, or an apple of this variety; also *ellipt.*

1860 R. Hogg *Fruit Man.* 5 Cox's Orange Pippin... Medium sized, roundish-ovate, and regular in its outline... A first-rate dessert apple. 1866 [see PIPPIN 2]. 1884 [see ORANGE *sb.*[1] 3 b]. 1920 E. A. Bunyard *Handbk. Hardy Fruits* 43 *Cox's Orange*..Dessert, November to March.. Raised in 1825, by Mr. Cox, a retired brewer of Colnbrook Lawn, near Slough, and introduced by Mr. Charles Turner, about 1850. 1934 *New Statesman* 15 Dec. 901/1 He can obtain a Cox from the fruit shop at Victoria Station. 1960 *Farmer & Stockbreeder* 12 Jan. 17/1 Cox's remain a good sale at firm values. 1962 *Times* 22 Oct. 11/4 An apple, even if it is a Cox's Orange, is not worth any great sum of money. *Ibid.* 25 Oct. 7/3 Trays of Coxes grown by Mr. G. N. Horne.

coxed (kǫkst), *ppl. a.* [f. COX *v.* + -ED.] Coxswained.

1939 Nickalls & Mallam *Rowing* vii. 116 Promotion to a coxed four can follow almost immediately. 1955 *Times* 15 July 3/7 The A.R.A. coxed four championships will be held at Henley-on-Thames on Saturday. 1963 *Times* 10 June 3/6 London took third place in the coxed fours.

co·xless, *a.* [f. COX *sb.* + -LESS.] Without a coxswain.

1906 *Daily Chron.* 18 June 6/4 Twirling, coxless, in impotent circles upon the bosom of the Thames. 1963 *Times* 10 June 3/6 Nottingham Union R.C. won the coxless fours event at the Ghent international regatta today.

Coxsackie (kuksæ·ki, kǫksæ·ki). Name of a town in New York State (see quot. 1949), used *attrib.* to designate any of a group of pathogenic enteroviruses or the diseases associated with them. So **Cox-, coxsa·ckie vi·rus.**

1949 G. Dalldorf in *Science* CX. 594/1 It has become desirable to name those viruses that are pathogenic for suckling mice and hamsters, and that have recently been associated with 'poliomyelitis'... It is suggested that the agent be called 'Coxsackie virus', since the first recognized human cases were residents of that New York village. Since a number of viruses may be involved, the term 'Coxsackie group of viruses' seems especially suitable. 1961 [see *ECHO VIRUS]. 1965 *Times* 3 Sept. 14/5 The coxsackie viruses of which there are over 30 serotypes..cause sore throats, sometimes associated with pain in the chest and abdomen. 1966 Wright & Symmers *Systemic Path.* II. xxxiv. 1196/2 Coxsackie viruses are the cause of various clinical syndromes. At least one of them—Coxsackie virus Type A7—has caused illnesses indistinguishable from poliomyelitis. *Ibid.*, Cases of Coxsackie meningitis.

coxswain (kǫ·kswēin, kǫ·ks'n), *v.* [f. the sb.] *trans.* To act as coxswain to (a boat); also *intr.* Hence **co·xswained** *ppl. a.*, having a coxswain; so of an event in which such boats compete.

1928 *Daily Express* 22 May 3/4 He..resumed the task of coxswaining the Brighton lifeboat. *Ibid.* 7 Aug. 12/6 In Heat 1 of the second round of the coxswained pairs. 1962 A. Silver *Lett. S. Butler* 39 At St John's the young man.. coxswained for the crew of the Lady Margaret Boat Club. 1963 *Times* 20 May 4/5 Coxswained fours have always been a big feature of provincial rowing.

coxy, *a.* (Later example.)

1927 F. B. Young *Portrait of Clare* 479 They wouldn't believe me; they said I was coxy about it.

coy, *v.*[1] Add: **6.** *trans.* To disguise or slight in a demure manner. *rare.*

1873 Trollope *Eustace Diamonds* I. xv. 198 She throws from her..all idea of coying her love. She would leap at his neck if he were there. 1874 —— *Way we live Now* I. xix. 124 Do not coy your love for me if you can feel it. 1880 —— *Duke's Children* III. viii. 88 She had coyed his love, so that he had been justified in thinking himself free from any claim in that direction.

coyish, *a.* (Later U.S. example.)

1833 H. Barnard in *Maryland Hist. Mag.* XIII. 322 A lovely grand-daughter..was present, and was pleased not to be very coyish.

coyishness. Delete † and add later example.

1869 R. D. Blackmore *Lorna Doone* III. xxii. 317 She would cling to me and nestle to me, being scared of coyishness.

coyote. Add: (Earlier U.S. examples.)

1824 W. Bullock *6 Months in Mexico* 119 Saw a cayjotte, or wild dog, which in size nearly approached the wolf. *Ibid.* 261 Mexico produces an animal which seems to connect the wolf, fox, and dog: it is called the cocyotie. *Ibid.* 311 Cages in which lions, tigers, wolves, cayatoo, and wild cats were confined. 1834 A. Pike *Sketches* (Boston) 14 (Th.), The little gray collotes [*sic*] or prairie wolves, who are as rapacious and as noisy as their bigger brethren. 1849 E. Bryant *California* (ed. 5) xix. 219 A species of jackal called here the *coyote*, frequently approached within a few rods of us.

transf. (Earlier and later examples.)

1872 S. Powers *Afoot & Alone* 277 Many slouching fellows..are really squatters or 'coyotes'. 1909 'O. Henry' *Roads of Destiny* xvi. 266 She's married to Benton Sharp, a coyote and a murderer. 1948 *New Mexico Q. Rev.* Summer 198 Often *coyote* is used as a synonym for *native*, and is applied to Indians and *mestizos* (mixed bloods), as readily as to plants.

b. *coyote wolf* (earlier example); **coyote diggings** (earlier and later examples); **coyote getter** (see quot.); **coyote hole** = *coyote diggings* (see also quot. 1906).

1850 *San Francisco Picayune* 31 Aug. 3/1 There are Coyoto Diggings..from which, at the depth of from 17 to 25 feet, $23,000 have been taken out in two days. 1948 *New Mexico Q. Rev.* Summer 199 In mining lingo of early California, derived from the Mexicans, 'coyote holes' or 'coyote diggings' were small drift tunnels. 1961 *New Scientist* 13 Apr. 17/1 What the Americans bluntly call 'coyote getters'. These small, mushroom-like appliances, fixed to a metal stake in the ground, are capable of firing a small-calibre cartridge loaded with sodium cyanide. 1851 *Sacramento* (Calif.) *Transcript* 1 May, He noticed one coyote hole where the miners had sunk the shaft through a strata..of ashes. 1906 *N.Y. Even. Post* 12 Sept. 7 Drilling coyote holes is the name applied by railroad contractors to drilling blast holes in grade running cuts through hills. 1850 L. H. Garrard *Wah-to-Yah* i. 13 We were serenaded by the Coyote wolf.

coz, var. *COS, 'cos *adv.* and *conj.*

cozymase (kōᵘzəi·mēⁱz, -s). *Biochem.* [a. G. *co-zymase* (Euler and Myrbäck 1923, in *Zeitschr. f. physiol. Chem.* CXXXI. 180), f. *CO-(ENZYME + ZYMASE.] One of the names of the co-enzyme nicotinamide-adenine dinucleotide (NAD).

1924 *Chem. Abstr.* XVIII. 2346 Co-zymase may be prepd. in impure form by boiling yeast and filtering. 1926–1946 [see *APOZYMASE]. 1962 *Lancet* 15 Dec. 1258/1 Reduced cozymase. 1964 Florkin & Stotz *Comprehensive Biochem.* (rev. ed.) XIII. vii. 41 The names 'cozymase', 'phosphocozymase'..should no longer be used.

cozzpot (kǫ·zpǫt). *slang.* [Perh. alteration of first syllable of COPPER *sb.*[4] + POT *sb.*[1] 9 d. Cf. *CRACKPOT, TOSSPOT.] A policeman.

1962 Parker & Allerton *Courage of Convictions* I. ii. 58 This cozzpot took me to the police station and in due course I was up in front of old man Henriques again. 1969 J. Ashford *Prisoner at Bar* x. 87 The cozzpots ain't givin' me a chance.

crab, *sb.*[1] Add: **4.** (Later examples.)

1968 J. R. Ackerley *My Father & Myself* xii. 139, I kept a stock of Blue Ointment handy for the elimination of crabs. 1970 *Observer* 12 Apr. (Colour Suppl.) 43 Green wings means someone's got gonorrhoea, and yellow wings for something else, like crabs.

7. d. The lifting-gear of a crane, travelling on rails and moving the load.

a 1877 Knight *Dict. Mech.* I. 644/1 One form of traversing-crane consists of a crab upon a carriage traveling upon rails on the beams overhead in a foundry. 1902 *Westm. Gaz.* 10 June 10/1 The total weight of the crane..is 474½ tons. The crab contains the whole of the mechanism for lifting and traversing, the speed of the traverse being about 26 ft. per minute. 1935 *Chambers's Encycl.* III. 540/2 The overhead traveller in a workshop or factory is a non-revolving crane; its essential parts are the main cross girders, the crab which moves to and fro along the cross girders on rails supported by them... The crab can..lift and shift the load from one part to any other part of the shop.

10*. [After G. *krebs* crab, unsold copy of a book.] A book returned unsold by a bookseller to the publisher.

1896 *Bookseller* 6 Mar. 278/1 The great bulk of the bookselling business [in Germany] with the publishers, is done by the 'on sale' system... These methods seem to work well in Germany, and the number of 'crabs', or books returned at the end of the year, is not often very great. 1960 Glaister *Gloss. of Book* 90/1 *Crabs*, a colloquialism for copies of a book returned by the bookseller to the publisher.

10.** *Naut. slang.* A midshipman, esp. a junior midshipman or naval cadet.

1916 *Chambers's Jrnl.* July 435/1 The sub, assisted by the senior 'snotties', had drilled the Crabs into a high state of discipline and efficiency. 1916 'Taffrail' *Carry On!* 43 To the senior Sub-Lieutenant..the newly joined Midshipmen are 'crabs' or 'warts', mere excrescences on the face of the earth.

11. crab-canon *Mus.* = *cancrizans *sb.*; **crab-eating** *a.*, that feeds on crabs (sometimes rendering L. *cancrivorus*); **crab-hole** *Austral.*, a hole burrowed by a land crab; so *crab-holed* adj.; **crab-pot (valve)**, in airships, a fabric valve with a sleeve which could be closed like a crab-pot; **crab-shell** (example); **crabwise** *adv.*, (moving) sideways or backwards like a crab; also *attrib.*

1908 *Strand Mag.* Jan. 30/2 Canons, so constructed that they would read the same backwards or forwards. For this reason they were called Crab Canons. 1959 Westrup & Harrison *Collins's Music Encycl.* 109/2 Crab canon (*canon cancrizans*) or retrograde canon. The part which imitates is written backwards, beginning with the end. Crab canon by inversion: the part which imitates is written backwards and upside down. 1847 Crab-eating [in Dict.]. 1883 *List Animals Zool. Soc.* (ed. 8) 82 *Procyon cancrivorus*, crab-eating raccoon. *Ibid.* 193 *Didelphys cancrivora*, crab-eating opossum. 1908 *Westm. Gaz.* 6 June 10/2 Stones are always met with in the stomachs of two species of Antarctic seals. The Crab-eating seal is one of these..swallowing them to act as grindstones, whereby the hard shells of the crabs may be properly broken up. 1932 S. Zuckerman *Soc. Life Monkeys* vi. 89 The length of the cycle in the common or crab-eating macaque is approximately the same as in other macaques. 1954 G. Durrell *3 Singles to Adv.* vii. 156 It was a broad, flat head with neat rounded ears and a dog-like muzzle. The creature's colouring was ash grey, but across the eyes was a wide black band...'A crab-eating raccoon.' 1966 R. & D. Morris *Men & Apes* i. 20 Crab-eating macaques in Bali are apparently treated with as much deference as the sacred monkeys of India. 1848 Mrs. Perry in Goodman *Church in Victoria* (1892) 72 Full of crab-holes, which are exceedingly dangerous for the horses...These holes are formed by a small land-crab, and then gradually enlarged by the water draining into them. 1891 R. Wallace *Rural Econ. Austral. & N.Z.* ii. 47 The surface [of the land] was closely pitted with crabhole-like water-hollows. 1903 'T. Collins' *Such is Life* 16 Price and Cooper, being cooks, had kindled an unobtrusive fire in a crab-hole, where three billies were soon boiling. 1956 S. Hope *Diggers' Paradise* xvii. 157 The ball may be lifted or dropped without penalty from wallaby scrapes, crab holes, ironstone outcrops and tractor marks. 1908 Mrs. A. Gunn *We of Never-Never* xii. 152 One hundred and thirty miles of sun-baked, crab-holed, practically trackless plains. 1918 W. E. Dommett *Dict. Aircraft* 16 *Crab-pot*, a fabric valve used for controlling the inlet of air to the ballonets of a non-rigid airship. 1950 *Gloss. Aeronaut. Terms (B.S.I.)* I. 53 *Crabpot valve*, a special type of fabric sleeve the operation of which is controlled by a hand-line. 1807 Beresford *Miseries* II. xviii. 185 An old *crab-shell*, which, in..his antiquarian fury, he shall consider as an inestimable treasure. 1904 *Daily Chron.* 6 May 8/1 There are barges in the way, and these have to be coaxed aside before the Adler can approach crabwise to the wharf. 1926 *Chambers's Jrnl.* 163/1 She went crabwise about the loch. *Ibid.* 224/1 A crabwise gait. 1927 R. A. Freeman *Certain Dr. Thorndyke* i. iii. 42 He began to advance, crabwise, across the deck in the manner of a wrestler attacking. 1934 C. Lambert *Music Ho!* v. 298 The innumerable inversions, augmentations, diminutions and crabwise canons of Schönberg's later works. 1963 *Times* 19 Apr. 5/5 It..swerved out of control and came crabwise down the middle lane and hit my lorry.

crab, *sb.*[4] *colloq.* [f. Crab *v.*[2] 2.] The action of crabbing or finding fault; an instance of this; an adverse criticism or objection.

1893 *Field* 11 Mar. 347/1 It will be said I am dreadfully 'on the crab', but I believe what I have written is only the simple truth. 1922 *Autocar* 10 Nov. 956 My only crab to them at present is that [etc.]. 1924 J. Buchan *Three Hostages* 25 The crab of the place was its neighbours. 1927 *Observer* 10 July 11 The only 'crab' we have against this is that cavalry of old effected most of their success by charging infantry.

crab, *v.*[2] Add: **2.** Also *absol.*

1891 F. W. Carew *No. 747; Autobiogr. Gipsy* xx. 228 Shice..alternately 'crabbed' and 'chy-iked' as the case might require. 1906 *Westm. Gaz.* 11 Aug. 1/2 The difference between us and you, said an American who had watched Mr. Chamberlain's Fiscal campaign, is that 'we boom, and you crab'. 1925 F. Scott Fitzgerald *Great Gatsby* vii. 127 The thing to do is to forget about the heat...You make it ten times worse by crabbing about it. 1968 *Listener* 4 Apr. 444/2 To no other people can the classic principle of 'first crib and then crab' have been so consistently applied.

b. To interfere with or obstruct the working, progress, or success of.

Cf. quots. 1812, 1890 in Dict.

1899 R. Whiteing *No. 5 John St.* xxi. 191 We was crabbed. ..The mugs might ha' wrecked the show. 1901 *N.Y. Independent* 12 Dec. (Cent. D. Suppl.), The use of foreign tires of course crabbed the deal. 1920 *Glasgow Herald* 18 Aug. 7 Posen contains a large German Irredentist minority, which might crab the Polish military defence. 1922 C. Sandburg *Slabs of Sunburnt West* 66 You're trying to crab my act. 1934 E. Bowen *Cat Jumps* 235 Seeming to crab Patsey's marriage. 1941 G. Heyer *Envious Casca* iv. 62 You've done all you can to crab Willoughby's play. 1955 *Times* 24 May 16/2 You made it your business to crab the performance.

crab, *v.*[3] Add: **3. b.** *Aeronautics.* To put (an aeroplane) in a position diverging from the straight course; to fly at an angle to the longitudinal axis. Also in other *transf.* uses (*trans.* and *intr.*) and with *back*, *in*, *on*. Also **crab** *sb.*[5], a divergent position.

1929 A. C. McKinley *Appl. Aerial Photogr.* 13 So that ..the necessary amount of crab can be obtained at which the airplane must fly. *Ibid.* 36 Adjusting the Camera for Crab.—On approaching the starting point of each strip the pilot will crab the airplane in order to make allowances for the wind. 1959 J. L. Nayler *Dict. Aeronaut. Engin.* 70 *Crab angle*, a colloquial term for the angle of yaw. 1962 *Punch* 17 Jan. 134/3 He knows which [TV] cameras can be crabbed (moved sideways). 1964 G. Lyall *Most Dangerous Game* iii. 25, I crabbed in towards him, holding the knife low. *Ibid.* xviii. 116, I crabbed back to my chair and sat down. 1966 'W. Haggard' *Power House* i. 12 Once skidding it [sc. a car] crabbed on helplessly. 1970 W. K. Kilford *Elem. Air Survey* iv. 89 Crabbing gives rise to loss of stereoscopic cover, since the area of overlap between two consecutive photographs is reduced.

cra·bber[2]. [f. Crab *v.*[2]] One who finds fault or criticizes adversely.

1909 *Daily Chron.* 14 June 6/4 They wore the determined look of the born 'crabber'. They were there to discover something wrong. *Ibid.* 27 July 4/4 We have fallen from the frying-pan of the boomers into the fire of the crabbers. 1936 'J. Tey' *Shilling for Candles* x. 118 There were one or two crabbers..people who wanted his job.

crab-eater. Add: **3.** A crab-eating seal, *Lobodon carcinophagus* (*crab *sb.*[1] 11).

1938 *Geogr. Jrnl.* XCI. 525 The two most abundant species of antarctic seal are the crabeater and the Weddell. 1959 *New Biol.* XXIX. 113 Rough estimates of seal populations within the Falkland Islands Dependencies have been made by R. M. Laws. He suggests..1,000,000 crabeaters.

crab-grass. 3. (Earlier U.S. examples.)

1743 J. F. Gronovius *Flora Virginica* 134 Panicum spicis alternis oppositisve [etc.]..Crab-grass. 1785 Jefferson *Notes Virginia* 69 Our grasses are Lucerne,..greensward, blue grass, and crab grass. 1833 B. Silliman *Man. Sugar Cane* 13 The Crab Grass (*Digitaria sanguinalis*) which comes up abundantly on new land, is cut up with hoes.

crab-stock. Add: **2.** *Pottery.* Used *attrib.* to denote parts of eighteenth-century English stoneware. Also *absol.*

1909 A. Hayden *Eng. Earthenware* vi. 212 A teapot enamelled in colours having what is known as a 'crabstock' handle, spout, and lid. 1924 Rackham & Read *Eng. Pott.* vi. 93 The handle and spout were moulded in the form of a short length of woody stem—the so-called 'crabstock'. 1956 G. Taylor *Silver* vii. 157 A tea pot of 1748..its crab-stock spout and handles.

crack, *sb.* Add: **1. c.** (Earlier example.)

1836 Dickens in *Morning Chron.* 26 Oct. 3/5 Green eid jist fetch him a crack over the head with the telescope.

d. In phr.: *to have* (or *take* or *give*) *a crack*, to make an attempt, or trial; to 'have a shot' or 'go'. *colloq.* (orig. *U.S.*).

1836 Hill *Yankee Story Bk.* (Weingarten). 1885 C. M. Yonge *2 Sides of Shield* II. i. 2 Uncle Regie, will you have a crack at the rabbits to-morrow? 1918 H. C. Witwer *Baseball to Boches* iv. 164 The medico says I'll be all right.. to take another crack at them Germans. 1922 E. Wallace *Flying* 55 xxxix. 236 I'd take a crack at some of them with Fifty and even with Meyrick, who is a smashing good horse. 1939 'N. Shute' *What happened to Corbetts* iv. 110 I'll have a crack at going on board. 1947 —— *Chequer Board* vii. 168 He said they wanted me, so I said I'd give it a crack. 1955 M. Banks *Commando Climber* viii. 146 Are you having a crack at the Slav Route? 1959 M. Hastings *Hour-Glass to Eternity* ii. i. 148 We'd like to have a crack at climbing the peak. 1959 J. Thurber *Years with Ross* vii. 131 I'm going to give Bergman a crack at that job. 1966 *Sunday Mail Mag.* (Brisbane) 2 Jan. 11/5 He returned to Australia determined to take another crack at the land.

2. b. The break (*of* dawn, *of* day). *colloq.* (orig. *dial.* and *U.S.*).

1887 *Outing* (U.S.) X. 7/1 At 'crack of day' as the sergeant of the guard expressed it, the stir of camp was started by waking up the cook. 1899 Dickinson & Prevost *Gloss. Cumberland* 146/2 *Crack o' day*, the first dawning before sunrise. 1923 C. E. Mulford *Black Buttes* ii. 27 You boys git what sleep you can. We'll round 'em up at the crack of dawn. 1929 J. B. Priestley *Good Compan.* ii. v. 375 'And when will it arrive at Middleford?'..'About the crack of dawn, I suppose.' 1948 W. S. Maugham *Catalina* xxix. 185 He had slipped away at crack of dawn.

5. Also, a sharp or cutting remark. *colloq.*, (orig. *U.S.*). Cf. *wisecrack.

1896 G. Ade *Artie* ii. 14 After that first saucy crack with the half [dollar] I laid low three or four hands. 1903 A. H. Lewis *The Boss* 120 This is exec'tive session, an' that crack about bein' a taxpayer is more of a public utterance. 1923 R. D. Paine *Comr. Rolling Ocean* viii. 129 Do you remember the day before that when he made that crack at you in front of Miss Crozier? 1924 W. M. Raine *Troubled Waters* xviii. 197 Make another crack like that and there'll be trouble right here in Cell Fifteen. 1930 Wodehouse *Very Good, Jeeves!* v. 127 Just one crack like that out of him ..and I should infallibly have done his upper maxillary a bit of no good. 1958 *Manch. Guardian* 7 June 4/6 An anti-British crack gets as greedy a laugh as an anti-American gibe in London. 1967 *Listener* 28 Dec. 846/3 Mr Davis's book..is devoid of 'personalities' in the malign sense, except for one snide (and unworthy) crack at Pope Paul VI on page 114.

7. c. *spec.* An opening between floor-boards or in a floor; esp. in phr. *to walk a* (or *the*) *crack*; also *fig.* *U.S.*

1825 J. K. Paulding *J. Bull in Amer.* vii. 81 When I had qualified myself by being able to walk a crack after swallowing half a gallon of whiskey. 1869 H. B. Stowe *Oldtown Folks* xxxvii. 483 Your minister sartin doos slant a leetle towards the Arminians; he don't quite walk the crack. 1875 —— *We & Neighbors* ix. 100 They don't come it round Jim. Any boy that don't toe the crack gets it. 1902 W. N. Harben *Abner Daniel* 73 He could walk a crack with a gallon sloshin' about in 'im.

d. A slight opening between a door and the door-post; similarly of a window.

1892 J. C. Harris *Uncle Remus & his Friends* 143 When he got little nigher, he tuck notice dat de front door wuz on de crack. 1898 M. Deland *Old Chester Tales* 237, I always think the door was open a little crack, and you could see out. 1911 R. D. Saunders *Col. Todhunter* xi. 161 Then the front door was opened on the crack. 1969 Gish & Pinchot *Lilian Gish* vi. 62 The panel was lowered a crack, and the morning light Griffith had so adroitly created touched her face.

e. *Mountaineering.* A vertical fissure, too narrow to admit the whole body.

1923 G. D. Abraham *First Steps to Climbing* v. 57 At times the way may lie up narrow clefts or cracks which only give room for the jamming of an arm or a leg. 1957 Clark & Pyatt *Mountaineering in Brit.* ii. 42 The party found itself at the foot of the upper pitch, which was tackled by means of a thin crack which widened into a chimney.

f. Phr. *to paper over the cracks*: to use a temporary expedient; to create a mere semblance of order, agreement, etc.

[1865 Bismarck *Let.* 14 Aug. (1876) 65 Wir arbeiten eifrig an Erhaltung des Friedens und Verklebung der Risse im Bau.] 1910 *Encycl. Brit.* XI. 871/2 Neither power was quite prepared for war, and..the convention of Gastein, to use Bismarck's phrase, 'papered over the cracks'. 1952 *Ann. Reg. 1951* 56 Mr. Bevan agreed to paper over the cracks for the period of the election. 1958 *Listener* 7 Aug. 213/3 Sir Malcolm did not succeed in papering over the cracks in the First Symphony.

19. Phr. *half a crack*: half a crown; two shillings and sixpence. *slang.*

1933 R. A. Knox *Body in Silo* xxiii. 237 So I hired the coat and hat and the rest of the outfit for half a crack from one of the artists. 1948 C. Day Lewis *Otterbury Incident* 85 A simple, ordinary coin of the realm, vulgarly known as half a crack or a demi-dollar.

crack, *v.* Add: **1. b.** (Earlier U.S. example.) Also with *down* (trans. and intr.).

1835 J. H. Ingraham *South-West* I. xix. 202 He would reload.., cock his beaver, take aim, and crack again. 1938 M. K. Rawlings *Yearling* iv. 31 I've helt back my shot... It goes agin me to crack down at sich a time. 1943 Hunt & Pringle *Service Slang* 25 *Crack down*, to shoot down (a Hun plane).

3. a. Also *colloq.*

1936 N. Coward *To-night at 8.30* I. 96 Get out of here before I crack you one. *Ibid* II. 71, I should like to crack you over the head with a bottle. 1954 F. Sargeson in C. K. Stead *N.Z. Short Stories* (1966) 8 Good Lord...And did he crack you?

c. *crack down*: to repress, to take strong measures against. Const. *on*, *upon*. *colloq.*

1940 Graves & Hodge *Long Week-end* xiv. 225 The police had 'cracked down hard' on the London night-clubs. 1942 D. Gilbert *Lost Chords* 94 She would crack down upon him with all the force of a woman unjustifiably scorned. 1942 J. B. Priestley *Black-out in Gretley* iv. 56 Our people have cracked down rather severely on it just lately. 1947 'N. Shute' *Chequer Board* 68 Keep them smart, 'n crack down on them if they're not dressed right. 1955 A. L. Rowse *Expans. Eliz. Eng.* i. 25 Rewarding good service..cracking down on negligence. 1968 *Globe & Mail* (Toronto) 17 Feb. 1/1 The Unemployment Insurance Commission will try to crack down on chiselers.

6. d. Phr. *to crack hardy* (or *hearty*), to put a good face on, to assume or maintain a bold bearing; see also quot. 1916. *Austral.* and *N.Z.*

1916 C. J. Dennis *Songs Sentim. Bloke* 120 *Crack hardy*, to suppress emotion; to endure patiently; to keep a secret. 1926 J. Doone *Timely Tips for New Australians*, *To 'crack hardy'*, to simulate courage. 1929 K. S. Prichard *Coonardoo* xv. 142 He smiled gratefully to her for 'cracking hardy', pretending she was not tired. 1949 F. Sargeson *I Saw in Dream* i. xiv. 177 Daley's got a big family to keep. ..So he cracks hearty.

7. b. quasi-*trans.*, with spoken words as object. To utter a crack (*crack *sb.* 5), to joke.

1957 H. Roosenburg *Walls came tumbling Down* (U.S. ed.) iii. 78 'I may be home before you,' she cracked. 'Just tell the Red Cross to come for me with an airplane.' 1958 *Woman's Own* 4 June 23/1 'Hello, Sister,' he cracked. 'What do you want to know about me? Name? Religion? Size in socks?'

8. (Earlier and later examples.) Also (in *pass.*), to be reputed (usu. in negative sentences).

1829 *Kentuckian* 28 May, He is not the thing he is cracked up for. 1835 Crockett *Van Buren* 175 Great men ..are not the things they are cracked up for. 1836 *Knickerbocker* VIII. 51 New-Orleans is not..half so bad a place as it is 'cracked up to be'. 1939 *War Illustr.* 14 Oct. p. ii/2 An article from a Paris correspondent cracking up the blue-lit nights of Paris. 1951 E. Bagnold *Loved & Envied* 234 The emotions have been found by then to be not all they are cracked up to be. 1969 'A. Gilbert' *Missing from Home* vii. 97 It's not always all it's cracked up to be.

9. b. Delete † *Obs.* and add later examples.

1937 R. Stout *Red Box* i. 4 It's a tough one, and I doubt if anyone could crack it but you. 1955 E. Hillary *High Adv.* viii. 148 If we didn't crack the route to the South Col pretty soon, we might as well go home. 1960 *Analog Science Fact & Fiction* Nov. 19/1 When the case of the Teleporting Juvenile Delinquents had come up he'd been assigned to that one too, and he'd cracked it. 1962 *Listener* 29 Nov.

931/2 A code message that will clear everything up when it is cracked. 1967 *Technology Week* 23 Jan. 79/1 We are cracking the code of the life process.

c. To surpass, break (a record).

1953 *Racing Times* 15 July 4/4, I was fortunate..to see him crack the one-mile record in the Futurity Stakes.

15. Also *freq.* with *up* (not restricted to *racing slang*). Cf. *crack-up.

1918 F. M. Ford *Let.* 6 Jan. (1965) 86, I wrote about half a novel in the Salient, but got tired of it when I cracked up. 1922 *Daily Mail* 17 Nov. 11 She..looked all over the winner.., but when the pinch came she cracked up with dramatic suddenness. 1945 T. Rattigan *Love in Idleness* i. 268 Try to get him to let up just a little. After all, we don't want him cracking up on us, do we?

22. b. Colloq. phr. *to get cracking*: to get started; to 'get a move on'. Cf. Get *v.* 32 b. Also with noun or pronoun (or other object) interposed between *get* and *cracking*.

1937 Partridge *Dict. Slang* 324/1 *Get cracking*, to begin work. 1938 'N. Shute' *Ruined City* vi. 121 If I could get this yard cracking again I'd be a very happy man. 1943 H. Bolitho *Combat Report* 73 Dickie yawned and said, 'Well, I must get cracking.' 1947 'N. Shute' *Chequer Board* 190, I got the solicitor cracking on it before I left England. 1949 S. Gibbons *Matchmaker* xviii. 221 Come on, let's get cracking, we're late now. 1950 A. L. Rowse *England of Elizabeth* v. 162 Norwich, which still had many void spaces.., is told to get cracking: if owners do not rebuild.. the corporation shall do so. 1957 E. Hyams *Into Dream* iii. vi. 234 'I'd better get cracking.' 'Very well. Take care of yourself.' 1969 *New Yorker* 12 Apr. 124/2 Before Dr. Latham can get cracking with his computer, someone at the Mission Control..will flip a switch.

23. *trans.* To decompose (heavy oils such as petroleum) by the application of heat and high pressure alone or by means of a catalyst so as to produce lighter hydrocarbons (e.g. petrol) of better quality and with a better yield than can be obtained by distillation. (Cf. **catalytic cracking*.) Hence **cracked** *ppl. a.*, **cra·cking** *vbl. sb.*

1868 B. Silliman in *Chem. News* 10 Apr. 171/1 By the process called 'cracking', heavy oils unfit for illumination are broken up into bodies of less density, from light naphtha to the heavier illuminating and lubricating oils. 1869 *Amer. Jrnl. Sci.* 2nd Ser. XLVII. 13 Rapid distillation 'cracks' the oil, because it necessitates increased temperature to force the vapors from the still. 1884 S. F. Peckham *Rep. on Petroleum* 179 The standard and prime oils, consisting largely of 'cracked' oils. 1896 B. Redwood *Petroleum* I. 317 The 'cracking process', whereby a considerable quantity of the oil which is intermediate between kerosene and lubricating oil is converted into hydrocarbons of lower density and boiling point, suitable for illuminating purposes. 1922 *Encycl. Brit.* XXXII. 80/1 Modern cracking dates back to the patent obtained in 1889 by the late Sir Boverton Redwood and Prof. Dewar. *Ibid.* 80/2 Cracked gasoline. 1930 *Economist* 12 July 87/2 At Tampico a new vapour-phase cracking plant is being erected for the production of high-grade gasoline. 1934 [see *anti-oxidant]. 1947 *Archit. Rev.* CI. 132/1 To make water gas, which may be enriched to the required calorific value by cracking gas oil in the plant. 1952 *Economist* 6 Sept. 581 Government's inquiry about degrading the cracking units. 1958 *Times* 2 June p. v/4 The oil feedstocks are thermally cracked in the presence of steam to make a fixed gas and heavy tar. 1959 *Times Rev. Industry* Aug. 70/2 Thermal 'cracking'..evolved to increase the quantity of gasoline obtainable from any given volume of crude oil. 1967 Bland & Davidson *Petrol. Processing Handbk.* iii. 62 Catalytic cracking and catalytic reforming.. have replaced their thermal counterparts in most of the present-day refineries.

24. *trans.* and *intr.* Of a door: to be slightly ajar; to leave slightly ajar. Cf. *crack *sb.* 7 d. Chiefly *U.S.*

1899 in Wentworth *Amer. Dial. Dict.* (1944) 139/2 Crack the door a little to let out the smoke. 1938 J. Rice *Somers Inheritance* iii. v. 135 The door leading into the Sunday School room was cracked open. 1964 *Spectator* 14 Feb. 205 Mr. Kennedy..made it his practice..to leave the door to his office cracked a little so that any personal assistant who felt the need to talk to him might walk right in.

IV. Phraseological use of the verb-stem.

25. *to cry crack*: to give up; to desist; to cry creak (Creak *v.* 5).

1888 'R. Boldrewood' *Robbery under Arms* II. xii. 204 We..never cried crack till we got to Nulla Mountain, where we knew we were pretty safe. 1922 Joyce *Ulysses* 300 He never cried crack till he brought him home as drunk as a boiled owl.

crackajack, var. *crackerjack *sb.* and *a.*

crack-down. *colloq.* Also **crackdown.** [f. *to crack down*, *crack *v.* 3 c.] An instance of 'cracking down', legal or disciplinary severity, repression.

1935 *Washington Post* 18 May, A threat of a 'crack-down' by the middle class group against those who put forward the legislation for abolishing public utility holding companies. 1941 *Economist* 3 May 587/2 With this case.. assuming large proportions, prophecies of a 'crack-down' are busily circulating in Washington. 1959 *News Chron.* 22 Oct. 1/4 A big national crack-down on wildcat strikers. 1962 *Economist* 22 Dec. 1203/1 January 1st will bring a major crackdown on 'expense account living'. 1970 *Daily Tel.* 4 Apr. 12 An enthusiastically prosecuted crack-down on every aspect of the illicit drug trade.

cracked, *ppl. a.* Add: **1. b. cracked cocoa** *U.S.* = *cocoa-nibs* (Cocoa 4).

1934 in Webster. 1939 J. P. Marquand *Wickford Point* xvi. 172 Supper..usually consisted of pale scrambled eggs and toast and cracked cocoa.

2. b. *cracked heel* (esp. in horses): see Scratch *sb.*[1] 2 a, Grease *sb.* 3.

1850 H. Beasley *Druggist's Gen. Receipt Book* 93 (*heading*) Ointments for..cracked heels. 1886 G. Fleming *Pract. Horse Keeper* 96 A predisposition to cracked heels is engendered by clipping the legs and pasterns in winter. 1898 F. T. Barton *Our Friend the Horse* 193 Cracked Heels... The result of irritation through sand, wet, etc. 1908 *Animal Management* (*War Office*) 66 Chapping...When confined to the hollow of the heel, the condition is called 'cracked' or 'greasy heel'.

5. (Further examples.) Also phr. *to be cracked about* or *on*, to be infatuated with.

1825 S. Woodworth *Forest Rose* 29 Poor fellow! He is a little crack'd I calculate. 1874 'S. Coolidge' *What Katy did at School* iii. 46 He's an elegant fellow. All the girls are cracked about him—perfectly cracked! 1960 *Spectator* 1 July 15/2, I brought myself to carry the English mistress's books—this was..usually done by girls who were 'cracked' on this woman or that. 1968 D. Hopkinson *Incense Tree* iv. 40 It was almost a social obligation to be 'cracked' on someone. 1968 [see next].

8. (See *crack *v.* 23.)

cra·ckedness. [See Cracked *ppl. a.* 5.] Unsoundness of mind, craziness.

1910 W. J. Locke *Simon the Jester* xiv, This was sheer crackedness of brain...When people begin to talk that way they are not allowed to go about loose. 1968 *Listener* 29 Aug. 278/2, I suppose all writers of children's classics have been cracked, or at least extremely weird. One respects their crackedness as a rare endowment.

cracker. Add: **4.** (Earlier Amer. examples.)

1766 G. Cochrane *Let.* 27 June (D.A.), I should explain to your Lordship what is meant by *crackers*; a name they have got from being great boasters; they are a lawless set of rascalls on the frontiers of Virginia, Maryland, the Carolinas and Georgia, who often change their places of abode. 1767 *N.Y. Mercury* 21 Sept., in *Mag. Amer. Hist.* (1878) II. 250 A number of people called Crackers, who live above Augusta, in the province of Georgia, had gone in a hostile manner to.. Okonee.

b. *attrib.*; **the Cracker State**, Georgia. *U.S.*

1808 *Balance* (Hudson, N.Y.) 6 Sept. 144 (Th.), [Verses by] a cracker planter. 1838 C. Gilman *Recoll. Southern Matron* xix. 130 The servant told us that a *cracker* man and woman were in the drawing-room. 1872 Schele de Vere *Americanisms* 659 Georgia..little deserves the name of Cracker State, by which it is occasionally designated. 1887 *Harper's Mag.* May 843/1 Numbers of lawyers would gather together and relate their observations of Cracker life. 1908 *Daily Chron.* 26 Aug. 7/4 Our cracker guides go everywhere in thin cotton trousers. 1910 *Washington Herald* 27 Nov. 9 Through November the 'Cracker State' has occupied the center of motordom's stage.

6. b. *attrib.*, as *cracker-motto*, *-paper*, *-poetry*, *-rhyme*.

1868 W. S. Gilbert *Bab Ballads* 54 Only find out who it is that writes those lovely cracker mottoes! 1906 *Westm. Gaz.* 8 Sept. 2/2 A tragedy in rhymed verse not up to the standard of cracker poetry. 1907 *Ibid.* 25 Apr. 2/1 The gaily tinted scraps of cracker-paper that children love. 1934 Dylan Thomas *Let.* 2 May (1966) 115 Cracker-rhymes that we..will imagine as the disregarded fruits of genius. 1942 'G. Orwell' *Critical Ess.* (1951) 128 The collection of cracker-mottoes at the end of *Man and Superman*. 1954 J. R. R. Tolkien *Fellowship of Ring* i. 45 The crumbs and cracker-paper, the forgotten bags and gloves and handkerchiefs.

c. An attachment to the end of a whip-lash by which a cracking sound can be produced. *U.S.*, *Austral.* and *N.Z.*

1835 Monett in J. H. Ingraham *South-West* II. 288 To the end of the lash is attached a soft, dry, buckskin cracker... So soft is the cracker, that a person who has not the sleight of using the whip could scarcely hurt a child with it. 1880 A. A. Hayes *New Colorado* (1881) x. 140 Each wagoner must tie a brand-new 'cracker' to the lash of his whip. 1890 'R. Boldrewood' *Col. Reformer* I. xviii. 110 Stockwhips garnished with resplendent crackers. 1907 W. H. Koebel *Return of Joe* 164 Fresh and efficient crackers swung continually at the ends of the stockwhips. 1966 'J. Hackston' *Father clears Out* 64 I'd plaited a whip specially for the occasion with a new green cracker on it.

8. c. A cracking plant (see *crack *v.* 23).

1951 [see *catalytic *a.*]. 1963 Williams & Jones *Liquid Fuels* i. 11 (*caption*) Fluid catalytic cracker. 1969 *New Scientist* 7 Aug. 280/2 The products of the crackers go to other plant for conversion into intermediate and final petrochemical products.

9. (Earlier examples.)

1739 in *New Eng. Hist. & Gen. Register* (1868) XII. 296 Wee haue..sent a box of Crakers to you. 1773 *Newport Mercury* 20 Dec., Advt. (Th.), Said Johnson bakes the very best sort of crackers. 1781 W. Moss *Essay Management & Nursing of Children* 108 Hard biscuit, commonly called crackers, are sometimes given; but they are heavy, owing to their being made without yeast and not fermented.

b. *attrib.* and *Comb.*, as *cracker-bag*, *-peddler*; **cracker-hash**, a sailors' hash of biscuits and meat; so **cracker-stew.**

1876 *Wideawake* (Boston, Mass.) 233/1 He picks up the cracker-bag and helps himself. 1897 Kipling *Capt. Cour.* v. 127 On his return the cocoa-tins and cracker-bags were counted out by the Frenchman's wheel. 1904 *Daily Chron.* 4 May 10/2 Cracker hash, a savoury mixture of salt meat and biscuit. 1924 R. Clements *Gipsy of the Horn* ii. 38 Sea-biscuits broken into pieces and baked with small morsels of beef or pork and called 'cracker-hash'. 1928 *Sunday Express* 8 Jan. 4 Memories that go back to the days of 'cracker-hash', 'lobscouse', and hard-case. 1962 W. Granville *Dict. Sailors' Slang* 36/1 *Cracker hash*, minced meat and biscuits. 1860 J. G. Holland *Miss Gilbert's Career* v. 76 'I'm going to be a cracker-peddler,' said Fred. 1909 *Daily Chron.* 14 Apr. 5/2 This is better..than cracker stew.

11. c. A pound(-note). *Austral.* and *N.Z.* *slang.*

1934 W. S. Howard *You're telling Me!* 300 What about money?..We haven't got a cracker. 1941 Baker *Dict. Austral. Slang* 20 *Cracker*, a £1 note. 1947 *Coast to Coast 1946* 37 That damned old foxskin won't be worth a cracker to you. 1960 N. Hilliard *Maori Girl* iii. ix. 240 I've got nothing, Harry: not a cracker.

cra·cker-ba:rrel. orig. *U.S.* A barrel containing crackers (sense 9); used *attrib.* of a plain or unsophisticated outlook or point of view, or of a person with such a view or views.

1877 *Harper's Mag.* Mar. 603/1 Down at the farther end of the room turning fragments out of a cracker barrel. 1905 *Springfield* (Mass.) *Weekly Republican* 15 Sept. 12 Dr. Hall..is not above sitting on a cracker barrel in a country grocery for a chat with old acquaintances. 1919 E. E. Cummings *Let.* 25 Aug. (1969) 61 In order that you may not sit on a cracker-barrel playing whist. 1933 J. T. Flynn *God's Gold* i. ii. 37 Politics, rum, riches, and religion —these were the favorite topics of American cracker-barrel debaters. 1938 *Time* 31 Oct. 26/3 On the air and in print the Burns character is that of a cracker-barrel philosopher. 1958 *New Statesman* 22 Feb. 244/3 The late Harold Ross..was always careful to keep the *New Yorker* 'a family magazine'...It's always kept close to the cracker-barrel, and remains a bastion of the commonsense American virtues. 1961 *John o' London's* 16 Nov. 548/2 Endless cracker-barrel philosophising. 1963 *Times Lit. Suppl.* 1 Feb. 76/5 Cracker-barrel Yankee wit.

cra·cker-box. *U.S.* [Cracker 9.] A biscuit-box. Also *attrib.* and *fig.* (cf. *cracker-barrel).

1857 *Harper's Mag.* July 149/1 Their instruments consisted of a couple of drums..and a Boston tin cracker-box. 1902 S. E. White *Blazed Trail* xxxv. 237 Entering, he would nod briefly and seat himself on a cracker-box. 1911 H. S. Harrison *Queed* xviii. 230 Republican cracker-box orators were trying somehow to make capital of the thing. 1932 E. Wilson *Devil take Hindmost* xxii. 231 With his.. alternations between inspiring religious eloquence and the crackerbox vernacular of the backwoods, he seemed out-spoken. 1945 B. MacDonald *Egg & I* (1946) i. 15 Mother tore the partitions out of the crackerbox house. 1956 M. Stearns *Story of Jazz* (1957) xiv. 159 He heard barnyard imitations by Jim Turner on the fiddle and Billy Nichols on the crackerbox.

cra·ckerjack, *sb.* and *a.* *colloq.* (orig. *U.S.*). Also **crackajack.** [A fanciful formation upon Crack *v.* or Cracker.] **A.** *sb.* **1.** Something that is exceptionally fine or splendid. Also, a person who is exceptionally skilful or expert.

1895 W. C. Gore in *Inlander* Nov. 62 He got a crackerjack when he bought that horse. *Ibid.*, As a pitcher he's a cracker-jack. 1896 *N.Y. Herald* 2 Apr. 7/4 There are so many crackajacks in the lot that..some good men will have to.. wait for a chance. 1897 *Outing* (U.S.) XXIX. 481/1 The old-fashioned dogs were heavier and slower than the latter-day crack-a-jack. 1898 *Boston Globe* 16 Aug. 7/3 This regiment is a crackerjack. 1909 'O. Henry' *Roads of Destiny* iii. 51 If the story was a cracker-jack he had Mesrour, the executioner, whack off his head. 1925 Wodehouse *Carry on, Jeeves!* ii. 45, I couldn't be expected to foresee that the scheme, in itself a cracker-jack, would skid into the ditch as it had done. 1927 H. A. Vachell *Dew of Sea* 262 The McCullough who was out with Prince Charlie in '45 challenged attention. 'Looks a crackerjack,' said Miss Angell. 1933 J. Buchan *Prince of Captivity* 219 I've got a crackerjack of an editor. 1963 *Listener* 14 Feb. 301/3 Our hero-worship of these crackerjacks in their leathers and goggles knew no moderation.

2. (Proprietary term.) A sweetmeat composed chiefly of popcorn and syrup. *U.S.*

1902 *Sears, Roebuck Catal.* 20/3 Cracker Jack...Price, per case of 100 packages 2.85. 1905 R. Beach *Pardners* (1912) i. 31, I bought a dollar's worth of everything, from cracker-jack to cantaloupe. 1947 E. W. Barkins *Dr. has Baby* 131 She dangled a red balloon, ate popcorn, peanuts and Cracker Jacks.

B. *adj.* Exceptionally fine or good; of marked excellence or ability.

1910 O. Johnson *Varmint* iii. 43 Say, by the way, look out—he's a crackerjack boxer. 1911 R. D. Saunders *Col. Todhunter* ix. 123 You've given me a cracker-jack talk on Missouri politics. 1920 C. H. Stagg *High Speed* vi, It was a crackajack piece of work for a chauffeur. 1956 *Times* 1 Aug., A highly mobile, well-equipped, crack-a-jack, small Regular Army. 1966 *Punch* 19 Jan. 101/3 These seventy-odd pieces of crackerjack journalism begin with Walter Lippmann's putting Mr. Rockefeller in the witness stand sometime in 1915.

cra·ckers, *pred. a.* *slang.* [f. Cracker *sb.*; cf. Cracked *ppl. a.* 5.] Crazy, mad; infatuated.

[1925 Fraser & Gibbons *Soldier & Sailor Words* 65 *To get the crackers*, to go off one's head. Mad.] 1928 *Daily Express* 21 Jan. 2/2, I shall go 'crackers' (meaning mad) if anything happens to Ted. 1928 E. Wallace *Flying Squad* xxv. 286 That old boy's crackers if ever there was one! 1938 G. Greene *Brighton Rock* vii. viii. 343 'You're crackers,' Dallow said. 1940 in Harrison & Madge *War begins at Home* viii. 183 This war is driving me crackers... I'm bored to death. 1941 N. Marsh *Surf. Lampreys* xiv. 207 She's gone completely crackers, it seems. 1943 'E. M. Delafield' *Late & Soon* x. 142 He's crackers about Primrose. 1956 L. McIntosh *Oxford Folly* 77 Only the thought of my thirty bob an hour keeps me from going crackers. 1959 *Daily Tel.* 5 Nov. 21/4 Liberal Party is 'crackers', says Ld. Morrison of Lambeth.

crackey, *int. U.S.* (Examples.) Also **crackee.**

1830 *Painesville* (Ohio) *Telegraph* 15 June 3/2 Oh! Crackee what luck! 1854 M. J. Holmes *Tempest & Sunshine* xii. 74 He gave vent to his joy by uttering aloud, 'Crackee, this is jest the thing!' 1913 G. Stratton-Porter *Laddie* ii. 52 'Whose money do you suppose that is?' 'Oh crackey! You can depend on a girl to see everything,' groaned Leon.

cracking, *ppl. a.* Add: **3.** (Earlier and later examples.) Also as *adv.*

1833 S. Smith *Lett. Major Downing* (1834) 17 If you don't hear of cracking work down there, that will make 'em stare, I'm mistaken. 1840 *Spirit of Times* (N.Y.) 7 Mar. 7 Long Island, where Decatur *was* to have made a cracking race. 1903 *N.Y. Sun* 29 Nov. 11 The Hunters Champion Steeple-chase resulted in a cracking good race. 1940 N. Mitford *Pigeon Pie* iii. 52 'What was she like?' 'Cracking bore, as usual.' 1957 I. Cross *God Boy* (1958) 110 Probably turning out to be a cracking-good saint.

cracking, *pr. pple.*, in phr. *to get cracking*: see *crack *v.* 22 b.

crack-jaw, *a.* Add: Also *transf.*

1920 M. Beerbohm *Let.* 18 May (1964) 249 Oh the tedious, crack-jaw, arid intellectual snobbery of *Diana* [*of the Crossways*]!

crackled, *ppl. a.* **a.** Add examples (as used in *Ceramics*: cf. Crackle *sb.* 3).

1876 A. W. Franks *Catal. Oriental Porc. & Pott.* 13 Crackled vases were called Tsui-khi-yao, under the southern Sung dynasty. *Ibid.* 15 Saucer. Very solid and coarse Chinese ware, crackled, and mottled with brown and yellow. 1922 *Edin. Rev.* Apr. 322 We may describe specimens of intentionally crackled ware of the Sung Ch'uan type as Ka-ware. 1957 *Encycl. Brit.* XVIII. 359/1 Our artists have found in greyish crackled porcelain a background for painting that inspires them to break new ground.

crackless, *a.* Delete *rare* and add further examples.

1883 'Mark Twain' *Life on Mississippi* 571 Solid blackness—a crackless bank of it. 1941 J. Masefield *Gautama* 22 The bowl, The bell-like, swan-like, chipless, crackless whole Desired piece. 1970 *Fremdsprachen* 44 After the concrete was cured, the slab appeared to be crackless to the naked eye.

crackling, *vbl. sb.* Add: **2. c.** Attractive women collectively; *a bit of crackling*, an attractive woman. *colloq.*

1949 Partridge *Dict. Slang* (ed. 3) 1023/1 *Crackling*... Usu. *bit of crackling*, a girl. 1958 M. Kelly *Christmas Egg* iii. 180, I never get those jobs with nice bits of blonde crackling thrown in. 1968 P. Dickinson *Skin Deep* iii. 32 'You know her?' 'I do, sir. Nice bit of crackling, she is.' 1970 A. Fowles *Dupe Negative* vii. 73 She was no lady... Sir Bernard, in his rags to riches climb, must have retained a preference for crackling.

3. b. (Earlier and later U.S. examples.) Also *fig.*

1834 D. Crockett *Life* xvi. 106, I looked like a pretty cracklin ever to get to Congress! 1835 A. B. Longstreet *Georgia Scenes* 24, I am perhaps..the best man at a horse swap that ever stole *cracklins* out of his mammy's fat gourd. 1843 W. T. Thompson *Major Jones's Courtship* ix. 36, I haint eat nothing but..sassingers, and cracklin-bread ever sense the killin commenced. 1846 J. J. Hooper *Adv. Simon Suggs* (1851) x. 133 'Ef them fellers aint done to a cracklin,' he muttered. 1946 E. B. Thompson *Amer. Daughter* 84 There was the rendering of lard that brought on crackling bread.

crackly, *a.* Add: Also *transf.*

1923 Kipling *Land & Sea Tales* 15 There was a roar of rather crackly laughter. 1969 *Listener* 12 June 835/2 It's all put together well enough, sprightly, black-and-white, with clever, crackly dialogue.

crackpot (kræ·kpɒt). *colloq.* [For *cracked pot* (cf. Crack-brain and Pot *sb.*[1] 7 a).] A crack-brain, a crazy creature, a crank. Also *attrib.*

1883 *Broadside Ballad* (Farmer), My aunty knew lots, and called them crack-pots. 1919 J. C. Snaith *Love Lane* xiv. 65 He felt that it would annoy this Toplofty Crackpot. 1927 H. M. Tomlinson *Gallions Reach* xxxiii. 254 We can't go doddering across Malaya behind an inspired crackpot following the Holy Grail, can we? 1934 A. Woollcott *While Rome Burns* 15 The true story of crackpot Marigold. 1944 L. Lariar *Man with Lumpy Nose* xiv. 130 Everybody knew that Herb was a member of the Crackpot Brooklyn youth organization. 1959 J. Cary *Captive & Free* 211 The public is used to grievance-mongers and despises 'em—they'll put him down for a crack-pot. 1959 G. Jenkins *Twist of Sand* v. 91 The High Command still thought it a crackpot idea, fraught with all kinds of difficulties and dangers.

crack-up (kræ·kʌp). [f. *crack *v.* 15.] Disintegration (under strain), collapse; a crash.

1926 Maines & Grant *Wise-Crack Dict.* 7/1 *Crack up*, wreck of an aeroplane. 1927 C. A. Lindbergh *We* ii. 53 The propeller came in contact with the ground... My first 'crack-up'! 1936 F. Scott Fitzgerald in *Esquire* Feb. 41 (*title*) Crack-up: a desolately frank document from one for whom the salt of life has lost its flavor. *Ibid.* Apr. 39/1 How an..optimistic young man experienced a crack-up of all values. 1936 L. C. Douglas *White Banners* xix. 393 Then came this crack-up in the stock market. 1937 *New Republic* 19 May 37/1 No..aircraft in the world..has suffered so many crack-ups. 1938 *Polit. Q.* Jan.–Mar. 25 The dangerous crack-up in our foreign situation. 1958 *Times Lit. Suppl.* 23 May 286/5 Berlin under air attack,.. and the final crack-up. 1963 S. Weintraub *Private & Public Shaw* vi. 178 Motorboats used to speed to the scene of a crack-up on the water.

crack-voiced, *a.* [Crack- 2.] Having a cracked or broken voice.

1874 Hardy *Far fr. Madding Crowd* I. vi. 67 The crack-voiced cock-pheasants' 'cu-uck, cuck'.

cracky, *a.* **2.** (Earlier and later examples.)

1850 C. Godley *Lett. Early N.Z.* (1951) 47 She is odd, some say *cracky*, and never sees anyone. 1924 'R. Crompton' *William—the Fourth* v. 75 She's cracky, I think. 1936 M. Huxley *Let.* 21 Feb. in *Lett. A. Huxley* (1969) 400 Probably you will think we have gone cracky.

cradle, *sb.* Add: **2.** Phr. *cradle-to-grave*, used *attrib.* (Cf. quot. 1709 in Dict.)

1943 *Time* 22 Mar. 13 Sir William Beveridge, author of the British 'cradle-to-grave' social security report. 1951 *Amer. Speech* XXVI. 39 He shifts easily from such levels as cant to slang or colloquialism...While such cradle-to-grave lexicography must be impressive to the uninitiate, it may also reflect an assurance born largely of scissors and paste. 1962 C. Walsh *Utopia to Nightmare* vi. 76 Everyone has cradle-to-grave security.

6. h. The 'rest' or support for a telephone receiver not in use.

1929 *P.O. Electr. Engin. Jrnl.* XXII. 193/1 With the P.O. type [of telephone set] a disconnection could only be produced when the microtelephone was replaced on the cradle in a very violent manner. 1953 J. Mortimer *Like Men Betrayed* x. 136 Kit put the telephone back in the cradle. 1959 D. Beaty *Cone of Silence* xiv. 158 The sharp click of the receiver returning to its cradle.

i. *Cricket.* A device used to deflect a ball thrown upon it in practising short-range fielding.

1934 *Times* 1 May 6/5 Slip catches were flying off the cradle, and there was fielding practice.

8. b. A frame placed round the neck of an animal to prevent its biting an injury or sore.

1831 Youatt *Horse* xviii. 323 The possibility of blemishing himself should be prevented by a cradle or wooden necklace, consisting of round strips of wood, strung together, reaching from the lower jaw to the chest. 1940 *Chambers's Techn. Dict.* 206/2 *Cradle*, a frame encircling the neck of a horse.

16. b. *cradle-robber.*

1920 S. Lewis *Main St.* xxxii. 389 Darned if this bunch of cradle-robbers didn't get hold of some young kids.

17. cradle-board, among N. American Indians a board to which an infant is strapped; also *attrib.*; **cradle-books** = Incunabula 2; **cradle-cannon** *Billiards*, a series of cannons with the two object-balls close on either side of a corner-pocket; **cradle Catholic**, one who is born into the Roman Catholic church; a Catholic 'from the cradle'; **cradle-heap, -hill** (examples); **cradle-hole** (example); **cradle-knoll** *U.S.*, a small knoll, as on a logging road; **cradle-rocker**, (*a*) = Rocker[1] 2 b; (*b*) one who rocks a child's cradle; similarly **cradle-rocking**; **cradle-snatcher**, one who weds, or is enamoured of, a much younger person, *slang* (orig. *U.S.*); so **cradle-snatch** *v.*, **cradle-snatching** *vbl. sb.*

1879 *Jrnl. Anthropol. Inst.* VIII. 341 Cranial modification adopted by the Chinooks and other Flat-head Indians.. may be confidently ascribed to the undesigned pressure of the cradle-board on a head of brachycephalic type. 1942 *Antiquity* XVI. 94 It [sc. a skull] belongs to a woman 30 to 35 years old, and shows cradle-board flattening. 1956 E. Wilson *Red, Black, Blond & Olive* i. 49 A baby on a 'cradle-board'. Wrapped up and strapped to this wooden back, the children can be laid down to sleep or stood up against the wall. 1902 *Daily Chron.* 7 Nov. 3/3 The whole of this second volume is devoted to *Incunabula*—the 'Cradle books' —the first fruits of the early presses. 1927 *Publishers' Weekly* 31 Dec. 2315/1 Black letter books and cradle books. 1907 *Westm. Gaz.* 6 Mar. 10/1 The 'cradle-cannon' has been responsible for another extraordinary break. 1952 R. Macaulay *Let.* 25 July (1961) 342 The very nice, intelligent, cradle-Catholic talks Controller at the B.B.C. 1965 *Guardian* 10 Feb. 8/3 She and her three brothers..were all cradle Catholics, her grandparents having come into the Church. 1830 Galt *Lawrie T.* I. 186 It was then but the mere blazed line of what was to be a road; stumps and cradle heaps, mud-holes and miry swails, succeeded one another. 1855 Haliburton *Nat. & Hum. Nat.* II. 374 The bye-road was so full of stumps and cradle-hills, it was impossible to drive in it. 1854 Thoreau *Walden* 316 Deep ruts and 'cradle-holes' were worn in the ice. 1897 R. E. Robinson *Uncle Lisha's Outing* v. 32 There were moss-covered cradle-knolls and mouldering trunks of the old trees whose uprooting had formed them. 1969 L. G. Sorden *Lumberjack Lingo* 28 *Cradle knolls*, small knolls or mounds of earth that require grading in the construction of logging roads. 1795 S. J. Pratt *Gleanings through Wales* I. vi. 53 The little cradle-rocker, was singing a lullaby to the suckling. 1891 Hardy *Tess* iii, The cradle-rockers had done hard duty for so many years, under the weight of so many children, on that flagstone floor, that they were worn nearly flat. 1907 *Daily Chron.* 28 Aug. 6/7 How many of the cradle-rockers would gladly help to emancipate their sisters, but cannot. 1878 *Design & Work* 16 Feb. 213/3 *Cradle-Rocking*... An American invention for rocking a cradle. 1891 Hardy *Tess* iii, The cradle-rocking and the song would cease simultaneously for a moment. 1938 D. Smith *Dear Octopus* iii. i. 98, I may be neurotic but I've never cradle-snatched my brother-in-law. 1967 J. Aiken *Ribs of Death* i. 11, I don't usually cradlesnatch. But there was something about you that made me think you were older. 1925 *N.Y. Times* 8 Sept. 2/2 'Cradle Snatchers' is concerned with the activities of three wives. 1965 R. Erskine *Passion Flowers in Business* xi. 142 Crispin asked me to dance. 'Cradle-snatcher,' said Miranda nastily. 1933 E. A. Robertson *Ordinary Families* viii. 189 Our jokes..were generally about cradle-snatching, because Dru was two years older than Basil. 1958 Osborne & Creighton *Epitaph G. Dillon* i. 26 So you've taken to cradle-snatching, have you. Not content with taking another woman's husband, you have to pick up a 'young gentleman' as well.

cradle, *v.* Add: **5. b.** *spec.* To replace (a telephone receiver) on its 'cradle' or 'rest'.

1956 R. Fuller *Image of a Society* viii. 211 Rose cradled the receiver with the mingled disappointment and relief of the thwarted telephonist. 1969 R. Stark *Blackbird* (1970) ix. 56 'I wouldnt have believed it,' Ken said, and cradled the phone.

cradler. b. (Earlier Amer. example.)

1766 Washington *Diaries* (1925) I. 227 Let into it with my whole force and two cradlers hired.

craft, *sb.* Add: **9. c.** An aircraft or spacecraft.

1881 W. D. Hay *300 Yrs. Hence* ix. 210 The traffickers in heavy goods, and such-like craft. 1909 *Westm. Gaz.* 2 Dec. 5/1 What could be more concise in its definition of the complete flying-machine than the word 'craft', which is commonly used to signify a ship of any description. 1935 C. Day Lewis *Time to Dance* 33 For no silver posh Plane was their pigeon..But a craft of obsolete design. 1969 *New Yorker* 12 Apr. 53/1 The Apollo craft itself will stay in orbit sixty-nine miles above the moon.

11. Short for 'art and craft' (*art *sb.* 17), as *craft-bowl*, *-work*. Also **craft-conscious** *a.*, conscious of the value of craftsmanship; **craft-consciousness**, awareness of belonging to a craft (sense 7); **craft-union**, a trade-union of men of the same skilled craft.

1903 *Daily Chron.* 7 Jan. 3/2 The Bow inkstand and craft-bowl. 1945 H. Read *Coat of many Colours* xxx. 145 Then we still need a distinction between what we may call a craft-conscious public, and a public merely out for instinctive enjoyment. 1927 *Observer* 29 May 13/6 The appearance of class-consciousness, or rather craft-consciousness, among business men. Of course professional men..had craft-consciousness long ago. *c* 1926 'Mixer' *Transport Workers' Song Bk.* 28 The 'apathies' and the 'carpers' Of craft union pupilage! 1928 *Britain's Industrial Future* iii. xiv. § 3. 155 The Craft Unions, which in general are the oldest, aim at combining men who pursue the same skilled trade, whatever industry may employ them. 1966 *Guardian* 7 Dec. 8/5 As some non-craft union leaders point out, the craft unions could encourage them by further relaxing their attitudes to dilutees. 1928 *Daily Express* 24 Feb. 5/2 Another institution..where games and craftwork are an important part of convalescent treatment.

craft, *v.* Add: **2.** Revived, esp. in the U.S., in sense 'to make or construct skilfully'.

1963 *Listener* 14 Feb. 290/1 Performances such as 'Lycidas' were essentially public events, monuments crafted out of a shared language. 1967 *Boston Sunday Herald Mag.* 26 Mar. 13/3 (Advt.), An exciting collection [of furniture]..crafted of butternut and pecan. 1970 *Yankee* Nov. 208 (Advt.), This beautifully crafted antique-pine replica.

crag, *sb.*[1] Add: **1. b.** *crag and tail* (earlier and later examples).

1815 J. Hall in *Trans. R. Soc. Edin.* VII. 201 The district..in the neighbourhood of Noble-house, exhibits a series of low hills, possessing the characteristic forms of craig and tail. 1960 L. D. Stamp *Britain's Struct.* (ed. 5) x. 90 The crag has a 'tail' partly of rock, partly of superficial deposits. This crag-and-tail structure is perfectly illustrated by the Edinburgh example where the gentle slope up to the Castle from the eastern side represents the 'tail'.

4. crag-bound *a.* = *crag-fast*; **crag-fast** *a.*, also of men (later examples).

1908 *Westm. Gaz.* 15 Oct. 9/4 It was presumed that the missing men might have become what is known as 'crag-fast'...A sound came back as though from..the Pillar...It suggested..that it was from someone crag-bound on the Pillar. 1936 L. MacNeice tr. *Aeschylus' Agamemnon* 35 Nor ran to splinters on the crag-bound coast. 1940 F. S. Chapman *Helv. to Himalaya* ii. 22 Soon we heard the shouts of what appeared to be a cragfast tourist.

craggedness. (Later U.S. example.)

1812 J. J. Henry *Camp. agst. Quebec* 181 A bleak and dreary coast and country, whose craggedness inspired disagreeable sensations.

cram, *sb.* Add: **2.** (Earlier examples.)

1810 M. Wilmot *Diary* 25 Jan. (1935) p. xxii, We all made the best of our way to the adjoining room, but a cram, with hoops, is the most ridiculous thing imaginable. *c* 1845 C. Brontë *Professor* (1857) II. xxi. 109 A garden made and provided for such crams.

4. a. (Earlier examples.)

1828 H. Alford *Jrnl.* 2 Dec. in *Life* (1873) ii. 36 At the lecture Evans gave a quantity of cram about the choruses in the Eumenides. 1832 J. S. Mill in *Monthly Repos.* VI. 658 Modern education is all *cram*—Latin cram, mathematical cram, literary cram, [etc.].

5. (Example.)

1912 T. Okey *Art of Basket-Making* vi. 38 The crams should be turned down just a little short of the stake along-side which they are to be inserted.

cram-. Add: **2. cram-shop**, a school run by a crammer (sense 2). *colloq.*

1926 *Contemp. Rev.* Nov. 614 The school becomes that unholy thing, a cram-shop. 1946 J. Cary *Moonlight* 242 The young man..made Ella promise to play [the piano] with him every afternoon when he could escape from what he called his cram-shop.

crambo. 1. b. (Earlier example.)

1811 *Wynne Diaries* 12 Sept. (1940) III. x. 340 They were obliged to dance reels and play at dumb Crambo.

cram-jam, *adv.* Now chiefly *U.S. dial.* [Emphatic combination of stems of Cram *v.* and Jam *v.*] Chock- or cram- (full). So **cram-jam** *v. trans.*

1880 *Punch* 25 Sept. 142/1 The house was cram-jammed from ceiling to floor by an audience that showed itself decidedly appreciative and strictly critical. 1903 *Dialect Notes* II. 296 *Cram jam full*, crammed full. 1905 *East Coast Visitor* 17 Aug. 7/3 I've seen these flats cram-jammed with fowl of all sorts. 1909 *Daily Chron.* 8 Mar. 3/2 That garden was simply cram-jam full of them. 1953 *Publ. Amer. Dial. Soc.* xix. 10 (Southern Appalachians) *Cram-jam-full*, very full.

cramp, *v.* Add: **4. c.** *trans.* and *intr.* To deflect or turn to one side. *U.S.*

1875 'Mark Twain' in *Atlantic* Mar. 286/2 A boat hates shoal water...*Now* cramp her down! Snatch her! 1883 —— *Life Mississippi* x. 99 Cramp her up to the bar! What are you standing up through the middle of the river for? 1924 W. M. Raine *Troubled Waters* iii, She tried to cramp to the left.

5. c. Phr. *to cramp one's style*: to restrict one's natural actions or behaviour.

[1819 Lamb *Let.* 7 June (1935) II. 250, I will never write another letter with alternate inks. You cannot imagine how it cramps the flow of the style.] 1917 A. Woollcott *Let.* 2 Sept. (1944) 26, I think the very fact of a censorship cramps one's style. 1919 *Punch* 9 Apr. 283 (*caption*) Cramping his style. 1923 *Saucy Stories* 1 Nov. 124/1, I always go out with Edith...Edith never cramps my style. 1928 W. S. Maugham *Ashenden* 21 If I get into any trouble, you will never be admitted into any of the allied countries for the rest of your life. I can't help thinking it would cramp your style. 1930 R. Lehmann *Note in Music* 44 He was so debonair and independent: it would take a lot to cramp his style. 1956 B. Goolden *Singing & Gold* viii. 179 He got a kick out of being with Daphne even if she cramped his style. 1963 J. Joesten *They call it Intelligence* xx. 190 Troll did not allow this misfortune to cramp his style.

crampet. Add: **4.** A wall-hook.

1901 Waterhouse *Conduit Wiring* 24 The first straight run or runs may be fixed up in position by means of saddles, clips or crampets, as the case may require. 1909 W. S. Ibbetson *Electric Wiring* x. 183 The conduit may be fixed to walls by means of saddles, crampets, or clips.

crampy, *a.* (Earlier non-medical example.)

1858 E. C. Gaskell *Let.* 10 May (1966) 504 We live in a queer pretty crampy house, at the back of a great farm house.

cranberry. Add: **3.** *cranberry bog*, *marsh*, *pie*, *sauce*; **cranberry-gatherer, -rake** *U.S.*, an implement used in gathering cranberries.

1807 in *Mass. Hist. Soc. Coll.* 2nd Ser. III. 94 They annually sell a hundred or two bushels of craneberries, which grow in great plenty in their craneberry bogs. 1948 *Sat. Even. Post* 4 Dec. 70/3 Caleb got lost and wandered down The bottom land to the cranberry bogs. *a* 1877 Knight *Dict. Mech.* 642/1 *Cranberry-gatherer*,..adapted to catch below the berries on the stalk, and collect them in a bag or box attached to the rake-head. 1748 J. Eliot *Field-Husbandry* i. 5 There are three kinds, viz. Thick Swamp, Boggy Meadow, and smooth, even, shaking Meadow; this last sort is called Cramberry Marsh. 1832 *Boston Transcript* 21 Jan. 1/1 She had a harsh face, like a cranberry marsh all spread with spots of white and red, as if she had the measles. 1924 C. C. Deam *Shrubs of Indiana* 278 We have records of cranberry marshes as far south as Warren County. 1832 L. M. Child *Frugal Housewife* 68 Cranberry pies need very little spice. 1869 Mrs. Stowe *Oldtown Folks* xxvii. 340 Cranberry pies, huckleberry pies. 1849 *Cultivator* VII. 52, I have used a wooden machine, made like a cranberry rake,..to gather my clover-seed this season. 1767 J. Adams *Diary* 8 Apr. in *Wks.* (1865) II. 206 Tufts.. determined to go over and bring [them]..to dine upon wild goose, and cranberry sauce. 1854 M. S. Cummins *Lamplighter* xix, She doesn't know anything about nursing, let her stick to her cranberry sauce and squash pie. 1948 *Reader's Digest* Mar. 147/2 We were winding up a huge supper of moose, roast grouse, cranberry sauce.

crane, *sb.*[1] Add: **3. f.** A moving platform on which a camera is mounted for the taking of angled 'shots'. So *crane-arm*, *-hook*, *-shot*, etc.

1937 *Cine-Technician* III. 100/1 One crane of Soviet make..with..8-metre raise, mounted on a pneumatic dolly. *Ibid.* 103/1 All the apparatus,..is well designed and made, the camera crane and model table at Mosfilm being notable examples. 1940 *Chambers's Techn. Dict.* 207/1 *Crane shot*, in motion-picture production, a shot taken with the camera mounted on a crane-like structure, to get vertical panning or a combination of vertical and horizontal panning. 1953 K. Reisz *Film Editing* ii. 173 Ladle on crane-hook standing under spout of cupola. 1961 G. Millerson *Telev. Production* iii. 24 Assisting the cameraman is an operator who pushes and guides (tracks) the mounting. Where the crane-arm is to be raised/lowered or slewed, he may also carry out these adjustments.

7. b. *crane-driver.*

1897 *Daily News* 26 Feb. 3/5 The well-known engineers.. were summoned..by..a crane driver in their employ. 1957 Manvell & Huntley *Film Music* iii. 71 The scene of a crane-driver singing on the dockside at five o'clock in the morning when the film opens was shot on location.

crane, *v.* Add: **3. b.** Of a camera mounted upon a crane: to alter range or direction.

1957 Manvell & Huntley *Film Music* ii. 34 The camera cranes across the hall as she waltzes up to the steps. 1960 N. Kneale *Mrs. Wickens in Fall* 151 The Camera cranes downwards from the tapestry..down to the helmet of a suit of armour. 1962 *Movie* Sept. 4/2 The camera cranes up and moves in.

cranial, *a.* Add: *cranial index* [Index *sb.* 9 b], the ratio of the width of the skull to its length (now usu. expressed as a percentage).

1868 *Anthropol. Rev.* VI. 391 A Dalecarlian has ·75, and an 'old Norseman' from Lough Larne C. Antrim ·73 as the cranial index. 1901 N. C. Macnamara *Hunterian Oration R. Coll. Surg.* 15 When the cranial index rises above 80, the head is called brachycephalic, a broad head; when it falls below 75 the term..long head is applied to it. *Ibid.* 15 Assuming the length of the cranium to be 100, the width is expressed as a fraction of it, and is known in the living subject as being the cephalic, and in the bare skull as the cranial index. 1927 Peake & Fleure *Priests & Kings* vi. 122 One skull..is somewhat broad-headed, having a cranial index of 80·9. 1969 A. H. Schultz *Life of Primates* xiii. 211 The breadth of the brain, expressed as a percentage of its length, which is the widely discussed 'cranial index' of anthropologists, varies individually in a large series of fully grown orang-utans.

craniate (krēⁱ·niět), *a.* and *sb.* *Zool.* [ad. mod.L. *crāniātus*: see Craniata.] **A.** *adj.* Having a skull or cranium. **B.** *sb.* A member of the Craniata.

1879 A. S. Packard *Zool.* viii. 401 The Vertebrates are divided into the skulless or *acraniate*..and the skulled or *craniate* (*Craniata*). 1888 *Encycl. Brit.* XXIV. 183/2 The fact that the nerve tube is dorsal..suggests that the Craniates' ancestor had a dorsal median nerve. 1900 G. C. Bourne *Compar. Anat.* I. 24 The craniate vertebrates. 1909 E. S. Goodrich in E. R. Lankester *Treat. Zool.* IX. i. 1 The great gap which separates the lowest known Craniate from *Amphioxus*. 1959 *Chambers's Encycl.* XII. 590/2 The most primitive extant craniates, the cyclostomes..possess no jaws. *Ibid.* 591/1 The primitive craniate condition.

crank, *sb.*[1] Add: **6.** *crank-arm*; **crank-case**, the case or covering in which the crank-shaft of a motor engine is enclosed; **crank-chamber** (see quot. 1902).

1889 P. N. Hasluck *Model Engin. Handybk.* ii. 14 The bed-plate is..cut away to allow the crank-arm and cross-head to pass. 1911 *Encycl. Brit.* XVII. 1005/1 In a turning piece, the perpendicular let fall from its connected point upon its axis of rotation is the arm or crank-arm. 1951 *Good Housek. Home Encycl.* 25/1 A flexible brush should be used to brush away any dust..at both ends of the crank arms. 1878 *Technol. Dict.* (ed. 3) 175/2 *Crank-case*, das Kurbelhelft, das Kurbelfutteral. 1900 *Autocar* 29 Sept. 932/2 Working upon a pin secured to the crank case, is a second and larger pinion wheel. 1904 *Motor Boat* 29 Sept. 228/1 Lamp oil is a rather imprudent thing to carry aboard, as some unwise friend may mistake it for engine lubricant and put a charge or two into the crank-case. 1920 *Cornhill Mag.* Sept. 311 For forgetting to put any oil in the crank-case, he was stopped fourteen days' pay. 1949 *Gloss. Aeronaut. Terms (B.S.I.)* 11 *Crankcase sump*, that portion of the engine in which lubricating oil is collected and led to the oil pumping system. 1902 A. C. Harmsworth et al. *Motors* vii. 135 The crank chamber, or base chamber, as it is usually termed, forms the base of the cylinder. 1904 *Motor Boat* 22 Sept. 215/2 Each cylinder has a bore of 90 mm...; both are cast together..and are mounted on a cast iron crank chamber. 1922 *Times* 20 June 8/5 The cylinders are cast in one with the crank-chamber, a practice forgivable in a low-priced chassis.

crank, *sb.*[2] **1.** Delete † *Obs.* and add later example.

1909 M. B. Saunders *Litany Lane* i. v, A glimpse..of grey west tower half hidden by high buildings, then re-appearing by some quaint crank of street or turning.

5. Delete *U.S. colloq.* and substitute *colloq.* (orig. *U.S.*). (Earlier and later examples.) Also *attrib.* and *Comb.*

1833 in J. P. Dunn *Indiana* (1919) II. 1124 Uncle Sam's 'Old Mother Bank' Is managed by a foreign crank. 1889 G. B. Shaw *London Music 1888–89* (1937) 264, I moved amidst cranks, Bohemians, unbelievers, agitators, and..riff-raff of all sorts. 1906 *Nature* 8 Nov. 25/2 A *crank* is defined as a man who cannot be turned. These men are all cranks. 1924 G. B. Stern *Tents of Israel* vii. 97 Danny remained at his crank school in Hampstead. 1933 Dylan Thomas *Let.* Oct. (1966) 40 Don't think I'm regaling you with some crank-ridden, pornographic notion. 1934 H. G. Wells *Exper. Autobiogr.* I. v. 261 The normal Fabian gathering had a real horror of the 'currency Crank', as it termed anyone who ventured to say that money has ways and tricks of its own. 1949 Koestler *Promise & Fulf.* ii. v. 274 More crank visitors. Among them a German who cures diseases by soul-waves and *mana*. 1961 M. Spark *Prime of Miss Jean Brodie* i. 7 It has been suggested again that I should apply for a post at one of the progressive schools... But I shall not apply for a post at a crank school. 1968 D. Hopkinson *Incense-Tree* iii. 29 As I went to a crank school, it followed that some of the parents of my friends should be cranks.

crank, *v.*[1] Add: **8. a.** To move or operate (a motor engine) by a crank. **b.** *intr.* To turn a crank, as in starting a motor engine. **c.** *fig.*

1908 *Smart Set* June 91 The chauffeur..jumped out, and began to crank her up. 1909 F. R. Jones *Gas Engine* 181 Small motors are 'cranked' or otherwise turned by hand. *Ibid.* 183 The cranking should be done immediately after the gas is turned on. *Ibid.*, Then crank up to a fair speed and close the switch. *Ibid.* 184 Until the motor has been cranked up to high speed. *a* 1910 'O. Henry' *Rolling Stones* (1916) 192 The proprietor..was cranking the cash register. 1920 V. W. Pagé *Automobile Starting*, etc. (ed. 6) 424 If the starting motor rotates but does not crank the engine. *Ibid.* 453 And still the starting motor makes no effort to crank the car. 1924 W. M. Raine *Troubled Waters* vi. 59 Silently she watched him crank the car and drive away. 1929 J. B. Priestley *Good Companions* i. v. 194 The car refused to start again. She cranked away until she was breathless and aching. 1956 H. Gold *Man who was not with It* (1965) xi. 92 Andy..starts to crank out the story of his life.

cranked, *ppl. a.* Add: **4.** *Aeronaut.* Of an aircraft wing or an aircraft with such a wing: see quot. 1959.

1939 *War Illustr.* 18 Nov. 295 It [sc. the Junker 87] has a distinctive 'cranked' wing. 1959 F. D. Adams *Aeronaut. Dict.* 52 *Cranked wing*, a wing whose spanwise axis abruptly departs from a straight line, as seen either in planform or in a vertical plane. A gull wing is one form of cranked wing. 1961 *Flight* LXXX. 396/1 The cranked delta has four lift engines in the centre.

crap, *sb.*[1] Add: **7. a.** *coarse slang.* Excrement; defecation. Also *Comb.*, as *crap-house*, a privy.

1898 in Wright *Eng. Dial. Dict.* 1925 J. Dos Passos *Manhattan Transfer* 20 You don't want to shovel crap.. all your life. 1937 Partridge *Dict. Slang* 189/1 *Crap*. 4 A defecation..Esp. *do a crap*. 1951 J. D. Salinger *Catcher in Rye* xvi. 140 There didn't look like there was anything in the park except dog crap. 1953 A. Baron *Human Kind* 178 To clean *their* crap up after them! 1958 E. Hyams *Taking it Easy* 191 As cunning as a crap-house rat.

b. *transf.* Rubbish, nonsense; something (occas. someone) worthless, inferior or disgusting. *slang.*

1898 Wright *Eng. Dial. Dict.* s.v., What crap's that y'er talkin? 1929 T. Wolfe *Look Homeward* (1930) xiv. 190 Little Maudie will fill up the column to-morrow with some of her crap. 1934 J. T. Farrell *Young Manhood* (1936) ii. i. 163 Today of all days giving us that crap that he has to help his old man. 1935 G. Ingram *Cockney Cavalcade* viii. 121 Surely the Yid wouldn't take away 'crap'? 1951 J. D. Salinger *Catcher in Rye* i. 5 Where I was born, and what my lousy childhood was like, and how my parents were occupied..and all that David Copperfield kind of crap. 1953 W. R. Burnett *Vanity Row* xiv. 100 Don't give me that crap. 1954 R. Fuller *Fantasy & Fugue* 115 All art is crap...I've given it up entirely in favour of leisure. 1958 K. Amis *I like it Here* i. 8 A lot of awful craps go to Spain. 1964 *Punch* 19 Aug. 267/1 And what a load of crap that was.

crap, *sb.*[4] *U.S.* **1.** Var. *craps.

1843 J. H. Greene *Espos. Gambling* 89 Ask many of the merchants, what has resulted to them in consequence of their clerks being decoyed to the crap's table. 1891 *Amer. N. & Q.* VIII. 41 There is a sort of gambling game, called *crap*, or 'shooting crap', much played by newsboys, bootblacks and negroes. 1930 E. Pound *XXX Cantos* xxv. 113 They mustn't shoot crap in the hall of the council.

2. *attrib.* and *Comb.*, as *crap-game*, *-house*, *-shooter*, *-shooting*, *-table*.

1890 J. P. Quinn *Fools of Fortune* 539 The 'clock', policy playing, 'crap' games, and the sale of lottery tickets run on as though there were no let or hindrance imposed by State law. *a* 1910 'O. Henry' *Trimmed Lamp* (1916) 239 A cigar store near the Bowery that was running a crap game in the back room. 1930 H. M. Tomlinson *All our Yesterdays* ii. ix. 204 When France and Germany play their great crap game, we don't matter. 1936 J. G. Cozzens *Men & Brethren* i. 50 Your friends may not know any better than to start a crap game here, but you do. 1888 *Missouri Republ.* Feb. (Farmer), When arrested he was in the negro craphouse kept by Alex. Wells. 1895 *Westm. Gaz.* 13 Aug. 8/1 Most of the orders come from the Southern negro crap-shooters'. 1926 J. Black *You can't Win* iv. 27, I soon came to know the poker players, crap shooters and dice sharks. 1885 *Boston Jrnl.* 6 June 1/8 Dallas, Texas...250 negroes are being tried for 'crap-shooting', a game played with dice. 1946 Morehead & Mott-Smith *Penguin Hoyle* 218 There is little to skilful play in crapshooting other than knowing the odds concerned in various bets. 1902 T. M. Young *Amer. Cotton Industry* 103 At the crap-table he [sc. the Negro] will often lose or win at a sitting a sum equal to the profits of his labour for a whole year. 1958 G. Greene *Our Man in Havana* iii. i. 112 The dice ratted at the crap-tables.

crap, *v.* *coarse slang.* [Cf. *crap *sb.*[1] 7.] **1.** *intr.* To defecate. So **cra·pping** *vbl. sb.*

1846 [see *dunny *sb.*[2] 1]. 1859 Hotten *Slang Dict.* 26 *Crapping case*, a privy, or water-closet. 1874 *Ibid.* 132 *Crap*, to ease oneself by evacuation. *Crapping case*, or *ken*, the water-closet. Generally called *crapping-castle*. 1953 A. Baron *Human Kind* 178 They'd crapped on the floor, in the same rooms they'd slept in.

2. *trans.* To talk nonsense to; to act or speak deceitfully to. *U.S.*

1930 J. T. Farrell in Wentworth & Flexner *Dict. Amer. Slang* (1960) 128/1 Don't crap me! You're still in love with her. 1958 S. Ellin *Eighth Circle* (1959) ii. xviii. 180, I don't want you to crap me...I want your honest opinion.

3. *U.S.* phrases: *to crap around*, to behave foolishly; to 'mess about'; to concern oneself with (something worthless); *to crap out* [? cf. *crap *sb.*[4]], to be unsuccessful, to lose; to fail; to withdraw from an activity, game, etc.

1933 Hemingway *Winner take Nothing* (1934) 221, I have passed at dice for three thousand dollars and crapped out for the six. 1937 J. Weidman *I can get it for you Wholesale* (1939) iv. 31 We've decided to show these guys that we mean business. No crapping around. 1941 B. Schulberg *What makes Sammy Run?* iv. 52 What good do you think it's gonna do you to crap around with stuff like that? 1942 Berrey & Van den Bark *Amer. Thes. Slang* § 7/7 Get out of order; break,..*crap out*. *Ibid.* § 27/6 Come to nothing,..*crap out*. *Ibid.* § 219/7 Have bad luck,..*crap out*. *Ibid.* § 262/2 Fail,..*crap out*. *Ibid.* § 649/7 Be defeated; lose,..*crap out*. 1952 S. Kauffman *Philanderer* (1953) iv. 64

CRAPE

Let's not crap around. Let's get to the business in hand. **1964** A. Wykes *Gambling* vi. 144 And so the shooter cannot 'crap out'.

crape, *sb.* Add: **3. b. crape-fern,** a New Zealand fern, *Leptopteris superba*; **crape hair,** artificial hair used by actors for false beards, moustaches, etc.; **crape-hanger** *U.S.*, (formerly) a person who hung up crape as a sign of mourning; hence *U.S. slang*, a kill-joy; a pessimistic person; cf. **crêpe-hanger*; **crape ring,** the innermost and faintest of the three 'rings' (Ring *sb.*[1] 8 b) of Saturn.

1882 G. M. Thomson *Ferns N.Z.* 94 *Todea superba*, Colenso. (Crape-fern.) **1889** G. Nicholson *Illustr. Dict. Gardening* IV. 50/1 *Todea*..Crape Fern..A genus comprising about five species of greenhouse Ferns, almost confined to the South temperate zone. **1944** *Whitcombe & Tombs Mod. Jun. Dict.* (ed. 7) 101 *Crape-fern*, a popular name for a beautiful fern which grows only in New Zealand; also called 'Prince of Wales's feathers'. **1951** M. Crookes *Dobbie's N.Z. Ferns* (ed. 4) 20 *L*[*eptopteris*] *superba* (superb). 'Heru-heru'. The most beautiful fern in New Zealand; sometimes called 'Prince of Wales Feathers', or 'Crape Fern'. **1866** W. Davidge *Footlight Flashes* xviii. 240 Engaged in the manufacture of a huge black beard..'Mary, give me the scissors, and some more crape hair.' **1885** J. K. Jerome *On Stage* i. 4 Crape hair and the rouge pot changed all this. **1966** J. S. Cox *Dict. Hairdressing* 41/2 Crape hair. **1920** H. C. Witwer *Leather Pushers* 294 Good Lord, I was never surrounded by so many crape hangers in my life! **1925** G. P. Krapp *Eng. Lang. in Amer.* I. 117 Who will know a generation hence..that a crape-hanger is a reformer? **1945** S. Lewis *Cass Timberlane* (1946) xxxvi. 259 Look at those old crape-hangers, Father Cass and Uncle Bradd. [**1850** W. Lassell in *Monthly Not. R. Astron. Soc.* 13 Dec. 21 It appeared as if something like a crape veil covered a part of the sky within the inner ring. **1851** —— *Ibid.* 14 Nov. 11 The ring, which I have usually called the 'crape veil'.] **1865** Brande & Cox *Dict. Sci.* I. 583/1 Crape Ring. **1928** W. M. Smart *Sun, Stars & Universe* vii. 130 In 1850 Bond and Dawes independently discovered a faint extension of Ring B towards the ball of the planet, known from its dusky appearance as the Crape Ring. **1957** *Encycl. Brit.* XX. 8/2 Saturn itself can be distinctly seen through the Crape ring.

crapper (kræ·pəɹ). *coarse slang.* [f. *Crap *v.* 1 + -er[1].] A privy.

1932 in *Amer. Speech* VII. 330. **1939** J. Steinbeck *Grapes of Wrath* xxiv. 405 They got ten crappers for the whole shebang. **1966** in 'L. Lane' *ABZ of Scouse* 23. **1969** C. Himes *Blind Man with Pistol* x. 106 Go to the crapper? What for? They weren't children, they didn't pee in bed.

crappy (kræ·pi), *a. coarse slang* (orig. *U.S.*). [f. *Crap *sb.*[1] 7 b + -y[1].] Rubbishy; inferior, worthless; disgusting.

1846 *Swell's Night Guide* 49 Which of us had hold of the crappy (sh–ten) end of the stick? **1929** M. Lief *Hangover* xviii. 281 That damn fool took all his money out of the savings bank and put it in some crappy oil stock in Wall Street. **1942** Berrey & Van den Bark *Amer. Thes. Slang* § 7/14 In poor condition..*crappy*. *Ibid.* § 30/4 Poor; mean; contemptible..*crappy*. *Ibid.* § 38/5 Ugly..*crappy*. *Ibid.* § 239/6 Trashy, *crappy*. **1951** J. D. Salinger *Catcher in Rye* xvii. 150 The show..was on the crappy side. **1966** *Melody Maker* 31 Dec. 9 The new BBC pop channel..will end up like Workers' Playtime, with some really crappy groups. **1970** *Weekly Guardian* 14 Mar. 13/5 Rents as high as £52 a month 'for crappy quarters'.

craps. *U.S.* [Of obscure origin, but cf. *crabs*, Crab *sb.*[1] 9.] A game of chance played with dice; *to shoot craps*: to play at this game. Also *Comb.* (see *crap *sb.*[4]).

1843 J. H. Greene *Expos. Gambling* 88 The game of craps..is a game lately introduced into New Orleans, and is fully equal to faro in its..ruinous effects. **1850** Mayne Reid *Rifle Rangers* I. i. 9 So I stepped inter a shanty whar they wur a playin craps. **1888** *St. Louis Globe Democrat* (Farmer), A party of colored deck hands were playing craps. **1902** T. M. Young *Amer. Cotton Industry* 103 The besetting weakness of the negro is gambling, and his favourite game is 'craps'. **1903** *N.Y. Sun* 12 Nov. 1 Two young toughs came along and started to shoot craps. **1922** C. E. Mulford *Black Buttes* ii. 20, I never saw a redhead yet that could shoot craps worth a damn. **1942** A. Huxley *Let.* 30 Dec. (1969) 485 All of which has taken us rather far from throwing dice; but the relation between craps and Reality is a real one.

b. (Also *crap*.) A losing throw or 'shoot' of 2, 3, or 12 in the game of craps.

1890 J. P. Quinn *Fools of Fortune* 277 The numbers 7 and 11 are called 'craps'. **1909** Webster s.v., The caster..loses if it [sc. his throw] is 2, 3, or 12 (called a crap). **1916** D. C. Thompson *In Russia* (1918) i. 3 A crap game was started. Just when you would start to reach for the money, after seeing '7' or '11' on the dice, the boat would give another roll and 'craps' would appear.

crapulosity (kræpiŭlǫ·siti). [f. L. *crāpulōsus* Crapulous *a.* + -ity.] Inclination to drunkenness or to gluttony.

? **1538** *Rule of Honest Life* sig. B 1ʳ, Eate without crapulosyte. **1887** Browning *Fust & his Friends* 108 Pride seized me at last as concupiscence first, Crapulosity ever: true Fiends, everyone.

‖ **craquelure** (kraklür). [Fr.] A network of small cracks in the paint or varnish on the surface of a painting. Also *transf.* and *fig.*

1914 C. Bell *Art* 169 He will put up with a cunning concoction of dates and watermarks, cabalistic signatures, craquelure, patina, [etc.]. **1934** *Burlington Mag.* Jan. 3/2 Certain areas of the picture..have been largely repainted; the craquelure ceasing abruptly. **1942** *Antiquity* XVI. 99 When the incrustation [on silver] is appreciably thick, there is considerable surface expansion and consequent craquelure. **1956** M. Swan *Paradise Garden* ii. 24 Marcus left the aesthetic aspect of a work of art unexpressed, but remembered sizes, colours, *craquelures* and iconography. **1963** P. H. Johnson *Night & Silence* ii. 6 A *craquelure* of rose upon his cheeks. **1969** *Times Lit. Suppl.* 6 Nov. 1272/3 The craquelure in the Westminster pictures goes through the signatures on both panels.

crash, *v.* Add: **1. d.** To pass by (a traffic-light) when it is on the point of changing to red or has already done so.

1937 L. MacNeice *Lett. from Iceland* iii. 32 Traffic and changing lights, crashing the amber. **1967** *Punch* 10 May 662/2 Motorists who saw the bullion robbery van crash the lights in Holloway Road that day.

2. b. Now freq. with adv. (Further examples.) Also *fig.*

1886 'Maxwell Gray' *Silence of Dean Maitland* III. iv, He longed for the..high stone roof to crash in and hide him from that terrible gaze. **1893** M. E. Mann *In Summer Shade* xxviii, Bob..crashed over and fell in his blood at his murderer's feet. **1903** R. Langbridge *Flame & Flood* v, She was prepared for some new development which must crash in on her ignorance. **1922** D. H. Lawrence *England, my England* (1924) 172 Seats had crashed over.

6. a. *intr.* Of an aircraft or its pilot: to fall or come down violently with the machine out of control. Also of a motor car, motor cycle, or train, or its occupant(s): to suffer damage in an accident. Also *fig.* of a business or its proprietor: to suffer financial ruin.

1910 R. Loraine *Diary* in W. Loraine *R. Loraine* (1938) 106 The machine leapt higher,..then—paff!—I came to earth, having stalled and crashed. **1912** *Manch. Courier* 1 Sept. 3/1 The machine crashed down into the meadow. **1915** *War Illustr.* 6 Nov. 288/2 Cpl. Bennett disabled the German machine, which crashed to earth. *a* **1918** McCudden *5 Yrs. R.F.C.* (1919) 122 One of our machines had crashed about a mile away and..both occupants were dead. *Ibid.* 164 The Hun went down and crashed. **1919** P. Bewsher *Green Balls* 231 He's getting very low! My God! Did you hear that! He's crashed! **1923** *Motor Cycling* 26 Sept. 657/2 The first casualty was..one of the favourites, who crashed on Bray Hill, his machine catching fire. **1928** F. E. Baily *Golden Vanity* xv, Secretaryships to magnates are so precarious. Moreover, if French crashed, who would give his ex-secretary a job? **1970** *Times* 22 Oct. 4/6 A light aeroplane which crashed near a caravan site. **1971** *Daily Tel.* 8 Feb. 2/6 A car crashed into a bus stop queue in Aigburth Road, Liverpool, yesterday.

b. *trans.* To damage or destroy (an aircraft) in landing; to cause damage to (a vehicle) in an accident.

1915 H. Rosher *Let.* 14 Mar. in *In R.N.A.S.* (1916) 69 Our aerodrome here is a beastly small one. I have had several narrow shaves already of running into things, and feel sure that before long I shall 'crash' something. **1919** A. E. Illingworth *Fly Papers* 26 'If you want a walking stick,' I said, 'crash a machine yourself—don't come pinching mine.' **1928** C. F. S. Gamble *North Sea Air Station* i. 44 This machine was crashed as a result of a forced landing. *Ibid.* ix. 136 The last-named officer crashed his machine on landing (owing to the bad condition of the ground).

7. a. *colloq.* (orig. *U.S.*). To break into (a place); to gain admission to (a place, party, etc.) without the proper credentials or an invitation; to gate-crash. Also *transf.*, to 'break into' or enter (a group, profession, etc.). So *to crash the gate*: to gate-crash.

1922 *Dialect Notes* V. 147 (Bryn Mawr College) We crashed the gate at a swell joint like some finale hoppers. **1924** G. C. Henderson *Keys to Crookdom* ii. 20 The grand larcenist..will burglarize a place quickly enough, but the petty thief is too timid as a rule to 'crash a joint'. **1937** *Rotarian* Nov. 11/1 To 'crash a party'. **1938** Wodehouse *Code of Woosters* vi. 106 Staring at me..as if I had been a ticket-of-leave man who had got in by crashing the gate. **1951** *N.Y. Herald Tribune* 12 Dec. 27/3 I'm glad to see [he]..crashed television successfully. **1953** R. Fuller *Second Curtain* v. 80, I hope you'll forgive me crashing your excellent party. **1958** *Economist* 11 Jan. 117/2 To crash a party—to come uninvited—was branded as inexcusable.

b. *intr.* With *in*, *into*.

1929 Dunning & Abbott *Broadway* III. 104 Some of Scar Edwards' playmates might try to crash in, looking for trouble. **1931** Runyon *Guys & Dolls* (1932) ii. 40 He hears rumours of the party, and just crashes in. **1942** O. Nash *Face is Familiar* 80 The prominent and respectable dignitary who..crashed into a thousand anthologies by remarking, There but for the grace of God go I. **1950** T. S. Eliot *Cocktail Party* III. 147 She told me you were giving a party... So I said, I really must crash in.

c. *slang.* To sleep, esp. for a single night or in an emergency. Cf. **crash pad* (b).

1969 *Win* 1 Sept. 21/1 The two people who had come for help agreed to take these guys home with them, turn them on, feed them, give them a place to crash, love them a lot and keep them close to them all the next day. **1970** *Guardian* 31 Mar. 11/1 The homeless one was sure that someone would always offer him a place 'to crash'.

crash, *sb.*[1] Add: **4.** *Hunting.* The outcry made by hounds when they find the game. Also *fig.*

1781 P. Beckford *Thoughts on Hunting* xiii. 167 Where are all your sorrows,..one halloo has dispelled them all.—What a crash they make! **1837** 'Nimrod' *Chace, Turf, & Road* 50 There is no crash now, and not much music... At the pace these hounds are going there is no time for babbling. **1919** Masefield *Reynard* 85 But the whimpering rose to a crying crash By the hollow ruin of Tineton Ash. **1949** C. E. Hare *Lang. Field Sports* (ed. 2) xi. 136 Hounds may *speak*..with:..Many cries: e.g. on *hitting off* the line on a *screaming scent* (a *crash*). This is more like a 'muffled roar'.

5. *Theatr.* **a.** (See quots. 1891 and 1947.) **b.** A noise of breaking; the glass, etc., used to imitate the sound of the breaking of windows, etc.

1891 Farmer *Slang* II. 206/2 *Crash*, the machine used to suggest the roar of thunder; a noise of desperate (and unseen) conflict; an effect of 'alarums, excursions' generally. **1921** R. M. Heath *A.B.C. of Production* 70 'Glass crash', ready off L. up stage. **1947** *Gloss. Techn. Theatr. Terms* 12 *Crash*, a hand-operated, magnified version of the watchman's wooden rattle. *Ibid.* 13 Thunder..Sometimes an ordinary wine or beer barrel..charged with cobble stones.., thus *Thunder Crash*. **1952** W. Granville *Dict. Theatr. Terms* 52 *Crash*, any 'crash' effect. Splintering glass is known as a *glass crash*... A door slam is a *door crash*; breaking crockery is a *china crash*.

6. *Aeronaut.*, *Motoring*, etc. The act, or an instance, of crashing. (See *crash *v.* 6 a.)

1917 *Sphere* 3 Mar. 190/1 This particular victim of a 'crash' had been compelled to lie abed..for several weeks. *a* **1918** J. T. B. McCudden *5 Yrs. R.F.C.* (1919) 167 It is the ambition of every youthful pilot to down a Hun in our lines—and then land a crash alongside. *Ibid.* 185 About the end of May I had my first crash, on a Bristol Scout that I was delivering to J— G—. **1923** *Motor Cycling* 26 Sept. 659/2 Lowe's retirement was due to a crash caused by a burst front tyre. **1929** *Star* 21 Aug. 8/2 Air crash that caused 7 deaths. **1970** *Times* 22 Oct. 4/6 The pilot, who was alone in the plane, was killed in the crash. **1971** *Daily Tel.* 1 Jan. 4/4 (*headline*) 70 feared dead in rail crash.

7. *attrib.* and *Comb.*, as **crash barrier,** a barrier erected to halt an aeroplane, car, etc., that goes off its intended course; **crash boat** orig. *U.S.*, a boat used to rescue those involved in a crash at sea; **crash cymbal,** a cymbal hung in such a way as to make a crashing noise when struck with a drumstick; **crash-dive,** (*a*) a sudden dive made by a submarine when surprised or in imminent danger; (*b*) a dive made by an aircraft, ending in a crash; so **crash-dive** *v. trans.* and *intr.* (also *transf.*); **crash-halt,** of a motor vehicle: an abrupt halt; also *transf.*; **crash-helmet,** a helmet worn, esp. by motor-cyclists, to protect the head; hence **crash-helmeted** *a.*, wearing a crash-helmet; **crash landing,** a landing involving damage to the aircraft; so (back-formation) **crash-land** *v. intr.*; also *fig.*; **crash pad,** (*a*) a shock-absorbing buffer for protection of passengers in aircraft, motor cars, etc.; (*b*) *slang*, a place to sleep, esp. for a single night or in an emergency; **crash programme,** a course of research, training, etc., undertaken with rapidity and intensive effort, e.g. in an emergency; **crash-stop** = **crash-halt*; **crash-tackle** *v. trans.* and *intr. Football*, to tackle with great vigour; so **crash-tackling** *ppl. a.*; **crash truck, wagon** orig. *U.S.*, an emergency vehicle equipped for aid after an aeroplane crash, etc.

1947 *Jrnl. R. Aeronaut. Soc.* LI. 823/1 Crash barriers, as they were first called, were introduced into the Royal Navy in 1938. **1970** *Daily Tel.* 28 Aug. 1/7 About 50 miles of the 680 miles of motorway in use have crash barriers. **1936** *Baltimore Sun* 3 June 24/5 The crash boat pulled the midshipman and Wood from the water. **1946** L. G. Green *So Few are Free* (1948) v. 81 Until war brought crash boats into Cape waters, the record for the run from Table Bay to Dassen was held by a harbour tug. **1927** *Melody Maker* Apr. 387/1 It is impossible to execute this properly on the crash cymbal, because, as this crash has to be damped out to get the proper rhythm, both hands are occupied. **1961** A. Berkman *Singers' Gloss. Show Bus.* 34 'Dance cymbals' is a general classification which includes: 'fast'.., 'splash'.., 'bounce', 'ride' or 'crash' (specially tapered cymbals). **1919** *Times* 22 Mar. 8/1 The submarine commander, if the flying boats were sighted, could do a 'crash dive'. **1928** C. F. S. Gamble *North Sea Air Station* xii. 177 As soon as those aboard the submarine saw the seaplane, they 'crash-dived' the boat. **1942** *Air News* Aug. 38/2 A legend became current that the Nipponese pilots would crash-dive their airplanes rather than permit their capture intact. **1943** *Times* 6 Dec. 4/5 The U-boat crash-dived and results were not observed. **1945** *Daily Express* 20 Apr., Luftwaffe pilots have adopted Japanese suicide tactics...Pilots crash-dived on Russian bridges over the Oder, destroying three of them. **1959** A. C. Clarke *Across Sea of Stars* 204 Davis brought the jeep to its usual crash-halt in the parking space. **1918** W. G. McMinnies *Pract. Flying* 228 *Crash helmet*, a specially-made flying helmet designed to save the pilot's head in case of a crash. **1923** *Motor Cycle* 20 Sept. p. xlvii (Advt.), 5,000 New Crash Helmets..ex-R.A.F. **1937** 'G. Orwell' *Road to Wigan Pier* ii. 28 The wooden crash-helmets—a comparatively recent invention—are a godsend..and so strong that you can take a violent blow on the head without feeling it. **1959** *Times* 30 Sept. 12/6 Girls in crash-helmets and jeans. **1959** *News Chron.* 26 Sept. 3/3 A crash-helmeted youth wheeling his motor-bike down the path. **1969** *Daily Tel.* 31 Oct. 15/2 Zoot-suited spivs and crash-helmeted police. **1941** P. Richey *Fighter Pilot* 27 We learnt that Killy had got his Hun after all: it had crash-landed near Macon. **1946** R. Lehmann *Gipsy's Baby* 156 The stage rocked madly on its trestles, John crash-landed beside her, bellowed in her ear. **1959** *Listener* 22 Jan. 163/1 A body moving in a Lunik-type path will not normally crash-land upon the Moon..unless it is given a fresh impulse at the critical moment. **1928** *Literary Digest* 12 May 74/2 A 'Chinese three-point

landing' is a crash landing. **1942** *Time* 23 Feb. 21/3 The pilot..glided in for a crash landing. **1939** *Collier's* 24 June 55/1 In minor scrapes with death, pilots..shatter their goggles as their heads bang into crash pads—doughnuts of resilient material. **1959** *Engineering* 27 Feb. 265/2 Instrument clusters under foam-filled crash pads. **1968** *Listener* 31 Oct. 566/3 The hang-outs and crash pads..resound with the uncontrollable chattering of sour high-school drop-outs. **1970** *Guardian* 28 Oct. 13/1, I have..lived 'underground', slept in 'crash pads' and taken my food on charity. **1947** Crowther & Whiddington *Science at War* 49 In these 'crash' programmes there was more difficulty in arranging to make the sets than the valves. **1957** *Economist* 21 Sept. 958/2 Once this crash programme of vaccination has been completed, the demand for vaccine will fall off sharply. **1958** *Optima* Mar. 51/2 This consultation *en masse* resembles a crash programme—it is calculated to solve a large-scale problem in the most effective manner in the shortest possible period of time. **1959** *Manch. Guardian* 6 July 2/3 We went on as fast as before, and we were lucky to have only one crash-stop when we met a tractor. **1963** I. Fleming *On H.M. Secret Service* xvii. 193 A girl..came to a crash-stop in front of Bond. **1960** *Times* 1 Feb. 15/1 As Jackson was crash-tackled he let fly from just inside the penalty area. **1960** *Times* 2 Nov. 17/4 Stowe crash-tackled with such fury Oundle could not go far. **1954** J. B. G. Thomas *On Tour* 11, I have had the pleasure of playing first with Sale, alongside the crash-tackling Claude Davey. **1943** *Sci. News Let.* 387 (*caption*) Crash Truck..ready to rush to the scene of a crash and shoot hundreds of gallons of water on a burning plane. **1938** *Time* 5 Sept. 51 (Advt.), Known to the Navy as 'Crash Wagons', these three White trucks represent the last word in modern emergency equipment. **1958** 'N. Shute' *Rainbow & Rose* v. 215 At three hundred feet she was still spinning, and I shouted to the men to get the crash wagon.

b. *attrib.*, passing into *adj.* Undertaken with rapidity or intensive effort; organized for an emergency. Cf. **crash-dive* and **crash programme* above.

1952 *Sun* (Baltimore) 28 Feb. 1/1 The construction of the bases was undertaken as a 'crash' job, that is, a job to be done with all possible speed. **1954** *Newsweek* 30 Aug. 46/1 Crash teams of doctors, nurses, and physical therapists for all important diagnosis and initial treatment cannot be sent to critical areas in sufficient numbers. **1958** *Economist* 6 Dec. 866/1 The minister has already made what he thinks is adequate provision in the crash training college scheme announced last August. **1970** *Observer* 10 May 27/7 Don't go on a crash diet. You'll only put it back on again.

crash, *sb.*[2] Add: **2.** The name of a tint in textile fabrics, the colour of unbleached cotton.

1927 *Daily Express* 2 Apr. 6 In shades of Peach, Bracken, Sunburn, Caramel, Gold, Crash, White. **1927** *Daily Mail* 12 July 1 Colours:..Silver, Blush, Crash, etc.

crasher (kræ·ʃəɹ). [f. Crash *v.* + -er[1].]

1. Something that crashes or makes a crash; a loud harsh blow or percussion. Also *fig.*

1863 W. B. Cheadle *Jrnl.* 30 June (1931) 162 The Assinboine shot the lamb in the head and he fell a tremendous crasher to the foot of the cliff below. **1887** T. Darlington *Folk-speech S. Cheshire*, *Crasher*, a lie. **1922** A. S. M. Hutchinson *This Freedom* II. iv, Flick, rat-tat! What a crasher! You can feel it echo! **1928** E. Phillpotts in *Sunday Dispatch* 16 Sept. 2 He told the fatal tale to me... It was..of course a crasher for Archibald.

2. = *gate-crasher.

1924 *Daily Express* 13 Nov. 3/4 'Crasher'—a man who comes in without an invitation. **1928** *Ibid.* 12 July 8/6 The procedure of these 'crashers' is to wait until a function is well started and then to stroll into the hotel [etc.]. **1928** *Sunday Dispatch* 15 July 1 The dance 'crasher' has a most insidious rival in the dance 'cadger', declares a prominent hostess. **1950** *New Yorker* 9 Dec. 31/1 His main job after the curtain goes up is handling crashers.

crashing, *ppl. a.* Add: **c.** *colloq.* Overwhelming; esp. in phr. *crashing bore.*

[**1928** R. A. Knox in *Sunday Dispatch* 30 Sept. 12/7 Your rejuvenated Fausting, talking like seventy when he looks a generation younger, would be shunned..as a crash bore. **1930** W. S. Maugham *Breadwinner* I. 18 *Judy.* What a bore the people are who went through it. *Patrick.* Crashing.] **1931** R. Campbell *Georgiad* i. 16 Piping Nancy-boys and crashing Bores. **1935** 'N. Blake' *Question of Proof* vi. 96, I asked for a crashing great emergency to turn up..and it has. **1953** X. Fielding *Stronghold* 108, I still consider the poem a crashing bore, a typical 'provincial' effort.

crashworthiness (kræ·ʃwɒ̄ːɹðinės). [f. *Crash *sb.*[1] 6 + Worthiness, after *airworthiness.] The quality in an aircraft or motor-vehicle that makes it safer in the event of a crash. So **cra·shworthy** *a.*

1948 *Jrnl. R. Aeronaut. Soc.* LII. 572 British Civil Airworthiness Requirements already contain certain 'crashworthiness' provisions. **1961** *Aeroplane & Astronautics* CI. 403/2 Improvements in 'crashworthiness' are called for, involving not only aircraft design but also fire-fighting and emergency equipment at airports. **1966** *New Statesman* 20 May 718/3 He explained what he thought could be done to improve auto design, particularly in respect to 'crashworthiness'—how to help passengers avoid injury or death in the event of a collision. **1966** *Economist* 2 July 4/3 American and British physicians now have the opportunity to be unpopular again by demanding that vehicles which carry people be sufficiently crashworthy to protect occupants against their own mistakes.

crate, *sb.* Add: **2*.** An aeroplane, especially one of obsolescent design. Also applied to other vehicles. *slang.* (orig. *Services*').

1928 *Daily Mail* 7 May 6/5 (citing U.S. air terms) *Hay Wire Outfit.*—A crude plane. *Crate.*—Somewhat better..but just an old crate'. **1934** Webster, *Crate.* Slang. *a.* An airplane. *b.* An automobile. **1935** C. Day Lewis *Time to Dance* 41 We'll fly this bloody crate till it falls to bits at our feet. **1937** F. E. Baily *Treat them Gently* iv. 86 His Rolls 'll give you more respect in the eyes of the reporters than my old crate would. **1940** N. Monks *Squadrons Up!* i. 22 Shooting up..pilots who were flying 'tomato crates'. **1957** *Times* 11 Oct. 11/6 You must travel in an antiquated two-engined crate which goes puttering over Central Asia at about 90 miles an hour. **1959** [see *aviate *v.*].

3. *crate-float.*

1903 A. Bennett *Leonora* ii. 48 The crate-floats, drawn by swift shaggy ponies and driven by men who balanced themselves erect on two thin boards.

crate, *v.* [f. the sb.] *trans.* To pack in a crate or framework. Hence **cra·ted** *ppl. a.*, **cra·ting** *vbl. sb.*

1871 E. Eggleston *Hoosier Schoolmaster* (1872) xi. 65 The boys thought 'twas funny to crate me. **1897** *Outing* (U.S.) XXX. 366/1 Crated and swathed in white clothes, it appeared like a mummy. **1900** Kipling in *Daily Express* 20 June 4/5 Neatly-piled boxes of rivets and a mass of crated ironwork. **1901** 'C. Holland' *Mousmé* 265 The heavier articles, a man came and crated for me. **1924** *Glasgow Herald* 19 Apr. 7 The old engine was hoisted on board and crated. **1963** *Ann. Reg. 1962* 525 The Soviet Government..has issued a new order, for the dismantling of the weapons which you describe as 'offensive', their crating and return to the Soviet Union.

crater, *sb.* Add: **2. b.** A raised bowl-shaped hollow on the surface of the Moon.

1860 [see Dict. sense 2]. **1868** J. N. Lockyer *Elem. Astron.* xvi. 94 A view of the crater Copernicus, one of the most prominent objects in the Moon...Outside the crater-wall..many smaller craters..are distinctly visible. **1964** *Yearbk. Astron. 1965* 113 The solution to the old question of whether the lunar craters have a volcanic, meteoritic or some other origin may be near at hand.

4. Also, the cavity formed by the explosion of a shell. Also *attrib.*

1855 E. B. Hamley *Story Campaign Sebastopol* xxvii. 282 A magazine..had been blown up by a shell—..no serious damage was done by this explosion, which left..a vast crater like a quarry. **1914** *Scotsman* 26 Dec. 8 Seamed with dug-outs, burrows, trenches,..and pitted with craters. **1919** P. Bewsher *Green Balls* 200, I can see dotted around the fields the great craters of the shell-holes. **1919** G. K. Rose *2/4th Oxf. & Bucks Lt. Infty.* ii. 24 The Somme 'craterfield'. **1926** *Encycl. Brit.* Suppl. I. 228/2 A 6-in. howitzer mine shell..produces a cylindrical crater about four yards across and 10 ft. deep.

6. crater-lake, a lake formed by the collection of water in the crater of an extinct volcano.

1879 *Encycl. Brit.* X. 250/1 In some cases, where ancient crater-lakes or internal reservoirs have been shaken by repeated detonations, and finally disrupted, the mud which has thus been produced issues at once from the mountain. **1885** A. Geikie *Text-book Geol.* (ed. 2) III. i. 224 Explosion lakes (Crater-lakes) of this kind occur in districts of extinct volcanoes. **1959** *Chambers's Encycl.* VIII. 307/2 Crater-lakes occupy the craters of extinct or dormant volcanoes (e.g. the Maare of the Eifel district).

crater, *v.* [f. the sb.] **a.** *intr.* To form a crater or hollow. **b.** *trans.* To obstruct or destroy (a road or terrain) by craters formed by mines or mine shells. So **cra·tered** *ppl. a.* Also *fig.*

1884 *Cassell's Fam. Mag.* 576/1 This arrangement is found to give a better light than the solid rod, which is apt to 'crater' or become hollow in its burning end. Compound rods..do not crater in this way. **1917** 'Contact' *Airman's Outings* 134 Tens of thousands of shells had pocked the dirty soil, scores of mine explosions had cratered it. **1921** *Times Lit. Suppl.* 8 Sept. 570/4 The Second [German Army] could not get beyond Albert on account of difficulties in the cratered area. **1922** *Glasgow Herald* 22 Mar. 10 A number of roads had been cratered and a certain number of bridges blown up. **1925** H. Acton in *Oxf. Poetry* 2 Our lives are cratered with great pocks and scabs. **1941** *Illustr. London News* CXCVIII. 223 (*caption*) Laying land-mines to delay advancing troops: cratering a road surface for depth-charge. **1943** C. Day Lewis *Word over All* 21 Yet words there must be, wept on the cratered present.

cravat, *sb.* Add: **1. c.** A scarf or necklet of lace, fur, etc., worn by women.

1903 *Westm. Gaz.* 26 Nov. 4/2 Things..of the lacy kind..little collar-cravats, &c. **1904** *Daily Chron.* 19 Mar. 8/2 Talk no longer of stoles, boas, and ruffles, but of cravats, for by that name the necklets sold for early spring wear prefer to be known. **1905** *Westm. Gaz.* 11 Nov. 13/2 The cravat effect of the ermine on the shoulders is charming.

d. *Surg.* (See quot. 1900.)

[**1884** W. Pye *Surgical Handicraft* II. viii. 78 The first and most obvious use of this [triangular or scarf] bandage is to simply tie it round..with a reef knot, it having been previously folded up into a cravat.] **1900** G. M. Gould *Pocket Med. Dict.* (ed. 4) 186 *Cravat*, a bandage made from a triangular cloth. **1914** E. L. Eliason *Pract. Bandaging* ii. 79 The centre of the body of the cravat is looped under the wrist and the two ends carried one around each side of the neck and tied together.

crave, *sb.* Add: **b.** *Scots Law.* A demand or claim addressed to the court. (Cf. Crave *v.* 1 c.) Also **cra·ving** *vbl. sb.* in the same sense.

1707 in C. D. Bentinck *Dornoch Cath. & Par.* (1926) vii. 252 They might deem it to be their duty to grant Mr Bowie's crave for an Act of Transportability. **1913** *Act* 2 & 3 *Geo. V* c. 28 Sched. ii. (23) The sheriff may, at any time after the expiry of the induciae, upon a written craving being endorsed on the initial writ, decern in terms of the crave of the initial writ. **1938** *St. Andrews Citizen* 26 Mar. 2/2 Sheriff-Substitute D— S— granted declarator in terms of the crave. **1959** Jowitt *Dict. Eng. Law* I. 534/1 In sheriff court practice in Scotland part of the initial writ is called the crave.

crawfish, *sb.* Add: **1. b.** *S. Afr.* An edible marine crustacean belonging to any of several genera of the families Scyllaridæ or Palinuridæ, especially the Cape crawfish, *Jasus lalandii*.

1853 L. Pappe *Edible Fishes Cape G. H.* 11 This crawfish [sc. the Cape lobster, *Palinurus lalandii*], peculiar only to the West Coast, and common to Table Bay, is easily caught. **1954** K. H. Barnard *S. Afr. Shore-life* 27 Farther east, instead of the Cape Crawfish, other species are found which are called *stridentes* or noisy Crawfishes..Gilchrist's Crawfish from the Agulhas Bank has the two short whips on each of the shorter feelers...The Port Elizabeth Crawfish..is a squat form. **1961** *Cape Times* 21 July 11/5 Our *kreef* which still appears as crayfish or crawfish on Cape restaurant menus, is the *langouste*. We changed our *kreef* from crawfish to rock lobster to please American taste. It seems they despise the small, river crawfish with which they are familiar.

crawfish, *v.* (Earlier example.)

1848 *Congress. Globe* Feb. 277/3 No sooner did they see the old British Lion rising up.., than they crawfished back to 49°.

cra·wfishing, *vbl. sb.* and *ppl. a.* Fishing for crawfish. Also *fig.*

1844 *Whig Battering Ram* (U.S.) 9 Aug. 4/1 Look out for such a specimen of crawfishing as the Locofocos alone can practice. **1873** *Winfield* (Kansas) *Courier* 31 July 2/2, I apprehend that..you have been guilty of what the boys would call 'crawfishing'. **1903** *Daily Chron.* 8 June 9/2 It is only natural that by crawfishing for a good reference this steward neglects the men who actually do the hard work.

crawk, *v.* [Imitative.] *intr.* To utter a hoarse sound, squawk; also *trans.* with *out.* Also as *sb.*

1845 *Round Preacher* vi. 94 Nbt many hours 'ud pass afore they'd crawk out for the loaves and fishes. **1889** R. T. Cooke *Steadfast* xxvi. 287 A cooped barnyard fowl, that crawks, and cheeps. **1915** D. H. Lawrence *Rainbow* iv. 106 There was a sharp crawk of expostulation as one of the hens shifted over. **1930** *Aberdeen Press & Jrnl.* 26 Sept. 6/3 The eagle retired, crawking out his discomfiture.

crawl, *sb.*[1] Add: **b.** A walk at a leisurely pace. In *beer-crawl*, *gin-crawl*, *pub-crawl*, a slow progress from one drinking-place to another. *slang.*

1883 *Bird o' Freedom* 7 Mar. 6/3 Phil Benjamin was taking his daily constitutional, which consists in what is in vulgar parlance designated 'a gin crawl'. **1902** *Daily Chron.* 3 Dec. 5/2 The cockney 'beer crawl'. **1905** *Ibid.* 28 Mar. 4/6 Glasgow's most fashionable Sunday parade, the 'crawl' on Great Western-road. **1915** T. Burke *Nights in Town* 273 We did a 'pub-crawl' in Commercial Road and East India Dock Road. **1927** *Blackw. Mag.* Sept. 336/2 During our evening gin crawls in the various camps. **1959** *Observer* 24 May 16/3 Heads of the dress firms will take the 100 expected buyers on individual 'pub-crawls'. **1961** M. Jones *Potbank* xiii. 50 No other men under forty..came into the pub, except perhaps..as one short stop in a crawl.

c. In full *crawl-stroke.* A high-speed swimming stroke, in which the swimmer, lying face-downwards, usually with the face submerged, makes alternate overhand arm-strokes assisted by the quick movement of the legs.

1903 Sinclair & Henry *Swimming* (ed. 4) iii. 89 A young swimmer named R. Cavill, who revolutionised all ideas about speed swimming for short distances by introducing a further modification of this style [sc. the 'Trudgen'], which was at once termed the 'crawl' stroke. **1906** Taylor & Gibson *Extra Dry* 32 Arthur Haddock..swam out with a rope in his teeth, using the Australian crawl. **1912** F. Sachs *Complete Swimmer* 144 The double over-arm and crawl racers. **1926** *Westm. Gaz.* 10 Sept., To the school children across the Atlantic the 'crawl' is as natural as the ancient 'breast' stroke to the average British child. **1930** J. Weissmuller (*title*) Swimming the American Crawl. **1934** J. T. Farrell *Young Manhood* (1936) xv. 340 He about-faced and swam the crawl stroke back to the pool edge. **1945** Baker *Austral. Lang.* ix. 179 The type of stroke known as the *Australian crawl*—apparently first introduced about 1901. **1957** Wodehouse *Over Seventy* xvii. 160 Perfecting my Australian crawl in the swimming-pool.

2. crawl space (see quot. 1963); also **crawlway.**

1951 P. D. Close *Building Insulation* (ed. 4) iii. 67 Floors over unheated crawl spaces should be insulated. **1963** *Gloss. Build. Terms* (*B.S.I.*) 9 *Crawl space, crawlway*, an under-floor space providing access to ducts..and of a height sufficient for crawling. **1965** R. McDowell *Hound's Tooth* (1967) viii. 78 He..squirmed into the crawlway. **1967** *Boston Sunday Herald* 26 Mar. III. 5/4 We live near the water, so our house was built with a waterproof crawl space.

crawl, *sb.*[2] **1.** Delete † *Obs.* (See quot.)

1961 F. G. Cassidy *Jamaica Talk* v. 95 Though *sty* is more common, *crawl* is still used today.

crawl, *v.*[1] Add: **1. c.** *intr.* To swim using the 'crawl' (see *crawl *sb.*[1] c).

1911 T. E. Lawrence *Lett.* (1938) 107 Using crawling and trudgeon strokes in swimming. **1913** —— *Home Lett.* (1954) 259 He..swam all over the Euphrates, 'crawling' about.

3. c. To behave sycophantically or abjectly. Freq. const. *to. colloq.*

1943 C. H. Ward-Jackson *Piece of Cake* 23 *To crawl*, to ingratiate oneself with, to make up to. **1966** A. Sachs *Jail Diary* xxi. 187, I suck up to the police... I don't actually crawl to them but I am so eager to win their approval. **1966** *Listener* 29 Sept. 445/1 Instead of maintaining an appropriately hostile stance..,Mr Khrushchev was 'crawling' to Washington. **1970** P. Y. Carter *Mr. Campion's Falcon* xxi. 155 I'm going to have to crawl...I want to grovel fantastically. I'm afraid I really do need help.

5. Further *transf.* examples; also *fig.*

1902 Kipling *Traffics & Discov.* (1904) 185 There's a whole switchboard full o' nickel-plated muckin's which I haven't begun to play with yet. The starboard side's crawlin' with 'em. **1905** *Strand Mag.* May 567/1 Shore and bushes near Wells and Blakeney are simply 'crawling' with pied fly-catchers and redstarts. **1916** 'Ian Hay' *First Hundred Thou.* 305 Country districts crawling with troops. **1924** Wodehouse *Bill the Conqueror* vii. 143 'He must be quite rich.' 'Crawling with money.' **1945** E. Waugh *Brideshead Revisited* 88 The country is crawling with Communists. **1950** W. Stevens *Let.* 27 Feb. (1967) 671 The place is crawling with books that we have no room for.

crawler. Add: **1. c.** One who swims the crawl-stroke. ? *Obs.*

1912 F. Sachs *Complete Swimmer* 149 The majority of crawlers will complete one length of a bath—usually about 30 yards—without breathing at all.

3. (Earlier and later Austral. and N.Z. examples.)

1847 'A. Harris' *Settlers & Convicts* i. 6 It [sc. the sight of Sydney from the harbour] gives, even from the distance, a presage of the stirring, downright earnest life..that so strongly characterizes the race that lives, and breathes, and strives around: a race with whom it is one of the worst reproaches to be *a crawler*. **1900** H. Lawson *Over Sliprails* in *Prose Wks.* (1948) 261 The rest were crawlers, mostly pub spielers and bush larrikins. **1912** R. B. Booth *Five Yrs. N.Z.* vii. 39 A rough lot of characters were..seen there..the wretched debauched crawler, the villainous-looking 'lag'. *c* **1926** 'Mixer' *Transport Workers' Songbk.* 82 Who is dinkum, who's a crawler? **1934** *Bulletin* (Sydney) 21 Mar. 40/3, I promised myself that if ever I met the crawler I'd flay him alive. **1957** 'N. Culotta' *Weird Mob* (1958) xiii. 203 Don't overdo the generosity. That will make you a 'crawler' which is nearly as bad as being a bludger.

4. A domesticated animal of slow movements. *Austral.*

1853 C. B. Hall in T. F. Bride *Lett. Vict. Pioneers* (1898) 218 On this run, out of 1,500 head of cattle, all had been driven off but about 30 'crawlers'. **1890** 'R. Boldrewood' *Col. Reformer* xviii, The herd..consisted almost wholly of the well-bred station 'crawlers', as the stockmen term them from their peaceable and orderly habits. **1941** Baker *Dict. Austral. Slang* 20 *Crawler*,..a cow, calf or bullock of peacable [*sic*] habits.

5. Chiefly *pl.* An overall garment put on a child in order that it can crawl and still keep its ordinary clothes clean.

1891 *Cassell's Family Mag.* 340/2 A child's quilt or 'crawler' is a useful nursery appendage. **1923** *Daily Mail* 11 Jan. 4 Children's Crawlers. **1923** *Weekly Dispatch* 18 Feb. 15 A new style of crawler has just been designed. **1924** A. Huxley *Little Mexican* 202 A little toddling creature in blue crawlers. **1937** M. Allingham *Dancers in Mourning* xiii. 171 You were in crawlers when I was kicking my heels about.

6. In full *crawler tractor.* A kind of tractor moving on an endless chain; also *ellipt.* So *crawler track.*

1922 *Encycl. Brit.* XXXII. 739/1 The crawler type..can pass over marshy land where the wheeled tractor would be mired. **1926** *Ibid.* Suppl. III. 796/1 The crawler tractor had one or two continuous-chain tracks on which it ran and which it carried with it. **1936** *Forestry* X. 88 The modern track-laying or 'Crawler' tractor with its great flexibility and speed is replacing horse-dragging. **1954** *Highway Engin. Terms* (*B.S.I.*) 45 *Crawler track*, a device, consisting of an endless chain of plates which bear upon the ground, used in place of wheels on vehicles or equipment intended for traversing soft or natural ground. **1955** 'N. Shute' *Requiem for Wren* 6 There were now four big diesel tractors on the place, one of them a crawler. **1959** *Times Rev. Industry* June 75/1 Largest single shipment of..crawler tractors since production began. **1962** *Flight Internat.* LXXXII. 209/1 The use of a 2,500 ton crawler vehicle to carry each Saturn..from assembly building to launch pad. *Ibid.*, All other launch equipment..will be transported to the pad by the crawler. **1968** *New Scientist* 26 Dec. 714/1 The problem of operating mounted ploughs appeared to be insuperable—one reason for the diminishing popularity of the crawler.

crawling, *vbl. sb.* Add: **b. crawling-board** (see quots.).

1901 *Daily Chron.* 25 Nov. 7/3 A crawling-board used on a roof in connection with a ladder. **1902** *Law Rep.* [1902] I.K.B. 494 A crawling board, which was a contrivance ordinarily used in the repair of roofs, and consisted of a wooden plank about 18 to 20 feet long and 10 inches wide, across which were nailed transverse pieces of wood.

crawling, *ppl. a.* Add: (Later example.)

1941 *Coast to Coast 1941* 127 His glance was venomous as he added: 'You crawling bastard, you!'

b. crawling peg *Econ.*, a system whereby the exchange rate can be frequently and marginally adjusted.

1966 J. H. Williamson *Crawling Peg* 2 The 'crawling peg' will be used to mean a system under which such par changes as occur are implemented slowly, in such a large number of small steps as to make the process of exchange-rate adjustment continuous for all practical purposes. **1970** *Times* 9 Feb. 20/2 Washington, therefore, can see distinct advantages in a one-way upward crawling peg.

crawlsome (krǫ·lsəm), *a.* [f. Crawl *v.* + -some.] Addicted to mean, worm-like behaviour. Hence **craw·lsomeness.**

1900 H. Lawson *Over Sliprails* 73 We'd be sure to..reckon that it was done out of nastiness or crawlsomeness, and feel a contempt for him accordingly. **1904** *Delta* (N. Queensl.) *Advocate* in *Westm. Gaz.* 21 Mar. 4/2 In the whole course of their crawlsome, insanitary lives.

cray[2]. Chiefly *Austral.* and *N.Z.* = Crayfish. Also *attrib.*

1906 R. E. Vernède *Meriel of Moors* viii. 49 'There iddn't everybody 'ud like to be clipped by a cray.'..'Bring the crays, Micky.'..Micky having shouldered the handkerchief full of cray-fish [etc.]. **1916** C. J. Dennis *Ginger Mick* 46 We'll 'ave a cray fer supper when I comes a-marchin' 'ome. **1933** *Bulletin* (Sydney) 13 Sept. 20/3 Every drawing of a cray I have studied for years past shows the convivial crustacean with huge claws like the Atlantic lobster. *a* **1939** 'R. Hyde' *Houses by Sea* (1952) 116 As fishers..boats that bob across the bay Setting their cray-pots in the island's shadow. **1966** T. C. Roughley *Fish & Fisheries Australia* 124 Cray-fishing is a dangerous occupation as the home of the crays is along the sandstone reefs of the mainland. **1970** *Australian* 14 May 6/1 There would seem little connection between the price of craytails in New York and the destruction of wildlife in Australia.

cray·fish, *v. Austral.* [f. the sb.] *intr.* To move like a crayfish; to act in a cowardly or scheming manner. Cf. Crawfish *v.*

1900 H. Lawson *Over Sliprails* 63 The steamer was just crayfishing away from a mud island, where she had tied up for more wool. **1930** K. S. Prichard *Haxby's Circus* xvii. 194 When he thought I was goin' to hit him, [he] crayfished..lay down on the floor, and said I couldn't hit a man when he was down.

cray·fishing, *vbl. sb.* [f. Crayfish *sb.* 3.] Fishing for crayfish; also (*Austral.*) acting in a cowardly or scheming manner.

1835 J. H. Ingraham *South-West* I. xvi. 174 Others were engaged in the delicate amusement of cray-fishing. **1931** V. Palmer *Separate Lives* 194 The truth, the whole truth, and no cray-fishing, so help me God. **1961** *Guardian* 6 Nov. 7/5 Crayfishing parties on the Dorn were occasions of beribboned village festivity. **1966** 'J. Hackston' *Father clears Out* 203 Such hedging and crayfishing about only postponed the dread business.

craze, *sb.* Add: **2. a.** Also in weakened sense: a capricious and usually temporary enthusiasm; *the craze* = (all) the rage (see Rage *sb.* 7 b).

1887 *Fortn. Rev.* XLII. 284 A quiet craze touching everything that pertains to Napoleon the Great and the Napoleonic legend. **1888** Mrs. H. Ward *R. Elsmere* I. iv. 86 A child of many crazes, eager for poetry one week, for natural history the next. **1911** W. Raleigh *Let.* 15 Dec. (1926) II. 374 I've been lecturing at Glasgow...Now that I'm not there I'm quite the craze. **1934** *Planning* II. xxxvi. 5 The pre-war craze for awarding gold medals to approved products. **1967** *Amer. Speech* XLII. 40 The sport has reached craze proportions.

crazia (kra·tsia). Pl. **crazie** (-*ie*). [It., ad. G. *kreuzer* Kreutzer.] A copper coin of Tuscany, the twelfth part of a lira.

1787 Beckford *Lett. Italy* (1805) I. 259 Four crazie only, somewhat less than three pence English. **1897** *Daily News* 5 July 3/6 Tuscany [postage stamps] 9 crazie on white, unused, 7*l.* **1963** *Times* 21 Feb. 14/5 Other property included a good range of the old Italian states issues of which a 60 crazie brick-red on azure of Tuscany sold at £340.

crazing, *vbl. sb.* **1.** (Further examples.)

1940 *Chambers's Techn. Dict.* 207/2 *Craze* or *crazing*, the minute hair cracks which sometimes appear on the surface of pre-cast concrete work or of artificial stone.—Fissuring of faulty coats of paint or varnish in irregular criss-cross cracks. **1951** H. M. Langton *Morrell's Synth. Resins* (ed. 3) v. 211 A defect of polystyrene is a tendency for mouldings and castings to develop fine cracks—known as 'crazing'. **1954** *Highway Engin. Terms* (*B.S.I.*) 29 *Crazing*, the cracking of a surface layer into small irregularly shaped contiguous areas. **1961** J. N. Anderson *Dental Materials* (ed. 2) xxiii. 242 Crazing appears when a denture has been repeatedly dried out..due to the alternate expansion and contraction of the resin.

crazy, *a.* Add: **4. a.** Further examples, esp. in *colloq.* senses 'extremely eager, enthusiastic', etc. Also *to go crazy*; *to be crazy about* or *for* (a person), to be infatuated with, to be in love with.

1798 Coleridge *Anc. Mariner* VII. p. 47 The Pilot's boy, Who now doth crazy go, Laugh'd loud and long. **1826** M. R. Mitford *Our Village* II. 271 All the world..were crazy to have their fortunes told. **1892** H. Dacre *Daisy Bell* 5 Daisy, Daisy, Give me your answer, do! I'm half crazy, All for the love of you! **1895** Susan Hale *Lett.* (1919) 294 They were crazy that I should talk about the celebrated people I have known. **1898** *Scribner's Mag.* Oct. 447/2 He's crazy over this girl. **1904** R. L. McCardell *Show Girl & Friends* 35 You know everybody is just crazy about octets? **1917** Wodehouse *Man with Two Left Feet* 18, I love you. I'm crazy about you. **1926** J. Black *You can't Win* iii. 19 Well, if you're so crazy about a job, I'll make one for you. **1929** M. Lief *Hangover* 235 He was crazy in love with her and one moonlit night he proposed to her. **1930** *Daily Tel.* 1 Dec. 21/1 At Royal Wimbledon..he went 'crazy', to use an inoffensive and common golfing phrase. His figures to the turn..a total of 30. **1932** Grace Perkins (*title*) Boy crazy. **1936** Bentley & Allen *Trent's Own Case* vii. 72 We were talking about Eunice Faviell..and the way so many men go crazy about her. **1949** Wodehouse *Mating Season* iii. 28 And the unfortunate part of it all is, Bertie, that I'm crazier about him than ever. **1962** J. Ludwig in R. Weaver *Canad. Short Stories* 2nd Ser. (1968) 252 A Toronto deb-type, gorgeous, crazy for Jimmy, hot and burning.

b. (Earlier example.)

1855 H. Melville *Piazza Tales* (1856) 425 Hopelessly infected with the craziest chimeras of his age.

c. *like crazy* = *like mad* (Mad *a.* 1 c). *colloq.* (orig. *U.S.*).

1924 P. Marks *Plastic Age* xxiv. 288 She has been going an awful pace, tearing around like crazy. **1957** J. Osborne *Entertainer* v. 35 There she was, beating them with her umbrella like crazy. **1964** *Punch* 24 June 929/1 They're wooing us like crazy. **1968** *Ibid.* 28 Feb. 309/1 Here were all those guys consuming like crazy and having to be regularly restocked.

d. *slang* (orig. *U.S.*). (*a*) Of music, esp. jazz: unrestrained, wild; exciting. (*b*) Hence as a term of approbation: excellent, admirable, satisfying. Cf. *cool *a.* 4 e.

1927 *Melody Maker* June 573/2 Leath has established her reputation as a..first-class artist in '*Crazy Words, Crazy Tune*'. **1933** *Ibid.* 2 Sept. 2 Arthur Roseberry Goes All 'Crazy'. **1935** *Hot News* Apr. 13/1 He is a crazy player, and a crazier singer. *Ibid.*, 'Jamaica Shout' is, I think, the craziest record. *Ibid.*, Where musicians are concerned..if I say a man is crazy you may be sure that I think he is very, very good. **1953** [see *cool *a.* 4 e]. **1956** *New Yorker* 8 Dec. 159 (Advt.), Hi Fi can turn your bachelor apartment into a Harem..and the crazy part is you don't have to be an engineer to operate H. H. Scott components. **1959** *Punch* 14 Oct. 319 The swing-cats sway, the hipsters tap their feet As Victor pounds his low-down crazy beat. **1962** J. Baldwin *Another Country* (1963) II. iii. 274 She laughed. 'Black Label [Scotch]?' 'Crazy.'

e. *crazy like* (or *as*) *a fox*: very cunning or shrewd. *colloq.* (orig. *U.S.*).

1935 G. & S. Lorimer *Heart Specialist* vi. 164 'Is she drunk or crazy and is she a giraffe?..' 'Crazy as a fox,' I said glumly. **1944** S. J. Perelman (*title*) Crazy like a fox. **1967** M. Procter *Rogue Running* xxv. 165 'Crazy,' Martineau mused. 'Crazy like a fox. And as hard to catch.' **1969** D. Bagley *Spoilers* x. 294 I'm crazy like a fox.. How much did I win from you..?

f. In the colloq. phr. *crazy, mixed-up* (*kid*, *person*, etc.), descriptive of one whose mind is driven distracted by conflicting and unresolved emotions. Also used of a confused situation.

1955 *Melody Maker* 30 July 6/2 Full of the most peculiar phrases..something else is a 'crazy, mixed-up creep'. **1956** *Punch* 28 Nov. 653 We live in an age in which the Common Man is apt also to be a crazy, mixed-up personality. **1957** J. Braine *Room at Top* xi. 106 As they say in the films, I'm just a crazy mixed-up kid.

4*. Used (after *crazy quilt*) to denote a garden walk or pavement of irregular pieces of flat stone or tile, esp. in *crazy paving*; so (back-formation) *crazy-pave* v. trans., *crazy-paved* ppl. adj.

1923 *Daily Mail* 13 Jan. 11 Stone walks, either crazy or rectangular. *Ibid.* 19 Feb. 6 Visitors will pass by old red brick and crazy paths to a sunk lawn. **1925** A. S. M. Hutchinson *One Increasing Purpose* III. xvi, Under the blue tile..of the crazy-paving just by the rain-water butt. **1927** H. C. Bailey *Mr. Fortune, Please* 173 Along the house was a terrace with crazy pavement in which saxifrage grew. *Ibid.* 223 Mrs. Pemberton's garden was a pleasant place of crazy paving and rock plants. **1952** E. R. Janes *Flower Garden* 36 Probably the best form of crazy paving is the broken paving stones obtainable in some towns when stone pavements are replaced by asphalt. **1960** *News Chron.* 22 Apr. 8/6 Jimmy Gold is..getting ready to crazy-pave his front garden. **1963** *Times* 25 Feb. 14/5 Almost every room in the house has exposed oak beam ceilings and there is a small crazy-paved garden.

5. *crazy-sensible*, *-wayed*, *-witty* adjs. (nonce-wds.); **crazy flying,** aerobatics performed near the ground; **crazy quilt** (for *U.S.* read orig. *U.S.* and add further examples); also *attrib.* and *fig.*

1922 *Flight* XIV. 372/2 The next item was the extraordinary exhibition of 'crazy' flying..on a jazzified Avro. **1940** N. Monks *Squadrons Up!* 162 To lift that gold cup for the crazy-flying event. **1888** *Boston Evening Jrnl.* 20 June, A Washington letter gives a clever sketch of the conversation which recently took place at the house of a matron who receives a hundred or two of people on the afternoon of 'her day'. 'Crazy quilt conversation', the hostess declared it to be. **1911** H. S. Harrison *Queed* 49 The present system is a mere crazy-quilt, quite unsatisfactory in a thousand ways. **1946** W. S. Maugham *Then & Now* xxii. 118 Machiavelli's emotions were as various as the colours of a crazy quilt. **1958** *Listener* 13 Nov. 765/1 Can anything but crazy-quilt legislation emerge from a chamber exposed to pressures such as these? **1963** *Economist* 28 Sept. 1121/3 The crazy-quilt redevelopment of Lower Manhattan. **1923** D. H. Lawrence *Let.* 24 Sept. (1962) II. 754 California is..sort of crazy-sensible. **1957** J. Kerouac *On Road* (1958) 164 So amazingly himself, all raging and sniffy and crazy-wayed. **1851** H. Melville *Moby Dick* III. xiii. 92 He's too crazy-witty for my sanity.

crazy, *adv. slang* (chiefly *U.S.*). [f. the adj.] Extremely; 'madly'.

1887 *Lantern* (New Orleans) 9 July 2/2 He was crazy drunk in a saloon one night. **1906** E. Dyson *Fact'ry 'Ands* i. 9 I'm just crazy gone on 'er. **1930** J. Dos Passos *42nd Parallel* xi. 175 And here I was crazy anxious. **1957** A. MacNab *Bulls of Iberia* xv. 176 Llorente is *rabioso*, crazy keen.

crazy (krē¹·zi), *sb. colloq.* (orig. *U.S.*). [f. the adj.] A mad or eccentric person.

1867 W. L. Goss *Soldier's Story* 30, I was addressed as 'old crazy' by my companions, and told to keep still. **1884** 'C. E. Craddock' *In Tenn. Mts.* 46 They take some crazies at the State's expense. **1945** G. Millar *Maquis* i. 6 There were many crazies in the organisation. It said much for the officers at the top that the crazies were permitted. **1969** *Guardian* 22 Feb. 2/6 There's no leadership at all. All this is being done by the street crazies.

creak, *v.* Add: **2. a.** Also *fig.*

1938 F. Scott Fitzgerald *Let.* 20 Jan. (1964) 564 The last pages..begin to creak from 116 on. **1962** *Listener* 15 Nov. 796/2 The Civil Service is already needing reform, and some people have said that even the limited extra responsibility..may make it creak.

creakily (krī·kili), *adv.* [f. Creaky *a.* + -ly[2].] In a creaky manner, with a creaky sound.

1901 E. F. Benson *Luck of Vails* xiii. 141 His mouth was dry, and his heart hammered creakily in his throat. **1909** *Daily Chron.* 27 Aug. 7/6 One treading softly..the other loudly, creakily, and importantly. **1920** *Blackw. Mag.* Oct. 525/1 A steamer creakily blowing off regular jets of steam.

creaking, *vbl. sb.* **a.** Delete † *Obs.* and add later example.

1923 *Chambers's Jrnl.* Oct. 638/1 The 'creaking' of the partridges still hung in the stagnant air.

creaking, *ppl. a.* Add: *a creaking* (also *creaky*) *gate*, in proverbial phrases (see quots.).

1776 T. Cogan *J. Buncle, Junior* I. vi. 239 But they say a creaking gates [*sic*] goes the longest upon its hinges. **1854** A. E. Baker *Gloss. Northants. Wds.* 266 'A creaking gate hangs longest on the hinges.' Used figuratively of an invalid, who outlives an apparently healthier person. **1880** S. Baring-Gould *Mehalah* II. vii. 131 Your mother..may live yet a score of years. Creaky gates last longest. **1944** A. Christie *Towards Zero* 62 But it seems I am one of these creaking gates—these perpetual invalids who never die.

cream, *sb.*[2] Add: **1.** *Devonshire, whipped cream* (earlier examples).

1673 Whipped cream [see Whipped *ppl. a.* 3]. **1825** J. C. Loudon *Encycl. Agric.* 991 *Devonshire cream*, is a term applied in the county of that name, sometimes to sour curd, and sometimes to sour cream. **1865** E. Burritt *Walk to Land's End* viii. 269 That most delectable of luxuries, Devonshire or clotted cream.

2. e. *spec.* A full-bodied mellow sherry. In full *cream sherry*. Cf. **Bristol cream*.

1964 *Wine & Spirit Trade Record* 19 May 653 (Advt.), A noble newcomer to the ranks of the famous, Diamond Jubilee Cream, the magnificent sweet cream sherry. **1965** *Guardian* 28 Nov. 2/8 A superb cream of considerable age. Bot. 21/6. **1969** *Observer* 21 Dec. 4/1 Of the complete range of Sherries on show five..each depicting one of the five main classifications—Fino—Amontillado—Oloroso—Amoroso—Cream.

f. The liquid rich in droplets or particles of the dispersed phase that forms a separate (esp. upper) layer in an emulsion or suspension when it is allowed to stand or is centrifuged; *spec.* the liquid rich in globules of rubber that forms a layer on the surface of latex in the manufacture of india-rubber.

1903 J. G. McIntosh tr. *Seeligmann's Indiarubber* 57 The hydrocarbide elements solidify on the surface into a sort of thick cream. **1914** H. Brown *Rubber* 71 The latex is diluted with water and is allowed to stand until the 'cream', consisting of the rubber globules, rises to the surface...The cream is afterwards converted into solid rubber by pressure [etc.]. **1934** H. N. Holmes *Introd. Colloid Chem.* viii. 78 Creams rise or sink according to the densities of the two liquids. If the two liquids have the same density they never cream. **1950** J. W. McBain *Colloid Sci.* ii. 21 In the cream the droplets are close together, and they may even..clump together, but it is important to note that they have not coalesced. **1965** *Trans. Inst. Rubber Ind.* XLI. 144 The latex in the creaming tank is then left undisturbed for about three weeks at the end of which separation is as complete as practical and the tank contains 24,000 gallons of cream..and 12,000 gallons of serum.

7. *cream-can, -cheque, -lorry, -stand, -truck* chiefly Austral. and N.Z.; **cream-cake** (earlier example); **cream cracker,** a crisp, unsweetened biscuit; **cream horn,** a pastry case shaped like a horn and filled with cream and jam; also *attrib.*; **cream ice,** an ice-cream; **cream-laid** *a.* (earlier and *ellipt.* examples); **cream puff,** a shell of puff pastry with a cream filling; also *fig.* and *transf.*, as (*a*) see quot. 1919; (*b*) something of small consequence; (*c*) an effeminate person; **cream soda** orig. *U.S.*, a carbonated drink of soda water; **cream tea,** afternoon tea which includes bread or scones with jam and clotted cream; **cream-ware** (examples).

1855 E. Acton *Mod. Cookery* (ed. 3) xxvi. 554 A delicious cream-cake. **1959** M. Shadbolt *New Zealanders* 16 Father mounted me..on the konaki beside the cream-cans. **1921** *Daily Colonist* (Victoria, B.C.) 30 Mar. 11/6 (Advt.), Small farm for sale...Cream cheque about $100 per month. **1959** M. Shadbolt *New Zealanders* 12 The cream-cheques were never large; money was always short. **1906** Mrs. Beeton *Bk. Househ. Managem.* v. 93 Biscuits..Cream Cracker 6*d.* per lb. **1937** 'G. Orwell' *Road to Wigan Pier* i. 15 You'll like a cream cracker with your cheese. **1962** E. O'Brien *Lonely Girl* vi. 67 We ate cold turkey and cream crackers. **1908** J. Kirkland *Mod. Baker* III. lxii. 349 Cream Horns. Roll out some puff-paste..and cut up in long strips...Wind each of these pieces of paste round a tin mould shaped like a cornucopia...Fill with..whipped cream. *Ibid.*, (*caption*) Cream horn or cornucopia tin. **1960** E. W. Hildick *Boy at Window* xiii. 99 Daintily nibbling a cream horn. **1849** Thackeray *Pendennis* (1850) I. xxvii. 263 What ice will you have—water ice or cream ice? **1854** in W. James *Order of Release* (1948) xv. 233 Gentlemen..placidly imbibing a cream ice. **1909** Ware *Passing Eng.* 97/2 *Cream Ice Jacks*, street-sellers of ½d. ices. **1857** Trollope *Three Clerks* II. xi. 239 An elegant little chamber..supplied with cream laid note paper, new pens, and the 'Times' newspaper. **1939** 'J. Struther' *Mrs. Miniver* 27 An invitation written..on lavishly stout cream-laid. **1936** M. E. C. Scott *Barbara & N.Z. Backblocks* 14 Send us bread by the cream-lorry. **1960** B. Crump *Good Keen Man* 103, I missed the cream lorry this morning. **1889** *Kansas Times & Star* 5 Oct., What they [sc. schoolboys] dote on most in pastry is cream puffs. **1902** A. Bennett *Grand Babylon Hotel* viii. 86 A cream puff was served..by way of sweets. **1919** Downing *Digger Dial.* 18 *Cream puff*, a shell-burst. **1923** T. E. Lawrence *Lett.* (1938) 399 Chocolate éclairs or cream puffs. **1938** *Time* 28 Feb. 64/3 His saccharine cinema roles and cream-puff publicity have all too closely linked the word 'beauty' with the name 'Taylor'. **1945** L. Shelly *Hepcats Jive Talk Dict.* 23 *Cream puff*, weakling. **1958** *Spectator* 1 Aug. 165/1 Hamlet can be acted by old and young, fat and thin, beef-cake and cream-puff, amateur and professional. **1970** R. Crawford *Kiss Boss Goodbye* II. viii. 112 You don't scare me... What do you think I am, a cream-puff or something? **1854** *Amer. Agriculturist* 20 Dec. 233/3 A recipe has been sold all over the country for making 'cream-soda'. **1935** L. MacNeice *Poems* 31 Drunk with steam-organs, thigh-rub and cream-soda. **1963** M. Duggan in C. K. Stead *N.Z. Short Stories* (1966) 116, I..sat by the cream stand out on the main road. **1964** P. M. Hubbard *Pict. Millie* viii. 74 We just bathe and moon about and eat cream teas. **1960** N. Hilliard *Maori Girl* I. i. 9 A wooden bridge..stands, its stringers creaking beneath every cream-truck. **1780** in D. C. Towner *Eng. Cream-Coloured Earthenware* (1957) iii. 10 (Advt.), To be sold..At the Derby Pot Manufactury, A large quantity of Earthenware..consisting of..Enamelled Cream Ware, and plain Cream tentable ware. **1865** E. Meteyard *Life of Wedgwood* I. vii. 285 By the close of the year 1761 Mr. Wedgwood has brought his cream-ware to a considerable degree of perfection. **1962** Creamware [see *Astbury].

b. Also (as in *cream-cake*) used to designate many other confections cooked with cream, or filled with whipped cream or Devonshire cream, as *cream bun*, *scone*, *slice*, etc.

1723 J. Nott *Cook's & Confect. Dict.* sig. L6, Cream Tarts..Eggs..Flour..Milk..Butter..Salt...Make your Tarts of Puff-paste...Pour in your Cream. *Ibid.* sig. L6ᵛ, Cream-Toasts, or Pain Perdu. **1747** H. Glasse *Art of Cookery* ix. 112 A Cream Pudding..Cream..Mace..Nutmeg..Eggs..Flour..Almonds..Rose-water. **1848** Thackeray *Van. Fair* I. iii. 20, I ought to have remembered the pepper which the Princess of Persia puts into the cream-tarts in the Arabian Nights. **1861** Mrs. Beeton *Bk. Househ. Managem.* 191 (*heading*) Cream sauce for fish or white dishes. **1877** *Cassell's Dict. Cookery* 160/2 Cream Biscuits..Cream Fritters. *Ibid.* 161/2 Cream Pancakes..Cream Pudding. *Ibid.* 162/1 Cream Sauce. *Ibid.* 162/2 Cream Toasts. **1888** L. Hargis *Graded Cook Bk.* 13 Amber Cream Soup...One pint cream, one pint milk, pinch of salt, pinch of cinnamon, three eggs. **1892** T. F. Garrett *Encycl. Pract. Cookery* I. 469/2 Cream Scones..will take from ten to twelve minutes to cook. **1894** Cream bun [sold in London]. **1897** *Hearth & Home* 2 Dec. 171/2 Devonshire Cream Buns. **1906** Mrs. Beeton *Bk. Househ. Managem.* 906 Cream Buns. **1915** M. Byron *Cake Bk.* 71–2 Cream Scones. **1951** *Good Housek. Home Encycl.* 360/1 A purée or cream soup garnished with lettuce. **1962** *Which?* Jan. 17/1 Cream soups are different from plain soups in that they contain a certain minimum of fat. **1963** *Times* 14 May 4/1 Milburn, whose zestful run up suggested a Bunter in quest of a cream bun.

cream, *v.* Add: **2. c.** Of an emulsion or suspension, esp. rubber latex: to form cream (*sb.*[2] sense *2 f); to separate into a layer of cream and another layer.

1903 [implied in *creaming *vbl. sb.* c]. **1914** H. Brown *Rubber* 71 Castilloa latex..creams very readily. **1926** *IRI Trans.* II. 229 Iceland moss extract will cause the latex to 'cream' in a few hours. **1934** [see *cream *sb.*[2] 2 f]. **1950** *Thorpe's Dict. Appl. Chem.* (ed. 4) X. 553/1 Latex creams very slowly and incompletely under the influence of gravity. **1960** A. W. Adamson *Physical Chem. Surfaces* ix. 389 Various O/W [sc. oil-in-water] emulsions, if allowed to stand, would separate or cream into emulsion-rich and emulsion-poor portions.

d. *trans.* To cause (latex) to cream; to bring about creaming in.

1938 C. F. Flint *Chem. & Technol. Rubber Latex* v. 204 It is possible to cream latex. **1943** H. Barron *Mod. Synthetic Rubbers* (ed. 2) x. 160 It is difficult to cream or centrifuge Buna latex because the particles are much smaller than normal rubber particles. **1965** *Trans. Inst. Rubber Industry* XLI. 136 In 1924 Traube patented the use of organic colloids for creaming natural rubber latex.

5. *to cream off* (later examples).

1905 *Daily Chron.* 13 Sept. 4/3 What is the function of the Volunteers?..To be creamed off into foreign service? **1957** *Times* 14 Feb., Ring roads must be built to cream off the heavy industrial traffic.

7. To prepare (fish, chicken, etc.) in a cream sauce.

1906, **1908**, **1935** [see *creamed *ppl. a.* 2].

8. To work (butter and sugar, yolk of eggs and sugar, etc.) into a creamy consistency.

1889 R. Wells *Pastry Cook & Confect. Guide* 30 Then start and cream the butter and sugar together. **1892** T. F. Garrett *Encycl. Pract. Cookery* I. 839/2 Cream ½ lb. of butter, add ½ lb. of caster sugar. **1906** Mrs. Beeton *Bk. Househ. Managem.* 916 Cream the butter and sugar together until thick and white. **1915** M. Byron *Cake Bk.* 132 Cream eight ounces of butter with eight ounces of sugar. *Ibid.* 139 Cream four ounces of castor sugar with six yolks. **1916** D. Canfield *Understood Betsy* (1922) vi. 109 You put the silver around, while I cream the potatoes. **1951** *Good Housek. Home Encycl.* 167/1 Simple operations such as creaming sugar and fat.

9. To treat (the skin) with a cosmetic cream.

1921 W. J. Locke *Mountebank* xi. 142 She corseted herself, creamed her face, set a *coiffeur* to work his will on her hair. **1922** Joyce *Ulysses* 748 Stripped at the washstand dabbing and creaming. **1927** *Daily Express* 14 Oct. 3/6 To have their ankles massaged, creamed, and drilled into the approved slender lines of grace and beauty. **1952** D. Ames *Murder, Maestro, Please* xvi. 111, I blacked out..without even creaming my face. **1967** 'M. Erskine' *Case with 3 Husbands* vi. 88 Ginney picked up a jar and began to cream her face briskly.

creamed, *ppl. a.* Add: **2.** *spec.* Of chicken, fish, etc.: cooked with cream, prepared in a cream sauce.

1906 Mrs. Beeton *Bk. Househ. Managem.* 1517 Creamed Lobster..Creamed Peas. **1908** *Practitioner* Sept. 369 Later on creamed chicken or fish may be given. **1935** 'R. Crompton' *William—the Detective* vii. 141 Dishes of creamed chicken for 'our little doggie brothers'.

3. Of butter: beaten to a cream. Cf. *cream *v.* 8.

1892 T. F. Garrett *Encycl. Pract. Cookery* I. 216/2 Creamed butter..is best adapted for making cakes and some kinds of pastry. **1906** Mrs. Beeton *Bk. Househ. Managem.* 1116 Creamed butter for sandwiches.

4. Formed or obtained as a result of creaming (*creaming *vbl. sb.* c).

1914 H. Brown *Rubber* 69 If the creamed globules are separated from the liquid..they coalesce to form a mass of rubber. **1953** Kirk & Othmer *Encycl. Chem. Technol.* XI. 822 Creamed latex of commerce is usually at a much higher solids concentration than centrifuged latex and is higher in viscosity.

creamer. Add: **c.** Also *cow-creamer* (see *cow *sb.*[1] 8). A cream-jug. orig. *U.S.*

1877 Mrs. E. S. Ward *Story of Avis* 237 That rose-curlew on the creamer is like a singing leaf, I think. **1893** 'O. Thanet' *Stories Western Town* 215 But she remembered the silver service, the coffee-pot,..the creamer,..the sugar-bowl. **1960** *House & Garden* Aug. 46/3 Creamer..8*s.* 2*d.* **1970** *Guardian* 17 Dec. 9/5 Coffee pot, sugar bowl, creamer.

creamery. Add: **1.** (Earlier and later examples.)

1872 *Rep. Vermont Board Agric.* I. 164 Mr. Bliss spoke of the creameries in New York. **1971** *Deerfield* (Wis.) *Independent* 28 Jan. 1 Farmers in those days brought their milk to the creamery by horse and wagon.

b. (Earlier and later U.S. examples.)

a **1877** *N.Y. Bulletin* (Bartlett), In the general features of the butter market there is no change. The fine creameries are considered well sold at 23 cts.;..Western creamery, 22 cts. **1909** *Public Ledger* (Philad.) 24 June 12/4 Butter..desirable grades of both solid-packed creamery and prints.

c. creamery butter, butter made at a creamery, as distinguished from that made at a private dairy; so **creamery print** (Print *sb.* 5).

1881 *Bradstreet's* 26 Nov. 349/2 Creamery butter sold at 30 @ 32c. **1894** *Field* 9 June 846/2 The average price realised for Irish creamery butter was 88*s.* f.o.r. at this creamery. **1935** *Economist* 7 Sept. 458/1 The Imperial Economic Committee estimates that British production of creamery butter amounted to about 63,300 cwts. in the calendar year 1930. **1909** *Public Ledger* (Philad.) 24 June 12/4 Pennsylvania and other nearby creamery prints.

creamily (krī·mili), *adv.* [f. Creamy *a.* + -ly[2].] With a creamy tint or surface; in a creamy or smooth manner; also *fig.*

1906 Frances Campbell *Dearlove* 24 Peaches like mother-o'-pearl creamily yellow. **1948** *Time* 30 Aug. 72 Linda's creamily venomous politeness at small-town parties. **1953** *Holiday* Dec. 44 Glossy limousines..glide uptown, purring creamily.

creaming, *vbl. sb.* Add: **c.** The coming together of the droplets or particles of an emulsion or suspension into a cream; the bringing about of this process, esp. in rubber latex by the addition of a creaming agent (rather than by centrifuging).

1903 J. G. McIntosh tr. *Seeligmann's Indiarubber* 56 Coagulation by creaming after doubling the volume of the latex with water. **1914** H. Brown *Rubber* 69 This behaviour, which is termed 'creaming',..is only shown by certain latices. **1937** *Ann. Rep. Progr. Rubber Technol.* i. 12 The latex to which the creaming agent is added is allowed to stand..for about 48 hours. **1946** Albright & Lees in J. Alexander *Colloid Chem.* VI. xvi. 349 Concentration of latex may be accomplished commercially by one of three important methods: namely, centrifuging, creaming, or evaporating. **1950** Kirk & Othmer *Encycl. Chem. Technol.* V. 711 One of the most used methods of comparing the stabilities of a series of emulsions is to observe the rate of creaming and/or sedimentation. **1954** Jirgensons & Straumanis *Short Textbk. Colloid Chem.* xvi. 382 Centrifugation usually produces only creaming, and the concentrated emulsion must then be treated chemically in order to induce coalescence.

crease, *sb.*[2] Add: **1.** *spec.* In trousers.

1897 G. B. Shaw *Our Theatres in Nineties* (1932) III. 247 The same sort of young man,..the same crease down his new trousers. **1939** Joyce *Finnegans Wake* 404 What a pairfact crease!

2. b. In ice hockey and lacrosse, the area marked out in front of the goal past which the players may not carry the puck or ball.
1897 E. T. Sachs in *Hockey & Lacrosse* 94 (*caption*) Goal crease. 1962 *Globe & Mail* (Toronto) 28 Dec. 33/3 Delvecchio scooped in Detroit's fourth goal with Bower trapped outside his crease from an earlier save. 1966 Evans & Anderson *Lacrosse Fundamentals* i. 22 There is usually a crease attack man assigned to play just in front of the circular crease with a 9-foot radius around the goal. 1968 J. Prouxboat *Pro Hockey* 119 The goalkeeper is granted a small protected area in front of his cage. This is called the crease. 1969 *Rules Box Lacrosse* (ed. 24) 3 The goal crease shall be in the shape of a circle having a radius of nine feet from the centre of a straight line drawn between the goal posts.
5. *Comb.*, as *crease-resistant* adj., applied to fabrics which by a process of manufacture are given greater ability to recover from creasing. Also *crease-resist*, *-resisting*, adjs.; *crease-resistance*.
1933 *Times* 16 Oct. 17/6 Made of yards and yards of crease resisting material. 1936 *Brit. Pat.* 449,043 1/1 It has been proposed to condense urea and formaldehyde to form intermediate condensation products and to apply such products to textiles for the purpose of obtaining crease resistant and recovery effects. 1947 *Brit. Pat.* 587,572 1/1 Textiles such as cloth..have..been treated with finishing compositions containing melamine-formaldehyde condensation products in order to impart crease-resistance, crush resistance [etc.]. 1957 *Textile Terms* (ed. 3) 33 Crease-resist finish. 1957 *Woman* 16 Nov. 25/4 Cottons and rayons..that have been made crease-resistant. 1960 *Guardian* 8 July 6/6 One can see..how crease resistance is obtained. 1970 *New Yorker* 1 Aug. 28/1 The spirited styling of his..crease-resistant jacket.

crease, *v.*² **3.** Also, to stun (a person); to kill; to exhaust physically; *to crease it*, to die. *slang* (orig. *U.S.*).
1909 R. A. Wason *Happy Hawkins* 46 'Where was you hit, Jabez?' I sez. 'I was creased the same as they crease a mustang.' 1913 Mulford *Coming of Cassidy* vii. 119 The drawing bar-tender..dazed from shock of a ball that 'creased' his head. *Ibid.* x. 167 They creased you! 1925 Fraser & Gibbons *Soldier & Sailor Words* 66 *To be creased*, fainted. Laid out. 1934 'D. Hume' *Too Dangerous to Live* x. 109 One of you has got to crease a watchman—just a tap to make him dumb for half an hour. a 1935 T. E. Lawrence *Mint* (1955) II. vi. 114 Cunningham threatened..to crease us tomorrow. 1936 J. Curtis *Gilt Kid* vi. 61 If he hit him hard, he'd probably crease him. ..He'd got tapped for that. 1959 J. Braine *Vodi* iv. 75 The lovable old character who can tell who's going to crease it before even the doctors do. 1962 D. Warner *Death of Bogey* i. viii. 15 Christ..you creased him. 1966 P. Shaw et al. *Lern Yerself Scouse* 60 I'm dead creased, I am very tired.

creasing, *vbl. sb.*² Add: **1. b.** (See Crease *v.*² 3.) *U.S.*
1823 E. James *Exped. Rocky Mts.* II. 328 As it was near daylight, we..[hoped] to have an opportunity to prove our skill in the operation of 'creasing'. 1833 T. Paxton *Stray Yankee in Texas* 38 The 'creasing' of a horse is a feat which..would electrify a northern jockey. 1923 J. H. Cook 50 *Yrs. on Old Frontier* 69 A third method of capture is by 'creasing'.

create, *v.* Add: **2. d.** To design (a costume); see *Creation 5 c. Also, to design and execute a scheme of interior decoration.
1930 *Daily Tel.* 7 Apr. 7/7 A brown mixture tweed suit..created by Schiaparelli. 1957 *New Yorker* 5 Oct. 33/3 The interior decorating firm that created the foyer.
5. *intr.* To make a fuss, grumble, 'go on' about (something). *slang.*
1919 *Athenæum* 8 Aug. 711/1 'Creating' is presumably an abbreviation of 'creating [or making] a fuss'. 1932 S. Gibbons *Cold Comfort Farm* xviii. 250 Creatin'! I thought 'e'd 'ave to be picked off the ceiling. *Ibid.* xix. 251 'E always creates at me something awful when I've finished telling 'im anything spiteful. 1955 'E. C. R. Lorac' *Ask a Policeman* viii. 92 The old lady came and created about her nephew living here. 1956 K. Farrell *Cost of Living* ii. 44 If my old man doesn't get 'is bit of fish at one, and that means one, he'll create. 1959 M. Hastings *Hour-Glass to Eternity* i. iii. 87 What does he do but come aboard and start creating about the loss of time!

creation. Add: **2. b.** The formation or flotation of a business company.
1898 *Westm. Gaz.* 25 Mar. 6/1 An Australian Brewery Creation. *Ibid.* 6/3 Engineering creations are becoming quite plentiful. This week will witness the flotation of the old-established business of Jesse Ellis and Co.
3. b. At Cambridge University before 16 Oct. 1926, the ceremony on Commencement Day in which the professors in the various faculties (or other officials for some degrees) recited the names of those who had been admitted doctors (doctors designate) during the past year and the senior proctor the names of those who had been admitted masters (inceptors).
1832 J. Romilly *Diary* 3 July (1967) 18 Com[mencement] Tuesday. Went at 8 to see the Creation of M.A.s. 1839 *Ibid.* 2 July 173 The greatest batch of Drs time immemorial for Creation. 1860 *Univ. Cambr., Method Proceed. Arts* 26 The Inceptors in every year become complete Masters of Arts by creation on the commencement day. 1902 *Student's Handbk. Cambridge* 341 The day appointed for the..process known as 'creation' in the case of Masters and Doctors in the various faculties. 1928 *Camb. Univ. Cal.* 102 The new Statutes have abolished the practice by which (i) Bachelors in Arts, Law, [etc.]..remained 'Bachelors designate' until the last week-day of December..(ii) Masters and Doctors only attained the full degree by 'Creation'.
4. b. Used in various phrases with *all* (see quots.), or as an exclamation; also *to beat*, *lick*, or *whip creation*, to surpass everything. *U.S. colloq.*
1825 J. Neal *Brother Jonathan* I. 257 Giving out his challenge..in a loud voice, for 'Indian hug'..or 'close hug', to all creation. 1830–33 S. Smith *Sel. Lett. Downing* (1834) 14 But when in all creation any of 'em will be finished I guess it would puzzle a Philadelphy lawyer to tell. [illegible] (Th.), He pulls like all creation, as the woman remarked when the horse ran away with her. 1843 T. C. Haliburton *Attaché* II. ix. 159 'Creation, man,' said Mr. Slick, 'I have done it..and you didn't know it.' *Ibid.* xii. 130 Creation! how sweet they smelt! he looked. 1848 'E. Kirke' *Among Pines* xiii. 237 Go to Wilmington and telegraph all creation. 1873 C. M. Yonge *Pillars of House* I. vi. 117 An English gentleman, who, Mr. Travis [an American] says, if he has money to spend, does whip creation. 1890 K. Whiting No. 5 *John St.* xiv. 148 Our luncheon isn't goin' to beat creation. 1917 H. T. Comstock *Man thou Gavest* 347 This out-of-door stuff costs like all creation.
5. b. *spec.* The first representation on the stage of a dramatic character or role; a dramatic character 'made' by a particular artist. (Cf. Create *v.* 2 c and F. *création*.)
1872 *Gentl. Mag.* Feb. 195 One of Mr. Irving's latest successes has been the 'creation' of Jingle in the version of 'Pickwick'. 1881 *Punch* 17 Dec. 288/1 The character of Maximilian..was doubtless written expressly for Robson, and, therefore, as one of that actor's 'creations' carries with it a Robsonian tradition.
c. A costume, etc., designed by an expert modiste. Also applied to other kinds of 'confection'.
1862 *Englishwoman's Domestic Mag.* May 45/2 We will now give you..the last creation, as they say here [*sc.* in France]—invention would be thought sufficiently grand in England. 1873 *Young Englishwoman* June 284/1 Gagelin has just launched out into the world of fashion his new creations for the coming season. 1878 *Cassell's Family Mag.* IV. 107/2 The Empress is another of this season's creations. 1884 *Pharmaceutical Jrnl.* 22 Nov. 405/1 A fresh, clear beautiful complexion..comforts the mind and eye far more than all the pearl-white and rose-pink creations. 1886 Kipling *Plain Tales fr. Hills* (1888) 20 It was what *The Queen* calls 'a creation'. 1898 *Daily News* 17 Sept. 6/2 A recent 'creation'—to use the curious technical word—consists of dark blue repp with heliotrope velvet let in. 1921 *13th Ann. Drapery etc. Exhib.* (Advt.), Our creations are for such as want..the correct things. 1930 'J. Dashwood' *Three Daughters* (1930) xvi, Judy, radiant in a tiara and a creation of Rosalie's.

creative, *a.* Add: **1. b.** Spec. of literature and art, thus also of a writer or artist: inventive (cf. Invention 3 b), imaginative; exhibiting imagination as well as intellect, and thus differentiated from the merely critical, 'academic', journalistic, professional, mechanical, etc., in literary or artistic production. So *creative writing*, such writing; also freq. in the U.S. as a course of study.
1816 Wordsworth *Thanksgiving Ode* 30 Creative Art..Demands the service of a mind and heart..Heroically fashioned. 1876 Geo. Eliot *Dan. Der.* II. iv. xxii. 73 A creative artist is no more a mere musician than a great statesman is a mere politician. 1900 W. B. Worsfold *Judgement in Lit.* iii. 25 Aristotle has once and for all characterized the process of creative literature, and distinguished such literature from all other branches of letters. 1903 A. Bennett *Truth about Author* iii. 29 It was eight years since I sat down as a creative artist. 1907 G. K. Chesterton in Dickens *Pickw.* p. viii, In creative art the essence of a book exists before the book..The creative writer laughs at his [illegible] *Authority of Bible* [illegible] would scorn [illegible] creation in the [illegible].
c. *creative evolution* [tr. Fr. *L'Évolution créatrice* (H. Bergson 1907)]: in Bergson's philosophy, the process of evolution of new forms regarded as not taking place according to natural laws but giving rise to genuine novelty; hence *creative evolutionist*. Also *creative synthesis*.
1909 *Proc. Arist. Soc.* IX. 41 The special title of the book 'Creative Evolution'. 1918 G. B. Shaw *Doctors' Delusions* (1932) 312 The Creative Evolutionists, with Butler and Bergson for their prophets. 1921 —— *Back to Methus.* p. lxxviii, Creative Evolution is already a religion, and is indeed now unmistakably the religion of the twentieth century. 1934 —— *Prefaces* 364/2 Myself..who believe in the religion of Creative Evolution. 1941 *Mind* L. 397 The concept of creative synthesis, according to which by the interaction of parts in a whole, there emerge 'new properties' qualifying the whole, which were not present in the constituent parts.
d. Extended uses (chiefly of *b).
1930 *Monotype Recorder* XXIX. 39 The fear that prices will drop still lower when trade in general seems to be poor can no longer have power over the creative salesman, because he realizes that salesmanship definitely stimulates and helps to *stabilize* trade. *Ibid.*, The remainder..fail to see the vital necessity for advertising and establishing a creative sales policy. 1933 Dylan Thomas *Let.* Oct. (1966) 45 Aren't *any* of the Creative Lifers men of action? 1936 *Word Study* Sept. 1/2 At a recent 'panel discussion' in one of our great universities several speakers aired their views on 'creative education'. 1937 'G. Orwell' *Road to Wigan Pier* xii. 230 In my spare time I want to do something 'creative', so I choose to do a bit of carpentering. 1941 *Time* (Air Exp. Ed.) 14 July 2/1 The creative pauses of Adolf Hitler. 1942 *Sphere* 27 June 409/1 The sub-editors here [at the B.B.C.] need to be more creative than in Fleet Street. 1958 *Spectator* 14 Feb. 197/1 'Creative' commercial jobs, such as advertising, designing, modelling, public relations, TV production, or on a 'glossy' news-magazine. 1969 *Times* 13 Dec. 9/7 Middleclass mothers who leave a child alone with a roomful of creative toys all day may produce 'C' stream children as often as the working mother.

creatively, *adv.* (Later examples of extended usage.)
1906 *Westm. Gaz.* 27 June 2/2 The use of a name which the world has agreed to consider creatively typical... Goldsmith's 'Tony Lumpkin' is immortal as perfectly expressing in name the character. 1930 *Proc. Brit. Acad.* XVI. 309 If any one ever read (in the current phrase) 'creatively', it was he [*sc.* Chaucer].

creativity (kriāti·viti). [f. Creative + -ity.] Creative power or faculty; ability to create.
1875 A. W. Ward *Eng. Dram. Lit.* I. 506 The spontaneous flow of his [*sc.* Shakespeare's] poetic creativity. 1926 A. N. Whitehead *Relig. in Making* iii. 90 The creativity whereby the actual world has its character of temporal passage to novelty. *Ibid.* 152 Unlimited possibility and abstract creativity can procure nothing. 1959 *Radio Times* 23 Jan. 3/1 He [*sc.* Burns] was a man of overflowing creativity—in so far as the phrase applies to his poetry.

creator. Add: **2. b.** One who 'creates' a dramatic character or role. (Cf. *Creation 5 b.)
1872 *Gentl. Mag.* June 715 Her representation had always this want of dramatic propriety for those who had witnessed the original creator of the part.
c. One who 'creates' or designs a costume, etc. (see *Create *v.* 2 d, *Creation 5 c).
1901 *Westm. Gaz.* 23 May 3/1 Everything seemed so absolutely simple, and yet so absolutely impossible to any other creator.

creature. See also *critter. **5.** *creature of circumstance*: see *circumstance *sb.* 4 a.
6. d. Also *sing.*
1885 *Encycl. Brit.* XVIII. 681/1 Pietro..resolutely denied himself every creature-comfort. 1959 *Camb. Rev.* 24 Oct. 58/1 With ingenuity and goodwill the creature-comfort problems of increasing the numbers of our societies quite considerably could be solved.

crèche. Add: Also creche, *erron.* crêche.
1. (Earlier and later examples.) In later use, a day nursery for babies and young children. Also *transf.*
1854 Mrs. Gaskell *Let.* 27 Oct. (1966) 320 She [*sc.* F. Nightingale] said..there should be crêches for the rich as well as the poor. 1860 C. M. Yonge *Hopes & Fears* I. ii. x. 388 A range of buildings round a court, consisting of day-schools, a home for orphans, a crèche for infants. 1875 [see Foundling 3]. 1909 *Westm. Gaz.* 27 Aug. 3/1 Why on earth should Ireland be treated simply as a crèche for nurturing a superfluity of priests and nuns for the Anglo-Saxon world? 1949 *Here & Now* (N.Z.) Oct. 17/1 Even.. creches, will not lift the load of fatigue from the mother with a sick child to nurse and a teething baby. 1953 N. Tinbergen *Herring Gull's World* xxiv. 229 The 'crèche' system as found in some penguins. 1967 *Times* 21 Sept. (Suppl.) p. vii/4 Children's facilities, all of which will be supervised, will include a creche.
2. A representation of the infant Jesus in the manger, with attending figures, often displayed at Christmas; = Crib *sb.* 1 b.
1792 *Wynne Diaries* 6 Jan. (1935) I. 97 We..saw at church the Crêche and the three Magi... These images were quite prettily made and amused the children. 1903 *Westm. Gaz.* 29 Dec. 2/1 The shepherds guarding their flocks, the star, and the stable of the inn at Bethlehem are all represented by toy scenery called a 'crèche'. 1963 G. K. Wilkinson *Guinea-Pigs* xii. 189 Christmas will soon be on us and I hope that you will arrange the Holy Crèche..in the church.

credal: see *creedal *a.*

credentialled (krĭde·nʃăld), *a.* Furnished with credentials.
1891 *Sat. Rev.* 10 Jan. 47 A person himself qualified and credentialled by almost every possible kind of passage through the Academic mill. 1928 *Sunday Dispatch* 5 Aug. 8 The credentialled spiritualist in a semi-scientific Bloomsbury 'college'.

credenza (krede·ntsă). [It.] A sideboard. (Cf. Credence *sb.* 7.)
1880 Shorthouse *John Inglesant* xxviii, Below and

beside these were credenzas and cabinets upon which luxury and art had lavished every costly device and material. 1951 *N.Y. Times* 14 Jan. 66 A small credenza, or breakfront storage cabinet. 1961 M. Levy *Studio Dict. Art Terms*, *Credenza painting*, a decorative painting on a buffet or sideboard.

credibility. Add: **b.** *spec.* In contexts of a defence policy based on the theory of the effectiveness of a nuclear deterrent.
1960 *Times* 11 Feb. 11/6 There would be no gain in the credibility of the deterrent if N.A.T.O. became an atomic Power. 1962 *Listener* 2 Aug. 161/2 Worry about the credibility of massive retaliation so long as America alone has the power to start it has been racking Nato for several years.
c. **credibility gap** (orig. *U.S.*), a disinclination to believe a person, a statement, etc., used esp. of the non-acceptance at face-value of official statements; a disparity between facts and what is said or written about them.
1966 *Guardian* 28 Dec. 6/1 Official American statements are no longer taken on trust... The phenomenon..is called the 'credibility gap'. 1967 M. McCarthy in *Observer* 30 Apr. 12 We do not recognise them, helmeted, in a bomber aiming cans of napalm at a thatched village. We have a credibility gap. 1968 *Globe & Mail* (Toronto) 28 June 23/8 Don't look now, but I'm afraid my credibility gap was showing a bit in a column I wrote the other day.

credible, *a.* Add: **1. c.** (See quot. 1963.) Cf. *credibility b.
1960 *Times* 11 Feb. 11/6 As a guarantee of European nuclear retaliation against a nuclear attack a N.A.T.O. deterrent would be highly credible. 1963 *Daily Tel.* 12 Jan. 13/8 'Credible', in the language of nuclear strategy, does not mean 'adequately frightful'. It means 'such as an enemy will think likely to be used'. 1966 Schwarz & Hadik *Strategic Terminology* 42 Credible first strike capability. *Ibid.*, The deterrent effect must also be credible to the allies who are to be protected by the threat.

credit, *sb.* Add: **13. d.** An entry in the record of a pupil or student certifying that he or she has qualified in some course of study. *U.S.*
1904 E. G. Dexter *Hist. Educ. U.S.* 288 On the basis of 'credits', one credit representing a subject pursued daily.. for one year in the secondary schools, forty-five credits is the usual requirement. 1963 R. I. McDavid *Mencken's Amer. Lang.* vi. 292 An English university student..knows nothing of frats, class days, rushes, credits, points, majors, semesters.
e. In full *pass-with-credit*. A kind of 'distinction' awarded in some examinations to examinees obtaining more than a certain percentage of the maximum marks in a subject.
1917 *Regul. Oxf. & Camb. Schools Exam. Board* 49 Candidates who attain a sufficiently high standard in one or more of the subjects comprised in the several Divisions will be awarded a pass-with-credit in each of those subjects. 1963 Barnard & Lauwerys *Handbk. Brit. Educ. Terms* 72 *Credit*, a mark above the average standard in the now obsolete School Certificate Examination.
f. *Theatr.*, *Cinemat.*, *Broadcasting*, etc. The acknowledgement by name, with details of the service rendered, of each individual contributor (actor, producer, etc.) to a production or the like, usually published in a programme or on the screen; freq. in *pl.* Hence *credit line*, *list*, etc.; *credit title* (see quot. 1933). Also *transf.*
1914 *Writer's Mag.* Feb. 44/2 The Photoplaywright should have equal prominence in credit line with the staff writer of the new story. 1922 *Moving Picture Stories* 7 July 22/2 The introductory titles give her screen credits for both the original story and the continuity. 1926 *Atlantic Monthly* June 765/1, I wrote a yarn that a clock..had 'stopped at the fatal hour'. This was reprinted, without credit and without investigation, all over that part of the country. 1933 C. Winchester *World Film Encycl.* 481/1 *Credit titles*, the names on the screen of the technicians responsible for the scenario, sound, art direction etc. 1934 H. N. Rose *Thes. Slang* vii. 50/2 Line below a picture giving the name of the photographer, *the credit line*. 1937 H. G. Wells *Brynhild* v. 59 A fellow who had something to do with building the sets, committed suicide, simply..because he was mortified by not being given a credit—credits they call 'em—on the screen. 1937 *Amer. Speech* XII. 100 In the early days [of wireless] there were advertisements, later commercial announcements, plain commercials or plugs, now often simply credits. 1941 A. W. Read *Let.* 5 June in *Amer. Speech* (1964) XXXIX. 255 The suspicion that members of the staff may think that I contributed the material in order to wiggle my way into a credit line in the DAE. 1952 W. Granville *Dict. Theatr. Terms* 53 *Credit list*, the list of acknowledgements of indebtedness to firms who have lent articles of furniture, electrical gear, draperies, etc., to dress the stage. *Credits*, the credit list. 1955 'G. Carr' *Corpse at Camp Two* ii. 23 On D.-G.P.'s 'credit' titles he appeared as 'Dem Nixon', which..was his professional name as a film cameraman. 1958 *Punch* 1 Jan. 80/1, I am old enough to remember a time when a television play began with a title, a stream of credits and a splurge of mood music and then proceeded to discharge its burden of drama. 1958 *New Statesman* 12 July 47/1 This superb piece of radio..is 'by' Ewan MacColl and Charles Parker; but one would like to give a long credit-list of all involved in it. 1962 *Guardian* 15 Feb. 6/4 The talks producers..do not get credit lines [in the *Radio Times*] for their work.
14. *credit balance* (*fig.* example), *side* (later *fig.* examples); **credit card** orig. *U.S.*, a card issued by an organization authorizing a named person to draw on its account or to make purchases on credit; formerly also a traveller's cheque; **credit-draper**, a 'travelling' draper who delivers goods on credit; **credit insurance**, insurance against bad debts; **credit letter**, a letter of credit (see sense 10 b); **credit man** *U.S.*, a clerk who has charge of the credit accounts in a business; **credit rating**, the assessment (e.g. by a hire-purchase company) of the ability and intention of a customer to meet his liabilities; **credit sale**, a method of purchase by delayed payment, the total cost of the goods usually being paid off within a stipulated time; **credit squeeze**, the restriction of financial credit facilities through banks etc.; also as *v.*; **credit transfer**, a method of payment by which a debtor's bank, etc., on his instructions, will transfer the amount owed to the account of his creditor, whether in the same or another bank; **credit-worthy** *a.*, delete † *Obs.*; now, qualified to receive commercial credit; hence **credit-worthiness**, the extent to which a person, firm, etc., is allowed such credit.
1942 C. S. Lewis *Screwtape Lett.* ii. 18 He still believes he has run up a very favourable credit-balance in the Enemy's ledger by allowing himself to be converted. 1888 E. Bellamy *Looking Backward* xiii. 200 An American credit card..is just as good in Europe as American gold used to be. [illegible]

creditable, *a.* Add: **3.** Capable of being ascribed to.
1904 *Rep. Librarian Congress* 32 Many documents creditable to that period can be judged to be so and assigned to their proper group only by internal evidence.

Créditiste (krediti·st). *Canad.* Also Crédi-tiste. [Fr.] An adherent of the Quebec wing of the Social Credit party in Canada, or one of their members of parliament. Also *attrib.*
1962 *Kingston* (Ontario) *Whig-Standard* 23 Oct. 13/1 Réal Caouette and his 25 Créditistes are the real reason. 1963 *Ibid.* 6 Mar. 1/7 Créditiste candidates captured almost half of their total of 45,000-odd votes in the entire area here on June 18. 1965 *New Statesman* 8 Oct. 506/3 The Quebec *créditistes*..seceded from the party. 1967 *Canad. Ann. Rev.* 1966 20 Five Créditistes and 2 Independents joined 13 New Democrats in voting against the government. 1971 *Daily Colonist* (Victoria, B.C.) 6 May 1/3 Réal Caouette announced proudly in the Commons Wednesday that the Ralliement Créditiste has readopted the name Social Credit in English, Crédit Social in French.

creditor. Add: **5.** *attrib.*, as *creditor-account*, *-nation*.
1899 *Westm. Gaz.* 8 Sept. 2/3 Creditor-account balances. 1897 *Daily News* 30 Nov. 4/6 This country, the great creditor nation of the world. 1919 J. L. Garvin *Econ. Foundations of Peace* 145 America's new position as a great creditor-nation. 1948 G. Crowther *Outl. Money* (ed. 2) ix. 328 The view that the so-called 'creditor' nations may be just as responsible as the 'debtors' for the disequilibrium in the exchanges.

Cree (krī), *sb.* and *a.* Also †**Cris, Kris.** [ad. Canad.-Fr. *Cris*, short for earlier *Cristinaux*, *Christinaux*, f. 17th-cent. Algonquin **kiristino* (cf. Fr. *Kiristinon*, etc.), mod. Algonquin *kinistino*, showing the regular replacement of earlier *r* by *n*.] **A.** *sb.* **1.** An Indian people of central N. America; a member of this people. **2.** The Algonquian language of the Cree people. **B.** *adj.* Of or pertaining to this people or their language.
1744 A. Dobbs *Acct. Countries adjoining Hudson's Bay* (caption, facing p. ii), Cris or Christinaux N[ation]. 1760 in T. Jefferys *Nat. & Civil Hist. Fr. Dominions* I. 1 (caption) Kris of the Lakes. *Ibid.*, Kris named also Christinaux and Killistins. *Ibid.* p. xv, The same with Killistinoes, Killistinos, Crigs, Cris, Crees, &c. *Ibid.* xv. 323 Though we conversed with these people in the Cree, or Cristinaux language, which is the usual medium of communication, they were Chepewyans or Rocky Mountain Indians. 1823 in R. H. Fleming *Minutes of Council of N. Dept. Rupert Land* (1940) 41. 1845 J. Bird *Let.* 8 Aug. in J. Howse *Gram. Cree Lang.* (1844) p. xv, The Rev. Mr. Smithurst has, by well studying your Grammar, been enabled to read the Communion Service to the Swampies, in Cree, a few days ago. *Ibid.* p. xvi, The knowledge of the Cree language. 1844 *Ibid.* p. viii, The grammatical system of the Crees (and Chippeways) is composed of the same philological elements as are found in the Grammars of cultivated languages. *Ibid.* 3 An analytical and grammatical view of the Cree dialect. 1932 Bloomfield *Language* iv. 72 The Algonquian family includes the languages of eastern and central Canada (Micmac, Montagnais, Cree). 1964 P. Worsley in I. L. Horowitz *New Sociol.* xxii. 379 Northern Saskatchewan..is inhabited by a sparse population of Cree and Chipewyan Indians. 1968 *Globe & Mail* (Toronto) 13 Feb. 3/ The children, Cree pupils in Grade 3, ..wrote stories that have been collected in book form by the Indian Affairs Branch.

Creed, *sb.*² *Telegraphy.* The designation of an automatic tape-printing machine invented by F. G. Creed (see quot. 1911). Usually *attrib.*, as *Creed printer*, *system*, *translator*.
1911 *Encycl. Brit.* XXVI. 521/1 The Creed system is a development of the Morse-Wheatstone system, and provides a keyboard perforator which punches Morse letters or figures on a paper strip by depressing typewriter keys. The slips are passed through an ordinary Wheatstone transmitter and actuate Wheatstone receiving apparatus which in turn controls a 'Creed receiving perforator'. This machine reproduces a copy of the original transmitting slip, which can be passed on to any other Wheatstone circuit or can be run through a 'Creed printer'. 1916 T. E. Herbert *Telegr.* (ed. 3) 363 The Creed receiving perforator. 1923 H. H. Harrison *Print. Telegr. Systems* 208 The Creed recorder. 1928 A. E. Stone *Text Bk. Telegr.* 184 The Creed Receiving Perforator. *Ibid.* 186 The Creed Printer. 1959 *Chambers's Encycl.* XIII. 550/1 The Creed reperforator produces a perforated tape similar to the original. This can either be used for a retransmission or fed into a Creed printer to produce a copy on tape in roman print.

creedal, credal, *a.* (Earlier and later examples.) In modern usage the more usual form is *creedal*.
1868 *Church Union* 11 Jan. (Cent. Dict.), Four columns of creedal and two editorials. 1893 *Westm. Gaz.* 9 June 3/3 A man altogether above racial or creedal prejudice. 1957 *Economist* 19 Oct. 191 The Commonwealth is peculiarly flavoured by the mid-twentieth century in having as its symbol of unity a person, and not only a flag or some creedal form.

creek, *sb.*¹ Add: **2. b.** (Earlier U.S. and Austral. examples; also, N.Z. examples.)
1622 in H. R. Shurtleff *Log Cabin Myth* (1939) 155 Creeks and Swamps as they call them..offer all aduantages to their ..enimys. 1638 in *Massachusetts Hist. Soc. Coll.* 3rd Ser. VI. 42 They..are dispersed securely in their plantations sixty miles along the coast, and within the land also, along some small creeks and rivers. 1793 J. Hunter *Hist. Jrnl.*

Port Jackson xxi. 516 In the afternoon a creek obliged them to leave the banks of the river, and go round its head, as it was too deep to cross. 1820 S. Marsden *Let.* 8 May (1932) iv. 249 There are two warm springs opposite each other, one on each bank of the creek, about ten feet above the level of the fresh water which runs between them. 1840 *N.Z. Jrnl.* I. 292/1 A fine creek of water..runs down from the mountains..and wastes itself in the salt water. 1852 G. B. Earp *Gold Colonies Australia* viii. 131 A 'creek', in New South Wales jargon, means a water-hole in the interior, and not an arm of the sea, as we understand it. 1857 R. Paul *Lett. Canterbury* iv. 65 In the Australian Colonies, as in America, brooks are called creeks. 1963 B. Pearson *Coal Flat* iv. 59 In a little clearing in the.. young bush by the bank of the creek was Mrs Seldom's little house.
5. Phr. *up the creek*: (*a*) in a tight corner, in trouble; *spec.* pregnant; (*b*) crazy, eccentric. *slang.*
1941 A. Miller in Kozlenko *100 Radio Plays* 22/2, I mean that if I'm killed you're up the creek. 1943 Hunt & Pringle *Service Slang* 68 *Up the creek*, lost, either on patrol or during a night out; off one's course. 1945 *Penguin New Writing* XXIV. 32 Lord, it's my mortar lance corporal. If he breaks we'll be right up the creek. 1960 H. Pinter *Caretaker* III. 71 You don't know what you're doing... You're up the creek! You're half off! 1961 H. E. Bates *Day of Tortoise* 52 I'm in trouble. I'm going to have a baby... I've had it. Good and proper. I'm up the creek. 1961 J. Heller *Catch-22* (1962) viii. 78 You really are up the creek, Popinjay. 1963 E. Lambert *Drip Dry Man* x. 51, I know a girl who thinks her bloke may have put her up the creek. 1969 I. Kemp *Brit. G.I. in Vietnam* vi. 134 'You okay?' asked Donovan... 'I thought you were properly up the creek.'
8. *creek-bed* (earlier and later examples), *-timber* (U.S.); **creek-bottom** *U.S.*, level ground beside a stream or brook; **creek-gum**, the Australian river red gum, *Eucalyptus camaldulensis*.
1852 *Southern Lit. Messenger* XVIII. 314/1 Ahead of us.. is a creek-bed, fringed far back into the hills with tree and shrub. 1936 I. L. Idriess *Cattle King* iv. 30 Often you can dig in a dry creek-bed and obtain soakage water. 1965 M. Shadbolt *Among Cinders* xx. 194 We walked..following a creek-bed. 1822 J. Woods 2 *Yrs. Eng. Prairie* 224 On the creek bottoms, coffee-berry, poplar, pecon, white walnut. 1857 F. L. Olmsted *Journ. Texas* 81 The soil of the creek-bottoms bears good cotton. 1898 Morris *Austral Eng.* 178 Creek Gum. 1930 V. Palmer *Men are Human* x. 89 The sun, visible through a gap in the creek-gums, hung fixed in the sky. 1790 J. Backus *Diary* in W. W. Backus *Mem. Backus Fam.* (1889) 93 Explored the bottom to the Creek timber land. 1836 D. B. Edward *Hist. Texas* 36 We find.. [the] prairie..relieved by creek timbers and solitary groves.

Creek, *sb.*³ [f. Creek *sb.*¹] **A.** *sb.* **1.** An Indian tribe, also called Muskogee, formerly inhabiting a wide region in south-eastern North America, now settled in Oklahoma; a member of this people. **2.** The language of this people, belonging to the Muskogean stock. **B.** *adj.* Of or pertaining to this people or their language. **C.** *sb.* A confederacy (later Nation) of several tribes and languages of which the Muskogee or Creek proper were the most numerous and prominent. **D.** *adj.* Of or pertaining to this confederacy.
1725 G. Chicken *Jrnl.* in N. D. Mereness *Trav. Amer. Col.* (1916), I hope your Honour will be speedy to me in your Expresses, Especially in your answer to me when you have heard from the Creeks. *Ibid.*, Do you understand the Creek Language? 1732 in *Cal. State Papers Amer. & W. Indies* 217 They [*sc.* the Spanish] have a party among the Creek Indians as well as we. *Ibid.*, Two of our Indian traders having been killed near the Creek nation in their way thither. 1736 [see Indian *sb.* 2 a]. 1775 Adair *Amer. Ind.* 2 A large body of the English Indian traders, on their way to the Choktah country, were escorted by a body of Creek and Choktah warriors. 1789 W. Bartram *Creek & Cherokee Indians* in *Trans. Amer. Ethnol. Soc.* (1853) III. i. 11 The Cricks, or, as they call themselves, Muscoges, or Muscogulges, are a very powerful confederacy. 1791 —— (*title*) Travels through North and South Carolina,..the extensive territories of the Muscogulges or Creek confederacy, and the country of the Chactaws. 1837 *Penny Cycl.* VIII. 146/1 Creek Indians were, at the beginning of the present century, one of the most powerful native tribes within the limits of the United States of America. 1868 S. W. & L. C. Perryman (*title*) Constitution and laws of the Muskokee or Creek nation, translated into Muskokee language. 1877 L. H. Morgan *Anc. Society* iii. 440 In Crow my husband's brother's wife is 'my comrade'.., in Creek my 'present occupant'. 1921 *Hist. Amer. Lit.* I. ii. i. 209 Gaily dressed Creeks, quite oriental in appearance. 1933 Bloomfield *Language* iv. 72 The Muskogean family includes, among other languages, Choctaw, Chickasaw, Creek, and Seminole. 1946 G. Foreman *Last Trek of Indians* 163 They were recorded as part of the Creek Nation, their chiefs participating in the Creek general councils. 1965 *Canad. Jrnl. Linguistics* Spring 145 The Yuchi..were conquered by the Creeks... All the Yuchi are bilingual, and there couldn't be two languages further apart than Creek and Yuchi.

creeklet. (Later examples.)
1884 T. Bracken *Lays of Land of Maori & Moa* 91 In a forest, far away, One small creeklet, day by day, Murmurs only this sad lay. 1943 K. Tennant *Ride on Stranger* ix. 90 A rum-coloured creeklet. 1969 E. W. Morse *Fur Trade Canoe Routes* ii. iv. 47 The Indians in earlier days may possibly have been assisted by more beaver dams on the creeklets.

creep, *v.* Add: **2. c.** *trans.* To introduce gradually; slowly to increase (an amount of light, volume of music, etc.). Const. *in.* Cf. *fade *v.*¹ 9.
1949 T. Rattigan *Harlequinade* in *Playbill* 57 They've crept in numbers two and three [*sc.* spotlights] too early. 1960 N. Kneale *Quatermass & Pit* iii. 97 Creep in music.
4. e. Of a liquid: to spread over or cover a surface as a thin film; *esp.* (of a salt solution) to rise on the sides of the containing vessel, depositing crystals of the salt; (of the dissolved salt) to be deposited in this way.
1888 J. A. Fleming *Short Lect. Electr. Artisans* (ed. 2) 208 Difficulty sometimes occurs from the gradual 'creeping' up of the salts around the stopper. 1900 *Nature* 4 Oct. 562/1 Dr. Trouton gave a short account of his experiments on the creeping of liquids, and on the surface tensions of mixtures. He has found that the tendency of certain liquids to creep up the sides of their containing vessels is due to such liquids being mixtures. 1902 W. R. Cooper *Prim. Batteries* 195 When evaporation of a salt takes place in a glass vessel, crystals form on the vessel near the surface of the solution; and..the crystals grow upwards and finally grow over the top of the vessel. This 'creeping', as it is termed, is avoided in Leclanché cells by dipping the tops of the glass pots into ozokerite or paraffin wax. 1909 *Cent. Dict. Suppl.* s.v. *Oil-thrower*, Oil creeping along the shaft from the journal is thrown off. 1957 *Encycl. Brit.* XIV. 186/2 If an open vessel containing helium II is suspended inside the vacuum flask,..the liquid creeps over the edge, and drops off the bottom of the vessel. 1958 J. J. Bikerman *Surface Chem.* (ed. 2) i. 89 The well-known phenomenon of creeping of solutions or crystal climbing probably involves the relation between surface tension and concentration.
5. b. *trans.* and *intr.* To rob (stealthily); to use stealth. *Criminals' slang* (orig. *U.S.*).
1914 in Jackson & Hellyer *Vocab. Criminal Slang.* 1928 M. C. Sharpe *Chicago May* (1929) xxxi. 259 Panelling, or creeping, only carries eighteen months, because the John went to the room of his own accord. 1955 P. Wildeblood *Against Law* iii. 119 You could creep that drum six-handed, with jelly and all, and she'd think it was mice. 1958 [see *creep *sb.* 1 e].
8. (Earlier example.) Also, to increase very gradually in length under excessive stress.
1872 W. S. Huntington *Road-Master's Assistant* (ed. 2) 29 The rails in creeping have a tendency to move towards the foot of the grade. 1899 J. A. Ewing *Strength Mater.* 24 When a load exceeding the elastic limit is applied the strain which occurs at once is followed by a continued 'creeping' or supplementary deformation. 1911 —— in *Encycl. Brit.* XXV. 1014/1 The elastic limit is the point..at which a tendency to creep is first seen. 1924 F. C. Lea in *Proc. Inst. Mech. Engin.* II. 1053 The problem is to find the safe stress at which the material will not change form or creep. *Ibid.* 1072 At 11.40 a.m. on the 11th the specimen had crept 0·2 millimetre, but at 11.40 a.m. on the 14th the creeping had ceased. 1955 *Oxf. Jun. Encycl.* VIII. 286/2 Under this smaller but continuous burden the metal gradually deforms or 'creeps', and eventually breaks.
10. Of soil, talus, etc.: to undergo creep (sense *7 a); to move imperceptibly *en masse*.
1889 *Geol. Mag.* VI. 257 The whole outer layer of the soilcap will..creep slightly downwards. *Ibid.* 260 The creeping of the soilcap through the action of frost. 1918 *Econ. Geol.* XIII. 609 In many places hillside surficial material seems to be creeping up instead of down—perhaps due to swelling by weathering of the surface portions of certain underlying beds. 1965 Hatch & Rastall *Petrol. Sedim. Rocks* (ed. 4) II. v. 77 If a rock fragment on a slope is moved by its own expansion or that of the pore-water it may be exposed to the pull of gravity. It will then creep slightly downhill.
11. Of a rubber tyre. (Cf. *creep *sb.* 9.)
1903 *Motoring Ann.* 300 It is claimed for the Collier tyre that it cannot possibly creep. 1908 *Westm. Gaz.* 25 Feb. 4/2, I understand that they have a great tendency to creep.
12. Of a belt or rope: to slip or slide backwards on the pulley. (Cf. *creep *sb.* 8.)
1922 F. V. Hetzel *Belt Conveyors* 124 A poorly made belt..will creep more and cause more wear than a good belt.

creep, *sb.* Add: **1. c.** †(*a*) A creeping fellow; a sneak. *dial. Obs.* (*b*) *slang* (orig. *U.S.*). A despicable, worthless, stupid, or tiresome person. Cf. Creeper 1 b.
a 1876 E. Leigh *Gloss. Cheshire* (1877) 52 *A Creep*.., a creeping fellow. 1886 Brierley *Cast upon World* xviii. 218 His whole get-up so suggestive of what in those days was called a 'creep', that I could not help regarding him with additional loathing. 1935 *Jrnl. Abnormal Psychol.* XXX. 362 *Creep*, a worthless person. 1938 *New Republic* 7 Sept. 129/1 The man..is nothing but a creep. 1951 [see *charge *sb.* 3 d]. 1954 Wodehouse *Jeeves & Feudal Spirit* i. 7 They were..creeps of the first water and would bore the pants off me. 1958 *Spectator* 9 May 588/3 A pathetic fat city creep comes making eyes at the daughter. 1960 H. Pinter *Room* 117, I get these creeps come in, smelling up my room. 1966 *Punch* 16 Feb. 241 'Maurice Thew School of Body-building'? That'll be that phoney creep upstairs.
d. A stealthy robber; a sneak thief; esp. one who works in a brothel. *Criminals' slang* (orig. *U.S.*). Cf. *creep joint* (*6).
1914 in Jackson & Hellyer *Vocab. Criminal Slang.* 1928 M. C. Sharpe *Chicago May* (1929) xxxi. 255, I have been a badger, pay-off, note-layer, creep, panel, and blackmailer. 1960 *Observer* 25 Dec. 7/6 A creep is a highly expert thief... He is so quiet that he can move about a house for hours without waking anybody.
e. Stealthy robbery; petty thieving; esp. in a brothel. So *at* or *on the creep*: engaged in stealthy robbery. Also *attrib. Criminals' slang* (orig. *U.S.*).
1928 M. C. Sharpe *Chicago May* (1929) xxxi. 257 She may decide to shift to the creep or panel game. 1931 C. Rimington *Bon Voyage Bk.* 89/2 *At the creep*, picking lady's skirt pocket while walking. 1938 F. D. Sharpe *Flying Squad* i. 15 Billy's at 'the Creep' means that Billy earns his living stealing by stealth from tills whilst a shop is momentarily unwatched, or from a warehouse. 1958 F. Norman *Bang to Rights* iii. 121 A geezer who got captured while he was out on the creep, he used to go out and creep about all over the place trying to find things to knock off. One of his favourite stokes was creep offices [*sic*] in the city ..during the lunch hour.
2. (Earlier example.)
1849 Dickens *Dav. Copp.* iii. 29 She was constantly complaining of the cold, and of its occasioning a visitation in her back which she called 'the creeps'.
4. (Earlier and later examples.) Also, an enclosure in which young animals may feed, with an entrance too small to admit the mother. So *creep-feed*, *-feeding*; *creep-feed* v.
1819 *Sporting Mag.* IV. 209, I have heard that they [*sc.* pheasants] of late have been snared in creeps, like hares. 1886 C. Scott *Sheep-Farming* 92 To fatten lambs rapidly, the utmost care must be given to their feeding. When folded on green crops they should feed in advance of the ewes, by having 'creeps' provided. 1924 *Chambers's Jrnl.* Jan. 49/2 The rabbits have their chosen creeps through the greenery below. 1950 *N.Z. Jrnl. Agric.* Jan. 63/1 Creep feeding. From the earliest age piglets will take extra feed. *Ibid.* Apr. 371/3 A creep in which the suckers can feed away from the sow is a necessity. 1960 *Farmer & Stockbreeder* 16 Feb. 50 The piglets..will already be creep-feeding. *Ibid.* 22 Mar. 136/2 A suitable creep-feed mixture would be 2 parts cracked beans, 1 part crushed wheat, 3 parts crushed oats, 2 parts ground barley. 1964 R. Jeffries *Embarrassing Death* ii. 18 He could see sheep creep-feeding in one field.
6. *attrib.* and *Comb.* (sense 10) *creep-rate*, *-resistance*, *-resistant* adj., *-strain*, *-stress*, *-test*, *-testing*; **creep curve**, a curve showing the rate, circumstances, etc. of creep (sense 10); **creep joint** *U.S. slang*, (*a*) a brothel or unwholesome apartment-house, esp. one where patrons are robbed; (*b*) a gambling-game operating in a different location each night; **creep limit, strength**, the maximum stress to which material, esp. metal, can be subjected without 'creep' (see sense 10).
1931 H. J. Tapsell *Creep of Metals* iii. 31 The form of the creep curve for materials at normal service temperature. 1928 M. C. Sharpe *Chicago May* (1929) ii, New York was full of creep joints at that time [c 1896]. 1930 *Amer. Mercury* Dec. 455/1 *Creep-joint*, a gambling house that moves to a different apartment each night. 1946 Mezzrow & Wolfe *Really Blues* i. 3 Earned my Ph.D. in more creep joints and speakeasies and dancehalls than the law allows. 1950 A. Lomax *Mr. Jelly Roll* (1952) 50 Creep joints where they'd put the feelers on a guy's clothes. 1940 *Chambers's Techn. Dict.* 208/1 Creep limit. 1931 H. J. Tapsell *Creep of Metals* iii. 31 Ultimate failure of a material would result however small the initial creep rate may be. 1937 *Jrnl. R. Aeronaut. Soc.* XLI. 378 Cast alloys have better creep resistance than forged. 1947 *Mech. Engin.* LXIX. 273 The scarcity of creep-resistant materials for blades and rotors working at high temperature. 1960 *Times* 16 Mar. (Canberra Suppl.) p. vii/1 The use of high-strength alloy creep-resistant steels for the tubes and supports. 1948 *Jrnl. R. Aeronaut. Soc.* LII. 2/2 Strain deviations..are regarded as 'creep strains'. 1929 Tapsell & Remfry (*title*) The 'creep' strength of a 'high nickel-high chromium steel', between 600° and 800° C. 1957 *Times Surv. Brit. Aviation* Sept. 27/4 American writers have claimed an increase in creep strength of 20 per cent. by vacuum-melting. 1934 *Jrnl. R. Aeronaut. Soc.* XXXVIII. 205 When selecting a light alloy material for aero engine design..it is important to consider the creep stress of the material concerned. *Ibid.* 408 Magnesium alloys under creep test. 1931 H. J. Tapsell *Creep of Metals* i. 3 The experimental refinements now found necessary in creep-testing.
7. *Geol.* (Cf. 3.) **a.** A slow, imperceptible movement *en masse* of soil, talus, etc., usu. downhill under the influence of gravity but freq. with other processes (such as successive freezing and thawing) contributing to the effect.
1889 *Geol. Mag.* VI. 260 The normal rise of the surface particles was about $\frac{1}{4}$ of the depth of the frozen soil, and their creep about $\frac{1}{10}$ of the same depth. 1897 W. B. Scott *Introd. Geol.* iv. 82 Each freezing causes the fragments to rise slightly..and each thawing produces a reverse movement; hence the slow creep down the slope. 1897 [see *soil-creep*, Soil *sb.*¹ 10]. 1938 C. F. S. Sharpe *Landslides* iii. 21 The general term creep may be defined as the slow downslope movement of superficial soil or rock debris, usually imperceptible except to observations of long duration. 1942 C. A. Cotton *Geomorphol.* (ed. 3) iii. 29 Evidence of creep may be seen where trees and posts have been tilted from the vertical. 1960 B. W. Sparks *Geomorphol.* iv. 47 Rock creep is a movement of jointed blocks, partly as the result of soil creep and partly as a result of sliding. 1966 R. Common in G. H. Dury *Ess. Geomorphol.* 53 Shearing stresses in the material of a slope cause creep when they exceed the 'fundamental' shearing resistance.
b. A slow displacement of strata or the earth's crust by expansion or contraction or under compressive forces.
1900 [see *crust-creep* s.v. *crust *sb.* 13 b]. 1903 T. M. Reade *Evol. Earth Struct.* ix. 134 The horizontal expansion ..will produce, by small increments and minor alterations, a creep, ending in an anticlinal fold. 1906 Chamberlin & Salisbury *Geol.* III. 312 Continental creep along the steep slope between the continental platforms and the oceanic basins. 1942 E. M. Anderson *Dynamics of Faulting* viii. 182 A general 'creep' of the surface towards the mountain axis is supposed by others to be an essential part of the process of orogeny. 1964 L. U. de Sitter *Struct. Geol.* (ed. 2) xi. 149 Very accurate geodetic surface measurements

across the San Andreas fault..have shown that slow creep at an annual rate of 1 cm does take place along the fault line.
8. The slip of the belt on the pulley drum, or wheel over which it runs.
[1888 *Lockwood's Dict. Mech. Engin.* 95 *Creeping*, the very slight loss of speed which results when drums are driven by rope gearing, due to the slipping of the rope.] 1909 W. C. Unwin *Machine Design* I. 448 (*heading*) Creep of belt.
9. A creeping motion between the rim of a wheel and a rubber tyre. Cf. *creep *v.* 11.
1908 *Westm. Gaz.* 5 Mar. 4/3 The rims of the R.W. wheel allow no creep with a properly inflated tyre.
10. The continuous deformation of a material (esp. a metal) under stress, esp. at high temperatures. Cf. *creep *v.* 8.
1924 F. C. Lea in *Proc. Inst. Mech. Engin.* II. 1066 At stresses slightly above this the creep was continuous and the bar broke. *Ibid.* 1072 It will thus stand a higher stress without creep. 1931 H. J. Tapsell *Creep of Metals* iii. 31 Creep may be defined as the deformation of a material occurring with time under and due to an externally applied stress whether the deformation be of the nature of plastic or of viscous flow. 1943 E. G. Couzens in R. S. Morrell *Synthetic Resins* (ed. 2) xvii. 544 Flow takes place in two stages—creep, which is greatest following elastic deformation, and cold flow. 1952 *Jrnl. Iron & Steel Inst.* CLXXI. 333/2 In these materials creep was found to be primarily a process of slip in the austenite crystals. 1956 *Gloss. Terms Concrete (B.S.I.)* 16 *Creep*, a slow inelastic deformation or movement of concrete under stress. 1957 *Brit. Commonw. Forest Terminol.* II. 51 *Creep*, the increase in strain with time due to elastic after-effect and plastic deformation. 1970 *Fremdsprachen* 44 An uncertainty about the effect of creep and shrinkage.

creepage (krī·pėdʒ). [f. Creep *v.* + -age.] Gradual movement; *spec.* leakage of electricity.
1903 *Electr. World & Engin.* 7 Nov. 777 (Cent. D. Suppl. s.v. *Oil-thrower*), Special oil throwers are provided to prevent the creepage of oil along the shaft. 1958 *Which?* I. ii. 22/1 We tested the motor for safety...This included brush, creepage and wiring compliance. 1966 *McGraw-Hill Encycl. Sci. & Technol.* VII. 158/2 These [insulators] are made of glazed porcelain, with a series of skirts to lengthen the creepage paths and provide maximum resistance to flash-over.

creeper. Add: **1. d.** *pl.* (*a*) The feet; (*b*) shoes with soft soles. Also *attrib.* (in *sing.*). Cf. *brothel-creeper. *slang* (orig. *U.S.*).
1889 Barrère & Leland *Dict. Slang* I. 280/1 *Creepers*.. (American), the feet. 1904 'No. 1500' *Life in Sing Sing* 247 *Creepers*, soft shoes worn by burglars, sneak-thieves and prison guards. 1924 G. C. Henderson *Keys to Crookdom* 402 *Creepers*, rubber-soled shoes. Also called sneaks. 1951 *Sunday Pictorial* 29 Oct., Fancy shoes with thick crepe-rubber wedge soles which are known to connoisseurs as 'creepers'. 1955 E. Blishen *Roaring Boys* iv. 210 He pointed to my shoes, which were new and crape-soled. 'They're creepers... Real up-to-the-minute yobo's thick-soled creepers.' 1961 M. Dickens *Heart of London* i. 67 The two-inch soles of their 'creeper' shoes.
2. c. (Earlier example.)
1847 W. B. Dickson *Poultry* 15 The Dwarf Fowl, or Creeper (*Gallus Bankiva*, *S. pumilio*, Temminck, *Le Coq Nain*, Buffon).
10*. A pupil in the tea-planting trade, esp. in Ceylon.
1893 *Field* 8 Apr. 510/3 'Creepers', as they are called, are constantly coming out to learn tea. 1894 *Standard* 2 Jan. 5 A 'creeper', it seems, is the technical term for a pupil whose parents pay a high premium to have him taught the art and mystery of tea-planting in Ceylon. 1921 Ld. F. Hamilton *Here, There & Everywhere* ii. 48 [In Ceylon] Planters are divided locally into three categories: the managers,..the assistants,..and the premium-pupils, known as 'creepers'. 1931 E. Sutton tr. *Fauconnier's Soul of Malaya* i. iii. 30 The conceit of these blasted little creepers!
10.** *Cricket.* A ball which keeps low after pitching.
1848 *Punch Almanack* May–June, Till some 'ripper' or 'creeper' gives the great wicket-keeper A chance. 1927 *Daily Tel.* 14 June 6/1 A 'creeper' from Larwood got rid of Twining. 1963 *Times* 17 May 4/5 Ormrod had to deal with three astonishing creepers from the lively Arnold.
11. b. **creeper bridge, rope**, a bridge or rope of twisted creepers stretched across a tropical river; **creeper chain** *Mining*, an endless chain fitted with grips or hooks for traction of mine-cars, etc.
1892 H. W. Hughes *Coal-mining* 383 No better appliance has been introduced for minimising the cost of conveying tubs about the heapstead than that known as the 'finger' or 'creeper' chain... It consists of an endless chain travelling under the tubs, provided at intervals with vertical projecting pieces of iron fastened to the links. 1894 *Westm. Gaz.* 16 Jan. 5/3 A creeper rope tied from bank to bank. 1909 *Ibid.* 30 Dec. 5/4 We finally managed to get another creeper bridge between the island and the opposite bank, and hauled the women and children to a place of safety.
Hence **cree·pered** *ppl. a.*, having (Virginia) creeper growing on the walls; **cree·perless** *a.*, without such a creeper.
1894 *Pall Mall Gaz.* 20 July 3/3 Down in the hollow is a glimpse of the creepered farmhouse. 1938 L. MacNeice *Earth Compels* 21 A chart of tropic Swamp and twilight Of creepered curtains. 1904 H. G. Wells *Food of Gods* i. ii. § 1 The little house was creeperless.

creepie. Add: **1.** Also *creepy stool*.
1892 Jane Barlow *Irish Idylls* vii. 178 Pat, set the ould creepy stool for Mrs. Doyne. 1903 W. B. Yeats *Hour-Glass* (1904) 3 A creepy stool near it. 1922 Joyce *Ulysses* 43 Fiacre and Scotus on their creepystools in heaven.
2. (Example.)
1854 *Trans. Penns. Agric. Soc.* 163 The variety of poultry exhibited..comprising, in the tribe of barn-yard fowls,.. the Frizzle; the Creely and the Creepy.

creeping, *vbl. sb.* **1. b.** Delete † and add later examples.
1924 C. Mackenzie *Heavenly Ladder* x. 145 The only thing he regretted about this Good Friday was his cowardice over the ceremony of creeping to the Cross. 1957 *Oxf. Dict. Chr. Ch.* 1411/2 *Veneration of the Cross*, a ceremony of the Latin Rite for Good Friday, sometimes also called Creeping to the Cross.

creeping, *ppl. a.* Add: **2. a.** Also applied to a flaw or crack in steel.
1902 *Daily Chron.* 1 May 6/3 From an examination of the broken parts a 'creeping' flaw was found in the cross-section. 1914 H. Brearly *Case-hardening Steel* 110 Such cracks, generally spoken of as 'creeping cracks', are not often found in brittle material.
b. *creeping Jesus*, a person who slinks about or hides himself from fear of being ill-treated; an abject, sycophantic, or servile person; one who is hypocritically pious. *slang.*
c 1818 Blake *Everlasting Gospel* in *Wks.* (1927) 137 If he [*sc.* Christ] had been Antichrist, Creeping Jesus, He'd have don eany thing to please us. 1827 —— *Lett.* in *Wks.* (1927) 1138 God keep you and me from the divinity of yes and no too—the yea, nay, creeping Jesus. 1871 G. P. R. Pulman *Rustic Sketches* (ed. 3) 88 Creeping-jesus, applied to a person who seeks to hide himself in pursuit of sport or otherwise. 'Jack crawled ääder the weeld ducks lik' a creeping-jesus.' 1934 R. Campbell *Broken Record* 56 The Zulus naturally despise the creeping Jesus type who sucks up to them. 1937 'G. Orwell' *Road to Wigan Pier* x. 194 The outer-suburban creeping Jesus. 1945 A. Huxley *Time must have Stop* xxiv. 226 That fool who believed in Gaseous Vertebrates, that creeping Jesus who tried to convert people to his own idiocies! 1966 'L. Lane' *ABZ of Scouse* 23 *Creepin' Jesus*, applied to a person who enjoys bad health or constant misfortune; somebody who solicits sympathy by wearing an air of patient martyrdom.
c. *creeping barrage* (Mil.): see *barrage *sb.* and add examples.
1916 H. W. Yoxall *Let.* 22 Sept. in *Fashion of Life* (1966) iv. 32 The creeping barrage which went in front of our assaulting lines was almost geometrically straight, and lifted each time to the second. 1919 G. K. Rose *2/4th Oxf. & Bucks Lt. Infty.* 129 Our methods of attack..consisted, broadly speaking, in the advance of lines of Infantry behind a creeping barrage. 1957 *Encycl. Brit.* VI. 663/2 The limitation of the standing barrage was that the curtain of shells did not move with the troops; and while a creeping barrage made short bounds of 50 to 100 yards, a jumping barrage made longer ones.
d. *creeping paralysis*: locomotor ataxia. Also *fig.*
1913 in Dorland *Med. Dict.* (ed. 7) s.v. *Paralysis*. 1925 W. Deeping *Sorrell & Son* i. 10 He remembered that he had won his M.C. by 'doing something' as a protest against the creeping paralysis of intense fear. 1926 H. J. Laski *Let.* 30 May (1953) II. 843 The miners are still out, and industry, as a result, is afflicted with a kind of creeping paralysis. 1932 *Discovery* Apr. 112/1 A single glance at the film gives information concerning creeping paralysis (disseminated sclerosis). 1964 G. Durrell *Menagerie Manor* v. 103 The creeping paralysis, a terrible complaint that attacks principally the New World monkeys.
4. b. *creeping wheat* (earlier examples); *creeping willow*, a shrub, *Salix repens*, native to Europe and Asia.
1819 W. C. Wentworth *Descr. N.S.W.* 91 The creeping wheat, however, may be sown in the commencement of February. *Ibid.* 92 To the farmer..who keeps large flocks of sheep, the cultivation of the creeping wheat is highly advantageous; since in addition to its yielding as great a crop as any other species of wheat, it supersedes the necessity of growing..food for the support of his stock. 1894 W. Robinson *Wild Garden* (ed. 4) xvi. 262 Dwarf willows.. such as the Creeping Willow in its various forms, and the Woolly Willow, a dwarf silvery shrub. 1952 A. G. L. Hellyer *Sanders' Encycl. Gardening* (ed. 22) 434 [*Salix*] *repens*, 'Creeping Willow', to 3 ft., Britain, Europe, Asia.

creepy-crawly, *a.* Add: Also *transf.* and *fig.*, sneaking, servile; (of feelings, etc.) full of eerie or uncanny suggestion; of or pertaining to creeping or crawling insects. Also as *sb.* (chiefly *pl.*), such an insect, animal, etc.; a creepy-crawly feeling.
1858 C. M. Yonge *Christmas Mummers* iv. 37 The pink scarf came back in his mind, as clear..as if the silver creepey-crawleys and long-barbed flowers had all arranged themselves into the letters *thief*. 1891 'L. Malet' *Wages of Sin* vii. iii, I'm ever so hungry, and there's cold creepy-crawlies running up my legs. 1902 *Daily Chron.* 5 Dec. 4/5 The creepy-crawly atmosphere of 'Wuthering Heights'. 1907 *Ibid.* 1 Apr. 4/4 His way is mostly the creepy-crawly way... There's nothing heroic, or splendid, or even dignified, about his methods. 1909 G. B. Shaw *Let.* 14 Apr. (1956) 152 We..are back again in Algeria, but under creepy-crawly circumstances... We are in a frightful place. 1923 D. H. Lawrence *Kangaroo* vi. 133 She had puppies—four darling queer little things—tiny little creepy-crawlies. 1960 *Woman* 5 Mar. 29/2 Mice, spiders, moths and other creepy-crawlies. 1966 Auden *About House* 24 The lair, maybe, Of creepy-crawlies or a ghost.

cremation. Add: **2.** *attrib.* and *Comb.*, as *cremation-burial*, *-cemetery*, *grave*.
1913 E. T. Leeds *Archaeol. Anglo-Saxon Settlements* iii. 58 The number of cremation burials occurring at Frilford is uncertain. 1934 *Essays & Studies* XIX. 150 The description of the last rites paid to Beowulf shows that his was a cremation-burial. 1955 Woolley *Alalakh* vi. 211 Cremation burial in Sq. M 13. 1907 H. M. Chadwick *Origin Eng. Nation* iv. 74 The cremation cemeteries at Croydon and Beddington are also perhaps inconclusive. 1960 P. H. Reaney *Orig. Eng. Place-Names* 102 Repton..situated above the right bank of the Trent in a district where cremation-cemeteries have been found. 1950 H. L. Lorimer *Homer & Monum.* vi. 346 Of two Early Geometric cremation graves found.. in the Agora one contained a pair of large fibulae.

crème (krɛ̄m, krē¹m). Also **crême.** [Fr., = Cream *sb.*²] **1.** A cream (Cream *sb.*² 2 a) or custard. So *crème brûlée*, one topped with caramelized sugar; *crème caramel*, a custard coated with caramel; *crème Chantilly* [cf. *Chantilly 3], whipped cream sweetened and flavoured with vanilla; *crème renversée*, a custard turned out of a mould.
1845 E. Acton *Mod. Cookery* xx. 442 Crême à la Comtesse..is a very delicate kind of sweet dish, which..may be rendered more recherché by a flavouring of maraschino. 1846 A. Soyer *Gastronomic Regenerator* 528 *Crème au Caramel*... Have three quarters of a pint of milk in which you have boiled an ounce of isinglass, pour it upon the caramel. 1865 M. B. Chesnut *Diary* 5 Apr. (1905) xx. 376 We keep a cookery book on the mantelpiece, and when the dinner is deficient we just read off a pudding or a *crème*. 1868 E. Acton *Mod. Cookery* xxiii. 484 The French make their custards, which they call *crèmes*, also in small china cups. 1886 M. Clark tr. *Brisse's 366 Menus* 163 Crème brûlée. Burnt cream. [*Ibid.* 324 Crème à la Chantilly. Cream à la Chantilly. 1888 Mrs. Beeton *Bk. Househ. Managem.* xxxi. 828 Caramel Pudding. (Fr.—Crème Renversée.)] 1906 *Ibid.* lxv. 1730 Crème Caramel renversée. Caramel Pudding. 1908 C. H. Senn *Menu Book* 273 [*Crème*] *Chantilly*, whipped double cream, with vanilla flavour. 1909 *Cent. Dict.* Suppl., *Crème brulée*, caramel or browned sugar with cream. 1912 H. H. Munro *Unbearable Bassington* xv. 271 Jerome and the girls don't want to eat any more *crème renversée*. 1914 C. Mackenzie *Sinister St.* II. iii. v. 592 Our crême caramel is a much showier sweet than anything they've got at the House. 1930 A. Bennett *Imperial Palace* lx. 478 Evelyn heard the order: vermicelli soup,.. cutlets,..crème caramel. 1935 *Punch* 9 Jan. 33/1 Oh, many a *crème* have I consumed. 1949 'C. Hare' *When Wind Blows* 153 Spooning into his mouth the last of the tasteless crème caramel which the club almost invariably provided by way of a sweet. 1958 R. Godden *Greengage Summer* ix. 95 Meringues with crème chantilly. 1959 *Listener* 6 Aug. 227/2 To make the *crème brulée* take 1 pint of double cream, 6 oz. of caster sugar, 4 egg yolks, and vanilla essence. 1970 Simon & Howe *Dict. Gastronomy* 143 *Crème pâtissière*, also known as confectioner's custard or baker's custard, this is a thick French custard or cream used to fill tartlets, cream horns or puff creams.
b. A name for various syrupy liqueurs, as *crème de menthe* (peppermint), *crème de vanille*, *crème de noyau*, *crème de cacao*. (Cf. Cream *sb.*² 2 e.)
a 1821 Keats *Cap & Bells*, in *Poet. Wks.* (1907) 472 The least drop of *crème de citron*, crystal clear. c 1870 in H. W. Allen *3 Saint James's St.* (1950) viii. 186/2 Liqueurs.. Creme de Noyau—10/-. 1877 *Cassell's Dict. Cookery* 382/2 Noyau, or Crême de Noyau, is a sweet cordial flavoured with bitter almonds. 1892 T. F. Garrett *Encycl. Pract. Cookery* I. 477/1 *Crêmes*, a French term applied to certain cordials and liqueurs, to indicate the cream-like smoothness of these manufactures. 1903 *Daily Mail* 11 Sept. 3/3 Crême de menthe, with its strong peppermint flavour, is the one almost exclusively favoured by ladies. 1930 E. Waugh *Labels* 26 Shady young men in Charvet shirts sit round the bar repairing with powder-puff and lipstick the ravages of grenadine and *crème de cacao*. 1930, 1958 [see *Alexander *sb.*¹]. 1961 I. Fleming *Thunderball* x. 112 A tall glass of his favourite drink—crème de menthe frappé with a maraschino cherry on top.
c. *crème de riz*, a fine rice-flour; ground rice.
1896 Mrs. A. B. Marshall *Cookery Bk.* (ed. 2) 15 (Advt.), Marshall's Crème de Riz..highly prized for Cakes, Puddings, Blancmanges. 1960 E. David *Fr. Prov. Cooking* 97 *Crème de Riz*, Ground Rice.
2. Phr. *crème de la crème*, the élite, the very pick of society.
1848 F. A. Kemble *Let.* 22 Jan. in *Rec. Later Life* (1882) 336 The..pretensions of an Austrian *crème de la crème* are comprehensible and consistent. 1860 *Once a Week* 28 July 119/2 The elders—the *crème de la crème*, or those initiated into the highest mysteries of the sect. 1867 S. W. Baker *Nile Tribut.* xvii. 451 The society in this district was not *crème de la crème*. 1898 B. L. Farjeon *Miriam Rozella* xi, Need I say that he and Lady Laverock move in the best society, and are *crème de la crème*? 1920 D. H. Lawrence *Lost Girl* i. 8 In his palmy days, James Houghton was *crème de la crème* of Woodhouse society. 1967 R. Shaw *Man in Glass Booth* vi. 70, I love you all for you are the finest of your kind, the crème de la crème Americaine.

Cremnitz (kre·mnits). Also **Kremnitz.** [f. *Kremnitz* (formerly also *Cremnitz*), the German name of Kremnica, a town of eastern Czechoslovakia (formerly Körmöczbánya, Hungary).] Used *attrib.* to designate a white lead pigment used as a paint base.
1874 A. A. Fesquet tr. *Riffault's Manuf. Colors* ii. 56 The Clichy or chemical white lead is less dense and possesses less body than the Kremnitz white. 1880 W. J. Muckley *Char. & Use Colours* 28 *Cremnitz White*, is sometimes known as Vienna White. It is very bright in appearance, even surpassing Flake White, but not so dense in body. It is a preparation of lead. 1934 H. Hiler *Technique of Painting* ii. 99 *White lead*, the most important pigment in general use in the fine arts at the present time. It is also called.. *Cremnitz White*.

CREMNOPHOBIA

cremnophobia (kremnofōᵘ·biă). *Path.* [mod. L., f. Gr. κρημνός overhanging cliff: see -PHOBIA.] A morbid dread of precipices or steep places.

1903 in DORLAND *Med. Dict.* (ed. 3). **1908** G. B. SHAW *Sanity of Art* 97 [Nordau] is started off by the termination 'phobia' with a string of Agoraphobia.., Belenophobia, Cremnophobia.

crenulate, *a.* Add: **2.** *Geogr.* Of a shoreline: having many small irregular bays formed by the action of waves on softer rock.

1919 D. W. JOHNSON *Shore Processes* vi. 278 Early in the youth of the shoreline the curves will be changed to sharply and irregularly crenulate lines by differential wave erosion. ..We may call a shoreline of this character a crenulate shoreline. **1937** WOOLDRIDGE & MORGAN *Physical Basis Geogr.* xxi. 349 All variations in rock hardness, and divisional planes, are picked out by the waves, and the line of the exposed portions of the shore becomes complex or crenulate. **1967** D. G. FRY tr. *Zenkovich's Processes Coastal Devel.* viii. 492 A crenulate coastline often develops in igneous rocks, where there is alternation of veins and dykes of different compositions, and of zones of jointing.

creodont (krī·odǫnt). *Palæont.* [f. mod.L. *Creodonta* (E. D. Cope 1875, in *Proc. Acad. Nat. Sci. Philadelphia 1875*, 446), f. Gr. κρέας flesh + ὀδοντ-, ὀδούς tooth.] A member of the Creodonta, a sub-order of extinct carnivorous mammals, which lived during the Palæocene, Eocene, and Oligocene epochs. Also *attrib.* or as *adj.*

1891 FLOWER & LYDEKKER *Mammals* 607 The more typical Creodonts appear ..to be ..closely related to the true Carnivora. **1903** H. JOHNSTON *Brit. Mammals* 115 No Creodont ..has ever been discovered which possessed more than three true molars. *Ibid.* 188 The Creodont Carnivores. **1968** A. S. ROMER *Procession of Life* xv. 245 The Palaeocene and Eocene epochs ..saw the rise of a variety of archaic flesh-eaters, often grouped (there were several families of them) as the creodonts.

Creole, *sb.* Add: **2.** A creolized language.

1879 L. HEARN *Creole Sk.* (1924) 54, I explique myself to her, and she tell me in Creole— [etc.]. **1958** C. F. HOCKETT *Course Mod. Ling.* xlix. 423 There are several examples of creoles in the Caribbean area, spoken largely by the descendants of escaped Negro slaves. **1962** *Listener* 22 Nov. 868/3 A number of people working on Creoles met in Jamaica in 1959, and agreed to adopt Robert Hall's distinction between Creoles and Pidgins: a Pidgin is a first-generation *lingua franca* ..spoken by everybody as a *second* language; when in subsequent generations it becomes the *first* language of a community, it is a Creole. **1965** *Tablet* 22 May 587/2 Haiti, it appears, may soon have its vernacular, Creole, in the Mass. They already have a Creole missal.. .Chants, responses, and readings in Creole are to be found in it.

creolism. Delete † *Obs.* and add later example.

1893 *Athenæum* 7 Oct. 484/1 Though he had been born in Misiones (Paraguay), yet, as the son of a lieutenant-governor of a department and removed to Spain at an early age, he did not suffer from the disabilities of creolism.

creolization. Add: **b.** The fact or process of being creolized.

1934 PRIEBSCH & COLLINSON *German Lang.* I. ii. 35 The complete Creolization of Afrikaans has been prevented by contact with the Dutch of the Bible and literature. **1958** P. GAMMOND et al. *Decca Bk. Jazz* xxi. 262 Lyttelton.. recorded three brilliant Creole-styled versions of Tin Pan Alley songs ..all ..of which, by a process of 'Creolization', became better jazz than many a so-called New Orleans tune recorded ..by revivalists. **1969** *Language* XLV. 659 Its incipient creolization provides scholars with a unique opportunity to observe the dynamics of linguistic change.

creolize, *v.* Add: **2.** (Example.)

1834 T. WENTWORTH *West India Sketch Bk.* II. 219 Those ..have become *creolized*, or removed from the physical characteristics of the African.

b. *Philol.* To make into a creolized language.

1958 C. F. HOCKETT *Course Mod. Ling.* xlix. 423 An artificial language ..can be creolized. **1964** *English Studies* XLV. 383 The English dialects were never creolized.

creolized (krī·oləizd), *ppl. a.* [-ED[1].] **1.** Naturalized in the West Indies or Louisiana.

1880 G. W. CABLE *Grandissimes* iii. 20 The most throughly Creolized Américain.

2. *creolized language*: a language which has developed from that of a dominant group, first being used as a second language, then becoming the usual language of a subject group, its sounds, grammar, and vocabulary being modified in the process.

1932 W. L. GRAFF *Lang.* p. xxxiv, *Creolized language*, a language of a civilized people, especially European, mixed with that of one or more savage tribes. **1933** BLOOMFIELD *Language* xxvi. 474 In some cases ..a subject group gives up its native language in favour of a jargon.. .When the jargon has become the only language of the subject group, it is a creolized language. **1949** *Archivum Linguisticum* I. II. 155 It is likewise undeniable in the formation of pidginised languages and of their developments into creolised languages.

crêpe. Add: (Earlier and later examples.)

1797 *Wynne Diaries* 13 Jan. (1937) II. 151 After having equipped ourselves, the Bride and me, in a Nuptial garment, (white crepe), we went to Lady Hamilton's. **1872** *Young Englishwoman* Oct. 540/1 These tunics are also made of white crêpe de chine. **1907** *Daily Chron.* 15 July 8/4 The crêpe de soie ..is of the plainer kind, without a printed border. **1909** *Westm. Gaz.* 6 Mar. 15/1 Crêpe eclair is a silken fabric. **1923** *Weekly Dispatch* 11 Feb. 15 This dinner gown of amber crepe Romain. **1923** *Daily Mail* 14 Feb. 6 A very pretty gown of blue crêpe romaine. *Ibid.* 19 Apr. 8 The bride's gown was of white crêpe perle. **1925** *Good Housekeeping* Apr. 142/3 The crêpe twist [of acetate silk]. **1925** E. SITWELL *Poor Young People* 3 Wear crêpe de Siam, barèges Isabelle. **1928** *Sunday Express* 8 July 3 Crepe malika, a lovely material,..is one of the few innovations which are finding favour with the most important dress artists. **1951** *Good Housek. Home Encycl.* 231/2 Dull-finished fabrics, such as crêpe suède.

2. In full *crêpe rubber*. Indiarubber rolled into thin sheets with a corrugated surface.

1907 *Brit. Trade Jrnl.* 1 Sept. in W. H. Johnson *Para Rubber* (ed. 2, 1909) 117 After the crêpe rubber has left the vacuum drier. *Ibid.* 118 The first packing of the crêpe in the box. **1909** W. H. JOHNSON *Para Rubber* (ed. 2) 114 The manufacture of crêpe rubber..consists in passing the freshly coagulated rubber through a washing machine. **1914** H. BROWN *Rubber* 75 The corrugated sheet known as crêpe.. . The crêpe rubber, after drying, is sometimes converted into blocks by submitting it to pressure in steel moulds. **1926** *Blackw. Mag.* Apr. 575/2 Coarser qualities of rubber are always made into crêpe. **1937** H. BARRON *Mod. Rubber Chem.* v. 55 Since the beginning of this century smoked sheet and pale crêpe have virtually monopolised the field of raw rubber as raw materials for manufacture. **1963** A. S. CRAIG *Rubber Technol.* iii, For crepe rubber production the slabs are given much more severe treatment, being passed.. through a series of mills the rolls of which are grooved and are turning at slightly different speeds.

3. A small, thin pancake. (Cf. CRISP *sb.* 3.) So **crêpe Suzette** (usu. in pl.), a pancake served in a hot sauce, often containing a liqueur.

[**1877** E. S. DALLAS *Kettner's Bk. of Table* 143 *Crêpe*.— The French for pancake. **1907** A. ESCOFFIER *Mod. Cookery* xx. 723 *Suzette Pancakes.* Make these from preparation A [basic recipe], flavoured with curaçoa and tangerine juice. Coat them ..with softened butter, flavoured with curaçao and tangerine juice.] **1922** C. H. SENN *Luncheon & Dinner Sweets* 63 Pancakes à la Suzette. (Crêpes Suzette.) **1924** A. E. M. FOSTER *London Restaurants* 87 Crêpe Suzette is another special dish. **1928** *Vanity Fair* Sept. 31/1 Crêpes Suzettes are pancakes raised by Cunard to a remarkable point of perfection. **1951** *Good Housek. Cookery Bk.* (1957) 304/2 Add the liqueur and brandy to the sauce, and replace the folded crêpes in the pan. **1961** *Guardian* 27 Dec. 2/4 Henri Charpentier, creator of Crêpes Suzette.., died at Redondo Beach, California, on Sunday, aged 81.

4. *Comb.*: **crêpe-hanger** *U.S. slang* = *crape-hanger* (*CRAPE *sb.* 3 b); **crêpe paper**, a thin crinkled paper resembling crêpe; **crêpe ring** *Astr.* = *crape ring* (*CRAPE *sb.* 3 b); **crêpe sole** [SOLE *sb.*[1] 2], the underside of a shoe covered with crêpe rubber; so **crêpe-soled** *a.*

1930 *N. & Q.* CLIX. 119/1 In an American novel, the writer refers to a man as a mere 'crêpe-hanger'. *Ibid.* 232/1 A crepe-hanger is the ultimate in depressing persons.. .The expression derives from the lugubrious undertaker's assistants who put up black decorations for a funeral. [**1895** *Montgomery Ward Catal.* 123/3 Crepe tissue paper.] **1897** *Sears, Roebuck Catal.* 333/1 Assorted imported tissue, 2 pieces crepe paper. **1903** K. D. WIGGIN *Rebecca* (1904) xiii. 140 The [lamp] shade..was of crinkled crêpe paper. **1915** *Chemists' Windows* 27 Crêpe paper or plain stiff materials have enough 'body' to remain in pleat. **1959** I. & P. OPIE *Lore & Lang. Schoolchildren* xii. 258 The Maypoles consisted of a pram wheel decorated with crepe paper and streamers. **1959** *Listener* 17 Sept. 429/1 The innermost ring, C, is more generally known as the Crêpe or Dusky Ring. **1926–7** *Army & Navy Stores Catal.* 654/1 The 'Glastonbury' Slipper Boot in buff sheepskins ..thin crêpe sole. **1936** G. GREENE *Journey without Maps* III. iv. 267 A pair of gym shoes with crêpe soles. **1941** V. WOOLF *Between Acts* 232 Crepe soles?.. They last much longer and protect the feet. **1935** HEMINGWAY *Green Hills Afr.* (1936) I. i. 22 We went quietly. I had on these crêpe-soled boots. **1953** H. CLEVELY *Public Enemy* x. 62 He ..put on ..a pair of crêpe-soled sandals.

crêpé, *a.* (Earlier example.) **crêpy,** *a.* Also **crêpey, crepey.** Resembling crêpe. (Later examples.)

1828 LYTTON *Pelham* I. xvi. 113 Her own grey hair *crêpé*, and surmounted by a high cap of the most dazzling *blonde*. **1941** C. McCULLERS in *55 Short Stories fr. N. Yorker* (1952) 320 The jockey ..scrutinized the room with pinched, crêpy eyes. **1946** E. TAYLOR *Palladian* vii. 69 The inside of her arm showed grey and crêpy. **1959** P. H. JOHNSON *Humbler Creation* xlii. 281 The skin of her throat seemed to her a little crepey, so she massaged it with cold cream.

crêpeline (krē̍·pəlīn). Also **crêpoline.** [Fr., dim. of *crêpe* CRÊPE.] A light thin material of silk, or silk and wool, used for women's dresses.

1873 *Young Englishwoman* Apr. 202/2 Crêpeline, a material only to be found at the Halle des Indes [in Paris]. **1882** CAULFEILD & SAWARD *Dict. Needlework* 93/2 *Crêpeline, Crêpon*, or *Crape Cloth*, a dress material, having a silken surface, much resembling crape, but considerably thicker. **1898** *Daily News* 10 Dec. 6/3 Another novelty is crêpoline cloth,..as soft as cashmere. **1909** *Westm. Gaz.* 9 Oct. 15/1 Crêpolines, silk combined with wool.

crépinette (krē̍pine·t). [Fr., dim. of *crépine*: see CREPINE 2.] Minced meat with sauce or farce, wrapped in pieces of pork caul.

1877 E. S. DALLAS *Kettner's Bk. Table* 143 *Crépinette*, a flat sausage enveloped in pig's caul. **1889** *Mrs. A. B. Marshall's Cookery Book* 123 Crépinettes à la Ferdinand. **1907** A. ESCOFFIER *Mod. Cookery* 462 Shape the *crépinettes* thus formed rectangularly. **1911** FRANCATELLI *Mod. Cook* 331 Crepinettes of Partridge à la d'Estaing. **1951** *Good Housek. Home Encycl.* 426/2 Meat and Ham Crépinettes.

crêping (krē̍·piŋ), *vbl. sb.*

1. The crimping or frizzing of hair.

1889 *Daily News* 30 Dec. 6/6 The mistake of too tightly crimping—or, to use the new word, crêping—the hair. **1966** J. S. COX *Dict. Hairdressing* 41/2 *Crêping* 1. Weaving and boiling hair to transform it into crêpe hair. 2. Crimping.

2. The production of crêpe rubber.

1909 W. H. JOHNSON *Para Rubber* (ed. 2) 123 A large size pair of breaking-up and crêping machines.

crepis (krī·pis). **1.** *Bot.* [mod.L. (Linnæus *Genera Plantarum* (1737) 240), ad. Gr. κρηπίς, Theophrastus's name for another plant.] A plant of the large genus of herbs so called, belonging to the family Compositæ and including a few cultivated species.

1822 LOUDON *Encycl. Gardening* Index 1383/2 Crepis, ..of easy culture. **1904** R. J. FARRER *Garden Asia* 246 Twinkling gold of crepis. **1919** —— *Eng. Rock-Garden* I. 243 No Crepis is worthy of admission to the rock-garden ..except the following: *C. incisa*..; *C. rubra*..; and *C. lagoseris*. **1948** A. G. L. HELLYER *Amateur Gardener* viii. 199 Most species of crepis are weeds and all have flowers like small dandelions.

2. *Biol.* [ad. Gr. κρηπίς base.] A sponge-spicule forming the central axis of a desma.

1900 E. A. MINCHIN in Lankester *Treat. Zool.* II. 134 Each desma is formed typically by secondary deposits of silica upon a true spicule termed the *crepis* or foundation. **1940** L. H. HYMAN *Invertebrates* I. vi. 299 The deposited silica at first follows the shape of the crepis. **1963** I. F. & W. D. HENDERSON *Dict. Biol.* (ed. 8) 124/1 *Crepis*, the fundamental spicule by deposition of silica upon which a desma is formed.

cresce·ndo, *v.* [f. the sb.] *intr.* To increase gradually in loudness or intensity.

1900 *Westm. Gaz.* 2 July 2/1 A faint whir crescendoes rapidly into the shrill whoop of a steam-siren. **1901** *Ibid.* 12 Nov. 2/1 The trolly-cars, with their booming note which crescendoes up the scale with increasing speed and diminuendoes with the slackening of it. **1903** R. LANGBRIDGE *Flame & Flood* xvi, A bubbling torrent of vituperation that crescendoed as she leapt in air..and decrescendoed, as.. she turned away. **1927** *Daily Express* 24 Oct. 10/3 'The season'..starts in November, crescendoes to its height in January and February, to die away in April. **1969** *Daily Tel.* 3 Feb. 12/3 As well as the nudity story which crescendoed up from Italy to swamp much of the Paris after-dark fashion, there is real news.

crescent, *sb.* Add: **6.** More fully, *crescent roll.* (Cf. *CROISSANT.)

1899 *Daily News* 23 Sept. 3/1 Crescent rolls and hot milk. **1951** E. PAUL *Springtime in Paris* v. 106 Coffee cups, saucers and crisp crescent rolls in our hands.

crescentric (krēse·ntrik), *a. rare.* [f. L. *crescent-em* with second element after CENTRIC *a.*] = CRESCENTIC *a.*

1851 H. MELVILLE *Moby Dick* III. i. 6 And swimming on, in one solid, but still crescentric centre. **1896** *Naturalist* Oct. 338 The beautiful crescentric mounds around York.

crescograph (kre·skŏgrɑf). [irreg. f. L. *crescĕre* to grow + -O + -GRAPH.] An instrument invented by Sir Jagadis Chunder Bose (1858–1937), Indian plant physiologist, for recording the rate of growth in plants.

1918 J. C. BOSE *Life Movements in Plants* I. 157 High Magnification Crescograph. *Ibid.* 169 Magnetic Crescograph. **1919** *Ibid.* II. 255 Balanced Crescograph. **1919** —— in *Proc. R. Soc.* B. XC. 365 (*heading*) Researches on growth and movement in plants by means of the high magnification crescograph. **1957** *Encycl. Brit.* III. 926/1 He [*sc.* Bose]..invented many delicate and sensitive instruments, such as his crescograph for recording plant growth, magnifying a small movement as much as 10,000,000 times.

crest, *sb.*[1] Add: **7. e.** Chiefly *Electr. Engin.* A point in a wave-form at which the varying quantity is a maximum. Hence **crest factor**, the ratio of the maximum value (*crest value*) of an alternating current or voltage to its root-mean-square value; **crest voltmeter**, any instrument for measuring the maximum value of an alternating voltage. (Cf. *PEAK *sb.*[2])

1875 *Encycl. Brit.* I. 111/1 [*Acoustics.*] The wave represented by the dotted line, which..has its crests. **1914** H. PENDER *Amer. Handbk. Electr. Engineers* 1297 Crest-factor or peak-factor is the ratio of the crest or maximum value to the r.m.s. value. *Ibid.*, The crest value of a sine-wave is $\surd 2$. **1916** *Trans. Amer. Inst. Electr. Engin.* XXXV. I. 115 The crest voltmeter is a direct-reading instrument, reading either the r.m.s. value of a sine wave having the same crest as a high voltage wave to which it is connected, or the true crest value, depending upon its calibration. **1961** *Listener* 9 Nov. 767/2 With an alternating current system, the insulation has to withstand the maximum value of the crest of the voltage wave; and that crest value is higher than the *useful* value, and so some of the expensive insulating capacity is wasted. In a direct current system the crest value of the voltage *is* the useful value, so there is no waste of insulation.

9. e. (See quot. 1954.)

1916 C. E. ALLEN *Machinery's Screw Thread Bk.* 3 The crest is the prominent part of the thread, of either the male screw or of the female screw. **1954** *Defs. for use in Mech. Engin.* (*B.S.I.*) 16 *Crest*, that part of the surface of a thread which connects adjacent flanks at the top of the ridge.

11. crest-line, (*a*) a series of ridges; (*b*) the sky-line of a ridge (cf. 7 a).

1890 LD. LUGARD *Diaries* (1959) I. ii. 96 The house..does not run parallel with the crest line, so..I am bound to make the Stockade skew-wise a bit. **1901** 'LINESMAN' *Words by Eyewitness* (1902) 49 From the encircling rim are darting innumerable spurts of flame..from the rifles of the men clinging like flies to the crest-line. **1915** *Blackw. Mag.* Jan. 9/2 Hardly had we rejoined the battalion, which was formed up behind a second crest-line.., when a tremendous shell fire began to fall.

cresting, *vbl. sb.* Add: **2.** Ornamental edging on a chair, settee, etc. Also *attrib.*, as *cresting rail.*

1908 P. MACQUOID *Hist. Eng. Furnit.* IV. viii. 241 Sometimes the whole of the splat was formed of crossed diagonal lines..and the panel of the cresting lightly carved in place of painting. **1941** *Burlington Mag.* June 187/1 The back [of the *Director* chair] was of a light and elegant design with graceful curves to the splat and cresting rail. **1963** *Times* 2 Mar. 11/7 Compared with *Director* designs, the Portuguese cresting rail is deeper, generally less pierced, and treated much more as a pediment.

crestless, *a.* Add: **2.** *gen.* Without a tuft, top, ridge, or the like.

1889 in *Cent. Dict.* **1908** A. S. M. HUTCHINSON *Once Aboard the Lugger* I. ii. 33 It welled, rose-deeply;..crestless, flinging no intoxicating spume. **1908** *Daily Chron.* 31 Aug. 7/2 Great shiny blue crestless jays flitted over the scrub.

Creswellian (krezwe·liăn), *a. Archæol.* [f. the name of *Creswell* Crags in Derbyshire (see quot. 1926) + -IAN.] Pertaining or belonging to a cultural period of the Mesolithic or late Palæolithic, represented only in Britain and roughly contemporary with the Magdalenian period. Also *ellipt.* as *sb.*.

1926 D. A. E. GARROD *Upper Palaeolithic Age* III. 194, I would suggest that this industry is sufficiently well characterized to deserve a name..to differentiate it..from the classical Magdalenian of France..[and] the true Upper Aurignacian. I propose tentatively 'Creswellian', since Creswell Crags is the station in which it is found in greatest abundance and variety. **1948** KROEBER *Anthropol.* (ed. 2) xvi. 661 Some authors..prefer to give local designations to these marginal local phases contemporary with the Magdalenian period of France. Such are Creswellian, Grimaldian and (late) East Gravettian for England, Italy and the Danube respectively. **1969** K. P. OAKLEY *Frameworks for dating Fossil Man* (ed. 3) II. ii. 167 Other Epi-Gravettian cultures developed contemporaneously in various parts of Europe, notably the Creswellian in Britain.

Cretaceous, *a.* Add: **B.** *sb.* (usu. with *the*). *Geol.* The Cretaceous system or period.

1851 LYELL *Man. Elem. Geol.* (ed. 3) xvii. 209 (*heading*) Upper Cretaceous. I. Maestricht beds and Faxoe limestone. **1906** CHAMBERLIN & SALISBURY *Geol.* III. 160 In the Black Hills, the Cretaceous has in some places a thickness of no more than 1000 feet. *Ibid.* 162 The Appalachian mountains,..which had been reduced to a peneplain by the close of the Cretaceous. **1910** *Encycl. Brit.* VII. 415/1 There is a very general unconformity and break between the Lower and Upper Cretaceous. *Ibid.*, With the opening of the Cretaceous in Europe there commenced a period of marine transgression. **1960** L. D. STAMP *Britain's Struct.* (ed. 5) xii. 139 These strata form..transition beds between the Cretaceous and the Tertiary.

Cretan (krī·tăn), *a.* and *sb.* [ad. L. *Crētānus*. The forms used in the various translations of the Bible are, in Acts ii. 11 *Cretes* (Middle English (*a* 1400, ed. A. C. Paues), Geneva and A.V.), in Titus i. 12 *Cretayns* (Tindale and Coverdale), *Cretyans* (Cranmer), *Cretians* (Geneva and A.V.); Reims and Douay have *Cretensians*, and R.V. *Cretans* in both places.] **A.** *adj.* Of or belonging to the island of Crete in the Mediterranean. **B.** *sb.* A native of Crete.

Cretan bull, the bull beloved by Pasiphæ. *Cretan carrot*, the plant *Athamanta cretensis*, used in medicine. *Cretan hemp*, bastard hemp.

[*c* **893** K. ALFRED *Oros.* I Hu Cretense & Atheniense, Creca leode, him betweonum wunnon.] **1579** NORTH *Plutarch's Lives* 276 This was but a fraude and a Cretan lye, to deceaue the Cretans with. **1596** SHAKES. *Tam. Shrew* I. i. 175 When with his knees he kist the Cretan strond. **1654** OGILBY *Virgil, Bucolicks* v. 24 *note*, That excellent Cretan Archer. **1797** *Encycl. Brit.* V. 532/2 In order to distinguish the true Cretans from strangers, they were named *Eteocretes*. **1819** SHELLEY *Peter Bell 3rd* III. xiii, Lunches and snacks so aldermanic. Where reigns a Cretan-tongued panic. **1820** —— *Œd. Tyr.* II. ii. 3 What though Cretans old called thee City-crested Cybele? **1821** T. CAMPBELL in *New Monthly Mag.* II. 438 (*title*) Song of Hybrias the Cretan. [**1829** LOUDON *Encycl. Plants* 506 *Origanum Dictamnus*, Dittany of Crete.] **1868** S. HEREMAN *Paxton's Bot. Dict.* 164/2 Cretan carrot. **1874** HARDY *Far fr. Mad. Crowd* xxv, He was perfectly truthful toward men, but to women lied like a Cretan. **1880** E. GLAISTER *Needlework* viii. 94 Another good style of pattern for tidies is to be found in the Cretan work..which..was brought to London three or four years ago. **1881** C. C. HARRISON *Woman's Handiwork* III. 157 Cutting Cretan embroidery into strips. **1902** *Encycl. Brit.* XXVII. 273/1 In the palace at Knossós..have been found a large number of clay tablets..inscribed with the indigenous Cretan character. **1915** *Edin. Rev.* Jan. 127 Compare the Cretan and Dionysiac Oreibasiai or Mountain-Rites. **1952** A. G. L. HELLYER *Sanders' Encycl. Gardening* (ed. 22) 97 *Celsia*.. *cretica*, 'Cretan Mullein', yellow, 3 to 5 ft., Medit. Region. *Ibid.* 344 *Origanum*.. *Dictamnus*, 'Cretan Dittany', pink, summer, 1 ft., Crete. *Ibid.* 502 *Valeriana*.. *Phu*, 'Cretan Spikenard', white, Aug., 2 ft., Caucasus. **1955** *Times* 6 Aug. 3/5 An important inscribed baked clay tablet..carries three lines of Cypro-Minoan script to which the date 1300 B.C. is attributed. It shows an affinity with the Cretan linear script found by Sir Arthur Evans at Knossos.

cretin. Now usu. pronounced (kre·tin). Add: Also in weakened sense (esp. in form *crétin*): a fool, one who behaves stupidly. Also *attrib.* and *transf.*

1884 W. JAMES *Coll. Ess. & Rev.* (1920) 270 Bodily commotions..may be experienced in their fulness by *Crétins* and Philistines in whom the critical judgment is at its lowest ebb. *a* **1930** D. H. LAWRENCE *Pornography* (1936) 75 The blood in the body stands still, before such *crétin* ugliness. **1933** J. JOYCE *Let.* 13 Aug. (1966) III. 282 The crétin of a concierge..has misdirected half my mail. **1961** I. JEFFERIES *It wasn't Me* ix. 123, I know I'm a moron and a cretin like you're always calling people.

cretinize, *v.* (Earlier example.)

1846 R. FORD *Gath. Spain* xxiv. 338 The Spanish Bourbons, when not 'cretinised' into idiots, are creatures composed to cunning and cowardice.

cretinoid (kre·tinoid), *a.* [See -OID.] Resembling a cretin or cretinism.

1874 W. GULL in *Trans. Clin. Soc.* VII. 180 (*title*) On a Cretinoid State supervening in Adult Life in Women. **1886** *Buck's Handbk. Med. Sci.* II. 20 Myxœdema..Cretinoid disease. **1966** *Jrnl. Clin. Endocrinol. & Metabolism* XXVI. 111/2 The appearance of the face was typical for a cretinoid child, with a large, protruding tongue and heavy features.

crevasse, *sb.* **2.** (Earlier and *fig.* U.S. examples.)

1814 H. M. BRACKENRIDGE *Views Louisiana* 179 The terrors excited by a *crevasse* or breaking of the levee. **1850** *Congressional Globe* App. 149/2 A moral crevasse has occurred: fanaticism and ignorance..have accumulated into a mighty flood. **1897** G. W. CABLE *Old Creole Days* 104 The Anglo-American flood that was presently to burst in a crevasse of immigration upon the delta.

Crèvecœur (krę·vkör). [Fr., = heart-break (see quot. 1909).] A variety (usually black) of the domestic fowl of French origin, resembling the houdan in body, but characterized by a comb consisting of two large coral-red horns. Also abbrev. **Crève** (**Crêve**).

1855 *Poultry Chron.* III. 285 The Crèvecœur fowls. *Ibid.* 286 The Crèvecœurs exhibited at Baker Street. **1873** L. WRIGHT *Bk. Poultry* xxvi. 415 The Crève is the bulkiest in appearance of all the French races. **1883** *Daily News* 3 Oct. 2/2 The pretty Houdans and Crèvecœurs. **1904** S. W. THOMAS in *L. Wright's Bk. Poultry* xxxi. 452/1 The spiral crest is gone, and so is the Crève comb. *Ibid.* 456/1 The very finest Crèves even now attain greater size than the largest Houdans. **1909** T. W. STURGES *Poultry Man.* xvii. 527 The Crève-Cœur..is one of the oldest of the French breeds.. They derive their name from the peculiar shape of the comb (Crève-cœur ..split hearts), the comb being like a V with a broad base and filled up half its height with a protuberant growth of flesh.

crevette (krəve·t). [Fr., = shrimp.] A deep shade of pink, shrimp-pink.

1884 *Cassell's Fam. Mag.* May 371/1 Blues, greens.. salmon-pink, and the deeper crevette, or shrimp-pink. **1890** *Daily News* 5 July 3/4 Costumes of the colours known as crevette and eau-de-Nil.

crevicing (kre·visiŋ), *vbl. sb.* (See quots. and CREVICE *sb.* 1 b.) Also *attrib.*

1851 *Alta California* (San Francisco) 17 July 2/2 The early adventurers in the gold-diggings required simply..a strong sheath-knife for 'crevicing'. **1876** J. MILLER *First Fam'lies* ix. 68 A lot of picks and pans, and tom irons, and crevicing spoons, that lay up against the wall. **1886** J. W. ANDERSON *Prospector's Handbook* 117 Crevicing—Collecting gold in the crevices of rock. **1888** C. D. FERGUSON *Exp. Forty-niner* ix. 129 It was all crevicing, that is, working the crevices in the rocks. **1948** P. JOHNSTON *Cities Calif. Gold Rush* 41/2 As crevicing was better up the fork, they broke camp, and moved to Zumwalt Flat.

crew, *sb.*[1] Add: **6. c.** *Aeronaut.* In full *air crew* (see *AIR *sb.*[1] III. 4). The persons manning an aircraft or spacecraft.

1917 'CONTACT' *Airman's Outings* 79 But the airman has experience of what the aeroplane crews must be going through, and his thought is all for them. **1960** *Aeroplane* XCVIII. 422/1 The crew chief, in fact, is the sixth member of the V-bomber flight crew, and is taken on all sorties which entail landing away from base. **1969** *Listener* 1 May 605/2 Eugene Cernan, one of the three-man crew of Apollo 10, due to be launched on 18 May.

7. crew (hair-)cut orig. *U.S.*, a closely cropped style of hair-cut for men (app. first adopted by boat crews at Harvard and Yale Universities); also *transf.* and *fig.*; also *crew-cropped* adj.; **crew neck, neckline** orig. *U.S.*, a round neckline of a garment, esp. a sweater, fitting closely to the throat as on vests worn by oarsmen; so *crew-necked, -shaped* adjs.

1938 HEMINGWAY *Fifth Column* (1939) 111 Wilson noted his crew-cropped hair. **1942** R. KING *Design in Evil* iv. 41 A steward..with a sparkling crew cut of chestnut hair. **1944** D. W. BROGAN *Amer. Problem* iv. 69 Crew-cut hair and brogues. *a* **1953** DYLAN THOMAS *Quite Early One Morning* (1954) 68 He is vigorously welcomed at the station by an ea·nest crew-cut platoon of giant collegiates. **1958** *Times* 20 Jan. 11/3 The mature version of the intellectual crew-cut is, with sad inevitability, the egg-head. **1962** *Listener* 18 Jan. 138/2 These new works sometimes convey nautical accents through the use of ship-shape and Bristol-fashioned wood, planed and crew-cut wood surfaces. **1940** *Time* 11 Nov. 76/2 Doe-eyed Lucille Ball..gets the affections of Richard Carlson, whose crew haircut makes him the first genuine-looking Princeton undergraduate in cinema history. **1946** 'P. QUENTIN' *Puzzle for Fiends* x. 98 A young man with a blond crew haircut. **1940** *Illustr. London News* CXCVII. 385 (Advt.), The windcheater..famous wind-resisting-pullover of close-textured cotton with fleecy inner surface. Crew neck. **1944** U. SINCLAIR *Presidential Agent* (1945) ii. 36 The President was lying in bed, wearing pyjamas..covered by a knitted blue sweater, crew-neck style. **1957** J. BRAINE *Room at Top* vi. 51 A yellow crew-neck sweater and a golf jacket. **1950** *Here & Now* (N.Z.) Dec. 25/1 The war was fought for the right of every man to wear crew-necked jerseys and every woman to wear black-lace panties. **1939** M. B. PICKEN *Lang. Fashion* 39/2 *Crew neckline*, round, high neckline, as in a sweater. **1935** *Amer. Speech* X. 193/1 Young girls *went nautical* last fall in dresses that were 'deeply pleated fore and aft'. The necks were *crew shaped* and the pockets were *anchored*.

crew (krū), *v.* [f. CREW *sb.*[1]] *trans.* and *intr.* To act as (a member of a) crew of a ship, aircraft, etc.; to assign to a crew. Hence **crewing** *vbl. sb.*, the work of such a crew, or of one of its members.

1935 'A. ANDREWS' *Blue Tunnyman* vi. 93 Of the many racing men for whom I crewed, no one taught me more than Mr. J. Paine..in a converted Bristol pilot cutter. **1944** 'N. SHUTE' *Pastoral* v. 107 Bad luck on the chaps that had to crew for him. **1947** *Daily Tel.* 7 May 5/1 He criticises the crewing of the aircraft.. .Air Cmdre. Brown attributes the accident to..bad crewing, the navigator being the only one of the four operational members who knew the route. **1955** *Times* 13 June 12/2 The Swallow class boat Blue Phantom, crewed by the Glanville twin brothers. *Ibid.* 16 June 8/2 Arrangements had been made for the crewing of the ship to ensure this. **1957** RAWNSLEY & WRIGHT *Night Fighter* i. 19 When it came to my turn to be crewed up it was..no surprise to me to learn that I had been allocated to the youngest pilot in the squadron. **1967** *Jane's Surface Skimmer Systems 1967–68* 30/1 It..can be crewed by personnel experienced in the operation of these craft.

crewel, *sb.* **3. b.** (Earlier examples.)

1598 FLORIO *Worlde of Wordes* 260/3 *Passamano d'accia*, statute lace, crewell lace. **1880** L. HIGGIN *Handbk. Embroidery* iii. 19 'Stem stitch' (wrongly called also, 'crewel stitch', as it has no claim to being used exclusively in crewel embroidery).

crewman (krū·mæn). [CREW *sb.*[1] + MAN *sb.*[1]] A member of a crew (senses 5 and 6).

1937 *Sun* (Baltimore) 2 July 1/6 Captain Gray..will share the 22½-ton Sikorsky flying boat with six other crewmen. **1950** J. DEMPSEY *Championship Fighting* 179, I never rowed in a crew. But I do know that crewmen have a rhythm or 'beat' to which they time their strokes. **1961** *Flight* LXXX. 284/1 A Mercury spacecraft carrying special instrumentation and a 'crewman simulator'. **1964** *English Studies* XLV. 4 The crewmen of the Tuscan vessel which carried him to Corsica.

‖ **criant** (kriaṅ), *a.* [a. F. *criant* crying, loud, pr. pple. of *crier* to CRY.] 'Loud', garish.

1876 'OUIDA' *Winter City* i. 9 A *criant* bit of furniture hurt her as the grating of a false quantity hurts a scholar. **1884** *Tablet* 24 May 805/1 There is nothing garish or *criant* in either of these glorious canvases.

‖ **criard** (kriar), *a.* Also in fem. form **criarde.** [Fr.] Shrill; 'loud'; garish (cf. prec.).

1840 THACKERAY in *Fraser's Mag.* July 120/1 His pictures are chiefly effects of sunset and moonlight; of too *criarde* a colour as regards sun and moon. **1889** G. B. SHAW *London Music in 1888–89* (1937) 260 His voice is high pitched and a little *criarde*. **1924** *Blackw. Mag.* June 774/1 Man and label were equally *criard*. *Ibid.* Aug. 237/1 The palace furniture was equally *criard*.

crib, *sb.* Add: **3. a.** N.Z. examples: now esp. a small house at the seaside or at a holiday resort.

1862 C. R. THATCHER *Dunedin Songster* No. 1. 5 The weather and time had so peppered that tumbledown crib, I declare. **1929** W. SMYTH *Bonzer Jones* xvii. 213 'Here's my crib,' he announced. **1947** 'A. P. GASKELL' *Big Game* 88 If it's fine George will be taking them up to his crib. **1962** *Guardian* 21 July 6/4 An index to social status in New Zealand..is possession of the seaside bach (in Southland, the crib). **1963** *Truth* (N.Z.) 24 Sept., 'You know my seaside summer bach?' 'Bach? What the South Islanders call a crib?' **1970** D. M. DAVIN *Not Here, Not Now* II. ix. 115 Then back to the crib again, set off the road in the bush.

d. *slang* (chiefly *U.S.*). A saloon, 'low dive', or brothel. Also *crib-house, -joint.*

c **1857** B. A. BAKER *Glance at N.Y.* 23 Let's take a drink; there's a crib open. **1882** *Sydney Slang Dict.* 3/2 *Drum*, or *crib*, house of ill repute. **1901** 'J. FLYNT' *World of Graft* 219 *Crib*, gambling dive. **1926** J. BLACK *You can't Win* (1927) xiv. 199 I'll make the cribs myself. I'm dynamite with them old bums in the cribs. **1930** J. DOS PASSOS *42nd Parallel* 320 The little lighted cribhouses. **1932** B. DE VOTO *Mark Twain's Amer.* vi. 124 The palaces blended with scores of dance halls..parlor houses, cribs. **1958** P. GAMMOND et al. *Decca Bk. Jazz* iii. 42 Forced into the dives and crib-joints of the red-light district of New Orleans.

6. b. Substitute for def.: *dial., Austral.*, and *N.Z.* Food, provisions; a light meal or snack; a piece of bread, cake, etc. Freq. *attrib.*

1825 JAMIESON Suppl. *s.v.*, Haste ye, and gi'e me ma.. *crib*, Guidwife. **1872** *N. & Q.* 4th Ser. IX. 47/1 The gift.. was generally a small cake..and was called the 'christening crib'—a *crib* of bread or cake being a provincialism for a *bit* of bread, &c. **1880** M. A. COURTNEY *Gloss. Cornwall* 15/2 *Crib*, a crust of bread; fragments of meat. 'Eat up your cribs.' **1904** 'G. B. LANCASTER' *Sons o' Men* 159 Sereld.. growled because someone had spilt tobacco-ash into his crib—which is bushman for dinner. **1908** *Westm. Gaz.* 13 May 6/1 Half an hour's 'crib' time [at Blackball, N.Z.] is also granted. **1926** K. S. PRICHARD *Working Bullocks* xi. 108 Red picked up his crib-bag. **1928** J. DEVANNY *Dawn Beloved* xxx. 273 He stopped..to hang up his towel and crib tin. **1942** A. L. ROWSE *Cornish Childhood* ii. 30 He used to take it to work with him and at crib-time (i.e. lunch-time) would entertain his fellows with it. **1947** A. VOGT in D. Davin *N.Z. Short Stories* (1953) 364 Ben went to work [in the bush] each day like the rest of the men, with his crib and oil-skin. **1954** *Coast to Coast 1953* 37 Jacques was holding out his crib. 'Time to eat.'..Crib over, the men rolled cigarettes. **1971** J. TURNER *Stone Dormitory* iii. 30 'Just come in for me crib. It's time.'..'It's ready for you, Tom,' she said, putting the bread and cheese and tea before him.

10. b. The enclosure for trapped fish in a pound-net. *U.S.*

1873 *Rep. U.S. Fish Commission* I. 264 The pound-nets.. have several parts, termed the 'leader', the 'heart', the 'pot', 'bowl', or 'crib', and the 'tunnel'. *a* **1884** KNIGHT *Dict. Mech.* Suppl. 231/1 *Crib* (Fishing), the bowl or pound of a Pound Net.

13. (Earlier examples.)

1816 [see *crib-dam]. **1867** *Harper's Weekly* 20 Apr. 252/4 The flood-gates of the 'crib' were opened.

19*. A complaint, grumble. *colloq.*

1943 HUNT & PRINGLE *Service Slang* 26 People have their own pet cribs.

20. crib-bite *v. intr.*, to have the practice or habit of crib-biting; **crib-biter**, also, a grumbler; **crib-breakwater** *U.S.*, a breakwater made of cribwork; **crib-bridge**, a bridge whose piers are formed of cribs (see CRIB *sb.* 13); **crib-cracker** *slang*, a burglar (see CRIB *sb.* 3 b); so *crib-cracking*; **crib-dam** *U.S.*, a dam formed of cribs; **crib-muzzle**, a muzzle worn by a horse to prevent crib-biting; **crib-rail**, a transverse member of the frame of a railway coach.

1844 Crib-bite [see WIND-SUCK *v.*]. **1860** HOTTEN *Dict. Slang* (ed. 2) 124 *Crib biter*, an inveterate grumbler; properly said of a horse which has this habit, a sign of its bad digestion. **1879** *Rep. Chief of Engineers, U.S. Army* II. 1588 (Knight), Crib breakwater. **1899** *Westm. Gaz.* 8 Dec. 2/1 What military engineers call a 'crib' bridge. **1879** *Punch* 3 May 201/1 A bludgeon as big As a crib-cracker's nobby persuader. **1883** G. R. SIMS *How the Poor Live* ii. 10 His talents as a 'cribcracker', and his adventures as a pickpocket. **1906** *Daily Chron.* 5 Dec. 6/6 The house is adequately protected against burglars and is proof against the amateur crib-cracker. **1852** *Punch* 9 Oct. 161/1 He..From cly-faking to crib-cracking turned. **1816** *Niles' Reg.* IX. Suppl. 164/2 These dams are built with timber, in the manner of crib dams, secured to the rocks below with iron bolts. *a* **1884** KNIGHT *Dict. Mech.* Suppl. 231/1 *Crib muzzle* (Manège), a muzzle used to correct the equine habit of cribbing. **1958** *Engineering* 14 Mar. 344/1 The body pillars, cantrail and cribrails are in 12 s.w.g.

crib, *v.* Add: **2. c.** To place (Indian corn, etc.) in a crib. *U.S.*

1719 J. HEMPSTEAD *Diary* (1901) 93, I went to Stonington and Stephen to Cribb the Corn. **1831** J. M. PECK *Guide for Emigrants* II. 151 The value of the crop, then, before it is cribbed, is one hundred and twenty-five dollars. **1874** E. EGGLESTON *Circuit Rider* i. 17 Now, boys, crib your corn. **1939** *These are our Lives* 256 The buyers would crib eight and ten thousand bushel of wheat and corn.

9. b. To complain, to grumble. *colloq.* Cf. *crib-biter.

1925 in FRASER & GIBBONS *Soldier & Sailor Words*. **1957** L. P. HARTLEY *Hireling* xi. 90 She calls on the neighbours, she's out half the time and doesn't answer the telephone, and when I start cribbing she just laughs.

cribellum (kribe·lǝ̆m). *Zool.* [L. *cribellum*, dim. of *cribrum* sieve.] An additional spinning organ, having numerous fine pores, situated in front of the spinnerets in certain spiders. Also **cri·bellate(d)** *adjs.*, having a cribellum.

1888 ROLLESTON & JACKSON *Forms Anim. Life* 524 In a few Spiders a chitinoid plate, the cribellum, lies in front of the spinnerets. **1926** T. H. SAVORY *Brit. Spiders* 97 The British cribellated spiders. **1951** LOCKET & MILLIDGE *Brit. Spiders* I. iii. 28 The presence of the cribellum is of importance in classification and the whole group is usefully divided into cribellate and ecribellate spiders. **1958** [see *CALAMISTRUM].

‖ **criblé** (krible), *sb.* (and *a.*). [Fr.: see CRIBBLED *ppl. a.*] A type of engraving on wood or metal (see quots.). Also as *adj.*, engraved in this way, = CRIBBLED *ppl. a.*

1879 *Encycl. Brit.* VIII. 437/1 A kind of wood engraving ..called the *criblé*.. .It means, *riddled with small holes*.. .The effect of light and dark is produced in this kind of engraving by sinking a great number of round holes of different diameters in the substance of the wood, which, of course, all come white in the printing. **1960** G. A. GLAISTER *Gloss. Bk.* 91/2 *Criblé*, minute punctures or depressions made in surfaces of wood or metal.. .Criblé backgrounds can be used to lighten borders which would appear too dark.

cribo (krəi·bo, krī·bo). [Origin unknown.] A large harmless snake, *Drymarchon corais*, found in tropical North, South, and Central America and the West Indies; also called *gopher snake* (*s.v.* GOPHER *sb.*[1] 4) and *indigo snake* (*s.v.* INDIGO C. 2).

1871 KINGSLEY *At Last* I. ii. 78 This snake..has no power against another West Indian snake, almost equally common, namely, the Cribo. **1918** W. H. HUDSON *Far Away* xii. 176 The Cribo of Martinique..kills and swallows the deadly fer-de-lance. **1957** A. H. & A. A. WRIGHT *Handbk. Snakes U.S. & Canada* I. 204 Mexican indigo snake.. .Other common names: Black gopher snake, blue bull snake, blue gopher snake, corais, cribo.

Crichton (krəi·tǝ̆n). The surname of James *Crichton* of Clunie (1560–85?), a Scottish prodigy of intellectual and knightly accomplishments; freq. qualified by *admirable*, it is used allusively for any person who excels in all kinds of studies and pursuits. Hence *Admirable Crichtonism*; also *Crichto·nian* adj.

The epithet which became traditional was first applied in Johnstone's *Heroes Scoti* (1603) 41 Iacobus Critonius Clunius, Musarum pariter ac Martis alumnus, omnibus in studiis, ipsis etiam Italis admirabilis. In English it appeared first in Urquhart's *Jewel* (1652) 112 The admirable Crichtoun..did ..present himself to epilogate this his almost extemporanean Comedie.

1812 T. AMYOT *Life of Windham* I. 139 [Windham] was the admirable Crichton of his age and country. *a* **1845** BARHAM *Ingol. Leg.* 3rd Ser. (1847) 264 Like a small boy at Eton, Who's not quite a Crichton. **1889** *Blackw. Mag.* CXLVI. 707/1 The..endless resource and Admirable Crichtonism of Robert Hazel. **1890** *Punch* 11 Jan. 15/2 *Algy*.. .You lucky dog, you possess all the accomplishments I lack! *Jim*...Oh, nonsense! Why, you're making me out a regular *Crichton*! **1900** *Academy* 3 Feb. 99/2 Matthew, the self-taught, listening intelligently to a German song while he is swimming, is a little too Crichtonian. **1914** 'IAN HAY' *Lighter Side School Life* ii. 47 In addition to all this, he must be an Admirable Crichton. **1914** J. M. BARRIE (*title*) The admirable Crichton. **1953** A. CHRISTIE *Pocket Full of Rye* iv. 30 Inspector Neele raised his eyebrows. 'The admirable Miss Crichton.' 'I find one must *know* how to do everything oneself.' **1959** *Listener* 5 Feb. 244/1 Other people, in some sort of Admirable Crichton situation, in fact exercise authority even though they are not in authority.

cricket, *sb.*[2] Add: **1. b.** The playing of the game of cricket.

1851 J. PYCROFT *Cricket Field* x. 200 Have me to bowl.. Box to keep wicket, and Pilch to hit, and then you'll see Cricket. **1857** HUGHES *Tom Brown* II. viii, Such a catch hadn't been made in the close for years...'Pretty cricket,' says the Captain. **1898** RANJITSINHJI *With Stoddart's Team* (ed. 3) x. 209 The dropped chances were the result of poor cricket on the part of the fieldsmen. **1904** P. F. WARNER *Recov. Ashes* viii. 150 The rain came down in torrents, and no cricket took place until 2.15.

c. Cricket as it should properly be played; hence *fig.*, fair play; honourable dealings with opponents or rivals; esp. in phr. *not cricket*; so *to play cricket*, to act fairly; to 'play the game'.

1851 J. PYCROFT *Cricket Field* xi. 210 We will not say that any thing that hardest of hitters..does is not cricket, but certainly it's anything but *play*. **1867** J. LILLYWHITE *Cricketers' Compan.* 13 Do not ask the umpire unless you think the batsman is out; it is not cricket to keep asking the umpire questions. **1900** *Westm. Gaz.* 5 June 2/2 We should be very much surprised if the Duke really thought that to dissolve would be 'cricket'. *Ibid.* 31 July 1/3 We believe that the feeling is very widespread that it would not be 'cricket' to get back to power again as the result of an appeal to the country. **1911** W. DE MORGAN *Likely Story* 313 It is scarcely fair play to make a merit of patience—isn't cricket, as folk say nowadays. **1922** *Daily Mail* 14 Nov. 10, I appeal to the Conservatives to do what is patriotic and honourable and to play 'cricket'. **1930** 'VAN DINE' *Scarab Murder* 20 It didn't seem cricket to leave the poor devil there. **1955** *Times* 21 July 5/7 When one was called to the Bar one was a public menace but one learnt what was and was not cricket during the period of pupilage.

2. *cricket-ball* (later example), *-cap, -club* (earlier and later examples), *-field* (earlier example), *-jacket, -match* (earlier and later examples), *-pitch, -player* (earlier and later examples), *-playing, -week*; **cricket-bag**, a long bag in which a cricketer's equipment is carried; **cricket-shoes**, shoes specially designed for wear on the cricket-field, usually having spiked soles.

1868 *Saint Pauls* II. 549 At Florence, Rome, or Naples, the unwonted spectacle of cricket-bags may startle the natives. **1904** P. F. WARNER *Recov. Ashes* xiii. 244 The rubber was won: the 'ashes' were in my cricket-bag. **1909** *Daily Chron.* 10 Feb. 5/3 A neighbouring cricket-ball factory. **1873** L. TROUBRIDGE *Jrnl.* in J. Hope-Nicholson *Life amongst Troubridges* (1966) ii. 9 A blue and white cricket cap on the back of his head. **1916** E. F. BENSON *David Blaize* iii. 35 The school eleven cricket-cap which he had won last week. **1731** *Daily Advertiser* 29 June, Eleven Gentlemen of Kent, belonging to Esquire Steed's Cricket-Club. **1887** F. GALE *Game of Cricket* 153 Cricket clubs are very much larger affairs than they used to be. **1787** in H. T. Waghorn *Cricket Scores* (1899) Frontispiece, Representation of the Noble Game of Cricket, as played in the celebrated Cricket Field near White Conduit House. **1827** E. NEALE *Living & Dead* 163 A dozen different cricket jackets. **1903** *Cricket* 30 Apr. (Advt.), Cricket Jackets. Navy, Melton, Trimmed Ribbon, or Cord. **1677** in T. Barrett-Lennard *Fam. Lennard & Barrett* (1908) 317 The crekitt match at ye Dick**. **1955** *Times* 9 May 15/1 One travels not only to see a cricket match as such. **1890** Cricket pitch [see PITCH *sb.*[2] 13]. **1654** in *N. & Q.* (1924) CXLVII. 325/1 Cricket players on ye Lord's Day. **1851** J. PYCROFT *Cricket Field* xi. 214 Pugilists have rarely been cricket players. **1700** *Post Boy* 30 Mar. in *N. & Q.* 10th Ser. (1904) II. 394/2 Gentlemen, or others, who delight in Cricket-playing. **1894** W. B. YEATS *Let.* 6 Aug. (1954) 234 Those vigorous fair-haired, boating, or cricket-playing young men. **1906** A. E. KNIGHT *Complete Cricketer* ix. 323 The Need of Cricket Playing. **1849** 'BAT' *Cricketer's Manual* (Advt.), Spiked soles for cricket shoes. **1908** W. E. W. COLLINS *Country Cricketer's Diary* ii. 27, I should doubt whether he ever owned a pair of cricket-shoes in his life. **1873** LD. W. P. LENNOX *Recoll. 1806–73* II. v. 106 Few gatherings are more delightful than the one that takes place annually at Canterbury during the cricket week. **1916** E. F. BENSON *David Blaize* xiii. 256 There's a cricket week at Baxminster, and they've asked me to play in two matches. **1968** R. V. BESTE *Repeat Instructions* xi. 121 A local festivity such as a cricket week.

cricket-bat. 1. = BAT *sb.*[2] 3 a.

1743 BULKELEY & CUMMINS *Voy. S. Seas* 130 The Indians had nothing in their Hands but a Club, like to our Cricket-Batts. **1860** [see CRICKET *sb.*[2] 2]. **1900** *Daily News* 9 June 6/5 Each has been engaged on his particular branch of the work of cricket bat making. **1964** W. L. GOODMAN *Hist. Woodworking Tools* 37 The cricket-bat maker's side-axe.

2. *cricket-bat willow* = *bat-willow* (see *BAT *sb.*[2] V).

1926 G. C. DRUCE *Flora of Bucks.* 314 Salix alba, var. caerulea. Cricket-bat Willow. This occurs as a female planted tree, sometimes 100 ft. high. **1933** *Discovery* Feb. 66/1 There is more than one botanical variety being grown under the name of 'cricket-bat willow'. **1955** G. GRIGSON *Englishman's Flora* 257 Cricket bats, from the Cricket Bat willow, which is *Salix alba* var. *coerulea*.

cricketer. In first quot. for 1770 read *c* 1742.

cricketess: see CRICKETRESS in Dict. and Suppl.

cricketing, *vbl. sb.* (Earlier and later examples.)

1771 in H. T. Waghorn *Cricket Scores* (1899) 74 After the cricketing there was a match at running. **1772** *Ibid.* 78 A cricket-match was played in Mr. Louch's cricketing-field. **1955** *Times* 16 July 2/5 One of the great events of the cricketing year.

cricketing, *ppl. a.* (Earlier example.)

1823 M. R. MITFORD in *Lady's Mag.* July 386/1 One club of cricketing dandies encounters another such club.

cricketress. (Later example of form *cricketess.*)

1963 *Times* 20 Apr. 3/7 (Advt.), Since '63 you've saved the guesses Of cricketers and cricketesses.

crickey, var. CRIKEY *int.*

crickle (kri·k'l), *v.* [Echoic.] *intr.* To make a sharp, thin sound; to make a succession of sharp sounds. Hence **cri·ckle** *sb.*

1883 STALLYBRASS tr. *Grimm's Teut. Myth.* III. 929 You hear him [*sc.* the wild hunter] bluster in the air, so that it 'crickles and crackles'. **1926** *U.F. Ch. Mission Rec.* Aug. 361/1 There ran a crickle of wind in the thatch.

crickle-crackle. Delete † *Obs. rare*[-1] and add later example.

1914 W. J. LOCKE *Fortunate Youth* xiii, I like to feel that He's in the wind or in the crickle-crackle of the earth.

‖ **cri de cœur** (kridəkör). Also **cri du cœur.** [Fr., lit. 'cry of' or 'from (the) heart'.] An utterance of distress or anguish; also *fig.*

1905 G. K. CHESTERTON in *Daily News* 1 July 6/4 After this innocent cri de cœur of mine, I thought at first there would be a fight. **1909** BEERBOHM *Yet Again* 125, I could no more 'dash off' this my *cri de cœur* than I could an elegy on a broomstick I had never seen. **1928** D. H. LAWRENCE *Lady Chatterley* vii. 88 She needed help, and she knew it; so she wrote a little 'cri du cœur' to her sister. **1950** A. L. ROWSE *Eng. of Eliz.* x. 414 The Bishop's pathetic *cri de cœur*.

crikey, *int.* Add: Also **crickey.** (Earlier and later examples).

1838 *Actors by Daylight* I. 24 Crikey, oh crikey! How very flat, stale and abominable Seem to me all the acting of this world. **1922** JOYCE *Ulysses* 223 Crickey, is there nothing for us to eat? **1924** D. H. LAWRENCE & SKINNER *Boy in Bush* 85 Crickey! Stop up another night! It'd make ye sawney. **1960** J. RAE *Custard Boys* I. i. 16 Crikey, I thought, he's tough.

crim, *sb. U.S.* and *Austral. slang* abbrev. of CRIMINAL *sb.* 2.

1909 in WEBSTER. **1953** K. TENNANT *Joyful Condemned* xxii. 212 Some of the fellow crims remarked on it in filthy language. **1970** *Tel.* (Brisbane) 29 June 4/1 (*headline*) Crims 'in turmoil'.

crime, *sb.* Add: **4.** *crime-drama, fiction, film, magazine, -mystery, play, -prevention, -proneness, -romance, -story; crime-promoting, -prone, -stained* adjs.; **crime-buster** *slang*, one who represses organized crime; **crime reporter**, a reporter (sense 2 b) who describes crimes and the trials of criminals; **crime sheet**, in the army, a list in which the names of offenders and their offences are entered; **crime wave,** a sharp rise in the incidence of crime; **crime-writer**, an author who writes about fictional crimes; hence *crime-writing*.

1952 *Manch. Guardian Weekly* 31 July 7/1 The tall prim crime-buster came forward. **1962** K. ORVIS *Damned & Destroyed* i. 9 A man who takes pleasure in being called crime-buster promises without delay to put vice on ice in Montreal. **1933** P. GODFREY *Back-Stage* ix. 119 An English school of crime-drama arose and flourished. **1924** C. S. MONTANYE in *Saucy Stories* Jan. 41/2, I write crime fiction for the magazines. **1936** *Daily Tel.* 16 Oct. 7/2 'Middle Class Murder' is first-class crime fiction. **1935** *Ann. Reg. 1934* II. 307 A cinema in Chicago where he was enjoying a 'crime film'. **1940** J. CARY *Charley is My Darling* xi. 63 Charley is telling stories out of some crime magazine. **1906** *Daily Chron.* 26 May 7/5 The many strange features of the case certainly make it one of the most remarkable crime-mysteries which have occupied the attention of the London police for some years. **1952** GRANVILLE *Dict. Theatr. Terms* 53 *Crime play*, a thriller featuring murder, theft, or crime generally. **1928** HECHT & MACARTHUR *Front Page* II. 71 The time to catch 'em is while they're little kids. That's the whole basis of my crime prevention theory. **1944** 'G. ORWELL' in *Horizon* Oct. 240 The distinction between crime and crime-prevention practically disappears. **1823** MILL *Autobiogr.* (1924) 273 The Catholic priesthood added.. the crime-promoting doctrine of indulgences. **1960** L. T. WILKINS *Delinquent Generations* 1 Children born during the war might be more crime-prone than others. *Ibid.* 8 The greatest 'crime-proneness' is thus found to be associated with that birth group. **1936** 'J. TEY' *Shilling for Candles* xix. 200 He..might be only a crime reporter, but he knew just as much about crime..as any police force. **1951** WODEHOUSE *Old Reliable* iv. 48 Bill in her time had been..crime reporter, sob sister, [etc.]. **1906** *Daily Chron.* 15 May 3/5 All the stock characters of..crime-romance are here. **1915** D. O. BARNETT *Lett.* 41 Spent the morning filling up 'crime sheets' with all their offences. **1917** EMPEY *From Fire Step* 150 The Sergeant-Major keeps what is known as the Crime Sheet. When a man commits an offence, he is 'Crimed'— that is, his name, number, and offence is entered on the Crime Sheet. **1898** *Westm. Gaz.* 14 July 2/1 She wondered.. what manner of hideous crime-stained countenance that paper hid. **1934** J. CARTER *New Paths in Book Coll.* 35 Mystery stories, crime stories, spy stories. **1936** 'G. ORWELL' *Keep Aspidistra Flying* xi. 295 Fools demanding crime-stories and sex-stories and romances. **1920** *Times* 21 Jan. 12/1 (*headline*) Crime Wave. Murder, robbery and theft. **1945** *Daily Mirror* 8 Dec. 1/4 Scotland Yard..made a new move in its war against the mounting crime wave. **1946** 'M. INNES' *What happened at Hazelwood* III. i. 169 Amateur crime-writers are just as painfully incompetent as amateur actors. **1959** *Times Lit. Suppl.* 24 Apr. 238/5 A category-debate about whether crime-writing ought to grow closer to the straight novel and away from the master-mind detective story.

crime, *v.* Delete *rare* and add further examples. Now esp. in army use.

1917 [see *crime sheet]. **1929** C. E. MONTAGUE in *Mercury Story Bk.* 178 You know, Sergeant, the sort of a squadroon it is where a man's never crimed. **1957** N. SQUIRE *Theory of Bidding* 213 East here has used his judgment.. He has been slightly optimistic but cannot be crimed. **1957** E. HYAMS *Into the Dream* 91 He'd crime a man as soon as look at him.

Crimean (krəimī·ăn), *a.* [f. *Crimea*, name of a peninsula lying between the Sea of Azov and the Black Sea, the chief seat of a war (1854–6) between Russia and Turkey (with its allies).] Of, pertaining to, or characteristic of the Crimea; *Crimean Gothic*, name given to an East Germanic language, supposedly a dialect or descendant of Gothic, which continued to be used in the Crimea down to the sixteenth century; *Crimean shirt*, a shirt worn by workers in the Australian and New Zealand bush (also *Crimea shirt*).

[**1591** G. FLETCHER *Russe Commonw.* XIX. 72[v] Some thinke that the Turkes tooke their beginning from the nation of the Chrim Tartars.] **1855** F. NIGHTINGALE *Let.* in C. Woodham-Smith *F. Nightingale* (1951) x. 231, I have now had all that this climate can give. Crimean fever. Dysentery. Rheumatism. **1855** (*title*) The illustrated Crimean war song-book of the allies. **1862** R. HENNING *Let.* 29 Aug. (1966) 96 He..had seen fit to array himself in a Crimean shirt which he put on *over* his other clothes. **1893** K. MACKAY *Out Back* (ed. 2) I. ix. 108 Crimean shirts, tight-cut moles, and light square-toed bluchers completed their costume. **1895** 'R. BOLDREWOOD' *Crooked Stick* iii. 80 A young man, whose Crimean shirt and absence of necktie denoted..the presumed abandon of bush life. **1899** M. BEERBOHM in *Sat. Rev.* 14 Jan. 45/1 An actor made up as a young 'swell', with glengarry, cheroot, and 'Crimean beard'! **1902** W. SATCHELL *Land of Lost* ii. 14 He wore a blue Crimea shirt open at the throat. **1913** J. D. JONES tr. *Loewe's Germanic Philol.* I. v. 16 Crimean Gothic can have been no real Gothic dialect, as it did not undergo the different changes common to East and West Gothic. **1927** M. M. BENNETT *Christison* x. 110 Christison gave Mickey a Crimean shirt. **1943** C. L. WRENN *Word & Symbol* (1967) 137 The famous 'Crimean Gothic'.. rests on no other foundation than that of a hastily-written list of some seventy words and phrases which Busbecq thought he heard from two men who had been in the Crimea. **1958** A. S. C. ROSS *Etymology* II. 74 Germanic..falls into three groups, the first containing only Gothic (texts of IV and VI c. A.D., also fragments of Crimean Gothic recorded 1560 A.D.), the other two being North and West Germanic. **1971** *Times* 1 Jan. 11/3 If some Baptists or Crimean Tatars or dissident writers tried to hijack an aircraft they would almost certainly be more severely dealt with than ordinary citizens.

‖ **crime passionnel** (krīm pasyonęl). Also **crime passionel.** [Fr.] A crime due to passion; *spec.* a murder resulting from jealousy.

1910 *Encycl. Brit.* X. 86/1 In cases of what is termed 'crime passionel', French juries..almost invariably find extenuation. **1912** J. MASEFIELD *Widow in Bye Street* VI. 84 Two lawyers talked statistics, '"Crime passionel" in Agricultural Districts'. **1930** *Punch* 7 May 529/3 We too find the literature of the *crime passionnel* a trifle soporific. **1943** M. F. RODELL *Mystery Fiction* v. 35 The *crime passionel*, the crime of jealousy committed on the spur of the moment, makes poor material. **1955** *Times* 13 July 6/5 The case of Mrs. Ellis has..demonstrated in an extreme degree the fundamental difference between French and English law in relation to the *crime passionel*. **1969** *Times* 17 Nov. 6/6 A woman has been murdered in a *crime passionnel*.

crimes (krəimz), *int.* Later modification of CRIMINE *int.*

1874 in HOTTEN *Slang Dict.* s.v. *Crikey*. **1891** FARMER *Slang*, Crimini, Criminey, or Crimes. **1929** E. RAYMOND *Family that Was* II. ii, Crimes! I'm moving in the Upper Ten, I think.

criminal, *a.* Add: **2. b.** *criminal court*, a court (first in Scotland) having jurisdiction over criminal prosecutions.

[**1597** *Lawes & Actes* Table, s.v. *Courtes*, All Courtes, Civill and Criminall, suld be fensed at elleven houres before noone.] **1678** MACKENZIE *Laws & Customes* 5 The decisions of our Criminal Court..do bind the same or succeeding Judges. **1788** *Desp. Paris* II. 53, I herewith send your Lordship the Resolutions of the Châtelet (one of the Criminal Courts). **1834** *Act 4–5 Will. IV* c. 36 § 1, That the Lord Mayor.., the Lord Chancellor.., and all the Judges..shall be and be taken to be the Judges of a Court to be called the 'Central Criminal Court'. **1910** *Encycl. Brit.* VII. 322/2 *Criminal Courts.* — (1) The lowest is that of the justice of the peace, sitting..to determine in a summary way certain specified minor offences.

c. *criminal code*, a system of jurisprudence to be applied in criminal cases.

1788 *Desp. Paris* II. 91 Great offers have been made to the Grand Conseil to induce that Tribunal to recognize the new Criminal Code. **1829** *Observer* 19 Apr. 1/4 It is in vain that we pass measures for modifying the severity of our Criminal Code. **1878** *Times* 18 July 10/3 The Criminal Code.—The course..taken with reference to the Criminal Code Bill, of referring it to a Commission..will give..the assurance of a most acute and careful investigation.

criminalistic (kri:minăli·stik), *a.* [f. CRIMINAL *a.* and *sb.* or CRIMINAL(ISM + -ISTIC; cf. next.] Of or pertaining to criminalism; with a tendency towards criminality.

1924 F. H. ALLPORT *Social Psychol.* vi. 136 The converse, that all who have such attitudes are criminalistic, is far from true. **1936** S. & E. GLUECK *Preventing Crime* i. 3 The more 'inflammables' (such as poverty, broken and distorted home life..and the like) that can be removed from the environment of childhood and youth, the less possibility is there of criminalistic conflagration. **1945** —— *After-conduct of Discharged Offenders* vii. 84 The acquisition of a certain degree of maturation..is significantly related to changes in criminalistic behaviour. **1967** A. BOUCHER *Best Amer. Detective Stories* 221 Locard was..one of the great initiators of modern criminalistic science.

criminalistics (kri:minăli·stiks), *sb. pl.* [ad. G. *kriminalistik* (Gross 1897), f. CRIMINAL *a.* and *sb.* + -ISTIC.] (See quot. 1949.)

1949 O'HARA & OSTERBURG *Criminalistics, The Application of the Physical Sciences to the Detection of Crime* p. x, The authors have decided, for the purposes of the present text, to use the name *criminalistics* in referring to the work of the police laboratory. This is not entirely a neologism. The words *Kriminalistik, criminalistique*, and *criminalistica* are in common use in continental Europe.. We shall define *criminalistics* as that science which applies the physical sciences in the investigation of crimes. **1961** L. RADZINOWICZ *In Search of Criminology* ii. 34 Too large a place was assigned to the technical aspects of criminalistics.

criminaloid (kri·minăloid). [f. CRIMINAL *sb.* + -OID.] A man with a tendency towards crime; a first offender as opposed to a habitual criminal.

1895 tr. *Lombroso & Ferrero's Female Offender* 308 A middle type between criminaloids and born criminals. **1909** *Westm. Gaz.* 5 Jan. 2/1 The common incarceration of offenders manufactures the criminal out of the juvenile, the criminaloid and the single offender.

crimmer, var. *KRIMMER.

crimp, *sb.*[3] Add: **1.** (Earlier examples.) Also *sing.*, a curl or (artificial) wave of the hair.

1867 A. D. WHITNEY *L. Goldthwaite* v. 97 I've brushed out half my crimps. **1870** L. M. ALCOTT *Old-Fashioned Girl* i. 1. It's too wet. Shouldn't have a crimp left if I went out such a day as this.

b. (See quot. 1863.) orig. *U.S.*

1863 H. S. RANDALL *Pract. Shepherd* (ed. 7) vii. 75 Regularity and distinctness of 'crimp'—that curved and graceful form and arrangement of the locks and fibers in the sheared fleece which indicate extreme pliancy. **1874** *Rep. Vermont Board Agric.* II. 410 Fineness of fiber can be judged by its appearance to the eye, by its feeling when touched and by its fineness of crimp. *Ibid.* 411 Style of wool is judged by its crimp; the number of crimps to an inch of very fine wool is from twenty-seven to twenty-nine. **1956** G. BOWEN *Wool Away!* (ed. 2) xii. 143 The wool..is remarkably heavy, with..a very pronounced crimp.

c. *to put a crimp in* or *into*: to thwart or block; to impair or interfere with. *U.S. slang.*

1896 ADE *Artie* xii. 106 They'll put a crimp in him if things come their way. **1911** H. S. HARRISON *Queed* xxv. 321 They never forgive a man who puts a crimp into the party. **1918** *Nation* (N.Y.) 7 Feb. 166/1 All plans passed through their hands, and they took particular pleasure in putting a crimp into the Fire God. **1939** WODEHOUSE *Uncle Fred in Springtime* vi. 80 My prestige in the home is already low, and a substantiated charge of being A.W.O.L. would put a further crimp in it. **1969** *New Yorker* 27 Dec. 26/3 Finally, a giant black panther leaps upon me and devours my mind and heart. This puts a terrific crimp in my evening.

crimper[1]. Add: **1. b.** A hairdresser. *slang.*

1968 D. GRAY *Died in Red* ix. 52 She was then put under a drier. Twenty minutes or so later she was again taken to the stylist (or crimper as they called themselves). **1968** 'E. TREVOR' *Place for Wicked* i. 6 He'd opened up as a crimper..decorating the salon and supervising the work himself.

2. (Earlier example.)

1877 E. S. PHELPS *Story of Avis* ii. 31, I think I could have patented a crimper that would make a simpler system of punctuation in your finger than this.

crimping, *vbl. sb.*[1] Add: **3.** *crimping board.*

1837 W. JENKINS *Ohio Gaz.* 158 A last factory, producing 14,000 lasts, 200 boot trees, and 200 crimping boards per annum. **1969** E. H. PINTO *Treen* 150 Crimping boards and their correspondingly grooved or serated rollers..were used for forming the minute crimpings, gatherings or ruckings, on 17th- and early 18th-century fabrics.

crimple, *sb.* Delete † *Obs.* and add later examples. Now *dial.* and *U.S.*

1844 *Lowell* (Mass.) *Offering* IV. 148 My paper is full, and I can only say ribbons, bows,..wimples, and crimples. **1862** C. C. ROBINSON *Dial. Leeds* 278 Where the breadth of muslin was narrow the 'crimple' was made by means of a penknife and the thumb. **1881** S. P. McLEAN *Cape Cod Folks* vi. 131 Teacher,..how shiny those crimples in your hair look, with that streak of sun lighting on 'em!

crimpy (kri·mpi), *a.* [f. CRIMP *sb.*[3] or *v.*[1] + -Y[1].] Having a crimped appearance; frizzy.

1888 [see CRIMP *sb.*[3] 1]. **1894** *Daily News* 24 July 5/7 The special petroleum wash dried more quickly..and left the hair more crimpy. **1923** E. BOWEN *Encounters* 91 A woman, with bright crimpy hair.

cri·msoning, *vbl. sb.* [f. CRIMSON *v.* + -ING[1].] Crimson colour or colouring.

1873 G. M. HOPKINS *Jrnl.* 3 Nov. (1959) 240 Balks of grey cloud searched with long crimsonings. **1898** —— *Poems* (1930) 147 When the Rose ran in crimsonings down the Cross-wood.

crimsony (kri·mzəni), *a.* [f. CRIMSON *a.* + -Y[1].] Somewhat crimson; resembling crimson.

1844 *Florist's Jrnl.* V. 17 The flowers differ in having the violaceous tint not *upon* the crimsony scarlet..but really *mixed* with the crimson in the texture of the flower. **1905** HOLMAN HUNT *Pre-Raph.* II. 70 With a crimsony lustre. **1909** M. B. SAUNDERS *Litany Lane* I. vi, The dim crimsony browns, the curious blue-greens.

crin (krin, ‖ kræ̃n). [Fr., = horsehair.] A fabric made from horsehair alone or combined with some other fibre; = CRINOLINE *sb.* 1 a; so *crin végétal*, vegetable horsehair (see VEGETABLE *a.* 7 a).

[**1858** SIMMONS *Dict. Trade*, *Crin*, the French name for horsehair.] **1875** *Encycl. Brit.* I. 565/1 The most important fibre is the *crin vegetal*, or vegetable horse hair, produced from the dwarf palm (*Chamærops humilis*), with which a vast portion of the uncultivated parts of the country [*sc.* Algeria] is covered. **1900** *Daily News* 9 June 6/6 The crin or horsehair (sometimes called crinoline) hat is poetic enough. **1900** *Traveller* 4 Aug. 112/2 A big floppy *crin* hat, trimmed with..black lace. **1909** *Westm. Gaz.* 3 June 8/3 There is a large hat of black crin. **1923** *Daily Mail* 9 Feb. 11 A similar shape of blue crin straw and satin ribbon. **1963** *Times* 3 June 11/5 A beret of crin sprinkled with chenille bobbles.

crinkled, *ppl. a.* Add: **crinkled (tissue) paper,** paper that is crinkled, made in various colours, and used for making paper flowers and for decorative purposes; = **crêpe paper.*

1895 *Army & Navy Co-op. Soc. Price List* 526 Crinkled tissue paper. For making Fancy Lamp Shades. **1921** *Dict. Occup. Terms* (1927) § 559 Crinkled paper machine operator. **1929** E. BOWEN *Last September* viii. 91 She used to plait hats out of crinkled paper.

cri·nkle-root. *U.S.* = PEPPERWORT 1 b.

1847 *Knickerbocker* XXIX. 377 Sassafras is 'coming good' now too in the woods; and so is 'crinkle-root'. **1899** H. VAN DYKE *Fisherman's Luck* iv. 74 Crinkle-root is spicy, but you must partake of it delicately, or it will bite your tongue.

crinkliness (kri·ŋklinės). [f. CRINKLY *a.* + -NESS.] Crinkly condition.

1927 R. A. FREEMAN *Certain Dr. Thorndyke* xv. 228 Mr. Wamploe's advice produced on Polton's countenance a smile of most extraordinary crinkliness.

crinkly, *a.* Add: **b.** Characterized by a succession of crinkling sounds.

1827 J. WILSON *Noctes Ambr.* in *Wks.* (1855) I. 330 Does my voice come from my heart in a crinkly cough, as if the lungs were rotten? **1911** HUGH WALPOLE *Mr. Perrin* iii, The white, crinkly sound of the silk of her dress against the table.

crinolined, *ppl. a.* Add: (Later examples.) Also *fig.*
1927 E. SITWELL *Rustic Elegies* 78 Like the crinolined waterfalls. 1930 —— *Coll. Poems* 222, I sat at my dressing-table—that chilly Palely crinolined water-lily. 1934 LD. BERNERS *First Childhood* 200 A crinolined damsel seated at the piano.

crio-. Add: **cri·ophore,** a statue or other representation of a figure carrying a ram (1909 in WEBSTER); so **crio·phoric, crio·phorous** *adjs.*
1921 G. A. F. KNIGHT *Nile & Jordan* xiii. 160 Criophorous sphinxes lined the avenues of Thebes and other cities. 1952 *Jrnl. Theol. Stud.* III. 90 The criophoric figures, such as those of Hermes, which anticipate later Christian representations of the Good Shepherd.

criollo (kriǫ·lo). [a. Sp. *criollo* native to the locality: see CREOLE.] A variety of cocoa tree, *Theobroma cacao,* native to Central America; also, a name for high-quality cocoa or cocoa beans. Also *attrib.*
1908 H. H. SMITH *Cacao Planting* 39 Strong growing forastero or calabicillo, bearing the finest criollo beans. *Ibid.*, Good buds from pure criollo stocks. 1929 *Encycl. Brit.* V. 947/1 The finest type of bean the criollo, is grown in Venezuela, Ceylon, Java, Samoa, Madagascar and Nicaragua; but in Ceylon, and other places, criollo is being replaced by forastero. *Ibid.* XXIII. 52/1 There are two grades of Venezuelan cacao—the *criollo* or native, and the *trinitario* or Trinidad. 1955 D. H. URQUHART *Cocoa* ii. 15 Cross-fertilization took place with the Criollo trees and when seedlings were raised from them they were no longer pure Criollo. *Ibid.* xiii. 156 Criollos and allied types were the cocoas with distinctive flavour.

crip. *U.S.* colloq. abbrev. of CRIPPLE *sb.*
1918 'A—No. 1' *Mother Delcassee of Hoboes* 43 Straight crip, actually crippled or otherwise afflicted. *Phoney crip*, self-mutilated or simulating a deformity. *Ibid.* 84 Going to play the 'phoney crip' act for a spell, Dallas Bob? 1942 BERREY & VAN DEN BARK *Amer. Thes. Slang* § 120/2 *Crip*, a crippled animal. 1950 HEMINGWAY *Across River* viii. 62 He only loved people, he thought, who had fought or been mutilated.. So I'm a sucker for crips, he thought.

cripes (kraips), *int.* Vulgar perversion of CHRIST in the exclamation *(by) cripes!* (Cf. *CRIMES *int.*)
1910 A. H. DAVIS *From Selection to City* xii. 107 'By cripes!' he gasped,..'I've lost th'..th' *cheques*!' 1911 L. STONE *Jonah* I. ix. 100 Stinky spat on his hands, and seized the wooden mallet. Cripes! he would show Pinkey which was the better man of the two. 1919 E. H. JONES *Road to En-Dor* i. 2 'What's the suggestion?' Alec asked. 'Spooking,' said I. 'Cripes!' said Alec. 1929 *Sunday Dispatch* 13 Jan. 2/4 You've sold me a pup! But, by cripes, I'll..let you have it back. 1930 J. B. PRIESTLEY *Angel Pavement* iv. 177 That's what it was—oh, cripes!—awful hole! 1937 N. MARSH *Vintage Murder* vi. 59 By cripes,.. I'm sorry. 1952 A. F. GRIMBLE *Pattern of Islands* 68 The captain goggled at me for a second, 'Cripes!' he said.

cripple, *sb.* Add: **1. b.** A cattle disease. Also in *pl. dial.* and *Austral.*
1897 *Penrith Obs.* 7 Dec. (E.D.D.), Ass t'coo doctor what ails a coo when it'll eat a body's kytle, er owt else but gerse —that's cripple. 1929 *Times* 1 July 15/6 Lack of minerals in pastures causes innumerable diseases, such as..'cripples' ..in Australia.
3. (Earlier and later examples.)
1675 *New Jersey Archives* (1880) I. 115 The great Swamp or Cripple which backs the said two Necks of land. 1832 J. F. WATSON *Tales Olden Time* 57 Through that cripple browsed the deer. 1942 *Sat. Even. Post* 5 Sept. 11/1 When they came to the cripple he sloshed straight through.

cripplingly (kri·pliŋli), *adv.* [f. CRIPPLING *ppl. a.* + -LY[2].] So as to cripple or disable.
1899 *Daily News* 18 Feb. 6/3 The new skirts are very long, cripplingly so. 1927 *Daily Express* 4 Nov. 3/4 The damage, the cost, the contamination are cripplingly unbearable. 1955 *Times* 1 July 16/3 The standard rate of income tax.. remains at a cripplingly high level.

crise. Delete † *Obs.* and add later examples. Also in various French phrases, esp. *crise de* (or *des*) *nerfs,* an attack of nerves, a fit of hysterics.
1921 W. J. LOCKE *Mountebank* xxi. 272 Reason enough for a *crise de nerfs*. Even I, who had nothing to do with it, found my equilibrium disturbed. 1922 M. ARLEN *Piracy* II. i. 72 'Virginia has got a *crise*,' Lois Lamprey commented. 1923 W. J. LOCKE *Moordius & Co.* vii. 96 Sometimes these *crises de nerfs* are dangerous. 1933 'G. ORWELL' *Down & Out* xx. 148 The cook usually had a *crise de nerfs* and a flood of tears. 1946 E. TAYLOR *Palladian* v. 49 When the time came for her to go to the party..there would be a painful *crise des nerfs*. 1962 GREGOR & NICHOLAS *Moral & Story* v. 133 The *crise de conscience* that faces the serious contemporary novelist. 1962 N. MARSH *Hand in Glove* ii. 44 I'm afraid your Pixie has created a parochial *crise*. 1962 *Punch* 23 May 808/1 When there is a *crise d'amour* it is very peaceful for me. 1963 A. HARTLEY *State of England* iii. 79 The Algerian *crise de conscience* that dissolved the authority of the French state. 1970 P. BAIR *Tribunal* III. v. 201 What then caused this *crise de conscience*? 1970 *New Yorker* 10 Oct. 178/2 She has been a splendid advertisement for the benefits of a happy marriage—conspicuously more relaxed, far less subject to those old *crises des nerfs*.

crisis. Add: **6.** *attrib.* and *Comb.*
1841 *Times* 11 May 5/1 It may disappoint the crisis-mongers to hear us say so. 1896 *Westm. Gaz.* 23 June 3/1 All the aspects of a crisis night. 1898 *Ibid.* 4 Jan. 2/2 A Tory Government was 'crisis proof'. 1898 *Ibid.* 26 Mar. 5/1 The 'crisis'-less years of the late Liberal Government. 1898 *Ibid.* 24 June 2/3 The Near and not the Far East..was the crisis-centre. 1900 *Ibid.* 11 May 2/2 A crisis-avoiding peace-compelling Government. 1903 *Ibid.* 3 Jan. 2/3 A Crisis Fund, amounting to nearly two millions. 1938 E. WAUGH in *Tablet* 23 July 112/1 The crisis-minded always maintain that the problems of their particular decade are unique and insuperable. 1938 *Punch* 10 Aug. 163/1 How many of these people are crisis-conscious? 1939 WYNDHAM LEWIS *Let.* 5 Oct. (1963) 266 In the crisis-days prior to the war. 1940 W. EMPSON *Gathering Storm* 65 The point is to join up the crisis-feeling to what can be felt all the time in normal life. 1960 *Times* 24 Oct. (Financial Rev.) p. viii/6 Switzerland..has been a normal haven for 'crisis' money. 1965 H. KAHN *On Escalation* xiii. 245 Crisis-management problems.

crisp, *a.* **6. b.** Add to def.: clean, neat; clearly defined.
1937 D. M. JONES *In Parenthesis* IV. 97 The sky overhead looked crisp as eggshell. 1961 *Listener* 21 Dec. 1069/1 The superbly detailed capitals, as crisp now as they were 150 years ago. 1963 *Ibid.* 24 Jan. 160/1 A rational, crisp, glass-and-rectangular-framework architecture.

crisp, *sb.* Transfer † *Obs.* to senses in Dict. and add: **6.** An overdone piece of anything cooked, fried, or roasted; usually in phr. *to a crisp.* Also *transf.* orig. *U.S.*
a 1852 F. M. WHITCHER *Widow Bedott Papers* (1856) xxix, One time, they'll burn their bread to a crisp. 1852 H. MELVILLE *Pierre* XII. ii. 263 Burn it till it shriveled to a crisp! 1899 W. C. MORROW *Bohem. Paris* 44 It was some-time before Haidon could realize that he was not burned to a crisp. 1911 H. S. HARRISON *Queed* xxi. 258, I became absorbed in a book I was reading, and Jim came back to find the bacon a crisp. 1959 A. HUXLEY *Let.* 29 Nov. (1969) 881 Temperatures are in the eighties—..the vegetation..is burnt to a crisp. 1968 K. WEATHERLY *Roo Shooter* 35 The hot weather continued and the feed had dried to a crisp.
7. In full *potato crisp.* A thin sliver of potato fried until crisp and eaten cold. Usu. in *pl.* of such food produced commercially.
1929 *Star* 21 Aug. 13/3 Potato Crisp Factory. 1935 H. NICOLSON *Let.* 3 Sept. (1966) I. 213 We went to Harry's Bar..and there was a Pekinese being fed with crisps. 1950 T. S. ELIOT *Cocktail Party* I. 12 Potato crisps? No, I can't endure them. 1953 'N. BLAKE' *Dreadful Hollow* 43 It [*sc.* a public bar] held only one occupant, who was eating out of a packet of crisps. 1961 'T. HINDE' *For Good of Company* iv. 41 She..sloped her little red mouth and let a fingerful of crisps into it.

crisp, *v.* Add: **1. b.** To fold (cloth) which has just been woven.
1892 in *Eng. Dial. Dict.* 1927 *Daily Tel.* 21 June 8 The cloth may be crisped (folded lengthwise), rolled or lapped.

crispbread (kri·spbred). [f. CRISP *a.* 5 a + BREAD *sb.*] A food made from crushed whole grains such as rye and wheat, prepared in the form of thin crisp biscuits.
1926–7 *Army & Navy Stores Catal.* 61/1 Ryvita (Crisp-bread)—pkt. 1/6. 1927 *Lancet* 11 June 1244/2 *Vita-wheat Crisp Bread*... The crisp bread has a pleasant texture and flavour. 1928 *Daily Mail* 25 July 8/5 (Advt.), This crisp-bread [Vita-Wheat] keeps you slim. 1951 *Good Housek. Home Encycl.* 427/1 Crispbread may be substituted for bread or toast at most meals, and is enjoyed with cheese.

crisper. Add: **b.** A container in a refrigerator in which vegetables and fruit can be kept crisp and fresh.
1960 *Housewife* May 92/2 A..moisture-seal crisper, in which to store your fruit and vegetables. 1960 *Sunday Express* 15 May 17/6 Full-width crisper for fruit and vegetables.

crispish (kri·spiʃ), *a.* [f. CRISP *a.* + -ISH[1].] Somewhat crisp.
1930 WODEHOUSE *Very Good, Jeeves!* vi. 142 When not pleased Aunt Dahlia, having spent most of her youth in the hunting-field, has a crispish way of expressing herself. a 1948 E. ANDERSON in B. James *Austral. Short Stories* (1963) 333 Pork..with its crispish, tannish, delectable-looking strips of crackling.

crispy, *a.* Add: **2. b.** *crispy noodles,* crisp fried noodles served with Chinese food.
1940 A. SIMON *Concise Encycl. Gastronomy* II. 55/2 'Crispy Noodles'. Roll this dough out very thinly and cut into strips as thin as spaghetti... Throw into boiling oil or frying fat, frying a delicate brown. 1961 B. ALDISS *Primal Urge* ii. 41 They ate their chow mein, sweet and sour pork and crispy noodles. 1969 *Guardian* 27 Dec. 13/3 He took..[a] job..as a waiter in a Chinese restaurant.. fetching and carrying No. 31 with crispy noodles and No. 13 with soft.

criss (kris). [Variant form of w. midl. dial. *cress* = CREASE *sb.*[2] (sense 3).] The curved top of the stand on which tiles are made.
1881 *Instr. Census Clerks* (1885) 87 Brick, Tile-Maker, Burner, Dealer, Criss Maker. 1921 *Dict. Occup. Terms* (1927) § 474 *Criss maker* (tile making); a carpenter who makes criss.

Crissake(s), var. *CHRIS(S)AKE. Also **c-.** orig. *U.S.*
1923 J. DOS PASSOS *Streets of Night* i. 13 Go ahead, Fanshaw, for crissake, we can't wait here all day. 1930 —— *42nd Parallel* 72 Say wake up Ike for crissake. 1939 C. R. COOPER *Designs in Scarlet* v. 74 Get screwball for Crissakes, get wacky! 1959 H. HOBSON *Mission House Murder* xii. 81 For Crissake, Ma.

criss-cross, *sb.* Add: **2. b.** A network of intersecting lines.
1881 C. DE KAY *Vision of Nimrod* x. 179 The country gleaming With silvery crisscross of canals. 1901 *Daily Express* 28 Aug. 4/6 A great boulevard..hemmed in all round by the criss-cross of narrower streets. 1928 A. BENNETT *Strange Vanguard* xxxii. 214 A criss-cross of streets dotted with a thousand towers. 1944 S. PUTNAM tr. *E. da Cunha's Rebellion in Backlands* i. 8 An irregular line of low hills..forms a confused crisscross over the entire breadth of the Campos.
c. *fig.* The state of being at cross-purposes.
1907 *Daily Chron.* 23 Feb. 3/1 The practice of one manufacturing country assisting another with the sinews of war was described by Mr. Zangwill as 'a topsy turvy criss-cross, and Gilbertian'. 1909 *Westm. Gaz.* 20 Sept. 2/3 The absurd criss-cross of the authorities who look after us at playtime.

criss-cross, *v.* Add: **2.** *intr.* To intersect or cross repeatedly; also *fig.*
1953 G. E. M. ANSCOMBE tr. *Wittgenstein's Philosophical Investigations* I. § 66, We see a complicated network of similarities overlapping and criss-crossing. *Ibid.*, The various resemblances between members of a family.. overlap and criss-cross in the same way.

crista (kri·stă). Pl. **cristæ.** [L., = crest.] A ridge or crest; *spec.* in various anatomical and zoological senses.
1849–50 J. WEALE *Dict. Terms Archit.*, *Crista*, a crest; the apex or highest part of a shrine. 1889 *Jrnl. Morphology* III. 300 The crista, a cord-like ridge running the full length of the dorsal surface of the capsule. 1940 *Proc. R. Soc.* B. CXXIX. 257 The possibility of a two-way response of the cristae. 1964 J. Z. YOUNG *Model of Brain* vi. 97 Each [statocyst] contains three main types of receptor: (1) a macula, vertically placed and carrying a statolith; (2) a ridge of cells with long hairs, the crista; [etc.].

cristobalite (kristǭ·bălait). *Min.* Also erron. **christobalite, crystobalite.** [ad. G. *cristobalit* (G. vom Rath 1887, in *Neues Jahrbuch für Min.* I. 198), f. the name of Cerro San Cristóbal, near Pachuca, Mexico, where it was first found: see -ITE[1].] One of the three main forms of silica (the others being quartz and tridymite), formed at high temperatures as *high-cristobalite* and changing at lower temperatures to a structurally related metastable polymorph, *low-cristobalite,* which occurs both massive (e.g. in opal) and as small, usu. octahedral, crystals.
1888 *Min. Mag.* VIII. 36 A New Mineral. Cristobalite... An analysis..gave 91 per cent. of silica and 6 per cent. of oxide of iron and alumina. 1920 *Brit. Museum Return* 144 Artificial minerals (cristobalite, fayalite, rhodonite, apatite, spinel..) from furnace slags. 1935 *Discovery* July 206/1 When ware is fired above a certain temperature a very large number of extremely fine crystals of a mineral (cristobalite) are formed. 1959 *Chambers's Encycl.* IX. 423/1 Both cristobalite (stable 1,470°–1,710° C) and tridymite (stable 870°–1,470° C) are known to occur in certain natural lavas. 1962 C. FRONDEL *Dana's Syst. Min.* (ed. 7) III. 287 In recent years..study has shown that opal..is a submicrocrystalline aggregate of crystallites of cristobalite.

crit. Restrict † *Obs.* to sense in Dict. and add: **2.** Short for CRITICISM 2, 3, CRITIQUE 1. *colloq.*
1908 D. H. LAWRENCE *Let.* 13 May (1962) I. 11, I could write crits.—but..who would have 'em? 1952 W. GRANVILLE *Dict. Theatr. Terms* 53 *Crit*, short for *critique* or the notice of a first-night performance. 1966 R. TRICKETT *Elders* vi. 78 Someone had written a practical crit. of Collier's poem with an interpretation of that rather obscure stanza. 1970 K. GILES *Death in Church* iii. 69 The inquest stuff won't be so bad, more like a send-up than a crit.
3. Short for *critical mass* or *size* (see *CRITICAL *a.* 7 b). *colloq.* (orig. *U.S.*).
1957 in *Gloss. Terms Nuclear Sci.* (Nat. Res. Council, U.S.) 39/1. 1958 [see *CRITICAL *a.* 7 b].

criteriological (kraitiᵊ:riǒlǫ·dʒikăl), *a.* [f. next + -ICAL.] Pertaining to criteriology; dealing with criteria. Hence **crite:riolo·gically** *adv.*
1936 *Downside Rev.* LIV. 110 A criteriological problem—the difference between sense perception and rational knowledge. 1938 R. G. COLLINGWOOD *Princ. Art* viii. 171 A science of thought must be 'normative', or (as I prefer to call it) 'criteriological', i.e. concerned not only with the 'facts' of thought but also with the 'criteria' or standards which thought imposes on itself. *Ibid.*, Forms of thought can be dealt with only 'criteriologically'.

criteriology. Add: **b.** The study of criteria, esp. as a branch of logic.
1934 *Theology* XXIX. 48 St. Thomas..places his theodicy before his criteriology. 1940 R. G. COLLINGWOOD *Ess. Metaph.* xi. 119 The methods it [*sc.* psychology] has developed in its history as a science of feeling preclude it from dealing with the problems of criteriology. It has nothing to say about truth and falsehood. 1955 *Jrnl. Philos.* LII. 564, I propose here merely the outlines of a criteriology.

critic, *sb.*[1] Add: **3.** *Comb.* (freq. in appositive use).
1680 LD. ROCHESTER *Poems* 16 A great Inhabiter of the Pit; Where Critick-like, he sits and squints. 1754 W. COWPER in W. Hayley *Life W.C.* (1803) I. 16 This simile were apt enough, But I've another, critic-proof! 1906 *Westm. Gaz.* 29 Sept. 14/2 There have been murmurs.. against the critic-dramatist. 1938 H. READ *Coll. Ess. Lit. Crit.* I. i. 17 When such a critic-poet attempts to probe down into such a fundamental question as the form and structure of poetry. 1965 *Canadian Jrnl. Linguistics* Fall 40 Critic-centred comments on the text.

critical, *a.* Add: **5. a.** spec. *critical path*: the most important sequence of stages in an operation, determining the time needed for the whole operation; freq. *attrib.*
1959 in *Jrnl. Industr. Engin.* (1962) XIII. 508/2 (*title*) Critical Path planning and scheduling. An Introduction. 1960 SAYER et al. in *Factory* July 75/1 The critical path technique forces management to recognize planning and scheduling as two distinct functions. For example, the planning of sequence of roof construction on a building job would be done as a separate critical path analysis operation. 1964 C. DENT *Quantity Surveying by Computer* vii. 114 The critical path schedule indicates to what extent an operation may be delayed before the job becomes 'critical', i.e. before the completion date must be put forward. 1964 K. G. LOCKYER *Introd. Critical Path Analysis* i. 2 In 1958, the E. I. du Pont de Nemours Company used a technique called the Critical Path Method (CPM) to schedule and control a very large project. *Ibid.* 6 This determining sequence is critical to the performance of the project, and is hence known as the *Critical Path*. 1970 *Daily Tel.* 13 Oct. 21/4 The critical path through a diagram..is the one which takes the longest, because it is on that route the time for completing the whole operation depends.
7. *critical damping,* damping which is just sufficient to prevent oscillations; *critical potential* = *ionization potential; *critical pressure,* the pressure required to liquefy a gas at its critical temperature; *critical state,* the state of a substance when it is at its critical temperature and critical pressure; *critical temperature* (examples); *critical volume,* the volume of unit mass of a substance at its critical temperature and pressure.
1869 T. ANDREWS in *Phil. Trans.* CLIX. 583 Below the critical temperature this distinction is easily seen to have taken place. 1879 *Encycl. Brit.* VIII. 732/1 Clerk Maxwell has calculated that the critical temperature for water should be about 434° C., the critical pressure about 378 atmospheres, and the critical volume about 2·52 cubic [*printed* cubit] centimetres per gramme. 1884 *Phil. Mag.* XVIII. 212 These cooling agents may be said to lower the temperature sufficiently to produce liquid oxygen, provided a pressure of the gas above the critical pressure, which is 50 atmospheres, is at command. 1899 T. O'C. SLOANE *Liquid Air* i. 20 When a gas is at the critical temperature and at the critical pressure also, the least increase of pressure or decrease of temperature will convert it into a liquid. When in this condition, ready to be a gas or a liquid, it is said to be in the critical state. 1908 K. EDGCUMBE *Industr. Electr. Meas. Instrum.* 35 It is easy to determine how nearly a given instrument approaches the point of critical damping, by noting how much the pointer overshoots a reading as it flies up to it. 1916 *Physical Rev.* VII. 687 To determine this critical potential more accurately, and.. to demonstrate whether or not ionization took place there, current-potential curves were taken. 1922 GLAZEBROOK *Dict. Appl. Physics* II. 373/2 In galvanometers of the Thomson type the damping due to induced currents is small, and if critical damping is desired the retardation by air friction must be capable of adjustment. 1931 *Rev. Mod. Physics* III. 347 Bohr's theory was so quickly supported by the experiments on critical potentials. 1933 in J. K. Henney *Radio Engineering Handbk.* § 7. 146 The smallest amount of damping which will cause the coil to come to rest with no oscillation whatever is called the critical damping, and the coil is said to be critically damped. 1947 *Sci. News* V. 164 No amount of compressing will liquefy it [*sc.* a gas] as long as it is hotter than this critical temperature.
b. *critical mass* or *size*: in Nuclear Physics, the minimum mass or size of fissile material required in a nuclear reactor, bomb, etc., to sustain a chain reaction.
1940 in M. Gowing *Britain & Atomic Energy 1939–1945* (1964) 391 For a sphere well above the critical size the loss through neutron escape would be small. 1941 *Ibid.* 402 For a low critical mass, the material must be as dense as possible. 1945 *Statements relating to the Atomic Bomb* (*H.M.S.O.*) 16 This estimate was very rough and the critical size was known only to a factor of three. 1946 E. S. C. SMITH et al. *Appl. Atomic Power* 202 No explosion occurs at all unless the mass of the fissionable material exceeds the critical mass. 1958 J. CLEUGH tr. *Jungk's Brighter than Thousand Suns* xii. 191 The determination of this 'critical size'—referred to simply as 'crit' in the Los Alamos jargon—had been one of the chief problems studied by the theoretical department.
c. *Nuclear Physics.* Of a nuclear reactor: maintaining a self-sustaining chain reaction; esp. in phr. *to go critical,* to reach the stage of maintaining such a reaction. Also *transf.*
1949 H. SOODAK in C. Goodman *Sci. & Engin. Nuclear Power* II. viii. 91 For a critical pile, the fundamental mode, j = 1, is constant in time whereas the higher modes decay. 1955 *Ann. Reg. 1954* 393 At Harwell the more important news of reactors was that 'Zephyr' became critical in February. 1955 *Sci. News Let.* 6 Aug. 83/1 The possibility of really putting atoms to work for the good of the world will go 'critical', to use a term applicable to the atomic reactor. 1957 *New Scientist* 26 Dec. 6/2 The prototype reactor went critical at the end of 1954. 1970 *Nature* 21 Nov. 704/1 Sizewell B will be the largest nuclear power station in Britain when the last of its four reactors goes critical in 1977.
9. *Comb.*, as *critical-minded* adj.
1899 W. JAMES *Talks Teachers Psychol.* i. 6 Where the disciples are not independent and critical-minded enough. 1956 A. TOYNBEE *Historian's Approach to Religion* x. 128 Philosophies have been apt to arise in..critical-minded, disillusioned social milieux.

criticality. Delete *rare* and add: **2.** *Nuclear Physics.* The state or quality of being critical (see *CRITICAL *a.* 7 c). Also *attrib.*
1950 *U.S. Atomic Energy Commission, AECD 3024* (*title*) Conditions for criticality in certain types of nuclear reactors, by D. S. Selengut. 1952 *Sci. News Let.* 20 Dec. 391/3 A pilot reactor..started operation and reached criticality. 1955 *Ibid.* 16 Apr. 248/1 During these experiments, known as criticality tests, the familiar blue glow of the irradiated fuel elements will be visible under the water. 1956 *Sunday Times* 14 Oct. 10/2 This miscalculation in the quantity of fuel needed to achieve 'criticality' has brought other problems.

critically, *adv.* Add: **7.** *Comb.*, as *critically-minded* adj.; also *absol.*
1906 *Daily Chron.* 1 May 3/3 The critically-minded might find fault with some of the drawing. 1938 *Mind* XLVII. 196 Has there ever been a philosopher against whom critically-minded successors have not been able to bring the charge of unconscious self-contradiction?

criticiasis. *Jocular nonce-wd.* [f. CRITICI(SM + -ASIS.] Criticism regarded as a disease.
1874 SWINBURNE *Let.* 5 Mar. (1959) II. 287 The awful dropsy of criticiasis of verbal emendation is dangerous when it becomes incurable.

criticist (kri·tisist). [f. CRITIC(ISM + -IST.] An adherent of the critical philosophy of Kant (cf. CRITICISM 2 c). Also *attrib.*
1889 in *Cent. Dict.* 1892 *Monist* II. 295 The conceptions of philosophy held by the criticist, the positivist, and the evolutionist. 1898 W. JAMES *Let.* 15 June (1920) II. 74 It confirms the 'criticist' views of the philosophy of history. c 1905 C. S. PEIRCE *Sel. Writings* (1940) 299 It is difficult to find a Criticist who does not hold to more fundamental beliefs than any Critical Common-sensist does.

critter (kri·təɹ). Also 9 **crittur.** Widespread dial. and jocular var. CREATURE; *spec.* an ox or cow; a horse; a chicken; a person (usu. disparaging).
1815 D. HUMPHREYS *Yankey in Eng.* 41 Cooking for the crew, and taking care of the dum critturs. 1827 A. SHERWOOD *Gaz. Georgia* 139 *Beast*, or *crittur*, for horse. 1834 S. SMITH *Sel. Lett. Downing* 86 Bears and wolves and sich kind of critters. 1834 C. A. DAVIS *Lett. J. Downing* 19 There he [*sc.* Capt. Finny] was, sure enough: the crittur had just come out of his bush-pasture. *Ibid.* 31 There was that crittur Mr. Van Buren. 1853 R. S. SURTEES *Sponge's Sp. Tour* xiv. 73 One of your poor pryin', inquisitive critturs, what's always fancyin' themselves cheated. *Ibid.* xxvii. 166, I saw the critter's great pecker steadily down in his plate. 1853 [see CREATURE 2 b]. 1855 D. G. ROSSETTI *Let.* 11 May (1897) 126 That magnanimous crittur seems to have restored him his confidence. 1856 GEO. ELIOT *Let.* 6 June (1954) II. 253 A charming little zoological curate here, who.. is most good-natured in lending and giving apparatus and 'critturs' of all sorts. 1862 G. DU MAURIER *Let.* Apr. (1951) 127 My talk is rather an institution among critters who are like to each other. 1923 'B. M. BOWER' *Parowan Bonanza* vi. 65 She..can sling a pack or rope a critter better than lots of men that draw wages for doing it. 1926 E. M. ROBERTS *Time of Man* (1963) 321 'You came a-horseback?' Jasper said. He was moved almost beyond speech by this expression of loyalty. 'A-horseback I reckon.' 'Yes, we rode our critters.' 1942 M. CAMPBELL *Cloud-Walking* 32 He dozed off till Sary's chickens crowed for morning... Then he stirred to light a fire in the cook room and go out to the lot to feed the critters about the place. 1956 'N. SHUTE' *Beyond Black Stump* 143 I'd have said it wasn't possible. I mean, to tame a critter like that.

crivens (kri·vĕnz), *int. dial.* or *slang.* Also **crivvens.** [Vulgar corruption of CHRIST; perh. infl. by *heavens* (HEAVEN *sb.* 6 d).] An exclamation of astonishment or horror.
1917 A. S. NEILL *Dominie Dismissed* viii. 101 Crivens! What a fine essay that wud mak! 1927 *Scots Observer* 2 July 11/2 'Crivens!' ejaculated Oddfish, 'Are they awa' again?' 1935 G. BLAKE *Shipbuilders* iv. 88 Holy crivvens, I nearly broke my flakin' back. *Ibid.* 89 Crivvens, boy, it's great to see one of the old crush!

crizzle, *v.* Delete '*Obs.* exc. *dial.*' and add: Hence **cri·zzled** *ppl. a.*; **cri·zz(e)l(l)ing** *vbl. sb.*
1624, etc. [in Dict.]. 1677 R. PLOT *Oxfordshire* ix. 253 The glasses made of these being subject to that..fault called Crizelling. 1937 *Burlington Mag.* Nov. 217/1 The interior decay has taken the form of what in seventeenth-century England was called 'crizzelling'. *Ibid.* 218/2 The bowl was.. extensively crizzled.

croak, *v.* Add: **5.** *intr.* (Further examples.)
1851 MAYHEW *London Lab.* I. 424/1 They go mouching along as if they were croaking. 1896 A. MORRISON *Child of Jago* xxviii. 272 Run, for Gawd's sake, or the woman'll croak! 1961 'J. WELCOME' *Beware of Midnight* ii. 33 Your old man has croaked and left you the lot.
b. *trans.* To kill; to murder; to hang. *dial.* or *slang.*
1823 P. EGAN *Grose's Dict. Vulgar Tongue*, *Croaked*, hanged. A flash term among keepers of prisons, who, speaking of a thief that was executed, observe, 'He was *croaked*.' 1848 *Ladies' Repository* VIII. 316/1 *Croak*, to murder. 1877 F. ROSS et al. *Gloss. Holderness*, *Crooak*,..N. and E., to kill. 1910 E. A. WALCOTT *Open Door* vii. 83 'I never done it!' he gasped. 'I never hurt nobody. Who's been croaked?' 1930 *Punch* 26 Feb. 236 It was fairly clear that he had been croaked. 1945 L. A. G. STRONG *Othello's Occupation* 123 Who croaked Enameline?

croaker. Add: **4.** *slang.* A doctor, physician; esp. a prison doctor. Now chiefly *U.S.* Cf. CROCUS 4.
1859 in HOTTEN *Dict. Slang* 26. 1889 BARRÈRE & LELAND *Dict. Slang* I. 281/1 One man who had put his name for the 'butcher' or *croaker*, would suddenly find that he had three ounces of bread less to receive. 1931 'DEAN STIFF' *Milk & Honey Route* vii. 71 Every doctor, known among the hobos as the 'croaker' or 'pill peddler'. a 1935 D. RUNYON *More than Somewhat* (1937) 12 She..goes..to get a croaker to see if my wounds are fatal. 1946 MEZZROW & WOLFE *Really Blues* vii. 95 The most he needed was some bicarbonate of soda and a physic, not a croaker.

croaking, *vbl. sb.* **2.** (Earlier example.)
1787 J. WEDGWOOD *Let.* 16 June (1965) 305, I rejoice to find that all croaking against the treaty is at an end everywhere.

Croat (krōᵘ·æt). [ad. mod.L. (pl.) *Croatæ* (F. *Croate*, G. *Kroat*), ad. Serbo-Croatian *Hrvat*, formerly pronounced (χrwát). Cf. CRAVAT *sb.* (from a later variety of pronunciation).] **a.** A native or inhabitant of the former Austrian province of Croatia, now forming part of Yugoslavia; one of a race descended from the people which occupied that country in the seventh century. **b.** A soldier of a former French cavalry regiment, composed mainly of Croats. **c.** The language of the Croats. Also *attrib.* or as *adj.*
1702 *Milit. Dict.*, *Croat*, properly the People of Croatia; but in *France* there is a Regiment of Horse so call'd... These *Croats* are commanded upon all desperate Service. 1749 FIELDING *Tom Jones* VI. ii, Brother, you are absolutely a perfect Croat; but as those have their Use in the Army of the Empress Queen, so you likewise have some Good in you. c 1790 *Encycl. Brit.* V. 555/2 The Croats derive their origin from the Sclavi. 1815 *Wynne Diaries* 28 Sept. (1940) III. 386 We passed a Croat Regiment, the women that follow'd it were perfect Gypsies. 1862 *Chambers's Encycl.* III. 324/2 The Croatians are warlike, but the name Croats is employed to designate light-cavalry regiments in the imperial army, in which Magyars and others are mingled with true Croatians. 1920 H. A. L. FISHER *Studies Hist. & Pol.* 207 Croat and Slovene newspapers. 1942 L. B. NAMIER *Conflicts* 48 In Yugoslavia the conflict between Croats and Serbs..offered the Nazis rich opportunities for political intrigue. 1959 J. REMAK *Sarajevo* iii. 42 Franz Ferdinand replied..ending with a sentence spoken in Croat.

Croatian (kroˌēⁱ·ʃiăn), *sb.* and *a.* [f. mod.L. *Croatia*, f. *Croatæ*: see prec. and -IAN.] **A.** *sb.* **a.** A Croat. **b.** The language of the Croats, belonging to the Balto-Slavic group. **B.** *adj.* Of or pertaining to Croatia or the Croats.
1555 EDEN *Decades* (Arb.) 290 The Bohemians, Croatians, and Sclauons. 1607 TOPSELL *Four-f. Beasts* 160 The Croatian Dog, resembling a Wolfe in haire and bignesse. 1748 JOHNSON *Van. Hum. Wishes* 20 The fierce Croatian, and the wild Hussar. 1822 [see WEND *sb.* 2]. 1837 *Penny Cycl.* VIII. 161/2 The Damascene plum furnishes the favourite drink of the Croatians. *Ibid.*, The Croatian language is a dialect of the Sclavonian. 1876 [see SLAV *sb.*]. 1885 *Encycl. Brit.* XVIII. 785/1 There are two main branches of Slavonic. The so-called Southern or South-Eastern branch embraces Russian, Ruthenian (in Galicia), Bulgarian, Servian, Croatian, and Slovenian. 1921 *Glasgow Herald* 8 Jan. 6 Their intrigues with the Croatian separatists. 1936 A. W. CLAPHAM *Romanesque Archit.* iii. 70 The native product of Croatian culture.

croc[2] (krǫk). Colloq. abbrev. of *crocodile.*
1. = CROCODILE *sb.* 1 a.
1884 C. B. LEWIS *Sawed-off Sketches* 133 'You'll be sure to agree with me,' muttered the old Croc. as he chewed him down. 1921 *Blackw. Mag.* Jan. 102/1 The 'croc' is the one African animal regarding which nobody has any idea of sportsmanship. 1925 *Ibid.* Sept. 419/1 With loud yells to scare away any crocs that might be lurking round. 1936 P. M. CLARK *Autobiogr. Old Drifter* x. 135 Leaving the corpses of many crocs lying about behind us. 1964 C. WILLOCK *Enormous Zoo* iv. 60 The head of a croc spear is detachable, like that of an old-fashioned whaler's harpoon.
b. = *CROCODILE *sb.* 1 c; esp. in phr. *mock croc,* imitation crocodile-skin.
1963 *Honey* Oct. 59 Boots—in mock-croc. *Ibid.* 69/4 Mock-croc bag. 1967 *Woman* 9 Dec. 11/2 Chic burgundy mock croc shoes.
2. = CROCODILE *sb.* 4.
1948 'J. TEY' *Franchise Affair* ii. 27 An ordinary sort of girl, after all. Not the sort you would notice in a croc. 1958 *Listener* 6 Nov. 722/2 Walking in a school croc.

crocean, *a.*[1] Delete † *Obs.* and add later example.
1900 F. THOMPSON *Sel. Poems* 21 The crocean and amethystine In their pristine Lustre linger on its coat.

Crocean (krōᵘ·ʃiăn, krōᵘ·siăn), *a.*[2] (and *sb.*). Also **Crocian.** [f. *Croce* (see below) + -AN.] Of, pertaining to, or characteristic of the Italian philosopher and statesman Benedetto Croce (1866–1952) or his idealistic 'philosophy of the spirit'. Also *sb.*, a follower of Croce or of his philosophy.
1921 HANNAY & COLLINGWOOD tr. *Ruggiero's Mod. Philos.* 357 This is the new conception of reality that emerges from the very heart of the Crocian philosophy. 1925 *Glasgow Herald* 16 Apr. 4 The French 'Symboliste' poets also are open to attack along Crocean lines. M. Pau Valéry..is not a Crocean. 1926 *Ibid.* 2 Apr. 5 The Crocean doctrine that art is expression and that all expression is art. 1958 *Listener* 10 July 61/1 A historian of the Crocean persuasion.

crochet, *sb.* **2.** Add: *cro̊chet-cotton, -hook, -pin.*
1846 C. MEE *Crochet Expl. & Illustr.* 2nd Ser. 151 Three reels of No. 12 crochet cotton will be sufficient. 1873 *Young Englishwoman* Feb. 91/1 Work with crochet cotton No. 30 in a chain the required length. 1849 *Family Friend* I. 78/2 The Penelope Crochet Hook is invented by Mrs. Warren. 1860 *Ladies' Companion* 37 A coarse Crochet Hook. 1881 *Instr. Census Clerks* (1885) 44 Crochet Hook Maker. 1909 *Daily Chron.* 8 Dec. 9/2 By putting the crochet-pin into the upper half of the stitch.

crocheting (krōᵘ·ʃi,iŋ, krōᵘ·ʃe,iŋ), *vbl. sb.* [f. CROCHET *v.* + -ING[1].] The making of crochet-work; crochet-work. Also *attrib.*
1883, 1891 [see CROCHET *v.*]. 1901 *Daily Chron.* 17 Sept. 7/1 She..at once took up her crocheting. 1921 *Dict. Occup. Terms* (1927) § 375 A crocheting machine to make lace.

crock, *sb.*[3] Add: **3.** (Earlier and later examples.) Now usu. a broken-down or physically debilitated person; an invalid; a hypochondriac. *colloq.* or *dial.*
1876 O. MADOX-BROWN *Dwale Bluth* II. v. 158 Hare sher cumes at learst... Th'little doiling crock! 1880 W. H. PATTERSON *Gloss. Antrim & Down* 25 *Crock, sb.* a derisive term for a person who fancies himself ailing or delicate. 1889 *Illustr. Bits* 13 July 4/2 You are getting a bit of a crock—failing fast, I should say. 1920 R. MACAULAY *Potterism* IV. iii. 149 Shall we be a race of clever crocks, or ..be robust imbeciles? 1922 C. E. MONTAGUE *Disenchantment* iv. 58 Chance..gave me the job of marching parties of crocks, total and partial, real, half-real, and sham, across.. to the place where the faculty did its endeavour to sort them. 1969 *Sci. Amer.* Feb. 69/2 Physicians..blame the patient by labeling him a 'crock'—medical slang for a neurotic complainer.
4. *slang.* An old, worn-out vehicle, ship, bicycle, etc.; esp. as *old crock.*
1903 KIPLING *Traffics & Discoveries* (1904) 123 But if those cruisers are crocks, why does the Admiral let 'em out of Weymouth at all? 1905 G. B. SHAW in *Grand Mag.* Feb. 116 An old crock of a 1904 six-cylinder car. 1914 A. BENNETT *Price of Love* xii. 242 I'm going to buy you a bike. I've had enough of that old crock I borrowed for you. 1935 H. G. WELLS *Things to Come* ix. 71, I understand you want all of these out-of-date crocks of yours..to fly again. 1959 I. & P. OPIE *Lore & Lang. Schoolch.* iii. 55 When boys see an antiquated machine [*sc.* bicycle] they shout:..'Sell that crock and buy a bike.'

crock, *v.*[2] Add: **c.** To impart colour or dye to other articles, to stain: said also of the colour.
1855 *Knickerbocker* XLV. 566 A pair of green gloves..had 'crocked-off' very generously to whatever was in contact with them. 1885 A. WATT *Leather Manuf.* 322 The clear colours do not 'crock' so easily, and the little that does come off is hardly noticeable. 1895 *Montgomery Ward Catal.* 3/1 This black is perfectly fast color and will not crock.

crock, *v.*[4] *colloq.* [f. CROCK *sb.*[3]] *intr.* To become feeble, collapse, give way, break down. Also *trans.*, to cause to collapse; to injure or disable. Often with *up.* Hence **crocked** *ppl. a.*[1], hurt, damaged, disabled; **crocking** *vbl. sb.*, collapsing, breaking down.
1846 J. T. & W. E. BROCKETT *Gloss. N. Country Words* I. 114 *Crock*, to grow little in bulk, to suffer decay from age. 1893 *Idler* Mar. 221 An oarsman who is likely to 'crock up'. 1896 *Westm. Gaz.* 12 Dec. 2/1 Smith has crocked his knee. 1900 *Ibid.* 17 Dec. 8/2 The northern player, who is less likely to get 'crocked' than the Richmond man. 1906 *Daily Tel.* 23 Aug. 9/7 He limped out to bat, after remaining in obscurity as a crocked player for half a day. 1926 *Spectator* 12 June 983/1 Dressing is accomplished quickly considering my crocked-up hand. 1928 *Observer* 19 Feb. 28/6 Slogging home against the present stream would safeguard a crew against crocking later on. 1960 *Times* 22 Oct. 8/6, I had 'crocked' my knee at hockey.

crocked (krǫkt), *ppl. a.*[2] *slang* (orig. *U.S.*). [Perh. f. *CROCK *v.*[4]] Drunk; intoxicated.
1927 in *New Republic* 9 Mar. 71/2. 1957 KEROUAC *On Road* (1958) 76, I had traveling money and got crocked in the bar. 1970 *Guardian* 9 Apr. 1/2 The curtain fell and the audience retired to get crocked.

Crockford (krǫ·kfǒɹd). **1.** (Usu. *Crockford's.*) The name of an exclusive gambling club opened in St. James's Street, London, in 1827 by William Crockford (1775–1844). Also *transf.*
1827 W. MAGINN *Whitehall* II. iii. 117 The Opera was crowded... Tattersall's was crammed—Crockford's crowded. 1838 H. C. ROBINSON *Diary* 27 Apr. (1967) 188 D'Orsay looked haggard as if he had been unlucky at Crockford's. 1867 'OUIDA' *Under Two Flags* I. vi. 136 They had brought ..dice for hazard..and were turning the unconscious Star and Garter into an impromptu Crockford's. 1966 J. CLEARY *High Commissioner* vi. 131 It's *the* gambling club. A sort of millionaire's Crockford's.
2. A colloquial designation of 'Crockford's Clerical Directory', a reference book for the clergy and the Church of England, first issued in 1860 by John Crockford (1823–65). (Occas. *Crockford's*.) Also *fig.*
1891 S. WEYMAN *New Rector* I. i. 9 In more distant vicarages..there were..anxious searchings of the 'Guardian' and Crockford. 1909 H. G. WELLS *Tono-Bungay* I. i. 16 There was an old peerage and a Crockford together with the books of recipes. 1947 [see *ALL C. 2 c]. 1970 K. GILES *Death in Church* i. 22 'I'll look it up in *Crockford's*.'..Harry wondered whether the compiler of the clerical Who's Who had..been related to the gambler.

crocky (krǫ·ki), *a.*[2] [f. CROCK *sb.*[3] + -Y[1].] That is a crock; broken-down, physically enfeebled.
1880 W. H. PATTERSON *Gloss. Antrim & Down* 25 *Crocky, adj.* fanciful about his health; hippish. 1906 *Westm. Gaz.* 18 Sept. 10/1 Among the parts of a crocky engine. 1907 A. CONAN DOYLE *Through Magic Door* 109 The crockiest of spectators had a better chance of life than the magnificent young athlete. 1920 R. MACAULAY *Potterism* IV. iii. 149 Crocky imbeciles.

crocodile, *sb.* Add: **1. c.** = *crocodile-skin* (see also quot. 1968).
1907 *Yesterday's Shopping* (1969) 389/2 The 'Gadabout' writing case...Crocodile, lined sheep..67/6. 1908 *Daily Chron.* 15 Aug. 3/2 A large crocodile letter-case. 1908 *Westm. Gaz.* 19 Nov. 4/2 All the upholstering is in crocodile. 1968 J. IRONSIDE *Fashion Alphabet* 237 Many so-called 'crocodile' accessories are in fact made from alligator skins, crocodile being particularly difficult to tan and preserve.
4. (Examples.) Also of a boys' school, etc.
1891 H. ATTERIDGE in *Little Folks* Nov. 326/1 He saw what boys sometimes call 'a crocodile'—a girls' school out for a walk. 1898 J. K. JEROME *Second Thoughts* 311 We came upon a girls' school walking two and two,—a 'crocodile', they call it. 1922 *Blackw. Mag.* Oct. 487/2 The crocodile of small boys in the streets. 1926 I. M. PEACOCKE *His Kid Brother* ii. 37 To walk in a 'crocodile' of orphans. 1950 F. A. SWINNERTON *Flower for Catherine* 107 One saw her leading the long lines of schoolgirls which are called 'Crocodiles'. 1968 M. BRAGG *Without City Wall* xx. 201 The crocodile rows of little children.
b. A long procession of moving objects close together. Also *fig.*
1912 H. G. WELLS *Marriage* ii. 55 She drove her little crocodile of primly sensible thoughts to their sane appointed conclusion. 1928 *Manch. Guardian Weekly* 17 Aug. 136/2 Those roads which..do not carry an endless and snorting crocodile of cars. 1930 R. PERTWEE *Pursuit* I. xi. 55 Transport would pile up before and behind you in a ceaselessly cursing crocodile.
5. *crocodile-like* adj.; **crocodile-bird,** the Egyptian black-headed plover, *Pluvianus ægyptius,* so called from its habit of eating the insect parasites of the crocodile, probably the trochilos of ancient writers; **crocodile shears,** shears used in cutting into lengths and removing the faulty ends of steel or iron bars; **crocodile squeezer,** a machine with a pivoted upper jaw, used in the process of removing impurities from metals by the application of pressure.
1868 A. C. SMITH *Attractions Nile* II. 255 *Charadrius spinosus*..in all probability the true 'crocodile bird' or *trochilus* of Herodotus. 1966 C. SWEENEY *Scurrying Bush* x. 142 The Egyptian 'Plover' (*Pluvianus aegyptius*) or crocodile bird has been the centre of controversy. 1621 in W. Foster *Eng. Factories India 1618–21* (1906) 347 These viprous, dessemblinge, and crockadillike currs. 1897 A. PAGE *Afternoon Ride* x. 61 A large iguana, almost crocodile-like in its proportion. 1884 W. H. GREENWOOD *Steel & Iron* 347 Puddled bars are..sheared hot either by crocodile or guillotine shears. *Ibid.* 301 The single alligator or crocodile squeezer has two broad flat jaws. *Ibid.* 302 The crocodile squeezer makes about 60 strokes per minute.

crocodile, *v. intr.* (Further examples.)
1936 M. FRANKLIN *All that Swagger* xii. 116 The school crocodiled abroad with its instructresses. 1960 *News Chron.* 15 Mar. 4/5 The girls crocodile in, a mistress at head and tail. 1969 *Guardian* 25 July 9/5 The diminutive school-girls crocodiling through the Commonwealth Institute.

crocodiling (krǫ·kǒdailiŋ), *vbl. sb.* [f. CROCODILE *sb.* or *v.* + -ING[1].] **1.** Walking in a crocodile (CROCODILE *sb.* 4).
1889 [see CROCODILE *v.*].
2. = *ALLIGATORING *vbl. sb.*
1932 *Paint Manufacture* II. 245/2 (*caption*) Common Stopper Defects. A. 'Crocodiling' of Hard Stopper Applied over Soft Primer. 1953 [see *ALLIGATORING *vbl. sb.*].

crocus. Add: **5.** *crocus-bed.*
1878 O. WILDE in *Irish Monthly* Apr. 211 The crocus-bed is a quivering moon of fire. 1891 'M. O'RELL' *Frenchm. Amer.* 60 A..crocus-bed effect.

Crœsus (krī·sǝs). The Latin form of the name of a king of Lydia (Gr. *Κροῖσος*) in the sixth century B.C., who was famous for his riches, used allusively in phrases, as *Crœsus' wealth, as rich as Crœsus,* and hence typically for 'a very rich person'.
1390 GOWER *Confessio Amantis* v. 4730 in *Compl. Wks.* (1901) III. 75 If the tresor of Cresus And al the gold Octovien, Forth with the richesse Yndien Of Perles and of riche stones, Were al togedre myn at ones. 1577 T. KENDALL *Flowers of Epigrams* fol. 21r, As riche as Cresus Affric is. 1578 G. WHETSTONE *Remembraunce of G. Gaskoigne* sig. B1, What auailes..King Cressus welth. 1650 TRAPP *Comm. Prov.* xxx. 8, I shall not envie the richest Crœsus or Crassus upon earth. 1707 [see RICH *a.* 2 a]. 1754 H. WALPOLE *Lett.* (1857) II. 389 A contest between two young Crœsus's, Lord Thanet and Sir James Lowther. 1883 M. E. BRADDON *Golden Calf* xii, Ida, left alone amidst all the fascinations of the chief shop in a smart county town, and feeling herself a Crœsus. 1924 C. MACKENZIE *Old Men of Sea* (1963) xix. 245 We could have sailed in the poor old *Able and Willing*, and been rich as Creases all our lives. 1931 J. T. ADAMS *Epic of America* vii. 185 There he was, rich as Crœsus, and dictating to the government.

crofterize (krǫ·ftəraiz), *v.* [f. CROFTER[1] + -IZE.] *trans.* To convert into a croft-tenancy. So **cro:fteriza·tion, cro·fterizing** *vbl. sb.*
1907 *Times* 7 Oct. 9/4 Nor do the Scottish Lowlands.. desire the crofterization of the Lowland counties. 1908 *Ibid.* 11 Mar. 11/5 Amendments which shall not interfere with the crofterizing of the Scottish Lowlands.

croisette. Delete *Obs.* and add later examples.
1906 *Daily Chron.* 11 Oct. 3/2 To pay to the Administration, in the form of a tax, a certain number of croisettes—brass rods in the form of a St. Andrew's cross. 1920 A. STRATTON *Engl. Interior* 72 French doorways of the Louis XIV. period, with surrounding architraves broken at the top with *croisettes* and surmounted by a frieze.

croissant (krwasan). [Fr. (see CRESCENT *sb.*).] = CRESCENT *sb.* 6.
1899 W. C. MORROW *Bohem. Paris* 139 The odor of hot rolls and croissants. 1928 R. MACAULAY *Keeping up Appearances* i. § 2 Foaming coffee and milk, the crusty roll, the little tender croissant. 1958 X. FIELDING *Corsair Country* 96 We're having coffee and croissants in the sort of square you would come across in any of the larger French Mediterranean seaports. 1970 *New Yorker* 6 June 64/2 Resident jockeys..were being offered..the French racing papers, with their coffee and croissants.

‖ **Croix de Guerre** (krwa,dǝ,gēr'). Also **croix de guerre.** [Fr., lit. 'cross of war'.] A French medal first awarded during the war of 1914–18 (see quot. 1922).
1915 *Daily Chron.* 4 Aug. 7/7 M. Poincaré conferred upon King Albert himself the French Croix de Guerre. 1917 T. E. LAWRENCE *Home Lett.* (1954) 345 The French Government has stuck another medal on to me: a croix de guerre this time. 1922 *Encycl. Brit.* XXXI. 892/2 The *Croix de Guerre*.—Established in 1915 to commemorate individual mentions in despatches during the war 1914–18. The cross was awarded to soldiers or sailors of all ranks..who were mentioned in orders of the day for an individual feat of arms... The ribbon..is green with narrow red stripes. 1944 H. G. WELLS '*42 to '44* 78 A girl in M.T.C. uniform with the gold leaf of the Croix de Guerre over her pocket. 1960 G. MARTELLI *Agent Extraordinary* ii. 34 Dressed usually in a neat dark blue suit only distinguished by the ribbon of the Croix de Guerre.

Crokerism (krōᵘ·kəriz'm). [f. the surname *Croker* + -ISM.] **1.** The political principles of John Wilson Croker (1780–1857).
1851 CARLYLE *New Lett.* (1904) II. 114, I was reading in the *Quarterly Review*:—very beggarly Crokerism, all of copperas and gall and human baseness. 1927 *Observer* 9 Oct. 16/4 When Randolph Churchill set out to revive his party, what did he do? He swept away Crokerism.
2. In U.S. politics, the political following and influence of Richard Croker (1841–1922), who made himself master (*c* 1888) of the Tammany organization and subsequently attained to great power in the government of the state of New York. Hence **Cro·kerist,** an adherent of Crokerism; also *attrib.*
1897 *Daily News* 4 Nov. 3/4 Evidence..that Crokerism had brought New York to almost the lowest possible point. 1900 *Westm. Gaz.* 27 Oct. 6/2 Mr. Roosevelt..urged..the killing of Crokerism in the State. 1901 *Daily Chron.* 3 Oct. 5/7 The supporters of Crokerism and corruption..are opposed by the advocates of good government with freedom from bosses. *Ibid.* 1 Nov. 6/3 He will..vote the whole Fusionist 'ticket' or the whole Crokerist 'ticket' as it is offered to him.

Cro-Magnon (kromæ·nyon, kromæ·gnǫn). Also **Cromagnon.** [The name of a hill of Cretaceous limestone in the Dordogne department of France in a cave at the base of which skeletons of *Homo sapiens* were found in 1868 among deposits of Upper Palæolithic age; it had previously been supposed that modern man did not exist in Palæolithic times.] Used, chiefly *attrib.*, to designate a group of mankind characterized by a long low skull, a wide face, and wide orbits having upper and lower borders close to one another and almost parallel; the stature is moderate or tall.
The group persisted in Mesolithic and Neolithic times, and some authorities consider that it survived in the Guanches of the Canary Islands. Similar skeletal features can be detected in certain modern European and north African peoples.
1869 P. BROCA in Lartet & Christy *Reliq. Aquit.* (1875) ix. 99 The general characters of the Cro-Magnon race. 1869 A. DE QUATREFAGES *Ibid.* 124 The Cro-Magnon skulls introduce new elements into the question of European origins. 1874 *Leisure Hour* 31 Oct. 697 Outline of Cro-Magnon skull as seen in front. 1882 *Amer. Antiquarian* Apr.–July 242 Cromagnon skulls in Bavaria. 1912 R. MUNRO *Palæolithic Man* 200 All the skeletons of the Cro-Magnon type found in the Grimaldi caves. 1927 HALDANE & HUXLEY *Anim. Biol.* xiii. 331 Cromagnon man. 1934 *Discovery* Mar. 79/2 All comparisons between individual moderns and individual Greeks or Cro-Magnons, are beside the mark in a discussion of the meaning of human progress as a whole. 1939 R. CAMPBELL *Flowering Rifle* II. 53 To sink these new Cromagnons out of sight.

crombec (krǫ·mbek). [Used in this form by Le Vaillant *Histoire naturelle des oiseaux d'Espagne* 1802; f. Du. *krom* crooked + *bek* BEAK *sb.*] A popular name for African warblers of the genus *Sylvietta.*
[1875–84 LAYARD *Birds S. Afr.* 303 Sylvietta rufescens... The 'Stomp-stertje' of the Dutch colonists, and the 'Crombec' of Le Vaillant.] 1901 STARK & SCLATER *Birds S. Afr.* II. 117 Sylviella [*sic*] pallida. The Zambesi Crombec. 1908 HAAGNER & IVY *Sk. S. Afr. Bird Life* 80 The Crombec (*Sylviella rufescens*), known to the farmers as the Stompstertje (Stump-tail), is ash-grey above and tawny-buff below. 1955 MACKWORTH-PRAED & GRANT *Birds E. & N.E. Afr.* II. 423 Crombecs or Stump-tails...Called Crombecs by the early Dutch settlers in South Africa because of their curved bills.

Crombie (krǫ·mbi). Also **crombie.** The name of J. & J. *Crombie* Ltd., a Scottish firm of cloth-makers, used to designate a type of overcoat, jacket, etc., made by them. Hence *Crombie-coated* adj.
1951 *Trade Marks Jrnl.* 17 Jan. 67/1 Crombie Product... Men's and boys' coats, suits, jackets..all made from piece goods wholly or substantially of wool, worsted or hair. J. & J. Crombie Limited, Grandholm Works, Woodside, Aberdeen, Scotland; Manufacturers and Merchants;—7th September, 1949. 1957 J. BRAINE *Room at Top* x. 95 A young man in a Crombie overcoat came through the door. 1957 *Times* 11 Sept. 3/1 Visiting Americans lured by the theatre's magic name come crombie-coated to tread heavily on Yeats's dreams. 1963 *Guardian* 16 Jan. 9/7 A man in a heavy black Crombie glared at the loud-speaker.

Cromer (krō·məɹ). [The name of a town on the Norfolk coast.] *Cromer Forest Bed*: the name of a series of deposits which outcrops on the coast at Cromer, comprising two freshwater beds which enclose the Forest Bed proper, an estuarine bed of clay containing the transported remains of trees and rich in plant and animal fossils.
From its position (see quot. 1964) the series was formerly thought to be pre-glacial, of Pliocene age, but it is now generally thought to have been deposited during the first (antepenultimate) interglacial, in the early Pleistocene, and its flora and fauna are taken as typical of this interglacial in Britain.
[1840 C. LYELL in *Phil. Mag.* 3rd Ser. XVI. 377 A general subsidence..must have taken place..in order to explain the submergence and burial of the trees of which the stools are found *in situ*; and this forest bed could not have been brought up again..to the level of low water, without a subsequent upheaval.] 1863 —— *Antiquity of Man* 511/1 Cromer forest bed. 1882 C. REID *Geol. Country around Cromer* iii. 8 The so-called 'Cromer Forest-bed', celebrated for the number and variety of the fossil mammals which it has yielded. 1902 *Encycl. Brit.* XXXI. 439/2 The latest Pliocene, or pre-Glacial, flora of northern Europe is best known from the Cromer Forest-bed of Norfolk and Suffolk, a fluvio-marine deposit which lies beneath the whole of the Glacial deposits of these counties. 1946 L. D. STAMP *Britain's Struct.* xiv. 156 The Cromer Forest Bed series which we mentioned as the youngest Pliocene beds in Britain are claimed by some as early Pleistocene. 1964 K. P. OAKLEY *Frameworks for dating Fossil Man* I. 102 The Cromer Forest Bed was for long regarded by British geologists as 'pre-glacial', for it is overlain by the oldest known glacial deposits in East Anglia... In 1950 Woldstedt published Thomson's pollen-diagram of the Cromer Forest Bed which showed that it was undoubtedly interglacial in character.

Cromerian (kromīᵊ·riăn), *a.* and *sb.* [f. prec. + -IAN.] **A.** *adj.* **1.** *Geol.* and *Palæont.* **a.** Epithet of the Cromer Forest Bed series (see prec.); hence, of, pertaining to, or characteristic of this series.
1900 F. W. HARMER in *Q. Jrnl. Geol. Soc.* LVI. 725 The so-called Forest-bed Series of the Cromer and Kessingland coasts..may be known as Cromerian. 1925 A. KEITH *Antiquity of Man* (ed. 2) I. xv. 304 The Cromerian beds tell us of the mild climate of the last phase of the Pliocene period. 1931 *Discovery* Mar. 85/1 The Cromerian flora is practically the same as that in East Anglia to-day. 1964 [see *1 b].
b. Epithet of a stage (and the corresponding age) in the Lower Pleistocene, and also of the first (antepenultimate) interglacial in Britain (generally identified with the Günz–Mindel Interglacial of Europe); hence, of or contemporaneous with this stage or interglacial.
1922 OSBORN & REEDS in *Bull. Geol. Soc. Amer.* XXXIII. 431 Certain surviving marine deposits of the north of Europe may be correlated with those of the Mediterranean, as follows: 1. Sicilian Stage of the Mediterranean = Cromerian Stage of the North Sea basin and Baltic. *Ibid.*, Along the German-Baltic coast Depéret recognizes marine faunal beds of Sicilian–Cromerian age. 1956 R. G. WEST et al. in *Phil. Trans. R. Soc.* B. CCXXXIX. 344 The Cromer Forest Bed or Cromerian Interglacial is considered to be equivalent to the..Günz/Mindel Interglacial of Europe. 1964 K. P. OAKLEY *Frameworks for dating Fossil Man* I. 109 In China the industry of Peking Man is associated with a fauna which was for long identified as Cromerian, whereas new evidence indicated that it is of Mindel II age... It is therefore most important to distinguish between a horizon of Cromerian *age* (*i.e.* dating from the Günz–Mindel Interglacial) and one containing elements of Cromerian *fauna.* 1967 D. H. RAYNER *Stratigr. Brit. Isles* xii. 378 The Cromerian stage refers to the Cromer Forest Bed and associated sediments of north Norfolk. 1968 R. G. WEST *Pleistocene Geol. & Biol.* xiii. 305 Compared with the succeeding interglacials the Cromerian vegetational history is distinct in the absence or scarcity of *Hippophaë* in the late-glacial zone. 1969 *Proc. Geol. Soc. Lond.* Aug. 152 It is recommended that for the Pleistocene and Holocene of the British Isles the following ages/stages be adopted as a regional scale... Pleistocene:..Anglian, Cromerian, Beestonian [etc.].
2. *Archæol.* Of, pertaining to, or characteristic of an ancient culture or people formerly supposed to be represented by remains (now not generally held to be artefacts) found in some Pliocene deposits near Cromer.
1922 OSBORN & REEDS in *Bull. Geol. Soc. Amer.* XXXIII. 424 Beneath the Forest Bed there has been discovered recently by J. Reid Moir a bed of giant flints of human manufacture, an industry which may be termed Cromerian. 1934 L. S. B. LEAKEY *Adam's Ancestors* v. 98 (*caption*) Two sides of a Cromerian flake tool. 1946 F. E. ZEUNER *Dating Past* vi. 185 The Cromerian industry..is particularly well-known from the so-called 'foreshore site', a flint spread exposed at low-water. 1964 K. P. OAKLEY *Frameworks for dating Fossil Man* II. vi. 219 All these flakes are in a highly battered condition. They have been compared with the very dubious 'Cromerian industry' in Britain.
B. *sb.* **1.** *Archæol.* One of the 'Cromerian' people.
1921 J. R. MOIR in *Jrnl. R. Anthrop. Inst.* LI. 387 These large flint masses represent the cores from which the ancient Cromerians obtained the raw material in the manufacture of their artefacts. 1925 A. KEITH *Antiquity of Man* (ed. 2) I. xv. 305 In 1920, Mr Moir discovered a 'working floor', a site used for the knapping of flint implements by early Cromerians.
2. *Geol.* The Cromerian stage or age or the Cromerian interglacial.
1957 J. K. CHARLESWORTH *Quaternary Era* II. xxix. 598 In East Anglia, by referring the Cromerian to the Pliocene and at the same time to an interglacial epoch, as is commonly done.., the glacial beginnings are thrust into the earlier horizons. *Ibid.* xxxii. 692 The Cromerian saw the European facies fully established. 1968 R. G. WEST *Pleistocene Geol. & Biol.* xiii. 305 The flora is very similar to that of the Cromerian in its paucity of exotic genera.

Cromwell (krǫ·mwel). [See CROMWELLIAN *a.* and *sb.*] **1.** In full *Cromwell shoe.* A shoe of the type supposedly worn by Oliver Cromwell, usu. having a large buckle or bow.
1879 M. E. BRADDON *Vixen* I. iv. 86 The girl..made rather a pretty picture... The tawny hair, black velvet frock..and broad-toed Cromwell shoes. 1952 C. W. CUNNINGTON *Eng. Women's Clothing Pres. Cent.* ii. 36 *Shoes*:..the Cromwell with two straps. 1963 A. GERNSHEIM *Fashion & Reality* II. 93 Buckled Cromwell shoes..were also popular [about 1911].
2. In full *Cromwell chair.* = *Cromwellian chair* (see next).
1868 C. L. EASTLAKE *Hints Household Taste* iii. 80 Perhaps the most satisfactory type [of dining-room chair] is that which is commonly known in the trade as the 'Cromwell' chair. Its form is evidently copied from examples of the seventeenth century. 1881 C. C. HARRISON *Woman's Handiwork* III. 191 A square Puritan 'Cromwell' in oak, severely plain save for its dark cushion in maroon plush. *Ibid.* 192 The large, square-seated 'Cromwell' chair. 1969 J. GLOAG *Short Dict. Furnit.* (ed. 3) 275 *Cromwell chair*, mid-19th century trade term for a dining-room chair, copied or adapted from a mid-17th century prototype.

Cromwellian, *a.* and *sb.* Add: **A.** *spec.* Designating a type of chair (see quots.).
1905 A. HAYDEN *Chats on Old Furnit.* iii. 96 There is a style of chair, probably imported from Holland, with leather back and leather seat which is termed 'Cromwellian'. 1948 H. GORDON *Old Eng. Furnit.* iv. 33 In the ten years or so of the Commonwealth régime..furniture was extremely plain, and the chair that is generally labelled 'Cromwellian' may well have been a Puritan model. 1970 G. SAVAGE *Dict. Antiques* 108/2 *Cromwellian chair*, a modern term to describe a plain chair of Spanish design slightly decorated with turning, with slung leather seat and back, the leather held in place with large-headed studs.

cronk (krǫŋk), *a. Austral. colloq.* [Cf. CRANK *a.*[3] 3 and 4.] Of a horse: unfit to run in a race, or dishonestly run as though unfit; said also of the race. Hence *gen.*, unsound, liable to collapse; also, obtained by fraud.
1891 N. GOULD *Double Event* xvii, He'd never ride another 'cronk' race, he vowed. 1892 *Bulletin* (Sydney) 12 Nov. (Funk), 'Cronk' financial institutions. 1893 *Herald* (Melbourne) 4 July 2/7 (Morris), The word 'cronk', Mr. Finlayson explained, meant 'not honestly come by'. 1900 H. LAWSON *On Track* 39 'It's always the way!'..'I knew the beggar would turn up!..And the only cronk log we've had, too!' *Ibid.* 157 Generosity isn't understood nowadays, and what the people don't understand is either 'mad' or 'cronk'. 1930 *Bulletin* (Sydney) 8 Oct. 20/4 Snip Sinker.. was mostly too lazy or 'ad a cronk wrist or a kink in his back when there were big guns on the board.

cronyism (krōᵘ·ni,iz'm). Also **croneyism.** [f. CRONY *sb.* + -ISM.] **a.** Friendship; the ability or desire to make friends. **b.** (Chiefly *U.S.*) The appointment of friends to government posts without proper regard to their qualifications.
1840 A. CONOLLY *Let.* July in F. MacLean *Person from England* (1958) 43 It *must* end in my going to Khokund, probably *via* Khiva with the Envoy thence, Yakoob Bai, with whom I have established great croneyism. 1922 W. DE LA MARE in *Boswell's Life of Johnson* p. xix, Johnson's oddities, his queer habits, his cronyism, his truculence, his wit, his frailties. 1950 *Collier's Mag.* 24 June 78/1 [He] sets a heap of store by the solemn vows of cronyism. 1952 *N.Y. Times* 17 Aug. 8E/1 The amount of politically entrenched bureaucracy that has earned for Mr. Truman's regime its sorry reputation for corruption, cronyism, extravagance, waste and confusion. 1968 *Guardian* 13 July 9/5 The Congress would be more amenable to the argument that 'cronyism' should give way to the next President's right to choose his own Chief Justice.

crook, *sb.* and *a.* Add: **A.** *sb.* **10. b.** In polo, an act of crooking an opponent's stick (see *CROOK *v.*[1] 6).
1935 *Times* 18 June 5/5 Captain Ansell..scored with a 60 yards penalty given for a foul crook.
13. For *U.S. colloq.* read orig. *U.S. colloq.* Now esp. a professional criminal or an associate of criminals. (Add earlier and later examples.)
1879 *Chicago Tribune* 6 Feb. 5/2 The *Times* still continues its attacks upon the Government officials in the interest of the Pekin and Peoria crooks. 1882 *Sydney Slang Dict.* 3/1 *Crook*, a thief and burglar. One who gets his living on the best. 1891 *The Sun* (N.Y.) 19 June 6/4 (Funk), The slang word 'crook' now bids fair to be recognized in the statutes and consequently to be adopted as good English in the courts of law. A bill regulating admissions to the prison at Marquette excludes, among other classes of individuals specified, those known to be 'crooks' in police parlance. 1896 *Westm. Gaz.* 17 July 2/1 A crook what kep a little crib Dad went to when things was too lively. 1903 *Daily Chron.* 3 Nov. 5/4 All the saloon-keepers, and gamblers, and crooks, and confidence men, in fact all the predatory elements of society are..working for a Tammany victory. 1909 *Ibid.* 19 June 3/2 The people here..are clever and rather interesting scamps. Were they on a slightly lower social level they would be called 'crooks'. 1949 [see *CROOKERY]. 1953 'M. INNES' *Christmas at Candleshoe* xiii. 150 'The fact is that a gang of crooks——' 'I beg your pardon?' Miss Candleshoe is wholly at sea. 'The fact is that a band of robbers is prowling about outside this house now.'
B. *adj.* **1. b.** 'Bent', stolen. *Criminals' slang.*
1900 *Sessions' Paper Central Criminal Court* CXXXII. 462, I bought it from you b— straight; I did not know it was *crook*.
2. *Austral.* and *N.Z.*
a. Of things: bad, inferior; out of order, unsatisfactory; unpleasant, dreadful.
1898 *Bulletin* (Sydney) 17 Dec. Red Page/2, *Krook* or *kronk* is bad. 1915 E. G. PILLING *Anzac Memory* (1933) iv. 62 Had a very crook night, sickness, cramp, dysentery. 1917 E. MILLER *Camps, Tramps & Trenches* (1939) ix. 52 The rifle issued to me was a crook one that fired high and left. 1918 *N.Z.E.F. Chrons.* 5 July 250/2 It's crook to stay for years. 1929 W. SMYTH *Girl from Mason Creek* xv. 163 'Cow of a job,' he muttered...'It's a bit crook for yer.' 1931 V. PALMER *Separate Lives* 271 It can't be helped now. When things go crook in the beginning [etc.]. 1934 A. RUSSELL *Tramp-Royal in Wild Australia* xvii. 106, I wasn't feeling too well at the time—too much crook water an' not enough decent tucker, I suppose. 1945 J. HENDERSON *Gunner Inglorious* iii. 18 A cigarette first thing in the morning before a cup of tea, tastes crook. 1947 'A. P. GASKELL' *Big Game* 32 Isn't it crook about Keith and Gordon [being killed]? 1958 'N. SHUTE' *Rainbow & Rose* 64, I never knew it [*sc.* the weather] to be so crook. 1968 K. WEATHERLY *Roo Shooter* 111 You know how the old wagon is crook in water.
b. Dishonest, unscrupulous, 'crooked'.
1911 L. STONE *Jonah* I. xi. 132 Yous don't think any worse o' me 'cause Lil's crook, do yer? 1916 'ANZAC' *On Anzac Trail* 44 Protesting..in lurid language against what they styled 'a crook trick'. 1929 C. C. MARTINDALE *Risen Sun* 173 When sport goes crook, what can remain wholesome? 1933 *Bulletin* (Sydney) 6 Sept. 8/2 They think.. that the system is 'crook'. 1936 F. SARGESON *Conversation with Uncle* 18 They said it [*sc.* pulling a race-horse] was a crook business right through. 1953 J. W. BRIMBLECOMBE in J. C. Reid *Kiwi Laughs* (1961) 178 His mentor had a crook deal put over him.
c. Irritable, bad-tempered, angry; esp. in phr. *to go crook* (*at* or *on*), to become angry (at); to lose one's temper (with); to upbraid, rebuke.
1911 L. STONE *Jonah* II. iv. 190 Yer niver 'ad no cause ter go crook on me, but I ain't complainin'. 1916 C. J. DENNIS *Songs of Sentimental Bloke* 78 An' there I'm standin' like a gawky lout. An' wonders wot 'e's goin' crook about. 1933 P. CADEY *Broken Pattern* xviii. 197 If Phœbe's gone crook at you..she's had some good reason for it. 1937 N. MARSH *Vintage Murder* vii. 70 See him when he goes crook!..His eyes fairly flashed. 1946 P. FREEDMAN in *Coast to Coast 1945* 136 Her ma's always going crook because I break the plaster. 1959 *Listener* 15 Jan. 115/2, I cut off his boot to stop the foot swelling. I remember he went crook on me: he said they were new, and I'd darn well have to buy him a new pair. 1964 P. WHITE *Burnt Ones* 295 When his mum went crook, and swore, he was too aware of teeth, the rotting brown of nastiness.
d. Ailing, out of sorts; injured, disabled.
1916 C. J. DENNIS *Songs of Sentimental Bloke* 88, I sneaks to bed, an' feels dead crook. 1916 *Oil Sheet* Dec. 7 Now I've just been vaccinated, and am feeling pretty 'crook'. 1934 A. RUSSELL *Tramp-Royal in Wild Australia* xxvii. 178 An' when Dick says he's crook, he's crook. He's out there alone, you know. 1937 N. MARSH *Vintage Murder* viii. 94 Letting him out just because he kidded he felt crook. *Ibid.* x. 116 'He was looking horribly crook.' 'Ill?' asked Alleyn cautiously. 'Too right, sir.' 1938 'R. HYDE' *Nor Years Condemn* 208 A crook knee and arm. 1952 N. DONNAN in *Coast to Coast 1951–52* 141, I got a crook hip, I can't do heavy work now. 1956 P. WHITE *Tree of Man* 55 'She's crook. It looks like the milk fever,' he said. 1960 B. CRUMP *Good Keen Man* 45 It saves you from getting the crook guts,

boy. *Ibid.* 49 He said he wasn't crook or anything. **1968** K. Weatherly *Roo Shooter* 28 If I don't go out at least five nights a week the cook thinks I'm crook and gets all worried.

crook, *v.*[1] Add: **1. d.** *to crook one's elbow* or *little finger*: to drink alcoholic liquor (esp. with implication of excess). *slang.*

1825 Jamieson Suppl. I. 271/2 *To crook the elbow*; as, She *crooks* her *elbow*, a phrase used of a woman who uses too much freedom with the bottle, q. bending her elbow in reaching the drink to her mouth. **1836** *Public Ledger* (Philadelphia) 2 Aug. (Th.), William Martin was fined for, as he quaintly expressed it, crooking his little finger too often. **1859** Bartlett *Dict. Amer.* (ed. 2) s.v., To crook one's elbow or one's little finger, is to tipple. **1875** Besant & Rice *With Harp & Crown* xix, The secretary..might have done great things in literature but for his unfortunate crook of the elbow. As he only crooks it at night, it does not matter to the hospital. **1924** J. Masefield *Sard Harker* III. 251 Sir James has sacked his old man for crooking his little finger: going on the jag, in other words.

6. *trans.* In polo, to catch hold of (an opponent's stick) with one's own stick, so as to interfere with his play.

1890 G. J. Younghusband *Polo in India* ii. 21 No player shall crook his adversary's stick unless he is on the same side of the pony as the ball, or immediately behind. **1898** T. B. Drybrough *Polo* xi. 268 'Crooking' means 'interposing' a stick between the ball and an adversary's stick..so as to 'hook' and arrest it. **1902** *Encycl. Brit.* XXXI. 820/2 They have no off side [in American polo], and it is not permitted to crook the stick of an adversary.

croo·kdom. [f. Crook *sb.* 13 + -dom.] The realm of crooks.

1921 *Glasgow Herald* 23 Apr. 4 The brilliant amateur investigator, whose uncanny intuition and superman brain have paralysed the ranks of 'crookdom'. **1929** *Daily Express* 7 Jan. 9 Story of a careless young athlete's adventures in crookdom.

crooked, *a.* Add: **1. b.** *crooked stick*: see Stick *sb.*[1] 12.

3. b. Delete (*U.S.* and *Australia*) and add earlier and later examples.

1864 Hotten *Slang Dict.* (ed. 3) 112 *Crooked*, a term used among dog-stealers, and the 'fancy' generally, to denote anything stolen. **1898** *Daily News* 27 Aug. 6/6 Telling him that he rather thought he had bought 'a crooked lot'. **1902** *Daily Chron.* 26 Aug. 6/6 In the event of his being found.. to be dealing in 'crooked' things, or refusing to give information as to where he got his stuff.

c. *Austral.* and *N.Z. slang* = *crook *a.* 2 c; esp. in phr. *crooked on*, angry at. Usu. pronounced (krukt).

1944 L. Glassop *We were Rats* I. viii. 48 Ya oughtn' ter feel crooked on things. I s'pose it's because Bertha's outa town? **1957** 'N. Culotta' *They're a Weird Mob* (1958) vi. 86 'Are you not ashamed of yourself?' 'Yeah, I'm real crooked on me.'

5. b. *crooked-neck(ed)* (*spec.* applied to a variety of squash: cf. Crook-neck). *U.S.*

1784 *Massachusetts Spy* 22 Apr. 1/1 Crooked neck squash. **1796–1801** Fessenden *Orig. Poems* (1806) 134 Like a nice crook'd neck'd squash on the ground. **1871** C. D. Warner *Summer in Garden* viii. 104 The summer squash..was nearly all leaf and blow, with only a sickly crooked-necked fruit after a mighty fuss.

crookedy (kru·kėdi), *a. rare.* [f. Crooked *a.* + -y[1].] Crooked.

1907 F. Campbell *Shepherd of Stars* xv. 170, I call it a very crookedy sort of tune. **1938** M. K. Rawlings *Yearling* xvii. 203 I've lost my boy. My pore crookedy boy.

crookery (kru·kəri). [f. Crook *sb.* 13 + -ery 2.] The state or condition of a crook; the dealings or behaviour of crooks; the world of crooks (see Crook *sb.* 13).

1927 *Observer* 15 May 15/2, I do not call it good 'crookery' when the situation can only be made by turning out the light. **1928** *Daily Express* 22 June 10/3 If a boy came to me and said: 'I have received a call to crookery.' **1949** 'J. Tey' *Brat Farrar* x. 81 A very..intelligent crook. On the highest level of crookery. **1962** 'E. Ferrars' *Busy Body* vi. 67 They probably want to use you as an alibi for Tom while Tom's up to crookery of some sort. **1967** M. Stand *Diana is Dead* vii. 104 'You seem to know a lot about crookery,' the sergeant said.

Crookes (kruks). The name of Sir William *Crookes* (1832–1919), English scientist, used attrib. or in the possessive to designate phenomena observed and apparatus invented by him. **Crookes** or **Crookes's (dark) space,** the dark space between the negative glow and the cathode of a vacuum tube, observed when the pressure is very low; also called *cathode dark space, cathodic dark space*. **Crookes** or **Crookes's glass,** a type of glass which protects the eyes from intense radiation, bright sunlight, etc. **Crookes** or **Crookes's layer,** (*a*) the layer of vapour underlying any mass or liquid in the spheroidal state, insulating it from the surface on which it rests; (*b*) = **Crookes (dark) space*. **Crookes** or **Crookes's radiometer:** see Radiometer 2. **Crookes rays,** = **cathode rays*. **Crookes's tube,** a highly evacuated tube in which stratified electric discharges can be observed.

1884 A. Daniell *Princ. Physics* 325 Let us now suppose that the particles recoiling from the heated surface do not meet other molecules, but impinge on the walls of the vessel. A layer of particles in such a condition is called a Crookes' layer. **1885** *Encycl. Brit.* XIX. 249/1 In Crookes's radiometer the free path is very long. **1889** *Cent. Dict.*, *Crookes's tubes.* **1892** G. F. Barker *Physics* 329 The layer of vapor which has to support the drop is called a Crookes layer. **1893** J. J. Thomson *Recent Res. Electr. & Magn.* 108 Next to this [*sc.* the negative electrode] there is a comparatively dark region..called sometimes 'Crookes' space' and sometimes the 'first dark space'. **1896** *Daily News* 29 May 5/2 The rays which produce the fluorescence are certainly the Crookes rays. **1902** *Encycl. Brit.* XXVIII. 47/2 The Crookes dark space. **1906** *Amer. Jrnl. Sci.* 4th Ser. XXII. 312 The extremely tenuous condition of the residual elementary gas or gases in a Crookes tube. **1910** *Hawkins's Electr. Dict.*, *Crookes' Effect*, the radiant effect produced in a vacuum glass tube in which the exhaustion has been carried to a high degree, when electricity is discharged through it between suitable electrodes. **1918** J. H. Parsons *Dis. Eye* (ed. 3) x. 184 Smoked or orange tinted (not blue) glasses.. are most efficacious when made with Crookes's glass. **1927** L. B. Loeb *Kinetic Theory of Gases* vii. 241 The Crookes radiometer, so often seen in opticians' windows, consisting of a set of mica vanes blackened on one face and mounted on an axis so that they are free to rotate inside a partially evacuated glass vessel and which rotate when radiation falls on them. **1931** *Glass* Dec. 509/1 Several Crookes glasses were shown. **1958** C. G. Wilson *Electr. & Magn.* xii. 365 At a pressure of 10^{-3} mm. Hg. or less, the Faraday Dark Space and Negative Glow disappear and the Crooke's [*sic*] Dark Space almost fills the tube.

Hence **Crookesian** (kru·ksiăn) *a.*, pertaining to Crookes or to instruments invented by him (*rare*).

1899 *Science Siftings* XVI. 117/2 The Crookesian scale-pan. *Ibid.*, Crookesian radiometer.

crookesite (kru·ksəit). *Min.* [ad. Sw. *crookesit* (A. E. Nordenskiöld 1867, in *Öfv. K. Vet.-Akad. Förh. 1866* XXIII. 366), f. *Crookes (the discoverer of thallium): see -ite[1].] A brittle, lead-grey selenide of copper, thallium, and silver.

1868 J. D. Dana *Syst. Min.* (ed. 5) II. ii. 40 Crookesite... Massive, compact; no trace of crystallization. **1946** J. R. Partington *Gen. & Inorg. Chem.* xvi. 431 The only minerals rich in thallium are crookesite (17 p.c. Tl, with Se, Cu, Ag) and lorandite.

crookish (kru·kiʃ), *a.* [f. Crook *sb.* 13 + -ish[1] 2.] Pertaining to or characteristic of a crook (Crook *sb.* 13) or crooked behaviour, dealings, etc.

1927 *Spectator* 17 Dec. 1083 At the Royalty, *The Crooked Billet*—terrifyingly, confusingly crookish, with a throat-cutting scene against which I warn nervous parents. **1934** E. Bowen *Cat Jumps* 89 Marianne's coupé had been run up on to the rough grass..with its lights out. Its air was lurking and crookish. **1937** —— *Coll. Impressions* (1950) 117 *The Alchemist*, with its crookish hilarity.

crook-necked, *a.* Having a crooked neck; spec. *U.S.*, applied to a variety of squash (cf. Crook-neck and *crooked *a.* 5 b).

a **1529** [see Crook *sb.* and *a.* C. b]. **1818** *Massachusetts Spy* 11 Nov. (Th.), Upwards of ten tons of the best crook-necked winter Squashes. **1945** *New Eng. Homestead* 27 Oct. 3/2 They are..the size of a summer crooknecked squash.

crool (krūl), *sb. rare.* [f. the vb.] The sound described under Crool *v.*; = Cooing *vbl. sb.* 1.

1938 W. de la Mare *Memory* 71 The monotonous crool of a dove.

croon, *v.* **2. a.** Add to def.: *spec.* to sing popular sentimental songs in a low, smooth voice, esp. into a closely-held microphone (see quot. 1959 s.v. *crooning* below).

1920 *Catal. Victor Records*, Standard Songs. 'Croon, Croon, Underneat' de Moon' (Clutsam). **1931** H. Arlen (title of song in the musical production *You Said It*) Learn to croon. **1933** *Punch* 2 Aug. 122/1 Bing Crosby the crooner ..croons to his feminine class and is crooned to in reply. **1940** *War Illustr.* 5 Jan. p. ii/2, I used to sup while Roy Fox's 'boys' played joyously with Les Allen 'crooning'.

3. *spec.* To sing (a song, etc.) in a low, smooth voice (cf. prec.).

1915 C. Lean (title of song in the musical production *The Blue Paradise*) The tune they croon in the U.S.A. **1925** H. D. Kerr (*title of song*) Croon a little lullaby. **1931** Durante & Kefoed *Night Clubs* 227 His band stressed the soft notes, and Rudy [Vallée] crooned his way right into the heart of the nation. **1932** *Amer. Speech* VII. 250 Bing Crosby plaintively croons that he has 'Found a Million Dollar Baby in the Five and Ten Cent Store'. **1933** *Fortune* Aug. 47/2 Their use of 'jazz' includes both Duke Ellington's Afric brass and Rudy Vallée crooning *I'm a Dreamer, Aren't We All?*

Hence **crooning** *vbl. sb.* and *ppl. a.* (later examples).

1923 B. James (*title of song*) Carolina Mammy. A real Southern mammy song—the crooning kind. **1927** *Melody Maker* Aug. 784/3 'Muddy Water' has a feature in a sweet crooning vocal introduction. **1929** *Ibid.* Dec. 1139/3 His crooning style of singing. **1931** *Musical Courier* (N.Y.) in *Oxf. Compan. Mus.* (1938) 1018/2 No jazz or cheap crooning stuff had a place in her repertoire. **1932** *Literary Digest* 30 Jan. 23/2 'You can't help thinking badly of any man who would degrade himself whining in that way..' he said of crooning. **1935** Wodehouse *Blandings Castle* v. 116 Everybody knows what Crooning Tenors are...They sit at the piano and gaze into a girl's eyes and sing in a voice that sounds like gas escaping from a pipe about Love and the Moonlight and You. **1959** *Chambers's Encycl.* XII. 570/2 In that type of vocal performance known as 'crooning' the lower range of the voice is chiefly used, and that more in the manner of conversation than of singing, though falsetto notes are often introduced. There is a noticeable gliding or sliding from one pitch to another and the intonation is often deliberately indefinite...Characteristic also is a certain oscillation or catch in the voice as it comes to rest momentarily upon a sustained sound.

crooner. Add: **b.** *spec.* A singer who croons (see *croon *v.* 2 a).

1930 *Vanity Fair* July 57 Just call them Crooners. **1932** Thorne Smith *Bishop's Jaegers* (1934) 314 That sound.. is made nightly by one of the nation's most popular crooners. **1933** [see *croon *v.* 2 a]. **1948** *Penguin Music Mag.* Feb. 25 The B.B.C. could start the campaign by refusing to make 'stars' of its crooners. **1954** *Granta* 6 Nov. 23/1 Dickie Valentine turns out, from his old cuttings, to be a crooner, as I had suspected.

crooningly (krū·niŋli), *adv.* [f. Crooning *ppl. a.* + -ly[2].] In a crooning manner.

1902 *Munsey's Mag.* Apr. 42 'Ah, no, ah, no,' she said crooningly, as if she comforted the child. **1928** D. L. Sayers *Unpleasantness at Bellona Club* iii. 31 A melody of Parry's formed itself crooningly under his fingers.

crop, *sb.* Add: **22. crop-bound** *a.*, (of birds) unable to pass food through the crop; **crop-duster,** an aircraft used for sprinkling insecticide, fertilizer, etc., on crops; a person who flies such an aircraft; **crop-dusting** (see *dusting *vbl. sb.* 1 b); **crop-end,** a piece of metal cut off a bar of rolled iron or steel to remove imperfections and to reduce the bar to standard length; **crop-head,** a crop-end cut from that end of a bar of iron or steel which is at the top during the process of cooling and where most of the imperfections occur; **crop-mark** *Archæol.* (see quot. 1956); also *attrib.*; so *crop-marking*; **crop movement** (see Movement 8); **crop-over,** in the West Indies, the end of the sugar-cane harvest on a plantation, and the accompanying celebrations; **crop-plant,** a plant cultivated for food; **crop-spraying,** the spraying of crops with insecticide or the like; also *attrib.* or *ppl. a.*

1854 *Poultry Chron.* I. 136 'Crop-bound' fowls. **1897** *Daily News* 29 Dec. 7/2 The bird..had become crop-bound, and in order to remove the obstruction an incision five inches long was made in the crop. **1939** *Collier's* 24 June 17/1 These are the crop dusters. **1966** *Punch* 8 June 832/3 The first cropduster to be produced in partnership, this nippy little machine can carry..liquid DDT. **1880** *Encycl. Brit.* XIII. 332/1 Cuttings, 'crop ends', and 'scrap' of various kinds, often not very largely inferior in value to the bar iron. **1884** W. H. Greenwood *Steel & Iron* xvi. 347 Cutting off the rough or crop-ends of puddled, finished, or other bars. **1903** *Sci. Amer.* Suppl. 21 Feb. 22687 The rough ends—'crop heads'—are cut off and are placed by an electric crane in a car for shipment to any part of the works. **1935** *Proc. Prehist. Soc.* I. 157 The two fields showing crop marks. **1947** J. & C. Hawkes *Prehist. Britain* vii. 161 Two of the most spectacular discoveries made by this crop mark method are the Bronze Age temple of Arminghall and the Roman town of Caistor-by-Norwich where every building was clearly planned in pale lines in the corn. **1956** J. K. S. St. Joseph in R. L. S. Bruce-Mitford *Recent Archaeol. Excavations in Brit.* 275 In spring and early summer, differences in colour, density or luxuriance of growth commonly develop in response to hidden differences in the soil. These 'crop-marks', as they are termed, reveal to an observer, often in the finest detail, buried remains of which no trace can be seen on the surface. **1937** *Oxoniensia* II. 13 Under corn or grass however it becomes covered with very distinct crop-markings, as can be seen from the air-photographs. **1909** *Westm. Gaz.* 14 June 12/1 The crop movement began very early last year, and the farmers were paid for their wheat and other products promptly. **1894** G. Robson *Missions United Presb. Ch.* 35 The grinding routine of slavery was relieved at 'crop-over' and Christmas-time by boisterous revels. **1906** *Westm. Gaz.* 2 June 11/2 Burrowing into the roots of grasses, crop-plants, and trees. **1958** *Listener* 28 Aug. 301/2 The herd-animals and crop-plants which were destined to form the main basis of modern food-production. **1956** *Farm Implement & Machinery Rev.* 1 Apr. 2146/1 Crop spraying is carried out to remove the injurious influences which adversely affect yields. **1959** *Daily Tel.* 15 Oct. 20/5 Three crop-spraying helicopters. **1970** *East African Standard* 23 Jan. 12/6 'What's crop-spraying?' asked his mother. 'Well, you fly low over cultivated fields and spray the crops with weed-killer solution from the aircraft,' explained Bill.

crop, *v.* Add: **6.** Also *intr.*, to cultivate land; to work as a farmer. Chiefly *U.S.*

a **1847** in H. Howe *Hist. Coll. Ohio* (1847) 357 He came down the Ohio to Cincinnati, and cropped the first season on Zeigler's stone house farm. **1903** *Dialect Notes* II. 310, I am cropping with Mr. Brown this year.

b. *trans.* To grow or rear as a crop.

1921 *Discovery* Feb. 48/1 The pest..remains in existence until potatoes are again cropped in the field.

cropper[2]. Add: **1.** *spec.* a shearing machine in iron and steel work; also, the workman who operates it.

a **1884** Knight *Dict. Mech.* Suppl., *Cropper*, a powerful

hand machine for shearing off bolts or rod iron. **1921** *Dict. Occup. Terms* (1927) § 279 *Cropper* (iron and steel rolling); forge cropper, mill cropper and shearer; a shearer or a hot sawyer who cuts off badly shaped ends of finished bars of iron or steel after rolling. **1937** *Times* 13 Apr. p. xxxviii/1 Portable hand-lamps, bolt croppers, insulated pliers.

cropper[3]. (Further examples.)

1874 Hotten *Slang Dict.* 133 *Cropper*, 'to go a cropper', or 'to come a cropper', *i.e.*, to fail badly. **1874** Trollope *Way We live Now* (1875) I. xxxviii. 241 He would 'be coming a cropper rather', were he to marry Melmotte's daughter for her money, and then find that she had got none. **1951** T. Rattigan *Who is Sylvia?* I. 230 We bachelors welcome competition from married men. We so much enjoy watching them come the inevitable cropper. **1963** *Times* 30 Jan. 1/7, I came a proper cropper, dearie, all black and blue I was.

cropper[4]. Add: (Earlier and later examples.) Also *cropper boy, work.*

1881 *Instr. Census Clerks* (1885) 40 Letter-Press Printing: ..Cropper Hand. **1888** Jacobi *Printers' Vocab.* 28 *Cropper*, a short term for the 'cropper' small printing platen machine. **1896** *Daily News* 30 Oct. 10/6 Compositor (young). Jobbing and cropper. **1901** *Daily Chron.* 3 Dec. 9/7 Printers.—Young man seeks Situation in machine room; good reference for Cropper work. **1903** *Ibid.* 4 Mar. 9/7 Printers.—Cropper boy wanted. **1921** *Dict. Occup. Terms* (1927) § 529 *Machine minder*, platen; cropper hand; has charge of platen machine.

cro·ppie. *U.S.* Also **croppy.** [App. f. F. dial. *crape* or LG. *krape*.] Any of several North American freshwater fishes (see quot. 1889). Cf. Crappie.

1856 *Spirit of Times* 20 Sept. 43/1 Rock and black bass, croppy, and the common sunfish. **1889** Farmer *Americanisms*, *Croppie*, a local name for a species of green bass found in Lake Minnetonka. **1892** *Gentlewoman's Bk. Sports* I. 73 We caught bass, croppie, sun perch and pickerel. **1920** S. Lewis *Main St.* 18 A fisherman..holding up a string of croppies.

cropping, *vbl. sb.* Add: **1. d.** *Metal-working.* The operation of cutting off the ends of an ingot, bar, etc., to remove the pipe and other defects (see also quot. 1904).

1904 Goodchild & Tweney *Technol. & Sci. Dict.* 139/2 *Cropping*, cutting the ends of bars, rails, etc.; especially cutting iron bars into lengths suitable for making into a fagot. **1930** *Engineering* 21 Mar. 385/2 Piping was avoided, so that no serious amount of cropping was needed. **1968** *Gloss. Terms Mechanized & Hand Sheet Metal Work* (*B.S.I.*) 7 *Cropping*, separation of a semi-complete or complete workpiece. *Ibid.*, Cropping tool.

4. *Comb.*, as **cropping shears** = **crocodile shears.*

1873 *Spon's Dict. Engin.* VI. 2122 Two pairs of cropping shears at 55 revolutions a minute. **1884** W. H. Greenwood *Steel & Iron* xvi. 348 The crocodile, cropping, or alligator shears.

croppy[2]. Add: Also *croppy-boy*.

c **1830** (*title*) The Croppy Boy. **1898** *Westm. Gaz.* 14 Jan. 2/2 What form of higher education you deem at once most suitable for Croppy Boys and least objectionable..to their Protestant overseers. **1949** D. M. Davin *Roads from Home* 59 A line of papists and croppy-boys and Galway rebels.

¶ (Earlier Australian examples in the sense of 'convict'.)

1800 J. Elder *Jrnl.* 25 Dec. in *Austral. Lit. Stud.* (1966) II. 215 An attack from the Irish Croppies. **1830** R. Dawson *Present State of Australia* viii. 299 He had a constable's staff, and considered himself..as a look-out constable for *croppy*, (as they always call the runaway convicts). **1848** H. W. Haygarth *Bush Life in Australia* i. 9 Mr. Longbow ..was..robbed..by the well-known 'croppies'—'Black Joe' or 'Irish Jem'.

‖ **croquis** (kroki). [Fr.] A rough draft; a sketch, study.

1805 C. James *Mil. Dict.* (ed. 2), *Croquis*, Fr. a rough sketch taken of any thing. **1888** *Athenæum* 7 Jan. 23/2 The volume will be illustrated with etchings and *croquis*. *a* **1895** F. Locker-Lampson in T. H. Ward *Eng. Poets* (1918) V. 525 He sent her this copy containing His comical little *croquis*. **1912** C. Mackenzie *Carnival* xxi. 248 So Ronnie sat there, making little *croquis* of Jenny with soft outlines elusive as herself.

cross, *sb.* Add: **10. a.** (Later examples.)

1920 A. Huxley *Limbo* 184 You must try and be strong and bear it bravely. We all have our cross to bear. **1963** A. Heron *Towards Quaker View of Sex* iv. 40 They must practise self-denial and 'bear their cross'.

18. a. Celtic cross: see *Celtic *a.* 2.

22. b. (Earlier and later examples.)

1872 *Young Englishwoman* Nov. 594/1 Bows of dark blue velvet cut on the cross. **1955** 'C. Brown' *Lost Girls* x. 107 The skirt was cut on the cross. **1968** J. Ironside *Fashion Alphabet* 79 Garments cut on the cross or bias have 'give' as the bias is stretchy.

c. *Theatr.* A movement from one part of the stage to another in acting.

1838 *Actors by Daylight* I. 214 He was..well versed in all the crosses and recrosses necessary to impose on the million. **1896** G. B. Shaw *Our Theatres in Nineties* (1932) II. 129 At the end of each of his first vehement speeches, he strode right down the stage and across to the prompt side of the proscenium on the frankest barnstorming principles, repeating this absurd 'cross'—a well-known convention of the booth for catching applause—three times.

d. *Boxing.* A blow that crosses over the opponent's lead. Also *transf.*

1906 E. Dyson *Fact'ry 'Ands* xvii. 233 Ther revolvin' arm..got home a left lead 'n' er right cross. **1938** D. Runyon *Take it Easy* 26 What she lays on his brow is a beautiful straight right cross. **1950** J. Dempsey *Championship Fighting* xxii. 144 The right cross, deadliest of all counter-punches, is used when a left-jabber becomes careless.

e. *Association Football.* A cross-pass.

1961 *Times* 29 Sept. 4/3 They quickly turned the screw, with three goals—by Pointer, side-footing in Douglas's cross. **1962** *Times* 12 Mar. 3/2 Greaves failed to stroke home one of his crosses. **1968** *Listener* 23 May 682/1 For a high cross a well-trained full-back..is good enough—though not for the kind of calculated low cross George Best engineered for Billy Foulkes's decisive goal against Real Madrid.

29. (Earlier and additional examples.)

1802 *Sessions' Paper* June 334/2, I got it on..the cross. **1883** 'Mark Twain' *Life on Mississippi* lii, If I would shake the cross and live on the square for three months. **1915** A. Conan Doyle *Valley of Fear* II. iii. 201 It's mum with me so long as I see you living on the straight...But, by gum, if you get off on the cross after this it's another story. **1917** —— *His Last Bow* viii. 293 There's a stool pigeon or a cross somewhere, and it's up to you to find out where it is.

cross, *v.* Add: **3. b.** *to cross one's heart*, to make the sign of the cross over one's heart, to attest the truth or sincerity of a statement, promise, etc.; freq. in phr. *cross my heart (and hope to die)*.

1908 S. Ford *Side-stepping with Shorty* xx. 314, I wouldn't touched [*sic*] another thing; cross m' heart, I wouldn't! **1922** 'K. Mansfield' *Garden Party* 24 'Promise not to tell.' They promised. 'Say, cross my heart straight dinkum.' **1926** R. Macaulay *Crewe Train* x. 184 'Let's both swear.' 'Cross my heart and hope to die. Now what about bed?' **1952** A. Wilson *Hemlock & After* iii. 167 Cross her heart, might she die if she sneaked.

5. a. Also, to place (one limb) *over* another.

1881 C. E. L. Riddell *Senior Partner* I. ii. 29 She crossed her soft white hands one over the other. **1886** 'Maxwell Gray' *Silence of Dean Maitland* II. i, Staring at the sky, with one leg crossed over the other. **1902** 'H. S. Merriman' *Vultures* xxv. 223 The captain..crossed one leg over the other.

c. *Telephony.* To make a connection between (telephone or telegraph wires of different lines or circuits); freq. used in *pass.* of accidental connections. Also *transf.*, implying a misunderstanding. Also **crossed** *ppl. a.*, **crossing** *vbl. sb.*

1884 *Telegraphic Jrnl.* 31 May 469/1 (*caption*) Crossed wires. **1910** H. Belloc *Pongo & Bull* iii. 56 'Don't mind me, Eddie, the wires were crossed.' And with this meaningless but sufficient phrase, he jammed the receiver down again. **1910** *Hawkins's Electr. Dict.* 100/2 *Crossing wires*, a temporary expedient when a defective section is found to exist in a telegraph circuit, for preserving the continuity of the circuit by crossing the wire over to a neighboring line till the fault is remedied. **1931** Wodehouse *If I were You* i. 12 'There's a lunatic at the other end of the wire who keeps calling me Little Bright Eyes.' 'I fancy the wires must have become crossed, m'lady.' **1932** —— *Hot Water* ii. 57 Can we by any chance have got the wires crossed?..It *was* the idea, wasn't it, that we should pile on to a pot of tea together? **1936** —— *Laughing Gas* vii. 78 An unforeseen crossing of the wires in the fourth dimension. **1958** *Listener* 11 Dec. 976/1 This crossing of the political wires had many repercussions in politics.

7. a. (Further examples.) Also *absol.*

1816 Jane Austen *Emma* II. i. 7, I..must..apologise for her writing so short a letter..in general she fills the whole paper and crosses half. **1819** Keats *Let.* 3 Oct. (1958) II. 221 Brown has a few words to say to you and will cross this. **1924** R. Macaulay *Orphan Island* xiv. § 2. 176 Miss Smith had the sloping, flowing hand of the ladies of her period, and often crossed and recrossed.

8. b. *spec.* In *Cricket*: (*a*) in fielding, to cross to the other side of the wicket at the end of an over, or when a left-handed batsman replaces a right-handed one at the crease or vice versa; (*b*) of a bowler: to go across at the end of an over in order to bowl from the opposite wicket, thus bowling two overs in succession (*no longer permitted*).

1867 G. H. Selkirk *Guide to Cricket-Ground* iv. 59 Avoiding the necessity of the field crossing over so frequently. **1877** C. Box *Eng. Game Cricket* 447 To cross over is to change wickets, which a bowler is permitted to do twice in an innings. **1883** [see Dict.]. **1908** W. E. W. Collins *County Cricketer's Diary* ix. 162 'He's not quite so good [a bowler] as I thought,' he confided..as we crossed over.

d. *intr. Biol. to cross over*: of segments of chromatids of homologous chromosomes: to interchange and recombine during synapsis; to undergo crossing-over (see *crossing *vbl. sb.* 11).

1915 T. H. Morgan et al. *Mechanism of Mendelian Heredity* iii. 59 Sex linked factors cross over from each other. **1916** *Jrnl. Genetics* V. 284 If..for every chromosome which crosses over in this definite way, another similar chromosome in another nucleus does not cross over at all, [etc.]. **1920** [see *cross-over 4]. **1949** Darlington & Mather *Elem. Genetics* iii. 45 The chromosomes, or rather their constituent chromatids, cross-over and separate in germ cell formation just as the genes segregate and recombine.

e. *euphem.* To die.

1930 'R. Crompton' *William's Happy Days* ix. 224 My dear, dear little four-footed friend..'E crossed over last week. **1935** N. Collins *3 Friends* xvii. 265 Just before she crossed over she mentioned your name.

14. d. *slang.* To cheat or double-cross; to act dishonestly in or towards; cf. Cross *sb.* 29. Also *intr.* (see quot. 1925).

1823 in P. Egan *Grose's Dict. Vulgar T.* *a* **1891** Henley & Stevenson *Deacon Brodie* III. iv, in Farmer *Slang* (1891) II. 218/1 What made you cross the fight, and play booty with your own man? **1925** *Flynn's* 10 Jan. 877/2 *Cross*, to squeal; to betray...To deceive; to cheat one's pals. **1938** G. Greene *Brighton Rock* II. ii. 86 It wouldn't have happened if we hadn't been crossed. A journalist thought he could put one over on us. **1960** 'W. Haggard' *Closed Circuit* xv. 179 He'd been using us; he'd crossed us; and he knew too much for safety.

cross, *a.* Add: **1. c.** (See *cross-wind.)

d. Also *cross-surge, -swell, -tide.*

1864 Dickens *Mut. Fr.* I. I. xiv. 133 In the cross-swell of two steamers. **1867** Smyth *Sailor's Word-bk.*, *Cross-swell*, this is similar to a cross-sea, except that it undulates without breaking violently. *Ibid.*, *Cross-tide*, the varying directions of the flow amongst shoals that are under water. **1891** Kipling *Light that Failed* xv. 311 A boisterous little cross-swell swung the steamer disrespectfully by the nose. **1899** —— *Five Nations* (1903) 9 'Twixt wrench of cross-surges or plunge of head-gale. **1903** *Q. Rev.* Apr. 486 Like vortices upon a surface of water swept by violent cross-tides.

e. *Cricket.* Applied to a bat held in a slanting position. Cf. *cross- A. 8.

1871 F. Gale *Echoes from Old Cricket Fields* 25 If you hit her you could only do it with a cross bat. **1891** W. G. Grace *Cricket* viii. 224 Playing with a straight bat is more likely to protect your wicket than playing with a cross bat. **1928** *Daily Express* 12 Nov. 3/4 One amazing cross-bat shot, head high..over cover's head to the boundary.

8. (Earlier examples.)

1812 J. H. Vaux *Flash Dict.* in *Mem.* (1819) II. 165 Any article which has been irregularly obtained..is emphatically termed *a cross article*. **1882** *Sydney Slang Dict.* 3/1 *Cross cove*, a thief. *Ibid.* 10/2 Four deaners for lush for the cross coves and their blowers. **1890** 'R. Boldrewood' *Miner's Right* II. xv. 62 He believed all the 'cross boys' of all the colonies were congregated here.

cross-. Add: **A. 4. a.** *cross-arm, -bracing, -walk*; **b.** *cross-lane, -wall*; **c.** *cross-reef.*

1725 *Manchester Rec.* (U.S.) I. 165 Southerly by the Towns land cal'd the Pasonage land as the Cross wall now stands. **1744** F. Moore *Voyage to Georgia* 30 The Garden is laid out with Cross-walks planted with Orange-trees. **1813** Jane Austen *Pride & Prej.* II. v. 53 Leading the way through every walk and cross walk. **1866** Thomas Howard *Hist. Inglewood Reefs* ix. 51 What is known as a cross reef, that is two reefs, the one intersecting the other. **1904** *N.Y. Even. Post* 16 May 7 The Government service answers very well as a cross-walk in getting over a trying period in a young man's life. **1906** D. V. Allen in P. Galvin *N.Z. Mining Handbook* 54 At the intersection of cross-reefs carrying sulphides. **1908** *Westm. Gaz.* 1 Aug. 7/2 He attempted to run back, but the cross-wall was immediately behind him. **1909** *Cent. Dict.* Suppl., Cross-arm. **1909** Webster, Cross bracing. **1925** *Bell Syst. Techn. Jrnl.* IV. 524 Individual wires mounted on separate insulators attached to cross-arms on poles. **1926** F. W. Crofts *Insp. French & Cheyne Myst.* x. 134 They reached the cross-lane at Earlswood. **1940** *Chambers's Techn. Dict.* 204/1 *Counter-bracing*, the provision of two diagonal tie-rods in the panels of a frame girder or other structure. Also called *cross-bracing*. **1951** *Archit. Rev.* CIX. 140/2 An uncompromising system of concrete cross-walls is used, based on a method of construction developed in Denmark during the war. **1958** C. Tomlinson *Seeing is Believing* (1960) 19 The wind eludes them Streaking its cross-lanes over the uneasy water. **1964** *Listener* 11 June 950/1 Each pier of each ring is led back to the others by brick cross-bracing, making up as it were the spokes of a set of colossal wheels. **1964** M. McLuhan *Understanding Media* xxii. 224 Witness the portent of the crosswalk, where the small child has power to stop a cement truck.

5. *cross-current* (examples); also *fig.* and *attrib.*; *cross-traffic.*

1598 Florio *Worlde of Wordes* 430/3 *Trauersa*,..a crosse currant of waters. **1891** Meredith *One of our Conq.* II. x. 254 It was a happy cross-current recollection. **1899** Morley in *Westm. Gaz.* 18 Jan. 5/1 There have been cross-currents, and it was impossible either inside the House of Commons or elsewhere that Sir William Harcourt could speak with the authority of a united party. **1899** *Westm. Gaz.* 26 Jan. 2/2 Cross-current politics. **1925** T. Dreiser *Amer. Trag.* (1926) I. xix. 140 Several precious moments were lost as the cross-traffic went by. **1948** [see *cross-town *a.*]. **1952** C. Day Lewis tr. *Virgil's Aeneid* XI. 242 In the middle of these cross-currents, when partisan feeling had reached High pressure..the envoys returned.

6. b. *cross-pollinate.* **d.** *cross-petition.*

1904 *Westm. Gaz.* 7 Dec. 7/2 The husband denied various acts described as cruelty by the wife, and cross-petitioned for judicial separation. **1920** *Chambers's Jrnl.* 13 Mar. 238/1 Budding and cross-pollenating. **1923** *Daily Mail* 28 Feb. 5 Her husband..cross-petitions for the dissolution of his marriage.

8. *cross-batted* (cf. *cross *a.* 1 e).

1955 Miller & Whitington *Cricket Typhoon* xii. 230 Caught at deep mid-on from a cross-batted swipe. **1955** *Times* 12 July 12/1 An astonishing cross batted slash through the covers. **1968** *Listener* 11 July 61/2 It is a passive, cross-batted reaction to the short ball from a fast bowler.

9. *cross-alliteration, -association* (example), *-protection, -resistance, -striation, -stroke, -suit, -summons, -tabulation.*

1899 W. James *Talks* x. 98 As cross-associations multiply and habits of familiarity and practice grow, the entire system of our objects of thought consolidates. **1923** *Daily Mail* 28 Feb. 5 In this case there are cross-suits. One, by the wife.., her husband..cross-petitions for the dissolution of his marriage. **1926** J. S. Huxley *Ess. Pop. Sci.* xviii. 282 The histological character of the cells had changed, cross-striations arising in them. **1927** *Daily Express* 17 Aug. 7

I hope that I shall live to see the day when motorists will be able to take out cross-summonses against careless pedestrians. **1934** Priebsch & Collinson *German Lang.* x. 375 Ingeniously adapting *b*, by a cross-stroke, e.g. *geƀan*..to designate the voiced labial spirant (v). **1938** R. Graves *Coll. Poems* p. xv, I found the strictly matching consonantal sequences of Welsh bardic poetry too crabbed for English, but modified them to cross-alliteration. **1949** R. K. Merton *Social Theory* 15 Paradigms, by their very arrangement, suggest the *systematic* cross-tabulation of presumably significant concepts. **1951** Whitby & Hynes *Med. Bacteriol.* (ed. 5) xii. 195 Sera prepared against either organism will ..afford cross-protection in mice. **1956** *Ibid.* (ed. 6) x. 135 *Tetracycline Derivatives*... Acquired resistance common, with cross-resistance to all members of group. **1961** *Lancet* 2 Sept. 521/1 Cross-resistance between different penicillins is not necessarily absolute.

11. a. *cross-handled.*

1896 *Daily News* 15 June 7/1 The familiar cross-handled baskets of the fruit. **1912** T. Okey *Art of Basket-making* vii. 74 The basket is to be covered as well as cross-handled.

B. cross-accent *Mus.* = Syncopation 3; **cross-border** *a.*, that forms a border across a fabric, etc.; so *cross-bordered* adj.; **cross-break,** a break across a lode of ore or strata of rocks; **cross cause,** (*a*) a hindrance (*Obs. rare*); (*b*) *Law*, a cause in which each of the litigants has a suit against the other; **cross-check** *v. trans.*, (*a*) *Ice Hockey*, to obstruct by holding one's stick across an opponent (Webster, 1934); (*b*) to check (an observation, theory, etc.) by an alternative method of verification; so as *sb.*; **cross-colouring** *Geol.*, colour-markings in strata caused by the introduction of extraneous matter by the action of water; **cross-correlation,** a correlation between two distinct series of measurements, events, etc., ordered in time or space; **cross-correspondence** *Spiritualism* (see quot. 1909); **cross-court** *a.*, of a stroke in tennis, rackets, etc.: hit diagonally across the court; so as *v. trans.*, to hit diagonally across the court; **cross-cousin,** one of two cousins who are the children of a brother and a sister respectively; also *attrib.* in *cross-cousin marriage*; **cross-cultural** *a.*, pertaining to or involving different cultures or comparison between them; **cross-dating,** the establishment of the date of one archæological site or level by correlation with another; also in dendrochronology (see quot. 1946); so **cross-date** *v. trans.* and *intr.*; **cross-fade** *v. trans.* and *intr.* (*Broadcasting* and *Cinemat.*), to 'fade in' one sound or picture while 'fading out' another; so as *sb.*; so **cross-fading** *vbl. sb.*; **cross-fault** *Geol.*, a fault which crosses the strike of the displaced strata; **cross-ferry** = Ferry *sb.*[1] 2, 3; **cross-flute,** a transverse flute (see Flute *sb.*[1] 1); **cross-hair** = Spider-line; **cross-influence,** interchange of influences or tendencies; **cross-kick** *v. intr.*, in football, to kick the ball across the field; also as *sb.*; **cross-legs,** (*a*) crossed legs; also quasi-*adv.*; (*b*) (*Obs. slang*) a tailor (cf. quot. 1602 s.v. Cross *a.* 1 b); **cross-license** *v. trans.*, 'to give a license to another to use (a patent or invention) in return for a similar license' (Webster, 1961); also *intr.* and as *sb.*; hence **cross-licensing** *vbl. sb.*; **cross-linguistic** *a.*, pertaining to or involving different languages or comparison between them; **cross-link** *sb.*, **-linkage** *Chem.*, a chemical bond, or an atom or short chain of atoms, which connects two (long) chains in a polymer or other complex molecule; so **cross-link** *v. trans.*, to connect by cross-links, to bring about cross-linking in; *intr.*, to form a cross-link (*with*); **cross-linked** *ppl. a.*; **cross-linking** *vbl. sb.*, the formation of cross-links; also, a network or system of cross-links; **cross-match** *v. trans.* (*Med.*), to test the compatibility of (the blood of a blood-donor and a recipient); also said of the individuals; so as *sb.*; hence **cross-matching** *vbl. sb.*; **cross-member,** a strut fastened across the length of the chassis of a motor-car, etc.; **cross-modulation,** electrical intermodulation; *esp.* the introduction into one signal of new frequencies from another, unwanted, modulated signal; **cross-mouth chisel,** a cylindrical boring chisel with a diametrical blade; also *cross-mouthed chisel*; **cross-pass,** a pass across the field in football; **cross-peen** (also **-pane, -pein**), a hammer in which the peen runs crosswise to the direction of the handle; **cross-ply** *a.*, designating a tyre in which the layers of fabric are laid with the cords at right angles across one another; also *ellipt.* as *sb.*; **cross-rail,** a horizontal rail of a door or other framework; **cross-reel** *v. trans.*, to wind (yarn) on a reel with a reciprocating movement; so *cross-reeling* vbl. sb.; **cross-rhythm** *Mus.*, the simultaneous use of more than one rhythm; **cross-rib,** (*a*) *Arch.* (see quots.); (*b*) in a side of beef, a sternal rib running crosswise to the body; **cross-saddle,** a saddle on which the rider sits astride; also as *adv.*, on a cross-saddle, astride; **cross-seizing** *Naut.*, a seizing in which a number of turns of rope cross an equal number in the opposite direction; **cross-shed** (see quot. *a* 1877); **cross-shot,** (*b*) *Lawn Tennis*, a shot that sends the ball diagonally across the court; **cross-sleeper,** a sleeper laid transversely across a tramway or railway track as a support for the rails; **cross-talk** (earlier and later examples); in wider use, any unwanted transfer of signals from one circuit, channel, etc., to another; also, in *Radio*, a reproduced signal due to waves that are not of the frequency to which the receiver is tuned; (*b*) altercation, repartee, back-chat; conversation; also *attrib.*; so **cross-talker**; **cross tube** (see quot. 1888); **cross-vigil** (see quot.); **cross-volley** *Lawn Tennis*, a volley that sends the ball diagonally across the court; **cross-weaving,** weaving in which the warp-threads are crossed in regular order; **cross-winding,** (*b*) the winding of yarn on a reel in such a way that the strands of one layer cross those of the previous layer at an acute angle; **cross-wire** (example).

1934 C. Lambert *Music Ho!* iii. 219 It is often suggested that jazz rhythm..ends by becoming monotonous through its being merely a series of irregular groupings and cross-accents over a steady and unyielding pulse. **1959** D. Cooke *Lang. Music* v. 265 Driving duple rhythm (with triplets in chaotic cross-accent). **1962** *Listener* 31 May 969/3 Cross-accents and polyrhythms frequently suggest a concerto-like opposition of the various instrumental sounds individually or in groups. **1894** T. W. Fox *Mech. Weaving* 156 If a cross-border machine is employed, a considerable saving in cards results in the manufacture of handkerchiefs ..with a border all round. *Ibid.*, Certain classes of fabrics, such as cross-bordered, swivel, and compound. **1909** *Westm. Gaz.* 30 Mar. 11/4 Low values caused by cross-breaks. **1696** J. Sergeant *Meth. Sci.* III. viii. 323 Multitudes of Cross-causes may intervene, hindering that Effect from following. **1768** Blackstone *Comm.* III. xxvii. 451 When there are cross causes, on a cross bill filed by the defendant against the plaintiff in the original cause, they are generally contrived to be brought on together, that the same hearing and the same decree may serve for both of them. **1940** *Economist* 24 Feb. 341/1 Certain cross checks operated by the authorities suggest that the extent of evasion of the Regulations.. has been commendably small. **1951** *N.Y. Times* 24 June E9/7 Experiments designed to cross-check important results. **1953** *Ibid.* 18 Jan. S3/6 An Associated Press cross-check.. showed them..in support of the..action. **1960** E. H. Gombrich *Art & Illusion* viii. 274 We now know that touch is only one of a whole battery of cross checks at our disposal. **1962** A. Nisbett *Technique Sound Studio* xiii. 225 This front-line audience can be of immense value to him in cross-checking his calculation or intuition. **1968** *Globe & Mail* (Toronto) 5 Feb. 20/3 Penalties..Watson (P) (cross-checking). **1901** *Science* 31 May 869/2, I scarcely dare assert that it might not be secondary cross-coloring. **1920** H. E. Howard *Territory in Bird Life* iv. 133 Here we have a direct relationship..which at first sight appears to be exclusive of cross-correlation. **1965** *Math. in Biol. & Med.* (*Med. Res. Council*) I. 37 The computer is being used to apply the mathematical techniques of autocorrelation and cross-correlation to the interpretation of the EEG in the treatment of temporal lobe epilepsy. **1970** *Nature* 26 Dec. 1299/1 A simple check for the effects of such coupling is to look for differences in the cross correlation between records from different aerials. **1904** J. G. Piddington in *Proc. Soc. Psychical Res.* XVIII. 294 (*heading*) Cross-correspondences between the trance-utterances and script of Mrs. Thompson and those of other mediums. **1909** O. Lodge *Survival of Man* xii. 182 Cross-correspondence—that is, the reception of part of a message through one medium and part through another—is good evidence of one intelligence dominating both automatists. **1938** H. F. Saltmarsh (*title*) Evidence of personal survival from cross correspondences. **1915** M. E. McLoughlin *Tennis as I play It* (1916) xi. 235 The net-man is in line with the angle of almost all cross-court shots. **1923** *Daily Mail* 30 June 11 He cross-courted both returns with his backhand. **1960** *Times* 10 June 18/6 He..hit his cross-court drive..to within inches of the line. **1889** E. Tylor in *Jrnl. Anthrop. Inst.* XVIII. 263 The child of the brother may marry the child of the sister. It seems obvious that this 'cross-cousin marriage', as it may be called, must be the direct result of the simplest form of exogamy. **1932** *Brit. Jrnl. Psychol.* Jan. 265 All ortho-cousins are forbidden, while cross-cousins are considered suitable mates. **1970** E. Leach *Lévi-Strauss* 121 A cross-cousin is a cousin of the type 'mother's brother's child' or 'father's sister's child'. *a* **1942** B. Malinowski *Sci. Theory Culture* (1944) iii. 18 There is the comparative method, in which the student is primarily interested in gathering extensive cross-cultural documentations. **1949** M. Mead *Male & Female* ii. 26 All people who have had the good fortune to learn several languages in childhood have a precious degree of..cross-cultural understanding. **1937** W. S. Glock *Princ. & Methods Tree-Ring Analysis* I. 16 Two ring sequences that cross-date were synchronous in formation and effectively duplicate each other in whole or in part... Cross-dating as generally practiced is the establishment of the time identity in ring groups in two different trees by means of very high and convincing structural correlation between them. **1939** G. Clark *Archaeol. & Society* v. 142 As a basis he studied living trees, cross-dating many trees of the same age to make sure of eliminating individual or purely topographical variations. **1946** F. E. Zeuner *Dating the Past* i. 11 *Cross-dating.* Having constructed a number of plots of individual trees one proceeds to 'cross-date' them. This is the term used by dendrochronologists for correlating the ring-series of one tree with that of another. **1950** G. E. Daniel *100 Yrs. Archaeol.* iv. 148 The synchronological technique of cross-dating. **1937** *Printers' Ink Monthly* Apr. 50/2 *Cross-fade*, where one section of sound (musical or otherwise) is faded in while another is faded out. **1940** *Publishers' Weekly* 5 Oct. 1405 Cross fade on cue. **1953** K. Reisz *Technique Film Editing* ii. 187 The dubbing editor should be able to cross-fade from the dialogue to the music track at any time. **1957** Manvell & Huntley *Film Music* iii. 78 The turning of the page at a crossfade or cut can be helped by music's power. **1931** T. H. Pear *Voice & Personality* vi. 61 The process called by radio-play producers 'cross-fading'. **1900** *Geogr. Jrnl.* XVI. 461 Both longitudinal and cross-faults.. in the Eastern Alps. **1964** L. V. de Sitter *Struct. Geol.* (ed. 2) xiv. 193 The simplest kind of fault is the normal cross-fault, perpendicular to the axis. **1900** *Westm. Gaz.* 2 May 10/1 Nine cross-ferries, two of which carry vehicular traffic. **1903** *Daily Chron.* 20 Feb. 6/7 Vehicular cross-ferry traffic was suspended. **1876** Stainer & Barrett *Dict. Mus. Terms* 172/2 Cross-flutes were known to the Greeks by the name plagiaulos (πλαγίαυλος), and to the Romans as *tibia obliqua*. **1902** *Westm. Gaz.* 2 July 2/3 How this the flute, and that the cross-flute wrought. *a* **1884** Knight *Dict. Mech.* Suppl. 322/1 A telescope fitted with a cross hair. **1917** J. H. McConkey *End of Age* 50 The cross-hairs of God's telescope of prophecy are centered upon it. **1931** G. Stern *Meaning & Change of Meaning* i. 13 The 'big' words are especially liable to sense-loans and cross-influences. **1942** A. Koestler in *Horizon* V. 390 The revolution in physics has certainly affected the artist..and similar cross-influences are easy to discover. **1927** *Daily Tel.* 10 Feb. 16/7 You must swerve.. or feint (or even cross-kick) at an unexpected moment. **1954** J. B. G. Thomas *On Tour* vi. 66 Dobbin..cross-kicked into the middle. **1960** T. McLean *Kings of Rugby* xi. 179 A crosskick by Brown. **1823** in *Spirit of Public Jrnls.* 1823 (1825) 59 Although the world has with one assent, agreed to consider the race of *cross-legs* as the most peaceable and innocent set of people. **1889** Wilde *House of Pomegranates* (1891) 39 Sitting down cross-legs, in a circle. **1912** W. de la Mare *Listeners* 9 Then from his crosslegs he gets down. **1921** D. H. Lawrence *Let.* in E. & A. Brewster *Reminisc. & Corresp.* (1934) 24 If only you crossed the spoon and fork in front to look like two cross-legs. **1964** M. Gowing *Brit. & Atomic Energy 1939–1945* vii. 208 He did not believe that any agreement between Governments for free cross-licensing was necessary. **1965** *Economist* 18 Dec. 1330/3 Part of the cross-licensing agreement. **1954** J. H. Greenberg in H. Hoijer *Lang. in Culture* I. 6 A matter which might be tested by cross-linguistic comparisons. **1964** R. H. Robins *Gen. Linguistics* 208 Cross-linguistic appeals to equivalents or earlier forms in other languages are wholly irrelevant. **1936** Blaikie & Crozier in *Ind. Engin. Chem.* XXVIII. 1159/2 The number of cross links in different soluble polymers will vary and be less than in the insoluble variety. **1937** Norrish & Brookman in *Proc. R. Soc.* A. CLXIII. 207 The production of these cross-linkages as determined by the formation of insoluble polymers appears to depend upon the electron attracting or repelling properties. *Ibid.* 207 Isoprene was used in an attempt to obtain a cross-linked polymer with styrene, since isoprene must possess this cross-linking property. *Ibid.* 219 Styrene will not form an insoluble copolymer with the ether, although a small amount of cross-linking must occur. **1941** P. J. Flory in *Jrnl. Amer. Chem. Soc.* LXIII. 3100/2 Gelation occurs when the cross-linking index *y* (equal to the number of structural units which are cross-linked per chain) is equal to unity. **1963** J. Osborne *Dental Mech.* (ed. 5) xi. 240 A cross-linking agent may be added to the monomer. On polymerization such a substance cross-links in at least two directions with methyl methacrylate. **1970** *Nature* 6 June 939/2 This results in cross-linking which binds together the polymer chains within the enlarged crystallites. *Ibid.*, Cross-linking polymers by *y*-irradiation is known to decrease crystallinity. **1930** *Ann. Surg.* XCI. 487 In five cases, in which the donor and the recipient were of the same group, there was agglutination when the bloods were cross-matched. **1937** Kracke & Garver *Dis. Blood* xxxix. 433 It is..advisable to cross match the prospective donor and recipient to eliminate any possibility of an untoward reaction... In performing a cross match either the slide or the test tube method..may be used. *Ibid.*, The latter [is ascertained] by cross matching (donor vs. recipient). **1961** *Lancet* 19 Aug. 381/1 Early assessment allows more time for accurate cross-matching, and makes emergency transfusion safer. **1922** *Autocar* 10 Nov. 982 Another feature is the mounting of the steering box on the front cross-member of the chassis. **1962** *Which?* (Car Suppl.) Oct. 138/1 We had to replace the front suspension cross member. **1933** J. M. Stinchfield in K. Henney *Radio Engin. Handbk.* viii. 204 Cross-modulation and modulation distortion in the r-f stages of a receiver. **1942** *Electronic Engin.* XV. 285 This method..prevents cross-modulation of the two input voltages due to a common cathode impedance. **1958** W. F. Lovering *Radio Communication* xiii. 319 The cross-modulation component is..most likely to occur when the interfering signal is that from a strong local station. *a* **1877** Knight *Dict. Mech.* I. 650/1 *Cross-mouth chisel*, a boring-chisel of a cylindrical form with a diametrical blade. **1896** *Daily News* 26 Sept. 3/5 Cross-mouthed chisels of hardest tool steel. **1929** *Daily Express* 7 Nov. 19/2 A cross-pass from right-winger T. Maskell was retrieved just beyond the far post by his opposite number. **1961** *Times* 25 May 4/2 Hitchens headed into the top corner on the bounce a long cross-pass from Armfield. *a* **1877** Knight *Dict. Mech.* II. 1647/2 (*caption*) Cross peen for coopers. **1957** R. Lister *Decor. Wrought Ironwork* I. 11 If the pane is placed in a position running relatively across this hole instead of running parallel to it, it is called a *cross-pane*. **1958** *Listener* 24 July 143/1 Do not buy a crosspein hammer that is too small. **1965** Gough & Udall *Radial Ply Tyres* 1 (*caption*) A conventional cross-ply tyre in section. **1968** *Listener* 18 July 95/2 Try not to mix radials with cross-plies. **1970** M. Edwards *Car Handyman* viii. 104 A cross-ply tyre is one in which the plies of the tyre, or the beads, cross over... Usually, if the marking is in inches, say 5·20× 10, then the tyre is a cross-ply. **1836** Dickens *Sk.*

Boz II. 95 A tent bedstead without hangings or cross-rails. **1880** *Spon's Encycl. Industr. Arts* II. 739 In the interior of the framework, is fitted a conical grid, having its apex downwards, and resting on a cross-rail at a short distance from the bottom. **1902** *How to make Useful Things* 13/2 The bottom cross-rail is..3½ in. less in length than the width of the end of the fowl-house. **1890** Nasmith *Mod. Cotton Spinning Mach.* xiii. 267 The hanks being reeled, they are, if cross reeled, dyed or bleached, and, if in leas, bundled. **1926** Whiteman & McBride *Jazz* xi. 231 Six hundred fox-trotters..automatically were dancing in cross rhythm. **1927** *Melody Maker* Sept. 845/3 Brahms..employed syncopation and cross-rhythms about a century before modern 'syncopated orchestras' were dreamed of. **1946** A. Hutchings in A. L. Bacharach *Brit. Music of our Time* xvi. 205 They lacked the vigour which comes from cross-rhythm, or counterpoint. **1858** *Dict. Archit.* (*Archit. Publ. Soc.*), *Cross rib* (Fr. *arc doubleau*), a rib from one pier or pillar across to its respond, square with the vault to which the rib belongs... Willis calls it the transverse rib, and it is often called the arch rib. **1902** R. Sturgis *Dict. Archit.* III. 289 The wall ribs (*formerets*) and cross ribs (*arcs doubleaux*) were..pointed. **1897** W. E. Norris *Clarissa Furiosa* xxxiii. 293 His daughter..would..ride to hounds in a cross-saddle. **1897** *Westm. Gaz.* 22 Dec. 4/2 The cross-saddle position assumed by women on 'bikes'. **1905** *Daily Chron.* 1 Aug. 3/3 One of our Royal Princesses is to be taught to ride cross-saddle. **1930** S. G. Goldschmidt *Fellowship of Horse* ix. 131 Cross-saddle riding for women is not making the progress it should. *Ibid.* 136 The prejudice against the cross-saddle. **1883** *Man. Seamanship for Boys* 109 A Cross Seizing is used when the rigging is turned in with the end up. *a* **1877** Knight *Dict. Mech.* I. 650/1 *Cross-shed*, the upper shed of a gauze-loom. **1894** T. W. Fox *Mech. Weaving* 225 *O* shows the lifting for an open shed, and *c* that for a cross shed. **1889** H. W. W. Wilberforce *Lawn Tennis* xii. 43 It may be a difficult cross-shot. **1902** *Westm. Gaz.* 1 July 4/3 His cross-shots to the left-hand corner swift and sure. **1841** *Penny Cycl.* XIX. 255/1 The use of cross-sleepers..needs little remark. **1888** *Encycl. Brit.* XXIII. 506/2 The rail was spiked through to a longitudinal timber laid on cross sleepers. **1887** *Soc. Telegr. Engin. Jrnl.* XVI. 433 The annoyance caused by induction, known commonly by the name of 'cross-talk'. **1909** 'I. Hay' *Man's Man* viii. 132 A carefully rehearsed 'cross-talk' dialogue between two knock-about artistes of the Variety firmament. **1910** *Hawkins's Electr. Dict.*, *Cross talk*, conversation over one telephone circuit overheard in the telephone of another circuit, when their wires run side by side. This fault is due almost entirely to electrostatic induction. **1917** 'I. Hay' *Carrying On* i. 18 Each bus is in charge of the identical pair of cross-talk comedians who controlled its destinies in more peaceful days. **1923** Wodehouse *Adventures of Sally* xv. 184 As brisk and snappy as any cross-talk between vaudeville comedians. **1930** *Times Lit. Suppl.* 27 Mar. 276/3 Some of the crosstalk of the American shop-girls is entertaining. **1932** P. E. Terman *Radio Engin.* xiii. 470 The first kind of cross-talk is produced by heterodyne detection of two signals having a frequency difference lying within the tuning range of the receiver. *Ibid.* 471 Cross-talk [of the second kind] is caused by the unwanted signal modulating the carrier wave of the desired signal. **1955** *Times* 27 July 8/2 There was a good deal of cross-talk on the origin of recent rumours about sterling. **1957** R. W. G. Hunt *Reproduction of Colour* xii. 163 Hence *cross-talk* (that is, interference between the luminance and chrominance signals) in band-sharing systems..is minimized. **1961** A. Wilson *Old Men at Zoo* vi. 306, I thought that I should scream if I had to live with this cross talk act for long. **1970** *Which?* Apr. 114 The crosstalk rating shows how well the stereo channels were separated, so that signals on one channel did not affect the other. **1907** *Daily Chron.* 1 May 6/4 Those pioneer cross-talkers, the Christy Minstrels. *a* **1884** Knight *Dict. Mech.* Suppl. 232/1 *Cross tube boiler.* In the usual vertical form, this boiler has one or more horizontal cross tubes..placed across the fire-box. **1888** *Lockwood's Dict. Mech. Engin.*, *Cross tubes*, the heating tubes in a steam boiler—usually applied to boilers of the vertical type. **1932** *Times Lit. Suppl.* 28 Jan. 55/3 The Cross-vigil (crosfigell), that is, praying for lengthened periods with the arms outstretched in the form of a cross. **1905** *Daily Chron.* 20 Mar. 3/3 The American cross-volleys which may bring the English players a little nearer the net in doubles. **1843** *Penny Cycl.* XXVII. 179/1 *Cross weaving.*—This term may be conveniently applied to those varieties of woven fabric in which the warp-threads..cross over or twist around one another, thus forming a plexus or interlacing independent of that produced by the weft. *a* **1877** Knight *Dict. Mech.* I. 650/2 *Cross-weaving loom*, a loom for weaving with a crossed warp. **1892** Nasmith *Students' Cotton Spinning* 360 Cross winding is resorted to when the hank is to be dyed. **1866** Cross-wire [see Spider-line].

cross-bar, *sb.* Add: **1. a.** *spec.* The horizontal bar of a bicycle frame; also, in *Football*, the horizontal bar between the goal-posts.

1857 T. Hughes *Tom Brown* v. 120 There it flies, straight between the two posts, some five feet above the cross-bar, an unquestioned goal. **1956** M. Golesworthy *Encycl. Assoc. Football* 36 The crossbar over the goalmouth was first introduced into Football Association rules in 1875. **1966** T. Simpson *Cycling is my Life* v. 38 My left leg was still fastened to the pedal by the toe-strap and then bent over the cross-bar with me lying across the front wheel. **1967** S. Beckett *Film* 32 Two bicycles ridden by men with girl passengers (on crossbar).

2. (Later examples.)

1937 R. H. Stetson in *Mélanges ling. et phil. offerts à J. van Ginneken* 355 In Continental usage the 7 has a cross-bar. **1957** N. R. Ker *Catal. MSS. containing Anglo-Saxon* p. xxxi, The form and position of the cross-bar [of ð] often vary.

cross-bred, *ppl. a.* Add: (Earlier and later examples of *sb.*) Also, (designating) wool from cross-bred sheep.

1880 Grant & Foster *N.Z.* iv. 42 The weight of these [fleeces] in the grease seemed to average about 5½ or 6 lbs., while that of the cross-breds were [*sic*] very little heavier. **1899** *Daily News* 7 Mar. 8/6 Fine crossbreds are also fairly firm, but the stronger descriptions, and English wools generally, are in small demand. **1941** [see *auto-sex* s.v. *auto-]. **1955** [see *come-back *sb.*[2] 4]. **1962** *Economist* 11 Aug. 548/2 New Zealand's crossbred wools have drifted lower again.

cross-breed, *v.* (Later examples.)

1932 *Discovery* Mar. 73/2 There was a certain amount of selection and cross breeding [of crops] towards the end of the nineteenth century. **1955** *Sci. News Let.* 23 July 50 That individual is separated from the rest, then bred and cross-bred to develop a pure-breeding strain. **1955** *Bull. Atomic Sci.* Sept. 242 FAO technicians are cross-breeding the rice strain Japonica..with the native Indica species of Southeast Asia and India.

cro·ss-conne:ct, *v. Electr.* [Cross- 6 a.] *trans.* To interchange the connections of (electric wires); to connect (each of a set of two or more wires or terminals) to a different wire or terminal of another set. Also in *ppl. a.* and *vbl. sb.* Hence **cro·ss-conne:ction,** the arrangement of wires in this way; **cro·ss-conne:ctor,** a device used to effect this.

a **1877** Knight *Dict. Mech.* III. 2512/2 To cross-connect wires is to interchange them, so that a current from one wire is shifted to another at one station and then back again at a farther station, to work around a faulty station. **1884** F. Krohn tr. *Glaser de Cew's Mag.- & Dyn.-Electr. Mach.* 261 The segments of the collector are internally cross connected. **1893** W. P. Maycock *Electric Lighting* II. vii. 205 The armature is cross connected to avoid the use of 4 brushes, the cross-connectors consisting of copper rings with two lugs. **1893** Sloane *Electr. Dict.* 157 *Cross-connecting board*, a special switch board used in telephone exchanges and central telegraph offices. Its function is, by plugs and wires, to connect the line wires with any desired section of the main switchboard. **1893** G. Kapp *Dynamos* viii. 176 A four-pole cylinder armature with cross-connections. **1910** *Hawkins's Electr. Dict.*, *Cross connected dynamo*, a dynamo having the coils of its armature connected to corresponding bars of the commutator. **1967** D. H. Hamsher *Communication Syst. Engin. Handbk.* xxi. 10 Cross connection is a method of..connecting a terminated cable pair to any one of a group of other cable pairs similarly connected. *Ibid.*, Weather-proof terminals with prewired and soldered conductors of plastic-insulated wires are designed to cross-connect circuits at any one location.

cro·ss-cou:nter. *Boxing.* [f. Cross- 5 + Counter *sb.*[5] 3.] (See Counter *sb.*[5] 3.) Hence as *v. trans.* and *intr.* Also *fig.*

1864 'Mark Twain' *Sketches of Sixties* (1927) 149 The skirmishers fight shyly up to each other, counter and cross-counter, feint and parry. **1889** E. B. Michell *Boxing* 166 The answer to this cross-counter is to deliver the right at the face of the counterer. **1897** *Encycl. Sport* (1901) I. 134/2 That most effective of manœuvres, though difficult of execution, the right hand cross-counter. *Ibid.* 137/1 Should he show a tendency to counter at the body or cross-counter at the head, you may find an opening for an upper-cut. **1907** W. De Morgan *Alice-for-Short* xxvi. 276 As this and much more cross-countered continually with the dialogue about Alice's squashy days, Mrs. Heath had good excuse for misunderstanding.

cross-country, *a.* Add: **b.** Applied to flying across the country. Also *absol.*, a cross-country flight.

1909 *Daily Chron.* 9 Sept. 1/6 The first cross-country flight of note was made by M. Farman on October 30, 1908. **1948** 'N. Shute' *No Highway* xi. 287 He..had hit a hill, like any pupil on his first cross-country.

cross-cut, *sb.* Add: **5.** Short for *cross-cut saw.*

1848 E. Atkinson *Otago Jrnl.* III. 46 Pit-saws; whit saws; and crosscuts, and files for ditto. **1853** 'P. Paxton' *Yankee in Texas* 89 Felling trees, handling cross-cuts, rolling blocks. **1942** L. Rich *We took to Woods* (1948) 55 Excellence on a two-man cross-cut has nothing to do with size and strength.

6. *Cinemat.* (See *cross-cut *v.* 2.)

cross-cut, *v.* Add: **2.** *Cinemat.* To subject (a film or films) to cross-cutting (see quot. 1933). Also *intr.* and *transf.* Hence as *adj.* and *sb.*

1933 A. Brunel *Filmcraft* 95 You may need part of a rejected take because you are cross-cutting that scene. *Ibid.* 97 After you have eliminated the scene numbers.. you can begin to make the simpler cross-cuts. *Ibid.* 156 *Cross-cut*, to alternate in editing two or more scenes—as when one has two close-ups of characters facing each other. **1957** Manvell & Huntley *Film Music* ii. 40 The two scenes of the dancing action were hastily lengthened and cross-cut to give a rhythm. **1958** *Times Lit. Suppl.* 5 Dec. 698/1 Haddock's gestures as he tells the story exactly match and are cross-cut with those of the Chevalier de la Hadoque slashing away at the ruffians boarding his ship. **1962** *Listener* 30 Aug. 328/1 A montage sequence..with shots of Eton, night-club frolics..cross-cut against back-street slums. **1967** *Spectator* 30 June 772/1 The author has developed an impressive, cinematic technique for telling this story, crosscutting rapidly from episode to episode.

cro·ss-cu:tter. *Forestry.* [f. Cross-cut *v.* + -er[1].] One who uses a cross-cut saw.

1902 *N.Z. Illustr. Mag.* Feb. 374/1 Having severed the head from the trunk, the crosscutter's work is considered accomplished, and they pass on to another tree. **1936** 'A' in A. M. Rust *Whangarei & Dist. Early Rem.* 163 Hundreds of bushmen, loggers, jackers, bullock-drivers, cross-cutters and rafters were employed.

cro·ss-cu:tting, *vbl. sb.* [f. Cross-cut *v.* + -ing[1].] **1.** The action of cutting across.

1805 R. W. Dickson *Pract. Agric.* I. 345 Repeated cross-cuttings with the plough and harrowings. **1896** *Daily News* 16 Dec. 8/4 He had done 326 feet of cross-cutting still in ore.

2. *Cinemat.* The action or process of alternating two or more sequences in editing a film. Also *transf.*

1938 G. H. Sewell *Amat. Film-making* x. 98 Such a device of alternating shots of two ideas is known as cross cutting. **1954** *Encounter* Aug. 52/2 *Intolerance*..set a standard of screen narrative by cross-cutting and contrapuntal action never since surpassed. **1958** *Times Lit. Suppl.* 21 Feb. 97/4 The manipulation of a rather complex plot is unsure, the cross-cutting inexpert. **1963** *Movie* Feb. 31/1 These angles, which are constantly cut together, have been selected in deliberate opposition to conventional cross-cutting.

cro·ss-dye, *v. trans.* (See quots.) Hence **cro·ss-dye** *sb.*, a colour used in cross-dyeing; **cro·ss-dyeing** *vbl. sb.*

1885 J. J. Hummel *Dyeing Textile Fabrics* 466 The cotton warp may be dyed black, brown, dark blue, drab, &c., before weaving, in which case only the woollen or worsted weft is dyed subsequently. The finished goods are then said to be 'cross-dyed'. **1901** F. Beech *Dyeing Cotton Fabrics* iii. 79 This hawking machine will be found useful..in dyeing cotton cloths with such dyes as..Cross-dye blacks. *Ibid.* v. 220 Before the introduction of the direct dyes the method usually followed..is that known as cross dyeing. **1961** Blackshaw & Brightman *Dict. Dyeing* 54 *Cross dyeing*, the dyeing of one component of a mixture of the fibres after at least one of the others has been dyed already. *Ibid.*, The criteria are that the dyed yarn shall not change in shade or bleed into other fibres when the latter are cross dyed.

crosse. (Examples.)

1867 *Ball Players' Chron.* 20 June 5/4 It is played with a 'crosse', each player carrying one. **1892** *Photogr. Ann.* II. 51 We remember at a lacrosse match taking a shot at the moment a goal was scored, and the 'crosses' were being shied into the air by the jubilant scorers. **1909** *Westm. Gaz.* 23 Sept. 12/3 French is a fine sturdy player, but he is inclined to use only one hand on his crosse.

crossect (krǫsˌse·kt), *v.* [f. Cross + L. *secāre, sectum* to cut.] *trans.* To divide transversely.

1860 Trollope *Castle Richmond* III. iv. 69 These had since been bisected and crossected, and intersected. **1927** *Glasgow Herald* 1 Sept. 8 The Exchequer crossects our income.

crossed, *ppl. a.* Add: **2.** (Earlier example applied to a letter.) Of a cheque: marked with two parallel lines (see quot. 1957 and Cross *v.* 7 c).

1834 J. B. Byles *Law of Bills* (ed. 2) x. 129 Crossed checks. It is common practice in the city of London, to write across the face of a check the name of a banker. **1836** Mrs. Gaskell *Let.* 12 May (1966) 8 None of your niminipimini notes, but a sensible nonsensical crossed letter. **1876** B. T. Bosanquet (*title*) Crossed cheques: the object of crossing, with suggestions as to the amendments required in the statutes relating thereto. **1877** [see Dict.]. **1957** *Encycl. Brit.* V. 413/1 The form [of a cheque] may be printed 'crossed'; two lines being printed across it. A cheque so crossed can only be cashed through a banker. **1966** *Listener* 25 Aug. 291 (Advt.), Order through your bookseller or send crossed cheque.

cross-fertilization. Add: Also *fig.*

1900 *Westm. Gaz.* 26 Jan. 3/3 Nearly all eighteenth-century art in France and England hangs ultimately upon him; and, by their cross-fertilisation, how much of modern art too! **1928** C. H. Dodd *Authority of Bible* ix. 196 Alexander's conquests had brought the old order to an end. There was a mingling of cultures, a cross-fertilization of East and West. **1944** J. S. Huxley *On Living in Rev.* p. vii, The compulsory cross-fertilization of ideas. **1970** *Daily Tel.* 22 Apr. 21/5 The companies are now starting on the cross-fertilisation process, exchanging information, identifying common projects.

cross-firing, *vbl. sb.* (Earlier example.)

1837 Dickens *Let.* 22 June (1965) I. 276 This cross-firing of notes makes me smile.

cross-head, *sb.* Add: **1. b.** Any beam across the top of a piece of mechanism.

1844 H. Stephens *Bk. Farm* I. 417 The draught shackle ..is held in its place upon the cross-head..by the draught-bolt. *Ibid.* II. 134 The handle..terminates in a crosshead. *a* **1884** Knight *Dict. Mech.* Suppl., *Cross Head*, a cruciform-shaped four-handled bar, at the upper end of a drill-rod or earth-auger. **1884** *Science* III. 314 Two side-screws, carrying the top crosshead. **1888** *Encycl. Brit.* XXIII. 151/1 Thus avoiding torsion of the polar axis at the expense of greatly increased length of the cross-head. **1901** Merwin & Webster *Calumet 'K'* xvi. 322 Another endless series of cups was carrying the wheat aloft. It went over the cross-head and down a spout.

2. (Later example.)

1964 *Friend* 4 Sept. 1067/2 The memorandum continues (and from this point, inserting only crossheads, we quote it in full): We cannot [etc.].

3. *Mining.* A heading running across a vein.

1877 R. W. Raymond *Statist. Mines & Mining* 197 At the point of connection the eastern limit of the south vein is defined by a cross-head.., and to the east of this cross-head no trace of the fissure has as yet been found.

4. cross-head brasses *pl.*, the brass bearings of the cross-head of a steam or other engine; **cross-head pin,** the pin by which the connecting-rod is attached to the piston-rod.

1865 *Daily News* 26 July 3/1 No. 80 torpedo boat..broke down owing to a defect in the crosshead brasses. 1887 D. A. Low *Machine Drawing* xii. 51 The cross-head pin need not be drawn separately. 1889 Hasluck *Model Engin. Handybk.* viii. 91 The hole in cross-head must be broached out till the cross-head pin will nearly fit it.

cross-head, *v.* (Later example.)
1908 'Ian Hay' *Right Stuff* iii. 42, I doubt now if I could write out twenty lines of 'Paradise Lost' without cross-heading them.

cro·ss-hea:ding. [Cross- 4.] **1.** *Mining.* A transverse heading (see quots.).
1883 H. M. Chance *Rep. Mining Methods in Anthrac. Coal Fields* 533 *Cross-heading*, a passage driven for ventilation from the airway to the gangway, or from one breast through the pillar to the adjoining working. 1904 Goodchild & Tweney *Technol. & Sci. Dict.* 140/2 *Cross Heading*, a drift or passage from one level to another for ventilating purposes.
2. = Cross-head *sb.* 2.
1898 G. B. Shaw *Our Theatres in Nineties* (1932) III. 378 The interview, the illustration and the cross-heading, hitherto looked on as American vulgarities impossible to English literary gentlemen, invaded all our papers. 1967 *Listener* 16 Nov. 634 It's my great opportunity to be listed under the magical cross-heading 'from Professor Ayer and others'.

crossing, *vbl. sb.* Add: **2.** (Earlier example of 'writing across other writing'.)
1822 Shelley *Let.* 18 June (1964) II. 715, I intend to indulge myself in plenty of paper and no crossings.
11. *crossing-keeper*; **crossing-gate,** a gate at a level crossing which is closed to road traffic when a train is due; **crossing-over** *Biol.*, the formation of a genotype exhibiting characters derived from both parents when the characters are known to be linked; hence also, the interchange of segments between chromatids of homologous pairs of chromosomes which causes this breakdown of linkage.
1929 *Star* 21 Aug. 7/2 The railway crosses the road in several places without crossing-gates. 1921 D. H. Lawrence *Sea & Sardinia* iv. 128 At a level crossing the woman crossing-keeper darted out vigorously with her red flag. 1912 T. H. Morgan & E. Cattell in *Jrnl. Exper. Zool.* XIII. 79 In certain combinations, the relation between linkage and breaking of the linkage ('crossing-over' as we shall call it) is shown at once. *Ibid.* 91 There were seven cases of crossing-over in color, all males, in a total of 872 males. 1937, 1949 Crossing-over [see *asynapsis].

cross-lining, *vbl. sb.* (Examples of the sense 'cross-fishing'.)
1897 *19th Cent.* Aug. 199 Cross-lining for trout has lately been prohibited. 1900 *Daily News* 16 Aug. 3/2 A large landowner, who has..succeeded in stopping cross-lining on Lough Corrib. 1907 *Westm. Gaz.* 13 Dec. 4/2 A daring spirit suggested the project of cross-lining.

cross lots, cross-lots, *advb. phr.* *U.S.* Also **cross-lot.** [See Cross *prep.* and Lot *sb.* 6 a.] By a short cut. Also *attrib.*
1825 J. Neal *Bro. Jonathan* I. 138 They could push on, a pooty, tedious, clever bit furder, cross lots—they could. 1874 'Mark Twain' & C. D. Warner *Gilded Age* xxi. 154 The cross-lots path she traversed to the Seminary. 1889 'Mark Twain' *Yankee* xx. 241 After three hours of awful crosslot riding he had overhauled his game. 1922 A. Brown *Old Crow* 469 They might even go over to Mountain Brook by the path "cross lots'.

cross-over. 3. Delete *U.S.* and add further examples. Also of a tramway.
1895 *Daily News* 15 Oct. 3/2 At the starting point are four crossovers to suit any arrangement of traffic. 1901 *Westm. Gaz.* 29 Nov. 10/2 The castings necessary for the crossovers on electric tramways. 1928 *Daily Express* 22 Nov. 11/1 The cross-overs available were at Beckenham Junction and Penge. 1967 C. J. Freezer *Model Railway Terminol.* 5/1 A train proceeding in its correct direction along the main line can run directly over the facing crossover, or must reverse to cross over a trailing crossover.
4. *Biol.* (*a*) An instance of the process of crossing-over (see *crossing *vbl. sb.* 11); (*b*) an individual having characters inherited by crossing-over. Also *attrib.*
1912 T. H. Morgan & E. Cattell in *Jrnl. Exper. Zool.* XIII. 91 The sum of the 'straight' males is 580, while that of the cross-overs is 292. 1916 *Jrnl. Genetics* V. 285 The cross-overs appear to occur in numbers exactly proportional to the distance apart of the factors concerned. 1916 *Genetics* I. 134 Of these thirteen cases which involved crossing over, twelve were crossovers in only one chromosome and were non-crossovers in the other. *Ibid.* 135 A crossover chromosome. 1919 R. C. Punnett *Mendelism* (ed. 5) xii. 144 It is upon the proportion of 'crossover' gametes as compared with 'non-crossover' gametes that the distances between the factors along the chromosomes have been determined. 1920 L. Doncaster *Introd. Cytology* 224 The American investigators call these exceptional combinations crossovers, since in the combinations of *Ab* and *aB*, *A* and *a* are regarded as having crossed over from their normal combinations and to have exchanged places. 1964 D. Michie in G. H. Haggis *Introd. Molecular Biol.* vii. 203 Ultra-fine genetic analyses using fast-breeding micro-organisms are capable of charting the distribution of cross-over events over exceedingly short chromosomal intervals.
5. *attrib.* or as *adj.* That crosses over; characterized by crossing over or having a part that crosses over another.
1893 'M. Gray' *Last Sentence* III. ii, White pinafore, crossover shawl, and velvet hat. 1905 *Westm. Gaz.* 8 July 13/2 The cross-over bodice. 1906 *Ibid.* 15 Feb. 4/1 A cross-over ring set with a large brilliant and a cabochon emerald. 1939 F. Thompson *Lark Rise* i. 8 They stood at corners in their big white aprons and crossover shawls. 1968 *Economist* 11 May 18/3 Both Governor Branigin and Senator McCarthy were (on different pretexts) soliciting Republican 'crossover' votes.
b. cross-over block, road (see quots.).
1893 Sloane *Electr. Dict.* 158 *Cross-over block*, a piece of porcelain or other material shaped to receive two wires which are to cross each other. 1888 *Lockwood's Dict. Mech. Engin.*, *Cross-over road*, a short diagonal line of rails on permanent way, provided with a pair of points or switches at each end, and connecting two parallel lines of rails together. 1893 *Athenæum* 8 July 68/1 'Crossings' imply something more than merely the gaps left in the rails for a cross-over road. 1896 *Daily News* 18 Dec. 8/2 He let the goods train on to the up main line, but did not pull over the cross-over road points for the goods train to go across to the down line.

cross-point, *sb.* Add: **3.** *pl.* The points of a railway cross-over.
1896 *Westm. Gaz.* 13 July 2/2 When the train has to pass over cross-points.

cro:ss-re·ference, *v.* [f. the sb.] *trans.* To provide with a cross-reference or cross-references; to refer *to* by a cross-reference. Hence **cro:ss-re·ferenced** *ppl. a.*; **cro:ss-re·ferencing** *vbl. sb.* and *ppl. a.*
1902 H. B. Wheatley *How to make Index* iii. 73 If a general heading be divided into sections, and each of these be clearly defined, they should be cross referenced, but not otherwise. *Ibid.* 74 Cross referencing has its curiosities.. 'Cattle *see* Clergy'. 1907 W. James *Let.* 18 May in R. B. Perry *Tht. & Char. W. J.* (1935) II. 506 You..stop him and cross-reference him and counter on him. 1914 *Cath. Encycl.* XVI. 88 Where several forms of the same name occur, all the references are grouped under one spelling to which the other forms are duly cross-referenced. 1922 *Daily Mail* 7 Dec. 10 It saves the unnecessary labour of cross-referencing. 1955 *Sci. Amer.* Jan. 95/1 Well thought out, clearly printed, skillfully cross-referenced. 1956 *Nature* 3 Mar. 402/2 A series of cross-referencing indexes. 1965 *Math. in Biol. & Med.* (*Med. Res. Council*) II. 47 Although seemingly elaborate, this kind of cross-referencing is really quite natural in view of the facts of human family structure.

cross-road. Add: **2. b.** *fig.* (usu. pl.). A point at which two or more courses of action diverge; a critical turning-point.
1795 S. J. Pratt *Gleanings through Wales* I. xv. 238 Join with me..that they may speedily be conducted from the *cross* roads of life. 1898 *Nat. Rev.* Aug. 908 To place him at the cross roads of starvation and revolt. 1915 *War Illustr.* 6 Mar. 54 Britons at cross-roads of Honour, Glory and Death. 1924 E. Y. Mullins (*title*) Christianity at the cross roads.
c. Phr. *dirty work at the cross-roads*: see *dirty *a.* 2 b.
3. b. Also *cross-roads*; *spec.* in *U.S.* with the implication of smallness, cheapness, etc.
1845 *Southern Lit. Messenger* XI. 586/1 The country merchant and cross-road store-keeper. 1848 E. Bryant *Calif.* (1849) 98 The place had something of the air of a cross-roads settlement. 1868 *Putnam's Mag.* June 715/1 Now and then an enterprising specimen of the breed set up a 'cross-roads grocery'. 1869 *Champaign Co.* (Illinois) *Gaz.* 26 May 1/2 Defer the payments we may have promised to Cross-roads politicians. 1905 *Forum* Apr. 485 To place an obstacle in the way of the cross-roads politicians. 1935 *Amer. Speech* X. 5/2 A small crossroads community in Vermont.

cross-ruff, *sb.* **2.** Also in *Bridge.*
1905 R. F. Foster *Compl. Bridge* 226 Do not let go of the lead until you have made all your trumps separately by the cross ruff on the red suits. 1907 C. S. Street *Good Bridge* vii. 103 The careless player exhilarated by the success of his devastating cross-ruff continues it once too often. 1926 M. C. Work *Auction Bridge Complete* II. ii. 330 The Declarer who forces his long trump hand (except to obtain an entry in a cross ruff) is generally playing the game of the adversary.

cross-ruff, *v.* (Stress variable.) [f. the sb.] *intr.* To play a cross-ruff. Also *trans.*, to play (a hand) using a cross-ruff. So **cross-ruffing** *vbl. sb.*
1905 R. F. Foster *Compl. Bridge* 226 Having no established suit..he plays to make his trumps separately by cross ruffing. 1918 —— *Foster on Auction* II. 354 Cross-ruffing an entire hand. 1934 L. H. Watson *Play at Contract Bridge* xvi. 149 If you are going to play a hand by cross-ruffing, you must follow it through. 1958 *Listener* 4 Dec. 965/2 We cannot cross-ruff our way to a high level contract.

cro·ss(-)se:ction, *sb.* Add: (Earlier and later examples.) *spec.* = Section *sb.* 4 a.
1835 A. Gray *Lett.* (1893) I. 52 A cross-section shows the same structure as the rattan. 1870 *Spon's Dict. Engin.* II. 389 A front elevation and cross-section of a boiler. 1874 *Ibid.* VIII. 2924 The converting department, shown in ground plan by Fig. 6996, and in cross-section by Fig. 6998. 1878 J. H. Beadle *Western Wilds* x. 143 Five men were twenty days felling it, the object being to have it sawed into cross-sections to be shipped eastward to Europe. 1933 *Archit. Rev.* LXXIV. 123 This pictorial cross-section.
b. *fig.* A typical or representative sample, group, etc., or an examination of this.
1903 *Independent* 22 Jan. 210/2 A narrow, timorous artificial treatment of some such limited subject or 'cross-section' of a subject as may be represented 'concretely'. 1904 Kipling *Traffics & Discoveries* 322 Cross-sections of remote and incomprehensible lives. 1909 H. G. Wells *Tono-Bungay* I. i. 5 You will ask by what merit I achieved this remarkable social range, this extensive cross-section of the British social organism. 1938 *Ann. Reg.* 1937 275 The New York *Annalist's* index of business activity, covering a wide cross-section of statistically measurable activity. 1952 E. Grierson *Reputation for Song* xxvi. 222 The jury, a fair cross-section of the community.
2. *Physics.* Used of the apparent area (measured in barns or millibarns) of a nucleus, atom, elementary particle, etc., as representing the probability of a specified interaction with another particle, etc. Symbol σ.
1921 *Sci. Abstr.* A. XXIV. 393 A method is described whereby the effective cross section of gas molecules is determined by means of slowly-moving electrons having a definite single velocity and definite path. 1938 R. W. Lawson tr. *Hevesy & Paneth's Man. Radioactivity* (ed. 2) vii. 75 The effective cross-section is equal to the mass absorption coefficient divided by 6×10^{23}, and multiplied by the atomic weight. 1955 R. D. Evans *Atomic Nucleus* 826 In the wave or field model, the 'cross section' for a particular interaction is the ratio between the rate of energy removal (power) and the incident intensity (power per unit area) per target particle. The cross section can be visualized as an area in the incident wave front... In the corpuscular model, the 'cross section' is the fraction of the incident particles which suffer the specified interaction, divided by the number of target particles per unit area of a thin target. 1960 *Gloss. Atomic Terms* (*H.M.S.O.*) 16 *Cross section*, the apparent size of a nucleus. This varies for many nuclei according to the reaction occurring (capture, fission or scattering) and the speed of the oncoming particle. 1962 *Gloss. Terms Nucl. Sci.* (*B.S.I.*) 30 *Cross section*, of a given nucleus or atom for a given radiation, that area perpendicular to the direction of the radiation which one has to attribute to the nucleus or atom to account geometrically for its interaction with the radiation; or, in other words, the number of those interactions per unit time divided by the radiation flux and the number of nuclei or atoms present. 1967 Condon & Odishaw *Handbk. Physics* (ed. 2) IX. viii. 214/2 The intrinsic probability that a reaction will occur is measured by its cross section... All conceivable events that occur when a projectile strikes a nucleus have their separate cross sections. One speaks of the cross section for elastic scattering,..for absorption,..etc. 1971 *Nature* 23 Apr. 523/2 Other processes, such as the $^{16}O(n, \alpha)^{13}C$ reaction, may contribute to the more point-like flashes, but their cross-sections are generally small.

cro·ss-se:ction, *v.* [f. the sb.] *trans.* To make a cross-section of; to cut into cross-sections. Hence **cro:ss-se·ctioning** *vbl. sb.*
1876 *Van Nostrand's Eclectic Engin. Mag.* XIV. 394/1 The engineer should have an intimate knowledge of the ground on the line of the work. This is obtained by means of cross-levels taken at right angles to the reference line of the work... This work in engineering parlance is cross-sectioning. *Ibid.* 399/2 The finished work can be cross-sectioned and plotted in the same manner as the original ground. 1890 D'Oyle *Notches* 52 They were going down to 'cross-section' the old railway survey which ran through our valley. 1897 *Outing* (U.S.) XXX. 126/2 Much of the ground is..conveniently bounded and cross-sectioned by roads. 1908 *Westm. Gaz.* 22 Aug. 14/1 The eye would then cross-section its words, reading the lateral parts in indirect vision.

cro·ss-se:ctional, *a.* [f. Cross-section *sb.* + -al.] Of or pertaining to a cross-section.
1874 *Spon's Dict. Engin.* VIII. 2931 The difference in cross-sectional area between the two ends of the ram is the area acted upon by the water to lift it. 1896 E. Atkinson tr. *H. du Bois's Magnetic Circuit* 60, *S* is the cross-sectional area of the bar-magnet. 1916 H. Barber *Aeroplane Speaks* 94 The cross-sectional dimensions must be correct. 1923 W. E. Gibbs *Clouds & Smokes* 137 The cross-sectional area is many times that of the fume. 1941 *Manch. Guardian Weekly* 14 Mar. 214/3 Material descriptive of conditions at present governing the lives of under-twenties (cross-sectional). 1958 M. Argyle *Relig. Behaviour* vi. 58 When age differences in attitudes are found in a cross-sectional survey, it is sometimes possible to account for them by historical factors. 1968 R. A. Lyttleton *Myst. Solar Syst.* iv. 114 Only about 10^{-8} of its overall cross-sectional area can be producing any reflection.

cross-stone. (Later examples.)
1851 [see Harmotome]. 1868 *Amer. Naturalist* I. 264 A boulder..containing large crystals of staurotide, or cross-stone.

cro·ss-tie. [Cross- 4.] A transverse connecting piece (of timber, etc.); *spec.* in *U.S.* = Tie *sb.* 7 b.
1813 *Niles' Reg.* III. 323/1 The three large ribs are preserved in their proper relative situations by fifty-four crossties. 1833 in J. R. Commons et al. *Documentary Hist. Amer. Industrial Soc.* (1910) I. 219 Begun laying cross-ties of plantation railroad. 1858 Simmonds *Dict. Trade*, *Cross-tie*, a railway sleeper; a connecting band in building. 1886 *Encycl. Brit.* XX. 244/1 The longitudinals are connected and kept to gauge by transoms or cross-ties at intervals. 1890 *Harper's Mag.* May 888/1 Across this ditch two old 'cross-ties' made a bridge to the railway. 1908 *Daily Chron.* 14 Nov. 8/6 A few variations in frame design..taking the form of auxiliary cross-ties and supplementary diagonals. 1948 *Clarke County Democrat* 23 Sept. 1/2 Cross ties and switch ties.

cro·ss-tongue. [Cross- 4 a.] A cross-grained tongue of wood used to give extra strength to a joint in woodwork. Hence **cro·ss-tongue** *v. trans.*, to provide with a cross-tongue.
1876 *Encycl. Brit.* IV. 489/2 Surfaces..formed of inch or inch and quarter boards joined with glue, and a cross or feather tongue ploughed into each joint. 1901 *J. Black's Carp. & Build., Home Handicrafts* 86 In the back and sides the grain of the wood runs vertically, the back being necessarily cross-tongued in two places. 1904 Goodchild & Tweney *Technol. & Sci. Dict.* 141/1 *Cross tongue*, a thin slip of wood with the grain at right angles to its length.

cro·ss-town, *a.* and *adv.* Also **'cross-town.** [Cross- 10 a.] **A.** *adj.* Lying, leading, or going across a town.
1886 [see Cross- 10 a]. 1894 *Congress. Rec.* 28 May 5413/1, I do not believe that on the L street, or, as it is called, this cross-town road, it is possible for a cable or electric motor to be successfully used. 1900 G. Bonner *Hard-Pan* i. 10 Then he hastened his steps, and a few blocks farther on boarded a cross-town car. 1948 T. Sharp *Oxf. Replanned* iv. 89 Two medieval streets which have also, in the absence of any other single cross-town route, to carry the whole of the cross-traffic of a city of 100,000 inhabitants. 1970 *New Yorker* 9 May 44/1 She gets on the crosstown bus alone.
B. *adv.* Across the town. *U.S.*
1906 'O. Henry' *Four Million* (1916) xvi. 165 The crowd in the gutter scattered, and the fine hansom dashed away 'cross-town. 1916 H. L. Wilson *Somewhere in Red Gap* 401 A regular old-fashioned horse-car going cross-town.

cross-tree. 2. Delete † *Obs.* (*nonce-uses*) and add later poetical examples.
a 1889 G. M. Hopkins *Poems* (1948) 265 And where I see the sweet cross-tree I in an instant would be gone. 1934 Dylan Thomas *18 Poems* 29 The blood that touched the crosstree and the grail Touched the first cloud and left a sign. 1956 E. Muir *Coll. Poems* (1960) 228 The archaic peoples in their ancient awe, In ignorant wonder saw The wooden cross-tree on the bare hillside.

cro·ss-wind. (See Cross *a.* 1 c.) Now commonly treated as a single word. Add later examples.
1920 L. Bairstow *Appl. Aerodynamics* iv. 141 *Cross-wind force*, the component perpendicular to the lift and to the drag of the total air force on an aircraft or any part thereof. 1935 *Discovery* Feb. 42/2 It now appears doubtful as to whether stabilising fins..are not a disadvantage, as presenting a greater surface to a cross-wind. 1959 *Manch. Guardian* 14 July 7/2 The pilot..used a cross-wind to lift the right wing.

crosswise, *a.* [f. the adv.] Placed or running across; transverse.
1903 *Westm. Gaz.* 10 Sept. 4/2 Its crosswise pelerine. 1927 *Observer* 4 Dec. 16/4 The cross-wise streets..are growing more and more canyon-like. *Ibid.*, The crosswise thrust of traffic at every block.

cro·ssword, cro·ss-word. [Cross- 4 c.] In full *crossword puzzle.* A puzzle in which a pattern of chequered squares has to be filled in from numbered clues with words which are written usu. horizontally and vertically, occas. diagonally. Also *attrib.* and *Comb.*
1914 *N.Y. World* 6 Dec. ('Fun') 7/2 Solution to last week's cross-word puzzle. 1924 (*title*) The cross word puzzle book. 1925 *Punch* 1 July 724 The allure of Epstein and Oxford trouserings has been for the few; the Cross-word Puzzle captivated the general. 1925 'Torquemada' (*title*) Cross-words in rhyme for those of riper years. 1927 *Observer* 3 Apr. 7 Particularly that spot known to crossword solvers as the acnestis. 1935 Mrs. H. Richardson *Parody* 5 This parody business is overrated, and he deserts it for a crossword puzzle. 1970 *Times* 28 Apr. 1/4 A national crossword competition is to be held in Britain this summer..with The Times crossword puzzles used for heats and finals.
b. *fig.*
1928 Galsworthy *Swan Song* II. iv. 141 Religion used to be red-hot politics, then it became caste feeling, and now it's a cross-word puzzle. 1959 R. Longrigg *Wrong Number* v. 63, I could see the first few rows of audience: a crossword puzzle of black dinner-jackets, white shirt-fronts, dark dresses, pale shoulders.

crotal[2] (krǫ·tăl). Var. Crottle; hence *attrib.* or *adj.* = of the colour of lichen, golden-brown.
1778, 1861 [see Crottle]. 1901 N. Munro *Doom Castle* iii. 23 Now the tartan's in the dye-pot, and you'll see about here but *crotal*-colour—the old stuff stained with lichen from the rock. 1907 *Westm. Gaz.* 26 Dec. 2/3 When Autumn wears her crotal gown. 1938 L. MacNeice *I crossed Minch* II. xii. 165 He showed me some of the stockings and tweeds —one very fine, thick tweed (two blue, two white and the crotal). 1968 *Guardian* 30 Mar. 10/4 The lichen which makes the rich, rust-coloured dye they call crotal.

crotale (krōᵘ·tal). [a. F. *crotale* (see Crotal).] A type of castanet used mainly in Latin-American music; also = Crotalum. Usu. *pl.*
1938 *Oxf. Compan. Mus.* 702/1 The *Crotales* (Fr.; made of wood or metal) are a variety of castanet. 1949 E. Pound *Pisan Cantos* lxxix. 79 Is there a sound in the forest of pard or of bassarid Or crotale or of leaves moving? 1964 S. Barrett in Norton & Spacey *Drums & Drumming Today* 30 The gentle tinkle of Crotales.

crotchet, *sb.*[1] Add: **13. crotchet letter,** one having a hook-shaped hair-line.
1887 *Script Letters for Perforating & Sewing*, Crotchet letters b v f r w.

crotchetiness. (Earlier example.)
1837 Mill *Let.* May in *Wks.* (1963) XII. 336 His crotchettiness, and his fussiness, and his go-between inclinations.

crouch, *sb.*[2] Add: **b.** *Athletics.* A method of starting in sprint races in which the runner crouches down on all fours. In full *crouch start.*
1913 S. A. Mussabini *Compl. Athletic Trainer* 196 The old-fashioned stand-up position enabled the runners to keep 'set' on their marks for a very much longer time than the present-day straining 'crouch' will let them do. *Ibid.* 217 Good level running from the modern 'crouch' start. 1931 F. A. M. Webster *Atheletes in Action* 17 No matter what type of race one is competing in, provided that it calls for a crouch start, the first consideration must be that of generating immediate momentum. *Ibid.* 81 The start is made in the normal 'Crouch' position used by sprinters. 1956 H. Abrahams *Olympic Games Bk.* Pl. facing p. 16 (*caption*) Start of the 100 metres final. T. E. Burke (U.S.A.)..has already adopted the 'crouch' start.

crouched, *ppl. a.* Add: **b.** *spec.* in *Archæol.* Of a burial: with the body in a crouching posture, usu. on its side.
[1898 Pitt-Rivers *Excav. Cranborne Chase* IV. 82 On the near side of the collection of interments are seen Nos. 1 and 2 crouched on the right side.] 1915 H. R. Hall *Ægean Archæol.* vi. 160 Primitive *Hockergräber* (crouched burials) have also been found at Tiryns. 1925 A. Keith *Antiquity of Man* (ed. 2) II. iii. 48 This 'crouched' burial had been made by the people who lived on the old land surface and worked the Neolithic flints. 1954 S. Piggott *Neolithic Cultures* ii. 49 A second barrow..contained a crouched skeleton.

crouching, *vbl. sb.* Add: (Later examples.) Cf. *crouch *sb.*[2] b.
1904 Graham & Clark *Pract. Track & Field Athletics* ii. 17 To-day practically every one uses the low or crouching start, which experience has proved beyond all question to be the quickest. 1912 E. H. Ryle *Athletics* vi. 89 There are two methods of starting—the old erect posture and the modern crouching or 'all fours' method. 1913 E. W. Hjertberg *Athletics in Theory & Practice* II. i. 85 The 'crouching' start has proved its superiority in every respect.

croup, *sb.*[2] Add: **3. croup-kettle** = **bronchitis kettle.*
a 1884 Knight *Dict. Mech.* Suppl., *Croup kettle*, a small kettle and alcohol lamp for quickly raising a steam for inhalation in cases of croup. 1889 'Mark Twain' *Yankee* xl. 516, I rousted out the croup-kettle myself; for I don't sit down and wait for doctors. 1961 *Brit. Med. Dict.* 790/2 *Bronchitis kettle*..; called also *croup kettle.*

croustade (krusta·d). [Fr., f. *crouste*, older form of *croûte* Crust *sb.*] A crisp piece of bread, fried or baked and scooped out to form a mould, to receive a filling of meat or other savoury; also, a hollowed shape of rice or pastry for the same purpose; (see also quot. 1845).
1845 Bregion & Miller *Pract. Cook* 41 *Croustades*, fried crusts of bread. 1846 Soyer *Gastron. Regen.* 160 Prepare the croustades as above, and make a good purée of fowl. 1865 'Ouida' *Strathmore* xi, Congregate at luncheon, and take *croustades* and conversation together! 1892 T. F. Garrett *Encycl. Pract. Cookery* I. vii. 481/1 The Croustade may be made of bread or paste of any kind. *Ibid.*, Line a dozen small Croustade-moulds with the rolled-out paste. 1960 *Farmer & Stockbreeder* (Suppl.) 16 Feb. 6/2 Croustade of spinach.

‖**croûte** (krūt). [Fr., = Crust *sb.*] A crust of bread, toasted or fried, served as a foundation for certain dishes; also = next.
[1841 Thackeray in *Fraser's Mag.* June 714/1 What do you think was our dinner for six persons?..Removes. Plompouding; croute de macaroni.] 1906 Mrs. Beeton *Bk. Househ. Managem.* lxii. 1657 Croûtes, blocks or shapes of fried bread, used as a basis for dressing salmis, whole birds, etc. 1907 G. A. Escoffier *Mod. Cookery* 782 Set these croûtes in a crown on a round dish, and garnish their midst with a rocky pyramid of plombière ice. 1951 *Good Housek. Home Encycl.* 394/2 It [*sc.* caviare] may also be..served spread on croûtes of fried bread.

‖**croûton** (krū·toṅ). Also **crouton.** [Fr., f. *croûte* Crust *sb.*] A small piece of toasted or fried bread used in soups and to garnish stewed dishes and minces. Also, any small piece used for garnishing.
1806 J. Simpson *Compl. Syst. Cookery* 31 Put a little anchovy essence, squeeze a lemon,..and garnish with croutons. 1846 Soyer *Gastron. Regen.* 60 Put some croûtons in the tureen, with twenty very small *quenelles de volaille.* 1892 T. F. Garrett *Encycl. Pract. Cookery* I. vii. 481/2 Croûtons of aspic jelly..are made in almost any shape. *Ibid.*, Croûtons for garnishing or soup. 1907 G. A. Escoffier *Mod. Cookery* 535 Border the dish with neatly-cut *croûtons* of pale jelly. 1921 *Contemp. Rev.* Sept. 374 A purée or cream soup with crackers or croûtons. 1967 C. O. Skinner *Madame Sarah* xii. 263 She'd taste a morsel or spoonful of every dish, every sauce, every crouton served.

crow, *sb.*[1] Add: **3. a.** (Further examples.)
1843 'R. Carlton' *New Purchase* II. 235 The *rara avis*—the white crow—a good President. [1851 *San Francisco Picayune* 3 Dec. 1/6, I kin eat a crow, but I'll be darned if I hanker after it.] 1877 *N. & Q.* 5th Ser. VIII. 186/1 A newspaper editor, who is obliged..to advocate 'principles' different from those which he supported a short time before, is said to 'eat boiled crow'. 1884 'Mark Twain' *Lett.* (1917) II. 443 Warner and Clark are eating their daily crow in the paper. 1885 *Mag. Amer. Hist.* XIII. 199 'To eat crow' means to recant, or to humiliate oneself. 1930 'E. Queen' *French Powder Myst.* xxiv. 196, I should merely be making an ass of myself if I accused someone and then had to eat crow. 1970 *New Yorker* 17 Oct. 39/1, I was going to apologize, eat crow, offer to kiss and make up.
d. Colloq. phr. *stone* (or *stiffen*) *the crows*: an exclamation of surprise or disgust. Esp. *Austral.*
1930 L. W. Lower *Here's Luck* xxvii. 242 'Stone the crows!' stormed Stanley. 1934 B. Penton *Landtakers* (1935) II. iii. 120 'Gawd stiffen the crows,' Bill commented bitterly. 1938 J. Moses *Nine Miles from Gundagai* 82 Stone the crows, what's up, mate? Has Australia got the blues? 1948 C. Day Lewis *Otterbury Incident* iv. 46 Cor stone the crows, 'ave a 'eart, young gents. 1953 J. C. Trench *Docken Dead* iii. 46 Cor stone the crows, he thought, this could go on till Christmas.
8. b. *N.Z. colloq.* A person who pitches sheaves to the stacker.
1888 J. Bradshaw *N.Z. of To-Day* ix. 171 When harvest came..he ought to have taken his place as 'crow' upon the stack. 1913 A. I. Carr *Country Work & Life in N.Z.* v. 11 A 'crow'..whose work consists of passing the fork-fulls thrown up by the carter to the stacker. 1956 J. Dare *Rouseabout Jane* xxiv. 185 When it came to stacking the corn, my job was to be 'crow'.
c. *slang.* A derogatory name for a girl or woman, esp. one who is old or ugly; freq. in phr. *old crow.*
1925 'H. H. Richardson' *Way Home* (1930) vi. 477 It makes me feel a proper old crow. 1938 Runyon *Take it Easy* 27 She is by no means a crow. In fact, she is rather nice-looking. 1957 R. C. Sherriff *Telescope* II. i. 56 *Mayfield.* There's an old lady named Miss Fortescue. .. *Ben* (*laughing*). Coo!—I know *that* old crow.
11. *crow-scaring*; **crow-bait** (for *U.S. colloq.* read *colloq.* (orig. *U.S.*) and add earlier and later examples); *spec.* an old or worn-out horse; **crow-blackbird** (earlier example); **crow-eater** (examples); also, a South Australian; **crow-fig,** the berry of the nux vomica tree; **crow-flight, -fly,** a direct course, a straight line (cf. sense 3 c in Dict.); also quasi-*adv.*; **crow-needle** (examples); **crow-pheasant,** a large bird of India and China, *Centropus sinensis*; **crow-pick** *v. trans.*, to inspect (coal) and free it from stones and rubbish; hence *crow-picker.*
1857 *Spirit of Times* 14 Feb. 382/1 He had a ole ball-face, bob-tail rip, jest' 'bout fit for crow-bait. 1860 *Marysville* (Calif.) *Appeal* 25 Mar. 2/1 For many moments did the teamster 'cuss' and belabor his crow-baits. *a* 1910 'O. Henry' *Trimmed Lamp* 73, I think I like your horses best. I haven't seen a crowbait since I've been in town. 1920 J. M. Hunter *Trail Drivers of Texas* 98 At this I..rounded up my 'crow bait' and pulled out for home. 1957 A. MacNab *Bulls of Iberia* xiii. 141 He rode out to do the *réjon* act on an ancient crowbait borrowed from the picadors' stable. 1778 J. Carver *Travels* 473 The crow blackbird..is quite black. 1899 *Daily News* 13 Sept. 7/5 'The land of the crow-eater' was at no time a convict settlement. 1902 J. H. M. Abbott *Tommy Cornstalk* 2 It may have been that, to the early South Australians, means of subsistence came not easily. At any rate they are called 'Crow-eaters'. 1934 Crow-eater [see *Bananaland]. 1967 *Courier-Mail* (Brisbane) 25 July 2 The 'Crow-Eaters' have bustled ahead and watched industrialisation transform their once sleepy-hollow State. 1778 Crow fig [see Nux vomica]. 1830 *Oxford Jrnl.* 30 Oct. 3 He struck her; which exasperated the poor woman so much as to induce her to poison herself with crow-fig. 1895 Bloxam *Chem.* (ed. 8) 760 Nux-vomica, or crow-fig, contains about 1 per cent. of strychnine. 1875 G. M. Hopkins *Let.* 20 Feb. (1935) 30 A long crow-flight is between us. 1885 *Science* 7 Aug. 108/2 We clambered over the hills and spurs in the usual crow-flight of the Karens. 1964 *Economist* 17 Oct. 258/1 The road..runs crow-flight straight. 1846 *Wesleyan Methodist Mag.* Jan. 53/1 It lies..east..at a direct distance, crow-fly, of about eighty miles. 1929 T. E. Lawrence *Home Lett.* (1954) 376 To get to Plymouth (only 300 yards crow-fly) is four and a half miles of bad road! 1733 W. Ellis *Chiltern & Vale Farming* xxxvii. 301 Crow-Needle, bears a white Flower, about half the height of the Corn. 1881 H. & C. R. Smith *Isle of Wight Words* 46 Crow-needles, *Scandix Pecten.* 1870 *Ibis* VI. 234 Among the bamboo-copses and gardens around Kiungchow-foo, and all other towns in Hainan, the Crow-Pheasant was abundant. 1878 P. Robinson *In my Indian Garden* 7 The crow pheasant stalks past with his chestnut wings drooping by his side. 1883 'Eha' *Tribes on my Frontier* 155 That ungainly object the coucal, crow-pheasant, jungle-crow, or whatever else you like to call the miscellaneous thing. 1964 A. L. Thomson *New Dict. Birds* 171/2 *C*[*entropus*] *sinensis*, commonly known in India as the Crow-pheasant, is a large black bird with chestnut wings. 1920 *Glasgow Herald* 13 May 6 To crow-pick each hutch as it passes the steelyard. 1921 *Dict. Occup. Terms* (1927) § 047 *Crow picker*; inspects shale in mine before it is loaded, to see that only clean shale is loaded. 1922 *Glasgow Herald* 12 July 10 Frae crawpickers that craw us O' hauf oor hardwon rakes;..Deliver us, O Lord! 1897 *Daily News* 15 Jan. 6/1 His first employment was crow-scaring. 1933 W. de la Mare *Lord Fish* 40 He had taken up crow-scaring at seven.

Crow (krōᵘ), *sb.*[4] and *a.* [See quot. 1935.] **A.** *sb.* **1.** A North American Indian tribe formerly inhabiting the regions of the Yellowstone and Wind rivers, now occupying a reservation in Montana; a member of this people. **2.** The language of this people, belonging to the Siouan stock. **B.** *adj.* **1.** Of or pertaining to this people or their language.
1801 P. Fidler in *Amer. Heritage Bk. of Indians* (1961) 324 (*legend on map*) Is.sap.poo. Crow mountain Indians. 1804 W. Clark *Jrnl.* 12 Oct. in Lewis & Clark *Jrnls. Lewis & Clark Exped.* (1904) I. 189 The Chien..or Dog Indians..[are] at war with the Crow Indians. 1812 J. C. Luttig *Jrnl.* 17 Sept. in *Jrnl. Fur-trad. Exped.* (1920) 78 Lecomte..asked them what Nation they were, they answered Crows. 1837 W. Irving *Capt. Bonneville* II. 29 Fitzpatrick..succeeded in prevailing upon the Crow chieftain to return him his horses. 1846 W. G. D. Stewart *Altowan* I. viii. 207 The language used was Crow. 1857 C. Kingsley *Two Years Ago* I. iv. 101, I got it in fair fight.. by a Crow's tomahawk in the Rocky Mountains. 1877 L. H. Morgan *Ancient Soc.* III. iii. 440 In Crow my husband's brother's wife is 'my comrade'. 1894 *Outing* (U.S.) May 89/1 Our name 'Crow' for this large and flourishing tribe of Indians..is a translation of their own totemic name Ab-sár-ra-ké, or Ap-sar-ro-ke. 1900 *Knowledge* 2 July 153/2 Among the other Atlantic stocks are..the Siouans, some of the most famous tribes of the latter being the Sioux or Dakotas, and the Crows. 1907 F. W. Hodge *Amer. Indians* I. 367/2 Crows (trans., through French *gens des corbeaux*, of their own name, *Absároke*, crow, sparrowhawk, or bird people). A Siouan tribe forming part of the Hidatsa group. 1935 R. H. Lowie *Crow Indians* 3 The Crow name for themselves is 'Apsāruke', which early interpreters mistranslated as 'gens de corbeaux', 'Crow (or Kite) Indians'. To me the word was explained as the name of a bird no longer to be seen in the country. The squaw-man Leforge defines it as 'a peculiar kind of forked-tail bird resembling the blue jay or magpie' which tradition assigns to the fauna of eastern Nebraska and Kansas at the time the Crow lived there. Apart from this fanciful localization, his and my data thus agree well enough. 1969 W. K. Powers *Indians of Northern Plains* 246 The Crow hold their annual Sun dance at Lodge Grass, Montana, in June, and the Crow Indian Fair and Rodeo at Crow Agency, Montana, in August.
B. *adj.* **2. Crow-type** or **Crow system,** etc.: a type of kinship terminology, typical of societies with matrilineages, in which sisters and female cousins are classified under three terms, one applied to sister and mother's sister's daughter, another to mother's brother's daughter (and brother's daughter), and the third to father's sister's daughter (and her mother and daughter).
1925 L. Spier in *Univ. Washington Publ. Anthropol.* I. II. 73 II. Crow Type. In this system the father's sister is an 'aunt' and her female descendants through females are 'aunts'! 1949 F. Eggan in M. Fortes *Social Structure* 122 They [*sc.* the Hopi] possess a majority of features associated with the classic Crow type. 1964 F. G. Lounsbury in W. H. Goodenough *Explor. Cult. Anthropol.* 351 A formal account of the Crow- and Omaha-type kinship terminologies. 1968 *Internat. Encycl. Social Sci.* VIII. 396 Their kinship systems [i.e., those of 'the tribes of the Prairie Plains'] were also 'classificatory', in that lineal and collateral relatives were merged in the terminology, but they utilized the lineage principle to provide a wide extension to the sytem. There were two subtypes: (*a*) the 'Omaha' system, associated with patrilineal descent, and (*b*) the 'Crow' system, associated with matrilineal descent.

crow (krōᵘ), *v.*[2] *S. Afr.* [Transliteration of dialectal Afrikaans *grau, grou*, f. *grawe*, Du. *graven*, with Eng. (k) representing Afrikaans (χ).] *trans.* and *intr.* To dig.
1853 F. Galton *Trop. S. Afr.* iii. 79 This method of digging is called in Dutch patois 'crowing' the ground; thus, 'crow-water', means water that you have to crow for, and not an open well, or spring. 1868 J. G. Wood *Nat. Hist. Man* I. xxx. 343 The Damaras..will sometimes 'crow' holes eighteen inches..in depth. 1896 H. A. Bryden *Tales S. Afr.* 47 With this last implement she can the more easily crow up their dinner.

crow-bar. (Earlier U.S. example.)
1748 in *Documents rel. Colonial Hist. New Jersey* (1883) 1st Ser. VII. 208 Men, armed with clubs, axes & crow bars, came, in a riotous & tumultuous manner.

crowd, *sb.*[3] Add: **1. b.** *spec.* A mass of spectators; an audience. (Cf. quot. 1613 under sense 1 in Dict.)
c 1863 E. Dickinson *Poems* (1968) II. 539 Their [*sc.* balloons'] Ribbons just beyond the eye—They struggle—some—for Breath—And yet the Crowd applaud, below. 1921 *Times* 1 Mar. 16/7 The crowd jeered at Hobbs owing to his slow movements in the field due to his recent injury. 1955 *Manch. Guardian* 30 Apr. 3/6 In recent years the crowd at Wembley Stadium has not seen the game of Rugby League played at its best. 1970 *New Statesman* 9 Oct. 454/3 The crowd was very similar both in behaviour and appearance to the audience that came to the Beaulieu jazz festivals. 1971 *Sunday Times* 31 Jan. 12/1 Tennis players lecture the line judge, appeal to the crowd.
c. A collection of actors playing the part of a crowd; freq. *attrib.*
1899 L. Wagner *How to get on Stage* 71 Sir Henry Irving when his two sons elected to go on the stage..said ..'I could only allow them to stand in the crowd at the Lyceum, to accustom them to the boards, and afterwards procure them an engagement in a touring company.' 1909 J. R. Ware *Passing Eng.* 100/1 What do *I* do? Oh, I go on with the crowd. 1935 J. Dell *Nobody ordered Wolves* v. 69 Her sole ambition had been to do crowd-work. 1936 *Archit. Rev.* LXXX. 192 If you are 'crowd', you go to one of the big communal dressing rooms. 1937 *Ibid.* LXXXII. 286 The interior is designed as a background to this 'crowd-scene'.
2. c. For '*U.S.* and *Colonies*' read 'orig. *U.S.*' Add further examples.
1840 *Congress. Globe* Apr., App. 376/2, I became satisfied that Democracy had but few charms for that crowd. 1883 Sweet & Knox *Through Texas* 13 He 'always went heeled, toted a derringer, and was a bad crowd generally'. 1897 Kipling *Captains Courageous* x. 218 They treat him as one of themselves. Same as they treat me. .. I'm one of the crowd now. 1933 D. L. Sayers *Murder must Advertise* ii. 33 The 'varsity crowd don't quarrel like the rest of them. *Ibid.* iii. 41 He used to tag round with that de Momerie crowd. 1939 *Chatelaine* Jan. 19/3 My bridge crowd was over the other night. 1971 *Woman* 23 Jan. 59/1 She was going through a particularly rebellious phase and seemed to be in with a wild crowd.
d. *colloq.* A military unit.
1901 *Westm. Gaz.* 31 Aug. 2/1 My crowd on this day were left flank advance guard. 1929 P. Gibbs *Hidden City* vi. 23 'What was your crowd?'..'East Kents. 8th Battalion.'
e. Colloq. phr. *to pass* (*muster*) *in a crowd*, not to fall so short of the standard as to be noticed; not to be conspicuously below the average (freq. with the implication of mediocrity).
1711 Swift *Jrnl. to Stella* 9 Feb. (1948) I. 185 Will she pass in a crowd? Will she make a figure in a country church? 1846 R. Ford *Gatherings from Spain* ix. 94 The rider's.. great object should be to pass in a crowd, either unnoticed, or to be taken for 'one of us'. 1853 Dickens *Bleak Ho.* xxvi. 259 They were mighty particular. You would pass muster in a crowd, Phil!
4. *crowd-control, -mind, -morality, -panic, -pleaser, -psychology, -suggestion*; *crowd-drawing, -pleasing, -pulling* adjs. See also sense *1 c and cf. *mass *sb.*[2]
1966 *Truth* (Brisbane) 9 Oct. 41/6 We call them bouncers but the U.S. has another term for those beefy bar room keepers of the peace. They are called crowd control engineers. 1971 *Times* 4 Jan. 1/3 The debate about crowd control and safety at football stadiums. 1848 J. R. Lowell *Fable for Critics* 37 He has faith...And this is what makes him the crowd-drawing preacher. 1923 H. G. Wells *Men Like Gods* III. ii. 265 Crowds and the crowd-mind have gone for ever. 1915 A. C. Curtis (*title*) Politics and crowd-morality, a study in the philosophy of politics. 1906 *Westm. Gaz.* 22 Jan. 2/2 Fewer processions with banners, fewer crowd-panics. 1943 *Gen* 16 Jan. 30/1 An up-and-coming fighter is a tearaway chap, a real crowd-pleaser. 1962 *Times* 26 Feb. (Canada Suppl.) p. xvi/4 One of the biggest crowd-pleasing sports in Canada is the rough and tumble stock car racing. 1924 W. B. Selbie *Psychol. Relig.* 204 The whole subject [*sc.* conversion] is an interesting branch of the study of crowd psychology. 1955 Keepnews & Grauer *Pict. Hist. Jazz* ii. 22 Brown..turned the term into a crowd-pulling asset by billing his group as 'Brown's Dixieland Jass Band'. 1924 W. B. Selbie *Psychol. Relig.* 157 The whole thing comes from crowd suggestion.

crowd, *v.*[1] Add: **2. c.** *trans.* and *intr.* To hurry. *U.S. colloq.*
1838 *Knickerbocker* XII. 506 Well, children, don't *crowd* the old man so; give him time. *a* 1861 T. Winthrop *John Brent* (1883) v. 43, I might perhaps make it a new story; but I crowd on now to the proper spot where this drama is to be enacted. *Ibid.* xix. 169 He crowded on, more desperately ..as a lover rides for love. 1876 *Rep. Vermont Board Agric.* III. 627 He is for ever crowding and rushing, so as to get some particular piece of work done by such a time.
3. Also, to push *back, down* (also *fig.*).
1830 *Massachusetts Spy* 14 July (Th.), He was carting timber, and stepped upon the cart tongue to crowd some sticks back with his feet. 1874 *Rep. Vermont Board Agric.* II. 732 You are crowding him down to a gold basis. *Ibid.* 764 The excavation was..stopped upon a clean pebbly bottom, into which an iron bar could be crowded down its length. 1880 G. W. Cable *Grandissimes* xl. 318 He crowded his hat fiercely down over his curls and plunged out.
7. c. Delete † *Obs.* or *arch.* and add later examples.
1908 E. J. Banfield *Confessions of Beachcomber* I. i. 36 Get away from this. Don't crowd a fellow. Go to a rock of your own. 1933 P. Godfrey *Back-Stage* iii. 40 'Crowding' and 'upstaging' are tricks of the selfish actor. To 'crowd' is to stand just close enough to another actor to prevent his making any gesture freely. 1963 'J. Le Carré' *Spy who came in fr. Cold* vi. 44 They crowded him in the dinner queue. Crowding is a prison ritual akin to the eighteenth-century practice of jostling. It has the virtue of an apparent accident, in which the prisoner's mess tin is upturned, and its contents spilt on his uniform.
f. *to crowd the mourners*: to exercise undue pressure; to push or hurry in an unseemly manner. *U.S. colloq.*
1842 *Spirit of Times* XII. 426 In the second mile, however, Fashion commenced 'crowding the mourners' by brushing down both straight sides. 1904 W. H. Smith *Promoters* xix. 282, I don't want to crowd the mourners at your end of the line. 1923 *Dialect Notes* V. 205 Keep ca'm now, an' don't crowd the mourners.
g. To approach (a specified age) closely; to verge on. *U.S. colloq.*
1943 *Newsweek* 22 Nov. 52 Reynolds, now crowding 60, would disclose no plans last week. 1960 *Guardian* 5 May 9/6 Groucho is crowding 70, though not very hard. 1961 F. Crane *Reluctant Sleuth* iv. 33 Bobo's maybe twenty-five. George is crowding sixty. 1969 *Guardian* 18 Aug. 9/6 Mae West..confessed to 'crowding sixty'.

crowded, *ppl. a.* Add: **1. b.** *fig.* Full of events or experience of life.
1791 T. O. Mordaunt in *Bee* 12 Oct. 179 One crouded hour of glorious life Is worth an age without a name. 1957 F. King *Widow* III. iii. 317 The busy and crowded years in India. 1971 *Times* 21 Jan. 10/4 He [*sc.* Compton Mackenzie] records a crowded old age in which flow of writing is as non-stop as social life.

crowdedness. (Later examples.)
1895 W. Schlich *Man. Forestry* III. 181 As long as the degree of crowdedness is not too great. 1920 M. Webb *House in Dormer Forest* I. vi. 65 The phrase pleased him because of its crowdedness. 1923 D. H. Lawrence *Birds, Beasts & Flowers* 67, I long to see its [*sc.* the world's] chock-full crowdedness. 1930 O. Lodge in *Aberdeen Press & Jrnl.* 9 Sept. 6/2 The great crowdedness of space.

crowdie, crowdy. 2. Delete ? *Obs.* and add later examples.
1938 L. MacNeice *I crossed Minch* II. xii. 167 For my tea I had..a large hunk of 'crowdy'. *Ibid.* 168 'Crowdy' is a kind of..crumbly cream cheese, pure white and with practically no taste. 1946 *Farmhouse Fare* (ed. 2) 274 In Aberdeenshire a delicious crowdie is made from buttermilk.

crow·-hop, *v.* [f. Crow *sb.*[1] + Hop *v.*[1]] *intr.* To hop like a crow; also *fig.* (see quot. 1897). Also as *sb.*, a hopping movement like that of a crow.
1897 *Chicago Tribune* 25 July 15/2 *Crow Hop*, to 'craw-fish'. 'Leedy has crow hopped out of the special session of the Legislature.' 1903 *Wide World Mag.* Apr. 548 The ways they try to throw their riders may be classed under three heads. The first is known as the crow-hop. 1907 S. E. White *Arizona Nights* I. xiv. 207 Sometimes we crow-hopped solemnly around and around the prostrate Schwartz. 1944 R. F. Adams *Western Words* 45 *Crow hop*, when a horse jumps about with arched back and stiffened knees at a pretense of bucking.

crown, *sb.* Add: **2. c.** *crown of thorns* (*starfish*): a poisonous starfish, *Acanthaster planci* or *A. ellisi.*
1964 *Medical Jrnl. Australia* 18 Apr. 592/2 The 'crown of thorns' starfish is to be found entwined in the branches of living coral, on which it feeds. 1969 *Sci. Jrnl.* Nov. 15/1 The starfish in question, *Acanthaster planci*, is better known as the 'Crown-of-Thorns' starfish since its upper surface is covered with prominent spines. *Ibid.* (*caption*) Starfish, known as the 'Crown-of-thorns', is wreaking havoc amongst coral reefs in the Indian and Pacific Oceans.
6. d. *Crown and Anchor*, a gambling game played with three dice each having faces bearing a crown, an anchor, and the four card-suits; the players place their bets on a board or cloth bearing similar figures.
1880 G. A. Sala *America Revisited* (1882) II. vi. 78 The Crown and Anchor booth at Greenwich Fair. 1903 [see *banker[2] 4]. 1917 A. G. Empey *From Fire Step* 125 The two most popular games are 'Crown and Anchor' and 'House'. 1935 *Punch's Almanack* CLXXXVIII. p. viii/1 The..new Dukes..played Crown and Anchor in the corridors. 1969 R. C. Bell *Board & Table Games* II. v. 84 The bets are the same as in Crown and Anchor.
19. b. The arched surface of a bowling-green. (Cf. **crown green.*)
1897 *Encycl. Sport* I. 128/1 In Lancashire each green has a 'crown' varying in rise and slope. 1904 S. Aylwin *Gentle Art of Bowling* iii. 15 Greens with a crown or rise in the centre..are common in many parts of England.
28. b. *Dentistry.* An artificial structure made to cover or replace the natural crown of a tooth.
1820 L. S. Parmly *Lect. Nat. Hist. Teeth* iii. 75 The pivot soon wears away the fang. Thus the artificial crown becoming loose, it drops out. 1885 I. E. & R. E. Clifford *Crown, Bar, & Bridge-work* 6 The crown fits over the root like a cap. 1963 J. Osborne *Dental Mechanics* (ed. 5) xxiii. 415 Crowns may be of two types, first those that cover the natural crown of the tooth... Second are those crowns that replace entirely the crown of the natural tooth.
31. b. Short for Crown-glass.
1854 C. Tomlinson *Cycl. Useful Arts* I. 761/1 Regarding glass as a chemical,..the various kinds have been distributed in the following manner:..2..English crown,..3..foreign crown. 1902 *Encycl. Brit.* XXV. 41/2 The experiments of the eminent Jena glass-makers with phosphate crowns and borate flints. *Ibid.*, A triple combination of ordinary crown and flint with a boro-silicate flint. 1966 *McGraw-Hill Encycl. Sci. & Technol.* IX. 351 (*caption*) Heavy flints, flints, light flints, extra-light flints, short crowns, crowns, borosilicate crowns.
c. The boring end of a diamond or similar drill.
1883 *Encycl. Brit.* XVI. 444/1 The working part of the drill consists of the so-called crown, which is a short piece of tube made of cast steel, at one end of which a number of black diamonds are fastened into small cavities.
d. A term used to designate the fineness of wire used in carding operations.
1884 W. S. B. McLaren *Spinning* ix. 211 The crown..is the number of wires in 1 inch *along* it.
35. crown-bark, Peruvian bark obtained from *Cinchona officinalis*; **crown-berry,** the Cape cranberry, *Dovyalis rhamnoides*; **crown borer,** a drill having a cutter equipped with diamonds or steel teeth for boring purposes; **crown-bud,** the flower-bud of a chrysanthemum shoot that forms after the plant 'breaks' or branches (*first crown bud*) or, if this is removed, the bud that forms on the secondary shoot (*second crown bud*); **crown cap** orig. *U.S.*, a metallic cork-lined stopper designed to be crimped over the top of a bottle; **Crown Derby,** the Derby (see *Derby 5) porcelain made from about 1784 to 1848, bearing a crown as an additional distinguishing mark; freq. *attrib.*; **crown-fern,** a New Zealand fern, *Blechnum discolor*; **crown fire,** 'a forest fire in which the crowns of the trees are ignited' (Webster, 1909); **crown gall,** a disease of plants caused by the bacterium *Agrobacterium tumefaciens* and characterized by tumours;

crown green, a bowling green which is higher at the middle than at the sides; **crown-jewels,** also *fig.*; **crown lens,** a lens made of crown-glass, chiefly used as a component of an achromatic lens; **crown-roast** *Cookery*, a roast of pork or lamb consisting of rib-pieces arranged so as to look like a crown; **crown rot,** a disease of rhubarb, caused by the fungus *Erwinia rhapontici*; **crown rust,** a disease of cereals and grasses, caused by the fungus *Puccinia coronata.*
1872 J. Yeats *Nat. Hist. Raw Materials of Commerce* II. 234 The pale bark contains most cinchonine, the yellow most quinine; Loxa or crown bark the largest proportion of quinidine. 1899 J. M. Maisch *Man. Org. Materia Medica* (ed. 7) 163 Loxa bark or crown bark, chiefly from *C. officinalis.* 1907 T. R. Sim *Forests & For. Flora Cape Gd. Hope* 132 'Crownberry' is in use at East London, and may have originated in the crown-like calyx of *D*[*ovyalis*] *rhamnoides.* 1962 Watt & Breyer-Brandwijk *Medicinal & Poisonous Plants S. & E. Afr.* (ed. 2) 1386/3 Dovyalis rhamnoides...Cape cranberry, Cranberry, Crownberry. 1902 *Encycl. Brit.* XXXI. 643/1 Attempts are being made to substitute a rotary 'crown' borer for the percussion drill in sinking wells for petroleum. 1900 A. Wynne in W. D. Drury *Bk. Gardening* v. 140 These growths form buds (termed crown buds)... If these buds form in July..they are taken out, and another shoot is made, which produces a 'terminal', or second crown, bud. 1962 *Amateur Gardening* 7 Apr. 15/2 Each shoot produces what are known as 'first crown buds'. 1928 *Collier's* 1 Sept. 47/1 Corks, bottle cappers, tubing, crown caps. [1850 J. Marryat *Coll. Hist. Pottery & Porcelain* ix. 181 The Derby porcelain is very transparent... The earliest mark is not known..subsequently, the mark was a D surmounted by a crown.] 1863 W. Chaffers *Marks & Monograms Pottery* 141 *Crown Derby.* A later mark than the preceding [*sc.* Derby-Chelsea]. 1872 Lady C. Schreiber *Jrnl.* (1911) I. 168 We found some charming Crown Derby custard cups and covers. 1900 E. Glyn *Visits Eliz.* 227 When he saw the best Crown Derby smashed on the floor. 1946 W. Martin *Flora N.Z.* iv. 77 At higher altitudes the larger, erect-growing Crown Fern (*B. discolor*) replaces them. 1960 B. Crump *Good Keen Man* 12 A skinny old sow trotted out of the crown-fern above me. 1938 Weaver & Clements *Plant Ecol.* (ed. 2) ii. 47 *Crown fires* race through the tops of the trees at a high rate of speed. 1900 (*title*) An inquiry into the cause and nature of crown gall (Arizona Agric. Exper. Station Bull. No. 33). 1950 *N.Z. Jrnl. Agric.* Mar. 229/3 Crown gall (*Bacterium tumefaciens*) in Chinese gooseberries is invariably associated with plants that have been propagated by root grafting or from cuttings. 1961 *Amateur Gardening* 9 Dec. 23 The growths or swellings which have affected your tree of Prunus tibetica are not canker but crown gall. 1902 *Encycl. Brit.* XXVI. 327/2 There are two kinds of green—one the crown, and the other the level. 1904 S. Aylwin *Gentle Art of Bowling* iii. 15 For a crown green No. 2 bias.. will be sufficient. 1851 H. Melville *Moby Dick* III. xiv. 98 At mid-day, with a blinding sun, all crown-jewels. 1895 G. B. Shaw in *Sat. Rev.* 2 Feb., Such crown jewels of dramatic poetry as Twelfth Night and A Midsummer Night's Dream. 1834 P. Barlow in *Phil. Trans.* CXXIV. 202 The crown lens must be made concave and the flint lens convex. 1845 T. Dick *Pract. Astr.* II. iv. § 6. 244 The predominating refraction of the crown lens disposed the achromatic rays to meet at a distant focus. 1961 R. Auerbach tr. *Boutry's Instrum. Optics* viii. 141 Much more compelling practical considerations will decide our choice, namely which of the four objectives with the crown lens in front..is the simplest to make and to mount. [1912 F. M. Farmer *New Bk. Cookery* 132 Roast crown of pork.] 1934 Webster, Crown roast. 1962 Crown roast [see *best A. 5 a]. 1967 *Vogue* June 132 Crown Roast uses two Best Ends. Your butcher will prepare it for you. 1924 W. A. Millard in *Bull. Agric. Dept., Univ. of Leeds* CXXXIV. 6 Although many minor diseases attack the Rhubarb crop, none is of any importance in Yorkshire excepting that which has thus come to be known simply as 'Rhubarb Disease'. This disease would be better described as Rhubarb Crown Rot, for it is the crown of the plant which is most generally attacked. 1952 E. Ramsden tr. *Gram & Weber's Plant Dis.* III. 355/1 No part of a plant with any sign of crown rot should be used for planting. 1899 G. Massee *Text-bk. Plant Dis.* 249 Crown Rust (*Puccinia coronata*, Corda). A widely distributed rust..met with on wheat, barley, rye, and many wild grasses. 1900 J. Percival *Agric. Bot.* VII. xlviii. 712 Two species of crown 'rusts' are known, namely *Puccinia coronifera*..and *P. coronata*. *Ibid.* 713 The upper cells of the teleutospores in both species are surmounted by a ring or crown of blunt teeth, hence the name crown 'rust'. 1933 *Discovery* Nov. 351/1 Some years crown rust (*Puccinia coronata* Corda) on oats amounted to a really serious menace. 1956 J. G. Dickson *Dis. Field Crops* (ed. 2) xii. 323 Crown rust caused by *Puccinia coronata* Cda. is distributed widely.

crown, *v.*[1] Add: **1. d.** *dial.* or *slang.* To hit (a person) on the head.
1746 P. Lock *Exmoor Scolding* (ed. 3) 6 Chell trim the, chell crown tha, chell vump tha. 1866 R. Hallam *Wadsley Jack* x. 49 Wi' that, sumboddy behint crahn'd me wi' a umbrella. 1919 R. Lardner *Real Dope* iii. 92 If he hadn't been so old I would of crowned him. 1948 A. Baron *From City from Plough* 156 Get off that box..before I crown you with this shovel. 1959 'O. Mills' *Stairway to Murder* xxiii. 234 'Someone crowned me, I take it?' The sergeant nodded. 'With the poker from our own hearth.'
15*. *Dentistry.* To put an artificial crown (see *crown *sb.* 28 b) on (a tooth).
1885 I. E. & R. E. Clifford *Crown, Bar, & Bridge-work* 6 Roots decayed too far for pivoting can be crowned. 1907 *Westm. Gaz.* 23 Oct. 9/1 The teeth were crowned. 1963 C. R. Cowell et al. *Inlays, Crowns, & Bridges* vii. 66 The mobility of any tooth to be crowned must be tested.

crowner[1]. Add: **2.** *U.S.* (Earlier and later examples.)
1815 *Massachusetts Spy* 31 May (Th. 146), This is the crowner, the cap-sheaf. 1922 Alice Brown *Old Crow* xxvii. 320 Isn't that a joke, Rookie? Charlotte would say it's the crowner.

crown-wheel. Add: **b.** *spec.* In the gears of motor vehicles.
1908 *Westm. Gaz.* 29 Dec. 4/1 The transmission [of a motor-car] being by chain,..crown-wheel, and differential gears. 1926 *Amer. Speech* I. 686/2 Automobile nomenclature. ... American: ring gear... English: crown wheel. 1963 R. F. Webb *Motorists' Dict.* 70 *Crown-wheel*, the largest bevel gear in the differential..housing.

crow's foot, crow's-foot. Add: **4. b.** A mark or symbol resembling a bird's foot.
1871 *Scribner's Monthly* II. 502 [A chart] adorned at this point by the crowsfeet that call for a chain of mountains.
c. *Textiles.* (See quots.)
1948 J. T. Marsh *Textile Sci.* xiv. 300 Creases should not be allowed to form or they will tend to become fixed in the cloth and form 'crows' feet marks'. 1957 *Textile Terms & Defs.* (ed. 3) 35 *Crowsfeet*, undesirable creases, particularly in crêpe fabrics, that prevent the production of a uniform surface appearance. 1962 J. T. Marsh *Self-smoothing Fabrics* ii. 8 A heat-setting process had a stabilising effect and prevented the subsequent formation of a multiplicity of little wrinkles or crow's-feet marks when the fabric was relaxed.

crow's nest, crow's-nest. Add: **3.** *N.Z.* (See quot. 1930.)
1930 L. G. D. Acland *Early Canterbury Runs* i. 5 To save trouble and disturbing the sheep more than was necessary, 'Crow's nests' were built on some plains stations —platforms on poles or in cabbage trees from which the shepherds could see where the sheep were. 1949 in S. S. Crawford *Sheep & Sheepmen of Canterbury* v. 39.

croy (kroi). *Sc.* [Formed from early Sc. *croys*, pl. of *cro* wattled enclosure for catching fish: cf. Crew[2], Croo, Cruive.] A structure in a river designed to hold back the water and restrict it to a certain channel; also = Cruive 4.
[1493 in *Acts of Lords Auditors of Causes & Complaints 1466–1494* (1839) 179/2 Johne Erskin..dois na wrang In the occupationn of the Croys of montross and fisching of the sammyn.] 1825 in Jamieson Suppl. *a* 1877 Knight *Dict. Mech.* I. 652/1 *Croy*, a mound or structure projecting into a stream, to break the force of the water on a particular part and prevent encroachments. 1880 *Scottish Naturalist* V. 258 It [*sc.* a cormorant] frequented a croy at Benchill fishing-station. 1908 *Baily's Mag.* May 380 Croys may be convenient for casting from. 1909 W. L. Calderwood *Salmon Rivers Scot.* 70 At Edradynate the system of croy-building has been carefully developed by Mr. H. W. Johnston, so that in one large pool alone..there are eleven croys.

croze, *v.*[1] (Earlier U.S. example.)
1849 *Rep. U.S. Comm. Patents* (1850) 386, I also claim the apparatus for chamfering and howelling and crozing.

‖**cru** (krü). Also **crû.** [Fr., f. *crû*, pa. pple. of *croître* to grow.] A French vineyard or wine-producing region; the grade of wine produced there. Also *attrib.*, *Comb.*, and *fig.*
1824 tr. *Jullien's Topography of Vineyards & their Products* p. xiii, A wine is called of such a *cru*, meaning a circumscribed spot in a vineyard; it is also used in a more extensive sense, as the *cru* of such a district, &c. 1833 C. Redding *Mod. Wines* iv. 67 *Cru.* This word is applied in several ways. It means a vineyard, a particular spot in a vineyard, any vine land generally. 1862 C. Tovey *Wine & Wine Countries* iv. 131 There are in Bordeaux Wines four estates, or growths, classed as first *crûs* or growths. 1867 'Ouida' *Under Two Flags* I. vii. 139 He drank the right *cru* and lived in the right set. 1939 F. M. Ford *Let.* 24 Jan. (1965) 309 Chateau Pavie is usually esteemed one of the big cru wines. 1951 R. Postgate *Plain Man's Guide to Wine* v. 92 The finest château bottled Premier Cru Classé of Médoc is connected by an insensible progression with the Bordeaux ordinaires. 1962 *Economist* 27 Jan. 334/1 Although most provincial papers are conservative, their conservatism is..not of the same *cru* as..the journalistic norm in Paris. 1963 *Times* 22 Jan. 12/3 Many of the vineyards which are classified as *crus bourgeois supérieurs* or *crus bourgeois*—that is, which did not appear among the first five growths in 1855 —have profited by the great advances in viticulture since then. 1966 P. V. Price *France, Food & Wine Guide* 172 Just below the classed growths come the *crus bourgeois*.. then the *crus artisans.*

crubeen (krūbī·n, krū·bin). *Anglo-Irish.* [ad. Irish *crúibín*, dim. of *crúb* claw, hoof, paw.] The foot of an animal; esp. a (cooked) pig's foot.
1847 W. H. Gregory *Paddiana* II. 95 Did ye ever see a pair of *crubeens* like them? 1907 G. B. Shaw *John Bull's Other Island* IV. 81 It [*sc.* the pig] put in the fourth speed wid its right crubeen as if it was enthered for the Gordn Bennett [car race]. 1922 Joyce *Ulysses* 144 Florence MacCabe takes a crubeen and a bottle of double X for supper every Saturday. 1970 M. Kenyon *100,000 Welcomes* xii. 96 'No crubeens', said Rafferty, fingering the groceries...'All I wanted was crubeens.'

cruche[2] (krūʃ). [Fr., = pitcher.] = Cruse.
1856 J. C. Robinson *Invent. Objects Mus. Ornamental Art* 26 Cylindrical Mug or Cruche. *Ibid.* 27 Globular brown glazed Flemish or Old English Cruche ('Gray-beard'). 1874 Lady C. Schreiber *Jrnl.* (1952) I. 446 Bought our cruche.

crucial, *a.* Add: **2.** (Further examples.) Freq. in trivial use = 'very important'.
1957 F. King *Widow* II. x. 245 That's the crucial time for me, like the first month of a baby. 1963 *New Statesman* 8 Feb. 195/1 What is crucial, of course, is that these books aren't very good. 1968 *Ibid.* 23 Feb. 241/2 Twice at crucial moments in this volcanic tragic comedy he asked us to advise him what to do. 1971 *Times* 19 Jan. 1 (*headline*) Leaders arrange to meet in private before today's crucial debate.

crucifixion. Add: **2. c.** (See quot.)
1917 A. G. Empey *From Fire Step* 149 The famous Field Punishment No. 1. Tommy has nicknamed it 'crucifixion'. It means that a man is spread-eagled on a limber wheel two hours a day for twenty-one days. During this time he only gets water, bully beef, and biscuits for his chow. You get 'crucified' for repeated minor offences.

crucify, *v.* Add: **2. c.** (Later examples.)
1921 D. H. Lawrence *Tortoises* 45 Why were we crucified into sex? 1940 H. L. Ickes *Secret Diary* 10 Aug. (1954) III. 297 He knew he would be crucified when he reached his own country. 1947 A. Miller *All my Sons* (1958) I. 81 *Ann.* Don't you hold anything against him? *Keller.* Annie, I never believed in crucifying people. 1960 *News Chron.* 15 July 3/3 If I was an hour or two late for filming, I was just about crucified. 1971 *Sunday Times* 24 Jan. 29/5, I do wish David Coleman wouldn't say 'Is the TV action replay *crucifying* referees?' I don't much mind what TV does to referees, but I do mind what it does to the language.
d. To subject to 'crucifixion' (see *crucifixion 2 c).

crucifying, *ppl. a.* (Later examples.)
1934 S. Spender *Vienna* ii. 19 And made invisible by crucifying suns Day after day. 1962 *Listener* 22 Nov. 885/3 We talked of a trend in twentieth-century music that is not fortifying but crucifying. 1964 *Times* 10 Dec. 18/3 The crucifying bottlenecks at the docks.

cruck[2] (krʋk). [Var. Crock *sb.*[5], Crook *sb.* 5 c.] One of a pair of curved timbers, forming with other pairs the framework of a house; = Crock *sb.*[5] Freq. *attrib.*
1898 S. O. Addy *Evol. Eng. House* ii. 17 A building erected in this way is now said to be 'built on crucks'. 1934 *Archit. Rev.* LXXV. 214/2 The foreign prototypes of the English 'cruck' house. 1948 J. Walton in *Antiquity* XXII. 179 The development of the cruck framework. *Ibid.*, The cruck buildings of northern England. 1949 K. S. Woods *Rural Crafts* IV. xi. 170 Two arched or slanting timbers, called crucks, or crutches,..form each of the gable-ends, and support the roof-tree. 1970 H. Braun *Parish Churches* viii. 101 The great halls of the Anglo-Saxons were formed of lines of wide timber arches called 'crucks'.

crud (krʋd). [Var. Curd *sb.*] **1.** = Curd *sb.* *Obs.* exc. *dial.*
1362–1626 [see Curd *sb.* 1 a and b]. 1788 [see Ream *sb.*[2] 1 a]. 1887 J. Service *Life & Recoll. Dr. Duguid* I. ix. 54 There were nae mair deidly engagements noo than the attack on..cruds and cream.
2. A despicable or undesirable person or thing; nonsense, rubbish. Cf. *crut[3]. *slang* (orig. *U.S.*).
1940 *Amer. Speech* XV. 212 A 'crud' is a fellow who is slovenly in his personal appearance and with his possessions. 1942 Berrey & Van den Bark *Amer. Thes. Slang* § 825/23 *Unattractive girl*,..crud, crumb, [etc.]. 1943 *Amer. Speech* XVIII. 153/2 *Crud*, food, usually unpalatable. 1955 'T. Sturgeon' in 'E. Crispin' *Best SF 2* (1956) 147 Would you say that..the writer of all this crud, believes..in what he writes? 1966 'L. Lane' *ABZ of Scouse* 23 *Crud*, rubbish. 1966 'K. A. Saddler' *Gilt Edge* i. 9 Can't stand the man. A real crud.
b. A real or imaginary disease. *slang* (orig. *U.S. Army*).
1945 *Reader's Digest* Apr. 109 We also have what the men call 'crud', a skin outbreak like ringworm. 1945 *Time* 13 Aug. 76 Jungle rot; New Guinea crud; the creeping crud: GI names for any and every kind of skin disease. 1947 *Amer. Speech* XXII. 304/2 *Crud* (*the*), this word..is used more commonly than any other single Army or Navy term for imaginary disease. 1948 *Ibid.* XXIII. 295/1 In the Pacific theater, *crud* was colloquial for all kinds of diseases, though it was most often used for the various forms of fungus infection. 1966 F. Shaw et al. *Lern yerself Scouse* 56, *I got Bombay crud*, I am suffering from looseness of the bowels.
c. An undesirable impurity, foreign matter, etc. (see quots.). *slang.*
1950 *Gloss. Terms Nuclear Sci.* (Nat. Res. Council, U.S.) 41/1 *Crud*, a slang term referring to an undesirable impurity or foreign material arising in a process. 1959 *New Scientist* 26 Mar. 696/1 'Crud' (Chalk River Unidentified Deposit), an impolite word applied to rust deposited on the fuel elements in the high radiation zones which exist in the heart of a reactor. 1962 in D. E. Barnes et al. *Newnes Conc. Encycl. Nuclear Energy* 160/2 In corrosion problems crud may be defined as particulate corrosion products..deposited on the surfaces of circulating water systems... The term is also applied to..many other undesirable residues..appearing in plant processes.

cruddy (krʋ·di), *a.* [Var. Curdy *a.*] **† 1.** = Curdy *a.* *Obs.* exc. *dial.*
1509–1678 [see Curdy *a.*]. 1882 Mrs. Chamberlain *W. Worcs. Words* 8 *Cruddy*, curdled; full of curds. 1937 E. J. Labarre *Dict. Paper* 65/1 *Cruddy paper*, i.e. mottled.
2. Dirty, unpleasant, unsavoury (see also quot. 1952). *slang* (orig. *U.S.*).
1949 D. Levin *Mask of Glory* 150 Get the cruddy end of the stick. 1952 Berrey & Van den Bark *Amer. Thes. Slang* (ed. 2) § 284/6 *Ill-tempered*,..cruddy. 1960 J. Kirkwood *There must be a Pony!* (1961) iv. 32 He had the largest collection of cruddy-looking suède shoes in the world. 1960 W. Sheed *Middle Class Education* (1961) 157 A cruddy wall of fastidiousness. 1962 K. Orvis *Damned &*

Destroyed 144 Is that what's running around in your cruddy mind? **1966** *20th Cent.* Spring 20/2 The company gets cruddier every time I come here. **1970** C. Wood *Terrible Hard* xiv. 205 We're slavin' our guts out 'ere and the cruddy Brylcreem boys are poncing about playing basket-ball.

crude, *a.* Add: **1. b.** *crude oil*, natural mineral oil. So *crude petroleum*.

1865 *Atlantic Monthly* XV. 389 Wagons laden with crude oil for the refinery. **1896** B. Redwood *Petroleum* I. 215 The crude oil of Upper Burma. *Ibid.*, The solid hydro-carbons present in crude petroleum. **1931** *Discovery* Nov. 350/1 Crude-oil rail traction is the successor to steam rail traction. **1970** *Times* 16 Apr. 14/5 The tar lumps are residues of crude oil.

c. *crude fibre*, the insoluble residue left when vegetable matter is boiled alternately in dilute acids and alkalis, corresponding roughly to its indigestible part.

1895 C. F. Cross et al. *Cellulose* 165 'Crude Fibre'.—'Rohfaser'. **1910** *Encycl. Brit.* V. 606/2 In the analysis of fodder plants..the residue obtained after successive acid and alkaline hydrolysis is the 'crude fibre' of the agricultural chemist. **1927** R. G. Linton *Anim. Nutrition & Vet. Dietetics* i. 9 As obtained by ordinary analysis, crude fibre is a mixture of cellulose, lignin, cutin, pentosans, etc., **1965** *Brit. Poultry Sci.* VI. 23/2 There is little detailed information concerning crude fibre digestion by poultry.

11. *Statistics.* Unadjusted; not corrected by reference to modifying circumstances; spec. *crude birth-, death-rate*, the total figures before adjustment.

1889 *Jrnl. R. Statistical Soc.* LII. 442 The merest tyro in statistics knows that crude gross numbers are of little value. **1896** *Lancet* 15 Aug. 479/1 The mean crude or uncorrected death-rate. *Ibid.* 479/2 The range of corrected death-rates is far wider than that of crude death-rates. **1945** *New Biol.* I. 30 A crude birth-rate is the annual number of births per thousand living persons. *Ibid.* 36 Determination of the standard mortality rate, as opposed to the crude death-rate, is a simple matter if we know the age composition of the population and specific mortality rates for each year of life. **1965** *Times* 13 Feb. 8/3 These are crude figures, with exports valued f.o.b. and imports c.i.f.

crude (krūd), *sb.* [f. the adj.] Crude oil (see prec., 1 b).

1904 *Encycl. Americana* XII, s.v. *Petroleum Industry*, The crude..might be found in paying quantities if artesian wells were sunk. **1921** J. E. Pogue *Econ. Petroleum* 79 The details of a complete refinery differ according to the type of crude employed. *Ibid.* 82 Asphaltic crudes, such as those of the Gold Coast. **1960** *Times* 11 Apr. 15/5 In 1959 French bottoms carried almost 90 per cent. of the crude reaching France from all sources. **1970** R. Johnston *Black Camels* v. 86 We might be thankful for that crude to feed the refinery.

crudification (krūːdifikēⁱ·ʃən). [f. Crude *a.* + -ification; cf. next.] The process or result of making something crude; an example of this.

1910 H. G. Wells *New Machiavelli* (1911) II. iv. 263 A series of hits and anecdotes and—what shall I call them?—'crudifications' of the issue. **1965** I. A. Richards in *Times Lit. Suppl.* 27 May 438/3 Reading..can also be an occasion for mind-breaking failure, for stultifying confusion, for crudification.

crudify (krū·difəi), *v. rare.* [f. Crude *a.* + -ify.] *trans.* To make crude.

1899 H. James *Notebooks* 15 Feb. (1947) 276 It will be pretty, though, in making that plain, not to crudify the statement of it.

cruel (krū·ĕl), *v. Austral. slang.* [f. the adj.] *trans.* To spoil; to destroy all chance of success with.

1934 *Bulletin* (Sydney) 1 Aug. 46/3 The game's right enough so long as mugs don't try and play a hand. When they come in that cruels the whole show. **1967** *Courier-Mail* (Brisbane) 17 May 1 He asked yesterday if the purpose of the debate was to 'cruel' his object. **1967** I. Hamilton *Man with Brown Paper Face* iii. 41 I've got a good job and I don't want to cruel it while everything's going for me.

cruellie, cruelly (krū·ĕli), *sb. colloq.* [f. Cruel *a.* + -ie, -y⁶.] A cruel joke, remark, comment, etc. Also *attrib.*

1959 *News Chron.* 6 July 3/4 The famous American 'cruellie' joke—example: 'But what did you think of the play, Mrs. Lincoln?'—is on the way out. **1959** *Guardian* 3 Nov. 7/3 'Some of [the greeting cards] are cruellies... They might say on the outside "*Stay the way you are*", and.. on the inside.."*Mean, Cruel, Thoughtless*"..' How about a real cruelly, then, with the gum flavoured with cyanide? **1961** 'J. Ross' *Last August* xiv. 168 Con fancies himself as a psychiatrist... He has the best collection of sick jokes and cruellies.

cruelty. Add: **1. b.** *attrib.*, as *cruelty-lust*; cruelty man (also with capital initials) *colloq.*, an N.S.P.C.C. or R.S.P.C.A. officer.

1925 D. H. Lawrence *Refl. Death Porcup.* 59 His cruelty-lust is directed almost as much against himself as against his victim. **1954** D. V. Donnison *Neglected Child* iv. 82 Workers..had been shut out of houses by people who suspected they had 'put the cruelty man on to them'. **1958** *Sunday Express* 27 Apr. 6/6 The Cruelty Man has been asking questions about Janie. **1970** P. Dickinson *Seals* iii. 68 Mr. Fasting..had larruped Jamie...Mother had wanted to send for the Cruelty Man.

cruise, *v.* Add: **1. b.** Esp. of an aircraft or automobile: to travel at cruising speed; of a taxicab: to travel about at random seeking business.

1915 *Sphere* 5 June 229/2 A craft [sc. Zeppelin] which can slip through the air with the speed of an express train and cruise about for thirty-six hours. **1930** 'A. Armstrong' *Taxi!* v. 49 A 'crawling' or 'cruising' taxi being one that meanders along the road..looking for fares. **1934** *Discovery* Dec. 350/2 Aeroplanes like the Handley Page 42 which cruises at a speed of 110 m.p.h. **1959** 'E. Peters' *Death Mask* i. 7 He didn't cross to one of the parked cars..nor halt to look round for a cruising taxi.

c. Delete '*rare*' and add further examples.

1971 *Sunday Times* 3 Jan. 54/8 (Advt.), Board a luxury liner to cruise the most colourful waters in the world. **1971** *Observer* 10 Jan. 37/3 (Advt.), Cruise the lovely Erne waterway.

2. *trans.* and *intr. Forestry.* (See quots.) Chiefly *U.S.*

1879 A. P. Vivian *Wanderings in Western Land* 53 Experienced men are sent out into the forests exploring, or to use their own term 'cruising'; their object being..to find suitable lumber for chopping. **1895** *Outing* (U.S.) XXVII. 218/2, I found he was off 'cruising' (i.e. hunting up good timber tracts). **1919** T. K. Holmes *Man from Tall Timber* 40 Si and me cruised a part of this timber before ever you fellers come down from Blainesburg. **1953** *Brit. Commonw. Forest Terminol.* I. 35 *To cruise*, surveying of forest land to locate merchantable timber and estimate its quantity..; the estimate obtained in such a survey.

cruise, *sb.* Add: **1. a.** *spec.* A voyage taken by tourists. Also *attrib.*

1906 'O. Henry' *Four Million* (1916) 91 The hibernatorial ambitions of Soapy were not of the highest. In them were no considerations of Mediterranean cruises. **1933** N. Coward *Design for Living* II. iii, That world cruise was a fatal mistake. **1937** R. Macaulay *I would be Private* 43 Dressed in all those cruise clothes. **1962** *Listener* 11 Jan. 90/2 Cruise-ships and charabancs and monstrous hordes of hikers are blots upon the landscape [in Greece]. **1971** *Sunday Times* 3 Jan. 72/1 Standards on board British cruise liners are generally high.

2. A survey or estimate of the amount of timber in a particular area. Cf. prec., sense 2. Chiefly *U.S.*

1911 J. F. Wilson *Land Claimers* viii. 112, I finished the cruise today. **1953** [see *cruise *v.* 2].

cruiser. Add: **1. b.** A yacht constructed or adapted for cruising, as distinguished from a 'racer'; also, a motor-vessel designed for pleasure cruises on the sea, or on rivers, canals, etc. See also **cabin cruiser*.

1879 in É. Bonnaffé *Dict. Anglicismes* (1920). **1888** *Encycl. Brit.* XXIV. 724/2 As to the number of yachts now afloat, cruisers as well as racers, the British yacht fleet..now numbers..3000 yachts. **1971** *Observer* 10 Jan. 37/1 (Advt.), Explore beautiful uncrowded waterways in 2- to 6-berth luxury cruisers. *Ibid.* 37/2 Explore the fascinating inland waterways on a real canal cruiser.

c. One who goes on a pleasure cruise.

1940 *Times* 9 Jan. 6/4 Shovel-board, with which cruisers are familiar on board our liners. **1961** *Guardian* 11 Jan. 5/4 Most 'cruisers' find the time all too short for what they want to do.

2. In science fiction, an aircraft or space-ship.

1923 E. R. Burroughs *Chessmen of Mars* vii. 70 The cruiser 'Vanator' careened through the tempest. **1958** T. Godwin in 'E. Crispin' *Best SF* 3 94 The cruisers carried the colonists to their new worlds.

b. A police-car that patrols the streets. *N. Amer.*

1929 *Sat. Even. Post* 7 Dec. 68/2 The cruisers are high-powered seven-passenger touring cars manned by a crew of four. **1958** *Ottawa Citizen* 28 May 7/5 To lift all speed regulations from police cruisers chasing law-breaking suspects. **1967** *Boston Sunday Globe* 23 Apr. 25/1 In Weymouth Patrolmen Richard McDonald and Ralph Campbell were injured when a car hit their cruiser.

3. (See quot. 1900.) Cf. *cruise *v.* 2. Chiefly *U.S.*

1893 *Scribner's Mag.* June 695/1 My first day's experience as a 'Cruiser' or 'Landlooker'. **1900** E. Brucken *N. Amer. Forests* 81 A peculiar class of people variously known as woodsmen, cruisers, landlookers, whose business it is to give information as to the existence of pine timber, its location, amount, value. **1921** *Daily Colonist* (Victoria, B.C.) 11 Oct. 9/3 Dave Vanstone..was on a timber cruising expedition with his two head cruisers. **1946** R. Peattie *Pacific Coast Ranges* 232 With his cruiser's eye, he could measure the quantity and the quality of the timber from the water's edge.

b. A long-legged boot such as timber-cruisers often wear. *U.S.*

1902 S. E. White *Blazed Trail* xvii. 125 Dressed in broad hats, flannel shirts, coarse trousers tucked in high-laced 'cruisers'. **1946** *Sat. Even. Post* 11 May 41/1 He was wearing Tillamook light cruisers.

4. *Boxing.* Short for *cruiser-weight* (see *5).

1928 *Daily Tel.* 28 Feb. 16 Poor heavy-weights. Gallant 'cruisers'. **1928** *Daily Chron.* 9 Aug. 11/2 Cuthbert Taylor (flyweight), John Garland (bantam), and Alfred Jackson (cruiser) all survived their preliminary tests.

5. *attrib.* and *Comb.*, as *cruiser-pinnace, -squadron*; *cruiser-built* adj.; **cruiser stern** *Naut.*, a type of ship's stern without an over-hang, the projecting part being under the water; **cruiser tank**, a tank (*tank *sb.*⁷) of intermediate weight designed for rapid movement; **cruiser-weight** *Boxing*, for professionals: a weight of more than 11 stone 6 lb. but not exceeding 12 stone 7 lb.; for amateurs: a weight of more than 11 stone 11 lb. but not exceeding 12 stone 10 lb.; light heavy-weight; a boxer of this weight; also *attrib.*

1902 *Westm. Gaz.* 7 May 5/1 Cruiser-built merchantmen. **1934** T. E. Lawrence *Let.* 8 June (1938) 806 The new cruiser-pinnaces. **1901** *Westm. Gaz.* 30 July 6/2 Cruiser squadrons. **1915** G. S. Baker *Ship Form* I. viii. 74 If the water line at the stern is kept too full it results in..eddy-making, and partly to avoid this a 'cruiser stern' has been adopted in many recent ships. **1950** P. F. Anson *Scots Fisherfolk* viii. 111 In most modern fishing vessels of the larger type..a 'cruiser-stern' is now almost universal. **1940** *Illustr. London News* CXCVII. 133 Armoured turrets of 'cruiser' tanks..can be automatically swung in any direction. **1964** C. Willock *Enormous Zoo* v. 98 The rhino-catchers were used to taking their transport across country at which a cruiser tank might have balked. **1920** *Boxing* 25 Feb. 96/2 Two cruiser-weights..engaged in a 15 rd. side stake match. *Ibid.* 13 Oct. 163/3 Carpentier..wants the world's cruiser-weight title. **1922** *Daily Mail* 11 Nov. 11 Jack Bloomfield, the cruiser-weight champion. **1923** *Ibid.* 10 Jan. 9 He will go for the cruiser-weight trophy.

cruiseway (krū·zwēⁱ). [f. Cruise *sb.* + Way *sb.*¹] An inland waterway intended chiefly for pleasure cruising.

1967 *British Waterways* (Cmnd. 3401) 3 The Government propose that forthcoming legislation should make special provision for two distinct groups of waterways. The first will be those which are to form part of the [British Waterways] Board's commercial division... The second..will be maintained primarily for powered pleasure craft. They can no longer usefully form part of a commercial transport system and they will be known as cruising waterways—or 'cruiseways'. **1968** *Guardian* 26 July 4/6 The Leeds–Liverpool canal is classified as a cruiseway for all but eight miles of its length. **1971** *Observer* 17 Jan. (Colour Suppl.) 10/4 Barbara Castle's 1968 Transport Bill, which designated a few miles of the [canal] system as cruiseways.

cruising, *vbl. sb.* (in Dict. s.v. Cruise *v.*). Add: **b.** *attrib.*, as (Cruise *v.* sense a) *cruising-ground, -shirt*; (*cruise *v.* sense 1b) *cruising-altitude, -height*; **cruising radius, range**, the maximum distance that the fuel capacity of a ship or aircraft will allow her to travel and return at cruising speed; **cruising speed**, the best economic travelling speed for a ship or vehicle, esp. an aircraft.

1951 *Gloss. Aeronaut. Terms (B.S.I.)* III. 19 *Cruising height (cruising altitude)*, a constant altimeter indication, in relation to a fixed and defined datum, maintained during a flight or portion thereof. **1958** *Listener* 16 Oct. 593/1 The pressurised cabin was subject to cyclic reversals of load as the aircraft climbed to cruising altitude. **1780** Capt. Hope in *Lett. S. Hood* (1895) 11 But could not learn what latitude their cruizing ground was in. **1851** Melville *Moby Dick* II. ix. 61 The..*Pequod* had slowly swept across four several cruising-grounds. **1951** Cruising height [see *cruising altitude* above]. **1956** 'N. Shute' *Beyond Black Stump* 237 They were off the ground and climbing up to cruising height. **1927** G. Bradford *Gloss. Sea Terms* 45/2 Cruising radius is calculated with two points of view—one, the vessel's capacity in miles without refueling; the other, her capacity to remain at sea expressed in days running at normal speed. **1922** M. Luckiesh *Bk. of Sky* xxi. 230 These huge flying-machines with a cruising range of a thousand miles. **1959** J. Braine *Vodi* vi. 94 One of those woollen shirts with a fastener at the throat—cruising shirts they called them. **1919** *Sphere* 26 Apr. 67/2 The single-engine machine is the faster at full and cruising speeds. **1928** *Daily Chron.* 9 Aug. 3/6 It will have a 'cruising speed' of 15 miles an hour. **1935** P. W. F. Mills *Elem. Practical Flying* iii. It is usual to describe performances in terms of both maximum and cruising speeds—the latter being the speed at which the aeroplane flies level with the engine running at the cruising revolutions recommended by the makers. **1958** *Listener* 30 Oct. 683/1 You want a car that has..a high cruising speed.

crumb, *sb.* Add: **1. c.** One of the irregularly-shaped and highly porous aggregates of particles found in soil having a crumb structure. (Cf. sense 3 b.)

1906 [see **crumb structure*]. **1914** T. L. Lyon et al. *Soils* (1920) vii. 109 The soil particles are not homogeneous as to size, and neither do all the particles function as simple grains, being gathered together in groups called granules, or crumbs. **1961** J. MacBean *Soil* iii. 34 Clay soils which are treated with lime to allow of the flocculation or grouping together of the single particles into crumbs..are warmer and more easily worked.

d. In rayon manufacture (see quots.).

1927 M. H. Avram *Rayon Industry* 259 Following the steeping operation the blocks of alkali-cellulose are shredded. In this operation, which is usually carried out in a machine called a shredder or disintegrator, the cellulose is reduced to very finely divided particles called crumbs. **1927** T. Woodhouse *Artif. Silk* vi. 54 The action of the internal parts of the kneader breaks up the [alkali-cellulose] sheets effectively into small particles similar to small breadcrumbs, and hence these particles are called 'crumbs'. **1959** *Chambers's Encycl.* V. 643/2 Viscose process... The crumbs are placed in churns where the action of carbon disulphide causes the alkali-cellulose to change to cellulose xanthate.

2. b. A body-louse. *U.S. slang.*

1863 O. W. Norton *Army Lett.* (1903) 175 Fortunately, I am not troubled with the 'crumbs' now. **1898** *Scribner's Mag.* XXIII. 440/1 Just then I felt something crawling on my neck. It was a crumb. **1925** J. H. Mullin *Adv. Scholar Tramp* iii. 46 If there is crumbs hoppin' around on me, I don't want to encourage 'em too much.

c. A lousy or filthy person; an objectionable, worthless, or insignificant person. *slang* (orig. *U.S.*).

1918 H. M. Rideout *Key of Fields* 236 A couple of crumbs want to kill you. **1930** Wodehouse *Very Good, Jeeves!* iii. 83 This old crumb would be the occupant of the bed which I was proposing to prod with darning-needles. **1959** H. D. Barton *Loving Cup* 236 He's an absolute crumb called Stuart Rowlandson. **1970** *Women Speaking* Apr. 5/1 If a man doesn't like a girl's looks or personality, she's a..crumb.

5. *crumb rubber*; **crumb structure** [tr. G. *krümelstruktur* (E. Wollny 1882, in *Forsch. auf d. Geb. d. Agrik.-Physik* V. 146)], the condition of soil when its particles are aggregated into crumbs (sense *1 c).

1956 *Gloss. Terms Rubber & Rubber-like Materials* (ASTM Spec. Techn. Publ. No. 184) 28 *Crumb rubber*. When vulcanized rubber is milled, it does not become soft and plastic but forms a type of material known as crumb or spring rubber. **1957** *Times* 20 Dec. 17/6 Output of reclaim and crumb rubber for the 12 months was approximately 7 per cent. higher. **1906** E. W. Hilgard *Soils* vii. 109 The word 'Krümelstructur' (crumb-structure), adopted by Wollny for this phenomenon, has both fitness and priority in its favor. *Ibid.* 110 Clay is most frequently the substance which imparts at least temporary stability to the crumbs and crumb-structure. **1926** Tansley & Chipp *Study of Veg.* vii. 116 The 'primary' inorganic particles of soil show a tendency to aggregate into 'compound particles'... This 'crumb structure'..is found in all good agricultural and good forest soils. **1960** L. D. Stamp *Britain's Struct.* (ed. 5) xi. 96 It is mainly the maintenance of this soil-structure, especially this crumb structure, which the farmer means when he talks about..'maintaining a fine tilth'.

crumble, *sb.* Restrict *rare* to senses in Dict. and add: **2. b.** *Cookery.* Food, such as bread or a mixture of flour and fat, in the form of crumbs; a dish made from such crumbs together with fruit, esp. *apple crumble*. Also *attrib.*

1947 M. Given *Mod. Encycl. Cooking* I. 727 Apple crumble. **1951** *Good Housek. Home Encycl.* 589/1 Canadian or 'crumble' topping for pies. **1958** J. Hawthorne *Myst. Blue Tomatoes* xiv. 92 For 'afters' to-day she made them all an apple crumble. *Ibid.*, Her crumbles were delicious. **1958** *Listener* 12 June 995/1 Rhubarb crumble pudding... Rhubarb crumble.

crumbs (krɒmz), *int.* Also **by crum(s), by crumbs.** [In phr. *by crum(s)*, a disguised oath.] An exclamation of consternation, dismay, etc.

1892 'Q' *Three Ships* i. 24 She'll not weather Gaffer's Rock. By crum! if she does, they may drive her in 'pon the beach, yet! *a* **1918** W. Owen *Poems* (1963) 60 *Vrach!* By crumbs, but that was near. **1922** 'R. Crompton' *Just—William* viii. 166 'Crumbs!' said William, 'Talk about bad luck!' **1922** F. Hamilton *P. J.: Secret Service Boy* vii. 277 O crumbs! Did he really say that I didn't mind wearing that rig-out? **1943** H. Pearson *Conan Doyle* iii. 52 Devil of a temper you've got, Doyle! By Crums, it's hardly safe to go out with you. **1956** S. Gibbons *Here be Dragons* iv. 74 Nothing like that. Crumbs! I should say not.

crumby, *a.* Add: **3.** *slang.* (Freq. **crummy.**) Lousy; filthy, dirty, untidy; inferior, shoddy, distasteful. Hence **cru·mbiness, cru·mminess.**

1859 Hotten *Dict. Slang* 27 *Crummy-doss*, a lousy or filthy bed. **1889** Barrère & Leland *Dict. Slang* I. 283/2 *Crummy* (army), dirty; applied amongst soldiers to a man's appearance. **1899** Kipling *Stalky & Co.* 73 'It's a crummy place..' They crawled out, brushed one another clean. **1931** G. Irwin *Amer. Tramp & Underworld Slang* 58 *Crummy*, verminous; undesirable; inferior or cheap. **1932** J. Dos Passos *Nineteen Nineteen* 3 Feeling crummy in the baggy civies, he walked slowly. **1949** H. E. Bates *Jacaranda Tree* vii. 66 The whole thing's a bit crummy. **1949** *Here & Now* (N.Z.) Nov. 27/2 The authentic crumminess of the downtown settings. **1956** R. Fuller *Image of Society* vi. 156 The place soon got to look crumby. **1958** *Spectator* 30 May 705/2 A crumby seducer. *Ibid.*, Crumby phoneyness. *Ibid.*, There is nowhere free from crumbiness and sex. **1969** I. & P. Opie *Children's Games* viii. 237 The game has been taken up by the physical training instructors under such crummy names as 'Poison Circle Tag'.

crumen (krū·men). *Zool.* [ad. L. *crumēna* purse.] The suborbital gland in deer and antelopes, secreting a waxy substance.

1875 W. H. Flower in *Proc. Zool. Soc.* 160 There was no suborbital gland or crumen [in a musk-deer]. **1883** [see Suborbital *a.*]. **1902** F. A. Beddard in *Camb. Nat. Hist.* X. xi. 299 There is no crumen or suborbital gland [in the musk-deer].

crumhorn, var. Cromorne, Krummhorn.

1694–6 [see Cromorne]. **1954** *Grove's Dict. Mus.* (ed. 5) II. 547/1 The body of a crumhorn is a slender tube, almost invariably of boxwood, with the lower end bent up in a hook-like curve. **1966** *New Statesman* 1 July 25/3 A few drones on shawms and crumhorns would have encouraged vocal techniques.

crummy, *a.* Add: **5.** (See *crumby *a.* 3.)

crump, *sb.*⁴ Add: **2.** The explosion of a heavy shell or bomb, or the sound of this; hence, the shell itself; *crump-hole*, a hole or crater made by a shell. *Soldiers' slang.*

1914 *Times* 10 Dec. 6/1 The heavy shell..ending in a loud 'crump' as it bursts on the ground. **1915** D. O. Barnett *Lett.* 180 Suddenly a yellow cloud leaped up three times as high as the tower itself..and after a bit there was the deuce of a crump. *Ibid.* 220, I got buried by a six-inch crump. **1915** 'Boyd Cable' *Between Lines* 254 There was some fancy driving past them crump holes in the road. **1917** P. Gibbs *Battles of Somme* 171 The enemy was 'lathering' the field of observation with every kind of 'crump' and shell. **1930** Blunden *Poems* 186 A crump at any moment May blow us to bits. **1961** *Guardian* 3 Apr. 5/4 The steady crump of falling bombs.

3. *Mining.* A violent burst in the floor, walls, or ceiling of a mine.

1925 *Pendleton Reporter* 7 Nov. 8/2 The accident was due to a 'crump'. 'Crumps' are caused through the floor of the mine rising owing to an accumulation of gases below it. **1927** *Command Paper* 2946 (Reports XI) 244 Crumps are..caused by..great cumulative stress set up by the folding and thrusting of the strata..which..causes violent roof-falls and up-thrusting of the floor of the mine. **1967** *Gloss. Mining Terms (B.S.I.)* VIII. 8 *Bump (crump)*, a sudden and heavy release of strain energy in the major body of rock surrounding a mine working, resulting in displacement of the strata.

crump, *v.*² Add: **4.** *Soldiers' slang.* **a.** *trans.* To bombard with heavy shells. **b.** *intr.* To fire heavy shells; also, to explode with a 'crump' (see prec., 2). Hence **cru·mping** *vbl. sb.*

1915 'Boyd Cable' *Between Lines* 254 We could hear the blighters crumpin' away back down the road behind us. **1916** *Blackw. Mag.* Jan. 125/1 You may imagine with what methodical solemnity the Bosche 'crumps' the interior of that constricted area. **1919** W. Deeping *Second Youth* xxiii. 196 Five-point nines were still crumping on the road ahead of them. *Ibid.*, The crumping ceased, and they moved on. **1923** Kipling *Irish Guards in Great War* I. 170 No. 1 Company of the Irish saw a platoon of Coldstream in front of them crumped out of existence. **1952** E. F. Davies *Illyrian Venture* vii. 119 Mortars of about two-inch and three-inch size were crumping irregularly. **1968** J. R. Ackerley *My Father & Myself* vii. 66 Shells began to whizz over and crump in the ravine behind.

crumpet. Add: **2.** Now usually a soft, round, doughy cake made with flour and yeast, cooked on a griddle or the like and usu. eaten toasted with butter. Cf. Pikelet¹. (Add later examples.)

1899 Wilde *Importance of being Earnest* I. 22, I had some crumpets with Lady Harbury. **1912** E. H. Ryle *Athletics* iii. 53 The usual indigestible concomitants of a heavy tea—buttered buns and crumpets—ought to be eschewed. **1930** D. L. Sayers *Strong Poison* ix. 111 Nothing goes so well with a hot fire and buttered crumpets as a wet day without and a good dose of comfortable horrors within. **1969** R. & D. De Sola *Dict. Cooking* 78/2 Crumpets are similar to..muffins but of porous consistency with surface holes.

4. *slang.* **a.** The head; esp. in phr. *balmy* or *barmy on* (or *in*) *the crumpet*: wrong in the head, mad: see *balmy *a.* 7, *barmy *a.* 2 b.

1891 [see *balmy *a.* 7]. **1897** W. S. Maugham *Liza of Lambeth* ix. 153 You're all barmy on the crumpet. **1909** H. G. Wells *Tono-Bungay* III. iii. 356, I heard my aunt admit that one of the Stuart Durgan ladies did look a bit 'balmy on the crumpet'.

b. A trivial term of endearment; also *old crumpet*.

1900 G. Swift *Somerley* 40 You're Ophelia, Scrubby; but don't you go winking at the johnnies in the stalls, you giddy little crumpet! **1920** *Punch* 21 Jan. 45/1 Don't, Percival, old crumpet. **1923** Wodehouse *Inimit. Jeeves* ii. 23, I say, old crumpet, did my uncle seem pleased to see you?

c. Women regarded collectively as a means of sexual gratification; *occas.* a woman; sexual intercourse. So *a bit* (or *piece*) *of crumpet*: a (desirable) woman; a 'bit of fluff'.

1936 J. Curtis *Gilt Kid* 75 Fancy staying up as late as this and not having no crumpet. **1958** M. Pugh *Wilderness of Monkeys* 37 'Not much crumpet here, tonight,' Maguire said, smiling lecherously at the military man's companion. **1959** S. Delaney *Taste of Honey* (rev. ed.) II. ii. 81 He's gone off with his bit of crumpet. **1961** D. Moore *Highway of Fear* i. 11 There's a delightful piece of crumpet..I'm anxious to set me eyes on again. **1961** *Spectator* 25 Aug. 269 Thigh-slapping accounts of luscious foreign crumpet. **1969** D. Lambert *Angel in Snow* iv. 63 Ansell..watched the couples wistfully. 'Plenty of crumpet here, you know. Why don't you chance your arm?'

crumple, *v.* Add: **5. b.** Also *fig.* (usu. *pass.*).

1865 G. Meredith *Let.* 11 Aug. (1970) I. 315, I fear that you are crumpled up with this accursed dyspepsia. **1916** A. Huxley *Let.* May (1969) 100 The only time they tried to do anything *strong*..they were absolutely crumpled up.

c. Also, to give way, collapse.

1898 *Westm. Gaz.* 17 May 2/2 It may be well that the Spanish defeat should not be too immediately overwhelming. It may help to keep Spain stable internally if she does not 'crumple up' at once.

7. *fig.* To deprive of strength and energy.

1892 Kipling *Barrack-room Ballads* 47 For the sickness gets in as the liquor dies out, An' it crumples the young British soldier.

crunch, *sb.* Add: **1. b.** A crisis; a decisive point, event, confrontation, etc.; a show-down; esp. in phr. *to come to the crunch*: to come to the point; to reach a show-down.

1939 W. S. Churchill in *Daily Tel.* 23 Feb. 14/6 Whether Spain will be allowed to find its way back to sanity and health..depends..upon the general adjustment or outcome of the European crunch. **1948** —— *2nd World War* I. I. xvii. 243 When the imminence of an attack on Czechoslovakia became clear, Beck demanded an assurance against further military adventures. Here was a crunch. **1957** *Economist* 28 Sept. 1002/1 What Sir Winston Churchill would have called the 'crunch' of the economic battle has arrived. *Ibid.* 19 Oct. 200/1 No one is anxious to be the spearhead of the next wages struggle; and..the crunch may not be reached until some time after the turn of the new year. **1960** *Times* 21 July 15/5 Even the holders of Government bonds turn out to be chiefly philanthropic institutions and trade unions when it comes to the crunch. **1963** 'W. Haggard' *High Wire* v. 52 When it came to the crunch de Fleury wasn't to be relied on. **1969** 'J. Fraser' *Cock-pit of Roses* xv. 111 Now the crunch. 'How do you know, Andrew?..' No reply.

crunchable (krɒ·nʃăb'l), *a.* [f. Crunch *v.* + -able.] Capable of being crunched or crushed.

1906 H. G. Wells *In Days of Comet* I. iv. § 3 The coal-cellar..opened, and diffused small crunchable particles about the uneven brick floor.

crunchingly (krɒ·nʃiŋli), *adv.* [f. Crunching *ppl. a.* + -ly².] In a crunching manner; with a crunching action or sound.

1849 A. J. Symington *Harebell Chimes* 24 While there crisp'd 'neath her feet The snow crunchingly. **1927** *Chambers's Jrnl.* 79/1 Carwardine stepped out on to the shingle, with no particular caution as to noise—quite crunchingly, in fact.

crunchy (krɒ·nʃi), *a.* [f. Crunch *v.* or *sb.* + -y¹.] Fit for crunching or for being crunched; crisp. So **cru·nchiness,** the quality of being crunchy.

1892 W. Besant in *Pictorial World* 6 Feb. 434/2 Showing molars of a whiteness and crunchiness both beautiful and awful. **1928** *Daily Express* 14 June 4 The ripe-corn flavour and delightful 'crunchiness'..make it unusually tempting. **1929** *Ibid.* 3 Jan. 5 The crispest and crunchiest of nuts from Brazil. **1960** *News Chron.* 12 Oct. 8/2 The crisp crunchiness of a breakfast cereal. **1971** *Woman's Own* 6 Feb. 3 (Advt.), Crisp and crunchy biscuits.

crunkle (krɒ·ŋk'l), *v.*³ [Echoic.] *intr.* To make a harsh dry sound, as by grinding the jaws. So **cru·nkling** *vbl. sb.*

1882 E. A. Floyer *Unexpl. Balūchistan* 362 The 'crunkling' noise of so many feeding together. **1900** *Westm. Gaz.* 5 Sept. 2/3 The crabs..crunkled loud and long.

crusader. Add: **b.** *attrib.* and *Comb.*

1845 E. B. G. Warburton *Crescent & Cross* II. xlvi. 238 Such a scene unchanged might that old Crusader-castle have witnessed, six hundred years ago. **1864** J. A. Grant *Walk across Africa* 143 The drums..are on the ground, in a line, each having a large white cross on its head—a strange Crusader-like custom. **1906** *Daily Chron.* 6 Apr. 3/5 The finest Crusader ruin extant. **1907** G. L. Bell *Desert & Sown* ix. 198 The great crusader fortress towards which we were going. *Ibid.* 205 The castle is the 'Kerak of the Knights' of Crusader chronicles. **1936** T. E. Lawrence (*title*) Crusader castles. **1968** C. Forsyte *Murder with Minarets* xviii. 119 On one arm of a bay stood what they took to be a Crusader castle.

crush, *sb.* Add: **2. d.** A person with whom one is enamoured or infatuated; an infatuation; so *to have* or *get a crush on*, to be enamoured of, take a strong fancy to. *slang* (orig. *U.S.*).

1884 I. M. Rittenhouse *Maud* (1939) 338 Wintie is weeping because her crush is gone. **1895** J. S. Wood *Yale Yarns* 153 Miss Palfrey..consented to wear his bunch of blue violets. It was a 'crush', you see, on both sides. **1913** *Dialect Notes* IV. 10 (Have a) *crush* (on), to be conspicuously attached to some one. **1914** G. Atherton *Perch of Devil* I. 31 Some of the younger married women..get a crush on some other woman's husband. *Ibid.* 186 To be jealous you've got to have a fearful crush. **1928** *Punch* 2 May 484/1 Gervase and Pontefract had had a quiet sort of masculine crush on Joyce for some time. **1929** Jelliffe & White *Dis. Nervous Syst.* (ed. 5) iii. 335 They tend to be aggressive, domineering and often play the man role with their school-mates, or 'crushes'. **1952** V. Gollancz *My dear Timothy* 212 It is common to make fun of schoolboy and schoolgirl 'pashes' and 'crushes'.

4. c. A funnel-shaped fenced passage along which cattle, sheep, or horses are driven for branding, dipping, etc. In full *crush-pen*.

1856 W. Roberts *Diary* 18 Dec. in J. H. Beattie *Early Runholding in Otago* (1947) vi. 43 There was no crush pen or drafting race. **1872** C. H. Eden *My Wife & I in Queensland* iii. 69 A crush, which is an elongated funnel, becoming so narrow at the end that a beast is wedged in and unable to move. **1890** Mrs. C. Praed *Romance of Station* ii, The 'crush', or branding lane. **1892** W. E. Swanton *Notes on N.Z.* ii. 124 The [unbroken] horses are put in a stockyard, and there roped or driven into a crush. **1895** *Chambers's Jrnl.* 702/2 A crush—that is, long lines of parallel fences just wide enough for one horse to pass at a time—was erected; they were driven into this long lane. **1931** T. A. Harper *Windy Island* (1934) III. iv. 225 The lean-to in its turn was divided into crush-pens and a large receiving pen. **1936** M. Franklin *All that Swagger* x. 91 Delacy erected trap yards and drafting crushes. **1942** *E. Afr. Ann.* 1941–2 105/1 Several hundred protesting cattle must be put through the 'crush' and jabbed with the big hypodermic needle.

d. A group or gang of persons; = Crowd *sb.*³ 2 c; *spec.* a body of troops; a unit of a regiment. *slang* (orig. *U.S.*).

1904 'No. 1500' *Life in Sing Sing* 247 *Crush*, a crowd. **1916** 'Boyd Cable' *Action Front* 151 You want to ask something about someone in the old crush [sc. regiment]. **1924** A. J. Small *Frozen Gold* i. 40 Any one of that crush would do murder for no more than that 500 dollars reward. **1927** *Observer* 12 June 10/3 The best recruiter is the man who is pleased with his 'crush'. **1931** R. Dark (*title*) Shakespeare—and that crush.

e. A drink made from the juice of crushed fruit; = Squash *sb.*¹ 7.

1919 H. W. Moore *On Uncle Sam's Water Wagon* 91 Strawberry crush. Wash and mash one pint of fresh strawberries..put three tablespoonfuls into a glass... Fill up the glass with any charged water. **1935** *Economist* 30 Mar. 734/2 The fruit-eating habit..has inclined the public favourably towards fruit juice beverages ('crushes', 'squashes', etc.). **1952** A. Baron *With Hope* 25 Have you seen that place along the front where they sell orange crush? **1959** P. Roth *Goodbye, Columbus* iii. 37 Smelling still of all the orange crush they'd drunk that weekend.

6. crush bar, a bar in a theatre, where the audience may buy drinks during the intervals of the entertainment; **crush barrier**, a barrier erected to restrain a crowd; **crush-pen** (see sense *4 c above); **crush-yard** *Austral.* and *N.Z.*, a yard leading to the crush (sense *4 c); also *fig.*

1954 *Granta* 24 Apr. 22/2 Shall I follow the deception of the crush bar into the first world of critical abuse? **1968** V. C. Clinton-Baddeley *My Foe Outstretch'd* ii. 50 In the second interval he made his way..round the back of the Grand Circle to the crush bar. **1909** *Westm. Gaz.* 17 Sept. 9/1 To prevent mishap several other crush barriers are erected. **1970** *Guardian* 24 Mar. 11/6 The stations are not equipped with proper crush barriers. **1888** 'R. Boldrewood' *Robbery under Arms* III. xvi. 247 This was the crush-yard and no gateway. I was safe to be hanged in six weeks. **1921** H. Guthrie-Smith *Tutira* xvi. 127 More would have been done but for the number of previously shorn sheep being mixed up with the woolly, making in necessary to fill the crush-yard more often. **1950** *N.Z. Jrnl. Agric.* Apr. 377/2 For handling large herds a crush yard is recommended.

b. *spec.* in *Geol.* with reference to compression, thrust, or shattering of rock, as *crush-belt, -breccia, -conglomerate, -line, -material, -movement, -plane, -rock, -structure, -zone.*

1893 Geikie *Text-bk. Geol.* (ed. 3) vi. I. ii. 703 Dykes of 50 or 60 yards in breadth are reduced, where one of these crush-lines crosses them obliquely, to a thickness of no more than four feet. **1895** G. W. Lamplugh in *Q. Jrnl. Geol. Soc.* LI. 564 Essentially, these crush-conglomerates are rocks made up of scattered fragments set in a slaty matrix. *Ibid.* 571 The rocks bordering on the crush-zones. *Ibid.* 574 The crush material is again revealed. *Ibid.* 578 Three..separate zones of the crush-structure. **1903** Geikie *Text-bk. Geol.* (ed. 4) I. II. II. vii. 164 Angular fragmentary rubbish..has subsequently been consolidated by some infiltrating cement (Fault-rock, Crush-breccia, Crush-conglomerate). **1903** *Trans. Edin. Geol. Soc.* VIII. 30 The previous investigators of Fassa Valley failed to recognise the presence of the innumerable crush-planes with extremely low hade. **1903** *Nature* 12 Feb. 359/1 This passage-zone had been the great crush-zone of the district. **1904** *Ibid.* 16 June 166/1 The post-Bala crush-movements. **1930** Peach & Horne *Geol. Scotl.* 62 The belt of sheared rocks and flinty crush-material. *Ibid.*, The flinty crush-rock weathers with a black or brown surface. **1937** *Discovery* Oct. 324/1 The flint crush belt that runs throughout the length of the Long Island.

crushable, *a.* (Earlier and later examples.)

1863 Trollope *Rachel Ray* I. vii. 130 Make yourself comfortable..; you can't crush me. Or rather I always make myself crushable on such occasions. **1927** *Daily Express* 12 Dec. 5 Pack as you buy is a useful rule for all but very crushable goods.

crushed, *ppl. a.* Add: **3.** *crushed strawberry* (examples). Similarly *crushed raspberry*.

1881 C. C. Harrison *Woman's Handiwork* III. 165 Wine color, pomegranate, Indian-red, crushed strawberry. **1897** [see Strawberry 6]. **1906** S. W. Bushell *Chinese Art* II. 40 The brilliant *sang de bœuf*..is now succeeded by its derivative of softer hue,..*peau de pêche* (peach-bloom) or crushed strawberry (*fraise écrasée*). **1931** R. Lehmann *Let. to Sister* 9 A patch of those crushed-raspberry, button chrysanthemums.

crusher. Add: **4.** (Earlier example.)

1835 *Sessions' Paper* Aug. 643 'Here are two *crushers*.'..I looked out of the window, and saw both the policemen.

b. A ship's corporal or policeman; a regulating petty officer. *Naval slang.*

1908 'L. Yexley' *Inner Life of Navy* ii. 10 Whenever the 'Assembly' sounded, the Police (usually called Corporals or 'Crushers') would run round the decks cutting..at the boys as they rushed up the ladders. **1909** J. R. Ware *Passing Eng.* 100/2 *Crushers*, ships' corporals, who are the rank and file of the master-at-arms. **1912** 'Aurora' *Jock Scott, Midshipman* xiv. 165 One of the 'crushers' has a 'down' on you. **1943** C. S. Forester *Ship* 85 A crusher is a member of the ship's police.

Crusoe (krū·so). One who is shipwrecked on a desert island, like the hero of Defoe's book. Also *attrib.* and *Comb.*, as *Crusoe life, -like* adj. and adv. Hence **Cru·soeing,** living like Crusoe.

1888 R. L. Stevenson in *Scribner's Mag.* Feb. 252/1 And then you might go Crusoeing, a word that covers all extempore eating in the open air. **1907** *Daily Chron.* 3 July 5/5 There he had built himself a habitation, Crusoe-like, out of brushwood. **1908** *Ibid.* 16 July 1/5 Blades of penknives were fashioned into needles, hair-combs were made from bush thorns, and altogether the men led a regular Crusoe life. **1926** *Chambers's Jrnl.* 104/1 No other island..has accommodated more Crusoes during the last three centuries than Chatham Island. *Ibid.*, There is another and fell aspect of Crusoeing, however—an aspect which most fiction-writers carefully ignore.

crust, *sb.* Add: **1. e.** A livelihood, a living. *Austral.* and *N.Z. slang.*

1916 C. J. Dennis *Songs Sentimental Bloke* 120 *Crust*, sustenance; a livelihood. **1944** G. McCarthy in *Coast to Coast* 149 He's still there... Makin' a crust too. **1949** E. De Mauny *Huntsman in Career* ii. 128 'What do you do for a crust?' 'I work on a newspaper.' **1969** *Coast to Coast 1967–8* 100 You'd have to work pretty hard for that, I'd reckon, and your old man too. What does he do for a crust?

5. c. For '*U.S.* and *Canada*' read 'orig. *U.S.*' (Earlier and later examples.)

1809 A. Henry *Travels* 146 The crust upon the snow cutting his legs..to the very bone. **1966** T. Armstrong et al. *Gloss. Snow & Ice* 13 *Crust*, a hard snow surface upon a softer layer.

7. b. Impudence, effrontery. *slang.*

1900 *Dialect Notes* II. 31 *Crust*, forwardness. **1923** Wodehouse *Inimitable Jeeves* xiv, The blighter had the cold, cynical crust to look me in the eyeball without a blink. **1954** —— *Jeeves & Feudal Spirit* xi. 97 Actually having the crust to come barging in here!

13. b. *spec.* in *Geol.* (see 4 b), as *crust-block, -creep, -fold, -fracture, -lag, -movement, -strain, -stress, -torsion.*

1897 *Geogr. Jrnl.* June 669 There are two primary and permanent kinds of crust-movements. **1900** *Ibid.* Jan. 48 The great Rocky Mountain-Andes fold,..the longest and most continuous crust-fold of the present day. *Ibid.* Oct. 457 That phenomena of crust-torsion were induced by any combination of crust-pressures. *Ibid.*, The original cause of crust-strains. *Ibid.* 460 Old crust-forms and crust-fractures, especially such as allow occasional intrusion and outlet of volcanic material, are determining factors in the distribution of the subsequent deposits. *Ibid.* 461 Gigantic crust-creep of overthrust masses. **1903** *Trans. Edin. Geol. Soc.* VIII. 177 The form of the sill-complex was capable of being re-moulded periodically in harmony with the localised crust-stresses. **1907** *19th Cent.* Aug. 220 The remarkable crust-movements exhibited over a wide area. **1926** *Chambers's Jrnl.* 598/2, I would..hazard also the suggestion that crust-lag may be a potent factor, in conjunction with shrinkage of the earth's crust, in the causation of earth-tremors and earthquakes. **1929** *Encycl. Brit.* II. 980/2 According to current views of the mechanics of mountain-folding, a crust-block of old and hard rocks is always present, which receives the pressure of the thrust causing the folding. **1934** *Nature* 15 Dec. 940/2 In one crust-block (west of the volcano), the tilting occurred in the same direction as before.

crusta (krɒ·stă). Pl. **crustæ.** The L. original of Crust *sb.*, used in some scientific senses and combinations. **a.** *Anat.* The ventral part of the cerebral peduncle. **b.** *Path.* = Crust *sb.* 3. **c.** *crusta fibrosa, c. petrosa*, the cement of a tooth (Cement *sb.* 4). **d.** *crusta lactea*, an eruptive disease of infants at the breast; milk-scab, milk-blotch. **e.** *crusta phlogistica* (see quot. 1890).

1806 *Med. Jrnl.* XV. 33 Crusta Lactea..sometimes proves a very severe..disease; in some families attacking every child at the age of a few weeks. **1820** R. Hooper *Med. Dict.* (ed. 4) 266/1 *Crusta*,..a scab. **1848** Dunglison *Med. Dict.* (ed. 7) 238/2 Crusta phlogistica. *Ibid.* 851/1 At the part where the enamel terminates at the cervix of the tooth, the *crusta petro'sa, cemen'tum*, or cortical substance, commences in an extremely thin stratum. **1856** *Crusta lactea* [see *milk-scab* s.v. Milk *sb.* 10 a]. **1876** T. Bryant *Pract. Surg.* (ed. 2) I. xiii. 537 The *crusta petrosa* is formed on the gradually elongating root through the agency of the dental sac or capsule which surrounds the forming tooth crown. **1876** *Quain's Elem. Anat.* (ed. 8) II. 555 The lower or superficial part [of the peduncular fibres]..consists almost entirely of white fibres, collected into coarse fasciculi and is named the crusta [earlier edd., crust] or basis, or the fasciculated portion of the peduncle. **1890** Billings *Med. Dict.* I. 353/2 *Crusta fibrosa*, crusta petrosa... *Crusta phlogistica*, inflammatory or buffy coat of blood. **1910** *Encycl. Brit.* IV. 395/2 The ventral part of each crus forms the crusta, which is..the great motor path from the brain to the cord. **1911** *Ibid.* XXVI. 501/1 Surrounding the dentine where it is not covered by enamel is the 'cement' or 'crusta petrosa', a thin layer of bone. **1949** F. W. Jones et al. *Buchanan's Man. Anat.* (ed. 8) v. 266 Cement, or crusta petrosa. This covers the dentine that forms the root of the tooth. *Ibid.* xiii. 1400 The basis pedunculi (crusta)..has a crescentic outline, the concavity of the crescent being dorsal and occupied by the convexity of the substantia nigra.

f. *Antiq.* A thin plate of embossed metal, etc., inlaid on a vessel, wall, or other object.

1842 Brande *Dict. Sci.*, etc., *Crusta*, in gem sculpture, a gem engraved for inlaying on a vase or other object. **1910** *Encycl. Brit.* XIV. 627/2 The *proscenium* of the Odeum was lined with *crustae*, or 'marble-veneering', under one inch thick. **1911** *Ibid.* XXIII. 484/1 The *crustae*, or plaques decorated in repoussé, which were mounted on smooth silver cups. *Ibid.*, Cups adorned with golden *crustae*.

crustaceous, *a.* **1. a.** (Earlier and later examples of *crustaceous lichens*.)

1856 W. L. Lindsay *Pop. Hist. Brit. Lichens* iii. 80 Without a soil prepared..by crustaceous Lichens, there could have arisen no higher vegetation. **1921** A. L. Smith *Lichens* iii. 72 Some crustaceous lichens have a persistently scanty furfuraceous crust.

crustal, *a.* Delete *rare* and add to def.: esp. of the crust of the earth or moon.

1883 A. Winchell *World-Life* (1889) II. iii. 402 The addition of crustal layers upon the exterior [of the moon]. **1912** [see *batholith]. **1924** J. G. A. Skerl tr. *Wegener's Orig. Cont. & Oceans* 14 An explanation of mountain building must take into account immense tangential crustal movements. **1958** *New Scientist* 23 Oct. 1099/2 This boundary or discontinuity between the surface crustal rocks and the mantle. **1970** *Nature* 28 Feb. 845/2 Crustal earthquakes may be caused by what is effectively a gravitational loading of the crust.

crustless (krɒ·stlĕs), *a.* [See -less.] Made without a crust; with the crust removed.

1927 *Daily Tel.* 11 May 18/3 (Advt.), Cheese factory (outside trust), of reputed 'Matterhorne' Gruyere (crustless), also in loaves, desires connection. **1971** *Ibid.* 21 Oct. 17/4 Pass the crustless bread pieces through one small egg beaten up with 4 fl. oz. milk.

crustose, *a.* Delete *rare* and add earlier and later examples; esp. designating a lichen having a thin, closely adhering thallus (now more common than *crustaceous* in this sense).

1879 W. A. Leighton *Lichen-flora of Gt. Brit.* (ed. 3) 509 Crustose, forming a crust. **1938** G. M. Smith *Cryptogamic Bot.* I. xv. 515 These lichens are always crustose and with the thallus forming an incrustation that adheres closely to the substratum. **1959** *New Scientist* 5 Feb. 294/1 In a crustose lichen there is an upper cortex and a rather poorly defined gonidial layer of algal cells. **1967** M. E. Hale *Biol. Lichens* i. 10 Lichens are traditionally classified into three growth forms: crustose, foliose and fruticose.

crut² (krɒt). *U.S. slang.* = *crud 2 a (in quot. 1940 = excrement). So **cru·tting** *a.*, = *cruddy *a.* 2.

1925 Hemingway *In our Time* (ed. 2) 66 That son of a crutting brakeman. **1937** —— *To have & have Not* III. vii. 130 You miserable little crut. **1940** —— *For whom Bell Tolls* xi. 150 We can eat goat crut in Gredos. **1955** J. P. Donleavy *Ginger Man* (1957) viii. 63 Two years in Ireland, shrunken teat on the chest of the cold Atlantic. Land of crut.

crutch, *sb.* Add: **6. b.** Chiefly *Austral.* and *N.Z.* (See quot. 1965.)

1916 *N.Z. Jrnl. Agric.* 20 Sept. 228 It is necessary to hold each lot of sheep in the bath for the time necessary to secure thorough immersion. This may be done..by the use of the crutch. **1953** B. Stronach *Musterer on Molesworth* viii. 55 We had two men on the 'crutch' pushing the sheep's heads under, and seven men at the race. **1965** J. S. Gunn *Terminol. Shearing Industry* I. 18 *Crutch*, a mallet-shaped instrument (like a crutch) used to push sheep under in a swimming dip. Improved dips, especially spray dips, have caused this tool to become obsolete.

crutch, *v.*¹ Add: **4.** To push (a sheep) into a dip with a crutch (see *crutch *sb.* 6 b). So **cru·tching** *vbl. sb.* Chiefly *Austral.* and *N.Z.*

1886 C. Scott *Sheep-Farming* 135 The hot water tank into which the sheep are put next morning has three divisions, in each of which they are well crutched. **1933** L. G. D. Acland in *Press* (Christchurch) 7 Oct. 15/7 When sheep are dipped, they are shoved under with a thing like an inverted crutch. This is called crutching. **1940** E. C. Studholme *Te Waimate* (1954) xiii. 117 One day whilst trying to 'crutch' (push under) some slippery-backed old ewe..Geoffrey fell in.

5. To cut off the wool or hair from the hind-quarters of (a sheep, dog, etc.). So **cru·tching** *vbl. sb.* Chiefly *Austral.* and *N.Z.*

1915 J. R. Macdonald *N.Z. Sheepfarming* xxv. 68 If crutching is followed, any wool that might hinder the lambs from sucking may be clipped. **1920** *N.Z. Jrnl. Agric.* 20 July 8 The keeping-lambs are crutched, branded, dipped, and placed out. **1941** *Nature* 5 Apr. 421/2 'Crutching'..consists of shearing the wool away from the area around the tail so that this part keeps clean. **1942** R. B. Kelley *Animal Breeding* vii. 76 We find it advantageous also to 'crutch' long-coated bitches. **1946** F. Davison *Dusty* vi. 65 Blowfly season was drawing near. Morrison and Tom were crutching the sheep, cutting away the soiled wool from under their tails, where they would be most likely to be blown. **1953** O. E. Middleton in C. K. Stead *N.Z. Short Stories* (1966) 189 But aside from shearing and crutching times, life was good at the station.

crux. 1. Add to def.: *crux ansata* = Tau 2 b (see quot. 1930).

1841 J. G. Wilkinson *Manners & Customs Anc. Egyptians* 2nd Ser. I. xiii. 341 The sign of life (or *crux ansata*) was compelled to submit to the unintelligible name of 'Key of the Nile'. **1896** [see *ankh]. **1930** E. A. W. Budge *Amulets & Superstitions* xviii. 340 It is wrong, too, to call the sign ☥, *crux ansata*, the 'handled cross', for whatever object the hieroglyph may represent, it was certainly not a cross or anything like it.

3. b. The chief problem; the central or decisive point of interest.

1888 *Law Times* LXXXIV. 293/2 There remained the point, which was the *crux* of the case, whether the defendant was under any duty towards the plaintiff. **1934** R. Benedict *Patterns of Culture* vii. 232 The crux of the matter is that the behaviour under consideration must pass through the needle's eye of social acceptance. **1944** J. S. Huxley *On Living in Rev.* 179 Culture, not 'race', is, again, the crux of the American problem. **1971** *Sunday Times* 31 Jan. 12/1 The crux, however, is accommodation. This is now widely agreed to be the main determinant of university and polytechnic expansion.

cruzeiro (kruzēᵊ·ro). [Pg., lit. 'a small cross'.] The principal monetary unit of Brazil, which superseded the milreis in 1942, = 100 centavos; also, a coin representing this.

[**1927** *Glasgow Herald* 30 Aug. 12 A decree of financial reform was promulgated providing for the stabilisation of exchange on the basis of a new monetary unit to be called the 'cruzeir'.] **1942** *Times* 8 Oct. 9/1 No date has yet been announced for the calling in of the note issue and the substitution of the new currency unit, the cruzeiro for the milreis. **1964** W. McCord in I. L. Horowitz *New Sociology* 427 She..earned a few cruzeiros a day. **1970** *Sci. Jrnl.* May 73/1 The cruzeiro is still far from being stabilized.

crwth (krūþ). Also **cruth.** The Welsh form of Crowd *sb.*¹

1837 *Penny Cycl.* VIII. 198/1 *Cruth*, or *Crwth*, a musical instrument of the violin kind, formerly much used in Wales. **1880** [see Crowd *sb.*¹]. **1938** *Oxf. Compan. Mus.* 243/2 It is said that in remote parts of Wales a violin is still called a crwth. **1942** E. Blom *Mus. in Eng.* ii. 20 A crwth or rebec or bass viol player may have indulged boldly in a double stop. *a* **1953** Dylan Thomas *Under Milk Wood* (1954) 20 He intricately rhymes, to the music of crwth and pibgorn.

cry, *sb.* Add: **16.** (Earlier and later examples.) Also *more cry than wool*.

c **1460–1809** [see Wool *sb.* 1 g]. **1952** V. W. Brooks *Confident Years* xviii. 197 Her diary was much more cry than wool and Mary MacLane was a startling figure only because the times were so colourless and mild.

cry, *v.* Add: **21. e.** Colloq. phr. *for crying out loud*, an exclamation expressing astonishment or impatience. orig. *U.S.*

1924 H. C. Witwer *Love & Learn* vi. 148 'For crying out loud' butts in Hazel impatiently. **1933** M. Allingham *Sweet Danger* v. 69 Well for crying out loud!..That's a nasty scrape. **1941** 'R. West' *Black Lamb* (1942) II. 156 For crying out loud, why did you do it?

cryable (krəi·ăb'l), *a.* [f. Cry *v.* + -able, after Laughable *a.*] That may be cried or wept over.

1897 *Daily News* 22 Mar. 9/4 Tragedy means a cryable play. **1908** S. E. White *Riverman* xxi. 188 What laughable and cryable mistakes she made only those who have encountered a like situation could realize.

cry-baby, *sb.* Add: (Earlier examples.) Also **cry-babby.**

1852 A. Cary *Clovernook* 274 You had better be still, cry-baby. **1854** M. J. Holmes *Tempest & Sunshine* xiii. 180, I wouldn't be such a cry-baby, anyway. **1858** A. Mayhew *Paved with Gold* I. iii. 51 Don't take on like that, for if the chaps see you they are sure to call you 'cry-baby'. **1922** Joyce *Ulysses* 532 Crybabby! Crocodile tears! **1969** I. & P. Opie *Children's Games* ii. 76 There was a girl..with the reputation of a cry-baby.

Hence **cry-baby** *v. intr. U.S. colloq.*

1902 O. Wister *Virginian* vii. 85, I am not crybabying to the judge. **1966** H. Kemelman *Saturday the Rabbi went Hungry* (1967) xxvii. 181 I'd be the last one to crybaby on it.

cryo- (krəiˌo), combining form of Gr. κρύος frost, icy cold (cf. Kryo-); as in **cryobio·logy,** the biology of materials cooled to temperatures lower than those at which they normally function; low-temperature biology; hence **cryobio·logist,** one who studies or is skilled in cryobiology; **cryobiolo·gical** *a.*, of or pertaining to cryobiology; **cryoglo·bulin** *Biochem.* (see quots.); **cryo·logy** (see quots.); **cryopedo·logy** (see quot.); **cryophi·lic** *a.*, applied to bacteria which flourish at low temperatures; **cry·ophyte** (see quots.); **cryoplana·tion** (see quot.); **cryopla·nkton,** plankton inhabiting snow and ice; **cryopu·mp,** a vacuum-pump which produces a very high vacuum by the use of liquefied gases; hence **cryopu·mping** *vbl. sb.*, the use of the cryopump; **cry·osar,** a switching device in computers (see quot. 1959); **cryosu·rgery,** surgery using instruments that produce intense cold locally; cryogenic surgery; hence **cryosu·rgical** *a.*; **cryoturba·tion** [cf. G. *kryoturbat* adj. (C. H. Edelman et al. 1936, in *Verh. van het Geol.-Mijnbouwkundig Genootsch., Nederland*, Geol. Ser. XI. 332)], any physical disturbance to the soil produced by the action of frost on water in the soil.

1960 H. T. Meryman in *Ann. N. Y. Acad. Sci.* LXXXV. II. 509 The future of cryobiology is exciting, permitting as it does the attainment of indefinitely suspended animation. **1961** *Lancet* 16 Sept. 657/2 At the cryobiology laboratories.. A. Rowe..demonstrated new apparatus for the..low-temperature storage of bone-marrow. **1962** *Business Week* 16 June 72/1 Cryobiology is the marriage of two separate sciences: cryogenics, or extreme low-temperature physics, and biology. *Ibid.*, Cryobiologists have come up with two ways to preserve cells by freezing. **1964** *Internat. Science & Technol.* June 58/2 The realm of cryobiology encompasses everything below the optimum temperatures at which life functions... The problems of cryobiology stretch all the way from trying to understand what happens to an animal, an insect, or a single cell when it is cooled (or later warmed) to techniques for preserving useful cells like blood or destroying undesirable cells such as those in the brain of a patient suffering from Parkinson's disease. *Ibid.* 66/2 The trouble is that these high rates of heat transfer occur too late—at temperature differences between specimen and fluid that are too low for cryobiological use. **1947** Lerner & Watson in *Amer. Jrnl. Med. Sci.* CCXIV. 413/1 The term cryoglobulin is..suggested to represent a group of proteins with the common property of precipitating (or gelifying) from cooled serum. **1965** *Oxford Mag.* 25 Feb. 235/2 He [sc. Bagratuni] perfected a technique for the assay of cryoglobulins, proteins separating from blood at temperatures below that of the body. **1947** *Jrnl. Glaciology* I. 35 'Cryology'. Shortly before the war this new word for the study of glaciology was coined in Central Europe...In America the word 'cryology' is coming into fashion to describe the study of refrigeration. **1961** L. D. Stamp *Gloss. Geogr. Terms* 140/1 At the International Association of Scientific Hydrology in Zurich, Meinzen referred to four divisions of hydrology—potamology, limnology, hydrology..and cryology (the scientific study of ice and snow). **1946** K. Bryan in *Amer. Jrnl. Sci.* CCXLIV. 639 *Cryopedology*, the science of intensive frost action and permanently frozen ground including studies of the processes and their occurrence and also the engineering devices which may be invented to avoid or overcome difficulties induced by them. **1942** C. S. Morris in *Dairy Industries* VII. 63 (*title*) Cryophilic bacteria as a cause of milk samples failing the methylene blue test. **1962** *Lancet* 5 May 955/2 Where stored blood is used, the greatest danger is its accidental infection with cryophilic bacteria, though this happens only once or twice for each million bottles issued. **1909** Groom & Balfour tr. *Warming's Oecology of Plants* xxxvii. 154 Closely allied to plankton, but of a subsidiary..nature, is the glacial community forming the cryophyte-formation, which is composed of microphytes that are periodically exposed to ice-cold water. **1960** N. Polunin *Introd. Plant Geogr.* xv. 490 It is perhaps best to refer to the plants growing on snow or ice as 'cryophytes'. **1946** K. Bryan in *Amer. Jrnl. Sci.* CCXLIV. 640 *Cryoplanation*, land reduction by the processes of intensive frost-action...Includes the work of rivers and streams in transporting materials delivered by the above process. **1932** Fuller & Conard tr. *Braun-Blanquet's Plant Sociology* xii. 289 *Cryoplankton*, protista inhabiting snow and ice. **1961** *New Scientist* 25 May 434/2 Cryopumps which involve the liquefaction of hydrogen or helium to produce high vacua over large volumes. **1963** *Ibid.* 11 Apr. 99/1 'Cryopumping'—freezing the air in a chamber. **1959** McWhorter & Rediker in *Proc. Inst. Radio Engin.* XLVIII. 1207/1 The cryosar is a new semiconductor device, intended primarily for high-speed computer switching and memory applications, which utilizes the low-temperature avalanche breakdown produced by impact ionization of impurities. The name of the device was derived from 'low-temperature (*cryo-*) switching by avalanche and recombination'. **1962** *Engineering* 5 Jan. 21/3 Cryogenic devices such as..the cryotron and the cryosar. **1962** *New Scientist* 26 July 213 (*heading*) Cryosurgery cures Parkinsonism. **1965** *Observer* 2 May (Colour Suppl.) 14 (Advt.), Union Carbide also manufacture ..cryosurgical equipment. **1946** K. Bryan in *Amer. Jrnl. Sci.* CCXLIV. 633 A recent coinage by Edelman, Florshutz and Jeswiet (1936) is 'cryoturbation'. **1954** *Proc. Prehist. Soc.* XX. 134 At a similar time cryoturbation took place where a clay and sand interface occurred near enough to the surface.

cryoconite, mod. form of Kryokonite (in Dict. and Suppl.).

cryogenic (krəiodʒe·nik), *a.* [f. Cryogen + -ic.] Of or pertaining to the production or use of very low temperatures.

1902 *Encycl. Brit.* XXX. 287 Within recent years several special cryogenic laboratories have been established. **1933** *Discovery* Mar. 70/1 The ultimate liquid of all low-temperature or 'cryogenic' work—liquid helium. **1960** *Aeroplane* XCVIII. 643/1 A frictionless spinning element as in the cryogenic gyro. **1960** *Guardian* 15 June 6/2 My husband earns his living as a cryogenic physicist. **1962** I. S. Cooper in *Jrnl. Amer. Med. Assoc.* CLXXXI. 600/1 The technique of cryogenic surgery consists of first applying moderate cold to identify the locus of abnormal neural activity and then intense cold to make a circumscribed lesion there. **1967** *Technology Week* 20 Feb. 27 (Advt.), Flight simulation, high energy fuels, system design,..and fabrication of cryogenic support equipment are only a few of the advanced aerospace research..activities currently in progress.

Hence **cryoge·nics,** that branch of physics which deals with the production of very low temperatures and their effects on matter.

1958 *Oxf. Univ. Gaz.* 2 Oct. 81/2 Research into He³ Cryogenics being carried out in the Clarendon Laboratory. **1962** [see *cryobiology* s.v. *cryo-].

cryohydric (krəioˌhəi·drik), *a.* [f. Cryohydrate + -ic.] Of or pertaining to a 'cryohydrate'; *cryohydric point* or *temperature*, the eutectic point or temperature of a solution of a salt in water.

1890 *Jrnl. Chem. Soc.* LVII. 361 A solution at the cryohydric point is the coldest of the same solvent and dissolved substance obtainable. **1902** G. S. Newth *Text-bk. Inorg. Chem.* (ed. 9) I. xiv. 155 Such a solution is known as a constant-freezing solution, or sometimes a cryohydric solution. **1902** *Encycl. Brit.* XXVIII. 569/1 The solution must.. become saturated with respect to both ice and salt, and this can only occur at the cryohydric temperature. **1940** Glasstone *Physical Chem.* x. 762 The terms cryohydrate and cryohydric point are obsolescent and will not be employed further. **1964** G. W. Castellan *Physical Chem.* xv. 299 The eutectic mixture was originally thought to be a compound. In aqueous systems, this 'compound' was called a cryohydrate; the eutectic point was called the cryohydric point.

cryoscopic, cryoscopy, mod. forms of Kryoscopic *a.*, Kryoscopy.

1900 *Rep. Brit. Assoc.* 167 The cryoscopic behaviour of substances possessing constitutions similar to that of the solvent. **1903** *Nature* 15 Jan. 263/1 The methods of exact cryoscopy. **1908** *Practitioner* Sept. 435 In differential diagnosis, he regards cryoscopic examination as of great importance. **1909** *Ibid.* Nov. 664 Kümmel is satisfied in such cases with the cryoscopy of the blood. **1964** N. G. Clark *Mod. Org. Chem.* xxiv. 505 The depression is a constant, characteristic of the particular solvent; it is called the molecular depression constant or cryoscopic constant.

Hence **cryosco·pically** *adv.*, by means of cryoscopy.

1949 E. P. Abraham in H. W. Florey et al. *Antibiotics* II. xxiii. 872 Its molecular weight, determined cryoscopically.

cryostat (krəi·ostæt). [f. *cryo- + -stat.] An apparatus for maintaining a very low temperature.

1913 *Chem. Abstr.* VII. 2327 A Helium Cryostat. *Ibid.*, Making use of the He cryostat O. [*i.e.* K. Onnes] has detd. the vapor pressure of liquid He. **1917** *Ibid.* 2976 Cryostat for temperatures between 27 °K and 55 °K. **1946** *Nature* 24 Aug. 272/1 The experiments can be made in an ordinary cryostat. **1969** Brown & Bertke *Textbk. Cytology* iii. 19/1 Recently a refrigerated unit to cut frozen sections (cryostat) has become available.

Hence **cryosta·tic** *a.*; spec. *cryostatic hypothesis* (see quot.).

1950 A. L. Washburn in *Rev. Canadienne Géogr.* IV. 111. 34 Progressive freezing from the surface downward to the permafrost table sets up a large hydrostatic pressure in the unfrozen material that is confined between these surfaces... Because of analogy with a hydrostatic press and because freezing is the initiating factor, the writer suggests that this concept be termed the cryostatic hypothesis.

cryotron (krəi·otrɒn). *Electronics.* [f. *cryo- + *-tron.] (See quot. 1956.)

1956 D. A. Buck in *Proc. Inst. Radio Engin.* XLIV. 482 (*heading*) The cryotron—a superconductive computer component. *Ibid.*, The cryotron, in its simplest form, consists of a straight piece of wire about one inch long with a single-layer control winding wound over it. Current in the control winding creates a magnetic field which causes the central wire to change from its superconducting state to its normal state. The device has current gain, that is, a small current can control a larger current; it has power gain so that cryotrons can be interconnected in logical networks as active elements. *Ibid.* 486/2 Cryotron circuitry. **1960** E. Delavenay *Introd. Machine Translation* ii. 21 Cryotrons also provide immense storage facilities on a very limited volume of matter. **1965** *New Scientist* 4 Nov. 331/3 Networks of cryotrons for storage..can be built up on one plane in one sequence of operations.

cryptæsthesia (kriːptĭsþi·ziă, -ĕsþi·ziă). Also **cryptesthesia.** [f. Gr. κρυπτ-ός Crypto- + αἴσθησις perception + -ia¹.] A supernormal faculty of perception, whether clairvoyant or telepathic.

1923 S. De Brath tr. *Richet's 30 Yrs. Psychical Res.* II. ii. 64 A single phenomenon which the magnetizers of a past age called 'lucidity' or 'clairvoyance' (*hellsehen*); which is now called telepathy...I propose to name it cryptesthesia. **1926** *Spectator* 9 Oct. 601/2 The dowser..is a person endowed with a subconscious supernormal cryptesthesia.

cryptanalysis (kriːptănæ·lisis). orig. *U.S.* [f. Crypto- + Analysis.] The art of deciphering a cryptogram or cryptograms by analysis. Hence **crypta·nalyst,** one who deciphers cryptograms; **cryːptanaly·tic, -ical** *adjs.*, of or pertaining to cryptanalysis.

1921 *Let.* 12 Feb. (Manly Papers, Univ. of Chicago), William F. Friedman. Cryptanalyst. **1923** W. F. Friedman (*title*) Elements of cryptanalysis. **1934** Webster, Cryptanalyst. **1937** *Reader's Digest* Sept. 52/2 She was responsible for awakening Col. Friedman's interest in cryptanalytical science. **1938** W. F. Friedman *Military Cryptanalysis* I. 5 Valid, or authentic cryptanalytic solutions cannot and do not represent 'opinions' of the cryptanalyst. **1957** W. S. Allen *Ling. Study Lang.* 11 The interest of such material for cryptanalysis and prehistory. **1958** *Antiquity* XXXII. 216 Cryptanalytic, epigraphic and linguistic aspects. **1966** D. Kahn *Codebreakers* i. 7 The only man in the Navy with expertise in three closely related and urgently needed fields: cryptanalysis, radio, and the Japanese language. *Ibid.* xiv. 435 The cryptanalysts of the German Foreign Office, who had long ago cracked the Polish diplomatic code. *Ibid.* 436 The cryptanalytic service of the German Foreign Office.

cryptic, *a.* Add: **1. b.** Mysterious, enigmatic.

1920 A. Christie *Mysterious Affair at Styles* (1921) iv. 60 'That difficulty will not exist long,' pronounced Poirot quietly. John looked puzzled, not quite understanding the portent of this cryptic saying. **1936** C. S. Lewis *Allegory of Love* i. 25 How irresistible is that cryptic knight who comes and goes we know not whence or whither, and lures the reader to follow as certainly as he lured the Queen and Kay. **1936** A. Christie *Cards on Table* xix. 188 He might have amused himself by making some cryptic remark to the doctor and noted the startled awareness in his eye. **1940** N. Marsh *Surfeit of Lampreys* (1941) xv. 225 'What did his lordship say?' 'His lordship is cryptic. He doesn't say much.' **1965** Gowers *Fowler's Mod. Eng. Usage* 115/1 *Cryptic* might usefully be reserved for what is purposely equivocal.., and not treated as a stylish synonym for *mysterious, obscure, hidden*, and other such words.

3. *Zool.* Of markings, coloration, etc.: serving for concealment; protective.

1890 [see *anticryptic *a.*]. **1933** *Discovery* Sept. 276/2 Bright animal colour..is frequently cryptic. *Ibid.* 277/1 The pattern of tigers, in fact of all cats, and also of zebras, is really cryptic. **1964** V. B. Wigglesworth *Life of Insects* x. 149 By far the most frequent method of concealment among insects living or resting in exposed situations is camouflage, or 'cryptic coloration'.

cryptically, *adv.* Add: **b.** *Zool.* Of coloration (cf. *cryptic *a.* 3).

1922 *Encycl. Brit.* XXX. 725/2 This revealing coloration ..is as a rule hidden by a cryptically coloured fore wing. **1953** N. Tinbergen *Herring Gull's World* x. 95 Finding a nest with its cryptically coloured eggs takes some time.

crypto (kri·pto). *colloq.* [The combining form Crypto- used as a separate word.] A person who conceals his adherence to a certain political group; *spec.* a crypto-communist. Also *transf.*

1946 T. Driberg in *Reynolds News* 10 Mar. 2/2 Labour MPs of various shades of opinion—not by any means only the Communist 'fellow-travellers' or so-called 'crypto's'. **1947** W. S. Churchill in *Hansard Commons* 5th Ser. CDXXXVII. 456 Pacifists or 'cryptos', or that breed of degenerate intellectuals. **1949** 'C. Hare' *When Wind Blows* 147 Is there any chance of our being able to find a crypto?..I mean, another man like Ventry, who can really play the thing but doesn't let on that he can.

crypto-. Add: **1. cryptobranch** (examples); **cryptomne·sia** [after Amnesia] (see quot. *a* 1901); hence **cryptomne·sic** *a.*; **cryptophyte,** (*b*) (see quots.); hence **cryptophy·tic** *a.*

1872 G. M. Humphry (*title*) Observations in myology, including the myology of Cryptobranch, Lepidosiren, [etc.]. *Ibid.* 1 The muscles and nerves of the Cryptobranch. *a* **1901** Myers *Hum. Pers.* (1903) I. p. xvi, *Cryptomnesia*, submerged or subliminal memory of events forgotten by the supraliminal self. *Ibid.* II. 136 'Cryptomnesia' (as Professor Flournoy calls submerged memory). *Ibid.* 140 This cryptomnesic automatism. **1916** C. E. Long tr. *Jung's Coll. Papers Analyt. Psychol.* 91 The rudimentary glossolalia of our case has not any title to be a classical instance of cryptomnesia. *Ibid.*, The cryptomnesic image arrives at consciousness through the senses. **1961** W. H. Salter *Zoar* x. 138 Latent memory (cryptomnesia) is therefore left as an alternative explanation to sheer chance-coincidence. [**1904** *Botanisk Tidsskrift* XXVI. p. xiv, C. Raunkiær gav en Meddelelse om biologiske Typer..karakteriserede ved Graden og Arten.. iv. Jordplanter, *Kryptofyter*. De overlevende Knopper..befinder sig nede i Jorden.] **1913** *Jrnl. Ecol.* I. 17 Cryptophytes include plants with their dormant parts subterranean. **1937** *Nature* 18 Dec. 1035/2 Cryptophytes, whose surviving buds, etc., are either beneath the soil or at the bottom of water. **1964** Gleason & Cronquist *Nat. Geogr. Plants* xviii. 229 Cryptophytes (hidden plants) are perennial herbs with their buds well below the surface. **1925** *Glasgow Herald* 23 May 4 The dense cryptophytic life [in the jungle] underneath the thick carpet.

2. a. *crypto-communist, -fascist* (also as *adjs.*); *crypto-Jew, -semite.*

1946 *Newsweek* 10 June 44/3 To..Ernest Bevin these extreme left-wingers with their demands for 'working class unity' appear as fellow travelers. He has denounced them as 'crypto-communists'. **1947** *News Chron.* 8 Apr., He is an extreme Left-Wing Socialist...In the Commons he is, of course, dubbed a 'fellow-traveller' and a 'crypto-communist'. **1961** *Times* 1 Dec. 15/1 A crypto-communist reporter. **1965** *Times* 23 Feb. 10/5 The right wing..attacked M. Beuve-Méry as a crypto-communist. **1937** C. Connolly in L. Russell *Press Gang!* 91 Ah, summer! There's a crypto-fascist for you! **1942** E. Waugh *Work Suspended* ii. 86 They're the new hush-hush crypto-fascist department. **1956** D. J. Enright *Bread rather than Blossoms* 26 A crypto-fascist looks for open war. **1892** Zangwill *Childr. Ghetto* I. 3 The Spanish crypto-Jews who had reached England viâ Holland. **1957** *Encycl. Brit.* XXI. 231/2 Portuguese crypto-Jews, that is, descendants of Jews whom the Inquisition had compelled to embrace Christianity but who remained Jews at heart. **1920** *Punch* 26 May 415/1 Giving dancing lessons to the daughters of profiteers, Crypto-Semites and other unpropitious persons. **1937** Wyndham Lewis *Blasting & Bombardiering* v. v. 280 This 'young American poet' was undoubtedly a crypto-semite.

cryptobiosis (kriːpto,baiōᵘ·sis). *Biol.* [f. Crypto- + *-biosis* as in *anabiosis.] (See quot. 1959.)

1959 D. Keilin in *Proc. R. Soc.* B. CL. 166, I should like to propose the term cryptobiosis, that is, latent life, for the state of an organism when it shows no visible signs of life and when its metabolic activity becomes hardly measurable, or comes reversibly to a standstill. **1965** *New Scientist* 28 Oct. 270/2 When an organism is in a state of cryptobiosis mechanical injuries are tolerated that would be immediately fatal in the normal state.

cryptobiotic (kriːpto,baiǫ·tik), *a.* [ad. G. *kryptobiotisch* (O. Kuntze *Phytogeogenesis* (1884) iii. 40), f. Crypto- + Gr. βιωτικ-ός pertaining to life.] **a.** (See quot. 1916.) **b.** = *cryptozoic *a.* 1. **c.** Of or pertaining to cryptobiosis.

1916 B. D. Jackson *Gloss. Bot. Terms* (ed. 3) 97/2 *Cryptobiotic*, Kuntze's suggested expression for those lowly organisms which appeared in geologic times, but have left no trace of their existence. **1934** S. F. Light in C. A. Kofoid *Termites* iii. 23 Associated with the wood-eating habit of the termites is their cryptobiotic mode of life. They live shut off from the light in enclosed passageways. **1948** T. E. Snyder *Our Enemy the Termite* (ed. 2) 240 *Cryptobiotic*, living a 'hidden life', a term applied to insects, such as termites, living a life where they are concealed in wood, underground, etc. **1965** *New Scientist* 28 Oct. 270/2 Some further properties of the cryptobiotic state.

cryptococcosis (kriːpto,kǫkōᵘ·sis). *Path.* [f. mod.L. *Cryptococcus* + -osis.] A disease of man and animals caused by the yeast-like fungus *Cryptococcus neoformans*; torulosis.

1938 *Texas Jrnl. Med.* XXXIII. 310 Systemic cryptococcosis. **1962** *New Scientist* 12 Apr. 46/3 A fungus disease called cryptococcosis, which usually attacks the [koala] bears' central nervous system. **1970** Jubb & Kennedy *Path. Domestic Anim.* (ed. 2) I. iii. 253/1 Cryptococcosis (European blastomycosis) is a subacute or chronic mycosis, caused by *Cryptococcus neoformans* (*Torula histolytica*),.. which shows, in man, dogs, and possibly other species as well, a marked predilection for the central nervous system.

crypto-communist, -fascist: see *crypto- 2 a.

cryptogenetic (kriːpto,dʒene·tik), *a. Path.* [f. Crypto- + -genetic.] Of a disease: of obscure or unknown origin. Also **cryptoge·nic** *a.*

1908 R. Park *Mod. Surg.* I. 87 Cryptogenetic or spontaneous septicemia is a term applied to those cases in which the port of entry of the germs is no longer visible—e.g., a hypodermic puncture—or cannot be positively determined. **1908** *Practitioner* Feb. 249–50 Cryptogenic pernicious anaemia...So-called 'cryptogenic' sepsis, or pyaemia. **1962** *Lancet* 27 Jan. 190/1 All cases of liver cirrhosis which cannot be ascribed to abuse of alcohol, occlusive processes of the bile ducts, or hæmochromatosis will be referred to as 'cryptogenetic cirrhosis of the liver'.

cryptogram. Add: **cryptogrammatic** *a.* (examples); **cryptogra·mmic** *a.* = Cryptogramic *a.*; **cryptogra·mmist** = Cryptogrammatist.

1895 *Daily News* 17 Jan. 6/4 To offer insoluble conundrums in cryptogrammatic ellipsis. **1962** C. L. Wrenn in Davis & Wrenn *Eng. & Medieval Studies* 314 The cryptogrammatic runes and oghams of the Hackness Cross. **1901** *Westm. Gaz.* 3 Jan. 2/3 By a bewildering..system of..word-counting he constructed a cryptogrammic theory. **1906** *Daily Chron.* 18 June 3/3 Prophets, cranks, cryptogrammists. **1929** W. J. Locke *Ancestor Jorico* vii. 90 In peace times the Admiralty had no use for an expert cryptogrammist.

cryptograph. Add: **2. b.** An enciphering or deciphering device.

1879 C. Wheatstone *Sci. Papers* 342 (*heading*) Instructions for the employment of Wheatstone's cryptograph. **1928** *Daily Express* 22 May 2/4 A new invention known as the cryptograph, which could automatically code messages and decode them.

cryptographer. (Later examples.)

1896 *Daily News* 3 Feb. 3/4 He has..been very successful as a cryptographer, and published..what is perhaps the only attempt at a scientific method of analysis of ciphers. **1959** *Chambers's Encycl.* IV. 280/1 The countries at war [in 1914–18] established numerous intercepting stations.. and also a staff of cryptographers.

cryptography. Add to def.: Also, the art of writing or solving ciphers.

1950 J. Kobler in *Collier's* 28 Oct., Perhaps the most brilliant feat in all the long history of cryptography was accomplished by our own cryptanalysts. **1957** *Encycl. Brit.* V. 930/1 The..quite important roles which poor cryptography or good cryptanalysis have played in international relations. **1966** [see next].

cryptology. Add: **b.** (See quot. 1966.)

1945 J. S. Galland (*title*) An historical and analytical bibliography of the literature of cryptology. **1957** *Encycl. Brit.* V. 923/1 Because of..the remarkable progress made in communications-electronics technology, cryptology has come to play a very important role in governmental communications. **1966** D. Kahn *Codebreakers* p. xvi, Cryptology is the science that embraces cryptography and cryptanalysis, but the term 'cryptology' sometimes loosely designates the entire dual field of both rendering signals secure and extracting information from them.

cryptomere (kri·ptomiᵊɹ). *Biol.* [ad. G. *kryptomer* (E. Tschermak 1904, in *Beih. z. Bot. Centralbl.* XVI. 11), f. Crypto- + Gr. μέρος part.] A latent genetic factor or characteristic. So **crypto·merism,** the possession of this factor or characteristic.

1906 R. H. Lock *Rec. Progress Study Variation* viii. 189 The best general name for the class of phenomena we are about to describe is perhaps latency of characters, or cryptomerism. **1909** W. Bateson *Mendel's Princ. Hered.* (ed. 2) v. 93 Factors which may thus exist without making their presence visible have been named by Tschermak 'Cryptomeres'.

cryptomeria (kriptomiᵊ·riă). [mod.L., f. Crypto- + Gr. μέρος part. So named because the seeds are hidden or enclosed by scales.] An evergreen coniferous tree (*C. japonica*) allied to the cypresses, a native of North China and Japan, and now extensively cultivated in England; the Japanese cedar; also, the wood of this tree.

1841 D. Don in *Trans. Linnean Soc.* XVIII. 166 (*heading*) *Cryptomeria*. Ord. Nat. Coniferæ. *Ibid.* 170 The wood in *Cryptomeria* is compact, and the fibrous tissue is composed of very slender vessels. **1852** R. Fortune *Journey Tea Countries China* xviii. 304 The beautiful *Cryptomeria*, or Japan cedar. **1863** R. Alcock *Capital of Tycoon* I. iv. 103 A long avenue of cryptomerias and pines. **1904** D. Sladen *Playing the Game* I. xii. 124 An avenue of tall cryptomerias. **1957** C. Brooke-Rose *Languages of Love* 45 A false ceiling of wide boards, not, alas, of cryptomeria but of ordinary stained timber. **1966** Boom & Kleijn *Glory of Tree* 24/1 The *Cryptomeria* is a very beautiful tree.

cryptovolcanic (kriːpto,vǫlkæ·nik), *a. Geol.* [ad. G. *kryptovulkanisch* (W. Branco and E. Fraas 1905, in *Abh. K. Preuss. Akad. Wiss.* 57), f. Crypto- + Volcanic *a.*] *cryptovolcanic structure*: a nearly circular area not associated with igneous rock but having features suggesting volcanic activity.

1921 W. H. Bucher in *Bull. Geol. Soc. Amer.* XXXII. 75 The Steinheim Basin..has thus far been the only known representative of this type of structure, for which Branco and Fraas proposed the term 'cryptovolcanic'. **1954** W. D. Thornbury *Princ. Geomorphol.* (1962) ix. 211 A rare but interesting type of dome is the cryptovolcanic structure. It is thought to be produced by the sudden release of volcanic gases at depth. *Ibid.* xx. 521 Some alleged cryptovolcanic structures may prove to be meteorite craters. **1969** C. Ollier *Volcanoes* iii. 33 Gosses Bluff in central Australia may be a cryptovolcanic structure.

cryptoxanthin (kripto,zæ·nþin). *Chem.* [ad. G. *krypto-xanthin* (R. Kuhn and C. Grundmann 1933, in *Ber. d. Deut. Chem. Ges.* LXVI. 1746), f. Crypto- + Xanthin.] A yellow carotenoid pigment, $C_{40}H_{56}O$, widely found in nature and important as a precursor of vitamin A.

1934 *Chem. Abstr.* XXVIII. 1043 Cryptoxanthin... This new xanthophyll..constitutes almost ⅓ of the total pigment of the calyxes and berries of *Physalis franchetti*. **1946** *Nature* 24 Aug. 269/2 Carotenoids, such as cryptoxanthin and β-carotene. **1960** Fox & Vevers *Nature of Animal Colours* v. 76 Other vitamin A precursors are α-carotene, cryptoxanthin (3-oxy-β-carotene), found in many plants as well as in egg-yolk, and echinenone.

cryptozoa (kriptozōᵘ·ă), *sb. pl. Zool.* [Back-formation from *cryptozoic *a.*] The group of cryptozoic animals.

1911 A. Willey *Convergence in Evolution* iii. 25 Phanerozoa and cryptozoa are two well-marked physiological groups. **1968** *Sci. Amer.* July 108/1 They are examples of the many small animals that lead hidden lives; they are some of the cryptozoa, the animals in hiding.

cryptozoic (kriptozōᵘ·ik), *a.* [f. Crypto- + Gr. ζωή life + -ic.]

1. *Biol.* Defining a class of fauna composed of animals living a concealed or hidden life (see quots.); also, belonging to this class.

1895 A. Dendy in *Rep. 6th Meeting Austral. Assoc. Adv. Sci.* 99 The Cryptozoic Fauna of Australasia...I use the word 'Cryptozoic' for want of a better. **1897** Parker & Haswell *Zool.* II. xiv. 601 Cryptozoic forms, which live under stones, logs of wood, etc., such as Land-Planarians, Peripatus, Centipedes, and Woodlice. **1953** R. F. Lawrence *Biol. Cryptic Fauna of Forests* ii. 32 Certain examples of cryptozoic forms may take advantage of the shelter provided by human habitations.

2. *Geol.* (Usu. with capital initial.) A name given to the Pre-Cambrian era, now esp. when this is contrasted with the whole of the later part of geological history, from the beginning of the Cambrian to the present day; of or pertaining to the Pre-Cambrian era. Cf. *Phanerozoic *a.*

1911 *Encycl. Brit.* XXII. 266/2 The name pre-Cambrian is the equivalent of the 'Algonkian'..; the terms..cryptozoic, eparchaic and others have also been applied to the same period. **1930** G. H. Chadwick in *Bull. Geol. Soc. Amer.* XLI. 48 Cryptozoic or Cryptobiotic (Precambrian). **1958** R. C. Moore *Introd. Hist. Geol.* (ed. 2) iv. 59 The first major division of earth history determinable from rocks exposed at the surface is the Cryptozoic Eon. *Ibid.*, It is convenient also to designate the Cryptozoic rocks as Precambrian. **1968** *Courier-Mail* (Brisbane) 19 Nov. 26/8 The Cryptozoic ('hidden life') extends a further 3,000 million years—and yet its rocks show almost no traces of life! **1969** Dunbar & Waage *Hist. Geol.* (ed. 3) vii. 164/1 In rocks so generally devoid of fossils what little indication we have of Cryptozoic environments and climates is derived from the sediments.

cryptozoon (kriptozōᵘ·ǫn). *Geol.* [f. Crypto- + Gr. ζῷον animal.] A Cambrian reef-forming fossil, believed to be the remains of colonies of calcareous algæ; cf. *stromatolite. Also *attrib.*

1883 J. Hall in *36th Ann. Rep. N.Y. State Mus. Nat. Hist.* Plate 6, There occurs a bed of limestone..the surface being nearly covered by closely-arranged circular or sub-circular discs which are made of concentric laminæ...I, therefore, propose the term cryptozoön as a designation for this peculiar form and mode of growth. **1914** *Smithsonian Misc. Coll.* LXIV. 98, I do not know a true Cryptozoön older than the Cambrian fauna. **1936** *Geogr. Jrnl.* LXXXVII. 429 Near Etah a Cryptozoon reef occurs. **1965** M. E. Wilson in K. Rankama *Precambrian* II. 312 Concentric *Cryptozoon*-like structures believed to be of algal origin. **1969** Dunbar & Waage *Hist. Geol.* (ed. 3) viii. 182/2 The name *Cryptozoon*, once given to all Lower Paleozoic stromatolites, is no longer used.

crystal, *sb.* and *a.* Add: **A. 4. c.** *fig.* Esp. a prophecy derived from crystal-gazing. Now *rare.*

1902 *Westm. Gaz.* 6 Nov. 2/3 The Cleveland by-election will always be memorable if only for the fact that the crystal has had a new form given to it. **1914** *Concise Oxf. Dict.* Add., *Crystal* colloq., view of the future thus obtained [*i.e.* by crystal-gazing], prophetic utterance. **1931** H. G. Wells *Work, Wealth & Happiness of Mankind* (1932) xii. 596 Favours, buttons, crystal and claptrap: these are the forces that bring the politicians of the great powers of the world to office.

9. d. *spec.* in *Electronics*, a crystalline piece of a semiconductor (such as germanium, silicon, or galena) used in a device on account of its properties of electrical conduction; *crystal detector*, a detector (sense *3 f) in which a crystal diode is employed to rectify a high-frequency current; a crystal diode

used as a detector; *crystal diode*, a semiconductor diode, esp. one comprising a semiconductor crystal with which a thin metal wire is in point contact; also called *crystal rectifier*; *crystal receiver, set*, a receiving set employing a crystal detector.

1907 G. W. Pierce in *Physical Rev.* XXV. 31 (*title*) On crystal rectifiers for..electric oscillations. *Ibid.* 50 Crystal rectifiers employed in the construction of alternating current measuring instruments. **1908** J. A. Fleming *Elem. Man. Radiotelegr.* 332 Crystal detectors. **1913** *Year-Bk. Wireless Telegr.* 419 Crystal Detector, a form of oscillation detector depending on the fact that certain crystals (*e.g.*, carborundum) allow current to pass through them more readily in one direction than in the other. **1917** *Wireless World* June 168 The balanced crystal receiver, by keeping signals at a reasonable strength, enables the operator to receive in conditions which would otherwise make work impossible. **1923** J. A. Fleming *Wireless Telegr. & Teleph.* 20 The most generally used crystal is now galena (sulphide of lead). **1923** Hawkhead & Dowsett *Techn. Instr. Wireless Telegr.* (ed. 3) 129 A good commercial crystal detector..should rectify with very small changes in potential. *Ibid.* 223 Crystal Receiver. **1924** E. T. Larner (*title*) Crystal sets. **1926** J. A. Fleming *Electr. Educator* I. 379/2 This detector consists of two crystals, zincite and chalcopyrites, in contact. **1943** C. L. Boltz *Basic Radio* xiii. 209 Some years ago thousands of people regularly used crystal sets for listening to broadcast programmes. **1948** *Gloss. Computer Terms* (M.I.T. Servomechanisms Lab. Rep. R-138) 5 Crystal diode. *Ibid.*, *Crystal rectifier*, a circuit component having almost unidirectional current-flow characteristics, usually consisting of a small piece of germanium in contact with a thin wire. **1955** *Sci. News Let.* 11 June 378/3 A crystal diode is a device similar to the transistor. **1959** *Chambers's Encycl.* XI. 475/1 The crystal detector..has found a new application in the field of radar where it is used at frequencies up to about 30,000 Mc/s. **1966** *McGraw-Hill Encycl. Sci. & Technol.* I. 359/2 Modern crystal rectifiers..represent the simplest and most sensitive of all rectifying devices.

B. 2. c. *crystal-clear* (examples); **crystal ball,** a crystal (sense 4 a) in the shape of a globe, also *fig.*; **crystal clock,** a quartz-crystal clock (see *quartz); **crystal-gazing,** concentration of one's gaze on a ball of rock-crystal or the like in order to obtain a telepathic or hallucinatory picture; also *fig.*; similarly **crystal-gaze** *v. intr.*, **crystal-gazer** (cf. *crystal-seer, -seeing* in Dict.); **crystal lattice** (see *lattice *sb.*); **crystal microphone,** a microphone which depends for its action on the piezo-electric activity of a crystalline substance; **crystal-pulling** [cf. G. *kristallisationsgeschwindigkeit* used of a similar process by J. Czochralski 1918, in *Zeit. f. physikal. Chem.* XCII. 219], a method of obtaining pure single crystals for use in a junction diode or a transistor by inserting a seed crystal in a melt of germanium or silicon and gradually withdrawing it; also *attrib.*; **crystal-vision,** crystal-gazing, or the picture seen by this means.

1855 Browning *Men & Women* II. 31 The sights in a magic crystal ball. **1964** J. Drummond *Welcome, Proud Lady* xii. 51 Did you see that in your crystal ball..or did Mrs. Dannhauser put you up to it? **1968** G. Butler *Coffin Following* i. 7 In the crystal ball the old woman could see it. *c* **1520** *Everyman* (1890) 898 Now the soule is taken the body fro Thy rekenynge is crystall clere. **1840** Lowell *Irene* in *Poet. Wks.* (1873) 3 Hers is a spirit deep, and crystal-clear. **1845** Browning *Dram. Rom. & Lyrics* 15 Glasses they'll blow you, crystal-clear. **1859** J. W. Carlyle *Let.* 20 Feb. in *Geo. Eliot's Lett.* (1954) III. 17 A crystal-clear, musical, Scotch stream. **1952** C. Day Lewis tr. *Virgil's Aeneid* xi. 245 The issue you ask our advice on Is crystal-clear and does not require our comments. **1937** *Discovery* Jan. 18/2 A crystal clock..will increase the accuracy of the gravity observations. **1950** W. de la Mare *Inward Companion* 78 And do you rap? Or crystal-gaze? **1898** A. Lang *Making of Religion* v. 95 The crystal-gazer. **1920** R. Macaulay *Potterism* iii. i. 108 Thought-readers, crystal-gazers, mediums and planchette-writers. **1889** *Proc. Soc. Psychical Res.* V. 507 Crystal-gazing. **1945** H. Read *Coat Many Colours* xxxi. 151 To ask people to begin looking for profound human emotions..in painted canvas or carved stone seems to me like asking them to indulge in a new kind of crystal-gazing. **1960** *Farmer & Stockbreeder* 12 Jan. 63/1 A suitably cynical note on which to end a session of crystal-gazing. **1926** R. W. Lawson tr. *Hevesy & Paneth's Man. Radioactivity* xxvi. 218 The crystal lattice has been traversed by about as many α-particles as there are uranium atoms contained in the lattice. **1933** *Chem. Abstr.* XXVII. 1578 A practical application of the observed phenomena is a new crystal microphone. **1962** A. Nisbett *Technique Sound Studio* 247 *Crystal microphone or gramophone pick-up.* This generates a signal by means of a crystal bimorph. **1952** *Proc. Inst. Radio Engin.* XL. 1339/2 (*caption*) High vacuum melting furnace showing crystal pulling mechanism. **1960** *Engineers' Digest* XXI. xi. 125/1 Dislocation free single crystals of germanium can be produced by a special method of crystal-pulling. **1962** Simpson & Richards *Junction Transistors* iii. 39 *Crystal pulling*...During this process material from the melt freezes on the seed, building up new rows of atoms whose crystal axes are the same as those of the seed. **1889** *Proc. Soc. Psychical Res.* V. 486 Recent experiments in crystal-vision. **1898** A. Lang *Making of Religion* v. 90 Crystal visions, savage and civilised.

crystalline, *sb.* Add: **6.** A light soft dress-material.

1903 *Daily Chron.* 25 July 8/4 Crystalline differs very little from mousseline de soie, for it is a thin fabric with a silky sheen upon it, and a very charming one for afternoon summer frocks. **1923** *Daily Mail* 8 May 14 Soft crepe finish crystalline.

crystallinity. Add: (Further examples.) Also, amount or degree of crystallization.

1937 *Discovery* Oct. 302/2 Regularity of molecular arrangement (*i.e.*, crystallinity). **1947** *Nature* 4 Jan. 30/1 A displacement of the electric strength temperature curve with decreasing crystallinity. **1958** *Times Rev. Industry* May 52/3 The long hydrocarbon chains..show a high degree of crystallinity.

crystallite. Restrict '*Min.*' to sense 1 in Dict. and add: **2.** (Later examples.) Now distinguished in sense from Microlith, -lite: a crystallite is of smaller size, does not polarize light, and cannot be referred to any definite species of mineral.

Occas. used (following Vogelsang) for 'embryonic crystals' in other substances (in quot. 1914 the word denotes a branched crystal or dendrite such as forms in freezing metal).

1910 *Encycl. Brit.* VII. 568/2 Crystallites may also be produced by allowing a solution of sulphur in carbon disulphide mixed with Canada balsam to evaporate slowly. **1914** *Jrnl. Inst. Metals* XI. 65 Most metals only exceptionally form simple polyhedral crystals, but by preference assume the form of branched crystallites. A theory of crystallites was proposed by Vogelsang, who made a special study of these structures in vitreous rocks,.. in blast-furnace slags, and in artificial preparations. He observed that by cooling solutions of sulphur in viscous solvents, minute..globules were obtained, which..gradually developed crystalline outlines, so that the typical branched crystallites were thus produced. He therefore..regarded crystallites as 'embryonic crystals'. **1926** G. W. Tyrrell *Princ. Petrol.* v. 81 Crystallites are embryo crystals, not yet organized to full crystalline status... Microlites are somewhat larger bodies which can be definitely recognised as minute crystals. **1953** L. E. Spock *Guide Study of Rocks* iv. 59 Typical fresh obsidian is glossy and black, owing its darkness to closely spaced crystallites. **1958** A. D. Merriman *Dict. Metallurgy* 49 *Crystallite*...The term is often used in reference to the tiny incipient crystals forming on the solidifying of metals and alloys. **1966** *McGraw-Hill Encycl. Sci. & Technol.* VII. 10/2 Crystallites..are the most rudimentary forms [of rock grains] and abound in glassy rocks in which rapid consolidation has arrested further growth.

b. An individual crystal or grain in a metal or other polycrystalline substance; also, a part of a quasi-single crystal in which the orientation is uniform and the structure homogeneous. (Orig. an application of the word in the prec. sense, as in quot. 1914.)

1914 *Jrnl. Inst. Metals* XI. 64 The 'crystal grains' of ordinary cast metals arise in this way, their boundaries being produced by the mutual interference of neighbouring crystallites. **1920** *Chem. Abstr.* XIV. 3006 The dissoln. of the eutectic..occurs at the boundaries of the crystallites even of 'pure' metals. **1938** J. Newton *Introd. Metallurgy* ii. 27 These small units are individual crystals of the metal or alloy and are called crystallites, crystal grains, or simply grains. **1962** Simpson & Richards *Junction Transistors* iii. 39 To minimize the formation of improperly oriented crystallites on the surface of the main crystal..the seed is slowly rotated as it is withdrawn. **1963** W. G. Burgers in J. J. Gilman *Art & Science of growing Crystals* xxii. 417/1 A normal piece of metal..is usually fine-grained and built up of crystallites (grains) often a few hundredths of a millimeter in diameter.

4. A minute part of cellulose or other polymers having the highly ordered structure characteristic of a crystal. Cf. *micelle.

1926 *Jrnl. Physical Chem.* XXX. 457 Cellulose, in its natural condition, consists of aggregates of crystallites as found in the bast fibres, seed hairs, and other plant tissue as well as in the animal body. **1931** *New Phytologist* XXX. 4 In such fibres as ramie the crystallites are arranged parallel to each other. **1932** *Proc. R. Soc.* B. CIX. 449 The cell-wall of *Valonia ventricosa* is found by X-ray methods to be built up of two main sets of cellulose chains which form crystallites crossing at an angle. **1946** E. I. Valko in J. Alexander *Colloid Chem.* VI. xxix. 595 A single macromolecule can..traverse disorganized regions, so that it belongs to different crystalline regions. As a consequence, the size of the organized regions, i.e., the 'crystallites', is not determined by the length of the macromolecule. **1948** Schmidt & Marlies *Princ. High-Polymer Theory & Pract.* ii. 54 In high polymers these crystallites are not all of the same size and do not possess well-defined faces or edges. **1965** Bell & Coombe tr. *Strasburger's Textbk. Bot.* (new ed.) I. i. 53 (*caption*) Diagram illustrating the micellar structure of cell walls...Linear macromolecules are grouped to form micellae or crystallites. **1966** M. L. Miller *Struct. Polymers* x. 493 Each of these crystallites was supposed to consist of a bundle of parallel chains,..with each crystallite so small that a single molecule passed through several crystals...A modern view regards highly crystalline polymers..as a single crystalline phase with defects.

crystallitic (kristăli·tik), *a.* [f. Crystallite + -ic.] Of the nature of a crystallite.

1859 A. Harker *Petrol.* vii. 97 The general mass of the glass is full of very minute crystallitic bodies. *Ibid.*, These are rocks..having the glassy base rich in crystallitic growths.

crystallization. 1. b. (Earlier example.)

1842 Mill *Let.* 8 Feb. (1963) XIII. 498 This will necessarily require..a gradual crystallization of many thoughts at present held in a state of suspension.

crystallized, *ppl. a.* Add: **3.** Of fruit, ginger, etc.: preserved by impregnation with sugar, and usually coated with sugar crystals.

1875 L. Troubridge *Jrnl.* 25 Dec. in J. Hope-Nicholson *Life amongst Troubridges* (1966) 133 Funny little presents they were—mostly boxes of choc. and crystallised fruits. **1882** [see Glacé *a.* 2]. **1900** *Lancet* 6 Jan. 49/2 The greengages of crystallised fruit..and candied citron peel, all owed their brilliant colour to a salt of copper. **1908** C. H. Senn *Dict. Foods* 109 *Petit fours*, F., is but the generic name for all kinds of very small fancy cakes usually highly decorated with fancy icing, crystallised fruits, and bon-bons. **1928** F. Hurst *President is Born* xxxix. 392 Crystallized ginger. **1936** Auden & Isherwood *Ascent of F6* I. iii. 50 Drinks and eats little but is fond of crystallized apricots. **1965** Hutton & Bode *Simple Sweetmaking* ix. 70 *Crystallized Fruits.* The aim in crystallizing fruits is to replace some of the water in the fruit with a strong sugar solution. **1970** A. Parker *Cooking for Christmas* i. 56 Young stem ginger for dessert comes preserved in syrup..; or an alternative is good crystallized ginger, in boxes or tins.

crystallo-. Add: **cry:stallobla·stic** *a.* [ad. G. *krista*^ll^*oblastisch* (F. Becke 1903, in *Compt. Rend. Congr. Géol. Internat.* (1904) II. 563), f. Gr. βλαστικ-ός budding] (see quots.); **cry·stallo-ceramie** = *cameo-incrustation* (see *cameo b); also, glassware of this kind; **cry·stallogram** (or **crysta·llo-**), a photographic record of the X-ray diffraction pattern presented by a crystal, hence of its structure.

1913 C. K. Leith *Struct. Geol.* (1914) 77 There is, in the schists, relative perfection of crystal forms, dependent on the character of the minerals...This mineral form and arrangement in schists is the 'crystalloblastic' structure of Milch and Grubenmann. **1939** A. Johannsen *Descr. Petrogr.* (ed. 2) I. 207 *Crystaloblastic* [sic], a crystalline texture due to metamorphic recrystallization. A characteristic of this texture is that the essential constituents are simultaneous crystallizations and are not formed in sequence, so that each may be found as inclusions in all the others. **1960** Turner & Verhoogen *Ign. & Metamorph. Petrol.* (ed. 2) xxii. 592 Fabrics resulting from this latter process are called crystalloblastic, a term introduced by Becke to cover structures of the crystalline schists but nowadays generally extended to include also fabrics of like origin and character in rocks resulting from contact or metasomatic metamorphism. **1821** A. Pellatt (*title*) Memoir on the origin, progress, and improvement of glass manufactures: including an account of the patent crystallo ceramie, or, glass incrustations. **1849** —— *Curiosities of Glass Making* 29 A patent was, some years since, taken out by the author..for ornamental incrustation, called 'Crystallo-Ceramie'...By this process, ornaments of any description—arms, ciphers, portraits, and landscapes of any variety of colour—are enclosed within the glass, so as to become chemically imperishable. **1960** *Times* 12 Nov. 9/3 Collectors group crystallo ceramie into five well-defined classes. **1942** *Electronic Engin.* XV. 188 From the crystallogram it is possible to gain much useful information such as the crystal size and the preferred or random orientation of the crystals.

crystobalite, var. *cristobalite.

csardas (tʃāːɹdāʃ, zāːɹdās). Also **czardas.** [a. Hungarian *csárdás*, f. *csárda* inn.] A Hungarian national dance.

1860 *Players* I. 114 Having made herself thoroughly mistress of the 'Czardas', the national dance of Hungary. **1883** 'Ouida' *Wanda* vi, They ended their dances with the Hungarian czardas. **1886** W. J. Tucker *E. Europe* 217 The bewildering postures and maddening antics of the Csárdás. **1888** E. Gerard *Land beyond Forest* II. 245 Whenever the csardas comes to an end there is a violent clapping of hands to make the music resume. **1944** W. Apel *Harvard Dict. Mus.* 198/1 *Czardas*, a Hungarian dance, usually consisting of a slow, pathetic introduction called *lassu*, and a rapid and wild dance called *friss* or *friska*. F. Liszt's Hungarian Rhapsody no. 2 is a well-known example. *Ibid.* 343/2 The Czardas..is said to be a 19th-century revival of the old verbunko.

‖**cuadrilla** (kwadri·lya). [Sp.: see Quadrille *sb.*²] A group or company; *spec.* the troupe or following of a matador.

1841 G. Borrow *Zincali* ii. viii. 354, I alone once robbed a cuadrilla of twenty Gallégos, who were returning to their own country, after cutting the harvests of Castile. **1893** Chapman & Buck *Wild Spain* v. 67 The *Espada*, or Matador, receives on the day from £120 to £200, including the services of his cuadrilla or troupe, which consists of two picadors, three banderilleros, and a cachetero. **1898** *Daily News* 11 Aug. 7/1 The crowd, blaming the bull, instead of the toreador and his cuadrilla, insisted on that animal being killed. **1932** Hemingway *Death in Aft.* iii. 26 Each matador ..has a cuadrilla, or team, of from five to six men who are paid by him and work under his orders. **1934** [see *banderillero]. **1967** *Economist* 25 Nov. 846/1 They were judged, as members of the *cuadrilla* (armed band), to be jointly responsible for all its acts.

‖**cuartel** (kwaɹte·l). Also **quartel.** [Sp.] A military barracks.

1832 W. B. Dewees *Lett. from Texas* (1852) 142 Seven hundred Mexicans were at the quartel or barracks. **1890** *Outing* (U.S.) Oct. 8/1, I saw you, too, at the *cuartel* at Tucson. **1941** Steinbeck & Ricketts *Sea of Cortez* (1958) 26 We imagined being held in some mud *cuartel*. **1967** K. Giles *Death in Diamonds* ii. 37 The patrolman..took O'Connors to the *cuartel*.

cub, *sb.*¹ Add: **2. c.** *Cub*, a junior member of the Scout Association (see Scout *sb.*⁴ 2 c). In full *Wolf Cub.*

1922 A. Poyser (*title*) The Cub Song Book. **1923** *Daily Mail* 11 June 16 Boy Scouts and Cubs furnished a guard of honour. **1964** M. Kelly *March to Gallows* xiii. 165, I shan't slip a wicked potion in it... Cubs' honour.

3. b. An apprentice or beginner; *spec.* an apprentice pilot on a steamboat. orig. *U.S.*

[**1840** *Ninawah* (Peru, Ill.) *Gaz.* 14 May 2/3 Awaiting the arrival of 'a cub' (a young speculator).] **1875** 'Mark Twain' in *Atlantic Monthly* May 567/1 The pilot not on watch takes his 'cub' or steersman..and goes out in the yawl. *Ibid.* 568/2 Nothing delights a cub so much as an opportunity to go out sounding. **1895** Kipling *Land & Sea T.* (1923) 72 I'll take him as my cub, for there's no denying he's a resourceful lad. **1966** *New Statesman* 30 Dec. 956/1 Every cub knows that the first rule of reporting is: never show your story to your subjects before it's on the street.

5. *cub-bear*; (sense *2 c) *cub-master, -mistress*; (sense *3 b) *cub-engineer, -pilot, -reporter.*

1834 H. Brackenridge *Recoll.* vii. 79 Some would rather pass for cub bears than be disappointed in their endeavours to attract attention. **1875** 'Mark Twain' in *Atlantic Monthly* Jan. 71/1 They..learned to disappear when the ruthless 'cub'-engineer approached. **1921** *Daily Colonist* (Victoria, B.C.) 25 Mar. 9/1 To run a troop and pack requires a certain amount of money, which it is hardly fair to ask the individual Scoutmasters and Cubmasters to find. **1927** *Daily Mail* 12 July 10/4 A Hastings Rover Cub-master. **1927** *Daily Tel.* 21 June 3/2 A child who had fallen into a mill stream..was rescued by a local cubmistress. **1970** J. Wainwright *Prynter's Devil* vii. 160 Shut up, if you can't do anything but make noises like an outraged cub-mistress. **1859** 'Mark Twain' in *Univ. Missouri Stud.* (1938) XIII. 57/2 Our friend Sergeant Fathom, one of the oldest cub pilots on the river. **1875** —— in *Atlantic Monthly* Feb. 217/1 (*heading*) A 'cub' pilot's experience; or, learning the river. **1899** J. L. Williams in *Scribner's Mag.* XXV. 277 (*title*) The cub reporter and the king of Spain. **1908** A. Ruhl *Other Americans* ii. 9 The mere gringo feels like a cub reporter at the office of a campaign committee. **1925** E. Wallace *King by Night* xli. 183 Bobby was a cub reporter on my newspaper in Sacramento.

Cuba libre (kiū·bă li·bre). [Amer. Sp., lit. 'free Cuba'.] A long drink containing lime juice and rum (see also quot. 1898).

1898 *Harper's Weekly* 20 Aug. 813/1 We..rode through the swamp, sometimes drinking a little 'Cuba libre' (water and brown sugar). **1937** Hemingway *To have & have Not* iii. vii. 133 'What's the lady drinking?'..'A Cuba Libre.' **1963** D. Cory *Hammerhead* v. 85 She had taken full advantage of a round half-dozen of Cuba Libres. **1964** *House & Garden* Nov. 106/2 *Cuba Libre.* Shake one measure of Bacardi rum with the juice of half a lime or lemon. Pour into a tall glass with ice and top up with iced Coca-Cola.

Cuban (kiū·băn), *a.* and *sb.* [See -an.] **A.** *adj.* Of or pertaining to Cuba. **B.** *sb.* A native or inhabitant of Cuba.

Cuban heel, a medium high, comparatively straight heel; so *Cuban-heeled* adj. Also (both forms) with small initial.

1829 *Foreign Q. Rev.* III. 414 The dependence of the Cuban proprietors on usurious merchants..is as dreadful as in other colonies. **1850** R. B. Kimball (*title*) Cuba and the Cubans; comprising a History of the Island. **1874** *Chambers's Encycl.* III. 349/2 The Cuban sugar-trade. *Ibid.* 350/1 'Cuba for the Cubans', is the watchword of the creoles. **1877** *Encycl. Brit.* VI. 679/1 Great sympathy had long been shown for the Cubans by the people of the United States. *Ibid.*, The yearly campaigns up to the present time have shown that in the eastern interior the Cuban patriots are practically invincible. **1885** R. L. & F. Stevenson *Dynamiter* 144 Story of the Fair Cuban. **1908** *Sears, Roebuck Catal.* 813/1 This Gunmetal Blucher Oxford has..light flexible soles, high Cuban heels and large eyelets. **1926** *Contemp. Rev.* Aug. 180 Among the mass of the Cuban peasants, the Cuban politician..strikes a responsive chord. **1936** 'J. Tey' *Shilling for Candles* vii. 86 She held up a foot and exhibited her very modest cuban heel. **1940** N. Marsh *Surfeit of Lampreys* (1941) xvii. 263 Which of those ladies wore cuban-heeled shoes? **1959** 'F. Newton' *Jazz Scene* iv. 71 The fashion for mixed Cuban-jazz music.

Cubanize (kiū·bănaiz), *v.* Also **-ise.** [See -ize.] *trans.* To claim a right of protection or partial control over (a weaker but independent state), as the United States is alleged to have done with regard to Cuba. So **Cu:baniza·tion.**

1922 *Q. Rev.* July 151 The various Yankee associations whose ultimate aim is the attraction of Mexico within the political orbit of the United States, and its 'Cubanisation' by treaty. **1924** in J. A. Hammerton *Countries of the World* xv. 1525/1 Among Cuba's other claims to fame may be placed the fact that its political status has originated a new verb—to Cubanise... It is a quasi-protectorate of America and the word was invented to express this relation.

cubbed (kɒbd), *ppl. a. poet. rare*⁻¹. [f. Cub *sb.*¹ + -ed².] With a cub or cubs.

1889 Tennyson *Demeter & Perseph.* l. 54, I envied human wives, and nested birds, Yea, the cubb'd lioness.

cubby. 1. (Earlier U.S. example.)

1868 *Congress. Globe* 2 June 2762/3 [Many of the national banks] keep a little cubby of an office, loan no money,.. and yet draw interest on their circulation.

cube, *sb.*¹ Add: **1. b.** Of sugar (examples). Also *attrib.*

1897 *Sears, Roebuck Catal.* 17/3 Sugar..Cubes. **1916** *Daily Colonist* (Victoria, B.C.) 8 July 11/3 The vessel's cargo consists of 110 tons of corn, 50 tons of bean coffee, 1,000 cases of cube sugar, [etc.]. **1935** *Discovery* Aug. 240/1 Mr Benn's host never went on an expedition without a large carton of cubes, which he handed out generously to those Mongolians whose tents they visited.

c. An extremely conventional or conservative person (cf. *square *sb.*). So **Cubesville** (after **Squaresville*), a group or set of such persons. *slang.*

1959 J. Osborne *Paul Slickey* ii. viii, He's strictly from Cubesville. **1960** Wentworth & Flexner *Dict. Amer. Slang* 133/2 *Cube*, a super 'square';..an ultraconservative; a thorough bore. **1961** *Woman* 11 Mar. 5/1 No need to feel cubesville (that's *worse* than being a square) if you don't follow Kookie patter; even many Americans reckon it odd! **1963** *Sunday Times* 8 Sept. 29/3 As one who left school in July, I feel qualified to write of current oddities mentioned by 'Old Squares' (cubes in teenage slanguage). **1963** *Telegraph* (Brisbane) 24 May 17/2 Square itself is old hat. Too many adults cottoned on..the phrase and proclaimed 'I'm a square' in self defence. You'll simply be A Cube. **1968** 'G. Bagby' *Corpse Candle* x. 133 When I sang it to him..he told me I was a complete fool. Daisy Bell was for the cubes.

cube (kū·be), *sb.*³ [Amer. Sp. *cubé* (also used).] One of several South American plants of the genus *Lonchocarpus*, having roots which contain rotenone, used as an insecticide.

1924 *Bull. U.S. Dept. Agric.* no. 1201. 6 In 1920, while collecting fishes in Peru, Dr. W. R. Allen procured a supply of the dried roots of 'cube'. **1930** *Sci. Amer.* Nov. 391 The cube plant now grows in a part of South America where the climate is similar to that of the Malay States. **1940** H. J. Holman *Surv. Insecticide Mat.* ii. 46 The bulk of the cube root and powder at present exported from Para and Manaos in Brazil is said to be derived from this species [sc. *Lonchocarpus urucu*]. **1960** Gunther & Jeppson *Mod. Insecticides* xii. 197 The name cubé is employed in South America to refer to the principal native species of *Lonchocarpus*, *L. nicou* and *L. utilis*, which were used for poisoning fish, and it has now been adopted in commerce to describe roots of any *Lonchocarpus* species which also display pronounced insecticidal properties.

cube, *v.* Add: **4.** To cut into small cubes.

1947 *Home Institute Cook Book* (N.Y.) 23 *Cube*, to cut into small cubes or solids of six equal square sides. **1951** *Good Housek. Home Encycl.* 354/1 Slice or cube and serve hot with melted butter. **1960** A. Wesker *Kitchen* 17 Cube stale bread for onion soup.

cubic, *sb.* Add: **2.** A cubist painting. *rare.*

1924 Galsworthy *White Monkey* I. ii. 14 The cubic called 'Still Life'.

cubically, *adv.* Add: In the form of a cube or cubes.

1855 J. R. Leifchild *Cornwall* 61 Rocks rising cubically. **1856** *Phil. Trans. R. Soc.* CXLVI. 488 Cubically and spherically isotropic bodies.

cubicle. Add: Hence *gen.* any small partitioned space; *spec.* a carrel in a library.

1926 *Bull. Amer. Libr. Assoc.* Oct. 297 Seminars, cubicles, and private studies will be provided for..advanced students, and visiting scholars. **1938** K. M. B. Cross *Mod. Public Baths* 61 Dressing cubicles are not required in these changing rooms, where their use is confined to children. **1960** H. Littlewood *Learning to Swim* ii. 19 Go to your cubicle or the changing room and make sure that you dry yourself thoroughly. **1961** T. Landau *Encycl. Librarianship* (ed. 2) 69/1 *Carrel*...Also called cubicle or stall. **1962** A. Nisbett *Technique Sound Studio* 246 *Control cubicle*, the soundproof room equipped with control desk. **1963** *Times* 22 Apr. 2/6 No one seems to know whether cubicles originated in England or America, and the cows don't care, because cubicles are the last word in cow comfort... Some of the cleanest cows to be seen these days..are those in cubicle installations. **1971** *Hampshire Gaz.* (Northampton, Mass.) 19 Feb. 9/4 It adorns the cubicle next to the one marked 'Men'.

2. *Electr. Engineering.* A chamber or compartment to hold a switch-gear apparatus.

1911 J. F. C. Snell *Power House Design* 347 The oil-break switches are contained within glazed brick cubicles. *Ibid.* 349 The operating gallery and cubicle gallery in the Bahia Blanca power house. **1927** *Daily Tel.* 14 Mar. 4 Switch-gear cells and cubicles constructed of moulded stone. **1930** [see *armour-clad *ppl. a.*]. **1968** F. Kertesz *Lang. Nuclear Sci.* (Oak Ridge Nat. Lab. TM 2367) 18 *Cubicle* referred to the power supply and control center of an individual calutron.

Cubism (kiū·biz'm). [ad. F. *cubisme*, f. *cube* Cube *sb.*] An important early twentieth-century revolutionary pictorial movement arising out of the rejection of traditional Western single-viewpoint perspective: in its first 'analytical' stages characterized by simple geometric forms which soon gave way to further complexes of interlocking semi-transparent planes. In its second major or 'synthetic' phase, flat abstract coloured shapes were assembled and clarified in such a way as to achieve a revisionary significance. Hence **Cu·bist** [F. *cubiste*], an artist who adopts one of the styles of Cubism; also *attrib.* and as *adj.* Also **cubi·stic** *a.*, **cubi·stically** *adv.*

'The word "Cubism"..dates from 1908 and was pronounced for the first time, according to M. Léonce Rosenberg, by a member of the Hanging Committee of the Salon des Indépendants. As a canvas by Georges Braque was being carried by, this person exclaimed, "*Encore des Cubes! assez de cubisme!*" A journalist seized on the *mot* and spread it abroad, and the painter concerned, together with his associates, accepted the nickname and confessed themselves Cubists' (Rutter *Evol. Mod. Art* 80).

1911 *Illustr. Lond. News* 21 Oct. 648/1 Paris is perturbed by the Cubism and the Cubists of the Salon d'Automne. **1911** *Lit. Digest* (N.Y.) 18 Nov. 914/1 The cubists take the blocks of the pavement as their medium for interpreting the external world. **1913** tr. *Gleizes & Metzinger's Cubism* 16 To understand Cézanne is to foresee Cubism. **1914** A. J. Eddy *Cubists & Post-Impress.* 72 Cubism is simply a systematic use of planes. **1915** W. H. Wright *Mod. Painting* 187 Those whose criterion is prettiness are naturally attracted to Whistlerian and Cubistic modes. **1917** W. J. Locke *Red Planet* x. 113 All their talk was of Hauptmann and Sudermann..and in art—Heaven save the mark—the Cubist school. **1920** R. Fry *Vision & Design* 186 It is interesting to consider his Cubist period, since Marchand's reaction to Cubism is typical of his nature. **1924** Galsworthy *White Monkey* II. ii. 133 [Painter to model] 'No, I shouldn't be treating you cubistically.' **1928** —— *Swan Song* III. xiii. 317, I remember the first shows in London of those post-impressionists and early Cubist chaps. **1936** A. H. Barr *Cubism & Abstract Art* 30 Cubism in the early days developed under the mixed influence of Negro Sculpture and Cézanne. **1948** [see *abstraction 8]. **1966** J. Griffin tr. *Fry's Cubism* 9 Cubism first posed, in works of the highest artistic quality, many of the fundamental questions that were to preoccupy artists during the first half of the twentieth century. **1970** *Oxf. Compan. Art* 293/2 Part of the object of the Cubists was to represent solidity and volume in a two-dimensional plane. *Ibid.*, Cubism is the outcome of intellectualized rather than spontaneous vision.

transf. **1914** A. J. Eddy *Cubists & Post-Impress.* 64 A form of dramatic representation that is essentially Cubist, Futurist, and Orphist in its expression. **1920** A. Huxley *Let.* 4 Mar. (1969) 182 Paris shd be amusing: I was there in January and had an entertaining time among the cubists of literature. **1926** W. J. Locke *Old Bridge* vi. 91 The.. German tourist and his cubistically attired wife. **1927** W. S. Vines *Movements* 3 Mr. Blunden is a case in point, this critic claiming him for the Georgians, while that one will allege that cubistic symptoms have characterised, if not marred, his later work. **1927** *Observer* 6 Mar. 21/3 A few [ladies' coats] display cubistic ideas, amusing to study in detail.

cubitus (kiū·bitǒs). *Ent.* [L.: see Cubit.]

†**1.** *Obs.* (See quot.)

1826 Kirby & Spence *Entomol.* III. 369 *Cubitus*, the fourth and elongated joint [in the first pair of legs in Hexapods], answering to the Tibia in the legs.

2. = Cubit 1 d. Hence **cu·bital** *a.*, **cubito-,** combining form.

1895 J. H. & A. B. Comstock *Man. Study Insects* iii. 64 The principal veins of the wing—are termed..the costa,.. the subcosta,..the cubitus, [etc.]. **1898** *Amer. Nat.* XXXII. 88 The fifth principal vein is the cubitus... Between the cubitus and the anal margin there are typically three veins. *Ibid.*, The two groups of wing-tracheæ thus formed may be designated as the costo-radial group and the cubito-anal group. *Ibid.* 234 The medio-cubital cross vein. This is a cross-vein extending from media to cubitus. **1918** J. H. Comstock *Wings of Insects* xxvi. 390 The third and last of the branched veins in flies is the cubitus. **1957** Richards & Davies *Imms's Textbk. Ent.* (ed. 9) I. 45 The cubitus.. divides into two main branches, the first cubitus..being convex and the second cubitus..concave. **1964** R. M. & J. W. Fox *Introd. Compar. Ent.* iv. 117 A complete cubital system is present in only a few insects...Posterior to the cubitus is a variable number of veins.

cuckold, *sb.*² **1.** (Later U.S. example.)

1821 T. Nuttall *Trav. Arkansa* ii. 58 The cornfields, at this season of the year, are so over-run with cuckold-burrs (*Xanthium Strumarium*)..as to prove extremely troublesome to woollen clothes.

cuckoo, *sb.* Add: **3.** (Further examples.) Now usu. *slang* for 'a silly person'.

1889 J. K. Jerome *Three Men in Boat* x. 152 Give us a hand here, can't you, you cuckoo; standing there like a stuffed mummy. **1921** H. C. Witwer *Leather Pushers* i. 4 On account of this cuckoo forgettin' he was a box fighter,.. we lose five other bouts. **1924** Galsworthy *White Monkey* I. ix. 77 'Don't worry, we'll dig up the just-right cuckoos, somehow.' 'A Chinese Minister would be perfect,' mused Fleur.

9. *cuckoo-echoing* poet. Comb.; **cuckoo-fish** (U.S. example); **cuckoo fowl** (see quot.); **cuckoo's mate** (examples); **cuckoo scab** *Austral.* and *N.Z.*, a skin disease of sheep.

1879 G. M. Hopkins *Poems* (1918) 41 Cuckoo-echoing, bell-swarmèd, lark-charmèd. **1884** J. C. Brevoort in G. B. Goode *Fisheries & Fishing Industries U.S.* I. 257 When freshly taken from the water they grunt quite loudly, whence their popular name of Grunter, or Cuckoo-fish. **1850** D. J. Browne *Amer. Poultry Yard* 56 The 'cuckoo fowl'..was so called from its barred plumage, resembling the breast of the cuckoo. [**1802** G. Montagu *Ornith. Dict.* II, s.v. Wryneck, Appearing at the same time with the Cuckow, it has been termed that bird's servant or attendant.] **1831** G. Montagu *Ornith. Dict.* (ed. 2) 123 Cuckoo's mate.—A name for the Wryneck. **1898** C. M. Yonge *John Keble's Parishes* xvi. 202 Wryneck..or Cuckoo's mate, squeaks all round the woods ..just as the cuckoo comes. **1955** D. A. Bannerman *Birds Brit. Isles* IV. 119 The approximate dates of its [*sc.* the wryneck's] arrival in Britain, where it is commonly known as 'the cuckoo's mate', have already been given. **1941** Baker *Dict. Austral. Slang* 21 *Cuckoo scab*, a skin disease on sheep on the back of the head and ears. **1951** L. G. D. Acland *Early Canterbury Runs* 372 *Cuckoo scab*, a skin disease sheep get at the back of their heads and on their ears. I have only noticed it among merinos and in the back country.

cuckoo (ku·kū), *a. slang* (orig. *U.S.*). [f. the sb.] Crazy, out of one's wits.

1918 *Wine, Women & War* (1926) 75 Wish my daughter would grow up like that... [fn.] Seen her since. Certainly must have been cuckoo! **1923** Wodehouse *Inimit. Jeeves* xvii. 241 He pottered about the room for a bit, babbling at intervals. The boy seemed cuckoo. **1928** *Collier's* 29 Dec. 28/1 When everything..failed to reduce Jack's bulk, I was nearly cuckoo with rage and fear. **1955** M. Gilbert *Sky High* vi. 76 Never asked for references?.. She must be cuckoo.

cuckoo-land, short for *cloud-cuckoo-land. Hence *cuckoo-lander.*

1916 A. Huxley *Burning Wheel* 49 For still in Cuckoo-Land they're labouring, With hopes undamped and undiscouraged hearts. *Ibid.* 50 And Cuckoo-Landers not a few shall prove. **1958** *Observer* 28 Sept. 17/1, I am afraid that your paper..is living in a cuckoo-land. Have you no idea what ordinary Cockney people..feel about this mass invasion [of coloured people]?

cuckquean, *sb.* (Later example.)

1922 Joyce *Ulysses* 15 A wandering crone..their common cuckquean.

cucumber. Add: **2. b.** Hence *cucumber-cool.*

1955 Auden *Shield of Achilles* iii. 75 In his New Jerusalem even chefs will be cucumber-cool machine minders.

3. b. Short for *cucumber-tree* (U.S.). Also *attrib.*

1797 F. Baily *Tour N. Amer.* (1856) 178 Elm, oak, cucumber, and other trees. **1835** A. Parker *Trip to West* 47 The timber consists of the various kinds of oak,.. cucumber, [etc.]. **1904** 'O. Henry' *Cabbages & Kings* x. 161 Johnny Atwood..prated feebly of cool water to be had in the cucumber-wood pumps of Dalesburg.

4. *cucumber-frame* (later example), *sandwich*; **cucumber-beetle, -bug, flea beetle** *U.S.* (see quots.); **cucumber mosaic,** one of a group of virus diseases that attack cucumbers and related plants; **cucumber-tree,** (*a*) (earlier example).

1841 T. W. Harris *Insects of Massachusetts* 101 These striped cucumber-beetles..notorious..for their attacks upon the leaves of the cucumber and squash. **1948** *Ada* (Okla.) *Even. News* 2 July 4/4 A powerful insecticide that will kill such stubborn pests as Cucumber Beetles. **1838** *Mass. Zool. Surv. Rep.* 100 The cucumber-bug..is called *Galeruca vittata*. At first sight it appears much like the potato-insect. **1877** *Rep. Vermont Board Agric.* IV. 154 The Cucumber Flea Beetle,..a little black beetle..sometimes attacks the raspberry. **1934** Wodehouse *Right Ho, Jeeves* xx. 251 It sounded as if Carnera had jumped off the top of the Eiffel Tower on to a cucumber frame. **1916** S. P. Doolittle in *Phytopathology* VI. 145 The cucumber mosaic disease shows most markedly on the fruits, the first sign being a yellowish mottling near the stem end. *Ibid.* Plate V (*caption*) Cucumber mosaic. **1923** W. F. Bewley *Dis. Glasshouse Plants* vii. 144 (*heading*) Symptoms of Cucumber Mosaic Disease. **1935** *Jrnl. Min. Agric.* XLII. 338 These three diseases [*sc.* green-mottle mosaic, yellow mosaic, yellow-mottle mosaic], collectively known as 'cucumber mosaic', are widespread. **1950** *N.Z. Jrnl. Agric.* Feb. 157/3 Cucumber mosaic..is a virus disease which may cause serious losses in cucumbers, marrows, pumpkins, and squashes. Symptoms consist of stunting of the plants and mosaic mottling of the foliage. **1896** E. Turner *Little Larrikin* xv. 171 The fates chose that he should be allotted to find a cucumber sandwich for his hostess's sister-in-law. **1899** Wilde *Importance of being Earnest* I. 5 Why all these cups? Why cucumber sandwiches? Why such reckless extravagance in one so young? **1967** *Listener* 23 Mar. 398/1 The kind of smile and soft tone of voice you would connect with cucumber sandwiches and a vicarage lawn. **1785** Jefferson *Notes on Virginia* 65 Cucumber-tree. Magnolia acuminata.

cucumiform, *a.* (Examples.)

1826 Kirby & Spence *Entomol.* IV. xlvi. 265 Cucumiform (Cucumiformis). Cucumber-shaped. **1838** *Penny Cycl.* XII. 270/2 The *Cucumiform Holothuriæ*, whose body is but little elongated, more or less fusiform, pentagonal, with tentaculiform suckers.

cucuy, cucuyo. Add: Also **cocuyo, cucujo.**

1836 W. E. Shuckard tr. *Burmeister's Man. Ent.* iii. 491 Among the natives, all these insects are called *Cucujos* or *Cucujii*. **1874** *Science Record* 485 The brilliant radiance of the cocuyo is emitted from its ventral region. **1952** E. N. Harvey *Bioluminescence* xiii. 461 Almost every traveler to the Caribbean region has mentioned luminous elaterids, which Spaniards call 'cucujo', and the French, 'taupin'.

cud, *sb.* Add: **4. cud-chewer,** a ruminant animal.

1927 Haldane & Huxley *Anim. Biol.* iv. 112 In cud-chewers like the cow and sheep.

cud, *v.* (Later *poet.* example.)

1966 *New Statesman* 1 Apr. 473/1 Cows..Cudding, watching, and knowing.

cuddle, *sb.* Add: **2.** *Comb.* **cuddle seat** (see quot.); **cuddle skirt,** a skirt made of thick, soft material.

1947 *Britannica Bk. of Yr. 1946* 840/2 *Cuddle seat*, a contrivance for carrying small children, consisting of a seat hung from a strap slung over the shoulder. It was introduced by Australian war brides. **1958** J. Cannan *And be a Villain* i. 8 She had put on her new oatmeal 'cuddle' skirt. **1961** *Sunday Express* 5 Mar. 11 Wool and cashmere cuddle-skirt.

cuddleable (kɒ·d'lăb'l), *a. colloq.* [See -able.] = Cuddlesome *a.*

1928 E. Wallace *Again the Three Just Men* 32 The Lord has given you kissable lips and a cuddleable body. **1928** *Daily Express* 18 May 11/1, I do not want a brainless doll, but I would like a jolly, lovable, cuddleable woman.

cuddlesome, *a.* Delete *nonce-wd.* and add examples.

1893 *Ardrossan & Saltcoats Herald* 1 Sept. 3 The crowd of cuddlesome darlings in nice frocks. **1923** *Daily Mail* 27 Feb. 7 A rattlesnake has produced a family of thirty babies. There is nothing of cuddlesome, infant softness about these tenfold triplets. **1965** F. Raphael *Darling* xviii. 83 The soft toys were cuddlesome as anything. **1970** *Homes & Gardens* June 65/1 There were cuddlesome lion cubs and strokeable deer.

cuddly, *a.* Delete *nonce-wd.* and add to def.: Such as invites cuddling; = Cuddlesome *a.*

1915 Galsworthy *Freelands* xxxvi, She laid her face beside his on the pillow...It made everything seem cuddly and warm. **1922** D. H. Lawrence *England, my England* (1924) 55 He put his arm round her and drew her a little nearer to him, in a very warm and cuddly manner. **1923** *Daily Mail* 16 Mar. 14 Cuddly toys appeal to babies. **1927** Wodehouse *Small Bach.* i. § 1. 6 'Are you trying to convey the idea that she is short and stout?' 'Oh, no, sir, not stout. Just nice and plump. What I should describe as cuddly.' **1963** *Sunday Express* 3 Mar. 15/4 A child became seriously ill after sucking mothballs which can be used to stuff a cuddly toy.

cuddy². Add: **1. b.** *spec.* The captain's cabin.

1917 'Taffrail' *Sub* 109 Breakfast with the skipper was better than breakfast in the gunroom. There were no tinned salmon fishcakes and watery porridge in the 'cuddy'.

cuddy³. Add: **1. c.** A (small) horse. Chiefly *dial.* and *Austral.*

c **1930** in R. Ward *Penguin Bk. Austral. Ballads* (1964) 201 A..stockman..Apostrophized his bloody cuddy. **1944** [see *curl *v.*¹ 1 c]. **1945** Baker *Austral. Lang.* iii. 71 *Cuddy*, [is] used in North English dialect for a donkey. In Australia we use it for a small, solidly built horse. **1969** C. Geeson *Northumberland & Durham Word Bk.* 72 *Cuddy*, a name for an ass or a small horse.

cudgerie (kɒ·dʒĕri). Also **cugerie.** The native name in Australia of the trees *Hernandia bivalvis* (see quot. 1886) and *Flindersia schottiana*, a large rain-forest tree with light-coloured wood.

1884 A. Nilson *Timber Trees N.S.W.* 135 Cugerie. Flindersia australis. **1886** F. M. Bailey *Queensland Woods* 67 *Hernandia*, Linn. *H. Bivalvis*, Benth. *Flora Austr.*, 'Cudgerie' or Greasenut. A tall tree with a smooth bark... The kernel [of the fruit] contains 64·8 per cent. of oil. **1901** —— *Queensland Flora* iv. 1316 Cudgerie or Grease-nut... Involucel enclosing the fruit..divided nearly to the base into 2 valves.

cue, *sb.*¹ **2. b.** (Later U.S. example.)

1831 P. Wingate in B. Peirce *Hist. Harvard Univ.* (1833) 219 We were allowed at dinner a cue of beer, which was a half-pint.

cue, *sb.*² Add: **1. c.** *Cinemat.*, *Broadcasting.* A signal for action to begin or end (see quots.). Also *attrib.* Also *spec.*, a mark on a film serving as a signal or direction to a film editor or projectionist.

1932 *Techn. Descr. Broadcasting House* (*B.B.C.*) 76/1 In certain announcing rooms..a cue light is installed, so that the announcer may..give a cue to, say, the conductor of an orchestra to commence. **1940** *Chambers's Techn. Dict.* 214/2 *Cue* (*Cinematography*), an indication, visual or aural, for action or speech on the part of someone, during continuity. **1948** *Brit. Stand.* 1492 6 The motor cue shall consist of circular opaque marks with transparent outlines... The change-over cue shall consist of 4 frames. **1962** A. Nisbett *Technique Sound Studio* vi. 109 All BBC studios are equipped with green cue lights. *Ibid.* 110 For talks, etc., cue lights are also used to help regulate the timing and pace of the programmes.

2. c. A stimulus or signal to perception, articulation, or other physiological response.

1931 *Brit. Jrnl. Psychol.* Apr. 342 The sensory cue of the image on the retina. **1940** *Ibid.* Jan. 228 Tactile and labyrinthine 'cues' may come to be interpreted by the blind with the characteristic immediacy which we associate with visual apprehension. **1956** *Language* XXXII. 274 (*title*) Acoustic cues for nasal consonants. **1963** *Amer. Speech* XXXVIII. 73 Since listeners identify members of minimal pairs with the aid of contextual cues, stress placement has no functional role. **1966** *Listener* 6 Oct. 503/1 This business of olfactory cues, as they are called, has been recognized in mammals too..but..overlooked in man.

5. *cue list, sheet*; **cue-bid** *sb.* and *v. trans.* and *intr.*, **cue-bidder, -bidding** *Contract Bridge* (see quots.).

1932 G. G. J. Walshe *Contract Bridge* I. 29 When a suit has been agreed upon between the two partners bids in other suits show features... These bids (Cue bids) are especially valuable. **1963** *Times* 5 June 16/3 He..refused either to cue-bid or to raise either of my suits. **1932** G. G. J. Walshe *Contract Bridge* III. 106 The following hands.. will illustrate the methods of the Cue-bidder. *Ibid.*, The original bidder should not..initiate Cue-bidding without control of the first lead of three suits. **1933** E. Culbertson *Contract Bridge Blue Bk.* xviii. 269 The 'cue-bidding' methods which concentrate mainly on showing Aces. **1927** *Melody Maker* Sept. 937/3 The present campaign for a better cue list for the provincial or smaller orchestra. *Ibid.* 937/1 The bogey of the Cue Sheet or Suggestion List. **1959** W. S. Sharps *Dict. Cinemat.* 88/1 *Cue sheet*, the complete schedule of cues for a film production. **1962** A. Nisbett *Technique Sound Studio* vi. 106 A constant check is kept on timing, using either a script or for unscripted programmes, a timed cue-sheet.

cue, *sb.*³ **5.** *cue-ball* (examples).

1881 H. W. Collender *Mod. Billiards* I. 36 The cue-ball is that with which the play is made. **1935** *Encycl. Sports* 82/2 *Cue ball*, ball belonging to the person who is at the table.

cue (kiū), *sb.*⁴ Colloq. abbrev. of Cucumber.

1935 *Daily Tel.* 7 June 21/3 'Toms' and 'Cues'. Home-grown tomatoes and cucumbers are in splendid condition. **1963** *Ibid.* 17 Dec. 10/2 (*heading*) 'Tommies and Cues'.

cue (kiū), *v.*² [f. Cue *sb.*²] *trans.* **a.** To provide or furnish with a cue. Also const. *in*, and *fig.* So **cued** *ppl. a.*

1928 *Melody Maker* Feb. 197/2 The 1st alto had melody cued-in. **1937** *Printers' Ink Monthly* Apr. 50/3 *Cue someone*, to give a signal indicating 'proceed with the pre-arranged routine'. **1957** R. Maney *Fanfare* vii. 95 Stimulated by the martinis, and cued by questions, I volunteered a lot of suggestions. **1959** *Times* 19 Nov. 16/5 He knew the music in the sense that he was able to dispense with the scores and cue his players from memory. **1961** *Listener* 20 Apr. 683/2 Not being cued by the State Department's press officers that this was indeed mighty stuff. **1962** *Ibid.* 4 Oct. 537/3 This scaffolding..served its purpose which was to introduce Esterhazy and to cue in his *apologia pro vita sua.* **1962** *Amer. Speech* XXXVII. 227 Velarized or emphatic stops are cued by frequency lowering of second formant. **1964** T. Rattigan *Heart to Heart* in *Coll. Plays* III. 426 Super captions. Cue announcer. *Ibid.*, Take out caption. Cue David. **1966** S. Jackman *Davidson Affair* ii. 17, I..saw the floor-manager's hand drop to cue me in, and turned to face the camera. **1970** *Radio Times* 30 Apr. 10/2, I want to explain in the programme that we've just heard from our Washington correspondent, and then cue you in live.

b. *spec.* To make an indicatory mark on (a film negative or a recording) (see *cue *sb.*² 1 c).

1938 G. H. Sewell *Amat. Film-making* iv. 47 The amateur is advised to..'cue' the negative. **1958** *N.Z. Listener* 26 Sept. 9/1 Discs are easier to cue and edit if the programme is complicated and needed in a hurry.

‖**cueca** (kwe·ka). [Amer. Sp., f. *zamacueca*, a S. Amer. Indian song and dance.] A South American dance.

1912 J. Masefield in *English Rev.* Oct. 383 And girls in black mantillas fit to make a Poor seaman frantic when they dance the cueca. **1917** E. Hague *Spanish-Amer. Folk-Songs* 103 (*title*) Cueca or Zamacueca (from Chile). **1928** *Daily Express* 14 Sept. 4 The cueca, like the tango, comes from South America. It is the national dance of Chili. **1960** *Times* 22 Jan. 13/7 He himself danced the *cueca* in Buenos Aires.

‖**cuenca** (kwe·ŋkă). [Sp., bowl, socket.] Used *attrib.* of the decoration of tiles: ornamented with sunken patterns surrounded by a raised outline.

[**1911** *Encycl. Brit.* XXVI. 973/1 About 1550..there arose the method *de cuenca* in which the parts of the design to receive different coloured enamels were stamped, slightly concave.., their edges alone being left in relief. **1925** B. Rackham in *Hannover's Pott. & Porc.* I. 550 *De cuenca* ('cell') tiles, the term..was invented as a designation by J. Gestoso y Perez, author of *Historia de los barros vidriados Sevillanos*, 1903.] **1939** A. Lane *Guide to Tiles* vii. 59 Patterns were sunk in the soft surface of the tile..in this.. *cuenca* (hollow) technique. **1960** R. G. Haggar *Conc. Encycl. Contin. Pott.* 477/1 Ceiling tiles were extensively made in Spain from the sixteenth century with coloured glazes in the *cuenca* technique.

‖**cuerda seca** (kwę̄rda seka). [Sp., lit. 'dry cord'.] (See quot. 1960.)

1911 *Encycl. Brit.* XXVI. 973/1 The tediousness of the process gave rise, about 1450, to what is known as the *cuerda seca* (or 'dry cord') method, in which narrow fillets at the edges of the separating interlacings were first stamped upon the tile itself and filled with clay and manganese; these being fired (thus forming a 'dry cord' or line) formed shallow compartments which were in turn filled with coloured enamel. **1939** *Burlington Mag.* Oct. 156/1 Tile-work glazed in the so-called '*cuerda seca* technique'. **1960** R. G. Haggar *Conc. Encycl. Contin. Pott.* 121/2 *Cuerda seca*, 'dry cord': a technique employed by Spanish tile makers in which lines were drawn upon the tile in a purple pigment mixed with a greasy substance which effectually separated the colours filled in on either sides of them, and disappeared during the firing process.

cuesta (kwe·sta). [Sp. *cuesta* slope:—L. *costa* (see Coast *sb.*).] A gentle slope or inclined plain, esp. one that ends in a steep drop; a hill or ridge with one face steep and the opposite side gently sloping. Orig. local *U.S.*; adopted in the second sense as a term in *Physical Geogr.* (see quots.).

1818 in *Amer. State Papers* (*For. Relat.*) (1834) IV. 298 A high ridge or mountain surrounds them all; and a cuesta..more or less rugged and precipitous. **1854** W. L. Herndon *Amazon* I. 96 The road..ascends a steep and rugged '*cuesta*'. **1896** *Nat. Geogr. Mag.* VII. 294 The plains belong to four great topographic categories, which in the rich Spanish nomenclature of the region may be termed *mesas*, *bolsons*, *plazas*, and *cuestas* (including *bajadas*)... *Cuestas* and *bajadas* are inclined plains, which can also be classed as declivities. **1899** W. M. Davis in *Proc. Geol. Assoc. Lond.* (1900) XVI. 76 The outer slope is so gentle that its inclination is hardly noticeable. Such an upland may be called a 'cuesta'. *Ibid.* 77 Finding no name in use for the forms here considered, I have..advocated the general adoption of the term cuesta...By the same natural extension of the original meaning that makes mesa apply to the whole of a tabular elevation, instead of only to its upper surface, cuesta may be made to apply to the entire body of the unsymmetrical linear elevation that is characteristic of certain denuded coastal plains. 1901 Jrnl. School Geogr. Oct. 295 This is just at the cuesta-like escarpment. 1939 Geogr. Jrnl. XCIV. 414 The Chiltern Hills are a simple chalk cuesta. 1941 C. A. Cotton Landscape Developed by Erosion x. 94 Homoclinal ridges grade into cuestas, which are developed on escarpment-forming strata of very gentle inclination. 1963 D. W. & E. E. Humphries tr. Termier's Erosion & Sedimentation iii. 66 It is sometimes difficult to distinguish these cliffs from nonmarine cuestas. 1970 R. J. Small Study of Landforms iii. 75 The most important landforms of such scarp-and-vale scenery are, of course, the cuestas themselves.

cuff, sb.¹ Add: **2. c.** Colloq. phrases: off the cuff (as if from notes made on the shirt-cuff) orig. U.S., extempore, on the spur of the moment, unrehearsed; also attrib. (with hyphens); on the cuff, (a) orig. U.S., on credit; (b) N.Z., beyond what is appropriate or conventional; excessive (phr. a bit on the cuff perh. infl. by rhyming collocation a bit rough); to shoot one's cuffs, see *SHOOT v.

1938 New York Panorama (Federal Writers' Project, N.Y.) vi. 157 Double talk is created by mixing plausible-sounding gibberish into ordinary conversation, the speaker keeping a straight face or dead pan and enumerating casually or off the cuff. 1941 Time (Air Exp. Ed.) 4 Aug. 1/1 Talking off the cuff to a group of civilian-defense volunteers he made them a little homily. 1944 Penguin New Writing XX. 130 In that scene, shot off the cuff in a shockingly bad light, there leapt out of the screen..something of the real human guts and dignity. 1948 Economist 3 July 17/2 Mr Truman's off-the-cuff comment. 1960 News Chron. 6 July 7/7 He was infuriated by Mr. Macmillan's refusal to give off-the-cuff answers.

1927 K. Nicholson Barker 149 On the cuff, a charged account. 1938 F. D. Sharpe Sharpe of Flying Squad 332 On the cuff, payment deferred. 1945 B. Macdonald Egg & I (1947) xi. 135 Money was not important at all. All business was transacted on the cuff.

1942 in Webber & Colvin Johnny Enzed in Middle East (1946) 8 A bit on the cuff, that sort of thing. 1944 J. H. Fullarton Troop Target xi. 85 That's a bit on the cuff, Dig.

d. The turn-up on a trouser leg. Chiefly U.S.

1911 T. Eaton & Co. Catal. Spring & Summer 117 Trousers have belt loops, cuff bottoms and full width. 1917 Ibid. Fall & Winter 369 Trousers have five pockets, belt loops and finished with cuff. 1931 W. Faulkner Sanctuary vi. 52 Scraping at his trouser-cuffs. 1947 Book Nine (Caxton Press, N.Z.) 23 He tapped [the cigarette] ash into his trouser cuff. 1968 Observer 10 Mar. 25/5 A technique which guarantees there won't be glass fragments (identifiable by spectrography) in the cuffs of the thief's trousers. 1969 Catal. J. C. Penney Fall & Winter 561 Slacks...Rugged corduroy fabric. Belt loops and cuffs.

4. cuff-edge, -link(s).

1922 Joyce Ulysses 7 Across the threadbare cuffedge he saw the sea hailed as a great sweet mother. 1897 Sears, Roebuck Catal. 425 Solid gold Cuff Links, plain polished and raised ornamentation. 1915 'Bartimeus' Tall Ship iv. 75 This liberal display of fine linen and flashing cuff-links. 1970 A. Cameron et al. Computers & O.E. Concordances 39 We had slugs made for print chains and finally made cuff links out of the slugs. 1971 N.Y. Times 21 Feb. 40 (Advt.), Boutique cufflink collection in swivel lucite top cases.

Cuff, sb.⁵ U.S. Abbrev. of *CUFFEE.

1755 J. Hempstead Diary (1901) 656 An Indian free-woman wife to Mr. Tilley's Negro Cuff died. 1814 H. M. Brackenridge Views of Louisiana 211 They chased a she bear into a hollow tree... The chopping was renewed; madam Cuff again appeared, and was saluted as before. 1855 Whitman Leaves of Grass (1860) 29 Growing among black folks as among white, Kanuck, Tuckahoe, Congressman, Cuff.

cuff, v.¹ Add: **4.** trans. To discuss, talk over (a tale, matter); also, to tell (a tale). dial.

1746 P. Lock Exmoor Scolding (1879) 56 Oll vor..cuffing a tale. 1854 A. E. Baker Northants Words I. 165 The personal appearance and behaviour of Miss H— was cuffed over at the ball. 1867 W. F. Rock Jim an' Nell (1896) cx, Let's cuff another tale. 1891 R. P. Chope Dial. Hartland 39 Cuff over, to talk over, discuss. 'Let's ha' a pipe an' cuff it auver.'

cuffed, a. (Further examples.)

1951 V. Nabokov Speak, Memory xv. 234 Cuffed hands of wood nailed to boles in the old parks of curative resorts pointed in the direction whence came a subdued thumping of bandstand music. 1960 T. Hughes Lupercal 36 Only a plump, cuffed citizen Gets enough quiet to hear God speak.

Cuffee, Cuffy (kʌ·fi). U.S. colloq. Also with lower-case initial. [A personal name formerly common among Negroes.] **a.** A Negro; also used as a generic name. **b.** A black bear.

1713 S. Sewall Diary (1879) II. 386, I press'd him, and came away with some hope; obliged Cuffee to call for him. 1824 J. Doddridge Notes 21 When the bear approached him, he sprang out and hallooed at him; but cuffee..jumped at him with mouth wide open. 1837 Southern Lit. Messenger III. 86 The song ceased, and the cuffee advanced in silence. 1844 'J. Slick' High Life N.Y. I. 74 Jest as I was a thinking this, the cuffy come into the room. 1862 'Artemus Ward' His Bk. (1865) 61 Praps I'm bearin down too hard upon Cuffy. 1959 A. Salkey Quality of Violence iv. 65 A friend to Boss-man and cuffee.

cuffer². dial. or slang. Also cuffa. [f. *CUFF v.¹ 4 + -ER¹.] A yarn or story.

1887 J. Farrell How He Died 65 You made me start to pitch you this most interesting cuffer. 1895 P. H. Emerson Marsh Leaves 156 He'll spin up a rare cuffa along with old Jenks. 1899 F. T. Bullen Idylls Sea xxv. 219 The time-honoured 'cuffer' or yarn was going its soothing round. 1923 Blackw. Mag. May 661/1 There's plenty of cuffers, as they're called, about mermaids, phantom ships, dripping corpses, and such like.

cuffless, a. Add: **2.** Of trousers: without turn-ups. Chiefly U.S. (Cf. *CUFF sb.¹ 2 d.)

1957 J. Frame Owls do Cry xxxvii. 167 An attendant with black suit and cuffless trousers. 1958 B. Malamud Magic Barrel (1960) 76 A green suit with cuffless trousers.

cufuffle, var. CURFUFFLE.

cuirass(e), sb. Add: **4. c.** In full cuirasse band. A band made of linen pressed in layers to protect a cycle tyre.

1906 Daily Chron. 28 Nov. 9/2 The winter and tropical tyre..consists of a smooth vulcanised cover, with cuirasse band put on top by hand. 1907 Ibid. 12 Oct. 9/4 The Paris cuirasse band...Unlike the Sphinx, the cuirasse becomes an integral part of the tyre, being attached inside the cover.

5. In full cuirass respirator. An apparatus covering the chest and providing artificial respiration.

1939 Rep. Med. Res. Council Breathing Machines 16 The Burstall jacket respirator (1938)..consists of a one-piece aluminium cuirass to enclose the trunk..from the neck to the waist-line...The London County Council cuirass respirator (1938) has been designed..to overcome many of the disadvantages of the original Burstall cuirass. 1955 Times 11 May 6/3 A cuirass respirator..which allows the 'iron lung' to be dispensed with after the first few weeks of poliomyelitis. The patient has full use of his limbs and the breathing cycle is maintained by a small pump standing by the bed. 1959 Times 25 Sept. 8/5 Later demonstrations included the working of the R.A.F. portable version of an 'iron lung'—the Monaghan cuirass—which is smaller and lighter and allows more mobility than the ground equipment.

‖**cuir ciselé** (kwir siz'le). [Fr., chiselled leather.] Used attrib. of a form of decoration on book-bindings in which a design is cut into the leather by means of a pointed tool.

1933 Library Mar. 337 (title) Some cuir-ciselé bookbindings in English libraries. 1938 Times Lit. Suppl. 12 Nov. 732/2 One [blind-stamped binding] is from southern Spain, dating from about 1480, with a cuir ciselé design.

Cuisenaire rod (kwizĕnē·ɹ). Also Cuisenaire's rod. One of a set of wooden rods of different length and colour according to the number they represent, invented by the Belgian educationalist Georges Cuisenaire as an aid in the teaching of arithmetic to children.

1954 Cuisenaire & Gattegno Numbers in Colour p. vi, For the child engaged in work with Cuisenaire's rods, the monotonous and uncertain feeling that is associated with counting is transformed into an exciting, intense experience of the active intellectual mind. 1962 Listener 5 Apr. 593/2 Dr Dienes's structural games are widely used, as are the Cuisenaire rods and the original apparatus devised by Stern. 1965 in P. Jennings Living Village (1968) 208 Children are introduced to number work by using Cuisenaire Rods. Each child has a box of ten compartments with wooden rods of different colours and lengths. Each colour stands for a number of units, and the rods increase in length from one unit to ten units.

‖**cuisine bourgeoise** (kwizin burʒwaz). [Fr.] Plain home cooking; chiefly applied to French food. Also fig.

1951 E. David French Country Cooking 170 Duck cooked with turnips is one of the classic dishes of the French Cuisine Bourgeoise. 1965 Economist 3 July 13/2 The immediate result of M. Defferre's crockery coming to bits in his hands is to leave General de Gaulle more than ever in charge of France's cuisine bourgeoise. 1969 Guardian 30 Aug. 3/2 French cooking has two branches, haute cuisine and the cuisine bourgeoise.

cuivré (kwi·vre), a. Mus. [Fr., pa. pple. of cuivrer to play with a brassy tone, f. cuivre, pop.L. *copreum, sb. use of neuter of L. cupreus of copper.] Brassy; used as a direction to play a brass instrument with a harsh, blaring timbre. Also absol.

[1927 Grove's Dict. Mus. (ed. 3) I. 767/1 Cuivré, the French term for indicating stopped notes on the horn.] 1931 G. Jacob Orchestral Technique iii. 32 The horn..is also capable of the most savage attack when roused, the stridency of which is enhanced (though the power actually reduced) by 'stopping' the bell with the hand or a metal mute and blowing hard. The former effect, known as the 'cuivré' is indicated by a cross. 1938 Oxf. Compan. Mus. 243/2 Cuivré...The term is found occasionally in music for brass instruments, especially the horn.

cuke (kiūk). Colloq. abbrev. of CUCUMBER.

1903 Postcard in Merriam-Webster files 1 Aug., Received to-day and sold. 3 Boxes Cukes $2.12½. M. Homburger. 1949 O. Nash Versus 96 Who coined these words that strike me numb?..The cuke, the glad, the lope, the mum. 1960 J. Symons Prog. Crime xiv. 88 You know what he's like, cool as a cuke.

Culbertson (kʌ·lbəɹtsən). The name of Ely Culbertson (d. 1955), an American authority on contract bridge, used chiefly attrib. and ellipt. of a system of bidding at contract bridge.

1929 Bridge World I. 40, I have been using what is known as the Culbertson System of bidding. 1931 A. Woollcott Let. 22 July (1946) iv. 79, I was gone only three months, but in that time..everybody had taken up Culbertson. 1932 Times Lit. Suppl. 12 May 349/3 Bidding based on the Culbertson principles. 1959 Listener 2 July 38/1 The Culbertson Four-Five No Trump convention. Ibid. 38/2 Culbertson bidders.

culching, cultching (kʌ·ltʃiŋ). vbl. sb. [f. CULCH, CULTCH.] The practice of strewing an oyster-bed with culch. Also attrib.

1894 Westm. Gaz. 3 Apr. 5/3 A Burnham cultching boat. 1904 Nature 17 Mar. 466/1 The process known as 'culching', that is, scattering the floor of the bed with rock, loose coral, and so on, to afford the necessary anchorage for the byssus of the young oyster.

cul-de-sac (s.v. CUL in Dict.). Add: **2.** (Earlier example.) Also fig.

1800 A. Paget Let. 10 May in Paget Papers (1896) I. 201 This [i.e. Palermo] is such a cul de sac that it would (be) ridiculous to attempt sending you any news. 1885 A. Dobson At Sign of Lyre 30 You tried the cul-de-sac of Thought; The montagne Russe of Pleasure. 1955 Sci. Amer. July 96/1 It has enlisted great talent in what appears to be the 'most hopeless cul-de-sac in the novel's history'. 1955 Times 23 July 6/3 The Constitution would then be a sham and a cul de sac, not a bridge to self-government later. 1968 Listener 25 July 105/1 My concentration then upon something which was past, something that could never be recovered, was a sort of cul-de-sac.

culicicide (kiū·lisisaid). Also culicide (kiū·lisaid). [f. L. culex, culicis gnat: see -CIDE 1.] An insecticide for destroying gnats and mosquitoes; also cu:licici·dal, culici·dal adjs.

1899 Brit. Med. Jrnl. II. 683/1 The authors made the following classification of culicidal substances: 1. Substances which kill the eggs. [Etc.] 1900 C. Christy Mosquitos & Malaria vii. 47 Culicicidal Substances...The substances which kill the mosquito itself may be divided into odours, fumes, and gases. 1900 J. J. Eyre tr. Celli's Malaria II. 205 The most efficient culicide is tobacco smoke. 1901 Practitioner Mar. 263 By fumigating the rooms occasionally with some such culicicide as the dried flowers of the chrysanthemum. 1949 M. F. Boyd Malariology II. lv. 1203 Carbolic acid crystals and gum camphor, equal parts, were melted together and volatilized by heat. This mixture has been referred to as Mimm's culicide.

culicifuge (kiū·lisifiūdʒ). [f. as prec. + -FUGE.] A substance applied to the body or to clothing in order to keep gnats and mosquitoes away; so cu:licifu·gal a.

1894 Gould Dict. Med. 343/2 Culicifuge, an agent that prevents the biting of mosquitoes; as oil of pennyroyal. 1900 J. J. Eyre tr. Celli's Malaria II. 212 A person can use culicifugal odours or scents on his body and clothing. 1946 P. F. Russell et al. Pract. Malariology xxv. 518 Many chemicals have been recommended as culicifuges to be applied to the skin to repel mosquitoes from biting.

culicine (kiū·lisəin, -in), sb. and a. Ent. [ad. mod.L. Culicini or Culicinæ, f. culex, culicis gnat: see -INE¹.] A mosquito belonging to the tribe Culicini of the sub-family Culicinæ; of or pertaining to a member of this group.

1911 A. Alcock Ent. Medical Officers iii. 61 Theobald, making the Culicines a separate family, has divided them into ten groups. 1921 G. H. Carpenter Insect Transf. 198 Less abundant in these countries than the Culicine gnats are the Anophelines. 1923 H. M. Lefroy Ent. 419 The Anophelines hang parallel to the surface..but the Culicines hang at an angle. 1929 R. Matheson Mosquitoes N. Amer. 36 The species of Culicine mosquitoes have widely varying larval habits. 1964 H. Oldroyd Nat. Hist. Flies vii. 78 Any reference to 'mosquitoes' in what follows may be taken as applying to both tribes, unless it is qualified by the adjectives anopheline or culicine.

cull, sb.³ Add: **1.** Delete † and add later examples.

1958 Times 20 May 4/3 An annual cull should be carried out..to limit further increases in the grey seal population. 1968 Times Lit. Suppl. 30 May 559/4 A list of the words and phrases I'd found, which may be of interest as showing an average daily cull from an intelligent newspaper.

2. Also, a bird drafted as inferior, and fig.

1919 H. L. Wilson Ma Pettengill viii. 253 It made him feel like a social cull or an outcast, or something. 1950 N.Z. Jrnl. Agric. Oct. 359/1 Many culls are unhealthy or diseased and are a potential danger to all other birds with which they come into contact.

3. a. (Earlier and later examples.) Also cull lumber.

1829 J. MacTaggart Three Yrs. Canada I. 245 The refuse wood is called culls, and brings an inferior price. 1867 Trans. Ill. Agric. Soc. 1865–66 VI. 647 Culls are a quality manufactured from winding, worm-eaten, shaky or dry-rot timber, badly manufactured, or less than sixteen (16) inches in length. 1897 F. C. Moore How to Build ii. 23 The 'cull' lumber should be put in the closets, storerooms, and upper or attic rooms. 1953 Brit. Commonw. Forest Terminol. I. 35 Cull, (a) an inferior plant rejected from nursery stock, (b) trees or logs that are of merchantable size but are rendered unmerchantable by defects. 1969 L. G. Sorden Lumberjack Lingo 30 Cull, rejected logs having little or no value.

c. Fruit rejected as being of inferior quality. (See also E.D.D.) Also attrib.

1937 Nature 7 Aug. 222/1 The utilization of farm wastes and by-products is being investigated; already cull citrus fruits in California are being processed for producing citric

acid, citrus oils and pectin. 1951 New Biol. X. 56 Malformed and scarred apples were discarded, and it became a matter of moment to reduce the proportion of 'culls' or waste. 1962 Times 31 Mar. 9/7 Some cull apples go, at a low price, into cider manufacture.

cull, v.¹ Add: **4.** To pick out (livestock, etc.) according to their quality. Also absol.

The earliest examples are Austral. and N.Z. but the word is now widely used in Britain and elsewhere.

1889 Williams & Reeves Colonial Couplets 9 I'd far sooner choose To be writing to you, than be culling the ewes. 1927 M. M. Bennett Christison xii. 125 Christison used to cull on clearly defined lines. At first coarse calves were culled. 1929 'M. B. Eldershaw' House is Built i. 10 There were a few lean, dejected cattle, the best of them having been culled out hours before [by buyers]. 1950 N.Z. Jrnl. Agric. Apr. 387/3 All ewes on this farm, which are crossbred sheep, are culled for quality and not for age. 1968 J. Gordon Beagle Guide 170 Cull, to eliminate unwanted hounds. 1969 Listener 27 Mar. 439/3 The battery boys 'cull' (or liquidate) their hens when they've laid for about a year. 1970 Kenya Farmer Feb. 15/2 A bigger cow..will..fetch a higher price at culling.

5. Forestry. (See quots.) N. Amer.

1904 S. E. White Blazed Trail Stories 49 A log is culled, or thrown out, when..it will not make good timber. 1905 Forestry Bureau Bull. (U.S.) No. 61, 9 Cull, to take out of a forest by selection a portion of the trees. 1953 Brit. Commonw. Forest Terminol. I. 35 To cull, (a) to reject inferior plants from nursery stock, (b) to deduct the effective portion of a merchantable log or piece of timber from the gross volume in scaling or measuring timber. (Canada.) 1953 H. L. Edlin Forester's Handbk. iv. 59 Once raised, the stocks should be culled—that is, any obviously stunted, mis-shapen or diseased specimens should be thrown aside.

culler. Add: **1. b.** (See quots.)

1849 Rep. Comm. Patents: Agric. (U.S.) 322 When the tobacco is taken down, the 'cullers' take each plant and pull off the defective and trashy ground and worm-eaten leaves. 1881 E. Ingersoll Oyster-Industry 243 Culler, one who picks over oysters, or culls out the worthless and smaller ones. 1953 Times Lit. Suppl. 20 Mar. 194/5 Mr. Thomson describes his every-day duties and experiences as a 'culler' employed by the Government of New Zealand to shoot destructive deer. 1969 L. G. Sorden Lumberjack Lingo 30 Culler, man who quickly graded and sorted, by picking out the culls or rejects, lumber being cut at the mill.

c. local. (See quot.)

1906 J. Hockaday in Vict. Hist. Cornwall I. 521/2 Many comparatively rough blocks [of slate-rock] are refused by the contract men, and these are passed on to men and boys called cullers, who are paid a fixed price, and make as much out of them as they can.

cullible, a. (Earlier example.)

1811 Shelley Let. 12 Jan. (1964) I. 45 Because men are & have been cullible, I see no reason why they shd. always continue so.

culling, vbl. sb.¹ **3.** (Further examples.)

1938 Times 14 Feb. 20/1 Losses incurred through death and culling are commonly in the region of 30 per cent. 1958 Times 1 Nov. 9/4 Litters are bred on more sparing lines than formerly, which means that there are fewer surplus hounds, and the job of 'culling' or drafting is all the more difficult. 1964 C. Willock Enormous Zoo vi. 102 The experimental shooting of hippo—culling is the polite conservation term for it—had begun.

cullum, var. *COOLUNG.

culm⁴, var. COMB sb.²

1940 in Chambers's Techn. Dict. 215/1. 1953 Word for Word (Whitbread & Co.) 17/1 Culms (or Coombes), the rootlets which are sieved from the malt at the end of the malting process; they are used for poultry and cattle food.

culmen. 1. Delete † Obs. and add later example.

1928 C. T. Onions in Times 19 Apr. 10/7 The Oxford English Dictionary is the culmen [of a series of lexicons].

culminate, v. **4.** Delete rare and add later examples.

1896 Earl of Rosebery in Westm. Gaz. 12 Sept. 5/1 This brings to a head and culminates all the nameless massacres in Asia Minor. 1904 Illustrated Bee (Omaha) 25 Sept., A romance extending over several years was culminated. 1927 Daily Express 25 Apr. 1/5 They decided that an immediate ceremony would culminate their childhood romance.

culmination. Add: **3.** The raising of the level of the land on either side of a river by allowing flood-water to deposit silt on it. [Cf. It. colmare vb.]

1838 F. Maceroni Mem. II. 62 The process of culmination is particularly successful if practised high up a river much liable to winter floods.

4. Geol. **a.** (Also culmination of pitch.) A part of a fold, esp. a nappe, where the strata were at their highest before they were eroded. **b.** An axis of a system of folds joining the highest parts of successive folds.

1927 L. W. Collet Struct. Alps II. i. 27 Windows generally originate..on culminations of pitch. Ibid., The windows of the Lower Engadine and of the Tauern are due to culminations of the substratum. 1942 M. P. Billings Struct. Geol. iii. 49 Culminations and depressions trend essentially at right angles to the trend of the folds; the folds plunge away from culminations toward depressions. 1944 A. Holmes Princ. Phys. Geol. xviii. 392 The nappes..are found to undulate up and down in an alternating succession of broad culminations and depressions. 1965 Ibid. (ed. 2) xxx. 1157 (caption) Mt Blanc (15,782 feet) the highest culmination of the Hercynian massifs. 1966 McGraw-Hill Encycl. Sci. & Technol. XIII. 408/1 In the culminations, where the higher strata have been removed by erosion, the structure of the lower strata may be seen in the deep valleys.

culotte (kiulo·t, ‖külot). [Fr., = knee-breeches; cf. SANSCULOTTE.]

1. Knee-breeches; also culotte courte. (Rare in Eng. use.)

1842 Barham Ingol. Leg. II. 11 Ripping the lace from his coat, And from what, I suppose, I must call his culotte. 1848 Thackeray Van. Fair xlv. 408 She said that it was only the thorough-bred gentleman that could wear the Court suit with advantage: it was only your men of ancient race whom the culotte courte became. 1877 Encycl. Brit. VI. 472/2 Towards the beginning of the 17th century, the high hose..were transformed into 'culottes', which were full and open at the knees. 1952 Granta 29 Nov. 6/1 We may hire the appropriate wigs and culottes from London costumiers.

2. (Usually in pl.) A divided skirt. Also attrib.

1911 in C. W. Cunnington Eng. Women's Clothing (1952) iii. 105 Satin culottes, 10/6. 1927 Delineator Mar. 15 The culotte skirt, which originally made its début as a novelty, is translated this spring into practical fashions. 1939 A. Keith Land below Wind ix. 148, I was still fastening my culottes, into which I had hastily jumped. 1939 M. B. Picken Lang. Fashion 40/3 Culotte, informal trouser-like garment having leg portions that are full and fall together to simulate a skirt. Worn as sports skirt. 1960 Harper's Bazaar July 55 A daring revival from the thirties—the culotte..wide pyjama-legged culottes. 1966 Vogue Dec. 87 Beautiful culotte dress. 1967 Guardian 5 Aug. 5/1 Culotte skirts..and culotte dresses abound.

3. The soft hair or feather on the back of the forelegs of a dog.

1928 in Funk's Stand. Dict. 1948 C. L. B. Hubbard Dogs in Britain 462 Culotte, the fringes of soft hair on the backs of the fore-legs of Pomeranians. 1968 H. Harmar Chihuahua Guide 235 Culotte, the feathery tail on the back of the forelegs of the Pekingese, Pomeranian and Schipperke.

cult, sb. Add: **2. b.** Now freq. used attrib. by writers on cultic ritual and the archæology of primitive cults.

1901 A. J. Evans Mycen. Tree & Pillar Cult 25 Aniconic Cult Images. Ibid. 77 Cult Scenes relating to a Warrior God and his Consort. 1903 Folk-lore Sept. 264 The image of the patron deity, usually a simple copy of the cult statue. Ibid. 269 Inscriptions found at various cult-centres. 1904 Hastings's Dict. Bible V. 118/1 The female Divinity must be represented by the female animal, in order to carry out the mythological tale or the cult-act. 1906 D. G. Hogarth in Proc. Brit. Acad. 1905–6 375 Small objects dedicated in that temple, among which are several cult-figurines of the Goddess. 1928 Peake & Fleure Steppe & Sown 104 Already in Early Minoan times the double axe had become, not only a symbol of authority, but a cult object. a 1930 D. H. Lawrence Apocalypse (1931) vii. 117 Cult-lore was the wisdom of the old races. 1950 H. L. Lorimer Homer & Monum. vi. 349 The earliest cult-image of the goddess. 1950 Scott. Jrnl. Theol. III. 368 The rôle of the king in the great cult-drama at the beginning of every new year. 1957 Antiquity & Survival II. 167/1 Near it a cult mask, made of clay, was still lying on the floor...In a further room, we discovered a unique cult-standard..made of bronze, with a tang to fasten it to a pole.

cultching: see *CULCHING vbl. sb.

cultellus (kʌlte·lŏs). Ent. Pl. cultelli (-əi). [L., dim. of culter knife.] Each of the lancet-like mandibles of blood-sucking Diptera.

1899 D. Sharp in Cambr. Nat. Hist. VI. vii. 443 Cultelli (mandibles of other anatomists). 1937 J. R. de la Torre-Bueno Gloss. Ent. 70 Cultellus (pl., cultelli), one of the blade-like lancets in piercing flies; the mandibles of some authors.

cultic (kʌ·ltik), a. [f. CULT sb. + -IC, perh. after G. kultisch.] Of or pertaining to a religious cult.

1898 Peritz in Jrnl. Biblical Lit. XVII. 117 Whether as divinity, devotee, or cultic official, woman shares cultic duties with man. 1925 G. B. Gray Sacrifice in O.T. 193 Though of course women reckoned their descent from Levi, they did not exercise the special cultic Levitical service. 1925 J. E. McFadyen in A. S. Peake People & Bk. 216 Gunkel maintains that, though they [sc. the psalms] originated in poetry composed for the cult, most of them no longer presuppose any cultic action. 1965 C. L. Wrenn in Bessinger & Creed Medieval & Linguistic Stud. 42 The runes must be explored in the light of evidences..that may have cultic significance.

Hence **cu·ltically** adv., in a cultic manner.

1953 Scott. Jrnl. Theol. VI. 343 By their Christological formulas they turn [Christ] into an idol to be cultically worshipped.

cultish (kʌ·ltiʃ), a. [f. CULT sb. + -ISH; cf. G. kultisch.] Of, pertaining to, or resembling a cult, esp. one regarded as eccentric or unorthodox. Hence **cu·ltishness.**

1926 Scribner's Feb. 2 He takes the 'cultish' and highbrow qualities out. 1948 K. Davis Human Soc. (1959) xx. 576 Chiropractic and naturopathy, offshoots of osteopathy, though much more backward and cultish. 1948 Time 16 Feb. 72 The Foursquare Church is doing its best to evolve from cultishness. 1959 Guardian 16 Oct. 10/3 Many will sympathise with Dr. Fromm's outspoken criticisms of the cultishness of the psychoanalytic movement. 1965 W. Lamb Posture & Gesture viii. 106 They all had their development in Central Europe during the 1920s and 30s and are liable now to look cultish and dated.

cultism. Add: **2.** [This sense owes nothing to the Sp. word.] The spirit, system, or practice of a cult or a cultic activity. So **cu·ltist**, a person practising cultism; **culti·stic** a.

1917 C. R. Payne tr. Pfister's Psych. Method II. xvi. 453 For the evangelical pastor, the aim in question is not a cultistic servitude for the confessional as a means of supernatural grace, but an ethical..purification purpose. 1933 Fortune Aug. 47/3 But the cultist will often go to preposterous lengths. 1934 Webster, Cultism. 1949 R. K. Merton Social Theory & Struct. (1951) xii. 314 Cultism, informal cliques..these and other techniques may be used for self-aggrandizement. 1955 Keepnews & Grauer Pict. Hist. Jazz xix. 243 Bop did have its share of dubious quirks and of inferior musicians and cultism. 1957 P. Worsley Trumpet shall Sound x. 214 The Kukuaik movement on Karkar Island began as an orthodox independent Christian body and later turned towards Cargo cultism. 1962 Sunday Express 25 Feb. 13/2 The police were sent..after 68 members of the super-race cult...The cultists massed on a beach. 1970 Daily Tel. 23 Oct. 4 A deranged group of cultists calling itself the 'People of the Free Universe' and threatening death to anyone 'misusing the natural environment'.

cultivar (kʌ·ltivāɹ). Hort. [f. CULTIV(ATED ppl. a. + VAR(IETY 6 b.] A variety that has arisen in cultivation. Also attrib.

1923 L. H. Bailey in Gentes Herbarum I. III. 113, I now propose another name, cultivar, for a botanical variety, or for a race subordinate to species, that has originated and persisted under cultivation. 1953 Rep. 13th Internat. Hort. Congress 1952 50 The Code provides..that the term 'cultivar' (abbreviated as cv.) be applied to those special forms which have originated or are maintained only in cultivation. Ibid. 51 When revival of the earliest published cultivar-name would cause confusion, it is to be listed as a synonym. 1961 New Statesman 19 May 808/2 What happened to tulips, roses, daffodils, is now happening to lilium. Relatively few species are brought together from remote places, crossed and recrossed, and give rise to thousands of cultivars. 1970 R. Gorer Devel. Garden Flowers 19, I have also followed the practice of using the term cultivar (abbreviated as cv.) to indicate variation that has arisen as a result of cultivation.

cultivate, v. Add: **2. b.** Biol. = *CULTURE v. c.

1888 Ann. Bot. II. 373 The spores, cultivated in suitable media, give rise..to a copiously branched and septate mycelium. 1891 G. S. Woodhead Bacteria x. 195 The bacilli, when obtained pure, and cultivated in fluid, grew out into very long threads. 1910 Jrnl. Amer. Med. Assoc. 15 Oct. 1381/1 Adult tissues and organs of mammals can be cultivated outside of the animal body. 1924 T. S. P. Strangeways Tissue Culture in Rel. Growth i. 11 The tissues of such an embryo can be readily cultivated, even if removed as much as fourteen days after the death of the animal. 1926 Proc. R. Soc. B. C. 273 If the undifferentiated limb-bud of the embryonic Fowl was cultivated in vitro, it underwent a considerable amount of progressive development. 1953 Sci. Progress XLI. 212 In organ culture, complete rudiments or fragments of organs are cultivated. 1963 Penso & Balducci Tissue Cultures in Biol. Res. vi. 145 Blood cells that can be most easily cultivated are macrophages or monocytes.

cultivated, ppl. a. Add: **2.** Of the voice or utterance: indicating refinement in its user.

1908 Westm. Gaz. 2 Jan. 8/1 The prisoner is a well set-up and well-dressed man with a cultivated voice.

cultivation. Add: **1. a.** Also attrib., as cultivation field, system; **cultivation bank, terrace**, a bank or terrace formed either naturally or artificially on a cultivated hillside; a lynchet; **cultivation mark** Archæol., a mark on the soil caused by cultivation in an earlier period; **cultivation paddock**, the part of an Australian farming estate used for the raising of crops.

1923 Geogr. Jrnl. May 356 Hitherto I have used this word [sc. lynchet] to describe the cultivation-banks of the Celtic system. 1893 F. Adams New Egypt 94 An open space in the desert, beyond the cultivation fields. 1924 Mawer & Stenton Introd. Surv. Eng. Place-Names viii. 156 Air photographs..reveal..trackways and cultivation-marks which would otherwise be overlooked. 1853 St. Julian & Silvester Prod. N.S.W. iv. 170 Few stations of any magnitude are without their 'cultivation paddocks'. 1902 Westm. Gaz. 13 Dec. 2/1 Posy..went over the fence into the cultivation paddock. 1923 Geogr. Jrnl. May 347, I think..that the date 650 B.C. is probably within a century or two of the date when the Celtic cultivation-system was inaugurated. 1911 Antiquary VII. 416/1 Within the camp, on the southern slope of the hill-crest, there exists a series of broad cultivation terraces. 1951 Field Archæol. (Ordnance Survey) (ed. 3) 62 Each of these strips was separated from its neighbours by a 'balk' of untilled land, and if the common field so divided lay on the slope of a hill a cultivation terrace or 'lynchet' formed on the lower boundary of each in the course of time.

2. b. (Later examples.) In wider use (cf. *CULTURE sb. 3 c).

1910 Jrnl. Amer. Med. Assoc. 15 Oct. 1379/2 (heading) Cultivation of adult tissues and organs outside of the body. 1926 Proc. R. Soc. B. C. 279 After the 17th day of cultivation the eyes undergo no further development and usually begin to degenerate. 1959 Chambers's Encycl. XIII. 653/2 Under certain conditions of cultivation the normal organization of the explanted tissue can be preserved and..it may undergo a remarkable degree of histological and even anatomical development. 1965 White & Grove Proc. Internat. Conf. Plant Tissue Culture 20 The initial pH of the nutrient media is important for the cultivation of excised roots.

cultual (kʌ·ltiuăl), a. [ad. F. cultuel, f. L. cultus CULTUS.] Of or pertaining to a cult or organized religious worship.

1906 Westm. Gaz. 17 Dec. 2/3 Catholic Cultual Associations. 1912 F. von Hügel Eternal Life 163 There is nothing necessarily superstitious in..Cultual Acts.

cultural, a. Add: **3.** Relating to civilization, esp. that of a particular country at a particular period; cultural anthropology, the branch of anthropology that is predominantly concerned with the cultural, as opposed to the physical, aspects of the evolution of man; cultural attaché, an embassy official whose function is to promote cultural relations between his country and the country in which he is staying; cultural revolution, a cultural and social movement in Communist China, begun in 1965, which sought to combat 'revisionism' and restore the original purity of Maoist doctrine; also transf.

1875 W. D. Whitney Life Lang. 172 All these widely-sundered tribes of men, found at the dawn of history in every variety of cultural condition. 1884 Science IV. 21/2 In its cultural development, China stands wholly for itself. 1898 Daily News 30 Aug. 5/1 The gigantic cultural problems awaiting solution in the Russian Empire. 1909 Westm. Gaz. 22 Apr. 5/2 The Professor [Doerpsfeld] says the excavations reveal several distinct cultural deposits. 1909 A. H. Keane Central & S. Amer. (ed. 2) I. 48 The southern extremity of the cultural zone. 1921 E. Sapir Language x. 221 How do the peoples of the given area divide themselves as cultural beings? what are the outstanding 'cultural areas'? 1923 A. L. Kroeber Anthropol. 6 Here..is the cause of the seeming preoccupation of social or cultural anthropology with ancient and savage and exotic and extinct peoples. 1934 H. C. Warren Dict. Psychol. 66/1 Cultural lag, slowness in adapting institutions or cultural habits to new or changing conditions or situations; the condition which ensues when certain elements of culture change more slowly than other elements. 1937 Times 31 Aug. 11/2 There is a proposal..that three 'cultural attachés' should be appointed to the more important embassies and consulates in foreign countries. 1938 R. G. Collingwood Princ. Art xi. 242 The groupings recognized by physical anthropology do not coincide with those of cultural anthropology. 1939 V. G. Childe Dawn Europ. Civ. (ed. 3) vii. 92 Though the temporary nature of the settlements excludes..stratigraphical chronology, the cultural sequence is well established. 1966 Economist 20 Aug. 709/1 Lin Piao..has loyally used the army as a guinea-pig for the 'cultural revolution' dose of salts with which Mao is now purging the whole country. 1967 Guardian 16 May 1/3 Chinese leadership..has certainly been prepared to accept serious economic and political losses..for the sake of the 'cultural revolution'. 1967 Economist 21 Oct. 274/1 They support Syria's 'cultural revolution' (the regime has just taken over control of its mainly religious private schools). 1970 E. McGirr Death pays Wages ii. 43 He got the Cultural Attaché to translate that for him.

4. Biol. Relating to the culture of micro-organisms, tissues, etc. (see CULTURE sb. 3 c in Dict. and Suppl.).

1900 Jrnl. Exper. Med. V. 259 The bacillus recovered by us from our several autopsies always showed the same cultural characters. 1969 S. T. Lyles Biol. Microorganisms v. 103 Cultural methods will depend both on the organism studied and the purpose for which the study is made.

culturally, adv. Add: Also spec. with reference to a particular form of culture or civilization.

1921 R. A. S. Macalister Text-bk. Europ. Archæol. I. v. 120 Communities may be culturally classified as hunters, pastors, and agriculturists, according to the means whereby the majority of their members obtain their livelihood. 1921 E. Sapir Language x. 228 Culturally identical with [the Hupa Indians] are the neighboring Yurok and Karok.

2. Biol. With reference to culture or a culture (sense 3 c).

1893 Daily Tel. 29 Sept. 4/6 A fatal case..is officially described as 'culturally indistinguishable from true cholera'. 1906 Practitioner Nov. 635 Culturally, it possesses no features by means of which it may be readily distinguished from other diplo- or streptococci. 1908 Ibid. Jan. 68 Ruppell found that there was a striking resemblance between the non-virulent strains of meningococci and ordinary gonococci, morphologically, culturally, and as regards their immunising power against virulent meningococci.

culture, sb. Add: **3. b.** culture pearl = *cultured pearl.

1921 Current Hist. July 623/1 Jewelers in London have been greatly perturbed over a new type of [Japanese] 'culture' pearls which is said to be so perfect that it cannot be distinguished from the natural article. 1937 Discovery Mar. 87/1 X-ray photographs of culture pearls. 1963 Times 12 Mar. (Austral. Suppl.) p. v/7 Culture-pearl farms.

c. (Earlier and later examples.) Also applied to the similar growth of plant and animal cells and tissues, and of whole organs or fragments of them. Also **culture medium**, a substance, solid or liquid, in or on which micro-organisms, tissues, etc., are cultured.

1880 G. M. Sternberg tr. Magnin's Bacteria II. i. 113 Cohn, in order..to get rid of the moulds,..employed the following culture-fluid. 1885 C. S. Dolley Technol. Bacteria Invest. I. ii. 59 Sterilizing the culture medium is accomplished ordinarily by the use of heat sufficient to kill all germs. 1890 Jrnl. Chem. Soc. LVII. 487 Experiments upon the culture of excised barley embryos on nutrient liquids. 1910 A. Carrel in Jrnl. Amer. Med. Assoc. 15 Oct. 1379/2 The plasmatic media were inoculated with many tissues or organs, of which all were found to multiply or grow. The cultures of the different tissues—as we shall call them—contain common characteristics. 1914 Jrnl. Exper. Med. XIX. 398 (heading) The effect of dilution of plasma medium on the growth and fat accumulation of cells in tissue cultures. 1938 R. C. Parker Methods Tissue Culture xvi. 208 The method of tissue culture has also served as a direct means of studying the further development of complex structures and organ rudiments. 1939 F. A. Knott Clin. Bacteriol. ii. 19 The culture medium chosen depends upon the variety of bacterium sought. 1953 Sci. Progr. XLI. 212 Earle and his collaborators..have devised methods for obtaining mass cultures of unorganised tissue from an initial cell suspension or a single cell. 1960 L. Picken Organization of Cells iii. 83 The presence of minute amounts of penicillin in the culture medium leads to the production of various abnormal forms of bacteria. 1965 White & Grove Proc. Internat. Conf. Plant Tissue Culture 9 Experiments with excised roots as organ cultures. Ibid. 28 Sterile seedlings of tomato..were grown in 50 ml of the inorganic solution of the standard root culture medium. 1970 tr. G. Le Douarin in J. A. Thomas Organ Culture ii. 17 The patella of a 9-day chick embryo developed cartilage after 3 days of culture.

5. a. (Further examples.)

1837 Emerson Jrnl. 24 Nov. (1910) IV. 371 It seems to me that the circumstances of man are historically somewhat better here and now than ever,—that more freedom exists for Culture. 1849 J. A. Froude Nemesis of Faith x. 85 The end of all culture is, that we may be able to sustain ourselves in a spiritual atmosphere as the birds do in the air. 1855 J. Conington Academical Study of Latin (1872) I. 212 That part of our culture which we have not worked out for ourselves, or received from contemporary nations, we owe almost wholly to Rome, and to Greece only through Rome. 1869 M. Arnold (title) Culture and anarchy. Ibid. ii. 49 The great men of culture are those who have had a passion..for carrying from one end of society to the other, the best knowledge, the best ideas of their time. 1871 Geo. Eliot Middlem. I. ix. 137 He wants to go abroad again..[for] the vague purpose of what he calls culture, preparation for he knows not what. 1916 E. Wharton Xingu i. 3 Mrs. Ballinger is one of the ladies who pursue Culture in bands, as though it were dangerous to meet alone. 1939 tr. H. Johst in C. Leiser Nazi Nuggets 83 When I hear the word 'culture' I slip back the safety-catch of my revolver. 1940 K. Mannheim Man & Society II. ii. 85 The crisis of culture in liberal-democratic society is due, in the first place, to the fact that the social processes, which previously favoured the development of the creative élites, now have the opposite effect. 1948 T. S. Eliot Notes Def. Culture ii. 41 Culture is not merely the sum of several activities, but a way of life. Ibid. 42 Group culture..has never been co-extensive with class. Ibid. 43 The primary channel of transmission of culture is the family.

¶ With distortion of spelling to indicate affected or vulgar pronunciation.

1931 Atlantic Monthly Feb. 149/2 They believe in 'cultchah'. 1959 Listener 15 Jan. 128/2 Italian Bouquet, an Epicurean Tour of Italy..an American hotchpotch of culcher and food.

b. (Further examples.) Also, the civilization, customs, artistic achievements, etc., of a people, esp. at a certain stage of its development or history. (In many contexts, esp. in Sociology, it is not possible to separate this sense from sense 5 a.)

1871 E. B. Tylor (title) Primitive culture. 1903 C. Lumholtz Unknown Mexico I. 117 A thrifty people whose stage of culture was that of the Pueblo Indians of to-day. 1921 R. A. S. Macalister Text-bk. Europ. Archæol. I. iv. 99 A language associated with a superior culture has always a tendency to swamp languages that have not such an advantage. 1921 E. Sapir Language x. 222 Historians and anthropologists find that races, languages, and cultures are not distributed in parallel fashion. 1942 Bloch & Trager Outl. Ling. Analysis 5 The activities of a society—that is, of its members—constitute its culture...Language, then, is not only an element of culture itself; it is the basis for all cultural activities. 1948 T. S. Eliot Notes Def. Culture i. 28 The culture with which primitive Christianity came into contact..was itself a religious culture in decline. Ibid. v. 93 A careful fostering [by Russia in its satellites] of local 'culture', culture in the reduced sense of the word, as everything that is picturesque, harmless and separable from politics such as language and literature, local arts and customs. 1953 A. K. C. Ottaway Education & Society i. 8 A single word to express 'the whole life of a community' is a special use of the word 'culture', which has been developed by the social anthropologists. 1954 S. Piggott Neolithic Cultures v. 123 Probably the word culture should be employed to define the collective and tangible outcome (pot-making, house-planning, tomb-building) of the material and spiritual traditions of a group of people. 1963 Brit. Jrnl. Sociol. XIV. 21 By 'culture' is meant the whole complex of learned behaviour, the traditions and techniques and the material possessions, the language and other symbolism, of some body of people.

c. Phr. the two cultures: see quot. 1956.

1956 C. P. Snow in New Statesman 6 Oct. 413/1 The separation between the two cultures has been getting deeper under our eyes; there is now precious little communication between them... The traditional culture..is, of course, mainly literary..the scientific culture is expansive, not restrictive. 1959 —— Two Cultures & Sci. Revol. 16 Those in the two cultures can't talk to each other..very little of twentieth-century science has been assimilated into twentieth-century art. 1961 Listener 16 Nov. 809/1 The lack of communication between scientists and non-scientists, which has been so much discussed recently in terms of 'the two cultures'. 1967 'W. Haggard' Conspirators ii. 14 He could explain things to laymen simply, despising ill-digested chatter about two cultures.

d. attrib. and Comb., as culture-condition, -instinct, -monger, etc.; culture-loving ppl. adj.; **culture vulture**, a rhyming collocation indicating a person who is voracious for culture.

1889 G. B. Shaw London Music (1937) 201 The race of culture humbugs. 1897 Mary Kingsley W. Africa 28 The present culture-condition of West Africa. Ibid. Pref. p. ix, Your superior culture-instincts may militate against your enjoying West Africa. 1905 Daily Chron. 15 June 3/1 The culture-loving Catholic Gael. 1909 Westm. Gaz. 2 Jan. 11/2 A modernised, constitutional, culture-loving Turkish State. 1931 A. Huxley Music at Night 226 Most professional intellectuals will approve of culture-snobbery (even while intensely disliking most individual culture-snobs). 1933 Amer. Jrnl. Sociol. XXXIX. 301 (title) The bureaucratic culture pattern and political revolution. 1938 Dylan Thomas Let. 23 Mar. (1966) 190 Advanced writing..sells very well over there [America], they're such culture-snobs. 1939 'M. Innes' Stop Press III. iii. 376 You couldn't have a more persistent culture-hound. 1947 in Wentworth & Flexner Dict. Amer. Slang (1960) 134/2 Everybody can't be a culture vulture. a 1953 Dylan Thomas Quite Early One Morning (1954) 67 See the garrulous others, also, gabbing and garlanded from one nest of culture-vultures to another. 1955 20th Cent. June 536 An attack on contemporary 'Culture'-mongers. 1960 Koestler Lotus & Robot 279 Literacy, culture-hunger and leisure-time are increasing even more rapidly than the birth-rate. 1964 Amer. Speech XXXIX. 46 The Caribs were very great travelers, and their words became culture words, words found in all the Indian languages on the Caribbean Sea. Ibid. 50 These pirates could have made a culture word of America. 1966 L. J. Cohen Diversity of Meaning i. 15 We also have the notion of a culture-word or culture-sentence, as when a historian of ideas is concerned with the meaning of the word 'mass' in seventeenth-century physics.

Also spec. in Anthropol. and Sociol. (In some contexts the meaning shades into an attrib. use of sense 5 in Dict.)

1901 Contemp. Rev. Mar. 455 The ancient 'culture-heroine'. 1903 Daily Chron. 11 June 3/1 The hero-tales and culture-legends of the prehistoric period of the Hebrews. 1907 A. C. Haddon in N. W. Thomas Anthropol. Ess. 183 The death dances were introduced into the Western Islands by two culture heroes from New Guinea. 1921 E. Sapir Language x. 223 That a group of languages need not in the least correspond to a racial group or a culture area is easily demonstrated. 1922 D. H. Lawrence Fantasia of Unconscious xi. 203 The woman is now the responsible party, the law-giver, the culture-bearer. 1931 H. J. Rose tr. W. Schmidt's Orig. & Growth Relig. v. xiv. 221 Leo Frobenius, a pupil of Ratzel, enlarged..the doctrine of 'culture-circles' (or 'spheres', Kulturkreise). 1933 Downside Rev. LI. 185 Russia is a culture-complex in itself, and Russia's problem is not ours. 1936 Mind XLV. 294 In the modern world, with its ever-increasing facilities for culture-contacts, a world-culture is in process of formation. 1945 Mind LIV. 78 The culture-hero has a vague complex status, part man, part demi-god. 1948 T. S. Eliot Notes Def. Culture iv. 70 Since..the scattering of Jews amongst peoples holding the Christian Faith, it may have been unfortunate..that the culture-contact between them has had to be within those neutral zones of culture in which religion could be ignored. 1949 M. Mead in M. Fortes Social Structure 27 Teacher, physician, nurse..each in turn represents some different form of culture conflict. 1951 R. Firth Elem. Social Organiz. iii. 81 Terms such as 'culture-contact'..were introduced to express the way in which new patterns of behaviour or types of relationship were acquired and incorporated into a primitive system. Ibid. 109 He is culture-bound in his desires as well as his activities. 1953 Proc. Prehist. Soc. XIX. 41 (title) The prehistoric culture-sequence in the Maltese Archipelago. 1957 Burlington Mag. Nov. 246/2 A theory of culture-circles whereby societies are classified as hunting, food-gathering, harvesting and horticultural. 1960 Listener 18 Aug. 244/1 The people of the host country appear to lack the normal conventions of social behaviour or to have a different and apparently illogical system... Most Europeans in Africa withdraw into their own community, and quickly equate their own way of doing things with their own superior material culture...This reaction..has been called 'culture shock'. 1962 D. Harden Phoenicians i. 24 To pick out what is Egyptian and Mesopotamian among finds and culture-traits in Phoenicia is not nearly so hard. Ibid. ii. 25 Their position on the land-route between the two great culture-areas of antiquity laid them open to constant political domination and cultural influences from each. 1969 Listener 30 Jan. 155/2 There was a curious naivety..in the frank description of how destructive, physically and socially, the culture-contacts with these remote peoples could be.

culture, v. Restrict rare to senses in Dict. and add: **c.** Biol. To maintain (bacteria or other micro-organisms, tissues, organs, etc.) under artificial conditions in a suitable nutrient medium so that they can multiply, grow, or develop.

1908 Practitioner Sept. 463 The ovary and tube (unopened) were despatched..with a request to see what organism could be cultured, and to make a vaccine. 1934 Huxley & de Beer Elem. Exper. Embryol. vii. 209 Fibroblasts of the fowl have been cultured in vitro for over 20 years..and show unchanged characters and an unchanged rate of growth. 1962 C. V. Harding et al. in A. Pirie Lens Metabolism Rel. Cataract 460 The lens had been cultured for two days in 199 with 23% rabbit serum ultrafiltrate. 1971 Nature 16 Apr. 472/2 We decided to isolate and culture the bacteria. Ibid. 474/1 A technique for culturing green organisms such as filamentous algae and moss protonema on an agar substrate.

cultured, ppl. a. Add: **1. b.** Developed under controlled natural conditions, esp. cultured pearl. Cf. CULTURE sb. 3 b in Dict. and Suppl.

1921 Current Hist. July 623/2 It is quite impossible to tell the natural pearl from the cultured pearl. 1930 Pop. Sci. Dec. 45/3 Like 'natural' pearls, the cultured product can be dissolved in acids. 1940 Chem. Abstr. XXXIV. 4700 An x-ray study of aragonite in natural and cultured pearls.

2. ¶ With distortion of spelling to indicate affected or vulgar pronunciation.

1929 Galsworthy Exiled I, Quate! 'Ow culchad! Hairs and grices! 1940 H. G. Wells Babes in Darkling Wood IV. i. 324 He likes treading out music with a pianola, for example, to the great disdain of the culchad Trotsky.

cultureless. (Additional examples.)

1941 'G. Orwell' Lion & Unicorn I. vi. 54 A rather restless, cultureless life, centring round tinned food, Picture Post, the radio, and the internal combustion engine. 1948 —— Coll. Ess. (1968) IV. 457 A genuinely classless society, which Mr. [T. S.] Eliot assumes would be a cultureless society.

culturology (kʌltiŭrǫ·lŏdʒi). Social Anthropology. U.S. [ad. G. kulturologie (W. Ostwald Philos. d. Werte (1913) xiii. 286); cf. CULTURE sb. 5 a, b, -OLOGY.] The science or study of (a) culture. Hence **culturolo·gical** a., of or pertaining to culturology; **culturo·logist**, a student of culturology.

1939 L. A. White in Amer. Anthropologist XLI. 571 Application of the viewpoint and principles of the philosophy of evolution is as essential to the solution of many problems in culturology as it is in biology or physics. 1949 Antiquity XXIII. 55 What White calls the 'culturological' approach. 1956 F. B. Steiner Taboo i. 17, I am not referring to American culturology. 1957 Contrib. Indian Sociol. I. 50 The culturologist may feel at ease but the sociologist is at a loss.

cum. Add: Also used as a combining word to indicate a dual nature or function.

1871–3 Trollope Eustace Diamonds (1873) I. xiii. 173 The Belgrave-cum-Pimlico life. 1913 Kipling Diversity of Creatures (1917) 172 Easy motor-bike-cum-side-car trips round London. 1939 O. Lancaster Homes Sweet Homes 44 The fervent mediaevalism..developed a philosophic-cum-economic tinge. 1959 Manch. Guardian 3 July 5/6 Three short..dinner-cum-cocktail dresses. 1959 Viewpoint July 33 The atmosphere of laboratory-cum-workshop.

Cumacean (kiūmēⁱ·ʃiăn), a. Zool. [f. mod.L. Cumacea (see def.), f. Cuma genus of crustaceans, f. Gr. Κῦμα (see CYME).] Of or pertaining to the Cumacea, an order of small sessile-eyed crustaceans resembling prawns, with a hard brittle carapace. Also as sb.

1879 Ann. & Mag. Nat. Hist. III. 56 Diastylis Josephinæ ..appears to be the commonest Cumacean inhabiting the deep waters between Faröe and Shetland. 1887 G. O. Sars in Challenger Rep., Zool. XIX. II. 4 Perhaps even some of the palæozoic forms placed among the Phyllocarida may have formed a direct transition to the Cumacean type. 1900 —— Crustacea of Norway III. p. v, The Cumacean fauna of Norway. 1902 Encycl. Brit. XXX. 479/2 According to Sars, the Sympoda (or Cumaceans), in spite of their sessile eyes, have closer affinities with the stalk-eyed orders. 1952 F. R. Allison in Scottish Naturalist LXIV. 40 Most planktonic animals, including..cumaceans..recorded in the fulmar's diet, have a vertical diurnal migration which brings them into the surface waters at night.

Cumæan (kiūmi·ăn), a. and sb. [f. Cumæ (L. Cūmæ, Gr. Κύμη), an ancient city on the Italian coast near Naples, founded by the Greeks in the 8th cent. B.C. + -AN.] **A.** adj. Of or pertaining to Cumæ, esp. famous for the Sibyl mentioned by Virgil in the 'Æneid'. **B.** sb. A native or inhabitant of Cumæ.

1731 J. Trapp Works Virgil II. 347 Have you been led thro' the Cumæan Cave, And heard th'impatient Maid divinely rave? 1803 C. Wilmot Let. 6 Mar. (1920) 168 Aeneas..landed at the Cumaean shore a little way from the Lake Avernus. 1870 Brewer Dict. Phr. & Fable 819/1 The Cumæan sibyl was the conductor of Virgil to the infernal regions. 1931 D. Randall-MacIver Gk. Cities in Italy & Sicily i. 6 The Cumaeans, says Pausanias, showed a small stone urn in the cemetery of Apollo which they said contained her [sc. the Sibyl's] bones. 1968 Encycl. Brit. VI. 889/1 The antrum (cave) famous in legend as the seat of the oracle of the Cumaean Sibyl.

cumarone, var. *COUMARONE.

Cumberland (kʌ·mbəɹlænd). **1.** The name of the English county used attrib. to designate a piquant sauce served esp. with cold meat.

1878 'Short' Dinners at Home 146 Cumberland Sauce for Game (Cold)...Cumberland Sauce (Hot). 1922 F. Hamilton P. J. i. 22 We shall be able to offer you a tolerably good supper, and I will not forget your favourite Cumberland sauce. 1959 House & Garden Dec.–Jan. 36/2 Cold spiced beef, ham, chicken or turkey, accompanied by..Cumberland sauce.

2. Used to designate the manner of cutting up a pig's carcase in which the ham is cut away and cured separately. ? Obs.

1905 W. H. Simmonds Practical Grocer III. 103 It pays to sell the ham separately and convert the rest of the side into 'Cumberland cut' bacon or 'Irish rolls'. 1907 Yesterday's Shopping (1969) 20/2 Hams...Cumberland.

Cumbrian (kʌ·mbriăn), a. and sb. [f. med.L. Cumbria, f. W. Cymry:— prehist. W. *kombrogi, pl. of *kombrogos lit. fellow countryman (W. bro:— *mrog- region).] **A.** adj. **1.** Belonging to the ancient British kingdom of Cumbria, which included Cumberland. **2.** Belonging to Cumberland, or its system of rocks; also, more widely, belonging to the Lake District and its fells. **B.** sb. **1.** A native of the ancient British kingdom of Cumbria. **2.** A native or inhabitant of Cumberland.

1747 T. Carte Gen. Hist. Eng. I. III. 211 Feuds and quarrels between particular chieftains and their clans were not the only grievance under which the Cumbrian and Straeth-cluyd Britains laboured. Ibid. 212 Whether Rydderch imagined, that he was any obstruction to the Cumbrians returning to their former allegiance. 1779 A. Butler Lives Saints (ed. 2) I. 140 Among the Straith-Cluid Britons, and the Cumbrians, the latter inhabiting the country from the Picts wall, to the Ribble in Lancashire. Ibid., The Cumbrians..were protected by Urien, lord of Rheged. 1780 T. West Guide Lakes Cumb., etc. (ed. 2) 6 The travelled visitor of the Cumbrian lakes. 1833–4 J. Phillips in Encycl. Metrop. (1845) VI. 584/2 The analogous arches of limestone, which begird the primary district of the Cumbrian lakes. 1837 Penny Cycl. VIII. 223/2 The Cumbrians have been undeservedly said to be litigious. Ibid., Cumbrian peasantry have various festive meetings, called the kirn, or harvest-home, sheep-shearing, merry nights, and upshots. 1882 Encycl. Brit. XIV. 40/1 Kentigern,..the restorer of Christianity among the Cumbrians. 1901 A. G. Bradley Lake District 4 The simple dalesman,..canny Cumbrian as he is. 1902 W. G. Collingwood Lake Counties I. ii. 38 There have been many Cumbrian poets, most of them, like Wordsworth at Grasmere, cottage folk. W. P. Haskett-Smith Ibid. II. vi. 259 Leaving out Langdale, which runs down into Westmorland, there are five purely Cumbrian dales. 1963 Times 25 Feb. 11/7 While the Cumbrian fight their battles with energy, there is a strong community spirit.

Cumbric (kʌ·mbrik). [f. med.L. Cumbria (see prec.) + -IC.] The former Celtic language of Cumbria. Also attrib. or as adj.

1953 K. Jackson Lang. & Hist. Early Brit. I. i. 6 We shall occasionally employ Primitive Cumbric (Pr. Cum.) for the Brittonic dialect of Cumberland, Westmorland, northern Lancashire, and south-west Scotland. 1954 —— in N. K. Chadwick Stud. Early Brit. Hist. 67 By Western British I mean the ancestor of Welsh and probably of the Celtic language of Cumbria, called Cumbric here, which seems to have agreed with Welsh in the main. 1967 Peeblesshire (R. Comm. Anc. Monuments Scotland) I. 3 The inhabitants..were responsible for the Brittonic or 'Cumbric' place-names of the region. 1970 B. M. H. Strang Hist. Eng. v. 283 The far north-west was brought under English rule in 1092, but it would be foolish to assume that all use of Cumbric immediately ceased.

cumdach (kū·dăχ). [ad. Ir. cumhdach, f. MIr. cumdach, f. OIr. cumtach.] An ornamental book-casket.

1887 M. Stokes Early Chr. Art in Ireland iv. 89 The first cumdach we read of..was made for the 'Book of Durrow', by the king of Ireland, Flann Sinna..who reigned between the years 877 and 916. 1908 Daily Chron. 30 Mar. 3/3 It is marvellous that any of those old manuscripts escaped the destructive raids of plundering Danes and Normans, especially as the cumdach, or cover, meant to preserve them from other injuries, was generally enriched with silver or gold, and sometimes inlaid with precious stones or enamels. 1960 G. A. Glaister Gloss. Book 93/1 Cumdach, a jewelled and elaborately decorated box used in late 9th-century Ireland for keeping manuscript books.

cumulate, v. Add: **2. b.** To combine (the entries of an index, catalogue, etc.) in successive issues.

1905 Readers' Guide Period. Lit. I. p. vii, The monthly numbers were cumulated, and quarterly and annual volumes were issued. 1931 A. Esdaile Man. Bibliogr. 301 The English Catalogue appears annually, and is 'cumulated' every five years. 1965 Amer. N. & Q. Mar. 106/2 Monthly issues..will be cumulated every third month.

cumulated, ppl. a. **1.** Delete †Obs. and add later example.

1926 Brit. Weekly 2 Sept. 452/4 The cumulated common sense of the Anglo-Saxon mind.

3. (See prec., sense *2 b.)

1905 (title) Readers' guide to periodical literature (cumulated). 1938 L. M. Harrod Librarians' Gloss. 53 Cumulated book catalogue. 1959 Indian Nat. Bibliogr. 1958 p. v, All these errors have been corrected in the Cumulated Volume.

cumulative, a. Add: **2. c.** cumulative error.

1887 Encycl. Brit. XXII. 707/2 [Surveying.] Cumulative error, not eliminable by circuiting, may be caused when there is much northing or southing..in the direction of the line. 1920 W. N. Thomas Surveying 509 Cumulative errors are those which tend always in the same direction, i.e. either to make the apparent measurements always too large or always too small...Cumulative errors are directly proportional to the number of observations. 1957 Kendall & Buckland Dict. Statistical Terms 74 Cumulative error, an error which, in the course of the cumulation of a set of observations, does not tend to zero.

5. (Earlier example.)

1853 J. S. Mill Lett. (1910) I. 173 One very strong recommendation of the plan of cumulative votes occurs to me.

cumulet (kiū·miŭlĕt). A high-flying variety of fancy pigeon.

1876 Bazaar, Exchange & Mart 12 Jan. 74/3 Pair splendid white eyed cumulets. 1910 A. H. Osman Pigeon Bk. xii. 131 During the past few years a breed that has been taken up with considerable keenness by exhibitors is the Cumulet. 1965 W. M. Levi Encycl. Pigeon Breeds 596 The Flying Cumulet is one of the ancestors of the Racing Homer and of the Tipper.

cumulo-. Add: **cumulo-nimbus.**

1887 Leisure Hour 570/2 Similar cumulus and cumulo-nimbus forms range in latitude from London to near Cape Horn. 1928 D. Brunt Meteorol. iii, Cumulo-nimbus is the thunder-cloud. It frequently takes the form of towers or anvils. 1970 D. Francis Rat Race iii. 39 Great heaps of cumulo-nimbus cloud were boiling up.

Cunarder. (Earlier U.S. examples.)

1850 Knickerbocker XXXVI. 574 The 'Cunarders'. 1874 'Mark Twain' & C. D. Warner Gilded Age lv. 494 He pointed out where the Cunarders lay when in port.

cundum, var. *CONDOM.

cunit (kiū·nit). Timber Industry. Also cunet. [f. C III (= a hundred) + UNIT.] (See quot. 1956.)

1953 Brit. Commonw. Forest Terminol. I. 36 Cunet, cunit, a unit of stacked wood containing 100 cu. ft. of solid volume within the outside dimensions of the stack. 1956 N.Z. Timber Jrnl. Oct. 54/2 Cunet or cunit, a unit of stacked wood containing 100 cu. ft. of solid volume within the outside dimensions of the stack...Also known as Cubic unit. 1956 in N. K. Wallis Austral. Timber Handbk. 348.

cunjevoi (kʌ·ndʒĕvoi). Austral. [Native name.] **1.** The popular name for the green arum or spoon lily, Alocasia macrorrhiza.

1889 J. H. Maiden Useful Native Plants Austral. 165 Colocasia macrorrhiza..Alocasia macrorrhiza..'Pitchu' of the aboriginals of the Burnett River, Queensland; 'Cunjevoi' of those of South Queensland. 1930 Bulletin (Sydney) 12 Nov. 26/3 The cunjevoi..grows in dense clumps on river banks and in moist brush forests of eastern and tropical Australia. 1964 Mod. Encycl. Austral. & N.Z. 290/1 Cunjevoi or Spoon lily...Its bulbs (normally poisonous) were cooked in a special way and eaten by aborigines. 1965 Austral. Encycl. I. 221/1 One of the commonest Australian species is the cunjevoi,..whose large fleshy rhizomes extend for several feet over the surface of the ground.

2. (Also -boi, -boy.) A common ascidian, the sea-squirt (see quots.). Abbrev. **cunjie.**

[1821 S. Leigh in W. S. Ramson Austral. Eng. (1966) 121 Conguwa, a kind of living fungus, which at certain Seasons they detach from the Rocks on the Sea Shore.] 1911 A. E. Mack Bush Days 109 Down at the sea's edge grew the cunjeboy, brown and red, upon the rocks. 1925 Illustr. Austral. Encycl. I. 342/1 Cunjevoi, popular name of a polyp (Cynthia praeputialis) common on the Australian coasts. It is often used as a bait for fishing. 1945 Sun (Sydney) 15 May, The organism known to local fishermen as congeboy, cungevoi, or congevoi. 1960 W. J. Dakin et al. Austral. Seashores (rev. ed.) xviii. 341 The cunjevoi itself..prefers the outermost rocks facing the ocean. 1965 Austral. Encycl. I. 274/2 Pyura stolonifera (cunjevoi) is occasionally used as bait but is never marketed as such. Ibid. III. 145/2 Cunjevoi,..the name used in New South Wales for the ascidian or sea-squirt. Ibid., A simple ascidian of the cunjevoi type may also be seen growing on wharf-piles or other harbour installations. 1966 Baker Austral. Lang. (ed. 2) xiv. 302 Cunjie, a cunjevoi, used for bait.

cunnilingus (kʌ:nili·ŋgŏs). [mod.L., f. L. cunnilingus one who licks the vulva, f. cunnus female pudenda + -lingus (lingere to lick).] Oral stimulation of the vulva or clitoris. Also **cu·nnilingue** v. trans. and intr., to practise cunnilingus (on).

1887 L. C. Smithers tr. Forberg's Man. Class. Erotology v. 122 A man who is in the habit of putting out his tongue for the obscene act of cunnilinging. 1897 Havelock Ellis Stud. Psychol. Sex I. iv. 98 The extreme gratification is cunnilingus,..sometimes called sapphism. 1905 Ibid. IV. 21 Cunnilingus was a very familiar manifestation in classic times;..it tends to be especially prevalent at all periods of high civilization. 1965 New Statesman 26 Mar. 492/3 Cunnilingus, in which the tongue is used. 1969 G. Legman Rationale of Dirty Joke viii. 566 A man comes home and finds the vicar cunnilinguing his wife. 1970 Guardian 6 Mar. 8/1 'The Beard'—a late night play at the Royal Court which shocked many..because of the mock display of cunnilingus in the last scene.

cunning, a. **6.** (Earlier examples.)

1844 'J. Slick' High Life N.Y. I. 220 'Why, that pair,' sez she,..a burying her hands..down in the pocket of her cunning apron. 1844 Dickens Mart. Chuz. xvii. 216 Tea and coffee arrived (with sweet preserves, and cunning teacakes in its train).

cunningness. 2. (Later example.)

1965 Listener 18 Mar. 420/2 His work..lacks the nudging cunningness of some 'art' photographers.

cunny (kʌ·ni). slang. [Prob. dim. of *CUNT; but cf. CONY sb. 5 b.] = *CUNT 1.

1720 D'Urfey Pills VI. 197 All my Delight is a Cunny in the Night, When she turns up her silver Hair. 1865 E. Sellon New Epicurean (1875) 11, I frigged and kissed their fragrant cunnies. 1879–80 Pearl (1970) 216 Your private parts, or cunny, Should not be let for money. 1891 Farmer & Henley Slang II. 230 Cunny-haunted,..lecherous. 1922 F. Harris My Life & Loves I. x. 208 She had limbs like a Greek statue and her triangle of brown hair lay in little silky curls on her belly and then—the sweetest little cunny in the world.

cunt (kʌnt). [ME. cunte, count(e), corresponding to ON. kunta (Norw., Sw. dial. kunta, Da. dial. kunte), OFris., MLG., MDu. kunte:— Gmc. *kuntōn wk. fem.; ulterior relations uncertain.] **1.** The female external genital organs. Cf. QUAINT sb.

Its currency is restricted in the manner of other taboo-words: see the small-type note s.v. *FUCK v.

[c 1230 in Ekwall Street-Names of City of London (1954) 165 Gropecuntelane.] a 1325 Prov. Hendyng (Camb. Gg. I. 1) st. 42 Yeue þi cunte to cunnig and craue affetir wedding. c 1400 Lanfranc's Cirurg. 172/12 In wymmen þe necke of þe bladdre is schort, & is maad fast to the cunte. c 1425

Castle of Perseverance (1904) 1193 Mankynde, my leue lemman, I my cunte þou schalt crepe. **1552** Lyndesay *Satyre* Procl. 144 First lat me lok thy cunt, Syne lat me keip the key. *a* **1585** Polwart *Flyting with Montgomerie* (1910) 817 Kis þe cunt of ane kow. *c* **1650** in Hales & Furnivall *Percy's Folio MS.* (1867) 99 Vp start the Crabfish, & catcht her by the Cunt. **1743** Walpole *Little Peggy* in *Corr.* (1961) XXX. 309 Distended cunts with alum shall be braced. *c* **1800** Burns *Merry Muses* (1911) 66 For ilka hair upon her c—t, Was worth a royal ransom. *c* **1888–94** *My Secret Life* VII. 161, I sicken with desire, pine for unseen, unknown cunts. **1934** H. Miller *Tropic of Cancer* (1935) 15 O Tania, where now is that warm cunt of yours? **1956** S. Beckett *Malone Dies* 24 His young wife had abandoned all hope of bringing him to heel, by means of her cunt, that trump card of young wives.
transf. and *fig.* *a* **1680** Ld. Rochester *Poems on Several Occasions* (1950) 28 Her Hand, her Foot, her very look's a Cunt. **1922** Joyce *Ulysses* 61 The grey sunken cunt of the world. **1928** D. H. Lawrence *Lady Chatterley* xvi. 296 If your sister there comes ter me for a bit o' cunt an' tenderness, she knows what she's after.
2. Applied to a person, esp. a woman, as a term of vulgar abuse.
1929 F. Manning *Middle Parts of Fortune* I. viii. 159 What's the cunt want to come down 'ere buggering us about for, 'aven't we done enough bloody work in th' week? **1932** 'G. Orwell' *Coll. Essays* (1968) I. 88 Tell him he's a cunt from me. **1934** H. Miller *Tropic of Cancer* (1935) 28 Two cunts sail in—Americans. **1956** S. Beckett *Malone Dies* 99 They think they can confuse me...Proper cunts whoever they are. **1965** V. Henriques *Face I Had* 69 'What d'you think you're doing, you silly cunt?' the driver shouts at her.
3. *Comb.*
1680 Anon. in *Rochester's Poems on Several Occasions* (1950) 36 Fam'd through the World, for the C--nt-mending Trade. **1868** *Index Expurgatorius of Martial* 32 A satire on Baeticus, who was a priest of Cybele, and a cunt-sucker. **1891** Farmer *Slang* II. 230/2 *Cunt-struck*, enamoured of women. **1923** Manchon *Le Slang* 97 *Cunt-hat*,..chapeau de feutre. **1965** F. Sargeson *Memoirs of Peon* ii. 28 We were all helplessly and hopelessly c...struck, a vulgar but forcibly accurate expression.

cup, *sb.* Add: **4.** Also, a depression in the skin forming a rudimentary eye in certain lower animals. Also *eye-cup* or *cup-eye*.
1900 F. A. Bather in E. R. Lankester *Treat. Zool.* III. 30 Each cup is coated at its base with pigment. **1927** *Glasgow Herald* 9 July 4 In some of the sea-worms..we start with diagrammatically simple 'cup-eyes',..and gradually pass to very elaborate 'cup-eyes'. *Ibid.*, A minute optic skin-cup. **1929** *Encycl. Brit.* XX. 628/2 Eye-spots are found in Medusae, starfishes, and some Annelid worms...The first step..is the sinking of the eye-spot into a pit-like depression, thus forming an eye-cup (optic cup). *Ibid.*, The cells situated at the back of the cup. **1940** Parker & Haswell *Text-bk. Zool.* (ed. 6) I. 249 [Phylum Platyhelminthes.] When most highly developed the eye..is still of very simple structure, consisting of a cup formed of one or more pigment-cells having sensory cells in close relation to it with processes (nerve-fibres) passing to the brain.
6. c. That part of a brassière which is shaped to contain or support one of the breasts. Also *attrib.* and *Comb.*
1938 'E. Queen' *Four of Hearts* (1939) ix. 129 She didn't have to wear a cup-form brassière. **1957** *Housewife* Sept. 104 Thinnest foam rubber curved in the cups..achieves a natural line by gently contouring the bosom itself. **1959** *Ibid.* June 28 Marquisette cup section underlined lace. *Ibid.*, A perfect contour bustline..B and C cups. **1959** *News Chron.* 13 July 6/3 Cup fittings are based on the difference between underbust and full bust measurements. **1970** *Times* 16 June 7/6 The prettiest and the most alluring and flattering bathing suits are halter-necked with a vertiginous..plunge in the front and very soft, unsupported cups.
8. a. Also *ellipt.* (In quots. 1952 and 1969, ellipt. in sense *12 b (ii).)
c **1760** *Mother Goose's Melody* (1785) 19 Take a cup and drink it up, Then call your Neighbours in. **1784** [in Dict.]. **1847** C. M. Yonge in *Mag. for the Young* Sept. 189 There is the kettle..all ready for tea!..Won't you sit down and have a cup, Amy? **1952** A. Wilson *Hemlock & After* I. iii. 51 Anyway, none of it would be your cup, darling. **1969** J. Elliot *Duel* I. iii. 68 He enjoys his little Royal Society dinners...Not my cup.
12. b. *cup of tea* (colloq. phr.): (i) used of a person.
1908 W. De Morgan *Somehow Good* xvi. 159 'It's simply impossible to help liking him.' To which Sally replied, borrowing an expression from Ann the housemaid, that Fenwick was a cup of tea. It was metaphorical and descriptive of invigoration. *a* **1909** in Ware *Passing Eng.* (1909) 101/1 Oh, don't yer though. You are a nice strong cup o' tea. **1939** N. Marsh *Overture to Death* xi. 120 Miss Prentice ..seems to be a very unpleasant cup of tea. *Ibid.* xxiv. 279 She was a cranky old cup of tea. **1940** A. Christie *One, Two, Buckle my Shoe* 123 Sounds quite like that old cup of tea who came to see Mrs. Chapman.
(ii) *one's cup of tea*: what interests or suits one.
1932 N. Mitford *Christmas Pudding* xiv. 211 I'm not at all sure I wouldn't rather marry Aunt Loudie. She's even more my cup of tea in many ways. **1933** P. Fleming *Brazilian Adventure* I. iii. 31 The desire to benefit the community is never their principal motive...They do it because they want to. It suits them; it is their cup of tea. **1936** Auden & Isherwood *Ascent of F6* II. iii. 96, I had an aunt who loved a plant—But you're my cup of tea! **1937** N. Coward *Pres. Indic.* III. v. 121 Broadway by night seemed to be my cup of tea entirely. **1948** 'J. Tey' *Franchise Affair* v. 54 Probably she *isn't* your cup of tea...You have always preferred them a little stupid, and blond. **1965** M. Spark *Mandelbaum Gate* v. 141 Freddy had stood in the doorway of the dark Orthodox chapel and, regarding the heavy-laden altar and the exotic clusters of coloured lamps hung round it, said, 'It's not really my cup of tea, you know.'
(iii) *a different cup of tea* (and similar expressions): something of an altogether different kind.
1940 N. Mitford *Pigeon Pie* xiii. 215 A Fred racked with ideals, and in the grip of Federal Union, was quite a different cup of tea from the old, happy-go-lucky Fred. **1946** 'S. Russell' *To Bed with Grand Music* i. 20 London in war-time..is a very different cup of tea from Winchester. **1957** *Listener* 5 Dec. 954/1 The outwitted villain..is quite another cup of tea.
13. a. *cup-eyed* adj.
1922 T. Hardy *Late Lyrics* 33 Cup-eyed care and doubt.
c. Additional combinations and examples; **cup-day** (earlier example).
1860 Mrs. Gaskell *Let.* 27 Aug. (1966) 631 It was Cup Day at Ascot. **1894** *Daily News* 26 Feb. 5/1 Those mighty cup-fighters, the Blackburn Rovers. **1895** *Ibid.* 21 Feb. 5/5 The Wednesday men are noted cup-tie fighters. **1901** *Westm. Gaz.* 22 Apr. 7/3 A typical 'cup-fighting' team. **1902** *Encycl. Brit.* XXIX. 329/2 The expression 'a cup horse' is understood to imply an animal capable of distinguishing himself over a long distance at even weights against the best opponents. **1905** *Daily Chron.* 14 Apr. 8/1 A special brand of play known as 'the Cup-tie game'. *Ibid.* 25 Dec. 3/4 Old Internationals and Cup-final players. **1908** *Pearson's Weekly* 5 Mar. Suppl. p. iii/3 We're playing a cup-tie! **1910** *Westm. Gaz.* 14 Mar. 14/2 The cup-holders were defeated in their first match. **1963** *Times* 10 Jan. 3/4 The..good humoured indulgence afforded Hospital cup-ties. **1968** *Listener* 23 May 681/3 There is something wrong with a game when one of its outstanding young exponents, the new Cup-winners' goalkeeper with Under-23 honours, says about it a few weeks before the Cup Final: 'The worst time of the week for me is between three o'clock and twenty to five every Saturday afternoon.'
d. cup-and-cone *Metall.*, designating a fracture in which one surface of the metal consists of a raised rim enclosing a flat central portion into which the other surface fits; **cup-and-ring,** designation of a type of marks found cut in megalithic monuments, consisting of a circular depression surrounded by concentric rings; **cup-and-saucer** *a.*, designation of a naturalistic style in the late nineteenth-century theatre, introduced by T. W. Robertson; **cup-cake** orig. *U.S.*, a cake baked from ingredients measured by the cupful, or baked in a small (freq. paper) cup; **cup-custard** (U.S. examples); **cup-fungus,** any discomycetous fungus having a cup-shaped ascocarp; cf. *cup-mushroom*; **cup-grease,** a kind of semi-solid lubricant; **cup-head,** a hemispherical head to a bolt; hence **cup-headed** *a.*; **cup hook,** a hook which is screwed into a wall, shelf, cupboard, etc., and used for hanging up cups, etc.; **cup-leather** (see quots.); **cup-mark, -marking,** a shallow cup-like depression found cut in rocks or stone monuments (see 5); also **cup-marked** *a.*; **cup-mouthpiece** (see quot.); **cup mute,** a kind of mute for a trumpet or trombone; so *cup-muted* adj.; **cup-plant** *U.S.* (examples); **cup-sculpture** = *cup-marking; **cup-shrimp** (see quot.).
1925 Cup-and-cone fracture [see *cuppy *a.* c]. **1967** A. K. Osborne *Encycl. Iron & Steel Industry* (ed. 2) 99/2 *Cup-and-cone*..A type of fracture occurring in tensile test pieces from steels possessing reasonable ductility, and containing no local abnormality where the necking occurs. **1867** J. Y. Simpson *Arch. Sculpt.* 2 Cup and ring cuttings. **1875** C. Maclagan *Hill Forts* Index, Cup and Ring Sculpturings. *Ibid.* 41 On one monolith..are some 'cup and ring markings'. **1900** *Daily News* 11 Oct. 6/1 A rude dial at West Kirby looks like an example of 'cup and ring stones'. **1919** *Proc. Soc. Antiq. Scot.* LIII. 23 The cup- and ring-marked stone which was found near this spot. **1963** S. Piggott in Foster & Alcock *Culture & Environment* iv. 64 The cup-and-ring carvings of Galicia have again been brought into relationship with those of Ireland. **1881** *Times* 27 Dec. 3/5 It [sc. Albery's *Two Roses*] has more than the merit, though it has hardly met with the popularity of the 'cup and saucer' comedies of the late Mr. T. W. Robertson. **1892** W. Archer in G. B. Shaw *Prefaces* (1934) 667/2 The scheme of a twaddling cup-and-saucer comedy. **1933** G. B. Shaw in *Shaw on Theatre* (1958) 222 The stuffiness of the London cup-and-saucer theatre. **1828** E. Leslie *Receipts* 61 Cup Cake. **1886** *Harper's Mag.* Dec. 134/2 Cousin Carry with her eternal cup-cake. **1887** M. E. Wilkins *Humble Romance* 271 Mis' Steele made some cup-cake to-day..She put a cup of butter and two whole cups of sugar in it. **1907** *Mrs. Beeton's All about Cookery* (new ed.) 216/2 Cup Cakes, Plain (American Recipe)..3 level cupfuls of flour, 1 cupful of sugar, ½ a cupful of butter, 1 cupful of milk..Bake in shallow tins or small cups. **1911** E. Ferber *Dawn O'Hara* viii. 109 There were little round cup cakes made of almond paste that melts in the mouth. **1957** J. Braine *Room at Top* viii. 82 The cakes were fresh..meringues, éclairs, chocolate cup-cakes. **1853** *San Francisco Whig* 28 July 1/4 (Advt.), Cup Custard. **1862** 'G. Hamilton' *Country Living & Thinking* 72 We had cup-custards at the close of our breakfast that morning. **1867** Mrs. Whitney *L. Goldthwaite* x. 223 Cup-custards, even, disappeared,—cups and all. **1910** *Encycl. Brit.* XI. 341/2 Owing to the shape of the fruit-body many of these forms are known as cup-fungi, the cup or apothecium often attaining a large size. **1960** R. W. G. Dennis (*title*) British cup fungi and their allies. **1966** F. H. Brightman *Oxf. Bk. Flowerless Plants* 150 The *Pezizales* or Cup Fungi have a spore-producing layer which develops within a more-or-less shallow cup. **1900** Archbutt & Deeley *Lubrication & Lubricants* v. 122 'Cup' greases are usually thickened with soap from either horse fat, cottonseed oil, or rape oil, saponified with lime. **1935** *Oil & Gas Jrnl.* 14 Nov. 66/2 Large quantities of soft cup greases..are still used for chassis lubrication. **1951** *Good Housek. Home Encycl.* 187/1 The groove of the frame should be freed..of earth and rust, and packed with..cup-grease. **1929** *Encycl. Brit.* III. 827/2 The cup-head or coach-bolt. **1895** *Montgomery Ward Catal.* 400/1 Brass Cup Hooks, Size ⅝ in. ¾ in. ⅞ in. 1 in. **1925** *Black. Mag.* Jan. 5/1 He put his pipe to rest in a cup-hook screwed at an angle in the window jamb. **1970** R. Jeffries *Dead Man's Bluff* xix. 180 A weight had been suspended by running string through a cup hook. **1889** *Cent. Dict.*, *Cup-leather*, a piece of leather fastened around the plunger or bucket of a pump. For a bucket it is sleeve-shaped, and for a plunger it is made with a solid bottom. **1904** Goodchild & Tweney *Technol. & Sci. Dict.* 143/1 *Cup leather*, a leather ring, produced by forcing a flat ring of leather into a mould. **1930** *Engineering* 25 July 95/3 They have rams..and..glands with triple cup leathers. **1884** *Proc. Soc. Antiq. Scot.* XVIII. 110 Edge of Rock with Cup-marks. **1919** *Ibid.* LIII. 22 The fracture on one side cuts across a cup-mark. **1867** *Ibid.* (1870) VII. 270 A Kist, with a cup-marked Cover. **1875** C. Maclagan *Hill Forts* 45 The cup-marked stone figured on Plate XI. **1935** *Proc. Prehist. Soc.* I. 150 At either end of this are standing stones, one of which is cup-marked. **1867** J. Y. Simpson *Arch. Sculpt.* 7 In the centres of the remaining six series of circles there are no cup-markings. **1877** W. Greenwell *Brit. Barrows* 341 A square piece of the same stone..which has a circular pit or cup-marking on each face. **1911** *Encycl. Brit.* XVIII. 947/1 *Cup-Mouthpieces.*—Brass wind instruments are played by means of cup or funnel-shaped mouthpieces, generally made of silver...The shallower the cup the more suitable it is for producing the higher harmonics. **1955** L. Feather *Encycl. Jazz* ii. 64 A variety of mutes, including..cup..mutes. **1961** A. Berkman *Singer's Gloss* 61 *Cup mute*, a cone-like mute with an added metal or fibre cup which reduces the volume considerably, producing a fine, pleasing tone. **1967** *Crescendo* May 8/2 'Boss Bambino' has bossa nova rhythm and cup-muted trombone. **1846** A. Wood *Class-bk. Bot.* (ed. 2) 336 *Silphium perfoliatum*. Cup-plant. **1870** *Amer. Naturalist* IV. 580 Another species of the same genus, called the cup plant (*Silphium perfoliatum*)..is common in the moist ravines. **1968** Peterson & McKenny *Field Guide to Wildflowers* 184 Cup-plant, *Silphium perfoliatum*. **1911** W. T. Calman *Life of Crustacea* 245 A smaller species..(*Leander squilla*), and another very similar species ..*L. adspersus*,..are said to be sold on some parts of the English coast as 'Cup Shrimps'.

cup, *v.* Add: **3. a.** (Further example.)
1940 Dylan Thomas *Portr. Artist* 117, I cupped a match to let them see my face in a dramatic shadow.
c. *Golf.* To lodge (the ball) in a 'cup' or depression of the ground. (See Cup *sb.* 5.) Usu. as *pa. pple.* or *ppl. adj.*
1896 W. Park *Golf* 95 A cupped ball gives room for playing one of the finest strokes in golf. **1905** H. Vardon *Compl. Golfer* 81 When the ball is really badly cupped. **1909** *Westm. Gaz.* 11 May 12/2 The cleek is only for use when the ball lies cupped.
5. *trans.* To make concave or cup-shaped; to form into a cup.
1909 G. Stratton-Porter *Girl of Limberlost* xv. 299 'Are you afraid she is going?' Elnora asked. 'If you are, cup your other hand over her for shelter.' **1911** *Encycl. Brit.* XXVII. 39/2 Power presses for working sheet-metal articles include those for cutting out the blanks, termed cutting-out or blanking presses, and those for cupping or drawing the flat blank into shape. **1954** Xan Fielding *Hide & Seek* 228 The despatcher..cupped his hand to my ear and shouted.

cupferron (kʌ·p,ferǫn, kiū·p-). *Chem.* [G. (O. Baudisch 1909, in *Chem. Zeitung* XXXIII. 1298), f. L. *cup-rum* copper + *ferr-um* iron + *-on*.] A brownish-yellow crystalline compound, $C_6H_5N(NO)ONH_4$, used as a quantitative precipitant for iron, titanium, zirconium, and certain other metals in acid solution and formerly as a reagent for copper.
1910 *Chem. Abstr.* IV. 557 *Cupferron*..forms complex salts with Cu and Fe, and may be used in the quantitative separation of Cu and Fe from nearly all other metals. **1939** *Thorpe's Dict. Appl. Chem.* III. 335/1 Nitroso-β-phenylhydroxylamine (cupferron) furnishes insoluble red and white complexes with ferric and cupric salts respectively. **1963** Skoog & West *Fund. Anal. Chem.* ix. 195 The ammonium salt of phenylnitrosohydroxylamine, better known as cupferron, is a somewhat more selective organic reagent.

Cupid. Add: **1.** *Cupid's bow*, designation of a shape or outline resembling the double-curved bow of Cupid. *Cupid's dart*, (*a*) the popular name for a variety of *Catananche*; Cupidone; (*b*) (also *Cupid's arrows*) (see quots. 1884, 1910).
[**1858** Lytton *What will he do with It?* III. vi. vi. 165 Evil passions had destroyed the outline of the once beautiful lips, arched as a Cupid's bow.] **1875** T. Seaton *Fret Cutting* 139 It gives the lip that shape called Cupid's Bow. **1884** E. W. Streeter *Precious Stones & Gems* (ed. 4) 292/3 The brilliant hair-brown needles of Rutile, penetrating the crystal in all directions, impart a curious appearance to the stone, and such specimens are often cut for brooches, under the name of Flèches d'Amour, or 'Cupid's arrows', or 'Venus's Hair-Stone'. **1904** B'ness von Hutten *Pam* iv. viii, He's a sweetly pretty youth..with a cupid's-bow mouth. **1910** *Encycl. Brit.* XII. 272/1 This form of the mineral [göthite] has long been known as onegite, and the crystals enclosing it are cut for ornamental purposes under the name of 'Cupid's darts' (*flèches d'amour*). **1910–11** H. Cescinsky *Eng. Furnit. 18th Cent.* II. 246 Figs. 249 and 250 are instructive in exhibiting the evolution of the true 'Cupid's bow' top rail. **1929** J. L. Hodson *Grey Dawn* I. i, Big brown eyes, cupid's bow mouth and broad forehead.

1930 L. H. & E. Z. Bailey *Hortus* 127/1 Catananche. Cupids-Dart...Herbaceous annuals and perennials with narrow leaves borne near base of stem and longstalked blue or yellow heads. **1935** E. J. Salisbury *Living Garden* viii. 127 The dry grassy banks of southern Europe have given us Cupid's Dart (*Catananche coerulea*)..brought to this country at the end of the sixteenth century. **1938** 'J. Bell' *Port of London Murders* ix. 167 She touched up her mouth, curving the cupid's bow well above the natural contour of her upper lip. **1962** R. Webster *Gems* I. x. 163 When the enclosed crystals are long hair-like needles of red or golden-coloured rutile the material is called rutilated quartz...Other more popular names such as 'Venus hair stone', 'Cupid's darts' and 'Flèches d'amour' are applied to the material. **1969** *Gloss. Terms Dentistry (B.S.I.)* 44 '*Cupid's bow' operation*, a procedure to re-adjust the vermilion border..of the upper lip into the classical Cupid's bow conformation.

cupola, *sb.* Add: **1. a.** *spec.* The dome of the building in which the French Academy meets (*la coupole de l'Institut*).
1900 *Westm. Gaz.* 21 June 1/3 It is only by a majority of one that M. Hervieu is called 'under the cupola'. **1920** *Edin. Rev.* Oct. 276 These guardians of the Cupola. *Ibid.* 279 On the 20th of March 1919, M. René Boylesve..was received under the Cupola by the poet, M. Henri de Regnier.
4*. *Geol.* A small dome-shaped projection on a batholith.
1911 R. A. Daly in *Proc. Amer. Acad. Arts & Sci.* XLVII. 69 Plutonic Cupolas...The juvenile gases tend to accumulate in any cupola-like irregularities in the roof. *Ibid.* 70 Round intrusive bosses or small stocks are characteristic cupola forms on large batholiths...It is evident that every such cupola increases as well as localizes the danger of true volcanic action. **1914** —— *Igneous Rocks & their Origin* vi. 102 These projections of the igneous mass have been called 'cupolas', after the analogous relation of an artificial cupola to the building of which it is a part. Many stocks are cupolas on batholiths. **1954** M. P. Billings *Struct. Geol.* (ed. 2) xix. 319 Cupolas are isolated plutonic bodies that presumably connect downward with the main batholith.

cupolated (kiū·pǫlēⁱtėd), *a.* Delete † *Obs.* $rare^{-1}$ and add example.
1924 *Contemp. Rev.* Jan. 78 Their carbuncled, cupolated tops.

cuppa (kʌ·pă). A form, freq. in modern times, of *cup o'*. Also used *ellipt.* for *cup o' tea*. *colloq.*
1925 Wodehouse *Sam the Sudden* vi. 42 Come and have a cuppa coffee. **1934** N. Marsh *Man lay Dead* xii. 211 Taking a strong cuppa at six-thirty in their shirt sleeves. **1942** E. Langley *Pea Pickers* 78 At Seymour he got out and begged us to have a 'little cuppa tea and a sangwidge, miss'. **1949** S. Gibbons *Conf. at Cold Comfort Farm* iii. 39 Come whoam, come whoam, you piece o' dirt. We've no more business here, and I lusts for me cuppa. **1959** 'A. Gilbert' *Death takes a Wife* ix. 119 Sit down and have a cuppa. **1968** M. Richler *Cocksure* viii. 46 'Good morning,' Joyce said. 'Coffee?' 'If it's no trouble I'd prefer a cuppa.' **1970** G. Greer *Female Eunuch* 117 Barbara Castle dealt with [it] by the disgusting expedient of having a cuppa with the women and talking it over heart to heart.

cupped, *ppl. a.* Add: **2.** Lying as if in a cup.
1929 W. Faulkner *Sartoris* I. ii. 44 He repeated above the cupped match. **1946** Dylan Thomas *Deaths & Entrances* 30 And the bird descended On a bread white hill over the cupped farm.

cupper². *Oxford University slang.* [f. Cup *sb.* + *-er⁶.] A series of intercollegiate matches played in competition for a cup. Freq. in pl.
1900 *Oxford Mag.* 31 Oct. 52/2 In the Cuppers we are drawn to play against New College. **1903** *Ibid.* 11 Feb. 214/1 Hockey.—..We are drawn against Magdalen in the second round of the 'Cupper'. **1928** *Observer* 18 Mar. 13/4 The final of 'the rugger cupper'. **1937** C. Day Lewis *Starting Point* 37 Next week I've got to go into training for Cuppers. **1961** *Times* 8 Mar. 18/1 The Oxford University Rugby cuppers final.

cuppiness (kʌ·pinės). *Metall.* [f. *cuppy *a.* c + -ness.] The state or condition of being cuppy (see *cuppy *a.* c).
1927 *Jrnl. Iron & Steel Inst.* CXV. 453 It may be contended that this 'cuppiness' can be somewhat obviated by taking very light drafts. **1932** E. Gregory *Metall.* ii. 54 Axial segregation gives rise to the formation of internal flaws and 'cuppiness' in wire-drawing. **1949** R. T. Rolfe *Dict. Metall.* (ed. 2) 65 Cuppiness or cupping..is due to excessive cold working before further annealing, the exterior skin..being..enabled to withstand a much greater deformation than the unsupported interior metal, in which failure first occurs.

cupping, *vbl. sb.* Add: **3.** *Metall.* The process of forming a depression in sheet metal by forcing a plunger into it when it is laced over a die, used either to fabricate articles or as a test of the ductility of the metal; so *cupping test*.
1921 *Jrnl. Inst. Metals* XXV. 441 A ductility testing machine is described in which the direct pressure necessary to distort the material is measured. The pressure to cause the cupping is weighed directly by means of a confined liquid and a gauge. **1927** *Jrnl. Iron & Steel Inst.* CXV. 926 (*title*) A cupping test for determining qualities of thin metal sheets. **1964** Gregory & Simons *Steel Working Processes* v. 130 Cupping is applied to such parts as containers, brake drums, dishes, pans, trays, beakers, etc.
b. *Metall.* Cuppiness; also, the flaws present in cuppy wire.
1925 A. T. Adam *Wire-Drawing* x. 199 Two examples of segregation, one in a thick mild steel bar and the other in a wire, both of which resulted in cupping, may be given. **1949** [see prec.]. **1952** *Jrnl. Iron & Steel Inst.* CLXX. 223/1 The rupturing of the segregate bands is analagous [*sic*] to the well-known 'cupping' in segregated cold-drawn wire. **1968** R. N. Parkins *Mech. Treatm. Metals* iv. 245 If too large a reduction is attempted, the wire may break or develop internal cracks associated with localized necking and referred to as cupping.

cuppy, *a.* Add: **b.** *Golf.* *cuppy lie*, the position of a ball when it lies in a 'cup' or shallow depression. Said also of the ball.
[**1886** H. Hutchinson *Hints Game Golf* 32 If it [sc. the ball] lie 'cuppy', a jerking stroke will be necessary.] **1892** [in Dict.]. **1901** W. J. Travis *Pract. Golf* (1903) iv. 46 It is better to..play each shot the same way—except in the case of a very cuppy lie. **1922** Wodehouse *Clicking of Cuthbert* 88 Mortimer..found his ball in a nasty cuppy lie.
c. *Metall.* Of drawn metal, esp. wire: having internal cavities that lead to a cup-and-cone fracture under sufficient tensile stress.
1925 A. T. Adam *Wire-Drawing* x. 197 'Cuppy' wire—*i.e.* wire which breaks either in drawing or in bending with a very distinct 'cup and cone' fracture. **1927** *Jrnl. Iron & Steel Inst.* CXV. 470 He desired to ask the author whether internal cuppy fractures were not liable to be produced by wrong manipulation in the drawing. **1958** A. D. Merriman *Dict. Metall.* 52/1 *Cuppy wire*...Wire which, though apparently sound, shows on a longitudinal section a series of well-developed internal fractures which open to the surface under bending stresses.

cuprammo·nia. Also erron. **cupra-ammonia.** [f. as next + Ammonia.] A solution containing a cuprammonium salt; a cuprammonium solution.
1862 H. Watts tr. *Gmelin's Hand-bk. Chem.* XV. 143 The solution of cellulose in cuprammonia is precipitated by a large quantity of water. **1927** M. H. Avram *Rayon Industry* 497 The cupra-ammonia cellulose process.

cuprammonium (kiūprămōᵘ·niǫm). Also erron. **cupra-ammonium.** [f. L. *cupr-um* copper + Ammonium.] A complex ion of copper and ammonia, $Cu(NH_3)_4^{++}$, obtained in solution by adding excess ammonium hydroxide to a solution of a cupric salt; it forms deep blue aqueous solutions which dissolve cellulose and are used industrially, esp. in the manufacture of some man-made fibres. So *cuprammonium process, rayon, silk*, etc. Also used *ellipt.* for *cuprammonium solution*.
1862 H. Watts tr. *Gmelin's Hand-bk. Chem.* XV. 142 Cotton immersed in aqueous biphosphate of cuprammonium becomes first gelatinous, then slippery, and forms an almost transparent gummy liquid. **1888** Cross & Bevan *Paper-Making* xvi. 217 The cuprammonium solution. **1907** Roscoe & Schorlemmer *Treat. Chem.* (ed. 4) II. 435 This dissolves to a deep pure blue liquid containing cuprammonium sulphate, $CuSO_4$, $4NH_3$, H_2O, which was first described by Stisser in 1693 as an *arcanum epilecticum* and afterwards termed *cuprum ammoniacale*. **1921** T. Woodhouse tr. *Foltzer's Artif. Silk* 29 Despeissis Artificial Silk (Cuprammonium Process). *Ibid.* 36 Cuprammonium Solution..is a solution of copper oxide in ammonia. It is a blue liquid known to chemists as Schweitzer's reagent. **1925** *Good Housekeeping* Apr. 142/2 Cuprammonium silk. **1927** M. H. Avram *Rayon Industry* 192 The cupra-ammonium solution of cellulose. *Ibid.* 497 Cupra-ammonium Rayon. **1939** *Thorpe's Dict. Appl. Chem.* (ed. 4) III. 355/1 'Schweizer's reagent' or cuprammonium..has the property of dissolving cellulose (cotton wool, linen, filter paper, etc.). **1960** Wood & Holliday *Inorg. Chem.* xiv. 347 The deep blue solution of cuprammonium sulphate deposits crystals..of composition, $Cu(NH_3)_4SO_4.H_2O$. **1964** N. G. Clark *Mod. Org. Chem.* xvi. 333 One commercial process, giving Cuprammonium Rayon, exploits the ability of an aqueous ammoniacal copper salt solution to dissolve cellulose.

cuprea (kiū·priă). [L., fem. of *cupreus* Cupreous *a.*] Used *attrib.* in *cuprea bark*, the coppery-red bark of the S. American tree *Remijia pedunculata* (and other species), one of the sources of quinine.
[**1882** *Jrnl. Chem. Soc.* XLI. 66 The singular bark described by Dr. Flückiger as *China Cuprea*.] **1884** *Pharm. Jrnl.* 23 Aug. 141/1 The alkaloid of cuprea bark. **1889** G. S. Boulger *Uses of Plants* ii. 100 *Remijia Purdieana*, Willd., and *R. pedunculata*, Trian., from Colombia, have of late years been imported in enormous quantities as sources of quinine etc., under the name of Cuprea Bark. **1939** [see *cupreine]. **1970** W. Solomon in S. W. Pelletier *Chem. Alkaloids* xi. 325 It is doubtful in this connection whether Cuprea bark (*Remijia pedunculata*), which was the sole natural source of cupreine, is now obtainable.

cupreine (kiū·pri,in). *Chem.* [f. *cuprea + -ine⁵.] A cinchona alkaloid, $C_{19}H_{22}N_2O_2$, contained in cuprea bark.
1884 Paul & Cownley in *Pharm. Jrnl.* 20 Sept. 222/1 This latter alkaloid is susceptible of being split into two other alkaloids, one of them being..quinine..the other an alkaloid..to which we will provisionally give the name of 'cupreine'. **1886** *Encycl. Brit.* XX. 185/1 Homoquinine has been shown..to be decomposed on treatment with caustic soda into quinine and a new alkaloid, cupreine. **1939** *Thorpe's Dict. Appl. Chem.* (ed. 4) III. 160/2 Cupreine..occurs together with other cinchona alkaloids..in the so-called 'Cuprea bark'..which is no longer collected commercially. **1966** *McGraw-Hill Encycl. Sci. & Technol.* XI. 193/2 The cinchona alkaloids..are quinoline derivatives; in quinine and quinidine, $R = CH_3O$; in cinchonine, $R = H$; and in cupreine, $R = OH$.

cupro-. Add: **cu:pro-ni·ckel,** an alloy of copper and nickel.
1905 *Kynoch Jrnl.* Jan.–Mar. 18 The bullet is sheathed with a cupro-nickel envelope. **1922** *Encycl. Brit.* XXX. 135/2 This jacket is made from cupro-nickel which..contains from 80% to 85% copper and from 15% to 20% nickel. **1948** M. Laski *Tory Heaven* xii. 172 A golden sovereign..changed into four cupro-nickel half-crowns. **1962** *Metal Industry* CI. 102/1 The term cupro-nickel is now more widely applied to any copper-nickel alloy containing less than about 50 per cent of nickel.

curate. Add: **2. b.** *curate's egg*: taken as a type of something of mixed character (good and bad).
Originating in a story of a meek curate who, having been given a stale egg by his episcopal host, stated that 'parts of it' were 'excellent' (*Punch* 9 Nov. 1895, p. 222).
1905 *Minister's Gazette of Fashion* Aug. 141/1 The past spring and summer season has seen much fluctuation. Like the curate's egg, it has been excellent in parts. **1962** *Oxf. Mag.* 22 Nov. 91/1 All the same it is a curate's egg of a book. While the whole may be somewhat stale and addled, it would be unfair not to acknowledge the merits of some of its parts.
5. *jocular.* A small poker (see quot.).
1891 *N. & Q.* 7th Ser. XII. 206/2 A 'curate' is a small auxiliary poker with a steel point, intended for use, in contradistinction to the elaborate fire brasses, which are only kept for show.
b. A cake-stand with two or more tiers. Also called *curate's comfort, delight, friend*.
1914 G. B. Shaw *Fanny's First Play* III. p. 216 He places the tray on the table. He then goes out for the curate... Juggins returns with the cakes. **1934** M. Harrison *Weep for Lycidas* I. 152 There were two of those curious stands, known as 'curate's delights', full of cakes. **1937** N. Coward *Pres. Indic.* v. iii. 180 There..were spread tea-tables..'Curate's Comforts', and large bowls of strawberries and cream. **1968** M. Allingham *Cargo of Eagles* xv. 163 A three tier 'curate's friend' cake stand.
6. In Ireland, a spirit-grocer's assistant.
1909 M. Hayden & Hartog in *Fortn. Rev.* Apr. 781 'Curate'..is the assistant to a 'spirit grocer', such as most grocers are in Ireland. **1914** Joyce *Dubliners* 184 These two gentlemen and one of the curates carried him up the stairs and laid him down again on the floor of the bar.

curating (kiūᵊ·rātiŋ), *vbl. sb.* [-ing¹.] Acting as curate; performing the duties of a curate.
1831 W. Cobbett *Two-penny Trash* Jan. 159 Non-residence, or stipendiary curating. **1907** E. H. Begbie *Vigil* ii. 23, I am to commence vicar. No curating. *Ibid.* v. 68 You have begun curating for me already.

curative. II. (U.S. example.)
1857 D. E. E. Braman *Inform. Texas* i. 15, I place great confidence in the frequent outward use of cold water, as a preventive and curative.

curb, *sb.* Add: **12.** In this sense the spelling is now usu. 'curb' in U.S., 'kerb' in U.K.
b. The body of curbstone brokers. *U.S.*
1903 *Nation* (N.Y.) 4 June 446 The Stock Exchange and the 'curb'..gave..plain evidence what their opinion was.
15. curb-market, -price, -stocks *U.S.* (cf. Curb-stone b and *kerb *sb.*); **curb service** *N. Amer.*, service by a shop, etc., to customers in cars at the street curb.
1900 S. A. Nelson *ABC of Wall St.* 10 The curb market, with its swarm of brokers. **1914** *N.Y. Herald* 17 Nov. 6/4 Prices again strong in curb market. **1930** *San Antonio (Texas) Light* 31 Jan., Closing Curb Prices. **1931** *Kansas City Star* 25 Aug., The hoppers sit on the curb in front of the drug store, honk and attack the curb service boys when they come out. **1938** *Archit. Rev.* LXXXIV. 137 (*caption*) A curb-service restaurant in Washington, D.C. **1962** *Canadian Jrnl. Linguistics* VII. 73 A culture pattern based on the automobile, with its motels, filling stations, and curb-service. **1915** *World's Work* (N.Y.) Oct. 641 Unlisted (Curb) Stocks.

curbing, *vbl. sb.*² **2.** (Earlier U.S. example.)
1838 J. Hall *Notes Western States* viii. 106 After removing the rich soil, a stratum of hard clay presents itself, then gravel, and then another layer of clay, all of which are so compact as to require no curbing.

curb-stone. Add: **a.** (Earlier example.)
1791 *Act* 31 *Geo. III* c. 65 § 124 Curb Stone from Eleven to Thirteen Inches wide and from Five to Seven Inches thick.
b. *U.S.* (Earlier example.) Also **curbstone agent, operator.**
1848 W. Armstrong *Stocks* 7 This class comprehends..all those petty operators and non-descripts, who have neither a local habitation or scarcely, a name, that are dignified by the title of curb-stone brokers. **1861** *Knickerbocker* LVII. 635 All sorts of brokers, from the leading houses down to the curbstone 'operator'. **1862** R. B. Kimball *Undercurrents* (1868) 321 It is rather a habit with the Curbstone operator when he gets severely winged, to go into the cigar business. **1884** *Cent. Mag.* Aug. 629/2 Peddling 'privileges' to small speculators through curb-stone agents.

curd, *sb.* Add: **2. c.** The edible 'head' of such brassicas as cauliflower and broccoli.
1916 W. F. Rowles *Food Garden* xi. 201 We may expect to cut the curds at the end of April. **1950** *N.Z. Jrnl. Agric.* Feb. 154/1 Most cauliflower crops..benefit from a side dressing applied shortly before the curd begins to form.

1951 *Good Housek. Home Encycl.* 393/2 It [sc. the cauliflower] has a compact white head (i.e. flowers, often called the curd). **1969** D. Bartrum *From Garden to Kitchen* 29 Broccoli, with their white, solid flower-heads (curds) are like a small cauliflower but a much hardier vegetable.

curd, *v.* **2.** *fig.* (Later poet. example.)
1887 G. M. Hopkins *Poems* (1918) 65 His thew That onewhere curded, onewhere sucked or sank.

curdle, *sb.* (Later example.)
1933 D. L. Sayers *Murder must Advertise* iv. 65 What disgusting stuff cauliflower could be—a curdle of cabbage!

curdler. Add: **2.** A story, etc., that curdles the blood (cf. Curdle *v.* 3 b).
1886 Baumann *Londinismen* 36/1 *Curdler*,..blood-freezer. **1966** *Listener* 21 July 103/3 [A play] too complicated to be a real curdler.

curdy, *a.* **3.** (Later examples.)
1859 Lever *Davenport Dunn* xxxvi, His curdiest salmon declined, his wonderful 'south-down' sent away scarcely tasted. **1892** H. G. Hutchinson *Fairway Island* i, We'll eat this [salmon] that had the tide-lice on him. He'll be fine and curdy.

cure, *sb.*¹ **9.** Delete † and add later examples. Also, a catch of fish so treated.
1883 A. Shea *Newfoundland* 7 The cure of the fish requires much care and judgment...The best cure is effected when the weather is variable. **1902** *Encycl. Brit.* XXXI. 143 The fish caught round the Newfoundland coast are generally of good quality, but the Labrador cure..is often very inferior. **1911** 'Viking' *Art of Fishcuring* xiv. 67 When salting the fish in the tubs it would not be advisable to put one day's fish down upon the top of the previous day's cure. **1957** *Fish Marketing in W. Europe* (O.E.E.C.) ii. 41 The demand for salted herring was declining, and..the consumer was showing more interest in soft cures, such as pickled, marinated and smoked.
11. [After F. *cure*, G. *kur, cur* (see Kursaal).] A period of residence at a health-resort, under medical regimen, in order to restore or benefit one's health. Also *Comb.*, as *cure-guest* (= G. *kurgast*), *-seeker*.
1887 *Time* Oct. 420 The month's 'cure' at Carlsbad. **1898** *Daily News* 22 July 5/1 One of the speakers was an old cure servant. **1905** *Westm. Gaz.* 11 Sept. 10/2 The number of cure-guests registered [at Carlsbad]. **1906** *Ibid.* 27 Aug. 8/1 Cure-seekers at Homburg. **1908** T. P. O'Connor *Campbell-Bannerman* 123 He rarely took the cure [at Marienbad]. *Ibid.*, The severe waters which the other cure-guests were taking. **1921** D. H. Lawrence *Let.* ?8 May (1962) II. 653, I..can't sit supping for ever at these inside Baden-Baden cure-springs. **1955** *Times* 5 July 8/4 [He] has left Berlin on his annual leave for an undisclosed destination 'to take a cure'. **1967** E. S. Turner *Taking Cure* 9 Taking the cure was usually a quest for healing waters.
12. The process of vulcanizing rubber (see also quot. 1923) or of hardening or curing plastic; also (with qualifying adj.), the degree of hardness produced.
1902 C. O. Weber *Chem. India Rubber* ix. 301 The pigments and other colouring matters contained in the india rubber...contain some impurity which is responsible for their discoloration, or perhaps the 'cure' has been too prolonged or carried out at too high a temperature. **1907** H. L. Terry *India-rubber* 32 Fine Para rubber..varies slightly in its properties and price according as it is 'Upriver hard cure' or 'Island soft cure'. **1908** H. A. Wickham *Para Rubber* 24 Extraction and cure of the rubber *latex*. *Ibid.* 29 The antiseptic smoke-cure. **1909** *Westm. Gaz.* 9 Nov. 12/1 Fine Hard Cure Para Rubber. **1922** H. E. Simmons *Rubber Manuf.* viii. 48/2 This variation in rate of cure or vulcanizing capacity. *Ibid.* 98/1 There are two general methods of vulcanization, namely, what is known as the 'cold cure' and the 'hot cure vulcanization'. **1923** B. D. W. Luff *Chem. Rubber* 19 In works practice, and indeed in technical literature, the term 'cure' is frequently employed instead of 'vulcanisation'. While this has the merit of brevity, it is unfortunately used also to denote the 'smoking' of wild or cultivated rubber in the course of its preparation. *Ibid.* 136 Hydrochloric acid gives a rubber having a slower rate of cure. *Ibid.* 137 The effect of alum in retarding the cure of the rubber. **1943** Simonds & Ellis *Handbk. Plastics* iii. 136 The state of cure of a laminated material can be determined with some degree of success by a water-absorption test. **1947** R. L. Wakeman *Chem. Commercial Plastics* xxvi. 786 Where concentrations of catalyst in the order of 1 per cent are used, heating to 175–260 °F effects cure after several hours. **1961** L. R. Mernagh in W. J. S. Naunton *Appl. Sci. Rubber* xii. 1062 Hot-air cures may be divided into open-air cures at atmospheric pressure and oven cures.

cure, *v.*¹ Add: **5. b.** *absol.* or *intr.* To effect a cure; often in *kill or cure* (see Kill *v.* 7 e).
1593 Shaks. *2 Hen. VI*, v. i. 99 Whose Smile and Frowne, like to Achilles Speare Is able with the change, to kill and cure. **1764–1875** [see Kill *v.* 7 e]. **1787** Cowper *Stanzas Bill Mortality* I. 27 No Med'cine, though it often cure, Can always baulk the Tomb. **1908** *Smart Set* Sept. 82/1 Butter-milk is good for it...Warranted to cure in thirty days or money refunded.
9. *intr.* To reside for some time at a health-resort, following a regimen for the benefit of one's health. See *cure *sb.*¹ 11.
1902 *Westm. Gaz.* 22 Aug. 2/1 Those who have come up to 'cure' at Davos. **1905** *Ibid.* 9 June 10/1 They 'cured' together on the balcony, and rowed together on the lake.
10. *trans.* To vulcanize (rubber); also, to harden (plastic) or otherwise improve physical properties during manufacture by chemical treatment.
1853 C. Goodyear *Gum-Elastic* I. vii. 102 Among many experiments for drying and curing the gum,..the inventor was much elated with the result of one. **1881** *Encycl. Brit.* XII. 841/2 The calendered sheets are generally cured between folds of wet cloth. **1902** C. O. Weber *Chem. India Rubber* ix. 299 'Dry heat cured' water-proof fabrics. *Ibid.*, 'Cold cured' cloth. **1907** H. L. Terry *India-rubber* 79 Goods cured by Dry Heat..are less likely to be damaged by copper than those which are cold cured. **1908** H. A. Wickham *Para Rubber* 29 The standard rubber known in commerce as 'fine Parà' is smoke-cured. *Ibid.* 32 The weight of the cured rubber should approximate very nearly that of the *latex* used. **1922** H. E. Simmons *Rubber Manuf.* viii. 48/2 They cured all of their samples at a temperature of 140 °C. **1947** R. L. Wakeman *Chem. Commercial Plastics* xxvi. 782 Furfuryl alcohol can be reacted with formaldehyde to yield a viscous mass which can be cured to a thermoset composition by application of heat. **1961** D. W. Huke *Introd. Natural & Synthetic Rubbers* v. 82 The early synthetic rubbers were much more difficult to cure than natural rubber. **1964** Oleesky & Mohr *Handbk. Reinforced Plastics* i. 8 The resin is fully cured and has become an infusible solid.
b. *intr.* To become vulcanized, undergo vulcanization or curing.
1922 H. E. Simmons *Rubber Manuf.* viii. 48/2 A rubber which cures an hour and forty-five minutes more quickly than plain or smoked sheets. *Ibid.* 49/1 A rapid curing rubber. **1923** B. D. W. Luff *Chem. Rubber* 136 Sulphuric acid gives a slow-curing rubber if used in slight excess. **1961** D. W. Huke *Introd. Natural & Synthetic Rubbers* v. 83 With sulphur and accelerators present the compounded rubber may start to cure while being processed.
11. *trans.* To harden (concrete).
1918 Hool & Johnson *Concrete Engineers' Handbk.* ii. 156 Where products are cured in this way, it is necessary that racks or cars be used. *Ibid.* 158 The curing rooms usually open into the molding department as conveniently as possible to the machines supplying the greatest number of products to be cured. **1921** Hatt & Voss *Concrete Work* II. 175 Demonstrate the relative strength of concrete when cured in the hot sun, in dry air, and in wet sand. *Ibid.* 179 The stone must be cured under a wet cloth. **1953** *Archit. Rev.* CXIII. 85 If these connections are grouted, the whole erection is held up while this is being poured and cured. **1970** *Fremdsprachen* 44 We took great care in curing the concrete, believing that by preventing escape of water from the slab, drying shrinkage would be very small.

curelessness (kiūᵊ·ɹlėsnės). The quality or condition of being cureless.
1892 Kipling & Balestier *Naulahka* xiii, Her heart torn with the curelessness of it all.

Curetonian (kiuᵊɹtōᵘ·niăn), *a.* and *sb.* Designation of the Syriac version of the Gospels discovered by the Rev. William *Cureton*, and edited by him from the MS. in 1858.
1861 Scrivener *Introd. Crit. N.T.* 236 The Curetonian Syriac. *Ibid.* 237 Such cases..are common to the Curetonian with the Peshito. **1904** F. C. Burkitt (*title*) Evangelion Da-Mepharreshe: The Curetonian Version of the Four Gospels. *Ibid.* II. 17 Where the photograph clearly agreed with the Curetonian against the Peshitta.

curettage (kiure·tā3). *Surg.* [Fr.: see Curette and -age.] The application of the curette; scraping or cleaning by means of a curette.
1897 *St. Thomas's Hosp. Rep.* New Ser. XXV. 84 Various supplementary measures have been added to the curettage. **1908** *Practitioner* Feb. 180 No more than a curettage of the growth was attempted. **1964** L. Martin *Clin. Endocrinol.* (ed. 4) viii. 252 But curettage is seldom desirable in adolescents in whom the condition usually rectifies itself spontaneously. **1965** J. Pollitt *Depression & its Treatment* v. 72 Powerful analgesics or premedication may be urgently required should threatened or incomplete abortion occur, or a dilatation and curettage be needed as a result.

curettement (kiure·tmĕnt). *Surg.* [See prec. and -ment.] = prec.
1908 *Practitioner* Dec. 787 The right ovary had been removed elsewhere five years ago, at which time curettement had also been done.

curf (kǭɹf). *local.* Also **carf, kerf.** [var. f. Carf, Kerf.] A cherty limestone found in one of the strata of the Portland beds of stone.
1839 *Civil Engin. & Arch. Jrnl.* II. 375/2 A middle or curf bed occurs only in the southernmost of the quarries. **1893** *Spon's Mechanic's Own Bk.* (ed. 4) 564 Then..the Bastard-Roach, Kerf, or Curf is reached. **1936** *Discovery* Mar. 76/2 Between the 'Whit Bed' and 'Base Bed' there is an irregular deposit of curf and flint, locally called 'Pericott'.

curfew. Add: **1. a.** Also *transf.* and *fig.*
1922 Joyce *Ulysses* 414 When the curfew rings for you. **1929** D. H. Lawrence *Pansies* 103 The curfew of our great day..the tocsin of this our civilisation.
b. In extended use: a restriction imposed upon the movements of the inhabitants of an area for a specified period.
1939 *Punch* 18 Oct. 435/1 The attempt..to get a nine o'clock curfew imposed on members of the Women's Land Army in training..to prevent them going out with soldiers. **1964** *Ann. Reg.* 1963 309 The new Prime Minister..imposed a dusk-to-dawn curfew in Jerusalem. **1970** D. Stuart *Very Sheltered Life* 70 There was an immediate curfew. Everyone had to be off the streets.
3. *curfew law, order.*
1897 J. Bryce *Impr. S. Afr.* xxi. 447 Cape Colony has a so-called 'curfew law', requiring natives who are out of doors after dark to be provided with a pass. **1931** *Daily Express* 16 Oct. 11/3 The curfew order of the university is that no cars may be used after nine without leave.

curfuffle, *sb.* (Earlier and later examples.) Also in forms *carfuffle*, *gefuffle, *ker-, kurfuffle. Now widely used as a colloquialism.
1813 G. Bruce *Poems* 65 An' Jeanie's kirtle, aye sae neat, Gat there a sad carfuffle. **1971** *Times* 9 Jan. 16/4 Since the predictable pre-April curfuffle, there has been the predictable summer and autumn hush.

curia. Add: **4.** *Hist.* **a.** Each of four electoral bodies in the Austrian constitution of 1861. **b.** Each of three bodies, representing respectively the nobles, knights, and towns, into which the members of the estates of Bohemia were divided in 1446.
1907 *Westm. Gaz.* 28 May 2/1 How false was the idea given by the old Curia Parliament of what were the feelings and the aspirations of the people of Austria. **1908** *Ibid.* 16 Jan. 2/1 He wished to see Bohemia divided into curias—Germans being governed by a German, and Czechs by a Czech curia. **1910** *Encycl. Brit.* III. 26/1 (Austria). *Ibid.* IV. 126/2 (Bohemia).

curiara (kūriā·ra). The native name in Venezuela and Colombia for a dug-out canoe.
1910 'H. J. Mozans' *Up the Orinoco* 174 The curiara is smaller than the bongo or falca. **1910** M. B. & C. W. Beebe *Our Search* 12 A network of narrow channels..allowed us to explore the far interior in our shallow curiara or dug-out. **1927** *Chambers's Jrnl.* 290/1 Their curiaras are very strong and carefully made.

curie (kiūᵊ·ri, ‖küri). [Named in honour of Pierre *Curie* (1859–1906), co-discoverer of radium.] **1.** Orig., a quantity of radon (radon 222, radium emanation) in radioactive equilibrium with one gramme of radium; later extended to denote an equivalent quantity of any of the decay products of radium. Now, a unit of radioactivity equal to $3{\cdot}7 \times 10^{10}$ disintegrations per second, freq. used *loosely* as a unit of quantity of any radioactive substance in which there is this degree of radioactivity.
1910 Rutherford in *Nature* 6 Oct. 430/2 It was suggested that the name Curie, in honour of the late Prof. Curie, should..be employed for a quantity of radium or of the emanation...The name Curie should be used as a new unit to express the quantity or mass of radium emanation in equilibrium with one gram of radium (element). **1931** M. Curie et al. *Rep. Internat. Radium-Standards Committee* in *Rev. Mod. Physics* III. 432 It is recommended that the use of the term curie be extended to include the equilibrium quantity of any decay product of radium. **1954** *Brit. Jrnl. Radiol.* XXVII. 243/2 Amount of radioactive material shall be expressed in curies (c). The accepted definition of the curie is:—The curie is a unit of radioactivity defined as the quantity of any radioactive nuclide in which the number of disintegrations per second is $3{\cdot}700 \times 10^{10}$. With this definition the curie is independent of the disintegration rate of radium. **1955** *Sci. Amer.* July 50/3 The most powerful modern atomic bomb should release no more than 10 billion curies. **1963** Jerrard & McNeill *Dict. Sci. Units* 37 The curie is too large for normal laboratory work where the radioactivity is generally of the order of millicuries. **1968** *Radiation Quantities & Units* (Internat. Commission on Radiation Units & Measurements) 6 In accordance with the former definition of the curie as a unit of quantity of a radioactive nuclide, it was customary and correct to say: 'Y curies of ^{32}P were administered..' It is still permissible to make such statements rather than use the longer form which is now correct: 'A quantity of ^{32}P was administered whose activity was Y curies.' **1970** *Sci. Jrnl.* Aug. 43/1 A single 1000 MWe reactor will therefore accumulate in its fuel many thousands of millions of curies of fission products.
2. Curie point or **temperature,** a temperature at which the type of magnetism exhibited by a substance changes; *spec.* that at which a ferromagnetic substance, on being heated, loses its ferromagnetism and becomes paramagnetic; also, an analogous temperature for a ferroelectric substance, at which it either loses its ferroelectricity or becomes ferroelectric.
1919 *Chem. Abstr.* XIII. 951 At about 1280° Fe undergoes a transformation with respect to its magnetic properties and this is referred to here as the Curie or A4 point. **1925** *Jrnl. Iron & Steel Inst.* CXII. 267 It will be observed that the line of magnetic transformation (the Curie points) in the γ-phase approaches the α to γ transformation curve. **1957** *Encycl. Brit.* XIV. 650/1 The temperature at which any ferromagnetic material loses its magnetism is known as the Curie point; it is 770 °C. for iron and 358 °C. for cobalt. **1960** *McGraw-Hill Encycl. Sci. & Technol.* III. 623/2 The transition between ferrimagnetism and paramagnetism is also marked by a Curie temperature. **1967** *Electronics* 6 Mar. 24/1 (Advt.), High skin temperatures demand an antenna with a high Curie point coupled with low loss characteristics at high frequencies. **1968** Condon & Odishaw *Handbk. Physics* (ed. 2) iv. vii. 119 The ferroelectric range of Rochelle salt is very narrow and that of the phosphates and arsenates is limited to low temperatures. Both crystal types..are piezoelectric above the Curie point.

curing, *vbl. sb.* Add: **2. b.** The process of rendering a substance harder or more durable (cf. *cure *v.*¹ 10, 11).
1853 C. Goodyear *Gum-Elastic* I. x. 165 (*heading*) Curing or tanning, commonly known as the acid gas process. **1881** *Encycl. Brit.* XII. 841/1 After which the true vulcanization, or 'curing' as it is termed, can be brought about in the usual way. **1918** Hool & Johnson *Concrete Engineers' Handbk.* ii. 156 Conditions for curing must be such that the product will not be rapidly dried. **1921** Hatt & Voss *Concrete Work* I. viii. 180 The process of keeping the product damp is called curing. It is done by sprinkling, immersion, and by the use of steam. **1930** *Engineering* 25 July 97/3 The sustained pressure during the curing process [of rubber]. **1964** Oleesky & Mohr *Handbk. Reinforced Plastics* i. 8 Thermosetting plastics..become increasingly infusable on heating. They undergo a chemical change which is not reversible. This reaction is called polymerization or curing.

3. *curing-room* (examples).
1933 B. Silliman *Man. Sugar Cane* 46 The line of the floor..deviates only six inches in the curing rooms. **1878** *Rep. Vermont Board Agric.* 79 In connection with..the manufacturing room was the curing room.

curio. Add: **b.** *curio-shop.*
1920 M. Beerbohm *Let.* 18 May (1964) 245 Your visits to the curio-shops. **1970** V. McKenna *Some of my Friends have Tails* 77 We had to film a scene in a curio shop—the type that sells objects made from skin and ivory and leather.

‖**curiosa** (kiūᵊri,ōᵘ·să), *sb. pl.* [neut. pl. of L. *cūriōsus* (see Curious *a.*).] Curiosities, oddities; *spec.* erotic or pornographic books (cf. *curious *a.* 16 b).
1883 *Sat. Rev.* 17 Mar. 350/2 Indicating at the end of his preface such miscellaneous *curiosa* as may be found in the mighty volume ensuing. **1920** M. Secker *Let.* Jan. in C. Mackenzie *Life & Times* (1966) V. 169 No bookseller would buy a single copy of the Rainbow even with my imprint unless he dealt in 'curiosa'. **1947** N. Marsh *Final Curtain* ix. 143 She's not..the type to pore over literary curiosa unless..they were curious in the specialised sense. **1959** J. Thurber *Years with Ross* iv. 68 Nash was a keen collector of human curiosa. **1970** E. McGirr *Death pays Wages* ii. 43 Somebody told the old chap that the British Museum Reading Room has a huge collection of *curiosa*.

‖**curiosa felicitas** (kiūᵊri,ōᵘ·sā feli·sitǣs). [L. (Petronius *Sat.* cxviii), lit. 'careful felicity'.] A studied felicity of expression.
1752 Chesterfield *Let.* 16 Mar. (1774) II. 229 The delicacy and *curiosa felicitas* of that poet [*sc.* Horace]. **1817** Coleridge *Biog. Lit.* II. xxii. 168 The frequent curiosa felicitas of his [*sc.* Wordsworth's] diction. **1886** F. Harrison *Choice of Books* I. iii. 61 Tennyson..has..a *curiosa felicitas* of phrase. **1908** L. Johnson in Yeats & Johnson *Poetry & Ireland* 22 The singular charm, the *curiosa felicitas*, of Celtic style. **1933** R. Tuve *Seasons & Months* iii. 79 The 'Curiosa Felicitas' which produces such 'handsom Turns and apt Expressions'.

curious, *a.* Add: **16. b.** Used as a euphemistic description of erotic or pornographic works.
1877 'Pisanus Fraxi' (*title*) Index librorum prohibitorum: being notes bio-biblio-icono-graphical and critical, on curious and uncommon books. **1925** A. Huxley *Those Barren Leaves* I. v. 55 The publications of the Purity League figure invariably under the heading 'Curious' in the booksellers' catalogues. **1934** H. G. Wells *Exper. Autobiogr.* II. viii. 529 That redoubtable suppressed *Life and Loves* of his..which is sought after by collectors of 'curious' books. **1947** [see *curiosa]. **1970** I. Montagu *Youngest Son* 240 My voracious approach to literature included the pages in the bookseller's catalogue labelled 'Erotica' or 'Curious', and I have yet to see an atom of evidence that pornography ever did anyone any harm.
c. Phr. *curiouser and curiouser*, more and more curious; increasingly strange.
1865 'L. Carroll' *Alice in Wonderland* ii. 15 'Curiouser and curiouser!' cried Alice (she was so much surprised, that for the moment she quite forgot how to speak good English). **1931** D. L. Sayers *Five Red Herrings* xv. 167 'I formed the opinion..that Mr. Gowan had..not departed from Kirkcudbright on the Monday evening..but that he had remained concealed in his own house.'..'Curiouser and curiouser,' said Wimsey. **1939** M. Allingham *Mr. Campion & Others* I. ix. 203 'Perhaps it wasn't empty then?' 'In that case it's curiouser and curiouser.' **1970** *Guardian* 31 Dec. 8/1 The ways of film companies become curiouser and curiouser.

cu·rious-mi·nded, *a.* [Curious *a.*] Having a curious or inquisitive or strange mind.
1928 T. E. Lawrence *Lett.* (1938) 599 He'd not likely kill an unarmed, solitary man (Arabs are very curious-minded). **1930** M. Mead *Growing up in New Guinea* i. 1 One of the most fascinating studies open to the curious minded. **1954** J. R. R. Tolkien *Fellowship of Ring* 62 The most inquisitive and curious-minded of that family.

-curist (kiūᵊ·rist). As the second element of compounds such as *mind-curist, sure-curist* (= one who dabbles in mind-cure, etc.).
1889 in *Cent. Dict.* **1907** *Practitioner* Apr. 580 Whether the engineer calls himself Christian Scientist, or Mind Curist, or Hypnotist, matters nothing.

curite (kiūᵊ·rəit). *Min.* [a. F. *curite* (A. Schoep 1921, in *Compt. Rend.* CLXXIII. 1186), f. the name of Pierre *Curie* (see *curie) + -ite¹.] An orange-red hydrated oxide of lead and uranium.
1922 *Nature* 12 Jan. 63/1 The mineral was found at Kasolo, Belgian Congo, and occurs along with curite and chalcolite. **1938** R. W. Lawson tr. *Hevesy & Paneth's Man. Radio-activity* (ed. 2) xv. 150 The mineral curite..contains no trace of ordinary lead. **1958** C. Frondel in *Bull. U.S. Geol. Surv.* No. 1064. 95 Curite is a secondary mineral formed with other hydrated uranyl or lead-uranyl oxides by the alteration of uraninite. The lead of the mineral seems to be radiogenic and derived from the uraninite.

curium (kiūᵊ·riǫm). [mod. L., f. the name of Pierre *Curie* (see *curie) and his wife Marie (1867–1934): see -ium.] An artificially produced, highly radioactive metallic element of the actinide series, formed when plutonium 239 is bombarded with alpha particles and when americium 241 is bombarded with neutrons. Symbol Cm, atomic number 96.
1946 G. T. Seaborg in *Chem. & Engin. News* 10 May 1197/3 For element 96, containing seven 5f electrons, we suggest 'curium', symbol Cm, after Pierre and Marie Curie. **1957** *Oxf. Mail* 9 Sept. 1/7 Curium is 1,000 times more active than ordinary plutonium, which is generally regarded as hazardous. **1964** *Times* 21 Dec. 6/6 The United States Patent Office has issued a patent for a new element, called curium, to Dr. Glenn Seaborg. *Ibid.*, Curium is believed to have valuable properties for space travel.

curl, *sb.* Add **3.** Phr. *in curl, out of curl*: said of hair which is kept curled, or which has gone straight. Also fig., as in *to go out of curl*: to lose one's activity and 'vim', to become limp.
1913 D. H. Lawrence *Sons & Lovers* i. 10 Tha'rt not long in taking the curl out of me. **1924** Galsworthy *White Monkey* I. ix, 'If *you* got pneumonia,' he said, 'I should go clean out of curl.' **1964** E. McCarthy *Frankly Feminine* 52 A 'perm-set'..will stay in curl for around eight weeks.
d. *Cricket.* The action of the verb (see *curl *v.*¹ 9); *spec.* = Break *sb.*¹ 5, Spin *sb.*¹ 2 c.
1833 J. Nyren *Young Cricketer's Tutor* 98 They had such a peculiar curl that they would grind his fingers against the bat. **1867** G. H. Selkirk *Guide to Cricket-ground* vi. 91 Cover point must be careful to allow for the 'curl' after grounding. **1871** *Baily's Mag.* June 168 His bowling.. being straight..with a nice curl from the leg across the wicket. **1888** A. G. Steel in Steel & Lyttelton *Cricket* iii. 170 In 1878 there was another..slow bowler named Allan... His bowling had a considerable amount of spin, but..the most extraordinary thing connected with it was the inward curl in the air towards the body of the batsman.
e. *Math.* The vector product (written curl F or ∇ × F) of the operator ∇ (see *del) with some given vector F; it gives a measure of the 'vorticity' or rotation at each point in the vector field F.
1873 J. C. Maxwell *Electr. & Magnetism* I. 28 To interpret the vector part of ∇σ..let us examine the vector $\sigma - \sigma_0$ near the point *P*. It will appear as in the figure.., this vector being arranged on the whole tangentially in the direction opposite to the hands of a watch. I propose (with great diffidence) to call the vector part of ∇σ the curl, or the version of σ at the point *P*. **1882** O. Heaviside in *Electrician* 18 Nov. 8/1 When one vector or directed quantity, B, is related to another vector, C, so that the line-integral of B *round* any closed curve equals the integral of C *through* the curve, the vector C is called the curl of the vector B. **1911** *Encycl. Brit.* XXVII. 964/1 If A represent the magnetic force at any point of an electro-magnetic field, the vector (∇A) will represent the electric current. In the general case it is called the curl, or the rotation, of A. **1943** Margenau & Murphy *Math. Physics & Chem.* iv. 148 The curl of the linear velocity of any point of a rigid body equals twice the angular velocity. **1965** J. B. Marion *Princ. Vector Anal.* ii. 83 A paddlewheel placed in a fluid will remain stationary.. where curl v = o. A field which everywhere has a vanishing curl is called an irrotational field. *Ibid.*, Both the divergence and the curl are encountered frequently in hydrodynamics and in electromagnetic theory but only infrequently in the mechanics of particles.
f. In surfing: see quot. 1962.
1962 T. Masters *Surfing made Easy* 64 *Curl*, the curved top of a breaking wave. **1965** Farrelly & McGregor *This Surfing Life* iv. 43 The semi-hollow wave..allows you to pick up speed in the top half, and when it breaks you can move down to the bottom half and ride underneath the curl, free of the white water. **1968** W. Warwick *Surfriding in N.Z.* 10/3 Paddle towards the peak and as it becomes critical turn your board, and come back with the curl.

curl, *v.*¹ Add: **1. b.** Phr. *to curl* (a person's) *hair*: to horrify, to frighten. *colloq.*
1949 'P. Wentworth' *Spotlight* xix. 119 And anything like the language—..I give you my word it was enough to curl your hair. **1958** *Ann. Reg.* 1957 186 Mr. Humphrey said that, unless the Government stopped taking so much out of the economy, there would be 'a depression that will curl your hair'. **1887** W. S. Gilbert *Ruddigore* I. 16 When he's excited he uses language that would make your hair curl. **1890** *Monthly Packet* Christmas no., 118 I'll choose a place that will make your hair curl to think of. **1963** V. H. Gielgud *Goggle-box Affair* iii. 31 The amount of overtime she and Miss Plain worked..would have made the T.U.C.'s hair curl.
c. *to curl the mo*, to succeed brilliantly, to win. So *curl- (kurl-) the-mo, curl-a-mo*, etc., attrib. phrs., excellent, outstanding. *Austral. slang.*
1941 Baker *Dict. Austral. Slang* 42 *Kurl*, good, excellent. Also, 'kurl-a-mo'. **1944** *Truth* (Sydney) 13 Feb. 4/3 Breasley saw Kintore donkey-lick a field of youngsters in the Federal Stakes, and had salt rubbed into his wound when the Lewis cuddy Valour curled the mo in the Bond Handicap. **1945** Baker *Austral. Lang.* vi. 126 *Curl-the-mo* was apparently first used to denote the self-satisfaction of a man who twirled the ends of his flowing moustache. It was then applied to anything meriting approval, was shortened to *curl*... A popular song 'Curl-the-Mo, Uncle Joe'—written in praise of Joseph Stalin, who has a large moustache. **1953** —— *Australia Speaks* iv. 97 There is not infrequent mention in the sporting columns of newspapers of *curl the mo mazuma*..a way of saying a lot of money. **1963** *Sunday Mirror* (Sydney) 20 Jan. 43/2 Gili, with Mulley apparently 'curling the mo' was possied behind them for his challenge. **1969** *Coast to Coast 1967–68* 86 He..lifts one of the brimming pilsener glasses: 'Come an' get it! It's curl-a-mo chico. Lead in the old pencil.'
4. In colloq. use (cf. sense *1 b).
5. e. *fig.* To shrink or writhe with horror, shame, etc.; esp. const. *up*.
1913 Galsworthy *Fugitive* III. i, It's..feeling people.. dislike your being there...I curl up all the time. **1923** E. Wallace *Capt. Souls* xlv. 248 So it got you, huh? I couldn't understand how a fellow like you could see it without curling up! **1940** Wodehouse *Quick Service* i. 14 I'm going to call at his office and look him in the eye..and watch him curl up at the edges. **1960** *Sunday Times* 22 May 17/6 He cheerfully admits to things which would make a good New Statesman-ite curl at the edges. **1967** S. Knight *Window on Shanghai* xii. 57 When I think what some parts of Shanghai must have been like *before* liberation, it makes me curl up!
9. *Cricket.* **a.** *intr.* Of the ball: to turn in after pitching; also, to turn in its flight before pitching. **b.** *trans.* Of the bowler: to cause (the ball) to curve in the air.
1833 J. Nyren *Young Cricketer's Tutor* 69 Delivering his ball straight to the wicket, it curled in, and missed the Duke's leg-stump by a hair's-breadth. **1888** Steel & Lyttelton *Cricket* ii. 54 Apart from breaking or curling, the ball may shoot or bump. **1900** P. F. Warner *Cricket in Many Climes* 83 He makes the ball curl in the air. **1904** *Westm. Gaz.* 21 May 3/1 When he first came to England,.. he had that 'curl-in-the-air ball' to a very marked degree. *Ibid.*, Trott shone as a baseball player, and it is to this that he owed his power of curling a ball.

curled, *ppl. a.* Add: **3. c.** Of wood: having a wavy or curly grain. Chiefly in *curled maple* (see Maple 2).
1778 in *Pennsylvania Archives* (1907) 6th Ser. XII. 860 A Curl'd maple Teatable. **1813** H. Muhlenberg *Catal. Plants* 84 (*Alnus undulata* or *crispata*) Waved alder or curled alder. **1855** *Trans. Mich. Agric. Soc.* VI. 528 Much of this [maple] timber is curled and some bird's-eyed. **1911** *Encycl. Brit.* XVII. 664/2 The most constant use of curled maple is for the stocks of fowling-pieces and rifles.

curlicue. (Earlier U.S. examples.)
1844 'J. Slick' *High Life N.Y.* II. xxii. 54, I writ out my name..and handed it over, curlecues and all. *Ibid.* xxvii. 155 We made a curlecue round both the ships.

cu·rlicue, *v.* [f. the sb.] *trans.* and *intr.* To bend elaborately or fantastically. Also *transf.* and *fig.* Chiefly in **cu·rlicued** *ppl. a.*
1844 'J. Slick' *High Life N.Y.* I. i. 11 A kind of picket fence made out of iron, all *curlecued* over on the sides. *Ibid.* II. xxii. 66 Then her arms went curlecueing over her head. **1947** E. Kazan *Notebk. for Streetcar Named Desire* in Cole & Chinoy *Directing the Play* (1953) 297 Blanche is a social type, an emblem of a dying civilization, making its last curlicued and romantic exit. **1963** *Punch* 25 Sept. 456/2 The pink curlicued cupolas.

curling, *vbl. sb.*¹ Add: **3. a.** *curling-pin* (example), *-tongs* (earlier examples).
1909 in A. Adburgham *Shops & Shopping* (1964) xxiii. 273 Curling Pins. **1763** *Boston Post-Boy* 12 Dec., Curling Tongs. **1816** Scott *Antiquary* I. x. 228, I hae the curling-tongs here to gie it a bit turn ower the brow. **1840** Thackeray in *Fraser's Mag.* XXII. 410/1 Mr. Fitch..gave a twist of the curling-tongs to his beard.
b. *curling rink.*
1833 [see *bear *v.*¹ 11 b]. **1926** *Daily Colonist* (Victoria, B.C.) 10 Jan. 6/5 The Portage annual bonspiel will commence on..February 16 according to the..committee of the Portage curling rink.

curl-leaf. [Curl *sb.* or *v.*] = Curl *sb.* 4.
1886 *Harper's Mag.* July 283/1 Foreign varieties and their hybrids are sometimes afflicted with the curl leaf.

curl-paper. (Earlier and later examples.)
a **1817** Anna Lefroy in Jane Austen *Volume the Third* (1951) 130 In the Dressing room..he had the..satisfaction of picking up a curl paper. **1924** C. Mackenzie *Old Men of Sea* xix. 325 Ciggyrettes! Oh dear!..I did once have a try at chewing ciggyrettes, but I reckon I might as well have started in chewing curl-papers. **1966** —— *Paper Lives* xv. 203 'Do any women still put their hair in curl-papers? I thought they used these metallic things.'..'I recall..seeing our housemaid with her hair covered in curl-papers.'

curly, *a.* Add: **3.** *curly maple* (see *curled *ppl. a.* 3 c).
1909 G. Stratton-Porter *Girl of Limberlost* xi. 218 In an expressed crate was a fine curly-maple dressing table. **1942** C. Weygandt *Plenty of Pennsylvania* 29 Curly maple is hard to come by.
b. *Cricket.* Curling (see *curl *v.*¹ 9 a).
1868 J. Lillywhite *Cricketers' Comp.* 54 Mr. Jupp.. played Southerton's 'curly' deliveries with consummate skill.
5. *curly-brimmed* adj.
1890 Conan Doyle *Firm of Girdlestone* xviii. 142 Curly-brimmed hat. **1953** 'N. Blake' *Dreadful Hollow* 203 Both wore dark overcoats and those curly-brimmed Homburg hats.

curly-wurly (kɒ̄·ɹli,wɒ̄·ɹli), *a. dial.* and *colloq.* [See Curlie-wurlie *sb.*] Twisting and curling.
1853 W. Cadenhead *Flights of Fancy* 187 Wi' a' their curly-wurly stanes. **1907** N. Munro *Daft Days* xix, The dusting of the stair-rauls and the parlour beltings—the curly-wurly places, as she called them.

curragh (kʊ·rāχ, kʊ·rā). *Ireland* and *Isle of Man*. [Ir. *corrach* marsh, Manx *curragh* moor, bog, fen.] Marshy waste ground; *spec.* the proper name of the level stretch of open ground in Co. Kildare, famous for its race-course and military camp.

1664 in *Lex Scripta Isle of Man* (1819) 144 Digg and take away Timber in and out of the Curraughes on the North Side. 1894 Hall Caine *Manxman* vi. i, The bog-bane to the rushy curragh, say I, Nancy. 1908 — in *M.A.P.* XX. 362 A widower living alone in some little mud cottage on the curragh.

currajong, var. Kurrajong.

currant. Add: **4.** *currant loaf, -wine* (later example); **currant-borer** (examples); **currant-worm** (examples).

1867 *Amer. Naturalist* June 223 The currant-borer moth (*Trochilium tipuliforme*) darts about the leaves on hot sunny days. 1961 R. South *Moths Brit. Isles* (ed. 4) II. 344 This species seems to have been introduced into North America, where its caterpillar is known as the 'currant borer'. 1922 W. G. R. Francillon *Good Cookery* (ed. 2) xxi. 385 *Currant loaf*...Cream the yeast. Add some of the milk... Beat in the butter, sugar, fruits and egg. 1933 L. G. D. Acland in *Press* (Christchurch, N.Z.) 16 Sept. 15/7 *Brownie*. Bread baked with currants and sugar..called..now, usually, currant loaf. 1850 C. M. Yonge *Langley School* xxvii. 249 They each had a glass of currant wine. 1867 *Amer. Naturalist* June 222 The *Abraxas? ribearia* of Fitch, the well-known Currant-worm, defoliates whole rows of currant bushes. 1886 *Harper's Mag.* Aug. 447 The natural history of the currant worm and moth.

currant jelly. [Currant 4.] A preserve made of the strained juice of boiled currants heated and mixed with sugar in a preserving-pan. Also *fig.* Also *attrib.*, as *currant-jelly dog*, a harrier.

1747 Mrs. Glasse *Cookery* 145 To make Curran Jelly. Strip the Currants from the Stalks, put them in a Stone Jar [etc.]. 1762 W. Gelleroy *Lond. Cook* 303 To make Currant Jelly. 1831 *Athenæum* 31 Dec. 852/1 Swallowing the bitter powder of instruction by enclosing it in the currant jelly of amusement. 1851 *Illustr. Lond. News* 11 Jan. 27/3 Those agents of woodcraft, called, in flippant parlance, 'currant-jelly dogs'. 1869 L. M. Alcott *Good Wives* (1871) v. 47 Fired with a housewifely wish to see her store-room stocked with home-made preserves, she undertook to put up her own currant jelly. [1923 H. Cox *Dogs & I* xvii. 145 Those which are contemptuously termed 'Red Currant Jelly Dogs'..are composed either of Beagle Harriers or..dwarf Foxhounds.]

cu·rranty, *a.* [See -y¹.] Full of currants.

1876 M. E. Braddon *J. Haggard's Dau.* ix, Certain rock-cakes, seedy and curranty.

currawong (kʊ·rāwɒŋ). *Austral.* [Aboriginal.] The native name in Australia for a bird of the genus *Strepera* (see quot. 1926).

1926 *Austral. Encycl.* II. 19/2 Birds of the other genus, Strepera, are variously known as bell-magpies or black magpies, as well as by the aboriginal name of 'currawong' in New South Wales and Queensland. 1936 F. D. Davison *Children of Dark People* 5 Currawongs and willy wagtails flitted among the bushes. 1964 *Telegraph* (Brisbane) 4 Mar. 3/5 Jock was a currawong—a type of magpie—and he sounded off just after dawn every day. 1970 *Southerly* XXX. 8 In season the currawongs in the camphor-laurels cry like tin-shears.

currency. Add: **4. c.** Formerly a name for native-born Australians, as distinguished from *sterling*, or English-born. (See 4 b and the quots. there marked *fig.*) Also *attrib.* and as *adj.*

1827 P. Cunningham *N. S. Wales* II. xxi. 53 Our Currency lads and lasses are a fine interesting race. 1828 *Ibid.* (ed. 3) 48 The Currencies grow up tall and slender, like the Americans. 1878 *Punch* 10 Aug. 60/1 We currency-folk have..been able to absorb your convict refuse without contamination from its criminal leaven. 1894 W. C. Dawe (*title*) The confessions of a currency girl. 1899 *Macm. Mag.* June 127/1 The boys when questioned would say: 'I'm not English; I'm Currency.' 1953 *Landfall* VII. 173 She spoke the King's English like a currency lass.

6. *currency crank, restriction*; **currency note**, paper money used as currency, esp. the £1 and 10*s.* notes first issued by the Treasury for circulation as legal tender during the war of 1914–18; a treasury note.

1931 H. G. Wells *Work, Wealth & Happiness of Mankind* (1932) ix. 363 General discussion [on currency] has been further burked by dubbing anyone who raised the question, a 'Currency Crank'. 1944 G. B. Shaw *Everybody's Political What's What* xi. 84 The Currency Crank is a nuisance in every movement for social reform. 1891 J. L. Kipling *Beast & Man in India* v. 105 A currency note for a thousand rupees. 1914 *Proclamation* 3 Feb. in *Jrnl. Inst. Bankers* (1915) XXXVI. 113 Payment for the order at its face value in coins or currency notes. 1920 *Discovery* May 145/1 Our over-issues of currency notes. 1922 *Encycl. Brit.* XXXI. 969/2 The 1914 Act..allowed an issue of £1 and 10s. currency notes by the Treasury. 1967 'R. Simons' *Taxed to Death* ix. 151 Several printed forms about currency restrictions.

current, *a.* Add: **3.** *current account*, an account kept by a customer at a bank to meet his current expenses; *current affairs, events*, those in progress, those belonging to the present time; *current goods* (see quot. 1948).

1846 Dickens *Pict. Italy* 66 A means of establishing a current account with Heaven, on which to draw..for future bad actions. 1875 H. Fisher *Opening*, etc. *Spec. Banking Accts.* 1 The opening, working, and closing of certain classes of Current and Deposit Accounts. 1899 *Westm. Gaz.* 1 Sept. 6/3 It is the depositor, rather than the current-account customer, who is victimised by this custom. 1951 R. W. Jones *Thomson's Dict. Banking* (ed. 10) 206/1 A current or running account is the active account on which cheques are drawn and to which credits are paid. 1920 Beerbohm *And Even Now* 60 Swinburne did, from time to time, take public notice of current affairs. 1955 'C. Brown' *Lost Girls* x. 111 We began each afternoon's session with a 'current affairs' talk. 1957 *B.B.C. Handbk.* 102 Up-to-date information on current affairs. 1850 *Harper's Mag.* June 122 (*headline*) Monthly record of current events. 1920 S. Lewis *Main Street* ix. 107 The Thanatopsis Club..have some of the *best*..current-events discussions. 1936 *Discovery* Nov. 355/2 The distinction between capital goods and current goods is..one of the most important in the whole of economics. 1948 G. Crowther *Outl. Money* (ed. 2) v. 129 Every year the community produces a certain total of goods and services; some of them are for immediate consumption, the rest are goods whose value will last beyond the immediate present. These two categories can be called current goods and durable goods. All services are naturally current goods.

current, *sb.* Add: **8. a. current meter** (see also quot. 1868). **b.** Various *Combs.*

1862 *Catal. Internat. Exhib.* II. xiii. 13 As these instruments have no break pieces or current reversers they cannot get out of order. 1866 R. M. Ferguson *Electr.* (1870) 185 A contrivance for this purpose is called a rheotome or current-break. 1868 W. D. Haskoll *Land & Marine Surveying* xi. 170 The current meter is useful also to ascertain the velocity of under currents. 1884 F. Krohn tr. *Glaser de Cew's Mag.- & Dyn.-Electr. Mach.* 207 The current closers and interrupters. *Ibid.* 272 The current-energised rotating helix. 1888 Bottone *Electr. Instr. Making* (1894) 192 The current reverser for the Wheatstone single needle telegraph. 1946 *Nature* 13 July 54/2 A system which gives the constant line voltage required for current-using devices. 1962 Corson & Lorrain *Introd. Electromagn. Fields* v. 179 A current-carrying conductor. 1962 Simpson & Richards *Junction Transistors* xiii. 295 The more usual practice is to define the feedback, solely by the way it is derived, as 'voltage' (parallel) or 'current' (series) feedback. 1964 R. F. Ficchi *Electr. Interference* x. 210 A current-limiting device in neutral circuits.

‖**currente calamo** (kɒre·nte kæ·lămo), *adv. phr.* [L., lit. 'with the pen running on'.] Extempore; without deliberation or hesitation.

The actual phr. is not recorded in cl. L.

1776 W. Mason *Let.* 25 Mar. in *Walpole's Corr.* (1955) XXVIII. 254 What I here send you was written yesterday *currente calamo*. 1857 Trollope *Barchester T.* II. xiii. 258 His letter..was written, *currente calamo*, with very little trouble. 1915 W. J. Locke *Jaffery* ix. 112 Then to work, and in another three months, *currente calamo*, the book would be written.

currently, *adv.* **2.** Add to def.: now, at the present time.

1947 J. Hayward *Prose Lit. since 1939* 11. 20 The same writer..is currently writing a section of the secret history of the war. 1969 A. Glyn *Dragon Variation* vi. 187 These were probably the two most exciting and original games [of postal chess] he was currently playing. 1971 *Daily Tel.* 8 June 10/3 They are currently to be seen at the Hampstead Theatre Club.

curricle. 2. (Earlier example.)

1752 H. Walpole *Let.* 5 Aug. (1903) III. 114 These mountains, where the young gentlemen are forced to drive their curricles with a pair of oxen.

curriculum. Add: **curriculum vitæ**, the course of one's life; a brief account of one's career.

1902 *New Internat. Encycl.* III. 21/2 Anciently biography was more of a mere *curriculum vitæ* than it is now. 1939 'M. Innes' *Stop Press* II. iv. 269, I don't know much about Benton's *curriculum vitæ*... He must have an orthodox.. academic record. 1941 Koestler *Scum of Earth* 59 His superiors..knew all about my professional travels from the *curriculum vitae* in their files, written by myself. 1954 *New Yorker* 25 Dec. 18/2 As for Mr. Lapidus's *curriculum vitae*, he was born in Russia fifty-two years ago, grew up in Brooklyn, graduated from the Columbia School of Architecture in 1927, and took a job with the well-known firm of Warren & Wetmore. 1971 *Time* 22 Mar. 14/2 Eddie's *curriculum vitae*..has been served up in plentiful quantity in the press.

curry, *sb.*² Add: **1. b.** *curry-sauce* (examples); **curry-paste** (examples).

1855 E. Acton *Mod. Cookery* (rev. ed.) i. 45 A large tablespoonful of Captain White's currie-paste. 1906 Mrs. Beeton *Bk. Househ. Managem.* xvi. 450 Add the stock, curry-paste, sliced apple. 1845 E. Acton *Mod. Cookery* viii. 201 Currie sauce, highly onioned, is frequently served. 1948 *Good Housek. Cookery Bk.* 280 Curry sauce..is much improved by the addition of 1 tbsp. cream immediately before use.

2. *to give* (a person) *curry*: see quot. 1941. *Austral. slang.*

1941 Baker *Dict. Austral. Slang* 21 *To give someone curry*, to abuse, reprove, express anger at a person. 1944 *Coast to Coast 1943* 113 I'd like him not to be writing! Wouldn't I give him curry! *Ibid.* 124 I'm going to give those old tarts a bit of curry to-night, Ron. 1945 Baker *Austral. Lang.* vi. 120 A man who attacks another is said..*to give him curry* or *curried hell*.

curse, *sb.* Add: **3. b.** = Cuss *sb.* 2.

1790 [see Rantipole *sb.* 1]. 1854 B. Young in *Jrnl. Discourses* I. 83 We have known Gladden Bishop for more than twenty years, and know him to be a poor, dirty curse. *Ibid.* 169 Why don't you do it, you poor miserable curses?

c. An angler's name for a very small gnat or midge.

1889 F. M. Halford *Dry-fly Fishing* vi. 116 'Curses', or black midges or gnats. 1899 *19th Cent.* Jan. 122 The monstrously minute 'curse'.

4. d. *the curse*: menstruation. *colloq.*

1930 J. Dos Passos *42nd Parallel* 147 She was afraid her period was coming on. She'd only had the curse a few times yet. 1933 'E. A. Robertson' *Ordinary Families* vi. 115 Ill luck..had added a premature last straw to my load of misery: I had the curse. 1960 *Woman's Own* 19 Mar. 15/1, I always think it such a pity when girls..call it 'the curse'. 1969 G. Greene *Travels with my Aunt* xii. 120, I forgot the damn pill and I haven't had the curse for six weeks.

5. curse-word = *cuss-word* (Cuss *sb.* 3).

1897 R. M. Stuart *Simpkinsville* vii. 225 The popular after-dinner 'curse word story' of the cloth would never have been tolerated in Simpkinsville.

curse, *v.* Add: **2. a.** Also const. *for*.

1922 H. Walpole *Cathedral* II. iv. 229 He cursed Foster for a meddling, cantankerous fanatic.

cursus. a. Add to def.: Also used *spec.* of a type of neolithic monument (see quot. 1963). (Earlier and later examples.)

1740 Stukeley *Stonehenge* viii. 35 The western branch.. continues curving along the bottom of the hill, till it meets, what I call, the *cursus*. *Ibid.* ix. 41 About half a mile north of Stonehenge, across the first valley, is the *cursus* or *hippodrom*, which I discovered, August 6, 1723. 1951 *Field Archæol.* (*Ordnance Survey*) (ed. 3) 18 Another feature which is now regarded as part of the henge complex is the extremely elongated earthwork of the type originally called a 'cursus' by Stukeley..because he believed that they were designed as racecourses. 1963 E. S. Wood *Field Guide to Archaeol.* II. 138 A class of still enigmatic neolithic monuments are the cursûs... These are long parallel banks with outside ditches, with squared or curved ends, running across country often for long distances.

d. The regular varying cadences which mark the end of sentences and phrases, esp. in Greek and Latin prose.

1904 H. A. Wilson in *Jrnl. Theol. Stud.* V. 387 Prof. E. Norden has traced the use of the 'cursus' in Classical writers, Greek as well as Latin. 1910 *Encycl. Brit.* VIII. 304/1 The *cursus* or prose rhythm of the pontifical chancery of the 11th and 12th centuries. 1910 A. C. Clark (*title*) The cursus in mediaeval and vulgar Latin. 1959 H. J. Rose *Outl. Class. Lit.* vi. 188 Three rhythms found their way.. into medieval Latin prose..where naturally accented syllables took the place of long ones. Hence the three 'runs' (*cursus*) as they were called.

curtain, *sb.*¹ Add: **1. d.** On a bonnet (earlier examples).

1788 E. Sheridan *Jrnl.* (1960) 138 Bonnets I see most generally worn and some with very deep Curtains, The Bonnet itself is small. 1861 C. M. Yonge *Stokesley Secret* ii. 31 Her lilac-spotted sun-bonnet..with a huge curtain serving for a tippet.

e. *pl.* A wrinkled effect resembling a draped curtain on a painted or varnished surface. *colloq.*

1922 M. Toch *How to paint Permanent Pictures* 79 A very heavy-bodied Linseed Oil,..was so viscous that it flowed down..and formed 'curtains', and teardrops. 1951 R. Mayer *Artist's Handbk.* iii. 136 *Streamlines*. The surface defect resembling drops of water running down a window pane is variously known by painters and paint technicians as frilling, curtains, runs, or tears. 1958 *Listener* 28 Aug. 323/1 Just flow them [*sc.* jelly paints] on a little more generously... You are not likely to have any trouble with runs or 'curtains'.

2. c. In various ellipt. or allusive uses: (i) = *curtain-call; (ii) the finale of a play, act, or scene; also *transf.*; (iii) = *curtain-fall.

1884 *Referee* 31 Aug. 3/3 *Written in Sand* was well received, and Broughton had to 'take a curtain'. 1885 *Ibid.* 15 Mar. 7/3 It is singular, considering how excellently French dramatists write, that they so frequently fail in getting a good 'curtain'. 1895 G. B. Shaw *Our Theatres in Nineties* (1932) I. 165 The doggerel tags before the final curtain. 1897 E. Terry *Let.* 19 June in *E. T. & Shaw* (1931) 219 The last week I've dragged myself through that long long part, and toppled down when it was Curtain on Thursday night. 1917 R. Firbank *Caprice* xii. 102 The other afternoon I 'offered my services' and obtained three curtains at a gala matinée. 1919 Wodehouse *My Man Jeeves* 146 Curtain of act one on hero..kidnapping the child. 1928 *Evening News* 7 Aug. 7/3 There were ten curtains after the second act and an enthusiastic reception when the curtain fell. 1928 *Daily Tel.* 4 Dec. 9/1 'Sapper' gives a decidedly original curtain to his dramatic murder tale 'The Hidden Witness'. 1965 *Listener* 9 Sept. 393/1 A lyrical outpouring, leading to a most effective curtain.

d. In *pl.* (also occas. in *sing.*), the end (cf. sense *2 c (iii)). *slang.*

1912 D. Lowrie *Life in Prison* vii. 82 There ain't much dope here now, an' it's curtains t' get nailed with it. 1918 Wodehouse *Piccadilly Jim* xi. 114 'What's wrong?' 'Curtains!..I've been fired.' 1937 C. Day Lewis *Starting Point* II. i. 135, I rather fancy potassium cyanide. You just chew a piece, and quick curtain. 1940 N. Monks *Squadrons Up!* 213 Once he gets the enemy lined up in that ring, it is curtains for the enemy. 1956 Wallis & Blair *Thunder Above* (1959) xii. 131 If the Party ever got on to it..it would be curtains for Kurt.

e. *curtain up*: the beginning of a performance.

1942 E. S. L. Robinson *Curtain Up* 9 The call-boy makes his rounds rapping like Fate at each dressing-room door... 'Curtain up.' 1968 *Guardian* 19 Feb. 6/1 Curtain-up is a month away. 1969 'S. Troy' *Swift to its Close* vi. 90 What are you going to do till curtain-up?

3. b. *Mil.* (In full *curtain of fire, curtain fire.*) A concentration of rapid and continuous artillery or machine-gun fire, etc., on a designated line or area, to prevent the advance or retreat of enemy troops, or to clear the way for the combatant's advance. Also, a concentration of fire to block the progress of aircraft.

1916 'Boyd Cable' *Action Front* 114 Shells began to batter at their parapet, and to prepare a curtain of fire along their front. 1920 D. A. Macalister *Field Gunnery* (ed. 4) vii. 157 During an attack..the batteries, acting in concert, establish the 'curtain of fire' or 'barrage'. 1922 *Encycl. Brit.* XXX. 98/2 The idea also was evolved of barrage fire, a curtain of bursting shell to be put up in the path of the raiders. 1943 T. Horsley *Find, Fix & Strike* 92 We..began our glide through the curtain of lead towards the inner harbour.

c. Short for *iron curtain (see *iron *sb.*¹); also with capital initial. Also used in similar metaphors, esp. implying restriction of information.

1945 *Sunday Empire News* 21 Oct. 2/2 (*heading*) A curtain across Europe. 1946 *Spectator* 13 Sept. 257/2 The Russians.. would admit their 'iron curtain', but pointed out that there was also the Anglo-U.S. 'uranium curtain'. 1949 [see *bamboo *sb.* 2]. 1950 M. Peterson (*title*) Both sides of the Curtain. 1953 *School & Society* LXXVIII. 129 (*title*) The language curtain. 1955 *Times* 21 July 6/4 The reaction at G.H.Q. East Africa has been to tighten even further its own security curtain. 1970 'W. Haggard' *Hardliners* iv. 37 A foreigner from behind the Curtain.

6. d. A contrivance consisting of wooden slats which can be rolled up: *spec.* one of a number of these used to form a dam or weir. Also *attrib.*, as *curtain-dam, -valve, -weir*.

1895 *Montgomery Ward Catal.* 608/3 Low Curtain Office Desk... Has lap joint, dust and knife proof curtain. 1903 Thomas & Watt *Improvem. Rivers* viii. 244 Curtain Dams... The Caméré curtain..consists of narrow horizontal strips of wood, hinged together, and capable of being rolled up by a chain. *Ibid.* 253 The space between the two rows [of shutters] was then filled with water by opening curtain-valves. 1927 E. Weymann *Dams* 586 The curtains are suspended from hooks on the face of the frames. 1929 *Encycl. Brit.* XXIII. 489/1 The curtain weir... In it wooden curtains that can be rolled up from the bottom were substituted for the needles in the Poirée weir.

8. *curtain-cord* (later examples); **curtain-call**, a call by an audience for an actor or actors to take a bow after the fall of the curtain (see 2); **curtain-fall**, the fall of the curtain at the end of an act or scene; the situation or tableau when the curtain falls; also *fig.*; **curtain line**, the last line of a play, act, or scene; also *transf.*; **curtain-pole** (example); **curtain-raiser** (read: orig. *slang*); also *transf.*; **curtain rise**, the rise of the curtain at the beginning of an act or scene; **curtain wall**, (*b*) see quot. 1901; also *curtain walling*; hence *curtain-walled* adj.

1884 'F. Leslie' *Let.* 15 July in W. T. Vincent *Recoll. F. L.* (1893) I. x. 176 You will find a room specially adapted for rehearsing curtain calls. 1919 Wodehouse *Damsel in Distress* xii. 144 He felt a wave of stage-fright such as he had only once experienced before in his life—on the occasion when he had been young enough to take a curtain-call on a first-night. 1863 A. D. Whitney *Faith Gartney's Girlh.* xvii, She drew the curtain-cord to let in the first sunbeam. 1939 T. S. Eliot *Old Possum's Pract. Cats* 14 The curtain-cord she likes to wind. 1900 T. E. Pemberton *Kendals* ix. 276 He must be forgiven and at curtain-fall live happily ever after. 1909 *Daily Chron.* 26 Jan. 5/6 'A Merry Christmas!' he shouts light-heartedly at curtain-fall. 1962 *Times* 27 Nov. 14/7 Within the confines of curtain-rise and curtain-fall. 1939 D. L. Sayers *In Teeth of Evidence* 200 'I will rest on my laurels'—that was a beautiful curtain line you gave him there. 1959 *Listener* 31 Dec. 1171/2 Conversations [in a novel] end with brave, ringing curtain lines. 1865 Geo. Eliot *Ess.* (1884) 206 Unctuous personages.. who soar above the curtain-poles without any broomstick. 1908 *Sears, Roebuck Catal.* 880/1 Heavy brass curtain pole rings. 1940 *War Illustr.* 26 Jan. 24 What has happened to date is the curtain-raiser to that aerial blitzkrieg which is still part of the stock-in-trade of the Nazi boasters. 1955 *Times* 27 July 2/6 There was a curtain raiser earlier this month when the case was put that the proposed scheme was ultra vires. 1969 *Australian* 24 May 36/6 The three Australian selectors..will watch Sydney Seconds..in the curtain-raiser before focusing on the main game. 1905 *Daily Chron.* 11 Feb. 6/2 Miss Tree sings, at curtain-rise, to very charming purpose. 1962 Curtain-rise [see *curtain-fall]. 1901 R. Sturgis *Dict. Archit. & Building* 731/1 *Curtain wall*. In modern construction, most often a thin subordinate wall between two piers or other supporting members; the curtain being primarily a filling and having no share—or but little—in the support of other portions of the structure. Thus, in skeleton construction, curtain walls are built between each two encased columns and..on a girder at each floor level. 1930 *Engineering* 1 Aug. 131/3 The curtain wall [of the Welland Ship Canal] is 3 ft. 6 in. thick and set back 10 ft. 6 in. from the upstream face. 1950 *Archit. Rev.* CVII. 221 On the ground (banking floor) these curtain walls are of glass blocks to give the maximum light without permitting passers-by to see inside. 1952 *Ibid.* CXII. 392 'Curtain wall' is a recent American term for a form of rigid skin walling. It is basically an extension of sheet cladding to cover wider spans... In a more developed form it includes the growing practice, particularly on slab blocks, of covering a complete elevation with subsidiary framing holding both cladding and windows. 1959 *Listener* 3 Dec. 976/2 Curtain-walled office-blocks. 1958 *Archit. Rev.* Jan. 7 The increasing use of curtain walling and similar systems. 1963 *Listener* 28 Feb. 371/1 Curtain walling has made it possible to turn the whole façade into a huge shiny texture. *Ibid.* 371/2 Curtain walling is being used to create simple geometric form at the expense of the spaces behind.

curtaining, *vbl. sb.* Add: spec. in *Painting*, the formation of 'curtains' (see *curtain *sb.*¹ 1 e). *colloq.*

1940 in *Chambers's Techn. Dict.* 217/1. 1953 in *Gloss. Paint Terms* (*B.S.I.*) 11.

curtal, *a.* Add: **4. b.** *curtal-sonnet* (see quot.).

a 1889 G. M. Hopkins *Poems* (1918) Pref. 6, Nos. 13 and 22 are Curtal-Sonnets, that is they are constructed in proportions resembling those of the sonnet proper, namely 6+4 instead of 8+6, with however a halfline tailpiece (so that the equation is rather $\frac{12}{2}+\frac{9}{2}=\frac{21}{2}=10\frac{1}{2}$.

curtate, *a.* Add: (Example in sense 'shortened, diminished in length'.)

1893 J. R. Harris *Stichometry* 18 The lines represent respectively a somewhat curtate half-hexameter and a similarly divided iambic trimeter.

2. *Economics* and *Statistics*. Shortened or limited according to some formula or rule; *esp.* counting or calculated for the number of full years in a period, to the exclusion of the odd fraction of a year.

1875 *Encycl. Brit.* II. 78 This formula gives the average number of *complete* years that persons of the given age will live..and makes no allowance for the portion of the year in which death occurs. The expectation thus found is called the curtate expectation. 1927 Bowley & Stamp *Nat. Income, 1924* 21 When the averages of these curtate groups are taken, that for women is often higher than for men. 1927 B. C. Hoskins *Insur. Lexicon* 134 If payments cease with the last payment preceding the death, the annuity is said to be 'non-apportionable', or 'without proportion', or 'curtate'. 1957 Kendall & Buckland *Dict. Statistical Terms* 75 If an assurance matures in 3 years 9 months, the curtate duration is 3 years.

curvaceous (kɒɪvē¹·ʃəs), *a. colloq.* (orig. *U.S.*). [f. Curv(e *sb.* + -aceous.] Curving, full of curves; *spec.* of a well-rounded female figure.

1936 *Screen Book Mag.* Feb. 61 The curvaceous lady [*sc.* Mae West] receives from Paramount just as many dollars per week for her scenario work as she receives for her acting. 1959 *N.Z. Listener* 21 Aug. 8/1 Legs that are strong, and curvaceous, and graceful, and a pleasure to the eye! 1959 H. Hobson *Mission House Murder* xviii. 119 Sharon; lissome and curvaceous in a revealing leopard-spotted bikini. 1965 *New Statesman* 30 Apr. 693/1 Curvaceous red paint marks..seemed to indicate a vague and unrealised intention to paint a female nude. 1970 *Interior Design* Dec. 759 (*caption*) A curvaceous upholstered armchair.

curvature. Add: **1. c.** A generalization of the notion of curvature applied to a space or manifold of four (or more) dimensions, first made in the theory of non-Euclidean geometry and further developed by Einstein in the general theory of relativity; the property of not being Euclidean or 'flat'. So *curvature of space-time*, etc.

1873 W. K. Clifford tr. Riemann in *Nature* 8 May 36/2 If we assume independence of bodies from position, and therefore ascribe to space constant curvature, it must necessarily be finite provided this curvature has ever so small a positive value. 1910 *Encycl. Brit.* XI. 727/2 Riemann's work contains two fundamental conceptions, that of a manifold and that of the measure of curvature of a continuous manifold. 1916 *Monthly Notices R. Astr. Soc.* LXXVI. 707 The mathematical interpretation of G is the *curvature* of the four-dimensional system of reference. 1920 A. S. Eddington *Space Time & Gravit.* x. 158 We thus get the idea that space-time may have an essential curvature on a great scale independent of the small hummocks due to recognised matter. 1920 R. W. Lawson tr. *Einstein's Relativity: Special & General Theory* 127 Half of this deflection is produced by the Newtonian field of attraction of the sun, and the other half by the geometrical modification ('curvature') of space caused by the sun. 1959 Spitz & Gaynor *Dict. Astron.* 399 The curvature of space existing in the vicinity of a massive body, like the sun, affects the course of a ray of light. 1966 Taylor & Wheeler *Spacetime Physics* iii. 175 The existence of this curvature destroys the possibility of describing motion with respect to a single ideal Euclidean reference frame that pervades all space. 1971 *Sci. Amer.* May 22/3 A gravitational-radiation detector built on this principle should be capable of measuring the gravitational waves (and hence the curvature of space-time) predicted by relativity theory.

curve, *a.* and *sb.* Add: **B. 1. c.** *Physics, Statistics*, etc. A graph or line drawn from point to point so as to represent diagrammatically a continuous variation of a quantity, either with time or with respect to some other quantity.

1854 *Amer. Jrnl. Sci. & Arts* 2nd Ser. XVII. 423 The same general law gives the intensity of the induced magnetism as a function of the exciting force... The curves which represent the law of induction for diamagnetic substances are surrounded..by the curves for magnetic substances. 1874 *Proc. R. Soc.* XXII. 27 There will be three curves—one expressing the relation between temperature and pressure for gas with liquid, another expressing that for gas with solid, and another expressing that for liquid with solid. 1884 *Pharmaceut. Jrnl. & Trans.* 26 July 77/2 The object of these curves was to show clearly some of the most important factors in the growth of crops. 1886 F. Galton in *Jrnl. Anthropol. Inst.* XV. 263 Section of surface parallel to *XY* is a true curve of frequency. 1899 Temperature curve [see Temperature 10]. 1909 K. Pearson *Problem Pract. Eugenics* 11 The curve for all possibly reproductive wives is amply verified by the curve for young wives. 1911 W. E. Dalby in *Rep. Brit. Assoc. 1910* 695 The curves of mileage, passengers carried, and goods carried increase regularly with the increase of capital. 1940 R. S. Woodworth *Psychol.* (ed. 12) 62 If we outline the shape of the distribution by a line joining the tops of adjacent columns we have the distribution curve. 1958 *Times Rev. Industry* Apr. 81/2 In the..Community industries the curve of annual steel production increase is flattening out. 1959 *Listener* 13 Aug. 237/2 The population curve has slowed down.

2. *spec.* The curving line of the female figure. Usu. in *pl.*

1862 *Harper's Mag.* June 45/1 The full round shape hid half its voluptuous curves in the shade of the dark-green riding-suit. 1906 *Dress* Nov. 26/2 Where the figure of the client is naturally too flat to meet this present demand for curves, the corsetière is apt to add a few little pinked-out silken frills. 1929 Wodehouse *Mr. Mulliner Speaking* i. 9, I can remember the days..when every other girl you met stood about six feet two in her dancing-shoes and had as many curves as a Scenic Railway. 1931 A. Christie *Sittaford Myst.* i. 8 What was the good of a woman if she didn't look like a woman? Papers said curves were coming back. About time, too. 1963 van Praagh & Brinson *Choreogr. Art* 174 The emancipated modern woman is athletic, well-groomed and chic with good proportions and fewer curves than in earlier periods.

4. (Examples.) Also *attrib.* and *fig.*

1879 *De Witt's Baseball Guide* 24 The great difficulty in curve pitching is to obtain the required command of the ball. 1887 *Outing* (U.S.) May 98/1 Assertions were rife in all quarters that the curve was a fallacy. 1912 C. Mathewson *Pitching* 14 Big League ball-players recognize only two kinds of pitched balls—the curve and the straight one. 1960 H. Seymour *Baseball* xxiii. 278 Fooling him with a curve on the outside. 1970 W. Smith *Gold Mine* xxxv. 92 Are we going to sit back and let them have a free run? No, sir! We are going to throw down our own curve ball!

5. *pl.* Round brackets; parentheses. *U.S.*

1928 M. H. Weseen *Crowell's Dict. Eng. Gram.* 169 *Curves*. This term is sometimes used to denote parenthesis marks ( ). 1961 R. B. Long *Sentence & Parts* xx. 475 Curves are usable to inclose quite clearly what occurs where commas would be confusing and dashes a little strong.

C. curve-billed (epithet of a N. American thrush); **curve-fitting**, the determination of the equation of the curve which, subject to any conditions such as the possible number of parameters, most closely represents the points on a graph or describes most accurately the relation between the variables they represent; **curve-plotting**, the graphic representation of a curve in a plan or diagram by means of points marked on co-ordinates.

1881 *Amer. Naturalist* XV. 217 The curve-billed thrush (*H[arporhynchus] curvirostris*). 1924 *Proc. Nat. Acad. Sci.* X. 79 (*heading*) The development of a frequency function and some comments [on] curve fitting. 1930 *Meteorol. Gloss.* (ed. 2) 51 Two main types of problems arise in connexion with curve fitting. We may either require to represent a variable quantity as a function of some independent variable such as time,..by drawing a curve to represent the functional relationship, or we may require to represent the frequency of occurrence of different values of a quantity, ..by means of a curve whose formula must be obtained. 1970 *Computers & Humanities* IV. 343 Through use of a curve-fitting technique, high-order polynomial expressions relating pitch to time (rhythm) are generated for each incipit. 1905 Ashe & Keiley *Electr. Railways* 6 Curve plotting is accomplished by means of a series of perpendicular and parallel lines..termed coördinates. 1945 *Jrnl. Franklin Inst.* CCXL. 278 Curve-plotting output units.

curve, *v.* Add: **3.** *trans. Baseball*. To throw or pitch (a ball) with a curve (see Curve *sb.* 4).

[1856 *Spirit of Times* 6 Dec. 229/1 It is questionable.. whether his style of pitching is most successful, many believing a slow ball curving near the bat, to be the most effective.] 1878 *De Witt's Baseball Guide* 35 With a view of settling the vexed question as to whether a pitcher can or cannot curve a ball, practical experiments were made. 1960 H. Seymour *Baseball* xvi. 177 The pitcher..dominated the game merely because he could curve a ball.

curved, *ppl. a.* Add: **b.** *curved fire*, gun-fire with an angle of elevation or departure exceeding that of direct fire.

The angle is variously specified: see quots.

1879 [see *indirect *a.*]. 1883 G. Mackinlay *Text Bk. Gunnery* 162 Indirect or curved fire. Fire from guns, with reduced charges, and from howitzers and mortars, at all angles of elevation not exceeding 15°. 1897 *Text Bk. Gunnery* (*War Office*) (ed. 3) II. iv. 249 When the curvature of the trajectory becomes considerable, as in High Angle and Curved Fire. 1907 O. M. Lissak *Ordn. & Gunnery* 358 Direct Fire is with high velocities, and angles of elevation not exceeding 20 degrees. Curved Fire is with low velocities, and angles of elevation not exceeding 30 degrees. 1917 W. H. Tschappat *Ordn. & Gunnery* 426 For convenience of discussion curved fire will be considered as firings with elevations between 15° and 40° and high-angle fire as firings with elevations above 40°. 1920 D. A. Macalister *Field Gunnery* (ed. 4) i. 19 *Curved Fire*.—Where the angle of elevation is not above 30°.

c. In *Baseball* (see *curve *v.* 3).

1905 *Chicago Daily News* 17 July 4/1 His curved ball was breaking much better for him too.

cu·rvesome, *a.* [-some¹.] = *curvaceous *a.*

1935 *N.Y. Even. Jrnl.* 7 May, Six burleycue lassies who daily peel off enveloping garments..to disclose their curvesome charms. 1938 M. Brinig *May Flavin* iii. 289 She was fairly tall and had a curvesome figure, and this was the best thing about her. 1940 *Time* 8 July 49/1 Mazeppa, traditionally played by a curvesome female.

cu·rvingly, *adv.* [-ly².] In a curving manner.

1923 *Weekly Dispatch* 13 May 7 Decorated with blue and white stripes that ran out curvingly, somewhat in swastika fashion. 1938 E. Thompson *Youngest Disciple* i. 9 It was a valley cut deep in the mountains, as by a scimitar, not straight but curvingly.

curvometer (kɒ̄ɪvǫ·mitəɪ). [f. Curve *sb.* + -ometer.] An instrument for measuring the length of a curve.

1902 *Encycl. Brit.* XXX. 578/1 More complicated curvometers or kartometers have been devised.

curvy (kɒ̄·ɪvi), *a.* [-y¹.] Having a curve or curves; full of curves, marked with curves. *Comb.*, as *curvy-brimmed* adj.

1902 *Westm. Gaz.* 23 Oct. 3/2 The collar and the quaint curvy applications on the skirt should be of guipure lace. 1965 G. McInnes *Road to Gundagai* xii. 204 Men in curvy-brimmed top hats.

cusec. Abbreviation, used in Engineering, etc., of '*cu*bic foot per *sec*ond'.

1913 Thomas & Watt *Improvem. Rivers* (ed. 2) I. i. 46 The abbreviation 'second-feet' which is frequently used in America instead of 'cubic feet per second', has the equivalent of 'cusecs' among the Anglo-Indian engineers. 1915 C. E. Housden *Is Venus Inhabited?* 37 We should need a continuous flow of 20,000,000 cusecs. 1960 *Times* 25 July 11/6 A rise in river-flow from three cusecs in dry weather to 55,000 in wet.

cush (kuʃ), colloq. shortening of Cushion *sb.*, esp. in sense 3 c.

1895 A. Roberts *Adv.* ix. 111 The red was under the cush. Alias hesitated at which ball to play. 1905 *Westm. Gaz.* 21 Oct. 4/1 It is like watching a game of billiards with wooden cushes and beechwood balls. 1947 D. Davin *Gorse blooms Pale* 97 And these bottoms and legs bouncing through the room. Cushioned against reality. It cannoned off their cush. 1965 'A. Hall' *Berlin Memorandum* viii. 83, I was placed in a slow drift for a right-angle..and brought the nose round full-lock with the kerb for a cush.

cush, *v.*, var. *cosh* (see *cosh *sb.*³). Hence **cu·sher.**

1923 F. L. Packard *Four Stragglers* I. ii, 'A bit of a "cushing" expedition, was it?'..'Just the usual bash on the head with a neddy.' *Ibid.* II. iv, That was the method of the 'cusher'.

cushag (ku·ʃəg). *dial.* Also **-og.** [Manx *cuishag vooar*, lit. 'big stalk'.] The common ragwort, *Senecio jacobæa*.

1887 Hall Caine *Deemster* xxiv, There's gold on the cushags yet. 1894 — *Manxman* II. xxii, Philip plucked the cushag. 1900 *Westm. Gaz.* 1 Oct. 2/3 The term 'weeds' is defined as including thistles, cushags, and common docks.

cushaw (kɒʃǭ·, kʊ·ʃǭ). *U.S.* Also **cashaw.** [perh. Algonquin.] A winter crookneck squash, or a variety of this.

1588 T. Harriot *Briefe & True Rep. of Virginia* sig. C4ᵛ, *Coscushaw*, some of our company tooke to bee that kinde of roote which the Spaniards in the West Indies call *Cassavy*. 1698 G. Thomas *Pennsylvania* 21 Cucumbers, Coshaws, Artichokes. 1705 R. Beverley *Hist. Virginia* II. iv. 27 Their Cushaws are a kind of Pompion, of a bluish green Colour, streaked with White, when they are fit for Use. 1868 W. N. White *Gardening for South* 214 The best variety [of squash] for family use is the Cashaw, a long, cylindrical, curved variety. 1924 J. W. Raine *Land of Saddle-Bags* 28, I asked an old man why he preferred 'cushaws' (a large crook-neck squash) to pumpkins.

cush-cush (ku·ʃkuʃ). [Native name.] A species of yam, *Dioscorea trifida*, native to South America and cultivated for its edible tubers.

1871 Kingsley *At Last* ii. 49 Great roots of yam and cush-cush. *Ibid.* 72 Wild cush-cush roots. 1929 Nicholls & Holland *Textbk. Trop. Agric.* (ed. 2) xv. 446 Cush-cush Yam of Trinidad or Yampi of Jamaica (*Dioscorea trifida*, Linn.). 1960 *Guardian* 8 Jan. 4/7 The 'couscouche' yam, the cush-cush of the British West Indies..are often made into fritters. 1969 *Oxf. Bk. Food Plants* 182/1 All [yams] are of Old World origin except the cush-cush yam *Dioscorea trifida* which is native to America.

Cushing's syndrome (ku·ʃiŋ). *Med.* [f. the name of Harvey Williams *Cushing* (1869–1939), American surgeon, who described the condition in 1932.] A syndrome of hypertension, obesity, metabolic disorders, etc., caused by hypersecretion of hormones by the adrenal cortex. **Cushing's disease**, Cushing's syndrome accompanied and caused by an adenoma of the basophil cells of the pituitary gland.

1934 *Proc. R. Soc. Med.* XXVII. 397 He regarded the case..as a typical one of Cushing's syndrome. 1936 H. D. Rolleston *Endocrine Organs* iii. 104 (*heading*) Cushing's syndrome. Synonym: pituitary basophilism. 1937 *Index-catal. Surgeon-General's Office* 4th Ser. II. 130/2 (*heading*) Basophilism, pituitary (Cushing's disease). 1962 J. Howkins *Shaw's Textbk. Gynæcol.* (ed. 8) xviii. 447 *Cushing's disease* must be distinguished from Cushing's syndrome. Cushing's disease is due perhaps to a basophil adenoma of the anterior pituitary which leads to hyperfunction of the suprarenal cortex. The main symptoms and signs are those due to a disorder of the suprarenal which are now referred to under the term Cushing's syndrome. 1964 L. Martin *Clin. Endocrinol.* (ed. 4) i. 32 It is generally accepted that the symptoms of Cushing's syndrome result ultimately from adrenocortical overactivity. *Ibid.*, The basophil adenoma..is only present in about 50 per cent of cases of Cushing's syndrome.

cushion, *sb.* Add: **2. a.** (Further examples.)

1954 *Economist* 30 Oct. 411/1 Tea shares are a tricky market; numerous small companies whose shares seldom change hands make it impossible for the stock jobbers to hold any cushion of stock. 1955 *Times* 6 July 5/5 Mr. Eric Fletcher..moved an amendment to enable a plaintiff to be entitled to costs on High Court scale when he recovered a sum of £250 or more. He said the introduction of this 'cushion' was essential to justice. 1965 *Listener* 17 June 886/2 Part of their training was to find themselves jobs and hold them with no ecclesiastical cushion to fall back on.

3. c. Also of a bagatelle board.

1837 D. Walker *Games & Sports* 89 There are likewise two small cushions placed against the sides.

d. Also, a body of air which supports an aircraft, hovercraft, etc.

1928 *New Republic* 15 Aug. 331/1 When the plane catches itself on a cushion of air at the end of a plunge, you feel heavy. 1960, etc. [see *air-cushion* s.v. *air *sb.*¹ B. II]. 1967 *Gloss. Terms Air-Cushion Vehicles* (*B.S.I.*) 5 *Cushion*, a volume of air under pressure enclosed between the bottom of an ACV and the supporting surface by rigid structure, curtains, skirts or any combination thereof.

e. A sweetmeat in the shape of a cushion.

1906 E. Nesbit *Railway Children* ix. 190 I'll give you some peppermint cushions for the little ones. 1921 L. Thorpe *Bonbons & Simple Sugar Sweets* 49 Satin Cushions... With a pair of scissors cut the mixture into small cushions and leave them until quite firm. 1970 J. Aiken *Embroidered Sunset* v. 89 Bars of coconut candy, mounds of chocolate drops, of peppermint cushions.

11. *cushion-cover*; *cushion-like* adj. (later example); **cushion plant**, a plant that grows in a dense cushion-like tuft (cf. sense 5 b in Dict.); **cushion-rider**, an early name for a hovercraft type of vehicle supported by a 'cushion' of air; so *cushion-riding* vbl. sb. and ppl. a.; **cushion-stitch** (example).

1881 C. C. Harrison *Woman's Handiwork* I. 61 The cushion-cover..has a ground of royal purple velvet. 1960 I. Jefferies *Dignity & Purity* xii. 183 She carried on enthusing about cushion covers. 1951 S. Spender *World within World* 258 Fields enclosed by cushion-like hedges. 1903 W. R. Fisher tr. *Schimper's Plant-geogr.* III. iv. 705 The type of cushion-plants..is represented in the alpine region of mountains of higher latitudes in both hemispheres..by a multitude of forms. 1911 *Encycl. Brit.* XXI. 764/1 In 'cushion plants' the leaves are very small, very close together, and the low habit is protective against winds. 1959 *Times* 13 Aug. 10/2 This year's Farnborough flying display will feature..a 'cushion-rider'. 1961 *Spectator* 14 July 53 A cushion-rider can be lifted vertically and then driven over the ground at speeds which reach into the take-off speeds of ordinary jet aircraft. 1960 *Aeroplane* XCIX. 771/1 Doubtless it has other applications and ideas for cushion-riding craft in mind. 1961 *Spectator* 14 July 53 A transition between cushion-riding and ordinary aerofoil lift. 1880 L. Higgin *Handbk. Embroidery* v. 47 Cushion Stitches are taken..so as to leave all the silk and crewel on the surface.

cushion, *v.* Add: **1. b.** *fig.* (Later examples.)

1958 *Times* 23 Jan. 7/2 Aircraft firms should diversify their activities so that their other work could cushion the fluctuations in aircraft requirements. 1962 *Listener* 19 Apr. 672/1 The trouble he has been at..to take action now to cushion the economy against a recession.

cushioned, *ppl. a.* Add: **1. b.** (Later example.) Cf. sense *6 below.

1960 *New Statesman* 30 Jan. 146/1 An earth-to-moon flight with a cushioned landing on the moon and return flight to the earth.

e. *fig.* Of the voice: soft and smooth, velvety.

1909 *Daily Chron.* 20 Jan. 5/6 Complaining for two hours against fate in that cushioned voice of hers. 1920 Galsworthy *In Chancery* II. xiii, There was comfort in her cushioned voice.

2. b. *fig.* Comfortable; protected.

a 1941 V. Woolf *Captain's Death Bed* (1950) 32 A luxurious, educated, cushioned career for life.

6. Of air: made into a 'cushion'.

1935 *Times* 4 Mar. 11/3 It has been held that lift rapidly decreases when the helicopter moves above the area of cushioned air created where the down-wash from the screws strikes the earth.

cu·shioning, *ppl. a.* [f. Cushion *v.* + -ing².] Forming a cushion.

1887 Tourgée *Button's Inn* 304 The soft..snow..may have formed a cushioning mass saving him from instant death.

cushiony, *a.* Add: *fig.* Easy, comfortable, 'soft'. Cf. *cushy *a.*

1908 *Daily Chron.* 25 Feb. 4/4 A lot of them have rare cushiony jobs.

Cushite (kʊ·ʃəit), *a.* and *sb.* Also **Kushite.** [f. *Cush*, name of an ancient country in the Nile valley + -ite.] **A.** *adj.* Pertaining or relating to an ancient people of eastern Africa, south of Egypt. **B.** *sb.* A member of this people; a sub-family of the Afro-Asian family of languages; also called **Cushitic, Kushitic** (kʊʃi·tik).

1836 N. Morren tr. *Rosenmüller's Bibl. Geogr. Central Asia* I. ii. 80 A Cushite was the same as 'a man of colour'. *Ibid.* 82 Each nation placed its Cushites or Æthiopians to the south..as known to them. 1846 *Cycl. Bibl. Lit.* I. 503/1 The name Cush was applied to tracts of country both in Arabia and Africa —..on the very probable supposition that the descendants of the primitive Cushite tribes, who had settled in the former country, emigrated across the Red Sea to the latter region of the earth. *Ibid.* 504/1 Part of the Cushite population immigrated to Africa. 1883 R. N. Cust *Sk. Mod. Lang. Africa* I. 126 Lepsius recognizes them as the modern representatives of the Kushites of the Old Testament, the Ethiopians of Herodotus. 1910 *Encycl. Brit.* XII. 894/1 All these Cushitic languages, extending from Egypt to the equator, are separated by Reinisch as Lower Cushitic from the High Cushitic group, *i.e.* the many dialects spoken by tribes dwelling in the Abyssinian highlands. 1933 Bloomfield *Language* iv. 67 The fourth branch of Semitic-Hamitic is Cushite, south of Egypt; it includes a number of languages, among them Somali and Galla. 1939 L. H. Gray *Found. Lang.* 366 The Kushitic sub-division occupies part of the western coast of the Red Sea. 1954 Pei & Gaynor *Dict. Linguistics* 117 *Kushitic*, a branch of the Hamitic sub-family of the Semito-Hamitic family of languages; it consists of a great many vernaculars, the principal ones being Somali and Galla. 1961 A. J. Arkell in R. Oliver *Dawn Afr. Hist.* 10 The Cushites clashed with.. Assyria. *Ibid.*, The pharaoh sacked the old Cushite capital of Napata. 1970 *New Eng. Bible Jeremiah* xxxviii. 7 Ebed-melech the Cushite, a eunuch.

cushla, var. *acushla.

1928 'Brent of Bin Bin' *Up Country* xiv. 243 And sure, Cushla-ma-chree, if you can't stand me I'll up and go away.

cushy (ku·ʃi), *a. colloq.* Also **cushey.** [Anglo-Ind., f. Hind. *khūsh* pleasant.] Of a post, job, etc.: easy, comfortable, 'soft'. Of a wound: not dangerous or serious.

1915 D. O. Barnett *Lett.* 44 The billets here are very good..and we have rooms to ourselves... It's all very cushey and nice. 1916 *Blackw. Mag.* Jan. 91/2 I've got a cushy wound. 1916 *Daily Mail* 1 Nov., He's got a cushy job. 1917 P. Gibbs *Battles of Somme* 146 All our men who have had the luck to get a 'cushie wound'. 1928 E. Waugh *Decl. & F.* I. iii, I was sent to Ireland on a pretty cushy job connected with postal service. 1938 Auden & Isherwood *On the Frontier* III. i, There're too many healthy young men slacking in cushy staff jobs! 1957 *Listener* 26 Dec. 1066/2 It was not a particularly cushy job and two of our men were killed in action. 1970 A. Sillitoe *Start in Life* 285 You were always on the lookout for a cushy billet. 1971 *Time* 18 Jan. 30/3 Something is not quite right even at the state's cushiest 'correctional facilities' (bureaucratese for prisons), some of which could pass for prep schools.

Hence **cu·shiness**, the state or condition of being 'cushy'.

1930 S. Sassoon *Mem. Infantry Officer* ix. 268 There were times when I felt perversely indignant at the 'cushiness' of my convalescent existence.

cuspate (kʊ·spe¹t), *a.* [f. Cusp + -ate.] Shaped like a cusp.

1896 F. P. Gulliver in *Bull. Geol. Soc. Amer.* VII. 401 In this paper will be considered those forelands which present a more or less sharply pointed form, whose two sides are bounded by shore curves... These are called cuspate forelands. 1941 *Proc. Prehist. Soc.* VII. 6 The forecourt walls may be..convex, when the forecourt is cuspate, cusp-shaped or heart-shaped. 1952 F. P. Shepard in *Bull. Amer. Assoc. Petr. Geol.* XXXVI. 1911 Prominent points or horns found extending into the bays and lagoons inside many barriers. These may be referred to as 'cuspate spits'.

cuss, *sb.* **2.** (Earlier example.)

1775 *Narraganset Hist. Reg.* (1885) III. 263 A man that.. was noted for a damn cuss.

cuss, *v.* Add: (Earlier examples.)

1815 D. Humphreys *Yankey in Eng.* 104 Cuss, curse. 1841 Dickens *Barn. Rudge* xxxi. 116 Am I to thank thee, Fortun', or to cuss thee — which? 1841 Thackeray *Gt. Hoggarty Diamond* (1849) xi. 143 Have him in..for, cuss me, I like to see a rogue. 1849 — *Pendennis* I. xiii. 115 Dammin and cussin up stairs and down stairs.

b. With *out*. (See quot. 1881.)

1881 *N.Y. Times* 18 Dec. in *N. & Q.* (1882) 6th Ser. V. 65/1 *Cuss out*, to subdue by overwhelming severity of tongue. 'He cussed that fellow out', *i.e.*, he annihilated him verbally. 1901 S. E. White *Westerners* xvi. 134 Clearly he could not 'cuss out' the delinquents as they deserved.

cuss-cuss, var. *kas-kas.

cussed, *a.* Add: (U.K. examples.) *spec.* Obstinate, pig-headed.

1864 Trollope *Can you forgive Her?* I. xvii. 133 It's the cussidest place in all creation. 1952 A. Grimble *Pattern of Isl.* 149, I should of course have made up my mind in all decency then to find the place for myself... But I was cussed.

cussedly, *adv.* Add: (Earlier U.S. example.) Also *spec.*, obstinately, pig-headedly.

1846 D. Corcoran *Pickings* 10, I never keeps low company, and you is so cussedly vulgar. 1959 *Economist* 21 Feb. 661/1 Blindly and cussedly anti-farmer.

cussedness. Read: *colloq.* or *slang* (orig. *U.S.*). (Later examples.)

1904 *Daily Chron.* 9 Sept. 3/2 The kind of man..who out of sheer 'cussedness'..would succeed where a dozen others would fail. 1905 E. M. Forster *Where Angels fear to Tread* v. 153 He could understand pure cussedness, but it did not seem to be that. 1922 Joyce *Ulysses* 605 He put it down to sheer cussedness. 1931 E. O'Neill *Mourning becomes Electra* I. iv. 90, I guess there's bitterness inside me—my own cussedness, maybe.

custard. Add: **2. b.** *custard cup* (examples), *pudding* (earlier example); **custard pie**, a pie containing custard; commonly used as a missile in broad comedy, hence used *attrib.* or allusively to denote comedy of this type; **custard powder**, a preparation in powder form for making custard by mixing it with milk; **custard tree**, the tree bearing the custard-apple.

1825 *Columbian Centinel* 5 Jan. 3/5 (Advt.), Dishes, Tureens, Custard Cups. 1843 Dickens *Christmas Carol* iii. 95 A custard-cup without a handle. 1832 L. M. Child *Frugal Housewife* 68 It is a general rule to put eight eggs to a quart of milk, in making custard pies. 1920 S. Lewis *Main Street* xvi. 198 Mr. Schnarken slipped a piece of custard pie into the clergyman's rear pocket. 1933 *Punch* 29 Nov. 609/3 [The show] is wanting in straight-cut wit, and it falls back too often on custard-pie. 1940 Graves & Hodge *Long Week-End* ix. 133 Charlie Chaplin..won enormous popularity..with his custard-pie comedies. 1852 H. Beasley *Druggist's Gen. Receipt Bk.* (ed. 2) 268 Custard Powder consists of sago meal, coloured with turmeric, and flavoured. 1951 *Good Housek. Home Encycl.* 433/2 Mock custards..can be made with cornflour or with the various proprietary custard powders on the market. 1728 E. Smith *Compleat Housewife* 95 To make a Custard Pudding. 1808 T. Ashe *Trav.* x. 85 [Custard Island abounds] with the papaw, which is vulgarly known by the name of the Custard tree.

custardy, *a.* Delete '*nonce-wd.*' and add further examples.

1901 G. Meredith *Let.* 31 Dec. (1912) II. 522 The Madeira apples were custardy and curious. 1961 M. Beadle *These Ruins are Inhabited* (1963) iii. 39 Dinner..ends with something custardy.

custodee. (Earlier example.)

1812 M. Edgeworth *Tales Fashionable Life* V. 221 Sir Terence O'Fay's exploits in evading duns,..tricking *custodees*.

custom, *sb.* Add: **6.** (sense 4) *customs officer, official, union*; (sense 5) **custom-built, -made** *adjs.*, built or made to order or to measure; so **custom-build** *v. trans.*; **custom smelter** *U.S.*, a smelter who treats rock or ore for customers.

1960 *Design* 29 Feb., A willingness to accept a new situation and to custom build the standards for it. 1925 *Art & Publicity* 36 (Advt.), Custom-built exclusiveness without excessive cost. 1955 T. Sterling *Evil of Day* vii. 77 A custom built Rolls Royce cloud. 1957 M. Sharp *Eye of Love* iii. 33 His good custom-built suit. 1855 *Chicago Weekly Times* 16 Jan. 1/2 [There] may be found a large and splendid assortment of custom made boots and shoes. 1959 *Observer* 8 Mar. 15/5 Because of the peculiar idiosyncrasies of faces they [*sc.* spectacle frames] often need to be custom-made. 1705 in *15th Rep. Hist. MSS. Commission* (1897) App. vi. 11 The Justices of the Peace are to assist the customs-officers. 1923 D. H. Lawrence *Birds, Beasts & Flowers* 29 But here, even a customs-official is still vulnerable. 1880 G. T. Ingham *Digging Gold* 268 There is at Galena a small custom smelter. 1963 *Times* 22 Apr. (Zinc Suppl.) p. ii/1 The custom smelters (smelters without their own mines). 1903 'Vigilans sed Æquus' *German Ambitions* iv. 55 The Hague *Courant*, which advocated a Customs Union with Germany. 1956 *Planning* XXII. 224 Three small nations—Belgium, the Netherlands, and Luxemburg —have formed Benelux, a customs union for nearly everything except agricultural products.

b. *attrib.* passing into *adj.* Designating articles made to measure or to order, or places where such articles are made, or people producing work of this kind; = *bespoke *ppl. a.* Also *fig.* Hence as advb., in combs., as *custom-fitted, -mixed, -tailored* adjs. Cf. *custom-built, -made above. Chiefly *U.S.*

1830 *Williams's N.-Y. Ann. Reg.* 163 There are no manufactories of cotton or woollen but such as are used for custom work. 1851 C. Cist *Sk. Cincinnati* 175 Fine and coarse work for foreign markets, and custom work for home consumption. *Ibid.* 176 Two-thirds of these [shoes] at least, are made here, wholesale, or at custom shops. 1895 *Montgomery Ward Catal.* 269/3 For higher priced clothing, we refer you to our custom tailoring department, where we make clothing to order in any size and style desired. 1903 *N.Y. Times* 26 Sept. 6 (Advt.), Custom tailors charge for suits like these $35. 1905 *Washington Star* 24 Nov. 5 (Advt.), Double or Single-Breasted Sacks, as perfect-fitting as the finest custom garments. 1943 J. P. Marquand *So Little Time* (1944) iv. 31 He made Jeffrey conscious of his own custom-tailored suit, of the shine on his brown low shoes and the crease in his trousers. 1955 T. Sterling *Evil of Day* xxi. 207 He had designed the murder for one woman and no other. It was a perfect custom fit. 1957 W. H. Whyte *Organization Man* xxiii. 299 A small area with 'custom' houses. 1959 *Sunday Express* 1 Feb. 19/4 His custom-tailored suit. 1961 M. Beadle *These Ruins are Inhabited* (1963) vi. 51 An awesome superstructure of custom-fitted plugs and adaptors. 1964 *Punch* 23 Sept. 456/3 Custom-mixed after-shave lotion. 1968 *Listener* 12 Sept. 331/1 The custom cars, whose flamboyant shapes are public property while their mechanical niceties are reserved for the initiated.

customer, *sb.* Add: **3.** Also *attrib.*

1832 *Chambers's Jrnl.* I. 276/4 When customer work failed, he was fain to work a piece upon speculation. 1955 J. G. Davis *Dict. Dairying* (ed. 2) 98 Though the cost of these seals is comparatively high they have an established 'customer-appeal'. 1964 *Times Rev. Industry* Feb. 19/2 The costs of customer placation (warranty costs, the costs of preventive and corrective service and maintenance, and the cost of maintaining spares stocks) must be added. 1969 *Jane's Freight Containers 1968–69* 416/1 The increase in distant-market customer-confidence.

b. In extended use: an applicant or client.

1896 *Westm. Gaz.* 5 Dec. 4/2 A. R. Downer is the latest 'customer' for Bredin. He last night telegraphed..expressing his willingness to run the new professional 350 yards for £50 a side.

5. Also used of a fox that affords good sport in the field.

1854 R. S. Surtees *Handley Cross* xxxiv. 268 A light-coloured fox beat him so often as to acquire the name of the 'old customer'. 1899 *Westm. Gaz.* 15 Feb. 9/2 Almost immediately a fox went away, and he proved to be a real customer.

cu·stomize, *v.* orig. *U.S.* [f. Custom *sb.* 5 + -ize.] *trans.* To make to order or to measure; to model or alter according to individual requirements. So **cu·stomized** *ppl.a.*, **cu·stomizing** *vbl. sb.*

1934 H. L. Mencken in *Words* Nov. 5/2 Obviously American..are such curious forms as..*to customize*. 1960 *Times* 14 Sept. 12/7 In the shops [in the U.S.], too, much time is saved with such contractions as 'Customized Drapes' which really reads much more dashingly than 'Curtains made to customers' specifications'. 1963 *Engineering* 31 May 740/1 A new and more economical means of obtaining a Grand Tourer was devised by 'customizing' an ordinary production sports car. 1965 *New Society* 2 Dec. 5/3 Not bigger and better engines—but more gadgets, more comfort, more customising. 1967 *Electronics* 6 Mar. 22/1 (Advt.), The high cost of 'customizing' is eliminated. 1970 *Daily Tel.* 26 Jan. 11/3 The cult of 'customised' transport—cars and motorcycles sculpted and plastered to express the personal fantasies of their owners.

cut, *sb.*² Add: **2. c.** (*a*) (Earlier example.)

1760 Sterne *Tr. Shandy* III. iv. 15 Pell mell, helter skelter, ding dong, cut and thrust..have they been trimming it [*sc.* a jerkin] for me.

3. b. *pl.* Corporal punishment, esp. of schoolchildren. *Austral.* and *N.Z. slang.*

1915 *Bulletin* (Sydney) 28 Oct. 47/1 'Six cuts yer give him,' roared the whiskers...The stick emphasized the last remark by a rapid descent on the meek one's shoulders. 1938 P. Lawlor *House of Templemore* xi. 123 'Urry or yer'l git th' cuts. 1945 F. Sargeson *When Wind Blows* ii. 14 [You] would get the cuts for sure. 1963 D. Adsett *Magpie Sings* 57 If anyone was careless enough to use the wrong peg, their coat, hat and bag could be thrown to the floor without fear of getting the cuts.

6. For 'Chiefly *U.S.*' read 'orig. *U.S.*' Also, a reduction in wages, supplies, services, etc. (Later examples.)

1921 *Daily Herald* 29 Apr. 1/4 The L.C.C. do not contemplate any immediate cuts in their tramway service. 1946 *Daily Tel.* 27 Mar., A statement of unusual gloom emanated from the Food Ministry..prophesying a fresh 'cut' in the soap and margarine rations. 1968 R. Harris *Nice Girl's Story* ii. 14 The gas..flickered blue and cold. 'There's some kind of a cut, I think...' 1971 *Daily Tel.* 18 Feb. 15/2 It is still not known how much next year's cut of 10,000 in the total of 180,000 assisted passages..will affect candidates from Britain.

10. a. (Earlier example.)

1833 *Gentl. Mag.* July 44/2 Beldham was great in every hit, but his peculiar glory was the cut.

12. b. Intentional absence from or deliberate omission to attend (an event). Cf. Cut *v.* 33 c.

1851 B. H. Hall *College Words* 90 *Cut*, an omission of a recitation. 1856 *Ibid.* (ed. 2) 147 *Cuts*. When a class [at Bowdoin College] for any reason become dissatisfied with one of the Faculty, they absent themselves from his recitation, as an expression of their feelings. 1915 *Dialect Notes* IV. 233 *Cut*, unexcused absence from class. 1919 W. T. Grenfell *Labrador Doctor* (1920) ii. 22 Attendance at chapel was compulsory, and no 'cuts' were allowed.

14*. *Cinemat.* A quick transition from one shot to the next (see also quot. 1940). Cf. *cut *v.* 21 e.

1933 I. Dalrymple in A. Brunel *Filmcraft* 174 Don't ignore the stunt or effect cut. 1933 M. Hankinson *Ibid.* 225 A cut is always made between the second and third sprocket-holes of a frame..on the action and parallel on the track.. because..the ordinary joining machine loses two sprocket-holes on each bit of film it joins. 1940 *Chambers's Techn. Dict.* 217/2 *Cut*, the junction between one strip of continuous film of motion-picture and the next. 1944 S. Cole *Film Editing* 10 In similar circumstances such a cut would be acceptable even without any sound at all. 1959 *Viewpoint* July 19 A straight 'cut' instead of the conventional 'fade' helped to achieve a startling visual jerk. 1961 G. Millerson *Telev. Production* 298 The cut is the simplest transition—an immediate change from one shot to the next.

16. a. *spec.* Short for *hair-cut, used esp. with defining word.

1951 N. Marsh *Opening Night* iv. 90 I'm a shoulder-length natural ash-blonde and I've had to have an urchin cut and go black. 1953 *Encycl. Brit. Bk. of Yr.* 639/2 Fashion produced the Pony-Tail and the Poodle-Cut, two hairstyles for women. 1960 *Sunday Express* 14 Aug. 12/4 The short cut..was *made* for me. *Ibid.* 23 Oct. 14/3 One guinea for a 'cut'.

21. b. A gramophone record or recording. Cf. *cut *v.* 23 d. orig. *U.S.*

1949 *Music Libr. Assoc. Notes* Dec. 42 A recording artist *cuts* a master and the recording executive may reject the *cut*. 1962 A. Nisbett *Technique Sound Studio* 247 *Cut*,..one of several separately recorded bands..on a disc. 1970 *New Yorker* 12 Dec. 182/2 These two cuts, along with.. 'Little Sadie', showed promise of saying something interesting.

23. † b. A slice of meat as a slight meal. Cf. *cold cuts. *U.S. Obs.*

1770 Washington *Diary* 9 June (1925) I. 383 Had a cold cut at Mrs. Campbell's. 1773 *Ibid.* 21 Feb. II. 102 [They] calld here, but would not stay dinner, taking a Cut before it. 1816 U. Brown *Jrnl.* 24 Sept. in *Maryland Hist. Mag.* (1916) XI. 230 At last comes to an Orniary [*sc.* Ordinary], fed & took a cut. 1827 *Cincinnati Enquirer* 15 Aug. 2/5 A cold cut at Utica.

c. A number of sheep or cattle cut out from the flock or herd. *U.S., Austral.*, and *N.Z.*

1888 Roosevelt in *Century Mag.* Apr. 860/2 As the animals of a brand are cut out they are received and held apart by some rider detailed for the purpose, who is said to be 'holding the cut'. 1907 S. E. White *Arizona Nights* I. vi. 112 The round-up Captain appointed two men to hold the cow-and-calf cut, and two more to hold the steer cut. 1933 E. Jones *Autobiogr. Early Settler* xxi. 93 [They] took a small cut of twenty or thirty sheep up to the river. 1953 B. Stronach *Musterer on Molesworth* x. 68 At last we got a small cut of our mob [of cattle] over [the bridge] and the rest was easy.

d. *slang.* Share (of profit, etc.); commission; = *rake-off. orig. *U.S.*

1918 H. C. Witwer *Baseball to Boches* ix. 363 If you get nailed we'll give your wife a cut of our winnin's! 1940 Wodehouse *Quick Service* xii. 101, I don't mind giving Howard Steptoe his cut, but..five hundred pounds has got to be earmarked for me. 1957 W. H. Whyte *Organization Man* 282 The real money would come from..the company's cut..of every dollar spent in the shopping center. 1970 *New York* III. 30 Nov. 28/3 The net proceeds of a $2 million stock offering after the underwriter had taken his cut.

27. Also, a portion of a field cut, or intended for cutting, at one time. *U.S.*

1765 Washington *Diary* 6 Nov. (1925) I. 216 Finishd sowing Wheat at the Mill—viz 19 Bushls. in ye large cut within the Post and Rail fence and 6 B. in ye small cut. 1770 *Ibid.* 14 Sept. 399 Morris at Doeg Run began to sow his third Cut of Wheat. 1855 G. N. Jones *Florida Plant. Rec.* (1927) 132 The Cotton in the lower most cut of prelow will avrige knee high, the next two cutes will not avrige quit wast high.

35. *cut-line* (example); (*b*) (see quot. 1912); (*c*) descriptive wording below an illustration; **cut-through**, an act of cutting through; *spec.* in *Rugby Football* (Webster, 1934). Cf. Cut *v.* 18 b.

1883 *Encycl. Brit.* XX. 210/1 Another white line across the front wall, termed the 'cut line', because the in-player, when serving, must first make the ball rebound from the front wall above this line. 1912 M. Drake *Eng. Glass-painting* 183 The panes composing it [*sc.* the panel] should be laid in their places on a sheet of paper and their outlines traced by a pencil run round their edges. This sheet of paper will serve the glazier as a 'cut-line' drawing when the panels are handed to him for re-leading. 1923 — *Doom Window* xxv. 290 Cartoons began to be completed, and Reinecke and Sophie now made the cut-line drawings. 1938 L. M. Harrod *Librarians' Gloss.* 54 *Cut line*, matter appearing below an illustration. More often called a 'caption'. 1964 H. Waugh *Missing Man* xi. 48 Betty Moore's picture ran two columns wide on the front pages with the cutlines describing her as the 'widowed beauty'. 1960 *Times* 28 Dec. 4/4 Leicester were in trouble again after a cut-through by Jeeps. 1962 *Ibid.* 2 Mar. 4/2 There was a fast, weaving cut-through by Watkins.

cut, *v.* Add: **2. b.** *to cut both* (or *two*) *ways*, to have a double or mixed effect; to have both favourable and unfavourable aspects or implications.

1605 [in Dict.]. 1854 J. C. Rutter *Let.* 23 May in M. Lutyens *Millais & Ruskins* (1967) 192 What you state about the Woman's Brain..might cut both ways..might not the irritant arise from want of consummation? 1866 [see Double-edged *a.*]. 1935 *Discovery* Oct. 313/1 Clever arguments cut two ways.

7. d. *slang.* To divide or share (spoils, profits, etc.); to receive (a share). Also *intr.* Cf. *cut *sb.*² 23 d.

1928 E. Wallace *Again 3 Just Men* x. 216 It was wicked ..that anybody should have so much money if he could not 'cut' his share. 1932 — *When Gangs Came* xxviii. 278 'The other fellows' had refused to 'cut'.

8. c. *to cut fine*: see *fine *a.* 7 g.

9. d. *to cut and cover*: to plough so that the furrow-slice is turned over on an unploughed strip. *U.S.*

1839 [see *61 below]. 1861 *Trans. Ill. Agric. Soc.* IV. 111 Mr. Mills is not in favor of any implement that 'cuts and covers'. Col. Harris..says that cutting and covering is practiced by some of the Scioto farmers.

13. b. Also of land, to yield as a crop.

1754 *Essex Inst. Hist. Coll.* XLIII. 90 The Pasturing good..and cuts enough to keep all the stock. 1840 J. Buel *Farmer's Comp.* (ed. 2) 211 One acre of good grass will cut three tons of hay, or keep a cow... Four acres of lean, poor grass will cut little more..than three tons of hay. 1872 *Rep. Vermont Board Agric.* 351 It would cut only hay enough to winter four cattle. 1923 R. D. Paine *Comr. Rolling Ocean* x. 169, I quit the sea for a spell to run my own place—she cuts thirty ton o' hay.

14. c. *transf.* = 56 d. *U.S.*

1903 A. Adams *Log Cowboy* ii. 13 Flood had the first pick, and cut twelve bays and browns.

15. a. *to cut across* (fig.).

1927 Carr-Saunders & Jones *Soc. Struct. Eng. & Wales* 83 But these other associations cut across industrial, occupational, and income classifications. 1960 R. Davies *Voice* from Attic 38 We exist as a class which cuts across all classes.

16. b. esp. *U.S.* with *trail*. Also *ellipt.*

1899 T. Hall *Tales* 19 One of the men dashes breathlessly in..with the exciting report that he has cut the raiders' trail. **1903** A. Adams *Log Cowboy* viii. 90 If you have no authority to cut this trail then you don't cut this herd. *Ibid.*, They were merely cutting (trail cutting) in the interest of the immediate locality.

19. a. (Further examples with *away*, *off*.)

1834 Dickens *Sk. Boz* (1836) 1st Ser. I. 92 The linen-draper cut off..leaving the landlord his compliments and the key. **1848** —— *Dombey* xv. 156 Mr. Toodle..rapped her on the back; and said..'Polly! cut away!' **c 1867** T. W. Robertson *Caste* I. 7, I did get leave, and I did cut away; and while away, I was miserable. **1932** A. J. Worrall *Eng. Idioms* 67 The prefect told the small boy to cut off.

b. (Earlier examples.) Also with *along*, *out*. *to cut round* (U.S. colloq.): to make a display; to act in a lively, gay fashion.

1797 B. Hawkins *Lett.* (1916) 126 He was driving a wagon at the time he was taken, and they cut out and took the horses with him. **1833** S. Smith *Major Downing* 139 What made us cut back so quick from Concord? **1834** D. Crockett *Life* 63, I saw a little woman streaking it along through the woods like all wrath, and so I cut on too. *Ibid.* 65, I took my eldest brother..and cut out to her father's house to get her. **a 1852** F. M. Whitcher *Widow Bedott P.* (1856) 91 They say she cut round and hollered and laffed and tried to be wonderful interestin'. **a 1859** in Bartlett *Dict. Amer.*, Instead of sticking to me as she used to do, she got to cuttin' round with all the young fellows, just as if she cared nothin' about me no more. **1864** Dickens *Mut. Fr.* I. iii. viii. 240 I'll cut back and ask for leave. **1879** F. R. Stockton *Rudder Grange* viii. 86 [The dog] was only cuttin' round because he was so glad to get loose. **1902** E. Nesbit *Five Children & It* ix. 237 You'll be late for your grub!..Cut along home. **1932** A. J. Worrall *Eng. Idioms* 68, I told him to cut out and buy some tea. **1949** M. Innes' *Journeying Boy* ii. 4 'And now you'd better cut along.' Captain Cox was a great believer in the moral effects of abrupt dismissals on the young. **1958** T. Williams *Orpheus Descending* iii. 88 *Lady.* So you're—cutting out, are you? *Val.* My gear's all packed. I'm catchin' the southbound bus.

c. To get up *behind* a vehicle. *U.S.*

1848 *Popular Songs* 30 Another calls out 'cut behind'. **1860** O. W. Holmes *Prof. Breakf.-t.* viii. 171 Here is a boy that loves to..chalk doorsteps, 'cut behind' anything on wheels or runners [etc.].

21. b. *imp.* (*slang*) = *cut out* (see *56 a below). Colloq. phr. *cut the cackle* (see *CACKLE *sb.* 3 a).

1859 Hotten *Dict. Slang* 28 Cut,..to cease doing anything. *Ibid.*, *Cut that*, be quiet, or stop. **a 1871** T. W. Robertson *Caste* i. 6 *Geo.* Well, then, eighteen months ago —*Haw.* Oh, cut that; you told me all about that. **1907** E. S. Field *Six-Cylinder Courtship* 55 'My dear fellow—' I began. 'Cut it!' he commanded. **1919** F. Hurst *Humoresque* 314 Come on, Herman, cut the comedy. **1971** ..time we were getting across to our hotel.

c. To outdo, excel. Cf. senses 53 e and 56 f in Dict. Chiefly *U.S.* in modern use (see quot. 1952).

1884 *Referee* 13 Apr. 1/4 George's performance..is hardly likely to be disturbed for a long time to come, unless he cuts it himself. **1897** *Penrith Obs.* 21 Dec. (E.D.D.), He went thirteen feet t'first lowp, but I cut him bi' three inch. **1952** B. Ulanov *Hist. Jazz* (1958) xxv. 351 'Cut' also means to best a soloist or band in competition.

d. *to cut a corner* or *corners*: to pass round a corner or corners as closely as possible; *fig.*, to pursue an economical or easy but hazardous course of action; to act in an unorthodox manner to save time; also, to act illegally.

1869 'Mark Twain' *Innoc. Abr.* (1870) xxiii. 171 He cuts a corner so closely now and then..that I feel myself 'scrooching', as the children say. **1894** Kipling *Day's Work* (1898) 315 It was at this point that he began to cut corners. **1909** M. Dyer *Candles in Wind* 57 Her husband's tendency to 'cut corners' when confronted with awkward facts. **1915** *The Cape* I. xx. 38 They turn out of side-streets at high speed, and cut corners in a dangerous manner. **1947** W. H. Whyte *Organization Man* 292 A disciplining force that helped them resist the temptation to cut corners. **1966** 'S. Ransome' *Hidden Hour* xix. 149 He could cut a sharp corner without letting it bother his conscience. **1966** 'S. Woods' *Enter Certain Murderers* iii. 41 If Dad had cut any corners, I think I'd have known about it.

e. (*Cinemat.*, *Radio.*) *trans.* To edit (a film, etc.). Also *intr.*, to make a quick transition from one shot to the next. *imp.* A signal to stop.

1913 [implied in *1 c]. **1916** E. W. Sargent *Technique Photoplay* (ed. 3) 184 You can cut to some single person who overlooks the crime and later tells the story. **1937** *Amer. Speech* XII. 100 Cut is used by [radio] production men as an imperative to halt a rehearsal. **1938** *Times* 7 Jan. 13/6 In front of the [television] producer sit the sound engineer controlling total output, and the sound mixer selecting and cutting it. **1947** D. Lean in O. Blakeston *Working for Films* 29 The scene should be cut like this. **1953** K. Reisz *Technique Film Editing* iii. xiv. 240 We cut to a closer shot of Pip. **1959** *Elizabethan* June 26/1 When the director wants to stop the camera he calls out 'Cut'. **1960** N. Kneale *Quatermass & Pit* 1. 11 Cut—to the excavation. *Ibid.* 12 Cut—to where a spadeful of clay is being swung down from the truck.

f. *trans.* and *intr.* To cut out (see *56 r below).

1938 Hemingway *Fifth Column* (1939) iii. ii. 88 Cut those lights! **1957** *Granta* 9 Mar. 19/1 Then I would lie down on my back watching them, hoping their engines wouldn't cut just then. **1958** 'N. Shute' *Rainbow & Rose* i. 34, I gave her a little throttle..and then cut it as she rolled on to the grass. **1970** D. MacKenzie *Kyle Contract* (1971) 12 He drove into his carport and cut the motor.

22. b. To dilute or adulterate. Chiefly *U.S.*

1930 J. P. Burke in *Amer. Mercury* Dec. 455/1 We don't cut booch any more. **1938** *Amer. Speech* XIII. 190/2 Other types of narcotics are cut. **1954** *Encounter* July 27/2 My wife..had a cup of coffee cut with bourbon ready for me. **1955** *Times* 9 Aug. 6/1 Most of the wine..when mixed or 'cut' with Algerian wine, provides a good deal of the ordinary *vin courant* of France. **1966** *Guardian* 31 Aug. 11/6 When I was 13 I knew how much quinine and sugar water you needed to cut heroin and sell it. **1969** *Sunday Globe* 21 May 23/4 Use bleach which has been cut with water and spread on the counter-tops.

23. *Sound Recording.* To record; to make (a record). orig. *U.S.*

1937 *Printers' Ink Monthly* Apr. 50/3 *Cut a disk*, to make a recording. **1948** *Newsweek* 19 July 38/1 Bernard Baruch cut a record of 'Yankee Doodle'. **1958** M. White in P. Gammond *Decca Bk. Jazz* xviii. 221 The Hot Five..cut several more sides for the same recording studios, where a number of very fine sides indeed were cut. **1962** *Melody Maker* 7 July 8 She cut five titles which will be released as part of the sound-track album of her film.

29. Also, to divide cards as a means of selecting one's partner, and *transf.*

1880 H. C. Adams *College Days at Oxford* 52 They cut for partners. **1937** A. Thirkell *Summer Half* i. 28 While they cut for partners, while they dealt,..his mind was in a turmoil. **1938** C. Morgan *Flashing Stream* iii. 221 Cut for drinks. *(They throw dice on a table.)* **1960** *Times* 8 Jan. *Penguin Hours* 4 Before each rubber those taking part 'cut' for partners and choice of seats.

31. a. (Earlier examples.) Also, of a cricket ball: to turn sharply after pitching; of a bowler: to make (the ball) turn sharply after pitching.

1816 W. Lambert *Instr. & Rules Cricket* 15 Aided by a turn or motion of the wrist, the Ball may be made to cut or twist, after it has grounded. **1833** J. Nyren *Young Cricketer's Tutor* 31, I do not remember to have seen Lambert cut a ball with the bat held horizontally. *Ibid.* 65 Peter Stewart..could cut the balls very hard at the point of the bat. **1851** J. Pycroft *Cricket Field* vii. 148 The balls may be made to rise over the wicket. **1960** I. Peebles *Bowler's Turn* 67 Geary and Macaulay cut and spun the ball at a sharp pace.

b. (Example.)

1875 'Stonehenge' *Brit. Sports* (ed. 12) iii. i. v. 691/2 The ball after contact with the ground has a tendency to shoot with a very slight rise in comparison with a ball that is not cut. *Ibid.*, If the ball is purely cut, stand well back from it.

c. (Examples.) Also *intr.*

1874 J. D. Heath *Croquet Player* 33 If the ball is to be 'cut' to the left, the right side of it must be struck. *Ibid.*, Considerable practice will be required before the player can cut perfectly. **1966** J. W. Solomon *Croquet* 95 To cut the rush to one side or the other, was for a long time to me a matter of luck.

33. c. (Further examples.)

1794 *Gentl. Mag.* Dec. 1085/1, I was told of men..who..cut chapel, cut gates, cut lectures, cut hall, cut examinations, [etc.]. **1930** W. S. Maugham *Cakes & Ale* iv. 49 She was prepared to cut an engagement in London.

33*. To finish. *N.Z. slang.*

1916 *Anzac Book* 49 ..cut it.. **1947** A. P. Gaskell *Big Game* 41 Is the beer all cut? *Ibid.*, Cut your beer—we're going. **1962** ...drink down. We must cut this bottle tonight.

40. (Earlier example.) Also as *attrib. phr.*

1704 *Boston News-Let.* 12 June 2/2 Capt. Vaughn rode by said Ship, but cut & run. **1929** *Daily Chron.* 23 Oct. 9/1 If it is the cut-and-run mood that has conquered him go home. **1945** *Hutchinson's Pict. Hist. War* 27 Sept. 1944–13 Mar. 1945 p. 70 The enemy was only capable of carrying out a cut-and-run attack by a force consisting of two or three battleships and a couple of carriers.

41. *to cut loose*. **c.** To begin to act freely; to start off; to commence an attack; to let one-self go. *U.S.*

1902 Ade *More Fables in Slang* (1902) 182 She would approach the Piano timidly and sort of Trifle with it for a while, and say they would have to make Allowances, and then she would Cut Loose and worry the whole Block. **1903** S. E. White *Westerners* xxiii. 227 They might buck some. **1911** H. Quick *Yellowstone N.* 209 Wolfville..sent ..he cut loose. **1923** R. D. Paine *Comr. Rolling Ocean* xxi. 'Goodness gracious. I should call this the deuce of a mess'. **1941** H. L. Mencken *Newspaper Days* (1941) 259 When he got down his first dozen mugs..he cut loose with an exultant yodel.

49. *to cut didoes* (DIDO), *dirt* (DIRT *sb.* 6 d), *eyes*, (*no*) *ice*, *the painter* (PAINTER[2] 2 b), *a splurge*, *a swath* (SWATH[1] 3 c), *to waste* (WASTE *sb.* 10 d): see the sbs. For *cut one's loss(es)*, *cut prices*, see *LOSS *sb.*[1], *PRICE *sb.*

52. **cut back. c.** *Cinemat.* To return to a previous scene by repeating a portion of that scene. (See *CUT-BACK *sb.* 2.) Also *trans.*

1913 E. W. Sargent *Technique Photoplay* (ed. 2) 91 The same device may be used to get rid of a dinner scene..We cut-back to some other action. **1916** *Ibid.* (ed. 3) 194 It is well to remember..that not all plays may be cut back. **1959** Halas & Manvell *Technique Film Animation* 337 Cut back so many frames..and these frames are shot again.

d. *trans.* To reduce or decrease (expenditure, etc.). Also *intr.*

1943 *Sat. Even. Post* 6 Nov. 112 If the Army cuts back a program, it will not need the steel for some other program. **1953** *Manch. Guardian Weekly* 7 May 3 Western Europe was cutting back its defence programme. **1958** *Engineering* 4 Apr. 433/1 Industry is continuing to cut back markedly on capital spending plans. **1965** *Listener* 2 Sept. 334/1 Germany, another country whose aid budget had been built up impressively by the early nineteen-sixties, has recently cut it back. **1971** *Daily Tel.* 14 Jan. 2/7 A spokesman for ICI said that recruitment of Ph.D. students had been cut back.

e. *intr.* In *Surfing*, to turn one's surf-board back towards the breaking part of a wave.

1963 *Surfing Yearbk.* 41/1 *Cutting back*, when a rider is getting too far ahead of the curl, and has to change his direction to get in a better position relative to the wave. **1969** *Observer* 3 Aug. 35/1 He can 'cutback', turning the board back toward the breaking wave.

53. cut down. a. (Later examples.)

1884 J. Hatton *Irving's Impr. Amer.* II. v. 86 A ship laden with corn was cut down and sunk by floating ice. **1952** *Oxf. Jun. Encycl.* VI. 174/1 As the old trees are cut down, seedlings spring up naturally to replace them.

g. Also *intr.*, freq. with *on*.

1939 H. W. Horwill *Anglo-Amer. Interpreter* 54, I am cutting down on my meat. **1945** E. Bowen *Demon Lover* 90, I got my hundred [cigarettes] this morning... I can't seem to cut down, somehow. Mary, have you cut down? **1962** A. Nisbett *Technique Sound Studio* xii. 220 Methods of cutting down on the labour have been suggested.

h. *to cut down to size*: to reduce to suitable dimensions; *fig.* to reduce to a true or proper level of importance.

1821 M. Wilmot *Let.* 20 Mar. (1935) 100 We..*cut down* silk stockings to Cat's size. **1904** G. B. Shaw *Lett. to G. Barker* 6 Dec. (1956) 45 The theme is a huge one; and it can't be cut down to Court size. **1927** Hemingway *Men without Women* (1928) 33 He wheeled his horse..towards..the far side of the ring where the bull would come out... 'Pic him, Manos,' he said. 'Cut him down to size for me.' **1959** *Listener* 2 July 12/2 We are in danger of forgetting our place, of getting ideas above our station. It is good to have Mr. Graves to cut us down to size. **1959** *Guardian* 17 Sept. 8/1 Production and distribution of films has been cut down to size and other interests have been greatly extended. **1969** J. Drummond *People in Glass House* xxxvi. 140 You've had thirty good years, perhaps it's time you were cut down to size.

54. cut in. c. Of coaches (earlier example). In later use also, to drive a motor-vehicle between two others which are passing each other in opposite directions; more recently, to drive a motor vehicle, cycle, etc., past another and move sharply in front of the overtaken vehicle. Also *transf.*

1820 *Kaleidoscope* 25 July 29/2 Amid the din Of drunken coachmen *cutting in*. **1925** *Don'ts for Motorists* 100 How to avoid accidents...Don't 'cut in'. **1926** *Weekly Disp.* 5 Sept. 1/1 He is supposed to have been reported for cutting in at the second and eleventh tees. **1931** *Highway Code* 10 Your reluctance to stop dead may tempt you to 'cut in' by threading your way between other vehicles. This is a frequent source of accidents. **1954** *Ibid.* 8 Never cut in, that is do not pull in sharply in front of a moving vehicle which you have just overtaken. **1955** R. Bannister *First Four Minutes* 21, I was unguarded against the man outside me who was cutting in.

d. *spec.* To have one's name added to a lady's dancing programme; also (orig. *U.S.*), to supersede a partner during a dance.

1857 G. A. Lawrence *Guy Livingstone* vi. 47 Keeping all her after-supper waltzes for him religiously, though half the men in town were trying to cut in. **1896** Ade *Artie* x. 91 He did n't want no one else to cut in. **1919** *Ladies' Home Jrnl.* Sept. 169 All the men want to cut in when she dances. **1920** F. Scott Fitzgerald *This Side Paradise* (1921) i. ii. 70 The dance began... Boys cut in on Isabelle every few feet.

f. To receive a share (of profits, booty, etc.); also *trans.* (orig. *U.S.*), to give (a person) a share; freq. with *on*. *slang.*

1890 R. D. Blackmore *Kit & Kitty* III. xiv. 192 The brothers..smiled a sour smile, as much as to say,—'You don't cut in for any of it.' **1924** R. Lardner *How to write Short Stories* (1926) 335 They'll cut you in on the big money. **1930** I. Goldberg *Tin Pan Alley* 210 For plugging certain numbers these leaders collect—'cut in'—on payments and royalties. **1950** G. Greene *Third Man* xiv. 122 Were you going to cut me in on the spoils?

g. *trans.* To connect (an electric circuit, etc.). Also *intr.* of a motor.

1910 [see sense *56 r].

h. *trans.* To insert (a scene) into a film sequence. Also *transf.*

1928 *Film Weekly* 29 Oct. 17/3 These shots I cut in with other and varied material. **1934** C. Lambert *Music Ho!* iv. 232 A picture of the mother crying was 'cut in' with a picture of a dripping kitchen tap. **1947** D. Lean in O. Blakeston *Working for Films* 29 Now where would you cut in the close-up of the banana-skin? **1962** A. Nisbett *Technique Sound Studio* ii. 43 Special [sound] effects are cut in as required. *Ibid.* vii. 128 We may wish to edit music..to cut in a retake.

55. cut off. h. *spec.* To deprive of communication by telephone or telegraph; to disconnect (a telephone).

1891 E. S. Ellis *Check 2134* v. 38 The company can't afford to be cut off this way. **1932** D. Whipple *Greenbanks* vi. 91 The telephone had its merits after all; Ambrose could be cut off. **1940** Auden *Another Time* 91 Stop all the clocks, cut off the telephone.

56. cut out. a. Freq. *fig.* in recent colloq. use: to stop doing or using (something); to leave off, do without, omit, drop: esp. in imper. phr. *cut it out*.

1903 Ade *People you Know* 82 Cut it out! **1905** 'H. McHugh' *You can search Me* i. 27 I've been speculating again after faithfully promising her to cut out all the guessing contests. So cut out the yesterday gag. **a 1910** 'O. Henry' *Sixes & Sevens* (1916) xviii. 209 To be frank with you, Whatsup, I've cut out the dope. **1914** C. Atherton *Perch of Devil* i. 137 If it were more the primal instinct ..so much the worse, the more reason to 'cut it out'. **1923** R. D. Paine *Comr. Rolling Ocean* vi. 110 Will you cut out the boozy stuff while you are ashore in Jamaica? **1930** Auden *Poems* (ed. 2) 54 Its no use raising a shout. No, Honey, you can cut that right out. **1937** R. Stout *Red Box* iv. 54 Llewellyn..was expostulating: 'Now, Dad, cut it out..you now listen a minute.' **1939** D. L. Sayers *In Teeth of Evidence* 202 'The great man himself. London's rising dermatologist.' 'Cut it out,' said Scales. **1970** M. Guyson tr. *Solzhenitsyn's First Circle* xlix. 366 'Cut it out!' said Pryanchikov, struggling violently. 'I'm sick of prosecutors and trials.'

d. (Earlier N.Z. and U.S. examples.)

1862 E. Chudleigh *Diary* 13 Feb. (1950) i. 28 On the run all day cutting out bullocks we succeeded in yarding about 60. **1867** J. T. Thomson *Rambles with Philosopher* xxvi. 149 We scampered away to the pasture grounds of his cattle, in order to cut out one of the nags for slaughter. **1869** *Overland Monthly* III. 126 Another rides in, selects a stray horse and 'cuts it out', by chasing it out with his horse.

l. Phr. *to have* (*all*) *one's work cut out*: see Work *sb.* 30.

q. To finish shearing. Also in extended use: *trans.*, to finish; *intr.*, to come to an end. *Austral.* and *N.Z.*

1890 *Melbourne Argus* 20 Sept. 13/6 When the stations 'cut out', as the term for finishing the shearing is. **1896** H. Lawson *In Days when World was Wide* (1900) 47 The cheque was spent that the shearer earned, and the sheds were all cut out. **1919** W. H. Downing *Digger Dial.* 18 *Cut out* (v.), cease. **a 1925** F. S. Anthony *Follow Call* (1936) 156 I've never been able to save a ten-pound note. **1934** T. Wood *Cobbers* ..gone. **1937** *Bulletin* (Sydney) 31 May 38/3 Tomorrow they would cut out the last of the sheep and the men would be paid off. **1941** Baker *Dict. Austral. Slang* 21 *To cut out*..to complete any task. **1948** *Landfall* II. 113 After the flax cut out in 1918, we milked cows. **1950** H. P. Tritton *Time means Tucker* iv. 29/1 The last sheep was shorn, the bell rang, the whistle blew and Charlton was cut out. **1963** A. Lubbock *Austral. Roundabout* 157 The great mines in Victoria..began to cut out..but the miners often remained in the district.

r. *trans.* To disconnect or switch off (an electric circuit, etc.). Also *intr.*, to switch off; to cease operating.

1910 *Chambers's Jrnl.* May 350/1 By means of a switch near the keyboard the organist can cut the motor in and out as desired. **1912** *Dial.* Aug. 356/2 The dynamo is cut out automatically. **1917** *Blackw. Mag.* May 804/1 We continued in a westerly direction, with one cylinder still cutting out. **1924** A. W. Judge et al. *Mod. Motor Cars* III. 75 Which causes the hammer to vibrate and to cut-out the battery circuit. **1926** H. H. U. Cross *Electric Lighting* (ed. 4) 264 When the gears are fully extended, the electro-magnet is cut out by a disconnecting switch. **1928** *Motor Manual* (ed. 27) 110 When the dynamo speed falls below a certain minimum the device cuts out or opens the charging circuit. **1930** *Daily Express* 16 Aug. 5/5 When aero engines were much more liable to cut out and force one down in isolated places. **1935** *Jrnl. R. Aeronaut. Soc.* XXXIX. 471 There is a danger of the engine cutting out during take-off.

57. cut over. c. (Earlier example.)

1867 J. Lillywhite's *Cricketers' Compan.* 12 You will..prevent your wicket being cut over in that part which takes all the batting out of you.

58. cut under. (Earlier U.S. example.)

1859 Bartlett *Dict. Amer.* (ed. 2), *To cut under*, to undersell in price. New York.

59. cut up. o. *trans.* With *caper*, *shines*, etc.: to behave in a mischievous or frolicsome manner. *U.S.*

1775 in *Narrag. Hist. Reg.* III. 263 A man that was in company there the evening before that cut up a caper. **1846** D. Corcoran *Pickings* 28 He was cutting up all kinds of extra shines, like those bere theatric lepers. **1847** [see *monkey-shine*, (*MONKEY *sb.* 18)]. **a 1848** *Knickerbocker* (Bartlett), A wild bull of the prairies was cutting up shines at no great distance, tearing up the sod with hoofs and horns. **1945** *Dex* 'Dindon'. **1903** A. D. McFaul *Ike Glidden* ii. 11 It was not the like that cut up the mischief this time. **1945** R. Lewis C. *Timberlane* (1946) xvi. 102 People recognizing you and staring at you cutting up monkey-shines!

p. To conduct or manage (a contest) fraudulently. *Sporting slang.*

1923 *Daily Mail* 16 Jan. 7 Georges Carpentier, M. Descamps, his manager, and M. Hellers, the manager of the coloured Boxer Siki, have been acquitted by the French Boxing Federation of having arranged and 'cut up' the fight in which Siki was declared victor. **1923** *Weekly Disp.* 13 May 7, I read in newspapers now that more than half the races under National Hunt rules are cut up, and that jockeys and trainers are not out to win.

q. *intr.* Of the surface of the ground: to become broken up irregularly.

1891 W. G. Grace *Cricket* iii. 20 Snow had fallen during the night..the wicket was cutting up badly. **1909** *Mid Flat Estate* 36 Six-ton loads went through..before the roads 'cut up' and were spoiled.

r. *to cut up* (*old*) *touches*: to gossip or reminisce. *U.S. slang.*

1931 D. Runyon *Guys & Dolls* (1932) viii. 180 They are cutting up old touches. *Ibid.* xiii. 281 Having a drink together..and..cutting up old touches of the time when they run with the Hudson Dusters together. **1941** *New Yorker* 1 Nov. 27/3 He and Dutch would get together and cut up touches.

60. cut-and-come-again, *sb. phr.* (See quots.)

1886 F. T. Elworthy *W. Somerset Word-bk.* 177 *Cut and come again*, a very prolific variety of kale or winter greens; much grown in cottage gardens. **1959** *Listener* 21 May 911/1 Some..are cabbage..to cook *cut-and-come-again* (or cottage kale).

61. cut-and-cover. Also *Agric.* (see *9 d). *U.S.*

1839 J. Buel *Farmer's Comp.* xiii. 113 The cut-and-cover practice is still worse as it leaves..two thirds of the soil, unpulverized by the plough. **1879** *Rep. Vermont Board Agric.* IV. 93 The old Dutch wooden plow..[was] used among the stumps and roots of the newish lands, with the plowman's ideal of cut and cover.

61*. cut-and-fill: the process or result of removing material from a place and depositing it near by (see quots.); also as vb.

1904 Chamberlin & Salisbury *Geol.* (1905) I. iii. 183 This is cut-and-fill. The sediment eroded from the curve which is concave toward the stream is shifted down-stream, while that deposited in the curve which is convex toward the stream is brought down from above. **1934** Webster, *Cut and fill*, to construct, as a stadium, by using material excavated from the center to form walls. **1940** *Chambers's Techn. Dict.* 218/1 *Cut-and-fill*, a term used to describe any cross-section of highway or railroad earthworks which is partly in cutting and partly in embankment. **1965** G. J. William *Econ. Geol. N.Z.* vii. 80/1 The existence of current-bedding, imbricate arrangement of pebbles, cut-and-fill bedding, and scour channels.

cut, *ppl. a.* Add: **2. b.** Of money: see **cut-money* (*12 a below). *U.S.*

1844 in C. Cist *Cincinnati Misc.* (1845) I. 61 As late as 1806..the business house in Philadelphia in which I was apprentice, received over one hundred pounds of cut silver.

3. *cut glass* (earlier example); (*transf.*, *colloq.*) an affected expression or mode of speech; also (with hyphen) *attrib.*

1843 H. Murray *Ireland* ..a handsome gift..glass *cut-glass* bottle. **1856** Mrs. Stowe *We & Neighbors* xxxiv. 300, I arranged it in my best cut-glass dish and covered it with foamy billow of whites of eggs. **1945** Dylan Thomas *Let.* 29 July (1966) 264 A position of positions—cut-glass for job or jobs. **1957** *Granta* 9 Mar. 20/1 All I can remember now is a cut-glass decanter stopper. **1962** John Creasey..**1962** 24 May 12/1 An impeccable cut glass accent. **1966** 'L. Lane' *ABZ of Scouse* 96 She talks cut glass.

6. *cut out*: see quot. 1932. *N. Amer.*

1819 *N. Amer. Rev.* VIII. 11 The Nottoway at Cut Bank Bridge. **1837** S. Cummings *Western Pilot* 66 You pass close by this cut bank of the bar. **1884** 'Mark Twain' *Huck. Finn* xv. 127 In about a minute I come a-booming down on a cut bank with..big trees on it. **1932** ..is undercut. **1867** *Medicine Hat* (Alberta) *News* 28 Jan. 1/6 The man whose he was riding went over the cut bank near the iron bridge. **1932** *Dialect Notes* VI. 228 Cut-bank. This word (variously spelled cutbank, *cut-bank*, and cut bank) is often used for the outer bank at the bend or a stream, the bank which the stream cuts into, leaving it often perpendicular. **1968** H. M. Patterson *Findlay's River* 26 On the right bank of the Parsnip [River] there is a high cutbank, a sand-and-gravel cliff.

c. *cut paper*, paper cut into a desired shape, usually for decorative purposes. Also (with hyphen) *attrib.*

1847 E. Brontë *Wuthering Heights* I. xiii. 253 A fireplace hung with cut paper dropping to pieces. **1891** Kipling *Light that Failed* (1900) 9 Here they are. **1926** *Times* 31 May 1/4 ..cut-paper hand-frills. **1930** ...of colour.. 1961 *Pract. Embroidery & Fabric Collage* i. 18 ..cut-paper..

9. *cut and laid*, *cut and lay*: see Lay *v.*[1] 6 b. Also *ellipt.*, a hedge made in this manner.

1919 Masefield *Reynard* 11. 73 Robin made Pip [*sc.* his horse] go smash through the cut and laid. **1937** *Daily Express* 21 Oct. 9 Neat-cropped grass fields split by cut and lays.

10. Of prices, etc.: reduced. Hence **cut-price**, **-rate** *adjs.*, having, or offered at, reduced prices; also *fig.* orig. *U.S.*

1889 in *Dict. Amer.* **1886** Boston *Jrnl.* (Farmer), The plain people who enjoy a spectacular, musical, and dramatic season at cut rates. **1897** *Sears, Roebuck Catal.* 684 No wholesale house can meet our cut prices. **1904** *Westm. Gaz.* 27 May 10/2 The policy of the 'cut rate' may continue. **1904** *Daily Chron.* 15 June 7/3 One out of every eighty of the cut-rate arrivals was ordered back to Europe. **1910** *Sat. Even. Post* 30 Sept. 76/2 Tricky cut-price operators..selling below living prices. **1930** R. Simmat *Personal Salesmanship* 85 If a salesman once..gives a cut price he will..be always expected to give a cut price. **1934** J. B. Priestley *English Journey* 17 A number of blatant cut-price shops, their windows crammed with goods, mostly inferior and dubious, and lined with placards. **1958** *New Statesman* 30 Aug. 238/3 Their decision to offer unsecured personal advances at, in effect, cut-rates. **1963** A. Ross *Australia 63* i. 19 M.C.C.'s batsmen were offered only cut-price bowling, and they savaged it accordingly. **1964** *Daily Tel.* 18 Jan. 2/5 It is a choice between 'quality' and cut-rate shopping.

b. Diluted; adulterated. (Cf. *cut *v.* 22 b.) Chiefly *U.S.*

1938 H. Asbury *Sucker's Progr.* 343 Suckers..paid exorbitant prices for cut and adulterated liquor. **1969** D. Bagley *Spoilers* vi. 180 'This is morphine.' 'Cut or uncut?' asked Follet. 'It's pure—or as pure as you can make the stuff in a slum like this.'

CUTTING

11. (Earlier and later examples.)

1743 W. Ellis *Mod. Husb.* Oct. v. 147 The great Improvement of curing cut down Wheat in the Field, is, known to the meanest Rustic. **1932** W. Faulkner *Light in August* (1933) vii. 143 The small figure in outdoor underwear. **1941** J. Cary *House of Children* iii. 10 We should see mountain children in their father's cut-down trousers. **1949** F. Maclean *Eastern Approaches* ii. iii. 200 A new, cut-down Ford station wagon.

12. a. **cut cloth**, **drop** (see quot. 1961); **cut-money** *U.S.* (see quot. 1822); **cut-rope** *Naut.*, = 'Painter'; **cut sheet** (rubber), rubber cut into sheets from a pressed block; **b.** *cut-flower*, *-glass* (see *3), *-leaf*, *-paper* (see *3 c), *-steel*; **cut-card**, applied to a type of relief decoration on silverware, etc., in which a thin sheet of metal is cut ornamentally and soldered to the surface; **c.** *cut-leaved* (earlier example).

1920 *Catal. Eng. Silversmiths' Work* (V. & A. Mus.) 18 Another method of decoration was that of so-called 'cut-card' work, the decoration being cut out from a separate sheet of metal. **1939** *Ossomenia* IV. 200 A candle-cup, of 1672, with fine 'cut-card' decoration. **1952** G. Taylor *Silver* v. 135 The Parisian goldsmiths..effected a monumental elegance ..due to an admirable harmony. ..Perhaps the most attractive..is cut-card work, a term used to describe flat patterns of sheet metal applied to the body of the object to be decorated. **1961** R. Mayes *Dict. Art Terms & Techniques* 103 *Cut-card work*, a decorative technique in metalwork in which a design cut out of a sheet of metal is superimposed on the surface of an object of the same metal, usually around a protuberance such as a handle or a finial. The cut-card work is used almost exclusively for silver. **1884** J. Hatton *Irving's Impr. Amer.* I. vi. 165 The well-known Hampton Court cloth (in 'Charles II') was so perfect..that it was regarded as a cut cloth, with 'raking' and water-pieces. **1933** P. Godfrey *Back-Stage* vi. 75 The drop scene, cut-cloths, and borders..became as obsolete as the 'aside' in acting. **1961** Bowman & Ball *Theatre Lang.* 89 *Cut cloth, cut-cloth*, a British term for a cut drop. *Ibid.* 90 *Cut drop*, a drop painted and then cut out so that the spectator sees a scene formed not only by this drop but also by whatever is placed behind it. **1902** *Westm. Gaz.* 4 Apr. 8/1 The cut-flower trade. **1970** W. L. Sidwell *Cowpea Cut Flowers for House* i. 9 The keen cut-flower gardener should expect to have blooms and foliage from his garden from March to October at least. **1897** G. D. Schoenth *Arboricult Flora U.S.* 261 *Robinia pseudacacia decaisnana*, Cutleaf Locust. **1923** E. F. Wyatt *Invis. Gods* 16 Nesbitt saw..cut-leaved birch flickered their light foliage. **1931** T. Miller *Gard. Dict.* I. s.v. *Lilac*, Lilac, with cut Leaves, falsely call'd, The Cut-leav'd Persian Jasmine...The Cut-leav'd Sort..having its older Leaves deeply cut in. **1822** in W. Littell *Statute Law of Kentucky* (1814) IV. 45 To pay out money into the public treasury. **1842** H. Moore *Eng. Prairie* 230 We found change at these towns very scarce; what there was, was mostly cut-money; that is, when change is wanted they often cut dollars, half-dollars, and quarter-dollars, into smaller pieces with an ax or chisel. **1824** W. N. Blane *Excursion* 257, I was obliged to cut a silver dollar, into quarters, and even into eighths; a practice so common in the Western States, that the cut-money as it is called..was the only change that could be had in Missouri. **1909** *Westm. Gaz.* 3 July 3/2 The cut rope [painter] of an old boat in apt to be very rotten. *Ibid.* 12 Aug. 2/2 Boys who..cut the rope. **1955** *Times* 5 Oct. 3/4 *Cut sheet* (see Sheet *sb.*[1] 4 b). **1907** H. L. Terry *India-Rubber* xii. 161 Cut sheet rubber. **1748** Smollett *R. Random* ii. xliv. 79 A fourth [sword] cut steel inlaid with gold. **1896** *Westm. Gaz.* 10 Dec. 3/2 A cut steel buckle. **1935** W. de la Mare *Connoisseur* (1926) 334 The cut-steel brooch of coloured gems.

cut-away, *a.* (*sb.*) Add: **b.** (Earlier example.)

1849 *Theatr. Programme* 23 July 58 A nondescript coat something between a Newmarket 'cut-away' and a shooting jacket.

2. Applied to a model or drawing of a piece of apparatus, etc., in which part is cut away so as to reveal the interior.

1946 W. L. Beauchamp et al. *Everyday Problems in Science* 653 (caption) A cutaway view of the clutch and the transmission gears of an automobile. **1959** *Motor Man.* (ed. 36) v. 110 This cutaway drawing..shows the internal construction of the Hydrosteer control valve unit.

3. *Cinemat.* As *sb.*, the process of making a quick transition to another scene. (Cf. *cut *v.* 21 e.) Also *attrib.*

1951 G. H. Sewell *Amat. Film-Making* (ed. 2) ii. 100 Between the items of a procession you can keep it moving by quick shots of the spectators in the crowd...When you are taking the original scenes you must remember to take some cut-away material as well. **1953** Reisz *Technique Film Editing* i. 26 The suspense..achieved by devices such as the quick cut-away to the shot of the..**1962** *Times* 15 Mar. 486/3 We had to use the cut-away shots of the..camera operators laughing just to show the trip was the..**1970** *New Yorker* 19 Sept. 103/1 In a cutaway shot so placed that it pierces your heart, the old man's eyes glitter.

cut-back, cutback, *sb.* [f. Cut *v.* 52.] **1.** *Hort.* A plant which has been pruned by cutting off shoots close back to the main stem.

1897 *Garden* 21 Aug. 141 The 'cutbacks' [*sc.* young vines] planted in March..**1906** *Gard. Chron.* ..[*sc.* cuttings]..**1966** *Gloss. Landscape Work (B.S.I.)* 19 *Cut back*, a tree or shrub which has been cut back to induce strong growth and subsequently grown on for one or more years.

2. *Cinemat.* A scene which is a return to a previous action.

1913 E. W. Sargent *Technique Photoplay* (ed. 2) 90 Originally the cut-back was used to close up a gap in the action, to obviate the actual showing of a crime [etc.]. **1916** *Ibid.* (ed. 3) 360 *Cut-back*, one or more returns to a previous action, either to avoid the showing of prohibited action, to raise the effect through contrast or to quicken the action. **1927** *Sunday Times* 27 Feb. 6/4 In the last ten minutes a 'cut-back' occurs, and the quarrel and killing are reconstructed. *Ibid.* 15 May 10/5 The discovery of the 'close-up' and the 'cut-back' has always been accredited to Griffith.

3. A conglomerate product, as bitumen that has been thinned by the addition of lighter oils.

1936 *Jrnl. R. Aeronaut. Soc.* XL. 547 A layer of sound deadening cement (a conglomerate of asphaltum, lead oxide, cork and rubber, called 'cut back' by the automobile trade). **1940** *Chambers's Techn. Dict.* 218/1 *Cut backs*, blends of asphaltic bitumen with various solvents, for use at comparatively low temperatures for road surfacing.

4. A reduction in expenditure, production, etc. orig. *U.S.*

1943 *Iron Age* 6 May 152 More than 90 per cent of prime contractors holding Army ordnance contracts are now operating..below capacity because of recent cut-backs in ordnance contracts. **1952** *Manch. Guardian Weekly* 10 Jan. 3 How to get more raw materials to Britain without causing such cutbacks in American civilian production. **1957** *Economist* 28 Sept. 1009/1 The swingeing cutbacks recently inflicted on China's economic plan. **1965** *New Statesman* 30 Apr. 673/2 Even if the 20 per cent cutback some have forecast did actually take place in their sector, this would probably mean a loss of not much more than 15,000 houses.

5. *Surfing.* The turning of the surf-board back towards the wave. Cf. *cut *v.* 52 e.

1965 J. Pollard *Surfrider* ii.19 Swing back into the wave while riding it and you do a 'cut-back'. **1970** *Surf* I. x. 9/2 Will also ease up tail pressure in turns and cutbacks.

cute, *a.* Add: **2.** (Earlier and later examples.) Now in general colloq. use, applied to people as well as things, with the sense 'attractive, pretty, charming'; also, 'attractive in a mannered way'.

1834 C. A. Davis *Lett. J. Downing* 214 I'm goin' to show you about as cute a thing as you've seen in many a day. **1857** 'Porte Crayon' *Virginia Illustr.* ii. 166 'What cute little socks!' said the woman. **1879** F. R. Stockton *Rudder Grange* vi. 61 [The flat] was so cute, so complete. **1880** A. A. Hayes *New Colorado* (1881) vii. 97 The way that Smart Aleck hollered when we swung round some of them 'cute' curves. **1900** *Daily News* 15 Nov. 6/5 A small and compact wooden house, what the Americans would call 'cute'. **1908** *Daily Chron.* 21 Apr. 3/3 American visitors who are used to wide rectangular streets are delightfully bewildered when I take them through sinuous byeways and tortuous alleys. They proclaim it 'just too cute and lovely'. **1941** A. Huxley *Grey Eminence* ii. 18 The tiny boy..looking almost indecently 'cute' in his claret-coloured doublet and starched ruff. **1945** —— *Time must have Stop* vii. 77 A French accent so strong, so indecently 'cute', so reminiscent of the naughty-naughty twitterings of a Parisian miss on the English musical comedy stage. **1960** P. Mortimer *Sat. Lunch with Brownings* 92 She's ever so cute—blue eyes. **1966** *Amer. Speech* XLI. 285 The style also gets cute at times, as when he writes that such adjectives as *washable* and *non-shrinkable* are 'among the most..not-to-be-got-along-withoutable adjectives'.

cuteness. (Later examples.)

1845 *Knickerbocker* Aug. (Bartlett), He had a pair of bright, twinkling eyes, that gave an air of extreme cuteness to his physiognomy. **1872** H. M. Stanley *How I found Livingstone* xii. 464 Very fine people and singularly remarkable for commercial ''cuteness' and sagacity. **1903** *Booklovers Mag.* Dec. (Advt.), The illustration gives but a faint idea of the beauty and cuteness of the calendar itself.

Cuthbert[2]**.** A slang name for a man who deliberately avoids military service; esp. in the war of 1914–18, one who did so by securing a post in a Government office or the Civil Service; a conscientious objector.

1917 *Evening News* 25 Jan. [in a cartoon by 'Poy']. **1919** *Mr. Punch's Hist. Gt. War* 225 As a set-off to the anti-'Cuthbert' campaign in the Press the War Cabinet has..declared that 'the whole Empire owes the Civil Service a lasting debt of gratitude'. **1933** J. Cary *Amer. Visitor* xvi. 182 All you Cuthberts are fit for is to dodge responsibility at the cost of other people's lives. **1935** A. J. Cronin *Stars look Down* ii. xii. 372 Not good enough for you, eh? Not fancy enough for Cuthbert?

cuticle. Add: **1. d.** The dead skin at the base of a finger-nail or toe-nail. Also *attrib.*

1907 *Yesterday's Shopping* (1969) 538/1 Cuticle knife, ivory handle. *Ibid.* 538/2 Cuticle scissors...Cuticle cream for softening the skin at base of nail. **1919** *Ladies' Home Jrnl.* Jan. 67/2 Cutex, the Cuticle Remover, comes in 35c, 65c and $1.25 bottles. **1962** *Woman* 1 Dec. 9/2 Every night, take care of your cuticles. *Ibid.*, Apply cuticle cream. **1966** *Vogue* Nov. 61/2 Everything needed for a manicure..cuticle remover, orange stick, emery board.

cutie (kiū·ti). *slang* (orig. *U.S.*). Also **cutey.** [f. Cute *a.* + -ie.] A cute person; esp. an attractive young woman. (In quot. 1768 the sense is 'a superficially clever person'.)

1768 in A. Hare *Georgian Theatre in Wessex* (1958) iv. 72 Let shallow Cuties, who, in Love with Sound, Care not a Pin if Action's never found. **1917** D. G. Phillips *S. Lenox* II. viii. 204 It was the bartender. 'Evening, cutie,' he said. 'What'll you have?' 'Some rye whiskey,' replied Susan. **1923** R. D. Paine *Comr. Rolling Ocean* viii. 130 Her friends thought she was a cutey for turning the trick. **1927** 'J. Barbican' *Confess. Rum-Runner* xiv. 149 He goes about with a high-stepping cutie who's ace-high on the face and figure. **1927** *Daily Express* 5 Dec. 13 His sweetheart, a 'cabaret cutie'. **1945** W. Plomer *Dorking Thigh* 20 Just like a young cutie Between the wars.

cut-in, *sb.* [See Cut *v.* 54.]

1. An act of cutting in: in senses of the verb (see Cut *v.* 54 in Dict. and Suppl.).

1883 [see Cut *sb.*[2] 35]. **1898** *Westm. Gaz.* 21 June 9/2 Herminius himself, with all his weight, is likely to have a cut in for the same race. **1920** F. Scott Fitzgerald *This Side of Paradise* (1921) i. ii. 62 The cut-in system at dances. **1931** I. L. Reeves *Ol' Rum River* 182 Others within the organization will think they are entitled to a cut-in should the brewery attempt at any time to make anything other than near-beer. **1939** *Chatelaine* May 35/3 When I see a 'debbie' who never dances more than halfway round the ballroom without a cut-in, I make this mental note:..she'll soon be among the 'young marrieds'. **1953** *Time* 23 Feb., It was a world where..*cut-ins* (giving a performer a share of a song's profits)..were standing operating procedure. **1958** *Listener* 6 Nov. 731/1 He's still going to get to the petrol pump by a neat cut-in.

2. *Cinemat.* A 'leader' inserted into a film sequence. In full *cut-in leader*. Also, an interposed scene.

1913 E. W. Sargent *Technique Photoplay* (ed. 2) ii. 15 All other leaders have been between scenes, but this is right in the middle, so it is known as a '*cut-in leader*', because it is cut into the scene. *Ibid.* vii. 50 The quoted leader is handy, but it should not be used too much, and there is a growing tendency to use two and even three cut-in leaders in one scene. A straight leader and one cut-in is about the limit. A leader or a cut-in, but not both should suffice. **1921** Lescarboura *Cinema Handbk.* 21 *Cut-in*, anything inserted in a scene which breaks its continuity. **1953** K. Reisz *Technique Film Editing* ii. 190 The cut-in of the crowd..was necessary to indicate the passage of time while the horses file out of the paddock. **1961** G. Millerson *Telev. Production* 302 The cut-in provides shock treatment. A cut-in to a field of thistles would be a dynamic introduction to a programme on 'The Weed Menace'.

3. A device for starting an engine by completing the electric circuit.

1921 *Motor Electrical Man.* viii. 109 In some switch-boxes the automatic cut-in and cut-out for the charging circuit is fitted. **1924** A. W. Judge et al. *Mod. Motor Cars* III. 74 A more common form of cut-in is that depending upon electro-magnetic action. *Ibid.* 76 There are many other forms of electrical, thermostatic, and mechanical 'cut-out' and 'cut-in' devices for the battery-dynamo and battery-ignition circuits.

cutlass, *sb.* **2.** **cutlass-fish** (examples).

1884 G. B. Goode *Nat. Hist. Aquatic Anim.* 335 The name 'Cutlass-fish', which is current for the same species [*sc. Trichiurus lepturus*] in the British West Indies. **1963** P. H. Greenwood *Norman's Hist. Fishes* (ed. 2) ii. 14 At the other extreme are fishes with long bodies, which may be..very much compressed, as in the..Cutlass-fishes.

cut-off, *sb.* Add: **1.** (Later example.)

1954 J. Southward *Mod. Printing* (ed. 7) II. xv. 212 Cut off is the amount [*sc.* of paper] severed from the webs to form individual copies of an edition.

2. For *Western U.S.* read 'orig. *U.S.*' Add earlier examples. Also, a lateral channel dug across a bend in a river (also *attrib.*).

1773 *Acts Gen. Assembly Georgia* (1881) 300 To make any such cut off as shall be thought necessary from River to River. *Ibid.*, In such Cuts off and Clearing. **1817** S. R. Brown *Western Gaz.* 222 It is about four miles across the several branches of the Pascagola..intersected by bayous and cut-offs. **1913** Thomas & Watt *Improvem. Rivers* (ed. 2) i. 27 When a bend has become almost a complete curve, the river breaks through the intervening neck of land and forms a cut-off. *Ibid.* ii. 337 The entire river was dammed at or near the upper part of a sharp bend, and a cut-off or lateral canal was dug across the bend from the pool thus formed. *Ibid.* 367 Cut-off walls resting on the river-bed. **1937** Wooldridge & Morgan *Physical Basis Geogr.* xii. 173 The abandoned loops form 'cut-offs', 'ox-bows' or 'mort-lakes' which, in time, become silted up.

b. (Earlier and later examples.)

1806 Z. M. Pike *Jrnl.* 28 Jan. in *Sources Mississ.* (1810) i. 64 Observed Mr. Grant's trackes going through it; found his mark of a cut off, (agreed on between us) took it, and proceeded very well. **1818** *Boston Weekly Messenger* 23 July (Th.), They pointed [it] out to him as being a nigh cut-off to the high road. **1908** *Westm. Gaz.* 31 Dec. 3/1 The Great Western's Ashendon to Aynho 'cut-off', which will provide that company with a new route to Birmingham, nineteen miles shorter than its existing one. **1924** W. M. Raine *Troubled Waters* iii. 29 Evidently she was taking the cut-off back to the ranch, unaware that the bridge had been washed out by the freshet. **1947** J. Steinbeck *Wayward Bus* i. 6 Those who came over the cut-off from San Juan de la Cruz.

3. a. (Further examples.) Also *attrib.*

1956 *Kenyon Rev.* XVIII. 417 Abrupt voice cutoff. **1966** M. A. K. Halliday in C. E. Bazell *In Memory of J. R. Firth* 152 Lexis seems to require the recognition merely of linear co-occurrence together with some measure of significant proximity, either a scale or at least a cut-off point. **1970** *Globe & Mail* (Toronto) 25 Sept. B10/2 The cutoff day for commitments in the current quarter is Wednesday. **1971** *Times* 23 Jan. 18/1 It is used here as a cut-off point between the poor and the rest of the community.

e. In a magazine rifle, a device which prevents the feeding of cartridges from the magazine into the chamber, and enables the rifle to be used as a single-loader.

1890 [see 3 c in Dict.]. **1898** *Daily News* 9 May 3/1 Magazine Cut-off. **1904** *Westm. Gaz.* 9 Dec. 7/2 As the Navy considered that a cut-off was necessary this..is fitted to all naval rifles. **1919** 'Boyd Cable' *Old Contemptibles* ix. 141 In a twinkling every man..had his rifle muzzle over the parapet, and his fingers busy with magazine and cut-off.

f. An automatic safety device for shutting off light, esp. the light of a cinema-projector.

1906 *Daily Chron.* 28 June 2/7 Automatic cut-off devices. **1917** C. N. Bennett *Guide to Kinematography* ix. 146 Before the condenser..is a safety device called the hand light cut-off. **1923** F. A. Talbot *Moving Pictures* ix. 119 The 'cut-off', an automatic safety shutter, mounted between the lamp and the film, which falls to intercept the light when the machine is at rest.

g. Chiefly *Electr.* A marked increase in the attenuation (or decrease in the amplification) of an oscillation when its frequency reaches some value, esp. of an alternating current or voltage by a filter, waveguide, etc.; usu. *attrib.*, as *cut-off frequency*, *point*.

1926 Franklin & Terman *Transmission Line Theory* v. 141 This frequency is called the cut-off frequency of the high-pass filter, for it marks the transition from pass to attenuated frequencies. **1930** *Bell System Techn. Jrnl.* IX. iii. 483 The cutoff points are taken as those at which the attenuation reaches a value 10 db greater than that at 1,000 cycles. **1930** *Discovery* Dec. 398/2 The directional filter offers very little attenuation to frequencies on one side of a certain frequency (cut-off frequency). **1959** *Chambers's Encycl.* XII. 735/1 The capacitative reactance of the air chamber in front of the diaphragm tends to raise the frequency at which this occurs, but results in a much sharper cut off at still higher frequencies. **1970** D. F. Shaw *Introd. Electronics* (ed. 2) xii. 272 The current gain of a transistor falls off at high frequencies...This frequency dependence is expressed accurately by the expression $\alpha_f = \alpha_0/(1+jf/f_\alpha)$ where..f_α is called the alpha cut-off frequency.

cut-off, *a.* Add: **2.** Shut out, excluded, remote (see Cut *v.* 55 h). Hence **cut-offness**, the state or condition of being cut off.

1894 M. Dyan *All in Man's Keeping* I. xii. 203 Would he never lose this cut-off feeling, this awful ache for comradeship? **1927** D. H. Lawrence *Let.* 3 Aug. (1962) II. 993 Our being cut off..is our ailment...I wish I saw a little clearer how you get over the cut-offness. **1939** J. Cary *Mr. Johnson* 92 The poorer, more cut-off people do not want roads. **1960** *Encounter* XV. 73 The cut-off-ness of the modern 'intellectual man' from the world.

cut-out, *sb.* Add: **b.** In an internal-combustion engine, a valve through which exhaust gases can escape without passing through the silencer.

1905 *Motor Cycle* 2 Jan. 6 An exhaust cut-out. **1906** *Daily Chron.* 17 July 3/3 Nothing is easier than to have a 'cut-out' in the exhaust pipe, worked by a wire from the steering column, and so produce this noise and eliminate the back pressure whenever you wish. **1907** *Public Opinion* 17 May 628/2 The attention of the Committee of the Royal Automobile Club has been called to the increase in the use of sirens and exhaust cut-outs by certain motorists. **1926** T. E. Lawrence *Seven Pillars* (1935) viii. xciv. 521 The cars, with closed cut-out, would..carry the trenches by surprise.

2. a. The space formed by a piece or section being cut out (as of a floor). *U.S.*

1851 A. O. Hall *Manhattaner in New Orleans* v. 30 Above the bar and post-office (the former..looked down upon through a wide cut-out in the floor) are the..reading-rooms of the merchants.

b. A railway or canal cutting. *U.S.*

1898 *Engineering Mag.* XVI. 116/1 The dredge by which the cut-outs were excavated and embankments constructed.

c. A figure cut out (or designed for cutting out) of paper, cloth, cardboard, wood, etc.; *spec.* in *Theatr.* (see quot. 1961). Also *fig.*

1905 Calkins & Holden *Art of Mod. Advertising* 10 The grocer must be supplied with attractive counter slips, 'hangers', window-cards, 'cut-outs', posters and other forms of lithographed matter. **1920** E. G. Lutz *Animated Cartoons* 84 This model, specifically spoken of as a 'cut-out', is pushed over the background under the camera and photographed. **1923** *Daily Mail* 1 Mar. 6 (Advt.), The famous Polly Pratt cut-outs for the kiddies. **1927** E. G. Lutz *Motion-Pict. Cameraman* 169 A series of cut-outs for a continued action. **1927** *Daily Express* 27 Sept. 5/2 Cut-outs are all the rage—cut-outs in wood and in cardboard painted. **1927** *Ladies' Home Jrnl.* Dec. 68/3 It would be easy enough to adapt for the purpose the various animal cut-outs and illustrations. **1927** *Home Notes* 17 Dec. 863/2 Embroidered Appliqué and Making Directions Supplied with Each Cut-Out. **1949** *Here & Now* (N.Z.) Oct. 14/2 The stage size is the same every night—props are at a minimum, and such customary essentials as drapes..and cut-outs are eliminated. **1956** *Essays in Criticism* VI. 372 The character of Harcourt, which the author has simply left as a cardboard cut-out. **1959** J. Masters *Fandango Rock* 328 The upper row of the Moorish arches made the familiar pattern of cut-outs in brown-red paper against a backdrop of blue. **1961** Bowman & Ball *Theatre Lang.* 90 *Cut out, cut-out, cutout*, a small flat, usually of profile board, cut to simulate trees, rocks, etc... Hence *cut-out scenery*. **1971** B. Callison *Plague of Sailors* 15 The mountain was still a jet black two-dimensional cut-out.

3. a. An act of cutting out cattle from the herd. *U.S.*

1874 J. G. McCoy *Cattle Trade* 81 In the beginning of the cut-out, a few gentle cows or working oxen are driven a short space from the round-up and held, to form a nucleus, to which those cut out gather. **1907** C. E. Mulford *Bar-20* xi. 120 In this contest Hopalong Cassidy led his nearest rival, Red Connors,..by twenty cut-outs. **1920** —— *J. Nelson* xxiii. 254 Sam saw no use of collecting infants only to have them turned loose at the cut-out.

b. A place where cut-out animals are collected.

1920 J. M. Hunter *Trail Drivers of Texas* 98 Our camp was the catch and cut-out for all the other horses.

c. The finish of shearing, end of shearing-time; completion of the shearing of a specified group of sheep (e.g. the rams) or of a particular flock, etc. *Austral.* and *N.Z.*

1900 H. Lawson *Over Sliprails* 33 It was within a couple of days of cut-out, so I told Mitchell—who was shearing—that I'd camp up the Billabong and wait for him. **1922** C. G. Turner *Happy Wanderer* 143 A 'cut out'—the finishing of a line of sheep or of the shed itself. **1926** J. Devanny *Butcher Shop* vii. 73 Another two days would have seen the cut out. **1959** H. P. Tritton *Time means Tucker* iv. 34/1 At the cut-out, when we were at the office getting our cheques.

4. A person acting as middle-man, esp. in espionage. *slang.*

1963 J. Joesten *They call it Intelligence* i. i. 44 A very important figure in espionage is the 'cutout'. This is a trusted middleman. **1966** M. R. D. Foot *SOE in France* iv. 94 A cut-out..is a means of establishing contact between two agents which..affords the minimum for the enemy security services to bite on. One agent passes a message in a simple code to the cut-out... The cut-out holds the message till approached by the next agent down the line. **1969** E. Ambler *Intercom Conspiracy* (1970) ii. 45 Through our cut-out I have made an offer for the shares.

cut-out, *a.* [See Cut *v.* 56 j and Cut *ppl. a.* 11.] Formed by cutting out a piece of paper, cloth, cardboard, etc.; of or pertaining to the piece cut out. Cf. *cut-out *sb.* 2 c.

1799 [see Cut *ppl. a.* 11]. **1873** *Young Englishwoman* Apr. 202/2 Make a deep flounce..and set this under the cut-out place. **1886** Kipling *Plain Tales* (1888) 88 Mrs. Hauksbee was expecting some cut-out pattern things in flimsy paper. **1898** *Westm. Gaz.* 4 June 7/2 The slate-coloured cut-out frames, on which the prints are mounted. **1927** E. G. Lutz *Motion-Pict. Cameraman* 162 Cut-out figures photographed in sequence. **1929** *Publishers' Circular* 22 June 73 Attractive cut-out figures, in full colours. **1970** *New York* III. 7 Dec. 61/2 The Tom Thumb finger-puppet theater with cut-out figures.

cut-over, *a.* and *sb.* [See Cut *v.* 57 and Cut *ppl. a.* 11.] **A.** *adj.* Cleared of timber by cutting. orig. *U.S.*

1899 *Westm. Gaz.* 6 Jan. 10/2 At least 90 per cent. of the cut-over lands [on the Pacific coast] are of absolutely no value for agricultural purposes. **1911** H. Quick *Yellowstone N.* xii. 338 The solitary guest which is the only thing that brings the haunch to the spit in the Minnesota cut-over forest. **1946** *Nature* 13 July 71/2 Experiments carried out in certain marked strips of cut-over forest to ascertain the results of allowing the areas to lie fallow for a varying period of years.

B. *sb.* **1.** *Fencing.* An offensive disengage executed over the opponent's blade.

1897 *Encycl. Sport* I. 389/1 To elude quarte made with the hand very low, riposte with a cut-over thus. **1969** T. Parker *Twisting Lane* 111 Going forward he made the first attack: a fast lunge and cut-over in Quarte.

2. An area on which the timber has been cut. *U.S. rare.*

1922 H. Titus *Timber* vi. 60 If we had known we could have gone north..into the hardwood cutover and made a go of it.

cuttee. Delete '*nonce-wd.*' and add further examples.

1821 P. Egan *Life in London* ii. v. 301 The '*cut direct*' comes with the *severity* of a paralytic stroke on the feeling of the poor *cuttee*. **1859** *Habits of Gd. Society* ix. 277 It does the cutter as much injury as the cuttee.

cutter, *sb.*[1] Add: **1. b.** *cutter-in*, (*a*) in *Whale-fishery*, one who cuts up a whale so as to remove the blubber (cf. Cut *v.* 54 b); (*b*) a motorist who cuts in between or in front of vehicles (cf. *cut *v.* 54 c); *cutter-out*, *spec.* one who separates cattle from a herd (see Cut *v.* 56 d) (*U.S.*, *Austral.*, and *N.Z.*).

1843 E. Dieffenbach *N.Z.* I. ii. 51 [The tongue] is a monopoly of the 'tonguer' or 'cutter-in'. **1928** *Daily Express* 5 June 9/1 The cutters-in, and the speed-at-any-price merchants who spoil travelling. *Ibid.* 26 June 10/2 The cutter-in and the speeder-up..are the causes of..accidents. **1910** C. E. Mulford *Hopalong Cassidy* iii. 28 Each of the cutters-out rode after some calf. **1920** —— *J. Nelson* xxv. 259 There was only one pair of ropers..and only three cutters-out. **1940** E. C. Studholme *Te Waimate* (1954) xvi. 136 The 'cutter-out' rode through [the cattle], and after spotting the beast he wanted, would follow it out to the edge of the mob.

2. e. *Cinemat.* One who cuts or edits a film (see *cut *v.* 21 e). Also *film-cutter*.

1921 D. Boughey *Film Industry* vii. 65 So interwoven are the functions of the editor, or cutter..that the two must work in perfect harmony. **1928** *Sunday Express* 18 Mar. 4/3 In America the film-cutter is a man with a sub-editorial mind developed to the *n*th degree, and film-cutting..is one of the highest paid professions. **1961** K. Reisz *Technique Film Editing* (ed. 9) ii. 185 In a newsreel unit the man in charge is generally referred to as the Editor, and the man who does the actual assembling in the cutting room, the Cutter. **1966** *Listener* 17 Mar. 383/2 That shows how much you owe to the cutter and the director when it comes to the screen.

4. c. *Sound Recording.* A tool which cuts a groove in a recording medium. In full *cutter stylus*.

1908 *Jrnl. R. Soc. Arts* LVI. 642/1 With a 12 inch disc, when the cutter is ½ inch from the edge, it will in one revolution describe a line on the record of a length approximately equal to the circumference of a circle of 11 inches diameter. **1962** A. Nisbett *Technique Sound Studio* iv. 87 The hot-stylus method..helps to smooth the wall as the cutter moves on. *Ibid.* 88 The cutter stylus ploughs its way through the surface of a disc.

5. b. *Cricket.* (*a*) A batsman who 'cuts' the ball (see Cut *v.* 31 a) or who is adept at making this stroke. (*b*) A ball that turns sharply after pitching (see *cut *v.* 31 a).

1851 J. Pycroft *Cricket Field* vii. 148 The balls may be regularly rising: in this case every one would like to see a good cutter at the wicket. **1955** *Times* 15 July 3/3 Palmer..is a particularly effective driver and cutter. **1960** I. A. R. Peebles *Bowler's Turn* 40 By way of variation he bowled a cutter. **1966** E. R. Dexter *Ted Dexter Declares* ii. 21 They gave me a bit of bowling and I started with medium-pace swingers and off cutters.

7. b. An animal yielding an inferior grade of meat. orig. *U.S.*

1905 *Chicago Daily News* 3 July 7/7 Canners and cutters were rather slow, as packers did not care to secure them. **1916** *Yorkshire Post* 23 Feb. 10/6 Pigs continue to rise in value, 6[d] extra per score being charged for cutters and sows. **1958** *Times* 1 July p. iii/4 Any line of demarcation between porkers and the rather heavier 'cutters' (near bacon weights) has practically disappeared. **1971** *Daily Hampshire Gaz.* (Northampton, Mass.) 12 Feb. 5/5 Prices per cwt. were:.. 24 cutters, up to $21.90.

8. cutter-dredge, -dredger, a river-dredge fitted with knives; **cutter-loader,** a machine that cuts coal from a coal-face and loads it on a conveyor.

1913 Thomas & Watt *Improvem. Rivers* (ed. 2) i. 100 The cutter dredge,..in which the material is loosened by a series of knives shaped so that they will not clog, fastened to a shaft and revolving close to the end of the suction pipe. **1940** *Chambers's Techn. Dict.* 218/2 *Cutter dredger*, a dredger of the sand-pump or suction type. **1948** *Something Done* (*Central Office of Information*) 19/2 Besides the Meco-Moore, two other British cutter-loaders are now in the advanced experimental stage.

cutter, *sb.*[2] **3.** (Earlier example.)

1803 'C. Caustic' *Poet. Petition against Tractorising Trumpery* 37 Then condescend to be my crony, And guide my wild Parnassian pony, Till our aerial cutter runs Athwart 'a wilderness of suns'!

cut-throat. Add: **1. c.** Used *attrib.* or *ellipt.* of several games of chance, esp. of a three-handed card game in which one or more players score individually rather than in partnership. orig. *U.S.*

1823 I. Holmes *Acct. U.S.* 353, I have seen at least twenty boys surrounding a billiard table, playing at a sort of game of chance they call cut-throat. **1868** *All Year Round* 31 Oct. 489/2 It is not uncommon, therefore, to see merchants (especially American) having a social game of 'cut-throat monte', 'eucre', or 'poker', with piles of gold before them. **1870** J. C. Duval *Big-foot Wallace* 247 As soon as I can learn to play poker and cut-throat loo. **1904** E. A. Tennant *ABC of Bridge* (ed. 4) 100 Cut-Throat Bridge. Though the above is the simplest way of playing three-handed Bridge, Cut-Throat has great attractions for some people. **1932** E. Waugh *Black Mischief* vii. 257 Anstruther, Legge and William were playing cut-throat bridge.

d. Used *attrib.* or *absol.* of a razor consisting of a blade set in a handle, as distinguished from a safety-razor.

1932 D. L. Sayers *Have his Carcase* i. 18 At the cost of a slight cut on her finger, she drew up an open cut-throat razor. **1957** J. Frame *Owls do Cry* 75 The old lather-up and the strop and the cut-throat razor for him. **1959** *Sunday Times* 22 Feb. 29/3 The advent of the safety-razor..happily put an end to my first clumsy essays with the open cut-throat.

5. b. In full *cut-throat trout*: see quots.

1891 *Cent. Dict.*, *Cutthroat trout*, the Rocky Mountain brook-trout. **1897** *Outing* (U.S.) XXX. 163/2 The father of all the Pacific trout, the black-spotted or 'Cut-throat' (*Salmo mykiss*) with the scarlet splotch on his lower jaw. **1946** *Mazama* Dec. 33/2 Steelhead trout and sea-run cut-throat trout provide capital sport for fishermen.

6. Later examples of *transf.* use.

1848 W. Armstrong *Stocks* 31 [Harlem rail road stock] is generally considered to be most essentially a 'cut-throat stock'. **1881** *Bradstreet's* 16 Apr. 237/4 The three great objects ..are..to avoid cut-throat competition. **1903** *Westm. Gaz.* 17 Jan. 3/1 The wide valley of the Po, where, in nine months of the year, the cut-throat *tramontana* is a frequent visitant. **1914** W. B. Yeats *Responsibilities* 35 A cloud blown from the cut-throat north. **1938** E. Waugh *Scoop* i. v, It's going to be a tough assignment from all I hear. Cut-throat competition.

cutting, *vbl. sb.* Add: **1. a.** (Earlier example in *Cricket*: cf. Cut *v.* 31 a.)

1827 in W. Denison *Sk. Players* (1846) 39 There would be comparatively no cutting to the point or slip.

d. *Cinemat.* The action of *cut *v.* 21 e. Also in *Sound Recording.*

1921 A. C. Lescarboura *Cinema Handbk.* (1922) 21 *Cutting*, editing a picture by elimination of useless or unacceptable film. **1936** A. Asquith in C. Brahms *Footnotes to Ballet* vi. 244 Photographing the same scene from different positions and changing them in the cutting. **1958** *Listener* 6 Nov. 752/2 The success of the total effect was due to the cutting and assemblage of the recordings under the composer's care.

3. b. A quantity that may be cut.

1902 S. E. White *Blazed Trail* xxvii. 191 'Its a fine country,' went on Thorpe so everyone could hear, 'with a great cutting of white pine.' **1957** *Brit. Commonw. Forest Terminol.* 55 *Cutting*, the amount of timber..that can be cut from sawn timber.

9. c. *cutting-in*: (*a*) See *cut *v.* 54 c.

1856 [see Cut *v.* 54 c]. **1925** *Don'ts for Motorists* 55 Cutting in is another evil practice. **1960** *News Chron.* 7 June 1/2 Bad overtaking and cutting-in—normal faults of holiday drivers.

(*b*) See *cut *v.* 54 d.

1920 F. Scott Fitzgerald *This Side Paradise* (1921) i. ii. 74 The reassured beaux and the eternal cutting in. **1928** *Daily Express* 14 Dec. 19 The American practice known as 'cutting-in'..consists..of any man who wishes to dance tapping the shoulder of another man who is already dancing and abducting his partner.

(*c*) The action of starting an engine by some device which closes the circuit. Also used of the engine itself, and *attrib.*

1924 A. W. Judge et al. *Mod. Motor Cars* III. 74 When the dynamo speed is low, the spring holds the contacts apart, but as soon as the 'cutting-in' speed is attained [etc.]. **1928** *Motor Manual* (ed. 27) 110 A dynamo has what is termed a definite 'cutting-in' speed, which means that at, say, 400 revolutions per minute, it begins to generate effective current, which, by the action of a device known as the cut-out, connects the dynamo to the battery and charges it.

d. *cutting-out*: (*a*) See Cut *v.* 56 j. Also *attrib.*, as *cutting-out scissors*, large scissors for cutting patterns from fabric.

1819 M. Wilmot *Let.* 21 Dec. (1935) 47 [My] pink dress which you were at the cutting out of. **1840** [see sense 9 a in Dict.]. **1873** *Young Englishwoman* Mar. 147/2 Cutting-out scissors, small scissors, and button-hole scissors. **1930** *Times Educ. Suppl.* 24 May 238/4 Well qualified..in Practical Needlework, with Cutting-out. **1936** R. Lehmann *Weather in Streets* i. ii. 40 Pins in her mouth, the cutting-out scissors in her hand.

(*b*) = sense 1 c.

1874 J. G. McCoy *Cattle Trade* v. 81 Whilst from six to ten cow boys hold the herd together the ranchman with one or two assistants separate such as are suitable. This process is termed 'cutting out'. **1877** R. E. Strahorn *Hand-bk. Wyoming* 35 Our artist has given a very fair representation of the 'cutting out' scene. **1884** W. Shepherd *Prairie Exper.* 34 This cutting-out goes on all the day long. **1890** 'R. Boldrewood' *Squatter's Dream* ii. 13 He's the best cutting-out horse. **1910** W. M. Raine *B. O'Connor* 75 I'm running this cutting-out expedition. **1920** C. E. Mulford *J. Nelson* xxv. 255 Selecting from their best cutting-out animals, saddles were hastily changed, [etc.].

e. *cutting-up*: (*a*) Boisterous, frolicsome, or silly behaviour. *U.S.*

1843 'R. Carlton' *New Purchase* ii. 209 Cutting up..consists in cracking nuts and jokes—racing one another and slamming doors—in upsetting chairs, and even kicking up carpets! **1883** J. A. Macon *Uncle Gabe Tucker* 162 De perlicemen nebber would 'a' let John de Baptis' do any sich cuttin'-up as dat.

(*b*) *Founding.* (See quot.)

1888 J. G. Horner *Dict. Mech. Engin.* 101 *Cutting-up*, the gashing of the broken edges or faces of a sand mould preparatory to adding fresh sand for mending-up.

10. *cutting cylinder*, *-edge* (also fig.), *-knife* (example), *-machine*, *-plate*; (sense *1 d) *cutting-bench*, *-print*; **cutting-bed** (example); **cutting-box** (earlier example); **cutting compound** = **cutting oil* (see also quot. 1963); **cutting contest** orig. *U.S.*, an informal competition of jazz musicians; **cutting horse, pony** *U.S.*, a horse or pony trained in separating cattle from a herd (cf. Cut *v.* 56 d); **cutting oil,** a preparation for the lubrication and cooling of the tool and the piece of metal being cut or worked in various machining operations; **cutting-room** (later examples); also, a room where a film is cut or edited (see *cut *v.* 21 e); **cutting session** orig. *U.S.* = **cutting contest*; **cutting stick,** a strip of wood or other material which receives the edge of the knife in the cutting cylinder of a paper-cutting machine as it severs each sheet; **cutting stylus** = *cutter *sb.*[1] 4 c.

1881 W. B. Carpenter *Microscope* (ed. 6) v. 229 The circular cutting-bed, instead of being fixed on the upper end of the cylinder, is made to *screw* upon it. **1936** P. Rotha *Documentary Film* ii. 77 Nothing photographed, or recorded on to celluloid, has meaning until it comes to the cutting-bench. **1744** W. Ellis *Mod. Husb.* Jan. vii. 69 The Chaff-cutter's Way was, to..put them upon some Cavings of Wheat..that he first placed at the Bottom of the long Cutting-box. **1910** *Metall. & Chem. Engin.* May 293/1 They mixed a small amount..with their cutting compound and..the tool did not blunt nearly so rapidly as before. **1963** R. F. Webb *Motorists' Dict.* 71 *Cutting compound*, an abrasive paste used for smoothing the paintwork of a car before the final polishing. **1946** Mezzrow & Wolfe *Really Blues* 372 *Cutting contest*, competitive get-together of performers. **1969** *Listener* 13 Mar. 358/3 Not to mention cutting contests in jazz, *Eisteddfodau*, and the choral and brass-band competitions which flourish in Northern industrial towns. **1909** *Daily Chron.* 11 June 7/5 Each colour is rolled out in long sheets, and passes through separate rollers upon which what is known as a 'cutting cylinder' is rotating continuously, one to each colour. **1929** *Encycl. Brit.* XIV. 166/2 In this process the scratched material is rolled into a continuous sheet, and led..under a revolving 'cutting cylinder'. **1964** *Gloss. Letterpress Rotary Print. Terms (B.S.I.)* 23 *Cutting cylinder*, the cylinder that holds the knives that cut the web into individual lengths or sheets. **1825** Loudon *Encycl. Agric.* § 501 It resembles a large..shovel, strongly prepared with iron on the cutting edge. **1831** *Mech. Mag.* 23 July 334 The cutting-edges must be parallel to each other. **1909** A. Berget *Conquest of Air* ii. v. 212 Angle of cutting edge, 7 degrees. **1927** Carr-

Saunders & Jones *Soc. Struct. Eng. & Wales* 207 No one supposes that it is possible to put a sharp cutting-edge on a leaden blade. **1966** *Rep. Comm. Inquiry Univ. Oxf.* I. 56 There is a cutting edge to our recommendation that Oxford should..remain of medium size. **1881** G. W. Romspert *Western Echo* 177 Each firm has particular horses trained for this business, and they are called 'cutting horses'. **1937** *Dial. Notes* VI. 618 A *cutting* horse is one especially trained to separate..a single animal from a whole herd. (I have seen one 'cutting' horse who could 'cut' a hen from a big flock of chickens.) **1925** F. Walton *Linoleum* 48 An octagonal cylinder, arranged with a number of cutting knives. **1876** *Encycl. Brit.* IV. 43/2 A rotary cutting-machine or 'ripper'. **1891** Chase & Clow *Stories of Industry* II. 43 The cutting-machine or shearing engine..shears the nap off close, leaving a smooth face to the cloth. **1917** J. R. Battle *Lubricating Engin. Handbk.* 300 The usual soluble cutting oil is made of a combination of oils..and is designed to permit its being mixed with varying amounts of water to form a stable cutting emulsion. **1955** *Oxf. Jun. Encycl.* VIII. 248/2 Today there are various brands of cutting oils, most of which are mixtures of mineral oil, soaps, and an emulsifying chemical which makes the oils mix readily with water. These keep the tool and work-piece cool during the cutting operations, and so prevent excessive wear and distortion of the work. **1876** Preece & Sivewright *Telegraphy* 195 The cutting-plate itself is in the form of a screw, and thus acts both as a drill and cutting-plate. **1887** F. Francis *Saddle & Mocassin* (Lentzner), I had been furnished with a trained cutting pony. **1902** A. MacGowan *Last Word* 435 A lady that 's been as able as any cowboy on the range..to manage anything, from a cuttin' pony as fine as silk, to the meanest buckin' bronc. **1953** K. Reisz *Technique Film Editing* 279 *Cutting-print*, the particular positive print which the editor assembles and on which he works. **1840** Dickens *M. Humphrey* iv. 101 A young hairdresser..had..a floor-clothed cuttin'-room upstairs. **1902** *Daily Chron.* 25 Oct. 7/6 At the rear is a cutting-room, where meat is cut up. **1918** R. Wagner *Film Folk* vi. 273 The cutting room of a studio is the slaughterhouse of vain ambition. **1936** 'J. Tey' *Shilling for Candles* ii. 23 Treating me like bits on the cutting-room floor. **1936** *Words* Oct. 6/1 Cutting room terms are especially exotic. **1959** *Guardian* 9 Nov. 5/7 The young technicians..crowded into the cutting-room to watch some..experimental commercials. **1959** *Jazz Review* May 12/1 A wild cutting session was in progress and sitting around the piano were twenty or thirty musicians. **1888** Wilson & Grey *Mod. Printing Mach.* iii. xxvi. 431 Small cutting machines are frequently turned by hand... Cutting-sticks, which fit into the table immediately under the knife, are generally made of hard wood. **1926** *Wireless World* 15 Sept. 399/2 This megaphone concentrated the sound waves on to a mechanical diaphragm to which was connected a cutting stylus resting on a revolving disc of soft wax.

cu·tting grass. [Cutting *ppl. a.*] **1.** Any of several grasses or sedges of Australia and New Zealand having sharp-edged leaves or stems, esp. *Gahnia psittacorum*.

1831 Hovell & Hume *Journey of Discovery to Port Phillip 1824–25* 61 They had the misfortune to encounter that species of long grass, which is known in the colony by the name of the 'cutting grass'. **1858** T. McCombie *Hist. Victoria* i. 8 Long grass, known as cutting-grass, between four and five feet high, the blade an inch and a half broad, the edges exquisitely sharp. **1894** *Age* 19 Oct. 5/8 (Morris), 'Cutting grass' is the technical term for a hard, tough grass about eight or ten inches high, three-edged like a bayonet, which stock cannot eat because in their efforts to bite it off it cuts their mouths. **1968** G. R. Cochrane et al. *Flowers & Plants of Victoria* 16/2 The leaves of many sedges will often cause deep cuts, and they are commonly referred to as sword grass or cutting grass.

2. = *cane-rat* (*Cane *sb.*[1] 10).

1934 *Times Educ. Suppl.* 31 Mar. p. iv/3 The Curator of Mammals was recently informed that some 'cutting grass' were being sent as a gift from West Africa..On arrival, they turned out to be three young great cane rats, or 'ground hogs'. **1960** *Times* 29 Sept. (Nigeria Suppl.) p. xxi/6 Roast cutting-grass.

cutty, *a.* Add: **5.** Capable of cutting, sharp. *spec.* (*N.Z.*) **cutty grass** (also **cuttigrass**) = *Cutting grass 1.

1903 Kipling in *Windsor Mag.* Sept. 364/1 We'll draw fine, freehand, tribal patterns on their backs with the cutty edges of mussel-shells. **1910** L. Cockayne *N.Z. Plants* vii. 109 Here is also the home of the sedge family, to which the so-called 'cutty-grasses' belong. **1920** J. Mander *Story of N.Z. River* ii. 29 Among the rushes and cuttigrass. **1927** W. H. Guthrie-Smith *Birds of Water, Wood & Waste* (ed. 2) 162 The nest is planted deep..in the heart of a bunch of cutty grass. **1957** *Landfall* XI. 213 The cutty grass sawed at her dress.

cut-up, *sb.* Add: **1. b.** orig. *U.S.* (*a*) The act of making practical or verbal jokes; clowning. *rare.* (*b*) A person who 'cuts up' or capers (see Cut *v.* 59 m and *o). Also *attrib.*

1843 'R. Carlton' *New Purchase* ii. 209 Art and tact..are requisite for the cut-up...If the affair is not done up to the point—it is teasing; if beyond—it is horse-play. *Ibid.*, The cut-ups were usually in wet weather. **1882** I. M. Rittenhouse *Maud* (1939) iii. 91 He and the Menagers are the greatest friends imaginable and such cut-ups. **1911** R. W. Chambers *Common Law* v. 157 Jests emanating from the boarding-house cut-up—a blonde young man with rah-rah hair and a brier pipe. **1923** R. D. Paine *Comr. Rolling Ocean* viii. 129 As a college cut-up he was the star comedian of the campus. **1936** Wodehouse *Laughing Gas* x. 107 Anything for a laugh is your motto. Well, good night, old cut-up. **1960** B. Keaton *Wonderf. World of Slapstick* (1967) 122 We cutups and cute kids only had one bad moment. **1969** *Listener* 16 Jan. 87/2 We have to take his word for most of what we're told in the book's final section; for the cut-ups and the rave-ups, for that wild weekend with 'Ronnie' Laing.

2. (Earlier example.)

1845 *Encycl. Metrop.* VIII. 749/1, 6,000 frames making cut-ups, &c., produce 1,960,000 dozens.

‖**cuvée** (küve). [Fr., lit. 'vatful' (*cuve*, see Cuve).] The contents of a vat of wine; a particular blend or batch of wine.

1833 C. Redding *Mod. Wines* iv. 68 *Cuvée*, the contents either of a cellar or vat at the vintage. **1883** *19th Cent.* Sept. (Advt.) (Stanford), Perrier Jouet's..Extra Dry [Champagne] Reserved Cuvee. **1911** *Encycl. Brit.* XXVIII. 724/1 These vattings, and indeed all blendings of any particular batch of wines, are termed *cuvées*. **1920** G. Saintsbury *Notes Cellar-Bk.* vi. 99 Some of the great shippers used to send out wines cheaper than their official *cuvées*. **1939** E. Ambler *Mask of Dimitrios* xiv. 301 The wine must be champagne..it must be a vintage *cuvée*. **1968** *Vogue* Dec. 40/2 Even the great *Cuvée* could not wash away the lump in my throat.

cuvette. Add: **4.** *Geol.* (See quot. 1929.)

1907 *Daily Chron.* 28 Oct. 3/3 The sand dunes and cuvettes round Lake Chad. **1910** *Encycl. Brit.* X. 598/1 There are 'short-synclines', '*brachysynclinaux*' or 'cuvettes'. **1929** L. J. Wills *Physiogr. Evol. Britain* ii. vi. 79 *Cuvette* is a convenient term for a basin in which sedimentation is going on..as distinct from a tectonic basin due to folding of pre-existing rocks.

Cuvierian (kiuvi͡ɪə·riăn), *a.* [f. *Cuvier* (see below) + -ian.] Of, pertaining to, or named after the French naturalist Georges Cuvier (1769–1832); characteristic of his methods or system of classification.

Cuvierian ducts, two short transverse venous ducts in the vertebrate fœtus and in fishes which return the blood from the cardinal veins to the sinus venosus of the heart; the common cardinal veins. *Cuvierian organs* (see quot. 1897).

1856 *Chambers's Jrnl.* 27 Sept. 207/2 A Cuvierian examination of the various articles. **1870** G. Rolleston *Forms Anim. Life* 146 It has been said that the 'Cuvierian organs', certain structures of..doubtful function attached to the stem of the respiratory tree or inserted upon the cloaca, are wanting in all Cucumariae with unequal tentacles. *Ibid.* 262 (*index*) Cuvierian ducts. **1897** Parker & Haswell *Zool.* I. ix. 394 In the Holothurians..the 'Cuvierian organs' are simple filiform glandular tubes..connected with the cloaca. **1905** H. S. Pratt *Vertebr. Zool.* 52 Joining each end of the transverse sinus venosus is a large and conspicuous duct or sinus called the Cuvierian duct. **1962** D. Nichols *Echinoderms* vi. 81 In aspidochirotes there are two groups of special defensive cuvierian organs branching from the bases of the respiratory trees. **1964** W. Coleman *G. Cuvier Zoologist* iii. 50 The second Cuvierian anatomical rule, the subordination of characters, may be considered as the epitome of the hierarchical arrangement of functions and therefore of organs.

cwm (kum). *Geol.* [Welsh *cwm* (cf. Coomb[2]).] A valley; in *Phys. Geogr.*, a bowl-shaped hollow partly enclosed by steep walls lying at the head of a valley or on a mountain slope and formed originally by a glacier; a cirque.

1853 Mrs. Gaskell *Ruth* I. vii. 170 Some 'Cwm', or hollow. **1882** Geikie *Text-bk. Geol.* iii. ii. ii. 407 Several hundred feet below, in the corrie or cwm at the bottom, lies the re-cemented glacier. **1933** *Geogr. Jrnl.* LXXXII. 202 The snow-patches are cwm-ice masses occupying deep scallops in an elevated position of the old erosion-surface. **1936** *Nature* 19 Dec. 1041/2 Its glaciers..widened their heads into cwms and gave to the basin its only fiord. **1951** *Times* 27 Nov. 5/7 While 'cwm' may occur..purely as a place-name..technically the word is restricted to the huge cauldron-shaped hollows found high up on heavily glaciated slopes. **1953** J. Hunt *Ascent of Everest* ii. 14 When Mallory saw it..in 1921, he named it the 'Western Cwm'. **1957** G. E. Hutchinson *Treat. Limnol.* I. i. 59 Such amphitheaters are called *cirques* in the French-speaking parts of the Alps, *Kars* in the German-speaking regions, *cwms* in Wales, and *corries* in Scotland. All four terms have achieved some degree of international usage, but the first seems to have been the most widely employed.

cyan (səi·ăn). = Cyan-blue.

1889 in *Cent. Dict.* **1957** R. W. G. Hunt *Reprod. Colour* iv. 37 The colours of the dyes used are cyan, magenta, and yellow. **1964** *Observer* 22 Nov. 23/1 Pinkish lights would make the picture greenish-blue or cyan, the engineers' word for sickly green.

cyanamide (under Cyan- 2). Add: Also, a salt of this compound, in which one or both of the hydrogen atoms are replaced by another element or radical; *spec.* = calcium cyanamide ($CaCN_2$), used as a fertilizer and as a source of other nitrogen compounds.

Cyanamide process, the production of calcium cyanamide by the reaction between calcium carbide (CaC_2) and nitrogen at a high temperature.

1897 *Jrnl. Chem. Soc.* LXXI. i. 460 Among the products of the change..we get sodium cyanide and sodium cyanamide, $C(NNa)_2$. **1914** J. Knox *Fixation Atmosph. Nitrogen* 88 The metallic cyanamides are derivatives of cyanamide, H_2CN_2. **1922** Partington & Parker *Nitrogen Industry* 188 The second important method of nitrogen fixation—viz., the cyanamide process. **1926** J. F. Crowley in E. Fyleman tr. *Waeser's Atmosph. Nitr. Industry* I. p. xiv, Cyanamide has been found to be a useful fertiliser for particular soils. **1965** P. A. S. Smith *Chem. Open-Chain Org. Nitrogen Compounds* I. vi. 251 Substituted cyanamides are for the most part named by simply adding the name of the substituent, as 'methylcyanamide'. *Ibid.* 253 Heavy metal cyanamides, however, have solubilities similar to sulfides and acetylides.

cyanicide (səiæ·nisəid). *Metallurgy.* [f. Cyan(ide + -cide 1.] Any substance present in an ore or in tailings which consumes the cyanide in the solution used for cyaniding gold or silver, so reducing its effectiveness as a solvent for those metals.

1894 *Jrnl. Soc. Chem. Industry* 31 Oct. 951/2 To overcome the cyanide-destroying qualities of the acid or iron salts present it is necessary to have recourse to neutralisation.., with or without a preliminary water washing to remove such soluble 'cyanicides' as may be present. **1904** Julian & Smart *Cyaniding* xxviii. 202 The principal cyanicide remaining in the Rand concentrates is the basic ferric sulphate. **1936** J. V. N. Dorr *Cyanidation Gold & Silver Ores* ix. 152 If precious-metal ores contain cyanicides, which preclude the use of cyanidation, flotation may be the only solution. **1966** *McGraw-Hill Encycl. Sci. & Technol.* VI. 233/1 Sulfides of copper, iron, antimony, and arsenic (cyanicides)..consume cyanide and oxygen, thus causing poor gold recoveries.

cy:anida·tion. *Metallurgy.* [f. *Cyanide *v.*: see -ation.] The process of cyaniding gold or silver ores; the extraction of these metals from their ores by means of the cyanide process.

1896 *N. Amer. Rev.* Apr. 479 The development of certain wet processes, cyanidation..has supplemented the work. **1915** T. K. Rose *Metall. Gold* (ed. 6) xiii. 281 The roasting of gold ores as a preliminary to chlorination or cyanidation. **1936** [see *cyanicide]. **1959** J. Newton *Extractive Metall.* vii. 431 Cyanidation is the most important for recovering gold, but it is often used in conjunction with other processes.

cyanide. Add: *cyanide gauze*, a gauze rendered antiseptic by impregnation with a cyanide, used in dressing wounds; *cyanide hardening*, case-hardening of iron or steel by the cyanide process; *cyanide process*, (*a*) a method of extracting gold and silver from ores by treatment with a dilute solution of sodium or calcium cyanide (or, formerly, potassium cyanide); (*b*) the fixation of atmospheric nitrogen by chemical reaction at high temperatures so as to form alkali cyanides or other cyanogen derivatives; (*c*) a process for case-hardening iron or steel by immersing it in molten cyanide and then quenching it in water or oil.

1895 *Arnold & Sons' Catal. Surg. Instr.* 726 Double Zinc-Cyanide Gauze. **1913** A. Bennett *Regent* ii. § 4 He did nothing but cover up the place with a bit of cyanide gauze. **1925** R. W. G. Hingston in E. F. Norton *Fight for Everest: 1924* iii. vi. 350 The following equipment is required:.. Cyanide gauze, 2 lb. **1921** *Jrnl. Iron & Steel Inst.* CIII. 435 Cyanide Hardening Due to Nitrogen... Steel case-hardened in cyanide contains much nitrogenised material, and nitrogen is believed to exert an appreciable influence on the properties of steel so treated. **1928** *Ibid.* CXVIII. 365 General practice in cyanide hardening is discussed, and reference is made to the Shimer process for case hardening and the cyanogen gas process. **1890** J. S. MacArthur in *Jrnl. Soc. Chem. Industry* 31 Mar. 270/1 By the cyanide process, ores containing lead, zinc, or earthy carbonates which cannot be worked to profit by chlorination, may be as..profitably treated as any other. **1926** E. Fyleman tr. *Waeser's Atmosph. Nitr. Industry* 486 (*heading*) The cyanide processes. **1940** R. H. Heyer *Engin. Physical Metall.* x. 341 The speed, low cost,..and cleanliness of the hardened surface make the cyanide process readily adaptable to production heat treatment. **1945** D. M. Liddell *Handbk. Nonferrous Metall.* (ed. 2) II. xi. 294 Outside of smelting and refining, the present-day metallurgy of gold and silver may be summarized under three heads: mechanical methods.., amalgamation, and the cyanide process. **1959** *Chambers's Encycl.* VI. 413/2 The cyanide process was developed by J. S. Macarthur and R. W. and W. Forrest of Glasgow and used at Johannesburg in 1890.

cy·anide, *v. Metallurgy.* [f. the sb.] *trans.* To treat with a cyanide. **a.** To treat (ores of gold or silver) with a dilute cyanide solution as part of the cyanide process for extracting the metal. **b.** To immerse (iron or steel) in molten cyanide in order to case-harden it. Hence **cy·anided** *ppl. a.*; **cy·anider**; **cy·aniding** *vbl. sb.*

a. 1894 *N. Brit. Daily Mail* 30 July 5 Cyaniding syndicates. **1895** M. Eissler *Cyanide Process* 39 The cost of cyaniding was as follows. **1896** *Daily News* 10 Dec. 2/7, 2,666 tons cyanided, producing 352 ozs. **1900** J. Park *Cyanide Process* 124 The many worries which the use of cyanide entails on even the successful cyanider. **1945** D. M. Liddell *Handbk. Nonferrous Metall.* (ed. 2) II. xi. 294 Chlorination of gold and hyposulphite leaching of silver ores were practically superseded by cyaniding by the year 1900. **1959** J. Newton *Extractive Metall.* vii. 431 Flotation may be used to produce concentrates that can be cyanided or amalgamated. **1965** G. J. Williams *Econ. Geol. N.Z.* viii. 109/2 In the late 1890's interest revived in the hope that the ore could be cyanided.

b. 1921 *Jrnl. Iron & Steel Inst.* CIII. 435 After annealing the cyanided bar is still hard to file. **1940** R. H. Heyer *Engin. Physical Metall.* x. 348 In certain carburizing and cyaniding processes nitrogen is introduced into the steel in addition to the carbon. **1966** *McGraw-Hill Encycl. Sci. & Technol.* XIII. 313/2 Both carburizing steels and medium carbon steels..are often cyanided.

cyanidin (səiæ·nidin). *Chem.* [a. G. *cyanidin* (Willstätter and Everest 1913, in *Ann. d. Chemie* CDI. 204), f. Cyan- + *-idin.] An

anthocyanidin (usu. isolated as the chloride, $C_{15}H_{11}O_6Cl$) which in combination with various sugars constitutes many of the most important natural anthocyanin plant pigments.

1914 *Chem. Abstr.* VIII. 336 A violet sol[utio]n..is hydrolyzed..giving 2 mol[ecule]s glucose and cyanidin chloride, $C_{15}H_{11}O_6Cl$. *Ibid.* 3422 If quercetin be reduced at 35° instead of at 0°, a small quant[ity] (0·5–4·0%) of cyanidin is formed. **1939** *Thorpe's Dict. Appl. Chem.* (ed. 4) III. 512/1 Cyanidin is the most common and widely distributed of the anthocyan pigments of plants, and occurs naturally in the form of various glycosides. **1966** J. B. Harborne in T. Swain *Compar. Phytochem.* xvi. 283 The fifth [*Plumbago* species], *Plumbago rosea*, is very distinct, having glycosides of kaempferol, quercetin, pelargonidin, cyanidin and delphinidin in its petals.

cyanize (səi·ănəiz), *v. Chem.* [f. Cyan- 2 + -ize.] *trans.* To convert into a cyanide, esp. as part of a process for fixing nitrogen. Hence **cy·anized** *ppl. a.*, **cy·anizing** *vbl. sb.*, **cy:aniza·tion.**

1881 *Chem. News* 29 Aug. 106/1 Cyanised Camphor... This compound, in alcoholic solution, is partly resolved into hydrocyanic and camphic acids. **1890** *Ibid.* 15 Aug. 82/1 The process consisted of passing the gas from a coal fire through large vertical cylinders..filled with a mixture of wood charcoal and potash... The cyanised charcoal was withdrawn at the bottom of the cylinder and dropped into..water containing powdered spathic iron ore in suspension, by which the cyanides were converted into ferrocyanides or prussiates. **1926** E. Fyleman tr. *Waeser's Atmosph. Nitr. Industry* 499 In order to determine whether pure nitrogen is necessary for cyanisation. *Ibid.* 501 The cyanising apparatus. *Ibid.*, The cyanised briquettes contained 19 per cent. of sodium cyanide. *Ibid.* 507 Mond cyanises briquettes of barium carbonate and charcoal in a ring furnace.

cy:anocoba·lamin. *Biochem.* [f. Cyano- + *cobalamin.] Vitamin B_{12}.

1950 [see *cobalamin]. **1961** *Lancet* 26 Aug. 482/2 It seems probable that the pharmacopœial term 'vitamin B12', at present restricted..to cyanocobalamin, will soon have to be extended to include also hydroxocobalamin. **1962** *New Scientist* 15 Feb. 390/2 A by-product of the streptomycin fermentation is cyanocobalamin (vitamin B_{12}), an important constituent of liver extract, used in treating pernicious anaemia.

cybernation (səibəɹnē͡ɪ·ʃən). [f. *cybern(etics *sb. pl.* + -ation.] The theory, practice, or condition of control by machines. Hence (as a back-formation) **cy·bernate** *v. trans.*, to control in this manner; **cy·bernated** *ppl. a.*

1962 D. Michael (*title*) Cybernation: the silent conquest. **1962** *Punch* 7 Feb. 231/2 Cybernation..is becoming a dirty word in America. **1962** *Catholic Gaz.* Nov. 320/1 (*heading*) The cybernated society. *Ibid.*, When the machines are controlled by the computers human operators become unnecessary and the society is cybernated... Many jobs have been partially or completely cybernated. **1963** J. A. T. Robinson *Honest to God* vii. 139 All the drives of modern secular society, whether collectivized or cybernated. **1969** *Northwest* (Sunday Oregonian Mag.) 14 Dec. 18/1 The major problems of the day—cybernation, the revolution in human rights and the threats of growing militarism.

cybernetics (səibəɹne·tiks). [f. Gr. κυβερνήτης steersman, f. κυβερνᾶν to steer (see Govern *v.*) + -ics.] The theory or study of communication and control in living organisms or machines. Hence (as back-formation) **cyberne·tic** *a.*, pertaining or relating to cybernetics. So **cyberneti·cian, cyberne·ticist,** one who is skilled in cybernetics.

Used in Fr. form *cybernétique* (= the art of governing) by A.-M. Ampère *Essai sur la Philos. des Sciences*, 1834.

1948 N. Wiener *Cybernetics* 19 We have decided to call the entire field of control and communication theory, whether in the machine or in the animal, by the name Cybernetics. **1951** *Jrnl. R. Aeronaut. Soc.* Oct. 624/2 All these machines represent developments in that part of what has been called the cybernetic revolution which is gradually taking over those operations in the fields of numbers, quantities, and data that are strictly clerical or mechanical. **1952** *Science News* XXIII. 77 The cyberneticists approach the problem of neural activity from a purely functional angle, and seek to model the activity of the brain as a whole on the electronic devices of modern communication systems and servo-mechanisms. **1958** *Listener* 18 Sept. 413 The claim of cybernetics is that we can treat organisms *as if* they were machines, in the sense that the same methods of synthesis and analysis can be applied to both. **1959** *Times* 11 May 6/6 Cybernetics is the study of man in relation to his particular job or machine with special reference to mental processes and control mechanisms. **1961** *Times Lit. Suppl.* 6 Jan. 2/4 It is all right for cyberneticians to make machines like men. **1961** J. Wilson *Reason & Morals* ii. 113 If men are machines, at least their behaviour suggests that they are cybernetic or self-regulating machines. **1962** *Listener* 1 Nov. 718/1 Cyberneticians, as the people who practise cybernetics now appear to call themselves, can build a larynx with which an injured man can speak. *Ibid.* 718/2 In education, too, cybernetics begins to intrude as electronic teaching machines make good the lack of human teachers. **1968** *Brit. Med. Bull.* XXIV. 197/2 The integration of cells, organs and systems..appears to be done on a cybernetic basis with feed-back processes..clearly interwoven at all levels. **1970** *Nature* 12 Sept. 1167/1 The cyberneticist's approach to the concepts of psychology is not, however, in evidence here.

cybotaxis (səibotæ·ksis). *Physics.* [f. Gr. κύβος Cube + τάξις arrangement.] An arrangement of molecules in a liquid (see quot. 1927). Hence **cybota·ctic** *a.*, of or pertaining to cybotaxis.

1927 G. W. Stewart & R. M. Morrow in *Physical Rev.* XXX. 233 The present paper..interprets the space array..not as caused by fragmentary crystals, but by a type of molecular arrangement wherein there is combined mobility of the component molecules and yet a recognizable space array. To this state is given the name 'cybotaxis', which means 'space-arrangement'. The adjective is 'cybotactic'. A new word is necessary in order to distinguish this state from that called 'crystalline'. **1941** *Nature* 22 Nov. 617/1 There are..two main types of conception of liquid structure..(1) group and (2) statistical conceptions... The most elegant development of the group type of conception is Stewart's theory of 'cybotaxis'. In this the groups are considered to be of transitory nature, continually forming and disappearing. **1948** *Sci. News* VIII. 96 Their views fit in well with Stewart's theory that liquids have a 'cybotactic' structure, viz., that there are small aggregates of molecules arranged as they would be in a crystalline particle.

Cycladic (siklæ·dik), *a.* [f. L. *Cyclades* pl., a. Gr. Κυκλάδες (νῆσοι islands) + -ic.] Of or pertaining to the Cyclades, a group of islands in the Aegean, lying in a circle round Delos; *spec.* designating, or pertaining to, the prehistoric civilization of these islands. Also *absol.*

1915 H. R. Hall *Ægean Archæol.* ii. 24 It has become a misnomer to call the island culture by a name connecting it in any way with the 'Mycenæan'. The word 'Cycladic' is now used, as the chief discoveries of this early stage of Greek civilisation have been made in the Cyclades. **1920** *Discovery* June 178/2 'Cycladic'..is sometimes substituted for 'Minoan' when one speaks exclusively of the island sites outside of Crete. **1921** *Brit. Museum Return* 60 Large beaked jug of later Cycladic style..From Melos or Thera. **1927** Peake & Fleure *Priests & Kings* 113 These [sub-periods] have been named in the same way, ranging from Early Cycladic I to Late Cycladic III. **1957** V. G. Childe *Dawn Europ. Civilization* (ed. 6) iv. 49 While it has been customary to assign most cemeteries to the Early Cycladic period (before 2000 B.C.), Åberg has shown that some graves must be Middle or even Late Cycladic.

cyclamate (si·klămē͡ɪt, səi·klămē͡ɪt). [f. *cyclo*hexylsulph*amate*, f. Cyclo- + Hexyl + *sulphamate*, f. *sulpham*(*ic* (f. Sulph- + Am(ide + -ic) + -ate[4].] A salt of cyclohexylsulphamic acid, $C_6H_{11}\cdot NH\cdot SO_3H$, esp. the sodium and calcium salts, which have been used as artificial sweetening agents. So **cy·clamated** *a.*, containing a cyclamate, having had a cyclamate added.

1951 *Jrnl. Amer. Pharm. Assoc.* (Sci. ed.) XL. 1 (*heading*) Studies on cyclamate sodium (Sucaryl sodium), a new noncaloric sweetening agent. **1965** *Observer* 17 Oct. 33/1 The use of cyclamate as an 'artificial sweetener'. *Ibid.*, Cyclamates fed to rats at levels of 5 per cent or 10 per cent of the diet reduced their rate of growth. **1969** *Daily Tel.* 21 Nov. 8/2 The decision to withdraw the ban was the result of the recommendation by a special scientific panel convened to determine the best way of making cyclamated products readily available to diabetics. **1970** *New Scientist* 1 Jan. 21/1 The cyclamate issue..is certainly stimulating research into ways of satisfying Man's sweet tooth. *Ibid.*, Cyclamates are only 30 times sweeter than sucrose and saccharin 300 times sweeter.

cyclamen. Add: **c.** The shade of colour characteristic of the red or pink cyclamen flower.

1923 *Daily Mail* 29 Jan. 1 In shades of Powder Blue.. Cyclamen,..Flamingo, Pink. **1926** *Spectator* 24 Apr. 750/2 Sleeveless cardigan, in various shades of cyclamen, rose marie and saxe. **1960** *Harper's Bazaar* Aug. 80/2 Cyclamen rayon satin.

cyclane (səi·klē͡ɪn). *Chem.* [f. *cycl(ic *a.* 7 or Cycl(o- + -ane.] Any saturated cyclic hydrocarbon.

1932 *Brit. Chem. Abstr.* 559/2 (*heading*) Raman spectra of cyclanes... Raman spectra of *cyclo*-propane, -pentane, -hexane, -heptane, and -octane and their Me derivatives show great similarity. **1939** *Thorpe's Dict. Appl. Chem.* (ed. 4) III. 532/1 They [*sc.* cyclenes] can..readily be derived from various simple derivatives of the saturated cyclic hydrocarbons (cyclanes).

cycle, *sb.* Add: **3. b.** *Physics*, etc. A recurring series of operations or states, *spec.* in internal combustion engines. Also, short for *cycles per second* (abbrev. C.P.S., cps., c/s), the unit of frequency of an oscillation (as an alternating current, a sound wave, etc.).

a **1884** Knight *Dict. Mech.* Suppl. 382/2 The complete cycle of motions in the Otto engine is accomplished only by two complete revolutions of the working shaft, or four strokes of the piston. **1887** *Encycl. Brit.* XXII. 479/2 Generally in heat-engines the working substance returns periodically to the same state of temperature, pressure, volume, and physical condition. When this has occurred the substance is said to have passed through a complete cycle of operations. **1893** T. O'C. Sloane *Standard Electr. Dict.* 175 *Cycle of alternation*, a full period of alternation of an alternating current. It begins properly at the zero line, goes to a maximum value in one sense and returns to zero, goes to maximum in the other sense and returns to zero. **1920** *Whittaker's Electr. Engineer's Pocket-Bk.* (ed. 4) 348 In the early days, when electricity was used only for lighting, frequencies round about 100 cycles were usual. **1929** A. F. Collins *Aviation* 148 After the power stroke is completed three more strokes must take place before there is another explosion stroke and, hence, another power stroke. Then the series of strokes, or cycle, as it is called, begins all over again, and this is what is meant by a four-stroke cycle engine. **1940** *Chambers's Techn. Dict.* 206/1 C.P.S., cps., c/s, abbrevs. for *cycles per second*, the usual measure of frequency. **1944** A. Wood *Physics of Music* iv. 42 The lowest frequencies used in the orchestra are those of the double bass and the bass tuba, which lie between 60 and 80 cycles per second... Tones of male speech embrace a range of from 120 to 8000 c.p.s. **1959** *Consumer Rep.* (N.Y.) Sept. 452/2 Only one [tweeter] provided adequate and relatively uniform power in the high-frequency range to beyond 20,000 cps. **1965** *Electronics Weekly* 10 Mar. 22/5 It operates from a standard single phase 50 c/s supply. **1967** *Electronics* 6 Mar. 325/1 Markus continues to use 'cycles per second' instead of 'hertz'.

c. *Geol.* *cycle of erosion, sedimentation* (see quots.).

1904 Chamberlin & Salisbury *Geol.* (1905) I. iii. 78 It has now been seen that by whatever method erosion by running water proceeds..the final result of subaërial erosion must be the production of a base-level... The time involved in the reduction of a land area to base-level is a cycle of erosion. **1921** L. D. Stamp in *Geol. Mag.* LVIII. 109 A 'cycle of sedimentation' comprises the deposits of a complete oscillation of the basin, each oscillation including a positive phase of marine invasion and a negative phase of regression. **1960** —— *Britain's Struct. & Scenery* (ed. 5) vii. 70 As the cycle of erosion progresses..features which are due to varying resistance of the rocks..gradually become eliminated as the surface is reduced to a monotonous level..plain.

12. *cycle-shop, -track*; **cycle-car,** a light motor-driven vehicle with three (rarely four) wheels; **cycle-clip** = *bicycle-clip.

1913 (*title*) The autocycle, side car and cycle car user. **1914** *Morn. Post* 9 Feb. 5 A Cyclecar Paperchase. **1954** 'N. Shute' *Slide Rule* 12 A little car built of motor bicycle components, then known as a cyclecar. **1939–40** *Army & Navy Stores Catal.* p. xxxi/1 Cycle clips. **1955** P. Larkin *Less Deceived* 28 Hatless, I take off My cycle-clips in awkward reverence. **1922** Joyce *Ulysses* 153 Rover cycleshop. **1916** —— *Portrait of Artist* iv. 182 Some jesuits were walking round the cycletrack. **1936** *Min. Transport Circ.* 454 (*Roads*) 2 Separate cycle tracks will often be justified solely on the grounds of public safety. **1963** *Ann. Reg. 1962* 449 Extensive use was made of cycle tracks and pedestrian footpaths for segregating traffic about other parts of the town.

cyclery (səi·kləri). *N. Amer.* [f. Cycle *sb.* 11 + -ery.] A bicycle shop.

1897 *Trans-Mississippian* (Council Bluffs, Iowa) 20 Apr. (Advt.), Council Bluffs Cyclery. **1899** J. F. Fraser *Round World on Wheel* xxxvii. 484 There is a cyclery—that's an American word—where machines are hired out at a shilling an hour. **1901** *Daily Colonist* (Victoria, B.C.) 1 Oct. 8/2 (Advt.), The stock, tools, etc. of the B.C. Cyclery & Supply Co. **1936** Mencken *Amer. Lang.* (ed. 4) 176 In Pasadena, Calif.,..there is a *hattery*, in South Pasadena a *cyclery*.

cyclette (səikle·t). [Fr.: see -ette.] A small (motor) bicycle.

1898 *Daily News* 15 Nov. 7/6 The one-mile motor cyclette race. **1923** *Daily Mail* 28 Feb. 6 The Paris–Nice Trial. How the motor cyclettes fared under strenuous conditions.

cyclic, *a.* Add: **1. d.** Aeronaut. *cyclic pitch control*, a method of controlling the direction of motion of a helicopter by varying the angle of the rotor blades during each cycle of rotation. So *cyclic pitch lever, stick.*

1944 H. F. Gregory *Helicopter* (1948) xiv. 164 The first and probably the most common [method of controlling helicopters]..is called cyclic pitch control. What it means is the change of pitch of a blade..as the blade moves in its cycle of rotation. **1959** F. D. Adams *Aeronaut. Dict.* 55/2 *Cyclic pitch stick*, a control stick for cyclic pitch control. **1962** *Flight Internat.* LXXXI. 865/2 According to the position of a pendulous 'cyclic-pitch' lever, the complete rotor disc and resulting lift vector can be tilted to control the flight path.

7. *Org. Chem.* Of a compound: having a molecular structure containing one or more 'closed chains' or rings of atoms; also, of or pertaining to such compounds. See also *alicyclic, *carbo-cyclic, *heterocyclic, **isocyclic* adjs.

1898 *Jrnl. Chem. Soc.* LXXIV. i. 637 It is extremely difficult to separate benzene from cyclic hydrocarbons boiling at a much lower temperature. **1913** *Bloxam's Chem.* (ed. 10) 544 The cyclic or closed-chain series. **1923** T. H. Pope tr. *Molinari's Org. Chem.* II. 616 Cyclic compounds. **1937** *Thorpe's Dict. Appl. Chem.* (ed. 4) I. 33/2 Cyclic acetals, of use in perfumery and as industrial solvents, are made by condensing dihydric alcohols with unsubstituted araliphatic aldehydes. **1961** L. F. & M. Fieser *Adv. Org. Chem.* ii. 51 The hydrocarbon is indeed known, and since it has the same number of carbon atoms as propane but is cyclic it is called cyclopropane.

cyclically (səi·klikăli), *adv.* [f. Cyclical *a.* + -ly[2].] In a cyclic or cyclical way; in cycles.

1895 *Athenæum* 1 June 710/1 Mr. Burstall commenced the reading of a paper 'On the Measurement of a Cyclically Varying Temperature'. **1939** *New Republic* 27 Sept. 210 Employment in this area has been decreased by thousands, not as a result of recession but in addition to the thousands of men cyclically displaced. **1961** P. J. Davis *Lore of Large Numbers* 60 A decimal whose digits recur cyclically must be a fraction.

cycling, *ppl. a.* Add: *cycling lizard*, a kind of lizard found in Australia (see quots.).

1937 *Discovery* May 137/1 The Racehorse or Cycling Lizard..runs at an incredible speed..and the movement of the back legs bears a remarkable resemblance to the motion of a cyclist. **1955** A. J. Ross *Australia 55* 46 The 'cycling' lizard, whose leg motions resemble the pedalling action of a man on a bicycle,..of which the Commonwealth Railway brochure speaks so enticingly.

cyclize (səi·kləiz), *v. Org. Chem.* [f. *cycl(ic *a.* 7 + -ize.] *trans.* To make (a compound) cyclic (*cyclic *a.* 7); *intr.* to undergo cyclization, become cyclic. So **cycliza·tion**, the rearrangement of the atoms in a molecule to form one or more 'closed chains' or rings; **cy·clized** *ppl. a.*

1909 *Chem. News* 16 July 24/2 The cyclisation of ketone acids always occurs at one of the carbon atoms near the ketone function. Cyclic chains containing C_5 and C_6 are most easily formed. **1933** *Jrnl. Chem. Soc.* ii. 1111 The cyclisation of aromatic compounds in which a cyclic system forms part of an unsaturated side chain. **1934** Webster, *Cyclize* v.t. **1936** *Nature* 7 Mar. 411/2 Open 'cyclised' polypeptides. **1942** G. Egloff et al. *Isomerization of Pure Hydrocarbons* vi. 170, 1,1,3,3-Tetraphenylbutene cyclized at 215° in the presence of 'Floridin'. *Ibid.* vi. 171 The other form, when treated in the same manner, cyclized into 1-benzyl-2,3-diphenylindane. **1949** H. W. Florey et al. *Antibiotics* II. xxviii. 958 Attempts to cyclize this compound. **1954** *Archit. Rev.* CXV. 356/1 Cyclized (or isomerized) rubber is mostly used for paints which are required to withstand steam and condensation and is thus most commonly used in kitchens and bathrooms. **1967** M. E. Hale *Biol. Lichens* viii. 117 A similar pathway, but with reduction instead of cyclization, could explain the synthesis of fatty acids.

cyclo-. Add: **cyclobra·nchian**, a mollusc of the sub-order Cyclobranchia; **cyclogi·ro, -gyro** [cf. *autogiro], see quots.; **cy·clogram** [-gram], the figure produced by a cyclograph; **cyclograph** (*b*) *Electr.* (see quots.); hence **cyclogra·phic** *a.*; **cy:clomorpho·sis** *Biol.* [ad. G. *cyclomorphose* (R. Lauterborn 1904, in *Verh. Naturh.-Med. Ver., Heidelberg* VII. 614), occas. used in English], the phenomenon in certain organisms, esp. planktonic animals, of undergoing recurrent seasonal changes in form; **cyclople·gia** *Path.* [Gr. πληγή stroke], paralysis of the ciliary muscle; hence **cyclople·gic** *a.*, producing cycloplegia; *sb.* a cycloplegic agent; **cyclorrha·phous** *a. Ent.* [f. mod.L. *Cyclorapha, Cyclorrhapha* (F. Brauer *Monographie der Oestriden* (1863) 34), Gr. ῥαφή seam (ῥάπτειν to sew)], of or pertaining to the Cyclorrhapha, a division of dipterous insects in which the adult emerges from the puparium through a circular seam; distinguished from *orthorrhaphous *a.*; **cyclostro·phic** *a. Meteorol.* [see Strophe], designating the force acting on a wind as a result of the curvature of its path, and also a (hypothetical) wind in which this force exactly balances that arising from the horizontal pressure gradient; **cy·clothem** *Geol.* [Gr. θέμα something laid down (see Theme)], see quot. 1932; **cy·clothyme** *a.* and *sb.* = **cyclothymic*; **cyclothy·mia** *Psychiatry* [Gr. θυμός mind, temper], a condition marked by cyclic alternations of mood from exhilaration to depression with a tendency, when aggravated, to manic-depression; hence **cyclothy·mic** *a.* and *sb.*, (a person) affected with this; (cf. *cycloid *sb.* 3).

1837 *Penny Cycl.* VIII. 248/2 Cyclobranchians. **1839** *Ibid.* XIII. 485/1 With regard to the marine species [of Limacineans], which Cuvier has approximated to these, M. de Blainville observes that they constitute his genus *Peronia* in his order of Cyclobranchians. **1933** *Nat. Advis. Comm., Techn. Rep. 474* (Nomencl. Aeronaut.) 12/2 *Cyclogiro*, a type of rotor plane whose support in the air is normally derived from airfoils mechanically rotated about an axis perpendicular to the plane of symmetry of an aircraft, the angle of attack of the airfoils being always less than the angle at which the airfoils stall. **1931** S. R. Roget *Dict. Electr. Terms* (ed. 2) 77/1 *Cyclogram*, a record obtained from a cyclograph. **1946** *Electronic Engin.* XVIII. 378 The Lissajous figure is perhaps the most important type of cyclogram display. **1931** S. R. Roget *Dict. Electr. Terms* (ed. 2) 77/1 *Cyclograph*, a name sometimes given to an instrument with an optical or electron-jet 'pointer' moving in two dimensions under control respectively of different variables. **1940** *Chambers's Techn. Dict.* 220/1 *Cyclograph*, an instrument in which a beam of light or cathode rays is made to move under the action of two controlling forces at right-angles to each other, thereby producing a closed figure (cyclogram) on the screen. **1933** R. A. W. Watt et al. *Applications Cathode Ray Oscillograph* i. 3 Braun also introduced the cyclographic method of using the tube. **1940** *Chambers's Techn. Dict.* 220/1 *Cyclogyro*, a rotorcraft depending for its lift on power-driven rotors rotating on horizontal axes. **1926** C. Wesenberg-Lund in *K. Dansk. Videnskab. Skr.* (*Natur. & Math.*) 8th Ser. XI. 127 The cyclomorphose [in *Daphnia cucullata*] has been thoroughly studied in many lakes. **1930** *Ibid.* 9th Ser. II. 37 In the seasonal variations or cyclomorphoses Lauterborn saw accommodations to variations in the external medium. *Ibid.* 38 The cyclomorphosis is a process of senility. **1939** *Q. Rev. Biol.* XIV. 137/1 The phenomenon of cyclomorphosis among plankton organisms is particularly well exemplified in fresh-water Cladocera, especially in the group *Daphnia cucullata* and *D. longispina*. **1961** S. C. Kendeigh *Animal Ecol.* vi. 60/1 An interesting phenomenon is cyclomorphosis, a seasonal change in body form that develops in many plankton organisms. **1902** *Jrnl. Amer. Med. Assoc.* XXXVIII. 1136/2 Difficulty in accommodating for hyoscin commenced in fifteen to eighteen minutes, and there was complete cycloplegia in thirty to forty-eight minutes. *Ibid.* 1135/2 (*heading*) Comparative values of cycloplegics... Atropin has been the chief reliance of ophthalmologists as a cycloplegic. *Ibid.* 1137/1 Tests possess sufficient accuracy to enable us to select the best out of the candidates for cycloplegic honors, namely hyoscin hydrobromate. **1956** *Nature* 17 Mar. 523/2 Variable..disturbances of vision, due to cycloplegic and mydriatic actions. **1957** *Encycl. Brit.* XXIII. 213/1 The use of cycloplegics (a kind of eye drops) is a conventional clinical method which attempts to determine the intrinsic relation of lens systems to length of eyeball when the accommodation is relaxed. **1889** *Cent. Dict.*, Cyclorhaphous. **1901** G. H. Verrall *Brit. Flies* VIII. 11 Cyclorrhaphous flies of not at all a leathery texture. **1961** J. E. Collin *Brit. Flies* VI. 1 It is now more usual to..adopt as the two main divisions the Nematocera and Brachycera with the latter subdivided into the Orthorrhaphous- and Cyclorrhaphous-Brachycera. **1916** Cyclostrophic [see *geostrophic *a.*]. **1959** R. E. Huschke *Gloss. Meteorol.* 151 The cyclostrophic wind can be an approximation to the real wind in the atmosphere only near the equator..or in cases of very great wind speed and curvature of the path. **1932** Wanless & Weller in *Bull. Geol. Soc. Amer.* XLIII. 1003 The word 'cyclothem' is..proposed to designate a series of beds deposited during a single sedimentary cycle of the type that prevailed during the Pennsylvanian period. **1957** *Encycl. Brit.* IV. 844/1 Coals are more important than limestones in identifying cyclothems because most cyclothems include only one coal whereas two or more limestones may occur. **1925** W. J. H. Sprott tr. *Kretschmer's Physique & Char.* xii. 208 We call the members of that large constitution-class, from which the schizophrenes are recruited, 'schizothymes', and those that correspond to the circular psychotics are called 'cyclothymes'. **1932** *Brit. Jrnl. Psychol.* Jan. 236 The mean age of the cyclothyme group is greater than that of the schizothyme. **1951** *John o' London's* 9 Nov. 724/3 Dickens was undoubtedly a cyclothyme. **1921** Glueck & Lind tr. *A. Adler's Neurot. Constit.* 187 Dementia præcox, paranoia and cyclothymia. **1929** P. Mairet tr. *A. Adler's Probl. Neurosis* 27 A cyclothymia beginning late in life. **1925** W. J. H. Sprott tr. *Kretschmer's Physique & Char.* xiv. 259 Among cyclothymic temperaments a certain mood-disposition usually goes with a certain psychic tempo. *Ibid.*, Cyclothymic psychomotility is distinguished by the natural quality of reaction and bodily movement, which is now quick, now slow. **1926** W. McDougall *Outl. Abnormal Psychol.* 353 Most of us..are liable to mild alternations of this kind, moods of 'excitement' and of depression... When the liability to such alternations is well marked, the personality is said to be of the cyclothymic type. *Ibid.* 356 Cyclo-thymics, in whom periods of energetic euphoria alternate with despondent impotence. **1965** J. Pollitt *Depression & its Treatment* iv. 59 In cyclothymic subjects, and those who have suffered from manic episodes, a large dose continued too long may produce hypomanic features.

2. *Org. Chem.* Also *ital.* (see quot. 1958). A prefix used in forming the names of some cyclic compounds.

1894 G. M'Gowan tr. *Bernthsen's Text-bk. Org. Chem.* (ed. 2) xv. 323 Their 'official names' are Cyclo-propane, Cyclobutane, etc. **1900** E. F. Smith tr. *von Richter's Org. Chem.* (ed. 3) II. 17 In accordance with the decision of the Geneva Conference, they take the name of the normal hydrocarbons with like carbon content, and add to the same the prefix 'cyclo'—e.g., cycloparaffins. **1904** *Jrnl. Chem. Soc.* LXXXVI. i. 413 It was not found possible to eliminate nitrogen from this compound and so obtain a *cyclo*octane derivative. **1911** *Encycl. Brit.* XXII. 31/1 Cyclo-hexanol, $C_6H_{11}OH$, is produced by the reduction of the corresponding ketone. **1925** A. W. Judge *Automobile Engines* i. 10 A mixture of 20 parts benzole and 80 parts cyclohexane will enable an engine to be run at 200 lb. per sq. in. compression pressure. **1944** L. F. & M. Fieser *Org. Chem.* 49 Cycloparaffins (cycloalkanes) bear a close resemblance to the paraffins. **1952** *Sci. News Let.* 24 Dec. 412/1 A new mold chemical, cycloserine, was..reported promising against tuberculosis. **1955** *Electronic Engin.* XXVII. 513 A special solvent, the main constituent of which is cyclohexanone. **1955** H. Welch et al. in *Antibiotic Med.* (N.Y.) I. 72 Cycloserine is the generic name for a new antibiotic produced by *Streptomyces orchidaceus*. **1957** *Nomencl. Org. Chem.* (I.U.P.A.C.) (1958) 18 The names of saturated monocyclic hydrocarbons (with no side chains) are formed by attaching the prefix 'cyclo' to the name of the acyclic saturated unbranched hydrocarbon with the same number of carbon atoms. **1958** Packer & Vaughan *Mod. Approach to Org. Chem.* ii. 40 They are called *cyclo*-alkanes or *cyclo*-paraffins and are named correspondingly, e.g. *cyclo*-pentane, *cyclo*-hexane. *Ibid.*, In American practice italics are not used for the cyclo, e.g. cyclopentane. **1963** *New Scientist* 9 May 321/2 The photochemical process that Japanese chemists recently developed for converting cyclohexane into caprolactam—which is used in the manufacture of nylon 6—is now in full scale production.

cyclo-cross (səi·klŏkrǫs). [f. Cycle *sb.* 11 + *cross* as in Cross-country *a.*] Cross-country bicycle racing.

1953 *Cycle Sport Ann.* 63 Winter winds..offer a challenge to the enthusiasm of the Cyclo-Cross expert. **1963** *Times* 4 Feb. 20 The national cyclo-cross championship.

cyclodialysis (səiklodəi,æ·lisis). *Surg.* [ad. G. *cyklodialyse* (Heine 1905, in *Deutsche Med. Woch.* 25 May 824/2), f. Gr. κύκλο-ς any circular body, eye + διάλυσις separating (see Dialysis).] An operation for relieving the tension of the eyeball in cases of glaucoma (see quot. 1964).

1908 W. L. Pyle tr. *Meller's Ophthalmic Surg.* xv. 204 Cyclodialysis, though able to diminish the intra-ocular pressure, cannot be called preferable or even equal to iridectomy. **1961** *Lancet* 22 July 166/2 A localised cyclodialysis plus air injection. **1964** S. Duke-Elder *Parsons' Dis. Eye* (ed. 14) xxi. 310 A communication between the anterior chamber and the suprachoroidal space, which also may lower the tension by cutting off some of the blood supply to the ciliary body, comprises the operation of cyclodialysis.

cycloid, *sb.* Add: **3.** *Psychiatry.* A person characterized by a tendency to alternate between exhilaration and depression. Also *attrib.* or as *adj.*

1925 W. J. H. Sprott tr. *Kretschmer's Physique & Char.* xi. 127 In an adverse situation the cycloid is either sorrowful or hot-headed, but he is never in the very least nervous. **1927** Henderson & Gillespie *Text-Bk. Psychiatry* ix. 190 The 'syntonic' or 'cycloid' personality, in which, if mental illness develops, it tends to be of the manic-depressive kind. **1929** F. H. Garrison *Hist. Med.* (ed. 4) 679 Cycloids are cheerful, temperamental, well adjusted to the business world, but liable to sudden alternations of exhilaration and depression. **1943** H. Read *Educ. through Art* iv. 76 The 'depressive cycloid' temperament is certainly our old friend the melancholic humour.

cyclone. Add: **1. e.** Used (freq. *attrib.*) of a machine in which a flow of gas or liquid is used to remove or separate solids, usu. by centrifugal force.

1898 *Daily News* 8 Feb. 3/5 The 'cyclone'—a great grey tube with ramifications to all the machines that saw or chip wood... A forced draught carries the chips through the..tube to the boiler house. **1930** *Engineering* 22 Aug. 221/3 The cyclone filter consisted of a cylindrical vessel fitted with tangential air inlets near the top..and a conical bottom for the reception and discharge of the deposited dust. **1962** *Gloss. Coal Preparation* (*B.S.I.*) 17 *Cyclone classifier*, a device for classification by centrifugal means of fine particles suspended in water. **1967** *Gloss. Materials Handling* (*B.S.I.*) iii. 6 *Cyclone*, a device imparting a rotary motion to the fluid stream thereby causing the entrained particles to be separated by centrifugal force and gravity.

2. cyclone cellar *U.S.*, a cellar intended to give shelter during a cyclone; also *fig.*

1887 E. Custer *Tenting on Plains* (1889) 652 Those women who take refuge..in their cyclone-cellar. **1904** G. H. Lorimer *Old Gorgon Graham* 125 This was one of those holy moments..when an outsider wants to pull his tongue back into its cyclone cellar. **1929** *Monthly Weather Rev.* (U.S. Weather Bur.) LVII. 338/1 People had seen the tornado approaching and had taken to storm caves—the well known 'cyclone cellars' of the West—and basements, where they were safe. **1946** *Reader's Digest* Mar. 135/1 In the winter it was snug and cozy, and in summer-time as cool and nice as our cyclone cellar.

cyclonish (səi·klō͡ᵘniʃ), *a.* Somewhat cyclonic (*fig.*).

1884 *Harper's Bazaar* in *Advance* (Chicago) 1 Jan. (1885), Giving the door a cyclonish bang. **1893** *Nation* (N.Y.) 15 July 32/2 She is altogether of the 'breezy', indeed cyclonish, western type—a good tempered girl with no end of go.

cyclonite (səi·klŏnəit). *Chem.* [f. *cyclo*-trimethylenetri*nitr*amine.] A colourless crystalline substance, $(CH_2\cdot N\cdot NO_2)_3$, used as a high explosive; also called RDX.

1923 *Chem. Abstr.* XVII. 2051 A description is given of the characteristics of..cyclonite..for use as 'booster'. **1939** *Thorpe's Dict. Appl. Chem.* (ed. 4) III. 535/2 Cyclonite..was first described under the name 'Hexogen' in 1899 by G. F. Henning. **1948** *Times* 10 Jan. 4/2 Cyclonite..has approximately the strength of dynamite. **1951** Kirk & Othmer *Encycl. Chem. Technol.* VI. 40 Cyclonite, therefore, may be considered at least the equal if not the superior of any other solid bursting-charge explosive available on a production basis. **1966** Millar & Springall *Sidgwick's Org. Chem. Nitrogen* (ed. 3) xviii. 599 The very powerful military high explosive Cyclonite or RDX..was first prepared by the action of nitric acid (99 per cent) on hexamine.

Cyclopean, -ian, *a.* Add: **1. b.** (See Cyclopia.)

2. (Earlier example.)

1822 M. Wilmot *Jrnl.* 26 Mar. in *More Lett.* (1935) 158 Fondi: remarkable for its gate, its Ciclopean wall, its tower

cyclophosphamide (səiklofǫ·sfăməid). *Pharm.* [f. Cyclo- + Phosph- (in *phosphoric acid*) + Amide.] A fine white crystalline powder used as a cytotoxic drug in the treatment of tumours and some kinds of leukæmia, and usu. given intravenously or orally; 2-[di-(2-chloroethyl)-amino-]-1-oxa-3-aza-2-phosphacyclohexane 2-oxide, $C_7H_{15}Cl_2N_2O_2P$.

1960 *Brit. Med. Jrnl.* 24 Dec. 1837/1 We report our experiences in the treatment of Hodgkin's disease and other tumours of lymphoreticular tissue with a cyclic nitrogen mustard phosphamide ester, cyclophosphamide. **1963** *Times* 16 Feb. 5/4 A drug, cyclophosphamide,..killed some of the cells thought to be responsible for rejecting foreign tissue such as the newly grafted kidney. **1966** *Lancet* 24 Dec. 1382/1 Cyclophosphamide was tried, unsuccessfully, in one patient, but this drug is known to be antagonised by dapsone. **1967** *Martindale's Extra Pharmacopoeia* (ed. 25) 818/2 The most troublesome side-effect of cyclophosphamide therapy is alopecia. **1969** *Nature* 1 Feb. 468/2 Cyclophosphamide shows promise as a method of removing the wool from sheep chemically. **1970** *Sci. Jrnl.* Mar. 63/1 Modern drugs such as methotrexate and cyclophosphamide are extremely valuable in the treatment of many cancers and especially leukaemias.

cyclorama. Add: **2.** *Theatr.* A large backcloth or wall, freq. curved, at the back of a stage, used esp. to represent the sky.

1915 H. K. Moderwell *Theatre To-Day* ii. 52 The 'sky-drops'..have been replaced..by an invention that makes a natural landscape possible. This is the 'Horizont', which we may name for the purpose a cyclorama. **1930** *Observer* 18 May 11/2 The stage will have unique features, including an artificial horizon or cyclorama. **1957** *Oxf. Compan. Theatre* (ed. 2) 173 The cyclorama is, in essence, a curved wall, or section of a dome, built at the back of the stage, and demanding one quality—an absolutely unbroken surface. **1958** W. T. O'Dea *Social Hist. Lighting* vii. 166 In 1902 the first attempts were made to produce coloured sky effects by coloured light only, the 'cyclorama' being installed by Mariano Fortuny at the Scala Opera House in Milan. The cyclorama, which is a smooth, light-coloured surface the full height and width of the stage, can also be used for optical projection of scenery and effects, but it must not be in the path of any illumination other than that deliberately cast upon it.

cy·clostyle, *v. trans.* To print (copies) by cyclostyle. Hence **cy·clostyled** *ppl. a.*

1897 *Westm. Gaz.* 3 July 2/3 Note the gradual rise from the cyclostyled circular to the printed appeal. **1928** *Daily Express* 5 Mar. 2/3 No candidate can be directly boosted by name in any matter printed, typed, or cyclostyled. **1957** *Times* 19 Aug. 2/7 The inquirer's first carefree letter to manufacturers will be answered by a flood of shiny prospectuses and cyclostyled price lists.

cyclotron (səi·klŏtrǫn). [f. Cyclo- + *-tron.] An apparatus for accelerating charged atomic particles by subjecting them repeatedly to a (usu. horizontal) electric field as they revolve in orbits of increasing diameter in a constant (usu. vertical) magnetic field.

1935 E. O. Lawrence et al. in *Physical Rev.* XLVIII. 495/2 An apparatus of the type developed by Lawrence and Livingston was used to produce a beam of high speed deuterons... Since we shall have many occasions in the future to refer to this apparatus, we feel that it should have a name. The term 'magnetic resonance accelerator' is suggested... The word 'cyclotron', of obvious derivation, has come to be used as a sort of laboratory slang for the magnetic device. **1936** *Discovery* June 193/1 Princeton University is..developing its high voltage apparatus by the installation of a cyclotron, which has proved the most satisfactory means of procuring the high-speed ions necessary for the study of nuclear transformations. **1942** *Endeavour* I. 40/1 The cyclotron..consists essentially of a large electromagnet with circular pole pieces, between which is fixed a shallow cylindrical vacuum-tight metal chamber. **1951** *Sci. News* XXII. 58 The cyclotron derives directly from the linear accelerator. The great length and numerous electrodes of the linear accelerator can be avoided if the particles are made to move in a spiral path by a magnetic field. **1966** [see *dee *sb.* 2]. **1969** Auden *City without Walls* 97 The High Priests of telescopes and cyclotrons Keep making pronouncements.

cylinder, *sb.* Add: **2. b.** A cylindrical record for a phonograph. Also *attrib.*

1891 'Mark Twain' *Let.* 28 Feb. (1920) 297 Ask them on what terms they will rent me a phonograph for 3 months and furnish me cylinders enough to carry 75,000 words. **1893** *Harper's Mag.* Jan. 214/2 It's just a phonograph...It don't seem to be exactly in order. Perhaps the cylinder's got dry. **1907** *Pearson's Mag.* Jan. (Advt.), You cannot get the best results from any talking machine without using Columbia Disc or Cylinder Records. **1956** R. Gelatt *Fabulous Phonograph* ii. 17 Edison cylinders could be shaved and used over and over again while gramophone cylinders had to be discarded much more quickly. **1967** *Amer. N. & Q.* Sept. 15/1 Booth's wax cylinder recording of Othello's speech to the Senate.

6. *to function* (or *click, hit, operate*, etc.) *on all* (or *four, six*, etc.) *cylinders*, of an internal-combustion engine: to be working at full power; hence *fig.*, to function properly, to be in good form; so *to miss on all* (or *four*, etc.) *cylinders*, to be working badly, to be in bad condition.

1912 C. Mathewson *Pitching in a Pinch* xii. 269 So the best infielder takes time to fit into the infield of a Big League club and have it hit on all four cylinders again. **1917** 'Contact' *Airman's Outings* 72 The needle on the rev.-counter quivered to the left as the revolutions dropped, and the engine missed on first one, then two cylinders. **1932** Wodehouse *Hot Water* ii. 38 He had tended..to undernourish his spiritual self. He had given it the short end, and it was missing, he knew, on several cylinders. *Ibid.* xv. 245 His smiling face, taken in conjunction with the bottle of wine which he carried, conveyed to Gordon Carlisle the definite picture of a libertine operating on all six cylinders. **1936** —— *Laughing Gas* xxi. 228 The old *preux chevalier* spirit was functioning on all six cylinders. **1951** M. McLuhan *Mech. Bride* 96/1 A sure sign that you are clicking on all cylinders. **1958** *Spectator* 22 Aug. 249/3 Only when he [*sc.* an actor] allows himself to play deliberately on all cylinders does he stage a dazzling display. **1960** [see *back-line, backline].

9. b. Add: *cylinder block* (see 6), *card* (see 5), *gun* (see 5), *head* (see 6), *machine, saw, seal* (see 4 a), *stove.*

1923 H. R. Ricardo *Internal Combustion Engine* II. vi. 142 When the induction system is cast in the cylinder block the whole of its internal surface is rough. **1812** *Deb. Congress* 12 June (1853) 2188 In the year 1762 cylinder cards were first made use of by Mr. Peel. **1892** W. W. Greener *Breech-Loader* 141 An old or true cylinder gun will not..put three pellets into a pigeon thirty yards distant. **1884** 'Mark Twain' *Huck. Finn* xxxii. 332 We blowed out a cylinder-head. **1895** Kipling *Land & Sea Tales* (1923) 205 Cylinder-head blown off. **1860** W. Blackwood *Let.* 28 Mar. in *Geo. Eliot's Lett.* (1954) III. 284 We have got the new cylinder machine working perfectly. **1962** F. T. Day *Introd. Paper* iv. 44 A popular name for the M.G. [machine glazed] or cylinder machine is the 'Yankee'. **1851** C. Cist *Cincinnati* 181 They are..fed to a cylinder saw, which cuts them into staves of the proper thickness and curve. **1887** *Scribner's Mag.* Jan. 80 The earliest printing-press was a seal, and the cylinder-seal may be said to have been an archaic rotary press. **1922** *Guide Babylonian & Assyrian Antiq.* (*Brit. Mus.*) (ed. 3) 82 Cylinder seal..of Syrian type. **1927** Peake & Fleure *Priests & Kings* 64 Some of the Egyptians used mace-heads and cylinder-seals almost exactly similar to those found somewhat later in Mesopotamia. **1898** E. N. Westcott *David Harum* 162 The proximity of wet boots and garments to the big cylinder stove.

cylindered (si·lindəɹd), *ppl. a.* [See -ed[2].] Having a cylinder or cylinders (of a specified number or type).

1899 *Daily News* 14 Sept. 7/5 The engine..is one of Mr. Drummond's latest type of four-wheel-coupled inside cylindered express locomotives. **1908** *Westm. Gaz.* 14 May 4/2 As a hill-climber the Argyll, with its 120 by 140 cylindered engine, has great claims. **1934** A. W. Judge *Automobile & Aircraft Engines* (ed. 3) i. 20 Variation in the mixture strength in the different cylinders in the multi-cylindered engines. **1952** H. Webster *Railway Motive Power* iv. 71 This was an inside cylindered engine of tractive effort 12,808 lb.

cylindra·ceo-. Used in *Zool.* as combining form of Cylindraceous *a.*

1822 J. Parkinson *Outl. Oryctol.* 195 A straight cylindraceo-tubular operculated shell. **1887** W. Phillips *Brit. Discomycetes* 161 Asci cylindraceo-clavate.

cylindrite (sili·ndrəit). *Min.* Also **kyl-.** [ad. G. *kylindrit* (A. Frenzel 1893, in *N. Jahrb. Min.* II. ii. 125), f. Gr. κύλινδρ-ος Cylinder (see quot. 1893) + -ite[1].] A blackish-grey sulphide of lead, antimony, and tin, known only from Bolivia.

1893 *Jrnl. Chem. Soc.* LXIV. ii. 576 At the Santa Cruz mine at Poopó, in Bolivia, a new mineral has been discovered. This the author has named kylindrite from the cylindrical form in which it occurs. **1896** A. H. Chester *Dict. Min.*, Kylindrite. **1929** H. A. Miers *Mineralogy* (ed. 2) ii. vii. 399 Among other sulpho-salts may be mentioned..cylindrite, the representative of a small but curious class of compounds containing tin together with lead, antimony, and sulphur. **1968** Embrey & Phemister tr. *Kostov's Mineralogy* 183 Cylindrite is found as cylindrical forms..and occurs in the Bolivian mines in paragenesis with franckeïte, pyrite, and sphalerite.

cymbal. 1. Add to def.: Also used singly and struck with a drumstick or the like.

1934 E. Little *Mod. Rhythmic Drumming* 17 The cymbal should be struck with that part of the stick at a point about half-way between the fingers and the tip. **1959** Westrup & Harrison *Collins Mus. Encycl.* 174 There are also two ways of using a single cymbal: (a) hitting it with a stick, hard or soft, in the manner of a gong, (b) performing a roll on it with timpani or side-drum sticks. **1961** J. Blades in A. Baines *Mus. Instruments* xiv. 340 In modern works in general, an additional cymbal suspended on a stand..is necessary.

cymbidium (simbi·diəm). *Bot.* [mod.L. (O. Swartz 1799, in *Nova Acta R. Soc. Scient. Upsala* VI. 70), f. Gr. κύμβη cup.] A tropical orchid of the genus so named, with a hollow recess in the lip of the flower.

1815 *Curtis's Bot. Mag.* XLII. 1751 (*heading*) Sword-leaved Cymbidium. **1895** *Daily News* 21 Mar. 2/5 There were three or four large groups that included..cymbideums. **1911** C. A. Harrison *Orchids for Amateurs* xx. 71 Cymbidiums do not like being disturbed more than can be helped. **1930** T. W. Briscoe *Orchids for Amateurs* vii. 114 Cymbidiums are strong growers. **1959** *Times* 2 May 9/3 If one has a greenhouse that can be kept just free of frost, cymbidiums will grow perfectly well. **1962** *Amateur Gardening* 31 Mar. 19/1 A good deal of the colour naturally came from the cymbidiums.

cymblin(g, cymling, see Simlin.

cymoscope (səi·mǫskō͡ᵘp). *Electr.* (*Disused.*) [See Cyma and -scope.] A wave-detecting device used in wireless telegraphy; any device that serves as a detector of electromagnetic waves.

1906 J. A. Fleming *Princ. Electric Wave Telegraphy* vi. 353 It seems desirable to possess a term..which shall connote all the forms of electric wave-detecting device, and the word *cymoscope* has been suggested by the author. **1924** *Harmsworth's Wireless Encycl.* 639/1 One of the earliest forms of cymoscope to be used was the Hertz tuned loop.

cymotrichous (səimǫ·trikəs), *a.* *Anthrop.* [f. Gr. κῦμα wave (see Cyma) + τριχ-, θρίξ hair + -ous.] Having wavy hair. Hence **cymo·trichy**, wavy-hairedness.

1909 A. C. Haddon *Races of Man* 3 These three varieties [of hair] are now termed leiotrichous, cymotrichous, and ulotrichous. **1924** *Ibid.* (ed. 2) 5 *Cymotrichy*, or smooth, wavy and curly hair. *Ibid.* 6 Some cymotrichous peoples have very hairy bodies. **1936** *Antiquity* X. 244 Three sorts of hair—woolly, wavy and straight—are carefully distinguished... They would indicate that ulotrichy, cymotrichy, and leiotrichy were simultaneously present.

cynghanedd (kəŋha·neð). *Prosody.* [Welsh.] An intricate system of alliteration and rhyme in Welsh poetry.

1849 T. Stephens *Lit. of Kymry* iv. 490 The rhythmical consonancy, termed *Cynghanedd*, was introduced at that time [*sc.* late 14th century], and has ever since formed an essential feature in Kymric poetry. [**1878** G. M. Hopkins *Let.* 5 Oct. (1935) 15 Certain chimes suggested by the Welsh poetry I had been reading (what they call *cynghanedd*).] **1886** —— *Let.* 6 Oct. (1938) 222 His employment of the Welsh *cynghanedd* or chime I do not look on as quite successful. **1962** *Times* 31 Jan. (Surv. Wales) p. x/5 *Cynghanedd*, that complicated system of consonantal correspondencies and internal rhyme.

cyno-. Add: **cynolo·gical** *a.*, of or pertaining to cynology; **cyno·logist,** one who is versed in cynology.

1926 A. Huxley *Essays New & Old* 201 No cynological philosopher has arisen to denounce the abject degeneracy of the Irish terrier. **1962** *Times Lit. Suppl.* 7 Sept. 662/2 They set up a 'Cynological Institute' in..Prague [in 1911]. **1948** C. L. B. Hubbard *Dogs in Brit.* iii. 17 A foundation upon which cynologists might work out the origin of breeds.

cypræa, cyprea. (Examples.)

1822 J. Parkinson *Outl. Oryctol.* 153 *Cypræa*, an ovate and vaulted univalve. **1834** C. M. Yonge *Let.* 4 July in C. Coleridge *C. M. Y.* (1903) iv. 123, I send..Jane a *cyprea*. **1913** B. B. Woodward *Life of Mollusca* vi. 97 The young Cowry shell (*Cypræa*) has a thin, sharp lip.

cypress[1]. Add: **4. cypress pine** *Austral.*, a tree of the genus *Callitris.*

1847 Leichhardt *Jrnl.* i. 13 It was covered with cypress-pine, and an Acacia. **1885** *Spons' Mech. Own Bk.* 131 Cypress pine (*Callitris columellaris*) is a plentiful tree in Queensland. **1936** F. Clune *Roaming round Darling* ix. 76 Box-trees and cypress-pines in plenty. **1967** A. M. Blombery *Guide Native Austral. Plants* iii. 108 Cupressaceae. A family..including the well-known native *Callitris* (Cypress Pine).

Cyprianic (sipriæ·nik), *a.* [ad. mod.L. *Cyprianicus*, f. *Cyprianus*.] Of, pertaining to, or characteristic of St. Cyprian (Thascius Cæcilius Cyprianus), bishop of Carthage, martyred A.D. 258.

1695 J. Sage (*title*) The Principles of the Cyprianic Age. **1696** G. Rule (*title*) The Cyprianick Bishop examined and found not to be diocesan, nor to have superior power to a parish minister, or presbyterian moderator. *a* **1861** W. Cunningham *Hist. Theology* Wks. (1863) I. 164 The Cyprianic bishop was very different from the modern one. **1916** J. R. Harris *Testimonies* i. viii. 77 We have an almost contemporary witness for the Cyprianic text. *Ibid.* xiv. 127 Justin starts with the first chapter of the Cyprianic Christology. **1920** *Christian World* 19 Aug. 8/4 We need for the recovery of Christian unity a man of Cyprianic grasp, though not of Cyprian's opinions. **1945** G. Dix *Shape of Liturgy* v. 116 The 'Cyprianic' doctrine of the sacrifice came to prevail in the West.

cyprid (səi·prid). *Zool.* An ostracod crustacean of the family Cyprididæ (see Cypris).

1913 H. M. Cadell *Story of Forth* i. 8 Other shale seams are almost entirely composed of minute entomostraca or cyprids.

Cypriot (si·priǫt), **Cypriote** (si·priō͡ᵘt), *a.* and *sb.* [ad. Gr. Κυπριώτης, f. Κύπρος Cyprus. Cf. Cyprian.] **A.** *adj.* Belonging to Cyprus. **B.** *sb.* A native or inhabitant of Cyprus; the Greek dialect (ancient or modern) of Cyprus.

1599 Hakluyt *Voy.* II. i. 309, I was credibly informed by a Cipriot a marchant of good wealth. **1639** Fuller *Hist. Holy Warre* v. xxix. 278 Take their names as I find them in the Catalogue of Stephen a Cypriot. **1750** *Universal Mag.* July 4/1 For a while he triumphed over the Tyrians, Sidonians, and Cypriots. **1797** R. Heron *Collect. Voy. & Trav.* 198 St. Spiridion, the Cypriot. **1837** *Penny Cycl.* VIII. 261/2 Almost every house has a garden, of which the Cypriotes are very fond. **1878** *Murray's Handbk. Trav. Turkey in Asia* (ed. 4) 179/1 The Inscriptions found..in Cyprus are in three languages: 1. Cypriote. *Ibid.* 182/1 Androcles..was present in the Cypriote fleet which supported Alexander. **1920** J. A. Robertson *Hidden Rom. N.T.* i. 20 The wealthy Cypriote Mnason, who entertained Paul. **1920** G. E. Buckle *Disraeli* VI. 300 The occupation of Cyprus ensured the fair treatment of the Cypriot Greeks. **1920** *Q. Rev.* July 38 This Cypriote School, with its grand Græco-Oriental seriousness, had many ramifications, one of which, the Palyrene..lasted down to the third century A.D. **1927** *Times* (weekly ed.) 25 Aug. 208/3 The Cypriots in Egypt. **1955** *Times* 6 May 4/3 The House was entitled to know what progress had been made in the discussions with Cypriots about their constitutional future.

cypripedium (sipripī·diəm). *Bot.* [mod.L. (Linnæus *Systema Naturæ* (1735)), f. Gr. Κύπρις Aphrodite + πέδιλον slipper, in reference to the inflated pouch formed by the labellum; cf. Cypripedin.] A member of the large genus of orchids so named. Cf. Lady's slipper 1.

1775 T. Blaikie *Diary Scotch Gardener* 4 June (1931) 37 In those woods grow great plenty of the *Cepre*p*edium*. **1813** *Curtis's Bot. Mag.* XXXVIII. 1569 In the Cypripediums which we have before figured..the two lower external petals are united together. **1873** *Young Englishwoman* Aug. 395/2 *Cypripedium*, or Slipper plant. **1890** W. Watson *Orchids* xxviii. 155 When Cypripediums show signs of bad health, they should be at once shaken free of soil. **1902** *Westm. Gaz.* 31 Dec. 10/1 Cypripedium orchids. **1904** R. J. Farrer *Garden Asia* 250 On the shelving banks of grass in open places [grows] the cypripedium. **1938** T. Hay *Plants for*

Connoisseur 47 The hardy terrestrial Cypripediums have long been highly esteemed for the border or rock garden. **1962** *Amateur Gardening* 14 Oct. 6/3 There are a number of cypripediums suitable for cool house conditions.

Cyrenaican (saɪrīnēⁱ·ikăn), *a.* [f. as CYRENAIC *a.* + -AN.] Of or pertaining to the region of Cyrenaica in north Africa, or its people. Also as *sb.*

1607 TOPSELL *Four-f. Beasts* 548 There are more kinds of mice in the Cyrenaican region. **1942** 'M. HOME' *House of Shade* v. 80 Suppose he had taken on the character of a Cyrenaican or a Tripolitanian from some southern oasis. **1963** I. W. ZARTMAN *Govt. & Pol. N. Afr.* (1964) v. 97 There is also a Cyrenaican Workers' Federation, but there is no nation-wide union. **1969** J. WRIGHT *Libya* xii. 133 The Cyrenaicans lost heavily during tough fighting in the spring of 1914.

Cyrillic, *a.* (Earlier examples.)

1842 *Penny Cycl.* XXII. 104/2 The use of the Cyrillic letters..remained in full vigour among those [Slavonians] who belonged to the Eastern Church. *Ibid.* 127/1 The Servians..make use of the Cyrillic alphabet. **1856** [see ROUMAN *sb.* 2].

cyste·ctomy. *Surg.* [f. CYST + *-ECTOMY.] Surgical removal of either the gall-bladder or all or part of the urinary bladder, or of a cyst.

1891 C. W. M. MOULLIN *Surgery* Index, Cystectomy. **1896** J. G. SMITH *Abdom. Surg.* (ed. 5) II. ix. 961 Weir of New York records three cases of partial cystectomy. In one he removed a triangular portion of the bladder measuring 2½ inches on the sides. **1962** *Lancet* 19 May 1048/2 Total cystectomy was necessary. **1967** S. BOYARSKY *Neurogenic Bladder* xxvi. 203/2 Three patients continued to have chronic cystitis with copious drainage of purulent material from the urethra. Cystectomy was performed in all 3 after attempts at local therapy with anti-bacterial irrigations.

cysteine (si·stiˌin, si·stiˌəin). *Biochem.* Also **-ein.** [ad. G. *cystein* (E. Baumann 1882, in *Zeitschr. f. physiol. Chem.* VIII. 302), f. CYST(INE + *-EINE.] An amino-acid, $HS{\cdot}CH_2{\cdot}CH(NH_2){\cdot}COOH$, which is a reduction product of cystine and a constituent of glutathione and many proteins.

1884 *Jrnl. Chem. Soc.* XLVI. A. 1382 Cystine and Cysteine...The hydrochloride of the new base cysteine is obtained by evaporating the filtrate. **1923** *Nature* 24 Feb. 274/1 The reduction of methylene blue by the sulphydryl compounds, reduced glutathione, cystein, and thioglycollic acid, is an autocatalytic reaction. **1961** *Lancet* 8 July 87/2 Two cysteine molecules readily give up their hydrogen atoms..to reunite as cystine. **1970** R. W. MCGILVERY *Biochem.* xvii. 382 Cystine enters metabolism as cysteine.

cysticotomy (sistikǫ·tŏmi). *Surg.* [f. CYSTIC (in *cystic duct*) + -TOMY.] Incision into the cystic duct.

1900 in DORLAND *Med. Dict.* **1901** W. W. SEYMOUR tr. *Kehr's Gallstone Dis.* II. 135 After 190 conservative gall-bladder operations (among them 37 cysticotomies..) I have lost..only 3 patients. **1908** *Practitioner* Dec. 826 A stone was extracted after the incision of the cystic duct (Cysticotomy). **1938** M. THOREK *Mod. Surg. Technic* III. xxxv. 1606 Cysticotomy. This operation consists of opening the cystic duct followed by immediate repair. The operation is only of historical interest.

cystinuria. (Later example.)

1963 *Lancet* 19 Jan. 127/1 In patients with cystinuria, lysine and arginine are poorly absorbed from the jejunum and ileum.

cystitic (sisti·tik), *a. Path.* [f. CYSTIT(IS + -IC.] Affected with cystitis.

1910 *Practitioner* July 44 Distended cystitic bladder gave way.

cysto-. Add: **cysto·scopy** [Gr. σκοπιά look-out, watch], examination of the bladder with a cystoscope; **cysto·stomy** [Gr. στόμα mouth], the formation of an opening into the bladder by incision.

1908 *Practitioner* Sept. 434 Cystoscopy is of great use in determining the character of the fluid issuing from each ureter. **1909** *Ibid.* Nov. 664 Cystoscopy and ureteral catheterization are necessary. **1938** *Nature* 24 Sept. 555/1 The radiological methods include..cystoscopy. **1910** *Practitioner* Mar. 378 As a general rule the operation of cholecystectomy shows a higher mortality than the cholecystostomy, but in Rimann's statistics the reverse is the case (cystectomy 3·4 per cent. and cystostomy 14·3 per cent.).

cytase (sai·tēⁱs,-z). *Biochem.* [f. CYT(O- + *-ASE.] Any of various enzymes found in some plant seeds which hydrolyse the hemicellulose constituents of cell walls and were formerly thought to dissolve the whole cell wall. Hence **cytasic** (səitēⁱ·zik) *a.*

1895 W. JAGO *Bread-making* 123 The enzyme, which thus dissolves the parenchymatous cell-walls of the endosperm, has received the name Cytase. Cytase is secreted by the embryo during germination. **1899** J. R. GREEN *Soluble Ferments* vii. 88 The existence of cytase has not been known for many years, our acquaintance with it dating back only to 1886, when it was discovered by De Bary. *Ibid.* xxi. 371 *Bacillus mesentericus vulgatus* has been shown to be possessed of diastasic, inverting, cytasic, and peptonising power. **1938** E. C. MILLER *Plant Physiol.* (ed. 2) xi. 802 The cytases are destroyed at a temperature of 60 °C. or above. **1956** *New Biol.* XXI. 12 The secretion of enzymes (cytases) attacking the cell walls of the endosperm. **1962** HOWARTH & WARNE *Lowson's Textbk. Bot.* (ed. 13) ii. 32 When the seeds, e.g. of lupin and date, germinate, an enzyme, cytase, hydrolyses the hemicellulose, forming sugars.

cytaster (səi·tæstəɹ). *Biol.* [a. G. *cytaster* (W. Flemming *Zellsubstanz* (1882) 379), f. as prec. + *ASTER 4.] **a.** = *ASTER 4. **b.** An aster-like structure which may be present in or induced in the cell when the nucleus is not dividing.

1892 A. A. CROZIER *Dict. Bot. Terms* 50/2 Cytaster, a series of achromatic rays extending from each pole of the nucleus in karyokinesis into the cytoplasm. **1901** E. B. WILSON in *Arch. f. Entwickelungsmechanik* XII. 541 In these eggs the first stage is..followed by the appearance of a variable number of vague clear spots in the cytoplasm which..become surrounded with radiating lines of granules and finally assume the form of asters. These, which are obviously identical with Morgan's 'artificial astrospheres', may be called cytasters, in contradistinction to the nuclear asters that are connected with the nucleus. **1925** —— *Cell* (ed. 3) 1129 *Cytaster*...An aster not associated with chromosomes; commonly employed as equivalent to 'accessory aster' or supernumerary aster. **1952** A. HUGHES *Mitotic Cycle* iv. 108 Cytasters induced in activated *Echinarachnius* eggs were examined. **1960** L. PICKEN *Organization of Cells* vii. 254 Any part of the cytoplasm can participate in the formation of scattered cytasters and of the normal amphiaster.

cytidine (səi·tidīn). *Biochem.* [ad. G. *cytidin* (P. A. Levene and W. A. Jacobs 1910, in *Ber. d. Deut. Chem. Ges.* 3152), f. as prec. + *-IDINE.] A nucleoside, $C_9H_{13}N_3O_5$, composed of cytosine and ribose and obtained by the hydrolysis of ribonucleic acid.

1911 *Jrnl. Chem. Soc.* C. I. 96 On partial hydrolysis of nucleic acid with ammonia, cytidine, $C_9H_{13}O_5N_3$, is obtained. **1958** *Oxf. Univ. Gaz.* 23 Apr. 894 An investigation into the metabolism of cytidine nucleotides, with special reference to cytidine diphosphate polyols.

cyto-. Add: **cytoche·mistry,** the chemistry of cells; *spec.* a branch of biochemistry using microscopical techniques for this study; hence **cytoche·mical** *a.*, pertaining to cytochemistry; **cytoche·mist,** one who studies cytochemistry; **cy:todiagno·sis,** diagnosis by examining the cell-contents of effusions into the serous cavities of the body (Dorland 1903); **cyto·gamy,** (*a*) the fusion of cells; conjugation, syngamy; (*b*) a kind of reproduction that sometimes occurs in *Paramecium* and some other ciliated Protozoa, in which two organisms undergo autogamy while in contact with each other but without any exchange of nuclear material; hence **cyto·gamous** *a.*, undergoing cytogamy; **cyto·gamont,** an organism in the process of cytogamy; **cytolysin** (səitǫ·lisin, səitolai·sin) [see *LYSIN], a substance that causes cytolysis; **cyto·lysis,** the dissolution of cells; hence **cytoly·tic** *a.*; **cytomorpho·sis,** the series of morphological changes undergone by cells during their life; **cytopa·thic, cytopathoge·nic** *adjs.*, of, pertaining to, or producing damage to cells; **cytopatho·logist,** one who studies cytopathology; **cytopatho·logy,** the pathology of cells; **cytoplasmic** *a.* (examples); **cytoto·xic** *a.*, toxic to cells; of or pertaining to a cytotoxin; **cytoto·xin,** any substance having a toxic effect on cells; **cyto·trophy** (see quot.); **cytotro·pism** = **cytotrophy*; **cy·tozyme** (see quots.).

1940 CASPERSSON & SCHULTZ in *Proc. Nat. Acad. Sci.* XXVI. 507 At the centre of the discussion was the nucleolus, which in its staining properties resembled some of the cytoplasmic components, and in some cases appeared..to be extruded into the cytoplasm. With the recent advances in cytochemical technique, these problems can be studied more critically. **1946** *Nature* 21 Dec. 917/1 (*heading*) Establishment of Cytochemical Techniques...It might be feared that his remarks will leave the cytochemist with a gloomy feeling. [**1899** A. GRAF in *Nova Acta Acad. Leopoldino-Carolinæ* LXXII. 280 Cytochemie.] **1905** GOULD *Med. Dict.* Suppl. 204/1 *Cytochemistry*, the chemistry of living cells. **1960** *Times* 26 May 3/2 Honours graduate required for the cyto-chemistry section of the Division of Pathology. **1962** *Sci. Survey* XI. 185 Another way of exploring the function of structures within cells is to establish their nature by combining electron microscopy with cell chemistry (cytochemistry). **1908** *Practitioner* Oct. 621 The method of cytodiagnosis..usually gives satisfactory results, especially in the case of pleural exudates. **1965** tr. *Smolka & Soost's Outl. & Atlas Gynaecol. Cytodiagnosis* 3 The development of fundamental cytology and cytodiagnosis in other medical fields has been furthered by the work of the American Society of Cytology. **1939** R. WICHTERMAN in *Nature* 15 July 123/1 Cytogamous paramecia, which are considerably smaller than typical vegetative ones, are very insecurely attached to each other at their anterior ends at the beginning of the process...The micro-nucleus of each cytogamont leaves its place near the macro-nucleus and then gradually increases in size. **1900** B. D. JACKSON *Gloss. Bot. Terms* 71/1 *Cytogamy*, the union of cells. **1906** M. HARTOG in *Cambr. Nat. Hist.* I. i. 33 This process [*sc.* syngamy] is called also 'conjugation' or 'cytogamy'. **1939** R. WICHTERMAN in *Nature* 15 July 123/1, I am therefore proposing a new term, *cytogamy*, for this phenomenon as distinct from autogamy in single individuals and true conjugation involving a nuclear transfer in joined pairs. **1961** MACKINNON & HAWES *Introd. Study Protozoa* iv. 292 The life-cycle of *Paramecium* is complicated. It consists of periods of growth interrupted by binary fission, which is its only means of multiplication, and periodic recourse to the curious sexual processes of conjugation, autogamy, and cytogamy. **1905** *Jrnl. Path. & Bacteriol.* X. 111 The cytolysins contained in [snake] venom. **1937** *Times Lit. Suppl.* 6 Mar. 173/2 Prominent among the mobile agents for defence found in the blood stream are the cytolysins. **1907** *Brit. Med. Jrnl.* 20 Apr. 923/1 There seems to be some toxic condition produced which overcomes natural resistance, resulting in cytolysis. **1946** *Nature* 21 Dec. 917/2 It is therefore necessary to kill the cell fragments, to remove diffusion difficulties by thorough cytolysis. **1904** *Keene's Bath Journal* 28 May 5/5 The 'cytolytic' milk of a cow has been tried. **1910** *Practitioner* Feb. 199 The virulence of the cocci was evidenced..by the cytolytic findings. **1908** C. S. MINOT (*title*) The problem of age, growth, and death; a study of cytomorphosis. **1968** BLOOM & FAWCETT *Textbk. Histol.* (ed. 9) ix. 219/1 The cartilage cells in a center of ossification undergo a regular sequence of changes referred to as the cytomorphosis of the cartilage cells. **1961** *Lancet* 29 July 248/2 A..degeneration of the embryonic cells, a cytopathic effect. **1956** *New Gould Med. Dict.* (ed. 2) 313/2 Cytopathogenic. **1959** *Brit. Jrnl. Exper. Path.* XL. 61 Experiments..to isolate cytopathogenic agents from sarcoidotic lesions. **1962** *Lancet* 26 May 1109/1 Cytopathogenic effects in tissue-cultures of sarcoid skin which are reminiscent of changes that might be associated with a virus. **1962** *Times* 26 June 3/2 A newly-created post of Cyto-pathologist. **1936** STEDMAN *Med. Dict.* (ed. 13) 281/2 *Cytopathology*, cellular pathology; morbid changes occurring in cells. **1957** *Encycl. Brit.* XV. 204/2 A major development of cytopathology is the study of 'new growths', among which cancer takes a leading place. **1889** *Cent. Dict.*, Cytoplasmic. **1920** L. DONCASTER *Study of Cytology* vii. 91 (*heading*) The cytoplasmic structures in spermatogenesis and oogenesis. **1952** G. H. BOURNE et al. *Cytology & Cell Physiol.* (ed. 2) i. 51 Phase-contrast microscopy reveals the presence of numerous cytoplasmic inclusions in the living cell. **1968** H. HARRIS *Nucleus & Cytoplasm* i. 8 The expression of the genetic information is effected by means of cytoplasmic regulatory mechanisms. **1907** *Practitioner* Aug. 191 The result of cytotoxic activity. **1943** Cytotoxic [see *ANTIRETICULAR *a.*]. **1961** *New Scientist* 19 Oct. 186/1 The so-called cytotoxic drugs used in the treatment of cancer. **1969** *Ibid.* 3 Apr. 30/3 There also seems to be a direct cytotoxic effect on the malignant cells. **1902** *Science* 2 May 697/2 In that they are destructive for the specific cells through which they have been produced, they are termed 'cytotoxins'. **1915** *Practitioner* Jan. 171 (*title*) Pituitary insufficiency and a pituitary antiserum or cytotoxin. **1964** M. HARRIS *Cell Culture & Somatic Variation* viii. 462 The possibility that cytotoxins and hemagglutins may be actually the same isoantibodies, demonstrated merely by two different techniques. **1901** G. N. CALKINS *Protozoa* 217 The phenomena of cytotrophy, or the mutual attraction of two or more cells. **1909** *Cent. Dict.* Suppl., Cytotropism. **1938** J. R. CARPENTER *Ecol. Gloss.* 76 *Cytotropism*, the coming together of cells of a frog's eggs in stage of early cleavage when artificially broken apart. **1927** HALDANE & HUXLEY *Anim. Biol.* viii. 166 A waxy substance called cytozyme produced by the breaking-up of cells. **1958** HARTMANN & GUENTHER tr. *Morawitz's Chem. Blood Coagulation* iv. 68 Fuld called the active 'zymoplastic agent' of the cells 'cytozyme', while I have referred to it as 'thrombokinase'.

cytochrome (sai·toˌkrōᵘm). [f. CYTO- + Gr. χρῶμα colour.] **1.** (See quot.)

1900 DORLAND *Med. Dict.*, *Cytochrome*, a nerve-cell having an ill-developed cell-body, in which the stained nucleus appears to be completely surrounded, and does not exceed in size the nucleus of a neuroglia-cell or a leukocyte.

2. Any of several closely related compounds, present in the cells of most aerobic organisms, which play an important part in cell respiration and consist of an iron-containing porphyrin attached to a protein; together they constitute the *cytochrome system*.

1925 D. KEILIN in *Proc. R. Soc.* B. XCVIII. 314 There is ample evidence that this pigment is not a simple compound, but a complex formed of three distinct haemochromogen compounds, the nature of which is not yet completely elucidated. I propose therefore to describe it under the name of *Cytochrome*, signifying merely 'cellular pigment'. **1948** *New Biol.* V. 126 Oxygen of the atmosphere enters the metabolism of the cell by combining with cytochrome, which then hands it on to other substances. **1953** FRUTON & SIMMONDS *Gen. Biochem.* xiv. 339 Cytochrome a_1 and a_2 were observed spectroscopically in certain bacteria that lack cytochrome a. **1957** *Endeavour* Oct. 196/1 [Lundegårdh's] suggestion.. was that the cytochrome system, which carries electrons from the substrates of the respiration system to molecular oxygen at the surface of the cell, might function as a carrier of anions towards the centre of the cell. **1965** BELL & COOMBE tr. *Strasburger's Textbk. Bot.* (new ed.) II. i. 282 An electron is transferred from the hydrogen to the cytochrome *c*, making its trivalent iron again divalent. **1969** *Times* 29 Apr. 13/1 Human cytochrome C is identical to the equivalent molecule in chimpanzees in all of its 104 units. **1970** AMBROSE & EASTY *Cell Biol.* vii. 228 Chlorophyll, haemoglobin, and the cytochromes all contain a common cyclic structure, called a porphyrin. **1970** *New Scientist* 16 July 119/1 Cytochrome c is a key molecule in the final stages of the 'burning' of foodstuffs for the provision of energy in living cells.

cytogenetics (səitodʒĕne·tiks), *sb. pl. Biol.* [f. CYTO- + *genetics* (s.v. *GENETIC *sb.*).] The study of cytology and genetics in relation to each other; *esp.* the study of the behaviour and properties of chromosomes as the constituents of cells that determine the hereditary properties of an organism. So **cytogene·tic**

(*b*) (cf. Dict., s.v. CYTO-), **cytogene·tical** *adjs.*, of or pertaining to cytogenetics; **cytogene·tically** *adv.*; **cytogene·ticist,** one who studies or pursues cytogenetics.

1931 E. B. BABCOCK in *Amer. Nat.* LXV. 5 (*heading*) Cyto-genetics and the species concept. *Ibid.* 7 Many such forms have been described as species and their hybrid nature has been discovered later by cytogenetic investigation. **1946** *Nature* 5 Oct. 461/2 The cell types of explanted tissues are cytogenetically fixed: they breed true to histological type. **1951** M. J. D. WHITE in L. C. Dunn *Genetics in 20th Cent.* xvi. 362 The cytogenetical investigations of the future. **1957** *New Biol.* XXIV. 31 To the cyto-geneticist the foraminifera present a range of material..rivalling the ciliates in diversity. **1960** *Times* 1 July 2/4 Cytogeneticist required for work in a clinical Cytogenetics group. **1970** *Watsonia* VIII. 1 Papers on the cytogenetics and experimental taxonomy of British plants. **1970** *Nature* 20 June 1177/2 Twenty infants..in whom the diagnosis of Down's syndrome had been confirmed cytogenetically were studied. *Ibid.* 19 Sept. 1271/1 During the past decade extensive cytogenetical studies have been carried out on man's normal chromosomes and chromosome variants.

cytokinin (səitokəi·nin). *Biochem.* [f. CYTO- + *KININ.] Any of numerous compounds which are present as growth regulators in higher plants and which promote cell division, inhibit ageing, and act with auxins to control the growth and development of the plant.

1965 E. MAEDA in *Plant & Cell Physiol.* VI. 653 (*heading*) Inhibition of lamina inclination by cytokinin in excised rice leaves. **1966** *Science* CLII. 726/3 Cytokinins occur in seeds, and they have a role in early stages of embryo growth and germination. **1968** *New Scientist* 5 Sept. 503/1 The cytokinins..are necessary for leaves to remain green and healthy. **1970** STREET & ÖPIK *Physiol. Flowering Plants* xi. 232 The sepals (and sometimes also the petals) synthesize auxin, gibberellin and cytokinin and export these hormones to the developing fruit.

cytology (səitǫ·lŏdʒi). *Biol.* [f. CYTO- + -LOGY.] The study of the cells of organisms; *exfoliative cytology*, the examination of cells that have been shed by an internal or external surface of the body as a means of detecting and identifying any tumours that may be present.

1889 [in Dict., s.v. CYTO-]. **1911** *Encycl. Brit.* XXI. 765/1 The elementary unit of plant structure..is the cell... Upon our knowledge of its minute structure or cytology, combined with a study of its physiological activities, depends the ultimate solution of all the important problems of nutrition,..variation, sex and reproduction. **1928** E. B. WILSON *Cell* (ed. 3) 17 The present work has been written by a student of embryology and cytology, with especial reference to the cell considered as the physical basis of heredity and development. **1949** *Ann. Internal Med.* XXXI. 661 (*heading*) A survey of the actualities and potentialities of exfoliative cytology in cancer diagnosis. **1952** C. P. BLACKER *Eugenics* x. 234 The second province of science (cytology) on which genetics is founded is concerned with the internal structure of the cell, especially of its nucleus. **1970** PASSMORE & ROBSON *Compan. Med. Stud.* II. xxviii. 9/1 While exfoliative cytology is a useful screening technique, a definitive diagnosis should be confirmed by more reliable means, such as biopsy, whenever possible.

Hence **cytolo·gic** (chiefly *U.S.*), **-lo·gical** *adjs.*, of or pertaining to cytology; also, in or on cells; **cytolo·gically** *adv.*, from the point of view of cytology; **cyto·logist,** one who studies cells.

1895 *Ann. Bot.* IX. 505 Boveri and with him many other cytologists believe that they [*sc.* the centrosomes] do actively direct and control the process of nuclear division. **1896** E. B. WILSON *Cell* ii. 70 A consideration of the forces at work in mitotic division..leads us into one of the most debatable fields of cytological inquiry. **1908** *Practitioner* Apr. 525 The infected region..should be bacteriologically and cytologically examined. **1922** R. C. PUNNETT *Mendelism* (ed. 6) 116 The cytologists have not provided us with any evidence of sex-chromosomes in plants. **1942** *Jrnl. Technical Methods* XXII. 75 Cytologic studies with the electron microscope. **1952** G. H. BOURNE *Cytol. & Cell Physiol.* (ed. 2) ix. 406 Apart from their capacity to induce malignancy, these agents have one cytological action in common. They are all capable of bringing about aberrations of mitosis. **1952** *Proc. Soc. Exper. Biol. & Med.* LXXIX. 252 (*heading*) Cytologic changes in rat adenohypophysis following administration of adrenocorticotrophin or cortisone. **1970** *Nature* 12 Sept. 1109/2 They are characterized cytologically by the presence of nucleoli of unequal size in most cells.

cytosine (səi·tosəin, -īn). *Biochem.* [ad. G. *cytosin* (A. Kossel and A. Neumann 1894, in *Ber. d. Deut. Chem. Ges.* XXVII. II. 2219), f. CYT(O- + -OS(E² + -INE⁵.] A crystalline pyrimidine base, $C_4H_5N_3O$, which is one of the constituents of nucleic acids and occurs in double-stranded DNA paired with the purine base guanine.

1894 *Jrnl. Chem. Soc.* LXVI. I. 631 Hydrolysis of nucleic acids...Cytosine, $C_{21}H_{30}N_{16}O_4 + 5H_2O$, is precipitated. **1954** *New Biol.* XVI. 15 In desoxyribose nucleic acid the purine is either adenine or guanine, the pyrimidine either cytosine or thymine. **1955** *Sci. Amer.* Oct. 70/3 DNA is a long-chain molecule constructed from four comparatively simple chemical units: adenine, thymine, guanine and cytosine. **1969** *Times* 8 Dec. 12/2 One possibility is that the cytosine arabinoside..destroys the normal cells in bone marrow.

cytotaxonomy (səi:totæksǫ·nŏmi). *Biol.* [f. *CYTO(LOGY + TAXONOMY.] **a.** Taxonomic classification of organisms using evidence from cytological investigation, esp. studies of chromosomes. **b.** The description of a particular organism or group of organisms, using this method. Hence **cy:totaxono·mic** *a.*, **cy:totaxo·nomist.**

1930 *Biol. Abstr.* IV. Index, Cytotaxonomy. **1937** *Amer. Jrnl. Bot.* XXIV. 126/1 The cyto-taxonomy of section Telephium of *Sedum* is discussed in the present paper. **1942** *K. Fysiogr. Sällsk. Lund Forhandl.* XII. 58 (*title*) Cytotaxonomic studies on boreal plants. **1954** A. LÖVE in *Rapp. 8me Congr. Int. Bot.* IX/X. 61 Cytotaxonomists..are able to deduce considerable understanding of the relationship of genera and even families from the study of chromosomes themselves. *Ibid.* 66 Cytotaxonomy is a very young branch of science, and its methods are still in a state of rapid development. **1964** DAVIS & HEYWOOD *Princ. Angiosperm Taxon.* vi. 193 Cytotaxonomy is the discipline which seeks to study variation and explain variational discontinuities and relationships in terms of cytology. **1971** *Watsonia* VIII. 286 A cytotaxonomic study of the collective species *Lamiastrum galeobdolon* in England and Wales showed two subspecies with different ploidy levels.

czar, tsar. Add: **b.** *transf.* A person having great authority or absolute power; a tyrant, 'boss'. orig. *U.S.*

1866 in Sperber & Trittschuh *Amer. Political Terms* (1962) 111/1 There wuz an immense crowd, but the Czar uv all the Amerikas didn't get orf his speech here. **1893** *McClure's Mag.* I. 375 He was being held up as 'The Czar'—a man whose iron heels were crushing out American popular government. **1899** *Daily News* 20 Apr. 5/5 It is unlikely that any successor will be found able to dominate..as 'Czar' Reed could do. **1931** E. LINKLATER *Juan in America* IV. vii. 338 Red-eye was a czar at whose anger the household grew faint. **1959** *Listener* 5 Nov. 784/1 The Czar—as we say—or President of the Motion Picture Producers' Association. **1970** *Guardian* 18 Apr. 10/6 Many [American] Presidents..establish a staff 'Czar' to cut down on 'unnecessary' memos and contacts.

czardas: see *CSARDAS.

Czech, *sb.* and *a.* Now usu. pronounced (tʃek). Add to def.: Also, the language of this people. Also = *CZECHOSLOVAKIAN *sb.* and *a.* (Later examples.)

1879 *Encycl. Brit.* VIII. 701/2 Czech, or Tsekh, is the national language of Bohemia, and is also largely spoken in Moravia and north-western Hungary. **1938** *Times* 24 May 12/5 My figure of 14,000 Czechs in Germany, which includes not only Czech-speakers but also bilinguals, does not refer to the small Czech-speaking proportion of the 186,000 Czechoslovak nationals resident in Germany. *Ibid.* 4 Oct. 7/1 Guarantees to the Czechs...His Majesty's Government would be prepared..to join in an international guarantee of the new boundaries of Czechoslovakia. **1957** *Encycl. Brit.* VI. 951/2 The native home of the Czechs today lies in the Czechoslovak republic..in the western parts of which.. they are the dominating and almost the sole population.

Czechize (tʃe·kəiz), *v.* [f. CZECH + -IZE.] *trans.* To make Czech in character, language, etc. So **Czechiza·tion.**

1920 *Glasgow Herald* 21 Sept. 8 If Magyarisation was bad, Czechisation is little better. **1927** *Contemp. Rev.* Feb. 173 Efforts to 'Czechise' place-names and to impose the Czech language. **1938** *Times* 26 Aug. 10/1 Increasing Czechization of industries and railways. **1941** A. J. P. TAYLOR *Habsburg Monarchy* xi. 179 Ruthenes and Slovaks can be polonised and magyarised—but Germans can't be czechised.

Czechoslovak (tʃekoslōᵘ·væk), *sb.* and *a.* **A.** *sb.* A native or inhabitant of Czechoslovakia. **B.** *adj.* Of or pertaining to Czechoslovakia or its inhabitants.

1917 L. B. NAMIER (*title*) The Czecho-Slovaks. An oppressed nationality. **1920** POKORNÝ & SELVER (*title*) The Czechoslovak Republic, its economical, industrial and cultural resources. **1929** (*title*) An anthology of Czechoslovak literature, selected and translated..by P. Selver. **1938, 1957** [see *CZECH *sb.* and *a.*].

Czechoslovakian (tʃekoslovæ·kiăn), *sb.* and *a.* = prec. Also, the language of this people.

1920 *Glasgow Herald* 7 Oct. 5 The efforts made by the Czechoslovakian Government towards the furtherance of peace. *Ibid.*, The debt of gratitude of Czechoslovakians towards the Allies. **1930** A. BENNETT *Imperial Palace* lxxi. 600 While they spoke in English they were thinking in French..or Czecho-Slovakian. **1952** E. F. DAVIES *Illyrian Venture* xi. 218 This was Checko, a big, good-looking Czechoslovakian pilot in the R.A.F.

D

D. Add: **I. 2.** *D-shaped* (examples); also *D-front, D link; D-fronted* adj.

1883 W. S. GRESLEY *Gloss. Coal-m.* 72 *D link*, a flat iron bar attached to chains, and suspended from a hemp rope to a windlass at surface. It is a loop in which one man is lowered and raised in an engine-pit. **1890** W. J. GORDON *Foundry* 135 A closed crucible with a D-shaped opening in one of its sides. **1895** *Westm. Gaz.* 22 Nov. 5/3 D-shaped and oval tubes. **1908** *Ibid.* 16 Nov. 4/2 A D-front limousine. *Ibid.* 19 Nov. 5/2 A 'D'-fronted landaulette.

II. 1. c. *D-layer, -region*: the lowest stratum of the ionosphere, occurring between 25 and 50 miles above the earth's surface, below the Heaviside or E-layer.

1930 APPLETON & RATCLIFFE in *Proc. R. Soc.* A. CXXVIII. 155 We therefore attribute the result of the small variation of the reflection coefficient with distance, to the influence of an absorbing zone (D region) of ionisation situated below the region (E region) in which the main bending takes place. *Ibid.*, Rays which travel to the more distant receiving stations have a longer path through the D region. **1935** *Nature* 8 June 953/2 Besides these two main regions (F and E), the existence of a so-called D or absorbing layer has been suggested. *Ibid.*, The appearance of echoes from the D layer is closely connected with the weakening of echoes from the E layer. **1955** *Sci. Amer.* Sept. 128/2 The lowest stratum of the ionosphere is called the D layer. Its electron density has not been measured accurately but is known to be low, because the layer does not reflect radio waves of one megacycle per second or higher frequency. **1968** G. M. B. DOBSON *Explor. Atmos.* (ed. 2) viii. 151 These radio waves will be absorbed as they come down through the *D* region.

4. *D notice*, short for *Defence notice* (see quot. 1967). Hence *D list*, a list of D notices currently in force.

The Services, Press and Broadcasting Committee was set up in 1912 as the 'Admiralty, War Office and Press Committee'. Its main function is to give guidance to the press, etc., about matters which, in the interests of national security, should not be publicly disclosed.

1961 *Times* 12 May 20/4 Mr. Lipton..asked the Prime Minister which Minister was responsible for preparing the D-list... *Mr. Macmillan*—Ministerial responsibility for D-notices rests on the Minister responsible for the subject covered by the notice. **1964** 'C. E. MAINE' *Never let Up* v. 41 No names were mentioned...The story had probably been put out with a 'D' notice, which meant that editors were asked to toe the security line. **1967** *Rep. Comm. Privy Counsellors* (Cmnd. 3309) I, A 'D' notice is a formal letter of warning or request, signed by the Secretary of a Committee known as the Services, Press and Broadcasting Committee, and addressed to newspaper editors, to news editors in sound broadcasting and television, [etc.]...Their purpose is to request a ban on the publication of certain subjects, indicated in the notices, which bear upon defence or national security.

III. 3. d., decent, esp. in *jolly d.*; D., Deputy, as D.A.A.G., Deputy Assistant Adjutant General; D.A.G., Deputy Adjutant General; D.A.Q.M.G., Deputy Assistant Quarter Master General; D., detective (*slang*); D., Dictionary, as D.A., Dictionary of Americanisms; D.A.E., Dictionary of American English; D.N.B., Dictionary of National Biography; D.O.S.T., Dictionary of the Older Scottish Tongue; D., dimensional, as 3-D, 3 D, three-dimensional; D., Distinguished, as D.C.M., Distinguished Conduct Medal (established 1862); D.F.C., D.F.M., Distinguished Flying Cross, Medal (established 1918); D.S.C., D.S.M., D.S.O., Distinguished Service Cross (established 1914), Medal (1914), Order (1886); D., District, as D.A., District Attorney (*U.S.*); D.C., District Commissioner; D.O., District Officer.

D.A., duck's arse (style of haircut); D. and C., dilatation and curettage; D. and P., d and p, developing and printing; db., decibel; D.B.E. Dame Commander of the British Empire (established 1917); D.B.S.T., Double British Summer Time; D.C., d.c., direct current; D.E.W., distant early warning; so

Dew line (see quot. 1956); D.F., direction(al) finding; D.I., Defence Intelligence; D.J., dinner jacket; disc jockey; D.M., D-*mark*, Deutsche mark; D.M.Z., demilitarized zone; D.N.A., de(s)oxyribonucleic acid (q.v.); D.O.A., dead on arrival (at a hospital, etc.); D.P., displaced person; D.Phil., Doctor of Philosophy; D.P.P., Director of Public Prosecutions; D.S.I.R., Department of Scientific and Industrial Research; D.Z., dropping zone.

d.d.d., D.D.D. (*dat, dicat, dedicat* gives, devotes, and dedicates); D.V. (earlier and later examples); d.d., D.D. (*dono dedit* gave as a gift).

See also (as main entries) *D-DAY, *D.D.T., *DORA, *D.T.

1949 N. SPAIN *Poison for Teacher* i. i. 19 Jolly d. of you to ask us. **1960** N. FAIRBROTHER *Cheerful Day* II. vi. 166, I *say*. Jolly *d.* It's *exactly* what I want. **1869** R. P. WHITWORTH *Comic Guide to Dunedin* 27 These..are gentlemen who give their talents and time to the paternal government for a certain modicium [*sic*] of payment, and are known as D.s. **1879** 'T. FREDUR' *Sk. Shady Places* ii. 16, I have a few friends among the D's (detectives), who give me the job to watch a house occasionally. **1916** J. B. COOPER *Coo-oo-ee* vii. 84 'The "D" started askin' wot's wot didn't he?' asked Sam. **1938** F. D. SHARPE *Sharpe of Flying Squad* xvi. 184 They [*sc.* crooks] very often know that a man is a 'D', as they call us, without being aware of his identity, because of the fact that he happens to be on the lookout. **1953** A. COOKE in *Manch. Guardian Weekly* 27 Aug. 7/2 One big studio has done a Technicolor 3-D movie of 'Kiss Me, Kate' which..has the effect of whisking the audience into the most privileged seat in a live theatre. **1954** *Ann. Reg. 1953* 381 The true 3-D films, involving left and right-eyed vision and the use of polarizing spectacles. **1969** *Listener* 4 Sept. 310/1 Now imagine that instead of the 3-D flesh and blood we are all composed of, we were 2-D people entirely confined to this 2-D world, the surface of the Earth, rather like flatfish. **1934** WEBSTER, D.A., District Attorney. **1959** *Times Lit. Suppl.* 30 Jan. 55/4 The author had worked at the House of Refuge (a very tough New York reformatory) and the New York Parental School (for truants) before he went to the D.A.'s office. **1951** *Sunday Pictorial* 29 Oct. (*caption*) The D.A. [haircut], so called because of its remarkable resemblance to a duck's rear. **1961** M. DICKENS *Heart of London* IV. 381 His hair, which was swept back in the popular D.A. hair-cut into a little drake's tail at the back. **1899** *Westm. Gaz.* 8 Nov. 2/1 Calling into consultation..the D.A.A.G. for Topography. **1960** *Blakiston's Illustr. Pocket Med. Dict.* (ed. 2) 187 D and C. Short for dilatation of the cervix *and* curettage of the uterus. **1968** J. FLEMING *Kill or Cure* i. 17 You ask him to perform an instant D. and C. on a young girl. **1924** *Brit. Jrnl. Photographic Alm.* 333 The appliances used in a.. D and P (developing and printing) establishment. **1945** *Daily Express* 5 May 1/2 All enemy forces..have surrendered to the 21st Army Group, effective 08.00 hours D.B.S.T. tomorrow. **1898** E. J. HOUSTON *Electr. Words* (ed. 4) 735/1 D.C., a contraction for direct current. **1924** *Times Trade & Engin. Suppl.* 29 Nov. 249/1 Recent developments at the Treforest power station included the installation of the Highfield transverter apparatus, enabling the transmission of d.c. at a high pressure over long distances. **1960** *Oxf. Univ. Gaz.* 19 Feb. 743/1 A variety of other samples, soils, British multiple coin-moulds, and Roman pewter, was also analysed by the d.c. arc method. **a 1912** W. T. ROGERS *Dict. Abbrev.* (1913) 53/1 *D.C.*, District Commissioner. **1950** 'N. SHUTE' *Town like Alice* ii. 39 The truck halted at the D.C.'s office and the subaltern went inside. **1914** *Daily Express* 10 Nov. 6/3 Our fellows were extraordinarily brave, and I think several of them should get the D.C.M. **1955** D.E.W. [see *DISTANT *a.* 8]. **1956** W. A. HEFLIN *U.S. Air Force Dict.* 165/1 *Dew line*, a line of radar stations at about the 70th parallel on the North American continent, financed by the American government but undertaken in cooperation with the Canadian government. **1957** P. FRANK *Seven Days to Never* ii. 53 The DEW line—the Distant Early Warning radar and interceptor net stretching from Alaska to Greenland. **1920** *Year Bk. Wireless Telegr.* 946 If the ship is equipped with D.F. apparatus every wireless station becomes a beacon on which a bearing can be taken. **1940** C. GARDNER *A.A.S.F.* 66 They'd managed to get Fécamp radio station shut down, so that German aircraft could no longer take D.F. bearings from it. **1919** *London Gaz.* 5 Dec. 15050/1 It is ordained that the award of the Distinguished Flying Cross shall entitle the recipient to have the initials D.F.C. appended to his name. *Ibid.*, The award of the Distinguished Flying Medal shall entitle the recipient to have the initials D.F.M. appended to his name. **1964** *Observer* 12 July 36/6 According to the *Security Gazette*, the Secret Service, MI5, is now known as DI5. **1966** J. PORTER *Sour Cream* ii. 20 We had several lectures from a languid D.I.5 man. **1961** A. BERKMAN *Singers' Gloss. Show Business Jargon* 19 *D.J.*, ..disk jockey. **1965** *Daily Tel.* 19 July 17/4 The BBC is plainly fascinated by the phenomenon of the disc jockey, now abbreviated to DJ. **1967** J. GARDNER *Madrigal* viii. 230 This Madrigal was young, slim and immaculate in a smoke-grey DJ with black silk lapels. **1960** *Amer. Speech* XXXV. 261 Some thousands of American troops defending the DMZ or Demilitarized Zone between North and South Korea. **1968** *Globe & Mail* (Toronto) 17 Feb. 4/3 The appearance of Communist tanks along the eastern end of the DMZ above Con Thien. **1903** E. GOSSE *Let.* 15 Jan. in E. E. Charteris *Life & Lett.* (1931) xix. 287 To see if anything has been overlooked by the *D.N.B.* **1930** *N. & Q.* 11 Oct. 257/2 Corrigendum for the D.N.B. **1954** G. DURRELL *Bafut Beagles* iii. 60 Sometimes I get palaver with the D.O., an' dat de tire me most of all. **1958** *Times* 6 Sept. 8/6 Even in these progressive times when authority is fast being transferred into African hands, the D.O., often under a new name and guise, and with new functions, may still be the administrative Jack-of-all-Trades, still tour his District, and still review cases in the Native Courts. **1958** T. FLINT *Emergency Treatment* (ed. 2) 9 Cases in which a spark of life is suspected..should not be classified as 'D.O.A.'. **1945** *Picture Post* 15 Sept. 18, I saw an exercise in the registration of D.P.s. **1895** *Edin. Univ. Cal. 1895–96* 136 Degree of D.Phil. Graduates who have taken the degree of Master of Arts in any Scottish University..in Mental Philosophy.. may offer themselves for the degree of Doctor of Philosophy (D.Phil.) in the same University after the expiry of five years. **1906** W. JOHNSTON *Graduates Univ. Aberdeen* p. xii, *D.Phil.*, Doctor of Philosophy. *Ibid.* p. xv, *Ph.D.*, Doctor of Philosophy. This form of contraction is used to distinguish the degree when obtained from a Continental University. *Cf.* D.Phil., which is the British form. **1938** *Times* 3 Feb. 14/2 A thesis on South African history, which brought him the degree of D.Phil. at Oxford. **1942** PARTRIDGE *Dict. Abbrev.* 34/2 *D.P.P.*, Director of Public Prosecutions. **1958** *Spectator* 27 June 827/1 What instruction, if any, I would give to the DPP should Archbishop Makarios visit this country. **1970** *It* 12–25 Feb. 19/1 With the large increase in crimes of violence and offences against property the time of the police and the DPP would have been better employed, in my submission, in those directions. **1917** *Flying* 19 Sept. 140/2 The first co-operative Research Association to be established under the auspices of the D.S.I.R. will probably be for cotton. **1964** *Guardian* 29 July 14/1 (*heading*) Tories plan a science shake-up. DSIR to disappear. **1887** *Times* 21 May 15/3 Brevet Major Archibald Hunter, D.S.O., from Supernumerary Captain, to be Captain. **1901** 'M. GRAY' *Four-Leaved Clover* i, He's got a D.S.O. You've got to deserve a D.S.O., mind you, before you get it. **a 1917** E. A. MACKINTOSH *War, the Liberator* (1918) 100 If you want a D.S.O. Or a small M.C. or so Don't go crawling rashly out When there's nobody about. **1930** *N. & Q.* 4 Oct. 245/2 There was a good sprinkling of D.S.O.'s and O.B.E.'s. **1843** C. BRONTË *Let.* 1 Aug. in Mrs. Gaskell *Life C. Brontë* (1857) I. xii. 296, I will continue to stay (D.V.) some months longer. **1918** D. V. [see *APOTROPAIC *a.*]. **1954** X. FIELDING tr. *Boulle's Bridge on River Kwai* (1956) IV. viii. 181 A second team was dropped on to a D.Z. some distance away.

b. *d* or *D* (*Physics, Chem.*) = diffuse: originally used to designate one of the four main series in atomic spectra, but now more frequently applied to electronic orbitals, states, etc., possessing two units of angular momentum.

1890 J. R. RYDBERG in *Phil. Mag.* XXIX. 335 K (D, 4) denotes the fourth line of the first diffuse series of the spectrum of potassium. **1922** A. FOWLER *Rep. Series in Line Spectra* iii. 15 Three of the chief series were recognised.. namely..Principal, Diffuse, Sharp...Rydberg's names.. are conveniently abbreviated to *P, D, S*. **1955** RICHTMYER & KENNARD *Introd. Mod. Physics* (ed. 5) vii. 245 The use by spectroscopists of the mysterious letters *S, P, D, F*.. to represent various values of *L*. *Ibid.*, The lowest *D* level. **1965** PHILLIPS & WILLIAMS *Inorg. Chem.* I. iv. 102 The withdrawal of the *d* electrons from the valence shell.

da. (Regional examples.)

1875 W. ALEXANDER *Sks. Life Ain Folk* 184 Da promis't to tak' me in o' 's bosie. **1922** JOYCE *Ulysses* 70 Waiting outside pubs to bring da home. **1936** 'N. BLAKE' *Thou Shell of Death* xiii. 232 Miss Judith grew up to be..the apple of her da's eye.

dab, *sb.*¹ Add: **3. b.** *pl.* Fingerprints. *slang.*

1926 N. LUCAS *London & its Criminals* i. 7 The finger-print system is without doubt the crooks' greatest enemy... The verifying of their 'dabs' soon brings their dossier to court. **1947** N. MARSH *Final Curtain* xvii. 263 Bailey's gone over it [*sc.* a tin] for dabs. **1957** K. FARRER *Gownsman's Gallows* xxii. 183 You'll get his photo and dabs by airmail today.

11. dab cricket, a children's pencil-and-paper game based on cricket.

1938 L. MACNEICE *I crossed Minch* vii. 98 While reviewing a novel, Mr. Mackenzie had played dab cricket... He explained to me his system...Consonants mean one; vowels mean two,..a double consonant following a double consonant means that the next man is out for nought. **1946** B. MARSHALL *George Brown's Schooldays* 115 He was going to continue his game of dab cricket.

dabbity (dæ·biti). *Sc.* [f. DAB *sb.*¹ (cf. 3. 4); cf. southern dial. *dabbit* (E.D.D.), small quantity.] A chimney-piece ornament.

1923 J. A. FLEMING *Sc. Pottery* i. 28 On the East Coast of Scotland the fisherfolk are famous for their gallery of 'wallie' dogs, lions, parrots, cats, etc. The Aberdonians call them 'dabbities'. **1961** *Times* 8 Apr. 11/6 'Chimney ornaments'. To the Scottish potter they were 'dabbities'.

dabitis (dæ·bitis). *Logic.* [L., = you will give.] The mnemonic term for that indirect mood of the first figure of syllogisms in which the major premiss is universal and affirmative, and the minor premiss and conclusion are particular and affirmative.

1599 BLUNDEVIL *Logike* 121 Celantes: Dabitis: Fapesmo: Frisesomorum. **1685** tr. *Arnauld & Nicole's Logic* viii. 35 These five modes are generally denoted by these words, Baralipton, Celantes, Dabitis, Fapesmo, Frisemorum. **1849** H. L. MANSEL *Artis Logicæ Rudimenta* 65 The five indirect moods of the first figure were called Baralip, Celantes, Dabitis, Fapesmo, Frisesmo.

dacca, dacha, varr. *DAGGA¹.

‖ **dacha** (dæ·tʃă). Also **datcha, datsha.** Pl. **da(t)chas, datche.** [Russ. *dácha*, pl. *dáchi*, lit. act of payment.] In Russia, a small house or villa for summer use, in the country near a town.

1896 *Edin. Rev.* Jan. 83 Russian officials run down to their *datchas*, or country houses. **1905** *Daily Chron.* 9 Sept. 4/5, I was privileged to visit one of these datche and to dine with the family. **1926** *Blackw. Mag.* Sept. 302/2 Colonel Napier.. had a dacha for the summer just outside the town. **1929** *Times* 14 Nov. 13/2 The settlers of German descent..are at present precariously housed in the 'Datchas', or summer chalets of the Moscow countryside. **1950** KOESTLER in R. H. S. Crossman *God that Failed* 66 Two-room flats..not to mention motor-cars and summer *datshas*. **1955** *Times* 8 Aug. 5/1 His principal guests—at a summer dacha 60 miles from Moscow..were the ambassadors and ministers accredited to the Soviet Government. **1971** *Guardian* 21 Jan. 3/3 Mr Krushchev..in the country dacha to which he had retired.

dachs (dæ·ks). Colloq. abbrev. (*sing.* and *pl.*) of DACHSHUND.

1886 G. MEREDITH *Let.* 24 July (1970) II. 821 The dog.. is the dearest little Dachs. **1892** B. POTTER *Jrnl.* 25 Oct. (1966) 291 The Duchess's *Dinah* is dead...The surviving dachs was noisy. **1960** M. SHARP *Something Light* viii. 77 *Some* sort of dog...We settled on dachs because they're so easy.

Dacian (dēⁱ·ʃiăn), *a.* and *sb.* Also 8 **Dacic.** [f. *Dacia* (see below) + -AN.] **A.** *adj.* Of or pertaining to Dacia, an ancient country or south-eastern Europe, or its people, or their language. **B.** *sb.* **1.** A member of this people. **2.** The language of this people.

a 1666 EVELYN *Diary* 26 Feb. an. 1645 (1955) II. 378 We came to the Forum Trajanum where his Columna stands yet intire, wrough[t] with admirable Bass-relievo & comprehending the Dacian War. **a 1773** A. BUTLER *Moveable Feasts* (1774) III. 134 The most illustrious Roman Conquerors often took Names..from Countries which they had subdued ..as the African,..the Germanic, Dacic, &c. **1776** GIBBON *Decl. & F.* I. i. 5 The Dacians, the most warlike of men, who dwelt beyond the Danube. **1818** BYRON *Ch. Har.* IV. cxli, *There* were his young barbarians all at play, *There* was their Dacian mother—he, their sire, Butcher'd to make a Roman holiday. **1837** *Penny Cycl.* VIII. 281/2 Domitian celebrated his pretended exploits against the Dacians by assuming the title Dacicus. **1847** *Howitt's Jrnl.* 12 June 325/2 A sentence or two will enable the reader to compare the modern Dacian with the ancient Roman. *Ibid.* 326/1 The adoption of the Cyrillian alphabet had the effect of carrying away..the association of Dacian words from their Latin sources. **1861** MAX MÜLLER *Lect. Sci. Lang.* IV. 116 We possess fragments of Dacian speech in the botanical names collected by Dioskorides. **1877** *Encycl. Brit.* VI. 758/2 He advanced to the Dacian capital. *Ibid.*, The Dacians come forward as one of the most powerful enemies of Rome. **1948** *Oxf. Jun. Encycl.* I. 413/1 Later began a series of invasions by Goths, Tartars, Huns and Magyars..., all of whom left their mark on the Dacian people. **1959** *Chambers's Encycl.* XII. 51/2 The remote ancestors of the Rumanians were the Dacians who inhabited lands north and south of the Lower Danube.

dacka, var. *DAGGA¹.

dacryo-, combining form of Gr. δάκρυον tear, as in **da:cryoadena·lgia** [ad. G. *dacryoadenalgie* (J. A. Schmidt *Über die Krankheiten des Thränenorgans* (1803) I. ii. 117)], pain in a lachrymal gland; **da:cryoadeni·tis,** inflammation of a lachrymal gland; **da:cryocysti·tis,** inflammation of a tear-sac.

1848 DUNGLISON *Med. Lex.* (ed. 7) 250/1 Dacryoadenalgia, Dacryoadenitis. *Ibid.* 250/2 Dacryocystitis. **1887** *Buck's Handbk. Med. Sci.* IV. 366/2 Neuralgia of the gland has also been observed, and the name 'dacryo-adenalgia' was given to it. *Ibid.*, Dacryo-adenitis, or inflammation of the lachrymal gland, is another affection which is seldom encountered. *Ibid.* 369/1 Acute dacryo-cystitis, or abscess of the sac, as it is termed, is a serious malady. **1908** *Practitioner* Feb. 288 Acute and chronic dachryocystitis [*sic*] are usually associated with epiphora. **1952** M. E. FLOREY *Clinical Appl. Antibiotics* xv. 458 A few results of the treatment of dacryocystitis have been recorded, but as the condition was usually chronic, distortion of the ducts often prevented chemotherapy alone from effecting a cure. **1964** S. DUKE-ELDER *Parsons' Dis. Eye* (ed. 14) xxxii. 512 Dacryo-adenitis occurs occasionally in general infections (mumps, influenza, etc.), sometimes leading to suppuration.

dacryon (dæ·kriǫn). *Anat.* [a. F. *dacryon* (P. Broca 1875, in *Bull. Soc. d'Anthropol. Paris* 2nd Ser. X. 361), a. Gr. δάκρυον tear.] In the cranium, the point of juncture of the lachrymal and frontal bones with the frontal process of the maxillary bone.

1878 R. BARTLEY tr. *Topinard's Anthropol.* I. i. 35 At the point where the posterior border of the ascending process joins the frontal and the os unguis, is the..dacryon. **1951** *Cunningham's Text-bk. Anat.* (ed. 9) 160 The upper end of the suture between the two bones is called the dacryon. **1959** *Chambers's Encycl.* I. 460/1 [Anthropometry.] The ophryon is now obsolete and the asterion, dacryon and inion are likewise falling into disuse.

dactyl. Add: **4. a.** A finger or toe. **b.** = DACTYLOPODITE. **c.** A part of the pretarsus of an insect.

1889 *Cent. Dict.*, *Dactyl*, a digit, whether of the hand or foot. **1946** *Nature* 9 Nov. 668/2 In ecdysis, any available rough surface is used to anchor the dactyl-claws of the walking legs. **1960** T. H. WATERMAN *Physiol. Crustacea* I. xvii. 564 In this crab autotomy never results after injury to the dactyl, the most distal segment of a walking leg.

dactylically (dækti·likăli), *adv. Prosody.* [f. DACTYLIC *a.*: see -LY².] With a dactylic rhythm.

1891 STEVENSON *Vailima Lett.* (1895) 85 'Ulufanua the isle of the sea', read that verse dactylically and you get the beat.

dactylo-. Add: In terms relating to the taking of finger-prints: **da·ctylogram** [Gr. γράμμα letter], a finger-print; **dactylo·grapher** [GRAPH(Y + -ER¹], one who takes or studies finger-prints (in quots. 1926 and 1931 the sense is 'typist'); **dactylo·scopy** [Gr. σκοπιά seeing], the examination of finger-prints; hence **da:ctylosco·pic** *a.*

1908 *Boston Transcript* 10 Oct., An interesting illustration of the practical value of the science of dactyloscopy. **1910** *Let. to J. A. H. Murray* 5 Mar., The dactyloscopic records of the Boston Police Department. **1913** DORLAND *Med. Dict.* (ed. 7), *Dactylogram*, a finger-print taken for purposes of identification. **1921** *Discovery* Oct. 259/1 You would not find two dactylograms alike, says Galton,..if you were to examine a series of 64,000,000. *Ibid.* [Poroscopy] is infinitely more fruitful in results than the one known by the name of dactyloscopy. *Ibid.* 259/2 In all cases of dactyloscopic analysis. **1926** F. M. FORD *A Man could stand Up* I. ii. 36 A dactylographer of respectability. **1931** H. G. WELLS *Work, Wealth & Happiness of Mankind* (1932) xii. 567 There would be stenographers and dactylographers swiftly available. **1935** *Discovery* Sept. 261/1 The print of a finger can be definitely characterised by the dactylographer. **1936** BENTLEY & ALLEN *Trent's Own Case* x. 122 Looking at these well-marked prints of a finger and thumb, he had no more than a vague recollection of the dactyloscopic terms to be used in describing them. **1970** *Reader's Digest* Dec. 207 In 1910, few policemen believed in dactyloscopy, the examination of fingerprints.

dad, *sb.*¹ Add: **2.** Used as a form of address to a person, not necessarily elderly, other than one's own father. *colloq.* (esp. in Jazz talk).

1959 J. C. HOLMES *Horn* 128 Here, dad, have a brew while I get these boys set up. **1960** *Time & Tide* 24 Dec. 1599/1, I think 77 *Sunset Strip* is real zoolie, dad. **1966** *Melody Maker* 30 July 8/3 Take that bit where everybody was called Dad...Altoist Bruce Turner..even called his wife 'dad'.

dad (deformation of *God*). Add further U.S. examples with verbs and pa. pples.

1834 W. A. CARRUTHERS *Kentuckian in N.Y.* I. 216 I'll be dad shamed if it ain't all cowardice. **1884** 'C. E. CRADDOCK' *Tenn. Mts.* I. 45 Dad-burn that..idle poultry. *Ibid.* III. 141 That dad-burned scoundrel. **1884** 'MARK TWAIN' *Huck. Finn* xxxiv. 354 It's de dad-blame' witches. **1901** W. H. HARBEN *Westerfelt* xiv. 195 'Don't act so dadratted foolish,' he said. *Ibid.* xxii. 300 Yes, dad burn it; you know she loves you. **1911** R. D. SAUNDERS *Col. Todhunter* vi. 84 I'll be dadblamed if I know what's goin to come of it all some day! **1927** *Hollis St. Theatre Progr.* (U.S.) 19 Sept., But who'd think where buildings are tall Business could be so dad-burned bum? **1944** T. D. CLARK *Pills, Petticoats & Plows* 156 There was a sentiment that 'a dad-blamed hog and a dad-gummed cow were the most aggravating things that ever made tracks on a piece of cotton land'. **1968** *Word Study* Feb. 7/2 'Darn it', 'dad gum it', 'heck'.

Dada² (da·da). [Fr. (*être sur son dada*, ride one's hobby-horse); title of a review which appeared at Zürich first in 1916, founded by Tzara (a Rumanian poet), Arp (an Alsatian German artist), and Huelsenbeck (a German poet).] Applied to an international movement in art and literature, characterized by a repudiation of traditional conventions and reason, and intended to outrage and scandalize. Hence **Da·daism,** the theory or practice of this movement. So **Da·daist(e** *sb.* and *a.*, **Dadai·stic, Dadai·stical** *adjs.* (Also with lower-case initials.)

1920 A. HUXLEY *Let.* 4 May (1969) 184 The three I have mentioned are almost wholly dada in style and sympathy. *Ibid.*, Most of the dadaist publications issue from a press named Au Sans Pareil. **1920** *Athenæum* 13 Aug. 221/2 The movement 'Dada'..has its headquarters in Paris, and its principal promoters are Francis Picabia and Tristan Tzara, neither of whom is of French nationality. *Ibid.*, Mr. Dent defines Dadaism as being a 'whole-heartedly æsthetic movement', in contradistinction to 'papaism' and 'nanaism'. *Ibid.* 222/2 [Guillaume Apollinaire's] most advanced, most nearly dadaistical poetry. **1920** *Times Lit. Suppl.* 2 Sept. 569/3 M. Mille still believes in *bon sens*, clarity and humour as valuable assets in art, a belief which has got him into serious trouble with the Dadaistes. **1920** *Glasgow Herald* 27 Nov. 6 Maeterlinck in his early days wrote verse not easily distinguishable from the work of the Dadaist. *Ibid.*, Dada means nothing... We gather some idea of the Dada position. **1923** A. HUXLEY *On Margin* 43 Dadaist literature always reminds me a little of..the inconsequent music of water. **1927** G. MURRAY *Class. Tradition* ix. 254 Does this mean that the words are only sounds and the poem a collection of sounds? That is clearly nonsense—a form of nonsense that is admired in certain French coteries, and is called 'dadaism'. **1934** C. DAY LEWIS *Hope for Poetry* 83 'Any work of art that can be understood is the product of a journalist' (Dada Manifesto). *Ibid.* 86 That was in fact the position that the Dadaists took up. 'Art is a private matter: the artist does it for himself.' **1935** D. GASCOYNE *Short Survey Surrealism* 25 The Dada spirit was something shared by a number of extreme individualists of various nationalities, all of whom were in revolt against the whole of the epoch in which they lived. **1947** A. E. BALAKIAN *Lit. Orig. Surrealism* 141 If the Dadaists sought to place themselves beyond the control of Beauty and its rules, it was because of their inner consciousness of something greater. **1956** C. GIEDON-WELCKER *Contemp. Sculpture* p. xiv, Dadaism created a metaphysic of banality by discovering the plastic vitality that emanates from nameless or unnoticed things. **1965** *Economist* 18 Sept. 1197/2 During the past year there has been an underground film festival featuring the latest 8 mm dadaistic experiment. **1965** H. RICHTER *Dada* 113 The idea of putting people in a position to exploit their mental and physical energies in a spirit of unbounded optimism and faith in themselves—this was the idea behind the wild and exuberant antics of Dada.

daddy. Add: **c.** = DOYEN 2.

1925 *New Yorker* 11 July 11 The Daddy of Sunday Painters. **1959** *Times* 26 Nov. 16/1 At full-back Uren, who must have been the daddy of the entire party, started falteringly.

2. Various slang uses (see quots.).

1859 HOTTEN *Slang Dict.*, *Daddy*, the stage manager.—*Theatr.* **1860** *Ibid.* [adds] *Daddy*, the person who gives away the bride at weddings. **1864** *Ibid.*, At mock raffles, lotteries, &c., the Daddy is an accomplice, most commonly the getter up of the swindle, and in all cases the person that has been previously arranged to win the prize. **1874** *Ibid.*, *Daddy*, the old man in charge—generally an aged pauper—at casual wards. **1886** *Graphic* 10 Apr. 399/2 The manager himself is sometimes known as the 'gorger', and 'daddy' is the stage-manager. **1901** 'M. FRANKLIN' *My Brilliant Career* xxi. 183 Joe Archer told me you ran into a clothes-line on race-night, and ever since then mother has kept up a daddy of a fuss about ours. *Ibid.* xxii. 194, I never felt such a daddy of a thirst on me before. **1941** BAKER *Dict. Austral. Slang* 22 *The daddy of them all*, the most notable or expert, the largest. **1962** *Woman* 31 Mar. 18/3 He [an Australian] had had his share of knockbacks. Including a daddy of a one from Belle herself. **1969** W. GARNER *Us or Them War* iv. 42 You graduate from taking little chances to taking big ones. This one was the daddy of 'em all.

3. Used in Jazz slang as a form of address. Also more generally applied to an older person, and (*U.S. slang*), a lover. Cf. *DAD *sb.*¹ 2; **sugar daddy*.

1926 C. VAN VECHTEN *Nigger Heaven* 285 *Daddy*, husband or lover. **1927** *Jrnl. Abnormal & Social Psychol.* XXII. 15 Come on, daddy, let's have some fun. **1935** WODEHOUSE *Blandings Castle* xi. 244 A two-timing daddy. **1948** *New Yorker* 3 July 28 The bebop people have a language of their own. They call each other Pops, Daddy, and Dick. **1957** J. KEROUAC *On Road* (1958) I. vii. 44 A waitress.. slightly hung-up on a few sexual difficulties which..I think you can manage, you fine gone daddy you. **1960** *Time & Tide* 24 Dec. 1599/1 He calls his colleagues the detectives, 'daddy', his clothes he refers to as 'threads'. **1962** *Amer. Speech* XXXVII. 34 This suggests to me that possibly the word *daddy*, meaning 'a male lover', may be a 'pure' Americanism. *Ibid.* 35 *Daddy* has a history beginning much earlier than 1935 as a term meaning simply 'lover' in Negro songs and blues. **1970** H. WAUGH *Finish me Off* (1971) 133 He wasn't pimping for her...He's my daddy and he plays it for five other girls.

da·ddy-o. Colloq. var. DADDY (various senses).

1949 *Music Libr. Assoc. Notes* Dec. 42 *Daddy-o*, friend, buddy. Originated with Negro musicians. **1952** R. ELLISON *Invisible Man* xxiii. 366 A group of zoot-suiters greeted me in passing. 'Hey, now, daddy-o,' they called. **1959** C. MACINNES *Absolute Beginners* I. i. 18 Oh yes it can, daddy-o. **1959** E. AMBLER *Passage of Arms* iv. 107 You Americans give away billions of dollars...Because big daddy-o wants to be loved. **1960** *Time & Tide* 24 Dec. 1599/3 The walls are crazy,..And the scene uncool for you, Daddy-o. **1963** 'A. GARVE' *Sea Monks* iii. 101 What you goin' to do..when you finished it, Daddy-O? *Ibid.*, D'you reckon we oughter let Daddy-O 'ave a tool like that? **1969** N. COHN *AWopBopaLooBop* (1970) vi. 56 Who calls the English teacher daddy-o?

Dædalic (dī·dălik), *a. Archæol.* Also **Dedalic.** [f. as DÆDALIAN *a.* + -IC.] Designating a Greek sculptural style of the 7th century B.C., done in clay, metal, and stone, mainly in Dorian areas.

1931 H. PAYNE *Necrocorinthia* xvi. 233 This is no place for a history of Daedalic sculpture. **1936** R. J. H. JENKINS *Dedalica* I. ii. 10 The term 'Dedalic' so used is not entirely a term of convenience, for the monuments in question are closely bound to each other stylistically, and represent unquestionably the Creto-Peloponnesian or Dorian sculptural tradition of the seventh century, when Dedalus was apparently active in the districts. **1962** R. W. HUTCHINSON *Prehist. Crete* xii. 340 Archaeologists have come to employ the term 'Dedalic' of the sculptural style characteristic of, though not confined to, Doric-speaking cities [in Crete] in the eighth and early seventh centuries B.C. *Ibid.* 342 The new Dedalic style affected not only figurines of clay or bronze and jewellery, but also had a notable effect on..the making of *pithoi* or large stone jars with moulded ornaments.

dafadar, daffadar, variant forms of DUFFADAR.

1849 E. B. EASTWICK *Dry Leaves* 69 A Dafadár of Captain Christie's corps. **1921** *Glasgow Herald* 25 Mar. 7 A force of one dafadar and ten sowars of mounted police. **1951** J. MASTERS *Nightrunners of Bengal* 337 The three-striped non-com was havildar (inf.) or daffadar (cav.).

daff (dæf). Colloq. abbrev. of DAFFODIL.

1915 C. MACKENZIE *Guy & Pauline* vi. 307 'We shall have all the daffs gone before we know where we are,' said the Rector. **1934** D. L. SAYERS *Nine Tailors* II. i. 69 You want a few more daffs. on the decani side, Mrs. Venables.

daffingly (dæ·fiŋli), *adv.* [f. *daffing*, pres. pple. of DAFF *v.*¹ + -LY².] Sportively.

1902 D. S. MELDRUM *Conquest of Charlotte* III. xix, 'Ah! he's married: that's crucial,' I said, daffingly. **1907** J. H. MCCARTHY *Needles & Pins* xx, 'I should have taken a great fancy to you,' he answered daffingly, 'if I had been a free man when we met.'

daffle (dæ·f'l), *v. dial.* or *colloq.* [f. DAFF *v.*¹ + -LE.] *intr.* To become silly, daft, or faltering; to act stupidly or inanely.

1796 W. H. MARSHALL *Rur. Econ. Yorks.* (ed. 2) II. 315 'He daffles', he wanders, or falters in his speech or conversation. **1853** R. S. SURTEES *Sponge's Sp. Tour* xxxvi, If your old man is done daffling with your draft I should like to have the pick of it.

Daffy (dæ·fi), *sb.*¹ Also **daff(e)y.** [The name of Thomas *Daffy*, an English clergyman of the seventeenth century.] Orig. in *Daffy's elixir*, a medicine given to infants, 'tinctura sennæ composita' (Dunglison), to which gin was commonly added; hence, a slang name for gin itself.

1680 *The True News: or, Mercurius Anglicus* no. 33 10–13 Mar. [2]/2 Whereas divers Persons have lately exposed to sale a counterfeit Drink called Elixir Salutis, the true Drink of that Name, having been long since published by Mr. Anthony Daffy, (and generally known by the name of Daffy's Elixir Salutis), [etc.]. **1681** [see ELIXIR *sb.* 2 b]. **c 1711** C. MORDAUNT *Let.* in Lady Hamilton *Mordaunts* (1965) iv. 79 Daphios Elixir for Chollick. **1768–74** [see ELIXIR *sb.* 2 b]. **1776** *Hibernian Jrnl.* 8–10 Apr. 179/3 (Advt.), Daffy's Elixir, just imported. **1821** *The Fancy* I. 304 While carrying on his new vocation as publican, Jack did not deny himself the use of drops of Daffy. **1828** *Sporting Mag.* XXI. 435 His predilection for daffey of late years grew upon him. **1838** DICKENS *O. Twist* I. ii. 15 [A little gin] to put into the blessed infants' Daffy. **1846** *Swell's Night Guide* 58, I takes the swell into the tape shop, took our daffies..and planted Flabby Bet on him. **1857** TROLLOPE *Barchester T.* xxiii, Not got a coral—how can you expect that he should cut his teeth? Have you got Daffy's Elixir? **1861** A. HALLIDAY in Mayhew *Lond. Labour* Extra vol. (1862) 430/2 When I goes in where they are a havin' their daffies—that's drops o' gin, sir. **1871** *London Figaro* 15 Apr. 10/2 [If the baby] should bawl persistently,..he would..bathe it in Mrs. Winslow's syrup, and thoroughly dose it with 'Daffy'. **1882** *Punch* 29 Apr. 193/2 A good many of them..had been partaking freely of *daffy*. **1967** E. BURTON *Georgians at Home* vi. 226 They [*sc.* Ward's pills] could hardly have done more harm or good than the snake-root and brandy.., Daffy's Elixir, usquebaugh, [etc.].

daffy (dæ·fi), *sb.*² *colloq.* (orig. *dial.*). Abbreviated form of DAFFODILLY. So **da·ffying** *sb.* (see quot. 1871).

1777 M. CUTLER in W. P. & J. P. Cutler *Life & Corr. M. Cutler* (1888) I. 63 Planted out in my borders in a great alley..early and late Daffies, and Peonies. **1871** *Leisure Hour* 25 Mar. 184/1 Another of our rustic treats..was going to gather daffodils...In Herefordshire this little festival was called Daffying. **1878** W. DICKINSON *Gloss. Cumberland* 25/1 *Daffy-doon-dilly, Daffy*, the daffodil. **1925** *Sunday at Home* June 536/2 Little winds just rose on purpose to stir the daffies. **1928** *Daily Express* 28 Mar. 3/3 This has been a bad season for 'daffies' owing to a recent spell of frost in Cornwall.

daffy (dæ·fi), *a. dial.* or *slang.* [f. DAFF *sb.*, *v.*¹; cf. *DAFFLE *v.*] = DAFT *a.*

1884 R. LAWSON *Upton-on-Severn Words* 14 *Daffy*, simple, soft. **1896** ADE *Artie* iii. 24 She'd make anybody daffy. **1902** *Munsey's Mag.* XXII. 491/2, I want to know what this new affair is. If I'm daffy, there's the reason. **1908** C. E. MULFORD *Orphan* xiv. 180 Old man Gordon was daffy on education, which is a good thing to be daffy over. **1922** *19th Cent.* Feb. 270 Guess the poor old devil's gone daffy. **1923** R. D. PAINE *Comr. Rolling Ocean* xv. 258 England was daffy over spy stories during the war. **1941** *Penguin New Writing* II. 87 Your eyes are all staring and aching..and they burn like hell. And it drives you nigh on daffy. **1959** I. & P. OPIE *Lore & Lang. Schoolchildren* x. 179 A person who is 'wanting in the upper storey' is..daffy. **1968** *Guardian* 28 Feb. 8/3 One of those charming fusions of the daffy benevolence of youth with the guilelessness of middle aged PROs.

dag, *sb.*¹ Add: **3. b.** *Comb.* **dag-boy, -cutter, -man, -pick** *v.*, **-picker** (see quots.). *Austral.* and *N.Z.*

1913 A. I. CARR *Country Work & Life in N.Z.* vi. 15 The dag cutter..has a seat handy and with a pair of shears cuts off all the wool he can, which he throws into another bin. **1933** *Bulletin* (Sydney) 8 Feb. 21, I work and whistle on my own..Dag-pickin' all day long. **1933** L. G. D. ACLAND in *Press* (Christchurch, N.Z.) 7 Oct. 15/7 The dags are afterwards gone through by a dag-picker or dag-boy, who cuts out any wool worth saving. **1958** *New Statesman* 23 Aug. 218/3 They were given jobs as..dag-men, which implied following around any one of 20,000 sheep and snipping off dung. **1965** [see *DAGGER *sb.*³ b].

4. A 'character', an extraordinary person, a 'tough' but amusing person (see also quot. 1941). *Austral.* and *N.Z. slang.*

1916 *Anzac Book* 47 Yes; 'Enessy was a dag if ever there was one! **1931** V. PALMER *Separate Lives* 222 Chook chuckled suddenly..'Ain't he a dag?' **1940** F. SARGESON *Man & Wife* (1944) 64 Struth he was a dag, Bill was. **1941** BAKER *Dict. Austral. Slang* 22 *Dag*, an amusing or eccentric person. Whence, *dag* adj., good, excellent: *a dag at*, expert at. **1945** *N.Z. Geographer* I. 35 He was a tough old dag, and no mistake. **1949** E. DE MAUNY *Huntsman in Career* i. 33 Scotty's a bit of a dag, isn't he? **1970** D. M. DAVIN *Not Here, Not Now* II. ix. 115 Gerald seemed to have become a bit of a dag since the old days.

dag (dæg), *sb.*⁵ *dial.* or *slang.* [perh. altered from DARG (one's) task.] A feat of skill; chiefly *pl.*, esp. in *doing dags* (see quots.).

1879 *N. & Q.* 5th Ser. XII. 128/1 'I'll do you (or your) dags.'—An expression used by children of young, and sometimes of older, growth, meaning, 'I'll do something that you cannot do.' **1886** F. T. ELWORTHY *W. Som. Word-Bk.* s.v., To 'set a dag' is to perform some feat in such a way as to challenge imitation...There's a dag for you—do it if you can. **1886** *Fun* (Farmer), He was very fond of what, in schoolboy days, we used to call doing dags. **1898** *Daily*

News 4 Oct. 6/3 What does your entertainment principally consist of?..Doing 'dags' to make the people laugh. **1902** *Windsor Mag.* June 114/1 Wearing it in your hat shows that you don't funk me. It's doing my dags to touch you.

dag, *v.*[1] **3.** Add to def.: The usual word in Australia and N.Z.

1889 WILLIAMS & REEVES *Colonial Couplets* 9 Dagging the hoggets, or drafting the rams. **1923** W. PERRY *Sheep Farming in N.Z.* vi. 73 The ewes should also be dagged.. before turning the rams in. **1965** J. S. GUNN *Terminology Shearing Industry* I. 20 *Dag*, wool mixed with dung, dirt, or other rubbish, hanging from the sheep. To 'dag' is to remove this wool with 'dagging shears'.

dageraad (dæ·gərā:d, -rā:t, ‖ dā·χ-, da·g-). *S. Afr.* Also **daggerhead, daggerheart.** [Afrikaans, f. Du. *dageraad* daybreak, the name of the fish being supposed to refer to its brilliant colouring.] The brilliantly coloured sea-fish *Chrysoblephus cristiceps* (family Sparidæ).

Derivation via Du. *dorade*, ad. Sp. and Pg. *dorado* DORADO (Boshoff & Nienaber *Afrikaanse Etim.*, 1967) does not account for the phonology of the S. Afr. word.

1853 L. PAPPE *Edible Fishes Cape Gd. Hope* 20 *Pagrus laniarius*..(*Dageraad*)..Strong conical teeth in the upper jaw, which..project from the mouth..those of the lower jaw much smaller. The whole fish is of a dark rose-colour... Lower jaw white; iris silvery; length 12 inches. **1906** *East London Dispatch* 3 July 3/3 This fish was probably what is known as a 'daggerhead' (pagrus laniarius). **1913** PETTMAN *Africanderisms* 135 *Dageraad, Pagrus laticeps*..is sometimes corrupted into Daggerhead, Daggerheart, etc. **1953** F. ROBB *Sea Hunters* viii. 122 Then we'll go inshore and fish daggerheads and soldiers. **1957** S. SCHOEMAN *Strike!* iii. 40 The dageraad is also called daggerhead along the Pondoland and Natal Coast.

dagga[1] (dæ·gă, da·χa). *S. Afr.* Also **dacca, dacha, dacka, dakha, dak(k)a.** [Afrikaans, f. Hottentot *dachab*.] A name for hemp, *Cannabis sativa*, used as a narcotic. Also applied to any indigenous plant of the genus *Leonotis*, called wild dagga, which is similarly used. Also *attrib.* and *Comb.*

[**1668** O. DAPPER *Kaffrarie* in I. Schapera *Early Cape Hottentots* (1933) 40 Zekeren krachtigen wortel, dien zy *dacha* noemen.] **1670** J. OGILBY *Africa* 583 A powerful Root, which they call *Dacha*; sometimes eating it, otherwhiles mingling it with Water to drink; either of which ways taken, causeth Ebriety. **1785** G. FORSTER tr. *Sparrman's Voy.* (1786) I. 145 *Bucku* (*diosma*) and wild *dacka* (*phlomis leonurus*) which are known both by the colonists and the Hottentots to be as efficacious as they are common. **1796** tr. *Le Vaillant's New Travels* III. 267 The people wished for tobacco and dacca (the leaves of hemp). **1822** W. J. BURCHELL *Trav. Interior S. Afr.* I. 366 The common hemp, called *dakka*, was here raised..as presents to the Bushmen, who smoke it instead of tobacco. **1835** J. W. D. MOODIE *Ten Yrs. S. Afr.* I. 41 Many of these people [*sc.* Hottentots] have..a pernicious habit of smoking a plant called 'dacha'. ..The 'dacha rookers' are held in great contempt by the tobacco smokers of their nation. **1894** C. H. DONOVAN *With Wilson in Matabeleland* vii. 140 It is exceedingly entertaining to watch these boys 'dakha-smoking'. **1910** J. BUCHAN *Prester John* vii. 119 He must have been a *dacha*-smoker, for he coughed hideously. **1912** *East London Dispatch* 28 June 9/6 The red Dagga, or 'Mfincafincane', of the Kaffirs. **1921** *Blackw. Mag.* Jan. 110/2 They are..much addicted to smoking a drug which is known as 'daka'. **1939** *Times Lit. Suppl.* 9 Sept. 531/3 Marihuana, otherwise.. ganja, dagga, and about a hundred other names. **1950** *Cape Times* 19 Sept. 12/7 Prison sentences for dagga smokers. **1953** P. LANHAM *Blanket Boy's Moon* vi. 131 There is..a huge trade carried on in South Africa in the growing, smuggling and peddling of dagga. **1969** *Sci. Jrnl.* Sept. 38/1 Some preparations of the [Cannabis] plant, like the *ganja* of India, are more intoxicating than marihuana... Others, like Indian *bhang*, Moroccan *kif*, and South African *dagga*, are nearly equivalent to marihuana.

dagga[2] (dā·ga). *S. Afr.* Also **daager, dagher, dargha.** [f. Zulu and Xhosa *daka* mud, clay, mortar.] A kind of mortar made of mud and cow-dung, often mixed with ox-blood. Hence as *vb.*, to smear with dagga.

1878 H. A. ROCHE *On Trek in Transvaal* xii. 251 A Kafir came to 'daager' or smear our floors. **1880** H. M. PRICHARD *Friends & Foes* 282 Kafir women..smear the walls and floor with 'dargha'. **1893** BLENNERHASSETT & SLEEMAN *Adv. Mashonaland* 32 We had heard..that 'daghering' and 'smearing' would be essential parts of our work. **1896** H. L. TANGYE *In New S. Afr.* v. 326 This lends some colour to the theory that the inhabitants adapted themselves..to the practice of the country and lived in dagher huts. **1899** W. H. BROWN *On S. Afr. Frontier* 63 The houses were built of 'dagga' (mud), brick, and corrugated iron. **1905** *Blackw. Mag.* Mar. 389/1 The room was floored with dagga—anthill earth brought to a high stage of hardness and mahogany-like polish by frequent dressings of bullock's blood and kraal manure. **1936** P. M. CLARK *Autobiogr. Old Drifter* vii. 92 A hut constructed of dagga—that is, ant-heap mixed with cowdung.

dagger, *sb.*[1] Add: **6. c.** = DOG-SHORE.

1838 *Civil Engin. & Arch. Jrnl.* I. 384/2 At 11 o'clock the dagger was knocked down, and the beautiful vessel..glided majestically into the river. **1896** *Strand Mag.* XII. 325 Being simultaneously released..these weights instantly fall, and..bring down the daggers, thus removing all obstacle to the passage of the ship down the ways.

da·gger, *sb.*[2] [f. DAG *sb.*[1] 3 + -ER.] *Austral.* and *N.Z.* **a.** *pl.* (See quot. 1945.)

1878 G. H. GIBSON *Southerly Busters* 179 I'm able for to shear 'em clean, And level as a die; But I prefers to 'tommyhawk', And make the 'daggers' fly. **1945** BAKER *Austral. Lang.* iii. 64 Handshears are known as daggers, jingling johnnies.

b. (See quot. 1965.)

1889 WILLIAMS & REEVES *Colonial Couplets* 28 He could do anything, he swore..Would take what came,..Be 'brander', 'rouse-about', or 'dagger'. **1952** *Arena* XXXI. 4 The dogs, the roussies and penners-up and daggers. **1965** J. S. GUNN *Terminology Shearing Industry* I. 21 In earlier times, but not often to-day, shedhands known as 'daggers' and 'dag boys' were given the job of 'dag picking' or recovering wool from the dags.

daggerhead, daggerheart: see *DAGERAAD.

daggett (dæ·gėt). Also **degote, degutt.** [ad. Russ. *dëgot'* tar.] A dark tar obtained by the distillation of the bark of the European white birch, and used in the preparation of Russia leather, and formerly as a local application for diseases of the skin.

1861 *Chambers's Encycl.* II. 104/1 [The birch] yields also the B[irch] Tar, or *Degutt*. [**1875** *Encycl. Brit.* III. 698/1 An empyreumatic oil, called *diogott* in Russia.] **1890** BILLINGS *Nat. Med. Dict.*, *Daggett*, birch-tar. **1900** DORLAND *Med. Dict.*, *Degote*, tarry oil of white birch. **1935** *Discovery* Oct. 299/2 Birch bark..by destructive distillation, yields an empyreumatic oil, known as Oil of Birch Tar, Dagget [etc.].

dagging, *vbl. sb.* Add: *spec.* the operation of cutting off the 'dags' or locks of dirty wool from (sheep). (Austral. and N.Z. examples.)

1898 H. B. VOGEL *Maori Maid* xix. 147 When the docking, and dagging, and cutting comes. **1915** J. R. MACDONALD *N.Z. Sheepfarming* xl. 120 Cleanliness, by way of proper dipping, dagging, and crutching are good preventive aids [against blowfly]. **1934** T. WOOD *Cobbers* viii. 99 Dagging, in polite terms, is the removal, by shears, of wool which is matted on the thighs by excreta.

daggy (dæ·gi), *a.* *N.Z.* [f. DAG *sb.*[1] 3 + -Y[1].] Of a sheep or wool: clotted with dags.

1923 W. PERRY *Sheep Farming in N.Z.* vi. 75 The ewes will be very daggy by the lambing season. **1940** G. MEEK in A. E. Woodhouse *N.Z. Farm & Station Verse* (1950) 153 A truck of daggy wool. **1950** *N.Z. Jrnl. Agric.* Oct. 310 One daggy sheep can stain quite a few clean sheep when penning up is in progress.

dagher: see *DAGGA[2].

Daghestan, Dagestan (dāgėstā·n). The name of a region of the eastern Caucasus, used *attrib.* and *ellipt.* of locally made rugs with geometric designs.

1904 W. D. ELLWANGER *Oriental Rug* vi. 61 The Daghestan rugs of Caucasia are only second in importance to those from Persian looms. **1913** W. A. HAWLEY *Oriental Rugs* xi. 199 The rugs which acquired the trade name of Daghestans are different from almost all other Caucasian pieces. **1931** A. U. DILLEY *Oriental Rugs* Pl. 51 (*caption*) Border motive became the property of Daghestan and Shirvan rugs.

Dago, dago. For *U.S.* read *slang* (orig. *U.S.*); add earlier and later examples. (Now a disparaging term for any foreigner.) Also *attrib.*

[**1723** BUMSTEAD in *New England Hist. & Gen. Reg.* (1861) XV. 199 The negro Dago hanged for fiering Mr Powell's house.] **1832** E. C. WINES *Two Years in Navy* (1833) I. vi. 145 These *Dagos* [of Minorca], as they are pleasantly called by our people, were always a great pest. **1858** *Knickerbocker* Jan. 7 And so, Bill, you served as a ingineer with these ere blamed dagos, you say. **1882** W. A. BAILLIE-GROHMAN *Camps in Rockies* 372, I waited until a lot of Dago emigrants passed. **1899** *Westm. Gaz.* 11 Dec. 3/2 Whilst licensed pilots take to drink, And Dago crews to prayer. **1902** [see *CHIN *sb.*[1] 1 e]. **1904** T. ROOSEVELT *Let.* 2 Sept. in H. F. Pringle *T. Roosevelt* (1931) 294 It will show these Dagos that they will have to behave decently. **1909** H. G. WELLS *Tono-Bungay* III. iv. 406 'E's a foreigner... That's what E is—a *Dago*! **1911** C. E. W. BEAN *'Dreadnought' of Darling* xxiv. 212 'Afghan' in the West is about as wide as 'Dago' on the coast. **1932** [see *ADJECTIVAL *a.* b]. **1934** N. MARSH *Man lay Dead* iii. 51 'Such indiscretion has doubtless been suitably chastised,' remarked the Russian... Charles Rankin..slipped his arm through Nigel's. 'Not a very delicious gentleman, that dago,' he said loudly. **1940** N. MITFORD *Pigeon Pie* iii. 41 There are Chinks and Japs and Fuzzy Wuzzies and Ice Creamers and Dagos, and so on. **1968** *Listener* 19 Dec. 819/2 England should have won. All that stopped us was that the dagos [*sc.* Paraguayans] got more goals than us.

2. The Spanish or Italian language.

1900 *Dialect Notes* II. 31 *Dago*, the Italian language. **1901** 'H. MCHUGH' *John Henry* 32 She said she was svelte. I suppose that's Dago for a shine. **1923** MRS. M. WATTS *Luther Nichols* 119 They were eternally being enjoined to say it in French, say it in German, say it in dago!

3. dago red, cheap red wine, esp. Italian. *U.S.*

1906 AITKEN & HILTON *Hist. Earthquake San Francisco* vi. 120 Casks of wine (real 'Dago red'). **1910** E. A. WALCOTT *Open Door* xii. 146 You know I'm.. pleased when the meal can be washed down only with diluted 'dago red'. **1963** *Freedomways* III. III. 410 Bad niggers stopped their drinking Dago red. **1966** J. DOS PASSOS *Best Times* iii. 83 As we poured down the dago red he would become mischievous.

Daguerrean (dăge·riăn), *a.* Also **Daguerreian, Daguerryan.** [See DAGUERREOTYPE.] Pertaining to Daguerre or the daguerreotype; photographic.

1843 J. EGERTON tr. *Lerebours's Treat. Photogr.* IV. XXV. 164 The Daguerreian phenomena. *Ibid.* 166 On the formation of the Daguerreian images. **1844** *Yale Lit. Mag.* IX. 381 His imitative bump is certainly 'large'—so large, in fact, that it becomes almost Daguerryan in its workings, in that its productions only want an appearance of life and health to counterfeit the original. **1851** C. CIST *Cincinnati* 161 Daguerrean rooms. **1878** W. ABNEY *Treat. Photogr.* ix. 63 The method of developing the Daguerrean image. *Ibid.* xxxiii. 264 Both of these eminent physicists employed the Daguerrean process with the greatest success in these researches. **1889** *Anthony's Photogr. Bull.* II. 5 A full fledged daguerrean artist.

dah (dā). [Burmese. See also *DAO, *DHA.] A short heavy sword, used also as a knife, especially in Burma; also *attrib.*

1832 J. BELL *Syst. Geogr.* IV. 588 A short sword called *dah*, having a blade of about a foot and a half in length. **1839** H. MALCOM *Trav. S.-E. Asia* I. II. iii. 247 Their chief tool, and one used for all manner of purposes, from the felling of a tree to the paring of a cucumber, is the dah. The handle is like that of a cleaver, and the blade like a drawing-knife. **1858** C. T. WINTER *Six Months Brit. Burmah* xiii. 104 The Burman has few agricultural implements...A cart, plough, dah (or sword-knife), and sickle are about all he requires. **1888** KIPLING *Departm. Ditties* (1890) 19 Amid the junglegrass..grinned and jabbered Little Boh Hla-oo and cleared the dah-blade from the scabbard. **1923** *Blackw. Mag.* Sept. 304/2 Thrust into his coloured sash were two silver-mounted dahs. **1947** 'N. SHUTE' *Chequer Board* v. 121 Many of them also wore their dahs, long straight steel blades with clumsy wooden handles.

dahabeeyah. Add: (Earlier and later examples in various spellings.)

1846 I. F. ROMER *Pilgrimage Temples of Egypt* I. 109 The Dahabieh and her crew were hired at a daily expense. **1876** *Western Morning News* (Plymouth) 2 Feb. 3/3 Three young English ladies..were sailing up the river in a dahabeah. **1905** E. F. BENSON *Image in Sand* i, A couple of dayabeahs moored to the bank were a blaze of Syrian awnings. **1916** J. B. COOPER *Coo-oo-ee* xvi. 244 The winding river, glowing with opal fire, and the dahabiehs, that was the Egypt of his dreams. **1936** E. M. FORSTER *Abinger Harvest* IV. 249 The Sahara where horsemen..are silently riding towards the Nile to intercept the dahabiyeh. **1952** C. P. BLACKER *Eugenics* 24 The party travelled in some style in a dahabeyah. **1963** *Times Lit. Suppl.* 11 Jan. 28/2 As his *dahabia* navigates the last of the cataracts.

dahil, var. *DAYAL.

Dahlgren (dā·lgren). Now *Hist.* [Name of J. A. B. *Dahlgren* (1809–70), U.S. naval officer.] In full *Dahlgren gun*. A cast-iron smoothbore gun invented by Dahlgren in 1850.

1861 *Times* 28 Dec. 6/6 They might..be relying on their new Dahlgren guns. **1862** G. B. MCCLELLAN *Disp.* 4 May in T. P. Kettell *Hist. Rebellion* (1865) 252 The rebels abandoned..four nine-inch Dahlgrens. **1867** H. LATHAM *Black & White* 86 Each [gunboat] was armed with two 15-inch Dahlgrens. **1876** VOYLE & STEVENSON *Milit. Dict.* (ed. 3) s.v., The Dahlgren 11-inch was once a formidable gun on the seas. **1910** *Sat. Even. Post* 30 July 27/3 They're big fellows—those Dahlgrens and Columbiads.

dahllite (dā·ləit). *Min.* [ad. G. *dahllit* (Brögger and Bäckström 1888, in *Vet. Akad. Stockh. Oefv.* XLV. 493). f. the names of T. and J. *Dahll*, Norwegian mineralogists: see -ITE[1].] A variety of hydroxyapatite containing carbonate.

1890 *Jrnl. Chem. Soc.* LVIII. 714 (*title*) Dahllite, a new Norwegian mineral. *Ibid.*, This new mineral, the only combination of a phosphate and a carbonate yet known, has been named dahllite, in honour of..Tellef and Johann Dahll. **1957** *Encycl. Brit.* II. 93/1 Carbonate-containing apatites related to fluorapatite and hydroxyapatite are known as francolite and dahllite, respectively.

Dahoman (dăhō^u·măn), *sb.* and *a.* Also **Dahomean, Dahomeyan, Dahomian.** [f. the name of the country *Dahomey* or the tribal name *Dahomeh*.] **A.** *sb.* **1.** A member of the people of the former kingdom of Dahomey, now part of the West African republic of the same name. **2.** The language of this people. **B.** *adj.* Of or pertaining to this state or people.

1793 A. DALZEL *Hist. Dahomy* I. i. 1 The Dahomans were formerly called Foys, and inhabited a small territory, on the north-east part of their present kingdom. *Ibid.* v. 21 The Dahoman Prince received the news of this victory with every possible demonstration of joy. **1820** J. MACLEOD (*title*) A voyage to Africa; with some account of the manners and customs of the Dahomian people. **1837** *Penny Cycl.* VIII. 285/2 The Dahomans..made their appearance on the coast. *Ibid.*, The captains of these forts..were now prisoners in the Dahoman camp. **1851** F. E. FORBES (*title*) Dahomey and the Dahomans, being the journals of two missions to the King of Dahomey. **1864** R. F. BURTON *Mission to Gelele* I. iii. 36 The Ffon, or Dahoman, a dialect of the great Yoruba family, has,..a G and a Gb, the latter..difficult to articulate. **1877** *Encycl. Brit.* VI. 766/2 The Dahomans have at several times penetrated along the beach towards the east as far as Badagry. **1883** C. A. MOLONEY *W. Afr. Fisheries* 50 It has been for some time conveniently contrary to the Dahomean fetish to cross water. **1941** *Jrnl. R. Anthropol. Inst.* LXXI. 49/2 Perhaps the Dahomean diviners did not know as much. **1957** *Encycl. Brit.* VI. 978/2 Of this population the majority was constituted by the Dahomeyans or Dahomi (in their own language called Fon

or Djedjé). **1959** *Chambers's Encycl.* IV. 349/1 In 1890.. French troops were heavily defeated by 9,000 Dahomeans. **1965** R. & D. MORRIS *Men & Snakes* ii. 27 When a Dahoman caught sight of one of the reptiles, he immediately prostrated himself on the ground.

dai: see *DAYE.

Dail Eireann (doɪl ē^ə·rən). [Ir. *Dáil Éireann* assembly of Ireland.] The lower house of the Parliament of the Republic of Ireland (before 1922, the Sinn Fein Parliament in Ireland). Also ellipt. **Dail.**

1919 *Times* 21 Jan. 9/4 A committee had been nominated 'to select Irish technical terms for the work of "An Dail Eireann",' the Gaelic appellation of the new Irish Parliament. *Ibid.* 22 Jan. 9/4 The first word in English was spoken when the Dail had been in session for about three-quarters of an hour. **1921** *Punch* 30 Mar. 254/3 The Government should enter into unconditional negotiations with Dail Eireann. **1923** *Contemp. Rev.* Sept. 332 Parliament has been sitting continuously ever since the last Dail elections. **1964** *Ann. Reg.* 1963 280 It was bitterly contested in the Dail.

dailiness. Delete *rare* and add examples.

1898 A. BENNETT *Man from North* xix, The drab dailiness of her existence in Carteret Street. **1899** *Academy* 14 Oct. 432/1 That which Mrs. Meynell has well called the 'dailiness' of life. **1906** *Daily Chron.* 22 May 6/6 All necessary implication of dailiness has long disappeared from the word 'journal'. **1906** E. V. LUCAS *Wand. in Lond.* i. 9 Perhaps a touch of grime is not unnecessary. Perhaps houses can be too clean for the truest human dailiness. **1935** W. J. BLYTON *Country Airs* viii. 74 Our homely, uncouth husbandry, with its divine humilities and dailiness.

daily, *a.* (*sb.*) Add: **A.** *adj.* **1. b.** *daily girl*, etc.

1921 *Dict. Occup. Terms* (1927) § 900 *Daily servant, daily girl*..; a non-resident general servant. **1933** A. BRYANT *Pepys: Man in Making* 434/2 Taylor, Pepys' 'daily woman'. **1939** D. L. SAYERS *In Teeth of Evidence* 36 The daily woman, Mrs. Crabbe, had been in the house till nearly dinner-time. **1971** *Woman* 13 Feb. 13/2, I gave Mrs. Candy, the daily help, a suit for her daughter.

3. *spec.* **daily breader** *colloq.*, one who earns his own living or 'daily bread'; **daily dozen** *colloq.*, used of physical exercises performed each day on rising (see quot. 1925).

1906 *Westm. Gaz.* 25 Sept. 1/3 In place of the neat villa of the daily-breader..it now journeys between rows of low, monotonous houses. **1930** J. A. R. MARRIOTT *How we Live* i. 8 Most of them live by their own exertions. They are in fact (to use the colloquial but expressive phrase) 'daily breaders'. **1942** N. STREATFEILD *Table for Six* 99 There I was catching a tube, a daily breader. **1919** *Red Cross Mag.* Dec. 30/3 It was while engaged in this work that he evolved what he calls his 'short-hand system' of setting-up exercises, or 'the daily dozen'. **1925** *Yale Univ. Obituary Rec.* 1349 Walter [Chauncey] Camp..devised..exercises for cabinet officers and civilian personnel in Washington out of which grew the 'Daily Dozen' series, which he introduced in all naval stations and in several aviation and army cantonments. **1936** WODEHOUSE *Laughing Gas* iv. 43 Ann is one of those girls who always look as if they had just stepped out of a cold bath after doing their daily dozen. **1965** W. LAMB *Posture & Gesture* ix. 121 Physical exercises, whether performed privately as a daily dozen, or in a class to the accompaniment of music, may be a pleasant and stimulating way of passing the time.

B. *sb.* (orig. *U.S.*) (Earlier and later examples.)

[**1823** D. WEBSTER *Priv. Corr.* (1857) I. 333, I am glad to see that you publish, in the Daily [*sc.* the *Boston Daily Advertiser*], your narrative.] **1832** J. K. PAULDING *Westward Ho!* I. xxi. 190 'Make out an estimate of the cost of establishing a paper.' 'A daily, sir?' 'Ay, a daily, if you wish.' *Ibid.* 191 I'll attack them in my Daily. **1846** [see WEEKLY *sb.*]. **1933** *Archit. Rev.* LXXIV. 176/2 The women's pages of the popular dailies. **1965** *New Statesman* 30 Apr. 679/1 The national daily he would like is still denied him. *Ibid.*, Two Fleet Street dailies as well as the *Sun* are..vulnerable.

2. 'Daily bread', food; livelihood.

1906 E. DYSON *Fact'ry 'Ands* xiv. 182 What's er bonzer like you doin' spreadin' sour paste fer yer daily? **1922** JOYCE *Ulysses* 57 Boland's breadvan delivering with trays our daily but she prefers yesterday's loaves.

3. A domestic cleaner or servant who does not live on the premises.

1933 MRS. C. S. PEEL *Life's Enchanted Cup* xix. 261 In my youth there were charwomen but the 'daily' is a new invention. **1953** 'P. WENTWORTH' *Watersplash* v. 31 Mrs. Deacon..was Miss Blake's daily and a very good cook. **1967** L. MEYNELL *Mauve Front Door* iii. 31 Most 'dailies' I have known have been disastrous. They come late; charge exorbitantly; drop ash all over the place.

4. *pl.* The first prints from cinematographic takes (*TAKE *sb.*), rapidly made for film producers or editors.

1934 *Tit-Bits* 31 Mar. 12/2 Every time a scene is successfully 'shot' it is called 'a take'; the whole of the day's 'takes' are then assembled and shown to the producer in a private projection room, but are then known as 'the rushes' or 'the dailies'. **1952** L. ROSS *Picture* (1953) iii. 106, I haven't had a chance to tell you how wonderful I feel the dailies (rushes) are.

Daiquiri, daiquiri (dəi·kiri, dæ·k-). Also **daquiri.** [f. *Daiquiri*, name of a district in Cuba.] A cocktail containing rum, lime, etc.

1920 F. SCOTT FITZGERALD *This Side of Paradise* (1921) I. iii. 115 Here's the old jitney waiter. If you ask me, I want a double Daiquiri. **1921** J. HERGESHEIMER *San Cristóbal de la Habana* 17 The moment..had arrived for a Daiquiri...I lingered over the frigid mixture of Ron Bacardi, sugar, and a fresh, vivid green lime. **1929** F. SHAY *Drawn from Wood* 181 Daquiri. **1935** *Lady Sysonby's Cook Bk.* 264 Daquiri (*Long drink*). The juice of 1 fresh lime, 1 teaspoonful of sugar. Fill your tumbler with dry crusted ice, and pour over Baccardi rum. Add sprigs of mint on the top. **1958** G. GREENE *Our Man in Havana* V. iv. 228 It was always a daiquiri I used to drink with him. **1967** L. JAMES *Chameleon File* (1968) x. 118 Wilson ordered the frozen Daiquiri which had been invented at La Florida.

dairy, *sb.* Add: **4.** *dairy cow, produce*; *dairy-fed* adj. (earlier and later examples); **dairy butter**, butter made at a private dairy; **dairy cream**, real cream as distinct from synthetic cream; **dairy factory** chiefly *N.Z.*, a factory with plant for the conversion of milk into butter or cheese; **dairy-farm** (examples); **dairy-farmer** (earlier and later examples); **dairy herd**, a herd of milch-cows; **dairy shorthorn**, a shorthorn bred primarily to yield milk.

1874 *U.S. Dept. Agric. Rep.* 1873 250 During 1873,.. 'good dairy' butter touched 40 cents as its highest extreme. **1656** R. VERNEY in M. M. Verney *Mem.* (1894) III. viii. 271, 4 Dairy Cowes..13 draught Bullocks. *c* **1830** *Farm-Rep. Glouc. Hill-Farm* 17 in *Brit. Husbandry* (1840) III, The twenty heifer-calves are bred to keep up the stock of dairy-cows. **1963** A. CLARKE *Coll. Plays* 291 A thousand Shorthorns and half as many dairy cows Are hers. **1962** L. DEIGHTON *Ipcress File* ii. 19 The notice that told customers not to expect dairy cream in their pastries. **1970** *Harrod's Summer Food News* (back cover), Fresh dairy cream, fresh eggs, pure sugar, fresh milk, and natural flavourings are used to produce these wonderful ice creams. **1888** J. P. DOWLING *Dairying in Australia* iv. 19 Dairy Factories—Co-operative and otherwise. **1888** R. M. MCCALLUM *Rep. Dairy Factories in N.Z.* 5, I have inspected a number of the dairy factories in the colony. **1950** *N.Z. Jrnl. Agric.* Jan. 21/3 There are two dairy factories in the area, a butter factory at Wairoa and a cheese factory at Nuhaka. **1959** A. MCLINTOCK *Descr. Atlas N.Z.* 57 Dairy factories are in the main owned cooperatively by the farmers. **1784** J. TWAMLEY *Dairying* 93 There is no branch in Husbandry seems of more importance..than the conducting and managing of Dairy-farms. **1807** A. YOUNG *Agric. Essex* II. 270 Dairy farms at Bumpstead and Hempstead, with much more grass than arable. **1895** *U.S. Dept. Agric. Yearbk.* 1894 295 The dairy farm should be carefully selected, all the requirements of the business being well considered. **1790** W. H. MARSHALL *Rural Econ. Midl. Counties* I. 354 A dairy farmer declares, that, one year, he lost forty pounds, by the mismanagement of his dairywoman. **1906** *Westm. Gaz.* 10 Oct., People often ask why English dairy-farmers have not adopted the co-operative methods. **1831** *Lincoln Herald* 21 Oct. 1/1 For dairy-fed porkers the price is at 4s to 5s. **1961** *Guardian* 28 Apr. 6/6 An organised..sales programme..for dairy fed meat and poultry. **1879** J. P. SHELDON *Dairy Farming* I. 5/2 Whilst a cow is kept in the dairy herd. **1928** *Daily Express* 3 Feb. 5/2 Instruction is given in the management of small dairy herds. **1842** *Ainsworth's Mag.* I. 44 The sale of dairy produce at the market town. **1930** T. S. ELIOT tr. *St. J. Perse's Anabasis* 59 The Stranger.. honoured with gifts of dairy produce and fruit. **1932** *Discovery* Feb. 58/2 Indeed with the most important of English breeds—the dairy shorthorn—the quality of calves when they enter the dairy herd seems still largely a matter of chance. **1957** *Encycl. Brit.* V. 47/1 Strains of Shorthorns have been selected for milk and butterfat production, as well as beef, and in the United States are called Milking Shorthorns; in Canada, Dual-Purpose Shorthorns; in England and Australia, Dairy Shorthorns.

dais. Add: Also **daïs.** Pl. **daises, daïses.** (The disyllabic pronunciation is now the more usual.)

1893 F. THOMPSON *Poems* 50 Underneath her azured daïs, Quaffing, as your taintless way is, From a chalice. **1898** H. NEWBOLT *Island Race* 69 The College Eight and their trainer dining aloof, The Dons on the daïs serene. **1907** R. M. BURROWS *Discoveries in Crete* i. 10 At one end of a pillared hall..there is a narrow raised daïs. **1927** in *Amer. Speech* (1929) V. 132 *Dais*—two syllables. **1967** A. C. GIMSON *Everyman's Eng. Pronouncing Dict.* 118/2 *Dais,-es* 'deiis (deis),-iz.

2. c. *Freemasonry.* (See quot. 1866.)

1866 *Masonic Eclectic* Sept. 371 *Dais*, the platform or raised floor in the East, on which the presiding officer is seated. **1925** A. HARDINGE *Life H. H. M. Herbert* I. 223 The crippling decisions of the Grand Master and the 'Dais' or board. *Ibid.*, The 'Dais' was consternated at the audacity of so young a brother.

daisy, *sb.* Add: **1. c.** Slang phrases: *under the daisies*, dead and buried; *to push up daisies*, *to turn one's toes up to the daisies*, to be in one's grave, to be dead.

1842 BARHAM *Ingol. Leg.* Ser. II. *Babes in the Wood* iv, Be kind to those dear little folks When our toes are turned up to the daisies. **1866** G. MACDONALD *Ann. Q. Neighb.* I. xi. 356, I shall very soon hide [my name] under some daisies. *a* **1918** W. OWEN *Poems* (1963) 65 'Pushing up daisies' is their creed, you know. **1928** S. VINES *Humours Unreconciled* xxi. 268, I think she's drinking herself under the daisies, so to speak. **1938** G. HEYER *Blunt Instrument* xiii. 252 'Where is the wife now?'..'Pushing up daisies...Died..a couple of years ago.' **1961** S. CHAPLIN *Day of Sardine* xiii. 245 Everybody goes to hell their own way and by the time you're privileged to be able to help a lost soul he'll be pushin' up daisies. **1970** *Guardian* 30 Dec. 9/6 In ten years time I think I should be pushing up daisies.

6. b. *U.S. slang.* (See sense 5.) Also as *adv.*

1886 [see sense 5]. **1887** F. FRANCIS *Saddle & Mocassin* x. 189 Well, if he can kick anything out of a Government mule, he's a daisy burro. **1892** *Harper's Mag.* Feb. 438/1 A passenger informed on him for having his coat unbuttoned. Daisy passenger, wasn't it? **1902** S. E. WHITE *Blazed Trail* xxxvii. 252 She's my daisy Sunday best-day girl. **1905** R. BEACH *Pardners* (1912) v. 130 The noose sailed up and settled over him fine and daisy. **1927** E. WALLACE *Mixer* i. 14 I'll introduce you to the daisiest night club in town.

7. *daisy-dotted, -flowered, -frilled, -peeping, -powdered* adjs.; **daisy fleabane** *U.S.*, any of several plants of the genus *Erigeron*; **daisy roots** *Rhyming slang*, boots; also *ellipt.*; **daisy-tree** *Austral.* and *N.Z.* (see quots.).

1848 A. GRAY *Man. Bot.* 206 *Erigeron annuum*, Daisy Fleabane. *Ibid.*, Narrow-leaved *E. strigosum*, Daisy Fleabane. **1872** *Rep. Vermont Board Agric.* I. 279 *Erigeron annuum* and *E. strigosum*, Daisy Fleabanes, acrid plants, mingle their coarse stalks quite too freely with the hay from newly seeded land. **1931** W. N. CLUTE *Common Names Plants* 131 'Kiss-me-and-I'll-tell-you' replied an attractive native of the Southern States when asked the name of that plant which people of colder climes know as the daisy fleabane. **1925** W. J. ARKELL in *Oxf. Poetry* 10 The daisy-dotted meadow. **1881** WILDE *Poems* 209 Your queen in daisy-flowered smock. **1924** E. SITWELL *Sleeping Beauty* x. 36 Her daisy-frilled frock. **1929** BLUNDEN *Near & Far* 41 Through spring's daisy-peeping wonder. **1820** 'JANUS WEATHERCOCK' in *London Mag.* Mar. 287/1 A dark-haired girl, 'amorous of mischief', curled on the daisy-powdered grass. [**1859** HOTTEN *Slang Dict.* 290 *Daisy recroots* (so spelt by my informant of Seven Dials, he means, doubtless, *recruits*), a pair of boots.] **1859** MATSELL *Vocabulum* 24/1 *Daisy-roots*, boots and shoes. **1873** HOTTEN *Slang Dict.* 366 *Daisy roots*, a pair of boots. **1879** *Macm. Mag.* XL. 501/2 I piped three or four pair of daisy-roots (boots). *Ibid.* 503/1 While waiting for my pal I had my daisies cleaned. **1943** *Gen* 25 Sept. 50/1 Your toes is poking out of your daisy-roots. **1898** MORRIS *Austral Eng.* 113/1 *Daisy Tree*, two Tasmanian trees, *Aster stellulatus*..and *A. glandulosus*. **1926** J. C. ANDERSEN in *Trans. N.Z. Instit.* LVI. 702/2 *Olearia*: daisy tree, daisy-tree, tree-daisy.

daisy-cutter. 2. *Cricket.* (Earlier and later examples.)

1857 *Bell's Life* 1 Nov. 7/1 The umpires called play, Grange being again on the defensive to the under-hand 'daisy cutters' of Sadler. **1963** *Times* 13 June 13/3 The ball that kept low had to be a daisy-cutter although a first-class cricket pitch has surely been sufficiently mown to lose all likeness to a flowery meadow.

dak. Now the usual spelling of DAWK *sb.*[2]

daka, dakha, varr. *DAGGA[1].

‖ **dakhma** (dā·kmă). Also **dokhma.** [Pers.] A tower of silence (see SILENCE *sb.* 2 c).

1865 [see SILENCE *sb.* 2 c]. **1908** *Westm. Gaz.* 4 Aug. 10/1 Neither God is pleased nor the dead are benefited by a showy expenditure on a Dokhma. **1912** H. G. RAWLINSON *Bactria* iii. 40 Alexander promptly ordered the *dakhmas*, or Towers of Silence, to be closed. **1957** *Encycl. Brit.* XXIII. 989/1 The dead..are to be left on the appointed places (*dakhmas*) and exposed to the vultures and wild dogs.

Dakin (dē^i·kin). The name of H. D. *Dakin* (1880–1952), British chemist, used in the possessive to designate a solution of sodium hypochlorite used as an antiseptic, or a dressing saturated in this fluid.

1920 MARTINDALE & WESTCOTT *Extra Pharmacop.* (ed. 17) I. 56 Dakin's (Stronger) Hypochlorite Solution. **1927** A. J. CLARK *Applied Pharmacol.* (ed. 2) 10 The chief chlorine compounds used for disinfection of the body are: (1) eusol and Dakin's fluid. **1928** EDMUNDS & GUNN *Cushny's Textbk. Pharmacol.* (ed. 10) 160 Dakin's solution is prepared by adding chlorinated lime to a solution of sodium carbonate. **1929** F. A. POTTLE *Stretchers* (1930) 149 Small eclat removed. Dakin dressing. **1963** *Brit. Pharmaceutical Codex* 1198 Surgical solution of chlorinated soda, Dakin's solution. **1964** J. J. WALSH *Understanding Paraplegia* xi. 78 The urinal should be..soaked for two hours in a suitable antiseptic solution such as 16% Dakins [*sic*] Solution.

Dakota (dăkō^u·tă), *sb.* and *a.* Also **Dacota, Dacotah, †Dahcota.** [Dakota (Santee dialect) *dakota*, lit. 'allies'.] **A.** *sb.* **1.** A North American Indian tribe inhabiting the upper Mississippi and Missouri river valleys, speaking a language of the Siouan stock; a member of this people. Also commonly called *SIOUX.

[**1804** LEWIS & CLARK *Jrnls.* (1904) I. 132 This Great Nation who the French has given the Nickname of Suouex, Call themselves *Dar co tar*...Those *Dar ca ter's* or Suoux inhabit..the country on the Red river.] **1810** *Ibid.* (1962) 508 As to Sioux Indians..the Indians call themselves Dacota. **1824** W. H. KEATING *Narr.* I. viii. 376 The Dacotas are a large and powerful nation of Indians. **1852** S. R. RIGGS *Gram. & Dict. Dakota Lang.* p. xv, The nation of the Sioux Indians, or Dakotas, as they call themselves, is supposed to number about twenty-five thousand. **1877** L. H. MORGAN *Anc. Soc.* II. ii. 86 In some tribes, as among the Dakotas, the gentes had fallen out. **1948** A. L. KROEBER *Anthropol.* (ed. 2) viii. 322 Among the Dakota each man was free to seek his own 'vision' or inspiration from supernaturals.

2. The language of this people.

1923 A. L. KROEBER *Anthropol.* v. 116 The number of words recorded in Klamath,..is 7,000;..in Dakota, 19,000. **1933** BLOOMFIELD *Lang.* iv. 72 The Siouan family includes many languages, such as Dakota, Teton, Oglala, [etc.].

B. *adj.* Of, pertaining to, or designating this people or their language.

1809 in D. Jackson *Lett. Lewis & Clark Exped.* (1962) 476 A Large Mantle, made of the Buffalow skin, worn by the Scioux, or Soue, Darcota Nation. **1841** J. BELL *Let.* 31 Aug. in J. D. Richardson *Compilation Messages of Presidents*

(1897) IV. 62 They are divided into bands..the generic name for the whole being the Dahcota Nation. **1846** C. LANMAN *Summer in Wild.* (1847) 56 Here it was I first saw.. Sioux or Dacotah Indians. **1852** S. R. RIGGS (*title*) Grammar and dictionary of the Dakota language. **1880** *Encycl. Brit.* XII. 831/2 The Dakota or Sioux nation is at present the most powerful of the Indian tribes in North America. **1881** *Harper's Mag.* Feb. 472/1 The story is told by..the author of a grammar and dictionary of the Dakota tongue. **1902** *Encycl. Brit.* XXXIII. 890/1 The Dakota Indians invented a chronological table, or winter count. **1959** E. TUNIS *Indians* 21/2 About thirty different Indian names are listed for the Dakota Indians, and a fair percentage of them can be translated to mean 'cutthroats'.

Hence **Dako·tan, -ian** *sb.* and *adj.*, (an inhabitant) of the Dakota Territory, or of the states of North or South Dakota.

1884 M. D. WOODWARD *Diary* 22 Mar. in *Checkered Yrs.* (1937) 32 The Dakotan of the next generation should be an educated person. **1889** *Cent. Dict.*, Dakotan, *a.*..*n.* **1898** *Monthly South Dakotan* I. 5 Here the judge first inculcated the principle of Dakotan jurisprudence. *Ibid.*, One hundred sorrowing Dakotans. **1914** W. H. RIVERS *Kinship & Soc. Organ.* 89 Professor Kroeber does not specify which kinds of grandfather and father-in-law are classed together in Dakotan nomenclature. **1947** *Amer. Speech* XXII. 249 Mr. Bruce Nelson, of Bismarck, N.D., tells me that *Dakotan* now prevails [over *Dakotian*] in both Dakotas. **1861** (*title*) Weekly Dakotian. **1877** L. M. MORGAN *Anc. Soc.* II. vi. 154 (*heading*) Dakotian Tribes. A large number of tribes are included in this great stock of the American aborigines.

Dalcroze (dælkrō^u·z). The name of É. Jaques-*Dalcroze* (1865–1950), a Swiss exponent of musical education through physical exercises, used *attrib.* and *ellipt.* to designate his system, the movements involved, or an institution where the method is taught. Cf. *EURHYTHMIC *a.*

1913 G. B. SHAW *Let.* 30 June in A. Dent *Shaw & Mrs. P. Campbell* (1952) 125 The Dalcroze school at Hellerau..is very interesting. **1920** D. W. BLACK in B. Russell *Pract. & Theory Bolshevism* I. iv. 65, I saw..some quite wonderful Eurythmic dancing, in particular an interpretation of Grieg's *Tanz in der Halle des Bergkönigs* by the Dalcroze method. **1920** D. H. LAWRENCE *Women in Love* (1921) xiv. 172 'Will you sing while I do Dalcroze?'..'While you do—?,' she asked vaguely. 'Dalcroze movements,' said Gudrun. **1964** *New Society* 13 Feb. 16/3 The period 1915–30 was a time for trying out experimental methods: Dalcroze eurythmics, the Winnetka technique, the project method, the Dalton Plan, the Junior Republics, the Montessori system.

Dalecarlian (dalėkā·ɹliăn), *a.* [f. *Dalecarlia* (see below) + -AN.] Of or pertaining to the province of Dalecarlia (Dalarna) in central Sweden or its inhabitants or their language. Also as *sb.*, a native or inhabitant of Dalecarlia.

1837 *Penny Cycl.* VIII. 290/1 The Dalecarlians are distinguished by their stature, courage, spirit of independence, and frankness of character. **1881** C. C. HARRISON *Woman's Handiwork* I. 95 Quaint Dalecarlian lace of an antique pattern has been brought to America by Swedish peasants immigrating to the far West. **1932** *Antiquity* VI. 429 A late Dalecarlian runic alphabet. **1959** *Times* 10 Jan. 9/6 The Dalecarlian poet Erik Axel Karlfeldt.

Dalek (dā·lek). [Invented word.] A type of robot appearing in 'Dr. Who', a B.B.C. Television science-fiction programme; hence used allusively. Also *attrib.* and *Comb.*

1963 *Radio Times* 26 Dec. 11/1 Dalek voices: Peter Hawkins, David Graham. **1966** *BBC Handbk.* 39 The main activity over the period in this 'merchandising' operation concerned the widely popular Daleks from the 'Dr. Who' series. Some sixty licences for the production of Dalek-inspired articles were issued. **1969** C. HODDER-WILLIAMS 98·4 ix. 49 Under what interesting new law do you propose to enforce this regime? Or have you hired the Daleks? **1971** *Radio Times* 30 Dec. 10/1 Who are the Daleks? Dr. Who's most dangerous enemies, written into his second adventure in 1963 by Terry Nation, who named them after an encyclopaedia volume covering DAL-LEK.

Daliesque (dā:li,e·sk), *a.* Also **Dali-esque.** [f. the name of Salvador *Dali*, Spanish painter (born 1904) + -ESQUE.] Resembling the style, or in the manner, of the paintings of Dali; surrealistic.

1941 *Time* 24 Feb. 40/2 A Daliesque-Italian-primitive trifle in which a monkey-like Satan deftly garners three damsels. **1958** *Observer* 26 Jan. 5/5 A patch over one eye on which somebody had painted a Daliesque white eyeball. **1968** *Punch* 24 Apr. 619/2 The present Bond (Lucky James?) seems less dominant—the Dali-esque melting pistol on the jacket may be symbolic.

dallop, var. *DOLLOP *v.*

Dally (dæ·li), *a.* *N.Z. colloq.* [abbrev. DALMATIAN.] Of Dalmatian origin. Also as *sb.*, a native of Dalmatia; a person of Dalmatian parentage.

1940 F. SARGESON *Man & Wife* (1944) 10 Two young Dallies who ran an orchard..had come out from Dalmatia. **1950** O. E. MIDDLETON in D. M. Davin *N.Z. Short Stories* (1953) 411 You can buy an awful lot of Dally plonk for four pounds. **1961** B. CRUMP *Hang on a Minute* 77 Henry.. got this dog off an old Dally scrub-cutter. *Ibid.*, A couple of days later they found the Dally dead on the hillside.

Dalmatian, *a.* and *sb.* Add: (Earlier and later examples.)

1581 W. ALLEN *Apol. Two Eng. Coll.* iii. f. 29ᵛ He [*sc.* Gregory XIII] hath made one [seminary] for the Dalmatians. **1654** T. BAYLY *End to Controversie* 11 Dalmatians, who immediately after their supper inducing, or mixing, the Heresies of Manchæus with the Catholike Religion, were overthrown, and supplanted by the Turks. *a* **1680** EVELYN *Diary* June an. 1645 (1955) II. 446 A prospect down the Adriatic as far as Istria & the Dalmatian side. **1788** [see SERVIAN *sb.*]. **1810** *Sporting Mag.* XXXVI. 61/2 Portrait of a Dalmatian dog. **1813** J. C. EUSTACE *Tour through Italy* I. vii. 163 This wonderful event..is attested by the ocular evidence of some Dalmatian peasants. **1881, 1897** [see PLUM PUDDING c]. **1911** *Encycl. Brit.* XXIII. 507/2 The Venetians..regained their power over the Dalmatians. **1936** A. W. CLAPHAM *Romanesque Archit.* iii. 63 That masterpiece of Dalmatian art the west doorway of the cathedral of Trogir (Trau). **1959** *Chambers's Encycl.* IV. 579/2 The Dalmatian..is large with a white coat covered with black or brown spots.

2. A Romance language formerly spoken by natives of Dalmatia.

1911 *Encycl. Brit.* XXIII. 507/2 Of these nine [Romance] languages, Dalmatian is now extinct. **1933** BLOOMFIELD *Lang.* iv. 61 The Dalmatian is extinct: one of the dialects, Ragusan, died out in the fifteenth century; another, Veliote, survived into the nineteenth. **1954** PEI & GAYNOR *Dict. Linguistics* 52 *Dalmatian*, an extinct Romance language, spoken formerly on the eastern coast of the Adriatic.

dalo (dā·lo). [Fijian.] = TARO.

1879 *Encycl. Brit.* IX. 156/2 The taro or dalo..is grown in ditches, by streams, or on irrigated ground. **1924** *Countries of World* XIX. 1874/1 The yam and the dalo, for which irrigation is necessary, are the staple roots.

Dalradian (dælrā·diăn), *a.* *Geol.* [f. *Dalrad-*, altered form of *Dalriada*, name of an ancient kingdom in Scotland and northern Ireland + -IAN.] Epithet of a series of metamorphosed sedimentary and volcanic rocks found in a belt running north-east from the west coast of Ireland through the southern highlands of Scotland, thought to be of early Cambrian age and in part probably pre-Cambrian; of or characteristic of this series. Also *ellipt.*

1891 A. GEIKIE in *Q. Jrnl. Geol. Soc.* XLVII. 75 It is well known that from the old kingdom of Dalriada, in the north of Ireland, a colony settled in Argyllshire, and gradually acquiring dominion over the whole of Scotland, gave that kingdom its present name. I would therefore propose that the term 'Dalradian' might be adopted..for the crystalline schists of the north of Ireland and centre and south-west of Scotland. (The adjective ought properly to be 'Dalriadan', with the accent on the second syllable; but I feel compelled to alter it into a form more consonant with English habits of pronunciation.) **1903** —— *Text-bk. Geol.* (ed. 4) I. II. ii. 192 The limestones of the Dalradian metamorphic series of Scotland (Islay), which may possibly be pre-Palæozoic. **1910** *Encycl. Brit.* XIV. 744/2 The other metamorphic areas of the north [of Ireland] present even greater difficulties... Hence it is useful to speak of them merely as 'Dalradian'. **1931** GREGORY & BARRETT *Gen. Stratigr.* i. 33 The Lennoxian, along the southern side of the Dalradian, is a band of slates, grits, and fine conglomerates composed of fragments of the Dalradian Schists. **1959** *Chambers's Encycl.* XI. 164/2 The Dalradian Series, which is of considerable thickness, contains a great variety of metamorphosed sediments including quartzites, grits, limestone, slates, phyllites, mica-schists and gneisses. **1962** *Times* 6 Feb. 3/7 His work on structures and metamorphism in the Dalradian of the Scottish Highlands. **1967** D. H. RAYNER *Stratigr. Brit. Isles* iii. 66 This genus is only known from beds of Lower Cambrian age..and the uppermost Dalradian strata are therefore Cambrian, but how much is uncertain.

Dalton[1] (dǭ·ltən). The name of the high school (at *Dalton*, Mass., U.S.A.) in which the educational plan so named (devised by Miss Helen Parkhurst) was first adopted in 1920, which consists essentially in dividing up the year's work into monthly 'assignments' which the pupils contract to carry through (with certain preliminary aids) on their own responsibility and with their own discipline. Hence **Da·ltonize** *v. trans.*, to manage or arrange by this educational method; **Da:ltoniza·tion**; **Dalto·nian** *sb.*[2], an advocate of the method; **Da·ltonism**[2], the method itself.

1920 *Times Educ. Suppl.* 18 Nov. 605/4 Mr. Ernest Jackman, headmaster of the Dalton High School, writes:—I am glad to answer the questions of various English educators regarding Miss Parkhurst's plan, now to be renamed at her desire, 'The Dalton Plan'. **1922** H. PARKHURST *Educ. on Dalton Plan* ii. 15 The Dalton Laboratory Plan provides that means by diverting his energy to the pursuit and organization of his own studies in his own way. *Ibid.* iv. 40 Demonstrating the superiority of the Dalton Plan from the point of view of economy. **1924** A. J. LYNCH *Individual Work & Dalton Plan* iii. 31 Review provision is made for six laboratories or subject-rooms corresponding with the six subjects that are Daltonised. *Ibid.* 34 The teaching of arithmetic under the Dalton Plan. *Ibid.* 47 Convinced Daltonians recognise at once that assignments are the heart and centre of the plan. *Ibid.* vi. 124 He could find no fault with any other Dalton teacher. **1927** A. HUXLEY *Proper Studies* 117 The first step in the Daltonization of a school consists in the abolition of class rooms and the substitution of specialist rooms. *Ibid.* 125 In a well-run Daltonized school the problem of discipline solves itself. *Ibid.* 133 These ancient seats of learning [*sc.* Oxford and Cambridge] were Daltonized long before Daltonism was invented. **1959** *Chambers's Encycl.* IV. 353/2 In a number of English schools, where a modified Dalton plan is still used, one finds a mixture of class-teaching and of individual assignments. **1964** [see *DALCROZE].

dalton[2] (dǭ·ltən). Also **Dalton.** [f. the name of John *Dalton* (see DALTONIAN *a.* and *sb.*).] A name for the atomic mass unit (see *ATOMIC *a.* and *sb.* A. 1), used chiefly in *Biochem.*; freq. used as a dimensionless unit of molecular weight.

1938 C. M. BEADNELL *Dict. Sci. Terms* 65/1 *Dalton*, mass unit, being the 1/16 of mass of O atom. **1967** *New Scientist* 27 Apr. 196/1 The size in molecular weight units was about five million daltons—in other words each molecule of DNA weighed as much as five million atoms of hydrogen. **1969** *Nature* 11 Oct. 150/1 All DNAs were sheared to a single-stranded molecular weight of about 400,000 Daltons by passage through a French pressure cell. **1970** *Ibid.* 28 Nov. 889/2 Thus it would be correct to write..'the molecular mass of protein X is 250,000 [*read* 25,000] daltons'; or 'the relative molecular mass (that is, molecular weight) of protein X is 25,000'...It would, however, be incorrect to say: 'the molecular weight of protein X is 25,000 daltons', for the dalton is a unit of mass, and molecular weight is dimensionless.

Daltonian, *a.* and *sb.* **A.** *adj.* (Earlier and later examples.)

1813 *Ann. Philos.* II. 32 (*heading*) On the Daltonian theory of definite proportions in chemical combinations. **1962** S. TOULMIN in *Quanta & Reality* 21 With his Periodic Table, Mendeléeff had shown that chemical atoms fell into well-defined families, having resemblances which could not be explained on Daltonian principles alone. **1962** M. P. CROSLAND *Hist. Stud. Lang. Chem.* iii. 264 Even Berzelius ..made use of Daltonian symbols in at least one publication.

dam, *sb.*[1] Add: **2. a.** (S. African examples.)

1857 D. LIVINGSTONE *Missionary Travels* v. 103 The industry of the Boers augurs well for future formation of dams and tanks. **1884** A. K. JOHNSTON *Africa* (ed. 4) xxiv. 391 Wherever there is a homestead, there is generally a water 'dam', with an orchard and garden.

4. d. A reservoir or tank, as of loam and brick construction, in which metal is collected for heavy castings.

a **1877** KNIGHT *Dict. Mech.* I. 674/2. **1880** *Encycl. Brit.* XIII. 299/2 The tymp arch usually projects a little forward from the earth wall, constituting the 'fore hearth', at the base of the front of which is the dam, a block of stone or mass of firebrick pierced by a vertical cavity..through which the molten pig iron is drawn off from time to time.

e. In full *rubber dam*. A sheet of soft rubber pierced with one or more holes and fitted in the mouth so as to protect the exposed tooth or teeth from saliva while a filling or other operation is done; also (without an article), rubber in the form used for this purpose. Chiefly *U.S.*

It was invented by S. C. Barnum, of New York, in 1864.

1872 L. P. MEREDITH *Teeth* 117 By the use of the rubber-dam inconvenience and unsuccessful operations may be avoided. **1908** G. E. HUNT in C. N. Johnson *Text-bk. Oper. Dentistry* ix. 146 For operations on the anterior six teeth a triangular shaped piece of dam will be found both efficient and economical. **1927** J. D. H. JAMIESON *Oper. Dentistry* iii. 30 It is used in the form of strips of rubberdam. *Ibid.* 33 Rubberdam is supplied in three grades of thickness. **1940** S. D. TYLMAN *Theory & Pract. Crown & Bridge Prosthesis* xxiii. 284 In those patients whose flow of saliva is copious, much time may be saved by the use of a rubber dam. **1969** R. E. MCDONALD *Dentistry for Child & Adolescent* x. 176/2 A few explanatory words and referral to the rubber dam as a 'raincoat' for the tooth or as a 'Halloween mask' will invariably allay the child's fear.

dam (dām), *sb.*[5] *Obs.* exc. *Hist.* Also **daum, dawm.** [Hind. *dām*.] An Indian copper coin of the value of one fortieth of a rupee.

1781 F. BALFOUR tr. *Forms of Herkern* 39 The sum of twenty one lacks. of dams. **1801** R. PATTON *Asiat. Mon.* 182 A crore of dams. **1871** in E. G. BALFOUR *Cycl. India* (ed. 2) II. 10. **1884** *Encycl. Brit.* XVII. 343/1 The gold coinage and the silver rupee are seldom seen [in Nepal], the ordinary currency consisting of the copper dāms and paisā, and the mōhar or half rupee. **1962** R. A. G. CARSON *Coins* 515 On occasional issues both of Sher Shah and his successor Islam Shah (1545–52) the coins were struck on a square flan. A new denomination, the dam, was struck in copper at a weight of 21·5 gm.

dam (form of *damn*): for recent examples see *DAMN *a.* and *adv.*

Dama: see *DAMARA.

damage, *sb.* Add: **5.** Esp. in phr. *what's the damage?* how much is there to pay?

1829 J. HUNTER *Hallamshire Gloss.* 29 'What is the damage?' This expression is equivalent to 'What expence have I incurred?' 'What must I pay?' **1852** [see Dict.]. **1875** J. G. HOLLAND *Sevenoaks* xxii. 303 What's the damage for the sort o' thing ye're drivin' at this morning? **1888** A. C. GUNTER *Mr. Potter of Texas* xi, 'What's the damage?' 'Damage?' echoes Lubbins, not understanding this Americanism. 'Yes, how much do I owe?'

damaged, *ppl. a.* Add: **b.** *damaged goods*: merchandise that has deteriorated in quality through unsaleability, exposure to the elements, etc. Also *fig.* of persons, esp. implying a woman's loss of virginity (see also quot. 1911).

1809 [see GOOD *sb.* 8 b]. **1815** J. SCOTT *Visit to Paris in 1814* xiii. 264 Damaged goods of every description, were

brought out for the Cossack-market. **1841** S. WARREN *Ten Thousand a Year* i, Didn't I feel like damaged goods, just then! **1872** *Porcupine* Aug. 330/3 Everything seemed to be done in some shops to steal money and pass off damaged goods. **1911** J. POLLOCK tr. *Brieux's Damaged Goods* III, They are not bad cases [of syphilis]; they are simply the damaged goods of our great human cargo. **1936** *Times Lit. Suppl.* 14 Nov. 925/3 When she came home, somewhat damaged goods in the world's eyes, her father was dead. **1952** *Essays in Criticism* II. 176 Donald Farfrae is not aware when he marries her that Lucetta is damaged goods.

Damara (dā·mărā, dæ·-). Also **Dama** (dā·mă). In full *Hill, Berg*, or *Mountain Damara*. One of a negroid people in the mountainous parts of South-West Africa, who have adopted the language of the Nama Hottentots.

1801 J. BARROW *Trav. S. Afr.* I. vi. 397 The Damaras are obviously the same race of people as the Kaffers. **1880** *Encycl. Brit.* XI. 731/2 *Damaraland*, a region of South-Western Africa..so called from the native race known..to the Cape colonists as Damara (Damra, or Dama). *Ibid.*, 85,000 are Herero proper, 30,000 Hill Damara, 3000 Bushmen. **1884** A. K. JOHNSTON *Africa* (ed. 4) xxvi. 463 These degraded Hill or Berg-Damaras name themselves *Houquain* or 'real men'. **1902** *Encycl. Brit.* XXXII. 736/2 The Bantu ..enslaving the Mountain Damaras. **1935** L. G. GREEN *Great Afr. Mysteries* x. 122 The Berg Damaras..living in the almost inaccessible mountain strongholds of South West Africa. **1947** M. OLDEVIG *Sunny Land* 53 There were also the Dama people, living in the mountains further to the north, who were capable of forging iron weapons—the Hilldamaras or Klipkaffirs of the present day. *Ibid.*, The Damas. **1956** A. G. MCRAE *Hill called Grazing* xii. 126 This derelict appeared to be a Berg Damara, a tribe with hardly any tribal cohesion.

damascene. Add: **B.** *sb.* **2.** Also *attrib.* and *Comb.*

1880 G. C. M. BIRDWOOD *Industr. Arts India* I. 141 A vast establishment of..damascene workers, chiefly for ornamenting arms. **1882** CAULFEILD & SAWARD *Dict. Needlework* 139/1 Damascene Lace... The difference between it and Modern Point lace..consists in the introduction into Damascene of real Honiton sprigs, and the absence of any needle-worked Fillings.

damasky (dæ·mǎski), *a.* [f. DAMASK *sb.* + -Y[1].] Of or pertaining to damask (sense 3).

1931 E. BOWEN *Friends & Relations* II. vii. 170 The dark-and-light damasky stripes of the lawn. **1958** W. SANSOM *Cautious Heart* ii. 27 We sat among the well-known damasky walls.

‖ **damassé.** Add: **B.** *adj.* Woven like damask.

1882 CAULFEILD & SAWARD *Dict. Needlework* 141 *Damassé*, a French term applied to all cloths manufactured after the manner of damask, in every kind of material. **1896** *Godey's Mag.* Apr. 443/2 White damassé mohair.

dambo (dæ·mbo). *Central Africa.* [Mang'anja *dambo* treeless grass-covered plain, open glade in the bush.] A grassy clearing.

1907 *Macm. Mag.* Jan. 194 Large open patches, or *dambos*, covered with fresh green grass and well-watered. **1916** *Cornhill Mag.* Mar. 385 Crossing those 'dambos' in the fierce heat of the day. **1954** *New Biol.* XVII. 13 Wherever the drainage is sufficiently retarded to give grey soils, seasonal water-logging is common and grassy 'dambos' result on the valley flats.

dame. 1. Delete †*Obs.*, and add: The title given to Benedictine nuns who have made their solemn profession (cf. DAN[1], DOM[1]); also, any fully professed nun.

c **1425** *Hampole's Psalter* Metr. Pref. 24 At a worthy recluse prayer cald dame Merget kyrkby. *c* **1590** in *Cath. Rec. Soc. Publ.* (1908) V. 192 Dame Isabel Whitehead an ancient religious woman. *a* **1700** *Ibid.* (1911) IX. 335 She leaving the world went over to the English Benedictine Dames of our Blessed Lady of Consolation. *Ibid.* 339 The R[d] Dames, Dame Magdalena, D. Augustina, D. Maria, and D. Clementia. **1795** in B. N. Ward *Dawn Cath. Revival* (1909) II. xxiii. 82 The three houses of English Dames at Paris. **1867** A. T. DRANE *Chr. Schools* II. iv. 179 Dame Mabel Wafre, abbess of Godstow. **1908** P. NOLAN (*title*) The Irish Dames of Ypres, being a history of the Royal Irish Abbey of Ypres.

2. c. A girl; a woman. Chiefly *U.S. slang.* Also *dial.* (see E.D.D.).

1902 *Commentator* (N.Y.) Jan. 104 Look to de frowsy dames erbout us. **1923** G. H. MCKNIGHT *Eng. Words* iv. 61 In the vocabulary of modern youth, chivalry is dead... A girl is a *jane*, a *dame*, a *moll*, [etc.]. **1928** *Punch* 12 Dec. 666/1 *Skid* is reputed to be a squire of dames ('dame' being apparently the American feminine of 'guy'). **1929** G. MITCHELL *Mystery of Butcher's Shop* xvii. 192 Here was this frightful dame named Bradley coming and invading the place. **1936** A. HUXLEY *Eyeless in Gaza* xxv. 353 Mr. Beavis..began to describe his researches into modern American slang...'I might say you had a dame complex, Anthony.' **1962** J. CANNAN *All is Discovered* iii. 77 I've never set eyes on the dame.

d. In modern pantomime, a comic character, that of a middle-aged woman, traditionally played by a man.

1902 in A. E. WILSON *Prime Minister of Mirth* (1956) iii. 53 Mr. Robey is different. To a stage which for years had been inhabited by pantomime 'dames', by the drink comedian, by the lodger and by the lodger's wife he came. **1925** M. W. DISHER *Clowns & Pantomimes* iii. 44 The wife of Noah..is sister to the dame of pantomime. **1933** G. ROBEY *Looking Back on Life* xi. 88 My most important pantomimes ..were in the big cities...I was often cast for the Dame. **1946** M. DICKENS *Happy Prisoner* ix. 189 He had been afraid they were going to guy her up like a pantomime dame. **1950** *Oxf. Jun. Encycl.* IX. 273/2 Pantomimes..all have their stock characters—the Fairy Queen, the Demon King, the Clown, the Dame (played by a male comedian). **1961** E. WILLIAMS *George* xxiii. 368 He was like a lively don who spends his vacs playing dame in some witty pantomime.

7. d. The title of women members of the Order of the British Empire; also *Dame Commander, Dame Grand Cross.*

1917 *Times* 2 June 6/1 The New Order of the British Empire...Degrees of classification for women,..(1) Dame Grand Cross, (2) Dame Commander, (3) Dame Companion. **1930** *Ibid.* 1 Jan. 14 Made a Dame of the Order of the British Empire for her services to aviation.

10. Also **dame's school.**

a **1817** JANE AUSTEN *Generous Curate* in *Volume the First* (1954) 73 A twopenny Dame's School in the village.

‖ **dame de compagnie** (dam də koṅpanʸi). Also, semi-anglicized **damdecompany.** [Fr.; lit. 'lady of company'.] A paid female companion. Also *transf.*

1784 H. MANN *Let.* 8 Oct. in H. Walpole *Lett.* (1858) VIII. 518 The duchess brought with her, as a *dame de compagnie*, a Frenchwoman. **1821** SHELLEY *Let.* 18 Feb. (1964) II. 265 The situation of *Dame de compagnie* is one indeed in which there is little to be hoped compared with what is to be feared. **1832** *Edin. Rev.* July 481 The female professor, late *dame de compagnie* to La Fayette. **1848** THACKERAY *Van. Fair* xxxiii. 289 Marry a drawing-master's daughter, indeed!—marry a *dame de compagnie*—for she was no better, Briggs. **1885** 'L. MALET' *Col. Enderby's Wife* II. iii, A nice, gentle, little person in grey, who put in an appearance at dinner—*dame de compagnie*, I suppose. **1897** *Sat. Rev.* 5 June 641 'Mees' became a 'damdecompany' to an old Contessa.

‖ **dame d'honneur** (dam donǭr). [Fr.; lit. 'lady of honour'.] A maid of honour, lady-in-waiting.

1805 C. WILMOT *Let.* 26 Aug. in Londonderry & Hyde *Russ. Jrnls.* (1934) II. 170 My name was given in to the Countess Protassoff, Dame of Honneur for a presentation. **1848** THACKERAY *Van. Fair* lv. 500 Some said she..had become a *dame d'honneur* to the Queen of Bulgaria. **1900** M. CORELLI *Master Christian* 261 Madame Bozier, who had been her first governess, and who now lived with her, as a sort of dame d'honneur.

damewort (dē^i·mwɒɹt). [f. DAME + WORT *sb.*[1]] A book name for the garden rocket, *Hesperis matronalis*; = DAME'S-VIOLET.

1776 WITHERING *Veget. Gt. Brit.* II. 403 Damewort, with a simple upright stem.

damfool (dæ·mfū·l). *colloq.* Also (*jocular*) **damphool, -phule.** [f. *DAMN *a.* + FOOL *sb.*[1]] A 'damned fool' (DAMNED *ppl. a.* 4); a person who behaves stupidly; *transf.*, a foolish thing or affair. Also *attrib.* or as *adj.*, foolish, stupid. Hence **damfoo·l** *v. trans.*, to treat as a fool. Also **damfoo·lery, damfoo·lishness.**

1881 'MARK TWAIN' *Lett. to Publishers* (1967) 135 Send me some of those damphool 'ready letter-writers'. **1882** J. J. JENNINGS *Theatr. & Circus Life* xxviii. 409 All the victim can do is to balance the account by putting experience on the debit side of the ledger and damphoolishness on the other. **1883** G. M. HOPKINS *Let* 28 Jan. (1955) 172 There is a great deal of nonsense about that set, often it sickens one..but still I disapprove of damfooling people. **1886** 'F. ANSTEY' *Fallen Idol* xi. 216 When I open the note I see in Greek characters, and forgif me that I rebeat such words to you at all, but I see written there—'Do not a damfool be!' **1898** 'O. THANET' *Heart of Toil* 100 Was that the damfool way he had talked himself? **1900** KIPLING in *Daily Express* 20 June 4/5 'Fighting heroically' in some damfool trap he's walked into with his eyes open! **1908** G. H. LORIMER *J. Spurlock* iv. 66 The grasshopper isn't the only damphool in the good old summer-time. *Ibid.* xii. 328 It's all damfoolishness. *a* **1909** in J. R. Ware *Passing Eng.* (1909) 103 Now, Hennery, I am going to break you of this damfoolishness, or I will break your neck. **1909** D. H. LAWRENCE *Let.* 8 May (1962) I. 53 She has lost all that damfoolery of faddishness about this, that and the other. **1913** W. J. LOCKE *Stella Maris* ii, It's perhaps the only tremendous thing in my damfool of a life. *a* **1953** DYLAN THOMAS *Quite Early one Morning* (1954) 51 Proclaim their right, as Englishmen, to look at the damfool place however they willynilly will. **1960** *Analog Science Fact/ Fiction* Oct. 31/1 This desperate bit of damfoolery.

damine (dē^i·min, -əin), *a.* *Zool.* [f. L. *dama* deer + -INE[1].] Belonging to or resembling the fallow deer, *Dama dama.*

1891 FLOWER & LYDEKKER *Mamm.* 323 Damine group of existing Deer. **1950** G. K. WHITEHEAD *Deer* x. 151 Fallow deer belong to the damine group.

dammar. Add: In full **dammar pine, dammar tree**, any tree yielding dammar resin.

1829 J. C. LOUDON *Encycl. Plants* 802 This genus [*sc.* *Agathis*] is formed of the Dammar Pines. **1832** W. ROXBURGH *Flora Indica* (ed. 2) II. 603 In the Bednore country, it is called the Dammar tree by the English, and blossoms during the hot season. **1846** LINDLEY *Veg. Kingd.* 229 Liquid storax is thought to be yielded by the Dammar Pine. **1880** C. R. MARKHAM *Peruv. Bark* 347 The *Canarum strictum*, or black dammer tree. **1959** J. C. T. UPHOF *Dict. Economic Plants* 12/1 *Agathis alba* Foxw. (syn. *Dammara alba* Lam.). White Dammar Pine.

dammed, *ppl. a.* Add later examples with *up*.

1899 *Westm. Gaz.* 17 May 8/1 In a dammed-up glacial valley. **1904** W. M. GALLICHAN *Fishing Spain* 115, I made a few casts with the fly over a dammed-up pool. **1946** *Nature* 12 Oct. 527/1 While, however, we have had a partial moratorium on the creations of fundamental science.., we have undoubtedly a new stock of dammed-up ideas.

dammit (dæ·mit), for *damn it*, esp. used in comparative phrases.

1908 E. WALLACE *Angel Esquire* xii, 'Outside as quick as dammit!' he cried. **1921** WODEHOUSE *Jill the Reckless* i. 31 When I'm alone with Barker—for instance—I'm as chatty as dammit. **1931** F. D. GRIERSON *Mystery in Red* vii. 106 You were as near rude to him as dammit. **1956** A. H. COMPTON *Atomic Quest* iii. 193 Dammit, give it stuff to spare. **1961** *Guardian* 24 Apr. 9/7 The score standing as near as dammit at two. **1971** *Sunday Times* (Colour Suppl.) 21 Feb. 43/3 Whenever I talk about Rizzo and brutality, I have to use the word 'allege', but dammit all, I have so many *facts*, baby, there ain't nothing alleged about it.

damn, *v.* Add: **5.** (Further examples.) Also, *damn* (one's) *eyes!*, used as an abusive expression.

1761 STERNE *T. Shandy* III. xii. 64 From the great and tremendous oath of William the Conqueror, (*By the splendour of God*) down to the lowest oath of a scavenger, (*Damn your eyes*). **1836** DICKENS *Let.* 20 Sept. (1965) I. 175, I will see them d—d before I make any further alteration. **1850** H. MELVILLE *White Jacket* II. xxvi. 170 What man-of-war's-men call a *damn-my-eyes-tar*, that is, a humbug. And many damn-my-eyes humbugs there are in this man-of-war world of ours. **1906** 'Q' *Mayor of Troy* xi. 151 D—n your eyes, it's *twins*—and both *girls*! **1912** KIPLING *As Easy as A.B.C.* 5 It's refreshing to find any one interested enough in our job to damn our eyes. **1922** JOYCE *Ulysses* 287, I was just passing the time of day with old Troy..and be damned but a bloody sweep came along and he near drove his gear into my eye. **1943** N. BALCHIN *Small Back Room* xiv. 200, I shall have to let go of the other wrench. Damn and blast. **1953** H. MILLER *Plexus* (1963) v. 175 Those things never happen to me. So you peddled candies in the Café Royal? I'll be damned.

damn, *sb.* Add: **2.** Phr. *not to give a damn*.

1895 J. L. WILLIAMS *Princeton Stories* 165, I don't give a damn for the girl. **1929** *Eugenics Rev.* July 86/2 See the happy moron, He doesn't give a damn. I wish I were a moron. My God! Perhaps I am! **1939** WODEHOUSE *Uncle Fred in Springtime* xviii. 263, I don't give a single, solitary damn. **1959** J. CARY *Captive & Free* i. 13 It was obvious, as one angry young woman remarked, that he didn't give a damn—and so they were enraged.

damn (dæm), *a.* and *adv.* Also **damn', dam', dam.** Clipped form of DAMNED *ppl. a.* (See also *DAMFOOL.) *damn all*: see *ALL A. 8 f.

1775 *Narragansett Hist. Reg.* (1885) III. 263 A man that.. was noted for a damn cuss. **1776** *Ibid.* (1882) I. 304 You damn old Tory Raskel. **1787** *Mirror* 164 Don't beef and butter go off damn soberly? **1882** in T. M. Healy *Lett. & Leaders* (1928) I. 150 T. P. quoted my answer as 'I'm damn glad'. **1897** C. M. FLANDRAU *Harvard Episodes* 4 You've gone through the whole damn thing yourself. *Ibid.* 218 'But he isn't bad, really bad.' 'No, certainly not; merely a damn fool.' **1901** MERWIN & WEBSTER *Calumet 'K'* iv. 62 My only order was, 'Clear the road—and be damn quick about it'. **1903** KIPLING *Five Nations* 199 I've known a lot o' people ride a dam' sight worse than Piet. **1918** W. J. LOCKE *Rough Road* xviii, It's all dam funny! **1919** H. J. LASKI *Let.* 14 Nov. (1953) I. 221, I won't say that this is the best of all possible worlds; but it's damn near it. **1928** D. L. SAYERS *Bellona Club* iii, If you understand that..you understand a damn' sight more than I do. **1928** E. WALLACE *Again the Three Just Men* 209 It's none of your dam' business. **1929** H. MILES tr. *Morand's Black Magic* iii. 48 A dam' cracker that I don't know ran into me. **1938** S. V. BENÉT *Thirteen O'Clock* 281 Lisa's got her damn-fool side. **1941** N. COWARD *Australia Visited* i. 6 We were in it once and for all and intended to damn well get on with it. **1945** C. S. LEWIS *That Hideous Strength* ix. 228 You're in a dam dangerous position already. **1959** 'O. MILLS' *Stairway to Murder* xiii. 138 It was a damnfool thing to do, and I realise it now. **1966** C. MACKENZIE *Paper Lives* viii. 116 These dam new towns with which the Accommodation people are infesting the countryside. **1970** N. MARSH *When in Rome* 80, I call it a damn poor show. Leaving us high and dry. **1970** D. STUART *Very Sheltered Life* 248 You one of those damn' Yankee reporters?

damned, *ppl. a.* Add: **4.** In the Southern U.S., a common epithet prefixed to *Yankee*.

1812 *Weekly Reg.* III. 45/1 Take the middle of the road or I'll hew you down, you d—d Yankee rascal. **1818** H. B. FEARON *Sk. Amer.* 210 His friend..said that there was 'nothing in America but d—d Yankies and rogues, and that it was not fit for a dog to live in'. **1833** H. BARNARD *Let.* 30 Apr. in *Maryland Hist. Mag.* (1918) XIII. 361 It is only surpassed by their hatred of the d—d Yankees. **1865** *N.Y. Even. Post* 28 Sept. 1/1 They swore to some men of a cavalry patrol camped across the river, that they would shoot the first d——d Yankee who tried to cross the bridge.

c. Substantival use of superlative in phr. *one's damnedest* (*damndest*): (*a*) the worst one can do, the utmost evil or harm possible; (*b*) one's very best effort (the usual recent sense).

1830 [see DAMNED *ppl. a.* 4 a]. **1846** J. J. HOOPER *Adv. Simon Suggs* xii. 144, I..tried my d—dst, but it wouldn't grind no way. **1891** H. HERMAN *His Angel* 176 Now do your damnedest at your peril. **1928** S. VINES *Humours Unreconciled* xviii. 237 She..had done her 'damnedest' to please him. **1932** S. GIBBONS *Cold Comfort Farm* xx. 268 Reuben..gave her thirty pounds with which to do her damndest. **1955** *Times* 31 Aug. 5/3 The frenzy of fanaticism invites choir and orchestra to do their damnedest. **1958** HAYWARD & HARARI tr. *Pasternak's Dr. Zhivago* I. vi. 183 Now, do your damnedest to get hold of a cab.

damning, *vbl. sb.* **2.** (Later example.)

1914 G. B. SHAW *Pygmalion* (1916) II. 131, I dont mind your damning and blasting..but there is a certain word I must ask you not to use.

‖ **damnosa hereditas** (dæmnōu·sā hīre·ditæs). [L., = 'insolvent inheritance'.] In Law, an unprofitable legacy; now widely applied to any inheritance, tradition, etc., involving more burden than profit.

1848 WHARTON *Law Lex.* 159/1 *Damnosa hæreditas*, a disadvantageous inheritance. **1889** S. ALEXANDER *Moral Order* III. ii. 307 Murder and lying and theft are a *damnosa hereditas* left us from a time when they were legitimate institutions. **1912** J. E. C. FLITCH *Modern Dancing* vii. 98 Lottie Collins left a legacy of style behind her which her successors probably found to be a *damnosa hereditas*. **1931** *Times Lit. Suppl.* 7 May 357/1 He is not himself immune from the effects of a *damnosa hereditas* common to the species. **1935** *Ibid.* 3 Jan. 1/3 He bequeathed his resentment in a Will so complicated..that it threatened..to become the *damnosa hereditas* of *The Times*. **1938** R. G. COLLINGWOOD *Princ. Art* x. 214 Hume did not recognize the difference; and his failure has been a *damnosa hereditas* for all subsequent philosophy. **1955** *Times* 10 May 4/2 The rule that an executor was not compelled to accept a *damnosa hereditas* did not provide a reliable guide.

damnum (dæ·mnŏm). *Law.* [L., = hurt, harm, damage.] A loss or wrong: see quot. 1862.

1828 *Reg. Deb. Congress U.S.* IV. I. 424 It is a loss which gives no legal title to indemnity; it is a *damnum*, but a *damnum*, as the law has it, *absque injuria*. **1862** J. W. SMITH *Man. Com. Law* 264 Damnum is such a damage, whether pecuniary or perceptible, or not, as is capable, in legal contemplation, of being estimated by a jury. **1969** R. F. V. HEUSTON *Salmond's Law of Torts* (ed. 15) xx. 719 A person who suffers *damnum* cannot recover compensation on the basis of *injuria* suffered by another.

Damoclesian, *a.* (Later example.)

1963 *Times* 13 May 17/1 In particular, the Damoclesian sword of a veto from General de Gaulle hangs over the whole proceedings.

damp, *sb.*[1] Add: **7.** *damp-proof* adj. (earlier and later examples); **damp-proof course** (earlier example); **damp-proofed** *ppl. a.*, rendered impervious to damp; **damp-proofing** *vbl. sb.* (also *attrib.*).

1870 *English Mechanic* XII. 262 *Damp-proof paper.*—Can any of your readers inform me where to obtain paper which will be unaffected by constant damp? **1876** *Encycl. Brit.* IV. 462/2 A damp-proof course, intended to prevent that rise of damp from the soil in the brickwork...This damp course is formed in various ways, as a layer of asphalt, or asphalt canvas, or some similar material. **1889** F. E. GRETTON *Memory's Harkback* 199 Provided your day be fine, and your shoes damp-proof. **1963** *Gloss. Build. Terms (B.S.I.)* 17 *Damp-proof membrane*, a damp-proof course within a floor or flat roof. **1962** *Times* 3 Dec. (Agric. Suppl.) p. iii/4 A damp-proofed floor. **1934** H. HILER *Notes on Technique of Painting* i. 38 Any good damp-proofing solution, such as shellac dissolved in petrol, etc. **1962** *Listener* 22 Mar. 534/2 You can get heat-resisting, mould-resisting, and damp-proofing paints.

damp, *a.* Add: **4.** As quasi-adv. in *damp-dry* v. trans. and intr., to dry to the state of being only damp.

1956 *N.Y. Times* 15 Jan. 65 (Advt.), It *automatically* fills, washes, rinses, damp dries, shuts itself off! **1960** *Daily Mail* 9 Feb. 8/8 Incidentally, a spin drier damp dries. **1961** M. BEADLE *These Ruins are Inhabited* (1963) iii. 39 If fed a shilling, this latter [*sc.* a gas heater] will yield enough warmth to damp-dry socks draped on the rungs of a chair.

damp, *v.* Add: **1. c., d.** (Earlier and later examples.) In wider use: to impose or to act as a resisting influence on (an oscillation or vibration of any kind) so that it is either progressively reduced in amplitude or, if the resistance is sufficiently great, converted into non-oscillatory return to an equilibrium position; also used with the oscillating body as obj. So *to damp out*: to damp, to extinguish by damping. So **damped** *ppl. a.*

1877 *Phil. Mag.* III. 482 The measurement of a resistance according to Weber's method of 'damped vibrations'. **1897** NICHOLS & FRANKLIN *Elem. Physics* III. xv. 168 The vibrations of a heavy, elastic body, such as a tuning fork, are but slightly damped. **1899** *Phil. Trans. R. Soc.* A. CXCII. 247 Any natural vibrations of the suspended system would be rapidly damped out. **1911** *Encycl. Brit.* XX. 348/1 This needle..must be so damped that when the current is cut off it returns to zero at once without overshooting the mark. **1922** GLAZEBROOK *Dict. Appl. Physics* II. 1040/2 The above methods of detection are appropriate for the reception of damped wave signals but not for continuous wave signals. **1927** E. G. RICHARDSON *Sound* iv. 121 All vibrations of solids of whatever type are damped by internal friction. **1929** E. MALLETT *Telegr. & Teleph.* xii. 320 The discharge current is oscillatory and of the 'damped wave' form. **1935** F. J. CAMM *Pract. Motorist's Encycl.* 82/2 The leaf-springs of the car's suspension system..damp-out a large proportion of the road shocks and vibrations suffered by the wheels of the car. **1940** *Illustr. London News* CXCVI. 576, I was very impressed by the springing, which damped-out all but exceptionally bad bumps on the road. **1950** *Engineering* 7 Apr. 397/1 Spring-loaded clamping devices heavily damped by hydraulic means to provide virtually solid clamping. **1957** *Encycl. Brit.* XXIII. 175/1 The low elasticity of the gut causes these high constituents [of the note] to be quickly damped. **1961** BICKLEY & TALBOT *Introd. Theory Vibrating Systems* iv. 37 We consider the typical case of a mass *m* controlled by a spring of stiffness *s* damped by a resistance proportional to the speed.

dampen, *v.* **3.** (Earlier and later U.S. examples.)

1827 J. F. COOPER *Red Rover* I. xv. 246 It is seldom that ..he is dampened with salt water. **1902** C. C. MUNN *Rockhaven* i. 9 We took a handful o' matches, an' dampenin' 'em, rubbed the ends round the eyes..o' the critter. **1906** H. D. PITTMAN *Belle of Bluegrass C.* xxi. 302 The heavy dews.. fell upon her, dampening her hair.

dampener. Add: Also *fig.*

1920 *Glasgow Herald* 5 Nov. 11 Socialism has received a decided dampener.

damper. Add: **2. c.** In an organ: a thumping-board (see THUMPING *vbl. sb.* b).

1879, 1881 [see THUMPING *vbl. sb.* b].

4. b. *Electr. Engin.* One of a set of short-circuited conductors in the pole faces of a synchronous electric motor or generator which resist any tendency of the machine to 'hunt', i.e. oscillate by running alternately faster and slower than the synchronous speed. Also *damper bar, winding.*

1906 A. RUSSELL *Altern. Curr.* II. 191 In order to prevent phase swinging, Hutin and Leblanc provided the field magnets with 'amortisseurs', or 'dampers', which tend to prevent any relative change between the positions of the magnetic field due to the armature and the field due to the field magnets. **1920** *Whittaker's Electr. Engineer's Pocket-Bk.* (ed. 4) 223 Care must be taken that the damper bars have not the same pitch as the armature slots as this might cause ripples in the c.c. pressure. **1934** WEBSTER, Damper winding. **1964** N. N. HANCOCK *Matrix Analysis of Electr. Machinery* xi. 198 The damper windings of synchronous machines are mechanically simple but electrically complicated devices. *Ibid.* 201 For synchronous motors it may be a wholly false assumption, since high resistance dampers may be used to obtain adequate starting torque.

c. Any device designed to damp mechanical vibrations; *spec.* a shock-absorber on a motor-car.

1929 NEWTON & STEEDS *Motor Vehicle* xxix. 332 Designers ..try to reduce the friction [in a laminated spring] to the minimum, and they introduce additional friction when it is required by external devices which are called 'dampers'. **1935** W. K. WILSON *Pract. Soln. Torsional Vibration Probl.* vii. 365 There is a definite setting for every damper at which the maximum reduction of vibration amplitude is obtained. **1952** A. W. JUDGE *Mod. Motor Engineer* (ed. 5) III. iii. 59 In most cases this damping action is improved by the use of dampers or shock absorbers fitted between the axles and the chassis frame. **1958** *Engineering* 7 Mar. 295/1 The car was tried without any suspension dampers at all. **1958** *Chambers's Techn. Dict.* Add. 971/1 Yaw damper suppresses directional oscillations in high-speed aeroplanes, while a roll damper does likewise laterally. **1961** BICKLEY & TALBOT *Introd. Theory Vibrating Systems* x. 122 In many mechanical systems friction is unwanted, and minimized, but in some cases vibration dampers are a feature of the design.

6. For '*Australia*' read 'Chiefly *Austral.* and *N.Z.*'. (Add earlier and later examples.)

1827 P. CUNNINGHAM *Two Years in N.S.W.* II. xxviii. 190 The farm-men usually bake their flour into flat cakes, which they call dampers, and cook these in the ashes. **1843** S. STEPHENS *Let.* 4 Sept. 169 (MS.), Flour, from which I make what we call 'dampers' in a frying pan. **1918** KIPLING *Land & Sea T.* (1923) 96 Wonderful hot cakes called 'dampers'. **1939** J. MULGAN *Man Alone* (1949) xiv. 138 He..then cooked a damper of flour and oatmeal. **1944** W. E. HARNEY *Taboo* (ed. 3) 37 You eat up, old men. I will wait for mine to cool off—hot dampers make me sick. **1964** F. CHICHESTER *Lonely Sea & Sky* 49 In order to bake 'damper', which is unleavened bread, we used to hang the oven high above the log fire and pile hot ashes on the lid.

6*. A till, a cash register; a drawer in which cash is kept. *slang.*

1846 R. L. SNOWDEN *Magistrate's Assistant* 344 To rob a till, to pinch a lob: or draw a damper. *Ibid.*, A till, a lob or damper. **1944** D. RUNYON *Runyon à la Carte* (1946) 104 Go over to his hotel..and get the night clerk to open his damper.

7. b. damper weight (see quot.).

1888 *Lockwood's Dict. Mech. Engin.* 102 *Damper weight*, a weight used to counterbalance that of the damper of a steam boiler in order to render it easy of adjustment.

damping, *vbl. sb.* Add: **b.** The action of damping an oscillation or an oscillating body (also *damping out*: see DAMP *v.* 1 c, d in Dict. and Suppl.); the resistance to an oscillation; also, the amount of this, as measured by the rate at which the oscillation diminishes in amplitude.

1870 *Phil. Mag.* XXXIX. 435 An oscillatory current.. sustains itself twice as long against the damping action of resistance as a comparatively steady current of the same maximum value. **1874** *Ibid.* XLVII. 296 The damping effect which the multiplier exerts on the swing of the needle. **1877** RAYLEIGH *Theory of Sound* I. iii. 57 The vibrations of a tuning fork properly constructed and mounted are subject to very little damping. **1879** *Encycl. Brit.* X. 51/1 By damping is meant the decrease of the extent of the oscillations of the galvanometer needle arising from the dissipation of energy through the resistance of the air, the action of currents induced in neighbouring metallic circuits, the viscosity of the suspension fibre, and so on. **1881, 1883** [in Dict.]. **1907** J. ERSKINE-MURRAY *Handbk. Wireless Telegr.* i. 28 Damping or decrease of amplitude is due to two causes—firstly, to dissipation of energy, as heat, in the local circuits; secondly, to radiation. **1912** *Q. Rev.* July 242 The damping-out of oscillations cannot be expected to be as rapid. **1945** *Electronic Engin.* XVII. 455 Resulting in better damping of the receiver diaphragm and consequently suppression of the effects of resonance. **1958** *Van Nostrand's Scientific Encycl.* (ed. 3) 476/2 The metal specimen is vibrated and the rate of damping-out of the vibrations observed. **1967** CONDON & ODISHAW *Handbk. Physics* (ed. 2) II. iii. 22/1 As the magnitude of the damping is increased, the rate of decay of the oscillations becomes greater.. until..the motion becomes aperiodic and is said to be critically damped.

2. *damping off*, the collapse of seedlings or cuttings caused by any of several parasitic fungi and encouraged by excessive moisture in the plant's environment. (Cf. DAMP *v.* 6.)

1899 G. MASSEE *Text-bk. Plant Dis.* 54 The term 'damping off' is applied to a disease of seedlings, characterised by the falling over and dying of the plantlets. **1928** F. T. BROOKS *Plant Dis.* i. 4 Parasitic attack of seedling plants at soil level often causes 'damping off'. **1970** LIEBSCHER & KOEHLER tr. *Fröhlich & Rodewald's Pests & Dis. Tropical Crops* 293 Symptoms of damping-off appear in seedbeds. **1971** *Daily Colonist* (Victoria, B.C.) 13 Mar. 38/4 The dreaded 'damping-off' disease, a soil-borne infection that attacks and rots the stems of small seedling plants just at soil level.

3. *attrib.*, as (sense *1 b) **damping capacity** [tr. G. *dämpfungsfähigkeit* (O. Foeppl 1923)], the ability of a metal or other solid to absorb vibrational energy and dissipate it as heat; **damping coefficient, constant,** or **factor,** any number representing the degree to which an oscillation is damped, usu. defined as the reciprocal of the time in which the amplitude decreases by a factor *e*; **damping winding** = *DAMPER 4 b.

1931 *Proc. Amer. Soc. Testing Materials* XXXI. II. 157 Engineering materials..are able to dissipate energy without failure when subjected to cyclic stresses below the fatigue limit...This property is known as 'mechanical hysteresis effect', 'internal friction' (of solids), or 'damping capacity'. The latter nomenclature was proposed by O. Foeppl in 1923, who offered the following definition: 'Damping Capacity' is the amount of work dissipated into heat by a unit volume of the material during a completely reversed cycle of unit stress. This damping capacity is measured in inch-pounds per cubic inch per cycle. **1953** S. P. TIMOSHENKO *Hist. Strength of Materials* xii. 378 The damping capacity of metals and its relation to fatigue strength. **1959** *Chambers's Encycl.* X. 326/2 Damping Capacity..is low in a sonorous metal such as bell-metal, and high in a 'dead' metal like lead. **1906** J. A. FLEMING *Princ. Electr. Wave Telegr.* i. 21 The Napierian logarithm of the ratio of any maximum current or ordinate to the next maximum in the opposite direction multiplied by twice the frequency, gives us the value of the damping coefficient. **1929** W. E. DALBY *Balancing of Engines* (ed. 4) viii. 259 The damping coefficients derived from the dynamic magnifier are more likely to approximate to the damping of the loaded bridge than damping coefficients determined from the residual oscillations. **1922** F. F. FOWLE *Stand. Handbk. Electr. Engineers* (ed. 5) xxiv. 1963 The damping constant of a circuit is a measure of the ratio of the dissipative to the reactive component of its admittance or impedance. **1906** J. A. FLEMING *Princ. Electr. Wave Telegr.* iii. 162 The damping factor *a* is a quantity the dimensions of which are those of the reciprocal of a time, whilst the logarithmic decrement is a mere numeric. **1920** *Whittaker's Electr. Engineer's Pocket-Bk.* (ed. 4) 224 Damping windings are also fitted on synchronous generators..to damp out pulsations due to variations in the armature ampere-turns.

damping, *ppl. a.* (Further examples.)

1928 S. P. TIMOSHENKO *Vibration Probl. Engin.* i. 22 These damping forces may arise from several different sources, such as air or fluid friction, internal friction of the material of the vibrating body, or friction between sliding surfaces. **1936** W. HEITLER *Quantum Theory of Radiation* i. 34 But because of the damping force the amplitude of the oscillator will decrease. **1958** C. G. WILSON *Electr. & Magn.* v. 137 These oscillations cease when all the energy of the coil is dissipated in overcoming the damping forces present.

dampish, *a.* **3.** (Later example.)

1963 *Times* 23 May 4/4 A dampish pitch and slow outfield at the start.

dampishly, *adv.* (Later example.)

1906 *Westm. Gaz.* 7 Apr. 2/2 The leading lady solemnly wipes the inside of her glass..and breathes into it—breathes, I fancy, rather dampishly.

damsel. Add: **III. 6. damsel-fish,** a small brightly-coloured fish of the family Pomacentridæ.

1905 D. S. JORDAN *Guide to Study of Fishes* II. xxii. 381 The *Pomacentridæ*, called rock-pilots or damsel-fishes, are exclusively marine. **1931** J. R. NORMAN *Hist. Fishes* xii. 246 Interesting examples of commensalism occur among the Pomacentrids or Damsel-fishes (*Pomacentridae*) of tropical coral reefs. **1968** J. E. RANDALL *Caribbean Reef Fishes* 189 Many of the damselfishes are highly territorial and pugnacious.

dan[2]. Add: Also **dan-buoy,** spec., one marking an area cleared by minesweepers, or indicating the position of sea-mines.

1916 *Chambers's Jrnl.* Oct. 665/2 Dan buoys, seen in the half-light,..are apt to be deceptive. **1943** *H.M.'s Minesweepers* v. 31/2 The Captain..gives him practice in dropping and picking up the dan-buoys which are used to mark the position of mines or the limits of a cleared area. **1962** *Times* 29 May 14/6 We sight the 'dan', or marker, buoys of our pots.

Hence **da·nner, da·n-layer,** a vessel used to lay dans.

1942 J. MASEFIELD *Generation Risen* 44 (*title*) The danlayers. **1943** *H.M.'s Minesweepers* vii. 49/1 Certain ships being detailed to act as dan-layers. **1946** J. IRVING *Royal Navalese* 60 *Danner*, a special vessel, following behind a force of mine-sweepers, which lays Dan-buoys to mark..the channel as it is swept. **1954** *Jane's Fighting Ships 1954–55* 195 Now rated as..inshore minesweepers.. but used mainly as danlayers.

Dan[4] (dæn). The name of one of the twelve tribes of Israel and of a town in its territory, taken to represent the northern limit of Israelite settlement in Old Testament times, and used in proverbial phrases to indicate a farthest extremity, esp. in phr. *from Dan to Beersheba.* (Cf. Judges xx. 1, II Samuel xxiv. 2, I Kings iv. 25.)

1738 SWIFT *Polite Conv.* I. 76, I remember, you told me, you had been with her from Dan to Bersheba. **1768** STERNE *Sent. Journ.* I. 85, I pity the man who can travel from Dan to Beersheba, and cry, 'Tis all barren. **1828** SCOTT *Jrnl.* 9 Apr. (1941) 222 The whole saving will not exceed a guinea or two for being cursd and damnd from Dan to Beersheba. **1905** H. G. WELLS *Mod. Utopia* ix. 268 The Utopians distinguished two extremes of this Kinetic class according to the quality of their imaginative preferences, the Dan and Beersheba, as it were, of this division.

dan[5] (dæn). [Japanese.] In Judo, a degree of proficiency; the holder of such a qualification.

1941 M. FELDENKRAIS *Judo* 166 All teachers..are of the first Dan in the beginning of their teaching career. **1954** E. DOMINY *Teach yourself Judo* i. 17 There are six *kyu* grades and ten *dan* grades. The word *kyu* means pupil and *dan* means degree or master. **1958** G. KOIZUMI *Judo* 44 Two third Dans, members of the grading panel, are required to recommend to First Dan. **1968** R. HARRIS *Nice Girl's Story* ii. 16, I often practised my judo. I..was, by this time, a 4th Dan.

Hence **dan-holder,** one who is, or possesses, a dan.

1941 M. FELDENKRAIS *Judo* 167 It is too easy for a higher Dan holder to beat a single opponent so he is opposed to a group. **1956** K. TOMIKI *Judo* i. 8 Of the *dan*-holders those from the 1st to the 5th wear a black belt or sash.

Danaert (dæ·nəɹt). Also **dannert.** The name of the German inventor of spring steel wire, usually barbed and in a spiral form, used in anti-tank and other forms of defensive warfare. Usu. *attrib.*, as *dannert wire.*

1945 *Penguin New Writing* XXVI. 31 Pillboxes had already been put up and a dannert wire apron fence encircled the pier arcade. **1948** *Hansard* CCCCXLV. 1535 There is concealed dannert wire in the grass around the camps. **1960** C. FITZGIBBON *When Kissing had to Stop* xi. 200 A score of bell-tents, surrounded by coils of dannert. **1961** *Wire Industry Encycl. Handbk.* 333/1 *Danaert wire*, the circles of barbed wire widely used in wartime for barbed wire entanglements. Also known as concertina wire.

danaid. Add: **b.** = *DANAINE.

1881 S. H. SCUDDER *Butterflies with Ref. Amer. Forms* 298 Tribe Festivi Fabricius [*Danaides* Boisd.] Danaids. **1892** W. L. DISTANT *Naturalist in Transvaal* 65 The female Hypolimnas being present with the Danaids. **1936** *Discovery* July 212/2 Danaids and various Nymphalids. **1937** H. D. PEILE *Guide Coll. Butterflies of India* 76 The females of this genus [*sc. Pareronia*] appear to be mimics of common blue and black Danaids. **1957** M. A. WYNTER-BLYTH *Butterflies of Indian Region* vi. 63 The Danaids are chiefly a tropical and sub-tropical family.

danaine (dæ·neˌin), *a.* and *sb. Ent.* [ad. mod. L. *Danainæ*, f. generic name *DANAIS: see -INE[1].] **A.** *adj.* Of or belonging to the subfamily Danainæ of the family Nymphalidæ of butterflies. **B.** *sb.* A member of this subfamily.

1897 E. B. POULTON *Theory Mimicry* in *Hope Rep.* (1901) II. 2 All of them possess a dark tip to the fore wing crossed by a white bar, as in the Danaine butterfly. *Ibid.*, The abundant black-and-white Danaines. **1913** *Oxf. Univ. Gaz.* 4 June 951/2 The rare Danaine butterfly *Amauris ansorgei.* **1930** R. A. FISHER *Natural Selection* vii. 161 This is probably true of many Danaines. **1945** E. B. FORD *Butterflies* v. 96 Professor G. D. Hale Carpenter has seen the male of the African Danaine *Amauris psyttalea* Plötz hover over the female and dust her with the scented powder with which the wing-pocket is filled in this insect. **1957** *Encycl. Brit.* XV. 517/2 The above facts point directly to the conclusion that there is some advantage in mimicking Danaine butterflies.

danais (dæ·neˌis). [mod.L. (P. A. Latreille 1807, in *Magazin für Insectenkunde* VI. 291), a. Gr. *Δαναΐς*.] A butterfly belonging to the genus so named, now usually included in the genus *Danaus.*

1878 P. ROBINSON *In my Indian Garden* 101 The coppery Danais flitted at ease about the shrubs. **1892** W. L. DISTANT *Naturalist in Transvaal* 65 Birds may..make an experimental dash at a *Danais.*

Danakil (dæ·nākil, dānā·kil), *sb.* and *a.* Also **Dankali** (dæŋkā·li). [Cushitic.] **A.** *sb.* **a.** A member of a widely spread Hamitic people of north-eastern Ethiopia. **b.** The language of this people. **B.** *adj.* Of or pertaining to this people, their language, or the areas they inhabit.

1875 *Encycl. Brit.* I. 61/1 Abyssinia..may be regarded as ..having..E., the territory of the Danakils. **1885** *Ibid.* XVIII. 778/2 The third or Ethiopic division [of the Hamitic Family] includes as its chief members the Beja or Bishárin, the Saho, the Dankali..the first two lying along the Red Sea north of Semitic Abyssinia, the others south of it, to the equator. **1902** *Ibid.* XXXII. 703/2 The native population [of French Somaliland] is either of the Danakil or of the Galla race. **1910** *Ibid.* I. 85/2 The Danakil lowlands have a hot, dry climate producing semi-desert conditions. **1934** L. M. NESBITT *Desert & Forest* 11 The territory inhabited by the Danakils is divided into two parts by the frontier separating Abyssinia from Eritrea. **1936** G. B. SHAW *Simpleton of Unexpected Isles* Pref. 4 The Abyssinian Danakil kills a stranger at sight. **1937** —— *Platform & Pulpit* (1962) 283 In Abyssinia a Danakil woman will not marry a man until he proves that he has at least four homicides to his credit. **1961** D. BATES *Fly-switch from Sultan* xviii. 102 Some splendid savages called the Danakil, who hovered around us eyeing our private parts with a predatory eye and a sharp knife. **1969** *Times* 16 Sept. (Somali Republic Suppl.) p. ii/6 Nearly half the population of the territory are Somalis of the Issa ethnic division who overflow into the Somali Republic and into Ethiopia, and the other half are Muslim *Dankalis* (or Afars) who also spread into Ethiopia.

dance, *sb.* Add: **7.** *dance-band, -floor, -frock, -rhythm, -step*; **dance-card,** a card bearing the names of (a woman's) prospective partners at a dance; **dance-director,** the person who, in musical comedies, arranges the dances; **dance-drama,** a rendering through dancing of a dramatic situation; **dance-hall** orig. *U.S.* (earlier and later examples); **dance hostess,** (*a*) a woman who holds a dance at her house, etc.; (*b*) a dancing-partner (sense *b*, *DANCING *vbl. sb.* b); **dance-house** (earlier and later examples); **dance programme** = **dance-card.*

1927 *Melody Maker* Aug. 739/1 It was his boast then that he would have a symphonic dance band. **1962** J. WAIN *Strike Father Dead* 67 The convention that lays it down that English dance-band singers must put on an American accent. **1895** J. L. WILLIAMS *Princeton Stories* 199 You will here meet several of those whose names you have on your dance-card, and you may make up your mind whether to remember that fact or not. **1922** JOYCE *Ulysses* 11 Old feather fans, tasselled dancecards, powdered with musk. **1932** WODEHOUSE *Louder & Funnier* 84 The dance-director is instructed to think up a lot of different business for the first encore. **1924** *New Republic* 26 Nov. 11 Spend a few hours in a New Mexico pueblo at the end of the day of one of their sacred dance-dramas. **1938** *Encycl. Brit. Bk. of Yr. 1938* 251/2 Dance dramas, marionette shows, musical comedies, [etc.]. **1958** *Times* 13 Aug. 5/3 Darrell's dance-drama *The Prisoners*..was not only topical..but also gripping. **1968** *Jrnl. Mus. Acad. Madras* XXXIX. 1 The dramatic poem of the Composer Nauka Caritram was also produced as a dance-drama. **1928** *Melody Maker* Feb. 171/3 The dance floor was crowded. **1959** M. SHADBOLT *New Zealanders* 32 There was even a crude dance-floor erected for the night a little back from the beach. **1904** *Westm. Gaz.* 18 Feb. 4/2 An accordion-pleated lace net is one of the prettiest dance-frocks I have seen for some time. **1858** *Mass. Acts & Resolves* 125 Any person who shall offer to view..any.. show, concert, or dance-hall exhibition of any description shall be punished by a fine. **1904** CONRAD *Nostromo* I. viii. 107 From the doors of the dance hall men and women emerged tottering. **1934** T. S. ELIOT *Rock* i. 40 Everythink useful for the people: dance 'alls, picture palaces, swimmin' baths. **1909** *Daily Chron.* 8 July 6/5 Lady Londesborough was one of the chief dance hostesses last night. **1934** F. B. YOUNG *This Little World* ix. 179 A young woman of a most undesirable class—a 'dance-hostess' (the word was vaguely familiar and unpleasant) in a London night club. **1961** A. WILSON *Old Men at Zoo* ii. 113 She's a dance hostess. **1848** *Western Boatman* (Cincinnati) June 133 That afternoon I wrote a letter to a friend of mine in Natchez, who was a woman that kept a dance-house. **1875** MRS. STOWE *We & Neighbors* xli. 375 He told me that he was in the constant habit of passing through the dance-houses, and talking with people who kept them. **1946** G. FOREMAN *Last Trek of Indians* 256 His forsaken wife, Comes-at-Rain, sprang through the window of the dance house. **1906** *Dialect Notes* III. 133 Got your dance-program filled up yet? **1913** C. MACKENZIE *Sinister St.* I. II. i. 149 The dance programme with Muriel's name fourteen times repeated. **1926** S. T. WARNER *Lolly Willowes* I. 42 A bunch of dance programmes kept for the sake of their little pencils. **1968** D. HOPKINSON *Incense-Tree* iv. 46 Dance programmes were usual—with the names of the dances on one side and space for the names of partners on the other, and a small pink pencil tied on with blue cord. **1880** GROVE *Dict. Mus.* II. s.v. Melody, In the matter of rhythm there are two things which play a part—the rhythmic qualities of language, and dance rhythms. **1947** A. EINSTEIN *Mus. Romantic Era* xvi. 290 A pathetic recitative changes suddenly into the most impudent dance-rhythm. **1920** S. LEWIS *Main St.* 380 The rude fiddling and banging dance-steps in the barn. **1936** *Discovery* June 186/2 For the different kinds of spirits [to be exorcised] different dances are held, each with its special dance-steps. **1962** *Times* 26 Apr. 8/1 Ready-made and established dance-steps.

dancer. Add: **6.** *slang.* (See quots.)

1864 HOTTEN *Slang Dict.* 117 *Dancer*, or *dancing-master*, a thief who prowls about the roofs of houses, and effects an entrance by attic windows, &c. **1930** E. WALLACE *Lady of Ascot* xiii. 120 There were active young men who called themselves dancers, and whose graft was to get into first-floor flats and get out quickly with such overcoats, wraps, and movables as could be whisked away in half a minute.

dancing, *vbl. sb.* Add: **b.** *dancing-class, -club, -list, -party, -teacher*; **dancing-partner,** (*a*) a person with whom one dances; (*b*) (see quot. 1921).

1870 MRS. STEPHENS *Married in Haste* xxxi. 172 Constance had never felt..pleasure in departing for her dancing classes. **1902** 'G. M. MARTIN' *Emmy Lou* 277 'There's to be a dancing club on Friday evenings,' she explained, 'and I'm invited.' **1918** A. BENNETT *Roll-Call* II. i. 217 They had belonged to two dancing clubs whose members met weekly in the saloons of the great hotels. **1871** MRS. STOWE *Pink & White Tyranny* xxi. 263 Her dancing-list seemed in a fair way to be soon filled up for the evening. **1920** *World's Pict. News* 27 Feb. 9/2 (*caption*) Gaby..with her famous dancing partner. **1921** *Dict. Occup. Terms* (1927) § 899 *Dancing partner*.., an expert dancer engaged by proprietor or manager of dancing hall or by individual dancer, to act as partner when required and to teach ball-room dancing. **1852** J. REYNOLDS *Pioneer Hist. Illinois* 52 They arrange all things necessary for the dancing party. **1889** *Kansas Times & Star* 13 Dec., The dancing party given by Mrs. Kirk Armour and her sister, Mrs. Ed Smith, at the Casino last night was the most brilliant social event of the early winter. **1841** J. F. COOPER *Deerslayer* xxiv. 112 His step as lofty as dancing-teachers and a light heart could make it. **1880** G. W. CABLE *Grandissimes* xliii. 336, I could be..a dancing-teacher.

dandelion. Add: **3. dandelion-clock** = CLOCK *sb.*[1] 8; **dandelion coffee,** a beverage prepared from dried dandelion roots; **dandelion greens** *U.S.*, fresh dandelion leaves used as a green food or herb.

1876 *Aunt Judy's Mag.* XIV. 628 He did not ask now why dandelion clocks go differently with different people. **1925** W. DE LA MARE *Broomsticks* 51 Buttercups and dandelion-clocks and meadow-sweet. **1950** 'P. WOODRUFF' *Island of Chamba* vii. 109 You could blow her away like a dandelion clock in the summer fields. **1852** S. MOODIE *Roughing it in Bush* 89, I met with an account of dandelion coffee published in the *New York Albion*, given by a Dr. Harrison, of Edinburgh, who earnestly recommended it as an article of general use. **1886** *Harper's Mag.* Sept. 578/2 If you'd asked pleasanter, I should just as soon told you that we use dandelion coffee. **1907** *Yesterday's Shopping* (1969) 518/2 Schweitzer's Dandelion Coffee. **1887** M. E. WILKINS *Humble Romance* 234 There were..two old women—one.. searching for dandelion greens among the short young grass. **1889** R. T. COOKE *Steadfast* xvii. 190 Dandelion greens is better 'n a doctor. **1945** *New Yorker* 26 May 60 With the shad came a dish of dandelion greens in a tart sauce. **1963** *New Good Housek. Cookbook* 452/2 Heap chopped, cooked dandelion greens and scallions on tomato slices.

dander, *sb.*[3] (U.S. examples.)

a **1800** *Spirit of Farmer's Museum* (1801) 278 An infant child..had ever since its birth, been grievously afflicted with a certain disorder in the head, called by the learned, 'the dander, or dandriff'. **1860** J. G. HOLLAND *Miss Gilbert's Career* viii. 131 A young man that..keeps the dander all off his coat collar..always makes a good husband. **1875** —— *Sevenoaks* v. 65 I've took more nor three quarts o' dander out iv 'is hide.

dander, *sb.*[4] For *U.S. colloq.* read *colloq.* (orig. *U.S.*) and add earlier and additional examples.

1831 H. J. FINN *Amer. Comic Ann.* 148 A general roar of laughter brought Timmy on his legs. His dander was raised. **1832** SEBA SMITH *Major Downing* 104 My dander began to rise, and I couldn't hold in any longer. **1834** C. A. DAVIS *Lett. J. Downing* 34 He was as wrathy as thunder—and when he gets his dander up, it's no joke, I tell you. **1847** HALLIWELL *Dict. Arch. & Prov. Words* I. 291/2 *Dander*, anger. **1861** G. DU MAURIER *Let.* Dec. (1951) 97 If you want to get on you must put the kicking straps on your dander and offend nobody. **1966** *Listener* 17 Mar. 395/1 Precocious manifestos raise the critical dander.

Dandie Dinmont (dæ·ndi di·nmǫnt). Also shortened to **Dandie.** [Name of a character in Sir Walter Scott's novel *Guy Mannering* (q.v. ch. xxii, 'Dandy Dinmont's Pepper and Mustard terriers', and Note C).] A breed of terrier from the Scottish borders, having short legs, long body, and rough coat. Also *attrib.*

[**1826** SCOTT *Jrnl.* 1 Apr. (1939) 145 Sometimes attending to the humours of two curious little terriers of the Dandie Dinmont breed.] **1848** *Sporting Life* 8 Jan. 246/1 The dog celebrated by Sir Walter Scott as the Pepper and Mustard, or Dandie Dinmont breed. **1859** 'STONEHENGE' *Shot-Gun* 77 The Dandie Dinmont..is an excellent rabbit dog. **1862** J. BROWN *Our Dogs* 29 From this dog descended Davidson (the original Dandie Dinmont) of Hyndlee's breed. **1875** MRS. STOWE *We & Neighbors* i. 7 A rough coated Dandie Dinmont terrier. **1894** R. B. LEE *Mod. Dogs* (*Terriers*) 287 He has never known one of his Dandies show the 'white feather'. **1925** *Chambers's Jrnl.* 772/1 The two so-called Dandies. **1952** R. LEIGHTON *Complete Bk. Dog* (rev. ed.) 241 When Dandies fight it is a serious matter. **1963** S. M. LAMPSON *Country Life Bk. Dogs* 122 The Dandie Dinmont has the distinction of being the only breed of dog named after a character in fiction.

dandy, *sb.*[1] (and *a.*). Add: **2.** Now usu. *a dandy.*

1822 *Pennsylv. Intelligencer* 3 Dec. (Th.), The reader will suppose this was a dandy of a thing, since it was on writing paper. **1887** *Harper's Mag.* June 160/1 'Death loves a shining mark', and she hit a dandy when she turned loose on Jim. **1897** S. HALE *Lett.* (1919) 319 Mrs. B. was a dandy, she didn't fuss nor worry. **1919** H. L. WILSON *Ma Pettengill* iv. 111 It was just one punch, though a dandy. **1968** D. HELWIG in R. Weaver *Canad. Short Stories* 2nd Ser. 376 We..sat..waiting for Barrow Man to light his fire. At nine-fifteen he did it. It was a dandy.

B. *adj.* **2.** Fine, splendid, first-rate. *colloq.* (orig. *U.S.*). Freq. in phr. *fine and dandy.*

1794 *Massachusetts Spy* 27 Aug. (Th.), My uncle Cuthbert blew out a prodigious puff of my dandy tobacco. **1842** W. BAGLEY *Let.* 22 June in N. E. Eliason *Tarheel Talk* (1956) iv. 128, I now have a real dandy suit of clothes & I step about New York just as if I was some great one. **1894** P. L. FORD *Hon. Peter Stirling* (1898) 163 'If I was as big as him,' said one, 'I'd fire all the peelers.' 'Wouldn't that be dandy?' cried another. **1908** C. E. MULFORD *Orphan* vi. 73, I got yore smokin', Orphant!..Here she is, right side up and fine and dandy! **1910** S. E. WHITE *Rules of Game* I. i, 'How's Mrs. Orde..?' he inquired. 'Mrs. Orde is fine and dandy.' **1926** 'R. CROMPTON' *William—the Conqueror* v. 83 Oh, how *dandy*! **1940** *War Illustr.* 5 Jan. 571/1 The troops had told the Dominions Secretary that the crossing had been 'dandy', and the General told us the same thing. **1940** O. NASH *Face is Familiar* 259 Candy is dandy But liquor is quicker. **1964** WODEHOUSE *Frozen Assets* ii. 31 I'm fine and dandy now, but before I saw you I was feeling extremely blue.

b. As *adv.* Finely, splendidly. *U.S. colloq.*

1908 S. E. WHITE *Riverman* xli. 323 'She's holding strong and dandy,' said Orde.., examining critically the clumps of piles. **1952** in Wentworth & Flexner *Dict. Amer. Slang* (1960) 140/1 She and her husband get along just dandy. **1963** O. NASH *Everyone but Thee & Me* 110 And, should Furnace or freezer act less than dandy, There's always a quaint old handy-man handy.

dandydom (dæ·ndidəm). [f. DANDY *sb.*[1] + -DOM.] The condition of a dandy; the world of dandies.

c **1850** in *Daily Chron.* (1902) 4 Nov. 3/2 It 'flustered the realms of dandydom'. **1885** *Society in London* 155 A glorified dragoon who has reached the apotheosis of old dandydom. **1899** *Strand Mag.* Mar. 273/2 He was..reduced to a state of dilapidated dandy-dom.

dandyfunk. *Naut.* [? f. DANDY *sb.*[1] II + FUNK *sb.*[2]] Hard tack soaked in water and baked with fat and molasses.

1883 in *Amer. Speech* (1959) XXXIV. 28. **1902** A. B. LUBBOCK *Round the Horn* iii. 94 Loring..proposed that we should make some dandyfunk for tea. *Ibid.* 95 The dandyfunk..steaming hot, a mixture between a cake and a pudding. **1903** C. PROTHEROE *Life Mercantile Marine* ix. 87 'Dandy-funk' is another dish..the substitution of a little molasses furnishing the excuse for another name. **1931** E. LINKLATER *Juan in America* I. iii. 37 They grew tender over memories of lobscouse and dandy-funk.

dandyishly (dæ·ndiˌiʃli), *adv.* [f. DANDYISH + -LY[2].] Like a dandy, in the manner of a dandy.

1868 *Good Words* 1 Nov. 699 Dandyishly dressed in spotless white linen. **1909** H. G. BARKER *Voysey Inheritance* IV. 172 Very dandyishly dressed, he still seems by no means so happy as his clothes might be making him.

dandy-wink. [f. DANDY *sb.*[1] II + WINK *sb.*[2]] A small windlass worked by short fixed levers, e.g. in a fishing-boat.

1883 W. C. RUSSELL *Round Galley Fire* 92 The dandy-wink is manned, the beam secured, and the net is then dragged in over the side.

danebol: see *DENNEBOL.

danegeld, -gelt. Add: Also *transf.* and *fig.*

1911 FLETCHER & KIPLING *School Hist. Eng.* ii. 39 It is always a temptation to an armed and agile nation, To call upon a neighbour and to say:—'We invaded you last night—we are quite prepared to fight, Unless you pay us cash to go away'. And that is called asking for Dane-geld. **1920** *19th Cent.* July 28 A policy dictated by fear should be interpreted in terms of *danegeld* payable by Great Britain to the Moscow Camarilla. **1955** *Times* 18 May 11/5 Amounted to little more than blackmail and the payment of *danegelt*, which has never yet satisfactorily settled any dispute.

dane gun. *W.* and *S. Afr.* Also **Dane gun.** [f. DANE + GUN *sb.* 3.] Any of several kinds of primitive firearms, orig. introduced by Danish traders.

1900 *Daily News* 25 Sept. 8/3 In the village were found 300 Dane guns, 40 rifles, three Martini-Henry carbines. **1901** *Daily Chron.* 31 Aug. 3/4 Three Ashantis, armed with their long dane guns. **1920** *Blackw. Mag.* Mar. 384/2 Every man or boy who could wield a spear or hold a dane gun. **1935** *Geogr. Jrnl.* LXXXV. 126 The dance that followed was accompanied by the firing of dane-guns. **1947** J. STEVENSON-HAMILTON *Wild Life S. Afr.* xv. 103 The native with his Dane gun, or Tower musket.

dang, *sb. slang.* [f. DANG *v.*] A damn, cuss.

1906 SOMERVILLE & 'ROSS' *Irish Yesterdays* 113 He wouldn't give a dang for them.

danger, *sb.* Add: **C.** *danger-area, -level, -point, -spot, -zone*; **danger angle,** (*a*) *Naut.* the angle enclosed by lines drawn from two known points to a point marking the limit of safe approach to a danger to navigation, so that a ship by steering a course keeping the two known points at a larger or smaller angle will avoid the danger; (*b*) *Gunnery* (see quot. 1918); **danger line,** a line, real or imaginary, representing the division between safety and danger; **danger money,** a payment made beyond basic wages for dangerous work; also *fig.*

1892 *Notes on Navigation H.M. Ships* (ed. 3) 13 The use of a danger angle in passing outlying rocks with land behind should also not be forgotten. **1902** *Encycl. Brit.* XXXI. 109/2 To avoid an unnecessarily wide détour in rounding points and shoals, extensive use is now made of both horizontal and vertical danger angles. *Ibid.*, The vertical danger angle enables similar results to be attained by measuring the vertical angle subtended by a known height. **1918** E. S. FARROW *Dict. Mil. Terms*, *Danger angle*, the angle which the tangent to the trajectory at the point of splash makes with the plane containing the point of splash and parallel to the horizontal plane through the muzzle of the piece in the firing position. **1929** *Star* 21 Aug. 12/4 The gas leak..which made a portion of New Bridge-street a 'danger area'. **1939** *News Review* 30 Nov. 10 Sandringham is a 'danger area' inasmuch as the East coast..might well be the route taken by German bombers on their way to London. **1955** A. L. ROWSE *Expansion Eliz. Eng.* i. 12 We get flashes of light upon the lurid scene from the comparative security of the towns..which swept like a sickle around the danger-area. **1935** *Discovery* Dec. 360/2 A simple timing device will indicate whether this approaches the danger-level. **1967** *Oxford Computer Explained* 31 *Danger level*, a predetermined level of stock, which when broken requires that any outstanding delivery previously requested be hastened. **1890** *Congress. Rec.* 5 June 5654/2, I believe the good sense of our law-makers will still hold us inside the danger line of peril. **1892** *Notes on Navigation H.M. Ships* (ed. 3) 8 The five-fathom line on most Admiralty charts is to be considered as a caution or danger line. **1902** *Monthly Weather Rev.* 3/1 The December floods of the Tennessee.. continued considerably above the danger lines for the first few days. **1953** L. P. HARTLEY *Go-Between* xiv. 167 Again I was lucky with the Psalms; the Sunday before there had been forty-four verses; this Sunday there were forty-three, seven below the danger line. **1942** W. H. BEVERIDGE *Social Insurance* 39 If an occupation is specially hazardous it should carry remuneration—'danger money'. But to give danger money only in the form of higher wages, that is to say, only so long as no accident has occurred, is of little value. **1953** B. BOLAND *Return* in J. C. Trewin *Plays of Year* IX. 341 There is nothing on this earth so dangerous as putting out a finger to touch another human being's life. ..We should get danger-money. **1958** *Economist* 15 Nov. 573/2 It means..adding a new regiment of Naafi volunteers, with their 'danger money', to the British men and women who will have to be protected. **1966** *Punch* 19 Jan. 90/1 It may be that the rage for allowances was originally inspired by the dockers, with their claims for danger money, dirty money,..and all the rest. **1835** J. A. ROEBUCK *Short Rev. Long Session* 11/1 When this excitement reaches danger-point the Lords will yield. **1897** *Daily News* 22 Apr. 6/3 The Macedonian difficulty, which is the real danger-point. **1910** *Westm. Gaz.* 6 Apr. 2/3 The axle is the danger-point in all heavy vehicles which are run at high speeds. **1933** *Archit. Rev.* LXXIII. p. lviii, Such nosings..are apt to wear smooth..and furnish a danger-point at a critical part of the tread. **1905** *Westm. Gaz.* 4 Oct. 3/2 The danger-spot in our new Treaty with Japan..is the provision for 'insuring the independence and integrity of the Chinese Empire'. **1939** L. MACNEICE *Autumn Jrnl.* xii. 49 Education..Trains us to keep the roads nor reconnoitre..the beauty-spots or danger-spots. **1907** *Westm. Gaz.* 5 Feb. 7/3 An alarming fire broke out in the City danger-zone soon after six o'clock last night. **1925** E. F. NORTON *Fight for Everest: 1924* 58 Anywhere beyond the Base Camp may be considered as the 'danger zone'. **1927** *Observer* 5 June 19/2 The chief new feature [of contract bridge] introduced in America has been what is known as the 'Vulnerable' or 'Danger Zone'. **1928** V. WOOLF in *Times Lit. Suppl.* 19 Jan. 34/1 Let us, as we approach the danger-zone of Hardy's philosophy, be on our guard. **1954** A. J. P. TAYLOR *Struggle for Mastery* xvii. 373 He and his advisers recognized that there was a 'danger-zone', an imaginary period when the British might suspect German designs and destroy her navy before it could hold its own. **1969** *New Yorker* 14 June 44/1 Ashe and Clark Graebner have long since entered the danger zone where any major mistake can mean the loss of the set.

D. Colloq. phr. (*to be*) *on the danger list*: (to be) dangerously ill (as of a patient in hospital).

1938 S. PUDER (*title*) On the danger list. A case history. **1950** C. MACINNES *To Victors* III. 346, I asked them what they thought of the men on the danger list. **1960** E. H. CLEMENTS *Honey for Marshal* xi. 184 The fellow wasn't on the danger list any more...He could have come up by ambulance. **1970** *Times* 16 Feb. 3/8 Five patients on the danger list were slightly better, the hospital said.

dangered, *ppl. a.* (Later poetic examples.)

1819 KEATS *Otho* I. i. Poems (1889) 423 This danger'd neck is saved. **1915** *Oxf. Mag.* 21 May 317/1 High Powers that love this dangered folk.

dangle, *sb.* Delete *rare* and add examples for sense: something that dangles.

1903 *Westm. Gaz.* 26 Mar. 8/2 We get dangles in the shops made and ready for our use, fashioned of silk cords and tassels. **1909** *Cent. Dict.* Suppl., *Dangle-money*, an early Chinese bronze coinage, so called from its resemblance to and former use as dangles of a musical instrument. **1909** M. B. SAUNDERS *Litany Lane* I. ix, A pair of long jet earrings representing funeral urns with cloths over them had replaced the usual golden dangles. **1909** WEBSTER s.v., A dangle of curls. **1937** PARTRIDGE *Dict. Slang* 207/2 *Dangle-parade*, a 'short-arm' inspection: New Zealand soldiers'. **1957** J. KEROUAC *On Road* (1958) I. vii. 44 On the wall was a nude drawing of Dean, enormous dangle and all.

dangly, *a.* [f. DANGL(E *v.* and *sb.* + -Y[1].] = DANGLING *ppl. a.*

1903 *Westm. Gaz.* 1 Jan. 3/1 Then often passementerie motifs dangle from bolero fronts of the short sac kind, for these attractive dangly things are more popular than ever. **1959** C. WILLIAMS *Man in Motion* iv. 47 She was wearing dangly ear-rings.

Danian (dēi·niăn), *a. Geol.* [ad. F. *Danien* (A. d'Orbigny *Prodrome de Paléont.* (1850) II. 290), f. L. *Dania* Denmark, f. *Danus* DANE: see -IAN.] Epithet of a stratigraphical stage or series (not represented in Britain) lying below the Montian stage, and held variously to be the highest division of the Cretaceous or the lowest of the Tertiary; of or pertaining to this stage or the period during which it was deposited. Also *absol.*

1873 *Q. Jrnl. Geol. Soc.* XXIX. 377 Cretaceous Period. Danian or Waipara Formation... This is the Cretaceo-Tertiary formation of Dr. Hector. *Ibid.* Index, Danian formation of New Zealand. **1882** GEIKIE *Text-bk. Geol.* 823 The uppermost division, or Danian, of the Continental chalk appears to be absent in England, unless its lower portions are represented by some of the uppermost beds of the Norwich Chalk. **1893** P. LAKE tr. *Kayser's Comp. Geol.* 314 In the North of France the succession of the Upper Cretaceous rocks closely resembles that of England; but still higher zones are represented, and these are united under the term Danian. **1925** *Countries of the World* XXX. 3072/1 A beautiful stone resembling marble is secured from phosphatic beds of Danian Age. **1960** L. D. STAMP *Britain's Struct.* (ed. 5) xii. 139 These strata form..transition beds between the Cretaceous and the Tertiary and are sometimes grouped with the one, sometimes with the other, sometimes separated as the Danian and the Montian. **1969** *Proc. Geol. Soc. Lond.* Aug. 152 The rocks commonly accepted as of Danian age should be included in the Cainozoic. **1971** *Nature* 19 Feb. 553/2 Impoverished Danian faunas are.. often separated by a bedding plane from Cretaceous faunas of similar sedimentary facies.

Daniell (dæ·nyĕl). [The name of John Frederic *Daniell*, English physicist (1790–1845).] **a.** Used *attrib.* or in the possessive to designate inventions of Daniell or their modifications, as **Daniell('s) battery** or **cell,** a cell in which the cathode is zinc in either dilute sulphuric acid or a solution of zinc sulphate and the anode is copper in a saturated solution of copper sulphate, with the zinc sulphate solution either floating on top of the copper sulphate solution or separated from it by a porous plate; also *Daniell's hygrometer.*

1840 *Phil. Trans. R. Soc.* CXXX. I. 191 Five cells of Professor Daniell's constant battery were charged and connected with the electro-magnetic machine. **1863** E. ATKINSON tr. *Ganot's Physics* VI. vi. 269 There are many sources of error in Daniell's hygrometer. *Ibid.* X. i. 609 The current produced by a Daniell's battery is constant for some hours. **1884** F. KROHN tr. *Glaser de Cew's Magn.- & Dyn.-Electr. Mach.* 233 If the instrument were always graduated with a Daniell cell. **1948** GLASSTONE *Physical Chem.* (ed. 2) xii. 934 In the familiar Daniell cell..the metallic zinc at the left-hand electrode passes into solution as zinc ions, liberating two electrons. **1955** H. L. PENMAN *Humidity* iv. 28 Daniell's hygrometer (1827), where the cooling is produced by forced evaporation of ether. **1957** *Encycl. Brit.* III. 216/1 The Daniell cell has an open-circuit voltage of about 1·08.

b. Used *ellipt.* for *Daniell cell.*

1871 *Engl. Mechanic* 430/2 The improved Daniell devised by me. **1878** *Encycl. Brit.* VIII. 93/2 The sawdust Daniell, invented by Sir Wm. Thomson (1858), is very convenient when portability is desired. **1922** GLAZEBROOK *Dict. Appl. Physics* II. 70/2 Kelvin's tray battery, which is a form of gravity Daniell, has been largely used in submarine telegraphy.

Daniglacial (dēiniglēi·ʃiăl, -ʃăl, -glēi·siăl), *a. Geol.* Also **Dani-glacial.** [f. L. *Dani-a* Denmark + GLACIAL *a.*] Epithet of the first division or 'sub-epoch' of the Late Glacial epoch in north-western Europe, when the ice-sheet of the last glaciation retreated from Denmark and halted near the tip of the Scandinavian peninsula; of or pertaining to this sub-epoch. Also *absol.* Cf. *GOTHIGLACIAL, *FINIGLACIAL *adjs.*

1912 G. DE GEER in *Compt. Rend. XI Congr. Géol. Internat. 1910* I. 253 As to the first of the late-glacial sub-epochs, which properly may be called the Dani-glacial one, or that part of the last ice-recession when the ice-border retired from the extreme limit of the last glaciation past Denmark..,its duration is not yet known. **1923** —— *Let.* 3 Nov. in W. J. Sollas *Anc. Hunters* (1924) xiv. 662, I hope it will be possible to date not only the commencement of the Gothi-glacial sub-epoch but of the Dani-glacial as well, and thus to determine the whole time which has elapsed since the maximum of the last glaciation. **1927** PEAKE & FLEURE *Hunters & Artists* i. 6 The first of these lines has been called the Daniglacial moraine, and the second the Gothiglacial, while the space between the two is known as the area of the Daniglacial retreat, and that between the second line and a third to be described is termed the area of the Gothiglacial retreat. **1957** J. K. CHARLESWORTH *Quaternary Era* II. l. 1525 The Daniglacial recession alone may have lasted 10,000–15,000 years. **1960** L. D. STAMP *Britain's Struct.* (ed. 5) xiv. 170 The Dani-Glacial was the period of the ice retreat from Denmark—when ice still covered the whole of Norway and Sweden.

Danish, *a.* Add: *Danish blue* (*cheese*): see quot. 1948; *Danish modern*: a modern style of furniture, characterized by simple clear lines, light woods, and lack of carved or painted decoration; *Danish pastry*: a yeast cake garnished with sugar, spice, nuts, icing, etc.; also *ellipt.*

1948 *Good Housek. Cookery Bk.* 383 *Danish Blue*, a soft white cheese with a blue mould veining, made in Denmark in imitation of Roquefort. **1953** R. FULLER *Second Curtain* I think I shall have just a little cheese. They often have Danish blue. **1955** D. BARTON *Glorious Life* 27 Their Danish Blue sandwiches and bitter arrived. **1948** A. H. RUTT *Home Furnishing* (ed. 2) xiii. 217 Finnish, Norwegian, and Danish Modern employ laminated wood and bentwood effectively. **1969** S. GREENLEE *Spook who sat by Door* xv. 133 He entered the panelled reception room, decorated in Danish modern. **1934** WEBSTER, Danish pastry. **1950** B. SCHULBERG *Disenchanted* (1951) v. 50 The waitress.. brought him..Danish pastry and coffee. **1963** 'M. ALBRAND' *Call from Austria* i. 12 'I could do with another cup of coffee.' 'And a Danish?' **1970** *New Yorker* 26 Sept. 33/2 She stopped at a delicatessen along the way to pick up a Danish pastry ring. *Ibid.*, When the Danish was finished.. she ran the palms of her hands along the sides of her chair.

Danite (dæ·nəit). *Obs.* exc. *Hist.* [f. *Dan*, the name of one of the sons of Jacob and of the tribe of Israel founded by him + -ITE[1]. Cf. *Genesis* xlix. 16 and 17.] **1.** A member of the Hebrew tribe of Dan.

1535 COVERDALE *Judg.* xiii. 2 There was a man at Zarga, of one of the kynreds of the Danites, named Manoah. **1667** [see HERCULEAN *a.* 2]. **1847** J. KITTO *Cycl. Bibl. Lit.* I. 514/1 The inability of the Danites to expel the Philistines and Amorites. **1968** *Encycl. Brit.* XIII. 119/1 The Levite, however, was persuaded by the Danites migrating northward to accompany them and he became the first priest of the famous sanctuary at Dan.

2. A member of an alleged secret order of Mormons supposed to have arisen in the early days of that sect to act as spies and suppressors of disaffection.

1838 *Test* (Rushville, Ill.) 12 Dec. 3/4 There, Patton, one of the bloodiest of the Danites, directed two of his bands. **1857** *Congress. Globe* 24 Feb. App. 289/3 They suppose that there is a secret society existing there, called Danites, Shanpips, or Destroying Angels. **1882** C. WAITE *Adv. Far West* 252 The Danites are expected to act as spies upon the federal officers and other gentiles; to watch the feelings and spirits of the saints, and to report the first indications of disaffection. **1905** *Daily Chron.* 22 June 3/1 The duties of the Danites, or 'avenging angels'. **1948** *Jrnl. Amer. Folk-Lore* Jan.–Mar. 20 See those dreadful Danites how they lynch many lives.

Dankali, var. *DANAKIL *sb.* and *a.*

dannert: see *DANAERT.

Dano- (dēi·no), used as combining form of *Danus* DANE, DANISH, = 'Danish and', as *Dano-German, -Irish*; **Dano-Norwegian** *sb.* and *a.*, (a modified form) of the Danish language used in Norway after its separation from Denmark, and now one of the two standard languages.

1931 *Times Lit. Suppl.* 19 Feb. 128/2 The Dano-German Münte. **1880** *Encycl. Brit.* XIII. 252/2 The many feuds between the Irish clans in which the Dano-Irish shared. *Ibid.*, During the independence of the Dano-Irish kingdom of Dublin. **1892** J. Y. SARGENT *Gram. Dano-Norwegian Lang.* p. vi, I have given..the Syntax of sentences simple and compound, so far as the Dano-Norwegian idiom differs from the English. **1911** *Encycl. Brit.* XIX. 818/1 The close of 1899 and the beginning of 1900 were occupied by a discussion..as to the adoption of the *landsmaal*..in place of the *rigsmaal* or Dano-Norwegian. **1933** BLOOMFIELD *Lang.* iv. 59 The present-day standard languages [in Scandinavia] are Icelandic, Danish, Dano-Norwegian, Norwegian Landsmaal, and Swedish. *Ibid.* xxvii. 484 The Norwegians modified their standard Danish in the direction of Norwegian speech-forms. This Dano-Norwegian *Riksmaal* ('national language') became the native speech of the educated upper class. **1961** L. F. BROSNAHAN *Sounds of Lang.* ix. 205 The *Riksmål*, which originated in the Dano-Norwegian speech of the educated population of the south-east.

‖ **dansant** (daṅsaṅ), fem. **dansante** (daṅsaṅt), *a.* [Fr., pr. pple. of *danser* to DANCE.] Accompanied by dancing, as in *thé dansant* (erron. *-ante*), an afternoon entertainment at which there is dancing and tea is served; *soirée dansante*, an evening party with dancing.

1819 M. WILMOT *Let.* 24 Oct. (1935) 28 Another Eve there was a *thé danceant* at old Prince Esterhazy's. **1841** *Punch* 28 Aug. 83/2 The elegant *soirées dansantes*, nightly held at the 'Frog and Fiddle'. **1845** *Ibid.* 26 July 52/1 Among the fashionable parties of the season we have observed the frequent announcement of a *Thé Dansante*, or a dancing tea. **1849** THACKERAY *Pendennis* I. vii. 71 What did Lady Snapperton do..at her *déjeuné dansant* after the Bohemian Ball? **1854** —— *Newcomes* xiv, Mrs. Toddle Tompkyns's *soirée dansante* in Belgrave Square. **1858** *Punch* 29 May 214/1 Two medical students..got up an impromptu *Thé Dansante.* **1872** E. BRADDON *Life in India* v. 151 The projection of balls and parties *dansantes*. **1945** J. BETJEMAN *New Bats in Old Belfries* 28 How restful to putt, when the strains of a band Announced a *thé dansant* was on at the Grand. **1971** *Times* 16 June 4/1 In the inelegant heart of London beside Piccadilly Circus it is still possible to go to a thé dansant.

‖ **danse du ventre** (daṅs dü vaṅtr). [Fr., lit. 'dance of the belly'.] = *belly-dance* (*BELLY *sb.* 17).

1893 *National Police Gaz.* (U.S.) 23 Dec. 2/2 The *danse du ventre* has reached New York at last, and has raised quite a storm. **1926** F. M. FORD *A Man could stand Up* II. i. 90 They had watched the Hun do the *danse du ventre*! **1967** G. FALLON *Rendezvous in Rio* xv. 129 She arched her back and started to gyrate her stomach in the fascinating *danse du ventre.*

danse macabre: see MACABRE *a.*

‖ **danseur** (daṅsör). [Fr., dancer.] A male ballet-dancer. So **danseur noble** [see NOBLE *a.*], the partner of a ballerina.

1828 J. EBERS *Seven Years of King's Theatre* iii. 66 They directed their endeavours towards forming an engagement with Albert, *premier danseur.* **1912** J. E. C. FLITCH *Mod. Dancing* ii. 32 Perhaps the greatest danseur who has ever lived—Gætano Vestris. **1943** K. AMBROSE *Ballet-Lover's Pocket-Bk.* 36 When the attributes of the ideal *danseur* are under discussion, it is first necessary to determine whether one is speaking of the male dancer as a *soloist*, or primarily as a partner to a *ballerina*: a *danseur noble.* **1947** *Ballet Ann.* I. 55 A *danseur noble* and perfect partner in classical adagio. **1959** *Observer* 24 May 19/5 Some of those portly senior *danseurs* were prudently left behind. **1961** *Times* 11 May 18/6 Mr. Royes Fernandez—a true *danseur noble.*

danseuse. (Earlier examples.)

1828 J. EBERS *Seven Years of King's Theatre* iii. 69 A carriage was sent for the accommodation of the *première danseuse.* **1840** A. BUNN *Stage* II. iv. 90 Those charming stanzas addressed to the fair *danseuse.*

Dantean, *a.* (Earlier example.)

1785 A. SEWARD *Let.* 25 Aug. (1811) I. 77 The Dantean Angel of Vengeance is diabolically insatiable.

So **Danteia·na** [see -ANA]; **Da·nteish** *a.* = DANTEAN *a.*; **Dante·scan** *a.* = DANTEAN *a.*; **Dantesque** *a.* (earlier and later examples); **Da·ntism,** the branch of study concerned with the works and life of Dante; **Danto·logist,** one versed in Dantology; **Danto·logy** = *DANTISM.

1936 *N. & Q.* CLXXI. 172/2 (*heading*) Danteiana. **1924** *Blackw. Mag.* 7 Jan. 24/1 A Danteish spot..had one redeeming feature. **1834** *Q. Rev.* LI. 23 Mr. Wright's double triplets..sound to our ears as little like the Dantescan harmony as Cary's blank verse. **1876** *Encycl. Brit.* V. 291/2 Dantescan commentators and scholars. **1813** J. FORSYTH *Remarks Excursion Italy* 66 Style..speckled even to affectation with *Dantesque* terms. **1924** H. DE SÉLINCOURT *Cricket Match* v. 126 He moved away with a look of deep dejection upon his Dantesque face. **1922** *Glasgow Herald* 23 Nov. 3 Since Witte's day Dantism has developed into something like an exact science. **1910** *N. & Q.* 30 July 83 It is as yet little known to Dantologists. **1903** *Daily Chron.* 7 Jan. 3/3 This..branch of Dantology. **1959** *Times Lit. Suppl.* 25 Sept. 548/3 There has been a remarkable flowering of Dantology under such scholars as Pézard, Gilson, [etc.].

danthonia (dænþōu·niă). *Bot.* [mod.L. (de Candolle & Lamarck *Flore Française* (1805) III. 32), irreg. f. the name of Étienne *Danthoine*, Fr. botanist + -IA[1].] A member of a large genus of tufted perennial pasture grasses so named, chiefly of Australia and New Zealand.

[**1863** *Proc. Calif. Acad. Nat. Sci.* II. 182 *Danthonia Californica* (California Oat Grass)..on borders of cultivated fields near the bay of Oakland. **1870** *Trans. N.Z. Inst.* II. 104 *Danthonia semi-annularis*, a variable grass found in nearly all soils and situations throughout the colony.] **1918** *N.Z. Jrnl. Agric.* 20 Mar. 135 On poor tussock country ..the use of danthonia invariably increases the carrying capacity. **1921** H. GUTHRIE-SMITH *Tutira* xix. 167 On hard hill-tops and narrow ridges and knobs, danthonia and microlena had grouped themselves in little companies. **1952** A. R. D. FAIRBURN *Three Poems* 59 In time of drought The danthonia shines like a flame. **1955** *Landfall* IX. 108 The paddock of dry danthonia grass.

Dantonist (dæ·ntŏnist). [f. the name of *Danton* + -IST.] A follower of Georges Jacques Danton (1759–94), one of the leaders in the French revolution. So **Dantone·sque** *a.*, resembling the style of Danton.

1849 THACKERAY *Pendennis* I. xix. 179 He avowed himself a Dantonist, and asserted that Louis the Sixteenth was served right. **1879** *Encycl. Brit.* IX. 606/2 Two parties.. the *Exagérés*, or Hebertists..and the *Modérés*, the Dantonists. **1899** A. H. BEESLY *Danton* 316 The indictment against the Dantonists. **1901** *Daily Chron.* 31 Dec. 3/1 He who had corrected the proofs of the Dantonist appeal for mercy. **1924** *Contemp. Rev.* Apr. 420 He can cut a knot with a phrase of Dantonesque audacity. **1928** *Sunday Express* 10 June 10/5 Camille Desmoulins was one of..the Dantonists.

Dan(t)zig (dæ·nzig, -tsig). Name of a city (now Gdańsk) near the mouth of the Vistula and of the district containing that city, used *attrib.* chiefly to designate kinds of timber grown in that district, as *Dantzig deal, fir, oak.* **Dantzig beer,** a black syrupy beer made at Dantzig; **Dantzig spruce,** beer made by adding a decoction of the buds or cones of spruce.

a **1592** GREENE *Fr. Bacon* (1594) sig. D3, All the westerne kings That lie alongst the Dansick seas. **1629** J. PARKINSON *Paradisi* II. liii. 524 Of Cowcumbers there are diuers sorts... The Dantsicke kinde beareth but small fruit, growing on short branches or runners: the pickled Cowcumbers that are vsually sold are of this kind. **1720** J. STEUART *Letter-Bk.* (1915) 122 Two small caves with best double strong Dantzig waters. **1843** *Ainsworth's Mag.* IV. 327 Having..laughed as frequently at the bottle of transparent 'Dantzic'. **1855** OGILVIE Suppl., Black-beer, or Dantzic. **1862** *Chambers's Encycl.* IV. 334/2 It [*sc.* the Norway spruce fir] is the.. Danzig Deal of the market. **1871** S. T. AVELING *Carpentry & Joinery* 9 The weight of a cubic foot of English oak generally considered seasoned is about 50 lb...Of Dantzic oak about 48. **1879** *Encycl. Brit.* IX. 223/2 The well-known 'Danzig-spruce' is prepared by adding a decoction of the buds or cones to the wort or saccharine liquor before fermentation. **1889** J. J. WELCH *Text Bk. Naval Archit.* 111 The wood employed [for decks] is generally Dantzic fir. **1945** [see *GOLDWASSER].

Danubian (dæniū·biăn), *a.* [f. med.L. *Danubius, Danuvius*, Gr. *Δανούβιος*: see -IAN.] Of or pertaining to, bordering on, the river Danube, or pertaining to the prehistoric cultures of the surrounding region.

Danubian principalities, (Hist.) the principalities of Moldavia and Wallachia, now forming part of Rumania. *Danubian reed*, the giant reed *Arundo donax*, cultivated in Australia.

1847 *Howitt's Jrnl.* I. 327/2 By these [devices] does Russia tyrannize over the Danubian principalities. **1848** THACKERAY *Van. Fair* lxvi. 600 Tyrolese glove-sellers and Danubian linen-merchants. **1875** *Encycl. Brit.* II. 700/1 The Danubian provinces. **1876** [see SERB *a.*]. **1878** GLADSTONE *Berlin Treaty* 27 The union of the Danubian Principalities. **1909** WEBSTER, Danubian reed. **1927** PEAKE & FLEURE *Priests & Kings* 138 Their civilization, which in many respects resembled that of Thessaly, is known as Danubian. **1941** AUDEN *New Year Let.* I. 26 The dazed uncomprehending stare Of the Danubian despair. **1957** G. CLARK *Archaeol. & Society* (ed. 3) iv. 116 A settlement of 'Danubian' (Neolithic) peasants on a loess patch near Cologne.

dao (dā·o). Also **dhao.** [Burmese.] = *DAH.

1876 VOYLE & STEVENSON *Milit. Dict.* (ed. 3), *Dhao*, a Burman tool or weapon (half chopper, half sword) used in clearing jungle and in cutting down trees. **1884** *Encycl. Brit.* XVII. 163/2 The only implements of tillage being the *dáo* or hill knife, and a *kodáli* or hoe. **1923** *Blackw. Mag.* May 569/2 They..are armed with..spear and *dao.*

dapping (dæ·piŋ), *vbl. sb.* [f. DAP *v.* + -ING[1].] Fishing by a method in which the bait is allowed to dip or bob lightly on the water. So **da·pper,** one who daps; **dap** *sb.*, the bait used in dapping; **da·pped** *ppl. a.*, dipped.

1799, 1867 [see s.v. DAP *v.*]. **1908** *Westm. Gaz.* 13 June 11/3 Heavy trout being captured with the natural insect on 'dapping' tackle. **1920** *Glasgow Herald* 17 July 4 A position from which he can lower his 'dap' on to the water without drawing the fishes' attention. *Ibid.*, The floating imitation of the dry-fly fisher or the dapper's living lure. **1928** *Daily Express* 28 July 4/6 The patient 'dapper', who thrusts his rod between the branches of over-hanging foliage and dangles a blue-bottle temptingly at the extremity of a 3x cast. **1933** 'R. CROMPTON' *William—the Rebel* viii. 163 His father..needed all his rods—either for 'dapping' or 'trolling' or 'casting'. **1960** *Times* 3 Dec. 9/4 It also rises freely in the daytime to a dapped grasshopper. **1967** *Daily Mail* 8 July 6/6, I am a beautiful dapper: My daddy-longlegs skimmed the surface of the lake like a corps de ballet. *Ibid.*, Levin, though he lacks the light touch needed for dapping, does..cast as to the manner born.

dapple, *sb.* **2.** (Later examples.)

1868 LOWELL *Let.* 4 Sept. (1894) I. 453, I should like to lie under a tree for a year, with no other industry than to watch the dapples of sunlight on the grass. *a* **1889** G. M. HOPKINS *Poems* (1918) 52 For earth..her dapple is at an end. **1916** D. H. LAWRENCE *Amores* 49 When the yellow dapples Of autumn tell the withered tale again.

daquiri: see *DAIQUIRI.

Darby. Add: **6. Darby and Joan club,** a club for elderly men and women.

1942 *Times* 18 Dec. 2/4 The Darby and Joan Club, which is believed to be the only one of its kind, was opened by Lord Soulbury at 16, Leigham Court Road, Streatham, yesterday. **1967** *Nursing Times* 18 Aug. 1083/3 For the gregarious elderly there are Darby and Joan clubs. **1970** D. CLARK *Deadly Pattern* v. 106 She often sang for the Darby and Joan club.

Dard (dāɹd), *sb.*[1] and *a.* **A.** *sb.* **a.** A member of any of several peoples of Dardistan, in the extreme north-west of the Indian sub-continent. **b.** Any of the Indo-Aryan languages of these peoples. **B.** *adj.* Of or pertaining to these peoples or their languages.

1873 G. W. LEITNER *Results Tour Dardistan* I. III. p. ii, He might have studied..my Dardu 'Vocabularies and Dialogues'..and then would have been enabled to have *added* something to our knowledge of..the Dard languages. **1875** F. DREW *Jummoo & Kashmir* xix. 442, I think that the Dārd character, at all events of the lower classes, is generally straightforward. *Ibid.* xviii. 422 The existence of the Dārds as a separate race, as well as something of their language, have..been facts within the reach of readers of travels. **1879** *Encycl. Brit.* X. 598/1 The Dards are described as decidedly Aryan in features. **1902** *Ibid.* XXVIII. 729/2 Eleven different languages, which have all been usually classed together under the name Dard. **1920** *Blackw. Mag.* May 620/1 The 'Indians' to whom Herodotus refers are none other than the Dards. *Ibid.*, In Ladakh to this day we find pure Dard settlements.

dard (dāɹd), *sb.*[2] *Bot.* Formerly also **dart.** [Fr., fruit-spur.] (See quots.)

1925 W. R. DYKES tr. *Lorette's System of Pruning* p. xxviii, The dart (*dard*) is an eye which has not grown beyond a certain point and which is surrounded by a rosette of two or three leaves. It is the sign of fruitfulness in the pear. *Ibid.*, In ordinary circumstances, the dart..swells and adds to itself a rosette of leaves, which gradually increase in number from three to five. Continuing to develop, the dart is finally crowned with seven leaves and is then known as a fruit-spur (*lambourde*). **1946** *Ibid.* (ed. 2) 5 We have.. adopted the original French spelling of dard and brindille, written in the first English edition dart and brindle—for both the latter have meanings totally unrelated to the objects referred to as dard and brindille. *Ibid.* 24 The dard is a bud which has not grown beyond a certain point [etc.]. **1951** *Dict. Gardening* (R. Hort. Soc.) II. 638/2 *Dard*, a bud on an apple or pear tree...It may develop into shoot if it receives too much sap, but usually with a steady supply of sap.. becomes a spur carrying a fruit bud.

dare, *v.*[1] **B. 5. a.** (Later examples.)
1886 *Harper's Mag.* Dec. 105/2 Jabe Pennell begun to hunt him an' dare him. **1908** L. M. Montgomery *Anne of Green Gables* xxiii. 254 All the silly things..were done.. because the doers thereof were 'dared' to do them. **1969** I. & P. Opie *Children's Games* ix. 264 Sometimes you're dared to go and tie the wifies' doors together.

dare, *sb.*[1] Add: **1.** (Later examples.)
1897 Kipling *Capt. Cour.* viii. 161 In a minute half the boats were out and bobbing in the cockly swells, but Troop kept the *We're Heres* at work dressing-down. He saw no sense in 'dares'. **1959** I. & P. Opie *Lore & Lang. Schoolch.* xi. 230 To give themselves guts when accepting a dare.
2. (Later U.S. example.)
1904 H. R. Martin *Tillie* 13, I would love to play in the evening if I had the dare.

dare-all. Add: Also *attrib.* or as *adj.*
1902 *Daily Chron.* 18 Mar. 3/2 Their dare-all Vikings came sailing over the sea to possess themselves of our homes. **1939** 'A. Bridge' *Four-Part Setting* iii. 29 This had developed in her..a dare-all and try-all attitude which didn't really belong to her character. **1942** Partridge *Usage & Abusage* 97/1 Venturesome journalists and dare-all writers should employ them with care and discretion.

dargha: see *dagga[2].

dariole. Delete *Obs.* and add: Now an individual dish of various kinds prepared in a dariole mould; a sweet or savoury custard, soufflée or jelly; *dariole* (*mould*), a very small tin mould shaped like a flower-pot.
1846 A. B. Soyer *Gastron. Regen.* 507 *Darioles.* Line (very thinly) a dozen small dariole moulds with paste.. flour..eggs..sugar... They will be as light as sugar and eat as delicate. **1890** *Mrs. Beeton's Cookery Bk.* 216/2 Oyster darioles. **1892** T. F. Garrett *Encycl. Pract. Cookery* I. 516/1 Receipts for..various Darioles will be found..the original principle being that of lining a Dariole-mould.. with thin paste and filling up with rich cream, or custard. **1895** *Army & Navy Co-op. Soc. Price List* 15 Sept. 308 Dariols, Plain. Do. Fluted. **1901** *Daily Chron.* 23 Nov. 8/4 Line some dariole moulds thinly with tomato aspic. *Ibid.*, Set a dariole on each slice of tomato. **1903** *Ibid.* 21 Mar. 8/5 Darioles à la Maraschino. **1930** *Daily Mail* 6 June 15/3 Dariole of salmon is made with flaked cooked salmon, set in a mould with aspic jelly, sliced hard-boiled eggs, and shrimps. **1965** *Listener* 16 Sept. 434/2 We then took six individual dariole moulds, the kind used for baking little castle cakes.

Darjeeling (dāɹdʒi·liŋ). The name of a town and district in Bengal, used attrib. or absol. to designate tea grown there.
1882 *Tea Cycl.* 294/2 A good field for Darjeeling tea will be found to exist in an active competition with China for the supply of Thibet. **1907** F. F. Thurstan *Few Facts concerning Tea* 5 *Darjeeling*, very flavoury, fine quality, colouty liquor, darkish, well made, tippy leaf; appearance no criterion. **1964** L. Deighton *Funeral in Berlin* i. 10 He poured boiling water on to the Darjeeling. 'You like Darjeeling?' **1967** N. Freeling *Strike Out* 16 Arlette poured out tea..Darjeeling tea.

dark, *a.* Add: **6. d.** *Phonetics.* Of an *l* sound: formed with the tip of the tongue against the teeth-ridge and the rest of the tongue placed as for the articulation of a back vowel. Of a vowel: articulated with the front of the tongue somewhat depressed and the back raised in the direction of the soft palate. Opp. *clear *a.* 13 b. Hence **darkness,** this quality of a sound.
1899 W. Rippmann *Elem. Phonetics* 42 When the tongue is raised a little further back we obtain lower, darker sounds..; when it is raised a little further forward we obtain higher, clearer sounds. **1906** —— *Sounds of Spoken Eng.* 47 The 'darkness' of the [l] is particularly noticeable when it comes at the end of a word. **1909** I. F. Williams *Phonetics for Scottish Students* x. 47 In forming l in English the tip of the tongue is raised to the teeth-ridge, and the part immediately behind the tip is somewhat hollowed. This hollowing makes the English l much darker than the French l, where the part behind the tip is arched. In Scottish the hollowing is much more considerable than in English, and the l still 'darker'. **1918** [see *clear *a.* 13 b]. **1942** *Amer. Speech* XVII. Suppl. 27 [ɑ] seems often to be a somewhat 'darker' or more retracted sound than the normal American variety. **1953** *English Studies* XXXIV. 250 To express 'vowel + glide' and not a 'dark' flavour of the following consonants. **1958** A. Cartianu et al. *Course Mod. Rumanian* 22 English dark *l* in words like: *middle, tell, almost* does not exist in Rumanian.
11. Freq. in *superl.* as an epithet for Africa and hence applied (chiefly joc. or ironically) to other places that are considered remote, uncivilized, etc.
1890, 1891 [in Dict.]. **1907** R. Dunn *Shameless Diary of Explorer* 68 He met the missionary in darkest Africa. **1915** L. Einstein *Let.* 14 Mar. in *Holmes–Einstein Lett.* (1964) II. 110 We had an interesting enough journey through darkest Europe to reach here. **1958** *Listener* 4 Dec. 956/2 Television comes to darkest Surrey. **1959** G. D. Painter *Marcel Proust* I. xii. 206 The beautiful Marie Nordlinger.. had arrived from darkest Manchester to study painting and sculpture in Paris. **1964** C. Willock *Enormous Zoo* i. 6 The Mitumbe hills..jagged, unfriendly and epitomizing darkest Africa. **1968** C. Cooper *Thunder & Lightning Man* ii. 27 They shunted him off to darkest Somerset.
11*. Of a theatre, etc.: closed.
1916 *Variety* 27 Oct. 12/1 The Star and Garter theatre, Hyde & Behman's local Columbia Circuit burlesque house, through being dark last week, when 'The London Belles' refused to accept the engagement, lost its share of the probable gross receipts. **1921** *Daily Colonist* (Victoria, B.C.) 30 Oct. 13/3 Last night the picture show was 'dark', and most of those people with families were spending the evening at home. **1953** *Economist* 28 Mar. 853/1 The owner or lessor of the theatre..is likely to cover his expenses..except in bad times when he has the burden of a 'dark' theatre. *Ibid.* 853/2 It could happen that as many as one-fifth of all the seats in the London theatres were not even on sale, because the theatres were dark.
13. a. *dark-glancing* (later example), *-shut*; **b.** *dark-browed* (earlier and later examples), *-faced*, *-haired* (earlier example), *-hearted* (example), *-leaved* (earlier example), *-toned*, *-visaged*, *-winged*.
1830 Tennyson *Poems* 87 Darkbrowed sophist, come not anear. **1952** E. Pound *Personae* 271 Come not anear the darkbrowed sophist. **1923** D. H. Lawrence *Birds, Beasts & Flowers* 171 They are royalty, dark-faced royalty, showing the conscious whites of their eyes. **1931** Blunden *To Themis* 22 Dark-glancing onward as he sings and guides. **1833** J. S. Mill *Let.* 25 Nov. (1910) I. 77 Dark-haired men with formidable moustaches. **1870** D. G. Rossetti *Let.* 15 Mar. (1965) II. 816 Where..the dark-hearted golden sunflowers shine? **1817** Keats *Epistle to G. F. Mathew* in *Poems* 55 Where the dark-leav'd laburnum's drooping clusters Reflect athwart the stream their yellow lustres. **1912** W. de la Mare *Listeners* 27 Laid in their dark-shut graves. **1934** *Burlington Mag.* Sept. 132/2 A fine relief in dark-toned wood. **1935** C. Day Lewis *Time to Dance* 32 Over the dark-toned earth. **1906** *Daily Chron.* 30 July 6/6 Another dark-visaged countryman of Reid, a typical specimen of the black Celt. **1925** E., O. & S. Sitwell *Poor Young People* 5 Where now a dark-winged southern wind soft grieves.
c. dark ages (often with capital initials), also (*a*) often restricted to the early period of the Middle Ages, between the time of the fall of Rome and the appearance of vernacular written documents; (*b*) (freq. in *sing.*) the period between the end of the Bronze Age and the beginning of the historical period in Greece and other Aegean countries; (*c*) *transf.*, a period of obscurantism or ignorance; *joc.*, an obscure or little regarded period before the present; **dark arches (moth),** a British noctuid moth of the genus *Xylophasia* (cf. *arch *sb.* 7); **dark chamber,** (*b*) (example); **dark current** *Electr.*, the current that flows in a photoelectric device when there is no light (or any other radiation capable of causing a current) incident on the photosensitive region; **dark-field** = *dark-ground; **dark glasses,** (*b*) spectacles with darkly-tinted lenses; **dark-ground,** applied *attrib.* to denote a type of illumination used in microscopy in which direct light is prevented from reaching the eyepiece and the only light seen is that scattered by the object, which appears as bright against a dark background; hence in *transf.* sense to denote any light-coloured matter on a dark background; **dark night (of the soul)** [tr. Sp. *noche oscura* (St. John of the Cross)], a period of spiritual aridity suffered by a mystic; also *transf.*; **dark smoke** (see quot. 1954).
1748 Smollett *R. Random* I. p. iv, In the dark ages of the world, when a man had rendered himself famous for wisdom or valour, his family..represented his character and person as sacred and supernatural. **1834** M. Edgeworth *Helen* iv. 46, I must go back..quite to the dark ages, the time when I knew nothing of my daughter's character but by the accidental lights which you afforded me. **1860** C. M. Yonge *Hopes & Fears* I. x. 370 What was natural science with the one, was natural history with the other. One went deep in systems and classifications, and thrust Linnæus into the dark ages. **1871** Geo. Eliot *Middlem.* I. ii. 20 We must have Thought: else we shall be landed back in the dark ages. **1876** F. Kilvert *Diary* 4 May (1944) 308 It was built in the Dark Ages of fifty years ago and was simply hideous. **1887** Kipling *Under Deodars* (1889) 58 Centuries ago—in the Dark Ages, before I met you, dear. **1907** G. Murray *Rise of Greek Epic* ii. 29 There lies between the prehistoric palaces of Crete, Troy, or Mycenae, and the civilization which we know as Greek a Dark Age covering at least several centuries. **1915** W. Leaf *Homer & Hist.* i. 34 The answer to the question lies somewhere in what, following Professor Murray, I have called the Dark Ages, the three or four hundred years which precede the first glimmer of authentic history in the eighth century. **1935** (*title*) Map of Britain in the Dark Ages (Ordnance Survey) [p. 5] This map covers that portion of English and Welsh history which falls between the years 410 A.D. and 871 A.D. **1943** F. M. Stenton *Anglo-Saxon Eng.* viii. 267 No other king of the Dark Ages ever set himself, like Alfred, to explore whatever in the literature of Christian antiquity might explain the problems of fate and free will. **1950** H. L. Lorimer *Homer & Monuments* viii. 461 Even in the Dark Age there must have been some degree of communication, as the common features of proto-Geometric culture show. **1952** Childe & Simpson *Anc. Monuments Scotland* 6 The 'Dark Ages'...Approximately fifth–eleventh centuries A.D. **1953** K. H. Jackson *Lang. & Hist. Early Brit.* 377 Dark-Age Latin. **1957** G. E. Wright *Bibl. Archaeol.* iv. 56/2 Shortly before 1700 B.C. a dark age settled over Egypt which was to last some one hundred and fifty years. This was caused by the invasion of Asiatics whom the Egyptians called *Hyksos*. **1832** J. Rennie *Consp. Butterfl. & Moths* 65 The Dark Arches..appears the end of June and beginning of July. **1921** Dark Arches [see *arch *sb.* 7]. **1951** Colyer & Hammond *Flies Brit. Isles* xxi. 267 *L[arvaevora] ferox*..has been bred from the Dark Arches Moth. **1958** W. J. Stokoe *Caterp. Moths* (ed. 2) I. 255 *The Dark Arches*, sub-family Agrotinae, Apamea monoglypha, occurring in all parts of the British Isles...It is very variable in its general colouring. **1860** C. M. Yonge *Hopes & Fears* II. iv. 66 'Where is she?' 'In the dark chamber, doing a positive of the cathedral.' **1914** *Astrophys. Jrnl.* XXXIX. 438 Two spurious currents..are found in the photo-electric cell...The second is what has been called a 'dark current', in the same direction as the light current. **1963** B. Fozard *Instrumentation Nucl. Reactors* vi. 67 Dark current is the current flowing in the cathode, or in the complete multiplier tube, with no illumination of the cathode. Dark current is due mainly to thermionic emission of electrons from the cathode and the early-stage dynodes. **1865** R. Beck *Achromatic Microscope* 34 In every kind of 'dark-field illumination' the light comes upon the object from below, but at such an oblique angle as never to enter the object-glass direct. **1966** D. G. Brandon *Mod. Techniques Metallogr.* 16 Features that can be observed by dark field illumination are always faintly visible by direct illumination...The same is not true of transparent biological specimens. **1927** Hemingway *Men without Women* (1928) 187 We were both tired of the sun...You could not sit outside the hut without dark glasses. **1965** R. Erskine *Passion Flowers in Business* v. 58 One would see later if dark glasses in the office would be a good thing. **1860** *Q. Jrnl. Microscopical Sci.* VIII. 207 (*heading*) On a dark-ground illuminator. **1949** H. C. Weston *Sight, Light & Efficiency* iv. 142 A well-known ophthalmologist recommended dark-ground reading matter for the use of partially-sighted children. **1965** W. J. Garnett *Freshwater Microscopy* (ed. 2) iii. 55 For the observation of microscopic freshwater life..much more can be achieved by the use of dark-ground illumination. *Ibid.*, It is important with all dark-ground work to use a strong source of light. **1864** D. Lewis tr. *St. John of the Cross's Wks.* I. 3 The dark night, through which the soul passes, on its way to the Divine Light. *Ibid.* 57 Faith, the dark night of the soul. **1913** C. Mackenzie *Sinister St.* I. II. xv. 409 An Half-hour with St. John of the Cross made him ask himself whether this were the dark night of the soul through which he was passing. **1927** J. S. Huxley *Relig. without Rev.* iv. 124 The sense of being forsaken...Mystics have called it 'the dark night of the soul', and describe it as an abandonment of the soul by God. **1951** G. Greene *End of Affair* II. i. 52 What do I know of phrases like 'the dark night' or of prayer? **1970** *Guardian* 22 Dec. 2/4 Governor Nelson Rockefeller..has emerged from the dark night of the soul that afflicts all politicians pondering the supreme sacrifice. **1954** *Beaver Committee Rep.* (*Cmd.* 9322), New legislation should prohibit the emission of dark smoke.. from any chimney...By 'dark smoke' we mean smoke of density equivalent to, or greater than, shade 2 on the Ringelmann Chart. **1958** *Times* 31 May 3/7 The Clean Air Act 1956 will be fully in force to-morrow, and it will be an offence punishable by fine to emit dark smoke from any chimney in England and Wales.
d. *Physics.* **dark lines** = *absorption lines* (*absorption 6); so *dark-line* attrib. or as adj.; **dark space,** one or other of two non-luminous regions (the *cathode or *Crookes or *first dark space* and the *Faraday or *second dark space*) in a vacuum tube traversed by an electric discharge; also *dark discharge*.
1802 W. H. Wollaston in *Phil. Trans.* II. 378 C, the limit of green and blue, is not so clearly marked as the rest; and there are also, on each side of this limit, other distinct dark lines. **1838** M. Faraday in *Phil. Trans.* I. 138 A purple stream..appeared on the end of the positive rod,.. but never joining the negative glow, there being always a short dark space between. *Ibid.* 139 The dark discharge through air..leads to the inquiry, whether the particles of air are..capable of effecting discharge from one to another without becoming luminous. **1878** *Encycl. Brit.* VIII. 64/1 The dark spaces that sometimes appear in the spark in gas at the atmospheric pressure. *Ibid.*, When the discharge takes place in highly rarefied gas, a dark space of this kind almost always separates the positive from the negative light. *Ibid.*, Pending further investigation, Faraday called it the dark discharge. **1879** *Ibid.* IX. 728/1 The dark lines of the spectrum of sunlight. **1895** S. P. Thompson *Elem. Less. Electr. & Magn.* (ed. 2) II. iv. 307 The kathode exhibits a beautiful bluish or violet glow, separated from the conductor by a narrow *dark space*. **1920** Dark space [see *cathode dark space* s.v. *cathode c]. **1928** W. M. Smart *Sun, Stars & Universe* vi. 83 Certain groups of lines which are known to belong to the arc spectra of the elements concerned.. together with their dark-line counterparts in the solar spectrum. **1958** [see *Faraday dark space*].

dark, *sb.* Add: **1. a.** *dark of the moon* (later U.S. examples).
1871 E. Eggleston *Hoosier Schoolm.* (1872) x. 87 But it must be rendered in the dark of the moon. **1889** Farmer *Americanisms* 193 *Dark Moon* or *Dark of the Moon*, the period between the moon's change from 'full' to 'new'. Also provincial in England. **1945** Tennessee Williams *Battle of Angels* II. i. 41 In the dark of the moon, beside a broken fence rail in some big rolling meadow.

6. *attrib.* and *Comb.* **dark adaptation,** self-adjustment of the eye to reduced intensity of light by means of an increase in the sensitivity of the retina; **dark-adapted** in *passive* and as *ppl. a.*, of an eye in which there is dark adaptation.
1909 E. B. Titchener *Text-Bk. Psychol.* I. xviii. 74 The immediate after-effect of general adaptation is always this contrary trend of vision:..if dark-adapted, [you are] now light-sighted. **1909** *Ibid.* xix. 80 When dark-adaptation has gone a certain distance. **1920** Dark adaptation [see *adaptation 2 b]. **1922** R. S. Woodworth *Psychol.* x. 225 Go into a dark room, and at first all seems black, but by degrees—provided there is a little light filtering into the room—you begin to see, for your retina is becoming dark-adapted. **1934** H. C. Warren *Dict. Psychol.* 68/2 *Dark-adapted eye*, an eye whose condition has been so modified by the withdrawal of general light stimulation that faint stimuli have become more effective. **1935** *Discovery* May 138/1 This increased sensitiveness of the retina is called Dark-Adaptation. **1946** *Nature* 31 Aug. 303/2 Perhaps it will be possible to find somebody who has more visual purple in the dark-adapted state. **1963** C. D. Simak *They walked like Men* ii. 11 By now my eyes had become somewhat dark-adapted and I could make out the formless shape of furniture.

darkfall (dā·ɹkfǭl). [f. Dark *sb.* + Fall *sb.*[1]] The coming on of dark; dusk, nightfall.
1897 'O. Rhoscomyl' *White Rose Arno* 292, I can have fifteen of them here by darkfall. **1929** A. Clarke *Pilgrimage* 41 At darkfall in a house where nobles throng. **1930** —— *Flame* 34, I hear them coming, For it is darkfall now.

da·rkling, *sb.*[2] [subst. use of Darkling *a.*] = Dark *sb.*
1903 *Westm. Gaz.* 13 Jan. 2/3 At darkling of the moon. **1909** H. G. Wells *Tono-Bungay* IV. i. 443 She carried some rugs for me through the shrubbery in the darkling. **1923** —— *Men like Gods* I. vii. 112 He..blundered by two couples of lovers who whispered softly in the darkling. **1963** A. Garner *Moon of Gomrath* xiii. 102 Once Anghalac sounds you may not know peace again, not in the sun's circle nor in the darkling of the world.

darl, colloq. abbrev. of Darling.
1930 K. S. Prichard *Haxby's Circus* xxviii. 329 'Oh, darl, don't you bother,' he begged. 'I hate you to get all het-up.' **1952** J. Cleary *Sundowners* i. 13 I'm so glad you're built like you are, darl. **1959** C. MacInnes *Absolute Beginners* 16 'Hi, darl,' she said. 'Hi, hon,' I answered. That's how we heard two movie stars address each other at a film we went to ages ago. **1970** J. Cleary *Helga's Web* iv. 72 'Darl——.' He hadn't called her that for several years: short for *darling*, there had been a time when he had called her nothing else.

Darling, *sb.*[2] The name of a river in western New South Wales used *attrib.* in the names of certain plants growing in its neighbourhood; also **Darling shower,** a local name for a dust-storm.
1898 Morris *Austral Eng.* 115/1 *Darling Pea*, an Australian plant, *Swainsonia galegifolia*. *Ibid.*, *Darling Shower*, a local name..for a dust storm, caused by cyclonic winds. **1899** *Westm. Gaz.* 20 Feb. 10/1 The Riverina breeds a particularly distressful variety of dust-storm known as 'Darling showers'. **1909** Webster, Darling lily, *Crinum flaccidum*. **1936** A. Russell *Gone Nomad* viii. 65 By this time, owing to the protracted drought, sand-storms or 'Darling showers', as we facetiously called them, had become an almost daily occurrence in the Red Country. **1945** Baker *Austral. Lang.* iii. 63 *Darling Pea*...One of Australia's poisonous plants; cattle eating it become afflicted with staggers and die: so a man wandering in gait or dazed in appearance is said to be suffering from Darling Pea. *Ibid.* x. 189 Cobar shower, denotes a duststorm. Darling shower, Wilcannia shower, and Bedourie shower are employed similarly. **1963** A. Lubbock *Austral. Roundabout* 11 It [*sc.* Patterson's Curse]..does not poison animals, as does the innocuous-sounding Darling Pea.

darling, *a.* Add: **c.** Sweetly pretty or charming, 'sweet'. *affected.*
1805 E. Cavanagh *Let.* 4 Oct. in Londonderry & Hyde *Russ. Jrnls.* (1934) II. 187 We all follow'd them out of doors across a Garden to a *darling* place. **1854** *Punch* Mar. 116 Isn't it the darlingest, sweetest, prettiest, little dear darling darling! Oh! did you ever!! **1858** Queen Victoria *Let.* 1 Apr. in R. Fulford *Dearest Child* (1964) 84 Darling Beatrice came..in short clothes with darling little stockings and..pink satin shoes—really too darling. **1906** *Punch* 21 Mar. 215/1 You have a darling little note-book..to match your frock. **1908** B. Harraden *Interplay* 373 It is perfectly darling of you to have chosen Hughie. **1937** D. Aldis *Time at Heels* v. 109 Your little boy did so beautifully...He was simply darling. **1952** M. McCarthy *Groves of Academe* (1953) i. 14 I've seen them with you in the Co-op, Dr. Mulcahy. They're *darling*. **1970** *New Yorker* 28 Feb. 54/2 Isn't it going to be darling!

darling, *v.* (Later examples.)
1942 T. Rattigan *Flare Path* II. i. 123 Wonderful the way you stage people darling each other. **1967** *Punch* 19 July 82/2 In the interval she seems to be in every bar and every corridor, darlinging everybody in sight.

darlint (dā·ɹlint). *dial.* and *joc.* = Darling 1.
1888 Kipling *Soldiers Three* (1890) 89 She cried loud, poor darlint, bein' mishandled. **1922** Joyce *Ulysses* 143 Quicker, darlint! **1930** E. Pound *XXX Cantos* x. 43 Siggy, darlint, wd. you not stop making war on insensible objects.

darn, *sb.*[2] *slang* (orig. *U.S.*). [Cf. next.] *by darn*, used as a form of asseveration. Also *not to care* (or *give*) *a darn*, not to care at all.
1840 C. F. Hoffman *Greyslaer* II. iv. 206 But, by darn, the capting's cleared out without speaking to one..but ourselves. **1850** W. K. Northall *Life & Recoll. Yankee Hill* 119 You may put down all our family...I don't care a darn. **1854** M. J. Holmes *Tempest & Sunshine* xxiv. 330, I don't care a darn how many Miss Betsy's I git. **1891** M. E. Ryan *Told in Hills* 92, I don't care a darn about the sheep just now. **1920** Galsworthy *Foundations* I, *Anne.* Which do you like to be called—John or James? *James.* I don't give a darn. **1957** I. Cross *God Boy* (1958) i. 8, I would love shooting and kicking out at my enemies, not giving a darn what they try to do to me.

darn, *adv.* and *a. slang* (orig. *U.S.*). [Arbitrary perversion of *damn *a.* and *adv.*]
A. *adv.* Extremely, intensely.
1789 Webster *Diss. Eng. Lang.* 385 The word (*dern*) is in common use in New England and pronounced *darn*. It has not, however, the sense it had formerly; it is now used as an adverb to qualify an adjective, as *darn sweet*; denoting a great degree of the quality. **1797** D. Mackintosh *Rational Ess. Eng. Gram.* 82 We say, dea'rn or da'rn, heark'en, heart', hearth'. **1869** Barnum *Struggles & Triumphs* (1871) 146 Darn glad to see you, by hokey; I came down here to have lots of fun. **1892** *Century Mag.* June 264 It was a darn good churn too. **1922** S. Lewis *Babbitt* ii. 18 You're so darn scared of the car that you drive up-hill with the emergency brake on! *Ibid.* v. 64 You're pretty darn near talking socialism! **1969** *New Yorker* 12 Apr. 62/3 We want to make darn sure we get there and back.
B. *adj.* 'Blessed', 'confounded'. Also *absol.*
1840 C. F. Hoffman *Greyslaer* III. iv. 141 'Jim, you've done the darn thing agin us to-night, and no mistake' said one. **1899** S. Hale *Lett.* (1919) 348 You know they are all here improving their minds, learning some darn thing or other. **1904** H. R. Martin *Tillie* 40 To fill out blanks answerin' to a lot of darn-fool questions 'bout one thing and 'nother. **1924** A. J. Small *Frozen Gold* iii. 87 I'm doing my darnest to drive you out of Cedar Falls. **1929** *Melody Maker* Jan. 15/1, I know a darn' sight more about the makings of a piano than I did a month ago.

darn, *v.*[1] Add: **c.** *transf.* To mend (a hole in a wall, road, etc.) by filling-in.
1801 W. Beattie *Entertaining & Instruct. Tales* (1813) I. 30 He staps wi' strae ilk navus bore, An' ilka crevice darns. **1850** 'H. Hieover' *Pract. Horsemanship* 146 There is a mode of keeping our present roads in order, that I have heard termed 'darning' them: i.e. if a part is seen somewhat lower than the surface, the unbroken pieces of granite are got up to it, and there left to be crushed by the wheels of carriages.
2. To ornament or embroider with darning-stitch.
1882 [see s.v. *darning *vbl. sb.* 3]. **1900** Day & Buckle *Art in Needlework* 108 The flower stalk is defined by darning the first row in a darker colour. *Ibid.*, The background is darned diaper fashion.

darn, *v.*[2] (= 'damn'). Earlier and later N. Amer. examples.
1781 *Pennsylvania Jrnl.* 20 June, In New England prophane swearing..is so far from polite as to be criminal, and many..use..substitutions such as *darn* it, for d—n it. **1809** A. B. Lindsley *Love & Friendship* 8 Darn my skin 'f you wouldn't dewe it. **1825** J. K. Paulding *J. Bull in Amer.* iii. 36 If I don't have him before the justice, darn my soul. **1922** S. Lewis *Babbitt* v. 51 Darn it, I thought you'd quit this darn smoking! *Ibid.* viii. 119 Every small American town is trying to get population and modern ideals. And darn if a lot of 'em don't put it across! **1968** *Globe & Mail Mag.* (Toronto) 17 Feb. 9/3 'Play, darn it!' he shouted to the open-mouthed pianist.

darnation. (U.S. examples.)
1798 *Aurora* (Philadelphia) 14 Aug. (Thornton, s.v. *Nation*), It seems as if the Irish are as incorrigible as the darnation Bostonians. **1825** S. Woodworth *Forest Rose* I. iii, Darnation take the garlic, I say. **1832** [see Tarnation *sb.*]. **1840** *Daily Pennant* (St. Louis) 9 May (Thornton, s.v. *Chip*), It wouldn't take so much..to make it a darntation sight riz-er. **1878** Mrs. Stowe *Poganuc P.* iii. 33 If I didn't hold on to him he'd have us all to the darnation in five minutes. **1924** in *Dialect Notes* V. 265.

darned, *ppl. a.*[1] Add: **2.** Formed, made, or ornamented with darning-stitch.
1881 C. C. Harrison *Woman's Handiwork* I. 86 Italian punto a maglia or darned netting. **1882** Caulfeild & Saward *Dict. Needlework* 145/2 The Darned Embroidery most practised in Europe has been chiefly worked upon cotton. **1895** *Montgomery Ward Catal.* 79/1 Darned or embroidered net, 72 in. **1909** *Westm. Gaz.* 27 Feb. 15/2 The theatre or afternoon dress is of darned filet over satin or cashmere. *Ibid.* 7 Aug. 15/1 Darned net, such as was worn about 1830, or even earlier, looks very well.

darned, *pa. pple.*, *ppl. a.*[2], and *adv.* (Earlier and later examples.)
(*a*) **1808** J. N. Barker *Tears & Smiles* 18 I'll be darned, sir, if I think this is the way. **1847** J. K. Paulding *Bucktails* II. ii. 33 I'll be darned but I guess I've lost my way. **1888** A. C. Gunter *Mr. Potter of Texas* xxii, He sinks back.. in amazed astonishment and mutters: 'Wall, I'm darned!'
(*b*) **1815** in *Amer. Speech* (1928) III. 231. **1834** C. A. Davis *Lett. J. Downing* 41 'Do you want another report?' 'Not by a darn'd sight,' says he. **1891** M. E. Ryan *Told in Hills* 123 She..was the 'darndest, cutest, little customer he ever saw'. **1904** W. H. Smith *Promoters* xviii. 269 Darndest fellow to take things up that way.
absol. **1844** 'J. Slick' *High Life in N.Y.* II. 233 There must al'ers be an eend tu every thing that's sweeter than common, that's the darndest of it. **1907** N. Munro *Daft Days* iv. 31 This is a funny dog...Isn't He the darnedst? **1960** *Farmer & Stockbreeder* 8 Mar. 147/1 The public are eating between four and five eggs each week and the Egg Board are doing their darndest to increase this figure still further.
(*c*) **1807** L. Beach *Jonathan Postfree* 23 Drove down old Squire Herdy's cattle—darn'd ugly creatures to drive. **1822** Woodworth *Deed of Gift* 45, I have taken a liking to you, 'cause you are so darn'd pretty. **1888** *Harper's Mag.* July 323/2 In Colorado the man who tells the first story has a darned poor show. **1922** S. Lewis *Babbitt* v. 57 Machine looks brand new now—not that it's so darned old, of course; had it less 'n three years. **1962** J. Ludwig in R. Weaver *First Five Years* 19 By God, she tells herself, this is a darned good face.

darner. Add: **2.** (Later example.)
1966 *Price List* (*Olney Amsden & Sons, Ltd.*) 29 *Needles*... Sharps, Crewels, Darners, etc.
b. A darning-machine.
1876 J. S. Ingram *Centenn. Exposition* x. 346 The Complete Darner. *Ibid.* 347 The top..could be removed for the insertion of a ball of darning-cotton into the swell of the 'darner'. **1897** *Sears, Roebuck Catal.* 324/1 The Magic Darner is a machine recently invented and patented for mending hosiery, silk, wool or cotton.
c. A darning-ball or similar device.
1895 *Montgomery Ward Catal.* 91/1 Egg Darners, with handle. **1909** *Daily Chron.* 25 May 9/1 Even the embroiderer who does not use a hoop will find it more convenient in working stockings to do the embroidery over a darner. **1966** *Price List* (*Olney Amsden & Sons, Ltd.*) 29 Wood Darner.. 1/6.

darning, *vbl. sb.* Add: **1. c.** Embroidering with darning-stitch; also = darning-stitch.
1882 Caulfeild & Saward *Dict. Needlework* 148/1 The embroidery is done in Satin stitch or in plain Darning. **1930** tr. *T. de Dillmont's Encycl. Needlework* 519 Close ground-work of darning and little wheels.
3. *darning-cotton, wool*; **darning-egg** = *darning-ball*; **darning-machine,** a machine for darning stockings, clothes, etc. (Knight *Dict. Mech.* Suppl. *a* 1884); **darning-needle,** (*b*) (see *devil's darning-needle* s.v. Devil *sb.* 25 b, c); **darning-stitch,** (*b*) a straight stitch used in embroidery to make a regular open-work pattern.
1811 Jane Austen *Let.* 18 Apr. (1952) 270 *She* is in want of chimney lights for Tuesday; —— & *I*, of an ounce of darning cotton. **1876** Darning-cotton [see *darner 2 b]. **1897** *Sears, Roebuck Catal.* 335/3 Our woman's tool set consists of..one pinking iron, one darning egg, [etc.]. **1925** E. Glasgow *Barren Ground* III. ii. 410 Dorinda..slipped her darning-egg into one of Nathan's socks. **1956** K. Hulme *Nun's Story* iv. 62 The lean self-abnegating face bent over a darning egg. **1881** C. C. Harrison *Woman's Handiwork* I. 27 Darning stitch is used to restore old embroideries. **1882** Caulfeild & Saward *Dict. Needlework* 148/2 Fillings for the centre of any designs that are not worked in Satin or Darning stitch. *Ibid.* 149/1 Darn the thick lines up and down in Point de Reprise or plain Darning stitch. **1934** M. Thomas *Dict. Embroidery Stitches* 72 Darning Stitch is really a long running or tacking stitch, the difference being that in darning used for decoration, only one thread of the material is picked up with each stitch. *Ibid.* 73 Darning Stitch..is also used in pattern darning. **1912** A. Bennett *Matador* 61 I've made this bouquet for you...You must excuse it being tied up with darning wool. **1932** D. C. Minter *Mod. Needlecraft* 178/2 Darning wool is better slightly finer than that of which the knitted garment was made.

‖ **darshan** (dā·ɹʃăn). [ad. Hindi *darśan*, f. Skr. *darśana* view, f. *drś* to see.] The sight of an august or holy personage.
1920 M. K. Gandhi in *Young India* 20 Oct. 3/1 In vain did Mrs. Gandhi and others plead with the crowds for self-control and silence... The answer was that they had come many miles to have *darshan* and *darshan* they must have. **1947** J. P. Chander *Teachings of Mahatma Gandhi* 140 *Darshan*. Love that is satisfied with touching the feet of its hero and making noise at him is likely to become parasitical. **1960** Koestler *Lotus & Robot* I. v. 152 To be in the mere presence of the guru conveys darshan, a spiritual enrichment and an intellectual lesson which need not be conveyed in words. **1962** *Times* 9 Oct. 9/1 Many of these people come simply to have *darshan* of the President. **1964** V. Naipaul *Area of Darkness* x. 251 She gave *darshan*—made an appearance, offered a sight of herself—only on important anniversaries. **1971** *Illustr. Weekly India* 4 Apr. 11/1 A Khatri cannot have *darshan* of the Kul Devi before *dev kaj*. *Ibid.* 23/2 An old man who only wanted to touch his feet and receive darshan was nearly killed by a lathi.

dart, *sb.* Add: **1. d.** A light pointed missile thrown at a target in the game called *darts*. Also *attrib.* and *Comb.*
1901 *Stationer, Printer*, etc. 1 June 322 Ring Boards, Dart Boards. Parlour Cricket. **1916** H. G. Wells *Mr. Britling* I. v. § 11 Dart-throwing and ring-throwing stalls. **1924** B. Gilbert *Bly Market* 18 Coconut Saloons. Shooting Galleries. Dart-Saloons. *Ibid.* 419 Darts, darts, darts, penny a dart. Over 50 wins the prize... No skill needed. **1929** *B'ham Post* 12 Jan., A peculiar thing I noticed was that two darts, such as they use in public-houses, were near the body. **1941** 'G. Orwell' *Lion & Unicorn* 15 A nation of stamp-collectors, pigeon-fanciers..coupon-snippers, darts-players. **1958** *Times* 29 Apr. p. x/4 In the past 30 years darts has become the most important of all public house games. *Ibid.* p. x/5 The darts themselves have changed during the past 20 years, from the comparatively light, wooden-stemmed type with feather flights to a shorter kind with much heavier metal body carrying plastic flights.

7. (Earlier Australian example and an Anglo-Irish example.) Also, (one's) fancy or favourite.
1882 *Sydney Slang Dict.* 3 *Dart*, object of attraction, or enticing thing or event, or a set purpose. **1895** in Morris *Austral Eng.* (1898) 115 'Fresh strawberries eh!—that's my dart,' says the bushman when he sees the fruit lunch in Collins-street. **1914** Joyce *Dubliners* 112 Suddenly..he thought of Terry Kelly's pawn-office... That was the dart! Why didn't he think of it sooner?

darter. 5. (Earlier U.S. example.)
1842 J. E. DeKay *Zool. N.Y.* IV. 20 The Tessellated Darter. *Boleosoma tessellatum*... It is usually seen at the bottom of clear springs or streams, lying for a while perfectly still, and then suddenly darts off with great velocity at its prey. This habit has acquired for it the popular name of *Darter*.

Dartmoor (dā·ɹtmŭ³ɹ, -mǭ³ɹ). Name of a district in Devonshire, applied *attrib.* (also *ellipt.*) to special breeds of ponies and sheep produced there.
1831 Youatt *Horse* iv. 59 The Dartmoor pony is larger than the Exmoor, and, if possible, uglier. **1837** —— *Sheep* vii. 252 The South Downs never succeeded well on the heath-clad hills of the Dartmoor sheep. **1902** *Encycl. Brit.* XXV. 189/2 Native ponies include those variously known as English, New Forest, Exmoor, Dartmoor. *Ibid.* 194/2 The Dartmoor is a hornless, longwool, white-fleeced sheep. **1906** *Westm. Gaz.* 24 Sept. 8/1 An intelligent Dartmoor, that would follow its master like a dog. **1937** Hull & Whitlock *Far-Distant Oxus* i. 24 In the last stall a tiny shaggy Dartmoor butted her with his head. **1955** *Times* 6 July 7/3 Miss Calmady-Hamlyn, of Buckfast, had both male and female champions in the Dartmoor pony classes for the second year running. **1970** *Observer* (Colour Suppl.) 26 Apr. 36/2 Dartmoors look rather like shaggy dogs.
b. Special Combs. **Dartmoor-clip** *v.* [*Dartmoor*, name of the convict prison near Princetown], to cut (a person's) hair very short as for a convict; **Dartmoor crop,** hair so cut; **Dartmoor granite** (see quot.).
1932 Kipling *Limits & Renewals* 197 He was run round to the barber an' Dartmoor-clipped for wearin' oily and indecent appendages. **1930** E. Raymond *Jesting Army* I. i. 11 He was the only officer on the ship with his hair shaved down to his skull in what was known as the 'Dartmoor Crop'. **1904** Goodchild & Tweney *Technol. & Sci. Dict.* 74/2 *Dartmoor Granite*, a greyish granite used in large work —e.g. London Bridge.

Darwin (dā·ɹwin). The name of Charles *Darwin* (see Darwinian *a.* 2), used *attrib.* (also *ellipt.*) to designate a race of tulips with tall stems and large self-coloured flowers.
1889 F. Darwin *Let.* 13 Apr. in E. H. Krelage *Drie Eeuwen Bloembollenexport* (1946) II. 553 Allow me to thank you for your courteous note in which you tell me of your wish to name after my father a new strain of tulips...I hope I shall see the 'Darwin Tulips' at Paris. **1891** *Gardeners' Chron.* 4 July 10/3 The Darwin Tulips are of Flemish origin. *a* **1916** H. H. Munro *Toys of Peace* (1919) 237 The Darwin tulips haven't survived the fact that most of the cats of the neighbourhood held a parliament in the middle of the tulip bed. **1922** F. M. Ford *Let.* 24 Oct. (1965) 145, 200 Darwin tulips. **1969** Hay & Synge *Dict. Garden Plants* 368/2 The old tulips of the Dutch flower painters..are usually not so vigorous as the Darwins or Darwin hybrids.
b. Used in the possessive: see *Darwinian *a.* 3.

Darwinian, *a.* (*sb.*) **1.** Delete † and add earlier and later examples.
1794 W. B. Stevens *Jrnl.* 3 June (1965) II. 161 The lines are truly Darwinian. **1797** Scott *Lett.* (1932) I. 62, I do not for example, think quite so severely of the Darwinian style, as to deem it utterly inconsistent with the Ballad. **1950** H. Darbishire *Wordsworth* i. 17 He is balancing his lines in the Darwinian manner with ornamental epithets.
2. (Earlier and additional examples.)
1860 T. H. Huxley in *Westm. Rev.* Apr. 566 The Darwinian hypothesis has the merit of being eminently simple and comprehensible. **1866** K. Stanley *Jrnl.* 22 Aug. in B. & P. Russell *Amberley Papers* (1937) I. 525 It was a very fine address, Darwinean in principle. **1884** W. S. Gilbert *Princess Ida* II, Darwinian Man, though well-behaved, At best is only a monkey shaved! **1971** M. S. Howard *Jonathan Cape* 313 The logical conclusions of Darwinian theory had yet to be drawn in simple terms to demonstrate the link between animal behaviour patterns and those of human society.
b. (Earlier examples. In quot. 1809, the sense is 'a follower of Erasmus Darwin'.)
1809 *Monthly Pantheon* Apr. 262 Hear this, ye..Darwinians. **1869** A. R. Wallace *Malay Archip.* I. iv. 61 This is..the first instance known of a 'flying frog', and it is very interesting to Darwinians. **1915** A. Huxley *Let.* Aug. (1969) 76 Laforgue was also a hearty Darwinian and liked the thought of being a developed beast.
3. Darwinian tubercle, a projection sometimes present on the edge of the human external ear (see quots.), believed by some scientists to be a relic of the pointed ear of quadrupeds; also called *Darwin's tubercule*, *Darwin's peak*.
1890 Billings *Med. Dict.*, *Darwinian tubercle*.., eminence on edge of helix of external ear, believed to correspond with end of pointed ear of apes, and therefore considered a vestige. **1920** I. F. & W. D. Henderson *Dict. Sci. Terms* 69/2 *Darwinian tubercle*, the slight prominence on the helix near the point where it bends downwards. **1957** *Encycl. Brit.* VII. 823/2 Round the margin [of the ear] in its upper three-quarters is a rim called helix, in which is often seen a little prominence known as Darwin's tubercle, representing the folded-over apex of a prick-eared ancestor.

Darwinism. 2. (Earlier example.)
1864 T. H. Huxley in *Nat. Hist. Rev.* Oct. 567 What we may term the philosophical position of Darwinism.

Darwinize, *v.* Add: Also *trans.* So **Da·rwinized** *ppl. a.*
1920 G. B. Shaw in *Public Opinion* 13 Aug. 160/2 It has restored faith in Providence to a Darwinised world. **1929** Blunden *Nature in Eng. Lit.* 14 The great mind which compares and sifts evidence until a new *De Rerum Natura* darwinizes us.

‖ **Dasein** (da·zəin). *Philos.* Also 9 **Daseyn.** [G., f. *da* there + *sein* being, a favourite word of Goethe's.] **a.** In Hegelian philosophy: existence, determinate being. **b.** In existentialism, esp. that of Heidegger and Jaspers: human existence, the being of man-in-the-world.
1846 J. D. Morell *Hist. Philos.* II. v. 141 This may appear clearer to the German scholar, if we say in Hegel's language, that *Sein* and *Nichts* form *Daseyn*. **1874** W. Wallace *Logic of Hegel* Proleg. p. clxxx, To bring a thing into *Daseyn* is to give it definite being. **1938** G. Reavey tr. *Berdyaev's Solitude & Society* ii. 56 Heidegger speaks with greater authority of the *Dasein* than of existence itself. **1956** F. Copleston *Contemp. Philos.* x. 160 As an empirical being which is 'there',..I am *Dasein*, object.

dash, *v.*[1] Add: **11.** (Earlier example.)
1800 T. Morton *Speed the Plough* II. ii. 28 But dash it, Lady Nelly, what do make thee paint thy vace all over we rud ochre zoo?
16. b. dash-pot (later examples and examples of extended application); **dash-wheel** (earlier example).
1878 in J. Dredge *Electr. Illum.* (1885) II. App. p. lxiv, The arm of the lever may be at right angles to the carbon, one end being weighted and the other attached to the core of a solenoid; the core may have a dash-pot action. **1902** *Encycl. Brit.* XXVIII. 86/1 The arc-lamp mechanism is provided with a dash-pot, or contrivance in which a piston moving nearly air-tight in a cylinder prevents sudden jerks in the motion of the mechanism. **1926** *Gloss. Terms Electr. Engin.* (*Brit. Engin. Standards Assoc.*) 50 *Dash-pot*, an appliance for preventing the sudden or oscillatory motion of any moving part of a piece of apparatus, by the friction of air or of a liquid. **1930** *Engineering* 21 Feb. 249/2 The proper function of the piston chamber is, however, that of a dash-pot. **1931** *Flight* 13 Nov. 1131/2 The dashpot is double acting, and there is a powerful check to spring recoil. **1936** *Gloss. Terms Railway Signalling* (*B.S.I.*) 16 *Dash pot*, a cylinder with a piston valve in which the escape of air or liquid is checked by the valve to assist in lessening shock. **1940** *Chambers's Techn. Dict.* 225/1 *Dash pot*, a device for damping out vibration; it consists of a piston attached to the part to be damped, fitting loosely in a cylinder of oil. **1959** *Times Rev. Industry* Mar. 36/1 The variable feed rate for cutting is controlled by a hydraulic dashpot. **1962** *Engineering* 15 June 793/1 The relay trip current setting is adjustable ..simply by altering vertically the position of the self-locking dashpots. **1839** Ure *Dict. Arts* 226 Put this mixture into the colour trough..and after two days wash in the dash-wheel.

dash, *sb.*[1] Add: **7. c.** Sometimes implying the use of strong language; hence as a mild substitute for *devil*.
1883 Ld. R. Gower *My Remin.* II. xxviii. 259 Who the D—h is this person..and what the Dash does he here? **1899** A. Nicholas *Idyl of Wabash* 64 A dreadful thought which if put in print would have contained a dash.
f. One of the two signals (the other being the dot) which in various combinations make up the letters of the Morse alphabet. Also *dash-and-dot*, more usually *dot-and-dash* (Dot *sb.*[1] 8).
1859 T. P. Shaffner *Telegr. Man.* 469 Whether the dots, spaces, and dashes be long or short, they should be uniform. **1873** F. Jenkin *Electr. & Magn.* xxii. § 4 Morse signals are sent by a simple key...A short depression or mere tap sends the short elementary signal technically called a *dot*; a longer depression sends the second elementary signal technically called a *dash*. **1882** Ogilvie Suppl., *Dash-and-dot*, consisting of dashes and dots; as, the dash-and-dot alphabet. **1916** J. Buchan *Greenmantle* xx. 268 The sound was regular and concerted—dot, dash, dot—dash, dot, dot..the longs and shorts of the Morse Code. **1942** *Electronic Engin.* XV. 36 On automatic sending it transmits S.O.S. three times followed by a long dash.
9. b. *to do one's dash* (Austral. colloq.), see quots. 1916 and 1966; *to have a dash* (*at*) (colloq.), to make an attempt.
1916 C. J. Dennis *Songs Sentimental Bloke* 121 *To do one's dash*, to reach one's Waterloo. **1923** Wodehouse *Inimit. Jeeves* iii. 37 The blighter's manner was so cold and unchummy that I bit the bullet and had a dash at being airy. **1930** —— *Very Good, Jeeves!* (1957) vi. 114, I supposed I had better have a dash at it and get it over. **1966** G. W. Turner *Eng. Lang. Austral. & N.Z.* vii. 152 It is tempting to wonder whether the expression 'he's done his dash' meaning 'he is played out', 'he has done all he can' is connected with the gold-miner's *dashing*.
11. (Earlier U.S. example.) **b.** A sprint. *U.S.*
1836 *Spirit of Times* (N.Y.) 20 Feb. 5/3 Or, I will make two races, for one thousand dollars each, give you fifty yards in a dash of one mile, and one hundred yards in a dash of two miles. **1895** *Chicago Tribune* 24 May 11/4 At least three of the rivals likely to meet in the 100-yard dash are said to be capable of tying the intercollegiate record. **1948** *P.C.C. Chron.* (Pasadena, Calif.) 31 Mar. 4/5 Anderson took a third in the open 100 yard dash. **1957** *Encycl. Brit.* XIX. 665/1 Distances up to and including 220 yd. are, in the United States, called dashes.
12. (Earlier and later U.S. examples.)
1868 *Rep. Comm. Patents 1867* (U.S.) I. 481/2 Carriage Boot..January 15, 1867. The apron is combined with a dash cover. **1911** J. C. Lincoln *Cap'n Warren's Wards* ii. 15 He says the buggy dash is pretty well scratched up.
b. Now esp. in motor vehicles; = *dash-board 1 b.
1902 Kipling in *Windsor Mag.* Dec. 13/2 Kysh's hands juggling with the levers behind the discreet backward sloping dash. **1906** *Daily Chron.* 14 Nov. 9/3 The coil and commutator, being fixed on the dash, are always in front of the driver. **1919** *Autocar Handbk.* (ed. 9) 253 A second lamp placed somewhere on the dash. **1929** *Daily Express* 30 Jan. 3/1 The car is fitted with electrical devices with lights on the dashfront to keep the driver informed when anything goes wrong with the lubrication or the ignition. **1944** *Coast to Coast 1943* 163 They got in the truck and by the light from the dash Black read the slip. **1966** M. Woodhouse *Tree Frog* xxvii. 205, I fitted the key into the truck's dash and backed off.
13. (Further examples.)
1796 in *Repert. Arts & Manuf.* (1797) VII. 290 Specification of the Patent granted to Mr. William Raley, of Newbald, in the East Riding of Yorkshire. November 10, 1796... O, the moving dashes or breakers. **1865** *Harper's Mag.* Mar. 541/2 Last summer Joe bought an old-fashioned dash churn. **1963** *Times* 18 May 11/5 *Meadar loinithe* is a plunging or dash churn.
14. dash-light, a light on the dash-board of a motor vehicle.
1926 *Catholic Mirror* June 47 'Don't they call this the dash light?' she queried, fingering the little nickel-plated illuminator. **1935** M. Eberhart *Cases of Susan Dare* 285 His mouth tightened in the little glow from the dashlight.

dash-board. Add: **1.** (Earlier examples.)
1846 *Rep. Comm. Patents* (U.S.) 81 An improvement in sleighs, principally connected with the dash-board, has been patented this year. **1851** *Gt. Exhib. Catal.* II. 635/1 Adjusting-iron for dash-lamps suitable for sweeps of carriage dash-boards.
b. In motor vehicles, the panel beneath the windscreen on which electrical instruments and controls are mounted. Also in aircraft.
1904 A. B. F. Young *Compl. Motorist* iv. 114 A cooling apparatus has now been fitted behind the bonnet and in front of the dashboard. **1925** W. Deeping *Sorrell & Son* v. § 1 Sorrell remained by the car. He liked the colour of it, and the compact brightness of the dash-board. **1942** *Electronic Engin.* XV. 9 A course meter on the dashboard gives a rough indication of distances.
2. b. In calico printing (see quot.).
1860 Ure *Dict. Arts* (ed. 5) II. 7 Dash wheels..were revolving wheels having dash-boards, which are much used in the washing processes necessary in calico printing.

dashed, *ppl. a.* **3.** Also *advb.*, deucedly, confoundedly.
1893 W. S. Gilbert *Utopia* II, How utterly dashed absurd.

dasheen (dæʃi·n). [Origin uncertain.] A cultivated variety (*Colocasia esculenta*) of the taro.
1899 *West Indian Bull.* I. 131 What is known in Trinidad as 'dasheen', also a *Colocasia*, is not, I think, grown in Jamaica at all. **1910** *U.S. Dept. Agric., Bur. Plant Industry Bull. No.* 164 27 It is not always easy to distinguish between the dasheens and the taros, for some of the dasheens have a tendency to throw sprouts from the tips of the tubers. **1913** W. H. Page *Let.* 26 Jan. in B. J. Hendrick *Life of W. H. Page* (1922) I. iv. 128 I'm going to grow dasheens. **1918** *Bull. Dept. Agric. Trinidad* XVII. 29 The name dasheen appears to be a corruption of the French phrase 'de la Chine'. **1953** P. Leigh Fermor *Violins of St. Jacques* 88 Living on dasheen and yam and breadfruit they had uprooted from the plantations. **1968** J. W. Purseglove *Tropical Crops* II. 587 Some..also provide temporary lateral shade, as do *Xanthosoma* spp. (tannias), *Colocasia* spp. (dasheen and eddoes).

dasher. 2. (Earlier U.S. examples.)
1846 *Rep. Comm. Patents* (U.S.) (1847) 233 What I claim as my invention..is the combination of the vertical dasher with the oscillating dashers. **1848** D. Drake *Let.* 7 Jan. in C. D. Drake *Pion. Life Kentucky* (1870) v. 93 The latter stages of the process [of churning], when the butter rises on the dasher.

dashiki (da·ʃiki). Also **dasheki.** [Of W. Afr. origin.] A West African type of shirt, sometimes worn symbolically by U.S. Negroes.
1969 *Observer* 26 Jan. 9/7 The revolution has spread to the Church: Negro ministers often wear *dashikis* or turtle-neck sweaters. **1969** *Sunday Times* 9 Mar. (Colour Suppl.) 40 The Harlem blacks want the dasheki, the short African jacket. **1970** 'J. Morris' *Candywine Devel.* xvi. 184 A sleek, oxblood brown singer who was flapping his *dashiki* shirt.

dashingness (dæ·ʃiŋnės). [f. Dashing *ppl.a.* 3 + -ness.] The quality of being dashing.
1934 E. Bowen *Cat Jumps* 91 Her dashingness, curtness, and air of experience. **1943** —— *Seven Winters* 31 Edwardian dashingness.

‖ **dasht** (dæʃt). [Pers. *dasht* desert, plain without water.] A name in Iran and some other parts of Asia for a desert, esp. a stony desert.
[**1875** *Encycl. Brit.* III. 223/2 The Dasht-Baha-rak is an extensive plain in this district, on which was formerly situated a large city, once the capital of Badakhshan.] **1901** *Mem. Geol. Surv. India* XXXI. 189 These floods, which, no doubt, were more frequent in former times, have

spread the pebbles over large areas in the desert, giving rise to the stony plains known by the name of 'dasht'. **1925** *Blackw. Mag.* Mar. 333/2 The coastal belt..was..primeval *dasht*—a dismal succession of open mud flats, patches of low tamarisk, pools of bitter water. **1950** W. B. Fisher *Middle East* xiii. 267 Firm sandy or stony stretches in which sand dunes (*rig*) may occur, giving a more 'normal' desert topography similar to that of Arabia or the Sahara. The term *dasht* is applied to this firm desert.

dassie (dæ·si, da·si). *S. Afr.* Also **dasje, dassi.** [Afrikaans: see Dassy.] **1.** = Das 2, Dassy.
1814 J. Pinkerton tr. *Thunberg's Acct. Cape of Good Hope* XVI. 129 The uppermost covers himself with the skin of a Dassi (*cavia capensis*), to keep out the cold and bad weather. **1828** J. Philip *Res. S. Afr.* I. ii. 27 Their Sonquaas (soldiers)..wander daily in the fields to catch dasjes, jackals, and other animals. **1835** Andrew Smith *Diary* 10 July (1940) II. 106 The dassie chews the cud. **1866** J. Leyland *Adv. S. Afr.* iii. 118 In these mountains I also saw the Dassi (Hyrax Capensis), or Rock Rabbit of the Cape. **1902** [see sense 2]. **1931** *Times Educ. Suppl.* 22 Aug. iv/3 Hyraxes, known also as 'Dassies', or 'Rock-rabbits', the conies of the Bible. **1946** *Cape Times* 28 Aug. 7/6 Eagles, Cape dassies, the mongoose. **1952** *Ibid.* 22 Nov. 4/6 A whole colony of dassies (rock-rabbits).
2. The sea-fish *Diplodus sargus* (family Sparidæ); black-tail.
1853 L. Pappe *Edible Fishes Cape G.H.* 22 *Cantharus Emarginatus*..Dasje..Rare in Table Bay, but more frequently caught in the several Bays to the East of the Cape. **1902** *Trans. S. Afr. Philos. Soc.* XI. 220 The Dasje might also with a little stretch of the imagination be likened to the rabbit or dassie, from its general shape, and this is the name by which it is known in Cape Town, Hout Bay, and Kalk Bay. **1930** C. L. Biden *Sea-Angling Fishes of Cape* xii. 199 Dassie are often caught that when taken from the water are silvery.

dassievanger (dæ·sivæŋəɹ, ‖ da·sifaŋəɹ). *S. Afr.* [Afrikaans, f. prec. + *vanger* catcher.]
Quot. 1867 identifies the bird with the popular name of 1 and the scientific name of 2.
1. The bateleur eagle, *Terathopius ecaudatus*. Cf. *berghaan.
1867 [see *berghaan]. **1889** H. A. Bryden *Kloof & Karoo* 273 A great black mountain eagle. We know him at once for a berghaan (cock of the mountain), or dassie-vanger (coney-eater). **1893** A. Newton *Dict. Birds* 132 Dassievanger (Coney-catcher), the Dutch name for an Eagle in South Africa, adopted by English residents.
2. A large, black bird of prey with a Y-shaped white mark on the rump, *Aquila Verreauxi*, found in S. and N.E. Africa and in Palestine.
1951 *Cape Argus* 20 Oct. (Mag.) 2/4 Most difficult of all birds to shoot, Hoesch found, was the black eagle or dassievanger.

dassy. (Earlier example.)
1846 H. H. Methuen *Life in Wilderness* iv. 79 A rock-rabbit, or dassy (*Hyrax Capensis*), was shot in the rocks, amongst which these animals always dwell.

dastoor, dastur(i, varr. Dustoor, dustoory.
1886 Kipling *Departmental Ditties* (ed. 2) 7 Roused his Secretariat to a fine Maratha fury, By a Hookum hinting at supervision of *dasturi*. **1909** M. Diver *Candles in Wind* i, Is the *zubbardusti* gentleman up there a toll-gate keeper that we should offer him *dasturi*? *Ibid.* viii, It was *dastur*; and there was no more to be said. **1926** *Blackw. Mag.* June 756/1 It was his custom—*dastur*—respected by his people and the British Resident.

dasylirion (dæsili·riǫn). [mod.L. (J. G. Zuccarini 1838, in *Allgemeine Gartenzeitung* 18 Aug. 258/1), f. Gr. δασύ-ς thick + λείριον lily.] A plant of the liliaceous genus of this name, indigenous to Mexico and the south-western U.S., having white bell-shaped flowers, cultivated as a greenhouse evergreen plant.
[**1858** *Curtis's Bot. Mag.* LXXXIV. tab. 5030 (*heading*) Bearded-leaved Dasylirium. **1866** Lindley & Moore *Treas. Bot.* I. 385/1 *Dasylirion*, a genus of *Bromeliaceæ*, consisting of Mexican plants with short stems, and densely crowded linear leaves which droop gracefully.] **1880** *Encycl. Brit.* XII. 262/1 The Dasylirions have stout woody stems and large heads of narrow leaves. **1933** *Punch* 11 Jan. 45/1, I simply yearn to look a dasylirion in the eye. **1951** *Dict. Gardening* (R. Hort. Soc.) II. 640/1 Dasylirions are excellent plants for subtropical bedding, their gracefully drooping leaves being ornamental at all times. **1963** W. Blunt *Of Flowers & Village* 29 The dasylirion has leaves as sharp as saws.

dasypeltis (dæsipe·ltis). [mod.L. (J. G. Wagler *Natürliches System der Amphibien* (1830) IV. 178), f. Gr. δασύ-ς thick + πέλτη small shield.] A small harmless egg-eating snake of the genus of this name found in central and south Africa.
[**1849** A. Smith *Illustr. Zool. S. Afr. Reptilia* Tab. LXXIII, On discovering that *Anodon* had been employed by conchologists, I adopted *Dasypeltis*, as proposed by Wagler.] **1887** *Encycl. Brit.* XXII. 194/2 A very peculiar genus of snakes, Dasypeltis, represented by three species only, is the type of a separate family. **1927** Haldane & Huxley *Anim. Biol.* xi. 223 The egg-eating snake, Dasypeltis. **1969** A. Bellairs *Life of Reptiles* I. iv. 119 Among the selective feeders are the egg-eating colubrids *Dasypeltis* and *Elachistodon*; many snakes are fond of birds' eggs.

datal, *a.* Restrict *rare* to sense in Dict. and add: **b.** Containing or including the date (as of a charter).
1837 T. D. Hardy *Rot. Chart.* 31 The Datal clause in Anglo-Saxon charters generally..precedes the names of the witnesses. *Ibid.* 34 William the Conqueror..also commemorated historical occurrences in his datal clauses. **1858** *Topographer & Genealogist* III. 120 Same seal and datal clause.

datcha, var. *dacha.

date, *sb.*[1] Add: **1. b.** *slang.* A foolish or comic person, esp. *soppy date*. (Usu. an affectionate term of abuse.)
1914 W. L. George *Making of Englishman* III. v. 302 These girls were used to the foreigner... I could fall into gallicisms now, and merely be called a 'date'. **1923** J. Manchon *Le Slang* 101 *You date!* que tu es drôle! **1935** G. Ingram *Cockney Cavalcade* iv. 55 A kid like that ought not to talk about love at her age..the soppy little date. **1959** I. & P. Opie *Lore & Lang. Schoolchildren* iii. 45 They say to him 'You're a soppy date.'
4. date-fish *U.S.*, a date-shell or piddock.
1838 *Knickerbocker* XI. 446 Each separate raisin therein embedded, bearing much resemblance to the date-fish in his rock. **1884** G. B. Goode *Nat. Hist. Aquatic Anim.* 707 Some cousins (*Zirphæa crispata*, *Platydon cancellatus*, etc.) are esteemed delicacies on the coast of California under the name of 'Date-fish'.

date, *sb.*[2] Add: **2. c.** An appointment or engagement at a particular time, freq. with a person of the opposite sex; a social activity engaged in by two persons of opposite sex. Cf. *blind date*. Also *attrib.* *colloq.*
[**1876** G. Meredith *Let.* 9 Mar. (1970) I. 512 If you do come I shall celebrate the event and make a date of it.] **1885** E. W. Howe *Mystery of Locks* 187 If he'll make a date with me, I'll exchange stories with him. **1896** Ade *Artie* vii. 65, I s'pose the other boy's fillin' all my dates? **1900** —— *Fables in Slang* (1902) 138 Her Date Book had to be kept on the Double Entry System. **1903** J. M. Forman *Journeys End* iii. 42, I must be going on. I've a date to keep. **1906** 'O. Henry' *Four Million* 176, I made a date for dinner this evening. **1916** C. J. Dennis *Songs of Sentimental Bloke* 121 *Date*, an appointment. **1919** W. S. Maugham *Moon & Sixpence* xlvii. 203 He walked away as if he'd remembered he had a date. **1923** A. Christie *Murder on Links* xxvi. 285 It still worried me that Bella hadn't kept her date with me. **1923** L. J. Vance *Baroque* xv. 147 I'll give you a ring and make a date. **1928** M. Baring *Comfortless Mem.* iii, Mr. Donne can't come; he's got a date. **1937** W. S. Maugham *Theatre* xxii. 209 He's got a date. Is the young ruffian having an affair with this girl? **1938** Wodehouse *Code of Woosters* ix. 206 Before parting, we had made a date for half-past four next day on the same spot. **1949** M. Mead *Male & Female* xii. 263 The girl in the perfect date-dress. *Ibid.* xiv. 282 The dress that is described as 'date bait'. **1958** *Times* 19 Feb. 3/3 The idea of *The Broken Date*, to give it its English title, is simple.
d. A person of the opposite sex with whom one makes or has made an appointment or engagement. *colloq.* (orig. and chiefly *U.S.*).
1925 *Amer. Speech* I. 102/2 My date was late last evening. **1943** Steinbeck *Once there was War* (1959) 125 On these terraces the soldiers come to sit about and to meet dates. **1951** J. D. Salinger *Catcher in Rye* vii. 60 Stradlater was in the back, with his date, and I was in the front, with mine. **1959** T. Griffith *Waist-high Culture* (1960) 32 In pairs we crowded into cars, our dates in our laps.
e. *Theatr. colloq.* A theatrical engagement or performance; a place where a performance is given, freq. as part of a tour. Also *transf.*, esp. (*U.S.*) a recording session.
1904 G. V. Hobart *Jim Hickey* vi. 101 God, please, if Danny is booked in Heaven won't You cancel his dates for a while and let him play here with me. **1933** P. Godfrey *Back-Stage* vi. 75 He then discusses the best provincial 'dates' to play before coming to London. *Ibid.* 115 He can book a series of touring dates with unsuccessful theatres. *Ibid.* xvi. 205 Smart watering-places, which in the holidays rank as No. 1 dates. **1936** N. Coward *To-night at 8.30* I. 92 *Bert.* She nearly got the bird second house. *Lily.* Too refined, I expect. For this date. *Ibid.* 93 This is as good a date as you can get. **1949** L. Feather *Inside Be-bop* ii. 16 Charlie's first small-band date, September 15, 1944. **1952** Wodehouse *Barmy in Wonderland* i. 17 He proposed to teach it a few simple tricks and get it dates on television.
7. *to date*: to the present time or moment. See Dict. and Up to date.
1936 *Nature* 21 Nov. 888/2 The results to date of the treatment..of cases of cancer..are detailed. **1940** *War Illustr.* 26 Jan. 24 It may be deduced that what has happened to date is the curtain-raiser to that aerial blitzkrieg. **1947** J. S. Huxley *Unesco* i. 16 The highest product of evolution to date. **1969** *Listener* 5 June 806/1 All this adds up to one of Dylan's most beautiful songs to date.
8. date-cancel *v. trans.*, to cancel by a written or stamped date; **date-letter,** a letter stamped upon gold or silver plate, pottery, etc., denoting the year of manufacture; cf. *date-mark*; **date-line,** also a line, or part of one, giving the date of issue of a newspaper or the date (and usually also place of origin) of a dispatch, letter, etc.; hence *v.* (usually in pa. pple., *date-lined*); **date-stamp,** a stamp with adjustable types, used in recording the date of posting or delivery of a letter or parcel, receipting a bill, receiving a book, and the like; also, the impression made by such a stamp; so **date-stamp** *v. trans.*, **date-stamping** *vbl. sb.*
1929–30 *Unemployment Book* 2 Date-cancel stamps immediately. **1863** W. Chaffers (*title*) Hall marks on gold and silver plate, with tables of annual date letters employed in the principal assay offices of England, Scotland and Ireland. **1935** *Burlington Mag.* June 288/1 Over-stamping on Sheffield-made candlesticks the London date-letter for 1775–6. **1888** *Missouri Republican* 24 Mar. (Farmer *Americanisms* s.v. *Fake*), The telegraph man, who has edited Mulhatton's yarns before, and knows a fake from a barn-door, by the date line alone. **1892** Date-line [see Dict.]. **1922** C. Sandburg *Slabs of Sunburnt West* 35 Turning among headlines, date lines, funnies, ads. **1942** W. Faulkner *Go down, Moses* 262 It [*sc.* a press report] was datelined from Joliet, Illinois, this morning. **1944** *Daily Express* 7 Sept. 1/5 A message..datelined Zurich to the German-controlled Scandinavian Telegram Bureau. **1959** *Elizabethan* June 28/1 The date line says Jenin, Thursday. **1970** *New Yorker* 3 Oct. 29/2 A recent article in the *Times*, datelined New Providence, commented that a 'developing crisis..has cast a pall over tourism here'. **1879** Trollope *John Caldigate* III. x. 139 They got hold of some young man at the post-office who knew how to fix a date-stamp with a past date. **1908** *Chambers's Jrnl.* Jan. 102/1 As in cases of delay in transmission or delivery, when the date-stamp shows the cause to be late posting. **1909** *Daily Chron.* 2 July 5/5 Mr. Buxton has undertaken that the date-stamps shall not in future bear the name of the firms. **1928** F. M. Ford *Last Post* vii. 168 Each egg wired to the bottom of its box, waiting till she had time to date-stamp it. **1930** C. Williams *War in Heaven* iii. 37 From date-stamp to waste-paper basket, from basket to files, from files to telephone Adrian pursued his investigation. **1959** *Clarendonian* Mar. 4 At a later stage two copies of each jacket come from the Bindery; one is date-stamped, ..and is then filed for future reference. **1886** Date-stamping [in Dict.]. **1908** *Chambers's Jrnl.* Jan. 102/1 The first process..to which letters are subjected is that of date-stamping. **1960** *News Chron.* 22 Sept. 9/7 A housewife faced a conference of food inspectors..to demand the date-stamping of all perishable foods.

date, *v.* Add: **2.** Also, to mark as being of a certain date or period; to render outdated or only briefly fashionable or appealing. Also *intr.*, to bear evidence of its or one's date or period; to be or become old-fashioned or outdated. *colloq.*
1895 *Westm. Gaz.* 13 Mar. 5/1 Every portion of a picture was 'dated' from every other portion of it. **1896** G. B. Shaw *Our Theatres in Nineties* (1932) II. 168, I have been led into this investigation of 'dating' by the fact that The School for Scandal..dated very perceptibly last Saturday night at the Lyceum in point of morals. **1901** *Daily Chron.* 17 May 6/6 In the four or five years it has been laid aside it [*sc.* a play] has not 'dated' in the slightest degree. **1911** in C. W. Cunnington *Eng. Women's Clothing* (1952) iii. 104 These hats would..never date their owners. **1915** H. G. Wells *Research Magnificent* 9 He had found the word 'Bushido' written with a particularly flourishing capital letter, and twice repeated. 'This was inevitable,' said White. ..'And it dates...Yes—this was early.' **1924** *Glasgow Herald* 6 Nov. 8 George Eliot still has her readers,..but she has begun to 'date' rather decidedly. **1925** *Sat. Rev.* 7 Nov. 1 Younger men say with contemptuous brevity that his views 'date'. **1927** E. Bowen *Hotel* ix. 103 The ornaments on Mrs. Kerr's dressing-table..'dated' her friend for her inevitably. **1928** Belloc *Convers. w. Angel* xxiv. 198 All the middle and early Victorians are already dating—except Macaulay. **1928** Galsworthy *Swan Song* III. xvi. 341 'I respected old Forsyte,' he said to his son...'He dated, and he couldn't express himself; but there was no humbug about him.' **1932** N. Coward *Cavalcade* II. iii. 80 'I was in "Mirabelle" then.'.. 'I was taken to see that.'..'*Taken* to see it, were you! That dates us a bit.' **1951** R. Knox *Stimuli* i. 4 Nothing..so dates our generation as this habit of talking about things being dated. **1957** *Times Lit. Suppl.* 20 Dec. 773/4 Almost all the films he discussed just ten short years ago have either badly dated or been forgotten.
d. To make or have a 'date' with (see *date *sb.*[2] 2 c); *spec.* to do so regularly. Freq. in *pass.* Also *intr.* and with *up*. *colloq.* (orig. *U.S.*).
1902 Ade *Girl Proposition* 70 Before he left that Evening he had himself all dated up for a return engagement. **1903** A. Kleberg *Slang Fables* 67 She was Dated to a chap with Uncounted wealth. **1917** E. W. Lardner *Gullible's Travels* iii. 86, I..dated her up to meet me down-town the next day. **1919** *Saucy Stories* Nov. 57/2 'What about the movies..?' 'All I can hand you is my thanks, I'm dated up!' **1924** *Ibid.* Feb. 30/1 What will..people..think when you are dated to show up with Miss Future Wife. **1924** 'W. Fabian' *Sailors' Wives* iv. 59 Warren Graves wants to date me for the Deuces Wild party Saturday night. **1928** *Collier's* 5 May 36/3 Dat fool gal datin' wid me and wawkin' off wid dat money man. **1938** D. Runyon *Furthermore* v. 95 These characters keep trying to date up the nurses. **1938** E. Bowen *Death of Heart* II. vi. 290 I'm afraid I've given you rather a miss... But it's been a thickish week, and I got all dated up. **1947** 'N. Shute' *Chequer Board* 62 The white troops found to their concern that every girl was dated up by a negro. **1958** S. Ellin *Eighth Circle* (1959) II. xvi. 172 'When will I see you again?' 'You can't!' she said in alarm. 'I mean, not this way, as if we were dating or something.' **1970** *New Yorker* 18 July 28/3, I was too busy with other social engagements. I was all dated up with the Greeks.
e. To assign *to* a specified date.
1913 E. T. Leeds *Archaeol. Anglo-Saxon Settlements* v. 86 This find can be dated by coins to about A.D. 290. **1957** G. Bibby *Testimony of Spade* xxiii. 371 The discovery.. of swords of the La Tène type together with Roman coins, of which the latest was dated to 54 B.C. **1971** *N. & Q.* CCXVI. 116/1 Both he [*sc.* Jordan] and Luick date the final /iu/ stage to the fifteenth century.

dated, *ppl. a.* Add: **3.** Belonging to or characteristic of a particular period; bearing evidence of its (or one's) date or period; old-fashioned, outdated. *colloq.*

1900 F. H. STODDARD Evol. Eng. Novel ii. 53 It is a dated society, and it is a dated woman, not the woman of all time. 1926 Atlantic Monthly June 769/1 Another newspaper sent out a man who, it happened, was 'dated' in his reportorial training. He..got the facts. 1944 Burlington Mag. Dec. 312/2 It is this meticulous realism which makes his figures appear so peculiarly 'dated'. 1954 KOESTLER Invis. Writing I. ii. 28 The orthodox 'proletarian' literature of the 'thirties appears to-day shallow and dated. 1962 Oxford Mail 19 Feb. 6/5 Much of the music has a dated air. 1970 Daily Tel. 17 Oct. 6/5 A great many I saw at the races were still wearing dated, conventional two-piece outfits—a bit too long, a bit too dowdy.

dateless, *a.* Add: **1. b.** Free from engagements or appointments. *U.S.*

1923 N.Y. Tribune 25 Apr., The young men at Northwestern University have agreed to join the young women of that institution in observing three dateless nights each week. 1944 Chicago Tribune 10 Dec. Grafic Mag. 4 Sometimes that mood indigo comes up briefly on a dateless Friday night.

4. (Earlier and later examples.) Also, foolish, 'clueless'.

a 1686 A. MARTINDALE Life (1845) iv. 79 Which he, being almost datelesse for age,..readily granted. 1848 MRS. GASKELL M. Barton II. vi. 98 Poor soul, she's gone dateless, I think, with care, and watching, and over-much trouble. 1854 N. & Q. 1st Ser. X. 211/1 'After he hit me o' th' heead I was dateless;' that is, I took no note of time. 1961 J. I. M. STEWART Man who won Pools v. 58 She'll be pretty dateless, won't she, as the wife of a man with hundreds of thousands of pounds? Ibid. xi. 121 He'd been pretty well flourishing a quarter of a million pounds at her like she was a dateless shopgirl.

5. *absol.* Cf. TIMELESS *a.* 2 b.

1894 Daily News 17 Sept. 4/6 Its [sc. the British Museum's] way of bringing itself..to date is to approach nearer and nearer to the dateless.

Hence **da·telessly** *adv.*

1956 K. CLARK Nude iv. 117 The standing woman is not so datelessly naturalistic. Her..complex pose seems to have been derived from an antique relief.

dating, *vbl. sb.* Add: *spec.* The act or process of becoming 'dated' (see *DATE *v.* 2, *DATED *ppl. a.* 3).

1896 [see *DATE v. 2]. 1936 C. CONNOLLY Rock Pool 12 The fault I am most conscious of..is that of dating...In this case my debt is with the nineteen-twenties.

b. *Archæol.* The determination of the age of an object, etc., found in archæological research. Also *attrib.*

1926 Encycl. Brit. I. 157/2 It is..possible..to lay down a chronological order by the sequence method of dating, based primarily on pottery. 1936 Discovery Nov. 362/2 The Bournemouth specimens were found in conditions which give no criterion for dating. 1950 F. E. ZEUNER Dating Past (ed. 2) i. 11 Dendrochronology may well become a dating-method in such areas.

c. The act or practice of making or having 'dates' (see *DATE *sb.*² 2 c). Also *attrib. colloq.* (orig. *U.S.*).

1939 G. GREENE Lawless Roads x. 249 Rating for dating. The famous sub-deb chart for getting along with boys. 1951 M. MCLUHAN Mech. Bride 72/2 She has to be so rigorously self-controlled during several years of dating. Ibid. 99/1 The precocious dating habits of middle-class children. 1952 W. SPROTT Social Psychol. 167 The 'dating' system in America. 1958 H. REILLY Ding Dong Bell (1959) viii. 78 He was around yesterday and again today. Rather heavy dating, considering that he and Carol have only known each other a month or so. 1959 Economist 30 May 821/2 The Americans are better since..they have already, at 17, several years of dating experience behind them. 1968 Globe & Mail (Toronto) 17 Feb. 53/4 (Advt.), Many more exciting dates, parties, and trips. Dating by computer!

dative, *a.* Add: **C.** *Comb.* **dative absolute,** in some inflected languages, a construction resembling the Latin ablative absolute, in which a substantive and participle in the dative case form an adverbial clause of time, cause, or coexistence; **dative-accusative** *a.*, having the functions of both the dative and the accusative case; *sb.*, such a grammatical form; **dative-object,** an object governed by the verb and in the dative case; **dative-phrase,** a phrase in which a preposition has a function equivalent to that of a dative case-ending in a language like Latin (cf. *case-phrase); **dative-verb,** a verb regularly constructed with the dative.

1870 F. A. MARCH Compar. Gram. Anglo-Saxon III. 152 Dative absolute.—A substantive and participle in the dative may make an adverbial clause of time, cause, or coexistence. 1918 M. CALLAWAY Syntax Lindisfarne Gospels i. 14 As before 1889, so in these later discussions, two views as to the dative absolute construction in Gothic are advocated. 1965 B. MITCHELL Guide to Old English v. 105 The dative absolute is used in imitation of the Latin absolute, e.g. gewunnenum sige 'victory having been gained'. 1933 L. BLOOMFIELD Lang. xxiv. 437 Nominative ye: dative-accusative you. 1940 C. C. FRIES Amer. Eng. Gram. 88 The six distinctive dative-accusative forms of pronouns (me, us, him, them, her, whom). 1959 M. SCHLAUCH Eng. Lang. i. 32 Prepositions ceased to 'control' more than one following case (a single form serving as dative-accusative, which for nouns had become identical with the nominative). 1927 E. A. SONNENSCHEIN Soul of Gram. 29 In English, in Greek, and occasionally in Latin the dative-object may become the subject... I was shown the way. 1940 C. C. FRIES Amer. Eng. Gram. 254 Nouns which formerly stood before the so-called impersonal verbs as dative-objects..now..functioned as subjects. 1927 E. A. SONNENSCHEIN Soul of Gram. 49 The dative-phrases of French and Spanish are simply developments of dative-phrases found in all periods of Latin. 1894 —— Gr. Gram. II. 168 (heading) Verbs taking the Dative (Dative Verbs).

datively, *a.* (Earlier example.)

1846 MONIER WILLIAMS Elem. Gram. Sanskrit ix. 160 Datively Dependent..those in which the relation of the first word to the last is equivalent to that of a dative.

dativo-. (Further example.)

1930 T. SASAKI Lang. of R. Bridges' Poetry 87 Use of the infin. in the dativo-locative function.

‖**dato** (dā·to), **datu** (dā·tu). Also **datoh, datoo, datto.** [Malay *datu* ruler, chief; *dato', datok* elder, title of respect or distinction; Tagalog *datò* ruling head of clan or tribe.] A landowner or chief in northern Borneo, the Philippine Islands, and some adjacent areas.

[1820 J. CRAWFURD Hist. Ind. Archip. III. VIII. i. 22 Native term for king in Javanese is Ratu, which is the same word that is written Datu in some other languages. Its literal meaning is grandfather, and by a slight inflection a senior or elder, from which last is taken its figurative meaning, a lord or chief.] 1841 J. BROOKE in E. Hahn James Brooke (1953) v. 68 The revenue will be collected by the three Datus, bearing the seal of the governor. 1867 SMYTH Sailor's Word-bk., Datoo,..a Malay term of rank, and four of whom form the council of the sultan of the Malayu Islands. 1897 Geogr. Jrnl. Jan. 36 The Galas district is nominally under the charge of the Dato. 1906 Daily Colonist (Victoria, B.C.) 13 Jan. 2/4 [In the Philippine Islands] W. J. Bryan yesterday was created a 'Datto' and saluted by fifty pieces of native artillery. 1925 Chambers's Jrnl. 46/1 The Malay point of view was well expressed by a certain Datoh, or landowner of good birth. 1960 S. RUNCIMAN White Rajahs III. iv. 219 The Resident and four Datus were lined up there to greet him.

datum. Add: **b.** *datum feature, level, mark, point, year.*

1954 Defs. Mech. Engin. (B.S.I.) 9 A datum plane, line or point establishes an exact geometrical reference as distinct from the physical reference provided by a datum feature. 1909 Daily Chron. 6 July 5/7 The price is below what one may call the datum level of £6,000. 1957 G. E. HUTCHINSON Treat. Limnol. I. ii. 165 The maximum depth will vary slightly with variations in water level, and ideally it should be referred to some independent datum level. 1926 Nat. Hist. Oxf. District 75 The datum mark of the Thames at Osney is 186 ft. above its outlet at the Nore. 1912 Aeroplane 5 Dec. 570/2 Their duty will be to enable battery commanders to fix 'Datum Points' in the sky. 1933 H. G. WELLS Shape of Things to Come IV. § 4. 359 The intrinsic quality of this book has been entirely overshadowed by its importance as a datum point in history. 1940 Economist 9 Nov. 581/1 The Treasury's choice of datum years does allow most industries the advantage of retaining part..of their..profits.

c. *Philos.* datum of consciousness, etc. (see quots.). Esp. *datum of sense* (cf. *SENSE-DATUM).

1846 W. HAMILTON Wks. T. Reid 749/2 The primary data of consciousness are..admitted..to be true. 1856 A. C. FRASER Ess. Philos. iv. 212 This notion of the Infinite is in fact an ultimate datum of consciousness. 1887 A. SETH Hegelianism iv. 118 That elementary statement must be originally made in virtue of..some immediate datum of experience. 1890 W. JAMES Princ. Psychol. II. xx. 252 'Can a doubleness, so easily neutralized by our knowledge, ever be a datum of sensation at all?' such an anti-sensationalist might ask. 1890 A. C. FRASER Locke 186 Locke thus reduces the entire certain knowledge of sensible things that man is capable of to one's present data of sense, and one's memory of past data. 1895 W. JAMES Coll. Ess. & Rev. (1920) 392 For our colleague Ladd,..the soul..furthermore performs a unifying act on the naturally separate data of sense. 1902 —— Var. Relig. Exper. 427 Mystical states merely add a supersensuous meaning to the ordinary outward data of consciousness.

d. *pl.* The quantities, characters, or symbols on which operations are performed by computers and other automatic equipment, and which may be stored or transmitted in the form of electrical signals, records on magnetic tape or punched cards, etc.

1946 Ann. Computation Lab. Harvard Univ. I. 11 Two card feeds for supplying the machine with empirical or other data. 1946 Math. Tables & Other Aids to Computation II. 97 The [IBM card] reader scans standard punched cards..and causes data from them to be stored in relays located in the constant transmitter. 1958 GOTLIEB & HUME High-Speed Data Processing i. 6 The machine as a whole can be considered to be a device which accepts data and instructions.., stores them.., operates on the data..and produces results by the output. 1960 E. DELAVENAY Introd. Machine Transl. ii. 21 A memory may be used to store data, or to store programme instructions. 1964 AFIPS Conference Proc. XXVI. 1. 219 Data is transferred to main storage as soon as two bytes are accumulated. 1967 Times Rev. Industry July 53/3 Up to 50 types of data can be analysed with one pass through the machine. 1969 P. B. JORDAIN Condensed Computer Encycl. 306 Data are recorded on the tape by magnetizing narrow lengthwise stripes (called tracks) in alternating directions. 1970 Sci. Jrnl. June 23/3 During each orbit data from the experiment is transmitted from the satellite to Fairbanks in Alaska and from there..to Oxford for initial processing. 1970 A. CHANDOR et al. Dict. Computers 99 Data is sometimes contrasted with information, which is said to result from the processing of data.

2. *pl.* Facts, esp. numerical facts, collected together for reference or information.

1899 W. W. F. PULLEN (title) Tables and data for the use of students in engineering laboratories. 1913 (title) Handbook of chemistry and physics; a ready-reference book of chemical and physical data. 1923 C. G. CONRADI Mech. Road Transport v. 51, I have concluded this chapter with three tables, giving the requisite data pertaining to various solids, liquids and gases..which may be employed as fuels. 1934 H. B. DWIGHT (title) Tables of integrals and other mathematical data. 1946 Nature 12 Oct. 519/1 Hexavalent chromium compounds were selected for this study as accurate X-ray and magnetic data for these compounds are available. 1958 MACDUFF & CURRERI Vibration Control vii. 165 Most of the data concerning shock and vibration on airplanes are classified. 1971 Daily Tel. 28 June 17 Desk research means collecting data from all published sources including government censuses, production figures, import and export statistics,..and trade publications.

3. Used *attrib.* and in *Comb.* in the pl. form, as *data bank, -handling vbl. sb., transmission*; **data link,** a telecommunications link over which data are transmitted, usu. to or from a data-processing centre; **data logger,** any instrument for making a recording, either continuously or intermittently, of the successive values of a number of different physical quantities; so **data-logging** *vbl. sb.*; **data processing,** the performance by automatic means of any operations on empirical data, such as classifying or analysing them or carrying out calculations on them; also *transf.*; so **data processor**; **data sheet,** a leaflet containing a summary of useful information on some subject.

1970 Daily Tel. 8 Oct. 9/4 Data banks containing comprehensive personal files covering criminal records, health records, income tax and so on, would be liable to abuse. 1964 Language XL. 214 Human beings are somehow specially designed to do this, with data-handling ability. 1968 Brit. Med. Bull. XXIV. 189/2 The computer is a machine which can perform any data-handling procedure. 1966 Guardian 4 May 20/3 Data links exist now. 1968 New Scientist 8 Feb. 312/2 From this central collection point the telephone data link will connect the various patients to the computer system. 1962 Aeroplane CIV. 34/3 A 500-channel data logger capable of scanning 100 channels per second. 1963 B. FOZARD Instrumentation Nucl. Reactors xiii. 157 Instrumentation of a nuclear power reactor is commonly undertaken on a very large scale and incorporates data-logging equipment. 1954 Instruments & Automation Dec. 1916/1 New 'Model CRC 102-A Electronic Computer' and its auxiliary..are designed for..data processing. 1956 Collier's Year Bk. 105/1 Data processing, in the broadest definition, means the handling of information by arithmetic rules and logic. It is performed by most types of business machines, the simple mechanical adding machines as well as the complex electronic card sorting systems or accounting machines. 1959 B.S.I. News Nov. 5/1 Punched cards and tapes used in automatic data processing. 1960 Times 22 Jan. 15/5 It..is designed for both mathematical calculations and commercial data-processing. 1966 R. L. GREGORY Eye & Brain iv. 46 Some of the data processing for perception takes place in the eye. 1954 Instruments & Automation Dec. 1905/1 The most advanced techniques for automatic control of plant, factory, and office operations..were represented..(4) electronic computers and data processors. 1960 E. DELAVENAY Introd. Machine Transl. ii. 19 The big I.B.M. data-processers..are..endowed with high operational speeds. 1898 Machinery (N.Y.) June 312/1 Four data sheets containing information on laying out gear blanks. 1971 Amateur Photographer 3 Mar. 31 This, No. 1 in a series of data sheets covering all major facets of photography, provides basic information on the general purpose black-and-white films on the market. 1946 Math. Tables & Other Aids to Computation II. 91 The design of data-transmission and servo units, the development of codes and related equipment for the rapid set-up of problems, and the conversion of continuous variables..into digital form.

¶Used in pl. form with sing. construction.

1807 W. IRVING Salmag. xviii. 366 My grandfather..took a data from his own excellent heart. 1902 A. S. TOMPKINS Hist. Rec. Rock Co., N.Y. 46 There is but little data to estimate Indian populations. 1931 H. F. PRINGLE T. Roosevelt p. viii, The author has preserved this data. 1955 [see biomedical s.v. *BIO-]. 1963 Daily Express 23 Sept. 3 He took out a patent but some of the data is missing. 1964, 1970 [see *1 d above]. 1965 J. ALLAN Speaking of Computers 5 Incidentally, by general usage data is now accepted as a singular collective noun. 1971 Computer Weekly 13 May 3/1 They have done little to analyse and interpret this data.

‖**daube** (dōb). Also in anglicized form **daub.** [Fr.] A braised meat (usu. beef) stew with wine, spices, etc. So *à la daube, en daube*: stewed or braised.

1723 [see *À LA b]. 1747 H. GLASSE Art of Cookery ii. 20 Beef à la Daub. You may take a Buttock..of Beef, lard it, fry it..put it into a Pot..stew it. 1877 E. S. DALLAS Kettner's Bk. of Table 371 A stew of fresh-water fish is called a matelote..of beef, a daube. 1927 V. WOOLF To Lighthouse I. xvi. 125 They were having Mildred's masterpiece—Bœuf en Daube. 1961 Spectator 7 July 11 Then came a daube of beef..with an unthickened but short sauce of wine and tomato purée.

Daubenton's bat (dobañtoñz). *Zool.* [f. the name of Louis Jean Marie *Daubenton* (1716–?1800), French naturalist + BAT *sb.*¹] A species of bat, *Myotis daubentoni*, of Europe and parts of Asia.

Orig. named *Vespertilio daubentonii* (1817).

1887 G. H. C. HAIGH in Zoologist XI. 293 Vespertilio daubentonii... The cry of Daubenton's Bat is very weak and shrill, sometimes prolonged into a sort of chatter. 1904 J. G. MILLAIS Mammals Gt. Brit. I. 86 Daubenton's Bat. Myotis Daubentoni... Daubenton's Bat has not infrequently been confused with Natterer's and the Whiskered Bats; it, however, possesses very much larger feet. 1960 Times 14 June 14/7 The Daubenton's bat was usually near the old mill dam.

dauermodification (dau·əɹmǫdifikē·ʃən). *Genetics.* Also ‖ **Dauer-, dauermodifikation** (pl. -en). [ad. G. *dauermodifikation* (V. Jollos 1913, in *Biol. Centralbl.* XXXIII. 233), f. *dauer* duration + *modifikation* MODIFICATION.] A character, usu. one artificially induced in an organism, which is inherited through the cytoplasm and tends to disappear after a few generations.

1938 Genetica XX. 126 A transitory change of the plasmatic nature in a basal part of the original individual, which change is not a common modification, neither a plasmatic mutation, because it is inherited in some degree for a few generations, but it disappears in succeeding generations..it may be classified as a case of 'Dauermodifikation' (permanent modifications). 1943 Hereditas XXIX. 424 The so-called dauermodifications..are inherited for some time and then disappear. 1944 Nature 5 Aug. 167/1 The carcinogens..induce in plants dauermodifications of limited or unlimited persistence. 1968 M. W. STRICKBERGER Genetics xiii. 259 Persistent abnormalities ('dauermodifikationen') can be induced in plants and animals by chemical and environmental treatment and carried for several generations by the female line.

daughter. Add: **2. b.** Used in pl. in the names of various women's societies, as *Daughters of Liberty* (Boston, 1769–70), *Daughters of the American Revolution* (1890), *of the Confederacy* (1894), etc. Also *sing.*, a member of one or other of these societies. *U.S.*

1769 Boston Gaz. 16 Oct. 1/3 And as true Daughters of Liberty, they made their Breakfast upon Rye Coffee. 1890 (title) Daughters of the American Revolution Constitution and By-Laws. 1894 Confederate Veteran II. 180/1 Daughters of the Confederacy are also organized. 1911 R. D. SAUNDERS Col. Todhunter i. 4 Working the Daughters of the Confederacy as a political proposition. Ibid. 7 Mrs. Todhunter, an ardent Daughter had gone early in the day. 1940 E. FERGUSSON Our Southwest 54 All our national and sectional Daughters are there and active, and even more patriotic Daughters of Texas' founders, fighters, signers, or early arrivals. 1962 Listener 18 Jan. 125/1 You are fed some dull crap by a sweet bunch of uptown Daughters of the American Revolution.

6. d. *Nuclear Sci.* A nuclide formed by the nuclear disintegration (either spontaneous or induced) of another nuclide. Orig. short for *daughter atom, element.

1926 [see *7 c]. 1933 O. H. BLACKWOOD et al. Outl. Atomic Physics xii. 222 At present we cannot tell whether the daughter of radium C″ is identical in all its properties with radium D. 1950 GLASSTONE Sourcebk. Atomic Energy v. 119/2 Since the daughter element also disintegrates, it is itself the parent of a daughter. 1956 A. H. COMPTON Atomic Quest 89 The average number of 'daughters' that each neutron generates. 1970 Nature 25 July 362/2 The decay product of natural uranium, thorium and their daughters.

7. *daughter-colony, -land, -nation.*

1901 National Rev. Nov. 347 The conduct of these daughter nations during our South African struggle. 1903 Westm. Gaz. 3 June 9/3 Everyone was too busy talking about their grand Imperial theories, and the duties of the mother-country, to bother about the dull little domestic facts that are worrying the daughter-land. 1905 Spectator 11 Feb. 205/1 The great self-governing daughter-nations. 1937 Discovery Aug. 229/2 The silver didrachm..of the Greek city states was introduced into the daughter-colonies and became the chief coin of Italy for nearly 400 years.

b. *Biol.*, etc. Applied to things having the relation of offspring of the first generation, or resulting from a primary division or segmentation. Cf. *daughter-cell* in Dict.

1876 Trans. Clinical Soc. IX. 137, I cut down upon the tumour so as freely to expose it, and then punctured it, when a quantity of clear water escaped, and with it two or three small daughter cysts. 1900 B. D. JACKSON Gloss. Bot. Terms 71/2 Daughter chromosome, a secondary chromosome, derived from division of the original. 1900 DORLAND Med. Dict. 188/1 Daughter-cyst, a small cyst developed from the wall of a larger one. 1924 J. A. THOMSON Science Old & New xliv. 257 A non-cellular organism multiplies by division, budding and spore-forming, and its daughter-units separate off. 1937 Nature CXL. 759 On this process [sc. crossing-over]..the later reduction in number and segregation to opposite daughter-cells equally depend. 1964 M. HYNES Med. Bacteriol. (ed. 8) xxix. 453 Daughter-cysts may be endogenous, i.e., developed within the primary cyst, or exogenous, i.e., buds penetrate the cyst wall and develop externally.

c. (sense *6 d) *daughter atom, element, product.*

1933 O. H. BLACKWOOD et al. Outl. Atomic Physics xii. 219 The atomic weight of radium is 226, and the daughter atom should be four units lighter because of the loss of an alpha particle. 1955 Gloss. Terms Radiology (B.S.I.) 6 The atom containing the original nucleus is sometimes called the parent atom and the resulting atom the decay product or daughter atom. 1926 R. W. LAWSON tr. Hevesy & Paneth's Man. Radioactivity viii. 87 A radio-element is in a state of radioactive equilibrium with its disintegration product, when the same number of atoms of the daughter element disintegrate as are formed in the unit of time. 1962 Newnes Conc. Encycl. Nuclear Energy 176/2 The decay products of a radioactive element are often referred to as its daughter products.

Daulian (dǭ·liăn), *a.* [f. mod.L. *Daulias* used as the generic name of the nightingale (*D. luscinia*), a. Gr. Δαυλίας lit. woman of Daulis, Procne, who was changed into a nightingale.] *Daulian bird*, an affected name for the nightingale.

1894 Daily News 22 Jan. 5/1 That tiny modest tome [sc. Keats's Poems of 1817], a brown Daulian bird in brown paper. 1904 Westm. Gaz. 9 June 12/1 Until three years ago the existence of the Daulian birds in Devonshire was strenuously denied. 1909 Ibid. 12 May 5/1 Londoners who want to hear the Daulian birds with complete certainty must now journey to Kew or Chingford.

dauncy (dǭ·nsi), *a. U.S.* and *dial.* [var. DONSIE *a.*] Sickly; delicate, not robust.

1846 E. W. FARNHAM Life in Prairie Land (1847) iv. 39, I shall give her enough to eat and wear, and I dont calculate she'll be very daunsey if she gets that. 1874 E. EGGLESTON Circuit Rider vi. 61 You look powerful dauncy, said the old man. 1880 W. H. PATTERSON Antrim & Down Gloss. 31 Donsy, Dauncey, sick, sick-looking. 1891 M. E. RYAN Pagan of Alleghanies v. 82 He ain't one o' yer skim-milk, dauncy ones. He's as stout as a young bull. 1912 I. S. COBB Back Home 148 He looks sort of dauncy and low in his mind. 1926 E. M. ROBERTS Time of Man (1927) iv. 152 Dorine said that he must be sick. 'He looks dauncy,' she said. Ibid. x. 377 'Pap he's sick today,' Joe said. 'I see that. He's dauncy all day.'

Davenport[2]**.** Used to designate china, earthenware, etc., made by a family of this name at Longport, Staffordshire, between 1793 and 1882.

[1829 S. SHAW Hist. Staffs. Potteries 2 The largest Potteries known, being Wedgewood's, Etruria;..Davenport's, Longport; Minton's, Stoke.] 1863 W. CHAFFERS Marks Pott. & Porc. 124 Davenport Stone China. Staffordshire. Longport. 1872 LADY C. SCHREIBER Jrnl. (1911) I. 156 We got nothing but a 'Davenport' plate. 1875 L. TROUBRIDGE Life amongst Troubridges (1966) 106 A bowl..marked with an anchor, this we thought at first was Chelsea, but know since that it is Davenport. 1954 G. SAVAGE Porcelain II. viii. 270 Much nineteenth-century Davenport porcelain survives. Ibid., Two Davenport plaques painted in the manner of Birkett Foster. 1965 A. CHRISTIE At Bertram's Hotel i. 10 The china, if not actually Rockingham and Davenport, looked like it.

davered (dēi·vəɹd), *ppl. a. dial.* [f. DAVER *v.*] Withered, faded, drooping.

1837 M. PALMER Dialogue Devonshire Dial. 6 Now, dear soul, her's like a daver'd rose. 1864 [see DAVER v. II].

So **da·verdy** *a. dial.*, dowdy, unkempt.

1906 GALSWORTHY Man of Property I. vii. 95 Even in the garden, that sense of things being pokey haunted old Jolyon; the wicker chair creaked under his weight; the garden-beds looked 'daverdy'. Ibid. II. iii. 148 That was how he liked 'em, all of a piece, none of your daverdy, scarecrow women! 1924 —— On Expression 7 What an expressive variant of the word 'dowdy'..is the word 'daverdy'..! Dowdy suggests the flannel petticoat, the thick, the dusty appearance; daverdy a sea-green, trailing, down-at-heeledness.

Davidic (dēivi·dik), *a.* [f. personal name *David* + -IC.] Of or pertaining to David as king of Israel, or as the reputed author of the Psalms. Also **Davi·dical** *a.*

[c 1000 Leechdoms (1866) III. 428 Of ðam dauiticum sealmum.] 1827 COLERIDGE Table-t. 10 Mar., I apprehend many of the Psalms to be Davidical only, not David's own compositions. 1865 Chambers's Encycl. VII. 819/1 The first [book]..contains the Davidic Jehovistic psalms. 1877 Encycl. Brit. VI. 841/2 We cannot well stop short of the admission that the Psalter must contain Davidic psalms. 1883 Ibid. XVI. 54/1 Amos foretold the redintegration of the Davidic kingdom. 1918 E. GRUBB Relig. Experience 186 The glories of the ancient Davidic monarchy. 1951 R. A. KNOX Stimuli vii. 15 The Saviour..to whose coming they looked forward, was to be a national Saviour, a Davidic ruler. 1957 G. E. WRIGHT Bibl. Archaeol. ix. 146/1 The southern kingdom, calling itself 'Judah', retained the Davidic dynasty.

daviesite (dēi·vizəit). *Min.* [f. the name of Thomas *Davies* (1837–92), British mineralogist: see -ITE¹.] (See quots.)

1889 L. FLETCHER in Min. Mag. May 171 Crystals of..an Oxychloride of lead (Daviesite), from Mina Beatriz, Sierra Gorda, Atacama, South America. 1951 in C. PALACHE et al. Dana's Syst. Min. (ed. 7) II. 58. 1963 M. H. HEY Index Min. Species, App. (ed. 2) 98 Daviesite...Delete..Syn. of hemimorphite.

Davis[1] (dēi·vis). The name of Dwight F. *Davis* (1879–1945), Amer. politician, used in *Davis Cup*: a cup presented by him and played for annually by international lawn tennis teams; also, the contest for this cup.

1901 Outing (U.S.) June 320/1 Another challenge has now been received from them..and a second attempt to 'lift' the Davis Cup will be made this season. 1936 'R. WEST' Thinking Reed xi. 374 He would secure front-row seats for the Davis Cup. 1950 Oxf. Jun. Encycl. IX. 309/1 In 1900 an American player called Dwight Davis presented the Davis Cup, which is competed for annually by many countries.

Davis[2] (dēi·vis). The name of Sir Robert H. *Davis* (1870–1965), used *attrib.* to designate: (*a*) apparatus invented by him to permit escape from a submarine; (*b*) a decompression-chamber for deep-sea diving.

1931 Hansard 10 June 1021 Six ratings..escaped from the wreck by means of the Davis submarine escape apparatus. 1936 R. H. DAVIS Deep Diving 6 The Davis submersible decompression chamber..was submitted to the Admiralty in 1929. Ibid. vi. 113 Divers could not do useful work there [beyond 204 feet] unless means could be found for improving the conditions of their ascent...The problem has now been solved by the invention of the Davis submersible decompression chamber. Ibid. xiv. 206 The procedure in case of emergency would be for two men to enter the locks wearing the Davis apparatus..to open the upper hatches and, floating through them, escape to the surface. 1940 E. ELLSBERG Men under Sea vii. 89 The Davis 'lung'..has always carried in addition to the chemical cartridge a small oxygen cylinder, from which fresh oxygen can be continuously supplied. Ibid. xxii. 305 A diver may work 20 minutes at a depth of 300 feet, come up in a total of 13 minutes' time to 60 feet, where he enters the Davis chamber, and then be hauled aboard. 1947 Sci. News IV. 12 The Davis submarine escape apparatus is the simplest form of diving apparatus. It consists simply of a counter-lung with a small oxygen cylinder and soda-lime canister, and a mouth-piece.

daw (dǭ), *a.* [Of obscure origin; Ir., Gael. *dath* 'colour' has been suggested.] Of a pale primrose colour, as the eyes of certain game fowl.

1856 W. B. TEGETMEIER Poultry Bk. xiii. 100 Black-breasted reds..have a fine long head; daw eyes. Ibid., The required 'daw eye'..is that which resembles the gray eye of a jackdaw. 1873 L. WRIGHT Bk. Poultry 277 There never was a Malay with red eyes; they are invariably pearl, yellow, or daw. 1913 W. BATESON Mendel's Princ. Heredity (ed. 3) 110 Malay fowls are peculiar in having a pale, yellowish white iris—the 'daw-eye' of fanciers.

dawg (dǭg). Colloq. var. DOG *sb.*

1898 J. D. BRAYSHAW Slum Silhouettes 125 Ev'ry markit mornin' yer ter be my dawg, ev'ry mornin' till yer like ter chuck it. 1939 O. LANCASTER Homes Sweet Homes 60 Beaten copper reminders that a man's best friend is his dawg (beloved of the golf-playing classes). 1966 'J. HACKSTON' Father clears Out 14 The dawg's 'ad 'is ribs stove in.

dawk, *sb.*² Add: **b.** *dawk-* or *dāk-bag*; **dawk-** or **dāk-wallah,** a letter-carrier.

1828 Asiatic Costumes 40 The dauk-wala is dispatched from the post-office every day with his bundle of letters. 1872 E. BRADDON Life in India vii. 260 The arrival at any village of the dāk-walla (letter-carrier) with a letter is an event to be remembered and talked of. 1923 Blackw. Mag. Nov. 678/2 My old dāk-wallah..had scented the battle from afar. 1926 Ibid. Nov. 587/1 An 'Urgent' dak bag arrived from the Agency with a letter from Baird. 1928 Ibid. Jan. 5/2 The correspondence came to an abrupt stop. Great soggy chunks of silence filled the incoming dāk-bag.

dawn, *sb.* Add: **1. b.** An opalescent colour resembling that seen in the sky at dawn.

1894 Daily News 11 Apr. 3/1 Palest pink and blue shot silk, called by the poetic name of 'Dawn', because it suggests the union of those colours in the early morning sky. 1927 Daily Express 21 Mar. 2 Colours include cedar, green, silver, new blue, dawn or bois de rose.

c. Phr. *came the dawn*: a cliché used to announce the break of day; hence *fig.*, used to indicate relief after a time of trouble, the dawning of understanding, etc.

1927 WODEHOUSE Meet Mr Mulliner v. 169 A benevolent glow irradiated the other's spectacles. 'Came the Dawn!' he murmured. 'Came the Dawn.' 1929 J. B. PRIESTLEY Good Companions III. v. 582 For her sake alone he..renounced wealth and fame. Love was his guiding star. Came the dawn. Yeogh!..What do you think you are—a little hero from Hollywood? 1948 C. DAY LEWIS Otterbury Incident iii. 29 As for Nick, you never saw such a 'Came-the-dawn' expression as he had on his face. 1967 'A. GARVE' Very Quiet Place I. iv. 60 It was fun staying up half the night..but, came the dawn, I was the one who had to..earn the rent. 1967 Guardian 18 Oct. 1 'Came the dawn.' In the days of silent films this caption introduced the sequence where the young lovers were united after a night of tropical storm.

3. *dawn-chill, -cloud, -flush, -mist, -wind; -lit adj.*; **dawn chorus,** the early-morning bird song; **dawn man,** an extinct primitive man; *spec.* (freq. with capital initials) the (fraudulently postulated) prehistoric type of man, *Eoanthropus dawsoni*: see *PILTDOWN; so *dawn woman*.

1899 A. WERNER Captain of Locusts 152 Holcroft shivered involuntarily in the dawn-chill. 1927 E. GREY Charm of Birds [i. 8 The robin's song is worth attention..and, though he may not open the Great Chorus at Dawn in May, he is the last to cease in the evening.] Ibid. iv. 70 In May..the great Dawn Chorus is at its fullest and best. 1966 Guardian 23 Mar. 3/2 The dawn chorus project which we carried out last spring produced the interesting result that skylarks apparently don't like singing in the rain. 1969 G. BLACK Cold Jungle xi. 160 The birds ought to be busy on their dawn chorus out in the Hebrides, with a new day practically settled in. 1901 KIPLING Kim xv. 383 Thence he vanished like a dawn-cloud on Jakko. 1906 Daily Chron. 30 June 4/6 A painter..saw a sunrise and put the dawn-flush into a picture. 1906 Westm. Gaz. 29 Oct. 2/3 Rare and transparent as the dawn-lit dew. 1912 R. BROOKE Grantchester in Poetry Rev. Nov. 507 Still in the dawnlit waters cool His ghostly Lordship swims his pool. 1913 Nature 2 Oct. 131/2 It is quite certain that they afford the

first evidence we have obtained of a hitherto unknown group of the Hominidae so fundamentally distinct from all the early fossil men found in Europe as to be worthy of generic distinction—a 'dawn-man' of a very primitive and generalised type. 1914 W. K. GREGORY in Amer. Museum Jrnl. XIV. 189 The Dawn Man of Piltdown. Ibid. 191/1 All agree that the Dawn Man dates at the very latest from the Old Stone Age. 1927 H. F. OSBORN in E. Eyre Europ. Civilization (1934) I. I. v. 80 We are descended from 'dawn-men' not from 'ape-men'. 1944 H. G. WELLS '42 to '44 190 The breeding season of the Dawn-Men may have been an annual affair. Ibid., The hardy steppe-bred Dawn-Woman of the early Solutrean. 1904 R. J. FARRER Garden of Asia xvi. 151 Across the broad landscape the dawn-mist lies in heavy, floating wreaths. 1887 KIPLING Departmental Ditties (1888) 35 The dawn-wind, softly, slowly, Brought to burning eyelids sleep. 1916 BLUNDEN Pastorals 35 And through green sprigs a little dawn-wind plains.

daxie (dæ·ksi). Colloquial pet-form of DACHSHUND. Cf. *DACHS.

1899 Daily News 28 Oct. 7/5 They (like Daxies again) delight in playing tricks. 1900 Ibid. 22 Dec. 6/3 Coercion distinctly disagreeing with daxies!

day, *sb.* Add: **6. a.** (The astronomical day is now reckoned from midnight to midnight.)

d. *of a day*: lit. lasting only a day, ephemeral; transitory, fleeting, fugitive.

1640 B. JONSON Under-Woods 234 A Lillie of a Day, Is fairer farre, in May, Although it fall, and die that night. 1746 WESLEY Serm. (1769) I. Pref. p. vi, I am a Creature of a Day, passing thro' Life, as an Arrow thro' the Air. 1818 KEATS Let. 3 May (1931) I. 153 My song should die away... Rich in the simple worship of a day. 1834 Rival Sisters 14 Man—the insect of a day. 1865 M. ARNOLD Ess. Crit. 1st Ser. Pref., Apparitions of a day.

e. A day noteworthy for its eventfulness, exertion, etc. *colloq.*

1926 HEMINGWAY Fiesta (1927) vii. 65, I say. We have had a day... I must have been blind [sc. drunk]. 1963 'W. HAGGARD' High Wire xii. 127, I expect you've had a day—I know I have. But there's one small thing still.

7. b. *one of those days*: a day of misfortune.

1936 P. FLEMING News from Tartary I. ix. 55 As we arrived at the inn, the building next to it..collapsed...It was one of those days. 1967 'S. WOODS' Case is Altered xiv. 166 Old Mr. Mallory was waiting to pounce on him, and it soon became obvious that it was going to be one of those days.

c. Used without a preposition or article. *U.S.*

1886 S. W. MITCHELL R. Blake 292, I saw a man at the Cape wharf day before yesterday, inquirin' about Mrs. Wynne. 1905 N.Y. Even. Post 20 May 4 Day before yesterday the President was again in a state of terrific determination. Ibid. 26 Sept. 6 Day after election people will want to know [etc.].

8. d. That period of the day allotted by usage or law for work; as, *an eight-hour day*. (See *EIGHT HOURS, *WORKING-DAY.)

1813, 1853 [see *WORKING-DAY]. 1850 Working Man's Friend & Fam. Instr. 14 Dec. 300/1 Being at the rate of 4s. 2d. per day of ten hours. 1870 Chambers's Jrnl. 10 Sept. 586/2 In government workshops,..by special act of Congress, eight hours has been constituted a legal day's work. 1880 C. MARVIN Our Public Offices (ed. 2) 121 [They] worked hard the whole of the seven hours of their official day. 1884 J. E. T. ROGERS Six Cent. Work & Wages xii. 327 It is plain that the day was one of eight hours. 1889 R. TANGYE One & All vii. 116 In 1871 a great agitation sprung [sic] up amongst the operative engineers at Newcastle-on-Tyne in favour of a nine hours' day. 1891–1948 [see *EIGHT HOURS].

9. c. *The Day* [tr. G. *Der Tag*]: a day on which an important event is expected to occur; esp. a day of military conflict or victory.

1914 O. SEAMAN in Punch 9 Dec. 470/1 [German Crown Prince loq.] Thank Father's God that I can say My constant aim was Peace; I simply lived to see the Day (Den Tag) when wars would cease. 1914 G. B. SHAW What I really Wrote (1930) ii. 30 When the German fire-eaters drank to The Day (of Armageddon) they were drinking to the day of which our Navy League fire-eaters had first said 'It's bound to come'. 1919 Ibid. xii. 321 Just as the lieutenants of the German and British navies..looked forward to 'der Tag' when the preparations would be brought to the test of warfare, the lieutenants of the United States navy are already looking forward..to 'The Day' when the British and American fleets shall fight for that power to blockade [etc.]. 1936 J. BUCHAN Island of Sheep xiii. 256 The reconnaissance is complete, gentlemen. Tomorrow is The Day. 1959 J. BRAINE Vodi iv. 68 My Dad..says we'll all have to fight for our country when Der Tag comes. That's German for the Day.

13. a. See also SEE *v.* 10 a.

b. *(in) this day and age*, (at) the present time; (at) the moment of speaking or writing.

[1917 A. WOOLLCOTT Let. 4 Dec. (1944) 41 You will receive a modest Christmas gift..of no conceivable use in this day and generation.] 1933 Week-end Rev. 7 Oct. 348/1 (title of film) This day and age. 1941 Time 13 Jan. 44/1 She knew that in this day and age a nun could be a scientist, if she were smart as well as conscientious. 1944 H. CROOME You've gone Astray xxi. 209 Do you mean to say that in this day and age..you're going to come the conventional? 1958 Spectator 18 July 116/2 The needs of this day and age. 1970 New Yorker 17 Oct. 39/2 What a comfort it was in this day and age to meet someone obliging.

19. day in (and) day out, every day for an indefinite number of successive days, continuously; **day off,** a day away from work, school, etc.; **day out,** a day away from home or one's lodgings; *spec.* a servant's free day; also *fig.*; **day to day** (later examples); hence *day-to-dayness*.

1828 W. CARR Dial. Craven (ed. 2) I. 102 'Day in and day out', all the day long. 1848 Punch 4 Nov. 182/2 The Servant-Girl's Idea of Life:—one long day out with 'the journeyman'. 1869 Ibid. 20 Mar. 111/2 Having made this a holiday with a view to having a 'day out', my landlady had not had notice to call me at any particular hour. 1883 B. HARTE Carquinez Woods ii. 51 It has been already intimated that it was his 'day off'. 1887 M. E. WILKINS Humble Romance (1890) 127 Sewing as she did, day in, day out. 1890 Peel City Guardian 4 Jan. 5/5 It was Fayle's day out, and he made the most of the chances offered. 1892 MRS. A. IRELAND Lett. G. E. Jewsbury p. xiii, Their fulfilment is wholly incompatible with a migraine or a 'day off'. 1893 Eng. Illustr. Mag. 488/2 The bus-driver spends his 'day off' in driving on a pal's bus, on the box-seat by his pal's side. 1898 A. E. T. WATSON Turf i. 17 It may not have been the animal's 'day out', it may do better later on. 1904 KIPLING in Windsor Mag. Dec. 10/1 Whatever 'e's done, let us remember that 'e's given us a day off. 1927 Public Opinion Jan. 56/3 The British Broadcasting Company will have to offer, day in and day out, a service. Ibid. Feb. 109/2 Work—day in day out—and not much money. 1933 S. JAMESON (title) A day off. 1933 Granta 26 Apr. 370/1 Those interested only in the day to day politics of the fall of the dollar and the Russian embargo. 1942 W. S. CHURCHILL End of Beginning (1943) 33 Even in the..United States the Executive does not stand in the same direct, immediate, day-to-day relation to the Legislative body as we do. 1948 F. R. LEAVIS Great Tradition IV. i. 200 That kind of self-sufficient day-to-dayness of living Conrad can convey. 1960 C. DAY LEWIS Buried Day vi. 116 One boy..was kicked around, jeered at or ostracised, day in day out for several years. 1963 Higher Educ. (Cmnd. 2154) xv. 220 The day-to-day conduct of policy must rest with the heads of institutions. 1966 G. N. LEECH Eng. in Advertising xv. 138 Lifebuoy Toilet Soap With Puralin gives day in–day out protection against B.O. 1971 Daily Tel. 1 July 1 (heading) Teachers' strike gives 400,000 pupils day off.

20. b. In various colloq. phrases, as *to make a day of it*: to devote a day to some pursuit, usu. one of pleasure; to spend the day in enjoyment or revelling (see MAKE *v.*¹ 18 c and cf. NIGHT *sb.* 4 e); *to make (one's) day*: see *MAKE *v.*¹; *if it's (or he's, etc.) a day*: of a period of time or a person's age, at least; *any day (of the week)*: at all times; without exception or doubt; cf. *every time* (*EVERY *a.* 1 e and *TIME *sb.* 18 b); *to call it (half) a day*: to consider that one has done a day's work; *fig.* to rest content, to leave off; *between two days U.S.*: overnight; *that'll (or that will) be the day* (app. orig. *N.Z.*): (*a*) that will be a day worth waiting for, experiencing, etc.; (*b*) (ironically) that is most unlikely; that will never happen; *those were the days*: an expression (nostalgic or ironic) of regret for time past.

1660 [see MAKE v.¹ 18 c]. 1731–8 SWIFT Polite Conv. (1963) 78 She's on the wrong Side of thirty, if she be a Day. 1763 BOSWELL London Jrnl. 28 July (1950) 327 Come..let us make a day of it. Let us go down to Greenwich and dine. 1777 [see IF conj. 1 a]. 1828 LAMB Let. Dec. (1935) III. 198 From this paradise, making a day of it, you go to see the ruins of an old convent at March Hall. 1829 G. GRIFFIN Collegians II. xxiii. 169 It's a long time since you an' I met..'Tis six years if its a day. [1833 J. NEAL Down-Easters I. ix. 134 He'll do it any day o' the week..let alone Saturdays—of course the speaker was a Marylander of Irish parentage.] 1838 J. C. NEAL Charcoal Sk. 140, I've a great mind to knock off and call it half a day. 1839 THACKERAY Catherine i, in Fraser's Mag. May 609/2 She's seventeen if she's a day, though he is the very first sweetheart she has had. 1840 DICKENS Old C. Shop xxxix, in Master Humphrey's Clock (1841) II. 11 Why you are a good deal better-looking than her, Barbara...You are, any day. 1840 R. H. DANA Bef. Mast (1854) xxviii. 177 Some rascally deed sent him off 'between two days'. 1843 Spirit of Times 4 Mar. 1/2 When a thing isn't 'worth a fig', one 'might as well call it half a day and quit'. 1885 [see NIGHT sb. 4 e]. 1889 'MARK TWAIN' Yankee 271 It would have been best for Merlin..to quit and call it half a day. 1902 CONRAD Youth 4 He was sixty if a day. 1903 A. D. MCFAUL Ike Glidden in Maine ii. 12 Hadn't been't he left town 'tween two days he'd be good way on the road to the pen'tentiary now. 1906 G. K. CHESTERTON Dickens viii. 188 Susan Nipper..is more of a heroine than Florence any day of the week. 1919 WODEHOUSE Damsel in Distress ix. 116 Albert rose, not unwilling to call it a day. 1922 S. LEWIS Babbitt xv. 193 'Remember how..we pinched the pants-pressing sign and took and hung it on Prof. Morrison's door? Oh, gosh, those were the days!' Those, McKelvey agreed, were the days. 1930 W. S. MAUGHAM Cakes & Ale 268 I've had my time and I'm ready to call it a day. 1931 A. L. ROWSE Politics & Younger Gen. 155 They would prefer a despotism of the civil service to a despotism of the law any day. 1934 J. T. FARRELL Young Manhood xvi, in Studs Lonigan (1936) 347 Jesus, those were the days, weren't they, Studs? 1936 J. B. PRIESTLEY They walk in City 240 He had to be there at nine..and then worked on until the Belvedere Trading Company..'called it a day'. 1941 BAKER N.Z. Slang vii. 50 That'll be the day!..a cant phrase expressing mild doubt following some boast or claim by a person. 1943 N. MARSH Colour Scheme vi. 101 He's a beaut. Wait till I get him. That'll be the day. 1951 —— Opening Night xi. 248 'If I've bungled,' Alleyn muttered, 'I've..bungled in a big way.'..Bailey astonished everyone by saying..'That'll be the day.' 'Don't talk Australian,' Mr. Fox chided. 1957 G. BELLAIRS Death in High Provence xiii. 149 Madeleine's sister is a great age, too. Eighty, if a day. 1957 J. BRAINE Room at Top xiii. 129 We'll call it a day...Don't think badly of me. 1960 H. PINTER Room in Birthday Party 107 'Maybe there are two landlords.'..'That'll be the day.' 1963 V. NABOKOV Gift iii. 185 A friend of his..complained that Carlsbad was no longer what it used to be. Those were the days! he said: 'you stand with your mug of water and there next to you is King Edward.' 1965 L. SANDS Something to Hide v. 83 'Got any free road-maps?' 'That'll be the day. Bob apiece.'

22. (Earlier example.)

1803 M. WILMOT Jrnl. 13 Apr. in Londonderry & Hyde Russ. Jrnls. (1934) I. 3 My precious Father..saw us safely into the two day Coach.

23. a, b. *day-beam* (earlier and later examples), *-fall, -hospital*; *day-old* adj. (also *sb.* = day-old chick, etc.).

1811 SHELLEY Let. 6 Jan. (1964) I. 38 The day-beam returning. 1952 C. DAY LEWIS tr. Virgil's Aeneid IV. 76 As Aurora was rising out of her ocean bed And the day-beam lofted. 1889 F. THOMPSON in Merry England XIII. 300 Who set upon her brow the day-fall's carcanet? 1960 T. HUGHES Lupercal 55 Waking, dragged suddenly From a choir-shaken height By the world, Lord, and its dayfall. 1843 Chambers's Jrnl. 30 Dec. 398/1 A kind of day-hospital, to keep the children from wandering idly abroad. 1951 Brit. Jrnl. Psychol. Aug. 307 The Day Hospital does not belong to the era of Individual Psychiatry. 1958 New Statesman 10 Jan. 34/1 It has long been recognised that given adequate out-patient facilities, day hospitals, occupation centres and hostels, many persons suffering from mental disorder need not enter a mental hospital. 1963 Guardian 16 May 4/6 The Marlborough day hospital, in London..had had no beds and has been working as a 'day' hospital where patients after attendance return to their homes and families at night; and as a 'night' hospital where patients can go to work during the day, returning for help in the evenings. 1907 Daily Chron. 8 July 4/4 Many poultry-keepers dispose of several thousand day-old chicks every season. 1911 R. BROOKE Let. Feb. (1968) 280 Every night I sit in a café near here..and read the day-old Times. 1928 Daily Tel. 11 May 19/4 Day-olds from reliable pedigree strains cost only 21s a dozen. 1930 MASEFIELD Wanderer of Liverpool 67 It was fine clear easterly weather with a day-old moon. 1959 B.S.I. News Apr. 18/2 Priority is being given to arrangements for the carriage of day-old chicks. 1960 Farmer & Stockbreeder 5 Jan. 109/3 Year after year crops of goslings are..in big demand as day-olds or 'growers'.

c. *day-drudge* (earlier example), *-nurse* (examples), *-trip, -tripper.*

1837 J. S. MILL in Westm. Rev. XXVII. 22 He [sc. Carlyle] possesses in no less perfection..the quality of the historical day-drudge. 1844 DICKENS Mart. Chuz. xxv. 309 The night-nurse..well beknown to Mrs. Prig the day-nurse. 1855 J. R. BESTE Wabash I. ii. 48 Our little Isabel amused the youngest children, and constituted herself their day nurse. 1903 A. BENNETT Leonora viii. 215 He had gone to London by a day-trip on the previous Thursday. 1967 C. DRUMMOND Death at Furlong Post viii. 107 A customer who had taken a day trip to Calais. 1897 Daily News 27 Sept. 3/5 The 'day-tripper' class of excursionists. Ibid., Day trippers by the Marguerite from the Thames. 1960 E. W. HILDICK Jim Starling & Colonel vii. 57 Day-trippers on their way to Blackpool.

f. *day-born.*

1849 THOREAU A Week 59 The Society Islanders had their day-born gods. 1903 Westm. Gaz. 26 Nov. 2/3 The day-born, hopeless longing dies.

24. day-and-night *attrib.*, throughout the day and night; **day-boarder** (example); **day-boy,** also *transf.* and *attrib.*; **day-bug** *Schoolboy slang*, = *day-boy*; **day-car, -coach** *U.S.*, any railway passenger carriage other than a sleeper; also *transf.*; **day care,** the supervision and care of young children during the day, esp. while their mothers are at work; freq. *attrib.*; **day continuation school,** a school for educating young workers released temporarily by their employers; **day dress,** = *day-gown*; **day editor,** the editor in charge of a newspaper during the day; **day-gown** (earlier example); **day-length,** the length of the day, esp. as it varies at different times of the year; also *attrib.*, *spec.* designating clothes of a suitable length for wear during the day (see also quot. 1949); **day nursery,** (*a*) a nursery used by children during the day (as distinguished from *night nursery*); (*b*) a nursery where children are cared for while their mothers are at work; **Day of Atonement** [tr. Heb. *Yōm Kippūr*], a Jewish fast day, observed from the ninth to the tenth of Tishri; **day-on** *Naut. slang*, one who does duty as officer of the day; **day release,** a system whereby employers allow employees days off from work for education; also *attrib.*; **day-wages**: so *day-wage attrib.*

1899 Westm. Gaz. 16 Mar. 10/2 The work will be carried out by day and night relays of men. 1964 S. DUKE-ELDER Parsons' Dis. Eye (ed. 14) xxvii. 436 Unless day-and-night nursing is available it is a wise precaution with many patients to tie their hands loosely to the bed at night. 1853 E. SEWELL Exper. Life iii. 26 A very tolerable school..where they were allowed..to attend as day boarders. 1914 Spectator 17 Oct. 516/2 We would train a portion of the men in what we may describe as 'day-boy' battalions... Up till the time of the Boer War..there were two battalions of London Militia who were always trained on the 'day-boy' system. The men lived in their own homes, and came to the depot each day for their recruit training. 1909 WARE Passing Eng. 105/1 Don't row with that fellow, he's only a day-bug. 1913 C. MACKENZIE Sinister Street I. I. vii. 103 When an older boarder called him a 'day-bug' Michael was discreetly silent. 1870 W. F. RAE Westward by Rail (1871) 50 This company build and run their own elegant sleeping coaches and palace day cars. 1904 Westm. Gaz. 26 Sept. 6/3 [Accident in Tennessee] Day-car and day-car were telescoped, buckled, and thrown over. 1964 Economist 4 Jan.

24/2 Adequately staffed day-care centres for children. 1873 Winfield (Kansas) Courier 11 Jan. 2/7 Elegant Day Coaches, [etc.]..are some of the modern improvements used on this Line. 1887 C. B. GEORGE 40 Yrs. on Rail xi. 226 A passenger on his way to the dining-car came out of the day coach into the ladies' end of my car. 1947 Shell Aviation News CXII. 4/1 Atlantic has applied to the Civil Aeronautics Board for permission to operate day coach services in the triangular area, New York, Washington and Pittsburg. 1919 Times Educ. Suppl. 25 Sept. 487/1 The day continuation school must sit between eight in the morning and seven in the evening. 1943 Ann. Reg. 1942 68 The Council..advocated..day continuation schools for young persons up to the age of 18. 1952 Oxf. Jun. Encycl. X. 142/2 A system of Day Continuation Schools was also included in the 1918 Fisher Act. All young people between the ages of 14 and 18 who had ended full-time schooling were to continue their education at such schools for a period of 320 hours each year. 1922 Liberty Dresses Spring 10 Day Dress, in crêpe-de-chine... Price 12½ guineas. 1873 W. MATHEWS Getting on in World xiv. 218 Mr. Brooks..acting as leading editor [of the New York Express], reporter, day editor, night editor, and even typesetter. 1877 Harper's Mag. Dec. 53/2 The day editor in charge. 1875 L. TROUBRIDGE Life amongst Troubridges (1966) x. 116 My old peacock cashmere evening, transformed into a day gown with long sleeves. 1920 GARNER & ALLARD in Jrnl. Agric. Res. XVIII. 582 The Stewart Cuban Mammoth tobacco which requires a day length of 12 hours or less to attain the blossoming stage has been grown commercially to some extent under an artificial shade. 1944 A. G. HATCHER in Mod. Lang. Notes Dec. 515 Among the many relationships expressed by noun combinations..the three-member compound is peculiarly modern..day-length dresses..bronze-finish lamp. 1949 New Biol. VII. 52 Day length neutral plants..flowered whatever the length of day. 1965 Harper's Bazaar Jan. 54/1 He will charge you about 20 or 25 gns. for a day-length dress. 1844 MRS. PARKES in Webster & Parkes Encycl. Domestic Econ. xxvi. i. 1187 Day nurseries should be prepared for the children by having the windows open early in the morning. 1850 Househ. Words II. 110/1 These institutions were to be Day-Nurseries for the children of the poor. 1884 Harper's Mag. Apr. 782/2 A 'Day Nursery and Temporary Home for Children', charging two cents a day to busy mothers. 1896 H. FRIEDERICHS In the Evening of his Days 70 (caption) The Day Nursery at Hawarden Castle. 1908 H. DE V. STACPOOLE Patsy ii, They were in the day nursery, which was also the schoolroom. 1955 Times 15 July 5/7 Under the scheme the boroughs would become responsible for maternity and child welfare, day nurseries, [etc.]. [1611 BIBLE Lev. xxiii. 27 On the tenth day of this seuenth moneth there shalbe a day of atonement.] 1819 L. ALEXANDER Hebrew Ritual 82 The Conclusion Prayer..concludes the service of the Day of Atonement. 1893 ZANGWILL Ghetto Tragedies 47 The great White Fast, the Day of Atonement. 1932 L. GOLDING Magnolia Street I. ii. 33 When she absented herself on the Feast of the New Year and the Day of Atonement, the understanding was she had a cold. 1914 'BARTIMEUS' Naval Occasions (1918) iv. 27 The Day-on flopped exhaustedly on to a Wardroom settee. 1945 Youth's Opportunity (Min. Educ. Pamphlet III) 12 Part-time day release by industry for the purpose of technical commercial and art courses. 1955 Times 23 May 6/2 In part-time day-release courses an effective proportion..of the students' time should be devoted to non-vocational work. 1965 J. MELVILLE There lies your Love i. 24 She was what was called a day-release student: a student who also had a job but was continuing her education. 1963 Times 10 May 6/5 The recent provisional agreement on pay increases for day-wage men and craftsmen had been endorsed.

dayabeah, var. DAHABEEYAH.

Daya(c)k, Dayakker, varr. *DYAK.

‖**dayal** (dā·yăl). *India.* Also **dhyal.** [See DIAL-BIRD.] = DIAL-BIRD.

1855 Orr's Circle Sci., Org. Nat. III. 307 The Dayal..which..is called the Magpie Robin by the English residents in Ceylon. 1893 NEWTON Dict. Birds 133 Dayal, or more correctly, it would seem, Dhyal (corrupted into Dial-bird), the Hindostani name commonly adopted by Anglo-Indians.

‖**dayan** (dayā·n). Also **dayyan.** Pl. **dayanim, dayans.** [Heb.] A religious judge in a Jewish community.

1880 Encycl. Brit. XIII. 687/1 Each congregation requires the services of a dayan or religious chief. 1891 M. FRIEDLÄNDER Jewish Religion II. vi. 463 The meat must not be used as food unless the animal has been examined by a competent person (Rabbi or dayyan) and declared..kosher. 1892 ZANGWILL Childr. Ghetto I. 4 The Dayanim, those cadis of the East End, administered justice. 1902 Daily Chron. 3 Jan. 5/1 The Rev. B. Spiers, the Dayan of the Beth Hamedrash of London. 1928 Daily Tel. 3 Jan. 10/4 Dayan Dr. Feldmann. 1928 Sunday Dispatch 30 Dec. 7/2 Problems of divorce and marriage brokerage as well as disputes between synagogues and questions of ritual are settled by the bench of five Dayans—or judges. 1957 L. P. GARTNER Jewish Immigrant in Eng. (1960) vii. 204 Rabbi Avigdor Chaikin..sat as a Dayan upon the Beth Din of the Chief Rabbi.

day-bed. Delete †*Obs.* and add later examples.

1940 Chambers's Techn. Dict. 225/2 Day bed, a low lounge with sloping head rest. 1969 L. J. CHIARAMONTE in Halpert & Story Christmas Mumming in Newfoundland 100 The girls sit down on a day bed (a single cot-type couch found in every kitchen).

day-dream. Add: **day-dream** *v. refl.*, to transport (oneself) imaginatively by means of a day-dream; **day-dreamer** (earlier example); **day-dreaming** *ppl. a.*; **day-dreamy** *a.* (example); also, having the quality of a day-dream.

1849 THACKERAY Pendennis I. xxviii. 272 This young day-dreamer built castles in the air for himself. 1899 Crampton's Mag. Jan. 109 Mr. Jones fell into a retrospective mood, day-dreaming himself back into the past. 1906 Daily Chron. 3 Aug. 8/5 Day-dreaming children can pass into this fairy world at will. 1921 T. S. ELIOT in Times Lit. Suppl. 31 Mar. 202/1 The day-dreamy feeling of Morris is essentially a slight thing. 1949 'M. INNES' Journeying Boy (1961) 170 There is no doubt, however, that Humphrey can day-dream himself into some rather alarming world of romantic adventure, with nervous consequences that are by no means desirable.

‖**daye** (dəi, dā·i). *India.* Also **dai, dhaye, dhye, dy, dyah.** [f. Hind. *dāi*, Pers. *dāyah*.] In N. India and Persia, a nurse; a wet-nurse; a midwife.

1782 India Gaz. 12 Oct. (Y., Suppl.), Dy (Wet-nurse) 10 Rs. 1810 T. WILLIAMSON East India Vade-Mecum I. 341 The Dhye is more generally an attendant upon native ladies. 1825 M. M. SHERWOOD (title) History of little Lucy and her dhaye. 1877 E. S. DALLAS Kettner's Bk. of Table 154 The English child born in India has a nurse called a da'i. 1883 C. J. WILLS In Land of Lion & Sun xxix. 326 The 'dyah', or wet-nurse, is looked on as a second mother. 1920 Outward Bound Oct. 82/1 The doctor and her least dangerous dai (nurse) clambered into the waiting barouche. 1927 Other Lands Apr. 89/2 The indigenous dais are..absolutely ignorant.

day labour. (Later examples.)

1839 DE LA BECHE Rep. Geol. Cornwall, etc. xv. 569 Though in some mines day-labour is also used under ground. 1850–52 J. M. WILSON Farmer's Dict. I. 498/1 The same class of men who perform all kinds of drudgery and day-labour in large towns. 1908 W. ST. CLAIR BADDELEY Cotteswold Shrine 76 At the same hour ended his day-labours. 1911 Rep. Labour & Soc. Cond. Germany III. vi–vii. 207 Half the labour is piece and the other half is day labour.

daylight. Add: **1. c.** (Examples with *through.*)

1712 [SWIFT] Law is a Bottomless-Pit III, in John Bull Still in his Senses vi. 25 I'll warrant ye, you shall see Day-light through them. 1922 JOYCE Ulysses 322 The Molly Maguires looking for him to let daylight through him.

2. b. *Photography.* The period during which film can be 'taken' by natural light; *spec.* (see quot. 1940). Freq. *attrib.*, occurring or performed during daylight (see also *6).

1889 E. J. WALL Dict. Photogr. 61 Daylight enlarging. 1940 Chambers's Techn. Dict. 225/2 Daylight, the average colour of sky and sun at noon, corresponding to a colour temperature of 6500 K. 1940 F. J. MORTIMER Wall's Dict. Photogr. (ed. 15) 258 Daylight enlargers.

3. d. (See quots.)

1930 Engineering 3 Jan. 14/1 The 'daylight', i.e. the distance [sc. in a hydraulic press] between the faces of the vertical rams in their highest positions and the surface of the work table. 1968 Gloss. Terms Mechanized & Hand Sheet Metal Work (B.S.I.) 7 Daylight. 1. On the press. The distance between the bed of the press and the face of the slide with the press at the top of its stroke and with the adjustment up. 2. On the die set. The distance between the inner faces of the die set with the press slide at the bottom of its stroke. 3. On the press tool. The distance between the closest points of a press tool with the press fully open and the adjustment up.

4. (Later U.S. examples.) Also in extended use of any vital organ. Also *to beat, scare*, etc., *the (living) daylight(s) out of* (a person), to beat, scare (a person) severely.

1848 E. BENNETT Mike Fink i. 10/1 We'll catch the fever and ager,..and that'll shake the day-lights out o' us. 1884 E. W. NYE Baled Hay 79 The driver bangs the mule, that is ostensibly pulling his daylights out. 1923 R. D. PAINE Comr. Rolling Ocean ii. 22 Putting seven of 'em in irons after they shot the daylights out of me left us mighty short-handed. 1944 E. CALDWELL Tragic Ground (1947) xiv. 163 If I could find a stick I'd grab it and beat the daylights out of you. 1951 E. TAYLOR Game of Hide-and-Seek II. i. 174 Though they scared the daylight out of me, I contempted them. 1955 F. YERBY Treasure of Pleasant Valley (1956) iii. 36 Didn't mean to hit him...Meant to throw close to him and scare the living daylights out of him. 1960 N. HILLIARD Maori Girl III. viii. 229 I'll go down there and belt the daylights out of him! 1960 R. RAE Custard Boys I. viii. 92 We'll get yer an' all the rest of yer gang, an' when we do we'll beat the living daylights out of yer. 1963 N.Z. Listener 13 Sept. 10/4, I hated him. He used to beat the daylight out of me. 1964 Illustr. London News 18 Jan. 102/3, I might have chuckled throughout 'The Suitor' if its chief actor did not happen to scare the living daylights out of me, as the current saying goes.

6. daylight factor (see quots.); **daylight-loading** *a.*, (of a film-spool, cartridge, etc.) adapted for loading by daylight without the use of a dark-room; **daylight-saving,** a method of securing a longer period of daylight at the end of the day, viz. by putting the clock forward (e.g. an hour). Cf. SUMMER TIME 2.

1915 First Rep. Dep. Comm. on Lighting (Cd. 8000) I. p. vi, The ratio of the actual value of the illumination to this enhanced value, expressed as a percentage, is termed the daylight factor, and is a measure of the lighting efficiency of the building at the point under consideration. 1930 Engineering 25 Apr. 550/1 'Daylight Factor', the ratio of the internal illumination to the illumination of a horizontal surface exposed to a hemisphere of sky. 1953 Terms Illum. & Photometry (B.S.I.) 14 Daylight factor. At a given point inside a building. The ratio of the illumination..at that point to that simultaneously existing..under an unobstructed sky. 1902 Photographic Catal., Rollable daylight loading Films. 1958 Newnes Complete Amat. Photogr. 257 Roll film tanks. There are two types: daylight loading and daylight developing. 1908 Hansard 4th Ser. CLXXXIV. 155 Daylight Saving Bill...Mr. R. Pearce..in moving the Second Reading said that the object of the Bill was to promote the earlier use of daylight in the summer. 1916 Ibid. 5th Ser. LXXXII. 321 The advocates of daylight saving are adept in securing the consent of one body of opinion on the ground that some other body has adopted it with enthusiasm. 1924 Punch 24 Sept. 338 Two more..weeks of daylight-saving. 1931 GALSWORTHY Maid in Waiting vi. 45 'Too early,' said Sir Lawrence, 'owing to Daylight Saving.'

Hence **day·lighter,** one of a body of men who sought to enforce the decrees of the Land League in Ireland by violence during the daytime.

1886 St. James's Gaz. 25 Nov. 11/2 Seeing the 'Day-lighters' she ran into the room where she knew the gun to be and closed the door.

day·lighting, *vbl. sb.* [f. DAYLIGHT.]

1. (See *DAYLIGHTER.)

1894 Daily Tel. 2 Apr., A case of 'daylighting' instead of moonlighting has been reported to the local police.

2. The process, degree, etc., of the illumination of buildings by daylight. Also *attrib.*

1937 Archit. Rec. Sept. 111/2 The problem of daylighting is therefore bound up with the problem of heating and it is..desirable to orientate the house so as to obtain the best results for both. 1958 House & Garden Mar. 53/1 Lectures..on..Daylighting and Architecture. 1959 Guardian 19 Dec. 2/4 A slab..200 feet high..would infringe daylighting rights in Denman Street. 1961 E. A. POWDRILL Vocab. Land Planning iii. 29 Daylighting standards have therefore been devised. Ibid. 32 It is not thought likely that the attainment of good sunlighting will normally demand a more generous spacing than daylighting.

day's work. Add: **b.** *all in the day's work*, something unusual but nevertheless taken as part of one's ordinary duty or routine. Freq. ironical.

[1738 SWIFT Polite Conv. i. 39 Will you be so kind to tie this String for me..? it will go all in your Day's Work.] 1820 SCOTT Monastery I. ix. 248 That will cost me a farther ride,..but it is all in the day's work. 1857 KINGSLEY Two Y. Ago III. iii. 91 All in the day's work, my boy. 1897 [see WORK sb. 33]. 1939 War Illustr. 220/2 He is not deeply impressed by his experience. 'It is all in the day's work' were his parting words to me. 1953 A. CHRISTIE Pocket full of Rye x. 60 This sort of thing seems ordinary enough to you, Inspector. All in the day's work.

daytimes (dēi·təimz), *adv. U.S.* [f. DAYTIME + advb. suff. -s.] In the daytime, during the day.

1854 M. S. CUMMINS Lamplighter xvii, Willie was very busy daytimes, but was always with them in the evening. 1936 L. C. DOUGLAS White Banners viii. 187 Her bed, daytimes, was strewn with illustrated magazines.

dazzle, *sb.* Add: **3.** The painting of large patches of colour on warships, etc., as camouflage in time of war. Hence **dazzle** *v. trans.*, to camouflage (a ship) in this way. Also *Comb.* in *dazzle-paint, -painted, -painting, -pattern.* Also *transf.*

1917 Admiralty Order 2 July (MS.), The 'Dazzle' painting of a ship with large patches of strong colour in a carefully thought-out pattern and colour scheme. 1919 Times 29 May 8/1 'Dazzle', to use the term employed by the camouflage department of the Admiralty. Ibid. 5 June 10/2 A 'dazzle' painted ship is on the whole more visible against sky and sea than the usual grey vessel, and this was its chief disadvantage. Furthermore, 'dazzle' painting was designed for short range, at which it is impossible to conceal a vessel against its background. 1919 Ibid. 9 June 6/4 Dazzle painting was never intended for use on 'ships of the line', but only for merchantmen..and war vessels working with them. 1919 Athenæum 11 July 583/2 The other service [sc. the Navy] invented 'dazzle'. 1920 Blackw. Mag. Oct. 94/2 A geological 'dazzle-painting' in ochre and red, brown, purple, and buff. 1920 Glasgow Herald 12 Mar. 7 A number of lantern slides were shown of ships 'dazzled' during the war. 1921 W. STEVENS Let. 5 Dec. (1967) 223 From time immemorial the philosophers and other scene painters have daubed the sky with dazzle paint. 1922 Glasgow Herald 28 Apr. 9 A 'tramp' steamer, 'dazzle' painted. 1928 C. F. S. GAMBLE North Sea Air Station xxii. 400 It was decided that all flying-boats should have their hulls 'dazzle-painted'. 1932 KIPLING Limits & Renewals 309 Then, area by area, she [sc. a sow] was painted with dazzle-patterns of greenish-yellow and purple-brown. 1948 C. DAY LEWIS Otterbury Incident i. 9 The superstructure was made of wood, and we'd dazzle-painted the sides.

4. *attrib.* Designating shoes, etc., in very bright or luminous colours.

1931 Star 8 May 7/4 Although the majority of women seem to prefer shoes with just two colours to match their frocks, 'dazzle' footwear are a good second. 1958 Economist 11 Jan. 94/1 Girls in tight jeans and dazzle socks. 1958 J. TOWNSEND Young Devils i. 11 The boy..his yellow dazzle socks flashing like twin beacons.

D-Day (dī·dēi). Also **D Day, D-day.** [*D* for *day.*] The military code-name for a particular day fixed for the beginning of an operation; *spec.* the day (6 June 1944) of the invasion of the Atlantic coast of German-occupied France by Allied forces. Also *transf.*, of non-military undertakings; later also used for *decimalization day* (e.g. in Britain 15 Feb. 1971, on which day decimal currency came into official use).

1918 Field Order No. 8, First Army, A.E.F. 7 Sept., The First Army will attack at H-Hour on D-Day with the object of forcing the evacuation of St. Mihiel salient. **1928** J. M. Saunders *Wings* 210 The word went out that 'D' day was to be Sept. 12. **1942** *Newsweek* 23 Nov. 27 A major Russian offensive long in preparation abiding the eventful D-day. **1944** *Times* 10 June 4/1 The Canadians landed on D Day at Bernières-sur-Mer. **1944** *Hutchinson's Pict. Hist. War* 12 Apr.–26 Sept. 342 By the end of D-Day plus two they had cleared their respective areas of dead and wounded. **1945** W. S. Churchill *Victory* (1946) 102 The total [of Germans] captured by the Allies since D-Day was 2,055,575. **1947** *Economist* 27 Dec. 1047 (*heading*) D-Day for the [Marshall] Plan. **1948** G. Mikes *How to scrape Skies* 21 You know that the pipes *will* burst one day but how exciting it is to spend first of all three pleasant days wrapping them up and then retiring to wait for D-day. **1950** [see *air-drop* s.v. *Air *sb.*[1] B. III. 2].

1963 *Rep. Comm. Inquiry Decimal Curr.* p. xiv, 'D-day'. Short for 'Decimalisation Day'. *Ibid.* xiii. 132 We hope.. that as many organisations as possible will change on 'D-day' [in South Africa]. **1970** *New Scientist* 5 Feb. 245/2 The Anti-Decimal Group will doubtless be busy between now and D-day. **1971** *Oxford Mail* 15 Feb. 1/2 D-day dawned with a minimum of fuss, and shoppers were taking the new coinage in their stride.

D.D.T. (didīti·). Also **DDT.** An abbreviation for *d*ichloro*d*iphenyl*t*richloroethane, a white, crystalline, chlorinated hydrocarbon used as an insecticide in the form of a powder, an aqueous emulsion, or a non-aqueous solution. Also *attrib.* and *Comb.*

1943 *Soap & Sanitary Chemicals* Dec. 117/1 An insecticide material of similar type is now being made in the United States by Geigy & Co., New York, under the name 'DDT'. **1944** W. S. Churchill *Memo.* 9 Mar. in *2nd World War* (1952) V. 612 The demand for the new insecticide, D.D.T., is urgent and increasing. **1944** *Chemical Age* 9 Sept. 246/2 All the DDT produced is at present reserved for the use of the armed forces. *Ibid.*, The DDT powder..was found effective. **1959** *New Statesman* 13 June 820/3 The mustard is to be powdered by helicopter with a DDT-type chemical. **1964** *Ann. Reg. 1963* 399 The accumulation of D.D.T. now going on in many people's fatty tissues might..be harmful in the long run. **1968** M. Pyke *Food & Society* viii. 118 The chlorinated hydrocarbon pesticides, of which DDT is by far the commonest, are very persistent in soil.

de. Add: **I. 4.** (Later examples.)

1865 Pusey *Eirenicon* 115 The poorer classes are not, for the most part, even acquainted with the distinction between what is to be believed to be *de fide* and what is popularly taught them as truth. **1951** *Essays in Criticism* I. 4 Modern critics..treat this as self-evident or *de fide*.

II. (Further examples.)

1778 H. Walpole *Let.* 4 July (1904) X. 271 The Congress has ratified the treaty with France, and intend to treat the Commissioners *de haut en bas*. **1868** *Good Words* 1 Aug. 516/1 A de-haut-en-bas-like drawling gait. **1882** W. R. Greg *Misc. Essays* ix. 181 But her *de-haut-en-bas* judgment of Macaulay is perhaps widest of the mark. **1885** A. Edwards *Girton Girl* I. ii. 39 Noble French family who.. dropped the 'de' from before their name, and settled here [*sc.* in Guernsey]. **1888** C. M. Yonge *Our New Mistress* v. 43 The names with a De before them are always the grandest. **1915** Mrs. H. Ward *Eltham House* vii. 119 Ask an officer to do without his uniforms. My frocks are just as much *de rigueur*. **1944** A. L. Rowse *English Spirit* 134 It seems to have been *de rigueur* to say that the Queen looked twenty. **1965** R. Ferguson *Woman with Secret* vi. 42 Her father thought he wasn't good enough for somebody with the 'de' in her name. **1966** *Listener* 17 Nov. 747/2 Hopkins has a distinct tendency to talk *de haut en bas* to his captive audience.

dea, deac. U.S. colloq. abbrevs. of Deacon *sb.* 1 c, d.

1821 *Massachusetts Spy* 28 Feb. (Th.), Deac Josiah Bridge. *Ibid.* 4 Apr., Dea. Ebenezer Read. **1907** *Springfield Weekly Republ.* 17 Jan. 11 Dea Wilson was among the foremost of the town's citizens. **1913** *Sat. Even. Post* 1 Feb. 34/3 Look at 'er, deac!

deacon, *sb.*[1] Add: **6.** (Earlier example.)

1851 J. S. Springer *Forest Life* 71 Directly over the foot-pole..and in front of the fire, is the deacon-seat. This seat constitutes our sofa or settee.

7. A very young or aborted calf, or its hide. *U.S. colloq.*

1873 *Chicago Tribune* 2 Jan. 6/2 Hides..deacons, 50 @ 65 ¢. **1898** E. N. Westcott *D. Harum* xvii, I guess you got a 'deakin' in that lot...That calf died, that's what that calf done. **1923** *Dialect Notes* V. 234 *Deacon*, a calf of veal age; the hide or skin of such a calf. 'That hide ain't worth much; it's only a deacon.'

Deacon, *sb.*[2] The name of Henry *Deacon* (1822–76), English industrial chemist, used attrib. or in the possessive to designate a process, for the manufacture of chlorine by the catalytic oxidation of hydrochloric acid.

1876 *Encycl. Brit.* V. 491/1 The production of chlorine by Deacon's process. **1921** *Dict. Occup. Terms* (1927) § 143 Deacon plant man, deacon process man. **1965** Phillips & Williams *Inorg. Chem.* I. xii. 446 Chlorine has been prepared by displacement with oxygen, in such exothermic reactions as..the Deacon process.

deactivate (dĭ,æ·ktivē[i]t), *v.* [f. De- + Activate *v.*] *trans.* To render inactive or less reactive; *spec.* to deprive of chemical action, to render less chemically reactive. So **de-activa·tion,** the process or result of deactivating; **dea·ctivator,** a substance which deactivates.

1904 *Sci. Amer.* Suppl. 20 Feb. 23523/1 He finds that the law of de-activation is still the same throughout this range of temperature. **1938** *Jrnl. Chem. Soc.* June 905 Just as the methyl substituent activates all positions,..so the carbethoxyl group deactivates all positions. **1939** *Nature* 29 Apr. 734/1 The stabilization of petrols by small quantities of organic compounds known as deactivators, which prevent oxidation, is an important advance. **1947** *Science News* V. 160 This drift activates or deactivates the aromatic nucleus. **1956** A. K. Osborne *Encycl. Iron & Steel Industry* 105/2 *Deactivation*, the process of prior removal of the active constituents, usually oxygen, from a corrosive liquid. **1958** *Economist* 20 Dec. 1051/1 (Advt.), Gum inhibitors and metal deactivators used in motor and aviation fuels to maintain quality during storage and use. **1970** J. Earl *Tuners & Amplifiers* i. 24 A switch for deactivating the stereo decoder on a weak stereo signal.

dead, *a.* Add: **A. 1. c.** spec. *dead head* (also *deadhead*), a faded flower head.

1960 D. Holman-Hunt *My Grandmothers & I* i. 7 Did you cut off *all* the deadheads? **1962** N. Marsh *Hand in Glove* v. 177 I'll go and snip the deadheads off the roses. **1970** C. Lloyd *Well-Tempered Garden* i. 48 The culling of dead heads is a ploy that figures persistently in the garden in summer.

2. c. In hyperbolical phrases expressing extreme fatigue or indisposition.

1813 Annabella Milbanke *Diary* (MS.), At home dead. **1894** *Pall Mall Mag.* Feb. 583 I'm nearly dead from being boxed up in the house all day. **1915** W. S. Maugham *Of Human Bondage* xliii. 208 You know, I'm simply dead. I don't think I can absorb anything more profitably. Let's go and sit down. **1962** J. Braine *Life at Top* xiv. 185 'It doesn't matter who started it now,' I yawned. 'Honestly, I'm dead on my feet, Susan.' **1970** P. Carlon *Death by Demonstration* xvi. 175 One job's enough. Come evening and I'm dead on my feet usually.

d. Of pain: dull and continuous, as opposed to sharp and sudden pain.

1863 T. B. Curling *Dis. Rectum* (ed. 3) 24 He complained of suffering from a dead, aching pain. **1894** Mrs. H. H. Gardener *Unoff. Patriot* 348 She only sat and stared, and was conscious of the dull dead pain.

3. c. Colloq. phr. *dead to the world*: unconscious or fast asleep; unaware of the external world.

1899 Ade *Doc' Horne* ii. 19 'Our host is dead to the world,' observed the actor... 'Let him rest,' said Doc'. **1906** E. Dyson *Fact'ry 'Ands* iii. 31 Heaven knows what blissful emotions were stirring softly in his bony breast, but he was 'dead to the world'. **1906** *Dialect Notes* III. ii. 133 *Dead to the world*, unconscious. 'He fell down and was dead to the world for a while.' **1955** E. Hillary *High Adventure* 166 He stumbled and fell slowly on to his face and lay there—dead to the world! **1957** G. Frick tr. *Yourcenar's Coup de Grâce* 56 A muffled sound of snoring rose from the great hall..where thirty exhausted lads lay dead to the world.

12. c. *Electr.* Of a circuit, conductor, etc.: carrying or transmitting no current; not connected to a source of electricity.

1903 A. H. Beavan *Tube, Train*, etc. xi. 134 The studs are 'alive' while the car is over them, and 'dead' as soon as it has passed. **1906** *Westm. Gaz.* 13 July 5/2 There was another stoppage..caused by a 'dead' car. **1929** D. Hammett *Dain Curse* (1930) xi. 109 The phone was there, but dead. **1937** David Jones *In Parenthesis* v. 112 Every telephonist with a dead instrument about his ears. **1968** *Globe & Mail* (Toronto) 3 Feb. 5/5, I tried to call the operator but the phone was dead.

14. b. *Acoustics.* Allowing little or no reverberation.

1907 *Science* XXVI. 879/2 The small room..when closed ..also serves to act as a dead air space between the larger room and the building wall. **1923** Glazebrook *Dict. Appl. Physics* IV. 694/2 A room considered to be right for speech may be just a little too *dead* for music. **1930** *Bell. Syst. Techn. Jrnl.* IX. 596 (*heading*) Reverberation Time in 'Dead' Rooms. *Ibid.*, With the advent of radio broadcasting and sound pictures very 'dead' rooms have been built. **1962** A. Nisbett *Technique Sound Studio* ii. 33 To do away with reverberation entirely and try to create entirely 'dead' studios. *Ibid.* 47 A section of the studio with almost completely dead acoustics, i.e. a 'dead-room'.

16. b. *slang.* Of a race-horse: not intended to win; fraudulently run in such a way that it cannot win; chiefly in *dead one, dead 'un.*

1864 *Baily's Mag.* June 121 A horse which has been regarded occasionally as a dead one has proved lively enough to beat the winner of the Two Thousand. **1868** *London Review* 11 July 38/2 The stable and owners might safely lay against what was technically a 'dead 'un' from the first. **1880** H. Smart *Social Sinners* v, Lord, what 'dead 'uns' he did back, to be sure! **1922** *N. & Q.* 12th Ser. XI. 206/2 *Dead meat.* Horses which are not out to win are so described.

c. Lacking resiliency or springiness; esp. of turf.

1870 *N.Y. Herald* 22 July 5/6 A dead ball was used, and again it was clearly demonstrated that this is the proper kind to play with. **1895** H. W. W. Wilberforce *Lawn Tennis* ix. 29 This form of game..arose from the very wet and dead state of the courts. **1909** P. A. Vaile *Mod. Golf* viii. 120 You will do well, should you have to choose [a driver] for yourself, to exercise moderation. Avoid too much spring. Don't have a 'dead' one. **1930** *Morning Post* 16 July 16/2 So well did Squires and Peach perform on the dead pitch that the Kent total of 317 was passed without the loss of another wicket.

18. b. Of a house: uninhabited. *slang.*

1879 J. W. Horsley in *Macm. Mag.* XL. 505/2 Me and the screwsman went to Gravesend, and I found a dead 'un (uninhabited house). **1896** A. Morrison *Child Jago* 231 On the look out for a dead 'un. **1922** *Daily Mail* 8 Aug. 2/2 We thought it was a 'dead' house, but we walked into a girl's room and she squealed.

c. *Mil.* Denoting an area which cannot be fired on from a particular point because of the nature of the ground, intervening obstacles, etc. (Cf. *dead angle* in D. 2.)

1899 *Westm. Gaz.* 9 Dec. 5/3 Besides the great advantage which we shall reap from the smashing power of these howitzers against field defences, we shall also find them most valuable to search out hollow or hidden ground 'dead' to other fire. **1900** *Daily News* 5 May 3/2 A high and rather steep hill, surrounded by a good deal of 'dead' ground. **1919** *Proc. Soc. Antiq. Scot.* LIII. 38 There is not a single piece of 'dead' ground in the whole fortress.

20. d. *Typogr.* That has been used or is no longer required, as copy after composition, or type ready for distribution or discarded.

a **1877** Knight *Dict. Mech.* I. 679/2 *Dead-letter*, type which has been used for printing, and is ready for distribution. *Dead matter.* **1898** J. Southward *Mod. Printing* I. xxiv. 154 The 'dead' letter..would, if of uniform face, constitute in itself a strong fount.

e. Of a cinema set: out of use. Cf. also quot. 1933.

1929 A. C. & C. B. Edington *Studio Murder Myst.* i. 7 The skeletons of 'dead' sets clothed in flowing veils of gray. **1933** P. Godfrey *Back-Stage* i. 20 Every stage accessory which becomes 'dead'—that is to say, which is not used again during the performance—must be cleared to below-stage.

21. (Illustrations of use in various games.) Cf. *Dead wood 1.

1875 *Encycl. Brit.* III. 407/1 (Baseball) A ball which hits the bat without being struck at, or the person of the striker or umpire, is a dead ball and out of play. **1876** *Ibid.* IV. 180/2 A 'dead bowl' is one knocked off the green, or against one lying in the ditch, or an illegally played bowl, and must at once be removed from the green. **1900** *Laws of Cricket* 4, 33a. If the ball, whether struck with the bat or not, lodges in a batsman's clothing, the ball shall become 'Dead'. **1902** *Encycl. Brit.* XXVIII. 426/2 So the game [*sc.* Rugby football] proceeds until the ball is once more 'dead'—that is, brought to a standstill. **1966** B. Johnston *Armchair Cricket* 97 A ball does *not* become dead when it strikes an umpire.

b. *Golf.* Of the ball: placed so near the hole that it can be holed with certainty at the next stroke. Also as *adv.*

1857 H. B. Farnie *Golfer's Manual* (1947) vii. 73 A ball is said to be dead..when it lies so close to the hole that the put is a certainty. **1881** R. Forgan *Golfer's Handbk.* iii. 30 If you can possibly win a 'half' [i.e. halve a hole] by running your ball 'dead' at the side of the hole,..then the cautious game is to be preferred. **1898** H. G. Hutchinson *Golf* (ed. 6) 83 Missing a four-inch put which your partner has left you..and receiving the cheery consolation, 'Never mind, partner, never mind—another time I'll try to lay you dead.' **1909** P. A. Vaile *Mod. Golf* v. 73 A man may lie 'dead' off a run up, but I am referring now to the well-lofted shot that falls 'plump' within an easy put of the hole and scarcely moves.

22. c. Of molten metal: thick and sluggish, either from insufficient melting or from having stood too long in a ladle. Cf. *Dead-melt *v.*

1884 W. H. Greenwood *Steel & Iron* xviii. 425 Too long exposure to the heat, or extreme 'dead-melting', produces a metal that runs dull and dead, affording ingots also of inferior quality.

d. Of ice: see quots.

1909 *Cent. Dict.* Suppl. I. 619/3 *Dead ice*, ancient ice retained in 'fossil glaciers' or elsewhere under the soil and not moving downward. **1937** Wooldridge & Morgan *Physical Basis Geogr.* xxii. 381 In Spitzbergen and elsewhere the ice has sometimes advanced over the low ground, but there has been no correspondingly rapid retreat. It has simply been left as 'dead ice', decaying by melting very slowly and without the production of large quantities of water. **1966** T. Armstrong et al. *Gloss. Snow & Ice* 13 *Dead ice*, any part of a glacier which has ceased to flow. Dead ice is usually covered with moraine.

24. c. *Cricket.* Of a bat: held in a defensive position with a slightly loose grip so that the ball strikes it and immediately drops to the ground.

1955 *Times* 13 July 8/6 But subsequently Bailey was simply Bailey, calm and unshakable, his whole defence built round the dead bat forward stroke. **1956** R. Alston *Test Commentary* 113 These days of dead-bat technique and over-cautious defence.

29. *dead load*, a load whose weight is constant and invariable; also *attrib.*

1888 in *Lockwood's Dict. Mech. Engin.* **1891** *Scribner's Mag.* X. 7 The greater engine-power will add to the dead load, thus still further diminishing the vessel's capability for carrying. **1930** *Engineering* 18 Apr. 503/2 To relieve the main girders of dead-load deflection and live-load stress. **1970** *Fremdsprachen* 43 The stratum..could be used for safe bearing pressures of 1,200 and 1,800 psf, for dead load and total load respectively.

30. *dead loss*: a complete loss; freq. *colloq.*, a person or thing that is totally worthless, inefficient, or unsuccessful; a complete failure; an utter waste of time. (Cf. quot. 1757 in Dict.)

1907 *Sears, Roebuck Catal.* 607/2 We seldom have two orders 'just alike' in every particular, consequently if the net was returned it would be a 'dead loss' to us. **1927** T. E. Lawrence *Let.* 27 Dec. in *To his Biographer, R. Graves* (1938) II. 144 This time it was a really good guard, and so I feel that the holiday has not been a dead loss. **1934** *Discovery* Nov. 317/2 Dead weight [on railways] means dead loss. **1951** 'J. Wyndham' *Day of Triffids* v. 82 Certain unmistakable *derniers cris*, some of them undoubtedly destined ..to become the rage of tomorrow: others, I would say, a dead loss from their very inception. **1956** D. M. Davin *Sullen Bell* 92 You think a dead loss like myself has no right to say it.

31. a. (Earlier examples of *dead certainty*.) Also *dead-earnest* in adjectival use.

1842 S. Kettell *Quozziana* 47, I saw, to a dead certainty, that if I should..be caught with my mouth open, I should be expected to say something. **1860** *Players* I. 154 'Done brown, to a dead certainty' said Buzzen to himself, as he went on eating. **1875** 'Mark Twain' in *Atlantic Monthly* Mar. 288/2 The grimmest and most dead-earnest of reading-matter. **1883** —— *Life Mississippi* xviii. 223 Ritchie's good-natured badgering was pretty nearly as aggravating as Brown's dead-earnest nagging.

b. *dead-on*: certain, unerring, exactly right (see quot. 1889).

1889 Barrère & Leland *Dict. Slang* I. 300/2 *Dead-on* (riflemen), straight on. A rifle-shot talks of the aiming being dead-on when the day is so calm that he can aim straight at the bull's eye instead of having to allow to the right or left for wind. He is said to be dead-on himself when he is shooting very well. **1959** *Punch* 17 June 815/1 She sang all night with pure, dead-on tone. **1966** 'K. Nicholson' *Hook, Line & Sinker* ix. 102 Don't you think a gesture like this is simply dead-on, when it comes to showing how with-it the Church is today?

d. *dead run*: a run at full speed without any let-up. *U.S.*

1889 K. Munroe *Golden Days* xii. 130 He..started on a dead run back over the trail. **1920** C. E. Mulford *J. Nelson* xii. 131 Striking into a dead run as he approached the rocky hump in the trail.

32. a. Also *dead and alive* (see Dead-alive *a.*); *dead and buried*; *dead and done* (*for, with*). All these phrases are also used *attrib.* (with hyphens). Hence **dead-and-goneness.**

1863 *All Year Round* IX. 473/1 The grave of Carthage, and other dead and buried cities of the Carthaginians. **1886** Baumann *Londinismen* 39/1 *Dead-and-done for*, rein futsch; *it had such a dead-and-done for look*, es sah so ganz erbärmlich aus. **1891** J. L. Kipling *Beast & Man in India* i. 7 Buddhism has been dead and done with in India proper for centuries. **1891** H. Herman *His Angel* ii. 40 The dead-and-goneness of emotional fervour. **1897** S. Erskine *Lord Dullborough* v, We..saw some six-months'-old playbills, announcing some dead-and-gone performance. **1909** *Westm. Gaz.* 10 Mar. 11/2 It is urged that the intellect of the Pollman is starved and himself broken on the wheel of a dead-and-done system. **1934** D. L. Sayers *Nine Tailors* II. iv. 158, I won't have you fretting yourself about that old business no more. All that's dead and buried. **1956** *Essays in Criticism* VI. 222 The dead-and-goneness of the past.

b. Also, (*as*) *dead as the* (or *a*) *dodo*, (*as*) *dead as mutton.*

1792 I. Bickerstaffe *Spoil'd Child* II. ii. 32 Thus let me seize my tender bit of lamb—there I think I had her as dead as mutton. **1838** [see Mutton 7]. **1904** H. O. Sturgis *Belchamber* iv. 51 The Radicalism of Mill..is as dead as the dodo. **1919** W. S. Maugham *Moon & Sixpence* ii. 10 Mr. Crabbe was as dead as mutton, but Mr. Crabbe continued to write moral stories in rhymed couplets. **1935** *Ann. Reg. 1934* II. 305 References appearing in the London newspapers to the effect that 'war debts are as dead as the Dodo' were cabled to the American press. **1960** *Guardian* 24 Mar. 11/1 Mr. Menzies..refused a request for a boycott..saying he had hoped this 'was as dead as a dodo'.

e. *to be dead on*: cf. Death *sb.* 16. *slang.*

1891 'S. C. Scrivener' *Fields & Cities* 22 These boys always were 'dead' on a rat, no matter what its size.

f. *to be dead nuts on*: see Nut *sb.*[1] 6.

g. Colloq. phr. (*I*, etc.) *wouldn't be seen* (or *found*) *dead in, with*: (I shall) have nothing to do with (something or someone); (I) hate, detest.

1915 Kipling *Debits & Credits* (1926) 29 'Wouldn't be found dead in *Hilarity*,' was Winchmore's grateful reply. **1931** T. R. G. Lyell *Slang, Phrase & Idiom* 671 No decent person would be seen dead with a specimen like that! **1933** A. G. Macdonell *England, their England* xiii. 222, I have to hang on to one [*sc.* a car] that my daughters say they wouldn't be seen dead in. **1937** M. Sharp *Nutmeg Tree* ix. 103 In the whole of France there wasn't a hat she would be seen dead in. **1966** A. E. Lindop *I start Counting* ix. 110 Do you think I'd be seen dead in gear like that?

h. Colloq. phr. *dead from the neck up*: brainless, stupid.

1930 J. Dos Passos *42nd Parallel* ii. 161 Most of the inhabitants were dead from the neck up. **1963** Wodehouse *Stiff Upper Lip* vi. 64 The sort of dead-from-the-neck-up dumb brick who wouldn't have thought of it.

B. 6. (Examples of *dead-meat*.)

1879 A. P. Vivian *Wand. in Western Land* 115 American dead meat can be delivered in perfect condition in English ports. *Ibid.*, The dead-meat trade is only in its infancy. **1897** *Westm. Gaz.* 18 May 2/3 One hideous monster was seen in the Park last week, puffing and rocking along and looking as much out of place as a dead-meat van in a Jubilee procession. **1908** *Ibid.* 22 July 4/3 If we are soon to get cheap beef and mutton it must be by developing the dead-meat trade.

7. *on the dead*: in dead earnest; honestly. *U.S. slang.*

1896 Ade *Artie* i. 7 On the dead, I don't believe any o' them people out there ever saw a good show. **1902** H. L. Wilson *Spenders* xxix. 340 Say, on the dead, Uncle Peter, I wish you'd come. **1903** A. H. Lewis *Boss* 184 But, on the dead! I'd like to learn how you..reconcile yourselves to things.

C. 2. (Further examples.) Esp. *dead broke* (see *Broke *ppl. a.* 3), *certain, easy, level, right, sure.* Now *colloq.*

a **1861** T. Winthrop *Canoe & Saddle* (1863) 280 Prairie-land lies dead level for leagues. **1871** J. Hay *Pike County Ballads* 10 He'd seen his duty a dead-sure thing. **1883** 'Mark Twain' *Life Mississippi* xxxix. 414 We'll cotton-seed his salad for him..that's a dead-certain thing. **1885** [see Go *v.* 48 e]. **1894** in E. R. Lamson *Yale Wit & Humor* 47 (*caption*) A Dead Easy Queen Caught His Eye. **1895** J. L. Williams *Princeton Stories* 166 You're dead right in saying he's too young. **1903** A. Bennett *Let.* 24 Aug. (1960) 96 She is dead right all through. **1904** W. H. Smith *Promoters* v. 92 For a dead easy mark in a business way, commend me to a preacher. **1908** G. H. Lorimer *J. Spurlock* i. 19 It was like having one of those mushy girls dead gone on you. **1922** D. H. Lawrence *Aaron's Rod* vii. 71 She liked him because of his dead-level indifference to his surroundings. **1923** 'B. M. Bower' *Parowan Bonanza* i. 15 'You're dead right, old girl,' Bill agreed. **1930** 'J. J. Connington' *Two Tickets Puzzle* xiv. 222 There's no great trouble in guessing who's mixed up in the business—that's dead easy. **1930** W. Gibson *Hazards* 12 He could always plane the deal Dead-level; ay, his work was always true. **1959** J. Braine *Vodi* i. 22 You're mardy. You're dead mardy. **1961** Simpson & Galton *Hancock* 43/2 Tony and Sid are dead bored. **1963** D. Lessing *Man & Two Women* 140 'That's right,' said Charlie, 'you're dead right.'

b. Slang phr. *dead to rights*: (*a*) completely, certainly; (*b*) red-handed; in the act; 'bang to rights' (*Bang *v.*[1] 10). (Cf. Right *sb.*[1] 14.) orig. *U.S.*

1859 G. W. Matsell *Vocabulum* 25 *Dead to rights*, positively guilty, and no way of getting clear. **1872** G. P. Burnham *Mem. U.S. Secret Service* p. v, A brief glossary of terms in the vernacular of criminals..*dead to rights*, caught with positive proof of guilt. **1881** *City Argus* (San Francisco) 2 July 4/4 Jimmy..was caught 'dead to rights', and now languishes in the city Bastile. **1889** Barrère & Leland *Dict. Slang* I. 301/1 I've got him dead to rights. **1947** 'A. A. Fair' *Fools die on Friday* 189 We've got her this time dead-to-rights.

D. 2. dead-ball line, in Rugby Football, a line behind the goal-line, beyond which the ball is considered 'dead' (sense 21 a); **dead-bird** (see quot. 1898); **dead-box,** a vehicle used for conveying dead bodies out of a mine; **dead-burned** or **-burnt** *a.*, of substances obtained by calcining refractory minerals such as gypsum or limestone: heated so strongly that vitrification occurred; of lime thus produced: that does not slake readily; **dead duck** *slang* (orig. *U.S.*), a person or thing that is useless, unsuccessful, bankrupt, etc.; **dead earth** *Electr.*, a complete or very low-resistance connection with the earth (see quots.); **dead finish** *Austral. colloq.*, (*a*) the 'limit' or extreme point (with regard to excellence, endurance, etc.); (*b*) any of several Australian trees or shrubs, esp. *Albizzia basaltica* or *Acacia farnesiana*; also, the thicket formed by such trees or shrubs; **dead-fold,** a sheep-pen; **dead furrow** *U.S.*, the last or finishing furrow left between 'lands' in ploughing; **Dead Heart** *Austral. colloq.*, the remote inland area of Australia; **dead horse** (see Horse *sb.* 18); **dead-house** (earlier example); (see also sense A. 18 b above); **dead leaf,** (*a*) the colour of a dead leaf; chiefly as *adj.*, = Feuillemorte *a.*; (*b*) *Aeronaut.* (see quot. 1918); **dead load,** (*a*) (see sense A. 29 above); (*b*) *pl.* (*U.S. colloq.*), great quantities; **dead marine** (see Marine *sb.* 4 d); **dead stick** *Aeronaut. colloq.* (orig. *U.S.*), (see quot. 1934); also *attrib.*, as **dead-stick landing,** a landing made with the engine 'dead'; **dead stock, deadstock,** (*a*) (see sense A. 20 b in Dict.); (*b*) (see Stock *sb.*[1] 53 a); **dead time,** (*a*) (see quot. 1909); (*b*) *Physics*, the period immediately after the registering of a pulse, a count, etc., when a detector or counter is not yet ready to register another pulse, etc.; **dead wagon** *U.S.*, a vehicle for conveying the dead; **dead white, dead-white,** (*a*) flat or lustreless white; (*b*) absolute white; pure white; also as *adj.*

1892 *Football Calendar 1892–93* 63 Not more than 25 yards behind the goal line, and parallel thereto, shall be lines, which shall be called the Dead-Ball Lines. **1905** *Westm. Gaz.* 30 Nov. 8/3 The necessity of lengthening the playing area to admit of the extensions behind the goals to the 'dead-ball' line. **1892** Stevenson & Osbourne *Wrecker* xxii. 349 Can't you give us 'a dead bird' for a good trade-room? **1898** Morris *Austral Eng.* 115/2 *Dead-bird*, in Australia, a recent slang term, meaning 'a certainty'. The metaphor is from pigeon-shooting, where the bird being let loose in front of a good shot is as good as dead. **1897** *Daily News* 12 May 5/7 He arrived at the pit's mouth in the dead-box, having fainted whilst below. **1939** *Iron & Steel Inst. First Rep. on Refractory Materials* 64 The addition of a small proportion of dead-burned magnesite to the chrome batch has been a common practice for many years. **1903** *Nature* 19 Nov. 64/2 Under favourable conditions gypsum actually breaks up at 63°·5, and forms insoluble anhydrite found in nature and identical with dead-burnt gypsum. **1904** Goodchild & Tweney *Technol. & Sci. Dict.* 151/1 *Dead burnt*, a term applied to lime which has become vitrified by fusion of calcium silicate in the limekiln. **1958** A. D. Merriman *Dict. Metall.* 54/1 This causes the lime to slake very slowly, and it is then referred to as 'dead burnt', in contradistinction to the pure lime. [**1829** *N.Y. Courier* 15 June 2/1 There is an old saying 'never waste powder on a dead duck'; but we cannot avoid flashing away a few grains upon an old friend, Henry Clay.] **1844** A. Jackson *Let.* 7 May in M. James *A. Jackson* (1937) xxiii. 481 Clay [is] a dead political duck. **1867** *New Mexican* 30 Mar. 2/2 The 'powerful' efforts of certain 'dead ducks' to prevent his appointment. **1888** *N.Y. Clipper* (Farmer), Long Branch is said to be a dead duck. **1958** 'A. Gilbert' *Death against Clock* 187 Once a chap's proved innocent..he's a dead duck to the Press. **1863** R. S. Culley *Handbk. Pract. Telegr.* vii. 105 Dead Earth. All the current passing through the fault ..No signal whatever beyond the fault. **1910** *Hawkins Electr. Dict.* 109/1 *Dead earth*, in telegraphy, a fault in the line involving a complete grounding or connection with the earth; a total earth. **1914** *Work* 26 Sept. 490/2 When cables are earthed intentionally the connection is complete, or a 'dead earth'. There are also 'partial earths' when a cable..leaks. **1881** A. C. Grant *Bush Life* xiv, 'He's the dead finish—go right through a man,' rejoins Sam, rather sulkily. **1885** H. Finch-Hatton *Advance Australia!* xvii. 272 On the western slopes, rose-wood, myall, dead-finish, plum-tree..all woods with a fine grain suitable for cabinet-making and fancy work. **1889** J. H. Maiden *Useful Native Plants* 355 *Acacia farnesiana*...Sometimes called by the absurd name of 'Dead Finish'. This name given to some species of *Acacia* and *Albizzia*, is on account of the trees or shrubs shooting thickly from the bottom, and forming an impenetrable barrier to the traveller, who is thus brought to a 'dead finish' (stop). **1902** J. H. M. Abbott *Tommy Cornstalk* 64 There are few colloquialisms more expressive of wearisome disgust, dissatisfaction and discontent than is 'Dead Finish'. It is almost synonymous with 'the Last Straw'. **1907** *Daily Chron.* 18 Mar. 4/4 There is a corporation which grows roses to compete with Nature's 'dead finish' trees. **1934** *Bulletin* (Sydney) 24 Jan. 21/3 They were made from myall, dead finish, ringed gidya and other fancy woods. **1959** C. & E. Chauvel *Walkabout* x. 69 If you go out that gate over there past the 'dead finish' tree and take the middle track you'll be right enough. **1897** L. Robinson *Wild Traits* vi. 168 A sudden change of diet from the frugal fare on the hill-turf and in the 'dead-fold' to that of lush cereals [etc.]. **1906** G. A. B. Dewar *Faery Year* 32 The dead-fold is formed of wattle hurdles bound about with swathes of straw. **1838** H. Colman *1st Rep. Agric. Mass.* 68 It [*sc.* the side hill plough]..avoids a dead furrow in the center. **1873** *Trans. Dep. Agric. Illinois* X. 94 The land between the rows should be plowed toward the trees, so as to have the 'dead furrow' in the center, to allow the water to pass off freely. **1894** *Irrigation Age* Jan. 34/2 With the discs straddling the dead furrow. **1906** J. W. Gregory (*title*) The dead heart of Australia. **1935** F. W. Jones in H. H. Finlayson *Red Centre* 8 That strange and undefinable attraction that the Dead Heart always has. **1945** *Salt* 2 July 23/1 Collective effort can radically alter the future of the so-called 'dead-heart' of Australia. **1812** J. J. Henry *Campaign against Quebec* 134 Many carioles..passed our dwelling loaded with the dead..to a place, emphatically, called the 'dead-house'. **1864** M. B. Chesnut *Diary* 27 May (1905) 311 Brushing scant locks which were fleecy white. Her maid would be doing hers, which were dead-leaf brown. **1896** *Daily News* 17 Oct. 6/5 A woollen skirt of a dead-leaf shade. **1905** *Westm. Gaz.* 21 Oct. 18/2 That delightful sort of golden browny shade that is really best described as dead leaf. *Ibid.*, This same peculiar dead-leaf colour. *Ibid.*, Some folds of dead-leaf-coloured crêpe de Chine. **1918** E. S. Farrow *Dict. Mil. Terms*, *Dead leaf*, in aviation, the term applied to an aircraft when its movement resembles that of a falling dead leaf. **1930** R. Lehmann *Note in Music* 35 The dead-leaf colour of the walls gave back a feeble reflection. **1869** 'Mark Twain' *Innoc. Abr.* lvii. 616 The old man's got dead loads of books. **1932** *Word Study* Jan. 3/2 The use of the phrase 'a dead stick' by some aviators. **1934** Webster, *Dead stick* (*Aviation*), a propeller that has ceased to revolve because the engine has stopped. —*dead-stick* adj. **1943** C. H. Ward-Jackson *It's a Piece of Cake* 24 *Dead stick*, engine stopped. **1946** B. Sutton *Jungle Pilot* 40 Poor Jimmy had had his motor stopped and was forced to make a 'dead stick' landing on the aerodrome. **1836** Dead stock [see Stock *sb.*[1] 53 a]. **1879** J. Scott *Farm Valuer* ix. 97 Interest is charged on the dead stock and the working cattle. **1958** *Times* 1 July p. i/7 His capital invested in livestock, deadstock and equipment. **1909** *Cent. Dict.* Suppl., *Dead time*, time during which the active work of accomplishing a purpose is not going on, although preparations for it may be in progress. Such, in pile-driving, is the time occupied in lifting the hammer. **1949** *Electronic Engin.* XXI. 455 There elapses about 10^{-4} sec., during which time a particle entering cannot initiate a count. This interval is the 'dead-time' for the counter. **1963** B. Fozard *Instrumentation Nucl. Reactors* v. 44 Geiger-Mueller counters have thus an inherent 'dead time' and are different in this respect from other types of ionisation detector. **1966** *Electronics* 17 Oct. 111 The sampling would be an integral part of the computer's program. It would occur many times during the wait or dead time of the tactical program. **1894** *Outing* (U.S.) XXIV. 7/1 Dead wagons, hospital ambulances and sanitary corps vehicles were the most prominent objects in the streets. **1794** R. Kirwan *Elem. Min.* (ed. 2) I. 327 Its colour white, two opposite faces silvery white, two others dead white, or yellowish. **1825** J. Nicholson *Oper. Mech.* 640 If it is to be finished flat, or, as the painters style it, dead white, grey, fawn, &c. **1857** G. Lawrence *Guy Liv.* xxx, The straight, beautifully-turned ankle, cased in dead-white silk. **1863** Mrs. H. Wood *Verner's Pride* xiv, The dead white of the roses was not more utterly colourless than Sibylla's face. **1920** R. Macaulay *Potterism* II. i. 67 Jane in a square-cut, high-waisted, dead white frock. **1922** D. H. Lawrence *England, My England* (1924) 110 She turned white—dead white.

dead-alive, *a.* Delete 'chiefly *U.S.*' before *dead-and-alive*, and add earlier and later examples.

1854 Thoreau *Walden* 166 He is dead-and-alive to begin with. **1901** 'M. Franklin' *My Brilliant Career* iii. 16 It's the dead-and-alivest hole I ever seen. **1908** E. J. Banfield *Confessions of Beachcomber* I. i. 9 The 'scrub fowl'..wastes no valuable time in the dead-and-alive duty of sitting. **1909** H. G. Wells *Tono-Bungay* I. ii. 67 This place.. gives me no Scope...It's dead-and-alive. Nothing happens. **1936** W. Holtby *South Riding* II. iii. 112 A bit dead-and-alive, isn't it? Other end of nowhere, eh?

dead beat, dead-beat, *sb.*[1] (*a.*) (Later examples.)

1927 *Motor Boat* 9 Sept. 226/3 The..Dead Beat compass ..returns after being displaced from its equilibrium position by one direct movement to the north pointing position. **1960** E. L. Delmar-Morgan *Cruising Yacht Equipt.* ii. 33 A light, dry card compass...The liquid-filled 'dead-beat' instrument has now taken its place.

dead beat, dead-beat, *sb.*[2] Add: (Earlier and later examples.) For '*U.S.*' read 'orig. *U.S.*' Also (orig. *Austral.*), a man down on his luck. Also *attrib.* Cf. *Beat *sb.*[1] 16.

1863 *Cornhill Mag.* Jan. 94 'Beau' Hickman [was] a professional pensioner, or, in the elegant phraseology of the place 'a deadbeat'. **1875** *Chicago Tribune* 13 Oct. 4/4 To go on a dead-beat spree. **1898** Morris *Austral Eng.* 115/2 *Deadbeat*. In Australia, it means a man 'down on his luck', 'stone-broke', beaten by fortune. **1902** W. Satchell *Land of Lost* iii. 18 This is the stranding-ground of the dead-beats of the world. **1909** Wodehouse *Mike* liv. 304 The Wrykyn team that summer was about the most hopeless gang of dead-beats that had ever made an exhibition of itself on the school grounds. **1912** E. Pugh *Harry the Cockney* xi. 114 He was a full private..attached to London's vast army of dead-beats..these miserable stricken creatures. **1958** *Spectator* 16 May 633/1 A company of British soldiers arrives in trucks, led by a deadbeat Temporary Major. **1959** 'J. Welcome' *Stop at Nothing* vi. 107 You don't want help from an old dead-beat like me. **1971** *Guardian* 18 Jan. 8/2 He didn't write me off as 'Oh, that dead-beat' when my name was mentioned.

dea·d-bea·t, *v. rare.* [f. the ppl. adj.] **1.** *trans.* To exhaust, wear out. *colloq.*

1868 M. E. Braddon *Run to Earth* II. ix. 169 If I ain't dead-beat, it's only because it isn't in circumstances to dead-beat me.

2. *U.S. slang.* **a.** *intr.* To sponge; to loaf. **b.** *trans.* To sponge on; to cheat.

1881 W. O. Stoddard *E. Hardery* 177 He's dead beated on you. **1888** *Boston Jrnl.* (Farmer), No party can dead-beat his way on me these hard times.

So **dead-bea·ten** *ppl. a.*, exhausted.

1875 J. G. Holland *Sevenoaks* i. 3 One by one—sick, disabled, discouraged, dead-beaten—they drifted into the poor-house. **1933** Masefield *Bird of Dawning* 220 He felt dead-beaten, yet saw endless things to be done.

dead-beatism. *U.S. slang.* [f. Dead beat *sb.*[2]] Worthlessness.

1869 J. H. Browne *Great Metropolis* 192, I have known men of fine talents..fall to the under plane of dead-beatism. **1882** *Congress. Rec.* 25 Jan. 615/1 [Are we] going to put a premium on judicial trumpery and dead-beatism?

dead-beatness. *rare.* [f. Dead-beat *a.* or *ppl. a.* + -ness.]

1. The quality or property of being aperiodic.

1898 E. J. Houston *Dict. Electr. Words* (ed. 4) App. B. 737/1 *Dead beatness*, possessing the property of aperiodicity. **1918** K. Edgcumbe *Industr. Electr. Measuring Instr.* (ed. 2) 154 A most important feature of all moving coil instruments lies in their dead-beatness.

2. Utter exhaustion.

1907 M. C. Harris *Tents of Wickedness* III. v. 299 The light came streaming in at the window that in the dead-beatness of last night everyone had forgotten to close.

deaden, *v.* Add: **4. e.** To make impervious to sound; = Deafen *v.* 3.

1901 R. Sturgis *Dict. Archit.* I. 751/1 *Deaden*,..to construct so as to be dead, in the sense of..impervious to sound, as a floor which has been made non-conducting. **1926** 'J. J. Connington' *Death at Swaythling Court* vii. 121 The kitchen is next the workshop and the walls are very badly deadened, so I could hear voices talking next door.

dead end, dea·d-e·nd. [Dead *a.* D. 2.] **1.** (See Dead *a.* D. 2.)

1886, 1889 [in Dict.]. **1960** D. & V. Nabokov tr. *V. Nabokov's Invit. to Beheading* xv. 147 Several times Cincinnatus found himself in a cul-de-sac, and then M'sieur Pierre would tug at his calves, making him back out of the dead end.

2. *fig.* Esp. a policy, course of action, etc., that leads nowhere; a 'blind alley'.

1922 M. Arlen *Piracy* III. xiv. 257, I felt..that there was a 'dead-end' at the end of my life. **1928** *Daily Tel.* 24 July 12/1 Young men..who are either working into a dead end or engaged in an industry that has a restricted future. **1934** *Discovery* Mar. 60/2 Once we came to a dead-end when we had no idea which way to turn. **1941** Auden *New Year Let.* III. 46 From the dead-ends of greed and sin.

b. *attrib.* or as *adj.* (*a*) That leads nowhere; having no possibilities for advance, promotion, etc.; (*b*) *dead-end kid*: a tough young person such as lives in back-streets or slums; also *transf.*

1928 *Observer* 15 Jan. 5 He deplores the fate of boys who get into dead-end employments. **1940** J. B. Priestley *Postscripts* 17 Overgrown, tormenting, cruel schoolboys—middle-aged 'dead end kids'. **1943** J. S. Huxley *TVA* xi. 86 No dead-end formula has killed the creative ability of the team. **1946** [see *dead-ending* below]. **1958** *New Statesman* 20 Dec. 869/2 It is not difficult to present France ..as Europe's Dead End Kid. **1963** *Economist* 20 Apr. 254/1 Workers shifting out of dead-end jobs and dead-end areas into 'approved' new spots.

3. *Electr.* (See quot. 1940.)

1922 Glazebrook *Dict. Appl. Physics* II. 670/1 A large increase in effective resistance also results from the attachment of a 'dead' coil having large mutual inductance to the section. **1925** P. J. Risdon *Crystal Receivers & Circuits* 10 The unused portion of the coil, although not directly in the circuit, is joined on to it, and produces an effect known as dead-end loss. **1940** *Chambers's Techn. Dict.* 226/1 *Dead end*, the unused portion of an inductance coil in an oscillatory circuit. *Dead-end effect*, the increase in effective resistance of an inductance coil due to currents circulating in the unused end-turns shunted by their self-capacity.

Hence **dead-e·nd** *v. trans.*, to bring to a dead end; *intr.*, to come to a dead end; **dead-e·ndedness,** the quality of leading nowhere or failing to advance; **dead-e·nding** *vbl. sb.* (see quot. 1946).

1921 *Blackw. Mag.* Nov. 641/2 Engineers..are not dead-ended so easily. **1946** A. Phelps *I couldn't care Less* x. 75 There were some..Americans who..made a habit of flying through the most appalling weather. We nicknamed them the 'Deadend Kids'...To this day the term 'Dead-ending' persists to describe flying through unusually bad weather. **1950** H. J. Massingham *Curious Trav.* iii. 53 The dead-endedness of modern life in which the rational faculty has overborne the poetic perception of reality. **1958** G. Usher *Death in Bag* xi. 105 The car..took an even narrower lane dead-ending in a concreted farmyard.

deadener. Add: **b.** *Logging.* (See quot.) *U.S.*

1905 *Terms Forestry & Logging* 34 *Deadener*, a heavy log or timber, with spikes set in the butt end, so fastened in a log slide that the logs passing under it come in contact with the spikes and have their speed retarded.

deadeningly (de·d'niŋli), *adv.* [f. Deadening *ppl. a.* + -ly[2].] In a deadening manner; so as to deaden.

1939 R. Murray *Good Pagan's Failure* iii. 150 The amazing material achievements to which we are almost deadeningly accustomed. **1962** I. Murdoch *Unofficial Rose* vii. 77 That honest simplicity..which made her for him so deadeningly structureless.

deadfall. Add: **1.** (Earlier U.S. examples.) Also *attrib.* with *trap.*

1829 *Massachusetts Spy* 8 July (Th.), In the act of getting in, the log or dead-fall fell upon his back and held him fast. **1843** Carlton *New Purchase* I. 2 We..were setting dead-falls and snares. **1902** S. E. White *Blazed Trail* xvii. 128 He had bound together..several of the oddly shaped pine timbers to form a species of dead-fall trap. **1959** J. D. Clark *Prehist. S. Afr.* vi. 132 The use of dead-fall traps.

b. *fig.* A trap, snare (cf. Trap *sb.*[1] 1 b).

1860 *Richmond Enquirer* 23 Nov. 1/8 (Th.), A continuance on the part of the Banks to issue specie would..catch us completely under the dead fall of Northern absorptive predominance. **1962** K. Orvis *Damned & Destroyed* xvi. 113 This rendezvous is a deadfall.

2. a. (Further examples.) Also *attrib.*

1898 *Westm. Gaz.* 12 Sept. 8/2 Travelling painfully over deadfall timber. **1968** R. M. Patterson *Finlay's River* 89 Then through deadfall and around small canyons they made their way down the west slope of the main range.

c. (Examples.) Also, see quots.

1837 A. Wetmore *Gaz. Missouri* 337 A small pot-house grocery or dead-fall of the village. **1867** *Harper's Mag.* June 131/1 In California, old Judge C— kept *a little dead fall*, as they call a rum-mill out there. **1903** A. Adams *Log Cowboy* xvi. 251 There's a deadfall down here on the river...that robs a man coming and going. **1937** D. Runyon *More than Somewhat* v. 100 The Bohemian Club is nothing but a deadfall.

dead-hand. Also **dead hand.** Add: **b.** *fig.* An oppressive and retarding influence. Cf. Mortmain sense c.

[**1871** *Scribner's Monthly* Nov. 19/1 The dead hand of Wesley has been stronger than the living hand of any pope.] **1935** *Discovery* Oct. 301/2 This cannot fairly be described as the 'dead hand' of the National Trust. **1955** *Times* 29 June 11/2 He would have fought the Government dead hand which fantastically enforces small papers ten years after the war. **1971** *Daily Tel.* (Colour Suppl.) 25 June 13/3 Eisenhower's dead hand on space was an obvious electoral issue for the two incoming presidential candidates to seize on.

2. *colloq.* An expert (*at* doing something).

1848 [see Dead *a.* A. 31 b]. **1862** G. O. Trevelyan *Interludes in Verse & Prose* (1905) 181 A young member of the Secretariat, a dead hand at a minute. **1888** 'R. Boldrewood' *Robbery under Arms* I. xv. 194 First-rate work it was, too; he was always a dead hand at splitting.

Hence **dead-ha·nded** *a.*, oppressively old-fashioned or out-dated.

1928 D. H. Lawrence *Lady Chatterley* xviii. 333 It was stupid, dead-handed higher authority that made the army dead.

deadhead, *sb.* Add: **3.** (Earlier U.S. examples.) Also *attrib.* and *transf.*

1841 *Spirit of Times* 23 Jan. 564/1 The house on Tuesday was filled as far as $300 could fill, barring 'the dead heads'. **1848** Bartlett *Dict. Amer.* App. s.v., Persons who drink at a bar, ride in an omnibus, or railroad car, travel in steamboats, or visit the theatre without charge, are called *dead heads*. **1863** *Rep. Maine Board Agric.* 15 The milch cow which barely pays the expense of keeping and care is a 'dead head' yielding no profit. **1869** W. H. Brewer *Rocky Mt. Lett.* (1930) 11 We had quite a train—some congressmen who have dead-head tickets over the road. **1887** C. B. George *40 Yrs. on Rail* ii. 32 Once in a while the conductor found it desirable to eject some would-be dead-head passenger while between stations. **1892** *Congress. Rec.* 31 May, App. 385/1 The free-delivery service is burdened by the collection and delivery of thousands of dead-head matter under the 'penalty-postage system'.

b. Used predicatively without article. *U.S. colloq.*

1874 'Mark Twain' & C. D. Warner *Gilded Age* xxx. 275 Senators and Representatives..always traveled 'dead-head' both ways. **1888** *Portland Transcript* 14 Mar. (Farmer), [Those letters] which had to do with the stage business and went dead-head.

c. A non-combatant accompanying a fighting force. *U.S. colloq.*

1864 T. C. Reynolds in J. N. Edwards *Shelby & his Men* (1867) xxvii. 470 The real fighting men did little injury, sneaks and dead-heads being the principal plunderers. **1867** J. N. Edwards *Shelby & his Men* xxi. 396 Accompanied by at least five hundred 'dead-heads', loafers, and amateur cavalry gentlemen.

d. A train, freight car, truck, etc., carrying no passengers or freight. Also *attrib. U.S. colloq.*

1938 L. M. Beebe *High Iron* 220 *Deadhead*,..empty passenger car. **1945** *Greeley* (Colorado) *Daily Tribune* 15 Dec. 1/7 The headon collision of a deadhead engine and a 37-car freight train. **1950** A. Lomax *Mr. Jelly Roll* 109 A deadhead, an empty mail car.

4. In full *dead head log*: a sunken or submerged log. *U.S.*

1902 S. E. White *Blazed Trail* lv. 380 He was enabled to catch the slanting end of a 'dead head' log whose lower end was jammed in the crib. **1905** *Terms Forestry & Logging* 34. **1907** *Black Cat* June 17 Numerous 'dead-heads' bobbed in the current like otters swimming with the stream.

5. A person who contributes nothing to an enterprise, activity, etc.; an unenterprising person.

1942 N. Streatfeild *I ordered Table* 215 It's an awfully sticky party this...I'm a deadhead to-night, but I can't help it. **1952** *Landfall* VI. 265 You can ignore him [God] too, as these Te Parenga deadheads do. **1964** *Economist* 4 Apr. 40/1 A 'dead head agency', a safe berth for unenterprising civil servants.

6. See *Dead *a.* A. 1 c.

dead-head, *v.* Add: **2.** *intr.* Of logs: to jam. *U.S. colloq.*

1922 H. Titus *Timber* viii. 79 Your hardwood will begin dead-heading in a hurry. *Ibid.* x. 89 If the raft goes to pieces and that one log dead-heads.

3. *intr.* To drive an empty train, truck, taxi, etc.; to travel in an empty vehicle. Also *trans. colloq.* (chiefly *U.S.*).

1911 *Daily Colonist* (Victoria, B.C.) 15 Apr. 3/3 Only O'Leary and the conductor..were in the car, which was deadheaded. **1929** *Folk-Say* I. 111 He [*sc.* a taxi-driver] has to deadhead all the way back. **1956** Wallis & Blair *Thunder Above* (1959) xi. 107 Kyle had flown up to Berlin.. as a check pilot and now had forty-eight hours before dead-heading back. **1962** K. Orvis *Damned & Destroyed* xii. 81 He hated to deadhead back an empty boat. **1970** *People* (Austral.) 26 Aug. 27/5 Another fireman was deadheading in the cab with us and he took over for me.

4. *trans.* To remove a dead flower or flowers from (a plant). Cf. *Deadheading *vbl. sb.* 2.

1952 C. E. L. Phillips *Small Garden* xi. 106 A great many herbaceous plants will go on and on if dead-headed. **1956** E. H. M. & P. A. Cox *Mod. Rhododendrons* 17 In a large collection,..it is impossible to dead head every plant. **1966** A. E. Lindop *I start Counting* vii. 96 The daffs were going off, and I dead-headed them and tied them down. **1970** C. Lloyd *Well-Tempered Garden* i. 48 One can distinguish broadly between those plants that are dead-headed with no further object than to tidy them up and those from which we are hoping to encourage another flowering.

dea·dheading, *vbl. sb.* [f. the vb.] **1.** *U.S. colloq.* = Deadheadism; also, the act or practice of being a deadhead.

1873 *Newton Kansan* 27 Feb. 3/4 Railroads occasionally complain of the deadheading, but no institution suffers so much from it as the press. **1903** G. C. Eggleston *First of Hoosiers* 263 Edward..objected..to all 'dead-heading' of the clergy, and to all 'discounts' made to preachers on the ground of their calling.

2. *Hort.* (See quot. 1954.)

1952 C. E. L. Phillips *Small Garden* vii. 62 The most elementary pruning is the 'dead-heading' of border flowers, by which we remove spent blooms in order to induce new ones. **1954** A. G. L. Hellyer *Encycl. Garden Work* 198/1 Faded flower heads may sometimes be removed...This..is often known as dead-heading. **1970** C. Lloyd *Well-Tempered Garden* i. 48 The first principle in dead-heading is always to cut back to something definite.

dead heat, *sb.* (Earlier examples.)

1796 *Sporting Mag.* VII. 260/2 The whole race was run *head* and *head*, terminating in a *dead heat*. **1823** 'J. Bee' *Dict. Turf* 94 'Dead heat', is when two winners come in nose to nose.

dead-heat, *v. intr.* Add: Const. *with.*

1902 *Daily Chron.* 21 May 3/5 Hitherto the London and North-Western have deliberately 'dawdled' over the thirty miles after Crewe, so as to only 'dead-heat' with their competitors. **1922** *Daily Mail* 22 Nov. 7 Chuck-a-Penny distinguished himself..by dead-heating with Eton and dividing the spoils.

de-adjectival (di,ædʒĕktəi·văl), *a.* [f. De- + Adjectival *a.*] Derived from an adjective.

1934 Priebsch & Collinson *German Lang.* iii. 225 De-adjectival verbs like *füllen* from *voll*, *heilen* from *heil*. **1962** B. M. H. Strang *Mod. Eng. Structure* vi. 97 Words that I shall call de-adjectival class nouns..are homonymous with adjectives.

dead letter. Add: **1. c.** *transf.* and *fig.* (See quots.)

1864 Hotten *Slang Dict.* 118 *Dead-letter*, an action of no value or weight; an article, owing to some mistake in its production, rendered utterly valueless. **1926** Fowler *Mod. Eng. Usage* 104/1 *Dead letter*..; the application of it [*sc.* the phrase] to what was never a regulation but has gone or is going out of use, as quill pens, horse-traction, amateur football, &c., or to a regulation that loses its force only by actual abolition (*the one-sex franchise will soon be a d. l.*), is a slipshod extension.

2. (Earlier examples.)

1703 in *Mass. Hist. Soc. Coll.* (1838) 3rd Ser. VII. 62 The other penny is lost in dead letters (remaining in the several Post Offices). **1737** *London Mag.* Jan. 54/1 John Jesse, Esq; Deputy Secretary of the Post-Office, succeeds the late Mr Williamson as Deputy Cashier; as does Mr John Barber, as Inspector of dead letters at the said Office. **1812** M. Edgeworth *Absentee* in *Tales Fash. Life* VI. xvii. 423 The letter went coursing after you...I took it for granted that it found its way to the dead-letter office.

3. *Typogr.* See *Dead *a.* A. 20 d.

dead-line. Add: **2.** (Earlier example.) orig. *U.S.*

1864 in *Congress. Rec.* 12 Jan. (1876) 384/1 The 'dead line', beyond which the prisoners are not allowed to pass.

b. *Printing.* A guide-line marked on the bed of a printing-press.

1917 F. S. Henry *Printing for School & Shop* xi. 183 If the chase is one that just fits the bed of the press, make certain that the type does not come outside of the dead-line on the press.

c. = Time-limit; esp. a time by which material has to be ready for inclusion in a particular issue of a publication. orig. *U.S.*

1920 *Chicago Herald & Examiner* 2 Jan. 10/4 Corinne Griffith..is working on 'Deadline at Eleven', the newspaper play. **1929** *Publishers' Weekly* 27 July 349 Deadline for *Poetry*'s $250 prize poem contest is September 1. **1948** *Daily Tel.* 31 May 6/5 The Security Council will not meet again until Wednesday, about 20 hours after the dead-line. **1958** *Woman's Jrnl.* Feb. 96/1 We wait till midnight... That's the deadline.

deadliness. Add: **2. b.** Dead accuracy. Cf. *Deadly *a.* 8 b.

1905 *Daily Chron.* 24 Aug. 6/4 A lack of deadliness in approaching.

deadlock, *v.* Add: *intr.* To come to a deadlock.

1903 *N.Y. Even. Post* 25 Nov. 6 The Legislature would have deadlocked over the vote had not Gov. Odell come to his rescue.

So **dea·dlocked** *ppl. a.*, brought to a deadlock.

1880 [see Dead-lock *v.*]. **1897** *Rev. Reviews* (N.Y.) Jan. 10/2 Mr. Cleveland makes it perfectly plain that the struggle going on in Cuba is a useless and ruinous one,—a *deadlocked* situation. **1931** Buck & Anthony *Bring 'em back Alive* 297 For several seconds we remained deadlocked, the animal making a perfect bedlam of the mess-room with his cries of rage. **1968** *Guardian* 24 Oct. 9/5 The deadlocked Vietnam peace talks.

deadly, *a.* Add: **8. b.** Characterized by dead accuracy.

1909 P. A. Vaile *Mod. Golf* v. 95 The peculiarity of the stymie stroke, played parallel with the ground, is its deadly direction.

dead man. Add: **3. b.** *dial.* or *slang.* A scarecrow.

1839 in G. C. Lewis *Gloss. Words in Herefordshire* 31. **1889** Barrère & Leland *Dict. Slang* I. 299/2 *Dead man*,.. (Popular), a scarecrow; a man made of rags.

4. (Example.)

1825 W. N. Glascock *Naval Sketch-Bk.* (1826) I. 11 Why don't they tuck-in those '*dead-men*' out of sight!

5. *techn.* Any of various objects buried in or secured to the ground and used as an anchorage or leverage (see quots.).

1840 W. T. Spurdens *Vocab. E. Anglia* (1858), *Deadman*, a piece of timber buried in the earth, to secure posts, or other timbers by. **1901** *Daily Colonist* (Victoria, B.C.) 15 Oct. 5/4 A deck hand..was killed by being struck on the head by a 'dead man', which is a post imbedded on a [river gravel] bar to haul the steamer over. **1930** *Engineering* 28 Nov. 667/3 The sand..was levelled by the scraper pulled along by tackle attached to a dead man. **1940** *Chambers's Techn. Dict.* 226/2 *Deadman*, the concrete, plate, or other anchorage for land ties. **1950** *N. Z. Jrnl. Agric.* Jan. 61/3 Where it is difficult to obtain ground hold for fence strainer posts, such as in swamps or rocky hill country, fences may be strained with a 'deadman', and the strainer post should be cast [of concrete] with wire holes suitably placed.

b. *Logging.* (See quot.)

1905 *Terms Forestry & Logging* 34 *Deadman*, a fallen tree on the shore, or a timber to which the hawser of a boom is attached.

dead men's bells (earlier example); also *dead man's bells.*

1818 *Edin. Mag.* Oct. 328/1 But dinnae pu' the dead men's bells. **1878** Britten & Holland *Dict. Eng. Plant-Names* 34 Bell, or Bells, Dead Man's, or Men's. **1960** Ary & Gregory *Oxf. Bk. Wild Flowers* 122/2 Other names for this common plant..are Dead Man's Bells (the plant is poisonous) and Fairy Thimbles.

dead man's handle. In an electric train, a controlling handle which must be held in position for the current to pass, so that the train is automatically brought to a standstill should the driver release his grasp through illness or accident. So *dead man's device, knob, pedal, treadle.* Also *transf.*

1908 *Internat. Libr. Technol. Electr.-Railway..Multiple-Unit Syst.* 11 Should an accident befall the motorman, causing him to release knob 14, the train will be automatically stopped. On this account knob 14 is sometimes called the dead man's knob, or handle. **1924** *Westm. Gaz.* 2 Dec., The Dead Man's handle..is a safety device for pulling up Underground trains. **1951** *Engineering* 13 Feb. 231/2 The usual dead-man's pedal..can be operated from both sides of the cab. **1958** *Times Rev. Industry* Sept. 48/2 The throttle is arranged as a 'dead-man's handle', and it is impossible to start the bus until the sliding doors have been closed. **1964** *Daily Tel.* 18 Feb. 20 The dead man's treadle.. should automatically stop a train if the driver becomes unconscious. *Ibid.*, A deadman's device.

dead-melt (de·dmelt), *v. trans.* To keep (metal) at a melting temperature until it is perfectly fluid and no more gas is evolved. Hence **dead-melted** *ppl. a.*, **dead-melting** *vbl. sb.*

1880 *Encycl. Brit.* XIII. 341/2 If cast immediately it is found that a much larger quantity of gas separates during solidification, rendering the steel porous, than is evolved if the metal is dead-melted, *i.e.*, allowed to remain melted for an extra half hour or more. *Ibid.*, The 'dead melting' effect of the extra time allowed in fusing steel for the molten metal to stand in the furnace after fusion is brought about is due [etc.]. **1884** [see *Dead *a.* A. 22 c]. **1952** C. A. Edwards *Struct. & Prop. Mild Steel* i. 15 Towards the end of the operation of making steel by the acid process..the oxygen content of the metal will be extremely low... A charge of this kind is known as 'dead melted' steel, which means that..solidification proceeds without the liberation of gas.

So **dead melt** *sb.*, the state of a melt when it contains no solid and is evolving no gas.

1929 W. Lister *Pract. Steelmaking* viii. 48 When the whole charge has come to a dead melt, but not before, commence to feed.

dead-oh (de·dōu), *a. Naut. colloq.* [f. Dead *a.* + Oh *int.*] **a.** = Dead drunk. **b.** Deeply asleep.

1889 Barrère & Leland *Dict. Slang* I. 300/1 *Dead-oh!* (naval), is said of a man in the last stage of intoxication. **1924** Masefield *Sard Harker* 160 He slept, as sailors say, 'dead-oh'.

dead-pan (de·dpæn), *a.*, *sb.*, *adv.*, and *v.* orig. *U.S.* Also **dead pan, deadpan.** [f. Dead *a.* + *Pan *sb.*[1] 6 c.]

A. *adj.* Of a face, look, etc.: expressionless, impassive. Of a person: having such a face. Also *transf.*, applied to speech, behaviour, etc.: detached, impersonal.

1928 *N.Y. Times* 11 Mar. VIII. 6/1 *Dead pan*, playing a rôle with expressionless face. **1929** *Variety* 17 Apr. 51/3 They clicked better at the Palace where the intimacy heightened the dead-pan comic's expression. **1939** I. Baird *Waste Heritage* i. 5 Matt's eyes lost their cold deadpan look. **1942** *Tablet* 19 Sept. 135/2 Mr. Attlee and Sir Stafford Cripps did their best to assume what in America is called a 'dead-pan' expression. **1947** *People* 22 June 2/6 Perhaps that accounts for Flynn's dead-pan immobility, his flat emotionless voice, and the air of bored indifference. **1949** *Times Lit. Suppl.* 5 Aug. 502/3 An official career far more eventful than most, and one oddly belying his mild demeanour and dead-pan wit. **1950** *Manch. Guardian Weekly* 23 Mar. 3/2 A dispatch from Washington reporting in dead-pan outline Mr. Acheson's Berkeley speech. **1953** L. A. G. Strong *Personal Remarks* 256 Malcolm Scott was severe, suety, impassive; what nowadays would be called dead-pan. **1957** *Sunday Times* 3 Mar. 3/3 For what is known as 'dead pan' humour no one can challenge Miss Jean Mann, whose facial expression gives no warning of the thrust to come. **1965** *Listener* 20 May 745/3 The superbly dead-pan warning: 'The publication of this book does not directly or indirectly imply that it can be regarded as authorized for use in churches.'

B. *sb.* An expressionless or impassive face, esp. one deliberately assumed; also, a person with such a face; the assumption of such a face.

Freq. treated as two separate words.

1933 'N. West' *Miss Lonelyhearts* (1949) 34 He practiced a trick used much by moving-picture comedians—the dead pan. No matter how fantastic or excited his speech, he never changed his expression. **1937** E. Linklater *Juan in China* xxi. 282, I told him it wasn't his..and all I got was a dead pan. **1939** N. Coward *Words & Music* II, in *Second Play Parade* 159 Now the wife of an Acrobat Is the 'Dead Pan' of the troupe. **1943** P. Cheyney *You can always Duck* iii. 53 This bar-tender is an interestin' sorta guy. A dead pan. Nothin' seems to worry him. **1951** E. Hyams *Sylvester* xx. 96 Eyes..set in round, flabby..faces, the dead-pans of a caste of men who..had given up their humanity. **1957** J. Braine *Room at Top* iv. 34 The millhand with the Alan Ladd deadpan.

C. *adv.* With a dead-pan face; in a dead-pan manner.

1933 Runyon in *Collier's* 28 Oct. 36/3 She does not scowl or anything else, but only looks very dead-pan. **1944** 'P. Quentin' *Puzzle for Puppets* xviii. 139 He said that completely dead-pan. For a moment I didn't take it in. Then I grinned. **1962** *New Yorker* 10 Mar. 157/1, 3 claims no authority and merely records, mostly deadpan, what in fact every Tom, Dick, and Harry is now doing..to the language.

D. *v. trans.* and *intr.* To speak, perform, behave, etc., with a dead-pan face or in a dead-pan manner.

1942 Berrey & Van den Bark *Amer. Thes. Slang* § 275/2 *Dead-pan*, to maintain an expressionless face. **1959** N. Marsh *False Scent* (1960) ii. 40 'After this,' she said slowly, dead-panning her voice to a tortured monotone, 'there is only one thing for me to do.' **1962** W. Knox *Little Drops of Blood* iv. 87 Moss dead-panned. 'I don't mind. Of course, I'll need to phone Miss Murdoch and disappoint her, but ——' he winked.

dead reckoning. Add: Also *Aeronaut.*

1917 Bosanquet & Campbell *Navigation for Aerial Navigators* i. 4 In aerial navigation..Dead Reckoning is the position arrived at as calculated from the estimated track and the estimated speed made good over the ground. *Ibid.* 5 These data enable us to find a Dead Reckoning position. **1935** C. G. Burge *Compl. Bk. Aviation* 477/1 Dead reckoning ..is a compromise between pilotage and navigation.

Dead Sea. b. (Earlier example.)

1817 Moore *Lalla Rookh* 222 Like Dead Sea fruits, that tempt the eye, But turn to ashes on the lips!

dead water. Add: **4.** (See quot.)

1904 Goodchild & Tweney *Technol. & Sci. Dict.* 151/2 *Dead water*, water which does not come into contact with the effective heating surface of a boiler.

dead weight. Add: **4.** *attrib.*, as **dead-weight debt,** a debt not covered by assets, such as the greater part of the British National Debt; **dead-weight (safety-)valve,** a safety-valve kept down by a heavy weight.

1827 [see sense 3 in Dict.]. **1894** *Westm. Gaz.* 7 May 3/1 Dead-weight expenses have almost reached the irreducible minimum. **1902** *Encycl. Brit.* XXXIII. 373/1 *Deadweight capacity*..is..the amount of deadweight which can be carried on the holds at load draught when the vessel is fully charged with coals and stores. **1904** Goodchild & Tweney *Technol. & Sci. Dict.* 151/2 Dead weight safety valve. **1905** *Daily Chron.* 16 May 4/4 There is dead-weight debt, and there is remunerative debt. **1909** *Ibid.* 29 Apr. 4/4 Having brought the dead-weight Debt down to the total at which it stood twenty years ago. **1927** T. Woodhouse *Artificial Silk* 16 A dead-weight safety valve. **1930** *Engineering* 10 Oct. 461/2 Should the steam stop valve on the boiler be closed..the deadweight valve is opened.

dead wood. Add: **1.** (Earlier and later examples.) Also (*U.S.*) in tenpins, a pin that has been knocked down and lies in the alley in front of those remaining.

1851 L. Clappe *Shirley Lett.* (1922) 84 If they ask a man an embarrassing question, or in any way have placed him in an equivocal position, they will triumphantly declare that they have 'got the dead-wood on him'. **1858** *Southern Lit. Messenger* XXVII. 351/1 He..sent his ball..straight to the left quarter of the Centre-Pin, and never left any dead wood on the alley. **1867** A. D. Richardson *Beyond Miss.* xi. 134 'The deadwood'—from the game of 'tenpins', in which a fallen pin sometimes lies in front of the standing ones so that the first ball striking it will sweep the alley. 'I have the dead wood on him' was used familiarly, meaning: 'I have him in my power.' **1947** *Time* 17 Mar. p. i, An automatic bowling pin spotter that sets up pins..removes dead wood..and even calls fouls! **1951** E. S. Gardner *Case of Borrowed Brunette* xix. 207 Well, they've evidently got the dead-wood on you now, Perry. They know that you took Eva Martell to that rooming-house.

b. *dead-wood fence*: a fence made of rough logs, fallen branches, etc. Chiefly *Austral.*

1813 J. Taylor *Arator* 208 They [*sc.* hogs] are the cause of dead wood fences, which render..labour unproductive. **1852** Mrs. L. A. Meredith *My Home in Tasmania* I. x. 157 A 'dead-wood fence', that is, a mass of timber four or five feet thick, and five or six high, the lower part being formed of..logs..and the upper portion consisting of the smaller branches skilfully laid over. **1959** in S. J. Baker *Drum* II. 103.

3. *fig.* A person or thing regarded as useless or unprofitable; a hindrance or impediment. Also *attrib.* orig. *U.S.*

1887 *Sci. Amer.* 1 Oct. 209/1 The commissioner [of patents] has made some effort..to cut the deadwood out of the examining and clerical forces left him as a legacy by his predecessor. **1903** *McClure's Mag.* July 326 No dead wood is taken into the concern unless it is through the supposed necessities of family or business relations. **1928** *Daily Express* 11 Aug. 9/5 These papers do not receive any advertising support from us unless they make a price which we consider is adequate when you cut out their dead-wood circulation. **1929** *Daily Tel.* 15 Jan. 6/3 Amalgamations of what have hitherto been competing concerns are being formed, the specific objects being to cut out any dead-wood which may be handicapping the smooth working of the machine, to promote efficiency by the pooling of brains and experience.

dead-work. 2. (Earlier and later examples.)

1839 *Penny Cycl.* XV. 246/1 All the underground work of mines in Cornwall..is of two distinct kinds: dead work, or that carried on in the rock or metalliferous deposit, for the purpose of trial and discovery; and productive labour. **1954** S. Piggott *Neolithic Cultures* ii. 39 The flint seam was at such a depth that a considerable amount of 'dead work' had to be done before the flint was reached.

de-a·erate, *v.* [f. De- + Aerate *v.*] *trans.* To remove air from. So **de-aerated** *ppl. a.*; **de-aerating** *vbl. sb.* and *ppl. a.*; **de-aeration, de-aerator.**

1791, 1830, 1878 [see De- II. 1]. **1919** H. G. Wells *Outl. Hist.* (1920) I. 17/2 The water has become deaerated and foul. **1940** *Chambers's Techn. Dict.* 227/1 *De-aerator*, a vessel in which boiler feed water is heated under reduced pressure in order to remove dissolved air. **1960** *Times* 12 Feb. 3/1 Condensing, Feed Heating and De-aerating plant.

dea ex machina: see *Deus ex machina.

deaf, *a.* Add: **7. deaf-aid,** a hearing aid; **deaf-ear,** (*c*) the ear-lobe of the domestic fowl.
1934 *Discovery* Nov. 324/2 The combined radio-gramophone and deaf-aid. 1939 *Nature* 15 Apr. 633/1 An efficient type of deaf-aid is operated by a pick-up coil, in which are induced currents from an energized cable hung round the auditorium or placed under the carpet. 1854 *Poultry Chron.* I. 225 The cock..should have large wattles, and a clear white deaf-ear. 1855 *Ibid.* III. 443 The importance of white deaf-ears seems however to have been overlooked by some of the competitors.

deafish, *a.* (Later examples.)
1899 G. MEREDITH *Let.* 11 July (1970) III. 1332, I am daily more and more a cripple, and am deafish. 1907 ZANGWILL *Jewish Trinity* 5 Unfortunately, the Mayoress of Middleton was deafish.

deal, *sb.*[2] **3.** Delete †(*obs.*). This sense (= HAND *sb.* 23 c) is still common.
4. b. (Earlier and later examples.) Now in gen. English use.
1863 J. SHERMAN in R. S. Thorndike *Sherman Lett.* (1894) 205 The war is prolonged, but little chance of its ending until we have a new deal. 1928 *Manch. Guardian Weekly* 10 Aug. 102/2 We are..dependent on Parisian sources for information about our latest deal with the French. *Ibid.* 104/1 [He] tries to explain the attitude of British Liberalism towards the naval deal. 1931 H. F. PRINGLE *T. Roosevelt* I. vi. 66 There had been deals and counterdeals between Tammany and anti-Tammany Democrats.
c. (*a*) *bad, raw* or *rough deal*, harsh or unfair treatment, swindling; (*a*) *fair* or *square deal*, equitable treatment, fair dealing. *colloq.* (orig. *U.S.*).
[1838 T. C. HALIBURTON *Clockmaker* 2nd Ser. 266 Six dollars apiece for the pictur's is about the fair deal for the price.] 1876 W. G. NASH *New England Life* ii. 30 That was a square deal, Mis Brown. 1912 J. SANDILANDS *Western Canadian Dict.* 37 *Raw deal*, a bare-faced swindle. 1927 LADY ASTOR in *Daily Tel.* 15 Nov. 9/3 Although we have got the vote we women have a long way to go before we get a positive square deal. 1928 *Daily Chron.* 9 Aug. 6/2 The men have had a fair deal. 1931 *Week-end Rev.* 18 Apr. 580/1, I do not believe that in ordinary life Martin would have had such a rough deal. 1938 E. BOWEN *Death of Heart* I. viii. 145 No outside people deserve the bad deal they get from love. 1940 E. C. BENTLEY *Those Days* viii. 237 The Opposition were quite content with this situation. If it was what is known nowadays as a raw deal, they did not mind. 1958 *New Statesman* 22 Feb. 219/1 The government in Djakarta refused to give Sumatra a fair deal.
d. *new deal, New Deal*, a new arrangement with a view to reform and betterment; *spec.* the programme of social and economic reform in the United States of America planned by the Roosevelt administration of 1932 onwards. Also *transf.* and *attrib.* Hence *new dealer, New Dealer*, one who advocates or supports a 'new deal'.
1834 in M. James *Andrew Jackson, Portrait of Pres.* (1937) II. xvii. 376 A new bank and a New Deal. 1863 [see sense 4 b above]. 1876 *Chicago Tribune* 7 Aug. 1/6 The 'New-Dealers' appointed a committee to confer with the old State Central Committee. 1909 H. JAMES *Princess Casamassima* (N.Y. ed.) II. xxxiv. 199 I'm one of those who believe that a great new deal is destined to take place. 1932 F. D. ROOSEVELT in *N.Y. Times* 3 July I. 8/7, I pledge you—I pledge myself—to a new deal for the American people. 1934 *Amer. Mercury* June 246/2 Fifty New Dealers who are helping the President make the country laugh. 1940 H. G. WELLS *New World Order* 61 The New Deal is plainly an attempt to achieve a working socialism and avert a social collapse in America. 1943 J. S. HUXLEY *TVA* 7 The Tennessee Valley Authority Act was one of the earliest New Deal measures, having been passed in May 1933—less than three months after Roosevelt took office. 1955 *Times* 3 Aug. 9/6 He ousts the 'new dealer', Malenkov, and re-directs Russian economy towards heavy industry. 1965 *Listener* 24 June 947/3 As a member of the New Deal team his position was in the highest degree anomalous.
e. *big deal*, an important business transaction; also *transf.*, something important, exciting, or satisfying; freq. used as an ironical exclamation. orig. *U.S.*
1928 Z. GREY *Nevada* xvi. 279 'Are you open to a big deal?' queried the rustler bluntly. 1949 A. MILLER *Death of Salesman* II. 100 My brother..pulled off a big deal today. I think we're going into business together. 1951 J. D. SALINGER *Catcher in Rye* i. 6 The game with Saxon Hall was supposed to be a very big deal around Pencey. *Ibid.* xviii. 163 The whole bunch of them—*thousands* of them—singing 'Come All Ye Faithful!' like mad. Big deal. It's supposed to be religious as hell. 1953 POHL & KORNBLUTH *Space Merch.* (1955) vii. 77 My bunk was all mine, twenty-four hours a day. Big deal. 1966 'S. HARVESTER' *Treacherous Road* ii. 22 So, I can charge an evening's entertainment to business expenses. Oh, big deal. Carry on.

deal, *v.* Add: **7. b.** To include (someone) *in* those to whom one deals cards for a game; freq. *fig.*, to include (a person) *in* an undertaking; to give (someone) a share or part. *colloq.* (orig. *U.S.*).
1942 BERRY & VAN DEN BARK *Amer. Thes. Slang* § 369/4 Give a share,..deal one in (on). 1965 J. M. CAIN *Magician's Wife* (1966) xiii. 101 You know everything, and yet you dealt yourself in. 1969 D. BAGLEY *Spoilers* iii. 72 He was on my original list, but he dealt himself in regardless and it would be too risky to leave him out now. 1969 W. GARNER *Us or Them War* xvi. 121 If they won't deal us in we may end up having to steal it.

dealer. Add: **5.** A jobber on the Stock Exchange.
1719 DEFOE *Anat. Exchange-Alley* 4 A young Dealer that has Money to lay out. 1837 *Penny Mag.* VI. 186/2 Dealers in bills purchase them either to get a commission, or in return for goods exported. 1870 *Gentl. Mag.* New Ser. V. 484 The dealers were almost unable to sell stock of any kind. 1890 *Cassell's Sat. Jrnl.* June 724/3 A jobber was engaged in 'banging' the market...Another dealer saw through the trick. 1902 *Encycl. Brit.* XXV. 23/2 Much of the work of the Stock Exchange account is carried out by a department of that institution corresponding to the bankers' clearing house. Its function is to bring into direct communication the ultimate buyer and the ultimate seller as represented by their respective brokers, thus eliminating, for the purposes of the settlement, the middleman known as the 'dealer' or 'jobber'. 1907 *Westm. Gaz.* 25 Mar. 9/3 Those dozens of other dealers are on the look-out for orders from brokers whose clientèle lies amongst the great body of the public. 1970 *Encycl. Brit.* XXI. 259/1 Dealers or jobbers in these securities take up their posts in this space every day, and dealers in other classes of securities similarly occupy the floor space allotted to their particular markets.

dealerdom (dī·lǝɹdǝm). [f. DEALER + -DOM.] The sphere or influence of a dealer or dealers; dealers collectively.
1921 W. DE MORGAN *Old Man's Youth* xxix. 287 Men have been so often taken at their own valuation, and have been worked up by dealerdom. 1963 *Spectator* 29 Mar. 393 The new Establishment of officialdom and dealerdom.

dealership (dī·lǝɹʃip). [f. DEALER + -SHIP.] The position, business, or privileges of a dealer; an authorized trading establishment.
1916 W. P. WERHEIM in *Sales Promotion by Mail* viii. 295 Getting more business, establishing more agencies and dealerships. 1964 *Economist* 24 Oct. 384/2 A row of dealerships along the same stretch of main road. 1967 *Boston Sunday Herald* 26 Mar. I. 45/2 (Advt.), Right now our dealership has openings for new and used car men. 1970 *Daily Tel.* (Colour Suppl.) 2 Oct. 9 A certain proportion became interested in real estate, car dealerships, restaurants, and so forth.

dealing, *vbl. sb.* Add: **5.** attrib., *dealing-book* (Stock Exch.).
1899 *Westm. Gaz.* 19 July 8/2 Dealing-books that have been innocent for weeks of more than two or three bargains a day have been filling rapidly. 1907 *Ibid.* 25 Mar. 9/3 There..is one man very much absorbed in his dealing-book.

dealkylation (di,ælkilēⁱ·ʃǝn). *Chem.* [f. DE- + *ALKYLATION.] The removal of an alkyl group or groups. So **dea·lkylate** *v. trans.*; **dea·lkylated** *ppl. a.*
1921 *Chem. Abstr.* XV. 865 Toluene is produced from xylene by the action of H..in the presence of Fe_2O_3 as a catalyst. Other similar dealkylations may be effected by the same method. 1937 *Jrnl. Amer. Chem. Soc.* LIX. 1418 Di-t-butylbenzene is dealkylated in the presence of phosphoric acid and benzene. 1946 *Nature* 30 Nov. 800/1 A broad programme on alkylation and dealkylation. 1968 S. PATAI *Chem. Amino Group* vi. 332 The formation of the major products—dealkylated amine and formaldehyde.

deamination (di,æminēⁱ·ʃǝn). [f. DE- + AMIN(E + -ATION.] The removal of an amino group or groups. Hence **dea·minate** *v. trans.*; **dea·minating** *ppl. a.*
1912 *Chem. Abstr.* VI. 1912 Deamination. Emulsions of liver..when incubated with asparagine, glycocoll or leucine, liberate NH_3. 1926 *Jrnl. Biol. Chem.* LXX. 140 Tyrosine is also deaminated during the fermentation. *Ibid.* 147 The bacillus is able to deaminate tyrosine. 1927 *Glasgow Herald* 4 June 4/2 The exogenous metabolism by which amino-acids are de-aminated. *Ibid.*, The preliminary process of splitting off the ammonia and forming a fatty acid is called de-amination. 1951 M. ABERCROMBIE et al. *Dict. Biol.* 65 *Deamination*, removal of amino (NH_2) group. In mammals occurs to many amino-acid molecules by action of deaminating enzymes in liver and kidney. 1962 DARDENNE & KIRSTEN in A. Pirie *Lens Metabolism Rel. Cataract* 419 Arginine and glutamic acid were rapidly deaminated. 1970 *Nature* 6 June 969/1 Nitrous acid causes deamination of primary amino groups attached to the ring structure of nucleic acid bases.

deaner, deener (dī·nǝɹ). *slang.* Also **dener, diener.** [Immediate origin uncertain, but prob. ultimately identical with DENARIUS, DENIER[3].] A shilling.
Before the advent of decimal currency esp. freq. in Australia and N.Z.
1839 H. BRANDON *Poverty, Mendicity & Crime, Deaner*, a shilling (country phrase). 1846 R. L. SNOWDEN *Mag. Assistant* 342 Shilling, Deaner, also twelver. 1851 MAYHEW *London Labour* I. 313/1 No, I'll give you a deuce o' deeners (two shillings). 1864 *Times* 12 Oct. 11/6 One woman said, 'Where's the "deaner"?' 1901 H. LAWSON *Joe Wilson & his Mates* II. 280 'Stumped?' inquired Jim. 'Not a blanky, lurid deener!' drawled Bill. 1904 *Daily Chron.* 27 Dec. 4/4 I've played it [sc. a cornet] six hour for a deaner and a haddock-bone disguised in batter as a bit o' cold, fried fish. 1907 *Ibid.* 26 July 4/7 In Ireland the 'denarius' lingers in the name for a shilling—a dener. 1908 *Ibid.* 5 Feb. 4/7 Here we come to classic reminiscences of the denarius! It is a 'bob', and 'owt deners' is a florin. 1922 *Contemp. Rev.* Sept. 367 A shilling..is frequently a 'deener' [in cockney speech]. 1930 V. PALMER *Men are Human* xxix. 273 I'd like a diener for every quart of whisky the old boy's sunk in his day. 1933 *Bulletin* (Sydney) 13 Dec. 31/1 O' course I waited, after spendin' the bob; 'e looked a bloke 'oo'd be good for more than a deener. 1937 A. UPFIELD *Mr. Jelly's Business* xiv. 146 If I 'ad lorst a deaner I'd have got hell. 1946 F. SARGESON *That Summer* 77 Could you give me the lend of a bob?..I'm on the beach myself, I said, but I can make it a deener.

deaness. 2. (Earlier example.)
1848 E. RUSKIN *Let.* 1 July in W. M. James *Order of Release* (1948) v. 115 A large party today at the Master of Pembroke's..and ditto in the evening at the Deaness of Christ Church (Mrs. Gaisford).

de-anglicize (di,æ·ŋglisaiz), *v.* [DE- II. 1.] *trans.* To remove English characteristics or influence from. Hence **de-anglicizaˑtion**; **de-a·nglicizing** *vbl. sb.*
1883, 1890 [see DE- II. 1]. 1894 D. HYDE in C. G. Duffy et al. *Revival of Irish Lit.* 117 (*title*) The necessity for de-anglicising Ireland. 1901 *Daily Chron.* 22 Aug. 6/4 Talking about Celticism and de-Anglicisation. 1905 *Westm. Gaz.* 27 Nov. 5/1 The Gaelic revival meant the de-Anglicising of Ireland. 1961 *20th Cent.* CLXIX. 70 Mencken had launched his de-Anglicization campaign in literature.

dear, *a.*[1] Add: **2. c.** (Later examples.) *Dear John*, as *sb.* (see quot. 1945); also *transf.* and *attrib.* Chiefly *N.Amer.*
1865 G. M. HOPKINS *Let.* 28 Aug. (1935) 1 Dear Bridges,—I left Manchester more than a month ago. 1908 D. H. LAWRENCE *Let.* 4 May (1962) I. 7 Dear Miss Jennings: With hot, boyish, impatience I looked for a letter from you. 1945 *Democrat & Chron.* (Rochester, N.Y.) 17 Aug. 17/2 'Dear John,' the letter began. 'I have found someone else whom I think the world of. I think the only way out is for us to get a divorce,' it said. They usually began like that, those letters that told of infidelity on the part of the wives of servicemen...The men called them 'Dear Johns'. 1947 *Amer. Speech* XXII. 187 It was a 'Dear John'. Quite a lot of the fellows had already had their 'Dear Johns'. 1957 W. CAMP *Prospects of Love* xvi. 103 There was a note from Fenny on the kitchen table. For the moment he enjoyed the irony of thinking it might be what the Americans called a 'dearjohn'. 1964 J. PHILIPS *Laughter Trap* (1965) I. ii. 13 Peter..had gone to war..in love with a girl named Elizabeth Schofield...He had received a 'Dear John' letter from Elizabeth, telling him she was married. 1970 *Post Office Subscriber Trunk Dialling: Dialling Instructions (London)* 1 Dear Customer, In 1966 we changed your old 3 letter and 4 figure numbering to all figures. 1971 *Guardian* 7 Jan. 11/8 The going is getting distinctively gritty for Pan Am, so much so that 1,876 of its staff are now walking round with a new year 'Dear John' from..the airline's president.
6. e. Of money: that can be borrowed only at a high rate of interest.
1878 [see MONEY *sb.* 3 c]. 1930 *Economist* 1 Feb. 239/2 A year of dear money, when Bank rate averaged a full one per cent. above the level of 1928.
C. *dear knows* (earlier and later examples).
1805 E. CAVANAGH *Let.* 4 Oct. in Londonderry & Hyde *Russian Jrnls. of M. & C. Wilmot* (1934) III. 190, I never seen such a good Lady..nor so generous I've reason to say dear knows. 1914–15 JOYCE *Portrait of Artist* (1916) v. 203 The dear knows you might try to be in time for your lectures. 1969 *Outlook* Mar. 226 The curate waded out into the sea of dear knows what but mostly rubbish.

dearness. Add: **2. b.** *attrib.*, as **dearness allowance, pay,** an allowance added to a basic salary, esp. in India, to cover an increase in the cost of living.
1955 *Times* 1 July 1/4 (Advt.), Pay. Rs. 1,000 to Rs. 1,600 p.m., according to qualifications and experience, plus dearness allowance. 1969 *National Herald* (New Delhi) 29 July 8/6 A part of dearness allowance was merged with basic pay last year. *Ibid.*, All employees whose basic pay and dearness pay do not total more than Rs. 620 will become eligible for overtime allowance.

deasil, *adv.*, *sb.* Add: Also **deisal, deisul.**
1897 *Daily News* 26 July 5/1 'Walking the deisul' round a person or place was lately, perhaps is still, a ceremony in the Highlands. 1945 *Archit. Rev.* XCVII. 49/3 English roundabouts run deisul, clockwise. 1947 AUDEN *Age of Anxiety* (1948) ii. 49 O Primal Age When we danced deisal.

death, *sb.* Add: **2.** *death-in-life*, life that lacks any satisfaction or purpose; living death. (Cf. quot. 1841 *s.v.* DEATHLINESS.)
1864 [in Dict.]. 1864–6 MRS. GASKELL *Wives & Daughters* (1866) II. xxv. 255 The Squire stood by in dumb dismay, touched in spite of himself by the death-in-life of one so young. 1907 *Folk-Lore* June 228 A seven years' death-in-life trance. 1934 'G. ORWELL' *Burmese Days* xxiv. 358 The sort of horrible death-in-life! The decay, the loneliness. 1955 E. BOWEN *World of Love* vi. 115 They conceived of no death, least of all death-in-life.
7. *to be the death of* (further examples); often hyperbolically.
1816 JANE AUSTEN *Emma* II. iii. 56 Oh! dear, I thought it would have been the death of me! 1863 'OUIDA' *Held in Bondage* I. iii. 63 A mill-wheel monotony would be the death of me. 1956 S. BECKETT *Waiting for Godot* I. 35 Estragon (convulsed with merriment). He'll be the death of me!
12. d. *to do* (a thing) *to death*, to overdo; to repeat too often or *ad nauseam*.
1882 H. D. TRAILL *Recaptured Rhymes* 112, I am also called Played-out and Done-to-death, And It-will-wash-no-more. 1886 H. BAUMANN *Londinismen* 43/2 *Done to death*,..überspannt geschrieben. *a* 1889 in Barrère & Leland *Dict. Slang* (1889) I. 320/2 Caricature of Academy pictures done to death in comic journals with utmost regularity for many past years. 1909 W. S. SPARROW *Hints on House Furnishing* II. iii. 134 Diapered patterns for wall-papers and carpets.. were 'done to death'. 1965 *New Statesman* 16 Apr. 605/1 It [sc. a tune] was mercilessly done to death by countless performers.

13. b. *to go one's death* (*on* or *upon*), to do one's utmost (for); to risk one's all (on). *Obs. U.S. slang.*
1833 D. CROCKETT *Sketches* 74 My little boys at home will go their death to support my election. *Ibid.* 173 You think they don't go their death upon a jig, but they do. 1835 A. B. LONGSTREET *Georgia Scenes* (1840) 199 I'll go my death upon you at the shooting match. 1878 *Scribner's Mag.* XV. 400/1 The consulship at Rio Janeiro is vacant, and being worth $6,000, he is moved to 'go his death on Rio'.
15. (Earlier and later examples.)
1788 W. COWPER *Let.* 3 Mar. (1904) III. 240, I have been in at the death of a fox. 1919 L. STRACHEY in *Athenæum* 13 June 454/1 Creevey..had a trick of being 'in at the death' on every important occasion:..he invariably popped up at the critical moment. 1933 N. COWARD *Design for Living* II. iii. 71 You have a tremendous sense of the 'right moment', Ernest. It's wonderful. You pop up like a genie out of a bottle, just to be in at the death!
16. (Earlier example.) orig. *U.S.*
1839 *Spirit of Times* 5 Oct. 368/3 [His] nose is so red that no musquito can stand the blaze of it. It's death upon gallinipers, too.
17. (Further examples.)
c 1384 CHAUCER *H. Fame* 502 But this as sooth as deeth, certeyn, Hit was of golde. *c* 1440 *Partonope* 6999 And in þis wise cristenyd was he, As siker as dethe, with-outen nay. 1598 B. JONSON *Ev. Man in Hum.* II. i, They would giue out..That I were iealous! nay, as sure as death, That they would say. 1766 GOLDSM. *Vicar* xix, As sure as death there is our master and mistress come home. 1831 S. E. FERRIER *Destiny* xlviii, Oh, as sure as death, then, that's just owning that you are going to be married. 1836 D. CROCKETT *Exploits & Adventures in Texas* x. 151 He hung on like grim death. 1837 T. HOOK *Jack Brag* III. v. 182 Jack..was holding on, like grim Death, by the companion. 1847 [see GRIM *a.* 4 b]. 1928 D. L. SAYERS *Unpleasantness at Bellona Club* ix. 112 If ever you see him again, Fentiman, freeze on to him like grim death. 1928 R. CULLUM *Myst. Barren Lands* xii. 106 'Think you'll ever get it?'..'Sure,' he said emphatically. 'Just as sure as —— death.'
b. (*a fate*) *worse than death*, a misfortune, situation, etc., regarded as being worse than death; *spec.* loss of virginity; rape (formerly *euphem.*, now usu. jocular).
[1653 D. OSBORNE *Let.* ? 28 Aug. (1903) 141 The Roman courage, when they killed themselves to avoid misfortunes that were infinitely worse than death. 1781 GIBBON *Decl. & F.* III. xxxi. 238 The matrons and virgins of Rome were exposed to injuries more dreadful, in the apprehension of chastity, than death itself.] 1810 JANE PORTER *Scottish Chiefs* III. iii. 68 But where was he who had delivered herself from a worse fate than death? 1832 F. TROLLOPE *Dom. Manners* (1949) 421 Then would a parliament of love ..doom to worse than death the spitter and tobacco chewer. 1894 RIDER HAGGARD *People of Mist* vi. 40 It is the custom of my mistress to carry a portion of this poison hidden in her hair, since a time might come when she must use it to save herself from worse than death. 1914 E. R. BURROUGHS *Tarzan of Apes* (1917) xix. 168 [The ape] threw her roughly across his broad, hairy shoulders, and leaped back into the trees, bearing Jane Porter away toward a fate a thousand times worse than death. 1926 *Sat. Rev.* 6 Feb. 166/1 The heroine, Countess Olga, is saved from a fate worse than death. 1949 R. HARVEY *Curtain Time* 22 If she hadn't got across, she would've had what they call a Fate Worse than Death. 1952 'C. BRAND' *London Particular* vii. 95 Madonna Lily was obviously marked up for a double dose of Worse than Death.
c. *like death* (*warmed up*), colloq. phr. indicating a feeling or appearance of extreme illness or exhaustion.
1939 *Soldiers' War Slang Dict.* 7/2 *To feel like death warmed up*, to feel ill. 1942 N. MARSH *Death & Dancing Footman* ii. 42, I look like death warmed up and what I feel is nobody's business. 1964 J. PENDOWER *Sinister Talent* xx. 185 It damned near killed me...I still feel like death warmed up. 1969 *Guardian* 14 Feb. 11/1 Much of the time I feel like death. I am in rather a bad temper.
d. *in the death*, in the end; finally. *slang.*
1958 F. NORMAN *Bang to Rights* 172 In the death this geezer got a repreave. 1962 R. COOK *Crust on its Uppers* iii. 42 In the death, though, it went the way of all the best bent jams. *Ibid.* xvi. 157 After all, you had to marry someone in the death.
18. a. *death-camp, -cart, -cult, -flower, -march, -note, -pill, -terror, -train*; also objective, as *death-control* (after *birth-control), *-dealer, -worship*.
1816 J. WILSON *City of Plague* II. iv. 103 Now may I ask whose pious care Hath plac'd these death-flowers here! 1841 BORROW *Zincali* II. III. ii. 60 The death-carts..went through the streets..picking up the dead bodies. 1897 *Daily News* 10 Dec. 7/2 During what was known as the 'death march' from Gundamuk to Peshawur..one battalion was..decimated..by cholera. 1903 *Daily Chron.* 20 Oct. 5/1 Said he, You death dealers, I'll stop you. 1904 R. J. FARRER *Garden of Asia* vi. 44 The glow of the death-flowers.. has faded. 1917 —— *On Eaves of World* I. iv. 71 The inevitable death-terror impels every human soul. 1920 MASEFIELD *Enslaved* 78 That devil's horn Its quavering death-note blew. 1922 JOYCE *Ulysses* 579 The deathflower of the potato blight on her breast. 1925 D. H. LAWRENCE *St. Mawr* 201 They had found their *raison d'être* in self-torture and death-worship. *a* 1930 —— *Etruscan Places* (1932) ii. 60 On the sculptured side of the sarcophagus the two death-dealers wield the hammer of death. 1938 AUDEN & ISHERWOOD *On Frontier* III. ii. 115 Your dreary death-cult is hardly likely to amuse a young lady. 1944 *Ann. Reg. 1943* 204 In the course of a three-day 'liquidation'..the Germans killed more than 1,000 Jews in 'death camps'. 1948 S. GILBERT tr. *Camus's Plague* III. i. 164 This system.. was really a great improvement on the death-carts driven by negroes. 1949 KOESTLER *Promise & Fulfilment* vi. 58 At that time the death trains had started rattling through Eastern Europe. 1955 *World Population & Resources* (*P.E.P.*) I. i. 9 Whereas death-control has always been actively fostered by Governments, birth-control has almost always been left to voluntary action. 1962 *Guardian* 16 Feb. 7/2 This is the authentic Jewish experience: herding, the death-march, endless hunger. 1963 I. FLEMING *On H.M. Secret Service* xvi. 175 Had Campbell got a death pill, perhaps one of the buttons on his ski-jacket or trousers? 1970 *Nature* 14 Feb. 595/1 Population growth, death control and the expansion of the labour force are proceeding rapidly.
b. *death-defying, -giving* adjs.
1878 GEO. ELIOT in *Macm. Mag.* July 168 What is martyrdom But death-defying utterance of belief? 1882 LONGFELLOW *In Harbor* 39 Life-giving, death-giving, which will it be? 1942 S. SPENDER *Life & Poet* 30 Tyrants who wish to freeze institutions into death-giving instead of life-giving forms. 1950 W. DE LA MARE *Inward Companion* 68 This death-defying acrobat.
d. *death-dark, -still, -white* adjs.
1906 *Westm. Gaz.* 19 Oct. 2/4 Wild lilies lie on a death-white heart. 1907 *Daily Chron.* 20 Aug. 3/2 Those death-still places which have no houses. 1921 W. DE LA MARE *Veil* 91 Rouse the Old Enemy from his death-still swoon. 1932 E. SITWELL *Bath* 81 They will play with..their death-dark negro slave.
19. death-bone *Austral.*, a bone pointed at a person and intended to cause his death (cf. *BONE *sb.* 1 e); **death camas(s)** = *death quamash*; **death cap** = *death-cup; **death-chair,** the electric chair; **death-cup,** the poisonous fungus *Amanita phalloides*; **death-feigning,** the feigning of death, esp. by an animal; **death-house,** (*a*) a place where a person dies; (*b*) *colloq.*, that part of a prison where persons awaiting execution are housed; also, the execution shed; **death-instinct** [tr. G. *todestrieb* (Freud 1920, *Jenseits des Lustprinzips* vi)], a destructive or self-destructive tendency postulated by Freud (cf. *DEATH-WISH); **death-or-glory** *attrib. phr.*, (*a*) (with capital initials) a regimental nickname (see quot. 1890); (*b*) *transf.*; **death quamash,** a plant of the western U.S., the bulb of which is poisonous to animals; **death-rate** (earlier example); **death-ray,** (chiefly in Science Fiction) a ray that causes death; **death-roll,** a list of the names of those who have been killed in an accident, battle, etc.; **death-tick** (earlier example).
1933 *Bulletin* (Sydney) 19 Apr. 20/3 Even the semi-civilised Binghi of to-day is not proof against the old belief in the death-bone. 1889 *Cent. Dict.* s.v. *Camass, Death camass*, the poisonous root of *Zygadenus venenosus*. 1937 *Range Plant Handbk.* (*Forest Service, U.S. Dept. Agric.*) W209 The more virulent species of death camas cause the majority of sheep losses. 1925 R. T. & F. W. ROLFE *Romance of Fungus World* iii. 37 *Amanita phalloides*, sometimes known as the Death Cap. 1949 *Oxf. Jun. Encycl.* II. 172/1 Many fungi are mildly poisonous, but only one British species, the Death Cap, is deadly poisonous. 1890 *N.Y. Tribune* 7 Aug. 2/1 Kemmler stepped into the death-chair. 1919 KIPLING *Lett. Travel* (1920) 269 Pinioned men in the death-chair before the current is switched on. 1904 *Westm. Gaz.* 6 Oct. 10/1 The 'death-cup' is very abundant in woods in this country. 1924 J. A. THOMSON *Science Old & New* vi. 33 The 'death-feigning' or 'playing "possum"' of various animals. 1959 SOUTHWOOD & LESTON *Land & Water Bugs* xii. 355 Pressure on the bug, as by forceps in the laboratory, causes 'death-feigning' immobilization. 1920 MASEFIELD *Enslaved* 8 The place was like a death-house save for cawings overhead. 1923 E. WALLACE *Missing Million* xxxviii. 293 He has three murders behind him, and the grey doors of the death-house in front of him. 1930 E. RICE *Voyage to Purilia* xiii. 168 The death-house—a huge edifice, crowded.. with condemned men awaiting execution. 1958 G. MIKES *East is East* 120 The Death House is a Chinese institution. People are sent there to die because the Chinese regard it as bad luck if people die at home. 1960 *Guardian* 2 May 1/7 There is less than 24 hours to go before Chessman leaves the death house..and goes to his..Maker. 1922 C. J. M. HUBBACK tr. *Freud's Beyond Pleasure Principle* vi. 38 The opposition between the ego or death instincts and the sexual or life instincts would then cease. 1961 J. A. C. BROWN *Freud & Post-Freudians* ii. 27 The Death instinct is a force which is constantly working towards death. 1890 FARMER *Slang* I. 199/1 The 17th [Lancers] are still well-known as the Death or Glory Boys, from their badge, which consists of a death's head, with the words, 'or glory'. 1959 W. K. RICHMOND *Brit. Birds of Prey* ix. 116 He is content to play second fiddle and leaves the death-or-glory stuff to his partner. 1962 *Guardian* 20 Dec. 6/1 Sir Roy Welensky's death-or-glory approach to political problems. 1884 MILLER *Plant-n.* 264/2 *Zygadenus venenosus*. 'Death Quamash', Hog's Potato. 1859 *Ann. Rep. Registrar-General* p. ii, The death rate was below the average. 1919 B. MUNN *Skeleton Man* xxvii. 90 Had the man once used his death rays he was watched carefully enough to have been caught..red-handed. 1947 CROWTHER & WHIDDINGTON *Science at War* 4 The situation was grave and the public had already begun to long for death-rays that would dispatch the strongest enemy at will. 1961 J. B. PRIESTLEY *Saturn over Water* vi. 85 You may have some ridiculous ideas..that we are..discovering fantastic gases or death rays out of science fiction. 1963 *Ann. Reg. 1962* 386 Unlikely rumours were also heard that it might be developed as a military death ray or anti-missile weapon. 1864 M. B. CHESNUT *Diary* 1 Aug. (1949) 425 Day after day we read the death roll. 1873 *Porcupine* XIV. 725/2 Two other actresses also appear on this week's death-roll. 1906 *Westm. Gaz.* 17 Oct. 10/1 The terrible disaster at Seaham in 1880, when the death-roll approximated to nearly 200. 1940 *Manch. Guardian Weekly* 5 Jan. 5 The full extent of the death-roll and damage will not be known for several days. 1853 H. MELVILLE *Cock-A-Doodle-Doo!* in *Harper's Mag.* Dec. 81/1, I might as well have asked him if he had heard the death-tick.

death-bed. b. Add: freq. (with derogatory implication) in *death-bed confession, repentance*; also *transf.*, of a belated change of conduct or policy.
1691–8 [in Dict.]. 1771 'JUNIUS' *Lett.* I. p. vi, A death-bed repentance seldom reaches to restitution. 1888 A. C. GUNTER *Mr. Potter of Texas* vi, Curse him and his death-bed confession! 1963 *Times* 2 May 17/3 It was deplorable that it should have needed unemployment of more than 900,000..to extract even this rather meagre deathbed repentance from the Government.

death-place (de·þplēⁱs). [f. DEATH *sb.* + PLACE *sb.*] The place where a person dies.
1830 MOORE *Byron's Life & Lett.* II. 778 His lost friend's melancholy deathplace, Missolonghi. 1901 *Spectator* 28 Sept. 430/1 Why..is the deathplace of Constantius given as Mopsucrenam? 1906 *Daily Chron.* 19 Dec. 4/5 Brocket Hall, Herts,..the death-place of two Premiers, Lord Melbourne and Lord Palmerston. 1908 *Westm. Gaz.* 16 Nov. 9/2 Violating the precedents which prescribe the death-place of the rulers of China. 1937 *Tablet* 23 Oct. 553/2 To follow Moses from birth- to death-place (neither of which will ever be ascertainable).

death-song. [DEATH *sb.* 18 a. Cf. G. *todesgesang*, earlier †*todtengesang*.] A song sung immediately before death, or to commemorate the dead. In U.S. *spec.* with reference to Indian customs.
1778 J. CARVER *Travels* 337 They are then bound to a stake..and obliged..to sing their death-song. 1843 *American Pioneer* II. 225 Four of the missionaries..were then led into the camp of the Delawares, where the death-song was sung over them. 1873 J. MILLER *Life amongst Modocs* ii. 21 They will not revisit their own camp..until it is first visited by their priest or medicine man, who chaunts the death-song. 1885 H. SWEET *Oldest Eng. Texts* 149 (*title*) Bede's Death-song. 1892 S. A. BROOKE *Hist. Early Eng. Lit.* I. 74 Like an Indian chief, he [sc. Beowulf] sang his death-song. 1947 B. A. DE VOTO *Across Wide Missouri* 84 Some of the Gros Ventres began to sing their death songs.

deathward, *adv.* Add: **B.** *adj.* Tending towards death.
1854 'G. GREENWOOD' *Haps & Mishaps* 43 O immortal stones,..mocking..the mournful mortality, the deathward throbbing, of the brows ye encircle! 1882 SWINBURNE *8 Yrs. Old* in *Tristram of L.* 257 Sounds of dying and dawning years, Now quickened on his deathward way. 1899 *Daily News* 17 Oct. 8/3 A pleasant-looking, neatly-dressed woman on the deathward side of fifty. 1937 J. M. MURRY *Necessity of Pacifism* v. 86 The individualized, by the very fact of its individualization, belongs to the deathward movement of society.

death-watch. 1. Substitute for second part of def.: *esp.* the death watch beetle, *Xestobium rufovillosum*, or *Trogium pulsatorium*, a book louse of the order Psocoptera.
1817 KIRBY & SPENCE *Introd. Ent.* II. xxiv. 383 A little wood-louse..—which on that account has been confounded with the death-watch—is said also..to emit a ticking noise. 1877 *Encycl. Brit.* VI. 132/2 Many of the *Malacodermata* are wood-borers; these include the Death-watch Beetles (*Anobium*). 1925 E. G. BLAKE *Enemies of Timber* vi. 136 The death-watch beetle..confines its activities exclusively to structural timbers, and is practically identical in its methods and habits with the common furniture beetle. 1955 *Times* 9 May 5/2 The death-watch beetle did not attack sound and permanently dry timber. 1964 N. E. HICKIN *Household Insect Pests* v. 56 One interesting habit shown by the species *Trogium pulsatorium* is the tapping noise which it produces by vibrating its abdomen against a material such as paper. It is probable that the first reference to the 'Death Watch' concerned this Book Louse and not the wood-boring beetle. 1969 R. F. CHAPMAN *Insects* xxviii. 575 The death watch beetle, *Xestobium*, produces tapping sounds by bending its head down and banging it against the floor of its burrow.

dea·th-wish. [tr. G. *todeswunsch*, f. *tod* death + *wunsch* wish.] A conscious or unconscious wish for the death of oneself or another; also *fig.*
1913 A. A. BRILL tr. *Freud's Interpr. of Dreams* v. 218 The death-wish towards parents is to be explained by reference to earliest childhood. 1929 P. MAIRET *Adler's Problems of Neurosis* vi. 84 After having triumphed over him and having had the death-wish granted by fate, he remained still unsatisfied. 1937 'C. CAUDWELL' *Illusion & Reality* 97 The same secret death-wishes are shown by these aristocrats if they turn revolutionary. 1947 *Partisan Rev.* XIV. 233 Even in America..the death-wish of the business community appears to go beyond the normal limits of political incompetence and geographical security. 1953 X. FIELDING *Stronghold* 153 A death-wish, a longing to be underground, to be in a tomb, buried. 1958 *New Statesman* 11 Jan. 29/1 Whatever handicaps Mr. Macmillan labours under as Prime Minister, they do not include the death-wish; he is still jauntily determined to win the next election.

deb (deb). orig. *U.S.* Also **debby.** Colloq. abbrev. of DÉBUTANTE; hence *debs'* (or *deb's, debbies'*) *delight*, an eligible or attractive young man in high society. Also *attrib.* or as *adj.*
1920 F. SCOTT FITZGERALD *This Side of Paradise* II. ii. 227 Dancing with mid-Western or New Jersey debbies. 1922 JOYCE *Ulysses* 437 Josie Powell that was, prettiest deb in Dublin. 1926 *Ladies' Home Jrnl.* July 26 One of my deb cousins makes a transcendent cocktail. 1927 *Sunday Express* 10 July 10 It appears that her comments on society have roused the ire of the great army of debutantes...How the angry 'debs' propose to deal with their victim..I cannot imagine. 1928 *Sunday Dispatch* 9 Dec. 11/2 The impossibility of parents doing any of the old kind of chaperonage in the hours kept by the present day (or night) 'debbies' during their present season. 1934 J. O'HARA *Appointment in Samarra* v. 135 Joe Montgomery could be classified under many headings...Well-dressed man. Debbies' delight. Roué. 1948 N. MARSH *Death in White Tie* iii. 33 Lord Robert half suspected his nephew Donald of being a Debs' Delight. 1940 M. DICKENS *Mariana* vii. 217 She presents her at Court, drags her to all the deb. dances. 1958 *Oxf. Mail* 1 July 6/6 His girl is a dim creation and Pauline Arden makes her unnecessarily 'debby'. 1958 *Daily Mail* 4 July 10/4 She danced with —— at a debby night spot clad in little more than a nylon nightie. 1966 J. BETJEMAN *High & Low* 56 The debs may turn disdainful backs On Pearl's uncouth mechanic slacks. 1970 *Daily Tel.* 5 Sept. 16 Somewhat debby (though none was in fact a deb), the girls must also be ready to display their social graces at Council cocktail parties.

debadge (dībæ·dʒ), *v.* [f. DE- II. 2 a + BADGE *sb.*] *trans.* To deprive of the badge which in the war of 1914–18 exempted a man from military service. Also **deba·dged** *ppl. a.*
1916 *Daily Mail* 23 Sept. 5/7 All single men under, say, twenty one, to be immediately debadged. *Ibid.* 1 Nov. 5/5 This practice of capturing debadged men has been brought prominently before the Man-Power Board.

debag (dībæ·g), *v. slang.* [f. DE- II. 2 a + BAG *sb.* 16.] *trans.* To remove the trousers from (a person) as a punishment or for a joke. Hence **deba·gging** *vbl. sb.*
1914 C. MACKENZIE *Sinister St.* II. III. vi, At Oxford..we should be out of sympathy with him, even up to the point of debagging him. *Ibid.* vii, 'We ought to debag him,' he cried. Appleby was thereupon debagged; but as..he continued to walk about trouserless and dispense hospitality without any apparent loss of dignity, the debagging had to be written down a failure. 1927 *Daily Express* 5 Oct. 3/3 If the Gun Room wishes to pay a tribute to one whom it loves, it de-bags the adored after dinner; if the Gun Room wishes to hurt the feelings of one it hates, it de-bags the hated one after dinner. 1958 B. NICHOLS *Sweet & Twenties* 118 A number of us chased Sir Robert down the moonlit High Street in an endeavour to debag him.

de-bamboo·zle, *v.* [f. DE- II. 1 + BAMBOOZLE *v.*] *trans.* To undeceive, disabuse.
1919 J. M. KEYNES *Econ. Conseq. of Peace* (1920) iii. 50 It was harder to de-bamboozle this old Presbyterian than it had been to bamboozle him. 1921 *Sat. Westm. Gaz.* 12 Feb. 18/2 Even with such an account before them, it is doubtful whether they [sc. visitors to Russia] will ever be 'de-bamboozled'. 1958 *Times Lit. Suppl.* 31 Jan. 54/5 Our rationalists and our materialistic demagogues alike, having found it relatively easy to debamboozle the masses of belief in God, love of neighbours, reason, etcetera, now find it extremely difficult to bamboozle them into believing or hoping anything.

deba·rbarize, *v.* [f. DE- II. 1 + BARBARIZE *v.*] *trans.* = UNBARBARIZE *v.*
1930 W. R. INGE *Christian Ethics* v. 223 The Catholic Church might..have..found a congenial task in debarbarising Christianity. 1963 *Guardian* 19 Feb. 4/2 Our chance to de-barbarise our society.

debark, *v.*[2] Delete *rare*, and add earlier and later examples. Hence **deba·rking** *vbl. sb.*
1742 W. ELLIS *Timber-Tree* II. 197 The Debarking of Oak in the common Season. 1963 R. R. A. HIGHAM *Handbk. Papermaking* v. 114 Any logs which have not been properly de-barked are picked out and returned by hand to the inlet side of the barker. 1965 G. MCINNES *Road to Gundagai* x. 166 The more enterprising de-barked the saplings. 1970 *Daily Tel.* 4 May 14 Last year two large chestnuts were deliberately killed by poison and debarking.

debark (dībā·ɹk), *v.*[3] [f. DE- II. 2 + BARK *sb.*[2]] *trans.* To perform an operation on (a dog) to prevent it from barking. Hence **deba·rking** *vbl. sb.*
1943 *Biol. Abstr.* XVII. 70 Dogs debarked by removal of the true vocal cords soon learned to bark again. 1959 *Times* 30 Sept. 8/4 A 'debarking' operation. *Ibid.*, A new operation for debarking dogs. 1961 *Daily Tel.* 19 Oct. 19/5 (*heading*) No de-barking. *Ibid.*, Two mongrels which annoyed neighbours with their barking are not to be 'debarked'.

débat: see next.

debate, *sb.*[1] Add: **2. c.** (Freq. in French form *débat.*) A type of literary composition, taking the form of a discussion or disputation, commonly found in the vernacular medieval poetry of many European countries, as well as in medieval Latin.
1841 T. WRIGHT *Latin Poems W. Mapes* 346 (*title*) Debate between the Body and the Soul. 1897 G. SAINTSBURY *Flourishing of Romance* v. 203 A form so popular with the French *trouvères* as the *débat*. 1903 E. K. CHAMBERS *Mediæval Stage* I. iv. 79 The *débat* is a kind of poetical controversy put into the mouths of two types or two personified abstractions, each of which pleads the cause of its own superiority, while in the end the decision is not infrequently referred to an umpire in the fashion familiar in the eclogues of Theocritus. *Ibid.* II. xxiii. 153 This *débat*-like theme is of course familiar in every branch of allegorical literature. 1933 R. TUVE *Seasons & Months* i. 12 The conflict between the vital and ascetic principles..seen in the *Owl and the Nightingale* débat. 1939 R. M. WILSON *Early M.E. Lit.* vii. 168 The *Thrush and the Nightingale*..is..a compromise between the debate and the lyric. 1963 M. D. LEGGE *Anglo-Norman Lit.* xiii. 335 The violent language used by both parties to the debate is, like the language of flytings, not to be taken seriously.

debating, *vbl. sb.* Add: **b. debating club** = *debating society*; **debating point,** a point which, though not necessarily essential to the matter in hand, furnishes a useful or interesting subject for debate; a proposition, contention, etc., used mainly to impress or disconcert one's opponent in a debate.
1830 C. C. BALDWIN *Diary* 6 Feb. (1901) 52 The Debating Club, composed of law students generally, hold a public debate. 1862 BAGEHOT *Coll. Works* (1965) II. 225 The House of Commons..became..the debating-club of fashion. 1927 G. B. SHAW *Doctors' Delusions* (1932) 135 Shallow petulances and *tu quoques* which have remained part of the vivisector's stock of debating points ever since. 1928 *Daily Tel.* 9 Oct. 6 'I do plead for the liberties of the people of England more than any of you do.' It was not a bad debating point, if such things had then been of any avail. 1965 *Listener* 2 Sept. 348/2 It may be no more than an interesting debating point, but in fact this eastern area was..little used or even visited by Indians.

debby: see *DEB.

de-beak (dībī·k), *v.* [f. DE- II. 2 + BEAK *sb.*[1]] *trans.* To remove the beak from. Hence **debea·king** *vbl. sb.*
1937 *Ohio Agric. Experiment Station Bimonthly Bull.* Jan.–Feb. 37 There is only one way to debeak [chickens] properly and effectively. Only the upper beak is removed. The beak is not cut off; it is torn off. 1955 *Sci. News Let.* 16 July 46/3 Debeaking turkeys reduces losses from cannibalism. 1959 *Observer* 8 Mar. 3/4 The ill-fated chicks..have to be..de-beaked to prevent them tearing one another to pieces. 1960 *Farmer & Stockbreeder* 15 Mar. 133/1 Once feather pecking and cannibalism become established as a vice in intensively kept layers the only cure is debeaking.

debel, -ell, *v.* (Later example.)
1897 H. N. HOWARD *Footsteps Proserpine* 12 Many mourned by man, by fate debelled.

debit, *v.* [5]Add: **3.** (*Gallicism.*) To put into circulation; to spread (news, etc.).
1879 T. F. SIMMONS *Lay Folks Mass Book* 366 In respect of those who debited these fables. 1916 G. B. SHAW *Pen Portraits & Reviews* (1932) 43 Mr Bennett talks shop and debits harmless tosh about technique for the entertainment of literary amateurs in a very agreeable and suggestive manner.

debitable (de·bitǎb'l), *a.* [f. DEBIT *v.* + -ABLE.] That can be debited.
1895 *19th Cent.* Oct. 650 Salaries and establishments out of, but debitable to, India. 1899 *Westm. Gaz.* 12 Oct. 5/3 The monthly maintenance is estimated at £11,000 sterling, all of which is debitable to England.

debiteuse (debitö·z, -tiū·z). *Glass-making.* [a. F. *débiteuse*, fem. of *débiteur* one who spreads (gossip), f. *débiter* to cut up, yield, MF. *débiter* to cut wood, sell retail, f. *de-* DE- I + *bitte* post on ship for fastening cables, f. ON. *biti* (see BITT).] An open, oblong, trough-like object made of refractory material and having a slit along the bottom, which in the Fourcault method of making sheet glass floats on the surface of the molten glass.
1922 *Encycl. Brit.* XXXI. 290/1 In drawing a sheet of glass a bait consisting of a narrow flat woven iron sheet of a length equal to the slit in the *débiteuse* is lowered within the lips forming the slit. 1948 W. C. SCOVILLE *Revolution in Glassmaking* 330 His 'debiteuse'..rested in a shallow tub of glass connecting with the melting furnace. 1969 R. PERSSON *Flat Glass Technol.* ii. 9 The glass sheet is drawn vertically from a slot in a refractory floater or debiteuse, which floats on the glass in the drawing chamber.

debonairness. (Later examples.)
1929 W. DEEPING *Roper's Row* x. 99 A notable claim to decency and debonairness. 1964 *Listener* 13 Aug. 248/1 His breezy debonairness depends on the convention of public-school illiteracy.

Debrett (dĭbre·t), *sb.* A colloquial designation of 'Debrett's Peerage of England, Scotland and Ireland', the first edition of which, issued in 1803, was compiled by John *Debrett* (*c* 1750–1822). Also *transf.*
1848, etc. [see *BURKE *sb.*]. 1849 THACKERAY *Pendennis* I. ix. 85 You cannot begin your genealogical studies too early; I wish..you would read in Debrett every day. 1873 [see WET-NURSE *v.* b]. 1910 BARONESS ORCZY *Lady Molly of Scotland Yard* iv. 82 If you have studied Debrett at all, you know..that the peerage is one of those old English ones which date back some six hundred years. 1970 *Guardian* 18 Aug. 1/5 The 1970 Debrett was published yesterday.

débridement (debridmaṅ). *Surg.* [Fr., lit. 'unbridling'.] The removal from a wound, etc., of damaged tissue or foreign matter (see also quot. 1842). Hence **debri·de** *v. trans.*, to perform débridement on.
1842 DUNGLISON *Med. Lex.* (ed. 3) 217/2 *Débridement*, the removal of filaments, &c., in a wound or abscess, which prevent the discharge of pus. In a more general acceptation, it means the cutting of a soft, membranous or aponeurotic part, which interferes with the exercise of any organ whatever. 1929 F. A. POTTLE *Stretchers* (1930) 149 Débridement about half the extensor muscles of thigh. Wounds of entrance and exit in knee joint debrided. 1967 S. TAYLOR et al. *Short Textbk. Surg.* xxxii. 473 Where there is..contamination of the wound by dirt and foreign bodies, full débridement must be carried out.

debrief (dībrī·f), *v. colloq.* [f. DE- II. 1 + *BRIEF *v.*[2]] *trans.* To obtain information from (a person) on the completion of a mission or after a journey. Usu. *pass.* So **debrie·fing** *vbl. sb.*
1945 *Picture Post* 31 Mar. 9/1 Crump and his men go through a medical test,..de-briefing and customs. 1945 *John o' London's* 16 Nov. 70/2 The R.A.F. use of the atrocity debrief. When airmen receive their orders for an operation they are said to be briefed for it—a quite legitimate extension of the legal term. But when they return to give their report, they do not just report, as one would think. They are debriefed. 1951 *Life* 17 Dec. 31/2 He went to the debriefing shed and made a routine report on the railroad bombing mission. 1964 *Observer* 13 Sept. 2/2 He [sc. a defecting Russian scientist] is now in the U.S., presumably being debriefed by American intelligence. 1964 J. PORTER *Dover One* xi. 138, I want everybody back at the station..for a de-briefing session. 1965 *Guardian* 22 Mar. 9/6 Leonov and Belyaev..will stay at the space station for several days to be debriefed and medically examined.

debris, débris. Add: **d.** = SLIME *sb.* 4, TAILING *vbl. sb.*[1] 2 b.
1871 *Cape Monthly Mag.* June 358 In the paucity of materials in the *débris* of pans worked for diamonds, I would have less difficulty in finding traces of these rocks. 1882 H. G. HANKS *Sec. Rep. State Min. Calif.* 283 *Debris*,..the silt, sand and gravel that flow from the hydraulic mines; called in miner's parlance, tailings, slums, and sometimes by the outlandish name of 'slickens'. 1902 D. WARD *Digest Criminal Cases Superior Courts Colony C.G.H.* 5/2 The accused, an employé, *not of De Beers*, but of a *débris* washer. 1967 *Gloss. Mining Terms* (*B.S.I.*) VIII. 12 *Dirt*, (*debris*),..any material, such as rock or clay, etc., associated with the mineral and extracted during mining operations.
2. *attrib.*, as **debris-cone,** a cone formed by the accumulation of volcanic ejecta, debris, etc.
1890 J. D. DANA *Char. Volcanoes* 113 The cone was found to be literally a debris-cone, not a lava-cone or cinder-cone in any part. *Ibid.* 171 Between 1880 and 1882 another debris cone began in the basin of Halema'uma'u. 1895 —— *Man. Geol.* (ed. 4) III. v. 285 The basin contained a debris-cone made of the fallen blocks, and not at all of ejected material.

de Broglie (dǝ bro·lʸi). The name of L. V. *de Broglie* (born 1892), French physicist, used *attrib.*, esp. in *de Broglie wave*, the wave which in wave mechanics is taken as accounting for or representing the wave-like properties of a material particle, esp. an elementary particle.
1922 *Nature* 17 June 781/1 An improved De Broglie photographic spectrometer was used. 1927 *Proc. R. Soc.* A. CXV. 208 (*heading*) The de Broglie Phase Wave in Generalised Space-Time. 1928 G. BIRTWISTLE *New Quantum Mechanics* xxiii. 189 The de Broglie wave length associated with the particle. 1952 R. E. MARSHAK *Meson Physics* iii. 97 The de Broglie wave length of the meson is much larger and comparable with the nuclear radius. 1966 D. G. BRANDON *Mod. Techniques Metallogr.* 38 The energy, E, of a single photon of light of wavelength λ is given by the de Broglie relation, $E = hc/\lambda = h\nu$, where ν is the frequency, c the velocity of light and h Planck's constant.

debt, *sb.* **5.** Add: *debt-collecting, -collector, -reduction*; **debt-raiser,** one who undertakes to raise money to pay off a debt; **debt-slave,** one who is in slavery for the redemption of debt; so *debt-slavery*.
1897 *Westm. Gaz.* 13 Apr. 2/1 But not by any means must it be supposed that the work of the County Courts is confined to debt-collecting. *Ibid.*, The small debt-collecting work which now largely occupies them. *a* 1852 in G. B. Hill *Talks about Autographs* (1896) 3 F.-M. the Duke of Wellington begs to inform Mr. Snip that he is neither the Marquis of Douro's steward nor Mr. Snip's debt collector. 1881 *Instr. Census Clerks* (1885) 121. 1899 *Westm. Gaz.* 19 May 3/1 Lord Salisbury may be an admirable political 'debt raiser'. 1905 *Daily Chron.* 1 Feb. 6/1 We have handed over £93,376 more towards debt reduction. 1895 MRS. GRINDROD *Siam* 40 People still sell themselves and their families to a wealthy chief, who will pay off their debts contracted through thriftlessness or gambling. These debt slaves give service for a specified term. 1962 *Daily Tel.* 23 Apr. 6 Mortgaging him as a debt-slave to a Pakistani employer already in Britain. 1895 F. A. SWETTENHAM *Malay Sketches* 229 The revolting practice of debt-slavery.

debug (dībʊ·g), *v.* [f. DE- II. 2 + BUG *sb.*[2]] **1.** *trans.* = *DELOUSE *v.*
1960 J. STROUD *Shorn Lamb* vi. 70 We'll..take them round to the Clinic, and..get them debugged there.
2. To remove faults from (a machine, system, etc.). So **debu·gging** *vbl. sb.*
1945 *Jrnl. R. Aeronaut. Soc.* XLIX. 183/2 It ranged from the pre-design development of essential components, through the stage of type test and flight test and 'debugging' right through to later development of the engine. 1959 *New Scientist* 26 Mar. 674/1 The 'debugging' time spent in perfecting a non-automatic programme. 1964 *Discovery* Oct. 51/3 This failure report plays a vital role in the process by which the scientist corrects or de-bugs his programme. 1964 T. W. MCRAE *Impact of Computers on Accounting* iv. 99 Once we have 'debugged' our information system. 1970 A. CAMERON et al. *Computers & O.E. Concordances* 49 Program translation, debugging, and trial runs of the concordance were performed at the University of Michigan Computer Center. *Ibid.*, By Christmas the program was debugged.

debunk (dībʊ·ŋk), *v.* orig. *U.S.* [f. DE- II. 2 + *BUNK *sb.*[4]] *trans.* To remove the 'nonsense' or false sentiment from; to expose (false claims or pretensions); hence, to remove (a person) from his 'pedestal' or 'pinnacle'. Also *absol.* Hence **debu·nker,** one who debunks; **debu·nking** *vbl. sb.* and *ppl. a.*
1923 W. E. WOODWARD *Bunk* i. 2 De-bunking means simply taking the bunk out of things. *Ibid.*, I'm a professional de-bunker. *Ibid.* 4 To keep the United States thoroughly de-bunked would require the continual services of..half a million persons. *Ibid.*, Just how do you go about your de-bunking operations? *Ibid.* 6 Recently we de-bunked the head of a large financial institution. 1927 *Daily Express* 21 Nov. 2/3 The Thucydidean school of what are known as 'debunking' historians. 1927 *Brit. Weekly* 29 Dec. 327/2 The somewhat ruthless process which in America is called 'debunking'—that is, pricking pretentious bubbles [etc.]. 1930 *Times Lit. Suppl.* 6 Mar. 174 The present fashion for 'debunking' great men. *Ibid.* 13 Mar. 217 He is not indeed a 'debunker', but he is as far from being a blind hero-worshipper. *Ibid.* 8 May 378 The aim of 'debunking' a reputation that has been swollen by the uncritical eulogies of contemporaries. 1934 *Municipal Engineering* 12 July 31/1 The London C.C. has decided to 'debunk' Waterloo Bridge, or, in other words, to take away the bunkum that has been attached to it. 1940 *Illustr. Lond. News* CXCVI. 758/2 In fact, he is a reverent man, who enjoys 'debunking' the 'debunkers', if that word may be taken now as acceptable and established English. 1948 *Sat. Rev.* 26 June 13/1 In dealing with military reputations, the author neither glorifies nor debunks. 1958 *Spectator* 13 June 777/1 It is his duty..to debunk the claims of the Fabians. 1960 *Guardian* 10 Dec. 5/3 No cynic, but a debunker.

debus (dībʊ·s), *v.* orig. *Army slang.* Also **debuss.** [f. DE- II. 1 + BUS *sb.*[2]] *intr.* To alight from motor transport. Hence **debu·s(s)ing** *vbl. sb.*
1915 G. ADAM *Behind Scenes at the Front* iv. 126 The battalion has 'embussed', as the orders now phrase it, 'at X, and will debuss at Y'. 1915 *Times* 12 Mar. 10/1 The words 'embuss' and 'debuss' have been consecrated in Staff orders. Many is the battalion which has received orders to 'embuss' at dusk at X, and 'debuss' at Z. 1917 *Ibid.* 28 Sept. 9/5 This Division had already had a very hard time...From the moment of 'de-busing' its life was made very difficult for it. 1958 P. SCOTT *Mark of Warrior* I. 20 The cadets arrived in trucks... They debussed in a tarmac-laid square. 1962 *Times* 26 Oct. (Suppl.) p. v/7 All are close to a car park and debussing station.

Debussyan (dǝbǖ·siǎn), *a.* and *sb.* [-IAN.] **A.** *adj.* Of or pertaining to the French composer Claude Debussy (1862–1918), his music or his style of musical composition. **B.** *sb.* An admirer or follower of Debussy.
1923 P. ROSENFELD *Mus. Chron. 1917–23* 147 There is no Debussyan fluidity, dreamfulness, satin. 1938 *Scrutiny* VII. 170 Any other composer of the time using the same more or less Debussyan idiom. 1959 *Radio Times* 16 Jan. 25/3 An individual mixture of Hungarian folk-song, Debussyan orchestration, and modal inflexions. 1959 R. RUSSELL in M. T. Williams *Art of Jazz* (1960) xviii. 209 The light polychrome orchestral palette of the Debussyians [*sic*].
So **Debussye·sque** *a.*, resembling the style of Debussy.
1946 R. BLESH *Shining Trumpets* (1949) x. 229 A rambling piece in unresolved Debussyesque harmonies. 1963 *Economist* 8 June 1008/1 This is not a Debussyesque hallucination.

début, *sb.* Add: Also *transf.*
1955 *Times* 6 May 14/3 Its considerable acuteness of observation make it a most promising *début* for its director. *Ibid.* 9 May 3/7 All [paintings] making their public London *début* in the Gardens. *Ibid.* 12 May 16/4 Lady Megan Lloyd George made her *début* on the Labour platform tonight. 1971 *Daily Tel.* 27 Apr. 20 The day's newcomer to the equity sections, Hilton's Footwear, made a rather disappointing debut.

débutante. Add: (Earlier and later examples.) Also, a young woman who has recently 'come out' (see COME *v.* 63 o); also *transf.* and *attrib.* See also *DEB.
1801 *Monthly Mirror* Aug. 134 The amiable *debûtante* was called for after the play, and crowns were thrown upon the stage. 1817 U. FOSCOLO *Let.* 20 July in H. C. B. Campbell *Journey to Florence* (1951) 14 The..Adonises, who from February until today, the end of the London season, have discussed the ten thousand *débutantes* of London Society. 1900 E. GLYN *Visits of Eliz.* 270 It was by far the prettiest débutante frock she had ever seen. 1903 *Daily Chron.* 20 June 8/4 The gauging of the skirt proclaims it a débutante among dresses. 1904 *Ibid.* 14 June 4/4 Lord Northampton's postponed ball for his débutante daughter. 1927 [see *DEB]. 1971 *Guardian* 3 Feb. 11/3 Her recollections as a debutante—the coming out balls and the charity balls and the hunt balls.

Debye (dǝbai·). The name of P. J. W. *Debye* (1884–1966), Dutch physicist, used *attrib.* to designate certain phenomena observed and principles enunciated by him, as *Debye effect* (see quots.); *Debye–Hückel theory*, a theory concerned with the inter-ionic forces in electrolytes; *Debye–Scherrer method*, a method for the identification of crystals by photographing the diffraction pattern formed by a beam of X-rays directed on to a powdered sample of the crystal under investigation; *Debye temperature*, a temperature characteristic of an idealized crystal lattice in Debye's theory of specific heats; also, a temperature calculated for a crystalline solid on the assumption that Debye's theory is a correct description of it; *Debye unit*, a unit of electrical dipole moment equal to 10^{-18} e.s.u. (approximately $3 \cdot 336 \times 10^{-30}$ coulomb metre).
1914 *Chem. Abstr.* VIII. 2845 A theoretical and mathematical discussion of the Debye effect, *i.e.*, the influence of rizing temp. on Röntgen ray interference phenomena. 1918 *Ibid.* XII. 2488 A mathematical paper dealing with the Bragg and Debye-Scherrer methods of X-ray spectrometry. 1930 *Chem. Soc. Ann. Rep.* XXVII. 327 The basis of the Debye-Hückel theory of the properties of electrolytic solutions is the idea of an ionic atmosphere. 1934 *Nature* 22 Sept. 459/1 The slopes of these lines yield the following moments in Debye units ($D = 1 \times 10^{-18}$ E.S.U.)...The average is 4·25 D. 1944 *Jrnl. Chem. Physics* XII. 289 A theory in which the Debye characteristic temperature θ was related to the curvature of the interatomic potential energy curve. 1947 *Thorpe's Dict. Appl. Chem.* (ed. 4) VIII. 228/2 Electric dipole moments are expressed in Debye units (represented D.); 1 D. is equivalent to 10^{-18} C.G.S. unit. 1955 H. B. G. CASIMIR in W. Pauli *Niels Bohr* 119 At temperatures well below the so-called Debye temperature Θ only lattice waves with a wave-length of many atomic distances are excited. 1957 *Encycl. Brit.* XXI. 35/1 When ultrasonic radiation passes through an electrolyte, the ions of one sign are affected differently by rapid changes in density from those of the other sign (Debye effect).

decade. Add: **1. b.** (i) A set of resistors (or capacitors or inductors) connected so as to provide a resistance of any integral value between one and ten times the lowest at a single setting of a switch, etc.; a decade is usu. one of several in a box which cover successive multiples of ten in value (so *decade box*, etc.).
1911 W. H. TIMBIE *Elem. Electr.* v. 121 (*caption*) Wheatstone bridge, 'decade' form. *Ibid.*, The coils composing the variable resistance are arranged in such an ingenious way that 10 variations in value in each 'decade' or row are made by the use of only 4 coils. *Ibid.* 122 One decade represents 1000-ohm steps, the next 100-ohm steps, etc. This gives a total range in the four decades of 9999 ohms, by one-ohm steps. 1952 TERMAN & PETTIT *Electronic Measurem.* (ed. 2) xiv. 615 In grouping condensers together to form decade units, it is desirable to employ the smallest possible number of condensers. 1952 F. K. HARRIS *Electr. Measurem.* vii. 229 There may be ten 1-ohm coils for the unit decade, ten 10-ohm coils for the ten decade, etc. *Ibid.* 231 The switching arrangement most common in decade resistance boxes..is the rotary switch having brushes which bear on contact studs.
(ii) A set of ten electronic devices each used to represent a digit in the counting of pulses. So *decade counter, system*, etc.
1948 *Sci. News* VII. 118 Each accumulator contains 10 decade counters and each decade counter 10 gadgets with the..name 'flip-flop'. *Ibid.* 119 The ten flip-flops within a decade counter represent from left to right the digits 0 to 9, while the whole decade represents one place of a ten-digit number. 1952 *Electronic Engin.* XXIV. 376 With a suitably larger power supply there is no limit to the number of decade units which can be used. 1958 *Engineering* 28 Mar. 389/3 The pulses are counted by a three decade system using multi-cathode discharge tubes which display in a digital form the measured voltage. 1963 B. FOZARD *Instrumentation Nucl. Reactors* viii. 88 The display for each decade consists of ten neon lamps arranged in a vertical column corresponding to the numbers 0 to 9.

decadent, *a.* Add: **2. b.** Said of other schools of literature and art characterized by decadence; *spec.* = *ÆSTHETIC *a.* 4. So as *sb.*
1894 M. BEERBOHM in *Yellow Bk.* II. 284 English literature..must fall at length into the hands of the decadents. 1906 R. BROOKE *Let.* Jan. (1968) 37, I contrive to keep the *mens insana in corpore sano* which is all the English decadent may hope for. 1942 WYNDHAM LEWIS *Let.* (1963) 324 The artist is labelled 'decadent' who departs from the Salon norm, or that of the Royal Academy, by the Hitlerite pundit of 'sanity'. 1958 *Times* 20 May 3/7 The last public appearance of Dadaism was that of an aesthetic outlaw under the Nazi ban and during the campaign of 1937 against 'Decadent Art'. 1961 M. LEVY *Studio Dict. Art Terms* 39 *Decadent movement*, a critical and disparaging term sometimes used loosely as a synonym for the Aesthetic Movement.

decadentism (de·kǎdĕntiz'm). [f. DECADENT *a.* + -ISM.] The distinctive qualities or spirit of the decadent school in art, literature, and music; decadent behaviour or characteristics.
[1886 *Athenæum* 24 July 117/1, I shall speak to you some other time about the *décadents* and the *décadentisme*, a malady of the hour.] 1895 *Westm. Gaz.* 27 May 2/1 The ego-mania of decadentism, its love of the artificial..and its exaggeration of the importance of art, have found their English representative among the Æsthetes. 1949 *Mind* LVIII. 114 Existentialism is the philosophy of decadentism.

decaffeinate (dīkæ·fīnēⁱt), *v.* [f. DE- II. 1 + CAFFEINE + -ATE[3].] *trans.* To remove the caffeine from, or to reduce the quantity of, caffeine in (coffee). So **deca·ffeinated** *ppl. a.*, **deca:ffeiniza·tion.**
1927 *Times* 24 May (Advt.), Even experts cannot detect that decaffeinisation makes a shade of difference in taste or aroma. 1934 WEBSTER, Decaffeinate, *v.t.* 1955 *Sci. News Let.* 8 Oct. 238/2 Makers of decaffeinated coffee who claim 97% or more of the caffeine has been removed from their products are correct. 1966 P. V. PRICE *France* 77 Decaffeinated coffee..may always be asked for by people who find that strong black coffee..interferes with their sleep.

decahydrate (dekǎhai·drĕt). *Chem.* [f. DECA- + HYDRATE *sb.*] A compound containing ten molecules of water of crystallization. So **decahy·drated** *a.*
1880 G. LUNGE *Sulphuric Acid* II. 6 The solubility of the decahydrated salt in water rises from 0° to 34° C. with the temperature. 1902 *Science* 24 Jan. 146/1 The equilibrium conditions were determined by the solid salt being in the form of the decahydrate.

decahydrona·phthalene. *Chem.* Also **deka-.** [f. DECA- + HYDRO- + NAPHTHALENE.] A colourless liquid, $C_{10}H_{18}$, much used in the paint industry as a solvent, with a molecular structure consisting of two fused saturated rings; cf. *DECALIN.
1877 *Jrnl. Chem. Soc.* II. 899 Tetrahydroisoxylene absorbs oxygen, while hexhydroisoxylene, like decahydronaphthalene, may be kept without alteration in the air or in oxygen. 1904 *Ibid.* LXXXVI. I. 987 Decahydronaphthalene, $C_{10}H_{18}$, obtained by hydrogenating tetrahydronaphthalene at 175°, is a colourless liquid with an odour of menthol. 1920 *Chem. Abstr.* XIV. 1751 The hydrogenation of naphthalene, converting this ring compd. into dekahydronaphthalene, $C_{10}H_{18}$. 1964 [see *DECALIN].

decal (dī·kǎl). [Abbrev. of DECALCOMANIA.] A transfer (TRANSFER *sb.* 3) produced by decalcomania.
1952 *Electronic Engin.* XXIV. 512 This firm is producing a book of decals or transfers. 1955 *Ibid.* XXVII. 513 These P.V.C. decals are intended for the identification marking of P.V.C. cables and are supplied in the form of ordinary (reverse) transfers. 1959 *New Yorker* 14 Nov. 53/1 The captain himself stopped the fellow's car at the gate as he was leaving, scraped the decal off with a razor blade. 1970 *Win* 15 Feb. 35/2 (Advt.), Put this decal in your car window and let the American flag freaks know where you stand.

decalage (dī·kǎlāʒ), ‖ **décalage** (dekalāʒ). *Aeronaut.* [Fr. *décalage* displacement, f. *décaler* to displace.] (See quots.)
1917 R. B. MATTHEWS *Aviation Pocket-Bk.* (ed. 5) XII. 267 *Decalage*, the difference in the angle of incidence between any two distinct aerofoils on an aeroplane; *e.g.*, the main plane and the tail; or more usually between the chords of the upper and lower planes of a biplane. 1933 *Flight* (Suppl. Aircraft Engin.) 27 Apr. 396d/1 The term aerodynamic decalage..defined as the angle that the no lift angle of the upper plane makes with the no lift angle of the lower plane, positive when the upper plane is at greater incidence than the lower.

decalescence (dīkǎle·sĕns). *Metallurgy.* [f. DE- II. 1 + CALESCENCE.] (See quot. 1954.) So **decale·scent** *a.*
1893 *Funk's Stand. Dict.*, Decalescence. 1910 E. F. LAKE *Composition & Heat Treatment of Steel* vi. 68 The temperature again falls uniformly until it reaches..1350° F., at which point the second change takes place..designed Ar 2. After the change has been completed at this point, it again lowers in temperature uniformly to the next point, or Ar 1... After this change takes place, it then gradually lowers to atmospheric temperature. These have been named the decalescent points. 1912 A. SAUVEUR *Metallogr. Iron & Steel* vii. 2 The notation Ar_1 and Ac_1 are used for the recalescence point and its reversal... The point Ac_1 has been called point of 'decalescence' by some writers. 1916 D. K. BULLENS *Steel* 417 For the majority of heat-treatment work, it is more important to know the location of the decalescent points than that of the recalescent points. *Ibid.* 425/2 Decalescence. 1954 *Gloss. Terms Iron & Steel* (*B.S.I.*) I. 11 *Decalescence*, absorption of heat without increase of temperature when the metal is heated through the critical points.

Decalin (de·kǎlin). Also **Dekalin.** [Trade mark in the U.S.] = *DECAHYDRONAPHTHALENE.
1920 *Chem. Abstr.* XIV. 1227 New turpentine substitutes ..Dekalin, $C_{10}H_{18}$...In the trade dekalin is not found pure, but always contains tetralin and is handled under the name tetralin extra. 1943 *Thorpe's Dict. Appl. Chem.* (ed. 4) VI. 352/2 The decalin so obtained was the pure *cis*-compound, whereas nickel at 160° yields chiefly *trans*-decalin. 1946 *Nature* 14 Sept. 365/2 The fuel used [in the torpedo] is a by-product of the coal distillation industry known as 'Dekalin'. 1964 N. G. CLARK *Mod. Org. Chem.* xix. 391 At 200° catalytic hydrogenation over nickel yields tetrahydronaphthalene ('tetralin'), while at 250°, decahydronaphthalene ('decalin') results.

decamethonium (de:kǎmĕþōᵘ·niǝm). *Pharm.* [f. DECA- + *METHONIUM.] A quaternary ammonium cation, $[(CH_3)_3N(CH_2)_{10}N(CH_3)_3]^{++}$;

also, the bromide or iodide salt of this, used as a muscle relaxant.

1949 G. ORGANE in *Lancet* 7 May 774/2 Decamethonium iodide (C 10) is a stable, neutral, non-irritant muscle relaxant, roughly five times as potent as *d*-tubocurarine chloride. 1963 *Lancet* 19 Jan. 130/1 With decamethonium muscle fasciculation was only slight, whereas with suxamethonium it was severe and long lasting.

decanate[1]. Delete †*Obs.* and add later example.

a 1963 L. MACNEICE *Astrol.* (1964) iv. 126 Certain stones are in 'sympathy' with certain decanates.

decant, *v.*[1] **c.** (Further examples.)

1915 J. BUCHAN *39 Steps* vii. 171, I was decanted at Crewe..and had to wait till six to get a train for Birmingham. 1925 WODEHOUSE *Carry On, Jeeves!* ii. 46 The nurse..got up with the baby and decanted it into a perambulator. 1959 T. S. ELIOT *Elder Statesman* II. 47 Let's hope this [conversation] was merely the concoction Which she decants for every newcomer.

decapacitation (dīkăpæsitēⁱ·ʃən). *Physiol.* [f. DE- II. 1 + *CAPACITATION b.] The process of removing the effects of capacitation; the action of decapacitating a spermatozoon.

1961 *Federation Proc.* XX. I. i. 418/2 (*heading*) A study of the decapacitation factor present in rabbit seminal plasma. 1969 *New Scientist* 31 July 234/1 Such sperm were said to have suffered decapacitation. 1970 *Nature* 11 July 182/2 To determine whether these compounds would behave like 'decapacitation factor' from seminal plasma, that is whether they would be absorbed by sperm and reverse the capacitated state, the following experiment was carried out.

So **decapa·citate** *v. trans.*, to deprive (a spermatozoon) of its ability to penetrate the *zona pellucida* of the ovum which it gains by capacitation; **decapa·citated** *ppl. a.*

1964 T. MANN *Biochem. Semen* (ed. 2) ii. 32 There is some evidence..that capacitated rabbit spermatozoa brought in contact with seminal plasma can be de-capacitated. 1970 *Nature* 11 July 182/2 We cannot say whether recapacitation represents the recovery of activity by decapacitated spermatozoa, or the activation of sperm not capacitated in the first place.

decaploid (de·kăploid), *a. Biol.* [f. DECA- + *-PLOID.] (See quot. 1940.)

1932 SANSOME & PHILP *Recent Advances in Plant Genetics* v. 165 In the first case the species with the highest chromosome number has ten sets as the basic number of the genus (decaploid). 1940 *Chambers's Techn. Dict.* 227/2 *Decaploid*, having ten times the haploid number of chromosomes. 1946 *Ann. Reg. 1945* 351 In the U.S.S.R. Zhebrak has worked on the properties of..decaploid wheats with seventy chromosomes. 1946 *Nature* 17 Aug. 239/2 Since the basic haploid number of *Artemia salina* is known to be 21, the present race must be considered as decaploid with a slight augmentation of the number of 105 tetrads.

decapod. Add: **A.** *sb.* **2.** A heavy-freight locomotive with ten driving wheels originating in the United States. Also *attrib.*

1888 *Scribner's Mag.* Aug. 183 Consolidation and decapod types of engines, which have four and five pairs of driving-wheels. 1903 *Westm. Gaz.* 25 Feb. 5/2 'Decapod'! New Hustling Locomotive for G.E.R. 1906 *Daily Mail* 17 Dec. 5/7 Messrs. Robert Stephenson and Co. of Darlington have just completed three..huge 'decapod' locomotives. 1947 L. M. BEEBE *Mixed Train Daily* 276 Decapods and other classifications with 250-pound engine pressure.

decapsulate (dīkæ·psiŭlēⁱt), *v. Surg.* [f. DE- II. 1 + CAPSULE *sb.* + -ATE[3].] *trans.* To remove the capsule of. So **decapsula·tion,** the removal of a capsule.

1907 *Practitioner* Oct. 471 The decapsulation of the normal healthy kidney. *Ibid.* Dec. 778 The renal artery of a cat, whose corresponding kidney had been decapsulated and fixed two months previously. 1962 H. L. KERN et al. in A. Pirie *Lens Metabolism Rel. Cataract* 386 The lenses were either decapsulated or mashed in the culture tubes. 1963 J. P. MERRILL in Strauss & Welt *Dis. Kidney* xiii. 457/1 Decapsulation of the kidney is of little value after acute renal failure has been present for several days.

decarbonization. Add: **b.** Removal of carbon deposit from inside an internal combustion engine.

1912 *Motor Manual* (ed. 14) vi. 232 Certain preparations in liquid form are sold for which it is claimed that, when injected in the cylinders, decarbonisation and thorough cleansing is effected.

decarbonize, *v.* Add: **b.** To remove carbon deposit from (an internal combustion engine). Also *absol.*

1915 R. D. PRICE *U.S. Patent 1,148,403*, Decarbonizing compound...My invention relates to a compound for use in removing the carbon deposits from the cylinders of gasolene engines and like internal combustion motors...I provide a decarbonizing composition consisting of the following ingredients. 1916 R. T. NICHOLSON *Book of Ford* vii. 93 Decarbonizing tools are sold, which can be inserted through the sparking-plug holes, and used to scrape the carbon out thus. *Ibid.*, Have nothing to do with chemical decarbonizing processes. 1925 *Morris Owner's Manual* x. 70 Materials required—for decarbonizing only—are the standard tool kit and a bottle of gold size. *Ibid.*, The head has to be lifted to decarbonize. *Ibid.* 71 When decarbonizing the Morris engine. 1930 *Engineering* 10 Jan. 44/1 This engine can be decarbonised and the valves ground in in less than an hour. 1967 E. RUDINGER *Consumer's Car Gloss.* (ed. 2) 30 Decarbonising, that is, scraping off the carbon..from the cylinder head and pistons.

decarbonizer (dīkā·ɹbŏnəizəɹ). [f. DECARBONIZE *v.* + -ER[1].] One who or that which decarbonizes; *spec.* see quot. 1921.

1890 in WEBSTER. 1921 *Dict. Occup. Terms* (1927) § 449 *Decarboniser* (sugar refining); attends a number of steam-heated cylinders..in which animal charcoal is decarbonised after revivification in char kiln.

decarboxylase (dīkaɹbǫ·ksilēⁱz, -ēⁱs). *Biochem.* [f. DE- + CARBOXYL + *-ASE.] Any enzyme which effects decarboxylation.

1940 E. F. GALE in *Biochem. Jrnl.* XXXIV. 399 When the organism is grown on the surface of broth agar, it has negligible decarboxylase activity. *Ibid.* 411 The production of decarboxylases may be the method by which the organism extends its range of existence, utilizing amino-acid decarboxylation when other substrates and methods of attack are no longer available. 1947 *Sci. News* IV. 64 The particular enzyme, containing B_1, was called pyruvic oxidase, sometimes decarboxylase.

decarboxylate (dīkaɹbǫ·ksilēⁱt), *v. Chem.* [back-formation from next.] **a.** *trans.* To remove a carboxyl group from. **b.** *intr.* To lose a carboxyl group, to undergo decarboxylation. Hence **decarbo·xylated, decarbo·xylating** *ppl. adjs.*

1922 *Jrnl. Biol. Chem.* L. 132 Are all strains of colon bacilli capable of decarboxylating histidine in our standard medium? 1930 *Biochem. Jrnl.* XXIV. 262 It is necessary to employ a decarboxylating reagent. 1946 *Nature* 27 July 132/1 The carboxyl group..is very unstable and in the presence of alkali readily loses carbon dioxide, forming a decarboxylated compound. 1959 A. ALBERT *Heterocyclic Chem.* vii. 269 Dehydracetic acid undergoes ring-opening to aceto-acetylacetone-3-carboxylic acid: this decarboxylates to acetoacetylacetone which cyclizes to 2:6-dimethyl-γ-pyrone. 1962 S. G. WALEY in A. Pirie *Lens Metabolism Rel. Cataract* 354 Only certain carboxylic acids..decarboxylate readily non-enzymatically. 1964 W. G. SMITH *Allergy & Tissue Metabolism* iii. 36 Mast cells can decarboxylate histidine at an appreciable rate.

decarboxylation (dīːkaɹbǫksilēⁱ·ʃən). *Chem.* [f. DE- II. 1 + CARBOXYL + -ATION.] The removal of a carboxyl group.

1922 *Jrnl. Biol. Chem.* L. 189 The decarboxylation of histidine is influenced by the presence of other amino-acids. 1940 [see *DECARBOXYLASE]. 1956 *Nature* 18 Feb. 333/2 The very high initial values for production of carbon dioxide in nitrogen..may be due to rapid decarboxylation.

decartel(l)ization (dīkāːɹtĕlaizēⁱ·ʃən). [f. DE- + *CARTEL(L)IZATION.] The abolition of the system of trade cartels. Hence **deca·rtellizer,** one who favours decartellization.

1947 *Times* 20 Oct. 5/6 At least three senior American officials have resigned from the decartellization branch. *Ibid.* 5/7 The socializers and decartellizers may expect to be displaced by experts in management and production. 1948 *Hansard* CCCCXLV. 338 The functions of the Division are supervisory and cover disarmament, decartelisation and the revival of peaceful industry. 1955 *Times* 4 July 9/7 Whereas only lately..decartelization in Germany was the order of the day, it is now again cartelization.

decastich. (Further example.)

1891 S. R. DRIVER *Introd. Lit. O.T.* viii. 376 A short poem (on the value of industry to the farmer), consisting of a decastich.

decasualize (dīkæ·ziŭălaiz), *v.* [f. DE- + CASUAL *a.* + -IZE.] *trans.* To remove the casual element from (labour). So **deca:suali·za·tion,** the abolition of casual labour.

1892 [see DE- II. 1]. 1907 *Ann. Rep. Universities' Settlement in E. London* 34 In its demand for nationalisation before means are taken to decasualise labour. 1910 *Fabian News* XXI. 16/2 Taxi-driver desires to decasualize his profession by acquiring regular clientèle. 1928 *Daily Tel.* 15 May 12/4 Bermondsey contains a large number of casual riverside labourers, whose decasualisation has been the steady aim of the Port authorities. 1961 *New Statesman* 5 May 698/2 The process of de-casualisation begun by Ernest Bevin. 1962 *Economist* 10 Feb. 497/1 Mr Cousins..spoke of the need to 'decasualise' labour relations in the docks.

decathlon (dekæ·þlǫn). [f. Gr. δέκα ten + ἄθλον contest.] In the modern Olympic games, a composite contest consisting of ten specific events. Also *attrib.* Hence **deca·thlete** *sb.*

1912 *Times* 16 July 13/1 The Decathlon was brought to a close to-day, the events to be decided being the pole jump, throwing the javelin..and 1,500 metres flat race. 1920 *Glasgow Herald* 23 Aug. 10 In the stadium proper most of the morning was given up to a succession of Decathlon events, including hurdles, throwing the discus, and pole jump. 1928 *Daily Tel.* 6 Nov. 19/4 The Springboks' Decathlon champion..broke down. 1968 *Evening Standard* 9 Aug. 30/5 Britain's leading decathlete. 1970 *Encycl. Brit.* VII. 154/2 Similar to the decathlon is the all-around championship of the U.S. Amateur Athletic Union.

decatise (de·kătaiz), *v.* [f. Fr. *décatir* to sponge or steam (cloth).] *trans.* To subject (cloth) to the action of steam in order to give it a permanent lustre or finish. Hence **de·catising** *vbl. sb.*

1907 *Textile Mercury* 16 Feb. 124/2 Fabrics of good quality should be decatised before dyeing. 1921 *Dict. Occup. Terms* (1927) § 384 Decatising machine minder. 1963 A. J. HALL *Textile Sci.* v. 224 A machine for decatising wool fabrics.

decaudation (dīkǭdēⁱ·ʃən). [f. DECAUDATE *v.* + -ATION.] Removal of the tail or 'tails'.

1897 *19th Cent.* May 805 Decaudation with mutilation is seen in *bike* for *bicycle*. 1927 *Daily Tel.* 9 Aug. 8/5 The decaudation and blanching and unstiffening of the waiter are another phase of the transformation which has abolished the frock-coat and the silk hat and women's hair.

Decauville (dəkōᵘ·vil). The surname of P. *Decauville* (1846–1922), French engineer, used *attrib.* or alone to designate a type of narrow-gauge railway invented by him.

1899 W. H. COLE *Light Railways* xiv. 245 The 2ft. rolling-stock on a French 'Decauville' line. *a* 1935 T. E. LAWRENCE *Mint* (1955) I. xxix. 98 We ran eight lorry loads of the Decauville rails. 1958 F. STARK *Alexander's Path* ix. 125 The little decauville running down from the sheds.

decay, *sb.* Add: **2. b.** *Physics.* The gradual decrease in the radioactivity of a substance; hence, the spontaneous transformation of a single atomic nucleus or elementary particle into one or more different nuclei or particles. Also *attrib.*

1897 RUTHERFORD in *Phil. Mag.* XLIV. 425 The intensity of the radiation varied widely, but in all cases the rate of decay was found to be in close agreement with theory. 1902 RUTHERFORD & ALLEN in *Phil. Mag.* IV. 708 The decay-curve for a copper wire exposed 210 minutes inside the laboratory. 1905 *Nature* 13 Apr. 574/1 Different samples gave for the half-period of decay from 52 to 55 seconds. 1931 G. GAMOW *Constitution of Atomic Nuclei* ii. 31 One of the most important characteristics of a decaying nucleus is its decay constant.., giving the probability of disintegration per unit time. 1938 R. W. LAWSON tr. *Hevesy & Paneth's Man. Radioactivity* (ed. 2) xxiii. 223 Uranium Z..may be a decay product of a uranium isotope other than ^{234}U. 1958 J. L. PERKIN in O. R. Frisch *Nucl. Handbk.* iii. 7 The decay of a nucleus via various excited levels of the final nucleus is shown diagrammatically. 1962 H. D. BUSH *Atomic & Nuclear Physics* iv. 81 The half-life of uranium I..is obviously too long to determine by measuring the decay of its activity. 1968 M. S. LIVINGSTON *Particle Physics* xi. 196 The most fundamental weak interaction is the decay of the neutron into a proton, an electron, and an antineutrino. 1969 *Times* 12 Mar. 4/7 The radioactive decay of uranium..has long been recognized as a means of fixing the ages of remote cosmological events.

c. A progressive diminution in the amplitude of an oscillation or vibration.

1906 J. A. FLEMING *Princ. Electr. Wave Telegr.* i. 15 Frictional resistance causes decay in the amplitude of the oscillations by dissipating their energy as heat. 1922 GLAZEBROOK *Dict. Appl. Physics* II. 111/1 The damping of the oscillations is determined by *b*..which is called the damping coefficient or the coefficient of decay. 1950 STEPHENS & BATE *Wave Motion & Sound* 357 This decay of amplitude is known as damping and the motion is referred to as damped harmonic motion. 1962 A. NISBETT *Technique Sound Studio* iii. 56 Some percussive instruments, such as tympani, continue to sound for some time, and have decay characteristics which are somewhat similar to that of reverberation.

decay, *v.* Add: **2. b.** Of an oscillation or vibration: gradually to decrease in amplitude, so that each swing is smaller than the one before. Also said of the amplitude of the oscillation.

1879 *Encycl. Brit.* VIII. 11/2 Sir W. Thomson investigated mathematically the discharge of a Leyden jar..and predicted that under certain circumstances the discharge would consist of a series of decaying oscillations. 1906 J. A. FLEMING *Princ. Electr. Wave Telegr.* 573 A very important matter in connection with practical electric wave telegraphy is the rate at which the wave amplitude decays during the emission of a wave train from the antenna. 1927 I. B. CRANDALL *Theory of Vibr. Syst.* i. 8 The natural oscillations may be made to decay very rapidly, or to disappear altogether, if the damping factor is made very large. 1944 A. WOOD *Physics of Music* ii. 23 Sound-waves are carrying energy more rapidly away from the fork, and the vibrations therefore decay more rapidly. 1959 *Chambers's Encycl.* I. 375/1 The amplitude of the swing about the final true position decays exponentially with time.

c. *Physics.* Of radioactivity: gradually to diminish in intensity; of a substance: to suffer a gradual decrease in its radioactive power, to undergo nuclear disintegration. Hence, of a radioactive substance, an atomic nucleus, or an elementary particle: to change or disintegrate *into* one or more different substances, etc.

1900 RUTHERFORD in *Phil. Mag.* XLIX. 177 The intensity of the 'excited' radiation falls to half its value in about eleven hours, or one decays 660 times faster than the other. 1913 —— *Radioactive Substances* viii. 339 The active deposit ..decays *in situ* and this results in an apparent decrease of the activity. 1942 J. D. STRANATHAN *'Particles' Mod. Physics* viii. 326 The half life period *T* of a radioactive substance is defined as the time required for one half of the active material present at any time to decay. 1958 W. K. MANSFIELD *Elem. Nuclear Physics* iii. 21 It is found experimentally that the probability of an unstable nucleus, known as a radioactive nucleus, decaying within a given time is constant. 1962 H. D. BUSH *Atomic & Nuclear Physics* iv. 80 Uranium X does not decay into a stable product but is

the parent of a chain of radioactive daughter products. 1968 M. S. LIVINGSTON *Particle Physics* iv. 72 In matter, when π^+ pions are slowed down by ionizing impacts, they decay into positive muons and muon neutrinos.

decayed, *ppl. a.* **1.** spec. in phr. *decayed gentlewoman.*

1851 HAWTHORNE *Ho. Sev. Gables* (1852) ii. 27 We might point to several little shops of a similar description..where a decayed gentlewoman stands behind the counter. 1921 G. B. SHAW *Lett.* (1952) 218 The celebrated decayed gentlewoman who had to cry laces in the street for a living but hoped that nobody heard her. 1961 J. GLOAG *Victorian Comfort* viii. 212 Impoverished widows and spinsters of the middle classes, who were officially described as 'decayed gentlewomen'.

Decca (de·kă). The name of a British company, used *attrib.* to designate air and sea radio-navigational systems developed by them.

1946 *Int. Meeting Radio Aids to Marine Navigation* II. III. xiii. 48 The Decca System of Radio Navigation consists of a number of transmitting beacon stations at fixed positions on shore, and special receiving equipment to be carried in the ship or aircraft by means of which the navigator can determine his position relative to these beacons. 1950 *Jrnl. Inst. Navigation* III. 330 The useful range of the Decca Navigator system is about 500 miles by day,..to about 300 miles by night. 1959 *Observer* 8 Feb. 4/5 A..flying aid invented in America and developed here—the 'Decca Navigator'; its technical description is a 'high accuracy hyperbolic navigation system'. This replaces the traditional string of radio beacons by a fine grid of radio signals sent out from four transmitters linked to make a Decca 'chain'.

deceased, *ppl. a.* Add: **1. c.** *transf.* Of a deceased person.

1906 *Times* 29 Aug. 11/2 London and North-Western stock was noticeably plentiful for delivery, and was said to have been sold heavily during the account on behalf of a deceased estate.

decelerate (dīse·lĕrēⁱt), *v.* [f. DE- + AC)CELERATE *v.*] *trans.* To diminish the speed of; to cause to go slower. Also *intr.* or *absol.*

1899 *Times* 30 Sept. 3/5 The 7.45 a.m. ex Exeter..is decelerated nine minutes. 1902 *Westm. Gaz.* 22 Oct. 2/1 Two years ago this timing was decelerated by 5 min. 1924 *Public Opinion* 26 Apr. 399/3 Pushing the third button decelerates the whole system. 1928 *Evening Standard* 18 Mar., There would be a catastrophe if you decelerated too suddenly. 1957 *Times* 3 Sept. 4/2 Their thrust reversal unit, designed to assist in decelerating aircraft after landing.

deceleration (dīselĕrēⁱ·ʃən). [f. as prec. + -ATION.] The action or process of decelerating.

1897 *Daily News* 20 July 5/2 As far as the Great Northern and Caledonian Companies are concerned, 'deceleration' has been the order of the day in making the summer arrangements. 1900 *Ibid.* 24 Mar. 5/6 These alterations and 'decelerations' affect only Chatham trains. 1922 *Field* 18 Feb., Our travel was one of smooth acceleration and deceleration. 1926 *Bulletin* 1 Dec. 5/6 Drive cautiously and avoid sudden acceleration or deceleration. 1930 A. BENNETT *Imperial Palace* iv. 17 Her accelerations and decelerations, her brakings, could hardly be perceived. 1955 *Roadcraft* (H.M.S.O.) vii. 70 There are two normal methods by which the speed of a motor vehicle may be reduced—(a) by the deceleration of the engine as the pressure on the accelerator is relaxed, and (b) by the application of the brakes. 1971 *Daily Tel.* 1 July 30/5 The cosmonauts could have died of a lack of blood supply to the brain caused by the strong deceleration during re-entry from space.

decelerator (dīse·lĕrēⁱtəɹ). [f. *DECELERATE *v.* + -OR.] An apparatus for reducing the speed of an engine.

1907 *Westm. Gaz.* 18 Nov. 6/3 The decelerator which automatically throttles the engine whenever the clutch is disengaged.

decelerometer (dīːselĕrǫ·mitəɹ). [f. *DECELERATE *v.* + -METER, after *ACCELEROMETER.] An instrument for ascertaining the deceleration of a moving body.

1924 *Sci. Amer.* Mar. 171/3 A decelerometer..is being used for making tests of the effectiveness of brakes. 1958 *Engineering* 21 Mar. 371/3 Recording decelerometers were found capable of giving a complete record of braking behaviour.

December. Add: **December moth** (see quots.).

1832 J. RENNIE *Consp. Butterfl. & M.* 38 The December Moth (*Pœcilocampa Populi*, Stephens) appears in December. 1907 R. SOUTH *Moths Brit. Isles* I. 113 (*heading*) The December Moth (*Pœcilocampa populi*). 1945 V. TEMPLE *Butterflies & Moths in Brit.* Pl. 80 (*caption*) Caterpillar of the December Moth.

decent, *a.* Add: **3. b.** *spec.* in mod. colloq. use (see quot.).

1949 R. HARVEY *Curtain Time* 63 Sometimes, if she knew one of the actors or actresses, she would knock at a door and call 'Are you decent?' (That old theatrical phrase startled people who didn't belong to the theatre, but it simply meant 'Are you dressed?')

5. b. Of a person: kind, accommodating, pleasant. *colloq.*

1902 E. NESBIT *Five Children & It* iii. 101 'Well,' said Cyril, 'if you ask me I think it was rather decent of her'—'Decent?' said Anthea; 'it was very nice indeed of her...' 1909 GALSWORTHY *Joy* III, Couldn't you just go up and give her a message..it would be most awfully decent of you. 1910 L. A. HARKER *Master & Maid* xvii. 255 Fellows had told him how cut up old Nick was when that chap died in his house, and Bruiser was a jolly sight decenter than old Nick. *Ibid.* xx. 308 He was a very decent chap, quite a man of the world. 1928 W. DEEPING *Old Pybus* ix. § 3 The pater has been rather decent. 1932 'N. SHUTE' *Lonely Road* vi. 125 That's really very decent of you. 1944 R. LEHMANN *Ballad & Source* 36 This is a ripping place, and they're being jolly decent to us.

decentralist (dīse·ntrălist). [f. DECENTRALIZE *v.*: see -IST.] One who believes in a policy of decentralization. Also *attrib.* or as *adj.*

1920 *Glasgow Herald* 18 Mar. 9 The struggle between the Centralists and the Decentralists or Regionalists in the matter of administration. 1921 *Q. Rev.* Apr. 398 The Centralists and Decentralists are about equal in numbers. 1941 *Commonweal* 20 June 204 For several years, we have had an evident decentralist movement. People have fled from the big cities, and landed outside. 1964 M. MCLUHAN *Understanding Media* I. i. 8 Automation technology..is integral and decentralist in depth.

decentralization. Add: Also *attrib.*

1898 *Daily News* 8 Sept. 5/1 The recommendations of Lord Lansdowne's decentralisation Committee. 1906 *Daily Chron.* 23 Jan. 5/2 The decentralisation schemes introduced by the late Government. 1908 *Ibid.* 12 May 6/4 With these larger and wider reforms the Decentralisation Commission has nothing to do. 1961 *Ann. Reg. 1960* 219 The decentralization reform..had to be modified.

decentralizer (dīse·ntrăləizəɹ). [f. DECENTRALIZE *v.*] = *DECENTRALIST.

1898 J. E. C. BODLEY *France* i. 36 The venerable savant, himself a decentralizer. 1963 *Economist* 19 Oct. 240/1 The 'decentralisers', who were ready to let Angola..and the rest move some way towards autonomy.

decentre (dīse·ntəɹ), *v.* Also **decenter.** [f. DE- + CENTRE *sb.* or *v.*] **1.** *trans.* To remove the centre from (in quot., CENTRE *sb.* 13).

1870 *Engineering* X. 485/1 In 1847, M. Beaudemoulin..made use of sand in decentering arches.

2. To place out of centre; to render eccentric; *spec.* in *Optics* (see quot. 1889). So **dece·ntrated** *ppl. a.*; **decentra·tion**; **dece·nt(e)red** *ppl. a.*; **dece·nt(e)ring** *vbl. sb.* and *ppl. a.*

1889 E. E. MADDOX *Clinical Use of Prisms* ii. 45 A lens is said to be normally centred when the optical and geometrical centres coincide..but to be decentred when, as in *Fig.* 21, the optical centre O, and the geometrical centre G, are apart. *Ibid.* 47 The word 'decentering' is generally applied to the removal of the optical centre away from the geometrical centre of the lens. *Ibid.* 49 The dislocation of the optical centre..to be obtained by decentration proper, that is by decentering the lens *in* its rim. 1909 *Cent. Dict. Suppl.*, Decentrated. 1946 A. HUXLEY *Let.* 26 Oct. (1969) 551 Too much concentration defeats its own end, and.. effective concentration has to work by a process of decentration. 1962 L. S. SASIENI *Optical Dispensing* v. 126 The right lens may be decentred out and the left lens decentred inwards. *Ibid.*, The amount of decentration produced by the asymmetry..is..likely to have appreciable prismatic effect. 1964 *Yearbook Astr. 1965* 96 The stresses..tend to de-centre the secondary, making it move off the optical axis.

decerebrate (dīse·ribrĕt), *a.* [f. DE- + CEREBRUM: see -ATE[2].] Deprived of the cerebrum; having the cerebrum removed or the brain-stem cut; also, resulting from this, as *decerebrate rigidity*, a state in which the limbs are extended and certain skeletal muscles rigidly contracted. So **decerebra·tion,** removal of the cerebrum, cutting of the brain-stem.

1897 C. S. SHERRINGTON in *Proc. R. Soc.* LXI. 244, I have found the 'long intra-spinal reflexes', like sub-cerebral rigidity ('decerebrate tonus',) locally abolished..by total severance of the sensory spinal roots belonging to their own region of terminal discharge. 1898 —— in *Jrnl. Physiol.* XXII. 319 In a communication to the Royal Society in 1896 I described under the name decerebrate rigidity a condition of long-maintained muscular contraction supervening on removal of the cerebral hemispheres. 1900 DORLAND *Med. Dict.* 192/2 *Decerebration*, the removal of the brain in performing craniotomy. 1902 *Encycl. Brit.* XXXI. 744/1 The decerebrate monkey exhibits 'cataleptoid' reflexes. Father Kircher's *experimentum mirabile* with the fowl and the chalk line succeeds best with the decerebrate hen. 1915 W. OSLER *Let.* 29 July in H. Cushing *Life* (1925) II. 484 It is a sort of psychical decerebration... I suppose it is the shock & strain. 1927 HALDANE & HUXLEY *Anim. Biol.* vi. 139 A 'decerebrate' animal, i.e. one in which the cerebral hemispheres have been removed, though unconscious, can to some extent adjust its standing posture. 1950 *Sci. News* XV. 19 An anaesthetised, or decerebrate animal. 1969 'M. INNES' *Family Affair* xvii. 187 Oswyn's virtually decerebrate, of course. 1970 J. CROSSLAND *Lewis's Pharmacol.* (ed. 4) ix. 181 Decerebration (which involves destruction of the brain above the level of the midbrain) releases this restraint and causes decerebrate rigidity.

decertify (dīsɜ̄·ɹtifəi), *v.* [f. DE- + CERTIFY *v.*] *trans.* To remove a certificate or certification from; *spec.* to remove certification of insanity from. Hence **de:certifica·tion.**

1918 *Oxf. Times* 11 May 7/3 The duty was cast on him of decertifying certain men if he thought they came within the terms of the order. *Ibid.*, I have decertified certain men. 1947 B. F. TUCKER *Guide to Nat. Labor Relations Act* 138 Decertification. The Board has power to revoke a certification already issued. 1959 *Times* 16 Mar. 10/2 The Newfoundland action in decertifying this union, which is affiliated to the powerful Canadian Labour Congress. 1960 *Sunday Times* 3 Jan. 11/4 The 1960s are likely to see the decertification of the majority of the remaining certified mental defectives. 1964 G. L. COHEN *What's Wrong with Hospitals?* viii. 177 They did all the pleasant jobs. I didn't dare open my mouth; I wanted to be de-certified.

deci-. Add: **1. decineper,** one tenth of a neper.

1931 *Telegr. & Teleph. Jrnl.* Mar. 119/2 The unit [of attenuation] adopted in France and Germany is the neper or decineper.

Decian (dɪ·ʃiăn), *a.* [f. *Decius* + -AN.] Of or pertaining to the Roman emperor Decius or his reign (A.D. 249–251), and esp. the persecution of Christians which took place under him.

1695 J. SAGE *Cyprianic Age* 92 He tells Cornelius Bishop of Rome, Ep. 59. That he was Proscribed in the Days of the Decian Persecution. 1717 N. MARSHALL *Wks. St. Cyprian* II. 13 The Decian persecution..began in Africa with a popular tumult. 1846 C. MAITLAND *Ch. in Catacombs* iv. 118 The Decian persecution at Carthage. 1869 T. W. ALLIES *Formation Christendom* II. xii. 323 At the eve of the great Decian persecution in 249. 1939 A. J. TOYNBEE *Study Hist.* VI. 203 The Decian persecution of the Christian Church.

decibar (de·sibāɹ). *Meteorol.* [f. DECI- + *BAR *sb.*[6] 2.] A unit of barometric pressure equal to one-tenth of a bar.

1910, 1914 [see *BAR *sb.*[6] 2]. 1967 *Oceanogr. & Marine Biol.* V. 90 The difference between the geostrophic circulation at 1750 decibars and that at 1000 decibars is negligible.

decibel (de·sibel). [f. DECI- + *BEL.] The usual unit (equal to one-tenth of a bel) used in comparing the power levels in two electrical communication circuits (or two parts of the same circuit) or the intensities of two sounds; freq. used to express a single power level or sound intensity relative to some reference level (stated or understood). Also used loosely in non-technical contexts. Abbrev. **db.**

1928 *Electrical Communication* VII. I. 33/2 If common logarithms are used, the reproduction is obtained in Decibels. 1929 W. H. MARTIN in *Bell System Techn. Jrnl.* VIII. 2 The Bell System has adopted the name 'decibel' for the 'transmission unit', based on a power ratio of $10^{\cdot 1}$...For convenience, the symbol 'db' will be employed to indicate the name 'decibel'. 1930 *Discovery* Dec. 398/2 The band-pass filter, which follows the low frequency modulator, allows the lower side-band to pass with an attenuation of six decibels. 1930 J. R. FIRTH *Speech* iv. 34 In order to measure differences of powers in units of sensation telephone engineers have adopted a logarithmic unit called the decibel (db.). 1937 *Nature* 28 Aug. 370 The First International Acoustical Conference...The 'phon'..was adopted as the unit in the subjective scale of equivalent loudness, while the use of the 'decibel'..was restricted to the scale of the associated energy or pressure level. 1948 *Punch Miscellany* 54/1 No one misses a single decibel of your conversation. 1953 L. F. BROSNAHAN *Some O.E. Sound Changes* iii. 28 The differences are measured in decibels, a decibel corresponding roughly to the change in loudness of a sound which is perceived as such. 1955 S. GIBBONS *Shadow of Sorcerer* iv. 37 A deep, fierce droning, louder by many decibels than the noise of wasps. 1958 *Engineering* 31 Jan. 133/3 Measurement at 800 c/s of the attenuation of a physical circuit with loss up to 10 db. 1959 K. HENNEY *Radio Engin. Handbk.* (ed. 5) xxviii. 23 Noise measurements are commonly expressed in decibels above a base which is approximately at 85 db below 1 mw. 1970 *Daily Tel.* 4 Nov. 12/4 Patrons are in for a fairly high rate of decibel battering from these two muscular and very loud players.

decidability (dīsəidăbi·lĭti). [f. DECIDABLE *a.* + -ITY, tr. G. *entscheidbarkeit.*] The quality of being decidable (see next).

1940 in W. V. QUINE *Math. Logic* 344. 1959 K. R. POPPER *Logic of Sci. Disc.* viii. 191 The questions I have raised..constitute the problem of decidability.

decidable, *a.* Add: **1. b.** *Logic* and *Math.* Of a statement or formula: capable, within the system to which it belongs, of being proved or disproved, or of being shown to be true or false, or to have some other property or not to have it; of theories, systems, etc.: having a solvable decision problem or computability problem.

1942 R. CARNAP *Introd. Semantics* xxvi. 163 *Ti* is decidable. 1950 W. V. QUINE *Methods Logic* (1952) 247 Elementary algebra is completable and mechanically decidable, while elementary number theory is not. 1958 [see *decision method* s.v. *DECISION 5]. 1963 G. T. KNEEBONE *Math. Logic* x. 279 In other cases..a system is to be reckoned decidable as long as both the derivable and the underivable formulae are recursively enumerable. 1964 E. BACH *Introd. Transformational Gram.* vii. 161 A decidable set of strings is a set such that after a finite number of steps it is possible to determine whether or not a given string is in the set.

deciduoma (dīsidiu,ōᵘ·mă). *Path.* [mod.L., f. DECIDUA + *-OMA.] An intra-uterine tumour probably caused by portions of the

decidua remaining after abortion; *deciduoma malignum*, a malignant and cancerous deciduoma.

1890 in BILLINGS *Med. Dict.* 1902 [see **chorio-epithelioma*]. 1907 F. J. MCCANN *Cancer of Womb* xi. 119 The deciduoma malignum consists as a rule of a small primary growth. *Ibid.* 122 Deciduoma may occur at any age during the child-bearing period. 1932 S. ZUCKERMAN *Soc. Life Monkeys* v. 79 The development of large masses of hypertrophied endometrium—deciduomata. 1967 R. A. WILLIS *Path. Tumours* (ed. 4) lxii. 1004 There is now universal recognition of its origin [*sc.* that of a chorion-epithelioma] from the chorionic epithelium, and of the erroneousness of the old name 'deciduoma malignum'.

decil, decile. Add: **2.** (Spelt **decile.**) *Statistics.* Any of the nine values of a variate which divide a frequency distribution into ten groups, each containing one tenth of the total population; also, any of the ten groups so produced.

1882 GALTON in *Rep. Brit. Assoc. 1881* 245 The Upper Decile is that which is exceeded by one-tenth of an infinitely large group, and which the remaining nine-tenths fall short of. The Lower Decile is the converse of this. 1907 G. P. WATKINS *Growth of Large Fortunes* ii. 18 Convenient relative numbers are the ratio of the upper decile, or the upper centile, to the median. 1934 *Brit. Jrnl. Psychol.* XXIV. 277 The group [of subjects] was formed by combining the highest and lowest deciles of a larger group, and there was thus too large a scatter. 1939 *Nature* 11 Mar. 439/1 Mr. Yule has selected the mean, median, upper and lower quartiles, interquartile range, and ninth decile. 1971 *Ibid.* 22 Jan. 224/3 The range of salaries indicated by the survey extends from \$24,500 or more in the highest decile to \$9,500 or less in the lowest decile.

decimal, *a.* and *sb.* Add: **A. 1. a.** (Further examples.) *decimal classification* (or *system*), a system of classifying library or archival material with a numerical notation subdivided as a decimal fraction allowing expansion after any figure; *spec.* the *DEWEY system.

1842 *Times* 23 Apr. 6/2 The standard measures are to be restored under the superintendence of three scientific gentlemen, who are inclined to introduce the decimal system. 1854 WILLIAM BROWN (*title*) Decimal coinage. A letter..to Francis Shand, Esq. chairman of the Liverpool Chamber of Commerce. 1858 TROLLOPE *Three Clerks* I. xi. 243 The much-vexed question of penny *versus* pound, as touching the new standard for the decimal coinage. 1876 M. DEWEY in *Public Libraries in U.S.A.* I. xxviii. 623 (*subheading*) A decimal classification and subject-index. 1885 [see *DEWEY]. 1930 *Library Jrnl.* 13 Dec. 1000 Why the Science Library adopted the Universal Decimal Classification. 1948 S. C. BRADFORD *Documentation* iii. 26 In 1883, a decimal classification was introduced in the Bodleian Library..and..is still in use. 1963 *Rep. Comm. Inquiry Decimal Currency* ii. 8 By far the most common decimal currency systems are two-place systems in which the major unit is divided into one hundred minor units or 'cents'. 1971 *Guardian* 22 Feb. 5/2 Bus services in many towns and cities went decimal yesterday.

decinormal (desinǭ·ɹmăl), *a. Chem.* [f. DECI- + NORMAL *a.*] Of a solution: having a concentration one-tenth of that of a normal solution; containing one-tenth of a gramme-molecule or gramme-equivalent of the dissolved substance per litre of solution.

1863 F. SUTTON *Syst. Handbk. Volumetric Anal.* 19 The decinormal solutions may be made either by weighing $\frac{1}{10}$ atom of test direct and diluting to 1000, or by diluting 100 parts of normal solution to 1000. 1898 *Rev. Brit. Pharm.* 32 Deci-normal silver nitrate. 1964 *Oceanogr. & Marine Biol.* II. 291 Water..acidified with decinormal hydrochloric or nitric acids.

decision. Add: **5.** *attrib.* and *Comb.*, as *decision-maker, -making, -taker, -taking, theory*; **decision method** = **decision procedure*; **decision problem** [tr. G. *entscheidungsproblem*] *Math.* and *Logic*, the problem of finding a decision procedure for a class of formulas; the *ENTSCHEIDUNGSPROBLEM; **decision procedure** *Math.* and *Logic*, an effective formal routine or mechanical method for deciding whether any selected formula of a given system, or a given class of formulas, is true or derivable within the system to which it belongs.

1938 S. CHASE *Tyranny of Words* xviii. 233 Mr. Baldwin was long the chief decision-maker for the British Empire. 1955 D. CHAPMAN *Home & Social Status* 2 Collaboration between researchers and 'decision-makers'. 1953 *Amer. Political Sci. Rev.* Mar. 7 The right of every voter to participate equally in the community's decision-making process. 1956 J. KLEIN *Study of Groups* xi. 148 The group moves into the decision-making phase. *Ibid.*, Progress in decision-making. 1960 *Times* 24 Mar. 2/2 An electronic laboratory (specialising in decision-making machines). 1965 M. FRAYN *Tin Men* v. 27 The decision-making faculties of a professional decision-maker like himself had to be nursed. 1948 A. TARSKI (*title*) A decision method for elementary algebra and geometry. 1958 FRAENKEL & BAR-HILLEL *Found. Set Theory* v. 297 A formalized theory *T* is decidable if there exists an effective, uniform method—a so-called decision method—of determining whether a given sentence, formulated in the vocabulary of *T*, is valid in *T*. 1939 *Jrnl. Symbolic Logic* IV. 1 (*heading*) On the reduction of the decision problem. 1954 I. M. COPI *Symbolic Logic* vii. 235 The decision problem for any deductive system is the problem of stating an effective criterion for deciding whether or not any statement or well-formed formula is a theorem of the system. 1945 W. V. QUINE in *Jrnl. Symbolic Logic* X. 3 No decision procedure is possible for the validity of polyadic schemata. 1950 —— *Methods of Logic* (1952) § 15. 82 A 'decision procedure'—i.e., a mechanical routine for deciding validity, implication, consistency, etc. 1957 *Technology* July 182/2 Any game of a finite kind which can be completed in a finite number of moves—and this includes simple games like noughts and crosses as well as draughts and chess—must have a decision procedure, even though we may not know for any particular game what this procedure is. 1962 *Times Lit. Suppl.* 24 Aug. 634/1 Wider ranges of privilege for economic decision-takers (and their wives and children). 1964 T. W. MCRAE *Impact of Computers on Accounting* iii. 98 Raw statistics, the backbone of the decision-taking process. 1964 J. Z. YOUNG *Model of Brain* iii. 28 'Signals' are the physical events in the communication channels that operate the decision-taking mechanism. 1961 *Jrnl. Acous. Soc. Amer.* XXXIII. 358/1 An algorithm based on statistical decision theory. 1964 T. W. MCRAE *Impact of Computers on Accounting* v. 120 Decision theory is probing the psychology of decision making, and attempts to provide an algorism for taking decisions.

decivilize, *v.* (Earlier U.S. example.)

1831 MRS. M. HOLLEY *Texas* (1833) 43 It sometimes happens that a white man from the States, who has become somewhat *decivilized* (to coin a word), is substituted.

deck, *sb.*[1] Add: **2. c.** *on deck* fig.: for U.S. read orig. *U.S.*, and add examples; also in sense 'alive'.

1867 *Ball Players' Chron.* 26 Sept. 5/4 Well, I went on deck and took up a bat. 1889 'MARK TWAIN' *Yankee* xxii. 274 Angels..are always on deck when there is a miracle to the fore. 1889 *Lisbon* (Dakota) *Star* 26 Apr. 4/2 A. H. Moore ..was kicked by a horse, a cow and a colt.., but is still on deck. 1946 K. TENNANT *Lost Haven* (1968) xix. 346, I couldn't bring it [*sc.* oakum] up here if Jame was on deck. She'd be down on me like a ton of lead. 1950 *Here & Now* (N.Z.) Dec. 12/1 If I am on deck when that time comes you will have a strong advocate for reinstatement in the service. 1966 *Times* 28 Feb. (Canada Suppl.) p. x/4 Dice games and three-card monte are not, so to speak, on deck but bingo flourishes.

3. b. By extension, any kind of floor or platform, as the floor of a pier or landing-stage, or the platform or roadway of a deck-bridge (see also quot. 1938).

1872 *Porcupine* XIV. 314/2 The decks of the three stages being swept pretty clear, by the devastating fire of the enemy. 1876 *Ibid.* XVIII. 330/1 Its deck is fairly rotting away. 1883 *Specif. Alnwick & Cornhill Rlwy.* 45 The girders are connected by a wrought-iron deck. 1910 A. WILLIAMS *Engin. Wonders of World* III. 282 The old suspension truss, which could then be removed piece by piece to make room for the upper deck. 1938 L. M. HARROD *Librarians' Gloss.* 56 *Deck*, one floor of a stack room containing the bookshelves, lifts, and workrooms. (American.) 1955 *Times* 9 May 18/2 Multi-deck car parks with direct access to the stores. 1961 *Daily Tel.* 19 Jan. 13/2 Other upper level walkways..include those on a 'deck' on the London Pavilion site. 1970 *Times* 9 Feb. 13/3 A service area and parking deck for 650 vehicles on the second floor.

c. *Aeronaut.* A main aeroplane surface, esp. of a biplane or multiplane; a wing.

1843 G. CAYLEY in *Mech. Mag.* XXXVIII. 275/1 Would it not be more likely to answer the purpose to compact it into the form of a *three decker*, each deck being 8 to 10 ft. from the other, to give free room for the passage of air between them? 1910 A. WILLIAMS *Engin. Wonders of World* III. 7/2 The biplane, with two 'decks' set one above the other. 1929 *Papers Mich. Acad. Sci. & Arts* X. 287 *Decks*, the wings of an airplane.

d. The floor of an omnibus or tramcar; *top* or *upper deck*, the upper floor or compartment of a double-decked vehicle.

1869 [implied in *DOUBLE-DECKED *a.*]. 1906 *Daily Chron.* 11 Sept. 7/1 The cars..have no upper deck, and carry only thirty-six passengers, as compared with accommodation for sixty-six in and on the double-deck pattern. 1966 P. MOLONEY *Plea for Mersey* 33 The best place to hear examples of all these quaint circumlocutions is the top deck of Liverpool's buses. 1968 *Listener* 26 Dec. 855/2 Two American soldiers sat on the lower deck of a bus smoking cigars.

e. *Aeronaut. slang.* The ground; *spec.* the landing-ground of an aerodrome.

1925 *Punch* 11 Nov. 521 *R.A.F. Officer* (unhorsed). Oh, he just stalled on top of the loop, did a roll and left me hangin' on the straps; then spun into the deck. 1941 *War Illustr.* 29 Aug. 93/1, I didn't see the bombs drop, but Mac, the rear-gunner, yelled over the inter-comm in a broad Scots accent—'There's one on the deck.' 1958 'N. SHUTE' *Rainbow & Rose* vii. 276 She spun her Moth into the deck.

f. The surface of a tape recorder above which the tape moves, together with its attachments such as the motor(s) and other mechanisms, the magnetic heads, and the circuits immediately associated with them, the whole being built as a single unit; any device for moving tape from one spool to another past magnetic heads; more fully as **tape deck*. Also, the corresponding part of a system for playing gramophone records.

1949 *Electronic Engin.* XXI. 149/3 At present the 'Tape-deck' recorder is available to manufacturers only who will fit the recorder into their own amplifying equipment. 1962 A. NISBETT *Technique Sound Studio* 249 Professional tape decks and some domestic decks have two further motors for the feed and take up spools. 1964 F. L. WESTWATER *Electronic Computers* iv. 67 Tape does not run continuously, but as required by the computer. It is mounted on what is called a Tape Deck. *a* 1965 *Manual of 'Ferrograph' Series 6* (*Model 631*) 4 The instrument resolves itself into three main units:—A. The mechanical deck carrying the motors, heads, reels, etc. B. The power unit and oscillator. C. The amplifier chassis containing the amplifier, monitor meter, etc. *Ibid.* 27 The mechanical unit is situated entirely on the hinged deck of the instrument. 1971 *Observer* (Colour Suppl.) 28 Feb. 33/2 (Advt.), The Garrard range..means you can tailor a deck to suit your needs... Perfect sound and record care for years.

5. b. A packet of narcotics; a small portion of some drug wrapped in paper. *U.S. slang.*

1922 E. F. MURPHY *Black Candle* (1926) I. v. 52 Small paper packages [of cocaine]..are called 'decks', and contain about a couple of sniffs. 1927 *Flynn's* 9 July 462/2 At night it was 'snow' that went over the counter..to poor devils who left behind them three dollars..for a deck. 1949 'J. EVANS' *Halo in Brass* (1951) iv. 29 A deck of nose candy for sale to the right guy. 1966 C. HIMES *Heat's On* iii. 27 When it's analysed, they'll find five or six half-chewed decks of heroin.

6. b. Part of a newspaper, periodical, etc., headline containing more than one line of type, esp. the part printed beneath the main headline. Also *attrib.*

1935 H. STRAUMANN *Newspaper Headlines* i. 28 These are first decks (and streamers) only. *Ibid.* iii. 87 The first three lines or 'decks' as they would be called in present-day journalism. 1965 L. H. WHITTEN *Progeny of Adder* (1966) 127 The eight-column headline told him of Pantelein's body being found. But it was the 'deck' headline that held him: *county coroner cites 'vampirism'.*

III. *deck-cabin, -cricket, -framing, -game, -passage, -passenger* (earlier U.S. example), *-scrubber, -stringer, -trumpet*; **deck-boy,** a boy employed on the deck of a vessel; **deck class,** a grade of accommodation entitling a person to deck-space only on board a ship; **deck-feather** (see quot.); **deck-hand** (earlier U.S. example); **deck-lander,** an aeroplane designed to be able to land on a ship's deck; **deck-light** (example); **deck-load** (earlier U.S. example); **deck-pot,** a pot used on whaling vessels to receive the scraps; **deck quoits,** a game played, chiefly on board ships, by throwing a rope quoit over a peg; **deck tennis,** a game played esp. on the deck of a ship by tossing a ring or quoit of rubber, rope, etc., back and forth over a net.

1900 *Westm. Gaz.* 22 Mar. 5/2 Prisoner said he was deck-boy on board the *Carisbrook Castle*. 1908 *Ibid.* 28 Aug. 12/1 The owner..sent his son, the deck-boy, down to the engineer. 1902 *Ibid.* 28 July 2/1 He..writes pictured post-cards at the deck-cabin table. 1953 A. SMITH *Blind White Fish* ii. 32 The second and third classes were more cosmopolitan;..all three could look down with equal disdain upon the deck class. *Ibid.*, The deck class passengers began to look around for sheltered niches in which to spend the night. 1969 J. H. VANCE *Deadly Isles* (1970) iii. 23 If he was lucky he might still find a berth available. If not, he'd go deck class, like the Polynesians. 1891 *Scribner's Mag.* X. 278 Deck cricket, quoits, and cock-fighting enliven the forenoons. 1879 *Encycl. Brit.* IX. 7/1 Deck feathers, the two centre tail-feathers. 1894 *Outing* (U.S.) XXIV. 396/1 Everything else, including the deck-framing and deck is of the same kind of material as those used in a regular battleship. 1896 S. A. BARNETT *Let.* Sept. in H. Barnett *Canon Barnett* (1918) II. xxxvii. 118 Of course there are the deck games. 1971 'A. GARVE' *Late Bill Smith* ii. 45 There was.. little provision for deck games. 1844 *Knickerbocker* XXIII. 88 On board of one of the steam-boats on the Mississippi, I encountered a deck-hand, who went by the name of Barney. 1928 *Daily Tel.* 18 Sept. 11/4 It [*sc.* the autogiro] should be able to act as a deck-lander on almost any ship. 1961 E. BROWN *Wings on my Sleeve* viii. 83 The Sea Hornet, a twin-engined deck-lander developed from the Mosquito, followed on the heels of the single-seater Hawker Sea Fury. 1849 N. KINGSLEY *Diary* 7 Mar. (1914) 7 Mate arrived today, Mr. Webb, put in deck lights & scuttle to house on deck. 1757 in *Essex Inst. Hist. Coll.* (1910) XLVI. 273 They hove overboard the Deck Load of Lumber. 1828 'C. SEALSFIELD' *Americans* ix. 105 The great difference of fare between a cabin and a deck passage..contributes to establish a distinction in this assemblage of people. 1883 'MARK TWAIN' *Life Mississippi* vi. 79 He only traveled deck passage because it was cooler! 1824 W. OWEN *Diary* 6 Dec. in *Indiana Hist. Publ.* (1906) IV. 57 Here the steerage, or as they are called deck passengers, sit, eat and sleep. 1872 E. EGGLESTON *End of World* xxviii. 187 He passed through to the place where the steerage or deck passengers are. 1904 *Sci. Amer. Suppl.* 5 Mar. 23551 The oil flows freely.. into the pots, while the refuse..is thrown into another receptacle, called the deck-pot. 1907 *Yesterday's Shopping* (1969) 1022/2 Deck quoits. As used on all steamships, and can be played in any sized room or lawn. 1971 'A. GARVE' *Late Bill Smith* ii. 64 The last time I was on a cruise a man died through over-exertion in a deck quoits competition. 1920 *Blackw. Mag.* Apr. 509/2 He..belaboured them methodically with a deck-scrubber. 1874 Deck stringer [see STRINGER 5 b]. 1874 THEARLE *Naval Archit.* 102 The deck-stringer plate. 1927 *Delineator* Mar. 9 They played deck-tennis and shuffleboard. 1932 L. GOLDING *Magnolia St.* III. vi. 530 The deck-tennis court was empty when Bella and Mick wanted to play. 1965 E. BROWN *Big Man* ix. 77 The other passengers..had declined..preferring the Sports Deck and their deck tennis. 1838 J. F. COOPER *Eve Effingham* I. ii. 52 A capital watch..and a deck-trumpet, in solid silver.

deck (dek), *sb.*[3] *colloq.* (orig. *Anglo-Indian*). Also **dekh.** [ad. Hind. *dekha* sight, *dekhnā* to see, look at.] A look, peep. Cf. *DEKKO.

1853 'PUNJABEE' *Oakfield* iv. 85 Some officer, stopping, as he passed by,..'just to have a dekh at the steamer'. 1886 YULE & BURNELL *Hobson-Jobson*, Deck. 1951 E. MILNE in J. Marriott *Best One-Act Plays of 1950–51* (1952) 99 Crickey, have a deck at Ronald Colman!

deck, *v.* Add: **6.** In *Lumbering*: to pile *up* (logs) on a skidway. *U.S.*

1901 *Munsey's Mag.* XXV. 392/1 Other men pile—technically, 'deck'—them [*sc.* logs] exactly as in the woods. 1905 *Terms Forestry & Logging* 35 Deck up.

deck-chair. [DECK *sb.*[1]] A folding chair, often with an adjustable leg-rest, used esp. on the deck of a ship as seating accommodation for passengers; a hammock chair. Also *attrib.*

1884 E. NESBIT *Let.* in D. L. Moore *E. Nesbit* (1933) v. 69, I sit here in my 'deck' chair. 1886 [see DECK *sb.*[1] III]. 1888 W. S. CAINE *Trip round World* i. 3 Ladies are grouped about in pleasant corners in easy deck-chairs. 1904 *Daily Chron.* 6 Oct. 8/1 At eleven o'clock soup and crackers..are served and little deck-chair groups are formed. 1926 *Spectator* 11 Sept. 372/2 To spend one's leisure lying on a deck-chair. 1937 J. BETJEMAN *Coll. Poems* (1958) 33 A tennis court, a summerhouse, deckchairs by the walnut tree. 1958 *Times* 23 Aug. 8/6 It is the weather and not the client that rules the days of the deck-chair attendant.

decker[2]. Add: **1.** See also DOUBLE-DECKER, *single-decker* (s.v. *SINGLE *a.* 17 b), THREE-DECKER, TWO-DECKER.

deckie (de·ki). *Naut. colloq.* Also **decky.** [f. DECK *sb.*[1] + -IE.] A deck-hand; = DECKER[2] 3 a. So **deckie-learner,** an apprentice deck-hand on a fishing-boat.

1913 *Q. Rev.* Apr. 435 This 'deckie'..has usually no more knowledge of seamanship than a ploughboy. 1934 'TAFFRAIL' *Seventy North* v. 113 He had passed through every grade—'deckie-learner', deck-hand..and finally skipper. 1935 'L. LUARD' *Conquering Seas* ii. 30 Alf making his first voyage as decky-learner. 1971 *Daily Tel.* 21 May (Colour Suppl.) 24/3 A young deckie sorts the arrivals [*sc.* fish] and throws them on to the ice.

decking, *vbl. sb.* Add: **2. b.** In extended uses (see *DECK *sb.*[1] 3 b).

1883 *Specif. Alnwick & Cornhill Rlwy.* 45 The superstructure consists of two wrought-iron plate girders..connected together by cross-bracing and by a decking of curved strips. 1897 *Daily News* 3 May 2/1 Piles, beams, and decking. 1898 *Ibid.* 4 Oct. 3/2 A huge decking..is being constructed in the river. *Ibid.* 12 Dec. 3/3 It was some time before the decking of the pier collapsed. 1900 *Westm. Gaz.* 29 Dec. 5/2 Some of the decking of St. Leonards pier was wrenched away by the waves. 1924 *Times Trade & Engin. Suppl.* 29 Nov. 252/3 The general design of decking is that of a reinforced concrete slab. 1954 *Archit. Rev.* CXVI. 129 Roof deckings are now well established in the vocabulary of present-day building materials. 1963 *Gloss. Build. Terms (B.S.I.)* 18 *Decking*, prefabricated units for the construction of a floor or roof. 1970 *Financial Times* 13 Apr. 13/7 Coated steel ..has found particular outlets in cladding, as well as roofing and decking.

3. In *Lumbering*: the action of piling logs on a skidway. Also *attrib. U.S.*

1901 *Munsey's Mag.* XXV. 392/1 A decking chain more than three hundred feet long is required to roll the logs to their places. 1902 S. E. WHITE *Blazed Trail* xi. 83 A shout of surprise or horror would have stopped the horse pulling on the decking chain.

deckle. 2. deckle-edged *a.* (examples).

1888 *N. & Q.* 7th Ser. V. 227/2 *Deckle-edged*. This term has lately been adopted in the advertisements of books to indicate that the edges of the paper have not been cut or trimmed, so that it is equivalent to..'rough edged'. 1899 T. VEBLEN *Theory of Leisure Class* vi. 163 A somewhat cruder type, printed on hand-laid deckel-edged paper. 1955 S. C. GILMOUR *Paper* vii. 65 Small sizes of deckle-edged notepaper are made in this way.

deckled (de·k'ld), *a.* [f. DECKLE + -ED[2].] Formed by a deckle; deckle-edged. Also *transf.*

1906 C. G. MCCRIE *Contemp. Portraits of Reformers* (dust-jacket), Vellum gilt, gilt top, deckled edges. 1947 D. M. DAVIN *Gorse blooms Pale* 150 The ash-tray was deckled and each time as my eye came to its degenerate line it winced with dissatisfaction.

decky, var. *DECKIE.

declai·m, *sb. rare*[−1]. [f. the vb.] = DECLAMATION.

1922 T. HARDY *Late Lyrics* 165, I went where my friend had lectioned The prophets in high declaim.

declaration. Add: **3. b.** *spec.* A declaration of love; a proposal of marriage.

1739–40 RICHARDSON *Pamela* (1740) I. 192, I am glad at my Heart, Madam, that I was *before-hand* in my Declarations to you. *Ibid.* 193 What Necessity was there for you to talk of your *former* Declaration? 1766 GOLDSMITH *Vicar* xvi. 144 My wife undertook to sound him..in the choice of an husband for her eldest daughter. If this was not found sufficient to induce him to a declaration, it was then fixed upon to terrify him with a rival. 1850 DICKENS *Dav. Copp.* xxxiii. 344, I went to Miss Mills's, fraught with a declaration. 1937 'E. M. DELAFIELD' *Ladies & Gentlemen in Victorian Fiction* iii. 89 The Victorian papa..was by no means invariably consulted before a 'declaration' took place.

4. b. (Earlier and later examples.)

1835 DICKENS *Let.* 18 Dec. (1965) I. 109 It will be unnecessary for me to remain here for the Declaration of the Poll on Monday. 1906 [see POLL *sb.*[1] 7 c].

8. b. In the game of Bridge, the naming of the trump suit or the declaring of 'no trumps' by any of the players.

1895 'BOAZ' *Laws of Bridge* 10 His partner must thereupon make the necessary declaration. 1905 in W. Dalton *'Saturday' Bridge* (1910) 12 If the dealer's partner make the trump declaration without receiving permission from the dealer, the eldest hand may demand: I. That the declaration so made shall stand. II. That there shall be a new deal. 1910 *Ibid.* 38 The declaration at Bridge affords an opportunity for the exercise of certain qualities which were never called into use in the game of Whist. *Ibid.* 39 The most expensive declaration..being No Trumps, when the value of each trick is twelve points.

c. *Cricket.* The closing, by the team batting, of an innings. (Cf. *DECLARE *v.* 11 b.)

1908 W. E. W. COLLINS *Country Cricketer's Diary* ix. 158 The Malvern boys had proved equal to the emergency.. after an apparently safe 'declaration' [by their opponents]. 1963 A. ROSS *Australia 63* i. 40 Benaud delayed his declaration.

declare, *v.* Add: **5. e.** *refl.* To declare one's love for another person; to propose marriage. Cf. *DECLARATION 3 b.

1840 DICKENS *Old C. Shop* viii, A concerted plot..having for its object the inducing Mr. Swiveller to declare himself in time. 1879 TROLLOPE *Eye for Eye* II. ii. 29 You should have thought of that before you declared yourself to her, Mr. Neville. 1907 M. E. BRADDON *Dead Love has Chains* ii. 40 People were beginning to say ill-natured things about us; and he must either declare himself, or must go away. 1965 C. D. BROAD in G. Cummins *Swan on Black Sea* p. xviii, This was his love for Catherine Mary Lyttelton; her tragic death..before he had declared himself.

11. b. *intr. Cricket.* To close an innings before the usual ten wickets have fallen; orig. 'to *declare* the innings at an end'. Also *trans.*

1889 *Cricketer's Guide* 7 On the last day of a double-innings match, or in a one-day match, the batting side may, at any time, declare their innings at an end. 1897 *Encycl. Sport* I. 245/2 *Declare*, to close an innings. 1901 *Daily News* 5 June 4/4 Warwickshire made 532 for four wickets, and then declared. 1955 *Times* 15 July 3/3 Barnett showed that he is still worth a few runs before Insole declared.

c. *trans.* and *intr.* In the game of Bridge, to name the trump suit, or to announce the intention to play 'no trumps'; in auction or contract bridge, to announce the number of tricks that one intends to make.

1895 'BOAZ' *Laws of Bridge* 10 The dealer, having examined his hand, has the option of declaring what suit shall be trumps, or whether the hand shall be played without trumps. 1899 'J. DOE' *Bridge Conventions* iii. 21 Before declaring ask yourself the question, Have we a better chance of making 30 than of losing 18? 1899 A. DUNN *Bridge* 27 The dealer should declare trumps 'on the top' of his cards. 1910 [see *BID *v.* 3 c].

13. *Racing.* To announce the withdrawal of (a horse) from a race for which it has been entered; said also *intr.* of the horse.

1847 *Weekly Times* 9 Oct., Stakes of 10 sovs each, 5 f[orfei]t, and only 3 if declared. 1897 *Westm. Gaz.* 5 Feb. 9/2 Two Grand National winners have just declared.

declarer. Add: **3. b.** In the game of Bridge, one who declares (see prec., 11 c).

1905 in W. Dalton *'Saturday' Bridge* (1910) 13 The declarer of the trump shall have the right to say whether or not the double shall stand. 1927 M. C. WORK *Contract Bridge* iii. 55 A double which is not 'free' may produce for the Declarer a game otherwise unobtainable. 1965 *Listener* 17 June 915/3 When Mrs Markus was declarer South led the two of diamonds.

‖ **déclassé** (deklase), *a.* and *sb.* Fem. **-ée.** [Fr., pa. pple. of *déclasser* DECLASS *v.*] **A.** *adj.* Reduced or degraded from one's social class; having come down in the world. **B.** *sb.* One who has been so reduced or degraded.

1887 *Fortn. Rev.* Aug. 227 It is only the *déclassé*, the ne'er-do-well, or the really unfortunate, who has nothing to call his own. 1905 *Spectator* 28 Jan. 144/2 Pamela..quits the company of artists and actresses, *declassés* and *divorcées*. 1921 *Glasgow Herald* 3 Aug., The attempt by a body of declassés to form the policy of the entire working-class of this country. 1921 *Times Lit. Suppl.* 29 Sept. 626/2 A girl of any family may, by force of circumstances, become *déclassée*. 1961 A. WILSON *Old Men at Zoo* ii. 97 Déclassé nations are very touchy.

declassify (dīklæ·sifəi), *v.* [f. DE- II. 1 + CLASSIFY *v.*] *trans.* (See DE- II. 1); *spec.* to remove (information, etc.) from the category of being 'classified' (see *CLASSIFIED *ppl. a.* c). So **declassifi·able** *a.*, **decla:ssifica·tion, decla·ssified** *ppl. a.*

1865 [see DE- II. 1]. 1946 *Nature* 14 Sept. 355/2 Much of the nuclear research material..is now declassifiable and provision has been made for revision of the rules for declassification of the secret documents. 1946 *Rep. Internat. Control Atomic Energy* IV. 53 This Committee was directed to report on a policy of declassification—that is disclosure—of scientific and technical material now classified as Secret. 1948 *Birmingham* (Ala.) *News* 21 Dec. 14/6 Forrestal asked Adm. Souers and Gen. Gruenther to 'declassify' it (make it non-secret) immediately. 1951 *Economist* 13 Oct. 864/1 No effective machinery is created for 'declassifying' secret information. 1953 *Encounter* Oct. 58/2 In the United States, the entire declassification procedure, by which work of potential military value is sifted and eventually tossed into the open, was devised by the scientists themselves. 1955 *Bull. Atomic Sci.* Apr. 145/3 Even the handling of declassified documents can cause trouble.

declination. Add: **11. declination axis,** that axis of an equatorial telescope which is at right angles to the polar axis, and to which is attached at one end the telescope and at the other the declination circle, so called because when the position of the telescope is changed by turning the declination axis there is an alteration in the declination of the object viewed; **declination circle,** (*b*) the graduated circle on an equatorial telescope which marks the declinations of the heavenly bodies; **declination compass** (see quot.); **declination magnet,** a magnet used in determining the magnetic declination and the magnetic axis.

1835 *Mech. Mag.* XXIV. 210/2 On these rollers turns the declination axis. 1888 *Encycl. Brit.* XXIII. 146/2 The equatorial in its simplest form consists of an axis parallel to the earth's axis, called the 'polar axis'; a second axis, at right angles to this, called the 'declination axis'; and a telescope fixed at right angles to the latter. 1905 *Westm. Gaz.* 17 Apr. 1/3 A large equatorial with a 26-in. photographic refractor at one end of the declination axis and a 30-in. reflector at the other. 1964 R. H. BAKER *Astron.* (ed. 8) iv. 108 The circle on the declination axis is graduated in degrees of declination. 1835 *Mech. Mag.* XXIV. 211, Y is the declination circle, fixed on the declination axis. 1888 *Encycl. Brit.* XXIII. 149/1 The declination circle is attached to the farther end of the declination axis. 1862 *Chambers's Encycl.* III. 461/1 The ordinary compass which must be used by making allowance for declination, is a declination compass. 1883 *Encycl. Brit.* XV. 238/1 The first step is to remove the torsion as far as possible from the suspension fibre by hanging to it a brass plummet E of the same weight as the declination magnet. After this weight has come to rest, it is replaced by the declination magnet. 1899 *Daily News* 3 Mar. 5/2 They have placed out here a declination magnet, a dip instrument for the inclination of the needle, and a deflexion instrument.

decline, *v.* Add: **13. d.** *Chess.* To refuse to take a piece or pawn offered in (a gambit).

[1833 W. LEWIS *Progr. Less. Chess* (ed. 2) 138 King's Gambit...The best move for the Black is to take your K.B.P.; but he may decline doing so.] 1847 H. STAUNTON *Chess-Player's Handbk.* IV. x. 347 (*heading*) The gambit declined. 1875 G. H. D. GOSSIP *Chess-Player's Man.* 705 The Queen's Gambit accepted and declined. 1899 E. E. CUNNINGTON *How to play Chess* 48 P–QKt4, offering the sacrifice of a P. to get an attack. Black need not take it (he may retreat the B to Kt3), in which case the Opening is called the Evans declined.

decli·vitously, *adv.* [f. DECLIVITOUS *a.* + -LY[2].] In a declivitous manner; on or down a steep slope.

1878 I. L. BIRD in *Leisure Hour* 5 Oct. 637/2 A good hotel declivitously situated. 1930 W. J. LOCKE *Town of Tombarel* vii. 215 The path lands you declivitously into the Place Georges Clemenceau.

declutch (dīklʌ·tʃ), *v.* [f. DE- II. 1 + *CLUTCH *sb.*[1] 6 a.] *intr.* To disengage the clutch of a motor vehicle; so **double-declutch** *v. intr.*, to release and re-engage the clutch twice when changing gear. Hence **declu·tching** *vbl. sb.*

1905 *Daily Chron.* 21 Mar. 7/4 Without once using a brake, changing his gear, declutching or slipping the clutch. 1906 *Westm. Gaz.* 16 July 5/1 The driver of the 'bus declutched at the top of the hill. 1925 *Morris Owner's Man.* 10 The expert driver will have recourse to double declutching. 1934 *Neuphilologische Mitteilungen* XXXV. 130 To *double-declutch*, to let the clutch in twice (instead of the more usual once) when changing gear. 1937 *Autocar Handbk.* (ed. 13) vi. 108 Syncromesh..rendered double declutching a refinement but no longer a necessity. 1963 *Which?* (Car Suppl.) Jan. 30 To make proper downward gear changes you had to double-declutch. 1966 'D. RUTHERFORD' *Black Leather Murders* i. 7 He double-declutched and rammed the gear-lever into bottom.

decode (dīkōᵘ·d), *v.* [f. DE- II + CODE *sb.*[1] 3 b or *CODE *v.* b.] *trans.* To decipher or translate (a coded message). Also *absol.* and *transf.* Hence **deco·ded** *ppl. a.*; **deco·der,** one who or that which decodes (see also quots. 1962 and 1964); **deco·ding** *vbl. sb.*

1896 *N. Brit. Daily Mail* 28 Aug. 5 The message was decoded. 1897 *Westm. Gaz.* 26 Jan. 2/1 If it seems desirable, I shall decode and publish at my own time. *Ibid.* 18 May 7/1 The decoded cables. 1897 *Daily News* 15 May 4/6 Formal evidence as to the de-coding of the Harris–Rhodes telegrams was given. 1897 *Times* 30 June 9 The cipher telegrams in which they and Mr. Rhodes are concerned, and which have now been decoded. 1920 *Glasgow Herald* 20 Aug. 8 The alleged decoded messages..contain ludicrous statements. As to whether it is due to incompetence in decoding,..we are not competent to judge. *Ibid.*, The decoder on his own showing was doubtful as to his interpretation. 1937 KOESTLER *Spanish Testament* II. iii. 263 Finally we landed up somewhere in the decoding department. 1956 *Spaceflight* I. 27/2 Received at the ground station, this signal is decoded automatically and interpreted as a series of line graphs. 1957 *Electronic Engin.* XXIX. 574 The decoders were produced in the laboratory at a cost of about £50 each. 1962 *Gloss. Autom. Data Processing (B.S.I.)* 83 *Decoder.* 1. A device capable of decoding a group of signals and generating other signals which may initiate an operation. 2. In data-processing equipment. A device with a number of input and output lines in which a specified combination of input signals causes a particular output line to give a signal. **1964** F. L. Westwater *Electronic Computers* (ed. 2) i. 2 Before it can operate, the computer must have numbers presented in the form which it can understand. This requires some kind of decoder or mechanism which will translate a number into a suitably recognisable electrical impulse. *Ibid.* vi. 99 Information on a card is normally decimal in character and has to be decoded into whatever binary code a computer employs. The decoding circuitry is..part of the computer design.

decohere (dīkohiə·ɹ), *v.* [f. De- II. 1 + Cohere *v.*] *trans.* To restore (a coherer) to its normal condition of sensitiveness. Also *absol.*, and *intr.* for *pass.* (Disused.)
1902 *How to make Useful Things* 5/1 The purpose of the tapper is to decohere the filings after they are affected by the etheric waves. *Ibid.* 6/1 The tapper keeps busily at work decohering in response to the continuously closing circuit caused by the waves. **1923** E. W. Marchant *Radio Telegr.* iv. 38 By using mercury, Lodge and others made coherers which decohered without tapping. **1924** S. R. Roget *Dict. Electr. Terms* 58/2 A coherer is said to 'decohere' when it resumes its normal feebly conducting state on tapping or disturbance, after it has been made conducting by electrical oscillations, etc.
Hence **decohe·rence**, **decohe·sion**; **decohe·rer**, a device for bringing a coherer back to its normal condition.
1899 Marconi in *Jrnl. Inst. Electr. Engin.* XXVIII. 275 This coherer forms part of a circuit..which circuit works a trembler or decoherer and a recording instrument. **1902** *Science* 21 Mar. 466 A short description of the single contact coherer used by him and an explanation of the so-called decohesion. **1902** *Young Engineer* I. 371/2 As the coherer is continuously rotating, decoherence follows immediately on coherence. **1903** *Sci. Abstr.* B. VI. 128 The self-induction of the telephone may have been sufficient to cause the decoherence. **1913** *Work* 4 Oct. 7/2 The decoherer circuit.

decoke (dīkōu·k), *v. colloq.* [f. De- II. 2 + Coke *sb.*] *trans.* = *decarbonize *v.* b. Also *transf.* Hence **deco·king** *vbl. sb.*
1928 *Daily Express* 19 May 13/2 My advice to all motorists who are puzzled about..the 'knocking' which invariably accompanies carbonisation, is that they should lose not an hour in getting the engine 'decoked'. *Ibid.* 31 May 3/7 Good car, but wants decoking. **1931** *Boys' Mag.* XLV. 169/1 The presence of the carbon will cause overheating, bad hill-climbing and loss of power, so when any of these things persist, it is time for the engine to be 'decoked'. **1958** M. Pugh *Wilderness of Monkeys* iii. 32 He found his pipe, decoked it with a spike intended to hold papers and was about to light it. *Ibid.* viii. 98 A meat skewer, which he normally used in decoking his pipe.

decoke (dīkōu·k), *sb. colloq.* [f. the vb.] = *decarbonization b.
1962 'A. Garve' *Prisoner's Friend* i. 35 Mr. Winter's car..it's in for a de-coke. **1967** *Observer* (Colour Suppl.) 28 May 35/1 Do your own de-coke for..half the normal garage charge.

‖**décollement** (dekolmaṅ). [Fr., f. *décoller* to unstick, disengage.] **1.** *Med.* The process of separating organs or tissues from surrounding parts; also, the state of being thus separated.
1842 Dunglison *Med. Lex.* (ed. 3), *Décollement*, the state of an organ that is separated from the surrounding parts, owing to destruction of the cellular membrane which united them; the skin is *décollée*; i.e., separated from the subjacent parts. **1908** *Practitioner* Sept. 455 The third danger zone is behind in the neighbourhood of the portal vein and vena cava. Injury to these vessels may be avoided by practising 'decollement' of the duodenum. *Ibid.*, The stage of decollement being completed, ablation may be proceeded with.
2. *Geol.* A process in which some strata are supposed to become partly detached from those underneath and slide over them, becoming folded.
1927 L. W. Collet *Structure of Alps* ii. viii. 129 The folds of the Jura must be considered..to be due to a phenomenon of '*décollement*'. *Ibid.* 139 Owing to the push exerted by the Alps in formation, the sedimentary rocks of the Jura have been detached on the Middle Muschelkalk, of which the salt beds played the rôle of a lubricant, and folded. This kind of folding is a '*décollement*'. **1935** E. B. Bailey *Tectonic Ess.* xi. 176 Complications have developed along a plane of *décollement* (ungluing) in the Trias. **1969** M. G. Rutten *Geol. W. Europe* x. 204 The French literature has become permeated almost completely by the idea of 'décollement' in the sense of gravity sliding since about 1950.

‖**décolletage** (dekoltaʒ). Also **decolletage, decoltage.** [Fr.: see Décolleté *ppl. a.*]
1. The low-cut neck of a bodice.
1894 *Season* X. 34/2 Into the breast pleats are placed fan-shaped gores..loosely smocked into the decoltage. **1896** *Westm. Gaz.* 14 Mar. 3/2 The sleeves and upper bodice are in creamy white chiffon, and the dark line round the décolletage of black jet. **1902** *Queen* 10 May 816/2 A soft fold of pink silk at the decolletage. **1933** Mrs. C. S. Peel *Life's Enchanted Cup* iv. 47 The decolletage was, in the case of young girls, decorously cut.
2. Exposure of the neck and shoulders by the low cut of the bodice. Also *fig.*
1894 *Sat. Rev.* LXXVIII. 596 Such art as that of John Oliver Hobbes, so typically feminine in its discreet décolletage, is not truth but effect. **1921** *Public Opinion* 8 July 37/3 A bold *decolletage* of shoulders still young and white. **1924** *Blackw. Mag.* Nov. 678/1 Skeletta's ever-increasing décolletage had a sort of embarrassed fascination for us. **1937** J. Laver *Taste & Fashion* xiii. 192 A considerable proportion of young women will be *décolleté*, for *décolletage* is a powerful weapon of attractiveness.

décolleté, *ppl. a.* Add: **c.** *fig.*
1890 *Forum* IX. 682 Of an American writer whose questionable book ran like wildfire into every reading home.., a publisher said: 'You cannot blame her; she was born *décolleté*.' **1952** S. Kauffmann *Philanderer* (1953) iii. 49 Before the wildness of the twenties, there was Victorianism; and before that there was the eighteenth century, very *décolleté*. **1966** *Listener* 20 Jan. 99/3 The glittering Mrs Cheveley, with her décolleté dresses and décolleté past.

‖**décolletée** (dekolte). [fem. pa. pple. of *décolleter*: see Décolleté *ppl. a.*] The low-cut neck of a bodice.
1907 *Daily Chron.* 2 July 10/5 Mohair braid round the décolletée and armpits in bretelles and as bands on the skirt. **1908** *Ibid.* 24 Aug. 7/5 Décolletée and sleeves are in Malines.

de:coloniza·tion. [f. De- II. 2 + Colonization *sb.*] The withdrawal from its former colonies of a colonial power; the acquisition of political or economic independence by such colonies. Also *transf.*
1938 M. J. Bonn *Crumbling of Empire* ii. 101 A decolonization movement is sweeping over the continents. An age of empire-breaking is following an age of empire-making. **1957** *Economist* 19 Oct. 213/2 Nor did the postwar return of the colonial powers reverse or halt the process of 'decolonisation'. **1960** *Guardian* 6 Oct. 10/4 Britain, as a liberal state, engaged in working out the logic of decolonisation. **1967** *Times* 30 Nov. 11/1 If the full glory of decolonization is to throw the alien power out neck and crop, this satisfaction can hardly be denied the National Liberation Front.

decolorize, -ourize, *v.* Add: **2.** *intr.* To lose colour; to become colourless.
1908 *Practitioner* Feb. 205 Characteristics of the gonococcus..Its characteristic half-moon shape, and the fact that it decolourises with Gram's method.
Also **deco·lo(u)rized** *ppl. a.*
1879 *St. George's Hosp. Rep.* IX. 509 Decolorised blood-clot. **1890** W. J. Gordon *Foundry* 177 Decolourized indigo. **1965** R. Cruikshank *Med. Microbiol.* (ed. 11) xlv. 646 In order to render the decolourised organisms visible..a contrast or counterstain is then applied.

de:compensa·tion. *Med.* [f. De- II. 2 + Compensation 1 c.] A state or condition of having lost compensation (sense 1 c), *spec.*, as in *cardiac decompensation, circulatory decompensation*, a failure of the heart to maintain adequate circulation, after a period of compensation. Hence **deco·mpensated** *ppl. a.*
1905 Gould *Dict. New Med. Terms* 208/1. **1938** H. A. Christian *Osler's Princ. & Pract. Med.* (ed. 13) 1037 Treatment of the Permanently Decompensated Cardiac. *Ibid.*, Sooner or later irremediable decompensation will come. **1953** *Faber Med. Dict.* 114/1 Cardiac or circulatory decompensation. **1961** *Lancet* 5 Aug. 314/2 We considered they had died from decompensated hypertensive cardiac failure. **1961** *Ibid.* 19 Aug. 429/1 Patients with enlarged hearts, cardiac decompensation, and diabetes.

decompress (dīkǫmpre·s), *v.* [f. De- II. 1 + Compress *v.*; cf. next.] *trans.* To relieve or reduce pressure. **a.** To subject (a diver, etc.) to decompression; to reduce the pressure of the air or other gas in. Also *absol.*
1905 *Jrnl. Physiol.* XXXIII. p. vi, We placed rats in 10–15 or 20 atmospheres of air.., and then decompressed them in about three seconds. **1909** G. W. M. Boycott *Compressed Air Work* i. 2 When decompressing from great depths or high pressures after short exposures, the saturation..is obviated by the first rapid drop in pressure. *Ibid.* 12 When a worker is suddenly decompressed by some such accident..he should immediately be placed in the air-lock. **1911** *Engineer* 10 Mar. 243/1 To decompress slowly but continuously. **1925** *Literary Digest* 27 June 24/1 When they have finished their labors, it is necessary that they be 'decomprest', that is, slowly restored to normal air conditions. *Ibid.*, The two air-locks..are used in putting the compress-air workers gradually under pressure and for decompressing them after they have finished work. **1962** [see *decompression 1 a].
b. *Surg.* To relieve excessive internal pressure in (a part of the body) by surgical means.
1914 A. P. C. Ashhurst *Surgery* xvii. 574 The most imperative indication is to 'decompress' the brain by removing some of the overlying cranium. **1954** E. L. Farquharson *Textbk. Oper. Surg.* xxv. 708 When the bladder has been chronically distended over a period of several days..it should be decompressed gradually. **1967** T. J. McNair *Bailey's Emergency Surg.* (ed. 8) lxxxviii. 970 Grossly distended intestine is decompressed by the method of Savage.

decompression (dīkǫmpre·ʃən). [f. De- II. 2 + Compression.] **1.** The process of relieving or reducing pressure. **a.** A reduction of the pressure of the air or other gas in an enclosed space; *esp.* (*a*) the process of subjecting a diver, etc., who has been in air under pressure to a gradual reduction in pressure in a special chamber until atmospheric pressure is reached; (*b*) in an aircraft, etc., a reduction of the air-pressure from that of the atmosphere to a lower value, *e.g.* as a result of a rupture of the cabin during high-altitude flight.
1905 *Jrnl. Physiol.* XXXIII. p. vi (*heading*) Estimation of the gas set free in the body after rapid decompression from high atmospheric pressures. **1906** *Westm. Gaz.* 27 Jan. 10/1 The men who controlled the airlocks, and were subjected to compression and decompression every few minutes, were in no case affected. **1939** H. G. Armstrong *Aviation Med.* xx. 335 The effects of sudden decompressions at less than one atmosphere. **1951** A. Grollman *Pharmacol. & Therapeutics* xxix. 688 The incidence of 'divers' bends' may also be reduced appreciably by using a mixture of helium and oxygen to replace the air in the pressure chamber used in decompression. **1962** J. Glenn in *Into Orbit* 41 You have one large handle for repressurizing the cabin with oxygen in case of a bad leak, plus another handle to decompress it..in case of fire. (Decompression—or getting rid of the oxygen—is the quickest way to put out a fire.)
b. *Surg.* The surgical relief of excessive internal pressure in a part of the body.
1906 *Trans. College Phys.* (Philadelphia) 3rd Ser. XXVIII. 73 (*heading*) Cerebral decompression; palliative operations in the treatment of tumors of the brain. **1949** I. Aird *Compan. Surg. Studies* xliii. 956 A distended bladder is prone to infection. If catheterization succeeds, gradual decompression is continued for five or seven days. **1966** E. L. Farquharson *Textbk. Oper. Surg.* (ed. 3) ix. 328 Subtemporal decompression was formerly employed as a method of relieving intra-cranial pressure.
2. *attrib.*, as *decompression symptom*; **decompression chamber,** (*a*) a chamber in which pressure is reduced gradually to that of the outside air; (*b*) a chamber in which a person or animal can be subjected to a reduction in the pressure of the air (or oxygen) below that of the atmosphere; **decompression sickness,** sickness resulting from the effects of too rapid decompression.
1932 *Daily Express* 28 Jan. 1/2 The minesweeper Tedworth..is equipped with special decompression chambers and the latest appliances for deep-sea diving. **1958** *Times Lit. Suppl.* 30 Oct. 564/2 He walks..into a decompression chamber to test his cameras for high altitude work on Mount Everest. **1961** *Lancet* 26 Aug. 484/1 Concern with..anoxia..led Haldane to expose himself to low oxygen-pressures in decompression chambers. **1941** *Observations on Decompression Sickness in Man* (*Flying Personnel Res. Comm.*, Farnborough) 2 The term decompression sickness is retained because it covers the whole range of phenomena that occur, while 'bends' refers only to one of many symptoms..while aeroembolism..refers to only one aspect. **1962** Decompression sickness [see *bend *sb.*⁴ 2 b]. **1971** *Sunday Times* 7 Mar. 45/7 Prolonged investigations by the Medical Research Council's experts on decompression sickness. **1906** *Jrnl. Physiol.* XXXV. p. vi, It would seem probable that any animal whose respired gases are conveyed to the cells without the intervention of a circulating liquid should be immune from decompression symptoms.

decompressive (dīkǫmpre·siv), *a.* [f. *decompress *v.* + -ive.] Of or pertaining to decompression (sense *1 b).
1910 H. Cushing in Osler & McCrae *Mod. Med.* VII. 457 Tumor-palliation (decompressive operations). **1931** H. Bailey *Emergency Surg.* II. xxv. 86 With the slight decompressive action of the injection the veil may be lifted sufficiently to enable us to decide on which side to trephine.

decompressor (dīkǫmpre·sǝɹ). [f. *decompress *v.* + -or.] An apparatus for reducing compression in a motor engine.
1919 *Chambers's Jrnl.* Dec. 830/2 The only levers needed to control this motor are one connected to a decompressor and clutch. **1923** *Daily Mail* 7 June 12 An ingenious decompressor to lower the engine compression..for easy starting.

deco·ncentrate, *v.* [See De- II. 1.] *trans.* To reverse or diminish the concentration of; *spec.* to dissolve (cartels or other large industrial groupings), to decentralize. Also *intr.*
1889 in *Cent. Dict.* **1958** *Times* 1 May 9/2 Why, it is asked, should Krupp be obliged to deconcentrate?

deconcentra·tion. [See De- II. 2.] The reversal or diminution of concentration; *spec.* the dissolution of cartels or other large industrial groupings; decentralization. Also *attrib.*
1889 in *Cent. Dict.* **1925** E. B. Wilson *Cell* (ed. 3) 545 The chromosomes are..in a state of deconcentration. **1955** *Times* 6 July 7/5 Some families..weathered the storm of deconcentration, and still control large business organizations. **1959** *Economist* 17 Jan. 204/1 Under the allied deconcentration laws Herr Alfried Krupp was obliged to sell his holdings in coal and steel.

decondition (dīkǫndi·ʃən), *v.* [f. De- II. 1 + *condition *v.* 9.] *trans.* To reverse or remove the conditioned reflexes of (someone); to undo the results of conditioning in (a person); (in quot. 1940, to lower the charged condition of). So **decondi·tioned** *ppl. a.*, **decondi·tioning** *vbl. sb.* and *ppl. a.*
1940 *Physical Rev.* 1 Dec. 993 The reapplication of potential showed that the tube had become partially deconditioned. *Ibid.*, The deconditioning of the tube was not due to any leak. **1941** *Spectator* 5 Sept. 232/1 Experiments in what is described as 'de-conditioning'. The idea is to reproduce warlike sounds and disturbing noises until the patients regain their normal reactions. **1959** *Times Lit.*

Suppl. 27 Mar. 181/3 Man is a transient being, subject to time, from which all his other 'conditionings' ensue. The need to 'decondition' man is..acknowledged in the Christian rite of baptism in which 'the world, the flesh and the devil' are abjured. **1960** Koestler *Lotus & Robot* ii. xi. 264 Zen started as a de-conditioning cure and ended up as a different type of conditioning.

decongestant (dīkǫndʒe·stănt), *a.* and *sb.* [f. De- II. 1 + Congest *v.* + -ant¹.] **A.** *adj.* = *decongestive *a.* **B.** *sb.* A decongestive agent.
1950 *Therapeutic Notes* Dec. 259 Benylin Expectorant is antispasmodic, decongestant, expectorant, [etc.]. **1952** *Official Gaz.* (U.S. Pat. Off.) 1 July 23/1 Decongestants, Digestants. **1958** Falconer & Norman *Drug, Nurse, Patient* xviii. 294 While nasal decongestants do have value, as a general rule there is a great tendency to overuse this type of medication. **1962** *Which?* Apr. 123/1 Decongestants constrict blood vessels and relieve the burning feeling and heaviness in the eyes. **1965** C. Andrewes *Common Cold* iv. 173 People then tend to take more of the decongestant spray. **1966** in G. N. Leech *Eng. in Advertising* xviii. 159 Never go to bed with a cold without decongestant Vick Vapour Rub.

decongestion (dīkǫndʒe·styən). [f. De- II. 2 + Congestion.] The relief of congestion. So **deconge·st** *v. trans.*, to relieve congestion in; **deconge·stive** *a.*, that relieves congestion.
1903 *Med. Record* 1 Aug. 167 Treatment..by electricity..has a decongestive local action. **1908** *Practitioner* Oct. 569 Heat relieves the pain, not by the old-fashioned theory of 'de-congestion', but by causing hyperæmia. **1945** *Ann. Reg. 1944* 49 The 'decongestion' of congested areas and the encouragement of reasonable balance between different regions. **1958** *Times* 8 Aug. 9/6 Suppose, for example, Devonshire County Council wishes to 'decongest' its main towns, while preserving green belts, beauty spots, and the best farmland.

decontaminate (dīkǫntæ·minēit), *v.* [f. De- II. 1 + Contaminate *v.*] *trans.* To remove contamination or the risk of contamination from (a person, area, etc.) affected by poison gas, radioactivity, etc.
1936 *Daily Tel.* 17 Oct. 13/5 Cars sped through the streets carrying men wearing gas-masks, whose task it was to decontaminate 'mustard gas' areas. **1938** *Protection of Your Home against Air Raids* 25 Mustard gas..'contaminates' clothing, or other objects exposed to it, making them dangerous to have near you or to touch until they have been 'decontaminated'. **1948** *Times* 5 Mar. 3/2 The vessels..were still so radio-active that the cost of decontaminating them..would have been greater than their value as scrap. **1958** *Observer* 12 Oct. 1/3 The package was elaborately decontaminated both by chemicals and radiation to prevent the placing of any live organisms on the moon's surface. **1971** *Nature* 5 Feb. 363/2 Breeders pay too little attention to the arduous task of decontaminating buildings where infected birds have been housed, before moving in fresh, healthy stock.
So **decontamina·tion,** the action or result of the verb; also *attrib.*
1935 *Discovery* July 186/2 The scheme of gas defence requires..that there should be an organisation for the decontamination of the areas affected. **1937** *Evening News* 15 Mar. 7/1 Decontamination squads, who would clear areas sprayed by mustard gas. **1955** *Times* 30 May 7/7 The techniques of decontamination..: a householder, on emerging from his cellar, should immediately take a broom and sweep down his roof. **1957** *Ibid.* 12 Oct. 6/1 Traces of radio-activity lingered on his hands and he was given a pair of surgical gloves to wear until to-morrow when the decontamination process can be completed. **1966** D. F. Galouye *Lost Perception* ii. 18 The original airport had been swallowed by a gaping crater where, even after two years, decontamination crews were still scouring the inner slope.

decontrol (dīkǫntrōu·l), *sb.* [De- II.] The removal of control; *spec.* the removal of government control, esp. of a control imposed in wartime or in an emergency. Also *attrib.* So **decontro·l** *v. trans.*, to free from control; **decontro·lled** *ppl. a.*, **decontro·lling** *vbl. sb.*
1919 G. H. Roberts in *Times* 14 Feb. 3/1, I am told that every one wants to get rid of Government control...The moment I can see those conditions safeguarded in respect of any commodity, decontrol will come. *Ibid.* 3/2, I do not propose to risk decontrolling any commodity while there is the slightest possibility of decontrol resulting in higher prices. **1919** *Observer* 23 Mar. 12/6 (*headline*) De-controlled. Tea after to-morrow: bacon next week. **1923** *Daily Mail* 27 Jan. 8 Rent decontrol. **1923** Ramsay MacDonald in *Hansard Commons* 5th Ser. CLX. 21 To decontrol a certain block of middle-class tenants..will do more harm than good. Look at what happens the moment decontrol comes. Either the rent goes up, or..the house will be sold. **1946** *New Statesman* 14 Sept. 182/2 The decision of the Price Decontrol Board to lift the controls on grains and dairy products. **1955** *Times* 27 Aug. 10/1 At least two million tons more domestic house-coal, it is considered, is needed before decontrol can be envisaged. **1957** *Ibid.* 23 Dec. 11/4 Of the tenants of our newly decontrolled flats some 85 per cent. have accepted and signed new agreements on the basis we offered. **1971** *Daily Tel.* 1 May 19/4 The decontrol of gold coins on April 1 has coincided with a slight increase in demand for gold in London and Zurich. **1971** *Ibid.* 23 June 19/4 The decontrolling of bank competition..suggests that the Government should also withdraw from supporting local authorities.

‖**décor** (dekǫr). Also **decor.** [Fr., ad. L. *decor* (see Decor).] The scenery and furnishings of a theatre stage. Also *transf.* and *fig.*, setting, surroundings.
1897 G. Saintsbury *Sir W. Scott* iii. 64 Scott had obtained part of the scenery..in his yachting voyage.., which also gave the *décor* for *The Pirate.* **1900** H. Harland *Cardinal's Snuff-Box* ii. 12 Not too much like a *décor de théâtre.* **1902** C. H. E. Brookfield *Random Reminiscences* x. 182 He was afraid some heavy piece of the *décor* had fallen upon me. **1931** *Times Lit. Suppl.* 29 Jan. 76/2 This author..seems to lavish all his art..on fitting to them [*sc.* his characters]..a telling *décor.* **1953** 'M. Innes' *Christmas at Candleshoe* xii. 135 Then he sees that this is another ghost—the ghost of some departed modest revelry, a tattered remnant of stage *décor.*
b. The decoration or furnishings of a room, building, etc.; the layout or method of display of an exhibition, etc.
1926 R. Macaulay *Crewe Train* ii. vi. 131 What with..the Russian ballet *decor* of Evelyn Gresham, and the..gunroom notions of Denham, the flat was a queer hotch-potch. **1927** *Observer* 24 July 18/3 Olympia last week marked a revolution in exhibition *décor.* **1953** E. Taylor *Sleeping Beauty* ix. 161 Those little silver cups which his mother seemed to think part of the décor of a schoolboy's bedroom. **1955** *Times* 11 July 12/5 A *décor* including washable walls of the useful material now commonly used for kitchen table tops.

decoration. Add: **1. a.** *Decoration day* (earlier and later examples). (See also quot. 1957.)
1871 *Michigan Gen. Statutes* (1882) I. 455 The thirtieth day of May, commonly called decoration day. **1877** *Independent* 24 May 25/1 Decoration Day dawned bright and beautiful. **1957** *Encycl. Brit.* VII. 126/1 *Decoration day*, a holiday..observed in the northern states..on May 30, originally in honour of soldiers killed in the U.S. Civil War, but subsequently also in honour of those who fell in later wars.
4. The composition placed in the head of a rocket firework which makes the display when the case explodes.
1873–4 W. H. Browne *Art of Pyrotechny* (1874) 23 The rocket is now ready for its cap or pot which is to contain the stars or other decorations that are to be used. *Ibid.* 30 We have another exceedingly beautiful decoration for our rocket heads, which is called golden rain.

decorticated, *ppl. a.* Add: *spec.* having had the cortex (Cortex 3 a) removed.
1927 G. V. Anrep tr. *Pavlov's Conditioned Reflexes* i. 14 This difference in the dynamic balance of life between the normal and the decorticated animal. **1933** *Brit. Jrnl. Psychol.* Oct. 145 A similar difference in the behaviour of normal and partially decorticated animals.

decortication. Add: *spec.* in *Surg.* (see quots.).
1900 Dorland *Med. Dict.* 193/1 *Decortication*,..removal of portions of the cortical substance of the brain. **1909** *Practitioner* Nov. 661 Where the surgeon finds during the course of a nephrectomy that the decortication of the sac is very difficult. **1936** Punch & Knott *Mod. Treatm. Dis. Respiratory Syst.* vi. 97 *Decortication.* The justification for this operation is based on the assumption that the lung cannot..obliterate the empyema cavity because it is bound down by..thickened and fibrous visceral pleura. The operation of decortication consists in freeing the lung of this inelastic coat. **1962** *Lancet* 5 May 946/1 This procedure was followed by a tension pneumothorax which eventually necessitated thoracotomy and decortication to produce good re-expansion of the lung.

decorticator. (Further examples.)
1879 *Encycl. Brit.* IX. 344/2 Child's decorticator, an implement..for simply rubbing and scouring the grain, or for removing the thin bran and germ previous to the operation of grinding. **1954** D. Unwin *Governor's Wife* iii. 58 The decorticator was a mile from the bungalow...The huge jaws devoured an endless belt of sisal leaves, chewing them up and spewing out flaccid heaps of blond fibre.

Decoudun (dikū·dɒn). [Named after Jules *Decoudun*.] A calender ironing machine of French invention, first made in England in 1876.
1889 *Laundry Management* p. xxi, The Decoudun Ironer...The decoudun varies in size and construction with each maker. A simple form consists of a polished metal bed (concave) and a heavy roller, fitted in a strong cast-iron frame. **1905** *Daily Chron.* 23 Feb. 9/2 Wanted to purchase, Decoudon Ironer, a Cudlipp Perfect preferred. *Ibid.* 31 Oct. 9/5 Laundry.—Wanted really good calender hands, for calender and Decoudun. **1911** in *Encycl. Brit.* XVI. 282/1. **1921** *Dict. Occup. Terms* (1927) § 918 Decoudun hand.

‖**découpage** (dekupāʒ). [Fr., lit. 'the action of cutting up or out', f. *découper* to cut up or out.] **1.** The decoration of a surface with an applied paper cut-out; an object produced by this technique.
1960 H. Hayward *Antique Coll.* 94/1 *Découpage.* Until the mid-1670's the mounts of folding fans were cut with open-work lace-like patterns copying the intricate designs of French and Flemish lace. **1961** *Times* 30 Aug. 11/1 The programme cover..is decorated with a *découpage.*
2. *Cinemat.* = *cutting *vbl. sb.* 1 d.
1963 *Times* 7 Mar. 8/6 Such is..the precision of the *découpage* that the mysterious tale unfolds in a completely fascinating way. **1963** *Times Lit. Suppl.* 15 Mar. 180/1 A post-production script following closely the final *découpage* of the film.

decouple, *v.* Restrict †*Obs. rare*⁻¹ to sense in Dict. and add: **2.** To make the coupling between (two oscillatory systems, or two modes of oscillation of a single system) very loose, so that there is little transfer of oscillations from one to the other; chiefly *Electr.* So **decou·pling** *vbl. sb.*
1938 *Wireless Engineer* Sept. 480 The time constants of the series and parallel type of filter, used for decoupling the A.V.C. bias from the grid circuit of the R.F. controlled valves, are examined. **1940** *Amateur Radio Handbk.* (ed. 2) 86/2 In some cases it is necessary to use decouplings which are effective for R.F. but which allow currents of speech frequency to pass. **1961** Harris & Crede *Shock & Vibration Handbk.* II. xxx. 27 The natural modes of vibration of a body supported by isolators may be decoupled one from another by proper orientation of the isolators. Each mode of vibration then exists independently of the others. **1962** Simpson & Richards *Junction Transistors* xv. 370 A collector decoupling filter, in which a resistor of value *R* is placed between the collector bias resistor and the power supply. **1964** R. F. Ficchi *Electrical Interference* x. 187 Another technique that is useful for decoupling power supply leads is the decoupling capacitor.
b. To muffle the sound or shock of (a nuclear explosion) by causing it to take place in an underground cavity.
1960 *Guardian* 15 June 1/2 He wanted to know whether 'de-coupling' (the way of muffling underground tests) was to be tested. **1960** *New Scientist* 24 Nov. 1369/1 There have been claims in the United States that an explosion conducted in a large spherical cavity is greatly muffled, or 'decoupled'. **1966** *Ibid.* 28 July 191/3 The decoupling effects of a nuclear explosion in a large sphere..are a mystery.

decoy, *sb.*² Add: **6. decoy ship,** one used to decoy enemy vessels.
1915 *War Illustrated* III. 262/2 Decoy ships flying a neutral flag. **1923** W. S. Churchill *World Crisis 1915* 290 Our two principal devices for destroying the German submarines were the Bircham Indicator Nets and the Decoy Ships, afterwards called the Q-boats. **1925** Fraser & Gibbons *Soldier & Sailor Words* 73 *Decoy ships*, a name for certain vessels (also known as 'Mystery Ships' and 'Q-Ships'), introduced in 1915.

decreation. Delete †*Obs.* and add later examples.
1918 P. T. Forsyth *This Life & Next* x. 97 It is creation not out of a chaos but a wreck. It is the recreation of a decreation. *a* **1930** D. H. Lawrence *Pornography & so On* (1936) 26 The excrementory flow is towards dissolution, de-creation, if we may use such a word.

decree, *sb.* **4. b.** *decree nisi* (examples).
1860, 1872 [see Nisi]. **1922** [see *absolute *a.* 5 c]. **1934** A. P. Herbert *Holy Deadlock* 163 She was a lonely decree-nisi mongrel of a woman.

decrement. Add: **2. b.** The ratio of the amplitudes of two successive cycles of a damped oscillation; also (more fully *logarithmic decrement*), the natural logarithm of this.
1879 *Encycl. Brit.* X. 50/1 Such needles have great advantages—where, for instance, the time of oscillation, the logarithmic decrement, or the extent of swing of the needle has to be observed. **1908** E. H. Barton *Text-bk. Sound* x. 577 Hartmann-Kempf found that the relation between logarithmic decrement and amplitude for three makes of tuning-fork is almost linear. **1913** [see *decremeter]. **1927** E. G. Richardson *Sound* iv. 121 To estimate the rate of decay, the ratio of the amplitudes in two successive periods is obtained, a quantity which is known as the decrement. It is the logarithm of this quantity..which usually figures in calculations. **1929** J. A. Ratcliffe *Physical Principles of Wireless* i. 14 The reciprocal of this quantity, $R\sqrt{\frac{C}{L}}$, is of great importance in the theory of oscillatory circuits; it is known as the decrement of the circuit, and corresponds to the logarithmic decrement of a mechanical oscillatory system. **1936** L. S. Palmer *Wireless Engin.* iii. 47 Circuits with small decrements have good selectivity. **1966** W. T. Thomson *Vibration Theory* ii. 48 A body vibrating in a viscous medium has a period of 0·20 sec. and an initial amplitude of 1·0 in. Determine the logarithmic decrement if the amplitude after 10 cycles is 0·02 in.

decremeter (de·krĭmitəɹ, dekre·mitəɹ). *Electr.* [f. Decre(ment + -meter.] An instrument which measures the logarithmic decrement of an oscillatory circuit.
1913 *Year-Bk. Wireless Telegr. & Teleph.* 294 The Marconi Company has brought out an instrument called a decremeter, which enables an approximate measurement of the decrement of a circuit to be made. **1931** Moyer & Wostrel *Radio Handbk.* iii. 112 The decremeter..may be considered as a wave meter in which the movable plates of the variable condenser carry a dial graduated in values of decrements...This device gives a direct reading of the decrement on a dial.

decrescendo. Add: (Earlier and later examples.) Also *attrib.* and as *v. intr.*
1806 in T. Busby *Dict. Mus.* (ed. 2). **1890** G. B. Shaw *London Music* (1937) 365 The Wagnerian interpretation of the *decrescendo* mark over the chord. **1903** R. Langbridge *Flame & Flood* xvi. 165 Her voice..crescendoed..and decrescendoed. **1968** *Language* XLIV. 87 A long decrescendo and downglide typically follow the peak of stress.

decretorial (dīkrĭtōə·riăl), *a.* Transfer †*Obs. rare* to sense 1 and add later examples of sense 2.
1909 G. Tyrrell *Christianity at Cross Roads* 179 He understood this destination as more than moral or decretorial, as an inherent potentiality of His spirit. **1921** *Glasgow Herald* 18 Apr. 7 The argument..which that gentleman plainly considers decretorial.

‖**decretum** (dīkrī·tǝ̆m). [L.] A decree; sometimes short for *Decretum Gratiani*.
1602 W. Watson *Decacordon* 163 The *decretum* of the order obserued in all elections. **1888** *Encycl. Brit.* XXIV. 639/1 The *Decretum* specially inculcated subjection of the wife to the husband.

decrypt (dīkri·pt), *v.* [f. De- II. 1 + *crypt* as in Cryptogram, Cryptograph; cf. It. *decriptare.*] **a.** To solve (a cryptogram) without knowledge of the key. **b.** To convert (a cryptogram) into plaintext by proper application of the key. So **decry·pted** *ppl. a.*, **decry·pting** *vbl. sb.*, **decry·ption,** the deciphering of a cryptogram.
1936 *Cryptogram* June 4 Those who succeeded in decrypting Message No. 1 of the last issue. **1938** R. Candela *Milit. Cipher of Commandant Bazeries* 111 Decrypting or Cryptanalysis is that discipline which teaches us how to attempt the restoration of ciphered texts to their intended meaning without the knowledge or use of the cipher or key involved. **1946** *Rep. Joint. Comm. Investigation Pearl Harbor Attack. 79th Congress 2nd Sess.* iv. 230 The message was decrypted and translated in rough form on December 6. *Ibid.*, The December 3 dispatch..was turned over on December 5, 1941, to the Radio Intelligence Unit for decryption and translation. *Ibid.* 231 No basis exists for criticizing the system which was set up for decrypting and translating the intercepted Japanese messages. **1957** R. Watson-Watt *Three Steps to Victory* liii. 318 Burn all codes save that needed to decrypt the final note. **1962** G. Jenkins *Grue of Ice* v. 72, I can get a bearing if a ship sends eleven letters. I proved it to the German Decryption Service. **1967** *Punch* 30 Aug. 327/1 The lower case was used for decrypted messages.

‖**dédale** (dedal). *rare.* [Fr.] = Dædal *sb.* 2.
1916 H. G. Wells *Mr. Britling* ii. ii. 260 The roads, he said, were not a means of getting from place to place, they were a *dédale*.

Dedalic, var. *Dædalic *a.*

dedecorous, *a.* For †*Obs.*⁻⁰ read *rare* and add later example.
a **1913** F. Rolfe *Desire & Pursuit of Whole* (1934) 59 He avoided them, because of the dedecorous way in which some of them had treated him.

dedendum (dīde·ndǝ̆m). *Mech.* [L., neut. gerundive of *dēdere* to give up, surrender.] **a.** 'That part of the tooth of a cog-wheel or gear which is inside the pitch-circle and is intercepted between the pitch-line and the circle which limits all the roots of the teeth and the spaces between them' (*Cent. Dict.* Suppl., 1909). Also *attrib.* **b.** (See quot. 1958.)
1901 *Internat. Libr. Technol.* II. (*Gear-Cutting*) §17. 7 A tooth is composed of two parts, the addendum, or part outside the pitch circle, and the part inside, which is called either the root or the dedendum. **1909** N. Hawkins *Mech. Dict.* 185/2 The dedendum circle is the circle within the pitch circle, to which the bottom of each tooth extends. **1925** Berard & Waters *Elem. Machine Design* x. 180 The bottoms of the grooves are limited by a root or dedendum circle, distant from the pitch circle by an amount called the dedendum. **1949** E. Buckingham *Analytical Mech. Gears* xii. 256 We shall use a constant axial pitch of 1 in...This would give us the following thread proportions:..Dedendum 0·3247 [in.]. **1958** *Chambers's Techn. Dict.* Add. 971/1 *Dedendum*, radial distance between the pitch and minor cylinders of an external screw thread; radial distance between the major and pitch cylinders of an internal thread. **1961** P. S. Houghton *Gears* (ed. 2) i. 6 The dedendum of spur, helical and worm gearing is the radial distance of the tooth measuring from the pitch circle to the bottom of the tooth space. For a bevel gear it is the depth of the tooth space below the pitch circle measuring along the back cone generator or edge angle.

dedes (dī·des). [Javanese.] Musk obtained from the rasse.
1817 [see Rasse¹]. **1843** *Penny Cycl.* XXVI. 406/2 This is the *Rasse* of the Javanese, who term the odoriferous secretion *dedes* or *jibet*.

dedicate, *v.* Add: **4. c.** *Forestry.* (See quots.)
1943 *Post-War Forest Policy* 50 in *Parl. Papers 1942–3* (Cmd. 6447) IV. 419 Woodland which is required for timber production should be *dedicated* to that specific purpose, and..woodland owners who so dedicate their land and also provide adequate assurances for subsequent good management should receive State assistance. **1953** H. L. Edlin *Forester's Handbk.* xxi. 348 In order to obtain financial assistance from the Forestry Commission, for any substantial area of privately owned woodland, it is necessary to Dedicate it. This implies that the owner undertakes to manage it on an approved plan, more or less in perpetuity.

dedicated, *ppl. a.* Add: **b.** Of a person: devoted to his aims or his vocation; single-minded in loyalty to his beliefs, or in his artistic or professional integrity.
1944 Auden *For Time Being* (1945) i. iii. 55 Having learnt his language, I begin to feel something of the serio-comic embarrassment of the dedicated dramatist. **1955** N. King *Shadow of Doubt* ii, in J. Trewin *Plays of Year 1954–55* XII. 527 *Laura*...Have you ever met a really...dedicated person? *Frank*. Yes. Yes, I think so. Particularly one during the War. He was a doctor. **1971** M. Russell *Deadline* xi. 130 I'm dedicated, Smutty...Let me show my true worth.

dedication. Add: **5. b.** *Forestry.* The assignment of land under certain conditions for the production of timber. Cf. *dedicate *v.* 4 c.
1943 *Post-War Forest Policy* 50 in *Parl. Papers 1942–3* (Cmd. 6447) IV. 419 The act of dedication on the part of the owner would include the following undertakings:—(1) To use the land in such a way that timber production is the main object. (2) To work to a plan..approved by the Forest Authority. **1946** *Q. Jrnl. Forestry* XL. 69 Dedication is a covenant under which both the owner of the land and the Government undertake certain obligations. **1950** *Ibid.* XLIV. 53 It is just two years since the proposed Covenant of Dedication was published, and although the Society had always been strongly in favour of dedication, the covenant, as drafted, was received with considerable misgiving and distrust.

dedifferentiation (dī:diferenʃi,ēi·ʃən). *Biol.* [ad. F. *dédifférentiation* (C. Champy): see De- II. 2.] The loss or reversal of differentiation (see Differentiation 1). Hence **de:differe·ntiate** *v. intr.*, to undergo such a process; **di:differe·ntiated** *ppl. a.*
1917 *Amer. Jrnl. Anat.* XXII. 188 Dedifferentiation or return to a more embryonic condition probably underlies all types of regeneration. *Ibid.* 189 Champy..has maintained that most of the cells in the body dedifferentiate in tissue culture. **1926** *Glasgow Herald* 18 Sept. 4 A good example of the dedifferentiated inert state of suspended animation is the 'brown body' of some of the moss-animals. **1926** J. S. Huxley *Ess. Pop. Sci.* 80 The sea squirt..can in unfavourable surroundings de-differentiate, or revert to an embryonic state. **1960** *New Biol.* XXXI. 90 Tissue cells undergo an apparent de-differentiation to form the young regenerate or the bud.

dedolent, *a.* Delete †*Obs.* and add later examples. Also **dedolant.** Both forms *rare.*
1893 'S. Grand' *Heavenly Twins* I. v. i. 3 She was sleeping soundly, not because she was dedolent but because she was exhausted. **1951** Auden *Nones* (1952) 34 A tanker sinks into a dedolant sea.

dedolomitization (dī:dǫlǒmitəizēi·ʃən). *Petrol.* [f. De- + Dolomitize *v.* + -ation.] The changing of dolomite into rock of another kind. So **dedo·lomitize** *v. trans.*, **dedo·lomitized** *ppl. a.*
1903 J. J. H. Teall in *Geol. Mag.* 4th Ser. X. 514 The cherty dolomites have been dedolomitised by the formation of magnesian silicates...But dedolomitisation has also been produced in another way...The carbonic acid freed itself more readily from the magnesia..converting the original dolomite into an aggregate calcite and periclase. **1930** Peach & Horne *Geol. Scotl.* 113 Another method of dedolomitization should be added to those enumerated by Teall. **1969** *Nature* 9 Aug. 607/2 The partially dedolomitized rocks have three components—calcite of the original limestone, unreplaced dolomite and calcite after dolomite.

deductible, *a.* Delete *rare* and add further examples (also **deductable**); *spec.* that can be deducted from one's tax or from one's taxable income.
1894 *Westm. Gaz.* 14 Aug. 2/1 Now tax is deductable either at the rate actually in force at the date of payment. **1913** *Standard* 3 Apr. 11/4 A dividend entrusted to an agent for payment on a date before April 6 is chargeable with the duty in force for the year 1912–13, and the duty is deductible notwithstanding that payment in individual cases may not happen to be claimed until after that date. **1925** *Glasgow Herald* 30 Jan. 9 The time within which..discount was deductable under the contract. **1954** I. Levin *Kiss before Dying* ii. ii. 89 We are..beginning the construction of a new gymnasium...Perhaps your father would be interested in making a contribution...Such contributions are tax-deductible. **1965** *Listener* 17 June 899/3 The all-in price for a London plane tree, including maintenance, is £40—not exactly cheap, but it is tax-deductible.

deduction. Add: **8.** *attrib.*, **deduction theorem** *Logic*, the rule or metatheorem that if within a system a formula B is derivable from a formula A, then 'If A then B' is a theorem of the system; the principle of conditionalization.
1941 O. Helmer tr. *Tarski's Introd. Logic* vi. 127 A general law..which..is known as the law of deduction (or the deduction theorem). **1951** *Mind* LX. 382 The Deduction Theorem holds for the calculus of causal propositions. **1962** W. & M. Kneale *Devel. Logic* v. 320 The principle of conditionalization (or 'deduction theorem')..was taken for granted by Aristotle.

deductive, *a.* Add: **1. c.** *deductive system* (Logic): a set of propositions or formulas, in which some are derived from others according to rules of proof, all such possible derivations being held to be included.
1910 Whitehead & Russell *Principia Math.* I. p. vi, A deductive system such as that contained in the present work. **1936** A. J. Ayer *Lang., Truth & Logic* ii. 40 Among the superstitions..is the view that it is the business of the philosopher to construct a deductive system. **1940** W. V. Quine *Math. Logic* 88 The highly explicit way of presenting formal deductive systems which is customary nowadays dates back only to Hilbert (1922) or Post (1921). **1951** *Mind* LX. 266 The calculus, or deductive system, of truth-functions.

deductivism (dīdɒ·ktiviz'm). *Philos.* [f. Deductive *a.* + -ism.] The preference for, use of, or belief in the superiority of, deductive as opposed to inductive methods; *esp.* either the doctrine that induction has no place in scientific method, or that induction in some manner requires justification by deduction. So **dedu·ctivist,** one who advocates deductivism; also *attrib.* or as *adj.*
1908 *Fabian News* June 56/2 The parallel sociological advance from metaphysics to deductivism. **1936** A. L. Rowse *Mr. Keynes* ii. 7 That excessive and unenlightening deductivism in economics. *Ibid.* 9 It is simply to recommend a more modest habit of mind to the deductivists. **1959** K. R. Popper *Logic of Sci. Disc.* i. 30 The theory of the deductive method of testing, or..the view that a hypothesis can only be empirically *tested*..might be called 'deductivism'. *Ibid.*, Duhem..held pronounced deductivist views. **1962** in E. Nagel et al. *Logic, Methodol., etc.* 265 (*title*) The Controversy: Deductivism Versus Inductivism.

dee, *sb.* Add: **2.** Either of the two hollow, D-shaped electrodes used to accelerate particles in a cyclotron.
1936 Lawrence & Cooksey in *Physical Rev.* L. 1131/2 The accelerating electrodes..are called duants, or dees. **1966** *New Scientist* 24 Nov. 464/1 In the traditional type of cyclotron, the particles are accelerated by being driven electrostatically from one to another of a pair of matched semicircular devices called dees.
3. Slang abbrev. of Detective *sb.* Cf. *D III. 3.
1882 *Sydney Slang Dict.* 3/2 Dee (D.), a detective policeman. **1895** *Brewer's Dict. Phr. & Fable* 339/2 Look sharp! the dees are about. **1943** *Penguin New Writing* XVIII. 72 Occasionally I'd see people we passed who'd pick them for dees. **1949** E. de Mauny *Huntsman in Career* 127 You've got to look out, if the dees come.

dee, *v.* Add: **b.** as *adj.* = Damned *ppl. a.* 4 a.
1889 Kipling *From Sea to Sea* (1899) 212 Dee fool. It's different in Upper Burma, where you get command and travelling allowances.

deedy, *a.* For *dial.* read: Chiefly *dial.* **1.** (Further examples.) Also, earnest, serious.
1895 Hardy *Jude* i. ii. 14 There! don't ye look so deedy! Farmer Troutham is not so much better than myself. **1938** F. D. Sharpe *Sharpe of Flying Squad* xxiii. 241 He said that he had seen four men in a pub in Bethnal Green in 'deedy' conversation. **1951** M. Kennedy *Lucy Carmichael* vi. v. 310 He was feeling genial and deedy, and..his imagination toyed with new campaigns. *Ibid.* vii. 326 The train came puffing in half empty, so he did not have to be so deedy on her behalf as he would have liked. **1959** *Listener* 19 Nov. 900/2 The opening scenes..were just a shade too fruity and deedy.

dee-jay, deejay (dīdʒēi·). *slang* (orig. *U.S.*). [Pronunciation of *D.J.* (see *D III. 3).] A disc-jockey.
1955 in L. Feather *Encycl. Jazz* (1956) 346/1. **1956** *Life* (U.S.) 19 Nov. 143/1 Some 1,200 radio disk jockeys..journeyed..to Nashville..to attend..'Dee Jay' (for disk jockey) convention. **1958** [see *A. and R.* s.v. *A III]. **1964** *Melody Maker* 28 Nov. 3 Deejays and promoters must stop being idiots.

de-emanate (dī,e·mănēit), *v.* [f. De- II. 1 + Emanate *v.*] *trans.* To deprive (a radio-active substance) of the power of emitting its characteristic radioactive gas or 'emanation'. Hence **de-e·manated** *ppl. a.*; **de-emana·tion.**
1902 Rutherford & Soddy in *Phil. Mag.* IV. 371 If thorium oxide be exposed to a white heat its power of giving an emanation is to a large extent destroyed. Thoria that has been so treated is referred to throughout as 'de-emanated'. *Ibid.* 575 Water-vapour exerts no influence..in de-emanating thoria. **1902** —— in *Jrnl. Chem. Soc.* LXXXI. 333 (*heading*) The de-emanation of thoria. **1904** Rutherford *Radio-activity* 216 Thus de-emanation does not permanently destroy the power of thorium of giving out an emanation. *Ibid.*, The de-emanated thoria was dissolved.

de-emotionalize (dī,ī̆mōu·ʃənăləiz), *v.* [f. De- II. 1 + Emotionalize *v.*] *trans.* To render emotionless. So **de-emo·tionalized** *ppl. a.*
1942 *Mind* LI. 84 The defence of the obsessional neurotic, namely de-emotionalised talk. **1957** J. S. Huxley *Religion without Revel.* (ed. 2) iii. 55 Hypostasised 'forces' and 'principles' are de-emotionalised refinements of personalised thinking.

de-emphasis (dī,e·mfăsis). [f. De- II. 2 + Emphasis.] A lessening or removal of emphasis; *spec.* in radio communications, a reduction in the relative strength of higher audio frequencies made in order to restore the original form of the signal.
1940 *RCA Review* Jan. 359 A de-emphasis circuit at the receiver. **1942** A. Hund *Frequency Modulation* ii. 221 Networks for a–f accentuation and deaccentuation are also known as circuits for a–f preemphasis and deemphasis respectively. **1957** K. W. Wittfogel *Oriental Despotism* 7 These methods, which use distortion and de-emphasis rather than open discussion. **1962** A. Nisbett *Technique Sound Studio* 254 Pre-emphasis of top (i.e. prior to transmission), with a corresponding de-emphasis at the receiver, helps to reduce the noise level still further. **1965** *Economist* 2 Jan.

27/3 The new policy aims were marked by major de-emphasis on air defence.

de-emphasize (dī,e·mfăsəiz), *v.* [f. De- II. 1 + Emphasize *v.*] *trans.* To remove emphasis from, or reduce emphasis on.
1938 *Time* 28 Nov. 22 One way to de-emphasize football is to hold the games in the morning. **1949** M. Mead *Male & Female* vii. 152 The dual character of the eliminative tract also provides a background for re-emphasizing and de-emphasizing sex differences. **1952** *Essays in Criticism* II. ii. 212 Without de-emphasizing feeling, he..emphasizes that poetic sensitivity is a correlative of a poet's knowledge. **1956** A. H. Compton *Atomic Quest* ii. 107 Ernest Lawrence..urged that this process be de-emphasized so that greater effort could be applied to the methods that gave greater promise of immediate production. **1961** A. West *Trend is Up* vii. 306 We're trying to make it a good college. We've de-emphasized football. **1970** *Economics* VIII. 145 The examination de-emphasizes technical detail.

deemstership (dī·mstəɹʃip). [f. Deemster + -ship.] The office of deemster in the Isle of Man.
1894 Hall Caine *Manxman* iii. iii, Let the Deemstership go to perdition. **1902** *Daily Chron.* 28 Mar. 5/2 When he retired from his deemstership of the Isle of Man in 1897.

deener: see *deaner.

deep, *a.* Add: **1. c.** *deep water*(*s*): see Water *sb.* 6 c in Dict. and Suppl.
d. *deep end*: the end of a swimming-pool at which the water is deepest; so in colloq. fig. phr. *to go* (*in*) *off the deep end*, etc.: to give way to emotion or anger; to 'let oneself go'.
1921 *Times Lit. Suppl.* 22 Dec. 853/3 Saint-Saëns rarely, if ever, takes any risks; he never, to use the slang of the moment, 'went in off the deep end'. **1922** Joad *Highbrows* vi. 186 She passed her life metaphorically at the end of a diving-board, ready at the slightest provocation to go in at the deep end. **1923** *Weekly Dispatch* 21 Jan. 5 Mr. Nicholas Hannen..plays the second fiddling husband admirably, except when, once or twice, he goes off the deep end a trifle too explosively. **1924** Galsworthy *White Monkey* ii. iv, Would it not be more in the mode, really dramatic—if one 'went over the deep end', as they said, just once? **1927** W. E. Collinson *Contemp. Eng.* 115 A very common phrase since the war is to go (in) off the deep end, an expression evidently taken from the deep end of a swimming-bath where the diving board is. **1934** F. W. Crofts *12.30 from Croydon* vii. 84 Yes, I've been. And found the old boy brimming over with wrath against you. And when he heard I was coming on the same job he fairly went off the deep end. **1963** T. Parker *Unknown Citizen* iii. 88 I'm not going to do what I've done before, go off the deep end, nothing like that. **1966** G. Greene *Comedians* i. iv. 113 Mr Smith trundled to the deep-end before he emerged. **1970** V. Gielgud *Candle-Holders* vii. 62 Simon Astley was too much like the instructor in a swimming-bath, who throws you in at the deep end.
2. b. Also, with numeral prefixed, having so many engagements or obligations.
1921 *Discovery* Sept. 242/2 He proposed to her, and she accepted him—if he would wait his turn; she was four deep already! **1935** Lady Fortescue *Perfume from Provence* 109 The lawyer was always engaged six deep whenever I called to see him.
c. *Cricket.* Of a fielder or fielding position: farther than normal from the batsman. (See also *deep field below.)
1867 G. H. Selkirk *Guide to Cricket Ground* ii. 25 Deep Cover, etc. When twenty-two are in the field the men that stand far out are designated by the names of the positions of the men they support, with the word 'deep' prefixed. **1897** K. S. Ranjitsinhji *Jubilee Bk. Cricket* ii. 54 When a man is put on the boundary behind extra-cover or cover, he is usually called deep-extra-cover or deep-cover.
3. c. *spec.* Of mining operations: far below the surface of the ground; so **deep-mined** *a.* (contrasted with *opencast*). *deep lead*: see Lead *sb.*² 6 b.
1839 *Penny Cycl.* XV. 239/2 Even in very deep shafts, when complete, daylight may be seen from the bottom. **1860** *Mining Gloss.* (ed. 2) 37 *Deep level*, the watercourse leading to the engine-shaft, being always the deepest adit in the mine. **1862** *Otago: Its Goldfields & Resources* 18 The deep sinking has not, so far, realised the sanguine expectations once entertained regarding it. **1899** *Daily News* 3 May 3/5 Deep-level mines and deep-level mining. **1921** *Dict. Occup. Terms* (1927) § 581 *Deep sinking labourer*; navvy or miner engaged in sinking deep holes or shafts. **1948** *Ann. Reg. 1947* 493, 4 million tons of deep-mined coal a week.
IV. a. *deep-well.*
a **1877** Knight *Dict. Mech.* I. 682/2 *Deep-well pump*, a pump specifically adapted for oil and brine wells which are bored of small diameters and to great depths. **1906** *Daily Chron.* 5 Sept. 3/3 To allow these deep-well waters..to run heedlessly to waste, is a policy of which a later generation of Australians may have bitter cause for complaint. **1963** *Gloss. Mining Terms* (*B.S.I.*) iv. 6 *Deep well pump*, any kind of pump delivering from a well, shaft or borehole.
b. *deep-bosomed* [Gr. βαθύκολπος], *-breasted, -faced, -grassed, -rutted.*
1851 Buckley *Iliad* 346 Trojan (dames) and deep-bosomed Dardanians. **1876** Pater *Greek Studies* (1895) 81 The deep-bosomed daughters of the Ocean. **1905** R. Garnett *Shakespeare* 56 Not all deep-bosomed earth's wide fruitfulness. **1909** *Westm. Gaz.* 14 Aug. 2/3 The brown-faced, deep-bosomed peasant women. **1935** W. G. Hardy *Father Abraham* 101 A tall, deep-breasted woman was standing in the door of the tent. **1908** *Westm. Gaz.* 23 Dec. 4/1 A number of golfers playing with deep-faced, round-headed drivers. **1906** *Ibid.* 21 May 1/3 The hedge on either side is just high enough to hide the deep-grassed meadows. **1836** T. Hook *G. Gurney* i. iii. 217 At the corner of the deep-rutted lane. **1899** *Daily News* 19 Oct. 6/1 Transports stopped in the deep-rutted roads.
c. deep-breathing *vbl. sb.*, the act of breathing deeply as a form of physical exercise; hence *deep-breathe* v. intr.; **deep-dish pie** *U.S.*, 'a pie, usually a fruit pie, baked in a deep dish and having no bottom crust' (Webster 1934); **deep-draft** or **-draught** *a. Naut.*, that displaces deep water; **deep drawing,** a kind of cold-working in which a sheet or strip of metal is subjected to considerable plastic deformation by being forced through a die, so producing hollow parts such as cylinders; **deep-etch, -etching,** a photo-engraving process in which the lithographic plate is slightly etched; hence **deep-etch** *v. trans.* and *intr.*; **deep field** *Cricket*, that part of the field which is near the boundary, esp. behind the bowler; also, a fieldsman or his position there; hence *deep-fielder*; **deep-(fat-)frying,** the frying of food in sufficient oil or fat to cover it completely; so *deep-(fat-)fry* v. trans. and intr., *deep-(fat-)frier, -fryer*; **deep kiss** (see quot. 1951); so *deep kissing* vbl. sb.; **deep litter,** a deep layer of litter (Litter *sb.* 3 b) used in poultry-houses, etc.; a method of keeping poultry in such conditions; also *attrib.*; **deep-milking,** the production of a good yield of milk; so *deep-milker*; **deep-rooter,** something which takes deep root; **deep-sinker** *Austral. slang*, (*a*) a drinking-glass of the largest size, so called from a fanciful resemblance to a deep mineshaft; (*b*) the drink served in such a glass; **Deep South,** the southernmost parts of the United States, usually taken to include the states of Alabama, Georgia, Louisiana, Mississippi, and sometimes South Carolina; also *attrib.*; **deep space,** a term for the regions of space that are either (*a*) beyond the solar system or (*b*) well outside the earth's atmosphere; also *attrib.*; **deep structure** *Linguistics* (see quot. 1965); **deep tank,** a tank for water ballast or fuel oil, formed by cutting off a part of a ship's hold; **deep therapy,** the treatment of disease by short-wave X-rays (see quots.); **deep trance,** a deep hypnotic state; also *attrib.*; hence *deep-tranced* used as *pa. pple.*
1964 *Punch* 26 Feb. 315/1 In the morning he deep-breathed. **1904** B. A. Macfadden *Building of Vital Power* 115 Through deep breathing conducted as an exercise, the lungs become larger. **1936** W. Faulkner *Absalom, Absalom!* vii. 217 He will raise the window and do deep-breathing. **1965** F. Sargeson *Memoirs of Peon* vi. 161 The rules were, nothing to drink except water; deep-breathing exercises and no tobacco. **1936** Wodehouse *Laughing Gas* ii. 29, I polished off the steak and put in a bid for deep-dish apple pie with a bit of cheese on the side. **1943** *Gen* 2 Jan. 33/2 Their [American] diet..consists of little but..deep-dish apple tart. **1908** *Westm. Gaz.* 30 Oct. 6/4 None of our inward water-borne traffic enters the Port in deep-draft vessels. **1925** *Forging, Stamping, Heat Treating* XI. 428/1 Several annealing and pickling operations..are necessary on extremely deep drawing. **1929** *Jrnl. Iron & Steel Inst.* CXX. 460 The force required and the stresses set up at various stages in the deep drawing of sheet metal. **1932** *Ibid.* CXXV. 559 The deep-drawing qualities of steel sheets. **1948** H. Missingham *Stud. Guide Comm. Art* ii. 141 *Deep-etch*, used on half-tone blocks where the white parts of the drawing would have a very small screen dot over them unless this instruction were given for their removal. **1960** G. A. Glaister *Gloss. of Book* 99/2 *Deep-etch process*, a method of making a lithographic plate by photo-mechanical means. It involves the printing on to metal of a photographic positive, and the very slight..etching of the plate. **1967** E. Chambers *Photolitho-Offset* 270 *Deep-etch*, lithographic platemaking process in which the work areas are slightly etched (about 0·003 in.) into the surface of the plate. Positives are used for printing to metal using special solutions, with the etches supplying space for the lacquer base of the ink-receptive printing image. **1897** *Process Photogram* Sept. 143 For a deep etching the solution would have to be renewed several times. **1955** in *Amer. Speech* (1956) XXXI. 84 They deep-fat-fry them. **1947** M. Given *Modern Encycl. Cooking* II. 1609 *Deep fat fryer*. Buy as a unit. **1926** S. E. Nash *Cooking Craft* vi. 35 French or Deep Fat Frying. **1870** *Times* 20 July 10/3 Mr. Francis was caught at deep field. **1900** *Westm. Gaz.* 22 June 3/1 There is scarcely a bowler nowadays who does not station at least one man in the deep-field. *Ibid.* 3/2 Our first two batsmen..succeeded in getting the ball between his deep fields. **1933** A. G. Macdonell *England, their England* vii. 114 Mr. Southcott..went away into the deep field, about a hundred and twenty yards from the wicket. **1870** *Times* 9 Aug. 3/4 Deep-fielders have now a better chance of compassing with effect the space allotted them. **1951** E. David *French Country Cooking* 18 A deep frier with a basket is necessary for chips. **1933** *Mod. Pract. Cookery* 195/2 Be very careful when deep-frying not to fill the pan too full. **1970** *New Yorker* 24 Oct. 43/2 It's cornmeal and onions, deep fried. **1932** E. Craig *Cooking with E. Craig* 113 Use 1 or 1½ lb...fat, lard or olive oil, for deep frying. **1948** A. C. Kinsey et al. *Sexual Behavior in Human Male* x. 369 Deep kissing is utilized as a prime source of erotic arousal...A deep kiss may involve considerable tongue contacts. **1951** Ford & Beach *Patterns Sex. Behav.* iii. 49 The 'deep kiss'..involves thrusting the tongue of one partner into the mouth of the other. **1946** Lippincott & Card *Poultry Production* (ed. 7) xii. 292 The usual procedure with deep litter is to start in the fall with two to four inches of dry litter, and add to this gradually until the floor is covered eight to ten inches deep. **1947** R. Seiden *Poultry Handbk.* 106 Deep-litter nests keep eggs clean. **1958** *Times* 19 Sept. 3/5 There had been a drop in egg production since they changed from batteries to deep litter. **1969** *Guardian* 25 Aug. 13/5 The sheep were all in deep litter. **1879** J. P. Sheldon *Dairy Farming* 5/2 It is important that she should be a 'deep milker'. **1833** *Ridgemont Farm Rep.* in *British Husbandry* (1840) III. i. 148 The large, heavy, slow-fattening but deep-milking cows of this country. **1879** J. P. Sheldon *Dairy Farming* 36/2 Deep milking..is a question of breeding and training. **1923** *Discovery* Sept. 244/2 The deep-milking propensity of some cows may be transmitted through the bull to the next generation of female calves. **1960** *Farmer & Stockbreeder* 8 Mar. 31/1 An opportunity to purchase well-framed, deep-milking Ayrshire cattle. **1898** *Westm. Gaz.* 19 Mar. 2/1 Her first cousin..is a deep-rooter, and must be looked after betimes. **1927** *Smallholder* 26 Mar. 105/2 The main-crop, deep-rooters should be sown at the end of May. **1886** F. Cowan *Australia: a Charcoal Sketch* 32 Long-sleever, Bishop Barker, and Deep-sinker, synonyms of Yankee Schooner. **1897** *Argus* 15 Jan. 6/5 (Morris), A tumbler—whether medium, small, or deep-sinker. **1936** 'R. West' *Thinking Reed* i. 23 He had thought himself right out of the illusions common to the Deep South. **1938** C. H. Matschat *Suwannee River* 285 A trip to the Deep South. **1940** C. McCullers *Heart is Lonely Hunter* (1943) i. i. 3 The town was in the middle of the deep South. **1941** W. S. Maugham *Writer's Notebook* (1949) 327 The vicissitudes of the war have brought her down to the deep South, but till then she had always lived in Portland, Oregon. **1952** *Spectator* 21 Nov. 710/2 A 'Deep South' American novel. **1967** *Times* 8 July 11/1 In the two years since the voting rights Act was passed the registration of Negro voters in the deep South has increased by 78 per cent. **1952** M. St. Clair in 'E. Crispin' *Best SF* (1955) 184, I am aware of an intense loneliness. It's a normal response to the deep space situation. **1954** *Jrnl. Brit. Interplanetary Soc.* XIII. 16 These specialized 'deep-space' rocket vehicles would, of course, be refuelled and serviced from satellite ships. **1960** F. Gaynor *Dict. Aerospace* 69 *Deep space*, a colloquial term for space beyond the outermost boundaries of our solar system. **1962** *New Scientist* 22 Mar. 676/2 Deep-space missiles..being fired far out into space and then falling back to approach the target at a steep angle and a high velocity. **1969** *Daily Tel.* 11 Jan. 1/4 They will also send back to Earth information about deep space during the 155 million-mile journey [to Venus]. **1965** N. Chomsky *Aspects Theory Syntax* i. 16 The syntactic component of a grammar must specify, for each sentence, a *deep structure* that determines its semantic interpretation and a *surface structure* that determines its phonetic interpretation. **1966** L. J. Cohen *Diversity of Meaning* (ed. 2) ii. 37 The basic rules generate so-called 'deep structures' which receive an interpretation from the semantic rules. **1909** *Cent. Dict. Suppl.*, Deep tank. **1935** *Jane's Fighting Ships* 113 No. 1 stokehold now converted into four oil-fuel deep-tanks. **1944** R. Phillips *Super-voltage X-ray Therapy* ii. 4 These descriptive terms—'superficial' therapy for voltages up to 130 kv., 'deep' therapy for the 200–400 kv. range, and 'supervoltage' therapy for 500 kv. upwards—have little to recommend them except brevity. **1967** *New Scientist* 29 June 760/1 Whereas the term 'deep therapy' formerly implied X-ray generators of several hundred kilovolts, it now implies accelerators of, effectively, several million volts. **1892** Deep trance [see Hypnosis 2]. **1927** W. de la Mare *Stuff & Nonsense* 48 A panther spies them there, Deep-tranced in speechless rapture. **1959** *Listener* 31 Dec. 1165/3 Deep-trance subjects.

deep, *sb.* Add: **3. d.** *Cricket.* Ellipt. for *deep field* (see *deep *a.* IV. c).
1906 Beldam & Fry *Gt. Bowlers & Fielders* 438 A fine example of the throw in from 'the deep'. **1924** H. de Selincourt *Cricket Match* (1928) iv. 89 One catch in the deep which he had held in a school cricket match..was still vividly remembered. **1954** J. H. Fingleton *Ashes crown the Year* 112 Lindwall having nobody in the deep. **1963** A. Ross *Australia 63* iii. 85 Graveney, chasing one in the deep, pulled a muscle and left the field.

deep, *adv.* Add: **1. d.** *Cricket.* In the deep field (see *deep *a.* IV. c).
1849 W. Lillywhite *Young Cricketer's Guide* 21 This fieldsman must stand deeper. **1857** Hughes *Tom Brown* II. viii, The batter..cuts it beautifully to where cover-point is standing very deep, in fact almost off the ground. **1891** W. G. Grace *Cricket* x. 265 Mid-on..is placed close in or deep according to the wish of the bowler.
3. c. (Further examples.)
1915 E. B. Holt *Freudian Wish* i. 5 The person does, in his dream, what he deep-down wishes to do. **1962** E. Bowen *Afterthought* 81 A slowly acquired deep-down knowledge.

deep-drawn, *ppl. a.* Add: **2.** Of metals: produced or worked by deep drawing; suitable for deep drawing (cf. *deep *a.* IV. c).
1925 *Forging, Stamping, Heat Treating* XI. 427 The importance of selecting the proper stock for deep drawn shells. **1926** *Jrnl. Iron & Steel Inst.* CXIII. 589 The manufacture of deep-drawn cylinders. **1959** *Motor Manual* (ed. 36) i. 7 The sheet is not pushed into the die by the punch, but is stretched into it, which is why the type of sheet steel known as deep drawn must be used.

deep-freeze, deep freeze. orig. *U.S.* [f. Deep *a.* + Freeze *sb.*¹] **1.** (Written *Deep-freeze.*) The registered American trade-name of a type of refrigerator capable of rapid freezing. Hence, a refrigerator or process in which food can be quickly frozen and stored almost indefinitely at a very low temperature. Also *attrib.*

1941 *Official Gaz. U.S. Pat. Off.* 8 July 260/2 *Deepfreeze.* For Refrigerating Apparatus and Parts Thereof. 1942 *Science Illustr.* Mar. 3/1 The Deepfreeze home locker looks like an up-ended white enameled barrel with an encased compressor housing at its side. 1951 *Good Housek. Home Encycl.* 218/1 'Polythene'..is particularly suitable for deep-freeze and quick-frozen foods. *Ibid.* 473/2 A refrigerator with a special deep-freeze compartment. 1957 W. H. WHYTE *Organization Man* 250 The chickens are stacked high in the deep freeze. 1958 *Sunday Times* 27 Apr. 22/6 The housewife with a deep freeze can keep them for six months. 1969 *News of World* 23 Nov. 9/5 They paid a deposit on a deep-freeze but then heard nothing more.

2. *transf.* and *fig.* 'Cold storage'; suspension of activity; suspended animation. Also *attrib.*

1949 L. C. WROTH in C. F. Bühler et al. *Standards of Bibliogr. Description* 114 Most historians won't bother. They are not interested in bibliography. For all that they may do about it, the meat will lie in the deep-freeze compartment until the end of time. 1950 *N.Y. Times Bk. Rev.* 8 Oct. 3/1 This can never be reprinted too often; it is a classic of deep-freeze. 1951 I. SHAW *Troubled Air* iv. 84 You must have led your life in deep freeze for the last twenty years. 1952 *Birmingham (Ala.) News* 6 May 26/3 A Swedish doctor has predicted that 'deep freeze' operations on human beings may not be far off. He said this is a technique in which heart action is purposely stopped by lowering the body temperature. 1957 *Daily Mail* 26 Sept. 9/5 The 'deep-freeze' girl who lay in a coma for 14 weeks. 1958 *Spectator* 6 June 722/1 It [sc. a suburb] is not alive either, of course; perhaps suspended animation is the best description, a kind of socio-economic deep freeze.

deep-freeze, *v.* orig. *U.S.* [f. prec.] *trans.* To subject to a deep-freeze process; to refrigerate; also *transf.* and *fig.* So **deep-freezer,** a deep-freeze; **deep-freezing** *vbl. sb.*; **deep-frozen** *ppl. a.*

1949 *Time* 18 July 34/1 Helen Hayes stuck a hand into her deep freezer. 1951 *Good Housek. Home Encycl.* 63/1 Deep freezer. *Ibid.* 438/1 Most fruits are suitable for deep freezing. 1953 A. HUXLEY *Let.* 8 Dec. (1969) 689 The product is deep-frozen the moment it leaves the pod. 1956 *Sci. News Let.* 3 Nov. 279/2 Deep-frozen shrimps from India may soon be making an appearance on dining tables in the United States. 1957 'B. BUCKINGHAM' *Boiled Alive* xxxiii. 249 If the cook deep-froze the body, he knew about the murder. 1959 *Listener* 5 Feb. 263/2 We all know the deep-frozen expert from the one who is humanly concerned to put something across. 1959 *Observer* 15 Mar. 16/3 Why should Gilbert and Sullivan be specially exempt?.. Perpetuity would deep-freeze them as well. 1959 *Woman* 2 May 33/1 With deep freezing and modern refrigerated transport, fish doesn't depend on the season nearly as much as it used to. 1970 *Daily Tel.* 28 Apr. 17 Leading manufacturers of refrigerators and deep-freezers..are producing space-saving models which are ideal for our small British kitchens. 1971 *Ibid.* 25 Jan. 10 No doubt these arguments are today accepted by all but a few deep-frozen ideologues on Labour's Left wing.

deepie (dī·pi). *colloq.* [f. DEEP *a.* + *-ie*, as in *TALKIE.] A three-dimensional cinematographic or television film.

1953 DILYS POWELL in *Sunday Times* 8 Feb. 9/7 Twentieth-Century Fox..has announced that all its important films are now to be what I will venture to call deepies. 1954 *Britannica Bk. of Year* 637/2 The normal kind of film was called a flat or flattie, the three-dimensional a deepie. 1966 *T.V. Times* (Australia) 7 Dec. 10/2 Three-D television is now the subject of experiment....This week, I discussed the future of 'deepies' with Mr B. Brownless.

deepish (dī·piʃ), *a.* [f. DEEP *a.* + -ISH¹.] Somewhat or rather deep.

1878 I. L. BIRD *Lady's Life Rocky Mountains* (1879) 197 We crossed a deepish stream on the ice. 1920 *Blackw. Mag.* July 86/2 It's only a deepish scratch. 1925 A. S. M. HUTCHINSON *One Increasing Purpose* I. xx. 128 Rather, rather deepish things. 1926 *Spectator* 6 Mar. 408/2 The Government was now in deepish water. 1959 *Times* 27 May 5/2 Dixon was caught at deepish mid-off.

deep-waterman (dīːpˌwǭ·təɹmæn). [f. *deep water* + MAN *sb.*¹] A sea-going vessel as opposed to a coaster.

1906 *Daily Chron.* 21 Mar. 6/4 Owners, not of 'deep-watermen' only, but of coasting craft as well. 1909 *Ibid.* 13 Aug. 3/1 What of the hundreds of fine deep-water men that still fly the Red Ensign? 1924 R. CLEMENTS *Gipsy of Horn* 57 This..constitutes the time-honoured decoration of a deep-waterman.

deer. Add: **4. b. deer-ball,** an underground fruit body of a fungus of the genus *Elaphomyces*; **deer-bleat** *U.S.*, an instrument serving to imitate the bleating of a deer; **deer-culler** *N.Z.* (see quot. 1947); so **deer-culling**; **deer-drive** (earlier U.S. example); **deer-fly,** any one of various flies which infest deer; also *attrib.*; **deer-horn,** (*a*) the material of a deer's horn; (*b*) *U.S.*, a large rough mussel of the Mississippi, *Trigonia* or *Unio verrucosa*, the shell of which is used for making buttons; **deer-stand** *U.S.*, a station for the shooters at a deer-drive; **deer-track,** (*a*) the marks of a deer's passage; (*b*) a route habitually taken by deer; **deer-yard** *U.S.* (earlier example).

[1640 PARKINSON *Theat. Bot.* 1320 *Tubera cervina.* The Deares underground balles or Mushromes are another sort of these Tuberaes.] 1854 MAYNE *Expos. Lex.*, Deer-Ball. 1866 LINDLEY & MOORE *Treas. Bot.*, *Deer balls*, a synonym of Hart's Truffles, Lycoperdon Nuts, and *Elaphomyces*. 1950 AINSWORTH & BISBY *Dict. Fungi* (ed. 3) 95 *Deer balls*, (Lycoperdon nuts or harts' truffles), *Elaphomyces* fruit bodies. 1852 MARCY & MCCLELLAN *Explor. Red River* 27 June (1853) vi. 50 The idea occurred to me of attempting to call them with a deer-bleat, which one of the Delawares had made for me. 1947 P. NEWTON *Wayleggo* (1949) 153 *Deer cullers*, men who shoot deer professionally for the Government or an acclimatization society. 1965 M. SHADBOLT *Among Cinders* xxvi. 260 He was a deer-culler when he wasn't writing. 1959 —— *New Zealanders* 113 He went deer-culling for a year. 1834 W. A. CARRUTHERS *Kentuckian* II. 108 Would you put this clamjamfry against a deer drive? 1853 J. BENWELL *Travels* 127 Dusky-looking deer-flies constantly alighted on our faces and hands. 1937 E. FRANCIS in *Public Health Reports (U.S.)* 22 Jan. 1 This name [sc. tularaemia] was given to the disease by the writer in 1920 after establishing the identity of..'deer-fly fever' in man. 1950 A. P. HERBERT *Independent Member* 276 The discomforts and dangers we should have to face ashore in Labrador—including black-fly, mosquitoes, deer-fly and dogs. 1843 'R. CARLTON' *New Purchase* I. xvii. 122 A powder horn, and its loader of deer-horn. 1880 *Encycl. Brit.* XII. 167/2 Deer-horn is almost exclusively used for handles by cutlers and walking-stick and umbrella makers. 1897 *Daily News* 1 Feb. 6/2 A saddle, probably Burgundian workmanship of 1400, composed of polished deer-horn plates. 1835 J. H. INGRAHAM *South-West* II. 137 After a farther ride of a mile..we arrived at the 'deer-stand'. 1787 W. ATTMORE *Jrnl. Tour N. Carolina* (1922) 24 One part of the Company go into the Wood..to trail for the Deer Tracks, and put the Dogs on the Scent. 1829 J. F. COOPER *Wish-ton-Wish* ix, Several times did I fall upon a maze of well-beaten deer-tracks, that as often led to nothing. 1859 J. CONWAY *Lett. from Highlands* 17 A pass, or deer-track, winding up through places apparently inaccessible. 1849 C. LANMAN *Lett. Alleghany Mts.* viii. 58, I discovered a large spot of bare earth, which I took to be a deer-yard.

deerlet. Add to def.: *spec.* the chevrotain.

1878 *Cassell's Nat. Hist.* II. 336 In the Water Deerlet of West Africa the external toes are smaller, whilst..each metacarpal..is independent of its neighbour, the Javan Deerlet differing in having the third and fourth fused into a 'cannon' bone. 1924 *Glasgow Herald* 8 Nov. 4 While camels chew the cud they resemble the old-fashioned chevrotains or deerlets in having only three chambers in their 'stomach' instead of the usual four.

deer-lick. *U.S.* (Earlier examples.)

1778 *Maryland Jrnl.* 2 June (Th.), I never saw a Deer-lick. Hunters have told me that Deer frequent those places for the mud. 1823 W. FAUX *Memorable Days Amer.* 234, I saw a deer-lick, at which I dismounted and took a lick.

deer-mouse. (Earlier and later examples.)

1840 E. EMMONS *Rep. Quadrupeds of Mass.* 62 *Arvicola Emmonsii.*...This beautiful animal..is known by the name of Deer Mouse. 1841 CATLIN *N. Amer. Indians* I. xxiv. 194 A small 'deer mouse', of which little and very destructive animals their lodges contained many. 1970 *Sci. Amer.* Feb. 58 The deer mouse (*Peromyscus maniculatus*), a small, white-footed species that is noted for its ubiquitous presence throughout North America.

deer-stalker. 1. (Earlier examples.)

1830–35 D. BOOTH *Analyt. Dict. Eng. Lang.* 257 In the woodcraft of former times, many devices for catching deer were employed by the deer stalkers. 1833 *Chambers's Jrnl.* 19 Oct. 298/1 One who really knew the mountains well, having been an incorrigible deer-stalker.

de-escalation (diˌeskălēⁱ·ʃən). orig. *U.S.* [f. DE- II. 2 + *ESCALATION.] The reversal of escalation. So **de-e·scalate** *v. trans.* and *intr.*

1964 *Christian Sci. Monitor* 28 Jan. 14/5 Just as the military must be able to escalate operations if the situation demands, they must be able to de-escalate also. 1965 H. KAHN *On Escalation* v. 91 Subsequent popular reaction, if the crisis de-escalates, may be an all-important influence. *Ibid.* xii. 230 In de-escalation, concessions and conciliation play something of the same role that demands and coercion play in escalation. 1965 *Observer* 29 Aug. 2/7 The Americans ..are making the suggestion that a process of de-escalation should be tried. 1967 *Spectator* 15 Sept. 293/3 A good case can be made for the thesis that the administration is seriously preparing to de-escalate the war. 1970 *Nature* 26 Sept. 1371/2 It is full of sound advice to administrators on how to de-escalate situations deliberately contrived to lead to confrontation. 1971 *Daily Tel.* 27 Jan. 7 It is perfectly possible to have de-escalation of the level of wage settlements and at the same time an increase in the real earnings of people.

deeshy (dī·ʃi), *a.* *Anglo-Irish.* [Orig. unknown.] Tiny; insignificant.

1825 T. C. CROKER *Fairy Leg. S. Ireland* I. 200 A deeshy daushy leather apron hanging before him. *Ibid.* 209 The three original diminutives are *tiny, dony*, and the Scottish *wee*...From the first and third they [sc. the Irish] form *weeny*, and by the use of the termination *shy*, they make *deeshy, doshy*, and *weeshy*. 1907 G. B. SHAW *John Bull's Other Island* II. 31 For shame, Patsy!..to be afraid of a little deeshy grasshopper.

deevy (dī·vi), *a.* *colloq.* Also **deevey, deevie, devey, devy.** [Affected alteration of *DIVVY *a.*] 'Divine'; delightful, sweet, charming.

1900 E. GLYN *Visits Eliz.* 11 Miss La Touche..said my hat was 'too devey for words'. 1904 E. F. BENSON *Challoners* xi, 'Martin,' she cried, 'you are too deevey!' 1905 *Punch* 8 Mar. 178/1 Do look at this sweet little monkey on the organ! Isn't he deevie! 1906 *Ibid.* 13 June 422/2, I had the most devy doll you can imagine. 1909 H. A. VACHELL *Paladin* i, The affair..was so appropriate, so obviously fashioned in heaven, so 'deevy'—a word coined in those days, and now regrettably become obsolete. 1930 V. M. SACKVILLE-WEST *Edwardians* i. 17 Tommy, you're going, aren't you? How too deevy! 1942 *English Studies* XXIV. 186 'Too deevy (divine) for words' is the flapper's reaction to what she moderately approves of.

Hence **dee·vily** *adv.*

1905 E. F. BENSON *Image in Sand* vii, How too deliciously eerie! How deevily mysterious!

de-excita·tion. *Physics.* [DE- II. 1.] The induced transition of an atom or other quantized system from an excited state to the ground state (or a lower excited state); the action of producing such a transition.

1964 *Proc. Physical Soc.* LXXXIII. 769 The theory of vibrational de-excitation in collisions should be most accurate for these simple molecular systems. 1968 M. S. LIVINGSTON *Particle Physics* xii. 217 Neutrinos are also not constituents of matter, but are radiated..in a variety of deexcitation transitions of nucleonic or particle states. 1970 *Nature* 17 Oct. 277/2 The de-excitation of a molecule and the appearance of a quantum of luminescent light is a discrete process.

So **de-exci·te** *v. trans.*, to cause the de-excitation of.

1964 *Proc. Physical Soc.* LXXXIII. 769 The theory.. predicts that light molecules will be effective in de-exciting vibrationally excited oxygen molecules.

defailance, -faillance. Delete †*Obs.* and add later examples.

1892 E. DOWSON *Let.* ?23 Nov. (1967) 254 Forgive my *défaillance*. I hope you didn't wait long; most contrite. 1902 BELLOC *Path to Rome* 194 This defaillance and breakdown which comes from time to time over the mind. 1938 HEMINGWAY *Fifth Column* (1939) III. i. 86 If you have a little defaillance I understand.

de-fat, defat (diˌfæ·t), *v.* [f. DE- II. 2 + FAT *sb.*²] *trans.* To remove fat or fats from. So **defa·tted** *ppl. a.*; **defa·tting** *vbl. sb.*

1923 DREYER in *Times* 15 June 11/6 If the bacillus is 'defatted'..it no longer retains the acid stain. *Ibid.*, Dr. A. C. Inman..began..to treat cases of human tuberculosis with the 'defatted' antigen. 1924 *Glasgow Herald* 15 Feb. 6 These 'defatted' vaccines. 1960 *Farmer & Stockbreeder* 29 Mar. 117 (Advt.), Quickly mixed powder; no waste. De-fats rubbers; prolongs their life. 1962 *New Scientist* 24 May 409/3 Methods involving simultaneous defatting and dehydration.

defatigable, *a.* Restrict †*Obs.* to sense 2 in Dict., and add later examples for sense 1.

1948 E. WAUGH *Loved One* (1949) 2 Then they lost interest. I did too. I was always the most defatigable of hacks. 1958 *Times Lit. Suppl.* 25 July 418/3 Mr Toynbee, by his own admission the most defatigable of marchers, has reverted to the written word.

default, *sb.* Add: **8.** *attrib.* Dealing with or connected with a default, as *default authority, interest, price.*

1897 *Westm. Gaz.* 29 July 5/3 If a client borrowed £100, say, and paid off £90, 'default interest' at the rate of one halfpenny per shilling per week..was at once charged on the £10 in arrear. 1908 *Daily Chron.* 13 May 7/7 While the county council is the default authority in case of the failure of the district council in sanitation, the Local Government Board is the default authority in case of the district council's failure to do what is needed in housing. 1909 *Westm. Gaz.* 9 Nov. 8/2 The Army authorities saying that if plaintiffs would not take the cattle the beasts would subsequently be issued at default prices.

defaulted (dĭfǭ·ltĕd), *ppl. a.* Restrict † *Obs.* to sense in Dict. and add: **2.** Not paid by reason of default.

1897 *Westm. Gaz.* 26 May 5/1 Nearly all the bonds issued of late by the Greek Government in respect of defaulted interest having found their way to London.

defaulter. c. (Earlier U.S. example.)

1806 WEBSTER, *Defaulter*, one who fails in payment, a debtor.

defeatism (dĭfī·tiz'm). [ad. F. *défaitisme*, f. *défaite* DEFEAT *sb.*: see -ISM.] Conduct tending to bring about acceptance of (the certainty of) defeat; a disposition to accept defeat.

1918 *Observer* 9 June 6/4 Irish Nationalists will henceforth support Pacifism, and that means defeatism. 1922 *Daily Mail* 13 Nov. 10 At an order from the leaders, the Labour Press has broken into a campaign of defeatism, lamenting that organised capital is still too strong to admit of any possible chance of the levy becoming law this time. 1926 C. L. GRAVES *Hubert Parry* II. 79 He was as far removed from foolish optimism as from 'defeatism'. 1928 GALSWORTHY *Swan Song* 124 To acknowledge the limitations of human nature was a sort of defeatism. 1942 F. G. HACKFORTH-JONES *One-One-One* vi. 65 Defeatism..must be guarded against; it was as dangerous an attitude of mind as the twin-headed monsters of Complacency and Wishful Thinking.

defeatist (dĭfī·tist). [ad. F. *défaitiste*: cf. prec.] One who advocates defeatism or accepts defeat. Also *attrib.* or as *adj.*

1918 *Observer* 9 June 8/6 The political creed of the party, ..is to support the Government in winning the war and in defeating the intrigues of Pacifists and Defeatists. 1918 *Times* 19 June 6/4 The Independent Nationalist Press..has applied the label 'defeatist' to those Nationalists who voted for a moderate policy at the Irish Convention. 1920 *19th Cent.* Mar. 556 The shop-stewards, too, in the great factories in Berlin and other towns were disloyal and 'defeatist'. 1921 N. ANGELL *Fruits of Victory* vii. 207 The repression of pacifist and defeatist propaganda during the War. 1921 *19th Cent.* Jan. 151 Throughout the Great War, as in the Japanese War, he was a defeatist. 1927 C. E. MONTAGUE *Right off Map* xxii, Lovel sat down to his separate table and map, but could not take his eyes from the abhorred defeatist. 1929 *Daily Tel.* 8 Jan. 10/3 The defeatist tactics of those who belittle his leadership.

defect, *v.* **2.** Delete 'Now *Obs.* or *rare*' and add: *spec.* to desert to a Communist country from a non-Communist country, or vice versa.

1955 *Times* 10 May 11/2 There must be many soldiers in the satellite armies who commit ideological sin in thought, but it is difficult to see how they could defect in action. *Ibid.* 27 Aug. 6/1 Dr. John..defected to east Berlin last year. 1959 *Ibid.* 28 May 15/5 A plot by a member of Parliament and a lobby correspondent to persuade a top Russian scientist to 'defect'. 1960 *Guardian* 16 Sept. 13/2 One of the two code clerks who defected to Russia.

defectivation (dife:ktivēⁱ·ʃən). *Linguistics.* [f. DEFECTIVE *a.* + -ATION.] The process or result of making, or becoming, defective.

1957 HJELMSLEV & ULDALL *Outl. Glossematics* 83 It involves syncretism, defectivation, and manifestation. 1963 *Language* XXXIX. 328 The intensive term of an opposition is also the one on which 'defectivations' and 'syncretisms' depend.

defective, *a.* and *sb.* Add: **A.** *adj.* **1. c.** *spec.* Mentally defective.

1898 *Amer. Jrnl. Sociology* Nov. 334 Numerous are the cases of idiotic women, the mothers of defective illegitimate children. 1899 *Act 62 & 63 Vict.* c. 32 § 1 A school authority ..may..make such arrangements..for ascertaining—(*a*) what children in their district, not being imbecile, and not being merely dull or backward, are defective, that is to say, what children by reason of mental or physical defect are incapable of receiving proper benefit from..instruction in the ordinary..schools. 1908 A. F. TREDGOLD *Mental Deficiency* viii. 123 A group of children existed who were so far defective that they could not be satisfactorily taught in the ordinary public schools, but who were not sufficiently defective to be certified as imbeciles or idiots. *Ibid.* 124 These defective children would suffer by association with imbeciles. 1933 L. P. CLARK *Amentia* p. xi, Though our compassion be great for the defective child, the need of understanding [etc.].

B. *sb.* **2. c.** A mental defective (see *MENTAL *a.*).

1899 *Pop. Sci. Monthly* Apr. 747 Laws preventing the marriage of defectives and of their immediate descendants would go far to stem the tide of harmful heredity. 1908 A. F. TREDGOLD *Mental Deficiency* viii. 127 A proportion of the urban defectives attending special schools are returned as cured to the ordinary schools. 1913 *Act 3 & 4 Geo. V* c. 38 § 1 The following classes of persons who are mentally defective shall be deemed to be defectives within the meaning of this Act:—(*a*) Idiots...(*b*) Imbeciles [etc.]. 1922 [see *MENTAL *a.*]. 1937 H. G. WELLS *Star Begotten* vi. 90 Institutions for defectives, lunatic asylums. 1964 M. CRITCHLEY *Developmental Dyslexia* xiii. 74 Any total series will probably include some children who are dullards, if not indeed defectives.

defeminize (dīfe·minəiz), *v.* [f. DE- II. 1 + L. *femina* woman + -IZE.] *trans.* To deprive of femininity. So **defeminiza·tion; defe·minized** *ppl. a.*

1900 *Amer. Jrnl. Psychol.* July 546 The most defeminized of these specimens, who are so prone to diminutives suggesting endearment. 1905 *Daily Chron.* 25 May 3/6 He thought this was 'monstrous and de-feminised'. 1907 *Ibid.* 9 Mar. 4/6 The so-called Feminism tends in reality to the 'defeminisation' of women. 1907 *Standard* 23 Mar., There was no need for women's suffrage, which would defeminise women.

defence, defense, *sb.* Add: **3. c.** (Earlier examples.) Also, in *Cricket*, the batting strength or batsmen collectively.

1828 G. T. KNIGHT in *Sporting Mag.* Mar. 338/2 The object is not to bring the batting down to the bowling.. not to diminish the means of defence, but to add to the powers of attack. 1830 M. R. MITFORD *Our Village* IV. 29 His hits were weak, his defence insecure. 1851 J. PYCROFT *Cricket Field* x. 203 Many a man..whose talent lies in defence, tries hitting. 1901 R. H. LYTTELTON *Out-door Games* iv. 81 Any reform of cricket law has for its object a levelling up of attack and defence—in other words, of batting and bowling.

e. *Psychol.* Behaviour which seeks to protect an organism from a real or apparent danger, esp. behaviour which seeks to resist the exposure of a neurosis. Also *attrib.*, as **defence mechanism,** a type of mental process, usually unconscious, that allows the ego or conscious mind to avoid conflict or anxiety, *e.g.*, by compensation, projection, or repression; also *transf.*

1909 A. A. BRILL tr. *Freud's Sel. Papers on Hysteria* (1912) iii. 51 The ideas of the 'defense' (abwehr) against an unbearable presentation, the origin of hysterical symptoms through conversion of psychic into physical excitement, the formation of a separate psychic group by an arbitrary act, leading to the defense—all these were..presented before my eyes. *Ibid.* 61 The assumption of a defense hysteria (abwehr hysterie) includes the requisite that at least one such moment has really occurred. 1913 *Amer. Jrnl. Insanity* 569 The psychosexual anaesthesia is nothing but a defense mechanism. 1917 D. HECHT in C. E. Long tr. *Jung's Analytical Psychology* (ed. 2) xiv. 424 By these means she remains at an infantile homosexual stage, which serves her as a defence. 1920 C. S. MYERS *Mind & Work* vi. 167 The true causes of one's emotional conduct are replaced by reasons which are invented subconsciously.. Such 'defence mechanisms', as they have been called, may come into play in any insoluble emotional situation. 1935 *Essays & Studies* XX. 131 Teachers, afraid of literature... put up a defence-mechanism of literariness. 1941 *Times Lit. Suppl.* 2 Aug. 372/2 The faith of modern Germany is, in psychological language, a 'defence' against reasoning. 1942 A. CHRISTIE *Body in Library* xiii. 115 The breaking of a piece of bad news nearly always sets up a defence reaction. It numbs the recipient. 1946 H. NICOLSON *Eng. Sense of Humour* 49 Freud was not alone in defining the sense of humour as a defence mechanism having as its main function the protection of the self against discomfort.

f. defence in depth, a system of defence comprising successive areas of resistance or mutually supporting fortifications.

1941 *Ann. Reg. 1940* 55 An elaborate system of 'defence in depth' was also formed all round the east and south coasts. 1949 KOESTLER *Promise & Fulfilment* II. ii. 205 Finally, the tiny area of Israel..gave them no opportunity for defence in depth.

5. Also, a defending force.

1916 'BOYD CABLE' *Action Front* 27 The defence, demoralised by that tornado of explosion, was pushed a good fifty yards further back.

c. The military resources of a country. Freq. *attrib.*

1935 C. R. ATTLEE in *Hansard Commons* 5th Ser. CCCII. 378 We talk about the co-ordination of defence; the co-ordination of the peace forces of the world is quite as important..as the co-ordination of the different services of this country. *Ibid.* 382, I am glad to hear that the Government are thinking of the question of having a Defence Minister. 1937 *Ann. Reg. 1936* I. 16 The Prime Minister still desired to keep the chief authority in defence matters in his own hands. *Ibid.* 17 The official title of the new Minister would be 'Minister for the Co-ordination of Defence'. 1937 F. P. CROZIER *Men I Killed* i. 20 These are the men who are planning 'Defence' at the risk of peace. *Ibid.* 21 To be accused of being anxious to avoid the wholesale slaughter of countless thousands..despite our £1,500,000,000 Defence Bill! 1966 *Listener* 6 Jan. 13/2 The diversion of money and productive resources to defence. *Ibid.*, Certain parts of industry have turned from civilian to defence production.

9. *attrib.*, as *defence area*; **defence bond,** a bond (BOND *sb.*¹ 10) issued by a government borrowing money for military defence (cf. sense 5 c above); also *defence loan bond*; **defenceman, defenseman** *N. Amer.*, in ice-hockey and lacrosse, a player in a defence position.

1940 *Economist* 6 July 10/2 The declaration of certain parts of the country to be defence areas in which movements of all persons are controlled—to be distinguished from protected areas where the restrictions apply to aliens only. 1940 *Ann. Reg. 1939* 118 The Government..issued.. defence loan bonds..bearing interest at 3½ per cent. 1941 *Ann. Reg. 1940* 46 The sale of..Defence Bonds..was going on apace. 1947 *News of World* 26 Jan. 1/2 A conversion offer is shortly to be made to holders of three per cent. Defence Bonds purchased between May 1 and Oct. 31 1940, which mature this year. 1895 *Athletic Life* Feb. 78/1 You will generally find it easiest to skate past a defence man if such tactics are necessary, by tiling straight at him and at the critical moment severing [*sic*] to his right hand side. 1940 T. STANWICK *Lacrosse* v. 41 Defense men who must play near the center stripe..should place themselves several yards behind their respective attack men. 1963 J. N. HARRIS *Weird World Wes Beattie* (1964) iii. 25 Rick Phelan was a hockey player....He was, in fact, the greatest prospective defenseman to come down the pike since the late Bill Barilko. 1966 EVANS & ANDERSON *Lacrosse Fundamentals* iv. 142 The defense man must not reach so far forward that he is off balance. 1970 *Toronto Daily Star* 24 Sept. 16/7 Brad Selwood, rookie Leaf defenceman, was sidelined after stopping a shot with his right hand in the second period.

defendant, *a.* and *sb.* **A. 2.** Delete ? *Obs.* and add later examples.

1896 *Daily News* 24 Nov. 5/7 Detailing my instructions to the defendant surgeon. 1907 *Westm. Gaz.* 6 Dec. 9/1 The defendant directors. 1924 *Pocket Oxf. Dict.* s.v., The defendant company.

defender. Add: **1. e.** *Sport.* The holder of a championship, cup, etc., who defends the title (opp. to *challenger*).

defensist (dīfe·nsist). *Russian Hist.* [f. DEFENCE, DEFENSE *sb.* + -IST.] One who, towards the end of the war of 1914–18, advocated the continuation of the war by Russia against Germany rather than the conclusion of a separate peace. Also *attrib.* or as *adj.* Hence **defe·nsism.**

1920 E. ANTONELLI *Bolshevist Russia* I. ii. 57 The sophistry of the self-styled 'liberal defensists'. *Ibid.*, Defensist in contradistinction to *defeatist*, meaning one who advocated continuing the war against Germany for the defence of Russia. 1932 M. EASTMAN tr. *Trotsky's Hist. Russian Rev.* I. xv. 303 'Our slogan..is pressure upon the Provisional Government..to induce all the warring countries to open immediate negotiations..and until then every man remains at his fighting post!' Both the idea and its formulation are those of the defensists. 1949 I. DEUTSCHER *Stalin* v. 131 The patriotic attitude or the 'defensism' of the Mensheviks. *Ibid.* 133 Stalin welcomed the semi-pacifist and semi-defensist Manifesto..The argument...implied that the defensists, Menshevik or even Liberal, acted in good faith.

defensive, *a.* and *sb.* Add: **A.** *adj.* **1. e.** *Cricket.* Of batting: characterized by cautiousness; having the protection of the wicket as the chief consideration.

1872 *Baily's Monthly Mag.* Aug. 167 Mr. Ottaway played a perfect defensive innings, while Mr. Grace was hitting. 1904 P. F. WARNER *How we recovered Ashes* i. 13 He can play a forcing or defensive game as circumstances demand. 1965 P. SHARPE *Cricket for Schoolboys* vii. 82 *Defensive stroke*,..stroke intended to protect wicket rather than score. 1967 C. J. GOODWIN *Coming in to Bat* i. 9 (*heading*) The defensive back stroke.

B. *sb.* **2. b.** *Cricket.* Defensive batting (see A. 1 e, above); blocking.

1851 J. PYCROFT *Cricket Field* iv. 58 The defensive was comparatively unknown: both the bat and the wicket, and the style of bowling too, were all adapted to a short life and a merry one. 1906 A. E. KNIGHT *Complete Cricketer* 344 To act on the defensive is to play with care, mindful of not getting out rather than of adding to the score.

defensor. Add: **4.** *Eccl.* An officer in charge of the temporal affairs of a church.

[1875 SMITH & CHEETHAM *Dict. Chr. Antiq.* I. 33/2 In Rome..the *Defensores* became by the time of Gregory the Great a regular order of officers.] 1905 F. H. DUDDEN *Greg. Gt.* I. 327 A certain defensor of the Milanese Church, named Valentinus.

deferral (dĭfə̄·răl). [f. DEFER *v.*¹ + -AL.] = DEFERMENT.

1895 W. J. LINTON *Love-Lore* 17 To-morrow will amend To-day's deferral. 1957 W. H. WHYTE *Organization Man* 130 The effect has been a large-scale deferral of dead-ends and pigeonholes for thousands of organization men. 1963 *Economist* 16 Mar. 1001/2 The new law ends this right of 'deferral' for so-called 'tax haven' income.

deferred, *ppl. a.* Add: *deferred payment*, payment that is postponed; *spec.* payment in instalments (see quot. 1951); also *fig.*; *deferred rate*, a cheaper rate charged for a telegram, cable, etc., which may be delayed in transit; so *deferred telegram*, one not for immediate delivery (see quot. 1908); *deferred stock* (U.S. example).

1796 *Deb. Congress U.S.* (1849) 792 A sum..will then become due and payable on the deferred stock. 1863 *Sydney Morning Herald* 25 Dec., Information regarding free selection, conditional purchase, deferred payment, or grazing rights. *Ibid.*, In the year previous to free selection and deferred payment, that is, during the year 1861. 1883 *Chambers's Jrnl.* 1 Sept. 545/2 The training undergone for the sake of the future recompense which forms the deferred payment. 1883 F. A. WALKER *Pol. Econ.* III. iii. 115 The Standard of Deferred Payments, usually called the Standard of Value. 1896 *Chambers's Jrnl.* 26 Sept. 610/1 It is this system of 'deferred payment', as it may be called, that enables the speculator to deal in shares although he may not possess any appreciable capital. 1898 MORRIS *Austral Eng.*, *Deferred payment*, a legal phrase. 'Land on deferred payment'; 'Deferred payment settler'; 'Pastoral deferred payment'. These expressions in New Zealand have reference to the mode of statutory alienation of Crown lands. 1907 *Westm. Gaz.* 18 June 4/2 The introduction of the deferred-payment system in connexion with the automobile business. 1908 *Ibid.* 3 Apr. 10/1 All deferred rate telegrams are posted instead of being wired. 1929 *Times* 2 Feb. 8/3 The cost of 'Urgent' telegram to be 1s., and 'Deferred' 6d., for 15 words. 1951 *Oxf. Jun. Encycl.* VII. 148/2 The other system of instalment selling is called 'deferred payments'. A written agreement is drawn up, as for a hire-purchase transaction, but the goods become the property of the buyer after payment of the first deposit.

defi: see *DEFY *sb.*

defibrillation (dīːfəibrilēⁱ·ʃən). *Med.* [f. DE- II. 1 + FIBRILLATION.] **1.** The separation of the fibres of tissues, *e.g.* cerebral tissues.

1935 H. J. WILKINSON tr. *Hultkrantz's Brain Preparations* i. 6 In the defibrillation method, spatulae and awl-shaped instruments are those most frequently used.

2. The stopping of fibrillation of the heart. Hence **defi·brillating** *ppl. a.*; **defi·brillator,** an apparatus used to control fibrillation of the heart.

1940 C. J. WIGGERS in *Amer. Heart Jrnl.* XX. 419 Defibrillation can be accomplished in the dog by the.. application of several comparatively weak shocks. 1956 *Blakiston's New Gould Med. Dict.* (ed. 2) 321/1 Defibrillator. 1958 *Electronic Engin.* XXX. 24 The complete circuit of the defibrillator is shown. 1961 *Lancet* 5 Aug. 293/2 Heart action is restored within a few minutes of releasing the aortic clamp. This results in perfusion of the coronaries with comparatively warm blood, and although ventricular fibrillation seems to be invariable at this stage it reverts easily with a defibrillating shock. 1967 *Spectator* 11 Aug. 159/3 Ventricular fibrillation may be controlled by a piece of electrical apparatus known as a defibrillator.

deficiency. Add: **1. e.** *Genetics.* = *DELETION 3.

1916 C. B. BRIDGES in *Genetics* I. 150 *Deficiency*...Tests showed..that a small section of the X in the bar region had become genetically non-existent! 1917 *Ibid.* II. 445 The general term 'deficiency' is used to designate the loss or inactivation of an entire, definite, and measurable section of genes and framework of a chromosome. 1956 *New Biol.* XX. 41 A deficiency arises when a chromosome is broken in two spots and the two end pieces join up to form a new chromosome from which the middle piece is missing... Large deficiencies, by which many genes have been removed from the cell, act as dominant lethals. Smaller deficiencies act as recessive lethals. 1969 G. W. BURNS *Sci. Genetics* xii. 220 Probably the best known disorder to be associated definitely with a chromosomal deficiency in man is the 'cri du chat' or cat-cry syndrome described by Le Jeune (1963).

2. deficiency disease, a disease caused by the lack of an essential or important substance in the diet; usu. = *AVITAMINOSIS; **deficiency payment,** a subsidy paid by the government to farmers to cover differences between market prices of agricultural produce and minimum prices guaranteed by the government.

1912 C. FUNK in *Jrnl. State Med.* XX. 341 (*title*) The etiology of deficiency diseases. 1914 *Lancet* 15 Aug. 460/1 If pellagra is a deficiency disease its relation to maize is similar to that of beri-beri to rice. 1933 *Discovery* July 234/1 The relief of deficiency diseases by the ever-increasing number of vitamins. 1968 M. PYKE *Food & Society* ii. 17 Expensive multivitamin tablets protect the well-fed people who can afford them against deficiency diseases they will never experience. 1932 *Act 22 & 23 Geo. V* c. 24 § 1 If, in any cereal year, the..average price of..wheat is less than the standard price, every registered grower shall..be entitled to receive..a payment (hereinafter referred to as a 'deficiency payment') representing..the difference.. between the said average price and the standard price. 1963 *Ann. Reg. 1962* 462 Britain's system of support for home agriculture (by means of deficiency payments). 1969 *Times* 13 Jan. 11/2 Although the primary intention at Oxford was to look at alternatives to the present deficiency payment system as a means of farm support, other problems kept intruding.

deficient, *a.* and *sb.* **B.** *sb.* Restrict † *Obs.* to senses in Dict. and add: **4.** = *DEFECTIVE *sb.* 2 c.

1906 F. THORESBY in *Westm. Rev.* Jan. 39 There are the deficients, i.e., those who from, or before birth, or by reason of their rearing, or both, never have..a fair start. 1927 CARR-SAUNDERS & JONES *Soc. Struct. Eng. & Wales* 213 [Authorities] vary notoriously...Some are active, while others close their eyes to the existence of deficients within their areas.

define, *v.* Add: **2. b.** Also *refl.*

1859 GEO. ELIOT *Lifted Veil* ii, I..saw the light floating vanities of the girl defining themselves into the systematic coquetry, the scheming selfishness, of the woman. 1888 MRS. H. WARD *R. Elsmere* xv, The slender figure suddenly defined itself against the road. *Ibid.* xxxii, The difficulties began to define themselves more sharply. *Ibid.* xliii, The half-coherent enigmatical sentences..began gradually to define themselves.

definiendum (dĭfiːnie·ndǒm). *Logic.* [a. L. *dēfiniendum* 'thing to be defined', neut. of gerundive of *dēfīnīre* DEFINE *v.*] That which is, or is to be, defined; the phrase of which a definition states or purports to state the meaning; in *Mathematical Logic*, the word or symbol (or the formula devised to contain the symbol) that is being introduced by definition into a system. (Cf. next.)

1871 [see *DEFINIENS]. *a* 1897 W. WALLACE *Lect.* (1898) 81 If definition that may be called which practically negatives the existence of the *definiendum*. 1910 WHITEHEAD & RUSSELL *Principia Math.* I. i. 11 We will give the names of *definiendum* and *definiens* respectively to what is defined and to that which it is defined as meaning. 1941 O. HELMER tr. *Tarski's Introd. Logic* vi. 150 The definiens is an arbitrary sentential function containing exactly the same variables as the definiendum, and containing no constants except primitive terms and terms previously defined. 1964 N. RESCHER *Introd. Logic* iii. 30 The expression whose meaning the definition attempts to explain..is called the *definiendum*.

definiens (dĭfi·nienz). *Logic.* [a. med.L. *dēfiniens* (not in Du Cange), pres. pple. of L. *dēfīnīre* DEFINE *v.*] The defining part of a definition; the phrase that states the meaning; in *Mathematical Logic*, the verbal or symbolic expression to which a word or symbol being introduced by definition into a system is declared to be equivalent. (Cf. prec.)

[*a* 1277 PETRUS HISPANUS *Summulae Logicales* (1947) 3.03 Unumquodque definiens in sua definitione sive definito.] 1871 T. M. LINDSAY tr. *Ueberweg's Syst. Logic* 173 The definiens has here a wider extent than the definiendum. 1910, 1941 [see *DEFINIENDUM].

definitely, *adv.* Add: **b.** *colloq.* Used as an emphatic affirmative: certainly; yes.

1931 A. A. MILNE *Two People* xvii. 329 'It would be disconcerting, wouldn't it?' 'Definitely,' said Reginald. 1933 WODEHOUSE *Heavy Weather* vii. 132 'So you wish to return to Tilbury House?' 'Definitely.' 'You shall.' 1959 *Sunday Times* 12 Apr. 32/4 Did they get more work done than at home? 'Definitely.' Would they recommend that the experiment is repeated another year? 'Oh, definitely.' 1967 L. J. BRAUN *Cat who ate Danish Modern* x. 88 'Do you design interiors around this theme?' 'Definitely!'

definition. Add: **4. c.** *definition in use*: a definition which does not provide an equivalent for the expression to be defined, but instead replaces the whole context in which that expression occurs by an equivalent not containing that expression; a contextual definition (cf. *CONTEXTUAL *a.* b).

1910 WHITEHEAD & RUSSELL *Principia Math.* I. iii. 69 Incomplete..symbols have what may be called a 'definition in use'...We define the use of ∇², but ∇² by itself remains without meaning. 1936 A. J. AYER *Lang., Truth & Logic* iii. 66 In a dictionary we look mainly for what may be called explicit definitions; in philosophy, for definitions *in use.* 1963 A. PAP *Introd. Philos. Sci.* ii. 30 Contextual definition (also called 'definition in use').

5. c. The degree of distinctness of the details in a photograph, film, television picture, etc.; so *high-, low-definition*, used to designate television systems using different numbers of scanning lines.

1889 E. J. WALL *Dict. Photogr.* 38 *Detail*, the definition of each minute part or parts of the material of a picture, whether in the negative or print therefrom. 1928 *Television* I. II. 7/3 It is claimed that much better definition and detail are obtained. *Ibid.* 10/1 A difference of phase of only one per cent. between the transmitter and the receiver is sufficient to spoil the definition of the received image. 1933 *Ibid.* VI. LXIX. 373/1 Other companies, too, may..be given similar opportunities of providing high-definition television apparatus for transmission experiments. 1935 *Discovery* Sept. 277/2 The pioneer work of Baird..with low and high definition scanning. *Ibid.* 278/1 The iconoscope camera, which is said to be capable of definition up to 500 lines. 1937 *Amateur Photography* (Newnes) 50 This explains the superb definition of the tiny cine-film pictures, in spite of their enormous enlargement when projected. 1943 *Gloss. Terms Telecomm.* (B.S.I.) 77 *High-definition television*, a system of television in which the number of scanning-lines exceeds 200 for each picture. 1946 *Nature* 20 July 88/2 In 1929..the B.B.C. decided to give Messrs. Baird Television, Ltd., facilities for experimental transmissions through the medium-wave London station. These transmissions, which were afterwards referred to as 'low-definition', employed 30 scanning lines and 12½ pictures per second. 1955 J. LIPINSKI *Miniature & Precision Cameras* ii. 34 It is not generally realized how serious a contribution camera shake makes in degrading the definition of a photograph taken with a miniature. 1969 M. J. LANGFORD *Adv. Photogr.* iii. 58 A high resolution lens may well form separate images of closely spaced lines, but unless the blacks and whites are also clearly definable by their contrast the impression of 'definition' will be poor.

definitional, *a.* Delete *rare* and add later examples.

1883 *Encycl. Brit.* XX. 49/2 Two distinct presentations are necessary to the comparison..but we cannot begin with such definitional differentiation. 1948 L. SPITZER *Linguistics* iii. 113 There is in the phrase *objet pour les yeux* something definitional and intellectual. 1963 J. LYONS *Structural Semantics* ii. 23 These are definitional statements, which will be amplified in what follows. 1970 *Economics* VIII. 126 We now have Y=C+I+X−M; or alternatively, Y+M=C+I+X which signifies a definitional equality between total supply and total demand.

Hence **defini·tionally** *adv.*, by means of a definition; as a definition.

1947 *Mind* LVI. 74, *f* might be a definitionally constructed, property determining a subclass of a class determined by a more abstract or less complex property. 1966 *Amer. Speech* XLI. 193 If all that can be said definitionally about *matter-of-factly adv* is 'in a matter-of-fact manner', then there is no reason for separate entry.

definitive, *a.* Add: **2.** Of an edition of a literary work, a textbook, etc.: authoritative; the most complete and authoritative to date.

1882 SWINBURNE *Let.* 27 Sept. in *N. & Q.* (1965) CCX. 304/2 Dr. Grosart..is about to publish what the French would call a 'definitive edition' of Daniel. 1887 *Amer. Jrnl. Philol.* VIII. 484 With the four volumes first mentioned the Goethe Society in Weimar begins the publication of the definitive edition of Goethe's works. 1928 T. S. ELIOT in E. Pound *Sel. Poems* p. vii, This book is, in my eyes, rather a convenient Introduction to Pound's work than a definitive edition. 1949 'G. ORWELL' *Nineteen Eighty-Four* I. iv. 44 Ampleforth..was engaged in producing garbled versions—definitive texts, they were called—of poems which had become ideologically offensive. 1959 *Spectator* 21 Aug. 235/1 That vague uneasiness one has come to feel in the presence of American 'definitive' biographies.

Defi·nitory, *sb.* [f. as DEFINITOR + -ORY¹.] Definitors (see DEFINITOR 1) collectively, or a council of these.

1898 FR. THADDEUS *Franciscans in England* 334 He lost no time in calling his Definitory together, and laying the matter before them. 1904 *Month* Mar. 245 At the Definitory itself he was chosen to be the first President of the renovated and perpetuated English Benedictine Congregation. 1960 J. B. DOCKERY *C. Davenport* i. 30 Provincial, Custos, and four Definitors constitute the governing body of a Province. These six are known as the Definitorium or Definitory.

definitory (dĭfi·nitŏri), *a.* [f. L. *dēfīnīre* to DEFINE, on type of adjs. such as AUDITORY: see -ORY².] Relating or belonging to definition.

a 1914 C. S. PEIRCE *Coll. Papers* (1932) II. § 271 Such propositions might be termed Definiform, or Definitory. 1928 W. A. PICKARD-CAMBRIDGE tr. *Aristotle's Topica* I. v. 102ᵃ In a word we may call 'definitory' everything that falls under the same branch of inquiry as definitions. 1935 *Mind* XLIV. 291 In ordinary discourse we are very seldom indeed conscious of the full definitory meaning of the complex terms that we use.

definitum (definəi·tǒm). *Logic.* [a. L. *dēfīnītum* 'thing defined', neut. of pa. pple. of *dēfīnīre* DEFINE *v.*] The thing or expression which a definition defines. Cf. *DEFINIENDUM.

[*a* 1277 PETRUS HISPANUS *Summulae Logicales* (1947) 5.10 Locus a definitione est habitudo definitionis ad definitum.] 1629 A. RICHARDSON *Logician's School-Master* 215 If the *definitum* bee larger than the *definitio*, then it outreacheth the limits of the thing. 1855 [see RESOLUTION 7 b]. 1944 G. E. MOORE in P. A. Schilpp *Philos. B. Russell* II. v. 197 It is only where..the sentence used to express the *definiens*..forms a part of the sentence which is the *definitum*, that one can be said to be giving a definition in spite of the fact that (1) is not fulfilled.

de·flagrating, *vbl. sb.* The action of the verb *to deflagrate*; used *attrib.* in **deflagrating spoon,** a metal spoon with a long handle, used for holding small quantities of materials that deflagrate.

1827 [see SPOON *sb.* 4 b]. 1867 BLOXAM *Chem.* 7 The experiment is most conveniently performed by heating the sulphur in a deflagrating spoon. 1936 G. H. J. ADLAM *Sci. Masters' Bk.* (2nd Ser.) 98 (*heading*) Deflagrating spoon.

deflate, *v.* Add: **1. b.** *intr.* for *pass.* Of an inflated object: to become emptied of the inflating gas; to 'go down'.

1902 *Daily Chron.* 2 Sept. 4/5 Mr. Spencer turning aside from the deflating balloon. 1925 *Glasgow Herald* 18 Apr. 9 When the bag deflated it formed a new bulkhead. 1971 *Country Life* 14 Jan. 59/3 The turkey cock deflated and rushed away, squawking for dear life.

2. *intr.* To 'climb down'; to lose spirit, confidence, etc.

1912 D. H. LAWRENCE *Let.* 5 Apr. (1962) I. 107 I'll write to Harrison. He seems inclined to deflate. On Tuesday he wrote me a cocky letter, yesterday, a sweet and friendly one. 1933 T. E. LAWRENCE *Lett.* (1938) 772 'Mr. Garnett' said the village postman importantly 'is gone to Spain.' 'Mr. Garnett is unfortunate' I replied..and the postman deflated. 1960 L. WRIGHT *Clean & Decent* 264 We may or may not deflate when a statistician tells us that of our neighbours on a London bus today, one in five *never* takes a bath.

b. *trans.* To reduce the size or importance of (a thing). Of a person's reputation, character, etc.: to depreciate, to 'debunk'.

1920 *Glasgow Herald* 17 Mar. 11, I rather wish not to inflate the currency of optimism, but I want to deflate the note of pessimism which is sometimes present. 1933 H. G. WELLS *Bulpington* iii. 113 Comfort, bathrooms, punctuality, duty, were all jumbled and deflated together under the blight of that word [*bourgeois*]. 1934 —— *Exper. in Autobiogr.* I. v. 247 Strong as is my disposition to deflate the reputation of Marx I have to admit [etc.]. *Ibid.* II. ix. 763 The belief in the possible world leadership of England had been deflated. 1940 J. BUCHAN *Memory Hold-the-Door* viii. 185 They were sansculottes who sought to deflate majestic reputations. 1940 *Manch. Guardian Weekly* 5 Jan. 2 The war's lack of intensity has deflated what military fervour there ever was. 1958 *Essays & Studies* XI. iv. 53 Lytton Strachey uses the tone of Gibbon in order to deflate the Victorians.

3. a. *trans.* To reduce the inflation of (a currency). Also *absol.*, to pursue a policy of deflation. **b.** *intr.* for *pass.* To be reduced by deflation.

1919 R. G. HAWTREY *Currency & Credit* 352 Every country will seek to keep pace with its neighbours. If one does not deflate its currency as quickly as the others the exchanges will turn against it. 1922 *Glasgow Herald* 7 Aug. 8 All of these costs..would require to deflate to pre-war standard to enable the sixpenny loaf to reappear. 1923 R. MᶜKENNA in *Daily Mail* 27 Jan. 3 If at this stage we made no further effort to deflate, trade would soon recover. 1926 *Westm. Gaz.* 30 July, To 'deflate' the franc from 240 or 200..to a level of only 25 to the £ is beyond the realms or practicability. 1963 *Ann. Reg. 1962* 15 Treasury belief that Britain could afford to deflate its way out of its balance of payments difficulties.

Hence **defla·ted, defla·ting** *ppl. adjs.*

1894 *Sat. Rev.* 8 Dec. 618/1 There are narrow edges to the rims on which a deflated tyre would rest. 1908 H. G. WELLS *War in Air* iv. 133 He left him in an extremely deflated condition, with all his little story told. 1931 —— *Work, Wealth & Happiness* (1932) xii. 616 Operations with the deflated armament firms, metallurgical industries and petroleum. 1933 *Mind* XLII. 266 The effect is to put Taine in his place (in the 'deflating' sense of these words). 1960 N. COWARD *Pomp & Circumstance* iv. 28 There's nothing more deflating than telling someone some exciting news and discovering that they already know it.

deflater: see *DEFLATOR.

deflation (dĭflēⁱ·ʃən). [See DEFLATE *v.*] **1.** The release of air from something inflated.

1891 [see DEFLATE *v.*]. 1968 PASSMORE & ROBSON *Compan. Med. Studies* I. xxix. 22/2 An imposed, maintained deflation of the lungs increases the frequency, or force, or both, of spontaneous inspiratory efforts.

2. *Physical Geogr.* [a. G. *deflation* (J. Walther 1891, in *Abhandl. d. math.-phys. Classe d. k. sächsischen Ges. d. Wissenschaften* XVI. 38).] The removal of particles of rock, sand, etc., by the wind.

1893 J. WALTHER in *Nat. Geogr. Mag.* IV. 176 We say of the wind that it 'sweeps' over the ground; for this word means nothing else than that the wind cleans the ground of all loose particles..that cover it. Translated into technical geologic language, it is called 'deflation', but that means nothing else than the every-day word 'sweep'. 1898 J. GEIKIE *Earth Sculpture* 20 The transporting action of the wind, or 'deflation' as it is termed, goes on without ceasing. 1910 LAKE & RASTALL *Text-bk. Geol.* 73 Erosion by wind divides itself naturally into two parts—removal by material or *deflation*, which of course comes under the heading of transport, and actual corrasion or wearing away of the rocks by the dynamical effect of moving sand. 1954 W. D. THORNBURY *Princ. Geomorphol.* xii. 302 Some geologists.. believe that deflation is a relatively insignificant process.. in the reduction of desert landscapes. 1970 R. J. SMALL *Study of Landforms* ix. 300 Deflation could gradually deepen the hollow until the water-table was exposed.

3. The action or process of deflating currency; an economic situation characterized by a rise in the value of money and a fall in prices, wages, and credit, usually accompanied by a rise in unemployment. Cf. *INFLATION, *DISINFLATION.

1920 R. G. HAWTREY in *Rep. Brit. Assoc. 1919* 252 To restore a depreciated unit to its normal gold value requires a measure of deflation. Deflation, which is a reversal of the process of inflation, must mean a decrease in the aggregate of money incomes. 1920 *Glasgow Herald* 11 May 10 The process of deflation likely to result from the new rights of the Federal Reserve system to discount on a graded scale. 1923 *Guernsey Star* 25 Jan., The primary ground on which a policy of gradual deflation is recommended is that it raises the exchange value of the pound sterling in relation to the dollar and hastens our return to the gold standard. 1956 *Ann. Reg. 1955* 227 Influential bankers and industrialists complained that the 'credit squeeze' had been overdone and warned against the danger of deflation.

4. *fig.* (Cf. *DEFLATE *v.* 2 a and b.)

1933 H. G. WELLS *Shape of Things to Come* II. § 12. 243 The mindless exaltation and the subsequent mindless deflation of American spiritual life. 1944 —— *'42 to '44* 157 Maybe his mental trouble is not hopeless. He may be cured by his deflation. 1958 G. J. WARNOCK *Eng. Philos. since 1900* xiii. 173 The contemporary philosopher's eye is characteristically cold and his pen, perhaps, apt to be employed as an instrument of deflation.

deflationary (dĭflēⁱ·ʃənări), *a.* [f. *DEFLATION + -ARY¹.] Of, pertaining to, or tending to deflation.

1920 *Glasgow Herald* 21 Aug. 7 The transition from an inflationary to a deflationary period in prices. 1923 J. M. KEYNES *Tract on Monetary Reform* iv. 152 Aurelian's deflationary zeal to restore the integrity of the coin excited an insurrection which caused the death of 7000 soldiers. 1929 *New Statesman* 1 June 231 The falling price level of recent years has been in large part the result of a world-wide deflationary movement which has inevitably reacted on credit and unemployment. 1957 G. E. HUTCHINSON *Treat. Limnol.* I. i. 125 There are a number of basins, the origin of which appears to involve wind action, either through its effect on the distribution of sand or through the deflationary or erosive action of wind acting on broken rock. 1959 *Ann. Reg. 1958* 478 The deflationary policies adopted by the Government in September 1957. 1962 *Listener* 5 Apr. 611/2 This makes him apply a deflationary technique to just those matters which have most strikingly affected British party life in the last century and a half. 1963 *Ibid.* 24 Jan. 179/1 Beneath the deflationary mockery of Beatrice and Benedick there is much human affection and goodwill.

deflationist (dĭflēⁱ·ʃənist). [f. *DEFLATION 3 + -IST.] One who advocates a policy of deflation. Also *attrib.* or as *adj.*

1921 *Glasgow Herald* 10 Feb. 4 If the deflationists have their way, the pound will be raised again to the value of 20s. 1922 *Edin. Rev.* July 194 The decision to get back to gold would divide the country into inflationists and deflationists. 1928 *Observer* 19 Feb. 18/2 The policy of the Suiyukai is positive and inflationist, while that of the Minseito is negative and deflationist.

deflator (dĭflēⁱ·tǫɹ). Also **-er.** [f. DEFLATE *v.* + -OR.] One who or that which deflates.

1896 *Westm. Gaz.* 2 Nov. 9/1 Some person..had strewn the road with a number of 'boot protectors', perhaps the most deadly deflator that could be constructed. 1964 *Economist* 12 Dec. 1205/2 The European common market deflaters.

deflectable (dĭfle·ktăb'l), *a.* [f. DEFLECT *v.* + -ABLE.] Capable of being deflected.

1893 in *Funk's Standard Dict.* 1925 *Contemp. Rev.* July 89 Woman..being less deflectable and in her nature more impressionable [than man].

deflectometer. Substitute for def.: An instrument for measuring the deflection or deformation of a body, esp. a metal bar, under stress. (Add earlier and later examples.)

1836 *Mech. Mag.* 6 Feb. 367/1 The effects produced [on the rails] by the passing engines and trains were minutely observed with this deflectometer. 1962 *Engineering* 23 Feb. 268/3 A series of deflectometers at the column positions was used to give an indication of the movement. A central deflectometer recorded the deflections in the new steel girder.

deflector. Add: **2.** *attrib.*

1940 *Chambers's Techn. Dict.* 230/1 *Deflector coil*, a coil so arranged that a current passing through it produces a magnetic field which deflects the beam in a cathode ray tube employing magnetic deflection. *Deflector plates*, electrodes so arranged in a cathode ray tube that the electrostatic field produced by a difference of potential between them deflects the beam. 1961 *Lancet* 30 Sept. 755/1 For protection of the area in front of the face, there was no significant difference between the performances of fabric 'filter masks' and paper or paper-insert 'deflector masks'. 1962 *Which?* Dec. 359/1 Eight of the typewriters had an erasure table which also acted as a deflector plate to stop the paper wrapping round the platen.

deflocculant (dĭflǫ·kiŭlănt). [f. as next + -ANT¹.] A deflocculating agent.

1930 *Engineering* 26 Dec. 814/1 Using alkaline deflocculants only, it would be possible to produce goods..superior to those obtained from the same mixes, hand-moulded. 1956 *Science News* XL. 66 Deflocculants are added in ceramic practice to give fluid suspensions containing high concentrations of solid matter.

deflocculate (dĭflǫ·kiŭlēⁱt), *v.* [f. DE- II. 1 + FLOCCULATE *v.*] *trans.* and *intr.* To undergo, or to cause to undergo, deflocculation. So **deflo·cculated** *ppl. a.*; **deflo·cculating** *vbl. sb.* and *ppl. adj.*

1907 *Sci. Amer.* 11 May 387 In the deflocculated condition ..graphite has a condition of fineness far beyond [etc.]. 1909 WEBSTER, Deflocculate *v.i.* 1940 *Geogr. Jrnl.* XCV. 67 The distinct turbidity of the White Nile was no doubt related to the deflocculating effect of its slightly alkaline water in contrast with that of the Blue Nile. 1956 *Sci. News* XL. 66 Dispersions of clays in water..brought about by deflocculating agents..are very sensitive to the addition of electrolytes. 1957 *Encycl. Brit.* X. 645/2 A soft, unctuous form results on treating carbon with ash or silica in special furnaces, and this gives the so-called 'deflocculated' variety [of graphite] when treated with gallotannic acid. 1963 D. W. & E. E. HUMPHRIES tr. *Termier's Erosion & Sedimentation* xi. 225 The marl is deflocculated and hydrolysis of the alumino-silicates occurs.

deflocculation (dĭflǫkiŭlēⁱ·ʃən). [f. DE- + FLOCCULATION.] The process by which floccules present in a liquid break up into fine particles, producing a dispersion; the action of deflocculating.

1904 *Nature* 7 July 238/2 The removal of the finest particles from the surface soil is attributed to deflocculation induced by the use of sodium nitrate. 1930 *Engineering* 18 July 61/1 The process of passing from sol to gel is termed coagulation, the reverse process is deflocculation. 1957 *New Biol.* XXIV. 37 The displacement of calcium by sodium from the clay particles in the flooded land caused them to contract and show the phenomena of 'deflocculation'.

defluvium (dĭflū·viŏm). *Path.* [L.] A complete shedding of some part (as the hair or the finger-nails) as a result of disease (see quots.).

1817 J. M. GOOD *Physiol. Syst. Nosol.* (1820) 501 Baldness. α Simplex. Hairs of the scalp of a natural hue: gradually dying at the bulbs, or loosened by relaxation of the integument. Defluvium capillorum. Sennert. Alopecia simplex. Sauv. 1899 G. T. JACKSON *Dis. Skin* (ed. 3) 93 The variety [of alopecia] called Defluvium Capillorum is that sudden and general fall and manifest thinning of the hair which comes on during or after some severe illness. 1906 *Practitioner* Nov. 692 All loss of hair is abnormal, and the popular belief in a physiological defluvium is false. 1910 J. N. HYDE *Pract. Treat. Dis. Skin* (ed. 8) 992 Onychomadesis (..Alopecia Ungualis; Defluvium Unguium; Onychoptosis).—Total and so-called intermittent shedding of the nails occurs, as in the partial form, in connection with systemic affections of a severe grade. 1930 W. J. O'DONOVAN *Hair* xii. 123 The defluvium of hair that coincides or follows severe general illness is a matter of common knowledge. 1940 L. MCCARTHY *Diagn. & Treatm. Dis. Hair* vi. 553 In cases of sudden loss of hair following a shock, in which the eyebrows and eyelashes are also involved, it may be impossible to say whether we are dealing with defluvium capillorum or with an alopecia totalis.

defocus (dĭfōᵘ·kǒs), *v.* [f. DE- II. 1 + FOCUS *v.*] *trans.* To put out of focus; also *intr.*, to go out of focus. So **defo·cus(s)ed** *ppl. a.*

1935 *Proc. Inst. Radio Engineers* XXIII. 1338 A defocused beam will cause..spot or pattern distortion. 1950 H. PENDER *Electr. Engineers' Handbk.* (ed. 4) II. 66 The apparent resolution is not appreciably impaired when the blue image is severely defocused. 1955 *Gloss. Terms Automatic Digital Computers* (B.S.I.) 9 The 'writing' beam is initially defocused so as to excite a small circular area of the screen. 1958 *Engineering* 21 Feb. 233/2 A more pleasing effect was obtained with the circular soft edged arrangement than with a de-focused primary beam. 1961 G. MILLERSON *Television Production* iii. 49 The director's only solution is.. to have cameras defocus when their shots are not being taken.

defoliant (dĭfōᵘ·liănt). [f. DEFOLI(ATE *v.* + -ANT¹.] A chemical used to cause defoliation.

1943 *N.Y. Herald Tribune* 7 Nov. 6 A cotton crop sprinkled with a specially prepared defoliant to make machine picking easier was harvested recently. 1956 *Nature* 3 Mar. 417/2 Research on the use of herbicides, arboricides and defoliants in East Africa. 1963 *Guardian* 1 May 8/5 The defoliants..are widely used by farmers. 1967 *Punch* 1 Mar. 302/2 The most powerful technological nation in the world has deployed its invention in massive weight; from 70,000-ton attack carriers, through flying headquarters equipped with computers, to defoliants for stripping the jungle of concealing vegetation.

defoliation. Add: **b.** The deliberate destruction of foliage (for military purposes).

1964 *Economist* 25 July 345/2 'Defoliation' of the jungle by tactical atomic bombs. 1965 *Times* 25 Mar. 10/1 Mr. Goldwater was demanding the nuclear defoliation of the Vietnamese jungles.

defoliator. (Examples.)

1887 *Amer. Naturalist* XXI. 580 Dr. Riley has published Bulletin No. 10 of the Division of Entomology, U.S. Dept. Agr., entitled 'Our Shade-Trees and their Insect Defoliators'. 1897 *Pop. Sci. Monthly* IV. 428 The leaf defoliators, as the rose chafer and flea beetle. 1936 *Forestry* X. 47 Repeated attacks by such forest insects as defoliators.

deformation. Add: **3.** Delete '*Physics*' and add further examples.

1900 *Proc. R. Soc.* LXV. 90 'Flow' or non-elastic deformation in metals. 1916 C. A. EDWARDS *Physico-Chem. Properties of Steel* xi. 125 One of the most useful properties possessed by metals is the facility with which they undergo plastic deformation when pressed, hammered, or rolled. 1953 AITCHISON & PUMPHREY *Engin. Steels* ix. 364 The deformation or warping is also partly attributable to the thermal stresses which are set up in the metal. 1967 H. J. STERN *Rubber* (ed. 2) xi. 490 Rapidly alternating stresses or deformations..give rise to the development of heat in the rubber. 1968 COULSON & RICHARDSON *Chem. Engin.* (ed. 2) II. iv. 152 Deformation of the drop is opposed by the surface tension forces so that very small drops retain their spherical shape.

b. *Geol.* The process by which a stratum, mass of rock, etc., undergoes change of form or structure by being compressed, faulted, folded, etc.; also, the result of this process. Also *attrib.*

1882 A. GEIKIE *Text-bk. Geol.* 312 Evidences of actual deformation within the mass of rock. 1904 CHAMBERLIN & SALISBURY *Geol.* (1905) I. 547 It is theoretically possible that deformation of the sub-crust may result from the internal transfer of heat without regard to external loss. 1937 *Q. Jrnl. Geol. Soc.* XCIII. 602 Such [sedimentation] fabrics in rocks of this type are never so well marked as deformation fabrics. 1955 *Sci. Amer.* July 40/1 There is little doubt that these four huge fracture zones resulted from some massive deformation of the earth's crust. 1963 E. S. HILLS *Elem. Struct. Geol.* iv. 77 It is therefore admissible to draw analogies between the mechanics of deformation of such rocks with the deformation of metals, both as to the effects in individual crystals and for the crystal aggregate as a whole.

deformational (dĭfǫɹmēⁱ·ʃənăl), *a.* [f. DEFORMATION + -AL.] Of or pertaining to deformation.

1903 *Nature* 12 Feb. 359/1 Several deformational movements had affected this district. 1965 G. J. WILLIAMS *Econ. Geol. N.Z.* vi. 66/2 Retrogressive metamorphism accompanying a deformational phase. 1970 *Physics Bull.* Apr 174/2 Physical properties such as deformational behaviour, strength characteristics,..and thermal properties.

deformeter (dĭfǭ·ɹmitəɹ). [f. DEFORM *v.*¹ + -METER.] An instrument for measuring deformation. Also *attrib.*

1927 G. E. BEGGS in *Jrnl. Franklin Inst.* CCIII. 380 The deformeter apparatus..consists of very precise means for producing known, but very small, displacements at sections where stress is sought, and optical means for measuring the deflections of the model at..load points. *Ibid.*, An indeterminate frame connected to the deformeter gauges and in position for deformation readings. 1946 *Nature* 20 July 91/2 Research work into structural analysis by the deformeter.

deframe (dĭfrēⁱ·m), *v.* [f. DE- II. 1 + FRAME *v.* 9.] *trans.* To remove (a picture) from its frame. So **defra·med** *ppl. a.*

1908 T. HARDY *Dynasts* III. II. iii. 78 A deframed old master. 1950 E. HEMINGWAY *Across the River* xv. 141 The portrait had been de-framed.

defreeze (dĭfrī·z), *v.* [f. DE- II. 1 + FREEZE *v.*] *trans.* = *DEFROST *v.* So **defree·zing** *vbl. sb.*

1901 *Westm. Gaz.* 22 July 10/2 A new process of 'defreezing' is now being tried. 1922 *Evening News* 20 Dec. 4/3 When these birds, only lightly frozen for the voyage, are 'defrozen'—the trade term. 1963 *Guardian* 4 Feb. 16/1 (*heading*) Electric charge to defreeze pipes. 1970 *Ibid.* 8 May 9/4 We recently defroze a rock hard salmon which left the sea..on August 16, 1969. It was superb.

defree·ze, *sb.* [f. the vb.] The act or process of unfreezing.

1962 *Daily Tel.* 3 Jan. 1/6 Describing yesterday as 'defreeze day', the Association gave warning of..a covering of water on roads already ice-bound. 1963 *Guardian* 30 Jan. 16/5 The 'de-freeze' of the main roads of Britain was almost complete.

defrost (dĭfrǫ·st), *v.* [f. DE- II. 2 + FROST *sb.*] *trans.* To unfreeze, remove the frost from; *spec.* (*a*) to unfreeze (frozen meat or other provisions); (*b*) to clear the frost from (e.g. the interior of a refrigerator, the windscreen of a motor vehicle or aircraft). Also *absol.* and *refl.*; occas. *intr.*, to become unfrozen. So **defro·sted** *ppl. a.*; **defro·sting** *vbl. sb.*

1895 *Daily News* 29 May 8/4 It was believed that Queensland defrosted beef could be brought into formidable competition with American chilled. 1895 *Australasian Pastoralists' Rev.* 15 Aug. p. viii, The difficulties hitherto attaching to the defrosting of Beef and Mutton. 1897 *Yearbk. U.S. Dept. Agric. 1896* 26 Each year there is visible improvement in the methods of defrosting meats in European markets. 1924 *Glasgow Herald* 19 Aug. 7 The electrical defrosting process experiments which were recently undertaken in Melbourne. 1937 'N. BLAKE' *There's Trouble Brewing* ix. 166 Nigel was standing near the door, his back to one of the refrigerators. 'Doesn't seem so cold as when I was here last.' 'No, sir,' said Carruthers. 'We're defrosting now.' 1951 *Good Housek. Home Encycl.* 232/2 An automatic de-frosting mechanism is incorporated in some machines. 1953 C. M. KORNBLUTH *Syndic* (1964) xix. 201 He defrosted some hamburger, fried it and ate it. 1957 *Daily Mail* 5 Sept. 11/5 The vegetables..now defrost in the car on the way from the home to the picnic-ground. 1957 *Economist* 9 Nov. 499/2 The heater provides a generous flow of warm air...The result is extremely efficient demisting and defrosting. 1958 *Listener* 17 July 81/1 The refrigerator that defrosts itself. 1960 *Housewife* Apr. 114/1 Simply defrost a packet of frozen chicken joints. 1965 J. CHRISTOPHER *Wrinkle in Skin* iv. 46 They had a mess of defrosted strawberries afterwards.

1970 *Daily Tel.* 28 Apr. 17 Both refrigerator and deep-freezer de-frost themselves continuously and automatically. 1970 *Which?* Sept 283/2 Most manufacturers recommend defrosting once a year.

defroster (dīfrǫ·stəɹ). [f. prec. + -ER¹.] A device for defrosting; *spec.* (see quot. 1930).
1927 *Refrigerating Engin.* Jan. 238 Defroster for packing houses and cold storage plants. 1930 *Popular Sci.* Dec. 74/3 This electrical 'defroster' warms the glass and prevents ice and snow from getting a hold on it. The device may be attached to the windshield of any car. 1937 *Sunday Express* 14 Mar. 13/4 The defrosters, unusual in England, created an impression. 1960 *News Chron.* 22 Sept. 13/4 The larger models [of refrigerator] feature an automatic or semi-automatic de-froster.

defterdar (deftəɹdā·r). Also 6–7 **teftadar, teftader,** 7–8 **tefterdar,** 9 **defturdar.** [Hind. *daftardār*, f. *daftar* DUFTER + *-dār* holder.] A Turkish officer of finance, *esp.* the accountant general of a province; also formerly, the Turkish minister of finance.
1589 HAKLUYT *Voy.* 205 Vnder him be three Subtreasurers called *Teftaders.* 1601 W. BIDDULPH *Trav. Four Englishmen* (1609) 88 *Defterdare,* that is Treasurer of Aleppo. 1615 G. SANDYS *Trav.* III. 211 Tendring to the *Teftadar* or Treasurer the reuenew of that *Sanziackry.* 1796 J. MORSE *Amer. Univ. Geogr.* II. 463 The first minister of finances is called Defterdar. 1836 E. W. LANE *Mod. Egypt.* I. 154 The Defturdár, having caused the Názir to be brought before him, asked him [etc.]. 1902 *Encycl. Brit.* XXXIII. 508/1 A complete budget of receipts and expenditure is drawn up by its *defterdar.*

defunctive, *a.* Delete †*Obs. rare*⁻¹ and add later examples. Also, becoming defunct; dying.
1920 T. S. ELIOT *Ara Vos Prec* 14 Defunctive music under sea Passed seaward with the passing bell. 1929 W. FAULKNER *Sound & Fury* 165 The lane..became defunctive in grass, a mere path scarred quietly into new grass. 1939 T. WOLFE *Web & Rock* I. 18 The day was drowsed with quietness and defunctive turnip greens. 1961 *Listener* 26 Oct. 681/1 The theatre has a whole barrel full of conventions for the purpose—slow curtains, quick curtains.. defunctive music preparing a slow ending.

defuse (dīfiū·z), *v.* [f. DE- II. 2 + FUSE *sb.*] *trans.* To remove the fuse from (an explosive device). Also *fig.*
1943 *Word Study* Oct. 7 A group of fliers defused a 1,000-pound bomb that had jammed in the racks when their plane was flying..over an Italian target. 1958 *Economist* 2 Aug. 352 Thought has to be given now, without delay, to the means of reducing the risks involved in this inevitable act of disengagement—of defusing it, in effect. 1960 *Times* 20 Jan. 6/1 Jack decides to save Barry by defusing the discovered bomb. 1967 *Listener* 1 June 708/2 Mr George Brown..says the Russians are anxious to 'defuse' the middle east. 1968 *Economist* 3 Feb. 11/2 The early release of the crew/alone would defuse this crisis, as things now stand. 1970 *Guardian* 9 May 1/3 President Nixon['s].. news conference..which he evidently hopes will defuse much of the students' anger.

defusion (dīfiū·ʒən). *Psychiatry.* [tr. G. *entmischung* (Freud *Das Ich und das Es* (1923) iv. 50), f. DE- II. 1 + FUSION.] A reversal of the normal fusion of the instincts; *spec.* a regression from the normal fusion of the life and death instincts.
1927 J. RIVIÈRE tr. *Freud's Ego & Id* iv. 57 Once we have admitted the conception of a fusion of the two classes of instincts with each other, the possibility of a 'defusion' of them forces itself upon us. The sadistic component of the sexual instinct would be a classical example of instinctual fusion..and the perversion in which sadism has made itself independent would be typical of defusion. 1945 J. C. FLÜGEL *Man, Morals & Society* viii. 101 The process of defusion, whereby the death instinct is forced to carry on its dread work without the saving grace of love. 1949 *Mind* LV. 83 Such processes as..the 'fusion' and subsequent 'defusion' of the various attitudes of the self. 1955 J. RIVIÈRE in M. KLEIN *New Directions in Psycho-Analysis* xiv. 360 They are both situations in which a measure of defusion of instincts has taken place. 1959 G. KENNEDY in S. HOOK *Psychoanalysis* 278 A partial defusion of these two normally fused sets of original instincts.

defy, *sb.* Delete †*Obs.* and add later examples. Also **defi.** [Perhaps a new formation on DEFY *v.*] Now chiefly *U.S.*
1888 B. P. BLOOD *Lion of Nile* in *Scribner's Mag.* Dec. 707 And all night long, roaring my fierce defy. 1897 *Harper's Mag.* Jan. 231 He sent out the last defy to the enemy in 1800. 1911 *Boxing* 9 Sept. 454/3 Marcel Denis launches forth with a defi to Young Brooks. 1945 *Sat. Rev. Lit.* 29 Dec. 10 Leonard..still hurls his defy into the teeth of his enemy.

degas (dīgæ·s), *v.* [f. DE- II. 2 + GAS *sb.*] *trans.* To remove unwanted gas from. Also **dega:sifica·tion; dega·sify** *v.*; **dega·sifying** *ppl. a.*; **degasser; dega·ssing** *vbl. sb.* and *ppl. a.*
1901 D. VON WAGNER *Man. Chem. Technol.* 12 If wood is heated above 150°, it is de-gasified. 1920 *Lit. Digest* 8 May 117 The first 'degassing chamber' was constructed in a cellar..in the shell-torn village of Cagnicourt, France. 1925 *Chem. Abstr.* 1938 (*heading*) Degasification and reactivity of carbonized fuels. 1928 *Brewer & Maltster & Beverageur* 15 Jan. 50 It is very difficult, once a water is de-gassed, to prevent it absorbing oxygen anew. 1934 WEBSTER, Degasser. 1936 *Nature* 26 Dec. 1091/2 The electrical vehicle used for transporting the degasser is operated by an 80-volt battery. According to recent experiments, an area contained by a circumference of about 300 meters and covering sewers.. can be degassed in 10–15 minutes. 1946 *Nature* 27 July 133/2 We believe that..(3) the type of source, as well as the degassing procedure followed in conditioning the fluoride before coating, affects the hardness [of magnesium fluoride films]. 1949 *Electronic Engin.* XXI. 423 Liquids and melted substances can be de-gassed by ultrasonic waves. 1953 *Jrnl. Iron & Steel Inst.* CLXXIII. 440/1 The author describes a method of desulphurizing and degassing steel in the ladle. 1958 *New Scientist* 7 Aug. 575/3 More information on the Dutch degasifying device..would..be valuable. 1965 *Times* 12 Mar. 24/1 (*caption*) This degasification station [at an oil-drilling rig] makes a strangely unreal picture.

de Gaullist, *a.* and *sb.* = *GAULLIST *a.* and *sb.* So **de Gaullism** = *GAULLISM.
1944 H. G. WELLS *'42 to '44* 63 It [*sc.* a novel] is, among other insanities, strongly de Gaullist. *Ibid.* 152 To build up a de Gaullist mentality in the infantile French mind. *Ibid.* 153 The de Gaullists in North Africa. 1946 W. S. CHURCHILL *Secret Session Speeches* 82 Any admixture of De Gaullist troops at the outset would destroy all hope of a peaceful landing. 1955 KOESTLER *Trail of Dinosaur* 199 English Labourites and Conservatives, French Socialists and de Gaullists, spoke from the same platform. 1963 *Daily Tel.* 4 Feb. 20/6 The political implications of 'de Gaullism' as they may affect Italy.

degauss (dīgau·s), *v.* [f. DE- II. 2 + the name of K. F. *Gauss,* German scientist (1777–1855). Cf. GAUSS.] *trans.* To protect (a ship) against magnetic mines by encircling it with an electrically charged cable (called the *degaussing belt* or *girdle*), so as to demagnetize it. Also in extended use, to remove unwanted magnetism from, to demagnetize. So **degau·ssed** *ppl. a.*, **degau·ssing** *vbl. sb.*
1940 *Times* 9 Mar. 6/5 Since the object of the device was to render ships non-magnetic, the verb to 'de-gauss'..was coined to indicate its function. *Ibid.*, The 'de-gauss-ing girdle'..consists of ordinary insulated electric cable, energized in a special way by an electric current. *Ibid.*, One of the officers responsible for its development expressed himself perfectly ready himself to take any properly de-gaussed ship over any number of magnetic mines. 1940 *Daily Tel.* 9 Mar. 1/4 A new word (de-gaussing) has been added to the wartime vocabulary of the Royal Navy as the result of the arrival of the German magnetic mine. It implies de-magnetising, or de-polarising a ship. 1940 *Manch. Guardian Weekly* 15 Mar. 201 A device which makes a ship incapable of affecting the magnetic needle in the mine. The apparatus is known as a degaussing girdle. It takes the form of an electric cable following the line of the hull round the ship's upper deck or bulwark. 1946 W. S. CHURCHILL *Secret Session Speeches* 25 The protective measures of the Admiralty—convoy, diversion, degaussing. 1954 C. P. SNOW *New Men* 13 He was working at Rosyth in one of the first degaussing parties. 1960 O. SKILBECK *ABC of Film & TV* 39 *Degaussing,* the process of removing residual magnetism from metallic objects. It is necessary frequently to degauss cutting room equipment. 1960 J. L. BERNSTEIN *Video Tape Recording* xii. 259 It is wise to degauss the heads prior to each recording session. 1968 W. H. BUCHSBAUM *Color TV Servicing* (ed. 2) v. 50 Degaussing a picture tube, just like demagnetizing the tape recorder head, is accomplished by generating a strong AC field around the degaussed object and then gradually reducing this field. 1969 *Daily Tel.* 21 July 10/8 Any colour receiver is sensitive to magnetic influences..and an elaborate degaussing process had to be applied to the first receivers that I used.

degeneracy. Add: **2.** *Physics.* **a.** A property of a quantized or an oscillatory system (see *DEGENERATE *a.* 3 a).
1928 SHEARER & DEANS tr. *Schrödinger's Coll. Papers Wave Mech.* 70 Multiplicity of the proper values corresponds to degeneracy in the theory of conditioned periodic systems and is therefore especially interesting for quantum theory. 1935 J. DOUGALL tr. *Born's Atomic Physics* v. 108 When the relativistic variability of mass is taken into consideration, the degeneracy of the hydrogen atom is certainly removed in part, but the motion is still simply degenerate. 1971 *Nature* 23 Apr. 495/3 Interaction between the free oscillations [of the Earth] and the Earth's rotation removes degeneracy in the spherical harmonics of the same degree.
b. A property of a system of particles or 'gas' (see *DEGENERATE *a.* 3 b).
1928 *Proc. Physical Soc.* XL. 327 An example of degeneracy is provided by the free electrons in a metal even at normal temperature. 1958 J. B. SYKES tr. *Ambartsumyan's Theoret. Astrophysics* xxxii. 544 Owing to the high ionisation inside the stars [*sc.* white dwarfs], the gas retains the properties of an ideal gas..up to densities of about 10^5 or 10^6 g/cm³; at higher densities a degeneracy sets in, at first of the electron gas, and later (for densities greater than 10^{4} or 10^{8} g/cm³) of the heavy particles.

degenerate, *a.* Add: **B. 3.** *Physics.* **a.** Of a quantized system: having two or more linearly independent eigenfunctions with the same eigenvalue; *spec.* having two or more states with the same energy; also applied to the eigenfunctions or the states. Also more widely, applied to any oscillatory system having two or more modes of oscillation with the same frequency, and to the modes themselves.
1923 H. L. BROSE tr. *Sommerfeld's Atomic Struct. & Spectral Lines* 564 We follow Schwarzschild and call the exceptional case considered degenerate. A degenerate case thus occurs..when the quantum conditions are not uniquely determined. 1929 CONDON & MORSE *Quantum Mech.* iv. 136 It is better to speak of a particular level as degenerate or non-degenerate, for there are mechanical systems in which some states are degenerate and others are not. 1940 F. SEITZ *Mod. Theory Solids* xii. 411 The six functions do not have the proper symmetry to have the same energy in a cubic crystal. Thus, the degenerate levels would split if interatomic interactions were taken into account. 1961 POWELL & CRASEMANN *Quantum Mech.* vi. 173 The eigenvalue *α* may be degenerate...Suppose, for example, that *α* is doubly degenerate, having linearly independent eigenfunctions ψ_1 and ψ_2 which may..be assumed to be orthonormal. 1968 R. C. STANLEY *Light & Sound* xvi. 315 If two or more normal modes [of vibration] formed by different reflection paths have the same resonant frequency..they are termed degenerate modes...In an irregularly shaped room fewer degenerate or near degenerate modes form.
b. Of a system of particles or 'gas' (such as the electrons in a metal or the interior of a white dwarf star): having properties which depart markedly from those of an ordinary gas as described by classical statistical mechanics, being described either by Fermi-Dirac or by Bose-Einstein statistics.
1928 *Proc. Physical Soc.* XL. 330 Thomas supposes the electrons in an atom to be degenerate in the sense of Fermi and Dirac. 1939 S. CHANDRASEKHAR *Introd. Study Stellar Struct.* x. 358 A completely degenerate electron gas is one in which all the lowest quantum states are occupied. 1951 J. DOUGALL tr. *Born's Atomic Physics* (ed. 5) viii. 265 Pauli and Sommerfeld (1927)..pointed out that the laws of classical statistics ought not to be applied to the electron gas within a metal, since it is bound to behave as a degenerate gas. 1954 D. TER HAAR *Elem. Statist. Mech.* iv. 95 In the interior of some stars such high densities will occur that notwithstanding the very high stellar temperatures the gas is degenerate. 1966 *New Statesman* 15 Apr. 534/1 These x-ray sources..might be made of 'degenerate matter'—matter so compacted that the atoms have collapsed down to the size of their nuclei. Such a star would be fantastically dense.

degenerate (dīdʒe·nĕrĕt), *sb.* [subst. use of the adj.] One who has lost, or has become deficient in, the qualities considered proper to the race or kind; a degenerate specimen; a person of debased physical or mental constitution.
1555 J. PROCTOR *Historie of Wyates Rebellion* f. 80 It is to be wished..that prouoked with so greate clemencie these degenerates reforme themselues. 1895 tr. *Nordau's Degeneration* I. iii. 18 In the mental development of degenerates, we meet with the same irregularity that we have observed in their physical growth...That which nearly all degenerates lack is the sense of morality and of right and wrong. 1901 H. ELLIS *Criminal* (ed. 3) iii. 51 Näcke..found the skulls of women..abnormal, and among degenerates generally..the stigmata of degeneracy are more common in women. 1919 M. K. BRADBY *Psycho-analysis* 17 The fact..is compatible with his being a genius or a degenerate, a scoundrel or a valuable citizen. 1952 W. J. H. SPROTT *Soc. Psychol.* viii. 142 The deplorable Jukes family, their dismal record of defectives and degenerates.

degenerate, *v.* **5.** Delete † *Obs.* and add later examples.
1893 J. PULSFORD *Loyalty to Christ* II. 131 The one seeking to regenerate, and the other to degenerate yet more and more the soul's nature. 1921 E. MACNEILL *Celtic Irel.* 17 Acquired habits..can degenerate and recreate a nation.

deglaciation (dīglēⁱʃiēⁱ·ʃən). *Geol.* [f. DE- II. 1 + GLACIATION 2.] The disappearance of ice from a previously glaciated region.
1895 J. D. DANA *Man. Geol.* (ed. 4) IV. iv. 969 Partial deglaciation, and the formation of a sheet of drift perhaps 20' in thickness, with occasional layers of soil embedded in the drift. 1947 R. F. FLINT *Glacial Geol. & Pleistocene Epoch* v. 64 By deglaciation is meant the uncovering of any area as a result of glacier shrinkage. 1961 *New Scientist* 25 May 444/3 It would be interesting to speculate upon the results of sudden deglaciation of the [Antarctic] continent.

deglamorize, deglamourize (dīglæ·mərəiz), *v.* Also **-ise.** [f. DE- II. 1 + *GLAMORIZE *v.*] To deprive of glamour. So **degla·morization; degla·morized** *ppl. a.*
1938 *Covina* (Calif.) *Independent* 22 Mar., The problem.. seems to be how the 'deglamorizing' of the 'whataman' idea among the dupes of the dictators is to be accomplished. 1941 *Sunday Jrnl. & Star* (Lincoln, Neb.) 2 Nov., Paulette Goddard recently became the first Hollywood glamor girl to de-glamorize herself in the interest of national defense. 1942 *Time* 27 Apr. 24/1 Deglamorization of the week was performed by an alert news photographer..who caught the handsome face of..Norma Shearer registering desolation. 1942 *Gen* 15 July 31/2 From the minute they join the depot, they find themselves undergoing a de-glamorisation course. 1962 *Times Lit. Suppl.* 26 Oct. 821/4 The deglamorized toughies of the 1960s. 1964 *Punch* 28 Oct. 657/3 It seems rather late to deglamorise war.

degote: see *DAGGETT.

degradation¹. Add: **1.** (Later example.)
1898 *Westm. Gaz.* 15 Jan. 5/1 Madame Dreyfus..sends to the Press a letter from her husband..written the day after the degradation.
4. c. In wider use in *Biochem.* and *Chem.*: a simplification of the structure of a molecule, brought about either naturally or artificially, in which it loses some constituent

atoms or is broken up into a number of simpler molecules. (Further examples.)
1900 *Jrnl. Chem. Soc.* LXXVIII. I. 650 (*heading*) Synthesis and degradation in the coumarone series. 1937 *Jrnl. Biol. Chem.* CXVIII. 781 The general conception has been that the enzymatic degradation of proteins proceeds in two distinct stages; i.e., the proteinases cleave the true protein into polypeptides and..the polypeptidases and dipeptidase complete the degradation to amino acids. 1958 L. SMALL in H. GILMAN *Org. Chem.* III. xii. 1027 The cyanogen bromide degradation..often succeeds with compounds that resist the Hofmann degradation..Other vigorous degradations are often employed to determine the fundamental structures present. 1949 *Nature* 15 Jan. 94/2 The identification of this degradation product affords support for the view that one point of attachment of the disaccharide grouping in glutoside is via a carbon atom directly linked to the indole nitrogen. 1954 *Biochem. Jrnl.* LVIII. 392/1 The terminal mononucleotide..was obtained by stepwise degradation of the polynucleotides listed in Table 1. 1958 [see *DEGRADE *v.* 5 d]. 1966 MORRISON & BOYD *Org. Chem.* (ed. 2) iv. 239 We may carry out a degradation: break the molecule apart, identify the fragments, and deduce what the structure must have been. 1968 *Times* 24 Oct. 7/8 Organic materials that accumulate in deep waters in the sea are relatively resistant to degradation by micro-organisms. 1970 [see *DEGRADE *v.* 5 d].
c. *Soil Science.* (See quot. 1958.)
1927 C. F. MARBUT tr. *Glinka's Great Soil Groups* 95 Holes made by burrowing animals were found filled with a brownish mass on which one could see the evidences of the later degradation which took place under the influence of the acid forest humus. 1958 *Yearbk. Agric. 1957* (U.S. Dept. Agric.) 756/1 *Degradation* (of soils), the change of one kind of soil to a more highly leached kind, such as the change of a chernozem to a podzol. 1965 B. T. BUNTING *Geogr. Soils* viii. 157 Invasions of podsols on the chernozem soils in climatic phases have caused degradation to gray wooded soils.
5. (Further examples.)
1958 *Engineering* 21 Mar. 358/1 The picture degradation normally experienced between successive generations of facsimile pictures is largely eliminated with the..machine. 1958 *Times Rev. Industry* June 26/1 Chemical reagents employed may easily lead to the degradation of the silicon. 1959 W. S. SHARPS *Dict. Cinemat.* 89/1 *Degradation,* the degree to which the quality of the film image at any stage is inferior to the original scene, or to the image at some previous stage. 1966 *Electronics* 17 Oct. 105 Indication to the aircrew of the amount of degradation of capabilities to allow assessment of mission capability.
6. (Later examples.)
1898 [see *AGGRADE *v.*]. 1909 W. M. DAVIS *Geogr. Ess.* xviii. 408 'Degradation'..is more properly associated with those leisurely processes..in which a graded slope is reduced to fainter and fainter declivity, although maintaining its graded condition all the while. 1922 C. A. COTTON *Geomorphol. N. Zealand* vi. 61 When..a stream cuts downward to establish or maintain grade, it is said to degrade; and the process is termed degradation.

degradative (dīgrē·dătiv), *a.* [f. *degradāt-*, ppl. stem of late L. *dēgradāre* + -IVE.] Causing degradation.
1940 *Jrnl. Research Nat. Bureau of Standards* Oct. 452 They involve the use of treatments which are known to have a degradative effect on the protein. 1950 *New Biol.* VIII. 101 Enzymes are capable of accelerating synthetic as well as degradative reactions. 1958 *New Scientist* 12 June 174/3 The degradative diseases, the pathetic conditions in which it is plain..that some part of the human frame has been worn out by too much use.

degrade, *v.* Add: **5. d.** *Chem.* and *Biochem.* To make (a molecule) simpler in structure; to split into a number of simpler molecules.
1935 TYSON & STILLER in Harrow & Sherwin *Textb. Biochem.* ii. 85 Zemplén has used a fundamentally different method to decide the point of union of the two hexose units in a reducing disaccharide. By degrading the reducing component till it no longer forms an osazone, he demonstrated..lactose..and other sugars. 1958 BALLANTYNE & WALKER *Dict. Named Effects & Laws* 192 *Wiesmann degradation.* An acyclic or cyclic α-hydroxy..is degraded to an aldehyde containing one less carbon atom by the action of a cold aqueous solution of sodium hypochlorite. 1963 *Union Carbide Stockholder News* Sept. 1/1 The detergents..can be quickly destroyed or degraded in sewage treatment systems to non-detergent-like products. 1970 *Nature* 26 Dec. 1311/1 Most body proteins are continually degraded and resynthesized. The protein content of various organs must represent the net balance between synthesis and degradation over some period of time.
8. Also occas. at Oxford University (now disused). Now at Cambridge, to take a specified examination when one is above the standing prescribed for it.
1906 *Oxf. Univ. Gaz.* 29 May 654/2 John O. Aglionby..was granted permission to degrade till Trinity Term, 1907, and to offer Modern History.
b. See quot. 1883; at Oxford University, to supplicate for a lower degree than that for which one originally entered.
1883 *Encycl. Dict.*, *Degrade,*..to take a lower degree than one is entitled to,..to descend to a lower class. 1921 *Oxford Univ. Registry Acc. Bks.* 4 Nov. (MS.), Wing, J. L...Degrading to B.Litt. £1.

degrade (dīgrēⁱ·d), *sb.* [f. the vb.] A piece of timber containing defects; also, the production of defects resulting in a lowering of the 'grade' or quality of the timber.
1922 R. C. BRYANT *Lumber* 242 This practice..has the disadvantage of producing a higher per cent of 'degrades', because the lumber as it leaves the kiln is bone-dry. 1953 *Oxf. Univ. Gaz.* 7 Aug. 1202/1 A thesis entitled 'A study of degrade in oak logs due to "ambrosia" beetles'. 1970 *Timber Trades Jrnl.* 21 Mar. 61/1 Strips ⅛ in. wide by 1 in. thick have been dried without excessive degrade at starting temperatures up to 66 °C..coupled with an initial humidity of 75%.

degraded, *ppl. a.* Add: **2. c.** *Soil Science.* Of soil; (cf. *DEGRADATION¹ 4 c).
1927 C. F. MARBUT tr. *Glinka's Great Soil Groups* 95 The well known degraded Tschernosem found in many parts of European and Asiatic Russia. 1965 B. T. BUNTING *Geogr. Soil* xv. 173 The westernmost area of modal chernozem is south of Wroclaw, but degraded chernozem occur as far west as Braunschweig.

degrain (dīgrēⁱ·n). [f. DE- II. 2 + GRAIN *sb.*¹] Used *attrib.* designating leather from which the grain has been removed. Hence **degrai·ned** *a.*
1925 *Civil Service Supply Assoc. Ltd. Catal.* 133 Men's Glove Department..Degrain Mocha Gloves. 1958 *Daily Express* 1 Nov. 8/1 Degrained Glove...Made from supple Degrain skins.

degras, dégras (degra·, de·grās). [F. *dégras,* f. *dégraisser* to remove grease from, with assimilation to *gras* fat.] **a.** The dark wax or grease obtained when fish-oils are rubbed into hides and recovered by expression and by washing the hides with alkali, as in the manufacture of chamois leather; the commercial product usu. contains added quantities of fish-oils and solid fats (cf. *MOELLON²), and is used in currying and fat-liquoring leather; also, a product made from fish-oils in imitation of this.
1882 *Encycl. Brit.* XIV. 390/2 This uncombined oil is washed out with a warm potash solution, and the fat so recovered, known as *degras,* forms a valuable material for the dressing of common leather by curriers. 1904 J. LEWKOWITSCH *Oils, Fats, & Waxes* (ed. 3) II. xvi. 1129 A large number of artificial dégras are now being prepared by blowing fish, liver, and blubber oils with air..thus imitating the natural process of oxidation the oils appear to undergo when skins are converted into chamois leather. 1908 H. R. PROCTER *Prin. Leather Manuf.* (ed. 2) xxv. 448 The residual oil..constitutes *moellon*...This is never sold for currying in its original purity; but, mixed with further quantities of fish oils, tallows, and sometimes wool-fat, it constitutes the ordinary degras of commerce. 1931 WILSON & MERRILL *Anal. Leather* xi. 382 The excess oil is pressed out after saturating the skins with water and constitutes the purest and best grade of moellon degras. 1937 BURTON & ROBERTSHAW in Atkin & Thompson *Procter's Leather Chemists' Pocket-Bk.* (ed. 3) xix. 326 Genuine degras (moellon) and sod oil were originally by-products of chamois dressing..; but are now frequently prepared from scrap skins or by direct oxidation of oils. 1958 A. KUNTZEL in R. O'Flaherty et al. *Chem. & Technol. Leather* II. xxviii. 431 It is a very valuable leather-greasing agent because of its high water compatibility and is known as degras or moellon.
b. *U.S.* The crude mixture of wax and fatty acids obtained by scouring wool or treating it with organic solvents, used industrially (*e.g.* as a source of lanolin and in the manufacture of lubricating greases) and as a substitute for degras (sense a) in leather manufacture; wool-grease, wool-fat.
1894 *Jrnl. Amer. Chem. Soc.* XVI. 555 In this country, the term 'degras' is generally applied to the grease or fatty matter recovered from the water in which wool has been scoured. 1925 F. S. HYDE *Solvents, Oils, Gums, Waxes* 124 Wool grease, the English 'Yorkshire grease', 'Suint', or American 'degras'—but not the true degras—is a dirty grease with a foul odor, obtained by scouring wool or sheep, and contains, besides fatty acids from soap employed in scouring, esters of palmitic and myristic acids, other waxy bodies, as well as mineral oil used in lubricating the wool. 1954 C. J. BONER *Lubricating Greases* iv. 149 Wool wax is ordinarily sold under the name of degras or wool grease.

degrease (dīgrī·s), *v.* [f. DE- II. 2 + GREASE *sb.*] *trans.* To remove grease or fat from. Hence **degrea·ser; degrea·sing** *vbl. sb.* and *ppl. a.*
1889 *Cent. Dict.*, Degrease. 1900 S. P. SADTLER *Handbk. Industr. Org. Chem.* (ed. 3) ix. 310 The only treatment of this kind, known technically as a degreasing process, is that with petroleum-naphtha. 1904 GOODCHILD & TWENEY *Technol. & Sci. Dict.* 153/1 *Degreasing,* the process of extracting fat or grease from wool and bones by means of benzene or other solvent. 1921 *Dict. Occup. Terms* (1927) § 158 s.v. *Extractor, Benzine degreaser, bone degreaser (benzine process), degreaser,* extracts grease (or tallow) from kitchen stuff, or butchers' scraps, by dissolving fat in petrol or benzine. *Ibid.* § 334 *Degreaser,* in fellmongery puts pelts into tanks..containing benzine..to remove grease. 1930 FIELD & WELLS *Electro-Plating* iv. 80 This principle is adopted by..'degreasers'. 1939 *Jrnl. Iron & Steel Inst.* CXXXIX. 168 The author compares the results of degreasing tests on a number of steel specimens covered with oil and fat..using petrol and three other unnamed degreasing agents. 1958 *New Scientist* 3 July 305/1 The zinc layer is removed by three minutes' 'pickling' (the steel being degreased at the same time). 1963 A. J. HALL *Textile Sci.* iv. 167 Wool fabric can also be degreased in a continuous manner by use of the I.C.I. trichloroethylene process. 1969 *Jane's Freight Containers 1968–69* 531/1 The complete container is degreased inside and out.

degree, *sb.* Add: **1. c.** *degree-cut* in gem-cutting: = TRAP-CUT.
1909 in WEBSTER.
6. c. *third degree*: see *THIRD DEGREE.
7. In legal use.
1880 *Encycl. Brit.* XII. 871/2 Each inn confers this status or degree [*sc.* of barrister] on its own members only. 1885 *Law Jrnl.* 13 June 364/1 That his Royal Highness..be called to the degree of the Utter Bar. 1959 JOWITT *Dict. Eng. Law* 601/2 *Degree,*..the state of a person, as to be a barrister-at-law, or a Master of Arts of a University.
10. c. *degree day,* a unit used to determine the heating requirements of buildings (see quots.). Also *attrib.*
1930 *Engineering* 11 July 34/1 The term 'degree-day' is used in the United States to denote the difference in external temperature between the daily mean and 65° F; we have no similar recognized term here. 1958 *Ibid.* 14 Mar. 322/3 The conception of Degree Days as a means of exercising control over heating installations and of forecasting fuel requirements is by no means a new one in this country. The London and Home Counties Coke Association started the first Degree Day Service in 1938. 1964 J. S. SCOTT *Dict. Build.* 94 *Degree-day value,* a figure which describes the relative coldness of a site. It is the sum of the number of days yearly by which the average temperature falls below 60° F. in Britain...In USA..the temperature chosen for degree-day charts is 65° F.
14. *degree of freedom*: see FREEDOM 10 in Dict. and Suppl.
15. *Comb.* (in sense 7 a), as *degree-day* (see also *10 c), *-fee, -level; degree-conferring, -granting* adjs.; *degree-giving* vbl. sb. and adj.
1903 *Westm. Gaz.* 11 July 2/2 The degree-conferring Universities of the United Kingdom. 1906 *Daily Chron.* 27 July 6/7 To make the Nottingham University College a degree-conferring University. 1882 J. ROWLEY *Diary* 9 May (1967) 12 A Degree day: 2 Mod. B.A., M.A. & 17 B.A. 1867 G. M. HOPKINS *Let. to Bridges* 5 Dec. (1935) 22 Would it be possible for you to see me on my way home? Certainly it would if the degree day does not fall too late. 1900 G. SWIFT *Somerley* 104 When the dance and degree-day were over. 1897 *Daily News* 23 Mar. 7/4 Many of the Degree-Fees have..been raised. 1909 W. JAMES *Mem. & Stud.* (1911) v. 102 'Civilization', with its herding and branding, licensing and degree-giving. 1946 *Nature* 19 Oct. 531/2 The difficulties..have led Lord Cherwell to propose the transfer of engineering training from the universities to degree-giving institutes of technology. 1895 *Westm. Gaz.* 27 July 5/1 It..had never been adequately recognized by degree-granting Universities. 1896 *Daily News* 20 Feb. 5/4 The larger degree-granting institutions. 1967 *Times* 28 Feb. (Canada Suppl.) 36 Some 60 degree-granting universities and colleges in Canada. 1956 *Nature* 18 Feb. 296/2 A National council to control the award of..degree-level qualification.

degreed, *a.* Add: **1.** (Later examples.) Also *absol.*
1905 *Daily Chron.* 11 July 3/2 The unfairness to the ratepayer,..to the class-teacher, to the non-collegiate, to the academical degreed. 1962 *Guardian* 27 Jan. 6/6 Degreed women are..very much the exception. 1967 *Ibid.* 28 June 6/5 Abstracts of papers by much-degreed ladies from the University of Saskatchewan.

degression (dīgre·ʃən). Restrict †*Obs.* to sense 1 and add: **2.** The decrease in the rate of taxation in a degressive scale.
1896 PALGRAVE *Dict. Pol. Econ.* II. 244/1 Graduated taxation therefore technically includes progression, degression, and regression. 1906 *Westm. Gaz.* 9 June 2/1 The income-tax so long planned in France..is English both in its taxation by schedules and in its exemptions of the less wealthy payers, though it effects the last-named result by a more complete scheme of 'degression' than is in operation here. *Ibid.* 31 July 5/3 If any change was made in the direction of degression by which everybody would first be taxed at the higher rate and then have to apply for abatements.
3. *Bibliogr.* See next, sense 2.

degressive (dīgre·siv), *a.* [f. L. *degress-,* ppl. stem of *dēgredī* to descend.] **1.** In taxation, of or pertaining to schemes in which the rate decreases successively on sums below a certain limit. Hence **degre·ssively** (Webster 1909).
1911 S. J. CHAPMAN *Outl. Pol. Econ.* xxxi. 379 Degressive taxation means that large incomes are taxed at a higher rate than smaller incomes, but not in a degree which involves as great a proportionate sacrifice for the former as for the latter. 1962 *Economist* 1 Dec. 939/2 This tax..might be made 'degressive' (i.e. declining in rate each year).
2. *Bibliogr.* (See quot. 1908.)
1908 F. MADAN in *Trans. Bibliogr. Soc.* IX. 53 (*title*) Degressive bibliography. *Ibid.*, The principle of degression..that is to say, the principle of varying a description [of a book] according to the difference of the period treated or of the importance of the work to be described. It may therefore be worth while to suggest four forms for the description of a book, showing by degressive changes what details may fairly be omitted in short descriptions. 1953 *Library* VIII. 20 Generalizations about degressive bibliography, a phrase increasing in popular esteem at the moment. 1963 *Times Lit. Suppl.* 10 May 348/3 It is flying in the face of the degressive principle to repeat..title-page wording in multi-volumed books.

‖ **dégringolade** (degrængolad). [Fr., f. *dégringoler* to descend rapidly.] A rapid descent; deterioration, decadence; change from bad to worse. Also as vb.
1883 *Sat. Rev.* 24 Nov. 648/1 The *dégringolade* of Tokka and the catastrophe of Obeid. 1895 G. B. SHAW in *Sat. Rev.* 27 July 109/1 Miss Lottie Collins..will soon find her

popularity degringolading from the summit on which the Tarara craze exalted it. 1906 W. DE MORGAN *Joseph Vance* xxxv. 346 This last is a short chapter, but is a record of a steady *dégringolade.* 1909 *Westm. Gaz.* 29 May 5/3 We.. share her distress in the ensuing dégringolade. 1926 *Spectator* 29 May 900/1 This *dégringolade,* this falling back into an undrained, unfenced, unploughed, unweeded prairie. 1926 *Nation* 24 July 460/1 The franc has this week looked the penny in the face, and has acquired a momentum in decline which suggests that a *dégringolade* is close at hand, unless prompt and effective measures are really taken. 1959 *Encounter* Sept. 69/1 The hero..underwent a convincing but totally unsensational *dégringolade,* taking, not to drugs or drink, but to an increasing sluggishness.

degum (dīgʊ·m), *v.* [DE- II. 2 + GUM *sb.*²] To deprive of gum; *spec.* (in the preparation of silk) to deglutinate. So **degu·mmed** *ppl. a.*; **degu·mming** *vbl. sb.* Also *fig.*
a 1884 KNIGHT *Dict. Mech.* Suppl. 249/2 *Degumming machine,* a machine used in treating silk before dyeing. 1887 [see DE- II. 2]. 1938 *Thorpe's Dict. Appl. Chem.* (ed. 4) II. 17/2 Generally silk is degummed as hanks of yarn. Degumming as cloth has..advantages. 1956 *Nature* 3 Mar. 429/2, 4·86 gm. of the radioactive cocoon fibres was cut into small pieces and degummed. 1959 N. MARSH *Singing in Shrouds* vii. 148 He treated his restive audience to a comprehensive de-gumming of *Hamlet* and *Macbeth.* 1960 A. E. BENDER *Dict. of Nutrition* 38/2 *Degumming agents,* used in refining of fats to remove mucilaginous matter consisting of gum, resin, proteins and phosphatides. 1963 A. J. HALL *Textile Sci.* ii. 42 There is about 20 to 30% of silk-gum associated with the silk fibres so that a weight loss of this order results from de-gumming—it makes the degummed silk so much the more expensive.

degut (dīgʊ·t), *v.* [DE- II. 2 + GUT *sb.*] *trans.* To remove the guts, contents, or essential elements of (in the senses of *gut* vb.).
1933 DYLAN THOMAS *Let.* Sept. (1966) 24 Open him [*sc.* Wordsworth] at any page: and there lies the English language...Degutted and desouled. 1948 WYNDHAM LEWIS *Let.* 25 Oct. (1963) 466 The publisher *who dulls these letters down* and deguts them will not only be doing a great disservice to Ezra but to the public of today and also tomorrow. 1960 *Sunday Times* 6 Mar. 6/8 We are certainly not going to de-gut the centre of London..to provide motorways and parking space. 1962 *Daily Tel.* 17 Dec. 8/8 The latest edition of 'That Was The Week That Was' [B.B.C. Television] was degutted as efficiently as any fish at Yarmouth! *Ibid.*, It was pleasant to watch the farmers degutting the unhappy Mr. Levin.

dehair (dīhē·ɹ), *v.* [f. DE- II. 2 + HAIR *sb.*] *trans.* To remove the hair from (a skin), to unhair. Hence **dehai·rer; dehai·ring** *vbl. sb.*
1902 *Mod. Amer. Tanning* I. 37 When [sulphide of sodium is] used in conjunction with lime a very satisfactory process is obtained which results in the skins being dehaired ..in much less time. 1921 *Dict. Occup. Terms* (1927) § 338 *Unhairer* (tannery); *dehairer*; (i) a beam man who spreads hides or skins on wooden beam; scrapes hair, by hand, with two-handled knife; (ii) passes hides or skins from lime pits, between unhairing machine, removing hairs. 1959 *Chambers's Encycl.* XIII. 456/2 The hide or skin..after de-hairing is immersed in liquors made from chromium salts.

de haut en bas: see DE II in Dict. and Suppl.

deho·rn, *v.* [DE- II. 2.]
1. *trans.* To deprive (an animal) of horns. Also *absol.* and *fig.* Hence **deho·rner,** one who, or an instrument which, dehorns animals.
1888 *Missouri Republ.* 15 Feb. (Farmer), Dehorning is performed when the calf is young, and the tips of the horns movable. 1888 Dehorn, Dehorner [see DE- II. 2]. 1889 FARMER *Americanisms* 197/1 The preferred age at which to dehorn is in the second year of the animal's life. 1895 *Montgomery Ward Catal.* 394/3 Cattle Dehorners...Newton's lately improved dehorning knife, revolving and sliding shear, each one making a draw cut. 1897 *Sears, Roebuck Catal.* (facing p. 33), The Adsit dehorning shears..makes a drawing cut completely encircling the horn, which it removes almost instantly. 1904 WILCOX & SMITH *Farmer's Cycl. Agric.* 339/1 The practice of dehorning cattle is generally recommended by the experiment station. 1907 *Sears, Roebuck Catal.,* This style Dehorner in the large size will clip any size horn from cattle of any age. 1914 *Boston Even. Transcript* 6 June 2/1 Four years ago they dehorned the speaker. 1920 F. B. HADLEY *Princ. Vet. Sci.* xiii. 260 When cattle are allowed to run together they should be dehorned. 1935 J. STEINBECK *To a God Unknown* 91 To borrow a dehorner for that long-horned bull. 1959 *Times* 27 Oct. 6/4 The high proportion of dehorned cows in the main cattle section. 1960 *Farmer & Stockbreeder* 26 Jan. 112/3 (Advt.), 6-volt calf dehorner. 1971 *Guardian* 20 Feb. 18/4 A cattle de-horning device.
2. *Forestry.* (See quots.)
1905 *Terms Forestry & Logging* 35 *Dehorn,* to saw off the ends of logs bearing the owner's mark and put on a new mark (Kentucky). 1957 *Brit. Commonw. Forest Terminol.* II. 59 *Dehorn,* to re-mark logs for change of ownership.
3. *Horticulture.* (See quot. 1954.)
1934 in WEBSTER. 1954 A. G. L. HELLYER *Encycl. Garden Work & Terms* 195/1 If..fairly large branches have to be removed..this should be done either right back to the main trunk, or to a suitably placed side branch. This latter process is known as dehorning.

de Horsey (dəhǭ·ɹsi). *Naut.* The name of Admiral Sir A. F. R. *de Horsey* (1827–1922), used to designate a rig for a cutter or launch consisting of a triangular or jib-headed fore-sail and a gaff mainsail. Hence **de Horsey-rigged** *a.*
1905 *Man. Seamanship* iii. 47 *Launches and Pinnaces.* The de Horsey rig, *i.e.,* one mast, a foresail set on the forestay, and a mainsail bent to a gaff. 1923 'BARTIMEUS' *Seaways* xii. 211 She's de Horsey-rigged and is propelled by oars in the absence of a suitable wind. 1953 J. MASEFIELD *Conway* (rev. ed.) IV. 249 The De Horsey-rigged cutters and sloop-rigged dinghies.

de:humi·dify, *v.* [f. DE- II + HUMIDIFY *v.*] *trans.* To reduce the degree of humidity of; to remove moisture from; to dry. So **de:humidifica·tion,** the removal of moisture; **de:humi·dified** *ppl. a.*; **de:humi·difier,** a device or substance for dehumidification; **de:humi·difying** *vbl. sb.* and *ppl. a.*
1921 *Sci. Amer.* 5 Mar. 188/2 Dehumidifier with dew-point control, showing the main parts of the controlling mechanism. 1932 *Engineering* 14 Oct. 458/1 The supply air has to be..dehumidified or dried. *Ibid.*, In this method of drying or dehumidification..the air..is passed through sprays of chilled water. *Ibid.*, The cooled dried air leaving the dehumidifier is warmed. 1933 *BBC Yearbk.* 102 This drying (or 'de-humidifying') of the air supply. 1933 *Archit. Rev.* LXXX. p. lxvi, The air conditioner comprises a slow speed fan and motor, heating coils, cooling and dehumidifying coils, air filter and humidifying spray for winter use. 1940 *Chambers's Techn. Dict.* 230/2 Dehumidified air. 1941 *Chem. Abstr.* 1674 Apparatus for humidifying, dehumidifying and cleaning air. 1947 *Jrnl. R. Aeronaut. Soc.* LI. 93/1 For tropical use, dehumidification is achieved by utilising the refrigerating system to cool the air below the dew point. 1958 *Engineering* 28 Feb. 274/2 Nevertheless, filtered and de-humidified air is supplied which is also heated or cooled as required. *Ibid.* 14 Mar. 352/2 Dehumidification..was very often omitted and 'conditioning' came to mean 'cooling'. 1969 *J. C. Penney Co. Catal.* Fall & Winter 870 (Advt.), 14-pint dehumidifier. Designed for the smaller home—prevents sweating pipes.

dehydracetic (dīhəidrăsī·tik), *a. Chem.* Also **dehy:droace·tic.** [tr. G. *dehydracetsäure* dehydracetic acid (A. Geuther 1866, in *Jenaische Zeitschr. f. Med. u. Naturwissen.* II. 410), f. *DEHYDR(O- + ACETIC *a.*] *dehydracetic acid*: a colourless, crystalline, cyclic compound, $C_8H_8O_4$, usu. obtained by heating ethyl acetoacetate.
1872 WATTS *Dict. Chem.* Suppl. 543 *Dehydracetic acid*... An acid said to be produced by the action of hydrochloric or carbonic acid on the sodium salt of ethyl-diacetic acid. 1911 *Jrnl. Amer. Chem. Soc.* XXXIII. 1119 The construction of a satisfactory structural formula for dehydroacetic acid has occupied the attention of chemists ever since its discovery. 1937 *Thorpe's Dict. Appl. Chem.* (ed. 4) I. 63/2 Ethyl acetoacetate loses alcohol on refluxing and forms dehydracetic acid. 1969 R. J. W. BYRDE in D. C. Torgeson *Fungicides* II. xii. 549 Dehydroacetic acid has proved of value as a postharvest treatment for fruit.

dehydrase (dīhəi·drēⁱz, -s). *Biochem.* [a. G. *dehydrase* (H. Wieland 1913, in *Ber. d. Deutsch. Chem. Ges.* XLVI. 3333), f. DE- II. 3 + HYDR- + *-ASE.] **a.** = *DEHYDROGENASE.
1914 *Chem. Abstr.* VIII. 3051 It is shown by means of a typical dehydrase, Schardinger's milk enzyme, that oxidase, reductase and mutase are 1 and the same enzyme. 1939 *Thorpe's Dict. Appl. Chem.* (ed. 4) III. 553/2 Citric acid dehydrase is present in the liver and in vegetable material acting on citric acid. 1959 N. CAMPBELL in E. H. Rodd *Chem. Carbon Compounds* IVB. viii. 942 Freudenberg also postulates a second process whereby catechins in the presence of dehydrases undergo condensation by dehydrogenation. 1961 [see *DEHYDRATASE].
b. = *DEHYDRATASE.
1953 *Adv. Enzymol.* XIV. 243 The usual English term 'dehydrase' for an enzyme dehydrating a substrate was changed to dehydratase, because *Dehydrase* in German.. means a dehydrogenating enzyme rather than an enzyme splitting off water. 1957 *Jrnl. Gen. Microbiol.* XVI. 480 The enzymic dehydration of tartaric acid to oxaloacetic acid, first established..for the *d*-isomer, occurs also with the *meso*- and *l*-isomers, and the attack on all three tartaric acids by bacteria of the genus *Pseudomonas* appears to occur principally by means of stereospecific dehydrases. 1961 [see *DEHYDRATASE].

dehydratase (dīhəi·drēⁱtēⁱz, -s). *Biochem.* [f. DEHYDRAT(E *v.* + *-ASE.] Any enzyme that catalyses the removal of a molecule of water from a substrate.
1953 [see *DEHYDRASE b]. 1961 *Rep. Comm. Enzymes Internat. Union Biochem.* vi. 34 The name 'dehydrase', which has been used for both dehydrogenating and dehydrating enzymes, will not be used. 'Dehydrogenase' will be used for the former and 'dehydratase' for the latter. 1966 *Proc. Nat. Acad. Sci.* LV. 869 Deamination of homoserine to α-ketobutyrate by homoserine dehydratase is the primary reaction in the catabolism of homoserine.

dehydrate, *v.* Add: **1. b.** *spec.* To remove the water from (foods), so as to preserve them and reduce their bulk.
1921 C. V. EKROTH in A. Rogers *Industr. Chem.* (ed. 3) II. 1158 One of the most important features of the food conservation movement since the outbreak of the war has been the practice of dehydrating fruits and vegetables. 1943 *Daily Tel.* 23 Oct. 4 As early as 1938 British food scientists began to study methods of dehydrating vegetables, meat and eggs and of storing them in that form.
c. *fig.* To render 'dry', lifeless, uninteresting, etc.
1957 *London Mag.* Nov. 73 They dehydrate Joyce, destroying both the devil and the child in him.
dehydrated *ppl. a.*, **dehydration** (examples in various senses of the verb).
1921 C. V. EKROTH in A. Rogers *Industr. Chem.* (ed. 3) II. 1156 Dehydrated or dried food has had practically all of its moisture removed. *Ibid.* 1170 The dehydration of cow's milk, partial or complete, produces, in the former case, the condensed and evaporated milk so familiar to us. 1943 *Harper's* July 166 *Life* shares with its elder brother, *Time,* ..a language which..revels in coining nifty, dehydrated words. 1944 *Evening Standard* 5 Dec. 6/4 It seems strange that words like flak..should escape censure by the austere.. while such bitter eloquence is directed against dehydrated, deprescribed, hospitalisation [etc.]. 1954 KOESTLER *Invis. Writing* ii. 26 Language, and with it thought, underwent a process of dehydration. 1955 *N.Y. Times* 6 Mar. I. 49/4 The Army is putting greater emphasis on dehydrated foods to meet the needs of swiftly moving, widely dispersed fighting units. 1958 *Times* 6 Feb. 11/2 A series of rather dehydrated arguments between a number of intellectual types.

dehydro- (dīhəi·dro), before a vowel also **dehydr-.** *Chem.* A prefix used in forming the names of some organic compounds, denoting (*a*) the loss of one atom of hydrogen (or freq. two), (*b*) the loss of a molecule of water (= ANHYDRO-); as **dehydrotriacetonamine; dehy:droasco·rbic acid,** an oxidation product, $C_6H_6O_6$, of ascorbic acid; **dehy:drocho·lesterol** [ad. G. *dehydro-cholesteryl* (A. Windaus et al. 1935, in *Ann. Chem.* DXX. 106)], the provitamin, $C_{27}H_{44}O$, of vitamin D_3, naturally formed in human and animal skin under the influence of sunlight.
1877 H. WATTS *Fownes' Man. Chem.* (ed. 12) II. 264 Triacetonamine, $C_9H_{17}NO$...Dehydrotriacetonamine, $C_9H_{15}N$. 1933 *Nature* 3 June 800/2 We found that by the action of a copper acetate solution upon ascorbic acid, a dehydro-ascorbic acid is formed. 1935 *Chem. Abstr.* XXIX. 7995 (*title*) 7-Dehydrocholesterol. *Ibid.*, 7-Hydrocholesterol.. with BzCl in C_5H_5N gives a dibenzoate...Heating..gives 58% of the benzoate..of 7-dehydrocholesterol. 1946 *Nature* 3 Aug. 169/1 The preparation from cholesterol (I) of 7-dehydrocholesterol (III), which on irradiation gave a highly antirachitic product (vitamin D_3). 1959 *New Biol.* XXIX. 35 Dehydroascorbic acid, the oxidized form of vitamin C, is reported as having been obtained..from sea-water. 1962 H. HEATH in A. Pirie *Lens Metabolism Rel. Cataract* 365 Present evidence would seem to indicate that it [*sc.* ascorbic acid] is actively secreted [by the ciliary body] in the nonionized, oxidized form, dehydroascorbic acid. 1965 *Nomencl. Org. Chem.* (*I.U.P.A.C.*) C. 42 Loss of two hydrogen atoms from a compound designated by a trivial name is denoted by a prefix 'didehydro-'...In common usage 'dehydro-' is often used in place of 'didehydro-'. For example the above compound [*sc.* 7, 8-didehydrocholesterol] is often termed dehydrocholesterol. 1968 R. S. CAHN *Introd. Chem. Nomencl.* (ed. 3) iii. 44 Removal of atoms is indicated in a few cases, e.g., dehydro (loss of 2H), anhydro (loss of H_2O).

dehydro·genase. *Biochem.* [f. DE- + HYDROGEN + *-ASE.] Any enzyme that transfers hydrogen from a substrate to an acceptor or performs an equivalent oxidation.
1923 *Chem. Abstr.* XVII. 1274 By the agency of dehydrogenases occurring in muscle, H. is abstracted. 1931 *Times Lit. Suppl.* 2 July 530/3 The temperature coefficients and energy exchanges of the citric acid dehydrogenase of cucumber seeds. 1946 *Nature* 17 Aug. 238/2 Different hydrogenase systems..have been studied by the methylene blue technique. 1958 FRUTON & SIMMONDS *Gen. Biochem.* (ed. 2) xii. 319 An important property of alcohol dehydrogenase (from yeast and from liver), and of other dehydrogenases, is that they catalyze not only electron transfer but also direct hydrogen transfer. 1961 [see *DEHYDRATASE].

dehydrogenated (dīhəidrǫ·dʒĕnēⁱtĕd), *ppl. a.* *Chem.* [f. DEHYDROGENATE *v.* + -ED¹.] Of a compound: deprived of some or all of its hydrogen.
1909 *Chambers's Jrnl.* Oct. 686/2 A direct conversion of the gas into alcohols and dehydrogenated alcohols.

dehydrogenation (dīhəidrǫ:dʒĕnēⁱ·ʃən). *Chem.* [f. DEHYDROGENATE *v.*: see -ATION.] The removal or loss of one or more atoms of hydrogen from a compound; = DEHYDROGENIZATION.
1866 ODLING *Anim. Chem.* 129 The oxidation or dehydrogenation of uric acid. 1950 *Sci. News* XV. 69 Such dehydrogenation reactions frequently occur at catalytic surfaces. 1956 *Nature* 17 Mar. 512/2 The structure of methyl lineolate dimer investigated by a dehydrogenation-oxidation procedure. 1964 N. G. CLARK *Mod. Org. Chem.* vi. 92 At high temperatures and in the presence of suitable catalysts, paraffins lose two atoms of hydrogen from adjacent carbons, with the production of a double bond and a molecule of gaseous hydrogen; such a process is known as 'dehydrogenation'.

deiamba (di,æ·mbă). Also **diamba.** [Native name.] Congo tobacco; = HEMP *sb.* 4.
1851 R. O. CLARKE in *Hooker's Jrnl. of Botany* III. 9 (*title*) Short notice of the African Plant Diamba, commonly called Congo Tobacco. 1861 R. BENTLEY *Man. Bot.* 637 Indian hemp is also used for smoking. The plant is also known under the name of *Diamba* in Western Africa. 1864 WATTS *Dict. Chem.* II. 309 Deiamba.

de-ice (di,əi·s), *v.* [f. DE- II. 2 + ICE *sb.*] *trans.* To remove or prevent the formation of ice on (parts of an aeroplane, machine, ship, etc.); to clear of ice. More usually as **de-icing** *ppl. a.* and *vbl. sb.*, removing or preventing ice-formation.
1935 *Jrnl. R. Aeronaut. Soc.* XXXIX. 794 An automatic de-icing device. 1937 *Ibid.* XLI. 731 The high pressure is mainly used for engine starting..and low pressure for de-icing. 1942 *Hutchinson's Pict. Hist. War* 18 Mar.–9 June 11 'De-icing' a Corvette...All hands set to work to strip away the icy covering. 1958 *Engineering* 28 Mar. 404/2 De-icing and de-misting of aircraft control cabin windows. 1961 W. VAUGHAN-THOMAS *Anzio* iv. 53 Parties were standing by to de-ice the mountain passes.

de-icer (di,əi·səɹ). [f. prec. + -ER¹.] A mechanical or other device to remove or prevent ice-formation, esp. on aircraft and on motor-car windscreens.
1932 *Product Engineering* Sept. 357/2 The installation is similar to that of the 'de-icer' without inflation tubes or compressed air equipment. 1934 *Shell Aviation News* No. 37. 20/2 Instal rubber de-icers on the propeller and pneumatic de-icers on the wings. 1941 [see *ANTICER]. 1958 *Times Rev. Industry* Aug. 30/1 Electroforms of..nickel are.. fitted on..aircraft propeller blades to guard the delicate de-icer elements against rupture.

deictic, *a.* Add: Also in *Grammar* and as sb.
1922 O. JESPERSEN *Language* xix. 383 The relation between a demonstrative pronoun or a deictic particle and genitival function. 1964 [see *ANAPHORA 1 b]. 1970 *Archivum Linguisticum* I. 6 The only exceptions to this restriction are the gen. dat. sg. fem. and the gen. pl. of the deictic *theser.*

deificatory (dīifikēⁱ·təri), *a.* Delete † *Obs.* and add later example.
1902 A. M. FAIRBAIRN *Philos. Chr. Relig.* II. II. ii. 474 He is not conceived as the subject of a deificatory process.

dei:oniza·tion. [f. next + -ATION.] The process of deionizing a substance or a space, or of becoming deionized; the removal or loss of ions.
1919 E. W. STONE *Elem. Radiotelegr.* (1920) viii. 157 To enhance both the extinction and ignition voltages..it is necessary to provide adequate means for the deionization of the arc. 1959 *Engineering* 20 Feb. 247/2 Silica is not removed by many demineralisation or deionisation treatments. 1962 M. G. DENAVARRE *Chem. & Manuf. Cosmetics* (ed. 2) II. 351 Deionization has become the most widely used method of purifying water in the cosmetic industry. 1967 H. COTTON *Adv. Electr. Technol.* xv. 641 There is a short interval of time required for the deionization of the discharge path.

deionize (di,əi·ŏnəiz), *v.* [f. DE- II. 1 + ION + -IZE.] *trans.* To deprive of ions or of ionic character; *esp.* (*a*) to cause the positive and negative ions in (a gas or a space) to recombine to form neutral atoms or molecules; (*b*) to take out of (a liquid, esp. water) the ionic constituents. Hence **dei·onized** *ppl. a.*
1906 G. MANN *Chem. Proteids* vi. 216 These compounds are therefore formed by the amphoteric amino-acids, simply adding CO_2, which thereby becomes de-ionised. 1919 E. W. STONE *Elem. Radiotelegr.* (1920) viii. 151 Let us assume that..the space between the electrodes is very quickly deionized. 1930 *Trans. Amer. Inst. Electr. Engin.* XLIX. 58/2 The layer of gas immediately adjacent to the surface becomes very quickly denuded of ions, and then ions which further reach the surface must diffuse through this layer of deionized gas. 1941 J. D. COBINE *Gaseous Conductors* iv. 102 The presence of the confining walls aids diffusion processes to deionize the entire volume of an ionized gas, for as the ions reach the walls they lose their charge. 1950 KUNIN & MYERS *Ion Exchange Resins* viii. 88 There are three methods for deionizing electrolyte solutions with ion exchange resins. 1956 *Nature* 28 Jan. 187/2 The resin was washed with de-ionized water. 1970 *Rep. Visitors Ashmolean Mus.* (Oxford) 1969 20 Mrs. K. Kimber..cleaned this by applying 'packs' of Sepolite..and deionized water to the unpolished surface of the marble.

deionizer (di,əi·ŏnəizəɹ). [f. prec. + -ER¹.] An apparatus for removing the ions from water or other liquids.
1956 *Nature* 17 Mar. 509/2 The portable deionizer..is supplied with exchangeable cartridges of the mixed resins. 1958 I. W. CORNWALL *Soils for Archaeologist* ix. 165 The best [water] is provided by percolation through a bed of the new ion-exchange resins in a de-ionizer made entirely of polythene. 1962 M. G. DENAVARRE *Chem. & Manuf. Cosmetics* (ed. 2) II. 351 The same water processed in a mixed bed deionizer shows less than 2 p.p.m. total solids.

deisal, deisul, varr. DEASIL in Dict. and Suppl.

Deiters (dəi·təɹz, ‖dai·tərs). The name of Otto Friedrich Carl *Deiters* (1834–63), German anatomist, used to designate various anatomical structures, as *cells of Deiters,* (*a*) the outer hair cells in the organ of Corti; (*b*) the astrocytes of the neuroglia; *nucleus of Deiters,* the lateral vestibular nucleus (NUCLEUS *sb.* 6 c) in the brain.
1867 *Quain's Anat.* (ed. 7) III. 767 Alternating with the outer ciliated cells [in the organ of Corti] are the cells of Deiters, which are fusiform and prolonged into a thread at each extremity. 1886 W. R. GOWERS *Man. Dis. Nerv. Syst.* I. III. 107 Fine fibres..form a network. At their intersections are peculiar cells consisting of a nucleus and small cell body ('glia-cells', 'cells of Deiters'). 1890 A. HILL tr. *Obersteiner's Anat. Cent. Nerv. Organs* 225 This region is.. known as the large-celled nucleus of the auditory nerve (Deiters' nucleus). 1964 J. Z. YOUNG *Model of Brain* Gloss. 335/2 *Deiter's* [sic] *nucleus,* nerve-cell region receiving signals from the organ of balance via the vestibular nerve. 1966 R. O. GREEP *Histol.* (ed. 2) xxxvi. 866/1 The supporting cells of the outer hair cells are the outer phalangeal cells, or Deiters' cells.

deity. Now freq. pronounced (dēⁱ·iti).

deixis (dəi·ksis). *Grammar.* [Gr. δεῖξις reference.] Indication, pointing out. Cf. DEICTIC *a.* in Dict. and Suppl.
1949 *Archivum Linguisticum* I. II. 181 His analysis of the basic personal pronouns in terms of deixis. 1964 R. A. HALL *Introd. Ling.* xxvi. 164 The function of pointing out is often called deixis. 1966 G. N. LEECH *Eng. in Advertising* xviii. 157 The definite article is characteristically employed in advertising in this sense of *absolute* deixis.

déjà vu (deʒa vü·). *Psychol.* [Fr., = already seen.] **a.** An illusory feeling of having previously experienced a present situation; a form of paramnesia. Also *attrib.*
1903 [see PROMNESIA]. 1906 *Jrnl. Abnormal Psychol.* Apr. 9 A feeling that the present is a part of past experience (*déjà vu*). 1914 tr. *Freud's Psychopathol. Everyday Life* xii. 323 My own experience of *Déjà vu* I can trace in a similar manner to the emotional constellation of the moment. 1937 KOESTLER *Spanish Test.* i. 21 He is unable to shake off a dream-like feeling that he has had this nightmare before, a feeling that the psychologists term *déjà vu.* 1949 M. DICKENS *Flowers on Grass* v. 107 The queer dreamlike warning came upon him again, that illusion of having experienced everything before. The *déjà vu phenomenon,* he knew it was called. 1953 H. H. PRICE *Thinking & Experience* iii. 80 We have all heard of the *déjà vu* illusion, and most of us have experienced it at some time or other. 1958 I. MURDOCH *Bell* xxv. 297 He had again the strange sensation of *déjà vu.*
b. *occas.* The correct impression that something has been previously experienced; tedious familiarity.
1960 *Times* 18 Feb. 15/2 Although better than her last novel, *Aimez-vous Brahms..*has a depressing air of *déjà vu.* 1962 *Listener* 11 Jan. 8a/3 Even when..some of them [*sc.* photogravures] are of well-known sites or monuments, the *déjà vu* is avoided by lighting or angle or texture.

déjeuner. Add: **b.** A breakfast service (see quot. 1875).
1774 *Descr. of Villa of H. Walpole* 105 A dejuné [ed. 1784, p. 74: dejeuné] of Seve china. 1869 LADY C. SCHREIBER *Jrnl.* (1911) I. 31 A very fine Chelsea Déjeuner which he sold for £90. 1875 E. METEYARD *Wedgwood Handbk.* 395 A déjeuner consists of a tray, one or more cups and saucers, occasionally a teapot, a cream jug, and a slop basin.
c. *attrib.*
1851 *Illustr. Catal. Gt. Exhib.* III. 726/1 Communion and déjeûne services. *Ibid.* 748/2 Papier maché—'standish', 'déjeûner tray', and bottle-stand. 1865 'OUIDA' *Strathmore* xii, In the breakfast-room every déjeûner delicacy was waiting. 1870 LADY C. SCHREIBER *Jrnl.* (1911) I. 65 A Sèvres déjeuner service. 1875 J. GRANT *One of Six Hundr.* x, A *déjeûner* service of splendid Wedgwood ware. 1899 *Westm. Gaz.* 5 July 1/3 Messrs. Christie sold yesterday the déjeuner-service of Napoleon I.

Dekabrist (de·kăbrist). Also **Deca-.** [a. Russ. *dekabrist,* f. *dekábr'* December.] One who took part in an uprising which occurred in St. Petersburg on 26 December 1825, on the accession of the Emperor Nicholas I.
1882 in OGILVIE (Annandale). 1885 *Encycl. Brit.* XIX. 649/1 The unfortunate conspiracy of the Dekabrists. 1903 *Daily Chron.* 22 Oct. 3/2 A reactionary policy led to a series of conspiracies—notably that of the Decabrists—which closed the door to enlightened government for many years to come. 1920 *19th Cent.* Sept. 420 *Russian Women* gives a description of the wives of the Dekabrists, aristocrats but our first revolutionists. 1925 *Glasgow Herald* 5 Mar. 4 Prince Wolkonsky, as became a descendant of a Dekabrist, was a Liberal.

Dekatron (de·kătrǫn). [f. Gr. δεκα- DECA- + *-TRON.] The proprietary name of any of several gas-filled multi-electrode counting tubes, each containing a central anode and a set of ten inter-connected cathodes. Also *attrib.*
1950 BACON & POLLARD in *Electronic Engin.* XXII. 173 (*title*) The Dekatron. A new cold cathode counting tube. *Ibid.*, A range of 'Dekatron' tubes has been designed which comprises..three types of multi-electrode cold-cathode tubes. 1951 *Electronic Engin.* XXIII. 85 These valves (Dekatrons) are gas filled tubes which have ten stable glow positions. 1958 *Engineering* 14 Feb. 201/3 The count is therefore shown on dekatron tubes in which the glow rotates through ten positions on each of three tubes which count the hundreds, tens and units respectively. 1963 B. FOZARD *Instrumentation Nucl. Reactors* viii. 77 These are multi-electrode gas-filled scaling tubes commonly arranged to give, for nuclear particle counting applications, a scaling factor per tube of ten (hence the name Dekatron).

dekh, var. *DECK *sb.*²

dekink (dīki·ŋk), *v.* [f. DE- II. 2 + KINK *sb.*¹] *trans.* To remove kinks from. So **deki·nking** *vbl. sb.*
1936 J. GRIERSON *High Failure* viii. 189 He would come down to help with the dekinking and rewinding of the aerial. 1957 *Observer* 7 July 2/5 A process for dekinking the hair [of Negro women].

dekko (de·ko). *slang* (orig. *Army slang*). Also **decko,** †**dekho.** [f. Hind. *dekho,* imperative of *dekhnā* to look: cf. *DECK *sb.*²] A look. Also as *vb.*, to look, look at.
1894 *Daily News* 8 Sept. 6/3, I had a 'dekho' round every 'house' in the place, but couldn't see you. 1917 A. G. EMPEY *From the Fire Step* 232 '*Dekko*', to look; a look at something. 1920 M. A. MÜGGE *War Diary of Square Peg* 219 *Dekko.*—Look! 1927 F. B. YOUNG *Portrait of Clare* VI. x. 744 He's promised to look in this evening, just to have a 'dekko' as he calls it, and see that you're all right. 1927 E. THOMPSON *These Men thy Friends* 63 'You haven't had a proper *dekko,* padre' he protested. 1933 *Punch* 16 Aug. 181/1 While they were hard-and-fast aground, James took a dekko, or look around. 1936 *Ibid.* 10 June 656/3 'Im and me was 'aving a last decko at the markwee when the housemaid comes out. 1958 *Observer* 11 May 15/1 Once I'd grabbed hold of the script and taken a good dekko at it, my worst fears were confirmed. 1961 J. STROUD *Touch & Go* xvi. 176 Phew! Dekko this, sir!

del (del). *Math.* [Short for DELTA, from its being represented by an inverted delta (sense 1).] A name of the symbolic operator ∇, defined as

$$\mathrm{i}\frac{\partial}{\partial x}+\mathrm{j}\frac{\partial}{\partial y}+\mathrm{k}\frac{\partial}{\partial z}\,.$$

1901 GIBBS & WILSON *Vector Analysis* iii. 138,

$$\mathrm{i}\frac{\partial}{\partial x}+\mathrm{j}\frac{\partial}{\partial y}+\mathrm{k}\frac{\partial}{\partial z}\,.$$

This symbolic operator ∇ was introduced by Sir W. R. Hamilton and is now in universal employment. There seems, however, to be no universally recognized name for it, although owing to the frequent occurrence of the symbol some name is a practical necessity. It has been found by experience that the monosyllable *del* is so short and easy to pronounce that even in complicated formulæ in which ∇ occurs a number of times no inconvenience to the speaker or hearer arises from the repetition. ∇*V* is read simply as 'del *V*'. 1924 C. E. WEATHERBURN *Adv. Vector Analysis* i. 4 ∇*V* (pronounced *del V*). 1962 CORSON & LORRAIN *Introd. Electromagn. Fields* i. 6 Gradient is commonly abbreviated as grad, and the operation on the scalar *f* defined by the term gradient is indicated by the symbol ∇, called del. 1964 S. SIMONS *Vector Anal.* v. 69 If we define a vector differential operator ∇ (called 'del' or 'nabla') by $\nabla = \mathrm{i}(\partial/\partial x) + \mathrm{j}(\partial/\partial y) + \mathrm{k}(\partial/\partial z)$, then it follows..that grad $V = \nabla V$.

delabializa·tion. *Phonetics.* [f. DELABIALIZE *v.* + -ATION.] The action of delabializing.
1907 H. M. CHADWICK *Origin Eng. Nation* iv. 66 The delabialisation in E 4..is confined to unaccented words. 1939 *Year's Wk. in Eng. Stud. 1937* 30 The actual phonetic processes involved in the 'delabialization'. 1959 *Brno Studies Eng.* I. 34 The delabialisation of *u.*

delafossite (delāfǫ·səit). *Min.* [a. F. *delafossite* (C. Friedel 1873, in *Comptes Rendus* LXXVII. 213), f. the name of G. *Delafosse,* French mineralogist (1796–1878): see -ITE¹.] A black oxide of iron and copper, $CuFeO_2$, found chiefly in crystalline form or as botryoidal crusts.
1873 *Jrnl. Chem. Soc.* XXVI. 1107 Finer specimens were subsequently found in the collection of the museum of natural history. The name delafossite is proposed for it. 1968 M. H. HEY in *Mineral. Mag.* XXXVI. 651 Two new microanalyses of delafossite from Nizhniĭ Tagil (the type locality) and Kimberly, Nevada, confirm the accepted formula $CuFeO_2$.

delanovite (delānǭᵘ·vəit). *Min.* Also **delanouite** (delæ·nuəit). [ad. G. *delanovit,* f. the name of *Delanoue,* French mineralogist: see -ITE¹.] A manganiferous clay of a rose-red colour.
1854 DANA *Syst. Min.* (ed. 4) II. App. 500 Delanovite, *Kenngott,* (Jahrb. k.k. Geol. Reichs., iv, 633).—A reddish amorphous earthy mineral...Kenngott places it near Montmorillonite. 1934 *Mineral. Mag.* XXII. 469 Substances named montmorillonite, as well as others named confolensite, delanouite, stolpenite, erite, severite, &c., show a great variety of composition.

‖ **délassement** (delasmañ). [Fr., f. *délasser,* f. *dé-* DE- + *las* weary.] Relaxation.
1804 M. WILMOT *Let.* 8 Aug. (1934) I. 122 The serpantine Mazes of two Polonaizes, which walk (for 'tis no more) is the only *délassement* one has. 1854 THACKERAY *Newcomes* xxii, Clive (who had taken a trip to Paris with his father, as a *délassement* after the fatigues incident on his great work)..declared the thing was rubbish. 1860 W. H. RUSSELL *Diary India* I. 56 We had the *délassemens* of many meals, and music, and whist, and songs at night. 1906 W. DE MORGAN *Joseph Vance* xxiv. 232 Illusion—hallucination—*délassement* of the senses—that sort of thing.

Delaware (de·lăwēᵊɹ). [f. the name of an American river (see quot. 1832).] **a.** A member of an Algonquian Indian people, formerly inhabiting the basin of the Delaware river. **b.** The language of this people. Also *attrib.*

1709 *Penn. Col. Rec.* (1852) II. 469 Chiefs of the Delaware Indians, settled at Peshtang above Conestogoe. 1721 *Doc. Col. Hist. N.Y.* (1855) V. 623 All the English..have..in Jersey & Pennsylvania, their own or home nations, called Delawares, [who] are exceedingly decreased. *a* 1762 S. NILES *Wars* in *Mass. Hist. Soc. Coll.* (1861) 4th Ser. V. II. 430 A large party of Delaware Indians. *a* 1772 J. WOOLMAN *Jrnl.* (1778) 201 None of them were quite perfect in the English and Delaware tongues. 1826 J. F. COOPER *Last of Mohicans* I. iv. 69 Look to a Delaware, or a Mohican, for a warrior! *Ibid.* 75 They spoke together earnestly in the Delaware language. *Ibid.* viii. 169 He spoke in Delaware. 1832 S. A. FERRALL *Ramble thro' U.S.* 46 The Lenni Lenape, or Delawares, as they were called by the English, from the circumstance of their holding their great 'Council-fire' on the banks of the Delaware river. 1877 L. H. MORGAN *Ancient Society* II. ii. 73 The number of gentes..varied in the different tribes, from three among the Delawares..to upwards of twenty among the Ojibwas. 1933 BLOOMFIELD *Lang.* iv. 72 The Algonquian family..includes the languages ..of New England (..Mohican, and so on, with Delaware to the south). 1965 *Canad. Jrnl. Linguistics* Spring 135 Some Algonquians..now speak..Delaware in Oklahoma.

delay, *sb.* Add: **1. c.** *spec. Electr.* (See quot. 1940.)

1930 *Bell Syst. Techn. Jrnl.* IX. 571 The transmission delay suffered by different portions of the frequency band must also be considered. This is necessary because..this delay tends to be different for different parts of the frequency band and the distortion produced is a function of the frequency-delay characteristics. 1937 W. L. EVERITT *Communication Engin.* (ed. 2) i. 23 *Delay Distortion*.. occurs when the phase angle of the transfer impedance with respect to two chosen pairs of terminals is not linear with frequency..thus making the time of transmission or delay vary with frequency. 1940 *Chambers's Techn. Dict.* 231/1 *Delay*, the time taken for a signal to travel from one end of an electrical communication system to the other, or along a part of such system. 1956 AMOS & BIRKINSHAW *T.V. Engin.* II. i. 20 The signal takes a finite time, usually termed delay, to pass through the amplifier. 1967 *Electronics* 6 Mar. 165/2 Circuit delay, meaning propagation delay due to the IC alone, is usually expressed on IC data sheets as t_p, but we reserve t_p for the propagation delay... We call this 'system logic delay'.

3. *attrib.*, as **delay action** = **delayed action*; **delay cable, line**, a device producing a desired delay in the transmission of an electrical or other signal, used esp. in computers.

1810 [see sense 1 a in Dict.]. 1879 *Man. Siege & Garr. Artill. Exerc.* II. 51 Delay Action for base of Battering Shell. 1900 *Daily News* 11 Apr. 5/6 Delay-action projectiles. 1928 in C. F. S. Gamble *N. Sea Air Station* xv. 280, 100-lb. bombs with 2½ seconds delay fuses. 1940 *Chambers's Techn. Dict.* 231/1 The surge..travels along the delay cable. 1941 S. R. ROGET *Dict. Electr. Terms* (ed. 4) 87/2 *Delay cable*, a cable of special characteristics by which a transmission line is connected to a surge measuring instrument to cause a time lag. 1947 *Electronics* Nov. 136/1 The system of storage used with delay lines depends upon timed distribution of the electrical impulses that represent the information. 1947 EMSLIE & MCCONNELL in L. N. Ridenour *Radar System Engin.* xvi. 667 The simplest delay line is a straight tube with parallel transducers at the end. 1958 *New Biol.* XXVI. 106 A typical storage system for a digital computer is what is called a 'delay line storage'. This is made up of tubes filled with mercury that carry pulses... These pulses travel up and down the mercury tube in a special order where the presence of a pulse stands for 1 and its absence for 0; then a whole collection of these os and 1s, say 00110110, may represent an instruction word in coded form, or a number on which the instruction will operate. 1962 N. H. CODLING in G. A. T. Burdett *Autom. Control Handbk.* viii. 41 A specialised cable type which finds applications in pulse-forming circuits for computers is the helically wound delay line. Delay cables have a very high inductance. 1964 C. DENT *Quantity Surveying by Computer* iii. 19 Delay line storage comprises units which store binary digits in the form of pulses, which are kept circulating in specially designed circuits called delay lines. 1965 *Economist* 13 Feb. 592/2 At the [colour TV] receiver the first two-colour signal is stored in a delay-line to await the arrival of the second, and the two are then combined. 1970 *Physics Bull.* July 306/2 The acoustic delay lines used in some of the earliest computers.

delayed, *ppl. a.*[1] Add: **b.** In specific collocations, as *delayed action*: an action that is delayed, esp. for a particular purpose; an arrangement or device that delays an action; freq. *attrib.*, as *delayed-action bomb*, a bomb that explodes some time after it has struck the target; *delayed-action fuse*, a fuse that delays the detonation of a charge until some time after the projectile has struck the target; *delayed drop, jump*: a parachute jump in which the parachutist delays longer than usual before pulling the rip-cord; *delayed neutron*: a neutron emitted by a nucleus which has been left in an excited state following the decay or fission of its parent nucleus; *delayed shock*: shock (see SHOCK *sb.*[3] 5) that appears a considerable time after the event(s) that produced it; also *transf.*

1892 *Chambers's Jrnl.* 560/1 A delayed-action fuse. 1909 *Westm. Gaz.* 2 Jan. 13/3 The thickest armour-plate can now be pierced by projectiles fitted with a delayed-action fuse. 1936 F. J. MORTIMER *Wall's Dict. Photogr.* (ed. 14) 175 *Delayed action*. Many shutters on modern cameras..are now made with a 'delayed action' adjustment. 1940 *Manch. Guardian Weekly* 10 May 362 The debate will prepare an explosion that will go off like a delayed-action bomb in about three weeks' time. 1942 *Aeronautics* June 62/1 Reaching their objective, dropped delayed-action bombs on their targets. 1958 *Listener* 20 Nov. 812/1 The electoral delayed action of those events as it works itself out will be even more difficult to detect and disentangle. 1962 G. LAWTON *John Wesley's English* 162 We find effective delayed-action mechanism such as: 'Indeed if thou canst save the soul of another, do; but at least save one, —thy own.' 1967 *Gloss. Sanitation Terms (B.S.I.)* 59 *Delayed action ballvalve*... The opening of the ballvalve is delayed until the water level in the cistern has fallen through a fixed distance. 1942 A. M. LOW *Parachutes* iv. 60 A delayed drop of 200 feet. *Ibid.*, The only really effective method of testing the matter was by a series of delayed jumps. 1946 W. F. BURBIDGE *From Balloon to Bomber* 43 The belief that a person lost consciousness during a parachute descent was widely held until delayed drops proved the contrary. 1941 M. GOWING *Britain & Atomic Energy 1939–1945* (1964) App. ii. 406 There are a few 'delayed' neutrons arising from products of the fission process. 1950 *Effects of Atomic Weapons* 240 More than 99 percent of the total number of neutrons accompanying the fission of uranium 235 or plutonium 239 are released almost immediately, probably within 10^{-8} second of the explosion. These are referred to as the *prompt neutrons*. In addition, somewhat less than 1 percent, called the *delayed neutrons*, are emitted subsequently. 1963 W. E. BURCHAM *Nuclear Physics* 703 A β-process may leave a product nucleus so highly excited that neutron emission is a predominant alternative to β^--decay. An emission of delayed neutrons, with a half-life corresponding to that of the preceding β-emitter, then ensues. 1938 DORLAND & MILLER *Med. Dict.* (ed. 18) 1272/2 *Delayed shock*, severe physical or mental disturbance, of which the symptoms occur a considerable time after the injury or mental impression is received. 1961 *Guardian* 29 May 12/6 The 'delayed shock' of the state's worst drought on record in 1959–60.

dele. Add: (Earlier example.) Also as *sb.*, an instance of the use of the sign so called.

1727–51 CHAMBERS *Cycl.* s.v. *Correction*, There are different characters used to express different corrections, *e.gr.* D or ẟ *dele* for anything to be effaced, or left out. 1821 DIBDIN *Bibliogr., etc. Tour* I. 129, I could discover..that.. he wished me to..leave him to his *deles* and *stets*! 1963 V. NABOKOV *Gift* iv. 258 To take but the word 'I', *ya* (formed in Russian script somewhat like a proof-reader's dele).

Hence **dele** (dī·li), *v.* = DELETE *v.* 2.

1705 S. SEWALL *Diary* 24 Dec. (1879) II. 150, I deled the Title, *In Obitum Crucis*. 1765 *N. Carolina Col. Rec.* (1890) VII. 81 To the *deleing* the Clause..we..agree. 1869 R. MORRIS *Spenser's Wks.* App. i. 689/2 The comma after ape should be deled.

delete, *v.* Add: **2. c.** To remove (a gramophone record) from the catalogue and thus no longer offer it for sale.

1937 *Gramophone* Oct. 190/1 H.M.V. are wisely persevering with their policy of announcing in advance which records they intend to delete from the Connoisseur's catalogue. 1949 *Ibid.* Oct. 90/1 The first two Bartók quartets, long deleted, are now eagerly sought after by collectors. 1966 *Melody Maker* 7 May 12/1 These justly famous tracks make a surprise reappearance because it was only in April that EMI deleted them.

3. *pass. Cytology.* Of a segment of a chromosome: to be lost from the chromosome. So **dele·ted** *ppl. a.* Cf. *DELETION 3.

1929 PAINTER & MULLER in *Jrnl. Heredity* XX. 296/1 Drawings of the chromosomes have been presented..the deleted *X* being indicated as '*X*—'. 1936 *Discovery* Sept. 269/1 It is usually a section in the middle which disappears, or is 'deleted', the two ends joining up to make a shortened chromosome. 1957 C. P. SWANSON *Cytol. & Cytogenetics* x. 368 At anaphase, the deleted portion is usually freed and does not undergo movement to one or the other of the poles.

deleteriousness. (Example.)

1812 SHELLEY *Let.* 29 July (1964) I. 316, I have *no* doubts on the deleteriousness of classical education.

deletion. Add: **2. b.** The action of deleting a gramophone record from the catalogue; a deleted record.

1937 *Gramophone* Oct. 190/1 The characteristic of last year's deletion list was the preponderance of French works. 1944 *Ibid.* Sept. 52/3 (*heading*) Decca deletions. 1954 *Gramophone Record Rev.* Oct. 574/2 From the many *Schwarzkopf* deletions I would single out the best complete version of the aria 'L'amero, saro costante'. 1969 *Gramophone* June 103/2 (Advt.), Classical, Shows and jazz LPs at reduced prices, mono and stereo deletions.

3. *Cytology.* The loss of a segment from a chromosome; also, the segment lost. Cf. *DEFICIENCY 1 e.

1929 PAINTER & MULLER in *Jrnl. Heredity* XX. 295/1 Deletions of *X* Chromosomes. *Ibid.*, We have obtained.. hyperdiploids of a different origin, based upon deletions of the *X*. 1938 E. B. FORD *Study of Heredity* iv. 92 In the instances so far mentioned, the fragment of chromosome which has broken away has succeeded in reattaching itself elsewhere. Sometimes, however, it fails to do so and is lost. This leads to 'deletion', in which some of the genes..find themselves unopposed by their allelomorphs. 1952 C. P. BLACKER *Eugenics* 245 A piece of a chromosome may break off, causing a deficiency or deletion. 1969 BROWN & BERTKE *Textbk. Cytol.* xxi. 480/2 The location where an interstitial deletion was 'extracted' is not represented by a gap in the chromosome.

deliberative, *a.* Add: **3.** *Gram.* Expressing deliberation or doubt.

1842 W. E. JELF *Gram. Gr. Lang.* II. iv. 501 The conjunct. after principal, opt. after historic tenses, has a deliberative force. 1858 J. CONINGTON *Wks. Virgil* I. 38/1 The deliberative conjunctive in Greek. 1893 HAYES & MASON *Tutorial Lat. Gram.* xlii. 238 The interrogative form of the jussive subjunctive is called the deliberative or dubitative subjunctive. 1930 J. F. MOUNTFORD *Kennedy's Rev. Lat. Primer* 160 The Subjunctive of Will (Volitive) expresses..what ought to be done as a matter of propriety or duty. When used in a question, this Subjunctive is called Deliberative.

delicate, *a.* Add: **8. b.** Also colloq. phr. *in a delicate condition* or *state of health*: pregnant.

1850 DICKENS *Dav. Copp.* xxvii. 287 Mrs. Micawber, being in a delicate state of health, was overcome by it. 1908 G. SANGER *70 Yrs. a Showman* xxx. 85, I was the more concerned as Mrs. Sanger was in a delicate condition.

IV. *delicate-looking* (examples); *delicate-minded.*

1853 MRS. GASKELL *Cranford* vi. 82 My mother was very pretty and delicate-looking. 1906 J. JOYCE *Let.* 12 Sept. (1966) II. 159 He wears spectacles, is delicate-looking. 1806 M. WILMOT *Jrnl.* 14 June (1934) II. 268 She opens into a very clear, very quick (?) judging, delicate minded, spirited Woman. 1891 M. BEERBOHM *Let.* 20 Aug. (1964) 21 Awkward for delicate-minded English people.

delicatessen (de:likăte·sən). orig. *U.S.* [G. *delikatessen*, Du. *delicatessen*, ad. F. *délicatesse* (see DELICATESSE).] Delicacies or relishes for the table; esp. *attrib.*, in *delicatessen shop, store*. **b.** *ellipt.* A delicatessen shop.

[1877 E. S. DALLAS *Kettner's Bk. of Table* 399 A house which abounds in foreign dainties of all sorts—Lingner's Delicatessen Handlung, 46, Old Compton Street, Soho.] 1889 *Kansas Times & Star* 7 Nov., Burglars broke into Blake's delicatessen store..and made a careful mess of the juicy stuff, canned and bottled. 1893 *Harper's Mag.* Apr. 660 They [sc. Germans in New York] maintained their delicatessen shops and pork butchers. 1893 W. D. HOWELLS *Coast of Bohemia* 261 The shop of an old German..who dealt in delicatessen. 1904 'A. DALE' [illegible] Tonight, Anna [illegible] dinner. 1904 *N.Y. Even. Post* 20 June 14 (Advt.), Our Modern Delicatessen Department on the fifth floor is prepared to furnish estimates for picnic luncheons. 1905 *Ibid.* 13 Nov. 7 Next week's opening of Mr. Conried's operatic delicatessen store on Broadway. 1908 *Daily Chron.* 8 Jan. 8/3 The German delicatessen shops which are now becoming such a feature in London life. 1916 *Daily Colonist* (Victoria, B.C.) 7 July 5/1 Home Cooked Meats and Delicatessen Goods in Great Variety. 1930 *Daily Tel.* 8 Apr. 9/6 The New York women..in their skyscraper kitchens, where everything can be done by electricity, with the delicatessen and..prepared food as a stand-by. 1960 *P.O. Directory: London Yellow Pages Classified (North)* 62/3 Delicatessens and Cooked Meats.

delicious, *a.* Add: **2. b.** With capital initial: designating a variety of eating apple of North American origin. Also as *sb.*

1903 BUDD & HANSEN *Amer. Hort. Man.* II. 70 The Apple. *Delicious*. Originated by Jesse Hiatt..Iowa.. a regular and heavy bearer. Fruit large, roundish conic, ribbed; skin medium thick..color yellow, washed with mixed red. 1932 *Discovery* July 220/1 Canada is proud of the Mackintosh Red, Delicious, Jonathan, [etc.]. 1959 A. H. MCLINTOCK *Descr. Atlas N.Z.* 42 The most popular export varieties are Sturmer, Delicious, Jonathan, Granny Smith, and Cox's Orange.

delictal (dili·ktăl), *a.* [f. DELICT + -AL.] Of, pertaining to, or of the nature of a delict.

1913 H. GOUDY in P. Vinogradoff *Ess. Legal Hist.* 208 Where..a delictal action was not strictly penal..it transmitted both actively and passively.

delight, *sb.* Add: **4.** *Lumps of Delight*: a former name for *Turkish delight* (see TURKISH *a.* 2 b).

1870 [see TURKISH *a.* 1]. 1875 L. M. ALCOTT *Eight Cousins* v, Phebe..crunched the 'Lumps of Delight' tucked into her mouth. 1894 *Daily News* 4 June 7/7 The Turkish, or rather Greek, sweetmeat known as Rahat Loukoums, or 'Lumps of Delight'.

delight, *v.* **1.** *absol.* Add examples.

1904 E. T. MEADE *Love Triumphant* v, I mean to go to London..to meet my equals. I shall delight, I shall delight. 1908 G. F. ATHERTON *Gorgeous Isle* iii, There had been much to delight and awe.

delignification (dīli:gnifikē[i]·ʃən). [f. DE- + LIGN(IN + -IFICATION.] The removal of lignin from woody tissue. So **deli·gnified** *ppl. a.*

1919 *Experiment Station Record* Feb. 160 Its action on the wood is described as one of delignification followed by digestion. 1943 *Chem. Abstr.* 10 Apr. 1877 Decomposition of delignified bast fiber. 1946 *Nature* 20 July 100/2 Jute fibres that have been subjected to intensive delignification. *Ibid.*, Completely delignified and partially mercerized jute fibres.

Delilah (dilai·lă). Also 6–8 **Dalilah**. The name of the woman who betrayed Samson to the Philistines (see Judges xvi), used allusively to mean a temptress or treacherous paramour.

1594 J. KING *On Jonas* (1597) xlvi. 630 You who esteeme ..to be the dearlings of the pleasure of Egypt, and be set vpon the knees of the Delilah of this world. 1614 T. ADAMS *Divells Banket* i. 6 If Dalilah inuite Sampson, ware his lockes; she will spoile the Nazarite of his hayres; there are many Dalilahs in these dayes. 1678 BUTLER *Hud.* III. ii. 1115 [Ye] Transform'd all Wives to Dalilahs, Whose Husbands were not for the Cause. 1768 H. BROOKE *Fool of Qual.* III. xiii. 30, I have no foreign Dalilahs, no secret amours. 1879 J. C. SHAIRP *Robert Burns* vi. 141 Other

Delilahs on a smaller scale Burns met with during his Dumfries sojourn. 1893 F. ADAMS *New Egypt* 63 This is about as far as the French Delilah dare at present go in the public incitements of her young Egyptian Samson. 1931 W. DE LA MARE *Seven Short Stories* 89 The gloved fingers, Delilah-like, had tapped again.

delineative, *a.* (Earlier examples.)

1841 H. S. FOOTE *Texas & Texans* I. 40 Others were perspicuously delineative of current events. [illegible] *Hist.* (1965) I. 260 In the absence of the delineative faculty..minds of this deep..class are apt to put up with reasons which lie on the surface.

delineator. Add: **3.** (See quots.) (*Tailoring.*) A pattern formed by rule; being expansible in the directions where the size may vary, as indicated by the varying lengths obtained by measurement. **2.** (*Surveying.*) A perambulator, or geodetical instrument on wheels, with registering devices for recording distances between points [etc.]. 1964 *Punch* 23 Sept. 442/3 The delineators are the small reflectors mounted on posts 132 feet apart [on U.S. roads].

delint (dīli·nt), *v.* [f. DE- + LINT[1].] *trans.* To remove the fibre from (cotton or similar seeds). Hence **delinter**. So **deli·nt** *sb.* (see quot. 1904).

1902 *Westm. Gaz.* 2 Jan. 7/2 A new process for delinting and hulling the cotton seed. 1904 L. L. LAMBORN *Cotton-seed Products* iv. 50 The cottonseed is run through the delinters, which pass to the condenser attached to the delinter. *Ibid.* 51 Linters, or delint, as the short fibre is also called, find extensive application in the arts. 1921 *Dict. Occup. Terms* (1927) § 119 Delinter; attends and feeds machine which removes short fibres adhering to cotton seeds, after long fibres have been detached in ginning process. 1922 *Encycl. Brit.* XXX. 591/1 The United States supplied large quantities of nitro-cellulose propellant for the Allies and for its own army, and used as raw material a considerable proportion of the shorter fibre 'waste' from the delinted cotton seed.

deliquescent, *a.* Add: **3.** (Extended uses.)

1937 [illegible] I. iv. 83 'Laissez Faire' was in its hey-day then; it is deliquescent now, though it lingers in the liking for 'Rugged Individualism'. 1947 T. H. WHITE *Elephant & Kangaroo* (1948) xvii. 144 A ghostly figure of the Virgin Mary...It was ghostly because Mrs. O'Callaghan had taken it into her head to give it a vigorous scrubbing..and this had taken off the paint. It had also taken off most of the left cheek, so that the Virgin now lived in her shadowy corner, chalk-white, leprous and deliquescent.

delish (dīli·ʃ), *a.* Colloq. abbrev. of DELICIOUS *a.*

1920 *Punch* 14 Jan. 38/1 Their music, I gather, is wholly delish. 1953 'N. BLAKE' *Dreadful Hollow* 51 Have a glass of port, won't you? It's rather delish. 1962 *Woman* 12 May 40/1 Making delish dishes for a picnic.

deliver, *v.*[1] Add: **8. d.** Colloq. phr. *to deliver the goods*: see *GOODS *sb. pl.* 1 a.

delivery. Add: **5. a.** (Earlier *Cricket* example.) Also, = BALL *sb.*[1] 4 c.

1816 W. LAMBERT *Instr. & Rules of Cricket* 12 Very few Bowlers run after delivery of the Ball. 1868 J. LILLYWHITE *Cricketers' Compan.* 55 Mr. Jupp..played Southerton's 'slow' deliveries with consummate skill. 1882 [see Dict.]. 1960 J. FINGLETON *Four Chukkas* 59 He thus played 'Snagg' to 368 deliveries.

9. *attrib.* and *Comb.*: *delivery cart, company, date, port, system, tube, van, wagon*; **delivery box, order** (see quots.); **delivery pipe**, a pipe through which liquids are ejected, *spec.* from a pump; also, a service-pipe; **delivery room**, a room in a hospital in which a mother is delivered (of a child).

1888 *Lockwood's Dict. Mech. Engin.*, *Delivery box*, the upper or delivery chamber of a series of two or three throw pumps, into which the liquid is lifted by the pistons and from which it is delivered. 1859 G. MEASOM *Illustr. Guide Lanc. & Carlisle Railw.* 125 (Advt.), Tradesmen's Delivery Carts and Vans. 1856 MRS. GASKELL *Let.* 29 Apr. (1966) 388, I have just sent you off a parcel by the Delivery Company. 1963 J. BRAINE *Life at Top* xi. 148, I started to worry about the Tiffield order again; ..Morrison would have to talk about delivery dates tomorrow. 1967 O. WYND *Walk Softly* i. 8 It's run like a precision machine. Delivery dates are met. 1882 H. BITHELL *Counting-house Dict.* 93 *Delivery order*,.. a document, entitling..the legal holder thereof, to the delivery of any goods..of the value of forty shillings, or upwards, lying in any dock, port, wharf, or warehouse. 1926 *Times Trade & Engin. Suppl.* 20 Nov. 245/1 Ownership in goods can be transferred by mere endorsement of a bill of lading or a delivery order. [illegible] 1888 *Lockwood's Dict. Mech. Engin.*, *Delivery pipe*..the series of pipes through which the liquids drawn up by pumping machinery are ejected. 1889 P. N. HASLUCK *Model Engin. Handybk.* 111 To connect the delivery-pipe union. 1895 *Daily News* 14 Sept. 5/1 It was noticed that this water had the power to dissolve the lead of the delivery pipes. 1720 Delivery port [see Dict. 4]. 1949 M. MEAD *Male & Female* iii. 62 In our own society, our images of the ..rituals of the delivery-room..overlay any realization of what a shock birth is. 1967 *Nursing Times* 8 Sept. 1199/2 Once labour is established, the mother should be in a delivery room where in a normal confinement she will have her baby without any further move. 1961 *Listener* 28 Dec. 1098/1 This means developing nuclear weapons and delivery systems so strong and so varied that no surprise attack could knock out the power to retaliate. 1967 *Guardian* 16 May 16/8 France..might..be interested..in the development of a European delivery system. 1968 *Listener* 19 Dec. 813/1 After the war, the Soviet Union had two defence priorities: the construction of nuclear weapons and delivery systems. 1879 NOAD & PREECE *Electricity* 221 The delivery tube conducts the gas into a graduated receiver. 1868 *Third Member for Birmingham* 22 Oct. (Advt.), William Keel's, delivery van. 1906 *Westm. Gaz.* 22 Feb. 11/1 For the purpose of manufacturing in England motor-cars, motor-omnibuses, and delivery-vans. 1907 *Ibid.* 21 Nov. 4/3 Delivered free by express motor delivery-van. 1960 *Library Assoc. Rec.* Aug. 260/2 *Delivery van*, a vehicle intended and adapted primarily for the transport of books in boxes or trays, and providing no facilities for the selection of books. 1882 G. W. PECK *Peck's Fun* 88 Smith took them [sc. the goods] out to put them in the delivery wagon. 1889 *Kansas Times & Star* 14 Dec., The sign painted on one side of the delivery wagon going to the various schools with supplies.

Della Robbia (de·lă rǫ·biă). [Name of a family of Italian painters and sculptors of the fifteenth century.] **a.** Used *attrib.* to designate the enamelled terra-cotta ware made by Luca Della Robbia and his successors. **b.** Any similar ware.

1787 P. BECKFORD *Let. fr. Italy* (1805) I. 307 Some curious specimens in Terra della Robbia. 1850 *Art Jrnl.* 1 Oct. 313/2 Many fine specimens of the Della Robbia ware are yet to be found in the Florentine churches. 1878 *Lloyd's Weekly* 19 May 5/4 [Standard], A round tablet in Della Robbia ware. 1886 *Encycl. Brit.* XX. 589/1 Though Luca was not the inventor of the process [sc. the production of terra cotta reliefs covered with enamel], yet his genius so improved and extended its application that it is not unnaturally known now as Della Robbia ware. 1961 M. BEADLE *These Ruins are Inhabited* (1963) i. 11 Della Robbia plaques.

delomorphic (dīlǫmǭ·ɹfik), *a. Anat.* [ad. G. *delomorph* (A. Rollett 1870, in *Centralbl. f. d. med. Wissenschaften* VIII. 338), f. Gr. δῆλος visible + μορφή form + -IC.] Having a definite form, *spec.* denoting certain cells of the gastric glands of the stomach. Also **delomo·rphous** *a.*

1881 J. N. LANGLEY in *Jrnl. Physiology* III. 272 He [sc. Rollett] found that in winter the gastric glands of the bat contained no delomorphous or border cells, but contained many in summer. 1890 BILLINGS *Med. Dict.*, *Delomorphous cells*, large parietal cells of secretory portion of peptic glands. 1891 W. D. HALLIBURTON *Chem. Physiol. & Path.* 633 Cells of a granular character called.. (Heidenhain), delomorphic cells (Rollett), or oxyntic cells (Langley).

deloul (dĭlū·l). Also **delool, delul, dolool.** [colloq. Arab. *delūl*, Arab. *dalūl*, lit. obedient.] A dromedary.

1830 J. L. BURCKHARDT *Notes on Bedouins* 260 What is called in Egypt and Africa *hedjein*, and in Arabia *deloul*, (both terms signifying the camels trained for riding). 1855 R. F. BURTON *El-Medinah* II. xx. 225 Bedouins bestriding hard-backed 'Deluls'. 1865 [see *HYGEEN]. 1875 *Encycl. Brit.* II. 242/1 The 'hejeen', or dromedary, sometimes also called 'delool', or 'tedool'.

delouse (dīlau·s), *v.* [f. DE- II. 2 + LOUSE *sb.*; cf. G. *entlausen*.] *trans.* To clear of lice. So **delou·sing** *vbl. sb.* and *ppl. a.*

1919 *Library Assoc. Rec.* Sept. 6 The inhabitants of each barrack went in turn to the delousing station on the other side of the island, where the delousing process took three days. 1921 *Glasgow Herald* 9 July 7 The fact that there were no adequate means of delousing these men. 1929 A. FORBES *Hist. Army Ordn. Serv.* III. ii. 46 Disinfectors for de-lousing clothing. [illegible]

b. *transf.* and *fig.* To free from something unpleasant. *slang.*

1942 BERREY & VAN DEN BARK *Amer. Thes. Slang* §335/2 *Delouse*, to get rid of displeasing companions. 1943 *Times Weekly* 18 Aug. 2 The road itself had been 'deloused' of mines. 1946 BRICKHILL & NORTON *Escape to Danger* xx. 179 His squadron was 'delousing' Fortresses as they came back from out of Holland.

de·lphically, *adv.* [f. DELPHIC *a.*: see -ICALLY.] In a manner characteristic of the Delphic oracle; enigmatically, obscurely, prophetically.

1927 *Observer* 15 May 17/2 More Delphically it [sc. a report] adds that for industrial purposes the general strike can be revocable only 'in the hands of men who know how and when to use it'. 1930 BLUNDEN *Leigh Hunt* viii. 106 Hunt had dipped into the future so delphically as to publish the hurried article announcing three 'Young Poets' who would do great things. 1960 *Guardian* 27 July 9/1 M. Nkayi merely read out a vague, almost oracular statement...He then fell delphically silent.

delphine (de·lfin). Also **-in.** [ad. DELPHINIUM.] = *DELPHINIUM b.

1909 *Cent. Dict.* Suppl. s.v. *Blue*, *Delphin blue*, a mordant coal-tar color of the oxazin type. 1923 *Daily Mail* 28 May 1 Shades: Lemon,..Champagne, Delphine.

delphinidin (delfi·nidin). *Chem.* [a. G. *delphinidin* (R. Willstätter 1914, in *Sitzungsber. preuss. Akad. d. Wissenschaften* 405), f. DELPHIN(IUM + *-IDIN.] An anthocyanidin which is usu. isolated as the chloride, $C_{15}H_{11}O_7Cl$, and which combined as glycosides forms many plant pigments.

1914 *Chem. Abstr.* VIII. 3421 The anthocyan of larkspur (delphinin) on hydrolysis gave 2 mols. dextrose, 2 mols. p-$HOC_6H_4CO_2H$ and 1 mol. delphinidin, $C_{15}H_{10}O_7Cl$. 1937 *Thorpe's Dict. Appl. Chem.* (ed. 4) I. 381/2 The anthocyanins ..fall into three groups represented by the pelargonidin, cyanidin, and delphinidin types. 1946 *Nature* 17 Aug. 240/1 A delphinidin flower pigment..in an acid cell sap is normally pink but is blue in the presence of excess aluminium. 1963 [illegible] xiii. 392 It is now known that violanin is a delphinidin-3-rhamnoglucoside combined in some way with p-coumaric acid.

delphinium. Add: **b.** A deep blue like that of the indigo-blue delphinium.

1923 *Daily Mail* 23 Apr. 6 Exclusive colourings, including..Delphinium, Saxe. 1927 *Observer* 9 Oct. 21 Nut Brown, Cocoa, Delphinium. 1970 *Guardian* 5 May 7/1 Mini dress..coffee, apple, rose or delphinium print on green. 1973 M. LEE *Dying for Fun* xlvii. 220 Pretty girls dressed in pastiche rags and delphinium-blue wellington boots.

delta. Add: **3. b.** *Electr.* In three-phase electrical equipment, the arrangement of the three windings in series in a manner represented by a triangle, each of the three wires of the circuit being connected to a junction of two windings; chiefly *attrib.*, as **delta connection**.

1902 *Encycl. Brit.* XXVII. 581/2 Any three-phase winding may be changed over from the star to the delta connection. *Ibid.* 592/1 If the three coils are closed upon themselves in a mesh or delta fashion. 1943 *Gloss. Terms Electr. Engin. (B.S.I.)* 19 *Delta connection*, the arrangement of the phase windings of a three-phase A.C. working, in which three conductors or windings are connected in series to form a closed circuit which may be represented diagrammatically by a triangle.

4. delta connection (see above); hence **delta-connected** *adj.*; **delta frequency, rhythm, wave**, the most slowly varying of the 'brain waves' recorded by an electroencephalograph, having a frequency of less than about three per second and normally present only during deep sleep; **delta plain**, the flat area of a delta; **delta plateau**, a raised delta plain; **delta-rays** or **δ-rays**, rays of low penetrating power consisting of slow electrons released by the passage of ionizing particles such as alpha-rays through matter; **delta wing**, a type of triangular swept-back aeroplane wing; so **delta-winged** *a.*, descriptive of aircraft with this type of wing.

[illegible] 1950 *Nat. Geogr. Mag.* Sept. 282/2 Among the latest to be tested against Father Time and the laws of aerodynamics is the new Air Force delta-winged plane. 1954 *Economist* 11 Sept. Suppl. 2/1 This year..Gloster Aircraft were able to put five delta-winged Javelins in the air at once.

deltiology (delti,ǫ·lŏdʒi). [f. Gr. δελτίο-ν, dim. of δέλτος writing tablet + -OLOGY.] The

hobby of collecting postcards. Hence **deltio·logist**, one who collects postcards.

1947 *N.Y. Times* 22 June VI. 20/2 (*heading*) The article on postcards..omits..the current passion for collecting postcards (now called by serious followers of the hobby 'deltiology'). 1959 BUTLAND & WESTWOOD *Picture Post Cards*, In recent years very few attempts have been made to cater for the Deltiologists. 1964 *Dalesman* Dec. 710/2 How many readers know the meaning of 'deltiology'? Certainly I was in the dark about it until a subscriber.. wrote to tell me all about her hobby, which is postcard collecting. 1970 *N.Y. Times Encycl. Alm.* 477/2 *Deltiologist*, a person whose hobby is collecting picture postcards.

deluginous (deliū·dʒinəs), *a.* [Fantastically f. DELUGE after *ferruginous, salsuginous*.] Like a deluge.

1835 G. DARLEY *Nepenthe* II. 60 He..enthralls Earth in deluginous ocean. 1923 *Glasgow Herald* 15 Nov. 8 The deluginous flooding of its markets. 1924 *Ibid.* 15 Apr. 8 When the Great War broke deluginous over the world.

delusion. Add: **3. b.** *delusions of grandeur*: a false belief concerning one's personality or status, which is thought to be more important than it is. Also *fig.*

1909 in *Cent. Dict.* Suppl. 1937 E. ST. V. MILLAY *Conversation at Midnight* I. 16 Take the electric refrigerator... Take mine...Delusions of grandeur, that's what it's got, all right; Thinks it's the *Queen Mary*. 1956 'M. INNES' *Old Hall, New Hall* II. i. 101 The University..had given them their impressive title while experiencing delusions of grandeur. 1967 P. MCGERR *Murder is Absurd* iii. 37 Man, you really have delusions of grandeur! You think you're going to audition Mark Kendall?

delustre (dīlv·stəɹ), *v.* Also (*U.S.*) **-luster.** [f. DE- II. 2 + LUSTRE *sb.*[1]] *trans.* Of textiles: to deprive of lustre or sheen. So **delu·string** *vbl. sb.*, **delu·stred** *ppl. a.*

1927 *Rayon Jrnl.* Apr. 13 (*title*) Delustering Rayon. *Ibid.* 13/1 The methods of delustering Rayon may be classified into three groups: the delustering of Celanese; the mechanical delustering of other varieties of Rayon (by abrasion); and the loading of Rayon to accomplish delustering. *Ibid.* 13/2 Celanese yarn..has been delustered without otherwise altering its composition of peculiar properties. 1929 *Daily Tel.* 22 Jan. 7/6 To get rid of that cold effect a special delustring process has to be gone through during the dyeing. 1934 B. L. HATHORNE *Rayon Dyeing & Finishing* ii. 49 The procedures used to produce delustered acetate fabrics vary with the type of fabric processed. 1950 '*Mercury*' *Dict. Textile Terms* 171/2 *Delustring*, chemical treatments for dulling the lustre of rayon yarns and fabrics in order to make them more nearly resemble silk. 1960 *Harper's Bazaar* Apr. 84 Dress of delustred oyster satin. 1963 A. J. HALL *Textile Sci.* ii. 61 Delustred acetate fibres. *Ibid.* 62 Delustring is hindered by lowering the temperature of the soap-liquor.

‖ **de luxe** (də lüks), *adj. phr.* [Fr., lit. 'of luxury'.] Luxurious, sumptuous; of a superior kind.

1819 Edition de luxe [see LUXE 2]. 1865 'OUIDA' *Strathmore* viii, I wonder governments don't tax good talk; it's quite a luxury, and they might add *de luxe*. 1885 Edition de luxe [see LUXE 2]. 1890 Trains de luxe [see LUXE 2]. 1908 *Westm. Gaz.* 6 June 5/1 We are conscious of something De luxe, but not oppressed by the sense of it. 1934 *Punch* 20 June 679/3 They will disclose Britannia, enthroned on the top of a *de luxe* model of one of those erections from which they mend tram-wires. 1949 E. POUND *Pisan Cantos* lxxvii. 52 Before the deluxe car carried him over the precipice. 1955 T. H. PEAR *Eng. Social Differences* viii. 182 Members of the upper economic strata..who patronise *hotels de luxe*. 1970 K. CHESNEY *Victorian Underworld* 336 These places were often little businesses engaged in a de luxe trade, glovers, bonnet makers, perfumers and so on.

delvauxene (delvǭ·ksīn). *Min.* Also **delvauxine.** [ad. F. *delvauxine* (A. H. Dumont 1838, in *Bull. de l'Acad. R. des Sci. et Belles-Lettres de Bruxelles* V. 298), f. the name of J. C. P. J. *Delvaux* de Fenffe (1782–1863), Belgian chemist: see -INE[5].] = next.

1844 DANA *Syst. Min.* (ed. 2) Suppl. 524 Delvauxene. Massive and earthy, with a yellowish-brown color. 1854 [see next]. 1965 *Mineral. Abstr.* XVII. 231/1 Delvauxine, destinezite, torbenite, richellite, and akaganéite at Richelle.

delvauxite (delvǭ·ksəit). *Min.* [ad. G. *delvauxit* (W. Haidinger *Handbuch d. Mineral.* (1845) iv. 512), f. *DELVAUX(ENE + -ITE[1].] A hydrated ferric phosphate.

1854 DANA *Syst. Min.* (ed. 4) II. v. 427 (*heading*) Delvauxene, *Dumont*. Delvauxit, *Haid.* 1864 *Chem. News* 24 Sept. 145/1 The formula $2Fe_2O_3.PO_5 + 24HO$ is assigned to delvauxite. 1963 *Mineral. Abstr.* XVI. 220/1 Amorphous coatings of delvauxite.

delve, *v.* **9.** (Earlier U.S. example.)

1838 C. GILMAN *Recoll. Southern Matron* xxix. 204 The poor mother..delving at her needle.

demagnetize, *v.* **1.** (Earlier example.)

1839 *Brit. Pat.* 8255 2 My said invented improvements consists [*sic*] in employing, as my prime movers, a series of electro-magnets, which are alternately..magnetized and demagnetized.

demagnetized (dimæ·gnetəizd), *ppl. a.* [f. DEMAGNETIZE + -ED[1].] Deprived of magnetic quality.

1876 PREECE & SIVEWRIGHT *Telegr.* 246 A demagnetised needle. 1945 [see *DOMAIN *sb.* 4 h].

demagogue, *v.* Delete *nonce-wd.* and add later U.S. examples.

1850 *Congress. Globe* 24 July App. 940 In Ohio, the master-spirits of the party, while demagoging upon the stump, have promised the people [etc.]. 1867 *Ibid.* 16 Mar. 146/1 There was a great temptation presented to members of the Republican party now to demagogue. 1876 *Congress. Rec.* 1 July 4338/1, I have not been one of those..trying to cut down a few dollars for the purpose of demagoguing before the county.

2. *trans.* To deal with (a matter) after the fashion of a demagogue.

1890 *Cincinnati Commercial Gaz.* 31 July, The President never thought of demagoging the matter by..spurning the goodly gift. 1897 *Congress. Rec.* 20 Feb. 2041/2 Here is a plain, common-sense question, not to be demagogued in any way.

demagoguery. (Later U.K. examples.)

1959 *Times Lit. Suppl.*, The now familiar recipe of aggressive demagoguery and muck-raking. 1966 *Listener* 10 Mar. 338/2 Wherever politicians rely overmuch on demagoguery or on cultivating the mystique of their own personalities, [etc.].

demand, *sb.*[1] Add: **7. demand curve**, a graph showing how the demand for a commodity or service varies with or depends on some other quantity; *spec.* one that shows how the demand, at any particular time, varies according to the price charged; **demand deposit** *U.S.*, a banking account from which the customer can withdraw funds without prior notification; **demand feeding**, the feeding of a baby when it cries, and not according to a timetable.

1936 *Economist* 11 Apr. 85/2 And how far is the [building society] movement prepared to cope with any future change, such as a downward turn in the 'demand curve' for owner-occupied houses? 1949 *Mind* LVIII. 199 The basis of the theory of demand is the demand-curve, which states a functional relationship between the price and the quantity of a commodity demanded by the aggregate of consumers. 1930 J. M. KEYNES *Treat. Money* II. xxiii. 7 Current Accounts in England and Demand Deposits in the United States roughly correspond to the Cash-deposits. 1947 L. TARSHIS *Elem. Econ.* III. xiii. 284 When someone writes a check, he instructs his bank to transfer a part of his demand deposit, or checking account, to another person. 1966 R. G. LIPSEY *Introd. Positive Econ.* (ed. 2) VIII. xlviii. 675 A demand deposit means that the customer can withdraw his money on demand. 1953 R. LIGHTWOOD in Gaisford & Lightwood *Paediatrics for Practitioner* I. xiii. 118 Some people find that 'demand feeding' obviates this [*sc.* early morning crying of a baby]. 1955 I. ASIMOV *Martian Way* (1964) 123 She followed the demand-feeding system or the 'if-you-want-it-holler-and-you'll-get-it' routine. 1970 *Radio Times* 16 Apr. 57/2 At about the same time that his [Dr. Spock's] book was published, a report came out attempting to prove that a baby could set his own feeding schedule ('demand feeding' as it has come to be known).

dema·ndingness. [f. DEMANDING *ppl. a.* + -NESS.] The quality or condition of being demanding.

1930 GWENDOLEN GREENE *Two Witnesses* i. 18 This conception he held to the last, never shrinking from its austere demandingness. 1949 M. MEAD *Male & Female* xv. 298 Conceited demandingness in a woman.

demantoid (dimæ·ntoid). [a. G. *demantoid*.] A green variety of andradite having a brilliant lustre.

1892 DANA *Syst. Min.* (ed. 6) 442 Demantoid is a grass green to emerald-green variety [of garnet] with brilliant luster. 1897 *Edin. Rev.* Oct. 346 The demantoid or Bobrovka garnet. 1957 *Encycl. Brit.* X. 29/2 The grass green demantoid is used as a gemstone and possesses high refractive and dispersive power. 1971 *Materials & Technology* II. viii. 519 By far the most valued garnet at the present time is the brilliant green demantoid.

demarcation. Add: **b.** (*b*) *demarcation dispute, rule*, etc., in reference to the precise scope and kind of work laid down by trade unions for their members in their rules, a dispute occurring between two unions where such rules appear to conflict.

1930 *Monotype Recorder* XXIX. 19 The amicable settlement of demarcation problems raised by developments in machines and processes. 1931 *Economist* 14 Feb. 340/1 In engineering, as in many other British industries, the burden of labour costs has been inflated unduly by rigid demarcation rules. 1940 R. POSTGATE *Verdict of Twelve* I. vi. 82 The enforcing of the rule book had already taken out of the hands of the members their power to start continual small strikes over 'demarcation disputes'. 1963 *Ann. Reg.* 1962 52 Demarcation dispute in the steel industry between the bricklayers' and steelworkers' unions over who should lay the dolomite blocks used to line furnaces. 1969 *Listener* 30 Jan. 157/2 David Haworth's *Observer* report on a demarcation dispute in Barrow may be read as a classic story of an employer setting two unions against each other.

demarcative (dimā·ɹkătiv), *a.* [f. DEMARC(ATION + -ATIVE.] Of, pertaining to, or characterized by demarcation: used esp. in *Linguistics*.

1955 M. REIFER *Dict. New Words* 88/2 *Demarcative function*, the use of phonemic and non-phonemic features.. to mark off the boundaries of words and morphemes. 1959 *Archivum Linguisticum* XI. II. 100 The demarcative phonological features can be termed prosodies of word delimitation. 1964 R. H. ROBINS *Gen. Linguistics* 166 The phonological features in question, juncturally analysed in American phonemics, are regarded in prosodic analysis as demarcative word prosodies.

demarcator (dī·maɹkē[i]təɹ). [f. DEMARCATE *v.* + -OR.] One who marks out boundaries.

1898 *Daily News* 25 Feb. 6/3 English demarcators and French got on together excellently well. 1900 *Ibid.* 22 Dec. 7/2 The demarcators failed to agree, and there is now a tremendous question between the two countries as to the lines drawn on the map by their respective experts.

démarche. Add: **b.** Esp. a diplomatic initiative, a political step or proceeding.

1678 [see Dict.]. 1940 *Ann. Reg.* 1939 210 This démarche ..plainly indicated to the Soviet leaders that Germany was prepared to concede to Russia more than Britain and France. 1955 *Times* 6 May 8/4 In his opinion British *démarches* against Athens radio broadcasts were unjustified. 1956 *Ann. Reg.* 1955 228 To induce the Bonn Government to make a joint démarche to the four Foreign Ministers.

demargarinated (dimā·ɹgărinē[i]ted), *a.* [f. DE- + MARGARIN + -ATE[3] + -ED[1].] Of an oil, having the stearin or solid part removed.

1920 W. CLAYTON *Margarine* 10 A deposit of 'stearin' ensues..., and if this is removed a 'winter' oil is obtained... Such winter oils are also said to be 'demargarinated'. 1939 *Thorpe's Dict. Appl. Chem.* (ed. 4) III. 410/2 The filtered [cottonseed] oil, known in commerce as 'winter oil' ('demargarinated' or 'destearinated' oil).

dematerialize, *v.*, **dematerialization** (further examples). **dematerializing** *vbl. sb.* (example).

1899 *Westm. Gaz.* 1 Apr. 2/3 'The administrators of our manufacturing and commercial firms and the commercial traveller are not necessarily condemned to a barren materialism...' Perhaps when he has de-materialised the commercial traveller he will try his hand on the Stock Exchange. 1914 C. MACKENZIE *Sinister Street* II. IV. iv. 928 These abrupt dematerializations of the underworld were really very difficult to deal with. 1935 *Burlington Mag.* Aug. 53/2 It is a 'dematerialisation' which afterwards El Greco realised in another way. 1956 R. M. LESTER *Towards Hereafter* ii. 30 The clothes began to take shape, and slowly filled out until the medium was sitting in her chair again, looking just as she had done before the dematerializing. 1960 M. CECIL *Something in Common* ii. 30 She..dematerialized backwards into her office.

deme, *sb.*[2] Add: **3.** *Ecol.* A local population of closely related plants or animals; also used as the second element of more precise terms, as in *ecodeme, gamodeme, topodeme*, etc. Hence **demolo·gical** *a.*

1939 GILMOUR & GREGOR in *Nature* 19 Aug. 333/1 In the course of work on the experimental delimitation of botanical groups, the need has arisen for a term which can be applied to any specified assemblage of taxonomically closely related individuals...We propose the term *deme* (from the Greek δῆμος) for this purpose, with appropriate prefixes to denote particular kinds of demes. 1954 *Genetica* XXVII. 150 The essence of the deme terminology is the construction of a series of category-terms by the addition of one or more virtually self-explanatory prefixes to the 'neutral' suffix '-deme'. *Ibid.* 151 Sometimes, however, demological units will coincide with recognised taxa. 1961 G. G. SIMPSON *Princ. Animal Taxon.* v. 176 The smallest unit of population that has evolutionary significance is a group of individual animals (of one species or subspecies) so localized that they are in easy and more or less frequent contact with each other...That minimal population unit is called a deme. 1966 R. L. SMITH *Ecology & Field Biol.* xxiii. 431 Collectively, individuals within each group make up a genetical population, or deme...Individuals that make up the deme are not identical. 1969 *Nature* 22 Nov. 750/1 The patterns of polymorphisms he found in Arizona indicate that within a species there are small units, or demes, of the order of a hundred or so mice that do not mix with other demes.

‖ **déménagement** (demenaʒmaṅ). [Fr.] The removal of household possessions from one place to another; moving house.

1875 LADY C. SCHREIBER *Jrnl.* (1911) I. 363 They were in the course of déménagement, going to Spa. 1892 C. M. YONGE *Let.* in C. R. Coleridge *C. M. Y. Life & Lett.* (1903) xii. 312 Gertrude had to be moved into the drawing-room... The *déménagement* will last for another week at least. 1937 LADY FORTESCUE *Sunset House* iv. 72 Bedlam had not been let loose; it was only the *déménagement* of the Italian family. 1952 M. STEEN *Phoenix Rising* xi. 245 It would have been rather tiresome..to have set up house..and be faced with *déménagement* at almost any moment.

dementalize (dime·ntăləiz), *v. Philos.* [f. DE- II. 1 + MENTALIZE.] To remove mind or mental characteristics from. So **deme·ntalized** *ppl. a.*

1936 R. G. COLLINGWOOD *Human Nature & Human Hist.* 24 Sciences of this type tend systematically to dementalize mind and convert it into nature. 1950 *Mind* LIX. 406 Can an adequate de-mentalized psychology ever be achieved?

‖ **démenti** (demaṅti). [Fr., f. *démentir* to give the lie.] = DEMENTIE. In modern use applied esp. to an official contradiction of a published statement.

1698–1883 [see DEMENTIE]. 1918 A. GRAY tr. *R. Grelling's The Crime* II. v. 228 The semi-official organ seeks to defend the reports of Pourtalès against the English *démentis*. 1921 *Contemp. Rev.* Jan. 103 My information contradicts the

dementis published by the *Deutsche Tageszeitung*. 1950 M. HAY *Foot of Pride* vii. 209 They..waited for the Vatican to publish a *démenti*.

‖ **dementia præcox** (dime·nʃiă prī·kǫks). *Med.* [mod.L. (A. Pick 1891, in *Prager Med. Wochenschr.* 8 July 312/2), used to render Fr. *démence précoce* (B. A. Morel, *Traité des dégénérescences* (1857), iv. 391), f. DEMENTIA + L. *præcox* (see PRECOCIOUS *a.*).] = *SCHIZOPHRENIA.

1899 *Amer. Jrnl. Insanity* July 116 'Dementia praecox' is Kraepelin's latest designation for the three types which it includes. *Ibid.* 118 The term 'Dementia Praecox', thus bringing together with katatonia a number of these variable forms..unifies them as belonging to one group. 1902 A. R. DEFENDORF tr. *Kraepelin's Clin. Psychiatry* 152 Dementia praecox is the name first applied by A. Pick, in 1891, to a group of cases..characterized by maniacal symptoms followed by melancholia and rapid deterioration. 1922 tr. *Freud's Introd. Lect.* xxviii. 384 Paranoia and dementia praecox, when fully developed, are not amenable to analysis. 1926 W. MCDOUGALL *Outl. Abnormal Psychol.* ii. 41 A boy presenting a clear picture of *Dementia Praecox* of the paranoid type. 1937 J. JOYCE *Let.* 30 Aug. (1966) III. 406 He was supposed to be an absolutely hopeless case of *dementia praecox*. 1970 N. MILFORD *Zelda Fitzgerald* 256 Dr. Meyer wanted Zelda to face her sickness squarely, not passively in the fixed terms of dementia praecox and schizophrenia.

Demerara (demĕrē[ə]·ră, demĕrā·ră). The name of a region of Guyana, used to designate a kind of (raw) cane-sugar, originally and chiefly brought from Demerara, the crystals of which have a yellowish-brown colour.

1848 *Sugar Question* II. 72 With a long price of 42s. for Demerara sugar, and more labour, the estates could go on. 1880 *Encycl. Brit.* XI. 251/1 The 'Demerara crystals' are very popular for their purity and saccharine strength. 1895 *Army & Navy Co-op. Soc. Price List* 106/1 *Sugars*.. Demerara, best, per 28 lb tin 5/3. 1895 *Young Woman* Dec. 98/2 Sprinkle them with a little Demerara sugar to form a crisp brown coating. 1901 *Brit. Med. Jrnl.* 4 May 1119/2 The West India Committee have caused several grocers to be prosecuted..for selling yellow crystals as Demerara sugar. 1959 *Which?* Oct. 134/2 Demerara sugar consists roughly of 98 per cent of sucrose, 0.7 per cent of other sugars and 0.5 per cent of moisture. In addition there are traces of some minerals and even smaller traces of a few vitamins.

demerit, *v.* Restrict † *Obs.* to senses 1, 2, 4 in Dict. and add: **2. b.** To lower in status by giving an unfavourable assessment of conduct. *U.S.*

1895 *Century Mag.* Oct. 843/2 He stands a fair chance of being demerited and punished until his hope of release before he is of age is almost extinguished.

demersal (dimɜ̄·ɹsăl), *a.* [f. L. *demersus*, pa. pple. of *dēmergĕre* to submerge: see -AL.] Of the eggs of fishes: sinking to the bottom of the sea, deposited at or near the bottom. Of fish, etc.: living near the bottom.

1889 *Nature* 13 June 159/2 The herring with its demersal eggs, fixed firmly to the bottom. 1911 *Ann. Rept. Sea Fish.* 1909 p. vi, in *Parl. Papers* XXIV, Demersal fish landed from each 'Area' of the North Sea. 1915 A. MEEK in *Rep. Dove Mar. Lab. Cullercoats* 14 The region where the demersal fry are mainly congregated. 1925 *Public Opinion* 16 Oct. 376/1 This increase was most marked in the bottom living or demersal fish. 1936 J. T. JENKINS *Fishes Brit. Isles* (ed. 2) 9 Practically all freshwater fish lay demersal eggs. 1959 *New Scientist* 7 May 1035/1 Demersal fishes (those caught near the bottom). 1968 F. R. H. JONES *Fish Migration* ii. 13 In the Pacific Ocean, however, demersal eggs are quite common among species whose Atlantic counterparts have pelagic eggs.

demethylate (dime·þilē[i]t), *v. Chem.* [f. DE- II. 1 + METHYLATE *v.*] To remove a methyl group from. Hence **deme·thylating** *ppl. a.*, **demethyla·tion.**

1911 *Chem. Abstr.* V. 2880 (*title*) Influence of caffeine on protein metabolism in dogs, with some remarks on demethylation in the body. 1933 *Ibid.* XXVII. 1624 (*title*) Demethylating the phenol ethers. 1940 *Nature* 3 Aug. 166/1 Demethylation definitely occurs in the animal body. 1967 I. L. FINAR *Org. Chem.* (ed. 5) I. xxvi. 682 Ethers that are sensitive to other demethylating reagents.

Demetian (dimī·ʃiăn), *a.* and *sb.* Also **Dimetian.** [f. *Demetia, Dimetia*, Dyfed in S.W. Wales (Cardigan, Pembroke, and Carmarthen) + -IAN.] Of or pertaining to Demetia in South Wales; also as *sb.*, the dialect of this region.

1841 *Anc. Laws & Inst. Wales* Pref. xii, The Dimetian or West Wales Code. 1877 H. HICKS in *Q. Jrnl. Geol. Soc.* XXXIII. 230, I propose now to divide the Pre-Cambrian rocks into two distinct series under the local names of Dimetian (*Dimetia* being the ancient name for a kingdom which included this part of Wales) for the lower, and Pebidian..for the upper series. *Ibid.* 231 The rocks which compose the Dimetian series are chiefly compact quartz schists. *Ibid.* 238 My endeavour has chiefly been to point out spots where the unconformity between them [*sc.* Pebidian rocks] and the Cambrians above, or the Dimetians below, is well marked. 1885 A. GEIKIE *Text-bk. Geol.* (ed. 2) 639 The so-called 'Dimetian' I regard as a granite which has invaded the Cambrian rocks. 1910 *Encycl. Brit.* XIII. 838/2 Three [legal] codes..called Venedotian, Demetian and Gwentian, are said to have been written down by Bleggwryd, archdeacon of Llandaff. 1913 J. M. JONES *Welsh Gram.* 8 Demetian, the dialect of Dyfed or South West Wales. *Ibid.*, Dialectal forms, chiefly Demetian and Powysian -*e*, begin to appear in the MSS. of the 15th century.

demi. Add: **B. 1.** *demi-lion* (later example).

1928 BLUNDEN *Jap. Garl.* 16 The inscrutable and dog-like grin Of demi-lions hedge me in!

demibranch (de·mibræŋk). *Zool.* [f. DEMI- + Gr. βράγχια gills.] = HEMIBRANCH (sense a).

1903 *Phil. Trans. R. Soc.* Ser. B. 150 A descending lamella and its corresponding ascending lamella together constitute a demibranch. The demibranchs are inner and outer, lying towards the visceral mass and the mantle respectively. 1946 *Nature* 12 Oct. 523/1 On percolation through the gills, the water passes into the interlamellar chambers..of the outer and inner demibranchs respectively.

‖ **demi-caractère** (dəmi,karaktḗr), *sb.* and *a.* *Ballet.* [Fr.] (Of) a dance which retains the form of the character dance but is executed with steps based on the classical technique; also applied to a dancer.

1776 *Publ. Advertiser* 2 Nov. 1/1 New Ballet Demi-caracteres, called Les Amusemens Champetre. 1828 J. EBERS *Seven Years of King's Theatre* iii. 72 The exquisite prettiness of the *demi-caractère* steps of Fanny Bias. *Ibid.* iv. 99 Fanny Bias was a dancer of the demi-caractère, perfect in..beautiful little half steps. 1950 *Ballet Ann.* IV. 70 A neck which is just too short for the demands of a classical ballerina. Hence she is at her best in *demi-caractère* rôles. 1957 *Times* 23 Aug. 11/3 A delightful *demi-caractère* dancer. 1963 VAN PRAAGH & BRINSON *Choreogr. Art* 177 Really talented demi-charactère dancers are able to fill roles in either of the other categories.

demidovite (demidǫ·vəit). *Min.* Also **demidoffite.** [ad. F. *démidovite* (N. Nordenskiold 1856, in *Bull. Soc. imp. des Naturalistes de Moscou* XXIX. I. 128), f. the name of Prince Anatolii Nikolaevich *Demidov* (1813–1870), Russian patron and traveller: see -ITE[1].] (See quot. 1955.)

1868 DANA *Syst. Min.* (ed 5) 403 Demidoffite [ed. 6, demidovite] occurs at Tagilsk, Urals, in mammillated crusts of a sky-blue color. 1955 M. H. HEY *Index Min. Species* (ed. 2) 214 Demidovite. Silicate and phosphate of Cu. *Ibid.* 401 Demidoffite...Syn. of Demidovite.

‖ **demi-glace** (de·mi,glās). *Cookery.* [Fr.; cf. GLAZE *sb.* 2.] Half-glaze; meat stock from which the liquid has been partially evaporated. Also *attrib.*

1906 MRS. BEETON *Bk. Househ. Managem.* x. 276 *Demi-glace sauce*. Ingredients — ½ a pint of Espagnole sauce..½ of a pint of good gravy, salt and pepper. 1951 *Good Housek. Home Encycl.* 438/2 *Demi-glace Sauce* (Half-glaze Sauce). Espagnole sauce to which has been added meat glaze or good gravy... Demi-glace is served with high-class meat dishes.

demilitarize, *v.* (Further examples.) Also, to place (a sovereign state) under an obligation not to maintain armed forces in a specified region. Hence **demi·litarized** *ppl. a.* Also **demilitarization** (later examples).

1905 *Westm. Gaz.* 4 Jan. 2/1 With Vladivostok demilitarised, and with Port Arthur and its entire hinterland confiscated. 1919 J. L. GARVIN *Econ. Foundations of Peace* xix. 485 If a demilitarised Germany were given a fair chance to recover..the powers of suasion and control by economic action would be paramount. 1934 H. NICOLSON *Curzon: Last Phase* 295 Ismet Pasha..made some passing reference to the possibility of establishing demilitarised zones in Thrace. 1936 *Economist* 14 Mar. 571/1 Owing to Herr Hitler's action last Saturday in reoccupying the demilitarised zone in the Rhineland the peace of Europe to-day is balancing..'on the razor's edge'. 1940 *Ibid.* 6 July 7/1 The French generals would still, it appears, resist an attack, but demilitarisation, not cession of territory or occupation, is what Italy asks of them. 1947 *Daily Tel.* 24 July, Mr. Marshall denied that the question of a 40-year Four-Power treaty guaranteeing German demilitarisation..cropped up on Monday. 1959 *Ann. Reg.* 1958 143 Israel and Jordan had agreed to try to carry out properly the July 1948 Agreement for the demilitarization of the Mount Scopus area. 1966 *Daily Tel.* 12 Oct., The draft of a treaty to keep space permanently demilitarised.

‖ **demi-mondaine** (dəmi,mondḗn). [Fr.; f. DEMI-MONDE.] A woman of the demi-monde.

1894 *Nation* (N.Y.) 12 July 29/1 The conclusion is irresistible that he has been fooled into believing *demi-mondaines* women of good society. 1898 W. J. LOCKE *Idols* xxi. 211 Her manner was that of the insolently luxurious demi-mondaine. 1922 *Daily Mail* 21 Nov. 6 Miss Mae Murray, most alluring of screen demi-mondaines. 1969 *Times* 25 Oct. p. iii/4 She presents a young, vivacious *demi-mondaine* who becomes steadily more tragic.

‖ **demi-monde.** (Earlier *attrib.* example.)

1859 G. A. SALA *Twice round Clock* 30 Countesses, actresses —*demi-monde* adventuresses.

demine (dimai·n), *v.* [f. DE- II. 1 + MINE *v.* 5.] In warfare: to remove mines from.

1945 *Manch. Guardian* 18 July 8/1 The time that it will take to 'de-mine' France is incalculable owing to the variety of local conditions.

demineralization (dimi:nĕrăləizē[i]·ʃən). [f. DE- II + MINERALIZATION.] The removal of salts, esp. from sea or brackish water; also, any abnormal loss of salts from the body. Hence **demi·neralize** *v. trans.*, to remove salts from; **demi·neralized** *ppl. a.*; **demi·neralizer**, an apparatus or installation used for demineralization; **demi·neralizing** *vbl. sb.*

1903 *Med. Record* LXIII. 621/1 The demineralization is proved by an increase in the mineral residue of the urine.. and by a correlative decrease in the mineralization of the blood. 1934 WEBSTER, Demineralize *v.* 1943 *Industr. & Engin. Chem.* Feb. 186/2 The precise composition of this demineralized effluent depends to some extent upon the quality of the raw water. *Ibid.*, The demineralizing process does not remove silica from water. *Ibid.* 192/1 The total cost of demineralizing this water is..27 cents per 1000 gallons. 1955 *Bull. Atomic Sci.* Oct. 286/3 Recent developments in the demineralization of sea water. 1956 *Nature* 28 Jan. 183/1 The clear end was cut off and dissolved in demineralized water and the metal then recovered by electrodeposition. 1958 *New Scientist* 6 Nov. 1199/1 The new water supply scheme for Guernsey will be of this form, and some 500,000 gallons of sea water will be demineralised each day. 1960 *Farmer & Stockbreeder* 1 Mar. 137/3 She [*sc.* a sow] has been demineralized, and her bones softened. 1960 *Times* 18 Aug. p. ix/4 A special de-mineralizer to remove substances that might be harmful to the boilers. 1962 *Lancet* 2 June 1168/2 The essential defect has been recognised as a failure in the calcification of new bone and cartilage, rather than a demineralisation of existing bone.

demiourgos (dimiau·ɹgǫs). A strict transcription of Gr. δημιουργός (see s.v. DEMIURGE).

1920 T. P. NUNN *Education; Data & First Princ.* 192 The 'pure' geologist still presses for the kind of understanding of the earth's structure that we might ascribe to the demiourgos who made it. 1924 W. B. SELBIE *Psychol. Relig.* 56 The familiar conception of the demiourgos in gnostic philosophy.

‖ **demi-pension** (dəmi,pāṅsyoṅ). [Fr.] The price of bed, breakfast, and one other meal at a hotel, etc.: usually in reference to France or other European country.

1951 N. MARSH *Opening Night* iv. 83 Your rental, *demi-pension*, here, is two [pounds]. 1959 *Listener* 15 Jan. 128/3 Cost? Demi-pension everywhere. Bread and wine make a good third meal.

demi-reputable (demi,re·piŭtăb'l), *a.* [f. DEMI- + REPUTABLE; see DEMI-REP.] Of doubtful reputation. Also **de:mi-reputa·tion.**

1897 W. J. LOCKE *Derelicts* xvii. 224 That fashionable demi-reputable world which had drawn him to his precipice. 1909 M. B. SAUNDERS *Litany Lane* x, People with doubtful histories, women of demi-reputation.

demi-semi. (Later examples.)

1901 *Contemp. Rev.* Mar. 358 England..no longer..employed the demi-semi-educated to educate the voters. 1906 *Daily Chron.* 1 May 3/1 Among the demi-semi-educated a laugh can always be raised by sitting down upon a silk hat! 1908 *Ibid.* 13 May 7/7 In the demi-semi-rural districts on the outskirts of towns. 1929 A. HUXLEY *Holy Face* 28 Her sage-green dress was only demi-semi-evening.

‖ **demi-sel** (dəmi,sel). [Fr., lit. 'half-salt'.] (See quot. 1946.)

1946 A. L. SIMON *Conc. Encycl. Gastronomy* IX. 13/2 *Demi-sel*, a French whole-milk soft cheese, about 4 oz. in weight, similar to *Gournay*, but with 2 per cent added salt. 1966 M. KELLY *Dead Corse* vi. 98 Why do you hanker for brown skin, Amazon pirate? You're lovely creamy white, like demi-sel.

demiss. 1. Delete † *Obs.* or *arch.* and add later examples.

1837 H. E. MANNING in J. R. S. Leslie *Life* (1921) 269, I wrote a very soft, demiss rejoinder. 1888 C. M. DOUGHTY *Arabia Deserta* I. 253 Not timid as the demiss Damascene Christians. 1903 J. BRYCE *Studies in Contemp. Biogr.* 53 By appearing too demiss or too unenterprising in foreign affairs.

demister (dīmi·stəɹ). [f. DE- II + MIST *sb.*[1] + -ER.] A device for clearing mist from the windscreen, etc., of a motor vehicle, aeroplane, etc. Also **demi·st** *v.*, **demi·sting** *ppl. a.* and *vbl. sb.*

1939 *Times* 28 Mar. 20/3 A lamp under the bonnet, and provision for a heater and demister. 1950 *Engineering* 3 Feb. 139/2 Built-in heating and demisting/de-icing equipment. 1957 *Economist* 9 Nov. 499 (Advt.), The heater provides a generous flow of warm air to the readily adjustable demister nozzle. The result is supremely efficient demisting and defrosting. 1963 *Times* 28 May 15/4 An air blower for demisting the rear window. 1966 T. WISDOM *High-Performance Driving* x. 108 Make sure the outside of the screen is clean and have the demister operating at full strength. 1968 C. BURKE *Elephant across Border* iii. 102 He..had to demist his glasses. 1968 R. PETRIE *MacLurg goes West* I. iii. 25 MacLurg turned on the blower to demist the car's windows.

demit (dimi·t), *sb. U.S. Freemasonry.* Also **dimit.** [f. DEMIT *v.*[2]] Written permission to leave a lodge, granted to a mason.

1856 R. MORRIS in A. G. Mackey *Encycl. Freemasonry* (1879) 221/2 A 'demit', technically considered, is the act of withdrawing, and applies to the Lodge and not to the individual. 1879 *Ibid.* 220/1 The granting of 'a dimit' does not necessarily lead to the conclusion that the Mason who received it has left the Lodge. He has only been permitted to do so. *Ibid.* 221/2 'A demit' is..an Americanism of very recent usage.

demi-tasse (dəmiˌtas, de·mitæs). Chiefly *U.S.* [Fr., lit. 'half-cup'.] A small coffee-cup; its contents. Also *attrib.* and as *adv.*

1842 *Galignani's Paris Guide* 17 To the latter [*sc.* the cafés] it is customary to retire immediately after dinner, to take a *demi-tasse* of coffee. **1870** J. Murray *Handbk. Paris* (ed. 4) II. 31 In the afternoon, when coffee is ordered, the waiter pours you out a small cup (demitasse). **1897** G. du Maurier *Martian* II. 49 Drinking our demi-tasse. **1906** 'O. Henry' *Four Million* (1916) 94 A roasted mallard duck.. with a bottle of Chablis,..a demi-tasse and a cigar. **1961** M. Beadle *These Ruins are Inhabited* (1963) v. 71 Coffee.. is served demitasse. **1966** *Daily Tel.* 18 Jan. 13/3 Add 2 gallons of hot, strong coffee and serve into demi-tasse cups.

‖ **demi-vierge** (dəmiˌvʸɛ̄rʒ). Also **demi-virgin.** [Fr. (*Les demi-vierges*, title of novel by M. Prévost, 1894).] A woman (esp. a young woman) of doubtful reputation or suspected unchastity, who is not a virgin except in the strict physiological sense of the word (Virgin 2). Also *fig.*

1908 *Westm. Gaz.* 14 Nov. 6/2 Scandal-mongering elders, vain, selfish, dissipated officers, and equally vain, selfish young ladies approximating closely to the demi-vierge. **1928** D. H. Lawrence *Lady Chatterley* ii. 17, I hope, Connie, you won't let circumstances force you into being a demi-vierge. **1937** *Times Lit. Suppl.* 27 Nov. 910/1 A more spiteful cat of a *demivierge* heroine has seldom adorned a novel. **1951** Koestler *Age of Longing* i. viii. 132 We call demi-vierges a certain category of intellectuals who flirt with revolution and violence, while trying to remain chaste liberals at the same time. **1953** R. Chandler *Long Good-Bye* xiii. 84 A couple of streamlined demi-virgins went by.

demnition (demni·ʃən). Chiefly *U.S.* Euphemistic pronunciation of Damnation 3 b.

1839 [see *bow-wow 3]. **1840** Poe *Business Man* in *Wks.* (1865) IV. 333 A democratic rabble is *so* obtrusive, and so full of demnition mischievous little boys. **1888** *Weekly Exam.* (San Francisco) 22 Mar. (Farmer), It was demnition hot.

Demo. Colloq. U.S. abbreviation of Democrat 2.

1793 J. Steele in *N.C. Hist. Comm. Publ.* (1924) XIII. 108 Mr. Smiley—a man who was very Popular in the State assembly, he is a great Demo(crat). **1804** Fessenden *Orig. Poems* 57 In vain each demo spouts and billows. **1805** —— *Democr. Unveiled* I. 117 Nothing did demos any good But syllogisms made of wood. **1948** Chesterton (Ind.) *Tribune* 28 Oct. 12/3 The program chairman kept peace between GOP and Demos.

demo (de·mo). Colloq. abbreviation of Demonstration; so **demo-disc** (see quot. 1963).

1936 J. Curtis *Gilt Kid* iii. 28 The anti-war demo last week. **1940** *New Statesman* 16 Mar. 372 Starting as a Fundamentalist he drifts steadily through the modern heresies until he lands up talking of Moses as a militant Unionist, the Exodus as a mass demonstration, and the events in Jerusalem as another demo. **1949** A. Wilson *Wrong Set* 100 Norman's out at the demo. At Trafalgar Square. **1961** *Guardian* 15 Sept. 10/6 She was fined £1 for obstruction in an anti-nuclear 'demo' this spring. **1963** *Time & Tide* 11 Sept. 8/3 One other method of becoming known is to send an agent a 'demodisc'—a gramophone record on which the singer or musician demonstrates his ability. **1967** *Listener* 3 Aug. 130/3 What had begun as a demo..came to look as fey as any folk festival. **1968** *Ibid.* 28 Mar. 410/1 The US Embassy provided a focus for an extreme Left 'demo' against 'fascist brutality' in Vietnam. **1970** *Daily Tel.* 19 Nov. 7/2 Only the other day she was a dedicated demo-girl, but now she has settled down to exemplary bourgeois domesticity. **1971** *Sunday Times* 21 Feb. 3/1 British Rail are running 31 'Demo special' trains today to bring an estimated 30,000 people to London to take part in the trade unions' protest demonstrations against the Government's Industrial Relations Bill.

demob (dīmǫ·b), *sb.* and *v.* Colloq. abbrev. of Demobilization and of Demobilize *v.* Also *attrib.*, as in *demob suit*, a suit issued to a soldier upon demobilization.

1920 *Glasgow Herald* 2 June 11 Some young soldiers.. who had been recently demobbed. **1921** *Ibid.* 5 Feb. 7 The unemployed demobbed men. **1922** W. J. Locke *Tale of Triona* v. 56 The impecunious demobbed. **1923** T. S. Eliot *Waste Land* ii. 11 When Lil's husband got demobbed, I said ..Now Albert's coming back, make yourself a bit smart. **1934** M. Weseen *Dict. Amer. Slang*, *Demob*, to demobilize; demobilization. **1945** *News Chron.* 17 May, No more Bevin boys will be directed to the pits, but those already there will have to remain until the time comes for them to be demobbed according to age and length of service, just as if they were soldiers. **1945** *Daily Mirror* 11 Aug. 3 (*headline*) It's 'total' demob. now—Many home by year's end. **1949** B. Boland in J. C. Trewin *Plays of Yr. 1948–49* 58, I know one of the places where I don't want to spend my demob. leave. **1949** G. Cotterell *Randle in Springtime* 282 The mockery of the black Homburgs and umbrellas, the demob suits, whose patterns one often recognised [etc.]. **1959** 'B. Mather' *Achilles Affair* 101 Jo had not long been demobbed from the Wrens.

democratizer. (Examples.)

1882 W. James *Will to Believe* (1897) 90 That element in reality..the rough, harsh, sea-wave, north-wind element, the democratizer—is banished. **1951** D. Riesman in *Amer. Scholar* XX. 268 Freud..used original sin as a democratizer of men.

‖ **démodé** (demōᵘ·de, ‖demode), *a.* [Fr., = Demoded.] Out of fashion.

1873 *Young Englishwoman* Apr. 199/2 Light prints, when démodée, make very economical jupons for summer wear. **1896** *Westm. Gaz.* 7 Mar. 3/2 A chance of reincarnation for some demodé white satin bodice. **1900** *Daily News* 16 Oct. 6/6 There is fashion in art, as in everything else, and the démodé painter soon passes into an obscurity which he does not deserve. **1928** *Observer* 26 Feb. 15/5 Ibsen, demodé as he is, seems to have the quality of the mountain that does not move. **1930** *Time & Tide* 4 Apr. 451 This does not mean that Mr. Birrell's mind is *demodé*. **1945** R. Hargreaves *Enemy at Gate* 61 Alarmed and incensed at the industry with which James [II] had refurbished the *démodé* mantle of 'divine right', [etc.].

demodulation (dīmǫdiulē¹·ʃən). *Electr.* [f. De- II. 3 + *Modulation.] The process of obtaining from a modulated wave or voltage a signal which has the same form as the signal originally used to modulate it; the recovery of modulation; cf. *detection (sense 3). Hence **demo·dulator**, a device or circuit used to effect demodulation.

1921 *Electrician* 15 Apr. 452/1 By the term 'demodulation' is meant the process of reproducing the original low frequency modulating wave from the carrier wave upon which it has been impressed. *Ibid.*, We may determine.. the components of voice frequency in the output circuit of the demodulator by considering the interaction of the various..frequencies. **1929** T. E. Shea *Transmission Networks & Wave Filters* I. i. 20 In radio systems, a 'carrier frequency' and 'two sidebands' are produced. For demodulation and recognition of signals the carrier frequency and one sideband may be suppressed. **1942** *Electronic Engin.* XIV. 728 If..it is desired to reduce the frequency modulation percentage of the broadcast signal..this can conveniently be done by passing a portion of the signal.. through a demodulator. **1962** Simpson & Richards *Junction Transistors* xviii. 460 The removal of the modulation from a carrier or i.f. signal is called demodulation or detection. **1963** G. Troup *Masers & Lasers* (ed. 2) ix. 162 Demodulation of microwave-modulated light has been achieved using travelling wave tubes.

demogra·phical, *a.* = Demographic. Also **demogra·phically** *adv.*

1902 *Daily Chron.* 7 Nov. 4/2 The application of what we might almost call demographical method to departmental studies of the French Revolution. **1904** *Biometrika* Jan. 100 Buénos-Ayres is a town which is altering demographically in two very sensible ways. **1952** C. P. Blacker *Eugenics* 175 Demographically enlightened people in Asiatic countries are appreciating the need for active measures to reduce fertility. **1965** *Math. in Biol. & Med.* (*Med. Res. Council*) II. 52 Since the Council of Trent (1545–1563), Catholic parish priests have been directed and required to register the demographical events taking place in the parishes under their jurisdiction. **1970** *Sci. Jrnl.* Apr. 84/2 Not that he neglects the demographical and social consequences of contraception.

demoiselle. Add: **2. c.** = *damsel-fish* (*damsel III. 6).

1884 G. B. Goode *Nat. Hist. Aquatic Anim.* 275 The Demoiselle and the Cichlid Families. **1926** C. W. Beebe *Arcturus Adv.* xii. 315 Out from this very coral rock in its path there would shoot a diminutive demoiselle, fins erect in righteous wrath. **1931** J. R. Norman *Hist. Fishes* xi. 208 A little coral-reef fish (*Amphiprion*) belonging to the family of Desmoiselles (*Pomacentridae*) has a ground colour of vivid orange. **1967** *N.Z. Listener* 29 Dec. 5/1 At a depth of 160 feet, diver Jeff Pearch is surrounded by a shoal of demoiselles as he swims behind the branches of a long-armed red sponge. **1968** J. E. Randall *Caribbean Reef Fishes* 189 The damselfishes (or demoiselles) are small reef fishes which are often very colorful.

demolition. Add: **3.** *attrib.*

1936 Mencken *Amer. Lang.* (ed. 4) vi. 289 In an Atlanta department-store the *News-Record* found..*demolition-engineers* who were once content to be house-wreckers. **1959** J. Austwick *Murder in Borough Library* iii. 21, I thought the houses in Clough Street were closed under a demolition order. **1971** A. Clarke *Mind to Murder* xii. 179, I went first to the demolition site, told the foreman I'd got to look for a room.

demon. Add: **2. d.** *fig.*, of an alcoholic drink. Also *attrib.*

1884 in *Africana Notes & News* (1961) 295 A good many of them would have..made their mark in the musical history of this country, had it not been for the demon—drink. **1887** [see Metheglin]. **1922** Joyce *Ulysses* 348 Had her father only avoided the clutches of the demon drink. **1936** Mencken *Amer. Lang.* (ed. 4) vi. 244 An Englishman.. never uses *rum* in the generic sense that it has acquired in the United States, and knows nothing of *rum-hounds*,..the *rum-trade*, and the *rum-evil*, or of the *Demon Rum*. **1948** Partridge *Dict. Forces' Slang* 54 *Demon vino*, Italian wine of the cheaper sort.

e. Applied to a being of superhuman or 'diabolical' energy, skill, etc. (cf. 3 a *spec.*); also to an action, etc.

1876 *Coursing Calendar* 21 A demon of a hare got up for Rose and Bar Girl. *Ibid.* 315 It was hard lines indeed for Mr. Watson to meet with such a demon of a hare for the decider. **1899** *Westm. Gaz.* 6 Feb. 3/1 He is a demon of accuracy. **1961** *Times* 4 July 11/4 'Demon' services were, by custom, reserved for male opponents.

f. *Cards.* (Also *Demon Patience.*) A simple, one-pack patience game, which rarely comes out. Also *Racing Demon*, Demon adapted for several players, each with his own pack, but played in competition.

1893 M. Whitmore Jones *Games of Patience* 3rd Ser. ix. 19 Demon Patience. **1900** 'L. Hoffmann' *Patience Games* 36 The Demon. **1918** H. G. Wells *Joan & Peter* xi. 388 A new card game, Demon Patience, a scrambling sort of game in which you piled on aces in the middle. **1919** K. Mansfield *Lett.* (1928) I. 245 You know how, when we get hungry, we are at last even unable to play Demon for wanting the hash-hammer to sound. **1936** 'P. Quentin' *Puzzle for Fools* xii. 95, I began to wonder whether she stole cards from a concealed pack...Her demon came out three times running. **1948** G. Greene *Heart of Matter* I. iii. i. 84 I've never played cards—except demon..and that's a patience.

3. a. *demon king* (examples); *demon star*, Algol [Arab., the demon: see Ghoul], the star β Persei. **d.** Instrumental, etc., as *demon-infested, -scooped, -stricken*.

1933 W. de la Mare *Fleeting* 169 Demon-infested rank morass. **1904** M. Beerbohm *Around Theatres* (1953) II. 59 In the 'seventies pantomime was flourishing still. Demon King and Fairy Queen..were familiar. **1940** 'N. Blake' *Whisper in Gloom* ii. 21. 148 Alec Gray is..a cheap snake, a proper young Demon King. **1942** R. Campbell *Flaming Terrapin* iii. 41 Sleep was a long dark tunnel demon-scooped Out of the Night's black rock. **1895** *Funk's Standard Dict.*, Demon star. **1928** *Daily Chron.* 1 Sept. 7/3 Algol, the Demon Star. **1936** *Discovery* June 187/2 The newly-made spirit-doctors proceed to exorcise the pepo..from the scores of demon-stricken people.

demon² (dī·mən). *Austral. slang.* [Appar. ad. *Diemen* (Van Diemen's Land, early name for Tasmania).]

1. A policeman.

1889 Barrère & Leland *Dict. Slang* I. 304/1 Demons (Australian), prison slang for police. 'The *demons* put pincher on me', I was apprehended. **1933** *Bulletin* (Sydney) 15 Nov. 38/1 Those 'demons', the fools, were chasing the wrong people, as usual. **1945** Baker *Austral. Lang.* v. 95 It was a natural development that police troopers should come to be called joes..although this use is not found often, demons and traps (1855) being more widely used. **1967** K. Giles *Death & Mr. Prettyman* ii. 61 'Tell the truth, Bert,' said the Australian, 'always help a demon in distress.'

2. (See quots.)

1909 J. R. Ware *Passing English* 107/1 *Demons*, old hands at bushranging: derived from men who arrive from Van Dieman's [*sic*] Land (Tasmania), some of whom are popularly supposed to have inaugurated bushranging in Australia. **1945** Baker *Austral. Lang.* ii. 42 A large number of synonyms for convict became current [in the early part of the nineteenth century], among them canary, transport, ..demon (a Van Diemen's Land convict), [etc.]. *Ibid.* 51 The following synonyms for bushrangers might be noted: rangers, white Indians and stickers-up. Demon was formerly used for an old hand at the game.

3. A detective.

1926 J. Doone *Timely Tips for New Austral.*, *Demon*, a slang word meaning a 'detective'. **1941** K. Tennant *Battlers* viii. 66 The showers were 'demons', or plain-clothes detectives. **1967** *Sunday Mail Mag.* (Brisbane) 23 July 5/3 To the Australian criminal a demon is a..detective.

demonically (dīmǫ·nikăli), *adv.* [f. Demonical *a.* + -ly².] In a manner befitting a demon; superhumanly.

1926 'L. Malet' *Dogs of Want* v. 124 He danced divinely. Perhaps it were more accurate to say demonically. **1962** I. Murdoch *Unoff. Rose* xviii. 270 Keeping pace demonically with her love for Felix.

demonish, *a.* Add: (Later examples; occurs commonly in D. H. Lawrence's works.) So **de·monishness**, devilishness.

1926 D. H. Lawrence *Plumed Serpent* iii. 55 A profound unbelief that was fatal and demonish. **1927** —— *Mornings in Mexico* 12 The penetrating, demonish voice. a **1930** —— *Pornogr. & So On* (1936) 85 Renoir didn't try to get away from the body. But he had to..rob it of..its natural demonishness.

demonomaniac (dīːmŏnomē¹·niæk). [f. Demonomania: see -ac.] One who believes himself to be possessed by a devil.

1891 tr. C. Lombroso's *Man of Genius* iii. i. 173 For the demonomaniacs of a hundred years ago..are now substituted the modern paranoiacs. **1930** Chesterton *New Jerus.* ix. 177, I do not say that psychologists admit the discovery of demoniacs; and if they did they would doubtless call them something else, such as demonomaniacs.

demonstration. Add: **8.** *attrib.*, chiefly in sense 5.

1883 [see sense 5 in Dict.]. **1899** *Westm. Gaz.* 16 June 4/1 The demonstration farm of the Northumberland County Council at Cockle Park. **1902** E. Lanteri *Modelling* i. The notes of which I made use for my demonstration-classes at the Royal College of Art. **1902** *Brit. Forestry: Dept. Cttee. Rep.* 9 in *Parl. Papers* (Cd. 1319) XX, That two areas for practical demonstration be acquired...We suggest that the Alice Holt Woods in Hampshire be made available as soon as possible to serve as a Demonstration Area. **1907** *Westm. Gaz.* 22 May 12/2 The Board of Education recognises gardening as a school subject...To each school is allotted a demonstration plot, which is used by the teacher for object-lessons. **1909** J. J. Findlay (*title*) The Demonstration Schools Record. **1943** J. S. Huxley *TVA* xx. 52 The TVA..set up five demonstration parks in the Valley. **1961** W. Vaughan-Thomas *Anzio* viii. 156 Hitler sent from Germany a special regiment, the Infantry Lehr Regiment, a demonstration unit composed of approved Nazis. **1970** J. Holt *Underachieving School* 186 It may very well be that in such schools we will have the kind of educational laboratory and demonstration centers which in your country, I guess, are supplied by many of the State schools.

demoralizingly (dīmǫ·rălaiziŋli), *adv.* [f. Demoralizing *ppl. a.* + -ly².] In a demoralizing manner; to a demoralizing degree.

1926 W. J. Locke *Old Bridge* iv. xii, It seemed..so demoralizingly vicious to drink a friend's cocktails or whisky behind closed doors.

De Morgan's laws (dəˌmǭ·gən). *Logic* and *Math.* [Named after the English mathematician Augustus *De Morgan* (1806–71), but known to logicians in the Middle Ages.] Two laws of the propositional calculus, viz. that the negation of a conjunction is logically equivalent to the alternation of the negations of the conjoined expressions, and that the negation of an alternation is logically equivalent to the conjunction of the negations of the alternated expressions; also, the analogous truths in the algebra of classes. Symbolically, $\sim(p.q) \equiv \sim p \vee \sim q$ and $\sim(p \vee q) \equiv \sim p . \sim q$. Also **De Morgan's theorem(s)**; **De Morgan** *absol.*

1918 C. I. Lewis *Survey Symbolic Logic* ii. 125. 3·4 and 3·41 together state De Morgan's Theorem. **1932** —— & H. Langford *Symb. Logic* ii. 33 These always follow from their correlates by some use of De Morgan's Theorem. **1950** W. V. Quine *Methods of Logic* (1952) i. 50. 53 De Morgan's laws are useful in enabling us to avoid negating conjunctions and alternations. **1957** P. Suppes *Introd. Logic* ix. 209 Equations (23) and (24) are De Morgan's laws. **1965** P. Caws *Philos. Sci.* xi. 68 What is called by De Morgan is seen to be equivalent.

Demosthenean, -ian, *a.* (Earlier examples.)

a **1739** C. Jarvis tr. *Cervantes's Don Quixote* (1742) II. ii. xv. 169 An enterprize worthy to employ..Ciceronian and Demosthenian eloquence. **1807** J. Mackintosh *Jrnl.* 15 Jan. in *Life* (1835) I. 323 He [*sc.* Fox] was the most Demosthenean speaker since Demosthenes. **1834** *Deb. Congress U.S.* 10 Mar. 843 This Demosthenian pouring-out of the shreds and patches of old Grecian orations.

Demosthenic, *a.* (Earlier example.)

1821 Shelley *Let.* 22 Oct. (1964) II. 360, I congratulate you on your demosthenic energy.

demote (dīmōᵘ·t), *v.* orig. *U.S.* [f. De- + Promote.] *trans.* To reduce to a lower rank or class. Hence **demo·tion.**

1893 in *Funk's Stand. Dict.* 489/2 The school children in Senator Wilson's district of Iowa 'use the word *demote* as an antithesis of *promote*.' **1901** *Smithsonian Rep.* (1902) 75 Promotion and demotion, i.e., advancement in 'age' (rank) by common consent in recognition of prowess, etc., with corresponding reduction in 'age' as the penalty for cowardice. **1919** *Daily Mail* 7 Oct. 5/4 Major-General Blank 'demoted' to be 'demoted' to brigadier. **1929** R. A. Fisher *Genet. Theory Nat. Selection* 226 The agencies controlling promotion or demotion [etc.]. **1946** C. Fry *Phœnix too Frequent* (1949) 40 Demoted first and then hanged! **1954** Koestler *Invisible Writing* 60 Every demotion, each step down the slopes of the pyramid, is final and irrevocable. **1955** *Times Lit. Suppl.* 11 Nov. 673/1 Montesquieu has been demoted to the humble status of a rather exaggerated figure of fun. **1957** *Listener* 28 Nov. 881/1 Premiers at the Great Central Hotel, Marylebone, now, alas, demoted to railway offices. **1958** *Economist* 2 Nov. 676/1 The demotion of the Secretary from the Ministry of Internal Affairs in Ghana to the Ministry of Communications in the course of a cabinet reshuffle. **1968** A. Cowan *Fortunately in England* 244 In any case, our sister Hall are prefects we can't very well demote them.

demotic, *a.* Add: **1. b.** Of or belonging to the popular written or spoken form of modern Greek. Also as *sb.*

1927 H. Nicolson *Some People* iii. 76 And from the bridge the second officer shouts demotic to the Company's agent. *Ibid.* vii. 179 He slowly approached Essad Pasha and spoke to him in demotic Greek. **1946** R. Castle *Simonmania* ii. 53 There are the moderate advocates of Demotic, and the advanced who are all for dropping more and more inflections. **1964** *Language* XL. 274 Here we encounter Demotic forms, which vary with normal demotic. **1964** J. T. Pring in D. Abercrombie et al. *Daniel Jones* 357 Now it is the partisans of demotic who are to the fore; and demotic has become the language of primary education.

2. Delete 'Somewhat *rare*' and add further examples. Also as *sb.*

1925 T. S. Eliot *Waste Land* 212 Mr Eugenides, the Smyrna merchant..Asked me in demotic French To luncheon at the Cannon Street Hotel. **1958** *Times Lit. Suppl.* 17 Jan. 30/1 Mr Amis's attitudes as a writer are not so demotic as his choice of terms seems to pretend. There is a suggestion of linguistic slumming, tweedy slanginess. **1961** *Listener* 30 Mar. 573/3 In fact advertising has, of necessity, to use simple, forceful, easily understandable words—'demotic' language, as Mr Whitehead might call it. **1967** R. Mabey *Class* 110 Not only did ITV develop the popular touch—it nurtured those diverse speech tones and accents which belonged naturally to the more demotic channel. **1968** A. Cowan *Fortunately in England* 162 An occasional class of boys would do their best to add a gaiety of their own by singing unusual and demotic versions of such songs as Men of Harlech. **1969** R. Heppenstall *Portrait of Artist as Professional Man* 101 To a north countryman of demotic antecedents like myself, Louis's background, manner and social attitudes were characteristically southern English upper-middle class. **1970** C. P. Snow *Last Things* 7 At that stage, he had a knack of speaking what he thought of as American demotic.

demoticist (dīmǫ·tisist). [f. Demotic *a.* 1 + -ist.] A student of demotic script. Also **demo·tist.**

1902 *Encycl. Brit.* XXVII. 726/2 Though demotic has not yet received serious attention at Berlin, the influence of that great school has made itself felt amongst demotists, especially in Germany, America, and England. **1913** *Year's Work Class. Stud.* 188 The use made in the syntax of the writings of the modern demoticists is quite fresh and very interesting. **1936** *Trans. Philol. Soc.* 50 It is very difficult to get on without them [*sc.* phrases from the purified language], as is admitted even by many of the present-day demoticists. **1964** J. T. Pring in D. Abercrombie et al. *Daniel Jones* 358 Most of the advocates of revised orthography have been demoticists.

demount, *v.* Add: **3.** *trans.* To remove from a mounting or place of support; to dismantle.

1958 in Webster. **1958** *New Scientist* 24 Apr. 29/3 [The inflamed hydrogen] carried off the singularly impressive eyebrows of a scientist demounting the autoclave.

demou·ntable, *a.* [f. De- II. 3 + Mount *sb.*² + -able; cf. F. *démontable* (1870 in Robert).] That can be taken from its mount (see Mount *sb.*² 9) or setting; that can be dismantled. Also as *sb.* Hence **demountabi·lity**, the property of being demountable.

1909 W. Lockwood *Vehicles of Air* xiv. 467 *Demountable*, said of a mechanism designed with special provision for ready taking apart and reassembling. **1926** R. Fraser *How it Flies* xx. 458 *Demountable*, a type of construction which permits a machine to be easily taken apart for transportation. **1928** Webster *Add.*, *Demountable*, said of a form of rim, for an automobile wheel, which with its tire can be removed from the wheel. **1930** *Engineering* 21 Feb. 248/1 A new form of a demountable Timken bearing. **1934** Webster, Demountability. **1941** *Nature* 19 Apr. 481/1 Irradiation was carried out by means of a demountable X-ray tube. **1943** J. S. Huxley *TVA* x. 75 The demountable house..is mass-produced in sections, which are then transported to the destination on special trailers, and erected on pre-cast concrete foundations built on the site. **1964** *Archit. Rev.* XCVI. 23/2 [These houses] were built of various types of prefabricated panels; but the degree of prefabrication varied from..demountable.. **1971** *Nature* 4 June p. ix (Advt.), Demountable assemblies with photomultipliers...Integral assemblies for highest resolution.

Dems (demz). Abbrev. of *Defensively Equipped Merchant Ship*; a gunner on such a ship.

1943 *Hutchinson's Pict. Hist. War* 27 Oct.–11 Apr. 266 Defensively equipped merchant ship gunners ('Dems') taking a refresher course. A high-angle 4-in. gun being demonstrated to a 'Dems' class.

demyelination (dīmaiːĕlinē¹·ʃən). [f. De- II. 3 + Myelination.] The removal or destruction of the myelin of nerve tissue. Also **demy·elinating** *ppl. a.*; **demyeliniza·tion.**

1932 Donald & A. Miller *Med. Dict.* (ed. 16) 362/1 *Demyelination, demyelinization*, destruction or removal of myelin. **1934** *Nature* 1 Dec. 851/1 The nerve degeneration, a demyelination of the fibres of the posterior roots, which is responsible for the incoordinated movements in rickets, is due to a deficiency of vitamin A in the diet. **1961** *Lancet* 9 Sept. 610/1 The Council are to set up a research group on demyelinating diseases. **1970** *Sci. Jrnl.* Apr. 38/2 Demyelination is caused by the injection in the crude [rabies] vaccines of a small basic protein derived from brain tissue and known to cause demyelinating encephalitis in experimental animals.

demystify (dīmi·stifai), *v.* [f. De- II. 1 + Mystery + -fy.] *trans.* To clarify (obscure, clouded, or irrational beliefs); to reduce or remove bewilderment, irrationality, etc., in (a person).

1963 *Times Lit. Suppl.* 11 Jan. 19/2 And so he has devoted himself to 'demystifying politics'. **1967** D. Cooper *Psychiatry & Anti-Psychiatry* v. 101 [The subject(s)] are demystified in terms of their phantastic hopes of gratification from the therapist as a parent-figure.

Hence **demystifica·tion**, the action of demystifying.

1964 *Economist* 16 May 734/2 An attempt at demystification and therefore of human liberation. **1967** D. Cooper *Psychiatry & Anti-Psychiatry* v. 101 [illegible] **1969** *N.Y. Rev. Books* 16 Jan. 16/3 Aron talks again and again of demystification [illegible]

demyth (dīmi·þ), *v.* [f. De- II. 1 + Myth *sb.*] Used occasionally as a synonym of *demythologize *v.*

1960 J. Macquarrie *Scope of Demythologizing* vii. 214 Some English writers on demythologizing..have adopted the habit of talking rather slickly of 'demything'. Apart from the fact that this word is a more than usual inelegant barbarism, it also betrays a fundamental misunderstanding of Bultmann's intention. If 'demything' meant anything, it would mean 'the elimination of myth'...'Demythologizing' means 'the elimination of mythology', and this is not the same as the elimination of myth. **1961** *Times Lit. Suppl.* 14 Apr. 236/3 All three books agree that some measure of demything (if a less barbaric rendering of the German than 'demythologizing' is permissible) of the world-view and other assumptions of New Testament writers is necessary.

demythicize (dīmi·þisaiz), *v.* [f. De- II. 1 + Mythicize *v.*] To remove the attribution of a mythical character to (a legend, etc.); = *demythologize *v.* Hence **demy:thiciza·tion.**

1951 K. Grobel in *Jrnl. Bibl. Lit.* 101 One may have to reject Bultmann's own attempt to demythicize. **1953** R. H. Fuller tr. *Bartsch's Kerygma & Myth* I. 224 Other remedies might easily be found..more efficacious than the demythicization of the Gospel. **1967** *Religious Studies* II. 171 It is especially in Greek culture that myth was submitted to a long and penetrating analysis, from which it emerged radically 'de-mythicized'.

demythologize (dīmiþǫ·lŏdʒaiz), *v. trans.* and *intr.* [f. De- II. 1 + Mythologize *v.*] To remove the mythical elements (from a legend, cult, etc.); *spec.* *Theology* (cf. G. *entmythologisierung* and quot. 1953): to reinterpret the mythological elements in the Bible. Hence **demytho·logizing** *vbl. sb.* and *ppl. a.*; **demythologiza·tion.**

1950 *Scot. Jrnl. Theol.* III. 39 What it means..to be initiated into the Kingdom of Heaven, and to have them as 'de-mythologised'. **1951** F. V. Filson tr. *Cullmann's Christ & Time* 13 A framework, of which we must strip the account in order to get at the kernel ('de-mythologizing' or 'myth-removal'). **1953** *Radio Times* 23 Jan. 19/1 A new word has entered into New Testament criticism since Bultmann began his exegesis: *Entmythologisierung*, best translated 'demythologizing'. Bultmann claims that there are many passages in the Bible, especially those about God and his dealings with man, that have the form of mythological expressions; and if they are to be intelligible in the twentieth century they must be reinterpreted. **1955** J. Macquarrie *Existentialist Theol.* viii. 176 A demythologized Bible..would require to use the same objective symbolic language, and because that language would be symbolic, it would be more accurate to speak of transmythologization than of demythologization. **1959** *Observer* 18 Jan. 20/7 Recently, there have been attempts to demythologise this highly popular saint [*sc.* St. Thérèse of Lisieux]. **1961** *Listener* 11 Jan. 68/1 Bultmann unflinchingly demythologizes, until he reaches that inner heart of Christian belief which he identifies with the *Kerygma*, the gospel truth. **1969** R. Bambrough *Reason, Truth & God* iii. 37 There is no incompatibility between a reformed, demythologized Homeric polytheism, a refurbished, demythologized Islam, and a refurbished, demythologized Christianity. **1971** *Nature* 12 Feb. 460/1 Thus, for Roszak, science, which began as a demythologizing force, by its own logic creates new and darker mysteries.

den, *sb.*¹ **4.** (Examples.)

1901 R. C. Maclagan *Games of Argyle* 22 Equal sides being chosen, a 'den'..sufficiently large to contain the whole of the side who are 'in' is fixed. **1959** I. & P. Opie *Lore & Lang. of Schoolchildren* viii. 150 *Den* or *den-y*. Occasionally used as truce term, from the 'den' or sanctuary of certain catching games. **1968** *Proc. Leeds Philos. Soc., Lit. & Hist. Section* XIII. ii. 56 There is the *den*, in which players line up to take their turn in striking.

den, *v.*¹ **2.** *to den up* (earlier and later examples).

1843 *American Pioneer* II. 171 In that climate [*sc.* of Canada] the bears usually den up in the winter, and lie in something of a torpid state. **1918** *Chambers's Jrnl.* Mar. 187/1 The brown bear usually 'dens up' early in the season.

denatality (dīnǣte·lĭti). [ad. F. *dénatalité*, f. De- II. 3 + Natality.] A declining birth-rate.

1939 *Tablet* 26 Aug. 269/2 The Daladier Government has begun to concern itself with this problem of denatality, which is far more serious even than that of the national defence. **1941** Koestler *Scum of Earth* ix. 98 Laziness, selfishness, alcoholism, denatality are supposed to be the clinical symptoms of decline.

denatant (dīnē¹·tănt), *a.* [f. De- + Natant *a.*] Of fishes: swimming with the current. Hence **denata·tion**, the movement of fishes in the direction of the current.

1915 Denatant, -ation [see *contranatant *a.*]. **1920** *Rep. Dove Mar. Lab. Cullercoats* 38 A consideration of our herring shoals from the point of view of denatation. **1959** [see *contranatant *a.*]. **1968** F. R. H. Jones *Fish Migration* ii. 15 Denatant means swimming, or drifting, or migrating with the current.

denationalism (dīnæ·ʃənăliz'm). [f. De-nationalize: see -ism.] The loss or deprivation of national character. Also **denationalist**, one who would deprive a people of its national identity.

1916 Stanford & Forsyth *Hist. Music* xvi. 305 The nationalists and the denationalists. **1962** *Guardian* 9 Mar. 9 International control of the Rhineland and the Saar meant denationalism of the people.

denationalization. Add: **2.** The action of removing (an industry, etc.) from national control and returning it to private ownership.

1921 *Official Index to Times* Mar.–June 479/1 Russia...Textile industries—denationalization recommended. **1959** *Daily Tel.* 13 Mar. 13/4 The 13 steel companies still awaiting denationalisation. **1971** *Observer* 14 Mar. 29/3 A policy of, say, nationalisation or denationalisation of an individual industry.

denationalize, *v.* Add: **2. b.** To transfer (an industry, etc.) from national to private ownership. Also *intr.*

1921 *Times* 17 Jan. 10/2 The object of the..agitation is not to improve the [telephone] service, but to get it de-nationalized, 'to get it handed over to private capitalists'. **1947** *Observer* 21 Dec. 5/4 It was no part of the Conservative programme to denationalise coal. **1965** *New Statesman* 7 May 714/1 All this fuss about steel would have been quite unnecessary if the Churchill cabinet had stood firm against backbench pressure to denationalise in 1951. **1970** C. N. Parkinson *Law of Delay* 62 The Conservatives tried timidly to go into reverse over steel and road haulage. But a process by which those or other industries should be nationalised and denationalised is not technically possible.

denaturalizer (dīnæ·tiŭrălaiːzəɹ, -tʃər-). [f. Denaturalize *v.* + -er¹.] One who or that which denaturalizes.

1832 *Crisis* 19 May 31/3 Man..the great denaturalizer of other animals, has exerted his deforming powers most remarkably..upon himself. **1905** *Westm. Gaz.* 28 Mar. 4/2 That ideal 'de-naturaliser' which would..solve the whole problem.

denaturant (dīnē¹·tiŭrănt, -tʃər-). Also **-ent.** [f. Denature *v.* + -ant¹.] A substance added to alcohol or other substances as a denaturing agent. (Cf. Denature *v.* 2.)

1905 *Westm. Gaz.* 4 Apr. 5/2 Alcohol similar in purity to methylated spirits before the denaturent is added. **1906** *Daily Chron.* 15 Sept. 3/4 The problem of a cheap, available and efficient denaturant. **1920** *Auto* 4 Mar. 248/2 A real denaturant, which will leave the fuel effective for its purpose, but destroy its qualities as a substitute for drinkable spirit. **1922** *Chambers's Jrnl.* 624/2 Alcohol of 95 per cent. purity was used without denaturents. **1946** *Rep. Internat. Control Atomic Energy* II. v. 23 It is possible, both for U235 and for plutonium, to remove the denaturant.

denaturate (dīnē¹·tiŭrē¹t, -tʃər-), *v.* [f. De- + Nature + -ate³.] = Denature *v.* 2. So **dena·turated** *ppl. a.*

1895 G. Lunge *Sulphuric Acid* (ed. 2) II. 14 Decomposing the salt for saltcake..is avoided by 'denaturating' the salt under official supervision. **1904** *Daily Chron.* 4 Aug. 3/5 A proper denaturating medium is not known. **1911** *Jrnl. Physiol.* XLIII. 15 SO_4 assists the agglutination of denaturated egg-albumen much more powerfully than Cl.

denaturation (dīnē¹:tiŭrē¹·ʃən, -tʃər-). [f. Denatur(e *v.* + -ation.] The action of denaturing; *spec.* in *Biochem.* (see quot. 1965 and *denature *v.* 2 b).

1882 [see Denature *v.* 2]. **1911** *Jrnl. Physiol.* XLIII. 25 Complete 'heat coagulation' of proteins..consists of two processes: (1) the union of the protein and hot water ('denaturation'), (2) subsequent agglutination and separation of the product. **1946** *Nature* 13 July 58/1 It has not been found possible to obtain more than an insignificant amount of the protein of apple-fruits in a soluble form except by treatments so drastic as to lead, inevitably, to some degradation as well as denaturation of the protein. **1956** *Ibid.* 10 Mar. 447/2 Recent studies on the reversible denaturation of proteins. **1965** M. Joly *Physico-chem. Approach to Denaturation of Proteins* 3 Denaturation is any modification of the secondary, tertiary or quaternary structure of the protein molecule, excluding any breaking of covalent bond.

denature, *v.* Add: **2.** (Later gen. examples.)

1907 *Chem. Abstr.* I. 2009 Denaturing acetic acid...The strength of acetic acid to be denatured is not specified. **1924** H. E. Fosdick *Mod. Use of Bible* v. 158 Unless it is willing to be denatured, religion cannot get on without this exciting aspect of its thought. **1951** E. Barker *Princ. Soc. & Pol. Theory* II. ii. 47 We only dream; and our dream is one which denatures the State and unspheres law. **1965** *Times Lit. Suppl.* 25 Nov. 1063/3 To omit the commas..is..to denature or homogenize the lines.

b. *Biochem.* To modify (a protein) by heat, acid, etc., so that it no longer has its original properties.

1925 *Jrnl. Biol. Chem.* LXIV. 370 When a protein is denatured by dilute acids and alkalies..an increase occurs in the acid- and base-binding powers. **1970** R. W. McGilvery *Biochem.* viii. 150 Some proteins, notably trypsin, are denatured in hot water, but return to their active form upon cooling.

Also **dena·turing** *ppl. a.* and *vbl. sb.*

1907 R. F. Herrick *Denatured or Industrial Alcohol* i. 2 A number of other substances are used as denaturing agents. *Ibid.* 3 Such denaturing was accomplished by mixing ten per cent of commercial wood alcohol with ordinary alcohol. **1971** *Sci. Amer.* June 48/2 The first step isolated the elastic fiber..by the use of a denaturing extractant such as a concentrated solution of guanidine and enzymes that specifically break down collagen.

denaturize (dīnē¹·tiŭraiz, -tʃər-), *v.* [f. De- + Nature + -ize.] = Denature *v.* 2. Hence **dena:turiza·tion**, denaturation.

1898 *Trans. Amer. Inst. Electr. Engin.* 139 (Cent. D. Suppl.). **1905** *Westm. Gaz.* 26 Jan. 2/2 The general idea is that if the alcohol be denaturised..it should be subject to no impost. **1969** *Physics Bull.* Jan. 8/2 Denaturization of key enzymes in the cellular structure.

denazify (dīnā·tsifai), *v.* [De- II + *Nazi + -fy.] *trans.* To (attempt to) detach (Nazis, or their adherents) from Nazi allegiance, or connection; also *transf.* Hence **denazifica·tion**, the detachment of Nazis from their allegiance; the removal of Nazis from official positions; also *attrib.*

1944 *Spectator* 28 July 72/2 Whenever the problem of.. de-Nazifying the young Nazis, is being seriously approached, some encounter or incident turns up to make the..hopelessness of the situation inescapable. **1944** *Sat. Rev. Lit.* 19 Aug. 10 His hope for a de-Nazification of Germany's killers. **1945** *Times* 16 July 3/4 The rapid progress which has been made in carrying out the allied denazification policies. **1955** *Ibid.* 26 July 7/2 Herr Gunter d'Alquen.. was ordered by a denazification court in an interim judgment to-day to pay 60,000 marks. **1957** *Economist* 28 Dec. 1136/1 The medals issued from 1939 to 1945 all carry the stigma of the swastika. They..must give place on their holders' breasts to denazified successors. **1958** *Times Lit. Suppl.* 30 May 292/2 Denazification was a failure and so was decartelization. **1970** *Daily Tel.* 21 Feb. 8/6 In this way the casual viewer, denied the information that the anti-Jewish remark was addressed to a Jew, while the anti-Nazi remark was written during the Nazi regime, is liable to come away with the impression that here is an anti-Semite attempting to denazify himself.

dendrite. Add: **3.** *Anat.* Any of one or more processes from a nerve cell which are typically short and extensively branched and which conduct impulses towards the cell body.

1893, 1899 [see *dendron]. **1900** W. S. Hall *Physiology* 535 The dendrites, or protoplasmic processes, resemble more closely in appearance the cell-body itself than does the axon. **1902** *Encycl. Brit.* XXV. 394/2 In the simplest cases the dendrites carry the sensory impulse to the nerve-cell. **1927** Haldane & Huxley *Anim. Biol.* i. 40 Shorter branched outgrowths (dendrites). **1962** *Lancet* 29 Dec. 1359/2 The cell bodies and dendrites of both cerebral and cerebellar cortex are indeed conspicuously rich in protein. **1967** [see *dendron].

dendritic, *a.* Add: **1.** Also in *Geogr.*

1898 I. C. Russell *River Devel.* vii. 204 The well-developed dendritic drainage in Lebanon valley. **1937** Wooldridge & Morgan *Phys. Basis Geogr.* xiv. 189 The ..surface will be resolved into slopes leading down to drainage channels. In areas of homogeneous rocks such a process gives rise to a simple, tree-like or dendritic, drainage plan.

dendrochronology (deːndroˌkronǫ·lŏdʒi). [f. Dendro- + Chronology.] The science of arranging events in the order of time by the comparative study of the annual growth rings in (ancient) timber. Hence **de:ndrochronolo·gical** *a.*, **de:ndrochronolo·gically** *adv.*, **de:ndrochrono·logist.**

1928 A. E. Douglass *Climatic Cycles & Tree Growth* II. 5 We are measuring the lapse of time by means of a slow-geared clock within the trees. For this study the name 'dendro-chronology' has been suggested, or 'tree-time'. This expression covers all the dating and historic problems ..as well as the study of cyclic variations and the distribution of climatic conditions. **1937** *Geogr. Jrnl.* LXXXIX. 407 A dendrochronologist is a man who investigates the climate of the past by measuring the rings of trees. **1937** *Proc. Prehist. Soc.* III. 321 The modern ways of pollen-analysis, geochronology and dendrochronology must be applied to settle geologically, once for all, the cultural correlations with Scotland. **1959** *Antiquity* XXXIII. 238 Undateable except by a rough and ready guesswork calculation from dendrochronological dates. **1969** *Nature* 15 Nov. 682/1 Radiocarbon measurements carried out on dendrochronologically dated wood samples.

dendrogram (de·ndrŏgræm). [f. Dendro- + -gram.] A branched diagram representing the apparent similarity or relationship between taxa, esp. on the basis of their observed overall similarity rather than on their phylogeny. Also in extended senses.

1953 E. Mayr et al. *Methods & Princ. Syst. Zool.* iii. 58 Figure 8..is not a phylogenetic tree, because it is not based on any information on fossil forms which might be ancestral connections of the various branches. Such a diagrammatic illustration of degree of relationship based on degree of similarity (morphological and otherwise) may be called a dendrogram. **1961** G. G. Simpson *Princ. Anim. Taxonomy* ii. 62 Opinions or claims regarding dendrograms vary between two extremes: that they have nothing whatever to do with phylogeny or that they do faithfully represent phylogeny. **1969** *Computers & Humanities* IV. 140 A matrix of correlation coefficients is computed and clustered and a dendrogram plotted.

dendrograph (de·ndrŏgrɑf). [cf. Dendrography.] An instrument used to measure the periodical variations in the size of tree trunks. Hence **de:ndrogra·phic** *a.*

1924 D. T. MacDougal (*title*) Growth in trees..dendrographic measurements. *Ibid.* 37 Standardization of the dendrograph for measurement of daily equalizing variations. ..The essential feature of the instrument is a rigid metal frame surrounding the trunk with one fixed contact-point, and the second consisting of the short bearing-arm of a lever, the long arm of which carries a pen which traces the record.

dendron (de·ndrǫn). *Anat.* [f. Dendr(ite + *-on* as in *neuron, axon*.] = *dendrite 3.

1893 E. A. Schäfer in *Brain* XVI. 136, I propose therefore to term the axis-cylinder or nerve-fibre processes *neurons*; and the protoplasmic processes *dendrons*. [*Note*] The latter have been known as 'dendrites', but the termination 'on' is..preferable to 'ite'. **1899** *Ibid.* XXII. 209 The protoplasmic processes are also referred to as the dendrites (His), dendrons (Schäfer) and dendritic processes. **1967** C. R. Noback *Human Nerv. Syst.* ii. 32/1 Neurons have two types of processes, viz., axons and dendrites (dendrons).

Deneb (de·neb). *Astr.* Also **Denab.** [ad. Arab. *ḏanab* (*ad-dajāja*) (hen's) tail.] The star α in the constellation Cygnus.

1867 Smyth *Sailor's Word-Bk.* 243 Deneb. **1933** *S.P.E. Tract* XXXVIII. 571 Denab (α *Cygni*). **1955** *Sci. News Let.* 30 July 74/1 Just below Vega, toward the east, is the figure of Cygnus, the swan. The most familiar stars in this group form a cross, and at the head, toward the northern horizon, is the bright Deneb.

dener, var. *deaner.

dene·rvate, *v.* *Med.* [f. De- II. 1 + Nervate *v.*] *trans.* To deprive an organ, tissue, etc. of its nerve supply. So **dene·rvated** *ppl. a.*; **denerva·tion²**, the act of denervating or the state of being denervated.

1905 *Jrnl. Physiol.* XXXIII. 395 Considerable caution must be exercised in comparing the response of the denervated muscle with that of the normal muscle. But the results of the experiments tend to show that the responsiveness to nicotine is increased by denervation. **1948** *New Biol.* IV. 123 The cutting of all nervous pathways to the mammary site... If under such denervated conditions the tissues were still observed to develop, [etc.]. **1962** *Lancet* 12 May 1024/2 Neurogenic hypertension, produced by denervation of the moderator nerves, required that both the nerves from the aortic arch and the carotid sinus be resected. **1963** R. P. Dales *Annelids* vi. 130 The locomotory rhythm of *Hirudo* is carried by the nerve cord and is not interrupted by denervating several segments part way down the body. **1968** *Brit. Med. Bull.* XXIV. 257/2 Electromyography is in routine clinical use for the detection of primary muscle disease and chronic partial denervation.

dengue. (Earlier U.S. examples.)

1828 *Charleston Courier* 15 July, The Dengue. This.. epidemic exists at this time in our city. *Ibid.* 9 Aug. **1830** *Amer. Beacon* (Norfolk, Va.) 9 Nov. 2/2 A rheumatic fever, pronounced by the physicians to be the celebrated Dengue of 1818, made its appearance.

Denia (dī·niă). The name of a town in south-eastern Spain, used *attrib.* and *ellipt.* to denote the products grown in its neighbourhood.

1845 Dodd *Brit. Manuf.* V. 102 The 'Lexias' produce a dry wine, the 'Denias' a sweet wine. **1904** *Westm. Gaz.* 12 Aug. 10/1 Quantities of new Denia grapes flood the market.

denial. Add: **6.** *Bridge.* A bid of another suit in order to show weakness in the suit bid by one's partner.

1916 R. F. Foster *Auction Bridge for All* xxi. 81 Instead of continuing his denial of the major suits, and trusting his partner to protect them..the dealer quit. **1927** G. Mott-Smith *Contract Bridge* viii. 117 A take-out may be a 'denial'. **1959** *Listener* 5 Mar. 434/3 On each occasion I was told that the answer should have been a denial. **1964** *Official Encycl. Bridge* 122/2 *Denial bid*, a bid that indicates lack of support for partner's bid (an obsolescent term).

denidation (dīnidē¹·ʃən). *Med.* [f. De- + L. *nidus* nest: see -ation.] The shedding of the superficial layer of the uterus, such as occurs during menstruation.

1874 J. H. Aveling in *Obstetr. Jrnl.* II. 212 Nidation has been likened to gestation. Denidation may be compared with parturition. **1935** *Nature* 16 Mar. 413/1 We may speculate whether some subtle distinction may not be drawn between an embryo and a mere blastocyst, between criminal abortion and permissible denidation.

denier³. Add: **4.** A unit of weight used to estimate the fineness of silk, rayon, or nylon yarn.

The unit is based on a length of 450 metres of yarn weighing 0·05 gramme.

1839 Ure *Dict. Arts* 1105 The first of these raw silks will have a *titre* of 20 to 24 deniers. **1858** Simmonds *Dict. Trade*, *Denier*,..in Italy, a small weight equal to about a grain, by which silk is weighed. **1887** *Colonial & Indian Exhib., Rep. Col. Sect.* 341 The general sizes [of silk] seem to be 16 to 20 deniers, but it will range from about 10 to 24 deniers, single thread. **1927** T. Woodhouse *Artificial Silk* 78 Finally the hanks are weighed..to ascertain the denier count. **1952** *Sunday Times* 15 June 8/5 Denier is the thickness of the yarn: the lower the denier number, the finer the stocking. **1957** *Times* 30 Sept. 11/3 The makers claim that these 15-denier 'Carefree' nylons will outlast several ordinary pairs. **1960** *Textile Terms & Defs.* (ed. 4) 53 *Denier*, the weight in grammes of 9,000 metres of a filament or yarn. The denier system is used as the standard count for filament silk as well as for rayon, cellulose acetate, nylon and other man-made fibres.

denigrate, *v.* Restrict 'Now *rare*' to sense 1.

2. (Later example.)

1952 *Daily Herald* 17 Nov. 4/1 Elements in this country which have always sought to denigrate the work of the United Nations. **1971** *Sci. Amer.* Sept. 237/2 They attempted to denigrate..our most crucial findings.

Hence **denigra·tory** *a.*

1955 *Times* 9 May 3/4 A revival of Shakespeare's *Richard III*, certainly the play most denigratory to the King. **1967** *Coast to Coast 1965–6* 179 Here Miss Silver-and-Green made an insulting shrug of great beauty, and an exquisite denigratory hand movement. **1969** *Daily Tel.* 13 Nov. 8/3 This book..is provocative and controversial..and, intentionally or not, is denigratory.

‖ **dénigrement** (denigrəmaṅ). [Fr.] *fig.* Blackening of character, disparagement.

1883 *Sat. Rev.* 21 Apr. 486/1 A criticism approaching to *dénigrement*. **1935** *N. & Q.* CLXVIII. 287/2 That tendency to—fundamentally futile—*dénigrement* which besets many biographers now-a-days. **1941** *Horizon* 14 Feb. 151 It shows something wrong with the whole conception of the book to treat everything in the same tone of flippancy and *dénigrement*.

denim. Add: (Later examples of *pl.* = overalls, trousers made of denim.)

1932 J. Dos Passos *1919* 3 He was a smiling ruddy man in blue denims. **1958** J. Lodwick *Bid the Soldiers Shoot* III. vii. 222 In the map pocket of my denims. **1959** *Manchester Guardian* 24 June 7/1 Figures in Bermuda shorts or rolled-up denims.

denitrification (dīnaiːtrifikē¹·ʃən). [f. De- II. 1 + Nitrification.] The reduction, esp. by bacteria, of simple inorganic nitrogen compounds; *spec.* the reduction of nitrates through several intermediates to gaseous nitrogen.

1883 *Jrnl. Chem. Soc.* XLIV. 230 The denitrification is effected by the organisms which are developed; for if the liquid is sterilised by heat..the liquid remains clear and the nitrate is not altered. **1932** Fuller & Conard tr. *Braun-Blanquet's Plant Sociol.* viii. 242 Very acid soils without herbaceous cover, in conifer forests, have a decided tendency towards denitrification. **1934** G. J. Fowler *Introd. Biochem. Nitrogen Conserv.* vi. 132 Denitrification, as the name implies, is the reverse of nitrification. **1964** M. Hynes *Med. Bacteriol.* (ed. 8) i. 8 Denitrification is effected by other bacteria which use nitrates or nitrites as hydrogen acceptors, reducing them to ammonia and thus perhaps impairing the fertility of the soil.

denitrify, *v.* Add: *denitrifying bacteria* (see quot. 1951).

1902 *Encycl. Brit.* XXVI. 56/2 Fresh manure abounds in denitrifying bacteria. **1932** Fuller & Conard tr. *Braun-Blanquet's Plant Sociol.* viii. 241 Over against the nitrate formers stand the denitrifying bacteria which destroy nitric acid. **1951** M. Abercrombie et al. *Dict. Biol.* 68 *Denitrifying bacteria*; soil bacteria which in absence of oxygen break down nitrates and nitrites with evolution of free nitrogen.

denizen, *sb.* **2. c.** (Later examples.)

1933 *Shorter Oxf. Eng. Dict.* p. vii, *Denizens* are borrowings from foreign languages which have acquired full English citizenship, *aliens* are words that retain their foreign appearance and to some extent their foreign sound. **1934** *S.P.E. Tract* XLII. 35 Most words when first borrowed are *aliens*, but if they survive they are gradually accommodated to the language which borrows them and become *denizens*.

‖ **denkmal** (de·ŋkmāl). Pl. **-mäler.** [G.] A monument, memorial.

1877 *Echo* 31 July 1/4 A forthcoming centenary or inauguration of a 'Denkmal'. **1927** *Spectator* 26 Nov. 928/1 We raise our monuments to the fallen in lines reminiscent of the *denkmal*. **1953** W. Stevens *Let.* 15 June (1967) 782 They [*sc.* plans] will create from this sense of things completely lost a kind of community Denkmal.

Denmark (de·nmāɹk). The name of one of the Scandinavian countries, used *attrib.* to designate special kinds of products, as in *Denmark satin* (see Satin *sb.* I. 1 b). Also *Denmark street* (see quot. 1934).

1836 Dickens *Sk. Boz* in *Morning Chron.* 24 Sept. 3/2 A pair of Denmark satin shoes. *Ibid.*, Nor were the Denmark satins a bit behindhand. **1875** [see Satin *sb.* I. 1 b]. **1882** Caulfeild & Saward *Dict. Needlework*, *Denmark Satin*, a kind of worsted stuff employed for the making of women's shoes, measuring 27 inches in width. **1934** S. R. Nelson *All about Jazz* i. 29 Much knowledge of human nature is to be found in Tin Pan Alley and its English equivalent, Denmark Street, Charing Cross Road. **1967** *Guardian* 8 July 12/8 What will happen to Mick Jagger..is not the only question..worrying the world of 'pop' music...Denmark Street has also been exercising its mind on the identity of some ancient Greek called Euterpe.

denn (den). [OE. (Kt.) *denn* woodland pasture; possibly the same word as OE. *denn* habitation of a wild beast (Den *sb.*¹); cf. MDu. *dann* forest, haunt of wild beasts.]

An OE. word, revived by place-name scholars, which survives chiefly in place-names (see A. H. Smith, *Eng. Place-Name Elements* (1956) I. 129), as *Halden, Tenterden*, both in Kent.

A woodland pasture, esp. for swine.

1936 E. Ekwall *Conc. Oxf. Dict. Eng. Place-Names* p. xv, Originally the *denns* belonged to the various *lathes* and were often situated far from the district to which they belonged. **1948** L. D. Stamp *Land of Brit.* iii. 46 The upland settlements had rights of feeding swine in sections or 'denns' of the great Wealden oak forests, hence such names as Tenterden and Biddenden.

dennebol (de·nəbǫl). *S.Afr.* Also **danebol.** [Afrikaans, f. *den* pine + *bol* ball, bulb.] A fir cone.

1909 *State* II. 768 She..began to break up a 'dennebol' between two stones. **1913** Pettman *Africanderisms* 139 *Danebol* or *Dennebol*..A fir cone. **1947** L. G. Green *Tavern of Seas* viii. 65 Sweets made of sugar, water, eggs, naartjie peel and dennebol pits. **1953** *Cape Times* 31 Mar. 16/3 Dennebols for your fires only sixpence for 50.

denominal (dīnǫ·minăl), *a.* and *sb.* *Gram.* [f. De- II. 3 + Nominal A. 1.] = Denominative *a.* 3, *sb.* 2.

1934 Priebsch & Collinson *Ger. Lang.* II. iii. 225 The suffix *-ēn*..(*a*) to form durative deverbals..(*b*) to form inchoative denominals. **1959** *Archivum Linguisticum* XI. 108 *Imprison*—denominal verb.

denominator. Add: **2.** *common denominator*, (*a*) a common multiple of the denominators of two or more fractions, i.e. a number that is a multiple of the denominator of each of them; (*b*) *fig.* something common to or characteristic of a number of things, people, etc.; also *attrib.*; so *least* or *lowest common denominator*, the lowest possible common denominator; also *attrib.* and *fig.*

1594 [see Common *a.* 16]. **1763** [in Dict.]. **1875** *Encycl. Brit.* II. 530/2 In practice it is usually the *least* common denominator that the fractions are compared by. **1898** *Westm. Gaz.* 6 May 2/3 His speech would have to be rendered by interpreters into each of the hundred tongues.., presuming no common denominator language can be used. **1910** [see *club bore* s.v. *club *sb.* 19 b]. **1937** *Discovery* Mar. 75/1 All the black-and-white processes have a common denominator. **1946** R. A. Knox *Epistles & Gospels* 56 He wants the would-be-clever people to cultivate..a lowest-common-denominator sort of mind. **1958** *Observer* 11 May 14/6 Searching among this raw material of art for the lowest common denominator on which to base a new style.

‖ **de nos jours** (dənoʒur), *adj. phr.* [Fr., lit. 'of our days'.] Of our time or lifetime; contemporary.

The phrase always follows the sb. which it qualifies.

[**1864** Browning *Dramatis Personæ* 47 Dîs aliter visum; or, le Byron de nos jours.] **1909** Beerbohm *Yet Again* 192, I admire the Demosthenes *de nos jours*. **1958** *Listener* 14 Aug. 249/3 The musical Robespierre *de nos jours*.

denotation. Add: Hence **denota·tional** *a.*

1948 B. Russell *Human Knowl.* 293 An example of a denotational definition is 'the tallest man in the United States'. **1957** in N. Frye *Sound & Poetry* 138 Word-repetition..repeats the..word..with the same denotational semantic spectrum. **1963** J. Lyons *Structural Semantics* iii. 38 The advantage of taking the set of colour-terms..is that they are readily shown to..cover a well-defined 'denotational field'.

denotative, *a.* Add: Also as *sb.* (from sense b).

1944 *Mind* LIII. 35 Roughly, empirical ties are the denotatives: demonstrative pronouns ('this', 'that'), relative adverbs ('here', 'now'), also symbols often not called linguistic, such as gestures (pointing), etc.

denotatum (dīnotē¹·tǿm). *Philos.* Pl. **-tata.** [a. L. *dēnotātum*, neut. pa. pple. of *dēnotāre* to Denote.] That which is denoted by some expression; *esp.* an existent object of reference. (Cf. *designatum.)

1938 C. W. Morris in *Internat. Encycl. Unified Sci.* I. 83 Where what is referred to actually exists as referred to[,] the object of reference is a denotatum. It thus becomes clear that, while every sign has a designatum, not every sign has a denotatum. A designatum is not a thing, but a kind of object or class of objects—and a class may have many members, or one member, or no members. The denotata are the members of the class. **1940** [see *designatum]. **1944** M. Weitz in P. A. Schilpp *Philos. B. Russell* 100 So to interpret the entities of physics means..that they are no longer the denotata of proper names..nor the denotata of descriptions. **1963** J. Lyons *Structural Semantics* iii. 37 The vocabulary of a language is not rightly regarded as a set of isolated symbols standing in a relation of correspondence with a set of discrete and prior concepts or denotata existing in the 'outside world'. **1968** Y. R. Chao *Language & Symbolic Systems* 67 The word *dog* means the animal dog. The word is said to *refer to*, or *denote*, the thing and the thing is the *referent* or *denotatum*. **1968** *Language* XLIV. 40 With adjectives, it is clear that *und* is obligatory when the coördinated nouns do not refer to the same sample of the denotatum.

denounce, *v.* Add: **6. b.** *absol.* or *intr.*

1837 E. Howard *Old Commodore* iii, He first petitioned, then remonstrated, and, foolish boy! at last denounced. **1888** Mrs. H. Ward *Robert Elsmere* xl, I went to confront, to denounce you!..I went to denounce..and the Lord refused it to me.

de nouveau: see De II.

de·nsener. *Metallurgy.* [f. Densen *v.* + -er¹.] A piece of metal used as a chill in foundries.

1930 *Engineering* 11 July 57/2 A method of attaining this end has developed into the use of chills or 'denseners', which are inserted in the mould wall against certain heavy portions of castings. These consist of suitably shaped pieces of metal, generally of cast iron or steel, placed in such a manner that when the iron is poured..the chills abstract the heat. **1933** *Jrnl. Iron & Steel Inst.* CXXVIII. 554 The author deals with the use of spiral mild-steel wires as denseners in castings.

densify, *v.* Add: Hence **de·nsified** *ppl. a.*

1900 *Westm. Gaz.* 16 Aug. 3/3 The earliest Greek manuscripts all read *Ponton Pileton*, the meaning of which is 'a densified sea'. **1944** *Electronic Engin.* XVI. 365/1 The laminated densified wood used in electrical apparatus. [**1957** *Brit. Commonw. For. Terminol.* II. 59 *Densify*, to increase the density (of wood), e.g. by compression or by impregnating with synthetic resin, as in the manufacture of improved wood.] **1960** *Farmer & Stockbreeder* 19 Jan. Suppl. 14/3 Simple lining material such as flat asbestos cement, or densified hardboard.

densitometer (densitǫ·mitəɹ). *Photogr.* [f. Densit(y 4 + -ometer.] An instrument for the measurement of photographic density.

1901 *Camera Mag.* 23 Oct. 14 The idea at the root of the 'Densitometer' is that if it is possible to ascertain the density of a negative in a uniform and accurate manner the difficulties of bromide printing are at once overcome. **1955** *Gloss. Terms Radiology* (*B.S.I.*) 54 *Densitometer*, an instrument for measuring photographic transmission and/or reflection density. **1959** *New Scientist* 27 Aug. 283/1 The density of photographic emulsions must be accurately controlled in order to achieve the best results in colour printing. Normally this quantity is measured by means of a densitometer which successively views selected points of the image, producing a value for each position. **1961** *Printing News* 16 Feb. 9 Densitometers—particularly of the reflection type—will not always agree on a numerical value for density.

density. 4. (Later example.)

1967 E. Chambers *Photolitho-Offset* 271 *Density*, the quantitative measure of the blackening of a photographic emulsion.

dent, *ppl. a.* Add: **3.** Also *ellipt.* (See also quot. 1909.)

1853 *Trans. Mich. Agric. Soc.* V. 125 The land..was planted..with the 'Indian Yellow Dent'. **1873** *Trans. Dep. Agric. Illinois* X. 77 The Dent Corns—White and Yellow Dent, Large White, and Yellow Dent. **1909** W. Bateson *Mendel's Princ. Heredity* 264 According as the seeds [of maize] are opaque or semi-transparent, the varieties are distinguished as 'Dent' or 'Flint'. **1950** *New Biol.* VIII. 37 *Dent* or *field corns*..are flinty with soft starch extending from the base to the tip of the kernels.

dental, *a.* Add: **1. b.** *dental chair* (U.S.); *dental floss* [Floss *sb.*² 2], *silk*, floss silk or similar fibrous material used to clean between the teeth; *dental surgery* (orig. *U.S.*), dentistry.

1826 L. Koecker (*title*) The principles of dental surgery. **1841** G. Waite (*title*) An appeal to Parliament, the medical profession and the public, on the present state of dental surgery. **1859** J. Tomes (*title*) A system of dental surgery. **1880** 'Mark Twain' *Tramp Abroad* xxiii. 222 About five hundred soldiers gathered together in the neighborhood of that dental chair waiting to see the performance. **1887** Dental surgery [see Surgery 1]. **1907** *Yesterday's Shopping* (1969) 507/2 Dental silk—reel, 0/3½; waxed 0/5. **1910** *Nat. Dent. Hosp. Gaz.* IV. xx. 42 Drill a few holes on each side of the fracture and thread them through and across with an ordinary piece of dental floss silk, finishing off with a knot on the lingual surface. **1922** Joyce *Ulysses* 126 He took a reel of dental floss from his waistcoat pocket and.. twanged it smartly between two..of his resonant unwashed teeth. **1955** *Oxf. Jun. Encycl.* XI. 98/2 The discovery of anaesthetics was of the greatest importance in the progress of dental surgery. **1971** *Times* 12 Feb. 13/7 The decision of Johnson and Johnson to reintroduce into this country dental floss.

c. Engaged in dentistry or dental work, as **dental mechanic**, an operative who makes and repairs artificial teeth; **dental nurse**, a nurse who assists a dentist (the term is also applied to a dentist's receptionist); hence *dental nursing* vbl. sb.; **dental surgeon** (orig. *U.S.*), a dentist (cf. *surgeon-dentist*); hence *dental surgeoncy*; **dental technician** = *dental mechanic.*

1840 *Amer. Jrnl. Dent. Sci.* I. 157 The objects of this Society [*sc.* the Amer. Soc. of Dental Surgeons] are to promote union and harmony among all respectable and well-informed Dental Surgeons. **1860** *Brit. Jrnl. Dent. Sci.* III. 232/2 We, the Court of Examiners, have diligently examined —— and have found him competent to exercise the art and science of a Dental Surgeon. **1881** *Jrnl. Brit. Dent. Assoc.* II. 9 The burden of a Dental Surgeoncy to a Hospital. **1916** *Daily Colonist* (Victoria, B.C.) 12 July 3/5 (Advt.), Highly skilled dentists and dental mechanics. **1921** *Act* 11 & 12 *Geo. V* c. 21 § 3 (1) The occupation of a dental mechanic. **1922** Joyce *Ulysses* 571 Dental surgeon Bloom with tweezers. **1938** H. R. Cullwick *Handbk. for Dent. Nurses* 8 The subject of the duties of a dental nurse has been sadly neglected by dental literature in England... The question of a dental nurse having a hospital training is a debatable one. *Ibid.* 11 Dental nursing as a career. **1961** *Evening Standard* 14 July 19/2 (Advt.), Dental Nurse experienced preferred. *Ibid.* 17 Aug. 15/1 (Advt.), Dental Technician... Expd. Orthodontics and simple gold work.

dentality. (Example.)

1877 H. Sweet *Handbk. Phonet.* 38 § 110, (th) is, like (r) formed by the point. The essential difference between them lies in the dentality of (th), which involves a more horizontal position of the tongue, which has to be stretched out to reach the teeth.

dentalium (dentē¹·liŏm). Pl. **dentalia.** [mod.L. (Linnæus *Systema Naturæ* (ed. 10, 1758) I. 785), f. L. *dentāl-is*, f. *dens, dent-em* tooth.] A tooth-shell of the genus so named (cf. Dental B. 4). Also *attrib.*

1864 *Proc. Zool. Soc.* 137 The value of the *Dentalium* depends upon its length. *Ibid.* 138 At one period, perhaps a remote one, in the history of the inland Indians these *Dentalia* were worn as ornaments. **1913** B. B. Woodward *Life of Mollusca* iv. 55 The *Dentalium*, again, buries in the sand, leaving only the apex of the shell protruding. **1926** *Daily Colonist* (Victoria, B.C.) 4 July 21/5 The effort to secure the dentalium at seacoast places, and its employment so widely by peoples far removed from the seacoast means that much more than merely ornamental value was placed upon it. **1931** *Antiquity* V. 431 Dentalium and pusiostoma shells are the most numerous. **1932** *Jrnl. R. Anthrop. Inst.* LXII. 267 Two skeletons..had head-dresses of dentalium shells. **1938** *Nature* 19 Feb. 306/2 They had necklaces, too, made of dentalium, alternating with 'twin' pendants, which seem to me to imitate the canines of deer. **1960** K. M. Kenyon *Archæol. in Holy Land* ii. 38 At 'Ain Mallaha..a burial closely resembling that of Mount Carmel, with a crown of *dentalia* shells, was made in a pit lined with plaster.

dentaria (dentē·riă). *Bot.* [mod.L. (J. P. de Tournefort *Institutiones Rei Herbariæ* (1700) I. 225), fem. sing. of L. *dentarius* pertaining to the teeth, so called from the tooth-like scales on the roots of the plant.] A plant of the cruciferous genus so named; = *pepper-root* (Pepper *sb.* 5), Pepperwort 16, Tooth-wort 3.
1818 *Mass. Hist. Soc. Coll.* 2nd Ser. VIII. 169 Among those, that flower in June, the most interesting are..the sarsaparilla, the dentaria or tooth-root,..and the mountain ash. **1886** M. Arnold *Lett.* (1895) II. 327 The dentarias too are beautiful.

dentelle. Add: **1.** (Earlier example.)
1847 C. Brontë *Jane Eyre* I. xv. 279, I..gave her.. cashmeres, diamonds, dentelles, &c.
3. *dentelle binding, border, tooling* in book-binding.
1890 *Catal. Exhib. Rec. Bk.-bindings Grolier Club* 11 *Dentelle Border.* A tooled pointed border with finely dotted or Gascon ornaments in imitation of lace. *Ibid.* 21 [Book] Blue morocco; doublé with blue morocco, large dentelle tooling. **1938** *Times Lit. Suppl.* 12 Nov. 732/3 Several of the far more attractive *dentelle* bindings with which the names of Padeloup and Derome are usually associated.

dentex (de·nteks). [mod.L. (G. Cuvier 1815, in *Mém. Mus. Hist. Nat. Paris* I. 456, f. *Sparus dentex*, the Linnæan name of the fish), f. L. *dentex, dentix* a kind of marine fish.] The common name of a sea bream, *Dentex dentex*, found in the Mediterranean and along the North African Atlantic coast; also used for other members of the genus.
1836 W. Yarrell *Hist. Brit. Fishes* I. 112 Duhamel.. mentions a *Dentex* that weighed no less than seventy pounds. **1880** Günther *Fishes* 389 Dentex—..Marine fishes rather locally distributed in the Mediterranean. **1925** *Countries of World* IV. 2673/1 The dentex and barracada pike. **1953** tr. *Cousteau's Silent World* 10 Below I saw a big blue dentex (bream) with a bitter mouth and hostile eyes. **1957** R. Campbell *Portugal* iv. 70 The Azores are as good for under-water fishing as the Berlengas...Dentex..grow to a great size there.

denu·clearize, *v.* [f. De- II. 1 + Nuclear *a.* + -ize; cf. Demilitarize *v.*] *trans.* To deprive of nuclear armaments; to remove nuclear armaments from. Chiefly as *ppl. a.*, esp. with *zone.* So **de:nucleariza·tion.**
1958 *Economist* 25 Jan. 290/1 The Polish suggestion of a 'denuclearized' area in central Europe. **1958** *Times* 20 Feb. 8/5 A denuclearized zone in central Europe would answer the vital interests of all European States. **1959** *News Chron.* 21 July 1/1 What he [*sc.* Mr. Krushchev] wanted from Scandinavia was its de-nuclearisation. **1960** *New Left Rev.* July–Aug. 1/2 A de-nuclearised, NATO-free West Germany. **1962** *Guardian* 15 Dec. 6/4 Mr. McNamara..seems to be doing more to de-nuclearise Britain than CND has yet achieved. **1963** *Economist* 11 May 547/1 If warfare could in some way be denuclearised.

denudant (dĭniū·dănt). [f. Denude *v.* + -ant[1].] That which denudes; *spec.* in *Geol.*, an agent or agency which removes disintegrated matter and lays bare the underlying rock or formation.
1894 J. Geikie *Gt. Ice Age* (ed. 3) 259 Its [*sc.* the boulder-clay's] chief denudant has evidently been running water.

denudation. Add: **2.** (Later examples.) Also *attrib.* So **denuda·tional** *a.*
1913 A. Holmes *Age of Earth* iv. 60 The application of denudational statistics to the measurement of geological time will be considered. **1928** Nordenskjöld & Mecking *Geogr. Polar Regions* 58 Old denudational surfaces or raised peneplains. **1949** *Proc. Geol. Assoc.* LX. 165 (*title*) The denudation chronology of the dip-slope of the South Downs. **1954** W. D. Thornbury *Princ. Geomorphol.* iii. 35 Some geologists have used the term denudation as if it were synonymous with gradation, but, as this term implies removal of material, it is hardly logical to include deposition under it. **1956** *Nature* 28 Jan. 166/1 The sequence of cyclic denudational landscapes. **1960** B. W. Sparks *Geomorphol.* i. 2 In its second sense, geomorphology is the study of the evolution of landscapes. Such study is often termed denudation chronology. **1960** L. D. Stamp *Britain's Struct.* (ed. 5) iii. 22 The cycle of denudation on the land and of sedimentation in the water is brought to a close by earth movements.

denumerability (dĭniū:mĕrăbi·liti). *Math.* [f. next: see -ity.] The fact or quality of being denumerable.
1935 *Bull. Amer. Math. Soc.* XLI. 644 In a metrizable space the Borel theorem holds without the restriction of denumerability of the original family. **1941** Courant & Robbins *What is Math.?* ii. 79 (*title*) The denumerability of the rational numbers and the non-denumerability of the continuum. **1963** R. R. Stoll *Set Theory* ii. 91 Using the preceding theorem, the denumerability of the positive rationals can be established.

denumerable (dĭniū·mĕrăb'l), *a. Math.* [f. Denumerate *v.* + -able. Cf. G. *abzählbar*, Fr. *dénombrable.*] Of a set: infinite but countable; capable of being put into a one-to-one correspondence with the set of finite integers or natural numbers; also more widely, either finite or countably infinite; *enumerable *a.*
C. S. Peirce used the term in a different sense.
[**1893** C. S. Peirce *Coll. Papers* (1933) IV. i. iv. 91 The dinumerable [*sic*] is to the innumerable as logarithmic infinity is to ordinary infinity.] **1902** A. N. Whitehead in *Amer. Jrnl. Math.* XXIV. 367 Imagination is very misleading, since it presents to us special aggregates which are denumerable. **1903** B. Russell *Princ. Math.* xxxvi. 296 A series of this type..is denumerable, that is, by taking its terms in a suitable order..we can give them a one-one correspondence with the finite integers. **1948** —— *Hum. Knowl.* iv. x. 345 Each atom may contain any one of a certain discrete denumerable series of amounts of energy. **1960** S. Körner *Philos. of Math.* iii. 63 The class of all rational numbers and the wider class of all (complex) algebraic numbers..are denumerable. **1964** E. Bach *Introd. Transformational Gram.* vii. 151 The 'smallest' kind of infinite set is a set which is denumerable or countable. **1968** *Language* XLIV. 571 The set of all sets of integers is non-denumerable; the set of all Turing machines is denumerable; each recursively enumerable set corresponds to a Turing machine; therefore there are sets of integers which are not recursively enumerable. **1968** P. A. P. Moran *Introd. Probability Theory* v. 218 It therefore has at most a denumerable number of discontinuities.
Hence **denu·merably** *adv.*
[**1893** C. S. Peirce *Coll. Papers* (1933) IV. i. iv. 91 The innumerable is not only dinumerably [*sic*] more than the dinumerable but is innumerably more.] **1932** M. H. Stone *Linear Transformations in Hilbert Space* i. 20 Special notations for sets of elements constructed from a finite or denumerably infinite collection..of closed linear manifolds will be found..helpful. **1964** E. Bach *Introd. Transformational Gram.* vii. 153 The rules will specify the possible and denumerably infinite distributions for each of the forms of the language.

denutrition. (Earlier and later examples.)
1868 W. James in *North Amer. Rev.* CVII. 325 The study of those facts of elementary life, which may be summed up under the name of nutrition and denutrition. **1896** *Daily News* 25 Jan. 5/5 Her nervous system was a good deal shaken in consequence of an accident when out riding. Denutrition has been one of the consequences.

deoch an doris (dǫχăn,dǫ·ris, dyǫχ-). Chiefly *Sc.* and *Irish.* Also **deoc(h)-an-doruis, deoch-an-dorris, deoch an dorus, doch-an-dorris,** etc. [Gael. *deoch an doruis*, lit. 'drink at the door'.] A parting drink, a stirrup-cup.
1682–91 J. Fraser *Chron. Frasers* (1905) 124 That ordinary farewell drink, a parting called Deoch i Dorrish, which, as it is said, Prior Dawson had invented. *Ibid.* 366 He must drink Doch in Dorris, a homely drow which the Highlanders takes [*sic*] with their familiars. **1810** W. Hickey *Memoirs* (1925) IV. 343 Sir George..insisted upon our taking the *Dukkin Dorreege* (I know not whether I spell it correctly. It means the parting glass, or glass at the door, in the Irish language). **1819** Scott *Br. Lamm.* II. iv. 85 The Lord Keeper, the Master, and the domestics, had drunk *dock-an dorroch*, or the stirrup-cup, in the liquors adapted to their various ranks. **1824** —— *Redgauntlet* I. ii. 22 This was a parting cup..and..fell under the exception of *Dochan dorroch.* **1839** D. M. Moir *Life Mansie Wauch* xxii. 283 To give Peter Farrel a dram by way of 'doch-an-dorris', as the Gaelic folk say. **1866** 'An Old Stager' *Stage Reminisc.* 13 Having partaken of *deoch an doruis* with our worthy host, we left the house. **1892** W. Ewing *Poems* 14 Then doch and dorus wis proposed. **1912** Lauder & Harper *It's Nice when you love a Wee Lassie* 3 We gave them as much as they wanted to eat, An' a wee deoch an' doris before their retreat. **1914** Joyce *Dubliners* 96 Let us have another one as a *deoc an doruis.* **1915** A. D. Gillespie *Lett. from Flanders* (1916) 240 An impromptu concert, which wound up with a song from me and 'wee doch-an-dorris' from Bankier. **1920** *Glasgow Herald* 17 Nov. 9 A deoch an doris or a kindly pledge in whisky. **1931** A. J. Cronin *Hatter's Castle* ii. iv. 199 I'm wi' ye, juist a wee deoch-an-dorris to keep out the cauld. **1970** *Guardian* 2 June 12/7 The Scot['s]..last drink, the *deoch an doris*, by ancient custom must be taken standing, and..need not be paid for.

deontic (diǫ·ntik), *sb.* and *a. Philos.* [f. Gr. δέον, δεοντ- (see Deontology) + -ic.]
A. *sb.* **1.** *pl.* [After *ethics, eudemonics*, etc.] (See quots.)
a **1866** J. Grote *Moral Ideals* (1876) vii. 102 A science of duty (deontics or deontology). **1906** J. S. Stuart-Glennie in *Sociol. Pap.* II. 250 The second order of ethical sciences.. form the contents of three classes of science—Economics, Deontics, and Juridics. **1929** *Mind* XXXVIII. 275 *Deontics*.., meaning..that which, either perfectly or in a definite degree, is obligatory upon all.
2. *sing.* (See quot.)
1926 *Mind* XXXV. 395 Ethical arguments..should be able to exhibit their 'deontic' in the same way as inference reveals its 'logic'.
B. *adj.* Of or relating to duty, obligation, etc.
1951 *Mind* LX. 1 (*title*) Deontic logic. **1955** A. N. Prior *Formal Logic* 221 There are 'laws of movement' of deontic operators analogous to those of ordinary modal operators.

‖**Deo volente** (dī·o vole·nti). Also (erron.) **Deo volens.** [L.] God willing; if nothing prevents (the fulfilment of a promise). Abbrev. D.V. (D III. 3).
1767 Gray *Let.* 6 June in *Wks.* (1884) III. 268 My intention is (*Deo volente*) to come to Cambridge on Friday or Saturday next. **1854** E. K. Kane *Jrnl.* 11 Sept. in *Arctic Explor.* (1856) I. xxvii. 356 *Deo volente*, I will be more lucky tomorrow. **1954** M. Lowry *Let.* 25 Jan. (1967) 360 Deo volens, they will more than get it back in the end. **1957** B. & C. Evans *Dict. Contemp. Amer. Usage* 131/1 The use of the phrase 'D.V.', the initials of *Deo volente*, or the English form *God willing*, as an interjection after an expressed intention is a verbal counterpart of knocking on wood and has about the same value.

deoxy-, *prefix. Chem.* [f. De- II. 3 + Oxy- 2.] A prefix used in forming the names of chemical compounds, indicating the loss of one or more atoms of oxygen; = Desoxy- (which is less common).
1871 *Jrnl. Chem. Soc.* XXIV. 404 Deoxycodeine, $C_{18}H_{21}NO_2$. **1949** *Biochem. Jrnl. Suggestions to Authors* 9 Spellings, etc., adopted..deoxy (prefix) *not* desoxy.

deoxyco:rticoste·rone. *Biochem.* Also **desoxy-.** [f. *Deoxy- + *Corticosterone.] A steroid compound, $C_{21}H_{30}O_3$, occurring in the adrenal cortex and also prepared synthetically.
1937 Steiger & Reichstein in *Nature* 29 May 926/1 This compound differs from corticosterone only by the absence of the fourth oxygen, and can therefore also be called desox-corticosterone. **1938** *Chem. Abstr.* XXXII. 578 The structure of the biol. active desoxycorticosterone from stigmasterol. **1941** *Ann. Reg. 1940* 344 Deoxycorticosterone was prepared artifically. **1952** G. Bourne et al. *Cytology* (ed. 2) xi. 478 From the adrenal cortex there have been extracted two main groups of physiologically active steroids. One group, exemplified by desoxycorticosterone, acts primarily on the salt balance of the body and regulates the excretion of Na and Cl by the kidney. **1956** *Nature* 28 Jan. 189/2 Comparisons are being made of the wool growth of adrenalectomized sheep maintained on different levels of cortisone and deoxycorticosterone. **1965** *New Scientist* 1 July 33/1 The true mineralocorticoid hormone is aldosterone, but there are compounds such as deoxycorticosterone and fluorocortisol that have potent related properties.

deoxyco·rtone. *Biochem.* [Shortened from prec.] = prec.
1949 *Lancet* 28 Jan. 159/1 Deoxycortone plus ascorbic acid..seems to be an effective substitute for the naturally occurring anti-arthritic steroid hormone. *Ibid.* 26 Nov. 993/1 The patients were treated with an intramuscular injection of 5 mg. Deoxycortone (desoxycorticosterone) acetate in 1 ml. of oleum arachis. **1963** *Brit. Pharmaceutical Codex* 233 Deoxycortone is a mineralocorticosteroid which causes sodium retention and potassium excretion. It is used ..in the treatment of Addison's disease.

deo:xyribonu·cleic a·cid. *Biochem.* Also **desoxy-.** [f. *Deoxyribo(se + Nucleic *a.*] A generic term for any of the nucleic acids which yield deoxyribose on hydrolysis, which are generally found in and confined to the chromosomes of higher organisms, and which store genetic information. Also called *deoxy-, desoxyribose nucleic acid.* Abbrev. D.N.A., DNA.
1931 Levene & Bass *Nucleic Acids* ix. 294 (*heading*) Desoxyribonucleic acids. **1938** *Jrnl. Chem. Soc.* 1722 Secondly there are deoxyribonucleic acids, in which the sugar component is d-2-deoxyribose and of which 'thymus nucleic acid' is the classical example. **1944** *Jrnl. Biol. Chem.* CLVI. 691 The key rôle of the desoxyribose type of these substances in the reproduction of inheritable characteristics has been emphasized by the identification of desoxyribose nucleic acid (DNA) as a major component of chromosomal nucleoprotein. **1947** *Thorpe's Dict. Appl. Chem.* (ed. 4) VIII. 622/1 Deoxyribonucleic acid has been identified in only a few viruses, for example those of psittacosis,..vaccinia,..influenza [etc.]. **1951** *Lancet* 16 June 1287/1 In normal animal and plant cells deoxyribonucleic acid (D.N.A.) is located entirely within the nucleus. **1953** *Sci. News* XXIX. 109 The structure proposed for the salts of deoxyribose nucleic acid consists of two chains coiled in the same direction about the same axis. **1955** *New Biol.* XIX. 9 DNA molecules are chains which can be broken down into nucleotides. Each nucleotide molecule can be broken down by hydrolysis into a molecule of phosphoric acid, one of desoxyribose (a five carbon sugar) and one of a purine (adenine or guanine) or of a pyrimidine (thymine, cytosine or 5-methyl cytosine, but in some phages thymine or 5-hydroxy-methyl-cytosine). **1955** *Sci. Amer.* Oct. 70/3 The chromosomes are made largely of DNA (desoxyribonucleic acid). DNA can exist in a great multitude of forms, all built of the same building blocks but with the units arranged in different sequences. The hereditary information carried by a DNA molecule is contained in the order of arrangement of these units. **1962** A. Huxley *Let.* 26 Dec. (1969) 945 He talked such nonsense.. about true intelligence residing only in the DNA molecule. **1964** *Listener* 19 Mar. 493/2 The spiral of deoxyribonucleic acid. **1970** *Nature* 5 Sept. 998/2 Demolition of the hallowed idea of the molecular biologist that RNA cannot synthesize DNA has prompted a reinvestigation of other ideas.
Hence **deo:xyribonu·clease** [*-ase], any enzyme that catalyses the hydrolysis of DNA, causing it to split up into smaller nucleotide units.
1946 *Jrnl. Biol. Chem.* CLXVI. 393 (*heading*) Streptomycin and desoxyribonuclease in the study of variations in the properties of a bacterial virus. **1965** Peacocke & Drysdale *Molecular Basis Heredity* vii. 69 If..one or other of the two nucleic acids has been eliminated from the material by digestion with a specific enzyme (deoxyribonuclease or ribonuclease), the remaining nucleic acid can be located. **1970** R. W. McGilvery *Biochem.* xx. 481

Degradation of nucleic acids begins with partial hydrolysis ..catalyzed by ribonucleases or deoxyribonucleases.

deo:xyribonu:cleopro·tein. *Biochem.* Also **desoxy-.** [f. *Deoxy-.] 'A nucleoprotein that yields a deoxyribonucleic acid on hydrolysis' (Webster 1961). Abbrev. D.N.A.P.
1944 *Jrnl. Biol. Chem.* CLVI. 692 Most desoxyribonucleoproteins contain very small amounts of tryptophane. **1955** *Sci. News Let.* 30 Apr. 280/3 The chemical has properties which place it in the class of biologically important substances termed desoxyribonucleoproteins...The substances are giant molecules which are found exclusively in the chromosomes and appear to be the carriers of cell heredity. **1961** *Lancet* 19 Aug. 436/2 One of these..was probably glycoprotein in nature, and was clearly different from the desoxyribonucleoprotein (D.N.A.P.).

deoxyri·bose. *Biochem.* Also **desoxy-.** [f. *Deoxy- + *Ribose.] Any of the sugars derived from ribose by the replacement of a hydroxyl group by a hydrogen atom; **deoxyribose nucleic acid**: see *Deoxyribonucleic acid.
1931 Levene & Bass *Nucleic Acids* i. 26 This reaction is now the general method for the preparation of 2-desoxy sugars. By this procedure 2-desoxyribose [etc.]..have been prepared. **1961** *Times* 18 Aug. 12/4 The sugar component of the nucleic acid of chromosomes is known chemically as deoxyribose. **1967** *Listener* 3 Aug. 142/1 Each nucleotide is composed of a base, a sugar called deoxyribose, and phosphate.

depalatalization (dīpæ:lātălaizē·ʃən). *Phonetics.* [f. De- II. 1 + Palatalization.] The loss of palatalization.
1952 *Archivum Linguisticum* IV. ii. 137 The depalatalisation of the unfamiliar sounds [ṇ] and [ʎ]. **1955** *Sci. Amer.* Aug. 79/1 We should not overlook the reverse phenomenon of depalatalization: the dropping of y before vowels in words where it used to be generally pronounced.

departed, *ppl. a.* Add: **4. b.** Also (esp. *joc.*) *the dear departed.*
1814 Jane Austen *Mansf. Park* iii. 56 Barely enough to ..enable me to live so as not to disgrace the memory of the dear departed. **1838** Dickens *Nickleby* iii. 20 The dear departed had never deigned to profit by her advice. **1868** L. M. Alcott *Little Women* (1871) xi. 137 Beth..sat making a winding-sheet, while the dear-departed lay in the domino-box. **1910** W. S. Houghton (*title*) The dear departed. **1964** E. Longford *Victoria R.I.* xxiv. 355 If the living Albert kept Queen Victoria and Disraeli apart, the dear departed brought them together.

department, *sb.* Add: **3.** Freq. in trivial use.
1824 T. Creevey *Let.* 23 Sept. (1903) II. 83 Lady —— has two maids here—one French and the other Italian, the latter of which presides over the bonnet department. **1966** *Listener* 20 Jan. 88/1 It will depend on two things: organization and performance in the field...There are limits to what may be expected of us in either department.
4. Freq. in French form.
1846 R. Ford *Gatherings from Spain* iv. 30 The French.. introduced their own system of *départements*, by which districts were neatly squared out. **1964** *Ann. Reg. 1963* 245 From 12–16 June he visited the Charente and Poitou regions and from 24–29 September the *départements* of Vaucluse, Drôme, Ain, and Rhône.
5. department store orig. *U.S.*, a large shop selling many different kinds of article. Cf. next.
1887 in F. Presbrey *Advertising* (1929) xxxv. 314 *Evening Wisconsin*..H. Heyn's Department Store. **1893** *Harper's Mag.* Apr. 659/2 They [*sc.* Brooklyn stores] compare favorably with the best and largest of the department stores of New York. **1910** H. G. Wells *Mr. Polly* i. 24 One of those large, rather low-class establishments which sell everything from pianos and furniture to books and millinery —a department store. **1953** *Manch. Guardian Weekly* 30 July 7 Department stores in up-and-coming cities around the world.

departmental, *a.* Add: **3.** Of a store: consisting of or comprising several departments.
[**1907** *Yesterday's Shopping* (1969) p. xi/2 A list of the Society's Departmental Price Lists..will be found on page xii.] **1911** R. Strong in W. B. Robertson *Encycl. Retail Trading* xii. 114 Small wonder that..the growth of a departmental store is sometimes phenomenal. **1924** *Times Trade & Engin. Suppl.* 29 Nov. 235/2 The departmental stores and the larger retail shops throughout the Western States. **1926** A. James *Commerce* i. 53 Departmental stores contain under one roof many departments. It is possible to purchase groceries, boots and shoes, furniture, clothing, toilet requisites, as well as listen to the latest jazz music from the orchestra. **1955** *Times* 14 May 12/1 Perhaps the departmental stores were the most consistent in their experience.

departmentalization. (Later examples.)
1933 *Jrnl. R. Aeronaut. Soc.* XXXVII. 346 In the beginning there was no departmentalisation of flying. The pioneers designed, financed, built, flew, crashed and repaired their own aircraft. **1940** *Economist* 27 July 125/1 The ultimate task of linking these 'Special' and 'Registered' accounts into an organic whole will demand more elastic methods. As a result of this departmentalisation, sterling is rapidly losing the homogeneity which is the hall-mark of a free currency. **1961** *Daily Tel.* 25 Feb. 8/2 If these negotiations fail, then President de Gaulle will have no alternative but to proceed with his interim plans for Algeria—departmentalisation, and the admission of more Moslems to office. **1962** *Listener* 5 July 19/2 Professor Haydn was arguing against departmentalization of history, the separate analysis of religious, political, and economic 'factors'.

departmentalize, *v.* Add: (Examples.) Hence **departme·ntalized** *ppl. a.*
1900 *Westm. Gaz.* 6 Dec. 9/3 The Bovril business was strikingly departmentalized, and..the managing-director ..still occupies the position he has held from the start. **1924** W. B. Selbie *Psychol. Relig.* 40 That tendency to departmentalize human nature from which modern psychology has in great measure helped to set us free. **1930** *N.Y. Times* 10 Aug. v. 2/2 Everything should be organized and departmentalized. **1931** W. Rose *Outl. Mod. Knowl.* p. xiii, Scientific investigation is thus becoming more departmentalised; each worker finds himself compelled to devote his mind and energies to a more restricted field. **1931** A. L. Rowse *Politics & Younger Generation* ii. 51 The normal continental type of a departmentalized social life. **1936** *Mind* XLV. 250 The feeble-minded..seem to be more departmentalized than normal children.

departure. Add: **8.** *departure lounge* (at an airport).
1948 *Airports & Air Transportation* May 404/1 Adjoining ..the first departure lounge, is a new staff bar. **1963** 'R. Erskine' *Passion Flowers in Italy* iv. 48, I spent it [*sc.* the time] in the Departure Lounge.

‖ **dépaysé** (depèize), *a.* [Fr., lit. '(removed) from one's own country'.] Removed from one's habitual surroundings; = *déraciné. Also as *sb.* Cf. Exile *sb.*[2]
1909 in Webster. **1924** *Glasgow Herald* 25 July 8 A golden eagle was discovered in the Clock Tower of the Houses of Parliament..a noble creature strangely dépaysé or out of its element! **1928** A. Waugh *Nor Many Waters* v. 208 She knew so well that drifting European life of the *dépaysé.* **1958** *Times Lit. Suppl.* 5 Apr. 182/2 In those unknown countries [*sc.* romance] he would have been as completely *dépaysé* as his own Comedie or Racine. *Ibid.* 2 July 447/1 Not yet so *dépaysé* that he had forgotten the language of the village pub.

dependability (dīpendăbi·liti). [f. Dependable: see -ility.] The quality of being dependable, reliability.
1901 F. T. Bullen *Sack of Shakings* 264 Next to the Trades in dependability..are the west winds of the regions north and south of the Tropics. **1922** *Daily Mail* 7 Nov. 8 Renowned for entire Dependability. **1928** *Daily Express* 21 Apr. 10/3 That essential solidity and dependability that every wise woman seeks in her life mate. **1960** *Farmer & Stockbreeder* 2 Feb. 69 Farmers everywhere give the 780 Special the biggest vote of confidence because they know its dependability is based on experience.

dependent, *a.* Add: **2.** *dependent differentiation*: see *Differentiation 1.

deperm (dīpɔ̄·ɹm), *v.* [f. De- II. 2 + Perm(anent *magnetism*).] *trans.* To demagnetize (a ship). Hence **depe·rming** *vbl. sb.* and *ppl. a.* Cf. *Degauss *v.*
1946 *Jrnl. Inst. Electr. Engineers* XCIII. i. 508/1 The process known as 'deperming' has been developed to remove or compensate the permanent longitudinal and athwartship magnetism. *Ibid.*, Deperming processes. *Ibid.* 514/2 The sequence of the flashes is of major importance since the stability of the deperm is governed by it. *Ibid.* 514/2 The model was then depermed. **1954** *Jane's Fighting Ships 1954–55* 59 Converted to special duty as mobile wiping/deperming units.

depersonalization (dīpɔ̄ɹsənălaizē·ʃən). [f. Depersonalize *v.* + -ation.] **1.** The action of depersonalizing or fact of being depersonalized.
1907 J. R. Illingworth *Doctr. Trin.* x. 191 Madness, prison, suicide may be the end, and all equally symbolise the destruction of proper personality, or, to use a modern term, the depersonalization of the man. **1912** F. von Hügel *Eternal Life* 69 A certain depersonalisation of his conception of this same Christ. **1929** *Times* 14 Aug. 6/3 That progress means the 'de-personalization' of the individual... **1929** J. Jenkins *Nature of Catholicity* iii. 74 The doctrine of transubstantiation, with its crass 'objectification' and hence its depersonalization of Christ's presence. **1953** H. Read *True Voice of Feeling* i. 14 We will consider..what was lost in this particular case of 'depersonalization' or 'de-poetisation'. **1958** *Economist* 4 May 64/2 It is fashionable to say that depersonalisation goes along with depersonalisation.
2. *spec.* in *Psychol.* [ad. F. *dépersonnalisation* (L. Dugas 1898, in *Revue philos.* XLV. 502).] A morbid state involving a loss of the sense of personal identity and a feeling of the strangeness or unreality of one's own words and actions; in extreme cases involving also an obsessive feeling of dissolution of the personality.
1904 *Amer. Jrnl. Psychol.* XV. 589 The peculiar feelings of strangeness and depersonalization. **1918** J. Ward *Psychol. Princ.* xv. 364 In some forms of so-called 'depersonalisation',..the individual doubts his own existence or denies it altogether. **1941** *Brit. Jrnl. Psychol.* XXXII. 113 This watching from without—known as depersonalization when reaching a high grade—is found in schizophrenia and also in melancholia. **1960** *Times* 15 Jan. 15/1 Three [participants in a test of isolation] showed evidence of depersonalization. **1963** H. Burn *Drugs, Med. & Man* (ed. 2) xxi. 211 The LSD intoxication stimulates an acute schizophrenia upheaval, marked by a feeling of depersonalization, of being withdrawn from reality.

depersonalize, *v.* Add: Hence **depe·rsonalized** *ppl. a.*, **depe·rsonalizing** *vbl. sb.*
1901 G. B. Shaw *Capt. Brassbound* iii. 277 (*stage direction*) Sir Howard (falling back on the fatalism of the depersonalized public man). **1919** C. W. Emmet in B. H. Streeter *The Spirit* v. 187 There is no substitution of the divine nature for the human, no depersonalising of man. **1950** *Mind* LIX. 419 The ego here is not a kind of depersonalised pure spectator.

depetal (dīpe·tăl), *v.* [f. De- II. 2 + Petal.] *trans.* To remove the petals from (a flower). So **depe·talled** *ppl. a.*
1936 D. E. Gascoyne *Man's Life is this Meat* 42 The depetalled flower. **1936** L. MacNeice *Let.* 16 Aug. in Auden & MacNeice *Lett. fr. Iceland* (1937) iii. 39 We like to dream of..a lazy music hour by hour depetalling the daisy. **1949** E. Bowen *Heat of Day* xi. 201 Children..engaged innocently on some act of destruction—depetalling daisies, puffing at dandelion clocks.

dephosphorylation (dīfɔsfɔ̄rilē·ʃən). *Chem.* [f. De- II. 1 + *Phosphoryl + -ation.] The removal of a phosphate group from a compound.
1931 *Jrnl. Biol. Chem.* XCII. 765 The function of such phosphorylations and dephosphorylations has..not been elucidated. **1956** *Nature* 11 Feb. 274/1 In algae it is considered that the oxidative assimilation of hexose follows the same glycolytic pathway as in higher organisms, and that these substances after dephosphorylation yield hexose phosphate. **1963** *Lancet* 5 Jan. 59/2 The free energy for sperm movement is provided by the dephosphorylation of adenosine triphosphate.
Also **dephospho·rylate** *v. trans.*, **dephospho·rylated** *ppl. a.*
1929 *Jrnl. Biol. Chem.* LXXXIII. 796 There occur.. others [*sc.* enzymes] capable of dephosphorylating nucleotides. **1954** *Biochem. Jrnl.* LVIII. 391/2 The dephosphorylated product was isolated by chromatography.

depicturement. (Earlier example.)
a **1866** J. Grote *Moral Ideals* (1876) xiii. 307 We read with interest the depicturement of the lives of others.

deplane (dīplē·n), *v.*[1] [f. De- II. 2 + *Plane *sb.*[5]] **a.** *intr.* To leave an aeroplane (after arrival at one's destination). **b.** *trans.* To remove from an aeroplane. So **depla·ning** *vbl. sb.* and *ppl. a.*
1923 *Blackw. Mag.* July 11/2 Dudley left me, saying.. that he was to 'deplane' [*sic*] as soon as possible. **1948** *Aeroplane* 2 Aug. 118/1 The passengers..were out of the machine, which had been towed away from the passenger deplaning area. **1948** in *Amer. Speech* (1952) XXVII. 72 A passenger under influence of liquor will be deplaned. **1967** L. James *Chameleon File* (1968) iv. 58 After clearing immigration control, he..watched the crowd of deplaning passengers.

depoliticize (dīpǒli·tisaiz), *v.* [f. De- II. 1 + Politicize *v.*] *trans.* To render non-political, to remove from the sphere of political activity or influence. Also *absol.* So **depo:liticiza·tion**; **depoli·ticized** *ppl. a.*
1928 *Daily Tel.* 11 Mar. 10/6 The 'depoliticisation' of the public services...The entire public services [in Jugoslavia] have come to be regarded as existing to enable political leaders to reward their followers for party services. **1937** *New Statesman* 4 Sept. 329/2 'Don't talk politics to us,' a bank clerk said to me, 'we [*sc.* the Germans] are a depoliticized people.' We know we can know nothing, and ..knowledge is dangerous here.' *Ibid.*, For it is the intelligent of all classes who are depoliticised. **1960** *Guardian* 19 Sept. 8/1 The distribution of irrigation water is easier to 'depoliticise' than, say, the status of Kashmir. **1963** *Economist* 3 Aug. 422/2 Spokesmen of the People's Party, aware of the political dynamite involved in any large-scale attack on the nationalised empire, have stressed that their aim is to 'de-politicise' rather than de-nationalise. **1970** *Guardian* 23 Dec. 3/7 To denationalise and depoliticise Jerusalem is.. the first practical step towards preparing a long-term plan for Jerusalem's future.

depo·lymerize, *v.* [f. De- II. 1 + Polymerize *v.*] **a.** *trans.* To break down (a polymer) into monomers or other smaller units. **b.** *intr.* To undergo this process. So **depo:lymeriza·tion**; **depo·lymerized** *ppl. a.*; **depo·lymerizing** *vbl. sb.* and *ppl. a.*
1893 *Funk's Stand. Dict.*, Depolymerization. **1909** Webster, Depolymerize *v.* **1924** *Jrnl. Textile Inst.* XV. T. 383 The formation of polarized marks can be ascribed to a process of twisting or gliding caused by lateral stresses in the fibre, which bring about a depolymerisation of the cellulose at these points. **1939** *Jrnl. Biol. Chem.* CXXVII. 253 The depolymerizing effect of pancreatin on the nucleic acid. *Ibid.* 257 The fully depolymerized substance has a molecular weight of 1341. *Ibid.* 258 Desoxyribonucleic acid, depolymerized by Feulgen's method.., is dephosphorylated by crude and by purified enzyme. **1962** H. Heath in A. Pirie *Lens Metabolism Rel. Cataract* 363 This has been attributed to the consequent lack of depolymerizing peroxide and hence the accumulation of large molecules of the mucopolysaccharide. **1962** *Lancet* 12 May 1007/2 Tomkins et al. have found that the highly polymerised (tetrameric) form of the enzyme, glutamic dehydrogenase, catalyses the oxidation of glutamic acid while the depolymerised (monomeric) form catalyses oxidation of alanine. **1964** N. G. Clark *Mod. Org. Chem.* xiv. 272 This liquid trimer is preferred to the low-boiling acetaldehyde, into which it slowly de-polymerizes under the reaction conditions. **1968** H. Harris *Nucleus & Cytoplasm* ii. 21 These phages induce depolymerization of the bacterial DNA. *Ibid.* 22 This phage does not depolymerize the bacterial DNA.

deport, *v.* Add: **5. b.** In Indian use, = *Detain *v.* 1. So **deporta·tion** = *Detention 1.
1909 J. Morley *Indian Speeches* 144 Great uneasiness is growing in the House of Commons as to the matter of deportation. You know what deportation means. It means that nine Indian gentlemen on December 13 last were arrested and are now detained. *Ibid.* 149 If he is one of these nine deported men, he is not put into contact with criminal persons. **1910** V. Chirol *Indian Unrest* vii. 99 The deportation struck just at that type of agitator whose influence is most pernicious because it is most subtle. **1914** Mrs. Besant *India & the Empire* 123 Harmless men like my friend Lajpat Rai, who is here, are deported or imprisoned without trial.

deportable (dĭpǭə·ɹtăb'l), *a.* [f. Deport *v.* + -able.] Liable to, or punishable by, deportation.
1891 Stevenson *Vailima Lett.* (1895) 100 This intervention would have been a deportable offence. **1957** T. P. Neill *Common Good* xi. 182/1 No immigrant is admitted if he cannot show that he is not liable to become a public charge, and if he does so within five years, he is automatically deportable.

deportee (dĭpǭəɹtī·). [f. Deport *v.* + -ee[1].] One who is or has been deported; *spec.* in Indian use, = *Détenu.
1895 *Westm. Gaz.* 13 Dec. 5/1 One party of fifteen deportees from Constantinople having been massacred. **1909** J. Morley *Recoll.* (1917) II. 309 The failure to tell the deportee what he is arrested for. **1910** V. Chirol *Indian Unrest* vii. 99 The grounds on which Government announced the release of these deportees last winter were even more unhappily chosen than the moment for the announcement. **1914** *Morn. Post* 3 Feb. 9/4 To intercept the Umgeni and give the deportees an opportunity of returning to their homes. **1920** *Glasgow Herald* 13 Mar. 9 The Dongola embarked yesterday at Devonport 100 Russian prisoners of war..., 9 Russian deportees, and an escort. **1939** A. Toynbee *Study Hist.* V. 124 The Milesian deportees were settled at the mouth of the River Tigris. **1959** *Times Lit. Suppl.* 6 Mar. 126/3 In that sparsely populated, primitive and depressed region..the deportees were more or less abandoned to fend for themselves.

deposit, *sb.* Add: **6.** *deposit account* (earlier U.S. example); (sense 3) *deposit bed, gold, mine.*
1851 C. Cist *Cincinnati* 89 Their average deposit account during that period was about eight hundred thousand dollars. **1849** C. Lanman *Lett. from Alleghany Mts.* i. 11 Heretofore the gold ore of Lumpkin county has been obtained from what is called the deposit beds. *Ibid.* 17 The deposit gold is extracted from the gravel by means of a simple machine called a rocker. **1833** H. Barnard in *Maryland Hist. Mag.* (1918) XIII. 346, I..hired a horse for 50 cents to go down to see the deposite mines, which are spread over the whole country.

depositional (dī:pŏzi·ʃənăl), *a.* [f. Deposition 8 + -al.] Of, relating to, or caused by deposition.
1900 *Geogr. Jrnl.* XVI. 461 Subordinate or local systems of crust-strains..are concentrated along old depositional and structural limits. **1902** [see *Aggrade *v.*]. **1967** *Oceanogr. & Marine Biol.* V. 120 Finer material is moved by longshore drift and currents to build up such depositional features as tombolos on the lee shore.

depot. Add: **4.** (Earlier example.)
1795 tr. *Moritz's Travels* 241 There was written on the sign: 'The Navigation Inn'; because it is the *depôt*, or storehouse, of the colliers of the Trent.
b. *Physiol.* The site of an accumulation or deposit of a substance (esp. fat) in an animal body. So *attrib.*, applied to any substance stored for eventual absorption by the organism, or to an action or process concerned with the deposition of such a substance.
[**1835–6** [in Dict., sense 1].] **1906** L. Hill *Rec. Adv. Physiol. & Bio-Chem.* xi. 288 The proteid metabolism..only begins to increase in the final stage of starvation when the reserve of depôt fat is almost exhausted. **1912** E. H. Starling *Princ. Human Physiol.* xi. 884 From the physiological standpoint the most important intracellular depôt of fat is in the liver. **1930** *Jrnl. Biol. Chem.* LXXXVII. 148 Fat obtained from the various depots. *Ibid.*, The total depot fat. **1936** *Nature* 21 Mar. 479/1 As we pass from depot fats of aquatic to those of land animals we find marked simplification in the mixed fatty acids. **1959** *Chambers's Encycl.* V. 601/1 The composition of the fat depots is practically identical with the fat in the food. **1961** *Lancet* 12 Aug. 345/1 A steroid preparation which is highly concentrated, and has a depot effect. *Ibid.* 9 Sept. 577/2 Daily injections.. were replaced by a single intramuscular depot injection. **1970** Passmore & Robson *Compan. Med. Stud.* II. vi. 16/2 There are two main types of insulin in clinical use, those with a rapid onset and short duration of action, and those whose action is slow in onset and lasts longer, the depot insulins.
Also **depot** *v. trans.*, to place in a depot.
1921 H. G. Ponting *Gt. White South* 274 When near the summit, Captain Scott told off four more of the party to depôt their surplus and return.

deprecating (de·prĭkēitiŋ), *ppl. a.* [f. Deprecate *v.* + -ing[2].] That deprecates or expresses disapproval or disavowal; deprecatory.
1871 L. W. M. Lockhart *Fair to See* xviii, A bright, but withal deprecating, smile on her lovely face. **1919** Wodehouse *Damsel in Distress* xi, Albert waved a deprecating hand. **1925** M. A. Lowndes *Some Men & Women* 199 With a queer, half-deprecating, half-humorous look on his handsome face.

depreciating, *ppl. a.* (Earlier Amer. example.)
1777 J. Adams *Wks.* (1854) IX. 463 There is so much injustice in carrying on a war with depreciating currency that we can hardly pray with confidence for success.

depressant, *a.* and *sb.* Add: **A.** *adj.* **b.** In ore and froth flotation: having the property of a depressant (sense *c).
1945 A. F. Taggart *Handbk. Mineral Dressing* xii. 32 Lime..present in excess in sulphide flotation..has a definite depressant effect on the copper sulphides other than chalcopyrite. **1959** *Chambers's Encycl.* X. 235/2 Alkalis are generally depressant in action on all sulphide flotations, preventing the collectors from working.
B. *sb.* **b.** Also **depressent.** A depressing influence.
1894 Stevenson & Osbourne *Ebb-Tide* i. v. 87 So strong a tonic to the merely weak, and so deadly a depressent to the merely cowardly. **1923** Kipling *Irish Guards in Gt. War* I. 121 There are times when Extreme Unction can be a depressant. **1961** *Times* 22 Mar. 18/3 To the ambitious athlete there is really no reason why standards should not be an incentive rather than a depressant.
c. In ore and froth flotation, an agent that causes certain of the solids to sink while not affecting the others.
1934 Webster, *Depressant*..*Ore Dressing*, a reagent that depresses. **1939** A. M. Gaudin *Princ. Mineral Dressing* xv. 393 Practical depressants for sulphide flotation include alkali hydroxides. **1951** *Engineering* 19 Oct. 492/1 Agents, known as depressants, are..used to promote wetting of.. constituents, such as pyrites, which would otherwise float with the coal. **1962** *Gloss. Terms Coal Preparation (B.S.I.)* 24 *Depressant*, a substance which when added to a pulp prevents a particular mineral or minerals from floating.

depressed, *ppl. a.* Add: **5. b.** *depressed area*, an area of economic depression.
1928 *Britain's Industrial Future* xx. 276 Already the local rates in the depressed areas have to shoulder too large a part of the burden of relieving unemployment. **1958** *Listener* 11 Sept. 371/2 There are no depressed areas now.
c. *depressed class(es)*, in India, persons of the lowest castes, 'untouchables'.
1931 *Economist* 17 Oct. 696/1 The other Minorities, notably the Depressed Classes and the Anglo-Indians, have still got to be fitted into the scheme. **1957** *Encycl. Brit.* X. 13/2 He [*sc.* Gandhi] undertook several more fasts in the interests of communal tolerance and the rights of the depressed classes.

depressent, var. *Depressant *sb.* b.

depressingness (dĭpre·siŋnės). [f. Depressing *ppl. a.* + -ness.] The quality of being depressing.
1923 *Glasgow Herald* 9 June 7, I wonder if he finds 'pervading depressingness' in these verses. **1927** *Chambers's Jrnl.* 3/2 More likely to stop a fellow from getting back to the mark, by its wretched 'depressingness'.

depression. Add: **5.** spec. *the Depression*, the financial and industrial 'slump' of 1929 and subsequent years. Also *attrib.*
1934 A. Huxley *Beyond Mexique Bay* 233 Since the depression, books on Mexico have been almost as numerous.. as books on Russia. **1935** 'J. Guthrie' *Little Country* xiii. 212 'I thought you had a baby.' 'No, darling,' said Carol. 'None of us are having them now. It's the depression.' **1935** *Punch* 19 June 719/1 All the wireless sets in Little Wobbly are pre-depression models. **1957** M. Sharp *Eye of Love* iii. 39 It was the Depression that had finished him off. **1963** H. Garner in R. Weaver *Canad. Short Stories* 2nd Ser. (1968) 37 An old Scots syndicalist I'd met on a road gang..in the early years of the depression.
6. b. *Psychol.* Freq. a sign of psychiatric disorder or a component of various psychoses, with symptoms of misery, anguish, or guilt accompanied by headache, insomnia, etc.
1905 *Psychol. Rev.* XII. 111 If these symptoms of depression—the motor retardation, the difficulty of apprehension and of association—become aggravated, one finds various forms of melancholia. **1934** H. C. Warren *Dict. Psychol.* 73/1 *Depression*..the pathological usage refers to a mood of pronounced hopelessness and overwhelming feeling of inadequacy or unworthiness. **1960** Koestler *Lotus & Robot* ii. viii. 202 Even patients with severe depression-psychosis..turned their heads slowly and worked up a mask-like smile. **1962** *Lancet* 2 June 1171/2 Even psychiatrists may profit from the reminder that 'events at the onset of depression..must be interpreted with caution for failure at work..or in a love affair may be early symptoms, rather than causes'.

depressive, *a.* Add: **2. b.** Involving or characterized by depression as a psychiatric illness. Hence as *sb.*, one who suffers from this condition. Cf. *Manic-depressive.
1905 *Psychol. Rev.* XII. 113 The German alienists are taking up again..depressive insanity, to which they rightly ascribe a great importance. **1951** R. E. Money-Kyrle *Psychoanalysis & Politics* iv. 84 Religion is a form of psychotherapy which promotes a belief in the existence of idealized good objects against persecutory and depressive guilt. **1962** *Lancet* 2 June 1171/2 An undiagnosed depressive illness is one of the commonest fatal mistakes in medicine. **1962** *Times* 13 Dec. 15/2 Antonio was an endogenous depressive. **1965** J. Pollitt *Depression & its Treatment* iii. 33 Guilt forms a nucleus for many delusional complaints of retarded depressives.

depre·ssurize, *v.* [f. De- II + *Pressurize *v.*] *trans.* To cause an appreciable drop in the pressure of the gas inside (a container, etc.), esp. to a natural level from an artificially sustained level (in quot. 1960, to subject to decompression: cf. *Decompress *v.* a). So **depre·ssurized** *ppl. a.*
1944 *Aviation* July 278/3 While in action, the cabin is de-pressurized and demand oxygen employed. **1947** *Jrnl. R. Aeronaut. Soc.* LI. 972/1 This altitude was the maximum at which it was reasonably possible for a fit pilot to retain sufficient control to make a normal descent, should the cabin for any reason become de-pressurised. **1958** *Daily Mail* 12 Nov. 4/5 There was a hissing sound as Larry helped him depressurise the airlock. **1960** Koestler *Lotus & Robot* vi. 188 They cannot go back into the pressure chamber to dissolve the bubbles by getting gradually depressurized. **1963** *Times* 24 May 12/7 An Aer-Lingus Viscount..made an emergency dive of 12,000 ft. today after a door unfastened and the passenger cabin became depressurized.

depriment, *a.* Delete †*Obs.* and add later *fig.* example. The word is still *rare.*
1939 *Mind* XLVIII. 282 We have the peculiar 'depriment' emotion of reverence in the contemplation of the moral law.

deprint (dī·print), *sb.* ?*Obs.* [f. De- + Print *sb.*] An offprint. Hence **depri·nt** *v.* (*Cent. Dict.* Suppl., 1909).
1885 Skeat in *Academy* 22 Aug. 121/3 Various terms, such as 'deprint', 'exprint', &c., have been proposed to denote a separately printed copy of a pamphlet. **1895** (*title*) Modern Language Notes: Deprints of Articles by George C. Keidel, Ph.D.

deprived, *ppl. a.* Add to def.: in modern use freq. applied to children who lack the benefits of normal home life, parental affection, etc.
1945 *Lancet* 3 Mar. 294/1 Many of them tell how bitter were their feelings as 'deprived children'. **1957** *Times Lit. Suppl.* 25 Oct. 637/1 The plot concerns a middle-aged spinster who was a 'deprived' child overshadowed by her brother and his tragic early death..in a search for the secret of her unhappy childhood.

deproletarianize (dī:prōulitēə·riănaiz), *v.* Also **deproletarize.** [f. De- II. 1 + Proletarianize *v.*] *trans.* To free of proletarian character or qualities; to cause to lose proletarian nature. Also *absol.* Hence **deproleta:rianiza·tion**; **deproleta·rianized** *ppl. a.*; **deproletariani·zable, deproleta·rian** *adjs.*
1954 *Encounter* Dec. 29/2 The net effect of capitalism, then, is in the long run to deproletarise. **1957** T. Kilmartin tr. *Aron's Opium of Intellectuals* iii. 72 In concrete terms, what would the triumph of a 'deproletarianised' working class involve? **1959** *Economist* 16 May 601 The slogan for the 1960s—more exciting perhaps than euphonic—should be to deproletarianise British society, both materially and psychologically. *Ibid.* 602/1 The social importance of what can be called the 'deproletarianisable' income groups also means..that the Conservatives should never again consider imposing restrictive controls on hire purchase. *Ibid.* 30 May 834/1 The signs of deproletarianization are as yet few and far between... A 'deproletarianized' family will surely expect to pay an economic price or rent for its house, not to have a council house provided for it. **1961** R. Williams *Long Revolution* 327 In the 'deproletarian' period since 1945. **1967** *Listener* 17 Aug. 205/1, I have never had anything but hatred and opposition for deproletarianising and back-to-the-land schemes.

deproteinize (dĭprōu·tinaiz), *v.* [f. De- II. 1 + Protein + -ize.] *trans.* To remove protein from, usu. as a purification measure in a process of chemical isolation. Hence **depro:teiniza·tion**; **depro·teinizing** *vbl. sb.* and *ppl. a.*
1956 *Nature* 4 Feb. 237/2 The plasma was deproteinized, and the resultant liquid..desalted. **1962** J. H. Kinoshita et al. in A. Pirie *Lens Metabolism Rel. Cataract* 405 The lens was homogenized in the deproteinizing mixture. **1963** F. Haurowitz *Chem. & Function of Proteins* ii. 20 This method finds extensive use in clinical laboratories in the deproteinization of biological fluids. **1967** *Oceanogr. & Marine Biol.* V. 347 Electron micrographs of cuticle deproteinized with hot potassium hydroxide.

depside (de·psaid). *Chem.* [ad. G. *depsid* (Fischer & Freudenberg 1910, in *Justus Liebig's Ann. d. Chem.* CCCLXXII. 35), f. Gr. δέψ-ειν to knead (see quot. 1910) + -ide.] Any of a group of organic compounds found principally in lichens (see quot. 1956).
1910 *Jrnl. Chem. Soc.* XCVIII. i. 266 It will no doubt be possible to prepare, by similar means, a large number of analogous substances from other hydroxybenzoic acids, and since many undoubtedly occur in nature, as, for example, tannin, it is considered advisable to classify them under the collective name *depside* (δέψειν tan). **1935** [see next]. **1956** E. H. Rodd *Chem. Carbon Compounds* III. xii. 784 Depsides are substances derived from two or more molecules of phenolcarboxylic acids by esterification of a carboxyl group of one molecule..with a hydroxyl group of a second. **1959** [see next]. **1967** M. E. Hale *Biol. Lichens* viii. 111 The depsides..make up the largest group of lichen substances.

depsidone (de·psidōun). *Chem.* [a. G. *depsidon* (Y. Asahina 1934, in *Acta Phytochimica* VIII. 34), f. prec. + -one.] Any of a group of compounds related to the depsides and

found almost exclusively in lichens, having the benzene rings joined by an ether linkage as well as by the ester linkage.
1935 *Chem. Abstr.* XXIX. 147 The remaining compds. of the orcinol group of Zopf have, in addn. to the depside linkage, a diphenyl ether linkage between the ortho OH groups with the consequent formation of a closed-ring structure. These compds. have been termed depsidones. **1959** *New Biol.* XXIX. 86 Nineteenth-century chemists investigated the lichen acids very thoroughly...They established that the acids are all either depsides, which are rather readily broken down by alkalis, or depsidones, which are not so broken down.

depth. Add: **I. 3. c.** *in depth*, profoundly; with deep insight or penetration. Hence (hyphenated) as an *attrib.* phr.
1959 B. C. Brookes in R. Quirk et al. *Teaching of English* v. 148 It takes the scientist perhaps twenty or thirty years to reach..that mature grasp of his subject which enables him to see it clearly both in depth and in relation to other disciplines. **1959** *Listener* 9 July 69/2 Dr. Waidson presumably had the choice of writing in depth about a few novelists or of skimming over as many authors as he could get into his survey. **1966** *Punch* 19 Jan. 83/3 Why haven't you asked my views on Sport? Not quite interviewing me in depth are you, Mr. Haverwood? **1967** *Electronics* XL. 35 (Advt.), He's backed by General Electric's total electronic capability—in-depth technical backup assistance. **1971** *Guardian* 20 Feb. 9/6 What I do is history plus reportage, an in-depth extension of my former journalism.
5. *depth of field, depth of focus*: see *Field *sb.* 16 c, Focus *sb.* 2 e in Dict. and Suppl.
12. b. Applied *attrib.* to an interview, approach, etc., that seeks to discover motives or attitudes that are not normally divulged, the results of which are used esp. as a basis for certain advertising techniques. Cf. **depth psychology* (see sense *IV).
1948 *Jrnl. Appl. Psychol.* Oct. 550 To orient ourselves to the problem and sketch in its broad outlines we began with a series of a hundred 'depth interviews' of television families. **1957** *Bookseller* 28 Sept. 1216/3 No real harm can be done if, using the results brought to the surface of the human subconscious and unconscious mind by the 'depth diggers', one astute manufacturer sells more of a commodity than a rival marketing a brand of equal quality and price. **1957** *Times Lit. Suppl.* 1 Nov. 661/2 Space-salesmen have now become students of the sub-conscious mind, and advertisements are governed by the 'depth-approach'. **1970** *Guardian* 10 Apr. 7/3 Skilled depth-interviewers.
IV. depth bomb, charge, a bomb capable of exploding under water; so **depth-charge** *v. trans.*, to attack with depth charges; **depth finder,** an apparatus for sounding the sea; *spec.* **sonic depth finder,** one in which the measurement is made by timing the echoes from the sea-bottom of sound waves transmitted from the ship; **depth-keeping** *vbl. sb.*, the maintenance of a submarine, fishing-net, etc., at a certain depth; **depth psychology** [tr. G. *tiefenpsychologie* (S. Freud *Das Ich und das Es* (1923) i. 17)] = *Psycho-analysis b; hence *depth-psychological* adj.; **depth psychologist,** one who practises or is skilled in depth psychology; **depth recorder,** a device for recording either how far below the surface of the sea it is or the depth of water below a vessel; so **depth-recording** *vbl. sb.*
1918 E. S. Farrow *Dict. Mil. Terms* s.v., Submarines.. are pursued and destroyed by dropping depth bombs from the observing aircraft or warship. **1944** *Jrnl. R. Aeronaut. Soc.* XLVIII. 219 Depth Bombs (D.B.)..have a very thin case and are detonated by hydrostatic fuse. **1917** *War Illustr.* 18 Dec. 361 Telegraph to seamen..who prepare to drop Depth Charges to destroy U-boat. **1920** *Blackw. Mag.* Mar. 315/2 The depth-charge thrower, which later came into common use..was as yet far from perfection. **1928** C. F. S. Gamble *N. Sea Air Station* 14 Lieutenant Williamson.. described how depth-charges (he said 'bombs exploding 20 feet under the surface of the water') might possibly be used to destroy submarines. **1918** *Daily Mail* 23 Sept. 2/4 From the captain of a U-boat..came to me the following description of what it is like to be depth-charged. **1940** *War Illustr.* 19 Jan. 628/3 That would find the submarine for us—and then we could depth-charge it. **1923** *Hydrographic Rev.* I. 72 Navy sonic depth finder recently developed at the Engineering Experiment Station, Annapolis, Maryland. **1948** R. de Kerchove *Internat. Maritime Dict.* 232/1 Echo sounder... Also called depth finder. **1916** Kipling *Tales of 'The Trade'* i. 6 Depth-keeping,..very difficult owing to heavy swell. **1923** *Blackw. Mag.* Oct. 527/1 Accurate depth-keeping being out of the question, I surfaced. **1959** H. Barnes *Oceanogr. & Marine Biol.* i. 26 Modern practice is tending more to the use of a depth-keeping device, either on the net itself or on the towing wire...The paravane principle is often employed. **1958** *Times Lit. Suppl.* 23 May p. xii/2 A sympathetic understanding of depth-psychological concepts. **1947** *Partisan Rev.* XIV. 528 Those commonplaces are too humdrum for the depth psychologist. **1957** *Observer* 27 Oct. 18/6 Motivational research took the place of market research, and the depth psychologists became the *eminences grises* of salesmanship. **1927** J. Riviere tr. *Freud's Ego & Id* i. 18 In the last resort the quality of being conscious or not is the single ray of light that penetrates the obscurity of depth-psychology. **1947** W. Empson *Seven Types of Ambiguity* (ed. 2) p. x, Some critics do not like to recognise this process because they connect it with Depth Psychology. **1960** H. Read *Forms of Things Unknown* viii. 133 Myth and dream, symbol and image—all the paraphernalia of depth-psychology—are conceived as shadow play. **1963** L. B. Lefebre tr. *Boss's Psychoanalysis & Daseinsanalysis* v. 87 The unconscious became so much the mark of psychoanalytic theory that psychoanalysis, and all doctrines derived from it, eventually became known as 'depth' psychologies. 'Depth' entered the picture because Freud..undertook to view mental phenomena in terms of a 'topographic' approach, and to regard the unconscious as a 'psychical locality'..'below' consciousness. **1911** *Encycl. Brit.* XIX. 295/1 Lord Kelvin's sounding machines..in the later form known as the 'depth recorder', where..results are obtained by the automatic record of the position of a piston forced upwards in a tube by..increased pressure. **1961** *Listener* 24 Aug. 269/1 After the war, echo-sounding and depth-recorders were introduced to fishing boats, and echoes were received of big shoals of fish. **1959** H. Barnes *Oceanogr. & Marine Biol.* i. 23 If the speed of the towing vessel can be accurately controlled, then the depth of the net for a given towing speed may be determined in a series of trial runs, using a depth-recording device attached to the net.

deputize, *v.* **2.** Delete 'usually in reference to musical engagements' and add examples of wider use.
1900 *Westm. Gaz.* 27 Sept. 5/2 Sir Samuel Scott being absent in South Africa, has cabled the neighbouring Tory candidate, Mr. Boulnois, to deputise for him. **1904** *Daily Chron.* 15 Sept. 3/6 He was called upon at various times to deputise for other high officials. **1906** *Sat. Rev.* 30 June 807 With the exception of Mr. Stead and Mr. T. P. O'Connor, the representatives of the press were deputising in the absence of their chiefs. **1915** *Observer* 31 Jan. 9/6 Mrs. Levison..deputising for Lady Jellicoe, read to the company a letter. **1929** D. G. Mackail *How Amusing* 472 The parlourmaid was deputizing for the housemaid. **1955** *Times* 5 Aug. 4/1 Evans, the Kent wicketkeeper for whom McIntyre deputized in the Leeds Test, will not play again this season.

deracialize (dīrē·ʃiălaiz), *v.* [f. De- + Racial + -ize.] *trans.* To remove racial characteristics or features from. So **deracializa·tion.**
1899 S. N. Patten *Devel. Eng. Thought* 365 Religious leaders, no longer deracialized by education, are dissatisfied with foreign platitudes and commonplaces. **1907** *Westm. Gaz.* 21 Feb. 2/2 The good service they have rendered by deracialising (so to speak) the elections. **1931** A. Keith *Ethnos* 27 Where Huxley went wrong was in believing that when Europeans belonging to separate racial stocks..were planted together..they became, if I may coin a term, deracialized. *Ibid.* 90, I am convinced that deracialization is possible.

‖ **déraciné** (derasine), *a.* [Fr.: see Deracinate *v.*] 'Uprooted' from one's (national or social) environment. Also as *sb.*
1921 *19th Cent.* May 770 The unseen Jew *déraciné* provides munitions of argument for the revolutionary group. **1926** J. Buchan *Dancing Floor* i. vi, She rides well, but her manners are atrocious. Lord, how I dislike these *déracinés*! **1931** in W. Rose *Outl. Mod. Knowl.* 751 To be delocalised is not to be déraciné. **1935** Auden & Isherwood *Dog beneath Skin* iii. iv, I'm quite déraciné, as they say in Bloomsbury. **1952** D. Davie *Purity of Diction in Eng. Verse* ii. 24 The typical *déracinée*, Fanny Price. **1964** R. Church *Voyage Home* v. 71 The dreadful self-consciousness of so many *déraciné* Americans, aping the hyper-civilized European decadents. **1967** *Listener* 22 June 832/1 Our 'dynamic' epoch has..produced a succession of *déracinés* ranging from Bartók to Stravinsky.

derate (dīrē·t), *v.* [f. De- II. 2 + Rate *sb.*[1]] *trans.* and *intr.* To diminish or remove the burden of rates (upon). Hence **dera·ting** *vbl. sb.* and *ppl. a.*
1928 W. S. Churchill in *Hansard's Parl. Deb.* Ser. v. CCXVI. 849 Twelve months will be required after that Bill is passed, to enable the new valuation to be made for the purposes of de-rating. *Ibid.* 869 The cost of the complete de-rating of agriculture is about £4,750,000. **1928** *Daily Express* 22 June 2/4 [He] asked the Government to expedite their derating scheme, which is expected to be worth £3,000,000 a year to the coal industry. **1928** *Daily Tel.* 17 July 11/4 [He] proposed to derate hospitals supported by voluntary subscriptions. **1955** *Times* 10 June 9/4 At the time they were de-rated there might have been some justification. *Ibid.* 15 June 3/1 Lord Meston said that the whole subject of derating should be reviewed.

deration (dīræ·ʃən), *v.* [De- II. 2 + Ration *sb.*] *trans.* To free (a rationed commodity) from rationing. Also **dera·tioning** *vbl. sb.*
1920 *Glasgow Herald* 25 Nov. 8 The Food Controller was able to announce..last night that sugar will be derationed as from Monday. *Ibid.* 15 One substantial instalment of decontrol I can announce to you to-night—the derationing of sugar. **1945** *John o' London's* 16 Nov., Goods become derationed, only because some bureaucracy rationed them. **1955** *Times* 16 May 4/6 Withdrawal of price controls..hand in hand with derationing. **1959** *Listener* 1 Jan. 8/1 A minor news item of 1958 that strikes the fancy as one thinks back is that coal was 'derationed'.

derationalize (dīræ·ʃənălaiz), *v.* [f. De- II. 1 + Rationalize *v.*] To deprive of reason. So **derationaliza·tion.**
1871 C. Hodge *Systematic Theol.* I. i. iii. 279 To call upon men to worship gravitation..is to call upon them to derationalize themselves. **1879** W. James *Coll. Ess. & Rev.* (1920) 129 Further considerations..may supervene and make relative or derationalise a mass of thought. **1938** H. Read *Poetry & Anarchism* iv. 48 The dialectical method ..leaves us with an art which is not so much art as an instrument of derationalization.

deratization (dīrætaizē·ʃən). [f. De- + Rat *sb.*[1] + -ization.] The expulsion or extermination of rats.
1914 *Standard* 13 Oct. in *N. & Q.* 11th Ser. X. 386 The Board of Trade communication gives a translation of the circular, in the course of which it more than once uses the curious word 'deratization', meaning, apparently, the clearing away of rats. **1921** *Glasgow Herald* 28 May 5 A discussion of 'Deratisation of Ships at British Ports'. **1929** *Stat. Rules & Orders No. 832* (*title*) The Public Health (Deratisation of Ships) Regulations. **1949** *Times* 9 Oct., The port health authority's 'measures against rodents' include 'deratization' and 'rat-proofing'.

Derby. Add: **1. b. Derby day** (earlier example); **Derby dog** (example).
1838 *Observer* 26 Aug. 2/2 During last Epsom races, on the Derby day we believe, [etc.]. **1867** *Punch* LII. 227/1 The Mystery of the Derby dog..the never-failing apparition of the Derby dog at Epsom.
d. Applied to any kind of important sporting contest; also *air Derby* (see also *aerial Derby*, s.v. *Aerial *a.* 5); *local Derby*, a match between two teams from the same district.
1909 *Daily Chron.* 17 June 5/6 The twenty-ninth Medway Barge Sailing Match, known locally as 'the barge Derby'. **1914** *Daily Express* 3 Oct. 3/1 A local Derby [*sc.* football match] between Liverpool and Everton. **1914** *Whitaker's Almanack 1915* 822/2 Air 'Derby' round London (94½ miles). **1919** *Sphere* 28 June 259 (*heading*) An air derby at 129 miles per hour. **1962** *BBC Handbk.* 37 It would still be right for local talent to be nursed and local derbies to be played.
2. b. A kind of sporting-boot having no stiffening and a very low heel (see also quot. 1968).
1901 *Daily News* 23 Feb. 6/4 The Prince Consort is represented..as wearing low-heeled, square-toed 'Derbies', with buckles on them. **1904** *Westm. Gaz.* 15 Apr. 10/2 Russia calf Derbys for shooting-boots. **1968** J. Ironside *Fashion Alphabet* 130 *Derby*,..the most common form of shoe. A tie shoe with eyelets and laces, the quarter and facings stitched *on top* of the vamp.
5. Denoting a variety of porcelain made at Derby, *esp.* a soft-paste porcelain made from about 1750, Crown Derby being a variety made from 1784. Cf. *Crown Derby* s.v. *Crown *sb.* 35.
1850 [see *Crown Derby* s.v. *Crown *sb.* 35]. **1868** C. L. Eastlake *Hints Household Taste* ix. 194 The qualities which distinguish old Chelsea, Derby, Worcester, and Plymouth china are well known to connoisseurs. **1869** Lady C. Schreiber *Jrnl.* (1952) I. 1. One small Derby group. *Ibid.* 4 A small Derby statuette of Neptune. **1875** W. F. Tiffin (*title*) A Chronograph of the Bow, Chelsea and Derby Porcelain Manufactories. **1885** *Encycl. Brit.* XIX. 641/2 The Derby under-glaze blue was remarkably fine. **1957** Mankowitz & Haggar *Encycl. Eng. Pottery* 70/2 The work of the earliest phase of Derby porcelain was nonphosphatic. ..The next phase of Derby was characterized by the use of pale delicate colours, [etc.].
b. Derby red = *chrome red* (Chrome 3).
1904 Goodchild & Tweney *Technol. & Sci. Dict.* 155/1 *Derby red*, a scarlet red pigment of good covering and staining power, sometimes used as a substitute for vermilion. **1937** *Thorpe's Dict. Appl. Chem.* (ed. 4) I. 550/1 *Austrian cinnabar.* Derby red, Chinese red.
6. *Derby* or *Derbyshire cheese*, a hard, pressed cheese made from partly skimmed milk, produced chiefly in the Derbyshire district. Also *ellipt.*
1902 *Encycl. Brit.* XXVII. 355/2 Derby cheese in its best forms is much like Leicester, being 'clean' in flavour and mellow. **1905** W. H. Simmonds *Pract. Grocer* III. 79 The true 'Derbyshire' cheese..is a small cylindrical or flat and thin cheese of pale colour, and generally of rich, buttery quality. 'Derby Goudas' are a variety shaped like the Dutch Gouda. **1955** J. G. Davis *Dict. Dairying* (ed. 2) 189 *Derbyshire or Derby cheese*, one of the oldest of our national cheese, but it was not until recently that a uniform method came to be adopted for its manufacture. **1970** *Listener* 12 Nov. 661 You can buy a special Derby cheese in good food shops in London that's got a green stripe in it—known as Sage-green Derby.
7. *Derby scheme*: in the war of 1914–18, a recruiting scheme initiated in October 1915 by the seventeenth Earl of Derby. Also *Derby recruit* and simply *Derby*, a soldier recruited under this scheme.
1915 *Times* 4 Dec. 9/6 Last week of the Derby Scheme. *Ibid.* 20 Dec. 9/2 Men who have been attested and classified under the Derby scheme. **1917** P. Gibbs *Battles of Somme* 177 Old English regiments with new men in them, including some of the 'Derby recruits'. *a* **1918** J. T. B. McCudden *Five Yrs. R.F.C.* (1919) 198 It was at that time [Feb. & Mar. 1917] that the 'Derby' scheme was operating. **1925** Fraser & Gibbons *Soldier & Sailor Words* s.v., Men of the 'Groups' of 'Derbies', awaiting their turn to be called up,.. wore armlets lettered 'G.R.' (General Reserve). **1927** W. S. Churchill *World Crisis, 1916–18* i. x. 239 It was evident that the Derby scheme could only be a palliative.

derealize (dīrī·ălaiz), *v. Philos.* [f. De- II. 1 + Realize *v.*[1]] To deprive of reality, make unreal. So **dere·alizer** (see quot. 1909); **dere·alizing** *ppl. a.*
1889 W. James in *Mind* XIV. 351 Corroborated, not de-realised, by the ultimate principle of my belief. **1904** —— *Meaning of Truth* (1909) iv. 109 We have no transphenomenal absolute ready, to derealize the whole experienced world by, at a stroke. **1909** —— *Pluralistic Univ.* ii. 49 Pluralism, in exorcizing the absolute, exorcizes the great de-realizer of the only life we are at home in. **1964** B. Frechtman tr.

Sartre's Saint Genet 161 And what if, by means of this makebelieve, he drew everything—trees, plants, utensils, animals, women and men—into a derealizing whirl?

deregister (dīre·dʒistəɹ), *v.* [f. DE- + REGISTER *sb.*[1]] *trans.* To remove from a register. Hence **de:registra·tion.**

1924 *Glasgow Herald* 19 Nov. 11 Disobedience would almost certainly have caused de-registration of the union under the Arbitration Act. **1925** *Ibid.* 23 Jan. 10 Mr Justice Powers indicated plainly that he would deregister the union if he did not receive a pledge that the men would obey the award. **1928** *Daily Express* 4 Feb. 8/7 The late Dr. Axham was de-registered for assisting Sir Herbert Barker. **1951** *Here & Now* (N.Z.) May 14/1 The compulsory union officialdom is necessarily afraid of any action that may lead to 'deregistration' and with that to the possible loss of their jobs. **1971** *Daily Tel.* 15 Feb. 1/1 All unions will be automatically registered with the new Registrar, and it will be up to unions to 'de-register' themselves. *Ibid.* 13 Apr. 6/3 De-registration or non-registration will lay the unions open to heavy taxation.

derequisition (dīrekwizi·ʃən), *v.* [f. DE- II. 1 + REQUISITION *v.*] *trans.* To convey (requisitioned land, property, etc.) to its original owner. Hence as *sb.* Also **derequisi·tioning** *vbl. sb.*

1945 *News Chron.* 7 May 3/1 In the North, Blackpool's vast Olympia building, just derequisitioned, has been booked. **1946** *Ann. Reg. 1945* 3 The de-requisitioning of factory and storage space. *Ibid.* 342 Our studio space remained restricted through delays in de-requisition. **1955** *Times* 9 July 11/7 The corporation had agreed to seek authority from the Minister to derequisition the plaintiffs' premises. **1957** *Times Lit. Suppl.* 13 Dec. 753/2 The fishing inn and its stream provide a lively contrast with the de-requisitioned mansion beside the gloomy lake.

dereserve (dīrizɜ̄·ɹv), *v.* [f. DE- II. 1 + RESERVE *v.*] *trans.* To deprive of exemption from military service, as in the wars of 1914–18 and 1939–45. Hence **dereserva·tion,** the action of the verb; **derese·rved** *ppl. a.*

1941 *Hutchinson's Pict. Hist. War* 22 Jan.–18 Mar. 247/1 We are reviewing the reserved occupations, and carrying out a great deal of dereservation. **1941** in *Amer. Speech* XVI. 277 Men in Class B who have been dereserved may be called up for civil defense in the near future. **1944** *Manpower* (*Min. of Information*) 54 We have seen..how the deferment of de-reserved men is now on an individual basis.

derestri·ct, *v.* [DE- II. 1 + RESTRICT *v.*] *trans.* To remove restrictions from; *spec.* to remove a speed limit on traffic in a specified road, area, etc. So **derestri·cted** *ppl. a.*; **derestri·ction** *sb.*

1935 *Punch* 29 May 634/3 Was it really necessary, for example, to add the word 'Derestriction' to our language in order to say that the speed-limit would no longer be enforced on certain roads? **1936** 'J. TEY' *Shilling for Candles* xxv. 270 Grant took full advantage of the lunchtime lull in traffic, and in derestricted areas excelled himself in the gentle art of speed with safety. **1936** *Times* 9 Jan. 7/2 In the opinion of the Minister there had been a *prima facie* case for derestriction. **1937** *Evening News* 15 Feb. 8/3 Mr. Belisha's henchmen were busy removing the derestriction signs from the lamp-posts. **1955** *Times* 24 June 4/7 The whole of the required two million tons to derestrict coal for the household market would have to come from America. **1960** *Ibid.* 9 Feb. 14/6 Few of the Ministry of Transport driving test routes include derestricted roads. **1966** J. MILES in T. Wisdom *High-Performance Driving* viii. 83 Since we've just passed a de-restriction sign, you can start motoring. **1971** *Daily Tel.* 19 May 17 De-restriction of the present broadcasting hours..is the least independent television can reasonably expect from the Government.

de rigueur: see DE II.

derisive. Add: **b.** That causes derision, ridiculous.

1896 *Westm. Gaz.* 25 Feb. 2/1 In thirteen years he has brought a paper costing money to keep it going and with a derisive circulation to the front rank of the world's journalism. **1923** *Daily Mail* 15 May 8 Germany has provided only a derisive amount to make good that cruel injury.

derisory, *a.* Add: **b.** = prec.

1923 *Westm. Gaz.* 19 Mar., In comparison with what it was hoped to do the result is derisory. **1923** *Daily Mail* 5 June 8 Of the total German payments for reparations France has received in cash or kind the derisory amount of £14,500,000, and England the equally preposterous amount of £5,700,000. **1971** *Oxford Times* 23 July 1/6 Both rejected the present rate offer as 'derisory'.

derivation[1]. Add: **1. d.** *Mus.* Borrowing, in an organ: see *BORROW *v.*[1] 2 c. So **deri·ved** *ppl. a.* = *BORROWED *ppl. a.* 2 b.

1905 T. CASSON *Pedal Organ* 22 It is true that they often call the borrowing by another name, such as 'transmission', 'derivation' and even 'duplication', but that is not straightforward.

6. c. *Transformational Gram.* (See quots.)

1957 N. CHOMSKY *Syntactic Structures* iv. 29 Given the grammar [Σ, F], we define a derivation as a finite sequence of strings, beginning with an initial string of Σ, and with each string in the sequence being derived from the preceding string by application of one of the instruction formulas of F. **1964** E. BACH *Introd. Transformational Gram.* ii. 15 A derivation..is a sequence of strings of symbols of which the first string is an initial string and in which every string follows from the preceding one by the application of a rule.

derivational, *a.* Add: (Later examples.) Also as *sb.*

1953 C. E. BAZELL *Linguistic Form* 76 All morphemes which may make part of a base but which may not be stems are derivational affixes, more briefly derivationals. **1964** E. BACH *Introd. Transformational Gram.* iii. 36 Our ability to reconstruct the 'derivational history' of this terminal string depends on a process of matching succeeding lines in the derivation. **1964** R. H. ROBINS *Gen. Linguistics* vi. 258 Both these types may be illustrated from English. *-hood* is a class-maintaining derivational suffix.

derivative, *a.* and *sb.* Add: **A.** *adj.* **4.** *Geol.* **a.** Of fossils: occurring in rocks other than those to which they are native. **b.** Of rocks: formed from materials derived from older rocks.

1871 C. LYELL *Student's Elem. Geol.* Index, Derivative shells of the Red Crag. **1894** J. GEIKIE *Gt. Ice Age* (ed. 3) 371 The shells which they occasionally contain are probably, in most cases, derivative—they do not occupy the positions in which the molluscs themselves lived. **1900** J. E. MARR *Sci. Study Scenery* ii. 9 The derivative class has been formed by accumulation of material..not having been in a state of fusion immediately before its accumulation. **1904** GOODCHILD & TWENEY *Technol. & Sci. Dict.* 155/1 Conglomerates, sandstones, shales, and clays are good examples of derivative rocks.

derivativeness. (Later examples.)

1927 M. SADLEIR *Trollope: Commentary* 374 Undoubtedly *The Three Clerks* was derivative..and to the fact of its derivativeness may be attributed its popularity. **1967** *Listener* 26 Jan. 129/1 The author himself made this derivativeness perfectly clear: 'I am struggling to think other people's thoughts after them.'

derive, *v.* Add: **11.** Freq. in mod. use, prob. at first as a gallicism.

1895 tr. *P. Bourget's Outre-mer* ii. 36 How all literature derives from him [*sc.* Shakespeare] in every English-speaking country. **1899** *Daily News* 28 Nov. 6/5 As a draughtsman he derives from Charles Keene. **1901** *Ibid.* 22 Jan. 5/4 The theory of the mediæval empire derives immediately from Rome. **1907** *Daily Chron.* 18 Oct. 4/6 Thackeray derived straight from Goldsmith. **1971** *Daily Tel.* 19 Nov. 13/3 Richard Rountree..is powerful in a role that must derive from those paragons of policemen Sidney Poitier used to play.

derived, *ppl. a.* Add: **d.** *derived fossils* (Geol.), fossils occurring in formations other than those to which they are native. Cf. *DERIVATIVE *a.* 4 a.

1869 *Geol. Mag.* VI. 259 We must in the first place determine whether there are any derived fossils in the bed. **1940** *Cement, Lime & Gravel* Dec. 26 (*title*) Derived Upper Llandovery fossils in Bunter pebbles.

derivometer (derivǫ·mītəɹ). Delete †*Obs.* and add later example.

1928 *Times* 15 Aug. 13/6 The afterpart of the cabin..contained the derivometer for measuring the airship's deviation from the straight course.

dermabrasion (dɜ̄ːɹmăbrē^i·ʒən). *Surg.* [Blend of Gr. δέρμα skin + ABRASION.] The removal of superficial layers of the skin with a rapidly revolving abrasive tool.

1954 BLAU & REIN in *Archives of Dermatology* Dec. 758 The abrasion..should remove similar numbers of pits and pilosebaceous units...This method we have called dermabrasion. **1958** *Times* 28 Mar. 7/6 Dermabrasion—a method recently introduced for the removal of certain blemishes and lesions of the skin..consists of removing the superficial layers of the skin by 'planing' the skin with a rapidly revolving wire brush. **1967** *New Scientist* 5 Oct. 41/1 No completely satisfactory way yet exists of getting rid of unwanted tattoos...Techniques such as excision and dermabrasion at best only obscure the pattern.

dermat-, dermato-. Add: **de:rmatogra·phia** *Path.*, = *DERMOGRAPHISM (s.v. *DERMO-).

1899 L. D. BULKLEY *Man. Dis. Skin* (ed. 4) xiii. 159 A name can be written on the skin with a blunt point (dermatographia, autographism). **1912** H. FRENCH *Index Diff. Diagnosis of Main Symptoms* 771 If letters or figures are marked out on the skin in this way, they appear as though they had been written in red, so that the condition has also been termed dermatographia.

de:rmatogly·phics. [f. DERMATO- + Gr. γλυφικός (see GLYPHIC *a.* and *sb.*).] The science or study of skin markings or patterns, esp. those of the fingers, hands, and feet; also, such skin markings themselves. Hence **de:rmatogly·phic** *a.*, **de:rmatogly·phically** *adv.*

1926 CUMMINS & MIDLO in *Amer. Jrnl. Phys. Anthropol.* IX. 471 (*title*) Palmar and Plantar Epidermal Ridge Configurations (Dermatoglyphics) in European-Americans. *Ibid.*, The term 'dermatoglyphics'..is used herein for the first time...It is proposed..as a designation of the division of anatomy embracing the surface markings of skin. *Ibid.* 475 Methods of formulation are discussed..in company with the dermatoglyphic features which they symbolise. **1928** STEDMAN *Med. Dict.* (ed. 10) 276/2 *Dermatoglyphics*, the surface markings (sharply sculptured ridges) of the skin, especially of the palmar and plantar regions. **1938** *Nature* 3 Dec. 1001/1 The new quantitative method from dermatoglyphics. **1963** *Lancet* 12 Jan. 114/2 The repeated demonstration of dermatoglyphic abnormalities in the autosomal trisomies..leads us to wonder whether each autosomal trisomy syndrome will prove to be dermatoglyphically distinctive.

de·rmatomy:osi·tis. *Path.* [ad. G. *dermomyositis* (H. Unverricht 1891, in *Deutsch. med. Wochenschr.* 8 Jan. 43/2), f. DERMATO- + MYOSITIS.] A disease of unknown origin characterized by œdema, dermatitis, and muscle inflammation.

1899 in *Index-Catal. Libr. Surgeon-General's Off.* IV. 161/1. **1905** *Med. Press & Circ.* CXXX. 207 Dermatomyositis, or acute polymyositis. **1958** WALTON & ADAMS *Polymyositis* i. 2 The terms polymyositis and dermatomyositis were used almost indiscriminately by the early authors. **1963** *Lancet* 5 Jan. 58/1 Skin lesions suggestive of dermatomyositis were never seen. **1970** J. BETHLEM *Muscle Path.* 39 If the disease is accompanied by skin involvement it is referred to as dermatomyositis.

dermatophyte (dɜ̄·ɹmătofəit). [f. DERMATO- + -PHYTE.] A pathogenic fungus that grows on skin, hair, feathers, etc. Hence **dermatophy·tic** *a.*; **de:rmatophyto·sis,** a superficial infection caused by such a fungus; **de:rmatophy·tid,** a secondary skin eruption caused by a dermatophyte or its toxic products.

1882 *Syd. Soc. Lex.* II, *Dermatophytic*, relating, or appertaining, to dermatophytes. **1885** H. VON ZIEMSSEN *Dis. Skin* 511 The various dermatophytes..are not so constituted as regards size, shape, and arrangement as to be microscopically differentiated from each other. **1894** F. T. ROBERTS *Theory & Pract. Med.* II. II. xcviii. 1144 This class [*sc.* Parasitic Diseases] includes all the affections produced by the various animal and vegetable parasites that infest the human skin. We shall therefore..describe, firstly, dermatozoic affections..and, secondly, dermatophytic diseases. **1894** R. QUAIN *Dict. Med.* 614/1 Epiphytic skin-diseases. Synon.: Tineae; Dermato-mycoses or Dermatophytoses; Ringworms. **1929** *Biol. Rev.* IV. 41 The name Dermatophytes is usually employed in a restricted sense, and is applied, not to all the fungi causing dermatomycoses, but only to those which cause superficial skin diseases or ringworms. **1934** *Brit. Jrnl. Dermatol.* XLVI. 139 A young man with a typical dermatophytosis of the feet and a dermatophytid of the hands was successfully treated. **1966** W. D. STEWART et al. *Synopsis Dermatol.* xv. 312 Dermatophytid is an eruption occurring secondary to a fungal infection. **1970** REBELL & TAPLIN *Dermatophytes* (ed. 2) 4 The dermatophytes may be thought of as a group of taxonomically related fungi with affinity for cornified epidermis, hair, horn, nails, and feathers. *Ibid.* 6 Most dermatophyte thalli regularly undergo fluffy degenerative changes in laboratory culture. **1971** *Nature* 5 Feb. 435/1 In industry..there were more than 5,000 spells of incapacity attributable to dermatophytosis from June 1967 to June 1968.

dermo-. Add: **dermogra·phia, dermo·graphism** *Path.*, an irritable condition of the skin, occurring esp. in cases of urticaria, in which lines drawn on it leave a reddish elevated mark.

1900 *Med. Rec.* (N.Y.) 4 Aug. 197/2 (*heading*) Dermographia and anidrosis. **1896** H. LELOIR in T. L. Stedman *Dis. Skin* 819 The dermographism was more marked upon the face than upon the rest of the body. **1908** *Practitioner* Feb. 252 From these individuals we get a history of attacks of urticaria, whilst they may even show dermographism. **1971** D. M. PILLSBURY *Man. Dermatol.* xv. 190 Physical agents, e.g., heat, cold, sunlight, scratching (dermographism) or heavy pressure may be responsible for urticaria in some individuals.

dermoid, *a.* Add: **B.** *sb.* A dermoid cyst.

1897 *Allbutt's Syst. Med.* III. 686 The intraperitoneal dermoids may be very numerous. **1906** *Practitioner* Nov. 664 These are the dermoids of the head and neck, or mediastinum; others, more complex ones, occur in the sexual organs. **1964** S. DUKE-ELDER *Parsons' Dis. Eye* (ed. 14) xxxi. 508 Sometimes a bridge of skin links the coloboma to the globe, or there is a dermoid astride the limbus at the site of the coloboma.

dern, *sb.* U.S. var. DARN (= damn). (Earlier examples.)

1853 *S. Lit. Messenger* XIX. 222/1 Cave said he did not care a dern for the oysters. **1874** E. EGGLESTON *Circuit Rider* (1895) 120, I tole him as how I didn't keer three continental derns fer his whole band. **1876** 'MARK TWAIN' *Tom Sawyer* vi. 56, I wouldn't give a dern for spunk-water.

dern, U.S. var. *DARN *a.*

1853 in *D.A.E.* **1876** 'MARK TWAIN' *Tom Sawyer* ix. 97 I'd druther [rather] they was devils a dern sight. **1883** —— *Life Mississippi* xviii. 219 'Where was you born?' 'In Florida, Missouri.' 'Dern sight better stayed there!' **1919** T. K. HOLMES *Man fr. Tall Timber* xviii. 220 The dern fool! Thirty thousand against thirty millions!

dern, U.S. var. *DARN *v.*[2]

1830 in *D.A.E.* **1848** J. J. HOOPER *Widow Rugby* in *Simon Suggs' Adv.*, etc. 128 Who dars to call me *hit*? Dern his old gray har, it shan't purtect him! **1883** 'MARK TWAIN' *Life Mississippi* xviii. 226 'Dod dern' was the nearest he ventured to the luxury of swearing. **1891** H. C. BUNNER *Short Sixes* 100 'Dern you,' said the keeper to Dr. Tibbitt.

derned, U.S. var. DARNED *a.* (= damned).

1843 *Spirit of Times* 11 Feb. 591/3 He said he would 'be *dod derned* if he'd go'! *Ibid.*, Report and be derned. **1873** J. H. BEADLE *Undevel. West* xxi. 405 [He] was rich afo' the war; derned poor now. **1898** H. S. CANFIELD *Maid of Frontier* i. 6 The derned rangers will have to make the same ride.

dernier, *a.* Add: **c.** *le* (or *the*) *dernier cri* [Fr., lit. 'the last cry']: the very latest fashion. Also in predicative use and (without article) *attrib.*

1896 *Westm. Gaz.* 10 Dec. 3/2 At a moment when cut-steel is *le dernier cri*. **1903** *Lady's Realm* XIV. 407/1 The shade *par excellence* for grey Father Thames is scarlet; in Paris this is the *dernier cri*. **1906** *Westm. Gaz.* 13 June 6/7 There was a time when the lisp was, so to speak, the 'dernier cri'. **1922** JOYCE *Ulysses* 249 Henry and James's wax smartsuited freshcheeked models, the gentlemen Henry, *dernier cri* James. **1928** *Daily Express* 16 June 5/5 That is why they wear hats and gowns that they are told are le dernier cri. *Ibid.* 9 Nov. 5/4 Any old necklace can be made modern and 'dernier cri'. **1960** *Times* 8 Jan. 13/6 The drabness of much *dernier cri* music.

d. *dernier mot*: the last word (see WORD *sb.* 25).

1834 J. S. MILL in *Monthly Repos.* VIII. 311 Propositions..are put forth with an air of authority,..as if they were the *dernier mot* of some great question. **1905** *To-Day* 8 Mar. 175/2 One would have thought this group the *dernier mot* of singing. **1937** S. SPENDER *Forward from Liberalism* 94 Private property..and inheritance, appeared to me..the *dernier mot* of legislation.

derrick, *sb.* Add: **2. f.** A structure erected over a deep-bored well, esp. an oil-well, to support the drilling apparatus. orig. *U.S.*

1861 *Daily Dispatch* (Richmond, Va.) 30 Apr. 1/4 Shanties, derricks, engine-houses, and dwellings, were at once enveloped in flames. **1865** *Times* 18 Oct. 9/5 All along the last few miles to Oil City the derricks have been becoming thicker and thicker. *Ibid.* 24 Oct. 6/1 The derrick used at oil wells..is a mere tall pyramid-shaped scaffolding, about 50 ft. high. **1883, 1885** [see RIG *sb.*[4] 3 a]. **1961** H. VAN DONGEN in K. Reisz *Technique Film Editing* (ed. 9) viii. 152 It is the precise spot chosen by the oil-prospectors for drilling, and it is here that the approaching derrick will settle down.

derricking (de·rikiŋ), *vbl. sb.* and *ppl. a.* [f. DERRICK *sb.* + -ING [1, 2].] **a.** *vbl. sb.* The action of operating the jib of a derrick-crane. **b.** *ppl. a.* Operating as a derrick.

1888 *Lockwood's Dict. Mech. Engin.* s.v. *Derrick*, Portable cranes..are provided with movable jibs, whose mechanism for derricking is made in various types. **1911** C. W. HILL *Electr. Crane Constr.* iii. 31 An electric motor..was provided to drive the lifting and derricking motions. **1940** *Chambers's Techn. Dict.* 236/1 *Derricking jib crane*, a jib crane in which the inclination of the jib..can be varied by shortening or lengthening the tie-ropes between post and jib. **1958** H. H. BROUGHTON *Electr. Cranes* (ed. 3) xvii. 236 Derricking motion. The movement of a crane jib when it is pivoting in a vertical plane is known as derricking or luffing. *Ibid.* 239 Hoisting rope is taken up which prevents the load from being lowered as it would be with ordinary derricking gear.

derrid, derride (de·rid, -əid). *Chem.* [a. G. *derrid* (M. Greshoff 1890, in *Ber. d. Deut. Chem. Ges.* XXIII. 3538), f. *DERR(IS: see -ID[4], -IDE.] **a.** An extract of derris as obtained by Greshoff (see quots.). **b.** (See quot. 1939[2].)

1890 *Pharmaceut. Jrnl. & Trans.* XXI. 559/2 *Derris elliptica.* This plant was stated to afford the most important of several drugs passing under the name of 'tuba root'..a resinous body with an acid reaction, to which the name 'derrid' has been given...In the root it exists together with a brown colouring matter called 'derris red', derived from the tannin of derris. **1900** *Jrnl. Chem. Soc.* LXXVIII. I. 109 Derride, $C_{33}H_{30}O_{10}$, prepared from the root of *Derris elliptica, Benth.*..is a pale yellow substance. **1915** J. D. GIMLETTE *Malay Poisons* v. 93 A poisonous resin called tubain by Leonard Wray and derrid by Greshoff. **1939** MEYER & KOOLHAAS in *Rec. Trav. Chim. Pays-Bas* LVIII. 211 The name *derride* has been assigned to the new substance in memory of Greshoff, who was one of the first to isolate a substance from derris root: this substance was not examined at the time and he termed it derride. **1939** *Jrnl. Chem. Soc.* 1100 Meyer and Koolhaas have suggested the name 'derride' for this substance [*sc. l*-elliptone], somewhat unfortunately, as it has already been used by Greshoff..for a resinous extract from *D. elliptica* root.

‖ **derrière** (deriēᵊɹ). *colloq.* [Fr.] = BEHIND C. *sb.*

1774 G. SELWYN *Let.* 13 Aug. in *15th Rep. Hist. MSS. Commission* (1897) 278 *S'il fait le fier*..I shall give him several spanks upon his *derrière*. **1814** in J. Agate *These were Actors* (1943) 31 He..exposed his *derrière* to his mistress. **1933** J. MARTIN-HARVEY *Autobiogr.* vii. 83 Another poor girl with an undisguisable '*derrière*'. **1935** G. B. SHAW *Let.* 17 Mar. in *Shaw & Mrs. P. Campbell* (1952) 308 You shouldnt have left that poor young lady..with a sore derrière by dealing a quite useless parting kick.

Derringer. Add: Also **Deringer** [the name of Henry *Deringer* (1786–1868)]. (Earlier U.S. examples.)

1853 G. D. BREWERTON *Overland with Kit Carson* (1930) 188 A preventive to interference..in the shape of Bowie knives, 'Derringers', and 'six-shooters'. **1854** J. R. BARTLETT *Personal Narr. Explor. Texas* I. iii. 48 My carriage driver carried a pair of Deringer pistols.

derris (de·ris). [mod.L. (J. de Loureiro *Flora Cochinchinensis* (1790) II. 432), a. Gr. δέρρις a leather covering, membrane (referring to the pod).] **a.** *Bot.* A member of the genus of woody tropical climbing plants so called, belonging to the family Leguminosæ.

1860 G. BENTHAM in *Jrnl. Linnean Soc.* IV. Suppl. 21 *Derris* has the flat pod of *Lonchocarpus*, varying likewise in consistency from thin and membranous to thick and almost woody. *Ibid.* 22, I adopted for the whole genus Loureiro's old name, *Derris*, because I could clearly identify it with his character. **1936** D. H. GRIST *Outl. Malayan Agric.* 182 Two species of derris are commonly cultivated in the Malay Peninsula. **1968** J. W. PURSEGLOVE *Tropical Crops* I. 257 Derris is propagated vegetatively by well-ripened stem cuttings.

b. A preparation, esp. the powdered root, of various species of *Derris* (and also of other genera containing rotenone), used extensively as an insecticide. Also *attrib.*

1890 [see *DERRID, DERRIDE]. **1919** N. E. MCINDOO et al. in *Jrnl. Agric. Research* XVII. 199 Derris, known widely as a powerful East Indies fish poison, was found to fulfill several of the requirements of a general insecticide; it acts both as a contact insecticide and as a stomach poison. *Ibid.* 200 Derris powder..was found to be efficient against dog fleas, chicken lice, house flies. **1932** R. F. FORTUNE *Sorcerers of Dobu* iii. 174 Derris, a vegetable poison in the roots of a tall liana, is used publicly for stupefying fish. **1936** *Discovery* Feb. 42/2 The successful derris treatments for Warble Fly damage. **1958** *Times* 25 Jan. 9/1 The derris will take care of the caterpillars and the BHC, of most of the other pests that afflict our plants. **1963** H. MARTIN *Insecticide & Fungicide Handbk.* i. 8 The other principal insecticide available prior to 1939 was derris, the ground root of species of the genera *Derris* and *Lonchocarpus*. **1971** *Daily Tel.* 3 Apr. 6/4 Don't forget the dustings of derris powder round their [*sc.* daffodil bulbs'] necks to scotch the maggots of the bulb flies.

derry[2] (de·ri). *Austral.* and *N.Z.* [app. jocular adaptation of *derry* in the refrain *derry down.*] A 'down' (DOWN *sb.*[3] 5); esp. in phr. *to have a derry on*, to be prejudiced against.

1896 *Argus* 19 Mar. 5/9 (Morris), Have you any particular 'derry' upon this Wendouree? **1897** D. MCK. WRIGHT *Station Ballads* 107, I ain't down on them if they ain't got a derry on me! **1900** H. LAWSON *Verses Pop. & Humorous* 185 It's orful when the p'leece has got a derry on a chap. **1915** *Bulletin* (Sydney) 15 Aug. 47/1 If you do, the Johns have a derry on yer ever after. **1918** *Chrons. N.Z.E.F.* 20 Dec. 245/1 He..[has] a particularly keen 'derry' on the parade-ground N.C.O. **1948** D. W. BALLANTYNE *Cunninghams* (1963) I. x. 58 She didn't like the Baptists though, had had a derry on that crowd ever since Hilda took her to an evening service.

derry[3] (de·ri). *slang.* [f. DERELICT *a.* and *sb.* + -Y[6].] A derelict building.

1968 BUSBY & HOLTHAM *Main Line Kill* v. 52 Red said it was time to kip in the derry...The derry was the second of a row of empty houses. **1969** *Guardian* 5 Aug. 4/8 Mary..lives with her husband, two Belgian boys, three English girls, and a young Frenchman in a 'derry'—a deserted house—in Chelsea.

dertrum (dɜ̄·ɹtrɒ̆m). *Zool.* Also **dertron.** [mod.L., ad. Gr. δέρτρον beak.] The extremity of the upper bill of a bird when in any way distinguished from the rest of the beak.

1889 in *Cent. Dict.* **1893** A. NEWTON *Dict. Birds* I. 33 Various parts of the rostrum have received special names: *culmen*, the dorsal ridge of the upper bill; *apex* or tip; *dertrum*, in which it often terminates. **1959** VAN TYNE & BERGER *Fund. Ornith.* 566 Dertrum. Extremity of the upper mandible (maxilla); the hook of the bill. **1964** G. R. MOUNTFORT in A. L. Thomson *New Dict. Birds* 93/1 Upper mandible... may be divided into the culmen, or dorsal ridge from the tip (dertrum) of the bill to the forehead, and the upper mandibular tomia, or lateral cutting edges.

derv (dɜ̄ɹv). [f. the initial letters of *diesel-engined road vehicle.*] Diesel oil for road vehicles. Also *attrib.*

1948 in PARTRIDGE *Dict. Forces' Slang 1939–1945* 54. **1952** *Economist* 15 Mar. 666/2 Suggestions that 'derv' fuel might be given preferential treatment in any increase in motor fuel duty. **1958** *Oxford Mail* 17 Jan. 6/4 The fuel which supplies the motive power for buses is fuel oil (or 'derv'). **1958** *Times* 17 July 10/6 The major fuel companies yesterday announced increases of ½d. a gallon from to-day in Derv, vaporizing oil, gas oil, and light fuel oil.

Derwenter. (Earlier example.)

1853 J. ROCHFORT *Adventures of Surveyor in N.Z.* ii. 16 Here we engaged a new steward..who I think was a Derwenter from what I have since seen of him.

derzie, -y, varr. *DURZEE.

desacralization (dīsæːkrăləizē^i·ʃən). [f. DE- II. 1 + SACRAL *a.*[2] + -IZ(E + -ATION.] **a.** The process of the ritual removal of a taboo. **b.** Secularization; the process of rendering something less sacred. So **desa·cralize** *v. trans.*

1911 *Encycl. Brit.* XXVI. 338/2 Temporary direct taboos, whether natural or acquired, may be removed by a process of desacralization or of purification. Thus, new crops are frequently taboo till the chief has partaken of them;..and the crops thus desacralized become free to all. **1942** H. WEBSTER *Taboo* i. 34 A state of taboo which has been formally imposed by the constituted authorities may be as formally lifted by an act of desacralization. **1959** W. R. TRASK tr. *Eliade's Sacred & Profane* 13 Desacralization pervades the entire experience of the nonreligious man of modern societies. *Ibid.*, The man who lives..in a desacralized world. **1964** M. MCLUHAN *Understanding Media* xv. 155 The visual desacralizes the universe and produces the 'nonreligious man of modern societies'. **1967** C. DAVIS *Question of Conscience* III. i. 183 Thus, nature, the State and society have all been desacralized.

desai (de·səi). *Indian Hist.* Also 7 **desie,** 9 **desaye, dessai.** [Marathi *deśāī.*] A native Indian revenue official, or petty chief.

1698 J. FRYER *Acc. E. India & Persia* 120 The Desie or Farmer, who squeezes the Countryman, as much as the Governor does the Citizen. **1800** WELLINGTON *Suppl. Desp.* (1858) II. 116 He has sent 300 horse to seize the dessays of the villages which you mention. **1835** J. BIRD tr. *Ali Mohammed Khan's Pol. & Stat. Hist. Gujarat* 408 (Y.), The Desayes..made a complaint at Court. **1883** *Pioneer Mail* 24 Jan. (Y.), The Desai of Sawantwari has arrived at Delhi on a visit. **1885** G. S. FORBES *Wild Life in Canara* 20 They were also instructed to rendezvous..under their Dessaies at any point which might be threatened.

desalinate (dīsæ·linēⁱt), *v.* [f. DE- II. 1 + SALIN(E *sb.* + -ATE[3].] *trans.* To remove salt from. Hence **desa·linated** *ppl. a.*; **desalina·tion,** the removal of salt.

1949 KOESTLER *Promise & Fulf.* I. xiv. 162 The strange experiment of de-salinating the soil of the barren area. *Ibid.* II. ii. 209 The blond, good-looking young man with his neutral, 'de-salinated' features. **1958** *New Scientist* 29 May 75 Desalination of desert waters. **1960** *Times* 15 Mar. 3/1 Among its present projects are..desalination studies. **1962** *Economist* 16 June 1138/1 Visions of deserts blooming under the touch of desalinated seawater. **1969** *Times* 2 May (Suppl.) p. v/1 Desalination techniques have advanced enormously since the Second World War.

desalt (dīsǭ·lt), *v.* [f. DE- II. 2 + SALT *sb.*[1]] *trans.* To remove salt or salts from. So **desa·lter,** a desalting apparatus; **desa·lting** *vbl. sb.* and *ppl. a.*

1909 in WEBSTER. **1945** *Times* 29 June, The Permutit sea water de-salting apparatus for producing drinking water from sea water without the use of heat. **1956** *Nature* 14 Jan. 77/1 Active work continues on the desalting of water. *Ibid.* 83/1 A simplified form of desalter, employing ion-exchange membranes. *Ibid.*, The solution to be desalted..is separated from the anode. *Ibid.* 4 Feb. 237/2 The plasma was deproteinized, and the resultant liquid and the two specimens of urine desalted. **1965** *Listener* 2 Sept. 342/1 Tenders to build a combined nuclear-power station and de-salting plant on the shores of the Mediterranean. **1968** *Times* 22 Nov. 14/8 Desalting water by a flash distillation process.

desaturate (dīsæ·tiurēⁱt), *v.* [f. DE- II. 1 + SATURATE *v.*] *trans.* To cause to become unsaturated, to make less saturated. So **desa·turated** *ppl. a.*; **desatura·tion.**

1911 *Engineer* 10 Mar. 243/1 The statement that the blood is desaturated in its passage through the lung requires proof. *Ibid.*, Desaturation of the blood in its passage through the lungs. **1931** *Brit. Jrnl. Psychol.* Jan. 285 At very low intensities, under ordinary conditions of vision..all sensations are completely de-saturated, and the spectrum is simply a band of grey. *Ibid.* 287 De-saturation is not, in fact, complete in the case of spectral colours...Even if there be no complete de-saturation, the problem of explaining a partial de-saturation is essentially the same. **1962** H. C. WESTON *Sight, Light & Work* (ed. 2) i. 17 White is thus the de-saturated 'colour' *par excellence*. **1970** *Amateur Photographer* 11 Mar. 62/3 Distant colours are desaturated by white light scattered by the atmosphere.

‖ **désaxé** (dezakse), *a.* Also **-axe.** [Fr.] Of the crankshaft of a motor-car: see quot. 1908.

1906 *Daily Chron.* 14 Nov. 9/3 The setting of the crankshaft desaxes, or out of line with the cylinders. **1908** *Westm. Gaz.* 4 June 4/2 An uncommon feature of the Metallurgique engine is the setting of the crank-shaft *desaxés*—that is to say, the centre of the crank is set slightly out of line with the centre of the cylinder. **1912** *Motor Manual* (ed. 14) iii. 66 Principle of Offset Cylinder, or Desaxe Crankshaft Setting. **1963** BIRD & HUTTON-STOTT *Veteran Motor Car* 100 It is the first known example of the *désaxé* crankshaft...*Désaxé* engines became quite the thing about ten years later.

descale (dīskēⁱ·l), *v.* [f. DE- II. 2 + SCALE *sb.*[2] 5.] *trans.* To remove scale from (metal or other surfaces). So **desca·ler,** a substance or device for descaling; **desca·ling** *vbl. sb.*, the process of removing scale; also *attrib.*

1932 W. H. HATFIELD in *Jrnl. West of Scotland Iron & Steel Inst.* (1931–32) XXXIX. 80/2 Descaling. The removal of the scale incurred in rolling, heat-treatment, and other work operations is a very essential part of the manufacture of corrosion-resistant steels...The descaling solution has a fairly long life. **1958** *Times Rev. Industry* Feb. 19/1 To maintain output during descaling operations. **1958** *New Scientist* 31 July 526/1 It involved rotating the evaporation tubes while the de-scaler was being applied. **1959** *Design* Oct. 59/2 The heater is easy to keep clean, and descale where the water is hard.

descend, *v.* Add: **3. b.** *to descend on* or *upon*: to visit unexpectedly; freq. applied to unwanted visitors.

1916 A. HUXLEY *Lett.* (1969) 98, I have at the moment staying with me in Balliol young Robert Nichols, who descended on me for a day or two. **1922** *Ibid.* 208 Aunt Ethel has wisely not divulged the fact that she is going to be at Como..otherwise Aunt N would have been sure to descend upon her. **1971** R. LEWIS *Error of Judgment* i. 12 What a trial it must be for her..to have an HMI descend on the college.

descendancy, -ency. Delete †*Obs.*, substitute *arch.*, and add later example.

1934 DYLAN THOMAS *Let.* 9 May (1966) 124 If a poem, in the John Donne descendency, is fairly good, they print it; if very good in the Tennyson descendency, they refuse to.

descended, *ppl. a.* Add: **2.** That has descended, fallen, or dropped.

1853 *Lancet* 29 Jan. 112/1 (*heading*) Excision from the inguinal canal of an imperfectly descended testicle. **1966** WRIGHT & SYMMERS *Systemic Path.* I. xxvi. 812/1 A testis that is still descending at about the end of the first year of life may fail to reach the bottom of the scrotum, remaining imperfectly descended.

descender[2]. Add: **b.** (Later example.) Also in *Printing* and *Palæography*, a descending stroke; a stroke which extends below the body of a letter.

1802 *Monthly Mag.* XIV. 70/1 Each small letter is to be without any tail-piece or descender. **1938** *Times Lit. Suppl.* 30 Apr. 304/3 Where it saves space..is by very nearly eliminating the 'descenders' of the lower-case. **1954** N. R. KER in R. M. Wilson *A. Riwle* (*Caius MS.*) p. xii, The end of the descender of *r*, as of other tailed letters, often turns sharply to the left.

descent. Add: **1. g.** The descent of Christ into hell.

1883 B. F. WESTCOTT *Hist. Faith* vi. 76 The eternal meaning of Christ's Descent, Resurrection, Ascension, Session in heaven, as set forth in our Creed. **1894** H. B. SWETE *Apostles' Creed* v. 56 The doctrine of the Descent had found a place in three synodical declarations. *Ibid.* 57 Cyril..assigns great importance to the Descent, making it one of his ten primary *credenda*. **1967** *Cath. Dict. Theol.* II. 163/2 The perspective runs on from the events of Bethany to what happens at the Descent, and from there to the final judgment of the world.

7. a. Also *attrib.*

1950 *Amer. Anthropologist* LII. I. 2 We may differentiate unilineal descent groups from a kinship system proper. **1951** R. FIRTH *Elem. Soc. Org.* i. 8 In such a small community less importance is attached to preserving a male descent-name than to marking the establishment of a new social unit. **1957** P. WORSLEY *Trumpet shall Sound* i. 18 The framework of Fijian social organisation was a system of agnatic descent-groups. **1958** G. LIENHARDT in Middleton & Tait *Tribes without Rulers* 105 Every tribe contains descent groups from many clans of both categories.

descloizite. (Examples.)

[**1854** A. DAMOUR in *Ann. de Chim. et de Phys.* XLI. 78 Je propose de lui donner le nom de *descloizite*, comme hommage à..M. Descloizeaux.] **1854** DANA *Syst. Min.* (ed. 4) II. 361 *Descloizite*..Trimetric..Lustre bright. Colour mostly deep black. **1951** J. R. PARTINGTON *Gen. & Inorg. Chem.* (ed. 2) xxi. 644 Vanadium is widely distributed, the principal ores being carnotite,..vanadinite,..descloizite, [etc.].

description. Add: **2. d.** *Philos.* Phr. *knowledge by description*: see *ACQUAINTANCE 1 b.

descriptionist. Add: **b.** One who professes to give a mere or pure description, free from evaluation, prediction, or explanation; also *attrib.*

1914 C. D. BROAD *Perception* ii. 91 The descriptionist view is liable to underrate it. **1950** *Mind* LIX. 423 Munro eschews all dealings with evaluation, but like so many other pure descriptionists in fact assumes a value attitude. **1953** C. E. BAZELL *Linguistic Form* viii. 103 One man is interested in city-structure, and another is interested in the description of cities...If a city is built in the form of a square, this is not more likely to escape the eye of the descriptionist than that of the structuralist.

descriptive, *a.* Add: **2.** *Math.* Of that branch of geometry in which the relations of lines, figures, and solids are studied, esp. in their projections on two planes. (Cf. F. *géométrie descriptive*, Monge 1794–5.)

1824–5 *Encycl. Metrop.* (1845) I. 312/2 A new species of geometry introduced..by Monge, during the period of the revolution, under the designation of descriptive geometry. **1841** T. G. HALL *Elem. Descr. Geom.* 19 The object of Descriptive Geometry is the invention of methods by which we may represent upon a plane..the form and position of a body which possesses three dimensions...The means by which descriptive geometry attains its object is the method of Projections. **1885** C. LEUDESDORF tr. *Cremona's Projective Geom.* 50 Projective Geometry..dealing with projective properties..is chiefly concerned with descriptive properties of figures. **1913** BLESSING & DARLING *Elem. Descr. Geom.* i. 1 The practical value of descriptive geometry lies in the knowledge gained in solving graphical problems which arise in engineering and architecture, and in making and reading working drawings. **1961** *New Scientist* 23 Feb. 489/1 Gaspard Monge—scientist, mathematician, father of descriptive geometry and founder of the first polytechnic school.

3. a. Consisting solely or principally of description; concerned with, or signifying, observable things or qualities, or what is the case rather than what ought to be or might or must be; not expressing feelings or valuations; relating to this type of meaning or interest. (Opp. *emotive, prescriptive, evaluative.*) Also (*rare*) as *sb.*

1885 W. JAMES in *Mind* X. 28 Our inquiry is a chapter in descriptive psychology. **1943** M. FARBER *Foundat. Phenomenol.* vii. 216 The distinctions [were] a matter of descriptive analysis. **1944** C. L. STEVENSON *Ethics & Lang.* ix. 210 In any 'persuasive definition' the term defined is a familiar one, whose meaning is both descriptive and strongly emotive. **1948** *Mind* LVII. 482 A special form of the descriptive fallacy—the fallacy of believing that all verbs are used to describe empirically distinguishable activities or states of affairs. **1951** *Ibid.* LX. 443 The traditional discussions of Free Will, confusing descriptive with prescriptive laws. **1952** R. M. HARE *Lang. of Morals* vii. 111 The descriptive properties which a particular strawberry had. **1955** J. L. AUSTIN *How to do Things* (1962) xii. 157 There is also a slide towards 'descriptives'. **1956** A. J. AYER *Probl. Knowl.* i. 7 A philosophic statement is not..like..a statement in any of the descriptive sciences..tested by observation. *Ibid.* ii. 80 There is the method of Descriptive Analysis.

b. *Linguistics.* Describing the structure of a language at a given time, avoiding comparisons with other languages or other historical phases, and free from social valuations; as in *descriptive grammar, linguistics*, etc. (Opp. *normative, prescriptive, historical*; cf. *SYNCHRONIC *a.*)

1927 *Mod. Philol.* Nov. 217 (*heading*) Descriptive linguistics. *Ibid.* 218 Today descriptive linguistics is thus recognized beside historical, or rather as precedent to it. **1933** BLOOMFIELD *Lang.* i. 18 Descriptive studies did not merge with the main stream of historical work. **1933** JESPERSEN *Ess. Eng. Gram.* i. 19 Descriptive grammar..aims at finding out what is actually said and written by the speakers of the language investigated. **1944** *Amer. Speech* XIX. 211 The later directions of linguistic research: dialect geography and descriptive analysis. **1947** E. H. STURTEVANT *Introd. Ling. Sci.* vi. 51 Descriptive linguistics forms the basis for historical linguistics. *Ibid.* 53 Most of our school grammars must be classed as descriptive. **1953** J. B. CARROLL *Study of Lang.* ii. 16 Several Greek grammarians, notably Dionysius Thrax, Apollonius Dyscolus, and Herodian, developed descriptive grammars of Greek. *Ibid.* 19 The European linguist who best formulated the methodology of descriptive linguistics..was Ferdinand de Saussure. **1961** H. A. GLEASON *Introd. Descriptive Linguistics* (ed. 2) xiii. 202 A descriptive grammar..is a systematically organized set of statements about the constructional patterns that characterize grammatical statements. **1966** J. J. KATZ *Philos. Lang.* ii. 8 The philosopher of language should..draw his linguistic information from the theory of language as developed in descriptive linguistics.

4. *Gram.* Of a clause: see quot. 1903. Of adjectives, etc.: assigning a quality; not primarily restricting the application of the expression modified. (Opp. *limiting.*) *descriptive adjective*: see quot. 1933.

1903 HALE & BUCK *Latin Gram.* IV. 260 The Volitive Subjunctive may be used..in Relative Clauses, determinative or descriptive..that is, telling *what kind* of person or thing is meant. *Ibid.* 302 A Descriptive Clause is necessarily a free one when it refers immediately to an antecedent *complete in itself*, e.g. a word denoting a person. **1927** E. A. SONNENSCHEIN *Soul of Gram.* i. 37 Descriptive genitives. **1933** BLOOMFIELD *Lang.* xii. 202 The adjectives are divided into two classes, *descriptive* and *limiting*, by the circumstance that when adjectives of both these classes occur in a phrase, the limiting adjective precedes and modifies the group of descriptive adjective plus noun. **1954** P. M. ROBERTS *Underst. Gram.* ii. 48 *A car'penter's hammer* ('a hammer customarily used by carpenters'—descriptive genitive). *Ibid.* iv. 93 Descriptive adjectives occur in three main sentence positions.

descriptivism (dĭskri·ptiviz'm). [f. DESCRIPTIVE *a.* + -ISM.]

1. *Philos.* The doctrine that value-words, and hence value-judgements, are equivalent in meaning to certain descriptive expressions; see also quot. 1961. (Cf. *DESCRIPTIVE *a.* 3 a.)

1961 WEBSTER, *Descriptivism*, a theory of ethics according to which only descriptive or empirical statements are meaningful. **1963** R. M. HARE *Freedom & Reason* ii. 17 The thesis..that moral judgements are a kind of descriptive judgements..is descriptivism.

2. *Linguistics.* The practice or advocacy of descriptive linguistics; the belief that the descriptive part of linguistics is fundamental. (Cf. *DESCRIPTIVE *a.* 3 b.)

1961 in WEBSTER. **1964** R. A. HALL *Introd. Ling.* lxxv. 442 A strict descriptivism has on occasion been preached. **1966** N. CHOMSKY *Cartesian Linguistics* 106 One of the most striking features of American descriptivism in the 1940s was its insistence on justification in terms of precisely specified procedures of analysis.

descri·ptivist. [f. as prec. + -IST.] An adherent or advocate of descriptivism. Also *attrib.* or as *adj.* So **descriptivi·stic** *a.*, **descriptivi·stically** *adv.* (Webster 1961).

1952 *Archivum Linguisticum* IV. I. 11 The eminent phonemicist and descriptivist W. F. Twaddell. **1960** R. M. HARE in J. O. Urmson *Conc. Encycl. West. Philos.* 141/2 Those views which hold that moral judgments are used to give some sort of information..are called 'descriptivist'. *Ibid.* 142/1 Stevenson's views did not..find favour with descriptivists. **1962** *Med. Ævum* XXXI. 149 There is some truth in the descriptivists' well-known claim: that a truly diachronic approach must be based on the comparison of descriptions of successive *états de langue*. **1963** R. M. HARE *Freedom & Reason* ii. 16 A naturalist is not the only sort of 'descriptivist'—if we may use this term for one who holds that value-words are simply one kind of descriptive word.

descriptor (dĭskri·ptǫ̆r, -əɹ). *Linguistics.* [a. L. *dēscriptor* describer.] An expression or sentence-element that has the function of describing.

1933 JESPERSEN *Syst. Gram.* 12 Recently V. Brøndal..has made a very bold attempt at a completely new system..*Relator* (R) and *Relatum* (r), *Descriptor* (D) and *Descriptum* (d). **1946** C. W. MORRIS *Signs, Lang. & Behavior* iii. 76 *Descriptors*, identifiors which describe a location. **1949** *Mind* LVIII. 27, I shall call the part of a sentence which performs the descriptive function of that sentence its 'descriptor'. **1953** W. J. ENTWISTLE *Aspects of Lang.* v. 157 The *descriptum* corresponds to quantity and has its pure expression in numerals; the *descriptor* with quality and is pure in adverbs. **1965** *Rev. Internat. Doc.* XXXII. 19/1 Any term, group of terms, or any other entity assigned to a document for the purpose of indexing it will be referred to as a descriptor.

descriptum (dĭskri·ptɒ̆m). Pl. **-ta.** [a. L. *dēscriptum*, neut. pa. pple. of *dēscrībere* to DESCRIBE.]

a. *Linguistics.* (Brøndal's term for) a sentence-element that has the function of numeration. **b.** *Philos.* The object for which a description, or the descriptive element of a sentence, stands; a thing given by description.

1933 [see *DESCRIPTOR]. **1936** *Mind* XLV. 37 We might distinguish these two varieties..as the theories (1) that the Self is a Descriptum, and (2) that it is a Logical Construction. **1938** H. REICHENBACH *Exper. & Prediction* § 25. 220 Existence is a quality not of individual things but of *descripta*. **1947** R. CARNAP *Meaning & Necessity* i. 32 The entity for which a description stands (if there is such an entity) will be called its *descriptum*. **1949** *Mind* LVIII. 29, I shall call that which is described by the descriptor, the 'descriptum'. The descriptum of an indicative sentence is what would be the case if the sentence were true; and of an imperative sentence, what would be the case if it were obeyed. **1953** [see *DESCRIPTOR].

deseam (dīsī·m), *v. Metallurgy.* [f. DE- II. 2 + SEAM *sb.*[1] 6.] *trans.* To remove surface defects from metal ingots, blooms, etc.), usu. by means of a gas-torch. So **desea·ming** *vbl. sb.*

1941 D. TAYLOR in *Junior Inst. Engineers Jrnl.* LI. 263 Steel conditioning. (The deseaming process in particular.) *Ibid.*, Ingots may be..ground and chipped or deseamed. **1952** *Gloss. Terms Welding and Cutting* (*B.S.I.*) 115 *Deseaming*, the removal of surface defects from cold ingots, blooms, billets and slabs by means of manual gas cutting.

desegregate (dīse·grigēⁱt), *v.* [f. DE- II. 1 + SEGREGATE *v.*] *trans.* To reunite (persons, classes, races, etc.) hitherto segregated; esp. (orig. *U.S.*) to abolish racial segregation in schools and other institutions. So **de:segrega·tion,** such reunion or abolition.

1952 *N.Y. Times* 14 Dec. E9/2 A 'statement of experts'..has been filed in behalf of the NAACP, citing the 'effects of segregation and the consequences of de-segregation'. **1953** *Life* (U.S.) 13 July 36/1 It is hoped that this decision of the court will help to desegregate the white movie theaters. **1958** *Listener* 12 June 966/1 He looked forward to a day when the schools [in Rhodesia] might possibly be desegregated. **1959** *Manch. Guardian* 8 Aug. 5/1 The statement demands eight years compulsory education for African children and immediate desegregation. **1959** *Times Lit. Suppl.* 28 Aug. 491/4 The citizens of Little Rock, after a year's closure of Central High, decided that even desegregated education was better than none. **1969** *Daily Tel.* 20 Oct. 20/7 A widespread movement by White parents to beat school desegregation in the Deep South.

desensitize (dīse·nsitaiz), *v.* [f. DE- II. 1 + SENSITIZE *v.*] *trans.* To reduce or eliminate the sensitivity of. *spec.* **a.** *Photogr.* To reduce the sensitivity to light of (a plate or film). **b.** *Med.* To render (a person or animal) insensitive to an allergen. **c.** *Psychiatry.* To free (someone) from a neurosis or complex. **d.** *Printing.* To treat (a lithographic stone or plate) with a solution which makes the areas not bearing an image repel printing-ink. Also *intr.*, to become insensitive (*rare*).

So **dese:nsitiza·tion,** the act or process of desensitizing; **dese·nsitizer,** a desensitizing agent; **dese·nsitizing** *vbl. sb.* and *ppl. a.*

1904 *Brit. & Col. Printer* 10 Mar. 15/2 Don't leave the plate too long out of the bath or it will desensitise. **1921** *Glasgow Herald* 9 Mar. 7 A single bottle contains enough solution for the desensitising of hundreds of small plates. *Ibid.* 6 Apr. 7 The discovery of phenosafranin as a desensitiser. *Ibid.* 11 May 7 The desensitising properties of pheno-safranine. **1922** *Brit. Jrnl. Photogr. Alm.* 368 The desensitising action. *Ibid.* 369 The transparent backing..acts as a powerful desensitiser. **1924** R. C. LOW *Anaphylaxis & Sensitisation* x. 197 One attack of dermatitis venenata does not produce an immunity or desensitisation to further attacks. *Ibid.*, Any attempt to desensitise patients by repeated applications of the irritant are [*sic*] not likely to be successful. **1935** HOWARD & PATRY *Mental Health* xiv. 318 The enucleation of 'sore-spots' or complex-determined topics, memories, associations and reactions which require desensitization and replacement by more wholesome resources and performances. **1937** *Sunday Dispatch* 16 May 17 Thereafter, a course of special injections (known as 'desensitisation') is given in gradually increased doses to accustom the system to the guilty factor. **1937** *Discovery* Oct. 299/1 The fixing of a photographic plate by desensitisation. **1940** A. L. M. SOWERBY *Wall's Dict. Photogr.* (ed. 15) 190 A desensitiser is a substance which, when applied to a sensitive emulsion, reduces its sensitivity to light. **1951** R. MAYER *Artist's Hand-Bk.* xii. 383 The portions of the stone that have not been drawn upon are so desensitized that ..if printing ink is smeared on these areas, it may easily be washed off. **1955** E. POUND *Section: Rock-Drill* xcii. 81 For 40 years I have seen..desensitization. **1955** *Sci. News Let.* 13 Aug. 106/3 Meanwhile, for the hay fever sufferer this season, shots to desensitize, or immunize, him against the pollens are the best that can be offered. **1959** *Chambers's Encycl.* I. 709/2 On the allergic hypothesis, desensitization of the patient to the allergens to which he is sensitive seems the most rational treatment to attempt. **1961** *Lancet* 19 Aug. 413/1 Bronchoconstriction and histamine release..were followed by desensitisation. *Ibid.* 413/2 The desensitising effects of increasing small doses of a specific antigen. **1967** E. CHAMBERS *Photolitho-Offset* xiii. 198 Whereas counter-etching is to clean the plate surface to hold the image, pre-etching follows counter-etching or surface treatment and is generally known as a desensitising etch. **1968** *Listener* 15 Feb. 205/2 The patient does desensitise with a tape-recorder alone, but one patient said that without the therapist she wouldn't continue treatment.

desequestrate (dīːsīkwe·strēⁱt), *v.* [f. DE- II. 1 + SEQUESTRATE *v.*] *trans.* To release from sequestration; to return to its owner. So **de:sequestra·tion.**

1959 *Times* 12 Jan. 8/5, 378 British firms and properties, and about 10,000 current bank accounts..will be desequestrated as soon as an agreement has been signed. **1959** *Daily Tel.* 13 Mar. 21/3 The first Briton to apply for desequestration papers. *Ibid.*, Desequestration will not take place until his claim has been formally admitted.

desert, *sb.*[2] Add: **5. a.** *desert-belt, -cave, -folk*; **b.** *desert-frequenting, -haunting, -worn* adjs.; **c.** *desert-brown, -grey, -long* adjs.; **desert boot** (see quot. 1948); **desert island,** an uninhabited, or seemingly uninhabited, and remote island; also *attrib.* and *fig.*, esp. (of equipment, cultural objects, or behaviour) suited to the social isolation and limited baggage allowance of a castaway on a desert island; **desert-lemon** *Austral.*, a rutaceous tree, *Eremocitrus glauca* (*Atalantia glauca*), bearing a small acid fruit; **desert oak** *Austral.* (see OAK 3 b); **desert pea** (see PEA[1] 3); **desert polish,** the polish imparted to rocks or other hard surfaces by the friction of the windblown sand of the desert; **desert rat** *colloq.*, a soldier of the 7th (British) armoured division, whose divisional sign was the figure of a jerboa, and which took part in the desert campaign in N. Africa (1941–2); **desert varnish,** a dark-coloured film composed of iron and manganese oxides, usually with some silica, deposited on exposed rocks in the desert and becoming polished by wind abrasion.

1913 KIPLING *Songs from Books* 142 For he knows which fountain dries, behind which desert-belt. **1948** PARTRIDGE *Dict. Forces' Slang* 54 *Desert boots*, brown boots reaching either halfway up the ankle or to just over it and tightly laced; they had crepe soles and, made of suede or of reversed calf, they did not need to be polished. **1964** *Listener* 12 Nov. 764/2 He was wearing suede desert boots. **1923** *Daily Mail* 5 Mar. 13 Nigger, Regal Blue, Grey, Desert Brown. **1885** W. B. YEATS in *Dublin Univ. Rev.* Sept., In gloom Of desert-caves. **1916** R. GRAVES *Over the Brazier* 15 Soft words of grace He spoke Unto lost desert-folk. **1872** 'MARK TWAIN' *Roughing It* (1882) v. 28 The desert-frequenting tribes of Indians. **1905** *Westm. Gaz.* 28 Aug. 10/2 This species (*Varanus griseus*) is a large, desert-frequenting lizard. **1906** *Ibid.* 1 Oct. 4/3 The kiang [*sc.* wild ass] is a desert-frequenting species. **1901** *Ibid.* 2 Jan. 2/1 Two stalwart sportsmen with..their desert-grey hounds gliding near them. **1894** R. B. SHARPE *Handbk. Birds Gt. Brit.* I. 112 From its pale coloration this Pipit might be considered a desert-haunting bird. **1607** TOPSELL *Four-f. Beasts* 13 They are driuen to a coast vnnauigable, where were many desart Islandes inhabited of wilde men. **1690** LOCKE *Govt.* II. § 14 The Promises and Bargains..between the two Men in the Desert Island, mentioned by Garcilasso De la vega,..are binding to them, though they are perfectly in a State of Nature. **1743** F. SHERIDAN *Let.* 16 Nov. in *Private Corresp. D. Garrick* (1831) I. 17 To something worse than the desert Island. **1856** C. M. YONGE *Daisy Chain* II. i. 337 It is like having all the Spaniards and savages spoiling Robinson Crusoe's desert island! **1922** C. E. MONTAGUE *Disenchantment* xiii. 175 All castaways together, all really marooned on the one desert island. **1930** F. BRETT YOUNG *Jim Redlake* III. v. 376, I always except the Meistersinger. I think I should choose it as my desert-island opera. **1939** *Mind* XLVIII. 156, I find that desert-island morality always rouses suspicion among ordinary men. **1942** *Radio Times* 23 Jan. 15/4 '*Desert Island Discs*'...Vic Oliver discusses with Roy Plomley the eight records he would choose if he were condemned to spend the rest of his life on a desert island with a gramophone for his entertainment. **1889** J. H. MAIDEN *Useful Native Plants Austral.* 8 *Atalantia glauca*,..'Native Kumquat', 'Desert Lemon'. **1932** AUDEN *Orators* III, Spare us the numbing zero-hour, The desert-long retreat. **1896** Desert oak [see OAK 3 b]. **1903** 'T. COLLINS' *Such is Life* 91 She had revelled in the audacious black-and-scarlet glory of the desert pea. **1929** K. S. PRICHARD *Coonardoo* 217 It was a good season, the desert pea scarlet under the mulga. **1967** A. M. BLOMBERY *Guide Native Austral. Plants* 101 (*caption*) *Clianthus formosus* (Sturt's Desert Pea). **1903** A. GEIKIE *Text-bk. Geol.* (ed. 4) I. III. II. i. 436 On the sandy plains of Wyoming, Utah, and the adjacent territories, surfaces even of such hard materials as chalcedony are etched into furrows and wrinkles, acquiring at the same time a peculiar and characteristic glaze ('desert-polish'). **1944** in *Shorter Oxf. Eng. Dict.* Add. (1956) 2487/3 As we stewed our tea—desert-rat style. **1945** W. S. CHURCHILL *Victory* (1946) 217 Dear Desert Rats, may your glory ever shine. **1958** *Times* 17 Apr. 11/4 The desert rat insignia will continue to be worn by all officers and men of the 7th Armoured Brigade Group. **1903** A. GEIKIE *Text-bk. Geol.* (ed. 4) II. 1425/1 Desert-polish or varnish. **1904** C. R. VAN HISE *Treat. Metamorphism* 547 In arid regions the hardened film has frequently been smoothed by the wind-blown sand, so as to present a polished surface. Such polished hardened films are known as 'desert varnish'. **1944** A. HOLMES *Princ. Physical Geol.* xiii. 270 The loose salts are blown away, but oxides of iron, accompanied by traces of manganese and other similar oxides, form a red, brown, or black film which is firmly retained. The surfaces of long-exposed rocks and pebbles thus acquire a characteristic coat of 'desert varnish'. **1970** R. J. SMALL *Study Landforms* ix. 294 Capillary rise is associated with the chemical breakdown of the interior of large boulders and the deposition of a hard crust of 'desert varnish'..on their surface. **1890** 'R. BOLDREWOOD' *Col. Reformer* ix, Sun-burned and desert-worn passengers.

desertic (dĕzɜ̄·ɹtik), *a.* [f. DESERT *sb.*[2] + -IC.] Characteristic of a desert. Also *fig.*

1936 *Nature* 19 Dec. 1042/1 With development of desertic conditions in North Africa and Arabia bringing increasing settlement near rivers, some degree of isolation and a high degree of long-continued local intermarriage developed. **1956** WYNDHAM LEWIS *Red Priest* xxxiii. 286 There is nothing awaiting me but a hideous desertic existence. **1963** D. W. & E. E. HUMPHRIES tr. *Termier's Erosion & Sedimentation* 406 *Hamada*..., an Arabic term used in the Sahara to describe a bare stony plain...It is a typical feature of the desertic climate.

deserving, *ppl. a.* Esp. in phr. *deserving poor.*

1801 *Observer* 29 Nov. 4/2 [The bill] proposed to enable Overseers to relieve the deserving Poor. **1971** B. INGLIS *Poverty & Industrial Revolution* i. 17 The deserving poor had been issued with small badges which served as a kind of begging licence.

de-sex, desex (dīse·ks), *v.* [f. DE- II. 2 + SEX *sb.*] *trans.* **a.** To castrate or spay. **b.** To deprive of the distinctive qualities of sex; also, to remove or minimize the sexual appeal of. So **dese·xed** *ppl. a.*

1911 *Experiment Station Record* XXV. v. 474 Measurements of desexed heifers show that in conformation they approach the type of a castrated male rather than that of a normal male. **1928** *Time* 16 Jan. 18 Those who show themselves to be morons should be de-sexed before they reach the reproductive stage. **1962** A. NISBETT *Technique Sound Studio* xii. 212 Small changes of pitch therefore tend not to change the apparent sex of a voice, but rather to de-sex it. **1963** *Economist* 28 Nov. 1338/1 The desexed version of a more uninhibited..dance.

desexualize (dīse·ksiuăləiz), *v.* [f. DE- + SEXUAL + -IZE.] *trans.* **a.** To deprive of sex or sexual characters; to deprive of the distinctive qualities of a sex. So **dese·xualized** *ppl. a.*

1894 *Idler* Sept. 195 The most highly cultured, mentally most richly endowed women I have known—not *desexualised.* **1919** M. K. BRADBY *Psycho-Analysis* 50 They regard all human energy as sexual, or 'libido', though Jung holds that it may become de-sexualised by turning from a sexual aim to non-sexual surrogate. **1926** *Spectator* 19 June 1038/2 The self-flatterer..desexualizes the human form that he may pride himself on his idealization of it. **1940** *Mind* XLIX. 352 Thus the 'conclusion' that every thirst for knowledge is a 'desexualised' form of sexual curiosity is not an unwarranted generalisation but either an inexcusable linguistic eccentricity or a fruitful linguistic modification. **1958** *Times Lit. Suppl.* 14 Feb. 90/2 Later adolescent pictures show a normal interest in sex, but both boys and girls become fashion-plate types, essentially desexualized. **1969** P. A. ROBINSON *Freudian Left* 207 Intensification of genital sexuality..left the body essentially desexualized.

b. To castrate.

1913 in DORLAND *Med. Dict.* (ed. 7).

desiccant, *sb.* (Later examples.)

1940 *Chambers's Techn. Dict.* 236/2 *Desiccants*, substances of a hygroscopic nature, capable of absorbing moisture and therefore used as drying agents. **1958** *Engineering* 14 Mar. 352/3 Air passing through the desiccant (activated alumina) is practically sterile when it leaves. **1967** *Times Rev. Industry* May 80/2 A hermetically sealed polythene film bag, with a quantity of desiccant enclosed within it.

desiccate (de·sikĕt), *sb.* [f. the ppl. adj.] A desiccated substance or product.

1926 *Science* 28 May 549 Some of my experiments were carried out with this desiccate. **1940** *Nature* 6 July 30/2 We decided to try desiccated 9–11 day chick embryos as a source of active principle...The desiccate was reconstituted by adding sterile distilled water.

design, *sb.* Add: **9.** *attrib.* and *Comb.*, as *design book, consultant, engineer*; **design-conscious** *a.* (see *CONSCIOUS *a.* 12).

1936 *Burlington Mag.* Nov. 235/2 The silversmiths appear to have had in their minds the forms depicted in German design-books. **1955** H. READ *Grass Roots of Art* (rev. ed.) vii. 137 The buying public..was becoming design-conscious. **1960** *Guardian* 30 Sept. 10/2 Plagiarism..is a rare headache to the design-conscious Scandinavian nations. **1954** H. READ *Anarchy & Order* 226 The attempt of certain artists to adapt themselves to the modern industrial system by calling themselves 'design consultants' has had no appreciable effect on the cultural situation. **1970** J. QUARTERMAIN *Man who walked on Diamonds* iv. 23 It was a design consultant's idea of hell. **1964** F. L. WESTWATER *Electronic Computers* iii. 49 Not infrequently, a design engineer will ask the logical designer to make alterations for various reasons.

designate, *a.* Add: *spec.* in the University of Cambridge.

1878 *Cambr. Univ. Cal.* 5 There are three days of general admission to the title of Bachelor Designate of Arts in every year. **1892** *Ordinances of Cambr. Univ.* 152 At the creation of Doctors of Law in every year, the names be arranged in order of seniority according to the seniority of the Doctors Designate as Masters of Law. **1925** *Cambr. Univ. Cal.* 85 Bachelors in Arts, Law, Medicine, Surgery, and Music remain 'Bachelors designate' until the 31st of December.

designatum (de·s-, de·zignēⁱtɒ̆m). *Philos.* Pl. **-ata.** [f. L. *dēsignātum*, neut. pa. pple. of *dēsignāre* to mark out, indicate (see DESIGNATE *v.*).] The object or class of objects, whether existing or not, that a sign designates. (Cf. *DENOTATUM.)

1938 [see *DENOTATUM]. **1940** *Jrnl. Religion* July 266 No appeal to revelation can deliver a man from responsibility for determining the designata and denotata of the words he uses. **1940** B. RUSSELL *Inquiry* vii. 109 The designatum of 'this' is continually changing. **1958** MEYER & WILKINSON tr. *Carnap's Introd. Symb. Logic* § 20 The color red is the designatum of the French word 'rouge'. **1964** *Language* XL. 266 The designatum of the whole expression is a Boolean sum of the designata of the members.

desirable, *sb.* Add: **2.** One who is desirable.

1669 H. MORE *Antid. Idolatry* viii. 93 The highest of all desirables, that is, God himself. **1853** R. S. SURTEES *Sponge's Sp. Tour* xvi, Certainly all parties concurred in placing him high on the list of 'desirables'. **1904** *Daily Chron.* 21 Sept. 8/5 Nor did the individual who spoke proudly of 'moving in a circle'—of 'desirables'—realise the vulgarity of the expression. **1905** *Westm. Gaz.* 21 June 8/2 The real undesirables..would have to be kept back and looked after until they became 'desirables'. **1908** *Ibid.* 26 June 2/1 Could not the undesirables be got rid of, and the desirables multiplied?

desiredness. (Earlier and later examples.)

a **1866** J. GROTE *Exam. Util. Philos.* (1870) xix. 327 There being rather want in the sense of *absence* of what should be than want in the sense of desiredness. **1920** A. C. PIGOU *Econ. of Welfare* v. x. 771 The marginal desiredness which money has for them.

desize (dīsəi·z), *v.* [f. DE- II. 2 + SIZE *sb.*[2] 2.] *trans.* To remove size from (textiles). So **desi·zing** *vbl. sb.*

1934 in WEBSTER. **1955** *Times* 13 July 1/7 (Advt.), Textile Chemist for work on sizing and desizing of natural and synthetic fibres. **1962** J. T. MARSH *Self-smoothing Fabrics* xi. 152 Zymolysis with the well-known products commonly employed for desizing.

desk, *sb.* Add: **3. c.** A specified section of a large organization, such as a newspaper office, government department, etc., responsible for a particular subject or operation. Freq. in *U.S.*, the department in a newspaper office where copy is edited. Cf. **city desk.*

1927 U. SINCLAIR *Money Writes* 18 The reporters who write up the sensational event—each one is hoping to attract the attention of the 'desk'. **1958** E. NEWBY *Short Walk in Hindu Kush* ii. 24 [At] the Foreign Office...I was interviewed by a representative of the Asian Desk. **1966** J. BINGHAM *Double Agent* vi. 87 At the table next to Henry Blundell sat little George Patterson, in charge of the East Russian desk...At the same table was Mike Parsons, who worked at the Czech desk. **1970** R. GADNEY *Drawn Blanc* vi. 61 They gave me a desk in Soviet Counter Intelligence, it's a big outfit now.

d. The reception desk or office of a hotel, office building, etc.; the person or persons on duty at the reception desk.

1963 D. B. HUGHES *Expendable Man* vi. 200 I'll tell the desk not to put through any more calls. **1966** G. LYALL *Shooting Script* xix. 150 Room 17, I think you said? And the desk knows I'm coming? **1970** P. BAIR *Tribunal* II. iii. 76 Ask the desk to ring through to Miss Jackson's room.

5. *desk calendar, diary, job, -light, -top*; **desk-bound** *a.*, obliged to remain at work at a desk; **desk copy** orig. *U.S.*, a free copy of a book, esp. one supplied for the personal use of a teacher; **desk-man,** (*b*) a man who works at a desk, *spec.* a journalist who works mainly at a desk; a white-collar worker; **desk-room** orig. *U.S.*, space for a desk rented in a business office.

1944 *Time* 24 Apr. 26 Few thought he would be desk-bound for long. **1962** *Listener* 28 June 1104/2 There were desk-bound jurists who had hatched out theories of crime 'as remote from reality as they are harmful'. **1907** *Yesterday's Shopping* (1969) p. xxxi/5 Desk Calendar Pads. **1922** F. SCOTT FITZGERALD *Beautiful & Damned* I. iii. 100 On a desk calendar he marked the days off. **1968** H. C. RAE *Few Small Bones* II. i. 71 He ruffled the pages of his desk calendar. **1942** *Amer. Speech* XVII. 121 The teaching profession avoids the appearance of receiving forbidden favors by asking its publishers not for *free* copies but for *desk* copies. **1962** *Publishers' Weekly* 23 Apr. 39/2 Professors when they request 'desk copies' frequently also order books ..for their own libraries. **1960** T. HUGHES *Lupercal* 52 Outstripping his desk-diary at a broad desk. **1963** L. DEIGHTON *Horse under Water* xxxii. 125 He flips through the desk and finds a nice leather desk diary. **1965** 'R. L. PIKE' *Police Blotter* (1966) vii. 105 He managed to get a soft desk job in the war. **1929** E. WILSON *I thought of Daisy* i. 29, I turned aside the adjustable desk-light..so that it lit only the farther wall. **1913** *Writer's Bull.* Oct. 101/1 The salaries paid..are..better than those paid to many 'desk men' in the offices of large newspapers. **1925** A. S. M. HUTCHINSON

One Increasing Purpose I. xxx. 181 The city desk-man's feeble stoop. 1961 Times 8 Feb. 13/7 Millions of deskmen from an inflated officialdom have been out in the fields helping the peasants. 1967 R. J. Serling President's Plane is Missing (1968) viii. 142 The IPS bureau chief was regarded as a superb deskman and a skilled writer. 1868 R. B. Kimball Undercurrents 9, I occupied an office—no, I had 'desk-room' in a basement office. 1870 J. K. Medbery Men & Myst. Wall Street 117 Many of the operators, as well as the smaller brokers,..have simply desk-room. 1926 Kipling Debits & Credits 337 Our War-side merely applied for desk-room in your lodges. 1930 D. Hammett Dain Curse xiii. 144 He..returned his feet to the desk-top. 1935 R. Stout League of Frightened Men xi. 197 What we are displaying on this desk-top is the soul of a man. 1960 Design July 56 Several basic broad desk tops, pedestals, panels. 1968 Daily Tel. 12 Nov. 22/6 Desk-top computers for use in homes..may be made possible by an invention described today.

desmoncus (desmǫ·ŋkǝs). Bot. [mod.L., f. Gr. δεσμός bond + ὄγκος hook.] A climbing plant of the genus of palms of this name, common in tropical America.
1899 J. Rodway Guiana Wilds 14 His head grazed by the formidable hooks which hung from the horrid desmoncus.

desolatingly (de·sǒlēˡtiŋli), adv. [f. Desolating ppl. a. + -ly².] In a manner that desolates or saddens.
1888 'L. Malet' Counsel of Perfection xiv. 323 These desolatingly encouraging persons. 1909 H. G. Wells Tono-Bungay I. iii. 81 A drab-coloured passage..not only narrow and dirty but desolatingly empty.

‖ **de son tort** (dǝ sǫn tor). Law. [Fr., lit. 'of his wrong'.] By his own wrongdoing, without authorization; spec. in phrases, e.g. executor de son tort (see Executor 3 b), de son tort demesne (see quot. 1835).
1670, 1767 [see Executor 3 b]. 1835 Tomlins Law-Dict. s.v. Tort, De son tort demesne, of his own wrong. 1874 Law Rep. Chancery Appeal Cases IX. 251 Responsibility may no doubt be extended in equity to others who are not properly trustees, if they are found either making themselves trustees de son tort, or actually participating in any fraudulent conduct of the trustee. 1959 Jowitt Dict. Eng. Law I. 573/1 Where B, the defendant, pleaded that what was alleged against him had been done by the order of C, his master, then A, the plaintiff, might reply that B had done it de son tort demesne sans que C, by commande, that is to say, of his own wrong without being ordered by C.

desorb (dīsǭ·ɹb), v. [Back-formation from next.] **a.** trans. To remove (a substance, etc.) from the surface upon which it is adsorbed. **b.** intr. Of a substance, etc.: to leave the surface upon which it is adsorbed (and pass into).
1924 Proc. R. Soc. A. CVI. 64 (heading) The energy required to desorb the gas-film. Ibid. 67 Some gases..are desorbed from tungsten, etc. with practically zero latent heat. 1937 Nature 10 July 48/1 Tswett conceived the idea of making the solution flow in one direction..through a column of the adsorbent, so that the adsorbed material might afterwards be desorbed, or 'eluted', by the same or some other more suitable solvent. 1964 New Scientist 26 Mar. 818/2 The free radicals were left to accumulate on the surface, from which they 'desorbed' back into the chamber. 1967 E. Page in E. A. Flood Solid-Gas Interface I. iv. 123 The desorption was accomplished by allowing the gas to desorb into a measured volume with the simultaneous application of heat.

desorption (dīsǭ·ɹpʃǝn). [f. De-, after Adsorption.] The liberation of a substance from the surface upon which it is adsorbed or from the liquid in which it is dissolved.
1924 Proc. R. Soc. A. CVI. 68 The desorption of a fresh film of carbon monoxide from tungsten. 1937 Nature 17 July 108/1 It seems possible that the existence of a sharp transition temperature may be associated with the desorption of gas from a film originally formed under poor vacuum conditions. 1958 R. E. Treybal Mass-Transfer Operations i. 3 If air is brought into contact with an ammonia-water solution, some of the ammonia leaves the liquid and enters the gas phase, an operation known as desorption or stripping.

desoxy-: see *deoxy-.

despair, sb. 1. counsel of despair: see *counsel sb. 2 c.

despecialize (dīspe·ʃǎlaiz), v. [f. De- + Special a. + -ize.] **a.** trans. To eliminate as a specialist or specialized vocation, subject of study, etc. **b.** intr. To pass from a specialized to a general condition. Hence **despecializa·tion**.
1896 F. L. D. Herbertson tr. P. de Rousiers' Lab. Quest. Britain II. iii. 190 Service is a very despecialized trade, at any rate in ordinary families, where a man-servant is an unknown luxury. Ibid. III. i. 253 The increasing despecialisation of the worker. 1898 Edin. Rev. Apr. 281 The whole tendency of the age of machinery has been to..de-specialise the average workman. 1903 S. N. Patten Heredity & Soc. Progr. 61 Emotion is the same force expressed as feeling, and in conscious beings is the index of the despecialization and regeneration acting within them. Ibid., The despecialized part is not restored, but a new specialization begins in the part affected by the emotion. 1967 Times 14 Nov. 11/5 A modest measure of despecialization would be to require all sixth formers to study four subjects at the same level. 1970 Nature 26 Dec. 1245/1 The report pours cold water on the idea that chemistry courses should be despecialized.

desperateness. 3. (Later example.)
1963 Times 30 May 17/4 The people had the will and desperateness—they held out.

de-spin, despin (dīspi·n), v. [f. De- II. 1 + Spin v.] trans. To counteract a spinning motion of, esp. of a satellite in space; to prevent from spinning. Hence **de-spun, despun** ppl. a.
1960 Aeroplane XCIX. 429/2 The H.T.P. jets..are intended to de-spin the satellite after separation. 1966 Electronics 3 Oct. 173 A Comsat-sponsored design study..is developing electronically despun microwave antennas for both NASA and the Air Force. 1967 New Scientist 30 Nov. 527/1 The new method provides a platform on top of the satellite that is not spinning. (A 'despun platform' is the name given to it by Hughes.) 1970 Sci. Jrnl. Jan. 18/1 To maintain permanent radio illumination of the Earth the horn aerial mounted at one axis is mechanically despun in the opposite direction to spin stabilization.

despisement (dīspai·zmĕnt). Delete †Obs., substitute arch., and add later examples.
1886 C. A. Briggs Messianic Proph. 358 Her days of oppression and despisement are over. 1887 Critic X. 251/2 An uncomfortable sensation of familiarity bordering on 'despisement'.

des(s)atine, varr. Dessiatine.
1884 Dessatine [in Dict.]. 1901 Daily Chron. 29 Aug. 5/1 The Tsar is said..to own in private property, mostly in the Baltic Provinces, a million dessatines of land.

dessay (de·sēˡ), repr. a form of pronunciation of (I) dare say (Dare v. III).
1931 S. Lawson Ricker Dust xviii. 317 The old woman was was [sic] waiting to live in it. Die in it too, he dessayed.

dessert. Add: Also 9 desert. **1. b.** Now also in British usage. (Earlier and later examples.)
1789 W. Maclay Jrnl. 27 Aug. (1890) iv. 138 The dessert was, first apple-pies, pudding, etc.; then iced creams, jellies, etc.; then water-melons, musk-melons, apples, peaches, nuts. 1833 H. Barnard in Maryland Hist. Mag. (1918) XIII. 379 The desert was pudding, and pies of all kinds. [illegible] 1864 A. V. Kirwan (title) Host & guest: a book about dinners, wines, and desserts. Ibid. xv. 215 A dessert should above all things be simple. Ibid. 219 Brandied fruits, compôtes and fruits preserved in syrup, are generally produced at a French dessert. 1935 W. Cather Lucy Gayheart I. iii. 21 He might have asked her and the boys to go out to the dining-car with him and have a dessert. 1966 Woman 24 Sept. Pullout 1 (heading) A starter. A main dish. A dessert.

2. dessert-knife (examples), **-plate, -spoon** (earlier examples); **dessert-service** (examples).
1765 J. Wedgwood Let. 6 July (1965) 35 Mr Grants dessert service will be sent today. 1769 Ibid. 20 Sept. (1965) 80, I have modeled you three or four sorts of desert plates. 1779 W. B. Stevens Jrnl. 6 June (1965) i. 86 He stabbed himself in three Places with a Desert Knife. 1808 Jane Austen Let. 27 Dec. (1952) 243 A whole tablespoon and a whole dessert-spoon, and six whole teaspoons. 1851 Mrs. Gaskell Let. Apr. (1966) 152 Miss Yates gives her a dessert service. 1911 Daily Colonist (Victoria, B.C.) 27 Apr. 3/1 (Advt.), Dessert Knives, Sheffield steel, fine handles. 1926–7 Army & Navy Stores Catal. 700/1 A Dessert Service consists of 12 Plates and 6 Dishes.

‖ **dessous** (dǝsu·). [Fr., = lower part; (pl.) underclothing (of women).] **1.** Underwear.
1901 'C. Holland' Mousmé 265 Mousmé had adopted that dainty dessous of Western woman. 1902 E. Glyn Refl. Ambrosine 237 The tiniest pink satin slippers peeped out of billows of exquisite dessous.

2. Phr. **dessous des cartes**: the underside of the cards as they lie face downwards when dealt; fig. a secret aim or object, explanation, etc. kept in reserve.
1756 Ld. Chesterfield Let. to Dayrolles 26 Nov., Misc. Wks. 1777 II. 435 There must be some dessous des cartes, some invisible wheels within wheels, which..I cannot guess at. 1880 Ouida [illegible] 276 Sir Walter and Arthur laughed at this dessous des cartes. 1885 'L. Malet' Col. Enderby's Wife III. iii, Just a little something behind, an explanation, you know, a dessous-des-cartes.

destabilize (dīstēˡ·bilaiz), v. [f. De- II. 1 + Stabilize v.] trans. To deprive of stability, to render unstable.
1924 in Webster. 1966 Guardian 24 Oct. 8/4 A first strike, which would destabilise the overall strategic situation. 1965 Phillips & Williams Inorg. Chem. I. ii. 52 It also appears to destabilize the d electrons slightly.
So **desta·bilizing** vbl. sb. and ppl. a.
1924 W. M. Hughes in These Eventful Years II. 285 The creation of a new political party..seems likely to have a destabilizing influence on Commonwealth politics. 1962 Economist 28 Apr. 266/2 Excessive and destabilising flows of short-term capital. 1962 W. B. Thompson Introd. Plasma Physics vi. 121 One where I is the total current enclosed within a cylinder of radius r, this de-stabilizing term is [etc.]. 1965 H. Kahn On Escalation xiii. 269 Ordinary technology might be almost as destabilizing as the..'doomsday machine'.

de-Stalinize (dī,stā·linəiz), v. [f. De- II. 1 + Stalin (adopted name of Iosif Vissarionovich Dzhugashvili (1879–1953), head of government of the Soviet Union) + -ize.] intr. To counteract the excesses of Stalinism or the influence of Stalin, as by reversing or amending his policies, removing monuments to him, and renaming places named in his honour; also trans., to affect in some way by de-Stalinization. So **de-Stalinized** ppl. a.; **de-Sta·linizing** vbl. sb.; **de-Sta·linizer**.
1957 Economist 21 Sept. 1013 The echoes of destalinisation and Hungary have reverberated among the Apennines. 1959 Times 7 Feb. 7/6 The GUM department store..where..escalators carry hopeful consumers through six floors of de-Stalinized delight. 1960 Guardian 20 Apr. 8/5 Russia decided to de-Stalinise in design. 1962 Times 6 Feb. 12/1 The so-called de-Stalinization policy. 1962 Listener 8 Mar. 404/1 Mr Barak's real offence was to be..too zealous a de-Stalinizer in Czechoslovakia itself. 1963 Daily Tel. 13 Sept. 12/2 The dismissal of the Prime Minister..and other 'de-Stalinising' changes. 1965 Spectator 5 Mar. 285/1 This giant-size statue of Stalin, stationed at the end of the Volga Don Canal in Volgograd, was demolished in the wave of destalinisation.

destocking (dīstǫ·kiŋ), vbl. sb. [f. De- II. 1 + Stock v.¹ 15 + -ing¹.] The reduction (by a retailer, etc.) of the quantity of stock held by the re-ordering of lower goods than are needed to replace those sold.
1959 Economist 31 Jan. 386/1 America ran into a destocking recession in 1949. 1963 Ann. Reg. 1962 474 The rate of stock-building was low and in the early months some destocking may have occurred. 1970 Daily Tel. 14 Jan. 16 This trend may be expected to continue as British industry swings round from de-stocking to building up its stocks.

destool (dīstū·l), v. [f. De- II. 2 + Stool sb.] trans. To remove (an African tribal chief) from authority. Cf. Stool sb. 1 e. Hence **destoolment**.
1929 R. S. Rattray Ashanti Law & Const. 82 Failure to accept such guidance and advice was a legitimate cause for destoolment. Ibid. 83 The person of a Chief was..invested with sanctity, just so long as he sat upon the Stool of his dead ancestors. This is the reason why there was a reaction the moment a Chief was destooled. 1964 Economist 25 Sept. 998/2 Assassination is a deplorable way of destooling a chief.

destroyer. Add: **b.** A type of small, fast warship armed with guns, torpedoes, etc., used for escort-work, attacking submarines, etc., orig. built for attacking torpedo-boats. Also attrib. and Comb., spec. in **destroyer-escort** (chiefly U.S.), a small destroyer; **destroyer-leader** U.S., a large destroyer.
Destroyer was the proper name of an American torpedo-boat in 1882 (see Leisure Hour, 1882, 674/2, quoting Brooklyn Eagle).
1893 Torpedo-boat destroyer [see Torpedo sb. 4]. [illegible] 1895 Daily News 27 July 3/1 The Rocket, another destroyer,..will not be ready for sea until Wednesday next. 1897 Pop. Sci. Monthly Nov. 139 It is becoming difficult to obtain the requisite area in screws of 'destroyers' without either resorting to an abnormal width of blade or to a large diameter. 1898 Kipling Fleet in Being ii. 25 [illegible] [illegible] (1945) 194 A destroyer leader. [illegible] 1962 Daily Colonist (Victoria, B.C.) 31 Oct. 1/1 The crippled Canadian destroyer-escort Kootenay will be towed back to Halifax for repairs.

destroying, vbl. sb. **b.** Naut. colloq. The action of serving in a destroyer.
1940 'Taffrail' Endless Story i. 13 Most destroyer officers, their detractors averred, went in for 'destroying' because they..needed the extra pay.

destroying, ppl. a. Add: **destroying angel** U.S. = *Danite 2.
1838 Peoria (Ill.) Register 24 Nov. 1/5 They had assembled under the title of different societies, called Danites, Gideonites, and the destroying Angels. 1857 [see *Danite 2]. 1872 'Mark Twain' Roughing It (1882) xii. 71 Half an hour later we changed horses, and took supper with a Mormon 'Destroying Angel'. 1943 B. De Voto Yr. of Decision 83 So in 1842 O. P. Rockwell, one of the Sons of Dan (the 'Destroying Angels' of ten-cent fiction), crept up to a window in Boggs's house and shot him.

destruct, v. Add: Delete †Obs. rare⁻¹ and add: Revived, chiefly in the U.S., in Rocketry. Also as sb. and attrib.
1958 Aero-Space Terms (Air University) 9/2 Destruct, the deliberate action of detonating or otherwise destroying a rocket missile or vehicle after it has been launched, but before it has completed its course. Said of friendly missiles esp. during test flights. 1958 Times 25 Sept. 10/2 The rocket..turned back..towards the shore; at this point it was destroyed (or 'destructed' as the official explanation puts it) by remote control. 1963 New Scientist 18 Apr. 139/1 A brief stop at the 'arming tower', where the 'destruct' equipment is installed. 1969 Guardian 21 Jan. 2/3 He had asked for a 'destruct' system to be installed in the electronics and code area of the ship.

destructful, a. Delete † and add later examples.
1881 Wilde Poems 223 Renovated By more destructful hands. 1895 E. W. Hopkins Relig. India (1896) xi. 270 She..will be..destructful of her husband.

destructional (dīstrʊ·kʃǝnǎl), a. [f. Destruction + -al.] Of or pertaining to destruction; formed by destructive agencies, spec. by denudation.
1900 R. T. Hill Physical Geogr. Texas Region (Topogr. Atlas U.S., Folio 3) 4 Destructional plains originate in the degradation (or planation) of older and higher surfaces down to a lower level. 1904 Amer. Jrnl. Sci. Jan. 38 The steep cliff is clearly in both cases a destructional surface from which material has fallen away. 1956 Gloss. Hydrologic Terms (U.N. Econ. Comm. Asia, ST/ECAFE/Ser. F, No. 10) 11/1 Destructional landscape, one resulting from action of water, streams, glaciers, winds, waves and animal organisms.

destructivity (dīstrʊkti·viti). [f. Destructive a. + -ity.] Destructiveness. Also **destru·ctivism**.
1902 Encycl. Brit. XXVII. 609/1 With the result that seismic destructivity can be accurately expressed in mechanical units. 1937 Times Weekly Ed. 19 May 5/1 It [sc. Marxism] demolishes the foundation of an individualistic society. On the Continent this is now called destructivism. 1963 Dylan Thomas Let. July (1966) 17 The..strangled destructivism of so much modern writing.

desubstantialize (dīsǝbstæ·nʃǎlaiz), v. [f. De- II. 1 + Substantialize v.] trans. To make less substantial; to take away reality from.
1940 Mind XLIX. 316 A being on the lower threshold of agency would by its timidity [illegible] desubstantialize its objective towards qualitative nonentity. 1954 [illegible] 176 We simultaneously desubstantialize that which we want to build up in the soul.

desuperheat (dīs(j)ū·pǝɹhīt), v. [f. De- II. 1 + Superheat v.] trans. To reduce the temperature of (superheated steam). So **desu·perheated** ppl. a., **desuperhea·ting** vbl. sb.
1931 Craig & Anderson Steam Power viii. 201 It is often necessary to desuperheat steam when the pressure and temperature..are too high. 1959 C. D. Swift Steam Power Plants xvii. 429 The desuperheating water must be brought into a fine mist in order to get a large and intimate contact surface with the steam to be desuperheated. Ibid., The element which [illegible] desuperheated steam temperature must be carefully located.
Hence **desu·perheater**, an apparatus which desuperheats steam.
1931 Craig & Anderson Steam Power viii. 201 A desuperheater is sometimes used in a feedwater heater. 1951 Engineering 23 Nov. 643/2 Steam..is passed through desuperheaters..which reduce the temperature to..above saturation.

desy·nchronize, v. [De-.] trans. The reverse of Synchronize v. (esp. sense 2 b). So **desy·nchronized** ppl. a.
[illegible]

detached, ppl. a. Add: **b.** Of persons, their conduct, etc.: characterized by detachment (see Detachment 4 b, and next).
1913 D. H. Lawrence Sons & Lovers xii. 322 He was pale and detached-looking. 1924 A. D. Sedgwick Little French Girl II. xi, She might be detached, and even callous; but she was not lax. [illegible] 'Cecil A Quiet Lives ii. 127 Gray's maturity appears in the detached clear-sightedness with which he could observe his own character.

detachment. Add: **4. b.** More widely, freedom or aloofness from ordinary concerns or emotional commitments.
1885 H. James in Harper's Mag. Feb. 342/2 [illegible] 1924 A. D. Sedgwick Little French Girl I. i, 'C'était un bien méchant homme,' Madame Vervier remarked in a tone of surpassing detachment. 1926 W. S. Churchill in W. R. Inge Lay Thoughts of a Dean 166 That sense of detachment and impartiality, that power of comprehending the other man's point of view. 1935 W. S. Maugham Don Fernando x. 201 This person seems to preserve a curious detachment: it would never occur to you that he was a mystic.

detail, sb. Add: **6.** attrib. and Comb., as **detail work**; **detail man** U.S. (see quot. 1961).
1928 Proc. 17th Ann. Meeting Amer. Drug Mfrs. Assoc. I. 96 With the detail man, we tell our message through the human ambassador. 1961 Webster, Detail man, a representative of a drug manufacturer who introduces new drugs to professional users (as physicians or pharmacists). 1964 New Statesman 12 June 906/3, I..became a drug pedlar, or more politely a medical representative. The American-owned company I joined preferred the term 'detailman', because it indicated the essential purpose of the work: to persuade the General Practitioner to prescribe the company's drugs by serving him with suitably tempting details. 1907 Daily Chron. 8 June 3/4 The parish councils would..be competent to relieve the controlling authorities of much detail work. 1922 H. Crane Lett. (1965) 97 The infinite and distasteful detail work I have been doing at the office. 1937 Burlington Mag. May 251/2 Inscription and detail-work are examples of the best Netherlandish engraving.

de-tail, detail (dī,tēˡ·l), v.² [f. De- II. 2 a + Tail sb.¹] trans. To deprive of its tail or tails.
1837 Marryat Snarleyyow III. i. 14 The dog had been detailed. 1900 G. Swift Somerley 148 His de-tailed coat looked like a ragged Eton jacket.

detailing, vbl. sb. Add: **b.** Fine Arts. = Detail sb. 4.
1931 C. H. Reilly in W. Rose Outl. Mod. Knowl. 976 Fine craftsmanship, such as the detailing of the Maison Carrée at Nîmes. 1958 Listener 20 Nov. 827/1 The timber detailing..seems to me not so satisfactory. 1960 House & Garden Dec. 31/1 Its warm colour..looks well with white detailing.

detain, v. Add: **1.** spec. To place (a political offender) in confinement.
1918 Rep. Comm. Rev. Conspiracies 86 in Parl. Papers (Cd. 9190) VIII, Such men are the leaders and organizers of the movement. They are now detained or their arrest is intended under Regulation III of 1818. 1940 J. Anderson in Hansard Commons 23 May 277, I have found it my duty, in the exercise of my powers under Regulation 18B of the Defence (General) Regulations, 1939, to direct that Captain Archibald Henry Maule Ramsay, Member of Parliament, be detained.

detainee (dītēˡnī·). [f. Detain v. + -ee¹; cf. Détenu.] A person detained in custody, usually on political grounds and in an emergency, without or pending formal trial.
1928 in M. H. Weseen Crowell's Dict. Eng. Gram. 118. 1940 Times 11 Dec. 9/4 There were other varieties of detainees with whom he had the greatest sympathy. 1957 Times 10 Sept. 6/6 The [Mau Mau] detainees were..employed on constructive development work. 1959 Listener 26 Feb. 359/1 Within hours of the return to the island [sc. Cyprus] of Sir Hugh Foot, the Governor, came the mass release of the detainees.

detect, v. Add: **5.** intr. To be engaged in work of detection; to act as a detective.
1926 D. L. Sayers Clouds of Witness iv. 105 Parker..was paid to detect and to do nothing else. 1930 M. Kennedy in D. L. Sayers Great Short Stories of Detection, Mystery & Horror (1931) 2nd Ser. 276 (title) Mr. Truefitt detects. 1957 A. Christie 4.50 from Paddington xviii. 178 'Good evening, Inspector Craddock.' 'Coming to detect in the kitchen?' asked Bryan with interest.

detection. Add: **3.** Electr. The process of obtaining a required electrical signal from a carrier wave or current that contains it; demodulation.
1906 J. A. Fleming Princ. Electr. Wave Telegr. vi. 353 (heading) Detection and measurement of electric waves. 1922 Glazebrook Dict. Appl. Physics II. 1040/2 All the above methods of detection are appropriate for the reception of damped wave signals but not for continuous wave signals. 1953 F. Langford-Smith Radio Designer's Handbk. (ed. 4) xxxvi. 1292 There are three main methods of performing the functions of F-M detection and A-M rejection in commercial F-M receivers. 1959 K. Henney Radio Engin. Handbk. (ed. 5) xii. 18 Detection is the process of removing desired information from a composite signal which conveys it through a communication or a processing system. 1962 Simpson & Richards Junction Transistors xviii. 460 The removal of the modulation from a carrier or i.f. signal is called demodulation or detection.

detective, a. and sb. Add: **A.** adj. **detective camera**, a term formerly used for a hand camera adapted for taking instantaneous photographs.
1881 Brit. Jrnl. Photogr. 28 Jan. 44/2 A form of the detective camera, in which the finding arrangement and the stock of slides are omitted, is in progress. 1882 Year Bk. Photogr. 27 Among novel apparatus we may mention..Mr. Bolas' so-called 'Detective Camera'. 1888 Brit. Jrnl. Photogr. 18 May 305 The subject of detective cameras is capable of considerable subdivision. 1893 [in Dict.].

B. sb. (Earlier examples.) Also attrib., as detective agency, anecdote, fiction, film, force, -inspector, novel, novelist, -sergeant, service; detective story (later examples).
1850 Dickens in Household Words 13 July 368/1 To each division of the Force is attached two officers, who are denominated 'detectives'. Ibid. 369/1 The two Detectives of the X division. 1852 —— Bleak Ho. xxv. 251 Detective Mr. Bucket.
1872 E. Crapsey Nether Side N.Y. 56 All the large commercial cities are now liberally provided with 'Detective Agencies', as they are called. 1959 Encounter XII. v. 30 The detective agency girls. 1850 Dickens in Household Words 14 Sept. 577/1 (title) Three 'detective' anecdotes. 1922 'Sapper' Black Gang ii. 28 What I'm going to tell you now..may seem extraordinary and what one would except in detective fiction. 1928 R. A. Knox Footsteps at Lock iii. 26 The Muse of detective fiction..cannot tell a plain unvarnished tale. 1958 Times Lit. Suppl. 6 June 316/5 This is one of the tales in which M. Simenon indulges his characteristic inquisitiveness about people while preserving the 'puzzle' convention of detective-fiction. 1911 C. N. Bennett et al. Handbk. Kinematography xiii. 100 Moreover, the bulk of modern motion picture detective films are of the Nick Carter and Sexton Blake variety. 1849 Alta California (San Francisco) 24 Dec. 3/3 The badge is of such a character that, when it becomes necessary to employ any of them [sc. policemen], as a detective force, they are [illegible]. 1850 Dickens Reprinted Pieces in Wks. (1858) VIII. 307 I'm an Officer in the Detective Force. 1888 A. C. Gunter Mr. Potter of Texas xx. Sergeant Brackett, of the British detective force. 1898 Westm. Gaz. 17 Nov. 7/2 Detective-inspector Egan said that he arrested the prisoner upon the charge. 1938 Times Lit. Suppl. 8 Oct. 642/2 Some long-suffering detective-inspector at Scotland Yard. [illegible] Cent. May 718 We note that the pilot of a detective novel is, in effect, an argument conducted under the guise of fiction. 1942 'N. Blake' in H. Haycraft Murder for Pleasure p. xxii, The detective novel proper is..read almost exclusively by the upper and professional classes. 1926 E. M. Wrong Crime & Detection p. xx, One temptation the detective novelist does well to avoid. 1850 Dickens in Household Words 13 July 368/2 The Detective Sergeant..fairly owned..that he could afford no hope of elucidating the mystery. 1969 Listener 24 Apr. 583/1 Interest centres on the man sent in to investigate, Detective-Sergeant Demosthenes de Goede. 1848 Mrs. Gaskell Mary Barton II. iii. 31 A well-known officer in the Detective service. 1905 G. K. Chesterton Club of Queer Trades iii. 96 The detectives in the detective stories. 1911 F. Swinnerton Casement ii. 75 'Douse the glim'..that old phrase in the detective stories he had read. 1963 Auden Dyer's Hand 153 The detective-story society is a society of apparently innocent individuals.

detectivism (dīte·ktiviz'm). [f. Detective + -ism.] The activities of a detective; detective work.
1894 Academy 23 June 514/2 An incredible piece of detectivism. 1896 Athenæum 15 Aug. 215 Literary detectivism of a high order. 1905 Book Lover 50/1 A sea story with a little detectivism and plenty of humour in it.

detector. Add: **3. c.** (Earlier example.) **e.** detector-pad. **f.** Any of various devices or circuits designed to carry out detection (sense *3).
1860 Encycl. Brit. XXI. 114/2 A 'detector' or common telegraphic galvanometer. 1894 O. Lodge Work of Hertz & Successors 29 We can easily see the detector respond to a distant source of radiation now..separated from the receiver, therefore, by several walls and some heavily gilded paper, as well as by 20 or 30 yards of space. 1898 Science Siftings 11 June 121/2 A Hertz-wave 'detector' resistance included in the circuit. 1902 Encycl. Brit. XXXIII. 232/2 The coherer, or detector, is inserted between the earth and the outer end of this last wire. 1924 Wireless Weekly 8 Oct. 744/2 Seven valves (all 'peanuts'), used respectively as detector, oscillator valve, three stages of intermediate frequency, second detector valve, and one stage of transformer-coupled note-magnification. 1928 Morning Post 6 Feb. 3/4 The nature of the circuit connected to the detector helps to determine the amplification. 1940 Illustr. London News CXCVI. 192/1 Imagine that two vehicles are converging on the crossing, the one on the T-shaped road being nearer the junction... The former would reach his detector pad first, and cause the lights to change in his favour. 1957 D. G. Fink Telev. Engin. Handbk. vi. 91 The sum of the luminance signal, the chrominance subcarrier, and the sound carrier is carried out at the second detector.

detemporalize (dīte·mpǒrǎlaiz), v. [f. De- II. 1 + Temporalize v.] trans. To make timeless in character; to remove the impression of passage of time from; to detach from a particular time.
1914 Philos. Rev. XXIII. 18 The very intensity of his vision detemporalizes all he sees. 1954 K. Tillotson Novels of Eighteen-Forties i. 97 In Wuthering Heights the past..is not only..a means of detemporalizing the action.

detend (dī,te·nd), v. [f. De- II. 1 + Tend v.²; cf. F. détendre to relax, slacken.] trans. To reduce the tension or intensity of.
1930 H. Read Wordsworth vii. 243 The philosopher begins with his intuitive apprehension and detends it—resolves it into that lower tension of expression which is prose. 1949 Koestler Insight & Outlook xx. 282 Self-transcending emotions de-tend the body as no action is required.

de-tension (dīte·nʃǝn), sb. A reduction of anxiety; a relaxation.
1949 Koestler Insight & Outlook xv. 58 The delight in laughing is one of the many variants of catharsis or de-tension.
Hence **de-te·nsion** v., **de-te·nsioning** vbl. sb. and ppl. a.
1952 A. Huxley Let. July (1969) 649 This process of remembering will induce a complete detensioning. 1958 Times 29 Sept. 5/4 A substitute 'detensioning' agent, something less harmful than smoking. 1960 New Statesman 10 Mar. 376/2 Macmillan is attempting to de-tension the conference and work for some face-saving compromise. 1966 Observer 3 Apr. 13/1 Detensioning had already led to the first Labour gain.

detent (dīte·nt), sb.³ Delete †Obs. and add to def.: the holding back or inhibition.
1907 W. James Mem. & Studies (1911) 256 They are forces of detent in situations in which no other force produces equivalent effects, and such is a force of detent only in a specific group of men. 1929 R. Bridges Test. Beauty iv. 150 A pinprick or a momentary whiff or hairbreadth motion freeth the detent of force.

‖ **détente** (detã·nt). [Fr., = loosening, relaxation.] The easing of strained relations, esp. in a political situation.
1908 Times 17 Aug. 5/4 A change in the European situation..had..set in... The characteristic feature of this transformation may be called a détente. 1921 Von Hügel in L. S. Hunter John Hunter 213 The profound impression of a comforting détente made upon my mind by my first contact with his spirit. 1928 Daily Tel. 29 May 9/5 To bring about an early détente in Turco-Persian relations. 1940 Ann. Reg. 1939 182 The very slight détente..which had followed the signature of the Franco-German accord..had brought her [sc. France] no real tranquillity. 1955 Times 6 June 7/7 The sense of détente is not to the liking of the party zealots who cry for more revolutionary fervour under the leadership of the party.

detention. Add: **1.** (Later examples, spec. of the confinement of a political offender.) Cf. *preventive detention.
1909 J. Morley Indian Speeches 146 There is no fixed limit of time of detention. 1920 Statem. Moral & Mat. Progr. India 1919 26 in Parl. Papers (Cmd. 950) XXXIV. 744 The continued detention of dangerous characters already under control or in confinement. 1940 J. Anderson in Hansard Commons 5th Ser. CCCLXI. 290 Regulation 18 B of the Defence Regulations was last night amended by the addition of a provision enabling me to order the detention of members of organizations which have had associations with the enemy.

c. At schools: keeping in as a punishment. Also concr. and attrib.
1882 Boy's Own Paper 3 June 574/3 Mr. Rastle..set them each twelve propositions of Euclid to learn by heart, and two hours a-piece in the detention-room, there to meditate over their evil ways. 1909 Wodehouse Mike xliii. 243 There is only one thing to be said in favour of detention on a fine summer's afternoon, and that is that it is very pleasant to come out of. 1931 'R. Crompton' William's Crowded Hours ii. 34 He was unable to answer two very simple questions that the Latin master asked him, and was given a detention.

5. attrib., as **detention barrack**, a military prison; **detention camp**, in the war of 1914–18, a camp in which aliens and others were kept under restraint; also applied to other places of incarceration; **detention centre**, an institution in which young offenders are detained for short periods.
1906 Act 6 Edw. VII c. 2 § 6. 5 A soldier sentenced to imprisonment..may be confined in a detention barrack. 1916 J. Buchan Greenmantle v. 62 The lieutenant discoursed a lot about prisoners and detention-camps. 1958 New Statesman 25 Jan. 93/1 The new governor, by..his dramatic gesture of a Christmas amnesty for 100 prisoners from the detention camps, has provided a much needed tonic. 1948 Act 11 & 12 Geo. VI c. 58 § 48 Detention centres, that is to say places in which persons not less than fourteen but under twenty-one years of age who are ordered to be detained in such centres under this Act may be kept for short periods under discipline suitable to persons of their age and description. 1961 Listener 19 Oct. 612/1 Mr. Sewell Stokes's talk on detention centres..paints a rosy picture of this comparatively new method of dealing with boys and young men of fourteen to twenty-one.

détenu. Add: (Further examples.)
1862 Times 8 July 12/4 If detention in its carriages is frequent, it is a reason for making the détenus as comfortable as possible. 1918 Parl. Papers (Cd. 9198) VIII. 105 (title) Report of Sir N. Chandavarka and Mr. Justice Beachcroft on Detenus and Internees in Bengal. 1920 H. V. Lovett Hist. Indian Nat. Movem. vii. 196 The Committee recommended the release of six only of all the above enumerated détenus. 1956 M. Collis Last & First in Burma xxiii. 209 He had passed through Uganda and was able to secure for the détenu some extra comforts. 1969 Hindusthan Standard (Calcutta) 5 Aug. 8/1 The High Court asking for production of the detenus.

detergency. Add: (Later examples.) Also, the process of cleansing a solid by means of a liquid, the action of a detergent.
1943 Industr. & Engin. Chem. Jan. 110/1 Detergency is not completed until the soil has been suspended in the detergent solution with sufficient stability so that it can be rinsed away rather than redeposited. 1950 Thorpe's Dict. Appl. Chem. (ed. 4) X. 803/1 Use of Launderometer... Detergency is measured either by the increase in the whiteness of the soiled cloth on washing..or by the dirt content of the detergent solution. 1959 Times 14 Sept. 2/6 Chemistry problems associated with detergency. 1965 R. D. & M. J. Vold Colloid Chem. v. 96 These effects may..contribute to effective detergency by helping to maintain the separated dirt in colloidal suspension.

detergent, a. and sb. Add: **B.** sb. (Further examples.) Now esp., any of various synthetic solids or liquids which are soluble in or miscible with water, which resemble soap in their cleansing properties, but which differ from it in not combining with the salts present esp. in hard water; also, any of various oil-soluble substances which have the property of holding dirt in suspension in lubricating oils; so **detergent oil**, an oil containing such a substance. Also attrib. and Comb.
1938 Encycl. Brit. Bk. of Yr. 1938 331/1 Word came of a new process for rapid saponification; the 'Igepals' are new detergents of German origin. 1941 Ann. Rep. Prog. Chem. 1940 102 A large number of synthetic substances used during recent years as detergents. 1941 Jrnl. R. Aeronaut. Soc. XLV. App. 141 [It] has a viscosity index of approximately 100—nearly twice that previously available in special detergent oils. 1951 Good Housek. Home Encycl. 45/2 Once a week use hot water and soapless detergent to remove all traces of grease. 1952 M. Laski in Observer 26 Oct. 5/2 Detergent-packets..almost always fall over and spill. 1957 G. I. Brown Introd. Org. Chem. xviii. 223 Synthetic detergents..are usually more soluble in water than soap is; they do not form a scum in hard water; they enable water to spread and penetrate more fully over or through an article being cleaned. 1957 Technology July 168/4 Detergents usually contain a surface active agent, alkylbenzenesulphonate, which influences surface adsorption and, incidentally, causes the familiar foaming (of doubtful utility). 1958 Engineering 21 Mar. 353/3 Detergent powders..which are bought by a large section of the population. 1958 Spectator 4 July 12/1 Like detergent foam in a millrace. 1958 Sunday Times 23 Nov. 28/3 He's..detergent-white and very cuddly. 1959 Engineering 16 Jan. 96/1 Improved performance has followed the use of heavy-duty and detergent oils. 1959 Which? Sept. 105/1 'Detergent' is often used—loosely—as a synonym for 'synthetic detergent'. Ibid., A good detergent is one which removes dirt easily from the clothes, holds the dirt in suspension in the washing-water, so that it can be rinsed away and will not get re-deposited on the clothes, and—in the process—does them no damage. 1960 V. B. Guthrie Petroleum Products Handbk. ii. 34 The use of detergents in oils has introduced additional demands on oxidation inhibitors. 1966 McGraw-Hill Encycl. Sci. & Technol. IV. 81/2 Soap..was the principal detergent until superseded in production in 1954 by synthetic detergents.

determa (dītā·ɹmā). Also 8 tetermer. [Native name.] A native wood of Guyana.
1769 E. Bancroft Nat. Hist. Guiana 80 The Tetermer Tree grows to near fifty feet in height. 1851 Illustr. Catal. Gt. Exhib. IV. I. 984/2 Determa, transverse and vertical sections, from River Demerara. 1858 Simmonds Dict. Trade, Determa, a native wood of Guiana, which is used for masts, booms, and planking for colonial craft.

determinable, a. Add: **B.** sb. Philos. [tr. G. das bestimmbare.] That which is capable of being given a more determinate form or of being more precisely specified; spec. (in W. E. Johnson's use) a general term or concept (e.g. colour) under which several specific terms or concepts fall (e.g. red, yellow, green). Also as adj.
1878 S. H. Hodgson Philos. of Reflection I. iv. 272 Maimon adds [in Die Kathegorien des Aristoteles: Propädeutik 248] that since the critical philosophy has already fixed the meaning of the expressions, matter and form, he will use expressions of his own instead of them, namely, the determinable and its determination. Ibid., Space is therefore the matter, the determinable (bestimmbare). 1906 W. James Let. 3 Apr. in R. B. Perry Thought & Char. W. J. (1935) II. 392 Taking 'experience' concretely..seems to me the only way in which to leave all its determinations real so far as they are attained, and at the same time to leave always a determinable..that provides for what is fertile and developable in the process. 1921 W. E. Johnson Logic I. xi. 174, I propose to call such terms as colour and shape determinables in relation to such terms as red and circular which will be called determinates. 1949 G. Ryle Concept of Mind ii. 44 Many disposition-concepts are determinable concepts. 1960 S. Körner Philos. of Math. viii. 167 Perceptual characteristics which in the philosophical literature are sometimes called 'determinables' or 'respects of likeness', such as 'colour', 'shape', etc., are all internally inexact.

determinacy. Delete rare and add later examples.
1953 C. E. Bazell Linguistic Form 108 It is the fact that different postulates tend to lead to similar solutions which guarantees the relative determinacy of phonemic as opposed to componential analysis. 1958 Listener 18 Dec. 1019/2 The determinacy which exists in events where large numbers of atoms are involved.

determinandum (dītǝ̄:ɹminæ·ndǒm). Logic and Philol. [a. L. dēterminandum 'thing to be determined', neut. of gerundive of dēterminā̆re to Determine.] That part of a proposition (or an expression) which is to be made determinate by a thought-process (or by another part of the expression).
1921 W. E. Johnson Logic I. i. 9 The determinandum is defined as what is presented to be determined or characterised by thought or cognition; the determinans as what does characterise or determine in thought that which is given to be determined. 1924 Jespersen Philos. Gram. xi. 146 At the outset indefinite and indeterminate,..the subject is..a determinandum which only by means of the predicate becomes a determinatum. 1961 [see *determinans].

determinans (dītǝ̄:ɹminæ·nz). Logic and Philol. [a. L. dēterminans, pres. pple. of dēterminā̆re to Determine.] The limiting or qualifying part of a proposition or expression.
1921 [see *determinandum]. 1961 Brno Studies III. 13 Syntagmatic relations of the type 'determinans—determinandum'.

determinant, a. and sb. Add: **B.** sb. **2. b.** Logic. = *conjunct sb. 5.
1887 J. N. Keynes Formal Logic (ed. 2) IV. i. 335 We may speak of the elements combined in a conjunctive term as the determinants of that term. 1892 W. E. Johnson in Mind I. 237 The separate constituents a, b, are called the determinants of the determinative synthesis a.b. 1949 Mind LVIII. 3 We find..a marked difference between the use of the word 'determinant' by symbolic logicians of the late 19th century, and the medieval use of its Latin equivalent. For such writers as Schroeder and J. N. Keynes a 'determinant' is any one of the elements combined in a logical conjunction.

c. Philol. A limiting or qualifying expression; esp., in a compound word, the element that limits or qualifies the meaning of the other element.
1869 [see sense 2 in Dict.]. 1934 Priebsch & Collinson German Lang. II. v. 258 Composite words consisting of a determinant word (Bestimmungswort) and a nucleus (Grundwort). The determinant may be an adjective..or a substantive. 1960 [see *determinatum]. 1962 A. Martinet Functional View Lang. ii. 51 Some determinants are lexical (an adjective like great), and others grammatical (an adjective like my, the article the, or the 'plural' moneme).

4. Biol. In Weismann's theory of heredity, each of the hypothetical units in the germ-plasm supposed to determine the character and development of a cell or group of cells (hence called a determinate) in the organism; also, a gene or other hereditary 'factor'.
1893 Parker & Rönnfeldt tr. Weismann's Germ-Plasm I. i. 57, I shall designate the cells or groups of cells which are independently variable from the germ onwards as the 'hereditary parts' or 'determinates', and the particles of the germ-plasm corresponding to and determining them, as the 'determining parts' or 'determinants'. Ibid. 59 A determinant is always a group of biophors. 1905 Westm. Gaz. 30 Mar. 2/1 Selection acts on the determinants and produces the variations once so plausibly attributed to the Lamarckian principle of acquired characters. 1920 Cambr. Bulletin Feb. 16 Germ-cell Determinants. 1944 C. D. Darlington in Nature 5 Aug. 165/1 How far are we justified in assuming the same kind of determinant in the cytoplasm where determinants are not fastened to the immediate products of their activity? 1971 Nature 19 Mar. 185/1 Factors in the cytoplasm of many eggs can influence the development of the cells, part of which they eventually constitute. This is most clearly seen in the case of the germ cell determinants.

determinantal (dītǝ̄:ɹminæ·ntǎl), a. Add: Consisting of or expressed as a determinant.
1867 C. L. Dodgson Elem. Treat. Determinants 12 The determinantal coefficient of any Element of a square Block is the Determinant of its complemental Minor. 1968 Fox & Mayers Computing Methods for Scientists & Engineers v. 95 Finding a zero of some function expressed in determinantal form.

determinate, ppl. a. Add: **C.** sb. **1.** Biol. (See *determinant B. sb. 4.)
1893 Parker & Rönnfeldt tr. Weismann's Germ-Plasm I. i. 58 Similar hereditary parts or determinates may be observed in butterflies, in which the colours on the wings often form very complicated lines and spots of slight extent but of great constancy.

2. Philos. (See quot.)
1921 [see *determinable sb.].

determinative, a. Add: **2. b.** Gram. determinative adjective, pronoun, etc. (see quots.); determinative compound = *tatpurusha.
1921 E. Sapir Lang. vi. 135 The words of the typical suffixing languages (Turkish, Eskimo, Nootka) are 'determinative' formations, each added element determining the form of the whole anew. 1924 H. E. Palmer Gram. Spoken Eng. II. 24 To group with the pronouns all determinative adjectives..shortening the term to determinatives. 1933 Bloomfield Lang. xiv. 235 One can..distinguish..determinative (attributive or subordinative) compounds (Sanskrit tatpurusha). 1961 R. B. Long Sentence & its Parts 486 The, a, and every are exceptional among the determinative pronouns in requiring stated heads.

determinatum (dītǝ̄:ɹminēˡ·tǒm). Logic and Philol. [a. L. dēterminātum 'thing determined', neut. of pa. pple. of dēterminā̆re to Determine.] A qualified or limited element of a phrase or proposition; the part determined.
1924 [see *determinandum]. 1960 S. Potter Lang. in Mod. World v. 67 In English the determinant (air) always precedes the determinatum (man).

determiner¹. Add: **1. d.** Biol. = *determinant B. sb. 4.
1909 W. Bateson Mendel's Princ. Heredity 79 Hitherto we have spoken of the determiner for such a colour as grey in rabbits and mice as 'dominant' over the colours lower in the scale, such as black or chocolate... We shall then speak of the determiner for grey as epistatic to that for black. 1922 R. C. Punnett Mendelism (ed. 6) 113 A female determiner incapable of carrying sex-limited factors. 1926 J. S. Huxley Ess. Pop. Sci. 236 The future symmetry of the embryo is not influenced by anything that one could call a 'determiner' in the cytoplasm. Ibid. 246 It then became an obvious task to discover more precisely where this structure-determiner was situated and how it acted.

3. Gram. A class of limiting expressions modifying nouns (see quot. 1933). Also attrib.
1933 Bloomfield Lang. xii. 203 Our limiting adjectives fall into two sub-classes of determiners and numeratives... The determiners are defined by the fact that certain types of noun expressions (such as house or big house) are always accompanied by a determiner (as, this house, a big house). 1961 R. B. Long Sentence & its Parts ii. 39 The most characteristic modifiers of nouns in positions in front of the nouns are (1) determiners, such as the and possessives, and (2) adjectives. Ibid. 40 Determiner modifiers, pronominal in function, normally precede adjectival modifiers. 1964 Language XL. 37 Traditionally, grammarians have recognized two kinds of determiners: the, usually called the definite determiner, and a, usually called the indefinite determiner. It is now common to regard other elements as determiners, or as parts of determiners... Further, it is convenient to regard all, any, and the like as part of the determiner (these items are frequently called predeterminers). And finally, pronominal genitives such as his and John's are shown to behave as determiners by transformational criteria.

determinism. Add: **2.** (Examples of extended uses.)
1944 G. Bateson in J. McV. Hunt Personality & Behavior Disorders II. xxiii. 714 The phrase 'economic determinism' has..become a slogan. Ibid. 716 It is this sort of cultural 'genetics' and cultural 'physiology' which I have tried to sum up with the phrase 'cultural determinism'. 1945 K. R. Popper Open Society II. xxii. 196 Sociological determinism [is the view that]..all our opinions..depend upon society and its historical state. 1950 C. D. Darlington in Darlington & Mather Genes, Plants & People p. xvii, Mendel directed his enquiries with a rigorous determinism. He assumed that every property of every seedling was determined by something that happened in its two parents. 1957 N. Frye Anatomy of Criticism 6 The fallacy of what in history is called determinism, where a scholar with a special interest in geography or economics expresses that interest by the rhetorical device of putting his favorite subject into a causal relationship with whatever interests him less.

deterministically (dītǝ̄ɹmini·stikǎli), adv. [f. Deterministic a. + -ly².] In a deterministic manner; inevitably.
1885 W. James Let. 5 Feb. (1920) I. 238 Since such activity has failed to be universally realized, it was (deterministically) impossible from eternity. 1935 Mind XLIV. 85 The relation of the part to the deterministically unfolding pattern of the whole.

deterrence. Add: Also spec., the reduction of the likelihood of war by the fear that nuclear weapons will be used against an aggressor; so graduated deterrence (see quot. 1966).
1955 Economist 5 Nov. 457 (heading) Graduated Deterrence? 1958 New Statesman 19 July 90/1 The argument about self-preservation and deterrence is trivial and thoughtless. At the moment the nuclear powers will probably not start war against each other except in the event of an accident. 1959 Times Lit. Suppl. 9 Jan. 17/3 The present state of passive deterrence (in which each side could retaliate devastatingly if attacked, so neither side can afford to attack) provides a framework of insurance. 1962 Observer 13 May 15/2 The 'deterrence only' school include most of the adherents of finite deterrence (or minimum deterrence)...If each side can be certain of inflicting terrible damage on the other with its strategic forces, then neither side will use them. 1966 Schwarz & Hadik Strategic Terminology 59 Graduated deterrence, a strategy threatening a whole range of countermoves, each of which is designed not so much to punish the enemy and to destroy his war-making capacity or to change the local balance of power, as to demonstrate the will to inflict greater punitive damage, to escalate the conflict, in the hope of convincing the enemy to refrain from pursuing his objective.

deterrent, sb. Add: Esp. the nuclear weapons of any one country or alliance; spec., the hydrogen bomb. Freq. the (great) deterrent. Also attrib. and Comb.
1954 Statement on Defence p. 4 in Parl. Papers 1953–54 XXII. 474 The primary deterrent, however, remains the atomic bomb and the ability of the highly organised and trained United States strategic air power to use it. 1954 New Statesman 14 Aug. 173/2 The theory and practice of the Great Deterrent is in fair way to becoming the theory and practice of the Great Bluff. 1955 Ibid. 26 Feb. 271/1 The Government has concurred in the Nato decision to rely on nuclear weapons and begun itself to manufacture a British H-bomb. It admits..that this leaves the United Kingdom well-nigh defenceless if the deterrent fails. 1957 J. Slessor (title) The great deterrent. 1958 Punch 27 Aug. 275/1 A lot of good that would do the human race, with..the stockpile of deterrents rising as high as the Empire State Building. 1959 Times Lit. Suppl. 9 Jan. 17/3 The theory of the H-bomb as a deterrent. 1959 Observer 14 June 16/2 Britain should also be prepared to give up her independent deterrent and stop the manufacture of nuclear weapons. 1962 Listener 29 Mar. 548/1 There is no sign..to suggest that the Great Powers will be able to abandon deterrent strategy in the forseeable future. 1963 Economist 12 Oct. 101/1 Restraint on German 'deterrent-mongers'.

detin (dī,ti·n), v. [f. De- II. 2 + Tin sb.] trans. To remove the tin from (scrap tin plate).
1909 in Cent. Dict. Suppl. 1923 Glasgow Herald 15 Aug. 7 The major portion of the old tins was disposed of after being detinned and pressed into billets. 1929 Times 21 Sept. 7/4 Scrap metal..suitable for detinning. 1960 E. S. Gyngell Appl. Chem. Engineers (ed. 3) xiii. 207 The well-known process used for detinning metal cans.

detonation. Add: **1. b.** In an internal combustion engine, the rapid and premature combustion of the mixture in the cylinder before it is ignited by the spark or the flame from the spark, producing a rattling or thumping noise (cf. *knock sb.¹).
1912 Motor Manual (ed. 14) vi. 201 With super-compression of the charge instead of a gradual expansion of the ignited gases a 'detonation' occurs. 1937 Autocar Handbk. (ed. 13) v. 91 Should the charge be fired off too early, detonation, or, as it is more popularly known, 'pinking', will result. 1968 Practical Motorist Oct. 205 The detonation causes a distinctive knock ('pinking') when the engine is under load.

detortion, -sion. Restrict 'Now rare or Obs.' to senses in Dict. and add: **3.** Biol. (Spelt detorsion.) Evolutionary reversion to a primitive linear anatomical organization in a species of organisms, esp. gastropod molluscs, whose intermediate ancestors underwent torsion through 180°.
1902 Encycl. Brit. XXX. 796/2 Not an absence of torsion, but an actual detorsion of the visceral commissure. 1930 G. R. de Beer Embryol. & Evol. vii. 54 That intermediate stages are not a priori impossible is proved by the fact that intermediate stages in detorsion in later Gastropods are not wanting. 1968 R. D. Purchon Biol. Mollusca i. 25 The penis is situated on the right side of the head..and has evidently been 'left behind' when the mantle cavity retreated posteriorly during the process of detorsion.

detour, sb. Now freq. pronounced (dī·tūᵊɹ). Delete 'now usually lit.' and add later examples.
1922 Joyce Ulysses 237 Better turn down here. Make a detour. 1923 E. O'Neill Hairy Ape 50 They make wide detours to avoid the spot where he stands in the middle of the pavement. 1966 'J. Hackston' Father clears Out 26 It's no use travelling by the compass if one has to make wide detours.

detour, v. Add: (Later examples.)
1937 D. Aldis Time at her Heels x. 225 Mary detoured to pick up her bag from the table. 1955 E. Bowen World of Love ii. 41 A stranger, who detoured to drop them at the Montefort gates.

b. trans. To send by a detour. U.S.
1905 N.Y. Even. Post 18 Sept., The Missouri Pacific is detouring its St. Louis–Kansas City trains over the Burlington Railway via Chillicothe. 1967 Boston Globe 18 May 12/2 (caption) Workman carries signs detouring motorists off Cambridge st. and Broadway.

detoxicate (dītǫ·ksikēˡt), v. Delete nonce-wd. and add later examples.
1906 Practitioner Nov. 590 Focalisation of the infection in the liver, with disturbance of its detoxicating mechanism. 1927 Observer 7 Aug. 3 It detoxicates the blood and keeps it clean and free-running. 1964 Economist 14 Mar. 1020/3 A toxic amine, which ordinarily is detoxicated in the liver.
Hence **detoxica·tion**, the action of depriving of poisonous qualities; **deto·xicator**, that which detoxicates; also **deto:xifica·tion** and **deto·xify** v.
1905 Gould Dict. New Med. Terms 213/1 Detoxification, the power of reducing the poisonous properties of a substance. Ibid., Detoxify, to deprive a substance of its poisonous attributes. 1906 Practitioner Nov. 586 A reducer and detoxicator of toxic substances. Ibid. 593 The detoxication of the poisons found in the blood. 1948 W. W. Pigman Chem. Carbohydrates xi. 509 Many phenols and alcohols when injected into animals are detoxified by conjugation as glucuronides. 1955 Sci. News Let. 9 Apr. 229/3 The vaccine for this is made by detoxifying the toxin, or poison, produced by the diphtheria germs. 1956 Nature 4 Feb. 219/1 Biochemical aspects of mechanisms of detoxication. Ibid. 25 Feb. 357/2 The conversion of insecticides to harmless metabolites or detoxification. 1961 New Scientist 7 Sept. 601/1 Many tumours do not have the ability of detoxifying drugs. 1963 Guardian 22 Jan. 8/5 You give the impression that detoxification of ordinary coal gas is neither a practical nor an economic proposition. 1967 New Scientist 16 Nov. 410/1 Resistance mechanisms often involve detoxication—that is, the resistant individuals are able to convert the insecticide into relatively harmless products. 1970 Daily Tel. 20 Jan. 14/1 This is Cocteau's second attempt at detoxication, the first having failed owing to lack of medical knowledge about withdrawal. 1971 Ibid. 2 Mar. 2/7 People with a serious drink problem may be detained in the 'detoxification centres' for 10 days while they receive medical treatment.

detractation (dītræktēˡ·ʃǝn). Delete †Obs. and add later example.
1888 H. S. Salt Shelley 141 Against Shelley he never uttered a word of detractation.

detrainment (dītrēˡ·nmĕnt). [f. Detrain v.² + -ment.] The action of discharging (persons, esp. troops) from a train, or of alighting from a train.
1899 Westm. Gaz. 9 Dec. 6/1 Three trains of mounted troops have been detrained at Arundel,..being covered in their detrainment by the New Zealand Mounted Infantry. 1916 Ld. E. Hamilton First 7 Div. 142 On October 11th the detrainment of the 2nd A.C. was completed. 1923 W. S. Churchill World Crisis, 1911–1914 I. xvi. 363 His detrainments at St. Omer, etc., were not completed till the 19th. 1928 Observer 18 Mar. 13/1 All detrainments of the assault divisions were made well back on a wide front. 1963 Times 24 May p. vii/4 London Transport has done everything it can to..explain why these detrainments are occasionally necessary.

‖ **détraqué** (detrake), sb. and a. Also fem. **détraquée.** [Fr., pa. pple. of détraquer to put out of order, derange.] **A.** sb. A deranged person; a psychopath.
1902 W. James Var. Relig. Exper. i. 7 From the point of view of his nervous constitution, [George] Fox was a psychopath or détraqué of the deepest dye. 1936 D. Barnes Nightwood iii. 80 Those who love a city..become the shame of that city, the détraqués, the paupers. 1969 Daily Tel. (Colour Suppl.) 25 Apr. 58/4 The détraquées—or potential nymphomaniacs—with their desires inflamed by concealment and repression.

B. adj. Deranged; crazy; psychopathic.
1925 A. Huxley Those Barren Leaves IV. v. 307 German romanticism, a little détraqué, turns..into expressionismus. 1930 Times Lit. Suppl. 13 Mar. 209/2 Here too are Augusta Leigh, refined, plaintive and détraquée, [etc.]. 1958 A. Wilson Middle Age of Mrs Eliot II. 255 She may be détraquée, as they say, but she's strictly U.

detribalization (dītrəi:bǎlǝizēˡ·ʃǝn). [f. next.] The process of becoming detribalized (see next).
1928 R. L. Buell Native Probl. Afr. II. 1064/1 Detribalization, Belgian Congo. 1941 G. Wilson (title) Essay on the economics of detribalization in Northern Rhodesia (Rhodes–Livingstone Papers No. 5). 1959 Times Lit. Suppl. 6 Mar. 131/3 It is their [sc. West Indians'] detribalization and assimilation into White life which has paradoxically made them more self-aware..of their African-ness.

detribalize (dītrəi·bǎlǝiz), v. [f. De- + Tribal a. + -ize.] trans. To render (a person) no longer a member of a tribe; to destroy the tribal habits of. So **detri·balized** ppl. a.
1920 Contemp. Rev. Sept. 397 Numbers of natives who, through living in close contact with the settlers have become quite detribalised. 1925 Glasgow Herald 25 Apr. 8 The African National Congress..is a very small body composed principally of detribalised natives. 1927 Public Opinion 272/1 White men have detribalised him. 1944 Archit. Rev. XCVI. 106/4 There is the Bantu tradition—the cultural unity of a tribal society which still persists or to which the detribalized African is still close, if only in time. 1955 Times 20 Aug. 7/4 It is not only the intelligentsia who are detribalized, but thousands of men, of all levels of education and intelligence, who have left their reserves to seek employment in Nairobi and elsewhere. 1969 Guardian 7 July 9/5 Tom Mboya chose..to detribalise himself.

detu·ne, v. [f. De- II. 1 + Tune v.] trans. **a.** To alter or adjust (an oscillatory system, as a resonant circuit or a microwave cavity) so that its resonant frequency no longer coincides with the frequency of some other oscillation with which it interacts. **b.** (See quot. 1963¹.) Hence **detu·ned** ppl. a.; **detu·ning** vbl. sb.
1924 S. R. Roget Dict. Electr. Terms 60/2 Detuning (or Mistuning), the alteration of the period of oscillation of a circuit from that of the waves being dealt with. 1958 Oxf. Mail 23 Aug. 3/6 A new Hillman Husky powered by a detuned version of the overhead valve Minx engine. 1962 Simpson & Richards Junction Transistors xiv. 358 Detuning also occurs in the external-gain control method described below. 1963 R. F. Webb Motorists' Dict. 75 Detune, to deliberately tamper with an engine to reduce its performance or efficiency... A high-powered racing engine that has been slightly 'detuned' will stand up to normal wear and tear better than a fully tuned engine. Detuning consists normally of fitting lower compression pistons, etc. 1963 G. Troup Masers & Lasers (ed. 2) vi. 105 Detuning the cavity centre pulls the oscillation frequency away from the molecular frequency. 1970 Daily Tel. 17 June 13/5 The engine is a detuned version of Rover's aluminium, 3½ litre V8.

deuce¹. Add: **2. deuce-game** (see quot. 1897); **deuce-set**, a set in which the sides are level, each having won five or more games.
1886 Cassell's Fam. Mag. Oct. 704/2 It also scores back to deuce points and deuce games. 1897 Encycl. Sport I. 621/2 Deuce-game, the game won, which makes the score in games level when each side has won more than five. 1908 Westm. Gaz. 16 Nov. 14/1 Losing the first game after a deuce set. 1969 New Yorker 14 June 67/1 Games are five-all. It is a so-called deuce set.

deurbanize (dī,ǝ̄·ɹbǎnǝiz), v. [f. De- + Urbanize v. 2.] trans. To deprive (a district) of its urban character. Also fig.
1924 Public Opinion 15 Feb. 154/3 Its work is to deurbanise the minds of British rulers. 1927 Observer 29 May 7 Can we de-urbanise England? Or will all our efforts to empty the towns merely result in urbanising the countryside?

‖ **deus absconditus** (dī·ǝs æbskǫ·nditǝs). [L., 'hidden god'; cf. Isaiah xlv. 15.] A god who is hidden from man. Also fig.
1932 Character & Personality Sept. 51 Where..in the annals of Victorianism is the sentimental God of the nineteenth century ever faced with a deus absconditus as in Luther's teaching? 1960 Auden Homage to Clio 21 Pray to a Deus Absconditus. 1965 S. Marcus Dickens: from Pickwick to Dombey vi. 217 The novelist has become a deus absconditus, asserting his godhead by his apparent indifference to his creation.

‖ **deus ex machina** (dī·ǝs eks mæ·kinā). Also (nonce-wd.) fem. **dea ex machina.** [mod.L., tr. Gr. θεὸς ἐκ μηχανῆς, lit. 'god from the machina' (the device by which gods were suspended above the stage in the Greek theatre): cf. Machine sb. 6.] A power, event, person, or thing that comes in the nick of time to solve a difficulty; providential interposition, esp. in a novel or play.
1697 J. Sergeant Solid Philos. 136 Nor is it at all allowable in Philosophy, to bring in a Deus è Machinâ at every turn, when our selves are at a loss to give a Reason for our Thesis. 1840 Barham Ingol. Leg. 135 This was, however, no less a personage than the Deus ex machinâ,—the illustrious Aldrovando himself. 1857 Trollope Barchester T. II. xv. 292 Doctor Gwynne was the Deus ex machina who was to come down upon the Barchester stage, and bring about deliverance from these terrible evils. 1873 Hotten Slang Dict. 223 Mary Ann, the title of the dea ex machinâ evolved from trades-unionism at Sheffield. 1889 A. E. Haigh Attic Theatre iv. 190 The mêchanê was used under various circumstances; but the most ordinary occasion for its employment was to introduce the 'deus ex machina' at the end of a play, when affairs had reached such a complicated condition that only divine interference could put them right again. 1896 W. Caldwell Schopenhauer's System ii. 62 If you assume in every brain a spirit like a sort of deus ex machinâ, you ought to concede a spirit to every stone. 1928 C. Dawson Age of Gods p. xiv, The dominant fashion is to look to the racial factor as the deus ex machina

of the human drama. 1937 M. Sharp *Nutmeg Tree* ix. 108 Her function at Les Sapins was that of a *dea ex machina*; and the make of her car would not bear examining. 1962 W. Nowottny *Lang. Poets Use* vi. 145 Fruitful Ambiguity, that popular *deus ex machina* of the contemporary critical scene. 1970 T. Hilton *Pre-Raphaelites* v. 133 The accepted interpretation, that he [sc. Ruskin] descended to the Brotherhood like some *deus ex machina*, ignores the real similarity of artistic purpose that bound him to the group.

deuteranomalous (diū:těrănǫ·mǎləs), *a.* *Ophthalmology.* [f. Deutero- + Anomalous *a.*] Of, pertaining to, or exhibiting partial deuteranopia.

1932 W. S. Duke-Elder *Textbk. Ophthalmology* I. xxv. 976 Holmgren (1880) reported 2 protanopes..and Reichert (1915) and v. Kries (1918) deuteranomalous cases. 1952 *Brit. Jrnl. Psychol.* Nov. 312 A study made upon..19 protanomalous and 32 deuteranomalous subjects.

deuteranope (diū·těrănōᵘp). *Ophthalmology.* [a. G. *deuteranope* (J. von Kries 1897, in *Zeitschr. f. Psychol. d. Sinnesorgane* XIII. 248), f. Deutero- + Gr. priv. ἀν- + ὤψ eye.] One who is green-blind. So **deuterano·pia**, green-blindness.

1901 Baldwin *Dict. Philos.* I. 274 *Deuteranopia*,..the name proposed by v. Kries for what was formerly called green-blindness. 1902 *Ibid.* II. 787 The wave-length which corresponds to the colourless sensation, *W*, is different for the two sorts of colour-defectives, proteranopes and deuteranopes. 1937 *Nature* 10 July 49/1 A diagnosis can be made from the tables of dichromasy in general, and of protanopia, deuteranopia, and tritanopia in particular. 1962 H. C. Weston *Sight, Light & Work* (ed. 2) i. 18 The green-blind are called 'deuteranopes'.

deuterated (diū·tĕrēᴵted), *a.* [f. *Deuter(ium + -ate² + -ed².] Containing deuterium; having had an atom of ordinary hydrogen replaced by one of heavy hydrogen (deuterium). So **deutera·tion**, the process of substituting a deuterium atom for an ordinary hydrogen atom in a molecule.

1947 *Nature* 11 Jan. 62/2 The Raman spectra of deuterated methyl halides. 1955 *Trans. Faraday Soc.* LI. 620 Data are given for all the partially deuterated species of methyl cyanide and methyl acetylene. 1956 *Nature* 31 Mar. 629/1 The effects of deuteration on biological activity. 1967 *New Scientist* 29 June 780/2 As such catalysts are normally used to promote reaction between hydrogen and organic molecules their efficiency in deuteration is not surprising. 1971 *Nature* 7 May 30/2 All the samples of lunar fines examined release CD_4 when treated with deuterated acids.

deuteric (diū·tĕrik), *a.* *Petrol.* [f. Gr. δεύτερ-ος Deutero- + -ic.] Pertaining to or resulting from the changes that take place in igneous rock during the later stages of consolidation of the magma (see quots.).

1916 J. J. Sederholm in *Bull. Comm. Géol. Finlande* IX. xlviii. 142 It would be advisable to discriminate between ..metasomatic changes which belong to a later period of metamorphism, i.e. are secondary in the strictest sense of the word, and those which have taken place in direct continuation of the consolidation of the magma of the rock itself. I propose to call the latter deuteric, as distinct from secondary changes. 1929 *Econ. Geol.* XXIV. 101, I thus conclude that the term deuteric is established as covering those metasomatic changes in igneous rocks caused by the reaction between the minerals already formed and the emanations..which have been given off from the same magma from which the rock itself crystallized. 1946 *Nature* 10 Aug. 206/2 In the deuteric or late stage there was a development of concentrations of titaniferous magnetite, apatite and other accessory minerals. 1967 K. D. Watson in P. J. Wyllie *Ultramafic & Related Rocks* IX. v. 319/2 Some of the calcite in the kimberlite at Bachelor Lake is probably deuteric.

deuterium (diūtɪᵊ·riəm). *Chem.* [mod.L., f. Gr. δεύτερ-ος second + -ium.] One of the isotopes of hydrogen, differing from the commonest isotope in having a neutron as well as a proton in the nucleus and present to about 1 part in 6000 in naturally occurring hydrogen (elemental and combined); also called *heavy hydrogen*. Symbols ^2H (also H^2), D.

1933 H. C. Urey et al. in *Jrnl. Chem. Physics* I. 513/1 We wish to propose that the names for the H^1 and H^2 isotopes be protium and deuterium. 1934 *Times Educ. Suppl.* 18 Aug. p. i/2 The newly discovered 'heavy water' or deuterium oxide in Antarctic snow. 1935 *Discovery* June 179/2 'Heavy water'..is rich in deuterium, the hydrogen isotope of mass 2. 1958 *Engineering* 31 Jan. 134/3 The deuterium or heavy-hydrogen content of seawater is calculated to be sufficient to sustain present energy consumption for a thousand million years.

deutero-. Add: **deu·terocone**, the inner or lingual cusp of an upper premolar tooth of certain mammals; **deu:teroco·nid**, the corresponding cusp of a lower premolar; **deu·terograph**, a duplicate written or printed passage; **deutero·merite**, a deutomerite; **deu·terotheme** (see quot. 1897); **deutero·toky** [Gr. τόκος bringing forth], that form of parthenogenesis in which the virgin female produces offspring of both sexes; amphitoky; **deu:teroto·xin** (see quots.); **deu:terozo·ic** (see quots.).

1892 W. B. Scott in *Proc. Acad. Nat. Sci. Phil.* 412 As early as the Puerco, however, we find that p⁴ in every known genus is complicated by the addition of a second cusp upon the inner or lingual side of the protocone, which may be called the deuterocone. 1922 W. K. Gregory *Orig. & Evol. Human Dentition* II. i. 104 A similarly situated, but better developed, basal cingulum in later mammals may be traced from the premolars, where it gives rise to the so-called 'deuterocone' or internal spur, backward to the 'protocone' of the molars. 1968 R. Zangerl tr. *Peyer's Compar. Odontology* 188 W. B. Scott thus proposed a special terminology for the description of the premolars, in which the cusps are simply numbered as protocone, deuterocone, tritocone, tetracone, and '-conid' respectively. 1892 W. B. Scott in *Proc. Acad. Nat. Sci. Phil.* 415 In the premolars, therefore, when a cusp occurs occupying the position taken by the metaconid in the molars, it cannot be regarded as homologous with that element, but rather with the deuterocone of the upper premolar and may consequently be called the deuteroconid. 1907 H. F. Osborn *Evol. of Mammalian Molar Teeth* viii. 199 There appears a cuspule on the inner side of the crown of the protoconid...(This is the deuteroconid of Scott's terminology.) 1894 R. B. Girdlestone (*title*) Deuterographs. Duplicate Passages in the Old Testament. 1896 *Expositor* Jan. 36 We can explain..the repeated occurrence in the same book of deuterographs. 1888 Deuteromerite [see *protomerite* s.v. Proto- 2 b]. 1897 W. G. Searle *Onomasticon Anglo-Saxonicum* p. xii, The Anglo-Saxon personal names may be divided into several classes. 1. The first and chief class consists of names that may be termed dithematic names, as they consist of two elements or themes, mostly monosyllabic, a first element or prototheme, and a second element or deuterotheme. 1937 H. G. H. Halvorson in *Harvard Univ. Summary Theses 1937* (1938) 271 (*title*) A study of Old English dithematic personal names: deuterothemes. 1895 *Cambr. Nat. Hist.* V. 141 It is a curious fact that the result of parthenogenesis in some species is the production of only one sex, which in some Insects is female, in others male; the phenomenon in the former case is called by Taschenberg Thelyotoky, in the latter case Arrhenotoky; Deuterotoky being applied to the cases in which two sexes are produced. 1965 F. A. E. Crew *Sex-determination* (ed. 4) v. 76 Deuterotoky or amphitoky in which both impaternate males and females are produced. *Ibid.* 77 Deuterotoky is found in several species of moths. 1900 Dorland *Med. Dict.* 200/1 *Deuterotoxin*, any one of the second of the four groups of diphtheria-toxins. 1904 Deuterotoxin [see *prototoxin* s.v. Proto- 2 b]. 1898 J. E. Marr *Princ. Stratigr. Geol.* vi. 59 Another suggestion was to split the Palæozoic age into an earlier Proterozoic and later Deuterozoic division. 1904 Goodchild & Tweney *Technol. & Sci. Dict.* 156/2 *Deuterozoic rocks*, the group of rocks comprising the Devonian Rocks, Old Red Sandstone, and Carboniferous System.

deuterogamy. Add: **2.** *Bot.* The condition in which fertilization by the fusion of gametes is replaced by other processes, as in some fungi, the higher algæ, and flowering plants.

1898 P. Groom in *Trans. Bot. Soc. Edinb.* 8 Dec. 140 Bearing in mind the analogy to the sexual act, we may describe these nuclear unions as being cases of deuterogamy. 1928 C. W. Dodge tr. *Gäumann's Compar. Morphol. Fungi* iv. 12 Copulation of the gametes is suppressed and replaced by many secondary processes... All these secondary processes are called deuterogamy (secondary pairing). 1952 F. L. Wynd tr. *Gäumann's Fungi* 60 Copulation therefore does not occur between true sexual cells, but between the gametangia. This secondary type of sexuality is called deuterogamy.

deuteron (diū·tĕrǫn). Also †**deu·ton.** [f. Gr. δεύτερ-ος second + *-on, after *proton.] The nucleus of an atom of deuterium, consisting of a proton and a neutron in combination.

1933 H. C. Urey et al. in *Jrnl. Chem. Physics* I. 513/2 Lewis (private communication) proposed the name deuton for the deuterium nucleus... It would appear to us that the name deuteron would follow the Greek definition better. 1933 *Times* 9 Dec. 9/3 In heavy hydrogen the nucleus is about double the weight of that of ordinary hydrogen. It is this nucleus of heavy hydrogen which is called the deuteron. 1934 *Physical Rev.* XLV. 612/1 Fortunately deutons have very long free paths in an atmosphere of deuterium. 1936 *Discovery* Feb. 36/1 Protons and deuterons are to be introduced into this magnetic field and directed under the accelerating chamber. 1942 *Ann. Reg. 1941* 352 Two processes of fission induced by deuteron bombardment. 1963 *New Scientist* 17 Oct. 159 Produced in the Institute's cyclotron by bombarding copper with deuterons.

deuton: see prec.

Deutsche mark, Deutschemark (doi·tʃəmā:ɹk). Also **Deutschmark.** [G., 'German mark'.] The monetary unit of the German Federal Republic, instituted in June 1948; also, a coin representing one Deutsche mark. Abbrev. *D.M.* or *D-mark*.

1948 *Sunday Times* 27 June 1/2 He will..take stock of the first week of economic life with the new Deutsche mark. 1948 *Economist* 27 Nov. 868/1 The future of the D-mark was a matter for considerable alarm. 1957 *Ibid.* 7 Sept. 817/2 A half-world that has to learn to live with the D-mark. 1958 *Spectator* 27 June 840/1 The Federal German Republic, whose gold and currency reserves have reached the record total of DM 23,926,000,000. 1959 *Times* 15 May 13/6 They are Germans on a *Drang nach Osten*..dutifully getting their Deutschmark's worth. 1969 *Listener* 22 May 723/1 It wasn't until the Deutschmark crisis that we began to suffer on our exchanges.

‖**deux-chevaux** (dö,ʃəvo). [Fr.] A two-horse-power motor vehicle; a small French car. Also *attrib.*

1962 *Harper's Bazaar* Dec. 76/2 The French Director-General who arrived..in a workaday Deux Chevaux. 1963 V. Canning *Limbo Line* xiv. 189 They used a *deux-chevaux* with the top rolled back. 1964 'J. Welcome' *Hard to Handle* ix. 93 A *deux chevaux* van was drawn up on a path outside the door.

‖**devadasi** (dēᴵvădā·si). [a. Skr. *devadāsī*, lit. 'a female servant of a god' (cf. Deva).] A nautch girl in a Hindu temple.

1817 tr. *Dubois' Manners & Customs of India* III. iii. 401 Next to the Sacrificers, the most important persons about the temple are the dancing girls, who call themselves *Devadasi*, Servants or slaves of the gods. 1886 in Yule & Burnell *Hobson-Jobson* 237/2. 1913 J. N. Farquhar *Crown of Hinduism* (1919) 397 To this day troops of dancing-girls, who are called *devadasis*, servants of the god, and who now and then do take part in the ritual, but whose real occupation is prostitution, are connected with most of the great temples of the South and West. 1931 *Times Educ. Suppl.* 8 Aug. 313/3 Legislation to abolish brothels and the system of dedication to temple uses of *devadasis*. 1958 E. Linklater *Position at Noon* i. 16 His sublime, inspired, and supernacular tart (or *devadasi*).

devalorize (dīvæ·lǒraiz), *v.* [f. De- + Valor + -ize: see Valorization.] *trans.* To lower the value of; to devalue. Hence **devaloriza·tion.**

1925 *Glasgow Herald* 1 Apr. 8 To face honestly the question of devalorising the franc. 1928 *Observer* 15 July 12/2 So that he and they may equally share the devalorisation. 1931 *Daily Express* 16 Oct. 8/4 The French currency had been devalorised by eighty per cent.

devaluate (dīvæ·liu,ēᴵt), *v.* [f. De- + Value *sb.* + -ate³.] *trans.* = *devalue *v.*

1898 W. J. Locke *Idols* xiii. 128 He is relying on his speech to-morrow to devaluate the evidence. 1924 *Glasgow Herald* 1 Mar. 8 The country has gone a great length in the direction of devaluating the vote by conferring it on men whose politics are as immature as themselves.

devaluation (dīvæliu,ēᴵ·ʃən). [f. prec. or next: see -ation.] The process of devaluing or fact of being devalued; *spec.* the reduction of the official value of a currency in terms of gold or of another currency. Hence **devalua·tionist**, one who advocates or supports devaluation.

1914 *Eng. Hist. Rev.* Jan. 140 The devaluation of the ancient Merovingian pence. 1921 *Glasgow Herald* 10 Dec. 8 Devaluation has an important bearing on Lord Inchcape's criticism of teachers' salaries. 1922 *Ibid.* 12 May 9 The probable further devaluation of the mark. 1935 *New Statesman* 1 June 793/1 Events in Belgium showed plainly the practicability of limited and controlled devaluation of the currency, and strengthened the hands of the devaluationists in France. 1961 D. Woodward tr. *Simenon's Premier* iii. 84 After the disastrous experiments made by previous governments..the only solution was a large-scale devaluation. 1969 *Daily Tel.* 14 Mar. 3 The 1968 balance of payments figures..must make the devaluationists think twice about the effectiveness of their remedy.

devalue (dīvæ·liu), *v.* [f. De- II. 2 + Value *sb.*] *trans.* To reduce or annul the value of; to deprive of value; *spec.* to effect the devaluation of (a currency). Also *absol.* Hence **deva·lued** *ppl. a.*

1918 *Guardian* 24 Oct. 847/3 The chemist has succeeded in devaluing the ruby. 1925 *Glasgow Herald* 29 Apr. 10 To return to the gold standard on the basis of a devalued sovereign. 1956 C. Wilson *Outsider* v. 133 Hegel..devalued life, failed to recognize that thought is only an instrument to 'more abundant life'. 1957 *Times* 18 Nov. p. iii/1 The pound would have to be devalued. 1965 *New Statesman* 23 Apr. 631/1 Wilson missed his chance to devalue last October.

devastating, *ppl. a.* Add: Freq. *fig.*, esp. in trivial or hyperbolical use: very effective or upsetting; astounding, overwhelming, 'stunning'. Cf. next.

1889 A. James *Diary* 4 Aug. (1965) 50 The somewhat devastating episode of July 18th when Harry after a much longer absence than usual presented himself, doubled by William! 1910 *Daily Chron.* 22 Mar. 1/3 It is expected that a devastating strike will be declared. 1924 G. W. Hillyard *40 Yrs. Lawn-Tennis* iii. 83 No man who ever walked on to a court was equipped with more perfect style, greater physical advantages, or more devastating strokes. 1925 *New Yorker* 8 Aug. 4/2 Not since the Tango provided luscious livelihoods for many svelte youths has so devastating a dance agitated the town. 1926 S. Jameson *Three Kingdoms* ii. 63 She was struggling with a devastating shyness. 1927 H. T. Lowe-Porter tr. *Mann's Magic Mountain* v. 378 Everything, whether in jest or earnest, was 'devastating', the bob-run, the sweet for dinner, her own temperature. 1933 E. Shanks *Enchanted Village* ix. 133 Oh yes, poor old Julian—I think, to be honest, that he's a devastating bore. 1936 R. Lehmann *Weather in Streets* i. 11 Oh, darling have you got to go? How *devastating*. 1957 I. Murdoch *Sandcastle* ix. 150 From the very depths of his being the knowledge came to him, suddenly and with devastating certainty. 1971 *Sunday Times* 28 Mar. 32/6 Ludovic Kennedy published his devastating analysis of the Christie murders and established the innocence of Timothy Evans.

devastatingly (de·vǎstēᴵtiŋli), *adv.* [f. Devastating *ppl. a.* + -ly².] In a devastating manner; so as to devastate: freq. in trivial or hyperbolical use (cf. prec.).

1905 *Daily Chron.* 28 Sept. 4/6 The devastatingly tidy housemaid. 1915 J. C. Powys *Visions & Revisions* 96 The stripping from human beings of their characteristic 'outer garments' makes them so dreadfully, so devastatingly, alike! 1927 *Observer* 15 May 8 A man in whom a natural violence of lust and temper is restrained by the unnaturally strict taboos of his childhood's environment, but breaks loose all the more devastatingly when his inhibitions are removed by drink. *Ibid.* 21 Aug. 5 The devastatingly betraying sentences of Theodore Parker. 1928 *Daily Tel.* 15 May 8 His conversation..is positive, narrow-minded, egregiously self-satisfied, impregnably commercial, but it is devastatingly entertaining. 1966 *Listener* 20 Oct. 578/3 The final sequence..is devastatingly lovely in its sadness.

develop, *v.* Add: **3. d.** *Chess.* Phrases: *to develop one's game*, to move (one or more men) from the original position into positions more useful for attack or defence; *to develop a piece*, to bring that piece towards or into a position where it is of greatest service.

1847 H. Staunton *Chess-Player's Handbk.* 22 In chess, this is attempted by the first player putting a Pawn en prise of the enemy early in the game, by which he is enabled more rapidly and effectually to develope his superior Pieces. *Ibid.* 147 Leaving him the option of exchanging Knights or of protecting his K.P., either of which would afford you an opportunity of developing your game. 1864 *Chess Player's Mag.* 52 Black, it seems, had no other chance of developing his game than to give up Rook for Knight at this point. 1952 E. Lasker *Chess Secrets* 41 He has developed one more piece than White.

e. To show the details of (a piece of work) in a drawing (cf. *development 7 d).

1888 [see *development 7 d].

f. To realize the potentialities of (a site, estate, property, or the like) by laying it out, building, mining, etc.; to convert (a tract of land) to a new purpose or to make it suitable for residential, industrial, business, etc., purposes.

1890 [see sense 3 a in Dict.]. 1901 *Times* 6 Nov. 5/1 Hogarth's house in Chiswick..will probably be purchased by a builder who will do what is called develop the property; we all know pretty well what that means. 1931 *Economist* 20 June 1312/1 It was unjust to apply this burden to 'developed' land, that is, to land the value of which has been raised by the energy or enterprise of its owner in putting it to good use. 1932 *Times Lit. Suppl.* 26 May 392/2 The tenants will be turned out as soon as possible and the whole estate 'developed' as a site for a factory.

5. d. *intr.* To come to light, become known. *U.S.*

1864 Webster 366/1 The plans of the conspirators develop. 1903 *N.Y. Even. Post* 17 Sept., A new feature of the shooting developed to-day, when it was discovered that [etc.]. 1927 *N.Y. Times* 29 Sept. 1/6 It developed that Beach had been pressed for money.

e. *trans.* To reveal, bring to light. *U.S.*

1889 *Kansas Times & Star* 8 May, A census of Kansas City's saloons develops the startling fact that there are about 1,000. *Ibid.* 16 July, A search for the pioneer bicyclist hereabouts developed Dr. Henderson, who proudly rode a wheel on our streets in 1880. 1932 T. J. Grayson *Leaders & Periods of Amer. Finance* v. 95 He did develop the fact that enormous lump sum appropriations had been made.

developable, *a.* Add: **c.** *Photogr.* Capable of being developed (see Develop *v.* 5 b).

1878 Abney *Treat. Photogr.* iv. 30 An exposure in the camera to produce a developable image would have to be very prolonged. 1907 *Westm. Gaz.* 26 Oct. 13/2 They take only a short time to produce a developable condition in the emulsion. 1958 *Amateur Photographer* 31 Dec. 393/2 The need for one quantum (the unit of energy) to strike each grain of silver to make it developable.

developer. Add: **c.** *Photogr.* An operative who develops photographs.

1899 *Daily News* 28 Apr. 5/2 We had two men on the train—our regular developer and his assistant. 1921 in *Dict. Occup. Terms* (1927) § 527.

d. An apparatus for developing a person's muscles, etc.

1900 E. Sandow *Strength* (rev. ed.) 41 The great value of the Developer lies in the fact that it serves to render the muscles pliable. *Ibid.*, Exercise with the rubber Developer affords a welcome change from work with the dumb-bells. 1907 *Westm. Gaz.* 17 Jan. 4/2 With regard to the use of grip dumb-bells and the developer, they are not intended for the abnormal development of any one muscle. 1971 *Woman* 13 Mar. 45 (Advt.), The world's most successful bustline developer.

e. With defining word prefixed: a person who develops or matures at a specified time or speed, as *late* or *slow developer* (see also quot. 1961).

1930 R. Lehmann *Note in Music* vi. 249, I was a very slow developer. By the time I started to wake up and think for myself, it was too late. 1936 A. Thirkell *August Folly* vi. 166 He is one of those late developers, trying to himself and to others. 1942 F. J. Schonell *Backwardness in Basic Subjects* iii. 53 They subscribe to the doctrine that dull children are frequently slow developers. 1953 R. Fuller *2nd Curtain* vi. 103 How frightful I must have been at thirteen.. I was a late developer. 1961 *Where?* III. 15 *Late developer*, a term used to describe a child whose mental capacity matures notably later than that of most children. It is often used of a pupil who shows increasing evidence of higher ability than was apparent at 11 plus. 1971 *T.V. Times* 11 Mar. 7/3, I have loved my 30's. I am a late developer.

f. One who develops (see *develop *v.* 3 f) a site, etc.; a speculative builder.

1938 N. L. North *Real Estate Selling* xvii. 257 The result was no sales, defaulted contracts, 'busted' developers. 1958 *Observer* 2 Mar. 8/4 As 'developers' and L.P.T.B. leap-frogged over each other..far into the Home Counties. 1971 M. Russell *Deadline* ix. 103 A developer who was.. putting in a bid for the airfield with a view to erecting.. a shopping precinct.

developing, *vbl. sb.* (see under Develop *v.*). Add: Also *attrib.*, as *developing cup, room, solution*; **developing circle** *Spiritualism*, a group of people who meet for the purpose of developing the latent psychic abilities they may possess; **developing (out) paper** (see quot. 1918).

1870 *Spiritualist* 15 Oct. 111/1 A Developing Circle... Mr. A. C. Swinton..has favoured us with the following details about the gradual development of mediumship in the circle held at his house. 1960 H. Edwards *Spirit Healing* i. 16, I received a number of messages from mediums stating that I possessed the healing gift. So I attended two church developing circles, where I again received encouragement to utilise my gift of healing. 1967 *Psychic News* 22 Apr. 8/3 In my view, would-be healers should sit in a developing circle to receive knowledge from competent spirit guides before attempting to treat patients. 1878 Abney *Treat. Photogr.* xvi. 118 The solution should be flowed back into the developing cup. 1884 T. C. Hepworth *Photogr. Amat.* 149 Wavy markings on a developed plate may generally be traced to dirty dishes or developing-cups. 1890 *Anthony's Photogr. Bull.* III. 166 Any one who has ever manipulated developing paper. 1918 *Photo-Miniature* Mar. 15 *Developing Out Papers* (*D.O.P.*), papers requiring to be developed, but of such sensitiveness that they can be printed by exposure to gaslight, yet developed by the same, placed at a greater distance. 1882 Developing room [see Dict.]. 1861 Developing solution [see Dict.].

developing, *ppl. a.* Add: Esp. in *developing country, nation*, a poor or primitive country which is developing higher economic, industrial, and social conditions.

1964 H. Myint (*title*) The economics of the developing countries. 1969 R. Blackburn in Cockburn & Blackburn *Student Power* 208 Bourgeois economists once talked about the economically 'backward' countries; then 'underdeveloped' was felt to be a kinder adjective. They now prefer to refer to poor capitalist countries as 'developing nations'. 1971 *Times* 17 Mar. 4/6 Many developing countries are facing growing unrest among young people.

development. Add: **3. d.** The act or process of developing (see *develop *v.* 3 f) a mine, site, estate, property, or the like; also, a developed tract of land. Freq. *attrib.*, esp. in *development work* (see also sense 11 below). Cf. *ribbon development.

1885 [in Dict., sense 11]. 1897 *Daily News* 4 Jan. 2/1 During the year an unusual amount of development work was done on the producing mines. 1898 *Westm. Gaz.* 21 Apr. 4/3 Advance moneys for development purposes. 1900 *Daily News* 11 June 2/1 Development operations have been carried out upon the Le Roi, No. 2 Property. 1904 Conrad *Nostromo* II. iii. 126 The new loan connected with railway development. 1936 P. Nash *Dorset* 6 All those courageous enemies of 'development' to whom we owe what is left of England. 1947 S. J. Truscott *Mine Econ.* (ed. 2) ii. 8 Sampling for the establishment of reserves is conducted primarily upon the exposures along the roads by which the deposit is explored and the property developed, this being development sampling. 1957 W. H. Whyte *Organization Man* 274 The little developments that encircle some towns. 1962 A. Christie *Mirror Crack'd* i. 16 All the people from the Development doing their shopping. *Ibid.* 17 Where once there had been meadows with cows, there was the Development.

7. d. The indication of the full details of a piece of work, esp. in a technical drawing.

1888 *Lockwood's Dict. Mech. Engin.* 107 *Development.* A drawing is said to be developed when certain working details are drawn in full. Thus a propeller blade is developed when the various transverse sections are shown. 1904 Goodchild & Tweney *Technol. & Sci. Dict.* 156/2 *Development*, giving full details of some part of a piece of work.

10. b. *Chess.* The disposition of the forces for attack or defence at an early stage of a game.

1864 *Chess Player's Mag.* 195 The difficulty..of meeting the many new developments of the attack, especially those resulting from 9 Q.Kt. to B.3rd, has led to a reaction in favour of the move Q.Kt. to R.4th. 1865 *Ibid.* 48 In order to fix a Pawn at his Q. fifth, and so obstruct the development of Black's cavalry. 1889 Freeborough & Ranken *Chess Openings* 13 There are two styles of development; the attacking and the defensive. In one the pieces are spread about to secure the greatest possible command of the board. In the other they are kept together mutually supporting or defending each other. *Ibid.* 14 When you cannot see your way to an attacking move, play a development move. 1970 B. Larsen *Larsen's Selected Games of Chess 1948–69* 181 Severe cases of neglected development are rare in master games.

11. development area, any of several areas designated by the Distribution of Industry Act of 1945 as places suffering or liable to suffer from unemployment, where new industries, etc., are encouraged. See also *3 d.

1945 *Act* 8 & 9 *Geo. VI* c. 36 § 1 In any area specified in the First Schedule to this Act (hereinafter referred to as a 'development area') the Board of Trade may for the purpose of facilitating the provision of premises needed for meeting the requirements of industrial undertakings (including requirements arising from the needs of persons employed or to be employed therein)..acquire..land for the provision ..of such premises. 1948 *Times Rev. Industry* Aug. 28/3, 117 new factory buildings and extensions were completed in the development areas. 1961 E. A. Powdrill *Vocab. Land Planning* iv. 65 Industrial estates should not necessarily be confused with the Government's programme for inducing industry to occupy sites in Development Areas.

developme·ntalism. [f. Developmental *a.* + -ism.] Belief in development or evolution; evolutionism.

1934 *Mind* XLIII. 393 Bartlett's notion of a 'Schema'.. incorporates developmentalism. 1936 M. T. Hodgen *Doctrine of Survivals* vi. 188 Developmentalism may well suffer defeat. 1947 C. S. Lewis *Miracles* xiv. 146 A certain degree of 'evolutionism' or 'developmentalism' is inherent in Christianity.

developmentalist. Delete *nonce-wd.* and add later examples. Also *attrib.*

1936 M. T. Hodgen *Doctrine of Survivals* iii. 72 All developmentalists were committed to the search for social origins. 1936 *Times Lit. Suppl.* 20 June 524/1 Questions for which the current developmentalist methodology has no answer. 1957 *Antiquity* XXXI. 215 Opposition to the independent developmentalists came from those who could demonstrate..the diffusion of certain traits.

‖**développé** (devlope). *Ballet.* Also **développée.** [Fr., pa. pple. of *développer* to stretch out, unwind, develop.] A movement in which one leg is raised and then fully extended.

1913 C. D'Albert *Dancing* 61 *Développé*...A temps, signifying the slow unfolding of the leg to its full length, whilst placing it in the position demanded. 1953 *Ballet Ann.* VII. 127/2 Some stunning développées growing out of incredible *pirouettes*. 1959 *Times* 10 Sept. 6/3 Miss Svetlana Beriosova..generously expressive in the sweep of her *développés*.

deverbal (dīvɔ̄·ɹbǎl), *sb.* and *a.* *Gram.* [f. De- + Verb + -al.] **A.** *adj.* Derived from a verb. **B.** *sb.* = *deverbative *sb.*

1934 Priebsch & Collinson *German Lang.* II. iii. 225 The suffix *-ēn* had two specific functions: (*a*) to form durative deverbals, e.g. *hangēn* 'to be hanging'. 1946 *Ibid.* (ed. 2) 256 The deverbal derivatives. 1959 *Archivum Linguisticum* XI. 108 Runner—deverbal noun.

deverbative (dīvɔ̄·ɹbǎtiv), *sb.* and *a.* *Gram.* [f. De- + Verb, after *denominative*.] **A.** *sb.* A word formed on or derived from a verb. **B.** *adj.* Derived from a verb.

1913 J. M. Jones *Welsh Gram.* 381 Other Aryan stem-forms, mostly deverbatives and denominatives. 1949 *Archivum Linguisticum* I. 167 The verbal base is either radical or thematic, the first being monosyllabic and primary, i.e. neither deverbative nor denominative. 1951 *Trans. Philol. Soc. 1951* 116 Secondary formations, perhaps deverbatives based on the common verbs. 1970 *Language* XLVI. 381 The three morphemes are mutually exclusive, and each of them is related..to large sets of nominal roots and deverbative nouns.

devey, var. *deevy *a.*

deviability (dī:viǎbi·lĭti). [f. Deviate *v.*: see *-bility.] Capability of being caused to deviate or of being deflected. So **de·viable** *a.*, that can be caused to deviate; deflectable.

1902 [see *beta 2 f]. 1904 S. Bottone *Radium* iii. 25 These experimental demonstrations of the deviability of the α and β rays. 1913 J. Cox *Beyond Atom* iii. 50 The α rays bear a positive charge... Though deviable by magnetic and electric fields, the amount of deviation is minute compared with that of kathode rays or β rays. 1938 R. W. Lawson tr. *Hevesy & Paneth's Man. Radioactivity* (ed. 2) ii. 33 The deviability of the α-rays in strong magnetic fields.

deviance (dī·viǎns). [f. Deviant *ppl. a.* + -ance.] Deviant state or quality; the behaviour or characteristics of a deviant. So **de·viancy.**

1944 G. Bateson in J. McV. Hunt *Personality & Behavior Disorders* II. xxiii. 722 The term 'deviance' implies standardization; and..deviance is allowed for and expected to occur in all cultures. 1954 W. J. H. Sprott *Sci. & Social Action* vi. 104 Deviance is the departure on the part of participants from culturally expected rules of conduct. *Ibid.* 106 Different societies put different pressures upon their members, and therefore every society provides its own deviancy hazards. 1956 J. Klein *Study of Groups* x. 134 If deviance is displayed not by one member, but a set of members sharing a set of norms, control by group pressure is less possible. 1968 Freedman & Doob (*title*) Deviancy: the psychology of being different. 1970 *New Society* 31 Dec. 1150 (Advt.), Crime, deviance and social sickness.

deviant, *ppl. a.* Restrict †*Obs. rare* to sense 2 and add: **1.** *spec.* Deviating from normal social, etc., standards or behaviour.

1935 M. Mead *Sex & Temperament* xviii. 302 Because it is believed marriage must be based upon contrasting personalities, deviant men often choose deviant women. 1944 G. Bateson in J. McV. Hunt *Personality & Behavior Disorders* II. xxiii. 722 A character structure which is normal among us may be deviant among the Kwakiutl. 1957 *Observer* 3 Nov. 10/5 Even among experimental groups, there is always a certain percentage of deviants who do not conform to standard norms, and in social terms these deviant groups are most necessary. 1959 B. Wootton *Social Sci.* x. 314 Elaborate studies of deviant behaviour have been..undertaken which make no attempt to establish the norms from which the subjects are presumed to be deviating.

deviant (dī·viǎnt), *sb.* [f. the ppl. adj.] **a.** Something that deviates from normal. **b.** = next, 1.

1927 *Jrnl. Amer. Med. Assoc.* 5 Feb. 376 Omental bands or deviants, when extensive, are prone to contain a great deal of fibrous tissue. 1928 M. Mead *Coming of Age in Samoa* (1929) xi. 169 These girls all represented the deviants from the pattern in one direction...At any time, they, like all deviants, might come into real conflict with the group. 1950 B. Wootton *Test. Social Sci.* vi. 126 The fact that there are always individual deviants shows that a strong-minded (or weak-willed) individual can substitute purely personal choices of his own for those favoured by the community in which he lives. 1952 M. McCarthy *Groves of Academe* (1953) iv. 74 Seasoned nonconformists and dissenters, sexual deviants, feather-bedders. 1957 [see prec.]. 1959 *Listener* 28 May 947/2 Our attitudes to offenders and other 'deviants'.

deviate (dī·viět), *sb.* [f. the vb.] **1.** A person who, or thing which, deviates; esp. one who deviates from normal social, etc., standards or behaviour; *spec.* a sexual pervert.

1912 *Pedagogical Seminary* XIX. 186 To analyze and diagnose mental deviates whose deviation has caused social maladjustment. 1940 *School & Society* 20 Apr. 503 (*title*) Contribution to the IQ controversy from the study of superior deviates. 1947 Ogburn & Nimkoff *Handbk. Sociol.* ix. 180 Group pressure tends to cut off extreme deviates. 1952 W. J. H. Sprott *Social Psychol.* i. 8 He was a 'social deviate'. *Ibid.* vi. 92 We do not expect uniformity, of course; there are plenty of eccentrics or 'deviates'.

2. *Statistics.* The value of a variate measured from some standard point of a distribution, usu. the mean, and usu. expressed in terms of the standard deviation of the distribution.

1925 R. A. Fisher *Statistical Methods* iii. 47 Table I. shows that the normal deviate falls outside the range ± 1·598193 in 10 per cent of cases. 1937 C. Burt *Backward Child* ii. 22 He works out the 'mean variation' by taking the deviates about the general average and then averaging in turn the deviates themselves. 1967 Condon & Odishaw *Handbk. Physics* (ed. 2) I. xii. 179/1 If X is normally distributed with parameters m and σ, then $Y = (X - m)/\sigma$, termed a standardized normal deviate, is normally distributed with parameters 0 and 1.

deviation. Add: **2. d.** *Statistics.* The amount by which one of a set of measurements, numerical observations, etc. differs from the arithmetical mean of the whole set; *standard deviation*, a common measure of the scatter or dispersion of a set of measurements, equal to the square root of the mean of the squares of the deviations.

1858 [in Dict., sense 2 a]. 1875 F. Galton in *Phil. Mag.* 4th Ser. XLIX. 35 Medium values will occur very much more frequently than extreme ones, the rarity of the latter rapidly increasing as the deviation slowly increases. 1894 K. Pearson in *Phil. Trans. R. Soc.* A. CLXXXV. 80 Then σ will be termed its standard-deviation (error of mean square). *Ibid.* 104 We have no less than three measurements deviating by more than four times the standard-deviation from the mean. 1925 R. A. Fisher *Statistical Methods* iii. 46 A deviation exceeding the standard deviation occurs about once in three trials. Twice the standard deviation is exceeded only about once in 22 trials. 1946 F. J. Schonell *Backwardness in Basic Subjects* (ed. 3) v. 80 This is at a point 1·6 standard deviations from the mean of the normal population group. 1968 J. H. Burn *Lect. Notes Pharmacol.* (ed. 9) 135 The standard deviation usually makes the scatter look greater than the average deviation makes it look. Thus for the cocaine figures, using the standard deviation we have 57·5 ± 23·1 mg. per kg., instead of 57·5 ± 16·2 mg. per kg. 1970 *Nature* 12 Dec. 1081/1 In VLF data handling we usually assume a normal distribution of deviations.

3. a. Add to def.: *spec.* the behaviour or characteristics of a deviate.

1912 [see prec., 1]. 1934 R. Benedict *Patterns of Culture* viii. 272 He [*sc.* the unsupported individual] may gradually achieve a..less tortured attitude toward his deviations. *Ibid.* 273 Much more deviation is allowed to the individual in some cultures than in others. 1960 Hinsie & Campbell *Psychiatric Dict.* (ed. 3) 683/1 Sexual deviation—such as homosexuality, transvestism, pedophilia, fetishism, sexual sadism. 1962 R. Cook *Crust on its Uppers* ii. 29 Every single person one knows seems to be mixed up in some kind of deviation or other.

d. *spec.* (*a*) *deviation of the complements*, in *Biochem.*, the prevention of the complements from acting upon the receptors. (*b*) *Embryol.* Divergence in the development of an animal from the ontogenetic stages of its ancestor.

1906 *Practitioner* Dec. 748 Another and much more important and serious cause of failure comes about by means of the phenomenon described by Neisser and Wechsburg, and known as the 'deviation of the complements'. 1930 G. R. de Beer *Embryol. & Evol.* xv. 102 The appearance of characters in the early stages of development is caenogenesis, and these characters which loom so largely in neoteny and deviation are flies in the Haeckelian ointment of recapitulation.

e. *Politics.* Departure or divergence from the principles, policies, or directives of a government or political party, used esp. of such actions in a Communist society. Also *transf.*, any deliberate divergence from prescribed rules or standards. So **devia·tionism**; **devia·tionist** *sb.* and *a.*

1930 W. H. Chamberlin *Soviet Russia* iii. 78 The Right Deviationists..favored a larger production of goods for immediate consumption. 1931 *Economist* 5 Sept. 425/1 That need may cause a further change in the plan as striking as the recent deviations from the pure doctrines of Communism. 1937 tr. *Lenin's Sel. Wks.* IX. 92 A slight syndicalist or semi-anarchist deviation would not have been terrible. *Ibid.* 126 A deviation is not yet a finished trend. A deviation is something that can be rectified. People have just wandered somewhat from the path, or are beginning to wander from the path, but they can still be put right. This, in my opinion, is what the Russian word *uklon* means. 1937 'G. Orwell' *Road to Wigan Pier* xiii. 253 To them, the whole Socialist movement is no more than a kind of exciting heresy-hunt—a leaping to and fro of frenzied witch-doctors to the beat of tom-toms and the tune of 'Fee, fi, fo, fum, I smell the blood of a right-wing deviationist!' 1938 H. G. Wells *Brothers* iii. 47 We know of their groups and their—what is your word?—deviations. 1940 *Economist* 31 Aug. 271/1 Lenin tried to reconcile his conscience to the existing degree of governmental control by talking of 'bureaucratic deviationism'. 1947 L. Hastings *Dragons are Extra* i. 23 No deviation from the Party line. 1952 *Scottish Jrnl. Theol.* V. 186 The Church cannot submit to any other, cannot flirt with any other. Deviationism is nothing less than unfaithfulness, adultery, harlotry. 1955 H. Hodgkinson *Doubletalk* 45 Deviationist is a Communist who, whether with good or evil intentions, strays from the path of the official party line. 1959 *Times* 7 Feb. 7/7 Such deviationist diversions as 'rock 'n roll'. *a* 1963 L. MacNeice *Astrology* (1964) v. 156 Here Kepler appears to be what orthodox astrologers might call a mystical deviationist.

deviceless, *a.* Add: Also [after Gr. ἀμήχανος], against which no device avails.

1851 M. Arnold *Lett.* (1932) 118 We are growing old, and advancing towards the deviceless darkness.

devil, *sb.* Add: **6. b.** (Earlier example.)

1780 T. Pasley *Private Sea Jrnl.* 5 Mar. (1931) 69 He has many good qualities, but as the old adage says, has not Divil enough in him.

7. a. (Earlier example of the Tasmanian 'devil'.)

1829 H. Widowson *Pres. State Van Dieman's Land* xviii. 180 The devil, or as the naturalists term it '*dasyurus ursinus*', is very properly named... It is as great a destroyer of young lambs as the hyena; and, generally speaking, is as large as a middling-sized dog.

8. Various other contrivances (see quots.).

a 1877 Knight *Dict. Mech.* I. 691/1 *Devil*, a machine for making wood screws. 1890 W. J. Gordon *Foundry* 72 A herculean metallic disk, grimly named the 'devil', armed with steel cutters on its circumference that takes off a pound of shavings at every revolution. 1901 *Farm, Field & Fireside* 13 Dec. 362/2 Large surfaces are dealt with by burning, an instrument called a 'devil' being generally employed by painters for 'burning off' doors, panels, etc.

9. b. *devils on horseback*, angels on horseback (see *angel *sb.* 7); also, a similar dish consisting of a prune or plum wrapped in a bacon-rasher and served on fried bread.

1909 in *Cent. Dict.* Suppl. 1963 R. McDouall *Cookery Book* 228 Devils on horseback are much the same [as canapés diane], except that a stoned prune is substituted for the chicken-livers.

11. Also, a dust-storm in South Africa.

1897 Baden-Powell *Matabele Camp* 284 A 'Devil' with its roaring pillar of dust and leaves, comes tearing by. 1900 *Daily News* 3 Apr. 3/1 The 'dust devils' that sweep across the blustering plain. 1901 *Westm. Gaz.* 16 Mar. 3/1 The 'devil' in South Africa will pick up boots and tins of sardines, even bottles of whisky and saddle bags.

17. *devil take the hindmost*: see Hindmost *a.* 1 b.

22. l. (Further examples.) Freq. shortened to *talk of the devil*; esp. used in reference to a person who appears unexpectedly when one is talking about him.

[1591 Lyly *Endimion* I. iii. 3, O that we had Sir Tophas.. in the midst of our myrth, *& ecce autem*, wyl you see the deuill?] 1666 G. Torriano *Piazza Universale* 134 The English say, Talk of the Devil, and he's presently at your elbow. 1721 J. Kelly *Sc. Prov.* 299 *Speak of the Dee'l, and he'll appear*, spoken when they, of whom we are speaking, come in by Chance. 1922 E. O'Neill *Anna Christie* (1923) I. 9 Speak of the devil. We was just talkin' about you. 1958 G. Greene *Our Man in Havana* III. iii. 136 'What's the matter, Hasselbacher?' 'Oh, it's you, Mr. Wormold. I was just thinking of you. Talk of the devil,' he said, making a joke of it.

n. (Further examples.) See also Drive *v.* 1 b, Needs *adv.* d, Spoon *sb.* 3 a.

1606 J. Day *Isle of Guls* sig. D4v, You were worse then the deuil els, for they say hee helps his seruants. 1661 A. Brome *Songs & Other Poems* 136 The Devil's ever kind to his own. 1806 [see Good *a.* 7 a]. 1837 F. Chamier *Arethusa* II. i. 13 Weazel was the only midshipman saved besides myself: the devil always takes care of his own. 1857 Trollope *Barchester T.* II. vii. 123 'Better the d—— you know than the d—— you don't know,' is an old saying. 1940 R. A. J. Walling *Why did Trethewy Die?* vii. 195 The devil looks after his own. 1960 *Times* 30 Jan. 3/3 They are probably wise to leave well alone on the devil-you-know principle.

23. c. *devil-driven, -haunted, -possessed* adjs.

1898 Kipling in *Spectator* 2 July 15/2 Afraid of the devil-haunted beach of noises. 1899 *Daily News* 20 Sept. 3/3 Then came the famous devil-possessed district. 1906 R. Whiteing *Ring in New* ix. 59 The devil-driven and purposeful way in which people do most things in this part of the world. 1922 W. B. Yeats *Trembling of Veil* II. viii. 108 The only Young Ireland politician who had music and personality, though rancorous and devil-possessed. 1926 M. Leinster *Dew on Leaf* II. i, Jack is lonely, wretched, devil-driven.

24. devil-crab, the velvet crab, *Portunus puber*; = *lady-crab* (Lady *sb.* 16); **devil dance,** the dance of a devil-dancer; **devil-devil,** (*a*) in Australian folk-lore, a devil, an evil spirit; (*b*) *Austral. slang* (see quot. 1933); **devil float,** a tool for roughening the surface of plaster; **devil liquor** (see quot. 1912); **devil-on-the-coals** *Austral. slang*, a small damper hastily baked in hot ashes; **devil-wood** (U.S. examples).

1871 C. Darwin *Desc. Man* II. ix. 269 When a Devil-crab (Portunus puber) was seen..fighting with a *Carcinus mænas*, the latter was soon thrown on its back. 1899 *Strand Mag.* June 655/2 Prickly devil-crabs. 1849 R. Caldwell *Tinnevelly Shanars* 19 The musical instruments..chiefly used in the devil-dance are the tom-tom..and the horn. 1901 Kipling *Kim* ix. 212 He had seen devil-dance masks at the Lahore Museum. *Ibid.* x. 248 Devil dances, and spells and charms. 1930 G. Knight *Intim. Glimpses Myster. Tibet* 29 The Devil Dances of Tibet..represent either some historical, legendary, or mythological event. 1900 H. Lawson *Over Sliprails* 108 Black Jimmie shifted away from the hut [of the dead woman]..for the 'devil-devil' sat down there. 1933 *Bulletin* (Sydney) 20 Dec. 34/1 The alternation of wet seasons or floodings on the one hand and of droughts on the other induces characteristic alternations of depressions and rises in heavy soils. These have received various local names, of which melonhole, gilgai, Bay of Biscay, devil-devil and crab-hole are the most frequently met. 1954 B. Miles *Stars my Blanket* xvii. 117 It's full of skulls, that cave is. Can't get a blackfella near it. They say the devil-devil's in it. 1939 Devil float [see *devil *v.* 6]. 1912 Thorpe *Dict. Appl. Chem.* (ed. 2) I. 149/1 The aqueous condensate obtained by cooling the waste gases [from ammonium sulphate manufacture] is a very noxious-smelling liquid, and is hence termed 'devil-liquor'. It contains sulphuretted hydrogen, pyridines, and similar substances, and hydrocyanic acid, and is also difficult to dispose of. 1930 *Engineering* 5 Sept. 311/3 The evaporation of the devil liquor or the aqueous condensates. 1862 A. Polehampton *Kangaroo Land* 77 Instead of damper we occasionally made what is colonially known as 'devils on the coals'. 1818 A. L. Hillhouse tr. *Michaux's N. Amer. Sylva* II. 153 *Olea Americana*..belongs exclusively to the Southern States, the Floridas and Lower Louisiana;..it has hitherto received no name from the inhabitants of the country, except on the banks of the river Savannah, where it is called Devil Wood. 1832 D. J. Browne *Sylva Amer.* 225 The wood..when perfectly dry—is excessively hard and very difficult to cut and split—hence is derived the name of Devil Wood. 1938 W. R. Van Dersal *Native Woody Plants U.S.* 175 Devilwood..often occurs in sandy soil.

25. b. devil's grip *colloq.*, pleurodynia; **devil's horse** *U.S.*, the praying mantis; †**devil's missionary** *N.Z.* (see quots.); **devil's picture(d)-books, picture-gallery** or **pictures** *colloq.* = Devil's books.

1888 *Amer. Jrnl. Med. Sci.* XCVI. 490 So agonizing was this pain that it was nicknamed the 'devil's grip' by a sufferer from the disease in Rappahannock County, Virginia, and this name became a common one there afterwards. 1924 *Ibid.* CLXVIII. 569 Devil's grip..is certainly due to an acute infection, and the recurrences seem like those of a protozoan infection. 1962 *Daily Tel.* 21 Nov. 1/3 Mr. Duncan Sandys..has Bornholm disease, it was said last night. The disease is also known as 'Devil's Grip'. 1883 Sweet & Knox *Through Texas* (1884) xliv. 629 Another of the most peculiar and interesting insects in Texas is called the 'devil's horse'. 1937 *Nature* 14 Aug. 264/1 Less graceful but fascinating are the devil's-horses, a showy grasshopper commonly six inches long which travels over the land by the million devouring vegetation as it goes. 1837 E. G. Wakefield *Brit. Colonization N.Z.* ii. 31 [The lawless renegade Englishmen] really deserve a name which has been given them—that of 'Devil's missionaries'. 1839 *Colonial Gaz.* XXIX. 474 [Low types, vagabonds, etc.,] devil's missionaries as they are appropriately called. 1786 Devil's pictur'd beuks [see Devil's books]. 1913 D. H. Lawrence *Sons & Lovers* i. 20 Morel never in his life played cards,.. 'the devil's pictures', he called them! 1927 W. E. Collinson *Contemp. Eng.* 31 My father..jocularly referred to the cards as the Devil's picture-gallery. 1964 A. Wykes *Gambling* vii. 161 Some old shipmasters won't allow 'the devil's picture-books' aboard.

devil, *v.* Add: **5.** To worry (someone) excessively; to harass, annoy, tease. Chiefly *U.S. colloq.*

1823 W. Faux *Mem. Days* 216 Go..tell our great Father, the President, how we are deviled and cheated. 1883 Sweet & Knox *Through Texas* (1884) iii. 47 They devilled the poor fellow almost to death. 1940 C. McCullers *Heart is Lonely Hunter* (1943) I. iii. 42 Sometimes it was fun to devil Portia. 1946 K. Tennant *Lost Haven* (1947) iii. 48 A man with any push would form a progress association and devil the shire council about the roads.

6. To scratch or score the surface of (plaster) to provide a rough surface for the next coat.

1939 *Archit. Rev.* LXXXV. 212 All these materials.. should be covered with a pricking-up coat, well scratched, a floating coat, 'devilled' with a 'devil float', and a finishing coat, hard trowelled.

devil-may-care. Add: Hence as *sb.*, a devil-may-care person, attitude, etc.

1857 G. W. Thornbury *Songs of Cavaliers* 120, I and some seventy devil-may-cares Rode to Bristol. 1928 *Manch. Guardian Weekly* 21 Sept. 230/3 An air of devil-may-care.

devil's-guts. Add: **c.** An Australian plant of the genus *Cassytha*. Also called **devil's twine.**

1889 J. H. Maiden *Useful Native Plants* 14 This and other species of *Cassytha* are called 'Dodder-laurel'. The emphatic name of 'Devil's guts' is largely used. It frequently connects bushes and trees by cords, and becomes a nuisance to the traveller. 1962 N. C. W. Beadle et al. *Handbk. Vasc. Plants Sydney Distr.* 133 *Cassytha* L., Devil's Twine. 1965 *Austral. Encycl.* II. 279/2 There are 12 indigenous species

[of *Cassytha*].. one species, *C. glabella* (devil's twine), being found in all the states.

deviltry. (Earlier U.S. example.)

1788 *Massachusetts Spy* 28 Aug. (Th.), His shoes were made of the leather of hypocrisy, tanned with the bark of presumption, and curried in the shop of deviltry.

devirilize (dīvi·rilaiz), *v.* [f. De- + Virile *a.* + -ize.] *trans.* To deprive of virility or manly qualities; to devitalize. So **deviriliza·tion**; **devi·rilized** *ppl. a.*

1901 *Amer. Jrnl. Psychol.* XII. 277 The devirilized effects of transcendental and idealistic habits of thought. 1920 W. J. Locke *House of Baltazar* v, These new women are out for the devirilisation of man.

devoice (dīvoi·s), *v.* [f. De- + Voice *v.* 8.] *trans.* = Devocalize *v.* Hence **devoi·cing** *vbl. sb.* and *ppl. a.*

1932 D. Jones *Outl. Eng. Phonetics* (ed. 3) v. 21 The voiced sound is commonly said..to become unvoiced or to be devocalized or better devoiced. 1950 —— *Phoneme* p. xv, Devoicing..voicing. 1951 Trager & Smith *Outl. Eng. Structure* 31 The voiced spirants show marked devoicing. 1964 E. Bach *Introd. Transformational Gram.* vi. 132 The North German pronunciation will result automatically from the general devoicing rule for word final consonants. 1965 W. S. Allen *Vox Latina* 69 A voiced velar plosive.. is devoiced before the *t*.

devoir, *sb.* Add: †**6.** A school exercise or piece of home-work. *Obs.*

c 1845 C. Brontë *Professor* (1857) I. xvi. 264 One day I gave, as a devoir, the trite little anecdote of Alfred tending cakes..to be related with amplifications. *Ibid.* 270, I made a report of the other devoirs. 1849 —— *Shirley* III. iv. 100 Were the faults of that devoir..grammatical errors, or did you object to the substance? 1857 Mrs. Gaskell *Let.* 13 Aug. (1966) 464, I should be glad if some one would look over the French dévoirs [*sic*], please.

devolution. Add: **9. b.** In Irish politics, with reference to a scheme proposed as a substitute for Home Rule.

1898 J. Redmond in *Hansard Commons* 4th Ser. LIII. 379 Liberal popular opinion in England is in the direction..of the diminution of the magnitude of the Home Rule question, from 1886 and 1893, down to some scheme of devolution and federalism. 1907 A. S. T. Griffith-Boscawen *14 Yrs. Parl.* 323 In the middle of the summer holidays [1905] the country had been startled by the promulgation by Lord Dunraven and his friends of a plan of 'devolution' of Irish Government, which was neither Unionism nor Home Rule, but a sort of half-way house, in which Irishmen of all persuasions were to live in happiness for ever after. 1907 *Daily Chron.* 9 May 4/6 The vehement language in which some of the Orange Members spoke of resistance to Mr. Birrell's Irish Council scheme might..almost justify fears of another 'Devolution War'. 1971 *Times* 19 Mar. 21/3 The form of devolution that is peculiar to that corner of the United Kingdom [*sc.* Northern Ireland] will have to be changed.

devolutionary (dīvǒl^i^ū·ʃənări), *a.* [f. Devolution + -ary¹.] Of, pertaining to, or characterized by devolution.

1896 *Daily News* 1 June 8/3 The symptoms of a coming devolutionary change. 1920 *Glasgow Herald* 17 Apr. 6 He has arresting ideas of the Clyde Valley as a devolutionary unit.

devolutionist (dīvǒl^i^ū·ʃənist). [f. Devolution + -ist.] One who believes in or advocates the principles of (political) devolution. Hence as *adj.*

1905 *Westm. Gaz.* 20 Feb. 1/2 Sir West Ridgeway, now a Devolutionist. 1920 *Glasgow Herald* 17 Apr. 6 Everyone is a convinced devolutionist to-day. 1968 *Economist* 11 May 9/2 The Tories' private plans seem more devolutionist than anything yet accepted by Labour.

Devon (de·vən). The name of a county in the south-west of England, used *attrib.* or as *sb.* to designate (*a*) a breed of cattle noted for the quality of their milk; (*b*) a breed of sheep.

1834 W. Youatt *Cattle* iii. 15 The skin of the Devon, notwithstanding his curly hair, is exceedingly mellow and elastic. 1837 *Penny Cycl.* VIII. 454 The North Devon oxen are famed for their docility. *Ibid.*, A cow bred from a North Devon by a Yorkshire bull. *Ibid.*, South Devon nott, brown face and legs, long wool, pure. 1844 H. Stephens *Bk. Farm* II. 172 The Devons may perhaps be classed among the Galloways and Angus. 1851 *Ibid.* (ed. 2) II. 144/1 Lord Farnham's Devon ox, slaughtered in Dublin in 1828. 1960 *Farmer & Stockbreeder* 1 Mar. 15/1 Main beef breeds include Shorthorns, Herefords, Devons, [etc.]. *Ibid.*, Other types [of sheep] include Cluns, Kerries, Devon Closewool and pure Down breeds.

b. *Angling.* The name of an artificial fly.

1924 *Blackw. Mag.* Apr. 492/1, I..fished out across the stream with a heavy Devon. 1960 *Times* 30 Apr. 9/4 Indifference sometimes shown by new run, settled-down salmon, to fly, prawn or devon pulled across their nebs.

Devonshire (de·vənʃəɹ). = prec.; used *attrib.* to designate various articles produced or animals reared in Devonshire, or characteristic of Devonshire, as *Devonshire cake, cider, pie, pottery, sheep, tea*; **Devonshire cream** (see Cream *sb.*² 1); **Devonshire slipper** (see quot. 1921); **Devonshire wainscot,** a species of moth, *Leucania putrescens* (see Wainscot *sb.* 4).

1747 H. Glasse *Art of Cookery* viii. 72 A Devonshire squab-pye. 1791 [see Regale *sb.*¹ 2]. 1807 [see Not *a.* and *sb.* 3 a]. 1825 H. Wilson *Mem.* III. 113 She gave us most excellent Devonshire cream, and hot Devonshire cakes. 1837 Youatt *Sheep* vii. 253 The Devonshire notts, or polled sheep used, forty or fifty years ago, to be at least middle-woolled, if not short-woolled sheep. 1845 Dodd *Brit. Manuf.* V. 176 Straw-plait... There are several descriptions of plait made in England—such as..the 'Devonshire'. 1858 Geo. Eliot *Scenes Cler. Life* I. 167 The respective merits of the Devonshire breed and the short-horns. 1881 *Instr. Census Clerks* (1885) 57 Devonshire Slipper Maker. 1893 Yonge & Coleridge *Strolling Players* xxxvi. 336 On the mantelpiece some Devonshire pottery. 1907 R. South *Moths Brit. Isles* I. 310 The Devonshire Wainscot (Leucania (Cirphis) putrescens). So far as the British distribution of this species is known, it seems to be confined to the coasts of South Devon and South Wales. 1909 *Bradshaw's Railway Guide* Apr. p. i (Advt.), Symons' prize medal Devonshire Cyder. 1921 *Dict. Occup. Terms* (1927) § 344 *Devonshire slipper maker*; a brown saddler who makes the fitting known as a slipper for one type of lady's side saddle. 1928 E. Waugh *Decline & Fall* II. vi. 199 The sale of brass toasting forks, picture post-cards and 'Devonshire teas'. 1959 *Listener* 8 Oct. 599/2 Another tasty and satisfying dish is Devonshire pie. 1959 R. D. Macleod *Key to Names Brit. Butterflies & Moths* 32/1 L[eucania] putrescens, Devonshire W[ainscot]. 1970 D. Clark *Sweet Poison* i. 29 Devonshire ice cream. Devonshire cider—Devonshire everything.

devulcanize (dīvʌ·lkănaiz), *v.* [f. De- + Vulcanize *v.*] *trans.* To restore (vulcanized rubber, etc.) to its former unvulcanized condition.

1899 *Eng. Mechanic* 19 May 316 Any one who can devulcanise indiarubber 'so as to make it soluble in benzine' will assuredly take out a patent for the process.

devy, var. *deevy *a.*

dew, *sb.* Add: **5. c.** *dew-grey* adj. **e.** *dew-damp, -drenched* (examples), *-soaked* adjs. **f.** *dew-lipped* (earlier example).

1899 A. R. Cowan *Hist. Kiss* 86 The grass, still dew-damp in the glade. 1906 T. S. Moore *Poems* 29 She.. ordered the dew-damp hair. 1885 W. B. Yeats in *Dublin Univ. Rev.* Apr. 56/2 And from the dew-drench'd wood I've sped. 1919 V. Woolf *Night & Day* xvi. 206 Her rather pale, dew-drenched look. 1932 W. Faulkner *Light in August* (1933) vii. 149 Against the dewgray earth..fireflies drifted. 1818 Keats *Endymion* II. 408 Just as the morning south Disparts a dew-lipp'd rose. 1941 W. de la Mare *Bells & Grass* 139 Dew-soaked shoes.

Dewalee, Dewali (dīwā·li). Also 7 **Dually,** 9 **Divali, Diwali, Dewallee.** [Hind. *diwālī*, ad. Skr. *dīpāvalī* (*dīpālī*) row of lights, f. *dīpa* light, lamp.] A Hindu festival with illuminations held on the day of the new moon in the month Aśvina or Kārttika.

1698 Fryer *Acc. E. India & P.* 110 The first New Moon in October, is the Banyans Dually. 1820 T. Coats in *Trans. Lit. Soc. Bombay* (1823) III. 211 The Dewallee, Deepaullee, or Time of Lights, takes place twenty days after the Dussera, and lasts three days. 1849 E. B. Eastwick *Dry Leaves* 84 The Diwālī happening to fall on this day, the whole river was bright with lamps. 1883 Monier-Williams *Religious Thought & Life in India* 432 The Dīvālī is celebrated with splendid effect at Benares. 1952 J. Masters *Deceivers* xiv. 142 It was the Dewali, the festival dedicated to lights and gambling, which fell always on the twentieth day after Dussehra. 1969 *Eve's Weekly* (Bombay) 20 Dec. 33/1 It is not Divali, but Christmas.

Dewar, dewar (diū·əɹ). The name of Sir James Dewar (1842–1923), British physicist and chemist, used *attrib.* or *absol.* to designate a double-walled vessel or flask, usu. of glass or copper, having the space between the walls empty of air to prevent conduction and convection of heat to and from the inner container. Cf. *vacuum-bottle, -flask*, and *-vessel* s.v. Vacuum 4.

1899 T. O'C. Sloane *Liquid Air* xi. 246 The Dewar vacuum bulb consists of a double or treble walled glass vessel, with the space or spaces between the vessels hermetically sealed and with a nearly perfect vacuum therein. *Ibid.* 248 Various shapes and modifications of Dewar bulbs. 1902 *Encycl. Brit.* XXX. 281/2 Vacuum or 'Dewar' vessels. 1930 *Engineering* 5 Sept. 314/1 A calorimeter consisting of a wide-necked Dewar flask, [etc.]. 1964 N. G. Clark *Mod. Org. Chem.* vii. 112 The reaction is carried out in a Dewar vessel. 1966 *New Scientist* 21 July 138/3 A replenisher that automatically maintains the level of liquefied gas in a dewar has been developed.

dewater (dīwǭ·təɹ), *v.* [f. De- II. 2 + Water *sb.*] *trans.* To remove the water from. So **dewa·tered** *ppl. a.*, **dewa·tering** *vbl. sb.* and *ppl. a.*

1909 in Webster. 1923 *Glasgow Herald* 10 Oct. 10 There is now more hope of getting through 'dewatered' roadways. 1929 *Times* 10 May 22/3 The Lagan Navigation Company shall..dewater such portion of the canal. 1930 *Engineering* 4 Apr. 436/1 These will be employed when de-watering of the culvert is necessary. 1960 E. L. Delmar-Morgan *Cruising Yacht Equip. & Navig.* vi. 81 This type of dewatering protective was used very widely on flooded machinery. 1962 *Times* 1 Feb. 3/4 Construction of and dewatering inside a cofferdam. 1969 *Washington Post* 22 Feb. B1/4 Water consumption drops and the tunnel can be, in the jargon of the engineers, 'dewatered'. That means drained. 1970 *Fremdsprachen* XIV. 135 The carbon slurry is drawn off the bottom of the column to one of the dewatering bins.

dew-bow (diū·bōᵘ). [f. Dew *sb.*, after *rainbow*.] An arch resembling a rainbow, occurring on a dew-covered surface.

1873 [see Dew *sb.* 5 b]. 1920 *Conquest* May 346/1 That curious phenomenon known as the dew bow. *Ibid.* 346/2 The..author may have seen a dew bow effect at his feet. 1960 *New Scientist* 31 Mar. 792/3 Dewbows..are rainbow effects of moonlight shining on dew-pearled grass.

dewdrop. Add: **b.** A glass bead resembling a drop of dew.

1880 *Harper's Mag.* June 31/1 'Grass-work' consists in the fastening of small glass beads or 'dew-drops' to the artificial blades.

Dewey (diū·i). The surname of Melvil *Dewey* (1851–1931), American librarian, applied *attrib.* to designate a decimal system of library classification developed by him. (See *decimal classification* s.v. *decimal A. 1 a.)

1879 *Library Jrnl.* IV. 139 The books on the shelves were numbered and arranged according to the 'Dewey system'. 1885 *Ibid.* X. 26 The scheme looks like a modification or improvement of the Dewey Decimal System. 1931 *Times Educ. Suppl.* 27 June 250/1 Classification of books—Dewey system. 1959 *Observer* 14 June 19/5 If anthropologists wrote with a little more levity, those of us who have to cook would be constantly at the shelves numbered 572 (Dewey classification) of the public libraries. 1970 E. McGirr *Death pays Wages* iii. 59, I suppose they have some sort of Dewey System master index.

Deweyism (diū·i,iz'm). [f. *Dewey* (see below) + -ism.] The tenets of the American philosopher and educationist John Dewey (1859–1952); pragmatism in philosophy or education. So **Deweyite** (diū·i,ait), a follower of Dewey or of his philosophy.

1904 W. James in *Mind* XIII. 464 How can a Deweyite discriminate sincerity from bluff? 1906 —— in *Jrnl. Philos., Psychol. & Sci. Methods* III. 337 The style of article that has usually discussed pragmatism, Deweyism, or radical empiricism. 1954 D. Riesman *Individualism Reconsidered* p. x, A renewed drive against 'Deweyism' and progressive education generally.

dewindtite (diwi·ntait). *Min.* [a. F. *dewindtite* (A. Schoep 1922, in *Compt. Rend.* CLXXIV. 625), f. the name of Jean Dewindt, Belgian geologist: see -ite¹.] A hydrated basic phosphate of lead and uranium (see also quot. 1956).

1922 *Jrnl. Chem. Soc.* CXXII. II. 305 Dewindtite, a new radioactive mineral...This occurs at Kasola, Katanga, Belgian Congo, as a canary-yellow powder intimately mixed with torbernite and a white, powdery material. 1956 *Amer. Min.* XLI. 921 The validity of 'dewindtite' as a separate species is in doubt, as it appears to be structurally identical to renardite, but chemically different.

dewlap. Add: **3.** Comb., *dewlap-deep* adj.

1916 A. Huxley *Burning Wheel* 28 Great oxen, dewlap-deep In meadows of lush grass. 1922 Blunden *Shepherd* (ed. 2) 21 Where milch cows dewlap-deep may wade.

Dew line, D.E.W. line: see *D III. 3.

dew-point. Add: **b.** *attrib.*, as *dew-point apparatus, depression, front, hygrometer, meter.*

1843 Dew-point hygrometer [in Dict.]. 1933 *Discovery* Dec. 360/1 The dew-point apparatus..is designed to discover the temperature to which a metal must be cooled before dew will deposit upon it. 1956 W. A. Heflin *U.S. Air Force Dict.* 165/1 *Dew point depression*, the lowering of the dew point by decreasing water content or pressure. 1958 *Engineering* 4 Apr. 422/1 A dew point meter for use with any gases that are not corrosive to brass or copper. 1960 *Aeroplane* XCIX. 624/1 The dew-point front is the boundary between moist 'maritime tropical' air..and dry 'tropical continental' air.

dew-pond (diū·pǫnd). [Dew *sb.*] A shallow pond, usually of artificial construction, occurring on downs where there is no adequate water-supply from springs or surface-drainage.

These ponds were originally thought to be fed by the condensation of water from the atmosphere, but it is now generally thought that dew plays little or no part in supplying the water.

1865 *Jrnl. R. Agric. Soc.* 2nd Ser. I. 273 *Dew-ponds.* These ponds are chiefly constructed on the highest ridges of the chalk range. 1877 H. P. Slade (*title*) A short practical treatise on dew ponds, the farmers' summer water suppliers. 1879 *Athenæum* 14 June 757/1 The Wiltshire farmers, having learned the value of 'cloud ponds' or 'dew ponds', have formed them at much expense on the tops of the hills. 1903 Kipling *Five Nations* 71 Only the dewpond on the height Unfed, that never fails. 1905 A. J. & G. Hubbard *Neolithic Dew-ponds* 2 The gang of dew-pond makers commence operations by hollowing out the earth for a space far in excess of the apparent requirements of the proposed pond. They then thickly cover the whole of the hollow with a coating of dry straw. The straw..is covered by a layer of well-chosen, finely puddled clay, and the upper surface of the clay is then closely strewn with stones. 1919 *Chambers's Jrnl.* Apr. 268/2 Every one must have heard of the famous dew-ponds of Sussex.

dewy, *a.* Add: **5. b.** Innocent and trusting; naïve.

1958 *Times* 20 Oct. 3/1 The street-walker..should surely not be played..like the dewy ingenue from *Stage Struck*. 1962 *John o' London's* 8 Feb. 139/2 Once as dewy-innocent as the great director.

6. b. *dewy-eyed*, also *fig.*: = sense *5 b above.
1938 'E. QUEEN' *Four of Hearts* (1939) i. 9 Hollywood agents, fat or thin, tall or short, dewy-eyed or soiled by life. 1960 *Guardian* 7 Nov. 6/6 He is not..dewy-eyed about young people, but he feels that promotion should come early.

dexamethasone (deksăme·pāsōᵘn, -zōᵘn). *Pharm.* [f. *dexa-* (blend of *hexa-* and *deca-* in *hexadecadrol* (see quot. 1958¹), representing the number 16 in the systematic names) + METH(YL + -*a* + *CORTI)SONE.] A synthetic steroid, $C_{22}H_{29}FO_5$, which resembles cortisone in its effects and is used in the treatment of some blood disorders and as an anti-inflammatory agent.
1958 *Arthritis & Rheumatism* Aug. 330 Preliminary results of..investigations on the effect of dexamethasone in rheumatoid arthritis are encouraging. 1958 *Ann. Rheumatic Dis.* Dec. 376 In December, 1957, four separate 16α-methylated derivatives of hydrocortisone were made available for clinical trial..16α-methyl 9α-fluoroprednisolone first received the generic name of 'hexadecadrol', but this was later changed to 'dexamethasone'. 1961 *Lancet* 12 Aug. 348/2 The therapy was changed from triamcinolone to dexamethasone. 1968 J. H. BURN *Lect. Notes Pharmacol.* (ed. 9) 95 Dexamethasone is a derivative of hydrocortisone, which is very much more active than hydrocortisone.

dexamphetamine (deksæmfe·tāmīn, -in). *Pharm.* [f. DEX(TRO- + *AMPHETAMINE.] The dextro-rotatory isomer of *AMPHETAMINE, used in the form of the sulphate and phosphate and having effects similar to those of amphetamine sulphate but more marked. Also called **dextro-amphetamine**.
1949 *Brit. Med. Jrnl.* 17 Dec. 1394/1 (*title*) Acute psychosis caused by dexamphetamine. 1952 *Martindale's Extra Pharmacopœia* (ed. 23) I. 516 The action of dexamphetamine sulphate is similar to that of amphetamine sulphate but it is effective in smaller dosage. 1958 *Spectator* 24 Jan. 103/2 A similar list of dangers can be made out for dexamphetamine (dexedrine). 1964 *Economist* 9 May 576/1 Dexamphetamine amylobarbitone alias 'purple hearts'.

Dexedrine (deksĕdrīn). Also **dexedrine**. [prob. f. DEX(TRO- + -*edrine* as in *BENZEDRINE.] The proprietary name of a preparation of dexamphetamine sulphate.
1942 *Trade Marks Jrnl.* 4 Nov. 458/1 *Dexedrine*, pharmaceutical preparations and substances for use in medicine and for veterinary use; sanitary substances;..Smith, Kline & French Laboratories. 1949 *Lancet* 21 May 885/1 I have used 'Dexedrine' (d-amphetamine sulphate) in the treatment of twenty-two cases of chronic sinus disease. 1955 *Sci. News Let.* 2 Apr. 232/3 Only brain stimulants such as caffeine and Dexedrine ('pep pill' ingredient) are able to keep tired aircrews working efficiently. 1958 [see prec.]. 1970 *Sci. Jrnl.* Feb. 7/1 Amphetamines, better known to most people as benzadrine [*sic*] and dexedrine—or as 'speed' in the drug subculture—have acquired a dirty name in the medical profession.

Dexter (de·kstəɹ), *sb.*² Also **dexter**. [Said to have originated from the name of a Mr. *Dexter*, who is credited with having established the breed.] One of a breed of small hardy Irish cattle originating from the Kerry breed. Also called *Dexter Kerry*.
1880 *Encycl. Brit.* XIII. 225/2 The variety known as the 'Dexter', a cross between the 'Kerry' and some unknown breed, is shorter and plumper than the pure 'Kerry'. 1864 *Country Gentlemen's Catal.* 17/2 If the capacity to make good beef is..a *sine quâ non* in the perfect dairy cow,..it must actually be conceded that..the little Dexter-Kerry..presents the nearest approach of any! 1890 *Daily News* 20 June 4/5 The best dexter animal in the show. *Ibid.* 22 June 9/1 The dainty Jersey or pigmy Dexter and Kerry. 1902 *Encycl. Brit.* XXV. 192/2 The Dexter breed..Until recently it was called the Dexter Kerry. 1953 D. O. C. P. MOULD *Ireland of Saints* ii. 23 The typical cow of Celtic Ireland was rather like the modern dexter.

dextral, *a.* Add: **1. c.** That uses the right hand in preference to the left; right-handed: see DEXTRALITY 2. Hence as *sb.*, a (predominantly) right-handed person.
1877 *Lancet* 1. 49 On Dextral Pre-eminence by William Ogle M.D. 1904 *Westm. Gaz.* 6 Aug. 11/3 As if fate had chosen to make it a dextral child. 1927 *Amer. Jrnl. Psychol.* XXXVIII. 327 People fall into four groups, pure or crossed dextrals, and pure or crossed sinistrals...In the pure dextral there is right-handedness and a dominant right eye. 1964 M. CRITCHLEY *Developmental Dyslexia* viii. 54 A fair number of dyslexics are unequivocal dextrals with no family history of left-handedness or ambidexterity.

dextran (de·kstræn). [a. G. *dextran* (C. Scheibler 1874, in *Zeit. Ver. Rübenz.-Ind.* XXIV. 321), f. DEXTRO- + -AN.] **a.** *Chem.* An amorphous, gummy substance produced by the fermentation of sucrose and some other organic materials, and now known to include many structurally related branched polysaccharides of glucose and to be produced only by a few micro-organisms, notably bacteria of the genus *Leuconostoc*. **b.** *Med.* A preparation of degraded, partially hydrolysed dextran in a suitable solvent (as saline solution), used as a substitute for blood plasma.
1879 *Jrnl. Chem. Soc.* XXXVI. A. 912 Beetroot Gum..is converted..after some days' boiling with a solution of potash into dextran. 1939 *Thorpe's Dict. Applied Chem.* (ed. 4) III. 567/2 Dextran, $(C_6H_{10}O_5)_n$, or viscose, is a gum which occurs in the unripe sugar beet...An animal dextran is found in the galls produced on elms by the louse. 1946 *Adv. Carbohydrate Chem.* II. 209 The term 'dextran' has been applied to carbohydrate slimes originating from cane syrup, fermenting vegetables and dairy products. 1948 *Daily Express* 24 Feb. 3/2 Synthetic blood..is known as dextran. 1963 *Brit. Pharm. Codex* 235 Dextran injection is used as a plasma substitute. It is not, however, a complete substitute. 1970 R. W. MCGILVERY *Biochem.* xxviii. 707 A column packed with beads of dextran, available under the trade name *Sephadex*, can be used to separate molecules of varying size.

dextrane. Delete entry and substitute: var. *DEXTRAN.

dextro-amphetamine: see *DEXAMPHETAMINE.

dezincification. Add: **b.** A type of corrosion that causes copper-zinc alloys to become soft and porous as a result of the leaching out of the zinc.
1898 *Engineer* 15 Apr. 363/3 The deterioration in the strength of Muntz metal bolts was due to a partial dezincification of the metal. 1939 *Jrnl. R. Aeronaut. Soc.* XLIII. 616 In untreated glycol brass was found to suffer attack by 'dezincification'. 1960 U. R. EVANS *Corrosion & Oxidation of Metals* xiv. 474 It is today common practice to add 0·02 to 0·06% arsenic to brass condenser alloys..; since this plan has been adopted, dezincification has ceased to be a serious menace.

dghaisa (dəi·sä). Also **dghajsa**, **dhaissa**. [Maltese.] A boat resembling a gondola, used in Malta.
1893 'S. GRAND' *Heavenly Twins* I. II. i. 214 A rainbow fleet of dghaisas..propelled by oarsmen who stood to their work. 1923 *Blackw. Mag.* Dec. 744/1 Not my business the fare—six dghaisa-man. *Ibid.* 750/2 Innumerable dghaisas skated about the surface of the harbour. 1940 M. DICKENS *Mariana* ix. 199 Wireless messages from Aunt Annabelle telling them to jump into a dhaissa and come ashore. 1964 *House & Garden* Dec. 28/1 The Maltese dghajsa..is first cousin to the gondola of Venice.

dha (dā). Also **dhar**. [Burmese.] **1.** A Burmese measure of length.
1821 P. KELLY *Univ. Cambist* (ed. 2) I. 115 The Dha, or Bamboo, consists of 7 royal Cubits; 1000 Dhas make 1 Dain, or Birman League, equal to 2 English Miles 2 Furlongs. 1888 *Encycl. Brit.* XXIV. 490/1 Burmah.—..dha 154 inches.
2. = *DAH, *DAO.
1859 J. W. PALMER *New & Old* 423 The Burmese dropped their..lances and dhars. 1920 *Blackw. Mag.* June 834/2 The two men were allowed to come on after they had deposited their guns and dhas.

dhal, **dhall**, variants of DAL.
1866 [see DAL]. 1923 *Nature* 12 May 626/2 The addition of oatmeal and dhall to the British ration.

dhaman (dā·mān). *India.* Also **dhamin**. [Hindi (Skr. *dharmaṇa*).] **1.** The rat-snake, *Ptyas mucosus*.
1878 P. ROBINSON *In my Indian Garden* 92 A pair of gorgeous dhaman snakes. 1927 *Chambers's Jrnl.* 2 July 495/1 The local natives all think that the dhaman or 'rat snake' is the female of the cobra. 1960 *Times* 16 Aug. 12/7 That's our resident dhamin—rat snake.
2. The grass *Cenchrus ciliaris*, used for fattening cattle. **b.** The dhamnoo, *Grewia tiliæfolia*, and other trees of the same genus.
1877 E. C. MOLONEY *Forestry W. Afr.* 248 Dhaman of Punjab. 1889 W. COLDSTREAM *Illustr. Grasses S. Punjab* Pl. xi, no. 11, (*caption*), *Dhaman* is..equivalent to the English clover. 1923 *Encycl. Brit.* XXXI. 212 The Gramineæ..known indiscriminately as 'dhaman' (*Grewia asiatica*, *G. tiliæfolia*, *G. vestita*). Wood is tough and elastic. 1932 PEARSON & BROWN *Commercial Timbers of India* I. 173 Dhaman timber as a beam is about 10 per cent. stronger than teak.

dhamnoo (dæ·mnū). [Hind.] An Indian tree, *Grewia tiliæfolia*, of the family Tiliaceæ, or the timber therefrom; = *DHAMAN 2 b.
1859 J. W. FOYLE *Illustr. Bot. Himal. Mts.* I. 104 *Grewia elastica*, called *dhamnoo* by the natives..affords timber which is highly valued for its strength and elasticity. 1964 J. LINDLEY *Veget. Kingd.* 372 The wood of Grewia elastica, called Dhamnoo, affords timber highly valued.

dhan (dān). *India.* [Hind. *dhān*, ad. Bengali *dhān*, rice, grain, *dhānā* fried barley, rice, etc. (Skr. *dhānā*).] Rice in the husk.
a 1815 W. ROXBURGH *Flora Indica* (1832) II. 201 Dhan the Bengalee name of the plant, and the unhusked rice, and Chaul the clean rice. 1858 SIMMONDS *Dict. Trade*, *Dhan*, a vernacular name in parts of Hindustan for rice. 1908 *Animal Managem.* 299 In Burma, unhusked rice (*dhan*), and in Madras, koolthi, is issued instead of gram or barley.

dhandh (dænd). Also **dandh**, **dhand**. [Sindhi] A lake or swamp of the territorial division of Sind in India.
1851 R. F. BURTON *Sindh* i. 11 Numerous channels..carry off the surplus water, feeding and refilling the lakes and dandhs, whose moisture has evaporated during the cold season. 1887 *Encycl. Brit.* XXII. 92/1 In the cold season the lakes or *dhandhs* are covered with wild geese, *kulang*, ducks, teal, curlew, and snipe. 1926 *Blackw. Mag.* Dec. 807/1 Any swamp or overflow water or tank or jhil in Sind is a dhand. 1928 *Ibid.* Mar. 389/1 In that year the dhand extended more than a hundred miles north and south, and ten to thirty miles east and west. *Ibid.* 389/2 Several canals took off from the dhand.

dhani (dæ·ni). Also **dhunny**. A kind of palm, the leaves of which are often used for making thatch in the tropics.
1936 *Chambers's Jrnl.* 9 Jan. 85/2 Bamboo matting, jungle-wood boards, and dhunny thatch make ready prey for flames. 1926 *Blackw. Mag.* June 731/2 She arrived soon after seven at the steamer anchorage, covered the flat islands of dhani palm and mangrove swamp. *Ibid.* Dec. 728/1 Huts..squalid and filthy and stinking, but for the most part newly roofed with dhani thatch against the rains. *Ibid.* 732/2 There was a sampan with a hooped covering of dhani leaves.

dhao, var. *DAO.

dharma (dā·ɹmă). Also **dhamma**, **dharm**, **dharmma**, **dherma**, **dhurm**. [Skr., = decree, custom.] In India, social or caste custom; right behaviour; law; esp. in Buddhism and Hinduism: moral law, truth. So **dharmakaya**, **dharmasastra**, **dharmasutra** (see quots.).
1796 W. JONES tr. *Inst. Hindu Law* p. viii, Our Menu with his divine Bull, whom he names as Dherma [ed. 3 (1863), *dharma*] himself, or the genius of abstract justice. *Ibid.* p. xi, A number of glosses or comments on Menu were composed by the Munis, or old philosophers, whose treatises, together with that before us, constitute the Dhermasastra, in a collective sense, or Body of Law. 1840 M. MÜLLER *Lect.* (ed. 2) 295 The..Dharma Sûtras..texts. 1859 M. MÜLLER *Hist. Anc. Sansk. Lit.* 115 ..Complete Sûtra work..the third (Sâmayâcarika or Dharma-sûtras), treating on temporal duties, customs, and punishments. 1850 R. S. HARDY *Eastern Monachism* i. 3 Expositions of the doctrines of Budha..are called bana, or the Word; and the system itself is called dharmma, or the Truth. 1862 M. MÜLLER *Chips* (1867) I. ix. 196 The second and third baskets [*sc.* canonical books of the Buddhists] are sometimes comprehended under the general name of *Dharma*, or law. 1884 S. BEAL *Buddhism in China* x. 100 Here we have the germ from which proceeded the idea of..formula of an invisible presence; the teaching and power of the law (*dharma*) represented the *Dharmakaya*, or law-body of Buddha. 1886 *Encycl. Brit.* XXI. 287/2 The *Dharma-sûtras*, or 'rules of (religious) law'. 1904 W. B. SELBIE tr. *Bousset's What is Religion?* 277 Inflections of the moral law or Dharma are the result of folly based on ignorance. 1938 D. T. SUZUKI *Mahayana Buddhism* ii. 46 The general explanation is that Dharmakaya is the permanent, undifferentiated, comprehending Truth, but the detailed explanation differs according to the different schools of Buddhism. 1958 J. KEROUAC *Dharma Bums* i. 5 An oldtime bhikku in modern clothes wandering the world..to turn the wheel of the True Meaning, or Dharma. 1961 *Listener* 31 Aug. 316/3 Everything was composed of these six elements, they were the origin of all things and were the *Dharmakaya*, the body and consciousness of Dainichi the Great Illuminator, the primordial and eternal Buddha. 1964 *Listener* 28 Mar. 548/1 By religion I am translating the Sanskrit word *dharma*..For Hindus it acquired a special meaning of sacred law, the law of the *Dharma-sastras* or law books..*Dharma* is eternal.

dharmsala (dā·ɹmsālă). Also **dharma sala**, **dhormsal**, **dhurmsal(l)a**, **durhm sallah**. [Hind. (Skr. *dharmaśālā*, f. *dharma* custom, decree, *śālā* house).] A building devoted to religious or charitable purposes, *esp.* a rest-house for travellers.
a 1805 in WELLINGTON *Dispatches* (Stanford). 1826 W. B. HOCKLEY *Pandurang Hari* I. 255, I then proceeded to the *durhm sallah*, or place where travellers put up, and where I had left the two females. 1838 *Penny Cycl.* XII. 233/2 A *Dharma Sala*, one or more sheds or buildings for the accommodation of the mendicants or travellers who are constantly visiting the math. 1920 *Chambers's Jrnl.* 134/1 Even the precincts of the regimental mosque and *dharmsala* were severely examined. 1922 *Blackw. Mag.* Oct. 530/1 At the door of the dharmsal I was met by some of the priests.

Dhimini, var. *DIMINI.

dhobeying, var. *DOBEYING.

dhobi. Add: **2. dhobie('s) itch**, ring-worm affecting the arm-pit and groin in hot moist climates; also, a form of contact dermatitis (see quot. 1967).
1890 BILLINGS *Med. Dict.*, *Dhobie's itch*, Tinea circinata tropica of scrotum, thighs, and perineum; so called in India. 1926 *Practitioner* Jan. 58 The name 'Dhobie itch' is used to denote several kinds of troublesome eczema generally first appearing on the covered hairy regions of the body. 1964 M. HYNES *Med. Bacteriol.* (ed. 8) xxvii. 417 Ringworm of the groin (dhobie itch). 1967 A. C. ALLEN *Skin* (ed. 2) vii. 241 (*caption*) Contact eczematous dermatitis (so-called 'dhobie itch') produced by contact with the ink from the betel nut used for marking laundry.

dhol (dōᵘl). [ad. Skr. *ḍhola*.] A large Indian drum. Also **dho·luk**, **dho·lkee**, drums similar to the dhol, but varying in size or shape.
1865 *Proc. R. Irish Acad.* IX. I. 116 (*Dhôl*)..(*Dhôlkee*).—Ordinary Drum and Little Drum. Both played by hand as

accompaniment to the voice, or struck with a stick when in concert with pipes or loud instruments. 1892 C. R. DAY *Mus. & Musical Instr. S. India* vii. 140 The Dhol..is a species of drum chiefly employed in the native bands that are usually heard at weddings and other festivities. *Ibid.*, The Dholkee is a smaller dhol, much used by women in the Deccan. *Ibid.*, The Dholuk and Dâk are drums somewhat similar to the dhol, but are generally rather larger and vary slightly in shape in different parts of India. 1954 *Grove's Dict. Mus.* (ed. 5) IV. 459/1 The typical drums are the *tabla* (right, and left hand, wedged in the crook of the knee) and the *dhol* (slung like a waif).

dhoon (dūn). Also **dhun**, **doon**, **dun**. [Hind. *dūn*, ad. Hindi *dūna* valley.] Any of the flat valleys lying parallel to the base of the Himalayas dividing the sub-Himalayan hills into two ranges; *spec.* the valley of Dehra.
1814 R. R. GILLESPIE *Let.* 29 Oct. in *Asiatic Jrnl.* (1816) II. 151/2 We met in the far-famed Dhoon, the Tempē of Asia; and a most beautiful valley it is. 1877 *H. M. Elliot's Hist. India* VII. 106 Khalilu-lla Khân..having reached the Dûn. 1879 MEDLICOTT & BLANFORD *Geol. India* II. xxii. 521 They consist of two ranges, separated by a broad flat valley, for which the native name 'dun' (doon) has been adopted in India. 1922 *19th Cent.* Jan. 46 In the lower valleys and hollows, or *dhuns*.

dhoona (dū·nă). [Hind. *dhūnā*.] A resin obtained from the Indian tree *Shorea robusta*.
1846 J. LINDLEY *Veget. Kingd.* 394 The dhoona or dammer pitch, generally used in India for marine purposes, and as incense. 1904 *Public Opinion* 20 June 582/1 The odour of dhoop (incense) and dhoona greeted our breath.

dhoop (dūp). [Hindi, Bengali *dhūp* incense, resin, gum.] An Indian plant, *Vateria indica*; also the pitch obtained therefrom.
1851 *Illustr. Catal. Gt. Exhib.* IV. I. 877/1 Piney resin of dhoop tree..from Canara. 1904 *Public Opinion* 20 June 582/1 The odour of dhoop [see *DHOONA].

dhormsal, var. *DHARMSALA.

dhrupad (dru·pad). [ad. Skr. *dhrupada* kind of dance.] A classical form of north Indian vocal music, consisting of a prelude and four sections developing various parts of the raga, usually sung in a slow tempo.
1848 B. A. PINGLE *Indian Mus.* (ed. 2) ii. 63 There are three different modes of rendering the Râgas or Râginis and Jûtias or Dhuns, in vocal as well as instrumental music..The principal among these are the Alapa or Joga and Dhrupada, [etc.]. 1921 H. A. POPLEY *Music of India* vi. 88 The *Dhrupad* is usually in slow time and in selected tâlas such as Kûtâla, Rûpaka tâla, Chantâla, and Dhîma tâla. 1959 A. N. SANYAL *Ragas & Raginis* i. The present work chiefly concerns the materials representing the classical forms, namely Dhrupads and Khyal forms. *Ibid.* (*index*) 175/1 Dhrupada (colloquial 'Dhrupad'). 1968 *Indian Mus. Jrnl.* v. 34 Translations of Sanskrit treatises, original treatises, dhrupada collections..and the like.

dhudheen, var. DUDEEN.
1853 'C. BEDE' *Verdant Green* viii. 67 A great consumption of tobacco was going on, not only through the medium of cigars, but also of meerschaums, short 'dhudheens' of evenest colour, and the genuine yard of clay. 1867 P. KENNEDY *Banks Boro* xxiv, Their hands went into their pockets more than once in search of the treacherously-consoling *dhudheen*. 1905 *Daily Chron.* 7 July 3/7 Removing his clay dhudheen from his lips.

dhuine-wassel, var. DUNIWASSAL.
1903 *Daily Chron.* 7 July 3/7 Scottish Divine wassels with aggressive feathers in front of their Tam o' Shanter caps. 1920 *Blackw. Mag.* Apr. 517/1 Kilted Hebridean dhuine-wassels.

dhun, var. *DHOON.

dhunny, var. *DHANI.

dhu stone (diū, dʒū). [ad. Welsh *du* black + STONE *sb.*] A type of dolerite found in Shropshire.
1879 G. F. JACKSON *Shropshire Word-Bk.* 117 Dhu-stone, basalt, of a black or very dark colour, quarried at Titterstone Clee. 1902 C. G. HARPER *Holyhead Road* II. ix. 99 Had his road not been one of the best in the world, constructed of hard dhu-stone—he could not have kept that excellent time. 1938 P. T. JONES *Welsh Border Country* vii. 79 The Clee Hills..are crowned with the hard 'dhu stone' that provide the valuable road-metal known as *dhu stone*. 1963 L. F. CHITTY in Foster & Alcock *Culture & Environment* vi. 172 Professor Shotton has found no evidence for the use of Clee Hill dolerite ('Dhu Stone') for stone implements.

dhyana (diā·nă). *Hinduism* and *Buddhism*. [Skr. *dhyāna*.] Profound meditation; the penultimate (seventh) stage of yoga (see quot. 1850).
1850 R. S. HARDY *Eastern Monachism* xxi. 255 By the power of nimitta the thoughts that prevent the exercise of dhyána will be restrained. *Ibid.* 270 The word dhyána is said to mean, 'that which burns up evil desire, or the cleaving to existence'. It is sometimes used in the sense of meditation, and at other times is allied to samádi; in some places it is a cause, and in others an effect. 1871 S. BEAL *Catena Buddhist Scriptures* I. 90 Next come the eighteen heavens...Their general collective title is 'the Worlds of the Brahmas', because of their purity. The distinctive title is 'the Heavens of the four Dhyânas', because all disturbing influences are removed from them, and those who dwell in them are employed in contemplation (dhyâna). 1902 W. JAMES *Var. Relig. Exper.* 401 'Dhyâna' is their [*sc.* Buddhists'] special word for higher states of contemplation. There seem to be four such stages recognized in dhyâna. 1962 A. HUXLEY *Island* vi. 76 The operations are called yoga, or dhyana, or Zen. *Ibid.* 79 *Dhyana* is contemplation.

dhye, var. *DAYE.

diabeah, var. DAHABEEYAH.
1864 J. A. GRANT *Walk across Africa* 366 Baker led us to his 'diabeah', or Nile pleasure-boat.

diabetic, *a.* (and *sb.*). Now usually pronounced (dəiăbe·tik).

diabetogenic (dəi:ăbītodʒe·nik, -bet-), *a.* [f. DIABET(ES + -O + *-GENIC.] Giving rise to or produced by diabetes.
1903 DORLAND *Med. Dict.* (ed. 3) 210/2 *Diabetogenic*, producing diabetes. 1938 *Encycl. Brit. Bk. of Yr.* 1938 227/2 The full diabetogenic effect—usually excretion of sugar and ketones in the urine. 1946 *Nature* 26 Oct. 589/1 The intravenous injection of high diabetogenic doses of alloxan..in the dog produces a very severe diabetic-uræmic syndrome. 1970 W. B. YAPP *Introd. Anim. Physiol.* (ed. 3) vi. 165 Probably the 'diabetogenic hormone' and 'ketogenic hormone' that have been described were simply impure preparations of it [*sc.* growth hormone].

‖ **diable** (diablə). Also **diabolo**, 'int. [Fr.] = DEVIL *sb.* 20 b.
1598 SHAKES. *Merry W.* I. iv. 70 O Diable, Diable: vat is in my Closet? 1765 STERNE *Tr. Shandy* VII. viii. 28 What's wrong now?—Diable!—a rope's broke! 1867 'OUIDA' *Under Two Flags* I. xiii. 202 They were fine men—diable!—they were fine men. 1937 LADY FORTESCUE *Sunset House* vi. 123, I feel it all in my arms and down to the toes of my feet, *diable*.

‖ **diable au corps** (diablokọr). [Fr., lit. 'devil in the body'.] = DEVIL *sb.* 6 b; restless or inexhaustible energy; devilment.
1869 G. B. SHAW *Our Theatres in Nineties* (1932) I. 197 Miss Lottie Collins..has still her Tararaboom-de-ay..and 'Her Diable au corps' manner. 1926 G. SAINTSBURY *Peace of Augustans* vii. 276 But in abeyance and subordination..the *diable au corps* has to be necessary in everything—Congreve and Sheridan are nearer to each other. 1959 *Manchester Guardian* 24 July 4/5 Skiffle has a diable au corps, which is denied to some profounder forms of art.

diablotin. Add: **2. a.** In the West Indies, the rare black-capped petrel, *Pterodroma hasitata*. **b.** A name in Trinidad for the guacharo.
1823 J. LATHAM *Gen. Hist. Birds* VII. 366 They [*sc.* Trinidad Goatsuckers] were served up without the heads or feet, under the name of Dumpy Ducks, or Diablotins. 1891 *Ibis* 6th Ser. III. 137 The Capped Petrel or Diablotin (*Œstrelata hasitata*) formerly bred on the tops of the mountains of Dominica. 1958 J. C. GREENWAY *Extinct & Vanishing Birds of World* 151 If, as is supposed, the diablotin nests somewhere in this vicinity [*sc.* Hispaniola], it must be considered to be in great danger in that part of its range. 1962 G. A. C. HERKLOTS *Birds of Trinidad & Tobago* 131 Guacharo or Oilbird or Diablotin.

diabolo (di-, dəiæ·bolo). [It., = devil.] The game of the devil-on-two-sticks revived under this name. Also, the wooden top with which the game is played. Hence **dia·bolist**, a player of the game.
The game consists in balancing and spinning a double-headed top on a string (which is supported on two sticks), throwing it into the air, and catching it again.
1907 G. K. CHESTERTON in *Daily News* 23 Sept. 12/1 The inventor of the modern Diabolo is M. Gustave Phillipart, a French engineer, well known as an authority on the automobile world. 1922 C. E. M. JOAD *Highbrows* iv. 135 They invent some quaint new game. 1907 *Westm. Gaz.* 25 Sept. 12/1 The Devil Game: Diabolo. *Ibid.* 586/2 It is not difficult to learn to spin the diabolo. 1908 *Daily Chron.* 28 Sept. 587/2 The diabolo game can be played by sides of almost any number. 1907 *Westm. Gaz.* 25 Sept. 12/1 The boy who has won the title of the diabolist. 1922 C. E. M. JOAD *Highbrows* iv. 135 They invent some quaint game of diabolo, or croquet and bowls, or roller-skating. 1964 W. L. GOODMAN *Hist. Woodworking Tools* 196 The curiously small, diabolo-shaped mallet.

diacetic (dəi,ăsī·tik), *a.* [See DIACETATE.] *diacetic acid*: = *acetoacetic acid*. So **diaceturia**, the presence of this acid in the urine.
1882 *Brit. Med. Jrnl.* 6 May 669/1 As the ferric chloride reaction points to the presence of a compound allied to ethyl-aceto-acetate, he [*sc.* A. Deichmüller] concludes that the compound is free diacetic acid. 1885 *Med. News* 3 Oct. 367/2 Diabetic patients suffering from diaceturia are liable..to develop coma and die. 1908 *Practitioner* Feb. 197 Urine did not contain acetone nor diacetic acid. 1953 E. P. JOSLIN *Diabetic Man.* (ed. 9) xxxii. 285 The simplest method for the detection of acidosis by urinary examination is Gerhardt's ferric chloride reaction for diacetic acid.

diacetylmo·rphine. *Chem.* [f. DIACETYL + MORPHINE *sb.*] An acetyl derivative of morphine, more commonly called *HEROIN. (Cf. *DIAMORPHINE.)
1875 *Jrnl. Chem. Soc.* XXVIII. 315 Isomeric diacetylmorphines. 1884 *Ibid.* XLVI. 613 Morphine dissolves easily in excess of acetic anhydride at 85°, to form diacetylmorphine. 1912 *Chem. Abstr.* VI. 2287 Descriptions and tests are given for diacetylmorphine hydrochloride. 1951 A. GROLLMAN *Pharmacol. & Therapeutics* iv. 89 In heroin (diacetylmorphine), acetyl (CH_3CO-) groups replace H in the OH groups at positions 3 and 6 of morphine. 1965 ALSTEAD & MACARTHUR *Clin. Pharmacol.* (ed. 21) xii. 299 *Diamorphine hydrochloride* (heroin). This is diacetylmorphine and its actions resemble those of morphine.

diachronic, *a.* Delete *nonce-wd.* and add: **2.** *Linguistics.* [tr. F. *diachronique* (F. de Saussure *a* 1913, in *Cours de linguistique générale* (1916) III. 120).] Pertaining to or designating a method of linguistic study concerned with the historical development of a language; historical, as opposed to descriptive or synchronic. Also *transf.*, in Anthropology, etc. Hence **diachro·nically** *adv.*; **dia·chrony**.
1927 *Mod. Philology* No. 218 De Saussure..outlines the relation of 'synchronic' to 'diachronic' linguistics. 1937 JESPERSEN *Analytic Syntax* xvii. 69 A view which is historically (diachronically) impeccable. 1938 R. H. LOWIE *Hist. Ethnol. Theory* xiii. 251 We..ought to study the changes going on before our eyes: a 'synchronic' approach must be combined with a 'diachronic' one. 1951 E. E. EVANS-PRITCHARD *Social Anthropol.* iii. 61 Social anthropologists generally study synchronic problems while historians study diachronic problems. 1957 R. W. ZANDVOORT *Handbk. Eng. Gram.* p. v, Contemporary and historical (or, in the terminology of modern linguistics, synchronic and diachronic) grammar are..best treated separately. 1959 W. BASKIN tr. *F. de Saussure's Course in Gen. Linguistics* i. iii. 81 Everything that has to do with evolution is diachronic. Similarly, synchrony and diachrony designate respectively a language-state and an evolutionary phase. 1963 *Canadian Jrnl. Linguistics* Fall 54 The synchronic and diachronic approaches to..language; the synchrony and diachrony of grammatical theory. 1969 C. L. WRENN *Word & Symbol* 13 So too must the whole cultural significance of a country be examined with a minute appreciation of its language seen both diachronically and synchronically. 1968 *Assoc. Teachers of Russian Jrnl.* XVII. 8 Diachronic study is concerned with the movement of a language through time, with the changes that occur in all its planes and the reasons for them.

diachronism (dəi,æ·krŏniz'm). [f. as DIACHRONIC *a.*: see -ISM 1.] *Geol.* The existence of a geological feature that transgresses palæontological zones (see quot. 1929).
1926 W. B. WRIGHT in *Rep. Brit. Assoc.* 1926 354 (*title*) Stratigraphical diachronism in the millstone grit of Lancashire. 1929 L. J. WILLS *Physiographical Evol. Brit.* xiv. 164 The sediments..do not represent the deposits of a given epoch of time...There is a great divergence between the lithological and chronological classification. This is another example of diachronism. 1966 D. T. DONOVAN *Stratigr.* v. 119 It is likely..that diachronism is the rule, not the exception, for formations.
2. *Linguistics.* A diachronic method of linguistic study; diachronic treatment.
1962 Y. MALKIEL in Householder & Saporta *Problems in Lexicography* i. 15 The fundamental dimension (diachronism versus synchronism).
So **diachroni·stic** *a.*; **diachroni·stically** *adv.*
1933 *Publ. Mod. Lang. Assoc. Amer.* XLVIII. 629 The diachronistic investigation of the great linguists of the nineteenth century. 1951 S. ULLMANN *Princ. Semantics* i. 36 It is not language that is synchronistic or diachronistic, but the approach to it, the method of investigation, the science of language. 1957 *Archivum Linguisticum* IX. 26 The term *langue* should be applied only to a language considered diachronistically.

diachronous (dəi,æ·krŏnəs), *a.* [f. as DIACHRONIC *a.*: see -OUS.] **1.** *Geol.* Exhibiting or characteristic of diachronism (see prec., sense 1); not of a uniform geological age.
1926 W. B. WRIGHT in *Rep. Brit. Assoc.* 1926 355 The Haslingden Flag Marine Band was not contemporaneous in the north-east and south-west..analogous phenomena are demonstrable in almost every highly fossiliferous formation. It is necessary therefore to use the term diachronous to describe a bed having such relations to the zonal succession. 1929 L. J. WILLS *Physiographical Evol. Brit.* xiv. 164 The grit facies must therefore transgress the time-zones..i.e., the grits are diachronous. 1966 D. T. DONOVAN *Stratigr.* v. 125 There is a recurring tendency among stratigraphers to be alarmed when a formation is discovered to be diachronous. 1969 *Proc. Geol. Soc. Lond.* Aug. 146 Once the marker points are adequately defined, subsequent demonstrations of facies-changes, diachronous boundaries, lacunæ or similar phenomena do not affect their validity.
2. *Linguistics.* = *DIACHRONIC *a.* 2.
1936 *English Studies* XVIII. 93 Since de Saussure and Bally we have learnt to distinguish diachronous and synchronous.

diaclinal (dəi,ăklai·nal), *a.* *Geol.* [f. Gr. διά through + κλίν-ειν to bend + -AL.] Of a valley, river, etc.: crossing a fold; passing through an anticline or syncline.
1874 J. W. POWELL in *Bull. Philos. Soc. Washington* I. 50 Still other valleys were found cutting across folds. These were called diaclinal valleys. 1875 — *Explor. Colorado River* xi. 160 Three varieties [of transverse valleys] are noticed: (*a*) diaclinal, those which pass through a fold, [etc.]. 1904 W. & E. E. HUMPHREYS tr. *Termier's Erosion & Sedimentation* v. 107 Valleys are classified by their relationship to the dip of the beds. These are: monoclinal or homoclinal rivers..; diaclinal..which cross folds.

diact (dəi·ækt). *Zool.* Shortened form of next.
1887 tr. *F. E. Schulze* in *Challenger Rep.*, *Zool.* XXI. 36 The two rays of a diact belong either to the same or to different axes. *Ibid.*, A..typical diact structure.

diactine (dəi,æ·ktin), *a.* *Zool.* [f. DI-² + *ACTINE 2.] Of a monaxon sponge spicule: having two rays, growing in both directions.

Also as *sb.*, a spicule of this type. Hence **dia·ctinal** *a.*
1888 SOLLAS in *Challenger Rep.* XXV. p. liii, Diactine (*diactina*).—A monaxon in which growth proceeds in both directions along the axis. 1900 E. A. MINCHIN in E. R. Lankester *Treat. Zool.* II. 116 A simple monaxon rod, which may be either diactinal (*rhabdus*), or monactinal (*style*). 1940 L. H. HYMAN *Invertebrates* I. vi. 297 Monaxons that develop by growth in both directions from a central point are named diactinal monaxons, diactines, or, briefly, rhabds.

diagenesis. (Later examples.)
1965 *New Scientist* 30 Dec. 893 Diagenesis—the chemical alteration that takes place in sediments, turning them into rocks for posterity. 1965 G. J. WILLIAMS *Econ. Geol. N.Z.* vi. 65/2 Barium and strontium were concentrated by low-grade diagenesis during the consolidation of 28,000 ft. of Triassic sediments.

diagnosis. 2. (Earlier example.)
1840 W. WHEWELL *Phil. Inductive Sci.* I. VIII. ii. 492 The Characteristick has been termed by some English Botanists the Diagnosis of plants; a word which we may conveniently adopt.

diagram, *v.* Delete *rare* and add further examples. Chiefly *U.S.*
1936 M. H. BRADLEY *Five-minute Girl* xii. 221 What else was she doing but winding the net about him, diagramming the ignominy of every exit? 1938 S. CHASE *Tyranny of Words* vii. 67 The..Ogden and Richards analysis can be diagrammed by a triangle. 1964 C. CHAPLIN *Autobiogr.* xxi. 354, I had read Major H. Douglas's *Social Credit*, which analysed and diagrammed our economic system. 1970 *Nature* 14 Nov. 685/2 These two reactions can be diagrammed according to two schemes.

diagrid (dəi·ăgrid). [f. DIA(GONAL *a.* + GRID.] A supporting structure consisting of diagonally intersecting ribs of metal, concrete, etc.
1943 *Westinghouse Wartime Engin.* Jan. 30 Construction engineers resorted to a new type of structure..known as Diagrid...The walls were built around the old plant, and the Diagrid roof framing erected. 1954 *Archit. Rev.* CXV. 424 A stressed concrete diagrid with in situ concrete joints cast between short precast units. 1963 *Engineering* 13 Dec. 764/2 The stress relieving of the diagrid in a French nuclear reactor.

diakinesis (dəi:ăkəinī·sis). *Cytology.* [mod.L., ad. G. *diakinese* (V. Häcker 1897, in *Biol. Centralbl.* XVII. 701), f. DIA-¹ + Gr. κίνησις motion.] The last stage of meiotic prophase, immediately preceding the disappearance of the nuclear membrane.
1902 E. B. WILSON *Cell* (ed. 2) 440 *Diakinesis* (διά, through), the segmented-spireme-stage, following the synapsis, in the primary oöcyte or spermatocyte, during which the chromosomes persist for a considerable period in the form of double rods. 1922 *Encycl. Brit.* XXX. 784/1 The nuclear wall is dissolved and the chromosomes separate as if repelled from one another: this process is known as 'diakinesis'. 1939 *Nature* 8 July 81/1 The number of bivalents attached to the nucleolus at zygotene to diakinesis. 1970 AMBROSE & EASTY *Cell Biol.* x. 326 The final stage of the first prophase of meiosis is sometimes called diakinesis (or 'moving apart').

dial, *sb.*¹ Add: **6. c.** (Further examples.)
1933 *Punch* 5 Apr. 384/3 The major hesitated, and then a grin lamped up his dial. 1958 L. A. G. STRONG *Light above Lake* xxi. 146 You should have seen the solemn dials on all the Gardas and officials.
d. On a telephone, a circular plate marked with letters, numbers, etc., above which is a disc which can be rotated by means of finger-holes to establish connection with another telephone. Cf. *DIAL *v.* 4.
1879 *U.S. Pat.* 222,458 Automatic Telephone-Exchanges. ..each station comprises..a dial instrument..to make intermittent breaks in the electric current, the number and character of which are successively indicated on a dial. 1903 *Daily Mail Yr. Bk.* 164/1 A moveable dial is attached to the instrument, and by manipulating this..the connection is completed and conversation becomes possible. 1914 W. ATKINS *Princ. Automatic Telephony* 8 To call, a finger is placed in the hole of the appropriate figure and the dial revolved until the stop-piece is reached. The dial is then let go and revolves back. 1921 [see *DIAL *v.* 4]. 1931 [see sense 8 b below].
e. (See quot. 1940.)
1922–3 *T. Eaton's Catal.* Fall & Winter 401/1 *Radio Supplies*... Dials and Knobs. 1928–9 *Ibid.* Fall & Winter 245/2 One knob 'selects' the stations. The dial is illuminated by a miniature light within and geared control gives fine adjustments. 1940 *Chambers's Techn. Dict.* 239/2 *Dial* (*Radio*), the mechanism for adjusting, and for indicating the adjustment, of the tuning controls. 1960 *Which?* May 100/1 BBC and ITA channel numbers could be read easily on most sets, though the DECCA dial could be misread.
8. b. dial telegraph (orig. *U.S.*), a telegraph having a dial marked with letters, numbers, etc., and operated in such a way that the needle on the dial at the receiving station copies the movements of that at the transmitting station; **dial (tele)phone**, a telephone operated by means of a dial; **dial tone** (orig. *U.S.*), = **dialling tone*.
1860 G. B. PRESCOTT *Telegraph* 160 The dial telegraphs are those in which a needle traverses a dial, upon the margin of which are placed the letters of the alphabet. 1886 Dial Telegraph [see sense 8 a in Dict.]. 1930 U. PARROTT *Strangers may Kiss* II. iii. 59 The first dial telephone exchange was opened. 1931 *Punch* 27 May 564/2, I like the dial telephone. Bless its funny little dial! *Ibid.*, One can do such a lot with the dial phone without getting into a hot exchange of snappy come-backs. 1923 *Bell System Techn. Jrnl.* II. II. 62 The subscriber will first remove his receiver from the hook and will hear the so-called..'dial tone', which indicates that the apparatus is ready to receive the call.

Dial, **dial** (dəi·ăl), *sb.*³ *Pharm.* [Proprietary name, first registered in Switzerland in 1918, a shortened form of *diallylbarbituric acid*.] A preparation of diallylbarbituric acid, used as a sedative.
1922 *Chem. Abstr.* XVI. 4299 Veronal, phenobarbital and dial condense readily in AcOH soln. 1931 *News Chron.* 19 Mar. 9/6 The opinion that dial, a preparation of veronal often taken by sufferers from sleeplessness, should be scheduled under the Dangerous Drugs Act. 1942 E. WAUGH *Put out more Flags* 142 She..returned to bed, took two tablets of Dial and slept, gently.

dial, *v.* Add: **4.** To manipulate a telephone dial (*DIAL *sb.*¹ 6 d) so as to establish connection with (another telephone). Also *intr.*, to manipulate a telephone dial. Also *transf.* See also *DIALLING *vbl. sb.* 2 b.
1921 *Conquest* Jan. 124/3 The operator's attention is obtained..by dialling 'o' usually. *Ibid.* 125/1 In order to dial a number, say 7, the finger is put in the hole above 7, and the dial is rotated to the stop and let go. *Ibid.* 126/3 The subscriber begins to dial by putting his finger in hole 5. 1928 *Daily Tel.* 28 Feb. 15 She goes to the 'central supply conveyor', at which, by dialing on an indicating switchboard, she lets the store know her requirements. 1930 *Punch* 21 May 579/1, I keep meeting people who are quite worn out with dialling all day. 1932 E. BOWEN *To North* xxv. 276 Very clumsily, slowly, she dialled a number. 1938 L. MACNEICE *Earth Compels* 52 Dial her number, None will reply. 1966 *Listener* 21 July 103/1 This one dialled the safe's vital combination.
b. As *imp.*, in *dial-a-bus*, etc., used chiefly of an agency or service that is accessible by telephone. (Freq. in *attrib.* collocations.)
1963 *Daily Tel.* 6 Sept. 23/5 Britain's first 'Dial-a-prayer' system comes into operation on Birmingham's subscriber trunk dialling telephones next Tuesday. 1966 *Economist* 19 Mar. 1127/3 The Boston-to-Washington system, 'dial-a-stop' and all, could be built and operating within ten years. 1969 *Nature* 15 Feb. 601/2 The later introduction of a development in transport called 'dial-a-bus' or 'telebus' is also recommended. 1969 *Daily Mail* 5 Apr. 5/1 A 24-hour dial-an-expert service has been set up to give advice to farmers. 1971 *Times* 8 Jan. 3/8 The organisation also offered the public a 'dial-a-chat' service, where anyone could telephone its central switchboard..and speak to one of its members.

dialect. Add: **2. b. dialect atlas, geography**: see quots. 1933; hence *dialect-geographer*, *dialect-geographical* adj.
1932 *Missouri Alumnus* Apr. 232/1 The American Council of Learned Societies is financing a Dialect Atlas of the United States and Canada. 1933 BLOOMFIELD *Lang.* iii. 51 Dialect atlases, collections of maps of a speech area with isoglosses drawn in, are an important tool for the linguist. 1948 *South. Folklore Q.* Dec. 231 The project was..inspired by the great European dialect atlases. 1926 *Germanic Rev.* I. 291 In 1898 Carl Haag introduced the term 'Kernlandschaften'..into the treatment of dialect geography. 1933 BLOOMFIELD *Lang.* xix. 321 The study of local differentiations in a speech-area, dialect geography, supplements the use of the comparative method. 1936 *Language* XII. 245 The dialect-geographers..have found variations. 1948 *Neophilologus* XXXII. 183 The results of the English dialect-geographical inquiry.

dialectic, *sb.*¹ Add: **2. c.** In more general use, the existence or working of opposing forces, tendencies, etc. Also in pl. form (const. sing. vb.).
1925 tr. *Bukharin's Historical Materialism* iii. 75 For Marx, dialectics means evolution by means of contradictions. 1939 *Nature* 21 Jan. 97/1 He [*sc.* J. B. S. Haldane] does not succeed in dispelling the fog that surrounds the uses of the term 'dialectic'. The mildest use seems only to imply that in any complex system of things, people, or thoughts, opposing forces or tendencies are at work, so that processes are likely to oscillate first one way then another between extremes. 1953 *Times Lit. Suppl.* 25 Dec., 'Dialectics', that magic word which nowadays too often gives an aura of perfection to the most outrageous obscurities and rhapsodizings..and not only among Marxists. 1965 *Listener* 25 Nov. 837/2 Schofield presents only one half of the dialectic and virtually ignores the other half, namely the counter-pressures which parents, and adult society in general, must bring to bear on rebellious youth. 1967 *Ibid.* 3 Aug. 140/3 Attempts to formulate the rules of dialectics usually result in arid scholasticism. Dialectics is indeed the grammar of Marxist thinking. *Ibid.* 14 Sept. 321 He [*sc.* J. K. Galbraith] suggested that for every apparently dominant force in modern capitalism, an equal and opposite force existed or could be conjured up: big trade unions to oppose monopolies, the state to oppose both, and so on, in an endless dialectic.

dialectic, *a.* and *sb.*² Add: **A. 1. b.** In Marxist theory used specifically in relation to materialism. Cf. next.
1892 E. AVELING tr. *Engels's Socialism Utopian & Scientific* 39 Modern materialism is essentially dialectic. 1926 M. EASTMAN *Marx, Lenin & Sci. of Revolution* I. ii. 24 That is the philosophy of 'dialectic materialism', the intellectual background of scientific socialism, and..the official state philosophy of the Union of Soviet Socialist Republics. 1952 W. J. H. SPROTT *Social Psychol.* x. 208 Marxists..have always accepted the reaction of ideas on..economic processes. This is, indeed, a field of 'dialectic' operation. 1961 P. USTINOV *Loser* viii. 138 Those impregnated with the spirit of dialectic materialism.

dialectical, *a.* Add: **1. b.** spec. *dialectical materialism*, the theory propagated by Karl Marx and Friedrich Engels according to which political events or social phenomena are to be interpreted as a conflict of social forces (the 'class struggle') produced by the operation of economic causes, and history is to be interpreted as a series of contradictions and their solutions (the thesis, antithesis, and synthesis of Hegelian philosophy). See DIALECTIC *sb.*¹ 2. So *dialectical materialist*.
[1891–92 G. PLECHANOW in *Neue Zeit* X. I. 278 Die dialektische Methode — das war das wichtigste wissenschaftliche Vermächtniss, das der deutsche Idealismus seinem Erben, dem modernen Materialismus, hinterlassen hat... Der moderne dialektische Materialismus weiss..besser als der Idealismus, dass die Menschen ihre Geschichte unbewusst machen.] 1927 D. KVITKO tr. *Lenin's Materialism* ii. 107 As far as dialectical materialism is concerned there does not exist a fixed immutable boundary between relative and absolute truth. 1934 *Discovery* Feb. 31/2 This conversion of the quantitative into the qualitative..is stressed by dialectical materialists today. 1934 V. ADORATSKY (*title*) Dialectical materialism. The theoretical foundation of Marxism-Leninism. 1936 M. PLOWMAN *Faith called Pacifism* 26 They pride themselves on being dialectical materialists, unencumbered by any ethical humbug. 1939 'G. ORWELL' *Coming up for Air* III. i. 185 A lot of stuff that nobody else understood, such as dialectical materialism and the destiny of the proletariat and what Lenin said in 1918. 1963 V. NABOKOV *Gift* iv. 237 Such methods of knowledge as dialectical materialism curiously resemble the unscrupulous advertisements for patent medicines, which cure all illnesses at once.

dialecticism. Add: **2.** *Philos.* Dialectical philosophy or practice.
1901 *Pop. Sci. Monthly* Jan. 298 A man..may be determined to do nothing not pronounced reasonable, either by his own cogitations (rationalism), or by public discussion (dialecticism), or by crucial experiment. 1936 H. READ *Surrealism* 81 Blake labours in hope that Enthusiasm and Life may not cease. In the whole of his writings I feel the presence of an instinctive dialecticism. 1948 *Theology* LI. 30 A world in which dialecticism and existentialism are the desperate and tortured answers that faith must grasp at.

dialler, **dialer**. Add: **2.** A device for dialling telephone-numbers automatically.
1969 *Daily Tel.* 3 Feb. 4/5 The first batch of 'card diallers'—numbers prestored on cards which automatically dial a number when 'punched'—is also due for field trials shortly. *Ibid.*, The automatic diallers are expected to benefit businesses mainly with a large volume of regular calls. 1971 *Sci. Amer.* June 90/3 As a prototype for a bubble mass-memory we have designed and built a 'telephone repertory dialer', a device for storing 50 to 100 frequently called telephone numbers.

dialling, *vbl. sb.* Add: **2. b.** The act or process of *DIAL *v.* 4.
1931 *Punch* 27 May 564/1 Those who think they can understand dialling but can't. 1965 C. FREMLIN *Jealous One* xxi. 169 After four or five minutes of fruitless dialling..she allowed herself to admit that there were no taxis available.
4. (sense *2 b) *dialling apparatus*, *code*; **dialling tone**, the sound produced by a telephone where automatic exchanges are in operation, indicating that the line is in order and the caller can start dialling.
1933 *Discovery* Apr. 132/1 Mechanical equipment is required to pilot the call through each successive stage from the dialling apparatus to the final selector. 1962 A. NISBETT *Technique Sound Studio* x. 176 Dialling is rather more of a performance than being on the receiving end of a phone call, simply because most dialling codes are much too long. 1922 *Telegr. & Teleph. Jrnl.* VIII. No. 84, 82/1 As soon as a connexion has been made with an idle final selector, a 'Dialling tone' is transmitted to the subscriber. 1927 *Daily Express* 28 Oct. 3/1 The dialling tone is heard as soon as you place the receiver to your ear—if your line is in order. 1966 A. PRIOR *Operators* v. 47 'Hello, hello!' But there was just the dialling tone.

dialogue, *sb.* Add: **1. c.** In *Politics*, discussion or diplomatic contact between the representatives of two nations, groups, or the like; hence *gen.*, valuable or constructive discussion or communication.
1953 *Times* 13 May, M. Mayer went on to speak of the 'dialogue' which was tending to establish itself between east and west. 1961 L. MUMFORD *City in History* vi. 181 He would have turned the urban dialogue into the sterile monologue of totalitarian power. 1962 *Listener* 29 Nov. 896/2 The Hungarian Communist party congress was another sounding-board for the increasingly acrimonious but still largely oblique dialogue between Peking and Moscow. 1963 *Ibid.* 31 Jan. 191/2 The essential working basis within the Community is provided by the 'dialogue' between the Commission and the Ministers. 1966 *Rep. Comm. Inquiry Univ. Oxf.* I. iv. 264 We would expect a continuing dialogue between the Hebdomadal Council and the Council of the Colleges on the working of Oxford's part in the Universities Central Council on Admissions. 1970 *Guardian* 25 Sept. 3/2 The new society, based on dialogue between all sections of society. 1971 *Times* 27 Feb. 13/5 May I comment on Professor Rahner's plea..for a strengthening of dialogue between religious believers and non-believers?

dialyse, **-ze**, *v.* Add: **b.** *spec.* in *Med.* To subject (blood) to dialysis (see *DIALYSIS 5 b).
1914 *Jrnl. Pharmacol. & Exper. Therap.* V. III. 277 An apparatus made of celloidin or other dialysing membrane. 1944 *Acta Medica Scandinavica* CXVII. 122 One might try to remove these substances from the blood by dialysis. For this purpose the blood must be dialysed..through a system of tubes or membranes outside the body, and then brought back again into the patient's body. 1963 *Lancet* 12 Jan. 82/2 The dialysing area of the particular hæmodialyser employed is given, together with the urea clearance achieved by its use.

dialyser, **-zer**. Add: **b.** *spec.* in *Med.* = *HÆMODIALYSER.
1944 *Acta Medica Scandinavica* CXVII. 123 (*heading*) The artificial kidney: a dialyser with a great area. *Ibid.* 125 Blood may be let into or out of the patient and into or out of the dialyser. 1966 DUNLOP & ALSTEAD *Textbk. Med. Treatment* (ed. 10) 755 This device allows periodic access to blood vessels so that they may be connected to a suitable dialyser.

dialysis. 5. In wider use: any process in which particles of different kinds are selectively removed from a liquid as a consequence of differences in their capacity to pass through a membrane into another liquid. (Further example.)
1950 KIRK & OTHMER *Encycl. Chem. Technol.* V. 15 The most important field for dialysis today is the recovery of caustic soda from industrial wastes and the refining of crude caustic.
b. *spec.* in *Med.* The process of allowing blood to flow past a suitable membrane on the other side of which is another liquid, so that certain dissolved substances in the blood may pass through the membrane and the blood itself be purified or cleansed in cases of renal failure, poisoning, etc.; the dialysis may take place outside the body in an artificial kidney or inside it using a natural membrane such as the peritoneum. Also, an occasion of undergoing this process.
1914 *Jrnl. Pharmacol. & Exper. Therap.* V. III. 276 We have devised a method by which the blood of a living animal may be submitted to dialysis outside the body, and again returned to the natural circulation. 1944 [see *DIALYSE *v.* b]. 1953 J. A. LUETSCHER in Smith & Wermer *Mod. Treatment* xix. 414 Potassium Intoxication...Efforts may be made to eliminate potassium from the body by dialysis (artificial kidney or intestinal or peritoneal dialysis) if the situation is urgent. 1966 DUNLOP & ALSTEAD *Textbk. Med. Treatment* (ed. 10) 750 Such patients may require several dialyses with the artificial kidney. 1968 *Listener* 11 July 42/1 The system of treatment is known as haemodialysis, or more simply dialysis. It is a way of cleaning the blood by external means. 1970 R. W. MCGILVERY *Biochem.* xxvii. 669 The accumulation [of urea in the blood] can be relieved by dialysis, now sometimes performed by simply flushing large volumes of fluid through the peritoneal cavity.

‖ **diamanté** (diămante). [Fr., pa. pple. of *diamanter* to set with diamonds, to make shine like diamonds.] Material to which a sparkling effect is given by the use of paste brilliants, powdered glass or crystal, etc. Also *attrib.*
1904 *Daily Chron.* 3 May 6/4 The duchess was in black with diamanté wings in her hair. 1909 *Ibid.* 20 Mar. 4/6 A long stole-like panel of crystal and diamanté embroidery. *Ibid.* 18 Aug. 4/5 Garnitures of pale blue silk embroidery and diamanté. 1923 *Weekly Dispatch* 29 Apr. 15 The gown being draped Greek fashion and trimmed with bands of diamanté. 1955 *Times* 20 May 12/5 A high-necked fitted bodice fastened with diamanté buttons.

diamidine (dəi,æ·midīn). *Chem.* [f. DI-² + AMIDE + -INE⁵.] Any compound having a structure based on two amidine, $-C(NH)(NH_2)$, groups; usually, a compound in which these groups are joined by a chain of carbon atoms or two benzene rings or both.
1931 *Jrnl. Chem. Soc.* 2992 We have..tested the hypoglycæmic action of a series of diamidines of basic fatty acids of varying length of chain. 1956 *Nature* 31 Mar. 604/2 An entirely new group of chemotherapeutic drugs, the diamidines. 1968 J. H. BURN *Lect. Notes Pharmacol.* (ed. 9) 120 The early stage of infection by either species is treated by injections of diamidines.

diamond, *sb.* Add: **1. d.** *pl.* Shares in a diamond-mine.
1905 *Daily Report* 22 Mar. 1/2 Lace Diamonds have been bought from Johannesburg. 1907 *Daily Chron.* 28 Oct. 1/7 In Mines diamonds declined. 1964 *Financial Times* 3 Mar. 19/4 Gold shares were irregular, while Diamonds were strong.
5. d. (Earlier example.)
1875 *Cincinnati Enquirer* 6 July 4/5 In the last seven innings the ball hardly got outside the diamond.
e. In a bicycle, the diamond-shaped frame of steel tubing. More fully *diamond frame*. (Now disused.)
1891 *Young Man* Apr. p. ii/2 (Advt.), Safety bicycle.—Diamond frame. 1897 *Outing* (U.S.) XXIX. 488/2 Those had canvas luggage-cases in the diamond of their wheels. 1898 *Cycling* 26 Valises which fit into the 'diamond' of the frame. 1917 *Cycling Man.* 2 Diamond frame, with horizontal top tube.
f. *Gliding.* (See quot. 1960¹.) Also *attrib.*
1960 *Times* 16 May 5/6, I was trying for my diamonds (a premier award in gliding)...I failed, but got my diamond heights over Newcastle. 1960 *Sunday Times* 6 Nov. 23/6 Diamond height had suddenly become a possibility. *Ibid.* 23/8 Diamond distance remains to be achieved. 1971 *Daily Tel.* 21 July 4/3 (*heading*) Burton third in gliding 'diamond'. *Ibid.*, The 'diamond distance' is considered one of the highest achievements in international gliding.
12. diamond dust (earlier example); **Diamond State** *U.S.*, the state of Delaware (see quot. 1934); **diamond stitch**, an embroidery stitch producing a diamond pattern (see quot. 1964).
c 1702 C. FIENNES *Journeys* (1947) III. ix. 239 The true diamond..cannot be divided nor cut but by some of it self diamond dust. 1866 *Galaxy* 15 Oct. 386 Without other significance than such..as attaches to the 'Diamond State', 'the Empire State', [etc.]. 1934 G. E. SHANKLE *State Names* 107 Delaware gets the nickname, the Diamond State, from the fact that it is small in size but great in importance. *c* 1926 'MIXER' *Transport Workers' Song Bk.* 10 You should also learn to crotchet, Diamond stitch and centre tuck. 1964 *McCall's Sewing* xiii. 243/1 *Diamond* or *chevron stitch*. This looks like the honeycomb stitch except that the thread between the rows is carried on top of the fabric instead of under the fabric.

diamond-back, *a.* and *sb.* Add: **A.** *adj.* (U.S. examples.)
1887 *Lippincott's Mag.* Sept. 456 Baltimore,..the home of the soft-shell crab, the diamond-back terrapin. 1894 G. W. CABLE *J. March* xxvii, Di'mon'-back rattle-snake hisself cayn't no mo' scare me 'n if I was a hawg. 1931 H. F. PRINGLE *T. Roosevelt* I. v. 58 People of the blue blood dined nightly on diamond-back terrapin. 1956 C. H. POPE *Reptile World* 219 The diamond-back rattlers..are the giants among rattlesnakes.
B. *sb.* **c.** A rattlesnake, *Crotalus adamanteus* or *C. atrox*, having diamond-shaped markings on its back. *U.S.*
1907 R. L. DITMARS *Reptile Bk.* xlv. 449 The flattened trails of the big Diamond-backs across the dry sandy roads. 1908 *Daily Chron.* 28 Aug. 7/4 The fangs of a big diamond-back are three-quarters of an inch long. 1956 C. H. POPE *Reptile World* 219 The western diamond-back (*Crotalus atrox*) is a reptile of dry country and deserts.

diamorphine (dəiămǭ·ɹfīn). *Chem.* [Shortened form of *DIACETYLMORPHINE.] = *DIACETYLMORPHINE.
1914 *Chemist & Druggist* LXXXV. 819/2 Diamorphinæ Hydrochloridum...The name 'diamorphine' cannot be regarded as satisfactory, as it gives no indication of the presence of the acetyl radicle on which the special properties of the compound depend. 1916 P. W. SQUIRE *Comp. Brit. Pharmacopœia* 551 Diamorphine Hydrochloride..is described officially as the hydrochloride of an alkaloid obtainable by the action of acetic anhydride on morphine. 1920 *Act* 10 & 11 *Geo. V* c. 46 § 8 The drugs to which this Part of this Act applies are morphine, cocaine, ecgonine and diamorphine (commonly known as heroin). 1965 [see *DIACETYLMORPHINE].

Diana. Add: **1. e.** *transf.*
1784 COWPER *Task* IV. 517 Nymphs were Dianas then, and swains had hearts That felt their virtues. 1897 *Westm. Gaz.* 25 Oct. 2/1 Some of these fair Dianas are clad in divided skirts. 1931 [see *ASTRIDE *a.*].

dianetics (dəiăne·tiks), *sb. pl.* [Alteration of DIANOETIC *sb.*] A system, developed by the American writer L. Ron Hubbard, that has as its aim the relief of psychosomatic illnesses by a process of cleansing the mind of harmful mental images. Hence **diane·tic** *a.*, of or pertaining to dianetics.
1950 *Time* 24 July, A new cult is smouldering through the U.S. underbrush. Its name: dianetics. Last week its bible, *Dianetics: The Modern Science of Mental Health*, was steadily climbing the U.S. bestseller lists. 1950 A. HUXLEY *Let.* 10 Dec. (1969) 634 We have been looking into dianetics. 1951 L. R. HUBBARD *Dianetics* p. xii, The main contributions of dianetic theory to the field of psychology and psychotherapy are seven fundamental assumptions. *Ibid.* i. 6 Dianetics..contains a therapeutic technique with which can be treated all inorganic mental ills and all organic psycho-somatic ills. 1954 J. CHRISTOPHER *22nd Cent.* 140 A group of pseudo-scientific mystical fanatics..who went in for a thing called dianetics. 1955 A. HUXLEY *Let.* 24 Oct. (1969) 769 Dianetic procedures were tried..but there was absolutely no recall. 1960 *Guardian* 1 July 9/4 A series of cod psychoanalytic scenes and 'dianetic' games are then evolved.

diapason, *sb.* Also pronounced (dəiăpēⁱ·sən).

diapasonal (dəi,ăpēⁱ·sənăl, -pēⁱ·z-), *a.* [f. DIAPASON *sb.* + -AL.] Of or pertaining to the diapason or melody. Also **diapaso·nic** *a.*
1928 *Daily Tel.* 31 Jan. 8 There was no 'diapasonal excess', which a distinguished Mus. Doc. has declared to be the abuse of modern organ-playing. 1928 *Observer* 8 Apr. 9/4 The trills, roulades, cascades, and diapasonic fertility of 'Zampa'.

diapause (dəi·ăpǭz). *Ent.* [f. DIA-² + PAUSE *sb.*] A period of retardation or suspension of development in some insects. Also *transf.*
1893 W. M. WHEELER in *Jrnl. Morphology* VIII. 68, I shall designate..the intervening resting stage as the diapause. 1934 *Times* 18 Sept. 17/2 The duration of the life-cycle depends on the absence or presence of a diapause in the adult stage. 1956 *New Biol.* XXI. 106 Many species [of Earthworms] survive unfavourable conditions by going into a state of diapause in which they roll themselves up into a tight ball in spherical earthen cells which are lined with mucus. 1969 R. F. CHAPMAN *Insects* xxxv. 717 Diapause is a delay in development which, although its effect is usually to facilitate survival during unfavourable periods, is not immediately referable to the adverse environmental conditions.

diaphanous, *a.* Add: **diaphanously** *adv.* (later example).
1969 A. GLYN *Dragon Variation* i. 4 A pot-bellied shah was reclining on cushions..while a diaphanously veiled maiden knelt before him, offering a cup.

diaphone¹ (dəi·ăfōᵘn). [f. DIA-¹ + PHONE *sb.*²] A low-pitched fog signal operated by compressed air and characterized by the 'grunt' which ends each note. Also *attrib.*.
1906 *Daily Colonist* (Victoria, B.C.) 20 Jan. 6/5 The fog alarm will consist of a diaphone, operated by compressed air. 1950 *Engineering* 29 Dec. 570/1 Three types of sound signal compared were a diaphone with a fundamental frequency of about 180 cycles p. second, [etc.]. 1970 *Motor Boat & Yachting* 16 Oct. 36/1 The fog horn was one of the diaphone type and I can't beat the description in Reed's—'a strong note of low tone ending with a well-defined "grunt"'.

diaphone² (dəi·ăfōᵘn). *Linguistics.* [f. DIA-¹ + PHONE *sb.*¹] All the different forms of a phoneme that collectively occur in all the dialects of a language (see quot. 1932).
1932 D. JONES *Outl. Eng. Phonet.* (ed. 3) xi. 52 The term diaphone is used to denote a sound used by one group of speakers together with other sounds which replace it in the pronunciation of other speakers. 1950 [see *DIAPHONIC *a.*²]. 1953 W. J. ENTWISTLE *Aspects of Lang.* iv. 114 The diaphones are also found in the speech of a single individual. 1961 KURATH & MCDAVID *Pronunc. Eng. in Atlantic States* iii. 101 (*heading*) The regional and social dissemination of the diaphones of stressed vowels.

diaphoneme (dəiăfōᵘ·nīm). *Linguistics.* [f. DIA-¹ + PHONEME.] The collective dialectal variants of a phoneme; = prec. Hence **diaphone·mic** *a.*; **diaphone·mically** *adv.*
1939 D. JONES in *Proc. Third Internat. Congr. Phonetic Sci.* 6 As long as the divergences are not wide enough to interfere with intelligibility, we get higher degrees of abstractions which might be termed diaphonemes. 1939 H. E. PALMER *Ibid.* 7 It might be appropriate to divide each degree of abstraction into..the phonemic and..the diaphonemic aspects of the abstraction. 1959 *Amer. Speech* XXXIV. 266 The core contrasts..can also be reached by way of the diaphonemic pattern, but a smaller set will emerge because of the phonetic integrity that is built into the diaphoneme. The criterion for assigning diaphonemically different entities to a single slot in the phonemic inventory is dialectal complementation among the contrasts.

diaphonic (dəiăfǫ·nik), *a.*² *Linguistics.* [f. *DIAPHONE² + -IC.] Of or relating to a diaphone or diaphones. Hence **diapho·nically** *adv.*
1932 D. JONES *Outl. Eng. Phonet.* (ed. 3) xv. 108 There exist..diaphonic variants; that is to say, there are English speakers who use a somewhat different diphthong. 1950 — *Phoneme* xxvii. 198 Overlapping of diaphones is..especially liable to happen when a sound lies near the limit of a diaphonic 'area'. *Ibid.* 202 The ʔ belongs diaphonically to *k* and *p* as well as to *t*. 1965 *Amer. Speech* XL. 227 The pronunciation (tagə)..is phonologically unlikely in the diaphonic variations before (g).

diaphorase (dəi,æ·fǫrēⁱs, -ēⁱz). *Biochem.* [a. G. *diaphorase* (Adler, von Euler & Günther 1938, in *Arkiv f. Kemi* XII. B. LIV. 4), f. Gr. διάφορ-ος different + *-ASE.] A flavoprotein enzyme reported to catalyse the reduction of dyes by diphosphopyridine nucleotide.
1938 *Brit. Chem. Abstr.* 1048 Diaphorase...Top and bottom yeasts have the same diaphorase activity. 1939 E. ADLER et al. in *Nature* 15 Apr. 641/2 We have used the name 'diaphorase' to designate the enzyme which catalyses the transport of hydrogen from dihydrocodehydrogenase I (CoH_2I) to acceptors like methylene/blue or cytochrome. 1966 S. P. COLOWICK et al. in Florkin & Stotz *Comprehensive Biochem.* XIV. i. 78 The biological significance of the diaphorase activity..is vague, at best, but it may prove useful in understanding the mechanism of hydrogen transfer. 1968 A. WHITE et al. *Princ. Biochem.* (ed. 4) xvii. 366 The diaphorase activity initially reported has proved to be the lipoyl dehydrogenase described above; diaphorase activity, therefore, is entirely artifactual.

diaphototropic (dəi:ăfōᵘtotrǫ·pik, -trōᵘ·pik), *a.* *Bot.* [f. DIA-¹ + PHOTOTROPIC *a.*] Of a plant, leaf, etc.: turning at right angles to incident light. So **di:aphoto·tropism**. Cf. DIAHELIOTROPIC *a.*, -ISM.
1901 D. T. MACDOUGAL *Pract. Text-bk. Plant Physiol.* viii. 127 Leaves..place their surfaces at right angles to the direction of the rays in response to their diaphototropism. *Ibid.* 138 The leaves of *Taraxacum* are diaphototropic, apogeotropic, and epinastic. 1929 J. C. BOSE *Growth & Tropic Movem. Plants* xiii. 133 (*title*) Dia-phototropism and negative phototropism. *Ibid.*, The seedling..assumed a diaphototropic position under intense and long-continued action of light.

diaphragm, *sb.* Add: **2. b.** A thin rubber or plastic contraceptive cap with a flexible metal rim which fits over the cervix.

1933 G. M. Cox *Clinical Contraception* ix. 120 The patient may be fitted with a diaphragm or vault pessary if she so desires. **1948** N. Mailer *Naked & Dead* (1949) iii. ii. 490 They wanted a baby, but now he cannot afford another one, and he is wondering if her diaphragm has been set properly. **1968** M. Richler *Cocksure* xvi. 96 He found himself buying tubes of vaginal jelly, diaphragms in all available sizes, prophylactics. **1970** *Contraceptives* (Suppl. to *Which?*) 47 There are basically three types of cap that can be used by women—diaphragm, cervical cap and vault cap. All fit in the vagina, over the cervix.

4. a. (Add examples of the use of the diaphragm in a camera.)

1878 W. Abney *Treat. Photogr.* xxix. 205 In the doublet lens the position of the diaphragm is important. **1892** *Photogr. Ann.* II. 38 The diaphragm case. *Ibid.* 39 A flare spot is..really the reflection of the diaphragm aperture. **1918** *Photo-Miniature* Mar., *Diaphragm shutter*, one working approximately in the position of the diaphragm in the doublet lens. Constructed of leaves or blades which open and then close the aperture in the exposure shutter.

diapir (dəi·ăpīᵊɹ). *Geol.* [f. Gr. διαπειραίνειν to pierce through.] An anticlinal fold in which overlying rocks are pierced by a mobile rock core. Also *attrib.* Hence **diapi·ric** *a.*; **di·apirism.**

1918 *Economic Geol.* XIII. 467 A peculiar characteristic..is the tendency of the Miocene beds, and especially the salt formation, to pierce through the overlying rocks at certain points on anticlinal axes, thus giving rise to a structure so common that Mrazec [in 1906 *Bull. Soc. Sci. Bucarest*] has coined the term 'diapir fold' to describe it. **1923** *Bull. Amer. Assoc. Petrol. Geologists* VII. 581 Mrazec..has written, in French, considerably in detail on the subject of diapirism, and has found this interpretation applicable to many folds. **1932** *Ibid.* XVI. 1062 The conception of diapiric structure, that is, folds in which the core unconformably penetrates overlying beds. **1942** M. P. Billings *Struct. Geol.* iii. 54 Piercing or diapir folds are anticlines in which a mobile core has been able to break through the more brittle overlying rocks. *Ibid.* xiv. 257 In Rumania..the salt, forced upward by orogenic pressure, has penetrated the sediments at the crest of the anticlines to form diapir folds. **1949** *Q. Jrnl. Geol. Soc.* CIV. 475 Diapiric upswelling and doming is, in fact, a regular phenomenon in the intrusion of granitic magma. **1956** *Ibid.* CXII. 263 (*title*) The Ardara granitic diapir of County Donegal. **1970** *Nature* 25 July 351 Deep seismic reflexion surveys have revealed diapiric structures in deep water. *Ibid.*, The diapirs described by Ewing *et al.* were salt domes.

diapsid (dəi,æ·psid), *a.* [f. mod.L. *Diapsida* (H. F. Osborn 1903, in *Mem. Amer. Mus. Nat. Hist.* I. 455), f. Dɪ-² + Gr. ἁψίς, ἁψιδ- arch.] Of a reptile skull: having two pairs of temporal arches, as in reptiles once grouped in the subclass Diapsida. Also as *sb.* Also **dia·psidan** *a.* and *sb.*

1903 *Mem. Amer. Mus. Nat. Hist.* I. 460 In all the Diapsidan types without exception the supratemporal fenestra is relatively small. **1933** A. S. Romer *Vertebr. Paleont.* vi. 128 It was once thought that all reptiles originally had two temporal openings, and hence forms with two openings (or two bars across the temple, which amounts to the same thing) were termed 'diapsid' (two-arched). **1956** —— *Osteol. Reptiles* ii. 473 It is very possible that all diapsids and diapsid derivatives have sprung from a common ancestor. **1969** A. Bellairs *Life of Reptiles* I. ii. 24 In many reptiles there are in fact two such openings in the skull on each side, one above the other, and hence an upper and a lower arch. This condition is known as diapsid.

diarchal (dəi,ā·ɹkăl), *a.* Also **diarchial, dyarchal, -ial.** [f. Dɪᴀʀᴄʜʏ + -ᴀʟ.] Of or pertaining to a diarchy; *spec.* (see *ᴅɪᴀʀᴄʜʏ).

1921 *Glasgow Herald* 11 Mar. 10 Among the Ministers recently elected under the new 'Dyarchal' system was an Indian who had been previously convicted on charges of conspiracy. **1921** *Evening Standard* 20 May 4/3 Mr...Sastri ..took a prominent part in the dyarchal reforms in India. **1924** *Observer* 3 Aug. 5/1 In Bengal, the Diarchal plan is in suspense. **1927** *Ibid.* 27 Mar. 11/2 The diarchial system had to be suspended. *Ibid.* 13 Nov. 18/2 The dyarchial system [in India]. **1932** *Times Educ. Suppl.* 1 Oct. 375/3 In a total population in the existing dyarchial provinces (excluding Burma) there are roundly 131,000,000 adults.

diarchic (dəi,ā·ɹkik), *a.* Also **dyarchic.** [f. Dɪᴀʀᴄʜʏ: see -ɪᴄ.] = prec. So **dia·rchical** *a.*

1920 *Edin. Rev.* Oct. 386 The diarchic character of the Swiss Government. **1921** *Spectator* 5 Feb. 161/2 The preposterous dyarchical constitution forced upon India by Mr. Montagu. **1926** *Contemp. Rev.* Feb. 240 He broke every rule of the diarchic principle. *Ibid.* Mar. 274 The dyarchic executives. **1963** *Times* 28 Jan. 9/3 The new dyarchic central government would presumably deal with 'national' issues.

diarchy. Add: Also **dyarchy.** (Further examples.) Also *spec.* the system of provincial government in India established by the Government of India Act of 1919 (9 & 10 Geo. V c. 101) (see also quot. 1917).

1885, 1886 [see Dʏᴀʀᴄʜʏ]. **1917** L. Curtis *Let.* 6 Apr. in *Papers Princ. Dyarchy* (1920) iii. 105 Self-government in any large country thus involves the operation of two authorities with separate mainsprings side by side. The word 'dyarchy' ..has been coined to denote this principle. **1919** *Let. Govt. India Const. Reform* 55 in *Parl. Papers* (Cmnd. 123) XXXVII, The only method by which this can be attained is one which involves the division of the functions of government between two different sets of authorities, a method which has been compendiously styled 'dyarchy'. **1920** H. V. Lovett *Hist. Indian Nat. Movement* vi. 163 In the Government of India there was to be no dyarchy, but the Indian element in the Viceroy's Executive Council was to be increased. **1959** *Times* 24 Feb. 11/2 The proposed constitution is essentially a diarchy. It comprises the two communal chambers, as well as very extensive communal self-government at municipal and local level. **1964** A. Swinson *Six Minutes to Sunset* i. 9 Montagu proposed..a system known as diarchy, whereby some functions of government should be taken over by the Indians, while others should be retained by the British.

diaschisis (dəi,æ·skĭsis). *Path.* [mod.L., a. G. *diaschisis* (K. von Monakow 1906, in *Neurologisches Centralblatt* XXV. 1030), ad. Gr. διάσχισις division, f. διασχίζ-ειν (see Dɪᴀsᴄʜɪsᴍᴀ).] A loss of functional activity in a region of the nervous system, resulting from injury to some other connected region of the system.

1915 H. C. Thomson *Dis. Nervous Syst.* (ed. 2) xlii. 401 Temporary attacks in which a patient has difficulty in speaking..may be caused by 'diaschisis' (Monakow), i.e. by a temporary loss of function produced by the shock of a lesion some distance away. **1942** *Brit. Jrnl. Psychol.* Apr. 284 The lowering of vigilance produced by cortical damage would seem to be comparable to the incidence of 'Diaschisis' postulated by von Monakow.

diascope (dəi·ăskōᵘp). [f. Dɪᴀ-¹ + -sᴄᴏᴘᴇ.] A magic lantern or projector by which the image on a transparent slide or the like is projected on to a screen, wall, etc. Cf. *ᴇᴘɪᴅɪᴀsᴄᴏᴘᴇ, *ᴇᴘɪsᴄᴏᴘᴇ¹. So **diasco·pic** *a.*

1911 [see *ᴇᴘɪsᴄᴏᴘᴇ¹]. **1937** *Nature* 27 Feb. 380/2 A classroom epidiascope..has a 500-watt lamp, and the change-over from episcope to diascope is instantaneous. **1951** *Engineering* 2 Nov. 564/1 An exhibition of 'Visual Aids to Industry'..includes diascopes.

diastematic, *a.* Delete ?*Obs.* and add later example.

1952 *Latin Liturg. MSS.* (Exhib. Guide, Bodl. Lib.) 11 An important step towards the development of a 'diastematic' notation was the method of scratching or tracing a line in the parchment.

diathermal, *a.* Delete †*Obs.* and add later *fig.* example.

1923 *Blackw. Mag.* Nov. 584/2 The Diplomatic Body's process of diathermal hibernation.

diathermic, *a.* Add: **2.** Of or pertaining to diathermy. Also **diathe·rmically** *adv.*, by means of a diathermic current.

1910 F. Nagelschmidt in *Proc. R. Soc. Med.* (Electro-Therapeutic Sect.) IV. 2, I have designed a special diathermic apparatus. **1910** *Brit. Med. Jrnl.* 12 Nov. 1547/2 During diathermic operations the blood vessels and lymph channels are securely sealed. **1929** C. C. Martindale *Risen Sun* 100 They were sure I liked heat, and pumped an incredible amount, diathermically, into me each day. **1949** E. B. Clayton *Electrotherapy & Actinotherapy* xvii. 284 All diathermic treatments to the head should be given with the patient reclining on a couch.

diathermy (dəi·ăþɜːmi). [ad. G. *diathermie* (F. Nagelschmidt 1909, in *Münch. med. Wochenschr.* 14 Dec. 2575/1), f. Gr. διά through + θερμός heat + -ʏ³.] The therapeutic passing of high-frequency electric currents through the body by means of external electrodes in order to generate heat within the body; *medical diathermy*, diathermy in which the tissues are warmed but not sufficiently to change their nature; *surgical diathermy*, diathermy in which there is sufficient heating to produce a local change such as destruction of tissue or coagulation of bleeding vessels.

1910 F. Nagelschmidt in *Proc. R. Soc. Med.* (Electro-Therapeutic Sect.) IV. 1 Diathermy is one effect of high-frequency currents of all kinds. **1910** *Archives Roentgen Ray* June 19 Diathermy is but one phase of d'Arsonvalisation, which, besides the production of heat, has other physiological effects dependent on the tension, length, form and frequency of the waves. **1911** *Brit. Med. Jrnl.* 14 Oct. 900/1 The now familiar diathermy apparatus. **1918** *Proc. R. Soc. Med.* (Electro-Therapeutic Sect.) XI. 39 (*heading*) Practical difficulties in performing surgical diathermy of malignant growths. **1929** *Times* 19 Apr. 16/3 Dr. F. D. Howitt visited Craigweil House and again applied the diathermy treatment to the King. **1930** D. McKenzie *Diathermy in Oto-Laryngology* i. 10 When we are giving medical diathermy we begin the current at zero and raise it gradually. **1957** B. O. Scott *Princ. & Pract. Diathermy* v. 55 Short-wave diathermy with its higher frequency as compared to long-wave diathermy is more convenient to apply to a patient since it can be used with air-spaced electrodes. **1966** *Listener* 13 Jan. 62/3 When a third subject was heated by diathermy —which is a kind of heat treatment sometimes given in hospitals—his speed of counting was increased too.

diatomite. (Further examples.)

1905 *Chambers's Jrnl.* Aug. 622/1 Diatomite is an excellent non-conductor of heat. **1939** J. G. D. Clark *Archaeol. & Soc.* ii. 51 Among the more specialized minerals the digging of which has benefited archaeology, one may mention diatomite, a freshwater silt used for insulation purposes. **1957** G. E. Hutchinson *Treat. Limnol.* I. i. 41 Deposits of diatomite are frequently found at the sites of such lakes.

diatonicism (dəiătǫ·nisiz'm). *Mus.* [f. Dɪᴀᴛᴏɴɪᴄ *a.* + -ɪsᴍ.] The quality or state of being diatonic; music the tonality of which is predominantly diatonic; a diatonic system.

1931 *Times Lit. Suppl.* 23 July 580/4 Returning from Schönbergian wanderings to a 'sincere diatonicism'. **1938** *Scrutiny* VII. ii. 173 Even his most complex departures from diatonicism can be referred back to some scale system or chordal structure. **1959** *Listener* 20 Aug. 297/1 Frank added-note diatonicism is allied to blatantly nautical-English melody.

diatonism (dəi,æ·tǫniz'm). *Mus.* [f. Dɪᴀᴛᴏɴɪᴄ *a.*: see -ɪsᴍ.] = prec.

1927 *Music & Lett.* July 325 An insipid diatonic scheme which lacks the primitive freshness of Beethoven's diatonism. **1928** G. Cooke *Theory of Music* 19 Recent attacks upon conventional and arbitrary Diatonism.

diaxon (dəi,æ·ksǫn). *Anat.* Also **-one.** [f. Dɪ-² + *ᴀxᴏɴ.] A nerve-cell having two axons.

1900 in Dorland *Med. Dict.*

diazo-. Substitute for first sentence of def.: A formative used in chemical nomenclature to denote the substituent $=\overset{+}{N}=\overset{-}{N}$ (as in *diazomethane*, CH_2N_2) and also the group —N=N— when one bond only is to a carbon atom of an organic group (as in *benzenediazochloride*, C_6H_5—N=N—Cl). † Formerly also used in naming compounds in which the group —N=N— has both bonds attached to carbon atoms in organic groups (now replaced by *azo-*) and compounds containing the diazonium group, $—N_2^+$. (*Diazo compound* is still used *loosely* to denote azo and diazonium compounds as well as true diazo compounds.) Add earlier and later examples.

1859 P. Griess in *Proc. R. Soc.* IX. 595 The new body, for which I propose the provisional name diazodinitrophenol, is soluble in alcohol and ether. **1908** J. C. Cain *Chem. Diazo-Compounds* xviii. 133 Hantzsch..adopted Blomstrand's formula for the diazo-salts. He preferred also to call these 'diazonium' salts. **1940** S. Miall *New Dict. Chem.* 169/1 The diazonium salts are by far the most important diazo-compounds. **1951** I. L. Finar *Org. Chem.* xxiv. 489 Diazo-compounds have the structure Ar—N=N—Y... Azo-compounds, on the other hand, have the structure Ar—N=N—Ar. **1951** C. R. Noller *Chem. Org. Compounds* xiii. 258 Compounds containing the characteristic group $>CN_2$ are known as aliphatic diazo compounds. **1959** *Chambers's Encycl.* IV. 690/1 Azo dye synthesis depends on two reactions discovered by Griess (1858): first, diazo compound formation..; secondly, adding this diazo component to an aromatic amino or hydroxy compound, called the coupling component. **1965** *Nomencl. Org. Chem.* (I.U.P.A.C.) C. 217 Compounds RN=NX are named by adding, after the name of the parent compound RH, the syllables 'diazo' joined to the designation of the atom or group X.

diazo (dəi,æ·zo), abbrev. of Dɪᴀᴢᴏᴛʏᴘᴇ. Cf. Dɪᴀᴢᴏ- in Dict. and Suppl.

1948 *Jrnl. Documentation* Dec. 184 An important advantage of the diazo system of reproduction..is that there is no need to decide beforehand what the eventual demand will be. **1953** *Library Assoc. Rec.* LV. 10 (*title*) Progress in reflex copying; Diazo and after. **1964** *Economist* 9 May 618/2 The most important of these [*sc.* sensitized papers] is diazo paper.

diazonium (dəiăzōᵘ·niǝm). *Chem.* [a. G. *diazonium* (A. Hantzsch 1895, in *Ber. d. Deut. Chem. Ges.* XXVIII. 1735), f. Dɪᴀᴢ(ᴏ- + *-ᴏɴɪᴜᴍ.] A formative used in chemical nomenclature to denote the ion $—N_2^+$ attached to a carbon atom in an organic compound (usu. an aromatic one, as *benzenediazonium chloride*, $C_6H_5N_2^+Cl^-$). Hence used *attrib.*, as *diazonium compound*, any of the compounds containing this group, some of which are commercially important, being prepared by diazotization and used in the manufacture of azo dyes and in chemical synthesis.

1895 *Jrnl. Chem. Soc.* LXVIII. i. 516 (*title*) Diazonium Compounds and Normal Diazo-compounds. *Ibid.*, The salts of diazobenzene with oxy-acids..contain the radicle N⋮NPh, which is strictly analogous to ammonium, behaving in all respects as a compound alkali-metal, and may therefore be termed 'diazonium'. *Ibid.* 517 Benzenediazonium mercurochloride, $PhN_2·HgCl_3$. **1908** [see *ᴅɪᴀᴢᴏ-]. **1938** G. H. Richter *Textbk. Org. Chem.* xxiv. 509 For most purposes the diazonium salts are not isolated; the solution is employed for whatever purpose is required...The pure *dry* diazonium salts are quite explosive. **1964** N. G. Clark *Mod. Org. Chem.* xxii. 460 On adding a diazonium salt solution to a solution of a phenol in aqueous sodium hydroxide, an azo-compound is formed...Compounds of this type are highly coloured, and are used as dyes. **1965** *Economist* 22 May 942/3 Ultra-violet light shone on the original, passes through the plain bits and decomposes a diazonium compound coating on the copy paper beneath.

dia:zotiza·tion. *Chem.* [f. Dɪᴀᴢᴏᴛɪᴢᴇ *v.* + -ᴀᴛɪᴏɴ.] The process of converting a

compound or a group of atoms into a diazo or diazonium compound or group; *esp.* the conversion of an aromatic primary amine into a diazonium compound by the action of nitrous acid in ice-cold acid solution.

1893 in *Funk's Stand. Dict.* **1902** *Chem. News* 18 July 32/1 This compound, on diazotisation in acetic acid in presence of sulphuric or nitric acid, behaves normally, and forms a dinitrodiazonium salt. **1939** *Thorpe's Dict. Appl. Chem.* (ed. 4) III. 580/2 These diazo-derivatives are generally..prepared from aromatic amines, and the process..is termed *diazotisation*. **1949** N. G. Heatley in H. W. Florey et al. *Antibiotics* iii. 152 The nitrite content of each is then estimated photometrically by the diazotization of the *p*-aminobenzoic acid. **1963** A. J. Hall *Textile Sci.* iv. 179 A number of disperse dyes are of a type which contain amino (NH_2) groups capable of diazotisation.

diazotize, *v.* Add: So **dia·zotizable** *a.*, capable of being converted into a diazo or diazonium compound or group.

1899 *Jrnl. Soc. Chem. Industry* 31 May 488/1 With diamines such as *m*- or *p*-phenylene diamine..compounds are obtained which are still diazotisable. **1963** A. J. Hall *Textile Sci.* iv. 181 The components of an insoluble azoic dye are a base having at least one diazotisable amino (NH_2) group in its molecule and a coupling component.

diazotype. Add def.: A method of photographic copying or colouring, using paper or textile materials impregnated with a diazonium compound. (Earlier and later examples.)

1890 *Photogr. News* 28 Nov. 934/2 Feer's process is not a primuline process, although it is a diazotype process. **1967** *Times Rev. Industry* May 117/1 (Advt.), Diazotype printing processes have endless permutations.

dibatag (di·bătæg). [Native name.] A species of antelope, *Ammodorcas clarkei*, found in Somaliland, having recurved horns ringed at the base.

1891 *Proc. Zool. Soc.* ii. 210 The new Gazelle...Somali name 'Debo Tag', which means 'carries tail high'. **1894** *Ibid.* ii. 318 The Dibatag is common enough where it is found at all, but it is very local in its distribution. **1945** F. Harper *Extinct & Vanishing Mammals of Old World* 675 Dibatag; Clarke's Gazelle...This..gazelle is confined to the interior of British and Italian Somaliland and southeastern Ethiopia. **1970** Dorst & Dandelot *Field Guide Larger Mammals Afr.* 234 Very shy and alert, Dibatag move around a great deal.

dibbler. Add: **3.** (Later examples.)

1967 *Times* 15 Nov. 7/8 An animal thought to have become extinct at some time in the past 83 years has come to light again in Western Australia—the rat-sized carnivorous marsupial popularly known as the dibbler. **1970** *Wildlife in Austral.* Sept. 67/1 The re-discovery of..the dibbler, the noisy scrub-bird and the shortnecked tortoise.

dibbuk, var. *ᴅʏʙʙᴜᴋ.

dibenzanthracene (dəi:benzæ·nþrăsīn). *Chem.* [f. Dɪʙᴇɴᴢ(ᴏ- + Aɴᴛʜʀᴀᴄᴇɴᴇ.] Any of several hydrocarbon derivatives of anthracene with a molecular structure consisting of five fused benzene rings and the formula $C_{22}H_{14}$; *spec.* 1, 2, 5, 6-dibenzanthracene, a colourless crystalline compound, present in coal tar and tobacco smoke, and having carcinogenic properties.

1918 *Jrnl. Chem. Soc.* CXIV. i. 494 (*title*) Synthesis of the isomeric hydrocarbons 1:2:5:6-dibenzanthracene and 3:4:5:6-dibenzphenanthrene. **1944** *Ann. Reg. 1943* 352 Ethyl mercuric chloride proved to be a polyploidogenic agent, dibenzanthracene facilitating its action. **1962** D. B. Clayson *Chem. Carcinogenesis* iii. 65 Magnus (1939) introduced 1:2:5:6-dibenzanthracene into mice by forced feeding and obtained lung tumours. **1964** R. Schoental in E. Clar *Polycyclic Hydrocarbons* xviii. 141 For example, 1.2,5.6-dibenzanthracene..is a potent carcinogen, but the 1.2,3.4-dibenzanthracene..is inactive.

diborane (dəibōᵊ·rēⁱn). *Chem.* [f. Dɪ-² + *ʙᴏʀᴀɴᴇ.] A borane containing two atoms of boron; *spec.* B_2H_6, a colourless poisonous gas having a disagreeable odour.

1926 *Brit. Chem. Abstr.* 227/2 Structure of ethane and diborane, B_2H_6. **1956** *Chem. & Engin. News* 6 Feb. 560/2 The names of some of the boron hydrides are: BH_3, borane (3), or simply borane; B_2H_6, diborane (4); B_2H_6, diborane (6); [etc.]. **1962** P. J. & B. Durrant *Introd. Adv. Inorg. Chem.* xvii. 516 Diborane, B_2H_6, is made by the action of a metallic hydride on a halide of boron. **1968** *Materials & Technol.* I. xii. 517 Diborane is of interest as a fuel for rockets to be launched under a water surface.

dibs² (dibz). [colloq. Arab. *debs* = Heb. *debaš* honey, wine syrup.] A thick sweet syrup made from grape-juice in Eastern countries; also, a similar syrup made from dates.

1757 B. Plaisted *Jrnl. Calcutta to Busserah* 112 The inspissated Juice of the Grape, called here Dibbs. **1841** E. Robinson *Bibl. Res. Palestine* II. 442 The finest grapes are dried as raisins; and the rest being trodden and pressed, the juice is boiled down to a syrup, which under the name of *Dibs* is much used by all classes wherever vineyards are found. **1864** W. K. Tweedie *Lakes & Rivers of Bible* 62 He will fetch fresh-baked bread, and a supply of dibs—a kind of honey made from grapes. **1946** N. Glueck *River Jordan* 96 Dibs, something like honey, usually made of dates.

dicaryon, var. *ᴅɪᴋᴀʀʏᴏɴ.

dice, *v.* Add: **1. a.** In extended use (*fig.*): to take great risks, esp. in phr. *to dice with death*. Freq. in Motoring contexts.

1941 Prince Chula Chakrabongse *Dick Seaman* xiv. 357 Racing motorists usually referred to driving in a race as either 'cracking' or 'dicing', the latter word having been derived from the journalists' former habit of writing about their being 'speed demons dicing with death'. **1969** *Observer* (Colour Suppl.) 23 Mar. 24/1 On the M1 or the M4 a lot of them try to dice with us...You'll get the bod who will come up behind you, an Aston Martin, a Sprite, a Mini Cooper, and they'll be flashing their lights to get by you. **1971** *Daily Tel.* 29 Jan. 3/8 [He] had been 'dicing' along the road with the driver of another car. They were trying to 'carve each other up' after a motoring incident a mile away.

d. To reject, throw away; to leave alone. *Austral. slang.*

1944 L. Glassop *We were Rats* i. i. 5 It's me name, but it's too cissy, so I dices it and picks up 'Mick'. **1953** K. Tennant *Joyful Condemned* xxii. 214 I'll dice it—for now.

dicentric (dəise·ntrik), *a. Biol.* [f. Dɪ-² + *-ᴄᴇɴᴛʀɪᴄ.] Of a chromosome or chromatid: having two centromeres. Hence as *sb.*, a dicentric chromosome or chromatid.

1937, etc. [see *ᴀᴄᴇɴᴛʀɪᴄ *a.*]. **1956** *Nature* 18 Feb. 325/1 Breakage of chromosomes is followed by reunion to give dicentrics. **1970** *Ibid.* 15 Aug. 707/2 Chromosome fragments and structural abnormalities (for example, dicentrics and acentrics) were rare.

dicey (dəi·si), *a. slang* (orig. *Air Force*). [f. Dɪᴄᴇ *sb.* + -ʏ¹.] Risky, dangerous; uncertain, unreliable.

1950 'N. Shute' *Town like Alice* x. 303 He..made a tight, dicey turn round in the gorge with about a hundred feet to spare. **1959** P. Capon *Amongst those Missing* 139 The river got a little dicey. I thought we'd wait for the moon. **1959** 'J. Bell' *Easy Prey* xi. 117 What d'you say we..get a bit more information? That sketch map looks dicey to me.

dichlor-, dichloro-. Add: **di:chlorodiflu:orome·thane**, a colourless, easily liquefied gas, CCl_2F_2, used as a refrigerant and as a propellant in aerosols, etc.

[**1930** *Engineering* 13 June 776/2 Dichlorofluoromethane as a refrigerant.] **1958** *Times Rev. Industry* Feb. 68/3 The liquid refrigerant used in the compression cycle may be dichlorodifluoromethane. **1964** N. G. Clark *Mod. Org. Chem.* ix. 171 Dichlorodifluoromethane..is used as a propellant in insecticide sprays, hair lacquers, shaving lather, etc.

dichoptic (dəikǫ·ptik), *a. Zool.* [f. Dɪᴄʜᴏ- + Oᴘᴛɪᴄ *a.*] Of certain dipterous insects: having the eyes widely separated.

1899 D. Sharp in *Camb. Nat. Hist.* VI. 440 When the eyes of the two sides meet in a co-adapted line of union the Insect is said to be 'holoptic', and when the eyes are well separated 'dichoptic'. **1949** *Handbks. Identification Brit. Insects* (R. Ent. Soc.) IX. i. 5 In many families [of Diptera] the males have the eyes approaching or meeting (holoptic) and the females have them well separated (dichoptic).

dichotomy. 1. b. Add to def.: Something divided into two or resulting from such a division; something paradoxical or ambivalent.

1942 *N.Y. Times* 28 Nov. 12/4 'An absolute dichotomy between science and reason on the one hand and faith and poetry on the other.' In other and simpler days people spoke of the conflict between science and religion, or the clash between the two, or the wide gulf between the two. **1950** *Listener* 28 Sept. 430/1 The questions of thinking in one language and writing in another, the doubtful dichotomy of east and west. **1957** 'J. Wyndham' *Midwich Cuckoos* xiii. 105 By a dichotomy familiar to us all, a woman requires her own baby to be perfectly normal, and at the same time superior to all other babies. **1966** *Listener* 3 Mar. 323/2 Their uncritical use of the 'Communist' versus 'free world' dichotomy.

dichromasy (dəikrōᵘ·măsi). *Ophthalmology.* [ad. mod.L. *dichromasia*, f. Dɪ-² + Gr. χρῶμα colour.] = *ᴅɪᴄʜʀᴏᴍᴀᴛɪsᴍ 2.

1909 in Webster. **1937** *Brit. Jrnl. Psychol.* Oct. 218 The tables of dichromasy in general, and of protanopia, deuteranopia and tritanopia in particular. **1969** [see next].

di·chromat, di·chromate, *sb.*² *Ophthalmology.* [f. Dɪ-² + Gr. χρωματ-ικός (see Dɪᴄʜʀᴏᴍᴀᴛɪᴄ *a.*): see -ᴀᴛᴇ¹.] A person affected with dichromatism.

1908 A. Duane tr. *Fuchs's Textbk. Ophthalmology* (ed. 3) ii. xi. 538 The man with partial color blindness recognizes only two colors in the spectrum, while for the man with normal sight at least three fundamental sensations may be assumed to exist...Hence, we call the former dichromates, the latter trichromates. **1937** *Brit. Jrnl. Psychol.* Oct. 218 'Hidden' figures which are visible only to the dichromate. **1959** *New Scientist* 30 Apr. 972/2 There are three types of dichromat, known as protanopes, deuteranopes and tritanopes respectively. **1969** *Nature* 26 July 414/2 For human dichromats the presence of a restricted region in the visible spectrum that cannot be discriminated from white light.. is considered to be diagnostic of dichromacy.

dichromatic, *a.* Add: **1. b.** *fig.*

1962 *Listener* 28 June 1128/3, I am not sure that he [*sc.* a novelist] presents his world with justice; his characterization is too dichromatic.

2. *Ophthalmology.* Of, pertaining to, or affected with dichromatism.

1897 Norris & Oliver *Syst. Dis. Eye* I. 654/2 Dichromatic color-blindness. **1905** J. W. Baird *Color Sensitivity of Peripheral Retina* 21 The color sensitivity of the periphery is so weak as to approximate in some degree the condition of the dichromatic, or even of the monochromatic retina. **1907** *Smithsonian Inst. Rep. 1907* 620 In dichromatic vision color perception is so limited that all of the shades perceived may be made by combining two of the spectral colors. *Ibid.*, The ordinary color blindness is dichromatic. **1959** *Chambers's Encycl.* V. 552/2 Subjects who have what is called two-colour (dichromatic) vision..could be of three varieties: red-blind subjects, green-blind subjects and blue-blind subjects.

Hence as *sb.*, = *ᴅɪᴄʜʀᴏᴍᴀᴛ.

1897 W. Thomson in Norris & Oliver *Syst. Dis. Eye* I. 600 We find people who can see only two colors in the whole spectrum. They are called dichromatics.

dichromatism. Add: **2.** *Ophthalmology.* Defective colour vision in which only two (rather than three) pure colours, in different combinations, are sufficient to produce a colour indistinguishable from any perceived.

1909 in Webster. **1910** *Proc. R. Soc.* B. LXXXII. 463 The defect which has caused the non-perception of certain red rays has not caused the dichromatism. **1959** *New Scientist* 30 Apr. 972/2 In dichromatism, the colour-blind subject requires only two primaries in order to match any spectral colour.

Dick, *sb.*¹ Add: **1. a.** *Tom, Dick and Harry*: see also Tᴏᴍ *sb.* 1 a; *clever Dick*: a clever or smart person; usu. ironical: a 'know-all'; also *attrib.*

1887 in H. Baumann *Londinismen* 27/1. **1895** J. T. Clegg *Works* I. 238 There's olez tuthri cliverdicks to smile At owt they thinken rayther caut-o'th'-road. **1933** J. B. Priestley *Wonder Hero* vi. 222 One o' these clever Dicks from London bought it. **1957** J. Braine *Room at Top* xxiv. 197 'Clever Dick,' she said. 'Think yer knows everythink, doncha?' **1957** *Observer* 29 Sept. 12/5 He wrote an article for the *Radio Times*, accompanied by a picture of two actors with funny hats and cutlasses, to disarm all clever-dick criticism. **1969** I. & P. Opie *Children's Games* iv. 154 There is bound to be some clever-dick who has hidden in a coal-hole and refuses to show himself.

3. b. The penis. *coarse slang.*

1891 in Farmer *Slang.* **1929** F. Manning *Middle Parts of Fortune* I. v. 95 Dost turn thysen to t'wall, lad, so's us 'ns sha'n't see tha dick? **1934** H. Miller *Tropic of Cancer* 68 That circumcised dick of his. **1963** J. T. Story *Something for Nothing* ii. 42 At a time when sex was being introduced into the school curriculum as something entirely new, they were already playing 'dicks and bums' with boisterous enjoyment. **1969** P. Roth *Portnoy's Complaint* 79 You might have thought that..my dick would have been the last thing on my mind.

dick, *sb.*⁶ *slang.* [? Arbitrary contraction of Dᴇᴛᴇᴄᴛɪᴠᴇ *sb.*] A detective; a policeman.

1908 J. M. Sullivan *Crim. Slang* 8 *Dick*, a cop, detective (Canadian slang). **1912** A. H. Lewis *Apaches of N.Y.* 95 Still, those plain-clothes dicks did not despair. **1924** *Amer. Speech* I. 151/2 'Dick' and 'bull' and 'John Law' have become established as names for the police. **1928** E. Wallace *Gunner* xxix. 234 They'd persuaded a couple of dicks—detectives—to watch the barriers. **1956** J. D. Carr *P. Butler for Defence* xiii. 140 Plain-clothes C.I.D. men..are currently known as bogeys, busies, dicks, and scotches.

dickcissel (diksi·sǝl). [Imitative of the bird's cry.] The black-throated bunting or little meadowlark, *Spiza americana*, of North America.

1886 R. Ridgway *Man. N. Amer. Birds* 452 Dickcissel... dark grey, becoming whitish on belly and lower tail-coverts. **1961** *New Scientist* 6 July 16/2 Nesting success of redwing blackbirds, dickcissels and other ground-nesting birds was very low on treated areas in Louisiana and Texas.

dicken, dickin (di·kǝn), *int. Austral.* and *N.Z. slang.* Also **dickon, dikkon.** [var. Dɪᴄᴋᴇɴs.] An interjectional exclamation expressing disgust or disbelief: come off it!, cut it out! Also const. *on.*

1894 *Bulletin* (Sydney) 5 May 13/3 'And did yer stouch him back?' 'No.' 'Dicken!' 'Swelp me.' **1906** E. Dyson *Fact'ry 'Ands* v. 53 'Well, what price goin' inter trainin' fer er livin' skelington?'. 'Dicken! Skeletons is low.' *Ibid.* viii. 98 To all iv which ther murmur is dickin. **1918** *Chrons. N.Z.E.F.* 10 Apr. 101 'Say, Kiwi, some coot told me you were a wingless bird.' 'Dicken, Yank, yer gotter have arms when there's a war on.' **1937** N. Marsh *Vintage Murder* xxi. 238 Aw dikkon, Mr. Alleyn...I suppose it's N.Z. digger slang 'Dikkon'. It's the same as if you'd say 'Come off it'. Used to hear it on the Peninsula. 'Aw dikkon, dig.' **1949** D. M. Davin *Roads from Home* iii. iii. 237 Dicken on that for a joke. **1950** *Evening Post* (Wellington) 6 Nov. 8/2 The accused said, 'Murder..dicken on that, me murder anyone!' **1970** *N.Z. Listener* 12 Oct. 13/5 'You don't lie and cheat the way my mother does.' 'Ah, dicken.'

Dickensian, *a.* Add: (Later examples.) Freq. in sense 'marked by conditions or features resembling those described by Dickens'.

1928 *Daily Express* 16 Apr. 10/2 It makes foreigners wonder whether the comfort and the good fare of the old Dickensian type of inn were purely mythical. **1955** *Times* 11 May 6/7 It..has reverted to the Dickensian standard of longer hours, less pay, and poorer conditions. **1962** *Daily Tel.* 6 Dec. 12/2 Fog as dense, if not as black, as anything that Dickensian London ever knew.

b. *sb.* An admirer or student of Dickens or his works.

1903 *Westm. Gaz.* 24 Nov. 2/3 A keen Dickensian like Mr. Pett Ridge. **1905** (*title*) The Dickensian, a magazine for Dickens lovers. **1964** *English Studies* XLV. 336 Dickensians will find it of interest.

So **Dickensy** *a.* (earlier and later examples); also **Dickensia·na** (see Aɴᴀ *suff.*); **Di·ckensite** = Dɪᴄᴋᴇɴsɪᴀɴ *sb.*

1859 Mrs. Gaskell *Let.* 9 Mar. (1966) 538 My story..is fated to go to this new Dickensy periodical. **1886** F. G. Kitton (*title*) Dickensiana, a bibliography of the literature relating to Charles Dickens and his writings. **1888** A. Lang in *Good Words* Apr. 233/1 One is a Dickensite pure and simple. **1905** *Daily Chron.* 12 July 4/7 A correspondent.. assures us that he is..a Dickensite. **1949** D. G. Smith *I capture Castle* vi. 75, I had hoped the lawyers' office would be old and dark, with a Dickensy old lawyer. **1958** *Listener* 18 Dec. 1046/1 There may soon be a special department in public libraries devoted to Dickensiana.

dicker, *sb.*² Add to def.: A deal, bargain; articles or commodities as a medium of exchange or payment. (Further examples.)

1831 *Boston Evening Transcript* 22 Dec. 1/1 His 'dicker' was begun, And by aid of solemn face, He closed a bargain soon. **1833** J. Neal *Down-Easters* I. v. 81 A dicker's a dicker I allays concate, where people's upon honor, but not where they aint. **1880** *Harper's Mag.* May 907/1 An old watch and shot-gun..that he had taken as 'dicker' on accounts. **1940** Wodehouse *Eggs, Beans & Crumpets* 57 It was his intention to..make a dicker with it [*sc.* a dog] by means of the slab [of cheese] which he had just purchased.

dicker, *v.* Add: (Further examples.) Also in extended use (*intr.*): to dither, vacillate, hesitate.

1802 *Port Folio* (Phila.) ii. 268 (Th.), *Dickering* signifies all that *honest* conversation, preliminary to the sale of a horse, where the parties very laudably strive in a sort of gladiatorial combat of lying, cheating, and overreaching. **1824** *Woodstock* (Vermont) *Observer* 15 June 4/5 (Advt.) (Th.), The subscriber has for sale the following property which he wishes to dicker for. **1891** Stevenson & Osbourne *Wrecker* (1892) iv. 63 To this romance of dickering I would reply with the romance..of art. **1947** D. M. Davin *Gorse blooms Pale* 78 Phyllis dickers a bit but of course she finishes by getting up and saying she's going to play. **1960** *Sunday Express* 6 Nov. 17/5 He withdrew it [*sc.* a play] angrily when A–R. dickered over the production date. **1961** *John o' London's* 25 May 589/2 Dickering on the edge of adulthood. **1962** *Sunday Express* 30 Sept. 4/5 The large stores who dicker and dither before they take a cheque. **1963** B. Pearson *Coal Flat* ix. 159 Henderson, though he dickered, usually came round to the majority opinion.

dicky, dickey, *sb.* Add: **3. c.** See *ᴅɪᴄᴋʏ-ʙɪʀᴅ, ᴅɪᴄᴋᴇʏ-ʙɪʀᴅ 2.

9. c. An extra seat at the back of a two-seater motor car which can be closed down when not in use.

1912 *Motor Manual* (ed. 14) iv. 139 On most two-seaters a light, detachable, rear single or dickey seat can be arranged for if specially desired. **1926** W. Deeping *Sorrell & Son* v, I'll take it round to a garage for you, sir. Luggage in the dicky?

d. In other extended uses.

1907 Masefield *Tarpaulin Muster* iv. 57, I..went to the leadsman's dicky, or little projecting platform, on the starboard side. *a* **1922** T. Burt *Autobiogr.* (1924) 94 The hauling-engine, called the 'dickey', was at the surface.

dicky-bird, dickey-bird. Add: **a.** (Earlier examples.)

1781 H. Walpole *Let.* 2 Jan. (1904) XI. 354 The Sphinx was a harmless dicky-bird in comparison. **1820** T. Creevey *Let.* 23 Jan. in *Creevey Papers* (1903) I. xiii. 296 Lady Jersey..is like one of her numerous gold and silver musical dickey birds.

2. Rhyming slang for 'word'. Also in shortened form *dick(e)y*.

1932 'P. P.' *Rhyming Slang* 15 Word...Dicky bird. **1936** 'M. Benney' *Low Company* ii. 48 Didn't say a dicky bird, the poor girl didn't. **1943** M. Harrison *Reported Safe Arrival* 61, I give yer me dicky. **1963** J. Prescot *Case for Hearing* vii. 108 Never said a dicky-bird about doing the place myself. **1970** A. Draper *Swansong for Rare Bird* ix. 76 George didn't say a dicky bird when I ambled in.

dicot, dicotyl. Abbreviations of Dɪᴄᴏᴛʏʟᴇᴅᴏɴ.

1877 J. D. Hooker *Let.* 8 Feb. in L. Huxley *Life & Lett. of J. D. H.* (1918) II. xxxix. 236, 1. Monocots. 2. Dicots. a. Angiosp. b. Gymnosp. **1877, 1890** [see Mᴏɴᴏᴄᴏᴛ, Mᴏɴᴏᴄᴏᴛʏʟ]. **1903** *Amer. Jrnl. Sci.* Dec. 416 The main development of the early Dicotyls and other plants constituting the best horizon markers took place in the late Jurassic. **1907** J. D. Hooker *Let.* 13 May in L. Huxley *Life & Lett. of J. D. H.* (1918) II. xxix. 22 You ask why 'in the British Flora of Mr. Bentham and myself I begin Dicots with Ranunculaceae'! **1967** *Kew Bull.* XXI. 342 Finally the monocots and dicots must be equally old, as is also apparent from the fossil record.

dicoumarin (dəikū·mărin). *Chem.* [ad. G. *dicumarin* (R. Fittig 1885, in *Ber. d. Deut. Chem. Ges.* XVIII. 2525), f. Dɪ-² + Cᴏᴜᴍᴀʀɪɴ.] Orig., any compound with a basic structure of two coumarin molecules joined together; now *spec.* the compound 3,3′-methylenebis-(4-hydroxycoumarin), $C_{19}H_{12}O_6$, a white powder first isolated from spoiled sweet clover as the cause of hæmorrhage in sweet clover poisoning of cattle and later prepared synthetically and used as an anti-coagulant.

1886 *Jrnl. Chem. Soc.* L. 47 Salicylaldehyde and succinic aldehyde yield a dicoumarin..very similar in chemical behaviour to coumarin. **1941** *Jrnl. Biol. Chem.* CXXXVIII. 526 The hemorrhagic agent..present in improperly cured hay made from the common sweet clovers..is the dicoumarin, 3,3′-methylenebis (4-hydroxycoumarin). **1943** *Acta Physiol. Scand.* VI. 28 The author has presented data on the prothrombin depressing effect in animals and man of a dicoumarin..first isolated from moulded sweet clover hay. **1947** tr. *Schmidt's Textbk. Org. Chem.* (ed. 5) ii. xi. 475 The clinical value of dicoumarin is enhanced by its lasting effect and the fact that it can be taken orally. **1965** [see *ᴀɴᴛɪᴄᴏᴀɢᴜʟᴀɴᴛ *a.* and *sb.*].

dicoumarol, dicumarol (dəikū·mărǫl). *Chem.* [f. prec. + -ᴏʟ.] = prec. (in its specific sense).

Dicumarol is a proprietary term in the U.S. (see quot. 1942).

1942 *Official Gaz. U.S. Pat. Off.* 28 July 721/2 Wisconsin Alumni Research Foundation...Dicumarol for anti-coagulants. Claims use since May 15, 1942. **1943** *Lancet* 14 Aug. 195/1 The work on the anticoagulant drug 3:3-methylene-bis-(4-hydroxycoumarin) was done in America;..the Americans have adopted 'dicoumarol' as its short name. *Ibid.* 28 Aug. 276/2 'Dicoumarol'..should be spelled 'dicumarol'. **1948** *Reader's Digest* Nov. 32/1 From the Mayo Clinic, where dicumarol was used postoperatively in 1686 cases. **1948** *Sci. News* VII. 63 Dicoumarol is now being used in medicine in conditions where it is desirable deliberately to decrease the clotting power of the blood. **1952** *Ibid.* XXV. 113 Dicoumarol is the only naturally occurring anti-vitamin. **1970** Jubb & Kennedy *Path. Domestic Anim.* (ed. 2) I. 361/1 Dicoumarol is the toxic factor responsible for sweetclover poisoning of livestock.

dictamnus (diktæ·mnǝs). [L.: see Dɪᴛᴛᴀɴʏ. Adopted as a genus name by Linnæus in his *Hortus Cliffortianus* (1737) 161.] A perennial herb of the genus so called, belonging to the family Rutaceæ, esp. the cultivated species *D. albus* (=Fʀᴀxɪɴᴇʟʟᴀ); or the sub-shrub *Origanum dictamnus* (= Dɪᴛᴛᴀɴʏ 1).

1551 [see Dɪᴛᴛᴀɴʏ 3]. **1603** Holland tr. *Plutarch's Mor.* 569 To finde out the herbe Dictamnus, for to feed on it. **1728** R. Bradley *Dict. Bot.* s.v., The true Dictamnus is usually housed in Winter. **1846** J. Lindley *Vegetable Kingdom* 470 Dictamnus is found in the south of Europe. **1922** *Glasgow Herald* 5 Aug. 6 The dictamnus is easily raised from seeds. **1962** *Amateur Gardening* 31 Mar. 4/2 Dictamnus or 'dittany'..is absolutely reliable once established.

Dictaphone (di·ktăfōᵘn). Also **dictaphone.** [f. Dɪᴄᴛᴀ(ᴛᴇ *v.* + *-ᴘʜᴏɴᴇ.] The proprietary name of a machine which records and subsequently reproduces for transcription words spoken into it. Also *attrib.*

1907 *Trade Marks Jrnl.* 15 May 840 *Dictaphone.* Philosophical instruments, Scientific Instruments, and Apparatus for Useful Purposes. Instruments and Apparatus for Teaching. Columbia Phonograph Company General..New York. **1907** *Daily Chron.* 3 July 3/5 The 'dictaphone', an adaptation of the phonograph. **1920** *Chambers's Jrnl.* 264/2 A familiar object in many business and editorial offices is the dictaphone, an instrument which records on wax cylinders letters or articles spoken into it. **1920** *Glasgow Herald* 22 May 5 Typing for half an hour an aggregate of 3991 words from dictaphone records. **1926** 'J. J. Connington' *Death at Swaythling Court* xvi, You know he's an expert on gramophones and dictaphones and all that kind of truck...He got a dictaphone record of a telephone message from Hubbard to his clerk. **1961** *Evening Standard* 14 July 20/2 Dictaphone Typist urgently required. **1967** *Times Rev. Industry* Mar. 41/1 One of Dictaphone's directors.. doubts whether a 'generic' name has any specific advantage 'for people may call it a Dictaphone, but they will not necessarily come to us'.

dictation. Add: **4.** *attrib.*, as **dictation speed**, the speed of speech (as) of a person dictating.

1957 H. Nicolson *Let.* 21 Feb. (1968) III. 332 'Has your husband joined you,' asked Vita at dictation-speed, 'or is he still in Formosa?' **1971** M. Russell *Deadline* v. 51 Larkin recounted the incident at dictation speed.

dictionary. Add: **1. c.** Colloq. phr. *to have swallowed the* (or *a*) *dictionary*: to use long or recondite words.

1934 'G. Orwell' *Burmese Days* ii. 29 Have you swallowed a dictionary?..We shall have to sack this fellow if he gets to talk English too well. **1966** M. Torrie *Heavy as Lead* x. 124 'The whole point is that my Society deprecates, as much as you do...' The voices began again, 'Aw, cut it out!' 'Put a sock in it!' ''Ev've swallered the dictionary!'

3. *dictionary-making* (later example), *meaning*, *word* (earlier example), *work*.

1794 W. B. Stevens *Jrnl.* 19 Nov. (1965) 206 He seems to be quandaryed (that's not a dictionary word I believe). **1837** Mill in *Westm. Rev.* XXVII. 19 A few phrases,..by adding up the dictionary meanings of which, we may hunt out a few qualities. **1854** W. C. Roscoe in *Prospective Rev.* X. 398 [Shakespeare] leaves his meaning to rest in great measure on the atmosphere that hangs about his language, rather than on its dictionary meaning and grammatical construction. **1887** *Trans. Philol. Soc. 1885–6* p. ix, The main difficulty in the Dictionary work is to trace the history of the development of the meanings of a word. **1929** C. I. Dodd *Apples & Quinces* ii. iv. 146 It was over the Dictionary work that Amanda made the acquaintance of Mr. Jasper Stafford. *Ibid.* v. 156 Amanda went back to Oxford and Dictionary-making.

Dictograph (di·ktǫgrɑf). Also **dictograph.** [f. Dɪᴄᴛ(ᴀᴛɪᴏɴ + -ᴏ + -ɢʀᴀᴘʜ, after *phonograph.*] The proprietary name of a device used as an internal telephone without mouthpiece or earpiece, by means of which speech in one room is picked up by a sensitive microphone and reproduced through a loudspeaker elsewhere in the building. Also *attrib.*

1907 K. M. Turner & W. Donnan *U.S. Pat.* 843,186 1/1 Our invention relates to what we shall term a 'dictograph', being a telephonic system or apparatus by which a person—for example, the manager of an office—may dictate letters to any one of his corps of stenographers without requiring them to leave their places at their own desks. **1907** *Times* 26 June 12/4 Mr. [K. M.] Turner of New York..had the honour of exhibiting the Dictograph before the King and Queen. **1912** *Times* (weekly ed.) 23 Feb., Indianapolis advices..say that a 'dictograph' which was concealed in the office soon after John McNamara's arrest enabled Government stenographers in the room below to take the daily conversations. **1921** *Trade Marks Jrnl.* 27 Apr. 852 *Dictograph.* Telephone Instruments and Apparatus (all included in Class 8). Dictograph Telephones, Limited, Dictograph House..London S.E. 1. **1926** *Spectator* 20 Mar. 520/1 Among inventions for great captains of industry, I looked with awe upon the Dictograph Telephone. **1962** *Punch* 3 Oct. 498/2 Those status-creating Dictograph telephones.

dicty (di·kti), *a. U.S. slang.* [Origin unknown.] **a.** Conceited; snobbish. **b.** Elegant, stylish, high-class.

1926 A. Niles in W. C. Handy *Blues* 22 Gay-cattin' 'roun' with dicty cats. *Ibid.* 33 *Dicty*, uppish and conceited. **1926** C. Van Vechten *Nigger Heaven* 12 'Winter Palace?' she inquired...'Naw..too many ofays and jig-chasers.' 'Bowie Wilcox's is dicty.' 'Too many monks.' **1944** A. Rose in G. Myrdal *Amer. Dilemma* xliv. 966 There are only a few dozen words and phrases that are uniquely Negro..such as 'dicty' which means trying to put on airs and act upper class without having the basis for doing so. **1959** R. Gant *World in Jug* 162 Nice boy, Julian..from one of those dicty schools and Oxford too, who had surprisingly turned to jazz.

dictyospore (di·ktiospōᵊɹ). *Bot.* [f. Gr. δίκτυο-ν net + Sᴘᴏʀᴇ.] (See quot. 1940.) So **dictyospo·ric** *a.*

1933 E. W. Mason *Annotated Account Fungi* (Imp. Bur. Mycol.) List ii. Fasc. 2. 12 In the classical species of this little understood genus the thallus breaks up into dictyosporic elements. **1940** *Chambers's Techn. Dict.* 243/1 *Dictyospore*, a multicellular spore divided into segments by both transverse and longitudinal walls. **1962** C. J. Alexopoulos *Introd. Mycology* (ed. 2) xviii. 394 Conidia are also named in accordance with their shape or structure. Thus we have *dictyospores*.., which possess both vertical and horizontal septa.

dictyostele (di·ktiostīl, diktəi·ǫstīl). *Bot.* [f. Gr. δίκτυο-ν net + Sᴛᴇʟᴇ 2.] In some ferns and dicotyledons, a stele which is so interrupted by leaf-gaps as to resemble a network of strands. Hence **dictyoste·lic** *a.*

1902 G. Brebner in *Ann. Bot.* XVI. 521 *Dictyostele*, a vascular tube with large overlapping leaf gaps, so that the whole structure becomes a network of vascular strands. **1911** *Encycl. Brit.* XXI. 736/2 It is better to call..a segment of a broken-up stele a *meristele*, the whole solenostele with overlapping leaf-gaps being called a *dictyostele*. **1934** Webster, Dictyostelic. **1955** *Ann. Bot.* XIX. 495 The vascular structure of the attenuated specimens was dictyostelic.

dicumarol: see *ᴅɪᴄᴏᴜᴍᴀʀᴏʟ.

Didache. Add: **2.** (With small initial.) The instructional or didactic element in early Christian theology, as distinct from 'kerygma' or preaching.

1936 C. H. Dodd *Apostolic Preaching* i. 6 For the early Church..to preach the Gospel was by no means the same thing as to deliver moral instruction or exhortation...It was by *kerygma*, says Paul, not by *didaché*, that it pleased God to save men. **1953** *Scottish Jrnl. Theol.* VI. 435 *Didache* was an expanded version of the *kerygma*, embracing Christian life and conduct, but including also instruction in the 'historical facts' as well as the 'saving facts', in the sacraments and eschatology. **1960** *Times Lit. Suppl.* 15 Apr. p. xii/1, The 'kerygma' is not of course opposed to the 'didache': the message must imply the teaching.

didakai, didakei: see *ᴅɪᴅɪᴄᴏɪ.

diddums (di·dǝmz). Also **didums.** [= *did 'em*, i.e. did they (tease you, etc.)?, with addition of plural *s.*] An expression of commiseration addressed to a child and jocularly to an adult; hence used as a meaningless term of address to children, etc. Hence **di·ddum** *v. trans.* (*rare*), to say 'diddums' to.

1893 E. F. Benson *Dodo* I. vii. 142 Women who were content to pore on their baby's face..saying 'Didums' occasionally. **1908** *Magnet* I. 1, Poor little thing! He misses his mammy, you know! Diddums! **1926** *Spectator* 1 May 795/1 On being 'diddumed' when sleeping..it [*sc.* a cat] opened its eyes crossly. **1928** *Internat. Jrnl. Psycho-Analysis* IX. 17 Where an hysteric will spend endless time consciously rejecting, say, an obscene word, the obsessional

will spend the same time in an agony of confusion over a simple nursery phrase, e.g. 'Diddums' or 'Ducky'. **1928** R. Macaulay *Keeping Up Appearances* viii, Sometimes they inquire..as to the past life in general of the object of their solicitude, as 'Was he?' or 'Diddums?' **1961** M. Kelly *Spoilt Kill* ii. 109 Mr Luke gave it to me...That was his way of saying diddums to baby. **1970** B. Turner *Another Little Death* iv. 31 'Diddums,' I said soothingly, and booted him in the backside to land headlong at her feet.

diddy (di·di). *slang* or *dial.* Also **deddy, diddey.** [Alteration of Tɪᴛᴛʏ³.] A woman's breast or nipple, or her milk; the teat of an animal.

1788 Grose *Dict. Vulg. Tongue* (ed. 2), *Diddeys*, a woman's breasts or bubbies. **1841** C. H. Hartshorne *Salopia Antiqua* 393 The cow's got a sore diddy. *Ibid.*, Gie th' lickle un a drop o' the diddy. **1876** T. E. Brown *Doctor* 48 Take a baby from the deddy Just when the mother's gettin' it reddy! **1922** Joyce *Ulysses* 610 Cuts off their diddies when they can't bear no more children. **1961** S. Beckett *Happy Days* i. 32 What's she doing? he says—What's the idea? he says—stuck up to her diddies in the bleeding ground.

didgeridoo (di:dʒĕrĭdū·). Also **didjeridoo, didjeridu, dijiridu**, etc. [Imitative.] A musical instrument of the Australian Aborigines, consisting of a long tube made from bamboo or a hollow sapling which is blown into to produce a resonant sound.

1924 F. T. Macartney in *Bulletin* (Sydney) 18 Dec., Didjeridoo—didjeridoo! A blackfellow blows through a length of bamboo To the regular beat of an ironwood stick. *Ibid.*, And all..Is dark while you hark to the didjeridoo. **1936** F. D. Davison *Children of Dark People* i. 10 He would make music on his didjeridoo. **1938** F. D. McCarthy *Austral. Aboriginal Decorative Art* 36 The chanting tubes or *didjeridu* are painted with red, yellow, white and black dashes on a red or plain field. **1944** W. E. Harney *Taboo* (ed. 3) 80 From the river bank came the droning of the didgeredoo. **1959** S. H. Courtier *Death in Dream Time* iv. 38 Only the didgeridoo, the native hollow-log trumpet, could produce that galvanising noise. **1965** *Sunday Mirror* 5 Sept. 23/1 A didgeridoo blown by Australian entertainer Rolf Harris. **1971** *Times Lit. Suppl.* 19 Nov. 1453/2 The everlasting drone of the dijiridu.

didicoi, didicoy (di·dikoi). *slang* or *dial.* Also **didakai, -kei, diddekai, diddicoy, didekei, -ki, -kie, -ky, didikai, -koi, didycoy.** [Romany.] A gipsy (see also quot. 1966).

1853 'E. R.' in M. Carpenter *Juvenile Delinquents* iv. 126 Gipseys, romaneys, didycoys, 'our people', as they call themselves. **1907** *Daily Chron.* 5 Oct. 6/2 Making raids on gipsy encampments with the object of getting them to send their young 'didekies' (children) to school. **1936** G. Greene *Journey without Maps* i. iii. 79 A didicoi..was the name they gave in Gloucestershire to gipsies. **1936** *Punch* 18 Mar. 321/3 Dappled with mire, By the didakai's fire. **1959** 'O. Mills' *Stairway to Murder* iv. 37 What's a man of your age and education doing wandering the country..? You don't strike me as a natural diddicoy type. **1960** W. Robertson *Shadow of Rope* xiii. 131 Them there diddicoys is wholly afeard o' the ma'sh. **1961** *Guardian* 23 May 5/5 These were the dreaded scrap-metal Didakeis. **1966** *Ibid.* 3 Nov. 4/6 'Didicoys'—the Irish tinkers and other nomads around London who far outnumber the true Romanies.

didn't: see Dᴏ *v.* 29.

dido². For *U.S. slang* read *dial.* and *U.S. slang.* Add earlier and later examples.

1807 J. R. Shaw *Life* ix. 140 A jolly Irishman, who cut as many didos as I could for the life of me. **1880** L. Parr *Adam & Eve* I. vi. 151, I thought..you'd be cuttin' up a Dido with everything. **1891** 'Q' *Noughts & Crosses* 76 There was a pretty dido goin' on atween the dree. **1919** H. Jenkins *John Dene of Toronto* (1920) v. 84 'Well, you can't,' snapped John Dene, 'receiver's off. Your boys have been playing dido all morning on my phone.'

didy (dəi·di). *U.S. colloq.* Also **didie.** [Infantile alteration of Dɪᴀᴘᴇʀ *sb.*] = Dɪᴀᴘᴇʀ *sb.* 2.

1902 in *Dialect Notes* II. 232. **1942** O. Nash *Good Intentions* 62 There are few spectacles..more untidy Than 1914 or something in a didy. **1945** B. Macdonald *Egg & I* (1947) xvi. 175, I gathered up my baby..and the didy bag.

die, *sb.*¹ Add: **1. a.** Also, a cube bearing other devices on its faces, or a solid with more or less than six faces (see quots.).

1874 *Macomb* (Ill.) *Eagle* 23 Nov. 1/5 'Now, gentlemen,' said she, 'we will throw poker dice.' **1910** *Encycl. Brit.* VIII. 176/2 Eight-sided dice have comparatively lately been introduced in France as aids to children in learning the multiplication table. **1927** W. E. Collinson *Contemp. Eng.* 32 Crown and Anchor is played by means of dice marked with crowns, anchors, hearts, etc. and a board similarly marked. **1960** R. C. Bell *Board & Table Games* v. 125 Games with two-sided dice. *Ibid.* 141 Three special dice are used marked with a crown, an anchor, a heart, a spade, a diamond, and a club.

2. b. (*g*) Colloq. phr. *no dice*: (it is or was) useless, hopeless, unsuccessful, profitless, etc.; nothing; 'nothing doing'. orig. *U.S.*

1931 D. Runyon *Guys & Dolls* (1932) 136 He is a guy I consider no dice. *a* **1939** R. Chandler *Trouble is my Business* (1950) 62 The old man's had a stroke...No dice there. *Ibid.* 222 No dice, sister. I'm putting the pressure on. **1943** P. Cheyney *You can always Duck* iv. 67 'She can come back here and go on driving a car.' 'No, sir,' I tell him. 'No dice. That dame has started bein' Mrs. Cara Travis an' she's goin' on bein' Mrs. Travis.' **1952** Wodehouse *Barmy in Wonderland* viii. 81, I was around at her bank this morning trying to find out what her balance was, but no dice. Fanny won't part. **1959** 'H. Howard' *Deadline* iv. 47 She was on her way back to report that it was no dice. **1959** M. Pugh *Chancer* 10 Nothing doing. I'm not going. No dice. **1968** *Globe & Mail* (Toronto) 10 July 1/4 'It's no dice as far as I'm concerned,' said one picket who made a derisive gesture.

die, *v.*¹ Add: **1. e.** *to die on* (someone): (*a*) to die in the presence of or while in the charge or care of (someone); (*b*) to cease to function for or be of use to (someone); to cease to interest.

1907 J. M. Synge *Aran Islands* i. 47 A farmer was in great distress as his crops had failed, and his cow had died on him. **1930** Kipling *Limits & Renewals* (1932) 243, I decided to drive in on a gust under the spitfire-sprit—and, if she answered her helm before she died on us, to humour her a shade to starboard. **1931** *Times Lit. Suppl.* 10 Dec. 1002/4 Carruthers drank all New York could give him for thirty-six hours and..the 'drink died on him'. Well, sometimes the novel dies on Mr. Waugh. That happened this time. **1934** A. Thirkell *Wild Strawberries* iii. 53 Let Weston know that the horn died on us this morning, so he'd better fix it up. **1936** J. Tickell *See how they Run* iv. 44 'I want to look after her while she's in England.' 'Suppose she died on you?' **1936** C. Day Lewis *Friendly Tree* vi. 80 That was one thing which had not died on her—the love of birds.

3. a. *to die hard* (earlier and later examples); *to die in one's boots* or *shoes* or *with one's boots on*: to die a violent death, *spec.* to be hanged; so *to die with one's boots off*: to have a peaceful or unspectacular death or end.

1694, etc. [see Sʜᴏᴇ *sb.* 2 d]. *a* **1700, 1712** [in Dict.]. **1784** *Gentl. Mag.* LIV. i. 19/1 The only solicitude too many of them discover is, whether the criminals *die hard*, according to the Tyburn phrase. **1873** J. H. Beadle *Undevel. West* xxii. 435 It will be said in Western dialect, 'They died in their Boots'. **1873** J. Miller *Life amongst Modocs* vi. 75 If you keep on slinging your six-shooter around loose..you will..die with your boots on. **1888** [see Hᴀʀᴅ *adv.* 3]. **1895** W. Rye *E. Angl. Gloss.* 21 Died with his *boots on*, viz. died a violent death. **1903** Masefield *Ballads* 22 So I'm for drinking honestly, and dying in my boots. **1946** B. Sutton *Jungle Pilot* 99 An aircraft which ends its career by dying with its 'boots off' and being deliberately burnt to ashes on the ground is a sight at once undignified and pathetic. **1959** *Listener* 6 Aug. 200/1 They died with their boots on; they hardly ever surrendered.

c. (*I*) *hope* (or *wish*) *I may die*, (*I*) *hope to die*, etc.: colloq. asseverations of the truth of what one says.

1865 Dickens *Mut. Fr.* II. iii. xi. 100 'Wish I may die,' cried Mr. Riderhood, with a hoarse laugh, 'if I warn't a goin' to say the self-same words to you.' **1899** R. Whiteing *No. 5 John St.* xxiv. 244, I see it in the piper. Wish I may die! **1912** Mulford & Clay *Buck Peters* xvii. 160 'There's a Witch's Ring right here on the range!' 'Nonsense!' 'Hope I may die! I'll show you, to-morrow.' **1926** [see *ᴄʀᴏss *v.* 3 b]. **1927** I. Gershwin *Let's Kiss & Make Up* 2, I didn't mean to start any scene to make you sigh. Hope to die! **1968** M. Allingham *Cargo of Eagles* xi. 133 'Off the record?' 'Never a word, may I die.'

7. c. Substitute for def.: Used hyperbolically to indicate extreme feelings of amusement, embarrassment, etc.; esp. in phr. *to die with* or *of laughing*. (Add further examples.)

1820 M. Wilmot *Let.* 12 Jan. (1935) 50 Once in a tender love speech, I thought I should have died, when Lady Grace..told him *audibly* he had turned over *two leaves*! *a* **1930** D. H. Lawrence *Phoenix II* (1968) 84 He looked like a positive saint: one of the noble sort, you know, that will suffer with head up and with dreamy eyes. I nearly died of laughing. **1949** D. Smith *I capture Castle* xi. 190 She knew some of the manaquins [*sic*] at a dress show—I could have died. **1969** Widdowson & Halpert in Halpert & Story *Christmas Mumming in Newfoundland* 162 He was dressed up, you know, and he was like an old shepherd. Well I nearly died.

14. (Later example.)

1928 F. T. Brooks *Plant. Dis.* xii. 201 Stone fruit trees often die back in association with the presence of various fungi.

die-back. Substitute for def.: The progressive dying of a shrub or tree shoot from the tip backwards, caused by disease or unfavourable conditions, esp. a disease of fruit trees. (Add further examples.)

1895 *Dept. Agric. Yearbk. 1894* (U.S.) 199 Die-back manifests itself by a number of striking characters. The foliage becomes very dark green, [etc.]. *Ibid.* 200 Finally a reddish brown resinous substance exudes on the twigs, forming the so-called die-back stain. **1910** G. Massee *Dis. Cultivated Plants & Trees* 430 Die back of willow shoots.. attacks the tips of willow shoots and causes them to die back. **1916** B. D. Jackson *Gloss. Bot. Terms* (ed. 3) 112/2 *Die-back*, of *Salix*, a disease due to *Diplodina salicina*; ~ of *Citrus*, some uncertain condition of health; ~ of *Prunus*, from *Naemospora crocea*. **1928** *Forestry* II. 27 Out in the forest die-back or canker due to frost is only likely to occur in particular areas. **1933** *Discovery* Feb. 66/2 *Phytophthora cambivora* has been isolated from dying beech in two separate woods in Somerset, and has been found to be the undoubted cause of the die-back. **1938** *Nature* 2 Apr. 612/1 Trees may be killed [by elm disease] in a single year, or the dieback may be more gradual. **1949** E. Hyams *Not in our Stars* i. 8 The Morello plantation has very bad die-back. **1957** *Country Life* 8 Aug. 248/1 Slight lightning discharges may do no more than cause dieback in a few of the topmost twigs. **1968** *Punch* 27 Mar. 466/3 The central brown stain from dieback disease.

die·-cast, *v.* [f. Dɪᴇ *sb.*¹ 6 c + Cᴀsᴛ *v.* 50.] *trans.* To cast (hot metal) in a die or mould. Hence **die-cast** *ppl. a.*

1909 *Daily Chron.* 7 Aug. 6/7 The frames are now die-cast instead of moulded. **1960** *Times Rev. Industry* Feb. 16/1 The first diecast radiator grilles were fitted.

die-casting, *vbl. sb.* [f. prec. + -ɪɴɢ¹.] The action of *ᴅɪᴇ-ᴄᴀsᴛ *v.*; also *concr.*, a piece of metal cast in a die or mould. Also *attrib.*

1911 *Machinery* (N.Y.) May 715/2 (*heading*) Die-casting and die-casting machines. **1930** *Engineering* 14 Feb. 196/3 The association has also continued its support of the researches in die-casting and die-casting alloys. **1942** J. N. Greenwood *Gloss. Metallographic Terms* (ed. 2) 17 *Die casting*, the casting of alloys in metal moulds or dies. There are pressure and gravity die castings, according to whether the metal is forced into the die or not. **1955** *Times* 6 July 16/7 A reorganization of their facilities for producing gravity die-castings and high pressure die-castings is contemplated.

diederik, diedrik (di·dǝrik, di·drik). *S. Afr.* Also **didric, diedrick.** [Imitative.] A small African cuckoo, *Chrysococcyx caprius* (see quot. 1790).

1790 E. Helme tr. *Le Vaillant's Trav. Afr.* I. 347 The green-golden cuckoo of the Cape...Perched on the tops of large trees, it continually repeats, and with a varied modulation, these syllables, *di, di, didric*, as distinctly as I have written them; for this reason I have named it the *didric*. **1853** *Edin. New Philos. Jrnl.* LV. 82 The pretty notes of the..*diedrick* further enliven the growing day. **1939** S. Cloete *Watch for Dawn* ix. 122 The intermittent calls of michi and diedrik. **1959** Gill & Winterbottom *First Guide S. Afr. Birds* (ed. 6) 109 *Didric* or *Diederik Cuckoo*, Diedrikkie; *Chrysococcyx caprius*.

die-hard, *sb.* and *a.* Add: **A.** *adj.* (Further examples.) Cf. sense *B. 2.

1922 *Weekly Dispatch* 5 Nov. 1/1 The splitting away of the great masses of the Conservative Party—of the Die-hard section. **1923** E. A. Ross *Russian Soviet Republ.* 395 The die-hard Tories. **1955** *Times* 19 May 10/1 There were also some diehard Mau Mau leaders who were doing everything in their power to prevent their followers from giving themselves up. **1971** *Internat. Affairs* (Moscow) Apr. 39/2 Diehard British chauvinism, rich with anti-Sovietism, was expressed back in October 1970, when the Conservative Party held its first conference..after it came to power.

B. *sb.* **2.** A person who is extremely conservative, stubborn, or irreconcilable, esp. on a political or other issue; *spec.* (*a*) one of those who were prepared to 'die in the last ditch' in their resistance to the Home Rule Bill of 1912; (*b*) one of those members of the Conservative Party who followed the leadership of the Marquess of Salisbury in 1922. Also *transf.*

1857 Hughes *Tom Brown* i. v. 114 They [*sc.* Rugby football players] are 'the fighting brigade', the 'die-hards' larking about at leap-frog..and playing tricks on one another. **1912** *Tatler* 9 Oct. 32a (*caption*) One of the 'Die-hards'. Lord Willoughby de Broke speaking recently at Ballyroney against Home Rule for Ireland. **1920** *Studies* June 197 The heroes of Marathon were conservative 'die-hards': the modern Athenian..was a Socialist or Spartacist. **1922** *Times* 18 Oct. 14/2 Diehards demand freedom. Lord Salisbury on Coalition. **1922** *Daily Mail* 25 Oct. 10 Mr. Chamberlain said: Politics have many vicissitudes. A few days ago I was orthodox; to-day I am a 'Die-hard'. *Ibid.* 3 Nov. 13 Die hard, the section of Conservatives led by Lord Salisbury who were hostile to Mr. Lloyd George's Premiership and Home Rule. **1927** A. T. Hagg *Labour Community Song Bk.* 6 The Tory Die-hards in their clubs They sing this plaintive song. **1957** *New Yorker* 29 June 60/2 The few million white Algerian French citizens, who the diehards declare must not be abandoned. **1963** F. T. Visser *Hist. Syntax* I. iv. 365 It would seem to have been a linguistic die-hard.

3. The Scottish terrier.

1900 *Westm. Gaz.* 9 July 3/1 To anyone in doubt as to a suitable dog to take up as household guard or companion the Scottish terrier, often called the 'Die-hard', or Aberdeen terrier, and the Dandie Dinmont, immortalised by Sir Walter Scott, are strongly recommended. **1921** *Melbourne Argus* 13 Sept. 5 The name 'Die-Hards' has been given to the well-known black or brindle Scottish terrier.

Hence **die-hardism** (dəi·hāɹdiz'm), the principles, tenets, or spirit of a die-hard.

1922 *Glasgow Herald* 8 Mar. 8 If the atmosphere of co-operation can be created, 'die-hardism' will be painlessly asphyxiated. **1923** *Ibid.* 28 July 8 There is a core of 'die-hardism' in all sincerely held convictions, and its language does not consist only of 'outworn shibboleths'. **1926** *Manch. Guardian Weekly* Sept. 181/1 Mr. Churchill has reversed his former die-hardism and struck out boldly. **1934** A. J. Toynbee *Study Hist.* III. 77 The most remarkable feat of Spartan 'Diehardism' was the attempt of the Royal Martyrs ..to reclothe the dry bones of the 'Lycurgean' system in flesh.

dieldrin (dī·ldrin). *Chem.* [f. the name of O. *Diels* (1876–1954), German chemist + *-drin*, after *ᴀʟᴅʀɪɴ: so called because the Diels–Alder reaction (the **diene synthesis*) was used in preparing it.] A crystalline solid, $C_{12}H_8Cl_6O$, white when pure and light brown in commercial preparations, which is an epoxide of aldrin and is used as an insecticide.

1949 *Dieldrin* (Interdepartmental Comm. Pest Control, Bur. Entomol. & Plant Quarantine, U.S. Dept. Agric.) 1 The coined generic name 'dieldrin' is announced for an insecticidal product..meeting the specifications set forth in the attached statement. **1950** *Adv. Chem. Series* (Amer. Chem. Soc.) i. 177 The most interesting oxidation of aldrin..

is the oxidation with per acids. This oxidative process.. produces 6, 7-epoxy-6, 7-dihydroaldrin, the compound now called dieldrin. **1956** *Nature* 17 Mar. 532/2 In connexion with the..Malaria Control Pilot Project, an area of about 300 square miles has been set aside for treatment with dieldrin. **1957** *New Scientist* 23 May 25/1 Dieldrin, used as a sheep dip in Australia as a protection against blowflies, was found to impregnate the wool fibres almost irremovably. **1969** *Times* 24 Mar. 4/5 There are severe restrictions on the use of the persistent insecticide dieldrin.

dielectric, *sb.* and *a.* Add: **B.** *adj.* **2. b.** *dielectric constant*: one of the physical parameters of a non-conducting medium, equal to the ratio of the electric displacement at any point in the medium to the displacement an identical charge distribution would produce in a vacuum, measured at the same point; the relative permittivity. (Some writers call this the *relative dielectric constant*, using *dielectric constant* for the ratio of electric displacement to electric field strength: cf. *PERMITTIVITY.)

1875 *Jrnl. Chem. Soc.* XXVIII. 38 (*heading*) Experimental determination of the dielectric constants of insulators. **1893** E. ATKINSON tr. *Ganot's Elem. Treat. Physics* (ed. 14) IX. iii. 734 In crystallised bodies the dielectric constant varies with the direction of the axes. **1936** HAUSMANN & SLACK *Physics* xxiii. 445 The dielectric constant..is very nearly unity (1·000586) for air at normal temperature and pressure. **1943** *Electronic Engin.* XVI. 206 The physical constant associated with a dielectric is called the dielectric constant. This is generally defined as the ratio of the capacitance of a condenser with the dielectric material between the electrodes to the capacitance of the electrodes alone when placed in vacuo. **1955** C. P. SMYTH *Dielectric Behavior & Struct.* i. 1 In the m.k.s. system the dielectric constant of free space is 8·854 × 10^{-12} farad per m. **1959** B. I. & B. BLEANEY *Electr. & Magn.* xvi. 454 The dielectric constant of a substance affords some valuable information as to the structure of its constituent molecules. **1967** M. NELKON *Fund. Physics* xxxvi. 673 Mica has a higher dielectric constant of 7.

die-link (dəi·liŋk). *Numism.* [f. DIE *sb.*^{1} 5 + LINK *sb.*^{2} 3 a.] A relationship established between two or more coins made or stamped by the same die. Hence as *v. trans.* (used *pass.*). Also **die-linked** *ppl. a.*, **die-linking** *vbl. sb.*

1941 C. H. V. SUTHERLAND in *Numismatic Chron.* 6th Ser. I. 99 The consideration of die-axes, and of die-linkings, which can be achieved only by the study of very numerous examples. *Ibid.* 104 The small thunderbolt accompanying the obverse portrait on the 'Seated Livia' series is not found on any other group; we cannot therefore hope for die-links between one group and another. *Ibid.* 105 A twin die-linked series, forming a parallel to the twin die-linked *Clementiae–Moderationi* series. *Ibid.*, The twin *Clementiae–Moderationi* series, being so closely die-linked. **1958** *Times* 11 Feb. 11/5 The hoard included coins of Edward the Elder..which were die-linked, 20 being from the same die of one moneyer. **1958** *Oxf. Univ. Gaz.* 7 Mar. 678/1 Such die-links serve to confirm that issues for both Titus and Domitian were produced in the same establishment.

dielytra (dəi,e·litrā). *Bot.* [mod.L. (Chamisso & Schlechtendal 1826, in *Linnæa* I. 556), f. Gr. δι- DI-^{2} + ἔλυτρον sheath.] A plant belonging to the genus once so called, now included in the genus DICENTRA.

1864 C. M. YONGE *Trial* II. vi. 122 Do you know this Dielytra? I think it is the prettiest of modern flowers. **1872** G. M. HOPKINS *Jrnl.* 29 June (1959) 220 Dielytras—in the full-blown flower there are at least four symmetrical 'wards'. **1900** J. M. ABBOTT in W. D. Drury *Bk. Gardening* viii. 269 Dicentras (Dielytras) are lovely spring- or early summer-flowering plants.

diene (dəi·in). *Chem.* [f. DI-^{2} + -ENE.] **a.** Any organic compound containing two double bonds between carbon atoms.

1917 *Chem. Abstr.* XI. 3031 (*title*) The melting points of hydrocarbons, especially dienes, having a system of conjugated double bonds. **1929** *Ibid.* XXIII. 3232 Dienes such as isoprene in relatively pure condition. **1961** L. F. & M. FIESER *Adv. Org. Chem.* v. 197 Butadiene is significantly more stable..than dienes with isolated (nonconjugated) double bonds.

b. *attrib.*, esp. in **diene synthesis,** an addition reaction in which two carbon atoms joined by a double or triple bond in an unsaturated compound become attached to the first and fourth carbon atoms of a conjugated diene, forming a ring of six carbon atoms in the resulting molecule; the Diels–Alder reaction; **diene value,** a number expressing the degree of unsaturation in fatty compounds.

1928 *Chem. Abstr.* XXII. 1144 (*heading*) Addition of 'diene' hydrocarbons. **1930** *Ibid.* XXIV. 356 (*heading*) The 'diene-syntheses'. An ideal method for the synthesis of organic compounds. **1957** *Encycl. Brit.* I. 549/2 The diene synthesis consists essentially of the addition of substances containing two conjugate double bonds, such as the butadienes, to substances such as maleic acid or quinones. **1936** *Rep. Prog. Applied Chem.* 433 Linseed oil itself shows a definite 'maleic' or 'diene value', equivalent to about 5–7% of conjugated triene glycerides.

diener, var. *DEANER.

diervilla (dī^ə·ɪvilā). *Bot.* [mod.L. (Tournefort 1707, in *Acta Acad. 1706* VII. 85), f. the name of *Dierville*, a French surgeon, who discovered a species of the plant in Acadia 1699–1700.] A plant of the genus so named, comprising deciduous shrubs having pink, crimson, yellow, or white flowers; some species are also called WEIGELA or *bush-honeysuckle* (see HONEYSUCKLE 3 b).

1806 B. MCMAHON *Amer. Gardener's Cal.* 293 Flowering shrubs..may now be planted with good success, such as.. St. Peter's-wort, Diervilla, roses, and all kinds of hardy deciduous shrubs. **1816** *Curtis's Bot. Mag.* XLIII. 1796 (*heading*) Yellow-flowered Diervilla. **1884** G. NICHOLSON *Illustr. Dict. Gardening* I. 475/1 The gracefully spreading form of Diervillas renders them remarkably well adapted for shrubberies. **1924** A. J. MACSELF *Flowering Trees & Shrubs* x. 134 Formerly known as Weigelas, a name indeed still familiarly used, the Diervillas are worthily held in high esteem. **1951** *Good Housek. Home Encycl.* 117/1 Other favourite flowering shrubs and trees that will grow in most soils are..Weigelias or Diervillas, Heaths, [etc.]. **1959** *Times* 31 Jan. 9/4 The forsythias, diervillas, pyracanthas.

dies. Add: **b. dies non.** Also *transf.*, a day that does not count or on which there is no activity.

1897 *Daily News* 19 Apr. 2/1 This has been almost a 'dies non' in the City, owing to the Stock Exchange being closed. **1937** *N. & Q.* CLXXIII. 369/2 Thursdays are a *dies non* in the players' week.

Diesel, diesel (dī·zəl). **a.** The name of Rudolf *Diesel* (1858–1913), German engineer, used *attrib.* to designate a type of internal-combustion engine invented by him, in which air alone is drawn into the cylinder, this air being so highly compressed that the heat generated ignites the fuel-oil when it enters the combustion space. Also *ellipt.*, a diesel engine; a locomotive, motor vehicle, etc., driven by a diesel engine (also *attrib.*).

1894 S. B. DONKIN *Text-bk. Gas, Oil, & Air Engines* Index 413 Diesel engine. **1894** —— tr. *Diesel's Rational Heat Motor* 53 The three vertical cylinders of the Diesel motor. **1928** J. K. ROBERTS *Heat & Thermodynamics* xiv. 284 The use of a Diesel engine results in considerable fuel economy as compared with an engine working on a 'constant-volume' cycle. **1930** *Engineering* 16 May 623/3 The first two Diesel locomotives in the world were built by The General Electric Company in 1917 and 1918. **1934** T. E. LAWRENCE *Lett.* (1938) 806 The new four-cylinder Gardner 60 bhp light Diesel is having its gear box fitted for test in 159 hull. **1957** [see sense b below]. **1966** J. BETJEMAN *High & Low* 15 Through the midlands of Ireland I journeyed by diesel. **1970** *Railway Mag.* Oct. 584/2 A number of class '47' diesels have been noted working under the wires in the London area recently.

b. *attrib.* and *Comb.*, as *diesel-driven*, *-engined*, *-hauled*, *-hydraulic* adjs.; **diesel-electric** *a.*, driven by electric motors powered by current from a generator, which in turn is driven by a diesel engine; also *absol.*, a diesel-electric engine, locomotive, or vehicle; **diesel oil,** a petroleum fraction used as fuel in diesel engines.

1905 H. ALLEN *Gas & Oil Eng.* 267 Test of a 500 B.H.P. Diesel Oil Engine. **1916** *Motor Ship & Motor Boat* 30 Nov. 379, 1200 h.p. Diesel-engined Auxiliaries. **1921** *Sci. Amer.* 2 Apr. 267 Advantages of the Diesel-Electric System over the Straight Diesel Drive. *Ibid.*, The advantage is decidedly with the Diesel-electric, since the engine runs continuously in one direction...A Diesel-electric operated boat can change from full speed ahead to full speed reverse in.. seconds. **1926** *Marine Engineer* Mar. 113/2 The primary valuation of a Diesel oil is obtained from the specific gravity and viscosity. *Ibid.* Apr. 145/2 Below are shown microphotographs of Diesel oil and boiler oil, before and after treatment in a..centrifuge. **1930** *Engineering* 16 May 624/2 The Diesel electric locomotive for the London and North Eastern Railway. **1935** *Hansard Commons* 18 June 236 Diesel oil used for the purpose of road vehicles. **1936** *Discovery* Mar. 89/1 Diesel-driven parcels vans. **1955** A. ROSS *Australia 55* iii. 43 The train is..diesel-hauled. **1957** *Economist* 30 Nov. 799/2 The 52 diesel-hydraulic locomotives ordered..bring the total number of main-line diesels ordered so far..to 282, of which all except 66 are diesel electrics. **1958** *Listener* 12 June 964/1 The experimental diesel car service introduced between Banbury and Bletchley. **1958** 'N. SHUTE' *Rainbow & Rose* i. 7 They've got a diesel-engined boat.

dieselize (dī·zəlaiz), *v.* orig. *U.S.* [f. *DIESEL, DIESEL + -IZE.] *trans.* To equip with a diesel engine or with diesel-electric locomotives. So **die:seliza·tion**; **die·selized** *ppl. a.*

1946 in *Amer. Speech* (1947) XXII. 158/2 The Burlington's main line passenger trains are largely dieselized. **1949** *Time* 30 May 38/1 He..hopes to cut costs more by dieselizing the entire [railway] line. **1950** *Engineering* 24 Mar. 328/1 In the U.S.A., where 'dieselisation' has gained..a foothold for long hauls. **1958** *Economist* 1 Feb. 381/2 The area needs an integrated system of road transport and dieselised railways. **1958** *New Statesman* 15 Nov. 668/3 Priority must be given to those schemes bringing quickest results, e.g., dieselisation before electrification. **1970** *Railway Mag.* Oct. 550/2 Some interesting combinations of locomotives could be observed on this train before dieselisation. *Ibid.* 553/1 The North Island system has been entirely dieselised for some years.

diet, *sb.*^{1} **7. diet-sheet,** a paper showing a daily diet, esp. for the inmates of an institution.

c **1863** FLORENCE NIGHTINGALE in C. Woodham-Smith *F. Nightingale* (1950) xvii. 397 Revised diet sheets for Troop-Ships. **1902** J. H. M. ABBOTT *Tommy Cornstalk* 182 We all lay in, or upon, our beds with our board-mounted diet-sheets in our hands. **1918** W. OWEN *Let.* 22 July (1967) 566, I see an old Diet Sheet on the table. **1967** *Vogue* Jan. 51/1 This is the Diet sheet. Vary meat, eggs, chicken and fish..to avoid boredom.

† **dietheroscope** (dəi,iþe·rŏskō^{u}p). *Obs.* [ad. It. *dieteroscopio* (G. Luvini 1874, in *Atti d. R. Accad. d. Scienze Torino 1873–74* IX. 389), f. Gr. δι- DI-^{2} + αἰθήρ ETHER + -O + -SCOPE.] An instrument for measuring variations in atmospheric refraction, usually consisting of a telescope having additional lenses or mirrors which bring two images of any object into the field of vision.

1877 *Q. Jrnl. Meteorol. Soc.* III. 414 The dietheroscope serves to show every little change of position, apparent or real, of any distant object. Its principal application is the measurement of the variations of atmospheric refraction. **1889** *Times* 21 Mar. 3/3 There are also Luvini's dietheroscope for observing the changes of atmospheric refraction optically; [etc.].

diethyl. Add: **2.** *diethylstilbœstrol* (or *-bestrol*), $C_{18}H_{20}O_2$, a synthetic compound having œstrogenic properties; = *STILBŒSTROL.

1938 E. C. DODDS et al. in *Nature* 5 Feb. 248/2 In this way 4:4'-*dihydroxy-αβ-diethylstilbene* (diethylstilbœstrol) was obtained. **1943** *Thorpe's Dict. Appl. Chem.* (ed. 4) VI. 272/2 Diethylstilbœstrol..confers the mating instinct on ovariectomised animals. **1944** *Ann. Reg. 1943* 359 Carcinoma of the prostate was successfully treated by oral administration of the synthetic œstrogen diethylstilbœstrol. **1969** *Daily Tel.* 18 Dec. 15/8 Heavy doses of the synthetic oestrogen diethylstilbestrol are being given to prevent pregnancy in rape victims.

dietitian (dəi,ĕti·ʃăn). Delete *rare* and add later examples. Also dietician.

1905 *Springfield Weekly Republ.* 29 Dec. 16 The dietitian's work consists in ordering food and preparing the dietary for the patients. **1906** *Daily Chron.* 14 Aug. 5/3 An ex-graduate of Harvard, and an experienced dietician. **1928** *Daily Express* 2 Aug. 2/3 An outline of each day's menu is supplied to the steward's department by the dieticians on board. **1960** *Act* 8 & 9 *Eliz. II* c. 66 § 1 Chiropodists, dietitians, medical laboratory technicians. **1971** *Daily Tel.* 18 June 4/5 His team includes dieticians, nutrition experts and engineers.

Dietl's crisis (dī·t'lz krəi·sis). *Path.* [f. the name of Joseph *Dietl* (1804–1878), physician of Cracow.] (See quots.) Freq. pl., *Dietl's crises*.

1897 *Internat. Clinics* IV. 94 The symptoms..—sudden, great abdominal pain, vomiting, distention of the belly, tenderness, and signs of collapse—were strongly suggestive of Dietl's crises. **1907** *Practitioner* Dec. 750 In such cases [of movable kidney] as present symptoms, the severity of these may present every gradation from a dull aching pain in the back to the acutest paroxysms of renal pain. To the latter the name of Dietl's crises is commonly applied. **1910** E. L. KEYES *Dis. Genito-Urinary Organs* xliv. 494 The tumor..is usually very painful and tender, and its growth is often attended by renal colic (Dietl's crises). **1953** A. W. BADENOCH *Man. Urol.* xii. 349 The so-called Dietl's crisis due to twisting of the pedicle producing torsion of the kidney is also very rare and I have not seen a case.

dif, diff, *sb.* and *a. colloq.* **A.** *sb.* Abbrev. of DIFFERENCE *sb.*

1896 ADE *Artie* 24 He said he didn't know her. 'What's the diff?' I says. **1909** J. R. WARE *Passing Eng.* 108/2 There is a great diff between a dona and a mush. You *can* shut up a mush (umbrella) sometimes. **1910** O. JOHNSON *Varmint* iv. 58 'What's the diff?' said the Tennessee Shad, yawning. **1914** S. LEWIS *Our Mr. Wrenn* vi. 62 It wouldn't make any dif. what they met as long as they was fighting together. **1919** *Punch* 28 May 417/1 But Lallie couldn't see the dif between a man and a novelist. **1967** 'H. CALVIN' *Nice Friendly Town* ix. 123 Here today gone tomorrow. What's the diff, duckie?

B. *adj.* Abbrev. of DIFFERENT *a.*

1908 *Punch* 10 June 416/3 They're not *really* a bit like themselves, but are something quite dif.! **1924** LAWRENCE & SKINNER *Boy in Bush* v. 61 'He's different from the rest of you, and his lingo's rotten.' 'He's not dif!' said Tom.

difference, *sb.* Add: **6. difference limen** or **threshold** [tr. G. *unterschiedsschwelle* (G. T. Fechner *Elemente der Psychophysik* (1860) I. x. 239)], the amount by which two stimuli or sensations must differ for the difference to be perceived; also, the degree of ability to perceive differences between stimuli.

1876 J. WARD in *Mind* I. 459 This 'relativity' between difference-threshold and stimulus [may] be due to something not in any sense psychical. **1895** E. B. TITCHENER tr. *Kuelpe's Outl. Psychol.* 160 With an area of contact of 1 mm. diameter, the difference limen on the index finger of the right hand was $\frac{1}{10}$ to $\frac{1}{20}$. **1902** W. JAMES *Var. Relig. Exper.* vi. 135 When one is sensitive to small differences in any order of sensation, we say he has a low 'difference-threshold'—his mind easily steps over it into the consciousness of the differences in question. **1953** C. E. OSGOOD *Method & Theory Exper. Psychol.* (1956) I. ii. 73 Absolute and difference thresholds are correlated in hearing, however, both varying concomitantly as functions of frequency.

different, *a.* Add: **1. c.** *colloq.* Out of the ordinary, special, *recherché*.

1912 D. F. CANFIELD *Squirrel Cage* iii, What a perfectly lovely couch...Why, it is so beautifully *different!* *Ibid.* xviii, To avoid being 'queer' and 'different' one had to play a good hand [at Bridge]. **1930** *Publishers' Weekly* 8 Feb. 709/1 They are always striving to write a piece of copy that will be 'different', that 'will hit the reader right between the eyes'. *Ibid.* 15 Mar. 1554/1 A 'different' book ad appeared in the Sunday, March the 9th, New York *Herald Tribune*. **1930** *Week-end Rev.* 7 June 467 Ireland this year! For a 'different' holiday, with all the charms of foreign travel and none of the disadvantages. **1965** *New Statesman* 7 May 712/2 The *Mail*'s bold, different typography.

differentiability (diferĕnʃiăbi·liti). [f. DIFFERENTIAB(LE *a.* + -ILITY.] Capability of being differentiated.

1909 in *Cent. Dict.* Suppl. **1925** G. A. BLISS *Calculus of Variations* v. 144 Hilbert's differentiability condition. **1947** *Mind* LVI. 48 The *f*'s must be single-valued, but no restrictions of continuity or differentiability are implied. **1964** E. A. POWER *Introd. Quantum Electrodynamics* vi. 78 The usual differentiability conditions of the calculus of variations are required.

differential, *a.* and *sb.* Add: **A.** *adj.* **3.** *differential analyser*, a machine designed for solving differential equations.

1931 D. BUSH in *Jrnl. Franklin Inst.* CCXII. 447 The differential analyzer. A new machine for solving differential equations. **1957** *Technology* July 188/1 The analogue laboratory contains a large electro-mechanical differential analyser. **1962** *Gloss. Terms Automatic Data Proc.* (*B.S.I.*) 8 *Differential analyser*, (1) an analogue computer using interconnected integrating gears to solve differential equations; (2) by extension, any computer designed primarily for solving differential equations.

4. b. *spec.* Applied to mechanisms devised for imparting differing velocities, e.g. to the two halves of the driving axle of a motor vehicle (so that the wheels revolve at different rates when turning a corner).

1888 [in Dict.]. **1902** A. C. HARMSWORTH *Motors & Motor-Driving* x. 213 The differential gear acts on the principle of the action of the pair-horse whippletree and equalising bar. **1903** *Sci. Amer.* 7 Feb. 91/2 Single-chain drives are more popular than ever, although..it looked as if the double outside chain drive to both rear wheels from a differential countershaft would supplant this form. **1904** A. B. F. YOUNG *Compl. Motorist* iii. 62 A separate shaft, parallel with the rear axle, called the differential shaft, driven by bevel gearing from the secondary shaft in the gear case. **1925** *Morris Owner's Man.* 26 If any adjustments to the differential bearings are required. *Ibid.*, This adjusting sleeve is easily accessible through an opening on the upper side of the differential carrier. **1967** E. RUDINGER *Consumer's Car Gloss.* (ed. 2) 32 The differential gears are in the middle of the rear axle.

B. *sb.* **3. b.** The difference in wages between one class of workmen and another (esp. between skilled and unskilled workmen) or between one industry, etc., and another; a difference in prices of similar products, services, etc.

1941 *Economist* 8 Feb. 178/2 The present economic differentials between one occupation and another and one status and another are, over a wide field, calculated to make people stay where they are. **1950** D. W. BROGAN *Era of F. D. Roosevelt* v. 100 Controversy arose over wage differentials between North and South. **1955** *Times* 17 June 4/4 This trend would lift the skilled differential and the future might show a relative improvement in payment for skill. **1959** *Punch* 19 Aug. 30/2 The rise of Oundle's fees to £435 puts the school within £25 of Eton...What does Eton do now? Does it preserve its status by leaping ahead.., or does it get the Headmasters' Conference to negotiate a national agreement on differentials? **1965** *New Statesman* 23 Apr. 634/2 The major companies have reacted with countrywide cuts of 1d., 1½d. or 2d. off certain grades [of petrol]. Some..cut-price companies have followed suit, to try to keep their 3d. differential.

4. A differential gear, *spec.* of a motor vehicle.

1902 *Daily Chron.* 30 June 6/2 He broke the differential of his 70 h.-p. Panhard car 50 kilometres from the finish. **1905** *Westm. Gaz.* 17 Feb. 8/2 A motor-boat requires no differential. **1930** J. B. PRIESTLEY *Angel Pavement* ii. 72 Wanted new plugs and mag. and brakes re-lining and something doing to the differential, and just cleanin' up a bit. **1959** *Motor Man.* (ed. 36) iv. 94 When a rigid rear axle is used, the differential naturally moves up and down in relation to the chassis whenever the wheels meet a bump.

differentiation. Add: **1.** *Biol.* Esp. *dependent differentiation* and *self-differentiation* (see quot.).

1926 J. S. HUXLEY *Ess. Pop. Sci.* xviii. 269 The future differentiation of a particular region at a particular moment may be of two kinds. It may be entirely independent of other organs, in which case we speak of self-differentiation, or it may depend upon some other organ, when it is called dependent differentiation.

b. *Philol.* The action of differentiating or fact of being differentiated: used of two adjacent phonemes, the meanings of synonyms, etc.

1925 P. RADIN tr. *Vendryès's Language* I. iii. 60 The two contiguous phonemes..take advantage of their differences to exaggerate them until they no longer have anything in common. This is the process of *differentiation* as opposed to accommodation. **1926** FOWLER *Mod. Eng. Usage* 114/1 *Differentiation*. In dealing with words, the term is applied to the process by which two words that can be used indifferently in two meanings become appropriated one to one of the meanings & one to the other. **1928** *S.P.E. Tract* XXXI. 323 Differentiation..consists in giving deliberate help to that natural process of language which is always tending to give different meanings to..synonyms. **1934** M. K. POPE *From Lat. to Mod. Fr.* II. i. 64 Differentiation is at times due to the same instinct, as for instance when plosive + plosive is differentiated to fricative + plosive. **1939** L. H. GRAY *Found. Lang.* iii. 69 Dissimilation..may be..either contiguous (differentiation) or incontiguous (dissimilation proper).

differentiator (difĕre·nʃiē^{i}təɪ). [f. DIFFERENTIATE *v.* + -OR.] One who or that which differentiates; spec. (*a*) *Biol.* an organ, part, etc., that stimulates or controls differentiation; (*b*) a device that produces an output proportional to the rate of change (with respect to time or another variable) of the input.

1889 *Cent. Dict.* s.v., The radicals of written Chinese serve as differentiators of the sense. **1894** *Times* 11 Aug. 11/1 It was..advisable to have a 'differentiator'..and this was difficult to construct, because any irregularities in a curve were magnified in its differential coefficients. **1924** *Discovery* June 77/1 Some influence..made the tissues of the host build themselves up in the special and orderly way of differentiation. It was a differentiator. **1969** P. B. JORDAIN *Condensed Computer Encycl.* 156 An ordinary transformer produces an output voltage proportional to the rate of change of the input current, but its use as a differentiator is not common.

differentness. Delete *rare* and add further examples.

1934 F. SCOTT FITZGERALD *Tender is Night* I. vi. 36 His manner was softer than at the studio, as if his differentness had been put on at the gate. **1966** G. N. LEECH *Engl. in Advertising* xxii. 200 These deviant styles derive their power mainly through their 'differentness'. **1967** *Economist* 19 Aug. 657 A demonstration of differentness? A badge of his unhappiness?

difficult, *a.* **2. a.** Delete *arch.* and add further examples (after F. *difficile*). Cf. DIFFICILE *a.* 3.

1904 *Westm. Gaz.* 20 Jan. 3/2 Lady Verona refers to her husband as 'rather difficult'. **1929** *Times* 2 Feb. 10/1 A letter from a 'difficult' customer. **1960** J. FINGLETON *Four Chukkas* 2 A touring team is infinitely better off without the 'difficult' player.

difficultness. Delete †*Obs.* and add later examples.

1934 C. S. FORESTER *Peacemaker* 19 He was experienced now in Mary's 'difficultness'. **1947** *Trans. Philol. Soc. 1946* 195 The growth in more serious shape of that difficultness—at times almost suggestive of nascent 'persecution-mania'—which prevented any happy relationships with almost all of his academic colleagues.

diffractometer (difrækto·mitəɪ). [f. DIFFRACT *v.* + -O + -METER.] An instrument for measuring diffraction; esp. an instrument used in diffraction analysis in Crystallography.

1909 in WEBSTER. **1926** *Proc. R. Soc.* B. XCIX. 275 Certain size changes of erythrocytes have been followed under the 'diffractometer'. **1954** *Philips Techn. Rev.* XVI. 123 The new instrument, which from now on will be termed an X-ray diffractometer..comprises..a basic diffraction unit..a Geiger counter goniometer, and an electronic circuit panel with automatic recorder. **1955** *Gloss. Terms Radiol.* (*B.S.I.*) 29 *Neutron diffractometer*, an instrument..for measuring with a neutron beam..the intensities of the diffracted beams at different levels. **1960** *Times Rev. Industry* May 31/1 Crystal structure can be examined by means of a diffractometer. **1961** *Engineering* 15 Sept. 333 Our instrument, known as the linear X-ray diffractometer, incorporates an analogue computer, which computes the crystal and detector settings for each of the many thousands of X-ray reflections that have to be measured for a complete determination of atomic structure. **1970** *New Scientist* 21 May 383/1 A new type of neutron diffractometer—the 'hedgehog' neutron diffraction spectrometer—has been developed.

diffractometry (difrækto·mĕtri). [f. prec.: see -METRY.] The measurement of diffraction; the use of a diffractometer.

1956 *Jrnl. Applied Physics* XXVII. 1215 (*heading*) Comparison of X-ray wavelengths for powder diffractometry. **1965** W. PARRISH *X-ray Analysis Papers* p. vii, Specialized topics such as..automatic single crystal diffractometry.

diffuse, *a.* Add: **2. f.** In Forestry, *diffuse-porous* adj., applied to woods in which the pores are scattered evenly throughout the growth ring.

1902 F. ROTH *First Bk. Forestry* III. 222 The diffuse-porous woods, like maple, yellow poplar, and cherry, where pores are usually very small and evenly scattered through the annual ring. **1928** *Forestry* II. 65 Under sub-alpine conditions woods which are normally diffuse-porous tend to become ring-porous. **1953** H. L. EDLIN *Forester's Handbk.* ii. 27 In some trees, those with diffuse-porous wood, they [*sc.* the vessels] are scattered evenly through the tissues.

diffused, *ppl. a.* Add: **2.** *diffused lighting* (see quot. 1926).

1926 *Gloss. Terms Electr. Engin.* (*B.S.I.*) 146 *Diffused lighting*, a system of lighting in which the luminous flux, after passing through a diffusing medium, reaches the area to be illuminated, in part directly and in part indirectly. **1933** *Discovery* Aug. 251/2 Strip-lighting which flood-lights the walls and gives a diffused lighting to the room.

diffuser. Add: **2. b.** *Engin.* A duct in a centrifugal pump, compressor, wind tunnel, etc., so shaped that it reduces the velocity of a fluid by increasing the cross-sectional area of flow.

1926 W. J. KEARTON *Turbo-Blowers & Compressors* vii. 186 The diffusers are of the 'free vortex' type, no guide vanes being fitted. **1946** J. G. KEENAN *Gas Turbines & Jet Propulsion* iv. 68 The diffuser is the name given to a ring of fixed guide vanes..surrounding the impeller. *Ibid.* 70 One object of the diffuser is to provide a smooth canalization of the flow from the impeller. **1959** D. D. WYATT in *High-Speed Aerodyn. & Jet Propul.* XII. E. § 33. 350 The jet diffuser..consists of a divergent shroud attached to the outlet end of the jet nozzle.

diffusion. Add: **2. e.** Formerly used as a semi-technical term in psychological writings: the arousal of a widespread response by a stimulus; the dissemination of nervous energy.

1859 A. BAIN *Emotions & Will* I. 10 Where feeling exists there must be a free diffusion of nervous energy over the brain and its outlying connexions. **1890** W. JAMES *Princ. Psychol.* II. xxiii. 373 There are probably no exceptions to the diffusion of every impression through the nerve-centres. **1918** J. WARD *Psychol. Princ.* I. iv. 79 This 'diffusion' or 'irradiation'..diminishes as we pass from the class of organic sensations to the sensations of the five senses.

3. b. *Anthropol.* The spread of elements of a culture or language from one region or people to another; also, the simultaneous existence of such elements in two or more places.

1871 E. B. TYLOR *Prim. Culture* I. i. 8 How good a working analogy there really is between the diffusion of plants and animals and the diffusion of civilization, comes well into view when we notice how far the same causes have produced both at once. **1937** R. H. LOWIE *Hist. Ethnol. Theory* xii. 223 Repudiating the notion that contact automatically precipitates diffusion. *a* **1942** B. MALINOWSKI *Sci. Theory Culture* (1944) iii. 17 The other dominant tendency of older anthropology laid primary stress on diffusion, that is, the process of adopting or borrowing by one culture from another various devices, implements, institutions, and beliefs.

diffusionist. Add: (Further examples.) Also *attrib.* or as *adj.* Cf. *DIFFUSION 3 b.

1926 *Encycl. Brit.* III. 566/2 As substitutes for the 'evolutionary' syntheses.., several 'diffusionist' schemes have been elaborated. **1931** *Times Lit. Suppl.* 5 Mar. 167/2 The diffusionist school..has done good work..by attempting the stratigraphical analysis of a given 'culture-complex'. **1934** *Nature* 12 May 707/2 Other schools, the 'diffusionists', the 'functionalists', [etc.]. **1946** J. S. HUXLEY *Unesco* ii. 46 The diffusionists on the one hand and on the other their opponents who believe in parallel and independent evolution of cultural patterns. **1951** E. E. EVANS-PRITCHARD *Social Anthropol.* iii. 46 The so-called evolutionary theories.. were attacked from two directions, the diffusionist and the functionalist. *Ibid.*, Those who became known as diffusionist anthropologists. *Ibid.* 47 Diffusionist anthropology is still predominant in America. **1961** L. MUMFORD *City in History* iii. 91 The old diffusionists, like G. Elliott Smith, who jumped too quickly at an answer, cast discredit on the question.

Hence **diffu·sionism** *Anthropol.*, the theory that all or most cultural similarities are due to diffusion.

1937 R. H. LOWIE *Hist. Ethnol. Theory* xi. 186 We.. concede..some of the strong points of German diffusionism. *a* **1942** B. MALINOWSKI *Sci. Theory Culture* (1944) iii. 21 The basis of diffusionism must be a correct identification of cultural realities plotted on the map.

dig, *v.* Add: **1. a.** *spec.* To make an archæological excavation (cf. quot. *c* 1755 in Dict.).

1740 W. STUKELEY *Stonehenge* x. 43 (*heading*) How the body is posited. What has been found in digging into these barrows. **1907** E. WHARTON *Fruit of Tree* I. iii. 32 As an archaeologist..I should really like to come here and dig. **1911** T. E. LAWRENCE *Lett.* (1938) 125 My orders are to.. bring out Woolley (new chief), and the stores and dig for three months. **1912** *Ibid.* 136, I would like to dig in the Persian gulf.

c. Hence, to understand. Cf. sense *6 c. *slang* (orig. *U.S.*).

1936 *N.Y. World Telegram* 6 Oct. 16/1 'You dig?' is a short cut for 'You understand?' **1952** B. ULANOV *Hist. Jazz Amer.* 344 The man who really 'digs' can more often than not describe the next development in jazz before the musicians have reached it. **1957** C. MACINNES *City of Spades* I. xi. 89 Twist now—you dig?

d. To have 'diggings'; to lodge. *colloq.*

1914 C. MACKENZIE *Sinister St.* II. III. xi. 717 Soon it would come to the point of declaring outright that he did not want to dig with him. **1919** W. T. GRENFELL *Labrador Dr.* (1920) iv. 64 Two or three classmates would 'dig' together.

e. To make incisions with action resembling digging.

1930 W. A. THORPE in *Connoisseur* Oct. 226/2 To produce relief motives the operator has to 'dig' with his wheel at a steep angle to the surface.

3. a. *spec.* To excavate archæologically.

1743 W. STUKELEY *Abury* xvi. 92 The very same appearances as I had so often seen, in digging the barrows about Stonehenge and Abury. **1912** T. E. LAWRENCE *Lett.* (1938) 134 The right way to dig a temple. **1949** W. F. ALBRIGHT *Archaeol. of Palestine* ii. 41 The initial plan to dig the great site systematically..had to be abandoned because of the prohibitive expense. **1968** R. L. S. BRUCE-MITFORD *Sutton Hoo Ship-Burial* 18 It was the richest treasure ever dug from British soil.

6. c. *slang* (orig. *U.S.*). (*a*) To understand, appreciate, like, admire; (*b*) to look at or listen to; to experience. Cf. sense *1 c.

1935 *Hot News* Sept. 20/2 If you listen enough, and dig him enough, you will realise that that..riff is the high-spot of the record. **1941** *Life* 15 Dec. 89 Dig me? **1943** M. SHULMAN *Barefoot Boy with Cheek* 90 Awful fine slush pump, I mean awful fine. You ought to dig that. **1944** C. CALLOWAY *Hepsters Dict.*, Dig v.—(1) Meet. (2) Look, see. (3) Comprehend, understand. **1944** M. ZOLOTOW *Never whistle in Dressing Room* iii. 52 When they see a pretty girl they shout, 'Dig the chick.' **1947** R. DE TOLEDANO *Frontiers of Jazz* p. x, I recognize it when I see it, the same as I dig good Jazz when I hear it. **1949** L. FEATHER *Inside Be-Bop* iii. 28 Dizzy didn't dig the band's kind of music and the band didn't dig Dizzy. **1958** *Punch* 8 Jan. 92/1 The lines of communication get tangled. In other words the people don't quite 'dig' you. **1958** *Listener* 29 May 912/1 He wants to 'dig' the whole of life, and is convinced that experience comes only to the irresponsible. **1958** *Punch* 25 June 853/3 Does the beat generation really dig such crazy old-world catch-phrases? **1959** C. MACINNES *Absolute Beginners* 60 If you *like* the other number, I mean like the looks of them, really dig them sexually. *Ibid.* 62 Everything you learned, you hadn't learned until you'd really dug it: i.e., made it part of your own experience. **1960** N. MITFORD *Don't tell Alfred* xviii. 192 Of course he's a man's man, you might not dig him like we do. **1969** *New Yorker* 29 Nov. 48/1, I just don't dig any of these guys. I don't understand their scenes.

10. dig down. c. *intr.* To pay money from one's own pocket. *U.S. colloq.*

1942 in BERREY & VAN DEN BARK *Amer. Thes. Slang* § 550/3. **1951** J. STEINBECK *Log from 'Sea of Cortez'* (1958) p. xxvi, She was a wise and tolerant pushover for any hard-luck story...Even when she knew it was a fake she dug down.

11. dig in. c. Colloq. phr. *to dig in one's feet, heels, toes*: to adopt a firm position; to keep resolutely or obstinately to one's decision, opinion, attitude, etc.

1933 *Punch* 16 Aug. 174/1, I am prepared to declare mosques open and to grace the inaugurations of new caravanserais, though I personally have no taste for ritual. But at that point I dig in my toes. **1941** L. A. G. STRONG *Bay* 179 One thing I had dug in my heels over was the church I went to. **1956** N. COWARD *South Sea Bubble* I. i. 8 You jumped at him before he had time to get his breath and now he's dug his feet in.

d. *intr.* or *refl.* To fix oneself firmly in a position; *spec.* (*a*) *Mil.*, to excavate a trench or the like in order to withstand an attack or consolidate a position; (*b*) *Cricket*, to consolidate one's position as a batsman.

1851 *Knickerbocker* XXXVIII. 183 [The crab] pinched, scratched, 'dug in', and held on. **1917** A. G. EMPEY *From Fire Step* 145 The machine-gunners went over with the fourth wave to consolidate the captured line, or 'dig in', as Tommy calls it. **1919** J. B. MORTON *Barber of Putney* xvi. 263 Word came back that they [*sc.* a platoon] were to go to a certain point and dig in. **1922** *Daily Mail* 21 Nov. 8 The most alarming of Sir Percival Phillips's disclosures is that our 'limpets' in Mesopotamia are digging themselves vigorously in. **1934** C. DAY LEWIS *Hope for Poetry* vii. 41 D. H. Lawrence dug himself in in the Unconscious. **1944** BLUNDEN *Cricket Country* iv. 49 Such a side is free from the solemn rule of 'digging in' which big cricket prescribes. **1949** *Manch. Guardian Weekly* 14 July 2/4 The policy of stimulating expansion rather than digging in to protect the status quo. **1959** *Times* 29 May 4/4 Watson was bowled by the second ball he received. But..Phillips dug in with gallant determination.

e. *intr.* To set to work earnestly and energetically; to work hard. *dial.* and *U.S. colloq.*

[**1877** F. ROSS et al. *Gloss. Holderness* 53/2 *Dig-intiv* it, lads, and you'll seean get it deean.] **1884** 'MARK TWAIN' *Huck. Finn* xxxviii, We got to dig in like all git-out. **1951** F. S. ANTHONY *Me & Gus* (1953) 26 We'll dig in like niggers, Mark, and show those old jokers over the fence how to smack up wood.

f. To begin eating, esp. heartily. *colloq.*

1912 *Dialect Notes* III. 574 Dig in and help us eat the rest of this turkey. **1952** A. BARON *With Hope* IV. ii. 119 Sit down and dig in. Your grub's getting cold.

13. dig out. a. *fig.* To obtain, get hold of, or get out by search or effort.

1864 [in Dict.]. **1877** *Gentl. Mag.* CCXL. 596 This last-named prince..had hidden himself in a cupboard in the midst of a roll of carpet, and was with difficulty dug out to be girt with the sword of Othman. **1887** *Harper's Mag.* May 884/2, I don't believe it is worth while to dig out the glasses. **1929** 'P. WILLIAMS' *Jacob's Ladder* xix. 269 It was Carolyn who..dug out two old volumes of eighteenth century pictures lying forgotten in a cupboard. **1930** *Daily Express* 30 July 3/7 England..may need more batsmen of the type who have to be dug out. **1968** *Globe & Mail* (Toronto) 3 Feb. B2/2 The Europeans..can dig out many reasons for their shortcomings.

14. dig up. a. Also, to obtain, find, search out. Cf. *13 a. Now *colloq.*

1611, 1861 [in Dict.]. **1895** *Century Mag.* Sept. 674/1, I heard he was tryin' to dig up a trade with a man who's got a mine over in the Slocan country. **1909** 'O. HENRY' *Options* 50 Ogden digs up a deck of cards, and we play casino. **1959** I. & P. OPIE *Lore & Lang. Schoolchildren* iii. 53 'Tell us news, not history.' 'Where did you dig that up?'

dig, *sb.*^{1} Add: **1. b.** *colloq.* An archæological excavation; an expedition for the purpose of an archæological excavation.

1896 A. J. EVANS in *Academy* 13 June 494/1 He showed me several clay bulls..obtained through his dig. **1908** *Chambers's Jrnl.* July 527/1 They [*sc.* tomb-hunters] speak of the different excavations as 'digs'. **1911** T. E. LAWRENCE *Lett.* (1938) 112 We are only two on this dig. **1940** 'M. INNES' *Secret Vanguard* vi. 56 The dig at Dabdab must be completed before the rains. **1957** K. M. KENYON *Digging up Jericho* 39 The first stages of a dig..start long before one actually gets down to excavating. **1969** *Times* 23 Jan. 13/1 The many archaeological digs that have been going on since the beginning of the Aswan Dam project.

6. *pl.* Lodgings (cf. DIGGING *vbl. sb.* 5); also occas. as *sing.* (and *Comb.*). *colloq.*

1893 *Stage* 11 May 16/2 'Being in the know' regarding the best 'digs' can only be attained by experience. **1905** *Varsity* 16 Nov. 79/1 An invitation from a friend in digs. **1908** A. S. M. HUTCHINSON *Once Aboard the Lugger* I. i. 27, I have heard that one can work far better by living near the hospital in digs. **1916** W. OWEN *Let.* 6 Nov. (1967) 414, I like this digs far better than the Queen's Hotel life. **1959** A. LEJEUNE *Crowded & Dangerous* vii. 78 His old digs.. where he lived when he used to work for us. **1959** *Times* 6 Mar. 12/6 Dig-hunting is a serious business...The good woman is offering you a room in her home for a whole year.

dig (dig), *sb.*^{3} *Austral.* and *N.Z. colloq.* Abbrev. of *DIGGER 2 f.

1918 *Chrons. N.Z.E.F.* 21 June 221/2 Be shrewd, sweet Dig. *Ibid.* 22 Nov. 198/1 'How far's the war, Dig?' was the first question we asked. **1933** *Bulletin* (Sydney) 18 Oct. 10/3 He gave his verdict: 'That's good, Dig.; that's strong.' **1946** E. G. WEBBER *Johnny Enzed in Italy* 4 The war has finished and a large number of Old Digs have bustled their way down various gangways. **1965** G. MCINNES *Road to Gundagai* ii. 25 Often they shouted at us..'Howsit up in the dress circle, dig?'

Digby (di·gbi). [The name of a seaport of Nova Scotia.] A dried or cured herring of a type caught at Digby. In full *Digby chicken* or *chick*.

1829 G. HEAD *Forest Scenes Wilds N. Amer.* 40 A small species of herring...They are extremely delicate, and are salted in great quantities every year. They have gained the nick-name of Digby chickens. **1862** *Chambers's Encycl.* III. 557/2 A variety of small herrings or pilchards, which are smoked and dried for export; they have a high flavour, and are known in trade as Digbies. **1883, 1887** [see CHICK *sb.*^{1} 4]. **1958** *Encycl. Canadiana* III. 267/1 Digby is noted..for fish, including cured herrings known as 'Digby chickens'.

digest, *sb.* Add: **1. b.** *spec.* A periodical composed wholly or mainly of condensed versions of articles, stories, etc., previously published elsewhere. Also *attrib.*

1922 (*title*) Reader's Digest. **1946** (*title*) The Literary Digest. A monthly magazine of popular literary interest. **1957** *Times Lit. Suppl.* 11 Oct. 604/4 The Chaplain's loving biography has been edited on the principles of 'condensed reading', as popularized by the 'digest' kind of magazine. **1958** J. CANNAN *And be a Villain* i. 6 Mad ideas they'd got from medical articles in Digests. **1967** G. STEINER *Lang. & Silence* 258 A time of fantastic intellectual cheapness,..the century of the book club, the digest, and the hundred great ideas on the instalment plan.

digester. Add: **4. d.** *Papermaking.* An apparatus in which wood, grass, etc., are turned into pulp by the action of hot water, chemicals, etc.

1898 *Sci. Amer.* LXXVIII. 185/3 The digester is filled to the top with chips..and the acid is then piped in. **1906** *Chambers's Jrnl.* Feb. 206/2 These pieces [of wood]..are fed into a 'digester', which in some cases is large enough to produce fifteen tons of pulp at one operation. **1962** F. T. DAY *Introd. to Paper* i. 13 The principle was to boil the wood with soda in digesters.

‖**digestif** (diʒestif). [Fr.] = DIGESTIVE *sb.* 1.

1908 *Daily Chron.* 15 Apr. 4/6 My husband thought a cigarette an excellent digestif. **1934** H. MILLER *Tropic of Cancer* 164 She will be obliged..to flop somewhere on the boulevard and sip her *digestif*. **1962** *Guardian* 13 July 8/1 Drambuie..ranks with Benedictine and Cointreau among the world's five top-selling digestifs.

digestive, *a.* and *sb.* Add: **A. 2. b.** *spec.* Designating a type of wholemeal biscuit. Also *ellipt.*

1876 *Off. Guide Cunard Steamship Co.* 158 (Advt.), Huntley & Palmers..Digestive Biscuits. **1894** L. HERITAGE *Cassell's New Universal Cookery Book* 1022 *Digestive Biscuits.*—Required: a pound of finely-granulated, or.. wholemeal [flour], half a pound of white flour [etc.]. **1925** ST. JOHN ERVINE *Anthony & Anna* I. 24 Which will you 'ave, sir—the charcoal or the digestive biscuits? **1935** G. BULLETT *Jury* xxiii. 234, I had to open a parcel of grocery to find the right kind of biscuits...Digestives were what she fancied. **1949** N. MITFORD *Love in Cold Climate* II. ii. 194 A few rather broken digestive biscuits. *Ibid.* 197 Weeks since I tasted digestives, my favourite food, too. **1955** *Observer* 19 June 7/5 Biscuits long known as *digestive* are being renamed *sweetmeal*.

digger. Add: **2. b.** Also *attrib.*

1881 H. W. NESFIELD *Chequered Career* vii. 75 Their manner of accosting me was simply their 'digger' style of humour. **1894** C. J. O'REGAN *Voices of Wave & Tree* 10 But that was digger nature.

d. In modern use, a member of a group of hippies who believe in a society where all food and possessions are shared freely and land is cultivated to feed the poor. Also *attrib.*

1967 G. LEGMAN *Fake Revolt* 19 Try to round up votes and get a bearded Digger or mock-saintly Provo elected mayor. **1967** *Economist* 15 July 217/2 A loosely knit group called the Diggers—taking the name from agrarian communists in the seventeenth century who cultivated waste lands to feed the poor—is operating pads for homeless hippies and dispensing free food and clothing, obtained by soliciting contributions from shopkeepers. **1968** *Guardian* 29 Apr. 7/5 The first Digger conference opened at London Anti-university over the weekend. **1969** *Ibid.* 27 Sept. 9/1 A leader of the London Diggers..describes his group as 'communal hippies, nonviolent basically and nonauthoritarian'.

e. A person who digs for archæological purposes. Cf. *DIG *v.* 1 a and 3 a.

1911 T. E. LAWRENCE *Home Lett.* (1954) 149 Thompson is not a digger, so the direction of that part would be my share. **1914** KIPLING *Lett. Travel* (1920) 256 Their dream (the diggers' dream always) is to discover a virgin tomb. **1960** *Times* 22 Feb. 14/1 Woolley will always be remembered as one of the most successful diggers ever engaged in field archaeology.

f. *colloq.* An Australian or New Zealander; *spec.* in the wars of 1914–18 and 1939–45, a soldier from Australia or New Zealand, esp. a private soldier; freq. as a term of address = *COBBER *sb.*^{2} Also *attrib.* Cf. *DIG *sb.*^{3}

1917 *Chrons. N.Z.E.F.* 5 Sept. 28/1 He ain't no digger; that's the colonel or the sergeant-major. *Ibid.* 14 Nov. 154/2 Two hefty diggers escorted the little lady to her home. **1917** E. MILLER in *Camps, Tramps & Trenches* (1939) xxiv. 192 A digger officer would have worded the message quite differently. **1918** *Digger* 13 On the wing with the jam, Digger. **1919** [see *AUSSIE *sb.* and *a.* 1]. **1919** W. H. DOWNING (*title*) Digger dialects. A collection of slang phrases used by the Australian soldiers on active service. **1923** D. H. LAWRENCE *Kangaroo* v. 98 We're mostly diggers back from the war. **1929** C. C. MARTINDALE *Risen Sun* 14 Where my experience of the Diggers really began was a little club in the Turl, to which hospital cases came. **1940** *War Illustr.* 16 Feb. p. iii/1, They are forming a new Digger Expeditionary Force, but old blokes like me are not wanted. **1941** *Illustr. London News* CXCIX. 534/1 They are a mixed lot, these 'diggers', some from offices, others from factories. **1948** R. FINLAYSON *Tidal Creek* i. 8 Put your bag under the seat, digger. **1963** *Evening Post* (Wellington, N.Z.) 26 Dec., Remembering the old diggers at this festive time of the year.

6. digger plough, a plough that breaks down the furrow-slice by means of a projecting wing or continuation of the mould-board; **digger's delight** *Austral.*, a species of speedwell, *Veronica perfoliata*, so called from the supposition that it grows only on auriferous soil.

1935 *Hutchinson's Techn. & Sci. Encycl.* I. 38/2 Modern mould-board ploughs fall generally into one or other of two classes: general purpose ploughs..and digger ploughs, which leave the soil flat and broken. *Ibid.*, In the general purpose plough the share is a pointed wedge...In the digger plough the share is wider and steeper and has a concave upper surface. **1960** *Farmer & Stockbreeder* 23 Feb. 64/2 Both competitors will use two ploughs—a free digger plough on stubble and a semi-digger on a one-year ley. **1878** W. R. GUILFOYLE *1st Bk. Austral. Bot.* 64 Digger's Delight...A pretty, blue-flowering shrub with smooth stem-clasping leaves. **1888** D. MACDONALD *Gum Boughs* 147 Such native flowers as the wild violet, the shepherd's purse, or the blue-flowered 'digger's delight'.

digging, *vbl. sb.* Add: **4. b.** Archæological excavation, or the site of such an excavation.

1911 T. E. LAWRENCE *Let.* 31 Mar. (1938) 101 Digging results will appear in *The Times*. *Ibid.* 23 May 106 She was really too captious at first, coming straight from the German diggings at Kalaát Shirgat. **1938** D. GARNETT in *Lett. T. E. Lawrence* 40 It was in this bungalow that he lodged his Arab friends..when he brought them from the Carchemish diggings to visit England.

6. digging-party (see PARTY *sb.* 7 and 8); **digging plough** = *digger plough* (see *DIGGER 6); **digging-stick,** a primitive digging implement consisting of a pointed stick, sometimes weighted by a stone.

1853 MRS. C. CLACY *Lady's Visit to Gold Diggings of Australia* iii. 19 But for our digging party entire, which consisted of my brother, four shipmates, and myself, no accommodation could be procured. *Ibid.* 32 Four other of our shipmates had also joined themselves into a digging-party. **1923** KIPLING *Irish Guards in Gt. War* I. 55 Some Garrison gunners threw three bombs at an enemy digging-party. **1891** R. WALLACE *Rural Econ. Austral. & N.Z.* xix. 272 Howards Digging Plough..turns over the furrow slices most perfectly and breaks what is then left at the surface by means of a projecting wing or continuation of the mould-board. **1865** Digging-stick [in Dict.]. **1947** I. L. IDRIESS *Isles of Despair* xii. 82 Women..carrying their food baskets and digging sticks. **1959** TINDALE & LINDSAY *Rangatira* ii. 199 The hoe and digging-stick method of cultivating the soil [by the Maoris]. **1960** K. M. KENYON *Archæol. in Holy Land* ii. 49 Cultivation of the ground was probably carried out by digging-sticks, pointed sticks weighted by stones, of which the evidence survives in the form of heavy stones pierced by a hole.

diggish (di·giʃ), *a. Children's slang.* [f. *DIG *v.* 6 c + -ISH^1.] Excellent, splendid.

1963 *Daily Mail* 22 Jan. 6/3 Schoolchildren produce their own variations. Mr. Peter Opie..tells me that instead of merely saying they dig something, they describe it as *diggish*.

diggy (di·gi), *a. colloq.* [f. DIG *sb.*^{1} 4 b + -Y^1.] Inclined to give sly digs.

1904 *Daily Chron.* 26 Mar. 3/3 General Frey is very 'diggy' against what we should have thought he would have found to be equally fine representatives of our Indian Army—the Sikhs. **1906** *Ibid.* 23 Jan. 3/1 Our official German critics..are always very 'diggy' towards Buller.

digit, *sb.* Add: **3. b.** Freq. used *attrib.*, esp. of parts and data in mechanical calculators, digital computers, etc., as *digit counter*, *pulse*, etc.

1921 C. D. LAKE *U.S. Pat.* 1,372,965 2/2 If the hole in the card represents the numeral 5, the register wheel will

be turned five digit spaces. **1946** *Nature* 12 Oct. 500/2 'Digit trays' for carrying pulse groups representing numerical data from one unit to another. **1948** *Proc. Symposium Large-Scale Digital Calculating Mach.* 34 Numbers are transferred from one unit of the ENIAC to another by digit trunks, which carry eleven lines, one for each digit of the number, and one for the sign of the number. *Ibid.* 271 If we now look at the digit pattern in the register at the beginning of digit time 1, we will find that all the digits will have been advanced by one digit position, excepting digit number 1, which has been deleted. **1955** R. K. RICHARDS *Arith. Operations in Digital Computers* vii. 193 'Digit counter' will be used to describe the individual multistable counting elements. **1959** *Economist* 20 June 1110/2 In the search for higher computer speeds, one obvious approach was to raise the 'digit frequency' or pulse rate of the machine. **1959** C. V. L. SMITH *Electronic Digital Computers* x. 238 Some means is needed to reduce the four levels of 1's essentially to two...This is accomplished by a circuit called a 'digit resolver'. **1963** GOULD & ELLIS *Digital Computer Technology* iv. 29 Single logic elements can be made to work at digit rates of say 100 megacycles per second. *Ibid.* vi. 50 It is not uncommon to speak of pulses which represent unit signals as 'digit pulses'. *Ibid.* 51 Such circuits are also of use where a deliberate delay of one digit time is required, and are commonly called digit delays.

digital, *a.* and *sb.* Add: **A.** *adj.* **4.** Of, pertaining to, or using digits [DIGIT *sb.* 3]; *spec.* applied to a computer which operates on data in the form of digits or similar discrete elements (opp. **analogue computer*).

1938 C. CAMPBELL *U.S. Pat.* 2,113,612 9/1 The emitter.. differs from the other emitters in that it has twelve digital conducting spots. **1945** J. ECKERT et al. *Appl. Math. Panel Rep.* 171.2R (*title*) Description of the ENIAC and comments on electronic digital computing machines. **1946** D. R. HARTREE in *Nature* 12 Oct. 500/2 [Computers] of the other class handle numbers directly in digital form...The American usage is 'analogue' and 'digital' machines. **1947** *Electronic Engin.* XIX. 178 Digital computing apparatus works out problems by methods which are basically the same as those which would be employed if an attempt were made to perform the calculations on paper. **1947** *Math. Tables & Other Aids to Computation* Oct. 359 We are engaged at the RCA Laboratories in the development of a storage tube for the inner memory of electronic digital computers. **1957** *Technology* July 182/2 The digital computer in normal use plays a purely deductive role in that it follows explicit instructions in operating on data that are fed into it. **1958** *Engineering* 28 Mar. 389/2 An in-line display of a digital form, which shows the unknown voltage to four significant figures. **1958** *IBM Jrnl. Res. & Devel.* II. 191/1 Voltage-to-digital conversion is accomplished by connecting a precision potentiometer to the output shaft of a servo-driven converter. **1964** *Times Rev. Industry* Mar. 58/2 A digital clock that presents the hours and minutes in figures. **1968** *Brit. Med. Bull.* XXIV. 189/1 Of the various types of computer which play a useful part in medical practice and research,.. the most important..is the electronic digital computer.

digitalization (di:dʒitălaizē^i·ʃən). *Med.* [f. DIGITAL(IS 2 + -IZATION.] The administration of digitalis or one of its active constituents to a patient or an animal so that the required physiological changes occur in the body; also, the state of the body resulting from this.

1882 in *Syd. Soc. Lex.* **1919** *Jrnl. Amer. Med. Assoc.* LXXIII. 1823/1 Only after a considerable number of days will a patient on this dosage come to be much below the grade of digitalization at which he started. **1951** [see *DIGITALIZE *v.*^{1}]. **1961** *Lancet* 9 Sept. 573/2 Prompt return of nausea precluded further digitalisation. **1970** PASSMORE & ROBSON *Compan. Med. Studies* II. viii. 2/1 The patient is then said to be fully digitalized...Thereafter, digitalization is sustained by the maintenance dose.

digitalize (di·dʒitălaiz), *v.*^{1} *Med.* [f. prec.: see -IZE.] *trans.* To subject to digitalization; to administer digitalis to. So **di·gitalized** *ppl. a.*^{1}, **di·gitalizing** *vbl. sb.*

1927 *Jrnl. Amer. Med. Assoc.* 10 Sept. 884 The general practitioner should not thoughtlessly digitalize his patient unless he has..facilities for determining the exact condition of his heart. **1951** A. GROLLMAN *Pharmacol. & Therap.* xviii. 351 If rapid digitalization is required, a so-called digitalizing dose may be administered at one time. *Ibid.* 353 The full digitalizing dose of digitoxin. **1961** D. M. DUNLOP *Textbk. Med. Treatment* (ed. 8) 565 On an average the total dose required to digitalize a fully grown adult is of the order of 1·5 to 2 g. (20 to 30 gr.) of digitalis powder..over a period of several days. **1970** [see *DIGITALIZATION].

digitalize (di·dʒitălaiz), *v.*^{2} [f. *DIGITAL *a.* 4 + -IZE.] *trans.* = *DIGITIZE *v.* 2. So **di·gitalized** *ppl. a.*^{2}

1962 *Aeroplane* 10 May 25/2 A digitalized radar terminal-area control system. **1965** *Math. in Biol. & Med.* (*Med. Res. Council*) I. 31 They already had a simple program worked out which would deal with digitalized outputs from recorders. **1967** COX & GROSE *Organization & Handling Bibl. Rec. by Computer* II. 45 The film may be the backing store and the picture digitalized for insertion.

digitally, *adv.* Add: **2.** By means of digits (DIGIT *sb.* 3); in digital form.

1957 N. CHAPIN *Introd. Automatic Computers* ii. 5 In addition to digital computers, there is an important group of computing machines known as analog computers because they operate with a physical analogy of a problem rather than with digitally expressed information about a problem. **1960** *McGraw-Hill Encycl. Sci. & Technol.* IV. 175/2 *Digital computer*. Any device for performing mathematical calculations on numbers represented digitally. **1964** *Data Acquisition & Processing in Biol. & Med.* II. 219 A digitally coded paper tape suitable for direct computer usage. **1970** *Nature* 12 Sept. 1123/2 The outputs of two receivers.. were combined, and the detected signal was sampled digitally every 50 ms over a period of 13 m 39 s.

digitize, *v.* Restrict *rare* to sense in Dict. and add: **2.** To convert (a continuously variable quantity) into a sequence of digits, generally for use in a digital computer, etc.; to represent in digital form. So **di·gitized** *ppl. a.*, **di·gitizing** *vbl. sb.*

1953 *Proc. Inst. Radio Engin.* XLI. 1456/2 It is..desirable to digitize the analog input signals as near to the system full scale as possible. **1957** *Electronic Engin.* XXIX. 568/2 Digitizing an analogue quantity makes the basic assumption that it may adequately be represented by a sufficient number of discreet intervals. **1959** *Times Rev. Industry* Aug. 24/3 All analogue quantities generated..are digitized. **1963** *Engineering* 8 Nov. 604/3 After digitising, the information is stored in 10 bit binary form. **1965** *Math. in Biol. & Med.* (*Med. Res. Council*) IV. 166 The slight discrepancy of the digitized wave. **1971** *Sci. Amer.* Feb. 81/3 We shall assume that our slab is a square and that it has been digitized into 64 *x*, *y* values or mesh points.

digitizer. Add: **2.** An instrument that converts continuously variable quantities into digital form, generally for use in a digital computer.

1953 *IRE Trans. Electronic Computers* II. III. 1/1 The digitizer..operates by 'reading' a coded representation of a given shaft position. **1956** *Electronic Engin.* XXVIII. 113/1 There are many ingenious ways of performing the analogue to digital conversion..and the commutator and digitizer types are probably the most useful. **1961** *Flight* LXXIX. 431/2 A digitizer converts the signals into a serial bit stream of binary data for transmission. **1970** *Computers & Humanities* IV. 244 A digitizer was used to measure and record the coordinates of bends in the outline.

digitorium (didʒitō^ə·riəm). [f. DIGIT *sb.* + -ORIUM.] A small portable keyboard used for exercising and strengthening the fingers in piano-playing; a dumb piano.

1876 in STAINER & BARRETT *Dict. Mus. Terms*. **1921** *Times* 15 Jan. 1/3 Wanted, Digitorium.—State size and price.

Digitron (di·dʒitrɒn). *Electr.* Also **digitron.** [f. DIGI(T *sb.* 3 + *-TRON.] The proprietary name of a cold-cathode character display tube.

1958 *Trade Marks Jrnl.* 18 June 622/1 *Digitron.* Electric discharge tubes...Ericsson Telephones Limited. **1960** N. MCLOUGHLIN et al. in *Electronic Engin.* Mar. 140/1 A new form of visual indication which has recently been developed is the Digitron, a cold-cathode character display tube. This tube, unlike the Dekatron, is not a counting device but is used to indicate the state of count or of switching position in the associated circuits. **1961** *New Scientist* 10 Aug. 350/1 The digitron..is essentially a neon lamp in which the cathode is formed in the shape of the required character—a number, a letter or a symbol.

diglossia (dəiglɒ·siă). *Philol.* [mod.L., ad. F. *diglossie*, f. Gr. δίγλωσσος bilingual + -IA^1.] (See quot. 1964.)

1959 C. A. FERGUSON in *Word* XV. 325 Diglossia. In many speech communities two or more varieties of the same language are used by some speakers under different conditions. *Ibid.*, The term 'diglossia' is introduced here, modeled on the French *diglossie*,..since there seems to be no word in regular use for this in English. **1964** E. PALMER tr. *Martinet's Elem. General Linguistics* v. 139 Linguists have proposed the term 'diglossia' to designate a situation where a community uses..both a more colloquial idiom of less prestige and another of more learned and refined status. **1965** *Language* XLI. 502 The concept of diglossia rests..on the observation that different languages or dialects enjoy varying social and political status within a community.

digonal (di·gŏnăl, dəigō^u·năl), *a. Cryst.* [f. DI-^{2} + Gr. γωνία angle + -AL.] Denoting an axis of two-fold symmetry.

1890 G. H. WILLIAMS *Elem. Crystallogr.* (ed. 2) iii. 47 Two other sets of axes are of use in the isometric system. One is the set of intersection-lines between the principal and secondary planes of symmetry...The first..are called the digonal. **1898** *Nature* 27 Jan. 309 Only digonal, trigonal, [tetragonal,] and hexagonal axes [of symmetry] are possible with crystals. **1911** A. E. TUTTON *Crystallography* x. 136 A digonal axis..is such that half a complete revolution about it restores the original appearance of the crystal. **1929** H. A. MIERS *Mineralogy* (ed. 2) i. 65 The crystals of this class..possess three digonal axes coinciding with the crystal axes, and no other symmetry.

digoxigenin (didʒɒksi·dʒĕnin, didʒɒksidʒe·nin). *Chem.* [f. *DIGOX(IN + *GENIN.] A hydrolysis product of digoxin, $C_{23}H_{34}O_5$.

1930 SYDNEY SMITH in *Jrnl. Chem. Soc.* 509 Digoxigenin separates from ethyl acetate in stout colourless prisms. **1956** *Nature* 11 Feb. 278/2 A higher dose of digoxigenin is required to produce the same degree of potentiation as digoxin.

digoxin (didʒɒ·ksin). *Chem.* [f. DIG(ITALIS + T)OXIN.] A crystalline glycoside, $C_{41}H_{64}O_{14}$, obtained from the leaves of the woolly foxglove, *Digitalis lanata*, and used chiefly in the treatment of heart-diseases.

1930 SYDNEY SMITH in *Jrnl. Chem. Soc.* 508 Digoxin, a new digitalis glucoside. *Ibid.*, This new glucoside, which has been named *digoxin*, resembles gitoxin in its sparing solubility in chloroform. **1956** [see prec.]. **1961** *Lancet* 9 Sept. 573/2 He was placed on maintenance digoxin. **1968** J. H. BURN *Lect. Notes Pharmacol.* (ed. 9) 36 Thus digoxin is given by mouth, or it is given intravenously.

digram. (Further example.)

1964 *Endeavour* Jan. 15/2 In English..'TR' is more frequent than 'TU', and Q is always followed by U. Such 'di-gram', and even some 'tri-gram', frequencies have frequently been counted.

diguanide (dəi·gwæ̆nəid). *Chem.* [f. DI-^{2} + GUANID(IN)E.] **a.** A colourless crystalline compound, $NH[C(NH)NH_2]_2$, which is a condensation product of two guanidine molecules with the loss of a molecule of ammonia; guanylguanidine.

1910 *Jrnl. Chem. Soc.* XCVIII. I. 896 Diguanide $NH[C(NH)NH_2]_2$, crystallises from absolute alcohol in glistening prisms, m. p. 130°. **1928** G. M. DYSON *Chem. Chemotherapy* v. 95 Several guanidine derivatives..can be considered pharmacologically inactive...Thus, guanyl urea (315)..is inactive and diguanide (316) in the form of its hydrochloride is inactive. **1946** *Nature* 16 Nov. 707/2 The diguanide system..provided structural features similar to those found in the earlier active pyrimidine compound '2666'. **1952** MORTON & HOGGARTH in E. H. Rodd *Chem. Carbon Compounds* IB. xiv. 935 Diguanide itself is a strongly basic substance.

b. Any compound derived from diguanide by replacement of hydrogen atoms by other atoms or groups; also, a diguanidine.

1911 *Jrnl. Chem. Soc.* C. I. 928 Diguanides of the type $NH:C(NHR)\cdot NH\cdot C(NH_2):NH$, where R = aryl, are readily obtained in the form of their hydrochlorides by heating dicyanodiamide with the hydrochlorides of aromatic amines in aqueous solution. **1961** *Lancet* 9 Sept. 563/2 The present compounds are not diguanides like synthalin. **1961** [see next]. **1966** *New Scientist* 24 Nov. 433/1 The longing of diabetics for a hypoglycaemic drug which could be taken orally..was realized ten years ago when the sulphonylureas and diguanides were introduced.

diguanidine (dəigwæ·nidin, -in). *Chem.* [f. DI-^{2} + GUANIDINE.] †**a.** = *DIGUANIDE a. *Obs.* **b.** Any compound containing two of the radicals $—NH\cdot C(NH)NH_2$, derived from guanidine by the loss of a hydrogen atom.

1879 *Jrnl. Chem. Soc.* XXXVI. 781 It is now found that this substance is a di-acid base, diguanidine, $NH:C(NH_2)\cdot NH\cdot C(NH_2):NH$. *Ibid.*, Diguanidine and its salts are colourless. **1927** *Brit. Pat.* 274,259, An improved process for the production of alkylene diguanidines. **1952** MORTON & HOGGARTH in E. H. Rodd *Chem. Carbon Compounds* IB. xiv. 931 This compound depresses the concentration of blood sugar, a property which is even more marked in certain diguanidines such as 'Synthalin'. **1960** *Brit. Med. Bull.* XVI. 251/1 Diguanidines..may be considered as being made up of two guanidine molecules joined by a chain of methylene groups. **1961** *Lancet* 30 Sept. 776/1 Synthalin with two guanidines joined by a chain of methyl groups is a diguan*idine*. In phenformin the two guanidines coalesce.. and it is a diguanide (or biguanide).

dihedral, *a.* (and *sb.*). Add: **1.** spec. *Aeronaut.* Designating the inclination, esp. upwards, of a wing or other surface to the horizontal, or the angle to the horizontal so formed; *dihedral board*, a device used in measuring the dihedral angle of the wings of an aeroplane.

1909 *Westm. Gaz.* 11 Mar. 4/1 The problem of automatic transverse stability is solved by giving the aeroplanes a dihedral angle or keels. **1916** H. BARBER *Aeroplane Speaks* iii. 108 Another method of securing the dihedral angle..is by means of the dihedral board. **1917** C. C. TURNER *Aircraft of To-Day* viii. 140 Inherent stability is secured..by a dihedral setting of the wings, i.e. sloping up to right and left from the body. **1960** C. H. GIBBS-SMITH *Aeroplane* I. ix. 38 They had at first rigged it with a dihedral angle, and found it flew badly in gusty winds.

B. *sb. Aeronaut.* A dihedral angle or inclination. Also *transf.*

1913 A. E. BERRIMAN *Aviation* iii. 37 The tail plane is invariably set at a lesser effective angle of incidence than the main wings. This arrangement is commonly described as the fore-and-aft dihedral. **1929** W. FAULKNER *Sartoris* v. 364 While in level flight, dihedral would be eliminated for speed. **1935** C. G. BURGE *Complete Bk. Aviation* 581/2 If the planes are inclined upwards towards the wing tips, the dihedral is positive; if downwards, it is negative. **1962** *Times* 25 May 18/5 Michelotti has..lifted the dihedral of the wings [of a car], setting twin headlamps beneath. **1969** J. M. BRUCE *War Planes of First World War* III. 134 This aircraft differed from the first in having no dihedral on the lower wings.

dihybrid (dəihəi·brid). *Biol.* [f. DI-^{2} + HYBRID *sb.*] A hybrid that is heterozygous with respect to two independent genes. Also *attrib.* So **dihy·bridism.**

1907 R. C. PUNNETT *Mendelism* (ed. 2) 34 Where the original parents differ in two pairs of characters, the case is termed one of dihybridism. *Ibid.* 62 In the dihybrid cases.. the two pairs of characters behaved quite independently, in so far as the process of segregation was concerned. **1918** BABCOCK & CLAUSEN *Genetics Rel. Agric.* v. 86 The F_2 segregates in accordance with normal dihybrid expectations. **1965** BELL & COOMBE tr. *Strasburger's Textbk. Bot.* (new ed.) II. ii. 334 Hybrids heterozygous for two characters are known as dihybrids. **1969** G. W. BURNS *Sci. Genetics* iv. 68 The F_1 individuals in this instance are referred to as dihybrid individuals because they are heterozygous for each of the two pairs of genes.

dihydro-. Add: **dihy:drostreptomy·cin,** a hydrogenated derivative of streptomycin with similar antibiotic properties.
1946 *Jrnl. Amer. Chem. Soc.* LXVIII. 1390/2 Streptomycin has been catalytically hydrogenated to dihydrostreptomycin. 1951 A. Grollman *Pharmacol. & Therap.* xxii. 456 Although large doses of dihydrostreptomycin induce the same vestibular disturbances as does streptomycin, it is much less neurotoxic. 1970 Passmore & Robson *Compan. Med. Studies* II. xx. 31/2 Dihydrostreptomycin..is now rarely used because it is more likely than streptomycin to damage the cochlear nerve and produce deafness.

di:hydroxya·cetone. *Chem.* [f. Di-[2] + Hydroxy- + Acetone.] A colourless crystalline compound, $CH_2OH·CO·CH_2OH$, isomeric with glyceraldehyde and having strong reducing properties.
1895 *Jrnl. Chem. Soc.* LXVIII. i. 496 Attempts to prepare dihydroxyacetone by the action of nitrous acid on diamidoacetone hydrochloride were not very successful. 1951 I. L. Finar *Org. Chem.* xviii. 358 Dihydroxyacetone..is not optically active, and hence, by the definition given above, is not a sugar. If..the proviso that a sugar is always optically active is rejected, then dihydroxyacetone will be a ketotriose. 1967 [see *glycerose]. 1969 Ottaway & Irvine tr. *Netter's Theor. Biochem.* xiii. 715 In the mitochondria a very active oxidation of glycerol phosphate takes place. The dihydroxyacetone phosphate which is formed diffuses into the cytoplasm,..and so acts as a transporter of hydrogen to the respiratory chain.

di-iodo-. Add: **di:iodoty·rosine,** an iodine-containing derivative of tyrosine, $C_9H_9NO_3I_2$, which is the precursor of thyroxine in the body and the compound into which iodine is first converted when absorbed by the thyroid gland from the blood.
1927 *Biochem. Jrnl.* XXI. 170 It is highly probable that thyroxine is formed in nature by the coupling of two molecules of diiodotyrosine with the loss of one side chain. 1949 *Blakiston's New Gould Med. Dict.* 297/2 *Diiodotyrosine*, $C_9H_9O_3NI_2$. A substance found in skeletons of corals, sponges, and other marine organisms. It..has been used in hyperthyroidism. 1964 L. Martin *Clinical Endocrinol.* (ed. 4) iii. 122 In goitrous cretins..the thyroid may be unable to trap iodide from the blood, or..to couple molecules of diiodotyrosine to form thyroxine.

dijiridu, var. *didgeridoo.

dikaryon (dəikæ·riǫn). *Biol.* Also **dicaryon.** [a. Fr. *dikaryon* (R. Maire 1912, in *Mycol. Centralblatt* I. 214), f. Di-[2] + Gr. κάρυον nut, taken as = nucleus.] A pair of unfused haploid nuclei of opposite mating type in a cell or spore which divide simultaneously when the cell divides; also, a dikaryotic cell, spore, or mycelium. Hence **dikaryo·tic** *a.*, containing a dikaryon; composed of dikaryons.
1913 W. B. Grove *Brit. Rust Fungi* i. 5 The two adjacent nuclei are said to be 'paired', and together they constitute a *synkaryon* or *dikaryon*. 1941 *Bot. Rev.* VII. 393 A dikaryotic hybrid may be defined as a hybrid made up of dikaryotic diploid cells, *i.e.* cells containing a conjugate pair of haploid nuclei. 1947 C. E. Skinner et al. *Henrici's Molds, Yeasts & Actinomycetes* (ed. 2) i. 9 The mycelium with paired nuclei of opposite 'sex' which divides conjugately is known as the dikaryon or is described as dikaryotic. 1971 *Nature* 30 Apr. 552/2 [They] report the successful growth *in vitro* of the corn smut (*Ustilago maydis*) dikaryon. *Ibid.* 552/3 It is the fusion of two of these non-pathogenic haploid cells which gives rise to the pathogenic dikaryotic mycelium.

dikaryophase (dəikæ·riŏfē¹z). *Biol.* Also **dicaryophase.** [f. prec. + Phase.] A phase in the life-cycle of a fungus when it contains dikaryotic cells. So **dika·ryophyte**, a dikaryotic mycelium.
1916 B. D. Jackson *Gloss. Bot. Terms* (ed. 3) 110/2 *Dicaryophase*, the stage ending in the production of teleutospores. 1932 *Proc. 6th Int. Congress of Genetics* II. 191 The haploid nuclei..remain associated in the relatively long, parasitic dikaryophase. *Ibid.*, New 'dikaryophytes'..have been found in nature. 1940 *Chambers's Techn. Dict.* 246/1 *Dikaryophase*, that part of the life-history of many Basidiomycetae in which the hyphae are made up of segments, each containing two nuclei.

dik-dik (di·kdik). Also **dikdik.** [Native name in E. Africa; also in Afrikaans.] Any of several small African antelopes.
[1870 W. T. Blanford *Geol. & Zool. Abyssinia* iii. 268 The '*Beni Israel*' or '*Om-dig-dig*', one of the smallest Antelopes known, abounds on the shores of the Red Sea.] 1883 F. L. James *Wild Tribes of Soudan* ix. 76 A *dik-dik*..made its appearance,..uttering the peculiar cry from which it has obtained its name. 1895 *19th Cent.* Sept. 489 Besides lion and rhinoceros there were..Gerenook, Dik-Dik (*Nanotragus Saltii*),..and many kinds of birds. 1906 *Daily Chron.* 23 Mar. 3/3 The dainty little Somali dik-dik. 1920 *Blackw. Mag.* May 660/1 There is nothing to be seen of wild life save occasional tracks of dik-dik. 1928 *Daily Express* 31 July 4 The dikdiks and smaller gazelles. 1970 *East African Standard* 23 Jan. 6/3 There have been bushbuck, dikdik and a porcupine born to inmates of the Orphanage. Another dikdik.. has been presented to the Orphanage.

dike, dyke, *sb.*[1] Add: **2. c.** A water-closet or urinal. *slang.*
1923 Manchon *Le Slang* 104 *Dike*, les cabinets; *to do a dike*, aller aux lieux, aux *gogues*. 1940 M. Marples *Public School Slang* 112 Other synonyms [for 'lavatories'] are *rears*, *lats*,..*dykes* (Oundle, 1920+), [etc.]. 1960 N. Hilliard *Maori Girl* 74 But the dyke's in the bathroom. 1965 G. McInnes *Road to Gundagai* v. 76 The outside (and only) lavatory, known locally as the dyke. 1967 J. Cleary *Long Pursuit* iii. 82, I learned..to respect her privacy. And I don't mean just when she went to the dike.
10. (sense 9 b) **dike-phase, -rock, -swarm** (see quots.).
1909 *Q. Jrnl. Geol. Soc.* LXV. 642 In our district..the main dyke-phase intervenes between the uprise of the Cruachan and the Starav Granites. 1954 G. W. Himus *Dict. Geol.* 44 *Dyke phase*, the closing episode in a volcanic cycle, characterized by the injection of minor intrusions, especially dykes. 1896 *Science Progress* IV. 476 A..difference of opinion exists as to the propriety of establishing an intermediate division between the plutonic and the volcanic types of structure. Such a division is roughly represented by the 'dyke-rocks' of Rosenbusch. 1903 A. Geikie *Text-bk. Geol.* (ed. 4) I. ii. ii. 197 Professor Rosenbusch [in *Mikrosk. und Physiogr. d. Massingen Gesteine* (ed. 2, 1887) II. 277]..groups the igneous rocks in three great sections: 1st, the deep-seated rocks..; 2nd, dyke-rocks (Gang-gesteine), which may have been injected as dykes and veins at a less distance from the surface (hypabyssal) [etc.]. 1965 G. J. Williams *Econ. Geol. N.Z.* xi. 167/2 Alkalic dyke-rock material which occurs in the alluvium of New River. 1909 *Q. Jrnl. Geol. Soc.* LXV. 646 A suggestion as to the relationship of the dyke-swarms to this focus. 1960 L. D. Stamp *Britain's Structure* (ed. 5) ix. 79 Into these cracks molten rock forced its way so that dykes were formed, so numerous that the expression 'dyke swarm' is often used.

dike (dəik), *sb.*[2] and *v.*[2] *U.S. slang* or *colloq.* [Of obscure origin; perh. a corruption of Deck *v.* 2.] (See quots.)
1851 B. H. Hall *College Words* 100 At the University of Virginia, one who is dressed with more than ordinary elegance is said to be *diked out*. 1871 Schele de Vere *Americanisms* 597 *Dike*, denoting a man in full dress, or merely the dress, is a peculiar American cant term as yet unexplained. To be *out on a dike* is said of persons, mainly young men, who are dressed more carefully than usual, in order to pay visits or to attend a party. 1899 B. W. Green *Virginia Folk-Speech* 115 *Dike*, to dress fine. 'You are diked up to-day.' 1902 S. Clapin *Dict. Amer.* 159 *Dike*, to attire oneself faultlessly for social purposes. *Diked out*, to be dressed up, with connotation of being in one's best clothes. 1923 M. Watts *Luther Nichols* 62 All right for you..comin' round here all diked out like Sunday.

dike, dyke (dəik), *sb.*[3] *slang.* [Of obscure origin.] A lesbian; a masculine woman. Also *attrib.*
1942 Berrey & Van den Bark *Amer. Thes. Slang* § 405/3 Masculine woman,..*dike, dyke.* 1959 F. Norman *Stand on Me* iii. 37 Nearly everyone is kinkey on dykes for some reason. 1964 E. Ambler *Kind of Anger* ii. 54 You know about that dike partner of hers? 1965 'E. McBain' *Doll* (1966) vii. 92 'Was your wife a dyke?' 'No.' 'Are you a homosexual?' 'No.'

diketone (dəikī·tōᵘn). *Chem.* [f. Di-[2] + Ketone.] Any compound in which there are two carbonyl groups, —CO—, each attached to two carbon atoms.
1896 *Jrnl. Chem. Soc.* LXX. i. 586 By the action of bromine water on quercitol, a diketone, $C_6H_8O_3$, is probably formed. 1946 *Nature* 10 Aug. 205/2 The glycol from butaldehyde gives mostly a keto-alcohol, further oxidation to the diketone being much more difficult. 1967 I. L. Finar *Org. Chem.* (ed. 5) I. v. 268 Butane-2, 3-dione,..$CH_3·CO·CO·CH_3$, is the simplest α-diketone.

dikey (dəi·ki), *a. slang.* [f. *dike *sb.*[3] + -y[1].] Having the appearance or characteristics of a lesbian.
1964 S. Bellow *Herzog* 99 Lucas warned me to look out for something dikey. 1966 C. Israel *Hostages* 155 The tall, dikey French woman. 1969 'J. Morris' *Fever Grass* ix. 83 Helen's gone dikey in her old age.

dikh (dik). *India.* Also **dik(k.** [ad. Hind. *diq.*] Trouble, worry, vexation.
1873 W. Heeley *Lay Mod. Darjeeling* (Y.), And if his locks are white as snow, 'Tis more from dikh than age! 1923 Kipling *Land & Sea Tales* 243 There has been great *dikh* in this case? *Ibid.* 246 There has been great *dikh-dari* [i.e. trouble-giving].

dikkon, var. *dicken, dickin *int.*

dikkop (di·kǫp). *S. Afr.* Also **diccop, dickop, dik-kop, dikop.** [Afrikaans, f. *dik* thick + *kop* head.] **1.** The stone-curlew.
1853 *Edin. New Philos. Jrnl.* LV. 83 The *dickop*..seem to emerge from their daylight concealment. 1858 A. W. Drayson *Sporting Scenes* 17 One or two of the bustard tribe are also found here, and are called the *diccop*, *coran*, and *pouw*. 1873 tr. *J. Verne's Meridiana* xvi. 148 The hunters shot..some '*dikkops*', whose flesh is very delicate eating. 1891 *Daily Graphic* 17 Aug. 5 Eleven snipe, one dikkop, one wild turkey, one blue crane. 1903 E. Glanville *Diamond Seekers* 116 They're playing dik-kop...The dik-kop drops his wing and shams hurt to lead you off. 1939 S. Cloete *Watch for Dawn* ix. 114 The moonlit nights when dikkops and the kievietjes screamed as they flew to water. 1947 J. Stevenson-Hamilton *Wild Life S. Afr.* xxxii. 270 The dikkops or stone curlews.—Two species of these are described in South Africa, the Cape Dikkop (*Burhinus capensis*), and the Water Dikkop (*Burhinus vermiculator*).
2. A form of blue tongue (see *blue tongue 1).
1870 T. Baines *Diary* 8 Dec. (1946) III. 762 One of whose horses was standing apart, suffering from the 'dikkop' form of horse sickness. 1959 *Cape Times* 28 Jan. 2/3 Dikkop, the dreaded sheep-killing disease, is raging at Murraysburg.

‖ **diktat** (di·ktæt). [a. G. *diktat* Dictate *sb.*] **a.** A severe settlement or decision, esp. one imposed by a victorious nation upon a defeated nation, a dictated peace; used *spec.* with reference to the Treaty of Versailles of 1919. **b.** A dictate, decree, or command; a categorical assertion.
1933 'A German Diplomat' *Hitler—Germany & Europe* (Friends of Europe, No. 2) 9 The treaty of Versailles..was not a 'Diktat' which artificially imposed a solution foreign to reality. 1940 *Time* 1 Jan. 47/1 If a Final Treaty is negotiated between victor and vanquished..at least a year after the Preliminary Treaty, or *Diktat*, is imposed. 1940 *Times Weekly* 7 Aug. 9/3 (*headline*) Axis plans for Rumania—popular resistance to 'Diktat'. 1941 *Mind* L. 290 He decreed that sense-data are not to have unnoticed characteristics...This *Diktat* of his would seem to imply that sense-data can have determinable shapes without having determinate ones. 1948 A. J. Toynbee *Civilization on Trial* v. 79 The psychological effect of the British *diktat* of A.D. 1842. 1959 *Listener* 22 Jan. 158/2 The Soviet draft was fundamentally different from the Versailles '*diktat*'. 1963 *Times* 27 May 13/4 Naked aggression against any state in Africa which does not accept the diktat of himself and his friends.

dilatation. Add: **1. c.** **dilatation and curettage** (or **curetting**) *Gynæcol.*, an operation involving dilatation of the cervix and curettage of the uterus, carried out for diagnostic purposes, to terminate a pregnancy, or to arrest irregular menstrual bleeding.
1906 D. B. Hart in Allbutt et al. *Syst. Gynaecol.* (ed. 2) 70 *Dilatation and curetting.* This is a very important procedure where malignant disease of the mucous membrane of the body is suspected. 1920 H. S. Crossen *Operative Gynecol.* (ed. 2) v. 268 (*caption*) The safe method of securing the necessary fixation of the cervix, for Dilatation and Curettage. 1964 *Jrnl. Obstetrics & Gynaecol.* LXXI. 668/2 Dilatation and curettage has a major place in the diagnostic and therapeutic armamentarium of the obstetrician and gynaecologist.

dilatometry (dəilĕtǫ·mĕtri). [f. Dilato(meter + -metry.] The measurement of expansion by means of a dilatometer.
1929 *Engineering* 13 Sept. 361/1 The very slow rates of heating or cooling, which can be used..in dilatometry. 1930 *Ibid.* 16 May 654/1 *Dilatometry.* The specimens made for measuring the permeability were also used for measuring the dilatation.

dildo[1]. Delete †*Obs.* and add earlier and later examples; *spec.* an artificial penis used for female gratification.
c 1593 T. Nashe *Choise of Valentines or the Merie Ballad of Nash his Dildo* (1899) 20 Curse Eunuke dilldo, senceless counterfet. 1598 Florio *Worlde of Wordes* 261/3 *Pastinaca muranese*, a dildoe of glasse. *Ibid.* 278/2 *Pinco*, a prick, a pillicock, a pintle, a dildoe. 1785 Grose *Dict. Vulgar T.* s.v., Dildoes are made of wax, horn, leather, and diverse other substances. 1886 Burton *Arab. Nts.* X. 239 Of the penis succedaneus,..which the Latins called phallus and fascinum, the French godemiché and the Italians passatempo and diletto (whence our 'dildo'), every kind abounds. 1952 Auden *Nones* 11 The nude young male who lounges Against a rock displaying his dildo. 1965 *New Statesman* 9 Apr. 570/2 Why does it matter so much to them whether lesbians use a dildo or not?

dildo[2]. Delete † *Obs.* and add earlier and later examples.
1672 W. Hughes *Amer. Physitian* 43 The Tree was long since called by the Spaniards, and by the Negroes that lived there, the Dildoe-Tree; and the English retain the same name still. 1926 Fawcett & Rendle *Flora of Jamaica* V. 279 C[*ereus*] *peruvianus*...Dildo. Dry parts of Jamaica on southern side. 1956 J. Hearne *Stranger at Gate* xix. 156 It's like getting a dildo thorn out of your foot.

dilex, var. *dillesk*, Dulse.
1883 J. B. Bloomfield in N. Okoshi *Fisheries Japan* (Fish. Exhib. Publ.) 27 A kind of seaweed called dilex, which they found upon the rocks.

dill, *sb.*[1] Add: **3. dill pickle** orig. *U.S.*, a pickled cucumber, gherkin, etc., flavoured with dill.
1906 'O. Henry' *Four Million* (1916) 157 He saw her beginning upon a huge Dill pickle. 1916 H. L. Wilson *Somewhere in Red Gap* v. 217 A big dill pickle, two deviled eggs, and a half of one of these Camelbert [*sic*] cheeses. 1971 J. Sangster *Your Friendly Neighbourhood Death Pedlar* ii. 38 She..pulled out a bottle. 'Dill pickle?'.. She laid one pickle on each plate.

dill (dil), *sb.*[6] *Austral.* and *N.Z. slang.* Also **dil.** [app. back-formation from *dilly *a.*[1]] A fool or simpleton; *spec.* one who is duped by a trickster.
1941 Baker *Dict. Austral. Slang* 23 *Dil*, a simpleton or fool. (2) A trickster's victim. 1949 *Evening News* 16 Feb. 4/6 Sydney has developed its own picturesque slang. They talk of..'a dill' (a weak character). 1957 'N. Culotta' *They're a Weird Mob* (1958) i. 13 Well don't stand there like a dill. *Ibid.* ix. 133 Joe said they were a 'lot o' dills'. 1961 P. White *Riders in Chariot* xv. 503, I am the same dill that always stuck around! 1969 *Telegraph* (Brisbane) 28 Aug. 2/7 At the start he felt a bit of a dill in a wig and robes.

1970 *N.Z. Listener* 12 Oct. 12/5 She acted like a dill. She shouldn't have antagonised us.

dilly (di·li), *sb.*[5] *slang* (orig. *U.S.*). [f. *dilly *a.*[2]] A delightful, remarkable, or excellent person or thing; freq. ironical.
1935 *Amer. Mercury* June 229/1 Ain't that a *dilly* (or *honey*)! 1950 'S. Ransome' *Deadly Miss Ashley* vi. 65 She looked as if she had had a rugged night of it; but she was easily repairable and still a dilly. 1951 R. S. Prather *Bodies in Bedlam* vii. 48 In order to get inside you have to pass three guards at three different gates. The first two aren't so tough, but the last one is a dilly if you don't have an appointment. 1958 R. Chandler *Playback* xix. 159 You're the most impossible man I ever met. And I've met some dillies. 1968 M. Kane *Walk of Devil* vii. 79 'Did you swing it, Jake?' Berger asked Winthrop..'It's a dilly,' Winthrop said. 1970 *Daily Colonist* (Victoria, B.C.) 1 Dec. 9/4 The new [oil] well, also in Norwegian waters, looks a dilly.

dilly (di·li), *a.*[1] *dial.* or *colloq.* (chiefly *Austral.*). [Perh. f. Daft *a.* + Silly *a.*] **a.** (See quot. 1873.) **b.** Foolish, stupid, mad.
1873 Williams & Jones *Gloss. Somersetshire* 11 *Dilly*,.. cranky, queer. 1906 E. Dyson *Fact'ry 'Ands* xvi. 214 Who should come sprintin' upstairs but me nibs, pale's er blessed egg, hair on end—fair dilly. 1908 J. Gunn *We of Never-Never* xiii. 168 Gone clean dilly, I believe. 1915 C. J. Dennis in *Bulletin* (Sydney) 15 Apr. 43/2 Ther's a Gawd 'Oo's leaning near To watch our dilly little lives down 'ere. 1942 Berrey & Van den Bark *Amer. Thes. Slang* § 151/10 Foolish; nonsensical; ridiculous,..*dilly*.

dilly (di·li), *a.*[2] *colloq.* [f. the first syllable of Delightful *a.* or Delicious *a.* + -y[6].] Delightful; delicious.
1909 *Punch* 26 May 362/1, I sent out the ordinary cards ..with 'Dancing' in one corner of the card, but in the *other* corner was 'Bare feet'. Wasn't it a dilly idea? 1922 C. E. M. Joad *Highbrows* iii. 103 Have you heard that new waltz, 'Luscious Love'? It's simply dilly.

dilly-bag. Add: **b.** *Austral. colloq.* Any small bag in which articles are carried.
1934 *Bulletin* (Sydney) 5 Dec. 48/1 He had seen young Willie leave the farmhouse carrying the tea and a dilly-bag of cakes. 1937 Partridge *Dict. Slang* 221/1 *Dilly-bag*, a wallet; a civilian haversack... In C. 20, often used by women for a small shopping-bag or for a general-utility purse-bag.

dilo (dī·lo). The Fijian name for the domba, *Calophyllum inophyllum*. Also *attrib.*, as *dilo oil, tree*.
1874 in Lindley & Moore *Treas. Bot.* Suppl. 1879 *Encycl. Brit.* IX. 156/2 The dilo.., the oil from its seeds being much used in the [Fiji] islands.., in the treatment of rheumatism. 1894 B. Thomson *S. Sea Yarns* 184 His arm was thick and knotted as yon dilo-tree.

dilruba (dilrū·bā). An Indian musical instrument having a long neck, three or four main strings played with a bow, and several sympathetic strings.
[1887 *Hindu Mus. & Gayan Samaj* III. 53 List of musical instruments in India... Dilarubâ.] 1921 H. A. Popley *Music of India* vii. 107 The *Dilruba* is very much like a sitār, but smaller; and instead of a bowl, it has a belly, covered with sheep-parchment. In shape it is something like the sārangī, and like that instrument it is played with a bow made of horsehair. 1969 R. Shankar *My Music* i. 38/1 Several other bowed stringed instruments not as popular as the *sarangi* are the *dilruba*, *esraj*, and *sarinda*.

dilutee (dailiutī·). [Irreg. f. Dilute *v.* + -ee[1].] An unskilled or semi-skilled worker who takes a place hitherto occupied by a skilled worker.
1918 *Glasgow Herald* 18 Dec. 7/6 It is being considered whether the dilutee should go and leave the skilled workman at his job, and whether the dilutee should be discharged to provide for the re-entry of the skilled worker who has been to the war. 1921 *Ibid.* 7 Mar. 12 The 50,000 dilutees whom the Government were particularly anxious to force upon the building trade. 1923 G. D. H. Cole *Trade Un. & Munit.* 135 Before any skilled men were taken for the army, all 'dilutees' of military age and fitness should first be removed. 1960 *Guardian* 4 May 18/2 Dilutees were not now needed because there was not enough work..for skilled men.

dilution. Add: **4.** The substitution of unskilled or semi-skilled for skilled workers.
1916 *Times* 17 June 7/5 They had no objection to the Admiralty introducing dilution all round, but..they must ask..that the position of the skilled workers would not be unduly harmed. 1919 *Daily Mail Year Bk.* 111/1 True, the dilution of labour includes the employment of a large contingent of unskilled men besides women. 1921 *Ibid.* 60/2 Dilution by taking in unskilled and partially skilled men, and particularly ex-Service men. 1940 *Economist* 24 Feb. 322/1 No decisions have been reached about the scale and conditions of dilution in general, female employment in particular, and retraining. 1964 *Times Rev. Industry* Feb. 71/2 They may..sponsor dilution (more jobs for women at equal pay and quicker apprentice training).

dim, *a.* Add: **4. b.** Applied to a person: not 'bright' intellectually; somewhat stupid and dull.
1892 Stevenson & Osbourne *Wrecker* xxi. 325 He's a very pleasing creature, rather dim, and dull, and genteel, but really pleasing. 1910 R. Brooke *Let.* 3 July (1968) 243 They were very dim, and said, couldn't *we* find an advance agent? 1916 M. Beerbohm in *Cornhill Mag.* June 719 The young writers of that era..strove earnestly to be distinct in aspect. This man had striven unsuccessfully...I decided that 'dim' was the *mot juste* for him. 1923 J. Trevelyan *Life of Mrs. H. Ward* x. 192 Teachers and many 'dim' people of various professions would find her as accessible as her strenuous hours of labour would allow. 1924 'W. Fabian' *Sailors' Wives* v. 63 The sexperts, which is a combination of sex and expert: I glued it together myself. Not so dim; yes? 1933 J. C. Masterman *Oxford Tragedy* xv. 229 The dim little research Fellow with clumsy manners and no conversation. 1950 *Listener* 7 Dec. 709/2 Constable and Turner were neglected and the dim and second-rate were crowned with triumphant laurels.
c. Of a thing, situation, etc.: dull, poor, undistinguished. *colloq.*
1958 B. Hamilton *Too Much Water* xi. 248, I personally had rather a dim war. *Ibid.* xii. 266 A rather dim situation for the M.C.C.
d. Colloq. phr. *to take a dim view*: see *view *sb.*

dim, *v.* Add: **2. c.** *to dim out*: to reduce the brightness of (street-lighting, etc.), esp. in time of war; to impose a 'dim-out' on (a city, etc.).
1942 *Amer. Speech* XVII. 204/2 The city was dimmed-out. 1945 *Daily Express* 20 Apr. 1/8 Street lighting need no longer be dimmed out.

dime, *sb.* Add: **2. a.** *pl.* Money; financial gain; freq. *the dimes*. *U.S. colloq.*
1843 *Spirit of Times* 21 Jan. 560/3 Times are hard, and dimes are scarce. 1845 S. F. Smith *Theatr. Apprenticeship* (1847) 7, I, in search of 'the dimes', acted plays in newly-built theatres. *a* 1861 T. Winthrop *J. Brent* (1883) xxvi. 233 Count out yer dimes, and I'll fill out a blank bill of sale. 1871 *Harper's Mag.* June 37/2 Among the excursionists there would be..such as travel to gather ideas rather than dimes.
b. (Earlier examples.) Hence *dime-novelish* adj., *dime-novelist*; **dime-store** *U.S.*, a shop in which the maximum price was originally a dime; also *attrib.* or as *adj.*, *spec.* designating a cheap and inferior article; cf. *five-and-ten* (*cent store*) (*five C. 2).
1859 (*title*) Beadle's dime song book. 1860 (*title*) Beadle's dime book of dreams..compiled from the most accredited sources for the 'Dime series'. 1861 *Vanity Fair* 26 Jan. 38/2, I invested in the dime editions of startling narratives. 1864 *N. Amer. Rev.* July 304 A Dime Novel is issued each month. 1865 A. H. Stephens *Diary* (1910) 424 A little primer-looking sort of a child's book. It was a dime novel. 1879 *Amer. Punch* Apr. 40/1 Written to order by the hundred, by a Dime novelist in New York. 1887 *Scribner's Mag.* July 120/1 It was a trifle boyish, and 'dime-novelish'. 1914 R. Herrick *Clark's Field* 7 The facts are not all dime-novelish, but very human and significant. 1928 Weseen *Crowell's Dict. Eng. Gram.* 188 *Dime store*, colloquial name for a store that specializes in articles selling for ten cents. *The five and ten* is a variant. 1931 *Kansas City Star* 23 Oct., A dime store in Emporia. 1938 *Newsweek* 31 Jan. 36/1 'Best buy' was a dime-store product, which cost 5 cents a gram. 1942 Berrey & Van den Bark *Amer. Thes. Slang* § 21/14 Cheap; paltry,..*dime-store*.

dimension, *sb.* Add: **3. c.** [a. F. *dimension* (J. B. J. Fourier *Théorie anal. de la Chaleur* (1822) ii. § ix. 154).] The power to which any one of the fundamental quantities or units is raised in the expression defining a derived quantity or unit in terms of them; also (in *pl.*), all the fundamental quantities in such an expression, each raised to its appropriate power, which together show how the unit of the derived quantity depends on the fundamental units; *method of dimensions*, dimensional analysis.
The 'fundamental quantities' are usually taken to be mass, length, and time, with the addition of one or more other quantities in certain cases (such as electrical and magnetic phenomena).
1864 *Rep. Brit. Assoc. 1863* 132 The value of a force is directly proportional to a length and a mass, but inversely proportional to the square of a time. This is expressed by saying that the dimensions of a force are LM/T^2. 1877 Ld. Rayleigh *Theory of Sound* I. iii. 47 From the necessity of a complete enumeration of all the quantities on which the required result may depend, the method of dimensions is somewhat dangerous. 1878 A. Freeman tr. *Fourier's Anal. Theory of Heat* ii. 128 Every undetermined magnitude or constant has one dimension proper to itself, and..the terms of one and the same equation could not be compared, if they had not the same exponent of dimension. We have introduced this consideration..in order to make our definitions more exact, and to serve to verify the analysis. *Ibid.* 129 The dimensions of x, t, v with respect to the unit of time are 0, 1, 0, and those of K, h, c are −1, −1, 0. 1911 *Encycl. Brit.* XXVII. 736/2 Velocity is of +1 dimension in length and −1 dimension in time. 1925 *Phil. Mag.* 6th Ser. L. 32 The dimensions of the viscosity, η, are $ML^{-1}T^{-1}$, of the density, ρ, ML^{-3}, whilst σ/ρ' is of no dimensions. *Ibid.* 31 Much information can be obtained concerning F by means of the method of dimensions. 1933 A. W. Porter (*title*) The method of dimensions. 1960 *McGraw-Hill Encycl. Sci. & Technol.* IV. 197/2 Quantities with the same dimensions can be expressed in the same units. 1964 H. S. Hvistendahl *Engin. Units* i. 7 For example, in the case of force, the term 'dimensions' is now generally understood to mean LMT^{-2}, and not merely 1, 1, −2. 1969 L. Young *Syst. Units Electr. & Magn.* i. 7 Area has dimensions of length squared; denoting 'dimensions of' by square brackets, we write $[A]=[L]^2$. 1970 *Nature* 29 Aug. 935/2 In SI units the constant $\mu_0=4\pi\times10^{-7}$ is indispensable in many formulae, if the dimensions are to balance.
5. dimension lines, straight lines usually having an arrow at each end, indicating the parts or lines to which the figured dimensions refer in a technical drawing.
1887 D. A. Low *Machine Drawing* 5 Dimension lines and centre lines are best put in of different colour. 1902 P. Marshall *Metal Tools* 18 The marking of dimension lines on metal surfaces is generally done with a steel scriber. 1927 G. E. Draycott *Technical Drawing* xii. 184 No dimension line should run closer than ¼" to a line to which it is parallel. 1961 Bellis & Schmidt *Archit. Drafting* xiii. 81/2 Use a 2H pencil for extension and dimension lines.

dimension, *v.* Restrict *rare* to sense in Dict. and add: **2.** *trans.* To mark the dimensions on (a working drawing, diagram, or sketch). Chiefly in *pa. pple.* or *ppl. a.* So **dime·nsioning** *vbl. sb.*, the action of marking dimensions; the dimension lines, etc., on a drawing.
1885 *Marine Engineer* 1 Apr. 27/1 Twenty-five large plates of fully dimensioned drawings. 1887 D. A. Low *Machine Draw.* 5 Many a good drawing has its appearance spoiled through being slovenly dimensioned. 1892 *Ibid.* 99 Rough dimensioned sketches. 1904 *Westm. Gaz.* 28 July 2/1 All parts being carefully illustrated by dimensioned drawings. 1907 *Install. News* Oct. 1/2 Diagrams are more valuable when dimensioned. 1927 G. E. Draycott *Technical Drawing* xiv. 219 (*heading*) Inking-in and dimensioning drawings. 1966 *McGraw-Hill Encycl. Sci. & Technol.* IV. 611/1 Working types of drawings may differ in styles of dimensioning. 1966 G. K. & H. J. Stegman *Archit. Drafting* ix. 211/1 A modular detail requires fewer small fractional dimensions than a detail dimensioned in the regular method.

dimensional, *a.* Add: **3.** Of, pertaining to, or involving dimensions of units or physical quantities.
1888 *Proc. Physical Soc.* X. 49 Prof. S. P. Thompson considered part of the difficulties of dimensional equations arose from the fact that no distinction was made between scalar and vector quantities. 1914 *Physical Rev.* 2nd Ser. IV. 369 The application of dimensional reasoning to mechanical problems is often useful in the interpretation of model experiments. 1964 H. S. Hvistendahl *Engin. Units* i. 10 The reason for the apparent ambiguity of the dimensions in such cases, is that our dimensional formulae do not distinguish the relative directions of the lengths combining as some power of L in those formulae.
b. *dimensional analysis*, a method of mathematical analysis in which physical quantities are expressed in terms of their dimensions (i.e. length, mass, time, etc.) so that conclusions may be drawn regarding their mutual relations; it is based on the fact that quantities added to or equated with each other must have the same dimensions.
1922 P. W. Bridgman *Dimensional Analysis* ii. 17 The purpose of dimensional analysis is to give certain information about the relations which hold between the measurable quantities associated with various phenomena. 1949 O. G. Sutton *Sci. of Flight* iii. 68 In analysing complex physical phenomena.., mathematicians..employ a dissection process known as *dimensional analysis*. 1957 *Encycl. Brit.* VI. 387/2 The principal use of dimensional analysis is to deduce from a study of the dimensions of the variables in any physical system certain necessary limitations on the form of any possible relationship between those variables. 1958 J. S. Scott *Dict. Civ. Engin.* 105 *Dimensional analysis*, in the model analysis of rivers, ships, ports..and other structures involving the flow of fluids, the use of dynamic similarity to enable the experimenter to see quickly what is the relationship between different variables.

dime·nsionally, *adv.* [f. Dimensional *a.* + -ly[2].] With regard to dimensions; in a dimensional way.
1888 *Proc. Physical Soc.* X. 44 We can transform both dimensionally and numerically by the relation $[K^{-1}\mu^{-1}] = v[LT^{-1}]$. 1927 I. B. Crandall *Theory Vibr. Syst.* ii. 56 Dimensionally, the *r* in this formula merely offsets other linear dimensions in the definition of μ and ρ. 1957 *New Biol.* XXII. 58 The choice of this dimensionally insignificant object for detailed study. 1969 L. Young *Syst. Units Electr. & Magn.* x. 126 To each constant of proportionality that is treated dimensionally there corresponds a base unit.

dimensionless, *a.* Add: **1. d.** Of a physical quantity or its unit: having no dimensions (*dimension *sb.* 3 c); of the nature of a pure number or ratio, and therefore having a value independent of the choice of units for other quantities.
1904 *Proc. Inst. Mech. Engineers* 248, *K* is a dimensionless quantity approximating to unity, and the working formula becomes $V = K\sqrt{(2P/\rho)}$. 1914 *Physical Rev.* 2nd Ser. IV. 345 The coefficients of a complete equation are dimensionless numbers, i.e. if the quantities *Q* are measured by an absolute system of units, the coefficients of the equation do not depend on the sizes of the fundamental units but only on the fixed interrelations of the units which characterize the system and differentiate it from any other absolute system. 1970 *Nature* 28 Nov. 889/2 It would..be incorrect to say: 'the molecular weight of protein X is 25,000 daltons', for the dalton is a unit of mass, and molecular weight is dimensionless.

dimer (dəi·məɹ). *Chem.* [f. Di-[2] + Gr. μέρ-ος share, part (after *polymer*).] A compound related to some other compound by having twice the number of atoms of the various elements in its molecule; usually, a compound the molecule of which is a double molecule formed by the joining together of two identical molecules.

1930 *Jrnl. Amer. Chem. Soc.* LII. 320 From the distillate was isolated a very small amount..of a crystalline solid having the analytical composition of tetramethylene carbonate. Molecular weight determinations indicated that it is a dimer: $\overline{COO(CH_2)_4OCOO(CH_2)_4O}$. 1951 S. Coffey tr. *Wibaut's Org. Chem.* vii. 105 If two molecules of the monomer combine, the polymer so formed is called a *dimer*, if three molecules combine, a *trimer*..and so on. 1964 J. W. Linnett *Electronic Struct. Molecules* iii. 50 Let us consider the dimer of CN, namely cyanogen, C_2N_2. 1965 *New Scientist* 29 Apr. 291/1 An excited molecule can join with a ground-state molecule to form a new double molecule (or dimer).

dimercaprol (dəi:məɹkæ·prǫl). *Chem.* [Abbrev. of *dimercaptopropanol*; the full form is also used occas.] = B.A.L. (see *B III).
1945 *Nature* 24 Nov. 617/1 The new compound 2 : 3 dimercaptopropanol (III) was made. 1947 *Chem. Abstr.* 4570 Antimalarials, dimercaprol, radiotherapy, and rutin are discussed. 1949 H. W. Florey et al. *Antibiotics* II. xxi. 815 Dimercaptopropanol (also known as 'British Anti-Lewisite' (BAL)). 1963 *Brit. Pharmaceutical Codex* 267 Dimercaprol is used in the treatment of acute poisoning by arsenic, mercury, gold, bismuth, thallium and antimony.

dimeric (dəime·rik), *a.* [f. Gr. διμερ-ής bipartite + -ic; in sense 2 formed directly on *dimer.] **1. a.** *Zool.* Bilateral; having a right and left side. **b.** *Bot.* = Dimerous *a.* b.
1897 L. H. Bailey in *Ann. Rep. Smithsonian Inst.* 455 A comparison of bilateral or dimeric animals with rotate or polymeric animals. 1900 in B. D. Jackson *Gloss. Bot. Terms* 78 *Dimeric, dimerous*, with two members in each part or circle.
2. *Chem.* That is a *dimer; having a molecular formula in which the numbers of atoms are twice those for some other compounds.
1929 *Jrnl. Amer. Chem. Soc.* LI. 2553 The products of such reactions..have..been assumed to be dimeric (and cyclic) for no other reason than that they were obviously not monomeric. 1938 [see next]. 1964 J. W. Linnett *Electronic Struct. Molecules* vii. 107 Dimerization to the tetrazane occurs..because, in the dimeric molecule, each of the nitrogen atoms has a formal charge of zero.

dimerize (dəi·mĕrəiz), *v.* [f. *dimer(ic *a.* + -ize.] **a.** *trans.* To render dimeric; to form a dimer of. **b.** *intr.* To undergo dimerization; to be converted into a dimer. Hence **dimeriza·tion**, the process by which two identical molecules join to form a single molecule; the action of forming a dimer; **di·merized, di·merizing** *ppl. adjs.*
a 1855 C. B. Mansfield *Theory of Salts* (1865) III. ii. 247 It does not at all follow..that therefore two molecules of the same body, when formed side by side, shall become polymerized or dimerized into a compound of double equivalent weight. 1931 *Chem. Abstr.* XXV. 5139 (*title*) Dimerization of isoprene. 1938 H. Gilman *Org. Chem.* I. iv. 442 Two types of addition of alkali metals take place. One is a simple 1, 2-addition..and the other has been termed a dimerizing addition... The dimerized addition products show the general reactions of Grignard reagents. 1938 *Jrnl. Chem. Soc.* 288 In the formation of a methylene-dicyclane by dimerisation of an open-chain diene the first step is..the production of an open-chain dimeric structure. 1952 *Chem. Rev.* L. 485 Fluoranthene dimerizes by a free-radical mechanism. 1959 Cram & Hammond *Org. Chem.* xv. 352 Olefins which are heavily substituted with halogen and other electron-attracting substituents dimerize to form cyclobutanes at high temperatures. 1964 J. W. Linnett *Electronic Struct. Molecules* iii. 45 Why does NO not dimerize to N_2O_2 since it has an odd number of electrons? 1965 *New Scientist* 30 Dec. 921/2 The thioketones..tend to dimerise spontaneously—that is, to form new compounds out of molecular pairs. 1970 *Nature* 28 Nov. 805/2 The bifunctional protein of the double mutant retains the ability to dimerize.

dimethyl. Add: **2.** *dimethylamine*, a colourless gas, $(CH_3)_2NH$, which liquefies at 7° C. and has an ammoniacal odour; *dimethyl phthalate*, a colourless liquid ester, $C_6H_4(COOCH_3)_2$, used esp. in insect repellents and as a plasticizer.
1852 Watts tr. *Gmelin's Handbk. Chem.* VII. 319 Dimethylamine..is formed as a hydriodate..by the action of ammonia on iodide of methyl. 1910 *Jrnl. Soc. Chem. Ind.* XXIX. 1398/2 Perfectly clear solutions of rosin are obtained by warming it with 30 per cent. of its weight of dimethyl or diethyl phthalate. 1944 Knipling & Dove in *Jrnl. Econ. Ent.* XXXVII. 479/2 Dimethyl phthalate had been found by the Orlando workers to be a good flea repellent. 1946 *Lancet* 13 Apr. 559/2 Dimethyl-phthalate (D.M.P.), a repellant used against the mosquito during the war. 1947 *Thorpe's Dict. Appl. Chem.* (ed. 4) VIII. 7/2 Dimethylamine has been employed as an accelerator in vulcanising rubber, and in the manufacture of detergent soaps. 1947 R. L. Wakeman *Chem. Commercial Plastics* xxii. 665 The number of plasticizers which can be used successfully with cellulose acetate is also restricted. Dimethyl and diethyl phthalates ..are often used.

Dimetian, var. *Demetian *a.* and *sb.*

dimidiately (dimi·diĕtli), *adv.* [f. Dimidiate *a.* + -ly[2].] In a dimidiate manner.
1857 T. Moore *Handbk. Brit. Ferns* (ed. 3) vii. 168 Pinnules obovate, obliquely ovate, or dimidiately subquadrate. 1887 W. Phillips *Brit. Discomycetes* 51 Cup subsessile, externally pruinose, dimidiately elongated, or obliquely contorted.

Dimini (dimī·ni). Also **Dh-**. [Gr. Διμήνι.] The name of a locality of north-eastern Greece used *attrib.* to designate a kind of pottery ornamented with spirals, found there by excavation.
1912 Wace & Thompson *Prehist. Thess.* 16 Painted pottery, Dhimini ware. 1925 V. G. Childe *Dawn Eur. Civiliz.* 69 The invaders [of Eastern Thessaly]..introduced a new pottery, Dimini ware, rather coarser than the older fabric.

diminished, *ppl. a.* Add: **1.** *diminished return* (cf. *diminishing *ppl. a.* 1 b).
1817 D. Ricardo *Princ. Pol. Econ.* ii. 61 On the best land, the same produce would still be obtained with the same labour as before, but its value would be enhanced in consequence of the diminished returns obtained by those who employed fresh labour and stock on the less fertile land. 1843 J. R. McCulloch *Princ. Pol. Econ.* (ed. 3) III. vii. 494 The greatest amount of capital and labour may be employed in fashioning raw produce and adapting it to our use, and in transporting it from where it is produced to where it is consumed, without a diminished return.
b. *diminished responsibility*: a state of mental disturbance or abnormality, not classifiable as insanity, but recognized in law as a ground for exempting a person from full liability for criminal behaviour.
1957 *Act* 5 & 6 *Eliz. II* c. 11 § 2 Persons suffering from diminished responsibility. Where a person kills..he shall not be convicted of murder if he was suffering from such abnormality of mind..as substantially impaired his mental responsibility. 1958 *Spectator* 31 Jan. 126/1 Suppose that at his trial Taper should plead diminished responsibility, alleging that Sir Reginald's speeches had driven him.. barmy. 1963 *Times* 6 Feb. 6/2 Found Not Guilty of murdering the girl but Guilty of manslaughter on the grounds of his diminished responsibility.
2. Delete 'now only in phr. from Milton' and add further examples.
1925 Belloc *Mr. Petre* i. 22 A diminished man. One who had 'had an accident'. 1932 A. Huxley *Brave New World* xii. 204 Bernard had to slink back, diminished, to his rooms. 1960 *Woman's Own* 3 Dec. 18/3 Amy hated..to have him look like that, diminished and remote.
4. c. *diminished chord*, a chord containing a diminished interval or intervals. **d.** *diminished seventh* (*chord*), a chord in which the interval between the outer notes is a diminished seventh.
1839 [see Seventh *sb.* 2]. 1920 P. C. Buck *Unfigured Harmony* (ed. 2) 12 Diminished sevenths are extremely useful, though more liable to be over-used than any other chord. 1926 A. Niles in W. C. Handy *Blues* 16 The device.. of ending up the tune on the diminished seventh chord. 1949 L. Feather *Inside Be-Bop* ii. 71 The diminished chord, for no apparent reason, is used less frequently in bebop than in earlier jazz. 1953 R. Vaughan Williams *Some Thoughts on Beethoven's Choral Symphony* 49 So is the diminished seventh of Bach's *Barabbas* a well-worn device. 1955 G. Abraham in H. van Thal *Fanfare for E. Newman* 14 Equally characteristic of the more mature Wagner is the lightening of the diminished sevenths.., leaving a purely string chord. 1958 C. Wilford in P. Gammond *Decca Bk. Jazz* ii. 32 The harmonies of ragtime..are made distinctive by very wide use of diminished chords.

diminishing, *ppl. a.* Add: **1. a.** spec. *diminishing glass*, an instrument which causes objects to appear smaller than they appear to the naked eye; *diminishing mirror*, a convex mirror in which the image is reduced in scale; *diminishing rod*, that part of the mechanism of a cotton-roving machine which gives the bobbins of roving their conical ends.
1665, 1816 [in Dict.]. 1861 Dickens *Gt. Expect.* II. xiv. 225 A private sitting-room..fitted up with a diminishing mirror. 1868 C. M. Yonge *Chaplet of Pearls* I. i. 2 They [*sc.* children] looked like a full-grown couple seen through a diminishing-glass. 1890 J. Nasmith *Mod. Cotton Spinning* x. 168 The slide in its reciprocal vertical movement causes, by means of the 'diminishing rod' or 'hangar bar', the upper cradle to oscillate in its centre. 1896 *Daily News* 28 May 2/2 Diminishing and magnifying glasses such as are used by artists.
b. Esp. in *the law of diminishing return*(*s*), in Economics, the principle that the expenditure of labour or capital beyond a certain point does not produce a proportionately corresponding return (see quot. 1938); also *transf.*
1815 E. West *Ess. Application of Capital to Land* 12 In the progress of improvement an equal quantity of work extracts from the soil a gradually diminishing return. 1848 Mill *Pol. Econ.* I. xii. § 2. 216 The general law of diminishing return from land would have undergone..a temporary supersession. 1883 F. A. Walker *Pol. Econ.* II. 23 The great comprehensive principle to which we give the name, 'the law of diminishing returns in agriculture'. 1930 J. S. Huxley *Bird-Watching* iv. 61 From the point of view of..the collector of birds seen and recorded, the law of diminishing returns has set in; only very seldom is he rewarded by the sight or sound of a species new or little known to him. 1938 F. Benham *Economics* ix. 130 We may now state the Law of Diminishing Returns. As the proportion of one factor in a combination of factors is increased, after a point, the marginal and average product of that factor will diminish. 1965 *Listener* 24 June 926/1, I, for one, am not prepared to accept that the law of diminishing returns applies to television. I do not believe that what is trivial and bad will drive out what is serious and good. 1970 *Times* 7 Jan. 9/8 The law of diminishing returns has begun to set in: we can't go on forever extending our hearts to shapely matrons.

diminuendo. Add: (Further examples.) Also *attrib.* or as *adj.*
1890 W. James *Princ. Psychol.* II. xviii. 71 Certain violin-players take advantage of this in diminuendo terminations. 1937 A. Huxley *Let.* 15 Dec. (1969) 429 Pardon this diminuendo, due to my not noticing that the ribbon had to change its direction! 1955 *Times* 9 May 3/4 There were 'expressive' ritenutos and diminuendos and protracted cadences. *Ibid.* 26 May 11/6 Labour's promise to do away with the 11 plus examination for selective secondary education was another scheme that suffered a *diminuendo*. 1958 A. J. Toynbee *East to West* xiii. 38 En route one changes, *diminuendo*, from South Australian broad gauge to Commonwealth standard gauge. 1959 *Encounter* Aug. 34/1 He was looking a bit diminuendo, and smiled rather nervously.
Hence as *v. intr.*, to become quieter or fainter; to grow less.
1901 *Westm. Gaz.* 12 Nov. 2/1 Their booming note crescendoes up the scale with increasing speed and diminuendoes with the slackening of it. 1905 *Daily Chron.* 27 Oct. 6/4 The wail of Niobe diminuendoes in the receding distance. 1962 *Guardian* 19 Oct. 11/7 He began to howl and diminuendoed down to a mutter.

diminution. Add: **10.** *Cytology.* [a. F. *diminution* (V. Herla 1895, in *Arch. de Biol.* XIII. 485).] The loss or expulsion, during the embryogenesis of certain organisms, of some chromosome material from the nuclei of cells that go to form somatic tissue.
1925 E. B. Wilson *Cell* (ed. 3) iv. 326 In these cases.. the process of diminution is somehow connected with the segregation of germ-cells from somatic cells. 1942 *Nature* 17 Jan. 67/2 In Sorghum we know that the chromosomes which undergo 'diminution' are in fact dispensable not only in parts of the plant but also in parts of the species. 1965 C. D. Darlington *Cytology* II. iii. 658 Coordinated reactions of centromeres, heterochromatin and cytoplasm are no doubt responsible for diminution.

dimit, var. *demit *sb.*

dimmer, *sb.* Add: *spec.* a device for reducing the brilliance of a light, esp. in a theatre, cinema, etc. Also *fig.* and *attrib.*
1905 *Chambers's Jrnl.* 26 Aug. 624/2 What is called a 'dimmer' is in use at some such places [*sc.* theatres] already. 1909 *Cent. Dict. Suppl.*, *Dimmer*, in *elect.*, an adjustable reactive coil used for reducing the amount of light of incandescent lamps. 1913 *Work* 14 June 212/1 The resistance of the dimmer coils. 1916 H. L. Wilson *Somewhere in Red Gap* ix. 378 It was a suit that the automobile law in some states would have compelled him to put dimmers on. 1926 H. T. Wilkins *Marvels Mod. Mechanics* 236 Levers on these dials operate the 'dimmers'. 1933 P. Godfrey *Back-Stage* i. 18 The electrician has been busy sliding the stops on his dimmer-board.

dimmer, *v.* Delete *nonce-wd.* and add earlier and later examples.
1873 C. G. Leland *Egypt. Sketch-Bk.* 22 He..looked over the top-rail at the beautiful Monte Christo, which was dimmering in the distance. 1914 Kipling *Lett. Travel* (1920) 269 Then the river dimmered up like pewter.

dimmish, *a.* (Further examples.)
1884 'Mark Twain' *Huck. Finn* xxxvi. 372 We was a standing there in the dimmish light. 1967 A. Laski *Seven Other Years* xiii. 183 An amiable but..'dimmish' heavy-weight of fifteen. 1969 J. Drummond *People in Glass House* xvii. 72 She was a sort of goose-girl type...Dimmish. But pretty.

dimoric (dəimǭ·rik), *a.* *Pros.* [f. Di-[2] + Mora[1] 3 + -ic.] Containing two *moræ*; having the length of two short syllables.
1901 [see Trimoric *a.*].

dimorphemic (dəimǫɹfī·mik), *a.* *Linguistics.* [f. Di-[2] + *morphemic *a.*] Containing, or belonging to, two morphemes.
1936 *English Studies* XVIII. 160 *Glazier* differs further from *glacier* in that it is dimorphemic. 1952 A. Cohen *Phonemes Eng.* iv. 102 The dimorphemic diphthongs [i-ə] and [u-ə] are manifested by the tendency to preserve the length of the first vowel.

dimorphotheca (dəimǭ:ɹfoþī·kă). *Bot.* [mod.L. (S. Vaillant 1722, in *Hist. et Mémoires Acad. Sci. 1720* 279), f. Gr. δίμορφος Dimorphous *a.* + Theca 2.] A plant of the South African genus of herbs or sub-shrubs so named, of the family Compositæ; the Cape Marigold (see Marigold 1 d).
1861 *Curtis's Bot. Mag.* LXXXVII. 5252 (*heading*) Grassy-leaved Dimorphotheca. 1866 Lindley & Moore *Treas. Bot.* I. 409/1 *Dimorphotheca*, the Cape Marigold. 1890 *Garden* XXXVIII. 180/3 Like many other handsome composites from the Cape, this Dimorphotheca had long ceased to obtain much notice as a garden plant. 1960 *Times* 10 Sept. 9/3 The dimorphothecas have also done well this year.

dim-out (di·maut). [f. phr. *to dim out* (*dim *v.* 2 c).] A reduction in the brightness or use of lighting, e.g. as a precaution against air-raids; in a theatre or cinema, the ending of a scene, etc., by a slow rather than fast diminution of lighting; the resulting partial darkness. Also *transf.* and *attrib.* Cf. *black-out.
1942 *N.Y. Times* 3 May 2E/5 In New York City the dim-out recalled the 'lightless nights' caused by the coal shortage in 1917. 1943 J. Steinbeck *Once there was War* (1959) 3 It is evening, and the first of the dim-out lights come on. 1944 *Times* 9 Nov. 2/3 One of the effects of the reduced gas-pressure at Manchester will be that the dim-out in the city will be more dim. 1947 'N. Blake' *Minute for Murder* v. 100 Darkness. Damn the dim-out. 1960 Lindsay & Crouse *Sound of Music* 7 Sister Sophia enters below the curtains and crosses the stage, carrying a large ring of keys. Dim Out. 1965 *B.B.C. Handbk.* 11 It would be naïve to suppose the new [television] channel was launched into an entirely friendly world. The black-out of the first night was succeeded by the dim-out of the first few months. 1968 O. Wynd *Sumatra Seven Zero* iv. 54 There didn't appear to be any activity at the hotel entrance, and there was a dim-out down there, as though all the staff had gone to bed.

dimps. Add: (Further examples.) Also **dampse, dempse, dimpse(s), dimpsey, dimpsie,** etc.
1810 *Monthly Mag.* XXIX. ii. 435/1 *Dumps, dimpse, dampse, dimmet*, twilight. D[evonshire]. 1867 W. F. Rock in W. W. Skeat *Nine Specimens Eng. Dial.* (1896) 47, I thort I glimpsed Jan slinge to tha rebeck i' the dimpse. 1885 F. T. Elworthy *Rep. Comm. Devonshire Verbal Provincialisms* 91 Just as the dempse was coming on. 1891 R. P. Chope *Dial. Hartland* 40 Twaz gittin' dimps avore us stairted. 1892 H. C. O'Neill *Devonshire Idyls* 25 But 'twas got quite dark all of a minute, and 'twere only in the dimpses when the maid shut her eyes. 1896 G. Chanter *Witch of Withyford* xii. 149 There be pixies in the dimpsey here. 1925 *Glasgow Herald* 1 Aug. 4 At the 'dimpsie' the colours of the flowers become more etherealised. 1950 L. A. G. Strong *Which I Never* 161 'Twasn't dark. 'Twasn't more than dimpsey. 1965 in P. Jennings *Living Village* (1968) 95 The Somerset village of Compton Martin..is on the lower north flanks of the Mendips...Local words include..*Dimpsey*: dark, twilit.

dim-wit, dimwit (di·mwit). *colloq.* (orig. *U.S.*). Also **dim wit.** [f. Dim *a.* + Wit *sb.*] A stupid or slow-witted person.
1922 *Dialect Notes* V. 141 She's the worst dim-wit on campus. 1925 *New Yorker* 20 June 15/2 An archduke, a sort of royal dim wit. 1943 [see *bird brain* s.v. *bird *sb.* 9]. 1944 *Penguin New Writing* XIX. 71 'What's the matter?' 'Cramp, you dimwit, cramp!' 1956 'J. Wyndham' *Seeds of Time* 217 He had an uncomfortable awareness of how many ways there were for even a dimwit to contrive a fatal accident.
Hence **dim-witted** *a.*, stupid, dull; **dim-wittedness**, the quality of being dim-witted.
1940 E. B. Mann *Troubled Range* (1941) xi. 135 Of all the damn' dim-witted stunts! 1948 'E. Crispin' *Buried for Pleasure* ix. 68 They say he's got 'a madman's cunning', which is their excuse for being too dim-witted to catch him. 1960 *20th Cent.* Oct. 359 Some Northerners..can give an impression of dim-wittedness.

dinah (dəi·nă). *slang.* [Corruption of Dona 2.] A man's sweetheart or favourite woman.
1898 [see Tart *sb.* 2]. 1909 J. R. Ware *Passing Eng.* 109/2 *Dinah*, a favourite woman; *e.g.* 'Is Mary your Dinah?'

Dinantian (dinæ·nʃiăn), *a.* *Geol.* [ad. F. *dinantien* (A. de Lapparent *Traité de Géol.* (ed. 3, 1893) II. 819), f. *Dinant*, a town in Belgium + -ian.] The name of the series of rocks in continental Europe deposited during the Lower Carboniferous period, and of the corresponding geological epoch. Also *absol.*
[1888 J. Prestwich *Geol.* II. i. 8 Belgium..Carboniferous ..Lower..Limestones of Dinant.] 1903 A. Geikie *Text-bk. Geol.* (ed. 4) II. vi. ii. 1051 The Carboniferous system of the European continent has been grouped by some geologists in three major divisions: 1st, the Lower (Culm, Dinantian), comprising all the Lower Carboniferous rocks up to the Millstone Grit, [etc.]. 1905 *Q. Jrnl. Geol. Soc.* LXI. 264 The designation Dinantian, as including the whole of the Carboniferous Limestone, seems hardly justified by the continuity or clearness of the sequence in the district which has suggested the name. 1959 *Chambers's Encycl.* III. 104/2 The Carboniferous System is well developed throughout Europe where the Dinantian, Namurian, Westphalian and Stephanian are recognized. 1969 T. N. George in *C.R. 6me Cong. Int. Stratigraphie & Géol. du Carbonifère, 1967* I. 193/1 British Dinantian rocks were deposited under conditions of sustained subsidence. *Ibid.* 197/2 It is still uncertain whether Vaughan's Avonian precisely equates with the earlier-named Dinantian.

dinar. Add: (Further examples.) Now the standard monetary unit of Iraq, Jordan, Kuwait, Tunisia, etc. (see quots.).
1811 P. Kelly *Universal Cambist* I. 346 Persia. Accounts are kept in Tomans of 50 Abassis..or 10,000 Dinars simple. 1931 W. F. Spalding *Tate's Money Manual* I. 54 The monetary unit [of Iraq] will be the gold dinar of 1000 fils. 1951 *Statesman's Year-Book* 1211 On 1 July, 1950, Jordan began to issue its own currency, the Jordan dinar, divided into 1,000 fils. 1958 *Economist* 8 Nov. 534/2 A new currency made its bow to the exchange markets this week. It is the Tunisian dinar, based on the French franc...The dinar has been given an equivalent of 1,000 French francs. 1962 *Statesman's Year-Book* 1186 The Kuwait Dinar (at par with the £ sterling) of 1,000 fils, replaced the Indian external Rupee on 1 April 1961. 1970 *Financial Times* 23 Mar. 12/2 The average per capita income [in Algeria] of 910 dinars.
b. [a. Serbian *dinar*, f. L. *dēnārius* Denarius.] The monetary unit of Yugoslavia (formerly of Serbia).
1882 *Statesman's Yearbook* 402 The Servian dinar is equal to one franc. 1907 *Macm. Mag.* Sept. 839 In Belgrade ..they gauged my ignorance of the number of nickel *paras* and *piastres* that go to the *dinar*, or franc. 1926 *Survey of Budget 1927–8* (Yugoslavia) 3 The changes in the economic conditions of this country, which were the consequences of the stabilisation of the dinar. 1927 *Economic Jugoslavia* 34 The National Bank..has paid up capital to the amount of 30 million dinars.

Dinaric (dinæ·rik), *a.* [f. *Dinara*, a mountain in Dalmatia + -ic.] Denoting a mountain range which extends in a south-easterly direction along the eastern side of the Adriatic, and a race of people inhabiting the coast of the northern Adriatic, characterized by tall stature, a very short head, dark wavy hair, and straight or aquiline nose. Hence as *sb.*, a member of this race.
1833 *Penny Cycl.* I. 387/1 The Dinaric Alps. 1862 *Chambers's Encycl.* III. 572/1 Dinaric Alps, that branch of the Alpine system which connects the Julian Alps with the western ranges of the Balkan. 1898 *Pop. Sci. Monthly* Oct. 732 The Adriatic or Dinaric race. 1900 Deniker *Races of Man* ix. 333 Dark, brachycephalic, tall race, called Adriatic or Dinaric, because its purest representatives are met with along the coast of the Northern Adriatic and especially in Bosnia, Dalmatia, and Croatia. 1924 *T. P.'s & Cassell's Weekly* 13 Sept. 658/2 The pyramid-headed man from the Dinaric lands. 1928 C. Dawson *Age of Gods* xvi. 370 There is..no evidence for the occurrence of the Dinaric type in the neolithic period. 1939 C. S. Coon *Races of Europe* v. 140 This is our first meeting with the Dinaric race. *Ibid.*, The stature was tall, as with modern Dinarics, and the long bones slender. 1944 A. Holmes *Princ. Physical Geol.* xviii. 397 Farther south the backward thrusts maintain their direction and mark the beginning of the Italian ranges which pass round the head of the Adriatic into the Dinaric Alps.

Dinas (di·nās). [f. *Dinas* Rock, in the Vale of Neath, Wales.] Used *attrib.* in **Dinas brick**, fire-brick made from **Dinas clay**, a kind of rock consisting almost entirely of silica.
1875 Hunt & Rudler *Ure's Dict. Arts* (ed. 7) I. 532 Stone bricks..are manufactured at Neath, in Glamorganshire...They are usually known as the 'Dinas bricks'. 1879 *Encycl. Brit.* IX. 844/1 Dinas clay, which is really nearly pure silica. 1880 *Ibid.* XIII. 294/2 Dinas brick, which perfectly resists the ordinary steel melting temperatures of coke-fired furnaces.

dincum, var. *dinkum *sb.* and *a.*

din-din (di·ndin). *colloq.* Also **din-dins.** [Childish or jocular reduplication of Din(ner *sb.*)] Dinner.
1905 E. M. Forster *Where Angels fear to Tread* ii. 47 'Din-din's nearly ready,' said Lilia. 1929 Wodehouse *Summer Lightning* xiii. 265 'Come along, Carmody. Din-dins.' Hugo had sunk into a chair. 'I don't want any dinner,' he said, dully. 1957 E. Taylor *Angel* v. 220 Must be time for dindins, madam. 1958 F. Norman *Bang to Rights* 113 If you get captured with one you'll get a few days no din-din.

dine, *v.* Add: **1. b.** *to dine out* (examples) (cf. Diner 1 b, Dining *vbl. sb.* 1 b); *to dine out on*: to be given hospitality at dinner partly or chiefly for the sake of one's conversation or knowledge about (a specific incident or topic, etc.).
1816 Jane Austen *Emma* II. vii. 120 They will not take the liberty with you; they know you do not dine out. 1835 Dickens *Let.* (1965) I. 67, I have received an Invitation to dine out to day. 1852 W. S. Landor *Let.* 11 Apr. in *N. & Q.* (1968) CCXIII. 414/1, I never dine out, or go into parties in the evening. 1923 W. S. Maugham *Our Betters* III. 170 Don't you remember that killing story about your father's death. You dined out a whole season on it. 1934 N. Marsh *Man lay Dead* xv. 268 In a couple of years you will be dining out on this murder. 1936 A. Christie *Cards on Table* vii. 66 If I were only to dine in houses where I thoroughly approved of my host I'm afraid I shouldn't dine out very much. 1951 J. C. Fennessy *Sonnet in Bottle* vii. 245 He does a very good imitation of it—he's dined out on it ever since. 1966 H. Yoxall *Fashion of Life* xi. 109, I dined out for the rest of this African journey on my acquaintance with the Princess's fiancé. 1970 'D. Halliday' *Dolly & Cookie Bird* v. 66 I've dined out on a few stories about her.

diner. Add: **2. b.** A restaurant, orig. and still occasionally one built to resemble a railway dining-car. *N. Amer.*
1935 *Amer. Mercury* July 311/2 The diner was back from the sidewalk, under a couple of big trees. 1960 *Weekend Mag.* 23 Apr. 42/2 My own test began at a roadside diner.. over coffee. 1965 *New Statesman* 5 Nov. 713/1 These diners are roughly equivalent to our pull-ups and the French *relais routiers* before they were turned into haunts of gastronomic fashion. The first 'diners' were old Pullman dining-cars sold off by railway companies and I was amused to see that the newest ones, specially built, attempt to imitate the dining-cars.

dineric (dəine·rik), *a.* *Physics.* [f. Di-[2] + Gr. νηρ-ός wet: see -ic.] Of or pertaining to the interface of two liquids.
1905 *Jrnl. Physical Chem.* IX. 546 The exponential formula has been found to describe a large number of dineric equilibria, especially organic liquids, over very wide ranges of concentration. 1915 *Ibid.* XIX. 275 To avoid circumlocution the surface separating two liquid phases will be called a dineric interface.

dinette (dəi·net, dəine·t). orig. *U.S.* [irreg. f. Dine *v.* + -ette.] **a.** A small room, an alcove, or part of a room set aside for meals. **b.** A set of articles of dining furniture, usu. compactly designed. Also *attrib.* **c.** A small restaurant.
1930 *Ladies' Home Jrnl.* Jan. 38/3 Two clumsy, heavy French doors between the living room and the dinette. 1931 *Sears Catal.* Spring 622 (*heading*) Stylish dinette at bigger savings. 1940 S. Lewis *B. Merriday* ii. 26 Fun, frolic, and 'fried dogs' at Dinty's dinette. 'Where all of Sladesbury's Bohemia hangs out.' 1942 *Archit. Rev.* XCI. 101/3 The usual worker's house [in Canada] has three bedrooms, a living-room, kitchen with 'dinette', bathroom, [etc.]. 1953 W. R. Burnett *Vanity Row* viii. 63 A little dinette, hidden by a beaverboard partition. 1955 J. Cannan *Long Shadows* iii. 35 The kitchenette is situated in close proximity to the dinette. 1957 W. H. Whyte *Organiz. Man* xxiv. 313 A wife was so ashamed of the emptiness of her living room that she smeared the picture window with Bon Ami; not until a dinette set arrived did she wash it off. 1962 *Radio Times* 29 Mar. 28/2 The divans and 'dinette' seats [in a caravan] are upholstered in deep foam rubber. 1970 *Globe & Mail* (Toronto) 26 Sept. 52/3 (Advt.), 20% off! Dinette sets. Has drop-leaf table, two well-padded chairs.

ding, *sb.*[1] Add: **2.** *Surfing.* (See quot. 1962.)
1962 T. Masters *Surfing made Easy* 64 *Dings*, dents or holes in surfboard. 1968 W. Warwick *Surfriding in N.Z.* 17/2 When repairing a ding or damaged area on your board, don't rush. 1970 *Surf* I. x. 33/1 Real ridiculous—we all got dings.

Dingaan('s) apricot (di:ŋgān(z) ē¹·prikǫt). [f. the name of *Dingaan*, chief of the Zulus from 1829 to 1840.] A name in Natal for the Kei-apple.
1853 E. Armitage in *Jas. Chapman's Trav.* (1868) II. 449 The Kei apple, or Dingan's apricot, invaluable for forming thorny fences and yielding a pleasant fruit. 1891 R. Russell *Natal* 31 The Dingaan apricot, or Kaw apple.

ding-a-ling. [Echoic.] = Ding *sb.*[2]
1894 E. L. Banks *Campaigns Curios.* 77, I was awakened by the ding-a-ling of the front-door bell.

‖ **Ding an sich, Ding-an-sich** (di:ŋ an zi·χ). *Philos.* Also **ding an sich.** [G.] A thing in itself (see Thing *sb.*[1] 14 e). Cf. *an sich.
1865 S. H. Hodgson *Time & Space* i. 25 The Ding-an-sich is that which cannot be reached or affected by consciousness. 1872 [see Noumenalize *v.*]. 1875 G. H. Lewes *Problems* II. 419 Metaphysicians have erected the idle mystery of the *Ding an sich*. 1897 *Mind* VI. 240 This Reality for us remains little more than a *Ding an Sich*. 1940 Auden *Another Time* 69 Do all clerks for instance Pigeon-hole creation, Brokers see the Ding-an-Sich as Real Estate?

dingbat (di·ŋbæt). *slang.* Also **ding bat, ding-bat.** [? f. Ding *v.*[1] + Bat *sb.*[2]]
1. *U.S.* In various uses (see quots.); esp. (*a*) a piece of money; *pl.* money; (*b*) = Thingummy (cf. *dingus); (*c*) a tramp or hobo.
1838 in *Amer. Speech* (1963) XXXVIII. 10 We can take a '*Quaker*' before we start—apply a '*Ding Bat*' [some kind of drink?] at Providence. 1861 in Bartlett *Dict. Amer.* (ed. 4, 1877) 177 It has been found necessary to expend the *dingbats*, to put something more substantial on the 'fly' [= in motion]. 1864 G. A. Sala in *Daily Tel.* 19 Oct., Little John, erst a hog-driver..and recently in trouble for manufacturing bogus 'dingbats'. *Ibid.* 1 Nov., I paid for my Kissingen in five-cent 'dingbat' or 'spondulick'—two of the many names given to the fractional currency. 1877 Bartlett *Dict. Amer.* (ed. 4) 177 *Dingbat*, a bat of wood that may be thrown (dinged); a piece of money; a cannon-ball; a bullet. 1895 *Dialect Notes* I. 387 *Dingbat*. Mr. Philip Hale, of the *Boston Journal*, has been collecting information..concerning this word. The following definitions appear:—(1) Balls of dung on buttocks of sheep or cattle. (2) Blow or slap on the buttocks. (3) Flying missile. (4) Squabble of words or pushing. (5) Money. (6) In some of the N.E. schools, the word is student slang for various kinds of muffins or biscuit. (7) Affectionate embrace of mothers hugging and kissing their children. (8) Term of admiration. 'They are regular *ding-bats*' (speaking of girls). 1918 'A-No. 1' *Mother Delcasse of Hoboes* 44 Stew Bum..Ding Bat..Fuzzy Tail..the dregs of vagrantdom. 1923 *Frontier* May 10 That blasted 'ding bat' of a Ford, as Stub calls it, just naturally stood on its hind legs..and turned a flip-flop. 1926 J. Black *You can't Win* vi. 65 If you was some kind of a rank dingbat you wouldn't have been invited down here. 1931 J. Thurber *Owl in Attic* ii. 78 It is sitting on a strange and almost indescribable sort of iron dingbat. 1944 F. Brown *Angels & Spaceships* (1955) 208 It was his dingbat. I mean, he made it and he thought he knew what it was.
2. *Austral.* and *N.Z.* In pl., esp. in phr. *to have the dingbats*, *to be dingbats*, to be mad, stupid, eccentric; also, to be a victim of delirium tremens; *to give* (a person) *the dingbats*, to inflict a feeling of nervous discomfort.
1918 *Chrons. N.Z.E.F.* 27 Sept. 109/2 'Ave you got the dingbats? 1926 J. Devanny *Butcher Shop* x. 96 George, 'e 'ad the dingbats. 1933 L. Acland in *Press* (Christchurch)

14 Oct. 15/7 *Dingbats.* Slang, of Australian origin, for delirium tremens. The dingbats, I believe, are really the snakes, weasels, etc., which a sufferer sees. **1937** Partridge *Dict. Slang* 221/2 *Dingbats,* eccentric; mad, gen. slightly: Australian military. **1943** F. Sargeson in *Penguin New Writing* XVIII. 71, I knew it would give me the dingbats if I just stayed on there waiting. **1945** *Southern Cross* (London) 15 Dec. 4/3 Even old George, used as he was to pink snakes, ding-bats, and spotted elephants, seemed a little surprised. **1949** *Landfall* III. 146 Your mother's dingbats. **1959** G. Slatter *Gun in Hand* iv. 42 Boozin' again! You'll end up with the dingbats, you will.

3. *Austral.* [Perh. f. Ding(o + Bat(man^{2}.] An army batman.

1919 in Downing *Digger Dial.* **1940** *Bulletin* (Sydney) 3 Jan. 35/3 There is a vast difference between a dingbat in the British Army and one in the A.I.F.

ding-dong, *adv., sb.,* and *a.* Add: **B. 1. b.** *fig.* Esp. (*a*) a heated argument; a quarrel; 'cut and thrust'; (*b*) a tumultuous party or gathering. *colloq.*

1922 Joyce *Ulysses* 266 Yes, she was back. To the old dingdong again. **1928** *Manch. Guardian Weekly* 19 Oct. 301/1 Accustomed to cut a good figure in the ding-dong of public argument. **1933** H. Belloc *Charles I* 351 A ding-dong of assertion and counter-assertion. **1935** G. Ingram *Cockney Cavalcade* ix. 142 I've been having a ding-dong with my old man. **1936** N. Coward *Hands across Sea* in *To-Night at 8.30* II. 18 Are you going to Nina's Indian ding-dong? **1956** 'J. Wyndham' *Seeds of Time* 84 You can't have a proper ding-dong with those quiet ones. **1961** Ashley Smith *East-Enders* vi. 93 The sons and daughters..coming up for a ding-dong which went on till far into the night.

dinge (dindʒ), *sb.*^{2} [f. Dinge *v.*^{2}, or back-formation from Dingy *a.*] Dinginess.

1846 E. D. Bancroft *Let.* 2 Nov. (1904) 12, I cannot get accustomed to the London dinge. **1854** Thackeray *Newcomes* xxxv, A noble dinge, a venerable mouldy splendour. **1860** —— *Round. Papers* (1863) 117 The dinge and wrinkles of their wretched old cotton stockings. **1916** Galsworthy *Five Tales* (1918) 249 His mood threw a dinge even over the children. **1968** J. R. Ackerley *My Father & Myself* xvi. 182 The dust and dinge of the cluttered house.

dinge (dindʒ), *sb.*^{3} *U.S. slang.* Also **dingy.** [f. Dingy *a.*] A derogatory term for a Negro. Also *attrib.* or as *adj.*, esp. with reference to a jazz style developed by Negro musicians. (See also quot. 1942.)

1848 *Ladies' Repository* Oct. 316/1 *Covess dinge,* a negress, sometimes called *dinge blowen*... *Dinge kinch,* a negro child. ..*Dinge,* a negro man. **1904** in 'No. 1500' *Life in Sing Sing* xiii. 247. **1909** 'O. Henry' *Roads of Destiny* 134 These dingies will cheat you out of the gold in your teeth if you don't understand their ways. **1933** E. Hemingway *Winner take Nothing* 43 That big dinge took him by surprise..the big black bastard. **1940** R. Chandler *Farewell, my Lovely* i. 9 'A dinge,' he said. 'I just thrown him out.' *Ibid.* 10 'You say this here is a dinge joint?'..'I told you it's a coloured joint.' **1942** Berrey & Van den Bark *Amer. Thes. Slang* § 32/8 *Dinge,* of negroes, dark or dusky in colour, dingy. *Ibid.* § 385/14 Negro,..*dinge, dingy. Ibid.* § 576/2 *Dinge,*..a negro musician. *Ibid.* § 579/1 *Dinge,* 'Negro vibrato' played with a very rapid, violent shake. **1958** V. Bellerby in P. Gammond *Decca Bk. Jazz* xvii. 205 The 'dinge' piano trill, deriving from the efforts of the early Negro instrumentalists to sing through their instruments, instinctively holding the rich overtones of Negro speech. **1969** A. Hunter *Gently Coloured* i. 4 A big buck nigger. A dinge. A spade. *Ibid.* 8 A dinge bit. It has to be.

dinger (di·ŋəɹ). *dial.* and *slang* (orig. and chiefly *U.S.*). [f. Ding *v.*^{1} 3.] Something superlative; a 'humdinger'.

1809 *Amer. Mag.* Nov. 1 This land of our dads..is a dinger at nailing the scads. **1892** *Leeds Merc. Suppl.* 22 Oct. 8/8 *Dinger,* anything of a superlative character, as in size, quality, &c. 'It's a dinger.' **1904** *Topeka* (Kan.) *Daily Capital* 1 June 4 The alfalfa crop this year is going to be a 'dinger'. **1939** Steinbeck *Grapes of Wrath* iv. 37 See how good the corn come along until the dust got up. Been a dinger of a crop.

dinghy. (Now the usual spelling.) Delete ‖ and add: **2.** (Earlier example.) Also, one with an engine.

1818 'A. Burton' *Adv. J. Newcome* iii. 176 The coofs hae stown awa the dingey. **1932** T. E. Lawrence *Lett.* (1938) 757 Our two Squadrons both sent us their dinghies, asking us to check the timing and tune them. **1935** *Ibid.* 855 We launched the Dinghy: the quietest and sweetest tick-over of any Dinghy yet! **1957** *Encycl. Brit.* XV. 878/2 (*heading*) The Motor Dinghy. This is an open boat..driven by an engine installed inside..or, more usually, by an outboard motor.

c. An inflatable rubber boat, esp. one carried on an aircraft for use in an emergency. In full, *rubber dinghy.*

1939 *War Illustr.* 14 Oct. 156/2 We alighted and, after signalling the men in the boat, blew up our rubber dinghy and pushed it out with a line to each end. **1942** *Jrnl. R. Aeronaut. Soc.* XLVI. 3 When folded the dinghy forms a cushion..which is strapped to the seat-type parachute. **1958** *Daily Mail* 15 Aug. 2/7 He radioed: 'Can see wreckage of an aircraft. Two tyres floating in sea. Several inflated dinghies on surface.'

dingo, *sb.* Delete ‖ and add: **2.** *Austral. slang.* A contemptuous term for a person: a cheat, scoundrel, traitor, coward.

1928 'Brent of Bin Bin' *Up Country* xvi. 182 The bitch has twice the guts of the old dingo. **1941** K. Tennant *Battlers* viii. 90 The bagmen are a mob of dingoes. **1948** V. Palmer *Golconda* ix. 67 That old she-dingo..wants us to believe this boy was loosing hell on them. *Ibid.* xxxi. 261 I'd be a hell of a dingo..if I didn't help you now.

dingo (di·ŋgo), *v. Austral. slang.* [f. the sb.] **a.** *intr.* To retreat, back out, act in a cowardly or treacherous manner; *to dingo on* (someone) (see quot. 1941). **b.** *trans.* To back out of; to shirk.

1935 *Bulletin* (Sydney) 29 May p. ii/4, I gave him a rather hot time for the first half; in the second round he 'dingoed', letting us through repeatedly, much to his team-mates' disgust. **1941** Baker *Dict. Austral. Slang* 23 *To dingo on,* to betray, let down, 'rat on' a person. **1952** J. Cleary *Sundowners* iii. 186 You ain't dingoing it, are you? You can't toss in the towel now.

dingus (di·ŋgəs). *colloq.* Also **dinges, dingis.** [f. Du. *ding* thing; cf. *dingbat.] A gadget, contraption, 'thingummy'.

1876 *Pioche* (Nev.) *Jrnl.* 23 Sept. 3/1 The latest thing in the way of a soul-warmer that the youths of Pioche have got is a dingis made thusly. **1882** G. W. Peck *Peck's Sunshine* 21 They pull out a dingus and three joints of fish-pole come out. **1898** *Empire* 27 Aug. (Pettman), 'Where d'ye find the animile?' 'Animal, Mr. Pike?' 'The dingus—the gentleman who lumbers round in space.' **1913** Pettman *Africanderisms, Dinges,* thing, almost universal in its application, things animate and inanimate in Dutch-speaking districts are all of them *dinges* if the speaker fails to recall their names. **1915** S. Lewis *Trail of Hawk* xxii. 203 That dingus in front is a whirling motor. **1928** *Blackw. Mag.* Jan. 30/1 Even an oiler, sent in an emergency to start such a homely inadequate dingus, can do no more. **1937** Robinson & Browne *How to be Perfect Husband* 126 To people who have been married for years this dingus should have a powerful appeal. **1939** 'N. Blake' *Smiler with Knife* 19 I'll just stick the whole dingus together again.

dingy, *a.* Add: **2. b.** Also, 'drab'; dull.

1920 H. G. Wells *Outl. Hist.* 201 Narrow and dingy-spirited specialists.

c. As an epithet in the vernacular name of certain butterflies and moths.

1832 J. Rennie *Butterfl. & Moths* 20 The Dingy Skipper ..appears about the end of May and middle of July. *Ibid.* 69 The Dingy (*M[amestra] furva*). *Ibid.* 142 The Dingy Wave..appears in June. **1876** *Encycl. Brit.* IV. 596 Plate xxix, *Hesperia tages* (Dingy Skipper). **1907** R. South *Moths Brit. Isles* I. 181 The Dingy Footman (*Lithosia griseola*). **1908** *Ibid.* II. 8 The Dingy Shears (*Dyschorista fissipuncta*). *Ibid.* 139 Dingy Mocha (*Ephyra orbicularia*). *Ibid.* 219 [The] Dingy Shell..is in the male..sprinkled and shaded with darker brown. **1955** E. B. Ford *Moths* v. 70 The Dingy Footman, *Eilema griseola* Hb...is usually of a light greyish shade. **1970** Higgins & Riley *Butterflies of Britain & Europe* 331 *Erynnis tages* Dingy Skipper.

dingy, var. *dinge *sb.*^{3}

dining, *vbl. sb.* Add: **1. b.** (Earlier example.)

1846 R. Ford *Gatherings from Spain* viii. 84 There we make our home—far from..distant dining-outs, visits, [etc.].

2. *dining alcove, area, chair, -hall* (examples), *-parlour* (earlier example), *recess*; **dining-car** (earlier and later examples); **dining-coat** *U.S.*, a dinner jacket.

1937 'M. Hillis' *Orchids on Budget* (1938) ix. 147 A large living-room with a dining-alcove. **1961** 'J. Welcome' *Beware of Midnight* ix. 102 The dining alcove in the kitchen where we were breakfasting. **1957** 'A. Vail' *Love me Little* xii. 90 He came into the dining *area* (I hate words like that). **1838** *Amer. Railroad Jrnl.* VII. 328 The introduction of dining cars. **1892** Kipling *Lett. Travel* (1920) 80 He knows when the train will..drop the dining-car. **1959** *New Statesman* 7 Nov. 610/2 The unofficial strike by railway dining-car workers in protest against the extension of Pullman car services. **1911** *Daily Colonist* (Victoria, B.C.) 27 Apr. 20/6 (Advt.), Dining Chairs,..with shaped head and three slats in back. **1970** *Country Life* 1 Oct. (Suppl.) 34 (Advt.), [A] set of 8 18th century Walnut Dining Chairs of unusual design. **1907** Lady Grove *Social Fetich* 152 'Tuxedo', 'dining coats', or 'dinner jackets'. **1667** J. Lauder *Jrnl.* Oct. (1900) 172 The dining hall, a large roome with a great many tables. **1815** *View N.Y. State Prison* 12 A corresponding room in the south wing is used as a dining-hall. **1870** 'F. Fern' *Ginger-Snaps* 237 Mrs. Fire-Fly.. swept into the dining hall in a train about six yards long. **1700** Pepys *Let.* 22 Feb. (1926) I. 291 Try whether you can recollect enough of my..dining parlour. **1906** M. H. B. Scott *Houses & Gardens* iv. 20 The introduction as an appendage to the hall of a dining recess. **1959** 'M. Halliday' *Thicker than Water* iv. 40 They would eat in the little dining recess in the kitchen; they seldom used the dining-room when they were on their own.

dink (diŋk), *sb.*^{1} *Austral.* [Origin unknown.] A ride or lift on the bar of a bicycle. Also *v. trans.*, to give (a person) such a lift.

1934 *Bulletin* (Sydney) 5 Sept. 20/2 The fortunate Melbourne schoolkid with a bike..is asked by his cobbers for a 'dink'. **1941** Baker *Dict. Austral. Slang* 25 *Double-dink,* to carry a second person on the top bar of a bicycle. It is also a noun. Exchangeable terms are 'dink', 'donk', and 'double-bank', both as verbs and nouns. **1948** *Coast to Coast 1947* 135 The lame one who used to let me dink him home on his bicycle.

dink (diŋk), *sb.*^{2} *U.S.* [Imitative.] A drop-shot in lawn tennis. Also *attrib.* So as *v. intr.* (see quot. 1942).

1939 J. D. Budge *On Tennis* 120 Some players resent their opponent's using the drop shot, or the 'dink' shot as they scornfully refer to it. **1942** Berrey & Van den Bark *Amer. Thes. Slang* § 717/2 *Dink,* a ball that drops just beyond the net. *Ibid.* § 717/3 *Dink,* to barely knock the ball over the net. **1959** *Times* 30 June 3/3 Some delicate touch shots cross-court and half-court—the dink as the Americans call it. *Ibid.* 1 Sept. 3/3 Drop shots, stop volleys and dinks were conspicuous by their absence. **1969** *New Yorker* 14 June 45/2 Nobody in his right mind, really, would try those little dink shots he tries as often as he does. *Ibid.* 61/2 He will dink. He spins his first serve in more.

dink, *sb.*^{3} *U.S. Mil. slang.* [Origin unknown.] A derogatory or contemptuous term for a Vietnamese person.

1969 *Eugene* (Oregon) *Register-Guard* 3 Dec. 2A/4 He also criticized U.S. military training, which he said permits references to the Vietnamese as 'gooks, dinks, or slopes'. **1970** *Guardian* 30 July 7/5 These are not people...They are dinks and gooks and slant-eyed bastards.

dink, *sb.*^{4} and *a.*^{2} Abbrev. of *dinkum *sb.* and *a. Austral.* (and *N.Z.*) *colloq.*

1906 E. Dyson *Fact'ry 'Ands* viii. 92 'Twasn't fair dink t'go outside ther firm. **1939** W. E. McKinlay *Ways & Byways of Singing Kiwi* i. 24 One of the Battalions being known as the 'Square Dinks' and another as the 'Fair Dinks'.

Dinka (di·ŋkā), *sb.* and *a.* [f. native name *Jieng* people.] **A.** *sb.* A member of a Sudanese people or their language; *collect.*, this people. **B.** *adj.* Of or pertaining to this people.

1861 J. Petherick *Egypt, Soudan & Cent. Afr.* xx. 345 The Dinka, an extensive negro tribe, inhabit the eastern interior. *Ibid.* 348 Five Dinka children. *Ibid.* 360 The Djibba speak a different language from the Dinkas. **1873** E. E. Frewer tr. *Schweinfurth's Heart of Africa* II. 501 List of mammalia observed during my travels..with their native names...Dinka: Agohk. Dyoor: Abworro. Bongo: Gumbi. **1900** *Daily News* 27 Feb. 5/7 Four days before reaching Bahr el Jaraf I was attacked by Dinkas. **1926** *Blackw. Mag.* Oct. 561/2 Wol explained in the Dinka dialect the reason and object of my sudden visit. *Ibid.* 562/1 The native Dinka never hurries unless he is being hunted. **1949** E. E. Evans-Pritchard in M. Fortes *Social Structure* 86 The adoption of a Dinka boy. *Ibid.*, Marriages of adopted Dinka into collateral major, and even minor, lineages undoubtedly take place sometimes. **1968** Y. R. Chao *Language & Symbolic Systems* 99 Dinka and Luo..have almost 1 million speakers each.

dinkel (di·ŋkĕl). [G.] A species of wheat, *Triticum spelta.*

1866 Lindley & Moore *Treas. Bot.* I. 409/2 Dinkel... *Triticum monococcum.* [**1884** tr. *A. de Candolle's Orig. Cultivated Plants* v. 364 European names [for spelt], on the contrary, are numerous...*Spelta* in Saxon, whence the English name, and the French, *épeautre*; *Dinkel* in modern German.] **1921** J. Percival *Wheat Plant* xxii. 325 Of the wheats with a brittle rachis Common Spelt or Dinkel is the race most extensively grown. *Ibid.* 327 A considerable amount of Dinkel grain is used in South Germany in soups. **1965** R. F. Peterson *Wheat* i. 14 (*heading*) *Triticum spelta*: spelt or dinkel wheat.

dinki-di, var. *dinky-di(e *a.*

dinkum (di·ŋkəm), *sb.* and *a. dial.* and *colloq.* (chiefly *Austral.* and *N.Z.*). Also **dincum.** [Origin unknown.] **A.** *sb.* **1.** Work; esp. hard work; a due share of work.

1888 'R. Boldrewood' *Robbery under Arms* v, It took us an hour's hard dinkum to get near the peak. **1891** S. O. Addy *Suppl. Sheffield Gloss.* 18 'I can stand plenty o' dincum.' This word is used by colliers at Eckington. **1900** Wright *Eng. Dial. Dict.* s.v., You have gotten to do your dinkum, soa you understand. **1941** Baker *Dict. Austral. Slang* 23 *Dinkum,* hard work or honest toil.

†**2.** An Australian; spec. an Australian soldier in the war of 1914–18 (see also quot. 1919^{2}). *Obs.*

1919 [see *Aussie *sb.* and *a.* 1]. **1919** Downing *Digger Dial.* 19 *Dinkums* (*the*), the 2nd Division. Also applied to the New Zealanders.

3. = *dinkum oil* (see sense B, and cf. *fair dinkum*).

1916 *Anzac Bk.* 56 A friend met me and asked if I had heard the latest dinkum. **1933** *Bulletin* (Sydney) 22 Feb. 34 First time I have a cert I'll pass the dinkum to you.

B. *adj.* Honest, genuine, real; as *adv.*, honestly. Phr. *fair* (or *square, straight*) *dinkum,* fair and square, honest; as *advb. phr.*, honestly, genuinely; *interrog.*, really?, is that so? Also *dinkum oil,* the honest truth, true facts. Cf. *dial.* '*fair dinkum!* fair play' (quot. 1900).

1894 *Bulletin* (Sydney) 5 May 13/3 'And did yer stouch him back?' 'No.'..'Fair dinkum?' 'Yes.' **1895** A. A. Grace *Maoriland Stories* 105 Well it ain't goin' to be *honest injun,* that's plain; not what I call *square dinkum.* **1900** Wright *Eng. Dial. Dict.* s.v., *Fair dinkum!* fair play! n. Lin[colnshire]. **1906** E. Dyson *Fact'ry 'Ands* vi. 64 'Meanin'?' he pointed to the centre of his breast, and his eyes were round with inquiry. 'Fair dinkum,' telegraphed Linda. **1911** L. Stone *Jonah* I. ix. 106 'I don't run after people I don't want,' said Pinkey, smiling through her tears. 'Fair dinkum?' cried Chook. **1916** *Anzac Bk.* 22/2 'Ere's some er the dinkum coc'nut ice the tart useter make. **1916** C. J. Dennis *Ginger Mick* 87 That's the dinkum oil frum Ginger Mick. **1919** V. Marshall *World of Living Dead* 36 This was the dinkum stuff—straight from one of the heads. **1921** *Spectator* 5 Feb. 169/1 This, as we Australians say, is 'dinkum'. **1922** [see *cross *v.* 3 b]. **1925** *Spectator* 21 Nov. 930/1 Every 'dinkum Bushman'. **1927** R. Rees *Life's what you make It* iii. 177 'What does dinkum mean?' 'You've been in New Zealand for months and months and you don't

know what dinkum means? It's good, excellent, superlatively fine, genuine, not faked. That's what dinkum is.' **1927** *Weekly Dispatch* 23 Oct. 2/5 Real dinkum Australians ..knew..that wombats can't fly. **1930** A. W. Groom *Merry Christmas* iii. 21 I'll strike a dinkum business deal with you—fair and square. **1934** *Punch* 24 Oct. 454/3 Indeed I am a bit wounded that you [*sc.* Melbourne] did not invite me to come and open the new century for you. I would have made a dinkum speech for you. **1937** N. Marsh *Vintage Murder* vii. 73 Give you a pain in the neck, dinkum, she would. **1942** C. Barrett *On Wallaby* iii. 43 Mobs of [goanas], mister. They was twelve foot long. Square dinkum. **1944** F. Clune *Red Heart* 42 The Simpson Desert..is the only fair-dinkum desert in Australia. **1944** J. H. Fullarton *Troop Target* ii. 18 Anyway there's no dinkum oil. Only latrinograms..it may be all hooey. **1957** 'N. Shute' *On Beach* iv. 111 He's dinkum, and she's not a bad sort. **1969** *Private Eye* 6 June 14 'Here's fifty pounds down.'..'Are you fair dinkum? I can do with the oscar.' **1969** *Sun* (Melbourne) 12 July 58/1 Fair dinkum, North's been waiting so long for its ship to come in, the pier's collapsed.

dinky (di·ŋki), *a.*^{1} and *sb.*^{1} *dial.* and *colloq.* (chiefly *Sc.* and *N. Amer.*). Also **dinkey, dinkie.** [f. Dink *a.*] **A.** *adj.* Neat, trim, dainty; small, tiny, trifling.

1788 E. Picken *Poems* 230/1 *Dinkie,* neat, handsome. **1858** M. Porteous *Real 'Souter Johnny'* 29 Ye'll observe yon dinkie pile In your ain cauf-lan'. **1880** Mrs. L. Parr *Adam & Eve* xxviii, You must leave me a dinkey little corner to squeeze into by. **1887** *Courier-Journal* (Louisville, Ky.) 1 Feb. 8/4 Jumping on a dinkey train while in motion. **1893** *Columbus* (Ohio) *Dispatch* 8 Apr., The British Artillery-men wore little dinky caps with a yellow band. **1896** Ade *Artie* xvii. 154 I'll come hot-footin' in here with my knee-pants and a dinky coat. **1904** 'O. Henry' *Cabbages & Kings* x. 169 A train of cars was waitin' for us on a dinky little railroad. **1905** E. Phillpotts *Secret Woman* I. i. 16 You're all angel yourself—all, to the dinky dimple there at the corner of your li'l mouth. **1915** *Punch* 20 Jan. 49, I shall have a couple of the dinkiest little wounded subs to show you. **1917** 'Contact' *Airman's Outings* 224 Winkle, the dinky Persian with a penchant for high life, has presented the family with five kittens. **1929** D. Coke *Monkey Tree* xvi, Miss Des Vaux asserted her superiority by saying that it was a 'dinky notion'. **1960** K. M. Wells *Cruising North Channel* 24 You will need a stove of sorts, something better than the dinky little two-burner alcohol contraption with which so many so-called cruising ships are fitted.

B. *sb.* Any small object or contrivance; *spec.* a small boat (perhaps a corrupt form of Dinghy) or a small locomotive. Chiefly *U.S.* (in spelling *dinkey*).

1849 *Pacific News* (S.F.) 27 Nov. 4/2 Picked up adrift, in San Pablo bay, a small copper Dinkey. **1874** *Kalama* (Wash.) *Beacon* 20 Jan. 4/2 The passenger train from Tacoma..passed the Des Chuttes bridge..an hour or two previous to the 'dinkey'. **1905** *Terms Forestry & Logging, Dinkey,* a small logging locomotive. **1905** G. S. Wasson *Green Shay* 195 They'll make out to cast loose their dinky all right. **1948** *Milwaukee Jrnl.* 18 July 6/3 The huffing and puffing steam dinkeys..still see service when traffic is heavy.

dinky, *a.*^{2} and *sb.*^{2} = *dinkum *a.* and *sb. Austral.* (and *N.Z.*) *slang.*

1941 Baker *Dict. Austral. Slang* 23 *The dinky,* the truth. Also adj., *dinky,* true.

dinky-di(e (di·ŋkidəi·), *a. Austral.* and *N.Z. slang.* Also **dinki-di.** [f. *dinkum *a.*] = *dinkum *a.*

1918 N. Campbell (*title*) The Dinky-di soldier and other jingles. **1928** A. Waugh *Last Chukka* 84 That was absolutely dinky die, my dear. **1938** P. Lawlor *House of Templemore* xviii. 194 'Dinky die?' asked Percy Andrews. **1952** J. Cleary *Sundowners* ii. 104 We're in for a dinky-di storm. **1956** S. Hope *Diggers' Paradise* 237 No one but dinky-di Aussies need submit a manuscript. **1969** *Australian* 24 May 18/3 Sinister karate chopping Japanese battling with true-blue, dinki-di locals.

dinner, *sb.* Add: **2.** *dinner-basket, -company* (example), *-dress* (examples), *-gong, -guest, -money, -pot, -roll, -service, -ware, -wine*; **dinner-bucket** *U.S.* = **dinner-pail*; **dinner-call** *U.S.*, a formal call upon one's host or hostess after a dinner party; **dinner-card,** (*a*) a card bearing an invitation to dinner (*Obs.*); (*b*) a card bearing a name and indicating a person's place at a dinner-table; **dinner-coat,** a dinner jacket; hence *dinner-coated* adj.; **dinner-dance,** a dinner followed by dancing; hence *dinner-dancing* vbl. sb.; **dinner-horn** *U.S.*, a horn used to announce dinner on a farm, etc.; **dinner jacket,** a dress-coat without tails worn in the evening as a less formal alternative to the swallow-tailed coat; hence *dinner-jacketed* adj.; **dinner-pail** *U.S.*, a pail in which a workman carries his dinner with him; hence in slang phr. *to hand, pass,* or *turn in one's dinner-pail,* to die; **dinner speech** *U.S.*, an after-dinner speech; so *dinner-speaking* vbl. sb.; **dinner-wagon** (earlier example).

1821–8 D. Wordsworth *Tour on Continent* in *Jrnls.* (1941) II. 318 The mother..hastening to her dinner-basket, chearfully presented me with her whole stock. **1939** F. Thompson *Lark Rise* i. 15 The leazer's water-can and dinner-basket. **1901** *Scribner's Mag.* XXIX. 404/2 Billy put on his coat, took his dinner-bucket. **1895** J. L. Williams *Princeton Stories* 263 It's two years now, and it's not good form to let a dinner call go more than two years in Princeton. **1754** *Connoisseur* 2 May 80, I received..a dinner-card from a friend, with an intimation that I should meet some very agreeable ladies. **1865** Dickens *Mut. Fr.* II. iii. xvii. 152 Mrs. Veneering..sends them every one a dinner-card. **1881** C. C. Harrison *Woman's Handiwork* ii. 125 Designs for dinner cards for Thanksgiving or Christmas. **1905** E. Wharton *House of Mirth* i. iv. 60 There would be notes and dinner-cards to write. **1907** M. C. Harris *Tents of Wickedness* i. iii. 35 His dinner-card lay on the side of the cloth next her, and she..glanced at it. 'Mr. Paul Fairfax'—so that was his name. **1922** F. Scott Fitzgerald *Let.* 31 Jan. (1964) 153 He looks like a sawed-off young tough in his first dinner-coat. **1929** Yeats *Let.* 2 Mar. (1954) vi. 758 To-night we dine with Ezra—the first dinner-coated meal since I got here. **1816** Jane Austen *Emma* II. vii. 119 Their love of society..prepared every body for their keeping dinner-company. **1901** *Lady's Realm* X. 613/1 From one dinner-dance to the next. **1910** *Westm. Gaz.* 15 Apr. 5/2 A dinner-dance—quite a small affair. **1965** *Observer* Suppl. 18 Apr. 46/1 Dinner-dancing after 11.30. **1815** *Belle Assemblée* July 274/1 Round dress..made to answer the double purpose of a morning or dinner-dress. **1897** M. Corelli *Ziska* xiii. 262 The Princess herself, attired in a dinner-dress made with quite a modern Parisian elegance. **1956** J. D. Carr *P. Butler for Defence* vi. 62 Helen, in her dark-blue dinner-dress, stood in the doorway. **1838** *Knickerbocker* XII. 227 How startling is the sound of the dinner gong! **1922** D. H. Lawrence *Aaron's Rod* xiv. 200 He did not notice the dinner-gong, and only the arrival of the chamber-maid..sent him down to the restaurant. **1811** L. M. Hawkins *Countess* I. xiv. 240 Mr. Sydenham, his son, and his charge, were to be dinner-guests. **1965** F. Sargeson *Memoirs of Peon* vii. 207 It was upon dinner-guest occasions that my gastropodous writhings were dispensed with. **1835** C. Gilman in *Southern Rose* 5 Sept. 2/1 The business was scarcely settled, when the dinner-horn sounded. **1849** *Congress. Globe* 10 Jan. App. 80/2 The dinner horn will be heard across broad fields, and will be answered by the keen appetites attendant upon honest labor. **1867** 'T. Lackland' *Homespun* iii. 290 From that time until the dinner-horn sounds, no tented field..ever furnished a busier..spectacle. **1891** M. E. Braddon *Gerard* III. vii. 208 Jermyn took up the loose pages, folded them carefully, put them in an inner pocket of his dinner jacket. **1894** *To-day* 17 Mar. 182/1, I see that the so-called 'dinner-jacket' is getting to be the regular wear at the theatres. **1924** Galsworthy *White Monkey* I. iv, Full fig, or dinner jacket? **1968** *Listener* 6 June 748/1 The struggle to rescue opera from the dinner-jacket brigade and to present it to ordinary people at reasonable prices. **1911** C. E. W. Bean '*Dreadnought' of Darling* i. 5 Any other dinner-jacketed, white-shirted, black-tied visitor in the room. **1936** 'M. Innes' *Death at President's Lodging* (1937) iv. 63 Round the high-table there stood, gowned and for the most part dinner-jacketed, the Fellows. **1835** Dickens *Sk. Boz* (1836) 2nd Ser. 14 Investing part of the day's dinner-money in the purchase of the stale tarts. **1942** A. P. Jephcott *Girls Growing Up* iii. 47 Ordinary timetable lessons are supplemented by..dinner money collections [etc.]. **1856** M. J. Holmes *Homestead* vi. i, The little 'tin bucket'..serves the treble purpose of dinnerpail, washbowl, and drinking cup. **1900** *Nation* LXXI. 323/2 He comes something short of ex-President Harrison's ability to see a 'spiritual significance' in the full dinner-pail. **1904** *N.Y. Even. Post* 19 Feb. 3 Thousands of men with their dinnerpails on their way to work. **1905** A. M. Binstead *Mop Fair* iv. 65 Evelyn Godolphin Mountprospect..passed in his dinner pail. **1922** Wodehouse *Clicking of Cuthbert* i. 14 A sliced ball, whizzing in at the open window, had come within an ace of incapacitating Raymond Parsloe Devine...Two inches, indeed, to the right and Raymond must inevitably have handed in his dinner-pail. **1964** —— *Frozen Assets* iii. 49 My godfather.. recently turned in his dinner pail and went to reside with the morning stars. **1775** P. V. Fithian *Jrnl.* (1934) II. 68 Tea..is boild in a common Dinner-Pot, of ten or fifteen gallons. **1871** Mrs. Stowe *Old Town Fireside Stories* v. 168 A gret iron pot as big as your granny's dinner-pot with an iron bale to it. **1833** *Chambers's Jrnl.* II. 32/2 You find Mrs B...flying about the dining-room,..marshalling glasses and dinner rolls. **1962** *Which?* May 144/1 We made plain dinner rolls, using ½ pint of water to 1 lb flour—a heavy dough. **1845** *Ainsworth's Mag.* VIII. 117 The furniture of the table ..reminds one of..a Russian dinner-service. **1968** *Guardian* 9 July 7/5 Their bargain dinner service..costs £4 5s. **1910** *Westm. Gaz.* 11 Apr. 8/3 Mr. W. W. Jacobs..said..Dinner-speaking was a gift which was never put into his stocking. **1852** *Harper's Mag.* VI. 89* That celebrated public dinner-speech. **1890** *Ibid.* Apr. 799/2 The modern dinner speech is a happy blending of sparkling banter, [etc.]. **1862** *Illustr. Catal. Internat. Exhib.* XI. No. 5719, A wainscot sideboard; dinner wagon, to correspond. **1895** *Montgomery Ward Catal.* 527/1 Pure white dinner ware..with gold decoration. **1905** *Daily Chron.* 2 May 7/1 The kitchen was strewn with smashed dinner-ware. **1961** *Times* 6 June 5/6 The dinner-ware is based on melamine crystal developed by Cyanamid. **1905** *Wine List of T. W. Stapleton & Co.* July, Sherry... Good Dinner Wine 36/-. **1920** G. Saintsbury *Notes on Cellar-Bk.* v. 78 Good Carbonnieux or Olivier..are admirable dinner wines.

dinoflagellate (dəinoflæ·dʒĕlᵉⁱt). *Zool.* [f. mod.L. *Dinoflagellata* (O. Bütschli 1885, in Bronn *Kl. u. Ordn. Thier-Reichs* I. ii. 907), f. Gr. δῖνο-ς whirling, rotation + L. *flagellum*: see *flagellate *sb.*] A member of the sub-class Dinoflagellata, which includes protistans having two flagella. Also *attrib.* or as *adj.* (*Cent. Dict.*, 1889).

1901 G. N. Calkins *Protozoa* 260 The dinoflagellate *Ceratium hirundinella.* **1934** *Discovery* Aug. 218/2 They have the characteristic flagella and horizontal groove of the Dinoflagellates. **1959** *New Biol.* XXIX. 43 A dense growth of phytoplankton, often dinoflagellates such as *Gymnodinium* or *Goniaulax.* **1967** P. A. Meglitsch *Invert. Zool.* iii. 42/2 Most of the dinoflagellates are marine although there are a number of fresh-water forms.

din't, dint (dint), contracted colloq. form of *didn't, did not* (see Do *v.* 29).

1961 *New Left Rev.* Jan.–Feb. 24/1 He din't see why I shouldn't. **1967** C. Drummond *Death at Furlong Post* iv. 39, I dint expect the pleasure of your cumpny. **1967** 'A. Blaisdell' *Something Wrong* (1968) xiv. 162 'I thought maybe she'd asked you,' said Sue...'I get you. But she dint.' *Ibid.*, I dint *know* her. **1968** *Punch* 14 Feb. 227/3 'Course, I 'ad to 'elp her with joining the letters up, din't I?

dioch (dəi·ǫk). Also **diock.** [?Native name.] An African weaver-bird of the genus *Quelea.*

1889 *Cent. Dict.*, Diock. **1905** G. E. Shelley *Birds Afr.* IV. i. 119 Quelea cardinalis...The Cardinal Dioch inhabits Eastern Africa from 7° S. lat. to 5° N. lat. **1930** G. L. Bates *Handbk. Birds W. Afr.* 495 The Red-headed Dioch is a bird of the Savannah Belt going all the way to Portuguese Guinea. **1964** *New Scientist* 18 June 736/1 The Black-Faced Dioch (*Quelea quelea*), commonly known as the Quelea bird.

diode, *a.* Add: **B.** *sb. Electr.* **a.** A thermionic valve of the simplest kind with just two electrodes, a cathode or 'filament' and an anode or 'plate'. **b.** = **semiconductor diode*: cf. *crystal diode* (see *crystal *sb.* 9 d). Also *attrib.* or as *adj.*

1919 W. H. Eccles in *Electrician* 18 Apr. 475/2, I propose to give the name 'diode' to a tube with two electrodes. **1921** —— *Contin. Wave Wireless Telegr.* i. 257 A bulb with two electrodes, namely, anode and cathode, is called a diode tube. *Ibid.* 306 This example shows plainly that two constants are required to define the chief properties of a diode. **1929** J. A. Ratcliffe *Physical Princ. Wireless* ii. 23 The diode usually consists of a straight wire filament which is heated by an electric current. **1943** C. L. Boltz *Basic Radio* x. 154 The diode's only use is as a rectifier. By adding another electrode we can increase the utility of the valve. **1944** *Electronic Engin.* XVI. 408, C charges through a diode valve. **1955** [see *crystal *sb.* 9 d]. **1958** *Times Rev. Industry* July 29/1 Impurities in single crystals in connexion with the production of transistors and diodes. **1959** K. Henney *Radio Engin. Handbk.* (ed. 5) ix. 7 Logic circuits using diodes may be considered another class of switching applications. **1970** D. F. Shaw *Introd. Electronics* (ed. 2) xi. 226 The rectifying property of a semi-conductor diode is a consequence of the asymmetrical conduction across the contact between a metal and a semi-conductor.

diœcy (dəi,ī·si). *Biol.* [f. Diœc(ism + -y^{3}.] = Diœcism.

1944 *Nature* 1 Apr. 392/1 Plants are not mobile, and hence sex separation, or diœcy as it is here called, is not an efficient method of control...Consequently, diœcy is relatively uncommon in plants. **1970** T. Dobzhansky *Genetics Evol. Process* xi. 386 Muller (1925) surmised that this is due to the preponderance of hermaphroditism (monœcy) among plants, and the separation of sexes (diœcy) among animals.

-dione, *suffix. Chem.* [f. Di-^{2} 2 + -one.] A suffix used to form the names of compounds containing two carbonyl groups, as in *menadione.

dionine (dəi·ǫnin). *Pharm.* Also **dionin.** [ad. G. *dionin,* a former proprietary name.] Ethylmorphine hydrochloride, used in the treatment of glaucoma, iritis, etc., and to alleviate coughing.

1899 *Jrnl. Chem. Soc.* LXXVI. i. 724 *Ethylmorphine hydrochloride* or *dionine*..is a white, odourless, finely crystalline powder,..used as an anodyne and narcotic. **1923** J. Joyce *Let.* 28 Mar. (1966) III. 73 Dr Borsch thinks it is useless to prolong the dionine treatment. **1949** M. A. Jennings in H. W. Florey et al. *Antibiotics* II. xxxi. 998 The use of the drugs 'dionin' and pilocarpine also tended to reduce the effective dose. **1963** A. H. Douthwaite *Hale-White's Mat. Med.* (ed. 32) 173 Dionine is ethyl morphine... Its action is like that of codeine.

Dionysic, *a.* Delete *rare.* ? *Obs.* and add further examples.

1882 A. Jessopp in *19th Cent.* May 728 A survival of the old belief in the Dionysic possession. **1920** D. H. Lawrence *Women in Love* xix. 279 But I hate ecstasy, Dionysic or any other.

dioptre (dəi,ǫ·ptəɹ). The usual spelling in Great Britain of Diopter in sense 5: a unit for expressing the power of a lens, equal to the reciprocal of its focal length in metres.

[**1876** *Compte-rendu: Congrès internat. des Sciences médicales 1875* (Brussels) 610 L'unité du système métrique, c'est le mètre: l'unité dioptrique..sera nommée *dioptre.*] **1882** E. Nettleship *Dis. Eye* (ed. 2) i. 15 Some system of numbering is required which shall indicate the refractive power of the lenses used for spectacles...In the *second system,* which is fast displacing the old one, the metrical scale is used; the unit is a weak lens of 1 metre (100 cm.) focal length, and known as a dioptre (D). **1907** *Practitioner* June 820 Patient was myopic to 2·0 dioptres. **1964** S. Duke-Elder *Parsons' Dis. Eye* (ed. 14) v. 51 A lens with a focal length of half a metre will be twice as strong as one with a focal length of 1 metre: the refractive power of such a lens is therefore 2 dioptres.

diorama. Add: **b.** A small-scale representation of a scene, etc., in which three-dimensional figures or objects are displayed in front of a painted background, the whole often being contained in a cabinet and viewed through a window or aperture in the front; hence, any small-scale model of a scene, building-project, or the like; also, a miniature

set (*set *sb.*^{1}) used in Cinematography and Television where a full-sized set or location would be impracticable.

1902 *Westm. Gaz.* 10 June 3/2 The most interesting feature of the Museum..is the diorama gallery, in which are shown about a dozen large tableaux of battles. **1926** *Ann. Report Imperial Inst.* 33 A certain number of 'dioramas' or modelled panoramas have been put in position in various courts. These models are electrically lit and are so placed that the exhibits relating to the particular industry or activity can be grouped around them. **1939** *Illustr. London News* 29 Apr. 715 The 'Perisphere', a huge sphere, 200 ft. in diameter, enclosing a diorama of a city of tomorrow. **1959** W. S. Sharps *Dict. Cinemat.* 90/2 *Diorama,* a small set used in place of a much larger one, usually in order to suggest location or time.

Dioscuric (dəi,ǫskiū°·rik), *a.* Also **dioscuric.** [f. Gr. Διόσκουροι, f. Διός, gen. of Ζεύς Zeus + κοῦρ-ος, κόρ-ος boy, son + -ic.] Of, pertaining to, or resembling the legend of the twins Castor and Pollux. Also **Dioscu·rian** *a.* Hence **Dioscurism** (dəi,ǫ·skiŭriz'm).

1903 J. R. Harris *Dioscuri* 42 Let us..examine a third case of twin saints in the Christian calendar, and test it.. for Dioscurism. *Ibid.* 47 We naturally enquire..whether there are any Dioscuric features about them. *Ibid.* 61 The popular religion was deeply tinctured with Dioscurism. **1956** E. E. Evans-Pritchard *Nuer Relig.* v. 129 These dioscuric descriptions of twins are common to many peoples.

diose (dəi·ōᵘz, -s). *Chem.* [f. Di-^{2} + -ose^{2}.] A generic term, analogous to triose, hexose, etc., for a 'sugar' containing two carbon atoms; the only possible one is glycolaldehyde, $CHO{\cdot}CH_2OH$.

1904 *Jrnl. Phys. Chem.* VIII. 509 Dioses, trioses, tetroses. **1948** W. W. Pigman *Chem. Carbohydrates* i. 17 Carbohydrates are usually classified according to the number of carbon atoms, *e.g.* pentoses, hexitols, heptonic acids. 'Monose', however, has been used as a short term for the preferred term 'monosaccharide', and also as a superfluous synonym for 'formaldehyde'. Similarly, glycolic aldehyde is called a 'diose', whereas the ending '-biose' denotes a disaccharide, as in 'melibiose' and 'gentibiose'.

diosgenin (dəi,ǫ·sdʒĕnin). *Chem.* [a. G. *diosgenin* (Tsukamoto & Ueno 1936, in *Jrnl. Pharm. Soc. Japan* (*Trans.*) LVI. 136), f. mod. L. *Dios*(*corea* (see below) + *genin.] A crystalline sapogenin, $C_{27}H_{42}O_3$, obtained chiefly from Mexican yams of the genus *Dioscorea* and used in the preparation of steroid hormones such as cortisone.

1937 *Chem. Abstr.* XXXI. 3493 (*heading*) Constitution of diosgenin. **1955** *Sci. Amer.* Jan. 59/3 Several groups sought a route to cortisone from the abundantly available natural sterols, particularly cholesterol, ergosterol and diosgenin. **1966** *New Scientist* 15 Dec. 619/1 Chemists..had revolutionized steroid therapy by synthesizing progesterone from the diosgenin found in the roots of the wild Mexican yam. **1966** E. Palmer *Plains of Camdeboo* xvii. 275 Our Elephant's Foot is *Dioscorea elephantipes* (*Testudinaria elephantipes*), and its genus has recently leaped to fame for certain of its members—among them our Elephant's Foot—contain diosgenin from which cortisone is manufactured.

diosphenol (dəi,ǫsfī·nǫl). *Chem.* [f. Dios(ma + Phenol.] A crystalline odoriferous compound, $C_{10}H_{16}O_2$, the main constituent of buchu leaf oil.

1880 F. A. Flückiger in *Pharm. Jrnl.* XI. 219/1 On submitting 35 kilograms of round buchu leaves to distillation (*Barosma betulina*), I obtained 180 grams of essential oil... An oily layer..concretes, and affords a crystallized mass of what we may call *diosphenol,* with allusion to Diosma, the original Linnean name bestowed on the buchu plants. **1934** *Jrnl. Chem. Soc.* i. 240 The very intense absorption of diosphenol is, of course, due to its being an enolised ketone. **1949** E. Guenther *Essential Oils* III. 371 The main constituent of buchu leaf oil is diosphenol (buchu camphor) $C_{10}H_{16}O_2$, a ketophenol.

dioxan (dəi,ǫ·ksăn). *Chem.* Also **dioxane** (-ēⁱn). [f. Dioxy-, Diox- + -an, -ane.] Any of three liquids, $C_4H_8O_2$, with a molecular structure consisting of a saturated ring of four carbon and two oxygen atoms; *spec.* 1,4-dioxan, used in large quantities as a solvent. Also, any of the derivatives of these compounds.

1912 *Jrnl. Chem. Soc.* CI. 1803 This compound rapidly loses dioxan. **1946** *Nature* 19 Oct. 553/2 Alginic acid diacetate swells, but does not dissolve, in water,..dioxan and glacial acetic acid at ordinary temperatures. **1947** *New Biol.* III. 124 A solution of Formvar in 1–4 dioxane is commonly used. **1951** A. Grollman *Pharmacol. & Therap.* xii. 230 A number of dioxane derivatives have been..demonstrated to have an adrenolytic but not a sympatholytic action. **1960** E. H. Rodd *Chem. Carbon Compounds* IV. xvi. 1525 The 1:3-dioxans are methylene ethers of 1:3-glycols and the 1:2-dioxans, of which little is known, are cyclic peroxides. **1965** Phillips & Williams *Inorg. Chem.* I. xv. 568 Advantages accrue on going to a solvent such as dioxan or dimethyl-formamide which have very low solvolysis constants.

dip, *v.* Add: **5.** *to dip snuff.* Also *absol.*

1849 *Knickerbocker* XXXIV. 117 The 'gude woman' sat in the corner 'rubbing snuff', or 'dipping'. **1864** J. T. Trowbridge *Cudjo's Cave* xxxiv. 332 For this excellent woman snuffed, 'dipped' and smoked. **1913** M. W. Morley *Carolina Mts.* 169 Nor is snuff taken after the manner of former generations of snuff-takers. Here the people 'dip'.

b. *intr.* or *absol.* To pick pockets. Also *trans. slang.*

[**1817** *Sporting Mag.* (Farmer), I have dipped into 150.. pockets and not found a shilling.] **1857** 'Ducange Anglicus' *Vulgar T.* 6 *Dip,* to pick pockets. **1925** *Brit. Weekly* 12 Mar. 573/2 If you don't want to get 'dipped' [*i.e.*, have your pocket picked], buy..small nuts and put them in your pocket with your cash. There isn't one of the boys can dip you then. **1929** *Detective Fiction Weekly* 2 Mar. 696/1 The first fourteen years I dipped I got grabbed eleven times. **1930** E. Wallace *White Face* vii. 89 You went over and you dipped him for his clock and pack. **1967** *Listener* 14 Sept. 325/3 Somebody is..going to dip his back pocket.

6. c. *Forestry.* (See quot.)

1877 T. Kelly in *N.Z. Country Jrnl.* I. 244 On the side the tree will most easily fall [the experienced bushman] dips it—that is, he cuts in a deep notch from the circumference towards the centre.

d. To lower (the beams of the headlights of a vehicle). Also *absol.*

1922 *Motor* 31 Oct. Suppl. p. xxxiv (Advt.), The A–L Anti-Dazzle Focus Headlight Attachment..No 'dipping'. No 'dimming'. **1928** '*Motor' Manual* (ed. 27) xi. 152 The type of headlamp which is so arranged that the beam can be dipped, swivelled, or both, at the will of the driver. **1936** *Discovery* Oct. 302/1 One effect of this beneficent discovery will be to render unnecessary the regulations for dipping and extinguishing headlights. **1959** '*Motor' Manual* (ed. 36) viii. 218 Do not engage in headlight battles. Always dip when another vehicle approaches. **1965** Priestley & Wisdom *Good Driving* ii. 20 A dipper switch..enables you to dip the beam. **1969** G. Black *Cold Jungle* x. 142 Headlights came at us. Rob dipped.

dip, *sb.* Add: **1. a.** (Earlier example in sense of 'a bathe'.)

1778 F. Burney *Evelina* I. xix. 122, I should advise you to take another dip.

g. *Pros.* An unstressed element in a line of alliterative verse. (G. *senkung.*) Cf. *lift *sb.*^{2}

1894 H. Sweet *Anglo-Saxon Reader* (ed. 7) p. lxxxviii, Each verse usually consists of four metrical elements, two lifts and two dips—that is, two strong- and two weak-stress elements. **1961** *Rev. Eng. Stud.* XII. 346 A metrical interpretation which..sets up metrical units ('lifts' and 'dips') which are defined in terms of stress relationships. *Ibid.* 347 Sequences of two unstressed syllables are counted as two 'dips', not one, if the two syllables belong to separate 'breath-groups'.

h. A receptacle from which a prize may be obtained by dipping; also *fig.*; *lucky dip* = Lucky-bag 1, or *bran-tub.

1915 'Bartimeus' *Tall Ship* vii. 129 Pennies..to be extracted at great personal risk from an electric dip. **1927** W. E. Collinson *Contemp. Eng.* 20 Here one could for a small sum put one's hand into the..lucky dive (dip) and draw out a prize. **1933** J. Thorpe *Happy Days* i. 28 Most of the sweet-shops at that time had 'lucky-dips'. **1934** [see *Blighty, blighty *sb.* c]. **1956** 'J. Wyndham' *Seeds of Time* 25 A sort of rigid, lucky-dip, take-it-or-leave-it system. **1967** P. Moyes *Murder Fantastical* xv. 219 Everything for the Lucky Dip is *wrapped* because of the bran.

5. b. *Mining.* Short for *dip-head* (see 11 in Dict.).

1877 *Encycl. Brit.* VI. 69/1 The drawing roads for the coal may be of three different kinds,—(1) levels driven at right angles to the dip,..(2) rise ways, known as jinny roads, jig-brows, or up-brows,..(3) dip or down-brows, requiring engine power. **1883** W. S. Gresley *Gloss. Coal-M.*, *Dip*.., a heading or other underground way driven to the deep... It is usual to drive a pair of dips about 10 yards apart every 180 yards or so. **1967** *Gloss. Mining Terms* (*B.S.I.*) viii. 11 *Dip,* an underground roadway driven downhill, usually following the inclination of the strata.

8. Also, a vat or tank in which sheep-dip is used. So *dip yard.*

1878 E. S. Elwell *Boy Colonists* 103 A large dip was built there. **1900** S. E. Blacke *Flights from Land of Bell-bird* i. 16 Then there were visits to the 'dip' beyond, where the sheep were washed. **1922** W. Perry *Sheep Farming in N.Z.* iii. 20 It is wise to build a dip. **1933** L. G. D. Acland in *Press* (Christchurch) 2 Dec. 15/7 *Yards* for dipping [sheep]..are generally called *dip yards.*

b. A grade of turpentine. (See quot 1884, and cf. *dipping *vbl. sb.* 2 b.)

1856, 1884 [see *virgin-dip* s.v. Virgin *a.* 17 b]. **1862** 'E. Kirke' *Among Pines* 167 I've four barr'ls of 'dip' and tu of 'hard'. **1896** *Pop. Sci. Monthly* Feb. 473 The *dip* or crude turpentine is emptied.

9. Also, any sauce or dressing; *spec.* the fat and juice left in the pan after meat has been cooked.

1846 Worcester, *Dip,*..sauce made of fat pork for fish. *U.S.* **1894** T. F. Robley *Hist. Bourbon Co., Kansas* 26 Some 'rashers' are cut from the 'flitch' of bacon and the grease tried out; eggs are fried, and 'dip' is made. **1931** 'N. Bell' *Life & Andrew Otway* viii. 345 'Lots of eggs..and loads of what cook used to call dip.'..'I shall eat the dip.. with a piece of bread. You know, rub it round the plate.' **1960** *Spectator* 10 June 848 Jam-butties one day and bread and dip the next.

b. *pl.* Dough-boys. *Austral. colloq.*

1859 D. Bunce *Trav. with Leichhardt* 161 Dr. Leichhardt gave the party a quantity of dough-boys, or, as we called them, dips. *Ibid.* 171 Dr. Leichhardt ordered the cook to mix up a lot of flour, and treated us all to a feed of dips.

c. A savoury mixture into which biscuits, etc., are dipped.

1960 J. Kirkwood *There must be a Pony!* (1961) viii. 61 We were up to our necks in dips: clam dip, cheese dip, mushroom dip. **1962** L. Deighton *Ipcress File* xxi. 143, I was loaded with anchovy, cheese dip, hard egg and salmon. **1962** *Woman's Own* 1 Dec. 49/2 Have a trolley of savouries and 'dips' ready to wheel in. *Ibid.* 50/2 Use as a dip with crisps or savoury biscuits.

10. (Later examples.)

1926 J. Black *You can't Win* iv. 35 No Missouri dip would take his roll, extract two fifty-dollar bills, and put the rest back in his pocket. **1936** J. Curtis *Gilt Kid* xxix. 279 'Ginger King,' he said, 'I'm going to nick you for a dip.' **1938** F. D. Sharpe *Sharpe of Flying Squad* i. 13 They have rich, picturesque names, such as..'Jimmy the Dip'. **1970** *Daily Tel.* 29 Apr. 4/6 New Yorkers who have had their pockets picked or handbags rifled on the city's Underground in recent years learned yesterday that the person responsible was probably a professional 'dip'.

11. dip compass = *dipping compass* (see Dipping-needle); **dip equator,** the magnetic equator (see Equator 3 b); **dip pen,** one that has to be dipped in the ink (opp. *fountain-pen*); **dip regulator** (see quot.); **dip-rod,** (*b*) = **dip-stick*; **dip-roller,** a form of roller used in printing-works for taking up ink; **dip-stick, dipstick,** a rod for measuring the depth of liquid; **dip-switch, dipswitch,** a switch that dips the beams of a vehicle's headlights.

1897 *Strand Mag.* Mar. 344/2 Its [*sc.* the Pole's] variable position was approximately determined by Sir James Clark Ross by help of the dip compass. **1883** *Encycl. Brit.* XVI. 164/2 The line of no dip is called the magnetic or dip equator. **1945** B. Macdonald *Egg & I* (1947) viii. 96 An ink bottle and a dip pen. **1949** 'N. Blake' *Head of Traveller* xi. 172 'What are you looking for?'..'My dip pen.' 'Stylo run out?' *a* **1884** Knight *Dict. Mech.* Suppl., *Dip regulator,* a device used in gas works for regulating the seal of the dip-pipes in the hydraulic main, and for drawing off the heavy tar from the bottom of the main without disturbing the seal. **1923** *Man. Seamanship* II. 201 Care should be taken that the correct amount of oil is in the oil pump at the bottom of the base chamber; an oil cock or dip rod is fitted for this purpose. **1934** *Times* 13 Feb. 10/4 The diprod, coil, Lockheed fluid tank on the dash..are all handy. *a* **1884** Knight *Dict. Mech.* Suppl., *Dip-roller* (Printing), a roller to dip ink from the fountain. **1927** *Observer* 4 Dec. 11/4 The near side of the engine [of a new Ford motor vehicle] carries the self-starter and the oil-filler and dip-stick gauge only, and has almost a European appearance. **1953** *Word for Word* (*Whitbread & Co.*) 18/1 *Dipstick,* an instrument used to measure the quantity of wort in the fermenting squares prior to fermentation. **1970** *Motoring Which?* July 92/1 Apart from the awkward dipstick of the 18/85, day-to-day servicing was easy on all three cars. **1952** *San Francisco Chron.* 15 June 13L, American: dimmer switch. British: dip switch. **1962** *Times* 13 Feb. 3/4 The floor buttons for dipswitch and screen-washers are too small. **1967** E. Rudinger *Consumer's Car Glossary* (ed. 2) 33 *Dipswitch,* foot-operated button or hand-operated lever with which the driver changes the headlamps from the main beam to the dipped beam.

b. In various attrib. uses of sense 5.

1839 Ure *Dict. Arts* 964 Where the coal-measures are horizontal, and the faults run at a greater angle than 45° to the line of bearing, they are termed dip and rise faults. *Ibid.* 968 The true dip-line of the plane which leads to the outcrop. *Ibid.* 974 On the dip side of the gallery. *Ibid.* 992 The subterraneous fire broke forth with two heavy discharges from the dip-pit. *Ibid.* 994 In the dip-mine a double tram-road is laid. **1877** *Encycl. Brit.* VI. 63/2 Galleries driven at right angles to these [*sc.* the dip head level and lodgment level] are known as 'dip' or 'rise headings', according to their position above or below the pit bottom. **1879** *Ibid.* X. 297/1 A quarry is usually worked to the dip of a rock, hence the strike-joints form clean-cut faces in front of the workmen as they advance. These are known as 'backs', and the dip-joints which traverse them as 'cutters'. *Ibid.* 303/1 Dip-faults will often be observed to deviate considerably from the normal direction of dip. **1882** A. Geikie *Text-bk. Geol.* 502 In general they [*sc.* joints of stratified rocks] have two dominant trends, one coincident, on the whole, with the direction in which the strata are inclined from the horizon, and the other running transversely at a right angle or nearly so. The former set is known as dip-joints, because they run with the dip or inclination of the rocks. **1887** P. McNeill *Blawearie* 24 The firm..had turned their whole force of men into the dip-workings, in order to exhaust the coal bordering on the march. **1900** *Geogr. Jrnl.* XV. 220 Its [*sc.* the Arun's] course was determined by the original dip-slope of the Wealden dome. **1957** *Encycl. Brit.* IX. 119/1 The dip-slip is the component of the net-slip measured directly down the dip of the fault plane. **1960** L. D. Stamp *Britain's Struct.* (ed. 5) v. 45 Valleys parallel to the dip of the rocks (dip valleys). **1961** J. Challinor *Dict. Geol.* 61/2 *Dip-slope,* a slope of the ground which is determined..by the dip of the beds. It is applied particularly to such a slope which ends upwards along the top of an escarpment and which is then opposed to the, usually steeper, scarp slope on the other side. **1965** G. J. Williams *Econ. Geol. N.Z.* xviii. 309/2 The amount of dip coal is difficult to estimate and it was suggested that 'the amount of workable dip coal is likely to be somewhat greater than the rise coal'. **1967** *Gloss. Mining Terms* (*B.S.I.*) viii. 11 *Dip face,* a coal face advancing downhill. *Ibid.*, *Dip workings,* workings lying to the dip of any designated point in a seam.

Dip, dip (dip). (Also with full point.) Colloq. abbrev. of Diploma *sb.* 2; so **Dip. AD,** Diploma in Art and Design; **Dip. Ed.,** Diploma in Education; see also *Dip. Tech.

1895 W. C. Gore in *Inlander* Nov. 64 *Dip,* diploma. **1963** *Times* 11 May 9/4 Only four of the 29 colleges awarded DipAD recognition will be allowed to offer diploma courses in all four areas of art and design. **1967** M. Drabble *Jerusalem the Golden* v. 96, I would be stealing the state's money, wouldn't I? By doing a Dip Ed without meaning to teach? **1969** Nairn & Singh-Sandhu in Cockburn & Blackburn *Student Power* 104 The problems of art education in this country centre around the enigmatic initials 'Dip. AD'. This stands for the degree now conferred by authority after four years' attendance at our leading colleges of art: the Diploma in Art and Design. **1971** 'P. Hobson' *Three Graces* i. 9, I can do my engineering dip. at the Tech.

dipeptidase (dəipe·ptidēⁱz, -s). *Biochem.* [a. G. *dipeptidase* (Grassmann & Haag 1927, in *Zeitschr. f. physiol. Chemie* CLXVII. 189), f. *dipeptide + *-ase.] Any enzyme that catalyses the hydrolysis of a dipeptide into its two constituent amino-acids but does not act on tripeptides or higher peptides.

1927 *Brit. Chem. Abstr.* A. 794/1 The dipeptidase hydrolyses all the dipeptides tested, but has no action on various natural proteins. **1939** W. B. Yapp *Introd. Animal Physiol.* i. 29 In the crab *Maia squinado* and in the marine snail *Murex anguliferus*..the gut contains the same four proteases as are present in vertebrates, namely proteinase, carboxy-polypeptidase, aminopolypeptidase, and dipeptidase. **1954** A. White et al. *Princ. Biochem.* xi. 233 Dipeptidases... specifically act only on certain dipeptides; an example is glycylglycine dipeptidase. **1960** E. L. Smith in P. D. Boyer et al. *Enzymes* IV. i. 8 *Dipeptidases.* As yet, none of these enzymes have been obtained in homogeneous form..however, it has been demonstrated that these enzymes do not attack tripeptides.

dipeptide (dəipe·ptəid). *Chem.* Also †**-id.** [f. Di-^{2} + *peptide.] Any compound composed of two amino-acid residues linked by a peptide bond.

1903 *Jrnl. Chem. Soc.* LXXXIV. i. 694 (*heading*) Derivatives of dipeptides and their behaviour towards pancreas ferments. **1955** *Sci. Amer.* May 37/1 A group of linked amino acids is known as a peptide: two units form a dipeptide, three a tripeptide and so on. **1965** G. W. Anderson in Florkin & Stotz *Comprehensive Biochem.* VI. vii. 258 When the term *dipeptide* is used, one ordinarily thinks of a compound formed of *alpha* amino acids which are joined by amide linkage of the carboxyl group of one acid with the amino of another. **1970** *New Scientist* 30 Apr. 217/3 The resulting dipeptide joins successively onto valine, then ornithine, then leucine.

diphasic, *a.* Add: (Later examples.) Also (*b*) marked by two phases or stages (e.g. of growth); (*c*) consisting of two phases of matter (solid, liquid, or gas.)

1900 Dorland *Med. Dict.* 206/1 *Diphasic,* doubly varied; said of electric currents of action in muscle. **1932** J. S. Huxley *Probl. Relative Growth* ii. v. 70 'Diphasic' mammals or birds such as arctic fox,..herons, owls, etc. **1936** *Discovery* Sept. 291/2 The diphasic wave that accompanies muscular and nervous activity. **1940** *Chambers's Techn. Dict.* 247/2 *Diphasic,* (of certain Trypanosomes) having a life-cycle which includes a free active stage. **1956** *Nature* 11 Feb. 280/1 Protozoa of this group [*sc.* Trypanosomidæ] being usually cultivated on complex diphasic media.

diphenhydramine (dəi:fenhəi·drămīn). *Pharm.* [f. Diphen(yl + *hydr-* (perh. from *hydrol) + Amine.] An antihistamine compound, $(C_6H_5)_2CH{\cdot}O{\cdot}CH_2CH_2N(CH_3)_2$, used in the form of its hydrochloride, a white powder with a bitter taste, in the treatment of allergic disorders. Also *ellipt.* for *diphenhydramine hydrochloride.*

1947 *Jrnl. Amer. Med. Assoc.* 27 Sept. 225/2 Diphenhydramine hydrochloride...The following dosage form has been accepted:..Kapseals Benadryl Hydrochloride: 50 mg. **1952** *Brit. Encycl. Med. Practice* (ed. 2) XII. 147 Synthetic antihistamine compounds..for the treatment of allergic disorders..include diphenhydramine. **1964** W. G. Smith *Allergy & Tissue Metabolism* vi. 69 In order to inhibit the activity due to histamine in the samples 2 to 3 mg. of diphenhydramine were added to the bath.

diphonemic (dəifonī·mik), *a. Linguistics.* [f. Di-^{2} + *phonemic *a.*] Applied to a sound that can be assigned to either of two phonemes. So **dipho·neme,** a sound of this kind.

1950 D. Jones *Phoneme* xx. 99 If a sound is assigned to two phonemes, it may be termed 'di-phonemic'. *Ibid.* 100 The sound might be treated as di-phonemic, i.e. assigned to one phoneme in some words and the other phoneme in other words. **1953** W. J. Entwistle *Aspects of Lang.* iv. 114 There may be diphonemes, i.e. phonemes which belong to two different orders, but these are set at a convenient distance in Hindustani and Japanese. *Ibid.* 115 Perhaps it would be simplest not to admit the existence of diphonemes (i.e. of single phonemes which have two values simultaneously) until more evidence is available. **1957** *Phonetica* I. 103 The existence of di-phonemic /h/ + /j/.

dipho:sphopy·ridine nu·cleotide. *Biochem.* [f. Di-^{2} + Phospho- + Pyridine + *nucleotide.] One of the names of the co-enzyme nicotinamide-adenine dinucleotide (NAD).

1938 *Biochem. Jrnl.* XXXII. 2235 Reduced coenzyme I (diphosphopyridinenucleotide) is oxidized extremely slowly by methylene blue and other oxidation-reduction indicators. **1955** *Sci. Amer.* Jan. 76/3 It turned out that the rickettsiae very definitely used the coenzymes known as DPN (diphosphopyridine nucleotide) and coenzyme A, both of which are essential for many oxidation reactions in animal cells. **1962** T. O. Sippel in A. Pirie *Lens Metabolism Rel. Cataract* 368 The oxidized and reduced forms of diphospho-and triphosphopyridine nucleotide were measured fluorometrically.

dipho:sphothi·amine. *Biochem.* Also **-in.** [Di-^{2} + Phospho- + *thiamine.] = *cocarboxylase.

1939 Barron & Lyman in *Jrnl. Biol. Chem.* CXXVII. 144 Lipmann's studies..on the action of diphosphothiamine as a catalyst for the oxidation of pyruvic acid have been confirmed by experiments. **1942** *Ann. Reg. 1941* 344 Important advances in vitamin research included the clarification of the functions of diphosphothiamin in carbohydrate metabolism. **1960** A. E. Bender *Dict. Nutrition* 32/2 Cocarboxylase is the diphosphate of vitamin B_1, alternatively known as thiamine pyrophosphate or diphosphothiamine.

diphtheroid, *a.* Add: **B.** *sb.* Any bacillus, esp. of the genus *Corynebacterium,* which resembles the diphtheria bacillus but is not pathogenic.

1908 *Practitioner* Jan. 138 The true diphtheria bacillus, and not one of these diphtheroids. **1949** H. W. Florey et al. *Antibiotics* I. v. 221 Diphtheroids, streptococci, and *L. acidophilus* were completely inhibited by that concentration. **1962** *Lancet* 5 May 933/1 The effect could also be elicited by..a diphtheroid.

diphyletic (dəifile·tik), *a. Taxonomy.* [f. Di-^{2} + Phyletic *a.*] Having two lines of descent; supposedly derived from two distinct sets of ancestors; also, of or pertaining to a classification of groups of organisms in accordance with the view that they have a diphyletic origin.

1902 *Nature* 25 Sept. 526 The 'Myriapoda', if a natural group, are diphyletic. **1903** *Amer. Nat.* Jan. 74 The diphyletic origin of the birds as represented by the Ratitæ and Carinatæ. **1912** *Rep. Brit. Assoc.* 578 Yet others claim a diphyletic origin for Man from the Apes. **1963** Davis & Heywood *Princ. Angiosperm Taxon.* ii. 46 A species (or unit of lower rank) derived by hybridisation between two ancestral species..is diphyletic by either definition.

diplanar (dəiplēⁱ·năɹ), *a. Math.* [f. Di-^{2} + Planar *a.*] Of or pertaining to two planes.

a **1865** W. R. Hamilton *Elem. Quaternions* (1866) 113 Any two quaternions (or quotients), which have different planes (intersecting therefore in a right line through the origin), may be said, by contrast, to be Diplanar.

dipleurula (dəiplū°·riŭlă). [mod.L., dim. of Dipleura.] A supposed bilaterally symmetrical ancestor of the echinoderms. Also, a stage in the development of some echinoderms.

1896 H. M. & M. Bernard tr. *Lang's Text-bk. Compar. Anat.* II. viii. 546 The radiate, but at the same time asymmetrical Echinoderm proceeds ontogenetically from a bilaterally symmetrical larva, the so-called Dipleurula. *Ibid.* 547 (*heading*) Metamorphosis of the Dipleurula Larva. **1900** F. A. Bather et al. in Lankester *Treat. Zool.* III. viii. 4 Zoologists have imagined a phylogenetic stage, the two-sided or *Dipleurula* stage,..more or less repeated in the *Dipleurula* larvae of recent Echinoderms. *Ibid.* 5 The simplest larval form among recent Echinoderms,..known as *Auricularia,*..differs from the *Dipleurula* in being bent upon its ventral surface. **1902** *Encycl. Brit.* XXVII. 620/1 We reach the conception that this supposed bilateral ancestor (or *Dipleurula*) may have become fixed. **1962** D. Nichols *Echinoderms* x. 121 It is not strange that early in life a common plan exists between those forms with larvae, or most of them. This stage has been termed the *dipleurula* ('little two sides'), a word which has unhappily been used to denote also a hypothetical common ancestor to the entire phylum.

diplo-, dipl-. Add: **diplacu·sis** *Path.* [Gr. ἄκουσις hearing], double hearing, the hearing of two notes when only one is produced, due to the hearing of a different tone in each ear, or to the arousing of two tonal sensations in the same ear; **di:plobaci·llus** (see quot. 1957^{1}); **diploge·nesis,** (*b*) the supposed change of germ plasm produced by changes due to environment, bringing about inheritance of acquired characteristics; **diplogra·phic** *a.* = *diplographical*; **diploha·plont** *Biol.*, an organism whose life-cycle embraces a diploid and a haploid phase; so **diplohaplo·ntic** *a.*; **di·plophase** *Biol.* [a. G. *diplophase*], the phase in the life-cycle of an organism when the nuclei are diploid; **di·plopore** *Zool.*, any of the pores that occur in pairs on the surface of the theca of certain cystoids (order Diploporita); also, a thecal canal that ends in one of these pores.

1890 Billings *Med. Dict.* 400/2 Diplakousis or Diplacusis. **1895** E. B. Titchener tr. *Külpe's Outl. Psychol.* 299 The abnormality (*diplacusis*) may be restricted to a single ear. **1970** J. V. Tobias *Found. Mod. Auditory Theory* I. x. 391 One possible explanation for diplacusis is distortion. **1901** *Jrnl. Exper. Med.* V. 213 The occurrence of acid-resisting diplococci or diplo-bacilli. **1908** *Practitioner* Feb. 203 The diplobacillus of Morax-Axenfeld. **1957** M. B. Jacobs et al. *Dict. Microbiol.* 78/2 *Diplobacilli,* bacillus-type bacteria occurring in pairs, as, for example, *Moraxella lacunata.* **1957** *Bergey's Man. Determinative Bacteriol.* (ed. 7) 419 Genus VII. *Moraxella*...Small rod-shaped cells which occur as diplobacilli **1896** *Nat. Sci.* Nov. 288 Cope's theory of Diplogenesis. **1823** J. Elmes *Mem. & Life C. Wren* 23 In 1647..he had a patent granted him for seventeen years, for a diplographic instrument for writing with two pens. **1921** *Bot. Abstr.* VI. 251 (*heading*) Genetical phenomena and taxonomy in haplonts and diplohaplonts in the vegetable kingdom. **1938** *Bot. Rev.* IV. 136 Diplohaplonts with unlike alternating generations are common among the Phaeophyceae. *Ibid.* 137 A lifecycle identical with that of the diplohaplontic Phaeophyceae with dissimilar generations has been described for *Stigeoclonium.* **1965** Bell & Coombe tr. *Strasburger's Textbk. Bot.* (new ed.) i. i. 36 In many of the more highly developed algae..the diploid zygote grows first into a multicellular diploid plant, the sporophyte. After the development of numerous diploid spore-mother cells meiosis takes place freely. Thus from the original nuclear fusion.. a multitude of haploid cells (gonospores) are produced. These organisms are termed diplohaplonts. **1925** E. B. Wilson *Cell* (ed. 3) 1130 *Diplophase,* that phase of the life-history, particularly in the antithetic alternation of generations, in which the nuclei are haploid, as in the sporophyte. **1965** J. Wilkinson tr. *Langeron's Outl. Mycology* (ed. 2) ix. 374 Meiosis intervenes between the diplophase and the haplophase, fertilization between the haplophase and the diplophase. **1899, 1962** Diplopore [see *haplopore* s.v. *haplo-].

diplococcus (s.v. Diplo-). Substitute: (diplǫkǫ·kǒs). *Bacteriol.* Pl. **diplococci** (-kǫ·kəi, -kǫki). [mod.L., f. Gr. διπλό-ος double + κόκκος grain, seed, adopted as a genus name by A. Weichselbaum 1886, in *Wiener med. Jahrb.* LXXXII. 483.] Any coccus that occurs predominantly in pairs, esp. one belonging to the genus of parasitic bacteria so called, which includes the pneumococcus, *D. pneumoniæ.*

1883 [in Dict. s.v. Diplo-]. **1911** M. Herzog *Text-bk. Disease-producing Microörg.* xxxii. 378 The diplococci show particularly often in the shape of candle-lights, hence the name *Diplococcus lanceolatus,* which means lancet-shaped. **1927** R. A. Kelser *Man. Vet. Bacteriol.* xi. 133 In the affected lung tissue Streptococcus equi commonly appears singly and in pairs. This characteristic led some investigators to consider the organism a diplococcus. **1949** Kelly & Hite *Microbiol.* vi. 68 Whereas the term diplococcus means any paired coccus, the same word begun with a capital letter refers only to the gram-positive paired cocci of the genus *Diplococcus.* **1969** S. T. Lyles *Biol. Microorg.* xx. 438 Another gram-negative diplococcus type is associated with the nervous system and produces meningitis.

Hence **diploco·ccal** *a.*, of, pertaining to, or caused by a diplococcus; **diploco·ccoid** *a.*, resembling a diplococcus or diplococci.

1903 *Med. Record* 2 May 705/1 Two cases of diplococcal septicaemia. **1908** *Lancet* 15 Feb. 484/1 A diplococcal infection of the throat. **1950** K. A. Bisset *Cytol. & Life-hist. Bact.* ix. 117 The two-celled structure [of *Neisseria gonorrhœæ*] is concealed by the gram-complex in the gram-positive species, and has been misinterpreted as 'diplococcal' in the gram-negative. **1910** *Practitioner* Apr. 489 The characteristic diplococcoid arrangement of the germs.

diplodocus (diplǫ·dǒkǒs, di:plodōᵘ·kǒs). [mod.L. (O. C. Marsh 1878, in *Amer. Jrnl. Sci.* 3rd Ser. XVI. 414), f. Gr. διπλό-ος double + δοκός a beam.] An individual of the extinct genus of gigantic herbivorous dinosaurs of the order Sauropoda, of which remains have been found in the Upper Jurassic of western North America. Also *allusively.*

1884 *Amer. Jrnl. Sci.* 3rd Ser. XXVII. 161 The skull of *Diplodocus* is of moderate size. **1890** *Guide Exhib. Gall. Dept. Geol. & Palæontol. Brit. Mus.* (*N.H.*) II. 9 The *Diplodocus,* an animal intermediate in size between *Atlantosaurus* and *Morosaurus.* **1905** E. R. Lankester *Extinct Animals* v. 204 Mr. Andrew Carnegie has presented to the Natural History Museum a complete reconstruction of the skeleton of a closely allied Dinosaur—the Diplodocus—which was excavated in Wyoming. **1905** *Westm. Gaz.* 6 July 2/1 Divo Pietro Aretino, the heroic literary blackguard, the diplodocus of the reptile press. **1927** Haldane & Huxley *Anim. Biol.* xi. 242 Diplodocus (herbivorous, gigantic, and semi-aquatic). **1928** G. B. Shaw *Intell. Woman's Guide Socialism* xlii. 162 Adam Smith's eighteen men are as extinct as the diplodocus. **1962** E. H. Colbert *Dinosaurs* vi. 111 In the long, attenuated sauropod *Diplodocus,* from western North America, the nostrils are on the very top of the skull.

diploicin (diploi·sin). *Chem.* [a. G. *diploïcin* (W. Zopf 1904, in *Ann. d. Chem.* CCCXXXVI. 60), f. *Diploicia* (see def.) + -in^{1}.] A crystalline depsidone, $C_{16}H_{10}O_5Cl$, isolated from the lichen *Buellia* (= *Diploicia*) *canescens.*

1904 *Jrnl. Chem. Soc.* LXXXVI. i. 1020 *Diploicin* forms sheaves of brown, thick crystals, melts at 225°, is only slightly soluble in alcohol, ether, glacial acetic acid, or benzene, has only a slight tendency to redden litmus paper, and gives no coloration with ferric chloride. **1946** *Nature* 26 Oct. 590/1 Diploicin..according to Burger and associates, possesses tuberculostatic activity *in vitro.* **1949** Florey et al. *Antibiotics* I. xiii. 574 Barry..found that when diploicin was treated with aqueous alkali it was converted to a sodium salt of the acid (II) which inhibited the growth of *Myco. smegmatis*..and of *Myco. tuberculosis*...Although it was not determined whether diploicin was actually antibacterial Barry stated that he was using the substance in a 'limited animal protection experiment'. **1967** M. E. Hale *Biol. Lichens* viii. 117 The first successful synthesis of a depsidone, diploicin.

diploid. Add: **B.** *adj. Biol.* [a. G. *diploid* (E. Strasburger 1905, in *Jahrb. f. wissensch. Bot.* XLII. 62): see *-ploid.] Of a cell or its nucleus: having two homologous sets of chromosomes, one from each parent, each containing the haploid number of chromosomes; also, of or pertaining to diploidy, (of an organism) having somatic cells that are diploid. Also as *sb.*, a diploid organism. Hence **di·ploidy,** the condition of being diploid; **di:ploidiza·tion; di·ploidize** *v. trans.* (see quot. 1930).

1908 [see *haploid a.]. 1914 G. N. Calkins *Biol.* ix. 209 Each has the number of chromosomes characteristic of the species (in modern terminology the diploid number). 1924 E. W. MacBride *Study Heredity* viii. 216 Other examples of such 'tetraploidy' (i.e. doubling the ordinary 'diploid' number of chromosomes) are known to occur amongst the progeny of hybrids between different species. 1928 *Amer. Naturalist* LXII. 57 Stronger evidence for diploidy of biparental males appears when three character differences, all affecting one structure, the wing, are involved. 1930 A. H. R. Buller in *Nature* 1 Nov. 687/1 The term *diploidisation* has been introduced here for the first time to designate the process by which a haploid cell is converted into a diploid cell or a haploid mycelium into a diploid mycelium by the formation of conjugate nuclei within the cell's or the mycelium's interior. A haploid mycelium of one sex may be said to *diploidise* a haploid mycelium of opposite sex. 1952 *New Biol.* XIII. 31 Most animals and many flowering plants are diploids. 1955 *Sci. Amer.* June 56/3 The body cells of a normal salamander are diploid. 1957 *Encycl. Brit.* XI. 486 Forty-eight is the diploid chromosome number in man... Haploid cells arise from diploid ones by a complex process known as meiosis. 1957 [see *haploid a.]. 1964 G. H. Haggis et al. *Introd. Molecular Biol.* x. 277 It is possible by exploiting a special phenomenon, known as sex-duction, to bring about a stable state of diploidy for a short length of the chromosome. 1968 [see *haploid a.]. 1971 *Nature* 7 May 49/2 If a diploid spermatozoon fertilizes a normal ovum, the resulting zygote would be triploid.

diplomatic, *a.* and *sb.* Add: **A. 3.** *diplomatic bag*, one containing the official mail of members of the diplomatic body; cf. *bag sb. 7 b; *diplomatic immunity*, the exemption from arrest, taxation, searches, etc., granted under international law to diplomatic personnel, their families and staff, when staying in a foreign country.

1911 *Encycl. Brit.* VIII. 299/2 In certain cases..the exterritoriality of ambassadors implies a fairly extensive criminal jurisdiction; in other cases the dismissal of the servant would deprive him of his diplomatic immunity and bring him under the law of the land. 1936 M. Ogdon (*title*) Juridical bases of diplomatic immunity. 1956 *N. & Q.* July 277 These papers were returned to Commander Bulloch via the United States Diplomatic Bag in January of 1895. 1970 R. G. Feltham *Diplomatic Handbk.* v. 52 The diplomatic bag is a sealed bag or container clearly marked as such and containing official documents and articles for official use, which a head of mission is entitled to receive and dispatch without it being interfered with. 1971 J. Sangster *Your Friendly Neighbourhood Death Pedlar* ii. 43 Do you need traveller's cheques,..a gun permit, diplomatic immunity?

diplomatize, *v.* Add: **1.** (Later U.S. example.)

1834 *Knickerbocker* Aug. 120 Ladies were invited..to prepare themselves for future honors..by becoming Latinized, Græcised, mathematicized, and at length diplomatized.

3. Delete *rare* and add later examples.

1898 *Daily News* 6 Aug. 7/3 Italian unity is owing in great part to the work of the people with Garibaldi, and Cavour did nothing but diplomatize the movement. 1906 *Daily Chron.* 24 Dec. 4/6 Count Nicholas Ignatieff who may be said to have 'diplomatised' the Russo-Turkish War of 1877.

diplont (di·plǫnt). *Biol.* [a. G. *diplont*, f. Diplo- + Gr. ὤν, ὀντ- being: see Onto-.] A sexual organism that is diploid at all stages of its life other than the gamete, which is haploid.

1925 E. B. Wilson *Cell* (ed. 3) vi. 492 From this organism [the haplont] arise the gametes..and by their union is produced the zygote from which arises a diploid, asexual spore-producing diplont (in plants the sporophyte), thus completing the life-cycle. 1943 L. W. Sharp *Fund. Cytol.* xi. 160 *Cladophora glomerata* is also reported to be a diplont, but it differs from *Codium* in having zoöspores. 1951 [see *haplont]. 1965 Bell & Coombe tr. *Strasburger's Textbk. Bot.* (new ed.) i. i. 37 Meiosis can be deferred right up to the formation of gametes. In this case, rare in plants, all the cells other than the gametes are diploid and the organisms are termed diplonts.

Hence **diplo·ntic** *a.*

1929 *Hereditas* XIII. 333 Diplontic sterility may be caused by species crosses, special genetic factors and by inbreeding. 1970 *Jrnl. Phycol.* VI. 122 The life cycle of *N[octiluca] miliaris* appears thus to be of the diplontic type.

diplotene (di·plotīn). *Biol.* [ad. Fr. *diplotène* (H. von Winiwarter 1900, in *Archives de Biol.* XVII. i. 70), f. Diplo- + *-tene.] The fourth stage of the prophase of meiosis (following pachytene), in which the four chromatids of each tetrad begin to separate into two pairs, each pair containing one chromatid from each of the two orginal paired chromosomes. Also *attrib.* or as *adj.*

1925 E. B. Wilson *Cell* (ed. 3) vi. 545 Probably, therefore, the diffuse stage should be regarded as a highly modified diplotene in which the duality of the early diplotene..in some manner persists throughout. 1929 *Jrnl. Genetics* XXI. 5 *Diplotene looping.* As soon as the double threads begin to open to form the loops characteristic of *diplotene*..it appears as though..real association has been taking place only between two chromosomes..for the third chromosome is seen to be lying free. 1931 *Nature* 9 May 711/1 At meiosis, it does not happen until pachytene (possibly at the moment at which the diplotene loops appear). *Ibid.*, This condition is fulfilled by the pairing of chromosome threads when they are still single, and their separation at diplotene when they have at last come to divide. 1951 G. H. Bourne *Cytol. & Cell Physiol.* (ed. 2) x. 435 After 3 days the oocytes entered the contraction phase, from which most of them had emerged by the 6th day, to pass into the pachytene phase; this was quickly followed by the diplotene. 1965 Bell & Coombe tr. *Strasburger's Textbk. Bot.* (new ed.) i. i. 34 In diplotene the pairing of the still extended chromosomes, completed in pachytene, is again lost.

dipole (dəi·pōᵘl). [f. Di-² + Pole *sb.*²]

1. A pair of non-coincident equal and opposite electric charges or magnetic poles (usu. but not necessarily close together); an object, esp. a molecule, atomic particle, etc., having such charges or poles; **dipole moment,** the product of the distance between the two charges or poles of a dipole and the magnitude of either of them; the electric or magnetic moment of a dipole.

1912 *Sci. Abstr.* A. XV. 184 Taking the view that in dielectrics the electrons when displaced are urged back by forces proportional to those displacements, and that *dipoles* of constant electric moment are present also, the author [*sc.* Debye] obtains the following expression. 1931 W. M. Deans tr. *Debye's Dipole Moment & Chem. Struct.* 15 The methods in general use for the determination of dipole moments depend on the measurement of the dielectric constant of the substance in question. 1934 *Nature* 17 Mar. 415/2 The permanent dipole of a molecule in a solution may be regarded as surrounded by an 'atmosphere' of dipoles of opposite sign produced partly by induction in the polarisable solvent molecules and partly by orientation of adjacent permanent dipoles. *Ibid.* 26 May 802/1 Few branches of physical chemistry can show a more rapid development than the study of dipole moments. 1956 *Sci. News* XLI. 51 The electrical properties of organic materials are critically influenced by whether or not they contain permanent molecular dipoles. 1957 *Endeavour* XVI. 188/1 Nuclei with spin quantum numbers greater than one-half usually possess an electric quadrupole moment as well as a magnetic dipole moment. 1965 A. Holmes *Princ. Physical Geol.* (ed. 2) xxvii. 989 The main part of the magnetic field is..like that of a powerful bar magnet (a 'dipole') placed near the middle of the earth. 1968 R. A. Lyttleton *Myst. Solar Syst.* iii. 97 The dipole-moment [of Mars] can be at most 1/3000th that of the Earth.

2. *Radio.* An aerial consisting either of two equal metal rods mounted close together in line or of a single rod, with the electrical connection made to the centre of the aerial and with a total length usually about half the wavelength to be transmitted or received; freq. *attrib.* (See also quots. 1947, 1960.)

1929 *Proc. Inst. Radio Engin.* Dec. 2207 (*caption*) Position of the dipole on the airplane... Securely stretched dipoles are most commonly used... Fig. 15 shows the installation of such a dipole antenna. 1930 *Telegr. & Teleph. Jrnl.* XVII. 31/1 These arrays generally consist of a number of dipoles..arranged in the form of a curtain. 1931 *Jrnl. R. Aeronaut. Soc.* XXXV. 761 The directional properties were investigated with a di-pole receiver. 1943 *Electronic Engin.* XVI. 197 Dipoles of length other than half wavelength. 1947 Siegert & Purcell in L. N. Ridenour *Radar System Engin.* iii. 82 'Window' is the British and most commonly used code name for conducting foil or sheet cut into pieces of such a size that each piece resonates as a dipole at enemy radar frequency. 1955 *Sci. Amer.* Mar. 38/1 A parabolic 'dish', either solid or made of a wire screen, reflects incoming radio waves to a focal point, where a small dipole or rod picks up the energy and converts it into an electric current which is then conveyed by a cable to a sensitive receiver. 1960 *Times* 16 Sept. 6/7 The idea of using a cloud of thin strips of metal foil to reflect radio waves originated during the war... Such a strip is tuned automatically to a wavelength twice the length of the strip—an aerial or reflector of this kind being known as a dipole. 1964 *Ann. Reg. 1963* 185 On 9 May..the U.S. released..some 400 million copper dipoles, or space needles, each about ¾ inch long.

dipped, *ppl. a.* Add: **1. c.** *transf.* Extended or carried below a surface or level.

1925 Penderel-Brodhurst & Layton *Gloss. Eng. Furniture*, *Dropped or dipped seat*, the seat of a chair having a concave upper surface between the two side rails. 1929 *Evening News* 18 Nov. 10 A wonderfully flared skirt dipped at sides.

d. Of the beams of the headlights of a vehicle: lowered.

1937 *Sunday Times* 10 Jan., Even in London, I always drive with my headlamps in the dipped position, except in extremely well-lighted thoroughfares. 1963 *Times* 28 Feb. 4/5 Birmingham's campaign for dipped headlights has 'sensationally' cut night accidents and casualties in the city. 1969 *Guardian* 23 June 7/5 In town traffic, dipped beams produce appreciable glare.

Dippel's oil. [f. the name of the discoverer J. C. *Dippel* (1672–1734), German alchemist.] Bone oil. In full *Dippel's animal oil.*

1819 W. T. Brande *Man. Chem.* vii. 429 When horn, hoofs, or bones, are distilled *per se*, a quantity of solid carbonate of ammonia, and of the same substance combined with empyreumatic oil,..are obtained; hence the pharmaceutical preparations called spirit and salt of hartshorn, and Dippel's animal oil. 1904 in Goodchild & Tweney *Technol. & Sci. Dict.* 1938 *Thorpe's Dict. Appl. Chem.* (ed. 4) II. 29/2 'Dippel's oil', sometimes also termed 'bone oil',..is obtained by the destructive distillation of bones in the preparation of bone charcoal. 1966 Millar & Springall *Sidgwick's Org. Chem. Nitrogen* (ed. 3) xxi. 619 Pyrrole..was first isolated in a pure condition by Anderson in 1858 who obtained it from a product of distillation of bone, bone oil or Dippel's oil.

dipper. Add: **1. d.** (Earlier and later examples.)

1889 in Barrère & Leland *Dict. Slang.* 1896 A. Morrison *Child of Jago* xi. 111 Such dippers—such pickpockets—as could dress well. 1968 M. Allingham *Cargo of Eagles* ii. 34 [The wallet] could have been pinched..and ditched there... Dippers often do that.

5. b. (Earlier example.)

1842 *Lowell* (Mass.) *Offering* II. 234, 236 (Th.), You all know the Dipper? Yes, it is in the Great Bear. The Little Dipper is in Ursa Minor.

c. In full *dipper dredge*. A type of dredging boat or machine (see quots.).

1877 *Encycl. Brit.* VII. 465/1 The dipper dredge consists of a barge, with a derrick-crane reaching over the stern, suspending a large wrought-iron bucket which brings up the dredged material. 1879 *Scribner's Monthly* Nov. 55/1 The channel has also been assisted somewhat in its development, by an Osgood dipper dredge. *a* 1884 Knight *Dict. Mech.* Suppl., *Dipper*, a form of dredging machine which has a large ladle on the end of a spar. 1959 *Chambers's Encycl.* IV. 634/2 Such a crude form of dipper dredge is still to be found in China.

7. b. = *dip *sb.* 8.

1891 R. Wallace *Rural Econ. Austral. & N.Z.* xvi. 247 The dipper..consists of a narrow well or trough of masonry. 1956 J. Gibbons et al. in D. L. Linton *Sheffield* 258 Installation of electricity, sheep pens and dippers.

c. As the name of various mechanical devices or instruments; *big dipper*, see *big a. B. 2.

1925 *Morris Owner's Man.* 28 In the bottom case are fitted the troughs for feeding oil to the connecting-rod big ends through the oil dippers which are fitted to same. 1928 *Daily Express* 28 Sept. 9 It should be made compulsory for all motorists to have dippers affixed to their headlights. 1951 N. Balchin *Way through Wood* iii. 46, I suppose he was driving dipped and it was one of those dippers that switches off the offside light. 1963 [see *headlight].

8. dipper switch = *dip-switch.

1935 *Times* 1 Oct. 8/4 The equipment includes such fittings as self-cancelling traffic indicators..and a foot-operated dipper switch. 1965 Priestley & Wisdom *Good Driving* ii. 20 A dipper switch..enables you to dip the beam.

dipping, *vbl. sb.* Add: **2. b.** A grade of turpentine. (Cf. *dip sb. 8 b.) *U.S.*

1832 D. J. Browne *Sylva Amer.* 232 The turpentine thus procured is the best, and is called pure dipping.

3. *dipping-machine, -tank, -trough, -vat*; **dipping-wheel** *U.S.*, a wheel consisting of revolving buckets or nets set in a river for catching fish.

1886 C. Scott *Sheep-farming* 145 The material best adapted for making the tub of a dipping machine is concrete. ..On a sheep farm the dipping-trough should be always the landlord's property, and a fixture. 1903 *Daily Chron.* 10 Feb. 6/4 The molten spelter, with which the dipping tanks were filled ready for the day's work. *a* 1884 Knight *Dict. Mech.* Suppl., *Dipping vat*, the trough containing fine glazing slip in which biscuit ware is dipped to be covered with the material which, baked on, forms glaze. *Ibid.*, *Dipping wheel*, a contrivance used in Southern rivers to meet local demand for fish. It is set in the stream so as to be turned by the current, and has a number of dip-nets which raise the fish and tumble them out at the axis in the manner of the scoop-wheel.

dipping, *ppl. a.* Add: Of the beams of headlights: lowered or capable of being lowered (cf. *dip *v.* 6 d).

1922 *Motor* 7 Nov. 744/2 (*caption*) Duco dipping headlights. 1957 *Act* 5 & 6 *Eliz. II* c. 51 § 3 No light shown by a vehicle, other than a dipping headlight, shall be moved by swivelling, deflecting or otherwise while the vehicle is in motion.

dippy (di·pi), *a.* *slang.* [Origin obscure; ?f. Dip *v.*] Mad, insane, crazy. Also const. *about*, *over*, in love (with). Also *absol.*

1903 C. E. Merriman *Lett. from Son* ix. 119 Her desolateness appeared to touch a hidden, sympathetic chord in my nature. Whatever the cause, I was dippy for fair. 1904 Ade *True Bills* 35 It is just as easy to love a girl who has the coin as it is to get dippy over the Honest Working-Girl. 1905 A. M. Binstead *Mop Fair* i. 6, I should be compelled to lament the fact that I had become hopelessly dippy. 1922 J. A. Dunn *Man Trap* xii. 167, I got lost there... Damned fool. No water! Too much sun! Went dippy and threw away everything. 1923 Wodehouse *Adv. Sally* x. 121 I'd be just as happy in two rooms and a kitchenette, so long as Fillmore was there. You've no notion how dippy I am about him. 1930 J. B. Priestley *Angel Pavement* vii. 353 The daughter she talks about seems to be completely dippy. 1938 F. S. Anthony *Me & Gus* ii. 11 A chap dippy over a girl. 1967 *Times Rev. Industry* Mar. 29/2 In past days the senile and the slightly dippy were clapped into institutions.

diprionid (dəiprəi·ǭnid), *a.* = Diprionidian *a.*

1888 Rolleston & Jackson *Anim. Life* 769 Specimens of diprionid Graptolites.

dipso, colloq. abbrev. of Dipsomaniac *sb.* and *a.*

1880 G. B. Shaw *Let.* 29 Nov. (1965) 36 How long do you suppose an average woman (dipso as described)..would last alive after the inability to retain food or drink set in? 1923 A. Huxley *Antic Hay* xiv. 217 Vicious young women. Lesbians, drug-fiends, nymphomaniacs, dipsos—thoroughly vicious, nowadays. 1935 E. Williams *Night must Fall* 30 Maniacs can't get far without cash..however dipso or nympho they may be. 1959 J. Braine *Vodi* xxv. 269, I didn't mean that you were a dipso... But I've never been out with you when we didn't go into a pub.

Dip. Tech., colloq. abbrev. of *Diploma in Technology.* Also, a person who holds this qualification.

1957 *Technology* Oct. 292/3 Recognition of Dip. Tech. courses. 1958 *Times* 19 Sept. 5/6 First Crop of Dip. Techs... The names are announced to-day of the first crop of students to be awarded the Diploma in Technology by the National Council for Technological Awards. 1958 *Economist* 15 Nov. 591/2 The universities have..agreed that the Dip Tech (created in 1956) shall rank as equivalent to first degrees in universities, and this means that Dip Techs should be able to work for higher research degrees in universities if opportunity offers. 1969 R. Layard et al. *Impact of Robbins* 73 There were 5,500 students on these courses, if we include those leading to the Diploma in Technology (Dip. Tech.), most of which have now been converted to degree courses.

Dipylon (di·pilǫn), *a.* and *sb.* *Gr. Archæol.* Also **Dipylum**, and with small initial. [L., a. Gr. δίπυλον, neut. of δίπυλος double-gated, f. δι- two + πύλη gate.] **A.** *sb.* A double gateway in which the two gates are placed side by side, *spec.* a gateway in Athens on the north-west side of the city. **B.** *adj.* Denoting or pertaining to the Dipylon of Athens; *spec.* designating a style of Greek pottery belonging to the Geometric period found during excavations near this gate, or a similar Bœotian ware of the same period and of similar style, or to the designs found on such pottery.

1835 *Penny Cycl.* III. 10/1 The direction of the wall from the Ilissus along the south and west sides of the city to the Dipylum is quite clear. 1896 *Daily News* 12 June 5/1 Through the modifications of the Dipylon period..the fashions of Greek ladies were always changing. 1902 *Encycl. Brit.* XXV. 572/2 We engrave an excerpt from a Dipylon vase. *Ibid.* 759/1 The discovery of the Dipylon Gate, the principal entrance of ancient Athens. 1911 W. M. F. Petrie *Revolutions of Civilisation* iii. 59 The rise of a new art began to dawn in the dipylon vases. 1950 H. L. Lorimer *Homer & Monum.* v. 164 The numerous indications that the Dipylon shield was wielded by a hand-grip dissociate it from the Minoan shield. *Ibid.* 165 The form of the Dipylon shield is simple, produced apparently by taking a hide cut to an oblong shape and stiffening it by a vertical stave down the middle and a horizontal one at top and bottom. 1955 *Archit. Rev.* CXVII. 295 A truncated dipylon gripped between canted buttresses.

diquat (dəi·kwǫt). *Chem.* [f. Di-² 2 + Quat(ernary *a.*] A quaternary compound with herbicidal properties (see quot. 1960).

1960 *Agric. & Vet. Chem.* I. 197/2 Diquat is one of a group of quaternary dipyridylium salts which possess quite unique properties. *Ibid.*, Diquat is thus a contact herbicide with a very rapid action, two main outlets being as a pre-harvest desiccant and as a contact weedkiller. 1962 *New Scientist* 22 Feb. 441/3 Diquat is quickly absorbed by weeds, killing rapidly on contact with their foliage. 1965 *Economist* 9 Jan. 130/2 A product like the paraquat/diquat 'chemical ploughing' agents (for squares: weedkiller).

Dirac (diræ·k). The name of Paul Adrien Maurice *Dirac* (born 1902), British mathematician and physicist, used *attrib.* to designate phenomena, theories, etc., in *Math.* and *Physics* discovered, described, or postulated by him. Also in *Fermi-Dirac.

1933 Harnwell & Livingood *Exp. Atomic Physics* 461 (Index), Dirac-Fermi statistics. 1934 *Physical Rev.* XLVI. 110/1 When the electron is very intimately connected with an attractive center, something happens to its spin which is not adequately described by the Dirac theory. 1935 *Ibid.* XLVIII. 284/2 The interaction of the heavy particles with the electrons (represented, for instance, by the Dirac 'density matrix'). *Ibid.* 435/1 The frequencies of the emitted light are determined by the eigenvalues of the Dirac equation. 1955 W. Pauli *Niels Bohr* 33 The coupling between one Boson field and one Dirac field. *Ibid.* 39 The Dirac equations for spinors are preserved. *Ibid.* 47 A Dirac spinor with four components can be decomposed into two irreducible parts. 1970 G. K. Woodgate *Elem. Atomic Struct.* iii. 36 The radiation field also is treated as a quantized system (Dirac method).

direct, *v.* Add: **1. c.** *absol.* (Later examples.)

1812 Shelley *Let.* 18 June (1964) I. 195 You may direct the Post Off. at Chepstow for if we are gone Eliza will be there. 1835 Dickens *Let.* 4 July (1965) I. 68 You may direct to me if you please at 18 York Place Fulham Road.

5. c. *trans.* and *intr.* To supervise and control the making of a film or the production of a play, etc.; to guide or train (an actor, etc.) in his performance. orig. *U.S.*

1913 F. W. Sargent *Technique Photoplay* (ed. 2) xxix. 162 *Director*, one who produces photoplays, directing the preparation and action. 1933 *Punch* 6 Dec. 638/2 Little Robert Lynen..had he been properly directed would have us all in tears. 1938 W. S. Maugham *Summing Up* 154 The remedy of course is for the author to direct his own play. 1967 *Listener* 12 Jan. 59/1 Ronald Neame has directed efficiently.

6. e. In a national emergency, etc.: to assign (workers) to a particular industry or employment.

1943 *Hutchinson's Pict. Hist. War* 17 Feb.–11 May 33 The keenness which was so wonderful a feature of the old volunteer Home Guard is no less marked in those who are known as the 'directed men'. 1945 *Yorkshire Post* 29 June, Men and women who were released from the Forces to undertake civilian jobs into which they were directed by the Ministry of Labour. 1946 *Lancet* 5 Jan. 19/2 The charge that the Government was seeking to direct medical labour.

direct, *a.* Add: **1. b.** *direct vision.* Also used *attrib.* to designate (*a*) in *Photogr.*, a type of view-finder in which the subject is viewed directly through a lens or sight; (*b*) in *Television*, a picture that can be viewed without magnification, or the apparatus for receiving such a picture. *direct wave* or *ray*, in *Radio*, the wave that passes from a transmitter to a receiver along the surface of the earth or (distinguished from *surface wave*) directly through the intervening air.

1911 B. E. Jones *Cassell's Cycl. Photogr.* 188 *Direct finder*, or *direct-vision finder*, a finder in which the view or object is inspected direct. 1915 —— *Cinemat. Bk.* iii. 19 A wire-frame direct-vision finder..is fitted in addition. 1937 *Discovery* Oct. 317/1 The receiver itself is a 23-valve superhet model with a direct vision television picture. 1952 Howe & Ducloux tr. *Kerkhof & Werner's Television* (Photo 36), Television picture with 625 lines, as reproduced by a direct-vision receiver. *Ibid.* 408 *Direct-vision tube*, picture tube intended for direct viewing without magnification. 1919 E. W. Stone *Elem. Radiotelegr.* (1920) ix. 198 A reflected wave from a station cannot arrive at the receiving station as soon as the direct wave which followed its normal path along the surface of the earth. 1943 C. L. Boltz *Basic Radio* ix. 142 Always there is a wave travelling along the surface of the earth. This is called the ground ray or direct ray. 1943 F. E. Terman *Radio Engineers' Handbk.* x. 674 The ground wave can conveniently be divided into..a surface wave and a space wave. The surface wave travels along the surface of the earth. The space wave is the result of..a direct wave and a ground-reflected wave.

2. c. Also applied to gunnery fire with an elevation not exceeding 15°.

1879 *Man. Artill. Exerc.* i. v. 24 Direct fire at masonry is either for demolition or for breaching.

6. d. *direct action*, (*b*) the exertion of pressure on the community through any action which is directly effective, such as strikes, sabotage, or demonstrations, as distinguished from action through constitutional processes; also *attrib.* Hence *direct actioner*, *actionist*, one who engages in direct action.

1912 J. R. MacDonald *Syndicalism* iv. 24 The Programme of Direct Action. 1919 *Times* 28 June 14/3 He had been a direct actionist for 35 years. 1920 S. & B. Webb *Hist. Trade Un.* (ed. 5) 672 The vast majority of Trade Unionists object to Direct Action..for objects other than those connected with the economic function of the Direct Actionists. ..Trade Unionists..are not prepared to disapprove of Direct Action as a reprisal for Direct Action taken by other persons or groups. 1920 *Edin. Rev.* Oct. 361 The root idea of direct action dates back in this country to Chartist days. In practice, direct action almost invariably implies either a sectional strike by a particular group or groups of labour, or a general strike by all groups of labour combined. 1931 S. Jameson *Richer Dust* iv. 86 The men were talking of 'direct action', and not taking much notice of their elected [trade-union] leaders. 1957 *Peace News* 18 Jan. 1/4 We have confined ourselves..too naïvely to public meetings. There is a kind of direct action which..can have most profitable results. 1958 *Observer* 7 Dec. 1/3 Members and supporters of the Direct Action Committee Against Nuclear War climbed a barbed-wire fence yesterday and 'invaded' a rocket base being built near Swaffham, Norfolk. 1961 *New Left Rev.* Mar.–Apr. 21/2 People who sneer at intellectuals and direct actioners. 1965 P. Arrowsmith *Jericho* xxvi. 285 Charles sat down on the ground. The other forty direct actionists followed his example. 1970 *Guardian* 4 Apr. 10/4 Why is direct action working where 'constructive', peaceful protest did not?

f. Of or pertaining to the work and expenses actually incurred during production as distinct from subsidiary work and overhead charges, i.e. to prime or initial costs or charges; also, applied to labour employed for the construction of works directly (without the intervention of a contractor).

1898 S. S. Dawson *Accountant's Compendium* 71/1 The prime cost..is the original or direct cost of same. 1903 *Encycl. Accounting* II. 263 These 'expenses' or charges are broadly divisible into 'direct' and 'indirect'. 1922 J. D. Hackett in *Management Engineering* Feb., Absence, such as is being considered here, applies mainly to direct or 'productive' labor and not to indirect or 'non-productive' employees. 1923 *Ibid.* May, *Direct labor*, work done exclusively in the making of a product, in contradistinction to subsidiary work also necessary for production. 1925 Ryall *Primer of Costing* 49 Direct labour may therefore be defined as—'Labour applied to a works order which can be measured and directly charged to that order or product'. 1930 *Daily Mail* 1 Aug. 7/7 The Office of Works is about to substitute direct labour for the present system of work done by contract.

g. Of a dye: not requiring a mordant; = Substantive *a.* 1 c.

1902 *Encycl. Brit.* XXVII. 558/1 Direct Colours.—The characteristic feature of the dyestuffs belonging to this class is that they dye cotton 'direct'. *Ibid.* 558/2 Wool and silk are dyed with the Direct Colours in the same manner as cotton. 1927 Horsfall & Lawrie *Dyeing of Textile Fibres* ix. 225 Deltapurpurine and Diamine Scarlets B and 3B are direct cotton colours in common use. 1966 R. C. Cheetham *Dyeing Fibre Blends* i. 8 The classes of dye in general use can be graded in a descending order of molecular size, ranging from aggregated acid and direct dyes to finely dispersed and levelling acid dyes.

h. *Metallurgy.* Designating a process by which wrought iron is obtained from the ore without the intermediate stage of cast iron.

1875 *Ure's Dict. Arts* (ed. 7) II. 941 Chenot's sponge may be obtained by an 'internal' or 'direct' method, in which the ore is reduced..by a hot current of carbonic-oxide gas. 1880 *Encycl. Brit.* XIII. 291/1 The 'direct' methods of Clay, Chenot, Yates, Blair, Snelus, Du Puy, Siemens, and others. 1884 W. H. Greenwood *Steel & Iron* xi. 212 The direct processes necessitate the use of purer and richer ores and fuels than the indirect processes. 1925 *Jrnl. Iron & Steel Inst.* CXII. 18 The difficulties that beset the commercial production of iron and steel direct from the ore in other direct processes.

i. direct-arc furnace, an electric-arc furnace in which the arc is formed between an electrode and the charge in the furnace; **direct-connected** *a.* = next (*a*); **direct-coupled** *a.*, (*a*) coupled without an intermediate transmission device; (*b*) in *Electr.*, applied to two circuits coupled by means of one or more resistors, capacitors, or inductors common to them both; also to an amplifier in which the signal path from one stage to the next contains only a resistor (or no circuit element at all), so that d.c. signals can be amplified; so *direct coupling*; **direct coupler** (see quot.); **direct current** *Electr.*, a current flowing in one direction only, as distinguished from an alternating current; abbrev. d.c., D.C.; also *attrib.*; **direct grant,** a grant of money paid directly to a school from public funds; freq. *attrib.*, esp. in **direct-grant school,** a school that receives a direct grant and in return observes certain conditions regarding the admission of pupils, etc.; also *ellipt.*; **direct injection,** a kind of fuel injection whereby fuel is supplied directly to the combustion chambers or their intakes in an internal combustion engine, without the use of a carburettor; usu. *attrib.*; so *direct-injected* adj.; **direct mail,** advertising matter or the like sent through the post to prospective customers; also *attrib.*; **direct method,** a method of teaching a foreign language through conversation, reading, etc., in the language itself without using the pupil's native language and without study of formal rules of grammar; also *attrib.*; **direct-reading,** applied *attrib.* to a measuring instrument calibrated in the actual quantity measured; **direct realism** *Philos.*, the theory that in perception we are directly aware of external or physical objects as they really are; **direct sowing** *Forestry* (see quot. 1891); **direct voice** *Spiritualism*, speech said to emanate directly from a disembodied spirit, without using a medium.

1921 J. N. Pring *Electric Furnace* xi. 202 In the direct-arc furnaces, such as those of Héroult and Girod, a zone of very high temperature is produced in the slag immediately below the electrodes. 1962 *Times* 2 Mar. 15/5 One of the Birlec three-ton direct-arc furnaces. 1901 *Feilden's Mag.* IV. 441/2 A direct-coupled electrically-driven fan. 1910 G. W. Pierce *Princ. Wireless Telegr.* 96 Direct coupled transmitting and receiving circuits. 1953 F. Langford-Smith *Radio Designer's Handbk.* (ed. 4) xii. 529 A direct-coupled amplifier is one in which the plate of one stage is connected to the grid of the next stage directly, or through a biasing battery or equivalent. 1962 Simpson & Richards *Junction Transistors* ix. 214 Another very useful direct-coupled amplifier is the Darlington compound connection. 1916 *Standard Rules Amer. Inst. Electr. Engin.* 97 *Direct coupler*, an apparatus which magnetically joins two circuits having a common conductive portion. 1907 J. Erskine-Murray *Handbk. Wireless Telegr.* ii. 43 This method of connection is now called 'direct coupling', as opposed to inductive coupling by means of a separate primary and secondary. 1910 *Hawkins's Electr. Dict.* 119/1 *Direct coupling*, connecting the shaft of a dynamo armature directly to the shaft which drives it. 1931 Moyer & Wostrel *Radio Handbk.* ii. 87 In direct capacity coupling, the coupling is made closer by reducing the capacity of the common condenser... In indirect capacity coupling, the coupling is made closer by increasing the capacity of the coupling condensers. [1886 *Jrnl. Soc. Telegr. Engin.* XV. 193, I am glad that people are beginning to use the term 'direct' when they mean a current which does not alternate.] 1889 E. J. Houston *Electr. Words* 165 Direct current. 1893 D. C. Jackson *Electro-Magn.* I. 92 A current constant in direction, but not necessarily so in value, is often called a Direct Current. 1893 *Jrnl. Soc. Arts* XLI. 623/2 The general merits of alternate-current, as against those of the direct-current systems. 1915 Hawkhead & Dowsett *Techn. Instr. Wireless Telegr.* 83 A direct current may be passed through a known resistance and the amount of heat generated may be measured. 1959 *Chambers's Encycl.* XI. 547/2 The term rectifier is used of any electrical device which can convert an alternating current into a source of direct current. 1945 *Guide to Educational System* (Min. Education) 58 *Direct-grant school*, school receiving grant direct from the Ministry of Education (*e.g.* nursery, special or grammar school). 1957 *Economist* 21 Dec. 1030/1 A number of excellent schools known in the jargon as 'direct grant'. *Ibid.*, To qualify for direct grant status, a school must not make a profit. 1958 *Times* 29 July 9/6 The direct-grant grammar school may be given a choice of accepting local authority rule or becoming entirely independent of the public system and public funds. 1970 *Guardian* 9 Apr. 12/3 Obviously the Conservatives will look on the 'direct grants' with tenderness. 1962 *Economist* 17 Mar. 1039/2 By making it reversible, lighter, faster and direct-injected, the company turned the diesel into a marine engine. 1935 *Jrnl. R. Aeronaut. Soc.* XXXIX. 802 Americans to give serious attention to the direct-injection engine, thus eliminating the carburet-

tor. 1930 H. Crane *Let.* 29 Dec. (1965) 360 The direct mail advertising business. 1959 *Times Lit. Suppl.* 23 Jan. 49/1 The work shown covers the realm of the poster, Press, direct mail, showcards, [etc.]. 1962 E. Godfrey *Retail Selling* x. 109 Direct mail advertising of this type can create a big response. *Ibid.* 110 Another useful direct mail prestige advertisement which many shops use is to send a Christmas card to all their regular customers. 1904 S. Bertelsen tr. *Jespersen's How to teach Foreign Lang.* i. 2 The method is by some called the 'new' or 'newer';..the 'direct' comes a little nearer. 1914 W. Owen *Let.* 3 Feb. (1967) 232 The majority of English Teachers have an execrable Accent, and what is worse, no notion of the *Direct Method.* 1917 H. E. Palmer *Scientific Study & Teaching of Lang.* iv. 72 In many cases..the Direct Method..resolves itself into the negative precept: there must be no translation. 1885 *Jrnl. Soc. Telegr. Engin.* XIV. 497 Ten years ago hardly anyone had even seen a direct-reading instrument. 1887 *Ibid.* XVI. 618 An instrument..which indicates the quantity to be measured without any adjustment on the part of the observer, is a 'direct-reading' instrument. 1956 *Nature* 17 Mar. 513/1 Electronic circuits for a direct-reading spectrograph. 1925 J. E. Turner (*title*) A theory of direct realism. 1956 H. H. Price in H. D. Lewis *Contemp. Brit. Philos.* 391 These three arguments..do at least refute the 'Direct Realism' of common sense. 1963 W. Sellars *Sci., Perception & Reality* 61 Direct realism and classical phenomenalism share..the 'phenomenalistic theme'. 1967 *Encycl. Philos.* VII. 78/1 Naive realism is the simplest form of direct realism. 1891 W. Schlich *Man. Forestry* II. ii. 41 Under 'direct sowing' is understood the formation of a wood by the sowing of seed directly on the area which it is proposed to stock. 1953 H. L. Edlin *Forester's Handbk.* x. 164 Many trials of direct sowing of various kinds of tree have been carried out in Britain, but no system has come into general use. 1926 A. Conan Doyle *Hist. Spiritualism* I. xiv. 330 Again, in reporting upon Mrs. Lord the Commission got the Direct Voice, and also phosphorescent lights after the medium had been searched. 1946 G. N. M. Tyrrell *Personality of Man* viii. xxv. 217 The 'direct voice' in which, it is alleged, a deceased person speaks from some isolated point in space. 1967 *Listener* 19 Oct. 488/2 The Society for Psychical Research has a standing offer of £1,000 to any medium who can produce physical phenomena such as materialisation, levitation or direct voice under test conditions.

dire·ctedness. [f. Directed *ppl. a.* + -ness.] The quality of being directed.

1922 J. Y. Simpson *Man & Attainm. Immort.* xii. 267 The specificity of action, the directedness, the working out of what looks like purpose. 1929 E. Bowen *Last Sept.* x. 117 She still had that quality of directedness.

directee (direktī·). [f. Direct *v.* + -ee.] One who is directed or is under direction.

1928 G. B. Shaw *Intell. Woman's Guide Socialism* lxx. 337 By paying the director more than the directee it creates a difference of class between them. 1938 *Mind* XLVII. 252 Some relation between the director and the directee which is explicitly to be proved absurd.

direction. Add: **1. e.** The art, technique, or an instance of directing a play, film, etc. (see *direct *v.* 5 c).

1938 C. Morgan *Flashing Stream* 36 The play was first performed at the Lyric Theatre, London under the direction of Mr. Godfrey Tearle. 1939 W. C. & H. S. Pryor *Let's go to Movies* 45 (*heading*) Direction. 1949 G. B. Shaw in E. J. West *Shaw on Theatre* (1958) 279 They are not concerned with direction as a fine art; but they cover the mechanical and teachable conditions which are common to all productions. 1959 *Oxf. Mag.* 12 Mar. 338/1 The pace and punch of his direction, camera-work and cutting are far better than anything he ever did in England.

11. direction-finder *Telecommunications*, a receiving device that determines from which direction radio waves come to it; so *direction-finding*; **direction indicator,** a device on a motor vehicle used to indicate the direction the driver intends to take; = *trafficator; **direction-post** (earlier example).

1913 *Year-Bk. Wireless Telegr. & Teleph.* 307 Direction-finding from ship-board. *Ibid.* 316 The uses of the direction-finder. 1919 R. Stanley *Wireless Telegr.* (ed. 2) II. 270 A direction-finding aerial system. 1920 *Discovery* May 131/2 The Marconi direction-finder not only receives wireless signals; it also indicates the direction of the sending station. 1921 L. B. Turner *Wireless Telegr.* 178 Direction-finding stations on land..are usually grouped in pairs at the ends of a suitable base line, so that the intersection of the two orientations determined gives the position of the source of signal, e.g. a ship or an aeroplane. 1935 *Discovery* June 155/2 The practical problems of direction-finding are closely connected with the vagaries of the waves returned from the ionosphere, and continued success has been attained in the development of direction-finders less subject to the errors produced by these waves. 1946 *Electronic Engin.* XVIII. 20 An automatic V.H.F. direction finder. 1966 M. R. D. Foot *SOE in France* x. 324 He tried to get in touch with Dowlen; but Dowlen was caught by direction-finders. 1937 *Times* 13 Apr. p. xxii/4 The makers of direction indicators and fog and anti-dazzle lamps. 1959 *Motor Manual* (ed. 36) vi. 183 Another change since the war has been the progressive displacement of the semaphore-type of direction indicator by flashing light signals. 1795 K. P. Moritz *Trav.* 142 Where there are cross-roads, there are direction posts, so that it is hardly possible to lose oneself in walking.

directional, *a.* Add: **4.** Of or pertaining to the direction or guidance of affairs.

1921 *Glasgow Herald* 5 Mar. 6 The responsibility of so improving his directional control of the conditions of working. 1922 *Ibid.* 21 Mar. 7 Their doctrine of one directional authority. 1928 *Daily Express* 26 Sept. 1 All directional and organisation expenses.

b. Of or pertaining to the direction of films, etc.

1949 *Here & Now* (N.Z.) Oct. 29/3 Camera, lighting, directional 'touches' are in their subordinate and proper place. 1962 *Movie* Sept. 25/1 The emotional effect of *Exodus* is the stronger for being free from directional nudges and promptings.

5. *Telecommunications.* Concerned with the record of directions indicated by radio signals received from a vessel, etc. Also, pertaining or relating to apparatus that transmits or receives radio signals, etc., more strongly in or from certain directions than others.

1914 R. Stanley *Wireless Telegr.* 179 To illustrate the effect of directional aerials at the sender and at the receiver. 1929 *Bell Syst. Techn. Jrnl.* VIII. 313 The use of directional receiving antennas is essential to satisfactory and economic results over such distances as the transatlantic radio path. 1930 *Aberdeen Press & Jrnl.* 26 June 7/1 An occasional directional message from the Cape Race wireless station was the only guidance they could pick up. 1935 *Economist* 30 Nov. 1062/2 Aerial navigation without wireless is to-day almost unthinkable. Night flying has only been made feasible by directional wireless. 1957 *BBC Handbk.* 60 To provide an effective signal in the area served by each programme, highly directional short-wave transmitting aerials are used. *Ibid.* 219 An efficient type of directional aerial. 1962 A. Nisbett *Technique Sound Studio* 255 The microphone is thus intensely directional.

directionality (dire:kʃənæ·liti). [f. Directional *a.* + -ity.] The state or quality of being directional; maintenance of direction.

1951 Parsons & Shils *Gen. Theory of Action* 405 An individual or group shows a consistent directionality in its selections. 1964 *Language* XL. 250 Linguistic relativity implies a one-way directionality. 1967 *Technology Week* 20 Feb. 34/3 Equipped with dipoles, the spacecraft was making cosmic noise measurements with no directionality. 1968 Chomsky & Halle *Sound Pattern Eng.* 386 There is..no reason to assume that the restriction has directionality, i.e., should be formalized to operate from left to right.

directive, *sb.* Delete †*Obs.* and add: *spec.* a general instruction how to proceed or act.

1902 *Encycl. Brit.* XXXIII. 647/1 The ecclesiastical régime..arrogates to itself the right of interfering by means of 'Directives' with the political life of nations. 1910 *Ibid.* II. 603/2 A few simple orders called 'directives' sufficed to set armies in motion with a definite purpose before them. 1941 *Economist* 15 Feb. 209/2 The actual handling of news items [by the B.B.C.] is subject to 'directives' or advice from Government departments. 1941 *Manch. Guardian Weekly* 14 Mar. 214/4 You would think it enough to call an order a direction. No, he [*sc.* the Civil Servant] has to make it 'a directive'. 1961 B. Fergusson *Watery Maze* ii. 52 A 'Directive' in Service parlance is a form of order, setting out one's terms of reference, defining one's responsibilities, and showing the extent and limitations of one's authority. 1971 *N.Y. Rev. Bks.* 22 Apr. 34/1 This authorization of September 1970 made official a practice which long preceded the issuance of the directive.

directivity (direkti·viti, dəi-). [f. Directive *a.* + -ity.] **1.** The quality or state of being directed by a vital force or power as distinguished from the physical forces, *spec.* as a theory of evolution.

1903 *Daily Chron.* Apr., That directivity which could not be explained without the power of the Deity behind it. 1907 *Hibbert Jrnl.* Oct. 150 Sheep, oxen, horses, and even geese may be feeding in the same grass-land. In each case the molecular compounds of the grass are identical, but they find themselves ultimately disposed in very different arrangements in the different animals. This can only be possible under the influence of directivity, but 'directing' differently in each creature.

2. The property or degree of being directional (see *directional *a.* 5). Also *attrib.*

1928 *Jrnl. Inst. Electr. Engin.* LXVI. ii. 955 General considerations of the directivity of beam systems. 1930 *Techn. News Bull., U.S. Bureau of Standards* Dec. 117/2 Directivity of reception or transmission, or both, will inform the pilot of the direction of danger. 1935 *Proc. Inst. Radio Engin.* XXIII. 357 Experiments with directivity steering for fading reduction. *Ibid.*, It has been noticed in the past that fading was affected by the directivity of the receiving antenna.

directly, *adv.* Add: **6. c.** Shortly; very soon; in a little while. *dial.* and *U.S.*

1851 H. Melville *Moby Dick* I. iii. 20 Supper?—you want supper? Supper'll be ready directly! 1882 'Mark Twain' *Stolen White Elephant* xvi. 268 When you say you will do a thing 'directly', you mean 'immediately'; in the American language—generally speaking—the word signifies 'after a little'. 1891 R. P. Chope *Dial. of Hartland, Devonshire* 40 *Drackly* or *dreckly*, directly; in the dialect this does not mean immediately, but shortly. 'I'll kom drackly; I mus' finish ot I'm 'bout fust.' 1936 M. Mitchell *Gone with Wind* xlvi. 824 Scarlett..leaned over the banisters. 'I'll be down terrectly, Rhett,' she called.

Directoire (dire·ktwāɹ), *sb.* and *a.* [Fr.; see Directory *sb.* 6.] **A.** *sb.* = Directory *sb.* 6.

1795 [see Directory *sb.* 6]. 1886 *Athenæum* 27 Mar. 432/1 The thanks of all playgoers are due to Mrs. Langtry for dressing the character of Pauline in the costume of the *Directoire.* 1937 J. Laver *Taste & Fashion* vii. 96 In France the Directoire and the Empire introduced a whole new class of speculators and adventurers.

B. *adj.* (Also with small initial.) **1.** Of, pertaining to, or resembling a style of dress prevalent at the time of the French Directory, characterized by its extravagance of design and its imitation of Greek and Roman costume. Also ellipt. as *sb.*, a hat of this style.

Spec. *Directoire knickers*, a fashion name (see quot. 1968).

1878 *Cassell's Family Mag.* 756 The Directoire..very nearly obscures the face. 1880 [see Redingote]. 1888 [see Empire *sb.* 8]. 1896 *Godey's Mag.* Feb. 214/2 An immense *directoire* bow of ribbon or chiffon. 1904 *Daily Chron.* 18 June 8/2 Such hats, with their picturesque brims, are known as Directoires. 1908 *Ibid.* 16 June 4/5 Worth has made a feature of Directoire gowns for Ascot. 1911 *Alfred Weeks's Sales Catal.*, Stockinette Directoire Knickers, slightly soiled, were 3/11, Reduced to 1/6¾d. 1938 E. Bowen *Death of Heart* i. iv. 75 Eddie's letter..she slipped up inside her woollen directoire knickers. It stayed just inside the elastic band, under one knee. 1951 E. Paul *Springtime in Paris* xiv. 260 The Carthaginian Queen, as a soloist, was permitted to wear a 'Directoire' gown. 1968 J. Ironside *Fashion Alphabet* 68 Directoire knickers..are straight knickers to the knee, with elastic at waist and knee.

2. Of, pertaining to, or resembling a style of furniture or objets d'art prevalent at the time of the French Directory.

1942 *Burlington Mag.* Nov. 286/2 Members of the Royal Family have contributed several unique French 18th century snuff-boxes,..the Duke and Duchess of Gloucester a directoire example chased and enamelled. 1955 A. West *Heritage* ii. 57 The pleasant directoire chairs and tables, which gracefully paid their frivolous tributes to the arts of Greece and Rome. 1962 *House & Garden* Dec. 63/1 An enormous bamboo Directoire bed. 1963 *Ibid.* Feb. 76/4 Directoire is the name given to the style of works of art produced in the last five years of the eighteenth century.

director. Add: **1. g.** One who directs a film or play, etc. (see *direct *v.* 5 c). orig. *U.S.*

1911 *Moving Picture World* 22 July 108/2 The director explains to the players the action of a..scene. 1914 R. Grau *Theatre of Science* 362 The world-famous director, D. W. Griffith. 1915 E. A. Dench *Making Movies* i. 1 Good actors, authors and photographers are indispensable, but unless they are guided by a talented director, results will be disappointing. 1933 *Punch* 30 Aug. 234/2 Mr. Herbert Wilcox, its [*sc.* the film's] director, has skilfully overcome the difficulties of adaptation. 1938 W. S. Maugham *Summing Up* 107 The director shouts 'curtain up'. *Ibid.*, I use the American word director rather than the English one, producer, because I think it better describes what should be the function of the person in question. 1944 *Sunday Times* 16 Jan. 2/6 In the meantime I pity the poor director, as we are beginning to call the producer. 1963 [see *directorial *a.* 1 b].

directorial, *a.* Add: **1. b.** *spec.* Of or pertaining to directors or the direction of films, etc.

1921 *Moving Picture Stories* 5 Aug. 26/3 Lloyd Ingraham is taking a month's vacation at the estate of David Kirkland, his directorial colleague. 1937 *Times* 25 Oct. 12/3 The directorial work was done by Mr. Clarence Brown. 1963 *Listener* 28 Feb. 392/3 BBC television has recently been..holding on to the directorial talent it has itself developed and at the same time luring in the best directors from independent television.

directory, *sb.* Add: **3. b.** = *telephone directory* (s.v. *telephone *sb.*); freq. *attrib.*, as **directory enquiries**, the section of the telephone-exchange which supplies callers with information about telephone numbers, etc.

1908 *Daily Chron.* 21 Sept. 4/6 Daily reports of all new and changed names for the Directory are forwarded to this department. *Ibid.*, The latest Directory information. 1922 Joyce *Ulysses* 697 Vulcanite automatic telephone receiver with adjacent directory. 1946 'S. Russell' *To Bed with Grand Music* v. 68 She hunted through telephone directories, questioned directory-enquiries. 1966 'S. Woods' *Enter Certain Murderers* ix. 157 Will you do something for me..? Get on to Directory Inquiries.

directrice. Delete †*Obs.* and add later examples.

1960 *Guardian* 25 July 4/3 The terrible Mlle Renée..who is the directrice of his salons. 1963 *N. & Q.* Mar. 112/1 Ibsen Club.—Founded November, 1909..with Miss Catherine Lewis as Directrice and Producer. 1969 *Daily Tel.* 13 Sept. 5/2 Madame Eve was also the training directrice of her attractive sales force.

dire·mpt, *v.* Delete †*Obs.* and add later example. Also **dire·mpted** *ppl. a.*

1885 W. James *Meaning of Truth* (1909) i. 5 Does not the word 'content' suggest that the feeling has already dirempted itself as an act from its content as an object? 1900 —— *Let.* 10 June in R. B. Perry *Tht. & Char. W. J.* (1935) I. 647 Once as streams of individual thinking, once as physical permanents, without the *immediately real* ever having been either of these dirempted things.

diremption. Add: **c.** *Bot.* An abnormal separation or displacement of leaves.

1869 M. T. Masters *Veget. Teratol.* 87 The term 'diremption' has sometimes been applied to cases where leaves are thus apparently dragged out of position. 1900 B. D. Jackson *Gloss. Bot. Terms*, *Diremption*, the occasional separation or displacement of leaves.

dirge, *v.* Add: **b.** To sing as a dirge.

1895 *Punch* 5 Oct. 162/2 They might all dirge in chorus the old duet of 'Again we come to thee, Savoy'.

c. *intr.* To utter a dirge.

1907 C. E. Mulford *Bar-20* xxi. 206 Shortly afterward the mournful cry of a whip-poor-will dirged out on the early morning air. 1921 *Chambers's Jrnl.* 211/2 The dead tops of the Gwynfrwyn trees were swaying and dirging dismally.

dirge(e, var. *durzee.

dirhem. (Later examples of the form *dirham.*)

1970 *New Yorker* 29 Aug. 48/2 There were ten dirhams in her pocket. 1971 *Ashmolean Mus. Rep. of Visitors 1970* 43 The purchase of a group of silver dirhams of the 'Arab-Sasanian' type.

dirigibility (di:ridʒibi·liti). [f. Dirigible *a.*: see -ility.] The quality of being dirigible; controllability.

1875 *Q. Rev.* CXXXIX. 137 One most important use of dirigibility would be in facilitating the descent, and in avoiding the many dangers to which the aeronaut, in his present helpless position, is so often exposed. 1902 *Westm. Gaz.* 25 Mar. 9/2 Proving the dirigibility of the aerostat. 1903 *Ibid.* 16 Nov. 7/3 Wireless dirigibility experiments. 1908 *B'ham Inst. Mag.* Jan. 254 The problem of..perfect dirigibility of dynamic flying machines.

dirigible, *a.* Add: **B.** *sb.* A dirigible balloon or airship. Also *transf.* (Cf. Fr. *dirigeable* sb.)

1885 *Sci. Amer.* 26 Dec. 405/1 He has devised two systems of dirigible balloons... The first of these, the independent dirigible, is of much the greater importance. 1907 *Daily Chron.* 26 July 5/6 He had been up in a captive balloon, but never before in a dirigible. 1910 *Daily Mail* 6 June 8/1 Night-flying dirigibles. 1927 *Glasgow Herald* 28 Apr. 15 A new dictionary of air terms has been compiled... So far as England is concerned, the word 'dirigible' will disappear and only 'airship' remain. 1953 [see *aqualung].

‖ **dirigisme** (diriʒiz'm). Also **dirigism.** [Fr. *dirigisme*, f. *diriger* to direct.] The policy of state direction and control in economic and social matters. Also *transf.* Hence **dirigiste, dirigistic** *adjs.*

1951 *Archivum Linguisticum* III. 220 Linguistic *dirigisme*, standards of correctness in a constantly evolving language. 1952 V. A. Demant *Religion & Decline of Capitalism* iv. 94 These are but a few of the reasons for the increasing *dirigisme* of economic life on the part of the state. 1957 *Times* 26 Feb. 4/3 Their [Sinn Fein] programme is a strange amalgam of bombast, Chauvinism, and dirigism. 1957 *Economist* 12 Oct. 16/2 The French hope that the new community will pursue a 'dirigiste', or at least a Keynesian policy regulating and guiding investment on a European scale, and ensuring that the Germans do not upset the whole scheme by deflating too much. 1959 *Times Rev. Industry* Dec. 75/3 A 'manifest crisis'..accompanied by dirigistic measures such as restrictions on production. 1967 *New Scientist* 9 Nov. 329/1 He warned his listeners against 'too much dirigism', reminding them of the USSR where crude political interference had forced men into politically neutral fields. 1971 Halsey & Trow *Brit. Academics* iv. 88 The liberal critique..fears..that 'dirigism' will supplant university autonomy.

dirk, *sb.* **2.** *dirk-knife* (U.S. examples).

1835 C. F. Hoffman *Winter in West* I. 212, I..gave the game the *coup de grace* with a dirk-knife which I had about me. 1843 'R. Carlton' *New Purchase* I. xx. 189 Tom [was] talking and laughing away like a fellow whittling poplar with a dirk knife!

dirndl (dɜ·ɹnd'l). [G. dial., dim. of *dirne* girl; cf. G. *dirndlkleid* peasant dress.] A style of woman's dress imitating Alpine peasant costume with bodice and full skirt; also **dirndl skirt,** a full skirt with a tight waistband.

1937 M. Hillis *Orchids on Budget* iii. 43 And an evening frock cut like a dirndl is amusing only if you don't have to wear it too often. 1938 *Times* 1 June 19/2 Nor must the ever popular dirndl be left out of the cotton catalogue. 1947 N. Cardus *Autobiogr.* 272 Painted dames sporting the Tyrolean dirndl. 1957 J. Braine *Room at Top* xxiv. 193 She was wearing a blue dirndl skirt and a white blouse. 1968 *Times Lit. Suppl.* 3 Oct. 1131/2 Austria was 'in' and English country house guests frolicked in dirndls.

dirt, *sb.* Add: **2. e.** A mean action, remark, etc. *U.S.*, *Austral.*, and *N.Z. slang.*

1893 S. Crane *Maggie* (1896) iii. 20, I got dis can fer dat ole woman, an' it 'ud be dirt teh swipe it. 1916 C. J. Dennis *Songs of Sentimental Bloke* 34 A bloke 'ud be a dawg to kid a skirt Like 'er. An' me well knowin' she was square. It 'ud be dirt! 1947 'A. P. Gaskell' *Big Game* 31 That was dirt, kicking mine [*sc.* my balloon] over the fence.

f. Scurrilous information or gossip; scandal. Also *attrib.*

1926 Hemingway *Sun also Rises* ii. 9 'Do you know any dirt?' I asked. 'No.' 'None of your exalted connections getting divorces?' 1934 E. Waugh *Handful of Dust* ii. 82 Good morning, darling, what's the dirt today? 1958 [see *angle sb.¹ 1 c]. 1959 P. Moyes *Dead Men don't Ski* iv. 41 'Come on, spill the dirt. What were they saying?'..'I always thought you didn't listen to gossip.' 1964 Wodehouse *Frozen Assets* v. 87 He doesn't think much of you... He thinks you fall short in the way of dishing the dirt.

6. c. *to eat dirt* (earlier example); also (*U.S.*), to make a humiliating confession or retractation.

1857 Kingsley *Two Y. Ago* II. v. 200 You have wonderfully changed your tone. Who was to eat any amount of dirt, if he could but save his influence thereby? 1885 *Mag. Amer. Hist.* Feb. 199/2 'To eat dirt' is to retract or 'eat humble pie'. 1891 Farmer *Slang* s.v. *Dirt*, *To eat dirt*,..to retract. 1903 Clapin *Dict. Amer.*, *Eat dirt*, to retract, to be penitent, the Yankee equiv. of 'to eat one's words'.

e. *to do dirt* (*to*), to harm or injure maliciously. (Cf. *dirty *a.* 2 d.) *slang* (orig. *U.S.*).

1893 S. Crane *Maggie* (1896) xiv. 111 Yer doin' me dirt, Nell! I never taut ye'd do me dirt. 1893 M. A. Owen *Voodoo Tales* 274 Ef I tek ter doin' dirt, den Ise willin' ter be jacky-me-lantuhn—an' sarve me right, too! 1913 E. C. Bentley *Trent's Last Case* vi. 137 They have been known to dynamite a man..who had done them dirt. 1926 J. Black *You can't Win* v. 57 You are the only human being I've met..that hasn't tried to do me some kind of dirt. 1929 D. H. Lawrence *Sex, Lit. & Censorship* (1955) 203 The so-called 'humour' is just a trick of doing dirt on sex. 1951 N. Balchin *Way through Wood* iv. 63 She doesn't do you dirt with characters like Bule—not serious dirt. 1951 J. C. Fennessy *Sonnet in Bottle* vii. i. 245 It was doing dirt to one of their own people. 1956 E. Pound tr. *Sophocles' Women of Trachis* 22, I'm telling you: do dirt to others but.. Don't weasel to me. 1959 'W. Haggard' *Venetian Blind* vi. 84 Dotties could do you dirt; they could remark..in public, that..you were living with the curate.

7. a. *dirt roof* (also *-roofed* adj.) *U.S.*

1881 *Rep. Indian Affairs* (U.S.) 121 Carpenter shop,..log, dirt roof. 1910 *Outlook* 2 July 483 A log cabin of two rooms, with a dirt roof. 1873 J. H. Beadle *Undevel. West* xxxiv. 734 To his joy he came upon a dirt-roofed log-house.

c. (sense 3 c) *dirt-washer, -washing.*

1869 S. Bowles *Our New West* ix. 179 The dirt-washers swept eagerly over the rich surface deposits. *Ibid.*, The old and simple dirt washing for gold was resumed.

d. dirt farmer *U.S.*, a practical farmer; one who farms his own land; so *dirt-farming*; **dirt-line,** a layer of dirt and debris accumulated on the surface of a glacier and imprisoned by the seasonal layer of snow; **dirt money** = *dirty money* (see *dirty *a.* 6 b); **dirt-poor** *a.* (orig. *U.S.*), extremely poor; **dirt road** (chiefly *N. Amer.*), an unmade road, having merely the natural surface; **dirt track,** an unpaved track or course; spec. (*a*) made of cinders and brick-dust for motor-cycle racing; (*b*) composed of soil, esp. as distinguished from turf, for flat-racing.

1924 H. Croy *R.F.D. No. 3* 148, A real dirt farmer, not..one of them city dudes. 1932 E. Wilson *Devil take Hindmost* xvi. 170 He himself has been a dirt farmer, not a white-collar farmer! 1955 *Sci. News Let.* 23 July 50/2 It appears that all visiting experts are high officials in Russian agriculture, not one being a 'dirt farmer'. 1920 *Boston Even. Transcript* 2 Oct. iv. 1/1 Doing its multiple duty of making dirt-farming a fine art. 1894 J. Geikie *Gt. Ice Age* (ed. 3) 30 The beds of snow..being usually marked off by a 'dirt-line' or crust formed of a mixture of dust, small grit, and occasional remains of insects. 1949 'M. Innes' *Journeying Boy* x. 119 Dirt money..the extra pay dockers and people get for doing something thoroughly nasty. 1937 *Time* 26 Apr. 41/1 Nearly blind and dirt-poor, Inventor Dave Mallory (Karloff) devises a burglar alarm worked by electric eyes. 1971 D. O'Connor *Eye of Eagle* xix. 128 They were dirt-poor; they could not bribe people. 1852 Mrs. Stowe *Uncle Tom's C.* vii. 62 Der's two roads to de river—de dirt road and de pike. 1931 G. T. Clark *Leland Stanford* iv. 85 Traveling by horse-drawn stage..over rough and dusty dirt roads. 1959 A. Fullerton *Yellow Ford* v. 40, I heard the noise of the car and a moment later saw it come bumping up the dirt road. 1959 N. Lofts *Heaven in your Hand* 109 The dirt road became a cobbled street. 1969 E. W. Morse *Fur Trade Canoe Routes* ii. v. 58 This little channel, still evident, leads from Bell Lake to the present dirt road. 1902 *Encycl. Brit.* XXIX. 335/2 Practically all flat racing in the United States is held on 'dirt-tracks', *i.e.*, courses with soil specially prepared for racing, instead of turf courses. 1924 Masefield *Sard Harker* i. 63 The road was not macadam but dirt-track, with soft going, after the first mile. 1928 *Daily Tel.* 20 Mar. 14/2 Motor-cycle racing on 'dirt tracks', a form of sport very popular in Australia. 1928 *Times* 2 July 6/7 Dirt Track Racing. 1949 Koestler *Promise & Fulfilment* xv. 168 A tiny two-seater training plane..which performed perilous hops from an improvised dirt-track runway in Jerusalem to Tel Aviv and back. 1961 W. Vaughan-Thomas *Anzio* viii. 162 One column struck south-east down the dirt track that led towards the Spaccasassi Creek.

dirt-eater. Add: *spec.* one of a class of 'poor whites' in some parts of the southern United States; = *clay-eater* (Clay *sb.* 9).

1840 C. F. Hoffman *Greyslaer* iii. xii. 223 Even Bettys, little fastidious as he was, recoiled from the fare which these 'Dirt Eaters', as the Indians called them, placed before him. 1866 *Ret. Agric. Soc. Maine* 46 It rests with you..whether you will take rank with the poor whites, the dirt-eaters of the South, or with the best classes of the north. 1940 R. O. Cummings *American & his Food* 87 Distinguished from other southerners by hookworm disease were dirt-eaters scattered in sand barrens and pine woods from South Carolina to Mississippi.

dirtiness. Add: **1. b.** The quality or state of an atomic or nuclear weapon having considerable radioactive fall-out. (Cf. *dirty *a.* 1 g.) *colloq.*

1956 *Life* 29 Oct. 44/2 To have any meaning, the ban would not be limited to the *size* of bombs..but to the amount of radioactivity they would be allowed to generate—their 'dirtiness'. 1961 *Guardian* 25 Oct. 2/5 Asked if the explosion..was of a 'dirty' bomb, he replied..'The question of "dirtiness" does not bear a direct relation to yield.' 1961 *Daily Tel.* 31 Oct. 22/5 Fall-out from last week's test is expected to reach Britain tomorrow. It may give an indication of the 'dirtiness' or otherwise of Russia's latest hydrogen weapons.

dirty, *a.* Add: **1. e.** Phr. *the dirty end* (*of the stick*), the difficult or unpleasant part (in a situation). (See also Stick *sb.*¹ 14 e.)

1924 'Sapper' *Third Round* vi. 167 I've been had for a mug. Somehow or other they've handed us the dirty end. 1930 J. Dos Passos *42nd Parallel* v. 71, I guess I always get the dirty end of the stick, all right. 1936 Wodehouse *Laughing Gas* v. 62, I mean what's downtrodden and oppressed and gets the dirty end of the stick all the time. That's me. 1957 *Listener* 31 Oct. 681/2 The Indians [in Malaya]..may be given the dirty end of the stick.

f. *transf.* *colloq.* Not streamlined; opp. *clean *a.* 13 c; spec. (*a*) of untidy or imperfect oarsmanship; (*b*) of the lines of an aircraft, used esp. of one with its landing gear unretracted.

1925 'Ian Hay' *Paid with Thanks* xvi. 213 Stroke,..you have still got a dirty finish. 1932 *Flight* 23 Sept. 891/1 One would imagine that what with the external wing bracing struts joining their lower ends to the hull on the chine, and with a planing bottom showing no flare or reversed curvature towards the chine, the boat would be extremely 'dirty' at certain running speeds. 1959 *Times* 13 Mar. 18/1 They went up quickly, in spite of some dirty bladework. 1962 *Flight Internat.* 1 Mar. 330/1 Stalling speeds clean and fully 'dirty' are almost 30 kt apart.

g. Of a nuclear weapon: having considerable radioactive fall-out. *colloq.*

1956 *Life* 29 Oct. 44/1 The H-bomb..is a relatively 'clean' bomb unless it is made 'dirty' (more radioactive) by using the tremendous heat of fusion to set off another fission process. 1958 *Observer* 13 Apr. 13/3 Those tests which are really hazardous to health—the tests of large 'dirty' H-bombs—cannot be held undetected. 1958 *New Statesman* 9 Aug. 162/3 He had to..admit that the US is now deliberately withdrawing A-bombs from its stockpile in order to make them 'dirtier'. They are then known as 'salted'.

2. (Further examples.) Spec. *dirty book*, a pornographic book; so *dirty bookshop*; *dirty joke, story*, a 'smutty' joke or story; *dirty weekend*, a sexually illicit weekend.

1912 R. Brooke *Let.* 7 Aug. (1968) 392, I shall..repeat poetry to you; you will repeat dirty stories to me. 1913 —— *Let.* 22 May (1968) 461 We must do some show together..where poetry & music &..satire and suffering & dirty jokes & triumphal processions shall be..mixed together. 1916 W. S. Maugham *Writer's Notebk.* (1949) 107 When at last he became more communicative it was..to show one a collection of dirty postcards. 1940 G. Marx *Let.* 16 Dec. (1967) 189 A master comic, not a dirty line or joke in the entire two and a half hours. 1944 'G. Orwell' *Crit. Essays* (1946) 125 The dirty post cards that used to be sold in Mediterranean seaport towns. 1959 J. Braine *Vodi* xxiv. 250 Each had some kind of talent, even if it were only for billiards or dirty jokes. 1960 *Guardian* 2 Nov. 11/2 For Lawrence to be confined always to dirty-book shops would be perhaps the greatest irony in literary history. 1963 P. Moyes *Murder à la Mode* xii. 208 You and Veronica were going off for what, in my day, used to be called a dirty week-end. 1965 K. Giles *Some Beasts No More* iv. 74 He was probably a messenger boy from the dirty book trade. 1969 D. Barron *Man who was There* i. 21 An angry German at the Karachi customs who was being accused of smuggling dirty books into the country. 1969 R. Quest *Cerberus Murders* xvii. 93, I get reasonably well paid—enough to enable me to..have a dirty weekend in Scarborough now and again. 1970 B. Spock *Decent & Indecent* 51 Fear of sexual impotence is its most obvious aspect; this is an important ingredient of a large proportion of all jokes and dirty stories.

b. Colloq. phr. *dirty work at the crossroads.*

1914 Wodehouse *Man Upstairs* 269 A conviction began to steal over him that in some way he was being played with, that some game was afoot which he did not understand, that—in a word—there was dirty work at the cross-roads. 1915 'Ian Hay' *First Hundred Thou.* xvi. 221 That..usually means dirty work at the cross-roads at no very distant period. 1927 R. Knox *Three Taps* v. 47 No question of accident..or of dirty work at the cross-roads? These rich men have enemies, don't they? 1960 A. Christie *Adv. of Christmas Pudding* 184 Good for you, old boy. Some dirty work at the crossroads—eh?

d. Also *absol.* in phrase *to do the dirty*: to play a dirty trick.

1914 *Daily Express* 13 Nov. 5/4 The Germans have been 'doing the dirty' on us by donning khaki and kilts to approach our trenches. 1915 D. O. Barnett *Lett.* 157, I hope our friends the 133rd will..do the dirty on their Prussian friends. 1929 J. B. Priestley *Good Comp.* iii. v. 607 Anyhow they did the dirty on yer. 1930 R. H. Mottram *Europa's Beast* xii. 282 If you've been doing the dirty on my friends. 1942 G. Kersh *Nine Lives Bill Nelson* iii. 15 It took me to do the dirty on him.

5. c. Of jazz music: having a slurred or rasping tone.

1927 *Etude* May 339 (*title*) More hot and dirty breaks. 1952 B. Ulanov *Hist. Jazz* (1958) x. 111 The brass smears and 'dirty' reed inflections then much favored by jazz musicians. 1955 L. Feather *Encycl. Jazz* vii. 277 His plaintive tone and style, sometimes called 'dirty' and often employing 'growl' effects. 1963 *Listener* 7 Feb. 264/2 The lady..had the 'dirty' tone and tough attack, while the man played straight and smooth.

6. a. Also *dirty-mindedness*; *dirty-looking* adj.

1817 W. Tucker *Family Dyer & Scourer* i. 7 A kind of dirty looking green. 1920 D. H. Lawrence *Lost Girl* xiv. 325 There was a strange mountain town, dirty-looking. 1915 F. M. Hueffer *Good Soldier* III. iv. 183 The sorts of dirty-mindedness that counsel in that sort of case can impute. 1935 'G. Orwell' *Clergyman's Daughter* v. 289 Just generalised suspicion. A sort of instinctive rustic dirty-mindedness.

b. dirty dog *slang*, a despicable or untrustworthy person; also with implication of lasciviousness; **dirty look** *colloq.*, a scowl, sour expression; **dirty money** (see quot. 1897); **dirty old man** *slang phr.*, used with the implication of lasciviousness; **dirty word,** (*a*) a vulgar or 'smutty' word; (*b*) a word made disreputable by what are regarded as its discreditable associations.

1928 S. Vines *Humours Unreconciled* xv. 204 Who's been calling me a dirty dog, I should like to know. 1959 I. & P.

OPIE *Lore & Lang. Schoolch.* x. 189 Less specifically he [sc. the tell-tale] is a..'dirty dog',..'spoil sport'. **1964** J. PORTER *Dover One* ix. 107 'Bit of a dirty dog, our Gordon Pilley,' said Sergeant MacGregor with a smirk. **1928** J. P. McEVOY *Showgirl* xi. 164, I sneaked a peep across the room and caught Jimmy giving me a dirty look. **1933** *Punch* 19 Apr. 444/1 It is therefore customary..to incline the head in the direction of the culprit, composing the features to an expression of great hatefulness but of even greater contempt. The correct adjustment of that expression constitutes the Dirty Look. **1934** WODEHOUSE *Right Ho, Jeeves* xi. 137 Deprived of Anatole's services, all he was likely to give the wife of his b. was a dirty look. **1897** S. & B. WEBB *Industr. Democr.* I. 313 When any class of work involves special unpleasantness or injury to clothing, 'black money' or 'dirty money' is sometimes stipulated for. **1960** *Sunday Express* 14 Aug. 1/1, 1,100 dockers..are claiming 'dirty money' for handling a cargo of red oxide. **1932** R. LEHMANN *Invit. Waltz* III. xiv. 240 Mum thinks he's harmless..In fact she was quite umbrageous with me when I called him a dirty old man. **1942** T. RATTIGAN *Flare Path* II. ii. 137 You don't think I waited up for you, do you? Keeping the women company, that's all. *Teddy.* Then you're a dirty old man. **1957** 'R. GORDON' *Doctor in Love* i. 10 I'm going to stay a bachelor. Changing imperceptibly from gay young to dirty old. **1971** D. CLARK *Sick to Death* i. 9 A man of my age on the look out for a lovely young lass puts me into the dirty-old-man class. **1842** LOVER *Handy Andy* xiv. 126 'Don't say popery,' cried the cook; 'it's a dirty word! Say Roman Catholic when you speak of the faith.' **1925** *New Yorker* 19 Sept. 6, I shall insert into the second act of each play one of the three remaining Dirty Words that haven't yet been pronounced on the stage. **1937** E. ST. VINCENT MILLAY *Conv. Midnight* i. 36 Charity is a dirty word, it's gone to bed Too often with philanthropists, it's *effutata*. **1957** D. KARP *Leave me Alone* xx. 298 To young people compromise is a dirty word. **1959** *Listener* 25 June 1094/1 Since then 'liberalism' has become a dirty word in Indonesia.

II. Used adverbially as an intensive: very, exceedingly. *slang.*

1920 GALSWORTHY *Foundations* III, 'E wants to syve 'is dirty great 'ouse. **1943** BAKER *Dict. Austral. Slang* (ed. 3) 26 *Dirty big*, used adjectively as an equivalent of 'bloody'. **1944** J. H. FULLARTON *Troop Target* ix. 67, I was lying flat on my back when a great dirty black Junkers dropped a big one. **1945** J. HENDERSON *Gunner Inglorious* xvi. 134 I'm going to grow a dirty big fig-tree outside my home. **1958** F. NORMAN *Bang to Rights* III. 124 There was one chap who had a dodgey foot, and he had to wear a dirty great boot and because of this they called him the Boot. **1970** D. M. DAVIN *Not Here, Not Now* IV. v. 311 The buggers are in now. They've got a dirty big majority already. **1971** D. CLARK *Sick to Death* iii. 65 Time for a dirty great pint.

dirty, *v.* Add: **1.** *fig.* Also const. *up.*

1953 W. P. McGIVERN *Big Heat* iii. 32 It might occur to her to blackmail you..with the threat of dirtying up your husband's name. **1959** 'P. QUENTIN' *Shadow of Guilt* v. 46 She, too, would be dirtied up by scandal. **1964** WODEHOUSE *Frozen Assets* i. 25 He's writing a novel... He keeps clutching his brow and muttering 'This damned thing needs dirtying up'..If it isn't the sort of stuff small boys scribble on fences, nobody will look at it.

b. To contaminate with radioactive matter (cf. *DIRTY *a.* 1 g).

1955 *Sci. News Let.* 26 Mar. 197/2 Great bangs that dirty the planet's atmosphere with radioactive debris, signaling atomic advances, may be used sparingly, hiding progress to competitive nations.

dirzie, var. *DURZEE.

dis (dis), *v. Printer's slang.* Also **diss.** Colloq. abbrev. of DISTRIBUTE *v.* 5. Hence **dis** *sb.*, type ready for distribution.

1889 BARRÈRE & LELAND *Dict. Slang*, *Diss* (printers), abbreviation for distribution, *i.e.*, printed off type—to be returned to its respective cases, and re-composed. **1899** J. SOUTHWARD *Mod. Printing* II. 168 It is not necessary for an operator to read the matter to be 'dis'd'. *Ibid.* 169 The type being ordinary 'dis', and in no way assorted. **1903** 'No. 7' *25 Yrs. in 17 Prisons* x. 96 There was 'pie' to the left of us, 'pie' to the right of us..and what had only taken a week to 'set up' took nearly a month to 'dis'. *Ibid.* xii. 115 For the first week or two I was put upon 'dissing'. **1970** *Brit. Printer* Aug. 69/1 Type is not dissed and some fairly elaborate machines are needed to cast refills for the cases.

dis (dis), *a.* Colloq. abbrev. of DISCONNECTED *ppl. a.* Hence, broken, not working (see also quot. 1925). Also as vb.

1925 FRASER & GIBBONS *Soldier & Sailor Words* 106 *Gone dis*, a colloquialism for mentally weak. The term comes from a telegraph or telephone signaller's phrase meaning disconnected. **1931** T. R. G. LYELL *Slang* 219 The poor old chap's brain's going dis. **1932** 'R. STRANGER' *Outl. Wireless* iii. 151 If the circuit is broken at any point, electron current ceases to flow, and we say that the circuit is 'dis'. **1932** *Telegr. & Teleph. Jrnl.* July 218/2 Her mistress had just gone out intending to report her line 'dis.' via a call office. **1937** *Wireless World* 26 Feb. 215/3 There's no warning whistle to tell him [sc. the wireless listener] the speaker is 'dissed'. **1969** P. DICKINSON *Pride of Heroes* 124 You've heard about me dissing my wirelesses? **1971** —— *Sleep & his Brother* v. 122 That dismal little switchboard in her room went dis... I found a hairpin which she'd dropped across the terminals.

disa (dai·sǎ). *Bot.* [mod.L. (P. J. Bergius *Descr. Plant. ex Capite Bonæ Spei* (1767) 348), of obscure origin.] A tropical African terrestrial orchid of the genus so named, with dark green leaves.

1844 *Curtis's Bot. Mag.* LXX. 4091 (*heading*) Horned-flowered Disa. **1890** WATSON & BEAN *Orchids* XXX. 235 A position in a house which suits cool Odontoglossums will be found agreeable to Disas. **1930** T. W. BRISCOE *Orchids for Amateurs* vi. 109 The Disas are terrestrial, with more or less tuberous roots. **1951** L. G. GREEN *Grow Lovely, growing Old* vi. 63 Disas and proteas lasted longer too, and cost less than garden flowers. **1966** E. A. C. L. E. SCHELPE *Introd. S. Afr. Orchids* 9 Only a few of the Disas would be recognised as orchids by the non-botanist.

disablement. Add: **3.** *attrib.*

1898 *Westm. Gaz.* 23 June 9/1 A crushing liability for death and disablement claims. **1898** *Daily News* 25 June 6/6 The first 26 weeks' disablement pay. **1920** *Act* 10 *Geo. V* c. 10 § 2 The rate of disablement benefit. **1961** *Lancet* 2 Sept. 539/2 A disablement resettlement officer (D.R.O.) from the local labour exchange attends the weekly assessment conference.

disaccharide (dəisæ·kǎrəid). *Chem.* Also †**-id.** [f. DI-[2] 2 + SACCHARIDE.] Any sugar that consists of two monosaccharide residues linked together.

1892 [see *BIOSE]. **1905** E. F. ARMSTRONG in *Proc. R. Soc.* B. LXXVI. 592 The enzymes which are capable of inducing the hydrolysis of disaccharides or bioses. **1938** *Thorpe's Dict. Appl. Chem.* (ed. 4) II. 298/1 The formulæ of the disaccharides have undergone revision on the basis of the accepted pyranose formula for glucose. **1954** *Ibid.* XI. 176/1 Disaccharides..can be split by acid hydrolysis or enzymes into their constituent hexose monosaccharides. **1970** R. W. McGILVERY *Biochem.* xxvi. 634 The infant living on the milk produced by his mother's mammary glands generates approximately 60 per cent of his ATP by oxidizing the fat in the milk and 40 per cent by oxidizing the disaccharide, lactose.

disadvantaged, *ppl. a.* Add: Later examples, esp. in *Sociol.*; also as *sb.*

1934 H. G. WELLS *Exper. Autobiogr.* II. viii. 572 This mannered ungraciousness towards disadvantaged people. **1949** R. K. MERTON *Social Theory* (1951) i. 76 Social mobility for the otherwise disadvantaged. **1962** *Amer. Speech* XXXVII. 18 A large mass of bottom-income, unskilled, manual, and service workers, including what our sociologist colleagues call the 'disadvantaged'. **1966** *Sat. Rev.* 16 Apr. 100/1 Schools in economically disadvantaged neighbourhoods. *Ibid.* 104/1 It's the familiar story of the disadvantaged child. **1967** *Boston Globe* 20 May 2/2 He said urban school teachers generally hold 'negative and fearful attitudes toward the disadvantaged and disordered youngsters in their schools'. **1970** *New Yorker* 26 Sept. 108/2 A revolt by the disadvantaged aimed at those holding power.

disaffiliation (di:sǎfilię̄ⁱ·ʃən). [f. DISAFFILIATE *v.*: see -TION.] The action of disaffiliating.

1926 *Amer. Review of Reviews* Nov. 473 The British Labor party confirms disaffiliation with the Communists. **1927** *Glasgow Herald* 9 Mar. 11 The ground upon which disaffiliation is urged. **1927** *Observer* 10 July 14/6 The announcement of the disaffiliation of the Union of Post Office Workers. **1955** *Times* 9 July 11/7 The defendant union were unwilling to expel him and were only concerned to do so by virtue of the award and the sanction (of disaffiliation) hanging over their heads. **1969** *Listener* 16 Jan. 87/1 The various modes of disaffiliation that have been adopted by the post-Hiroshima young.

disagreeable, *a.* **3. b.** (Earlier U.S. examples.)

1827 J. F. COOPER *Prairie* xii, We are disagreeable about his camping on the prairie, instead of coming in to his own bed. **1836** *Knickerbocker* VIII. 151 Had I not become accustomed to such dangers, I should have felt very disagreeable.

disambiguate (disæmbi·giu,ę̄ⁱt), *v.* [f. DIS- 8 + AMBIGU(OUS *a.* + -ATE[3].] *trans.* To remove ambiguity from.

1963 *Language* XXXIX. 175 A speaker can disambiguate parts of a sentence in terms of other parts. **1967** *Ibid.* XLIII. 619 When necessary, the Greek spelling is disambiguated by an appended phonetic transcription.

disambiguation (di:sæmbigiu,ę̄ⁱ·ʃən). [f. as prec. + -ATION.] Removal of ambiguity; also, the result of such removal.

1827 G. BENTHAM *Outl. Logic* 88 Disambiguation—where it is to fix the sense of an *ambiguous* term. This operation has been termed *distinction* by some Logicians. *a* **1832** J. BENTHAM *Logic* in *Wks.* (1843) VIII. 249/1 Disambiguation is distinction applied to words. **1967** R. H. ROBINS *Short Hist. Linguistics* vi. 138 Limitation of a word's field of reference or its disambiguation as the result of specific collocations. **1968** *Proc. Arist. Soc.* Suppl. XLII. 135 We already know the correct syntactical or semantical disambiguation of his utterance. **1970** *Language* XLVI. 100 Pairs of sentences can easily be found in which disambiguation rests upon comma intonation alone.

disamenity (disǎme·niti, -ī·niti). [f. DIS- 9.] A disadvantage or drawback (of a locality, etc.). Freq. in *pl.*

1924 *Glasgow Herald* 19 Apr. 4 The disamenities of warm countries in the summer time. *Ibid.* 29 Nov. 4 The disamenities attendant on the growth of civilisation. **1938** *Rep. Brit. Assoc. Advancem. Sci.* 1937 383 The increased social expenditure and disamenities of social life (such as road congestion, housing shortage, loss of open spaces) in expanding areas. **1965** *Listener* 13 May 711/1 The political disamenity, dislocation, and social disturbance suffered both by the inhabitants and the immigrants. **1970** *New Scientist* 13 Aug. 342/2 Two other rapidly growing sources of disamenity..are air travel and tourism.

disanalogy. Delete †*Obs.* and add later examples.

1948 M. BLACK *Lang. & Philos.* (1949) i. 10 Risk is always involved in neglecting the disanalogy between the things compared. **1959** P. F. STRAWSON *Individuals* v. 171 These analogies and disanalogies. **1966** J. J. KATZ *Philos. Lang.* iii. 35 There is a certain disanalogy here.

disappear, *v.* Add: **1.** Also with advb. expressions introduced by prepositions.

1837 E. HOWARD *Old Commodore* I. iv. 111 Richard Stubbs..disappeared down the Jacob's ladder. **1842** LEVER *J. Hinton* lii, A mounted party..entering one of the gates of the city, disappeared from our sight. **1881** MRS. RIDDELL *Senior Partner* xxii, The boy disappeared into the retirement of the back room. **1888** MRS. H. WARD *R. Elsmere* v, Almost every year he disappeared to France. *Ibid.* x, Rose caught a gray dress disappearing up the little stairs. **1922** JOYCE *Ulysses* 427 He disappears into Olhousen's the pork butcher's.

3. *trans.* To cause to disappear.

1897 *Chem. News* 19 Mar. 143 We progressively disappear the faces of the dodecahedron. **1949** *Amer. Speech* XXIV. 41 The magician may speak of disappearing or vanishing a card.

disappoint, *v.* Add: **2.** Also *absol.*, to cause disappointment.

1842 LEVER *J. Hinton* ix, The Duke has disappointed so often, that he is sure to go now. **1881** MRS. RIDDELL *Senior Partner* xxxii, I shall look out for you at Waterloo at a quarter to five, and trust you will not disappoint. **1927** *Daily Tel.* 3 May 16 Innuendo [sc. a racehorse] has disappointed in the past when fancied. **1966** *Listener* 13 Oct. 549/3 Ormandy's CBS album of the Berlioz *Requiem*.., of which I had high hopes, disappoints.

disappropriation. (Later example.)

1964 *Listener* 19 Nov. 797/2 This demands an act of *disappropriation*, a disowning of the self,..and the substitution of the passive, perfectly detached will centred on God.

disarmer. Add: **b.** An advocate of disarmament. So *nuclear disarmer* (see *NUCLEAR *a.*).

1906 *Westm. Gaz.* 30 July 7/1 England is so richly provided with naval equipment that she can easily afford to masquerade as a 'disarmer'. **1960** *Guardian* 21 Sept. 8/5 One can be a pacifist and a unilateral disarmer, prepared to accept the consequences for oneself and one's country.

disarmingly (disǎ·ɹmiŋli), *adv.* [f. DISARMING *ppl. a.* + -LY[2].] In a disarming manner; so as to disarm opponents (usu. *fig.*).

1901 'L. MALET' *R. Calmady* VI. iv. 525 An expression of disarmingly innocent penitence. **1905** *Daily Chron.* 11 Mar. 7/2 Hascombe smiled disarmingly. **1920** *Blackw. Mag.* Dec. 714/2 A name..which she..disarmingly admitted there was small prospect of her ever otherwise being able to change. **1928** *Daily Tel.* 21 Aug. 8/5 He is a business man himself, speaking in terms which are disarmingly straightforward.

disassemble (disǎse·mb'l), *v.* Restrict † *Obs. rare*[−0] to sense in Dict. and add: **b.** To take to pieces, to take apart. (The opposite of *ASSEMBLE *v.*[1] 2 b.) So **disasse·mbly,** the act or process of disassembling.

1922 *Short Stories* Feb. 41/2 This generating plant was partly disassembled. **1930** *Sci. Amer.* Nov. 389 The entire 50-foot line can be assembled or disassembled in six minutes. **1954** *New Biol.* XVI. 36 The assembly and disassembly of protein molecules in tissue fluids has been shown..to be continually taking place. **1958** *Times Lit. Suppl.* 11 Apr. 192/1 The reconstruction of disassembled paintings. **1958** *New Scientist* 1 May 31/1 (*diagram*) Disassembly cell. **1967** KARCH & BUBER *Offset Processes* x. 462 Remove the screws from the bars and disassemble.

disassociation. Add: **b.** = *DISSOCIATION 3.

1934 H. C. WARREN *Dict. Psychol.* 82/1 *Dissociation*,..redundant syn. *disassociation*. **1936** W. TEMPLE *Church & Teaching* ii. 30 Psychology..is suffering badly from what it has itself taught us to call the disassociation of personality.

disaster, *sb.* Add: **3.** *attrib.*, as **disaster area,** an area in which a major disaster has occurred; also *fig.*

1960 *Times* 14 Sept. 8/1 President Eisenhower designated sections of the state [of Florida] a major disaster area. **1969** M. DRABBLE *Waterfall* 244, I was merely a disaster area, a landscape given to such upheavals. **1970** *Guardian* 25 Nov. 2/1 Pakistani officials..described the difficulties they had experienced in bringing aid to the disaster area.

disazo- (disæ·zo). *Chem.* [f. Gr. δίς twice + AZO-.] A combining form used in organic chemistry to denote the presence in the molecule of a compound of two azo groups. Also used *attrib.* as *disazo.*

1886 E. F. SMITH tr. *von Richter's Chem. Carbon Compounds* 465 Such compounds can also be obtained by a second introduction of two molecules of a diazo-compound into phenols.., and are called disazo-derivatives. **1891** ROSCOE & SCHORLEMMER *Treat. Chem.* (ed. 2) III. III. 353 Disazobenzene..is obtained from the corresponding amido-disazobenzene..by diazotizing and boiling with alcohol. **1913** BLOXAM & LEWIS *Chem* (ed. 10) 719 Disazo-dyestuffs..contain the ·N:N· group twice, and are of three kinds. **1930** E. FYLEMAN tr. *Curtis's Artificial Org. Pigments* i. 6 In the patent literature a number of disazo-dyestuffs are also mentioned, but none of these have attained any importance. **1966** MILLAR & SPRINGALL *Sidgwick's Org. Chem. Nitrogen* (ed. 3) xvii. 588 The azo dyes form the largest class of organic dyes...Monoazo, disazo, trisazo, and polyazo dyes are distinguished.

disbursal (disbə̄·ɹsǎl). [f. DISBURSE *v.* + -AL.] The act of disbursing, disbursement.

1589 in *Rep. Hist. MSS. Comm., Var. Coll.* (1904) III. 49 Divers disbursalles..as for some bookes. **1895** *Manchester Guardian* 30 Sept. 6/8 Accommodation..upstairs for the disbursal and reception of goods. **1898** in *Westm. Gaz.* 9 Mar. 2/3 All moneys received shall be duly acknowledged, and their disbursal shall be controlled by the Mansion House committee.

disburse, *v.* Add: **1. d.** (Later example.) Also *fig.*

1789 *Loiterer* 21 Mar. 6 Any intention to defraud, or any inability to disburse. **1920** E. POUND *Umbra* 111 Disburse Can she, and wake Such firm delights.

disc, *sb.* (Now the usual form in Great Britain.) Transfer the entry from DISK and add: **2. d.** A phonograph or gramophone record. Also *attrib.* and *Comb.*

1888 *Leisure Hour* 209/1 A disc about eleven inches in diameter can, it is said, contain four minutes' talk. **1907** *Sound Wave & Talking Machine Record* Dec. 60/2 The world today always associates Edison's phonograph with a cylinder apparatus, but the first phonograph we look at in this patent is a disc (he called it a disk) machine. **1919** A. SEYMOUR *Good-bye-ee!*, A gramaphone [*sic*] record with the picture of a foxterrier on the disc. **1929** *Sunday Dispatch* 20 Jan. 16 A fine disc by his orchestra. **1941** *B.B.C. Gloss. Broadc. Terms* 9 *Disc recording.* (1) Process of registering sound by electro-mechanical means in the form of lateral corrugations in a spiral groove on a disc coated with, or composed of, plastic material. (2) (Also *Disc.*) Disc on which sound has been so recorded. (3) (By extension.) Programme material so recorded. **1951** *Ann. Reg.* 1950 386 A whole classical symphony could be recorded on the two sides of a 12-inch disc without any break in the course of the movements. **1958** *Listener* 18 Dec. 1049/2 The puerile pop singer who becomes the demi-god of the frenzied disc-addicts. **1962** *Melody Maker* 7 July 4 His discs have sold in millions all over the globe...He doesn't make hit discs today. **1962** A. NISBETT *Technique Sound Studio* vii. 134 *Disc editing.* Discs are edited by dubbing (i.e. copying), and extensive professional equipment is required. **1968** *Listener* 27 June 845/3 The only complete opera-in-English available on disc in this country.

e. Short for *identity disc* (see *IDENTITY).

a **1918** W. OWEN *Poems* (1931) 104 Let my inscription be this soldier's disc.

f. In full *parking disc.* A device used in **disc parking.*

1960 *Economist* 9 Apr. 193/3 A parking disc is a device a driver himself displays on his vehicle to show the time he arrived or the time he ought to leave a parking area, or both times. **1970** *Oxford Mail* 13 July 6 Whilst there is adequate warning, on entry into the city, that disc parking is operative it does seem ridiculous that an ample supply of discs is not readily available from garages, stores and traffic wardens.

7. a. *slipped disc* (see *SLIPPED *ppl. a.*).

8. b. *disc brake* (hence *-braked* adj.), *-brooch*, *-crank*. **c.** *disc-cutter*, *-fan*; also in the names of agricultural machines (orig. *U.S.*), as *disc-cultivator*, *drill*, *-harrow* (examples; hence as *v. trans.*), *plough*; **e.** also instrumental and similative, as *disc-adjusting*, *-capped*, *-like* adjs.; **f. disc area** *Aeronaut.* (see quots.); **disc-clutch** (examples); **disc electrode** (see quot. *a* 1884); **disc-jockey** orig. *U.S. slang*, a person who selects and introduces gramophone records for transmission on radio or television; abbrev. D.J., *DEE-JAY; also as *v. intr.*; hence **disc-jockeying** *vbl. sb.*; **disc loading** *Aeronaut.* (see quots.); **disc parking,** a parking scheme for vehicles whereby each must display a disc on which is shown the time when the vehicle was parked.

1898 *Cycling* 54 Disc-adjusting bearings. **1919** H. SHAW *Text-Bk. Aeronaut.* xi. 146 The propeller during rotation marks out a circular area which is known as the 'disc area'. **1962** *Gloss. Aeronaut. Terms* (*B.S.I.*) v. 16 *Disk area*, the area of the circle described by the tips of the blades. **1904** M. MACLEAN *Mod. Electr. Pract.* IV. 241 Electric disc brake.—This is a brake that is not very largely in use [on trams] in this country. **1928** *Flight* 20 Dec. 1325/2 The multiple disc brake offers unlimited possibilities for having large braking surfaces. **1950** *Autocar* 15 Dec. 1257 At the moment the disc brake, like most innovations, is more expensive than the system which it seems to supplant. **1964** J. GRIFFITHS *Your Car* x. 105 *Disc Brakes.* These are a comparatively recent development on popular cars...The modern type was developed on racing cars, but it originated with the brakes used for the landing wheels on large aircraft. **1959** G. FREEMAN *Jack would be Gentleman* x. 221 Prosser with his disc-braked car is able to leave his braking that little bit longer. **1937** *Burlington Mag.* Feb. 99/1 The first Kentish jewelled disc-brooches. **1958** D. WHITELOCK *Changing Currents in Anglo-Saxon Studies* 21 Mr Bruce-Mitford's article on the late Saxon disc-brooches. **1928** *Daily Express* 17 Aug. 2/7 The system of milk distribution by means of disc-capped bottles. **1906** *Daily Chron.* 14 Nov. 9/3 The enormously increased popularity of the multiple disc or 'plate' clutches. **1909** *Westm. Gaz.* 11 Feb. 4/2 A multiple disc-clutch. **1888** *Lockwood's Dict. Mech. Engin.*, *Disc crank*, or *crank disc*, or *crank plate*, a crank of circular outline in which the metal is so disposed that the varying motion of the connecting-rod is suitably balanced. **1894** *Irrigation Age* Jan. 34/1, I have found one of the best tools that we have yet used to be the disc cultivator. **1835** URE *Philos. Manuf.* 198 A toothed pinion..gives each disc-cutter a quick rotatory motion on its centre. **1897** *Sears, Roebuck Catal.* 159/3 The disc drill when polished will cut through sod and trash. **1907** L. H. BAILEY *Cycl. Amer. Agric.* I. 207 The disc drill is also used very extensively in many sections of the country. *a* **1884** KNIGHT *Dict. Mech.* Suppl., *Disk electrode*,..an electrode for telegraphic instruments in which the connection is secured by the contact of the peripheries of two disks. **1884** F. KROHN tr. *Glaser de Cew's Magn. & Dyn.-electr. Mach.* 104 A kind of voltaic battery in which only one metal was employed, the disk-electrodes of which were rendered active by polarisation. **1903** *Westm. Gaz.* 9 Dec. 8/2 The air..is drawn out by a disc-fan. *a* **1884** KNIGHT *Dict. Mech.* Suppl., *Disk harrow*, a harrow..carrying a number of sharp-edged and concave disks. **1891** R. WALLACE *Rural Econ. Austral. & N.Z.* xix. 270 The Disc-Harrow..with its saucer-shaped discs strung on two axle-shafts. *Ibid.* xxii. 309 After the land has been disc-harrowed. **1907** L. H. BAILEY *Cycl. Amer. Agric.* I. 385 The revolving disc harrow or plow, with its concave discs moving obliquely through the soil. **1948** *British Birds* XLI. 29 This area of waste ground had been disc harrowed a few days before. **1960** *Farmer & Stockbreeder* 15 Mar. 98/1 A more recent development is the powered disc-harrow. **1941** *Variety* 13 Aug. 36/3 Gilbert is a disc-jockey who sings with his records. *Ibid.* 51 (*headline*) Art Green disc-jockeys from Manhattan Beach. **1942** *Time* 6 July 67/1 Some stations merely hired 'disk-jockeys' to ride herd on swing records, in the traditional milk-man's matinee style. **1955** *Times* 22 July 9/4 We will do well to try to emulate the disc-jockeys and to put into our voices the warm, gay humanity which they lavish on Mrs. Soup and Mrs. Gravel. **1971** *N.Y. Times* (Entertainment section) 9 May 17/4 All the other disk jockeys had names like Brown and Green. **1941** *Variety* 13 Aug. 36/3 (*headline*) Disc-jockeying in Fifth Avenue window. **1955** N. FITZGERALD *House is Falling* x. 164 I've had to take a spot of disc-jockeying on commercial programmes to make ends meet. **1893** J. TUCKEY tr. *Hatschek's Amphioxus* 137 A disc-like thickening of the hypoblast. **1956** *Nature* 11 Feb. 277/2 Red cells retain their disk-like form. **1934** *Jrnl. R. Aeronaut. Soc.* XXXVIII. 508 (W/A) is the air column loading in lbs. per square foot of cross-section of the downwardly displaced column of air. In conformity with helical propeller practice this may be referred to as the disc loading. **1962** *Gloss. Aeronaut. Terms* (*B.S.I.*) v. 16 *Disk loading*, the thrust of the rotor divided by the disk area. **1960** *Daily Tel.* 2 Apr. 1/1 Provision of disc parking on the lines of the system used in Paris. **1970** Disc parking [see sense *2 f above]. **1881** *U.S. Patent* 9603, Disk plow. **1907** Disc plow [see *disc harrow* above]. **1940** *Chambers's Techn. Dict.* 251/2 *Disc plough*, a plough which cuts a furrow slice by means of a sharp-edged steel disc, of saucer-like shape, set obliquely to the ground surface.

disc (disk), *v.* Chiefly *U.S.* and *N.Z.* Also **disk.** [f. the sb.] To cultivate with a disc cultivator or disc harrow. So **di·scing** *vbl. sb.*; *discing machine* (see quot. *a* 1884).

a **1884** KNIGHT *Dict. Mech.* Suppl., *Disking machine*,..a steam-cultivating implement to be drawn by an engine over sod or plowed sod. **1917** *Nat. Weather & Crop Bull.* Mar. 4/2 Cutting corn-stalks, disking, and plowing have commenced. **1945** B. MACDONALD *Egg & I* (1946) iii. 45 When the garden..had been ploughed, disked, harrowed and dragged. **1950** *N.Z. Jrnl. Agric.* Mar. 227/1 The soil should be disced and worked down to a fine tilth. *Ibid.* 314/1 Stalks should be ploughed or disced in..after the tobacco harvest. **1952** M. H. HOLCROFT *Dance of Seasons* x. 75 After the ploughing came the discing, a confused and bumpy distribution of clods which marred the beauty of the furrows. **1960** *Farmer & Stockbreeder* 15 Mar. 107/3 The amount of discing needed varies of course with the seasons.

discard, *sb.* Add: **1. c.** *gen.* The fact of being discarded; dismissal. Also, the act of dismissing or abandoning.

1782 L. L. DALRYMPLE *Jrnl.* 12 Oct. in E. V. Mason *Jrnl. Young Lady Virginia* (1871) 29 Nancy had an admirer lately...He got his discard yesterday. **1906** *Daily Chron.* 29 Sept. 5/4 'The Chief's' sudden discard of South Africa and adoption of Protection under the name of Tariff Reform.

2. Delete *rare* and add examples.

1926 *Jrnl. Iron & Steel Inst.* CXIII. 76 The charts show the results of the analyses..throughout the section of the ingot, after the removal of the top and bottom end discards. **1943** E. H. THOMPSON *A.L.A. Gloss. Library Terms* 47/2 *Discard*, a book officially withdrawn from a library collection because it is unfit for further use or is no longer needed. **1949** R. T. ROLFE *Dict. Metallogr.* (ed. 2) 62 *Crop, crop end or discard*, the end (or ends) of an ingot cut off and discarded, as containing the central pipe or other defects. **1955** W. W. DENLINGER *Complete Boston* II. 65 The litter should be looked over carefully for possible defectives and discards. **1967** *Times Rev. Industry* Aug. 36/1 Using the discard of coal preparation plants to make lightweight concrete or building blocks.

b. Fig. phr. *(in)to the discard*: into oblivion or disuse.

1905 *Smart Set* Oct. 14/1 I'm much obliged to the lady; but she goes to the discard, too. **1927** H. E. FOSDICK *Pilgr. Palestine* 260 One surely does not mean to sweep into the discard as spiritually futile the elaborate symbolism of Eastern worship. **1944** J. S. HUXLEY *Living in Revol.* ii. 27 The old concept of economic man has gone into the discard.

discarnate, *a.* Restrict †*Obs. rare* to sense 1 and add: **2.** Divested of the flesh or the body, disembodied.

1895 *Westm. Gaz.* 5 Nov. 2/3 Any *à priori* belief in a discarnate existence of one's personality. *a* **1901** MYERS *Hum. Pers.* (1903) II. 274 We cannot simply admit the existence of discarnate spirits as inert or subsidiary phenomena. **1901** J. H. HYSLOP in *Proc. Soc. Psychical Research* XVI. 216 We ought to expect *a priori* that a discarnate memory should be defective in its communications from a transcendental world. **1920** *Public Opinion* 9 July 34/3 It is rash to claim that a given phenomenon..must..be due to discarnate influences. **1922** E. PHILLPOTTS *Grey Room* iv. 106 This death-dealing ghost, or discarnate but conscious being.

So **discarna·tion,** disembodiment.

1901 *Contemp. Rev.* Aug. 221 The discarnation of personality in death.

discharge, *v.* **1. c.** Delete † and add later example.

1899 *Daily News* 12 June 3/4 Kennerley was not aware that the firearm was loaded, and it discharged in his face.

discharge, *sb.* Add: **3. b.** Applied *attrib.* to a tube or an electric lamp containing a gas or metal vapour in which an electric discharge can be produced between two electrodes. Cf. **gas-discharge.*

1898 *Phil. Mag.* 5th Ser. XLVI. 296 Hertz made the rays travel between two parallel plates of metal placed inside the discharge-tube. **1935** *Discovery* Apr. 111/2 The latest drifting signs make use of small electric discharge tubes instead of filament lamps. **1936** *Nature* 14 Nov. 836/1 The new discharge bulb lamps are already in use on the Continent...In appearance they are like the ordinary 'pearl' lamp but they have no filament. They contain a small quartz mercury vapour discharge lamp..and they are corrected for colour. **1936** *Discovery* Dec. 365/2 The methods of determining with accuracy the masses of swift charged atoms and molecules produced in the discharge tube. **1959** *Listener* 12 Mar. 454/1 About 1930, the discharge lamp emerged as a competitor of the tungsten filament lamp for street lighting.

dischargeable, *a.* Delete *rare* and add later example.

1897 *Daily News* 20 July 5/6 The notes are dischargeable on August 1st, 1900.

dischargee (dis,tʃaɹdʒī·). [f. DISCHARGE *v.* + -EE.] A person who has been discharged.

1894 *Scottish Rev.* July 58 Government finds place for its deserving dischargees in its public service. **1952** *N.Y. Times* 28 Sept. 23 Dischargees without combat service.

disciple, *sb.* **3.** Add to def.: The name was suggested by Alex. Campbell of Lexington, Kentucky, in 1832. (Add earlier examples.)

1834 J. M. PECK *Gaz. Illinois* 203 A new sect [was] recently organized by a union of 'Reformed Baptists' and 'Christians' who call themselves 'Disciples'. **1835** J. MARTIN *Gaz. Virginia* 76 The precise distinction between the regular Baptist and the Reformers, called the disciples of Christ, not being in all cases drawn.

discipline. 2. Delete *arch.* and add later examples.

1942 *Spectator* 27 Feb. 204/1 The distribution of academic disciplines in which they [sc. candidates for the Foreign Office] had specialised. **1958** G. J. WARNOCK *Eng. Philos. since 1900* xiii. 172 It is only quite recently that the subject-matter, or rather the tasks, of philosophy have come to be clearly distinguished from those of other disciplines. **1962** *Lancet* 13 Jan. 113/1 Sir Leonard Parsons..had been the first to draw into the paediatrics of his time other disciplines such as biochemistry and immunology.

8. (Later example.)

1816 JANE AUSTEN *Let.* 9 July (1952) 457 Her illness must have been a very serious one indeed...Tell your Father I..most sincerely join in the hope of her being..much the better for her present Discipline.

disclass, *v.* (Further examples.)

1916 G. B. SHAW *Pygmalion* 196 Her father..formerly a dustman, and now fantastically disclassed, had become extremely popular in the smartest society. **1934** —— *Prefaces* 626/2 The baron's cadet..cannot avail himself of the public elementary and secondary schools because such a step would disclass the man of family.

discli·max. *Ecol.* [f. DIS- 9 + *CLIMAX *sb.* 4 b.] (See quots.)

1936 F. E. CLEMENTS in *Jrnl. Ecol.* XXIV. 265 Disclimax...The most frequent examples of this community result from the modification or replacement of the true climax, either as a whole or in part, or from a change in the direction of succession. *Ibid.* 266 Selective cutting not infrequently initiates disclimaxes, as may likewise the similar action of other agents such as fire or epidemic disease. **1938** WEAVER & CLEMENTS *Plant Ecol.* (ed. 2) iii. 86 Disturbance climaxes or *disclimaxes* are nearly always the result of disturbance by man or domesticated animals. **1962** H. HANSON *Dict. Ecol.* 110 *Disclimax*, an enduring climax community altered by disturbance by man or domesticated livestock, e.g., a grassland which has replaced a deciduous forest.

disco (di·sko). Colloq. abbrev. (orig. *U.S.*) of *DISCOTHÈQUE. Also, the equipment for playing records at a discothèque.

1964 *Playboy* Sept. 55/3 Los Angeles has emerged with the biggest and brassiest of the discos. **1970** R. GARRETT *Run Down* i. 29 I've watched the smut bookshops, the neon palaces, the gambling dens, the discos. **1972** *Oxf. Mail* 7 Jan. 2 (Advt.), Velvet Sound's Disco for hire.

disco-. Add: **di·scoplasm, di·scotriæne** (see quots.).

1900 W. MYERS tr. *Ehrlich & Lazarus's Histol. Blood* 51 The discoplasm loses its power of retaining the hæmoglobin, and gives it up to the blood plasma in ever increasing quantity. **1913** DORLAND *Med. Dict.* (ed. 7), *Discoplasm*, the structural part of a red blood-corpuscle. **1888** SOLLAS in *Challenger Rep.* XXV. p. lvii, Varieties of the Triæne... *Discotriæne*...The cladome is a disc in which separate cladi are not distinguishable; and the axial rods representing them extend but a short distance from the cladal origin. *Ibid.*, The discotriæne combines in itself the characters of the desma and the ordinary spicule.

discography (diskǫ·grǎfi). [f. *DISC *sb.* 2 d + -OGRAPHY. Cf. Fr. *discographie.*] A catalogue raisonné of gramophone records; a list of the recordings of a single composer or performer; also, the study of recordings. Hence **discogra·phical** *a.*, pertaining or relating to discography; **disco·grapher,** one skilled in discography.

1935 *Melody Maker* 14 Dec. 8/3, I have seen all kinds of discology, discography, and long lists of records. **1936** R. D. DARRELL *Gramophone Shop Encycl. Recorded Music* p. vii, A presentation of each composer's discography work by work, with complete recorded versions of each composition in its original form. **1941** *Jazz Information* Nov. 22/1 Important as these items are to discographers, they were merely sidelines with James P. **1946** *Jazz Mag.* III. i. 5/2 Eric S. Tonks' new complete 'Discography of Jazz' on sale...Tonks has spent some time..on his discographical details, that these records hardly represent a short 'History of Jazz'. **1946** R. VERAKKER in A. McCarthy *PL Yearbk. Jazz* 140 Those to whom discography appears a needlessly involved science. **1952** [see *DISCOPHILE]. **1955** R. BLESH *Shining Trumpets* (ed. 2) xi. 323 The assembling of discographical information and biographical material in America. **1957** *Times Lit. Suppl.* 29 Nov. 714/3 Its scholarship is the 'discography' (a curious discipline analogous to bibliography). **1966** *Guardian* 8 Mar. 12/6 Leonard Petts..has spent some time compiling a discography of Sir Winston Churchill for the British Institute of Recorded Sound...[Churchill] must have left his words behind in odd corners not yet found by the discographers.

discohere (diskohī·ɹ), *v.* [DIS- 6.] **1.** *Electr.* = *DECOHERE *v.*

1899 *Standard* 8 May, The current causes a small hammer to strike the coherer and to cause the filings to discohere, so that the circuit may be completed and broken at will.

2. *fig.* To cease to cohere.

1954 W. FAULKNER *Fable* (1955) 31 [The malleable mass] has relinquished, dis-cohered..flowing..back to its own base anonymity.

discombobulate (diskǫmbǫ·biulę̄ⁱt), *v. U.S. joc.* Also **discombobberate** and other variants. [Prob. jocular alteration of *discompose* or *discomfit.*] *trans.* To disturb, upset, disconcert. So **discombo·bulated** *ppl. a.*; **discombobula·tion,** upset, embarrassment.

1834 *Sun* (N.Y.) 21 Mar. 2/3 May be some of you don't get discombobracated. **1838** J. C. NEAL *Charcoal Sk.* 14 While you tear the one, you'll discombobberate the nerves of the other. **1839** *Spirit of Times* 16 Mar. 24/2 Finally, Richmond was obliged to trundle him, neck and heels, to the earth, to the utter discombobulation of his wig. **1916** R. FIRBANK *Let.* 11 Feb. (1964) 178, I put my own discombobulation first to lead up unnoticeably to yours. **1943** *Sat. Rev. Lit.* 23 Jan. 4/1 President Roosevelt's sarcastic reply, when asked as to the wisdom of raising an army too large to be supplied from the home-front, in terms of 'discombobulating the domestic economy'. **1957** M. MILLAR *Soft Talkers* ii. 21 It seems as though we were getting all discombobulated for nothing. **1962** R. P. BLACKMUR in E. Hubler et al. *Riddle of Shakespeare's Sonnets* 138 The hues attract, draw, steal men's eyes, yet penetrate, discombobulate, amaze the souls or psyches of women. **1970** 'E. QUEEN' *Last Woman* i. 17, I don't want you people to be in any way discombobulated.

discommodious (diskǫmō̆·diəs), *a.* Delete †*Obs.* and add later examples.

1897 *Blackw. Mag.* Nov. 593/2 A [fashion] discommodious for warm weather. **1928** *Ibid.* Feb. 161/2 Confound and blister that blasted German and all his damned discommodious works.

disconcert, *v.* **2.** Add: Also *absol.*

1908 *Smart Set* Sept. 47 She was conscious of a baffling reserve, a poise that disconcerted.

disconcertingly (diskǫnsə̄·ɹtiŋli), *adv.* [f. DISCONCERTING *ppl. a.* + -LY[2].] In a disconcerting manner.

1898 *Westm. Gaz.* 5 Apr. 2/2 The Japs have found it disconcertingly expensive. **1900** H. G. WELLS *Love & Mr. Lewisham* xi, He regarded Lewisham critically and disconcertingly over gilt glasses. **1906** GALSWORTHY *Man of Property* I. i, He had eyes disconcertingly inattentive at times.

disconfirm (diskǫnfə̄·ɹm), *v.* [DIS- 6.] *trans.* (To tend) to show the falsity of invalidity of (a hypothesis, etc.); to count against. (Opp. CONFIRM *v.*) Hence **disconfi·rming** *ppl. a.*

1936 R. CARNAP in *Philos. Sci.* III. 425 We may, if we wish, call a sentence disconfirmed in a certain degree if its negation is confirmed in that degree. **1943** *Jrnl. Symbolic Logic* VIII. 122 Certain observation sentences confirm or disconfirm a hypothesis...The concepts of confirmation, disconfirmation, and neutrality..have just been loosely characterized. **1945** *Mind* LIV. 2 A statement is also applicable in principle, if it is possible to describe the kind of data which would confirm or disconfirm it. *Ibid.*, Applicability..to the confirming or disconfirming character of experiential data obtained in the test of a hypothesis are often made. **1949** A. PAP *Elem. Anal. Philos.* xiii. 336 A statement is meaningful if and only if one can describe evidence which would confirm it and evidence which would disconfirm it. **1968** P. M. POSTAL *Aspects Phonol. Theory* viii. 161 All such assertions are, however, easily disconfirmed.

disconfirmation (diskǫnfəɹmę̄ⁱ·ʃən). [DIS- 9.] The process or result of disconfirming.

1937 O. NEURATH in *Philos. Sci.* IV. 276, I..proposed that we employ the terms 'confirmation' and 'disconfirmation'. **1943** [see *DISCONFIRM *v.*]. **1945** *Mind* LIV. 2 The verifiability theory of meaning..of confirmation and disconfirmation..one more comprehensive than those of conclusive verification and falsification. **1963** *Language* XL. 261 Critics of transformational theory take these experimental results as disconfirmation.

disconformable, *a.* Restrict †*Obs.* to sense in Dict. and add: **2.** *Geol.* Containing or constituting a disconformity.

1905 A. W. GRABAU in *Science* 27 Oct. 534/2 We need a term to express the relation where two formations thus conformable in their bedding but comprising between them a time break of greater or less magnitude. To speak of such strata as unconformable, without qualifying the term,..suggests that the older strata have suffered folding and erosion before the deposition of the later...We might speak of such formations as disconformable. **1925** N. E. ODELL in E. F. Norton *Fight for Everest: 1924* 295 The true nature of this important junction..it is difficult to tell, but it has the appearance of a disconformable one. **1970** *Nature* 12 Sept. 1127/1 The glacial beds..are disconformable on Aztec siltstone.

disconformity. Add: **2.** *Geol.* An unconformity between two parallel, approximately horizontal sets of strata, the lower set having undergone erosion but not deformation before the upper set was deposited.

1906 A. W. GRABAU in *Bull. Geol. Soc. Amer.* XVII. 569 Generally this kind of overlap implies some erosion of the underlying concordant formations, thus producing a disconformity. **1949** W. J. MILLER *Introd. Physical Geol.* vi. 137 If..two sets of beds separated by an erosion surface have their stratification surfaces practically parallel, there is a more or less hidden unconformity which may be called a disconformity. **1965** G. J. WILLIAMS *Econ. Geol. N.Z.* i. 3/1 The limestones were followed in many areas by disconformities and unconformities.

discontinuity. Add: **2.** *attrib.*, as **discontinuity layer,** a layer of water in a lake or the sea in which the temperature changes rapidly with depth from that of the water above it to that of the water below; a thermocline separating an epilimnion from a hypolimnion.

1911 E. M. WEDDERBURN in *Trans. R. Soc. Edin.* XLVII. 625 Attention has been drawn to the very rapid temperature changes which are found at the discontinuity layer. **1931** *Discovery* June 199/2 In general the presence of the 'discontinuity layer' tends to prevent the free mixing of the layers [of water in the sea]. **1942** T. H. JENKIN in *Jrnl. Animal Ecol.* XI. 252 The term 'Discontinuity Layer' is retained to cover the layer of very variable thickness in which there is a sufficient temperature gradient to prevent any direct mixing with the cool hypolimnion below. **1964** *Oceanogr. & Marine Biol.* II. 125 At depths of hundreds of metres, perhaps below the discontinuity layer, light is reduced to 50–15% of surface man.

discophile (di·skofəil). Also **discophil.** [f. *DISC *sb.* 2 d + -PHIL, -PHILE.] An enthusiast for and collector of gramophone records.

1940 *Commercial Appeal* 14 June 1/4 Many discophiles might have preferred one of the less hackneyed earlier symphonies. **1952** *Times Lit. Suppl.* 15 Aug. 514/2 The little world of discography and the discophil has been expanded internationally and for some years has needed a comprehensive encyclopædia of..the electric recordings of classical music. **1959** *Sunday Times* 17 May 20/8 While learned discophiles deal with the relationship between the Ambrosian and Gregorian Chant, humbler discophiles may now make their own programmes. **1962** *Times Lit. Suppl.* 12 Jan. 20/4 The discophile's cult of the vocal record.

discothèque, -theque (di·skotek). [a. Fr. *discothèque*, after BIBLIOTHÈQUE. Cf. *DISC *sb.* 2 d.] A club, etc., where recorded music is played for dancing.

[**1951** SACKVILLE-WEST & SHAWE-TAYLOR *Record Guide* 32 On the analogy of 'bibliothèque', the French have taken to using the very attractive 'discothèque' for 'record library'; it is a pity that this formation is impossible in English.] **1954** *New York Times* 3 Oct. 36/1 Germain-des-Prés *discothèque* night clubs..where phonograph discs are played for dancing. **1960** *Atlantic Monthly* Sept. 48/1 We go after dinner, to a Left Bank *discothèque.* *Ibid.* 46/1 Since this is a *discothèque*, jazz is blaring from the walls and record sleeves are scattered about. **1964** *TV Guide* (U.S.) 12-18 Dec. 19/1 A discothèque is a small, intimate night club that plays recorded music for dancing and dancing, and those dances include dancing the frug, the monkey, and the Watusi a delight because they move with the beat. **1965** F. RAPHAEL *Darling* xxii. 105 They danced in smoky discothèques. **1970** *Morning Star* 6 Jan. 3 Mr. Moon, his wife and two friends had attended the opening of a discothèque at the Red Lion public house on Sunday.

discount, *sb.* Add: **6.** (In sense 2 b) *discount market*; (in sense 4) applied *attrib.* to a shop that sells goods below the normal retail price; also *discount selling, trading.*

1858 G. CLARE *Money-market Primer* xii. 135 The very important function performed in the economy of the discount market by the middleman, through whom..nearly the whole of the discount class of business is transacted. **1922** W. F. SPALDING *London Money Market* vii. 106 The discount market is one of the most important sections of the London Money Market. **1933** B. ELLINGER *This Money Business* iii. 20 The Discount Market in London has been in existence for about seventy years. **1949** *Consumer Rep.* Aug. 343/1 Buying..nationally-advertised goods from discount houses at savings up to 20%. **1958** *Oxf. Mail* 30 Jan. 6/3 Service in discount shops [in the U.S.] is poor. **1960** *Guardian* 10 June 3/1 The grocery concern which..is relatively demanded by trade associations. *Ibid.* 7/3 A Lincolnshire grocer..is to build a discount store in Manchester. **1962** *Listener* 13 Dec. 994/2 The discount house, which sells all these things by self-service at very low prices, has grown phenomenally in recent years in the United States, but there are still only half a dozen of them in this country. **1964** *Times Rev. Industry* Apr. 72/3 One supermarket chain in California..has switched over entirely to discount trading, including leasing departments in discount stores. **1970** *Times* 23 Sept. 14/6 Strictly speaking a Discount Store is an enormous single storey self-service store..which sells a huge range of goods..at below normal retail prices.

discounted, *ppl. a.* (Later example.)

1964 *Times Rev. Industry* Feb. 98/2 The principles of discounted cash flow and sinking fund accounting.

discounter. Add: **2.** One who is involved in discount trading.

1921 A. WALL *Analytical Credits* v. 126 It is essential for him to be known in the trade as a discounter. **1964** *Economist* 1 Feb. 411/1 The discounters—a terribly misleading word for retail traders who take only a modest profit. **1964** *Times Rev. Industry* Apr. 72/3 Retailers of other merchandise, such as shoes, jewellery, [etc.]..have also sought to cut their losses by leasing departments from discounters.

discover, *v.* Add: **8. d.** *spec.* To bring to public notice, make famous or fashionable.

1908 *Hay Maw's Mag.* Sept. 114/1 It is interesting just here to note that while editor of the Westminster, Mr. Macdonald 'discovered' Ralph Connor (Rev. Dr. Gordon), the celebrated Canadian novelist. **1936** M. BARING *Daphne Adeane* i. 5 She was merged in the ranks of the unnoticed, till she was suddenly 'discovered'. **1958** *Times Lit. Suppl.* 5 Sept. 503/3 In a very short time she had producers..vying with each other for the honour of 'discovering' her. **1963** J. FLEMING *Death of Sardine* iii. 41 One day, when Trigoso Praia, or *Plage*, was 'discovered' the road might be an important promenade.

discrete, *sb.* (Later example.)

1967 *Electronics* 6 Mar. 116 Integrated circuits will be turning up routinely in new products throughout 1967. The big switch from discretes is on.

discretionary, *a.* Add: **1.** (Additional examples.)

1960 *Life* 1 Feb. 53/1, 34 million families..control most of the $64 billion 'discretionary income' (money left after necessary expenditures) in the country. **1961** *Times* 10 June 11/4 American sociologists have invented the term 'discretionary time' to cover the hours left over from eating, sleeping, and earning a living. **1964** HALL & HOWARD *Church in Social Work* 285 In mid-1958 boys earned an average of 9s. 1d. and girls £5 6s., or about £3 of which was available for discretionary spending.

discriminability (diskri:minǎbi·liti). [f. DISCRIMINABLE *a.* + -ITY.]

1. The quality of being discriminable.

1901 E. B. TITCHENER *Exper. Psychol.* I. ii. 189 Clearness implies a maximal discriminability or separability from other processes. **1954** BROWN & LENNEBERG in Saporta & Bastian *Psycholinguistics* (1961) 489/2 A total of six hundred..was..determined at discriminability. **1961** *Jrnl.* ...that the colour. **1966** *Times Lit. Suppl.* 15 Nov. 1058/1 Tarski's doubts about the effective discriminability of a class of logical truths. **1970** *Sci. Amer.* Mar. 62/1 It is clear..that increasing the discriminability of the items by incorporating an additional attribute into each item would increase performance.

2. The capacity to discriminate; discriminative power.

1931 *Brit. Jrnl. Psychol.* XXII. 38 The ability..to set up different responses to A and A'..falls short of the limits of temporary discriminability.

discri·minable, *a.* Delete †*Obs. rare* and add examples.

1946 *Mind* LV. 131 Feelings..are discriminable elements. **1964** GOULD & KOLB *Dict. Social Sci.* 447/2 A great deal of human behaviour..cannot be directly associated with specific, discriminable organic states. **1970** *Jrnl. Gen. Psychol.* Jan. 85 The middle of the list is least discriminable.

discriminant, *a.* and *sb.* Add: **A.** *adj.* **3.** *discriminant function*, in *Statistics*, A function (usually linear) of a set of variables which is used to allocate items to any of several classes with minimum probability of misclassification, each variable taking one value for each item.

1936 R. A. FISHER in *Ann. Eugenics* VII. 179 Discriminant functions. When two or more populations have been measured in several characters, $x_1, ..., x_s$, special interest attaches to certain linear functions of the measurements by which the populations are best discriminated. **1948** *New Biol.* IV. 43 The technique of discriminant functions is a means of using all the measurements of two or more measurable attributes for the distinction of two sub-classes of a population. **1962** *Lancet* 15 Dec. 1258/2 To combine the information from several observations on each case, a discriminant function, was calculated previously.

B. *sb.* **2.** That which enables one to discriminate; a distinguishing feature; *spec.* a discriminant function.

1920 *Expositor* May 338 The formula ought not to be used as a discriminant between one Gospel and another. **1958** *New Scientist* 16 Oct. 1083/2, I should like to suggest to the 'discriminant' (or 'for the factors responsible for variance). **1962** *Lancet* 15 Dec. 1258/2 In order to calculate the discriminant, the value being known for each patient in this series. *Ibid.* 1259/2 Over half the patients..have negative discriminants.

discriminate, *v.* Add: **3. a, b.** *spec.* To exercise racial discrimination (cf. *DISCRIMINATION 1 c).

1866 A. JOHNSON *Speech* 27 Mar. in H. S. Commager *Documents Amer. Hist.* (1935) II. 16/2 Congress can repeal all State laws discriminating between whites and blacks in the subjects covered by this bill. **1899** B. T. WASHINGTON *Fut. Amer. Negro* vi. 190 We find the Negro forgetting his own wrongs, forgetting the laws and customs that discriminate against him in his own country. **1936** *New Statesman* 8 Aug. 190/2 When discriminating racially, popular opinion lays emphasis on the Negro's colour. **1968** *Listener* 3 Oct. 427/2 If you now extend for ever with the expectation of being discriminated against, the chances are you won't ever be disappointed.

discrimination. Add: **1. a.** Also with *against.*

1949 *Times* 10 Sept. 5/6 Any sort of discrimination against the trade of any country or countries...Discrimination has always tends to be practised against exports from 'hard-currency countries'. **1951** J. R. WINTON *Dict. Econ. Terms* 70 Generally speaking the dollar shortage..discrimination has been widely practised..against the U.S.A.

c. *spec.* The making of distinctions prejudicial to people of a different race or colour from oneself; racial discrimination.

1866 A. JOHNSON *Speech* 27 Mar. in H. S. Commager *Documents Amer. Hist.* (1935) II. 16/2 Thus a perfect equality of the white and coloured races is attempted to be fixed by Federal law in every State of the Union over the vast field of State jurisdiction covered by these enumerated rights. In no one of these can any State ever exercise any power of discrimination between the different races. **1889** T. WASHINGTON *Fut. Amer. Negro* vi. 148 Let the very best educational opportunities be provided for both races; and add to this an election law that shall be incapable of unjust discrimination. **1906** *Ann. Amer. Acad. Pol. & Soc. Sci.* XXVII. 111. 208 Travel skipped were obliged to secure the discrimination in southern society. **1958** *Ann. Reg.* 1957 158 Direct talks between South Africa..and India and Pakistan..on the dispute over alleged discrimination against Indians in the Union. **1971** *Times* 5 Mar. 14/7 Mr Maudling could still say that 'In our Immigration Bill is not as a matter of discrimination, unless you're looking for discrimination'.

d. *Psychol.* In a learning situation, the identification of a stimulus and the choice of an appropriate response. Freq. *attrib.*

1896 W. JAMES *Princ. Psychol.* I. xiii. 494 The reaction may be withheld until the signal has consciously been discriminated as this or that. **1902** TITCHENER *Exper. Psychol.* I. ii. 62 ...discrimination-time, associations-time) and then discrimination-time. **1908** E. B. TITCHENER *Primer Psychol.* 260 In the discrimination reaction, he [sc. the subject] moves only when he has apprehended some one or two or more familiar stimuli. **1954** *Discrimination learning*. **1958** E. R. HILGARD *Theories of Learning* v. 130 Skinner believes that his arrangement for obtaining discrimination is superior to that usually used. *Ibid.*, The discrimination experiment is always complicated by these 'irrelevant' cues. **1962** C. E. OSGOOD *Method & Theory Exper. Psychol.* III. xvi. 352 What is it in behavior that enters to check this overboard generalization of response? The process is called discrimination. **1968** J. H. FLAVELL *Developmental Psychol. J. Piaget* ii. 56 Discrimination is the complement of generalization, as students of learning have long known. **1967** NEIMARK & ESTES *Stimulus Sampling Theory* iv. 430 The classic character of discrimination theories..is based upon the number of elementary processes invoked. *Ibid.* 431 Any comprehensive account of learning must provide an adequate account of stimulus generalization and discrimination. **1970** L. ELKINTON tr. *Le Ny's Learning & Memory* 57 The maze is a more complex discrimination situation in which there are possible alternative responses. *Ibid.* 58 In both types of situation (discrimination apparatus and maze), the reaction..is a partially new reaction.

2. *Electr.* **a.** (See quot.)

1931 S. R. ROGET *Dict. Electr. Terms* (ed. 2) 89/2 *Discrimination*, an instrument with an earlier system of charging of an electric supply which automatically switches over to a different meter when a predetermined maximum demand has been exceeded.

b. Any circuit the output of which depends on some property of the input that departs from a fixed level or value; *esp.* one that converts a frequency-modulated signal to an amplitude-modulated one by modulating the output amplitude according to the deviation of the input frequency from a fixed (carrier) frequency.

1935 *Proc. Inst. Radio Engineers* XXIII. 1126 An automatic frequency control system will consist of two distinct circuits; a frequency discriminator or frequency sensitive circuit that generates a bias varying with changes of the intermediate-frequency signal carrier frequency..and a frequency control circuit. **1942** *Electronics* Sept. 44 The frequency discriminator which..has been used in various forms in automatic frequency control systems. **1949** *Nature* 4 Jan. 26/1 In order to be able to read a frequency-modulated signal, the receiver uses a device known as a 'discriminator', which in effect converts it into an amplitude-modulated signal, which may then be detected by any of the usual methods. **1962** SIMPSON & RICHARDS *Junction Transistors* xvii. 457 The most common type of frequency-modulation detector, the discriminator, was developed originally for vacuum-tube receivers. **1964** F. J. M. FARLEY *Elem. Pulse Circuits* iv. 75 The circuit 120 is an example of a discriminator, that is, a circuit which transmits only the upper portion of a pulse. In common with other discriminator circuits..it may be used..to select signals of amplitude greater than a predetermined threshold value.

discriminatory, *a.* Delete *rare* and add later examples.

1941 F. MASON *Essay on Racial Tension* x. 88 The way we make against her [sc. the Negro] in this country..if we fail to discriminate by discriminatory devices which were defensible as in the interests of the African. **1962** *Times* 7 Apr. 12/3 The fact that companies will have to pay on their gains at the higher corporation tax level..was thought to be discriminatory. **1971** *Guardian* 8 Mar. 6/4 Asians..have not forgotten the similar 'discriminatory' Immigrants Act of 1968.

discursivity (diskǫɹsi·viti). *Philos.* [f. DISCURSIVE *a.* + -ITY.] = DISCURSIVENESS.

1940 *Mind* XLIX. 447 Kant..never completely abandoned his view of the discursivity of the human intelligence. **1946** *Ibid.* LV. 52 Kant..never suggests that the possession of..discursivity). **1950** *Ibid.* LIX. 17 Such experience [is..the understanding's] thoroughgoing discursivity.

discussant (diskǫ·sǎnt). orig. *U.S.* [f. DISCUSS *v.* + *-ANT[1].] One who discusses, esp. one who takes part in a set discussion.

1927 *Jrnl. Amer. Med. Assoc.* 4 June 1818 The discussants were Drs. Sydney Kuh and Francis Cerry. **1966** D. C. HAYS *Readings in Automatic Lang. Processing* p. v, A discussant in the case of group discussions, how can..authority be predicted from what the discussants have said to one another in the course of a few hours of conversation? **1966** *Economist* 10 Sept. 1004/2 Besides the speakers there were five discussants who did not participate in the deliberations of the workshops..whose function was to belabor controversial points of each paper after summary oral presentation.

discussion. Add: **3.** (Earlier example.)

1853 C. BRONTË *Villette* I. vi, By the small dainty messes Miss Marchmont's cook used to send up..to the discussion of which we used to devote half our hours... She always said I could not be able to do without my dinner.

6. *discussion-circle, group, programme.*

1947 C. DAY LEWIS *Starting Point* III. iii. 169 When the Neale class gets together, it's just one devastating discussion-circle from morning to night. **1942** H. J. LASKI *Let.* 31 July (1953) 1365 A discussion group to thrash out the principles of governmental re-organisation. **1940** GRAVES & HODGE *Long Week-End* xiii. 187 The B.B.C...appointed Education Officers..whose duty it was to organize 'Discussion Groups'. **1958** *New Statesman* 15 Mar. 331/3 One radio feature which is not yet practicable on TV is the B.B.C.'s monthly international discussion-programme, *Radio Link.*

disease, *sb.* Add: **1.** (Revived in modern use with the spelling *dis-ease.*)

1909 *Daily Chron.* 17 Sept. 3/1 Perhaps he..kept dark the apprehensions of his artist soul, communicated his ease not his dis-ease. **1920** A. S. M. HUTCHINSON *This Freedom* II. ii. 166 They were in a curious dis-ease whose occasion was not but the vision of a smutty youth. **1960** *Times Lit. Suppl.* 22 Mar. 1960 She had a curious dis-ease in meeting socially doctors whom also she met professionally. **1960** *Guardian* Mar. 28/1 The fear and dis-ease which underlie the more obvious nostalgias.

4. *disease-producing* adj.

1903 *Westm. Gaz.* 27 Aug. 5/1 The invasion of the body by disease-producing organisms. **1931** J. S. HUXLEY *Uniqueness of Man* iii. v. 98 Disease-producing bacteria.

diseasing, *ppl. a.* Restrict †*Obs.* to sense 1 and add example of sense 2.

1919 W. MAYNELL *dad. Sarah & War* vii. 68 There's no disgraceful and diseasing drunkenness, but only this divine intoxication.

diseconomy, dis-economy (disǐkǫ·nōmi). *Econ.* [DIS- 9.] The opposite of economy; an absence of economy; *spec.* an increase in costs arising when a business organization exceeds an 'optimum size'.

1937 *Economica* IV. 441 This fall may be caused by..large scale dis-economies. **1954** *Encounter* Dec. 30/1 Orthodox 20th century economics..has nevertheless contained certain 'dis-economies' of large-scale operation which ensure that a firm has an optimum size, growth beyond which again becomes less efficient. **1960** *Times Rev. Industry* Apr. 31/1 Advantages of mass production have been saddled us with the 'diseconomies'..of mass storage. **1969** *Daily Tel.* 17 Nov. 18/6 Early loss of recently-trained people to overseas companies..can be described as the dis-economies in a narrowly national sense.

disengaged, *ppl. a.* Add: *spec.* in *Theatr.* colloq. Unemployed.

1937 P. GODFREY *Back-Stage* v. 70 It takes many years before the superfluous actors and actresses will admit to themselves that the professional terms as 'liberty' and 'disengaged' are no longer applicable to them in a temporary sense.

b. Uncommitted; opp. *ENGAGED *ppl. a.* 4.

1958 *Oxf. Mag.* 6 Mar. 340/2 Oppressed by the facile and disengaged conformism of the present age. **1960** *20th Cent.* Apr. 303 The accent to-day is on the personal, the disengaged, the apolitical and the private sense.

disengagement. Add: **f.** A withdrawal of military forces; a renunciation of military or political influence in a particular area.

1957 *Economist* 28 Dec. 1113/2 Even if..the sun shines brightly on the new disengagement..and..the troops are be pulled back from their forward positions. **1958** *Spectator* 31 Feb. 159/1 A policy of disengagement in Central Europe. **1958** *Observer* 23 Feb. 5/3 Topic No. 1:..was 'Disengagement', by which was meant some plan whereby Russian and American troops would pull back to the 1945 armistice line in Europe. **1966** SCHWARZ & HADIK *Strategic Terminology* 66 *Disengagement*, design for diminishing international tensions by creating a neutral or demilitarized zone between the armed forces of two antagonistic powers or groups of states.

disequilibrium. Add: **b.** *spec.* in *Econ.*

1923 J. M. KEYNES *Tract Monetary Reform* iv. 160 When the disequilibrium was purely seasonal, this was an unqualified advantage. *Ibid.* 161 The post-war method is a most rapid and powerful corrective of real disequilibria in the balance of international payments. **1948** G. CROWTHER *Outl. Money* (ed. 2) x. 331 Any excess of Saving over Investment produces disequilibrium—a shortage of purchasing power to buy the products of industry, falling prices and unemployment. **1962** *Listener* 4 Oct. 511 The kind of balance of payments disequilibria which..

diseuse (dī·zǫ̈z). [Fr., fem. = talker.] A female artiste who specializes in monologue.

1896 *Cosmopolitan* XX. 344 She is only a concert hall singer (or diseuse, to use a newly-coined and specific title). **1923** *Daily Chron.* 20 June 8/4 Mme. Anna Thibaud, the celebrated diseuse. **1927** M. BARING *Tinker's Leave* xxv. 274 Bielor played songs, and Zarova, a well known diseuse sang. **1955** *Times* 1 Jan. 7 Another Italian was an artist of the *diseuse*, and her inviolably sympathetic.

disfellowship, *v.* (Earlier U.S. example.)

1831 *Troy* (N.Y.) *Watchman* 3 Sept. (Th.), They were disfellowshipped by the church.

disfiguringly (disfi·giuriŋli, -fi·gəɹ-), *adv.* [f. DISFIGURING *ppl. a.* + -LY[2].] In a disfiguring manner; so as to disfigure.

1911 MRS. H. WARD *Case of R. Meynell* i, Certain lines on the forehead..showed themselves disfiguringly. **1923** S. BARING-GOULD *Early Remin.* ii. 23 Telegraph stations planted..disfiguringly, on the summit of church towers.

dis·function, disfu·nctional *a.*, varr. *DYSFUNCTION, *DYSFUNCTIONAL *a.*

1951 PARSONS & SHILS *Gen. Theory of Action* 35 Functional in one context but disfunctional in another. **1952** KOESTLER *Arrow in Blue* xviii. 169 Yet, as far as I know, I have never had any glandular disfunction. **1959** P. KIRBY *Freud* viii. 220 Freud thought only reason could resolve this pull between social functions and disfunctions. **1969** *Indian Med. Jrnl.* V. 72 The disfunction in the phylum expressed in social interactions resultant from man's neurosis.

disguise, *sb.* **5.** See also **BLESSING vbl. sb.* 4 c.

disgust, *v.* **1. b.** (Later example.)

1873 L. TROUBRIDGE *Life amongst Troubridges* (1966) 46 He is the music master and oh how I do loathe and disgust him and his lesson.

dish, *sb.* Add: **2. a.** *transf.* and *fig.* (Earlier and later examples.) *spec.* An attractive person, esp. a woman. Now only in informal use.

1599 SHAKES. *Much Ado* II. i. 283 Heere's a dish I loue not, I cannot indure this Lady tongue. **1606** —— *Ant. & Cl.* II. vi. 134 He will to his Egyptian dish againe. *Ibid.* v. ii. 275, I know, that a woman is a dish for the Gods, if the diuell dresse her not. **1929** D. HAMMETT *Dain Curse* (1930) xix. 217 He turned his wet lips to me and..said 'What a swell dish you are.' **1938** J. CURTIS *They drive by Night* xxiii. 263 So you're Queenie, are you? And a nice little dish you are. **1945** P. CHEYNEY *I'll say she Does!* v. 141 She's a swell dish—a lovely piece of frail that one. **1955** PRIESTLEY & HAWKES *Journey down Rainbow* 64 The purple band..marched on to the field, accompanied by drum-majorettes—every one a dish, as they say. **1958** P. G. WILSON *Middle Age of Mrs. Eliot* 68 That man I've been talking to is rather a dish, just too him he's a bottom-pincher. **1959** H. HOBSON *Mission House Murder* vi. 19 'Is she a good-looker?'..'What a dish, my dear.'..It's gorgeous. **1965** *Spectator* 1 Jan. 12/3 She was..a dish who always looked at it as if he were a dish.

c. Slang phr. *one's dish*: something exactly suited to one's tastes, requirements, abilities, etc.; one's 'cup of tea'. *orig. U.S.*

1918 H. C. WITWER *From Baseball to Boches* 106 They [sc. grenades] ought to of been my dish, seen' what a notorious pitcher I was. **1927** *Daily Express* 28 Nov. 4/1 Dice do not..tell me..Some men are not my dish. **1930** *Amer. Mercury* v. 114 Did you lose your nerve And cloud your conscience because it wasn't Your dish really? **1955** *Punch* 14 Jan. 42/3 Thus they are aimed to provide the right dish. **1957** WODEHOUSE *Over Seventy* iii. If a mystery story is not my dish.

4. b. A microwave reflector or aerial with a concave surface.

1948 *Gloss. Terms Waveguide Technique* (*B.S.I.*) 11 *Dish*, a reflector the surface of which is part of a sphere or of a paraboloid of revolution. **1956** *Electronic Engin.* XXVIII. 539 The actual radiator consists of a 30ft diameter dish. **1957** *Ann. Reg.* 1956 333 The great radio-telescope at Jodrell Bank in Cheshire, with its 250-ft. reflecting 'dish'. **1960** *Listener* 14 Jan. 18 Once the dish is pointed with a new telescope employs dipoles parabolic instead of dish. **1962** *Times* 20 Dec. 1/1 The 1967 *Ibid.* 2 Oct., a grid 1 dish of carbon fibre material.

5. Also, more commonly, the condition of wheels having such concavity; the amount of such concavity.

1810 (in Dict.). **1837** W. B. ADAMS *Carriages* 98 Some wheels..get more dish than others. **1844** H. STEPHENS *Bk. Farm* II. 1154 The third or front wheel may be fixed without dish. **1888** *Engl. Mech.* 6 Jan. 375/1 The dish is considerable, amounting to 2 inches in the 5-foot wheel. **1891** *Fur, Fin & Feather* Mar. 167 The left hind wheel of his wagon is out of dish.

9. b. *dish-washing* (earlier and *attrib.* examples).
1889 R. Cooke *Steadfast* i. 16 The song of the kettle as it piped away on the shortened hook, where it kept hot for dish washing. **1897** *Outing* (U.S.) XXX. 124/2 A dish-washing machine. **1963** *Which?* 6 Feb. 46/1 Dishwashing machines are comparatively rare.

c. Consisting, or having the form, of a dish; dish-shaped.
a 1823 D. Wordsworth *Second Tour Scotl.* in *Jrnls.* (1941) II. 344 Porters in white hats with dish-crowns close to the head. **1874** S. J. Thearle *Naval Archit.: Wood & Iron Shipbuilding* 72 The hollow or dish keel is a variety of the flat keel system. **1893** *Kennel Gaz.* Aug. 214 Jess III is spoilt by her dish face. **1960** *Times* 2 Jan. 9/1 The working springer tends to be smaller, lower slung, often with a dish face. **1962** *Listener* 19 July 112/1 A dish aerial at Goonhilly Down. **1968** *Times* 20 Dec. 6/8 A special dish telescope 20ft. in diameter.

10. dish-cross, -rim, -ring (see quots. 1908, 1931); **dish-lift**, = Dumb-waiter 2; **dish-mop** orig. *U.S.*, a small mop used for washing dishes, etc.; **dish-pan** *U.S.*, a pan in which dishes, etc., are washed; hence **dish-pan hands**, an inflamed or sore condition of the hands caused by washing-up or by the use of cleaning materials in housework; **dish-rag** chiefly *U.S.* (examples); also *transf.*, the dishcloth gourd; **dish-towel** (earlier example).
1785 *Daily Universal Register* 1 Jan. 3/2 Dish crosses with lamps, 14 oz. to 20 oz. each. **1908** B. Wyllie *Sheffield Plate* 75 'Dish-crosses' or 'spiders'..served two purposes: either to keep a hot dish from marking the polished tables..or to support a spiritlamp which kept the contents of the dish above hot. **1908** H. N. Veitch *Sheffield Plate* 136 These dish-crosses..are suitable for both round and oval dishes. **1859** Sala *Twice round Clock* 81 With every modern convenience and improvement: with bath-rooms..dish-lifts, [etc.] **1920** *Contemp. Rev.* June 890 He walked to the dish-lift and listened intently. **1897** *Sears, Roebuck Catal.* 98/1 This dish mop is made of cotton and is securely fastened to handle. **1913** E. H. Porter *Pollyanna* v. 47 Nancy, hurrying with her belated work, jabbed her dish-mop into the milk pitcher. **1955** E. Coxhead *Figure in Mist* viii. 229 She picked up a dish-mop and began to scour a..saucepan. **1872** *Newton Kansan* 5 Sept. 4/5 Put your corn, while hot, in a dish-pan. **1942** W. Faulkner *Go Down, Moses* 176 Ash began to beat on the bottom of the dish-pan with a heavy spoon to call them to breakfast. **1944** B. Hutchison *Hollow Men* xiii. 187 'And me'—she chuckled at this—'sunk without trace in domesticity, teaching school, dish-pan hands [etc.].' **1839** *Southern Lit. Messenger* V. 329/2 When he landed he lay there as limber as a dish-rag. **1872** 'Mark Twain' *Roughing It* 44 It really pretended to be tea, but there was too much dish-rag and sand in it. **1890** *Cent. Dict.* s.v. *sponge-gourd*, Vegetable sponge or dish-rag. **1904** *N.Y. Tribune* 22 May, A novel enterprise, that of raising dishrags, is being exploited by a number of Southern California horticulturists. **1939** Dylan Thomas *Map of Love* 21 The horrid Woe drip from the dish-rag hands. **1774** in B. Wyllie *Sheffield Plate* (1908) 71 Dish-rims. **1908** B. Wyllie *Ibid.* pl. lxxi, Revolving Dish Stand for round or oval dish. Possibly this is what was meant by a 'dish-rim'. **1931** E. Wenham *Domestic Silver* v. 75 [Dish-crosses] represent to the English what the misunderstood dish-rings mean to the Irish, namely, a stand on which to place a hot dish to prevent it from marking the table. **1869** Mrs. Stowe *Oldtown Folks* 275 Aunt Lois..found her dish-towel freezing in her hand.

dish, *v.*[1] Add: **1.** (Later examples with *out*.)
1930 Brophy & Partridge *Songs & Slang of Brit. Soldier* 116 Cooked food was 'dished out' by the orderly men of the day. **1958** Hayward & Harari tr. *Pasternak's Dr. Zhivago* I. v. 133 Food was dished out and the used plates stacked in the hand-worked service-lift.

2. With *out*. (Later examples.) In modern use: to distribute; to give or hand *out* (often with the pejorative implication of a lack of care or discrimination). So *to dish it out* (U.S. colloq.): to deal out punishment; to fight hard.
1934 *Black Mask Mag.* Oct. 13/1 You take [money] away fast enough, but you don't like to dish it out. **1939** I. Baird *Waste Heritage* vi. 82 Gabby could take it and he could dish it out. **1939** *War Illustr.* 21 Oct. p. ii/1 The drivel that is dished out between news bulletins. **1955** A. Huxley *Genius & Goddess* 36 Tripe and hogwash dished out by the moulders of public opinion. **1959** *Camb. Rev.* 2 May 471/2 They dish out the gestetnered flysheet which does duty for it, free of charge. **1966** B. Kimenye *Kalasanda Revisited* 11 Offices..which dish out highly-coloured literature extolling the wonders of their country.

6. (Earlier example.)
1846 R. Ford *Gath. Spain* vii. 69 The Andalucian horse..is given to *dishing* with the feet.

dish-cloth. Add: **2. dishcloth gourd,** the gourd or the plant of any of the species of *Luffa*, esp. *L. cylindrica*, of which the spongy inner portion of the fruit may be used as a cloth.
1900 L. H. Bailey *Cycl. Amer. Hort.* II. 948/2 *Luffa*.. Dishcloth Gourd. Vegetable Sponge..*L. cylindrica*..The commonest Dishcloth Gourd. **1963** J. Organ *Gourds* vii. 93 British seedsmen on the whole restrict themselves to *L. cylindrica* which is usually listed as *Luffa*, Dishcloth gourd, or Vegetable Sponge.

dished, *ppl. a.* **b.** (Later examples.)
1878 *Trans. Ill. Dept. Agric.* XIV. 210 The head was short and fine, with a dished face and rather thin jowls. **1908** *Animal Managem.* 21 The face..'stag-faced' or 'dished'. **1958** *Daily Mail* 19 Aug. 4/8 The least we can do..is to adopt 'dished' steering wheels on all cars. **1962** *Which?* (Car Suppl.) Jan. 4/2 Dished, when used for a steering wheel, means that the centre hub is lower than the rim.

dishing, *vbl. sb.* (Further examples.)
c 1806 D. Wordsworth *Tour Scotl.* in *Jrnls.* (1941) I. 281 We grew impatient for our dinner..but we had to wait at least another half hour before the ceremony of dishing up was completed. **1929** W. Lang *Sea-Lawyer's Log* iii. 29 The Duty Watch are busily engaged in the Barrack Room washing and drying their table utensils (a process known in the Service as 'dishing up'). **1955** *Times* 4 May 13/5 The dishing out of Constitutions which did not satisfy the aspirations of these African peoples.

dishoard (dis,hōə·ɹd), *v.* [Dis- 6.] *trans.* To release or bring out of a hoard. So **dishoa·rding** *vbl. sb.*
1926 D. H. Robertson *Banking Policy* v. 48 Automatic Stinting which in each case exactly cancelled the intended Dis-lacking involved in the process of Spontaneous Dishoarding. **1933** —— in *Economic Jrnl.* XLIII. 401 The converse operations to Saving, Lacking and Hoarding may be called Dissaving, Dislacking and Dishoarding. Hoarding (Dishoarding) may be alternatively defined as acting in such a way as to decrease (increase) the velocity of circulation of money against output. **1955** *Ann. Reg. 1954* 258 More food was available in the towns, partly as a result of 'dishoarding' by peasants more ready to surrender their produce as consumer goods became more plentiful. **1963** *Times* 11 Feb. 6/7 Now that industry is 'dishoarding' labour.

dishonest, *v.* **3.** (Later example.)
1922 Joyce *Ulysses* 383 He would ever dishonest a woman whoso she were..if it so fortuned him to be delivered of his spleen of lustihead.

dishouse, *v.* **1.** Add: also, to deprive of a habitation. Chiefly in **dishou·sed** *ppl. a.* (also *absol.*), **dishou·sing** *vbl. sb.*
1900 *Daily News* 4 Jan. 4/1 The evil of dishousing altogether would be substituted for the evil of living in places unfit for habitation. *Ibid.* 8 Jan. 7/1 Such considerations as the fate of the dishoused. **1922** Weekly *Times* 2/3 To secure sites for the dishoused. **1922** *Ibid.* 15 Dec. 5/1 The dishousing of the inhabitants of the slums. **1921** *Contemp. Rev.* Oct. 564 Large closing orders are out of the question because of the dishousing they would occasion.

dishu·man, *a.* *nonce-wd.* [Dis- 10.] Unhuman.
1920 D. H. Lawrence *Lost Girl* vi. 94 Whether she would ever be able to take to his strange and dishuman element, who knows?

dishu·maniza·tion. *nonce-wd.* [f. Dishumanize *v.* + -ation.] The act or process of dishumanizing.
1867 W. James *Novels & Tales* XVII. Perfect..London was a terrible place to die in; doubtless not so much..by conscious cruelty or perversity as under the awful doom of general dishumanisation.

dish-washer. Add: **2.** (Examples.) orig. *U.S.*
1867 *Rep. Comm. Patents 1865* (U.S. Pat. Off.) I. 538 [The] dish-washer..consists of two disks placed one above another [etc.]. **1889** *Kansas Times & Star* 9 July, The patent new dish washer washes a bushel of dishes in a few minutes. **1921** *Daily Colonist* (Victoria, B.C.) 2 Apr. 2/6 (Advt.), Lessen the labors of your wife. Electric washing machines, electric dish washers, [etc.]. **1956** N. Shute *Beyond Black Stump* 21 A new dishwasher stood where the old one had stood. **1958** *Woman's Own* 3 Mar. 11/4 Laundry goes straight into the washing machine and so we will dread the washing up with the electric dish washer. **1971** *Which?* Mar. 69/1 Refrigerators have a simpler mechanism than automatics or dishwashers.

di·sh-wa·tery, *a.* [See -y[1].] Resembling dish-water. Also *fig.*
1890 *Columbus* (Ohio) *Dispatch* 13 Feb., The review [of a political quarrel] would be in its tenor dishwatery. **1910** W. James *Mem. & Studies* (1911) xi. 284 Mawkish and dish-watery. **1928** *Sunday Express* 22 Jan. 9/4 Then came the soup. Great greasy tins of a dish-watery liquid.

dishy (di·ʃi), *a.* *slang.* [f. *dish *sb.* 2 a + -y[1].] Very attractive.
1961 *Sunday Tel.* 19 Mar. 9/1 He encountered the dishy St. Tropezienne on his holiday plane. **1961** *Sunday Express* 10 Dec. 10 The manufacturer..dreamed up that dishy little number you're mad for. **1962** J. Gardner *Liquidator* iii. 50 'Mm, is that him?' said the girl, all velvet. 'He's dishy.' **1966** G. N. Leech *Eng. in Advertising* xxiii. 198 The dreamiest, dishiest fabrics ever. **1970** *Evening Standard* 17 Mar. 35 (Advt.), I like Peter...And his dishy new sports car.

dishybilly. Joc. corruption of Dishabille 1.
1922 Joyce *Ulysses* 418 See her in her dishybilly. Peels off a credit.

disillude, *v.* (Further example.)
1894 G. B. Shaw *Let.* 23 Apr. (1965) 427 Sergius in the play..when disilluded, declares that life is a farce.

disimpale (disimpē·l), *v.* [Dis- 6.] *trans.* To unfix from something pointed. Also *fig.*
1906 *Daily Chron.* 13 Feb. 5/4 Sir Henry..was not only disimpaled by the dome of ..? **1930** W. de la Mare *Desert Islands* 185 Though the tale-tellers, who..their fingers..sharpest of tenter-hooks, he is bound to disimpale him in the end.

disimperialism (disimpiə·riăliz'm). [Dis- 9.] The reversal of imperialism; the acquisition of independence by former imperial territories.
1959 J. Strachey *End of Empire* xiv. 214 The steady march of a voluntary, or at least largely non-violent, process of dis-imperialism. **1962** *Economist* 6 Jan. 14/2 Rhodesia, where Britain faces its most painful test of disimperialism.

disimprove, *v.* **b.** (Example.)
1904 *Jrnl. Educ.* Dec. (1960) II. 75 My voice has disimproved through want of practice and from too much smoking.

disincentive (disinse·ntiv). [Dis- 9.] A source of discouragement, esp. to economic progress or development. Also as *adj.*
1946 *Daily Tel.* 2 Nov. 4/2 In view of the urgency of the need for maximum production any disincentive is a serious consideration. **1954** *Times* 18 Mar. 9/2 It is admitted also that P.A.Y.E. presents the disincentive effects of taxation in the promptest possible manner. **1957** *Technology* July 168/1 Taxation, Mr. Thorneycroft maintained, was undoubtedly a disincentive. **1963** *Higher Educ.* xiv. 212 in *Parl. Papers 1962–3* (Cmnd. 2154) XI. 639 We think it probable that it [sc. the financing of students by loans] would have undesirable disincentive effects. **1964** *New Society* 27 Feb. 8/3 Commuting is not a disincentive for some of the men to take active part in local organizations.

discommodate, *v.* (Later example.)
1922 Joyce *Ulysses* 431, I don't think you need over excessively disincommodate yourself in that regard.

disinfest (disinfe·st), *v.* [Dis- 6.] *trans.* To rid (a person, building, etc.) of infesting insects, vermin, etc. So **di:sinfesta·tion**, the process of disinfesting.
1920 *Lancet* 1 Oct. 681/1 (*heading*) Disinfection and disinfestation in the field. *Ibid.* 681/2 The term 'disinfestation' or 'to disinfest' was employed during the war to differentiate the two processes. **1935** *Discovery* Apr. 116 Hydrogen cyanide, or HCN, is the material most generally used for fumigation or disinfestation. *Ibid.* 116/2 The task of thoroughly disinfesting slum property. **1942** *Times* 21 Sept. 5/3 Typhus, which is louse-borne,..we overcome by salutary measures of disinfestation. **1959** *Times* 7 Dec. (Agric. Suppl.) p. vi/2 The use of radiation..could be usefully applied for the disinfestation of products handled in bulk. **1966** *Listener* 10 Mar. 351/2 The irradiation method has great economic potential in..the disinfestation of dried and smoked fish, the disinfestation of dried fruits, (etc.). **1970** H. Braun *Parish Churches* xii. 225 Disinfestation is nowadays possible, but requires a desperate amount of scraping away of ruined timber before the beetle can be reached and destroyed.

disinflation. Add: **2.** The reversal of a state of monetary inflation; the return to a state of equilibrium from an inflationary state; a policy to check or reduce inflation. So **disinfla·tionary** *a.*
1947 *Economist* 22 Mar. 424/1 One test..of a disinflationary (or at least a non-inflationary) policy should be that it does not stimulate the creation of additional money. **1948** G. Crowther *Outl. Money* (ed. 2) iii. 108 In the course of the year 1947..it was necessary to find a word to describe a retreat from inflation which did not go so far as deflation; 'disinflation' was invented to serve the purpose. **1955** *Times* 17 Aug. 7/5 This might make sense if we had heavy unemployment and a favourable trade balance, but it can hardly be expected to make an effective contribution to disinflation. **1957** *Times* 17 Dec. 14/3 A high Bank rate..although disinflationary in intent, will lose much of its effectiveness if it is not backed by price stability.

disinhibition (disinhibi·ʃən). [Dis- 9.] (See quot. 1927.)
1927 G. V. Anrep tr. *Pavlov's Conditioned Reflexes* iv. 67 We are now afforded some justification for regarding disinhibition, as we did a short while ago, as being the 'inhibition of an inhibition'. **1937** *Brit. Jrnl. Psychol.* July 29 Such disinhibition, as Pavlov calls it, has been found associated with the various types of cessation or diminution of activity we have studied. **1964** J. Z. Young *Model of Brain* xvi. 284 Such a mechanism recalls the suggestion that enzymes exist in an inhibited form and that demand brings them into action by disinhibition.

disinsectize (disinse·ktaiz), *v.* [f. Dis- 6 + Insect *sb.* : see -ize.] *trans.* To remove insects from, esp. from an aircraft. Hence **di:sinsectiza·tion, disinse·ctizing** *vbl. sb.*, the removal of insects.
1937 *Times* 11 Nov. 11/6 Further study of disinsectizing agents [for aircraft] was decided on. **1947** *Shell Aviation News* CXI. 8/1 The International Sanitary Convention of Aerial Navigation 1933 and 1934..lays down aircraft disinsectization procedure. **1959** *Times* 9 Nov. 9/4 No completely satisfactory method of disinsectizing aircraft while in flight can be recommended. *Ibid.*, The two prongs of defence are vaccination and disinsectization of aircraft.

disintegrate, *v.* Add: **1. c.** To cause (a substance or an atom or nucleus) to undergo disintegration.
1920 Rutherford in *Proc. R. Soc.* A. XCVII. 394 Possibly the actual energy required to disintegrate the atom is small compared with the energy of the α-particle. *Ibid.* 395 No evidence has been obtained to show that helium can be disintegrated by the swift α-particles. **1932** Cockcroft & Walton in *Ibid.* A. CXXXVII. 229 We describe experiments which show that protons having energies above 150,000 volts are capable of disintegrating a considerable number of elements. **1942** J. D. Stranathan *Particles* v.

183 With these energetic particles it has become a relatively easy matter to disintegrate almost any atom.

2. b. Of a nucleus or particle, or a radioactive substance: to undergo disintegration (see *disintegration sense a, below); to decay (see *decay *v.* 2 c).
1904 F. Soddy *Radio-Activity* viii. 122 In the case of the disintegrating atoms the cause of the disintegration is at present unknown. It proceeds at a definite rate, a fixed fraction of the total atoms disintegrating in the unit of time, without hindrance or acceleration by any agency known... The internal energy of the chemical atom becomes for the first time knowable when it disintegrates. **1926** R. W. Lawson tr. *Hevesy & Paneth's Man. Radioactivity* i. 4 We now define a substance as being radioactive when the atoms of which it is composed disintegrate spontaneously. **1942** *Ann. Reg. 1941* 351 Those mesons born practically at rest disintegrate in the stratosphere into electrons. **1962** H. D. Bush *Atomic & Nuclear Physics* iv. 73 Which nuclei disintegrate in a particular time interval is a matter of chance. **1963** S. Tolansky *Introd. Atomic Physics* (ed. 5) xix. 321 Yukawa..postulated that the meson can spontaneously disintegrate, being in this sense radioactive.

disintegration. Add: **a.** spec. in *Nuclear Physics*, a process which a nucleus may undergo, spontaneously or under bombardment, in which it either emits one or more particles and becomes a different nuclide or else splits up into two or more smaller nuclei; also, the decay of an elementary particle; an instance of such a process. Also freq. *attrib.*
1903 Rutherford & Soddy in *Phil. Mag.* V. 446 The first stage in the disintegration of thorium is not directly into the emanation. *Ibid.* 583 Radioactive change can only be of the nature of an atomic disintegration. **1904** F. Soddy *Radio-Activity* viii. 121 The disintegration theory [of radioactivity] regards the property as due to a fixed proportion of the total number of atoms which are undergoing disintegration. *Ibid.* xii. 172 Consider the case of a disintegration series in which there is a parent element, A, disintegrating at an excessively slow rate. **1914** H. G. Wells *World set Free* i. § 1, He set up atomic disintegration in a minute particle of bismuth. **1932** Cockcroft & Walton in *Proc. R. Soc.* A. CXXXVII. 229 (*heading*) The disintegration of elements by high velocity protons. **1933** *Discovery* June 179/2 This third hydrogen has been produced artificially in the Cavendish Laboratories in Cambridge, England, and at Princeton by the nuclear disintegration process. **1942** *Ann. Reg. 1941* 351 Fermi's theory of β-ray disintegration. **1942** Stranathan *Particles* viii. 348 Most atoms emit a γ-ray photon and pass to their normal energy states long before they undergo a subsequent nuclear disintegration. **1945** *Electronic Engin.* Sept. 668/1 The enormous energy which may in certain circumstances be obtained from a chain of atomic disintegrations. **1946** *Nature* 14 Sept. 373/3 The disintegration-rate..correctly specifies the strength of a radioactive source. **1949** *Nucleonics* Dec. 49 The curie shall be defined as that quantity of any radioactive species (radioisotope) undergoing exactly 3.700×10^{10} disintegrations per second. **1954** H. Semat *Introd. Atomic & Nuclear Physics* (ed. 3) xi. 362 A more accurate determination of the half-life of the radioactive disintegration of the neutron.

c. disintegration constant, a measure of the rate of disintegration of a radioactive substance.
1926 R. W. Lawson tr. *Hevesy & Paneth's Man. Radioactivity* iii. 33 The half-value thickness is related to the absorption coefficient in the same way as the half-value period to the disintegration constant. **1962** H. D. Bush *Atomic & Nuclear Physics* iv. 75 Measurement of the change of activity with time enables the disintegration constant to be found.

disi·ntegratively, *adv.* [f. Disintegrative + -ly[2].] In a disintegrative manner, in a way that causes disintegration.
1874 *Rep. Med. Off. P.C.* 6 in *Parl. Papers* (C. 1068) XXXI. 355 A force..acting disintegratively upon organic matter.

disinterest, *sb.* **2.** Delete †*Obs.* and add later examples.
1896 *Sat. Rev.* Suppl. Christmas 4/2 We here see Morris working, with entire disinterest, at his work, and caring above all things for fine workmanship. **1905** *Globe* 19 Sept., The American Press..reproached Japan with her want of disinterest.

3. Delete *rare* and add later examples. (Cf. next.)
1900 *Pilot* 1 Sept. 283/1 The general reader may, without confessing to a heart of stone, feel a certain disinterest in weather and Mrs. Bouveries, however nice and however ill. **1904** *Sat. Rev.* 9 Jan. 34 The whole election was a model of quiet disinterest. **1940** *Economist* 19 Oct. 480/2 Peace may bring back the old disinterest and the old candour in mutual criticism which springs perhaps from friendship but does nothing to advance it. **1962** *Guardian* 9 Apr. 6/3 The days of service disinterest are over. The RAF has begun to think seriously..about the future of space. **1965** M. Morse *Unattached* i. 17 The general reaction..was a mixture of curiosity, disinterest, fear, and embarrassment.

disinterested, *ppl. a.* **1.** Delete † ? *Obs.* and add later examples. (Often regarded as a loose use.)
1928 in C. F. S. Gamble *N. Sea Air Station* xiii. 222 Being disinterested with the rest of the proceedings, I opened the file and began to read the theory of Wave Transmission. **1928** *Daily Express* 21 June 11/4 She is listless and disinterested. **1928** *Sunday Dispatch* 8 July 15/2 The English public is disinterested in its theatre. **1946** 'S. Russell' *To Bed with Grand Music* vii. 95 She was disinterested rather than credulous. **1960** I. Bennett *Delinquent & Neurotic Children* viii. 299 Parents disinterested and failed to cooperate. **1960** *Guardian* 2 Mar. 7/2, I am always annoyed when anybody refers to me as a religious poet. Could a poet be disinterested in religion? **1970** *Daily Tel.* (Colour Suppl.) 15 May 20/4 Oxford, after three successive defeats, are almost entirely disinterested in the Boat Race.

disinterestedly, *adv.* Add: **2.** Without interest or concern; unconcernedly.
1941 *Penguin New Writing* IX. 9 We lean over the stern, smoking, spitting, staring disinterestedly at the creaming green of the wake. **1955** A. Ross *Australia 55* 37 Girls, who chain-smoke or read disinterestedly.

disintoxicate, *v.* Delete ?*Obs.* and add later *fig.* examples. Hence **disintoxica·tion**, removal of the effects of intoxication. Also *fig.*
1927 *Scots Observer* 20 Aug. 9/5 'Spiritual life,' says Mr. Santayana, 'is not a worship of values whether found in things or hypostatised into supernatural powers. It is the exact opposite; it is *disintoxication* from their influence.' **1930** *Advt.*, Natural mineral water of Contrexéville. Diuretic and for disintoxication. **1935** *Mind* XLIV. 94 The lesson of that tragedy is acceptance of life, not through intoxication and excess, but through disintoxication, submission, conciliation, self-limitation. **1936** *Scrutiny* IV. iv. 381 The next stanza is the only one in the poem to be completely disintoxicated and disenchanted. **1963** Auden *Dyer's Hand* 27 In so far as poetry, or any other of the arts, can be said to have an ulterior purpose, it is, by telling the truth, to disenchant and disintoxicate.

disinvestment (disinve·stmĕnt). *Econ.* [Dis- 9.] The consumption, realization, or reduction of investment; a diminution of capital goods.
1938 J. M. Keynes in *Times* 7 Oct. 10/1 The balance-of-trade position and the net disinvestment in this country's foreign assets..also needs particular attention. **1940** *Economist* 7 Dec. 704/2 The inflow of the metal [gold] provided a vehicle for disinvestment and was a positively deflationary force. **1948** G. Crowther *Outl. Money* (ed. 2) v. 160 In the early part of a depression the gradual liquidation of these stocks [of raw materials] is an important cause of 'dis-Investment'. When it is finished..one reason for the decline in Investment is removed. **1971** *Sunday Express* (Johannesburg) 28 Mar. 15/4 Also noticeable was some disinvestment by institutions.

‖**disinvoltura** (disinvǫltū·rā). [It., f. *disinvolto* unembarrassed, f. *disinvolgere* to unwind (*volgere* to wrap).] Self-assurance; lack of constraint.
c 1847 F. A. Kemble *Records of Later Life* (1882) III. 322 Acted to the life by a woman, who moves with more complete *disinvoltura* in her men's clothes than most men do. **1899** M. Beerbohm in *Sat. Rev.* 6 May 555/1 When they [*sc.* the English] are drunk, they do, indeed, talk, and do attain even a certain degree of *disinvoltura*. **1947** A. Einstein *Mus. Romantic Era* xvi. 265 Bellini..with his elegant insouciance, with his melancholic *disinvoltura* of intellectual and physical customs.

disjunct, *a.* Add: **B.** *sb.* *Logic.* One of the components in a disjunctive proposition; also, a disjunctive proposition (see quot. 1948 and Disjunctive *a.* 2).
1921 W. E. Johnson *Logic* I. iii. 30 In the disjunctive function 'Not-both *p* and *q*', *p* and *q* are disjuncts. **1922** —— *Ibid.* II. x. 211 We shall take..the implicants and disjuncts to stand for particular propositions. **1948** H. Reichenbach *Elem. Symbolic Logic* v. 194 We can regard the corresponding combination '*F* v *G*' as one class, which is the disjunct of the classes '*F*' and '*G*'. **1954** I. M. Copi *Symbolic Logic* ii. 12 The two statements so combined are called disjuncts (or alternatives). **1962** W. & M. Kneale *Devel. Logic* iii. 160 A disjunctive proposition..was said to involve a complete opposition..of its disjuncts.

disk. The current spelling in Great Britain is now *disc, q.v.

dislocation. Add: **1. e.** *Crystallography.* A displacement of the lattice structure of a crystal.
1934 G. I. Taylor in *Proc. R. Soc.* A. July 368 The block after the unit slip, or 'dislocation' as we may call it, has passed through from left to right. *Ibid.*, The passage of a positive dislocation across a crystal from left to right produces the same effect as the passage of a negative one from right to left. **1955** *Sci. Amer.* July 81/1 We think of a dislocation as a line running through a crystal (although it is really a region of small but finite cross section). Around such a line is a region of energy higher than in the rest of the crystal. **1958** *Van Nostrand's Sci. Encycl.* (ed. 3) 526/2 Dislocations are important in determining the mechanical and electrical properties of solids, and play an important part in solid state physics. **1969** *New Scientist* 15 May (Feature Section) 6/2 The crystal merely provides a kind of 'space' in which various well-defined families of defect 'particles' exist. The most notable of these elementary defects are vacancies..; interstitials..; and dislocations.

dislove, *v.* (Later example.)
1922 Joyce *Ulysses* 50 The foot that beat the ground in tripudium, foot I dislove.

disloyalist. (Earlier U.S. examples.)
1863 in *Boston Sunday Her.* 24 May 1/3. **1870** *Congress. Globe* 7 July 5310/3 The county of Monroe [in Missouri] was the place where disloyalists fleeing from other counties took shelter all the time.

dismal, *a.* Add: **7.** *Dismal Desmond*: a toy dog with drooping ears; also *transf.*, a gloomy person; *Dismal Jimmy* (colloq.): a gloomy person.
1926–7 *Army & Navy Stores Catal.* 891/3 Dismal Desmond. An amusing, lovable puppy. **1934** A. Christie *Parker Pyne Investigates* 66 The gentleman selling Dismal Desmonds does not know what to make of it. **1938** E. Bowen *Death of Heart* II. iii. 219 A Dismal Desmond dog sat on the bed. **1939** 'M. Innes' *Stop Press* II. vii. 319 A blob of a nose, a mottled snout, a lachrymose eye, a..drooping ear—the creature is called a Dismal Desmond and is known in every nursery. **1941** A. Christie *N or M?* i. 5, I wasn't conscious of looking a Dismal Desmond. **1968** 'R. Raine' *Night of Hawk* x. 52 Larry's a character and a half. Don't let the dismal Desmond look fool you. **1927** *Melody Maker* Sept. 931/2 We are not prophets nor dismal Jimmies. **1940** H. G. Wells *All Aboard for Ararat* iii. 92 It's up in all the offices now; the Dismal Jimmy stuff is barred.

dismayingly (dismē^i·iŋli), *adv.* Delete †*Obs.* and add later examples.
1911 M. & J. Findlater *Penny Monypenny* III. iv. 316 It was now dismayingly hot. **1926** *Chambers's Jrnl.* 337/2 She would have found it dismayingly difficult to conduct the most juvenile class.

dismiss, *v.* Add: **3. c.** *Cricket.* To put (a batsman or side) out (usu. *for* a score).
1875 *Field* 22 May 501 Ten runs later Mr Longman was dismissed, and sundry changes were made in the bowling. **1892** *Times* 22 July 7 Afterwards Gunn saw the rest of his side dismissed, and took out his bat for a faultless 98. **1912** A. Brazil *New Girl at St. Chad's* vii. 111 The St. Hilary side was dismissed for sixty-seven. **1933** D. L. Sayers *Murder must Advertise* xviii. 306 The Brotherhoods were dismissed for 155, and the Pym Eleven gathered themselves together from the four corners of the field.

dismoded (dismō^u·ded), *a.* Anglicization of *démodé *a.*
1898 *Blackw. Mag.* Nov. 693/2 A tune dismoded, common. **1907** *Ibid.* Sept. 428 His ambition was the ambition now wholly dismoded, to make scholars and gentlemen. **1917** *Daily Mail* 25 Aug. 2/4 The works of Frith, Leighton, and other dismoded veterans. **1922** *Blackw. Mag.* June 806/1 The men of genius who are its peculiar glory seem dismoded to the anarchs who write in hopeless competition with them.

dismountable (dismau·ntăb'l), *a.* [f. Dismount *v.* + -able.] Capable of being dismounted. Of a gun or cannon: capable of being removed from its carriage for transport.
1711 *Fingall MSS.* in *10th Rep. Hist. MSS. Comm.* (1885) App. v. 168 The garrison had a battery..which..was not dismountable, by reason of its lowness. **1900** *Engineering Mag.* XIX. 789/2 A gun made dismountable, that it may be easily carried by men or animals. **1903** *Westm. Gaz.* 17 Nov. 9/1 The gun is provided with..dismountable shields. **1960** *Farmer & Stockbreeder* 12 Jan. 81/3 About 50 black-and-white steers are early weaned in these home-made dismountable pens along one side of the dry cows' shed.

dismutation (dismiutē^i·ʃən). *Chem.* and *Biochem.* [Dis-.] Disproportionation: usually, disproportionation involving the simultaneous oxidation and reduction of a compound in a biological context.
1926 *Chem. Abstr.* XX. 929 The great biological significance of dismutation instead of simple oxidation is emphasised. **1938** *Nature* 4 June 992/1 The principle of the hydrogen shift (dismutation), which governs the mechanism of alcoholic fermentation, has been carefully studied. **1954** *New Biol.* XVI. 65 The semiquinones probably disappear by dismutation into quinone and quinol. **1970** R. W. McGilvery *Biochem.* xiv. 270 Reactions such as this in which the oxidation of half of the substrate is counterbalanced by reduction of the other half are called dismutations.
Hence (as a back-formation) **dismu·te** *v. intr.*, to undergo dismutation.
1947 *Science News* V. 91 In neutral or slightly acid soils, tervalent manganese..dismutes forming manganese dioxide and bivalent manganese. **1957** G. E. Hutchinson *Treat. Limnol.* I. xv. 802 This material dismutes, the reaction being equivalent to $Mn_2O_3 \rightarrow MnO_2 + MnO$.

Disneyesque (di:zni,e·sk), *a.* [f. the name of Walter Elias *Disney* (1901–66), American cartoonist + -esque.] Having the characteristics or resembling the style of an animated cartoon made by Walt Disney or his company. So **Di·sneyland**, the name of a large amusement park near Los Angeles, devised by Walt Disney, applied *transf.* to any fantastic or fanciful land or place; a never-never land.
1939 Auden & Isherwood *Journey to War* ii. 63 Lady Precious Stream utters some piercing, Disneyesque sounds. **1956** A. Huxley *Adonis & Alphabet* 170 The Mother of the greeting cards inhabits a delicious Disneyland, where everything is syrup and Technicolor, cuteness and schmalz. **1961** *Guardian* 3 Nov. 8/3 There are moments when the animal kingdom threatens to turn into Disneyland. **1963** P. H. Johnson *Night & Silence* xxxiii. 237 Already all over America these pretty, Disneyesque buildings proliferated.

disome (dai·sō^um). *Biol.* [f. Di-[2] + *-some[4].] Any pair of homologous chromosomes. Hence **diso·mic** *a.*, relating to or characterized by a disome; usu. (of a haplont), having one extra chromosome which is homologous with a chromosome of the normal haploid set; also as *sb.*, a disomic organism, etc.

1921 A. F. Blakeslee in *Amer. Naturalist* LV. 259 The homologous chromosomes therefore form tetrasomes, to use a new term, instead of disomes as in normals or trisomes as in triploid plants. *Ibid.*, The following terms are suggested to designate sets with numbers of chromosomes from 1 to 12: monosome, disome, trisome, [etc.]. **1924** —— in *Proc. Nat. Acad. Sci.* X. 114 The ratios actually found resemble those in disomic rather than in trisomic inheritance. **1946** *Nature* 21 Sept. 418/1 Genetical tests based on the difference between tetrasomic and disomic ratios. **1954** *Genetics* XXXIX. 339 PWT strains, whether originating as disomics or as heterocaryotic ascospores, might be expected to have a wild phenotype. **1965** Fincham & Day *Fungal Genetics* (ed. 2) v. 120 Although aneuploidy in the nuclei of vegetative mycelium of Neurospora has not been shown to occur..the occurrence of n + 1 (or disomic) ascospores is well established.

disorderly, *a.* Add: **B.** *sb.* A disorderly person.
1852 G. C. Munday *Our Antipodes* III. ii. 80 A brace of disorderlies in handcuffs. **1855** *Illustr. Lond. News* 21 July 74/1 One of the drunk and disorderlies. **1905** *Daily Chron.* 9 Oct. 5/3 The Pope was obliged to threaten the disorderlies with expulsion from the Vatican.

disordination (disǫ:ɹdinē^i·ʃən). (Later examples.)
1896 *Pop. Sci. Monthly* Feb. 511, I would propose the term disordination, the etymological opposite of coordination. **1897** *Educ. Rev.* XIII. 52 The phenomena of hypnotic disordination or disruption of consciousness.

disorient, *v.* Delete †*Obs.* and add later examples, esp. in *diso·riented*, *diso·rienting* ppl. adjs.
1931 A. L. Rowse *Politics & Younger Generation* 21 Others..emerged from it [*sc.* the war] shaken and disoriented. **1950** W. Lowry *Let.* Jan. (1967) 189, I was trading on your..disorienting yourself from your own orbit. **1951** R. M. Williams *Amer. Society* (1952) xiv. 535 Individuals are likely to..that..until the person is literally disoriented. **1957** J. Kerouac *On the Road* (1958) 274 Where disoriented people have to go to be near a specific elsewhere. **1964** A. W. Gouldner in I. L. Horowitz *New Sociol.* 203 A disorienting normlessness.

disorientate, *v.* **b.** (Later examples.) Hence *disorientated* ppl. a.
1928 F. Brett Young *My Brother Jonathan* 582 His freedom disorientated him. **1959** *Times Lit. Suppl.* 16 Oct. 598/2 A somewhat disorientated person, who suddenly discovers that an affair has been taking place in the neighbourhood between a strange blonde girl and someone who has apparently assumed his own identity. **1960** *Times* 13 May 16/2 The talented yet disoriented adolescent. **1965** *Nursing Times* 5 Feb. 184/1 She was confused and disorientated because of time, place or person.

disorientation. Add: **2.** Also, a confused mental state, often due to disease, in which appreciation of one's spatial position, personal identity, and relations, or of the passage of time, is disturbed.
1902 A. R. Defendorf tr. *Kraepelin's Clin. Psychiatry* 15 Disorientation..is possibly a special form of a moderate clouding of consciousness. **1957** *Times* 4 Oct. 13/5 Exposure to noise levels of about 150 decibels may result in disorientation, nausea and vomiting. **1962** Henderson & Gillespie *Text-bk. Psychiatry* (ed. 9) vi. 107 Simple inattention..may cause partial disorientation.

disowner (disō^u·nəɹ). [f. Disown *v.*] One who disowns.
1805 J. Smith *Perm. Mess. Exodus* ix. 126 The disowners of God.

disparition. Delete †*Obs.* and add later examples.
1922 Joyce *Ulysses* 689 The disparition of three final stars, the diffusion of daybreak. **1940** V. Woolf *Writer's Diary* (1953) 336 This disparition of an echo.

dispatch, *sb.* **12.** Add: *dispatch-carrier*; **dispatch-boat** (earlier example); **dispatch money** (see quot. 1923); **dispatch note**, a memorandum required to be made in addition to the customs declaration for foreign parcel post; **dispatch-rider**, one who rides on horse-back, bicycle, or motor-cycle to carry dispatches; so **dispatch-riding**; **dispatch-vessel** = *dispatch-boat*.
1794 *Deb. Congress U.S.* 20 Mar. (1849) 1482 The President..[shall] be authorized to employ, as dispatch-boats, such of the revenue cutters of the United States, as the public exigencies may require. **1905** *Daily Chron.* 10 Jan. 3/1 A dispatch-carrier for the Foreign Office. **1878** *Law Jrnl. Rep.* Q.B. XLVII. 513/1 If the Court shall be of opinion that..despatch-money..is to be paid per working day or day of twelve hours. **1923** De Hart & Bucknill *Maclachlan's Merchant Shipping* (ed. 6) 428 Despatch money is a payment which the Shipowner agrees to make for time saved out of the lay-days. **1892** *Post Office Guide* Oct. 383 Parcels for Foreign Countries must..be accompanied by a Despatch Note. **1968** *Ibid.* 113 Despatch notes can be obtained beforehand from any post office which accepts parcels for abroad. **1890** *Daily News* 16 Oct. 7/2 A report brought from the north by dispatch riders. **1914** Wrybong. **1951** *Oxf. Jun. Encycl.* IV. 267/1 The army dispatch-rider..proves invaluable for messages which cannot be sent by any other method. **1907** *Daily Chron.* 10 Jan. 3/1 An interesting dispatch-riding test from Newcastle and London to Manchester. **1809** *Deb. Congress U.S.* 21 Feb. (1853) 432 There may be time for the despatch vessel to go to France and return.

dispatcher. Add: **1.** *spec.* (*a*) *N. Amer.* = *train-dispatcher* (see Train *sb.*[1] 22 b); (*b*) (see quot. 1954).
1878 A. Pinkerton *Strikers* xviii. 219 The conductor..promptly passed the dispatcher. **1931** H. F. Pringle *Th. Roosevelt* II. 441 Orders were telephoned to the chief dispatcher of the Pennsylvania Railroad in New Jersey. A special train was to be made up. **1953** D. Maclean *Eastern Approaches* II. viii. 299 The dispatcher beckoned to us and Vivian and I..made fast the static lines to the special hooks on the inside of the fuselage. **1954** A. W. Fielding *Made to Sink* 227 The 'plane's despatcher—the sergeant responsible for launching the personnel and containers out of the body of the machine.

dispense, *sb.*[1] Restrict †*Obs.* to senses in Dict. and add: **3. b.** In full *dispense bar*. A bar in a club or hotel for the use of staff.
1934 C. L. Anthony *Touch Wood* I. 27 A bar! They called it the dispense bar. **1961** *Daily Tel.* 7 Apr. 25 At the club..no intoxicating drinks were to be supplied after 2 a.m., except in the dispense bar.

dispenser. Add: **3. b.** Also, one who dispenses a commodity.
1881 W. Carleton *Lightning-rod Dispenser* 3 This rail-road smash reminds me..Of a lightning-rod dispenser that came down on me one day. **1925** *Daily Tel.* 13 May 20 Soda water dispenser required, to supervise six fountains.

5. A container that dispenses an appropriate measure of a commodity (usu. with defining word).
1947 Steinbeck *Wayward Bus* 104 He passed the sugar dispenser politely to her. **1959** *Brit. Pat.* 694,333. This invention concerns a new and improved dispenser for..dispensing..a household dispenser. **1959** *Engineering* 27 Feb. 267/1 Ice-cream dispensers have been installed in a number of factories. **1960** *Woman's Own* 2 Jan. 43/3 Metal detergent dispenser. **1970** *New Scientist* 30 Apr. 244/1 Looking forward to hitting Dubois City (population 300) with its liquid ice dispenser.

dispermy (dai·spəɹmi). *Biol.* [f. Di-[2] + Gr. σπέρμα seed + -y[3].] The entrance of two spermatozoa into a single egg. Hence **di·spermic** *a.*
1896 E. B. Wilson *Cell* 335 Dispermy. **1905** *Rep. Brit. Assoc.* 413 Dispermic eggs. **1956** *Nature* 3 Mar. 429/1 The egg was presumably dispermic. *Ibid.*, This evidence supports the postulate of dispermy. **1965** *New Scientist* 1 Apr. 35/2 The condition known to biologists as dispermy, in which two sperms fertilise one egg.

dispersal. Add: **b.** *concr.* in *Aeronaut.* One of several stations, situated at scattered points round an airfield, in which aircraft are parked in order to minimize losses from air attack. Also *attrib.*
1940 *Flight* 7 Nov. 1/1 The machine is shown at its dispersal point near the aerodrome, where, when not in use, it rests in order to minimise the effect of any attack on the aerodrome. **1942** T. Rattigan *Flare Path* iii. She only saw it..at dispersal on the ground, the runway for a..dispersal. **1943** *Daily Tel.* 24 June 6 At Maupertuis airfield bombs burst in the dispersal area. **1949** P. Brickhill & C. Spitfires over Malta iii. 81 Sitting in dispersal, we discussed..**1956** 'N. Shute' *Rainbow & Rose* i. 22 There's a..taxiing to dispersal and handing over to the ground crew took a bit of time. **1966** J. Beedle 43 *Squadron* xiii. 147 The last of the four Hurricanes was ready to taxi back to dispersal.

dispersant (dispə̄·ɹsănt). [f. Disperse *v.* + -ant[1].] An agent which causes a substance to form a dispersion in the surrounding medium, or which helps to maintain an existing state of dispersion; a dispersing agent.
1944 C. W. Gregory in *Petroleum Refiner* Dec. 504/1 To complete the petroleum dictionary, it seems advisable to coin..Dispersant—An agent for imparting dispersing powers to a dispersing agent or medium (noun). **1948** *Oil & Gas Jrnl.* 8 Jan. 58/1 Dispersants are defined as additives which prevent the deposition of sludge or varnish. **1968** *New Scientist* 5 Dec. 560/1 A dispersant..C.1.25 has been fitted with a boom under each wing for the spraying of a 'dispersant' on oil slicks. **1971** *Sunday Times* (Johannesburg) (Business Section) 28 Mar. 15/3 (Advt.), Governments can today be supplied with oil slick dispersants which will minimize toxic hazards.

disperse, *v.* Add: **2. a.** *spec.* To scatter or station (ships, aircraft, etc.) at separate points in order to minimize losses from air attack. Also *intr.*
1941 *Manchester Guardian* 24 Apr. 8/3 We took the decision to disperse plant over the countryside..and dispersed them..were carried out on a very daring basis. **1944** *Return to Attack* (Army Board, N.Z.) 12/2 During the day vehicles were 'dispersed' 150 to 200 yards apart which meant that no attractive target was presented to enemy aircraft. **1948** *Daily Tel.* 9 Dec. 6 The fleet is well dispersed..The dropping of the 'atomic bomb' flash will be reduced by the spread of the ability of the fleet to disperse in the face of that all-powerful weapon.

9. *Chem.* The verb-stem is used *attrib.* in **disperse phase, system** (see *dispersed*); hence **disperse·oid** = *disperse system*.
1915 E. W. Washburn *Princ. Physical Chem.* xxv. 361 If we imagine any phase within a given system to be separated into smaller and smaller particles, then as the size of these particles gradually decreases the surface of contact between this phase and its neighbors will correspondingly increase and the effects due to the nature of surface tension..will gradually become more apparent, and these surface forces will eventually begin to be an important factor in determining the properties of the phase. Such a system is called a disperse system. Whenever this situation exists to an appreciable extent..we have what is called a disperse system or a dispersoid. **1927** Crocker & Matthews *Theoret. & Exper. Physical Chem.* 273 *Disperse phase*, the discontinuous constituent of a colloidal solution corresponding to the solute in true solution. *Disperse system*, any colloidal solution. A two-phase system with greatly developed surface. **1954** S. C. Blackton *Dual* vi. 136 Dusts..will here be regarded as belonging to that section of the colloidal state named aerosols, and consisting of solid disperse phase in gaseous dispersion medium. **1939** *Thorpe's Dict. Appl. Chem.* (ed. 4) III. 292/1 In general the amount of solid substance dispersed or peptized varies with the amount of solid phase present, reaching a maximum for medium quantities. This relationship was termed the..solid-phase rule, but..disperse-phase rule is a more suitable name. **1962** *McGraw-Hill Encycl. Sci. & Technol.* IX. 585/2 Condensed dispersoids and fine mechanical dispersoids generally tend to flocculate or agglomerate.

dispersed, *ppl. a.* Add: **c.** *dispersed phase, system*, = *disperse phase, system* (see prec.).
1915 E. W. Washburn *Princ. Physical Chem.* xxv. 362 The degree of dispersion of a dispersed phase is usually defined as the ratio of its surface to its volume. **1949** S. Clayton *Theory Emulsions* 1 That liquid which is broken up into globules is termed the dispersed phase. **1924** *Thorpe's Dict. Appl. Chem.* (ed. 4) III. 279/1 Such dispersed systems can be classified into..(a) mechanical suspensions, (b) colloidal solutions, (c) molecular solutions.

dispersedness. (Later examples.)
1867 *Atlantic Monthly* LXXX. 524 It..gives added cohesion to a great institution whose topographical dispersedness is surpassed only by its enormous enrollment. **1917** A. S. Pringle-Pattison *Idea of God* viii. 139 The finite facts in their dispersedness and multiplicity seem to be utterly contingent.

dispersion. Add: **6.** So *dispersion* = the degree of scatter of values in a set of observations.
1867 *Jrnl. R. Statist. Soc.* LX. 865 Notice that no term measuring the dispersion of weights enters in the formula; but that the approximation was made on the assumption that the weights were equal. **1930** *Engineering* 13 June 756/1 The observations were characterised by a large dispersion, which was attributed to the numerous factors involved. **1963** R. Folard *Instrum. Nucl. Reactors* vii. 70 A commonly used measure of dispersion..is the standard deviation, which is defined as the square root of the mean of the squared deviation values.

6*. *Physical Chem.* A type of intimate mixture in which one substance is present in a large number of separate small regions distributed throughout another, continuous, substance; examples are emulsions (one liquid in another) and aerosols (a solid or a liquid in a gas); also, the state of being so distributed. Also *attrib.*, as **dispersion medium**, a substance containing another substance 'dispersed' in it.
1913 [see *dispersed *ppl. a.* c]. **1919** E. Hendrick *Chem. Everyday Life* 74 Soap is a colloid, and when we get a little of it in a great deal of water we have it in dispersion. **1924** A. Findlay *Physical Chem.* 173 The colloid sol..consists of finely divided particles (the dispersed phase) distributed throughout a dispersion medium. **1927** Crocker & Matthews *Theoret. & Exper. Physical Chem.* 274 *Dispersion medium*, the continuous constituent of a colloidal solution, corresponding to the solvent in true solution. **1944** *Petroleum Refiner* Dec. 504/2 The property of maintaining insoluble matter (sludges and contaminants) in suspension in the oil so that they will not settle out. **1957** *Encycl. Brit.* VI. 31/2 A foam is a gaseous dispersion (usually of air) in a liquid continuum. *Ibid.*, Pastes are concentrated dispersions of fine solid particles in a liquid continuum. **1958** *Times Rev. Industry* May 69/1 The main dispersion will probably use 500 gallons of emulsion paint in an hour.

7. dispersion hardening *Metall.*, a process of ageing produced by heating at high temperatures.
1932 *Metallurgist* VIII. 110/1 The hardening depends on the degree of dispersion of the particles, and it is accordingly sometimes known as dispersion hardening. **1934** H. O'Neill *Hardness of Metals* vi. 195 'Dispersion hardening' and 'precipitation hardening' are good labels to employ for the phenomenon in general.

dispersu·ade, *v.* *nonce-wd.* [f. Dis- + Persuade *v.*] 3.
1952 Auden *Nones* 17 Nothing..can..dispersuade the Furies.

dispireme (daispai·rīm). *Cytology.* Also **-em.** [f. Di-[2] + *spireme.] (See quot. 1896.)
1890 W. Turner *Cell Theory* 31 In the..dispirem stage, the chromatin threads thicken and shorten. **1896** E. B. Wilson *Cell* 336 *Dispireme*..that stage of mitosis in which each daughter-nucleus has given rise to a spireme.

displaced, *ppl. a.* Add: **b.** *displaced person*, one removed from his home country by military or political pressure, esp. a non-German compelled to work in Germany in the 1939–45 war, and thereafter homeless. Abbrev. D.P.

1944 *Sat. Even. Post* 22 July 14/1 The Refugees..or, as they term these people here..the Displaced Persons. **1945** *Broadcaster* (U.S.) June 7 The real difficulty was and is the care of the slave laborers, men, women and children the Germans had imported from all over Europe to do their work for them. These we call Displaced Persons and for brevity refer to them as DP's. **1946** *Ann. Reg. 1945* 64 The huge number of 'displaced persons' who were still wandering about. **1961** L. Mumford *City in History* vi. 198 The very migrations of prisoners, slaves, refugees, displaced persons, all widened the bonds of human association.

displacement. Add: **2. d.** (Earlier example.)
1873 J. C. Maxwell *Electr. & Magn.* I. i. 60 The electric polarization of an elementary portion of a dielectric is a forced state into which the medium is thrown by the action of electromotive force, and which disappears when that force is removed. We may conceive it to consist in what we may call an electrical displacement, produced by the electromotive force.

e. displacement law, any of three laws in *Physics*: (i) that the wavelength at which a black body radiates most energy is inversely proportional to its absolute temperature; (ii) (see quot. 1922); (iii) that when an atom emits an alpha particle its atomic number decreases by two, and when it emits a beta particle its atomic number increases by one.
1904 T. Preston *Theory of Heat* (ed. 2) vi. 600 This important equation expresses the fact that if radiation of a particular wave-length corresponding to a definite temperature is adiabatically altered to another wave-length, then the temperature changes in the inverse ratio. This is known as Wien's displacement law. **1922** *Encycl. Brit.* XXXII. 561/1 Perhaps the most comprehensive connexion of spectra with the periodic table is established by the 'displacement law' of Kossel and Sommerfeld..The displacement law states that, when an element is ionized, the enhanced series have the same type of complexity as the arc series produced by the element to the left (i.e. in the preceding group) in the periodic table. **1926** R. W. Lawson tr. *Hevesy & Paneth's Man. Radioactivity* xiii. 114 In this way we obtain the following displacement law: The emission of α particles by the element results in the production of an element situated two places lower down in the natural sequence of the elements, whereas when the disintegration element emits β rays, an element situated one place higher in the system. **1927** Rutherford in *Proc. Physical Soc.* XXXIX. 360 This generalisation, known as the displacement law, first put forward in general form by Russell, Fajans and Soddy, gives us at once the atomic number and chemical properties of each element in the radioactive series, and thus fixes the ordinary physical and chemical properties of each radioactive element, as well as its atomic weight. **1968** M. S. Livingston *Particle Physics* ii. 16 Soddy and others established the displacement law.

f. *Psychol.* The substitution of one idea or impulse for another, as in dreams, obsessions, etc.; the unconscious transfer of intense feelings or emotions to something of greater or less consequence; also = *sublimation 5 c.
1913 A. A. Brill tr. *Freud's Interpr. Dreams* vi. 286 The result of this displacement is that the dream content no longer resembles the core of the dream thoughts, and the dream reproduces only a distorted form of the dream-wish. *Ibid.* vi. 299 The dream activity makes use of the displacement of psychic intensities up to the transvaluation of all psychic values. **1955** H. C. Abraham tr. *K. Abraham's Clin. Papers* III. vii. 189 We should transgress the rule if we attempted to make an arbitrary separation between displacement and non-alternative..object choice. **1962** Henderson & Gillespie *Text-bk. Psychiatry* (ed. 9) vii. 118 Displacement is a term given to the transfer of the emotional setting of one idea to some other apparently insignificant idea. **1964** Gould & Kolb *Dict. Social Sci.* 204/1 'Fiddling' with objects such as pencils is a masturbation substitute or displacement. *Ibid.* 204/2 In social psychology the concept *displacement* has been used with respect to the hostility expressed toward minority groups.

g. Applied *attrib.* to an activity or behaviour pattern occurring outside its normal context and arising from a conflict of impulses (see quots.).
1947 E. A. Armstrong *Bird Display & Behaviour* ii. 36 Toying with odds and ends may be regarded as..providing emotional outlets. They are 'displacement' or 'substitute' activities, common enough in human life. *Ibid.* viii. 106 'Displacement' activities will be regarded as a special type within the larger class of false activities. **1951** N. Tinbergen *Study Instinct* v. 114 It is a well-known fact that displacement activities often occur in a situation in which the fighting drive is thwarted..drive to escape are both activated. *Ibid.* 116 Herring gulls and many other birds..feed their young displacement ..If it is not at all improbable that this..is displacement feeding. *Ibid.* 117 If male sticklebacks are forced to nest very closely together, they will show nearly continuous displacement digging. *Ibid.* viii. 210 In man learned patterns, like lighting a cigarette, handling keys or handkerchief, &c., often act as displacement activities. **1957** *New Scientist* 26 Sept. 42 Displacement activities usually result from a conflict between two strongly activated but antagonistic drives, or from strong motivation, usually sexual, in the absence of the necessary stimulus required for the release of the consummatory act. **1962** *Listener* 1 Aug. 169/1 Many such irrelevant actions are known in animals. Dr Tinbergen has called them 'displacement activities'. They are seen in conflict situations, where an animal's automatic behaviour mechanism cannot work smoothly in the given environment. *Ibid.* 169/2 Human displacement activities are not easily catalogued.

3. (Later example.)
1935 C. G. Burge *Complete Bk. Aviation* 263/1 Displacement, the amount of fluid displaced by the gas in an aerostat.

4. *Bot.* Abnormality in the position or form of a leaf or organ.
1869 M. T. Masters *Veget. Teratol.* 89 Instances of displacement of leaves arising from suppression.

display, *v.* Add: **5. c.** *Ornith. intr.* To engage in or use display (see *display *sb.* 2 b).
1902 *Zoologist* 4th Ser. VI. 197 The displaying bird crouched, upon which the pairing took place. **1936** *Nature* 27 June 1057/2 Birds of many species pair before they display and often pair for life. **1953** Bannerman *Birds Brit. Isles* II. 188 A male coal tit displaying to an apparently disinterested female with her back turned to the ardent suitor. *Ibid.* 292 To watch a male gold-crest displaying to the lady of his choice.

display, *sb.* Add: **1. c.** The presentation of radar echoes or signals on the screen of a cathode-ray tube; a visual presentation of data from a computer, whether by means of a cathode-ray tube or some other device; also, a device or system used for this. Also *attrib.*
1945 *Electronic Engin.* XVII. 684 If the target is out of sight the Radar display panel is used. *Ibid.* 716 Photographs of the cathode-ray tube face showing..the main display. **1946** *Ibid.* XVIII. 165 The I.F. rather than the video is fed from the main console to the display units. **1947** L. J. Haworth in L. N. Ridenour *Radar System Engin.* vi. 173 Target range is displayed as a horizontal coordinate and the display is expanded in the vertical dimension. *Ibid.*, The signals from a given target are concentrated in two 'pips'..the basis of range and azimuth position. **1958** *Listener* 30 Oct. 691/1 New displays with much higher power, on which the display was sometimes covered with layer displays. **1962** R. S. Ledley *Digital Computer & Control Engin.* xxii. 736 Another scheme is to use an electroluminescent surface. Wires buried in the surface can be controlled by the computer to generate voltages, and hence light, as desired for picture displays. **1964** Huskey & Korn *Computer Handbk.* xviii. 21 The number of bits..has been extended to its limit..the sign of the display is the number..**1967** *Technology Week* 20 Feb. 22/1 A Univac developmental effort is in progress..to discover whether future command and control display systems for manned space flight should use all-digital techniques. **1969** *Times* 24 Jan. 21 Advanced alpha-numeric display system, with a..display unit as a source of the..display.

2. b. *Ornith.* A specialized pattern of behaviour used by birds as a visual means of communication, often in conjunction with characteristic calls. Also *attrib.*
1901 *Zoologist* 4th Ser. V. 344 Whether it was a conscious display or not..the bird seemed to be enjoying the..reflex action. **1910** *Ibid.* 4th Ser. XIV. 201 One may go further..to call them 'sexual' display..that is, in order to better adapted to show off the beauties of their plumage. **1914** *Proc. Zool. Soc.* 524 The Display Ceremonies..seem so very like the 'display' during courtship. **1923** *Brit. Birds* XXVII. 34, I had an opportunity of seeing the display of the male. **1937** *Ibid.* XXX. 274 It seems that 'display-building'..is being regarded as a manifestation of sexual excitement..shown by the Great Crested Grebe. **1938** *Ibid.* XXXII. 86 The display flight was similar to that of related species. **1949** *Ibid.* XLII. 110 He immediately displayed the display attitude, exposing the throat at very close quarters and not attempting to touch one. **1953** Bannerman *Birds Brit. Isles* I. 23 A description of the spring display of the hooded crow has been given by the Misses Baxter and Rintoul, who watched a bird jumping a little distance from her..ruffling display. *Ibid.*, The aerial display by a pair of Golden Eagles..looking for 'figure-of-eight' flights. **1964** A. L. Thomson *New Dict. Birds* 203/2 When two or more incompatible behaviour..occur..and some instances of this have been specialized in evolution for a communicatory function and this may be designated 'display'.

5. (sense 4) *display-ad* (colloq.), *-face*, *-heading*; **display-case**, = Case *sb.*[2] 7 a) in which items are displayed for inspection; also **display cabinet**; **display hand**, (*a*) one who sets up display-type; (*b*) a pyrotechnist employed chiefly to assist in firework displays; **display-letter** (example); **display lighting**, lighting used to illuminate objects, buildings, etc., on display; **display-type** (examples); **display window** orig. *U.S.*, a large shop-window in which merchandise is displayed.
1929 Mencken *Amer. Lang.* v. 160 Want-ad, display-ad. **1930** A. Bennett *Imperial Palace* xxxvi. 250 Display-ads of the kind you're putting now in London dailies. **1933** *Connoisseur* Nov. 328/1 Two Chippendale display cabinets in the Chinese taste. **1965** M. Spark *Mandelbaum Gate* iii. 73 The crib-figures..were spread about on the glass top of a display cabinet. **1851** *Ibid.* 16 Dec. 2/6 (*Advt.*), A glass counter and display-case full of women's things, gloves, belts, cosmetics. **1929** 'N. Shute' *Trains like Alice* 328 A flower piece..in a glass display case. **1896** *Daily News* 3 Dec. 12/7 (Advt.), Compositor.—First-class jobbing and Display Hand seeks situation. **1921** *Dict. Occup. Terms* (1927) § 148 *Display hand*..; pyrotechnist: assists at firework display, lighting fuses [etc.]. **1907** 'Mark Twain' in *N. Amer. Rev.* Jan. 4 According to the display-heading—'Rich Woman Fell Down Cellar'. **1919** R. Skinner *Making Type Work* 33 Two Many Capital Letters in a Display Heading Confuse the Eye. **1894** W. B. Wood *Recoll. of Stage* xiii. 431 Professional displays of character and form of newspaper headlines, titlepages, and headings or displayed advertisements. **1856** *Daily News* 1 Dec. 12/7 (Advt.), Compositor..**1951** M. Hessinger *Student's Guide Commerc. Art* ii. 80 Display faces..comprise the larger sizes of letters used for newspaper headlines, titlepages, and headings or displayed advertisements. **1869** *Boston Herald* 15 Mar. 4/1 Most of the chief stores in the city..a display window. **1890** *Ibid.* 13 Nov. 4/1 The store was..The display window..a yard long. **1950** *Ann. Reg.* 1949 465 Restrictions were still imposed on shop-window displays, advertisements, lighting. **1865** *Boston Herald* 15 Mar. 4/1 Most of the chief stores..contain one or more display windows..typical of the display-window. **1967** Karch & Buber *Offset Processes* 536 *Display type*, the largest and specially designed type faces used to attract attention. **1934** Webster, Display window. **1945** Steinbeck *Cannery Row* xi. 47 The final climax came with the front of Holman's bootery broken out and the party trying on shoes in the display window. **1964** C. Buchanan *Traffic in Towns* 67 Conventional shopping streets with display windows facing pavements.

b. *Mus.* Designating a piece of music that specially displays the performer's skill or virtuosity.
1959 *Listener* 9 July 76/3 This is a display piece; it shows off an orchestra's virtuosity. *Ibid.* 17 Dec. 1093/2 Hindemith's thirty-four-year-old Concerto for Orchestra, a rollicking piece of display music. **1961** *Ibid.* 9 Nov. 789/3 Such an outburst of the Zerbinetta..unhappily preserves the most regrettable features of the old display aria.

disponible, *a.* Add: (Examples.) Cf. F. *disponible*. Also *absol.* as *sb.*
1899 G. Meredith *Let.* 31 May (1970) III. 1328 You are more disposable and should decide to come and stay. **1909** H. James *Novels & Tales* III. p. vi, He [*sc.* Turgenev] saw them, in that fashion, as *disponibles*, saw them subject to the chances, the complications of existence. *Ibid.* v. p. xviii, That extremely disponible figure of Christina Light whom I had ten years before found left on my hands at the conclusion of 'Roderick Hudson'. **1912** Conrad *'Twixt Land & Sea* (1914) I. 11 He gave me the names of all the disponible ships with their tonnage and the names of their commanders. **1961** R. C. Knight tr. *Bonnard's Gr. Civ.* III. i. 17 Euripides is open to every influence..[He] is always disponible. He does not know how to forget his feelings and other people's..when any situation touches him too deeply. **1965** *Punch* 24 Mar. 447/3 One's picture of the upper-class Parisian..ardent, informed, *disponible*, never in the way or out of it.

disposable, *a.* Add: **2. b.** *disposable income* (see quots.).
1948 *in Amer. Speech* (1951) XXVI. 292/2 'Disposable income' then, represents the money actually available for consuming power, for savings and for investments. **1950** T. H. Marshall *Citizenship & Social Class* 49 Those whose disposable income and capital do not exceed (420 and £500 respectively..'Disposable' means the balance after considerable deductions have been allowed for dependants, rent, ownership of house and tools, and so forth. **1971** *Daily Tel.* 23 July 17 Savings have been running at 9 p.c. of total personal disposable income.

3. Applied to an article designed to be thrown away after one use. Hence *absol.* as *sb.*
1943 L. E. Holt *Care & Feeding of Children* (ed. 16) i. 5 The disposable paper diapers are a great convenience and involve relatively little expense. **1951** A. A. Woodman *Sunday Express Baby Bk.* iii. 34 There are several brands of disposable napkins. **1960** *Which?* Apr. 81/2 It [*sc.* a vacuum cleaner]..could easily be used with disposable bags. **1965** *Nursing Times* 5 Feb. 181/2 Wall cupboards..for the storage of disposables have been included. **1966** *Sunday Times* (Colour Suppl.) 25 July 8/1 A lot of ward equipment is disposable now—things like catheters or blood drips..Bed pans and bottles are made of paper-mache, to be disposable. **1970** *Guardian* 9 Apr. 11/2 Visitors to the Design Centre's exhibition of disposables..have been getting..steamed up about its implied refuse problems.

disposedly, *adv.* Add: In later instances echoing the 'high and disposedly' of quot. 1610 in Dict., = with lofty dignity.
1904 R. J. Farrer *Garden of Asia* 187 Go-betweens conduct the negotiations, high and disposedly as Queen Elizabeth. **1927** J. H. McCarthy *Needles & Pins* vii. The company..marched up the aisle very disposedly. **1934** Wodehouse *Leave it to Psmith* ix. § 5. 195 Through the belt of rhododendrons a figure..made their whittle, moving high and disposedly in the direction of the back premises.

disposition. Add: **11.** *Comb.* in *Philos.*, as *disposition-concept, -term, -word*.
1936 R. Carnap in *Philos. Sci.* III. 440 Let us consider the question whether the so-called *disposition-concepts* can be defined. *Ibid.* 448 In the case of a disposition-term, the reduction cannot be replaced by a definition. **1949** G. Ryle *Concept of Mind* ii. 44 Many disposition-concepts are determinable concepts. **1958** *Aspects of Translation* 54 One may use the word 'disposition' for this kind of property, and may say..that dispositional statements..**1965** P. Caws *Philos. Sci.* viii. 54 Carnap..reinterprets the applicability of disposition-words.

dispositional, *a.* Delete *rare* and add examples. Hence **dispositiona·lity**; **dispositionally** *adv.*
1908 W. R. B. Gibson *Probl. Logic* xii. 129 We might refer to them as dispositional possibilities..To say that 'S can be P' is to say that S is dispositionally capable of P. **1909** W. M. Urban *Valuation* ii. 53 Actualisation of dispositional tendencies by acts of presumption, judgment, and assumption. **1921** R. M. Jones *Later Periods of Quakerism* II. xxiv. 970 In spite of these dispositional tendencies she rendered a remarkable service to the Society of Friends. **1949** G. Ryle *Concept of Mind* ii. 43 When we describe glass as brittle, or sugar as soluble, we are using dispositional concepts. To possess a dispositional property is not to be in a particular state, or to undergo a particular change; it is to be bound or liable to be in a particular state, or to undergo a particular change, when a particular condition is realized. **1956** A. J. Ayer *Probl. Knowl.* iv. 137 The word 'remember' is used dispositionally, and to say that one can properly be said to remember things that one is not at the moment thinking of. **1963** *Listener* 21 Nov. 831/1 One way of a dispositional verb like 'can'. **1963** A. P. Ayer *Introd. Philos. Sci.* ii. 31 The so-called dispositional predicates..designating a property to react in a characteristic way to a certain kind of stimulus. **1964** *Amer. Philos. Q.* I. 218 Goodman also construes dispositionality to be problematic or mysterious.

DISPOSSESSED

dispossessed, *ppl. a.* Add: Also *absol.*
1901 *Munsey's Mag.* Apr. 411/2 Throughout Ireland, on the whole, Protestants are the possessors, Catholics the dispossessed. 1909 *Englishwoman* Apr. 250 Woman rebelled because she belonged to the classes of the dispossessed. 1944 H. Laski *Love on Super-Tax* i. 5 Spring as it came to the needy and the dispirited, to the fallen and the dispossessed.

dispropo·rtionate, *v.* *Chem.* [Back-formation from next.] *intr.* To undergo disproportionation.
1934 in Webster. 1946 *Nature* 9 Nov. 673/1 The monoethoxy compound..is unstable at room temperature, disproportionating rapidly into SiF_4 and $(C_2H_5O)_2SiF_2$... The triethoxy compound is the most stable of the three, having no tendency to disproportionate even at temperatures near its boiling point. 1950 A. F. Wells *Structural Inorg. Chem.* v. 144 Many lower halides disproportionate into a mixture of metal and higher halide. 1964 J. W. Linnett *Electronic Struct. Molecules* iii. 51 The ion contains an odd number of electrons and, in this environment, does not disproportionate to O_2 or O_2^-, both of which contain an even number of electrons.

disproportiona·tion. *Chem.* [f. Disproportion *sb.* or *v.* + -ation.] A transfer of atoms or valency electrons between two or more identical molecules or ions, resulting in molecules or ions containing the same elements in different proportions or in different oxidation states.
1929 E. E. Reid *Coll. Org. Chem.* xxxiv. 550 Reactions of this sort have received the name 'disproportionation'. 1939 *Jrnl. Amer. Chem. Soc.* LXI. 2769/2 The identification of tri-*p*-tolylmethane..and a polymeric residue as the products of disproportionation of tri-*p*-tolylmethyl..makes it possible to suggest a mechanism for the reaction. 1946 *Nature* 9 Nov. 673/1 The disproportionation proceeds to about one third of completion, at which point an equilibrium is set up. ..Disproportionation occurs during the fluorination. 1960 *Times Rev. Industry* Sept. 55/1 Disproportionation of the monochloride into metallic aluminium. 1969 *Nature* 19 Apr. 257/1 Disproportionation of $MSiO_3$ compounds into M_2SiO_4 olivine yields SiO_2.

disqualification. **1.** (Examples in Sport and Motoring.)
1887 Bury & Hillier *Cycling* xi. 297 Furnishing false, misleading or incomplete information shall be a ground for disqualification [for cycle races]. 1899 J. D. Champlin *Young Folks' Cycl. Games & Sports* 695 Any contestant..is liable to disqualification—subject to the discretion of the referee. 1935 *Encycl. Sports* 226/2 Disqualification..may ensue from a number of causes and may be permanent or temporary. 1954 F. C. Avis *Boxing Ref. Dict.* 30 Disqualification, the dismissal from a contest through flagrant or persistent fouling. 1971 *Times* 21 Apr. 6/5 The disqualification [from driving] was set at six months. *Ibid.*, A driver..appealed successfully, against the disqualification order.

disqualify, *v.* **b.** (Examples in Sport and Motoring.)
1887 Bury & Hillier *Cycling* xi. 299 It shall be the duty of each competitor to see that he starts from his proper mark, and in default he may be disqualified for the race in question. 1897 *Encycl. Sport* I. 139/1 He [sc. the referee at boxing] can disqualify a competitor in the event of a foul. 1898 *Ibid.* II. 221/2 If a horse or his jockey jostle another horse or jockey, the aggressor is disqualified. 1906 *Times* 31 July 5/5 Partisans..assisted him to the finish, and the judges had no alternative but to disqualify him. 1963 Bloodgood & Santini *Horseman's Dict.* 63 An animal may be disqualified for a variety of reasons because either with, or without intent to defraud, it has been incorrectly described; because of carrying the wrong weight; of being doped, etc. 1971 *Times* 1 Apr. 4/4 He also pleaded guilty to driving a bus while disqualified. 1973 *Times* 19 Apr. 1/2 Mr. [Stirling] Moss..was disqualified for six months and fined £35 by magistrates at Thame, Oxfordshire, for two offences.

disquieten (diskwai·ĕt'n), *v.* [f. Dis- 6 + Quieten *v.*] = Disquiet *v.*
1921 *Glasgow Herald* 9 Mar. 10 Her condition is disquietening. *Ibid.* 23 Sept. 7 In view of disquietening reports as to their activities in the Near East. 1928 L. Rossiter *Sex Age* viii. She turned, bewildered and disquietened, from the sickening turmoil.

disquietingly (diskwai·ĕtiŋli), *adv.* [f. Disquieting *ppl. a.* + -ly^2.] In a disquieting manner; also used to introduce a statement of a fact considered disquieting.
1901 'L. Malet' *Sir R. Calmady* iv. vii. 346 The creaking of a board, the rustle of a curtain,..were..disquietingly replete with possible meaning. 1922 *Public Opinion* 23 June 580/1 A tendency to change strangely and disquietingly. 1928 *Britain's Industr. Future* (Lib. Ind. Inq.) iv. xxvii. 398 In some industrial centres there is a disquietingly high percentage of registered unskilled workers between the ages of eighteen and twenty-five. 1970 *Nature* 28 Nov. 818/1 Disquietingly, no major known human teratogen..has been identified by prospective epidemiological analyses.

Disraelian (dizrē·liăn), *a.* [f. the name of Benjamin *Disraeli*, first Earl of Beaconsfield (1804–1881), Conservative statesman and prime minister.] Pertaining to or characteristic of Disraeli or his opinions, measures, or writings.
1880 T. W. Reid *Politicians of To-Day* I. 38 Perhaps no better word than 'impartiality' could be found to describe that peculiar quality of the Disraelian mind and character. 1901 *N. Amer. Rev.* Feb. 252 The great Disraelian Myth, which has changed the most un-English of all our Prime Ministers into an almost sacramental Symbol of Patriotism. 1927 *Observer* 11 Dec. 15/2 A devout Disraelian Tory. 1927 H. Miles tr. *Maurois's Disraeli* 296 Even abroad the altogether Disraelian boldness of this coup was extolled. 1962 *Daily Tel.* 12 Dec. 10/2 Seeking to give the Disraelian One Nation ideal more substance.

disregard, *sb.* Add: **2.** (See quot. 1940.)
1940 *Manch. Guardian Weekly* 8 Mar. 192 What Mr. Elliot inelegantly but conveniently called 'disregards'—..forms of income which are to be disregarded when the means of an applicant for supplementary pension are assessed. 1955 *Times* 15 June 3/6 There was a case for saying that the disregards in relation to pensions were not adequate. 1959 *Times* 21 June 14/4 The regulations dealt with..those disregards which related to earnings. The National Assistance Bill was necessary to give powers to deal with the other disregards.

disrelate (disrĭlē·t), *v.* [See Disrelated *ppl. a.*] *trans.* To sever the connection between, cause to have no connection.
1864 J. Owen *Duty Saints Persev.* xv. 353 True Believers who only are the members of Christ disrelate themselves to him. 1874 ... 25 Nov. 2/3 Something analogous to a sense of prudery has caused the author slightly to disrelate the two contestants for Lois. 1946 *New Republic* 29 Apr. 630 The vertical character of Thomas' vision tends to disrelate the components of immediate experiences.

disremember, *v.* (Earlier and later examples.)
1815 in *Dict. Hist. Amer. Industr. Soc.* (1910) IV. 25, I belonged to the Society about fifteen months; as to the constitution I disremember, the rules I recollect. 1834 M. Edgeworth *Tour in Connemara* (1950) 93 He had some outlandish name that was gone out of his head—that he disremembered. 1885 G. M. Hopkins *Poems* (1918) 52 Quite Disremembering, dismembering all now. 1928 D. Byrne *Destiny Bay* xv. 313 Either in Ohio or Illinois, I disremember which. 1938 M. K. Rawlings *Yearling* i. 13 'Do you mind what we said we'd do, full moon in April?' 'I disremember.'

disrobing, *vbl. sb.* (Later examples.)
1903 'A. M'Neill' *Egregious Eng.* 58 The bare business of robing and disrobing taken up pretty well half her waking day. 1912 L. A. Harker *Mr. Wycherly's Wards* x, Neither of them cared a whit for Jane-Anne and her disrobings.

disrupture, *sb.* (Earlier Amer. example.)
1785 Jefferson *Notes Virginia* (1787) 27 The evident marks of their disrupture and avulsion from their beds by the most powerful agents of nature, corroborate the impression.

diss, var. *dis *v.*

dissava (disā·vă). Also dissauva, dissave, dissauva. [Sinhalese *disāwa*.] A governor of a district of Ceylon.
1681 R. Knox *Hist. Rel. Ceylon* 35 He gives order to his Dissaua's or Governors of the Countreys to..choose out Boyes. *Ibid.* 50 Next under the Adigars, are the Dissauva's. 1720 Defoe *Capt. Singleton* 294 The King of the Country..sent down a Dissauva, or General, with an Army. 1803 R. Percival *Ceylon* 258 The Dissauvas, as long as they hold their office, are allowed by the king a certain portion of land for their services. 1859 J. E. Tennent *Ceylon* II. 92 The dissave of Oovah..placed himself at the head of the insurrection.

dissa·ve, *v.* Also dis-save. [Dis- 6.] *intr.* To spend more than one's income, by drawing on savings or realising capital. So dissa·ving *vbl. sb.*; dissa·ver, one who dissaves.
1936 J. M. Keynes *Gen. Theory Employment* vii. 82 No one can acquire an asset..unless..someone else parts with an asset of that value..otherwise there must be dis-saving an equal sum. 1946 *Economist* 13 Apr. 602/2 Nothing has been allowed for dis-savings, which must loom ever larger as increasing supplies of goods create opportunities for spending. 1950 *Daily Tel.* 6 Jan. 4 Federal Reserve report refers to people who spend more than they earn as 'dis-savers'. 1952 *Times* 1 Dec. 7/2 The dissaving which estate duties do for them when they die. 1954 Newlyn & Rowan *Money & Banking* ix. 191 African holders of temporarily idle balances tend to dissave in an effort to maintain their standards of consumption. 1971 *Daily Tel.* 23 July 17 In addition to..the encouragement to buy on credit there is a possibility of a substantial dis-saving.

dissected, *ppl. a.* Add: **1.** *dissected map* (earlier example).
1807 Southey *Lett. from Eng.* vi. 69 They have..dissected maps which they combine into a whole.
2. b. *Physical Geogr.* Formed by the dissection of a once flat plateau or plain.
1904 W. M. Davis *Elem. Physical Geogr.* ix. 278 An excellent illustration of a well dissected upland is found in the Ozark plateau of southern Missouri. *Ibid.*, The maturely dissected surface has much less strength of relief than the plateau of West Virginia. 1949 W. G. Moore *Dict. Geogr.* 51 *Dissected plateau*, a plateau into which a number of valleys have been carved by erosion; its origin as a plateau is patent, however, when the tops of the mountains and ridges are seen to be level against the skyline, showing that they once formed part of a continuous surface. 1965 A. Holmes *Princ. Physical Geol.* (ed. 2) xix. 573 (*caption*) A maturely dissected plateau of the Gondwana cycle being invaded by the valley flats of the present cycle.

dissection. Add: **6*.** *Physical Geogr.* The breaking up by erosion of a flat surface such as a plateau or plain into hills, or flat uplands, and valleys.
1909 in *Cent. Dict. Suppl.* 1937 Wooldridge & Morgan *Physical Basis Geogr.* xiii. 177 If the active deepening of the valleys is continued after the stage of mature dissection, the ridge-crests will remain sharp. 1968 R. W. Fairbridge *Encycl. Geomorphol.* 15/1 Deep canyon dissection is followed by secondary headward erosion of smaller tributaries.

dissector. Add: **b.** A dissecting instrument.
a 1860 A. E. Smith *Med. Student* (1862) 8 He perpetually carries a Dublin dissector under his arm. 1920 *Practitioner* July 118 If the appendix is concealed under massive granulations, careful search with swab, blunt dissector and the occasional use of scissors will rarely fail to bring it into view.

dissel-boom. Also disselboom. (Earlier and later examples.)
1822 W. J. Burchell *Trav. S. Afr.* I. viii. 150 The pole (disselboom) is ten feet long, having at the end a strong iron staple. 1930 R. Campbell *Adamastor* 65 In my last trek be thou the Star To whom I hitch my disselboom.

dissemblable, *a.* Add: Hence *absol.* as *sb.*
1928 V. Woolf *Orlando* iv. 161 What a phantasmagoria the mind is and meeting-place of dissemblables!

disseminated, *ppl. a.* Add: Of a disease: dispersed or spread throughout an organ, a tissue, or the whole body.
1876 *Brit. Med. Jrnl.* 25 Nov. 675/2 (*heading*) A case of disseminated insular sclerosis. 1906 *Lancet* 17 Nov. 1351/1 A man suffering from Disseminated Sclerosis which resembled tabes dorsalis. 1932 *Discovery* Apr. 112/1 A single glance at the film gives information concerning..paralysis (disseminated sclerosis). 1963 H. Burn *Drugs, Med. & Man* (ed. 2) xviii. 181 Pemphigus and disseminated lupus erythematosus are cured by cortisone. 1964 S. Duke-Elder *Parsons' Dis. Eye* (ed. 14) xvii. 214 Disseminated choroiditis. In this type, small areas of inflammation are scattered over the greater part of the fundus behind the equator. *Ibid.* xxiii. 349 Acute disseminated encephalomyelitis.

disseminule (dise·miniul). *Bot.* [Irreg. f. Dissemin(ate *v.* + -ule.] Any part of a plant that serves to propagate it, such as a seed or spore.
1904 F. E. Clements in *Bot. Surv. Nebraska* VII. 52 Disseminules designed to pass through a resting period are often brought into conditions where they germinate at once. 1936 Fuller & Conard tr. *Braun-Blanquet's Plant Sociol.* i. 17 The numerical superiority of shoots and disseminules of certain species is certainly important here. 1960 N. Polunin *Introd. Plant Geogr.* iv. 97 In nature only a small proportion of the plant bodies which become dispersed, and which may conveniently be termed disseminules (diaspores), actually become established and effect migration.

dissimilate, *v.* (Further examples in *Philol.*)
1935 G. K. Zipf *Psycho-Biol. of Lang.* (1936) 93 When *dh*..was dissimilated..to *d*. 1946 E. A. Nida *Morphol.* 287 Vowels may assimilate and dissimilate.

dissimilatory (disi·milătŏri), *a.* *Philol.* [f. Dissimilate *v.* + -ory.] Pertaining to or produced by dissimilation.
1901 H. Oertel *Lect. Stud. Lang.* iv. 232 The 'dissimilatory' loss of a repeated consonant is in no way similar to the syncope (haplology) of one of two similar syllables. *Ibid.* 234 Obscurity prevails with regard to the causes of dissimilatory substitution of similar sounds, such as *l* for *r*, or *n* for *l*. 1936 [see *assimilatory *a.*]. 1962 *Amer. Lang. Notes* Dec. 546 *Danforth* is a good example of dissimilatory loss of *r*. 1962 *Archivum Linguisticum* IV. 35 It could also be argued that the dissimilatory process took the short cut from **morm* to **borm*.

dissociated (disōu·ʃi,eitĕd), *ppl. a.* Add: *Psychol.* Characterized by the disjunction of associated mental connections or the disaggregation of consciousness; *dissociated personality*, a pathological state of the mind in which two or more distinct personalities exist in the same person. (Occas. used as active verb.)
1890 [see *dissociation 3]. 1911 I. H. Coriat *Abnormal Psychol.* 7 When an experience or complex has become dissociated, it tends to act automatically, and cannot be controlled by the will. 1912 B. Hart *Psychol. Insanity* iv. 47 That the continuous stream of her thought had been interrupted by the sudden appearance of a 'dissociated system of ideas'. 1918 C. S. Myers *Present-day Applic. Psychol.* 35 Irène had undergone a severe shock owing to the death of her mother... becoming 'dissociated', of doing the thing and not knowing why we do it. *Ibid.* 70 It links up proverbial knowledge of human nature with the phenomena of 'dissociated personality'. 1922 *Encycl. Brit.* XXXII. 200/1 As a little girl of three, 'Doris Fischer' was thrown down violently by her drunken father, and so sustained a psychic fracture, which 'dissociated' her into 'Margaret' and 'Real Doris'. 1924 W. B. Selbie *Psychol. Relig.* 291 It is not necessary to presuppose a secondary personality or a dissociated consciousness. 1940 *Brit. Jrnl. Psychol.* July 11 Children..who were apparently too dissociated for any expression of their energy in phantasy or cognition. 1957 R. L. Munroe

Schools of Psychoanalytic Thought iii. xi. 483 Another way by which dissociated impulses may find partial discharge is through the process of sublimation.

dissociation. Add: **2.** *dissociation constant*, the product of the concentrations of the dissociated ions in a solution divided by the concentration of the undissociated molecule when equilibrium has been reached.
1891 *Jrnl. Chem. Soc.* LX. i. 257 The author..communicates the determination of the dissociation constants of some 60 organic substances of acid character. 1955 J. G. Davis *Dict. Dairying* (ed. 2) 318 The degree of dissociation is thus a measure of the strength of the acid or alkali, a very strong acid like hydrochloric or a very strong alkali like caustic soda being practically completely dissociated in dilute solution. The degree of dissociation is expressed in terms of the dissociation constant.
3. *Psychol.* The process or result of breaking up associations of ideas.
1890 W. James *Princ. Psychol.* I. 506 What is associated now with one thing and now with another tends to become dissociated from either... One might call this the law of dissociation by varying concomitants. 1899 J. M. Baldwin *Handbk. Psychol.* (ed. 2) 218 The part played by dissociation is evident. If there were no such breaking up of representations, imagination would be simply memory. 1925 E. & C. Paul tr. *Janet's Psychol. Healing* I. xi. 676, I regard a memory, and especially a fixed idea,..as a system comprising a number of associated psychological phenomena... I have attempted to break up this system, to demolish it stone by stone; this is what I call the dissociation of a fixed idea. 1969 S. H. Bartley *Princ. Perception* (ed. 2) v. xii. 326 Dissociation brought about by local anaesthesia begins with effects upon the smallest nerve fibers and ends with the largest.
b. The disintegration of personality or consciousness; the state in which a person suffers from dissociated personality.
1897 E. Parish *Halluc. & Illus.* 71 If we..seek for some quality common to all the various states in which hallucinations occur, we shall find that their most striking characteristic is the dissociation of consciousness. 1906 M. Prince *Dissociation of Personality* iii. 22 A dissociation of the mind, known as a state of hysteria or 'traumatic neurosis'... Sometimes the mental dissociation produces a complete loss of memory. 1908 *Brain* July 257 Cerebral dissociation..is at least one of the essential features of the hypnotic state. 1922 *Encycl. Brit.* XXXII. 200/1 Other cases of dissociation [e.g. the 'Watseka Wonder']. 1935 *Brit. Jrnl. Psychol.* Oct. 176 Many of the shamanistic phenomena which have been described..can be explained by supposing varying degrees of dissociation. 1948 W. McDougall *Introd. Social Psychol.* (ed. 29) 84 Abnormal states of the brain, of which the relative dissociation obtaining in hysteria, hypnosis, normal sleep, and fatigue, is the most important. 1963 Langner & Michael *Life Stress & Mental Health* xv. 409 Such symptoms as fainting or amnesic periods (as well as alcoholism and drug addiction) are considered evidence of withdrawal and escape behavior.

dissoconch (di·sŏkǫŋk). *Zool.* [f. Gr. δισσός double + Conch.] The shell of a mollusc in the veliger stage; also, the shell of an adult bivalve.
1888 Jackson in *Proc. Boston Soc. Nat. Hist.* XXIII. 543 In the oyster..this [sc. the earlier shell] shell is not single, but double-valved, and, therefore, deserves a distinct name, as it precedes the dissoconch or true shell.

dissogony (disǫ·gŏni). *Zool.* Also -geny. [ad. G. *dissogonie* (C. Chun 1888, in *Bibliotheca Zool.* I. i. 64), f. Gr. δισσός double + -γονία offspring.] The phenomenon found among the Ctenophora, in which there are two periods of sexual maturity in the same individual, one in the larval and another in the adult form.
1896 R. Lydekker *R. Nat. Hist.* VI. 477 In at least one species (*Eucharis multicornis*) sexually mature larvae, or larvae which are capable of reproduction as such, also occur; these, when completely developed, become once more capable of reproduction as adults;—a method of multiplication which has been called dissogony. 1906 *Camb. Nat. Hist.* I. 419 To this series of sexual phenomena the name 'Dissogony' is given. 1911 *Encycl. Brit.* XVI. 214/2 The larva becomes sexually mature and lays eggs; it then loses its generative organs and develops into the adult, which again develops reproductive organs (dissogony). 1940 L. H. Hyman *Invertebrates* I. viii. 677 All the ctenophores are hermaphrodite, and many have two periods of sexual maturity, one in the larva and a final one in the adult, with a degeneration of the gonads between. This phenomenon, known as dissogeny, has been termed dissogony.

dissolve, *v.* Add: **7. b.** *Cinemat.* and *Television.* To cause (a picture) to become faint or fade away (*into* another); similarly *intr.* (cf. 13). Cf. Dissolving *ppl. a.* b. Hence as *sb.*, the act or process of dissolving a picture; a dissolving scene in a cinema film; a piece of apparatus with the aid of which this is produced; dissolver, an apparatus for dissolving a picture; also *attrib.*
[1845 *Poster for Assembly Rooms, Nottingham*, Dissolving Views & the Chromotrope.] 1912 F. H. Richardson *Motion Picture Handbk.* (ed. 2) 375 The lamps of a dissolver should each one be connected just as though the other one did not exist. *Ibid.* 377 To construct a home-made dissolver shutter. *Ibid.* 378 A dissolving effect with one lens is an impossibility. *Ibid.*, Your dissolver lens must be matched. 1915 *Ibid.* (ed. 3) 604 The upper dissolver lamp must be supplied with amperage equal to the projection machine arc. *Ibid.* 606 Dissolving moving picture. *Ibid.*, Many operators who run two machines dissolve one reel into the next. 1918 H. Croy *How Motion Pictures are Made* vii. 176 The second means of accomplishing a fade picture is by means of the dissolving shutter... The dissolving shutter is a mechanical device which, while the shutter is revolving, is closed by a blade slowly passing over the opening until it is entirely closed. *Ibid.* 178 The so-called 'dissolve', by which the figures of the scene gradually disappear while those of a succeeding scene slowly take their place. 1954 *Encounter* Aug. 52/1 The fade and the dissolve as means of visual transition from scene to scene. 1959 Halas & Manvell *Technique Film Animation* xix. 171 It [sc. the work book] also shows how each scene..is to be punctuated, whether by a straight cut, a fade or a dissolve. 1960 N. Kneale *Quatermass & Pit* i. 14 Dissolve..to a brass name-plate. *Ibid.* iii. 86 Slow Dissolve..to the Rocket Group laboratory. 1970 *New Yorker* 31 Oct. 132/3 Their sexual bouts lead to quick dissolves.

dissonantal (di:sŏnæ·ntăl), *a.* [f. L. *dissonānt-ia* Dissonance + -al.] = Dissonant *a.*; employing or characterized by dissonance.
1949 R. Blesh *Shining Trumpets* (1949) i. 8 The ultra-modern polyphonal and dissonantal school of today. 1948 E. Sitwell *Notebk. on Shakes.* v. 29 The rocking up and down of the dissonantal o's ('Crowne', 'Toe', 'top'). 1951 *Eng. Studies* XXXII. 68 Alliterative assonance—sometimes called para-rhyme, or dissonantal rhyme—is of course not unknown in English poetry before the twentieth century.

di·ssonate, *v. rare.* [f. L. *dissonātus* (see Dissonate *a.*) and -ate^3.] **a.** *intr.* 'To be dissonant or harsh; said of sounds' (*Cent. Dict. Suppl.* 1909). **b.** *trans.* To make dissonant.
1927 *Observer* 9 Oct. 14 From Adam de la Hale's sharp (and rising) fourth to Purcell's flat (and dissonating) seventh. 1961 *Times* 4 July 13/1 The piece is pleasant-sounding, 'dissonated' by some violin harmonics held over moving chords.

dissuader. Add: Also, a thing which dissuades.
1855 M. Arnold *Lett.* (1895) II. 216, I relied on a dissuader from you. 1923 G. Collins *Valley of Eyes Unseen* 29 The other came on, and received a like dissuader from my right.

distance, *sb.* Add: **5. f.** *Boxing.* Striking distance.
1806 *Sporting Mag.* XXVII. 130/1 He [sc. a boxer] was an excellent judge of his distance. 1808 *Ibid.* XXXII. 35/1 Dogherty commenced at a rally within distance with determined courage. 1816 F. W. J. Henning *Recoll. Prize Ring* 41 At last Rooke got within distance, and lashed out his left, catching Joe an ugly one on the ear. 1896 in J. C. Trotter *Boxing* 125 When in the event of a competitor being knocked down, his opponent shall retire out of distance. 1897 *Encycl. Sport* I. 139/1 When out of distance, the left hand may be brought up to the proper position... The boxer..must immediately get well out of distance. *Ibid.* 135/2 After the..blow..well out of distance.
11. b. *spec.* in *Boxing*, the scheduled length of a contest.
1949 L. Harvey *Training & Self Defence* 112 If..you can stay the full distance of a contest..work to a schedule during the fight. 1966 *Times* 24 Nov. 6/6 A fearless fighter who had..gone the distance three times with the famous negro, Al Brown. 1968 *Times* 30 Sept. 15/1 This greater power..would not be possible if the bout goes the distance.
12. *distance-piece* (examples); *distance-receptor Physiol.*, a sense organ, such as the eye or ear, that is responsive to stimuli from distant sources.
1930 *Engineering* 4 July 9/3 The introduction of a distance piece between a piston and its crosshead. 1962 *Times* 27 Apr. 11/7 The four toothed blades were separated by a distance-piece. 1906 C. S. Sherrington *Integrative Action Nerv. Syst.* ix. 324 It is in the leading segments that we find the 'distance-receptors'. 1927 Haldane & Huxley *Animal Biol.* vii. 300 Since they [sc. gastropods] are free-moving, they have need of distance-receptors. 1959 Marshall & Hughes *Physiol. Mammals* viii. 187 Exteroceptors..may be divided into: telereceptors or distance receptors, cutaneous receptors, and chemoreceptors. [1971 *Nature* 26 Feb. 495/1 Circumstantial evidence for the bursicles functioning as distance chemoreceptors came from a study of the escape response of the marine snail *Tegula funebralis* to the starfish *Pisaster ochraceus* Brandt.]

distance, *v.* Add: **1. a.** Also *fig.* distancing *vbl. sb.* and *ppl. a.* (later examples).
1950 ... II. 470 The distancing ('Verfremdung') which Brecht desires is complemented by his concreteness. 1949, 1956 [see *alienation 1 d]. 1958 R. Williams *Culture & Society* i. v. 87 The method..of documentary record..has a slightly distancing effect. 1961 W. J. Harvey *Art of G. Eliot* v. 116 George Eliot sometimes fails to achieve an adequate distancing of her experience. 1963 *Listener* 3 Jan. 21/2 Would it not be more true to say that the extreme formality..distances the hatred for contemplation? 1963 *P.M.L.A.* LXXVIII. 296/1 Primarily Chaucer's narrative method serves to distance us from the characters. 1967 *Medium Ævum* XXXVI. 284 In *The Parlement of Foules* and in *Troilus and Criseyde* Chaucer takes great pains to distance himself as narrator from the action of the poem, and this hardly seems to be an ironic distancing.

distanced, *ppl. a.* **1. a.** (Recent examples of *fig.* use.)
1949 Koestler *Insight & Outlook* xix. 267 Interest becomes disinterested, as it were, distanced, almost completely detached from the aims of the original drive. 1962 W. Nowottny *Lang. Poets Use* iv. 79 We take in the language of the last quatrain, so distanced from the language of common life, without recoil.

distant, *a.* Add: **8.** *distant early warning* (abbrev. D.E.W.) *line*, a radar system installed in North America for the advance detection of missile attack; *distant-water* attrib.
1955 *Times* 18 Aug. 6/3 Two task forces, totalling 3,000 men, now on their way to the Arctic regions of Canada, will be employed in erecting more than 50 radar stations there as part of the distant early warning (D.E.W.) line. Seven radar stations are known to be working in Greenland, besides those incorporated in the two newer warning lines across Canada, named Mid-Canada and Pine Tree. 1958 *Times* 21 July 9/6 The Distant Early Warning line in the far north is similarly run by civilians on private contract. 1958 *Listener* 3 July 6/1 The total catch of the distant-water fleet. 1960 *Times* 5 Feb. 5/1 The British distant-water fishermen are as vitally involved as anyone in the fishery limits dispute. 1962 J. Tunstall *Fishermen* ix. 218 In Hull all the ships are distant-water trawlers, whereas in the other ports there are more near and middle-water ships. *Ibid.* x. 228 By 1953 the restriction committee was called by the amusing title of 'The Distant Water *Development* Scheme'.

distearin (daisti·ărin). *Chem.* [ad. F. *distéarine* (Berthelot 1854, in *Ann. de Chim. et de Phys.* 3rd Ser. XLI. 221), f. Di-^{2} + Stearin.] A fat, the diglyceride of stearic acid. (Cf. Stearin 1.)
1866 H. Watts tr. *Gmelin's Hand-bk. Chem.* XVII. 117 (*heading*) Distearin. 1873 C. H. Ralfe *Outl. Physiol. Chem.* 48 Stearic acid forms with glycerin 3 compounds, Monostearin, Distearin, and Tristearin. 1879 [see Monostearin]. 1956 *Nature* 21 Jan. 146/2 These..experiments indicated that..the distearin was preferentially precipitated by urea.

distemperer. Delete *rare* and add later examples.
1881 *Instr. Census Clerks* (1885) 52 Colourer. Decorator. Distemperer. 1901 *Daily Chron.* 10 Sept. 9/7 Painter, Distemperer.—Good brush hand wants Work.

distensile (diste·nsail, -il), *a.* Delete †*Obs. rare*, add examples, and add to def.: Also, capable of distending or causing distension.
1879 *St. George's Hosp. Rep.* IX. 337 Over the whole of this tumour could be felt well-marked distensile pulsation. 1902 *Encycl. Brit.* XXXI. 560/2 As a result of these fibrous changes there is interference with the blood current, since the vessels become unyielding yet frangible, instead of distensile and elastic, tubes. 1907 *Practitioner* Nov. 621 This distensile force would have been amply sufficient to cause dilatation of the ventricular cavity. 1910 *Ibid.* Apr. 482 Forcible inflation under pressure may also be done by compressing the india-rubber tube between the distensile bulb and the proximal glass tube.

distinctive, *a.* Add: **1. b.** Applied *spec.* in *Linguistics* to a phonetic feature that is capable of distinguishing one meaning from another. Also *distinctive-feature*, applied *attrib.* to a theory or system propounded by R. Jakobson of classifying linguistic elements in terms of their distinctive phonetic features.
1927 Bloomfield & Bolling in *Language* III. 129 Normally we symbolize only phonemes (distinctive features) so far as we can determine them. 1933 Bloomfield *Lang.* v. 77 Part of the gross acoustic features are indifferent (non-distinctive), and only a part are connected with meanings and essential to communication (distinctive)... A feature that is distinctive in one language, may be non-distinctive in another language. 1942 Bloch & Trager *Outl. Ling. Analysis* 38 Divide all phonetic differences observable in the language into two kinds: *distinctive differences* or *contrasts*, capable of distinguishing one meaning from another; and *nondistinctive differences*, never used for this purpose. 1952 R. Jakobson et al. *Preliminaries to Speech Analysis* 3 The distinctive features are the ultimate distinctive entities of language since no one of them can be broken down into smaller linguistic units. The distinctive features combined into one simultaneous bundle form a phoneme. 1964 *Language* XL. 221 A problem wide open for experimentation is the psycholinguistic validation of 'distinctive-feature analysis'. 1965 *Ibid.* XLI. 168 A positional and articulatory distinctive-feature description has been more thoroughly exploited than any other in analyzing clusters. 1965 N. Chomsky *Aspects of Theory of Syntax* i. 55 Jakobsonian distinctive-feature theory. 1968 Chomsky & Halle *Sound Pattern Eng.* 64 We take 'distinctive features' to be the minimal elements of which phonetic, lexical, and phonological transcriptions are composed.

distinguished, *ppl. a.* **4.** Add: *Distinguished Conduct Medal* (abbrev. D.C.M.), a distinction, instituted in 1862, awarded to British warrant officers, non-commissioned officers, and men for distinguished conduct in the field. *Distinguished Service Order* (abbrev. D.S.O.), an order of distinction for British naval, military, and air force commissioned officers, instituted 9 Nov. 1886. Also *Distinguished Service Cross* (D.S.C.), *Flying Cross* (D.F.C.), *Service Medal* (D.S.M.), *Flying Medal* (D.F.M.): see quots.
1862 *Circ. Mem.* 233 25 Nov. in *Army List* (1863) Jan. 330 No retrospective action will be given to the warrant in question, so far as 'distinguished conduct' medals without Annuity or Gratuity are concerned. 1914 *Times* 19 Oct. 8/5 His Majesty has been pleased to approve of the establishment of a medal, to be called the Distinguished Service Medal, to be awarded to chief petty officers, petty officers, men, and boys of all branches of the Royal Navy, to non-commissioned officers and men of the Royal Marines, and to all other persons holding corresponding positions in his Majesty's Service afloat, for distinguished conduct in war

in cases where the award of the Conspicuous Gallantry Medal would not be applicable. His Majesty has further approved of the award of the Conspicuous Service Cross (to be designated in future the Distinguished Service Cross) to all officers below the rank of Lieutenant-Commander in addition to the officers previously eligible for this Decoration. 1918 *Times* 3 June 7/5 'The Distinguished Flying Cross', to be awarded to officers and warrant officers for acts of gallantry when flying in active operations against the enemy..'The Distinguished Flying Medal', to be awarded to non-commissioned officers and men for acts of courage or devotion to duty when flying, although not in active operations against the enemy.

distinguo (disti·ŋgwo). [L. = I distinguish.] A distinction in thought or reasoning.
1895 G. Saintsbury *Last Vintage* (1950) 176, I do not propose to deal with some correspondence which followed..as to the Scotchness of Mr. Gladstone's alleged practice of intimating dissent (or at any rate a sort of *distinguo*) by the syllable *ma*, which it seems was not intended for the heat of an English sheep, but for the Italian equivalent of 'but' itself. 1920 *Contemp. Rev.* Mar. 364 When one hears that in so unqualified a way, certain distinguos at once present themselves to the mind. 1945 R. G. Collingwood *Idea of Nature* 11 Renaissance cosmology had avoided this conclusion by a *distinguo*. The world of nature as it appears to our senses was admitted to be unknowable; but behind this world of so-called 'secondary qualities' there lay other things..knowable because unchanging.

distomatosis (dai:stōumătōu·sis, -stǫm-). *Path.* and *Vet. Sci.* [mod.L., ad. F. *distomatose* (A. A. Florance 1866, in *Diss. Faculté Méd. Strasbourg* XLI. 2), f. as Distomatous *a.*: see -osis.] = next.
1894 G. Fleming tr. *Neumann's Parasites & Dis. Dom. Animals* ii. iv. 516 The term *Distomatosis*..or *Distomiasis*..designates any affection due to Distomes. *Ibid.* 517 The distomatosis of herbivora has been known for a very long time under numerous designations, which have been applied almost exclusively to the ovine species. 1894 *Jrnl. Compar. Med. & Vet. Archives* XV. 162 An epizootic of distomatosis. *Ibid.* 173 Three cases of pulmonary distomatosis in Kansas cattle. 1955 Gaiger & Davies *Vet. Path.* (ed. 4) xxxiii. 649 (*caption*) Distomatosis in liver of ox.

distomiasis (distomai·ăsis, -stǫm-). *Path.* and *Vet. Sci.* [mod.L., ad. F. *distomiase* (Wiame 1862, in *Ann. de Méd. Vét.* 22), f. Distoma: see *-iasis.] (See quot. 1961.) = *liver-rot* (Liver *sb.*^{1} 7). Cf. *Fascioliasis.
1894 [see prec.]. 1907 *Westm. Hosp. Rep.* XV. 60 Cerebral distomiasis displays itself with the irritative and paralytic phenomena of a cerebral neoplasm. 1961 *Brit. Med. Dict.* 449/1 *Distomiasis*, any infection by trematodes or flukes; the term is now obsolete since the worms formerly in the genus *Distoma* are now correctly placed in other genera. The term still survives in medical textbooks, with many qualifying adjectives.

distort, *v.* **4.** Delete †*Obs. rare* and add later examples.
1970 *New Scientist* 12 Nov. 342/3 The way in which the teeth distort is shown in Figure 3. 1962 *Which?* (Car Suppl.) Oct. 135/2 On the hill, the jack mounting points distorted slightly.

distortion. Add: **3. b.** *Psychol.* The alteration of repressed or unconscious elements before they appear in the conscious mind.
1910 *Amer. Jrnl. Psychol.* XXI. 304 The distortion in the dream-making is thus a means of evading the censor. 1925 A. & J. Strachey tr. *Freud's Coll. Papers* III. 361 Obsessional thoughts have undergone a distortion similar to that undergone by dream thoughts before they become the manifest content of a dream. 1957 R. L. Munroe *Schools of Psychoanalytic Thought* iii. xi. 482 The child whose personal experience was especially unfortunate..is especially prone to..distortion.
4. *Electr.* A change in the wave-form of a signal by an electronic device such as an amplifier or during transmission from one point to another, usually impairing the quality of its reproduction. Hence distortionless *a.*, not producing any distortion; also, not affected by distortion.
1887 O. Heaviside in *Electrician* 3 June 80/1 The attenuation..increases so fast with the frequency, thus leading to a most prodigious distortion in the shape of irregular waves... Now the distortion and the attenuation, though different things, are intimately connected. *Ibid.* 24 June 143/2 'Distortion'..I chose myself as preferable to 'mutilation' and similar words. Its meaning is obvious. Make current-variations in a certain way at one place. If the current-variation at another place are similar, no matter how much attenuated they may be, there is no distortion. 1892 — *Electr. Papers* II. 129 The distortionless system..brings the speed of the current into full view. 1914 A. B. Rolfe-Martin *Wireless Telegr.* 117 It is found that, owing chiefly to dielectric losses in the condenser, the distortion has a still greater weakening effect on radiating power. 1922 *Jrnl. Inst. Electr. Engineers* Apr. 397/2 For satisfactory transmission of speech the circuit should also be practically 'distortionless'. 1937 *Proc. Inst. Radio Engineers* XXV. 321 It is possible to obtain distortionless reception of considerably higher levels of modulation if the two side-frequencies..are transmitted. 1962 Simpson & Richards *Junction Transistors* xii. 276 Class B amplifiers must be used in push-pull arrangements if distortion-free amplification is to be obtained. *Ibid.* xiii. 298 This is readily seen by considering any such distortion voltage as being produced by a fictitious generator added within the closed loop of a distortionless amplifier.

distractibility (distræktĭbi·lĭti). [f. Distractable *a.* + -ity.] The condition of being distractible; inability to give prolonged attention to a task or object; esp. in *Psychol.*, the tendency to have the direction of one's attention easily changed by chance stimuli.
1902 A. R. Diefendorf tr. *Kraepelin's Clin. Psychiatry* 16 Distractibility is more marked in chronic nervous exhaustion. 1923 *Jrnl. Appl. Psychol.* VII. 358 This same group also showed a larger percentage of 'irritability', 'distractibility', ..and marked 'slowness of reaction' than did the other two groups. 1940 *Mind* XLIX. 65 The fluctuations and distractibility of attention are of obvious importance to applied psychologists. 1962 Henderson & Gillespie *Text-bk. Psychiatry* (ed. 9) vi. 109 Distractibility is a disorder of the attention in which the patient gives attention to every passing stimulus.

distraction. Add: **2. c.** Applied *attrib.* to behaviour of birds that is intended to distract the attention.
1943 *Trans. Linn. Soc. N.Y.* VI. 248 The instinctive response in face of danger to eggs and especially to young of drawing attention to the adult and away from the offspring, is usually termed 'injury-feigning'. It would be better to call it 'nest-protecting display' (Murphy 1926) or distraction display. 1950 *Brit. Birds* XLIII. 7 The pseudo-sleeping figure occurs commonly in the Oyster-catcher's complex series of distraction-behaviour patterns. 1954 Fisher & Lockley *Sea-Birds* viii. 173 But gannets, petrels and other sea-birds..distraction displays are absent. 1961 Bannerman *Birds Brit. Isles* X. 213 Although distraction or 'injury feigning' displays are uncommon, they are sometimes elaborate, particularly by birds with chicks. *Ibid.*, 'Distraction flight' in which the cock or hen rises from the nest or near the chicks with fluttering flight.

distraughtly (distrǭ·tli), *adv.* [f. Distraught *a.* + -ly^2.] In a distraught manner.
1926 *Chambers's Jrnl.* June 383/1 She strove, strove distraughtly to shriek, to move. 1958 S. Spender *Engaged in Writing* 32 Botolf's long..hair had been shoved distraughtly..across his head.

distress, *sb.* Add: **2. d.** (Earlier examples.)
1803 *Sporting Mag.* XXI. 21/2 Fifth Round... Firby began to exhibit symptoms of distress. 1836 *Spirit of Times* 12 Feb. 13/1 She showed some symptoms of distress and the backers of the field thought there was still a chance.
5. (sense 2 c) *distress call, light, message, signal, signalling*; **distress committee**, a committee set up to help people in distressed circumstances; **distress work**, work provided for people in distress.
1913 *Year-bk. Wireless Telegr. & Teleph.* 319 Accumulation..enabling the ship to issue distress calls. 1929 *Times* 9 Dec. 1/1 (*heading*) Sea search after distress call. 1905 *Daily Chron.* 21 Sept. 1/7 The establishment of distress committees. 1911 *Encycl. Brit.* XVIII. 818/1 Distinguishable distress-lights. 1921 *Discovery* Apr. 92/2 A distress message is preceded by a signal consisting of three dots, three dashes, and three dots sent as one sign, and repeated at short intervals. This signal is allotted to as the S.O.S. signal. 1873 *Porcupine* 8 Sept. 269/1 It is necessary for all craft to carry a gun for use in making distress signals. 1913 *Year-bk. Wireless Telegr. & Teleph.* 318 Distress Signalling. 1905 *Daily Chron.* 13 Nov. 4/4 The distress work is not provided except in emergencies.

distressed, *ppl. a.* Add: **a.** Applied *spec.* to a person living in impoverished circumstances.
1844 C. Ridley *Let.* (1958) xv. 180 Little round mats..which she works at all the evening as if she were a distressed gentlewoman. 1932 R. Lehmann *Invit. Waltz* iii. iv. 202 It's all right, Aunt Sybil, I swear. Don't look so like a distressed gentlewoman. 1935 E. Bowen *House in Paris* I. ii. 19 A 'distressed lady'..jotting down what she spent. 1967 *Daily Mail* 27 Nov. 10/1 A field on the other side of Oxfordshire belonging to the Distressed Gentlefolks Aid Association. 1971 *Daily Tel.* 16 July 7/4 Distressed gentlefolk, daughters of a Harley Street surgeon whose legacy..could not or would not foresee inflation and a dearth of husbands.
b. *distressed area*, a region where there is much unemployment.
1928 *Britain's Industr. Future* (Liberal Ind. Inquiry) I. ii. 30 The continued depression of these industries entails the unemployment, on a large scale, of many workers..and raises acute problems in a number of distressed areas. 1936 *Discovery* Nov. 355/2 The development of newer industries is vital to the recovery of our distressed areas, which remain the one black spot in the otherwise remarkable position of Great Britain. 1942 *Times Weekly* 2 Dec. 15 If their 400 hat factories became derelict they would be a distressed area.
c. Upon which a distress is levied.
1896 *Westm. Gaz.* 24 June 3/1 The Government had never sent any inquiry..whether..between land distressed and land not distressed.
d. Of furniture: see quot. 1940. orig. *U.S.*
1940 J. Judson *What Every Woman should know about Furniture* 33 If you're a real antique lover you can even buy pieces with 'distressed' finish—that is, with simulated marks of age and wear. 1966 *Daily Tel.* 4 Oct. 15/5 'We're coming to a lot of "distressed" finish,' said the American designer solemnly... The current vogue in America is for furniture that has been 'bashed' or scarred to make it look like English antique. 1967 *Boston Sunday Globe* 23 Apr. 23 (Advt.), Traditional lamps..rendered in distressed wood finishes. 1970 *Cabinet Maker & Retail Furnisher* 16 Oct. 129/1 Heavily-distressed oak pieces..has 12 full-working drawers with distressed fronts.

distributary, *a.* and *sb.* Add: **A.** *adj.* **2.** *spec.* in *distributary canal, channel, river.*
1926 *Spectator* 24 Apr. 756/2 To excavate additional distributary canals and field channels. 1954 W. D. Thornbury *Princ. Geomorphol.* xxii. 584 Shoestring sands may be ..river channel fillings, delta distributary channel fillings, [etc.].
B. *sb.* **b.** A river branch which flows away from the main stream without returning to it, as in a delta; a similar branch of a glacier.
1863 J. Fergusson in *Q. Jrnl. Geol. Soc.* XIX. 328 One consequence of any such alteration in the course of the main stream is, that the initial or terminal oscillation of any tributary or distributary is continually altering its position. 1881 *Times* 16 Aug. 3/6 [A river's] breaking up into distributaries as it approaches the sea. 1954 W. D. Thornbury *Princ. Geomorphol.* vii. 173 A river crosses its delta through a number of channels known variously as distributaries, mouths, or passes. 1957 G. E. Hutchinson *Treat. Limnol.* I. i. 76 The possiblity that the glacier that filled the basin had a western distributary which cut the present outlet.

distribution. Add: **1. b.** (Earlier example.)
1793 tr. A. R. J. Turgot (*title*) Reflections on the formation and distribution of wealth.
2. b. The occurrence of linguistic elements in a language, in terms of their characteristic position or context (see quots.).
1933 Bloomfield *Lang.* v. 81 Non-distinctive features occur in all manner of distributions. 1951 Z. S. Harris *Meth. Struct. Ling.* ii. 15 The distribution of an element is the total of all environments in which it occurs. 1953 C. E. Bazell *Ling. Form* 6 And if distribution is (as will generally be granted) the surest clue to semantics, it is neither a semantic unit, nor the basis of semantic analysis. 1964 R. A. Hall *Introd. Ling.* v. 26 By distribution, we mean the conditions under which the various elements (allophones, allomorphs, etc.) occur.
3. c. *Statistics.* The way in which a particular measurement or characteristic is spread over the members of a class.
1854 *Amer. Jrnl. Sci. & Arts* 2nd Ser. XVII. 396 The very cause which determined the distribution of the atomic weights according to a numerical law. 1886 *Jrnl. Anthrop. Inst.* XV. 351 (*title*) The comparative distribution of Jewish ability. 1895, etc. [see *frequency distribution* s.v. *Frequency 6 b]. 1959 *Chambers's Encycl.* XIII. 150/1 Further examples..are distributions according to income, volume of trade, height and weight. 1971 *Nature* 18 June 416/3 The NSF survey gives the following distribution of graduate students by area of science..: physical sciences 18 per cent.., social sciences 21 per cent [etc.].
9. *attrib.*, as **distribution board** *Electr.*, an insulated panel carrying terminals, fuses, etc., for controlling a number of subsidiary electrical circuits; **distribution map** (see quot. 1951).
1907 *Installation News* Apr. 11/2 The ordinary type of distribution board where the switches and fuses are enclosed under one cover. 1933 *Archit. Rev.* LXXIV. 202 Most buildings are now wired on the 'distribution board' method which collects all fuses together. 1947 J. & C. Hawkes *Prehist. Brit.* vii. 177 For all work of this kind.. the prehistorian uses distribution-maps. In these the find-spots..are plotted on a map. 1951 *Oxf. Jun. Encycl.* IV. 254/1 In recent years new kinds of maps have been developed to show such things as the distribution of rainfall, temperature, population, crops, and other things of importance and interest. These are called distribution maps, or statistical maps.

distributional, *a.* Add: (Further examples.) Hence distribu·tionally *adv.*
1933 A. M. Woodbury in *Ecol. Monogr.* III. 168 (*heading*) Distributional Units. 1936 *Amer. Speech* XI. 171/1 Mr. Zipf has succeeded in defining distributional factors and in demonstrating their significance in language. 1947 *Int. Jrnl. Amer. Linguistics* July 171 This procedure allows meanings to be catalogued in terms of distributionally defined units. 1951 Z. S. Harris *Meth. Struct. Ling.* ii. 16 Two utterances or features..distributionally equivalent. 1953 C. E. Bazell *Ling. Form* 83 The view of traditional grammar that *he eats* and *he is eating* stand in immediate opposition is fully confirmed by distributional criteria. 1958 C. F. Hockett *Course Mod. Ling.* xix. 174 The..problems below are in what may be called pure distributional analysis. 1963 J. Lyons *Structural Semantics* ii. 11 The lexeme is a formal unit of grammatical analysis, established distributionally.

distributism (distri·biŭtiz'm). [f. Distribute *v.* or Distribut(ive *a.* + -ism.] The theory or practice of the 'distributive state' (see next); = Agrarianism 1. So distri·butist *a.* and *sb.*; also distri·butivism (*rare*).
1915 E. Barker *Pol. Thought in Eng.* viii. 223 Mr. Belloc's Distributivism. 1918 *Irish Theol. Q.* Jan. 102 Industrialism *versus* the distributist state—return to agriculture and handicrafts. 1926 G. K. Chesterton *Outl. Sanity* i. 30 Disquisitions throwing doubt on the detailed perfection of a Distributist State. *Ibid.* vi. 224 Several things that may bring us a stage nearer to Distributism. *Ibid.*, I do not claim that all Distributists would agree with me. 1933 T. S. Eliot *Use of Poetry* iv. 73 Wordsworth then proceeds to expound a doctrine which nowadays is called distributism. 1953 *Universe* 31 July 6/4 It is Distributist to advance schemes of real co-operation among owner-producers. 1964 *English Studies* XLV (Suppl.). 214 It was Chesterton the Distributist and advocate of 'three acres and a cow'.

distributive, *a.* Add: **1. b.** Of, pertaining to, or designating a political system or state in which personal property is owned by the largest possible number of people.

1912 Belloc *Servile State* 6 A reaction towards a condition of well-divided property or the *Distributive State*. *Ibid.* iii. 50 The State..was an agglomeration in which the stability of this *distributive* system (as I have called it) was guaranteed by the existence of co-operative bodies. 1919 G. B. Shaw *Matter with Ireland* (1962) 212 An economic Utopia which Mr Chesterton and Mr Belloc call the Distributive State. 1925 R. H. Tawney *T. Wilson's Disc. Usury* 19 Elizabethan England..is still, in a convenient modern phrase, a Distributive State.
2. b. *distributive fault*, a fault in which the displacement is distributed among several parallel planes at short distances from one another instead of being confined to a single plane.
1904 Chamberlin & Salisbury *Geol.* (1905) I. 494 Sometimes the faulting is distributed among a series of parallel planes,..thus giving rise to a distributive fault. 1955 W. L. Russell *Struct. Geol. for Petroleum Geologists* v. 113 Step and distributive faults consist of a number of parallel breaks, each with a throw in the same direction.

distributively, *adv.* **d.** (Earlier examples.)
1773 E. Bentham *Introd. Logick* ii. i. 39 The predicate of the former (Six) is affirmed of all the Planets taken collectively: that of the latter (opake Bodies) is affirmed distributively. 1834 Whately *Elem. Logic* (ed. 5) III. § 11. 215 The middle term is used in one Premiss collectively, in the other, distributively.

distributivity (distri:biŭti·vĭti). [f. Distributive *a.* + -ity.] The character or quality of being distributive.
1947 H. Reichenbach *Elem. Symb. Logic* ii. § 11. 47 The laws of commutativity, associativity, and distributivity supply the reason that these manipulations closely resemble algebraic transformations. 1952 W. V. Quine *Methods of Logic* i. 54 The law which makes this possible is known as the law of distributivity of conjunction through alternation. 1965 Patterson & Rutherford *Elem. Abstract Algebra* iii. 73 Associativity of multiplication in $M_{p,p}$ is a consequence ..of distributivity of multiplication over addition for real numbers.

distributor. Add: **b.** *spec.* (i) An electric cable from which lines run to the premises of individual consumers.
1901 C. H. Wordingham *Central Electr. Stations* xviii. 186 The system of distributors should be as flexible as possible, since it is always preferable to lay them under the footways and near the buildings. 1929 G. W. Stubbings *Underground Cable Systems* vi. 71 The pressure drop in a distributor is fixed by the statutory limitations; in a feeder it is governed by economic conditions only. 1967 H. Cotton *Adv. Electr. Technol.* iii. 101 The main consideration in the design of distributors is the voltage drop.
(ii) A device in an internal combustion engine that passes the current to each sparking plug in turn so as to maintain the correct firing order of the cylinders; also *attrib.*
1905 T. H. Hawley *Motor Ignition Appliances* v. 29 The high-tension discharge from B, instead of going to the sparking plug direct, must first find the path open through the four-point distributor D. 1925 *Morris Owner's Man.* iv. 44 The distributor, the distributor rotating arm, and—at the driving end of the magneto—the end cover, should.. be removed, wiped clean, and polished with a fine dry cloth. 1964 J. Griffiths *Your Car* xi. 119 The distributor itself is simply a kind of switch, with a rotating arm that conveys the high-voltage supply from the coil to each of the sparking plugs. *Ibid.*, The contact breaker, condenser, and rotor arm are all part of the distributor unit, and are revealed by lifting off the distributor cap.

district, *sb.* Add: **3. e.** (Earlier examples.) Also, *spec.* the area served by a maternity hospital or a midwife for home confinements. Colloq. phr. *on the district*: see quot. 1933.
1854 C. M. Yonge *Castle Builders* v. 64 The Miss Shaws.. have undertaken to get..a district appointed for us to visit. *Ibid.* v. 66 Several excellent persons..had attempted visiting and instructing the poor in his [sc. the vicar's] district. 1933 Partridge *Slang To-day & Yesterday* ii. iii. 191 'Each of the London teaching hospitals undertakes the care of the parturient poor in its own district.'..A student engaged on his three or six months' course of such midwifery is said to be 'on the district'. 1961 *Observer* 7 May 35/3 The different conditions which exist in maternity hospitals and on the district. 1964 G. L. Cohen *What's Wrong with Hospitals?* v. 87 'I'm me own boss on the district,' Nurse Bailey..expounded the advantages of domiciliary work. *Ibid.* 109 Midwives on the District recommend bed-rest but cannot enforce it.
f. A territorial division of the Methodist communions comprising a number of circuits.
1831 J. M. Peck *Guide for Emigrants* 258 There are three districts, the Illinois, the Kaskaskia, and the Wabash districts, over each of which is a presiding Elder. 1839 [see Circuit *sb.* 6]. 1885 [see sense 6 in Dict.]. 1970 B. Drewery in S. G. F. Brandon *Dict. Compar. Relig.* 440/2 The Methodist constitution..forms series of concentric circles, from the local chapels organised into 'circuits', each with a team of Ministers under the 'Superintendant', the Districts under 'Chairman', and the central authority of the annual Conference of 650.
g. With capital initial: short for the London *Metropolitan District Railway*; also *pl.* shares in this railway. Also *attrib.*
1886 [see Gas *v.* 1]. 1898 *Westm. Gaz.* 29 Nov. 8/1 We cannot find any sufficient reason for the recent rise in Districts. 1902 *Ibid.* 29 May 7/3 To travel on the District from Ealing to the Mansion House and back, third-class, will in the future cost 8d. 1909 *Ibid.* 16 July 10/3 Districts were also good in tone at 17⅞. 1911 W. Owen *Lett.* (1967) 80 Transported to the other side of London by those wretched District trains. 1959 *Chambers's Encycl.* XI 500/1 The District owned the south side of the Inner Circle from Mansion House to South Kensington... District trains reached Uxbridge over the Metropolitan.
6. *district-visitor* (earlier example), *-visiting*. **district heating**, a method of supplying heat or hot water from a single source to a number of separate buildings or a whole district; **district nurse**, a nurse who serves a rural or urban district; so *district nursing*; **district officer**, a representative of the Government in a colonial district; **district school** (earlier and later U.S. examples); **district system** *U.S.*, a system of electing members to the House of Representatives by electing one member for each district of a State (see sense 3 d in Dict.).
1908 A. G. King *Pract. Steam & Hot Water Heating* xxiii. 288 District Heating. This type, if it may be so termed, of steam and hot-water heating owes its inception to..Mr. Birdsall Holly, of Lockport, N.Y. 1913 A. M. Greene *Elem. Heating & Ventil.* xi. 279 District heating or heating from a central station..has been extended to heat towns or portions of towns. 1943 *Archit. Rev.* XCIII. 102/3 The district heating pipes by which a house and a whole estate.. can be centrally heated. 1970 *Daily Tel.* 7 Dec. 8/4 District heating costs more to install than individual heating systems but running costs are substantially less. 1883 *Cassell's Family Mag.* Apr. 314/1 *District Nurses.*—Some seven or eight years ago, an Association was formed in London for providing a body of skilled and trained nurses to nurse the sick poor at their own homes. 1894 Mrs. H. Ward *Marcella* II. iii. iv. 313 Marcella's Association allowed its District Nurses to live outside the 'home' of the district on certain conditions. 1957 *New Yorker* 12 Jan. 28/3 Terrified of infections and vaccines, she barred the door to the district nurse. 1883 *Cassell's Family Mag.* Apr. 314/2 She..receives training in the practice of district nursing for a period of six months. 1956 in A. Pryce-Jones *New Outl. Mod. Knowledge* 220 So much can now be provided by way of meals, home helps, and district-nursing services. 1861 *Statem. Mat. & Moral Progr. India 1859–60* iv. xiv. § 6 in *Parl. Papers* [265] XLVII, Appeal cases, exceeding in value 1,000 Rupees, do not come before the District Officers. 1931 A. R. Radcliffe-Brown *Method in Social Anthropol.* (1958) i. iii. 92 An officer of one of the African colonies..was asked if it would be a good thing to give a training in anthropology to those who would ultimately become district officers. 1793 F. Asbury *Jrnl.* 7 Mar. (1852) II. 186 Building a house for conference, preaching, and a district school. 1854 M. J. Holmes *Tempest & Sunshine* viii. 111 He handed him five hundred dollars, telling him..to send her for two years to the district school. 1946 Partridge & Bettman *As We Were* 19 Only in the backward regions of the country does the district school still survive. 1816 *Deb. Congress U.S.* 20 Mar. (1854) 214 Under the district system,..the weight of Pennsylvania, great as she is, dwindled down to a solitary vote. 1902 *Encycl. Brit.* XXXIII. 581/2 The House of Representatives is composed of members elected by popular vote... Each State is at liberty under the Constitution to adopt either the 'general ticket' system, *i.e.*, the plan of electing all its members by one vote over the whole State, or to elect them in one-membered districts (the 'district system'). 1854 C. M. Yonge *Castle Builders* vi. 78, I read to him about your plans for district visiting. 1935 *Scrutiny* IV. 116 No amount of observation of the district-visiting kind..will produce a convincing substitute for adequate response to the quality of the working-class life. 1850 Kingsley *A. Locke* II. xii. 183 Katie had never heard of her before—'some district visitor' or other?

district, *v.* Add: Orig. and chiefly *U.S.* (Earlier example.)
1792 *Mass. Acts & Laws* (1895) 184 Resolve for districting the commonwealth, for the purpose of choosing federal representatives.

disturb, *v.* Add: **1. d.** *refl.* To put oneself out by moving, etc. (e.g. in order to assist a person).
1831 T. L. Peacock *Crotchet Castle* iii, The stranger was rising up, when Mr. Crotchet begged him not to disturb himself. 1888 Mrs. H. Ward *R. Elsmere* xiii, 'Can I find anything for you?' he said springing up. She hesitated a moment, then..she said..: 'Pray don't disturb yourself. I know exactly where to find it.'

disturbed, *ppl. a.* Add: **b.** *spec.* in *Psychiatry*, emotionally or mentally unstable or abnormal; also (orig. *U.S.*), designed for or occupied by disturbed patients.
1904 T. Johnstone tr. *Kraepelin's Lect. Clin. Psychiatry* iii. 25 A few weeks before his admission he had some attacks of apprehension, and then became disturbed, ill-balanced and absent-minded. 1935 W. Seabrook *Asylum* xiv. 220 A disturbed hall as seen through the eyes of a disturbed patient. 1960 *Guardian* 9 Nov. 6/5 The standards she set were too high for the child, who became disturbed and began to pilfer. 1964 G. L. Cohen *What's Wrong with Hospitals?* viii. 164 Only by segregating those considered dangerous in a 'disturbed ward'—the erstwhile refractory section—can hospitals give the rest a parcelled-out freedom. 1970 *Guardian* 21 May 26/2 The centre..has kept its belief that disturbed people could respond to..a small community ..where they would never be regarded as patients.

disvalue (disvæ·liu), *sb.* Restrict †*Obs.* to sense in Dict. and add: **2.** *Philos.* Negative value. Hence disva·luable *a.*, having negative value; bad, evil, or noxious.
1925 C. D. Broad *Mind & its Place* xi. 504 Intrinsically good states..may have a great instrumental disvalue. 1929 J. Laird *Idea of Value* p. xvi, In the technical usage of philosophy, economics, aesthetics and certain other such studies, 'value' is usually taken to mean both value and disvalue. It is therefore contrasted with non-value, or with that which is neither good nor bad. 1942 *Mind* LI. 185 Having value sometimes means having positive as opposed to negative value, and sometimes means being either valuable or disvaluable. 1963 J. N. Findlay *Meinong's Theory* (ed. 2) ix. 291 When things are given as actual or non-actual, we experience their values or disvalues by serious feelings and unserious desires.

disyllabic, dissyllabic, *a.* Add: Also as *sb.*
1934 Jespersen *Essent. Eng. Gram.* iv. 41 *Fire* and *hire* are disyllabics as early as in Shakespeare. 1935 Curme *Gram. Eng. Lang.* II. 341 The dissyllabics, *often* and *early.*

ditch, *sb.*^{1} Add: **1. c.** Calcutta, so called in allusion to the Mahratta Ditch (see Mahratta 3). *slang.*
1886 Yule & Burnell *Hobson-Jobson* 246/2 *Ditch*; and *Ditcher*, disparaging sobriquets for Calcutta and its European citizens.
d. The trench or piece of ground immediately surrounding a bowling green.
1861 *Chambers's Encycl.* II. 289/1 If a bowl..strikes the jack, and then rolls into the ditch, it reckons as if on the green. 1886 T. Taylor *Rules of Bowling* 16 When the jack is run into the ditch by a bowl in the regular course of the game. 1902 *Encycl. Brit.* XXVI. 328/1 There is no excuse for short play on the part of the first players; their bowls would be far better in the ditch. 1962 *Bowls* (*Know the Game*) 3/2 The green must be level and surrounded by a ditch and bank.
2. c. (*a*) *Naval slang.* The sea; (*b*) *R.A.F. slang.* The English Channel or the North Sea.
1922 *Man. Seamanship* I. i. 33 A smart seaman would not talk officially of the sea by a favourite slang expression 'the ditch'. 1925 Fraser & Gibbons *Soldier & Sailor Words* 78 'He fell into the ditch', *i.e.* overboard. 1945 Partridge *Dict. R.A.F. Slang* 23 *The Ditch*, the sea; especially the English Channel. (Adopted from nautical slang.)

ditch, *v.*^{1} Add: **6.** (Earlier example.)
1816 *Sessions' Papers* May/June 248/2 When I got up, they cried..ditch him, and I was immediately thrown into the ditch.
b. *slang.* (*a*) *trans.* To bring (an aircraft) down into the sea in an emergency. (*b*) *intr.* To come down into the sea in an emergency. Cf. *ditch *sb.*^{1} 2 c. Hence ditching *vbl. sb.*
1941 *Times Weekly* 15 Oct. 19 The pilot..must 'ditch' his aircraft in the sea, near enough to a ship for him to be picked up. 1941 in A. G. Smith et al. *Investigations Behaviour Aircraft* (1957) 23/2 Hudson ditching trials. 1943 Redding & Leyshon *Skyways to Berlin* xv. 100, I..got ready to ditch the plane. *Ibid.*, We had had ditching practice. *Ibid.* 101 There is a standard procedure in ditching. 1944 *Times* 20 Mar. 5/7 The crews run over the 'ditching' procedure to be followed should they have to come down on the sea. 1958 *Daily Mail* 15 Aug. 2/5 If an airliner is ditching, it should give a Mayday call. 1962 D. Slayton in *Into Orbit* 23 Most flyers have had to learn some kind of ditching technique in case they make a forced landing over water.
c. *transf.* and *fig.* To defeat, frustrate; to abandon, discard; to jilt. (See also quot. 1899.) *slang* (orig. *U.S.*).
1899 'J. Flynt' *Tramping with Tramps* (1900) 393 *Ditch*, or *be ditched*, to get into trouble, or to fail at what one has undertaken. To be 'ditched' when riding on trains means to be put off, or to get locked into a car. 1911 *Springfield* (Mass.) *Republican* 31 Aug. 1 Its enactment into law would have ditched them in their present reciprocity campaign. 1921 T. Wolfe *Let.* 13 Nov. (1956) 22, C. is going to see a girl in Brookline that he met in the bank last summer and it seems the Ohio girl has been ditched. 1923 *Saucy Stories* 1 Mar. 74/1, I was gonna ask you to hop over to a dance... Can't you ditch the other guy? 1924 G. C. Henderson *Keys to Crookdom* 104 After they have committed the hold-up, they 'ditch' the stolen motor vehicle. 1926 J. Black *You can't Win* vii. 85 We got 'ditched' off our train at Port Costa, and crawled into a hay car for the night. 1946 'S. Russell' *To Bed with Grand Music* ii. 27 Let's ditch the others and go and dance privately. 1948 L. A. G. Strong *Trevannion* 186 The one who scuppered his parents and ditched his wife. 1950 T. S. Eliot *Cocktail Party* ii. 117 It isn't simply the end of an illusion In the ordinary way, or being ditched. 1958 P. Kemp *No Colours or Crest* viii. 175 Davis..was struggling to carry the heavy wireless set; I shouted to him to ditch it and save himself.

ditcher. Add: **2.** (Earlier U.S. example.)
1846 *Rep. Comm. Patents* (*U.S. Pat. Off.*) (1847) 256 Having thus fully described my improved ditcher what I claim therein as new..is [etc.].
3. (Earlier example.)
1877 A. Arnold *Through Persia* II. xiv. 251 Huge iron vessels, long and narrow, built for the Suez Canal, and locally known as 'ditchers'.
4. A bowl which runs or is driven off the green. Cf. *ditch *sb.*^{1} 1 d.
1886 T. Taylor *Rules of Bowling* 17 A bowl which runs off the green..and which has not previously touched the jack, is called a 'ditcher', and must be immediately removed to the bank. 1948 G. T. Burrows *All about Bowls* xiv. 86 *Ditcher*, a bowl that runs on into the ditch or is driven there or is turned quickly into the pebbles, for reasons of strategy.
5. A resident in Calcutta (cf. *ditch *sb.*^{1} 1 c). *slang.*
1886 [see *ditch *sb.*^{1} 1 c]. 1909 *Westm. Gaz.* 17 June 8/1 The existence of the unfortunate ditchers left to themselves in the off season. 1923 *United Free Ch. Miss. Rec.* June 248/1 Residents in Calcutta are sometimes referred to..as 'Ditchers'.

dite, *sb.*[2] Also **dit.** Phr. *not to care a dit(e)*: not to care at all. Cf. DOIT 2.
1907 *Westm. Gaz.* 7 Sept. 13/1 'Don't care a dite,' Sylvia said despondently. 1920 *Blackw. Mag.* Oct. 488/2 'I suppose your major won't mind that?' 'Not a tuppenny dit.'

diterpene (dəitɜ̄·ɹpīn). *Chem.* [f. DI-[2] 2 + TERPENE.] Any terpene with the formula $C_{20}H_{32}$; also, any of the simple derivatives of such a hydrocarbon, differing in their substituents or their degree of saturation from the diterpenes proper. Hence **dite·rpenoid,** a term used in place of *diterpene* when this is restricted to compounds of the formula $C_{20}H_{32}$ only; also *attrib.* or as *adj.*
1902 F. J. Pond tr. *Heusler's Chem. Terpenes* 431 Very many diterpenes are known, but they have never been thoroughly characterized. 1949 *Q. Rev.* (*Chem. Soc.*) III. 36 A full understanding of the chemistry of the diterpenoids has been developed only in the last twenty years. *Ibid.* 37 (*heading*) Diterpenoid resin acids. 1953 E. H. Rodd *Chem. Carbon Compounds* II. xii. 489 The diterpenoids (C_{20}) and triterpenoids (C_{30}) are mostly obtained from plant or tree gums and resins which, unlike the essential oils, are not volatile in steam. 1956 *Chem. & Ind.* 3 Nov. 1275 (*heading*) The preparation of an intermediate for the synthesis of bicyclic diterpenes. 1956 I. L. Finar *Org. Chem.* II. viii. 324 Phytol, $C_{20}H_{40}O$,..is an acyclic diterpene.

dithematic (dəiþimæ·tik), *a.* [f. DI-[2] + THEMATIC.] Of a word: containing two significant themes or stems. Also as *sb.*
1916 E. Weekley *Surnames* 26 These Teutonic names were originally all dithemetic [*sic*]. 1922 —— in *N. & Q.* 12th Ser. XI. 52/2 Some old Teutonic dithematic. 1927 *Englische Studien* 10 Nov. 76 In the course of the tenth century dithematic names, such as *Wulfstan, Æthelstan* and *Ælfred*, are used almost exclusively among the higher classes of society.

dither, *v.* Add: **1.** (Further examples.) Also in gen. colloq. use: to vacillate, to act indecisively, to waver between different opinions or courses of action.
1908 'I. Hay' *Right Stuff* i. 6 If there is a viva-voce, be sure to speak up and give your answers as though you were sure of them...The one thing the examiners will not thole is a body that dithers. 1923 H. C. Bailey *Mr. Fortune's Practice* iii. 81 All newspapers are run by madmen, but the 'Watchman' merely dithers. 1927 *Manch. Guardian Weekly* 16 Dec. 463/1 While Governments dither and talk limply of disarmament and peace large numbers of normally inarticulate citizens grow increasingly restive. 1930 J. B. Priestley *Angel Pavement* vii. 359 'I don't know what on earth you're trying to say,' she told him...'Oh, don't dither so much, silly.' 1932 C. Williams *Greater Trumps* x. 168 She re-ordered her thoughts; this was mere dithering. 1938 E. Bowen *Death of Heart* ii. v. 255 The lady..was already dithering round a table of new novels. 1948 'N. Shute' *No Highway* i. 27, I don't think it [*sc.* a tailplane] had any continuous movement—it wasn't dithering all the time. 1959 *Times* 14 Dec. 13/4 She was the first producer we had ever had who never dithered about which was Up Stage and which Down.
2. To confuse, perplex, make nervous (esp. in *pass.*). So **di·thered** *ppl. a.*, confused, perplexed; also (*Austral.*), drunk.
1919 Masefield *Reynard* 98 He's done: he's dithered. 1932 N. Lindsay *Cautious Amorist* v. 70 Dithered we [w]as already by the booze we had ashore. 1936 D. G. Smith *Call it a Day* ii. ii. 75 It's these girls in the shops. They just dither you. 1948 V. Palmer *Golconda* xvii. 140 I've seen him so dithered by printed words he didn't know whether it was this week or next. *Ibid.* v. 32 They have a right to know what the prospects here actually are. At present they are dithered by rumours.

dither, *sb.* Add: **b.** A state of tremulous excitement or apprehension; chiefly in phr. *all of a dither*; also, vacillation; a state of confusion. *colloq.* or *dial.*
1819 'P. Bobbin' *Sequel to Lanc. Dial.* 6 (E.D.D.), I'm aw on o' dither, if th' wynt bo sturs a twig. 1891 C. Wordsworth *Rutland Words* 11 Those children keep me in the dithers, they do. 1899 Watts-Dunton *Aylwin* xii. 331 The sight o' both on us..might make the poor body all of a dither if she was very ill. 1929 J. B. Priestley *Good Comp.* iii. ii. 500 They'll rehearse all right...When it comes to the night, all of a dither. 1931 E. Sackville-West *Simpson* iii. xvii, She quickly pulled herself together, feeling that such a state of dither would not, if she showed it, illustrate her name. 1939 N. Marsh *Overture to Death* xxi. 243 Eleanor was thrown into a dither by finding us there together. 1957 S. Jameson *Cup of Tea for Mr. Thorgill* ii. 31 Always in a dither of enthusiasm and misplaced devotion—and what a bore that is! 1958 *Sat. Rev. Lit.* 31 May 8/3 She came up with Stanley Baldwin and his policy of delusion and dither, which left England nearly helpless against Hitler. 1970 M. Pereira *Pigeon's Blood* xi. 127 Such brains are usually characterised by two things: the speed with which they can reach vital decisions; and the speed with which they can grasp how to implement such decisions. A total absence of dither, if you like.

dithery (di·ðəri), *a.* orig. *dial.* [f. DITHER + -Y[1].] Dithering, trembling.
[1866 E. L. Linton *Lizzie Lorton* III. ix. 245 A puir lile diddery doddery horphan.] 1887 T. Darlington *Folk-Speech S. Cheshire* 172, I went queite sick an' dithery. 1931 J. Cannan *High Table* xi. 175 Learned and famous though he was, he was all dithery over having two young nobodies to dine. 1942 J. D. Carr *Seat of Scornful* xiii. 135 She'd already sounded all wild and dithery, but now was worse.

dithing (di·ðiŋ), *vbl. sb. dial.* Also **dithying,** *ppl. a.* [Cf. DITHER *v.*] Quivering, trembling.
1818 R. Wilbraham *Gloss. Cheshire* 14 *Dithing*, a trembling or vibratory motion of the eye, from dither or didder. 1913 Masefield *Daffodil Fields* 66 Red dithying sparks flew..from the fire's heart. 1916 Blunden *Pastorals* 34 Such gusts of rain..dashed on the belfry tower..And roused the dithying goblin on the vane.

dithizone (dəiþəi·zōᵘn). *Chem.* [ad. G. *dithizon* (H. Fischer 1929, in *Zeitschr. f. angew. Chem.* XLII. 1025), f. *diphenylthiocarbazone*.] The crystalline compound diphenylthiocarbazone, $C_{13}H_{12}N_4S$, which is used as a reagent for the estimation and separation of lead and other metals.
1929 *Brit. Chem. Abstr.* A. 1412/2 (*heading*) Detection of heavy metals by means of 'dithizone'. *Ibid.*, Dithizone forms coloured complex compounds with many metals. 1956 *Nature* 31 Mar. 620/1 Phosphate adsorption followed by dithizone extraction. 1957 G. E. Hutchinson *Treat. Limnol.* I. xv. 820 The zinc extractable by dithizone from the waters of Japanese mountain lakes varied from 1·3 to 5 mg. m.⁻³. 1958 *Oxf. Univ. Gaz.* 23 Apr. 880 Some new metal-dithizone complexes.

dithyrambically (diþiræ·mbikăli), *adv.* [f. DITHYRAMBIC *a.*: see -ICALLY.] In or as in dithyrambs; with dithyrambic or 'lyrical' expression.
1891 Symonds *Biogr.* (1895) II. 332 Tell me if you would like me to write what I think about their excellence—not dithyrambically, as here, but soberly as art requires. 1905 *Spectator* 11 Mar. 371/1 M. Santos-Dumont writes interestingly, if dithyrambically, of the future of the airship.

ditto, *v.* Add: **b.** To say or do the same as another person; to agree.
1894 H. Gardener *Unofficial Patriot* 299 They are sulking in their tents and we are dittoing in ours. 1901 *Westm. Gaz.* 22 Mar. 2/1 No, Mr. Balfour knew nothing of Lord Lansdowne's communication. 'Nor I,' dittoed Lord Cranborne. 1922 Joyce *Ulysses* 621 Quite so, Mr Bloom dittoed.

di·ttograph, *v.* [f. the sb.] *pass.* To be repeated by dittography. Hence **di·ttographing** *vbl. sb.*
1897 *Expositor* June 409, x. 22*c* is certainly 'dittographed' from *v.* 22*a*. 1906 S. R. Driver *Jeremiah* 349 The ה at the end..is simply dittographed from the following [word]. 1943 C. L. Wrenn in *Trans. Philol. Soc.* (1944) 20 The -*i* of *maegsibbi* seems obviously to be a mere dittographing of the *i* of *affectui*.

Divali, var. *DEWALEE.

divan. Add: **3.** (Further examples.) Now usually a low bed or couch with no back or ends.
1840 Dickens *Old C. Shop* xi, The bed being soft and comfortable, Mr. Quilp determined to use it, both as a sleeping place by night and as a kind of Divan by day. 1919 W. S. Maugham *Moon & Sixpence* xxix. 129, I had a divan in my sitting-room, and could very well sleep on that. 1954 'V. H. Collins' *One Word & Another* 47 A *divan* is an upholstered piece of furniture...It can be used as a *bed* (often called *divan-bed*) in a bedroom, or as a *settee* in the day and a *bed* at night. 1965 M. Forster *Bogeyman* x. 175 The cover over her divan was red and white striped cotton.
7. *divan-bed, -cover, -seat, -sofa.*
[1919 F. Hurst *Humoresque* 99 A leather-and-oak 'davenbed' had obviously and literally been dragged to the least conspicuous corner.] 1933 *Discovery* Aug. 254/2 There is a divan-bed opposite the window. 1932 D. C. Minter *Mod. Needlecraft* 217/2 Box-shape divan cover. 1898 G. B. Shaw *Philanderer* ii, There are circular recesses at each side of the fireplace, with divan seats running round them. 1893 Divan sofa [see SETTEE[2] a].

dive, *v.* Add: **1. c.** Of a submarine: to submerge.
1872 tr. *Verne's Twenty Thousand Leagues under Sea* (1874) ii. iv. 168, I quite approved of the *Nautilus* entering it [*sc.* the gulf]. Its speed was lessened: sometimes it kept on the surface, sometimes it dived to avoid a vessel. 1902 *Encycl. Brit.* XXXII. 575/2 An officer..dived with her [*sc.* a submarine] in water about 16 ft. deep. 1955 *Oxf. Jun. Encycl.* VIII. 49/2 When a submarine is to dive beneath the surface, its buoyancy is reduced by allowing water to enter large tanks..inside the hull.
d. *Aviation.* To descend or fall precipitously with increasing momentum.
1908 H. G. Wells *War in the Air* iv. § 5 He could feel the airship diving down, down, down. 1914 Rosher *In R.N.A.S.* (1916) 37, I switched on and off, and dived down through the opening to 1,000 feet. 1916 H. Barber *Aeroplane Speaks* 136 *Dive*, to descend so steeply as to produce a speed greater than the normal flying speed. 1959 *Chambers's Encycl.* I. 115/1 There also exists a diving altitude above which the jet comes into its own.
Hence **di·ve-bomb** *v. trans.*, to attack with bombs at a low level after diving. Also *transf.* So **dive-bomber** (= G. *sturzkampfflugzeug*), a dive-bombing aircraft; **dive-bombing** *vbl. sb.* and *ppl. a.*
1935 *Evening News* 11 July 7/1 In dive-bombing, which is the most accurate form of aerial attack on surface targets yet devised, the aircraft is aimed bodily at the target in the course of an almost vertical dive, which is maintained for several thousands of feet. 1936 *Jrnl. R. Aeronaut. Soc.* XL. 720 Dive bombing is limited to low altitude and suffers from insufficient penetration, although the aiming is probably good. 1937 *Flight* 4 Nov. c/1 (*caption*) Great Lakes dive bombers of the U.S. Marine Corps. 1939 *Times* 29 Sept. 10/4 The North Sea Attack. Failure of German Dive-Bombers. 1940 *Ibid.* 8 July 3/4 The aircraft..delivered a dive-bombing attack. *Ibid.* 23 July 2/4 Patrolling off the South Coast, three Hurricane pilots spotted 16 Me. 110s flying line astern to dive-bomb a convoy. 1958 A. J. Toynbee *East to West* xxxi. 94 They [*sc.* hawks] ignored his impertinence and dive-bombed us thick and fast. 1971 P. C. Smith *Stuka at War* iv. 38 The dive-bombers blasted a coastal gun battery on the Isle of Wight.
5. (Earlier example.)
1748 Smollett *R. Random* I. xiii. 102 Walking a few paces, [he] dived into a cellar, and disappeared in an instant.

dive, *sb.* Add: **1. b.** *Aviation.* A precipitate descent. (Cf. *nose-dive and *DIVE *v.* 1 d).
1914 Rosher *In R.N.A.S.* (1916) 13 When in the air, he bawls in your ear, 'Now when you push your hand forward, you go down, see!' (and he pushes your hand forward and you make a sudden dive). 1915 *War Illustr.* 27 Feb. 46/2 The excitement of the dive,..and the swift upward leap of the machine. 1936 *Discovery* Mar. 73/2 The pilot cannot pull the nose of his aeroplane up so quickly that he stalls it with the subsequent danger of a steep dive or spin. 1970 D. L. Brown *Miles Aircraft since 1925* 111 He opened the throttle wide and put the nose down into a steep dive.
c. Of a submarine: submerging, submersion.
1915 W. E. Dommett *Submarine Vessels* iv. 42 When preparing for a dive, the..valves are tried. 1962 G. Weller *All about Submarines* (1963) iii. 36 On *Hunley's* first dive, the flames of her lantern flickered low after only a half-hour...On the next trip the submarine stayed down five times as long.
3. For 'In *U.S.*' read '*colloq.* (orig. *U.S.*)'. (Add further examples.)
1871 *N.Y. Herald* 6 July 8/2 One of the gayly decorated dives where young ladies..dispense refreshments to thirsty souls. 1885 *Referee* 10 May 3/3 A grand entrance takes the place of the tavern, which is relegated to down below, and is called a 'dive'. 1892 Stevenson & Osbourne *Wrecker* viii, I visited Chinese and Mexican gambling-hells, German secret societies, sailors' boarding-houses, and 'dives' of every complexion of the disreputable and dangerous. 1897 *Daily News* 17 Apr. 3/1 From highway into byway they go; now up into tottering garret, then down into dim dive. 1910 *Westm. Gaz.* 25 Jan. 4/1 This dingy 'dive' can boast of many glorious memories. 1940 Auden *Another Time* 112, I sit in one of the dives On Fifty-Second Street. 1958 *Spectator* 4 July 8/3 The degenerate dives of Berlin.
4. *attrib.* and *Comb.* **dive brake** (see quot. 1962); **dive-keeper** *U.S.* (see Dict. sense 3): examples.
1940 C. Gardner *A.A.S.F.* 238 The 87's, with their dive-brakes on, came down vertically to about 600 feet. 1954 *Economist* 11 Sept. (Suppl.) 3/1 The Hawker factories producing Hunters contain two or three hundred complete and half-complete machines waiting for their new dive brakes. 1962 *Gloss. Aeronaut. Terms* (*B.S.I.*) v. 5 *Dive brake*, any device primarily used to increase the drag of an aircraft at will. 1887 *Chicago Tribune* 4 May 3/1 Consternation has seized the divekeepers. 1910 S. E. White *Rules of Game* i. xvi, One of the saloon keepers at Twin Falls...This dive-keeper..had offered transportation.

dive-bomb, -bomber, -bombing: see *DIVE *v.* 1 d.

divergent, *a.* Add: **2. b.** *Psychol.* Of thinking, reasoning, etc.: of a kind that produces a wide variety of possible answers to a problem. So **diver·ger,** one who thinks in this way.
1956 J. P. Guilford in *Psychol. Bull.* LIII. iv. 274 Production factors fall into two groups—convergent-thinking factors and divergent-thinking factors. Such a distinction seems not to have been emphasized in prior literature on thinking. 1966 L. Hudson *Contrary Imaginations* iii. 38 With new tests, it seems vital that we should avoid question-begging if we possibly can. For this reason, I propose to name them technically. The 'High IQ' I shall call a converger; the 'High Creative', a diverger, and the two styles of reasoning, convergent and divergent, respectively. 1970 *Nature* 25 July 420/2 If you are better at conventional IQ tests than at 'open-ended' tests..you are a converger; if not, you are a diverger.

diversification. Add: **c.** *Econ.* The spread of investment over a variety of enterprises, or the production of a variety of different articles, services, etc., often as a safeguard against the effects of fall in demand for a particular product.
1939 S. R. Dennison *Location of Industry* i. ii. 38 Industrial diversification will result in the development of subsidiary and complementary industries, and also of entirely new lines of manufacture. 1944 A. Cairncross *Introd. Econ.* ii. vi. 76 Such diversification makes the firm less vulnerable to sudden changes and allows it to remain a going concern where smaller, less diversified concerns would be forced to give up business. 1971 *Brit. Printer* Jan. 104/1 A stone-layer, named Hans Rüegger, venturing into an early example of diversification, added a printing shop to his business.

diversify, *v.* Add: **1. c.** *Econ.* To introduce or use diversification in (see prec.). Also *intr.*, to engage in diversification.
1939 S. R. Dennison *Location of Industry* i. iii. 72 As industry becomes diversified, as the needs of consumers grow.., then the market exerts a stronger attraction. 1944 A. Cairncross *Introd. Econ.* ii. vi. 76 Firms may seek to spread their risks by diversifying their output, or markets, or sources of supply, or processes of manufacture. 1967 *Times Rev. Industry* Apr. 39/1 Westland Engineers itself diversified when it bought up Unique Balance Company, making sash balances for windows. 1973 *Guardian* 6 Mar. 1/1 An explosives firm at Great Oakley, near Harwich, has decided to 'diversify' into potato chips. Its factory was threatened with closure unless a new product was introduced.

diversion. Add: **1. c.** An alternative route by-passing a road that is temporarily closed.
1955 *Times* 19 Aug. 5/1 One of the hazards for the motorist is finding a way round the many places where the road is up with the aid of, sometimes in spite of, what has become almost a Viennese institution—the Umleitung or diversion. 1958 'A. Gilbert' *Death against Clock* xii. 167 When you put up your road blocks, then I leave to use the diversion.
6. *attrib.* and *Comb.*: **diversion-cut,** a channel made to divert impure water past a reservoir; **diversion weir,** a weir erected to divert water from a river to the head of an irrigating canal.
a 1877 Knight *Dict. Mech.*, *Diversion-cut*, a channel to divert past a reservoir a stream of impure or turbid water which would otherwise flow into the reservoir. A by-wash. 1893 *Amer. Rep. U.S. Geol. Surv.* 1892–3 III. 231 One of the latest..diversion weirs constructed in this country is that built at the head of the Turlock and Modesto canals.

diversionary, *a.* Delete *rare*⁻¹ and add later examples.
1957 P. Worsley *Trumpet shall Sound* vii. 242 Diversionary action was sometimes taken. 1958 *Times Lit. Suppl.* 21 Nov. 677/2 The book tells the story in great detail ..explaining..why the diversionary bombing failed. 1960 *Times* 13 Apr. 6/3 Waddington is not only a V-bomber base but a diversionary airfield for civil aircraft.

diversionist (divə̄·ɹʃənist, dai-). [f. DIVERSION + -IST; cf. Russ. *diversant*.] In Communist usage: a saboteur; also, one who conspires against the government. Also *attrib.* or as *adj.* Hence **diver·sionism,** the activity of a diversionist.
1937 *Daily Tel.* 28 Aug. 12/3 A woman railway worker..has been sentenced to be shot..for 'putting sulphuric acid in the water bottles' of three trains bound for Moscow.. Her subtle-washer and fellow 'diversionist'..was sentenced to 10 years' imprisonment. 1949 F. Maclean *Eastern Approaches* I. ii. 24 For some years past numerous politicians and others had met with their fate, variously branded as 'Trotskists', 'wreckers', 'Fascist spies', 'diversionists' and so on. 1951 Koestler *Age of Longing* I. vi. 110 We have proofs that Nadesha Filipovna was one of the leaders of this diversionist conspiracy. 1955 *Reporter* 16 June [illegible] indulged in right-wing diversionism and petty-bourgeois wrecking? 1955 *Times* 1 July 10/6 He pleaded Guilty to political crimes and diversionist activity, but denied collaborating with the Gestapo during the war.

diversity. Add: **4.** **diversity factor** *Electr.* (see quot. 1943).
1905 *Fabian Tract* cxix. 6 When we speak of a good diversity factor we mean that the generating station is so happily situated that it meets a regular and constant maximum demand for diverse purposes...A continuous 'diversity factor' makes a good 'load factor'. 1925 *Standardization Rules Amer. Inst. Electr. Engin.* 17 *Diversity Factor*, is the ratio of the sum of the maximum power demands of the subdivisions of any system or parts of a system to the maximum demand of the whole system or of the part of the system under consideration, measured at the point of supply. 1930 *Sci. Abstr.* B. XXXIII. 38 Nomograms for determining the diversity factor of a network having a number of loads of different magnitudes and durations. 1943 *Gloss. Terms Electr. Engin.* (*B.S.I.*) 90 *Diversity factor*, the ratio of the sum of the maximum demands of the several consumers or loads to their maximum simultaneous demand.
5. *Radio.* Applied *attrib.* to a system of reception in which, in order to reduce the effects of fading, a signal is received simultaneously by several channels, the signal or combination of signals of best quality being automatically selected.
1930 *Proc. Inst. Radio Engin.* XVIII. 1738 The R.C.A. receiving stations now employ the diversity system, usually with three antenna groups for a single unit. 1938 *Admiralty Handbk. Wireless Telegr.* II. R. § 54 *Diversity reception*. This is the name given to various schemes that may be utilised at a receiving station...If a signal fades at one place, it may be quite strong at another a few hundred feet away. 1966 *Electr. Comm.* XXXVI. 123/1 Techniques of diversity reception have advanced rapidly...Two or more [radio] signals are separately received at different frequencies or over different paths.

diverter. Add: Also **divertor. b.** *Electr.* Any of several devices that provide an alternative path for an electric current or for magnetic field lines; *spec.* (*a*) a resistance placed in parallel with the series winding of a series- or compound-wound d.c. motor or generator in order to alter the size of the current through the windings; (*b*) a device that protects a transmission line or associated equipment from surges by providing a path to earth for the surge current.
1891 J. W. Urquhart *Dynamo Constr.* xvii. 242 A device, having for its object the controlling of the magnetic intensity was formerly employed in the case of magneto machines. It consisted in placing a 'divertor'..of soft iron partially across the limbs of the magnet, so as to magnetically short-circuit the magnet. 1909 Hawkins & Wallis *Dynamo* (ed. 5) I. xvi. 542 Where it may be necessary to alter the amount of compounding..a diverter is employed just as with a series-wound dynamo. 1934 J. Henderson *Automatic Protective Gear* v. 63 When the straight-through fault current reaches an abnormal value..the diverter relay operates instantaneously. 1946 *Nature* 23 Nov. 742/1 Their development was first stimulated by the requirements of surge diverters (lightning arresters) for overhead power transmission lines. 1965 B. J. Cory *High Voltage Direct Current Converters & Systems* viii. 213 Diverters are installed to protect this item from overvoltages that would arise on account of arc-quenching. 1968 J. F. Whitfield *Electr. Installations Technol.* viii. 167 Terminal voltage is adjusted by means of the diverter resistance..which carries part of the current, thus reducing field current and terminal voltage.

divertible, *a.* (Examples.)
1904 W. James in *Atlantic Monthly* July 103/2 He who had the keenest eye for instances and illustrations, and was least divertible by casual side-curiosity, would score the quickest triumph. 1928 *Manch. Guardian Weekly* 31 Aug. 178/3 [illegible] will be diverted into the parallel valley of the Atchafalaya River.

diverticulectomy (daivəɹtikiŭle·ktŏmi). *Surg.* [f. DIVERTICUL(UM + *-ECTOMY.] The excision of a diverticulum.
1926 Young & Davis *Young's Pract. Urology* II. xvi. 335 *Diverticulectomy*. This term is applied to operations for the removal of vesical diverticulum. 1928 J. S. Horsley *Operat. Surg.* (ed. 3) xxx. 837 Approach to the bladder for resection of the bladder wall, diverticulectomy, for stone or for drainage is frequently indicated. 1961 *Lancet* 30 Sept. 744/1 Vesical diverticulectomy.

diverticulitis (daivəɹtikiulai·tis). *Path.* [f. DIVERTICUL(UM + -ITIS.] Inflammation of a diverticulum.
1909 in Dorland *Med. Dict.* 211/1. 1923 R. Knox *Radiogr. & Radio-Therap.* (ed. 4) I. 364 (*heading*) Diverticulitis of the colon. [illegible] 1969 N. S. Painter in B. C. Morson *Dis. Colon, Rectum & Anus* xvii. 201 When diverticula are causing no symptoms or only minimal symptoms the disease is called diverticulosis and when one or more diverticula are inflamed the condition is called diverticulitis.

diverticulosis (daivəɹtikiulōᵘ·sis). *Path.* [mod.L., a. G. *diverticulosis* (F. de Quervain 1914, in *Deut. Zeitschr. f. Chirurgie* CXXVIII. 81), f. DIVERTICUL(UM + -OSIS.] The presence of diverticula, esp. in the intestine.
1917 *Amer. Jrnl. Roentgenology* IV. 281/2 Both carcinoma and diverticulosis represent important causes of pelvic colon obstruction. 1948 A. S. Till in H. Souttar *Textbk. Brit. Surg.* I. xi. 352 (*heading*) Diverticulosis of the small intestine.

divertimento. b. Substitute for def.: (*a*) A composition designed primarily for entertainment, esp. a suite of movements for a chamber ensemble. (*b*) An arrangement of, or fantasia on, airs from an opera, etc. (Further examples.)
1948 H. Ulrich *Chamber Music* (1966) v. 122 Various versions of the divertimento differed widely in the number of movements, but all had two features in common: their light, cheerful content and their derivation from the three-movement *sinfonia*. 1955 *Times* 14 May 4/5 We would have been glad to have heard one of Mozart's earlier symphonies, or larger divertimenti. 1958 Auden & Kallman *Magic Flute* 58 And even those Divertimenti which He wrote to play while bottles were uncorked.

divertor, var. DIVERTER.

divide, *v.* Add: **1. c.** Phr. *divide and rule* (occas. *govern*) [tr. L. *divide et impera* (also used)]: a statement of the policy of not allowing subject peoples or factions to make common cause.
1602 W. Watson *Decacordon* iii. 69 According to Machiavels rule of *divide & impera*. 1605 J. Hall *Meditations* i. 109 For a Prince, that he may have good successe against other rebels or forraine enemies, it is a sure axiome, Divide and rule. 1870 Brewer *Dict. Phr. & Fable* 231/1 *Divide and Govern*. Divide a nation into parties, or set your enemies at loggerheads, and you can have your own way. [illegible] 1936 P. Fleming *News from Tartary* v. iii. 198 There has really been no need for the Chinese to put their immemorial colonial policy of *Divide et impera* into practice; nature has done it for them. 1948 B. Stevenson *Home Bk. Proverbs* (1959) 101/1 'Divide et impera'..was the motto of Philip of Macedon and of Louis XI of France, in dealing with his nobles. It was the traditional motto of Austria. Polybius, Bossuet, and Montesquieu used it, but it is generally ascribed to Machiavelli. 1962 *Listener* 26 Apr. 718/2 True to their traditional 'divide and rule' policy British diplomats tried to deepen the differences between the Kenya African National Union and the Kenya African Democratic Union.
8. a. Also with *up*.
1914 E. Cannan *Wealth* v. 82 Even the pasture was divided up with the small exceptions which we see in the 'commons' of the present day.

divide, *sb.* Add: **2.** *the Great* (*Continental*) *Divide*, the watershed of the Rocky Mountains; *fig.* a dividing or boundary line; *spec.* the boundary between life and death.
1868 *Congress. Globe* 14 July 4068/1 The doctrine of political equality forms the great 'divide' between parties now as heretofore. 1869 W. J. Palmer *Surv. across Continent* 171 The great Continental Divide at Arkansas Pass. 1872 J. H. Tice *Over Plains* 214 [Tales] of those who long since 'have gone over the Divide'. 1907 C. E. Mulford *Bar-20* xxiii. 226 Snip! goes his bill an' th' snake slides over th' Divide. 1908 —— *Orphan* xi. 139 If he was killed, he would have company across the Great Divide. 1909 *Daily Chron.* 16 Sept. 1/2 He was good to Ruth, and she, too, loved him. But between them still was 'the great divide'. She could not forget that he had bought her for a string of nuggets. 1955 C. S. Lewis *Let.* 25 June (1966) 263 Instead of saying the Great Divide came between the Middle Ages and the Renaissance, I said..that it didn't. 1965 *Listener* 16 Sept. 414/2 This is the divide between Barth and Aquinas, it is the divide between the conservative evangelical tradition and liberal theology, it is the divide between Biblicism and the Bishop of Woolwich.

divided, *ppl. a.* Add: **1. d.** *divided skirt*: see SKIRT *sb.* 1.

dividend. Add: **3. b.** (Later *fig.* example.)
1965 *Listener* 16 Sept. 402/2 Nothing in fact will pay better dividends in the long run than a determined effort to discover what is actually going on in the Health Service.
5. **dividend-stripping** (see quot. 1959); hence *dividend-stripper*.
1958 *Times* 22 Apr. 10/3 Nothing in the Budget created more concern among many Conservative backbenchers than the retrospective effect of the proposal by the Chancellor of the Exchequer to make the tax evading practice of 'dividend stripping' illegal—with effect from October, 1955. 1958 *Punch* 25 June 839/2 What Odhams and the *Daily Herald* had done was not indeed exactly the same as what the dividend-strippers had done, and Mr Houghton may have been right in saying that the object of that exercise was not to avoid taxation. 1959 *Times* 8 Apr. 17/2 Dividend-stripping (which is essentially a finance company operation for offsetting a dealing loss against tax reclaimed from dividends accumulated by the company which is the subject of the deal) was stopped some time ago.

di·videndless, *a.* [See -LESS.] Without dividends.
1899 *Westm. Gaz.* 26 Jan. 8/1 If the Hyderabad-Deccan Company were in its infancy instead of having laboured on through thirteen years of a dividendless career. *Ibid.* 2 Mar. 8/1 The dividendless stock of the District Railway. 1920 *Glasgow Herald* 4 Sept. 10 It may be noted that shareholders again go dividendless. 1938 *Times* 18 Aug. 16/2 Not only they but also the Preference stockholders..go dividendless.

divider. Add: **9.** A partition or screen; *spec.* a piece of furniture or the like dividing a room into two parts; freq. *room-divider*.
1959 D. Barton *Loving Cup* 95 Alastair slid back a panel in a walnut room divider and brought out brandy and glasses. *Ibid.* 96 Climbing plants trailing up the room divider. 1960 *House & Garden* May 59 Divider (with glass shelves), £30. 1967 E. Short *Embroidery & Fabric Collage* iii. 80 A feature of modern open-plan living-rooms is often the room divider, screening the dining area or study from the main sitting room. 1970 *Harper's Bazaar* Oct. 17/2 The restaurant is spacious with raffia'd dividers between the tables.

divine, *a.* Add: **5. b.** Freq. in trivial use.
1818, 1826 [in Dict.]. 1870 L. M. Alcott *Old-Fashioned Girl* (1874) iii. 42 Your foot is perfectly divine in that boot. 1960 R. Daniel *Death by Drowning* iv. 45 I've just bought a divine swim suit.
8. *divine-looking* adj.
1937 F. Scott Fitzgerald *Let.* 8 Oct. (1964) 18 I'm glad Stanley is divine-looking; sorry Andrew is repulsive.

divinely, *adv.* Add: **2.** Also in trivial use: excellently, extremely well.
1822 [in Dict.]. 1927 L. Mayer *Just between us Girls* vii. 43 Honestly those nobilities *can* make love divinely. 1928 E. Waugh *Decline & Fall* i. ix. 98 He plays just too divinely. 1929 E. Bowen *Last Sept.* xvii. 227 You do dance divinely. 1930 H. Wolfe *Uncelestial City* iii. 124 My dears, too grim, too psycho-analytically ungodly, and altogether too divinely Jim.

diving, *vbl. sb.* Add: **b.** *diving-suit*; **diving-board,** a board projecting some distance over the water, from which a swimmer dives; **diving-plane,** a horizontal rudder on a submarine for steering the vessel up or down; also a similar device fitted to instruments used in oceanography.
1893 Sinclair & Henry *Swimming* iv. 108 A spring diving-board is generally used for running headers. 1908 J. London *Let.* 26 Oct. (1966) 264, I supply it with.. diving-suits and boats. 1915 Kipling *Fringes of Fleet* 29 A mine and chain had jammed under her [*sc.* the submarine's] forward diving-plane. *c* 1938 L. Mumford *City Development* (1946) 112 The curve of a diver's body as he leaps from the diving-board. 1943 C. S. Lewis *Perelandra* ii. 27 If they have a scientific civilisation they may have diving-suits. 1959 H. Barnes *Oceanogr. & Marine Biol.* i. 31 (*caption*) Small high-speed Plankton Indicator... All models except the last have their own diving planes. 1959 *Jane's Fighting Ships 1959–60* 414 She incorporates several novel features including hydro-wings or diving planes fitted to the conning tower 'fin'.

diving-bell. Add: **b.** The air-filled web in which the water-spider lives under water.
1854 [see WATER-SPIDER]. 1961 *Listener* 7 Dec. 986/2 A water-spider beside its 'diving-bell', a bubble of air contained in a web.

divinity. Add: **6.** **divinity fudge** *U.S.*, a type of home-made fudge.
1913 E. H. Glover *'Dame Curtsey's' Bk. Candy Making* 34 Divinity Fudge. Three and one-half cups of granulated sugar, one-half cup of 90 per cent corn syrup, [etc.]. 1970 *New Yorker* 5 Sept. 36/3 My mother stayed out of the kitchen as much as possible, except for making divinity fudge perhaps once a year.

‖**divisa** (divī·sā). [Sp., device, emblem (cf. DEVICE).] In *Bull-fighting*: coloured ribbons denoting the breeder of a bull.
1932 R. Campbell *Taurine Provence* ii. 45 To seize the divisa (or cocarde) and present it to the lady of one's affections is the summit of gallantry. 1967 McCormick & Mascareñas *Compl. Aficionado* ii. 52 An attendant placed a short barb with the *divisa* of the ganaderia in the toro's *morillo*.

‖**divisi** (divī·si). [It., lit. 'divided', *pa. pple. pl.* of *dividere* to DIVIDE.] A musical direction (see quot. 1876).
1740 Grassineau *Mus. Dict.* 65 *Divisi*, signifies, divided into two or more parts. 1876 Stainer & Barrett *Dict. Mus. Terms* 135/2 *Divisi*,..a direction that instruments playing from one line of music are to separate and play in two parts. 1929 *Melody Maker* Mar. 310/1 Six first violins (playing three *divisi* parts). 1940 G. Jacob *Orchestral Technique* (ed. 2) x. 100 In quiet passages use 'divisi' rather than double-stops.

division. Add: **1. e.** *Hort.* The propagation of perennial plants by splitting clumps into parts capable of rooting themselves.
1805 T. A. Knight *Rep. Committee Hort. Soc. Lond.* 5 Almost every plant, the existence of which is not confined to a single summer, admits of two modes of propagation; by Division of its Parts, and by Seed. 1841 J. Loudon *Ladies' Compan. Flower Garden* 86/1 Plants are said to be propagated by division when they are taken up and separated into portions. 1915 T. W. Sanders *Pop. Hardy Perennials* i. 24 By division, the simplest of all methods, we can easily obtain strong plants of any perennial. 1971 E. Coxhead *One Woman's Garden* vii. 54, I started with one plant and by division now have three.
f. *Biol.* The spontaneous separation or breaking up of a cell into two or more approximately equal parts that constitute daughter-cells, usu. involving division of the nucleus (if any) followed by the breaking up of the cell as a whole; freq. as *cell division*. As a mode of reproduction of simple organisms usu. termed FISSION or *schizogony*.
1880 *Encycl. Brit.* XII. 13/1 Four types of Cytogenesis may be distinguished: (1) Rejuvenescence; (2) Conjugation; (3) Free-cell formation; and (4) Division. 1896 E. B. Wilson *Cell* ii. 45 In the multicellular organism all the tissue-cells have arisen by continued division from the original germ-cell, and this in its turn arose by the division of a cell pre-existing in the parent-body. 1901 T. H. Morgan *Regeneration* 149 The breaking up of lumbriculus or of a planarian into pieces that form new individuals is a typical example of division. 1920 L. Doncaster *Introd. Study Cytol.* xv. 247 The differentiation of the germ-layers is not conditioned by differential nuclear division, but by unequal division of the cytoplasm. 1940 Parker & Haswell *Text-bk. Zool.* (ed. 6) I. ii. 89 Multiplication [of *Paramecium*] takes place by binary fission (*D*), the division of the body being preceded by that of both nuclei. 1964 G. H. Haggis et al. *Introd. Molecular Biol.* i. 13 A remarkable bubbling of the cytoplasm occurs, and this becomes furiously active in the later stages of cell division. After division the movement dies away and the daughter cells spread out, [etc.]. 1970 Ambrose & Easty *Cell Biol.* i. 20 Cell division is the way in which reproduction occurs in most simple unicellular organisms.
10. c. Also *pl.* The parade of a ship's company according to its divisions.
1915 'Bartimeus' *Tall Ship* iii. 54 Nine o'clock, sir; all ready for divisions. *Ibid.* 55 A moment later the bugle overhead blazed forth 'Divisions'. 1947 B. Mason in D. M. Davin *N.Z. Short Stories* (1953) 338 It was Sunday, and we began to make ourselves 'tiddly' for Divisions. 1971 *Daily Tel.* 18 Jan. 10/3 'Divisions', or the muster and inspection of the ship's company each morning in harbour, long regarded as sacrosanct, now seldom occurs more than once a week.
e. A section of a railway line. Also *attrib.* *U.S.*
1858 W. P. Smith *Gt. Railway Celebrations* 98 The opening of the Western Division of the O. and M. Railroad. 1887 C. B. George *40 Yrs. on Rail* xii. 254 My plan.. is to have a book to be called the division book kept by each company. 1891 C. Roberts *Adrift Amer.* 71 Every line in the United States is divided into divisions of various lengths...Each division is under the supervision of a man who is called a division road master. *Ibid.* 96 Wallace was a division terminus. 1892 A. C. Gunter *Miss Dividends* 10 This citizen soldier..had been one of the division engineers of the Union Pacific Railway.
f. Any of the two or three grades of imprisonment to which certain misdemeanants could be sentenced with a view to separating hardened criminals from less serious offenders. *Obs.* exc. *Hist.*
1865 *Act* 28 & 29 *Vict.* c. 126 § 67 In every Prison to which this Act applies Prisoners convicted of Misdemeanor, and not sentenced to Hard Labour, shall be divided into at least Two Divisions, One of which shall be called the First Division; and..a Misdemeanant of the First Division shall not be deemed to be a Criminal Prisoner within the Meaning of this Act. 1898 *Act* 61 & 62 *Vict.* c. 41 § 6 Prisoners..not sentenced to penal servitude or hard labour, shall be divided into three divisions...Where a person is..sentenced to imprisonment without hard labour, the court may direct that he be treated as an offender of the first division or as an offender of the second division. If no direction is given by the court, the offender shall..be treated as an offender of the third division. 1918 A. Huxley *Let.* 30 Apr. (1969) 150 His trial takes place to-morrow, the appeal, which wont, I imagine, do any good except perhaps to change his six months from second to first division. 1947 *Rep. Commissioners of Prisons 1945* 64 in *Parl. Papers* 1946–7 (Cmd. 7146) XIV. 155 The provisions of this Bill [*sc.* the Criminal Justice Bill, 1938], so far as concerns the prison system, proposed..the abolition of sentences of Penal Servitude and imprisonment with Hard Labour, and of the Triple Division of Offenders.
g. In the Civil Service, the technical designation of the several grades of clerks.
1876 *London Gaz.* 12 Feb. 638/2 A Lower Division of the Civil Service shall be constituted. It shall consist of Men Clerks and of Boy Clerks, engaged to serve in any Department of the State to which they may, from time to time, be appointed or transferred. *Ibid.* 639/1 Promotion from the Lower to the Higher Division of the Service shall not be made without a special certificate from the Civil Service Commissioners. 1892 S. Savill *Civil Service Coach* 2 The salaries of Clerks in the Second Division..shall commence at 70*l.* per annum. 1898 *Guide Employm. Civil Service* 15 Second Division Clerkships...The Second Division forms at present the rank and file of the permanent Civil Service. 1959 *Chambers's Encycl.* III. 605/1 There is a Civil Service Clerical Association and a First Division Association (the latter representing the administrative grade and taking its name from the old name for that grade).
h. In Association Football, a group of teams in competition, usu. forming part of a league.
1899 Oakley & Smith *Football* iii. 170 A Second League has been formed, and the number of clubs admitted to the First League or Division has been enlarged. 1908 'Bedouin' *Scottish League Football* 24 The Second Division of the League was created in 1894–5. 1912 A. Bennett *Matador of Five Towns* 9 If Knype drop into the Second Division..it 'll be all up with first-class football in the Five Towns! 1966 *Listener* 20 Jan. 88/1 Ipswich Town topped the First Division for a season without ever seeming of international calibre. 1971 *Times* 13 Apr. 7/1 Leicester, in essence, are a first division club.

divisional, *a.* Add: **2.** *Divisional Court*: a court constituted by two or more judges of the High Court to try cases from one of the divisions (see DIVISION 10 a), when such cases are not suitable to be heard by a single judge.
1873 *Act* 36 & 37 *Vict.* c. 66 § 40 Such causes and matters as are not proper to be heard by a single judge shall be heard by divisional courts of the high court, which shall for that purpose exercise all or any part of the jurisdiction of the high court. 1928 *Daily Mail* 3 Aug. 10/6 In recent years the trend of decisions is very different and reached its culminating point in the Divisional Court this year. 1939 Bishop & Sachs *ABC of Divorce Practice* 4 The President of the Probate, Divorce and Admiralty Division, with other Judges attached thereto are the permanent Judges of the Divorce Division and a Divisional Court..is constituted by two of these Judges or by any two Judges of the High Court, sitting together.

divisionism (divi·ʒəniz'm). [f. DIVISION + -ISM.] The practice of painting with pure colours, and of achieving the effect of mixed colours by the juxtaposition on the canvas of contrasting colours instead of by mixing the desired shade on the palette. Hence **divi·sionist,** a painter who follows this practice; also *attrib.* or as *adj.*
1901 L. Villari *Giovanni Segantini* 53 His 'divisionist' drawings. *Ibid.* 73 In this replica he began to apply the system of divisionism. 1920 *Times Lit. Suppl.* 18 Nov. 745/4 Cézanne, with his insistence upon the volumes of objects, rebelled against the disintegration of form which was the logical consequence of divisionism. 1921 *Edin. Rev.* Apr. 304 The most enthusiastic Divisionist discovered there were limits beyond which his theories could not be put into practice. 1926 F. Rutter *Evol. Mod. Art* 41 A Divisionist painter desiring a grey..places on his canvas little touches, say, of pure violet in juxtaposition to little touches of a yellowish green. 1949 Koestler *Insight & Outlook* 391 The great influence which Seurat's system of 'divisionism' exerted on painters of his generation. 1970 *Oxf. Compan. Art* 318/2 Notable precursors of Divisionism were Watteau and Delacroix. 1971 *Listener* 4 Mar. 274/1 Pointillist or divisionist in detail.

divorce, *sb.* **4.** (Further examples.)
1905 *Macm. Mag.* Nov. 57 Is it an ill thing that the newspapers should publish detailed reports of divorce-suits? 1945 Divorce case [see *AT *prep.* 40]. 1955 M. Gluckman *Custom & Conflict in Afr.* iii. 79 Social factors and not only personal disharmonies may control divorce-rates in Western society. 1960 *Spectator* 23 Sept. 444 Where people aren't brittle and promiscuous and divorce-prone.

divorceless. Delete *nonce-wd.* and add earlier and later examples.
1595 Chapman *Coronet* st. iv in *Ovid's Banquet* sig. F1ᵛ, And in th' vntainted Temple of her hart Doth the diuorceless nuptials celebrate Twixt God and her. 1914 H. G. Wells *Englishman looks at World* xv. 207 There was a time in this country when our marriage was a practically divorceless bond.

divot, *sb.* Add: **2.** *Golf.* A piece of turf cut out with a club by a player in making a stroke.
1886 H. G. Hutchinson *Hints on Golf* 9 With an iron club an unskilful player is more likely to cut fids of turf—*golficè*, 'divots'—out of the green. 1890 —— *Golf* x. 272 A divot well replaced is, in most conditions of the ground, as a divot that has never been cut. 1935 O. Nash *Primrose Path* (1936) 105 The wretched golfer, divot-bound.

Divvers (di·vəɹz). *Oxford University slang.* [f. first syllable of DIVINITY + *-ER[6] + -S (representing the pl. *moderations*).] Formerly, divinity moderations, the first public examination in Holy Scripture.
1905 *'Varsity* 30 Nov. 109/3 Those who are in for 'Divvers' should make sure of knowing all about St. Stephen's Speech. 1913 *Isis* 8 Feb. 184/2 Honour Mods. and Divvers behind him and Groups before. 1949 L. A. G. Strong *Maud Cherrill* 12 An obsolete examination known to generations of undergraduates as Divvers. 1969 G. Smith in *Lett. of A. Huxley* 11 H. passes the Examination in Holy Scripture ('Divvers'), which he has previously failed.

divvy, *sb.* and *v.* Add: (Earlier and later examples.) Freq. *to divvy* (*up*), to divide (up).
1872 in *Amer. Speech* (1952) XXVII. 77 A fraudulent 'divy'. 1877 L. Grover in Goldberg & Heffner *America's Lost Plays* (1940) IV. 203 We divvy a cool $20,000. 1880 A. A. Hayes *New Colorado* (1881) xi. 156 The two men, unsuspicious of danger were 'divvying up' the spoils in the middle of the road. 1884 J. Hay *Bread-winners* x. 150 'You surely do not intend—' 'To strike Saul for a divvy?' 1926 Kipling *Debits & Credits* 156, I pinched 'em an' divvied with Macklin. 1936 'R. Hyde' *Passport to Hell* xv. 227 They didn't divvy up with the spoils like our crowd would have done. 1938 J. Rice *Somers Inheritance* iv. vii. 264 It says here his money is to be divvied up. How'd you like a cut-in? 1957 *Times Lit. Suppl.* 20 Dec. 766/5 The problem of 'divvying-up' the spoils of automation.

divvy (di·vi), *a. slang.* [f. first syllable of DIVINE (A. 5 b) + -Y[1]; cf. *DEEVY *a.*] Extremely pleasant, 'divine', 'heavenly'.
1903 *Daily Chron.* 1 Aug. 8/1, I heard one of them say that 'the dimpy was divvy', and this, when translated, meant that a certain dinner party was divine. 1924 F. M. Ford *Some do Not* i. ii. 48 The only divvy moment was when you stood waiting in the booking-office for the young man to take the tickets.

Diwali, var. *DEWALEE.

dix (dis). [Fr., = ten.] The lowest trump in Bezique and other card games; also, in Pinocle, a certain score of ten points, or the trump entitling the player to it.
1908 R. F. Foster *Pinochle* 5 In two-hand, any player holding the dix may exchange it for the trump card immediately after winning a trick. 1952 E. Kempson et al. *Hoyle Up-to-Date* 151 A player holding a dix may count it merely by showing it upon winning a trick.

‖**dix-huitième** (dizwityɛm), *a.* [Fr., = eighteenth.] Of or belonging to the eighteenth century. Also as *sb.*
1920 *Times Lit. Suppl.* 26 June 538/1 In a charming and characteristically dix-huitième manner. 1926 A. Huxley *Essays New & Old* 87 The more sophisticated find their *dix-huitième* in the original French documents..or in the ironic pages of Mr. Lytton Strachey. 1926 —— *Jesting Pilate* i. 166 The *dix-huitième* parody of Chinese art. 1936 F. R. Leavis *Revaluation* iii. 68 The post-war cult of the *dix-huitième*.

dixie[1] (di·ksi). Also **dechsie, dixey, dixy.** [Hind. *degchi, -cha*, a. Hindi *degachi, -chā*, Punjabi *dekachi, degāchi, -chā*, ad. Pers. *degcha*, dim. of *deg, dig* iron pot, kettle, cauldron.] An iron kettle or pot, used by soldiers for making tea or stew.
[1879 Mrs. A. E. James *Ind. Househ. Managem.* 40 Six dechsies and covers. *Ibid.* 45 A few *dechsies* (copper pots).] 1900 *Westm. Gaz.* 29 Mar. 8/1 On halting at Klip Drift we immediately got down our dizies [*sic*] and made tea for all. 1900 *Daily News* 10 July 3/2 The 'billy' is what Tommy calls a 'dixie'. 1901 *Westm. Gaz.* 26 June 8/1 With much difficulty water was procured from a spruit over a mile away, and the 'dixey' boiled. 1916 H. G. Wells *Mr. Britling* ii. iv. § 14 They shelled us again next day and our tea dixy was hit. 1916 *Anzac Bk.* 41/2 Tea made in the stew dixie, and tasting more of dixie and stew than of tea. 1942 C. Barrett *On Wallaby* iv. 71, I returned with..a dixie much the worse for wear.

Dixie[2] (di·ksi). *U.S.* [Origin obscure: see Mathews *Dict. Amer.*]
1. The southern United States; the South. Also *Dixie('s) Land, Dixieland.*
1859 D. D. Emmett *Dixie's Land*, Away! away! away down South in Dixie. *Ibid.*, In Dixie Lann whar I was bawn in, Arly on one frosty mawnin. 1859 *N.Y. Herald* 4 Apr., [Bryant's Minstrels are giving] Dixie's Land, another new Plantation Festival. 1861 G. P. Putnam (*title*) Before and after the battle; a day and night in 'Dixie'. 1864 W. Pittenger *Daring & Suffering* 35 That coat..I wore all through Dixie. 1866 C. H. Smith *Bill Arp* 139 I'm a good Union reb, and my battle cry is Dixie and the Union. 1901 W. Pittenger *Gt. Locom. Chase* 101 Now I will succeed, or leave my bones in Dixie. 1903 *N.Y. Times* 10 Dec. 5 Nearly 400 exiles from Dixie Land gathered at the annual dinner of the Southern Society. 1948 *Daily Ardmoreite* (Ardmore, Okla.) 15 July 1/8 The 20-minute show Dixieland put on..was more liberally sprinkled with rebel yells.
b. The music or words of the song of 'Dixie', written by D. D. Emmett in 1859.
1860 E. Cowell *Diary* (1934) 231 The irrepressible 'Dixie' predominates, but sentimental ditties are in high favor. 1904 *Minneapolis Times* 23 June 8 The orchestra in a Georgia theatre quieted a panic-stricken crowd by playing 'Dixie'. 1911 H. S. Harrison *Queed* xxi. 261 From far away floated the strains of 'Dixie', crashed out by forty bands.
c. *Dixieland*: used *attrib.* or quasi-*adj.* to designate a type of jazz music originally played in the southern United States, characterized by a strong rhythm of two beats to the bar and collective improvisation. Also *ellipt.* Hence **Di·xielander,** one who plays Dixieland jazz.
[1919 *Punch* 16 Apr. 293/1 'The Original Dixie Land Jazz Band has arrived in London,' says an evening paper.] 1927 *Melody Maker* June 553/1 A dance band..need not be merely the type of 'Dixieland' jazz band. 1934 S. R. Nelson *All about Jazz* i. 25 The Dixielanders are not so démodé as one would think. *Ibid.* vi. 141 'The Jazz Band' was a 'Dixieland' combination playing modernized 'Dixieland' style. 1937 *Étude* Dec. 835/1 *Dixieland*, the original, New Orleans jazz as developed by the famous 'Dixieland Five'. 1939 W. Hobson *Amer. Jazz Music* 210 Dorsey plays a strong bass-part foundation in the style known as 'Dixieland'. 1946 *Amer. Jazz* No. 1. 13/1 The results are probably nearer to the earliest Dixieland (containing a coloured element) than anything waxed in recent years. 1950 A. Lomax *Mr. Jelly Roll* (1955) iii. 126 The all-white Original Dixieland Jazz Band of 1917 (by chance the first band to record jazz) is generally reckoned the originator of 'Dixieland'. *Ibid.* iv. 179 The white New Orleans Dixielanders. 1970 *Melody Maker* 12 Sept. 35/7, I strongly suspect it will still be possible, somewhere or other, to listen to live Dixieland in 1999.
2. (See quot. 1873[2].) Also *attrib.*
1873 J. H. Beadle *Undevel. West* xxx. 660 'Dixie wine' as the Mormons call it, is rather strong and pungent. *Ibid.* 661 All that part of Mormondom south of the rim of the Great Basin is called Dixie, and extends some distance into Arizona. 1894 *Irrigation Age* Jan. 38/1 The famous 'Dixie Land', comprising the counties of Millard, Washington and Beaver. 1942 W. Stegner *Mormon Country* 345 Dixie sleeps peacefully all winter with hardly a Gentile intruder.

Dixiecrat (di·ksikræt). *U.S. colloq.* [f. *DIXIE[2] + DEMO)CRAT 2.] A member of a group of southern United States Democrats who seceded from the Democratic Party in 1948 because they opposed its policy of extending civil rights; a States' Rights Democrat. Also *attrib.* Hence **Dixiecra·tic** *a.*; **Di·xiecratism.**
1948 *Birmingham* (Ala.) *News* 22 May 5 Truman finds some Dixiecrats supporting him on A-veto stand. 1948 *Tuscaloosa* (Ala.) *News* 14 July 4/1 States Rights' walkout delegates..will..select a 'Dixiecratic' nominee for President. 1948 *N.Y. Times* 5 Sept. 5/2 The States Rights Democrats have Bill Weisner, telegraph editor of the Charlotte News, to thank for putting Dixiecrat into the American vocabulary. Mr. Weisner was writing a headline on a story about the States Rights Democrats which would not fit. Dixiecrats would. 1948 *Economist* 18 Sept. 464/1 Democrats will be able to express their sense of outrage by voting for the Dixiecrats—the..name given to the rebellious States Rights' Democrats. *Ibid.*, The Dixiecrat candidate. 1949 *Tuscaloosa* (Ala.) *News* 17 Dec. 4/3 Byrnes' conversion to Dixiecratism.

dizoic (dəizōᵘ·ik), *a. Zool.* [f. Gr. δι- DI-[2] + ζῷον animal + -IC.] Producing two young; applied to a spore producing two sporozoites.
1901 [see POLYZOIC *a.* 1 b]. 1903 [see MONOZOIC *a.* 2].

dizygotic (dəizəigǫ·tik), *a. Biol.* [f. DI-[2] + ZYGOTE + -IC.] Of twins: derived from two separate ova (and therefore not identical).
1930 *Nature* 15 Nov. 766 Dizygotic twins are..on the whole inferior to their brothers and sisters in the physical measurements, but this is certainly not the case with monozygotic twins. 1952 *New Biol.* XII. 27 Ordinary fraternal or dizygotic twins are due to the independent fertilization of two separate ova and may be of the same or of different sex. 1966 *Lancet* 24 Dec. 1401/2 The frequency of monozygotic twins ranges from 1·9 to 6·9 per 1000 births, but of dizygotic twins from 2·7 to 32·2 per 1000.

dizzy, *a.* Add: **3. b.** (Further examples.)
1878 J. H. Beadle *Western Wilds* xxxv, Dance houses and saloons multiplied and 'dizzy doves' gave an air of abandon to the streets. 1888 *Texas Siftings* 29 Sept. (Farmer), Professional beauties or maidens, commonly called dizzy blondes. 1889 *Kansas Times & Star* 4 Nov., Many of the local clergy last night warned the church members..against a 'Dizzy Blonde' company coming to one of the theaters soon. 1938 G. Heyer *Blunt Instrument* ix. 166 The dizzy blonde herself. 1945 [see *DOG *sb.* 15 q].
c. Startling, astonishing, vivid. *slang.*
1896 Ade *Artie* xvii. 158 They was out there in them dizzy togs cuttin' up and down the track. 1897 *Daily News* 10 Aug. 5/2 Four straw hats with 'dizzy bands'. 1923 R. D. Paine *Comr. Rolling Ocean* v. 84 When she limped into Brest a week overdue, the admiral called it a dizzy miracle.

dizzyingly (di·zi,iŋli), *adv.* [f. DIZZYING *ppl. a.* + -LY[2].] In a dizzying manner; amazingly.
1925 A. S. M. Hutchinson *One Incr. Purpose* iii. xvi. 335 Not hurtingly now, but, mysteriously, dizzyingly. 1952 Dylan Thomas *Let.* 21 Nov. (1966) 383 Forgive me for not answering your dizzyingly kind letter long before this. 1957 *New Yorker* 26 Oct. 71/1 This was a dizzyingly far cry from the small provincial company that London had hoped might be willing to accept him. 1964 *Rev. Eng. Stud.* XV. 315 But were the ordinary people who watched mysteries really so dizzyingly other-worldly?

Dizzyite (di·zi,əit). [f. *Dizzy*, nickname of Benjamin Disraeli (see, e.g., *Punch* 3 Apr. 1852, p. 139) + -ITE[1].] A follower or admirer of Disraeli (cf. *DISRAELIAN *a.*).
1903 W. Meynell *Disraeli* I. 61 Dizzyites..must marvel that one who received this close confidence could afterwards be jaunty at the expense of the dead woman whom Disraeli 'so truly loved'. *Ibid.* II. 483 If Disraeli bore his traducers no grudge, it would be superfluous indeed for true Dizzyites to bear them any.

djati (dʒā·ti). Also **d'jatti, jati.** [Malay *jāti.*] The teak tree, *Tectona grandis.*
1908 *Westm. Gaz.* 9 June 5/3 Java teak, known as d'jatti. 1940 E. J. H. Corner *Wayside Trees of Malaya* II. Pl. 215 (*caption*) Teak, *Jati, Tectona grandis* in the Residency Grounds, at Penang.

djebba, djibba(h), varr. JIBBAH. Occas. worn by women as a type of smock.
1896 *Daily News* 23 Oct. 2/1 They had turned their tattered djebbas inside out. 1904 *Daily Chron.* 27 July 8/2 The djibbah is produced in full, warm reds, purples, blues, and orange hues. 1909 H. G. Wells *Ann Veronica* vii, A purple djibbah with a richly embroidered yoke. 1927 *Spectator* 17 Dec. 1081/1 The books..are sold by ladies in djibbas. 1963 *Punch* 27 Feb. 302/2 A girl going into the senior school [at Roedean]..could wear her djibbah until she left at eighteen.

D.N.A., DNA, see *DEOXYRIBONUCLEIC ACID.

do, *v.* Add: **A. 2. c. γ.** (Further examples of *don't* = does not.)
1670 in *Coll. Rhode Isl. Hist. Soc.* (1902) X. 102 Evidence of..River being more than 11 Miles Long but how Much More dont say. 1774 P. V. Fithian *Jrnl.* (1900) 202 A Sunday in Virginia dont seem to wear the same Dress as our Sundays to the Northward. 1813 J. K. Paulding *John Bull & Br. Jon.* ii. 9 The old saying that a man don't know when he is well off. 1835 R. M. Bird *Hawks of Hawk-Hollow* I. xi. 143, I wonder she don't sing; for a speaking voice, she has the richest *soprano.* 1862 O. W. Norton *Army Lett.* (1903) 120 It don't take ten thousand acres here to support one family. 1918 A. Huxley *Let.* 28 June (1969) 157, I only hope that this letter will reach you, though your loss will not be very great if it dont. 1946 K. Tennant *Lost Haven* (1947) i. 15 A man what don't profit from all a woman's telling and hiding the bottles ain't worth the trouble.
3. d. *colloq., dial.*, and *U.S.* **done.**
1847 in D. Drake *Pioneer Life Kentucky* (1870) iii. 63 The weavil..'done' great injury to that grain. 1848, etc. [see E.D.D.]. 1849 N. Kingsley *Diary* (1914) 56 Anna done the fair thing last night. 1850 *Ibid.* 117 [We] worked in the old place and done middling well. 1873 'Mark Twain' & Warner *Gilded Age* xxxiii. 307, I think it done him good. 1924 W. M. Raine *Troubled Waters* xxi. 226 The little boss done right not to take that Cheyenne bid for the doggies. 1969 *Listener* 4 Sept. 312/3 After what they've done to me, I never could forgive them. And I never done anybody any harm.
B. 6. e. (Further examples const. *with.*)
1820 *Edin. Rev.* XXXIII. 93 They are so happy that they know not what to do with themselves. 1858 Bulwer-Lytton (*title*) What will he do with it? *c* 1920 D. H. Lawrence *Phoenix II* (1968) 116 And what are you doing with yourself these days, Mr. Noon?
f. *to have done it*: to have acted extremely foolishly; to have made a mess of things; *that does* (or *did*) *it*: that is (or was) the last straw. *colloq.*
1837 Dickens *Pickw.* xxxv. 391 Well, young man, now you *have* done it. 1842 S. Lover *Handy Andy* xlii. 320 By the powers, you *have* done it this time! 1857 Dickens *Dorrit* i. viii. 62 'You've done it,' observed Tip; 'you must be sharper than that, next time.' 1883 G. M. Hopkins *Let.* 28 Sept. (1938) 164, I began to fear I had, as people say, 'done it this time'. 1914 'E. Bramah' *M. Carrados* 82 'Now you've done it,' commented Mr. Carlyle. 1930 Belloc *New Caut. Tales* 45 His father made a fearful row. He said 'By Gum, you've done it now!' 1946 W. F. Brown *Through Windows* xiii. 64 Again I did not go and pleaded the same reason. That apparently did it, as they say—she stopped asking me to lunch parties.
g. *it is not* (freq. *isn't*) *done*: it is forbidden by custom, opinion, or propriety; it is bad form. *colloq.*
1879 E. Gosse in Charteris *Life & Lett.* (1931) 126 We haven't the originality to think of dying. It's never done here, in our set. 1911 R. W. Chambers *Common Law* i, 'You know,' he said, 'models are not supposed to come here unless sent for. It isn't done in this building.' 1926 E. M. Dell *Black Knight* i. viii, 'Oh, but you couldn't—you couldn't—live there by yourself!' protested Joyce. 'It isn't done, Ermine. It wouldn't be fitting.' 1928 *Observer* 29 Jan. 22/1 Undergraduates regard the conduct of the night of December 13 as the sort of thing that 'isn't done'. 1932 *Times Lit. Suppl.* 5 May 318/4 A first-class book made up of things which emphatically are 'not done'. 1963 A. Heron *Towards Quaker View of Sex* 56 When it is 'not done' to discuss sexuality—as in many Western sub-cultures. 1971 P. Worsthorne *Socialist Myth* viii. 188 That kind of behaviour simply is not done, simply is not cricket.
11. a. (Further examples.) Also, to decorate, furnish. With person as obj. (*colloq.*) = to operate on, attend to, etc.
1898 A. E. T. Watson *Turf* i. 21 Almost all these horses have their own boys, who ride at exercise, and, as the phrase goes, 'do' them, that is to say, groom and attend to them in their stables. 1901 *Daily Chron.* 16 Oct. 5/2 The [vaccinated] man who..has been 'done in the leg'. 1902 Mrs. C. S. Peel *How to keep House* iv. 42 Do flowers, write *menus*, do house accounts and see housemaid. 1919 D. Ashford *Young Visiters* (1951) v. 33 A small but handsome compartment done in dark green lether with crests on the chairs. 1929 P. Guedalla *Missing Muse* 147 The explorers ..went below and began to do their hair. 1960 C. Day Lewis *Buried Day* 116 Confirmation was habitually referred to as 'getting done'. We were 'done', according to our age group, in batches—like loaves. 1960 *Housewife* Apr. 29/2 She has 'done the flowers' at innumerable society dinners. 1962 R. Hyman *Mod. Dict. Quots.* 146 Can I do you now, sir?—Mrs. Mop (Itma, B.B.C. Radio Programme, 1939–1949).
d. Also *colloq.*, to act or behave in a manner characteristic of (a specified person, etc.).
1934 *Tit-Bits* 31 Mar. 12/3 'To do a Gaynor' means to smile upwards through eyes swimming with tears, a tribute to Janet Gaynor's ability to switch on the 'sunshine through the tears'. 'To do a Garbo', on the other hand, means to be proud, aloof, and unbending. 1943 N. Balchin *Small Back Room* 28 We do all the work and then R. B. sails in and does a God Almighty on us. 1960 L. Cooper *Accomplices* i. vi. 58 He's a fiend about it—not that he does a McCarthy or rants. 1963 *Times* 20 Apr. 9/4 Already those responsible for premises or institutions under threat of closure and persons haunted by the now inevitable word redundancy are saying 'They're doing a Beeching on us'.
e. Also colloq. phr. *done to the wide* or *the world*: absolutely done for, defeated, etc. Also *slang*, to beat up; to defeat; to finish up; to kill. Cf. sense 45 b below.
1780 *Sessions' Paper* 611/2 He..got one of our cutlasses, which was drawn;..and said, 'D—n my eyes, here is one of Akerman's bloody thieves, let us do him first.' 1794 *Sporting Mag.* III. 260/2 Much skill was displayed by both the combatants...Dame Fortune..at length favoured the tin-man, who, in the language of the schools, *did his man.* 1796 Grose *Dict. Vulgar T.* (ed. 3), *Do*,..to overcome in a boxing match; witness those laconic lines written on the field of battle, by Humphreys to his patron—'Sir, I have done the Jew.' 1905 Conan Doyle *Return of S. Holmes* 218 'You've done me,' he cried, and lay still. 1922 *Daily Mail* 6 Dec. 11 He came again after appearing 'done to the world' more than once. 1925 Fraser & Gibbons *Soldier & Sailor Words*, *Done to the wide*, utterly beaten. 1948 L. A. G. Strong *Trevannion* 3 If I do Sid, I'm to have a go at Sailor Berridge. 1954 M. Procter *Hell is City* ii. ii. 47 I'll do you if it's my last act in life. I'll swing for you with pleasure. 1959 *Encounter* Aug. 33/2, I..told him..I'd do him if I ever saw his face again. 1959 I. & P. Opie *Lore & Lang. Schoolch.* x. 195 'I'll clout you', 'I'll do you', or 'I'll do you up' (a threat of sinister implication in London S.E.).
f. Also *to do in the eye* (see *EYE *sb.*[1] 2 k).
g. (Earlier and later examples.) Also, to achieve; to travel at (a certain speed, etc.).
1808 *Sporting Mag.* XXXIII. 146/2 The Captain did the first mile in five minutes and a second. 1919 C. Mackenzie *Early Life Sylvia Scarlett* ii. i. 273 Good engine this. We're doing fifty-nine or an unripe sixty. 1919 G. B. Shaw *Augustus does his Bit* 228 The old cars only do twelve miles to the gallon. Everybody has to have a car that will do thirty-five now. 1963 M. Procter *Moonlight Flitting* i. 6 'That's a Rolls-Royce, isn't it?' 'Yes. It's practically new. Only done about a thousand.' 1966 J. Miles in T. Wisdom *High-Performance Driving* vii. 70 A Ford Zodiac and a caravan..doing 25 m.p.h. and taking up a lot of road. 1967 'L. Bruce' *Death of Commuter* vi. 62 What do you reckon to do to the gallon?
h. (Earlier and later examples.) Also, to attend (an entertainment).
1817 Lady Granville *Let.* June (1894) I. 119 We shall then meet them at Basle, do the Rhine, stay two or three days at Brussels, and home. 1857 J. F. Maguire *Rome* 8 Some of the latter evidently went to the Pope's Chapel as they had gone the previous night to the Opera, to hear the music, or to 'do' it, as they would the Coliseum. 1890 E. Dowson *Let.* 4 Mar. (1967) 139, I rather want..to do a St. James's Ballad Concert. 1932 D. L. Sayers *Have his Carcase* xviii. 239 We could go and do a show together. 1951 M. Kennedy *L. Carmichael* ii. i. 85 People over a wide area would ..dine and do a show, as they do at Stratford. 1955 *Times* 25 Aug. 12/3 During siesta the only activity comes from tourists 'doing' St. Peter's, the Colosseum, and the Trevi Fountain before sunstroke or tea. 1966 Auden *About House* 20 This unpopular art..Cannot be 'done' like Venice Or abridged like Tolstoy.
m. To arrest; to catch or seize hold of; to charge; to convict. *slang.*
1784 *Sessions' Paper* Jan. 221/1 He stepped on one side of me and said, 'You have not done me yet.' I immediately pursued him. 1812 J. H. Vaux *Flash Dict.*, *Done*, convicted; as, he was *done* for a *crack*, he was convicted of house-breaking. 1936 G. Ingram *Muffled Man* vi. 91 Blow me if one of your tribe [*sc.* policemen] don't go and do me, and I get found a quid. 1963 *Guardian* 23 Feb. 4/4 'This is a murder charge. There is no certainty that you will be done for murder.'..He did not say that Kelly would only be 'done' for robbery and not murder. 1968 'R. Simons' *Death on Display* iii. 44 I'm goin' straight. Last time I was done was two years ago, and I ain't been tapped on the shoulder since.
n. To look after or provide food for (an animal). *dial.* and *N.Z.*
1890 S. S. Buckman *Darke's Sojourn* viii. 72 Nobody can't be expected to do a flock on no vittles. 1916 *N.Z. Jrnl. Agric.* 20 Sept. 174 Ewes have been..'well done by' during the winter. 1923 W. Perry *Sheepfarming in N.Z.* viii. 116 Besides the usual pasture, roots and green feed given to the flock it is often found profitable to 'do' the show sheep especially well. 1950 *N.Z. Jrnl. Agric.* Oct. 347/2 The successful management of a sheep run calls for..an appreciation of the carrying capacity of tussock land to 'do' sheep well without either overstocking or understocking.
o. To provide food, etc., for (a person); to treat or entertain (*well*). Also, *to do oneself well*: to make liberal provision for one's creature comforts; *to do* (a person) *proud*: see PROUD *a.* 10 b. *colloq.*
1897 *Punch* 23 Oct. 185/1 The nightmare of an artist who does himself not wisely but far too well, at an unnecessary supper. 1902 *Daily Chron.* 16 Aug. 3/4 For ten francs a day one is done well there. 1902 *Westm. Gaz.* 25 Aug. 2 His Majesty has been to Westminster Abbey, and the Crystal

Palace,..and Madame Tussaud's—really we think that on the whole we have done him very well. *Ibid.* 22 Oct. 3/3 The man who had done himself fairly well on everyday cooking. **1928** *Daily Express* 7 Sept. 1/1 They do you well, with plenty of eggs, cream, [etc.]. **1940** G. D. H. & M. Cole *Murder at Munition Works* iv. 55 The Chief Constable will bear me out that they do one quite well there.

p. To spend completely. *Austral.* and *N.Z. slang.*

1928 *Bulletin* (Sydney) 15 Feb. 35/1 He grumbled: '..I'd just as soon ha' done me brass on goats.' **1931** V. Palmer *Separate Lives* 218, I did my last frog [*sc.* franc] on a feed at the estaminet to-night. **1959** G. Slatter *Gun in Hand* v. 71 Could you lend me a few bob to put on?..She was always doing her money cold on the donks. **1969** 'A. Garve' *Boomerang* i. 24 Right now I've done my money, but as soon as I can raise the fare I'll be getting back.

15. b. Also, to make shift to live *on* (a limited income).

1924 R. Macaulay *Orphan Island* xviii. 237 'Is that a good living wage?' he asked her; and she answered that they could just do on it, no more, with what she herself earned.

16. *to do or die* (later examples). Hence *do or die* as adj. phr., expressing determination not to be deterred by any danger or difficulty.

1809 T. Campbell *Gertrude of Wyoming* III. xxxvii. 71 To-morrow let us do or die! **1863** L. M. Alcott *Hospital Sk.* i. 13 The head..fermented with all manner of high thoughts and heroic purposes 'to do or die',—perhaps both. **1879** *Boy's Own Paper* 18 Jan. 2/2 Never soldier went into action with a more solemn do-or-die feeling than that with which I took my place on the field that afternoon. **1902** *Westm. Gaz.* 1 Mar. 4/1 She dips the first pen into the ink with a do-or-die expression. **1907** *Daily Chron.* 26 Sept. 4/6 The dominant motive with all was hatred of the foreign yoke, and the 'do-or-die' determination to shake it off. **1947** *People* 22 June 7/4 His hands grip the handlebars tightly, his face is grim with a do-or-die look. **1971** *Rand Daily Mail* 27 Mar. 23/6 Highlands Power..face a do-or-die effort from the Pretoria team.

b. (Examples.) Also *trans.*, to copulate with. Phr. *to do it* [It *pron.* 9], used *colloq.* in the same sense.

1913 D. H. Lawrence *Sons & Lovers* II. vii. 162 Do you think we *spoon* and do? We only talk. **1922** Joyce *Ulysses* 724 Not that I care two straws who he does it with. **1954** R. Bissell *High Water* xvii. 181 Them island girls they'd rather do it than eat. **1959** A. Sinclair *Breaking of Bumbo* II. x. 106 You don't *do her?* And you eat in *Chelsea?* There's something queer about you. **1967** V. Canning *Python Project* viii. 157 Some service-man..did your mother in Cyprus..and then..made an honest woman of her.

20. b. Hence *trans.* To do for, suffice for, satisfy (a person). *colloq.*

1846 *Congress. Globe* 20 July 1118, I have just enough [money] to do me to the end of the session. **1880** *Congress. Rec.* 22 Jan. 491/1, I should like to have ten minutes, but it will do me just as well in the morning. **1925** W. Deeping *Sorrell & Son* xiii, 'What's it to be, Do? An orange cocktail?' 'Yes, that will do me.' **1928** Galsworthy *Swan Song* I. iv. 25 Leicester Square would do me all right. **1951** J. B. Priestley *Festival at Farbridge* 325 That'll do me. Not choosy. **1965** F. Sargeson *Memoirs of Peon* i. 12 The reverse side..did me to draw and paint on.

29. (Later example of *didn't*.)

1939 Joyce *Finnegans Wake* 198 Didn't you spot her?

b. Colloq. phr. *no, you don't*: you will not be allowed to do what you intend; I shall prevent you from doing (something implied).

1884 *Boy's Own Paper* 4 Oct. 2/2 'No, you don't!' muttered Soady, starting his melody. **1912** P. Nash *Let.* 21 Aug. in Bottomley & Nash *Poet & Painter* (1955) 48, I started a new outdoor drawing—but the devil or someone said 'no you dont', and at half hour intervals I was interrupted by heavy rain. **1926** B. A. McKelvie *Huldowget* ii. 27 'No, you don't,' exclaimed Collishaw, and he caught the girl's arm and gently drew it aside. **1952** W. G. Hardy *Unfulfilled* 176 But then Barty started toward Pam, his face blazing, his fist up and Peter jumped forward. 'No, you don't.'

30. b. (Later example.)

1930 D. L. Sayers *Strong Poison* ix. 112 'Ev another crumpet, do, Mr. Bunter.

d. Colloq. phr. *don't —— me*: do not use the word —— or mention the name of —— to me.

1829 [see Dear *v.* 3]. **1870** Trollope *Vicar of Bullhampton* xli. 263 'But, Mrs. Brattle—' 'Don't Mrs. Brattle me, Mr. Fenwick, for I won't be so treated.' **1892** C. M. Yonge *Cross Roads* x. 112 'Emmie!' 'Don't Emmie me!'

31. b. *done*: used in *U.S. dial.* (chiefly *Southern*) as a perfective auxiliary or with adverbial force in the sense 'already; completely'.

1827 A. Sherwood *Gaz. Georgia* 139 *Done said it*, for has said it. *Done did it*, for has performed, or done it. **1836** *Spirit of Times* (N.Y.) (1846) 22 (Th.), He had done gone three hours ago. *Ibid.* 94 I'd done got the licker, and I was satisfied. **1853** 'P. Paxton' *Yankee in Texas* 114 Of Alabama origin..is that funny expression, 'done gone', 'done done', implying 'entirely gone', and 'entirely done'. **1854** M. J. Holmes *Tempest & Sunshine* ii. 24 I've done let my best horse and nigger go off with a man from the free States. **1873** J. H. Beadle *Undevel. West* xix. 356 People have done forgot they had any Injun blood in 'em. **1887** *Century Mag.* Nov. 96/1 'You done had supper?' she asked. **1917** H. T. Comstock *Man thou Gavest* 300, I done told Burke I—I was going to prove myself. **1938** M. K. Rawlings *Yearling* vii. 65 You couldn't see 'em. There ain't none left. They've done left here, jest like the Injuns. **1945** E. T. Wallace *Barington* 18, I don't know what you need with another boy. You done got four.

34. c. *nothing doing*: *lit.* nothing being done, or transacted; no business on foot; hence (*slang* or *colloq.*) an announcement of refusal of a request or offer, failure in an attempt, etc.

1827 De Quincey in *Blackw. Mag.* Feb. 211/2 Complaining 'that there was nothing doing'. **1858** *Leisure Hour* 25 Mar. 186/2 There's nothing doing now. **1870** *Porcupine* 26 Mar. 503/3 A friend of mine hailed an outfitter the other day, 'How is business?' 'Nothing doing.'

1910 *N.Y. Even. Post* 13 Dec. 7 Spottford offered the porter a dime. The negro waved it aside and said: 'Nothing doing; my price is a quarter at least.' **1915** 'I. Hay' *First Hundred Thou.* xx. 302 'Na pooh!'..also means, 'Not likely!' or 'Nothing doing!' **1928** *Boston Even. Transcript* 30 Mar. 15/7, I looked in the dictionaries. 'Nothing doing!' **1930** W. S. Maugham *Gent. in Parlour* x. 46 Then my girl asked me to marry her... I told her there was nothing doing. **1937** —— *Theatre* xii. 107 He can hardly expect me to ask him to come and sleep in here...Nothing doing, my lad. **1947** *People* 22 June 1/1 It was suggested that she should come incognito. Nothing doing.

38. do for ——. a. (Later examples.)

1914 B. Stoker *Dracula's Guest* 21 He..got..the name of an old woman who would probably undertake to 'do' for him. **1936** A. Christie *Cards on Table* xiv. 136 The superintendent's researches..led him..to Mrs. Astwell—who 'did' for the ladies at Wendon Cottage.

b. (Earlier example.) Now freq. in *pass.* Also *done-for* adj.

1740 *Sessions' Paper* 9 July 190/2 D—mn you, I'll *do* for you. **a 1817** Jane Austen *Persuasion* (1818) xi. 279 Give Anne your arm...She is rather done for this morning. **1847** J. S. Robb *Squatter Life* 128 They found Sam holding the straw figure in his arms, and looking in a state of stupor at the horse; he thought his master was 'done for'. **1950** J. Cannan *Murder Included* i. 12 These doomed and done-for ladies and gentlemen.

39. do to ——. b. Colloq. phr. *to do something* (or *things*) *for* or *to*: to improve; to render more pleasing or attractive.

1942 D. Powell *Time to be Born* (1943) ii. 43 The bathing-suit would have to be..very carefully cut indeed to 'do things for her'. **1960** *Daily Express* 21 Jan. 1/3 A beret always seems to *do* something to generals. **1961** *Sunday Express* 12 Nov. 18/7 Here it is—a suit that does things for a woman. **1961** *Guardian* 13 Dec. 6/6 A coffee flavoured liqueur..really does something to the ice-cream.

40. do with ——. c. *I* (*you*, etc.) *could do with*: I could make use of or profit from; I should be glad to have; I need. *colloq.*

1783 R. Benson *Let.* 3 Nov. in T. W. Thompson *Wordsworth's Hawkshead* (1970) 335, 1 cd. also do with some Apples. **1859** Geo. Eliot *A. Bede* II. xxv. 197 Well, I could do wi 't, if so be we want to get rid on't. **1936** J. B. Priestley *They walk in City* vi. 140 Ah'll pay up, ay, an' Ah'll give it to t'lad. He could do wi' it. *Ibid.* xii. 376, I could do with a cup of tea and a smoke. **1941** *Punch* 18 June 584/2 'If ever there was a man who could do with a little assistance—' 'Just independent perhaps.' **1955** M. Gilbert *Sky High* xiv. 205 You look as if you could do with a wash and brush up.

45. do down. b. To overcome, master, get the better of, bring to grief; to 'do' (sense 11 f in Dict.). *colloq.*

1911 H. S. Walpole *Mr. Perrin & Mr. Traill* viii. 154 He saw nothing but a spiteful and malignant world trying, as he phrased it, to 'do him down'. **1922** D. H. Lawrence *England, my England* 257 Poor Fanny! She was such a lady, and so straight and magnificent. And yet everything seemed to do her down. **1923** *Daily Mail* 12 Mar. 6 Sir Arthur Griffith-Boscawen..said they had been done down by what had been rightly called an act of treachery. **1936** L. A. G. Strong *Last Enemy* 13 To get your rights from Fosdyke was to cheat him, since he had never meant you to have them...It was a real injury to stop him from doing you down. **1958** *Economist* 29 Nov. 784/1 School for them is the great game of doing down, or being done down by, the teacher.

46. do in. b. To spend completely (cf. sense *11 p). *slang* (chiefly *Austral.* and *N.Z.*).

1889 *Referee* 19 May 2/1 A young fellow..rushes to 'do in' every spare fiver or tenner that comes into his possession. **1909** T. H. Thompson *Ballads about Business* 27 I'd..never make for home again until I'd 'done it in'. **1930** V. Palmer in *Bulletin* (Sydney) 19 Feb. 51/1 Now he's done his money in.

c. To bring disaster upon, do a great injury to, ruin; often, to murder, kill. *slang.*

1905 *Daily Chron.* 22 May 6/3, I heard people tell her to do me an injury, throw glasses at me, and 'do me in'. **1906** *Ibid.* 11 Dec. 4/4 It seems funny that the first blooming order I got in Enfield I should be done in. **1914** G. B. Shaw *Pygmalion* III in *Nash's Mag.* Dec. 308 My aunt died of influenza: so they said...But it's my belief they done the old woman in. **1918** W. J. Locke *Rough Road* vi, If you engage a second-rate man..who isn't used to this make of car, he'll do it in for you pretty quick. **1919** J. B. Morton *Barber of Putney* xiv. 235 'Yes,' said Graves. 'That's what did my nerves in. Still sleep bad.' **1928** Galsworthy *Swan Song* I. ix. 66 That house had 'done in' her father. **1963** *Listener* 4 Apr. 585/2 These were professional killers who 'did in' John Regan.

d. To exhaust, wear out; freq. *done in*, exhausted. *colloq.*

1917 S. McKenna *Sonia* v. 245 Loring mopped his forehead. 'I feel absolutely done in,' he murmured. **1955** E. Hillary *High Adventure* 143 For the first time I really feel a bit done in.

49. do out. c. (Later examples.)

1881 A. B. Evans *Leicestersh. Words* 139 Ye're ollus a-doin' out the house of a Saturday! **1910** A. Bennett *Clayhanger* III. i. 326 Once a week..his room was 'done out'. **1955** J. Cannan *Long Shadows* iii. 44 'E's not arriving till..this afternoon but I did the room out yesterday.

e. (Earlier and later examples.)

1825 T. Creevey *Creevey Papers* (1963) xii. 209 Mrs. Taylor and I having done Mylord and Mylady out of £3 apiece at Ecarté. **1840** R. Barham in *Bentley's Misc Mar* 269 Rebuked 'em For unhandsomely doing him out of his Dukedom. **1929** H. S. Walpole *Hans Frost* II. i. 118 Your aunt's so unselfish, she'd do herself out of anything.

50. do over. b. To cheat, swindle, get the better of. *slang.*

1781 G. Parker *View of Society* II. 43 And now, Hostler, can't you tell me how you have *done 'em over?* **1789** —— *Life's Painter* v. 44 His huntsman was his prime-minister..who could, at any time, *do him over*, as they phrased it, for half-a-crown or half-a-guinea. **1930** 'A. Armstrong' *Taxi* xii. 164 'Pinching (or knocking) a job off the rank', means cruising near a rank head and snapping up a fare who would otherwise have taken a cab from the standing. To 'do the rank over' has the same meaning. **1939** H. Hodge *Cab, Sir?* xv. 219 'Doing a man over', or 'doing the rank over', is..'wrongfully..taking away the fare from any other driver who..appears to be fairly entitled to it'.

c. To disable, wear out, tire out. *colloq.*

1789 W. Dunlap *Darby's Return* 13 For while we were watching, like sportsmen for plover, The linen took fire—and did us all over. *Ibid.* 14 We sneak'd into town;—very fairly done over. **1837** Dickens *Pickw.* xxxviii. 417 He's in a horrid state o' love; reg'larly comfoozled, and done over vith it. **1853** 'P. Paxton' *Yankee in Texas* 96 [The dogs] were completely done over and used up.

d. To handle (a person) roughly. *Austral.* and *N.Z. slang.*

1866 *Maungatapu Murders* 17 Since we are going to do these people [*sc.* their murder/robbery victims] over..I think we had better prevent him from doing us any harm. **1953** A. Upfield *Murder must Wait* ix. 81 'Done over properly, wasn't he?' 'From appearances, yes. Mitford must be a rough place.'

e. To copulate with; to seduce. *slang.*

1873 Hotten *Slang Dict.* 145 *Done over*..also means among low people seduced. **1961** R. Amato in C. K. Stead *N.Z. Short Stories* (1966) 233 All the sailors..want to marry the girl they've done over. **1961** *John o' London's* 3 Aug. 163/2 A truly Moravian rape-scene in a ruined church, with Cesira and Rosetta both done over by a screeching pack of Moroccan *goums*.

f. To decorate, refurbish; = *make over* (Make *v.*[1] 92 d).

1905 E. Wharton *House of Mirth* I. i. 10 It must be pure bliss to arrange the furniture just as one likes...If I could only do over my aunt's drawing-room I know I should be a better woman. **1908** *Smart Set* Sept. 84/1 If only somebody would 'do over' Browning into English. **1941** J. P. Marquand *H. M. Pulham* xxxiv. 386 We ought to keep this as the spare room and do the nursery over.

52. do up. c. Also, to dress up. *colloq.*

1897 M. Kingsley *Trav. W. Afr.* 21 Here and there in the street you come across a black man done up in a tweed suit, or in a black coat and tall hat. **1946** R. Lehmann *Gipsy's Baby* 10 The younger ones could not be said to be dressed, in the accepted sense. They were done up in bits of cloth, baize or blanket.

e. (Earlier and later examples.) Also, to ruin other than financially; to get the better of; to settle or finish; to beat up.

1785 Grose *Dict. Vulgar T.*, *Done up*, ruined by gaming, and extravagances, (*modern term*). **1835** Coleridge *Table-t.* I. 5 It is not easy to put me out of countenance,.. yet once I was thoroughly *done up*, as you would say. **a 1846** B. R. Haydon *Autobiogr.* (1927) III. xvi. 317 Lord Elgin saw Knight was done up, and done up was the whole clique. **1854** A. E. Baker *Gloss. Northants Words* I. 192 *Done-up*, ruined in circumstances. 'They can't go on much longer, they're quite *done up*.' **1887** *Lantern* (New Orleans) 30 Apr. 2/2 The idea of this gang jumping on J. C. Matthews and doing him up. **1894** *Harper's Mag.* LXXXIX. 389/1 They lame Bob Griffiths fer life. And then they do up Buck. Shoot a hole through his spine. **1904** W. H. Smith *Promoters* ii. 54 The thing to do is to do up your competitor. **1906** 'O. Henry' *Four Million* 121, I have many times told you those Dagoes would do you up. **1962** 'R. Simons' *Killing Chase* vi. 77 Some of the boys did me up last night.

55. *Comb.* in *attrib. phr.* as *do-as-you-please*, *DO-IT-YOURSELF.

1923 D. H. Lawrence *Kangaroo* ii. 24 The sense of do-as-you-please liberty. **1963** *Listener* 10 Jan. 73/2 The opportunities for enjoying the kind of do-as-you-please holiday that more and more Russians..now take for granted every year.

do, *sb.*[1] Add: **2. b.** *esp.* An entertainment or show (see quots in Dict.); a party; hence (orig. *jocular*), a military engagement, raid, or other 'show'.

a 1824 J. Briggs *Remains* (1825) 243 Such individuals should have their feast (or *do*, as it is called). **1915** 'Boyd Cable' *Between Lines* 110 We are about the first Terrier lot to be in a heavy 'do' in the forward trenches. **1925** Fraser & Gibbons *Soldier & Sailor Words* 78 *Do*, an event. A stunt. An attack, etc. *E.g.*, 'When is the do coming off?'; 'The Somme do'; 'The Havrincourt do', etc. **1955** *Times* 18 May 14/2 Miss Margaret Herbison broadcast on behalf of the Labour Party last night a talk which she described as a 'family do'. **1958** M. Kerr *People of Ship St.* ix. 108 Her family has a 'do' every year on the anniversary of the day her mother's father died. *Ibid.*, Christmas 'dos' are especially important.

c. Usu. in *pl.* Dealing, treatment; esp. in phr. *fair do's*. *colloq.* (orig. *dial.*).

1859 Hughes *Scour. White Horse* vi. 122 Only seemed to want what they called 'fair doos'. **1862** C. C. Robinson *Dial. of Leeds* 282 'A shabby dew', says a man who has had twopence given him for getting a waggon-load of coals in. 'A fairish dew', says another who has got a shilling and a lot of victuals away with him for the same. **1941** L. A. G. Strong *Bay* 168 Come on, Doctor. Fair do's. **1951** 'A. Garve' *Murder in Moscow* i. 15 There's no 'nobs' there; it's fair do's for everybody. **1952** A. Stuart in J. C. Trewin *Plays of Year* V. 114 It's fair dos. I need you and you need me. **1953** S. Beckett *Watt* i. 8 The lady now removing her tongue from the gentleman's mouth, he put his into hers. Fair do, said Mr Hackett. **1960** *She* Dec. 74/3 If *she* wants to go out agree to take her this year providing she will stay at home the next, to please you. That would be 'fair do's'.

4. An injunction to do (something specified). Only when coupled with *don't*: see Don't.

1902 'Stancliffe' (*title*) Golf do's and don'ts. **1920** G. C. Bailey *Complete Airman* 190 Between this and the next lesson the pupil should be encouraged to think well over the 'dos' and 'don'ts' of his trip. **1962** Y. Olsson in F. Behre *Contrib. Eng. Syntax* 91 There are certain do's and don'ts that he should keep in mind.

5. A success, a 'go'; esp. in phr. *to make a do of it*. *Austral.* and *N.Z. colloq.*

1902 *N.Z. Illustr. Mag.* V. 381/1 Your poor, rough, back section. It has been forfeited twice, and you'll never make a do of it. **1945** *N.Z. Geographer* I. i. 36 High-country men..take to the back-country because they can make a better 'do' of it than on a mixed farm down-country. **1947** D. M. Davin *Gorse blooms Pale* 91 Another go at making a do of things with his wife.

do-all. Add: (Later example.) Also *attrib.*

1816 Jane Austen *Emma* I. i. 19 A something between the do-nothing and the do-all. **1964** G. L. Cohen *What's Wrong with Hospitals?* i. 21 Medical advance soon prevented the do-all doctor from subsisting alone. He leaned increasingly on hospital resources.

doaty, var. *DOTEY.

doaty, *a.*, var. Doty *a.*

dob, *v.* Add: **2.** To betray, inform against. Const. *in*. *Austral. slang.*

1959 Baker *Drum* (1960) 105 *Dob in*, to betray, to focus blame on another; esp. *to dob someone in*. **1963** E. Thompson *Lawyer & Carpenter* vi. 100 Any time you run outside the law, I won't dob you in, either. **1964** *Punch* 18 Nov. 750/2 Those Canberra wowsers have really dobbed us in this time. **1970** M. Kelly *Spinifex* vii. 113 You won't dob me into the demons.

dobber[2] (dǫ·bəɹ). Chiefly *dial.* [f. *dob*, dial. var. *dab* a lump (Dab *sb.*[1]); cf. Dabber 2.] A large marble.

1874 E. Waugh *Chimney Corner* (1879) 116 He's as numb as a clay dobber. **1875** Nodal & Milner *Gloss. Lancs. Dial.* 107 *Dobber*..a large marble. **1934** J. O'Hara *Appt. Samarra* vii. 186 There was a game of marbles called Dobbers, played with marbles the size of lemons. You played it in the gutter..throwing your Dobber at the other fellow's. **1961** *Times* 17 Apr. 12/6 Bending down to bowl along the pitch to the ring of marbles with what we called a 'dobber'.

dobby, dobbie. Add: **3.** *dobby-weave* attrib.

1961 *Guardian* 30 Jan. 5/5 Some attractive suitcases..have 'dobby-weave' linings.

4. In full *dobby-horse*. A wooden figure of a horse. Cf. Hobby-horse 2, 4 b.

1879 J. C. Clough *B. Bresskittle* 9 (E.D.D.), Theer were shows an..dobby horses. **1886** R. Holland *Gloss. Chester* 101 *Dobby-horse*, a hobby-horse. An imitation horse which figures in the play performed by the 'Soulers'. **1933** E. K. Chambers *Eng. Folk-Play* 214 Cheshire with a 'Dobby-Horse'..and..there are similar rites with animals other than a Horse. **1937** J. B. Priestley *I have been here Before* II. 63 He says we all go round and round like dobby horses. **1968** D. Braithwaite *Fairground Architecture* iii. 46 Early horse roundabouts, known as 'Dobbies',..were being produced at King's Lynn certainly in the early 1870's. *Ibid.* 158/1 Steam-driven 3-abreast Dobby Horses.

dobe, 'dobe (dō^u^·bi). Colloq. shortenings of Adobe. *U.S.*

1883 E. W. Nye *Baled Hay* 135 The dobe pig pens..are not true to nature. **1897** *Outing* (U.S.) XXIX. 582/1 The afternoon monsoon..howls..as if it would tear the stubborn little 'dobe shanties off the earth. **1904** 'O. Henry' *Cabbages & Kings* iv. 73 Grass huts, 'dobes, five or six two-story houses. *Ibid.* vi. 104 A 'dobe house in a dirty side street. **1951** Steinbeck *Burning Bright* I. 1 The close-cut barley stubble stood in bunches with the black 'dobe earth between.

Dobermann (dō^u^·bəɹmæn). [Name of Ludwig *Dobermann*, a nineteenth-century German dog-breeder of Thuringia. See also *PINSCHER.] In full *Dobermann Pinscher*. A kind of German hound with a smooth coat and docked tail.

1917 *Policeman's Monthly* Jan. 5/3 Nowadays four breeds of dogs are being used for police purposes: the Continental Sheepdog, the Airedale Terrier, the Doberman Pinscher and the Rottweilers. **1922** R. Leighton *Compl. Bk. of Dog* viii. 119 The Germans pride themselves upon their sheepdogs, no less than upon their Boxers, their Dobermann Pinschers, their Saufangers, and Teckels. **1928** *Sunday Express* 24 June 11 A German named Dobermann, of Apolba, Thuringia..spent his life experimenting with different breeds, and at last, in the middle of the last century, he produced specimens of the 'Dobermann Pincher'—and died before disclosing his secret. **1959** *Times* 16 Jan. 1/7 Home Wanted for 18 month Dobermann Pinscher bitch.

dobeying (dō·bi,iŋ). *Naval slang.* Also **dhobeying.** [f. Dhobi + -ing[1].] Washing clothes, doing laundry.

1929 in F. C. Bowen *Sea Slang* 37. **1940** *Illustr. London News* CXCVII. 160 (*caption*) Destroyer-men doing their 'dobeying' on deck. **1955** *Times* 2 June 2/6 Some of the early plays..are the stuff of their lives; swabbing and dobeying being more natural than all the heroics.

dobie, doby, also **doba, dobbey, dobbie,** varr. *DOBE. Also *attrib.* and *Comb. U.S.*

1838 M. Whitman *Let.* 12 Mar. in *Overland to Pacific* (1936) VI. 294, I had walled the cellar with dobies. *Ibid.* 30 Oct. 327 The erection of a Doby house. **1842** J. Williams *Tour to Oregon* (1921) 84 Some of them build their houses with what they call 'dobbeys', made of mud, in the shape of brick. **1845** J. Clyman *Diary* 12 July (1928) 168 A strong doba or mud walled fort. **1854** *Fort Benton Jrnl.* 10 Oct. in *Montana Hist. Soc. Contrib.* (1940) X. 2 Men commenced Dobbie making. *Ibid.* 13 Oct. 3 Had Dobbie makers make up their mud into bricks. **1865** [see Adobe]. **1885** *Outing* (U.S.) VII. 52/1 Half sand and the other half 'doby'-mud. **1891** C. Roberts *Adrift Amer.* 86 These bricks..although they are spelt 'adobes', I always heard..spoken of as 'dobies'.

doc (dǫk). Colloq. abbrev. of Doctor *sb.*

?1850 E. Z. C. Judson *Three Years After* vii. 29/2 'Where's the Doc?' asked the old lady. **1854** R. Glisan *Jrnl. Army Life* 24 Nov. (1874) 149 Don't you think, Doc., ague makes a fellow powerful weak? **1887** F. Francis *Saddle & Mocassin* viii. 146 Anyhow, Doc Gilpen the Marshal jumped him [= took him to task]. **1904** G. H. Lorimer *Old Gorgon Graham* 173 Doc was cribbing those powerful Sunday evening discourses from a volume of Beecher's sermons. **a 1918** W. Owen *Coll. Poems* (1963) 72 The Doc.'s well-whiskied laugh. **1920** C. E. Mulford *J. Nelson* vi. 60 I'd like to shake hands with th' coyote that lugged th' Doc off to fix that laig. **1964** A. Christie *Caribbean Mystery* xxiii. 230 Molly's practically herself again. The doc says she can get up to-morrow.

‖**Doccia** (dǫ·tʃiä). The name of a town near Florence, Italy, used *attrib.* of a variety of porcelain produced there.

1857 J. Marryat *Hist. Pott. & Porc.* (ed. 2) xiv. 331 The early pieces of Doccia porcelain show a close imitation of the white Oriental. **1900** F. Litchfield *Pott. & Porc.* vii. 139 Doccia china is generally found in parts of table services. **1960** R. G. Haggar *Conc. Encycl. Continental Pott.* 138/1 Doccia was a hybrid porcelain akin to hard-paste, grey like most Italian porcelain in colour, badly turned out, often showing fire cracks and other technical faults.

Docete (dō^u^·sit). [Anglicized form.] = Docetist.

1781 Gibbon *Decl. & F.* II. xxi. 241 The Gnostics, who were distinguished by the epithet of *Docetes*. **1788** *Ibid.* IV. xlvii. 537 The *Docetes*, a numerous and learned sect of Asiatics. **1894** J. R. Illingworth *Pers. Hum. & Div.* i. 11 Had Christ been..a mere appearance as with the Docetes.

Docetism. (Earlier example.)

1846 Geo. Eliot tr. *Strauss's Life of Jesus* I. i. v. 274 To say..that the food which he took did not serve for the nourishment..of his body by real assimilation..would strike every one as Docetism.

doch-an-dorris, doch-an-dorroch, doch and dorus, etc., varr. *DEOCH AN DORIS.

dock, *sb.*[3] Add: **4. b.** *transf.* with preps., *in* (or *out of*, etc.) *dock*, in hospital, undergoing treatment; (of a vehicle) laid up for repairs. *colloq.*

1785 Grose *Dict. Vulgar T.* s.v. *Dock*, He must go into dock, a sea phrase, signifying that the person spoken of, must undergo a salivation. **1848** in *Amer. Speech* (1935) X. 40/1 *Hauled into dock*, sick at home. **1919** *Athenæum* 11 July 582/2 'Dock', hospital, is..probably from 'in dock'. *Ibid.* 22 Aug. 791/2 While 'in dock' (*i.e.* in hospital) one lay upon 'biscuits'. **1939** 'G. Orwell' *Coming up for Air* I. i. 11 The old car..was temporarily in dock. **1960** *News Chron.* 16 Feb. 6/5 He's just out of dock after the old appendix. **1963** 'R. East' *Pin Men* ii. 52 If Father's car hadn't been in dock.

c. In full *scene-dock* (see Scene 13).

1898 Sachs & Woodrow *Mod. Opera Houses* III. Suppl. I. 24 At each side of the stage the counterweight boxes practically form enclosing walls with a number of openings leading to a series of 'scene' docks on each side. The arrangement of these side docks..is essentially of French origin, and they afford a very ready means for the disposal of scenery which has to be quickly removed from the stage. *Ibid.* 34 There is a dock for each sequence of 'traps', so that the 'wings' belonging to each 'entrance' can always be kept in the dock opposite it. **1952** W. Granville *Dict. Theatr. Terms* 158 *Scene-dock*, usually shorted, by stage carpenters, to *dock*. A stowage space at the back, or side, of the stage.

7. *dock-head* (earlier Amer. examples); **dock-glass,** a large wine-glass originally designed for wine-tasting.

1920 G. Saintsbury *Notes on Cellar-Book* i. 9 The large, slightly pinched-in 'dock-glass', half filled, suited it as indeed it does almost any wine. **1953** *Word for Word* (Whitbread & Co.) 18/1 *Dock glass*, a goblet holding exactly a quarter of a pint, used originally by Excise Officers for wine tasting in the docks. **1657** *Rec. Early Hist. Boston* (1877) II. 142 To sett up a building att the west end of the house..by the dock head. **1736** *Ibid.* (1885) XII. 139 The Watch House at the Dockhead.

dock, *sb.*[4] Add: **dock brief,** a brief handed direct to a barrister in court, who has been selected by a prisoner, standing in the dock, to defend him. (Cf. Docker[3].)

1909 *Daily Chron.* 30 Apr. 6/7 The 'dock brief'..is the only exception to the rule that briefs must come through a solicitor. **1928** *Daily Tel.* 10 Jan. 9 Barristers who are not anxious to accept dock briefs are entitled to leave the court on hearing a prisoner ask for one.

dock, *v.*[1] Add: **2. b.** To make a deduction from (a person's pay) as a fine, subscription, etc.; also with the person as object. *colloq.* (orig. *dial.*).

1822 Cobbett *Weekly Reg.* 13 Apr. 81 Hence arose numerous schemes for *docking* you in this quarter. **1891** C. Wordsworth *Rutland Words.* s.v., Mr. A— has docked his men as last Saturday, I suppose. **1891** *Harper's Mag.* Nov. 888/2 Each man was 'docked', or charged, seventy-five cents a month for medical services. **1901** Merwin & Webster *Calumet 'K'* vii. 128 Every man that drops anything into the bins gets docked an hour's pay. *Ibid.*, I guess we won't take the trouble to dock you. **1937** V. Bartlett *This is my Life* xi. 170, I should find my salary docked or stopped altogether.

dock, *v.*[2] Add: **5.** *trans.* To join (a space vehicle) to another in space; also *intr.*, to become joined. Const. *with*. Freq. in **docking** *vbl. sb.* Hence **docked** *ppl. a.*[3]

1960 A. W. Nelson in *Amer. Rocket Soc. Paper* No. 1493–60. 9/1 The two are in essentially identical orbits and only a few hundred feet or so apart. Now the final 'parking' or 'docking' must be accomplished. **1961** W. Schroeder *Terminal Guidance Scheme for Docking Satellites* (Amer. Rocket Soc. Paper 1952–61) 5/1 While such a solution requires the minimum expenditure of fuel, it is unsatisfactory because of the time required to dock. *Ibid.* 1/1 It appears extremely likely that in the very near future a number of missions will require the docking of two vehicles in space. *Ibid.* 2/2 The propulsion system consists of two canted engines on the aft or docking end..and a single main engine on the nose. **1962** K. W. Gatland *Astronautics in Sixties* xi. 344 Although initally radar and other homing devices may be relied upon to achieve linkage of the space-craft,..experiments will be performed..to see if manual control of the docking procedure is possible. **1965** *New Scientist* 23 Dec. 851/2 Five such rendezvous and docking experiments in space are planned in 1966. **1966** *Times* 6 June 1/6 The target satellite's protective shroud hanging on to the docking collar. **1969** *Daily Tel.* 17 Jan. 1/2 Tass did not say whether an airlock connection allowed the crew to move from one docked capsule to the other, internally. **1969** *Guardian* 22 July 1 Astronauts Neil Armstrong and Edwin Aldrin soared up and away from the moon's surface to dock with the command module. **1971** *Times* 26 Apr. 1/6 Soyuz 10 was docked with Salyut at 02.47 B.S.T. yesterday. **1971** *Times* 21 June 1/2 If his background is in docking, that is an expert in the orbital problems of bringing the space station and the module together at a certain time and place, he will be invaluable.

docker[3]. (Earlier and later examples.)

1889 in Barrère & Leland *Dict. Slang.* **1891** *Strand Mag.* II. 89/1 Many young advocates do a brisk trade in what are termed 'dockers'. **1928** *Daily Tel.* 17 Jan. 10/6 A dock brief...Sometimes it is called a 'docker'.

Dockland, dockland (dǫ·klænd). A name, originally journalistic, for the districts about the London docks.

1904 *Daily Chron.* 21 Sept. 6/2 Clarkson-street School..is situated in the heart of Dockland. **1922** *Weekly Dispatch* 19 Nov. 8 For him there is glamour even in the mean streets of dockland. **1922** *Daily Mail* 12 Dec. 7 The Dockland Mission, formerly known as the Malvern Mission, in Canning Town. **1929** *Times* 7 Feb. 9/4 His work there won the whole-hearted love and devotion of Dockland.

docksman (dǫ·ksmæn). [f. *docks* (see Dock *sb.*[3] 5) + Man *sb.*[1]] A man employed at a dock or docks; *spec.* see quot. 1921.

1921 *Dict. Occup. Terms* (1927) § 745 *Docksman*, one of a team of men who open and shut dock gates by means of capstan. **1929** *Daily News* 25 July 5/6 The former Cardiff docksman.

dockyard. Add: **dockyard man, dockyard-man,** a man permanently employed in a Government dockyard; also (*colloq.*) **dockyard matey.**

1801 Nelson *Let.* 31 July (1845) 433 The Vessels should be..ready for the Dock-yard men to be put on board. **1829** W. N. Glascock *Sailors & Saints* III. 95 The 'Dock-yard-Maties'—a class of men, whose hostility, and turbulent insolence to naval officers..is proverbial. **1833** Dock-yard matey [see Matey]. **1837** Dickens *Pickw.* ii, Soldiers, sailors,..and dockyard men. **1906** *Outlook* 20 Nov. 495/2 Dockyardmen who are in danger of losing that *otium cum dignitate* which they have regarded as their perquisite. **1909** *Pall Mall. Gaz.* 12 Apr. 3/2 Naval men, or dockyardmen, which practically amounts to the same thing, raised Torpedo-boat No. 99 after she was sunk off Berry Head. **1918** 'Taffrail' *Little Ship* ii. 26 The dockyard-maties had slapped on the service gray paint over coal-dust and dirt alike. **1942** N. Monsarrat *H.M. Corvette* i. 11 Aft, the Torpedoman was arguing..with a welder, a Clydeside dockyard-matey.

doctor, *sb.* Add: **6.** Also, a wizard or medicine-man in a primitive tribe.

1858 *Compendium of Kaffir Laws & Customs* 123 Doctors are not entitled to fees, except a cure is performed, or the patient relieved.

b. Esp. in the West Indies, S. Africa, and W. Australia, a cool sea-breeze which usually prevails during part of the day in summer *colloq.*

1740 [in Dict.]. **1844** *Knickerbocker* XXIII. 46 [In St. Augustine] we were beginning the summer custom of gathering every morning to meet the 'doctor' (sea-breeze) on

the square. **1856** F. Fleming *Southern Africa* iv. 62 The South-easter, from blowing all pestilent vapours and effluvia out to sea..has obtained the local epithet of '*the Doctor*'. **1861**, etc. [see *Cape doctor*, *CAPE *sb.*[2] 4].

c. One who mends or repairs; esp. with a qualifying word; also *transf.*; *spec.*, see quot. 1899. *colloq.*

1899 *Daily News* 2 Mar. 9/1 The owner, nervous about a vessel, wants a further insurance, and the 'doctor' procures it for him...The 'doctor' is a broker who deals particularly with the overdue vessels. **1922** A. Haddon *Green Room Gossip* iv. 97 Miss Platt is a professional play-reader and 'play doctor'. **1938** M. Lane *E. Wallace* IV. i. 319 Du Maurier was a skilful 'play doctor', and the final script..bore only a family resemblance to the drama which Edgar had written at top speed.

d. Colloq. phr. *what the doctor ordered*, transf. and fig.: something beneficial or desirable; *you are* (freq. *you're*) *the doctor*: you are the expert; it is for you to decide.

1907 'O. Henry' *Trimmed Lamp* (1915) 95 You are the lady doctor. **1914** *Dialect Notes* IV. 71 She thought Ezry was jest *what the doctor ordered*. **1920** R. Macaulay *Potterism* I. iv. 50 Very well, mother. You're the doctor. [**1922** Joyce *Ulysses* 257 And what did the doctor order today?.. I think I'll trouble you for some fresh water and a half glass of whisky.] **1948** G. Vidal *City & Pillar* I. i. 16 The waiter brought her a drink. 'Just what the doctor ordered,' she said, smiling at him. **1950** Wodehouse *Nothing Serious* 61 'You admired my little friend?' 'She is what the doctor ordered.' **1962** 'H. Howard' *Double Finesse* ix. 106 You're the doctor. Go ahead and open it. **1969** C. Young *Todd Dossier* 120 The thing to do now was relax..and forget about it. 'Okay, you're the doctor,' Charlie said.

7. d. *Electro-plating*. **e.** *Photogravure*. Also *doctor blade*.

1886 A. Watt *Electro-Deposition* xiv. 184 The pad, or 'doctor', as it is sometimes called, is dipped in the gold solution and applied to the part to be gilt. **1926** C. N. Bennett *Elem. Photogravure* xviii. 117 Rapid rotary gravure [was] made definitely commercial, by the introduction of what is known as 'doctor wiping'. *Ibid.* 118 One side of the doctor is clamped firmly in a 'carriage' or holder, while the free, or 'wiping', edge is caused to press upon the printing cylinder at a slight angle. **1930** H. M. Cartwright *Photogravure* vii. 92 The cylinder receives ink from the inking or 'furnishing' roller...The ink is removed from the cylinder surface by means of the doctor. **1943** *Gloss. Terms Electr. Engin.* (*B.S.I.*) 101 *Doctor*, an anode covered with fabric or sponge saturated with the plating solution and applied locally to the article to be plated, which is made the cathode. **1961** T. Landau *Encycl. Librarianship* (ed. 2) 119/2 *Doctor blade*, a flexible knife which removes surplus ink from engraved plate in intaglio printing, *e.g.* photogravure. **1967** Karch & Buber *Offset Processes* 536 *Doctor blade*, a 'knife' of rigid plastic or thin sheet-metal which presses against the gravure press cylinder, and which wipes away ink from the surface of the cylinder.

9. (Earlier example.)

1860 C. M. Yonge *Hopes & Fears* I. II. iv. 229 The doctor ..and all her other radiant fabrications of..feathers.

11. A ship's cook (earlier U.S. examples); the men's cook at a station or camp (N.Z. example).

1821 *Massachusetts Spy* 1 Aug. (Th.), The cook, at sea, is generally called doctor. **1835** J. H. Ingraham *South-West* I. vi. 69 All [the crew] neatly dressed in white trousers and shirts, even to the sable 'Doctor' and his 'sub'. **1902** W. S. Walker *Zealandia's Guerdon* v. 55 'Cook-shop for all hands'. ..Close by this building is the 'doctor's'—cook's—residence.

13. **doctor's gum,** the West Indian tree *Rhus metopium*; also, see quot. 1887.

1858 [see Hog gum]. **1887** C. A. Moloney *Forestry W. Afr.* 279 Hog or Doctor's gum, Gamboge tree (*Symphonia globulifera*).

doctor, *v.* Add: **2. c.** To castrate (an animal).

1902 F. Simpson *Cats* ii. 30 It is necessary..to have your male cat doctored when he arrives at years of discretion. **1958** *Amer. Speech* XXXIII. 165 *Doctor*, to castrate [in Australian stockmen's language]. **1960** T. Clarke in J. Pudney *Pick of Today's Short Stories* XI. 80 If that cat's a Tom,..I'd better get it doctored. **1970** C. Kersh *Aggravations of Minnie Ashe* v. 63 He had been paid 'good English money' to doctor our cat.

doctorand (dǫktǫrænd). Also in L. form **doctora·ndus** (pl. -i). [G., ad. med.L.] A candidate for a doctor's degree.

1912 R. S. Rait *Life Med. Univ.* ii. 32 When our young English Doctorand received the permission of his Rector to proceed to his degree. **1921** *Edin. Rev.* Jan. 72 Increasing numbers of *doctorandi* sought admission to his laboratory.

documentalist (dǫ:kiume·ntălist). [f. Documental *a.* + -ist.] A person engaged in documentation (see *DOCUMENTATION 3 b).

1939 *Nature* 23 Sept. 560/1 The Section devoted to the Division of Work between Libraries and Documentalists. **1948** S. C. Bradford *Documentation* 11 The documentalist is enabled to put before the creative specialist the existing literature, bearing on the subject of his investigation, in order that he may be saved from the dissipation of his genius upon work already done.

documentary, *a.* Add: **4.** Factual, realistic; applied esp. to a film or literary work, etc., based on real events or circumstances, and intended primarily for instruction or record purposes. Also *ellipt.* as *sb.*

1930 P. Rotha *Film till Now* I. ii. 65 The Documentary or Interest Film, including the Scientific, Cultural and Sociological Film. **1932** *Cinema Q.* I. i. 67 Documentary is a clumsy description, but let it stand. The French who first used the term only meant travelogue. **1932** *Film in National Life* (*Rep. Comm. Educ. & Cult. Films*) viii. 115 § 174 A deliberate documentary film must be a transcript of real life, a bit of what actually happened, under approximately unrehearsed conditions. **1934** *Punch* 26 Dec. 720/1 Most documentary films seem to hinge upon the exposition of some staple industry. **1935** R. Spottiswoode *Gram. of Film* 288 The documentary as he defines it is still flourishing. **1936** *Times Lit. Suppl.* 25 Jan. 72/3 The documentary film—or, *tout court*, 'documentary'. **1941** [see *ACTUALITY 4 b]. **1947** J. Hayward *Prose Lit. since 1939* 32 'Mass-Observation', whose intriguing 'documentaries' of the British people at work and play contain the crude substance of innumerable novels, biographies, and essays. **1957** V. J. Kehoe *Film & T.V. Make-Up* i. 17 Some producers do not like the smoothness of the face created by the use of make-up. They strive to achieve what is termed a *documentary* effect..by the lack of make-up on men (even at times, on women). **1957** *Listener* 18 July 103/1 Mr. Owen's 'documentary', as he calls his attractive book, reveals him as an acute observer. **1962** *Observer* 8 July 20/4 Henry Cecil's light legal documentary fiction.

Hence **docume·ntarist,** one who makes documentary films.

1953 *New Statesman* 10 Oct. 419 A documentarist of sensation. **1959** *Times* 3 Nov. 15/3 The most ruthless of documentarists. **1961** *Times* 1 June 6/4 It is a film about film-making...A simple, puppyish documentarist has moved into the 'pad' of a group of junkies. **1962** *Times* 26 Apr. 8/5 The Swedish documentarist..in his new film.

documentation. Add: **3. b.** The accumulation, classification, and dissemination of information (see quot. 1948); the material so collected.

[**1905** *Publ. Office Internat. de Bibliographie* LXIX (*title*) L'Organisation rationnelle de l'information et de documentation en matière économique. Rapport présenté.. par P. Otlet.] **1927** D. P. Myers in *Libraries* XXXII. 107 (*title*) International documentation, its classification and purpose. **1938** *Rep. World Congr. Universal Documentation, 1937* 32 The documentation centre at the Science Library. **1940** *Proc. Brit. Soc. Internat. Bibliogr.* I. II. 11 The co-ordination of documentation. **1948** S. C. Bradford *Documentation* 9 Documentation is the process of collecting and subject classifying all the records of new observations and making them available, at need, to the discoverer or the inventor. **1967** *FID News Bull.* XVII. 70/1 The advent of a new discipline of science called *documentation*,..*scientific information*, *information science*, etc., is a commonly recognized fact.

dod- (Dod *sb.*[1]), used as an intensive with verbs and pa. pples. *U.S. colloq.* See also *DODGAST *v.*, *DOD-ROT *v.*

1829 *Yankee* Apr. 131/2 Dod burn the boy! what's he arter? **1835** A. B. Longstreet *Georgia Scenes* 231 I'll be dod blamed if I do. **1883** 'Mark Twain' *Life on Miss.* 220 He ejaculated: 'Well, I'll be dod-derned.' *Ibid.*, 'Dod-dern' was the nearest he ventured to..swearing. **1908** C. E. Mulford *Orphan* xiii. 164 'Dod-blasted postage stamp of a pelt,' he grumbled.

dodderer (dǫ·dərəɹ). [f. Dodder *v.* + -er[1].] One who dodders; one infirm in body or mind; a feeble or inept person.

1907 G. B. Shaw *John Bull's Other Island* III. 65 We're not all foostherin oul doddherers like Mat. **1917** W. J. Locke *Red Planet* xi. 131 Do you think I'm a blind dodderer? **1926** *Public Opinion* 23 Apr. 420/1, I am constantly meeting ponderous dodderers who are sure civilisation is in rapid decay. **1930** *Punch* 14 May 551, I see now that I have been a difficult old dodderer upon this subject.

doddering, *ppl. a.* (see under Dodder *v.*). Add: Now freq. mentally feeble or inept; futile, footling. Hence **do·dderingness.**

1908 *Fabian News* XIX. 82/2 Mr. Justin McCarthy, in his rather doddering introduction, explicitly warns us against Mr. Sheehy-Skeffington's portraiture of Davitt as an anti-clerical politician. **1915** Wodehouse *Something Fresh* iii, The amiable dodderingness which marked every branch of his life. **1921** G. B. Shaw *Back to Methuselah* IV. III. 203 He was a doddering old ass. **1926** *Brit. Weekly* 23 Sept. 519/3 Your puir, toom, dodderin', fushionless kirk.

doddery (dǫ·dəri), *a.* [f. Dodder *v.* + -y[1].] Apt to tremble or totter, from age or infirmity; shaky.

1866 [see *DITHERY *a.*]. **1919** J. Buchan *Mr. Standfast* xvii, When he got on his feet he was as doddery as an old man. **1921** *Chambers's Jrnl.* May 325/1 The old man..seemed to have become very doddery as he descended from the buggy. **1963** H. Calisher *Textures of Life* ii. 35 Its thick-walled ambling waste of space and doddery pediment.

doddle (dǫ·d'l), *sb.*[3] *colloq.* [?f. Doddle *v.* 2.] Something that is easy or requires little effort; a 'walk-over'; (see also quot. 1937).

1937 Partridge *Dict. Slang* 229/2 *Doddle*, money very easily obtained. **1966** M. Woodhouse *Tree Frog* xv. 122 If the climb had reached any level of difficulty higher than Moderate, which is the Climbers' Club's polite way of labelling a gumshoe doddle, we'd have died. **1970** A. Draper *Swansong for Rare Bird* vi. 40 Probation was a doddle really, and it didn't make much difference to me.

dodecaphonic (dō^u^:dikăfǫ·nik, dode:kăfǫ·nik), *a.* *Mus.* [Gr. δώδεκα twelve + Phonic *a.*] Of, relating to, or using, the twelve-note system, a method of composition formulated by Arnold Schönberg. Hence **dode·caphonism** (also dō^u^:dikæ·fōniz'm), **dode·capho:ny** (-fō^u^:ni, also dō^u^:dikæ·fōni), the twelve-note system; **dodecaphonist** (dode·kăfōnist, dō^u^:dikæ·fōnist), a composer using this system.

1950 L. O. Symkins in *Étude* Sept. 12 (*title*) Schœnberg's new world of dodecaphonic music. **1951** *Musical Q.* XXXVII. 95 We dislike Schœnbergism, atonalism, dodecaphonism, and the rest. **1952** *Ibid.* XXXVIII. 604 Schollum's Second Sonata is in its combination of linearity, Romantic expressiveness, and dodecaphony a mixture of styles. **1953** *N.Y. Times* 2 Feb. 18/8 Like the majority of dodecaphonists, he presents music governed by strict logic and organization. **1955** *Oxf. Compan. Mus.* (ed. 9) 698/1 All the twelve notes of the octave are employed in every composition, and all the notes are treated in such a way as to enjoy an equal footing...Theorists of the Schönberg school prefer to call his method not a chromatic one but 'Dodecaphonic'. **1955** *Times* 11 May 7/6 Dallapiccola's music, written before his conversion to dodecaphony, never failed to heighten the emotion. **1970** *Daily Tel.* 20 May 16/5 No amount of explaining could trace dodecaphony in Tippett's 2nd Symphony.

dodecyl (dō^u^·disil, dō^u^,di·səil). *Chem.* [f. Dodec(ane + -yl.] The univalent hydrocarbon radical $C_{12}H_{25}$—, derived from dodecane by the loss of one hydrogen atom.

1889 Muir & Morley *Watts' Dict. Chem.* II. 413/2 Dodecyl alcohol $C_{12}H_{26}OH$. **1963** A. J. Hall *Textile Sci.* vi. 294 Thus two very popular synthetic detergents in use today are dodecyl benzene sulphonate [etc.].

dodgast (dǫ·dgast), *v.* *U.S. colloq.* [f. *DOD- + *gast*, prob. for Blast *v.* 10. Cf. *DOD-ROT *v.*] In imprecatory and expletive use: to 'confound', 'curse'. Chiefly in **do·dgasted** *ppl. a.*

1888 *Detroit Free Press* (Farmer), It's a dodgasted funny thing,..but it's a fact. **1908** C. E. Mulford *Orphan* ix. 103 What can we do when our cayuses are so dod-gasted tired? **1909** *N.Y. Observer* 2 Sept. 319/2 Well, dodgast you, get in the stern there. **1914** W. J. Locke *Fort. Youth* xxi, It's a pity, sonny—a dodgasted pity!

dodge, *v.* Add: **13.** *to dodge Pompey*: (*a*) to evade work (*Naval slang*); (*b*) see quot. 1930 (*Austral. slang*).

1929 F. Bowen *Sea Slang* 38 *Dodging Pompey*, avoiding work on shipboard. Originally a naval phrase entirely. **1930** Billis & Kenyon *Pastures New* iii. 46 Browne detailed the laws passed, not to encourage the overlander, but rather to counteract his habit of stealing grass—'dodging Pompey', as it was known. **1961** F. H. Burgess *Dict. Sailing* 73 *Dodging Pompey*, skulking, or avoiding work by the use of any semi-legitimate excuse.

dodge, *sb.*[1] Add: **2. b.** *on the dodge*: engaged in crooked or dishonest proceedings.

1904 'O. Henry' *Heart of West* (1912) xi. 214 I've been on the dodge for a month, and I'd like to rest up. **1920** J. M. Barrie *Kiss for Cinderella* I. 26 If you wanted to get into Buckingham Palace on the dodge, how would you slip by the policeman?

Dodgem (dǫ·dʒĕm). Also **Dodge-Em,** and with small initial(s). [f. Dodge *v.* 4 + 'Em.] Esp. in *pl.*, a fairground amusement consisting of a number of small electrically-powered cars steered about in an enclosure. Also *attrib.*, as *dodgem car*.

1921 *Sci. Amer.* 15 Oct. 269/1 A device known as the 'Dodge-Em' is a clever piece of..work. Small cars, fitted with steering wheels, are placed on a polished..floor. A trolley connects each little car with an electrically charged mesh..overhead...The steering is only relative and it requires extreme ability to dodge the other fellow's car. **1937** *Night & Day* 23 Sept. 27 The big outfits—the Giant Racers, Over the Falls, the Helter-Skelter, the Dodge 'Ems. **1945** *Archit. Rev.* XCVII. 53 The sparks flying from the Dodgems. **1947** 'N. Shute' *Chequer Board* x. 278 Among the hurly-burly of his swings and roundabouts and flip-flops and dodgem cars. **1952** N. Coward *Relative Values* I. ii. 28 What on earth's a Dodgem?..One of those little motor cars you go on in Margate and bang into everybody. **1968** D. Braithwaite *Fairground Architecture* i. 20 The transport revolution has been reflected in a succession of devices from..'Dodg'em Cars'..to present day 'Space-Cruisers' and 'Hurricane Jets'. *Ibid.* vi. 103 'Dodg'ems' have been the most consistently popular ride on the fair-ground.

dodger, *sb.* Add: **2.** (Earlier examples.)

1831 J. M. Peck *Guide for Emigrants* II. 152 Dodgers are masses [of corn meal] like small loaves of bread, prepared in a similar manner [i.e. with water or milk], and baked in the spider or skillet. **1832** Mrs. F. Trollope *Dom. Mann. Amer.* I. 83 Hoe cake, johnny cake, waffle cake, and dodger cake.

b. A sandwich; bread, food. *Austral.* and *Services' slang.* Also *dial.* (cf. Dodge *sb.*[2]).

1919 Downing *Digger Dial.* 19 *Dodger*, bread. **1925** Fraser & Gibbons *Soldier & Sailor Words* 79 *A dodger*.. a sandwich. **1941** Baker *Dict. Austral. Slang* 24 *Dodger*, food of any kind. Also, a 'yunk of dodger', a slice of bread. **1957** 'N. Culotta' *They're a Weird Mob* (1958) 51 Smack us in the eye with another hunk o' dodger. **1966** 'L. Lane' *ABZ of Scouse* II. 27 *Dodger*, bread or sometimes crude cake. 'Gimme a slice er dodger.'

5. On a ship: a screen to afford protection from spray, etc. Also, a protective shield used on a rocket, etc.

1898 C. Hyne *Capt. Kettle* x. 260 Under shelter of the dodgers on the upper bridge. *Ibid.* 262 Kettle hung on behind the canvas dodgers at the weather end of the bridge. **1942** Masefield *Generation Risen* 42 Sprays are freezing on the dodger. **1956** *Jrnl. Brit. Interplan. Soc.* XV. 126 A dodger or shield had been placed at some distance from the thrust-balance. **1959** P. McCutchan *Storm South* vi. 75 Wynton took..shelter..behind a canvas dodger secured to the weather mizzen rigging.

dodger (dǫ·dʒəɹ), *a.* *Austral. slang.* [Orig. unknown.] Good, excellent.

1941 in Baker *Dict. Austral. Slang* 24. **1953** D. Stivens *Gambling Ghost* 1 Instead of having to risk a knock on the Pearly Gates everything was dodger.

dodgy, *a.* Add: (Later examples.) Also (*colloq.*) of things: difficult, awkward, tricky.

1896 *Daily News* 16 Oct. 6/3 The pious purpose perhaps justified the dodgy means. **1898** G. B. Shaw *Mrs. Warren's Prof.* I, Take care of your fingers: theyre rather dodgy things, those chairs. **1916** D. H. Lawrence *Let.* 13 Jan. (1948) 67 The roads are too dodgy to be grasped. **1959** *Times Lit. Suppl.* 7 Aug. p. xii/2 Docketing and definition are dodgy businesses. **1960** H. Pinter *Room* 108 It'd be a bit dodgy driving tonight.

dodo. Add: *transf.* and *fig.* An old-fashioned, stupid, inactive, or unenlightened person. Phr. (*as*) *dead as the* (or *a*) *dodo*: see *DEAD *a.* 32 b.

1874 [in Dict.]. **1886** *Pall Mall Gaz.* 11 Feb. 1/1 The old dodo at Scotland Yard, roused into a state of feverish activity..yesterday converted itself by a tremendous effort into a gigantic turkey-cock. **1922** F. Scott Fitzgerald *Let.* 18 June (1964) 164 Tom Boyd wrote me that Bridges had been a dodo about some Y.M.C.A. man. **1950** A. Wilson (*title*) Such darling dodos.

dodrat, var. *DOD-ROTTED *ppl. a.*

1897 *Outing* (U.S.) XXX. 173/2 This is the dodrattest place I ever struck.

dod-rot, *v.* *U.S. colloq.* [f. *DOD- + Rot *v.*] = *DODGAST *v.* Hence **do·d-rotted** *ppl. a.*

1842 *American Pioneer* I. 347 'Dod rot'em,' said the old hunter, 'I would not let them have a bushel.' **1885** 'C. E. Craddock' *Prophet Gt. Smoky Mts.* vi. 121 'Dod-rot that critter,' exclaimed the sheriff, angrily. **1887** *Century Mag.* (Farmer), 'You ketch us with yer dodrotted foolin',' says he. **1911** R. D. Saunders *Col. Todhunter* vii. 100 That dod-rotted old lady is a-movin' Heaven and earth to make a match.

doe. Add: **3.** *doe-eye*; *doe-eyed* adj.

1933 Dylan Thomas *Let.* (1966) 75 A broad creature, not ..to be confused with the slim, doe-eyed apparition of your green book. **1959** 'J. R. Macdonald' *Galton Case* (1960) ii. 17 The doe-eyed girl from the badminton court. **1959** *Manch. Guardian* 2 July 5/6 A certain tendency to doe-eyed sentimentality. **1963** N. Freeling *Because of Cats* iv. 64 Big wide doe-eyes nodded, yes.

dœgling (dö·gliŋ). Also **dogling.** [Native name in the Faröe Islands.] The bottle-nosed whale, *Hyperoodon ampullatus*, which yields *dœgling train oil*.

1866 H. Watts tr. *Gmelin's Handbk. Chem.* XVII. 180 Doegling Train Oil. From the doegling (the *Balæna rostrata* of Chemnitz, *Hyperodon* of later zoologists), a kind of dolphin. **1890** Roscoe & Schorlemmer *Treat. Chem.* (new ed.) III. II. 483 The glyceride of this acid forms, according to Scharling, the principal part of the dœgling train oil. **1947** *Antiquity* XXI. 87 The Bottle-nosed Whale or Dogling.

doek (duk). Also 9 **douk.** *S. Afr.* [Afrikaans, = cloth (Duck *sb.*[3]).] A cloth, especially a head-cloth. Cf. *KOPDOEK.

1798 Lady Anne Barnard *Jrnl.* 22 May in *S. Afr. Century Ago* (1924) II. 226, I offered her four schellings or a *doek*, viz, a handkerchief; she preferred the last. **1853** W. R. King *Campaigning in Kaffirland* iii. 19 Gaily dressed in startling cottons, with gaudy *douks* or bandanas on their woolly heads. **1944** M. de B. Nesbitt *Road to Avalon* viii. 66 The women in neat white print dresses and 'doeks'—coloured kerchiefs wrapped around their heads. **1956** N. Gordimer *Six Feet of Country* 43 Now she had on her head a woollen doek again, instead of one of the maids' caps Ella provided for her to wear.

doer. Add: **5.** An eccentric, a 'character', esp. in phr. *hard doer*, a 'hard case'. *Austral.* and *N.Z.*

a 1885 in *Penguin Bk. Austral. Ballads* (1964) 78 Many more hard doers, all gone to Kingdom Come. **1916** 'Anzac' *On Anzac Trail* 122 He is a rule, a 'hard doer'. **1919** Downing *Digger Dial.* 19 *Doer*, a person unusually humorous, reckless, undisciplined, immoral or eccentric. **1928** *Bulletin* (Sydney) 29 Feb. 21/1 Give me the real hard-doer, give me the decent chap. **1929** K. S. Prichard *Coonardoo* xxxi. 300 You know Monty,..one of the hardest doers in the Nor'-West. **1947** *Landfall* I. 166 Struth..a real hard doer! **1959** S. H. Courtier *Death in Dream Time* iii. 28 Laurie could have been joking. He was that kind of doer, you know.

dog, *sb.* Add: **1. f.** *the dogs*: a greyhound race meeting (see also *17 a). *colloq.*

1927 *Daily Mail* 28 July 7/4 'Going to the dogs' has..lost..its old suggestion of a descent to dissipation and ruin. Since greyhound racing at the White City..came into existence the expression has suggested a good adventure. **1928** A. P. Herbert *Trials of Topsy* 52 (*heading*) Going to the Dogs. Well Trix darling at last I've been to these *contagious* greyhounds. **1936** W. Holtby *South Riding* I. v. 54 Pretty little painted sluts minced..off to the pictures or dogs. **1959** *Economist* 13 June 1016/3 He..failed his Bar examinations because he preferred horse-racing, the 'dogs' and dancing.

3. b. (Further examples.)

1836 Dickens *Sk. Boz* 2nd Ser. 200 He had been a sad dog in his time. **1844** [see Lucky *a.* 1 a]. **1847** Dickens *Dombey* xxvi. 266 Well! we *are* gay dogs, there's no denying. **1900, 1910** [see *GAY *a.* 2 a].

d. = Watch-dog b. (*Schoolboys' slang.*)

1870 *Chambers's Jrnl.* Oct. 676/1 The boys withdrew..to read the forbidden prints, three taking their turn at a time, whilst three more 'played dog'—that is, stood ready to bark a warning should a pion be seen approaching. **1959** I. & P. Opie *Lore & Lang. Schoolch.* xvii. 373 In Kirkaldy watch-dog [*i.e.*, a boy keeping lookout] becomes either 'watchie' or 'dog'.

e. An informer; a traitor; esp. one who betrays fellow criminals. *U.S.* and *Austral. slang.*

1846 *Nat. Police Gaz.* (U.S.) 21 Feb. 210/2 Dick White has been playing the 'dog', and he and the 'coppers' are now within ten minutes of the house. **1888** 'R. Boldrewood' *Robbery under Arms* I. v. 69 Are you going to turn dog, now you know the way in? **1901** E. Dyson *Gold Stealers* xix. 231 'Tell me how you come to be in the Stream drive that night.' Dick..answered nothing. 'Come on, old man, I won't turn dog.' **1969** *Telegraph* (Brisbane) 11 Oct. 1/1 A 'dog' is the term applied by prisoners to fellow-prisoners who turn informer.

f. Something poor or mediocre; a failure. *U.S. slang.*

1936 *Metronome* Feb. 21/4 *Dog*, something [*i.e.* a song] that's kicked around. **1952** *N.Y. Times Bk. Rev.* 10 Aug. 8/3 '[The book will have] a record-breaking sale.' 'Yes, unless of course the book turns out to be a dog.' **1968** L. O'Donnell *Face of Crime* ix. 118 I'd be a fool not to take advantage. I had a real dog on my hands. **1970** *New Yorker* 15 Aug. 65/1 Audiences are in a mess...They don't know what they want... So many movies are *dogs*.

g. A horse that is slow, difficult to handle, etc. *slang.*

1944 P. Kendall *Service Slang* 5/2 *Dog*, affectionate term for cavalryman's horse, also called a job. **1945** Baker *Austral. Lang.* ix. 175 A *dog* is a horse difficult to handle. **1955** T. Rattigan *Separate Tables: Table by Window* iii, Is it going to be dry at Newbury?..Walled Garden's a dog on heavy going. **1958** J. Hislop *Start to Finish* xii. 132 A 'dog' means a horse which cannot be relied upon to do his best..a horse may be a 'dog' because there is something wrong with him.

13*. *pl.* Short for *dog's meat*: feet. *Rhyming slang.*

1924 Wodehouse *Leave it to Psmith* x. 211 You'll pick up your dogs and run round as quick as you can make it. **1939** M. Dickens *One Pair of Hands* x. 169, I feel more like goin' to bed and sleeping for a week than prancing round the ballroom on me poor dogs. **1939** Steinbeck *Grapes of Wrath* vi. 56 We ain't gonna walk no eight miles..to-night. My dogs is burned up.

13.** *pl.* Sausages. Cf. *hot dog* (*HOT *a.*). *slang.*

1925 Fraser & Gibbons *Soldier & Sailor Words* 80 *Dogs*, sausages. **1959** I. & P. Opie *Lore & Lang. Schoolch.* ix. 163 Sausages are 'bangers'..or 'dogs'.

15. e. Also ellipt. *a hair of the dog*, a drink.

1936 M. Mitchell *Gone with the Wind* x. 207 Do you think..Miss Pittypat would be having any brandy in the house? The hair of the dog——. **1966** A. E. Lindop *I start Counting* xviii. 224 George took Len off for a hair of the dog. **1967** N. Fitzgerald *Affairs of Death* ix. 152 What you need, Frank, is a good stiff hair of the dog.

m. *to keep a dog and bark oneself*.

1583 B. Melbancke *Philotimus* 119 It is smal reason you should kepe a dog, and barke your selfe. **1738** Swift *Polite Conv.* i. 10, I won't keep a Dog, and bark my self. **1965** J. Porter *Dover Two* xi. 147 'What time is it?' There was a clock right opposite him on the dining-room wall but Dover didn't believe in keeping a dog and barking himself. 'Just gone nine, sir.'

n. *to see a man about a dog*: used colloq. as a vague excuse for leaving or absenting oneself.

c 1867 D. Boucicault *Flying Scud* IV. i in Nicoll & Cloak *America's Lost Plays* (1940) I. 221 Excuse me Mr. Quail, I can't stop; I've got to see a man about a dog. **1939** D. L. Sayers *In Teeth of Evidence* 38 I've got to get back to London to see a man about a dog. **1963** *Amer. Speech* XXXVIII. 175 *See a man about a dog* was a Prohibition euphemism for 'buying liquor', whereas several contemporary students [at Kansas University] recognized it as a circumlocution for 'visiting a rest room'...At Johns Hopkins, the phrase served as an excuse for leaving the scene.

o. In various locutions involving an unpleasant circumstance or event unfit even for a dog.

1887 Baumann *Londinismen* 43/1 It isn't fit to turn a dog out. **1943** *Amer. Speech* XVIII. 46 Other examples of translated Yiddish being adopted by non-Yiddish-speaking people are, 'It should(n't) happen to a dog!' [etc.]. **1964** J. Porter *Dover One* i. 12 The Assistant Commissioner shuddered gently as he thought of all the messes you could get into in a kidnapping case. It wasn't the sort of job you'd wish on a dog.

p. *to put on dog*: to assume pretentious airs. *colloq.* Hence *dog* (ellipt.), pretentiousness, 'side'.

1871 L. H. Bagg *Four Years at Yale* 44 *Dog*, style, splurge. To put on dog, is to make a flashy display, to cut a swell. **1889** W. D. Howells *Hazard of Fortunes* I. 267 He's made the thing awfully *chic*; it's jimminy; there's lots of dog about it. **1915** Kipling *Fringes of Fleet* 36 Ah! That's the King of the Trawlers. Isn't he carrying dog, too! Give him room! **1924** W. J. Locke *Coming of Amos* xii, I don't want to put on dog, but the Lord didn't give me physical strength for nothing. **1926** —— *Old Bridge* II. v, Young Blake puts on dog and condescends to take the order. **1940** Wodehouse *Eggs, Beans & Crumpets* 48 An editor's unexampled opportunities for putting on dog and throwing his weight about. **1950** W. Stevens *Let.* 20 Feb. (1967) 670 Sweeney is completely without side or dog. **1962** 'A. Gilbert' *No Dust in Attic* xiv. 190 Matron put on a lot of dog about the hospital's responsibility.

q. *like a dog's dinner*: used of someone or something dressed or arranged in an ostentatiously smart or flashy manner. *colloq.*

1934 'C. L. Anthony' *Touch Wood* II. ii. 66 Why have you got those roses in your hair? You look like the dog's dinner. **1936** J. Curtis *Gilt Kid* v. 58 The geezer..was dolled up like a dog's dinner with a white tie and all. **1945** *Penguin New Writing* XXIII. 42 A dizzy blonde all dressed up like a dog's dinner. **1954** J. Trench *Dishonoured Bones* II. iii. 57 Tarting up my house and the gardens like a dog's dinner.

r. *like a dog with two tails*: very pleased, delighted.

1953 J. Trench *Docken Dead* v. 65 She's like a dog with two tails. **1954** P. H. Johnson *Impossible Marriage* III. x. 273 Ned came in..looking scared. He was not at all like a dog with two tails. **1969** B. Cobb *Scandal at Scotland Yard* iii. 34 Bagshaw didn't bear the resemblance to a dog with two tails which I had expected.

16. b. *dog-hound* (later examples).

1896 *Sportsman* 10 July 4/2 In beagles, the Cheshire won in the class for..dog hounds. **1955** *Times* 14 July 5/4 The Duke of Norfolk saw the doghound championship awarded to Distaff, a 1952 entered hound from his own pack.

17. a. *dog-basket*, *-couple* (later examples), *-food*, *-pack*, *-show* (earlier examples), *-train*, *-whistle* (earlier example). Also in ref. to greyhound racing, as *dog-race*, *-racer*, *-racing*, *-track*.

1842 Mrs. H. M. Stanley *Let.* 22 Sept. in N. Mitford *Ladies of Alderley* (1938) 46, I walked to Northwich to order a dog basket & other trifles. **1843** *Ainsworth's Mag.* III. 147 With his dog-couples slung across his shoulders. **1939–40** *Army & Navy Stores Catal.* 999/2 Dog Couples, medium, for spaniels, setters and pointers. **1907** *Athenæum* 3 Aug. 119 The whale-meat taken as dog-food was poisoning the animals. **1951** M. McLuhan *Mech. Bride* 77/1 The ultimate absurdity of this attitude gets frequent expression in those dog-food ads. **1925** Masefield *Trial of Jesus* 6 The sneered lips of the dog-pack in men snarling. **1927** F. Brett Young *Portrait of Clare* II. xi. 201 The huntsman and his whips had clattered over from the kennels with the dog-pack. **1864** *Chambers's Jrnl.* 502/2 Betting more than you can afford upon a dog-race. **1865** *Ibid.* 657/2 They are also fond of dog-racing. **1875** *Ibid.* 254/1 Excluded from enjoying the pleasures of bull-baiting, the Lancashire rough falls back on dog-racing or some similar sport which admits of betting. *Ibid.*, Manchester..being the headquarters of the rabbit-courser;..and the colliery districts generally, of the dog-racer. **1928** *Manch. Guardian Weekly* 10 Aug. 113/4 The Dog-racing bill. **1859** *Newcastle Courant* 1 July 3/1 (*heading*) Sporting dog and poultry show. **1861** *Times* 4 Apr. 12/2 (*heading*) Fancy dog show at Birmingham. **1863** in *N. & Q.* (1963) Mar. 106/1 The International Dog Show. **1928** *Observer* 25 Mar. 16/6 The Ministry of Health has decided that Wimbledon must put up with a dog-track, however much the Council and inhabitants may resent it. **1958** *Economist* 11 Oct. 155/1 A racecourse, a dog track, and thirty-nine pubs help the men to get through the £A2,000 a year that many of them have been earning. **1897** Kipling *Capt. Courageous* v. 121 He told them of mail-carrying in the winter up Cape Breton way, of the dog-train that goes to Coudray. **1842** *Ainsworth's Mag.* II. 222 Their word..is now less influential than a dog-whistle.

b. *dog-driver*, *-driving*, *-fancier* (earlier examples), *-fancying*.

1806 *Sporting Mag.* XXVII. 194/1 (*heading*) The dog-fancyer. **1821** P. Egan *Life in London* II. iii. 221 The dog-fancier in the corner..sidled up to the Swells. **1845** *Ainsworth's Mag.* VII. 5 I'm the only honest man in the dog-fancyin' line. **1895** Kipling *2nd Jungle Bk.* 148 The boy knows something of dog-driving. **1898** *Daily News* 17 Jan. 8/5 The Admiral..described how the two saved the life of their dog-driver..when he 'was rapidly freezing'. **1910** H. G. Wells *Hist. Mr. Polly* ix. 301 Drowning superfluous kittens, dog-fancying as required.

c. Further examples, chiefly *poet.*

1922 Joyce *Ulysses* 404 Swineheaded..or doghaired infants occasionally born. **1928** E. Sitwell *5 Poems* 4 Beneath my dog-furred leaves you see The creeping strawberry. *Ibid.* 10 And dark green dog-haired leaves of strawberries. **1929** —— *Gold Coast Customs* 22 The dog-whining dawn light. **1931** W. de la Mare *7 Short Stories* 134 He looked at me..with those dog-bright eyes. **1932** Auden *Orators* III. 85 A dog-hated dustman.

e. *dog-English*.

1938 F. M. Ford *Let.* 16 Mar. (1965) 290 He will at least write comprehensible dog-English. **1961** *N.Y. Times Bk. Rev.* 22 Jan. 6/4 They have been translated into a kind of academic dog-English.

18. a. dog-clutch, a device for coupling two shafts in the transmission of power, one member having teeth which engage with slots in another; **dog-eat-dog,** *phr.* used esp. *attrib.* of a ruthlessly competitive attitude (with allusion to the proverb 'dog does not eat dog': cf. quot. 1858 *s.v.* 15 m in Dict.); **dog-end** *slang*, a cigarette-end; **dogface** *U.S. slang*, a soldier, esp. an infantryman, in the U.S. army; **dog handler,** a person who is in charge of a dog or dogs, esp. a police dog; **dogman** *Austral.* (cf. Dog *sb.* 7 a), a man who gives directional signals to a crane-operator on a building-site, often riding on the goods lifted by the crane; **dog-paddle** *colloq.*, a stroke, or way of swimming, like a dog's; **dog-robber** *slang*, (*a*) in *pl.*, civilian clothes worn by a naval officer on shore leave; (*b*) a navy or army officer's orderly; **dog-sled** (earlier example); **dog tag** *U.S. slang*, a soldier's identity disc; **dog-team**, a team of dogs used for drawing a sled; **dog-town** (earlier example); **dog-trials** *pl.* (*N.Z.*), a series of tests of the skill of sheep-dogs in tending sheep; **dog tucker** *Austral.* and *N.Z.*, mutton used as food for working dogs (see quot. 1933).

1907 *Westm. Gaz.* 18 Nov. 6/3 The road-wheels are mounted on the..steel valves, leaving the enclosed driving-shafts free to transmit the power, through the medium of dog-clutches, to the hubs. 1930 *Engineering* 14 Feb. 198/2 The spindle is connected to the handwheel by a dog clutch. 1951 G. H. Sewell *Amat. Film-Making* (ed. 2) iii. 44 A dog-clutch on the camera motor mechanism engages with the main spindle of the magazine. 1933 'D. Stivens' *Milk & Honey Route* xv. 169 He knows and lives the justice of the jungle as well as he knows and lives the dog-eat-dog code of the main stem. 1959 N. N. Holland *First Mod. Comedies* xiv. 168 The impression we get is of a dog-eat-dog world. 1961 *Daily Tel.* 20 Feb. 21/4 Dog-eat-dog among the Lumumbists. 1963 M. Levinson *Taxi* vi. 72 No woman can call herself weak if she is prepared to throw herself pell-mell into the 'dog-eat-dog' kind of driving that goes on in the West End. 1964 R. Jeffries *Embarrassing Death* iii. 21 You don't want to be nice for this job..it's dog eat dog. 1935 H. Nevill *Sneak Thief on Road* 153 'Hard-up?' 'Dog-ends,' said Yank. 'Dust, funny mixings, ten a penny, cigarette ends out of the gutter.' 1941 *Punch* 10 Sept. 238/3 Our sojourning place was bare of everything but dust and cigarette-butts ('dog-ends', these last are called). 1955 P. Wildeblood *Against Law* iii. 18 The ensuing 'dog-ends' are unpicked, re-rolled and smoked again. 1941 *Time* 13 Oct. 2 Ordinary soldiers are called 'dog-faces' by the devil-dog Marines. 1943 Steinbeck *Once there was War* (1959) 143 There are too many little religious rules and prejudices [amongst the Arabs in north Africa] that an unsuspecting dogface can run afoul of. 1958 *Newsweek* 20 Oct. 42 No dogface who dug one (sc. a foxhole) will ever forget his blistered hands and aching back. 1968 R. Jeffries *Traitor's Crime* i. 9 The civilian Ware..was changing a fan-belt on a dog handler's van. 1971 *Sunday Express* 25 Apr. 17/6 Dog handler Mr. Robert Green..receives £720 from her estate. 1962 R. Clark (*title*) The Dogman and other poems. 1970 *Sunday Mail Mag.* (Brisbane) 10 May 5/2 Most Sydney men will not raise their eyes beyond ladies' leg level except to see one other phenomenon—the daring dogmen on the skyhooks. 1904 R. Thomas *Swimming* 428 How did Beowulf swim? I should say the human stroke..popularly but incorrectly known as dog paddle, which was the European stroke to about the year 1500. 1928 *Daily Express* 25 June 4/5 Try to push off from the side, performing the kick with a 'dog-paddle' arm stroke. 1898 W. P. Drury *Tadpole of an Archangel* 202 He was an absent-minded..young giant in..a dogrobber suit. 1946 G. Hackforth-Jones *Sixteen Bells* ii. 61 ... 'dog-robbers'... 1958 M. Dickens *Man Overboard* iii. 36 Then he..changed into dog robbers and went into the town to get drunk. 1967 *Everybody's Mag.* ... 1859 ... dog-sled ... 1936 ... dog tag ... dog-town ... 1934 *Press* (Christchurch) ... dog-trials ... 1933 ... dog tucker ...

b. dog's age *slang* (orig. *U.S.*), a long time; **dog's breakfast** *slang*, a mess; **dog's chance**, the poorest chance; cf. *dog-chance* s.v. 18 a in Dict.; also (N.Z.) *dog's show*.

1836 *Knickerbocker* VII. 17 That blamed line gale has kept me in bilboes such a dog's age. 1916 H. L. Wilson *Somewhere in Red Gap* v. 175 ... 1919 T. K. Holmes *Man fr. Tall Timber* v. ... 1937 Partridge *Dict. Slang* 231/2 *Dog's breakfast*, a mess: low Glasgow (–1934). 1959 *Times* 29 Apr. 10/4 He can't make head or tail of it. 'It's a complete dog's breakfast.' 1963 *Times* 22 Feb. 12/3 The warders..are very angry and have rejected the latest War Office offer as totally unacceptable. They feel the offer is a bit of a dog's breakfast. 1902 *Captain* VII. 542/1 They all felt that Adderman's wouldn't have a dog's chance when Ardenwood College had got fairly going. 1939 J. B. Priestley *Let People Sing* 50 Don't suppose I've got a dog's chance really, but I have to keep on trying. 1957 I. Cross *God Boy* (1958) vi. 46, I had to admire Bloody Jack for sitting on there even though he didn't have a dog's show of getting any fish.

c. dog-winkle, the marine gastropod *Nucella lapillus*.

1853 Forbes & Hanley *Hist. Brit. Mollusca* III. 386 This whelk [sc. *Purpura lapillus*] is called Dog-periwinkle on many parts of the coast. 1856 P. H. Gosse *Man. Marine Zool.* II. 129 *Purpura* (Lamk.). Purple, or Dog-winkle. 1901 *Westm. Gaz.* 16 Dec. 3/1 The Tyrian purple of the ancients can be obtained from the common dog-winkle (*Purpura lapillus*). 1950 ... *Gt. Brit.* (ed. 2) 196 *Nucella lapillus* (*Purpura*). Dog Winkle.

dog, *v.* Add: **1. a.** Also with *out*.

1843 'R. Carlton' *New Purchase* II. 180 We'll dog out the rats now. 1877 E. Ross et al. *Gloss. Words Holderness* 55 Dog-out-o, to obtain by persistent importunity. 1936 M. Allingham *Flowers for Judge* ix. 140 Someone murdered him very neatly indeed..Our astute friends..dogged that much out all right.

dogan (dōᵘ·găn). *Canadian slang.* Also **dogun** (-gən). [Perh. f. *Dogan*, an Irish surname.] An Irish Roman Catholic.

1854 *Hamilton* (Ontario) *Gaz.* 15 May 2/7, I would be overly liberal if I estimated their number as a couple of Dogans! 1933 'P. Slater' *Yellow Briar* (1934) ii. 22 Many a time I got a smart clout on the lug and was told to take that for a dirty little dogan.

Dogberry². (Earlier example.)

1846 R. Ford *Gath. Spain* xx. There is no absurdity, no inconceivable ignorance, too great for the local Spanish 'Dogberries', who rarely deviate into sense.

dog-biscuit. Add: **b.** *Military slang.* = *biscuit 1 b (b).

1925 Fraser & Gibbons *Soldier & Sailor Words* 79 Dog biscuit, an old Army term for the Army mattresses.

dog-box. Add: **c.** *transf.* A type of railway goods wagon or compartment in a railway carriage. *Austral.* and *N.Z. slang.*

1917 L. Miller *Diary* 19 Jan. in *Camps, Tramps & Trenches* (1939) i. 2 Our crowd moved off from headquarters in the railway goods wagons known as 'dog-boxes'. 1945 Baker *Austral. Lang.* ix. 196 The use of dog-boxes to describe small carriages on N.S.W. railways.

dog-brier. (Later U.S. examples.)

1840 C. F. Hoffman *Greyslaer* II. xi. 42 A spot where some huge rocks, covered only with dog-briers let down the light..into the forest.

dog-collar. Add: **1.** (Later examples.)

1864 *Ainsworth's Mag.* VI. 352 A dagger, a green velvet bonnet..and a small dog-collar. 1939–40 *Army & Navy Stores Catal.* 999/1 In ordering Dog Collars, the size of dog's neck must be given.

2. *spec.* A derogatory or jocular term for the clerical collar. (Further examples.)

1861 *Temple Bar* I. 386 The clerical High-Church neighbour who wears a stiff-starched dog-collar instead of a cravat. 1931 *Tablet* 21 Feb. 254/1 Jewish Rabbis..as well as Free Church pastors, are often seen wearing what the profane call 'dog-collars'. 1965 J. Porter *Dover* iv. 45 His dog collar gleamed whitely in the darkness of the hall.

b. *attrib.*

1906 *Daily Chron.* 17 Oct. 8/4 The latest ornaments for the throat are of the dog-collar pattern. 1904 *Ibid.* 3 Nov. 5/1 Diamond dog-collar brooch.

dog-dayed, *a.* *poet. nonce-wd.* Of the dog-days (cf. Dog-days *sb. pl.* 3).

1934 Dylan Thomas *18 Poems* 30 There from their hearts the dogdayed pulse Of love and light bursts in their throats.

dogeate (dōᵘ·dʒeᵻt). [f. Doge + -ate¹.] = Dogate.

1925 J. C. Powell in J. Buchan *Nations of To-Day, Italy* 5 In the Dogeate of Pietro Orseolo.

dog-fight. Add: **2.** *transf.* A general disturbance or mêlée; *spec.* a 'scrap' between aircraft.

1880 'Mark Twain' *Tramp Abroad* II. 197 No information about prize fights or other dog fights, horse races, walking-matches..or other sporting matters. 1917 Kipling *Diversity of Creatures* 208 The pitiless Whips were even then at the telephones to herd 'em up to another dog-fight. 1919 A. E. Illingworth *Fly Papers* 79 The battle develops into a 'dog-fight', small groups of machines engaging each other in a fight to the death. 1928 F. E. Baily *Golden Vanity* xiv. 201 No dividend, reserve fund wiped out, and a dog-fight at the annual general meeting. 1938 C. F. S. Gamble *N. Sea Air Station* xxii. 400 It was decided that all flying-boats should have their hulls 'dazzle-painted'..so that a pilot could, in a 'dog-fight', know at a glance who was in a particular machine. 1937 C. Burt *Boys' Book of Flying* xvi. 190 Inevitably occasions arise in aerial warfare—as in a 'dog-fight', where several machines are milling together in whirling confusion. 1939 *Times* 3 Nov. 8/6 During the dogfight the troops and others on the ground took cover, and the only damage done was one or two holes in the roofs of barns. 1958 *Economist* 13 Dec. 967/1 ... the dogfight that at present passes for industrial relations at BOAC. 1970 *Globe & Mail* (Toronto) 26 Sept. 35/6 On our performance so far, we're in a dogfight for fourth.

Hence as *v. intr.*, to fight or 'scrap'. Also *trans.* (*rare*).

1929 T. E. Lawrence *Lett.* (1938) 648, I live in barracks (i.e. we dog-fight promiscuously). 1934 V. M. Yeates *Winged Victory* i. 17 It was too risky to stop and fight... Dog-fighting was an amusement for rather nearer home. *Ibid.* vi. 51 They would dive and zoom..but they would never dogfight Camels. 1941 E. C. Shepherd *Milit. Aeroplane* 14 If the formation [of bombers] gets broken, the single machine may have to 'take evasive action', but it will not attempt to dogfight.

dogged, *a.* Add: **3.** Esp. in colloq. phr. *it's dogged as does it*: persistency and tenacity win in the end.

1864 M. B. Chesnut *Diary* 6 Aug. (1949) 429 'It's dogged as does it,' says Isabella. 1867 Trollope *Chron. Barset* lxi, There ain't nowt a man can't bear if he'll only be dogged... It's dogged as does it. 1896 *Daily News* 27 June 8/1 ... 'It's dogged as does it.' 1942 N. Marsh *Death & Dancing Footman* x. 195 ... 'It's dogged as does it,' said Chloris.

dogger³ (dǫ·gəɹ). *Austral.* [f. Dog *sb.* + -er¹.] One who hunts dingoes.

1934 *Bulletin* (Sydney) 31 Oct. 20/4 Many of the doggers, prospectors and blockholders who make the connection (i.e. travel) long distances by camel. 1935 H. H. Finlayson *Red Centre* i. 15 The dogger and prospector follow the explorer. 1946 F. Davison *Dusty* xviii. 204 Ask an old dogger of the western stations. 1965 E. Troughton *Furred Animals of Austral.* (ed. 8) 237 A Queensland 'dogger' who found it impossible to trap or poison an old-man dingo responsible for killing hundreds of sheep.

doggery. 4. (Earlier example.)

1835 D. P. Thompson *Timothy Peacock* 140 (Th.), A sort of dogging doggery, or sailor's hotel, situated near the wharf.

dogging, *vbl. sb.* (see under Dog *v.*). Add: Grouse-shooting using dogs to rouse the birds, as distinguished from 'driving'. Also *attrib.*

1886 Walsingham & Payne-Gallwey *Shooting* I. 86 Wet weather is always bad for 'dogging'. 1894 [see *dogging vbl. sb.*, Dog *v.*]. 1909 *Westm. Gaz.* 10 Aug. 4/2 On the 'dogging' moors..the actual shooting will begin as soon as it becomes legal. 1928 *Daily Tel.* 26 June 13/5 The shooting extends to 16,000 acres and is an excellent dogging moor.

b. *Austral.* The hunting of dingoes.

1934 A. Russell *Tramp-Royal in Wild Australia* iv. 37 Tuck had told me what he had made on his dogging expeditions. 1935 H. H. Finlayson *Red Centre* xiv. 142 More profitable is 'dogging'.

doggishly, *adv.* (Later example.)

1928 D. H. Lawrence in E. Rickword *Scrutinies* v. 62 The Forsyte trying to be freely sensual. He can't do it... He can only be doggishly messy.

doggishness. (Later examples.)

1905 A. Bennett *Tales 5 Towns* ii. 109 He had seldom felt less doggy... It seemed to him that dogginess was not the glorious thing he had thought. 1928 D. H. Lawrence in E. Rickword *Scrutinies* v. 66 Next time I'll get properly married and do my doggishness in my own house.

doggo (dǫ·go), *adv. slang.* [Of obscure origin: prob. f. Dog *sb.*] *to lie doggo*: to lie quiet, to remain hid. Also occas. with other verbs.

1893 Kipling *Many Inventions* 250, I would lie most powerful doggo when I heard a shot. 1916 H. Rosher *In R.N.A.S.* 141 ... lie doggo ... 1918 H. A. Vachell *Some Happenings* iv. You'll play doggo and keep out of sight. 1924 *Blackw. Mag.* Sept. 352/2, I stayed doggo in the scrub. 1925 F. Maclean *Back to Bokhara* iii. 136 Lying doggo with an expression of angelic innocence when he came to see if she was in bed and asleep.

dog-gone. Add: (See also *Sc. Nat. Dict.* s.v. *dag*.) **B. 2.** Also as *adv.* and *quasi-sb.*

1871 E. Eggleston *Hoosier Schoolmaster* 40 She was so dog-on stuck up. 1911 R. D. Saunders *Col. Todhunter* 95 You was so dog-gone proud of the blue coat. 1933 E. Caldwell *God's Little Acre* xvii. 266 That will be my ship coming in, and I don't give a dog-gone for the name you call it. *Ibid.*, When I get a load of it, I'll know dog-gone well my ship came in.

C. (Additional Scottish examples.)

1893 T. Stewart *Miners* 203 Trade's sae dagont dull. 1908 J. Lumsden *Doun i' th' Loudons* 244 That dagon'd buffer o' a wife.

doggy, *sb.* Add: **3.** An officer's servant or assistant. Cf. *dog-robber (b). orig. *Services' slang.*

1909 J. R. Ware *Passing Eng.* 113/2 *Doggie*,..officer's servant, especially cavalry. 1916 'Taffrail' *Carry On* 18 The captain's 'doggie'..midshipman A.D.C.—is relating certain shipwrecking experiences. 1921 *Blackw. Mag.* July 50/2 Is the Admiral going to have a doggie? 1952 A. Grimble *Pattern of Islands* ii. 53 My function would be to act as doggie—that is, clerical assistant and odd-job man—to the District Officer.

doggy, *a.* Add: **4.** Dashing, stylish, smart. Cf. *dog *sb.* 15 b. *slang.*

1889 E. Dowson *Let.* 28 July (1967) 95, I found it difficult to choose between the half-dozen..charming photos of Mignon... The 'doggy' ones..styled the best. 1891 Farmer *Slang* II. 301 *Doggy*,..stylish. 1905 [see *doggishness]. 1932 A. J. Worrall *Eng. Idioms* 6, I like your tie, it is very doggy.

5. Of Latin: debased, corrupt. (See Dog *sb.* 17 e.) *colloq.*

1898 *Daily News* 3 Jan. 6/3 He spoke Latin! patristic Latin of the doggyest order.

dog-house. Add: In literal use now chiefly *U.S.* (Later examples.)

1879 F. R. Stockton *Rudder Grange* vii. 77, I had no dog-house as yet. 1898 S. Hale *Lett.* (1919) 338 Behind the dog-house there is a warren of..four small animals. 1970 *Daily Progress* (Charlottesville, Va.) 21 Mar. C2/6 Going to paper a wall?..Build a doghouse? Lay a patio?

b. *fig.*, *in the dog-house*: in disgrace, out of favour. *slang* (orig. *U.S.*).

1932 *Editor* 6 Feb. 110/2 Dog-house—in disfavor. 'My moll caught me tryin' to make that twist, so I'm in the dog-house now.' 1940 Wodehouse *Quick Service* xix. 231 Already he was solidly established in the doghouse as the result of..that letter. 1955 Priestley & Hawkes *Journey down Rainbow* ii. 34 And the men, so often 'in the dog house,' ... 1958 J. Cannan *And be a Villain* iv. 83 He said to me, 'I'm in the dog-house over this.' 1963 P. H. Johnson *Night & Silence* ix. 55 He'd been getting bad grades, he was in the dog-house as it was.

2. A small structure, esp. a hut, of a shape suggesting a kennel. Usu. *colloq.* or *slang.*

1898 H. E. Hamblen *Gen. Manager's Story* 43 I'll have to drop off a flag, or they'll git our doghouse [sc. caboose, on a freight train]. 1949 *Manch. Guardian* ... the 'dog-house' ... The former is a check station where simulated flights are conducted, the latter is a protuberance that houses instruments on the rocket's otherwise smooth skin. 1961 F. Moyes *Sunken Sailor* v. 74 A dog-house which gave shelter to the helmsman as he stood at the wheel. 1961 *Gloss. Terms Glass Ind.* (*B.S.I.*) 14 *Dog-house*, a small extension of a glass furnace, into which the batch is fed. 1969 *Guardian* 1 July 11/6 ... a small blister on the smooth skin of booster rocket, used to house instruments.

b. *slang.* A double-bass.

1923 in G. McKnight *Eng. Words* iv. 45. 1933 H. T. Webster in *Forum* Dec. ... 1952 B. Ulanov *Hist. Jazz in Amer.* (1958) xxv. 349 ... no 'doghouse' for bass fiddle ...

dogie, dogy (dōᵘ·gi). *U.S.* [Of obscure origin.] A motherless, neglected, or under-nourished calf on a cattle range. Also *attrib.*

[1888 *Cent. Mag.* Oct. (Farmer), They were mostly Texas doughies—a name I have never seen written, it applies to young immigrant cattle.] 1892 *Outing* (U.S.) Feb. 358 A queer, pot-bellied little dogy (a calf prematurely weaned by the death of its mother and developed into a runt). 1903 A. Adams *Log of Cowboy* vii. 86 Before you could say Jack Robinson our dogies..were running in half a dozen different directions. 1911 H. Quick *Yellowstone Nights* ... 1920 J. M. Hunter *Trail Drivers of Texas* 130 ... 1962 E. B. Atwood *Regional Vocab. Texas* 57 ... the usual Texas term is *dogie*..A dogie, according to various informants is 'small', 'undernourished'..or of 'inferior breed'.

dog-leg, *a.* Add: Further examples of gen. use; *dog-leg fence* (later examples); *dog-leg hole*, in *Golf.*

1903 T. Collins' *Such is Life* 75 The wild cattle having walked over the dog-leg fence. 1904 S. Rudd *Sandy's Selection* 50 A cultivation paddock enclosed with a dog-leg fence. 1907 *Daily Chron.* 17 Aug. 7/2 There are two or three very fine specimens of the dog-leg kind of hole. 1909 *Westm. Gaz.* 30 Apr. 4/2 There is the occasional variety of the 'dog-leg' hole, where the player has to go round an angle. 1935 *Bulletin* (Sydney) 16 Jan. 21/1 ... the dog-leg fence and huts of stringybark. 1959 *News Chron.* 14 Dec. 4/3 ... 1962 *Times* 27 Nov. 13/2 Gone are the 'dog leg' corners of the earlier windscreen. 1967 *Listener* 9 Feb. 199/2 The dog-leg bend in the branch duct follows the wall above where the ward widens.

2. Applied to an inferior quality of tobacco. Also *ellipt.* as *sb.* *U.S.*

1858 *Nat. Intelligencer* (Washington) 10 July 3/3 A large quantity of 'dog-leg' tobacco ... thrown into the tub. 1863 'E. Kirke' *My Southern Friends* iii. 48 ... 1868 *Congress. Globe* 18 Mar., App. 187/1 Watching the neighbors pitch horseshoes for dog-leg tobacco. 1894 M. E. Ryan *Pagan of Alleghanies* ii. 25 Then the black-and-tan man treated himself to a fresh chew of 'dog-leg'.

dog-legged, *a.* Add: **b.** Of a fence (see Dog-leg *a.*). *Austral.* and *N.Z.*

1900 H. Lawson *Over Sliprails* 95 A spidery dog-legged fence. 1935 N. R. McKenzie *Gael fares Forth* 88 The 'dog-legged' [type of fence]..was apparently a purely Canadian type..made with stakes of varying length and the trunks of small trees.

dogoned, var. Dog-goned *a.*

1868 *All Year Round* 19 Sept. 353/2 He looks the dogondest cove ever since Jim Ford left. 1872 E. Eggleston *End of World* xxiii. 158 Clark township don't want none o' 'em, I'll be dog-oned if it do.

do-good. Delete †*Obs.* and add: In modern use (orig. *U.S.*) = *do-gooder. Also *attrib.* or as *adj.* Also **do-goo·ding** *ppl. a.* and *vbl. sb.*; **do-goo·dism**; **do-goo·dy** *a.* All these derivatives carry the disparaging sense of *do-gooder.

1923 *Nation* 21 Nov. 569/2 There is nothing the matter with the United States except..the parlor socialists, up-lifters, and do-goods. 1936 *Sat. Rev. Lit.* 14 Mar. 5/1 'Scientific and do-gooding people from all over America' had foregathered to a Child Recovery Conference. 1951 *Manch. Guardian Weekly* 14 June 3/1 Mr Truman..has been made aware that..anything that comes out of the State Department bears..a stigma indicating vague incompetence, sympathy to the Communists, and general 'do-goodism'. 1958 *Times Lit. Suppl.* 15 Aug. p. x/2 And the committee would, I am sure, be a snuggery of all the no-goods and do-goods whom I have spent half my life successfully avoiding. 1958 S. Ellin *Eighth Circle* (1959) II. xiii. 141 A decision that reeked of the genteel do-goodism, of the compulsive idealism that..infected people of..[his] class and type. 1958 *Listener* 11 Dec. 1007/3 His tyrannous wife..and her do-goody brother. 1959 *Guardian* 28 Aug. 3/3 The present-day young are less interested in vague do-gooding..than in reacting to a human situation. 1968 *New Statesman* 11 Oct. 469/1 Not in a do-good social worker way..but in a fundamental Christian way. 1969 *Listener* 23 Jan. 105/2 It contains all the do-good elements one can pack into 24 minutes: attractive, young and, of course, black war-widow, her husband having died a hero's death in Vietnam, starts life in California with her attractive young black son.

do-goo·der. orig. *U.S.* [f. *to do good* (see Good *sb.* 5 a) + -er¹. Cf. prec.] A well-meaning, active, but unrealistic philanthropist or reformer; one who tries to do good. Hence **do-goo·dery.**

1927 *St. Louis Post-Dispatch* 18 Jan. 14/5 The dogooder..is all the hokum, all the blather and all the babble of the modern so-called 'social movement'. 1943 *Time* 25 Oct. 32/1 Go find yourself out something about Charles P. Taft before dismissing him as famed do-gooder. 1954 Koestler *Invis. Writing* 250 The prim feeling of virtue of do-gooders. 1957 *Sunday Times* 12 May 8/6 [He] does not write as a sentimental do-gooder, or with primarily political motives. 1959 'M. Innes' *Hare sitting Up* I. i. 20 An out-and-out do-gooder, full of an exalted love of..humanity. 1961 *Guardian* 25 Mar. 5/3 The boy is trapped..by do-goodery. 1968 *Daily Tel.* 11 Apr. 17/4 Amateur 'do-gooders' could slip up when certain medical attention was needed, a coroner said yesterday. 1968 N. Marsh *Clutch of Constables* i. 13 Let us examine my philanthropy. Or rather, since I have no distaste for colloquialism, my dogoodery.

Dogra (dōᵘ·grā). Also **Dogri, Dogur.** [f. *Duggar* (see def.).] The name of a people inhabiting the Duggar district in north-west India; a member of this people; also (usu. *Dogri*), their language. Also as *adj.*

1845 H. M. Elliot *Suppl. Gloss. Indian Terms* 238 *Dogur*, a tribe scattered over various tracts of the North-West of Hindoostan. 1870 *Proc. R. Geogr. Soc.* XV. 13 The unhappy results of the incessant feuds waged between the mountaineers (who are to a man Mahomedans) and the Dogra troops of the Kashmir Rajah. The atrocities practised by the Dogras are a disgrace to a feudatory of the British crown. 1888 Kipling *In Black & White* (1889) 91 Sikhs, Pathans, Dogras—they're all alike, these black vermin. 1926 *Blackw. Mag.* Nov. 593/1 Dogras being high-caste Hindus are very particular. 1948 D. Diringer *Alphabet* vi. 373 Dogra or Dogri, a dialect of Punjabi, spoken by one and a half million people in the Jammu State and its neighbourhood. 1957 *Encycl. Brit.* VII. 503/1 Dogra, an inhabitant of the Duggar tract...The original home of the Dogra people was situated between the lakes of Siroensar and Mansar.

do·g's body, dog's-body, dogsbody.

1. See Dog *sb.* 18 b. See also quot. 1924.

1818 'A. Burton' *Adv. J. Newcome* ii. 76 I'll get you the Dog's-body Squeezer. 1858 [see Dog *sb.* 18 b]. 1924 R. Clements *Gipsy of Horn* 38 Sea-biscuits soaked into a pulp with water and sugar, are known as 'dogs-body'.

2. A junior person, esp. one to whom a variety of menial tasks is given; a drudge, a general utility person. *colloq.*

1922 T. E. Lawrence *Lett.* (1938) 365 I'll have got used to being a dog's body. 1925 Fraser & Gibbons *Soldier & Sailor Words* 80 *Dogsbody*, a Midshipman. A humorous semi-sarcastic colloquialism for any junior officer, R.N. *E.g.*, 'He's only a dogsbody sub'. 1928 *Daily Express* 3 Apr. 13/2 A midshipman is known..in the service as a 'snottie'... If he is a junior midshipman he is also a 'dog's body'. I defy anyone to be accurate and sentimental about a snottie who is a dog's body. 1932 H. Simpson *Boomerang* ix. 184 His post of dog's-body..left him no leisure at all. 1934 M. Hodge *Wind & Rain* II. i. 57 Oh, no. *I* don't understand anything! I'm just 'a dog's body'. 1942 M. Dickens *One Pair of Feet* vii. 116 People who told me I should be a house-parlourmaid 'on Privates [*sc.* private wards]' had over-estimated. I was Dogsbody. 1950 L. A. G. Strong *Which I Never* v. 169 As Assistant Stage Manager and general dog's body, she was grossly overworked and supremely happy. 1955 H. Spring *These Lovers fled Away* vii. 205 My status was never defined. I was everybody's dog's-body. *Ibid.* xvii. 481 Introduced to a secretary whose dog's-body she was to be. 1967 *Listener* 14 Sept. 332/3, I was a sort of general dogsbody to begin with—assistant stage-manager, and what have you.

dog's-ear, *sb.* **2.** (Earlier example.)

1840 R. H. Dana *Bef. Mast* iv, The first [man] on the yard goes to the weather earing, the second to the lee, and the next two to the 'dog's ears'.

dog's-ear, *v.* (Later examples of *dog-ear*.)

1903 *Westm. Gaz.* 8 Jan. 2/1 She dog-eared her book. 1940 R. Stout *Over my Dead Body* xi. 150 He..dog-eared a page and closed the book.

dog's-eared, *ppl. a.* (Later example of *dog-eared*.)

1895 M. Pemberton *Impregnable City* xv. 303 Dog-eared lilies.

dog's-tail. **1.** (Later example.)

1961 R. W. Butcher *Brit. Flora* II. 957 The crested dogstail..is a very common plant in grassland throughout the British Isles.

dog-tooth, dog's tooth. Add: **3. b.** A broken check pattern used esp. for men's and women's suitings.

1948 *Evening Standard* 22 Mar. 6/3 A suit..in fine dog's tooth worsted suiting. 1958 *Vogue* Jan. 41 The boldly checked tweed coat..in brown dog's tooth checked over stone. 1958 *Spectator* 27 June 831/3 He displayed a certain raffish elegance in his long, dark jacket and dog-tooth trousers. 1960 *Times* 22 Feb. 13/1 A skirt of black and white dogstooth check.

dog-whelk. Substitute for def.: A member of the marine mollusc family Nassariidae, esp. *Nassarius reticulatus*; also = *dog-winkle* (s.v. *dog *sb.* 18 c). (Add later examples.)

1901 E. Step *Shell Life* xv. 257 The Dog-whelks (*Nassa*) are almost as well known as the Purples. 1960 J. A. C. Nicol *Biol. Marine Animals* xi. 467 Another interesting case is the dog-whelk *Nucella lapillus*. 1968 N. F. McMillan *Brit. Shells* 50 ... Nassariidae, or dog-whelks, are scavengers. *Ibid.*, *Nassarius reticulatus*..Dog whelk.

dog-wolf. Add: **2.** = Wolf-dog 2.

1907 *Kennel Encycl.* I. 159 At a dog show at Spa, in 1882, several 'Dog-wolves' were exhibited, and the animals so termed in the catalogue.

dogy: see *dogie.

dohickey, var. *doohickey.

‖ **doigté** (dwate). *Fencing.* [Fr., f. *doigt* finger, Digit *sb.*] Fingering; the use of the fingers (see quots.). Also *fig.*

1889 Pollock et al. *Fencing* i. 51 The proper use of the thumb and fingers, called *doigté*, is of the greatest importance in fencing. 1931 F. C. Reynolds *Book of Foil* viii. 131 *Finger play or 'doigté'*. In describing the various movements of attack and parry, stress has been laid on the importance of directing the point mainly by the action of the thumb and forefinger. *Ibid.* 210 *Doigté*, the control of the point by the action of the thumb and forefinger. 1958 *Spectator* 4 July 27/2 Something missing in him as a politician, some lack in *doigté*.

doily, *sb.* or *a.* Add: **2. b.** A small ornamental mat made of paper, linen, etc., used on a plate beneath sandwiches, cakes, etc.

1864 *Hist. North-Western Soldiers' Fair* 89 [Donations] 2 cake doylies. 1905 P. Pennington *Woman Rice Planter* (1913) 197 You fill a basket, put a dainty doily over it. 1936 W. Holtby *South Riding* iv. ii. 218 The cheese cakes and lemon tarts lay on frilled lattice of doilies. 1954 J. Betjeman *Few Late Chrysanthemums* 95 Beg pardon, I'm soiling the doileys With afternoon tea-cakes and scones. 1958 *Times* 12 Nov. 3/4 His [*sc.* Magritte's] little painting of a figure cut out of a paper doily stuck on the same card as another paper-doily figure between two very solid rocks.

doing, *vbl. sb.* Add: **1. a.** Esp. in colloq. phr. *to take a bit of* (or *lot of*, *some*) *doing*: to require all one's efforts; to be difficult to do.

1936 *Discovery* Dec. 382/1 The pain and swelling combined to knock Almudena completely out for some time. And that took a lot of doing. 1964 L. Nkosi *Rhythm of Violence* II. ii. 31 Kitty... Did everything go according to plan? *Jimmy*. Took a bit of doing, but ... we pulled it off. 1969 *Time* 28 Feb. 49 His long-suffering wife..and their six kids put up with him, which takes some doing.

c. A scolding; a beating, thrashing, beating-up; a severe monetary loss. *dial.* and *colloq.*

1880 W. H. Patterson *Gloss. Antrim & Down* 31 *Doing off*, a scolding. 1888 G. Bidwell *Forging his Chains* xxxviii. 418 Punishment for violence against any prison authority..strokes of the flesh-cutting cat-o'-nine tails,..besides the effective preliminary 'doing' by the warders. 1897 W. S. Maugham *Liza of Lambeth* xi. 209 She 'as give yer a doin'; an' look at yer eye! 1909 J. R. Ware *Passing Eng.* 113/2 I've had a bad doing this week—lost thirty quid. 1923 *Hansard Commons* 13 July 1782 The hon. Member for Dundee..has given me a tremendous doing. 1959 *Times* 10 Mar. 5/5 The cops gave me a doing. 1968 B. Turner *Sex Trap* xiii. 167 'For God's sake, man! You'd get three years if you give him a doing,' she exclaimed.

2. (Further examples.)

1818 Keats *Let.* 13 Jan. (1958) I. 205 Lawk! Molly there's been such doings. 1923 Kipling *Irish Guards in Gt. War* I. 216 The talk in the camps turned on their doings—everything connected with the front line was 'doings'. 1949 E. Pound *Pisan Cantos* lxxx. 92 Stewing with rage Concerning the landlady's doings with a lodger unnamed.

4. *pl.* Materials for a specified adjunct of a dish or meal; also, a made or fancy dish. *U.S. local.*

1838 E. Flagg *Far West* II. 72 'Well, stranger, what'll ye take: wheat-bread and chicken fixens, or corn-bread and common doins?' by the latter..being signified bacon. 1843 'R. Carlton' *New Purchase* II. xl. 58 A snug breakfast of chicken fixins, eggs, ham-doins, and corn slapjacks. 1859 *Knickerbocker* LIII. 317 Tell Sal to..have some flour-doins and chicken-fixins for the stranger. *Ibid.* 318 Instead of 'store-tea' they had only saxifrax tea-doins, without milk. 1880 J. C. Harris *Uncle Remus* i. 18 Sposen' you drap roun' termorrer an' take dinner wid me. We ain't got no great

doin's at our house, out I speck de old 'oman..kin sorter scramble roun' 'en git up sump'n. 1908 *Dialect Notes* III. iv. 306 *Doing(s)*,..prepared dishes, especially fancy dishes.

b. *U.S. colloq.* Lace, trimming, ornaments, etc., of a dress.

a 1847 W. T. Porter *Quarter Race Kentucky* 84 [The girls] came pourin out of the woods..,fixed out in all sorts of fancy doins, from the broad-striped homespun to the sunflower calico. 1856 *Knickerbocker* XLVII. 406 Pretty girl that in the black fixings and white arrangements, with blue doings.

c. Applied to any concomitant, adjunct, or 'etcetera', or anything that happens to be 'about' or to be wanted. orig. *War slang.*

1919 *Athenæum* 25 July 664/2 'Doings', practically anything: 'Pass the doings.' 1925 Fraser & Gibbons *Soldier & Sailor Words* 80 *Doings, the*, a word with every kind of meaning and application. *E.g.*, In quarters, 'Pass the doings', might mean bread, salt, a pack of cards, or anything at hand. 'I'll have a drop of the doings' (*i.e.*, whatever drink there is going). 'Here comes Jerry with the doings' (*i.e.*, an enemy aeroplane sighted). 1927 W. E. Collinson *Contemp. Eng.* 101 Doings for gravy, salt, sugar, pepper, milk or anything wanted at a particular time for a particular purpose. *a* 1935 T. E. Lawrence *Mint* (1955) II. xvi. 144 Cook chuckled, snatched somebody's blacking-tin and with three swift passes of a boot-brush painted his doings jet-black. 1938 G. Greene *Brighton Rock* v. i. 193 Her skirt drawn up above her knees she waited for him with luxurious docility... 'You've got the doings, haven't you?' 1938 S. V. Benét *Thirteen o' Clock* 248 He's got a little doin's of a black moustache.

do:-it-yourse·lf. [See Do *v.* 11 and Yourself *pron.* 3.] The action or practice of doing work of any kind by oneself, esp. one's own household repairs and maintenance, usu. as opposed to employing someone else to do it. Also *transf.* and *fig.* Freq. *attrib.* or as *adj.* Abbreviated as *D.I.Y.* Hence **do-it-himse·lf, do-it-yourse·lfer, do-it-yourse·lfism**, and various other nonce derivatives.

[1616 T. Draxe *Bibliotheca Scholastica* 163/1 If a man will haue his businesse well done, he must doe it himselfe. 1693 W. Penn *Some Fruits of Solitude* (ed. 2) 66 Have but little to do, and do it thy self. *a* 1845 Barham *Ingol. Leg.* (1905) 288 If it's business of consequence, *Do it yourself!*] 1925 D. Beard (*title*) Do it yourself. A book of the big outdoors. 1949 *Here & Now* (N.Z.) Oct. 17/3 Husbands who have been brought up in the do-it-all-myself tradition of the previous generation. 1952 *Time* 30 June 45/3 Do-it-yourself has brought similar gains, and market shifts, to other industries. 1954 *N.Y. Herald Tribune* 21 Mar. iv. 17/5 There are other reasons, too, for the tidal wave of do-it-yourselfism. 1954 *N.Y. Times* 5 Sept. F1/7 To the do-it-yourselfer, plywood is as essential as paint, tools, plastics and ordinary lumber. 1955 *Pract. Householder* Dec. 203/1 A central pool such as a 'D.I.Y. Club' from which..tools can be hired is the obvious advantage. 1957 *Times* 12 Nov. (Canada Suppl.) p. xv/4 Canadian males are great devotees of do-it-yourself. 1957 *Financial Times Ann. Rev. Brit. Industry* 51/1 A country planning a nuclear power programme has a choice of two routes: do-it-yourself and the ready-made. 1958 *Spectator* 4 July 13/2 The builder is shoving up his do-it-himself boxes. 1959 *Manch. Guardian* 13 Aug. 5/4 The..do-it-yourself kit, complete with all necessary screws, bolts, nuts and instruction chart. 1959 V. Packard *Status Seekers* (1960) 76 Trimming hedges or repairing screen doors or other delights of the do-it-yourself. 1965 *Spectator* 5 Mar. 293/1 The whole of the artistic world has been debauched by the hogwash of the do-it-yourself vogue. 1966 *Guardian* 16 May 6/4 Do-it-yourselfers have been enthusiastic about..adhesive vinyl tiles. 1966 *Listener* 24 Nov. 781/2 This chamber-music is a universe away from the stuttering *simplesse* of do-it-yourself Hemingway. 1969 *Daily Tel.* 4 Oct. 11/1 For families or parties of young people especially, who don't mind a little do-it-yourselfing, this sort of holiday is usually memorable. 1970 F. McKenna *Gloss. Railwaymen's Talk* 34 *Do-it-yourself kit*, diesel-age term for steam locomotives. The arrival of the diesel locomotive..consigned the steam locomotive to the level of a do-it-yourself kit. 1970 *Guardian* 30 July 9/6 A DIY inflatable dome kit..for £30.

‖ **dojo** (dōᵘ·dʒo, dōᵘ·ʒo). [Japanese.] **a.** A room or hall in which judo is practised.

1942 P. Longhurst *Jiu-Jitsu* (new ed.) 7 The *dojos*, or Judo gymnasia, would be littered with casualties. 1964 'J. Munro' *Man who sold Death* iii. 27 Hakagawa's dojo, the gymnasium which was his classroom.

b. In full *dojo mat.* A padded mat on which judo is practised.

1966 P. O'Donnell *Sabre-Tooth* iii. 41 The *dojo*, the padded mat spread in the centre of the long combat room. 1967 'J. Munro' *Money that Money can't Buy* i. 11 Their practice sessions on the dojo mat. *Ibid.* xiii. 162 In one corner..was a dojo—a judo practice mat.

dol (dōᵘl). [f. L. *dolor* pain: see Dolour.] A unit of intensity of pain.

1947 J. D. Hardy et al. in *Jrnl. Clinical Investigation* XXVI. 1158/1 A scale of pain intensity is proposed, the unit of which is called a 'dol', composed of 2 just perceptible steps in discrimination of stimulus intensity. 1957 [see *dolorimeter]. 1962 J. D. Hardy in Keele & Smith *Assessment of Pain in Man & Animals* 194 Above 8 dols, however, the intensity of pain no longer increases rapidly with skin temperature. 1964 Keele & Armstrong *Substances producing Pain & Itch* iii. 32 On this scale 'ceiling' pain intensity has a value of 10½ dols.

‖ **dolce vita** (doltʃe vītă, dǫ:ltʃī vī·tă). [It., lit. 'sweet life'.] A life of luxury, pleasure, and self-indulgence. (Freq. preceded by *the* or *la*.) Also *attrib.*

[1960 *Times* 7 Dec. 15/5 *La Dolce Vita* is a film..'making up an apocalyptic fresco of seven nightmarish nights and seven sobering dawns'.] 1961 *Guardian* 30 Mar. 7/1 The grander villas, where the Roman playboys..pursued their Dolce Vita. 1961 *20th Cent.* Mar. 311 A headstrong and muddled woman..who combined convent education, *la dolce vita* and gambling. 1962 *Listener* 6 Sept. 367/2 His plans to abandon his *dolce vita* existence. 1970 *Times* 14 Oct. 2/6 A totally alien way of life—a dolce vita of Bicester, 1970 style.

dole, *sb.*¹ Add: **6. a.** *spec.* (usu. *the dole*). The popular name for the various kinds of weekly payments made from national and local funds to the unemployed since the war of 1914–18. Phr. (*to be* or *go*) *on the dole*: (to be or start being) in receipt of such unemployment relief; also *transf.* and *fig.*

1919 *Daily Mail* 11 June 8/4 You won't draw your out-of-work dole of 29s. this week. 1923 L. A. Harker *Master & Maid* (new ed.) xx. 265 If only I'd danced an Irish jig I believe I could have got the whole of them to increase the dole. 1925 *Westm. Gaz.* 26 Mar., 3,000 Aliens on the Dole. 1928 *Britain's Industrial Future* (*Lib. Ind. Inq.*) 277 To speak of Unemployment Benefit as 'the dole' is to misrepresent the facts. 1933 W. Greenwood (*title*) Love on the dole. A tale of the two cities. 1937 *Daily Herald* 21 Jan. 4/2 Beef producers must not consider themselves as 'on the dole for ever', but must use the assistance to reorganise the industry. 1955 *Times* 29 July 5/6 The nation was on the dole, and had been for 10 years.

8. dole-cupboard (see quot. 1910¹); **dole-drawer**, one who receives the dole (sense *6 a).

1910 *Encycl. Brit.* VII. 634/2 The livery cupboard..was often used in churches to contain the loaves of bread doled out to poor persons... They were then called dole cupboards. 1910 V. Tree *Let.* 9 Oct. in *Castles in Air* (1926) i. 37, I am rather inclined to sell our Dole cupboard we bought together. 1926 *Good Housekeeping* July 188/2 Profiteers, dole-drawers and music-hall artists—in fact, the only people who have any money to-day. 1938 R. G. Collingwood *Princ. Art* xv. 333 An audience of wage-earners or dole-drawers.

doler. Add: Also with *out.*

1965 *Listener* 24 June 953/1 Aaron is the speaker of smooth words, the doler-out of truth in prudent instalments.

doless, *a.* Also **doeless, do-less.** For *Sc.* and *U.S.* read *dial.* and *colloq.* (Add further examples.)

1876 T. E. Brown *Doctor* 12 A doeless sort of a woman. 1891 R. Mulholland *Haunted Organist* 229 It's the poor do-less pair we'd be only for our Ailsie, that's han's an' feet to us both. 1936 M. Mitchell *Gone with Wind* lv. 949 He's just a piddling, do-less, good-for-nothing. 1949 M. Mead *Male & Female* xv. 324 Too lazy, too do-less, to make an effort.

‖ **doli capax** (dǫ:li kæ·pæks), *adj. phr. Law.* [f. L. *doli*, gen. sing. of *dolus* (see Dole *sb.*³) + *capax* capable.] Capable of having the evil intention necessary for the commission of a crime. So **doli incapax**, incapable of having such an intention: usu. applied to a person under the age of fourteen.

a 1676 Hale *Hist. Placit. Cor.* I. 22 He might or might not be guilty according to the circumstances of the fact that might induce the court and jury to judge him *doli capax*, *vel incapax*. *Ibid.* 26 An infant under the age of fourteen years and above the age of twelve years is not *primâ facie* presumed to be *doli capax*, and therefore regularly for a capital offense..he is not to be convicted or have judgment as a felon. 1797 T. E. Tomlins *Jacob's Law-Dict.* II (s.v. *infant*), Also, under fourteen, though an Infant shall be *primâ facie* adjudged to be *doli incapax*; yet if it appear to the court and jury that he was *doli capax*, and could discern between good and evil, he may be convicted. 1880 *Encycl. Brit.* XIII. 3/1 After fourteen an infant is *doli capax*. 1961 *Times* 22 Feb. 11/3 Their condemnation of the *doli incapax* rule (no liability without established knowledge of wrongfulness)..has not been answered.

dolichocephal (dǫ:likose·făl). [ad. mod.L. Dolichocephali *sb. pl.*] A dolichocephalic person. Also *attrib.* or as *adj.*

1876 tr. *Peschel's Races of Man* 55 If the index [of breadth] sinks below 74, we speak of 'dolichocephal', narrow or long skulls. 1900 tr. *Deniker's Races of Man* ix. 316 In Sweden and Denmark they were dolichocephals and meso-cephals. 1901 [see *brachycephal]. 1930 C. G. Seligman *Races of Africa* vi. 132 Tall dolichocephals, with a stature about 67 inches.

dolichomorphic (dǫ:likomǭ·ɹfik), *a.* [f. Gr. δολιχός long + μορφή form + -ic.] Having disproportionately long bodily members, as the head and neck.

1930 R. L. Sutton *Long Trek* 162 In bodily conformation, the typical Mbulu is dolichomorphic. The head is considerably longer than it is wide, the neck is long, the chest is phthisical.

dolina, doline (dǫli·nă, dǫli·n). *Geol.* [ad. Russ. *dolina* valley, plain.] A depression or basin in a karstic limestone region, esp. one that is relatively extensive and funnel-shaped.

1882 A. Geikie *Text-bk. Geol.* III. II. ii. 355 The ground may there be found drilled with vertical cavities (*swallow-holes, sinks, dolinas*), by the solution of the rock along lines of joint that serve as channels for descending rain-water. 1922 *Geol. Mag.* LIX. 406 The funnel-shaped hollows which are so frequently met with on the surface of the karst are termed ponors. Many of these are largely filled with red earth, formed by the decomposition of the limestone, and are then known as dolinas. 1937 *Discovery* July 214/1 Huts clustered round a very wide dolina in the shelter of the highest elevation. 1942 O. D. von Engeln *Geomorphology* xxii. 572 The doline, of which the swallow hole and sink hole are variants, is the initial and fundamental unit of karst topography. The type doline has the form of a funnel top and occurs in the Adriatic karst. 1962 *Nature* 16 June 1037/2 Vertically walled, knife-edged arêtes and pinnacles of bare limestone standing like battlements around dolines covered by montane forest of mossy aspect.

doll, *sb.*¹ Add: **2. b.** A dummy used by a ventriloquist, or by a puppeteer.

1893 R. Ganthony *Pract. Ventriloquism* iv. 138 Hand Dolls to fit on hand (à la punch)... Knee Dolls 30 inch high, dressed any character. 1950 *Oxf. Jun. Encycl.* IX. 476/2 A conventional ventriloquist's dummy is a wooden doll with large glassy eyes, enormous shutter mouth, and teeth like tombstones. 1957 *Encycl. Brit.* XIV. 906/2 To operate, put the doll on the hand like a glove. 1967 *Listener* 23 Nov. 679/1 Voices that belong not to human beings but to ventriloquists' dolls.

3. (Earlier example.) Also in more general sense: a woman; a girl; esp. a very beautiful or attractive woman; also *occas.*, a pleasant or attractive man. Now *slang.*

The nineteenth-century examples are not clearly separable in meaning from those in Dict.

1778 F. Burney *Evelina* I. xxiii. 197 As to the women, why they are mere dolls. 1846 *New Swell's Guide to Night Life* 29 (Partridge, *Dict. Slang* Suppl.), Soldiers and their Dolls. 1851 [see Toff]. 1903 R. L. McCardell *Conversations of Chorus Girl* 53 At Vassar..a bunch of society dolls..are teaching me high kicking. 1923 G. H. McKnight *Eng. Words* iv. 61 In the vocabulary of modern youth, chivalry is dead... A girl is..a *chicken*, a *doll*, [etc.]. 1931 D. Runyon (*title*) Guys and dolls. 1961 'B. Wells' *Day Earth caught Fire* ix. 133, I don't dig you, doll. Look we're all entitled to so much water. 1961 in Webster *s.v.*, He is tall, handsome, and muscular. In short, he's a doll. 1967 I. Marder *Paris Bit* iii. 47 He came up to me..and shook hands warmly... 'Max, doll! How *are* you?' 1971 *Scope* (S. Afr.) 19 Mar. 139/1 You don't have to do it, doll.

5. doll-house (see *doll's house); **doll-land**, the realm of dolls.

1871 *Monthly Packet* Oct. 392 The inhabitants of Doll-land would never have recognized the difference between these home manufactures and the best Russian sables! 1959 *Chambers's Encycl.* XIII. 709/2 An interesting adjunct of doll-land is the doll's house.

doll (dǫl), *sb.*³ *Horse-Racing.* Also **dole.** [? var. Dool¹.] A hurdle used as a barrier (see quot. 1942).

1942 Berrey & Van den Bark *Amer. Thes. Slang* § 732/5 *Doles, dolls*, barrier hurdles across gallops to close them to horses and horsemen, or to mark turns on a course or gallop. 1958 J. Hislop *From Start to Finish* vi. 49 The best going is always nearest to the bushes or dolls marking out the gallop.

doll (dǫl), *v.*² *colloq.* [f. Doll *sb.*¹ 2.] *trans.* To dress *up* finely or elaborately; to deck *up*. Also *intr.* (for *refl.*) and *transf.* So **do·lled(-up)** *ppl. a.*, **do·lling(-up)** *vbl. sb.*

1906 *Even. Standard* 31 Aug. 3/3 The time fellows spent in dolling up before taking a wheel. 1916 H. L. Wilson *Somewhere in Red Gap* ix. 378 Jeff said he'd also doll up in his dress suit and get shaved. 1917 C. Mathewson *Sec. Base Sloan* xix. 261 He was..all dolled up in fancy togs. 1921 *Public Opinion* 9 Sept. 252/1 Katurah dolled herself up a little but not too much. 1921 H. C. Witwer *Leather Pushers* iv. 93 He..gives himself a swift dollin' up before the mirror. 1923 E. O'Neill *Hairy Ape* v. 46 All dis is too clean and quiet and dolled up. 1927 M. Eiker *Over the Boat-Side* 269 She had been exquisitely costumed. If she ever did run into Reverdy, she hoped it would be some time when she was dolled. 1928 *Daily Express* 11 Sept. 5/5 The look of bitter contempt and disgust that glittered out of the dolled-up servitor's eyes. 1928 Galsworthy *Swan Song* II. xiii. 216 He supposed his fellow-guests were 'dolling up' (as young Michael would put it) for this ball. 1932 E. Wilson *Devil take Hindmost* iv. 29 The new architecture..is a revelation of our helplessness under the industrial system. ..When we try to doll it up a little, the result is..the New School. *Ibid.* xxii. 234 A dolled-up blonde had called at his office. 1955 'N. Shute' *Requiem for Wren* iii. 67 She could put on her Number Ones and doll herself up smartly. 1959 J. Fleming *Miss Bones* ii. 18 Change the frames over.. doll them up—you know the sort of thing that makes them sell.

dollar. Add: **3.** (See also *almighty *a.* 2.)

c. Colloq. phrases (orig. and chiefly *U.S.*): *bottom dollar*, see *bottom *sb.* 19; (*it is*) *dollars to doughnuts* (or *buttons*, etc.), (it is) almost assured; a certainty; (*like*) *a million dollars*, see *million.

1884 G. W. Peck *Peck's Boss Book* 130 It is dollars to buttons that..she will be blown through the roof. 1890 *Texas Siftings* 8 Nov. 6/3 It is dollars to a doughnut..That some one will start a fire. 1904 *Boston Herald* 8 Aug. 6 It is dollars to cobwebs that every such person will be disappointed. 1904 *Utica* (N.Y.) *Observer* 29 June 6 They talk of fire drills;..it is dollars to doughnuts that not an excursion boat in New York harbor ever had one. 1932 *Atlantic Monthly* Mar. 390/2 It is dollars to doughnuts not a soul will see him. 1936 J. Curtis *Gilt Kid* xiii. 131 If he were seen it was dollars to doughnuts that he would be arrested.

6. *dollar-bill, -earner, -earning* adj., *-hunt, -hunter, -hunting* sb. and adj., *-mark* (earlier U.S. example), *-note* (earlier U.S. example);

dollar area, the area comprising countries where the American dollar is used as currency or as a basis for exchange, or whose sterling balances with British banks are freely convertible into dollars; **dollar-a-year man** *U.S.*, a man who serves the government for a nominal salary; also *transf.*; **dollar country**, a country in the dollar area; **dollar diplomacy** orig. *U.S.*, a foreign policy that seeks to further the financial and commercial interests of a country (*spec.* of America) abroad and often to extend its influence in international relations by means of these interests; **dollar gap**, the excess of a country's (*spec.* Britain's) receipts or imports from the United States or other countries in the dollar area over its payments or exports to those countries; **dollar imperialism** = *dollar diplomacy*; **dollar-sign** *U.S.* = *dollar-mark*; **dollar spot**, a discoloured patch caused by disease, as on an animal or on turf; the disease itself; **dollar store** *U.S.*, a shop in which all or most of the articles are priced at a dollar or less.

1946 *Hansard Commons* 19 Nov. 688 Will the hon. Gentleman give an assurance that no goods will be purchased from the dollar area which can be purchased from the sterling area? 1918 *Lit. Digest* 11 May 11/1 While the 'dollar-a-year men' undoubtedly did good work, the delay and friction in our war-machine became glaringly evident. 1952 *Times Lit. Suppl.* 31 Oct. 702/3 This business man first turned banker; and then he turned Secretary of the Navy, and later of Defence, to serve his country as a dollar-a-year man. 1774 N. Cresswell *Jrnl.* (1925) 21 A considerable sum in Four, Three, Two, One, Two-thirds, One-third and 'One-Ninth' of a Dollar Bills is struck in these Bills of Credit. 1813 *Kingston* (Ontario) *Gaz.* 7 Sept. 3/3 Two hundred and fifty pounds in Dollar bills. *c* 1828 A. Lawrence *Diary & Corr.* (1855) 87 Dollar-bill was found in your chamber on the morning you left home. 1883 Dollar-bill [in Dict.]. 1917 H. T. Comstock *Man thou Gavest* 195 He..tucked the letter and dollar bill in the breast of his shirt. 1947 *Hansard Commons* 8 July 2048 We are drawing nearly half our supplies from dollar countries. 1953 *Ann. Reg.* 1952 338 The more northerly 'dollar countries' of Venezuela, Colombia, the Dominican Republic, Cuba, and Mexico. 1910 *Harper's Weekly* 23 Apr. 8 An attempt is made..to outline..what is meant by the term 'Dollar Diplomacy', as it has come to be commonly applied to certain of the activities of Secretary Knox..in Honduras, in Liberia, [etc.]. 1940 *Economist* 31 Aug. 277/2 The Nazis reinforce Argentina's distrust of American leadership by whispering campaigns on 'dollar diplomacy'. 1958 *Times* 17 July 3/3 There were three conflicting forces...The first was international Communism, the second aggressive nationalism, and, third, and more recent—dollar doctrine and diplomacy. 1958 *New Statesman* 22 Feb. 224/3 Tourism is..our biggest single dollar-earner. 1962 *Harper's Bazaar* Aug. 13 Scotch whisky—Britain's biggest dollar-earning export. 1948 *Ann. Reg.* 1947 92 A year when the British people were introduced to a battle of the dollar gap. 1957 *Times* 18 Nov. (Annual Financial & Commercial Rev.) p. i/1 Sterling came under such exceptional pressure in the current year, however, largely because of..the re-emergence of the dollar gap. 1892 Stevenson & Osbourne *Wrecker* ix. 138 Of all forms of the dollar-hunt, this wrecking had by far the most address to my imagination. 1848 Mill *Pol. Econ.* II. iv. vi. 309 The life of the whole of one sex is devoted to dollar-hunting, and of the other to breeding dollar-hunters. 1900 *Daily News* 25 Sept. 4/6 The dollar-hunting Americans. 1964 *Daily Tel.* 12 Feb. 14 'Dollar imperialism' is condemned for disorganising or even preying on the liberation of Britain's empire. 1847 *Boston Weekly Mail* 23 Jan. 3/6 The dollar mark..is only applied, properly, to the United States coin, or currency, of that name. 1831 *Deb. Congress U.S.* 22 Feb., App. p. cxxxix, Taking the issues of one, two, and three dollar notes, in the Eastern States as a guide. [1857 *Hist. Mag.* I. 186 In 1784,..Jefferson in the memorial which proposed the dollar as the American money-unit, employed the $ sign.] 1895 *Montgomery Ward Catal.* 223/3 Linen Markers Rubber Type Outfit...Consists of five printers' alphabets,..punctuation marks, dollar sign, [etc.]. 1920 S. Lewis *Main St.* x. 115 The dollar-sign has chased the crucifix clean off the map. 1912 J. R. Mohler et al. tr. *Hutyra & Marek's Spec. Path. Dis. Dom. Animals* I. vi. 826 The number of the dollar spots may be considerable, even exceeding one hundred. 1922 C. R. Edmonds *Dis. Anim. S. Afr.* 340 About the fortieth day the 'plaques' or dollar spots first appear in the skin. 1926 *Bull. U.S. Golf Assoc. Green Section* May 129 Small [brown-]patch is generally limited to about the size of a silver dollar, from which it has been commonly referred to as 'dollar spot'. 1935 *Gardeners' Chron.* 23 Feb. 129/1 Investigation of diseases of turf of lawns and greens has been prosecuted in the U.S.A. for many years, and the commonest diseases and causal organisms are well known, viz., Brown-patch, due to Rhizoctonia Solani, and Dollarspot, due to an unnamed species of Rhizoctonia. 1956 *Dict. Gardening* (R. Hort. Soc.) (ed. 2) IV. 2169/2 Dollar Spot is due to the fungus *Sclerotinia homoeocarpa*. 1872 *Harper's Mag.* June 132/2 There are ever so many good girls in factories..and in 'dollar stores' too.

Dollardom (dǫ·ləɹdǫm). [f. Dollar + -dom.] A place where the people's main aim is to amass dollars; also, the inhabitants of such a place; rich Americans collectively.

1852 *Lantern* (N.Y.) I. 109/2 My dear Dollardom, listen to a suggestion of Diogenes. 1852 *Hamilton* (Ontario) *Gaz.* 20 May 3/2 The mercurial Demon of Republicanisms, hath been urging the matrons and maidens of Dollardom to the perpetration of strangely preposterous pranks. 1920 *Blackw. Mag.* Aug. 139/1 Manhattan Dollardom trooped into the Ritz-Carlton carrying flasks.

do·ll-ba:by. *U.S.* [Doll *sb.*¹ 2.] = Doll *sb.*¹ 2 and 3; in quot. 1807, a small image of the human figure on which clothes are displayed. Also *attrib.*

1807 Jefferson *Writings* (1898) IX. 83 The dresses of the annual doll-babies from Paris. 1853 J. G. Baldwin *Flush Times Alabama* 42 She never had more than a thimble-full of brains in her doll-baby head. *Ibid.* 292 The little girls..had been petted by their fathers and mothers like doll-babies. 1896 *Harper's Mag.* XCII. 808/2, I keep on looking just the same frivolous doll-baby. 1965 H. Gold *Man who was not with It* iv. 34 The doll-baby lashes fell shut.

dollop (dǫ·ləp), *v. colloq.* and *dial.* Formerly also **dallop.** [f. the sb.] *trans.* To serve, put or give *out* in large quantities; to cover (something) with a large quantity. Also *fig.* (See also quots. *a* 1825 and 1860.)

a 1825 R. Forby *Vocab. E. Anglia* (1830) I. 88 *Dallop*, to paw, toss, and tumble about carelessly. 1860 Hotten *Slang Dict.* 130 *Dollop*, to dole up, give up a share. 1960 R. Daniel *Death by Drowning* i. 17 Clare dolloped a spoonful of trifle on her plate. 1961 W. Buchan *Helen all Alone* 168 Close-grown shrubs, all dark and dolloped with snow. 1967 *Listener* 28 Sept. 401/2 Is there either anthropological or literary virtue in mere transcription from tape, actuality dolloped out hot from a machine?

do·ll's house. Also **doll-house, dolls' house.** [Doll *sb.*¹ 2.] A miniature toy house made for dolls; also *transf.* and *fig.*, applied esp. to a diminutive dwelling-house. Also *attrib.*

1783 Lady E. Fenn *Cobwebs to catch Flies* I. 90, I have a chest at home, in my doll's house. 1840 Dickens *Let.* 31 July (1969) II. 109, I don't believe there is anywhere such a perfect little doll's house as this...the garden flourishing, the road lively, the rooms free from creeping things. 1842 —— *Amer. Notes* I. iii. 113 Like articles of furniture for a pauper doll's-house. 1847 C. Brontë *J. Eyre* I. iv. 47 Some picture-books and doll's house furniture. 1853 Dickens *Bleak Ho.* vi. 57 A habitable doll's house..would set the boy up in life. 1855 *Fraser's Mag.* July 60/2 Each of these doll-house rooms is crowded with prints, [etc.]. 1865 Dickens *Mut. Fr.* II. iv. v. 198 'We live on Blackheath, in the charmingest of dolls' houses...' But Bella started up... 'I want to be something so much worthier than the doll in the doll's house.' 1882 H. F. Lord tr. *Ibsen's Nora* p. v, The play now given to us as *Nora* is called in Norwegian *Et Dukkehjem*. To a public unused to Ibsen's surprises, *A Doll's House* is a misleading title. 1903 G. B. Shaw *Man & Superman* Ep. Ded. p. xi, Doña Juana, breaking out of the Doll's House and asserting herself as an individual. 1926 W. Deeping *Sorrell & Son* xxiv, Christopher spent a week-end with Thomas Roland in his doll's house at Chelsea. 1959 [see *doll *sb.*¹ 5].

dolly, *sb.*¹ Add: **2. c.** A girl or woman, esp. a young, attractive one. *colloq.*

1906 E. Dyson *Fact'ry 'Ands* xiv. 181 Now I wouldn' turn it [*sc.* beer] down fer ther toffest Dolly on ther block. 1942 Berrey & Van den Bark *Amer. Thes. Slang* § 382/2 A female, esp. a girl or young woman,..*dolly*. *Ibid.* § 427/3 Cute girl,..*dolly*. 1968 *Observer* (Colour Suppl.) 21 Jan. 5/1 Every bird, dolly, housewife, and career girl in Britain today owes something to the ladies on the left. 1968 *Daily Mirror* 20 Aug. 9/3 He is very gone on girls, is always falling wildly in and out of love with dishy dollies.

3. b. *Cricket* (*colloq.*). (*a*) A donkey-drop (see *donkey 3 b); (*b*) a very easy catch (cf. *dolly *a.* sense b).

1906 A. E. Knight *Complete Cricketer* 344 *Donkey-drop* or *dolly*, a slang term for slow, dropping bowling. 1956 N. Cardus *Close of Play* 78 But he might give Mr. Champain an over or two of his 'dollies'. 1969 P. Dickinson *Pride of Heroes* 40 Sir Ralph bowls me a dolly, watches me cart it for six, and records his admiration.

4. h. A small platform on wheels or rollers, used as a truck or conveyance; *spec.* a mobile platform on wheels on which a film- or television-camera can be moved about; (see also quot. 1955). orig. *U.S.*

1901 Merwin & Webster *Calumet 'K'* vi. 104 Other gangs were carrying them [*sc.* planks] away and piling them on 'dollies' to be pulled along the plank runways to the hoist. *Ibid.* xiii. 246 And every stick that leaves the runway has got to go on a dolly. 1929 *Bookman* Feb. 622 'Moving in' or 'trucking up' means wheeling the camera (on a small rubber tired wagon or 'dolly') closer to the characters. 1937 *Electronics* June 14/3 The camera is mounted in a mechanically controllable mount, and this in turn may be placed on a movable platform or 'dolly' so that it can be moved about over the set. 1948 *Time* 21 June 1/2 With a heavy truck and a tractor pulling, and one of these dollies under each corner, the derrick..was moved 3½ miles. 1955 *Amer. Speech* XXX. 91 *Dolly*, a retractable support for a semi-trailer when it is not connected to a tractor. 1961 G. Millerson *Television Production* iii. 27 Two small cranes (or power-operated dollies) and two pedestal cameras will meet the reasonable demands of even an elaborate drama production. 1970 *Commercial Motor* 25 Sept. 101/2 Converter dollies for use with standard haulage semi-trailers.

i. *S. Afr.* A lure for fish (see quot. 1957).

1930 C. L. Biden *Sea-angling Fishes of Cape* 136 The lure or 'dolly' as the fishermen term it. 1957 S. Schoeman *Strike!* 117 Professional fishermen catch thousands [of snoek] on 'dollies', which are cigar-shaped pieces of lead.. with a large barbless hook attached. A few thin strips of shark skin or pork rind or even a piece of red rag or silver paper is tied to the shank of the hook.

j. *Electr. Engin.* (See quot.)

1940 *Chambers's Techn. Dict.* 257/2 *Dolly*, the operating member of a tumbler switch; it consists of a small pivoted lever projecting through the outer cover.

5. *attrib.* (sense *4 h) *dolly camera, -pusher, shot*; **dolly-bag** = *Dorothy bag*; **dolly mixture**, a mixture of tiny coloured sweets of various shapes; also *transf.*, *fig.* and *attrib.* (Cf. also *dolly *a.* c.)

1926 R. H. Mottram *Crime at Vanderlynden's* 29 She carried the day's takings clasped to her breast, in a solid little leather dolly-bag. 1958 *Birmingham Even. Mail* 27 Mar. 6/6 Dollybags swinging from the arm. 1958 M. Dickens *Man Overboard* ii. 30 The cameraman..climbed up to his perch on the big dolly camera. 1957 R. Hoggart *Uses of Literacy* vi. 167 The 'dolly-mixtures' pleasures of a constant diet of odd snippets, of unrelated scrappy facts, each with its sugary little kernel of 'human interest'. 1959 I. & P. Opie *Lore & Lang. Schoolch.* ix. 166 Other current sweet-shop favourites appear to be the same as thirty years ago, in fact bull's eyes, jelly babies, and dolly mixture have entered schoolchild language as descriptive nouns. *Ibid.* 169 A chap who has got duck's disease is..labelled..dolly mixture (after a species of very small sweet). 1956 H. Grisewood in A. Pryce-Jones *New Outl. Mod. Knowledge* III. 423 These various sorts of *structores* and *imaginarii* were the dolly-pushers, the cameramen..of the ancient world. 1933 in C. Winchester *World Film Encycl.* 481/2 'Dolley shot', a moving camera shot.

dolly, *a.* Add: (Further examples.) Also, stupid, foolish.

1886 Baumann *Londinismen* 43/2 *Dolly*, dumm. 1922 Joyce *Ulysses* 759 Grinning all over his big Dolly face.. didnt he look a balmy.

b. In games, esp. Cricket, denoting a very easy catch, shot, etc. *colloq.*

1895 *Eng. Dial. Dict.* s.v., Such a dolly-catch. 1903 Jessop in H. G. Hutchinson *Cricket* v. 130 These simple 'dolly' catches are much more difficult to hold than those from hard drives. 1909 *Westm. Gaz.* 2 Mar. 12/2 Till then he had played a grand game, and had saved many clever shots, only in the end to be beaten by a 'dolly one'. 1955 *Times* 11 July 4/3 Lane-Fox..failed to get to the pitch of the ball and cocked up a dolly catch.

c. Usu. applied to a girl: attractive; fashionable. *colloq.*

1964 *Telegraph* (Brisbane) 6 Apr. 8/1 Take note, girls... Our London men report that you haven't really been given top-of-the-pops praise by your boyfriend unless he has called you a Dolly Bird. 1967 *Sun* 22 Feb. 6/6 *Dolly*, *groovy*, *cool*, *hippy*, fashionable. 1967 *Punch* 6 Dec. 872/2 It studies tradition, 'dolly girls', protest, pop, film stars, pop art and the American influence. 1970 'D. Halliday' *Dolly & Cookie Bird* vi. 89 The Chinaman..really was dolly, with a long yellow face. 1971 'R. Crawford' *Badger's Daughter* I. i. 13 You'll have to take..that dolly-bird you hide in Romford with you.

dolly, *v.* Add: **d.** To prettify; to doll *up* (see *doll *v.*²).

1958 *Listener* 13 Nov. 768/2 You go home and dolly them [*sc.* the songs] up a bit. 1961 C. H. Douglas-Todd *Popular Whippet* 37 At all events, 'dollied up' or not, a tea-chest makes a most effective indoor kennel. 1966 *Economist* 12 Nov. 709/3 Last minute painting and dollying up is done with the flag poles in position. 1968 R. V. Beste *Repeat the Instructions* xiii. 139, I didn't imagine you were the sort of spiv who would take a bribe however it was dollied up.

2. *intr.* (*colloq.*) To move a film- or television-camera *in* or *up* towards the subject, or *out* away from it. Cf. *dolly *sb.*¹ 4 h. Also *fig.*

1939 L. Jacobs *Rise of Amer. Film* xxii. 442 The traveling of sound is like panning or dollying of the camera. 1957 R. Longrigg *Switchboard* II. 123 Dolly up a bit, Steve... Dolly *up*, you clot. 1961 *Guardian* 13 Oct. 5/2 As if this were the opening of a..television programme, you 'dolly up' to its owner. 1965 *Listener* 11 Feb. 238/3 The camera..remains static and does not 'dolly in' for a close-up, or alternatively 'dolly out'.

dollying, *vbl. sb.* (Earlier example.)

1848 Mrs. Gaskell *M. Barton* II. x. 138 He had been engaged in 'dollying' and a few other mischievous feats in the washing line.

Dolly Varden. (Further examples.)

[1841 Dickens *Barn. Rudge* xix. 41 As to Dolly [Varden], there she was again..in a smart little cherry-coloured mantle, with a hood of the same drawn over her head, and.. a little straw hat trimmed with cherry-coloured ribbons.] 1872 *Harper's Weekly* 25 May 407/4 Was ever any new costume more criticized than the new 'Dolly Varden'. 1872 E. Wordsworth *Let.* May in E. Romanes *C. M. Yonge* (1908) ix. 139 My hostess looking more like an old French *marquise* than ever in a red and black Dolly Varden dress, with pink skirt. 1876 *Yreka* (Calif.) *Union* 3 June, The first spotted trout..were given the name of Dolly Vardens by Elda McCloud. 1946 *Trail Riders Bull.* Oct. 5/1 Dr. George Rae..landed four Dolly Vardens..during a special 'time-out' for the anglers.

‖**dolma** (dōᵘ·lmă). [Turkish, f. *dolmak* to fill, be filled.] A Turkish dish of several ingredients (see quots.).

Widespread in various E. European countries. Pl. forms reflect local variations.

1889 *Cent. Dict.*, *Dolma*, a Turkish dish made of vine-leaves, egg-plant, gourds, etc., stuffed with rice and chopped meat. 1935 M. Morphy *Recipes of All Nations* 767 The *dolmas*..consists of uncooked mutton finely chopped with kidney fat, a little soaked bread, and sometimes uncooked rice, the whole highly seasoned and spiced. A teaspoon of the mixture is wrapped in either a blanched vine leaf, fig leaf, or cabbage leaf, and these are placed in a saucepan and braised. 1954 D. Dodge *Lights of Skaro* i. 17 A wooden tub full of *dolmas*, cool in their wrapping of pickled vine leaves.

1962 'W. Haggard' *Unquiet Sleep* iv. 29 The delicacies craved by the Attic palate even in London—*halva*, *foul medames*, anari cheese, *dolmades* rolled in vine leaves, tinned. 1962 *Listener* 24 May 931/1 Among these you are certain to find *dolmathes*—vine leaves stuffed with various mixtures..and served in a tomato, or egg and lemon, sauce.

dolman. Add: **4.** In full, *dolman sleeve.* A sleeve that is much wider at the arm-hole than it is at the wrist.
1934 C. M. Brown et al. *Clothing Constr.* (ed. 2) xxxvi. 278 The dolman sleeve is usually set into a very low armseye. 1939 M. B. Picken *Lang. Fashion* 43/2 Cap-like wrap or coat with the dolman sleeve. 1941 *Amer. Speech* XVI. 68/1 Summer sleeves show that Dolmans—the very draped and opulent sleeves—are back.

‖ **dolmus, dolmush** (dǝᵘ·lmʊʃ). [Turkish *dolmuş* a vehicle or boat which departs only when all the seats are taken, f. *dolmuş* filled.] A shared form of public transport (in Turkey), esp. a taxi. Also *attrib.*
1957 I. Orga *Young Traveller in Turkey* ii. 24 *Dolmuş* means shared taxi...It's cheaper to travel *dolmuş*. 1962 I. Fleming *When I grow Rich* iii. 44 Driving about all over Istanbul..in a packed *dolmus* (or shared taxi). *Ibid.* v. 87 They embarked in a *dolmus* rowing-boat, with two other people waiting to be rowed across. 1967 J. Rathbone *Diamonds Bid* xii. 108 We caught a dolmush, a shared taxi, outside the New Mosque.

do·lomite, *v.* [f. the sb.] = Dolomitize *v.*
1913 W. B. Lewes *Oil Fuel* 32 The carbon dioxide dissolved in the brine under enormous pressure would slowly become absorbed in actions upon the mineral matter present, such as forming carbonates and soluble bicarbonates of magnesia, thus dolomiting the calcareous deposits.

dolool, var. *Delool.

dolorifuge: see -fuge.

dolorimeter (dǝᵘlǝri·mɪtǝɹ). [f. L. *dolor* pain (see Dolour) + -i- + -meter.] An instrument for the measurement of pain or sensitivity to pain.
1949 *Sat. Even. Post* 3 Dec. 29/3 The instrument they developed has been christened a 'dolorimeter'. 1957 J. Eysenck *Sense & Nonsense in Psychol.* vi. 238 A psychological torture instrument, the dolorimeter, which measures pain in units called dols.

doloroso (dǫlǫrǝᵘ·sǝᵘ), *a.*, *adv.*, and *sb.* *Mus.* [It. = Dolorous *a.*] As a direction to the performer: plaintive(ly), pathetic(ally). As *sb.*, a passage played in this manner.
1806 Busby *Dict. Mus.* (ed. 2), *Doloroso*,...a term by which we understand that the movement before which it is placed is to be performed in a soft and pathetic style. 1947 A. Einstein *Mus. Rom. Era* xi. 143 The combination of *doloroso* and *agitato*..is typical.

dolphin. Add: **6. f.** (See quot.) *U.S.*
1905 *Terms Forestry & Logging* 35 *Dolphin*, a cluster of piles to which a boom is secured.
9. *dolphin-drawn, -torn* adjs.
1914 W. B. Yeats *Responsibilities* 32 Paintings of the dolphin-drawn Sea-nymphs in their pearly waggons. 1965 *Eng. Studies* XLVI. 383 If Keats did not invent the dolphin-drawn float himself he may have found this picture in..the numerous illustrations of Homer. 1932 W. B. Yeats *Words for Music* 2 That dolphin-torn, that gong-tormented sea.

dolphined (dǫ·lfind), *a.* *poet.* [f. Dolphin + -ed².] Containing or having dolphins.
1934 Dylan Thomas *18 Poems* 19 The shades Of children ..cry to the dolphined sea.

Dom³ (dǝᵘm). [a. Hind. *Ḍōm*, f. Skr. *Ḍōma*, *Ḍōmba*.] A member of a menial Hindu caste. Also *attrib.* or as *adj.*
1828 H. H. Wilson in *Asiatick Researches* XVII. 4 Nābhāji..was by birth a *Dom*, a caste whose employ is making baskets and various sort of wicker work. 1860 H. M. Elliot's *Races N.W. Prov.* I. 84 Ramgarh and Sahankot, on the Rohini, are also Dom forts. *Ibid.* 85 The Magahya Doms of Champāran are a race of professional thieves. 1872 M. A. Sherring *Hindu Tribes* 400 The Dom is generally considered by Hindus to be the type and representative of all uncleanness. 1891 H. H. Risley *Tribes of Bengal: Ethnogr. Gloss.* I. 243 When a man has been ejected from his own caste for living with a Dom woman. 1937 H. W. Tilman *Ascent of Nanda Devi* i. 7 The Doms..propitiate the local gods and demons.

domain, *sb.* Add: **3. d.** spec. *Austral.* (with capital initial). The name of a park in Sydney, Australia, popular for speech-making. Freq. *attrib.*
1860 E. Turner *Little Larrikin* xiv. 163 His ideas were as sweeping and extravagant as those of a Domain orator out of work. 1904 A. B. Paterson *Rio Grande's Last Race* 124 It's grand to be an unemployed And lie in the Domain. 1941 Baker *Dict. Austral. Slang* 24 *Domain cocktail* (or *special*), a lethal concoction of petrol and pepper which reputedly once had a vogue among deadbeat drinkers in the Sydney Domain. *Domain dosser*, a loafer or down-and-out who frequents the Sydney Domain. 1943 K. Tennant *Ride on Stranger* xxiii. 261 He, too, could see his old comrade, Mervyn Leggatt, across the Domain advocating the direct opposite to his own policy.

4. d. *Math.* An algebraic system with two binary operations defined by postulates stronger than those for a ring but weaker than those for a field; *esp.* (more fully *integral domain*), a commutative ring in which the cancellation law holds for multiplication of non-zero elements and (with most writers) which has a unit element for multiplication.
1896 *Bull. Amer. Math. Soc.* III. 102 The idea of the adjunction of its roots..to the totality of numbers heretofore in use, viz. the rational numbers, is now developed, and the properties of rational functions of the elements of this new 'domain', are taken up. 1904 F. Cajori *Theory of Equations* xiii. 134 A set of numbers is called a domain of rationality or simply a domain, when the sums, differences, products, and quotients of any numbers in the set (excluding only the quotients obtained through division by 0) always yield as results numbers belonging to the set. All rational numbers constitute such a domain. 1937 A. A. Albert *Mod. Higher Algebra* ii. 27 The most important type of integral domain is the field. 1941 Birkhoff & MacLane *Surv. Mod. Algebra* iv. 97 By a unique factorization domain (sometimes called a 'Gaussian domain') is meant an integral domain in which (i) any element not a unit can be factored into primes, (ii) this factorization is unique to within order and unit factors. 1958 Zariski & Samuel *Commutative Algebra* I. i. 22 An important class of unique factorization domains is given by the so-called euclidean domains or rings admitting a division algorithm. 1965 J. J. Rotman *Theory of Groups* iv. 66 A domain is a commutative ring with unit that has no divisors of zero.
e. *Math.* The set of values that an independent variable of a function can take; the graphical representation of this set; the set comprising all the first elements of the ordered pairs constituting some given set.
1902 *Encycl. Brit.* XXVIII. 545/2 The idea of a 'variable' is that of a number to which we may assign at pleasure any of the values which constitute a definite aggregate, called the 'domain' of the variable. 1914 A. R. Forsyth *Theory of Functions of Two Complex Variables* iii. 57 A restricted portion of a field of variation is called a domain; the range of a domain being usually indicated by analytical relations. 1937 Michell & Belz *Elem. Math. Analysis* I. i. 60 The parts of the plane in which N can lie are commonly called the domain of N and the same word is used for the corresponding aggregate of pairs of values of x, y which correspond to N. 1955 D. A. Quadling *Math. Analysis* iii. 14 A function which is defined by means of a formula may have its domain restricted by the character of the formula itself.. The domain of the function 1/x cannot include the number 0. 1967 M. R. Kinsolving *Set Theory & Number Systems* ii. 21 One might say that the domain of R is the set of all first elements appearing in the pairs (a, b) constituting R.
f. *Logic.* The class of all terms that bear a given relation to any term (see quot. 1903).
1903 B. Russell *Princ. Math.* I. ii. 97 All referents with respect to a given relation form a class. It follows that all relata also form a class. These two classes I shall call respectively the domain and the converse domain of the relation; the logical sum of the two I shall call the field of the relation. *Ibid.* 98 If paternity be the relation, fathers form its domain, children its converse domain, and fathers and children together its field. 1955 A. N. Prior *Formal Logic* III. ii. 277 Both the domain and the converse domain of the null relation are the null class.
g. *Math.* An open connected set of at least one point.
1906 W. H. & G. C. Young *Theory of Sets of Points* ix. 178 The set of all the points inside a triangle is called a triangular domain, or the interior of the triangle, and is a simple case of a region. *Ibid.*, The points of a domain always form an open set. 1957 E. T. Copson *Introd. Theory of Functions of Complex Variable* (rev. ed.) iii. 15 If we add to a domain its limiting points, the resulting set is called a *closed region*.
h. *Physics.* In ferromagnetic materials, a region which behaves as an elementary magnet, all the atoms or ions in a region having the axes of their permanent magnetic moments aligned in the same direction. Also *attrib.*
1926 E. C. Stoner *Magnetism & Atomic Structure* xiii. 295 The size of the domains may be much smaller than the size of the actual crystalline 'grains'. 1944 *Electronic Engin.* XVII. 144/3 The assumption of a molecular field, and the sub-division of a ferromagnetic into spontaneously saturated domains, form the foundation of the modern domain theory of ferromagnetism. 1945 *Rev. Mod. Physics* XVII. 15 In a demagnetized specimen..the directions of magnetization of the individual domains are distributed at random among the possible easy directions. 1966 Phillips & Williams *Inorg. Chem.* II. xiii. 26 Ferromagnetic substances commonly exhibit a domain structure, in which each domain is permanently magnetized. 1966 Carey & Isaac *Magnetic Domains* i. 13 In such a case regions may not exist in the crystal where the direction of the magnetization changes from one domain to the next; these regions are called domain boundaries or domain walls.
i. *Linguistics.* (See quot.)
1933 Bloomfield *Lang.* xv. 247 The substitute replaces only forms of a certain class, which we may call the *domain* of the substitute. 1964 *Language* XVIII. 14 A suprasegmental phone has a Domain, defined as the type of sequence of segmental phones which it covers. 1968 P. M. Postal *Aspects Phonol. Theory* iii. 41 Each rule 'realizing' morphophonemic structures as phonemic ones must be defined exclusively on a domain consisting of strings of morphons.

domaine-bottled (domēⁿ·n,bǫ·t'ld), *a.* [f. Fr. *domaine* estate + Bottled *ppl. a.* 2.] = *estate-bottled* (see *estate *sb.* 14). So *domaine-bottling*.
1960 *Spectator* 21 Oct. 627/3 A very special Chambolle Musigny and a particularly aristocratic Clos de Vougeot, both domaine-bottled. 1961 *Guardian* 21 Nov. 15/3 A certain vogue for offering so-called domaine-bottled Burgundies. 1967 E. Penning-Rowsell in C. Ray *Compleat Imbiber* IX. 195 The demand for authenticity, including..domaine and château bottling. 1970 *Observer* (Colour Suppl.) 6 Dec. 74/3 The 1967 domaine-bottled Montrachet.

domal, *a.* Add: **3.** *Phonetics.* (See quots.) Hence as *sb.*
1919 C. R. Lanman in *Festgabe Adolf Kaegi* 101 The derivative, domal..is usable as adjective or as substantive (domals). *Ibid.*, The best English equivalent for *mūrdhanya*, 'produced at the dome (of the palate)', is clearly domal. 1933 Bloomfield *Lang.* vi. 98 Apical articulation in the domal position (the tip of the tongue touching almost the highest point in the roof of the mouth). *Ibid.*, In Sanskrit..postdentals..and domals..are distinct phonemes. 1943 K. L. Pike *Phonetics* vii. 123 A *domal* sound (i.e. *cerebral*, or *cacuminal*) implies that the tongue tip articulates somewhere behind the alveolar arch. 1962 B. M. H. Strang *Mod. Eng. Struct.* iii. 30 Positions of articulation..domal (tongue tip to dome of palate).
4. Of, pertaining to, or shaped like a dome.
1928 E. R. Lilley *Geol. Petrol. & Natural Gas* xi. 274 The Bald Hill Dome is one of the largest domal structures that are common in northeastern Oklahoma. 1949 W. D. Thornbury *Princ. Geomorphology* ix. 208 Most large domal structures are of rather ancient geologic age. 1972 *Nature* 19 Feb. 538/1 A broad domal uplift of central Kenya of about 300 m in the late Miocene was succeeded by a major uplift of 1,400 m.

dome, *sb.* Add: **4. c.** *Geol.* Any of various kinds of geological structure resembling a dome in shape (see quots.).
[1833 C. Lyell *Princ. Geol.* III. xxi. 269 Suppose the five formations to lie in horizontal stratification at the bottom of the sea; then let a movement from below press them upwards into the form of a flattened dome, and let the crown of this dome be afterwards cut off.] 1900 *Rep. Geol. Surv. Louisiana* [illegible] The dome of the salt is situated on the [illegible] of the island. [illegible] 1909 J. P. Iddings *Ign. Rocks* I. [illegible] 1930 *Engineering* 10 Jan. 39/3 A sulphur salt dome in Louisiana has yielded 9,000,000 tons of sulphur. 1956 C. M. Nevin *Princ. Struct. Geol.* (ed. 5) iii. 26 A dome is a roughly symmetrical upfold, the beds dipping in all directions, more or less equally, from a point. 1958 *Nature* 2 Apr. [illegible] The productive domes are fairly typical of the oil fields, being of small extent but giving large yields. 1962 A. Holmes *Princ. Physical Geol.* vi. 73 Domes and basins represent the limiting cases in which the beds dip in all directions, outwards from, or inwards towards, the centre of the structure. *Ibid.* xvi. 349 Salt domes are curious structures occurring in great numbers along the Gulf Coast of the United States. 1962 E. A. Vincent tr. *Rittmann's Volcanoes* i. 20 The slowly extruded lava piles up into a dome over the mouth of the vent.
d. The head. *slang.*
1891 in Farmer *Slang* II. 305/2. 1928 C. Sandburg [illegible] 1959 [see *bonce 2]. 1969 [see *bonce 2].
5. c. *dome of silence*, the trade name of a type of castor (Castor² 2) fitted to furniture; also *fig.*
1924 *Trade Marks Jrnl.* 11 June 551 Domes of Silence... Metal Castors for Furniture. 1925 A. Christie *Secret of Chimneys* xvii. 170 Those boots of yours aren't exactly domes of silence, are they, Bill? 1947 G. Greene *19 Stories* 161 One of those cases of circumstantial evidence, in which you feel the jurymen's anxiety—because mistakes have been made—like domes of silence muting the court. 1960 *Woman* 13 Feb. 3/4 Domes of silence, small metal fittings for the legs of tables or chairs.
f. In full *dome fastener.* A press-stud consisting of a rounded portion which clips into a socket, used esp. as a fastener for gloves.
1920 *Daily Chron.* 14 Mar. 6/4 (Advt.), Kid Gloves..with 2 or 3 dome fasteners. 1926 *Daily Colonist* (Victoria, B.C.) 17 Jan. 7/1 (Advt.), Dome Fasteners, Black or White, assorted sizes. 1966 G. W. Turner *Eng. Lang. in Austral. & N.Z.* viii. 172 Press studs are called *domes* [in New Zealand].
6. *dome-headed* a., having a large, well-rounded head; **dome-light**, a dome-shaped lamp.
1910 H. G. Wells *Mr. Polly* vii. 158 A certain high-browed gentleman living at Highbury..This dome-headed monster of intellect dwells alone. 1912 W. de la Mare *Memory* 75 A sage, dome-headed, grey, Who looked a child. 1956 *Archit. Rev.* CXIX. 354/4 The metal combined curbs and linings made for use with glass domelights. 1962 K. Orvis *Damned & Destroyed* iii. 23 She sat back under the winking domelight.

domestic, *a.* and *sb.* Add: **A.** *adj.* **2. a.** In special collocations: *domestic economy* (see Economy 2 a); *domestic science*, the study or knowledge of household management, comprising cookery, laundry, needlework, etc.; = Housewifery 1 a; *domestic service*, the condition or occupation of a household servant (cf. Service¹ 1 a); *domestic slave*, (*a*) a household slave, esp. as distinguished from a predial slave (see Predial *a.* 3 a); (*b*) = Odalisque; (*c*) (see quot. 1799); *domestic slavery*, the condition of a domestic slave; also *fig.*; *domestic workshop*, a workshop in a private dwelling-house.

1778 [see Economy 3 a]. 1797 F. M. Eden (*title*) The State of the poor, or an history of the labouring classes in England,..in which are particularly considered, their domestic economy, with respect to diet, dress, fuel, and habitation. [illegible] What do we understand by 'Domestic Economy' or 'Domestic Science' as we often prefer to call it now? 1905 *Daily Chron.* [illegible] a domestic economy school. [illegible] 1911 'Bartimeus' *Tall Ship* iii. 37 Cooking and laundry, and hygiene—domestic science it's called. 1936 *Discovery* Jan. [illegible] The greatest amount of dissatisfaction was found among teachers of domestic science, art and music. [illegible] 1840 H. Reeve tr. *de Tocqueville's Democracy in America* IV. iii. 59 [illegible] domestic service is looked upon by masters in the same light. 1933 Mrs. C. S. Peel *Life's Enchanted Cup* xviii. 239 Two of these..were Ministry of Reconstruction Committees; one to enquire into the housing of the working classes..the other to enquire into conditions of domestic service. 1776 Gibbon *Decl. & F.* I. ii. 49 In the free states of antiquity, the domestic slaves were exposed to the wanton rigour of despotism. 1799 *Lady's Mag.* June [illegible] odalisk, or domestic slave. 1799 M. Park *Trav.* (ed. 2) xxii. 287 The domestic slaves, or such as are born in a man's own house, are treated with more lenity than those which are purchased with money. [illegible] 1878 *Act* 41 *Vict.* c. 16 [illegible] domestic workshops.
6. *domestic-minded* adj.
1889 W. B. Yeats *Let.* Feb. in *Lett. to K. Tynan* (1953) 85 The poor domestic-minded swindler! 1960 *Times* 4 Mar. 13/6 The Afrikaner, being at heart a generous and very much a domestic-minded person.
B. *sb.* **4.** Also *sing.*; *spec.* (*a*) plain cotton cloth; (*b*) *U.S. colloq.*, a kind of cigar.
1817 M. Austin *Let.* 11 July in *Austin Papers* (1924) 317 The Domesticks I expected, have Not arrived [illegible] 1864 G. A. Sala *My Diary in America* II. xvi. 375 Those eminently nasty rolls of tobacco called in New York 'domestics'. [illegible] *Domestic*, a plain cotton cloth; it may be grey, coloured, or with a check pattern.

domesticate, *v.* Add: **2.** (Earlier example.)
1741 Richardson *Lett. Impt. Occasions* [illegible] It domesticates her, as I may say.
4. (Later examples.)
1818 Keats *Let.* 16 Dec. (1958) II. 4 With Dilke and Brown I am quite thick—with Brown indeed I am going to domesticate—that is we shall keep house together. 1850 Thackeray *Pendennis* II. xxi. 210 He became a good deal under the influence of his uncle's advice, and domesticated in Lady Clavering's house.

domesticity. Add: **3.** A domestic or homely expression or idiom. *rare.*
1899 Wilde *Importance of being Earnest* i. 28 Jack is a notorious domesticity for John!

domett. Add: Also **domette.**
[illegible] Men's Overshirts, plaid domette flannel. 1901 *Daily Chron.* 13 Nov. 8/3 Close-grained cloth coats, lined with fur, or with satin, or sandwiched with domette between it and the cloth. 1910 *Practitioner* Mar. 167 A many-tailed bandage of domette. [illegible] *Army & Navy Stores Catal.* [illegible] Bandages... Domette, 3 in. × 6 yds.

dominance. Add: **2.** *Biol.* The phenomenon whereby one of a pair of alleles present in a genotype is expressed in a phenotype while the other allele is masked; the state or property of being dominant (*Dominant *a.* 7). Also *attrib.*, as *dominance modifier* = *dominigene.
1902 W. Bateson *Mendel's Princ. Hered.* 129 (*heading*) The facts in regard to dominance of characters in peas. [illegible] 1916 J. Wilson *Man. Mendelism* 24 If the pairs in which the parents differ are clearly defined at the hybrid generation as regards dominance and recessiveness. 1918 Babcock & Clausen *Genetics Rel. Agric.* v. 69 The condition of dominance..is determined by the fact that in the hybrid that character is expressed to the exclusion of its contrasted character. 1925 T. H. Morgan *Evol. & Genetics* 151 (*heading*) The dominance of the wild type genes. 1929 R. A. Fisher in *Amer. Naturalist* LXIII. 555 The real difficulty..is not so simple as that the selective action upon dominance modifiers is so small that there has not been time for it to have had any appreciable effect. 1937 T. Dobzhansky *Genetics & Origin of Species* vi. 172 It must be kept in mind that dominance versus recessiveness is a matter of degree. 1968 M. W. Strickberger *Genetics* vi. 99 This phenomenon, by which one trait appears and the other does not, even though the factors for both are present, is called dominance.
3. *Ecology.* The prevalence or predominance of one or more species in a plant community.
1923 G. E. Nichols in *Ecology* IV. 14 A distinction is frequently made between associations in which the position of dominance is shared by two or more species..and those in which a single species is dominant. 1932 Fuller & Conard tr. *Braun-Blanquet's Plant Sociol.* iii. 33 The measurement of the cubic volume of standing timber by the forester may be considered as a determination of dominance.

dominant, *a.* and *sb.* Add: **A.** *adj.* **6.** *Forestry.* Overtopping other trees; said esp. of those trees in a forest which have their crowns free to light on all sides.
1893 J. Nisbet *Sk. Trees Woodland Crops* 21 Four classes of stems become distinguishable, viz., (1) predominating, (2) dominant, (3) dominated, and (4) suppressed. 1908 [illegible] *Dominant*, a tree which has raised its crown above the level of the surrounding trees. 1930 *Indian Forest Rec.* XV. 1. 2 Dominant Trees, including all trees which form the uppermost leaf canopy and have their leading shoots free.
7. *Biol.* [tr. G. *dominierend* (Mendel 1866, *Versuche über Pflanzenhybriden* in *Verh. d. Naturforsch. Ver., Brünn 1865* IV. 10).] Of a hereditary character: appearing to the exclusion of another character in a heterozygous organism containing alleles for them both. Hence of an allele or gene: expressed to the exclusion of another allelic gene. *Const.* to, *over.*
1900 W. Bateson in *Jrnl. R. Hort. Soc.* XXV. 58 In the case of each pair of characters there is one which prevails over the other. This prevailing character Mendel calls the dominant character, the other being the recessive character. 1924 C. C. Hurst [illegible] Mendel found the first-named character dominant over the other. 1925 T. H. Morgan *Evol. & Genetics* 151 The genes that behave as dominant have been found to be largely recessive to the genes already present in the original type which are said, therefore, to be dominant to the new genes. 1953 W. Braun *Bacterial Genetics* i. 3 When a gene derived from one parent differs from the corresponding gene contributed by the other parent, the effect of only one of the two genes may [illegible] the detectable effects of the other. Such genes are called dominant. 1965 Bell & Coombe tr. *Strasburger's Textbk. Bot.* (new ed.) ii. 332 It is possible for one gene to suppress the functioning of the other, i.e. one is dominant while its allele is recessive.
8. *Ecology.* Designating or pertaining to the predominant species in a plant community.
1923 [see *dominance 3]. 1950 *Jrnl. Ecol.* XXXVIII. 108 Plant associations have been classified according to the 'dominant' species of the vegetation; that is, those which occur in a high proportion of quadrats.
9. *dominant wavelength*, the wavelength or hue of a colour that determines the other colours with which it will match.
1933 W. de W. Abney *Res. Colour Vision* 415/2 Dominant wave-lengths. 1957 R. W. G. Hunt *Reproduction of Colour* viii. 79 Subjective Term. Hue; Objective Term: Dominant Wavelength. *Ibid.* 193 Where θ is a correlate of the hue (in terms of dominant wavelength).
B. *sb.* **3.** *Biol.* A dominant allele or character; an individual in which a particular dominant allele is expressed.
1900 W. Bateson in *Jrnl. R. Hort. Soc.* XXV. 58 In this generation the numerical proportion of dominants to recessives is..as three to one. 1905 R. C. Punnett *Mendelism* [illegible] there are dominants which breed true to the dominant character [illegible] 1913 *Off. Univ. Gaz.* [illegible] proved to be a Mendelian dominant, and [illegible] recessive. 1930 R. A. Fisher *Genet. Theory Nat. Selection* iii. 50 If..the mutant genes are dominant just as often as they are recessive, selection will be far more effective in eliminating the disadvantageous dominants than in eliminating the disadvantageous recessives. 1965 [illegible] *Introd. Human Genetics* [illegible] Dominants that cannot be transmitted therefore cannot be distinguished from environmentally induced traits. 1967 *Listener* 6 Apr. 450/1 A predisposition towards depression is probably an inherited characteristic—biologically inherited, that is, as a Mendelian dominant, and not just acquired from the family environment.
4. *Ecology.* The chief constituent of a plant community.
1913 Tansley & Adamson in *Jrnl. Ecology* I. 83 It is difficult to resist the conclusion that beech is the natural dominant in the higher members of woods. 1916 [see *association 12]. 1933 *Forestry* VII. 122 There is little difference between the various stem-classes—dominants, co-dominants, and sub-dominants—as regards the distribution of form-classes. 1938 Weaver & Clements *Plant Ecol.* (ed. 2) iv. 91 The visible unity of the climax is due primarily to the dominants or controlling species.

dominative, *a.* Add: **1. c.** *Psychol.* Tending to dominate; exhibiting domination (see quots.).
1937 H. H. Anderson in *Jrnl. Soc. Psychol.* VIII. 335 (*title*) An experimental study of dominative and integrative behavior in children of preschool age. *Ibid.* 338 Dominative behavior included the following: 1. Verbal demands to secure materials. 2. Forceful attempts to secure materials. 1953 A. K. C. Ottaway *Education & Soc.* vii. 137 'Dominative' behaviour is typified by autocratic methods and the attempt to dominate the will of others.

dominee. S. African var. Domine *sb.* 2 a.
1950 *Cape Times* 20 Sept. 16/2 The dominees of the Ring. 1959 *Ibid.* 14 Apr. 1/8 Two dominees, members of the Northern Transvaal Regional Synod. 1966 A. Sachs *Jail Diary* xvi. 136 He really is a very fine man our Dominee. He is a very good Christian, but he is a warm person.

dominie. Add: **3.** In full *dominie apple.* A variety of large apple. *U.S.*
1817 W. Coxe *Fruit Trees* 115 The Domine was imported from England. 1876 J. Burroughs *Winter Sunshine* vii. 158 If they were the dominie apples..he certainly would [hasten his sermon].

dominigene (dǭmi·nidʒin). *Biol.* [ad. G. *dominigen* (R. Goldschmidt 1935, in *Zeitschr. f. indukt. Abstammungs- u. Vererbungslehre* LXIX. 74), f. *domini-erend* Dominant *a.* + *-gen* *gene.] Any gene that modifies the dominance of another gene.
1938 R. Goldschmidt *Physiol. Genetics* ii. v. 107 As a short term for such modifying genes, Goldschmidt (1935b) has proposed the word dominigenes. *Ibid.*, Landauer (1933) found a dominigene for the frizzle character of fowl. 1940 *Nature* 9 Mar. 390/2 Two linked complementary factors for N-type are favoured by one of us..while a dominigene is the pivot of another hypothesis. 1952 C. P. Blacker *Eugenics* x. 242 Dominance can be modified by certain genes in the genic milieu for which the term *dominigenes* has been proposed.

dominion. Add: **2. b.** Esp. a country outside England or Great Britain under the sovereignty of or owing allegiance to the English or British crown; *spec.* †(*a*) *pl.* the English possessions in America; †(*b*) the principality of Wales; (*c*) any of the larger self-governing nations in the British Commonwealth; also *attrib.* *The Old Dominion* (earlier and later examples); also, *the Ancient Dominion.*
1606 *First Charter Virginia* in H. W. Preston *Doc. Illustr. Amer. Hist.* (1886) 9 The said several Colonies and Plantations,..they being of any Realms, or Dominions under our Obedience. 1623–4 *Act* 21 *Jac. I* c. 3 Within this Realme or the Dominion of Wales. 1682 *Acts of Assembly Virginia* (1727) I. 142 His Majesty's Subjects, being in this his Majesty's Dominion of Virginia. 1700 *Act* 11 & 12 *Will. III* c. 12 § 1 Commanders in Chief of Plantations and Colonies within his Majesties Dominions beyond the Seas. *Ibid.* c. 19 § 7 Any Prisons..belonging to any County of this Realm, or the Dominions of Wales. 1705 F. Makemie *Persuasive* (*Dedication*), To..Her Majestys Governor of the Ancient Dominion of Virginia. 1778 *Cal. Virginia State Papers* (1875) I. 311, I should not see the old Dominion this winter. 1789 W. Maclay *Deb. Senate* (1880) 14 The member from the ancient dominion. 1850 W. H. Foote *Sk. Virginia* 393 The Smiths came to Virginia to commence log colleges in the 'Ancient Dominion'. 1901 *London Gaz.* 4 Nov., The following addition shall be made to the Style and Titles at present appertaining to the Imperial Crown of the United Kingdom and its Dependencies;..after the words 'of the United Kingdom of Great Britain and Ireland', these words, 'and of the British Dominions beyond the Seas'. 1907 *Times* 1 Mar. 10/2 They had found themselves in complete agreement with the Premiers and Ministers of Defence of the King's Dominions across the seas. *Ibid.* 22 Apr. 14/2 The Prime Ministers of the self-governing Dominions. 1910 A. H. Forbes (*title*) History of the British Dominions beyond the Seas (1558–1910). 1912 A. B. Keith *Respons. Govt. in Domin.* III. 1313 Since the Colonial Conference of 1907 Dominion is a technical term for the self-governing Colonies. 1922 *Daily Mail Year Bk. 1923* 84/1 The terms..offered to Ireland the full dominion status of Canada. 1931 *Act* 22 *Geo. V* c. 4 § 1 In this Act the expression 'Dominion' means any of the following Dominions, that is to say, the Dominion of Canada, the Commonwealth of Australia, the Dominion of New Zealand, the Union of South Africa, the Irish Free State and Newfoundland. 1949 *Newsweek* 1 Aug. 15/1 This constituted 'outside interference' in the affairs of the Old Dominion. 1952 *Times* 21 Feb. 3/1 The words 'British Dominions beyond the Seas' had disappeared from the Proclamations in the United Kingdom, Australia, and New Zealand...At the same time Canada had substituted 'Canada' for 'Dominion of Canada'. 1953 *Times* 8 Dec. 9/2 Her Majesty will visit three of the seven fully sovereign nations—it is significant that their only generic title of 'Dominion' is already obsolescent—which make up in the narrower sense the oversea Commonwealth. 1959 *Times* 18 Apr. 7/3 The Labour Party pamphlet suggested that they should be offered 'Dominion Status'. That is to say, they should become independent, but should be able by the act of self-determination to re-enter the Commonwealth.

domino. Add: **3. d.** (Earlier example.) Also *subst.* (see quot. 1873); *it is domino* (*with*), it is all up (with), it is the end (of), it is finished (for). *slang.*
1862 B. Brierley *Tales & Sk. Lancs. Life* 26 What dost think abeawt Sebastypol bein' takken?..Aw'll bet thi a quart o'ale ut it's domino wi' it neaw. 1864 Hotten *Slang Dict.* (ed. 3) 123 *Domino*, a common ejaculation of soldiers and sailors when they receive the last lash of a flogging. 1873 *Ibid.* (ed. 4) 147 A domino means either a blow, or the last of a series of things, whether pleasant or otherwise. 1894 J. T. Clegg *David's Loom* xxi. 245 It'll be domino for me neaw. 1898 *Daily News* 10 Feb. 7/5 The young delinquent sullenly declared that James struck him first, whereupon he 'gave him domino for himself'. 1911 A. Bennett *Hilda Lessways* v. ii. 330 I've..paid the cheque! So it's domino, now! 1927 *Chambers's Jrnl.* 45/1, I thought it was domino with me and my little schemes.

e. *pl.* The teeth. *slang.*
1828 W. T. Moncrieff *Tom & Jerry* II. v. 53 Sluice your dominos—vill you?..Drink, vill you? don't you understand Hinglish? 1857 A. Mathews *Tea-Table Talk* II. 122 The poor destitute gentleman was still diligently seeking his lost *dominos*. 1913 *Pedagogical Seminary* XX. 436 To drink is to sluice the dominoes.
f. In full *domino paper.* Paper printed with a design from a wood-block and coloured, used as wallpaper, etc. (see quots.). (Cf. Fr. *domino*, *papier dominoté.*) Also ‖ **dominotier,** a maker of such paper.
1839 W. Chatto *Treat. Wood Engraving* ii. 59 In France the same kind of cuts [*sc.* wood-cuts], probably stencil-coloured, were called 'dominos'...The word 'domino' was subsequently used as a name for coloured or marbled paper generally, and the makers of such paper..were called 'dominotiers'. 1924 N. McClelland *Hist. Wall-papers* 20 The industry which gave the *Dominotiers* their name was the making of 'domino papers', which consisted principally of marbleized papers and again of others with little figures and grotesques, crudely printed from wood-blocks and coloured by hand. These 'dominos' were made in Rouen and in other cities..of France. 1926 Sugden & Edmondson *Hist. Eng. Wallpaper* 27 In France, 'domino' papers are regarded as the real forbears of paper-hangings. *Ibid.* 28 'Domino' papers were usually small—16½ in. by 12½ in.—and all the earliest were 'marbled'.
g. *pl.* The keys of a piano (see also quot. 1889). *slang.*
1889 Barrère & Leland *Dict. Slang* I. 303/1 *Domino thumper* (theatrical), a pianist. 1891 Farmer *Slang* II. 306/1 *Dominoes*,..the keys of a piano. 1895 J. T. Clegg *Works* I. 169 Aw con play 'God save the Queen' wi two fingers, iv aw happen to catch th' reet dominoes to start off.
h. *to make* (*the*) *domino*: to go out at the game of dominoes; also *fig.*, to anticipate the end; to finish first.
1890 'Berkeley' *Dominoes & Solitaire* 11 Sometimes each hand constitutes a game in itself; and when this is so, the player who makes 'domino' wins. 1892 C. Santley *Student & Singer* (ed. 3) ii. 24, I did not notice the bar's rest before the 'Amen', and performed a solo, which called forth some witty remark from Benedict about the future career of the singer who made the 'domino'. 1912 'Jar' *Dominoes* 3 A player, when he has played all the dominoes from his hand, is said to 'make domino'.
i. *Mus.* An error in performance (cf. quot. 1892 for sense *3 h). *colloq.*
1946 *Penguin Music Mag.* Dec. 50 One can get away with a 'domino' once—even thrice, but then someone starts to say: 'Poor Blank, he is beginning to slip.'
j. *fig.* Used of a theory that a political event or development in one country, etc., will lead to its occurrence in others; also *transf.* Freq. *attrib.*
1954 D. D. Eisenhower in *N.Y. Times* 8 Apr. 18/1 You had broader considerations that might follow what you might call the 'falling domino' principle. You had a row of dominoes set up, and you knocked over the first one, and what would happen to the last one was the certainty that it would go over very quickly. 1965 *New Statesman* 19 Feb. 277/1 There was as much domino talk ('With the collapse of South Vietnam, Laos..would speedily be swallowed..') then as now. *Ibid.* 277/2 Even if..the domino theory works out in practice. 1965 *Guardian* 16 Dec. 9/7 The domino phenomenon...if one African country pulls out of the Commonwealth then..there is a very real possibility the others will go out in succession like dominoes. 1966 *Ibid.* 6 Sept. 8/4 The departure of Zambia would not be a freak exception, but the fall of the first domino. Tanzania would follow forthwith. 1971 *Times* 3 May 12/3 There is a 'domino theory' about the possible relaxation of drug abuse legislation in Europe and North America... By the 'domino theory', once a lead has been given, others may be encouraged.

dompt, *v.* For † *Obs.* read *rare* and add later examples. Hence **do·mpting** *ppl. a.*
1912 Galsworthy *Inn Tranq.* 258 What is grievous, dompting, grim, about our lives is that we are shut up within ourselves. 1928 V. G. Childe *Most Anc. East* v. 115 We see a group representing a hero dompting two lions.

don, *sb.*¹ Add: **1. c. Don Juan** (earlier and later examples); also *attrib.*; hence **Don Juanery** *sb.*, **Don Juane·sque** *a.*; **Don Quixote** (examples); also *attrib.*
[1734 Fielding *Don Quixote in England* Introd., The Audience, I believe, are all acquainted with the Character of Don Quixote and Sancho. I have brought them over into England, and introduced them at an Inn in the Country.] 1848 Thackeray *Van. Fair* xxii. 190 Don't trifle with her affections, you Don Juan! 1870 D. G. Rossetti *Let.* 15 Mar. (1965) II. 817 He is a complete Don Quixote in every way. 1890 G. B. Shaw *Let.* 16 Dec. (1965) 278 An Irish Don Juan who will eventually compromise Socialism by some outrageous scandal. *Ibid.*, Those who take the Don Juan view of me. 1898 W. Graham *Last Links* 33 Byron's manner was tinged with a vein of Don-Juanesque recklessness. 1900 A. Conan Doyle *Gt. Boer War* x. 167 His long thin figure, his gaunt Don-Quixote face. 1902 *Pall Mall Gaz.* 4 Jan. 6/3 This Don Quixote of a society has made an assault upon the most solid of windmills. 1925 D. H. Lawrence *Refl. on Death Porcupine* 182 It's Don Juanery, sex-in-the-head, no real desire, which leads to profligacy and squalid promiscuity. 1926 W. J. Locke *Old Bridge* ix. 138 Her father was a Don Juanesque clerk in a factory. 1963 Auden *Dyer's Hand* III. 106 B..tries to be a Don Juan seducer in an attempt to compel life to take an interest in him.
6. (Example of the shortened form.)
1897 *Daily News* 16 Mar. 8/3 Two detectives..saw the prisoners playing Don.

donah: see Dona 2.

donate (dǝᵘ·nĕ¹t), **donat** (dǝᵘ·næt), *sb.* [ad. med.L. *dōnātus, -a*, pa. pple. of *dōnāre* to give.] Name given to members or associates of certain religious orders (see quots.).
1804 L. de Boisgelin *Anc. & Mod. Malta* I. ii. iii. 239 Those who were desirous of being received into the confraternity, and to become brothers de stage or donats in the order, presented themselves before the brother who was to receive them, fell on their knees, and placing their hands on the missal which the brother held open, pronounced the following words. 1858 G. Bowyer *Ritual of Profession of Knights & Religious Ladies* p. xiii, A Donat, or Confrère, is a member of the Order, and participates in the divine offices, benefits, prayers, Masses, and pious works of the Order. He is called Confrère, as being a member of the Confraternity of Donats. The Donats are also called in the statutes Fratres de Stagio, or Fratelli di Staggio. 1886 *Encycl. Brit.* XXI. 174/1 There was also an affiliation of religious ladies (*dames*) and of *donats* or honorary members. 1886 M. Geudens *Life of St. Norbert* p. xxvii, The donates, or oblates, who offer themselves to the Order, make their vows also; but they are not perpetual. 1902 Bedford & Holbeche *Order Hosp. St. John Jerus.* 207 In addition to Members there should be Honorary Associates and Donats—..the Donats being persons who, from an appreciation of the works of the Order, had contributed to its funds. 1911 *Encycl. Brit.* XXIV. 13/1 Affiliated brethren (*confratres*) and 'donats' (*donati*, i.e. regular subscribers..to the order in return for its privileges). 1925 C. S. Durrant *Flem. Mystics & Eng. Martyrs* 45 Here [at Diepenveen] he saw about a hundred veiled nuns with no small number of lay sisters, donates, and servants.

Donatism, Donatist, donative, donatory. The pronunciation of the first syllable is now generally (dǝᵘ).

doncher (dǝᵘ·ntʃǝɹ). Also **doncha, dontcha, dontcher.** Colloq. representation of *don't you*, esp. in phr. *doncher know.*
1893 E. F. Benson *Dodo* II. xv. 309 It's an arful bore reading books, dontcherthink, what? 1897 Hall Caine *Christian* iv. ii. 367 'Oh, let us, Glo,' cried Betty. 'I'd love it of all things, doncher know!' 1897 M. Corelli *Ziska* i. 31 My mother has taken to 'studying character', don'cher know. 1908 A. Huxley *Let.* 29 June (1969) 28 We have as you so aptly put it a great many aw, ehm new *men* dontcher know. 1913 Kipling *Diversity of Creatures* (1917) 274 You don't work any of your English on me. 'So glad to see you, doncher know—an' ta-ta!' 1957 [see *Dick *sb.*¹ 1 a]. 1962 *Guardian* 31 Oct. 7/2 Don't want to spend me evenin' watching plays about incest and rape, doncher know? 1969 L. J. Chiaramonte in Halpert & Story *Christmas Mumming in Newfoundland* 96 Why dontcha come around with us?

done, *ppl. a.* Add: **1. c.** Colloq. phr. *the done thing*: the accepted, correct, or fashionable action or mode of behaviour; = Thing *sb.*¹ 15 a.
1922 C. E. Montague *Disenchantment* iv. 57 Others were anxious lest the taking of steep and thorny paths..should come to be 'the done thing'. 1940 Harrisson & Madge *War begins at Home* iv. 88 Interview results..show more people in favour of..the 'done thing' than written results, where people are..more candid. 1953 N. Fitzgerald *Midsummer Malice* ii. 32, I expect he made a pass at you. He still thinks it is the done thing in the theatre.

Donegal (dǫnĕgǭl, dǫnĕgǭ·l). The name of a county in the north-west of Ireland; used *attrib.* or *ellipt.* to designate something produced in or peculiar to the county, esp. a type of tweed or a kind of coarse, knotted carpet.
1903–4 *T. Eaton Catal.* Fall & Winter 17 Women's Trainless Suit, made of Oxford grey Donegal tweed. 1905 *Westm. Gaz.* 5 Aug. 10/2 There are some charming new tweeds this season. The Donegals are good. 1909 J. Joyce *Let.* 27 Oct. (1966) II. 257, I send you seven or eight yards of Donegal tweed to have a new dress made from. *Ibid.* 17 Nov. 264 Irish tweeds, Donegals and suitings. 1946 J. B. Priestley *Bright Day* vi. 188 A floppy Donegal tweed hat. 1963 *House & Garden* Feb. 79/3 Donegal carpets. Hand-knotted carpets were first made in County Donegal in 1898. *Ibid.* 79/4 Those Axminster and Savonnerie carpets of the eighteenth century, whose colour and design the modern Donegal often emulates. 1968 *Guardian* 2 May 7/3 Very soft Isle of Bute tweeds and Donegals.

doner: see Dona 2.

dong, *v.* Add: (Further example.)
1954 J. Masters *Bhowani Junction* xxxiv. 291 A coppersmith bird donged with maddening persistence among the bushes in the garden.
2. *trans.* To hit, punch (esp. *Austral.* and *N.Z.*); to force by reiterated noise, speech, or effort. Cf. Ding *v.*¹ 2 a. *colloq.*
1889 E. L. Linton *Through Long Night* I. i. xv. 243 She had to be dinged and donged into obedience. 1928 Blunden *Undertones of War* 291 The drum-tap dongs my brain To a whirring void. 1930 *Bulletin* (Sydney) 7 May 21/1, I done me block an' donged 'im proper. 1937 N. Marsh *Vintage Murder* vi. 66 It was certainly a high-class way of murdering anybody...Dong him one with a gallon of champagne. 1959 I. & P. Opie *Lore & Lang. Schoolch.* x. 196 'Dong him on the dome' (head). 1960 N. Hilliard *Maori Girl* III. x. 133 I'll dong you if you say it any more. 1961 P. White *Riders in Chariot* xi. 410 'I will dong you one,' shouted Hannah, 'before you tear this bloody fur.'

Dong (dǫŋ), *sb.*² [Coined by E. Lear 1877 in *Laughable Lyrics*, 'The Dong with a luminous nose'.] A fabulous creature represented as having a luminous nose; also *transf.*
1927 E. Bowen *Hotel* xi. 131 Their cigarette ends glowing and fading preceded them like a pair of luminous noses, and equidistant spots of fire advertised that other pairs of Dongs were promenading solemnly. 1954 L. MacNeice *Autumn Sequel* 49 Egdon gives to Snark and Dong His kind attention still.

dong (dǫŋ), *sb.*³ *coarse slang* (chiefly *U.S.*). [Origin unknown; perh. f. prec.] = Penis.
1930 in *Amer. Speech* V. 390. 1939 J. Steinbeck *Grapes of Wrath* xvi. 245 Tell 'em ya dong's growed sence you los' your eye. 1961 Partridge *Dict. Slang* Suppl. 1124/2 Three Canadian terms for 'penis'..are *dink, dong, hammer.* 1969 P. Roth *Portnoy's Complaint* 18, I was wholly incapable of keeping my paws off my dong.

dong, *sb.*⁴ *Austral.* and *N.Z.* *colloq.* [f. *dong *v.* 2.] A heavy blow, a punch.
1941 Baker *Dict. Austral. Slang* 24 *Dong*, a blow, esp. with the fist. 1965 *Telegraph* (Brisbane) 5 July 8 *Dong*, poke (punch).

donga. For *S. Africa* read Chiefly *S. Africa*, and add further examples. (See also quot. 1966.)
1948 H. Drake-Brockman in B. James *Austral. Short Stories* (1963) 2nd Ser. 102 There were..brilliant flowers in the dongas after it had rained. 1964 *Sunday Truth* (Brisbane) 9 Aug. 20/5 The Government Administration [of New Guinea] builds and supplies homes for its employees. These are called 'dongas'. 1966 Baker *Austral. Lang.* (ed. 2) viii. 177 *Donga*, any gully or depression in which men could settle themselves in order to loaf...Also used both in Tobruk and New Guinea for a makeshift shelter...Not long after the end of World War II, the word acquired use among New Guinea's white population for a house.

Dongola, dongola² (dǫŋgǭ·lā, dǫŋgǝᵘ·lā). [f. *Dongola*, the name of a province of the Sudan.] A type of leather resembling kid, made from goat, sheep, or calf skin. Freq. *attrib.*
1889 *Shoe & Leather Review* 10 Oct. 31/3 In twenty-eight minutes the pair of shoes were finished—a neat and perfect pair of Dongola kids. 1895 *Montgomery Ward Catal.* 270/1 Leather Clothing, commonly called Dongola Goat, is made of oil-tanned and dressed Rocky Mountain sheepskin. *Ibid.* 508/1 Ladies' Fine French Dongola Button Shoes. *Ibid.* 508/2 This shoe is made from a very fine grade of satin finish dongola. 1897 C. T. Davis *Manuf. Leather* (ed. 2) xxiii. 376 It is the purport of this section to treat, not of the more modern tannages, but of the earlier brands that were for a number of years..regarded as standard—we refer to the 'Dongola' and preceding tannages. *Ibid.*, The different processes through which goat-skins pass in their conversion into Dongola. *Ibid.* 424 This blacking is for..dongola. 1926 *Daily Colonist* (Victoria, B.C.) 3 Jan. 36/3 (Advt.), Ladies' Black Dongola Kid Pumps.

Donizettian (dǫnitse·tiăn), *a.* and *sb.* [-ian.] **A.** *adj.* Of, pertaining to, or characteristic of the Italian operatic composer Gaetano Donizetti (1797–1848) or his music. **B.** *sb.* An admirer or adherent of Donizetti.
1848 Thackeray *Van. Fair* iv. 29 The eternal Donizettian music with which we are favoured now-a-days. 1955 E. Dent in H. M. van Thal *Fanfare for E. Newman* vi. 103 The opera is..rich in melody of the finest Donizettian type. 1962 *Times* 2 Feb. 13/2 Mahlerites are likely to owe their allegiance to the..gramophone records conducted by Dr. Bruno Walter.., Donizettians to Mme. Callas and Miss Schwarzkopf.

donk² (dǫŋk), colloq. abbrev. of Donkey.
1916 J. B. Cooper *Coo-oo-ee* xvi. 241 The men from Ironbark were arrested at the appearance of the long narrow lorries, and the 'donks', or donkeys, in them. 1919 *Athenæum* 8 Aug. 727/2 Mules were 'donks'. 1922 'R. Crompton' *More William* (1924) xii. 206 Look out for the donk, you ole ass. 1957 Blunden *Poems of Many Years* 288 The old East Window..Where Jesus used to perch On that plump donkey bound for Egypt, yes, That was a work of art,—the donk no less.

donkey. Add: **1. b.** Colloq. phrases: (*a penny*, etc., *more and*) *up goes the donkey*, used with allusion to the cry of a travelling showman (see quot. 1889); *to talk the hind leg*(*s*) *off a donkey* (see *talk *v.*).
1841 *Punch* I. 41/2 The report of Sir John Pullon, 'as to the possibility of elevating an ass to the head of the poll by bribery and corruption' is perfectly correct, provided there is no abatement in the price. Let him canvass again, and.. if he will only stand 'one penny more, up goes the donkey!' 1889 Barrère & Leland *Dict. Slang* (1897) 304/2 'Three more and up goes the donkey', that is, three pennies more and the donkey will go up the ladder. This phrase, used by mountebanks to denote that the performance will begin when the sum required is complete, is often said mockingly to a braggart to imply disbelief. 1913 M. Beerbohm *Let.* 11 Mar. (1964) 222 They earnestly hope to raise the sum of £500; after which 'up goes the donkey', I suppose. 1970 *Brewer's Dict. Phr. & Fable* 336/2 *Two more and up goes the donkey*, an old cry at fairs, the showman having promised his credulous hearers that as soon as enough pennies are collected his donkey will balance himself on the top of a pole or ladder.
2. b. A simple card-game played with special cards.
1920 *Isis* 3 Mar. 6/1 Don't play 'donkey' with him: he is the champion of the Giler. 1959 G. Avery *James without Thomas* xi. 195 They hastily changed to Donkey.
3. a. *donkey-carriage* (earlier example), *-chair* (example), *-ride*, *-riding* (example); *donkey-like* adj.

1816 Jane Austen *Let.* 9 July (1952) 459 We set off in the Donkey Carriage for Farringdon. 1868 C. M. Yonge *Let.* 8 Oct. in C. Coleridge *C. M. Y.* (1903) 298 She had been out for a long turn in a donkey chair. 1899 *Westm. Gaz.* 18 July 3/3 This zebra–donkey combination appears to exhibit a more than donkey-like obstinacy. 1962 I. Murdoch *Unofficial Rose* xxvi. 250 The more owlish and donkey-like he felt himself becoming. 1855 D. G. Rossetti *Let.* 1 July (1965) I. 261 A donkey-ride at Clevedon. 1894 G. du Maurier *Trilby* I. ii. 154 Swings, peep-shows, donkey-rides. 1825 H. Wilson *Mem.* II. 86 Fanny..doated on donkey-riding.
b. **donkey-drop** *colloq.*, in cricket, tennis, etc.: a slow ball bowled or hit so that it travels in a high curve; **donkey jacket**, a thick jacket worn by workmen as a protection against rain, cold, etc., and later in more general use as a fashionable garment; **donkey-lick** *Austral. slang*, (*a*) *v. trans.*, to defeat easily (e.g. in a horse-race); (*b*) *sb.*, treacle or golden syrup; **donkey's breakfast** *slang*, (*a*) a straw mattress (see also quot. 1901); (*b*) a straw hat; **donkey's** or **donkeys' years** (occas. *ears*, with punning allusion to the length of a donkey's ears and to the vulgar pronunciation of *ears* as *years*) *colloq.*, a very long time; **donkey-sled** *U.S.* (see quot.); **donkey-work,** the hard or unattractive part of an undertaking.
1888 A. G. Steel in Steel & Lyttleton *Cricket* iii. 128 Are *you* going to bowl your donkey-drops? I'll hit them all out of the ground. 1906 Donkey-drop [see *dolly *sb.*¹ 3 b]. 1927 *Daily Tel.* 14 June 9/1 He is content to play 'donkey drops' back into court, awaiting the right ball to hit. 1929 *Morning Post* 4 Oct., Members of the City Corporation wanted to know at yesterday's meeting at the Guildhall what a donkey jacket is... Mr. Gower explained that the jacket was one with leather shoulders and back. 1959 J. Braine *Vodi* xxii. 237 He stood there..in his navy-blue donkey jacket and black corduroys. 1961 *Sunday Express* 19 Nov. 1/4 Mr. Osborne, wearing a fur-lined check donkey jacket over a light-weight suit, arrived 10 minutes early for the wedding. 1944 *Truth* (Sydney) 13 Feb. 4/3 Breasley saw Kintore donkey-lick a field of youngsters in the Federal Stakes. 1945 C. Mann in *Coast to Coast* 23 The filly would donkey lick them in the second. 1953 Baker *Australia Speaks* iii. 81 *Donkey lick* or *cocky's delight*, treacle or golden syrup. 1901 W. C. Russell *Ship's Adv.* iii, Explaining.. that the term *donkey's breakfast* signified the bundle of straw which sailors who are reckless of their money ashore carry on board ship with them as a bed. 1909 J. R. Ware *Passing Eng.* 114/2 When a gent puts a donkey's breakfast a-top of his nut. 1916 'Anzac' *On Anzac Trail* 7 We slept on the usual 'donkey's breakfast', of course. 1935 'J. Guthrie' *Little Country* vi. 131 City men in the bowler hat and in the straw hat that was called a donkey's breakfast. 1916 E. V. Lucas *Vermilion Box* lxxvii. 86 Now for my first bath for what the men call 'Donkey's ears', meaning years and years. 1927 H. S. Walpole *Great Trad.* xvi, I was at the wedding, you know,..'aving worked for Miss Janet and her sister donkey's years. 1928 S. Vines *Humours Unreconciled* ix. 121 He hasn't talked emotionally to me for donkey's years. 1955 J. I. M. Stewart *Guardians* i. 8 It was donkey's years since he had been in an English train. 1961 *Observer* 19 Mar. 3/3 American influence and financial participation have been strong here for donkeys' years. 1905 *Terms Forestry & Logging* 35 *Donkey sled*, the heavy sled-like frame upon which a donkey engine is fastened. 1920 *Nat. Rev.* Apr. 145 Most of the donkey-work of this preposterous League has fallen on British shoulders. 1928 *Sunday Dispatch* 8 July 20/5 It would never do for a player to bat and not take his share of the donkey-work afterwards! 1940 *Manch. Guardian Weekly* 15 Mar. 213 As Parliamentary Secretary to the Ministry of Health she has had to do most of the donkey work in debate.

‖ **donnée** (done). Also **donné.** [Fr., fem. pa. pple. of *donner* to give.] The subject, theme, or motif of a story, play, etc.; a datum; a basic fact, assumption, etc.
1876 H. James in *Atlantic Monthly* XXXVIII. 693/2 A silly young girl and a heavy, overwise young man who *don't* fall in love with her! That is the *donnée* of eight monthly volumes. I call it very flat. 1878 —— *Fr. Poets & Novelists* 136 The *donnée* of 'Le Père Goriot' is typical. 1885 O. Wilde in *19th Cent.* XVII. 806 There is a mention of a bodice for Eve, but probably the *donnée* of the play was after the Fall. 1896 W. James *Let.* 11 June (1920) II. 37 B.'s moral atmosphere is anyhow so foreign to me, a lewdness so obligatory that it hardly seems as if it were part of a moral donnée at all. 1920 T. S. Eliot *Sacred Wood* v. 57 It is perhaps the craving for some such *donnée* which draws us on toward the present mirage of poetic drama. 1925 *Glasgow Herald* 29 Aug. 4 In the same tale *données* from classical myth are also to be encountered. 1945 Auden *Coll. Poetry* 127 The flushed assault of your recognition is The *donnée* of this doubtful hour. 1956 *Essays in Criticism* VI. 73 Because she accepts sensation as a *donné*, Mrs Woolf accepts the flow of linear time. 1960 J. Bayley *Characters of Love* iv. 213 To move right away in discussion from the text and from the actual Jamesian *donnée*. 1966 *Listener* 14 July 65/1 One can see how these raw *données* could be cooked into a film—the boy's scales and sonatinas, the dockland sirens, the café radio.

donnishly (dǫ·niʃli), *adv.* [-ly².] In a donnish manner.
a 1913 F. Rolfe *Desire & Pursuit of Whole* (1934) xvi. 163 Conversation spluttered donnishly. 1924 R. Macaulay *Orphan Island* vii. 76 Mr. Thinkwell signed to him, donnishly, with his hand, to stop talking. 1954 *N.Y. Times Bk. Rev.* 21 Nov. 6 Pater's treatment..flirted donnishly with pagan graces.

Donnybrook, donnybrook (dǫ·nibruk). [The name of *Donnybrook*, a suburb of Dublin, Ireland, once famous for its annual fair.] A scene of uproar and disorder; a riotous or uproarious meeting; a heated argument.
1852 *Blackw. Mag.* Nov. 645/2 The Irish patriots insist on having a Donnybrook to themselves. 1887 Kipling *Plain Tales* (1888) 69 Hindus and Mohammedans together raised an aimless sort of 'Donnybrook'. 1915 *Lit. Digest* 17 Apr. 863/2 A campaign which the New York *World* called a 'Donnybrook'. 1964 *Word Study* Feb. 1/1 The lexicographical donnybrook provoked by *Webster's Third New International Dictionary*. 1966 *Economist* 12 Feb. 600/3 Imagine the Donnybrook there would be in France or Italy.

donor. Add: **b.** *spec.* A blood donor (see *blood *sb.* 19).
1910 *Johns Hopkins Hosp. Bull.* XXI. 67/1 The serum of both donor and donee is capable of agglutinating the corpuscles of the other. 1936 *Brit. Med. Jrnl.* 28 Mar. 651/2 The patient's blood should in any case be grouped...If possible suitable donors can be chosen and warned to be in readiness. 1962 'J. Le Carré' *Murder of Quality* xviii. 178 'What was Stella's blood group, do you know?' 'Mine's B. I know that. I was a donor at Branxome. Hers was different.'
c. A person, alive or dead, from whom an organ or tissue is removed for surgical transplantation; also, an animal treated in this way.
1918 *Jrnl. Med. Res.* XXXVIII. 35 In a number of cases the second lobe of thyroid from the third guinea-pig (the second donor) was transplanted into a control guinea-pig. 1930 *Physiol. Rev.* X. 549 The experiments.. were mainly carried out in guinea pigs and rats and.. thyroid gland and cartilage were transplanted simultaneously from donor to host in each case. 1955 *Jrnl. Clin. Invest.* XXXIV. 331/2 In two cases the kidneys came from living donors. 1963 *Brit. Med. Jrnl.* 14 Sept. 645/2 (*heading*) Renal transplantation in man: a report of five cases, using cadaveric donors. 1971 *Daily Tel.* 26 July 3/8 Doctors should only be allowed to remove an organ if the donor has given his consent in writing or if the nearest relative that it is practicable to contact, has given consent.
d. In artificial insemination: one from whom the semen is taken. Also *attrib.*
1947 [see *artificial insemination* s.v. *artificial *a.* 5]. 1950 *Ann. Reg.* 1949 30 The Church could not condemn [artificial insemination] if the husband were the donor. 1958 *New Statesman* 18 Jan. 58/3 (*heading*) Donor Babies. *Ibid.* 25 Jan. 98/1 To afford a new legally regulated status to A.I.D. would..establish the donors as a socially approved class of persons.
2. *Chem.* and *Physics.* An atom, molecule, etc., that loses a constituent part to something else; *esp.* (*a*) an atom, etc., that gives up a valency electron pair to another atom, so forming a co-ordinate bond with it; (*b*) in a semiconductor, an impurity atom which has a higher valency than the majority of the atoms and can give up a valency electron to the conduction band of the crystal; *donor bond*, a chemical bond which can be regarded as having had some of its strength transferred to another atom.
1927, etc. [see *acceptor 3]. 1946 *Trans. Faraday Soc.* XLII. 100 Here we have to consider both the electron work function of the donor and the electron affinity of the acceptor. 1952 C. A. Coulson *Valence* ix. 252 This shows that the central C-C bond is an acceptor, acquiring bond order at the expense of the two end donor bonds. 1964 Guirard & Snell in Florkin & Stotz *Comprehensive Biochem.* XV. v. 173 At the stage of purity studied, the transaminase has a rather broad specificity for both amino group donors and acceptors. 1966 *Chem. in Brit.* II. 164/2 For triplet energy transfer from donor to acceptor to occur, the donor triplet must lie close to, or above that, of, the acceptor. 1971 *Sci. Amer.* July 34/2 If there are more acceptors than donors, current is carried by holes (positively charged) and the material is designated a *p*-type semiconductor.
3. *attrib.*, as **donor country** (see quot. 1959). (See also senses 1 c and d above.)
1959 *Britannica Bk. of Year* 546/1 Another political coinage [in 1958] was donor country, a country giving financial or material aid to an underdeveloped nation. 1961 *Ann. Reg.* 1960 471 At least one donor country was realizing that aid could easily go down the drain.

donsie, donsy, *a.* and *sb.* **A.** *adj.* **2.** Delete ? *Obs.* and add earlier and later examples. Also, poor, dreary, low-spirited; sickly, feeble (cf. *dauncy *a.*). *Sc.*, *north.*, and *U.S. dial.*
1720 A. Ramsay *Familiar Epistles* 10 Has thou with Rosycrucians wandert? Or thro' some doncie Desart danert? 1805 *Lancaster* (Pa.) *Jrnl.* 2 Oct. (Th.), Citizen Lafferty must have a 'doncy' opinion of the cause, when he is afraid to bet even. 1835 J. D. Carrick *Laird of Logan* 273 Sic an unco wastrie in the way of claiths..made me a thocht donsy. 1853 *Yale Lit. Mag.* XVII. 223 (Th.), [She brought some letters] to my room, to keep me from feeling 'donsy'. 1880 [see *dauncy *a.*]. 1917 J. L. Waugh *Cute McCheyne* 108 My faither was sawney an' donsie.

don't. Substitute for def.: colloq. contraction of *do not* (see Do *v.* 29 and *29 b); also (now vulgar) contraction of *does not* (see Do *v.* A. 2 c. γ in Dict. and Suppl.). Hence as *sb.* (usu. *pl.*), a prohibition; the reiteration of 'don't'. Also (*rare*) as *v. intr.* (Add earlier and later examples.)
1670, etc. [see Do *v.* *A. 2 c. γ, 30 d]. 1738 [see Do *v.* 28]. 1741, 1831 [see Do *v.* A. 2 c, γ]. 1874 Hardy *Far from Madding Crowd* vi. 76 'Whose shepherd is he?' said the equestrian... 'Don't know, ma'am.' 'Don't any of the others know?' 1902, etc. [see *do *sb.*¹ 4]. 1919 H. L. Mencken *Amer. Lang.* vi. 210 *Don't* has also completely displaced *doesn't*, which is very seldom heard. 'He *don't*' and 'they *don't*' are practically universal. 1929 D. H. Lawrence *Pansies* 48 (*poem-title*) Dont's. 1944 *Living off Land* vi. 125 In looking after a snake bite patient.., remember the three 'don'ts'.
b. don't-care *sb.*, one who does not care (see Care *v.* 4 a); a careless, unconcerned, or indifferent person; so **don't-care-a-damnativeness** (or **-itiveness**) (*slang*), carelessness, unconcern; **don't-carish** *a.*; **don't-carishness.**
1841 F. A. Kemble *Let.* 11 Oct. in *Rec. Later Life* (1882) II. 134, I am grown old and stupid and sleepy and don't-carish. 1841 Thackeray in K. Meadows *Heads of People* II. 169 That reckless do n't-care-a-damnativeness which leads a man to disregard all the world. 1841 *Daily Picayune* (New Orleans) 24 Feb. 2/3 He..had..all the nonchalance and dont-care-a-dam-itiveness of De Bar. 1864 *Daily Tel.* 29 Sept., There must be a deeper and more philosophic reason for this don't-carishness. 1905 *Daily Chron.* 10 Aug. 3/7 If she is at all weak she soon becomes a 'don't care', and a street-door gossip. 1924 W. J. Locke *Coming of Amos* ii. 16 No nose on earth could so express the don't-care-a-damnativeness of disdain. 1955 Auden *Shield of Achilles* i. 29 Bloodshot images of rivers screaming, Marbles in panic, and Don't-Care made to care.
c. don't know, a person who does not know (something) or who has not reached a decision or opinion on a particular subject, esp. in answering a questionnaire or the like; hence (*nonce-wd.*) **don't knowist** (see quot. 1908).
1888 *Pall Mall Gaz.* 18 Jan. 11/2 The gospel according to 'Don't-know' is at present unfit to supersede the Synoptics. 1908 G. B. Shaw *Sanity of Art* 57 'Positivists' or 'Dont Knowists' (Agnostics). 1940 Harrisson & Madge *War begins at Home* xiv. 381 Their figures differ from ours in showing a lower number of Don't knows. 1959 *News Chron.* 28 Sept. 6/1 There has now been enough old-fashioned give-and-take to make the Don't Knows start thinking. *Ibid.* 6/5 The Don't Knows, the drugged, and the mindless. 1970 *Amateur Photogr.* 11 Mar. 29/1 All the questions are completely answered; there are none of those curious 'don't knows' that usually round off the totals.

dontcha, dontcher, varr. *doncher.

donut: see *doughnut.

doodad (dū·dæd). Chiefly *U.S.* [Origin unknown; ? cf. Dad *sb.*² 2.] A 'fancy' article (of dress), a 'thingummy'; esp. a trivial or superfluous ornament.
1905, 1912 in H. Wentworth *Amer. Dial. Dict.* (1944) 173/2. 1920 S. Lewis *Main Street* xxiv. 298 Have a nice square house, and pay more attention to getting a crackajack furnace than all this architecture and doodads. 1928 *Daily Express* 24 July 8/4 Men detest flounces...'More fakements and doodads! Why on earth cannot the woman keep things simple?' 1934 J. O'Hara *Appointment in Samarra* (1935) ii. 40 Decorated with Santa Claus and holly doo-dads. 1961 *Guardian* 29 Mar. 9/7 Houses with a drop of the old half timbering and the old Georgian doodads. 1966 D. Enefer *Painted Death* viii. 76 An open lacquered box with hair clips and other doodads.

doodah (dū·da). *slang.* Also **do-da, dooda.** [From the refrain *doo-da(h)* of the plantation song 'Camptown Races'.] **1.** Phr. *all of a doodah*: in a state of excitement; dithering.
1915 H. Rosher *In R.N.A.S.* (1916) 97, I had lunch with the R—s and five daughters (swish, I was all of a doo-da!). 1918 *Chambers's Jrnl.* May 299/1, I feel all of a doo-dah, all of a wonk. 1928 S. Vines *Humours Unreconciled* xiii. 165 It was the evenings, clearly, that made her 'all of a religious doodah'. 1929 J. B. Priestley *Good Comp.* ii. iii. 307, I don't care if a man's been fifty years in the business, there's the same old thrill comes back. Opening night—all of a doodah! 1952 Wodehouse *Pigs have Wings* i. 28 Poor old Clarence was patently all of a doodah.
2. = *doodad.
1928 D. L. Sayers *Unpleasantness at Bellona Club* v. 45 D'you mind stickin' all those dark-slides into one pocket and a few odd lenses and doodahs into the other? *a* 1935 T. E. Lawrence *Mint* (1950) iii. i. 166 The old lady next me in the underground wore a flippant skirt, all doo-dahs. 1945 *Salt* 26 Feb. 16/1 A bit of a kid done up, in trousers with do-da's danging on 'em. 1957 H. Croome *Forgotten Place* 68 They make little plastic doodahs to use in electrical machinery.

doodle, *sb.* Add: **2.** A doodle-bug. *U.S.*
1887 *Harper's Mag.* July 276/1 She wondered how the nice, fat little round 'doodles' were getting on in their tin can under the house; she never had had such a fine box of bait. 1939 *These are our Lives* (U.S.) 157 They not knowing any more than a doodle in the woods what she was saying.
3. doodle-bug a. *U.S.*, a tiger-beetle, or the larva of this or various other insects.
c 1866 G. W. Bagby *Old Virginia Gentleman* (1910) 48 Try to tame a catbird, call doodle-bugs out of their holes. 1876 'Mark Twain' *Tom Sawyer* viii. 76 Doodle-bug, doodle-bug, tell me what I want to know! 1903 S. E. White *Forest* vii. 81, I never hesitate to offer them [*sc.* trout] any kind of a doodle-bug they may fancy. 1944 *Democrat* 7 Dec. 2/2 Doodlebugs are those little insects which fan out or blow out little conical shaped holes in sandy or dusty places. *fig.* 1908 C. E. Mulford *Orphan* 163 You blamed doodle-bug, yu! 1918 —— *Man fr. Bar-20* xviii. 193 'You're a fine pair of doodle-bugs'... 'Don't you know an opportunity when you see one?' 1938 'E. Queen' *Four of Hearts* (1939) i. iii. 34 She'll tell you more about these doodlebugs than they know themselves.

b. As a nickname also applied to the German pilotless plane or flying bomb of the war of 1939–45; ellipt. *doodle.*

1944 *Times* 22 June 2/1 The first fighter pilot to shoot down what the R.A.F. men call a 'doodlebug' was Flight Sergeant Maurice Rose, of Glasgow. **1944** *Aeronautics* Aug. 27/1 The most elaborate German automatic weapon, the flying bomb,..doodle-bug, or power-bomb. **1944** *N. & Q.* 9 Sept. 122/1 *The doodle-bug.*—The new weapon with which our enemy has attacked us has borne a number of names since its appearance over Southern England on 13 June...The public in London has now generally accepted doodle bug, doodle bomb or simply doodle. **1969** T. Parker *Twisting Lane* 79, I left school in 1944, just after the doodle-bugs finished.

4. An aimless scrawl made by a person while his mind is more or less otherwise applied. So **doo·dle** *v. intr.*; also *fig.* (partly by assoc. with Dawdle *v.*), to idle; **doo·dling** *vbl. sb.* and *ppl. a.*; **doo·dlingly** *adv.*; also **doo·dler** *sb.*, one who doodles.

1937 R. M. Arundel *Everybody's Pixillated* p. ix, A 'doodle' is a scribbling or sketch made while the conscious mind is concerned with matters wholely unrelated to the scribbling. **1937** *Manch. Guardian* 5 May 8/4 Doodling is fidgeting about pictorially with a pen or pencil at odd moments to pass the time... In Australia..[caterers] have provided special menu cards with plenty of space on them in the hope of luring doodlers into doodling on these instead of on the tablecloth. **1937** *Lit. Digest* 26 June 19/3 'But everbody doodles.' So Gary Cooper, as Longfellow Deeds, in 'Mr. Deeds Goes to Town', defended himself. He wasn't crazy because he drew squares and circles on scraps of paper —he was just 'doodling'. **1938** *Life* 14 Nov. 7/2 It's a doodle picture. I've seen it dozens of times on telephone pads and what not. **1942** *Punch* 25 Feb. 158/1 Mr. Clement Attlee, the Premier's deputy, industriously drew doodles of intricate pattern. **1944** *Times* 17 Mar. 5/4 It is nevertheless a thrill to the humble 'doodling' addict to discover that Marshal Stalin himself is of the brotherhood. **1947** Auden *Age of Anxiety* (1948) i. 25 On memories stuffed With dead men's doodles. **1948** 'J. Tey' *Franchise Affair* xviii. 217 Robert sat doodling on..blotting-paper. A herring-bone pattern. **1951** E. Mittelholzer *Shadows move among Them* iii. viii. 326 He left it a doodled smear on the Penguin. **1953** C. Day Lewis *Italian Visit* iii. 39 Lightning sketches, Symbolic doodlings, hour by hour set down Haphazardly. **1955** H. Spring *These Lovers fled Away* ix. 128 When we had changed from a main line train and were doodling across country. **1959** H. Gardner *Business of Criticism* ii. ii. 120 When a writer's first drafts, scraps of memoranda, and 'doodles' have been preserved, we may possibly have a limited success in tracing the workings of the creative imagination. **1960** *20th Cent.* Mar. 233 Poetry is not the free unfettered self-expression of the doodler. **1961** *Times* 23 Nov. 15/2 Mr. Gwyn Thomas lets a witty pen run doodlingly on. **1967** *New Scientist* 12 Oct. 102/1 At boring committee meetings, in common with others who doodle, I can escape with relief into the insulated world of mathematical abstraction. **1970** H. Braun *Parish Churches* xviii. 219 Geometrical 'doodles' made with a mason's compasses are medieval.

doodle-bug: see *doodle *sb.* 3.

doodle-doo. Playful shortening of Cock-a-doodle-doo.

1785 Grose *Dict. Vulg. Tongue* (s.v. *Doodle*), *Doodle doo*, or cock a doodle doo, childish appellation for a cock, from its note when crowing. **1904** Barrie *Peter Pan* (1928) v. 133 *Hook* (*slowly*). Cecco, go back and fetch me out that doodle-doo. **1907** D. O'Connor *Peter Pan Picture Bk.* 53 He was silenced by a shrill and piercing cock's-crow from the cabin. 'Someone must bring me out that doodledoo,' roared the Captain.

doofer (dū·fəɹ). *slang.* Also **doofah, doovah, doover.** [Prob. alteration of *do for* in such phrases as *that will do for now.*] = Thingummy (see also quots.).

1937 Partridge *Dict. Slang* 234/2 *Doofer*, half a cigarette: workmen's [slang]. **1941** Baker *Dict. Austral. Slang* 25 *Doover*, any object, thingummy. **1945** —— *Austral. Lang.* viii. 153 Other words of particular note [in the 1939–45 war] were..*doover, doovah* or *doovah-dah* used as a general utility term for a thingumebob, specifically applied to any shelter, especially a humpy, and to a hospital bottle for urination, whence, *doover-joey*, a male hospital nurse. **1945** *Salt* 2 July 43/2 Idly I lean against the parapet of our doover. **1945** C. H. Ward-Jackson *Piece of Cake* (ed. 2) 24 *Doings*, used in place of any noun that cannot be recalled on the spur of the moment...Variants are 'Doofer' and 'Dooshanks'. **1969** 'S. Troy' *Swift to its Close* vii. 110 Mr. Turvey returns 'ome. Comes again to the Hangar in time for the doofah. **1970** P. Dickinson *Seals* iv. 95 This is a very fancy doofer indeed...It transmits along one wavelength and receives along another.

doohickey (dū·hiki). *colloq.* (orig. and chiefly *U.S.*). Also **dohickey, doohicky.** [f. *doo(dad + *hickey.] Any small object, esp. mechanical; a 'thingummy' (see also quot. 1928).

1914 *Our Navy* (U.S.) Nov. 12 We were compelled to christen articles beyond our ken with such names as 'dohickeys', 'gadgets' and 'gilguys'. **1925** Fraser & Gibbons *Soldier & Sailor Words* 81 *Doo hicky*, an airman's term for any small, detachable fitting. **1928** *Sunday Dispatch* 30 Sept. 10/2 He offered to run me down in the old doohicky—his latest, though second-hand car. **1949** R. Chandler *Little Sister* vii. 45 A pencil..broke its point on the glass doohickey under one of the desk legs. **1967** A. Lurie *Imaginary Friends* iv. 44 Just unhitch that dohickey there with a wrench.

doolally (dū·læli), *a. slang* (orig. *Services*'). [Spoken form of *Deolali* (Marashtra, India) + Tap *sb.*[4]] In full *doolally tap.* Characterized by an unbalanced state of mind.

1925 Fraser & Gibbons *Soldier & Sailor Words* 75 *Deolali tap* (otherwise *doolally tap*), mad, off one's head. Old Army. **1936** J. Curtis *Gilt Kid* vii. 76 What's the matter with that bloke? Doolally? *Ibid.* xviii. 181 'What's up with you for Christ's sake, kid? Come doolally tap?' Scaley was getting worried...His pal seemed to have gone mad. **1936** F. Richards *Old-Soldier Sahib* iv. 74 Time-expired men sent to Deolalie from their different units might have to wait for months before a troop-ship fetched them home...The well-known saying among soldiers when speaking of a man who does queer things, 'Oh, he's got the Doo-lally tap,' originated, I think, in the peculiar way men behaved owing to the boredom of that camp. **1943** Hunt & Pringle *Service Slang* 28 *Doolally*, very drunk or temporarily insane, without distinction.

Doolan, doolan (dū·lăn). *N.Z. slang.* [Prob. f. the Irish surname *Doolan.*] A Roman Catholic; an Irish Catholic.

1940 F. Sargeson *Man & Wife* (1944) 19 It sounded pretty awful to me, that sort of praying. Because I'm a Doolan myself, and Mrs Bowman [who was doing the praying] was always down on the churches. **1947** D. M. Davin *Gorse blooms Pale* 186 She'll have me a doolan yet, Father. **1963** B. Pearson *Coal Flat* iv. 63 They had called him 'the Doolan bugger'—*Doolan* on their tongues meant an Irish Catholic.

dool-owl. *dial.* [f. *dool* (Dole *sb.*[2]) + Owl *sb.*] An owl (as a symbol of gloom); in quots. *transf.*, a dull, depressing person.

1928 D. H. Lawrence *Lady Chatterley* (1930) xi. 196 When I look at women who's never really been warmed through by a man, well, they seem to me poor dool-owls after all. **1929** —— *Pansies* 144 The upper classes...Such bloomin' fat-arsed dool-owls.

doom, *sb.* Add: **11.** *doom-laden* adj.

1938 *Times* 24 Mar., Mr. Gennadi Rozhdestvensky gave this doom-laden symphony its full eloquence.

Doomie (dū·mi). *R.A.F. colloq.* [f. Doom *sb.* 4 + -ie.] A name given to an imaginary prophet of doom or giver of warnings.

1945 *Tee Emm* (Air Ministry) V. 151 (*caption*) Doomie says: Doomie-buoy says Don't shoot me. **1952** M. Tripp *Faith is Windsock* xiii. 204 The era which produced Chads, Doomies and Gremlins. **1962** *Guardian* 30 Nov. 12/4 Pinwheel was his first name, but..he was adopted by other RAF units under the name of Doomie—and Doomie he remained until the British Army adopted him (*circa* 1943) with the new name of Chad.

doomsday. Add: **3. doomsday machine** (see quot. 1961); also *doomsday bomb.*

1960 H. Kahn *On Thermonuclear War* x. 489 Technology, 1969..Doomsday Machines. *Ibid.* 500 The last two items in Table 65 are probably the most important, even if not very probable. Foremost are Doomsday Machines. I am not predicting that they will be built... It is most unlikely that either the Soviet Union or the U.S. would build such machines. **1961** *New Scientist* 26 Oct. 230/1 The Doomsday Machine is a hypothetical weapon which is capable of destroying all human life. **1965** B. Russell *Autobiogr.* (1969) III. iv. 211 Cobalt would be necessary for the Doomsday Bomb. **1968** *Observer* 31 Mar. 25/4 The idea of a nuclear 'Doomsday machine', capable of destroying all life on earth, is not *technically* absurd. **1970** A. Dipper *Hard Trip* xiv. 200 It is unutterably evil, and dangerous as a doomsday bomb.

doomy (dū·mi), *a.* [f. Doom *sb.* + -y[1].] **a.** Of a person: depressed, sad. **b.** Of a thing: depressing, weird.

1961 G. Smith *Business of Loving* xi. 219 The gaffer..was a doomy old nut who once or twice had talked about jagging it in. **1967** A. J. Marshall in L. Deighton *London Dossier* 142 The City of London...It's a doomy, forbidding place to me. **1968** *Crescendo* Jan. 22/2 Even the irrepressible Mr. Brookmeyer produces a somewhat doomy arrangement of 'Willow Weep For Me'.

doonga (dū·ngă). *India.* Also **dunga.** [ad. Hind. *dongā.*] A flat-bottomed dug-out with a square sail.

1905 *Westm. Gaz.* 3 Jan. 2/1 A line of stretchers was winding in and out past the dungas—all laden. **1922** *Chambers's Jrnl.* 759/2 There is the house-boat, the *doonga*—a species of long canoe, with a sloping rush roof.., in which you live. **1925** *Ibid.* 77/1 At Srinagar Durrant hired a large doonga-houseboat.

door. Add: **4. b.** One of two boards or metal plates attached to the ends of a trawl-net.

1911 *Encycl. Brit.* XXVII. 219/2 The trawl boards, or as they are frequently called 'doors', are of deal. **1928** Russell & Yonge *Seas* xiii. 276 The net is hauled by two wire warps, one attached to each 'door'. **1961** *Times* 10 June 11/5 The modern trawl, with its bobbined foot rope, extra sweep between doors (French trawl).

7. a. *door-catch, -chain*; *door-handle* (earlier example); *door-knob* (examples); *door-latch* (examples); *door-window* (later example). **c.** *door-wise* adv. (earlier and later examples).

1897 *Sears, Roebuck Catal.* 44/2 Screen Door Catches, enameled iron with stop. **1913** Masefield *Daffodil Fields* 74 A door-catch clacked. **1836** Dickens *Sk. Boz.* 1st Ser. (1837) 81 The door-chain was softly unfastened. **1967** *Gloss. Terms Builders' Hardware* (*B.S.I.*) iv. 14 *Door chain*, a device for preventing a door from opening beyond a limited distance, comprising a chain attached at one end to a fixing plate secured to the door frame and at the other to a stud. **1832** F. Trollope *Domestic Manners* II. xxvi. 104 A delicate silver knocker and door-handle. **1847** *Rep. Comm. Patents 1846* (U.S.) 63 Improvements in door-knobs. **1905** *Daily Chron.* 19 July 4/7 How many door-knobs are there in this happy island? **1678** *New Castle* (Del.) *Court Rec.* (1904) 362, 3 Iron door Latches. **1924** J. M. Murry *Voyage* xvii. 304 She fumbled blindly at the door-latch. **1907** *Westm. Gaz.* 31 Aug. 2/3 He bounded out through the wide-open French window. ..Vincent slammed the door-window behind as he jumped. **1769** *Monthly Rev.* XL. 372 A rude arch curtained doorwise. **1898** G. B. Shaw *Arms & Man* i. 4 The window is hinged doorwise.

8. door-casing, -facing, -trim *U.S.* = Door-case; **door chimes** [Chime *sb.*[1] 2], a chiming mechanical or electrical device acting as a door-bell; **door-pin** (later example); **door-stone** (later U.S. example); **door-to-door** *attrib.*, (of canvassing, selling, investigating, etc.) done methodically at one house after another; (of a journey) from actual start to actual finish.

1887 M. E. Wilkins *Humble Romance* 2 He lounged smilingly against the door-casing, jingling his scales, and waiting for the woman. **1962** A. Nisbett *Technique Sound Studio* x. 173 Door chimes and hand-bells. **1845** W. G. Simms *Wigwam & Cabin* I. 99, I had been hewing out some door-facings for a new corn-crib and fodder-house. **1877** *Congress. Rec.* 26 Nov. 705/1 This man..was sitting up in the door with his feet on the door facing. **1889** W. B. Yeats *Wanderings of Oisin* 90 Raise the door-pin with alarm. **1893** M. A. Owen *Voodoo Tales* 209 The aunties searched under every doorstone for 'tricks'. **1902** *Harper's Mag.* May 1004/1 When I arrived at the house my son, the editor, had just returned from a door-to-door sale of the *Mosquito.* **1929** 'G. Daviot' *Man in Queue* x. 128 You can be an ex-service man..out of a job...I don't want any door-to-door business. **1934** *Planning* II. xxxvi. 14 Apart from certain short-distance traffic which can be more cheaply and efficiently sent by road on account of the door-to-door facilities and absence of double handling, the main competition between road and rail is over certain long-distance traffic. **1943** *Our Towns* (*Women's Group on Public Welfare*) ii. 12 The door-to-door salesmanship which is one of the pests of town life. **1963** D. Ogilvy *Conf. Advert. Man* v. 96 When I was a door-to-door salesman I discovered that the more information I gave about my product, the more I sold. **1905** *N.Y. Even. Post* 30 Dec. 12 (Advt.), The corridors, floors, stairways, doortrims and walls are of marble.

doorman (dō^ə·ɹmæn). **1.** A farrier's assistant.

1896 *Daily Chron.* 25 Aug. 9/5 Farrier—Young man wants Job as doorman and jobbing. **1897** *Daily News* 10 May 11/5 The present scale of pay is 5*s.* a day for doormen and 6*s.* for firemen. **1901** *N. & Q.* 9th Ser. VIII. 184/2 [Advt. in provincial newspaper] Wanted, doorman, able to nail well.

2. = Doorsman.

1904 *Daily Chron.* 26 July 6/4 A publican whose doorman ejected a customer. **1905** *Westm. Gaz.* 27 June 9/1 Employment as porter, odd-man, lift-man, or door-man. **1927** *Blackw. Mag.* Nov. 701/1 At this moment entered the discreet white-clad doorman. **1959** *Times* 18 June 19/5 Mehmet Hassan..doorman..was sentenced to 21 months.

door-mat. Add: **b.** *fig.* Applied to a person upon whom people 'wipe their boots'.

1861 Dickens *Gt. Expect.* I. xii. 207 She asked me and Joe whether we supposed she was door-mats under our feet, and how we dared to use her so. **1899** *Westm. Gaz.* 15 Apr. 2/3 'Door-mat' duty was never the portion of the one who was to be four times Prime Minister. **1915** Mrs. H. Ward *Eltham House* x. 172 It's no good playing doormat. You've got to make people afraid of you. **1930** *Observer* 20 Apr. 5/5 She is not such a nullity and 'doormat' as Miss Byron. **1959** M. M. Kaye *House of Shade* xii. 163 A nice kind sugar-daddy of the adoring door-mat type. **1961** Wodehouse *Service with Smile* vi. 90 On these occasions he ceased to be a human doormat whom an 'Oh, Clarence!' could quell.

2. *slang.* **a.** A beard or moustache. **b.** = *door-step b.

1909 J. R. Ware *Passing Eng.* 115/2 *Door-mat*, the name given by the people to the heavy and 'unaccustomed beards which the Crimean heroes brought home from Russia in 1855–56...By 1882 the term came to be applied to the moustache only. **1935** A. J. Cronin *Stars look Down* i. i. 15 Here..have a doormat, do.

door-step. Add: **b.** *slang.* A thick slice of bread.

1885 *Eng. Illustr. Mag.* June 604/2 'Doorsteps', I found, were thick slices of bread spread with jam. **1924** Lawrence & Skinner *Boy in Bush* iv. 55 Everybody..chewed huge doorsteps of bread. **1933** W. de la Mare *Lord Fish* 88 The door-step [proved to be] a slab of bread with a scrimp of margarine. **1959** *Times* 5 Nov. 13/6 There is..nothing exclusive about the childish use of..'door step' for a thick piece of bread. **1969** *Listener* 17 Apr. 533/3 Won't you slice me a doorstep please?

c. *attrib.* and *fig.* (phr. *on the* (or *one's*) *door-step*, near by).

1906 *Daily Chron.* 4 Jan. 4/1 Dr. Cooper's fight is in every respect a 'doorstep' affair. **1908** *Ibid.* 20 Feb. 3/5 All the prisoners concerned in the 'doorstep' campaign. **1909** *Westm. Gaz.* 25 Oct. 9/2 We still want doorstep workers. **1909** *Daily Chron.* 30 Dec. 3/4 The Christmas-boxes that custom decrees, are as follows...This..includes only the doorstep tributes. **1949** Blunden *After Bombing* 17 Japan's young children, staring shy..From mother's back Or door-step-side. **1952** *New Statesman* 29 Mar. 370/3 Most visitors brought doorstep sandwiches and huge home-made pasties to eat with their pint-pots of tea. **1957** A. Huxley *Let.* 8 Apr. (1969) 823 The ultimate revolution..is here on our doorstep. **1958** A. White tr. *Colette's Claudine in Paris* iv. 30 Just a few yards from here, there's a delightful flat, and we'd be practically on each other's doorsteps. **1959** *Listener* 26 Nov. 945/3 Mr. Conrad Aiken has called it '..a

classic right on the doorstep'. **1963** *Times* 1 Feb. 6/3 Lord Champion said hire-purchase commitments were often entered into through a stupid desire to keep up with the Joneses. This feeling was exploited by doorstep salesmen. **1970** J. Porter *Rather Common Sort of Crime* ii. 20 A wide experience of doorstep salesmen had taught her to examine life's doorstep offers with the utmost care.

door-yard. (Earlier and later examples.)

c **1764** in T. D. Woolsey *Hist. Disc.* (1850) 54 The Freshmen..are forbidden to wear their hats..in the front dooryard of the President's or Professor's house. **1913** R. Frost *Boy's Will* 9 How drifts are piled, Dooryard and road ungraded. **1941** T. S. Eliot *Dry Salvages* i. 7 The rank ailanthus of the April dooryard.

doosed, doosid: see Deuced *a.*

dooted, var. Doted *ppl. a.* 2.

doovah, doover, varr. *doofer.

dop, *sb.*[3] *S. Afr.* [Afrikaans.] **1.** In full *dop brandy.* Cape brandy, made from grape-skins.

1889 '*Argus' Ann. & S. Afr. Directory* 338/1 Thus we obtain 'Cognac', 'Dop', and 'Cango' brandy from spirits of wine. **1895** W. C. Scully *Kafir Stories* 18 Jim..got his daily number of tots of poisonous 'dop' brandy. **1896** *Johannesburg Weekly Times* 8 Aug. 8 Several samples of whiskies, Cape dop, and Cape brandy were examined. **1896** Baden-Powell *Matabeleland Camp.* 347 We just had sufficient 'dop' (Dutch brandy) to give everybody a tot in which to drink her health. **1901** *Westm. Gaz.* 11 May 5/1 A bottle of 'Dop'—or Cape gin. **1910** 'R. Dehan' *Dop Doctor* (1913) xiii. 98 'Dop', being the native name for the cheapest and most villainous of Cape brandies, has come to signify alcoholic drinks in general. **1921** *Chambers's Jrnl.* 647/1 The Cape 'dop' bottle brought oblivion to his tortured mind. **1950** L. G. Green *At Daybreak for Isles* i. 4 Dop brandy was to be had for sixpence a bottle.

2. A tot, especially of wine as given to farm labourers in the Western Province of the Cape.

1950 *Cape Times* (Week-end Mag.) 17 June 5/4 The pay of the *mailer* is good...Added to this, there is an occasional *dop* from both shebeener and customer. **1961** *Cape Argus* 8 Aug. 2/4 He wanted to buy a 'dop'.

dopa (dō^u·pă). *Chem.* and *Biochem.* Also **DOPA.** [a. G. *dopa* (B. Bloch 1917, in *Arch. f. Dermatol. und Syphilis* CXXIV. 132), f. the initial letters of the formative elements of *dioxyphenylalanine*, a former name of the compound.] 3,4-Dihydroxyphenylalanine, $C_9H_{11}NO_4$, a crystalline amino-acid which occurs naturally (not as a constituent of proteins) and is used in the treatment of Parkinsonism; in man it is a precursor of noradrenaline and of melanin, being formed by the oxidation of tyrosine in the nerves and adrenal medulla.

1917 *Jrnl. Chem. Soc.* CXII. i. 675 When a frozen section of..skin is treated with a 1‰ solution of 3:4-dihydroxyphenylalanine (termed 'dopa'), oxidation and condensation occur with the formation of a..pigment (*dopamelanin*).. owing to the action of an..enzyme, *dopaoxydase.* **1923** *Nature* 3 Nov. 675/1 The presence of 'dopa' (3,4-dioxyphenylalanin) in the cocoons of night-butterflies and saw-flies causes spontaneous formation of melanine when water is admitted. **1956** *Ibid.* 3 Mar. 430/2 The mechanism of formation of melanin pigments in insects may be as follows: Phenylalanine → tyrosine → dopa → dopaquinone.. → dopachrome, [etc.]. **1962** H. Burn *Drugs, Med. & Man* xi. 115 There are..substances which can pass through the blood-brain barrier..easily. One of these is the substance out of which noradrenaline is made. It is called 'dihydroxyphenylalanine, and is known as dopa for short. **1968** J. W. T. Dickerson in Davison & Dobbing *Appl. Neurochem.* ii. 61 The conversion of tyrosine to DOPA is probably mainly a non-enzymic one. **1970** *Nature* 4 Apr. 21/1 During the past decade a new approach to Parkinsonism has evolved, culminating in the introduction of L-dopa. **1970** *Daily Tel.* 8 Sept. 3/3 General practitioners in the London postal area can now prescribe L-Dopa for patients with Parkinson's disease.

dopamine (dō^u·pămīn). *Biochem.* [f. *dopa + Amine.] The immediate precursor of noradrenaline in the body, found esp. in nervous and peripheral tissue and formed by decarboxylation of dopa; 3,4-dihydroxyphenylethylamine, $C_8H_{11}NO_2$.

1959 Senoh & Witkop in *Jrnl. Amer. Chem. Soc.* LXXXI. 6222/1 This investigation concerns itself with related hydroxylation mechanisms of 3,4-dihydroxyphenethylamine ('dopamine'), a biogenic amine of key importance in metabolism. **1962** H. Burn *Drugs, Med. & Man* xi. 115 When dopa enters the brain the substance dopamine is..formed. **1965** *New Scientist* 18 Mar. 719/1 Three of the key monoamines are noradrenaline, dopamine and 5-hydroxytryptamine. **1969** *New Scientist* 30 Oct. 234/2 It is thought that if dopamine is absent, there is no way of checking the excitatory action of the ACh in the caudate nucleus, and that this.. results in tremor and rigidity of the muscles. **1971** *Nature* 7 May 54/1 The principal biochemical defect associated with Parkinsonism is a marked decrease in the dopamine and serotonin content of the brain.

dopant (dō^u·pănt). [f. *dop(e *v.* + -ant[1].] The substance used in doping a semiconductor.

1963 Seidman & Marshall *Semiconductor Fund.* vi. 86 The melt consists of intrinsic polycrystalline germanium or silicon mixed with suitable dopants such as phosphorus, which produces an n-type melt, or gallium, which produces a p-type melt. **1969** *New Scientist* 13 Mar. 569/2 The concentrations..of dopants used during the fabrication of the circuit. **1970** *Physics Bull.* Apr. 146/2 The electrical role of implanted dopants in semiconductors.

dope, *sb.* Add: **1.** (Further examples.)

1876 *Territorial Enterprise* (Virginia City, Nev.) 13 Feb., Nothing was known of the mysteries of 'dope'—a preparation of pitch which, being applied to the bottom of the shoes, enables the wearer to glide over snow softened by the warmth of the sun. **1947** R. Peattie *Sierra Nevada* 218 They treated the running surface of the skis with 'dope', corresponding to our modern downhill wax. **1954** E. Eager *Half Magic* v. 97 Today he..had a double hot fudge dope.

b. A preparation, mixture, or drug which is not specifically named (see quots.); = Stuff *sb.*[1] 6. *slang* (orig. *U.S.*).

1872 *Chicago Tribune* 24 Dec. 4/4 He..bids us beware of the sugar, for it is full of flour and sand;..of the milk, for it is compounded of dope. **1900** Flynt & Walton *Powers that Prey* 186 Give me some more o' that dope there—quick —I—I—*am*—dyin'. **1901** Ade *Forty Mod. Fables* 188 Give me some perfumed Dope that will restore a Peaches and Cream Complexion. **1912** F. A. Talbot *Moving Pictures* 53 (*caption*) Two small barrels..made all the film base, or 'dope', as it is called, in 1891. **1915** A. D. Gillespie *Let.* 4 July (1916) 222 The hay fever is better now, more because the season is passing than by reason of the doctor's 'dopes', I think. **1915** E. Poole *Harbor* 60 Joe's father vaccinated about a score of children that week. The 'dope' he used was mailed to him by a drug firm. **1924** Webster Add., *Dope*...5. *Photogr.* A varnish used to facilitate retouching, block out portions of a negative, etc. **1928** *Sunday Dispatch* 8 July 9/5 'Dope finish' [is a slang expression] for face paint and powder.

c. A varnish applied to the cloth surface of aeroplane parts, in order to increase strength and to keep them taut and air-tight. Also, a liquid preparation applied to air-ship coverings, to increase gas-tightness.

1912 *Aeroplane* 19 Dec. 607 Cellon...The Fabric Dope used by the leading British and Continental Aeroplane and Hydro-aeroplane Builders. **1916** H. Barber *Aeroplane Speaks* 142 *Strut, Dope*, a strut within a surface, so placed as to prevent the tension of the doped fabric from distorting the framework. **1917** *Times* 1 June 9/5 The King and Queen ..went on through..the seaplane department, and the 'dope' room. **1918** A. Barr in A. J. Swinton *Aeroplane Handbk.* (1920) 144 The Germans at the beginning of the war used acetate dope and covered it with transparent oil varnish. **1930** *Flight* 17 Jan. 134/2 Dope is a solution of cellulose esters in various solvents and diluents. **1967** B. Robertson *Aircraft Markings* 14 The French had used a sky blue dope early in the war at the time when French uniforms were of horizon blue.

d. A substance added to petrol or other fuel, etc., to increase its efficiency; an additive.

1930 *Engineering* 11 July 31/1 A lubricant such as oleine ..or oleic acid, when added to a fuel containing metallic dope, had the effect of partially inhibiting the detonation-delaying action of the dope. **1937** *Times* 13 Apr. p. xix/4 Dopes both for petrol and for lubricating oils are now well known.

3. a. *colloq.* (orig. *dial.*). A stupid person, a simpleton, a fool. Also (*U.S. slang*), a person under the influence of, or addicted to, some drug (see sense *3 b).

1851 *Gloss. Words Cumberland* 8 *Dope*, a simpleton. **1866** E. L. Linton *L. Lorton* II. i. 17 A 'downo-canno dope'—which meant a spiritless simpleton. *Ibid.* III. ii. 39 She was..'a dozened lile dope'. **1909** *Sat. Even. Post* 1 May 5/1 He's an old dope, which he don't look like, or he's on. **1925** Fraser & Gibbons *Soldier & Sailor Words* 81 *Dope*, ..a fool. **1925** N. Venner *Imperfect Impostor* ii. 17, I am just the humble dope; the clay pigeon; the simple-minded confederate. **1928** E. Wallace *Flying Squad* xiv. 121 How many times have I told you not to frighten that girl, you poor dope? **1938** R. E. Sherwood *Idiot's Delight* i. 17 That wasn't an officer, that was a porter, you dope! **1948** *Chicago Daily News* 11 June 16/4 Cold tea sold to night life dopes for brandy at 75 cents a throw. **1957** J. Braine *Room at Top* 48 'Who's Alice?' I asked. 'You've met her, you dope.' **1959** P. Capon *Amongst those Missing* 108 Silly dope, he can't go on dodging the Court for ever.

b. 'Opium, especially the thick treacle-like preparation used in opium-smoking' (Cent. Dict. Suppl. 1909); hence applied to stupefying drugs and narcotics in general, or to alcoholic drink. *slang* (orig. *U.S.*). (See also sense *5.)

1889 *Kansas Times & Star* 8 Oct., The oldest of the trio, an Irishman from County Cork, was very hilarious...The 'dope' made him 20 years younger and very pugnacious. **1891** T. W. Knox in H. Campbell *Darkness & Daylight* (1892) xxviii. 570 The opium used for smoking—called by the smokers 'dope'—is an aqueous extract of the ordinary commercial gum. **1894** J. L. Ford *Lit. Shop* ix. 130 Opium-joints—those mysterious dens in which..the fumes of the burning 'dope' cloy the senses. **1896** Ade *Artie* viii. 75, I would advise you to stop smokin that double X brand of dope. **1896** *Chautauquan* Oct. 60/1 In San Francisco, large confiscations of 'dope' are made nearly every week. **1920** *Outward Bound* Oct. 38/1 The death of a fascinating actress from an overdose of 'dope'. **1922** *Public Opinion* 5 May 420/3 The cabarets in Constantinople are a meeting place for all the world's dope purveyors.

fig. **1937** *Ess. & Stud.* XXII. 146 Poetry was comparatively popular because it was dope, though a mild and fairly harmless dope like cigarette-smoking.

c. A medical preparation administered to a race-horse for the purpose of affecting its performance.

1900 *Westm. Gaz.* 30 Oct. 7/3 It is administered in capsules, given in a gelatine of varying thickness according to the required time when the 'dope' is desired to take effect. **1913** *Badm. Mag.* Jan. 88 A dope proper may be administered [to a horse] as a powder laid on the tongue, as a drink—usually given in old ale—or by the hypodermic syringe.

d. [Perh. arising from the ambiguity of the abbrev. *coke* = (*a*) cocaine, (*b*) Coca-Cola.] Coca-Cola or some other carbonated drink. *local U.S. slang.*

1915 *Printer's Ink* 23 Sept. 46/2 The propensity of a large portion of those who regularly drink Coca-Cola to call for their favorite drink as 'dope' or 'coke' or 'koke'. **1940** C. McCullers *Heart is Lonely Hunter* (1943) ii. i. 83 If she had any money she bought a dope or a Milky Way at Mister Brannon's. **1947** F. G. Patton in *55 Short Stories from New Yorker* (1952) 224 Want a dope?..I tried to call them cokes once.

4. Information, esp. on a particular subject or of a kind not widely disseminated or easily obtained; (a statement of) facts or essential details; also, information, a statement, etc., designed to gloss over or disguise facts; flattering or misleading talk. *slang* (orig. *U.S.*).

1901 'H. McHugh' *John Henry* 77 I've known Tommy for a long time, so he feels free to read his dope to me. **1904** W. H. Smith *Promoters* iii. 72 Unless you cover your dope with a sort of angel-of-light coating that would fool Gabriel himself, you'll never get there. **1905** R. Beach *Pardners* (1912) i. 26 He handed me the dope: 'In re Olive Troop Morrow *vs.* Justus Morroe'. **1910** S. E. White *Rules of Game* (1913) ii. iii. 126 Gosh! I get sick of handing out dope to these yaps. **1914** Jackson & Hellyer *Vocab. Criminal Slang* 8 The most popular slang term in use today in the unregenerate world—'dope'..signifying 'news', 'intelligence', or 'meaning'. **1915** Froest & Dilnot *Crime Club* xii. 290 Don't pull any of that dope on me. **1917** A. G. Empey *From Fire Step* 103, I was to send the dope to Cassell and he would transmit it to the Battery Commander as officially coming through the observation post. **1919** *Detective Story Mag.* XXVIII. i. 13, I suggested it to a detective, but he laughed at me and said the article was nothing but 'dope'. **1920** *Public Opinion* 9 July 43/3 Your patent dialectic dope By gulps we take with zest. **1920** Wodehouse *Damsel in Distress* xx. 235, I could help you there. I've got the thing down fine. I've got the infallible dope! **1921** *19th Cent.* May 748 He does not quite believe that the Bolshevik leaders themselves believe in their doctrines. He strongly suspects that on their part it is mainly 'dope'. **1922** H. L. Foster *Adv. Trop. Tramp* x. 140 Run up to Bolivia and get the dope on this affair. **1935** Auden & Isherwood *Dog beneath Skin* i. ii. 34 He had all the dope about the Army Contracts trial. **1945** A. Christie *Sparkling Cyanide* iii. x. 136, I shouldn't dream of..denying it. You've obviously cabled to America and got all the dope.

5. *attrib.* and *Comb.*, as (sense *3 b) *dope-addict, -dream, -merchant, -seller, -smuggler, -smuggling, -traffic, -trafficking*; **dope-book** *U.S. slang*, a book of information on any subject; also *fig.*; **dope-fiend** *slang*, orig. *U.S.*, a drug-addict; **dope-pedlar** (also **-peddler**), **-runner**, a seller of illicit drugs; hence *dope-peddling, -running* vbl. sbs.; **dope-ring**, a group of people engaged in obtaining, selling, and using narcotics; **dope-sheet** *U.S. slang*, a sheet of paper bearing information or instructions, esp. written or printed information about race-horses; **dope-shop**, a place where dope (sense *1 c) is applied to aeroplanes, etc.; **dope-stick** *slang* (examples are *N.Z.*), see quot. 1918.

1933 D. L. Sayers *Murder must Advertise* xi. 188 A dope-addict's dream. **1909** *Cent. Dict. Suppl.*, *Dope-book*, a miscellaneous collection of racing information. **1918** *Amer. Mag.* Aug. 38/2 The one that invents the dope book on the female race, and the bird that holds a patent on the complete understanding of human nature. **1908** U. Sinclair *Metropolis* 137 What will people think..seeing you sitting there like a man in a dope dream? **1934** H. G. Wells *Exper. Autobiogr.* II. ix. 821, I had expected to find a new Russia stirring in its sleep..and I found it sinking deeper into the dope-dream of Sovietic self-sufficiency. **1896** *Sun* (N.Y.) Dec., 'A dope fiend'...A victim of the opium habit. **1914** R. H. Davis *With Allies* (1915) 158 With the desperation of a dope fiend clutching his last pill of cocaine. **1963** *Punch* 20 Mar. 425/2 Mr. Trocchi's dopefiends. **1921** *Outward Bound* Apr. 44/2 The policy of the 'dope merchant'..would issue in sleep, death and fossilisation. **1923** *Jrnl. Amer. Inst. Crim. Law & Criminol.* Aug. 292 No other group comes closer to the bootlegger or the dope peddler. **1933** D. L. Sayers *Murder must Advertise* xii. 217 You must be the dead spit of some habitual dope-peddler. **1937** A. J. Cronin *Citadel* 388 He's nothing but a sleek dope peddler. **1959** *News Chron.* 19 Aug. 6/3 The bookies, the dope pedlars and bootleggers. **1959** P. Townsend *Died o' Wednesday* x. 177 Malosti's current livelihood, the dope-peddling organisation. **1959** F. Usher *Death in Error* ix. 143 Nothing worse than a little mild dope peddling goes on. **1929** D. Hammett *Dain Curse* (1930) iii. 24 What do you think he is?..Head of a dope ring? **1930** *Daily Express* 6 Nov. 19/4 Gangsters, dope rings, tough guys. **1957** P. Moore *Science & Fiction* 100 Veterans whose sole mission in life was to uncover dope-rings led by unscrupulous Martians. **1933** D. L. Sayers *Murder must Advertise* xiv. 238 He said you were a dope-runner. **1941** A. Christie *Evil under Sun* xi. 206 The result of her getting mixed up..with the dope-running stunt. **1953** K. Tennant *Joyful Condemned* xxxi. 303 You used to know all about his dope-running. **1930** *New Statesman* 27 Dec. 356/2 He was using the money to finance a gang of dope-sellers. **1903** Ade *People you Know* 111 When he arrived at the Track he gave up for a Badge and a Dope-Sheet and a couple of Perfectos. **1959** Halas & Manvell *Technique Film Animation* xix. 189 His work book..must next be elaborated on a different set of printed sheets,

generally known as camera-exposure charts or, in the U.S.A., as dope sheets. **1933** *Aeroplane* 26 Apr. 732/1 When the machine reaches the end of the line it is ready to go straight to the dope shop. **1937** D. & H. Teilhet *Feather Cloak Murders* ii. 43 Wonder if your friends..know you're so companionable with a dope smuggler, eh? **1941** A. Christie *Evil under Sun* xi. 215 The whole case hinges on dope smuggling. **1969** M. G. Eberhart *Message from Hong Kong* xii. 184 If it's dope smuggling it's too big for me. **1918** *N.Z. at Front 1918* 132 Cigarettes! smokes, fags, weeds, dope-sticks—they are known by many strange names. **1930** W. Smyth *Wooden Rails* xi. 168 Pass me those dope-sticks, there's a dear. **1933** J. B. Priestley *Wonder Hero* iii. 95 He wondered which of the queer fellows..hanging about street corners were in the dope traffic. **1933** D. L. Sayers *Murder must Advertise* xv. 249 The crookedness of dope-trafficking. **1954** J. Symons *Narrowing Circle* xii. 47 Dinmock had begun as an agent of the Barcini brothers, who got a seven-year stretch for dope trafficking.

dope, *v.* orig. *U.S.* [f. Dope *sb.*]

1. *trans.* To administer dope to (a person, a horse); to stupefy with a drug; to drug.

1889 Barrère & Leland *Dict. Slang* s.v., Doping is the stupifying men with tobacco prepared in a peculiar way... Nine out of ten saloons in the slums employ doping as a means to increase their illicit revenue.—*American Newspaper.* **1891** Farmer *Slang* II. 309/2 *Dope v.* (American), to drug with tobacco. Also *Doping* = the practice. **1900** *Westm. Gaz.* 23 Oct. 8/2 They urge a liberal investment on the American horse, and confidentially impart the information that the animal is 'doped'. **1902** H. L. Wilson *Spenders* xxxv. 429 That guy that doped me, he wa'n't satisfied with my good thirty-dollar wad. **1915** C. S. Jones *Hohenzollern* 169 The King (who, if we are to believe his fair companion, used to be liberally doped) would awake from a trance at a *séance* to find his *inamorata* in his arms. **1919** H. L. Wilson *Ma Pettengill* ii. 62 To say nothing of doping him with asperin and quinine and camphor and menthol and hot tea and soothing words. **1919** G. Page *Veldt Trail* xix. 200 'They must have kept him [*sc.* a horse] pretty short of food.'..'Or doped him,' suggested Birkdale. **1922** *Blackw. Mag.* May 632/2 He ladled into his nose sufficient snuff to have doped an ordinary man. **1955** *Times* 8 June 4/5 He had heard of greyhounds being doped, but not to make them run faster.

fig. **1908** G. H. Lorimer *J. Spurlock* iii. 41, I was so doped with my siren song that I steered straight for the rocks. **1928** G. B. Shaw *Intell. Woman's Guide Socialism* l. 218 Boycotting the Churches as mere contrivances for doping the workers into submission to Capitalism.

b. *intr.* To take or be addicted to drugs.

1909 in *Cent. Dict.* Suppl. **1926** N. Lucas *London & its Criminals* xviii. 253 Doping is so prevalent in the East that the traffic in drugs appeals naturally to the Eastern crook. **1933** D. L. Sayers *Murder must Advertise* v. 85 Did he dope?

2. To treat with an adulterant, etc.; to 'doctor'.

1898 *Let.* in *Congress. Rec.* App. 223/1 They will run their flutter mills and mixers, and dope the flour to suit themselves. **1906** in Asher & Heal *Send no Money* (1942) 95 You would like to know just where you are at before you 'dope' that kind of stock with a strange mixture. **1913** *Sunday Times* (Trenton, N.J.) 2 Mar. 1 Alternative offered to the water drinkers of Trenton: Typhoid if the water isn't 'doped' with hypochlorite of lime; an itch if it is.

b. *Electr.* To add an impurity to (a semiconductor) to produce a desired electrical characteristic.

1955 F. E. Terman *Electronic & Radio Engin.* (ed. 4) xxi. 753 The degree of doping is commonly expressed in terms of the conductivity (or resistivity) of the resulting material. **1956** L. P. Hunter *Handbk. Semiconductor Electronics* vii. 7 The melt is doped with a donor (arsenic) so as to produce a low-resistivity N-type emitter. *Ibid.* vi. 26 A uniformly doped crystal will be grown. **1962** Simpson & Richards *Junction Transistors* iii. 43 The distance that the barrier extends into the semiconductor is also smaller in the case of a heavily doped semiconductor. **1970** *Sci. Amer.* Feb. 22/3 By suitable masking and 'doping' techniques, which selectively altered the electrical behavior of small regions, several score transistors could be created on each wafer.

3. To smear, daub; *spec.* to apply 'dope' to (the outer fabric of an aeroplane or the like).

1868 *Putnam's Mag.* II. 363 With their snow-shoes thoroughly 'doped', the crowd resort to some suitable place for the contest, which begins with a grand dash. **1902** *Dialect Notes* II. 233 *Dope, v.* tr. 1. To smear, or lubricate. 2. To put salve on a wound. **1916** *Sphere* 25 Mar. 321/1 The wings are finished,..and the whole 'doped' and ready. **1917** *Times* 1 June 9/5 Some hundreds of girls were engaged in 'doping' the fabric which covers the planes, rudders and ailerons. **1922** *Encycl. Brit.* XXX. 59/1 [Airship.] Cotton.. appears to have some advantages owing to its great uniformity of contraction when doped. **1928** C. F. S. Gamble *N. Sea Air Station* 10 The envelope was 'doped' with aluminium paint on its top side and with yellow pigment on the lower portions. **1935** H. L. Davis *Honey in Horn* 12 There was a salad of lettuce whittled into shoestrings..and doped with vinegar and bacon grease. **1941** *Illustr. London News* CXCIX. 198 (*caption*) A rigger dopes fresh patches of fabric over the muzzles of the guns to assist in streamlining and speed until the aircraft opens fire.

4. *to dope out.* **a.** To make out; to find out, discover; to get the truth about.

1906 'O. Henry' *Four Million* (1916) 161 All the same, I believe it was the hand of Fate that doped out the way for me to find her. **1913** E. D. Biggers *7 Keys to Baldpate* ii. 31 Ther's something I haven't quite doped out. That is —who's trespassing, me or you? **1914** R. H. Davis *With the Allies* (1915) 10 We would study the morning papers and ..from them try to dope out the winners. **1918** C. Wells *Vicky Van* xv. 180, I dope out all this has to be so. **1919** Wodehouse *Damsel in Distress* ii. 30 Nature had it all doped out for me to be the Belle of Hicks Corners. **1922** *Short Stories* Feb. 102/2 'How'd you dope it out, Kid?' asked one. 'Tell us how you could do such good detective work.' **1926** J. Black *You can't Win* xxiii. 367, I had all the criminal lawyers in San Francisco doped out like race horses by this time. **1943** N. Marsh *Colour Scheme* v. 94 Uncle James dopes it out that it's been Questing's idea to get this place on his own.

b. To work out; to get hold of.

1906 'O. Henry' *Four Million* (1916) 191 He's doped out a fifty-dollar bill, anyway. **1919** *Detective Story Mag.* XXVIII. i. 6 He might have doped out a corking yarn about how the Phantom..made everybody believe that the explosion killed him. **1922** *Short Stories* Feb. 175/2, I didn't have time to dope out any plot till I got back to Denver.

Hence **doped** *ppl. a.*; **do·ping** *vbl. sb.* and *ppl. a.* Also **do·per**, one who dopes; one who administers or takes drugs.

1889 [see *dope *v.* 1]. **1900** *Daily News* 14 Nov. 8/4 'Doping' meant the administration to a horse of certain medical preparations, with the object of either stimulating or retarding the animal's progress in a race. **1903** *Westm. Gaz.* 15 Sept. 5/2 The Jockey Club propose to pass a rule forbidding doping in any shape or form. **1903** *Sun* (N.Y.) 23 Nov. 12 A 'doped' cigar was given to him in a pool and billiard room, and it had pretty near the same effect on him as knockout drops. **1913** *Daily Mail* 11 Mar. 7/5 'Doped' Athletes... It is well known..that the Russian skaters take such stimulants [as strychnine]. **1913** *Badm. Mag.* Jan. 89 A doped animal will run till it is done, mad drunk with the drug. *Ibid.*, A regular doper always tries his horses at home. **1913** *Aeroplane* 13 Mar. 303/1 The shiny surface of well 'doped' wings. **1914** *Racing Calendar* 23 July, I was unable to obtain evidence of the presence of a doping agent. **1921** *Blackw. Mag.* Apr. 535/2 She rejoined the remainder of the passengers.., having in her hands a doped handkerchief. **1922** *Ibid.* Jan. 124/1 He might publish in his doped and venal press the felicitations that were showered upon him. **1923** Wodehouse *Inimitable Jeeves* xiv, To tell me we ought to cook Harold's food ourselves to prevent doping. **1926** *Chambers's Jrnl.* 17 July 515/1 So the old reprobate's dopings had told at long last. **1926** [see *dope *v.* 1 b]. **1935** *Discovery* July 208/2 The 'doping' of the sample water with 'O.T.' is carried out automatically and with perfect precision. *Ibid.*, The colour of the doped solution is intermediate between those of the two permanent solutions. **1955** *Times* 7 June 3/3 A Harley Street specialist was one of three men who were alleged at the Central Criminal Court yesterday to have been concerned in a conspiracy concerning the doping of greyhounds. **1956, 1970** [see *dope *v.* 2 b].

dopester (dō^u·pstəɹ). *slang* (orig. *U.S.*). [f. Dope *sb.* + -ster.] **1.** One who collects information on, and forecasts the result of, sporting events, elections, etc. Cf. *dope *sb.* 4.

1907 *Cosmopolitan Mag.* Feb. 363/1 As we talked on a corner not long ago, a Dopester..stepped up to us. **1916** *All-Story Weekly* 25 Mar. 283 The dopesters had looked up those two coast fights, and that ended the betting. **1921** *Daily Colonist* (Victoria, B.C.) 2 Apr. 10/3 America is in line for the bulk of the Davis Cup matches this summer according to the dopesters who put their pencils to work right after the recent draw. **1964** *Economist* 25 Apr. 396/1 The inside dopesters, squeezing the latest gossip about intra-party machinations out of politicians.

2. One who sells, uses, or is addicted to, drugs.

1938 J. Rice *Somers Inheritance* i. vi. 45 A dopester seldom drinks and most drinkers have not yet taken to dope.

dopey (dō^u·pi), *a. slang* (orig. *U.S.*). Also **dopy.** [f. *dope *sb.* 3 b.] **1.** Sluggish or stupefied, with or as with a drug.

1896 *Sun* (N.Y.) Dec., A man who acts as if under the influence of the poppy drug is said to be dopy. **1900** Ade *More Fables in Slang* (1902) 179 A Young Man with Hair who played the 'Cello. He was so wrapped up in his Art that he acted Dopey most of the time. *a* **1909** Buck *Handbk. Med. Sci.* VI. 686 (Cent. Dict. Suppl.), The patients are said to be 'dopey'; they are markedly prostrated, indifferent to their surroundings, and want only to be left undisturbed. **1914** R. Brooke *Let.* Aug. (1968) 607 The general uneasiness..seems to take all the strength out of me. I feel 'dopey'. **1921** B. Matthews *Ess. English* 112, I began to hear men assert that they felt *dopy, i.e.* sluggish, as though they had taken an opiate. **1924** A. J. Small *Frozen Gold* iii. 97 Henderson was still a bit dopey [after a blow on the head]. **1931** L. Golding *Magnolia St.* iii. vi. 546 'Bella, you look all dopy! What's wrong?'..Her eyes took on a far focus. **1932** J. B. Priestley *Dangerous Corner* ii. 56, I took three of those tablets I have to make me sleep and now I feel absolutely dopey. **1957** E. Eager *Magic by Lake* 71 The four children..went on being dopey and droopy and sleepy all afternoon when they *did* get up.

2. Stupid, 'dumb'.

1896 *Cincinnati Enquirer* 2 Aug. 2/1 There is an impression of truth to the rather 'dopy' proposition that makes it worthy of newspaper space. **1939** S. V. Benét *Tales before Midnight* 150 'He said I was a little angel.' 'Was he dopey!' said her elder, blightingly. **1947** N. Marsh *Final Curtain* v. 81 Am I dopey? I tell you I *heard* her. **1957** J. Braine *Room at Top* 119 It's quite unmistakable, that look—a sort of dopy joyfulness. **1957** I. Cross *God Boy* (1958) xv. 118 'I'm off home—I don't feel like it,' I said. 'Don't be dopey,' said Joe. **1963** H. Garner *Best Stories* 227 Step began to laugh. 'That dopey foreman. He didn't bother to check with me.'

dopiness (dō^u·pines). *N.Z.* [f. *dopey *a.* + -ness.] A deficiency disease of sheep.

1932 *Discovery* Nov. 357/2 Recent investigations in New Zealand testify to..the value of top dressing of pastures with lime and superphosphate where sheep are affected with dopiness diseases. **1934** L. J. Wild *Soils & Manures in N.Z.* (ed. 3) v. 72 The sheep disease called 'dopiness' occurs notably in the Mairoa district. **1950** *N.Z. Jrnl. Agric.* July 33/3 Some green, succulent feed just before and at lambing assists in the prevention of diseases associated with lambing —dopiness (twin lamb disease or ante-partum paralysis) and milk fever.

‖ **doppelganger, dopple-, -gänger** (dǫ·pəlgæ:ŋəɹ, -ge·ŋəɹ). [See Double-ganger.] = Double-ganger 1.

1851 M. A. Denham in *Denham Tracts* (1895) II. 79 Hell-hounds, dopple-gangers, boggleboes. **1879** C. L. Dodgson *Euclid & Mod. Rivals* i. ii, Are their Doppelgänger available? **1907** N. Munro *Daft Days* xxviii, Miss Macintosh is surely your doppelganger. **1940** M. Lowry *Let.* 7 May (1967) 30 It may well be that you will observe my little doppelgänger poltergeist soul hoisting a drink in a bar in them parts. **1952** C. Day Lewis tr. *Virgil's Aeneid* x. 228 Not knowing that what so thrilled him was only a *doppelgänger.*

Dopper[3] (dǫ·pəɹ). *S. Afr.* Also **dopper.** [Afrikaans, of uncertain origin.] The sobriquet of a member of the 'Gereformeerde Kerk in Suid-Afrika', a strictly orthodox Calvinistic denomination, commonly regarded as being old-fashioned in ideas, manner, and dress. Hence **Do·pperdom** (*rare*), the Doppers collectively.

1850 N. J. Merriman *Jrnl.* 9 Dec. in *Kafir, Hottentot, & Frontier Farmer* (1854) 96 These Doppers are a sort of Dutch Church puritans. **1859** *Queenstown Free Press* 4 May (Pettman), The *Doppers* would not be satisfied with an angel from heaven. **1881** [see Dopper[2]]. **1898** W. Harcourt *Let.* 29 Aug. in A. G. Gardiner *Life W. H.* (1923) II. xxiii. 461 It is with a view to this that A. M. [*sc.* A. Milner] wants a display of more force..to 'convince Dopperdom that *England means war*', if Kruger does not do our bidding. **1900** *Daily News* 13 Feb. 7/2 The burghers being chiefly of the 'dopper' or back-country class. **1926** *Brit. Weekly* 27 May 158/1 The charming Dutch girl lived with her old Dopper father on a nearby farm. **1958** *Cape Times* 27 Dec. 3/6 The old dopper town of Krugersdorp.

doppie (dǫ·pi). *S. Afr.* [dim. of *dop *sb.*[3]] A grape-skin.

1948 *Cape Times* (Week-end Mag.) 24 Jan. 12/7 (Advt.), A Grape Crusher and two Wine and Doppies Pumps manufactured by Consani's for the Wine Industry. **1955** J. Packer *Valley of Vines* i. 13 If you fell into a tank among the *doppies*—the skins—and the wine, you died.

Doppler (dǫ·pləɹ). Also **doppler.** The name of C. J. *Doppler* (1803–53), Austrian mathematician and physicist, used *attrib.* or in the possessive to designate an effect first explained by him in 1842 (in *Abh. d. k. böhm. Ges. d. Wiss.* (1843) 5th ser. II. 465–82) and other phenomena related to it or caused by it, as *Doppler broadening* (of spectral lines), *Doppler('s) principle*; **Doppler effect,** the effect on sound, light, or other waves of relative motion between the source of the waves and the observer: the observed frequency of the waves is higher or lower than the emitted frequency according as the source (or the observer) is moving towards or away from the observer (or the source); **Doppler shift,** the change of frequency resulting from the Doppler effect. Also in equipment or procedures utilizing the Doppler effect, as *Doppler navigation, radar.*

1871 E. Atkinson tr. *Ganot's Physics* (ed. 5) v. i. 165 When a sounding body approaches the ear, the tone perceived is somewhat higher than the true one; but if the source of sound recedes..the tone perceived is lower. The truth of this, which is known as Doppler's principle, will be apparent from the following considerations. **1905** E. C. C. Baly *Spectroscopy* xii. 382 In order to see if the rapidly moving particles in the streamers caused any displacement in the position of the lines in the spectrum on what is known as the Doppler effect. **1926** H. Macpherson *Mod. Astron.* 34 The application of Doppler's principle to the measurement of the rotation and atmospheric motions of the Sun. *Ibid.* 119 The Doppler principle was first applied to the study of stellar motions as far back as 1868. **1927** A. S. Eddington *Stars & Atoms* 75 Owing to the Doppler effect a moving atom absorbs a rather different wave-length from a stationary atom. **1945** *Electronic Engin.* XVII. 583 The sound of an aeroplane flying low is heard approaching, passing overhead with the attendant 'Doppler' effect, or apparent change in engine speed, then dying away into the distance. **1953** *Ibid.* XXV. 113/1 The doppler frequency of a 3·5 cm radar detecting a target of 400 m.p.h. radial velocity. **1955** *Sci. Amer.* Sept. 136/3 Special sensitive equipment makes it possible to measure the amount of this Doppler shift even though the track may last only a fraction of a second. **1956** *Nature* 18 Feb. 299/2 This spectral line enables not only the spatial distribution of hydrogen in the galaxy to be studied, but also the velocity by the Doppler displacement. **1958** *Oxf. Mag.* 8 May 418/1 Atomic beams..were used to reduce the 'Doppler width' of spectral lines. **1959** J. L. Nayler *Dict. Aeronaut. Engin.* 84 A Doppler radar unit is used to obtain a running record of the speed of guided missiles. **1960** *Gloss. Terms Telecomm.* (*B.S.I.*) 266 *Doppler radar*, a form of radar in which the radio Doppler effect is used to determine the radial component of the velocity of a target relative to the radar system. **1963** G. Troup *Masers & Lasers* (ed. 2) iii. 38 The processes of spontaneous emission broadening, Doppler broadening and pressure broadening. **1966** *Economist* 16 Apr. 253 Marconi..has supplied a complete airadio system (comprising communications, approach navigation, Doppler navigation and automatic direction-finding equipment). **1966** *McGraw-Hill Encycl. Sci. & Technol.* I. 164/1 Airborne radar has been used for determining drift through the Doppler principle. **1967** *Listener* 27 Apr. 544/3 In the case of the red shifts in the light from the extra-galactic nebulae we interpret these as a doppler effect. **1967** *Times Rev. Industry* Feb. 51/3 Air India has placed an order worth nearly £250,000 with the Marconi Company for Doppler navigators to fit its fleet of Boeing 707s.

Hence **Doppler-shift** *v. trans.*, to cause a Doppler shift of.

1971 *Nature* 15 Jan. 158/2 The frequency of the light undergoing Rayleigh scattering by molecules is Doppler shifted by thermal motions.

dopy, var. *dopey *a.*

dor, *sb.*[1] **4. dor-bug** (earlier example).

1833 A. Greene *Life & Adv. D. Duckworth* I. 86 It's a dorbug!

Dora (dō^ə·rā). Also **D.O.R.A.** A jocular personification of the '*D*efence *o*f the *R*ealm *A*ct', the name being an acronym forming a familiar feminine proper name. The Act was first passed in August 1914 and provided the British Government with wide powers during the 1914–18 war.

1917 G. B. Shaw *London Mus. 1888–89* (1937) 382 The sixpence that went as tax to the Government, which might have stopped the performance by virtue of Dora, and didn't. **1918** 'I. Hay' *Last Million* vi. 68 We are up against official secrets again. A lady called Dora: you will become well acquainted with her. **1921** *Punch* 13 Apr. 293/2 To judge by his description, *Dora's* daughter [*sc.* the Emergency Powers Act of 1920] will be not a whit less drastic in her action than the old lady herself. **1920** P. Guedalla *Missing Muse* 194 These imbecile restrictions, which are the last legacy of 'Dora' to her grateful heirs. **1933** Lloyd George *War Mem.* I. vi. 177 This was the third edition of d.o.r.a., and was designed to give very greatly increased powers to the Authorities to secure munition production. **1952** H. Nicolson *King George V* viii. 112 In spite of D.O.R.A., the principles of Magna Carta were affirmed.

Dorcas[1]. Add: (Earlier example.) So *Dorcas meeting.* Also *ellipt.* Hence **Do·rcas** *v. intr.*, to work for a Dorcas society (*colloq.*).

1832 F. Trollope *Dom. Manners* I. xi. 154 Yet..spite of the old women and their Dorcas societies, atheism is awake and thriving. **1863** A. D. Whitney *F. Gartney* xix, There was great putting of heads together at the 'Dorcas', about it. **1866** Geo. Eliot *F. Holt* I. v. 120 Some prime Miss,.. in which Felix was no more interested than in Dorcas meetings. **1894** S. Baring-Gould *Queen of Love* II. xx. 40 Aunt Beulah will not be home for an hour. She is Dorcasing. **1900** A. Upward *Eben. Lobb* 39 Getting ready for a Dorcas that came to our house once a quarter to have tea and work for the heathen.

Dorcas[2] (dǭ·ɹkăs). [mod.L. (a. the specific epithet of *Gazella dorcas*, once used as a generic name in its own right), ad. Gr. δορκάς deer, gazelle.] In full *Dorcas gazelle.* A small gazelle, *Gazella dorcas*, found in northern Africa and western Asia.

1821 J. E. Gray in *London Med. Repos.* XV. 307 Fam[ily] 4. Antilopidæ...Gazelle, Dorcas. Antilope Dorcas. **1860** *Proc. Zool. Soc.* 415 Additions made to the Menagerie...1 Dorcas Gazelle, *Gazella dorcas.* **1893** R. Lydekker *Horns & Hoofs* 180 The last of the Asiatic gazelles in which the females are horned is the Dorcas gazelle. **1932** H. C. Maydon et al. *Big Game Shooting in Afr.* vii. 177 We had passed a belt that held Addra Gazelle, and now Dorcas and Rufifrons were very common. *Ibid.* 181 Herds of Addra and Dorcas Gazelle, and an occasional Ostrich wandered amid the ravines. **1970** Dorst & Dandelot *Field Guide Larger Mammals of Afr.* 242 Dorcas Gazelle live in small herds up to 20 head.

‖ **doré** (dore), *a. Metallurgy.* [Fr., lit. 'gilded'.] Containing gold.

1887 J. A. Phillips *Elem. Metall.* (ed. 2) 757 Doré silver, when it contains copper, is cupelled with lead..before parting. **1933** Liddell & Doan *Princ. Metall.* x. 242 If bullion contains predominantly silver with a little gold, it is known as 'doré bullion'.

Doric. Add: **B.** *sb.* **3.** *Typogr.* (See quots.)

1857 *Spec. Printing Types* (*H. W. Caslon & Co.*), Pearl Doric, No. 2...Brevier Doric, No. 1...Nonpareil Doric, No. 1...Pearl Doric, No. 1. **1888** C. T. Jacobi *Printers' Vocab.* 33 *Doric fount*, a particular kind of sans-serif type used for display work. **1900** T. L. De Vinne *Pract. Typogr.* I. 325 Specimen No. 3, usually called doric, is really a combination of a thick-faced roman and antique. **1954** *Archit. Rev.* CXVI. 119/1 Ionic..seems to have been invented by the type-founders in contrast to *Doric*, an early name for sans [serif].

‖ **dorje** (dǭ·ɹye). [Tibetan.] A representation of a thunderbolt in the form of a short double trident or sceptre, held by lamas during prayers.

1882 *Encycl. Brit.* XIV. 501/1 In the chief of these temples is preserved the famous *Dorjé* of Buddha, *i.e.*, the *Vajra* or Thunderbolt..the symbol of the strong and indestructible, which the priest grasps and manipulates in various ways during prayer. From this *dorjé*, according to one etymology at least, comes the name of the Himalayan sanatarium *Dorjiling* or Darjeeling. **1939** M. Pallis *Peaks & Lamas* xiii. 148 Others held a handbell..in the left hand and a *dorjé* or thunder-bolt sceptre in the right, the pair which symbolizes the marriage of Wisdom and Method.

dorm[2] (dǭɹm). Colloq. abbrev. of Dormitory *sb.* 1.

1900 *Dialect Notes* II. 17 The student..lives in the *dorm* (dormitory). **1904** Wodehouse *Gold Bat* xiv. 153 It went into Rigby's dorm. So it must have been a chap in that dorm. who did it. **1927** A. MacDonald *Dorty Speaking* iii. 22, I found Midge in the dorm when I went up to change for tea. **1936** A. Huxley *Eyeless in Gaza* vi. 65 It was against the school rules to go up into the dorms during the day. **1957** *Times Lit. Suppl.* 15 Nov. p. ii/1 Much of it is a sort of smouldering defence of the original Tom Brown ritual: dorm feasts, fagging, hero worship, [etc.]. **1969** C. Davidson in Cockburn & Blackburn *Student Power* 342 Female students are permitted to determine how strict or 'liberal' their dorm hours might be.

dormancy. Add: *spec.* of seeds and plants (see quot. 1929).

1911 *Bot. Gaz.* LII. 455 Fawcett obtained similar results with many kinds of weed seeds, the percentage of germination being increased, and the 'period of dormancy' shortened by freezing again and again. **1916** *Amer. Jrnl. Bot.* III. 99 Dormancy in plants is common in three organs, seeds, spores, and buds. **1929** Weaver & Clements *Plant Ecology* vi. 113 When a seed does not germinate immediately upon leaving the parent plant, it is said to be in a state of dormancy. Dormancy is not confined to seeds, however, but is also characteristic of many offshoots such as rhizomes, bulbs, tubers, etc. **1959** *New Scientist* 3 Sept. 338/3 Gibberellic acid..can also break dormancy so that freshly lifted potatoes immediately sprout when dipped in this substance and dormant seeds germinate. **1971** *Nature New Biol.* 16 June 195/1 Professor P. F. Wareing..reviewed the environmental and hormonal factors controlling seed dormancy.

dormeuse. 2. (Earlier example.)

1808 M. Wilmot *Jrnl.* 16 Aug. (1934) iii. 363 We..set off in the Dormeuse 4 horses abreast & two before.

dormitory, *sb.* Add: **1. b.** In universities and colleges: a building in which students reside; a hall of residence; a hostel. *U.S.*

1865 *Atlantic Monthly* XV. 551 He worked with them, studied with them,..slept in the same dormitory. **1892** *Univ. of Chicago Quart. Cal.* 9 Students are advised to make their residence in the dormitories. **1903** *N.Y. Even. Post* 7 Oct. 7 There are four regular dormitories or halls for women at Oxford. **1913** J. K. Lord *Hist. Dartmouth Coll.* 487 In 1899 the old home..was converted into a small dormitory for twenty men. **1964** G. B. Schaller *Year of Gorilla* (1965) x. 258 Through the generosity of Makerere College she was permitted to live in the girls' dormitory on the campus while I finished my work.

c. A small town, or a suburb of a large town, containing residences of those who work in the metropolitan area. Esp. *attrib.*

1923 *Westm. Gaz.* 6 Feb., The steady expansion of London's working centres and the ever-increasing sprawl of its surrounding dormitories. **1930** *Times* 12 May 9/5 Moreover, it is said, West Fulham is but a London dormitory—a place where voters sleep, and have their interests elsewhere. **1930** *Times Lit. Suppl.* 18 Sept. 736/1 Self-contained residential or 'dormitory' towns. **1949** *Here & Now* (N.Z.) Nov. 25/1 Otaki has a number of Dunlop-Rubber and Rimutaka-tunnel workers, but otherwise is a dormitory suburb. **1955** *Times* 30 June 5/1 The choice before the nation was either dormitory development, cutting into the green belts, or country town development well outside the large towns. **1961** L. Mumford *City in History* xvii. 549 This daily shuttling between dormitory and work-place. **1962** *Punch* 30 May 812/3 Orpington is a dormitory suburb. **1971** *Daily Tel.* 3 July 9/2 Wilmslow's commuter dormitories.

dornick[2]. *U.S. dial.* Also **darnick.** [Cf. Ir. *dornog* handful, small stone.] A pebble, stone, or small boulder.

1840 *Daily Pennant* (St. Louis) June 18 (Thornton), That ar man he tooks up a dornick, and made a heap of cavortins. **1869** 'Mark Twain' *Innoc. Abr.* xxxv, Darnick from the tomb of Abelard and Heloise. **1878** J. H. Beadle *Western Wilds* xii. 185 He gathered a dornick, and was just drawin' back to send the strange dog where they's no fleas. **1899** Ade *Fables in Slang* 124 A big White House, with..white-washed Dornicks in front of it. **1942** C. Morley *Thorofare* ii. xxxii. 147 You should have threw a dornick at me the other evening and put me in my place.

doronicum (dǫrǫ·nikǝm). [mod.L., ad. mod. Gr. δωρονεῖκον, ad. Arab. *darānaj, darūnaj.* Adopted by Linnæus in his *Systema Naturæ* (1735) as the name of a genus.] A perennial herb of the genus so named, belonging to the family Compositæ, and sometimes cultivated in gardens; Leopard's bane.

1607 Topsell *Four-f. Beasts* 247 That same herbe (called *Doronicum*) and of the Grecians, *Doronieu.* **1728** R. Bradley *Dict. Bot.* I [s.v. Doronicum Vulgare], The most common Doronicum of our Gardens hath divers Leaves rising from the Root. **1785** T. Martyn tr. *Rousseau's Lett. Elem. Bot.* xxvi. 397 *Doronicum* or *Leopard's-bane*, a wild plant of the Alps, and now common among the perennials of the garden, has the scales of the calyx in two rows. **1892** S. R. Hole *Bk. about Garden* 27 That group of iris (germanica) and doronicum reminds us of the cohorts of the Assyrians, 'all gleaming with purple and gold'. **1926** G. Jekyll *Colour Schemes for Flower Garden* (ed. 6) 27 Tulip Chrysolora of fuller yellow, yellow Wallflowers, the tall Doronicum, and..several patches of yellow Crown Imperial. **1970** C. Lloyd *Well-Tempered Garden* iii. 310 Doronicums.. will grow anywhere.

Do·rothy bag. [f. the feminine proper name *Dorothy.*] A woman's handbag gathered at the top by a drawstring and slung by loops from the wrist.

1907 *Yesterday's Shopping* (1969) p. xxxii/3 Dorothy Bags. **1909** *Westm. Gaz.* 23 Aug. 9/1 A 'Dorothy' bag, containing a quantity of jewellery. **1923** *Daily Mail* 15 Jan. 15 An effective Dorothy bag for evenings. **1962** *Harper's Bazaar* Dec. 7/1 The matching dorothy-bag purse has an expanding frame. **1970** *Times* 22 Dec. 7/4 Languid silk tassels..decorate black satin Dorothy bags.

Dorothy Perkins (dǫ·rəþi pɜ·ɹkinz). Also simply **Dorothy.** [Personal name.] A popular variety of rambling rose which bears clusters of double pink flowers.

1903 W. Paul *Rose Garden* (ed. 10) 262 *Dorothy Perkins*, an erect growing variety [of *Rosa wichuraiana*] of great beauty. **1908** J. H. Pemberton *Roses* 314 Dorothy Perkins (Wich.) Jackson & Perkins, 1901.—Shell pink, white centre. —Very vigorous pillar. **1912** H. H. Thomas *Complete Gard.* 110 Liberties may be taken with the Dorothy Perkins class of rose that would lead to disappointment with other kinds. **1913** —— *Rose Bk.* 13 *Dorothy Perkins.*—The most popular of all wichuraiana roses, bearing large clusters of big, double, rich pink blossoms. **1915** *Star* 5 Aug. 2/6 Suburban gardens round London are aflame with blooming 'Dorothys'. **1941** L. MacNeice *Plant & Phantom* 59 A trellis of Dorothy Perkins roses.

dorp. Delete †*Obs.* Still in use in South Africa, also for a small town.

1835 A. Steedman *Wanderings & Adv. Interior S. Afr.* I. 103 Beaufort, therefore, generally presents a scene of activity, arising from the number of farmers, who on various accounts have frequent occasion to visit the *Dorp.* **1852** C. Barter *Dorp & Veld* iv. 26 Maritzburg..an English town rising out of the ruins of a Dutch *dorp* or village. **1902** *Daily Chron.* 23 Apr. 7/3 The most remote 'dorp' has not been too far placed beyond the reach of the fertilising stream. **1902** J. H. M. Abbott *Tommy Cornstalk* 13 The dorp represented the Bush township. **1920** [see *backveld]. **1934** R. Campbell *Broken Record* vii. 162 The radios of God have always spoken out of dorps like Medina and Nazareth. **1954** D. D'Ewes *Mydorp* ii. 18 Uncle Robert and Aunt Kate..had left Mydorp..to settle themselves in a dorp down the river.

dorsal, *a.* (*sb.*) Add: **2. c.** *Phonetics.* (Of sounds) made with the back of the tongue.

1933 Bloomfield *Lang.* vi. 101 English has no dorsal spirants, but they occur in many languages. **1963** *Amer. Speech* XXXVIII. 216 The dorsal sound found in the German *ach* is very common in Armenian.

B. *sb.* **3.** *Phonetics.* A dorsal sound.

1964 E. Palmer tr. *Martinet's Elem. Gen. Ling.* ii. 50 A dorsal may also be alveolar.

Hence **dorsa·lity.**

1951 *Archivum Linguisticum* III. 135 Plosive release, tension and dorsality.

Dorset (dǭ·ɹsėt). The name of a county in the south-west of England, used *ellipt.* or *attrib.* in designation of things produced in or originally peculiar to the county.

1747 H. Glasse *Art of Cookery* xxi. 164 Golden-dorset. *Ibid.*, The Golden Dacket Dauset. **1824** J. Wight *Mornings at Bow St.* 4 There was toast and prime Dorset, and muffins and crumpets. **1836** Dickens *Sk. Boz* 2nd Ser. 24 Tubs of weekly Dorset, and cloudy rolls of 'best fresh'. **1891** R. Wallace *Rural Econ. Austral. & N.Z.* xxvi. 360 The English breed..the Dorset—a close-coated, white-faced horned breed, which produces two crops of lambs within the year. **1897** H. G. Wells *Plattner Story* 205 He wanted us to eat Dorset butter. **1940** *Chambers's Techn. Dict.* 259/1 *Dorset Down*, one of the Down breed sheep with fine close wool which is used largely for Cheviot quality woollens and for hosiery yarns. **1955** [see *blue-vinn(e)y* s.v. *blue *a.* 13]. **1957** Mankowitz & Haggar *Eng. Pott. & Porc.* 75/1 *Dorset clay*, Poole Clay..is generally known in Staffordshire as blue clay. **1960** *Farmer & Stockbreeder* 22 Mar. 63 Mr. K. Hebditch's Dorset Down flock. **1970** *Guardian* 28 Feb. 11/1 The Dorset horn sheep is the only breed which can produce more than one set of lambs a year.

dorsi-. Add: **dorsico·rnu**, the posterior grey column or horn of the spinal cord; hence **dorsi-co·rnual** *a.*; **do·rsiflex** *v. trans.*, to bend (a part of the body, esp. the foot or toe) towards its dorsal surface; **dorsifle·xion,** (*b*) *Anat.*, flexion or bending towards the dorsum or dorsal surface; **dorsi-ve·ntral** *a.*, **-ventra·lity** = Dorso-*ventral, -ventrality.*

1890 *Buck's Handbk. Med. Sci.* VIII. 528 The myelic cornua are strictly dorsal and ventral,..permitting the adjectives dorsicornual and ventricornual. **1908** *Practitioner* Oct. 561 The ankles can be dorsiflexed, until the dorsum of the feet touches the shin. **1902** D. J. Cunningham *Text-bk. Anat.* 251 If posterior or dorsal surfaces be approximated by the process of bending, then the flexion becomes *posterior* or *dorsi-flexion*, as at the knee- or wrist-joints. **1967** G. M. Wyburn et al. *Conc. Anat.* vi. 186 Dorsiflexion of the foot is limited by the wedge-shaped trochlear surface of the talus. **1882** Dorsi-ventral [see Dorso-, Dors-]. **1959** Southwood & Leston *Land & Water Bugs* 51 All Amyoteinae have an abdomen capable of great volumetric increase, accomplished by dorsi-ventral expansion. **1895** F. W. Oliver tr. *Kerner's Nat. Hist. Plants* II. 697 The oophyte is a lobed band-like thallus with marked dorsi-ventrality.

dorsum. 2. Delete (*nonce-use*) and add later example.

1905 W. G. Holmes *Justinian & Theod.* I. 27 The moat.. follows the trend of the ground as it rises on either side from the beach to the dorsum of the peninsula.

dory, *sb.*[2] Add: **b.** *attrib.* and *Comb.*, as *dory-becket, -fisherman, -fishing, -man, -mate, -roding*; *dory-modelled, -shaped* adjs.

1897 Kipling *Capt. Cour.* 154 Fiddling helplessly with a dory-becket. *Ibid.* 163 Men had met one another before, dory-fishing in the fog. *Ibid.* 178 It's yours and welcome, Harve, because we're dory-mates. **1923** T. S. Eliot *Waste*

Land 32, I had in mind the 'longshore' or 'dory' fisherman, who returns at nightfall. **1962** *Times* 6 Jan. 9/7 We would pass the dorymen hauling their line. **1890** K. Munroe *Dory Mates* 25 He delighted in being called his father's 'dorymate'. **1919** *Hist. Amer. Lit.* II. II. x. 9 The reader asks resentfully what they are doing in this dory-modelled *galère*, painted green below with a border of blue. **1897** Kipling *Capt. Cour.* 52 A tiny anchor..and some seventy fathoms of thin brown dory-roding. **1897** *Outing* (U.S.) XXX. 386/2 The boat..is dory-shaped, nine feet long.

‖ **dos-à-dos** (dozado), *adv. phr.* and *sb.* [Fr.] **A.** *adv. phr.* Back to back. **B.** *sb.* A seat, carriage, or the like, so constructed that the occupants sit back to back. **C.** In Bookbinding, *attrib. phr.* (see quot. 1952). Cf. *DO-SE-DO.

1837 J. F. Cooper *Recoll. Europe* I. 41 Some one kindly told him that they no longer danced *dos-à-dos*. **1859** *Habits of Good Society* xvi. 349 A liberal supply of ottomans, *dos-à-dos*, and sofas. **1882** H. De Windt *Equator* 119 The street cab of Batavia is a 'dos-à-dos' literally so called, as the passenger sits with his back to the driver's, thus forming a mutual support. **1952** J. Carter *ABC for Bk. Collectors* 66 *Dos-à-dos binding*, a style of binding, mostly used for devotional works, in which two volumes are bound back-to-back with a common lower board. **1970** *Sotheby's Catal. Books* 9 Nov. 28 Two modern calf gilt fitted boxes of dos-à-dos form.

dosage. Add: **1. b.** Used similarly in radiotherapy of X-rays and other ionizing radiation: see next.

1893 A. S. Eccles *Sciatica* 56 Care must be taken not to exceed the dosage either in strength or duration. **1912** Bythell & Barclay *X-ray Diagnosis & Treatment* 121 It is necessary to be extremely cautious until some idea is obtained as to the amount of dosage the skin will stand. **1918** R. Knox *Radiogr. & Radio-Therap.* (ed. 2) II. 424 The various systems of measuring the X-ray dosage. *Ibid.* 511 The most difficult question in radium treatment is that of dosage. **1928** *New Statesman* 28 July 510/1 The most careful and experienced practitioner may sometimes cause an X-ray burn after dosages which he has used..without injury on hosts of occasions. **1938** R. W. Lawson tr. *Hevesy & Paneth's Man. Radioactivity* (ed. 2) xxiv. 258 The maximum daily dosage of γ-rays. **1959** *Lab. Invest.* VIII. 173 This lack of certainty..of the scope of autoradiographic method for radiation dosage measurement prompted us to investigate the problem.

dose, *sb.* Add: **1. b.** A given quantity of X-rays or other ionizing radiation, esp. considered in relation to a person receiving it; a quantity of ionizing radiation received or absorbed at one time or over a specified period (e.g. in radiotherapy or the irradiation of plants); **dose rate**, the rate at which the dose is increasing. Also *attrib.*

absorbed dose (or simply *dose*): the quantity of ionizing radiation absorbed, measured (in rads) by the energy absorbed per unit mass of material; *exposure dose* (or *exposure*): the quantity of ionizing radiation to which anything is exposed or subjected, measured (in roentgens) by the ionization it produces in a given mass of air.

1912 Bythell & Barclay *X-ray Diagnosis & Treatment* 117 Heavy doses may occasionally produce a strong skin reaction. **1918** R. Knox *Radiogr. & Radio-Therap.* (ed. 2) II. 424 An erythema dose is one which causes slight erythema to appear within fifteen to twenty-one days. *Ibid.* 428 If the total dose is to be administered in several sittings. *Ibid.* 514 Exposures, with large quantities of radium in well-filtered doses, may be given up to twenty-four hours. **1947** *Radiology* XLIX. 283/2 Lower dose rates could not be used, as the period of fertility of mice is only eight months. *Ibid.* 352/1 Four dose levels (13, 4·3, 1·15, 0·115r) were empirically chosen. **1950** *Britannica Bk. of Year* 682/2 *Dose*, the amount of radio-active contamination received by a person, implement, or other object employed on or used in atomic energy research or utilization. **1955** *Bull. Atomic Sci.* 213/3 The biological effects of radiation are measured by the dose received, that is, by the energy absorbed by unit volume of the tissue from the radiation. **1956** *Nature* 17 Mar. 531/1 At very low dose-rates..the radiation times required would be inconveniently long. **1959** *Times* 7 Dec. (Agric. Suppl.) p. vi/5 Doses of radiation in the range 8,000–10,000 rads have been found to be sufficient for commercially acceptable sprout suppression in potatoes of several varieties. **1963** *Clin. Dosimetry* (Nat. Bur. Stand. Handbk. 87) 38/2 Numerous names were examined as a replacement for exposure dose, but there were serious objections to any which included the word dose. There appeared to be a minimum of objection to the name *exposure* and hence this term has been adopted by the [International] Commission [on Radiological Units and Measurements]... The elimination of the term 'dose' accomplishes the long-felt desire of the Commission to retain the term dose for one quantity only—the absorbed dose. **1969** *New Scientist* 24 Apr. 177/1 With high energy X-rays the dose at a depth below the surface is significantly greater than that on the surface skin. **1970** Passmore & Robson *Compan. Med. Stud.* II. xxxiii. 3/2 Such fall-out is estimated to have resulted in an average yearly dose of 2·4 mrads in the period 1954–9.

2. (Additional example.)

1894 P. L. Ford *Hon. Peter Stirling* 150 'He snubbed me,'..explained Miss De Voe, smiling..at the thought of treating Peter with a dose of his own medicine.

b. An unpleasant experience.

1847 E. Brontë *Wuthering Heights* I. 56 You have reason in shutting it up...No one will thank you for a dose in such a den! **1939** H. G. Wells *Holy Terror* I. ii. 41 Seems he don't like the idea of this new war that's coming...*We* had a dose.

c. *a dose of salts*: a dose of aperient salts; also *transf.* and *fig.* with *like*: very rapidly.

1837 *Crockett's Almanac* 3 I'll go through the Mexicans like a dose of salts. **1953** E. Simon *Past Masters* II. iii. 84 She went systematically through the residents, in the phrase of Monro, like a dose of salts. **1961** Dobie & Sloman *Tinker* II, If you think you're getting tired or anything, tell me and we'll be back down that bloody thing like a dose of salts. **1961** Wodehouse *Service with Smile* ii. 27 He boxed three years for Oxford... And went through the opposition like a dose of salts. **1968** J. Wainwright *Edge of Extinction* 31 If we don't hold 'em they'll go through this city like a dose of salts.

d. An occurrence of venereal disease.

1914 *Dialect Notes* IV. 105 *Dose*, venereal disease. **1922** Joyce *Ulysses* 151 Some chap with a dose burning him. **1952** B. Hamilton *So Sad, So Fresh* xix. 122 The cream of the joke is, she gave me the worst dose I've ever had. **1968** B. Turner *Sex Trap* xi. 97 She's riddled with pox. I know four blokes who've copped a dose from her.

do-se-do (dōᵘ·zidōᵘ). orig. *U.S.* Also **do-si-do.** [Corruption of *DOS-À-DOS.] A figure in square dancing in which two people pass around each other back to back and return to their places. Hence as *v. intr.* Also *transf.*

1929 D. Scarborough *Can't get Red Bird* v. 105 First couple out, round up four; Swing your pardners in a do-se-do. *Ibid.* 106 Haul six and get fixed; Do-se-do like picking up sticks. **1934** *Amer. Ballads & Folk Songs* (1957) 415 Bunch the heifers to the middle, circle stags and do-se-do. **1962** R. Nettel *Folk-Dancing* 147 *Do-si-do.* Partners face each other and walk forward passing right shoulders, then step sideways (back to back) and return to places walking backwards and passing left shoulders. **1969** C. Armstrong *Seven Seats to Moon* viii. 91 Tony Thees on J's trail in the Chevvy began to try to shake off Barry Goodrick in the Ford... They did tricky lane shifts, fast turns and do-si-dos around some blocks.

dosemeter: see *DOSIMETER.

dosh (dɒʃ). *slang.* [Origin unknown.] Money.

1953 H. Clevely *Public Enemy* xviii. 114 He hadn't enough dosh on him. **1959** I. & P. Opie *Lore & Lang. Schoolch.* ix. 155 Money..is referred to as..moolah, dosh (common), sploosh, [etc.]. **1970** M. Kenyon *100,000 Welcomes* xvi. 139 'America! The money 's in America!'..'Tis true. The Yankees have the dosh all right.'

dosimeter. Add: Also **do·semeter.** *spec.* A recording device to measure ionizing radiation, esp. one worn by a person exposed to potentially harmful radiation.

1944 (*title*) Photographic Film as a Pocket Radiation Dosimeter. (U.S. Atomic Energy Commission, MDDC-1065). **1957** *Financ. Times Ann. Rev. Brit. Ind.* 85/3 Dosemeters may be installed..to register..the flux of fast neutrons..at key points. **1959** *New Scientist* 5 Nov. 872/1 Normally dosimeters employing dyes will not register any perceptible effect on a sample 1 mm. thick if the radiation intensity is below about 1,000 to 10,000 roentgen. **1969** *Times* 16 July 4 (*caption*) Dosimeter access flap.

dosimetry. (Examples.)

1944 *U.S. Atomic Energy Comm. Rep. AECD-2278* (1948) (*title*) Photographic neutron dosimetry to date. **1960** *New Scientist* 2 June 1398/1 The dosimetry experiment is concerned with the..question as to precisely what dosages of radiation caused these observed effects. **1961** *Times* 17 Apr. 3/3 Appointments in the following fields:.. Radiation Dosimetry. **1968** S. J. Fleming in *Proc. 2nd Internat. Conf. Luminescence Dosimetry* 266 (*title*) The colour of spurious thermoluminescence in dosimetry phosphors.

dosology, dosometer: see Dosiology, Dosimeter.

doss, *sb.*[2] Add: **1.** (Earlier examples of spelling *doss.*) Also with suffixed *adv.*

1846 *Swell's Night Guide* 77 She stalled a lushy swaddy to a doss t'other darky. **1847** G. W. M. Reynolds *Myst. London* III. xxv. 71/2 May she be faithful to thy doss. **1892** R. Carrick *Romance Lake Wakatipu* iv. 16 [The bed] was accounted a luxury..compared with the doss-down the digger in pursuit of his calling was accustomed to. **1943** J. B. Hislop *Pure Gold & Rough Diamonds* 117, I thought it a great labour-saving idea and a great place for a doss-out. **1956** E. Blyton *Myst. Missing Man* xvii. 130 Only an old fellow who wants a doss-down somewhere.

doss, *v.*[2] Also with *down.*

1896 *N.Z. Alpine Jrnl.* II. ix. 169 Hodgkins and I 'dossed down' by the side of it. **1898** J. D. Brayshaw *Slum Silhouettes* 4, I was dossin' dahn at Shorty's. **1899** J. Bell *Shadow of Bush* iii. 11 There is a spare bunk in the wharé for one..and the other can doss down somewhere. **1924** *Chambers's Jrnl.* Jan. 31/2 We made a pretence of dossing down. **1932** *Daily Express* 25 June 17/6 If he wants to be on his way at daybreak he dosses down with his face to the east.

dossier. Now commonly with pronunc. (dɒ·si‚e, dɒ·sye). Add later examples: freq. used in the sense 'a bundle of papers or information about a person'.

1884 *Pall Mall Gaz.* 13 June 11/2 In neatly-docketed cabinets round his office stood the *dossiers* of all the criminals with whom he has had anything to do for the past eight years. **1912** H. Belloc *Servile State* ix. 176 A series of *dossiers* by which the record of each workman can be established. **1920** 'Sapper' *Bulldog Drummond* xii. 300 Here's his dossier...'Ditchling, Charles. Good speaker; clever; unscrupulous. Requires big money; worth it. Drinks.' **1939** M. Spring Rice *Working-class Wives* ii. 25 Questionnaires filled in by women of a better..position... Such dossiers would have served as 'controls'. **1955** *Bull. Atomic Sci.* Apr. 129/3 A file check of government dossiers. **1967** A. S. Neill *Talking of Summerhill* x. 125, I guessed they had rung up our Home Office to ask for my dossier.

dossy (dɒ·si), *a. slang.* [Cf. Sc. *doss* neat, spruce, *dossie* small, neat, well-dressed person.] Stylish, smart. Hence **do·ssily** *adv.*

1889 Gilbert *Brigands* in *Standard* 9 Nov. 3 (E.D.D.), We are dossy and neat From head to our feet. **1900** *Daily News* 31 July 8/2 What with the ladies' bonnets and blokes' dossy hats. **1903** 'Marjoribanks' *Fluff-hunters* 42 A dossy Sloane Street milliner. *Ibid.* 95 A dossily dressed girl.

Dosto(y)evskian (dɒsto(y)e·fskiăn), *a.* Also **Dostoievskian.** [f. the name of Feodor Michaelovich *Dostoevsky* (1821–1881), Russian novelist.] Of, or characteristic of, Dostoevsky or his works.

1925 *Sat. Rev. Lit.* 12 Dec. 404 The Dostoievskian emphasis upon purification through suffering. **1929** *London Aphrodite* IV. 316 We are forced back into intellectual vicious-circles of self-scorn, and that is too dostoevskian. **1931** S. Beckett *Proust* 62 A fine Dostoievskian contempt for the vulgarity of a plausible concatenation. **1949** Koestler *Promise & Fulfilment* xv. 170 The Götterdämmerung of British rule in the Holy Land was not in the Wagnerian, but rather in the Dostoievskian style. **1959** *Times* 21 Mar. 9/5 The Dostoyevskian appreciation of the spark of the divine in every human soul. **1961** *Times* 30 Mar. 15/2 Jesus is shown as a despised Dostoevskian epileptic.

dot, *sb.*[1] Add: **4. b.** *to a dot*: exactly, precisely.

1728 Fielding *Love in Several Masques* II. ix. 27 *La. Trap.* Are you blind? they are both alike to a Tittle. *Sir Pos.* To a dot. Her Hand to a dot. **1839** *Spirit of Times* 9 Nov. 428/3 There were a large number of horses in attendance.., and amongst them were some who had the 'go along' in them to a 'dot'. **1854** M. J. Holmes *Tempest & Sunshine* xv. 215 That was one of Tempest's capers to a dot. **1866** *Congress. Globe* 18 June 3235/3 He understands it to a dot. **1881** *Ibid.* 20 Apr. 356/1 That is the question. That is it to a dot. **1887** A. W. Tourgée *Button's Inn* 189 That'll suit me to a dot. **1924** Kipling *Debits & Credits* (1926) 312 You have it!.. That's him to a dot.

c. *the year dot* (i.e. a date too old to be particularized), very long ago. *colloq.*

1895 W. P. Ridge *Minor Dialogues* 166, I reckon *he* was born in the year dot, that 'orse was. **1928** E. Wallace *Again Sanders* v. 109 He was constantly rediscovering obvious things, or revivifying theories that had been decently interred in the year dot. **1956** 'A. Gilbert' *Death came Too* xii. 132 It's..the wife who poisons the husband, not some confederate he met in Cuba in the year dot. **1966** G. E. Evans *Pattern under Plough* xiii. 128 'That's been the same since the *Year Dot* when *Owd Hinery* were an infant'.. and..he explained that Owd Hinery was the Devil; 'and as "dot" comes before 1, the *Year Dot* was somewhere right back at the beginning'.

d. *on the dot*: (lit. of the clock-face), punctually, at the precise moment. Also in similar phrases.

1909 in *Cent. Dict.* Suppl. **1923** H. Crane *Let.* 21 June (1965) 137 From the dot of five till two in the morning. **1931** A. Christie *Sittaford Mystery* xiii. 189 They have no idea what a curse they are to everybody with their punctuality, and everything done on the dot of the minute. **1953** W. R. Burnett *Vanity Row* vii. 58 She's always been very scrupulous about settling her bill on the dot. **1958** R. Stout *Champagne for One* (1959) vi. 75 At six, on the dot as always, Wolfe entered.

e. *pl.* Originally, the notes on sheet music; hence, written or printed music. *slang.*

1927 *Melody Maker* June 586, I will give you the 'dots' for them. **1956** K. Baker in S. Traill *Play that Music* i. 22 When speaking of jazz, I mean that kind of music that is all spontaneous, fully extemporized, in other words—no 'dots'. **1968** *Crescendo* Apr. 38/2, I know of not one other guitarist in my home county..capable of playing an arrangement..in a manner that could be termed 'doing the dots justice'.

5. e. *Morse telegraphy.* (See *DASH *sb.*[1] 7 f.)

1838 *Ann. Electr., Magn., & Chem.* III. 146 The numbers consist of nothing more than dots made on the paper, with suitable spaces intervening. Thus ... ... would represent 325. *Ibid.*, The alphabetical signals are made up of combinations of dots and of lines of different lengths. **1859**, etc. [see *DASH *sb.*[1] 7 f].

f. *Television.* A picture element in colour television consisting of one of the three primary colours; also, one of the corresponding areas of phosphor on the inside of the tube, which when struck by a beam of electrons fluoresce a particular colour; *dot-sequential system*, a system of colour television in which dots of the three primary colours are formed in succession as the picture is scanned.

1937 *Discovery* Nov. 329/1 This amount of definition is determined by the number of dot elements into which the picture is arbitrarily divided. **1951** *Britannica Bk. of Year* 617/2 A dot-sequencial system..in which the colour is changed for each picture element or dot. **1957** *Encycl. Brit.* XXI. 913/1 The superposition of the colour signal on the brightness signal signifies that areas of the order of a picture element (dots) are reproduced with a repetition rate of only 15 per second. **1959** K. Henney *Radio Engin. Handbk.* (ed. 5) xxii. 64 The shadow-mask holes and dots are so positioned that the electron beam from, for example, the green gun can strike only green-emitting dots. **1970** *Physics Bull.* Nov. 515/2 Ciné recordings are made directly from the TV screen picture, with limitations set by phosphor dot size.

8. dot-and-dash *a.*, also *transf.*; and as *v.*

trans.; **dot-etching** [cf. 4, quot. 1821], in photolithography, a method of modifying the colour values of a half-tone negative or positive; **dot-map**, 'a statistical map indicating, by means of dots, the relative frequency of distribution of certain statistical data over a geographical region' (Webster, 1934).

1901 G. B. Shaw *London Mus. 1888–89* (1937) 396 This was a more sensible system, and less harshly crushing to the singer, than the dot and dash system of using trumpets and drums. **1901** W. T. Spencer in *Academy* 28 Sept. 266/2 Dickens..sat back in his chair, dot-and-dashing telegrams from Fancy-land. **1906** *Daily Chron.* 6 Nov. 3/3 They live in 'dot-and-dash-land', in a world of broken utterances, implied confidences, and vague memories. **1948** F. H. Smith *Photographs & Printer* 158 The development of dot-etching techniques for retouching screen negatives and positives has also helped the trend toward reducing the number of printings used. *Ibid.*, The screen negatives may if necessary be retouched to correct tone and colour values by dot-etching. **1960** G. A. Glaister *Gloss. of Book* 346/1 *Retouching*, the hand-correcting of colour separations in the photoengraving and photo-lithographic processes. This is known in America as *dot-etching*. **1968** *Gloss. Terms Offset Lithogr. Printing* (B.S.I.) 19 *Dot etching, dot reduction*, the chemical removal of silver from the edges of a half-tone dot image. **1923** K. G. Karsten *Charts & Graphs* lvi. 663 Dot-maps are often most easily made with colored map pins or map tacks. **1939** *Geogr. Jrnl.* XCIII. 274 The printing of the dot-map of the population of Australia gives some wrong impressions in detail.

9. Slang or dial. phr. (*to go*) *off one's dot*: (to go) out of one's senses. Cf. *DOTTY *a.* 2.

1890 *Yorkshireman* 35 (E.D.D.), I have gone Completely off my dot. **1919** H. S. Walpole *Secret City* III. x. 391 She says he's just goin' off his dot. **1929** D. H. Lawrence *Pansies* 141 And you have to act up like they do Or they think you're off your dot.

dot, *v.*[1] Add: **6.** To hit, strike; esp. in phr. *to dot* (*a person*) *one. slang.*

1895 W. P. Ridge *Minor Dialogues* 166 I'll dot *you* one.. if you don't keep that mouth of yourn ——. **1896** W. W. Jacobs *Many Cargoes* 239 Put your dooks up...I'm going to dot you! **1912** A. N. Lyons *Clara* xxi. 237 Some of us might dot you one. **1936** Wodehouse *Laughing Gas* xx. 222 This brace of thugs arguing and disputing as to which should have the privilege of dotting me. **1951** J. B. Priestley *Festival at Farbridge* 348 Any monkey tricks an' I'll dot yer one.

dotard. Add: **A.** *sb.* **3.** (See quot.)

1884 G. B. Goode *Nat. Hist. Aquatic Anim.* 58 The Harbor Seal. *Phoca vitulina.* The young are there [*sc.* in Newfoundland] also called 'Rangers', and when two or three years old..receive the name of 'Dotards'.

dote, *sb.*[1] Add: **4.** Decay in wood. (Cf. Dote *v.*[1] 4.)

a **1877** Knight *Dict. Mech.* I. 564/1 *Clear-stuff*, boards free from knots, wane, wind-shakes, ring-hearts, dote, sap. **1905** *Terms Forestry & Logging* 35 *Dote*, the general term used by lumbermen to denote decay or rot in timber. **1968** *Gloss. Terms Timber Preservation* (*B.S.I.*) 9 *Dote*..is commonly applied to timber that is slightly affected by decay and is not acceptable for certain purposes.

dotey (dōᵘ·ti). *Anglo-Ir.* Also **doaty, doty.** [Related to Dote *v.*[1]] A term of endearment, esp. for a child. Also *attrib.* or as *adj.*

1892 E. Lawless *Grania* II. III. iii. 30 Arrah, hush, my dotey! Be easy, now, there's a good child. **1898** in *Eng. Dial. Dict.* s.v., Come here, doaty, and give me a kiss. **1919** G. B. Shaw *Heartbreak House* I. 4 *Nurse Guinness.* Never mind him, doty. *Ibid.* 9 Remember that I am Lady Utterword, and not Miss Addy, nor lovey, nor darling, nor doty. **1922** Joyce *Ulysses* 257, I looked so simple in the cradle they christened me simple Simon.—You must have been a doaty, Miss Douce made answer. **1930** S. Beckett *Whoroscope* 2 My squinty doaty! **1936** 'N. Blake' *Thou Shell of Death* xiii. 230 Ah, a dotey little love she was. **1967** S. Beckett *No's Knife* 82 She'll say to me, Come, doty, it's time for bye-bye.

dotted, *ppl. a.* Add: **1. b.** Of engraving: executed by dots instead of lines; stippled. Cf. F. (*manière*) *pointillée.*

1802 *Monthly Mag.* XIV. 59/2 Both the prints are well engraven in the dotted manner. **1821** [see Dict. 2]. **1897** R. E. Graves in *Dict. Nat. Biogr.* L. 58/2 He [*sc.* Ryland] adopted the 'chalk' or dotted manner of engraving. **1908** A. M. Hind *Hist. Engraving* 290 We have already noted dotted work in plates of Giulio Campagnola. **1938** *Burlington Mag.* Mar. 150/1 The importance of Cologne for the production of dotted prints.

c. *dotted line*, a line of dots or small dashes; *spec.* (on a document): one to indicate the space left for signature (and therefore acceptance, etc., of its terms). Hence *to sign on the dotted line*, to agree fully or formally.

1772–84 [in Dict.]. **1891** J. A. Ewing *Magn. Induction in Iron* iv. 82 A cyclic process of magnetisation was gone through, the results of which are shown by the dotted lines. **1919** A. C. Lescarboura *Behind Motion-Picture Screen* 416 So it generally proved that the really big man who was in a position to sign the order 'on the dotted line', could not and would not spare the time for such a trip. **1921** Wodehouse *Indiscretions of Archie* xvi. 184, I spoke to him as one old friend to another..and he sang a few bars from 'Rigoletto', and signed on the dotted line. **1937** *Daily Herald* 5 Feb. 19/7 When the [football] talk was of transfer and things, and when Childs..had just signed on the dotted line for Luton. **1950** D. Gascoyne *Vagrant* 7 Just fill in (in block letters) on the dotted line your name And number. **1958** *Times* 13 Oct. 11/6 Far from being content to 'sign on the dotted line' he [*sc.* King George VI] insists on knowing why. **1968** M. S. Livingston *Particle Physics* x. 182 These are illustrated in Weisskopf's presentation by solid or dotted lines. **1971** *Sunday Express* (Johannesburg) (Homefinder) 28 Mar. 3/3 It is enough to warrant the expense of going to have a long, lingering look before you sign on the dotted line.

2. b. *spec.* Of moths.

1843 Humphreys & Westwood *Brit. Moths* I. 125 (*caption*) Graphiphora Baja (the dotted clay). **1845** *Ibid.* II. 17 (*caption*) *Cleora teneraria* (the dotted carpet). *Ibid.* 72 (*caption*) *Ptychopoda lividata* (the single dotted wave). **1907** R. South *Moths Brit. Isles* I. II. 187 The dotted footman (*Pelosia muscerda*). *Ibid.* 214 The dotted rustic (*Agrotis* (*Pachobia*) *simulans*). **1958** W. J. Stokoe *Caterp. Brit. Moths* (ed. 2) I. 305 The small dotted buff.. *Petilampsia minima. Ibid.* 339 The dotted chestnut.. *Dasycampa rubiginea. Ibid.* 389 The dotted fan-foot..*Zanclognatha cribrumalis. Ibid.* II. 188 The dotted border..*Erannis marginaria.*

dotter, *sb.* Add: **2.** A device in which a pencil dots an oscillating target fixed to a gun when fired without ammunition, used in training gunners to take aim.

1903 *Daily Chron.* 25 June 4/5 Neither Captain Percy Scott nor his dotter were on view. **1906** *Ibid.* 5 June 2/3 Admiral Percy Scott's dotter and aiming apparatus.

dotty, *a.* Add: **2.** (Earlier and later examples in sense 'feeble in mind, silly'.) *colloq.* or *dial.*

14.. in J. Glyde *New Suffolk Garland* (1866) 213 Ale mak many a mane to have a doty poll. **1888** E. Laws *Hist. Little England* 420/1 *Dotty*, silly from age; senile. **1896** *Evesham Jrnl.* 28 Nov. (E.D.D.), The Council hardly knows if he has not gone 'dotty'. **1936** Bentley & Allen *Trent's Own Case* vii. 73 The Englishman looked sick and a bit dotty. **1948** R. Lehmann *Note in Music* (ed. 2) IV. 130 Quite wrapped up in herself—with something pretty rum staring out of her eyes. A bit dotty, perhaps. **1971** *Sunday Times* 16 May 32/8 Taya Zinkin, with a kind of dotty inevitability, ends as an authority on India.

dottiness. b. *colloq.* Eccentricity, feeble-mindedness, craziness.

1934 Wodehouse *Thank You, Jeeves* xii. 162 Dotty, beyond a question. And who knew but what that dottiness might not run in the family? **1959** P. Bull *I know the Face* viii. 135, I was in the witness-box or court throughout and was drugged to the point of dottiness.

doty, *a.* Add: Also **doaty.** Hence **do·tiness.**

1885 *Spons' Mech. Own Bk.* 167 'Doatiness': a speckled stain found in beech, American oak, and others. **1948** R. de Kerchove *Internat. Maritime Dict.* 209/2 *Doaty*, said of the condition of timber when stained with yellow and black spots.

doty, var. *DOTEY.

Douai (dū·ē[i], dɑu·ē[i]). Also **Douay.** The name of a town in northern France used *ellipt.* or *attrib.* to designate the English translation of the Bible completed at Douai in the early 17th century and used in the Roman Catholic Church.

1837 *Dublin Rev.* Apr. 476 To call it any longer the Douay or Rhemish version is an abuse of terms. **1847** Webster, *Douay-bible*, an English translation of the Scriptures, sanctioned by the Roman Catholic Church. **1864** J. H. Newman *Apol.* I. 16 Words have been running in my head, which I find in the Douay version thus. **1867** G. M. Hopkins *Let.* 15 Aug. (1956) 41, I must check them by the Douay. **1957** *Oxf. Dict. Chr. Ch.* 419/1 *Douai-Reims Bible*, the traditional version of the Bible among English-speaking RCs...The work was begun at Douai, but owing to the migration of the college to Reims in 1578, the NT was completed in that city and published there in 1582. The OT, which did not appear till 1609, was published at Douai. **1971** 'S. Woods' *Serpent's Tooth* 42 She repeated the oath in a nervous voice, clutching the Douay Bible that had been provided for her.

double, *a.* Add: **A. 1. a.** (Additional examples.)

1611 Coryate *Crudities* 352 The Italian when he vttereth any Latin word wherein this letter *i* is to be pronounced long, doth alwaies pronounce it as a double *e*, *viz.* as *ee.* **1666** [see Yolk *sb.*[1] 1]. **1836** Dickens *Sk. Boz* 1st Ser. (ed. 3) 81 The chief pastime of the children..had been..to knock loud double knocks at the door. **1840, 1871** [see Door 1 b]. **1843** *Ainsworth's Mag.* III. 153 She..assailed his nerves by means of the thundering double-knocks of postmen. **1866** [see Knock *sb.*[1] 1]. **1873** *Young Englishwoman* Sept. 438/2 Round eggs..[may] contain a double yolk. **1906** Galsworthy *Man of Property* II. xii. 256 The only thing against her was that she had not a double name. **1951** *Catal. of Exhibits, South Bank Exhib., Festival of Britain* 135/1 Double sink, stainless steel. **1953** E. Simon *Past Masters* I. iii. 33 There was a big stove, two double sinks. **1961** *Guardian* 1 Feb. 6/4 A splendid double sink with a double drainer. **1968** J. Fraser *Evergreen Death* xix. 162 A man parks his car on a double yellow line and we can have him.

c. *double axe* (later examples).

1894 *Jrnl. Hellenic Stud.* XIV. 123 Fifty or more 'small heads of oxen, with a double axe between their horns, cut out of gold plate'. **1957** V. G. Childe *Dawn Europ. Civilization* (ed. 6) ii. 28 After Middle Minoan III the single-bladed axe was ousted in Crete by the two-edged variety—the Double Axe—known also to the Sumerians and elevated to become a fetish or symbol of divine power.

4. (Additional examples.)

1854 C. M. Yonge *Castle Builders* xxii. 348 Kate..continued it [*sc.* knitting] steadily when the double wool was a great deal too hot to be pleasant. **1873** *Young Englishwoman* May 247/2 Berlin Wool-work Border..it may be worked in single or double wool.

b. *double bassoon* (earlier example); *double reed*: see Reed *sb.*[1] 8 a.

1876 Stainer & Barrett *Dict. Mus. Terms* 137/1 *Double bassoon*, the deepest-toned instrument of the Bassoon family.

5. *to live* (or *lead*) *a double life*: to sustain two different characters in life, esp. one virtuous and respectable, the other immoral or blameworthy. Often of a married man who keeps a mistress.

1888 R. L. Stevenson in *Scribner's Mag.* Jan. 123/2 He began..to dream in sequence and thus to lead a double life. **1892** I. Zangwill *Childr. Ghetto* (1893) viii. 83 Esther led a double life, just as she spoke two tongues. **1907** *Times* 19 Dec. 9/4 The woman must have been murdered by a man who was leading a double life...The prisoner had been leading a double life. **1924** E. Wallace *Sinister Man* xxxv, She had never imagined that this gawk of a girl..could lead what was tantamount to a double life. **1953** L. P. Hartley *Go-Between* xi. 134 Since Marcus's return I had become vaguely aware that I was leading a double life.

6. *double act*: a performance by two entertainers; the entertainers themselves; also *transf.*; *double agent*: a spy who works on behalf of mutually hostile countries, usu. with actual allegiance only to one; *double aspect* [Aspect *sb.* 9]: in *Metaph.*, the two forms under which a reality may appear; also *attrib.*, as *double-aspect theory* (see sense C. 2 below), a philosophical theory, drawn from Spinoza, that mind and body (or matter) are the same thing viewed from two different aspects, subjective and objective; = *identity-hypothesis*; *double bar*: a species of finch found in Australia; *double bill*: see *BILL *sb.*[3] 8 c; *double bind* (see quot. 1962); so *double-binder*, a person whose action results in a double bind; *double blank*: a domino with both halves of its face blank; *double bluff*: see quot. and Bluff *sb.*[2] 3; *double boiler*: a saucepan consisting of two pots, the upper one containing the food to be cooked, and the lower one containing water which is heated; *double bond* Chem. [*BOND *sb.*[1] 13 e]: a chemical bond in which the two atoms 'share' two pairs of electrons rather than one pair; *double chair*: † (*a*) a light pleasure carriage having two seats *Obs.*; (*b*) a love-seat; *double chin*: a chin with a fold of flesh under it (cf. *double-chinned*, quot. 1387 s.v. Double *a.* C. 1); *double chorus*: see Chorus *sb.* 5; *double coal*: a superior kind of coal (the application varying locally); *double concerto* (see quot. 1842); *double consciousness*: see Consciousness 7; *double cream*: cream with a high fat-content; *double cube* Archit.: a room of which the breadth is equal to the height and the length is twice the breadth; also *attrib.* or as *adj.*; *double dagger*: see Dagger *sb.* 8, Diesis 2; *double date* U.S. colloq.: a 'date' (*DATE *sb.*[2] 2 c) involving two couples; *double decomposition* Chem.: the simultaneous decomposition of two compounds in a chemical reaction accompanied by the formation of two other compounds; *double dot* Mus.: see quots.; *double drummer* Austral.: a noisy type of cicada; *double elephant*: see Elephant 10; *double exposure* Photogr. [*EXPOSURE 1 e]: (*a*) an accidental exposure of the same plate or film twice; (*b*) the deliberate superimposition of a second image on an exposure already made; (*c*) *fig.*; *double fault* [Fault *sb.* 5 c]: two consecutive faults at lawn tennis, etc.; *double feast*: see Feast *sb.* 1 b; *double feature* [*FEATURE *sb.* 4 a (*c*)]: a cinema programme containing two full-length films; *double fertilization*: see quots.; *double figures* (rarely *double figure*): a total or score, esp. of runs at cricket, higher than nine and less than one hundred; *double fleece* Austral. and N.Z.: see quot. 1933; so *double-fleecer*; *double frame*: (*a*) *Typogr.* = Frame *sb.* 11 c; (*b*) *Cinemat.* and *Television*: see quot. 1959; *double-glazing* [Glazing *vbl. sb.* 1]: the action of furnishing a window with two layers of glass to reduce the transmission of heat, sound, etc.; two layers of glass fixed in a window; *double helix*: a pair of parallel helices intertwined about a common axis: the postulated structure of the DNA molecule; *double indemnity* U.S. (see quot. 1948); also *attrib.*; *double jeopardy* Law: the placing of a person in jeopardy twice for the same offence, against which there is a common-law immunity; *double knitting*: (*a*) a type of

knitting which is tubular and closed at both ends, used for ties, belts, borders of cardigans, etc.; (*b*) a thick knitting-wool made by doubling the yarn; also *attrib.*; hence *double-knit, -knitted* adjs.; *double negation* in Logic, a statement containing two negatives which, by mathematical analogy, thereby becomes positive in meaning; *double nelson* (see quot. 1889); *double O* (U.S. slang) [from the resemblance to a pair of eyes]: an intense look; *double oxer*: an oxer with a guard-rail on each side; *double play* in Baseball, a play by the defence in which two runners are put out successively by throws of the basemen; *double pneumonia*: pneumonia affecting both lungs; *double room*: a bedroom for two people; *double salt*: a salt which is composed of two simple salts and which when crystallized has physical properties different from its components but which in aqueous solution behaves as a mixture of them; *double saucepan* = *double boiler*; *double saw*(*-buck*) [*sawbuck] *U.S. slang*: (*a*) twenty dollars; a twenty-dollar note; (*b*) a twenty-year prison sentence; *double shuffle*: see Shuffle *sb.* 5; *double-spacing*: see *double-spaced* s.v. sense C. 1 and Spacing *vbl. sb.* 1 a; *double spread*: short for *double-page spread*; *double standard*, a rule, principle, judgement, etc., viewed as applying more strictly to one group of people, set of circumstances, etc., than to another; applied specifically to a code of sexual behaviour that is more rigid for women than for men; *double stem* Ski-ing: a position adopted for slowing down by making a point inward angle, i.e. by spreading the rear ends of the skis and bringing the front points together; *double summer-time* (see quot. 1962 and cf. Summer-time 2); *double u*: (*b*) *colloq.*, short for W.C. = water-closet; *double vision*: diplopia; *double wedding*: a wedding of two couples at the same time; *double window*: see quot. 1877.

1905 A. Bennett *Tales of Five Towns* II. 246 He wants me to tour with him..and do a double act. **1952** W. Granville *Dict. Theatr. Terms* 64 *Double-act*, two vaudeville artistes, cross-talk comedians, or singers, e.g. the famous Layton and Johnson team of the 1920's. **1959** *Times* 29 May 4/2 When their double act was done, having produced 87 runs..it suited Kenyon..that he soon had Gloucestershire in again. **1935** R. P. Blackmur (*title*) The double agent. **1941** Koestler *Scum of Earth* 79 The sensational trial..had revealed an amazing scene of plots, intrigues, spies, and double-agents. **1960** *News Chron.* 19 Feb. 3/4 A young Dutchman..said he was a double agent. He had joined the Germans only to get to Britain and there serve his country. **1865** J. Grote *Moral Ideals* (1876) 267 Conscientiousness..has a double aspect, outwards and inwards. **1870** S. H. Hodgson *Theory of Practice* I. i. 3 It was maintained that..the whole world of phenomena..had a double aspect, subjective and objective, was at once a mode of consciousness and an existing thing. **1909** A. J. North *Nests & Eggs of Birds Austral.* II. 279 Stictoptera bichenovii... This Finch, the 'Double-bar' of Sydney bird dealers, is another instance of a species being found in coastal as well as the inland districts of Queensland. **1933** *Bulletin* (Sydney) 13 Sept. 39/1 The grove of lemon-trees where, year after year, the double-bars had nested. **1959** J. Wright *Generations of Men* 212 The garden where her pretty diamond-sparrows and double-bars and finches nested. **1956** G. Bateson et al. in *Behavioral Sci.* I. 253/2 He [*sc.* a schizophrenic] has special difficulty in handling signals of that class whose members assign Logical Types to other signals... The hypothesis which we offer is that sequences of this kind in the external experience of the patient are responsible for the inner conflicts of Logical Typing. For such unresolvable sequences of experiences, we use the term 'double bind'. **1962** *Listener* 6 Dec. 949/2 Serious troubles can arise when.. a mother's normal life becomes subject to promptings from her unconscious...The child of such a parent finds himself repeatedly caught in a 'double bind', that is in a situation in which he is given simultaneous but mutually contradictory cues, so that whatever he does will be wrong. **1960** *Arch. Gen. Psychiatry* III. 359/2 The emotional importance of the double-binder to his 'victim'. **1801** J. Strutt *Sports & Pastimes* IV. ii. 240 One of them is a double blank. **1868** Sala *Notes & Sk. Paris Exhib.* iv. 34 The houses..gave to the outskirts of Paris an odd affinity to a city built of dominoes set on end. The double-sixes and double-fours, with here and there a double-blank in the shape of a dead-wall. **1927** Wodehouse *Small Bachelor* i. 20 It has been well said of Sigsbee H. Waddington that, if men were dominoes, he would be the double-blank. **1919** J. Buchan *Mr. Standfast* iii. 64 His device was apparently the Double Bluff. That is to say, when he had two courses open to him, A and B, he pretended he was going to take B, and so got us guessing that he would try A. Then he took B after all. **1879** A. D. Whitney *Just How* 260 Cut up and boil and mash..in a bain-marie, or double boiler. **1950** T. S. Eliot *Cocktail Party* I. i. 37, I suppose there must be a double boiler: Isn't there one in every kitchen? **1889** G. M'Gowan tr. *Bernthsen's Org. Chem.* i. 49 The assumption that the affinity which becomes free at each of two carbon atoms, upon abstraction of the hydrogen, is employed in creating a 'double bond' between them. *Ibid.* i. 50 By this term 'double bond' is not, however, to be understood a closer or more intimate combination. The olefines, on the contrary, are more readily oxidized than the paraffins, being thereby attacked at the point of the double bond. **1903** Walker & Mott tr. *A. F. Holleman's Text-bk. Org. Chem.* I. § 129. 150 The double bond must not be regarded as a mere doubling of the single one. **1944** *Hackh's Chem. Dict.* (ed. 3) 286/2 *Double bond*, a condition which exists in unsaturated compounds where two single valence bonds connect two atoms. **1964** N. G. Clark *Mod. Org. Chem.* ii. 14 The pair of valency bonds linking adjacent carbon atoms together..is referred to as an ethylenic or olefinic bond. When this type of linkage occurs between other atoms, it is simply termed a double bond. **1795** C. Pettigrew *MS. Let.* 19 Sept., I think it will be best to send the Double Chair. **1833** *Maryland Hist. Mag.* (1918) XIII. 338 Dr. Smith..and Drs negro boy left Salisbury with two easy riding horses and a double chair. **1904** P. Macquoid *Hist. Eng. Furniture* ix. 220 Double chairs or love-seats. **1934** *Burlington Mag.* Oct. 163/1 An oak chair dated 1672..of unusual width, but hardly wide enough to be described as a double chair. **1832** Double chin [see Chin *sb.* 1]. **1958** Double chin [see *ASTRAKHAN b]. **1803** J. Plymley *Agric. Shropsh.* 54 Coal, called the double-coal. **1839** Ure *Dict. Arts* 962 A section of the Quarrelton coal..showing the overlapped coal and the double coal. **1879–81** G. F. Jackson *Shropsh. Word-bk.*, *Double-coal*, a good coal for manufacturing purposes, much used. **1920** W. Gibson *Coal in Gt. Brit.* 207 The Seven Feet Coal..is the chief coal, but below it the Double and Bench coals are workable. **1842** J. F. Warner *Universal Dict. Mus. Terms* 26/1 We distinguish a concerto or concert for one instrument alone from a double concerto, i.e. a concerto for two instruments together, (concerto doppio,) or indeed for several instruments together. **1958** *Listener* 30 Oct. 706/3 The double Concerto in D minor. **1877** E. S. Dallas *Kettner's Bk. of Table* 303 We have double cream put in to sauces and soups. **1888** Mrs. Beeton *Bk. Househ. Managem.* xxxii. 898 For whipping and making sweets it is usual to ask for *double* cream, that is thick cream that has stood on the milk for twenty-four hours instead of twelve. **1936** Lucas & Hume *Au Petit Cordon Bleu* 161, 1 gill double cream. **1959** *Listener* 2 July 39/2 Whip the double cream until fairly stiff. **1776** A. Young *Tour in Ireland* (1780) 265 He has built, besides other rooms..a drawing one..., a double cube of 25 feet, being 50 long, 25 broad, and 25 high. **1930** H. Nicolson *Diary* 5 July (1966) I. 51 Down to Wilton with Vita... Go with Pembroke to the Palladian bridge and look back on the house all lit up with the Van Dykes showing in the Double Cube (Room). **1969** *Guardian* 19 June 13/8, I thought..about Inigo Jones's superb double-cube room. **1969** P. Dickinson *Pride of Heroes* 38 Beyond the hall..was the Chinese Withdrawing-room, a double cube. **1931** *Amer. Speech* VI. 204 Double date. **1952** S. Kauffmann *Philanderer* (1953) ix. 140 The ultimate triumph ..was more often accomplished on double dates than otherwise. **1955** M. Millar *Beast in View* xiv. 171 She had met Evelyn..on a double date with one of John's fraternity brothers. **1866** H. E. Roscoe *Less. Elem. Chem.* vi. 56 The decompositions here effected may serve as the type of a very large number of chemical changes classed as double decompositions. **1903** H. C. Jones *Princ. Inorg. Chem.* xxvii. 323 Sulphates can also be formed by double decomposition or metathesis. **1957** G. E. Hutchinson *Treat. Limnol.* I. x. 668 Krogh concludes that basaltic rocks in general take up CO_2, calcium carbonate and silicic acid being formed by double decomposition. **1801** Busby *Dict. Mus.*, *Double-Dot*, or *Dotted-Dot*, the Double-Dot consists of two points, one following the other. **1959** *Collins Mus. Encycl.* 201/1 The double dot, first suggested by Leopold Mozart in 1756.., indicates a prolongation of the normal length by three-quarters. **1927** *Austral. Encycl.* I. 269/2 In the Sydney district..the Double Drummer (*Cyclochila australasiae*). **1952** *Chambers's Shorter Eng. Dict.* Suppl., *Double drummer*, a large brown and orange cicada, remarkable for the large, swollen drums or covers to its sound-producing organs. **1892** W. E. Woodbury *Encycl. Photogr.* 223 *Double exposure*, an error often made by amateurs in unconsciously exposing the same plate on two occasions. **1911** D. S. Hulfish *Cycl. Motion-Pict. Work* II. ii. 91 The making of ghosts by double exposures. **1912** F. A. Talbot *Moving Pictures* xx. 225 The fairy..having been photographed only during the second exposure, appears at first very indistinctly. The result of the double exposure is shown in the illustration; and the gradual appearance of the fairy may be followed very easily. **1939** *Amer. Speech* XIV. 271 The good pun makes a double exposure on the mind. **1958** M. L. Hall *Newnes Compl. Amat. Photogr.* iv. 54 It is common nowadays for even simple cameras to have a shutter-film wind interlock which prevents blank negatives or double exposures. **1958** *Observer* 16 Feb. 13/6 The treatment [of a film] seems a bit outmoded (all those soulful close-ups and double exposures). **1909** *Cent. Dict.* Suppl., Double fault. **1921** A. W. Myers *20 Yrs. Lawn Tennis* 128 A universal 'Oh!' echoed round the arena when Wilding served a double fault. *Ibid.* 156 He served half a dozen double faults and two foot-faults. **1955** *Times* 2 July 2/7 A sad double fault gave Trabert all the assurance that he might have needed. **1934** Webster, Double feature. **1945** T. Williams *Glass Menagerie* (1948) I. i. 28 *Tom*: I'm going to the movies. ..There's a wonderful double feature down at Leowe's State. **1909** W. Bateson *Mendel's Princ. Heredity* xv. 270 The seed of maize is formed by a *double fertilisation.* It consists of two parts, an embryo, and an endosperm...The embryo is formed by the union of one nucleus of the egg-cell with one from the pollen-tube, and the endosperm is similarly formed by the union of the united polar nuclei with another from the pollen-tube. **1916** B. D. Jackson *Gloss. Bot. Terms* (ed. 3) 118/2 *Double fertilization*, in angiosperms, when one male cell from the pollen-tube fuses with the egg nucleus, the other with the upper polar nucleus, and this last with the lower polar nucleus. **1959** Foster & Gifford *Compar. Morphol. Vascular Plants* xix. 515 The participation of each of the two male gametes in a fusion process is uniquely characteristic of angiosperms, and is usually designated by the expression 'double fertilization'. **1860** J. E. Eardley-Wilmot *Remin. T. A. Smith* 251 Nor was Lord F. Beauclerk fortunate enough to mark a double figure in either innings. **1875** *Cliftonian* IV. 93 No one scored double figures. **1884** *Boy's Own Paper* Summer No. 26 Watch the ball, keep your temper, and don't be afraid; For that is the way double figures are made. **1894** *Times* 25 May 11/3 Mr. Mitchell for once in a way failed to reach double figures. **1933** *Press* (Christchurch, N.Z.) 7 Oct. 15/7 A sheep that is missed at one shearing and comes in the next has a *double fleece.* He is called a *double-fleecer.* **1904** *N.Z. Illustr. Mag.* X. 48/1 Sheep annually evaded the shearing muster and remained among the scrub..to develop into 'double-fleecers'. **1921** H. Guthrie-Smith *Tutira* xxiii. 224 We lived on..the fat wild sheep and double-fleecers. **1904** Goodchild & Tweney *Technol. & Sci. Dict.* 171/1 *Double frame*, a composing frame usually made of deal and holding two pairs of cases at the same time. **1959** Halas & Manvell *Technique Film Animation* 338 *Double frame*, one animation drawing photographed for two frames instead of one. **1943** *Fortune* Mar. 182 Double-glazing is quite a good insulator. **1957** *Housewife* Sept. 23/2 The north and south walls consist almost entirely of Plyglass double-glazing. **1958** *Chambers's Techn. Dict.* 974/2 *Double-glazing*, glazing with two panes separated by spacers and a layer of dehydrated air which prevents misting. **1960** *House & Garden* May 69/2 Double-glazing prevents heat loss. **1971** D. Devine *Dead Trouble* vii. 65 He added a sun lounge and installed central heating and double glazing. **1954** Crick & Watson in *Proc. R. Soc.* A. CCXXIII. 89 (*heading*) Detailed configuration of the double helix. **1962** T. Dobzhansky *Mankind Evolving* ii. 37 If the double helix separates into two single threads each can re-form an exact copy of the original double structure. **1968** J. D. Watson (*title*) The double helix. **1968** *New Scientist* 19 Sept. 592/1 The symbol of the molecular biological age is without doubt the 'double helix' of DNA. **1924** J. B. Maclean *Life Insurance* xiv. 257 Double indemnity benefits require but a few words. **1930** A. H. Mowbray *Insurance* xi. 170 Examination of the causes of accidents which will entitle the insured to double indemnity will disclose that these events are so spectacular ..as generally to be given considerable publicity. **1948** J. B. Maclean *Introd. Life Insurance* I. xiii. 219 'Double Indemnity'..is a provision for payment of double the face amount of the policy if death is the result of an accident. **1969** J. Weidman *Centre of Action* (1970) viii. 91 It's for her own good. With a double-indemnity clause, which comes to twenty-eight dollars a year extra, she'll be financially independent at seventy-seven. **1910** W. W. Willoughby *Constitutional Law of U.S.* I. §184. 439 It was held that by an act of Congress in 1902, the immunity from double jeopardy for crime as provided in the Constitution had been extended to the Philippines. **1969** M. L. Friedland *Double Jeopardy* i. 3 The history of the rule against double jeopardy is the history of criminal procedure. No other procedural doctrine is more fundamental or all pervasive. **1970** H. Waugh *Finish me Off* (1971) 194 If we charge him and the jury lets him off, there's no second chance. That's double jeopardy. **1895** *Montgomery Ward Catal.* 291/1 Fascinators, hand made, double knit of Shetland floss. **1964** *Observer* 12 July 8/4 The double-knit jersey revolution, which has gathered in phenomenal profits for a handful of bright ladies' knitters. **1907** *Yesterday's Shopping* (1969) 243/1 Welsbach [gas] mantles..double-knitted. **1855** Mrs. Gaskell *North & South* I. xii. 146 Mrs. Thornton..liked Mrs. Hale's double knitting far better. **1911–12** *T. Eaton & Co. Catal.* Fall & Winter 243/7 Beehive Double Knitting or Petticoat Yarn is a soft thick knitting yarn made of a fine quality of wool. **1938** M. Thomas *Knitting Bk.* 165 *Double Knitting.* A Tubular Fabric constructed on two knitting pins is worked as follows. **1960** *Farmer & Stockbreeder* 29 Mar. (Suppl.) 9/1 3 oz Emu Romany double-knitting or Scotch double-knitting wool. **1970** *Guardian* 24 Mar. 9/2 Regency Bainin double-knitting, for instance (used for our Aran tunic patterns) is available in twenty-four shades. **1883** F. H. Bradley *Princ. Logic* I. v. 131 (*heading*) The principles of identity, contradiction, excluded middle, and double negation. **1888** B. Bosanquet *Logic* I. vii. 324 The conclusion thus obtained..may be *bond fide* arrived at through the double negation I have described, and may be at first unsupported by the direct observation. **1961** I. M. Copi *Introd. Logic* (ed. 2) ix. 282 Using the Principle of Double Negation (D.N.), which asserts that *p* is logically equivalent to ∼∼*p*. **1969** F. I. Dretske *Seeing & Knowing* ii. 57 Logicians are not tempted to abandon the rule of double negation. **1889** W. Armstrong *Wrestling* 233 Probably the most dangerous move in Lancashire and Cornwall and Devon wrestling..is what is called the 'Double Nelson'...To get behind an opponent, place both arms under his, and clasp your hands round the back of his neck and thus bend his head forward till his breast-bone almost gives way. **1903** J. J. Miller *Scottish Sports* 127 So Ingram slipped on a double-Nelson, pinned him down for the requisite 30 seconds, and then politely assisted him to rise. **1917** R. W. Lardner *Gullible's Travels* (1926) ii. 48 So then I and Bishop knocked the street-car service and President Wilson and give each other the double-O. **1957** R. A. Heinlein *Door into Summer* (1960) i. 12 The cashier came over and leaned on my table, giving the seats on both sides of the booth a quick double-O. **1907** *Daily Chron.* 12 Nov. 4/4 Double-oxers, stone-walls..and broad ditches. **1958** *Times Lit. Suppl.* 31 Jan. 63/2 Such technicalities of the chase as..double oxers. **1867** *Ball Players' Chron.* 6 June 2/3 A double play by Willard and Shaw.. caused the Lowells to retire for a blank score. **1880** N. Brooks *Fairport Nine* ii. 36 A double play for the 'White Bears'..and not a run scored. **1968** *Washington Post* 4 July C1/7 The Yankee second baseman..grabbed the ball, stepped on second and threw to first base for a double play. **1968** *Globe & Mail* (Toronto) 10 July 26/6 McCovey then bounced into a double play. **1892** W. Osler *Princ. Med.* 525 Double pneumonia presents no peculiarities other than the greater danger connected with it. **1929** *Encycl. Brit.* XVIII. 100/2 Usually pneumonia affects one lower lobe but it may extend to the whole lung or even to parts of both lungs (double pneumonia). [**1926** F. Kilbourne *Dot & Will* (1929) 193, I would change to a single room in the hotel which was a little cheaper than the double one..we had.] **1931** *Times* 1 June 10/3 A large double room and private bathroom. **1849** D. Campbell *Inorg. Chem.* 176 These double salts are known as manganese alums. **1948** Glasstone *Physical Chem.* (ed. 2) x. 807 When a double salt can exist as a solid phase the behavior on evaporation depends on whether the compound is stable in contact with water or not. **1906** Mrs. Beeton *Bk. Househ. Managem.* iv. 59 The double saucepan is especially useful for making porridge and gruel. **1961** *Guardian* 24 Mar. 12/6 Unless you have a very low heat on the top of your stove..it is really best to use a double saucepan. **1850** *Knickerbocker* XXXVI. 297 Send me the two double 'saw-bucks'. **1925** *Writer's Monthly* June 486/1 *Double sawbuck*, a twenty dollar banknote. **1926** Maines & Grant *Wise-Crack Dict.* 7/2 *Double saw*, twenty dollar bill. **1929** 'C. Walt' *Love in Chicago* 25 'What'd it net yuh, State Street?' I asked. 'A little over a double saw-buck, 'n' I stuffed it all on Hip-Bones, 'n' she ain't come in fur 'er oats yet.' **1936** L. Duncan *Over Wall* i. 21, I learned quickly that a dollar-bill was a fish-skin;..a twenty a double-saw. **1945** L. Shelly *Jive Talk Dict.* 24 *Double sawbuck*, a

twenty-year jail sentence. **1948** *Time* 17 May 87/1 Any tout or hustler around the track can usually work Eddie for a 'double sawbuck'. **1950** H. E. Goldin *Dict. Amer. Underworld Lingo* 61/1 *Double-saw, double-sawbuck*, a prison sentence of twenty years. **1899** J. London *Let.* 7 Jan. (1966) 11 Surely the double-spacing could not have led to a mistaken estimate of length. **1956** F. C. Avis *Bookman's Conc. Dict.* 88/1 *Double spread*, text matter or, more usually, an advertisement stretching across the whole of two facing pages. **1951** E. Paul *Springtime in Paris* iv. 90 Without a robust double standard, the admittedly loose women play a losing game. **1962** *New Statesman* 16 Nov. 698/2 The greatest temptation into which the politically committed can be led is that of the double standard. **1963** Auden *Dyer's Hand* 3 In relation to a writer, most readers believe in the Double Standard: they may be unfaithful to him as often as they like, but he must never, never be unfaithful to them. **1968** R. Amberley *Incitement to Murder* vi. 176 He no doubt follows a double standard. One for business and one for everyday life. **1968** S. Hynes *Edwardian Turn of Mind* vi. 177 The point about the double standard is made—that it is unjust of the husband to demand greater fidelity of his wife than he offers her. **1936** O. Schniebs *Skiing for All* iii. 33 The double-stem (snow-plow position) is the brake in skiing. **1943** *Times Weekly* 18 Aug. 5/2 Double summer-time ended early on Sunday, when clocks were put back an hour. **1962** E. Bruton *Dict. Clocks & Watches* 61 *Double summer time*, introduced in Britain during the Second World War for economy. In winter the clock was one hour in advance of Greenwich mean time and in summer, two hours in advance of it. **1914** C. Mackenzie *Sinister St.* II. IV. ii. 859 The double-u is just next your bedroom. **1889** G. A. Berry *Dis. Eye* 504 The *diplopia* or double vision to which the condition gives rise. **1922** *Encycl. Brit.* XXX. 975/1 Paralysis of the muscles of the eye, producing diplopia or double vision. **1771** Smollett *Humph. Cl.* III. 255 Every thing is now prepared for our double wedding. The marriage-articles for both couples are drawn and executed. **1864** C. M. Yonge *Trial* II. vi. 118 There was a proposal to join forces, and have a double wedding..the two schoolfellows and two young friends. **1949** D. Smith *I capture Castle* ix. 129, I accepted him and Rose and I arranged to have a double wedding. **1819** M. Wilmot *Let.* 26 Nov. (1935) 31 We are..living at Vienna..with stoves and double windows in our rooms. *a* **1877** Knight *Dict. Mech.* I. 731/1 *Double-window*, one having two sets of sash, inclosing a body of air as a non-conductor of heat and to deaden noise. **1908** Kipling *Lett. of Travel* (1920) 133 The double windows are brought up from the cellar. **1949** D. Macardle *Children of Europe* xiii. 205 The cold of winter is so intense in Hungary that people who can afford it have double windows.

C. 1. *double-aspected, -bearded, -bottomed* (later example), *-columned, -curved, -dotted, -flowered* (example), *-mouthed* (later example), *-sighted* (examples), *-spaced, -tracked, -triggered* (examples), *-walled*; **double-brooded** (example); also of birds; **double-coated,** having two coats; **double-fronted** (later example); **double-threaded,** of a screw (also *fig.*).

1876 *Mind* I. 357 This double-aspected Whole may be taken as the *larger circle* including either of the two aspects. **1933** R. Tuve *Seasons & Months* iv. 158 Janus double-bearded (one of them forked) holds a nondescript object that is either bread or horn. **1833** B. Silliman *Man. Sugar Cane* 60 His apparatus is composed of a double bottomed copper boiler, covered by a dome. **1932** E. Step *Bees, Wasps, Ants, &c., Brit. Isles* 195 Another double-brooded species.. is *Claudius rufipes.* **1953** D. A. Bannerman *Birds Brit. Isles* I. 117 In the opinion of the above authority the species [*sc.* the citril finch] is..double-brooded. **1922** R. Leighton *Compl. Bk. Dog* xvii. 271 [The Cairn Terrier] must be double-coated. **1935** *Discovery* July 190/2 Some very promising results have recently been shown by sponsors of the Brewster process, in which an imbibation printing of yellow is superposed on a toned double-coated film. **1958** T. L. J. Bentley in M. L. Hall *Newnes Complete Amat. Photogr.* vi. 91 With the double-coated films current in the mid-thirties acceptable prints could be obtained with camera exposures varying by as extreme a range as 2000 to 1. **1965** *Times* 30 Aug. 12/1 They become the first company to offer sheet steel coated both sides with a p.v.c. film. The double-coated steel is called Stelvetite 'R'. **1967** Karch & Buber *Offset Processes* v. 192 The paper usually used for a flat is 80 lb. double-coated goldenrod stock. **1840** C. Brontë *Let.* in Mrs. Gaskell *Life C. B.* (1857) I. 212 Recording all their sayings and doings in double-columned close-printed pages. **1861** Double-columned [see Columned *ppl. a.* 3]. **1959** *Brno Studies* I. 137 Quoted from double-columned cheap edition. **1937** *Burlington Mag.* May 258/2 The recent discovery in Shang-Yin tombs of double-curved knives. **1837** Double dotted [see Dotted *ppl. a.* 3]. **1955** H. Van Thal *Fanfare for E. Newman* v. 69 A frustrated counterpoint in jerky double-dotted rhythm. **1902** *Westm. Gaz.* 22 Oct. 12/1 A perfectly formed double-flowered ox-eye daisy. **1965** *Bucks Examiner* 3 Sept. 13/2 (Advt.), A modern double-fronted Detached Bungalow. **1952** C. Day Lewis tr. *Virgil's Aeneid* IX. 203 The double-mouthed pipe tweedles for addicts. **1735** J. W. *Creed Expounded* II. ix. 253 Unless they are double-sighted Folk, who see, what other People can discern nothing of. **1846** Geo. Eliot tr. *Strauss's Life of Jesus* II. II. vi. § 79. 139 What..double-sighted beings, must Moses and Jesus have been, if they mixed with their cotemporaries without any real participation in their opinions and weaknesses. **1956** F. C. Avis *Bookman's Conc. Dict.* 87/2 *Double-spaced*, that style of typescript in which the inter-linear spacing equals the depth of a line of typescript. **1963** D. Heyes *12th of Never* (1964) i. 8 He then continued typing to the bottom of the double-spaced page. **1909** *Westm. Gaz.* 18 Mar. 4/1 The propeller is..made up of two portions of a double-threaded screw. **1910** *Daily Chron.* 28 Jan. 6/3 'A Will in a Well' is a double-threaded mystery story. **1937** E. Muir *Coll. Poems* (1960) 81 The double-threaded river That runs through life and death and death and life, Weaving one scene. **1887** C. B. George *40 Yrs. on Rail* v. 91 Accidents are reduced to a minimum, owing to good management and to the double-tracked roads. **1967** *Times* 23 Oct. 9/4 The management should.. stop spending precious capital on converting double tracked lines to single track. **1839** Z. Leonard *Adv.* (1904) 70 In a hurry, the one that was accustomed to the single trigger caught up the double triggered gun. **1630** *Orders for River of Thames* in Binnell's *Descr. Thames* (1758) 66 No fisherman..shall..use or exercise any Flue, Trammel, double-walled Net, or hooped Net. **1871** Double-walled [see Walled *ppl. a.* 1 b]. **1903** *Dublin Rev.* July 169 The double-walled hydrogen vessel. **1965** G. McInnes *Road to Gundagai* v. 82 It was an earthenware double-walled beehive filled with water.

2. a. *double-curve, -flow, -motor, -reduction, -spiral, -standard, -zero*; **double-aspect theory,** etc. (see sense A. 6 above); **double-base powder, propellant** (see quots.); **double-beat sluice** (see quot.); **double-bubble fuselage,** etc., the fuselage, etc., of a double-decked aircraft; **double-gate table,** a gate-table with two hinged movable legs to support leaves; **double-pole switch** (*Electr.*) (see quot. 1940); **double-tone ink** (see quots.).

1879 W. James in *Mind* IV. 330 The 'double-aspect' school postulate the blank form of 'One and the Same Fact'. **1909** *Hastings' Encycl. Relig. & Eth.* II. 757/1 Ward.. discusses..the Neo-Spinozism of the 'double-aspect' theory. **1931** G. F. Stout *Mind & Matter* 82 This is the so-called double-aspect theory; according to it mind and matter are different sides or aspects of the same thing. **1960** *Discovery* Oct. 62/2 The much discussed 'double-aspect' hypothesis of mind and brain. **1951** W. Ley *Rockets, Missiles & Space Travel* vii. 172 The propelling charge in that rocket was a double-base powder containing..nitrocellulose..nitroglycerin..and diphenylamine added as a stabilizer. **1960** F. Gaynor *Dict. Aerospace* 75 *Double-base propellant*, a solid propellant which consists largely of nitroglycerine and nitrocellulose. **1931** F. M. Du-Plat-Taylor *Reclam. Land fr. Sea* 72 Double-beat or compensated cylindrical sluices. **1947** *Jrnl. R. Aeronaut. Soc.* LI. 142/1 The cottage-loaf design of hull—or the double-bubble section as it applied to the landplane—was a very good form for the pressurised cabin. *Ibid.* 174/2 The Brabazon fuselage was a 16-ft. circle, and the Saunders-Roe boat had a beam of approximately 16 ft...The vices of the double-bubble fuselage did not then seem immediately apparent. **1959** J. L. Nayler *Dict. Aeronaut. Engin.* 116 *Double-bubble fuselage*, a two-decker airliner (or fuselage) with a cross-sectional shape like a figure eight. **1927** Peake & Fleure *Hunters & Artists* viii. 126 The 'double-curve' Ofnet skulls may.. show us a stage in the evolution of broad-headedness. **1950** H. L. Lorimer *Homer & Monum.* v. 284 From about 600 onwards the double-curve bow appears in connexion with mythological or heroic beings. **1930** *Engineering* 15 Aug. 189/2 There is..a double-flow low-pressure turbine in tandem with a high-pressure machine. **1940** *Chambers's Techn. Dict.* 260/2 *Double-flow turbine*, a turbine in which the working fluid enters at the middle of the length of the casing and flows axially towards each end. **1961** *Aeroplane* C. 394/1 Operational airline experience with the double-flow engine (this term is used here to cover both the British by-pass types and the U.S. and Russian turbofans) began just a year ago. **1908** *Daily Report* 5 Sept. 8/2 A 3 ft. 6 in. oak double-gate table. **1910** *Chambers's Jrnl.* 24 Dec. 55/2 A double-motor aeroplane. **1920** R. E. Neale *Whittaker's Electr. Engin. Pocket-Bk.* (ed. 4) 419 For the control of c.c. motors, double-pole switches and fuses are commonly used. **1940** *Chambers's Techn. Dict.* 261/2 *Double-pole*, said of switches, circuit-breakers, etc. which can make or break a circuit on two poles simultaneously. **1951** *Archit. Rev.* CIX. 62/1 The two-meter unit for both power and lighting circuits has one 60 ampere switch and one 30 ampere double-pole switch. [**1922** *Encycl. Brit.* XXX. 950/1 The demands of large users of continuous-current power..are best met either by geared generators (steam turbines driving continuous-current generators through double helical reduction gearing)..or [etc.].] **1957** *Ibid.* XXI. 355/2 Later he [*sc.* Charles Parsons] followed this up by a 'double-reduction' gearing which admitted of a still greater difference in speed of rotation between the propeller and the turbine. **1962** *Economist* 10 Nov. 605/2 By re-rolling the plated steel—the double-reduction process—..American producers are making 'thin tin'. **1963** R. F. Webb *Motorists' Dict.* 80 *Double reduction gears*, a method of increasing the number of gear ratios by the fitting of a second two-speed gearbox separated from the normal box. **1928** Peake & Fleure *Steppe & Sown* 96 Double-spiral ornaments made of copper wire. **1867** J. Laing *Theory of Business* iv. 46 The double-standard system..causes one of the two metals to be treated as bullion. **1963** *Times Lit. Suppl.* 29 Mar. 214/4 He was also a double-standard man. **1964** E. A. Nida *Toward Sci. Transl.* viii. 158 The double-standard capacity of new literates who can decode oral messages with facility but whose ability to decode written messages is limited. **1904** Mitchell & Hepworth *Inks* xi. 178 'Art shades'..are now much in vogue...A half-tone block printed in one of these inks..appears as if produced by two printings. Such inks have been described..as *double-tone inks.* **1954** J. Southward *Mod. Printing* (ed. 7) II. xix. 251 Doubletone inks are an adaptation of coloured inks designed for printing illustrations. These inks..impart the effect of more than one colour or shade. **1914** E. Pound *Let.* 19 Jan. in *Lett. J. Joyce* (1966) II. 327 He has exactly twice as much sense as the common american editor, a sort of double zero leaning toward the infinitesimal. **1964** A. Wykes *Gambling* ix. 214 The American double-zero wheel..*does* have a definite pattern.

3. *double-berth, -board, -damn* (later example), *-dot*; **double-check** *v. trans.*, to check (something) twice, or in two ways, in order to minimize the chances of inaccuracy; **double-date** *v. intr.* (*U.S. colloq.*), to make or participate in a 'double date'; **double-declutch** *v. intr.*, see *DECLUTCH *v.*; **double-dig** *v. trans.* (see quot.); so *double-digging*; **double-dink** *Austral.* = *DOUBLE-BANK *v.* 2; **double-fault** *v. intr.*, in Lawn Tennis and Squash Rackets, to serve two consecutive faults; hence *double-faulter*; **double-iron** *v. trans.*, to shackle with irons on both legs (cf. *double-ironed* in 4 in Dict.); **double-shuffle** *v. intr.*, to perform a double shuffle (Shuffle *sb.* 5); **double-space** *v. trans.* (see *double-spaced* (*C. 1)).

1966 *Times* 27 June 10/3 Strike-bound vessels have been double-berthed. **1874** *Rep. Vermont Board Agric.* II. 512 My plan was to double board and cleat the main body of the barn, having a basement or cellar under the whole barn. **1958** *Amateur Photogr.* 31 Dec. 2/2 (Advt.), You may.. automatically double-check your focus with your range-finder at the same time. **1969** E. Lathen *When in Greece* ii. 13 For several hours, he and Leonard double-checked specifications. **1946** P. Goodin *Clementine* xxi. 194 We'll have lots of fun—we'll be double-dating! **1951** J. D. Salinger *Catcher in Rye* vi. 50 I'd double-dated with that bastard a couple of times. *Ibid.* vii. 60 We once double-dated, in Ed Banky's car, and Stradlater was in the back, with his date, and I was in the front, with mine. **1952** H. Waugh *Last seen Wearing* (1953) 12 Marlene and Peggy were double-dating that night with a couple of boys from Carlton College. **1897** G. B. Shaw *Let.* 7 Oct. (1965) 810 Damn the printer..blast him..double-damn him! **1933** *Jrnl. R. Hort. Soc.* LVIII. I. 117 The soil should be double dug i.e. two spades deep. **1842** Double digging [see *BASTARD *a.* 8]. **1941** Baker *Dict. Austral. Slang*, Double dink. **1942** E. Langley *Pea Pickers* (1958) 26 We went double dinking on his white mare. **1897** *Outing* (U.S.) XXIX. 377/2 Two mink had double-dotted the course of the brook. **1921** A. W. Myers *20 Yrs. Lawn Tennis* 136 Dixon double-faulted in the eleventh game and lost it. *Ibid.* 78 The brilliant server and smasher became a double-faulter and a snatcher at lobs. **1922** W. T. Tilden *It's All in the Game* 118 Vincey took the first point on Dave's net but double-faulted away the next. **1927** *Daily Express* 6 June 1 Tilden..double faulted. **1961** *Times* 17 Jan. 14/7 At 8—all there came five empty hands with Amin, put out off the wood, getting in again and then double-faulting above the line. **1897** P. Warung *Tales Old Régime* 42 Here, guard! double-iron this man. **1909** M. B. Saunders *Litany Lane* I. i, Toeing, tipping, double-shuffling, hopping. **1922** Joyce *Ulysses* 513 He.. doubleshuffles off comically. **1958** J. Kerr *Please don't eat Daisies* 60 The necessity for double-spacing the script.

4. a. *double-based, -glazed*; **double-screened** (see quot. 1921); **double-sided,** (*a*) that can be or has been used on both sides, cf. *DOUBLE-FACED *a.* 1 c; (*b*) = Double-faced *a.* 1. **b.** *double-running.*

1954 K. W. Gatland *Devel. Guided Missile* (ed. 2) i. 34 Most of these, like cordite and ballistite, contain nitroglycerine—in which case they are known as 'double-based' propellants. **1962** Simpson & Richards *Junction Transistors* viii. 192 The uni-junction transistor or double-based diode. **1910** *Westm. Gaz.* 26 Mar. 6/2 Side windows and skylights, all of which are double-glazed, in order..to guard against changes of temperature. **1939** *Archit. Rev.* LXXXVI. 29 The double-glazed windows light the administrative offices. **1956** *Nature* 21 Jan. 111/2 Much can be done by the suitable design of houses to reduce the artificial heat load required, by double-glazed windows. **1931** L. F. Pesel *Eng. Embroidery* I. 20 Such linens are not really satisfactory, and make this double-running embroidery difficult. **1963** *Times* 1 June 11/7 The early double-running stitch gradually being augmented by coral-stitch, satin-stitch [etc.]. **1905** *Daily Chron.* 4 May, Double-screened Nuts. **1921** C. E. Evans *Hints Coal Buyers* (ed. 2) 56 Double Screened coal, indicates coal that has been screened at the Colliery, and screened also over two open screens in the spout at the Dock Tip, that is to say, 'Double Screened' at time of shipment. **1907** *Captain* XVIII. p. xxvii (Advt.), The Best Disc Records are Double-sided 8¾ in. **1934** *Mind* XLIII. 270 Christian philosophy is undeniably double-sided, exhibiting a rational and a religious aspect. **1936** *Burlington Mag.* Sept. 136/1 The double-sided panel from Valenciennes. **1956** *Nature* 25 Feb. 391/1 A piece of double-sided corrugated paper.

double, *sb.* Add: **3. i.** Esp. in phr. *at the double.* Also *fig.*

1865 *Chambers's Jrnl.* 213/1 Intellect not only marches, but marches at the 'double'. *Ibid.* 470/1 Ellsworth detailed twenty men.., and went at 'the double' down Pennsylvania Avenue. **1883** *Army Regulations* II. x. 242 A certain number of movements are to be performed at each drill at 'the double'. **1961** New Eng. Bible *Acts* xxi. 32 He immediately took a force of soldiers with their centurions and came down on the rioters at the double.

m. Substitute for def.: An actor or singer who takes two parts in the same piece; also an understudy or substitute. (Earlier and later examples.)

1808 S. W. Ryley *Itinerant* I. iv. 89 When the company is *thin*, and *one* actor is obliged to do *two* parts, we call that a *double.* **1818** *Sporting Mag.* II. 14/1 It would be impossible to find what the players call '*a double*' for Mr. Stephen Kemble. **1928** *Sunday Express* 8 Apr. 4 Two 'doubles' were employed..for some small scenes in which Miss Thorndike.. could not appear. **1960** O. Skilbeck *Film & TV Working Terms* 42 *Double*, one who impersonates an artist (usually the star) in a shot either because of danger, or because they have superior ability in some required form.

n. Further instances (e.g. = double agent, double bedroom, double event, double snipe, two 'tots' of whisky, two centuries scored by a batsman in one match).

1845 *Ainsworth's Mag.* VIII. 213 A very gentlemanly [Londoner]..armed with one of Purdey's first-rate doubles [*sc.* guns]. **1891** N. Gould *Double Event* xxvi, Messrs Isaacs and Moses..were always ready to lay the double. *Ibid.*, If he loses the Derby we may go for a recovery in the cup. But..Ike is confident he will win the double. **1902** *Harmsworth London Mag.* June 438/2 The men who play both cricket and football well..the best 'first-class doubles' who were at Oxford or Cambridge. **1918** E. Wallace *Down Under Donovan* xvi. 210 Mr. John President may yet pull off a double. **1920** *Field* 2 Oct. 488/2 There were several 'doubles' (a fish on each of the two hooks used on the line) of red gurnet and bream. **1920** G. Burrard *Notes on Sporting Rifles* 33 A hammerless ejector double rifle is the best and quickest to reload...Next come hammerless non-ejectors, and then hammer rifles, but a double is a *sine qua non*. **1921** *Spectator* 19 Mar. 357/1 A few snipe rose. We got four of them, two being 'doubles'. **1922** J. SYRETT *Alf* 99 'You've 'ad a lot of doubles to-night, Mr. Powell,' Flo remarked...'Don't want to go 'ome screwed again to-night, does yer?' **1929** *Star* 21 Aug. 17/1 When he reached his second hundred of the match, for no batsman before has twice done the 'double' in Test Matches. **1931** *Times* 16 Mar. 2/7 Large doubles [*sc.* bedrooms] now available. **1951** KOESTLER *Age of Longing* II. v. 264 Georges would.. send him another double on the house. **1953** WODEHOUSE *Ring for Jeeves* i. 16 A double, dear lady, is when you back a horse in one race and if it wins, put the proceeds on another horse in another race. **1957** A. GRIMBLE *Return to Islands* iii. 62 Her chance of pulling off the 'double', which is to say, first the miracle of her homing, and then the crowning marvel of her safe entry into harbour. **1959** 'M. DERBY' *Tigress* i. 18 Keep an eye on her. Start off by assuming that she *is* a double. **1963** *New Yorker* 29 June 46/3 (Advt.), The famous hotel Astor Singles from $9, doubles from $14. **1969** 'A. HALL' *Striker Portfolio* viii. 93 Being not only a potential defector but a double, he had broken down. **1971** 'A. GARVE' *Late Bill Smith* i. 34 We've done very well to be left with only two singles and two doubles. That's if they all embark, of course.

o. = *double feast* (see FEAST *sb.* 1 b).

c **1690** in *Month* (1882) Jan. 122 And his feast kept as a duble annualy upon yᵉ 2nd of Octobre. **1759** CHALLONER *Let.* 4 May in E. H. Burton *Life* (1909) II. xxiii. 7 He..will come over to receive his consecration here: and therefore I should be obliged to you if you would obtain for him ..a license to have this performed on any *double*. **1763** *Divine Office for Laity* IV. 229 The Transfiguration of our Lord. A greater Double. **1850, 1885** [in Dict., sense 3 n].

p. *Mus.* = VARIATION 14.

1806 BUSBY *Dict. Mus.* (ed. 2) *Double*, a word which in the old music carries the same sense as that which we now give to the term *variation*. **1962** *Listener* 26 July 153/2 'Ornamental variations' disappoint us when, as in Couperin's airs with 'doubles', the embroidering fantasy seems less developed than in previous work.

q. *Bridge.* A call by a bidder's opponent involving doubling of the score for tricks bid and made with a bonus to the declarer if he makes overtricks and an increase of the penalty if he fails to make his contract.

1903 in 'L. Hoffmann' *Card & Table Games* (ed. 3) 276 Something fresh is always cropping up; owing, perhaps, to the Declaration or to the Double. **1905, 1927** [see *DECLARER 3 b]. **1958** *Listener* 2 Oct. 541/2 North's double conventionally asked his partner to make some unlikely lead.

r. Double-screened coal.

1931 *Times* 16 Mar. 19/7 Lanarkshire [coal]..trebles.. doubles..singles.

s. *Darts.* A throw on the narrow space enclosed by the two outer circles of a dart-board; the space itself.

1935 *Encycl. Sports* 221 Victory goes to the player who reduces his total exactly to nothing with a 'double'. **1936** R. CROFT-COOKE *Darts* vi. 36 *Double Top*, 40, of course, i.e. double 20. Many players..start on the double 20, score on the treble, and leave themselves the double on which to get out. **1959** *Chambers's Encycl.* IV. 381/1 The players must generally begin and finish on a double.

6. Slang phr. *to come the double*, to act in a treacherous or evasive manner (example is *Austral.*); *to put a double on* (a person), to double-cross. See also quot. 1914.

1888 'R. BOLDREWOOD' *Robbery under Arms* II. xiii. 209, I didn't know myself that your Kate had come the double on you. **1914** JACKSON & HELLYER *Vocab. Criminal Slang* 29 *Double*, a conspiracy to deceive or defraud a victim; the 'double-cross'. Example: He got the double. **1923** E. WALLACE *Missing Million* xvii. 143 And I ask you..if you would think a girl who could write as this young lady wrote to me, would put a double on me as she did.

double, *v.* **1. e.** Add: (Further examples.) Also *absol.*, to play two parts. In extended use, to become or act as a double agent (cf. *DOUBLE *a.* 6 and *sb.* 3 n).

1918 H. CROY *How Motion Pictures are Made* v. 124 A young man, doubling for a leading lady in a bit of hazardous fire jumping. **1928** *Daily Express* 8 Apr. 4 Picturegoers should look out for the portions of the film in which Miss Thorndike was 'doubled' by other actresses. **1933** P. GODFREY *Back-Stage* ii. 24 The various Scottish thanes have to double and treble—soldiers, murderers, messengers, and apparitions. **1949** G. B. SHAW *Sixteen Self Sketches* vii. 39 The appointment of art critic to The World, which Archer was for the moment doubling with his regular function of dramatic critic, was transferred to me. **1959** *Times* 8 June 13/3 The umbrella, which can double as a sunshade. **1962** *Listener* 8 Mar. 253/1 His adventures at a small German court are 'doubled' and interwoven with the autobiography of a professor's cat. **1965** R. SEGAL *Crisis of India* iv. 204 Travelling traders, who frequently doubled as money-lenders and so could dictate terms to their debtors. **1968** 'B. MATHER' *Springers* xv. 160, I was already in a Red cell. I doubled for the Russians right from the beginning.

f. *Chess. trans.* To place two pawns or two rooks one behind the other on the same file.

1750 'A. D. PHILIDOR' *Chess Analysed* 4 He chuses rather to let you take his (Bishop)..tho' he suffers to have his Knight's Pawn doubled by it. **1806** *Chess made Easy* (ed. 5) 71 One must always strive to hinder the adversary from doubling his rooks. **1891** R. B. SWINTON *Chess for Beginners* ix. 72 Sometimes it is worth while to effect an exchange of pieces, only to cause your opponent to double his Pawns in taking one of yours. **1958** H. GOLOMBEK *Instructions to Young Chess Players* ii. 18 When two pawns of the same colour are in the same file they are called 'doubled'.

g. *Bridge. trans.* and *intr.* To declare a double (*DOUBLE *sb.* 3 q).

1894 'BOAZ' *Pocket Guide to Bridge* 6 The effect of doubling is that the value of each trick is doubled. **1898** 'L. HOFFMANN' *Card & Table Games* (ed. 2) I. 312 The main elements of novelty in Bridge..may be classed under the following heads:—..3. Licence to each party alternately to double and re-double the normal value of tricks. **1902** J. B. ELWELL *Bridge* 111 *Going over*.. The effect of 'over', 'over', etc., is that the value of each trick point is doubled, quadrupled, etc. **1906** *Bridge Pocket Book* 13 After the trump declaration has been made by the dealer or his partner, their adversaries have the right to double. **1909** *Strand Mag.* Jan. 71/2 The fourth player will be in a fine position, either to double the forced call or to overcall it. **1912** F. IRWIN *Fine Pts. Auction Bridge* 56 You can either double the two hearts or go to 'two no-trumps'. **1928** A. WAUGH *Nor Many Waters* ii. 74, I called, 'Three No Trumps.' And the man on my left doubled. **1965** *Listener* 20 May 758/2 North was hoping to play in Two Hearts doubled.

h. *intr.* To play two (or more) musical instruments. So *to double* (*on*): to play (an instrument) in addition to one's main instrument.

1927 *Melody Maker* May 421/1 Fred Livingstone..belongs to the class that doubles on both saxophones and clarinet. *Ibid.* June 551/3 Miss Ivy Read leads on the piano and is supported by Miss P. Pax (violin), Miss Brightwell (banjo doubling saxophone) and Miss Sibruk (drums doubling 'cello). **1934** S. R. NELSON *All about Jazz* iii. 68 It is usual to find only one [violin] in the smaller bands, except where the saxophones both double on this instrument. **1955** L. FEATHER *Encycl. Jazz* ii. 64 A clarinetist would double on tenor sax.

4. c. *to double up* (earlier and later examples). Also *intr.* In Betting, to double the stakes.

1789 W. DYOTT *Diary* July (1907) I. 63, I was very unpleasantly situated, being obliged to double up with a jolly ensign, or to take lodgings in town. **1940** WODEHOUSE *Eggs, Beans & Crumpets* 29 You doubled up when you won, thus increasing your profits by leaps and bounds. **1952** *Times* 21 Nov. 8/3 In favour of giving students a reasonable spell of living in college, without making them 'double up' on the staircases. **1958** J. K. GALBRAITH *Affluent Society* xiii. 148 People cannot afford to own or rent their own homes and must double up. **1970** R. GADNEY *Drawn Blanc* vii. 82 'Doubling up again, Donnelly?' someone said... The roulette wheel was spun once more.

6. a. (Later example.)

1852 E. RUSKIN *Let.* 26 Apr. in M. Lutyens *Effie in Venice* (1965) II. 301 Very fine looking Russians..wrapped up in immense cloaks doubled with furs.

11. (Later example with *on*.)

1888 'R. BOLDREWOOD' *Robbery under Arms* III. xv. 229 How did you find out Warrigal's doubling on me?

12. *trans.* Short for *DOUBLE-BANK *v.* 2. *N.Z.*

1947 *Book Miscellany* (Christchurch, N.Z.) IX. 33 After tea, he doubled me to the station. **1963** N. HILLIARD *Piece of Land* 57 A bike came past: a big boy doubling a girl on the crossbar.

‖ **doublé** (duble), *a.* [Fr., = lined.] **a.** Covered *with*, folded over. **b.** Of a book lining: made with a *doublure*. **c.** Plated with gold or silver.

1848 THACKERAY *Van. Fair* iv. 25 A sort of tent, hung round with chintz of a rich..India pattern, and *doublé* with calico of a tender rose-colour. **1890** *Catal. Exhib. Rec. Bk.-bindings Grolier Club* 11 When the inside of the cover is lined with leather it is termed *doublé*. *Ibid.* 27 Garnet morocco, with ornament in mosaic and gold; doublé, blue morocco, with border of foliage and flowers. **1901** 'L. MALET' *Hist. R. Calmady* III. x. 258 The noble house of Fallowfeild, *doublé* with all the gold of all the Barkings.

double-bank, *v.* Add: **b.** (Later examples.) Also, to double; to repeat (something); *spec.* to drive (motor vehicles) in two lines abreast; to park (a motor vehicle) alongside another stationary vehicle (cf. *DOUBLE-PARK *v.*). So **dou·ble-banking** *vbl. sb.*

1903 R. J. CLOW *Pillar of Salt* vii. 87 'I will not call twice at the same place during the year.'..'"Double banking" is a poor game,' continued the husky man. **1929** K. S. PRICHARD *Coonardoo* xxvi. 255 She..stuck it when the watches [over the cattle] were double-banked for two or three nights afterwards. **1933** P. GODFREY *Back-Stage* viii. 104 This often happens when two or more first nights synchronize, and the newspaper's regular critic has to be double- or treble-banked from other members of the staff. **1936** A. THIRKELL *August Folly* i. 14 Owing to the great depth of the shelves the books were double banked. **1948** PARTRIDGE *Dict. Forces' Slang 1939–1945* 60 *Double banking*, the practice, severely discouraged, of having two parallel lines of traffic going in the same direction. **1954** *Highway Code* 11 Do not let your vehicle stand in the carriageway.. alongside a standing vehicle, thus causing 'double-banking'. **1958** *Times* 1 May 4/2 Meters were less ugly than double and triple banking of cars all over the place indiscriminately. **1961** *Guardian* 10 Oct. 18/2 Mr. Macmillan has extended his double-banking of Ministers. **1963** J. VAIZEY *Education in Class Society* iii. 10 In order to accommodate the children coming out of the sixth forms, we shall have to adopt expedients like double-banking.

2. *intr.* To ride two on a horse, bicycle, etc. Also *trans.*, to ride (a horse, bicycle, etc.) thus. So **double-banking** *vbl. sb.* Chiefly *Austral.* and *N.Z.*

1888 'R. BOLDREWOOD' *Robbery under Arms* I. xix. 267 'We must double-bank my horse..for a mile or two.'..He jumped up, and I mounted behind him. **1926** T. E. LAWRENCE *Seven Pillars* (1935) v. lix. 338 The man made camel-less could double-bank another, riding two-up, in emergency. **1928** 'BRENT OF BIN BIN' *Up Country* (1966) vi. 67 Bert and Tim..double-banked on the horse, and returned to the high ground near the river. **1930** *Bulletin* (Sydney) 8 Jan. 20/4 His sister..somehow got him on to a horse..and double-banking to hold him on, she took him to Dr. Pirie's surgery at Liverpool. **1940** B. O'REILLY *Green Mountains* (1941) 67 Rose and I 'double-banked' on a fat round pony. **1945** BAKER *Austral. Lang.* iii. 72 It was possibly through the agency of the horse that Australia acquired..*double-bank*..employed mainly today when a cyclist gives another person a ride on the bar of his bicycle.

double-banked, *a.* Add: **c.** *transf.*

1929 F. C. BOWEN *Sea Slang* 40 *Double banked*, sleeping two in a cabin. **1964** *Times Rev. Industry* Feb. 57/3 The first 'double-banked' motor the firm has made..(mounting two sets of five cylinders each radially round a common drive shaft). **1968** *Economist* 17 Aug. 53/1 A double-banked defence network of anti-missiles.

double-barrel, *a.* (Earlier example.)

1807 Z. M. PIKE *Acct. Exped. Mississippi* (1810) 8 Apr. 240 Visited the treasurer, who showed me the double-barrel gun given by governor Claiborne.

sb. **b.** A hyphenated surname. (Cf. DOUBLE-BARRELLED *a.* 2.)

1952 A. POWELL *Buyer's Market* iii. 178 The double-barrel..has really no basis whatever, beyond the surname of a remote ancestor.

double-barrelled, *a.* Add: **1. b.** Of a telescope.

1955 *Sci. News Let.* 21 May 324/2 A double-barreled telescope that can record a golf ball's flight eight miles away will go to work for the Air Force to track guided missiles.

2. (Earlier and later examples.)

1777 *Maryland Jrnl.* 9 Sept. (Th.), The event of this double-barreled scheme has been, that the colonel and his party are defeated. **1912** W. OWEN *Let.* 24 July (1967) 151 Your sleek Thomas, Hopkins, Dixon,..and the rest of these double-barrelled guns, whose double-barrelled names I refuse to write. **1938** *Spectator* 21 Jan. 75/2 The two minor groups are generally allowed..to nominate one candidate in the double-barrelled constituencies. **1959** J. P. HUGHES *How you got your Name* vi. 103 In surnames the double-barrelled form does not appear before the eighteenth century. **1965** T. REESE *Bridge Conventions* 44 *Double-barrelled Stayman*, an extension of the Stayman Convention whereby both two clubs and two diamonds in response to 1 NT are conventional.

double bed. A bed to accommodate two persons. Also *attrib.* So **double-bedded** *a.*, having a double bed or two single beds.

1798 JANE AUSTEN *Let.* 24 Oct. (1952) 21 We have one double-bedded and one single-bedded room. **1839** E. A. HITCHCOCK *Jrnl.* (1930) 25 There were five double beds. **1844** *Ainsworth's Mag.* V. 407 He and his friend slept in a double-bedded room. **1910** *Bradshaw's Railway Guide* Apr. 1022 Inclusive terms..from 15/– per day or 4½ guineas per week for double bedrooms. **1912** J. JOYCE *Let.* 22 Aug. (1966) II. 310, I have taken a double-bedded room at 21 Richmond Place. **1915** 'H. J. JAMES' (*title*) Double bed dialogues. **1925** G. BURRARD *Big Game Hunting* 280 A good warm rug or blanket, preferably double-bed size. **1931** *Times* 16 Mar. 22/1 Large double bed room. **1958** HAYWARD & HARARI tr. *Pasternak's Dr. Zhivago* I. vi. 182 They went into another room which had..a wide double bed.

double(-)blind, *a.* Applied to a test or experiment conducted by one person on another in which information about the test that may lead to bias in the results is concealed from both the tester and the subject until after the test is made; orig. used of tests for determining the efficacy of drugs.

Quots. 1937, 1948 both refer to a double-blind test.

[**1937** *Jrnl. Amer. Med. Assoc.* 26 June 2178/2 The data consisted of the patients' judgments regarding changes in pain. These data were secured in a manner relatively free of bias by the use of the 'blind test'. **1948** *Amer. Heart Jrnl.* XXXVI. 529 The study was conducted by the 'blind' method. The materials for injection..were unknown to the observer as well as to the subject.] **1950** *Amer. Jrnl. Med.* IX. 146/1 The 'internal-evaluation' was made by skilled questioning under conditions of the 'double blind test' in which neither the physician nor the patient knew at the time whether the evaluation related to the placebo or khellin. **1954** *Proc. R. Soc. Med.* XLVII. 197 The only safe way to obtain unbiased opinions from either of them [*sc.* doctor or patient] is to make them express their opinions without knowing whether the patient received an active drug or not. This is known in America as a double blind test. **1961** *Lancet* 19 Aug. 423/1 Statistics and certain concepts, such as double-blind trials, are on everyone's mind to-day. **1968** [see *EXTRA-SENSORY *a.*]. **1970** *Sci. Amer.* Mar. 62/3 To demonstrate that no cheating was involved, the experiment was repeated on a double-blind basis: neither the investigator nor the subject knew what the dot pattern contained in advance. **1971** *Nature* 12 Mar. 113/2 It should be noted that this study was not double blind and hence is subject to observer bias.

b. *ellipt.* or as *sb.*

1960 J. R. WILSON *Double Blind* i. 12 In an ordinary blind trial the patient does not know which substance he is receiving...There may be reasons why this information is better withheld even from the doctor himself. This is known as a double blind. **1960** *N.Y. Times* 14 Aug. 77/3 After the first few cases of improvement with griseofulvin therapy were noted, a research method known as the 'double blind' was used. **1961** *Spectator* 3 Feb. 165/1 The well designed trial avoids this unconscious conspiracy by a system known as the double-blind.

dou·ble-cross, double cross. [f. DOUBLE *a.* + CROSS *sb.*] **1.** [CROSS *sb.* 29.] An act of treachery to both parties (orig. in gaming or sport) esp. by pretended collusion with each; more widely, betrayal of the other party in a (dishonest) transaction.

1834 W. H. AINSWORTH *Rookwood* III. iv. ii. 244 (*title of poem*) The Double Cross. **1848** *Sporting Life* 4 Mar. 4/2 All bets are off. It has..been 'rumoured', that a double cross was intended. **1874** HOTTEN *Slang Dict.*, *Double cross*, a cross in which a man who has engaged to lose breaks his engagement, and 'goes straight' at the last moment. **1887** *Referee* 21 Aug. 1/3 (Farmer), A double cross was brought off. Teemer promised to sell the match, and finished by selling those who calculated on his losing. **1896** ADE *Artie* ix. 79 Every time I see him over at the city hall he's whisperin' to one o' them red-necked boys and fixin' it to give somebody the double-cross. **1910** W. M. RAINE *B. O'Connor* 214 'Think you're getting the double-cross?' asked Leroy. **1920** A. E. W. MASON *Summons* xii, There was always a certain amount of money for the man who would work the double cross. **1930** H. G. WELLS *Autocracy Mr. Parham* I. i. 9 Espionage had never been so universal, conscientious, and respected, and the double cross of Christian diplomacy ruled the skies from Washington to Tokio. **1959** 'H. HOWARD' *Deadline* x. 80 It had to be someone on Lloyd's payroll who was working the double-X. **1970** E. MCGIRR *Death pays Wages* iv. 89 Part of the deal is that I guarantee you against a double-cross or hi-jacking.

Hence **double-cross** *v. trans.*, to give (a person or persons) the double-cross; loosely, to cheat; also *absol.*; **double-crosser**; **double-crossing** *vbl. sb.* and *ppl. a.*

1903 ADE *People you Know* 153 Although he had been double-crossed and put through the Ropes, he still had a Punch left. **1904** 'O. HENRY' *Cabbages & Kings* x. 161 'Twas thus I was double-crossed by the Tropics through a family failin' of goin' out of the way to hunt disturbances. **1910** W. M. RAINE *B. O'Connor* 240 Nothing like being on the spot to prevent double-crossing. **1915** WODEHOUSE *Something Fresh* i, A wealthy uncle who subsequently double-crossed them by leaving his money to charities. **1926** J. BLACK *You can't Win* iv. 42 This song is a favorite among negroes when in great trouble, such as..being double-crossed by a friend. **1927** *Observer* 10 July 17/5 This was apparently part of a deep-laid plot, for the Nationalists now consider they have been double-crossed. **1927** *Vanity Fair* Nov. 132/4 A 'rat' or a 'heel' is a double-crosser or a worthless person. **1928** *Hearst's International* Aug. 156/3 Are you going to be a dirty thief and a double-crosser? **1928** *Collier's* 18 Aug. 6/4 'You're a double-crossing rat,' I said. **1940** E. POUND *Cantos* lxv. 142 Congress has double XX'd me. **1942** C. S. LEWIS *Broadcast Talks* I. i. 11 Think of a country where..a man felt *proud* of double-crossing all the people who had been kindest to him. **1960** Double-crosser [see *BACK-STABBER].

2. [CROSS *sb.* 28.] A cross between two hybrids each obtained by crossing two separate inbred lines, e.g. in the production of maize.

1920 *Wallaces' Farmer* 18 June 1604/2 To gain the maximum results from inbred strains it is necessary to bring together four strains in what is called a double cross. **1934** *Discovery* Feb. 52/2 The so-called double-cross technique so largely used in breeding maize in America and used to a lesser extent with swine and poultry in several countries. **1967** BRIGGS & KNOWLES *Introd. Plant Breeding* xviii. 226 Jones, in 1918, proposed the use of the double cross to produce hybrid varieties of corn. By double cross is meant the hybridization of two single crosses and the use of the F_1 as the commercial crop. *Ibid.*, Double cross hybrids.

3. Shortened form of *double* CROSS-STITCH.

[**1934** M. THOMAS *Dict. Embroidery Stitches* 63 *Double cross stitch*, a canvas stitch, consisting of a cross stitch set diagonally with another cross stitch set straight and worked over the first.] **1960** G. LEWIS *Handbk. Crafts* 38 Frequently these two are the only stitches used to the neglect of the many others which would greatly enrich many pieces of work, such as doublecross.

double-deck. [See DOUBLE *a.* C. 2 and DECK *sb.*] Used *attrib.* in designations of structures, vehicles, etc., having two platforms, floors, or planes one above the other. So **double-decked** *a.* [DOUBLE *a.* C. 1]; **double-decking** *vbl. sb.*

1869 *Trans. Ill. State Agric. Soc. 1867–69* VII. 460 Sheep arrive here from the west in single decked cars, but leave in double decked ones. **1894** W. T. STEAD *If Christ came to Chicago!* 186 Mr. Pullman has devised an admirable double-deck car. **1903** A. H. BEAVAN *Tube, Train, Tram, & Car* xv. 212 Electric omnibuses..double-decked. **1906** [see *DECK *sb.*[1] 3 d]. **1910** A. WILLIAMS *Engin. Wonders World* II. 49/2 The makers claim that..it [*sc.* the Scherzer Rolling Lift Bridge] could be made double-decked. *Ibid.* 267/1 The double-deck floor accommodates four elevated railway tracks. **1917** C. C. TURNER *Aircraft of To-day* ii. 34 The 'double-deck' type of rectangular planes. **1930** *Times Educ. Suppl.* 17 May p. iv/3 The building will have double-deck lifts, which will permit the taking on and discharging of passengers at two floors simultaneously. **1955** *Times* 9 May 18/1 A double-deck self-parking garage. **1956** WALLIS & BLAIR *Thunder Above* (1959) i. 6 The huge double-decked airliner. **1959** *Manch. Guardian* 29 Jan. 6/5 The proposed double-decking of the first mile of the Great West Road. **1960** BAKER & JOHNSON *Dict. Highway Traffic* 55 *Double-deck road*, a collective term for a high-level road and a low-level road that both follow the same route, one over the other. **1970** *Globe & Mail* (Toronto) 28 Sept. 17/3 A gang of youths captured a double-deck bus.

double-decker. Add: **a.** (U.S. example.)

1835 *Western Monthly Mag.* June 339 The Washington is a splendid double decker, calculated to carry three hundred tons.

b. Delete *U.S.* and substitute orig. *U.S.* Also, a double-decked omnibus. (Examples.)

1867 *Terr. Enterprise* (Virginia, Nev.) 19 July 3/1 A 12-mule double-decker prairie schooner. **1878** *Design & Work* IV. 324/3 [Quoting U.S. newspaper] The car is a double-decker. **1887** *Harper's Mag.* Sept. 557/2 The street-cars are double-deckers, with seats upon the roof as well as within. **1895** *Popular Sci. Monthly* Apr 757 The 'double decker' or two-story cars. **1908** *Westm. Gaz.* 22 Apr. 4/2 The car of the future must be a double-decker. **1955** *Times* 6 May 19/5 Our policy of substituting double-deckers for single-deckers was continued.

c. In various technical uses: a loaf baked with a smaller upper portion; a woman's dress consisting of two skirts or flounces, one above the other; a miners' cage made in two storeys; a double-decked aeroplane; two beds, one above the other; a bridge or road built on two levels; a multi-layer cake or sandwich; 'a tenement-house having two families on one floor' (*Cent. Dict.* 1889).

1877 W. WRIGHT *Hist. Big Bonanza* 301 Those [*sc.* cages in the shaft of a silver mine] with two platforms are called 'double-deckers'. **1899** DICKINSON & PREVOST *Cumbld. Gloss.* 88/1 s.v. *Curn keak*, *Double-decker*, a pastry cake baked in the oven, having a layer of currants inside it. **1902** *Fortn. Rev.* June 1008 The English 'double-decker' is a fearful and wonderful production that errs on the side of heaviness. **1902** *Westm. Gaz.* 30 Oct. 3/1 The double-decker costume sketched has each of its flounces bordered with..grey and white squirrel. **1902** *Daily Chron.* 12 Nov. 8/6 A number of workmen were being brought up out of the mine..in the double decker cage. **1917** C. C. TURNER *Aircraft of To-day* viii. 137 Chanute made 700 glides in his 'double-decker' without an accident. **1934** WEBSTER Double-decker [sandwich].

d. *attrib.*

1867 [see *b]. **1881** *Harper's Mag.* Jan. 206/1 A good specimen of a 'double-decker' engine. **1906** E. DYSON *Fact'ry 'Ands* xvii. 227 She pays out double-decker 'am 'n' beef bull's-eyes. **1930** *Economist* 22 Feb. 405/2 The proposal of the Royal Commission for a double-decker rail and road bridge was rendered out of date. **1930** *Aberdeen Press & Jrnl.* 20 Mar. 7/6 Thirteen people were injured in an alarming double-decker 'bus smash. **1946** KOESTLER *Thieves in Night* 144 It is a kind of double-decker sandwich. There is a crusty top layer of apparent arrogance. **1959** *Listener* 13 Aug. 263/1, I also take a few sandwiches of the double-decker variety, that is three slices of bread and two fillings. **1963** *Times* 29 Jan. 14/4 The upswing of the past 12 months, unlike the earlier ones, has been a 'double-decker' affair.

double dummy: see DUMMY *sb.* 2 (in Dict. and Suppl.).

double Dutch: see DUTCH B. *sb.* 2 b.

double eagle. [f. DOUBLE *a.* + EAGLE *sb.* 5.] **1.** A gold coin of the value of twenty dollars. *U.S.*

1849 *Laws of U.S. concerning Money* (1910) 508 For all sums whatever, the double eagle shall be a legal tender for twenty dollars. **1859** BARTLETT *Dict. Amer.* (ed. 2) 185 There are also double eagles of twenty dollars. **1872** E. EGGLESTON *End of World* xxvi. 179 He..piled the double-eagles like a fortification in front of him. **1886** *Harper's Mag.* Dec. 36 Then the spokesman took a golden double-eagle from his belt. **1902** S. E. WHITE *Blazed Trail* viii. 61 His wages were twenty-five dollars a month, which his van bill would reduce to the double eagle. **1948** P. JOHNSTON *Lost & Living Cities Calif. Gold Rush* 41/1 A shower of bright double eagles clattered musically to the floor.

2. An ensign or armorial bearing, esp. in Germany.

1861 [see DOUBLE *a.* 1 c]. **1934** H. MILLER *Tropic of Cancer* (1948) 279 Goethe was..stamped with the German trade-mark, with the double eagle. **1936** *Burlington Mag.* Sept. 111/2 Besides the owl-jugs proper there exists another.. in the form of a double-eagle. *Ibid.* 112 (*caption*) Double-Eagle Jug, with the arms of von Khuenburg of Carinthia.

double-e·dgedness. The quality or condition of being double-edged.

1901 J. JASTROW *Fact & Fable in Psychol.* 165 With peculiar obliviousness to the double-edgedness of his remark, he writes [etc.]. **1908** *Dublin Rev.* Oct. 281 He did not realize the double-edgedness of epigram when delivered in the wrong time and place, to the wrong person.

‖ **double entente** (dūbl aṅtāṅt). [Fr.] = DOUBLE ENTENDRE.

1895 'H. S. MERRIMAN' *Sowers* (1896) xx. 163 He had a deft way..of planting a *double entente*. **1909** *Westm. Gaz.* 24 July 4/3 Our journalists have used more often the incorrect phrase *double entendre* than the French critics the phrase *double entente*, which is the term that our writers ought to employ. **1934** *Punch* 19 Dec. 692/1 There are of course occasions for laughter, the loudest laughter being very naturally given to those '*double ententes*' which are readily offered by the association of boy scouts and girl guides in camp. **1938** H. G. WELLS *Apropos of Dolores* iv. 154 There's a lot of this double-entente in French, I know. **1957** B. & C. EVANS *Dict. Contemp. Amer. Usage* 142/2 There is no use pointing out that *double entendre* does not exist in French and that the proper phrase for a double meaning, one of them usually indelicate, is *double entente*.

double entry: see ENTRY 9 b.

double event. [EVENT *sb.* 2 e.] Orig. in *Racing*, applied to the winning, by a horse, competitor, or team, of two races or matches at the same meeting or in the same season; hence *gen.* applied to two occurrences, acts, or performances of any kind. Also *attrib.*

1846 'SYLVANUS' *Pedestrian & other Reminiscences* xxiv. 221 The laying on a 'double event', when one of them remains in the bettor's power, having his victim bound in the..chain of certain..loss. **1863** *Illustr. London News* 566/3 They are not anxious to back anything for the 'double event'. **1872** *Gentl. Mag.* Dec. 696 He then 'backed himself for the double event' and went in for 'second schools', the last bar to a B.A. taking up mathematics. **1886** EARL OF SUFFOLK et al. *Racing* (1889) xiv. 268 The double-event betting. **1888** *Peel City Guardian* VI. 2/3, I was not codding, dear old boss, when I gave you the tip...Double event this time...Jack the Ripper. **1891** N. GOULD (*title*) The Double Event. **1898** *Cycling* 62 Most of the special burning oils are satisfactory, provided that 'double-event' oils are eschewed. **1899** *Chambers's Jrnl.* 25 Nov. 823/1 The reasons for a display of disgust at a 'double event' [*sc.* the birth of twins]. **1915** KIPLING *Fringes of Fleet* 30 It was a simple calculation of comparative speeds and positions, and when it was worked out she decided to try for the double event.

double-face. Add: **b.** (Later example.)

1910 W. B. YEATS *Green Helmet* 29 You say that, you double-face.

c. = DOUBLE-FACED 1 b. Also as *sb.*, a fabric so finished.

1873 *Young Englishwoman* Sept. 442/2 The wide sash of double-face ribbon. **1966** *Guardian* 6 Sept. 6/6 A raglan coat, of white/nutbrown doubleface.

double-faced, *a.* Add: **1. a.** (Additional examples.)

a **1877** KNIGHT *Dict. Mech.* I. 726/1 *Double-faced*, a term applied to an architrave, or the like, having two faces. **1927** JESPERSEN *Mod. Eng. Gram.* III. xi. 215 Some phrases are.. double-faced. **1963** VISSER *Hist. Syntax Eng. Lang.* I. ii. 99 In Old English the number of double-faced or amphibious verbs was far inferior to that of transitive verbs.

c. Of a gramophone record: having a recording on each side (cf. *double-sided* (s.v. *DOUBLE *a.* C. 4 a)).

1936 *Amer. Speech* XI. 5 The author has made ten double-faced phonograph records.

dou·ble-head, *v.* orig. *U.S.* [Cf. DOUBLE-HEADER b.] *intr.* Of a train: to run with two engines. Also *pass.*, to be drawn by two engines. So **double-heading** *vbl. sb.*

1904 *Delineator* Sept. 374 A heavy freight train had double-headed up the mountain, and at the summit the leading engine had been cut off to run down ahead of the train. **1959** *Times* 5 Oct. p. iii/6 The latest electric locomotives [in Switzerland] can haul 600 tons..over the steep gradients..without any double-heading. **1970** *Railway Mag.* Oct. 541/1 The flying passage of a northbound express double-headed by 'Black Staniers'.

double-headed, *a.* Add: Of a train: running with two engines. Of an electric locomotive (see quot. 1905).

1902 *Daily Chron.* 21 May 3/5 So heavy a train..doubtless ..will always be 'double-headed'. **1905** *Ibid.* 12 Jan. 8/5 The locomotive is double-headed, and controllable at either end, so that no turning is necessary. **1966** *Times* 3 Jan. 8/5 A double-headed express with 10 coaches.

double-header. Transfer *U.S.* to sense a and add: **b.** (Earlier and later examples.) orig. *U.S.*

1878 A. PINKERTON *Strikers* 216 'Double-headers', or freight trains composed of a larger number of cars than the single train, and drawn by two engines. **1971** *Sunday Express* (Johannesburg) 28 Mar. 24/3 Steam from a ruptured locomotive boiler scalded the flesh off two railmen when their double-header jumped the rails.

c. In baseball, lacrosse, etc., the playing of two games in succession on the same day. *N. Amer.*

1896 *Cincinnati Enquirer* 30 July 2/2 In case rain should stop to-day's or to-morrow's games double headers will have to be played the next day. **1967** *Boston Herald* 8 May 16/4 Leading the New York Yankees past Kansas City, 8–3, for a split of their Sunday doubleheader. The Athletics won the opener, 4–1. **1968** C. DRUMMOND *Death & Leaping Ladies* iv. 72 The Louisiana Lancers had..won two more double-headers on Wednesday and Thursday. **1970** *Globe & Mail* (Toronto) 26 Sept. 39/5 The Detroit Tigers won the first game of the doubleheader.

d. *Logging.* (See quot.) *U.S.*

1905 *Terms Forestry & Logging* 35 *Double header*, a place from which it is possible to haul a full load of logs to the landing, and where partial loads are topped out or finished to the full hauling capacity of teams.

e. *Gambling.* A double-headed coin. *Austral.* and *N.Z.*

1948 V. PALMER *Golconda* iv. 26 'What's the trouble?'.. 'Dirty work. That dago, Joe Comino, trying to ring in a.. double-header. Macy Donovan was keeping the [two-up] ring.'

double image. [f. DOUBLE *a.* + IMAGE *sb.*] **1.** An optical appearance or counterpart of an object seen double in certain circumstances, esp. as a result of an affection of the eyes like diplopia.

1880 W. JAMES *Coll. Ess. & Rev.* (1920) 172 The *other eye* ..is kept covered during the experiments to prevent double images. **1890** —— *Princ. Psychol.* I. xi. 425 The increased visibility of optical after-images and of double images, which close attention brings about, can hardly be interpreted otherwise. *Ibid.* II. xx. 226 To most of us the whole [visual] field appears single, and it is only by rare accident or by special education that we ever catch a glimpse of a double image. **1966** C. L. THOMSON *Your Sight* (ed. 4) ii. 20 The person may occasionally see a double image, due to the failure of the two eyes to merge their images properly, a condition termed diplopia.

2. *Art.* (See quots.)

1939 WEBSTER Add., *Double image* (Surrealism), a representation of an object which is at the same time without deformation a representation of a different object. **1958** M. L. WOLF *Dict. Painting* 86 *Double image*, a popular device in primitive and surrealist art; any image that allows two irreconcilable interpretations simultaneously.

3. Used *attrib.* in various technical senses.

a **1877** KNIGHT *Dict. Mech.* I. 726/2 *Double-image micrometer*, suggested by Roemer about 1678; brought into use by Bonguer about 1748. *a* **1884** —— *Dict. Mech.* Suppl. 267/2 *Double image prism*, a prism of Iceland spar giving a double image of the object of complementary tints, and also used by revolving the images to measure the angle of crystals examined under the microscope. **1940** *Chambers's Techn. Dict.* 261/1 *Double-image tacheometer*, a type of tacheometer, used with a horizontal subtense bar, which gives two images of the bar in the field of view.

double-jointed, *a.* (Stress variable.) Having joints that permit a much greater degree of movement of parts of the body than is normal. So **double-joi·ntedness.**

1831 J. ROBY *Trad. Lancs.* 2nd Ser. I. 103 The knave is.. of an incredible strength, being..double-jointed. **1912** *Lancet* 19 Oct. 1077/2 A boy..with..double-jointedness... The joints were very loose, and the child took particular pleasure in forming almost circles by locking the index and middle finger of each hand. **1930** 'GREENHORN' *Tinker, Tailor* vi. 131 His double-jointed, coffee-coloured hand sent the knife hissing on its deadly way. **1961** C. H. BARNETT et al. *Synovial Joints* IV. iii. 249 Anatomists can only plead ignorance of the minor structural differences that must distinguish the normal joint from the double jointed variety.

double-minded, *a.* (Later examples.)

1961 NEW ENG. BIBLE *James* i. 8 A man of that kind must not expect the Lord to give him anything; he is double-minded, and never can keep a steady course. *Ibid.* iv. 8 You who are double-minded, see that your motives are pure.

double-park, *v. trans.* and *intr.* orig. *U.S.* [f. DOUBLE *adv.* + *PARK *v.* 2 b.] To place or leave (a vehicle) parallel to another vehicle parked near the side of the road. Hence **double-parker**, one who parks his vehicle in this way; **double-parking** *vbl. sb.*

1931 *Kansas City Star* 12 Aug., Manhattan and Emporia each has a local ordinance forbidding double-parking of motor cars. **1932** *Ibid.* 5 Apr. 18 Practically all of Commercial street was used by double parkers. **1936** *Amer. City* Jan. 95/1 It was a frequent occurrence for an exasperated motorist in desperation to double-park his car. **1951** I. SHAW *Troubled Air* i. 26 The police after you? Have they finally got you for double-parking? **1959** *Observer* 14 June 17/6 This forces commercial vehicles to double-park. **1959** *Encounter* Nov. 47/2 The cars are double-parked so thickly.. that a moving vehicle can scarcely manœuvre. **1962** *Economist* 3 Feb. 412/1 Parked, and double-parked, cars restricted movement throughout central London. **1970** B. KNOX *Children of Mist* i. 10 The..duty car double-parked briefly outside the main entrance to let him out.

double-quick, *a.* and *adv.* (Later examples.)

1918 *Blackw. Mag.* Mar. 293 Down go his hands on the joystick in double-quick time. **1959** J. BRAINE *Vodi* xix. 223 If we were married and I made just one mistake in business..she'd be off double-quick.

doubler². Add: **5.** *N.Z. slang.* (See quot.)

1871 C. L. MONEY *Knocking about N.Z.* x. 141 [We] were served out with a 'doubler', or two lots of grog in one.

doublet. Add: **2. c.** (Example.)

1896 T. L. DE VINNE in Moxon *Mech. Exerc.* I. p. xviii, Typographic peculiarities have been followed, even to gross faults, like *doublets*.

d. A story or saying which occurs in two different biblical contexts, and hence is regarded as derived from distinct sources.

1891 F. P. BADHAM *Formation of Gospels* 11, I subjoin tables of the doublets in S. Matthew, S. Mark and S. Luke. **1899** J. C. HAWKINS *Horae Synopticae* II. § iv. 64 The 'doublets', or repetitions of the same or closely similar sentences in the same Gospel, are of great value in supplying hints as to the sources and composition of the Gospels. **1906** F. C. BURKITT *Gospel Hist.* 14 One of the really striking features about the narrative in Genesis..is the number of Doublets, *i.e.* stories told twice over. *Ibid.* 163 At the first glance they are real doublets; *i.e.* different accounts of the same event drawn from different sources. **1927** A. H. MCNEILE *Introd. N.T.* 64 Isolated sayings in *Mark* which occur in more or less similar forms in two passages...These are often called 'doublets'.

4. c. A pair of associated lines occurring close together in a spectrum.

1897 J. R. RYDBERG in *Astrophys. Jrnl.* VI. 235 The constituents of the doublets and triplets of the nebulous series are built up after exact rules,..while the constituents of the sharp series, so far as we know, are simple lines. **1926** R. W. LAWSON tr. *Hevesy & Paneth's Man. Radioactivity* v. 54 Just like the *D*-line, the strongest rhodium line also consists of a doublet. **1961** *Adv. Spectrosc.* II. 228 The structure which is stable at low temperatures shows a doublet in the infra-red..as well as in the Raman spectrum.

dou·ble-take, double take. orig. *U.S.* [TAKE *sb.*] A delayed reaction to a situation, sight of a person, etc., rapidly following an earlier inappropriate reaction; esp. a procedure in comedy, etc., in which an actor at first reacts unexpectedly or inappropriately to a given situation and then, as if more fully realizing the implications, reacts in an expected or more usual manner. Also, a second, often more detailed, look. Hence **double-take** *v. intr.*, to act in such a manner.

1938 *Chatelaine* Jan. 50/2 In case you aren't familiar with the 'double-take'—it's an exaggerated reaction to surprise. **1941** H. MACINNES *Above Suspicion* vii. 55 A young man had come out of the hotel door; he halted as he heard Frances' voice, and looked at her, giving what Hollywood has perfected as the 'double take'. **1942** BERREY & VAN DEN BARK *Amer. Thes. Slang* 595/4 *Double take*, Edward Everett Horton's specialty—first he beams at what he thinks is a compliment, then he realizes the truth and 'takes' it the second time, but not so happily. **1957** *New Yorker* 23 Nov. 43/1 The only person who recognized the former President during our stroll together happened to be a friend of ours..who performed a double take worthy of the late Oliver Hardy. **1958** L. FLETCHER in Cerf & Carmel *24 Favorite One-Act Plays* 125/1 Duffy has been examining lunch, but double-takes suddenly on above. **1958** *N. & Q.* Feb. 65/2 The Antonio-Aquilina scenes have to be read with a 'double-take': the comedy is obvious, the ugliness must be discovered lying just below the comic surface. *Ibid.* 66/1 The play is, in effect, an extended double-take. **1959** P. MOYES *Dead Men don't Ski* xix. 260 The Colonel did a sort of double-take, as though he'd seen the others for the first time. **1961** A. WILSON *Old Men at Zoo* i. 21, I said, 'Ah!' Then I did a quick double-take, and added, 'What reporter?' **1962** *John o' London's* 10 May 460/3 Mr. Ustinov sits there..blinking, double-taking. **1969** E. MCGIRR *Entry of Death* ii. 22 The old fellow momentarily double-took, but being quick-witted darted a cunning look at the saloon bar. **1970** —— *Death pays Wages* ii. 33 'It'll be the good doctor,' said Oakley rising. Piron did a double-take, for a dazzling brunette entered.

dou·ble-talk, double talk. orig. *U.S.* [TALK *sb.*] **a.** Deliberately unintelligible speech; speech that is a mixture of real and invented words; gibberish. **b.** Verbal expression intended to be, or which may be, construed in more than one sense; deliberately ambiguous or imprecise language; used esp. of political language that is subject to arbitrary national or party interpretation. Hence **double-talker**, one who uses such language; **double-talk** *v. intr.*; **double-talking** *vbl. sb.* and *ppl. a.*

1938 *New York Panorama* (Amer. Guide Ser.) vi. 156 Of late a humorously conceived system of language corruption called *double talk*..has made itself felt. *Ibid.* 157 *Double talk* is created by mixing plausible-sounding gibberish into ordinary conversation, the speaker keeping a straight face or *dead pan* and enunciating casually or *off the cuff*. **1941** *Time* 16 June 61/1 Thirteen recorded versions of this pandemic double-talk ballad are available. **1945** H. I. PHILLIPS *Private Purkey's Private Peace* xxii. 129 We got the right slant on bullies, greaseballs, double talkers, supermen, and dopes. **1948** AUDEN *Age of Anxiety* vi. 125 And all species of space respond in our own Contradictory dialect, the double talk Of ambiguous bodies. **1950** *Amer. Speech* XXV. 190 Back in Tsarist times Lenin and his associates inaugurated this double-talk and double-writing—especially the latter—in order to deceive the Tsarist censors and police. **1952** C. DAY LEWIS tr. *Virgil's Aeneid* IX. 203 You'll find no Atridae here, no double-talking Ulysses. **1956** *Ann. Reg.* *1955* 45 The 'directive'..was itself a masterpiece of the familiar technique of double talk. Terms such as 'free elections', 'free contacts'..meant different things in East and West. **1957** *New Statesman* 19 Oct. 1/2 The Archbishop of Canterbury is the best double-talker since the Delphic oracle shut up shop. **1958** D. EWEN *Compl. Bk. Amer. Mus. Theater* 230 'Melody in Four F.', a tongue-twister relating in double talk the adventures of being conscripted into the army. **1959** *Listener* 4 June 969/2 This has meant..a certain amount of double-talk, many carefully imprecise statements of intention. **1960** *Guardian* 23 Dec. 5/3 Britain's confused and double-talking attitude to the Cyprus question. **1961** *New Left Rev.* Mar.–Apr. (*front cover*), The Labour Party has double-talked its way around the issue. **1968** *Globe & Mail* (Toronto) 13 Jan. B5/1 Spiralling costs..have inflated the price of B.C. Hydro and Power Authority's Columbia and Peace River power and flood control dams—just how much nobody is certain and Mr. Bennett has persistently doubletalked on this subject.

double-team, *v. U.S.* **1.** [TEAM *sb.* 5.] *intr.* To combine two teams into one.

1843 W. T. NEWBY *Diary* 27 Sept. in *Oreg. Hist. Q.* (1939) XL. 235 We crawsed..a smawl streem. Dubeld teamed & crawsed the hill. *Ibid.* 28 Sept., We continued up the creek.. dubel teaming & passing the fork in 4 miles. **1934** H. VINES *Green Thicket World* 111 Often they had to double-team.

2. [TEAM *sb.* 3.] *intr.* To bring double force to bear *on* (or *upon*) a person.

1860 *Congress. Globe* 12 Jan. 424/2 In respect to the Senator's allusion to 'double-teaming' upon him..I do not exactly agree with my friend from Mississippi. **1865** M. B. CHESTNUT *Diary fr. Dixie* (1905) 346 Grant had double-teamed on Lee. **1904** T. WATSON *Bethany* (1920) 197 On the next day we double-teamed on one section of his army.

b. With *it*: to act in combination.

1884 'MARK TWAIN' *Huck. Finn* xix. 183 'Old man,' said the young one, 'I reckon we might double-team it together.'

dou·blethink, double-think. [Coined by 'George Orwell' (see quot. 1949) from DOUBLE *a.* 5 + THINK *sb.*] The mental capacity to accept as equally valid two entirely contrary opinions or beliefs.

1949 'G. ORWELL' *Nineteen Eighty-Four* I. iii. 37 His mind slid away into the labyrinthine world of doublethink. To know and not to know, to be conscious of complete truthfulness while telling carefully constructed lies, to hold simultaneously two opinions which cancelled out, knowing them to be contradictory and believing in both of them, to use logic against logic, to repudiate morality while laying claim to it, to believe that democracy was impossible and that the Party was the guardian of democracy. **1953** *Encounter* Nov. 26/1 He will react..either with straight abuse or with devious double-think. **1957** T. KILMARTIN tr. *Aron's Opium of Intellectuals* 119 How can one condemn the Soviet Union, since the failure of the Bolshevik enterprise would be the failure of Marxism and therefore of history itself? This is an admirable piece of philosophical double-think, typical of our latter-day intelligentsia. **1959** *Daily Tel.* 13 Nov. 12/2 They ask for increases in wages which are plainly impossible; or they pretend they want a shorter working week when they really want more overtime. Their followers know double-think when they see it, as well as the employers. **1969** *New Scientist* 2 Oct. 18/1 This symposium exhibited a form of intellectual doublethink that could pay lip service to global starvation one minute, and assume Britain would always be able to import most of her food the next.

double time. 1. *Mil.* (See DOUBLE *a.* 4 c.)

2. *Mus.* [TIME *sb.* 12.] Double the time specified or previously used. *Spec.* in *Jazz*: see quot. 1961. So **double-time** *v. intr.*

1877 W. S. GILBERT *Sorcerer* (1957) I. 217 Time was when this old heart would have throbbed in double-time at the sight of such a fairy form! **1939** D. BAKER *Young Man with Horn* i. 81 He..went into double time on the cymbal. *Ibid.*, He stayed right there double timing one-handed on the cymbal. **1961** A. BERKMAN *Singers' Gloss. Show Business Jargon* 19 *Double time*, doubling up the rhythmic beat without changing the singer's melody line.

doubleton (dɒ·b'ltən). *Card-playing.* [f. DOUBLE *a.*, after SINGLETON².] In Whist and Bridge: two cards only of one suit, in a player's hand. Also, one card of a doubleton.

1906 *Westm. Gaz.* 11 Aug. 14/1 B's ten of hearts is nearly sure to be a lead from the best of three—it cannot be a singleton... Similarly, if it is a doubleton [etc.]. **1922** *Evening News* 10 Apr. 8/6 Do not lead from a doubleton (a suit of two cards only) about which you have no information. **1927** *Observer* 20 Mar. 25 When he holds a singleton or weak doubleton of a suit. **1944** *Times* 17 May 6/3 Provision was made for revaluation after the first round of bidding to count three points for a void, two for a singleton, and one for a doubleton. **1960** T. REESE *Play Bridge with Reese* xxxiv. 118 Could the Queen of diamonds be doubleton?

doubletree (dɒ·b'ltrī). *U.S.* [f. DOUBLE *a.* + TREE *sb.*, after U.S. *single-tree* = SWINGLE-TREE.] The cross-piece to which the swingle-tree of a carriage, plough, etc. is attached.

1847 [see SINGLE-TREE]. **1952** STEINBECK *East of Eden* xiii. 121, I saw him..take the double-tree off the hay rake.

double wall. In full *double wall knot*: see WALL-KNOT. Hence **double-wall** *v.*

1801 J. J. MOORE *Brit. Mariner's Vocab.* sig. M1ᵛ, Double wall Knot. **1808** D. LEVER *Yng. Sea Officer's Sheet Anchor* 5 To Double Wall this Knot...The knot will appear.. having a double wall, and single crown. **1841** R. H. DANA *Seaman's Man.* vii. 37 Make the single wall slack, and crown it...Thus made, it has a double wall. *Ibid.* 38 It may be double walled by next passing the strands under the walling on the left of them.

doubling, *vbl. sb.* Add: **1. c.** *Mus.* The use by a single player of two (or more) musical instruments. Also *attrib.* (Cf. *DOUBLE *v.* 1 h.)

1926 WHITEMAN & MCBRIDE *Jazz* ix. 198 Most of the players perform on many different instruments...Doubling is, then, the main strength of the jazz orchestra. **1927** *Melody Maker* May 435/2 There is no doubt that doubling is an advantage, but a player should certainly not try to play a second instrument until he is an absolute master of the one he normally uses. *Ibid.* Aug. 809/2 A violinist anxious to learn a 'doubling instrument' quickly and easily. **1955** KEEPNEWS & GRAUER *Pict. Hist. Jazz* iii. 40 The excess of instruments on these bandstands indicates how much 'doubling' a musician was expected to do. **1968** *Melody Maker* 23 Nov. 18 Most of those at the colleges were drummers who had switched to vibes as a doubling instrument.

d. The action of DOUBLE *v.* 1 d.

1931 G. JACOB *Orchestral Technique* ix. 81 Unison doublings do not help much..except in the case of high trumpet parts being doubled in unison by clarinets. **1947** C. GRAY *Contingencies* viii. 146 The various doublings and *divisis* which are now the bread-and-butter of every orchestral composer.

doubloon. Add: **b.** *pl.* Money. *slang.*

1908 'O. HENRY' *Voice of City* (1916) 80 He's left his whole cargo of doubloons to a microbe. **1924** WODEHOUSE *Leave it to Psmith* i. 28 Aunt Constance keeps an eye on the doubloons and checks the outgoings pretty narrowly. **1959** P. BULL *I know the Face* v. 76, I..was anxious to lay my hands on anything that brought in the doubloons.

doubt, *sb.*¹ Add: **1. a.** *spec.* Uncertainty as to the truth of Christianity or some other religious belief or doctrine. Freq. *pl.* and occas. personified. (Cf. quots. *c* 1400, 1708, 1779, and 1850 in Dict.)

1855 BROWNING *Men & Women* I. 216 What have we gained then by our unbelief But a life of doubt diversified by faith, For one of faith diversified by doubt. **1915** G. K. CHESTERTON *Poems* 98 John Grubby, who was short and stout And troubled with religious doubt. **1924** C. MACKENZIE *Heavenly Ladder* xxiv. 296 It was all right so long as I said Mass myself; I had no doubts then. **1934** H. G. WELLS *Exper. Autobiogr.* I. iv. 188, I had not yet been confirmed... I suggested that I might have 'doubts'. **1960** P. MORTIMER *Saturday Lunch with Brownings* 109 For the first time in his life..he had Doubts. **1971** *Daily Tel.* 8 Apr. 10/6 (*heading*) 3 per cent. of church-goers have doubts.

b. Phr. *to give* (an accused person) *the benefit of the doubt*: to give a verdict of Not Guilty where the evidence is conflicting; to assume his innocence rather than guilt. Hence in wider use, to incline to the more favourable or kindly decision, estimate, or the like.

[**1844** C. NAPIER *Let.* 21 Feb. in W. Napier *Life* (1857) III. 48, I shall therefore..give him the benefit at your request.] **1848** *Bell's Life* 9 July 2/3 If he thought he was out, it must suffice; but he ought to have been quite certain, or..to have given the batter the benefit of the doubt. **1860** T. INMAN *On Myalgia* 104 We should more frequently give our patients the 'benefit of our doubts', and abstain from attempting to cure an inflammation [etc.]. **1892** [in Dict.]. **1961** P. USTINOV *Loser* xi. 259 He deserved the benefit of the doubt, for old times' sake. *Ibid.* xiii. 284 Perhaps, he now thought, he had just been a microcosm of a world addled by a desperate malady...No, he deserved no benefit of any doubt.

doubting, *ppl. a.* (See also THOMAS 1.)

douceur. Add: **1.** (Further examples.) Revived in sense 'something pleasant or agreeable'. So *douceur de* (*la*) *vie* or *de vivre*: the pleasure or the sweet things of life.

1700 CONGREVE *Way of World* IV. 57 My morning thoughts, agreeable wakings, indolent slumbers, all ye *douceurs*, ye *Someils du Matin adieu*. **1938** W. S. MAUGHAM *Summing Up* 294 They will not know the easy, sheltered life which makes many who were at their prime before the war look upon those years as did the survivors of the French Revolution when they looked back on the Ancien Régime. They will not know the *douceur de vivre*. **1952** W. PLOMER *Museum Pieces* 130 As if he were lamenting a *douceur de vivre* of which he had caught the last fragrance. **1958** I. MURDOCH *Bell* xxvi. 305 A curious relationship grew up between Michael and Dora, something undefined and wistful which had for Michael a certain ease and *douceur*. **1962** *Times* 13 Feb. 11/3 These institutions are sometimes a bit short of *les douceurs de la vie*. **1962** *Listener* 6 Dec. 979/2 Yet another reminder of that *douceur de vivre* which in 1939 vanished for ever. **1963** *Punch* 3 July 4/2 A certain softness, ..easy-going placidity, *douceur de vie*. **1965** 'J. DARCY' *Killing in Hats* iii. 59 Between the *douceur de la vie* which was about to pass, and the tantalising future.

3. (Later examples.)

1880 J. T. WHEELER *Short Hist. India* III. iv. 354 The commander-in-chief of the Bengal army..reckoned on receiving a handsome douceur. **1922** *Classical Rev.* XXXVI. 31/2 The Emperor..adds that a douceur will be necessary in order to get him to consent to the loan. **1965** N. GULBENKIAN *Pantaraxia* xiv. 305 On the few occasions when I have a bet, a winning one, my wife gets a little nonsense and all my servants get a little *douceur*.

douche, *sb.* Add: Comb. *douche-bag*, *-can*, *-glass*.

1908 *Practitioner* Oct. 579 At first I used an ordinary enamelled tin douche-can. *Ibid.*, It is an advantage..to be able to watch the limb which is being congested, and so I now use a cylindrical douche-glass. **1934** H. MILLER *Tropic of Cancer* 109 Over the bedstead hangs a douche-bag which he keeps for emergencies. **1966** *Observer* (Colour Suppl.) 20 Mar. 41/2 A few belts, a tweed beret and a douche bag were all that was left of Eva Braun's envied wardrobe.

dough, *sb.* Add: **2. b.** Money. *slang* (orig. *U.S.*).

1851 *Yale Tomahawk* Feb. (Th.), He thinks he will pick his way out of the Society's embarrassments, provided he can get sufficient dough. **1896** ADE *Artie* ii. 12, I pulled in the dough and picked up the cards. **1917** [see *BOODLE¹ 2 b]. **1919** *War Slang* in *Athenæum* 22 Aug. 791/2 'Dough' denotes money, but more especially the weekly pay. **1942** WODEHOUSE *Money in Bank* (1946) xxvii. 241 She's got more dough than you could shake a stick at. **1943** *Coast to Coast 1942* 59 It might mean that I'll get a chance of makin' some dough. **1944** C. A. LAWRENCE *Narrowing Wind* 60 You're in the dough and you kin afford to live there better 'n I kin. **1955** *Times* 3 Aug. 5/4 I'm going back to business and make myself a little dough.

attrib. **1904** *N.Y. Even. Post* 7 Nov. 3 This is Tammany's regular annual 'dough day'—that is, the day on which the district leaders come to Tammany Hall for election day funds. **1906** *Ibid.* 24 Oct. 4 In the country, election day without some sort of 'dough-bag' is an unheard-of thing. No 'dough-bag' means no votes.

5. *dough-bait*; *dough-like* adj.; **dough-cake,** (*a*) a cake made of dough; (*b*) *dial.* a simpleton, a fool; **dough-head** (example).

1904 *Westm. Gaz.* 19 Nov. 3/1 On the bank..men and boys..are fishing with quill-floats and dough-bait, the least artistic form of sport. *c* **1750** M. PALMER *Devonshire Dialogue* (1839) 33 How unvitty and cat-handed you go about it, you dough-cake. **1839** F. TROLLOPE *Dom. Manners* (ed. 5) 272 It won't convene for me to be mixing doe cakes and Johny cakes all day. **1844** LEE & FROST *10 Yrs. Oregon* xxii. 290 Becoming quite hungry we got out some flour, and baked some dough cakes. **1892** *Encycl. Cookery* I. 525/1 Small Dough Cake...Large Dough Cake. **1921** W. DE LA MARE *Memoirs of Midget* 8 Stuffing himself out with bread-and-dripping or dough-cake. **1854** M. J. HOLMES *Tempest & Sunshine* xv. 222 He inwardly accused them all of being 'doughheads'. **1928** A. B. CALLOW *Food & Health* 25 The 'indigestibility' of very new bread is due to its dough-like consistency.

dough-boy. Add: **2.** *U.S. slang.* †**a.** = ADOBE (cf. *DOBE, *DOBIE). *Obs.*

1856 in *Mont. Hist. Soc. Contrib.* (1940) X. 74 The Carpenters at Work getting the Doughboy tools ready. *Ibid.* 75 The men mixed their mud in preparation for making doughboys. *Ibid.* 82 The Carpenters made..three doughboy moulds.

b. *colloq.* An infantryman in the U.S. Army. Also *attrib.*

There have been several conjectures as to the origin of this sense (*e.g.* quots. 1887 and 1940).

[**1865** *Harper's Weekly* 25 Nov. 741/2 The battered excursionists..were designated as 'Bummers', 'Do-Boys', 'Raiders', etc.] **1867** in E. Custer *Tenting on Plains* (1887) xvi. 516 Wasn't I glad I was not a doughboy? **1887** *Ibid.*, A 'doughboy' is a small round doughnut...Early in the Civil War the term was applied to the large globular brass buttons on the infantry uniform, from which it passed..to the infantrymen themselves. **1898** *Scribner's Mag.* Aug. 133 Horse dealers..led their ponies up and down before.. dough-boy officers. **1904** *N.Y. Times* 13 June 6 A disgusting practice which reduced a bold cavalier to the level of a 'doughboy' at once. **1918** 'I. HAY' *Last Million* ix. 134 The true exile in this war is the American born Doughboy. **1920** E. W. BOK *Autobiogr.* (1921) xxxi. 293 The amazing part of the 'show'..was the American doughboy. **1940** O. L. SPAULDING in *Dict. Amer. Hist.* II. 163/1 The word 'doughboy'..can be traced with certainty as far back as 1854...The contemporary explanation then was, that the infantrymen wore white belts, and had to clean them with 'dough' made of pipe clay. **1966** A. LOOS *Girl like I* ix. 164 During World War I, she dressed as a doughboy in olive drab.

dough-face. 1. Substitute for def.: A mask made of dough. Also *transf.* (Further examples.)

1809 *Deb. Congress U.S.* 23 Feb. (1853) 1509 It is something like dressing ourselves up in a dough-face and winding-sheet to frighten others. **1820** *Massachusetts Spy* 22 Mar. (Th.), They saw their dough faces in the glass and were frightened. **1883** E. EGGLESTON *Hoosier School-Boy* xviii. 120 Two boys from the neighborhood, who had joined the party, agreed to furnish dough-faces for them all. **1940** C. MCCULLERS *Heart is Lonely Hunter* (1943) II. i. 98 One boy had..put on a dough-face bought in advance for Hallowe'een.

2. (Earlier examples.)

[**1820** *New Brunswick Times* 13 Apr. (Th.), [John Randolph of Roanoke] said, 'I knew these would give way. They were scared at their own dough faces...We had them.'] **1830** *Boston Transcript* 6 Dec. 2/3 The protecting duty will be repealed, if the anti-tariff party can get enough dough faces to join them.

doughnut. Add: (Now in general use in English; also in form **donut.**) Freq. made in the shape of a thick ring. **2. a.** *colloq.* or *slang.* Applied to various objects with a shape resembling the toroidal shape of a doughnut, as a motor-car or aeroplane tyre or a ring-shaped float (see quots.). In *Math.*, a torus.

1925 FRASER & GIBBONS *Soldier & Sailor Words* 82 *Dough nuts*, a Navy name for the 'Carley Floats', life-saving rafts of circular shape, carried on board ships of war. **1930** *Amer. Mercury* Dec. 455/2 *Doughnut*, an automobile tire. 'We clout ten doughnuts an' call it a day.' **1931** *Flight* 13 Feb. 145/1 The 'doughnut' wheels with which the Air-work School machines are being equipped. **1936** M. ALLIS *Eng. Prelude* xxviii. 202 An occasional stone doughnut of a sheepfold, tells of solitary shepherd vigils. **1955** *Times* 31 Aug. 9/4 The Air Ministry are also interested in quickly inflatable 'doughnut' buoyancy bags, which fit round the wheels like a horizontal lifebelt and can be inflated to about three times the size of a motor car tire. **1959** M. GARDNER in *Sci. Amer.* Mar. 148/2 107 We..then stretch and bend the cylinder into a torus...Every pair of antipodes on this diabolic doughnut will total 17.

b. *Nuclear Engin.* A toroidal vacuum chamber placed between the magnet poles of a betatron or synchrotron, in which electrons or protons are accelerated; applied also to similarly shaped vessels in other devices (see quot. 1958).

1941 *Physical Rev.* LX. 51/1 The central 7 cm. hole of the doughnut was formed by pushing the centers of the dished faces together and picking out the glass. **1942** *Ibid.* LXI. 94 To cut down bombardment of the inside of the vacuum doughnut..electrons are injected at 15 or 20 kilovolts for only a brief interval. **1942** *Electronics* Sept. 84/2 Prof. Kerst, inventor and designer, is shown preparing to place the doughnut vacuum tube between poles of the electro-magnet. **1949** *Times* 12 Aug. 2/3 The essential function of the synchrotron..is achieved by firing a burst of electrons..into an evacuated 'donut'. **1958** *New Statesman* 6 Sept. 266/3 In Zeta, the heavy hydrogen particles, forming a 'plasma', are contracted into a beam which travelling round in the 'doughnut' or 'torus' (like a giant motor-tyre) is 'pinched'.

3. *Comb.*, as *doughnut-shaped* adj.

1941 [see *BETATRON]. **1962** *Sci. Survey* VII. 108 The effect..may be produced by toroidal (or doughnut-shaped) magnetic fields encircling the sun.

Douglas¹ (dɒ·glăs). [The name of David *Douglas* (1798–1834), Scottish botanist.] In full *Douglas fir, pine*, or *spruce*: a large coniferous tree, *Pseudotsuga menziesii* (also known as *P. taxifolia*) or *P. glauca*, native to western North America.

[**1837** A. B. LAMBERT *Descr. Genus Pinus* III, (*heading*) Pinus Douglasii. Trident-bracted fir. *Ibid.*, I gladly adopt the name of *P. Douglasii*, in honour of the indefatigable botanist to whom I am indebted for the specimens from which I have been enabled to complete my description and plate of the species.] **1856** A. W. WHIPPLE in *Rep. Explor. Route to Pacific* (*U.S. War Dept.*) III. I. 79 *Douglass spruce*.. would afford a better material for railroad ties. **1868** F. WHYMPER *Alaska* 44 Where the Douglas pine, spruce, and hemlock had grown. **1873** G. M. GRANT *Ocean to Ocean* x. 259 The hillsides and the country beyond support a growth of splendid spruce, black pine, and Douglas fir. **1884** C. S. SARGENT *Rep. Forests N. Amer.* 209 *Pseudotsuga Douglasii.* ..Red Fir. Yellow Fir. Oregon Pine. Douglass Fir. **1931** *Discovery* Sept. 285/1 Methods of forcing the oil into the wood of refractory species, such as larch and Douglas fir, are being investigated. **1937** *Evening News* 11 Feb. 8/4 All the Office of Works stands are being constructed of Douglas Fir and Western hemlock from British Columbia. **1945** R. W. SERVICE *Ploughman* 156 The shack stood in a clearing, engulfed by Douglas pines three hundred feet high. **1947** R. PEATTIE *Sierra Nevada* 148 It is called Douglas fir or Douglas spruce indiscriminately. **1970** H. L. EDLIN *Collins Guide to Tree Planting & Cultivation* xix. 286 The green or Oregon Douglas fir, *P. menziesii*, comes from British Columbia and the neighbouring regions of Alaska, Canada, and the United States. *Ibid.*, Where there is less room, the best kind to plant is the blue or Colorado Douglas fir, *P. glauca*.

Douglas² (dɒ·glăs). *Anat.* The name of James *Douglas* (1675–1742), Scottish physician, used in the possessive or with *of*-adjunct to designate various anatomical structures described by or named after him; *Douglas's pouch* (or *pouch of Douglas*), a pouch of the peritoneum between the uterus and the rectum; the recto-uterine pouch.

1864 W. SHARPEY et al. *Quain's Anat.* (ed. 7) I. 251 The deficiency thus resulting in the posterior wall of the sheath of the rectus muscle is marked superiorly by a well-defined lunated edge, whose concavity looks downwards towards the pubes—the semilunar fold of Douglas. **1878** W. TURNER *Introd. Human Anat.* xii. 841 The interval between the uterus and rectum..is the utero-rectal pouch, or pouch of Douglas. **1880** *Lancet* 25 Sept. 494/2 Tappings through the peritoneum..for accumulation of fluid in Douglas's pouch. **1907** *Practitioner* Apr. 472 A boggy, ill-defined, and very tender swelling present in Douglas's pouch. **1954** L. T. MORTON *Garrison & Morton's Med. Bibliog.* (ed. 2) 107 Douglas described the peritoneum in detail; his name is perpetuated in the 'pouch', 'line', and 'fold of Douglas'. **1965** L. B. AREY *Developmental Anat.* (ed. 7) xviii. 329 The pelvic cœlom is thereby subdivided into two blind bays, the recto-uterine pouch (of Douglas) and the vesico-uterine pouch. **1970** *Cumul. Index Med. 1969* X. 5621/1 Puncture of Douglas' pouch in cytodiagnosis of ovarian neoplasms.

Douglas³ (dɒ·glăs). *Economics.* The name of Major Clifford Hugh *Douglas* (1879–1952), British engineer, used *attrib.* to designate his plan for 'social credit'. Hence **Dou·glasite** *a.* and *sb.*

1933 H. G. WELLS *Bulpington of Blup* ix. 346 A lady who wanted to help forward the Douglas Scheme. *Ibid.* 369 She took up..the Douglas Plan. **1938** D. GARNETT in T. E. Lawrence *Lett.* 866 Ede had asked Lawrence's opinion of the Douglas Credit Scheme. **1939** H. G. WELLS *Holy Terror* I. iii. 73 It included everything from..single-taxers to Douglasites. *Ibid.* 75 An earnest..little man..who had swallowed the Douglas gospel whole. **1962** *Listener* 6 Sept. 357/3 Orage..became the most persuasive exponent of Douglasite economics the Social Creditors ever had.

Doukhobor (dū·kŏbǭɹ). Also **Dukh-.** Pl. **-ors** or **-ortsy.** [Russ. *Dukhobór*, pl. *-bóry*, also *-bórets*, pl. *-bórtsy*, spirit-wrestler (see SPIRIT *sb.* 23 c).] A member of a Russian religious sect which originated in the 18th century, many of whose members emigrated to Western Canada in the late nineteenth century after persistent persecution. Also *attrib.*

1876 J. B. TELFER *Crimea* I. x. 112 The Douhobortsy fully believe the Scriptures to be the revelation of God. **1886** *Encycl. Brit.* XXI. 82/2 The 'Dukhobortsy' communities (warriors of the Spirit)..are renowned as colonizers. **1897** [see SPIRIT *sb.* 23 c]. **1899** *Westm. Gaz.* 24 Apr. 3/2 Some of the Russian Doukhobors who have been expelled for refusing to participate in war. **1899** *Daily News* 30 May 5/2 Several settlements of the persecuted sect of the Doukobohrs [*sic*] are established there [*sc.* in Canada]. **1921** R. M. JONES *Later Periods of Quakerism* II. xx. 836 The persecution of the Doukhobors in Trans-Caucasia. **1931** *Times Lit. Suppl.* 17 Dec. 1021/4 We learn that when the Dukhobor colonists marched in force to the local seat of justice to protest..'every man, woman and child in the procession was stark naked'. **1957** J. S. HUXLEY *Relig. without Revol.* vi. 135 The anti-social but extremely religious Doukhobors. **1968** WOODCOCK & AVAKUMOVIC *Doukhobors* i. 19 The name of 'Doukhobor'..was first used in anger and derision by one of their opponents, Archbishop Amvrosii Serebrennikov of Ekaterinoslav. It means 'Spirit Wrestlers', and it was intended by the archbishop, when he invented it in 1785, to suggest that they were fighting *against* the Holy Ghost. *Ibid.*, There is a central, constant element in Doukhobor Christianity.

Doulton (dōᵘ·ltən). The name of John *Doulton* (1793–1873), used *attrib.* and *absol.* to designate pottery made at the works instituted by him.

1878 L. JEWITT *Ceramic Art* vi. 100 This ware is called 'Doulton ware', or 'sgraffito ware'. **1900** F. LITCHFIELD *Pottery & Porcelain* vii. 178 The artistic pottery may be divided into three classes,..namely, Doulton Ware, Lambeth Faience, and Impasto. **1909** H. G. WELLS *Tono-Bungay* II. iv. 229 The India-rubber plant in a Doulton-ware pot. **1966** G. BURNETT *Dead Account* iv. 27 A ceiling-high fixture..full of expensive books and Doulton and..glass bric-a-brac.

douma, var. *DUMA.

dourine (dū·rin). [ad. F. *dourin*.] A contagious disease of horses transmitted by copulation and caused by the parasite *Trypanosoma equiperdum*.

1882 Power & Sedgwick *Lex. Med.* II, *Dourine*, the Arabic name of *Mal de coït*. 1897 M. H. Hayes *Veterinary Notes for Horse Owners* (ed. 5) xxiii. 470 Dourine..was first observed in Germany by Ammon in 1796, and has since..spread to..Germany, Austria, Russia, Italy, France, Algeria, Syria and America.. It is unknown in Great Britain. 1903 *Ibid.* (ed. 6) 510 Dourine is a specific disease which at first appears as an inflammation of the surface of the genital organs, and which causes grave alterations in the nervous system of the attacked animal. *Ibid.* 511 Mares are more liable to acute dourine than stallions. 1963 Jubb & Kennedy *Path. Domestic Animals* II. 506/2 The symptomatology of dourine can be divided into genital, cutaneous, nervous, and general manifestations which occur separately or concurrently.

douro (dū·ro). Also **duro.** [Fr., ad. Sp. *duro*.] A former Spanish coin.
1870 Lady C. Schreiber *Jrnl.* 12 Apr. (1911) I. 105, I got for a duro a small specimen of the embossed tile. 1872 *Ibid.* 26 May 145 We gave a douro and a half (6/3). 1905 *Daily Chron.* 21 June 5/4 That the Moorish Government should pay 9,000 douros by way of compensation. 1905 *Westm. Gaz.* 12 Sept. 2/1 No, my daughter, a douro, that is sufficient. Another sou would be excessive. 1908 *Ibid.* 21 July 5/1 The new law provides that all these 'Seville' douros shall be confiscated. 1925 *Chambers's Jrnl.* June 383/1 He proposes forty douros as a fair price.. A douro is equal to five francs. 1962 R. A. G. Carson *Coins* 317 In 1848 a new decimal coinage created new denominations... In silver the denominations were the douro and its half of 20 and 10 reals respectively.

dourouculi (dūrukū·li). Also **dorocouli, douracouli, douro(u)coli.** [a. S. Amer. Indian name.] Any of several S. and Central American monkeys of the genus *Aotus*, characterized by large, staring eyes, long, non-prehensile tails, and nocturnal habits; the night-monkey or owl-monkey.
1842 *Ann. Mag. Nat. Hist.* X. 256 The two species of Douroucouli are evidently distinct. 1861 *Proc. Zool. Soc.* 111 The following additions were announced to have been made to the Menagerie.. 1 Douroucouli Monkey. 1891 Flower & Lydekker *Mammals* 714 The Douroucoulis. 1894 H. O. Forbes *Primates* I. 166 The Douroucoulis. 1897 *Q. Rev.* Oct. 414 The Douroucoulis or Night Apes are truly nocturnal animals.. The group ranges from Costa Rica and Nicaragua to the south of Paraguay. 1902 F. E. Beddard *Mammalia* 556 The Douroucouli Monkeys. 1929 *Times* 17 May 18 (caption) A pair of douroucoulis, or nocturnal owl-faced monkeys, one of the feline, the other of the three-banded species, recently acquired by the Zoo. The douroucouli is native to Guiana, Brazil, and Venezuela. 1964 *Listener* 23 July 123/1 Finnicky small mammals, such as marmosets and douroucoulis.

douser. Add: *Cinemat.* (See quots.)
1921 A. C. Lescarboura *Cinema Handbk.* 21 *Douser*, the manually operated door in the projecting machine, which intercepts the light before it reaches the film. 1940 *Chambers's Techn. Dict.* 262/1 *Douser*, the automatic screen which cuts off the light falling on to the film from the projector arc, when it is not passing intermittently through the gate.

dove, *sb.* Add: **1. d.** = *dove-colour* (5 b).
1895 *Bow Bells* 29 Mar. 322/1 Sortie-de-bals..are almost always in neutral tints—dove, gray, or fawn. 1903 *Daily Chron.* 21 Nov. 8/4 Blouses in dove with dove charmingly.
2. e. = *dove-marble* (*5 b below). Also *attrib.*
1805 *Times* 7 Nov. 4/4 Vein, Dove, and fine Statuary chimney-pieces. 1872 *Rep. Vermont Board Agric.* 667 The first [sc. marble] to be mentioned is the 'Dove', it being of a dove color.
f. *Politics.* A person who advocates negotiations as a means of terminating or preventing a military conflict, as opposed to one (cf. *hawk *sb.*¹ 3) who advocates a hard-line or warlike policy. Also *attrib.* or quasi-*adj.* and *transf.*
1962 Alsop & Bartlett in *Sat. Even. Post* 8 Dec. 20/1 The hawks favored an air strike to eliminate the Cuban missile bases... The doves opposed the air strikes and favored a blockade. 1964 *New Yorker* 10 Oct. 108 Not one of them, whether a 'dove' or a 'hawk', took much stock in the notion of 'overkill'. 1966 *Guardian* 10 Jan. 9/8 The Republicans are themselves divided into two groups: the liberal Javits, or doubting dove wing; and the Gerald Ford, or hawk wing, which wants a 'total win' in Vietnam. 1966 *Listener* 21 July 93/2 The term 'hawks and doves'..came into circulation by Charles Bartlett, President Kennedy's great journalistic confidant, in the course of an apparently inspired account of what took place in the President's own National Security Council at the time of the Cuban missile crisis. *Ibid.* 6 Oct. 488/2 For the South Vietnamese there are no nice clear-cut issues, no hawk or dove solutions. 1967 *Boston Sunday Herald* 30 Apr. III. 5/3 It is unfair for the Administration and the hawks to try to compromise the patriotism of the doves. 1971 *N.Y. Rev. Books* 17 June 19/1 A perceptive columnist and long-time dove.
5. b. *dove-soft, -white, -winged adjs.*
1923 E. Sitwell *Bucolic Comedies* 44 And the miller's daughter Combs her locks, Like running water Those dove-soft locks. 1871 Swinburne *Songs before Sunrise* 66 Now, to stroke smooth, the dove-white breast of love. 1967 *New Statesman* 28 July 109/2 Their dove-white cars sped by heavy black armour contorted like paper-clips. 1867 G. M. Hopkins *Wr. Deutschland* (1918) st. 3, My heart, but you were dove-winged, I can tell.
b. *dove-coloured* (earlier and later examples); **dove-flower** (examples); **dove-marble,** marble of a dove-colour; **dove orchid** or **orchis** = *dove-plant*; **dove tree,** *Davidia involucrata* and its varieties.
1727 E. Dorrington *Hermit* iii. 227 A grave Gentlewoman..dress'd in plain Dove-colour'd Cloaths. 1876 J. S. Ingram *Centennial Exposition* xi. 361 A very fine dove-colored or mottled marble was shown. 1831 *Curtis's Bot. Mag.* LVIII. 3116 (heading) *Peristeria elata*. Lofty Dove-Flower. 1951 *Dict. Gardening* (R. Hort. Soc.) III. 1529/1 *P[eristeria] elata*. Dove or Holy Ghost Flower. 1872 *Rep. Vermont Board Agric.* 675 The first mills at Swanton were wholly employed in the manufacture of grave-stones from the dove-marble. 1918 *Chambers's Jrnl.* May 311/2 The 'dove' orchid, or Espíritu Santo flower of Central America. 1852 C. M. Yonge *Two Guardians* viii. 141 Those tropical plants.. the dove orchis or the zebra-striped pitcher-plant. 1931 A. Osborn *Shrubs & Trees for Garden* xxv. 324 Davidia. Chinese Dove Tree. 1970 H. L. Edlin *Collins Guide to Tree Planting & Cultivation* 226 When a dove tree is in bloom in May, these white bracts stand out in a bold display, as though a flock of white doves were alighting amid its bright-green foliage, and this explains the English name.

dove, occasional pa. t. of Dive *v.* (N.Amer. examples.) See also E.D.D.
1855 Longfellow *Hiawatha* viii. 96 Straight into the river Kwasind Plunged as if he were an otter, Dove as if he were a beaver. 1857 *Canad. Jrnl. Industry Sci. & Art* II. Sept. 351 In England when a swimmer raises his heels..head foremost, into the water he is said to *dive*, and is spoken of as having *dived*... Not so however, is it with the modern refinements of our Canadian English. In referring to such a feat here, it would be said, not that he dived, but that he *dove*. a 1940 F. Scott Fitzgerald *Last Tycoon* (1949) v. 101 He dove in after her. 1970 *Toronto Daily Star* 24 Sept. 17/4 Forest Hill struck first when Mike Brown dove on a loose ball.

doveish, var. Dovish *a.*

dovekie. Add: (Also and now normally), the little auk (*Plautus alle*).
1917 T. G. Pearson *Birds of Amer.* I. 31 The little Dovekies or 'Sea Doves' breed along the coasts of Greenland. 1928 W. B. Alexander *Birds of Ocean* xxi. 257 Little Auk or Dovekie (*Alle alle*). 1954 Fisher & Lockley *Sea-Birds* i. 17 Among the auks the dovekie and the Brünnich's guillemot from the north join the puffins, razorbills and guillemots in ocean wanderings.

dovelike, *a.* (Later examples.)
1930 R. Campbell *Adamastor* 71 With dove-like voices call the distant fifes. 1968 *Listener* 2 June 816/3 Even a journalist from the dove-like *New York Times* was seen wielding a sub-machine-gun during the Tet offensive.

Dover's powder. *Pharm.* [Name of Thomas *Dover* (1660–1742), English physician.] A preparation of opium and ipecacuanha (*pulvis Doveri*) used as an anodyne diaphoretic, etc.
[1774 *Pharm. Coll. Regii Med. Edin.* 158 Pulvis sudorificus, sive Doveri.] 1802 *Edin. New Dispensatory* (ed. 6) III. xxvi. 525/1 Compound Powder of Ipecacuanha, commonly called Dover's powder. 1834 *Boston Med. & Surg. Jrnl.* 23 Apr. 174 A grain and a half of Dover's powder..has sometimes caused extreme anxiety for the safety of children under eight months old. 1887 *Buck's Handbk. Med. Sci.* V. 325/1 Dover's Powder.. Powdered Opium 10 parts, Ipecac. 10 parts, Sugar of Milk 80 parts. 1951 A. & E. F. Goodman *Pharmacol. & Therapeutics* xxiv. 492 Opium is often added in order to further allay coughing by depressing the center, the well-known Dover's powder being a favorite prescription for this purpose. 1971 D. Clark *Sick to Death* v. 92 You can get it [sc. ipecacuanha] in Dover's Powders.

dovetail, *v.* **3.** (Earlier example.)
1813 *Analectic Magazine* II. 111 The various compartments of the dialogue dove-tailed into each other.

dovish, *a.* Also **doveish.** Delete †*Obs.* and add later examples in sense of *dove *sb.* 2 f.
1966 *Listener* 6 Oct. 488/1 They tend to take up strong positions, 'hawkish' or 'dovish'. 1967 *Guardian* 14 Oct. 6/5 This sudden upsurge of dovish voices in the Republican ranks. 1968 *Ibid.* 16 May 9/1 He does not take a particularly dovish line himself on Vietnam. 1971 *Ibid.* 23 June 2/1 'Dovish' sponsors of the initial proposal..conceded that it was meaningless after the Administration succeeded..in making the nine-month withdrawal contingent upon a full release of all prisoners.

Dow: see *Dow-Jones.

dowd, *sb.*¹ Add examples of recent currency, which appears to be due to a new back-formation from Dowdy *a.*
1899 *Westm. Gaz.* 30 Oct. 2/2 She's a dowd to-day. 1904 *Ibid.* 23 June 4/2 Only a duchess may dare to be a dowd just now. 1930 *Time & Tide* 13 Sept., To confound the shallow-pates who maintain that the suffragist must be a dowd, the leader of the W.S.P.U. appeared on platforms clothed in Paris frocks. 1959 *John o' London's* 24 Nov. 629/1 A homely and devoted dowd like some of the later translations by Victorian writers.

dowdy, *sb.*¹ *Obs.* or *dial.* = Dowd *sb.*¹ (see quots.).
1778 F. Burney *Evelina* II. ii. 35 'Perhaps Lady Howard may be able to lend you a cap.'.. 'Do you think I'd wear one of her dowdies?' 1880 W. H. Patterson *Gloss. Antrim & Down* 32 *Dowdy cap*, same as dowd. 1905 Wright *Eng. Dial. Dict. Suppl.* 93/1 *Dowdy*. Hmp. The linen bonnet worn by women when working in the field.

dowelled, *ppl. a.* (Earlier example.)
1805 *Times* 7 Nov. 4/2 Excellent dowelled flooring.

dower, *sb.*² Add: **4. dower-chest,** (*a*) = *wedding-chest* (Wedding *vbl. sb.* 4 b); (*b*) *U.S.* = *hope chest* (*Hope *sb.*¹ 5); **dower-land** (earlier U.S. example).
1881 C. C. Harrison *Woman's Handiwork* III. 142 Carved dower-chests from Spain and Italy. 1921 *Daily Colonist* (Victoria, B.C.) 19 Oct. 17/1 Carved Oak Dower Chest. 1922 *Daily Tel.* 12 June 20/1 Antique walnut cabinets, dower chest. 1935 N. Venner *Imperfect Impostor* iv. There was an old oak dower chest, curiously carved. 1957 *Daily Tel.* 14 June 3/1 A fine old carved dower chest of the Henry VIII. period. 1769 Washington *Diary* 16 May (1925) I. 325 Rid over my dower Land in York.

dowitcher (dau·itʃəɹ). [Iroquois.] Any of several long-billed waders (genus *Limnodromus*) of North America, belonging to the family Scolopacidae and resembling the sandpiper, esp. the red-breasted snipe (*Limnodromus griseus*).
1841 *Spirit of Times* 9 Jan. 529/3 The mellow attenuated trill of the soaring dowitcher. 1872 Coues *Key N. Amer. Birds* 252 Brown-back. Dowitcher. 1888 Lees & Clutterbuck *B.C. 1887* xvii. 182 The long-billed dowitchers are very much like a large snipe, of a plain cinnamon colour. 1934 L. W. Hutchison *Notes in Kings College Sun* xiii. 136 They had been shot..and were Dowitchers, a species of Arctic wildfowl. 1965 *New Scientist* 21 Oct. 165/3 Sandpiper-like American waders such as..the dowitchers..form a substantial proportion of the visitors.

Dow-Jones (dau dʒōu·nz). The names of C. H. *Dow* (died 1902) and E. D. *Jones* (died 1920), American economists, used *attrib.* to designate an index of the relative price of American securities based on the current average rates of an agreed select list of industrial and other stocks. Also in shortened form *Dow*.
1908 *Ticker* Jan. 38/1 The Dow-Jones System of Averages is simply a method of calculating the average price of 20 active railroad stocks and 12 industrial stocks. 1922 W. P. Hamilton *Stock Market Barometer* i. 7 The Dow-Jones average is still standard, although it has been extensively imitated. 1957 *Encycl. Brit.* XVIII. 474/1 The Dow Jones averages..were based on 20 industrials, 20 railroads and 15 utilities. 1965 N. Strand *Marketing Dict.* 43 Various ways of measuring the trend of securities prices on the N.Y. Stock Exchange, the most popular of which is the Dow-Jones average of 30 industrial stocks. *Ibid.* 45 In the case of the Dow-Jones industrial average, the prices of the 30 stocks are totaled and then divided by a divisor which is intended to compensate for past stock splits and dividends and which is changed from time to time. 1964 *Financial Times* 25 Feb. 3/1 As the Dow Jones Industrial Average approached the 800 level lively profit-taking was attracted, which limited the market's rise to mostly fractions. 1968 *Globe & Mail* (Toronto) 17 Feb. B1 Among the 30 Dow stocks, 18 declined, 7 advanced and 5 were unchanged. *Ibid.* B 8 The Dow-Jones industrial average was down 0·9 points to 932·45. 1970 *Daily Tel.* 1 June 19/1 Prices plummeted to the 'low' of 631 on the Dow in response to genuinely poor economic prospects.

down, *sb.*¹ Add: **3. down-feather.**
1606 Shaks. *Ant. & Cl.* III. ii. 48 The Swannes downe feather That stands vpon the Swell. 1884 Martin & Moale *Handbk. Vertebrate Dissection* II. 97 Lying beneath the contour feathers are down feathers (*plumulae*). 1959 C. J. Hylander *Feathers & Flight* i. 10 Down feathers are the first body covering of baby birds.

down, *sb.*³ **2. b.** An act of throwing down, as in wrestling. In American and Canadian football: see quots.
1840 W. G. Simms *Border Beagles* 134 He downed him; a fair upset down. 1882 in P. H. Davis *Football* (1911) 470 They must give up the ball to the other side at the spot where the fourth down was made. 1893 W. K. Post *Harvard Stories* 41 After three downs Spofford dropped back. 1897 *Encycl. Sport* I. 425/1 *A down*.—The term used to indicate the number of attempts made to advance the ball. Each side has three tries in which to advance the ball five yards. The end of each try, *i.e.* when the ball is held by the opposing side, is a down. As soon as the five yards have been gained it is first down again. 1927 *Observer* 11 Dec. 25/3 After the kick-off the side that has the ball must gain ten yards in a maximum of four 'downs'. 1959 *Times* 30 Nov. (Canada Suppl.) p. xxi/1 Canada has the downs system, by which a team must make 10 yards in three downs, or turn over the ball. 1968 *Globe & Mail* (Toronto) 28 Sept. 18/5 Fifteen of Montreal's first downs were earned on the ground to six for the Riders.
5. Chiefly in phr. *to have a down on*: to dislike, regard unfavourably, be ill-disposed towards. orig. *Austral.*
1856 W. W. Dobie *Recoll. Visit Port-Phillip* v. 84 The bushranger had been in search of another squatter, on whom 'he said he had a down'. 1862 C. R. Thatcher *Canterbury Songster* 10 I've got no 'down' on Travers. 1874 M. Clarke *His Natural Life* (1875) I. ii. vii. 237 He never ceased to.. find fault with him... It was evident that Mr. Frere had a 'down' on the Dandy. 1878 R. B. Smyth *Aborig. Victoria* I. 129 Blacks never like a quarrel to be of long standing:..nothing would make a man more miserable than to think that some of his tribe had a 'down' on him. 1904 *Daily Chron.* 8 Dec. 8/1 Why this 'down' on an always useful, sometimes dainty, garment? 1916 *Chrons. N.Z.E.F.* 27 Dec. 199 They've got a down on arrogance and swank. 1928 S. Vines *Humours Unreconciled* xiii. 179 Somebody'd got a down on him. 1947 W. S. Maugham *Creatures of Circumstance* 151 She had a down on Lady Kastellan and didn't care what she said about her.
7. A 'down' train or coach. *rare.*
1884 [see Up *sb.* 3].
8. The position or action of a dog lying down in response to an order to do so.
1948 E. H. S. Longhurst *Dog Training* 158 Tests for obedience classes.. Recall from Sit or Down (dog to be recalled by handler when stationary). *Ibid.* 159 Article to be given to the handler as he leaves the ring for the 'down'.

down, *a.* Add: **1. c.** (Earlier examples.)
1840 H. Cockton *V. Vox* v. 27 They met the 'down coach'. 1845 in J. R. Planché *Extravaganzas* (1879) III. 48 Opening of the Down Line to the 'Bee and Orange' Station. 1846 —— *Bee & Orange Tree* v. 28 The Fairy Atmospheric down train descends rapidly and then enters the Garden.
d. Of a payment: see Down *adv.* 12.
1926 *Automotive Industries* 16 Sept. 451 The dealer gets less total money on the whole transaction but permits the unpaid balance after the down-payment to be greater. 1930 *Sam Acheson* (Texas) *Light* 31 Jan. 14/6 Small down payment, balance like rent. 1959 J. Braine *Vodi* x. 138 The money for the first inescapable hundred down-payments. 1970 *New York* III. 16 Nov. 4/2 He had to pay $10,000 for the down payment on his house.
e. *down trip*, an unpleasant or depressing hallucinatory experience induced by the drug LSD. *U.S. slang.*
1968 [see *Freak *n.* 4]. 1968 J. Hudson *Case of Need* iii. i. 172 She was real depressed. She took a couple of down-trips, real freaks, and it shook her up.

down, *adv.* Add: **3. b.** *Theatr.* = *down stage *adv.*
1893 Wilde *Lady Windermere* i. 12 Duchess of Berwick (*Coming down C., and shaking hands*). *Ibid.* iv. 153 Lord Darlington (*Comes down C.*). Lord Darlington (*Coming down back of table*). a 1916 H. James *Compl. Plays* (1949) 191 Enter Noémie, Nioche and Lord Deepmere, rapidly down from centre. *Ibid.* 194 Her father has to come down to rejoin before I know. (*Coming down, left centre.*) Whose is it? 1939 N. Coward *Play Parade* 17 An ordinary bus sign R.C. (right centre) down on footlights.
5. *down for the count*: see *count *sb.*¹ 1 c.
14. b. (So many points, etc.) behind one's opponent in a game; opposed to *up*.
1896 *Outer Spec.*; 1897 *Encycl. Sport* I. 472/2 A player is said to be down when his opponent has won one or more holes more than he has. 1907 H. H. Hilton *Golfing Remin.* 103 At the fourteenth hole he was one down. 1959 *Times* 21 May 4/1 He had another hard [golf] match, but was, I think, never down.
15. Later examples, in the phrase *down to date* (after Up to date).
1889 *Cent. Dict.* II. 1461/2 *Down to date, up to date*, to the present time. 1892 'Mark Twain' *Amer. Claim.* 211 I've got part of him down to date, anyway. 1897 —— *Following the Equator* xxv. 244 He was down to date with them, too. 1901 *Daily Chron.* 1 Nov. 3/1 Are our methods of fire-fighting down to date? 1903 *Westm. Gaz.* 5 June 2/3 The most down-to-date dictionaries... 1937 *Daily Tel.* 19 Nov. 22 The down-to-date traveller discovered..that without the aid of aeroplanes it was only just possible to equal Fogg's record.
25. c. *down charge*: the order given to a setter or pointer in training to drop when the game rises and the shot is fired. So as *sb.* and *v.*; **down-charging** *vbl. sb.*
1833 *New Sporting Mag.* V. 259/1 Some sportsmen..make him down charge when the bird is missed. *Ibid.* 260/1 Call out directly, 'Down charge!' 1848 W. N. Hutchinson *Dog Breaking* ii. § 21. 14 Your setter..should make the young dog lie down (for the 'down charge'). 1859 'Stonehenge' *Shot-gun & Sporting Rifle* ii. i. 119 He springs up the birds, calling out 'Down charge' at the same moment in a loud voice. 1880 Ld. Walsingham et al. *Shooting* I. 54 His obedience to 'down charging' being frequently enforced. *Ibid.* 334 Provided the dog is fairly cured of chasing, taught to 'down-charge', find, return, and keep at heel.
27. down on, c. *to be down on*: to be opposed to; to show or express disapproval of. Chiefly *U.S.*
1850 *Alta California* 5 Aug., Here the factory girls appear to be down on the male help. 1864 *Daily Californian* 10 Sept., To refuse payment of the trifling bills of a few journals who had latterly perhaps been down on them. 1874 'Mark Twain' *Sk. New & Old* (1875) 205, I was down on sich doin's. 1883 —— *Life on Miss.* 268 If there's one thing an alligator is more down on than another, it's being dredged. 1897 W. James *Var. Relig. Exper.* xix. 335 Some persons..glory in saying that they 'are down on' this or that. 1903 *McClure's Mag.* Nov. 92 I'm down on grafting mayors and councilmen and legislators. 1961 'They're down on her for some reason or other,' he told himself. 'Angry, most likely, because she's hung on to her boy.'
27*. down along: in, or to, the West Country. Also *attrib.*, and *sb.* (= the West Country). *dial.* and *slang* (see quot. 1929).
1871 Kingsley *At Last* i, Their faces lighted up at the old pass-word of 'Down-Along'; for whosoever knows Down-Along, and the speech thereof, is at once a friend and a brother. 1899 *Westm. Gaz.* 12 Jan. 3/3 The people of the 'down-along' country are slow in putting thoughts into words. 1905 E. Phillpotts *Secret Woman* iii. xiii, Henceforth I shall come down-along once a year to visit you. 1907 *Westm. Gaz.* 18 Dec. 4/2 In his 'Devonshire Characters'.. there is a true tang of 'Down-along' in every page. 1929 F. C. Bowen *Sea Slang* 40 *Down along*, sailing coastways down Channel.
28. (Further examples.) Also *transf.*
a 1828 J. Bernard *Retrosp.* 37 This curious class of mammalia, the 'Down-Easter' as it is often called. *Ibid.* 240 He had lately quitted 'Down East', and was coming South. 1825 [see *away *adv.* 11 b]. 1829 *Mass. Spy* 25 Nov. (Th.), 'Where the deuce is Dennis [a town]?' 'Oh, down east.' 1830 S. Smith *Major Downing* 1, Some of the down-east antiquaries. 1835 J. H. Ingraham *South-West* I. xv. 161 Miserable-looking sloops and schooners, compared to which, our 'down-easters' are packet ships. 1843 *Ainsworth's Mag.* IV. 426 This old lass lived 'down east' near Chiselhurst. 1924 R. Clements *Gipsy of Horn* 84 If this ship was a 'down-easter' she'd be flaunting a main-royal. 1945 J. C. Colcord *Sea Lang.* 68 *Down East*, a general term for Maine and the Maritime Provinces of Canada... A 'down-easter' may be either a person or a vessel hailing from that region. 1948 W. Stevens *Let.* 4 Apr. (1967) 582 We are too close to that severe Down East that is Labrador.
30*. down to earth: back to reality. Also freq. **down-to-earth** *adj. phr.*, interested in everyday affairs; not affectedly superior; realistic; ordinary. So *down-to-earth(n)ess*.
1930 Wodehouse *Very Good, Jeeves!* ii. 29, I had for some little time been living, as it were, in another world. I now came down to earth with a bang. 1932 *Canadian Forum* Feb. 193/3 This book is full of such 'down to earth' observations. 1934 M. Hodge *Wind & Rain* i. ii. 27 She's awfully 'down to earth', you know. 1943 J. S. Huxley *Evol. Ethics* viii. 64 The general moral principle of evolutionary purpose can come down to earth in the concrete task of achieving minimum planning. 1956 *Essays by Divers Hands* XXVIII. 80 There is a gaiety and a down-to-earthness in much of his verse. 1957 L. E. Jones *Edwardian Youth* ii. 55 Julian [Grenfell] had a passion for red-blooded down-to-earthness. 1958 A. W. Fielding *Corsair Country* viii. 163 So I call for the bill. Which brings us down to earth with an ugly thud. 1961 *Daily Tel.* 12 May 24/2 Dame Irene Ward's hope that the committee..will include 'some down-to-earth people who know something about what goes on'. 1969 *Listener* 6 Feb. 189/2 This down-to-earthness most literally strikes home in the anchoring of the play to an actual place.
34. down-dropping.
1900 Kipling in *Daily Mail* 1 May 4/5 A whispering Guardsman, half of whose larynx had been put out of commission by a down-dropping bullet. 1936 W. Stevens *Opus Posthumous* (1957) 62 The same down-dropping fruit in yellow leaves.
35. *downside*.
1926 D. H. Lawrence *David* xiii. 100 The downside of the loaf, as it were. 1966 *Listener* 8 Sept. 337/2 A downside in foreign confidence.

down, *prep.* Add: **1. c. down cellar**: in the cellar or basement. *U.S.*
1805 *Pocumtuc Housewife* (1906) 47 Put in the soap-grease barrel down cellar. 1855 M. Thomson *Doesticks* x. 84 A patent medicine pedlar, with a..conservatory down cellar. 1870 'F. Fern' *Ginger-Snaps* 141 When we place a young plant down cellar and shut out light and sunshine. 1871 Stowe *Oldtown Fireside Stories* 50 Ef ye should be down cellar, and the candle should go out, now? 1877 E. S. Ward *Story of Avis* 141, I wonder if it wouldn't help you out to go down cellar and stir the cream yourself. 1884 H. Paul *Linden on Saugus Branch* 131, I rushed down-cellar to get our lantern.
2. d. down the course: said of a horse which is trailing some distance behind the leaders in a race.
1920 A. E. W. Mason *Summons* xx. 202 All our horses were down the course... They weren't running in their form to-day. 1957 *Daily Mail* 21 Jan. 9 Certain horses which ran second or third in the great 'chase at Aintree were 'down the course' this week at Birmingham. *Ibid.* 5 Mar. 9 What about the big and powerful stables..whose horses are down the course one day and up the next, according to the betting?
e. down home: at one's home, in one's native land or region; also as *sb.*, one's homeland; hence (usu. *attrib.* or *adj.*) used to designate something, esp. jazz music or blues, that is down-to-earth and unpretentious. *colloq.* (orig. *U.S.*)
[1828 J. H. Newman *Jrnl.* in *Autobiogr. Writings* (1956) v. 212 After a week's stay at Highwood, I went down home to Brighton.] 1931 *Amer. Speech* VII. 49 'Down home' [in eastern Idaho] is Utah. 1938 *N.Y. Amsterdam News* 12 Mar. 17 Almost primal emotions, hangover from the old 'down home' rent strut' days. 1958 J. C. Holmes *Horn* (1959) 85 Maybe later pick up gigs with a downhome band. 1960 *Melody Maker* 7 May 14/3 A measure of down-home funk.
f. down the line: in various *lit.* and *fig.* senses (see quots.). Also (with hyphens) *attrib.*: from one end to the other, at every point (see also quot. 1959).
1908 [see Line *sb.*² 26 b]. 1958 [see *bolt *sb.*¹ 15]. 1959 *Chambers's 20th Cent. Dict. Suppl.*, *Down-the-line*, of a ballet-dancer, inconspicuously placed, unimportant. 1962 Listener 1 Mar. 364/2 To others the risk is rather that consultative arrangements down the line may reinforce industry's very British predilection for cosy little get-togethers. *Ibid.* 15 Mar. 469/2 The view of many present-day managers..is that they would want to overhaul our methods all down the line. 1962 *Economist* 9 June 1000/1 Mr Yarborough described himself as a 'down-the-line supporter' of President Kennedy. 1967 *Listener* 31 Aug. 264/1 The authorities are trying to substitute, for tight central guidance, disciplines that will ensure some down-the-line control.

down, *v.*² Add: **1.** *to down tools*: to cease working, to go on strike. Hence *down-tools* is used attrib. to designate such action. Also *fig.*
1898 *Westm. Gaz.* 7 Apr. 6/3 The men..have ruined their position by..suddenly downing tools. 1915 *Daily Express* 4 Mar. 1/5 The 'down-tools' movement seems to have arisen spontaneously out of the engineers' dissatisfaction at not securing the twopence an hour increase when they asked for it. 1923 *Daily Mail* 3 Mar. 13 Yesterday some hundreds of the men 'downed tools'. 1958 *Times* 22 June 11/6, I have been compelled..to employ the lazy and the inefficient at wages negotiated by their trade union, if not imposed under a down tools threat. 1958 *Oxf. Mail* 23 Jan. 1/2 He was 'sincerely sorry' to leave his job as Minister of Agriculture. 'I hate downing tools in the middle of a job.' 1958 *Punch* 15 Jan. 109/1 More trouble seems to be brewing for the B.B.C., over complaints from manufacturers that after each half-hourly news bulletin workers down tools for five minutes to discuss it. 1969 *Listener* 1 May 614/1 The employer should not be able to impose sudden arbitrary action on the workers, the only remedy for which may be a down-tools strike.
b. *fig.* To overthrow, to get the better of.
1808 H. S. Canfield *Maid of Frontier* 43 We will have trouble in finishing the gang after Hartfield is downed. 1909 *Westm. Gaz.* 26 July 1/2 He's some chance of 'downing' his political opponents. 1904 F. Lynde *Grafters* xviii. 233 'A determination to make my brag good.' 'To down the ring, you mean?' 'Yes; to down the ring.' 1947 W. Deeping *Sorrell & Son* i. § 5 The thing is not to let every swine down you. 1946 K. Tennant *Lost Haven* (1947) 7 He never missed an opportunity of downing Thorne, but Lost Haven continued to support the new store. 1970 *Globe & Mail* (Toronto) 25 Sept. 31/3 Luke Walker pitched a seven-hitter for his 15th consecutive victory as the Pirates downed Montreal Expos 8–0 last night.
c. To drink down.
[1860 O. W. Holmes *Prof. Breakf.-t.* 52 Give a fellah a fo'-penny bun in the mornin', an' he downs the whole of it.] 1922 C. E. Mulford *Tex* ix. 145 Silently he poured out a drink and downed it mechanically. 1949 D. M. Davin *Roads from Home* iii. 4 John downed the two [drinks] that were waiting for him. 1954 Koestler *Invisible Writing* 321, I downed the sherry. *Ibid.*, [He] downs the rest of his beer and calls for more.

down and out, *adj. phr.* orig. *U.S. colloq.* [Down *adv.* 5, Out *adv.* 19 c.] Completely without resources or means of livelihood; absolutely 'done'; also *transf.* (see quot. 1934). Also *absol.* ('the down and out') and as *sb.*
1889 *Kansas Times & Star* 28 Nov., The bowers, saloon-keepers and sports will meet..to provide a turkey feast for the 'down and outs' in their line. 1901 'H. McHugh' *John Henry* 31 Say! I was down and out. 1910 'O. Henry' *Trimmed Lamp* (1916) 181 I'm down and out. 1917 J. Farnol *Definite Object* vi. 49, I don't want 'em to think I'm floatin' around with a down-an'-outer. 1921 H. Walpole *Young Enchanted* III. vi, Everybody over forty is tired and down and out, and everybody under thirty has swelled head. 1923 H. L. Foster *Beachcomber in Orient* x. 215 Nowhere in my travels had I ever found a city so full of the down-and-out as was Singapore at that particular moment. 1924 W. B. Selbie *Psychol. Relig.* 87 The down-and-outs converted there. 1928 H. Walpole *Wintersmoon* II. v, Next to Wildherne was a down-and-out with holes in his boots. 1928 Galsworthy *Swan Song* III. vi. 259 'You've never been down and out, I imagine, Mr. Forsyte?' 'No,' answered Soames. 1933 'G. Orwell' (*title*) Down and out in Paris and London. 1934 *Amer. Speech* IX. 31 *Down and out* refers to the opening lead at an ace followed by the lead in the order of their rank, of the next lower cards in the same suit. 1958 *Times Lit. Suppl.* 10 Jan. 15/2 After leaving school he emigrated to Canada..—a world of outsiders, down-and-outs, drunks, sensualists, homosexuals and eccentrics. 1968 T. Parker *People of Streets* 31 The last contribution is to the poor people really, the ones who they call the down-and-outs. 1961 Billy Costello, down-and-out dosser, twenty-four years old.
Hence **down-and-out** *v. trans.*, to 'do for', destroy; **down-and-outer,** one who is 'down and out'; **down-and-outness.**
1909 *Springfield Weekly Republ.* 4 Mar. 2 Compliments from political enemies follow the most distinguished down-and-outer of his day into the seclusion of private life. 1926 'Boyd Cable' *Air Men o' War* 186 That machine-gun upstairs is a certain invitation to sudden death and the German gunners to down and out us. 1922 H. L. Foster *Trip Tramp.* ii. 20 The down-and-outers of whom the old sea-captain had spoken. 1926 *Blackw. Mag.* Apr. 455/2 In the process of investigating the reason for their down-and-outness. 1926 *Boston Sunday Herald* Apr. 27/1 Two down and outers, looking for some rich people to marry, find each other.

down-beat (dau·nbīt), *sb.* [Down *adv.* 38.] A downward beat; *spec.* in *Mus.*, (the downward stroke of a conductor's baton or hand, indicating) the first or most heavily accented note of a measure; = Thesis 1.
1876 [see Up *a.* 4]. 1879 Grove *Dict. Mus.* I. 95/2 We make the down beat for the strong accents, and raise our hand for the others. 1891 [see Thesis 1]. 1937 *John o' London's* 7 May 217 To convert the energy of a down-beat in flapping flight into that forward thrust which gives to birds their speed through the air. 1946 P. Rosenfeld tr. *R. Schumann's Music & Musicians* 169 Music itself sought to return to its origins where the laws of downbeat did not yet oppress it. 1955 G. Abraham in H. van Thal *Fanfare for E. Newman* 13 Each wind instrument enters on a downbeat.

down-beat (dau·nbīt), *a.* orig. *U.S.* [cf. prec.] Pessimistic, gloomy, sombre; relaxed, unemphatic.
1952 *N.Y. Times Mag.* 6 Jan. 10 The visitor to Europe may be..distressed by the down-beat mood of the people. *Ibid.* 16 Mar. 22 No type of film is more chancy..than the one that is loaded with misery and ends on a note of despair. ..Such pictures have, in recent years..been tagged 'down-beat' films. 1955 *N.Y. Herald-Tribune* 19 Sept., That pictorially memorable march up the twilit hills of a dusty Southern town has an inexplicably plodding and down-beat air about it. 1958 *Times* 9 May 13/7 The checks among the men continue in down-beat tones. 1958 *Observer* 30 Nov. 16/7 It [sc. a play] had a nice natural down-beat ending. 1966 *Listener* 10 Jan. 88/2 In the rest they were crushed with typical continental down-beat deliberation.

down-calving, *a.* [f. Down *adv.* 34.] (See quot. 1886.)
1886 F. T. Elworthy *W. Somerset Word-Bk.* 205 *Down-calving*, in calf, and near the time of calving. (Very com.) 25 Down-calving cows and heifers.—Local advertisement of sale. 1931 *Daily Express* 25 Oct. 9/4 The challenge cup for the best down-calving cow. 1933 J. M. Murray *Community Farm* (1933) 47 We bought six down-calving shorthorn heifers.

down-comer. Add: **b.** Also for conveying gas. (Earlier, later, and *attrib.* examples.)
1888 *Lockwood's Dict. Mech. Engin.* 115 *Down-comer*, or *down-take*, the vertical pipe which conducts the waste gases from the top of a close-mouthed blast furnace into the blast main. 1920 *Chambers's Jrnl.* Jan. 60/1 At the extremities of the two drums are large tubular connections, the uptake being at the front end (where the hottest temperatures prevail), and the downcomer outside the furnace at the rear end, which is practically cool. 1950 *Engineering* 30 June 722/2 Fed from the steam and water drum through downcomer tubes.

down-coming, *ppl. a.* Coming down or downwards.
1851 H. Melville *Moby Dick* III. xlix. 308 Starbuck and Stubb, standing upon the bowsprit beneath, caught sight of the down-coming monster. 1862 *Harper's Mag.* July 167/2 They reached a spot where they passed the down-coming train in safety. 1922 Joyce *Ulysses* 437 He disappears into Olhousen's the pork butcher's, under the downcoming rollshutter. 1951 L. MacNeice *Ten Burnt Offerings* 50 Dust of down-coming hours. 1968 G. M. B. Dobson *Exploring Atmos.* (ed. 2) vi. 107 The intensity of the down-coming, cosmic radiation increases with height.

down-country, *sb.*, *adv.*, and *a.* [Down *a.* and *prep.*]
The phrase, which is current in North America, New Zealand, South Africa, etc., is to be distinguished from *down-country* in Dict. s.v. Down *sb.*³ 4.
A. *sb.* The flat part of a country (as opposed to the hilly regions). In the United States: see quots. 1823 and a 1870.
1823 J. F. Cooper *Pioneers* v, To them the road that made the most rapid approaches to the condition of the old, or, as they expressed it, the down countries, was the most pleasant. a 1870 R. M. Chipman *Notes on Bartlett* 129 *Down-country*, used in the interior to denote on or toward the seaboard; occasionally, the seaboard, or the land nearer a river's mouth. 1904 'G. B. Lancaster' *Sons o' Men* 192 In the down-country. 1933 Acland in *Press* (Christchurch) 30 Sept. 15/7 *Down country*, used (chiefly by people in the hills) to describe the localities near town or on the plains. They also speak of *down country* people or sheep. 1940 E. C. Studholme *Te Waimate* (1954) xii. 113 At Waimate we have our own down-country and Merinos or half-breds on the hills.
B. *adv.* In or to the flat part of a country; in or to the part of a country on the plains.
1874 A. Bathgate *Colonial Experiences* x. 135 A dozen or more horses with their pack-saddles empty..were returning down country. 1879 W. J. Barry *Up & Down* xxiv. 263, I sent her down-country in the coach. 1945 *N.Z. Geographer* I. i. 36 They can make a better 'do' of it than on a mixed farm down-country.
C. *adj.* Situated in, belonging to, or relating to the part of a country on the plains.
1896 A. Bryden *Tales S. Afr.* iii. 68 You know I don't..sport tall yarns for the benefit of down-country folk at Kimberley. 1901 Kipling *Kim* xiii. 353 Who ever heard of these Sahibs coming into the hills without a down-country cook? 1950 *N.Z. Jrnl. Agric.* Oct. 358/1 The breed of sheep on Glenaray [high-country sheep-run] is Romney type..and it is attractive to the down-country farmer when no longer suitable for the run.

down-draught. Add: **1.** (Further examples.)
1907 *Daily Chron.* 25 Oct. 8/5 It was maddening that these harsh down-draughts of the smoke should come to help the enemy. 1961 *Manch. Guardian* 4 Sept. 1/6 The sports plane was apparently caught by a down-draught.
b. *attrib.* or as *adj.* Designating a furnace, carburettor, etc., employing a downward draught of air or gas.
1906 C. Moore *Handbk. Pract. Smithing & Forging* ii. 6 These down-draught hearths are now being adopted in many of the modern works. 1935 *Jrnl. R. Aeronaut. Soc.* XXXIX. 503 A centrifugal fan delivers compressed air to a Stromberg down-draught carburettor. 1959 *Chambers's Encycl.* VI. 130/2 Intermittent kilns may be of the rectangular or round down-draught type. 1969 *Motor Manual* (ed. 36) iii. 51 Carburettors may be up-draught, horizontal or down-draught, according to the direction in which the main mixture stream is fed into the engine.
2. (Later example.)
1850 A. M'Gilvray *Poems* 58 Wives, and wives' friends are..a d—d down-draught, If they be poor.
b. A ne'er-do-well; a profligate. *dial.*
1832 *Aberdeen Shaver* Jan. 125 He is..nothing better than a down-draught, or ne'er-do-weel. 1849 C. Brontë *Shirley* xxii, They were chiefly 'downdraughts', bankrupts, men always in debt and often in drink.
3. The drawing or displacing of water by an object as it sinks.
1899 F. T. Bullen *Way Navy* 24 The down-draught of the anchor had sucked him after it almost to the bottom.

downed (daund), *a.* [f. Down *v.*¹ or Down *sb.*² + -ed².] Covered or lined with down.
1906 *Westm. Gaz.* 14 May 2/3 The poor hare..been sitting comfortably on a downed nest. 1939 Dylan Thomas *Map of Love* 22 Pack back the downed bone.

downer (dau·nəɹ). Colloq. var. Down *sb.*³ 5.
1915 C. Mackenzie *Guy & Pauline* i. 46, I knows better than go for to contradict you when you've got a downer on anyone. 1936 S. Sassoon *Sherston's Prog.* III. 224 He asserted that I'd got 'a downer' on some N.C.O.

downface, *v.* [Down *adv.*] *trans.* To contradict, controvert; to browbeat; to outstare; = *to face down* (see Face *v.* 3 a.). (Cf. *down-faced* s.v. Down *adv.* 38.)
She downfaces us that you've got the key of the padlock in a letter. 1922 Joyce *Ulysses* 323 He'd try to downface you that dying was living. 1936 E. Bowen *Death of Heart* I. ii. 102 Daphne's person was sexy, her conversation irreproachably chaste. She would downface any remark by saying 'You are awful,' or by turning away her face.

down-fold. *Geol.* A synclinal fold or depression. So **down-folded** *a.*, (of strata) dipping on each side towards a common axis.
1906 H. J. Mackinder *Britain & Brit. Seas* xi. 157 To the very top it consists of down-folded beds. *Ibid.* 60 Geographical valley and geological downfold here coincide with a precision that is rare. 1926 *Glasgow Herald* 15 Oct. 7 The rest of the region divides by arches or anticlines, but from the down-folds (synclines). 1954 J. F. Kirkaldy *Gen. Princ. Geol.* ii. 25 Stratified rocks..are often seen to be folded into up-folds (anticlines) and down-folds (synclines). 1970 *Sci. Amer.* Feb. 37/2 The blocks of faulted crust in the lower part of the experiment are tilted..in the direction opposite to what would be expected from the downfolds.

down-glide. *Phonetics.* A downward glide (see Glide *sb.* 2 b).
1890 H. Sweet in J. T. Hatfield et al. *Curme Vol. Linguistic Studies* 94 The up-glides and down-glides of the voice. 1930 G. E. Fuhrken *Standard Eng. Speech* xvi. In *that's his son* the down-glide coincides with the end-consonant in *son*. 1965 *Language* XLI. 475 In stressed syllables the long-voice, high, mid, and low are actualized as rapid downglides.

down grade, down-grade, *sb.* Add: (Earlier examples.) Orig. *U.S.* Also as *adj.*
1858 *Harper's Weekly* 31 July 483/3 A train thundering along a down-grade. 1872 'Mark Twain' *Roughing It* xvii. 141 A 'down grade', a flying coach, a fragrant pipe and a contented heart—these make happiness. 1872 *Newton Kansan* 19 Sept. 4/3 Greeley has had a down grade. 1876 J. Miller *Life amongst Modocs* vi. 76 He [sc. the stage-driver]..says, 'Hope, I am on the down grade, and have to crack the brake!' and sank down and died. And so it is that the 'down-grade', an expression born of the death of the old stage-driver, has gone into the language. 1900 Meredith & Webster *Calumet 'K'* xi. 199 They'd all strike like a freight train rolling down grade.

down-grade, *v.* [f. the sb.] *trans.* To lower in grade, rank, status, estimation, or the like. So **down-grading** *vbl. sb.*
1930 *Sunday Times* 12 Oct. 18/4 No further down-grading of London schools should take place until the basic principles of grading have been considered. 1964 *Labor Herald* (San Francisco) 8 Dec. 6 Aircraft union to fight downgrading pay cut drive. 1953 A. Baron *Human Kind* 97 There was a special examination to down-grade men. 1955 *Times* 27 Aug. 6/6 The dispute arose from a wartime agreement under which some N.U.R. wagon shop men were upgraded to fitters. Because of the dismissal of 15 fitters, the local branch of the Amalgamated Engineering Union claimed that the N.U.R. men should be downgraded once more. 1958 *Times Lit. Suppl.* 14 Feb. 89/3 A Pordenone has been downgraded to an Amalteo. 1959 G. D. Mitchell *Sociology* vii. 120 Non-manual employees whose subjective status is working class..have down-graded their non-manual occupations. 1964 F. Bowers *Bibliogr. & Textual Criticism* v. iii. 147 A corresponding down-grading of the Folio must be made. 1969 *Times* 24 Feb. 12/5 A scientific committee..recommended that the manned flight activities should be downgraded because they 'exaggerated' one aspect of space activity.

downhill, *sb.* Add: **3.** In Skiing: a downhill race.
1960 *Times* 22 Jan. 16/3 The British women's ski running championships began here to-day with the downhill.

Downing Street. [Name of Sir George *Downing* (c 1624–1684), British diplomat.] A short street in London running out of Whitehall towards St. James's Park and containing the Foreign Office and the official residence (No. 10) of the prime minister; hence used as a synonym for the Government (or the prime minister, or Foreign Office) of the day.
1781 A. Storer *Let.* 1 Mar. in *15th Rep. Hist. MSS. Comm.* (1897) App. VI. 467 Even though Lord North reject my application, or neglect the good-natured interference of those friends, who..have no small weight in Downing Street. 1831 Ld. Palmerston *Let.* 1 Mar. in H. L. Bulwer *Palmerston* (1870) II. viii. 48 The French..need not wish to have truer or warmer friends than they now have in Downing Street. 1849 Thackeray *Pendennis* I. xxxi. 308 Look! here comes the Foreign Express galloping in. They will be able to give news to Downing Street to-morrow. 1858 *Leisure Hour* 18 Nov. 728/1 The decrees and counsels of Downing Street will be heard simultaneously in Pekin or Canton. 1920 K. Jones *Fleet St. & Downing St.* 330 Thus would Fleet Street and Downing Street at last understand one another. 1920 E. H. Begbie *Mirrors of Downing St.* 7 The private opposition he [sc. Lloyd George] encountered in Downing Street. 1971 B. Graham *Spy Trap* xvi. 112 Dmitrov..instructed him to watch for a memorandum from Downing Street.

down lead. *Radio.* [Down *adv.* 38.] A wire that connects an elevated aerial or part of an aerial to a receiver or transmitter; a lead-in.
1913 *Work* 23 Aug. 413/2 The aerial..should be at least 100 ft. to 150 ft., including down leads. 1925 *Harmsworth's Wireless Encycl.* 729/1 Where the down lead enters the house care must be taken to secure perfect insulation. 1952 E. A. Laport *Radio Antenna Engin.* i. 38 The antenna consists of a large elevated capacitance area with two or more down leads that are tuned individually.

down-looking, *a.* (Earlier U.S. examples.)
1788 *Maryland Jrnl.* 9 May (Th.), Lindsey, a down-looking fellow, had on a new flaxen shirt. 1800 *Aurora* (Philad.) 23 July, A number of sneaking down-looking fellows, who occasionally assembled in a group.

downness (dau·n,nės). *rare.* [f. Down *a.* + -ness.] The fact or condition of being down; lowness.
1890 W. James *Princ. Psychol.* II. xx. 150 Rightness and leftness, upness and downness, are again pure sensations differing specifically from each other, and generically from everything else. 1927 *Scots Observer* 1 Oct. 2/2 A friend who positively finds his happiness in responding to human downness.

down pipe, downpipe. [Down *a.* 1 a.] A pipe leading downward; *spec.* a pipe to carry rain-water from a roof to a drain.
1858 [see Down *a.* 1 a]. 1904 Goodchild & Tweney *Technol. & Sci. Dict.* 172/2 *Downpipe*, a rainwater pipe. 1960 *Times* 1 Aug. 10/1 Even the gutters and downpipes are of copper. 1962 *Which?* (Car Suppl.) Oct. 139/2 Connection between exhaust manifold and down pipe [was] found to be slack. 1970 *Globe & Mail* (Toronto) 28 Sept. 24/3 (Advt.), Aluminum eaves, downpipes and upper facings.

down-point, *v.* [f. Down *adv.* + *point *sb.*¹] *trans.* To lower the value in points of (something rationed). So **down-pointing** *vbl. sb.*
1946 *News Chron.* 5 Mar. 1 The trade would welcome the downpointing of women's coats and costumes. 1947 *Sunday Express* 7 Dec., Unless the Board of Trade downpoint the garments, hundreds of thousands of them will be left in warehouses. 1948 *Times* 26 May 4/3 Sheets over 81 in... in width will be down pointed from seven to six coupons each.

down-range, downrange, *adv.* Chiefly *U.S.* [Down *prep.* 3.] In a position along the course of a missile, space-vehicle, or the like. Also *attrib.*, designating a station or observation-point thus placed.
1952 *N.Y. Times* 9 Mar. 4/1 'Down-range' stations at Jupiter Inlet on the Florida coast and on Grand Bahama Island already provide instrumentation 200 miles from the missile launching sites on Cape Canaveral. 1953 *Monsanto Mag.* July 3 They seized upon the light house as a reference point. From it they worked 'down range' to establish a series of stations from which to observe the flight of missiles. *Ibid.*, A typical down range station is a little electronic city set up to track missiles, predict weather, receive telemetry signals from missiles. 1962 G. Cooper in *Into Orbit* 28 We had already worked out rescue and recovery procedures for picking up the Astronaut down-range. 1966 *Times* 28 Feb. (Canada Suppl.) p. xii/3 Some experts believe it can fire a projectile 160 miles downrange.

downrightly, *adv.* (Later example.)
1947 Dylan Thomas *Let.* 5 June (1966) 310 And everybody downrightly refused to believe that *that* was the amount you got.

downshift, *sb.* and *v.* **A.** *sb.* A change to a lower gear on a car. **B.** *v. intr.* To change to a lower gear.
1959 *Observer* 1 Mar. 21/5 Downshifts are not entirely smooth but most driving is done in top. 1961 *Engineering* 27 Oct. 530 Accelerator kick-down gives downshift to second. 1961 'I. T. Ross' *Old Students* (1963) ii. 26, I downshifted and concentrated on the road.

downslope, down-slope (see below), *sb.*, *adv.*, and *a.* [f. Down *a.*, *prep.*: cf. Downhill.]
A. *sb.* (dau·nslōᵘp). A downward slope.
1908 *Westm. Gaz.* 5 June 2/1 We were away on the long down-slope—into California. 1914 D. H. Lawrence *Prussian Officer* 157 He had come in full view of the downslope. 1964 *Economist* 30 May 986/3 A frightening downslope.

B. *adv.* (dau·nslōᵘ·p). Down a slope; downhill.
1928 'M. Chapman' *Happy Mountain* iii. 22 The pent roofs of tar-paper that shielded the graves' earth from washing down-slope in the rains. 1938 C. F. S. Sharpe *Landslides* Plate II a (caption) Trees tilted downslope by creep tend to return to vertical position during growth. 1944 A. Holmes *Princ. Physical Geol.* x. 147 There are various kinds of mass movements of surface materials downslope. 1971 *Nature* 9 July 88/1 The adult animal..could be swept downslope by turbidity currents.
C. *adj.* (dau·nslōᵘp). Directed down a slope; downhill.
1938 C. F. S. Sharpe *Landslides* iii. 21 The general term creep may be defined as the slow downslope movement of superficial soil or rock debris. 1964 *Oceanogr. & Marine Biol.* II. 31 A turbidity current is essentially a down-slope current over the bed of the sea..which is driven primarily by excess density. 1971 *Nature* 5 Feb. 399/1 The largest component of downslope creep movement is thought to be due to seasonal changes in soil water content.

Downsman (dau·nzmǣn). [Down *sb.*¹ 2.] A native or inhabitant of the (Sussex) Downs.
1906 *Academy* 20 Jan. 63/1 The Downsman in the city May not his home forget. 1921 S. Leslie *Manning* 44 Morning after morning in the grey mist the shepherds and downsmen could hear the bell of their vigilant pastor. 1927 *Observer* 5 June 6/3 [He] founded the Society of Sussex Downsmen. 1963 *Daily Tel.* 25 Oct. 17/7 The Society of Sussex Downsmen have protested..against a proposal that five acres of downland..should be used as an amusement area.

downsome (dau·nsǫm), *a. colloq.* and *dial.* [See Down *adv.* 18, Down *a.* 3.] Inclined to be down or dispirited.
1888 F. R. Stockton *Dusantes* iii, When you left us at 'Frisco we felt pretty downsome. 1894 Blackmore *Perlycross* viii. 61 Then I just looked in at the *Bush*, because my heart was downsome.

down South, *advb. phr.* [Down *adv.* 2, 4.]
a. *U.S.* In or into the States south of the Mason–Dixon line. Also as *adj.*
1834 C. A. Davis *Lett. J. Downing* 25 Though I tell'd 'em down south my father was an Irishman,..I am as clear a Yankee..as the Major itself. 1862 'E. Kirke' *Among Pines* i. 12 Old Abe he's gwine to come down Souf. *Ibid.* iii. 60 Away down South in Dixie. 1884 'Mark Twain' *Huck. Finn* xxxiii, There was plenty other farmer-preachers like that,..down South. 1905 A. H. Rice *Sandy* 23, I lived down South, clean off the track of ever' thing. 1963 *N.Y. Times* 1 Dec. 42 This winter for your Down-South vacation fly Eastern.
b. *N.Z.* In or into the South Island or its southernmost provinces, Otago and Southland.
1867 Lady Barker *Let.* May in *Station Life in N.Z.* (1870) xviii. 137 (*heading*) A Journey 'Down South'. 1873 —— *Station Amusements in N.Z.* vi. 93 All the first-class pastureland 'down South', as [Otago] was called, had been taken up long before. 1873 Trollope *Austral. & N.Z.* II. xxvi. 444 'A railway for you gentlemen down south!' says a northern member. 'Certainly,—but on condition that we have one here, up north.' 1949 P. Newton *High Country Days* vii. 69 Lofty, in happy mood, and unfailing in his praise of his native province [sc. Southland], delighted young Wallace with tall stories of 'down south'.

dow·nspout. Chiefly *N. Amer.* [Down *a.* 1 a.] A pipe conveying rain-water from a roof, etc., to the ground or to a drain, etc.
1896 *Rules governing Printing of Specifications U.S. Patent Office* (rev. ed.) 49/2 Down-spout. 1912 M. Nicholson *Hoosier Chron.* xxxiv. 592 Mr. Harwood had to put a new downspout on the kitchen. 1968 *Globe & Mail* (Toronto) 3 Feb. 41/5 (Advt.), Fascia, eavestrough, downspouts, shutters. 1969 *Morning Star* 26 Aug. 2 Special non-drying paint is being put on the upper end of school downspouts in an attempt to stop vandalism.

Down's syndrome. [Name of J. L. H. *Down* (1828–96), English physician.] = *mongolism.
1961 G. Allen et al. in *Lancet* 8 Apr. 775/2 Some of the undersigned are inclined to replace the term 'mongolism' by such designations as 'Langdon-Down anomaly', or 'Down's syndrome or anomaly' or 'congenital acromicria'. 1961 *Ibid.* 21 Oct. 935/1 Our contributors prefer Down's syndrome to mongolism because they believe that the term 'mongolism' has misleading racial connotations and is hurtful to many parents. 1965 H. E. Sutton *Introd. Human Genetics* v. 37 The condition known as trisomy 21 syndrome or mongolian idiocy (sometimes referred to as Down's syndrome) had long been an enigma. 1965 *New Scientist* 2 Dec. 632/1 The chromosomal damage leading to Down's syndrome (as the 'mongol' condition is now described).

down stage, down-stage, *adv. Theatr.* [Down *prep.* 3.] At or towards the front of the stage. Also *attrib.* or as *adj.* Cf. *up-stage.
1898 L. Merrick *Actor-Manager* 41 It [sc. her train] would fall 'down stage'. 1933 P. Godfrey *Back-Stage* i. 15 He examines these things for himself before returning to his position on the down-stage centre mark. 1936 N. R. Smith *All Star Cast* 38 Verity came down-stage to the table. 1939 L. Hellman *Little Foxes* i. 11 Downstage, right, are a high couch, a large table, several chairs. 1968 *Listener* 18 July 72/1 Rather apart from the others..was Mr. Okara, advancing down stage to meet me with outstretched hand.

do·wnstart. [f. Down *adv.* + *start* after Upstart *sb.*] (See quot. 1949); also, one who pretends to be of lower origin than he is. Also *attrib.*
1898 G. B. Shaw *Sixteen Self-Sketches* (1949) viii. 44 My father was an Irish Protestant gentleman of the downstart race of younger sons. 1921 —— Pref. to *Immaturity* in *Prefaces* (1934) xxiii. 627/1, I was a downstart and the son of a downstart. 1949 —— *Sixteen Self-Sketches* ii. 7 The Downstart, as I call the boy-gentleman descended through younger sons from the plutocracy, for whom a university education is beyond his father's income, leaving him by family tradition a gentleman without a gentleman's means or education, and so only a penniless snob. 1960 G. Mikes *How to be Inimitable* 21 Quite a few people assert that they are of lower origin than they..are... The place of the upstart is being taken by the downstart.

down-state, downstate. *U.S.* [Down *prep.* 3.] The part of a State outside a large city, esp. the southern part. Also as *adv.* and *adj.* Hence **down-sta·ter,** an inhabitant of downstate. Cf. *up-State* (Up *prep.*² 6).
Used in various parts of the U.S. with varying local significance.
1909 *Daily Maroon* (Chicago) 2 Oct. 1/4 Springer, a husky full-back from down-state. *Ibid.* 12 Oct. 1/1 The down-staters have always supported their men loyally. 1932 W. Faulkner *Light in August* iii. 58 She had gone to visit her people downstate. 1942 *Amer. Speech* XVII. 30 (*heading*) Pronunciation in downstate New York. 1947 *Chicago Tribune* 22 June II. 5/6, I have had a great many letters from downstaters who want to correct the faults. 1970 *Globe & Mail* (Toronto) 25 Sept. 1/1 The New York Telephone Co., for the second day, turned on its 250 emergency generators in the downstate area.

dow·nstrea·m, *adv.* (*a.*, *sb.*). Also **down stream, down-stream.** [Down *prep.* 3.] **a.** Down the stream, in the direction of the current, towards the mouth of a river. Also *fig.*
1706 in *Rec. Plymouth* (Mass.) (1889) I. 26 Bounded by said river..down stream unto the upper end of the meadow. 1864 [see Down *prep.* 3]. 1865 H. B. Stowe *House & Home P.* 316 There is a general tendency to let all sorts of old forms and observances float down-stream. 1869 Blackmore *Lorna D.* vii, Even an otter might float downstream. 1878 R. L. Stevenson *Inland Voy.* 196 Canoeing was easy work ..to keep the head down stream. 1929 Belloc *Joan of Arc* iii. 62 The French forces lay downstream. 1957 P. Caraman *Henry Morse* vii. 66 Across the river, about a mile downstream, a path ran from the bank.
b. as *adj.* and quasi-*sb.* Also *fig.*
1842 *American Pioneer* I. 70 Steam-boats seem almost to say, we will do your up-stream business for nothing, if you will give us your down-stream business. 1891 W. Morris *News from Nowhere* (ed. 2) ii. 8 Even the up-stream bridges ..are scarcely daintier, and the down-stream ones are scarcely more dignified and stately. 1934 J. L. Myres in E. Eyre *European Civilisation* I. ii. 115 There were certainly casual immigrants from down-stream. 1954 R. St. B. Baker *Sahara Challenge* iii. 39 The down-stream rush of water is decreased. 1963 *Economist* 21 Sept. 1042/2 Much larger 'downstream' investment.

down-street, *adv.* (*a.*). *colloq.* and *dial.* [Down *prep.* 3.] Down the street; in, into, or toward the lower part of a town, etc.; (see also quot. 1962). Also *attrib.* or as *adj.*, and as *sb.*
1828 M. R. Mitford *Our Village* III. 199 He began..to look out for a wife, up street and down...The down-street lady was a widow. 1852 H. B. Stowe *Uncle Tom* xxviii, I believe I'll go down street, a few moments, and hear the news. 1865 W. S. Banks *Provincial Words Wakefield* 76 *Up Street*, means the upper part of the town... 'Dahn street' is used in a similar manner. 1876 W. Wright *Hist. Big Bonanza* (1877) 365, I was told down street..that there was a regular row in one of the shebangs up this way. 1888 B. Lowsley *Gloss. Berkshire Words* 73 *Down-strit*, the opposite direction in the main road through a village from *up-strit*. 1962 *Amer. Speech* XXXVII. 158 One goes *down city* only if he is going to the business center; if he is going no farther than the nearest grocery store or shopping center, his destination is *downstreet*.

down-swept, *a.* [Down *adv.* 35.] Swept or curving downwards.
1927 *Daily Express* 13 Oct. 2 Much lower [car] bodies.. [are] achieved by a down-swept chassis. 1933 C. Day Lewis *Dick Willoughby* vi. 55 Her eyes observed him keenly under the down-swept lashes. 1937 *Times* 11 Dec. 4/7 The frame is of downswept box section shape. 1963 *Times* 15 Jan. 12/2 From the..front bumpers to the downswept tail.

dow·n-swing. [Down *adv.* 38.] A downward swing; *spec.* (*a*) the descending movement of a golf-club about to hit the ball, following the back-swing; (*b*) a period of low or declining activity in trade, business, etc.
1899 [see *back-swing* s.v. *back- A. II]. 1922 Wodehouse *Clicking of Cuthbert* viii. 192 Do shout 'Boo!' at him when he is starting his down-swing! 1935 *Economist* 2 Feb. 250/2 Mr Crump..regards the crisis in the main as the normal downswing of the trade cycle. 1946 G. von Haberler *Prosperity & Depression* ii. 21 (*heading*) The Downswing. The reverse of the upswing. 1959 *Spectator* 11 Sept. 336/2 The sufferer should be told..about this..fluctuation in his malady, so that he doesn't get unduly discouraged during a down-swing. 1966 *Economist* 18 June p. xxx/1 After two years of popularity, clearing bank shares appear to be entering the downswing of their familiar cycle. 1970 *Nature* 21 Nov. 703/2 The present epidemic, hopefully, seems to be on the downswing.

down timber. *N. Amer.* A heap or tract of fallen trees, brought down by wind or storm or other natural agent.
1881 W. O. Stoddard *E. Hardery* 263 There was plenty of old 'down timber' to be cut up, and cleared away. 1895 H. S. Somerset *Land of Muskeg* 170 (*caption*) Horses in down timber. 1948 *Sierra Club Bull.* (San Francisco) Mar. 111 Up interminable slopes of down timber, where the small logs repeatedly broke under the weight of man and pack. 1951 R. P. Hobson *Grass beyond Mountains* 136 Windfalls or down timber lay piled up between the trees.

Downtonian (dauntōᵘ·niăn), *a. Geol.* [f. *Downton*, name of a locality in Herefordshire + -ian.] Of, pertaining to, or designating a stratigraphic stage or series in Europe placed at the top of the Silurian system or the bottom of the Devonian, and the age or epoch during which it was deposited. Also *absol.*
1879 C. Lapworth in *Ann. & Mag. Nat. Hist.* 5th Ser. III. Table facing 455 Silurian System: Upper Division (Downtonian). 1910 Lake & Rastall *Text-bk. Geol.* xx. 334 The Downtonian series varies little. Wherever it occurs, whether on the Welsh borders, in the Lake District or in the south of Scotland, it is essentially an arenaceous deposit and was formed in shallow water. 1938 A. K. Wells *Outl. Hist. Geol.* vii. 66 Both Downtonian and Dittonian represent successive stages in the transition from marine to continental conditions. *Ibid.* 67 The thickness of the Downtonian Stage diminishes. 1955 G. G. Woodford tr. *Gignoux's Stratigr. Geol.* iv. 110 Similar sediments are seen on the edges of the Baltic Shield...This Baltic Downtonian is attached to the Silurian by Scandinavian geologists. 1967 D. H. Rayner *Stratigr. Brit. Isles* v. 140 Molluscs..are often abundant in the lower Downtonian beds. 1971 B. W. Sparks *Rocks & Relief* viii. 295 In this account the Downtonian will be included in the Devonian.

down town, down-town, *adv.*, *a.*, and *sb.* Also **downtown.** orig. *U.S.* and chiefly *N. Amer.* [See Down *prep.* 2 b, 3.] **A.** *adv.* (See Down *prep.* 2 b.) Cf. Up-town *adv.*
1835, 1883 [see Down *prep.* 2 b]. c 1909 D. H. Lawrence *Collier's Friday Night* (1934) iii. 60 Well, my duck, I looked for you downtown. 1952 S. Kauffmann *Philanderer* (1957) vi. 91 The man who had seemed a dignified young wonder on 135th Street..looked like a pompous and overweening young ass downtown. 1968 *Globe & Mail* (Toronto) 17 Feb. 24 (Advt.), Tickets available downtown at box office prices exclusively at Moody's in the Colonnade.
B. *adj.* Of, pertaining to, or situated in the lower, or more central, part of a town or city. Cf. Up-town *a.*
1836 P. Hone *Diary* (1889) I. 200 The value of down-town property. 1852 *Harper's Mag.* V. 413/2 The down-town men..slip uneasily through the brick and mortar labyrinths of Maiden-lane. 1870 J. K. Medbery *Men & Myst. Wall Street* 67 On these securities therefore the down-town banks make call loans. 1883 [see Down *prep.* 3]. 1891 *Congress. Rec.* 28 Jan. 1906/1 The second ward of the city of New York..is what is called a down-town ward, a business ward. 1902 *Westm. Gaz.* 12 May 2/1 Men do not doff their hats in the down-town elevators which brought her up to the big office..near the top of one of the high down-town buildings. 1911 *Daily Colonist* (Victoria, B.C.) 13 Apr. 7/5 Pedestrians in the down-town district on Friday were interested in the sight of a veritable procession of lorries loaded with high grade Pianos. 1939 C. Morley *Kitty Foyle* 175 His necktie was sort of downtown and Bourse-looking. 1952 B. Ulanov *Hist. Jazz* (1958) iv. 33 Squalid little clubs in downtown New York. 1958 *Times Lit. Suppl.* 10 Jan. 21/2 The account of the wild sermon in down-town Kingston, vividly conjuring up the crucifixion. 1966 *New Scientist* 14 July 103/2 Mr. Crosland writes about *downtown* schools, meaning schools in the centres of cities. Pure American, of course, and a very useful term too. 1968 *Globe & Mail* (Toronto) 17 Feb. 1/8 A downtown Ottawa office building. 1969 *Australian* 24 May 24/2 (Advt.), Your home in downtown Sydney.
C. *sb.* The lower or business part of a town or city.
1851 H. Melville *Moby Dick* i, Its extreme down-town is the Battery. 1905 *N.Y. Even. Post* 4 Mar. 5 One of the diversions of downtown yesterday was watching the sure movements of a steeplejack. 1955 R. Blesh *Shining Trumpets* (ed. 3) vii. 160 New Orleans' downtown is the old quarter north of Canal Street. 1968 *Globe & Mail* (Toronto) 13 Jan. 45/1 (Advt.), At the subway stop.. 19 minutes from downtown.
Hence **dow·n-tow·ner,** one who lives in or frequents the down-town part of a city.
1830 J. F. Watson *Ann. Philadelphia* 244 They were the Achilles and the Patrocles of the 'downtowners'. 1887 *Courier-Journal* (Louisville, Ky.) 8 May 12/5 Jay Gould has set down-towners to eating snails.

downtrend. [Down *adv.* 38.] A downward tendency, esp. in economic matters.
1926 *Dry Goods Economist* 11 Dec. 28 Retail trade is large, and no pronounced down-trend has yet developed in that quarter. 1940 *Economist* 10 Feb. 250/1 (*heading*) Downtrend in farm prices. 1960 *Times* 5 Feb. 16/3 In 1959 the seasonal downtrend was not reversed until the second week in March.

down-turn, downturn, *sb.* [Down *adv.* 38.] A turning downward; a decline, esp. in economic or business activity.
1926 *National Provisioner* 19 June 40 Most fat cows and heavy heifers lost around 25c and in instances the downturn on better grades was even greater. 1940 *Economist* 10 Feb. 250/1 The great upturn [of prices] was basically caused by drought, and the great downturn by the breaking of the drought. 1947 Dewey & Dakin *Cycles* xvi. 232 The corrective economic downturn following World War II. *Ibid.* 233 The downturn in the 54-year price rhythm was hardly even

noted. 1955 *Ann. Reg. 1954* 168 He asked for measures which could be used if the economic 'downturn' became a recession. 1957 *Economist* 7 Sept. 760/1 The economy has already demonstrated this year that it can suffer a mild downturn in manufacturing. 1964 L. S. Hultzén in D. Abercrombie et al. *Daniel Jones* 87 In some idiolects this non-finitive text shape has an arrested down-turn or slight up-turn. 1970 *Daily Tel.* 14 Jan. 19 The down-turn in stocks suggests the possibility of a lower level of importing.

down-turn, downturn, *v.* [Down *adv.* 33.] *trans.* To turn downwards.
1909 R. Kane *Sermon of Sea* xx. 325 The rusty green and knitted surface of the sod are gradually cloven and down-turned. 1922 Joyce *Ulysses* 691 With solicitation, bending and downturning the upturned rug-fringe.
So **down-turned** *ppl. a.* (see Down *adv.* 35).
1880 *Rules on Specifications U.S. Patent Office* 25/2 The following are single words..downturned. 1909 *Westm. Gaz.* 21 June 2/3 Curve those lips..To a dismal down-turned crescent. 1942 T. S. Eliot *Little Gidding* 10, I fixed upon the down-turned face That pointed scrutiny.

down under, *adv.* [Down *adv.* 4.] At the antipodes; in Australia, New Zealand, etc. Also *attrib.* and *sb.* (after a prep.).
1899 [see Under *adv.* 2 f]. 1908 *Daily Chron.* 31 Oct. 4/6 The Bishop of London..does not take a very active interest in the 'down under' section of his see. 1909 *Westm. Gaz.* 11 Jan. 12/1 The same could happen 'down under' if New Zealand..were bracketted with New South Wales and Queensland. 1916 *Anzac Bk.* 145/2 Macaulay's prophecy concerning the man from 'down under' sitting on the ruins of London Bridge. 1922 *Daily Mail* 9 Dec. 11 The steeple-chaser Kinlark, a gift to the Prince from 'down under'. 1928 *Daily Express* 17 Feb. 4 Mr. Collins, the captain of the last visiting Australian team, was a stipendiary steward 'down under'. 1936 *Discovery* Jan. 10/2 Sydney, which proudly boasts of the finest natural harbour in the world, welcomes you to that vast Dominion 'down under'. 1941 L. A. G. Strong *John McCormack* x. 173 The second concert in Auckland brought to an end John's first visit Down Under. 1971 *Time* 14 June 2/2 Down under may be the place to make money, but up over is the place to spend it!

downward. Add: **A.** *adv.* **3.** *Comb.* (Further examples.)
1865 G. M. Hopkins *Poems* (1948) 134 A block of copse Close-rooted in the downward-hollowing fields. 1922 Joyce *Ulysses* 377 Exterior splendour may be the surface of a downwardtending lutulent reality. *Ibid.* 656 Slow erosions of peninsulas and downwardtending peninsulas. 1943 L. B. Lyon *Evening in Stepney* 14 Fuel for the downward-bending The sod-quickening fire. 1964 J. C. Catford in D. Abercrombie et al. *Daniel Jones* 35 A downward-forward displacement of the hyoid bone..can easily be observed.

dow·nwarp. *Geol.* [Down *adv.* 38.] A broad surface depression; a syncline. So **dow·nwarping,** the local sinking of the earth's surface to form such a depression.
1917 *Prof. Papers U.S. Geol. Surv.* xciii. v. 110/2 From Gibson the axis of the downwarp has a general north-northeast trend to the great bend of Chaco River. 1917 *Jrnl. Geol.* XXV. 146 The downwarping of certain layers near the top of the Lockport dolomite. 1936 *Geogr. Jrnl.* LXXXVII. 165 This resistance was not however sufficient to prevent another downwarp in Southern India. 1950 F. E. Zeuner *Dating Past* (ed. 2) vii. 280 A slow down-warping to the north is going on. 1955 Brown & Dey *India's Min. Wealth* (ed. 3) xviii. 663 The tectonic downwarp in front of the Himalayas. 1963 D. W. & E. E. Humphries tr. *Termier's Erosion & Sedimentation* v. 108 Local downwarping or uplift can lead rivers to reverse their courses either completely or in part. 1965 A. Holmes *Princ. Physical Geol.* (ed. 2) xxx. 1173 The primary importance of upwarping and downwarping in providing the conditions for gravitational tectonics on a spectacular scale.

dow·nwash. *Aeronaut.* [Down *adv.* 38.] The downward deflection of an air-stream by an aerofoil or other body; *downwash angle* (see quot. 1919). Hence (*rare*) **downwash** *v.*
1915 *Rep. & Mem. Advisory Comm. for Aeronautics, No. 196* v. 125 It appeared possible that the advantage of such increased lift might introduce a concomitant disadvantage in an increased 'downwash' in the neighbourhood of the tail planes. 1919 W. B. Faraday *Gloss. Aeronaut. Terms* 21 *Downwash angle*, the angle through which the airstream relative to the aeroplane is deflected by an aerofoil or other body. Measured in a plane parallel to the plane of symmetry. 1922 *Encycl. Brit.* XXX. 19/2 The reaction upon the rear plane is therefore inclined backward by the angle through which the air has been 'downwashed' by the leading plane. 1931 *Flight* 20 Nov. 1148/1 The amazing control of the Autogiros, old as well as new, is very largely due to the downwash from the rotor blades. 1966 D. Stinton *Anat. Aeroplane* v. 58 To generate a given lift at a given airspeed an aerofoil of long span has to impart a smaller downwash to the air it meets than an aerofoil of shorter span.

down wind: see Wind *sb.*[1] 18 a.

downy, *a.*[2] Add: **1. b.** *downy mildew*: a disease of plants caused by parasitic fungi of the order Peronosporales.
1886 *Bull. Bot. Div. U.S. Dept. Agric.* II. 7 (*title*) Fungous diseases of the grape vine. I. The downy mildew. 1909 B. M. Duggar *Fungous Dis. Plants* x. 152 The downy mildew of the grape is one of the most important disease-producing organisms among the Peronosporaceæ. 1950 *N.Z. Jrnl. Agric.* Feb. 157/1 Downy mildew (*Perono-plasmopara cubensis*)..appears as more or less angular, yellowish spots on the leaves, on the undersides of which a scanty white downy growth develops. 1970 *Times* 14 July 10/6 The aim was to produce a variety resistant to downy mildew.
3. c. downy woodpecker, a small species of North American woodpecker, *Dendrocopos pubescens*.
1808 A. Wilson *Amer. Ornithol.* I. 153 Downy Woodpecker. *Picus Pubescens*...is the smallest of our Woodpeckers. 1872 *Rep. Vermont Board Agric.* 321 The Downy Woodpecker (*Picus pubescens*, Linn.) a black and white bird, usually not over six inches long. 1948 *Pacific Discovery* Mar.–Apr. 18/1 A harsh *spick!* note tells of a downy woodpecker in the neighborhood. 1964 J. Bull *Birds N.Y. Area* ii. 289 The Downy Woodpecker nests in open woodland, orchards, suburbs, and city parks. 1971 *Islander* (Victoria, B.C.) 13 June 13/3 There are the..downy woodpecker, flicker woodpecker, [etc.].

downy (dau·ni), *sb. slang.* Also **downey.** [f. Downy *a.*[2]] A bed; so *to do the downy*, to lie in bed. Cf. Downy *a.*[2] 2 (and quot. 1605 for sense 4).
1846 *Swell's Night Guide* 117/2 Dab, a letter, doss, downey, bed. 1847 W. T. Porter *Quarter Race Kentucky* 58 The candidate yawned, looked at his bed,..finally..seating himself upon 'the downy'. 1854 'C. Bede' *Further Adv. V. Green* vii. 59 This'll never do..! Cutting chapel to do the downy! 1858 Trollope *Three Clerks* I. ix. 181 I've a deal to do before I get to my downy. 1868 —— *He knew he was Right* (1869) I. xlv. 347 The Colonel was lodged safe in his downey. 1906 E. Dyson *Fact'ry 'Ands* ix. 118 Er pair iv boots was stickin' out conspicuous et one end iv ther downy. 1968 *Gloss. Brit. Argot* (Paramount Pictures), *Do the downy*, remain in bed.

dowry. Add: **5.** Comb. *dowryless* adj.
1925 *Glasgow Herald* 31 Dec. 4 The 'Army man'..outraged the conventions of Breffne by marrying a dowryless maiden for love. 1949 M. Mead *Male & Female* xv. 303 She came to be chosen in marriage for her dowryless self alone.

doxographer (dǫksǫ·grăfəɹ). [f. mod.L. *doxographus* (Diels 1879), f. Gr. δόξα opinion + -γραφος writer: see -er[1].] A writer who collects and records the opinions or *placita* of the Greek philosophers. Hence **doxogra·phic, doxogra·phical** *adjs.*, of or pertaining to the doxographers; **doxo·graphy,** a collection of philosophical opinions.
1892 J. Burnet *Early Greek Philos.* 371 By the term *doxographers* we understand all those writers who relate the opinions of the Greek philosophers. *Ibid.* 374 The doxography [of the *Lucullus*] has come through the hands of Kleitomachus. *Ibid.* 375 Short doxographical summaries are to be found in Eusebios [etc.]. 1908 J. Adam *Relig. Teachers of Greece* xiii. 267 We have doxographical testimony to show that Diogenes pronounced the soul to be imperishable. *Ibid.* 268 The doxographers sometimes ascribe to him [*sc.* Democritus] the doctrine of a single world-soul or Deity identical..with the aggregate of fiery atoms in the world. 1919 *Jrnl. Hellenic Stud.* 180 The Greek doxographers know of no astronomer before Thales. 1937 *Mind* XLVI. 248 Nor was he..blind to the possibility that doxographic statements going back to Theophrastus may sometimes originate in an Aristotelian misconception or misrepresentation. 1952 *Jrnl. Theol. Stud.* III. 123 The problem how far [Basil] knew the philosophers from their actual works and how far he was dependent on anthologies and doxographical manuals. 1952 G. Sarton *Hist. Sci.* I. x. 239 We have only fragments and the sayings of doxographers, indirectly and poorly transmitted. *Ibid.*, Doxographic books of the latter [*sc.* Theophrastos] are known only indirectly through later extracts. *Ibid.* xii. 327 We really need a new critical edition of all the Ctesias fragments and of the doxography relative to him.

doyen. 2. (Delete last sentence of definition; see *doyenne[2].)

‖**Doyenne**[1] (dwaye·n). Also **Doyenné.** [ad. F. *doyenné*, in full *poire de doyenné*, lit. 'deanery pear'.] In full *doyenne pear*. A variety of pear, esp. *Doyenne du Comice*, a large yellow late-fruiting pear, a favourite for cultivation.
[1666 Evelyn *Kal. Hort.* (ed. 2) 107 Pears. The Squib-pear, Spindle-pear, Doyoniere, Virgin.] 1731 Miller *Gard. Dict.* (1733) s.v. *Pyrus* 35 Le Doyenné, *i.e.* The Deans Pear. 1822 Loudon *Encycl. Gardening* (1824) § 4437 Doyenné. *Synonym.* Dean's pear. 1858 *Trans. Ill. Agric. Soc.* III. 338 Fine crops of White Doyenne and Bartlett. 1860 R. Hogg *Fruit Manual* 181 Fruit small, roundish-obovate, or Doyenné-shaped. 1872 *Rep. Vermont Board Agric.* 109 A yearly crop of the finest Doyenne pears, in size, colour and quality, of any I have ever seen. 1958 *Listener* 20 Nov. 853/2 The best [varieties of pears] I think are Williams' Bon Chrétien, Marie Louise, Doyenne du Comice, [etc.].

‖**doyenne**[2]. [See Doyen 2.] A female doyen; the leading or senior woman in a group, society, etc.
1905 A. Bennett *Sacred & Profane Love* II. i. 104 That stately dowager, that impeccable *doyenne* of serious English fiction. 1926 A. Huxley *Let.* 10 Aug. (1969) 272 Anita Loos..is the doyenne of Hollywood. 1958 *Times* 24 Nov. p. xvi/5 The *doyenne* of this group is that assiduous and deservedly popular writer Mazo de la Roche. 1959 P. Bull *I know the Face* i. 13 Miss Elsie Fogerty, the doyenne of The Central School of Dramatic Art.

dozer[2] (dōu·zəɹ). Also **'dozer.** Colloq. shortening of *bulldozer.
1942 *Infantry Jrnl.* (U.S.) Oct. 51 The blade of the dozer is not much good. 1949 *Landfall* III. 174 Come trucks and lux and tools of trade Motor tyres a dozer-blade. 1958 *Daily Mail* 16 July 6/6 These modern devices have such names as the tractor shovel, the crawler-loader, the dozer. 1959 'J. R. Macdonald' *Galton Case* (1960) xxvi. 204 You could of knocked me over with a 'dozer. 1971 *Engineering* Apr. 52 The whole caterpillar-built range including lift trucks, track and wheel loaders, and dozers.

dozy, *a.*[1] Add: Also **dosey, dozey. 1. a.** (Further examples.) Also (*colloq.*), (mentally) sluggish; stupid; lazy.
1836 in R. McNab *Old Whaling Days* (1913) 451 The 4th mate..has been Sick and dozey ½ the time. 1924 Galsworthy *White Monkey* i. vii. 50 Soames directed his gaze at the pink face of dosey old Mothergill. 1959 J. Braine *Vodi* iii. 51 'The swine,' Dick said. 'You're dozy,' Liphook said. 'They were damned decent, really.' 1961 J. Maclaren-Ross *Doomsday Bk.* vii. 75 What's funny, you dozey berk?
2. (Earlier and later examples.)
1872 Schele de Vere *Americanisms* 464 *Dozy* and *dozed* are said in Pennsylvania of timber beginning to decay and unfit for use, while the decay is yet hardly perceptible, but the timber already brittle. (S. S. Haldeman.) 1923 D. H. Lawrence *Kangaroo* viii. 166 Fatuous letters from friends in England..as dozy as ripe pears in their *laisser aller* heaviness.

dozzle (dǫ·z'l), *sb. Metallurgy.* Also **dozzler.** A hollow refractory brick fitted to the top of an ingot-mould to provide a reservoir of molten metal, which flows downwards to fill cavities in the ingot; = Feeder 9 a.
1923 J. W. Hall in Harbord & Hall *Metallurgy of Steel* (ed. 7) II. i. 4 With this the fire-brick head, called in Sheffield a 'dozzler', can be lifted off. 1932 E. Gregory *Metallurgy* ii. 51 A short fireclay cylinder, known as a *core* or *dozzle*. 1954 *Gloss. Terms Iron & Steel* (*B.S.I.*) II. 8 *Dozzle*, a preheated hollow refractory brick used to provide a feeder head for small ingots.

drab, *sb.*[2] and *a.* Add: **B.** *adj.* **c.** In comb. with other names of colours.
1894 R. B. Sharpe *Hand-bk. Birds Gt. Brit.* I. 12 Sides of neck and under surface of body drab-grey. 1905 *Westm. Gaz.* 15 May 10/2 The rather soft fur of the underparts is drab-brown.
C. *sb.* **1. b.** *S. Afr.* (*pl.*) The long feathers on the part of the wing of a female ostrich near to the junction with the body. (In quot. 1896 *drab* is adj.)
1881 A. Douglass *Ostrich Farming in S. Afr.* xi. 68 The little white belly feathers should have been replaced by blacks or drabs. 1896 R. Wallace *Farming Ind. Cape Col.* xi. 235 Drab, long, and medium were about 10s. per lb. lower. 1913 C. Pettman *Africanderisms* 167 *Drabs*, corresponding growth from the female.
c. *fig.* A dull or lifeless appearance or character.
1903 *Daily Chron.* 31 Dec. 5/1 Despite the fact that so many of his works wore a drab, still those who knew him best recognised that the drab was the colour of his experience. 1909 *Westm. Gaz.* 4 Feb. 1/3 It is the one sustained note of colour in the dreary drab of Irish life.

draba (drē[1]·bă). *Bot.* [mod.L. (J. J. Dillenius in Linnæus *Systema Naturæ* (1735)), ad. Gr. δράβη a kind of cress.] A plant of the genus of herbs so named belonging to the family Cruciferæ, found in temperate and arctic regions, and cultivated as hardy annual, biennial, and perennial alpine plants. (See also *whitlow-grass* (Whitlow b).)
1629 Parkinson *Paradisi* ci. 390 Candy Tufts..is named by some, *Draba*, or *Arabis*, as Dodonæus, but *Draba* is another plant differing much from this. 1777 W. Curtis *Flora Lond.* I. Plate 133, Vernal Draba or Whitlow Grass. 1895 S. H. Vines *Students' Text-bk. Bot.* 603 Latiseptal silicula of Draba. 1968 *Q. Bull. Alpine Garden Soc.* XXXVI. 69 There are many magnificent drabas.

drabbet. (Earlier example.)
1819 J. Baxter *Jrnl.* 16 Feb. in *Amer. Speech* (1965) XL. 198 Bot Jack a piece of Bleu Cloth for a Coat 3½ yds at 3/6 & myself a pair of Trowsers called Drabbett.

drabble, *v.* Add: **2.** Also in extended use.
1903 Kipling *Five Nations* vii, Across the sad valleys all drabbled with rain. 1923 *Chambers's Jrnl.* 89/2 Thews who ..drabbled graybeards in their blood.

drabi (dræ·bi). [See quot. 1920.] A muleteer.
1900 *Pioneer Mail* 16 Mar. (Y.), The mule race for Drabis and grass-cutters was entertaining. 1920 *Chambers's Jrnl.* 296/2 (Indian frontier) The lot of the muleteer (or *drabi*, as he is generally called, this being the native rendering of the English word 'driver') is never a very easy one. 1920 *Blackw. Mag.* Nov. 569/2 (Mahsudland) Mule drabis. 1927 E. Thompson *These Men, thy Friends* 125 A fool of a *drabi*, who had left his mules and stumbled off, seeking India that lay to the east.

drably (dræ·bli), *adv.* (see under Drab *sb.*[2] and *a.*). Add: Also *fig.* Without brightness or colour, dully, uninterestingly.
1905 *Westm. Gaz.* 19 Sept. 10/1 Few guess that the dahlia ..has had a drably unromantic origin. 1918 *Cornhill Mag.* June 616 The desirability of expressing thoughts fully and truly in words..is too drably presented to the child. 1927 *Sunday Express* 1 May 9 Their novels look drably old-fashioned. 1956 *Ann. Rep. Smithsonian Inst.* 216 The sun's interior must be drably uniform.

drac(h), drack, colloq. abbrevs. of Drachma 1 b.
1935 L. Durrell *Let.* in *Spirit of Place* (1969) 31 They.. said, with uncertainty—24 dracs—1/-. *Ibid.* 33 There is a good peasant wine... It costs 6 dracks—3[d] per bottle. 1936 'T. B. Marle' *Candid Escort* i. 11 'Can you give me five drachs?' he asked. 1969 R. Airth *Snatch!* ii. 19, I stood outside the court-room in Athens with eight dracks in my pocket. 1971 P. Dickinson *Sleep & his Brother* iii. 54 I was sitting by the yacht-basin in Iraklion wondering where my next drach was coming from.

dracocephalum (drækose·fălǝm). *Bot.* [mod.L. (Linnæus *Genera Plantarum* (1737) 173), f. Gr. δράκων dragon + κεφαλή head, in reference to the shape of the flower.] An annual or perennial herb of the genus so called, belonging to the family Labiatæ and native to temperate Asia and Europe; = Dragon's head 2.
1840 J. Loudon *Ladies' Flower-Garden Ornamental Annuals* 226 Dracocephalum...The Dragon's Head... There are several other kinds of annual Dracocephalums, but they differ very little from each other, and they are none of them common in British gardens. 1904 R. J. Farrer *Garden Asia* 237 The pallid gleam of a dracocephalum. 1957 *Q. Bull. Alpine Garden Soc.* XXV. 95 A Dracocephalum..has flowers of a deep purple-blue.

dracone, Dracone (dræ·kōun). [f. L. *draco*, *-ōnem*, ad. Gr. δράκων dragon.] A large flexible container for transporting liquids, towed on the surface of the sea.
[1957 *Times* 26 Nov. 7/4 A model, which was christened Draconella at her launching, is 67 ft. long and 3 ft. in diameter.] 1959 *Times* 27 Apr. (Suppl.) p. xi/4, The most spectacular of the new flexible containers are the long sausage-shaped nylon and neoprene bags called Dracones. 1963 *Daily Tel.* 20 Dec. 18/3 The dracone..has been used in Greece to bring fresh water to islands and for the transport of fuel. 1967 *New Scientist* 2 Nov. 289/1 The Dracone, a towable 'barge' made of nylon fabric coated with synthetic rubber for transporting petroleum products, drinking water, etc.

draconiform (drăkǫ·nifǭɹm), *a.* [f. L. *dracon-*, *draco* Dragon: see -form.] Resembling a dragon in shape.
1888 J. C. Dunlop *Prose Fiction* I. 450 The accessory emblem of a draconiform monster.

Dracula (dræ·kiŭlă). The name of the king of the Vampires, invented by Bram Stoker in the novel of this name (1897), used allusively to denote a grotesque or terrifying person, etc.
Prince Vlad of Wallachia, who died in 1476, was also known as 'Dracula' (spelt in various ways, e.g. Ladislaus Drakula, Wladislaus Dragwlya). *Times Lit. Suppl.* 22 Oct. 1971, p. 1336.
[1897 B. Stoker *Dracula* iv. 51 The last I saw of Count Dracula was his kissing his hand to me; with a red light of triumph in his eyes, and with a smile that Judas in hell might be proud of. *Ibid.* xxvii. 383 We looked back and saw where the clear line of Dracula's castle cut the sky.] 1938 M. Allingham *Fashion in Shrouds* xvi. 279 If he's Dracula himself we'll get him and hang him. 1953 R. Lehmann *Echoing Grove* 266 A real killer in disguise she must be—a real female Dracula. 1971 'A. Gilbert' *Tenant for Tomb* xv. 252 Now that Lady Dracula is unlikely to be released..how about our considering our own plans?

draegerman (drē[1]·gəɹmæn). *N. Amer.* Also **dregerman.** [f. the name of A. B. Dräger (1870–1928), German scientist, inventor of a type of breathing apparatus + Man *sb.*[1]] One of a crew of men trained for underground rescue work. So *Draeger crew*.
1918 R. Drummond *Minerals & Mining, Nova Scotia* 343 Draeger men were soon ready but exploration was somewhat impeded by the jamming of the cage at the bottom of the shaft. 1936 *Ottawa Jrnl.* 20 Apr. 1/1 One of the Stellanton Draeger crew explained to bystanders. *Ibid.* 21/1 Stellanton's draegermen—a crew of rescue men trained to enter a colliery after an explosion. 1958 *Times* 25 Oct. 6/7 The dregermen..carry about 45 lb. of oxygen equipment on their backs.

draft, *sb.* Add: **2. a.** (Further military examples.) Spec. (esp. in U.S.), selective conscription. Also *attrib.* and *Comb.*
a 1772 J. Woolman *Jrnl.* (1774) v. 87 The military officers ..agreed on a draught [ed. 1900: agreed on draft]. 1780 Joseph Jones *Let.* 2 Dec. (1889) 58 Let an exemption from draft, or even militia duty out of the State, be offered by the law to every person who recruits a soldier for the war. 1878 J. H. Beadle *Western Wilds* 532 All the really valuable survivors of the volunteer army had returned to civil life; ..the draft-sneaks and worthless remained. 1918 'Captain X' *Our First Half Million* 20 Certain newspapers in discussing the draft..used an unfortunate word. It was 'conscript'. 1918 W. Owen *Let.* ? 25 Mar. (1967) 542 Two companies of A 4 (boys) on Draft Leave. 1931 J. T. Adams *Epic of America* xi. 338 There was no draft, there were only.. volunteers. *Ibid.* xiii. 384 No reliance was placed upon volunteering..and a universal draft act was passed. 1961 *Listener* 17 Aug. 240/1 These sermons were levelled at a nation of professional draft-dodgers. 1968 *Globe & Mail* (Toronto) 17 Feb. 1/4 The head of the Toronto Anti-Draft Program last night predicted a considerable increase in the numbers of draft-dodgers coming to Canada in the wake of new draft deferment rules for graduate students. 1968 *Listener* 6 June 733/2 Students..were protesting against the war by turning in their draft cards. 1971 *Times* 22 Feb. (Canada Suppl.) p. v/1 Nor are more than a few of them draft-dodgers from over the United States border.
b. Also, a person drafted for military service.
1916 'Boyd Cable' *Action Front* 110 I've seen one-half the battalion wiped out and built up with drafts. *a* 1917 E. A. Mackintosh *War, the Liberator* (1918) 23, I know the drafts are good men, I know they're doing well.
c. = *draft-ewe* (Draft *sb.* 7).
1844 H. Stephens *Bk. Farm* III. 1107 Drafts are ready for sale in September. 1886 C. Scott *Sheep-Farming* 107 Ewes and gimmers at 34/4½, being 3/- a head more than the drafts.
d. *U.S.* An act of drafting (see *draft *v.* 1 c).
1948 *Economist* 8 May 763/1 Many believe that General Eisenhower would yield to a draft. 1948 [see *draft *v.* 1 c]. 1950 *Manch. Guardian Weekly* 2 Nov. 2 He chose to yield to Governor Dewey's 'draft' and accept the nomination.
3. c. (Earlier example.)
a 1817 Jane Austen *Northanger Abbey* (1818) II. xi. 219 Giving ready-monied actual happiness for a draft on the future, that may not be honoured.
7. d. draft tube, † **box** [Draught *sb.* 23], an air-tight tube or enclosure that at its upper end receives the discharge from a turbine and at its lower end extends below the level of the water in the tail-race; (orig. simply *draft*).
1840 Parker & McKelvey *U.S. Patent 1658*, The percussion and reaction wheel..is inclosed in a box or case, which is denominated a 'draft'..the mouth of said draft-dipping into the water and being..below the level of the water in the tail-race. 1849 *Jrnl. Franklin Inst.* XLVIII. 402 The water passes from the wheels or rims into two air-tight chambers..called 'draft boxes', from which it passes into two air-tight iron tubes..called 'draft tubes', which..discharge the water beneath the surface of the lower level. 1876 *Ibid.* CI. 285 A turbine provided with this draft tube can be placed between both levels that constitute the fall, and produce the same result as if the turbine were placed at the bottom of the fall. 1927 *Sel. Engin. Papers Inst. Civil. Engin.* xxxiv. 3 Draught-tubes for hydraulic turbines may be either straight, or curved so as to divert the direction of flow through 90 degrees between the turbine and the tail-race. 1958 *Engineering* 28 Mar. 414/2 The draft tubes for two more units have been completed.

draft, *v.* Add: **1. a.** (Further military and stock-farming examples.) Also (chiefly *U.S.*), to conscript.
a 1772 J. Woolman *Jrnl.* (1774) v. 86 Orders came..to draft the militia. 1860 G. Duppa in S. S. Crawford *Sheep & Sheepmen of Canterbury* (1949) v. 46 Draft out rams as flock is dipped and keep them in a separate flock. 1862 O. W. Norton *Army Lett.* (1903) 102 We want them drafted if they won't volunteer. 1863 S. Butler *First Year in Canterbury Settlement* x. 153 You will see fit to draft out all the lambs that are ready for weaning. 1889 Williams & Reeves *Colonial Couplets* 9 Dagging the hoggets, or drafting the rams. 1940 E. E. Cummings *Let.* 10 Nov. (1969) 158, I avoided the American army, by visiting France with a Norton Harjes ambulanceunit, merely to have myself drafted later and serve six months at Camp Devens. 1946 F. Davison *Dusty* viii. 80 He and Harry were bringing a mob in to the yards to draft out some ageing ewes. 1947 'N. Shute' *Chequer Board* iii. 56 When I got drafted they.. put me into a construction unit. 1953 M. Scott *Breakfast at Six* (1960) iv. 34 Presently Larry was summoned to help draft the sheep. 1968 *Globe & Mail* (Toronto) 17 Feb. 4/8 A representative of the U.S. Office of Education said more than 150,000 prospective graduate-school students probably would be drafted.
c. *transf.* Esp., to force or persuade (a reluctant or allegedly reluctant person) to become a candidate for office. *U.S.*
1861 J. G. Holland *Lessons in Life* iii. 41 An old man.. was drafted into the grand jury. 1927 *Observer* 4 Dec. 20/7 Mr. Smith will be nominated. Mr. Coolidge will be 'drafted' by acclaim of the Convention. The Republicans will win. 1948 *Chickasha* (Okla.) *Daily Express* 4 July 1/1 Nevertheless he could be drafted, if a definite draft movement took place at Philadelphia.

draftee (drɑfti·). *U.S.* [f. Draft *v.* + -ee[1].] A conscript. Also *transf.*
1866 'F. Kirkland' *Pictorial Bk.* 162 The young draftee appeared a little bewildered. 1952 Steinbeck *East of Eden* xlii. 420 The draftees wouldn't look at their mothers. 1953 *Manch. Guardian Weekly* 12 Feb. 7 The American 'draftee'..is subject to these interferences. 1964 *Economist* 1 Feb. 412/1 The third willing draftee is Mr Richard Nixon. 1969 I. Kemp *Brit. G.I. in Vietnam* iv. 74 My companion, Bob Horton, was a draftee a little younger than myself.

drafter. Add: **3.** A draught-horse; also, a horse used for drafting (animals).
Cf. 1876 F. K. Robinson *Gloss. Whitby*, *Draughters*, waggon-horses.
1906 *Springfield Weekly Republican* 7 Feb. 2 (Advt.), A nice lot of well-broken useful horses, consisting of all classes from the nice, pleasant driver to the large, strong, rugged drafter. 1935 *Bulletin* (Sydney) 3 Apr. 20/1 Old horses seem to be the best camp drafters.

drafting, *vbl. sb.* (Further *attrib.* examples.)
1856 W. Roberts *Diary* 18 Dec. in J. H. Beattie *Early Runholding in Otago* (1947) vi. 43 There was no crush pen or drafting race. 1863 S. Butler *First Year in Canterbury Settlement* x. 154 The sheep are in the small yard C (which is called the drafting-yard). 1882 Armstrong & Campbell *Austral. Sheep Husbandry* xv. 177 A second gate hung on the inside of the race will act as a drafting gate, and, when not in use, will, when closed, leave the race secure. 1916 *N.Z. Jrnl. Agric.* 20 Sept. 229 Drafting gates 3 ft. wide. 1922 W. Perry *Sheep Farming in N.Z.* iii. 26 It will facilitate dipping if the entrance to the dip is at the end of the drafting race. 1950 *N.Z. Jrnl. Agric.* July 7/3 *Drafting race*: This race is long and narrow, just wide enough for the sheep to pass through in single file. Its purpose is to divide a mixed flock of sheep into any required number of smaller mobs, and for this purpose the race is fitted at its exit with one or more drafting gates, which guide the sheep into several drafting pens. 1956 J. Dare *Rouseabout Jane* xxi. 167 We had to move the whole lot [of sheep] to the drafting yards. 1959 A. Upfield *Bony & Black Virgin* xxi. 196 Nothing there but a well and drafting yards.

drag, *v.* Add: **1. b.** Also *fig.*, esp. in phr. *to drag one's feet* (orig. *U.S.*), to delay deliberately, hold back deliberately.
1946 *Life* 20 May 69/1 The Soviets are frankly 'dragging their feet' in making the European peace in order to prolong chaos. 1948 *News Chron.* 16 Sept. 1/3 He indignantly denied that the Government was 'dragging its feet' as it had been suggested in the American Press. 1950 *Hansard Commons* 28 Mar. 197 It is widely thought..in America that the British Government are lacking in zeal for the whole plan—'dragging their feet' is, I believe, the American expression. 1970 *Times* 24 Mar. 12/1 Many local authorities drag their feet. They wait for their sewage works to become ..overloaded. 1970 *New Scientist* 4 June 480/1 Many authorities are dragging their heels in setting up the zones.
e. *trans.* To take or escort (a person) to a particular place, event, etc., esp. against his will. *colloq.*
1924 P. Marks *Plastic Age* 136 No freshman was allowed to attend the Prom, but along with the other men who weren't 'dragging women' Hugh walked the streets and watched the girls. 1925 W. Deeping *Sorrell & Son* xxxviii. 385 'Sorry to drag you off like this.' 'Do you think I mind? —It was I who dragged you away.' 1952 J. Cannan *Body in Beck* vii. 142 In the evening I was dragged to an Olde Tyme Dance in the Town Hall.
f. To pull *on* or *at* (a cigarette); to inhale (cigarette smoke). *colloq.* (orig. *U.S.*). (Cf. Pull *v.* 12 b.)
1919 H. Leverage *White Cipher* viii. 121 He waited and dragged at the cigarette. 1926 L. H. Nason *Chevrons* (1927) x. 306 Eadie dragged on the cigarette. 1957 H. Croome *Forgotten Place* xi. 139 He lit one cigarette from the butt of another and dragged at it nervously.
6. (Example of use with *on*.)
1836 Dickens *Let.* 23 Jan. (1965) I. 120, I did set to work yesterday, and dragged on as well as I could.
9. b. *Austral.* and *N.Z. slang.* (See quots.)
1939 in Partridge *Dict. Slang* Suppl. (1961) 1071/2. 1941 Baker *N.Z. Slang* v. 39 From the New Zealand shearing sheds came those effective expressions *to drag the chain* and *swing the gate*, phrases applied to the slowest and the fastest shearer in a shed respectively. 1941 —— *Dict. Austral. Slang* 25 *Drag the chain*, to be slow, to be inferior, to 'tail' the field in any work or contest. 1959 G. Slatter *Gun in my Hand* 91 Stop dragging the chain and have one with me.
10. Delete † *Obs.* and add later examples.
1936, 1938 [see *dragging *vbl. sb.*].
b. *trans.* To arrest. *Criminals' slang.*
1924 E. Wallace *Room 13* ii. 31 After they dragged you I did some hard thinking. 1928 —— *Gunner* xxii. 185 If you particularly want him dragged, you'll tell me what I can drag him on.
13. (From Drag *sb.* 6 a.) To follow the line of scent of (an animal); to trail.
1773 Washington *Diary* 22 Dec. (1925) II. 133 Went out after Breakfast with the Dogs, dragd a fox an hour or two, but never found [it]. 1786 *Ibid.* 9 Feb. III. 12 Never got a fox afoot, tho I dragged one to Mr. Robt. Alexander's Pocoson.

drag, *sb.* Add: **1. e.** A motor-car. *Criminals' slang.*
1935 R. T. Hopkins *Life & Death at Old Bailey* x. 269 When the car thief knocks off a drag (car) from some West End car park. 1947 *Sci. News* IV. 50 There he was, ready with a 'drag' to transport you both at speed. 1960 *Observer* 25 Dec. 7/6 A stately great drag..with a smart chauffeur at the wheel.
f. (See quot. 1954.) Also *attrib.*, as *drag race, racer, racing, strip*. orig. *U.S.*
1954 *Amer. Speech* XXIX. 95 *Drag*, a race between two cars to determine which can accelerate faster. The race is over a given distance, with few exceptions a quarter of a mile. *Ibid.*, There are different types of drag racing: (1) drags from a dead stop; (2) drags from a rolling start. *Ibid.*, *Drag strip*, a straight course used in drag racing, usually an abandoned air strip. 1962 *Ibid.* XXXVII. 273 An establishment where youthful drivers congregate to plan illegal activities such as highway drag-races. 1964 *Guardian* 9 Jan. 3/2 An international drag festival is to be held in Britain...Drag racing was first seen in Britain in September when two American cars gave exhibition bouts. 1967 *Airfix Mag.* June 356/2 Any one of four separate versions can be built from the kit namely a stock, custom, saloon racing, or drag racing car. 1971 *Sunday Express* (Johannesburg) 28 Mar. 7/1 Gerry..has donated a grandstand for Margo's Rainbow dragstrip. 1971 *Capital Times* (Madison, Wis.) 15 June 29 The start of a drag race. *Ibid.*, There have been reports of drag racing by youths on country roads. *Ibid.*, The drag racers are gone.
3. d. (Further examples.) Also, an annoyance, a bore; a dull or boring person. (Cf. quot. 1813 for sense 7 a in Dict.)
1923 Manchon *Le Slang* 109 *Drag*,..difficulté, embarras, partic[t] une chose qui vous éprouve, vous épuise. 1936 'F. O'Connor' *Bones of Contention* 157 As sure as God 'tis a drag. 1946 Mezzrow & Wolfe *Really Blues* viii. 106 They may have been a drag and a headache to their mothers. 1954 L. Armstrong *Satchmo* viii. 126 Life can be such a drag one minute and a solid sender the next. 1959 C. MacInnes *Absolute Beginners* 42 Old Vern..is such a drip-dry drag that no one would ever take *him* for the male of the establishment. *Ibid.* 130 The whole thing was becoming something of a drag. 1968 *Listener* 29 Feb. 265/2, I know so many people that before they took it [*sc.* LSD] were such a drag, and when they took it, they really opened up.

e. A street, road; esp. in phr. *the main drag*. *slang* (now chiefly *U.S.*).
1851 Mayhew *London Lab.* I. 232/1 French news is generally liked in a fashionable drag. *Ibid.* 248/2 Another woman..whose husband has got a month for 'griddling in the main drag' (singing in the high street). 1905 [see *backgate]. 1914 Jackson & Hellyer *Vocab. Criminal Slang* 30 *Drag*, a main thoroughfare in any community; the main street...'The boys are pivoting on the main drag', i.e. begging on the street. 'The muffs are cruising on the drag tonight', i.e. soliciting on the street. 1931 'Dean Stiff' *Milk & Honey Route* 204 *Drag*, hobo term for the main street of the town, as distinguished from the *main stem*. 1936 J. Curtis *Gilt Kid* xix. 188 If he could find the main drag and jump a bus before the bogies got him, he should be able to make a clean getaway. 1962 K. Orvis *Damned & Destroyed* v. 38 You didn't just wander in off the drag to buy coffee. 1965 J. P. Carstairs *Concrete Kimono* ix. 79 We drove through..the main drag of Babaki.
f. The slow-moving portion of a cattle-herd which is being driven. Hence *drag-driver*. *U.S.*
a 1861 T. Winthrop *John Brent* (1883) viii. 71 Relieved from their drags, the herd frisked away with unwieldy gambolling. 1888 T. Roosevelt in *Century Mag.* Apr. 862/1 The rest [of the men] are in the rear to act as 'drag-drivers', and hurry up the phalanx of reluctant weaklings. 1920 J. M. Hunter *Trail Drivers of Texas* 44, I went up the trail twice, and drove the drag both times. *Ibid.* 151 All the men were in front of the cattle except myself, the drag driver, and the cook. *Ibid.* 172 We left the drags together in another herd. 1924 W. M. Raine *Troubled Waters* x. 101 I'm plumb fed up with the dust of the drag driver.
g. Feminine attire worn by a man; also, a party or dance attended by men wearing feminine attire; hence *gen.*, clothes, clothing. *slang.*
1870 *Reynolds's Newsp.* 29 May 5/5 We shall come in drag. 1870 *London Figaro* 23 June 3/4 Not quite so low.. as going about in 'drag'. 1887 *Referee* 24 July 3/1, I don't like to see low coms. in drag parts. 1927 *Sunday Express* 13 Feb. 5/5 A drag is a rowdy party attended by abnormal men dressed in scanty feminine garments, singing jazz songs in high falsetto voices. 1942 M. McCarthy *Company she Keeps* (1943) iii. 80 A kind of masquerade of sexuality, like the rubber breasts homosexuals put on for drags. 1959 C. MacInnes *Absolute Beginners* 27 My Spartan hair-do and my teenage drag and all. 1959 J. Osborne *World of Paul Slickey* ii. x. 80 You would never have the fag Of dressing up in drag You'd be a woman at the weekend. 1960 *20th Cent.* Mar. 255 Bad Taste, exemplified by.. Henry Kendall in drag. This is by no means the first time that Mr. Kendall has appeared to reverse his sex. 1966 *Listener* 23 June 918/3 Laurence Olivier, doing his Othello voice and attired painstakingly in Arab drag. 1967 *Spectator* 14 July 54/1 The gear shops flip their decor as often as they do the pop tunes blaring out the newest hits as you try on the latest 'drag'. 1968 R. Baker (*title*) Drag, a history of female impersonation on the stage.
h. Influence, 'pull'. *U.S. slang.*
1896 Ade *Artie* xii. 105 He knows I've got a drag in the precinct. *Ibid.* xvii. 160 If you've got any drag with him. 1923 Hemingway *In our Time* (1926) 193 We had a big drag with the waiter because my old man drank whisky and it cost five francs, and that meant a good tip.
6. c. (Earlier example.)
1803 W. Taplin *Sporting Dict.* II. 486 A *train scent*, (that is, a drag across the country).
7. d. *Angling.* A dragging motion on a fishing-line; also *concr.*, a device in a fishing reel.
1907 *Westm. Gaz.* 29 Nov. 3/1 The drag cannot be overcome—where the current fished into is far stronger than the current fished over. 1937 Hemingway *To have & have Not* i. i. 23, I felt his drag. He had it screwed down tight. You couldn't pull out any line.
e. *Aeronaut.* and *Hydrodynamics.* The force resisting the motion of a body through a gas or a liquid; *esp.* the resistance along the line of flight to the motion of an aircraft, etc.
1909 A. Williams *Engin. Wonders of World* III. 12/1 To prevent the resulting drag slewing the aeroplane round, the warping mechanism is linked up with the rudder. 1918 W. E. Dommett *Dict. Aircraft* 19 The horizontal component of the air pressure on a wing or aerofoil is known as the drag. 1931 *Flight* 1 May 384/2 And how little headway have made such conceptions as induced drag, profile drag and span loading. 1935 P. W. F. Mills *Elem. Pract. Flying* i. 4 The lift and drag forces act upwards and backwards respectively. 1948 *Sci. News* VII. 25 For an aerofoil it is necessary to introduce also the *induced drag*, i.e. the part of the total resistance which depends entirely on the lift. 1948 V. L. Streeter *Fluid Dynamics* iv. 67 Any body passing through a real fluid experiences a resisting force, called drag, which depends upon the form of the body and its surface roughness. 1952 *Economist* 20 Dec. 852/2 The pilot, when he lifted the nose-wheel of the aircraft off the ground, did so at a sharper angle than usual. The result of this was to give high 'drag', that is, to increase the resistance of the air to the passage of the aircraft. 1971 *Physics Bull.* Mar. 157/2 Perturbations of the orbit of an artificial satellite by the earth's oblateness and atmospheric drag.
f. *slang.* An inhalation of (cigarette) smoke; the act of smoking a cigarette. (Cf. *drag *v.* 1 f.)
1914 Jackson & Hellyer *Vocab. Criminal Slang* 30 *Drag*, an inhalation of smoke, tobacco or opium. 1920 F. Scott Fitzgerald *This Side of Paradise* (1921) i. ii. 58 The ponies took last drags at their cigarettes and slumped into place. 1926 L. H. Nason *Chevrons* (1927) x. 305 A long drag and a cloud of smoke rolled out into the aisle. 1957 C. MacInnes *City of Spades* i. v. 28, I lit up, took a deep drag, well down past the throat, holding the smoke in my lungs. 1962 *Coast to Coast 1961–62* 132 We stopped beside a little trickle of water for ten minutes' break and a drag.
g. *Cricket.* Back spin imparted to the ball by the bowler.
1920 E. R. Wilson in P. F. Warner *Cricket* 84 The two other spins which can be put on the ball are what have been called the drag (or back spin) and top spin. 1922 W. W. Armstrong *Art of Cricket* i. 45 The ball on which drag has been put and which never seems to arrive as soon as it is expected.
h. A slow type of dance, or the music for this; also (*slang*), a dancing party. *U.S.*
1923 in Wentworth & Flexner *Dict. Amer. Slang* (1960) 160/1 (*title of song*) Shoeshiner's Drag. 1928 *Melody Maker* Feb. 178 (Advt.), 'Rain' is a slow drag number. *Ibid.* 179/3 'Sugar', played in a nice drag rhythm. *Ibid.* 183/2 A fascinating *legato* drag rhythm. 1952 B. Ulanov *Hist. Jazz* (1958) x. 115 The cotillion orchestra and polite quartet that accompanied high society drags. *Ibid.* xviii. 220 The records they made ('Harlem Fuss' and 'Minor Drag') caused quite a stir.
i. *Mus.* A drum-stroke consisting of two or more grace-notes preceding a beat.
1927 *Melody Maker* Aug. 807/2 A rudimentary beat—the open drag. *Ibid.*, You must try this drag rhythm with the stick and the brush. 1931 G. Jacob *Orchestral Technique* vii. 71 The drag..may contain more than the two preliminary grace-notes (which really amount to an infinitesimally short roll). 1934 E. Little *Mod. Rhythmic Drumming* 13 *The Crush Roll.* Known variously as the 'Crush', 'Press' or 'Drag' Roll, this is a 'fake' beat which has found its way into drumming for a very good reason...The stick (held fairly loosely) is 'crushed' on to the drum head, and allowed to 'bounce' a number of times. 1961 J. Prebble *Culloden* i. 13 The sticks of the Main Guard came down on the skins in the drag and paradiddle of the General.
9. *drag-rope* (U.S. examples); **drag-fold** *Geol.*, a small fold in a bed that forms part of a larger fold or a fault; esp. one with the appearance of having been formed by shearing when stronger or more massive beds on each side of the folded bed moved relative to each other; **drag-line,** (*a*) *Geol.* each of a series of fainter glacial striations forming a fringe on the lee-side of an older set and produced when one glacier crosses the path of another; (*b*) an excavator having a bucket which is pulled towards the machine by a wire rope; also, the wire rope itself; also *attrib.*; **drag-mill** = Arrastre; **drag-saw** (examples); **drag-seine** *U.S.*, a haul-ashore seine (*Cent. Dict.* 1890 s.v. *seine*[1]); hence **drag-seining** *vbl. sb.*; **drag-shoe** = Shoe *sb.* 5 f; **drag strut** *Aeronaut.*, a strut designed to strengthen a wing against forces arising from drag.
1911 *U.S. Geol. Surv. Monogr.* lii. v. 123 A common type of fold is a drag fold.., by which the formation becomes locally buckled along an axis lying in any direction in the plane of bedding. 1937 *Geogr. Jrnl.* XC. 124 The repeated imbricate faults, shear-plan[e]s, and drag-folds. 1942 M. P. Billings *Struct. Geol.* xii. 221 The lineation results from the parallel arrangement of the crests of minute drag folds formed by the sliding of different layers over one another. 1965 G. J. Williams *Econ. Geol. N.Z.* xix. 342/2 The only folding has been in the form of drag-folds near major faults and gentle compactional folds in the lower Tertiary sediments. 1886 T. C. Chamberlin in *7th Ann. Rep. U.S. Geol. Surv.* (1888) 201 It clearly shows the older set by the drag-lines on their lee sides. 1919 C. G. Raht *Romance of Davis Mts.* 328 The intake canal was dug with..drag lines. 1922 *Glasgow Herald* 28 Sept. 7 Drag-line excavators. 1940 *Chambers's Techn. Dict.* 263/2 *Drag-line excavator*, a mechanical excavating appliance consisting of a steel scoop bucket which is suspended from a movable jib; after biting into the material to be excavated, it is dragged towards the machine by means of a wire rope. 1950 *Engineering* 17 Nov. 369/1 A drag shovel, a drag line, a grab crane. 1956 *Planning* XXII. 56 The largest type of draglines, which cost nearly £750,000, may remove ore at the rate of up to a million tons a year, depending on the depth. *a* 1884 Knight *Mech. Dict.* Suppl. 271/2 *Drag mill*, another name for the arrastra. 1783 in R. T. Durrett *Centenary of Louisville* (1893) 147, 4 pair drag ropes (damaged). 1856 E. E. Hale *If, Yes, & Perhaps* (1868) 147 The sled is fitted with two drag-ropes, at which the men haul. 1907 'O. Henry' *Heart of West* 148 'He bears a drag-rope.' 'Get him and saddle him as quick as you can.' 1868 *Iowa Agric. Soc. Rep. 1867* 220 Drag-saw, for cutting logs into fire-wood. 1893 *Spons' Mech. Own Bk.* (ed. 4) 355 The log is..brought under a drag-saw. 1945 B. MacDonald *Egg & I* (1946) i. iii. 50 The drag-saw barked and smoked. 1888 Goode *Amer. Fishes* 179 The method chiefly practiced by the colonists of New England was that of drag-seining. 1844 J. Backhouse *Narr. Visit to Mauritius & S. Afr.* vii. 138 The drag-shoe is not used on these occasions, lest the wheel should start out of it. 1935 C. G. Burge *Compl. Bk. Aviation* 592/2 The main planes would tend to fold back. This is prevented by dividing the space between the front and rear spars in each plane into rectangular panels by means of 'drag struts'. 1964 A. C. Kermode *Aeroplane Struct.* (ed. 2) ix. 171 To prevent the backwards or forwards movement of the wings, the spars in the old-fashioned conventional structure were usually braced together by a system of struts and wires, the struts being called compression or drag struts.

dragée. Add: In modern use not restricted to sweetmeats serving as a vehicle for drugs; often a sugared almond. Also *attrib.* and *transf.*
1853 C. Brontë *Villette* III. xxi. 48 He was fond of bonbons..and..would give his 'dragées' as freely as he lent his books. 1905 A. Bennett *Sacred & Profane Love* iii. iii. 249 Alice wanted to buy him some sweets...I asked him if he would like *dragées*. 1958 *Observer* 14 Sept. 11/3 Make-up in delicate *dragée* tones.

dragged, *ppl. a.* (See Drag *v.*) (Further examples.)
1831 Seba Smith *Life J. Downing* (1834) 118 The poor Huntonites seemed to be a most dragged out. 1938 J. Steinbeck *Long Valley* 138 I'm kind of dragged out. 1962 K. Orvis *Damned & Destroyed* vii. 49 'I'm dragged,' she said. 'Real dragged.' 1963 *Sunday Express* 13 Oct. 5/7 'She is having "dragged" walls—the latest technique in distressed paintwork.' 'Distressing' is a decorator's term for applying a top coat of the paint so that the tone of the undercoat shines through. 1965 *House & Garden* Feb. 47/2 The bedroom..has walls of pale aquamarine dragged paint.

dragging, *vbl. sb.* (Further examples.)
1812 [see Drag *v.* 10]. 1876 *Encycl. Brit.* IV. 403/2 The bristles..are sorted according to thickness by a process called 'dragging', which consists in passing them through a kind of comb, which retains those that are too stout to go between the teeth. 1882 *Sydney Slang Dict.* 32 *Dragging down*, or *pulling down*, stealing articles from shop-doors. 1936 J. Curtis *Gilt Kid* ii. 18 I'm a screwsman and not on the dragging lark. 1938 F. D. Sharpe *Sharpe of Flying Squad* 330 *Dragging*, stealing from vans. 1961 M. Levy *Studio Dict. Art Terms*, *Dragging*, a method of applying pigment with little or no vehicle by dragging it lightly over the tacky surface of a painting, in order to produce a broken effect.

dra·ggle-tai:ledness. [-ness.] Draggle-tailed condition or character.
1889 E. F. Knight *'Falcon' on the Baltic* iv. 62 The outrageously bad taste and gaudy draggletailedness of English girls of the same degree. 1904 *Westm. Gaz.* 2 Nov. 1/3 The terrible draggletailedness of some of the women.

draggy, *a.* Add: **b.** Boring; conventional; uncongenial; unpleasant. *colloq.* (orig. *U.S.*).
1922 S. Ford *Trilby May crashes In* iii. 43 We were both prepared to be thrilled, I expect, but we soon found that an early rehearsal is rather a draggy affair. 1964 *Punch* 4 Nov. 683/2 The 'draggy old days'. 1967 *Listener* 17 Aug. 205/2 We'll just have to ask my boss and his draggy wife to our rave. *Ibid.* 23 Nov. 668/3, I know it's draggy having the au pair feeding with us; but one has to be madly democratic if one wants to keep them. 1971 A. Hunter *Gently at Gallop* iii. 26 Charlie was only a brewer, remember—draggy rooms made him feel comfortable.

drag-net. b. (Later examples.)
1906 *Daily Colonist* (Victoria, B.C.) 5 Jan. 6/5 The Police Drag-Net.—Quite a grist has been gathered to the police mill for the magistrate's sitting this morning. 1928 Hecht & MacArthur *Front Page* ii. 73 They're throwing a drag-net around the whole North Side. 1958 *Oxford Mail* 1 Aug. 1/9 Police spread a 'dragnet' across the Jezreel Valley in a bid to recapture those who had got away.

dragon[1]. Add: **7. c.** *dragon china* (example of use in full form).
1853 E. M. Sewell *Experience of Life* ix. 80 Tea came, and..the wide cups of dragon china.
10. c. A very powerful armoured tractor.
1926 *Glasgow Herald* 8 Apr. 11 The tanks, dragons, light and heavy guns, cookers, etc. 1927 *Sunday Express* 1 May 7 Just as these 'tankettes' will largely supersede the infantry, so will the 'dragons' supersede horse-teams for bringing up the guns.
18. (Further examples.)
1903 *Folk-Lore* Sept. 293 A dragon-horse carrying on its back a scroll. 1937 *Burlington Mag.* Oct. 162/1 The *ch'i-lui*, also called dragon-horse, is known to us from classic writings.
19. a. *dragon-killer* (*fig.* example), *-scale* (example). **c.** *dragon-guarded* adj.; *dragon-ridden* (example).
1901 *Daily News* 22 Feb. 6/3 Their places of captivity stand for dragon-guarded castles. 1914 W. B. Yeats *Responsibilities* 32 In a dragon-guarded land. 1963 *Times* 17 May 24/2 (Advt.), Every quarter it reviews, comprehensively and authoritatively, the latest developments in pure and applied science. It is the best dragon-killer sixpence can buy. 1922 W. B. Yeats *Seven Poems* 8 Now days are dragon-ridden. 1885 —— in *Dublin Univ. Rev.* Apr., Until afar appear the gleaming dragon-scales.
21. dragon's teeth, also the colloquial name given to the cone-shaped anti-tank obstacles used in the war of 1939–45 (see also quot. 1971).
1943 Hunt & Pringle *Service Slang* 28 *Dragon's teeth*, a form of anti-tank obstacle. 1944 *Times* 28 Nov. 4/2 Extensive minefields, road blocks, dragons' teeth, tank ditches, [etc.]. 1971 *Oxf. Univ. Gaz.* 18 Feb. 671/1 'Dragon's teeth', that is to say, sharp hinged teeth which..protrude from the ground but can be made to sink into it for a car to pass.

dragon-fly. Add to def.: (See also quots. 1917 and 1937.)
1917 R. J. Tillyard *Biol. Dragonflies* i. 1 In his subdivision of the Class of Insecta, Linnaeus placed all the Dragonflies known to him in the single genus *Libellula*, forming the family *Libellulidae* of the..Order Neuroptera. His pupil Fabricius, in re-arranging the Orders of Insects.. constituted the Dragonflies as a separate Order under the name of Odonata, because of the form of their mandibles... The name Odonata persisted, and has been adopted by all subsequent writers. 1937 C. Longfield *Dragonflies Brit. Isles* 6 A somewhat unfortunate situation has arisen by the double use of the word 'dragonfly'. For the past sixty years or so it has been used as the popular name for the whole Order of the *Odonata*. However, the first use of the word..only designated the large species then known to naturalists of the Sub-order *Anisoptera*, in contradistinction to the 'damsel-flies' or *Zygoptera*. 1970 *Age* (Melbourne) 22 June, Dragon-flies together with the smaller damsel-flies make up the order Odonata.

dragonize, *v.* **2.** (Example of *absol.* use.)
c 1853 Mrs. Gaskell *Let.* (1966) 857 When Meta is at home she is to dragonize & prepare French lessons.

dra·g-out. *U.S. slang.* A violent fight, or one who engages in such a fight. (See also *knock-down *sb.* 2.)
a 1859 *Southern Sketches* (Bartlett), He's a rael stormer, ring clipper, snow belcher, and drag out. 1870 *Nation* 30 June 411/2 The number of encounters,..knock-downs, drag-outs, [etc.]..in which the Representative..has been engaged.

dragster (dræ·gstəɹ). orig. *U.S.* [f. *drag *sb.* 1 f + -ster.] (See quot. 1955.)
1954 *Amer. Speech* XXIX. 95 Dragster is one of the classifications in official acceleration competition. 1955 *Britannica Bk. of Year* 490/1 *Dragster* (drag + roadster), a 'hot rod', or car constructed from spare parts and designed to exhibit the maximum of engine efficiency and the minimum of elegance. 1961 *Engineering* 18 Aug. 200/1 His new dragster..is the first British-designed version of an American sprint car. 1967 *Times Educ. Suppl.* 13 Oct. 762 (*caption*) Some of the boys with a 'dragster' they built. 1971 B. Callison *Plague of Sailors* i. 43 One of those ostentatious James Bond-type dragsters that was all chrome-plated wheel spokes.

drail (drē[1]l), *v.*[2] *U.S.* [f. Drail *sb.*] *intr.* To fish with a drail.
1636 A. Shurt *Let.* 28 June in *Mass. Hist. Soc. Coll.* (1863) 4th Ser. VI. 570 Richard Foxwill..spake with a boate of ours (draylinge for mackrell). 1873 *Rep. U.S. Bureau Fisheries* I. xiv. 248 The usual method of taking them [*sc.* bluefish] with the line is by drailing or trolling. 1888 Goode *Amer. Fishes* 180 It is not known when the custom of drailing for mackerel was first introduced.

drain, *sb.* Add: **1. e.** *Colloq. fig. phr.*, *to go* (etc.) *down the drain*, to disappear, get lost, vanish; to deteriorate, go to waste.
[1925 F. Lonsdale *Last of Mrs. Cheyney* 10 It is she who closes the drain on them as they go down it on the Thursday.] 1930 W. S. Maugham *Breadwinner* i. 52 All his savings are gone down the drain. 1933 W. Chetham-Strode in *Famous Plays of 1933* 449 We're all so down the drain no one's got anything. 1951 Koestler *Age of Longing* 234 Others, dozens of Leontiev's colleagues, had..collapsed or made fatal mistakes and gone down the drain. 1952 'J. H. Chase' *Double Shuffle* ix. 184 We had paid out good money to get those policies, and we couldn't afford to let them go down the drain. 1958 *Listener* 7 Aug. 196/1 A poor devil who goes down the drain before a temptation that he finds too strong. 1960 H. Pinter *Dumb Waiter* in *Birthday Party* 130, I thought these sheets didn't look too bright. I thought they ponged a bit...I told you things were going down the drain. 1961 *Ann. Reg. 1960* 471 It appeared that at least one donor country was realizing that aid could easily go down the drain.
f. Colloq. phr. *to laugh like a drain*, to laugh loudly.
1948 Partridge *Dict. Forces' Slang* 109 *Laugh like a drain*, to chuckle 'consumedly'; laugh loudly, especially at someone's discomfiture. (Ward-room and also Army officers'.) 1957 M. Sharp *Eye of Love* iv. 48 'What did you call me?' asked Harry Gibson—and laughed like a drain. 1958 S. Gibbons *White Sand & Grey Sand* 222, I shall laugh like a drain if she's world-famous in another five years or so. 1966 'K. Nicholson' *Hook, Line & Sinker* xv. 174 Old Hester would laugh like a drain if she could see us singing hymns over her.
5. *drain-pipe* (*attrib.* and *fig.* uses, esp. *pl.*, narrow, tight-fitting trousers); *drain-tile* (examples).
1886 Baumann *Londinismen* 45/1 *Drain-pipe*, Schülersprache: Makkaroni. 1903 *Daily Chron.* 4 Aug. 5/1 Sooner or later those dreadful drain-pipe structures the Charing-cross and Cannon-street bridges will have to be rebuilt. 1915 A. D. Gillespie *Let.* 17 Sept. in *Lett. from Flanders* (1916) 304 We have..trench-mortars, drain-pipe mortars, rifle grenades, and..every kind of shell. 1950 *Strand Mag.* Mar. 83/1 'Drain-pipes haven't caught on,' says Eddie. 'The money won't run to them. We can't wear them for work.' 1954 [see *Edwardian *sb.* 3]. 1955 M. Berger-Hamerschlag *Journey into Fog* xviii. 248 But even Jeff entered the class in a new outfit of that macabre character, with drainpipe trousers, both sides braided with black silk. 1960 *Times* 7 Dec. 17/4 (*heading*) The drainpipe not to be despised. *Ibid.* 17/5 What matters is..to show them that a 'drainpipe' bag with a few clubs is not to be despised. 1851 J. Brown *Forester* (ed. 2) i. 46 We are now putting a drain-tile into the ditches and filling them up. 1940 *Chambers's Techn. Dict.* 264/1 *Drain tiles*, hollow tiles laid end to end without joints, to carry off surface or excess water.

drainage. Add: **4.** *drainage-canal*, *-line*; **drainage-basin,** the area of land drained by a river and its tributaries; a catchment area; = Basin *sb.* 12.
1882 *Nation* 13 July 33/1 The topography of its immediate banks and that of its drainage-basin..are fully set forth. 1885 A. Geikie *Text-bk. Geol.* (ed. 2) iii. ii. ii. 352 The proportion of mineral matter in river-water varies with the season...Its amount and composition depend upon the nature of the rocks forming the drainage-basin. 1965 A. Holmes *Princ. Physical Geol.* (ed. 2) xvii. 469 A main river and all its tributaries constitute a river system, and the whole area from which the system derives water and rock-waste is its drainage basin. 1900 *Westm. Gaz.* 20 Sept. 8/1 The opening of the drainage canal has given Chicago an excellent supply of pure water. 1966 *McGraw-Hill Encycl. Sci. & Technol.* 434/1 Drainage canals are deeply cut to facilitate the drainage of surrounding land. 1882 A. Geikie *Text-bk. Geol.* 922 The permanence of drainage-lines is one of the most remarkable features in the geological history of the continents.

drainboard (drē[1]·nbōəɹd). orig. and chiefly *U.S.* [f. Drain *v.* + Board *sb.* 1.] = *draining-board.
1905 *Sears, Roebuck Catal.* (ed. 115) 542/2 Nickel plated brass strainer and 24-inch enameled iron drain board with end piece. 1945 B. MacDonald *Egg & I* (1946) 95 Along one wall were a sink and drain-boards. 1956 'N. Shute' *Beyond Black Stump* i. 24 He went into the kitchen and unloaded five small trout..on to the steel drainboard.

draining, *vbl. sb.* Add: **4.** *draining-pen* (Sheep-farming); **draining-board,** a grooved and sloping board on which utensils are put to drain after they have been washed; so *draining-table*.
1906 *Westm. Gaz.* 30 Apr. 4/2 Few sculleries are equipped with a draining-board. 1933 *Archit. Rev.* LXXIV. 42/3 (*caption*) A typical gas-operated built-in refrigerator installed under the draining board. 1919 *N.Z. Jrnl. Agric.* 21 Apr. 224 Information is furnished..regarding the best method of finishing off the concrete floor of a sheep-dip draining-pen. 1965 J. S. Gunn *Terminology Shearing Industry* i. 25 In the older 'dips' and also in modern 'sheep showers', this is an adjoining draining pen to which sheep go after treatment with the 'wash' or 'dip'. 1895 *Army & Navy Co-op. Soc. Price List* 230/2 Plate Rack and Folding Draining Table Combined (Patent).

drainless, *a.* Add: **2.** Not provided with drains.
1902 H. Rumbold *Recoll. Diplomatist* I. 38 They were badly lighted and worse paved, drainless and malodorous.

drake[2]. **2. drake-fly**: restrict † to sense in Dict. and add: (*b*) a may-fly, used in angling.
1833 T. Hook *Parson's Dau.* i. v, A dab at killing trout; drake-fly, wasp-fly, or stone-fly, all one to him. 1927 H. Williamson *Tarka the Otter* iv. 57 The summer drake-flies.. hatched from their cases on the water and danced over the shadowed surface.
b. Used *attrib.* before the names of birds of the duck kind to denote the male of the species.
1889 [see Dict. 2]. 1907 in *Zoologist* (1908) Apr. 124 A drake Shoveler seen on the river at Eaton.
c. *Comb.* with *drake's*. **drake's tail,** (used of) unruly hair at the back of the head. Cf. *duck-tail.
1938 M. K. Rawlings *Yearling* iii. 24 The hair grieved him..it grew in tufts at the back. 'Drake's tails', his mother called them. 1960 C. Day Lewis *Buried Day* iv. 80 It horrified me..when I first caught sight of the back of my neck..to find that I had a drake's-tail of hair.

drama. Add: **1. b.** *Theatr.* = Melodrama 1.
1895 G. B. Shaw *Our Theatres in Nineties* (1932) I. 106 After the exasperatingly bad acting one constantly sees at the theatres where high comedy and 'drama' prevail, it is a relief to see even simple work creditably done. 1947 —— *Shaw on Theatre* (1958) 277 To him drama meant melodrama, its technical sense on the stage.

Dramamine (dræ·mămīn). *Pharm.* Also **dramamine.** The proprietary name of an antihistamine compound used as a drug to prevent nausea.
1949 Gay & Carliner in *Science* 8 Apr. 359/1 The dose of Dramamine was 100 mg every 5 hr and before retiring. Dramamine prevented seasickness in all but two of the 134 men. 1950 *Trade Marks Jrnl.* 7 June 516/1 Dramamine. Pharmaceutical preparation consisting of dimenhydrinate for use in the prevention and treatment of motion sickness, nausea and vomiting, and as antihistaminics. G. D. Searle & Co...United States of America. 1958 B. Hamilton *Too Much of Water* ii. 23 Edgar, beginning to feel liverish and sleepy from the dramamine tablet he had punctiliously taken an hour before sailing. 1962 I. Murdoch *Unofficial Rose* xxxvi. 344 He had provided himself with plenty of Dramamine for the trip. 1970 *New Scientist* 5 Mar. 445/2 Dramamine and hyoscine are found effective in stilling the internal turbulence.

dramatic, *a.* Add: **A. 1.** (Further examples.) *dramatic soprano*: see quot. 1961.
1710 C. Gildon *Life T. Betterton* p. vi, The Graces of Action and Utterance come naturally under the Consideration of a *Dramatic Writer*. 1826 *Blackw. Mag.* XIX. 197 (*title*) On cant in dramatic criticism. *Ibid.* 198/1 Justice and honesty require, that the bread of a performer, or the character of an author, shall not be sacrificed to the dull sport or the heedless haste of paragraphs in the newspapers. These are now almost the sole vehicles of dramatic criticism. 1864 *Round Table* 2 Jan. 43/3 Dramatic critics in New York. The use of the term 'Critic' in this heading is a misuse of English, but as the persons of whom we have now to write lay claim to the honorable appelation, we venture to treat it thus badly—on this occasion. *Ibid.* 43/3 Out of this mass of moral and physical corruption has come for years the greater portion of 'Dramatic Criticism' in New York. 1879 C. E. Pascoe *Dramatic Notes* 29 That competent dramatic writer, Mr. Charles Reade. 1907 W. Raleigh *Shakespeare*, *Index* 229/2 Irony, dramatic. 1926 Fowler *Mod. Eng. Usage* 295/2 *Dramatic irony*, i.e. the irony of the Greek drama...The surface meaning for the dramatis personae, & the underlying for the spectators. 1942 Partridge *Usage & Abusage* 167/2 Dramatic irony is that which consists in a situation—not in words;..when the audience in a theatre or the reader of a book perceives a crux, a significance, a point, that the characters concerned do not perceive. 1946 *Penguin Music Mag.* Dec. 44, I heard a female, who called herself a dramatic soprano, screech and bawl. 1961 A. Berkman *Gloss. Show Business Jargon* 19 *Dramatic soprano*, the strongest of the female voices, with a range up to about high C.
B. *sb.* **2.** (Further examples.) Also *transf.* and *fig.*
1796 G. Colman in R. B. Peake *Mem. Colman Family* (1841) II. viii. 253 They who are experienced in dramatics will, I trust, see that I have made a fair 'extenuation' of myself. 1936 L. C. Douglas *White Banners* xvi. 338 There was no need to dramatize it..; the strange chronicle furnished its own dramatics. 1957 P. Kemp *Mine were of Trouble* ix. 181 Running forward—I realize now, of course, that this was the most puerile dramatics—I seized the flag and ran back with it.

dramaticule. (Later example.)
1967 *Listener* 3 Aug. 148/1 His [*sc.* Beckett's] new play, *Come and Go*, can spare only 121 words of dialogue for its three women moving through their grim permutation-game, and so Beckett has the grace to call it 'a Dramaticule'.

dramatism (dræ·mătiz'm). Add: **2.** Dramatic quality.
1880 J. H. Ewing *Let.* 16 Mar. in H. Eden *J. H. Ewing* (1896) 222 Her writing is glorious—Imagination limited—Dramatism—nil! 1901 *Daily Chron.* 14 Nov. 3/1 The infamous stage management of the thing rather aided than detracted from its unspeakable dramatism.

dramatize, *v.* Add: **1. b.** (Later example.)
1900 *Daily News* 28 May 4/1 The glorious language in which Milton sang, Shakespeare dramatised, Richard Baxter prayed, and George Whitfield thundered.
2. *refl.* To behave melodramatically.
1934 H. G. Wells *Exper. Autobiogr.* II. viii. 620 My mother dramatized herself, indeed, but so artlessly that I rebelled against that.

dramatizing, *vbl. sb.* and *ppl. a.* Add: Also *fig.*
1888 Mrs. H. Ward *R. Elsmere* II. xviii. 107 Oh, to fall at her feet, and ask her pardon before parting for ever! But no—no more posing; no more dramatizing. 1934 H. G. Wells *Exper. Autobiogr.* II. viii. 620 Accident threw me in my receptive years mostly among non-dramatizing systematic-minded people.

Drambuie (dræmbū·i, -biū·i). [Proprietary term.] A whisky liqueur manufactured in Scotland.
1893 *Trade Marks Jrnl.* 20 Sept. 923 Drambuie, a Liqueur. James Ross, Broadford, Skye. 1935 A. L. Simon *Wines & Liqueurs* 22 *Drambuie*, a Scotch liqueur, golden in colour, with the flavour of Whisky and Honey. 1962 G. Thomas *Keep* i. 25 You took a few drinks. Why not? Anybody selling monumental masonry in..towns..where the only concession to gaiety is a striped shroud, deserves a bath in drambuie, no less.

dra·m-shop. Chiefly *U.S.* [Dram *sb.*[1] 3 b.] A shop or bar where spirituous liquor is sold in drams or small quantities.
1725 *New-England Courant* 8–15 Feb. 1/2 Certainly the Devil had brought himself as well as his Hogs to a fine Market, when he was thus expos'd to Sale in a Dram-Shop. 1761 J. Adams *Diary* in *Wks.* (1850) II. 122 Taverns and dram-shops are therefore placed in every corner of the town. 1839 [see Dram *sb.*[1] 6]. 1849 F. Parkman *California & Oregon Trail* 445 We passed..Boone's grocery and old Vogle's dram-shop. 1875 Holland *Sevenoaks* i. 2 A row of stores and dram-shops and butchers' establishments. 1951 R. Campbell *Light on Dark Horse* iii. 62 The Indians.. were quite capable of boycotting his dramshop.

‖**Drang** (dræŋ). [G.] Pressure; urge, strong desire; esp. in *Drang nach Osten* (lit. pressure to the east), a former German imperialistic policy of eastern expansion; also *transf.*
1906 W. James *Let.* 28 June in R. B. Perry *Tht. & Char. W. James* (1935) II. 471 The wide difference between your whole *Drang* in philosophizing and mine would give me a despairing feeling. 1909 V. Yovanovitch *Near-Eastern Problem* iii. 21 (*heading*) The 'Drang nach Osten' or the Pan-German Peril. *Ibid.* 22 The economic interests of the hour are paramount, and the real peril against which the Europe of the twentieth century must fight is the *Drang nach Osten*—in other words, the tendency of the Germans towards universal economic supremacy. 1914 H. H. Munro *Beasts & Super-Beasts* 153 It was, perhaps, a desire to outdistance all possible competition that influenced the management of the *Daily Intelligencer*..to transfer its offices.. to Eastern Turkestan...All took part in what was popularly alluded to as the *Drang nach Osten*. 1939 H. Nicolson *Diary* 9 Feb. (1966) 391 Berlin engineered the story about the *Drang nach Osten* in order to divert our attention from an impending drive towards the west. 1940 *Economist* 12 Oct. 450/1 A German offensive—of which the occupation of Roumania and the establishment of airfields within bombing distance of Istanbul is a first manœuvre—may well develop, following the old lines of the *Drang nach Osten*. 1941 A. Huxley *Grey Eminence* viii. 171 Crusading, for Wallenstein, was merely an excuse for the *Drang nach Osten*. 1951 M. Lowry *Let.* 25 Aug. (1967) 252 A philosophy and psychological *drang*. 1967 *Economist* 18 Mar. 996/2 It is impossible not to be uncomfortably aware of the omens that support your thesis of an economic *drang nach osten* in Yorkshire.

‖**drápa** (drā·pă). Pl. **drápur.** [ON. *drápa*, prob. f. *drepa* to strike (cf. Drepe *v.*).] A heroic laudatory poem of the Old Norse period.
[1843 G. W. Dasent tr. *Rask's Gram. Icelandic* xxi. 217 After [the Toglag] the Knútsdrápa, and several other laudatory poems were arranged.] *Ibid.* xxii. 223 Laudatory poems were of two kinds; a short eulogy..was called flockr

DRAPE

..: The longer and more stately kind of these poems were the drapur. **1913** W. A. Craigie *Icel. Sagas* i. 12 It is recorded of one man, Stúf the Blind, who was himself a poet, that he could recite more than thirty long encomia (called *drápur*) and as many shorter ones (*flokkar*). **1927** E. V. Gordon *Introd. Old Norse* 297 The *drápa* had a refrain of two or four lines every two, three, or four stanzas, and usually there were several refrains. **1959** A. G. Brodeur *Art of Beowulf* vi. 137 The tale of Hygelac's fall, which we know to be solidly historic, may have been derived..from poems of the type represented in Scandinavia by the *drápa*. **1968** G. Jones *Leg. Hist. Olaf Tryggvason* 9 The well-known Hallfred..wrote a drápa or encomium on him.

drape, *v.*^{1} Add: **5.** To place (oneself) against or on an object or another person, *esp.* in drunken unsteadiness. *colloq.*

1943 Hunt & Pringle *Service Slang* 28 *Draped*, the worse for drink—hanging on to lamp-posts or one's friend, i.e. draped around anything available. **1958** E. Dundy *Dud Avocado* I. i. 29 The lamp-post against which I was limply draped. **1959** *Punch* 9 Dec. 559/1 Paul Drake comes in and drapes himself across the arm of an over-stuffed chair. **1960** *Times* 4 Mar. 8/1 He..draped himself round a Belisha Beacon in a thoroughly drunken fashion.

drape, *sb.*^{1} Add: **c.** *pl.* Curtains. Chiefly *N. Amer.*

[**1895** *Montgomery Ward Catal.* 13/1 Drapery Silk...Suitable for throws, sash curtains, mantel drapes, etc.] **1908** *Sears, Roebuck Catal.* 885/2 A strong, well made Nottingham Lace Curtain,..one of the most stylish and attractive drapes one could possibly desire for the parlor window. **1934** J. T. Farrell *Young Manhood of Studs Lonigan* (1936) II. xviii. 373 He looked at a rose-green pottery lamp set on the table near the heavy blue velvet drapes. **1936** J. G. Cozzens *Men & Brethren* I. 117 The long drapes drawn together across the front windows. **1952** Granville *Dict. Theatr. Terms* 65 *Drapes*, curtains of velvet or fabric used in place of scenery in revues or in repertory productions in 'Little' theatres. **1962** *Listener* 2 Aug. 191/1 In America.. they [sc. Venetian blinds] are frequently used with unlined 'drapes' (at the side of the window only). **1968** *Globe & Mail* (Toronto) 17 Feb. 49/4 (Advt.), Drapes for sunroom and house. **1970** *New Scientist* 7 May 269/1 The cost of moving (including items such as legal fees, new drapes, etc) is £500.

d. A suit of clothes. *slang* (orig. *U.S.*). Also *attrib.*, as *drape suit*, a suit consisting of a long jacket and narrow trousers.

1945 L. Shelly *Jive Talk Dict.* 9/2 *Drape*, suit. **1951** [see *gear *sb.* 5 e]. **1952** A. Wilson *Hemlock & After* III. i. 201 An endless horizon of drape suits. **1957** M. Swan *Brit. Guiana* I. viii. 133 He was a..man of thirty-two, wearing gaberdine drapes and a bow-tie. **1958** E. Hyams *Taking it Easy* 238 Street-corner youths whose drape-suits and side-burns were evolving into a pastiche of Edwardian fashion. **1969** *Listener* 10 July 58/3 They wore drape-jackets with velvet cuffs and their brothel-creepers were in immaculate condition.

e. *drape forming* (see quot. 1964). So *drape technique*, etc.

1958 *Times Rev. Industry* Aug. 57/2 Machine for thermo-forming heavy industrial parts..uses drape..techniques. **1964** Wordingham & Reboul *Dict. Plastics* 56 *Drape forming* (drape vacuum forming), a method of shaping a thermoplastic sheet material in which the sheet is clamped into a frame, heated and a male mould pushed into the sheet to give positive mechanical stretching. Intimate contact is obtained by applying a vacuum.

drapery, *sb.* Add: **5.** (Further examples.) Also, usu. in *pl.*, curtains. *N. Amer.*

1895 *Montgomery Ward Catal.* 15/1 Drapery prints. For furniture coverings or draperies. *Ibid.* 351/3 Brass drapery chains. For looping back and holding in place curtains, portieres, etc. **1938** G. T. Buswell et al. *Daily-Life Arith.* III. iii. 74 Mrs. Wilson made window draperies. **1967** L. J. Braun *Cat who ate Danish Modern* ii. 18 'There's been a delay on the draperies; the fabric manufacturer discontinued the pattern.' 'Could the photographer shoot from an angle that would avoid the missing drapes?'..'*Never* call draperies *drapes*.' **1968** *Globe & Mail* (Toronto) 17 Feb. 46/8 Extras include..broadloom in living and dining room, draperies in living and dining room.

6. drapery drudge, man, an artist employed by another artist to paint the drapery in a composition.

1861 Drapery drudge [in Dict.]. **1894** H. Gamlin *G. Romney* 90 Unlike Reynolds and Gainsborough, he employed no drapery men.

draping (drē^{i}·piŋ), *ppl. a.* Hanging in graceful or 'artistic' folds.

1898 *Daily News* 7 May 8/4 A stiff collar on which are drawn folds of some softly draping stuff. **1903** R. Langbridge *Flame & Flood* xx, Vases with the draping honey-suckle.

draught, *sb.* Add: **24.** Phr. *to feel the draught*: see *feel *v.* 6 b.

47. c. (Earlier and later examples.)

1835 Dickens *Sk. Boz* (1837) 2nd Ser. 39 A pot of the real draught stout. **1971** *Daily Tel.* 13 May 13/6 Draught beer..is brewed from hops, malt and yeast and is served either by tap or by hand (suction) pump directly from the barrel.

48. c. draught-excluder, a device for excluding draughts; **draught-proof** *a.*, fitted so as to be proof against draughts; hence **draught-proof** *v. trans.*; **draught-screen,** a screen for keeping off draughts; **draught-scroll,** a scroll for regulating the draught of the roving on a spinning-mule; **draught-tube** (see *draft tube* s.v. *draft *sb.* 7 d).

1859 G. Measom *Illustr. Guide Lanc. & Carlisle Railways* 118 (Advt.), Draft excluders. **1895** *Army & Navy Co-op. Soc. Price List* 15 Sept. 187 Patent Draught Excluder. This simple invention consists of a roller covered with plushette, which revolves between two brass brackets when the door is opened or closed. **1909** *Lady's Realm* July 271/2 A most effectual draught excluder. **1908** *Westm. Gaz.* 17 Nov. 4/2 When closed it is entirely draught-proof. **1929** *Evening News* 18 Nov. 5/1 It seats a pilot and two passengers in an enclosed draught-proof cabin. **1960** *Farmer & Stock-breeder* 12 Jan. 107/3 Use a draught-proof surround for the first 4–5 days. **1960** *House & Garden* Oct. 119/1 Sound draught-proofed doors and windows. **1965** E. Gundrey *Foot in Door* xii. 90 Three burly men called on an old lady and offered to draughtproof doors. **1907** N. Munro *Daft Days* xv. 129 She got him in behind the draft-screen on the landing of the stair. **1922** F. J. Niven *Justice of Peace* x, Behind the draught-screen was the sound of soap-lather and water. **1968** *Listener* 20 June 798/3 He had a wooden draughtscreen..and that was elaborately painted. **1894** C. Vickerman *Woollen Spinning* 233 The form of the draft-scroll has to be varied in diameter at different points to suit the twine at different portions of the draft.

Dravidian (drǟvi·diăn), *a.* and *sb.* [f. Skr. *drāviḍa* pertaining to *Draviḍa*, name of a province of southern India. (See Tamil etym.)] **A.** *adj.* Of or pertaining to a non-Aryan people found in southern India and Ceylon, or their languages.

1856 R. Caldwell *Compar. Gram. Dravidian Lang.* 527 This shepherd people..gradually merged in the mass of the Drâvidian race. **1871** E. Balfour *Cycl. India* (ed. 2) II. s.v. *India* 39/2 That geographical distribution of the Kol and Dravidian languages. **1902** *Encycl. Brit.* XXX. 419/2 The languages spoken [in the Madras Presidency]—all of the Dravidian family—are Telugu in the north-east, Tamil in the south, Malayalam in the west, and Canarese in the central plateau. **1959** [see *ashram]. **1961** Burrow & Emeneau (*title*) A Dravidian etymological dictionary.

B. *sb.* **1.** A member of this people or linguistic group.

1856 R. Caldwell *Compar. Gram. Dravidian Lang.* 527 The builders of the cairns had settled in India earlier than the Drâvidians. **1871** E. Balfour *Cycl. India* (ed. 2) II. s.v. *India* 42/1 The uncivilized Dravidian speaking tribes are genuine Dravidians who have in a great measure escaped the culture which the more exposed tribes have received. **1884** D. Aubrey *Lett. fr. Bombay* 149 Every scheme appears to have been tried to draw the Hindoo, the Iranian, the Jain, the Dravidian to Christ. **1919** H. G. Wells *Outl. Hist.* 79/1 The Himalayas etc. divided off the Dravidians from the Mongolians. **1924** A. Huxley *Little Mexican* 58 Two expatriated Hindus and a couple of swarthy meridional Frenchmen, who might pass at a pinch as the Aryan compatriots of these dark Dravidians. **1928** C. Dawson *Age of Gods* iv. 82 The Dravidian was pictured as a mere jungle-dwelling savage.

2. Any of the group of languages spoken by this people.

1856 R. Caldwell *Compar. Gram. Dravidian Lang.* 48 The Drâvidian vocabularies have borrowed largely from the Sanscrit..The Sanscrit, in some instances, has not disdained to borrow from the Drâvidian. **1862** *Jrnl. Amer. Oriental Soc.* VII. 297 We should have expected sound philological method, if anywhere, in the comparison of Dravidian and Sanskrit, considering the accessibility of the material. **1871** [see B. 1 above]. **1928, 1934** [see C below].

C. *Comb.*, as *Dravidian-speaking*.

1871 [see B. 1 above]. **1928** C. Dawson *Age of Gods* iv. 83 The modern Dravidian-speaking Brahui. **1934** *Discovery* Feb. 44/1 The important distinction is that between the Aryan and Dravidian-speaking peoples, which broadly set off northern India against the south.

Dravidic (drǟvi·dik), *a.* [f. *Draviḍa*: see prec. and -ic.] = prec. adj.

1888 *Amer. Antiquarian* X. 59 They first entered India, became mingled with the Dravidic race, and afterward were driven out.

draw, *v.* Add: **14.** (Earlier example.) In *Bowls*, to cause (a bowl) to travel in a curve to a chosen spot on the green. Also *intr.* (with the bowl as subject).

1843 'A Wykehamist' *Pract. Hints Cricket* 13 A leg-ball between these lines should be drawn or played under-leg. **1868** [see Shot *sb.*^{1} 11 b]. **1897** *Encycl. Sport* (1901) I. 126/1 Every bowl..is shaved down very carefully and evenly on one side..the effect being to cause the bowl..to incline and turn (or draw) towards the reduced side. **1902** *Encycl. Brit.* XXVI. 329/1 In drawing..the object is to draw as near as possible to the jack. **1910** *Ibid.*, III. 348/1 Should the jack be driven towards the side boundary, it is legitimate for a player to cause his bowl to draw outside of the dividing string. **1962** *Bowls* ('Know the Game' Series) 24 The drawing shot which finishes resting against the jack is sometimes called the 'dead draw'. **1962** *Times* 25 Aug. 2/6 He drew shot brilliantly.

32. (Earlier *Cricket* example.) Also, to bowl out a batsman.

1833 *Sporting Mag.* V. Cricketers' Reg. 13 The stumps were drawn. **1861** *Once a Week* 10 Aug. 182/1, I drew his wicket with..my second ball.

38. (Earlier *Cricket* example.)

1871 'Thomsonby' *Cricketers in Council* 59 To 'draw' a match by refusing to play the ten minutes necessary for finishing it is, in our opinion, a very paltry proceeding.

52. b. Also *fig.*, to be unsuccessful, to fail (in a search); to be in vain.

1825 *Sporting Mag.* XVI. 25/1 One hundred sovereigns is a very pretty 'find' in any man's pocket, and particularly so in one which is sometimes drawn a blank. **1874** Lady C. Schreiber *Jrnl.* (1911) I. 270 Drew blank the only curiosity shop. **1914** *Sphere* 19 Dec. 296/1 She was worrying over Miss Titmus's probable annoyance at drawing a blank from her godchild. **1939** *Punch* 1 Nov. 484/1 Enquiries at the two houses either side of him have drawn blank. **1969** *Woman* 19 Apr. 36/2 Ask the health visitor to introduce you to other mothers..if you..draw a blank at the clinic.

75. *to draw level*. Also *transf.*

1932 *Punch* 27 Apr. 465/3 Other nations had drawn level with us. **1955** *Times* 1 Aug. 3/6 The South Africans have come from behind and surprised the favourites by drawing level in a Test rubber.

82. draw in. d. (Later example without *inf.*)

1813 Jane Austen *Pride & Prej.* III. xiv. 246 Your arts..may..have made him forget what he owes to himself and to all his family. You may have drawn him in.

f. (Earlier example.)

1840 R. Barham in *Bentley's Misc.* Mar. 274 As the evenings begin To close, or, as some people call it, 'draw in'.

draw, *sb.* Add: **1. c.** *Cricket.* A leg stroke in which the batsman deflects the ball so that it passes between the wicket and his legs. Also, a fieldsman placed so as to field balls so hit.

1836 *Nottingham Rev.* 30 Sept., Caught out at the draw through the ball being to the 'leg'. **1846** W. Denison *Cricket: Sketches of Players* 16 How many 'bats' have been compelled to yield up their wickets just when they fancied they had made a fine 'draw'. **1849** *Boy's Own Bk.* 78 As the ball [when drawn] generally gets away between long-stop and leg, it is advisable to place as 'draw' the fieldsman that can best be spared. **1857** [see Draw *sb.* 1]. **1893** R. Daft *Kings of Cricket* v. 88 Tom Hearne..was more successful with the old-fashioned 'draw' than any batsman I can remember.

d. Short for *draw-poker* (see Draw- a). *U.S. colloq.*

1857 *Phœnix* (Sacramento, Calif.) 20 Sept. 3/2 This mongrel, David, recently lost a sum of money, playing 'draw'. **1876** J. Miller *Life amongst Modocs* x. 133 The man..took a quiet game of 'draw' with the boys at the Howlin' Wilderness, and won at once the title of Judge. **1891** *Scribner's Mag.* X. 278 A small game of draw shortens the dying hours. **1945** *New Yorker* 14 Apr. 21 'Dealer's choice,' said Kelly. 'Draw or stud. Fifty-cent ante on draw.'

e. A puff on a pipe, cigarette, etc., a smoke. Chiefly *dial.* and *U.S.*

1876 'Mark Twain' *Tom Sawyer* 138, I could smoke this pipe all day, but he'd keel over with just two draws on a pipe. **1881** A. Wardrop *J. Mathison's Courtship* 24 I'll jist licht my pipe, an' ha'e a bit draw. **1895** 'G. Setoun' *Sunshine & Haar* 253 After making himself quite presentable, sat down for a 'draw'. **1908** A. M'Ilroy *Burnside* v, 'You'll tak' a draw,' the host would say, taking the pipe from his mouth and handing it to his guest. **1933** P. MacDonald *Myst. Dead Police* i. 7 There's nothing like a draw to quiet a chap down. **1969** *Flamingo Mag.* (E. Afr.) x. 45/4, I lit the hemp and had a draw of it.

2. b. The act of drawing a revolver in order to shoot. *U.S.*

1857 T. H. Gladstone *Kansas* v. 54 With my hand upon the pistols..he didn't stand out long. But I felt pretty bad ..till I got the draw on him. **1903** C. T. Brady *Bishop* i. 9 He had the reputation..of being the quickest man on the draw..in the Territory. **1908** C. E. Mulford *Orphan* iii. 37 And they would have gotten it, too, only I beat them on the draw. **1947** *Chicago Tribune* 22 June (Comics) 9 She might beat me to the draw!

5. (Earlier examples.)

[**1825** J. Neal *Bro. Jonathan* I. 50 Everybody was glad when he was beaten; everybody reckoned a draw-game, as a victory over him.] **1856** Mongredien in C. Tomlinson *Chess-player's Ann.* 134 With a view to a 'draw', by bringing the Black Pawn on to a Rook's file. **1860** E. M. Cowell *Diary* 17 June in M. W. Disher *Cowells in America* (1934) 117 Of course every one knows that the fight between Sayers and Heenan was undecided—*a draw*—and both have belts given to them. **1863** *Illustr. London News* 22 Aug. 191/1 Surrey *v.* England, at the Oval..ended in a 'draw' yesterday.

7. (Earlier and later U.S. examples.)

1786 *Maryland Jrnl.* 3 Nov. (Th.), A draw is placed over the deepest water, for permitting vessels to pass and repass. **1889** Morse *Amer. Geogr.* 181 The draw..is designed to require the strength of two men only in raising it. **1902** G. H. Lorimer *Lett. from Self-made Merchant* 21 Our schooner was passing out through the draw at Buffalo.

b. A drawer. *U.S.*

1692 in *Connecticut Probate Rec.* (1904) I. 463, I giue to Elizabeth Thomson..one table with a draue in it. **1748** *N.H. Probate Rec.* III. 565, I give..my chist of draws to my dafter Lidea. **1775** *Essex Inst. Hist. Coll.* XIII. 188 You know I can take a Draw at a time and lay them in the same manner into Dr Gardners. **1829** in W. L. Mackenzie *Lives Butler & Hoyt* (1845) 50 That celebrated receptacle of Chancery papers..the draw or bushel-basket..of his venerable predecessor. **1862** Lowell *Biglow P.* 2nd Ser. III. 108 Once git a smell o' musk into a draw An' it clings hold. **1898** E. N. Westcott *David Harum* 143 They're in the draw there. **1929** in Wentworth *Amer. Dial. Dict.* (1944) 178/2 The draw sticks. **1971** *Amherst* (Mass.) *Record* 28 July 15/1 Wanted to Buy. Two draw file and adding machine.

9. Also, a shallow valley containing a stream. *U.S.*

1882 W. A. Baillie-Grohman *Camps in Rockies* xii. 340 Among the rough and steep chains of mountains full of 'draws', 'pockets', and gulches. **1935** W. Cather *Lucy Gayheart* II. xi. 216 In the draws, between the low hills, thickets of wild plum bushes were black against the drifts. **1953** J. Masters *Lotus & Wind* xx. 253 There was a chance they'd miss the inflow of this draw. **1959** N. Mailer *Advts. for Myself* (1961) 137 The trail rose for a few hundred feet, and then dipped into an empty draw.

10*. *Founding.* A cavity inside a casting produced by the shrinking of the metal during solidification; a shrinkage cavity.

1907 McWilliam & Longmuir *Gen. Foundry Pract.* xxii. 189 In this class of work, 'draws' are often met with, which constitute another class of liquid shrinkage. **1925** *Foundry Trade Jrnl.* XXXII. 552/2 Many defects described by the practical moulder as draws are in fact blows, or in some cases a combination of the two. **1962** J. G. Tweeddale *Metall. Princ. Engineers* vi. 181 Draws are very troublesome in castings.

11. a. *draw-down* (later examples); *draw-in*, also in mod. usage, a roadside space out of the way of traffic where vehicles, esp. buses, may make temporary stops; cf. *layby; *draw-off* attrib., esp. in *draw-off tap* = *draw-tap* (Draw- a). **b. draw-out** (see quot.).

1943 J. S. Huxley *TVA* ix. 55 Ingenious new methods of temporary draw-down of water-level. **1965** G. J. Williams *Econ. Geol. N.Z.* xv. 238/1 Low-permeability holes where draw-down of aquifer pressures during discharge is substantial. **1968** *Gloss. Terms Offset Lithogr. Printing* (*B.S.I.*) 27 *Drawdown*, a method of comparing or examining inks by scraping samples down a sheet or slab to produce thin graded films. **1939** *Nature* 20 May 850/1 'Lay-byes' and 'draw-ins' should be made on every few miles of highway. **1954** *Gloss. Highway Engin. Terms* (*B.S.I.*) 26 *Draw in*, a part of the highway set aside for Public Service Vehicles to draw out of the traffic lanes to pick up and set down passengers. **1909** *Westm. Gaz.* 28 Sept. 3/2 The draw-off taps at the sink..fitted on single pipes. **1951** *Good Housek. Home Encycl.* 222/1 Only a trickle of water comes from draw-off points. **1959** *Times* 14 Mar. 9/7 Wassail bowls fitted with silver draw-off taps. **1960** G. A. Glaister *Gloss. of Book* 110/1 *Draw out*, a printing fault caused when the roller pulls out a loose type.

draw-. Add: **draw-box**: restrict † to sense in Dict. and add (*b*) = *drawing-frame* (Drawing *vbl. sb.* 6 a); **drawcard** = *drawing card* (see *drawing *ppl. a.* 4); **draw-cord** = *draw-string*; **draw-frame** = *drawing-frame* (see Drawing *vbl. sb.* 6 a); **draw-hoe** (see Hoe *sb.*^{2} 1 b); **draw-horse** (U.S. example); **draw leaf,** a leaf of a draw-table; so **draw-leaf table** = *draw-table*; **draw-nail** (see quot. 1960); **draw-sheet** (examples); (*b*) *Printing* (see quot. 1928); **draw-string** (earlier, later, and *attrib.* examples); **draw-tab,** a theatre curtain which can be pulled across the front of the stage; **draw-table,** an extending table, a table with additional pieces which can be drawn out to extend the length; **draw-top (table)**, see quot. 1904.

1909 *Cent. Dict.* Suppl., *Draw-box*, a set of three or more pairs of rollers attached to combing- and certain other machines for attenuating, or drawing out, the sliver. **1940** *Chambers's Techn. Dict.* 264/2 *Draw-box.* This consists of two or more pairs of fluted rollers between the doffer and the coiler of a carding engine. **1959** *Times* 16 Feb. 3/7 O'Neill, who is now the drawcard of Australian cricket. **1969** *Australian* 24 May 35/10 Yardley left St George-Budapest three seasons ago to become a top drawcard with Tranmere Rovers. **1971** *Sunday Times* (Johannesburg) (Business Section) 28 Mar. 14/1 The biggest single drawcard for overseas tourists is London. It is experiencing an unprecedented hotel boom. **1840** W. G. Simms *Border Beagles* 376 [I] have nothing to do but tie a few threads and lay a draw-cord through the end-loops of the net. **1935** *Burlington Mag.* Aug. 92/2 Only alternate threads were controlled by draw-cords. **1936** *Ibid.* Mar. 145/2 The scale-harness is used in order to economize in the number of the drawcords on which the pattern is arranged. **1897** W. S. Taggart *Cotton Spinning* II. i. 2 The full cans of sliver are taken from the card and put behind the draw-frame, so that the sliver can be passed up in the direction of the arrows through holes in the guide-plate A. **1822** [see Hoe *sb.*^{2} 1 b]. **1961** *Amateur Gardening* 23 Dec. 13 The draw hoe is pulled towards the operator and is ideal for the larger weeds, earthing up potatoes or making seed drills. **1845** S. Judd *Margaret* I. xvii. 160 Near Hash stands the draw-horse on which he smooths and squares his shingles. **1932** *Daily Express* 2 July 7/6 Draw-leaf table, size 5 ft. × 3 ft. when extended. **1955** R. Fastnedge *Eng. Furn. Styles* i. 9 Tables of joined construction..and indeed those with a draw leaf (first introduced about 1505). **1958** Osborne & Creighton *Epitaph G. Dillon* I. 11 A draw-leaf table with dining chair. **1702** Draw-nail [in Dict.]. **1960** R. Lister *Decor. Cast Ironwork Gt. Brit.* 226 *Draw nail*, a pointed rod for driving into a pattern to extract it from its mould after ramming. Also called *draw spike*. **1870** Z. P. Veitch *Handbk. Nurses* I. 9 Great care should be taken to arrange a macintosh and draw-sheet in such a way that no blood or other discharge can possibly reach the under sheet. **1928** H. Jahn *Dict. Graphic Arts Terms* 75 *Draw-sheet*, the sheet drawn on over the make-ready on a press. **1960** *Guardian* 1 July 5/4 Mum replaces the draw sheet. **1845** C. M. Kirkland *Western Clearings* (1846) 127 It was Miss Celestina Pye, and she certainly had no draw-strings in her lips just then. **1872** E. Eggleston *End of World* xiii. 89 She pulled out the folds of the chintz curtain, hanging on its draw-string half-way up the window. **1909** *Westm. Gaz.* 10 May 5/2 A belt of thick satin ribbon put through wide embroidered loops like a draw-string. **1955** J. Cannan *Long Shadows* iii. 51 The striped dress with the drawstring. **1957** *New Yorker* 16 Nov. 150/2 It's a wind-resistant..overblouse, cut straight to a drawstring waist. **1958** *Vogue* Sept. 133 Drawstring blouses solve the problem of tuck-in blouses and skirts that are constantly parting company. **1959** 'M. Erskine' *House of Enchantress* iii. 36 She had..a mouth pursed like the opening to a draw-string bag. **1965** C. D. Eby *Siege of Alcázar* (1966) i. 40 The city walls ran in an unbroken line across the neck of the Tagus loop like a drawstring on a sack. **1957** J. Osborne *Entertainer* I. 12 Ordinary, tatty backcloth and draw-tabs. **1904** P. Macquoid *Hist. Eng. Furn.* iii. 96 An oak draw-table of 1560. **1925** Penderel-Brodhurst & Layton *Gloss. Eng. Furn.* 55 *Draw table*, a term generally associated with heavy tables like the refectory table. The top is divided into three leaves, the two end ones lying under the centre leaf. **1938** *Burlington Mag.* Dec. 276/1 Celebrated drawtable from Hinton Abbey near Bath. **1904** P. Macquoid *Hist. Eng. Furn.* iii. 93 With the invention of the draw-top a revolution took place in tables...The top was in three pieces, the lower leaves drawing out and being supported by long armed brackets; the upper leaf dropped into its position, and so the table elongated to double its length. **1955** R. Fastnedge *Eng. Furn. Styles* i. 20 The draw-top table..wherein a subsidiary and additional leaf might be drawn out at each end of the table top.

drawback, *sb.* Add: **2. b.** *Bookselling.* A rebate of the paper tax given under certain conditions to the King's printers and the Oxford and Cambridge University presses. *Hist.*

1796 (*title*) The Poems of Ossian...Printed for A. Strahan and T. Cadell: And sold by T. Cadell Jun. and W. Davies. ..MDCCXCVI (Drawback.) **1797** (*title*) An Historical Essay on the Ambition and Conquests of France...London. Printed for J. Debrett..1797. (Drawback.)

5. (Examples.)

1843, etc. [see *false core* (*false *a.* 17 a)]. **1875** [see Core *sb.*^{1} 8]. **1960** Laing & Rolfe *Man. Foundry Practice Cast Iron* (ed. 3) iv. 90 When patterns are bedded in the floor, there is rarely more than one joint. Many devices are used to avoid a second, one method being the use of drawbacks.

draw-bar. Add: **1.** Also of other vehicles.

1904 *Westm. Gaz.* 30 Mar. 2/1 Two omnibuses..and four light wagons, all..connected by spring draw-bars. **1950** *Engineering* 17 Nov. 388/3 On wet heavy land, a tractor with rubber tyres..will give a drawbar performance. **1954** *Gloss. Terms Agric. Mach.* (*B.S.I.*) 8 *Drawbar*, a member fitted to a tractor for the attachment of hauled implements.

2. orig. *U.S.* (Examples.)

1670 *Groton Rec.* (1880) 36 A gat or a sufficient pair of draw barrs to [be] Kept and maintained at the end [of] Natha[niel] Lawrences field. **1811** *Massachusetts Spy* 3 Apr. (Th.), There were a pair of draw bars about twelve or fifteen yards from [his] door. **1836** *Southern Lit. Messenger* II. 162 On every side I was met by gates, drawbars, and gaps. **1884** R. Jefferies *Life of Fields* 85 The gates beside the lane were not gates at all, but double draw-bars framed together.

3. *Glass-making.* (See quot. 1948.)

1948 *Amer. Ceramic Soc. Bull.* XXVII. 356/1 *Draw bar*, submerged clay block used to define the position of sheet glass during drawing. **1970** *Glass Technol.* XI. 95A The refractory draw bar contains sealed cavities providing buoyancy in molten glass.

draw-boy. Add: **2.** An article exposed for sale in a shop window at a very low price to attract customers. *slang.*

1864 Hotten *Slang Dict.* 125 *Draw-boy*, a cunning device used by puffing tradesmen.

drawer^1. Add: **7. b.** *Printing.* = Tympan 4.

1896 De Vinne *Moxon's Mech. Exerc. Printing* 410 The pasting down of the vellum on the inner side of the tympan (now known as the drawer).

8. *drawer-off* (in various trades: see Dict. Occup. Terms, 1921).

1908 *Westm. Gaz.* 9 Oct. 3/1 'Drawers-off' in saw-mills.

drawer^2. **c. drawerful** (later examples).

1919 G. B. Shaw *Heartbreak House* I. 23 He has a whole drawerful of Albert Medals. **1962** *Sunday Express* 8 July 15/3 Stockings..by the drawerful.

drawing, *vbl. sb.* Add: **1. b.** (Earlier U.S. examples.)

a **1852** F. M. Whitcher *Widow Bedott P.* (1883) v. 18 She sent to borrer somethin or other—a loaf o' bread—or a drawin' o' tea. **1855** T. C. Haliburton *Nat. & Hum. Nat.* II. 350 'I believe,' she said, 'I have a drawing of tea left,' and taking from the shelf a small mahogany caddy, emptied it of its contents.

4. Also with *-in*.

1843 *Lowell* (Mass.) *Offering* III. 215 One of the writers.. has had an opportunity to work at 'drawing in' (that is, drawing the threads through the harnesses). **1844** G. Dodd *Textile Manuf.* iii. 101 Next succeed the various processes of 'winding', 'warping', 'beaming', 'drawing in', &c. **1927** T. Woodhouse *Artificial Silk* xi. 118 The weaver's beam is taken to the drawing-in department, where all the threads are drawn through the mails of the healds..and then passed through the dents of a weaving reed. **1960** *Textile Terms & Defs.* (*Textile Inst.*) (ed. 4) 57 *Drawing-in*, the process of drawing the threads of the warp through the eyes of the healds and the dents of the reed. The operation thus includes that of reeding.

5. *drawing up* (earlier example).

1816 Jane Austen *Emma* I. v. 72, I have seen a great many lists of her drawing up.

6. a. drawing account, an account from which money can be drawn, a current account; see also quot. 1962; **drawing-string** (earlier U.S. example).

1835 [see Depositor 2]. **1920** H. Crane *Let.* 30 July (1965) 41 A drawing account at the bank at my own disposal in addition to a good percentage commission on everything I sell. **1962** S. Strand *Marketing Dict.* 223 *Drawing account*, a company's system of credit permitting withdrawals by salesmen against future commission earnings. **1832** J. P. Kennedy *Swallow Barn* II. xviii. 225 Faces shortened as if with drawing-strings.

b. *drawing-box*; **drawing board** (later examples); colloq. phr. *back to the drawing board*: used after the failure of some enterprise, an invention, etc.; **drawing-paper** (earlier example).

1965 *New Statesman* 9 Apr. 561/1 The super-fast giants.. are now coming on to the drawing boards. **1965** *New Yorker* 6 Nov. 122 A fiery mushroom cloud, translatable by the most cretinous moviegoer as...'Back to the drawing board, you plucky amoebas!' **1967** E. Short *Embroidery & Fabric Collage* iv. 108 Small pieces can be stretched on a drawing board; for large hangings the floor may have to be used. **1968** *Listener* 2 May 562/3 The squeaking [of some newly designed chairs] as we rolled about was too much for the sound people. So again ordinary chairs replaced them. Back to the drawing-board, Byron. **1800** M. Edgeworth *Parents' Assistant* (ed. 3) IV. I. 160 Lady Augusta's apartment, in which her writing-desk, her drawing-box, and her piano-forte stood. **1873** C. M. Yonge *Pillars of House* II. xxiii. 264 John Harewood returned, bringing with him what Alda took for a dressing-case, and Cherry for a drawing-box. **1798** Jane Austen *Let.* 18 Dec. (1952) 41, I have been forced to let James & Miss Debary have two sheets of your Drawing paper.

drawing, *ppl. a.* **4.** Delete † *Obs.* and add later examples, esp. in sense 'that attracts a large crowd, an audience, etc.'; esp. in *drawing card*, an attraction.

1887 in *Amer. Speech* (1950) XXV. 32/2 It proves as good a drawing card as ever. **1894** W. T. Vincent *Rec. F. Leslie* I. ix. 155 A salary adequate to his 'drawing' capacity. **1906** *Westm. Gaz.* 8 Sept. 12/3 The income..varies exceedingly, depending upon the subject taught,..as well as the personal drawing-power of the teacher. **1930** *Daily Express* 6 Nov. 15/4 The biggest drawing card Europe has ever known was Georges Carpentier. **1959** *News Chron.* 9 Dec. 3/1 More than any other guest singer on the Covent Garden list, Maria Callas has the biggest drawing power and means the biggest business.

drawing-room^1. Add: **1. c.** *U.S.* Formerly, a section or carriage of a railway-train more luxurious or more private than usual. Also *attrib.*

1867 *Commerc. & Financ. Chron.* V. 347/2 A new and magnificent sleeping and drawing-room car of the Pullman patent has been..placed on the Michigan Central road. **1873** 'Mark Twain' & Warner *Gilded Age* 264 A lady timidly entered the drawing-room car. **1882** [see Dict. 3]. **1931** E. Linklater *Juan in Amer.* II. vi. 101 There were many so-called drawing-room compartments on the train. **1936** L. C. Douglas *White Banners* xviii. 393 Uncle Thomas will still eat, and buy two tickets so he can travel in a Pullman drawing-room.

3. b. Used allusively to qualify a version of a story, etc., fitted by its observance of the proprieties for the society of the drawing-room.

1877 *Porcupine* 20 Jan. 676/3 It was desirable to modify the language a little, and the drawing-room version ran thus. **1885** A. Edwards *Girton Girl* II. ii. 22 An actor in a similar bit of drawing-room comedy. **1887** Kipling *Under Deodars* (1888) 61 Where..did you pick up the *Chanson du Colonel*? It isn't a drawing-room song. It isn't proper. **1898, 1906** [see *ballad 2]. **1900** G. B. Shaw *Prefaces* (1934) xxix. 709/2 Love is the one subject that the drawing room drama dare not present. **1904** *Sat. Rev.* 9 Apr. 456/1 An infinity of barren drawing-room comedy or drawing-room comedy-drama. **1909** *Daily Chron.* 23 Sept. 1/3 The Prime Minister's Birmingham version of the Budget struck me..as having been intended for what I may call drawing-room use. **1915** Rosher *In R.N.A.S.* (1916) 82 What do you think of this story, the latest from the trenches? It's not quite a drawing-room one! **1959** J. Braine *Vodi* xv. 200 The neatly-folded scarf gave him a drawing-room comedy smartness. **1966** *Listener* 15 Dec. 893/3 He contented himself within his own natural resources—classical Lieder, the standard oratorios and the best of the 'drawing-room ballads'.

Hence **draw·ing-roo:my** *a.*, characteristic of the drawing-room, as being over-refined, insipid, etc.

1906 *Daily Chron.* 18 May 9/3 Miss Evelyn Millard's Desdemona was a disappointment. She was terribly drawing-roomy. **1907** *Ibid.* 18 June 6/4 The sentiment was sometimes rather drawing-roomy. **1947** N. Cardus *Autobiogr.* I. 132 Grieg's rather drawing-roomy 'Un Rêve'.

draw-knife. Now the more usual form of *drawing-knife*. (Further examples.)

1868 R. Taylor *Past & Present of N.Z.* xiii. 292 So destructive did the natives find the [introduced] mouse that they called it the *toro naihi*, or scythe,..literally 'drawknife', a very sharp instrument used by whalers in cutting off blubber from the fish. **1948** E. J. Stowe *Crafts of Countryside* xx. 121 Straight draw-knives are common and can be bought at any ironmonger's shop, but the curved ones are uncommon and are usually made by the blacksmith. **1959** *Times* 30 May 9/5 A farrier's rasp is a much safer instrument to use than a draw-knife. **1969** E. H. Pinto *Treen* 391 The origin of the draw-knife or shave..has never been traced.

drawl, *v.* Add: **1.** *intr.* (Later example.)

a **1918** W. Owen *Poems* (1963) 63 The blind-cord drawls across the window-sill.

3. Also freq. with quoted words as obj.

1842 Lever *J. Hinton* xxx, 'Them chaps always recover,' drawled out the doctor in a dolorous cadence. **1878** R. L. Stevenson *New Arab. Nts.* II. vi. 66 'Ye—es,' drawled Northmour.

drawn, *ppl. a.* Add: **1.** (Additional examples.) *drawn butter*, melted butter; *drawn-in*, (of a rug or mat) made of small cuttings of material drawn through a canvas foundation; **drawn-on** (see quot. 1940).

1747 H. Glasse *Art of Cookery* ii. 25 Have a little good drawn Gravy in a Bason. **1826** in *Doc. Hist. Amer. Industr. Soc. N.Y.* (1910) I. 299 Cod fish and potatoes, with drawn butter and eggs. **1879** Drawn butter [see Butter *sb.*^{1} 1 d]. **1880** 'Mark Twain' *Tramp Abr.* 281 Your brains.. sloshing around in your head same as so much drawn butter. **1901** *Harper's Mag.* CII. 661/2 Her mother had only drawn-in rugs, which Ellen had watched her make. **1940** *Chambers's Techn. Dict.* 265/2 *Drawn on*, said of a book

cover which is attached by gluing down the back; if the end-papers are pasted down it is said to be *drawn on solid*. **1956** F. C. Avis *Bookman's Conc. Dict.* 88/2 *Drawn-on cover*, a separate cover glued onto a stabbed inside with flat back; as seen in most magazines.

b. *Bot.* (See quot. 1900.)

1880 W. J. May *Greenhouse Management* 142 The plants are kept as near the glass as possible, so that they do not become drawn. **1900** B. D. Jackson *Gloss. Bot. Terms* 82/1 *Drawn*, applied to attenuated shoots, diminished and etiolated, often increased in length. **1962** *Listener* 26 Apr. 750/1 Do not let the seedlings become weak and drawn.

7. More widely, made or formed by drawing (see Draw *v.* 5 b). (Further examples.)

1930 *Engineering* 12 Dec. 755/3 Fourcault took out his first patents in 1904, and in 1913 a machine produced a substantial quantity of drawn sheet glass at the factory at Damprémy. **1933** *Archit. Rev.* LXXIV. 130/1 Arc lamps now fitted with drawn-wire bulbs. **1940** *Chambers's Techn. Dict.* 265/2 *Drawn-wire filament*, an incandescent lamp filament, made by a wire-drawing process. **1962** *Gloss. Terms Glass Ind.* (*B.S.I.*) 31 *Drawn glass*, glass, usually in sheet form, made by a continuous mechanical drawing operation. **1960** *Times* 21 Sept. 9/5 Carbon steel tubes (seamless and drawn-welded).

8. (Earlier example.)

1830 Advt. in A. Adburgham *Shops & Shopping* (1964) iv. 38 Whalebone drawn bonnets.

dray, *sb.*^{1} Add: **1. b.** A sled used in dragging logs in the woods. Also *attrib. and Comb. U.S.*

1902 S. E. White *Blazed Trail* vii. 52 A number of pines had been felled out on the ice, cut in logs, and left in expectation of ice thick enough to bear the travoy 'dray'. **1905** *Terms Forestry & Logging* 36 *Dray*, a single sled used in dragging logs. One end of the log rests upon the sled. **1969** L. G. Sorden *Lumberjack Lingo* 36 *Dray*, two runners with a bunk in the center to haul logs out of the woods; a single sled used in dragging logs.

3. b. Any two-wheeled cart. *Austral.* and *N.Z.*

1833 C. Sturt *Exped. S. Australia* I. p. xlix, They send their produce to the market..receiving supplies for home consumption, on the return of their drays or carts from thence. **1846** A. Whisker *Memo Book* 9 Jan. (MS.) I. 21 One of the Drays broke Down about 1 mile from the camp with shot and rum on it. **1872** C. H. Eden *Wife & I in Q'sland* ii. 31 A horse-dray, as known in Australia, is by no means the enormous thing its name would signify, but simply an ordinary cart on two wheels without springs. **1926** J. Doone *Timely Tips for New Australians* Gloss., *Dray*: In Australia this word denotes the springless type of cart generally being equipped with a tipping attachment.

4. *dray-load* (later Austral. and N.Z. examples), *-track*; **dray-cart** = sense *3 b (earlier N.Z. example); **dray-road** (see quot. 1905).

1848 *Handbk. N.Z.* v. 97 Dray-carts drawn by bullocks are chiefly used. **1866** M. A. Barker *Station Life in N.Z.* (1870) vii. 49 It was preceded by two dray-loads of small rough-hewn stone piles. **1901** 'M. Franklin' *My Brilliant Career* xiv. 118 With stacks of love to all at home, and a whole dray-load for yourself, from your loving sister, Sybylla. **1959** J. Pascoe *N.Z. Sheep-Station* 19 Then the 'clip' of wool is baled and driven across the Rakaia in dray-loads to be taken by trucks to the wool sale down country. **1845** E. J. Wakefield *Adv. in N.Z.* II. vi. 159 The proprietors constructed a dray-road up the steep side. **1905** *Terms Forestry & Logging* 36 *Dray road*, a narrow road, cut wide enough to allow the passage of a team and dray. **1907** *Westm. Gaz.* 14 Dec. 14/1 For four hot and weary days I had tramped along lonely and disused dray-roads and bridle-paths that led from a little mining township in the northern part of New South Wales to the coast. **1921** H. Guthrie-Smith *Tutira* xx. 188 It [sc. the pack-team] also, in olden days before the advent of a dray-road, played an important part in station activities. **1859** F. Fuller *Five Years' Residence in N.Z.* viii. 149 Improvements..[to] his Run.. [consist of] partial cuttings for the dray-track where required to be made. **1866** J. Murray *Descr. Province of Southland* iii. 19 When such bridges, culverts and crossings as cannot be dispensed with are made, the track which is often on the line of an old bridle path, becomes a 'dray track'. **1944** F. Clune *Red Heart* 81, I don't need any dray-tracks to find what I'm looking for.

dray, *v.* (under Dray *sb.*^{1}). Add: (Earlier N.Z. examples.) Also *to dray in* (U.S.).

1858 *Richmond-Atkinson Papers* (1960) I. vii. 423, I have arranged with Johnny Jones that he shall have a large shepherd's house..and woolshed,—to be drayed up so as to have no waste timber. **1859** F. Fuller *Five Years' Residence in N.Z.* viii. 149 Such tracks may become available.. for draying down the squatters' produce. **1905** *Terms Forestry & Logging* 36 *Dray in, to*, to drag logs from the place where they are cut directly to the skidway or landing.

Hence **dray·ing** *vbl. sb.* Also *attrib.*

1857 *Lyttelton* (N.Z.) *Times* 13 May 8/1 Stock owners have been enabled to complete their draying operations with ease. **1906** 'O. Henry' *Four Million* 248 A single gentleman connected with the draying business. **1906** —— *Rolling Stones* (1915) 13 You can get me a bunch of draying contracts. **1942** [see *bobbing *vbl. sb.*^{1}].

dray-horse. Add: (Later examples.) So *dray-horse* v. (*rare*).

1857 R. B. Paul *Lett. fr. Canterbury, N.Z.* i. 19 One of Mr. Russell's men then mounted a dray-horse. **1896** 'Mark Twain' *Let.* (1917) II. 637 You have been dray-horsing over the same tiresome ground for a year. **1906** 'O. Henry' *Four Million* 4 She lifts Tobin's hand, which is own brother to the hoof of a dray horse. **1916** J. B. Cooper *Coo-oo-ee* xix. 298 Damn ye, the dray horse wouldn't go in the shafts of that cart.

dreadful, *a.* (*adv.* and *sb.*). Add: **C.** (Earlier examples.)

1873 [see *penny dreadful* s.v. Penny 11]. **1888** C. M. Yonge *Our New Mistress* iv. 38 One of those cheap tales—'dreadfuls', I believe they call them—that one got at the station.

dreadnought, *a.* and *sb.* Add: Also **dread-naught. B.** *sb.* **2.** A fearless person.

[**1694** Motteux tr. *Rabelais's Pantagruel* v. xxxvi. 179, I am as stout as Hercules...My name's William Dreadnought.] **1827** M. Wilmot *Jrnl.* 3 July in *More Lett.* (1935) 268 Our Chezy dreadnought has received him into her house. **1832** Scott *Redgauntlet* (Waverley ed.) I. i. 6 To recollect that the author himself..was one of those juvenile dreadnoughts, is a sad reflection to one who cannot now step over a brook without assistance.

3. (Freq. with capital initial.) The name of the first British battleship (launched on 18 Feb. 1906) of a powerful type superior in armament to all its predecessors; hence, any of a class of battleships having their main armament entirely of big guns of one calibre. (Now disused.) Also *attrib.* and *transf.*

[**1587** Drake *Desp.* 27 Apr. in Hakluyt *Voy.* (1904) VI. p. xiv, A great leake sprange uppon the Dreadenoughte.] **1906** *Outlook* 20 Oct. 495/2 The Atlantic Fleet will consist of three *Dreadnoughts* and five of the *Canopus* class. **1908** *Westm. Gaz.* 14 Aug. 2/2 The mysterious *Dreadnoughts* which are being built in this country for the Brazilian Government. **1909** *Daily Chron.* 23 Mar. 1/1 Our Dreadnought strength and our strength in pre-Dreadnought ships, in comparison with those of Germany. **1914** *Daily Express* 26 Nov. 2/4, 7 Dreadnought Zeppelins: Airships built for the Invasion of Britain. **1915** *Ibid.* 23 Jan. 1/5 Vessels of the Dreadnought era. **1959** *Chambers's Encycl.* IV. 634/2 All maritime powers adopted the general design and it became customary to define battleships as dreadnoughts and pre-dreadnoughts. These definitions generally lapsed as the pre-dreadnought ships became obsolete.

4. *N.Z.* The name of a strain of wheat.

1916 *N.Z. Jrnl. Agric.* 20 Apr. 306 In Canterbury and North Otago the following wheats have given a good account of themselves from the point of view of yield: Red Marvel (French), Dreadnought (French). **1959** J. M. McEwan *Wheat Varieties of N.Z.* 14 Dreadnought has been grown in New Zealand from 1910 and is believed to be of English origin.

dream, *sb.*^{2} Add: **1. b.** Colloq. phr. *like a dream*: easily, effortlessly, without difficulty.

1949 R. Stout *Second Confession* (1950) xi. 87 The engine ..starts like a dream, warm or cold. **1961** *Guardian* 28 Nov. 16/2 The Piccadilly one-way system..worked 'like a dream' throughout the day.

3. c. An ideal or aspiration; *spec.* a national aspiration or ambition; a way of life considered to be ideal by a particular nation or group of people. Freq. with defining adj. prefixed, as *the American dream* (see *American A. 1 a).

1931 *N. & Q.* CLX. 107/1 If, in the course of centuries, the Russian dream comes true the history of Australia..may seem, to students belonging to a Communist society, just as primitive, curious and exciting as to us appear the struggles within the Heptarchy. **1936** M. Mitchell *Gone with Wind* xi. 214 He was still a young girl's dream of the Perfect Knight. **1937** L. Bromfield *Rains Came* Dedication, For all my Indian friends..but for whom I should never have..understood the Indian Dream. *Ibid.* I. xxxiii. 144 A ruler who would cherish the dream and carry it a little way farther along the way to fulfilment. **1937** Kipling *Something of Myself* vi. 149 Rhodes..said to me apropos of nothing in particular: 'What's your dream?' I answered that he was part of it.

4. a. *dream-consciousness, -picture, -play, -poem, -process, -sequence, -state*. **b.** *dream-country, -experience, -figure, -image* (examples), *-imagery, -kingdom, -landscape, -language, -life* (examples), *-stuff, -wish*. **c.** *dream-alliance*. **d.** *dream-interpretation*. **e.** *dream-awakened, -crossed, -fed, -ridden* adjs. **f.** *dream-heavy* adj.

1951 S. Spender *World within World* 310 Reader-writer walk together in a real-seeming dream-alliance leading into gardens inhabited by Stephen Daedalus and Marcel. **1899** W. B. Yeats *Wind among Reeds* 35 Unknown spears Suddenly hurtle before my dream-awakened eyes. **1917** D. H. Lawrence in *Seven Arts* Mar. 443 The thin, transparent membrane of her sleep, her overlying dream-consciousness. **1919** M. K. Bradby *Psycho-analysis* vii. 88 There is often no clear difference in the dream consciousness between idea and act, subject and object. **1925** W. de la Mare *Two Tales* 15 This vision was only of his dream-consciousness. **1895** Hardy *Far fr. Mad. Crowd* (new ed.) p. vi, The horizons and landscapes of a merely realistic dream-country. **1930** T. S. Eliot *Ash Wednesday* 20 The dreamcrossed twilight between birth and dying. **1910** W. James *Coll. Ess. & Rev.* (1920) 510, I woke again two or three times before day-break with no dream-experiences. **1885** W. B. Yeats in *Dublin Univ. Rev.* July 137/1 Where dream-fed passion is and peace encloses. **1919** M. K. Bradby *Psycho-analysis* x. 123 If we fail to discover what a dream figure symbolises, later dreams are likely to supply the clue. **1938** L. MacNeice *Modern Poetry* ii. 46, I required poetic diction and dream-figures. **1897** W. B. Yeats *Secret Rose* 2 The enchantment of his dream-heavy voice was in her ears. **1869** G. M. Hopkins *Jrnl.* 23 Dec. (1959) 194 The dream-images also appear to have little or no projection. **1956** W. Mellers in A. Pryce-Jones *New Outl. Mod. Knowledge* 345 He believed art to be a dream-image, lifting us above sordid actualities. **1929** T. S. Eliot *Dante* 65 A certain *habit* in dream-imagery can persist throughout many changes of civilization. **1913** A. A. Brill tr. *Freud's Interpret. Dreams* ii. 102 If the method of dream interpretation here indicated is followed, it will be found that the dream really has meaning. **1943** *Mind* LII. 78 No dream-interpretation, no symptom analysis occurs, as it were, *in vacuo*. **1925** T. S. Eliot *Poems 1909–1925* 96 Eyes I dare not meet in dreams In death's dream kingdom. **1936** *Burlington Mag.* Sept. 106/2 We detach our attention from this dream-landscape. **1935** L. A. G. Strong in *Amer. Mercury* Aug. 436/1 [James] Joyce..is, throughout *Work in Progress*, assaying a dream language. *Ulysses* dealt with day. *Work in Progress* deals with night. **1851** D. G. Mitchell (*title*) Dream life: a fable of the seasons. **1874** M. Clarke *His Natural Life* (1875) II. III. v. 143 It seems to me..that I have lived somewhere before, and have had another life—a dream-life. **1909** W. James *Coll. Ess. & Rev.* (1920) 487 The will to *personate* may fall outside of the medium's own dream-life. **1940** 'G. Orwell' *Inside Whale* 122 The idea is to give the bored factory-girl..a dream-life. **1899** 'Mark Twain' *Man corrupted Hadleyburg* (1900) 214 A dimly connected procession of dream-pictures. **1907** *Daily Chron.* 11 Nov. 3/4 A surprising Arabian Nights dream-picture. **1950** E. H. Gombrich *Story of Art* xxvii. 442 The experiment of painting dream-pictures was certainly worth making. **1897** G. B. Shaw *Let.* 4 July (1965) 779 It sounded at once serious and inexplicable, like a dream-play. **1942** L. B. Namier *Conflicts* 15 There was something singularly unreal and depressingly second-hand about this dream-play of French history. **1964** C. S. Lewis *Discarded Image* iv. 63 Every allegorical dream-poem in the Middle Ages records a feigned *somnium*. **1944** *Mind* LIII. 179 The imagery-world into which we pass ..is just the way the dream-process..presents itself. **1938** W. de la Mare *Memory* 84 Her dream-ridden eyes. **1959** *Encounter* Oct. 53/1 A legend that has the stylised simplicity of a dream-sequence in a Hollywood musical. **1899** *Westm. Gaz.* 12 Aug. 3/1 A waking and momentary dream state. **1951** M. McLuhan *Mech. Bride* 10/1 Such patterns can only persist in a dream state. **1899** 'Mark Twain' *Man corrupted Hadleyburg* (1900) 281 The Egyptian..rose, filled the planet with sound and splendor, then faded to dream-stuff and passed away. **1916** D. H. Lawrence *Amores* 27 The dream-stuff is molten and moving mysteriously. **1936** *Burlington Mag.* Aug. 85/1 His dream-wishes should be fulfilled in reality.

g. dream-boat, dreamboat *colloq.* (orig. *U.S.*), an exceptionally attractive or pleasing person or thing (= Dream *sb.*^{2} 3 b); *spec.* an extremely attractive member of the opposite sex; **dream-book,** a book containing interpretations of dreams; **dream-child,** a child seen in a dream; an imaginary child; so *dream-son*; **dream-time,** (*a*) *Austral.* = *alcheringa; (*b*) the time for dreams, when the fancy is allowed to run freely; **dream vision,** a conventional poetic form, freq. used by medieval poets, in which the author recounts an alleged dream, the subject of which is often open to allegorical interpretation; also, a poem presenting this form; **dream-work** *Psychol.* [tr. G. *traumarbeit*], the process by which dreams transmute their latent content into their manifest content in order to conceal their real meaning from the dreamer.

1947 Berrey & Van den Bark *Amer. Thes. Slang* Suppl. 1, *Liked person.* Cheezle peezle, dreamboat, [etc.]. **1949** in Wentworth & Flexner *Dict. Amer. Slang* (1960) 162/1 [Ava Gardner] will star opposite James Mason, who she says is a 'dreamboat'. **1951** C. M. Kornbluth in *Galaxy Sci. Fiction* Apr. 141/1 Other cars were showing up, all of them dreamboats. **1951** T. Rattigan *Who is Sylvia?* II. 245, I thought you'd be quite old and staid and ordinary and, my God, look at you, a positive dream boat. **1957** *Life* 29 Apr. 137/1 (*caption*) A dream boat for hot-rodders is a chromed roadster like this one. **1960** *Woman's Own* 10 Sept. 63/2 You've been a dreamboat, and so has Robin. **1793** J. Lackington *Mem.* (rev. ed.) xxxix. 415 Here you may find an old *bawd* inquiring for 'The Countess of Huntingdon's Hymn-book'..and Dolly for a Dream-book. **1803** M. L. Weems *Let.* 27 Aug. (1929) II. 272 To that list you may add..Some dream books, dreaming Dictionaries and above all, some Pilg. Progress. **1909** J. Barlow *Irish Ways* 17 There are fair-sized country towns, whose shops might be thoroughly ransacked without bringing to light any literary wares of more account than a dream-book. **1923** P. Colum *Castle Conquer* x, I bought ear-rings and brooches, dream-books and fortune-books, buckles and combs. **1822** Lamb in *London Mag.* V. 21 (*title*) Dream-Children; a reverie. **1903** *Westm. Gaz.* 28 Mar. 2/1 He's only my Dream-child. Some women have to be content all their life with Dream-children. **1926** M. Leinster *Dew on Leaf* 114 Dream-son be all that I shall ever know. **1963** Auden *Dyer's Hand* 510 Prince Hal will remain his [Falstaff's] dream-son and boon-companion. **1910, 1965** Dream time [see *alcheringa]. **1943** W. E. Harney *Taboo* (1944) 42 Years before, in the 'dream time'..water poured in from the east and flooded the country. **1937** Blunden *Elegy* 60 But the brain Fights dream-time in vain. **1906** R. K. Root *Poetry of Chaucer* iv. 65 Its general form as a poem of the dream-vision type associates the *Parliament of Fowls* with the essentially mediæval, French models of Chaucer's earlier period. **1929** L. Pound in Malone & Ruud *Stud. Eng. Philol. in Honor of F. Klaeber* 235 One is tempted..to dwell upon the popularity of the Dream-Vision form in the Middle Ages as bearing relation to the dream inspiration of poetry. **1947** H. S. Bennett *Chaucer & 15th Cent.* iii. 34 While he retains the dream-vision, he uses it in a new (if not novel) way to bring out the pathos of his story. **1957** C. Muscatine *Chaucer & Fr. Tradition* iv. 115 [In the *House of Fame*] Chaucer.. again adopts the dream vision as a frame. **1965** *English Studies* XLVI. 15 Chaucer's *Book of the Duchess* and his other dream-visions. [**1913** A. A. Brill tr. *Freud's Interpret. Dreams* vi. 262 The dream which we recollect upon wakening would thus only be a remnant of the total dream-work. **1938** *Brit. Jrnl. Psychol.* XXVIII. 294 Dream-work..enables a compromise to be reached between the satisfaction of the repressed urges and the need for sleep.

h. *attrib.*, passing into *adj.* Such as one dreams of or longs to have; ideal; perfect.

Some examples in senses 4a–g are not distinguishable from this use.

1896 E. Turner *Little Larrikin* xii. 129 The dream-cottage was dearer to him than all the beautiful houses he

owned. **1903** *Westm. Gaz.* 6 July 10/1 Mr. Gibson was not slow to grasp the resemblance between his dream-girl and the real. **1911** J. London *Let.* 30 May (1966) 347, I am building my dream-house on my dream-ranch. **1931** J. Cannan *High Table* xv. 220 Home at last to my Dream Girl. **1958** K. Goodwin in P. Gammond *Decca Bk. Jazz* xiii. 149 He fronted an all-star 'dream' band for a lengthy season at the famous Birdland niterie. **1959** D. Eden *Sleeping Bride* v. 40 She planned her dream home. **1960** *Guardian* 14 Apr. 12/4 A dream London in which there are fewer cars. **1961** *Listener* 7 Dec. 966/2 Housewives are offered dream houses with dream kitchens. **1967** Wodehouse *Company for Henry* iv. 66 We got engaged. The family put up a considerable beef..because I wasn't everybody's dream girl. **1971** *Guardian* 19 Jan. 5/6 This is not a dream car just built for a motor show.

dream, *v.*^{2} Add: **10.** *to dream up* (occas. *to dream out*): to picture (something) in one's mind; to think up, devise, invent.

1930 E. Pound *XXX Cantos* v. 21 And all of this, runs Varchi, dreamed out beforehand In Perugia. **1935** *Punch* 4 Sept. 262/2 The man who has a clearly formed ambition, who has dreamed *out* an ideal which his whole personality [etc.]. **1941** *Life* 3 Mar. 23/2 Ambassador Winant is about as far from the conventional picture of a..diplomat as Franklin Roosevelt could have dreamed up. **1942** *Time* 23 Mar. 60/2 Pondering Stanford's lack of a liberal arts school, Professor Dodds dreamed up one which would avoid the failings of most liberal arts colleges. **1950** *Manch. Guardian Weekly* 16 Nov. 3 A slick political trick, such as might have been dreamed up by a bright Chicago ward-heeler. **1958** *Listener* 30 Oct. 680/1 This compulsory 'cooling-off period' is not something we have just dreamed up. **1964** M. McLuhan *Understanding Media* v. 59 Plato.. failed to notice that Athens was a greater school than any university even he could dream up.

drea·mlessness. [-ness.] Dreamless condition.

1905 E. F. Benson *Image in Sand* xiii, Something that had mingled with sleep, but was previous to her deep dreamlessness. **1928** J. W. Vandercook *Black Majesty* i. 2 The black peasants of Haiti have slipped ever deeper into stupor and dirt and dreamlessness.

dreamy, *a.* Add: **3. b.** Perfect, ideal; delightful, beautiful. *colloq.* (orig. *U.S.*).

1941 *Life* 27 Jan. 79/2 Subdebs use a great many adjectives...The list merely begins with..*dreamy* and *super*. **1952** S. Kauffmann *Philanderer* (1953) xiv. 233 'Let us find a cool and lovely garden restaurant and have a slow, exquisite dinner...' 'O.K., Russ... Sounds dreamy.' **1953** M. Dickens *No More Meadows* viii. 307 She said she had a date with a dreamy boy. **1958** *Spectator* 10 Jan. 38/2, I have a Ford Edsel, the first to be imported here, it's dreamy, all electric push-buttons. **1959** 'N. Blake' *Widow's Cruise* 67 The water's absolutely dreamy. And I bet you're a super swimmer.

4. *dreamy-eyed* (examples).

1892 'Mark Twain' *Amer. Claim.* 17 A dreamy-eyed young fellow. **1958** E. H. Clements *Uncommon Cold* i. 29 She looked devoted and *very* intellectual...And sensitive. Dreamy-eyed.

drear, *sb.* Transfer †*Obs.* to sense in Dict. and add: **2.** A dreary person. *colloq.*

1958 M. Allingham *Hide my Eyes* x. 101 'What an evil-eyed old drear,' he remarked. **1966** J. B. Priestley *Salt is Leaving* v. 61 He was just a miserable little drear.

drear, *a.* **a.** and **b.** (Later examples in trivial use.)

1962 *Movie* Nov. 34/3 The Israelites are mainly drear, and the little fun comes from the performances of Anouk Aimee as the queen of Sodom and Stanley Baker as her brother. **1968** A. Clarke *Darkened Room* x. 127 I've bought you..a little towel—I think hospital ones are normally rather drear.

dreary (drī^{ə}·ri), *sb. colloq.* [f. the adj.] A dreary person.

1925 N. Coward *Hay Fever* I. 39 Damn, damn! It's those drearies. **1936** H. G. Wells *Anat. Frustration* xiv. 165 The parade of donnish and scholastic drearies.

dreck (drek). *slang.* Also **drek.** [a. Yiddish *drek* (G. *dreck*) filth, dregs, dung, f. MHG. *drec*:— Gmc. *þrekka- repr. by OE. *þreax* rubbish, rottenness, ON. *þrekkr*, OFris. *threkk*. Ult. origin uncertain but connection with Gr. σκατός dung, στεργάνος privy, L. *stercus* excrement is generally accepted.] Rubbish, trash, worthless debris.

1922 Joyce *Ulysses* 511 Farewell. Fare thee well. *Dreck!* **1947** *Horizon* Feb. 90 The anonymous countryside littered with heterogeneous *dreck*. **1965** E. Lacy *Double Trouble* v. 58 *Drek* your dolls are!..I wouldn't stick my customers with such junk! **1966** E. West *Night is Time for Listening* i. 13 'You *are* dreck,' she said. 'I hope you are killed.' **1967** O. Hesky *Time for Treason* v. 38 Meat better than the usual *drek* we get.

dredge, *sb.*^{2} **4. dredge corn** (see quot.).

1917 *Stat. Rules & Orders 1182* 2 in *Parl. Papers* XXVI. 402 For the purposes of this Order, 'Dredge Corn' shall mean a mixture of cereals, whether or not grown together, containing more than one cereal as a main constituent.

dredging, *vbl. sb.* **c.** *dredging-bucket*.

1872 *Porcupine* 29 June 198/3 Set the dredging-buckets to work, and scooped it all away.

dreelite (dre·-, drī·ləit). *Min.* Also **dréelite, dreeite.** *Obs.* exc. *Hist.* [ad. F. *dréelite*, named by P. A. Dufrénoy (1835, in *Ann. chim. phys.* LX. 102) after E. de *Drée*: see -lite and -ite^1 2 b.] The former name of a supposed sulphate of barium and calcium (see quots. 1888, 1950).

1836 *Amer. Jrnl. Sci.* XXX. 380 Dreelite, a new Mineral Species. **1888** *Jrnl. Chem. Soc.* LIV. 33 The mineral termed dreelite by Dufrénoy is shown by the angles of the cleavage planes and by the optical characters to be identical with barytes...Probably..the percentage of calcium sulphate given by Dufrénoy was due to impurities. **1896** A. H. Chester *Dict. Min.*, Dreeite, Dreelite. **1950** M. H. Hey *Index Min. Species* 370 *Dréelite*,..a mixture of baryte, gypsum, etc.

dreep, drepe, *v.* Also **dreap, dreip. 2.** Delete † and add: *Obs.* exc. *dial.* (see also quot. 1825). (Add later examples.)

1807 R. Tannahill *Soldier's Return* 47 Sers! how your tail, an' wings are dreeping! **1825** Jamieson Suppl., *Dreip*, ..to walk very slowly; as, 'There she comes *dreepin*'.'.. To do any piece of business slowly, and without any apparent interest. **1894** R. Reid *Poems, Songs, & Sonnets* 240 Never herriet mavis dreept sae lane and chill. **1941** L. A. G. Strong *Bay* viii. 179 A shuffling, dreeping old crone.

dreep (drīp), *sb. dial.* [f. Dreep *v.*] **1.** A wet, dripping condition; (see also quot. 1887).

1844 W. Jamie *Muse of Mearns* 103 They danced till in a dreep wi' sweat. **1887** Jamieson Suppl., *Dreep*, sb., drip, dripping, as from a roast, from the eaves, &c.: also, the eaves; and where drops from the eaves fall on the ground, as, 'Ye mun bide within your ain *dreep*.'

2. An ineffective, spiritless, or lugubrious person; a 'drip'.

1927 *Spectator* 5 Nov. 171 What can you expect of 'Sammy dreeps', 'dozened idiots' or 'glaikit stirks'? **1940** in *Sc. Nat. Dict.* (1952) III. s.v., That wumman's jist a dreep. I canna thole her. **1942** 'P. Wentworth' *Danger Point* xxii. 130 'Will you give me your impression of the girl.'..'Oh, a long, thin dreep. No guts. The sort that whines.' **1970** 'D. Shannon' *Unexpected Death* (1971) vii. 99 She was, he thought, a dreep.

dreepy (drī·pi), *a. dial.* [f. Dreep *v.* or *dreep *sb.* + -y^1.] Drooping, droopy, spiritless.

1892 R. O. Heslop *Northumberland Words* 253 She's but a poor dreepy creetur. **1927** *Observer* 6 Nov. 15/2 A weak, dreepy-drippy sort of face, obviously belonging to a man with about as much personal force as there is in a penn'orth of suet.

dreikanter (drəi·kantəɹ, -kæntəɹ). Pl. **dreikanter, -ers.** [a. G. *dreikanter*, lit. three-edged thing, f. *drei* three + *kante* edge (see Cant *sb.*^{1}) + *-er* -er^1.] An angular, faceted pebble the surface of which has been cut by wind-blown sand; *esp.* one with three facets.

1903 *Nature* 10 Dec. 143/1 A collection of wind-worn pebbles of quartz and quartzite from an old raised beach near Waverley, North Island, New Zealand,..was exhibited by the president. They have been cut by the sand driven by the wind into the characteristic *Dreikanter*. **1938** A. K. Wells *Outl. Hist. Geol.* xiii. 147 The quartzite and other pebbles in the Keuper in this region are not dreikanters. **1944** A. Holmes *Princ. Physical Geol.* xiii. 258 Such wind-faceted pebbles, which often resemble Brazil nuts,..are known as dreikanter or ventifacts. **1946** F. E. Zeuner *Dating Past* Pl. x. (*caption*) A 'dreikanter', a quartzite boulder facetted by blown sand. **1970** R. J. Small *Study of Landforms* ix. 301 Such abrasion is capable of..shaping individual boulders and small stone into 'ventifacts', notably the well-known 'dreikanter' with their three faceted sides.

drenching, *vbl. sb.* Add: **b. drenching-gun,** a device for giving a medicinal drench to animals.

1950 *N.Z. Jrnl. Agric.* Sept. 239/1 Where a drenching 'gun' is used. **1961** B. Crump *Hang on a Minute* 158 On a table in the middle of the floor was a drenching-gun. **1966** *Punch* 21 Sept. 446/2 Today's stockman must be proficient with hypodermic syringe and drenching gun.

Dresden (dre·zdən). The name of a town in Saxony, used *attrib.* or *absol.* to designate a variety of white porcelain made at Meissen near Dresden, or an object made of this, characterized by elaborate decoration and figure-pieces in delicate colourings. Hence (often *attrib.*) used to designate anything of a delicate or frail prettiness.

1735 E. Finch *Let.* 29 May in E. Burton *Georgians at Home* (1967) iv. 156 To gett either old or Dresden China. *Ibid.*, Exchang'd for a sett of Dresden. **1750** H. Walpole *Let.* 23 June (1941) IX. 106 On the cabinet stood a pair of Dresden candlesticks. **1752** [see China *sb.*^{1} 3 b]. **1753** Hanway *Trav.* II. 226 Fourteen apartments filled with China and Dresden porcelain. **1756** [see Porcelain 1 a]. **1850** Thackeray *Pendennis* I. xxxviii. 372 Wherever you sate down there were Dresden shepherds and shepherdesses convenient at your elbow. **1895** *Montgomery Ward Catal.* 9/1 Extra Fine Paris Sateens...New artistic designs in small Dresden china effects. **1899** J. M. Ward *One Poor Scruple* I. v. 65 Madge's delicate, Dresden-china, little figure. **1905** W. Holman Hunt *Pre-Raph.* I. 49 Etty was cramped by a taste for Dresden-china prettiness. **1908** *Daily Chron.* 28 Aug. 1/1 His own Dresden-china doll of a wife. **1912** C. Mackenzie *Carnival* iv. 38 Her honey-coloured hair and Dresden cheeks fascinated the impressionable child with all the wonder of an expensive doll. *Ibid.* xv. 155 You do look like a Dresden shepherdess with your heart-shaped face and slanting eyes. **1917** T. S. Eliot *Prufrock* 33 The Dresden clock continued ticking on the mantelpiece. **1962** G. K. Hunter *John Lyly* iv. 233 Phebe, the Dresden-china shepherdess. **1970** D. Devine *Illegal Tender* i. 5 How bored she was with that face! Fair hair, big blue eyes, little pouting mouth, pink and white complexion. Like a doll. Yet Chris approved. 'A Dresden Shepherdess,' he'd called her.

dress, *v.* Add: **7. a.** *spec.* To make or provide clothes for (an undressed doll); to put clothes on (an undressed doll).

1727 Ld. Hervey *Let.* 7 Nov. in *Lett. Lady M. W. Montagu* (1966) II. 88 Your pretending to be young enough to take a Pleasure in dressing Babys. **1800** Jane Austen *Let.* 1 Nov. (1932) I. 24 My Mother is very happy in the prospect of dressing a new Doll. **1803** M. Wilmot *Let.* 31 July in *Russ. Jrnls.* (1934) I. 26, I have heard people.. regret that..they had not collected specimens of the fashion of every year by dressing a Doll as every flash of fashion flitted by. **1825** H. Wilson *Memoirs* 22 Romping with her lovely children, dressing their dolls, and teaching them to skip. **1830** M. R. Mitford *Our Village* IV. 213 Good Mr. Norris..dressed his little daughter's doll. **1844** Disraeli *Coningsby* IV. iv. 44 Miss Millbank..was sitting at a round table covered with working materials, apparently dressing a doll. **1864** C. M. Yonge *Trial* I. ii. 23 Ethel, are they too big for Mary to dress some dolls for them? **1968** B. Drew (*title*) Let's dress a doll.

c. (Further examples.) Esp. *to dress for dinner*.

1741 Richardson *Pamela* III. xxx. 184 The three Gentlemen rode out, and returned just Time enough to dress before Dinner. **1771** Franklin *Autobiogr.* in *Writings* (1905) I. 237 One does not dress for private company as for a publick ball. **1782** L. L. Dalrymple *Jrnl.* (1871) 34 We are dressing for dinner. **1841** F. A. Kemble *Let.* 28 Mar. in *Rec. Later Life* (1882) II. 67 It is close upon time to dress for dinner. **1865** Trollope *Belton Est.* (1866) I. iii. 62 We'd better get ready for dinner now. I always dress, because papa likes to see it. **1922** W. S. Maugham *On Chinese Screen* xlix. 193 He always dressed, and..expected the three boys to wait at table. **1931** V. Woolf *Waves* 129 She has not dressed, because she despises the futility of London.

d. *to dress up* (further examples); also *intr.*, of children: to attire onself in a costume or in various clothes as a game; *to dress down*: to wear clothes less formal than would be expected; to dress informally.

1857 Trollope *Barchester T.* III. v. 83 Them two walloping gals, dressed up to their very eyeses. **1870** L. M. Alcott *Old-Fashioned Girl* v. 65, I ain't going to dress up for nothing; I look so lovely, some one must admire me. **1876** [see Nine *sb.* 6 b]. **1888** Mrs. H. Ward *R. Elsmere* I. I. iv. 85 He could see her dressing up with him on wet days, reciting King Henry to his Prince Hal. **1903** *Little Folks* Feb. 115/1 We began to dress up...Humphrey had on my white flannel pyjamas with a red sash...Violet had on the lace window-curtain. **1914** C. Mackenzie *Sinister Street* II. III. vi. 618 You'd be the first to laugh..if I dressed up..half-a-dozen of my friends in velvet jackets. **1959** D. Campbell *Evening under Lamp-Light* 21 Let's play grown-ups. Let's dress up in their clothes. **1960** *Harper's Bazaar* July 50 A predilection for dressing down at casual dinner parties. **1963** *Sunday Express* 17 Feb. 22/4, I want to be recognised as a good actress. That's why I dress down, rather than dress up. **1968** D. E. Allen *Brit. Tastes* vii. 175 On Tyneside the different income-groups differ less sharply than elsewhere in the styles and standard of clothing that they purchase, as if the better-off deliberately chose to dress 'down'.

f. *intr.* Of a male: to allow the sexual organs to be on one side or the other of the fork of the trousers.

1966 *Guardian* 18 Mar. 10/4 You..find an amusing piece by John Morgan on Carnaby Street—'We are "dressing" in the middle this year, man,' a pop singer explains. **1967** *New Statesman* 31 Mar. 450/2, I detected some sag on the right-hand side of the trouser front and got the fitter to pin it back. 'No no no!' said Roy...'Mr Silver dresses to the left.' The fullness on the right was critical. **1969** *Guardian* 31 July 6/1 All those little male problems like dressing to the right or left.

8. d. To fill (a theatre, etc.) by means of complimentary tickets. Cf. Paper *v.* 4 b.

1896 G. B. Shaw *Our Theatres in Nineties* (1932) II. 234 The theatrical deadhead gets his ticket on the implied condition that he 'dresses the house'. If he comes in morning dress, or allows the ladies who accompany him to look dowdy, he is struck-off the free list. **1933** P. Godfrey *Back-Stage* ix. 131 A large number of complimentary tickets is distributed to keep the stalls 'dressed'. **1961** Bowman & Ball *Theatre Lang.* 116 *Dress the house*,..to assign seats to an audience with artful spacing so that the theatre appears to be more crowded than it really is.

13. a. Also, to season (food, esp. a salad).

1795 tr. *Moritz's Travels* 240 The sallad, for which they brought me all the ingredients, I was always obliged to dress myself. **1796–7** Jane Austen *Pride & Prej.* (1813) II. xvi. 188 These two girls had been..dressing a sallad and cucumber. **1942** C. Spry *Come into Garden, Cook* ix. 114 It [sc. an American-type salad] may contain grape-fruit, orange, pineapple, grapes, peaches, and so on. Dressed with mayonnaise, it is often finished off with a sprinkling of ground nuts. *Ibid.* 115 In the happy days when..one might have little melons to serve as a first course, I have filled them with cubes of their own flesh mixed with diced cucumber and dressed them with a thin cream dressing.

j. *Type-founding.* To finish (types or lines of type) after casting, by grooving and smoothing them and adjusting their height and alignment.

1683 Moxon *Mech. Exerc., Printing* xxi. ¶ 2 This pair of Dressing-sticks will serve to Dress Brevier, Long-Primmer, and Pica. **1839** W. Nichol in T. C. Hansard *Print. & Type-founding* 231 The letters are then set up in a long stick, and again dressed. 1888 *Encycl. Brit.* XXIII. 699/2 The types are then dressed and the picker takes them in hand.

k. *intr.* To weigh (a specified amount) on removal of the skin and offal.

1872 J. G. Bourke *Jrnl.* 25 Nov. (MS.), A black tailed deer which dressed about..200 lbs. 1895 [see sense 13 a in Dict.].

dress, *sb.* Add: **4. a.** *dress allowance, -case, designer, -designing, -pattern, -protector, shop, show, stand* (examples); **dress agency,** an agency, shop, etc., that buys clothes privately and resells them; **dress-basket,** a travelling case for a woman's dresses; **dress-conscious** *a.*, designating a person who is sensitive and particular about clothes; **dress-form** chiefly *U.S.* (see quot. 1909); **dress house** (now *rare*), = Brothel *sb.* 3; **dress-improver** (earlier example); **dress length,** a piece of material sufficient to make a dress; **dress-parade, dress parade,** a display of clothes by mannequins (see also sense 4 b); also *fig.*; **dress-preserver,** (*a*) = **dress-shield*; (*b*) 'a leather-covered iron frame extending from the step of a carriage upward over the rim of the wheel, designed to prevent mud or water from being thrown into the carriage' (*Cent. Dict.* Suppl. 1909); **dress reform,** a movement to make dress more practical; so *dress-reformer*; **dress rehearsal,** a rehearsal of a play in costume, esp. the final rehearsal before the first public performance; also *transf.* and *fig.*; (cf. quot. 1793 s.v. Dressed *ppl. a.*); **dress sense** (see *Sense *sb.*); **dress-shield,** a piece of waterproof or other material fastened under the arms of a woman's bodice to protect it from perspiration; **dress-weight,** (*a*) a small lead weight placed in the hem of a dress, etc.; (*b*) cloth of a weight suitable for making into dresses. **b.** † *dress-box*; *dress cane, cloak, -clothes, -coat* (earlier examples), *-glove, -party, -shirt, -shoes* (examples), *-sword* (earlier example); **dress-carriage,** a carriage reserved for state or semi-state occasions; **dress-circle** (earlier example); **dress-parade, dress parade** *Mil.*, a formal parade in which officers and men wear dress-uniforms; also *fig.*; (see also sense *4 a above).

1931 W. Holtby *Poor Caroline* iii. 76 She inspected the garments for sale in a Court Dress Agency, wondering who wanted to buy tarnished tinsel slippers, and stained georgette frocks. 1965 R. Ferguson *Woman with Secret* ix. 59 A Belgian woman who ran a dress agency. 1907 A. Bennett *Grim Smile of Five Towns* 83 A woman who had a generous dress allowance. 1967 *Guardian* 16 Oct. 4/3 A girl..who used two years' dress allowance to buy a pony. 1894 *Country Gentlemen's Catal.* 152 Solid leather portmanteaux, ladies' dress baskets. 1911 'K. Mansfield' *German Pension* 43 A dress-basket neatly covered in a black tarpaulin. 1812 *Dramatic Censor for 1811* Feb. 99/1 The space now alloted to the *basket* to be one entire anti-room to the dress-boxes. 1849 Dickens *Dav. Copp.* xxiv. 383 We resolved to go downstairs to the dress-boxes. 1836 —— *Sk. Boz* I. 303 George..carried a dress cane. 1897 *Daily News* 8 Feb. 5/4 Many of these are what one might call semi-state carriages, but are known as Dress Carriages. 1897 *Westm. Gaz.* 9 Dec. 3/2 The travelling 'dress-case' that combines dressing-bag and trunk. 1899 *Ibid.* 3 Aug. 3/2 These..low broad dress-cases, that, if necessary, can go under a railway carriage seat. 1825 *News* 30 Jan. 36/1 In the dress circle there was not one lady. 1885 E. Sheridan *Let.* 10 July in *Jrnl.* (1960) ii. 62 She was examining my dress cloak and ask'd whether I thought it fit to *dance* in. 1890 C. M. Yonge *More Bywords* 161 She ran to fetch her dress cloak for Jane. 1814 C. S. M. Bury *Diary* 21 May (1838) I. 312 All the gentlemen..looked beautiful in their dress clothes. 1831 Dress-clothes [in Dict.]. 1929 Wodehouse *Mr. Mulliner Speaking* 258 He was prepared to stand or fall by his dress-clothes. 1959 N. Mailer *Advts. for Myself* (1961) 117 He was changing into his dress clothes by the time I followed him to the dormitory. 1767 J. Habersham *Lett.* (1904) 61, I want a dress Coat. 1819 M. Wilmot *Let.* 17 Dec. (1935) 40 Servants dress'd in rich embroidered dress coats. 1918 'Alpha of the Plough' *Leaves in Wind* 146 He is as dress-conscious as a milliner. 1958 *Spectator* 31 Jan. 130/3 Why is the bowler hat regarded by one whole class of very dress-conscious young men as untouchable? 1901 *Westm. Gaz.* 12 June 3/2 The tendency of the dress designer is just now to study old pictures and prints. 1903 *Ibid.* 31 Dec. 3/1 Dress-designing was never in a more advanced stage of development than now. 1893 *T. Eaton & Co. Catal.* Spring & Summer 97/2 Hall's patent dress forms. 1909 *Cent. Dict.* Suppl., *Dress-form*, a frame, sometimes of wire, in the form of a woman, used in making dresses. 1916 L. I. Baldt *Clothing for Women* viii. 162 For the purpose of designing, it will be necessary to have a dress form, on which a close-fitting lining previously fitted to the person for whom the garments are to be designed, has been placed. 1836 Dickens *Sk. Boz* II. 274 A strange chaos of dress-gloves, boxing-gloves, caricatures, albums. 1823 Dress-house [see **accommodation house*]. 1870 W. Acton *Prostitution* (ed. 2) ii. 9 The description of brothels called dress houses was much more prevalent a few years ago than..at present. *Ibid.* 13 The keepers of the old dress houses were mostly females of extreme avarice. 1872 *Young Englishwoman* Oct. 548 (*caption*) Horsehair dress improver. 1873 *Ibid.* Apr. 202/1 A dress length of 8 metres of the best quality costs 58 francs. 1889 *Young Ladies' Jrnl.* 1 Jan. 21/1 (Advt.), A full dress-length of beautiful cloth. 1907 *Daily Chron.* 16 May 5/7 The wife of a native chief who was given a dress length by Lord Kitchener. 1847 *Army Regulations* (U.S.) 91 All company-officers and men will be present at dress-parades. 1870 L. M. Alcott *Old-Fashioned Girl* xvi. 271 The dress-parade is over, and I'm ever so much obliged to you..for..showing me how to make the best of things. 1873 Dress parade [in Dict.]. 1948 S. Gilbert tr. *Camus's Plague* ii. vi. 116 Daily, round about eleven, you see a sort of dress parade of youths and girls. 1960 *Times* 14 Sept. 12/6 We were..bundled off to a dress parade, my wife agog for the latest fashions...However, we found ourselves deposited in front of the local military academy just in time for their annual inspection. 1825 H. Wilson *Mem.* III. 386 He is going to the Duke of Devonshire's dress party. 1844 *Lexington* (Ky.) *Observer* 25 Sept. 1/3 Just received..striped Chusans, in dress patterns. 1895 *Chicago Tribune* 6 Apr. 1 An Eton dress pattern for 10 cents is displayed on page 16. 1907 *Yesterday's Shopping* (1969) 840/2 Dress Preservers. 1931 J. Cannan *High Table* xvii. 251 Mrs. Logan, already so hot that she couldn't think why dress preservers had gone out of fashion. 1951 M. Kennedy *Lucy Carmichael* i. v. 35 Have you sewed in dress preservers?.. You know how you perspire. 1864 *Rep. Comm. Pat. 1862* (U.S.) I. 506 *Dress protector*...This article is intended for the use of infants, to protect their clothes and the dresses of their nurses. 1897 H. G. Wells *Plattner Story* 241 Little purchases:..dress-protectors, tape, and a pair of Lisle hose. 1876 *N.Y. Tribune* 28 Sept., No man can involuntarily throw one leg over the other without a shortening of what the recent Dress Reform convention calls the garmenture. 1889 *Kansas Times & Star* 16 Mar., Anna Jenness Miller's dress reform disciples now number about 400 here. *Ibid.* 7 Mar., Mrs. Jenness Miller, dress reformer, is back in New York from her Western lecture tour. 1828 J. Ebers *7 Yrs. of King's Theatre* iv. 81 On the preceding evening, a dress rehearsal was given. 1854 A. C. Mowatt *Autobiogr. of Actress* 134 The night before that on which the play was to take place we had a dress rehearsal. 1897 *Encycl. Sport* I. 563/2 After several undress and dress rehearsals the master may venture to ask a field to meet him. 1917 *Strand Mag.* Dec. 538/1 If you'll have one dress rehearsal, I'll promise to leave you in peace for the duration of the war. 1925 *Surv. Internat. Affairs 1920–23* i. 46 The work of the Brussels Conference..served as a 'dress rehearsal' for the First Session of the Assembly of the League of Nations. 1963 *Times* 24 May 5/1 Green's play had a slightly dress-rehearsal look about it. 1926 E. Glyn *Love's Blindness* iii. 37 She had, naturally, that dress sense,—that perception of chic. 1943 G. Battiscombe *C. M. Yonge* viii. 101 Most women..have some rudimentary feeling for dress, but Charlotte was born without dress-sense. 1884 *Queen* 9 Feb. (Advt.), You have just ruined that new dress under the arms because you did not have a Canfield Dress Shield. 1890 *Ladies' Home Jrnl.* June 18/4 The Canfield Seamless Dress Shield. 1905 *Daily News* 26 Sept. 5/1 Sewn on the inside of the bodice were two rubber dress-shields. 1892 *Harper's Mag.* Dec. 159 Maybe it's a dress-shirt shield. 1806 in *Doc. Hist. Amer. Industr. Soc.* III. 73, I had not long worked for him before I got on to light dress-shoes. 1828 J. Ebers *7 Yrs. of King's Theatre* 294 Drab pantaloons..and dress shoes. 1930 E. Waugh *Labels* vii. 171 The dress shops..were advertising their end of the season sales. 1971 C. Whitman *Death Suspended* v. 94 Neville was in advertising and Isobel ran a dress shop. 1930 *London Mercury* Feb. 320 They had had to cancel two fittings and three dress shows. 1965 A. Christie *At Bertram's Hotel* xii. 113 The idea of patronising a dress show of any kind would not even have occurred to her. 1904 M. Corelli *God's Good Man* 337 She had a figure which was the envy of all modellers of dress stands. 1912 *Cassell's Penny Bk. Dressmaking* 6/1 The Dress-Stand.—There is one accessory which all home dress-makers should endeavour to procure, i.e. a padded dress-stand...Adjustable models..are obtainable that will screw and unscrew, upon which garments may be fitted for various figures. 1833 R. Dyer *9 Yrs. of Actor's Life* vii. 112 Presenting me with a dress sword, with an inscription expressive of their regard. 1899 in A. Adburgham *Shops & Shopping* (1964) xxii. 261 Dress weights. 1939–40 *Army & Navy Stores Catal.* 637/2 An all wool dress weight fabric. 1950 H. McCloy *Through Glass Darkly* (1951) iii. 25 The lead dress-weights our grandmothers wore in the hems of their long skirts. 1959 *Times* 12 Jan. 11/5 Range of Terylene and cotton mixtures..used at present for lingerie..and dress-weights.

dressage (dre·sāʒ). [Fr., lit. 'training', f. *dresser* to train, drill.] The training of a horse in obedience and deportment; the execution by a horse of precise movements in response to its rider. Also *attrib.* and *fig.*

1936 *Field* 24 Oct. 1015/2 In recent years there has been considerable interest shown in the training and dressage of hacks..as witnessed in the complicated dressage test held at the recent Olympic Games. 1938 *Times* 17 May 8/7 The first international dressage tests to be held in this country will form one of the principal features of the International Horse Show. 1957 *Times* 14 Oct. 4/5 Miss Willcox and High and Mighty..duly won the open event with a dressage of 95·66. 1962 R. S. Summerhays *Elem. Riding* (ed. 4) xxvi. 145 Dressage is a series of well-defined phases or exercises in the training of a horse. 1963 *Times* 22 Apr. 8/5 Mr. Harold Wilson's Lloyd-Georgian deftness in all problems of party dressage.

dressed, *ppl. a.* Add: (Further examples.) Also with *up*.

1612 W. Strachey *Virginia* (1849) ii. 184/1 Apron or any kind of dressed leather. 1798, 1815 [see Undressed *ppl. a.* 8 b, 7]. 1862 G. Borrow *Wild Wales* I. ix. 92 Whenever I go to Chester, and a dressedup madam jostles against me, I shall call her carn-butein. 1870 'Mark Twain' *Sk. New & Old* (1875) 244 He murdered them 'in his splendid dressed-stone mansion'. 1891 R. Wallace *Rural Econ. Austral. & N.Z.* ii. 54 The best dressed mutton..was selling at 2½*d.* per lb. 1906 L. Stanfield (*song title*) All dressed up like a hippodrome horse. 1908–9 *T. Eaton & Co. Catal.* Fall & Winter 310/1 Fine dressed dolls, beautifully dressed. 1912 G. Whiting (*song title*) When you're all dressed up and have no place to go. 1940 W. Shewring *Topics* ix. 100 Isocrates and Demosthenes..are types of the pretentious bad artist—'all dressed up and nowhere to go'. 1963 'Han Suyin' *Four Faces* 54 She bent to smile at the dressed total face in the mirror. 1971 *Timber Trades Jrnl.* 21 Aug. 16/1 Prior to adoption of the new standards sizes for dressed lumber have been the same regardless of the moisture content.

dresser¹. Add: **1. b.** A dressing- or toilet-table; a bureau. *N. Amer.*

1895 *Montgomery Ward Catal.* 605/1 Dresser, well made, of hardwood...Has a good 20 × 24 German beveled mirror and 3 large drawers. 1900 F. R. Stockton *Afield & Afloat* 263 'There,' said he, opening a dresser drawer. 1906 'O. Henry' *Four Million* 180 Dulcie took a last look at the pictures on the dresser..and skipped into bed. 1924 H. Croy *R.F.D. No. 3* 103 She bent over the 'dresser' to examine herself in the mirror. 1927 M. de la Roche *Jalna* xiv. 159 The tiny light of the candle, reflected in the mirror on the dresser, only faintly illuminated their faces. 1968 *Globe & Mail* (Toronto) 3 Feb. 12 (Advt.), Triple Dresser...$280. 1970 S. Ellin *Man from Nowhere* iii. 13 Just tell me which dresser you want.

dresser². Add: **1. b.** *Type-founding.* An operative who finishes type after casting.

1683 Moxon *Mech. Exerc., Printing* xxi. 187 The Letter Dresser hath..his Letters Set up in Composing sticks. 1846 Dodd *Brit. Manufacturers* VI. 45 The long frame, filled with a single line of type, is removed to the dresser. 1888 *Encycl. Brit.* XXIII. 699/2 The dresser..slips them into a long stick..and..cuts with a plane a groove in the bottom.

2. b. *Theatr.* One who helps to dress an actor or actress (see quot. 1870).

1844 J. Cowell *30 Yrs. among Players* ii. xii. 81/2 The domestic paraphernalia of the housekeeper and ladies'-dresser. 1851 W. K. Northall *Before & Behind Curtain* 117 Carpenters, scene-shifters, gas-men, dressers and supernumeraries, all partook of the general joy. 1870 O. Logan *Before Footlights* 77 All theatres of any importance have 'dressers'. These help the players in change of dress, and fold up and put away their stage clothing after the piece is over. 1914 G. K. Chesterton *Wisdom of Father Brown* iv. 83 An aged servant or 'dresser', whose broken-down face and figure..contrasted queerly with the glittering interior of the great actress's dressing-room. 1970 K. Giles *Death in Church* v. 148 Her dresser had very pleasant memories of good Sergeant Honeybody.

dressing, *vbl. sb.* Add: **1. b.** *dressing up* (see Dress *v.* 7 d in Dict. and Suppl.); also *attrib.*

1852 C. M. Yonge *Two Guardians* xiii. 229 What difference can my dressing up..make to any one? 1864 Mrs. Gaskell *Wives & Daughters* (1866) I. xiii. 152 Molly had a private dressing-up for the Miss Brownings' benefit. 1868 L. M. Alcott *Little Women* 14 Meg..was as much a child as ever about 'dressing-up' frolics. 1944 M. Paneth *Branch Street* 92 We got hold of a big box full of old dressing-up material. 1950 A. Wilson *Such Darling Dodos* 130 The strange old hats and frocks in the dressing-up box in the nursery. 1957 R. Hoggart *Uses of Literacy* ii. 58 The game of dressing-up—trailing round the streets in grown-ups' cast-off clothes. 1964 D. Gray *Devil wore Scarlet* xi. 105 A huge trunk, which was evidently the 'dressing-up' box. It was filled with old clothes of every sort. 1968 P. Dickinson *Skin Deep* vii. 142 A good old-fashioned attic, where people have been putting things..to get them out of the way, old wickerwork cots and dressing-up clothes and iron bedsteads.

c. *Mil.* Proper alignment of troops. Cf. Dress *v.* 4.

1792 [see *Attention 5]. 1802 C. James *Mil. Dict.* s.v., Dressing of a battalion after the halt, is to bring all its relative parts in a line with the point..towards which it was directed to move. 1889 *Infantry Drill* i. 9 He will take up his dressing in line by moving..till he is just able to distinguish the lower part of the face of the second man beyond him. 1966 *Listener* 3 Mar. 317/3 'Come on, wake up. Get your dressing.' The familiar barks jerked their limbs into making the mechanical actions.

3. Also with *down* (see Dress *v.* 9).

1876 *Coursing Cal.* 223 Blucher was much faster in the stretches than Folly, who got a genuine dressing down. 1893 W. K. Post *Harvard Stories* 70 The poor man got such a dressing down that Randolph presented him with full forgiveness. 1925 E. F. Norton *Fight for Everest, 1924* 21 The following morning, when they were coming up for a second dressing I thought I would add a little dressing down on my own account. 1956 A. L. Rowse *Early Churchills* iv. 73 Sir Winston..gave Captain Thornhill a dressing down in open court.

5. a. dressing-comb, a comb used for dressing (see Dress *v.* 13 b) the hair. **b. dressing-forceps,** forceps used in applying and removing surgical dressings; **dressing-house,** a house for dressing ore; **dressing-station,** a place at which wounds are dressed.

1790 *Pennsylvania Packet* 19 Apr. 4/2 John Murduck..has likewise for Sale..Dressing, rake, and tail combs. 1881 C. C. Harrison *Woman's Handiwork* i. 57 Crewels..combed into fluffiness by a coarse dressing-comb. 1826 A. C. Hutchison *Pract. Obs. Surg.* (ed. 2) 180 By dilating the meatus urinarius with a common pair of dressing forceps. 1879 *St. George's Hosp. Rep.* IX. 772 Passing a pair of dressing-forceps through the joint to the lowest part of its outer aspect. 1872 *Rep. Vermont Board Agric.* 628 The same track takes it [*sc.* the ore] to the dressing house at the foot of the hill. 1894 Dressing-station [in Dict.]. 1915 A. D. Gillespie *Let.* 21 Mar. in *Lett. from Flanders* (1916) 61 A man who gets hit, not dangerously, with a dressing station handy, and a doctor to attend to him at once.

e. *Printing* (see Dress *v.* 3 b): *dressing-bench, -block, -hook.* **f.** *Type-founding* (see *Dress *v.* 13 j): *dressing machine, plane, stick.*

1683 Moxon *Mech. Exerc., Printing* xix. 183 The Dressing-Bench. *Ibid.* ix. 31 The Dressing-Block..is to run over the Face of the Form, and..to be gently knock't upon.., that such Letters as may chance to stand up higher than the rest may be pressed down. *Ibid.* xx. 184 The Dressing-Hook. 1888 *Encycl. Brit.* XXIII. 699/2 A machine, which produces types that do not require rubbing or dressing...The casting machine and the dressing machine are..mounted on a common frame...The letters pass through a channel one by one into the dressing machine. 1695 in H. Hart *Century of Typography* (1900) 55 Utensils for Printing. 4 Dressing Planes. 3 Dressing Blocks. 1683 Dressing-stick [see *Dress *v.* 13 j].

dressing-case. (Earlier and later examples.)

1790 *Pennsylvania Packet* 10 May 4/3 A few curious morocco dressing cases for gentlemen and ladies. 1922 J. M. Murry *Things we Are* 81 He wanted to know about the dressing-case; what colour it was, even, and how many bottles it contained. 1939–40 *Army & Navy Stores Catal.* 877/2 Dressing Case, in Morocco leather.

dressing-gown. (Later examples.)

1896 'Iota' *Quaker Grandmother* xxiv. 280 She put the key into her dressing-gown pocket. 1909 *Granta* 11 June, I rose, put on a dressing-gown..and went upstairs. 1971 'E. Ferrars' *Stranger & Afraid* x. 154 She's had a habit for years of getting up quietly and going wandering about the house in her dressing-gown.

dressing-table. Add: (Earlier and later examples.) Also *attrib.*

1692 J. Verney *Let.* in M. M. Verney *Mem. Verney Family* (1899) IV. xii. 468 Dressing table plate. 1899 *Daily News* 29 July 8/4 The 'dressing-table' effect, as these white muslins or lace over colour produce, is in great favour. 1960 *Farmer & Stockbreeder* 8 Mar. Suppl. 3/1 The dressing-tables had chintz frills.

dress-make, *v. intr.* Also **dressmake.** (In Dict. *s.v.* Dress-making *vbl. sb.*) (Add earlier and later examples.)

1884 S. O. Jewett *Country Doctor* ii. 6 She might dress-make or do millinery work. 1953 E. Coxhead *Midlanders* iii. 74 But, Connie, you know I can't dressmake.

dress-maker. Add: Also **dressmaker.** (Earlier and later examples.) Also *attrib.*, designating clothes, etc., made by a dress-maker or resembling such garments.

1803 *Morning Herald* 12 Feb. 1/2 To Milliners and Dress-makers.—Wanted immediately, two Persons who have lived as First Hands in respectable private Houses of Business. 1904 *Westm. Gaz.* 28 Apr. 4/2 Robes that really look like quite expensive dressmaker frocks. 1907 *Ibid.* 5 Jan. 13/1 Perhaps I should not say tailor suit, but dress-maker cloth suit, for those charming draped bodices..are..more the province of the dressmaker than the tailor. 1944 *New Yorker* 7 Oct. 54/2 Simple clothes softened with a bit of dressmaker detail. 1946 *Woman & Beauty* Feb. 120/1 Black crêpe-de-chine. Lovely for dressmaker suits. 1968 J. Ironside *Fashion Alphabet* 80 *Dressmaker suit*, a suit made by a dressmaker, usually softer than a tailor-made and using very little tailor's canvas.

b. *dress-maker's dummy* = Dummy *sb.* 5 a.

1949 D. G. Smith *I capture Castle* ii. 13 A dressmaker's dummy of most opulent figure with a wire skirt round her one leg. 1960 D. Holman-Hunt *My Grandmothers & I* i. 22 Against one wall..stood a dressmaker's dummy.

dress-up (dre·sʌp). [f. Dress *v.* 7 d.] The act of dressing up, esp. in one's best clothes; an occasion, gathering, etc., which demands formal dress. Also *attrib.* or as *adj.*

1865 A. D. T. Whitney *Gayworthys* II. vii. 132 'Will you go to meeting, Gershom?' Joanna asked him...'I guess not, ..the dress-up takes down the devotion, rather, for me.' 1878 Mrs. Stowe *Poganuc People* 52 As to that little dress-up affair over there,..I don't think any real harm has been done. 1887 'Pansy' *Little Fishers* iii, They are dress-up clothes. 1909 G. B. Shaw *Press Cuttings* 20 Arf of it's only ousemaidin; and tother arf is dress-up and make-believe. 1954 E. Pangborn *Mirror for Observers* (1955) ii. v. 141 It's always trousers nowadays except for evening dress-up, unfortunate for fat girls. 1965 *New Society* 8 July 12/2 The dress-up dinner party is nearly dead.

dressy, *a.* Add: **1. b.** *transf.* and *fig.* Excessively elaborate.

1864 Bagehot *Coll. Works* (1965) II. 366 A dressy literature, an exaggerated literature, seem to be fated to us. 1941 Blunden *T. Hardy* iii. 40 Hardy..was always oddly fond of heavy, dressy expressions.

2. (Earlier and later examples.) Also, formal; requiring full dress.

1785 E. Sheridan *Let.* 5 July in *Jrnl.* (1960) 59, I have got a Celbridge for the Honour of Ireland, these are for walks or Church, as a more dressy one I brought from London a white Persian Hat. 1902 in C. W. Cunnington *Eng. Women's Clothes* (1952) ii. 52 *Shoes* are more worn than boots for dressy occasions. 1945 'L. Lewis' *Birthday Murder* (1951) ii. 30 Victoria had planned to put on a red dress, but..chose a short, dressy black one instead.

Dreyfusard (drẽ·fusāɹ, -ǟɹd). Now *Hist.* [Fr., f. the name of *Dreyfus* + *-ard* -ard.] A defender or supporter of Captain Alfred Dreyfus (1859–1935), a Frenchman of Jewish descent who was convicted of treason in 1894 and declared innocent in 1906. Also *attrib.* or as *adj.*

1898 *Westm. Gaz.* 15 Sept. 5/2 This, of course, is a sop offered to the Dreyfusards. *Ibid.*, The opinion is expressed among Dreyfusard leaders that the President is rushing to his ruin. 1905 *Book Lover* Apr. 4/2 His affiliations have not been even *Dreyfusard* in France or Bryanite in the United States. 1919 W. B. Yeats in *Irish Statesman* 23 Aug. 212/1 It was amidst Socialist and Dreyfusard controversy that he discovered his belief. 1967 *Listener* 30 Mar. 435/2 Halévy was a passionate Dreyfusard.

dribbly (dri·bli), *a.* [f. Dribble *sb.* + -y¹.] Tending to dribble; dribbling; characterized by dribbles.

1909 R. Brooke *Let.* (1968) 172 A pair of dribbly dotards. 1951 Dylan Thomas *Let.* 12 Apr. (1966) 356 There in that English sink, intolerable, dribbly, lost. 1958 *Times* 18 Nov. 13/4 The..nature of pigment, the sort of mud-pie fascination that comes..of its liquid and dribbly aspect. 1967 D. Pinner *Ritual* xiv. 141 Those lovely dribbly dogs and those even lovelier dribbly policemen!

driblet, *sb.* Add: **4.** *Comb.* **driblet cone,** a cone produced by the successive ejections of small quantities of lava; a hornito.

1888 J. D. Dana in *Amer. Jrnl. Sci.* CXXXV. 32 The projectile process in the basalt-volcano..makes not cinder-cones, but dribblet-cones, 15 to 40 feet high, out of the projected masses, the falling driblets becoming plastered together. *Ibid.*, Such driblet-cones are of all angles from 30° to 90°. 1939 A. K. Lobeck *Geomorphol.* xix. 675 Where gases sputter out through the side of the dome, a *spatter* or *driblet cone* may be built up 10 to 12 feet above the ground. 1969 C. D. Ollier *Volcanoes* v. 58 Spatter may be erupted through a crack in the surface of a pahoehoe flow, and build up a small cone or spire of scoria and driblets. This is called a hornito or driblet cone. The term is usually restricted to small features perhaps 5 m high.

dried, *ppl. a.* Add: **1.** (Further examples); *spec.* of foods: deprived of moisture so as to be capable of being preserved for a long time, often in the form of powder.

1596 Shakes. *1 Hen. IV* ii. iv. 271 Dried neat's tongue. 1709 J. Lawson *New Voy. Carolina* 29 Loblolly made with Indian Corn, and dry'd Peaches. 1737 J. Brickell *Nat. Hist. N. Carolina* (1743) 319 When they [*sc.* Indians] go to War or their Hunting Matches..they generally carry with them..Bread, Indian Corn, dried Fruits, of several sorts. 1753 C. Gist *Jrnl.* (1893) 81 Got some corn and dried meat. 1895 *Montgomery Ward Catal.* Index. Dried fruits. 1907 H. W. Wiley *Foods & their Adulteration* ii. 115 *Dried eggs.* The rapid drying of fresh eggs is perhaps an unobjectionable method of preservation. *Ibid.*, Dried products are sometimes made from decayed eggs. 1909 G. F. Still *Common Disorders of Childhood* iv. 59 Another firm prepares a dried milk to which cream and milk sugar have been added before desiccation. 1918 *Lancet* 24 Aug. 245/1 The rôle of dried milk in infant feeding [etc.]. 1933 *Chem. Abstr.* 1919 Dried eggs may be held in storage for several months with no loss of vitamin D in the yolks. 1941 *Ann. Reg.* 1940 6 To purchase Turkish dried fruits..annually for the duration of the war. 1944 *Daily Tel.* 11 July 3 The housewife's greatest standby is Dried Eggs.

driedness (drai·dnės). [f. Dried *ppl. a.* + -ness.] Dried condition. Also *dried-up-ness*.

1907 Galsworthy *Country House* iii. ii. 231 The strange yellow driedness of his face. 1923 *United Free Ch. Miss. Rec.* June 248/1 Dried-up-ness is of the essence of the thing.

drier, dryer. Add: **1. b.** (Examples.) Also with *up*.

1881 *Instr. Census Clerks* (1885) 64 Woollen cloth manufacture:..Drier. *Ibid.* 82 Cardboard and pasteboard making:..Dryer. 1889 F. E. Gretton *Mem. Harkback* xii. 207 As many [hop] plantations are being gathered in about the same time, the drier has his hands full all at once. 1916 Blunden *Harbingers* 24 When the dryer in his oast Had loaded up his lattice-floors. 1941 M. Treadgold *We couldn't leave Dinah* vi. 99 She took full possession of the sink and Mick leaned up against the table, waiting to fulfil his function as Dryer-up-in-Chief with the dish-cloth. 1944 A. Thirkell *Headmistress* vi. 135 Captain Hornby was an excellent drier-up and Elsa was secretly much impressed by his demanding a washleather to finish off the spoons.

2. Delete (*dryer*). Also, a substance mixed with oils, inks, etc., to make them dry quickly.

1882 *Encycl. Brit.* XIV. 677/2 By slow degrees a proportion of 'dryers' is added [to linseed oil]. 1940 *Thorpe's Dict. Appl. Chem.* (ed. 4) IV. 91/1 The modern method is to heat the oil by means of closed steam coils..while the driers are employed in the soluble form. 1951 R. Mayer *Artist's Handbk. Materials* iii. 157 Driers or siccatives are metallic salts combined with materials such as oils or resins which mix with the usual paint and varnish ingredients. 1958 T. Landau *Encycl. Librarianship* 113/2 *Drier*, a substance used in printing inks to expedite drying by evaporation.

3. b. *spec.* A hair-drier.

1937 Hemingway *To have & have Not* iii. xviii. 253 He put a net over it wet and put me under the dryer. 1955 *Punch* 27 Apr. 531/1 Clamped down under a drier in the Hairdressing Salon. 1961 *B.S.I. News* Sept. p. 27 The term 'hair dryer' includes the flexible cord and connectors supplied with the dryer. The dryers are classified into the three following types: *a.* Earthed dryers *b.* Double-insulated dryers and *c.* All-insulated dryers. 1961 A. Wilson *Old Men at Zoo* ii. 106, I was under the drier when this girl started on about it.

drift, *sb.* **2. d.** Substitute for def.: A slow variation of the characteristics or operation of an electric circuit or device. Also *attrib.* (Add examples.) orig. *U.S.*

1889 in *Cent. Dict.* 1892 *Jrnl. Franklin Inst.* CXXXIV. 482 This magnification varies with the sensibility and is largest when the instrument is most sensitive. In galvanometer work this phenomenon is known as the 'drift' of the needles. 1931 *Physical Rev.* XXXVII. 396 This drift is normally very constant, so that an input current as small as 10^{-16} amp can be measured by noting the *change* in the rate of drift which it causes. 1932 *Rev. Sci. Instruments* III. 420 The degree to which the circuit is balanced is determined by noting the back and forth drift which a high sensitivity galvanometer will make, due to natural battery voltage fluctuations. 1962 Simpson & Richards *Junction Transistors* xiii. 310 The chief problem in the design of amplifiers to pass d.c. is to reduce slow changes in the operating conditions of the transistors to a minimum. These changes..are called drifts. 1963 B. Fozard *Instrumentation Nucl. Reactors* xi. 133 More commonly some form of automatic drift-correction circuit is used. 1970 J. Earl *Tuners & Amplifiers* ii. 31 This..is in receipt of a stabilised voltage (to avoid tuning drift).

e. *Aeronaut.* The horizontal deviation of an aircraft from its course in consequence of winds and air-currents; also, the rate or amount of such deviation.

1891 O. Chanute *Aerial Navigation* 26 How much of this new pressure..opposes forward progress, and may be denominated drift. 1896 H. S. Maxim in J. Means *Epitome of Aeronaut. Ann.* (1910) 113 The aeroplane..in which the drift will go the greatest number of times into the lift will be considered the most satisfactory. 1909 *Westm. Gaz.* 25 Feb. 4/1 The daring plan of jumping off a hill 45 ft. high with the object of ascertaining the proportion between lift and drift. 1916 H. Barber, *Aeroplane Speaks* 2 There's Drift, my horizontal component, sometimes, though rather erroneously, called Head Resistance. *Ibid.* 57 The Drift..must be overcome by the Thrust in order to secure the necessary velocity to produce the requisite lift for flight. 1920 *19th Cent.* July 145 The object of camber is to obtain the maximum lift, and to reduce the drift. 1958 'N. Shute' *Rainbow & Rose* ii. 41 We had about fifteen degrees of drift. 1968 G. D. P. Worthington *Airline Instrument Flying* xiii. 191 If drift has been correctly allowed for and time and rate of turn are correct, the aircraft should be on the correct track when it comes out of the turn.

f. *Astr.* Orig., any group of stars having a random distribution of velocities; usu. applied to a group of stars with an apparent systematic motion towards some point in the sky superimposed on their motion within the group, some or all of the nearest stars belonging to one or other of two such groups with opposite directions of motion.

1906 A. S. Eddington in *Monthly Notices R. Astron. Soc.* LXVII. 35, I define a 'drift of stars' to be a system of stars whose velocities relative to some system of axes are quite haphazard. *Ibid.* 40 Kapteyn..found this same disagreement, and pointed out the explanation—that there are really two drifts of stars. 1928 E. A. Fath *Elem. Astron.* (ed. 2) xv. 236 The motions of the stars in any drift are not along parallel lines...But when the group is taken as a whole it shows a decided group motion toward its vertex. 1938 W. M. Smart *Stellar Dynamics* iv. 106 In most regions of the sky it is found that one drift (drift I) is usually more prominent than the other. 1968 D. S. Evans *Observ. Mod. Astron.* iv. 137 The two drifts which describe the motion of stars relatively near the Sun are both considered to be in motion relative to the Sun.

g. In the migration of birds: the influence of wind currents.

1918 Baxter & Rintoul in *Ibis* VI. 255 We do not consider that the *direction* of the wind, apart from its *force*, stops birds migrating...We believe that the deviation from the direct route is largely, perhaps mainly, due to drift. 1960 E. Ennion *House on Shore* v. 66 Their capture..has gone a long way to establish the reality of 'Drift'. 1971 *Nature* 30 Apr. 580/2 Waders and waterfowl..frequently correct for drift in winds of moderate speed.

h. *Motor-racing.* A controlled slide.

1955 *Motor Trend* Jan. 33, I steer into the drift, accelerating slightly. 1957 S. Moss *In Track of Speed* v. 61 He was a great exponent of the 'drift'. He was always fastest out of a bend, having discovered the art of pointing the car to the ensuing straight before he was fully round. 1957 *Life* (U.S.) 15 Aug. 83/1 When Fangio puts his Maserati or Ferrari into a corner in a four-wheel drift (i.e., with all wheels skidding).

3. b. *Philol.* (See quot. 1921.)

1921 E. Sapir *Lang.* vii. 165 The drift of a language is constituted by the unconscious selection on the part of its speakers of those individual variations that are cumulative in some special direction. 1926 *Language* II. 133 We find the same drift at work from analysis to synthesis, from monosyllabism to polysyllabism. 1961 L. F. Brosnahan *Sounds of Lang.* vii. 155 These great trends, or *drifts*.., are some of the most interesting phenomena of historical linguistics.

8. c. A large mass of flowering plants (see quot. 1966).

1908 G. Jekyll *Colour in Flower Garden* iii. 24 The word 'drift' conveniently describes the shape I have in mind, and I commonly use it in speaking of these long-shaped plantings. 1915 C. Mackenzie *Guy & Pauline* i. 38 Planting a drift of..deep yellow primroses. 1965 *Listener* 23 Sept. 461/2 All of these bulbs can be planted in drifts of hundreds at a time. 1966 J. Berrisford *Wild Garden* viii. 94 Plants and bulbs should be set informally in..'drift-planting'—that is in ovals of irregular outline rather like cloud formations with one or two outliers a little way off so that they look as if sown by the wind.

9. Also *fig.*

1897 *Daily News* 3 May 4/1 They are not beggars..; they are merely human drift—men who live on 'nuffin'.

10. b. drift-peat, a deposit of peat associated with a glacial drift.

1894 J. Geikie *Gt. Ice Age* (ed. 3) 308 The drift-peat and timber that underlie the Carse-deposits of the 45 to 50-ft. level.

14. b. (See quot. 1811.)

1811 T. Davis *Agric. Wilts.* (ed. 2) 269 *Drifts*, the rows in which underwood is laid when felled. 1915 J. Buchan *Thirty-Nine Steps* vii. 165, I found shelter below an overhanging rock..where a drift of dead brackens made a tolerable bed. 1927 *Forestry* I. 33 The normal procedure is to cut the coppice..in drifts 3 feet wide and from 4 to 6 feet apart. The young trees..are planted up the middle of the drifts.

19. a. (sense *2 g) *drift-migrant, -migration.*

1952 *Scottish Naturalist* LXIV. 12 If a drift-migrant could continue on the wing indefinitely it would pass from one weather system to another. 1959 *Listener* 22 Jan. 160/1 The average drift-migrants that make a landfall are not necessarily lost. 1960 *Brit. Birds* LIII. 325 (*heading*) Autumn 'drift-migration' on the English east coast.

c. drift-angle, (*a*) *Naval Arch.*, the angle of lee-way (see Lee-way); (*b*) *Aeronaut.* (see quots. 1951 and 1967); **drift-bottle,** a bottle used for the charting of ocean-currents; **drift-indicator** *Aeronaut.*, a device for indicating drift (see sense *2 e, above); **drift-netter** (examples); so *drift-netting*; **drift-pile** *Canad.*, a pile of drift-wood in a river, etc.; **drift plate** or **sight** *Aeronaut.* = **drift-indicator*; **drift-weed,** (*a*) also *fig.*

1882 W. H. White *Man. Naval Arch.* (ed. 2) xiv. 621 The angle between this tangent and the keel-line, or 'drift-angle', (*angle de dérive*) as it is termed, gradually increases. 1906 E. L. Attwood *War-Ships* (ed. 2) 237 At the point P, where OP is drawn perpendicular to the centre line of the ship, there is no drift angle, as the tangent to the circle through P is the centre line of the ship. 1935 C. G. Burge *Complete Bk. Aviation* 264/1 *Drift-angle sight*, an instrument for determining the drift-angle. 1951 *Gloss. Aeronaut. Terms* (*B.S.I.*) III. 6 *Drift angle*, the angle, at any instant, between the longitudinal axis of an aircraft and the track. 1967 *Gloss. Terms Air-Cushion Vehicles* (*B.S.I.*) 6 *Drift angle*, the angle, in the horizontal plane, between the longitudinal axis of an ACV and the instantaneous true direction of motion relative to the local meridian. 1909 *Cent. Dict.* Suppl., Drift-bottle. 1912 Murray & Hjort *Depths of Ocean* v. 261 In order to study the currents, drift-bottles have..been used, in which are enclosed slips of paper with directions to the finder. *Ibid.* 262 (*caption*) Results of Dr. Fulton's drift-bottle experiments in the North Sea. 1966 *McGraw-Hill Encycl. Sci. & Technol.* IX. 252/1 Even slow surface currents have been observed and measured by the drift of debris or drift bottles. 1919 H. Shaw *Text-Bk. Aeronaut.* xiii. 169 There are two types of drift indicator, one of which is designed for..indicating leeway over the surface of the ground.., while the other is intended to indicate whether or not the machine is flying head to wind. 1889 *Cent. Dict.*, Drift-netter. 1913 *Q. Rev.* Apr. 438 Its rapid extension so alarmed the drift-netters that an agitation was started against it. *Ibid.*, Drift-netting is carried on by both steamers and sailing vessels. 1963 *Times* 18 May 11/1 Already there are signs that the prohibition of drift-netting will result in increased river catches. 1927 A. P. Woollacott *Mackenzie & his Voyageurs* 37 Log-jams and drift-piles are numerous and troublesome. 1968 R. M. Patterson *Finlay's River* 103 They found the river very swift and badly obstructed by driftpiles—dangerous piled-up jams of dead and uprooted trees. 1935 C. G. Burge *Complete Bk. Aviation* 475/1 There are many instruments in use to-day for determining this [*sc.* windage], the oldest being the drift plate..and the most modern..the tail drift sight. 1940 'N. Shute' *Old Captivity* vii. 218 The pilot sat hunched at the wheel,..glancing from time to time down through the drift sight. 1906 Somerville & 'Ross' *Irish Yesterdays* 223 She belonged to the drift-weed of the household. 1909 *Daily Chron.* 14 June 9/2 These dreadful drift-weeds of the great city.

drift, *v.* Add: **1. c.** Also (*colloq.*), to go away, get out; to come or go casually; to wander; freq. with adverbs, as *to drift around, by, in, out*; *to drift apart*, of a man and a woman: gradually to lose mutual affection, etc.

1864 'Mark Twain' in Harte & 'Twain' *Sketches* (1926) 122 She can calculate on my drifting around in the course of an hour or so. 1874 —— & Warner *Gilded Age* I. ix. 119 A week drifted by, and all the while the patient sank lower and lower. 1876 —— *Tom Sawyer* xxxii. 274 Villagers filed through Judge Thatcher's house..and drifted out raining tears all over the place. 1898 M. Deland *Old Chester Tales* 43 Peter, with a pretty girl on his arm, drifted in out of the windy and rainy darkness. 1903 *Red Book* July 278/1 Ralph and I had drifted apart years before. 1908 [see *Drifter e]. 1910 Wodehouse *Psmith in City* vi. 46 Let us drift aside into this teashop. 1925 [see *Drifter e]. 1942 R. Chandler *High Window* (1943) ix. 70 'Beat it,' he said. 'Drift.' *Ibid.* xix. 134 So speak your piece and drift away. 1945 *Coast to Coast 1944* 3 Early spectators were drifting in. 1954 'N. Blake' *Whisper in Gloom* i. vi. 83 Well, I'll be drifting. 1958 Hayward & Harari tr. *Pasternak's Dr. Zhivago* ix. 255 Tonya and I have never drifted apart and this year of work has brought us even closer together.

2. b. To allow or cause (a fishing-net or -line) to be borne by the current. Also *absol.*

1850 N. Kingsley *Diary* 21 Apr. (1914) 118 [We] drifted the seine across the river and floated down with the current. *Ibid.* 119 They drifted once more and made up the number of 51 salmon. 1907 *Westm. Gaz.* 29 Nov. 3/1 By casting or 'drifting' the dry-fly on a long line down stream to the fish. 1963 K. Mansfield *Anglers' Dict.* 51 For bass the best tactics are to anchor up-tide from some buoy, quay or wharf which bass are likely to visit and drift the line down to them. 1971 *Angling Times* 10 June 3 Colin drifted his peeler crab bait around the rocks on float tackle.

c. To drive (cattle or horses) slowly (see quot. 1893). Also with *in*. *U.S.*

1893 *Funk's Stand. Dict.* 555/3 *Drift*, to drive cattle slowly, letting them feed as they go. 1903 A. Adams *Log of Cowboy* iv. 51 The Rebel and Blades were following, to drift in what cattle we had held on our left. *Ibid.* vii. 85 We were drifting them back towards the trail. 1920 J. M. Hunter *Trail Drivers of Texas* 50 [They said] they would drift the horses along with two outfits instead of four.

drifter. Add: **b.** Also, a heavy, mounted percussion drill driven by compressed air, used in mining for horizontal working. Also *drifter drill*.

1918 R. Peele *Mining Engin. Handbk.* 1112 Drifters. 1921 *Engin. & Mining Jrnl.* 27 Aug. 340/1 Modern rock-drilling appliances of the hammer type fall into three well-marked groups, namely, the drifter..; the plugger..; and the stoper. 1946 G. J. Young *Elem. Mining* (ed. 4) xiv. 539 For work at the face 3½-in. drifter drills..are required. 1951 S. Higham *Introd. Metalliferous Mining* vii. 88 When rapid advance is desired, rigged drifters are used.

c. Also, a fishing-vessel or other small boat used by the Royal Navy, esp. in wartime, for patrolling, conveying stores, etc.

1916 *In Northern Mists* xxiv. 97 Like all these North Sea craft, trawlers, drifters, and the rest, she is built with bows nearly twice as high out of the water as the after-part. 1917 *Times* 19 May 6/4 Light cruisers..raided the Allied drifter line in the Adriatic and succeeded in sinking 14 British drifters. 1923 *Man. Seamanship* II. 19 In home waters capital ships are generally supplied with a drifter. 1961 Granville & Kelly *Inshore Heroes* x. 108 In addition to H.M.S. *Midge*, the Coastal Forces' base, to which the M.L.s were attached, there was the base for trawlers, drifters, armed yachts and other auxiliary craft employed on anti-submarine or mine-clearance work.

d. An object which is allowed to float freely in the sea to determine ocean-currents; a drift-bottle.

1897 T. W. Fulton in *Scot. Geogr. Mag.* XIII. 637 Surface currents of the North Sea...Drifters...were used, one-ounce wide-mouthed bottles. 1900 *Geogr. Jrnl.* XV. 275 On the voyage from Iceland to Jan Mayen in 1896 twenty drifters were thrown overboard.

e. A man following an aimless, irresolute, or vagrant way of life.

1908 *Daily Chron.* 28 Sept. 4/7 The drifter drifts to California, and brings up there because..he can drift no further. 1922 *Short Stories* Feb. 25/1 The trampers ain't all hoboes, some of 'em being just drifters. 1925 W. Deeping *Sorrell & Son* xxviii. 275 For months he had had a sense of drifting, and his character was not that of a drifter. 1971 *Guardian* 14 June 9/1 A drifter with nothing but the clothes he stands up in.

f. A wind causing snow to drift.

1922 R. J. Flaherty *My Eskimo Friends* 8 Then came snow, the winter's first big 'drift'r', and for three days there was no land or sea or sky. 1924 A. J. Small *Frozen Gold* ii. 49 A confused..blur of whining blizzards, roaring drifters, flat calm cold-snaps. 1970 *Globe & Mail* (Toronto) 28 Sept. 19/8 Winds ranging from a drifter to 15 knots.

drifting, *vbl. sb.* Add: **2.** *spec.* in *Mining* (see Drift *v.* 7). Also *attrib.*

1853 *Harper's Mag.* VI. 447 The shafting and drifting gets only the copper which is in the immediate course of those operations. 1882 *Rep. Prec. Metals* (*U.S. Bureau of Mint*) i. 70 This mine has been worked by the drifting method. *Ibid.* iii. 635 A..portion of the gravels or 'wash'..is removed by..drifting, from underneath worthless or comparatively barren ground. 1897 *Sears, Roebuck Catal.* 50/2 Drifting picks, adze eye, oil finish. 1927 R. Peele *Mining Engineers' Handbk.* (ed. 2) x. 500 All types of rock drill are used for drifting and crosscutting. 1952 M. H. Haddock *Basis Mine Surveying* vii. 259 Drifting in coal-mining has nearly always been between parallel strata with the frequently assured certainty of locating the deposit sought.

drift-wood, driftwood. Add: **1.** Esp. wood carried down by a river. *U.S.*

1785 Washington *Diary* 3 Aug. (1925) II. 396 It would probably be frequently choaked with drift wood, Ice, and other rubbish. *Ibid.* 25 Jan. (1925) IV. 79 The river there is..full of small islands occasioned by drift wood lodging on the rocks. 1821 J. Fowler *Jrnl.* 22 Oct. (1898) 26 The men Waided over and gathered drift Wood for the night. 1848 J. F. Cooper *Oak Openings* I. iii. 47 The drift-wood choked the channel. 1850 H. C. Watson *Camp-fires Revol.* 67 That exposed our boats to being all the time tangled in the drift-wood and bushes. 1884 'Mark Twain' *Huck. Finn* vii. 50 The river was coming up pretty fast, and lots of drift-wood going by on the rise.

2. *fig.* Also *attrib.*

1840 *Daily Picayune* (New Orleans) (D.A.E.), Kendall will cut a 'swell' at our next drift-wood training. 1865 J. G. Holland *Plain Talks* 224 A lodging place for political drift-wood. 1907 *Daily Chron.* 23 July 4/6 The metropolis attracts human driftwood from all quarters.

drifty, *a.* Add: **2. b.** Flowing.

1897 *Daily News* 1 May 8/2 The dress..is so soft and drifty. 1962 *Sunday Express* 30 Dec. 17/5 Long drifty skirts of voile.

drill, *sb.²* Add: **6.** (Later example.)

1967 *Lebende Sprachen* XII. 136/2 *Drill* or *language laboratory drill*, a series of exercises devised for giving practice in teaching or testing a particular skill. It may be in the form of a two-, three-, or four-phase drill.

b. The agreed or recognized procedure, esp. on formal occasions. *colloq.*

1940 D. Wheatley *Faked Passports* xxii. 265 Our Generals had so little imagination that the drill was always just the same and..the Germans got quite used to it. 1943 Hunt & Pringle *Service Slang* 28 *The drill*, the correct way of doing a job is always referred to in this manner, or as the 'right drill'. 1957 *Listener* 5 Dec. 925/2 He would come to dinner, swallow his soup, and then fall fast asleep. The servants knew the drill and kept his other courses warm.

7. b. *drill-hall, -ship*; **drill-book,** a manual of instruction in military or other drill; also *attrib.* and *transf.*; **drill order** (see *Order *sb.*).

1846 *United Services Mag.* ii. 235 The French drill-book. 1868 *All Year Round* 11 July 108/1 He walks in a fine drill-book style. 1900 *Daily News* 15 May 3/3 The Queensland Mounted Infantry contingent seem to have engrafted sufficient drill-book into their common-sense methods. 1906 W. Wood *Enemy in our Midst* vii. 73 On a certain assumption which was that an enemy would work according to drill-book and rule-of-thumb. 1937 G. Fairbanks (*title*) Voice and articulation drillbook. 1878 *Chambers's Encycl.* III. 671/1 Drill-halls, in which drill can be carried on comfortably in any kind of weather, are now common. 1891 *Scribner's Mag.* X. 565 Entertainments are also given in the drill hall every Friday evening. 1933 L. A. G. Strong *Sea Wall* x. 120 Jerry held..a grand tournament of all his pupils, hiring for the purpose a local drill-hall. *a* 1865 Smyth *Sailor's Word-Bk.* (1867) 264 *Drill-ships*, a recent establishment of vessels in which the volunteers composing the Royal Naval Reserve are drilled into practice. 1948 *Jane's Fighting Ships 1947–48* 50 *Zetland* was assigned to Solent Division R.N.V.R. as drillship.

drill, *sb.³* For *Cynocephalus* read *Mandrillus*. Also *attrib.*, as *drill baboon, monkey*.

1898 *Westm. Gaz.* 15 Feb. 11/3 The drill monkey, the most costly and rare of its kind in the gardens. 1905 *Daily Chron.* 29 Apr. 7/1 The finest drill baboon ever seen in confinement is in the Bellevue Gardens.

drill, *sb.⁵* (Further examples.)

1757 in E. W. Cunnington et al. *Dict. Eng. Costume* (1960) 253/2 Dressed in..a white drill Frock. 1876 *Monthly Pkt.* June 570 Boys, on admission, to be..supplied with..2 suits of brown drill. 1899 E. W. Hornung *Amateur Cracksman* 223 A girl in a white drill coat and skirt. 1918 H. G. Wells *Joan & Peter* xiii. 664 Both the pink gingham and the white drill had been tried on. 1934 'G. Orwell' *Burmese Days* vi. 96 Dressed in khaki pagri-cloth shirt, drill shorts and a pigsticker topi. 1967 *Listener* 3 Aug. 156/2 Hugh Griffith in khaki drill and a pith helmet.

drill, *v.³* **1. a.** Add to def.: *spec.* to shoot with a gun. *colloq.*

1808 E. S. Barrett *Miss-Led General* i. 11 It would be a terrible affair to *us*..if we should be drilled with a bullet. 1833 [in Dict.]. 1912 E. C. Bentley *Trent's Last Case* vi. 136 Thirty thousand men..would have jumped at the chance of drilling a hole through the man. 1930 *Amer. Mercury* Dec. 455/2 Go drill the mutt. He's strictly stool. 1936 M. Mitchell *Gone with Wind* xii. 223, I can drill a dime at fifty yards.

Drinamyl (dri·nămil). *Pharm.* Also **drinamyl.** [f. D(ext)r(o- + (*amphetam)in(e + Amyl².] The proprietary name of a preparation of dexamphetamine and amylobarbitone, used as a stimulant. Cf. *purple heart* s.v. *Purple *a.* and *sb.*

1950 in *Trade Marks Jrnl.* 13 Dec. 1053/2. 1952 *Martindale's Extra Pharmacopœia* (ed. 23) 516 Drinamyl...Tablets each containing d-amphetamine sulphate 5 mg. and amylobarbitone 32 mg. For control of mental and emotional distress. 1961 *Guardian* 23 Mar. 3/1 Drinamyl, a Schedule 1 poison known in the trade as 'purple heart'. 1968 J. Blackburn *Young Man from Lima* iii. 32 Two drinamyl tablets had restored his self-confidence. 1969 *Daily Tel.* 5 Mar. 25/6 A drug addict who took drinamyl and was formerly an addict of heroin.

dring (driŋ), *v. dial.* [Presumably local form of Thring *v.*] *trans.* and *intr.* To crowd, press, squeeze. Also *fig.* So **dri·nging** *vbl. sb.*

1810 *Monthly Mag.* XXIX. i. 435/1 *Dringing*, crowding. 1825 J. Jennings *Observ. Dial. W. Eng.* 34 *To dring*, to throng; to press, as in a crowd; to thrust. 1867 W. F. Rock *Jim an' Nell* 37 A thousan' happy fancies dring. 1922 Joyce *Ulysses* 723 I'd have to dring it into him. 1923 *Chambers's Jrnl.* Apr. 309/2 'I dring through it,' he said, head up to whatever chanced.

drink, *v.¹* Add: **8.** (Further examples.) Also with *away, up*.

1701 Defoe *Trueborn Eng.* ii. 31 Drink their Estates away, and Senses too. 1884 'Mark Twain' *Huck. Finn* xliii. 437 It's likely pap's..got it [*sc.* money] all away from Judge Thatcher and drunk it up. 1964 *Penguin Bk. Austral. Ballads* 123 Between them they drank every cent.

10. e. *to drink up*, to finish one's drink. (Cf. sense 2 in Dict.)

1919 C. Mackenzie *Early Life Sylvia Scarlett* i. v. 158 Drink up and have another. 1927 Hemingway *Fiesta* xiii. 150 Drink up, Harris. 1938 G. Greene *Brighton Rock* vii. ix. 347 Drink up. We'd better get on with the good work. 1949 'N. Blake' *Head of Traveller* ii. 40 Well, drink up, Jack, and we'll make a four-hand.

11. *to drink like a fish* (earlier examples).

1640 Fletcher & Shirley *Nightwalker* iv. sig. H2ᵛ, Give me the bottle, I can drink like a Fish now, like an Elephant. 1701 Farquhar *Sir H. Wildair* ii. 14 Drink like a Fish, and swear like a Devil. 1802 C. Wilmot *Let.* 2 Jan. in *Irish Peer* (1920) 24 Not forgetting I was not at Home, and consequently drinking like a Fish. 1837 [see Fish *sb.¹* 1 c].

12. e. Also in other analogous uses.

1746 P. Francis tr. *Horace's Sat.* i. iii. 24 He drank the Night away Till rising Dawn. 1836 Dickens *Sk. Boz* 2nd Ser. 90 The musicians exhibit unequivocal symptoms of having drunk the new year in.

15. (Later examples.)

1967 C. Ray *Compleat Imbiber* IX. 66 A vintage which was (as the wine merchants say) 'drinking very nicely now'. *Ibid.*, I sipped the wine, which drank like velvet. 1969 *Guardian* 23 May 9/3 Every one of these wines will drink well now: most of them will improve by keeping.

drink, *sb.* Add: **3. c.** Colloq. phr. *to have drink taken*, to have drunk alcoholic liquor; to be intoxicated or suffering from the effects of drink; hence in various *ellipt.* uses.

[1914 Joyce *Dubliners* 125 How easy it was to know a gentleman even when he has a drop taken.] 1924 Kipling *Debits & Credits* (1926) 186, I saw 'em, sir, come out..not drunk, but all—*all* havin' drink taken. 1930 'Sapper' *Finger of Fate* 286 When men of Denton's calibre get into

the condition of 'drink-taken', such trifles as the presence of other guests in the house do not deter them from being offensive. 1963 'A. Gilbert' *Ring for Noose* iii. 41 You're sure he's dead, not just drink taken?

6. For *U.S. humorous* read *colloq.* (orig. *U.S.*). Also, the sea. (Add earlier and additional examples.) Always preceded by *the*.

1832 J. K. Paulding *Westward Ho!* I. 121 Sing dumb, or I'll throw you into the drink. 1844 *Daily Picayune* (New Orleans) 24 Mar. 2/2 There never would have been any Atlantic ocean if it hadn't been for the Mississippi, nor never will be after we've turned the waters of that big drink into the Mammoth Cave! 1844 Dickens *Mart. Chuz.* xxxiii. 396 'He'd spill 'em in the drink:' whereby the Capting metaphorically said he'd throw them in the river. 1882 M. E. Braddon *Mount Royal* II. iv. 79, I was coming across the Big Drink as fast as a Cunard could bring me. 1941 *New Statesman* 30 Aug. 218/3 *The Drink*, the sea. 1942 T. Rattigan *Flare Path* i. 102 Down 'e goes into the drink turning and twisting. 1944 *Penguin New Writing* XX. 130 A British pilot was being pulled out of the drink. 1960 L. Meynell *Bandaberry* xiv. 183 [He] had fished us out of the drink just, and only just, in time.

8. a. *drink habit, party, table, time, tray*; also with *pl.* of sense 5 a, as *drinks party*, etc. **c.** *drink-sodden* adj.

1890 W. Booth *In Darkest Eng.* ii. vii. 243 The disorganised, sweated, hopeless, drink-sodden denizens of darkest England. 1890 W. James *Princ. Psychol.* II. xxviii. 685 The drink-habit is only a symptom of their disease. 1904 H. G. Wells *Food of Gods* iii. iii. 240 The drink-sodden wretchedness of the painted women at the corner. 1960 H. Edwards *Spirit Healing* xi. 92 The drink habit passed away. 1961 R. Jeffries *Evidence of Accused* v. 55 They had arrived at drink-time. 1962 E. Salter *Voice of Peacock* xii. 127 A funeral party and a drinks party on the agenda. 1963 *Harper's Bazaar* Feb. 15/1 Coiled mats of silver plate on copper..look good on the drinks trolley. 1963 L. Meynell *Virgin Luck* vi. 156 Who were all those people at the drink party? 1964 E. Bowen *Little Girls* iii. vii. 237 He had removed the tea tray, brought in the drink tray. 1966 *Observer* (Colour Suppl.) 13 Nov. 40/2 The Cocktail Hour, commonly known as drinks time, is a mysterious 6–8 p.m. limbo. 1966 *Listener* 24 Nov. 763/3 A middle-aged couple ensconced in a lovely home: the drinks tray, stage left, in constant use. 1966 J. B. Priestley *Salt is Leaving* vi. 70 Alan..followed her to the drinks table. 1967 L. Meynell *Mauve Front Door* xv. 214 The dramatist sends one of his characters to the always lavishly well-stocked drink table. 1970 C. Wood *Terrible Hard* iii. 36 Perhaps they shouldn't have given that drinks party so soon. 1971 'D. Halliday' *Dolly & Doctor Bird* ix. 119 Johnson moved across to the drinks table, and ..began to pour three neat doubles.

drinker. Add: **3.** A drinking-trough.

1947 *All-Pets Mag.* May 5 Keeping..plenty of cool water in drinkers for them [*sc.* canaries]. 1950 *N.Z. Jrnl. Agric.* Nov. 447/2 The only satisfactory place for drinkers other than specially designed and expensive systems is outside the floor space of the [hen] house. 1960 *Farmer & Stockbreeder* 19 Jan. Suppl. 37/2 If the brooder has fitted food troughs and drinkers.

drinkery. (Earlier U.S. examples.)

1840 J. P. Kennedy *Quodlibet* 222 The Sergeant took a small frame house next door to Sim Traver's Refectory,—or rather as Sim called it, his Drinkery. 1845 T. J. Green *Texian Exped.* xix. 368 We wended our way up town, and called into the first open 'drinkery'.

drinkie (dri·ŋki). *colloq.* Also **drinkie-pie, drinky.** [f. Drink *sb.* + -ie, -y⁶.] An affected, childish, or jocular form of Drink *sb.* 5.

1947 *Horizon* Oct. 88 'How about a little drinkie-pie?' Gwen said. 1953 D. Parry *Going up—Going down* v. vi. 241 We might go and have a lovely drinkie. 1962 J. Cannan *All is Discovered* i. 18 We can have our little drinky while we choose. 1966 R. Braddon *Committal Chamber* iii. 32 'There,' she smiled, passing him a glass of sherry. 'Drinkies.'

drinking, *vbl. sb.* Add: **4. a.** *drinking-fountain* (examples), *funnel* (N.Z.); **b.** *drinking club, -place, -saloon* (examples), *-shop*.

1967 E. Paul *Jewels in Jeopardy* i. 11 Rather dubious entertainment in the Soho area, strip-tease clubs, drinking clubs and so on. 1860 Dickens *Uncomm. Trav.* xiv. 216 A drinking fountain..to freshen its thirsty square. 1882 Drinking fountain [see Fountain 2]. 1968 A. Munro in R. Weaver *Canad. Short Stories* 2nd Ser. 260 The drinking fountain surrounded by little puddles of water. 1927 T. E. Donne *Maori, Past & Present* vii. 76 (*caption*) Drinking funnel for tohunga when tapu. 1955 H. J. Phillipps *Maori Carving Illustr.* 9/2 Four beautifully-carved drinking funnels are on exhibition. 1853 'P. Paxton' *Yankee in Texas* 143 Buffaloes have a regular drinking-place. 1870 'F. Fern' *Ginger-Snaps* 91 The man who..takes that child to bar-rooms and drinking places. 1855 R. Glisan *Jrnl. Army Life* (1874) 172 The lobbies and drinking saloons at the capitol. 1875 Mrs. Stowe *We & Neighbors* xli. 377 Finally we alighted before a plain house in a street full of drinking-saloons. 1855 *N.Y. Herald* 6 Nov. 5/3 A multitude of drinking shops have already been closed. 1891 Kipling *Light that Failed* xv, Dick entered the drinking-shop which was one source of her gains.

c. drinking-habit, addiction to alcoholic liquor; the drink habit; **drinking-nut** (see quot. 1909); **drinking up,** the finishing of a drink; **drinking-up time,** a short period after the legal closing-time in a public house which is permitted by law for the consumption of drinks bought before it.

1899 W. James *Talks to Teachers* viii. 64 They talk of the smoking-habit and the swearing-habit and the drinking-habit. 1909 *Chambers's Jrnl.* Apr. 256/1 A stalwart native ..with..a dozen 'drinking-nuts' (young cocoa-nuts), the ice-creams of the Pacific. 1960 *Guardian* 30 Nov. 2/3 The proposal for an extra quarter of an hour for 'drinking up'. 1961 *Times* 29 Mar. 7/3 'Drinking up' time of 10 minutes. 1968 'H. Carmichael' *Slightly Bitter Taste* x. 182 We stop serving at three o'clock and then there's ten minutes drinking-up time before we turn the key in the lock.

drinky (dri·ŋki), *a. colloq.* and *dial.* [f. Drink *sb.* 3 + -y¹.] Tipsy; drunk.

1846 J. J. Hooper *Adv. Simon Suggs* 44 But then he was 'drinky and played careless'. 1846 —— *Taking Census* in *Adv. Simon Suggs, &c.* 166 When a little 'drinky', he was wont to exhibit very fair horsemanship. 1871 G. P. R. Pulman *Rustic Sketches* (ed. 3) 92 *Drinky*, rather drunk. 'He was drinky, but not rigglar drunk.' 1874 Hardy *Far fr. Mad. Crowd* II. xii. 143 I've been drinky once this month already. 1901 'M. E. Francis' *Pastorals of Dorset* 52 I've seen a man as was a bit drinky-like throw off his hat and tread on it. *Ibid.* 236 He do seem to be a bit drinky.

drinky, var. *Drinkie.

driography (drai̯ǫ·græfi). [f. Dry *a.* + -graphy, after Lithography.] A lithographic printing process which dispenses with the use of water as a barrier to prevent ink from settling on non-printing surfaces. Hence **driogra·phic** *a.*

1970 *Publishers' Weekly* 14 Sept. 57/2 Traditionally, water has been required to keep non-image areas free from ink. Driography, which requires special inks, is based on fiddling with the adhesive and cohesive properties of ink. *Ibid.*, These driographic inks will also work on regular offset processes with dampening systems and conventional plates. 1971 *Brit. Printer* Feb. 62/2 The trick in driography has been to find two surfaces that are sufficiently different in 'wettability' to dispense with water completely. 1971 *Penrose Ann.* LXIV. 125/1 Dry Plate is a registered 3M brand name. The firm has adopted the words 'driography' and 'driographic' for generic description of the system its components and the presses, papers and inks involved.

drip, *v.* Add: **5.** *Naval slang.* To complain, grumble.

1942 *Gen* 1 Sept. 13/1 When she [*sc.* a Wren] grumbles she's guilty of 'dripping'.

drip, *sb.* Add: **2. b.** *Med.* The continuous slow introduction of fluid into the body (esp. intravenously) involving its passage drop by drop through a chamber; also, the fluid so introduced or a device for this. Also *attrib.*

1933 *Amer. Jrnl. Med. Sci.* CLXXXV. 701 The chill should be removed from the milk and the drip given continuously throughout the day and night. 1949 *Anesthesia & Analgesia* XXVII. 57 The bottle containing the solution is attached to a Murphy drip setup, and the solution injected by the anesthetist. 1961 *Lancet* 29 July 240/2 An intravenous normal saline drip was put up. 1968 J. H. Burn *Lect. Notes Pharmacol.* (ed. 9) 3 Noradrenaline is sometimes given as an intravenous drip.

3. b. *fig.* Nonsense; flattery; sentimental drivel. *slang* (orig. *U.S.*).

1919 *Wine, Women & War* (1926) 306 Drool about duties by Navy egg, regular and hard boiled. Usual R.O.T.C. drip. 1924 P. Marks *Plastic Age* xxiv. 297 That freshman, Larson, showed me a theme..that Kempton had corrected. It was full of errors that weren't marked, and it was nothing in the world but drip. 1946 'B. Gray' *Mr. Ball of Fire* i. 10 'We'll have nothing of the sort,' interrupted Joy, putting a welcome stop to this drip.

c. A stupid, feeble, or dull person; a fool; a bore. *slang.*

1932 G. & S. Lorimer *Men are like Street Cars* v. 114 He's no drip...Ted's a darn good egg. 1936 N. Marsh *Death in Ecstasy* xviii. 215 What about that little drip Claude? 1938 J. Cary *Castle Corner* 279 Ah, ye dirty devil, and what sort of a drip are ye to be dropped in a medical hall. 1951 I. Brown *I break my Word* 123 We now more often call a feeble, foolish creature a drip. 1951 J. Cannan *And all I Learned* xi. 197 Of all the wet drips! 1959 I. & P. Opie *Lore & Lang. Schoolch.* xv. 326 Someone considered over-affectionate is said to be soppy, sloppy, gormless, a drip, or a clot.

d. *slang.* A grumble, complaint. (Cf. *Drip *v.* 5.)

1945 'Tackline' *Holiday Sailor* v. 53 Some of you chaps, you're on the drip the whole bloomin' time. Nothing but moan, moan, moan. 1948 'N. Shute' *No Highway* xii. 293 He had a bit of a drip last time because we kept him in the dark. 1970 *Guardian* 20 Aug. 1/2 One of the accused, Able Seaman Edward Kirkbride, said he remembered someone saying: 'I am going to have a drip (complaint).'

8. drip coffee, coffee prepared by allowing boiling water to percolate through ground coffee (cf. *Dripped *ppl. a.*); also **drip coffee-pot, coffee-maker,** a percolating coffee-pot; **drip culture,** a hydroponic method of plant culture in which the nutrient solutions are supplied automatically by a drip-feed mechanism; **drip-feed,** a method of feeding, lubrication, etc., whereby liquid is supplied a drop at a time; freq. *attrib.*; **drip mat,** a small mat placed under a glass, cup, etc., to catch drips; **drip painting,** a method of painting by which the colour is dripped on to the canvas or other medium (not applied directly with brush or palette knife); also, a painting so produced; **drip-point** *Bot.* = **drip-tip*; **drip-proof** *a.*, protected from drips; **drip-tip** *Bot.* (see quots.).

1895 G. A. Sala *Life* II. 382 The renowned 'drip' coffee..is so strong that it is said to stain the saucer into which it is poured. 1897 *Outing* (U.S.) XXIX. 574/1 He..produced a jar of coffee and the drip coffee-pot. 1964 *Which?* Feb. 53/1 With most of the drip coffee makers..a major disadvantage was the time it took to pour the boiling water through the ground coffee. 1923 *Amer. Jrnl. Bot.* X. 559 The relative dry-weight values of the plants grown in the drip cultures are in every instance much superior to the corresponding values for the simple cultures. 1952 C. E. Ticquet *Successful Gardening without Soil* 176/1 Drip culture. 1907 *Westm. Gaz.* 18 Nov. 6/3 The neat four-sight drip-feed lubricator. *Ibid.* 27 Dec. 4/2 No adjustment of the drip-feed points is necessary, as the pumps send a continuous stream of oil to the various parts. 1916 *Motor Cyclists' A.B.C.* 115 Lubrication (Automatic Drip Feed). 1964 *Which?* Jan. 13/1 There were 14 drip-feed heaters—the paraffin flowed to the burner from an inverted fuel container. 1971 *Manch. Guardian Weekly* 24 Apr. 5 Contaminated hospital drip-feed solutions. 1953 *Word for Word* (*Whitbread & Co.*) 18/2 *Drip mats*, small glass or absorbent mats on which glasses are placed to prevent staining tables and counters. 1958 *Listener* 27 Nov. 888/1 The exhibition shows, as Pollock's imitators never have, the range of drip painting. 1960 *Times Lit. Suppl.* 9 Sept. p. xlvii/5 The drip paintings of 1949–50 (probably the peak of Pollock's achievement). 1900 B. D. Jackson *Gloss. Bot. Terms* 82/1 Drip-point, drip-tip. 1936 F. A. Westbrook in H. Pender *Electr. Engineers' Handbk.* (ed. 3) iv. xvi. 14 Drip-Proof Motors. This type of motor has the upper halves of the heads closed so that materials, including liquids, cannot fall into it. 1960 *Farmer & Stockbreeder* 15 Mar. 171 (Advt.), B.S.D. motors (drip-proof or totally-enclosed fan-cooled). 1897 J. C. Willis *Flowering Plants* I. 154 The drip-tip..or acuminate leaf-apex. 1952 P. W. Richards *Tropical Rain Forest* iv. 87 The exaggerated acumen or drip-tip so often seen in rain-forest trees has long attracted attention.

drip-dry (dri·pdrai·), *v. intr.* Of certain synthetic or chemically treated fabrics: to dry when hung up to drip, without subsequently requiring wringing or ironing; also *trans.*, to dry (a garment, etc.) in this manner; also *absol.* Hence as *adj.*, that will drip-dry; as *sb.*, such a garment or fabric. Also *transf.* and *fig.*

[1916 *Daily Colonist* (Victoria, B.C.) 16 July 3/5 (Advt.), Do not wring or squeeze but hang to drip dry in the open air.] 1953 *Woman* 4 Apr. 34/3 With this permanently pleated nylon nightie..you just..rinse out, drip dry, and you don't even think of ironing. 1954 *Sci. News Let.* 16 Oct. 252/1 It will not shrink, fade or wrinkle and it drip-dries without pressing. 1956 *World Tennis* Sept. 51/2 Dacron shorts can be drip-dried. 1957 *Woman* 16 Nov. 25/3 Cottons and rayons with a special finish—that is ones that have been made crease-resistant, drip-dry or non-iron. 1958 *Ibid.* 22 Feb. 60/2 (Advt.), The rayon which 'drip-dries' at least as well as the best 'no-iron' cottons. 1958 *Sunday Express* 23 Feb. 14/3, I wouldn't exchange the sweet scent of fresh ironing for all the drip-dry cottons from Manchester. 1959 *Woman* 4 Apr. 43/3 Drip-dries are usually resin-finished fabrics. 1959 [see *Drag *sb.* 3 d]. 1960 *House & Garden* Mar. 92/2 Plates will drip-dry without a smear. 1968 *N.Y. Rev. Books* 7 Nov. 3/2 His greying drip-dry shirts.

dripped (dript), *ppl. a.* [f. Drip *v.*] That has been allowed to drip or percolate, esp. (*U.S.*) of coffee so made.

1884 F. E. Owens *Cookbook* 307 Dripped coffee. 1909 'O. Henry' *Options* (1916) 50 But if you're ever in the Middle West just mention my name and you'll get foot-warmers and dripped coffee.

dripping, *vbl. sb.* **4.** Add: **dripping crust,** a pastry crust made with dripping; **dripping toast,** toast spread with dripping.

1747 H. Glasse *Art Cookery* viii. 75 A Dripping Crust...Beef-dripping..work it up well into..Flour. 1906 Mrs. Beeton *Bk. Household Managem.* xxxi. 883 *Dripping crust* (for plain pies and puddings). 1921 W. de la Mare *Crossings* 14 Maybe you'll come and have a bit of dripping toast to your tea.

drippy, *a.* Add: **2.** Drivelling, sloppily sentimental; having the characteristics of a 'drip' (see *Drip *sb.* 3 c); 'wet'. *colloq.* (orig. *U.S.*).

1952 *N.Y. Herald Tribune* 9 Apr. 29/1 'Isn't it too dreamingly drippy,' said Miss Temple. 1962 K. Orvis *Damned & Destroyed* ix. 66 It was kind of drippy, me not coming back. 1963 *Times* 6 Feb. 13/4 Mr. Richard Pasco as the drippy hero. 1967 O. Norton *Now lying Dead* iv. 63 Men get so drippy when they're over-civilized, don't they?

dripstone. 2. (Earlier example.)

1792 W. Bligh *Voyage to South Sea* ii. 21, I also directed the water for drinking to be filtered through dripstones.

drisheen (drĭʃī·n). [ad. Ir. *drisín* intestine.] A kind of sausage made from sheep's blood, milk, and seasoning.

1910 P. W. Joyce *English as we speak it in Ireland* 251 Drisheen is now used in Cork as an English word, to denote a sort of pudding made of the narrow intestines of a sheep, filled with blood that has been cleared of the red colouring matter, and mixed with meal and some other ingredients. 1914 Joyce *Portrait of Artist* (1916) ii. 99 Mr Dedalus had ordered drisheens for breakfast. 1928 *Daily Express* 9 Mar. 11/7 Drisheen is a kind of sausage made..mainly from sheep's blood and milk. It is the Irish cousin of the Bury pudding. 1939 Joyce *Finnegans Wake* (1964) 164 Correspondents..will keep on asking me what is the correct garnish to serve drisheens with. Tansy Sauce. Enough. 1968 T. Fitzgibbon *Taste of Ireland* 79/1 Drisheen, traditional name for a black or blood pudding in County Cork.

drive, *v.* Add: **3. d.** *trans.* and *intr.* To drive (bees) into a new hive (see quots.).

d. (Earlier *Cricket* examples.) See also quot. 1867, and *off-drive* (OFF *sb.* 3), *on-drive* (ON *sb.*). Also (*colloq.*), a forcible punch or the like.

1836 E. JESSE *Angler's Rambles* 301 He would..shew..the exact length which he could cover in a forward drive. **1839** *Sessions Paper* X. 16 May 33 Tighe then said, 'Give her a drive,' and they went off. **1845** 'N. FELIX' *Felix on Bat* I. iii. 16 A tall long-reached man could make a fine forward drive. *Ibid.* I. iv. 20 The difference between the forward play and the 'drive' being only in the additional force you must employ. **1867** G. H. SELKIRK *Guide to Cricket Ground* iii. 42 The ball is always endeavoured to be hit with this part of the bat [*sc.* about five inches from the bottom of the blade], which is called the 'drive'. **1898** S. R. CROCKETT *Standard Bearer* xxxiii. 294 He gied him aye the ither drive wi' his nieve.

g. An organized effort to gain a particular end, esp. to raise money; an intensive campaign or effort. orig. *U.S.* Also (*U.S.*), the sale of goods or stocks at a low price.

1889 'MARK TWAIN' *Yankee* xxii. 245 She was making the honest best drive at it she could. **1890** *Ann Arbor Reg.* 1 Mar. (Advt.), Ladies, we are going to give you a Benefit and it will be the drive of the season. **1893** *Chicago Tribune* 1 July 36 A Big Cut in Waists A Big Drive in Corsets. **1921** E. L. BOGART *War Costs* 210 The four-weeks drive which ushered in the First Loan. **1928** *Britain's Industrial Future* (*Lib. Ind. Inq.*) III. xix. 258 The remarkable American 'drive' for the sale of Liberty Bonds. **1930** *Nation* 4 Oct. 360/1 The opposition of the peasantry to the 'drive' for collective agriculture assumed its most dangerous form last winter. **1933** *Granta* 19 Apr. 358/1 Anyone who has read accounts of the expulsion of the German Jews from the legal and medical professions may have wondered why no similar drive was being made against Jews in university and school appointments. **1941** *Times Weekly* 30 July 15 The export drive of the automobile industry.

h. *whist-drive*: see WHIST *sb.*[2] b; so *bridge-drive* (*BRIDGE *sb.*[2]). (Cf. *drive-whist* under sense *9.)

i. Energy, intensity, persistence, initiative, determination to achieve one's purpose.

1908 KIPLING *Lett. to Family* iii. 25 'Hustle' does not sit well on the national character..'Drive', a laudable and necessary quality, is quite different. **1909** G. B. SHAW *Let.* 31 Dec. (1956) 161 Stella has not the drive, the zest for Hypatia. **1924** R. FRY *Transformations* (1926) 173 Roman art could become expressive when its exponent had some other qualities than Roman efficiency and drive. **1944** J. S. HUXLEY *On Living in Rev.* xv. 195 Such men seem to lack the drive and confidence needed for public life. **1955** E. HILLARY *High Adventure* 148 It seemed to us at Camp IV that there was a certain lack of drive. **1965** W. LAMB *Posture & Gesture* iii. 46 People who show drive and welcome work (or at least get down to doing it) *look* less fatigued.

j. *Mil.* A forceful advance or attack.

1911 *Encycl. Brit.* XXVII. 207/2 Lord Kitchener commenced his first drive. **1918** E. S. FARROW *Dict. Mil. Terms* 189 The drive of the Allies was costly.

k. A thrill; exhilaration, esp. resulting from the use of narcotics. *U.S. slang.*

1931 G. IRWIN *Amer. Tramp & Underworld Slang* 67 *Drive*, a thrill. Formerly that exhilaration derived from narcotics; now, any temporary pleasure or uplift of spirit. **1949** N. ALGREN *Man with Golden Arm* I. 58 *Sure* I like to see it hit. Heroin got the drive awright—but there's not a single to a ton.

2. a. Also, a broad path in a wood.

1880 R. JEFFERIES *Hodge & his Masters* I. viii. 188 There comes..the low, dull, rushing roar of hundreds of hoofs..There is a block in the treacherous 'drive'. **a 1887** — *Field & Hedgerow* (1889) 304 The fawns fed away..to one of the broad green open paths or drives. **1894** W. ROBINSON *Wild Garden* (ed. 4) ix. 94 (*heading*) Woodland drives and grass walks.

3. b. (See quots.) Also *attrib.*

1899 J. BELL *In Shadow of Bush* xiv. 83 The smaller trees..had been 'scarfed', or cut partly through in readiness, and skilfully, so that each, when struck, might again in its turn strike and bring down another. The noise of a fall or drive of this kind is like thunder. **1940** B. O'REILLY *Green Mountains* II. 76 In the felling of rain forest, much chopping may be saved..by the use of the 'drive' system. **1948** R. ST. B. BAKER *Green Glory* xviii. 183 This fiendish practice is known as a 'drive'. The axeman will lay into the trees in such a manner that each will fall on its neighbour, when the pressure from above is set off by a big 'drive' tree on the uphill side of all the others.

4. (Earlier Austral. example.) Also *N.Z.*

1857 in E. A. Cooke et al. *Fresh Evidence from Early Goldmining Pubns.* (1966) 11 A main drive shall be carried along the course of such lead or gutter. **1880** H. LAPHAM *We Four* 18 There was a deserted 'drive' up on the spur. **1900** M. BOYD *Our Stolen Summer* x. 93 Up ladders, down ladders, along narrow drives they went. **1912** C. PURNELL *Modern Arthur* 17 Then rushed the miners from their hollow 'drives'.

6. The means or mechanism by which something is driven; esp. a device by which power is transmitted from one part of a motor vehicle to another; a driving-gear. Freq. with defining word prefixed, as **belt drive* (*BELT *sb.*[1] 6), **chain drive* (*CHAIN *sb.* 19). orig. *U.S.*

1901 MERWIN & WEBSTER *Calumet 'K'* xiv. 263 He's putting in three drives entirely different from the way they are on the plans. **1902** A. C. HARMSWORTH et al. *Motors* 191 A very smooth and silent drive without the spreading or bursting action of the bevels. **1912** *Motor Man.* (ed. 14) 74 Three speeds and a reverse are provided, with direct drive on top speed. **1960** *Analog Science Fact/Fiction* Dec. 45/2 Something crunched heavily under their stern at the exact instant the drive cut out. **1961** *Listener* 7 Sept. 353/3 It [*sc.* a telescope]..has a most efficient drive, so that the object under study is held firmly in the field of view.

b. In a motor vehicle with automatic transmission: the position of the selector lever in which the gears are automatically changed as required.

1963 *Which?* (Car Suppl.) Oct. 116/2 The selector mechanism was a little inclined to snatch at *Drive*. **1965** PRIESTLEY & WISDOM *Good Driving* v. 40 *D for Drive.* In this position, which is usual for all driving conditions, the transmission starts in first gear and automatically changes up into second and then top gear. *Ibid.* 41 With the engine running, *Drive* is selected, the handbrake released. **1967** *Times* 31 Mar. 3/7 The coroner said the selector lever must have been in 'drive'. **1970** D. MACKENZIE *Kyle Contract* (1971) 13 He drove out of Palamos..He shifted into drive and settled back.

7. *Psychol.* Any internal mechanism which sets an organism moving or sustains its activity in a certain direction, or causes it to pursue a certain satisfaction; a motive principle; any tendency to persistent behaviour directed at a goal; *esp.* one of the recognized physiological tensions or conditions of need, such as hunger and thirst. **b.** Any type of persistent behaviour or disposition that would lead to the attainment of a certain goal.

[**1888** *Mind* XIII. 165 *Trieb* (for which there is no good single equivalent in English).] **1918** R. S. WOODWORTH *Dynamic Psychol.* ii. 42 The drive is a mechanism already aroused and thus in a position to furnish stimulation to other mechanisms. *Ibid.* iii. 65 As the individual grows up, his actions are more and more controlled by inner drives. **1931** J. C. FLÜGEL in W. Rose *Outl. Mod. Knowl.* 367 It is a question rather of simultaneous 'drives' or wishes. **1931** C. J. WARDEN et al. *Animal Motivation* I. 14 By a drive we mean an aroused reaction tendency which is characterized primarily by the fact that the activity of the organism is directed toward or away from some specific incentive, such as food, water, animal of the opposite sex, etc. **1955** C. BRENNER *Elem. Textbk. Psychoanalysis* ii. 27 A drive, then, is a genetically determined, psychic constituent which, when operative, produces a state of psychic excitation or..tension. This excitation or tension impels the individual to activity, which is also genetically determined in a general way, but which can be considerably altered by individual experience. **1958** W. STARK *Sociology of Knowl.* 234 It is not a true instinct, but what we have labelled a drive, i.e. not a tendency rooting in the body but rather an attribute of the whole personality. **1962** R. FINE *Freud* (1963) vi. 93 The individual finds it extremely difficult to recognize his unconscious emotional drives.

8. *Electr.* (See quot. 1940.) Cf. *DRIVER 2 j.

1940 *Chambers's Techn. Dict.* 267/1 *Drive*, generally, the alternating voltage applied to the grid of an amplifying thermionic tube. Specially, the master oscillator circuit and its immediately subsequent amplifying stages in a transmitter using independent drive. **1959** K. HENNEY *Radio Engin. Handbk.* (ed. 5) xviii. 36 The drive for modulator tubes which are driven into grid current must have very good output-voltage regulation. **1962** SIMPSON & RICHARDS *Junction Transistors* xii. 287 The input impedance of a *CE* power stage capable of supplying 5 watts will vary typically from 80 ohms at very low drive to perhaps 15 ohms at full drive.

9. *attrib.*, as (sense 2 a) *drive-gate*; **drive-whist** *U.S.*, progressive whist.

1903 *Westm. Gaz.* 17 Jan. 1/3 The words were painted in large letters of gold on the drive gate. **1927** F. B. YOUNG *Portrait of Clare* 127 For answer he pointed to a white drive-gate, on the left of the road. **1888** *San Juan Prospector* (Del Norte, Colo.) 15 May 3/1 Mr. Grover Allen entertained the Young Married Folks at drive whist last Monday evening.

drive-. Add: **drive-belt,** a belt that transmits torque to a mechanism, etc.; **drive-off,** (*a*) *adj.* (of a ship) from which a motor vehicle can be driven; (*b*) *sb.* the act of setting off in a motor vehicle; **drive-on** *a.*, (of a ship) on to which a motor vehicle can be driven; also *drive-on/drive-off*; **drive-wheel** (U.S. example); **drive-yourself** *a.* (orig. *U.S.*), designating a motor vehicle let out on hire and driven by the hirer; also *ellipt.* as *sb.*

1959 *Which?* Aug. 87/2 The drive belts [on the spin drier of a washing machine]..slipped, affecting the speed with which the drums rotated. **1967** *Lebende Sprachen* XII. 136/2 *Drive-belt*, the belt on a tape recorder for turning the reels. **1960** *Times* 20 Jan. 9/6 Specially built drive-on/drive-off Transport Ferry ships carry loaded lorries. **1964** R. PETRIE *Murder by Precedent* vii. 103 They were to draw in the main canteen for order of drive-off. **1963** 'W. HAGGARD' *High Wire* iii. 32 Those drive-on car ferries. **1887** C. B. GEORGE *40 Yrs. on Rail* ii. 28 The drive-wheels will slip. **1921** *Daily Colonist* (Victoria, B.C.) 2 Apr. 2/1 (Advt.), The First 'Drive Yourself' Auto Livery in Canada and it's a great success. Rent a Dodge, Overland, Chevrolet or Ford and try it yourself. **1929** D. HAMMETT *Red Harvest* xviii. 179, I found Dick Foley in a hired drive-yourself Buick. **1959** G. M. WILSON *Shadows on Landing* viii. 88 The drive-yourself car with David at the wheel. **1962** W. H. MURRAY *Maelstrom* viii. 109 He would hire a 'drive-yourself' at Tarbert.

drive-in (drai·vin), *a.* (and *sb.*). orig. *U.S.* [f. DRIVE *v.* 5 c.] Designating a restaurant, cinema, bank, etc., into or up to which a customer can drive his car and, without leaving it, have a meal, see a film, effect a business transaction, etc. Hence as *sb.*, such a restaurant, cinema, etc.

1930 *San Antonio* (Texas) *Light* 31 Jan. 14/5 (Advt.), Drive-in drink and sandwich, with living room and bath.

1930 C. BEATON *Diary* Dec. in *Wandering Years* (1961) 189 The diehards hang on..working at 'drive-in' quick-lunch counters. **1937** *Amer. Speech* XII. 320/1 At roadside *drive-ins*, where one may secure curb service while sitting in his car. **1941** *Time* 14 July 66 (*caption*) U.S.'s Biggest Drive-In Theater. **1950** *Archit. Rev.* CVIII. 408/1 Drive-in banks, drive-in restaurants, drive-in cinemas, drive-in shops..are new environment elements. **1955** PRIESTLEY & HAWKES *Journey down Rainbow* 24 Drive-in movie theatres. **1959** *Listener* 5 Feb. 251/2 Britain's first 'drive-in' bank (a branch of the Westminster Bank in Liverpool). **1960** *Daily Express* 30 Aug. 7 Britain's first open-air cinema—Americans call them 'drive-ins'—is planned for Sandown Park racecourse next year. **1970** *Globe & Mail* (Toronto) 26 Sept. 50/1 (Advt.), Manager early 40's, fully experienced in drive-in restaurant..management.

driven, *ppl. a.* Add: **3.** *driven well* (U.S.), a tube-well.

a 1877 in KNIGHT *Dict. Mech.* **1923** M. WATTS *L. Nichols* 5 A driven well..went dry in periods of prolonged drought.

driver. Add: **1. b.** A horse trained to be driven in harness. *U.S.*

1876 *Rep. Vermont Board Agric.* III. 168 Stylish, enduring roadsters, trotters and gentlemen's drivers, standing from fifteen to fifteen and one-half hands high. **1902** A. D. McFAUL *Ike Glidden* viii. 61 This is a pretty good driver you've got here. *Ibid.* 66 All prosperous people there keep a 'driver' and a 'trader'.

2. b. Also in Trotting. Phr. *in the driver's seat*, in a controlling position; in charge.

1923 *Nation* 18 July 49 He has swung blithely into the driver's seat and cheerfully undertaken to run the nation without knowing how. **1948** *N.Y. Times* 18 Apr. v. 2/1 Leo Durocher, after an involuntary year of exile, is back in the driver's seat. **1963** *Weekly News* (Auckland) 8 May 53/1 Two drivers were sprawled on the track after one of the worst smashes at a trotting meeting in the North Island this season. **1965** G. McINNES *Road to Gundagai* xi. 189 'A divine Australian voice'..with a hint of iron in it that made you feel he was in the driver's seat. **1968** *Globe & Mail* (Toronto) 15 Jan. 20/3 Wes Coke, the young driver from Petrolia, rang up three winners on the matinee card.

f. Short for *driver-ant*.

1897 M. KINGSLEY *W. Africa* 626 Bad language, such as I am accustomed to when a lord of creation gets drivers on him. **1966** C. SWEENEY *Scurrying Bush* vi. 84 Small, reddish-brown drivers clung to my toes.

g. *Cricket.* A batsman who drives or is skilled in driving (see DRIVE *v.* 8 b).

1906 A. E. KNIGHT *Complete Cricketer* 344 The batsman who is proficient at the stroke is..a good 'driver'. **1921** P. F. WARNER *My Cricketing Life* iii. 64 Lionel Palairet..was also a fine driver on both sides of the wicket.

j. That part of an electronic circuit designed to supply the input signal power required by the last stage of a power amplifier, esp. a transmitter or receiver; an electrical device, as a valve or transistor (*driver transistor, tube, valve*), immediately preceding the output stage; the stage as a whole immediately preceding the output stage (called also *driver circuit, stage*).

1924 S. R. ROGET *Dict. Electr. Terms* 68/2 *Driver*, an expression sometimes used for a source of oscillations in radio telegraphy, particularly in connection with testing operations. **1938** G. E. STERLING *Radio Man.* (ed. 3) v. 213 The driver is ordinarily a class *A* 'power amplifier' stage. **1948** GLASOE & LEBACQZ *Pulse Generators* iv. 124 The driver circuit is not an oscillator in the same sense as the circuit used in television, and may more properly be referred to as a 'regenerative pulse generator'. **1948** I. A. GREENWOOD et al. *Electronic Instruments* vii. 179 Since the voltages at the potentiometer arms come from relatively high-impedance sources, precise impedance-changing circuits (drivers) are necessary to reproduce these voltages across the resolver stator windings without loading the potentiometers. **1948** J. A. PIERCE et al. *Loran* vii. 211 The pulse-output circuit and exciter driver amplify and shape the pulse that drives the local transmitter. *Ibid.* ix. 295 [This] requirement is met by the use of a c-w driver stage that amplifies the output of the 90-kc/sec generator to a level of about 20 watts. **1949** F. G. GARRATT tr. *Deketh's Fund. Radio-valve Technique* xvii. 229 The operation of valves in this manner..involves the flow of grid current and the source of grid-signal voltage has to supply a certain amount of power. To meet this, therefore, the Class-B output stage is preceded by a so-called *driver valve* capable of supplying a certain power. **1962** SIMPSON & RICHARDS *Junction Transistors* xi. 246 The fourth method, direct coupling, may be used advantageously..in the coupling of a driver to a power stage. **1970** J. EARL *Tuners & Amplifiers* ii. 51 The majority of power amplifiers have their push-pull output transistors driven direct from a pair of driver transistors.

k. *Acoustics.* A device in a horn loudspeaker that converts electrical energy into sound.

1927 *Wireless World* 16 Nov. 666/1 It is a case of arranging for the most convenient method of coupling between the driver..in this case between the diaphragm and the air within the horn, so that the imprisoned air shall be acted upon to the best advantage. **1950** H. S. KNOWLES in K. Henney *Radio Engin. Handbk.* (ed. 4) xvi. 747 The driver unit is coupled to the horn by the sound chamber. **1967** BADMAIEFF & DAVIS *How to build Speaker Enclosures* ii. 19 Every enclosure alters in some way the performance of the driver placed in it.

l. *Sheep-shearing.* (See quots.) *Austral.* and *N.Z.*

1933 L. G. D. ACLAND in *Press* (Christchurch) 14 Oct. 15/7 *Driver*, a leather strap on the handle of a pair of shears, which fits over the back of the shearer's hand. **1959** H. P. TRITTON *Time means Tucker* iv. 31/1 There is a band of leather from the heel of one blade [of shears] to the back of the other. This is the 'driver', and its purpose is to prevent the hand from slipping forward onto the blades. **1965** J. S. GUNN *Terminol. Shearing Industry* I. 25 *Driver*, a leather strap on hand shears. This fits firmly round the handle and over the back of the shearer's hand, thus allowing more drive to be given to a blow while preventing the hands from slipping over the blades.

driveway. Add: **b.** (Earlier U.S. example.) Also, a private carriageway for a motor vehicle alongside, in front of, or leading to a house, garage, or other building; a drive. Chiefly *N. Amer.*

1870 *Congress. Globe* 2 Feb. 966/3, I doubt as to the policy of allowing this railroad to go along exactly in the track of where we propose to have a public drive-way. **1935** M. M. ATWATER *Murder in Midsummer* xvi. 146 Harold was in the drive-way beside the house, about to start his car. **1945** *Chicago Tribune* 13 May VII. 1/1 Beyond the driveway that runs before her front gate, the greens start popping out of the moist earth. **1952** *Manch. Guardian Weekly* 28 Feb. 12 We twist into a driveway and have arrived. **1965** PRIESTLEY & WISDOM *Good Driving* ix. 65 It is often safer..to use a side-turning. A side-road or driveway..will do. **1966** *Globe & Mail* (Toronto) 9 May 7/3 [I] shovelled gravel to make a new driveway. **1971** 'A. BLAISDELL' *Practice to Deceive* iv. 59, I always left the driveway light on for her.

c. A passageway for the conveyance of hay, grain, etc., into a barn.

1839 *Mass. Agric. Rep.* 1838 80 The building should be so placed that the barn floor could be laid upon the beams, and the drive-way be into the end directly under the roof. **1868** *Rep. Comm. Agric.* (*U.S. Dept. Agric.*) 242 Where it is practicable, it is best to have the drive-way for drawing in hay, grain and corn fodder enter the gable end. **1949** *Pacific Spectator* Spring 226 The upper floor, right on a level with the driveway, had a big haymow on the left.

d. A scenic highway. *Canada.*

1909 *Gow Ganda* (Ont.) *Tribune* 1 May 1/1 The advisability of spending three hundred thousand dollars in building a driveway and boulevard along the Canadian side of the Niagara River. **1927** M. DE LA ROCHE *Jalna* x. 114 They were gliding slowly along an ocean driveway in Rosamond Trent's car. **1958** *Saturday Night* (Toronto) 27 Sept. 7/1 More trees, parks and driveways as handouts from the taxpayers of Canada.

driving, *vbl. sb.* **3. a.** For 'in a carriage' read: 'in a carriage, motor vehicle, etc.' (Add further examples.)

1858 C. M. YONGE *Christmas Mummers* i. 11 Harry Mayne was perched on the driving-seat. **1882** G. W. PECK *Peck's Sunshine* 24 If he don't put on an old driving coat and go out on the road occasionally. **1895** *Montgomery Ward Catal.* 518/1 Men's river or driving shoes. **1897** *Sears, Roebuck Catal.* 228/1 Light weight driving gloves. **1911** *Madame* 20 May 318/2 No one who has not lived in the canton for three months shall be allowed a driving licence. **1926** *Amer. Speech* I. 686/1 American..rear view mirror. English..driving mirror. **1932** KIPLING *Limits & Renewals* 135 Phil, alone in the car..shifted into the driving-seat. **1933** *Punch* 13 Dec. 662/1 Reasonable plea for a driving-test was put forward by Lord Howe. **1936** *Times* 10 Jan. 12/2 His waistcoat, containing his savings and his driving licence. **1961** *Evening Standard* 14 July 19/2 (Advt.), Driving Instructor reqd..Motor School. **1962** N. MARSH *Hand in Glove* vi. 199, I mean his driving gloves. They're heavy leather ones with string backs. **1966** 'A. HALL' *9th Directive* xxiv. 224, I slid the driving-seat back a notch. **1966** B. KIMENYE *Kalasanda Revisited* 98 As soon as they were alone, she asked if she could take driving lessons. **1966** T. WISDOM *High-Performance Driving* xvi. 136 The driving test..in my mind is not a real *driving* test at all. It is just a traffic test. *Ibid.*, The driving-school customer is interested only in passing the test. **1968** *Autocar* 18 Apr. 26/1 Driving simulators are still a novelty. **1969** V. CANNING *Queen's Pawn* ii. 8 There was a pair of comfortable women's driving shoes down by the control pedals.

b. driving-band, (*a*) (see Dict.); (*b*) (see quot. 1893); **driving-iron** (example); **driving-stick,** a stick with which cattle, etc., are driven.

1886 F. C. MORGAN *Handbk. Artillery Matériel* (ed. 3) vi. 56 The driving band in the newest pattern of shells is made broader, and has cannelures round it. **1893** LLOYD & HADCOCK *Artillery* ix. 230 *Driving bands.* All projectiles, except case shot, are fitted with a band of soft metal, which is cut into by the rifling of the gun, and thus forms the medium for translating the rotatory motion to the projectile. *Ibid.* 231 'Cannelures' are cut in the driving band to make room for the copper which is displaced by the rifling when the projectile is fired. **1916** 'BOYD CABLE' *Doing their Bit* v. 70 One girl..is turning the copper driving bands. **1890** Driving iron [see IRON *sb.*[1] 4 e]. **1869** J. G. FULLER *Flower-Gatherers* 147 He took Jack's long driving-stick. **1926** D. H. LAWRENCE *David* xii. 88, I plant this driving-stick in the soft earth.

driving, *ppl. a.* Add: **1.** *driving force* or *power*, the force or power by which an engine or vehicle is driven; motor force or power; freq. *fig.*

1856 [in Dict.]. **a 1877** KNIGHT *Dict. Mech.* I. 948/2 The great weight of this wheel is..to carry the machine over the one half of its period in which the driving-power is absent. **1905** *Sketch* 26 July 38/1 The driving power [*sc.* of a dirigible balloon] is supplied by two 50 horse-power Bucket motors. **1909** F. HARRIS *The Man Shakespeare* 369, I always think of him as a ship over-engined; when the driving-power is working at full speed it shakes the ship to pieces. **1911** 'I. HAY' *Safety Match* i. 14 Brian Vereker would make a noble figurehead..but..the figurehead would require a good deal of imported driving-power behind it. **1927** M. PUPIN *New Reformation* 214 A cosmic stream of solar energy from which everything that lives and breathes on this terrestrial globe derives its driving-force, just as the mill on the mountain side derives its driving-power from the mountain stream. **1931** J. S. HUXLEY *What dare I Think?* vii. 246 The religious driving-force of a great many intelligent people is going to waste. **1965** *New Statesman* 30 Apr. 691/2 Hofmeyr was not only often acting prime minister but almost always the main driving force in the administration. **1967** CONDON & ODISHAW *Handbk. Physics* (ed. 2) II. iii. 23 For a given magnitude of driving force, resonance in the velocity amplitude occurs at minimum impedance. **1970** *Times* 9 Mar. 7/1 Someone persuaded the man known as the 'Driving Force' behind *Woman's Own* and *Modern Woman* to see me.

b. Energetic, dynamic, forceful. (Cf. *DRIVE *sb.* 1 i.) orig. and chiefly *U.S.*

1835 J. H. INGRAHAM *South-West* II. 92 They become thorough, driving planters. **1946** R. BLESH *Shining Trumpets* (1949) xi. 257 Mitchell's cornet playing throughout is of the utmost clarity and compressed, driving simplicity. **1952** B. ULANOV *Hist. Jazz Amer.* (1958) xix. 235 Lester is also a summary example of driving, vigorous tenor saxophone.

drizzle, *v.* Add: **4.** *intr.* To pick the gold thread out of tassels or embroideries into which it was woven; so *drizzler, drizzling* (also *attrib.*).

1884 tr. *Bauer's Posthumous Mem.* II. i. 117 Prince Leopold..diligently and indefatigably drizzled. *Ibid.* 118 This 'drizzling'..was invented in Paris..during the reign of King Louis XVI...The most fashionable ladies of the court felt no compunction in asking the gentlemen of their acquaintance for cast-off gold and silver epaulettes. **1896** *Godey's Mag.* Feb. 177/2 Drizzling—which was nothing more or less than picking the gold thread out of old gold tassels, braid, [etc.]. *Ibid.* 178/1 One of the Countess's principal grievances against the Prince seems to have been that he was a confirmed drizzler; she says Leopold would sit by her hour after hour diligently and indefatigably drizzling...The tall Prince..bending over his elegant drizzling-box of tortoise-shell. **1969** E. H. PINTO *Treen* 310 The value of the material in gold lace led to later destruction of much of it for reclamation of the gold. In France, in the late 18th century, *parfilage* or unravelling became a social and profitable pastime. Early in the 19th century, the destructive addiction came to England under the name of drizzling.

drogue. Add: **3.** *Aeronaut.* A truncated cone of fabric with a hoop at the larger end, used for various purposes: (*a*) a brake or anchor for aircraft, esp. seaplanes (see quot. 1919); (*b*) such a cone towed behind an aircraft as a target for gunnery practice; (*c*) a **wind-cone*; (*d*) an auxiliary braking parachute in an ejection-seat mechanism; (*e*) part of an air-craft-refuelling device (see quots. 1949 and 1966); (*f*) (see quot. 1962[1]). Also *attrib.*

1919 W. B. FARADAY *Gloss. Aeronaut. Terms* 59 *Drogue*, a fabric bag arranged to tow with its mouth open, thereby resisting passage through the water. **1931** *Flight* 13 Mar. 234/1 The manufacture of drogue targets. **1932** *Nuttall's Stand. Dict. Suppl.*, *Drogue*, an open fabric bag used to show the direction of the wind at an aerodrome. **1933** *Flight* 29 June 628/1 The target is called a 'drogue' and is the same size as the fuselage of a 'Bulldog'. **1934** *Exmouth Jrnl.* 12 May 8/2 (Advt.), At every air port, you see it—the flyer's drogue. It points the right direction for air traffic to take when starting or landing. **1941** [see *air-to-air* adj. s.v. *AIR *sb.*[1] B. III 1]. **1947** *Times* 30 Aug. 2/1 The drogue steadied his upward rush, and eight seconds later the seat parachute opened. **1949** *Flight* 11 Aug. 177 The tanker aircraft trails a 65ft. fuel hose which terminates in a conical drogue. **1951** *Engineering* 30 Mar. 368/2 To..slow down the seat after ejection, a 2 ft. diameter drogue parachute..was employed. **1951** *Jrnl. Brit. Interplanetary Soc.* X. 300 The most efficient braking device is the parachute drag brake (also known as 'drogue chute' or 'parabrake'). **1962** J. GLENN et al. in *Into Orbit* 244 *Drogue*, a small parachute used to stabilize the descent of a spacecraft during re-entry. **1962** *Times* 21 Feb. 10/2 The drogue parachute opened to retard the capsule further and then the main parachute billowed out to lower Colonel Glenn into the sea. **1966** *New Scientist* 20 Jan. 142/1 Apparatus used [for refuelling a helicopter in flight] was of the standard kind—a drogue containing the cup at the end of the tanker's hose and a probe from the nose of the helicopter which is flown into the cup and locked there.

drogulus (drǫ·giŭlǝs). [Coined 'on the spur of the moment' by A. J. Ayer perh. by subconscious association with DRAGON + L. *-ulus* as in DRACUNCULUS.] An entity whose presence is unverifiable, because it has no physical effects. Also *transf.*

1957 A. J. AYER in Edwards & Pap *Mod. Introd. Philos.* 608 Suppose I say 'There's a "drogulus" over there,' and you say 'What?' and I say 'Drogulus,' and you say 'What's a drogulus?' Well I say 'I can't describe what a drogulus is, because it's not the sort of thing you can see or touch, it has no physical effects of any kind, but it's a disembodied being.' **1959** L. S. PENROSE in *New Biol.* XXVIII. 98, I had difficulty in finding a suitable name for the activated complexes produced in these experiments. On showing one of them to Professor A. J. Ayer, I inquired whether it perhaps might be a 'drogulus'...He replied that it was undoubtedly a 'drogulus'.

droit[1]. Add: **1. b.** *droit(s) du* (or *de*) *seigneur* (‖drwadüsenyör), an alleged custom of mediæval times by which the feudal lord might have sexual intercourse with the bride of a vassal on the wedding-night, before she cohabited with her husband. Also *transf.*

[**1784** 'F. G. DESFONTAINES' (*title*) Le Droit du Seigneur. Comédie en trois actes,..mêlée d'ariettes.] **1825** H. WILSON *Mem.* III. 168 Lord Frederic Beauclerc..declares himself willing to..marry us, privately, by special licence, provided you agree to grant him les droits du seigneur. Worcester by dropping a made stitch at intervals; **drop-stroke**, in tennis, rackets, badminton, etc., a stroke that causes the ball or shuttlecock to drop sharply after crossing the net or striking the wall; **drop-tank** *Aeronaut.*, a (fuel-)tank that can be detached and dropped in flight; **drop-title**, a title which is set comparatively low on the page; **drop-volley**, a volleyed drop-stroke; **drop-wrist**, = *wrist-drop* (WRIST 5 d) (cf. quot. 1893 s.v. DROPPED *ppl. a.* 1 a).

1835 *Liberator* (Boston) 5 Dec. 196/5 The travellers on whom I bestowed your *drop cakes*. **1879** M. E. BRADDON *Vixen* II. ix. 151 Trimmer's drop-cakes..are always capital. **1904** S. A. MUSSABINI *Mannock's Billiards* I. 336 The plain 'drop cannon' which is employed to gather the balls between the two top pockets. **1909** *Westm. Gaz.* 2 Feb. 12/3 He unexpectedly missed a rather wide drop-cannon from hand. **1934** H. ADDISON *Textbk. Appl. Hydraulics* viii. 137 (*heading*) Falling surface or drop-down curves. **1940** *Chambers's Techn. Dict.* 267/2 *Drop-down curve* (Hyd. Eng.), the longitudinal profile of the water surface in the case of non-uniform flow in an open channel, when the water surface is not parallel to the invert. **1951** *Catal. of Exhibits, South Bank Exhib., Festival of Britain* 36/2 Gas cooker..with drop-down door. **1928** *Daily Mail* 31 July 1/2 Settee has drop end. **1960** *Measurement of Spectacles* (*B.S.I.*) 18 Dimensions of drop-end sides. **1921** *Newcastle-upon-Tyne Med. Jrnl.* Apr. 118 (*title*) Tendon fixation of drop foot. **1924** *Jrnl. Amer. Med. Assoc.* LXXXII. 30/1 (*caption*) Shoe modified for the drop-foot brace. **1925** *Ibid.* LXXXV. 1927/1 Drop-foot is the inability to dorsiflex the ankle joint or foot, and is the result of any agent that impairs or abolishes the muscular power in the anterior group of leg muscles. **1962** D. V. & F. DAVIES *Gray's Anat.* (ed. 33) 1212 Paralysis of all the dorsiflexor and evertor muscles of the foot..producing a 'drop foot'. **1925** *Glasgow Herald* 15 Jan. 8 A mass of dough, kneaded, drop-forged, or otherwise assembled into a solid entity. **1895** *Montgomery Ward Catal.* 555/2 Cold drawn steel tubing, with steel drop forged crown. **1897** *Westm. Gaz.* 7 Jan. 7/2 Cleeks and irons made of drop-forged steel. **1957** *Economist* 7 Dec. 897/1 The drop forgers and specialised ironfoundries. *a* **1884** KNIGHT *Dict. Mech.* Suppl. 277/2 *Drop forging*, one made in that form of press in which the blow is by impact instead of by mere pressure. **1897** *Outing* (U.S.) XXX. 278/2 The hammers that are pounding out the drop-forgings. **1909** *Engineer* CVII. 277 Drop forgings are cheaper and more accurate than hand forgings. **1925** *Jrnl. Iron & Steel Inst.* CXI. 527 Recent developments in drop-forging practice. **1959** *Ibid.* CXCI. 94/3 Calculating the power required in drop forging. **1895** *Montgomery Ward Catal.* 555/1 This machine..has our patent duplex drop frame for ladies' use. **1898** *Westm. Gaz.* 18 Nov. 9/1 In Professor Lilly's triangulated frame will be found an attempt by a skilled engineer to overcome the 'drop'-frame difficulty. **1906** *Bazaar, Exch. & Mart.* Suppl. 12 Oct. 1481/2 Gentleman's cycle, 23 in. drop frame. **1928** *Daily Tel.* 16 Oct. 7 Another feature demanded by lady drivers is a drop-frame for the divisional window between the front and rear seats. **1898** *Cycling* 91 The drop-framed safety. **1934** WEBSTER, *Drop-front* adj., of a desk, having a part of the front formed by a hinged lid or cover which may be lowered to form a writing table. **1951** *Good Housek. Home Encycl.* 74/2 Two types of door, the drop-front and the side opening, are available. **1895** *Montgomery Ward Catal.* 383/2 Drawer drop handles. **1898** J. P. ARKWRIGHT *Cabinet-Making* 84/2 Brass drop handles..look well. **1940** *Antiquity* XIV. 69 Three angons were actually pushed through the drop handle of the larger bowl. **1967** *Gloss. Terms Builders' Hardware* (*B.S.I.*) IV. 11 *Drop handle*, any handle pivoted to a plate or spindle so that it falls by gravity to a vertical position when not in use. **1937** M. ALLINGHAM *Dancers in Mourning* xix. 243 He then turned the bike sideways, showed off the drop handlebars with the special grips. **1895** *Montgomery Ward Catal.* 264 Our improved high arm in the new drop head case. **1903** *Sears Catal.* (ed. 113) 420 We recommend the drop head cabinet for the reason that it serves as a protection for the head, and when closed you have in it a handsome stand or table. **1932** *Autocar* 28 Oct. 7 (Advt.), Romney 2-seater Drop-head Coupé. **1934** *Times* 16 Oct. 7/3 This all-weather car has a patented drop-head which can be folded back by one person without leaving the car. **1936** J. G. COZZENS *Men & Brethren* II. 191 Bill Jennings sat at the battered drophead desk. **1951** *Bookman's Gloss.* (*R. R. Bowker Co.*) (ed. 3) 54 *Drop initials*, initial letters as tall as two or more lines of text, which lines are indented to allow room. **1963** *Punch* 9 Oct. 528/2 Elegant drop-initials three inches deep. **1967** J. SEVERSON *Great Surfing* Gloss., *Drop-knee*, a type of turn where both knees are bent—the trail leg crossed behind the lead leg with the trail-leg knee dropped closer to the board than the knee of the lead leg. **1970** *Surf '70* (N.Z.) 37/2, I found I had to change from a drop knee turn to a turn with both knees bent and my weight behind my legs. **1895** *Montgomery Ward Catal.* 609/2 Men's drop leaf desk. **1966** A. W. LEWIS *Gloss. Woodworking Terms* 45 The hinge is recessed into the face of the job, e.g. on bureau falls and on drop-leaf tables. **1861** *Harper's Mag.* Nov. 815/2 The dark, bright Spanish woman, on the alabaster shade of the drop-light. **1902** *T. Eaton & Co. Catal.* Spring & Summer 218/4 Electric drop-light shade. **1904** K. C. THURSTON *John Chilcote* viii, The drop-light from the ceiling being directly above his head. **1847** C. LANMAN *Summer in Wilderness* xxvi. 158, I..with a drop-line have taken, in twenty minutes, more trout than I could eat in a fortnight. **1925** *Duo-Art Service Manual* 14 Drop-off screw. **1952** *Time* 21 Apr., *Drop-off point*, place where copies of *Time* come off a plane, train or truck. **1958** *College English* XX. 16/2 There is no drop-off in volume. **1959** *Times* 31 Jan. 6/1 The rocket engines are fuelled by a mixture of liquid oxygen and ammonia representing about half the drop-off weight of 31,275 lb. **1897** STEPHENSON & SUDDARDS *Textbk. Orn. Design Woven Fabrics* iv. 49 This placing or 'dropping' of one diamond below another..gives the essence of the drop pattern. **1898** KENNEDY & COHEN in W. A. Morgan '*House*' *on Sport* 420 Saunders..relied on his..drop service. **1962** *Times* 8 Jan. 3/7 His cramping drop-service..just gave him the edge. **1905** *Daily Chron.* 16 Feb. 8/5 A drop-shelf, with chains attached to one of the panels. **1908** *Captain* Aug. 453/1 They never practice its [*sc.* the lob's] antithesis, the drop-shot. **1927** *Observer* 20 Mar. 27/3 Mixing up deep drives and clever drop-shots. **1963** *Times* 19 Jan. 3/3 He reached the first drop shot only at its second bounce. **1907–8** *T. Eaton & Co. Catal.* Fall & Winter 243/3 Child's iron and brass cot,..drop sides with brass top rails. **1959** *B.S.I. News* Apr. 18/1 Suitable forms of fastening devices for the drop side have been specified, with the intention of preventing the child from lowering it. **1962** *Engineering* 28 Dec. 839 Dropsides and tailboards for lorries have been made from bonded metal. **1963** *Which?* Jan. 13 A drop-side cot is probably used more continuously than any other piece of nursery furniture. **1946** *Bull. Amer. Meteorol. Soc.* XXVII. 162/1 Droppable radiosondes..can be dropped by parachute; they emit radio signals...The accuracy of the resulting..data is comparable with similar data from the usual radiosondes...These 'dropsondes' should be quite valuable. **1951** in *Meteorol. Abstr.* (1952) III. 6 (*title*) Dropsonde observations. **1963** *New Scientist* 14 Nov. 396/3 A 'dropsonde' to measure low densities and temperatures. **1968** G. M. B. DOBSON *Explor. Atmos.* (ed. 2) ii. 19 The instrument is carried up on a small rocket to a height of about 70 km and there released, when it descends on a large parachute...Such instruments are sometimes known as drop-sondes. **1890** in *Amer. Mail Order Fashions* (1961) 12 Ladies' Black Spun Silk Hose, drop stitch. **1905** *Smart Set* Oct. 9/2 Kind o' openwork, like a lady's drop-stitch sock. **1923** *Daily Mail* 29 June 1/2 (Advt.), French Lisle Thread Stockings..Dropstitch design. **1897** *Encycl. Sport* I. 621/2 *Drop-stroke*, a stroke by which the ball is made to drop dead, just clearing the net. **1898** COCKAYNE & GOWER in W. A. Morgan '*House*' *on Sport* 254 A 'drop' stroke [in rackets] is a return so soft that it hardly comes off the front wall at all. **1946** TAYLOR & ALLWARD *Spitfire* 102/2 By carrying a belly drop tank the range could be increased to around 2,000 miles. **1961** B. FERGUSSON *Watery Maze* xvi. 397 Drop tanks full of water were flown by a Skyraider from H.M.S. *Albion*. **1971** *Air Pictorial* XXXIII. 135/1 In-flight refuelling capability and up to six 600-gal. external drop-tanks can probably double this figure. **1893** WISE & SMART *Bibliography of Ruskin* I. 189 There is no title-page, the 'drop-title' on page 1 reading 'Memorandum of Association of the Guild of St. George'. **1907** *Westm. Gaz.* 9 Nov. 12/3 The drop-volley..was one of her favourite strokes. **1927** *Daily Express* 4 May 13/7 She would leap forward and summarily cut short the rally with a deft drop-volley. **1901** *Daily News* 3 Jan. 6/4 A door with window, and on either side of the latter a drop window. **1926** *Glasgow Herald* 26 June 9 The drop-windows permit of ready means of ventilation on warm days. **1860** R. J. JORDAN *Dis. Skin* 18 Painters..become affected with 'lead-colic' and 'drop-wrist', from having to do with white lead. **1902** H. J. STILES in D. J. Cunningham *Textbk. Anat.* 1203 The condition known as '*drop-wrist*', the result of paralysis of the extensor muscles of the forearm.

drop-in, *sb.* and *a.* [DROP *v.* 27.] **A.** *sb.* **1.** An unexpected or casual visit or visitor. *colloq.*

1819 [see DROP *sb.* 12 c]. **1942** BERREY & VAN DEN BARK *Amer. Thes. Slang* § 363/5 *Drop-in, happen-in*, a chance social call. **1970** M. MILLAR *Beyond this Point* (1971) iii. 32 Some of the people were drop-ins who had no real interest in the case.

2. (See quot. 1940.) *U.S. slang.*

1937 D. RUNYON in *Collier's Weekly* 16 Jan. 9/3 Nicely-Nicely commences to figure this is about as soft a drop-in as anybody can wish. **1940** *Amer. Speech* XV. 117/2 *Drop-in*, something which is easy; easy money. So-called because a fat *mark* may sometimes 'drop in' to a confidence-game without being steered.

B. *adj.* **1.** Designed to drop into position; *spec.* of a seat (see quot. 1960).

1921 *Spectator* 16 Apr. 485/2 Above this French window is a drop-in fanlight. **1960** A. QUINTON in G. Lewis *Handbk. Crafts* 232 Any type of 'drop-in' seat: i.e. one which can be lifted from its frame and has no springs. **1962** L. S. SASIENI *Princ. Optical Dispensing* viii. 195 There are three main types of these supplementary fronts: 'clip-on' fronts, 'grab' fronts, and 'drop-in' fronts.

2. At which one may 'drop in' (see DROP *v.* 27 b). *colloq.*

1958 *Listener* 27 Nov. 860/1 Mr. Walter Reuther's United Auto Workers runs night schools, 'drop-in' centres for retired members, recreation halls. **1962** *Guardian* 24 July 5/1 We were entertained..by the Federation of Lace Makers to a drop-in (bar ouvert toute la journée) view of the laces used in the show.

droplet. Add: *droplet infection*, infection conveyed by fine droplets of mucus sprayed into the air when a person opens his mouth to speak, cough, etc.

[**1904** *Westm. Gaz.* 20 Sept. 7/2 In the act of loud speaking, fine droplets of mucus are sprayed from the mouth into the air...It has been shown that by reading aloud for half an hour B[acillus] prodigiosus may be disseminated from the mouth to a distance of 24 ft. in front.] **1907** J. C. M'VAIL *Prevention Infect. Dis.* 275 When a patient is out of doors in a solitary place droplet infection may on occasion be disregarded. **1932** *Times Educ. Suppl.* 17 Sept. 358/1 Droplet infection was responsible for the spread of sore throat.

drop-letter. Add: **2.** (See quots.) Cf. *drop-initial* s.v. *DROP-.

1894 *Amer. Dict. Printing* 149/1 *Drop letters*, two-line letters, the top being as high as the top of a line of an advertisement or of reading matter, the remainder dropping down to the next line. This expression is not used in America, the equivalent being a two-line letter. **1950** *Dict. Printing Terms* (*Porte Publ. Co.*) (ed. 5) 53/2 *Drop-letter*, a large letter used for ornamentation at the beginning of a chapter or article.

drop-out. Also **dropout.** [*DROP *v.* 28*.] **1.** *Rugby Football.* A drop-kick made from within the defending side's twenty-five-yard line in order to restart play after the ball has gone dead.

1882, etc. [see DROP *sb.* 12 c]. **1905** *Westm. Gaz.* 12 Dec. 9/2 From the drop-out Cambridge began an attack. **1960** E. S. & W. J. HIGHAM *High Speed Rugby* xiii. 157 The kick-off, drop-out, penalty and free kick.

2. A person who 'drops out' (see *DROP *v.* 28* b), esp. from a course of study or from society; also, the act of withdrawing. Also *attrib.* *colloq.* (orig. *U.S.*).

1930 *Sat. Even. Post* 1 Mar. 110/2 The drop-outs are usually those with inferior mental capacity. **1960** *Times* 21 Nov. (Canada Suppl.) p. xviii/4 The bored students—mostly boys—and the 'drop outs'. **1962** *Guardian* 19 Jan. 8/5 The brilliant woman..becomes an embarrassing statistic in the academic 'drop-out' rate...Dr. Bunting was a 'drop-out' for six years when her children were young. **1966** *Listener* 29 Sept. 454/3 The older teachers are resigned to the low standards and the high rate of drop-out. **1967** *New Statesman* 15 Dec. 838/2 An international gathering of misfits and drop-outs, smoking pot and meditating in the Buddhist temples. **1970** *New Society* 5 Feb. 231/3 He seems to imagine that, with the exception of the drop-outs, the working class has been entirely absorbed in 'co-operative economic production and consenting political citizenship'. **1971** *Brit. Med. Bull.* XXVII. 5/1 Experience suggests a high drop-out rate for those being treated.

3. *Photogr.* The elimination of highlight dots from part of a half-tone negative or plate; also, a half-tone having such an area eliminated, or the area itself. Also *attrib.*

1948 FLADER & MERTLE *Mod. Photoengraving* p. xxi/2 *Dropout*, a highlight halftone negative or printing plate; 'dropping-out' is the elimination of highlight dot formations. **1967** KARCH & BUBER *Offset Processes* v. 179 The drop-out halftone..is made where it is desirable to eliminate the highlight background of an illustration. **1968** *Gloss. Terms Offset Lithogr. Printing* (*B.S.I.*) 11 *Drop-out*, the areas of a picture which are clear of printing image. *Ibid.* 17 *Drop-out mask*, a photographic mask of sufficient density and contrast to separate non-printing areas from the first printing tones.

4. In tape-recording, a momentary decrease in the amplitude of the recorded signal due to a flaw in the tape; also, such a flaw.

1955 *Jrnl. Audio Engin. Soc.* III. 31 Tape squeal, sticking, and level variations have been caused by deposits on the head due to the shearing of imperfections in the oxide. Because of the effect of level variation in the computer field, the latter phenomenon has generally been referred to as 'dropout'. **1958** *Engineering* 7 Mar. 310/2 When the signal amplitude falls below this limit, faulty reproduction of the signal occurs. This is called 'drop-out'. **1962** A. NISBETT *Technique Sound Studio* iv. 85 Where there is a slight flaw, such as a drop-out, making a simple joint (and removing the flawed scrap of tape) will save time. **1967** D. F. ELDRIDGE in C. B. Pear *Magn. Recording* iii. 101 Once the first permanent dropout is formed, additional dropouts in increasing numbers will follow rapidly.

droppable (drǫ·păb'l), *a.* [f. DROP *v.* + -ABLE.] Capable of being dropped; fit to be dropped or discarded.

1908 W. DE MORGAN *Somehow Good* xli. 446 Sally held tight to her groundless opinion long enough for the previous question to be droppable, without effrontery. **1965** *Language* XLI. 392 An informationless operator..would be droppable.

dropped, *ppl. a.* **1. a.** Add to def.: *spec.* in *Rugby Football*, designating a goal scored by a drop-kick.

1896 [in Dict.]. **1933** *Times* 4 Dec. 5/6 Norris increased Cheltenham's lead with a splendid dropped goal from a distance of 30 yards. **1971** *Times* 15 Feb. 9/2 Bushell put Sidcup ahead with a superb dropped goal from 30 yards.

c. *dropped head*: a head (see HEAD *sb.* 13) at the beginning of a chapter, etc., placed lower down the page than the first line on ordinary pages.

1904 in GOODCHILD & TWENEY *Technol. & Sci. Dict.* 176/1. **1906** E. JOHNSTON *Writing* xvi. 343 It would not be necessary for the first page of a chapter to have the ordinary dropped head and blank upper space.

d. *dropped handlebars* = *drop-handlebars* (see *DROP-).

1913 *Boy's Own Bk. Outdoor Games* 218 A warning regarding the use of 'dropped' handle-bars will not be out of place. These are only necessary in racing, and while sacrificing comfort are not conducive to fast riding. **1970** *Morning Star* 18 Aug. 4 One sees young lads on bicycles in a busy city bent over dropped handlebars on which the brake levers are right at the end of the bars.

e. *dropped scone* (see SCONE 1).

dropper. Add: **1. c.** One who passes counterfeit money, cheques, etc. (cf. *DROP *v.* 16 b). *slang.*

1938 F. D. SHARPE *Sharpe of Flying Squad* xxix. 297 These [cheques] are then passed on to other members of the gang..known as 'droppers'. Their job is to present the cheques at the banks. **1959** 'C. HARE' *Best Detective Stories* 236 The functionary whose mission it is to put forged currency into circulation is known technically as a dropper.

d. One who delivers goods, liquor, etc., from market or store to retailers; a 'shop-dropper'. *local Austral.* and *N.Z. colloq.*

1949 F. SARGESON *I Saw in Dream* II. xv. 255 [The Police] reckoned they'd got her [sly-grogging] this time, because they'd found out a dropper had been through the town a few nights before on his lorry. **1957** *Courier-Mail* (Brisbane) 26 Nov. 2 A shop owner has only to telephone an order to a 'dropper' and within a few hours it is delivered to his door.

4. (Later examples.)

1898 *Elizabeth & her German Garden* 37 Either you or the dropper-in will say something..better left unsaid. **1941** V. WOOLF *Between Acts* 48 Uninvited, unexpected, droppers-in, lured off the high road. **1969** J. COOPER *How to stay Married* 44 The droppers-in will be so embarrassed that they'll apologise and make themselves scarce.

5. e. *Hort.* A young bulb of certain bulbous plants, *esp.* a small bulb developed at the apex of a downward shoot growing from the base of the parent bulb.

1900 B. D. JACKSON *Gloss. Bot. Terms*, *Dropper*, the young bulb of a tulip, not of flowering size. **1907** *Ann. Bot.* XX. 429 The 'Droppers' of *Tulipa* and *Erythronium*. **1929** A. D. HALL *Bk. Tulip* 22 Occasionally also it will be noticed..that a stolon has started away from the base of the old bulb, turned downward and formed a bulb at the extremity...These bulbs are called 'droppers' and differ in no respect from other offsets. **1951** *Dict. Gardening* (R. Hort. Soc.) IV. 2163/1 A peculiar type of natural propagation is seen in the 'droppers' which are sometimes formed.

f. A vertical member of a fence or the like; *spec.* a light lath used between the main uprights of a fence to keep the wires spaced. Chiefly *Austral.*, *N.Z.*, and *S. Afr.*

1904 'G. B. LANCASTER' *Sons o' Men* 93 [He] had prayed for slotted droppers [in the fence]. *a* **1935** G. L. MEREDITH *Adventuring in Maoriland* vii. 68 Between the posts the wires are stapled to 'droppers', consisting of about three by one battens. **1950** *N.Z. Jrnl. Agric.* Nov. 458/3 The materials are likely to be more easily obtained than wooden droppers. **1967** *Coast to Coast 1965–6* 150 'D'you know what a Mallee gate is, Bob?' 'Yes, it's a short loose panel, just droppers and wires.'

dropping, *vbl. sb.* Add: **5. b.** *pl.* The waste material cast off from a machine in certain processes of textile manufacture.

1902 W. I. HANNAN *Textile Fibres of Commerce* 115 The primary impurities from each of the two processes of opening and scutching are known as the droppings.

8. (sense 6) *dropping(s) board, pit*; **dropping field, -point, zone,** a place prepared for the dropping of supplies, troops, bombs, etc., from aircraft.

1916 *N.Z. Jrnl. Agric.* 21 Aug. 100 Those who are not prepared to pay regular attention to cleaning are advised not to have dropping boards. **1950** *Ibid.* June 531/3 Droppings boards or wired-in droppings pits are coming back into use again [in hen-houses]. **1942** *Times Weekly* 9 Sept. 2 He had to walk two miles back to the 'dropping' field. **1947** J. MULGAN *Report on Experience* 80 They ran to a schedule and knew the dropping-points as intimately as their aerodromes. **1945** *By Air to Battle* (*H.M.S.O.*) 39 It was left to the battalion commander..to share with the pilot the responsibility of choosing the dropping zone from the air. **1956** J. TICKELL *Moon Squadron* viii. 83 In ten minutes, we would be over the DZ or dropping zone. **1968** A. J. JACKSON *Blackburn Aircraft* 469 The rear doors were removed for a demonstration over an Army dropping zone at Amesbury, Wilts.

dropsy. Add: **A.** *sb.* **4.** Money, *esp.* paid as a tip or as bribery. Cf. *DROP *sb.* 18 d. *slang.*

[**1616** T. DRAXE *Bibliotheca Scholastica* 33/1 He hath the siluer dropsie.] **1930** 'GREENHORN' *Tinker, Tailor* xi. 253 [He] lived only for tips, or 'dropsy' as it's called in the vernacular. **1934** P. ALLINGHAM *Cheapjack* ix. 94, I always thought he took the dropsy. If you'd made it half-a-dollar he might have taken it. **1955** P. WILDEBLOOD *Against Law* 103 A nice bit of dropsy to a copper usually does the trick. **1969** J. GARDNER *Compl. State of Death* ii. 18 'The clients immediately think if you're a wop you'll be a push-over for dropsy.' 'Dropsy?' 'Payola...'

drosera (drǫ·sĕrä). *Bot.* [mod.L. (Linnæus *Systema Naturæ* (1735)), f. Gr. δροσερός dewy.] An insectivorous herb of the genus so called, = SUNDEW; also, the dried and powdered plant, formerly used as a remedy for respiratory diseases.

1801 SMITH & SOWERBY *Eng. Bot.* XIII. 867 Every *Drosera* hitherto discovered grows in boggy situations. [**1812** J. STOKES *Bot. Mat. Med.* II. 189 Drosera.] **1846** J. LINDLEY *Veget. Kingd.* 433 The common Droseras are rather acid. **1907** *Yesterday's Shopping* (1969) 510/3 Homoeopathic pilules or tinctures...Drosera. **1922** *Chem. Abstr.* 3730 *Drosera* is used in medicine principally as a sedative in the treatment of whooping cough and bronchitis. **1928** MARTINDALE & WESTCOTT *Extra Pharmacop.* (ed. 19) 854 Drosera *not* found specific. **1962** *Amateur Gardening* 3 Mar. 8/2 Droseras do not require much warmth, but plenty of moist fresh air.

drosophila (drǒsǫ·filă). *Ent.* [mod.L. (C. F. Fallén *Geomyzides Sveciæ* (1823) 4), f. Gr. δρόσος dew + φίλος loving.] A fruit-fly of the genus so called, much used as an experimental subject in the study of genetics.

1829 J. CURTIS *Guide Arrangement Brit. Insects* 235 Drosophila. **1877** *Encycl. Brit.* VII. 256/2 The destruction of..various culinary plants by *Psila*,..*Drosophila*, &c. **1910** T. H. MORGAN in *Science* 22 July 120/1 (*title*) Sex limited inheritance in Drosophila. *Ibid.*, In a pedigree culture of *Drosophila* which had been running for nearly a year through a considerable number of generations, a male appeared with white eyes. The normal flies have brilliant red eyes. **1932** J. S. HUXLEY *Probl. Rel. Growth* ii. 63 The increase in total size of Drosophila caused by low temperature is accompanied by a decrease in wing-size. **1938** *Ann. Reg. 1937* 346 A new technique of artificial fertilisation of Drosophila opened up fresh vistas in genetic research. **1962** G. CREMER-BARTELS in A. Pirie *Lens Metabolism Rel. Cataract* 443 Pteridines..from the red eyes of a drosophila fly. **1969** *Times* 28 Mar. 6/8 Experiments with the fruit fly *Drosophila*.

drosophyllum (drōᵘsǒfi·lǒm). *Bot.* [mod.L. (H. F. Link 1806, in *Neues Jrnl. f. d. Botanik* I. II. 51), f. Gr. δρόσος dew + φύλλον leaf.] An insectivorous plant of the genus of plants so named, belonging to the family Droseraceæ and consisting of a single species, *Drosophyllum lusitanicum*, found in southern Spain, Portugal, and north Africa.

1875 C. DARWIN *Insectivorous Plants* xv. 335 The glands of Drosophyllum, without being stimulated, continually secrete, so as to replace the loss by evaporation. **1894** F. W. OLIVER tr. *Kerner's Nat. Hist. Plants* I. 154 It is not surprising to find *Drosophyllum* covered at the same time with remains of besmeared dead bodies. **1901** *Daily News* 5 Jan. 9/1 The lecturer..told how the drosophyllum went in for a sticky exudation that acted not as a deterrent, but as an actual death-trap. **1909** *Daily Chron.* 18 Feb. 7/5 Portuguese farmers use the fly-catching drosophyllum instead of mechanical flytraps to clear their houses of insects. **1942** F. E. LLOYD *Carnivorous Plants* viii. 102 *Drosophyllum* exercises its own proper power of digestion.

drot, *U.S.* var. DRAT *int.* and *v.*

1834 W. G. SIMMS *Guy Rivers* II. 100 Drot the man..who hasn't the courage to get in a passion. **1846** J. J. HOOPER *Adv. Simon Suggs* ii. 20 Drot it! what do boys have daddies for, any how? **1884** 'MARK TWAIN' *Huck. Finn* xix. 184 Drot your pore broken heart.

‖**dróttkvætt** (drōᵘ·ʰt,kvæit). Also 8 **droquæt**, 9 **drottkvædi, -kvæði.** [ON. *drótt-kvæði*, f. *drótt* king's household + *kvæði* poem.] Court metre, a complex verse-form employed by the scaldic poets of early Scandinavia.

1779 J. STRUTT *Chron. Engl.* II. III. 253 The droquæt..was most generally used, each verse of which consisted of six syllables. **1843** G. W. DASENT tr. *Rask's Gram. Icelandic* v. 218 Dróttqvæði has three trochees, or properly spondees in each line; but dactyls may also be used in the two first places. **1870** G. BAYLDON *Elem. Gram. Old Norse* v. 115 The metre usual in laudatory poems is called *Drótt-kvæði*..or 'heroic verse'. **1881** *Encycl. Brit.* XII. 623/1 The change in the phonesis of the language is well illustrated by the new metres as compared with the old Icelandic *Drott-kvædi* in its varied forms. **1911** *Ibid.* XXVII. 1046/1 In Icelandic poetry there was a highly artificial verse-system known as court-verse (*dróttkvaett*), which consisted of alliterative groups of two lines each, arranged in staves of eight lines. **1927** E. V. GORDON *Introd. Old Norse* 295 The favourite metre of the skalds was *dróttkvætt* (also called *dróttkvæðr háttr*). **1951** G. TURVILLE-PETRE *Heroic Age of Scandinavia* xv. 165 The most popular of the scaldic verse-forms was called the Court Metre (Dróttkvætt)...The line of the Court Metre contained three stresses, and these were generally distributed between six syllables.

drought. Add: **2.** *absolute drought, partial drought* (see quot. 1963).

1881 G. J. SYMONS in *British Rainfall 1880* 112, (1) 'Absolute Droughts', or all periods of 14 or more consecutive days absolutely without rain; and (2) 'Partial Droughts', or all periods of 28 days or upwards in which the total fall was less than a quarter of an inch. **1899** *Daily News* 12 June 7/2 With all the dry weather we had last year there was not one case of an absolute drought in London. **1963** *Meteorol. Gloss* (Met. Office) (ed. 4) 83 An 'absolute drought' is a period of at least 15 consecutive days, to none of which is credited 0·01 in., or 0·2 mm, or more of rainfall. A 'partial drought' is a period of at least 29 consecutive days, the mean daily rainfall of which does not exceed 0·01 in., or 0·2 mm.

5. *drought-proof, -resistant, -resisting, -stricken* adjs.

1936 I. L. IDRIESS *Cattle King* xxxvi. 314 He hoped in time with bores to make it drought-proof. **1952** *New Biol.* XIII. 41 Drought-resistant tetraploid types. **1916** *Nature* 15 June 333/2 Special drought-resisting wheats. **1927** W. G. KENDREW *Climates of Continents* (ed. 2) 54 Drought-resisting bushes. **1881** W. D. HAY *300 Years Hence* ii. 34 Drought-stricken Indian districts. **1911** E. M. CLOWES *On Wallaby* ii. 28 Especially in drought-stricken districts.

drown, *v.* **1.** Delete (Now *unusual*) and add further examples.

1924 A. D. SEDGWICK *Little French Girl* I. viii. 70 She had the sensation of drowning yet of keeping calm while she drowned. **1928** *Manch. Guardian Weekly* 31 Aug. 152/1 Public opinion is like the crowd that watches a man drown while convinced that something ought to be done. **1970** *Which?* June 171/1 Become exhausted and drown without lifejacket.

fig. **1902** A. SYMONS *Poems* I. 119, I sicken with a wild desire, I drown in sweetness. **1957** S. SMITH *Not waving but Drowning* 13, I was much too far out all my life. And not waving but drowning.

drowned, *ppl. a.* Add: **2.** *spec.* in *Physical Geogr.*: designating a valley or other landform that is partly or wholly under water as a result of a (permanent) change in the relative levels of land and sea (or lake).

[**1889** W. M. DAVIS in *Nat. Geogr. Mag.* I. 213 The antithesis of this is the effect of depression, by which the lower course may be drowned, flooded or fjorded.] **1902** LD. AVEBURY *Scenery of England* iv. 125 When the land is sinking..the drowned river-valleys make the coast irregular and complicated. **1902** *Geol. Atlas U.S.* LXXXIII. 17/2 Both types belong to the class known among physiographers as drowned shores, a designation which signifies that they are more deeply submerged now than they were at some shortly preceding epoch. **1908** *Ibid.* CLVIII. 1/3 A subsidence of the coast has transformed the lower portions of the old river valleys into deep marine channels...A shore line exhibiting these characteristics is termed a 'drowned coast'. **1937** WOOLDRIDGE & MORGAN *Physical Basis Geogr.* xxi. 354 The drowned river valleys or *rias* of South-western Ireland. **1957** G. E. HUTCHINSON *Treat. Limnol.* I. i. 11 The northern part of Lake Victoria, with its numerous drowned valleys. **1963** D. W. & E. E. HUMPHRIES tr. *Termier's Erosion & Sedimentation* v. 126 Superficially similar phenomena, sometimes classed as fjords (but simply 'drowned valleys'), occur in regions which have never been glaciated.

drowse, *v.* Add: **2.** Also with *away, off.*

1886 W. W. STORY *Fiammetta* ii. 39 He..now and then drowsed away into a half sleep. **1908** *Smart Set* Sept. 101/2, I must have drowsed off.

drubbing, *vbl. sb.* Add: Also *transf.* and *fig.*

1884 G. MEREDITH *Let.* 31 Dec. (1970) II. 755 He got well licked [at football]. A swim in the Baths afterward braced him, for victory or another drubbing. **1955** *Times* 24 May 11/3 The Communists, who are still licking their wounds after the drubbing they got in 1950. **1959** *Spectator* 21 Aug. 215/1, I shall be surprised, though, if the Establishment does not take another drubbing in the City over Harrods.

drug, *sb.*[1] Add: **1. b.** *spec.* Now often applied without qualification to narcotics, opiates, hallucinogens, etc., esp. *attrib.* and *Comb.*, as *drug-abuse, -addict, -addiction, -dependence, -evil, -fiend* (*FIEND 4 c), *-habit, -peddler, -peddling, -pusher, -pushing, -taker, -taking, -traffic.*

1883 W. BLACK *Yolande* II. xiv. 255 One of the results of using..those poisonous drugs, is that the will entirely goes. **1899** *Chemist & Druggist* LV. 1010/2 Defendant entered the house to be cured of the habit of taking drugs, and it was alleged that he had absented himself without leave, and obtained cocaine for injection. **1902** *Daily Chron.* 7 Nov. 5/6 Two remedies to the drug-evil were suggested by the Bishop of Kensington. **1905** *Ibid.* 2 May 5/7 The drug-taking Chilcote. **1906** R. BROOKE *Let.* 3 Apr. (1968) 48, I have to read stealthily.., a practice akin to that of secret drug-taking. **1907** *Daily Chron.* 3 Sept. 7/7 A drug-taker appropriated a bottle of drugs from a Brighton chemist's shop. **1907** *Chemist & Druggist* LXX. 107/2 The drug habit. Mrs. Florence Iggulden..died last week from morphine-poisoning. **1916** *Ibid.* LXXXVIII. 19/1 Narcotic drug traffic. **1916** *N.Y. Evening Post* 7 Jan., The Drug Addict, the Physician, and the Law. **1917** *Amer. Review of Reviews* LVI. 435 (*title*) Drug Addiction and the Harrison Law. **1920** [see *ADDICT *sb.*]. **1922** C. E. MONTAGUE *Disenchantment* xv. 199 Drunkards, thieves, liars, sorners, drug-takers. **1925** H. G. WELLS *Christina Alberta's Father* I. vi. 156 Artists' models and drug-fiends. **1930** 'E. QUEEN' *French Powder Mystery* xxxviii. 313 A gang of drug-peddlers operating..in this city. **1945** R. KNOX *God & Atom* ix. 127 If religion, as Lenin said, is the opium of the people, he himself has done his best to make drug-fiends of us all. **1949** E. PARTRIDGE in *Good Housekeeping* June 13/1 Some years ago, the League of Nations instituted an inquiry into the drug-traffic. **1959** 'F. NEWTON' *Jazz Scene* 293 Entertainment, petty crime, prostitution, drug-pushing and the like. **1966** *New Scientist* 29 Dec. 713/1 The idea of an evil shadowy figure corrupting our youth by 'pushing' drugs is largely nonsense. *Ibid.*, It is generally accepted that the addict is a psychopath before taking up the drug. **1967** E. & M. A. RADFORD *No Reason for Murder* xv. 105 How are we concerned? We've no case of drug peddling. **1967** *Observer* 10 Sept. 17/7 Areas of..drug-pushers, prostitution and delinquency. **1968** *Listener* 14 Nov. 639/1 Some drug dependences arise from medical causes, as with a diabetic who must have his insulin. **1970** *Times* 28 May 7/5 Pot-smoking is widespread in spite of dire warnings about the dangers of 'drug abuse' repeatedly broadcast by the armed forces radio. **1971** *Time* 7 June 54/3 The dream that drugs are a short cut to truth and beauty.

2. *a drug in the market* (earlier and later examples); now more usually *a drug on the market.* Also *transf.*

1833 *Knickerbocker* Mar. 157 Lace veils are a drug in the market. **1893** *Funk's Stand. Dict.* s.v., A drug on (or in) the market. **1921** GALSWORTHY *To Let* I. i. 10 Well, they wouldn't confiscate his pictures, for they wouldn't know their worth. But what would they be worth, if these maniacs once began to milk capital? A drug on the market. **1922** JOYCE *Ulysses* 192 Genius would be a drug in the market. **1944** W. S. MAUGHAM *Razor's Edge* iv. 125 He can't bear his feeling of being a drug on the market.

3. *drug-counter, -jar, -pot, -vase*; **drug clerk** *U.S.*, an attendant in a drug-store; **drug-fast** *a.* [*FAST *a.* 1 h] = *drug-resistant adj.; so *drug-fastness*; **drug-resistant** *a.*, resisting the effects of a drug or drugs; so *drug-resistance, -resisting* adj.

1849 *Whig Almanac 1850* 25/1 A drug clerk at $1,000. *a* **1910** 'O. HENRY' *Rolling Stones* (1916) 102 The drug clerk looks sharply at the white face half concealed by the high-turned overcoat collar. **1959** *Listener* 30 July 173/1 Flaubert's drug-counters and the inexhaustible inventories of Zola. **1926** *Jrnl. Infectious Dis.* XXXIX. 243 The fact that bacteria become tolerant or drug fast to various germicides in vitro immediately leads to the question of specificity. *Ibid.*, Thirty drug fast strains were tested. *Ibid.* 237 (*title*) Drug-fastness in its relation to the resistance of certain organisms toward familiar germicides. **1945** *Trans. Faraday Soc.* XLI. 363 In the case of phenol itself and of several substituted phenols, little or no adaptation or drug-fastness is in fact apparent. **1931** G. E. HOWARD (*title*) Early English drug jars. **1960** H. HAYWARD *Antique Coll.* 102/1 *Drug jar*, pot or jar intended for use on apothecaries shelves. **1961** *Antiquaries Jrnl.* XLI. 9 (*caption*) Drug-jar with inscription from Winchester. **1903** R. L. HOBSON *Catal. Eng. Pott. in Brit. Mus.* 137 English Delft Ware..Drug Pot, barrel-shaped. **1910** J. F. BLACKER *ABC of collecting Old Eng. Pott.* x. 78 Zachariah Barnes..was noted as the maker of wall-tiles and druggists' jars, or drug-pots. **1932** *Brit. Med. Jrnl.* II. 668/1 (*title*) Drug resistance, with special reference to trypanosomiasis. *Ibid.*, The practical significance of the capacity of trypanosomal infections to become drug-resistant, or drug-fast, for the treatment of

the infections in man..was at once recognized. **1961** *Lancet* 5 Aug. 309/2 Few new drug-resistant cases will appear. *Ibid.* 310/1 Drug resistance is already a significant problem in many areas. **1951** KOESTLER *Age of Longing* 242 Developing new and better drug-resisting strains. **1933** *Burlington Mag.* Mar. 108/1 South Kensington has recently acquired a drug vase.

drug, *v.*[2] Add: **3.** *intr.* To take or be in the habit of taking drugs; *esp.* to indulge in narcotics.

1893 *Funk's Stand. Dict.* s.v., She has drugged all her life. **1968** [see *DRUGGER 3].

drugger. Add: **3.** One who takes narcotic, etc., drugs; a drug-addict. *colloq.*

1941 J. CARY *Herself Surprised* lxxxi. 204 When first I knew Lizzie I thought she was a drugger or a drinker or worse. **1968** H. R. F. KEATING *Inspector Ghote hunts Peacock* vii. 98 Your precious Peacock..was nothing but a low-down little drugger. I may smoke because I need it for my work, but she just drugged to make herself lower than she was.

dru·g-store. orig. *U.S.* and chiefly *N. Amer.* Also **drugstore, drug store.** [f. DRUG *sb.*[1] + STORE *sb.* 12 a.] A pharmacy or chemist's shop, often also dealing extensively or mainly in other articles, as toilet requisites, stationery, magazines and newspapers, light refreshments, etc. Also *attrib.*, esp. as **drug-store cowboy,** a braggart, loafer, or good-for-nothing; a person who is not a cowboy but is dressed like one.

1810 *Washington* (D.C.) *Chron.* 17 Nov. 2/2 (Advt.), Cash Drug Store. **1819** H. MCMURTRIE *Sketches of Louisville* 137 There are..in Louisville..three drug stores. **1859** *Brit. Colonist* (Victoria, Brit. Columbia) 12 Dec. 2/5 (Advt.), Drugstore Government street, between Yates and Johnson. **1871** *Cincinnati Commercial* 30 Aug. 3/2 A gentleman..stopped at a corner drug store and asked for a glass of soda water. **1883** SWEET & KNOX *Through Texas* xv. 230, I felt a sort of drug-store taste in my mouth. **1889** [see HYPO *sb.*[1]]. **1903** *N.Y. Even. Post* 24 Sept. 8 It hardly pays to keep the [soda] fountains going in the drugstores. **1908** *Westm. Gaz.* 18 Aug. 3/1, I took them to a chemist's or I beg pardon: a drug-store. **1925** *College Humor* Feb. 57/1 (*heading*) With the Drug Store Cowboys. **1931** W. A. MCADOO *Crowded Years* iii. 34, I gave him fifty cents to buy some medicine for me at a drugstore. **1932** E. WAUGH *Black Mischief* iv. 150 Bunting strung..across the Boulevard Amurath, from Levantine café to Hindu drug store. **1935** C. W. THURLOW CRAIG *Paraguayan Interlude* xxi. 247, I have seen men carry guns this way; they were generally new-comers and quickly became objects of derision. 'Drug-store cowboys', they are called. **1936** M. DE LA ROCHE *Whiteoak Harvest* xix. 219 Almost every day they telephoned to the drug store for a pint of delicious ice-cream. **1957** P. FRANK *Seven Days to Never* iii. 102 She married..a marijuana-smoking drugstore cowboy. **1965** A. NICOL *Truly Married Woman* i. 2 He got down the new bottle of patent medicine which one of his friends who worked in a drug store had recommended to him. **1968** *Globe & Mail* (Toronto) 17 Feb. 51 (Advt.), Pharmacist manager for busy downtown drug store. **1969** *Daily Tel.* (Colour Suppl.) 24 Jan. 13/4 There is provision for a small drugstore selling coffee, books and records.

drum, *sb.*[1] Add: **1. b.** (Further examples.) See also *side-drum* s.v. SIDE *sb.*[1] 27, and sense 1 d below.

1804, etc. [see BASS *a.* 3 b in Dict. and Suppl.]. **1879** GROVE *Dict. Mus.* I. 466/2 The Tenor-drum is similar to the side-drum, only larger, and has no snares. **1888** tr. *Riemann's Catechism Mus. Instr.* v. 100 The tenor-drum used for rolls (tamburo rullante) has likewise no snares (strings of catgut) and sounds therefore dull and gloomy (though much higher than the big-drum). **1889** *Brit. Bandsman* Sept. 280/1 The big drum..was strapped to a tent pole. **1894** KIPLING *Seven Seas* (1896) 200 Oh, 'ark to the big drum callin'. **1923** J. M. MURRY *Pencillings* 239 One of those multi-musical Italian wanderers, with a big drum on their shoulders. **1940** G. JACOB *Orchestral Technique* (ed. 2) vii. 72 Such things as the tenor drum..need not be spoken of in detail.

d. *to beat* or *thump the* (*big*) *drum*(*s*): to make loud or ostentatious advertisement, protest, or the like.

1611 [in Dict.]. **1907** [see THUMP *v.* 1 a]. **1930** *Church Times* 4 July 3 Even the Bishop of London hesitates when the Protestant drum is loudly beaten. **1930** J. A. WILLIAMSON *Short Hist. Brit. Expansion* (ed. 2) II. VI. v. 255 The Conference of 1911 met under the shadow of war, and there is no doubt that this was realized by its members. The effect was not a public beating of the drum, but the exact contrary. **1961** *Lebende Sprachen* VI. 100/1 His old woman really beat the drum [= scolded him]. **1967** *Boston Sunday Herald* 26 Mar. I. 30/3 The second big objection to the Weston proposal is that it involves a flagrant violation of the home rule principle for which the reform element has beaten the drums so loudly in recent years.

g. *Austral. slang.* A warning, a piece of information, esp. a racing tip; freq. *the drum*, the facts; true or reliable information.

1941 BAKER *Dict. Austral. Slang* 25 *Drum*,..a racecourse tip. e.g., 'get (give) the drum' about a horse. **1944** L. GLASSOP *We were Rats* xvi. 88 I've got the drum from a friend in the Seventh Div. Headquarters in Melbourne. We're going to Bombay. *Ibid.* xxxiv. 193 I'm givin' ya the drum now. **1960** S. H. COURTIER *Gently dust Corpse* viii. 116 Ready to give me the drum about Cullerman yet? **1961** H. R. F. KEATING *Rush on Ultimate* i. 13, I know now all right..but I didn't a week ago. Not until Humphrey gave me the drum. **1968** D. O'GRADY *Bottle of Sandwiches* 7 Gave us the drum on where to get hold of the particular rifles we had our eyes on.

6. g. The cartridge-holding receptacle of a machine-gun; also, the contents of one of these.

1888 [in Dict.]. **1890** W. J. GORDON *Foundry* 29 To the breech is fixed a drum with 104 bullets. **1916** 'BOYD CABLE' *Action Front* 198 Can you fill the cartridges into these drums while I shoot? **1928** *Daily Tel.* 24 Apr. 12/6, I gave him a drum and he went down underneath me.

†**9. d.** A street. *Obs. slang.*

c **1789** ?G. PARKER *Sandman's Wedding* in J. S. Farmer *Musa Pedestris* (1896) 64 Just as he turned the corner of the drum, His dear lov'd Bess, the bunter, chanc'd to come. **1851** MAYHEW *London Lab.* I. 217/2 We..slink into the crib (house) in the back drum (street). **1889** BARRÈRE & LELAND *Dict. Slang* I. 332/1 *Drum* means also a street, a road...It may have come directly from the English gypsy *drum* (old form *drom*), which is, truly, from the Greek δρομός, a road.

e. *slang.* A house, lodging-place, or other building; esp. (*a*) *U.S.* a drinking-place, saloon, night-club; (*b*) a brothel, low dive; (*c*) a room or flat.

1846 *Swell's Night Guide* p. iii, The..cracksman, who would screw a drum. **1851** MAYHEW *London Lab.* I. 418/1 Suppose I want to ask a pal to..have a game at cards with some *blokes* at *home* with me, I should say..'Splodger, will you..have a touch of the *broads* with me and the other heaps of *coke* at my *drum*.' **1859** G. W. MATSELL *Vocabulum* 28 *Drum*, a drinking-place. **1867** J. GREENWOOD *Unsentimental Journeys* xxvi. 204 'Come along; I shall be a pot to your pot.' 'Where shall we go?' 'Oh! to the old drum, I suppose.' **1872** G. P. BURNHAM *Mem. U.S. Secret Service* p. v, *Drum*, a bad house, boarding-place, or small tavern. **1899** R. WHITEING *No. 5 John St.* I. ii. 15 'That's my drum two doors beyond.' His drum was better to look at. **1903** A. H. LEWIS *Boss* 32 He said the Dead Rabbit was a drum for crooks! **1908** K. MCGAFFEY *Sorrows of Show Girl* 234 We came to a door which the gee threw open and said, 'This is your stateroom.' Honest, I never saw such a drum. **1938** D. RUNYON *Furthermore* xii. 231 The bar in Good Time Charley's little drum in West Forty-ninth Street. **1966** L. SOUTHWORTH *Felon in Disguise* iii. 53 They probably checked at the Probation Office as soon as they left my drum. **1967** K. GILES *Death in Diamonds* vi. 118 'You get that way running a drum,' grunted Crook...'A drum?' 'The house that ain't a home,' said Harry.

f. = SWAG *sb.* 10; *to hump one's drum*: see HUMP *v.* 2. *Austral.* and *N.Z. slang.*

1866 W. STAMER *Recoll. Life Adventure* I. 304 Our ci-devant millionaire..'humping his drum', [would] start off for the diggings to seek more gold. **1872** C. H. EDEN *My Wife & I in Queensland* i. 17 They all chaffed us about our swags, or donkeys, or drums, as a bundle of things wrapped in a blanket is indifferently called. **1889** M. ROSS *Compl. Guide Lakes Central Otago* 44 'Time's up!' is called, the 'drum' (the local for swag) hoisted on, and the final ascent begins. **1933** *Bulletin* (Sydney) 18 Jan. 20, I sees a bloke comin' along the road from Winton with 'is drum up.

g. *slang.* A tin or can in which tea, etc., is made; so **drum-up,** a making of tea; the preparation of a meal.

1919 *Athenæum* 8 Aug. 728/1 I've some sugar. If you get some tea and hot water we'll have a drum up. **1931** F. GRAY *Tramp* ix. 104 He will bring his 'billy-can' or 'drum' to the door of a cottage or mansion and beg a little hot water to make his tea. *Ibid.*, A 'drum' is an old tin, say a half-pound coffee tin; two holes are made at the lip so that the string may be threaded to carry and control it. **1970** F. MCKENNA *Gloss. Railwaymen's Talk* 35 *Drum*, a tea-can.

10. (Further examples.) Now *Obs.*

1805 [see HURRICANE *sb.* 2 b]. **1883** L. TROUBRIDGE *Jrnl.* May in J. Hope-Nicholson *Life amongst Troubridges* (1966) 163 A dinner-party and after a 'drum'—lots of people we knew.

12. a. *drum-dance, -kit, -polka.*

1934 I. W. HUTCHISON *North to Rime-ringed Sun* ix. 93 The strange 'drum-dance'. **1936** *Discovery* June 186/2 A drum-dance is held [in one Kenya tribe] at which the possessed victim is danced to exhaustion. **1934** E. LITTLE *Mod. Rhythmic Drumming* (Advt. on back cover), 'Ajax' drum kit. **1965** G. MELLY *Owning-Up* v. 47 There were always a great many very steep steps to drag the drum kit up. **1849** *Theatrical Programme* 11 June 23 The Celebrated Drum Polka, with Solos by Mess. Kœnig, Collinet, etc. **1909** T. HARDY in *English Rev.* Apr. 4 Who now recalls those crowded rooms..Where to the deep Drum-polka's booms We hopped in boisterous style?

13. drum brake, a type of brake in which brake-shoes fixed to the vehicle are pressed against a brake-drum fixed to the wheel; **drum camera,** a camera in which the film is mounted on a drum or cylinder which rotates during exposure (recording the motion of an object as a trace on the film); **drum-drying,** an industrial drying process involving drum-shaped containers; **drum-fire** (see quot. 1918); also *transf.* and *fig.*; **drum-fish** *v. intr.* (*U.S.*), to fish for drum-fish; so *drum-fishing*; **drum winding,** an armature winding in which the conductor is wound from end to end on or in the surface of a cylindrical core; so **drum-wound** *a.* Also *DRUM-MAJORETTE.

[**1949** FRAZEE & BEDELL *Automotive Fundamentals* II. vi. 355 The similarity of the band clutch to certain types of brakes has led to its being called a 'drum brake' by some designers, although this name is not entirely accurate.] **1950** NEWTON & STEEDS *Motor Vehicle* (ed. 4) xxvii. 470 The vast majority of brakes are friction brakes and these may be sub-divided into: (1) Drum brakes and (2) Disc brakes. **1959** *Motor Manual* (ed. 36) v. 132 The internal-expanding drum brake has been practically universal on cars. **1962** *Which?* (Car Suppl.) Apr. 49/2 With drum brakes, the brake shoes are pressed outwards against a drum on the wheel hub. **1934** *Photographic Jrnl.* LXXIV. 389/1 (*heading*) High-speed drum camera. **1951** W. D. CHESTERMAN *Photogr. Study Rapid Events* vi. 80 Drum cameras are of two kinds. In the type which are more generally constructed, the film is attached to the outside or the inside surface of a rapidly rotating cylindrical drum, and the images are formed successively on the film by a variety of optical and mechanical means. In another type..the film is wound on a stationary drum, and the optical images are formed in rapid succession on the film surface by rotating optical parts. **1946** *Nature* 10 Aug. 194/1 Spray-drying, flash-drying and drum-drying have been developed with considerable success. **1917** *Times* 20 Aug. 6/4 The artillery duel increased to drumfire. **1918** E. S. FARROW *Dict. Mil. Terms*, *Drum fire*, a common name given to the artillery barrage or curtain of fire. Continuous bombardment, like the rolling of drums. **1955** *Bull. Atomic Sci.* Mar. 98/3 We expose ourselves quite unnecessarily to a drumfire of criticism. **1818** in J. R. Commons et al. *Doc. Hist. Amer. Industrial Soc.* (1910) I. 203 Sent the boat a Drum fishing and caught 5 Drum. **1855** *Knickerbocker* XLVI. 499 So highly enjoyed is drum-fishing among our bail-fishers. **1904** *Booklovers Mag.* III. 625/2 Senator Quay..was discovered..knee-deep in the surf at Atlantic City, drum-fishing. **1893** HAWKINS & WALLIS *Dynamo* vii. 117 One loop of the drum winding is the exact equivalent of the two loops of the ring winding. **1952** K. C. GRAHAM *Small Commutator Motors* ii. 32 A drum winding..is the general type of armature winding used in modern armatures. **1893** W. P. MAYCOCK *Electric Lighting* II. vii. 196 The armature is drum wound. **1904** R. M. WALMSLEY *Electricity in Service of Man* II. i. 756 A method of arranging the connections of drum-wound armatures..consists in winding and insulating the coils separately before placing them on the core.

14. *drum-and-trumpet history*: see HISTORY *sb.* 2.

drum, *v.* Add: **5.** Also, to solicit orders; to canvass. *U.S.*

1839 C. F. BRIGGS *Adv. Harry Franco* I. xiii. 90 Augustus ..had drummed in Arkansas, and collected in the lithograph cities of the west. **1882** *Congress. Rec.* 315/1 The merchants ..have many thousands..drumming for business in every town. **1901** *Chambers's Jrnl.* Dec. 827/1, I was 'drumming' for one of the two great houses which divided the wool and the hides of the Argentine.

7. In more general use, to dismiss or expel someone in disgrace. Const. *out.*

1968 A. COOKE *Talk about Amer.* xxxvi. 215 Shut your eyes and you can see Mark Twain running a newspaper only a few blocks away from Sutro till he was drummed out of town. **1970** *New Yorker* 12 Sept. 47 Confound it, Moxley! Any more of that and I'll have you drummed out of the Sierra Club!

8. b. *Austral. slang.* To inform or warn (a person); to give (someone) 'the drum' (see *DRUM *sb.*[1] 1 g). Also with *up.*

1919 V. MARSHALL *World of Living Dead* 30 He impressed upon me the exact location of the maternal abode, and proceeded to 'drum me up' with the message. **1959** BAKER *Drum* (1960) vi. 49 Look, mate, let me drum you about something. **1969** D. NILAND *Dead Men Running* iii. 86 Jesus, don't bite me, son. I was only gonna drum you.

9. b. To ring or knock on the door of (a house) to ascertain whether it is unoccupied before attempting a robbery; hence, to reconnoitre, with a view to a robbery. Also *intr.*, to steal from an unoccupied house, etc. *slang.*

1925 N. E. LUCAS *Autobiogr. of Crook* vii. 105 Crooks go 'drumming' in pairs, dressed as clerks or messengers... Should they find an office left unoccupied during the lunch hour they quickly and skilfully 'turn it over'. **1933** C. E. LEACH *On Top of Underworld* x. 138 *Drumming a place*, ringing or knocking to see if occupants are at home. **1936** J. CURTIS *Gilt Kid* ii. 23 'My God, you got the gaff weighed up good.' 'Not half. A bloke drummed it for me and put me wide.' **1962** *Observer* 25 Feb. 21/2 They were both making a steady living at drumming (housebreaking). **1967** J. WAINWRIGHT *Talent for Murder* 114 Peters was hanging around outside..the Wallaces' residence...He was..a police officer 'keeping observations' or a criminal 'drumming the joint'. **1970** 'B. MATHER' *Break in Line* vi. 77, I bet Chatterjee's been drumming every room in the joint.

13. *to drum up*: to make tea in a billy-can or the like; also, to prepare a meal under rough-and-ready conditions (out-of-doors). Cf. *DRUM *sb.*[1] 9 g. *slang.*

1923 KIPLING *Irish Guards in Great War* I. 58 The Irish 'drummed up', which is to say, stewed their tea or rations. **1931** F. GRAY *Tramp* ix. 104 Now comes the great event of the day. He is prepared and is going to 'drum up'. **1935** 'G. ORWELL' *Clergyman's Daughter* ii. 104 After getting to Bromley they had 'drummed up' on a horrible, paper-littered refuse dump. *Ibid.* 106 They..'drummed up' in thickets where firewood and water were handy, and cooked strange, squalid meals. **1962** *Listener* 13 Sept. 382/2 Only at midday was there a break, when fires were lit and water boiled..to make tea. This was known as 'drumming up'.

drumhead. 5. (Further examples.)

1899 *Westm. Gaz.* 22 Dec. 10/1 What he calls 'drum-head letters', written by soldiers at the front before and after battle. **1908** *Daily Chron.* 25 May 7/7 A drum-head service held in the camp of the Essex Imperial Yeomanry.

drumlinized (drʌ·mlinəiz'd), *ppl. a.* *Physical Geogr.* [f. DRUMLIN: see -IZE.] Formed into or covered with drumlins, or with structures shaped like them.

1907 *Bull. N.Y. State Museum* CXI. 402 Smooth ridging or ribbing making one doubtful whether to map the area as moraine or drumlinized drift. **1934** *Geogr. Jrnl.* LXXXIV. 145 The drumlinized tract on the west side. **1970** *Nature* 2 May 441/2 The bench surfaces are scoured into a drumlinized micro-topography of 1–10 cm relief.

drumlinoid (drʌ·mlinoid), *a.* and *sb.* *Physical Geogr.* [f. as prec. + -OID.] **A.** *adj.* Resembling a drumlin in shape. **B.** *sb.* A long, low hill of drift resembling a drumlin but of a less regular form; also, a mass of rock to which glacial action has given a drumlin-like appearance.

1898 *Geol. Survey Canada, Ann. Rep. 1896* IX. 182F High regular drumlinoid hills. **1901** A. P. BRIGHAM *Text-bk. Geol.* xiv. 270 The drumlinoid (having the form of a drumlin) is a rock structure superficially modified by ice movement. **1904** *Sci. Amer. Suppl.* 16 Jan. 23447 Some of the principal ridges present drumlinoid profiles. **1953** *Canad. Geographer* III. 19 In some places, individual drumlinoids may extend over a distance of several miles, in others they are very much shorter and resemble drumlins in form. **1968** R. W. FAIRBRIDGE *Encycl. Geomorphol.* 295/1 A 'drumlinoid'..is a spindle-shaped hill of till or a series of 'streaks' of till not developed into the typical drumlin shape.

dru:m-majore·tte. orig. *U.S.* [f. DRUM-MAJOR + -ETTE.] A female drum-major (DRUM-MAJOR 1 c); a girl who leads or takes part in a parade or the like, twirling a baton, etc.

1938 *Life* 10 Oct. 3/1 (*heading*) Drum majorettes are latest in ballyhoo. **1940** R. CHANDLER *Farewell, My Lovely* xxi. 162 The girls wore white silk blouses and drum majorettes' shakos. **1955** J. B. PRIESTLEY in Priestley & Hawkes *Journey down Rainbow* v. 84 The purple band..marched on to the field, accompanied by drum-majorettes. **1957** *New Yorker* 5 Oct. 36/1 There was the Drum and Bugle Corps of St. Rocco's Church in Newark, led by a drum majorette in blinding scarlet, gold, and blue, her knees and sabre flashing. **1971** *Graphic* (Durban) 7 May 1/2 There will be a display by drum majorettes. **1971** *Glasgow Herald* 22 June 11/4 (*caption*) The Swedish drum majorettes marching in George Square, Glasgow.

drummer. Add: **2.** For *U.S.* read orig. *U.S.* and add further examples.

1915 LD. REDESDALE *Memories* I. xiii. 287 The boarding house chiefly used by 'drummers'—travellers of English commercial houses. **1941** BAKER *Dict. Austral. Slang* 26 *Drummer*,..a commercial traveller.

b. A thief (see quots.). *slang.*

1856 MAYHEW *Gt. World of London* I. 46 Those who hocus or plunder persons by *stupefying*; as '*drummers*', who drug liquor. **1859** HOTTEN *Dict. Slang* 34 *Drummer*, a robber who first makes his victims insensible by drugs or violence, and then plunders them. **1960** *Observer* 25 Dec. 7/6 Nobody wanted to know the drummers, those squalid daytime operators who turn over empty semi-detached villas while the housewives are out shopping. **1962** 'J. BELL' *Crime in our Time* III. 51 They knock at the doors of houses to discover if the owners are at home. The police call them 'drummers', because they drum on the doors in this way.

c. *Austral.* and *N.Z. slang.* (See quots.)

1898 *Bulletin* (Sydney) 17 Dec., *Drummer*, the laziest and therefore the slowest shearer in a shed. **1941** BAKER *N.Z. Slang* v. 39 Just as *ringer* is used for the most expert shearer, *drummer* is used for the worst. **1949** P. NEWTON *High Country Days* v. 55 Hewett, the learner, and 'drummer' of the gang, with seventeen to his credit had also shown an improvement. **1959** H. P. TRITTON *Time means Tucker* v. 42/2 It's not every man that is drummer in four sheds running.

d. *Austral.* and *N.Z. slang.* A swagman or tramp. Cf. *DRUM *sb.*[1] 9 f. ? *Obs.*

1933 L. G. D. ACLAND in *Press* (Christchurch) 14 Oct. 15/7 *Drum*, swag, obviously from the shape. Hence *drummer*. **1945** BAKER *Austral. Lang.* v. 102 *Bender* (1885) and *drummer* (*circa* 1890) were once popular terms for tramps of slightly better class than the sundowner.

drumstick. Add: **2. e.** *Cytology.* An appendage of the nucleus of a polymorphonuclear leucocyte, composed of sex chromatin and characteristic of females.

1954 DAVIDSON & SMITH in *Brit. Med. Jrnl.* 3 July 6 A solitary chromatin nodule..becomes separated off from the main nuclear lobes in a proportion of the neutrophils... This characteristic 'drumstick', which is just visible at × 90 magnification, has to be distinguished from other appendages which occur in both sexes. **1964** L. MARTIN *Clinical Endocrinol.* (ed. 4) vi. 204 The drumstick consists of an ovoid chromatin mass of 1·5 μ diameter connected to the nucleus by a thin chromatin thread. **1966** W. A. DAVIDSON in K. L. Moore *Sex Chromatin* iii. 69 The drumsticks which have been identified in animals appear to be similar in form and size to those in man.

drunk, *ppl. a.* and *sb.*[2] Add: **1. a.** *drunk and disorderly*: the official form of a charge in police-court procedure (cf. DISORDERLY *a.* 2 b); so quasi-*sb.*, a drunk and disorderly person; the offence of being drunk and disorderly; *drunk and incapable*: see INCAPABLE *a.* 5.

1855 [see *DISORDERLY B. *sb.*]. **1874** GREENWOOD & MARTIN *Magisterial & Police Guide* 610 *margin*, Offences. Drunk and disorderly persons. **1893** T. MARRIOTT *Constable's Duty* 76 Offence Defined...Every person who..is found to be drunk and disorderly. **1922** JOYCE *Ulysses* 602 Fined ten bob for a drunk and disorderly.

B. *sb.* **1.** (Earlier and later examples.) Also, a state of drunkenness.

1779 W. SMITH *Let.* 8 June in *15th Rep. Hist. MSS. Comm.* (1897) App. VI. 430 They [*sc.* American soldiers] call a month's pay, which is 53s. 4d. paper money, but *three drunks*. Rum, 30 dollars a gallon. **1839** C. F. BRIGGS *Adv. Harry Franco* II. vii. 78, I have kept money enough to have a good drunk. **1846** *Spirit of Times* 18 Apr. 92/2 In an hour after he put down two gallons more to get up the drunk. *a* **1909** G. E. EVANS *Coll. Verse* (1928) 266 Cows in various states of drunk were scattered all around. **1952** M. TRIPP *Faith is Windsock* iii. 55 'I went on a seven-day drunk.' 'Like muck you did.' **1966** H. NIELSEN *After Midnight* (1967) xi. 135 She was sleeping off a drunk in the bedroom.

2. (Earlier example.)

1852 J. W. CARLYLE *Let.* 18 Oct. (1949) xvi. 236 When I got up at my usual hour (*six o' clock*), I reeled about like 'a drunk' (as Mazzini would say).

drunkometer (drʌŋkǫ·mitəɹ). *U.S.* [f. DRUNK *ppl. a.* + -OMETER.] A device for measuring the alcoholic content of a person's breath.

1934 *Jrnl. Amer. Med. Assoc.* 12 May 1597/1 Exhibit of automatic demonstration of the use of the 'drunkometer', a device for detecting drunkenness by testing the subject's breath. **1955** *Times* 20 July 6/1 Lord Mancroft said that three machines, the drunkometer, the alcohmeter, and the intoximeter had been developed in the United States. **1960** *New Statesman* 23 Jan. 95/1 Since the drunkometer has been used in New York, convictions for drunken driving have increased from 30 to 85 percent.

druther (drʌ·ðəɹ). U.S. dialectal alteration of (*I, you,* etc.) *would rather.* Hence **dru·ther(s), ru·ther(s),** a choice, preference.

1876 [see *DERN *a.*]. **1895** *Dialect Notes* I. 388 Bein's I caint have my druthers an' set still, I cal'late I'd better pearten up an' go 'long. **1896** 'MARK TWAIN' *Tom Sawyer, Detective* ix. 74 'Any way you druther have it, that is the way *I* druther have it. He ——'. 'There ain't any druthers *about* it, Huck Finn; nobody said anything about druthers.' **1941** W. A. PERCY *Lanterns on Levee* (1948) xxii. 292 'Your ruthers is my ruthers' (what you would rather is what I would rather). Certainly the most amiable and appeasing phrase in any language, the language used being not English but deep Southern.

dry, *a.* Add: **1. c.** (Earlier and later examples.)

c **1893** [see WET *a.* 2 a]. **1932** J. S. HUXLEY *Prob. Rel. Growth* ii. 71 The wet-season and dry-season forms of certain tropical butterflies. **1964** C. WILLOCK *Enormous Zoo* vii. 113 A Rhodesian dry season track which was rarely used by vehicles.

11. a. For in *U.S. political slang* read orig. in *U.S. political slang*, and add further examples. Also, of a person who favours prohibition; hence quasi-*advb.* in phrases *to go* or *vote dry.*

1870 *Scribner's Monthly* I. 63 Dry or wet, Mr. Dort? Indifferent, eh? Adolph, a hock-glass. **1887** *Courier-Journal* 7 Feb. 1/7 Athens, in which the State university is located,..is a dry town. **1888** *Detroit Evening Jrnl.* 20 Feb. (Farmer), If a county has voted on local option, and has gone dry. **1904** *N.Y. Even. Post* 3 Oct. 6 If every town and city in Vermont should vote 'dry' at the next election. **1908** [see WET *a.* 16 c]. **1916** *Literary Digest* (N.Y.) 1 Jan. 4/2 About as much 'dry' territory 'going wet' as there was of 'wet' territory 'going dry'. **1944** W. R. SCOTT *Revolt on Mount Sinai* xxii. 179 Many members who long had voted dry accepted the election result as a mandate from the people. **1971** *Scotsman* 20 May 20/8 If the people of Kirkintilloch could be consulted on the issue of whether they should remain 'dry' or 'wet' it was difficult to see whether they should not also be consulted on the question of whether they should enter the Common Market.

18. b. Of acoustics: lacking in warmth or resonance.

1961 *Listener* 2 Nov. 715/2 When an orchestra broadcasts in a television studio, where the acoustic is likelier to be drier. **1962** A. NISBETT *Technique Sound Studio* ii. 53 Some modern music is now written for drier acoustics than those which sound best for Beethoven.

C. 2. a. *dry-feed, -wash* vbs., *dry-washing*; **dry-clean** *v. trans.*, to clean (clothes and other textiles) without using water; also *intr.* and *transf.*; hence *dry-cleanable* adj., *dry-cleaned* ppl. adj., *dry-cleaner, dry-cleaning* vbl. sb. and ppl. adj.; **dry-dyeing** (see quot. 1904).

1817 W. TUCKER *Family Dyer & Scourer* i. 20 For dry cleaning Clothes of any Colour. **1897** *Chambers's Jrnl.* 25 Sept. 620/2 A firm of so-called 'dry-cleaners' of wearing apparel. *Ibid.*, In this dry-cleaning process the goods were immersed in a vessel of benzine. **1899** *Westm. Gaz.* 9 Mar. 3/1 A dry cleaning cloth ball. *Ibid.* 20 Apr. 3/3 The present extraordinary perfection of dry-cleaning. *Ibid.*, A good gown, dry-cleaned, is a much better thing than an inferior new one. Of course I do not pretend that all dry-cleaners are equally good. *Ibid.* 12 Oct. 3/2 How well they may dry-clean at home by the use of benzoline. **1908** *Daily Report* 27 Aug. 6/4 The conditions under which women and girls work in dry-cleaning establishments, where benzine is largely used. **1930** *Engineering* 12 Dec. 760/1 The advantages which dry cleaning [of coal] had to offer, as compared with wetwashing, were essentially financial ones...Had it been decided to wash the coal in the ordinary way instead of dry cleaning it, an additional 13 ovens would have been necessary. **1957** *New Yorker* 16 Nov. 150/2 Dry-cleanable colors include black and deep gold. **1958** Dry-cleaned [see *AUTOMAT 3]. **1970** 'D. HALLIDAY' *Dolly & Cookie Bird* viii. 111 Blood was spreading over..his cream jersey suit. It would never dry-clean. **1970** *Encycl. Brit.* VII. 709/2 Dry cleaning is the process of cleaning fabrics with liquids other than water. *Ibid.*, The dry-cleaned garment maintains its original shape and feel. For many years the dry-cleaning process was based on the use of highly flammable solvents. **1904** *Westm. Gaz.* 8 Sept. 4/2 Dry-dyeing is simply dyeing with aniline dyes soluble in spirit. **1907** *Daily Chron.* 8 July 4/4 New systems of dry-feeding young and adult stock. **1901** KIPLING *Kim* viii. 207 The halts for prayers (Mahbub was very religious in dry-washings and bellowings when time did not press). **1962** K. ORVIS *Damned & Destroyed* xiv. 93 He began worriedly dry-washing his hands.

3. dry bath *slang*, a search of a prisoner when he has been stripped naked; **dry battery** *Electr.*, a battery of dry cells; **dry-blower** *Austral.*, (*a*) a gold-miner; (*b*) used as a term of opprobrium; (*c*) (see quot. 1964); **dry-blowing** *Austral.* (see quot. 1894); **dry-bob** (see also BOB *sb.*[3] 1 and 2); **dry brush** (see quots.); freq. *attrib.*; **dry camp** *U.S.*, a camp or halt where there is no water; **dry cell** *Electr.*, a voltaic cell in which the electrolyte is contained in an absorbent material or is in the form of a paste, thus preventing spilling of the contents; **dry diggings,** (*a*) (orig. *U.S.*) gold-diggings away from a river or stream; (*b*) in South Africa, diamond-diggings at which the diamondiferous material is disintegrated by exposure to the atmosphere; **dry-dike** = *dry-stone dike* (see DIKE *sb.* 6 b); so *dry-diked* adj., *dry-diker*; **dry end,** that end of a paper-making or drying-machine from which the material emerges dry; **dry farming** chiefly *N. Amer.*, farming without a good supply of water; so *dry farm* sb. and vb., *dry farmer*; **dry-fly** *a.* and *v.*, substitute for def.: used to describe a method of fishing in which an artificial fly floats lightly on the water; (add earlier and later examples); **dry fly,** an artificial fly used in this type of fishing; **dry hopping** (see quot. 1956); **dry ice** orig. *U.S.*, solid carbon dioxide; **dry joint** *Electr.*, a soldered joint with faulty electrical continuity; **dry mounting,** a method of mounting photographs (see quot. 1958); **dry offset** (see quots.); **dry-plate clutch** *Mech.*, a plate clutch which operates without lubrication; **dry-point settlement, village,** one which is not liable to flooding; **dry run,** (*a*) *U.S.* a dry creek or arroyo; (*b*) *colloq.* (orig. *U.S.*), a rehearsal, test, 'dummy run'; hence as *v. trans.*, to rehearse, practise; **dry shampoo** (see quot. 1966); **dry shaver,** an electric or other razor for use without soap and water; **dry-ski** *a.*, designating a school, etc., for indoor training in ski-ing; **dry skid,** a skid of a motor vehicle on a dry surface; so **dry-skid** *v. intr.*; **dry spell,** a period of dry weather (see quot. 1920); **dry spinning,** a method of spinning natural or artificial fibres (see quots. 1904 and 1957); hence *dry-spin* vb. trans., *dry-spun* adj.; **dry-stone** *a.* (earlier example); **dry suit,** a type of diving suit, usu. made of sheet rubber, which uses the principle of air-insulation to protect the diver from cold, and under which warm clothing can be worn; **dry valley,** a valley in which the original stream or river has disappeared; **dry wall** (earlier example); hence **dry-wall** *v. trans.* and *intr.*, to build a dry wall (around); *dry-waller, dry-walling*; **dry wash** *N. Amer.*, the dry bed of an intermittent stream.

1933 G. INGRAM *Stir* v. 93 The warder..said he'd give him a dry bath just to see. **1965** *New Statesman* 30 July 452/3 Two or three times a week the Heavy Mob rushed into our cells and gave us a 'dry bath', which adequately describes the search of a man who is standing 'starkers' in the middle of his cell. **1885** *Electrician* 10 Jan. 174/1 (*heading*) Conversion of liquid into dry batteries. **1895** *Queenslander* 7 Dec. 1069 Every other man you meet in Coolgardie..is either a lord, a colonel, a captain, a doctor, an expert, an agent, a sharebroker, or a sharper; all the rest are dudes, drunkards, and dryblowers, professional liars, and loafers. **1935** *Bulletin* (Sydney) 20 Mar. 10/4 In the early days of the Westralian goldfields it gained me many friends amongst the isolated dryblowers to whom I passed on my weekly copy. **1936** F. CLUNE *Roaming round Darling* xix. 196 There we met Dryblower Tom Mitchell, awaiting the return of his camel-team, to go prospecting. **1964** R. WARD *Penguin Bk. Austral. Ballads* 208 A 'dry-blower' is a crude device, made of hessian and wooden saplings, and used in arid areas for separating gold from the ore. **1894** *Argus* 28 Mar. 5/5 (Morris), When water is not available, as unfortunately is the case at Coolgardie, 'dry blowing' is resorted to. This is done by placing the pounded stuff [*sc.* alluvial ore] in one dish, and pouring it slowly at a certain height into the other. If there is any wind blowing it will carry away the powdered stuff; if there is no wind the breath will have to be used. **1911** H. P. BOWIE *Laws Jap. Painting* iv. 66 Dry twig or old firewood line..is generally used in the robes of old men and produced by what is called the dry brush; that is, a brush with very little water mixed with the *sumi*. **1958** M. L. WOLF *Dict. Painting* 88 *Dry brush*, in Chinese art, a painting technique in which the ink is used sparingly with a minimum of moisture in the brush; known natively as *kan pi*. **1959** HALAS & MANVELL *Technique Film Animation* xix. 220 If the area common to the drawings is..less than two-thirds, then dry brush effects can be used to join the two objects. **1869** J. R. BROWNE *Adv. Apache Country* 128 We made a dry camp till morning. **1873** J. H. BEADLE *Undevel. West* xxviii. 615 We..find a pool with water enough for our horses, and to fill our jugs, as we must make a 'dry camp' to-night. **1887** *Outing* (U.S.) X. 4/2 We halted on an open place..and went into dry camp. **1920** J. M. HUNTER *Trail Drivers of Texas* 312 The round up boss..called for two or three men..to make

DRY

what is called a 'dry camp'. 1893 P. Benjamin *Voltaic Cell* xv. 309 Dry cells are best adapted to circumstances where current is intermittently needed and then only for a short time. 1936 *Discovery* Sept. 285/2 Causing a current from a battery of dry cells to pass through the fine wire. 1848 *Californian* 14 Aug. 2/3 In one part of the mine called the 'dry diggins', no other implements are necessary than an ordinary sheath knife, to pick the gold from the rocks. 1853 *Househ. Words* VIII. 321/1 The dry diggings at least furnish equal proof of energy and industry. 1858 *Brit. Colonist* (Victoria, Brit. Columbia) 11 Dec. 1/2 These are in fact a species of dry diggings. 1858 W. Howitt *Land, Labour & Gold* I. xi. 126 Next came the dry diggings; these were far enough from the stream to be free of its drainage. 1862 E. Hodder *Memories N.Z. Life* 222 There are two principal kinds of diggings: river diggings..and dry diggings, in the conglomerate and gravel accumulated on the slope of the mountains. 1873 F. Boyle *To the Cape for Diamonds* 123 Four 'dry diggings':—New Rush..Old De Beers, Dutoitspan, and Bultfontein. 1889 K. Munroe *Golden Days* x. 111 The dry diggings were those of hill-sides, or in gulches containing no steady supply of water. 1899 G. Lacy *Pictures of Travel* 173 The 'dry diggings' are thirty miles to the south-east of Pniel. They are so called because the gems are not found in river-wash, but in dry tufa, which has apparently never been in contact with water. 1910 J. Hart *Vigilante Girl* xxiv. 326 It had been a 'dry diggings', and the skeleton line of a long flume ran thread-like along the mountains. 1907 *Macm. Mag.* Jan. 196 The platform.. was some fifty feet above the valley, and the stones on its face, which was almost perpendicular, appeared to be irregularly dry-dyked. 1905 *Spectator* 11 Feb. 211/1 In the Boer War the 'dry dikers' of a certain East Yorkshire regiment used to be asked to volunteer to build 'sangars'. 1894 J. Dunbar *Notes Manuf. Wood Pulp & Papers* 39 The temperature of the cylinders was too high at the dry end of the machine to produce the desired result. 1927 T. Woodhouse *Artif. Silk* iii. 26 The delivery-end or 'dry-end' of a pulp drying machine. 1962 F. T. Day *Introd. Paper* iv. 36 The beginning of the paper making machine is described as the 'wet end', whilst the other end of the machine, which consists of drying cylinders and paper finishing calenders, is called the 'dry end'. 1919 E. Hough *Sagebrusher* xxxiii, A few scattered dry farms, edging up close to the river in the valley far below. 1952 D. F. Putnam *Canadian Regions* 376/2 The extra size of the dry farms. 1971 L. Davidson *Smith's Gazelle* iii. 50 The sheep could graze among the table rock and the wheat could be dry-farmed on most of the rest. 1912 R. A. Wason *Friar Tuck* iii. 36 Next came the dry farmer. 1919 H. L. Wilson *Ma Pettengill* v. 155 The forlorn shack of a dry-farmer. 1908 *Sci. Amer.* 22 Aug. 120 'Dry farming' consists in so preparing the soil in semi-arid regions that it will catch what little annual rainfall there is, and store it within reach of the roots of the plants to be grown. 1878 J. W. Powell *Rep. Lands Arid Region* 78 A company of Danes..have obtained a meagre subsistence by dry farming. 1906 *Nature* 26 July 304/2 The scientific aspect of what has been designated in the United States as 'dry-farming' consists in utilising to the best advantage all the water that falls in semi-arid regions. 1936 F. Clune *Roaming round Darling* xiii. 113 An Experimental Farm was started to carry out investigations with regard to cereals under dry-farming conditions. 1961 L. D. Stamp *Gloss. Geogr. Terms* 165/2 Dry farming implies a specialized technological treatment of land to overcome the short supply of water. 1846 G. P. R. Pulman *Vade-Mecum of Fly-Fishing* (ed. 2) vii. 84 If the dry fly be widely different in these respects, the fish will be *surprised*. 1897 *Westm. Gaz.* 7 Apr. 3/1 Dry Fly fishing is..using a dry and floating fly instead of a wet and sunk one. 1913 *Q. Rev.* July 66 Dry-fly fishing for sea trout is still in its infancy. 1890 Dry hopping [see Hopping *vbl. sb.*[2] 2]. 1956 *New Biol.* XXI. 14 Dry hopping is the practice of adding hops to the barrel of finished beer before it leaves the brewery. It gives added aroma and to some degree additional biological stability. 1925 *Off. Gaz. U.S. Patent Office* 28 Apr. 850/2 Dryice Corporation of America..*Dry Ice*, Carbon Dioxide (CO_2) in Solidified Forms, Mixtures, and Compounds. 1930 *Engineering* 17 Oct. 504/2 The oxidation of carbon-monoxide from electro-metallurgical and calcium-carbide furnaces to carbon dioxide for producing 'dry ice', or solid carbon-dioxide. 1938 *Archit. Rev.* LXXXIV. 119 Dry-ice refrigeration unit accommodating approximately 180 half-pint cartons of milk. 1968 *Times* 17 Oct. 18/6 The traps were baited with solid carbon dioxide, or dry ice, which evaporated at the rate of about three litres a minute. 1933 H. J. B. Chapple *Television* ii. 38 Unless [soldering is] thoroughly understood..a set fails to function owing to weak or dry joints. 1940 *Amateur Radio Handbk.* (ed. 2) iii. 42/2 The best and safest flux is pure resin, although it..needs greater care if dry joints are to be avoided. 1960 *P.O. Telecomm. Jrnl.* XII. 92/2 As 'dry' or imperfectly soldered joints may not show up until after the equipment has been in use for some time..any operator employed on soldering should have had adequate practice. 1903 *Photogram* X. 320/1 Wet or dry mounting. 1958 V. Drumm in M. L. Hall *Newnes' Complete Amat. Photogr.* xxxv. 329 The best method of mounting photographic prints is with the use of dry-mounting tissue in a hot press. The tissue is impregnated with shellac which melts at approximately 160 °F. 1958 T. Landau *Encycl. Librarianship* 113/2 *Dry offset*, printing by letterpress to a rubber cylinder from which impressions are taken on to paper. 1967 Karch & Buber *Offset Processes* ii. 30 *Dry Offset*. A shallow-etched relief plate transfers the image to a rubber blanket on a cylinder. The press needs no moistening rollers. 1927 *Observer* 15 May 22 A..dry plate clutch. 1928 *Motor World* 9 Mar. 162/2 There is an enclosed dry plate clutch, a three-speed unit gear-box. 1920 M. Aurousseau in *Geogr. Rev.* X. 228 We have two special cases of arrangements governed by water supply—the extreme conditions giving rise to what we will term *wet point villages* and *dry point villages*. 1946 L. D. Stamp *Britain's Struct.* xv. 173 For a thousand years these scattered 'dry-point' settlements remained typical of the heart of what is now Greater London. 1845 J. Palmer *Jrnl. Trav. Rocky Mts.* 28 Sept. (1847) 61 We took up a dry run for one or two miles, thence over a ridge to a running branch. 1893 *Harper's Mag.* Apr. 697/2 Arroyos, or dry runs,..collect the storm waters. 1941 *Amer. Speech* XVI. 165/1 *Dry run*, to practice; a dress rehearsal. 1943 *Sat. Evening Post* 27 Nov. 12 She had to locate his pulse, get her watch ready and make a couple of dry runs. 1944 *Air News Yearbk.* II. 18 There is long, hard planning, endless training, repeated dry runs..behind undertakings of this magnitude. 1958 *Spectator* 1 Aug. 167/1 A comedy series..needs the whole benefit of dry runs, ample film facilities, innumerable scriptwriters. 1968 D. Marlowe *Mem. Venus Lackey* i. 25, I had mentally dry-runned the sexual act a thousand times since puberty. 1970 *New Yorker* 10 Oct. 159/1 Since there might be difficulty locating the Commander's home, it would be advisable to make a 'dry run' sometime before the call. 1890 *Hairdressers' Weekly Jrnl.* 14 June 383/2 (Advt.), Niagara Foam..an American dry shampoo..20/- per gallon. 1913 *Queen* 24 May (Advt.), When you want your hair to look extra nice and bright..just treat it to a dry shampoo with.. Hair Powder. 1928 *Ibid.* Feb. 271/2 Dry Shampoo. Carbonate of Potash 1 oz. Water 32 ozs. Saponine ½ oz. Industrial Spirit 32 oz. Perfume, as desired. 1966 J. S. Cox *Dict. Hairdressing* 49/2 *Dry shampoo.* (1) A shampoo in powder form which is applied to the head as a powder, massaged in and then brushed out...(2) A shampoo composed of industrial methylated spirit or isopropyl alcohol and water with the addition of a foaming element such as saponin. 1937 *Night & Day* 1 July 4/2 (Advt.), If you drive a car you'll shave with a Schick dry shaver. 1963 *B.S.I. News* Feb. 33 This revision will specify the requirements for mains-operated dry shavers. 1954 *Springfield* (Mass.) *Daily News* 10 Nov. 22 The dry ski class..will begin tonight. 1957 *Times* 25 Nov. 11/4 Dry-ski schools which help muscles to become more flexible and teach beginners the basic movements. 1958 A. Hocking *Epitaph for Nurse* xi. 197 Taking a gravelly sharp corner at an almost reckless speed she got into a dry skid. 1961 I. Fleming *Thunderball* ii. 23 The Bentley dry-skidded to a stop in the gravel. 1887 Dry spell [see Spell *sb.*[3] 5 b]. 1920 *British Rainfall 1919* 27 A Dry Spell is a period of fifteen or more consecutive days no one of which is a 'Wet Day'. 1961 *Amer. Speech* XXXVI. 267 A sharp distinction is observed in Colorado between the relatively long *drouth* and the relatively short *dry spell*. 1864 A. J. Warden *Linen Trade* v. i. 697 Wet spinning differs chiefly from..dry spinning in having the spinning frame furnished with a receptacle for holding water. 1904 Goodchild & Tweney *Technol. & Sci. Dict.* 177/2 *Dry spinning*, flax may be spun wet or dry, the latter giving a softer and more spongy yarn. 1921 T. Woodhouse tr. *Foltzer's Artif. Silk* iv. 23 Spinning with the aid of water,..has been replaced by a system of dry spinning. 1957 *Textile Terms & Defs.* (*Textile Inst.*) (ed. 3) 94 *Dry spinning* is the process in which a solution of the polymer is extruded into a heated chamber to remove the solvent, leaving the solid filament. *Ibid.* 39 Dry-spun. 1963 A. J. Hall *Textile Sci.* ii. 75 Solutions which can be *dry spun*—that is, into warm air to evaporate off the solvent and leave solidified filaments behind. *c* 1702 C. Fiennes *Through Eng. on Side-Saddle* (1888) II. iii. 83 You scarce see a tree and No hedges all over yͤ Country, only dry stone walls. 1955 R. & B. Carrier *Dive* iv. 118 Rubber suits have been designed to protect the diver from the effects of cold water. There are two basic types..the 'wet suit' and the 'dry suit'. *Ibid.* 120 The next best thing is..a sealed sheet-rubber dry suit worn with one or more suits of long underwear or wool sweaters underneath. 1971 B. Graham *Spy Trap* xii. 84 Crabb had bought a new black Pirelli dry suit. 1898 Dry-valley [see *blind *a.* 11 c]. 1927 C. C. Fagg in *Proc. & Trans. Croydon Nat. Hist. Soc.* IX. 94 The development of the dry valleys of the Chalk belongs to the more recent phases of the Denudation of the Weald. *Ibid.* 96 (*caption*) The Dry Valley Systems. 1961 L. D. Stamp *Gloss. Geogr. Terms* 166/1 The origin of the dry valleys of the Chalk. 1778 G. White *Let.* 3 July in *Selborne* (1789) I. xli. 236 *Lathræa squamaria*, tooth-wort..on the dry wall opposite *Grange-yard*. 1883 *Encycl. Brit.* XVI. 450/1 The materials..may be built up alone (*dry walling*) or with the aid of mortar or hydraulic cement. 1886 F. T. Elworthy *W. Somerset Word-Bk.* 219 Dry-waller. 1914 G. Jekyll *Colour Schemes for Flower Garden* (ed. 3) ix. 86 An earth bank four and a half feet high, dry-walled on both sides. 1922 *Daily Mail* 10 Nov. 8/5 A dry-waller has to be born, not made. So old William told me when I found him dry-walling. 1872 J. G. Bourke *Jrnl.* (MS.) 25 Nov. (D.A.E.), There is a dry wash on this road. 1926 Mulford *Hopalong Cassidy's Protégé* ix. 110 A bridge spanned a dry-wash, dry most of the year. 1962 *Bad Lands of Red Deer River* (Board of Trade, Alberta) 25 The pieces..are lying scattered at the base of the cliffs or in dry washes where they have been carried by run-off.

dry, *sb.* **1. a.** Add to def.: the dry season (chiefly *Austral. colloq.*).

1897 [see Wet *a.* 2 f]. 1908 Mrs. A. Gunn *We of Never-Never* vii. 88, I—I—thought you'd reckon that travellers' water for the Dry came before your rooms. 1938 X. Herbert *Capricornia* (1939) vii. 74 The Dry! the good old Dry—when the grasses yellowed, browned, died to tinder. 1955 J. Cleary *Justin Bayard* xi. 172 Thinking of coming down there later in the Dry. 1968 K. Weatherly *Roo Shooter* 35 As the dry progressed and the heat remained constant, they stopped breeding.

2. b. *Austral.* A desert area; waterless country.

1909 *Bulletin* (Sydney) 21 Jan., A seventy-five mile dry. 1938 *Observer* 30 Oct. 11/4 The swaggie's..billy-can..is carried full of water, so that if 'on the wallaby' over a long stretch of 'the dry' (waterless country) he can..be sure of his..'billy tea'.

c. A dry wine, cocktail, etc. (see Dry *a.* 8).

1953 A. Upfield *Murder must Wait* xxiii. 205 We settled for a half bottle of gin and a few bottles of dry. 1958 'J. Welcome' *Run for Cover* vi. 108 'Good evening, Herbert. A "Dry" please.'..Herbert's dry martinis..were as pale as ice.

5. A prohibitionist; a person who opposes the use of alcoholic liquors. *colloq.* (orig. *U.S.*).

1888 in A. Randall-Diehl *2,000 Words & Defs.* 1896 *Chicago Record* 11 Feb. 6/5 Even though there might be some precincts where the 'wets' outnumbered the 'drys'—yet the whole county would go dry. 1920 *Eye Opener* (Calgary) 7 Feb. 1/3 The drys cannot pretend much longer that Alberta is 'prohibition'. 1930 *Daily Express* 6 Nov. 2/1 An active 'Dry'. 1965 Wodehouse *Galahad at Blandings* i. 13 The woman who runs the school is a rabid Dry and won't let her staff so much as look at a snifter. 1970 *Encycl. Brit.* XVIII. 610/2 The 'drys' assumed a considerable degree of power within both the Democratic and Republican parties.

6. The process of drying.

1957 *Economist* 16 Nov. 579 (Advt.), Soft, dry towelling that gives you a good, clean dry every single time.

b. *Theatr.* The act of 'drying up' on the stage (see Dry *v.* 5 d and *2 d).

1945 M. Agate *Madame Sarah* ii. 22 She..adopted the English custom of the stage-manager keeping an eye on the book from the prompt-corner in case of a 'dry'. 1960 *News Chron.* 14 Oct. 10/6 When no spark is struck..the effect is as embarrassing as a theatrical 'dry'.

dry, *v.* Add: **1. c.** *absol.* To dry crockery, cutlery, etc., after washing up.

1935 A. J. Cronin *Stars look Down* I. viii. 59 'Shall I dry for you, mother?' She shook her head, dried the dishes herself. 1949 D. Smith *I capture Castle* x. 166 Neil and Simon helped with the washing up...Ivy washed and we all dried. 1967 G. North *Sgt. Cluff & Day of Reckoning* vi. 51 She piled dishes into his hands to be carried to the kitchen: 'It'll save time if you dry for me.'

2. c. *to dry straight*: to come right eventually. *colloq.*

1897 W. J. Locke *Derelicts* xxii, I shall miss you terribly—at first—but it will all dry straight, Yvonne. 1936 Wodehouse *Laughing Gas* xxvi. 278 Cheer up, Joseph. Things will dry straight one of these days.

d. Suddenly to forget or fail to speak one's words in a play or other performance. *colloq.*

1934 *N.Y. Herald Tribune* 2 Sept. vii. 10/2 Thumbing the pages for certain theatrical terms, we find..*fluff* (but not *dry* or *dry up*). 1941 Brahms & Simon *No Bed for Bacon* xxi. 253 I'm sorry I dried at the beginning. 1953 L. A. G. Strong *Hill of Howth* i. 8 A colleague of mine once dried in the middle of a scene. 1955 J. Coates *Linda* xv. 170 She dried in the middle of a speech. Beryl prompted her and she went on. 1967 M. Shulman *Kill Three* iii. viii. 147 'O. K., Allan,' said the director into his microphone. 'If she fluffs badly or dries we'll go straight to Three.'

e. *to dry down*: of paint, to become dry.

1958 *Listener* 28 Aug. 323/1 Some complain that the jelly paints..dry down with a poor gloss. 1959 *Ibid.* 9 Apr. 651/1 The oil-based paints..may dry down with a patchy finish on standard hardboard.

4. c. dry out. *intr.* Of a drug-addict: to undergo a course of treatment designed to break dependence on the drug; of an alcoholic: to undergo a similar course of systematic disintoxication. Also *trans.*, to cure (a drug addict or an alcoholic) in this way. So **dry-out** *sb.*; **drying-out** *vbl. sb.* and *ppl. a.*

1967 *Guardian* 8 Feb. 7/3 They are not only making firmer contact with the addicts..but also giving some of those they have 'dried out' a purpose. *Ibid.* 7/5 The painful process of the detoxification ward, the 'dry-out'. 1969 *New Scientist* 13 Mar. 554/1 The removal of alcohol from the blood by using the artificial kidney may be found to be the safest (and cheapest) way of 'drying out' alcoholics in a state of acute intoxication. 1969 *Maclean's* Aug. 55/2 Too often a drinker 'dries out' and his case is closed. 1970 R. Haughton *Love* v. 143 A boy or girl would be pulled through a 'bad patch' (and the 'patches' of a drug-addict 'drying out' are very bad indeed). 1970 E. Tidyman *Shaft* (1971) iii. 41 By eight, she had undergone..the drying-out procedure in private institutions. 1971 *Daily Tel.* 2 Mar. 2 Drunks arrested by police in future may have to spend a compulsory three days in a 'drying out' centre.

5. b. Also *absol.* = sense *1 c above.

[1932 S. Gibbons *Cold Comfort Farm* vii. 92 She..flicked the reminders of dinner off the table with Adam's drying-up towel.] 1959 *House & Garden* Dec. 34/1 (*heading*) D for drying-up Essentially the masculine task. 1962 I. Murdoch *Unofficial Rose* ii. 28, I was just wondering if he'd mind drying up while we're at church. 1966 'K. Nicholson' *Hook, Line & Sinker* ix. 101 He seemed preoccupied while drying-up.

d. (Further examples); *spec.* = sense *2 d above (cf. quot. 1884[2] in Dict.). Also *trans.*, to cause (someone) to forget his words in a play or the like.

1853 *San Francisco Comm. Advertiser* 9 Dec. 2/4 She defied his Honor..and giving assurance of a disposition never to 'dry up', was carried down below to cool off. 1923 *N.Y. Times* 9 Sept. vii. 2/1 *Dry a man up*, to give the wrong cue, or to say something aside to disconcert a fellow-actor, and so cause him to dry up. 1928 F. Scott Fitzgerald in *Sat. Even. Post* 21 July 8/3 'Oh, dry up!' retorted Basil. 1933 P. Godfrey *Back-Stage* iii. 34 When an actor fails to remember his lines and the scene comes to an unpremeditated stop he 'dries up'. *Ibid.* 36 We didn't dry up! No, we kept that scene going! 1934 H. N. Rose *Thes. Slang* xii. 83/2 What's the idea of trying to dry me up in the last number? 1967 *Times* 10 May 3/8 (*headline*) Insurance to stop actor 'drying up'. 1969 *Listener* 31 July 140/3 Why is the advertising drying up? Who has stopped (or never even started) advertising in these six?

dryas (drəi·æs). *Bot.* [L.: see Dryad. First used as a genus-name by Linnæus, *Genera Plantarum* (1737) 148.] An evergreen subshrub of the genus so called, belonging to the family Rosaceæ and found in cold or alpine regions of the Northern Hemisphere. Cf. Avens.

1798 Smith & Sowerby *Eng. Bot.* VII. 451 We have cultivated the *Dryas* with success under a north wall. 1872 Hogg & Johnson *Wild Flowers Gt. Brit.* VII. 556 Dryas Octopetala...Synonymes.—*Dryas depressa*,..*White Dryas*, English. 1927 *Glasgow Herald* 26 Mar., A thin carpet of Arctic-Alpine vegetation such as a white-flowered dryas and dwarf-willows. 1936 D. McCowan *Animals of Canad. Rockies* ii. 18 Dryas spreads its lovely floral carpet in the sun. 1955 *Times* 4 July 10 That *was* Dryas, my dear fellow, for already your mountain avens is far back down the pass.

dry dock, dry-dock. Add: **b.** *in dry dock* (fig.): inactive, unemployed; in quarantine; in hospital. *colloq.*

1927 W. E. Collinson *Contemp. Eng.* 58 Those who have been in close contact with the infected patient may have to remain in quarantine or dry-dock. 1929 H. A. Vachell *Virgin* iii. 55 June found herself in dry dock, and likely to remain there, when her services were most in demand. 1930 J. Dos Passos *42nd Parallel* 153 Janey I'm in dry-dock girl.

dry goods. (Earlier and later N. Amer. examples.)

1657 J. Alrichs *Let.* 25 May in *Pennsylvania Archives* (1877) V. 285 Some of the dry goods, entirely scattered about, were wet and injured by the quantity of water in the ship. 1921 *Daily Colonist* (Victoria, B.C.) 26 Mar. 6/4 Fire.. worked havoc in the $40,000 stock of groceries, drygoods, shoes and general stock carried by Messrs. Malpass & Wilson. 1968 *Globe & Mail* (Toronto) 17 Feb. 31 Sellers and buyers of produce, hardware, dry goods and what-not.

drying, *vbl. sb.* Add: **2.** *drying-rack*; **drying day**, a sunny, windy day on which washing dries quickly.

1906 E. Nesbit *Railway Children* vii. 143 'I'll wash your Indian muslins at once.'..'It's a nice drying day—that's one thing.' 1971 R. Rendell *One Across* x. 79 Mrs. Blackmore's Monday wash was flapping on the line.. 'Lovely drying day!' 1884 T. C. Hepworth *Photogr. Amat.* xiv. 124 The skeleton drying-rack which I recommend. 1934 *Archit. Rev.* LXXV. 45/1 (*caption*) Kitchenette with gas-ring, sink, drying-rack.

dryness. Add: **4.** The condition of being 'dry' (see Dry *a.* 11 a) or without alcohol; prohibition.

1910 'Mark Twain' *Speeches* 430 When the others drink I like to help; otherwise I remain dry. This dryness does not hurt me. 1920 *Contemp. Rev.* July 79 'Dryness' in America is enormously increasing the consumption [of sugar] there. 1927 *Observer* 24 July 9/2 President Coolidge will..run as a staunch champion of 'dryness'. 1944 W. R. Scott *Revolt on Mount Sinai* xxii. 175 The platform..ordered the drys to stop measuring candidates for Congress..by their degrees of dryness.

dry-nurse, *sb.* Add: **1. a.** Phr. *at dry nurse* (cf. Nurse *sb.*[1] 2 a).

1839 R. Barham in *Bentley's Misc.* VI. 640 Neglecting the poor little dear out at dry-nurse. 1849 C. Brontë *Shirley* III. ix. 206 Mrs. Horsfall had him at dry-nurse.

Dryopithecus (drəi:opiþi·kɒs, -pi·þikɒs). [mod.L. (E. Lartet 1856, in *Compt. Rend. Acad. Sci.* XLIII. 221), f. Gr. δρῦς tree + πίθηκος ape.] A genus of fossil anthropoid apes of the Miocene period in France. So **dryopithecine** (-pi·þisin) *a.*, pertaining to this genus; **dryopithecoid** (-piþi·koid) *a.*, resembling this genus.

1862 *Geologist* V. 428 The shaft of the supposed humerus of the *Dryopithecus*, from the miocene of the South of France. 1863 C. Lyell *Antiquity of Man* xxiv. 499 Fossil apes...One of these, the Dryopithecus of Lartet, a gibbon or long-armed ape, about equal to man in stature, was obtained in the year 1856 in the upper miocene strata at Sansan, near the foot of the Pyrenees. 1912 A. Keith *Human Body* iv. 58 The very earliest of the large fossil anthropoids which have yet been discovered is the kind now named Dryopithecus. 1937 *Nature* 20 Feb. 326/2 If *Australopithecus* is not literally a missing link between the older dryopithecoid group and primitive man, what conceivable combination of ape and human characters would ever be admitted as such? 1939 *Ibid.* 2 Sept. 451/2 A primitive dryopithecine stage. 1957 *Antiquity* XXXI. 191 The Dryopithecine canine [tooth] was too large and specialized.

dry-up (drəi·ɒp). The action of drying up (see Dry *v.* 5).

1873 J. H. Beadle *Undevel. West* 711 The plowman returns to his work without waiting for a 'dry-up'. 1891 Farmer *Slang* II. 334/2 *Dry-up* (theatrical), a failure. 1939 A. Thirkell *Before Lunch* v. 136 The two elder ladies rapidly finished the wash-up and the dry-up. 1940 N. Coward *Australia Visited* iv. 22 It is well over a year now since I have written anything...This..dry-up of talent is extremely painful.

dso. Add: Also **dzo.** Cf. Zho.

[1866 H. A. Jaeschke *Tibetan & Eng. Dict.* 33/1 *Dzo*, hybrid of yag and common cow.] 1897 *Geogr. Jrnl.* July 36 They also breed herds of *dzo*, a very valuable hybrid between the cow and yak...The male *dzo* is used for ploughing. 1960 'S. Harvester' *Chinese Hammer* vii. 71 Some dzos..highly prized as draught animals and for a high fat content in their milk. 1969 C. D. Darlington *Evol. Man & Society* 625 The herdsmen stumping the hills probably came from Mongolia..bringing yaks, dzos (yak-ox hybrids) and sheep.

D.T. (di·ti·). Also **D.T.'s** (di·ti·z). Colloq. abbrev. of Delirium tremens.

1858 'Megathym Splene' *Almæ Matres* 33 The disease called D.T. (heaven forbid I should write it in full). 1861 B. Hemyng in Mayhew *London Labour* Extra vol. (1862) 224/2, I shall soon get D.T. and then I'll kill myself in a fit of madness. 1880 G. R. Sims *Ballads of Babylon* 79 A titled churl Who had just got round from a bad *d.t.* 1907 *Daily Chron.* 30 Aug. 7/3 'D.T.'s' without drunkenness. 1910 A. H. Davis *From Selection to City* ix. 82 He started to shake as if he had the d.t.'s. 1955 R. King *No Paradise* ii. iv. 168 Men, in the last stages of d.t., rend the darkness with their ravings. 1969 Kessel & Walton *Alcoholism* (ed. 3) iii. 34. Delirium tremens—DTs—generally begins two to five days after stopping very heavy drinking.

dual, *a.* (*sb.*) Add: **A. 3.** In specific collocations: *dual carriageway*, a road with separate carriageways, divided by a central strip, for up and down vehicular traffic; *dual control*, control exercised by two parties or persons jointly; in *Aeronaut.*, the duplication of the pilot's controls for instructional purposes; similarly in a motor vehicle; freq. *attrib.*; hence *dual-controlled* adj.; *dual ignition*, ignition in a motor-vehicle engine by two independent currents from a battery; *dual personality*, two distinct personalities in one individual; *dual-purpose* adj., serving two purposes or bred for two purposes; *spec.* of cars, capable of carrying people and goods; *dual-standard* adj., pertaining to or capable of transmission or reception of television programmes using either of two different picture-densities (see quot. 1961).

1933 *Proc. Inst. Municipal & County Engineers* LX. iv. 274/2 The intention of such roads was to so design them as to enable them to carry very fast traffic safely. The only method of doing that was by providing dual carriageways with central strips which could be so treated as to make them a delightful amenity. 1957 I. Murdoch *Sandcastle* viii. 128 A dried-up grass verge separated her from the dual carriageway. 1968 Dual carriageway [see *blub *sb.*]. 1884 *Illustr. London News* 23 Feb. 170/2 The Dual Control...was the proximate cause of our troubles in Egypt. 1913 *Captain* Sept. 1069/2 Military influence is seen in the two or three-seated machine with dual control. 1914 *Aeroplane* 15 July 60/2 Even if the 'Herring Pond' is crossed previously by a single or dual control machine. 1944 H. C. Dent *Education Act, 1944* 21 The compromise on the 'Dual Control' of schools is one of the outstanding triumphs of the Act. 1959 *Which?* May 37/2 All lessons should be on a dual-control car. 1930 *Daily Tel.* 24 July 10/4 The machine was dual-controlled. 1909 *Westm. Gaz.* 22 July 8/3 Some system of what is known as dual ignition for the modern high-class.. motor-car. 1917 A. G. Clark *Text Bk. Motor Car Engin.* II. 73 Having in view the high efficiency of the modern magneto, it may be accepted that the dual ignition will only be fitted on the more expensive and high-powered cars to facilitate starting up. 1905 *Strand Mag.* Apr. 451/2, I feel my dual personality rather puzzling. 1917 A. Waugh *Loom of Youth* i. ii. 29 In the corps as Officer Commanding..and as a clerk in orders...His dual personality embodied the spirit of 'the Church Militant'. 1935 B. Russell *Relig. & Sci.* v. 140 In cases of what is called dual personality..what seems to outside observation to be one person is, subjectively, split into two. 1958 'Lancaster' & Poling *Strangers in my Body* iii. 128 This twenty-five-year-old married female patient was referred for psychological examination with diagnosis of dual personality. 1914 *Scotsman* 8 Oct. 9/1 The number of 'dual-purpose' boats, equipped to burn both coal and oil, includes 38 battleships,..and 21 light cruisers. 1925 J. T. Cole in G. F. Finlay *Cattle Breeding* ii. xxxii. 431 The 'dual-purpose' Shorthorn type imported from England in the first half of last century. 1958 *Observer* 10 Jan. 5/4 Police action against owners of 'dual purpose' cars has increased sharply. 1958 *Church Times* 21 Nov. 1/2 He saw St. John's—at present a dual-purpose hall which is also dual-owned, the sanctuary belonging to the diocese and the hall to the local authority. 1961 *Which?* Nov. 301/2 A dual standard (sometimes called switchable) set, which operates at present on the 405-line system, but which is already partly converted and might need only a new tuning unit to make it capable of receiving both 405 and 625-line transmissions. 1965 *B.B.C. Handbk.* 111 A new television studio in Glasgow and a modified one in Bristol are equipped for dual-standard working.

B. *sb.* **3.** In chess problems, a choice in White's continuations. Hence **dual** *v.* in *pass.*, to admit of a dual solution.

1875 *City of London Chess Mag.* 116 The Black Pawn on the K·R file is here omitted as superfluous, and the above-mentioned duals entirely eliminated. 1903 *Daily Chron.* 11 July 8/7 Your problem is dualled by 22 17. 1906 A. C. White *Tours de Force* p. xxxiii, If White is allowed a choice of continuations, such a choice is called a dual, or multiple, continuation. There are two kinds of duals: Absolute duals ..and minor duals. 1966 *New Statesman* 13 May 705/2 That beautiful piece by Sikdar..had to be disqualified because a dual was discovered and couldn't be cured.

dual (diū·ăl), *v.* [f. the adj.] *trans.* To convert (a road) into a dual carriageway.

1959 *Times* 23 Dec. 4/2 Before 1958–59, 20 miles [of the Great North Road] were already dualled and during the year a further 24 miles were completed. 1961 *Times* 15 Dec. 8 (*caption*) Trunk road to be dualled in the current programme. 1969 *Oxford Times* 18 Apr. 17 (*heading*) Northern by-pass to be dualled to Banbury Road.

duant (diū·ănt). *Physics.* [f. L. *duo* two + -ant[1].] In an accelerator = *dee *sb.* 2; in an electrometer = *binant.

1930 *Science* 10 Oct. 376/2 Semicircular hollow plates in a vacuum not unlike duants of an electrometer. 1936 [see *dee *sb.* 2]. 1936 *Nature* 22 Feb. 316/2 The cyclotron at present has a limited voltage amplification due, in part, to a space charge built up within the duants. 1938 [see *binant].

duar, var. Douar.

1899 *Daily News* 6 Nov. 7/2 At the duar of Charifin a man stole a donkey. 1908 *Westm. Gaz.* 16 Apr. 5/2 The duar or village of the Kaid of the Ouled Buziri.

dub (dɒb), *sb.*[5] *Criminals' slang.* [Cf. Dub *v.*[3]] A key, especially one used for picking locks. Hence **du·bsman** (or abbrev. **dubs**), a turnkey, gaoler.

a 1700 B. E. *Dict. Cant. Crew*, *Dub*, c. a Pick-lock-key. 1789 G. Parker *Life's Painter* 139 A bunch of young dubs by her side. 1821 D. Haggart *Life* 31 We seized him, took the dubs, bound, and gagged him. 1839 Ainsworth *Jack Sheppard* ii. xii, Oh! give me a chisel, a knife, or a file, And the dubsmen shall find that I'll do it in style! 1887 Henley *Villon's Good Night* (Farmer), You coppers' narks, and dubs, What pinched me when upon the snam. 1923 *Chambers's Jrnl.* 6 Oct. 716/1, I pulled the dub of the outer jigger from his suck.

dub, *sb.*[6] *slang* (orig. *U.S.*). [Perh. related to Dub *v.*[1] 11, Dubbed *ppl. a.* 4] One who is inexperienced or unskilful at anything; a duffer, fool.

1887 *Courier-Journal* (Louisville, Ky.) 20 Jan. 6/4 Dem dubs is goin' to git it in de neck. 1896 Ade *Artie* i. 4 What kills me off is how all these dubs make their star winnins. 1902 H. L. Wilson *Spenders* xxx. 353 People can talk all they want to about your bein' just a dub—I won't believe 'em. 1905 *Smart Set* Oct. 18/1 I've made up my mind that I ain't goin' to keep on bein' a common dub all my life. 1911 H. Quick *Yellowstone Nights* 43, I was coming on pretty well for a dub. 1916 'Boyd Cable' *Action Front* 8 The Schmidt customer crowd didn't need to know a thing about me being here unless he was dub enough to tell 'em. 1923 J. Manchon *Le Slang* 111 *Dub*, *tennis*, un joueur médiocre. 1931 T. A. Harper *Windy Island* (1934) II. iii. 122 He was not exactly a dub at Latin and maths. 1943 K. Tennant *Ride on Stranger* xix. 213 Quinlan wasn't such a dub as he looks. 1949 O. Nash *Versus* 40 The unassuming dub Trying to pick up a Saturday game In the locker room of the club.

dub, *v.*[3] Restrict † *Obs.* to sense 1 and add earlier and later examples to sense 2.

1753 J. Poulter *Discoveries* 33 If the Seger is dub'd, that is, the Door lock'd or bolted. 1781 G. Parker *View of Society* II. i. 69 *Dub the Jigger* is, in other words, *shut the door*. 1958 F. Norman *Bang to Rights* i. 12 Everybody in the nick had already been dubbed up for the night.

dub, *v.*[4] *slang.* [Origin obscure.] *intr.* To pay *up*; so *to dub in*, to make a contribution.

1823 'J. Bee' *Slang* 72 *Dub up*, to pay at once. 1839 *Comic Almanack 1840* 36 'Come, dub up!' roars a third; and I don't mind telling you..that I..took out the sovereign and gave it. 1845 *Punch* Oct. 147/1 He has been compelled to 'dub up' out of his own pocket. 1846 *Swell's Night Guide* 79 A stranger may gain admission by the rum cull introducing him, and dubbing a tanner to the chairman. 1852 G. C. Mundy *Our Antipodes* I. v. 181 The juniors are compelled to *dub up*. 1923 Blunden *Christ's Hospital* 199 Five or six boys 'dub in' for a pot of strawberry jam or treacle. 1959 I. & P. Opie *Lore & Lang. Schoolch.* x. 195 The demand to hand over has elsewhere been heard expressed in the words: 'Cough it up', 'Dish it out', 'Dub up' (North Country).

dub, *v.*[5] [Shortened form of Double *v.*] *trans.* To provide an alternative sound track to (a film or television broadcast), especially a translation from a foreign language; to mix (various sound tracks) into a single track (see quot. 1959); to impose (additional sounds) on to an existing recording; to transfer (recorded sound) on to a new record. Also with *in*, *on*. So **dubbed** *ppl. a.*; **du·bbing** *ppl. a.* and *vbl. sb.*

1929 *N.Y. Times* 13 Oct. ix. 8/6 *Dubbing*, the process of re-recording from film to film, or from film to wax, or from wax to film, or from wax to wax. 1930 *Electronics* Nov. 373/2 These people are then re-photographed in silent close-ups, and then foreign players 'dub' in the same lines. 1930 Pitkin & Marston *Art of Sound Pictures* 270 *Dubbing*, a method of doubling the voice on the screen after the photographing of the picture. 1931 B. Brown *Talking Pictures* xi. 275 Dubbing on disc is no more difficult than on film. *Ibid.* 288 Dubbing from sound track to disc was frequently employed. 1938 *Encycl. Brit. Bk. of Yr.* 421/1 'Dubbed' versions, bearing sound tracks in the native languages. 1939 *Times* 25 Mar. 10/4 Queen Mary..went into what is called the 'dubbing' theatre, where sound is recorded. 1944 *Ann. Reg. 1943* 344 The Overseas Dispatch Department..handled some 300 copies of 80 different films..some 'dubbed' and some with foreign commentaries or foreign subtitles. 1952 *Record Year* 124 The Michelangeli set (GX 61004–7) is a dubbing, so unsuccessful as to preclude judgment on the merits of the performance. 1959 Halas & Manvell *Technique Film Animation* xix. 209 The final set of tracks required for dubbing are laid, and at the subsequent dubbing session these are run simultaneously and balanced together and amalgamated on to one single sound track. 1962 *Movie* Sept. 6/3 Watching a film he has previously made with the director for whom he is now working in the dubbing room. 1966 *B.B.C. Handbk.* 38 More international sound tracks for programmes were provided to allow foreign commentaries to be dubbed on.

‖ **dubbeltjie** (dö·bəlkʸi). *S. Afr.* Also **dubbeltjiedoring** (-do:rəŋ); **duiweltjie** (döū·vəlkʸi) and (formerly) **dubbeltje(doorn)**.

1. [Afrikaans, of uncertain origin.] One of several South African weeds, esp. *Tribulus terrestris* or *Emex australis*; also the spiny, angular burr of any of these weeds.

1795 tr. *Thunberg's Trav.* (ed. 2) I. 148 The seed-vessels of the *rumex spinosus* (dubelties). 1827 T. Phillips *Scenes in Albany & Caffer-Land* vi. 100 Our dogs were soon disabled by a prickly seed which gets into their feet..called a *dubbeltje*. 1912 *East London Dispatch* 13 Sept. 7/4 The every-increasing spread of the 'Dubbeltje Doorn' weed. 1949 *Cape Argus* (Mag. Section) 2 Apr. 2/5 A sheet of yellow dubbeltjie blossom.

2. [Afrikaans, f. Du. *dubbeltje* double stiver, double penny.] A penny; in *pl.*, money, pence. Cf. Doublejee, etc.

1822 W. J. Burchell *Trav. Interior S. Afr.* I. iii. 78 The only current coin, are English penny-pieces, which here pass for the value of two pence, and are called *dubbeltjes*. 1833 S. Kay *Trav. Caffraria* ii. xi. 283 At last he scarcely had *dublejees* (pence) sufficient to carry him back to the Colony. 1949 L. G. Green *In Land of Afternoon* (1950) vii. 102 English coins became legal tender at the Cape in 1806, and the heavy 'cartwheel' penny pieces bearing the head of George III soon became known as 'dubbeltjes', as they were worth two pence.

dubbin: see Dubbing *vbl. sb.* 4. Hence **du·bbin** *v. trans.*, to apply dubbin to; **du·bbined** *ppl. a.*, treated with dubbin.

1825 Jamieson, *Dubbin*, the liquor used by curriers for softening leather, composed of tallow and oil. 1855 J. Davies in *Trans. Philol. Soc.* 230 *Dubbin*, a kind of paste used by shoemakers. 1896 J. K. Snowden *Web of Old Weaver* vii. 77, I was seeking a pot of dubbin on the top shelf of her cupboard. 1897 C. T. Davis *Manuf. Leather* (ed. 2) xxx. 463 Dubbin them again and lay them by for a day or two. 1899 M. Cobbett *Bottled Holidays* 147 An omnibus odorous of freely-dubbined boots. 1909 C. E. Benson *Brit. Mountaineering* ii. 30 Dubbin..keeps them [*sc.* boots] supple, and prevents the leather perishing. *a* 1918 [see *blanco]. 1967 *Listener* 30 Nov. 702/2 Tommy grew up in the Gorbals, dubbining his football boots with bacon fat.

dubby (dɒ·bi), *a. colloq.* and *dial.* [f. Dub *v.*[1] + -y[1].] Blunt; short, dumpy.

1825 Jennings *Obs. Dial. West Eng.* 35 *Dub, dubbed, dubby*,..blunt; not pointed; squat. 1872 F. W. Robinson *Wrayford's Ward* III. 208 A nose that young ladies, I believe, call 'dubby'. 1904 Fowler & Felkin *Kate of Kate Hall* xxiii, A dubby piece of blacklead that couldn't write two consecutive words.

Dublin (dɒ·blin). The name of the capital of the Republic of Ireland, used *attrib.* in **Dublin (Bay) prawn**, the Norway lobster, *Nephrops norvegicus*.

1911 *Encycl. Brit.* XVI. 838/2 The Norway lobster..is less esteemed for food than the common species. In London it is sold under the name of 'Dublin prawn'. 1949 H. Smith *Master Bk. Fish* ix. 171 Dublin Bay prawns (*Langoustines*). ..All recipes given for prawns may be used for Dublin Bay prawns. 1951 E. David *French Country Cooking* 31 Dublin Bay prawns in fritters. 1963 *Times* 19 Jan. 10/6 The Norway Lobster or the Dublin Prawn, which, as scampi, we now purchase in the anonymous form of frozen packets of shelled 'tails'. 1971 *Nature* 29 Jan. 298/1 It is the tail of this animal, also known as the Dublin Bay prawn although it is really a close relative of the common lobster, which constitutes scampi.

Dubliner (dɒ·blinəɹ). [f. *Dublin* (see prec.) + -er[1].] A native or inhabitant of Dublin.

1900 *Westm. Gaz.* 11 Apr. 4/1 The ceremony..which Dubliners hope the Queen will perform. 1914 J. Joyce (*title*) Dubliners. 1961 *Observer* 22 Jan. 31/1 The other, a Dubliner I should judge by her accent, was the girl who gave her baby away to Jane Russell.

Dubonnet (dübone). [The name of a family of French wine-merchants.] The proprietary name of a sweet French aperitif; also, a glass of this wine.

1913 *Trade Marks Jrnl.* 25 June 1022 *Dubonnet*...The Best Appetizer in the World...A Wine. La Société Anonyme Dubonnet..7, Rue Mornay, Paris, France; Merchants. 1919 W. S. Maugham *Moon & Sixpence* lvii. 256 Ask *Monsieur* if he will not drink a little glass of Quinquina Dubonnet. 1930 H. Craddock *Savoy Cocktail Bk.* i. 62 *Dubonnet cocktail.* ½ Dubonnet, ½ Dry Gin. 1960 L. Cooper *Certain Compass* 149 What will you have? A Martini, Dubonnet, whisky?

dubs (dɒbz). *local.* [Short for *doubles.*] A term used in various senses in the game of marbles (see quots.).

1823 E. Moor *Suffolk Words* s.v., A player knocking two out of the ring cries 'dubs!' to authorize his claim to both. 1882 M. H. Foote *Led-Horse Claim* iv. 62 'What is it the boys say when they play marbles?'..'Fend dubs?' Hilgard suggested. 1896 *Dialect Notes* I. 220 In Missouri..dubs means, not doublets, but that the player has blundered, and by crying 'dubs' is entitled to play again. 1941 Baker *Dict. Austral. Slang* 26 *Dubs*, marbles which are placed in a ring in a game of marbles.

dubs, dubsman: see *dub *sb.*[5]

dubu (du·bu). [Native name.] In full *dubu-house*. In eastern Melanesia, a men's house or communal dwelling.

1917 *Nature* 27 Dec. 335/2 The men's club-houses, or *dubus*, have now almost died out. 1924 'R. Daly' *Outpost* xii. 118 At the end of the main street, in front of the dubu-house, there was a raised platform.

ducat. Add: **2. b.** (Also **ducket.**) A ticket, esp. a railway-ticket or ticket of admission. Cf. Docket *sb.*[1] 7. *slang.*

1871 J. H. Banka *State Prison Life* 493 Railroad ticket,.. ducket. 1874 Hotten *Slang Dict.* 152 *Ducket*, a ticket of any kind. Generally applied to pawnbroker's duplicates and raffle cards. 1879 *Macm. Mag.* Oct. 501/2 So I took a ducat (ticket) for Sutton in Surrey. 1935 *Sat. Even. Post* 26 Oct. 9/2 I'll slip you a workin'-press ducat. *Ibid.* 9 Nov. 76/3 It'll sell plenty ducats. Big publicity. 1970 *Guardian* 30 Nov. 8/2 My wife and I had a couple of ducketts to see the Marxes' Broadway musical, 'Animal Crackers'.

Duce (du·tʃe). Also **duce.** [It., = leader.] A leader; spec. *Il* or *The Duce*, title assumed by Benito Mussolini (1883–1945), the creator and leader of the Fascist state in Italy.

1923 B. Quaranta di San Severino *Mussolini* p. xv, Three..important elements account for the success of the 'National Fascista Party'..above all, the personality of Mussolini himself, the 'Duce', as he is called. 1928 *Observer* 12 Feb. 12 Signor Mussolini makes no allusion to theocracy in his preface to Major Barnes's book. Nevertheless, it is obvious that the book is acceptable to the Duce. 1930 *Economist* 25 Oct. 748/2 The Austro-German 'duce's' [*i.e.* Hitler's] anxiety..to deny that the damage has been done by the Nazis. 1937 A. Huxley *Ends & Means* v. 34 They yet persist in taking the same roads as are taken by the Duces and Fuehrers. 1937 [see *bilateral *a.* c]. 1965 'C. Hibbert' *Garibaldi & his Enemies* i. iv. 57 The sight of the 'Duce' [*sc.* Garibaldi] sitting astride his horse. 1971 R. Collier (*title*) Duce! The rise and fall of Benito Mussolini.

duchess. Add: See also *duchesse. **2. b.** Hence, a girl or woman, *spec.* one's mother or wife (cf. *dutch *sb.*[2]); also as a term of address to a woman.

1906 E. Dyson *Fact'ry 'Ands* xiii. 170 Er sorrer—eh, what? 'Tain't er little duchess, is it? 1909 J. R. Ware *Passing Eng.* 119/2 *Duchess*, mother—invariable title given between familiar friends when the mother of either is being asked after. 'How's the Duchess, Bob?' 1923 J. Manchon *Le Slang* 112 *The* (or *my*) *old duchess*..., ma vieille, ma femme. 1953 K. Tennant *Joyful Condemned* xxviii. 272 'Hold your noise, Duchess,' Jimmy commanded. 1967 L. Forrester *Girl called Fathom* xii. 147 Start talkin', Duchess. We're gonna toss what you got into the computer ..and see what comes out.

3. b. A size of writing paper (see quot.).

1923 H. A. Maddox *Dict. Stationery* 24 *Duchess*, a fashionable small broad size of private note paper, measuring 5⅝ in. or 6 in. by 4⅜ in.

duchesse (dɒ·tʃes, ‖ düʃes). [Fr., = duchess.] **1.** A kind of chaise-longue, consisting of two armchairs facing each other, with a stool connecting them. So *duchesse brisée* (see quot. 1937).

1794 *Cabinet-Maker* (*A. Hepplewhite & Co.*) (ed. 3) 16 (*caption*) Duchesse and Confidante. [1803 T. Sheraton *Cabinet Dict.* 337 Duchess, a kind of bed, composed of three parts, or a chair at each end and a stool between them.] 1936 'R. West' *Thinking Reed* i. 12 André..poured out a glass of Evian and sat back on the duchesse sofa. 1937 *Burlington Mag.* June 286/2 Our chaise-longue..consists of a *bergère* and a *bout-de-pied*. In this form it is usually called *duchesse brisée*. 1955 R. Fastnedge *Eng. Furnit. Styles* ix. 200 The *duchesse*..was formed by two facing 'Barjier chairs..with a stool in the middle'.

2. *duchesse dressing-chest, -table*, a dressing-table with a swing-glass; also *ellipt.*; *duchesse toilet-cover, -set*, a cover for a dressing-table, or a set of covers usually consisting of one long runner, one smaller, and two very small mats.

1863 M. B. Chesnut *Diary from Dixie* 10 Sept. (1905) xv. 240 The bride had a *duchesse* dressing-table, muslin and lace. 1881 C. C. Harrison *Woman's Handiwork* iii. 210 The Duchesse Dressing Table...Dear to woman's heart is the convenient little 'Duchesse'..with its snowy draperies. 1895 *Army & Navy Co-op. Soc. Price List* 15 Sept. 1102/2 Duchesse Toilet Covers. 1906 *Lady's World* Nov. 156/1 Very dainty Duchesse toilet sets can be made by placing some of these floral squares on a foundation of coloured silk. 1922 *Daily Tel.* 12 June 20/1 Burr walnut duchesse tables and washstands. 1930 *Ibid.* 5 Apr. 21/7 Duchesse dressing chests. 1957 J. Frame *Owls do Cry* i. xii. 54 The tweezers ..stayed in exactly the same place on the duchesse.

3. (See quot. 1878.)

[1873 *Young Englishwoman* July 338/2 The bodice and open Duchess sleeves are also trimmed.] 1878 *Cassell's Family Mag.* IV. 168/2 For balls..there are the Duchesse and Marquise sleeves. The Duchesse covers two-thirds of the arm, and is finished off with frills..and lace.

4. *duchesse satin, satin duchesse*, a very soft, heavy kind of satin.

1878 *Cassell's Family Mag.* IV. 365/1 Duchesse satin is the widest, and at the same time far the best. 1884 *Pall Mall Gaz.* 9 Aug. 9/1 A long train of cream duchesse satin. 1894 *Times* 19 Sept. 10/4 Charged..with stealing a large quantity of silk and satin duchesse. 1895 *Army & Navy Co-op. Soc. Price List* 15 Sept. 1095/2 Coloured Satins. Duchesse. 1968 J. Ironside *Fashion Alphabet* 248 *Duchesse*, a highly lustrous, heavy, firm, yet soft satin.

5. *Duchesse lace*, a variety of Brussels pillow-lace, worked with fine thread in large sprays.

1882 Caulfeild & Saward *Dict. Needlework* 166/2 Duchesse Lace is worked with a finer and different thread to that of Honiton. 1902 *Encycl. Brit.* XXX. 110/2 'Duchesse' and Bruges lace are the chief pillow-made laces. 1926–7 *Army & Navy Stores Catal.* 1110/3. Duchesse and lace-trimmed blinds.

6. *duchesse potatoes*, mashed potatoes mixed with egg and either baked or fried in small cakes or used as a decoration (see quots.).

[1930 R. Wakefield *Toll House Recipes* 172 Duchess potatoes.] 1947 I. B. Allen *Food for Two* 106 Pipe duchesse potato around the edge. 1948 L. Marion *Be your own Chef* xii. 209 Duchesse Potatoes...When you have a compact mash of potatoes..add two tablespoons of grated cheese and beat in a yolk of egg..melt a tablespoon of butter in the frying-pan..place your little pieces in it and turn them on both sides, until they are of an even golden colour. 1951 *Good Housek. Home Encycl.* 481/1 Mashed or Duchesse potatoes may be piped into fancy shapes or borders.

duck, *sb.*[1] Add: **2.** (Further examples); *like a duck in thunder, like a (dying) duck in a thunderstorm*: having a forlorn and hopeless appearance; *like water off* (or *from*) *a duck's back, like* (or *as*) *a duck* (*takes*) *to water*: easily, readily; *does* (or *will, would*) *a duck swim?*: a colloquial phrase of enthusiastic acceptance or confirmation.

1785 'P. Pindar' *Lyric Odes for 1785* 21 Gaping upon Tom's thumb, with me in wonder, The rabble rais'd its eyes, —like ducks in thunder. 1802 C. Wilmot *Let.* 31 Jan. in *Irish Peer* (1920) 35 On asking him what fault he had to find with her, he look'd 'like a Duck in Thunder', and made me instinctively wave [*sic*] the investigation. 1822 Scott *Peveril* I. xi. 269 Closed her eyes like a dying fowl—turned them up like a duck in a thunder-storm. 1824 *Blackw. Mag.* XVI. 347 The thing passed off like water from a duck's back. 1840 Dickens *Old C. Shop* ii. 81 Mr. Swiveller.. observed that last week was a fine week for the ducks, and this week was a fine week for the dust. 1842 S. Lover *Handy Andy* iv. 35 'What do you say..will you dine with me?' 'Will a duck swim?' chuckled out Jack Horan. 1863, etc. [in Dict.]. 1867 A. D. Richardson *Beyond Mississippi* xiv. 177 He takes to them as instinctively as a young duck to water. 1871 [see Water *sb.* 1 f]. 1885 *Boy's Own Paper* 23 May 542/3 'Perhaps you would not object to drinking the queen's health?' Would a duck swim? 1891 L. T. Meade *Sweet Girl Graduate* xvi. 133 'Do you really think that Maggie Oliphant cares for Mr. Hammond?'..'Cares for him!.. Does a duck swim? Does a baby like sweet things?' 1893 [see Take *v.* 74 e]. 1906 'O. Henry' *Four Million* (1916) 82 'With you, Dempsey?' she stammered. 'Say—will a duck swim?' 1917 J. C. Bridge *Cheshire Proverbs* 72 He winks and thinks like a duck i' thunner. 1933 A. Christie *Lord Edgware Dies* xxii. 183 You did look for all the world like a dying duck in a thunderstorm. 1960 C. Day Lewis *Buried Day* vi. 107, I had taken to vice like a duck to water, but it ran off me like water from a duck's back. 1971 'A. Gilbert' *Tenant for Tomb* i. 8 The Bear Lady said brightly that it was a nice day for the ducks.

3. b. Often *a duck of a . . .*; and applied to things as well as persons.

1819 M. Wilmot *Let.* 26 Nov. (1935) 31, I shall presently throw my letter into *the long drawer* at the top of my *duck of a* secretaire. 1841 *Punch* 18 Sept. 112/1 If our remarks were made with an affectionate eye to the young ladies of the satin-album-loving school, we should assuredly style this 'a duck of a picture'. 1884 W. L. Rede *Sixteen String Jack* i. ii, Oh, isn't he a duck of a fellow? 1891 Farmer *Slang* s.v., A duck of a bonnet.

c. With hypocoristic suffix *-s*. Used as a familiar form of address.

1936 J. Curtis *Gilt Kid* viii. 81 She crossed her legs. Her thighs were white and shapeless. 'Got a fag, ducks?' 1958 E. Hyams *Taking it Easy* 200 Talked like you'e did, ducks. 1958 *Times* 1 Oct. 11/6 One is waited on; called 'sir', not 'ducks'. 1963 *Times* 13 Feb. 11/4 The comfortable northern friendliness of the expression 'ducks' as employed by comfortable northern females to all and sundry—to warm the heart towards the species.

4. b. A fellow, 'customer'. *U.S. slang.*

1857 *Phoenix* (Sacramento) 11 Oct. 4/1 No such 'duck' as this could *nab* the 'Ubiquitous'. 1872 'Mark Twain' *Roughing It* xlvii. 331 Are you the duck that runs the gospel-mill next door? 1903 A. Adams *Log of Cowboy* ix. 125, I can't quite make out this other duck, but I reckon he's some big auger. 1904 W. H. Smith *Promoters* v. 100 As you said, Goldsby, Slosher's a slick duck.

c. (See quots.) *U.S. slang.*

1938 *Amer. Speech* XIII. 156/1 *Duck*, a flying boat. 1942 E. Colby *Army Talk* 76 *Ducks*. Air Corps men..give the name to amphibian airplanes capable of landing on water or ground.

11. b. *duck-shooter, -shooting.*

1792 Scott *Let.* 10 Sept. (1932) I. 22, I have turned a keen duck shooter, though my success is not very great. *Ibid.* 30 Sept. 26, I have quite given up duck-shooting for the season. 1859 J. Conway *Lett. from Highlands* viii. 73 A day's duck-shooting. 1945 C. Mann in B. James *Austral. Short Stories* (1963) 72 Even old duck-shooters have now almost got used to them.

12. duck-dive, a vertical dive down into water by a swimmer; hence as *v. intr.*, to make such a dive; **duck('s) arse, ass** (also **anatomy, behind**) *slang*, a style of haircut in which the hair at the back of the head is shaped like a duck's tail (cf. D.A. s.v. *D III. 3); **duck's** (or **ducks'**) **disease** *colloq.*, a facetious expression for shortness of leg; also *duck-disease*; **duck-shover** *Austral.* and *N.Z. slang*, a cab-man who does not wait his turn in the rank, but touts for passengers; also *transf.* (see quot. 1941); so *duck-shove* vb. intr. and trans., *duck-shoving* vbl. sb.; **duck soup** *slang* (orig. and chiefly *U.S.*), something requiring little effort; a person easy to overcome or cheat; a 'cinch'; **duck-tail, ducktail** *colloq.*, (*a*) *S. Afr.*, a young hooligan or 'teddy-boy'; (*b*) = *duck's arse* (see above); **duck-walk**, a duckboard track (see *duckboard).

1969 N. Cohn *AWopBopaLooBop* (1970) vi. 55 He looked like another sub-Elvis, smooth flesh and duck-ass hair. 1928 S. Vines *Humours Unreconciled* viii. 103 Mr. Sheepshanks..soon got his host expanding a theory of the 'duck-disease', as he facetiously called the shortness of leg from which the Japanese were suffering. 1942 G. Mitchell *Laurels are Poison* ix. 99 She..began to come upstream in a series of duck-dives, testing the depth of the water. 1953 L. Charteris in J. Merril *Off Beaten Orbit* (1959) 114 Any good swimmer can duck-dive. 1969 'I. Drummond' *Man with Tiny Head* xv. 176 Nigel took a deep breath, duck-dived and swam under water. 1951 *Sunday Pictorial* 29 Oct., The D.A. therefore stands for Duck's Anatomy—or some such word. 1960 Wentworth & Flexner *Dict. Amer. Slang* 165/2 *Duck's ass*,..a boy's haircut. 1961 J. I. M. Stewart *Man who won Pools* iv. 48 His girl had..made him quit that Duck's Behind for a straight sleeking back with oil. 1925 Fraser & Gibbons *Soldier & Sailor Words*, *Ducks' disease*, short-legged. 1960 B. Marshall *Divided Lady* i. vi. 28 Plinio, the barman with duck's disease, came running up. 1960 *Spectator* 4 Nov. 677 As the leader of the [New Zealand] Opposition complained, the Government has 'dodged and duckshoved' the issue. 1969 *Sunday Mail* (Brisbane) 31 Aug. 5/1 Some Cabinet Ministers said that most local authorities were 'duck shoving' on the State's litter problem. 1898 Morris *Austral Eng.*, Duck-shover. 1908 *Westm. Gaz.* 18 Jan. 3/2 A swanker and a duck-shover. 1941 Baker *Dict. Austral. Slang* 26 *Duckshover*, one who adopts unfair business methods. 1870 *N. & Q.* 4th Ser. VI. 111 'Duck-shoving'..is the term used by our Melbourne cabmen to express the unprofessional trick of breaking the rank, in order to push past the cabman on the stand for the purpose of picking up a stray passenger or so. 1896 *Otago Daily Times* 25 Jan. 3/6 (Morris), 'Duck shoving', a process of getting passengers which operated unfairly against the cabmen who stayed on the licensed stand and obeyed the by-law. 1912 A. H. Lewis *Apaches of N.Y.* iv. 84 'Them Gophers are as tough a bunch as ever comes down the pike.' 'Tough nothin'!' returned Slimmy: 'they'll be duck soup to Ike.' 1929 D. Hammett *Red Harvest* xxvi. 257 it was a juicy row, while it lasted—no duck soup for the coppers at that. 1966 Ogilvy & Anderson *Excurs. Number Theory* i. 4 The number 307 comes out, in binary notation, to be 100110011 which would not have the convenience of 307 at the grocery store. perhaps, but is duck soup for the Computer. 1966 B. E. Wallace *Murder in Touraine* xvii. 152 Now all he had to do..was to..avoid a guard: it was going to be duck soup if you were the right kind of duck. 1955 D. Keene *Who has W. Lathrop?* (1966) xi. 129 The blond youth was in this up to his ducktail haircut. 1959 *Chambers's 20th Cent. Dict.* Add., *Ducktail*, the white Teddy Boy of South Africa. 1960 *Guardian* 28 Mar. 1/2 He [*sc.* Dr. Verwoerd] described South Africa's overseas critics as 'the ducktails (Teddy boys) of the political world'. 1961 *Personality* 16 May 27, I have long since ceased to use the label 'Teddy boy' and now think entirely in terms of 'ducktails'. 1961 *Listener* 7 Sept. 343/2 The '*stilyagi*' with their tight trousers and duck-tail hair-dos. 1968 *N.Y. Rev. Books* 7 Nov. 3/3 He was buying a bottle of hair-oil..to soothe his ducktail. 1915 *Blackw. Mag.* Apr. 449/1 Where there are no duck-walks, we employ planks laid across the mud. 1917 *War Illustr.* 28 Apr. 239/2 The 'duck-walk' is laid for easy crossing of difficult surfaces.

duck (dɒk), *sb.*[4] orig. *U.S.* Also **D.U.K.W., dukw.** [*DUKW* is a combination of factory serial letters designating certain features of the vehicle.] An amphibious vehicle (see quots.).

1943 *War Illustr.* 15 Oct. 301 'Ducks' they are called in soldier slang, and it is easy to see why. In the first place there is something duck-like about these queer motor-barges-cum-trucks which are as much at home on the sea as on the land; and then their factory serial letters placed together spell 'Dukws'. 1944 *Hansard Commons* 2 Aug. 1466 The marvellous American invention, the 'Duck', spelt D.U.K.W., is a heavy lorry which goes at between 40 and 50 miles per hour along the road, and can plunge into the water and swim out for miles to sea in quite choppy weather. 1945 [see *amphibious *a.* 2 b]. 1945 *Manch. Guardian* 18 July 5/2 Officially known as 'Dukws'—a combination of the factory serial letters D for boat, U for lorry body, and KW for lorry chassis—they quickly became known in the Army and Navy as 'Ducks'. 1958 *Observer* 14 Sept. 1/1 Supply craft might stand offshore and disgorge their cargoes into ducks and other smaller craft.

duck, *v.* Add: **2. b.** To back out, withdraw; to make off, abscond; to default. *colloq.* (orig. *U.S.*).

1896 Ade *Artie* ii. 9, I think I'll have to duck on that present. 1900 —— *Fables in Slang* 42 Having delivered herself of these Helpful Remarks she would Duck. 1910 W. M. Raine *B. O'Connor* 249 Coming through the cañon Del Oro in the night, he ducked; I believe he's in Mexico now. 1911 H. S. Harrison *Queed* xxi. 270 It's about over. And now I must pay for my fun—duck back to the office. 1917 H. L. Wilson *Ruggles of Red Gap* ii. 26, I duck out every morning before she's up. 1919 Wodehouse *Damsel in Distress* xvi, He saw me, too, and what do you think he did? Ducked down a side-street, if you please. 1968 *Globe & Mail* (Toronto) 17 Feb. 7/3 A wealthy bachelor, he ducked away from a question on the extent of his financial resources.

c. *trans.* To get away from, to avoid, dodge (a person or thing). *colloq.* (orig. U.S.).

1896 Ade *Artie* vi. 55 He was with a lot o' them Prairie avenue boys, and purty soon he ducks 'em and comes over an' touches me for two cases. 1926 J. Black *You can't Win* vii. 80 We'll get a passenger train out of Cheyenne, kid, if we can duck Jeff Carr. 1928 *Daily Tel.* 6 Nov. 13/3 (American Election) Both sides, he says, have 'ducked' the problems of Labour and foreign policy. 1936 'R. Hyde' *Passport to Hell* 155 Soldiers who had ducked the church parades since the beginning of the War fell out of the lines to pray there. 1959 M. M. Kaye *House of Shade* vi. 79, I should like to duck the whole situation by getting roaring drunk. 1963 *Listener* 21 Feb. 351/1 His peculiar play certainly ducked the questions of subjective and objective judgments that he chose to raise in *Radio Times*.

6. *trans.* and *intr.* In the game of Bridge (see quots.).

1905 R. F. Foster *Complete Bridge* 263 Ducking..is refusing to part with the command of your own suit, and is usually resorted to in situations in which no finesse is possible. *Ibid.* 264 The dealer sees that it is impossible to catch the K, Q, 10 of spades, so he ducks the suit by leading a small card. *Ibid.* 266 As there is no card in dummy's hand but the club ace that will bring the suit into play, that card must be kept as a re-entry until the third round, and the first two rounds of the suit must be ducked. *Ibid.* 312 *Ducking.*—Refusing to win tricks when able to do so. 1928 *Daily Express* 27 Aug. 4 You can frequently make the most of a suit by deliberately losing the first trick. This method of play, called 'ducking', is founded not only on the law of average probabilities but also on the expectation that the cards are normally distributed.

duck-bill, *sb.* Add: **e.** *duck-bill (scraper)*, *Archæol.*, a scraper (sense 4 e) flattened like a duck's bill.

1911 J. Chambers *Stone Age Lake Lothing* 11 Here are disc-shaped and duckbill scrapers. 1959 J. D. Clark *Prehist. S. Afr.* viii. 206 Another form of scraper made on a blade is known as a 'duck-bill'.

duckboard (dɒ·kbōəɹd). [f. Duck *sb.*[1] + Board *sb.*] Usually *pl.* In the war of 1914–18, a slatted timber path laid down on wet or muddy ground in the trenches or in camps; also in wider use (*spec.* see quot. 1940). Also *attrib.* Hence **du·ck-boarded** *a.*, furnished with duckboards.

1917 *War Illustr.* 17 Mar. 109 Walking wounded are helped along the duck-boards that flank the light railways. *Ibid.* 14 July 467 They..flung duck-board bridges over the Douve river. 1920 G. K. Rose *2/4th Oxf. & Bucks Lt. Infty.* 31 For four miles the path lay along a single duck-board track, capsized or slanting in many places. *Ibid.* 72 The day was spent in..rebuilding dug-outs or laying fresh duckboards (wooden slats to walk on in the trenches). *Ibid.* 73 A duck-boarded communication trench. 1925 Fraser & Gibbons *Soldier & Sailor Words*, *Duckboard glide*, a common term for after-dark movements along the trenches, when secrecy and quietness was essential...*Duckboard harrier*, a despatch 'runner' or messenger, whose duty took him along the duckboards in the trenches. 1926 *Glasgow Herald* 9 Jan. 9 Certain underground stations are feeling the effects of the flood, particularly that of Mirabeau, where duck-boards have been laid upon the platform. 1932 E. Bowen *To North* xix. 196 Lady Waters, in grey knitted wool, standing out on a duckboard. 1940 *Chambers's Techn. Dict.* 270/2 *Duck board*, a board which has slats nailed across it at intervals and is used as steps in repair works on roofs.

duck egg: see Duck's egg.

ducket. Delete 'obs.' (see *ducat 2 b).

ducking[3]. (Later U.S. examples.)

1904 *N.Y. Times* 10 May 4 (Advt.), Splendid selection of duckings from which to make them [*sc.* awnings]. 1920 J. M. Hunter *Trail Drivers of Texas* 177 The silver..was placed in duckin' sacks.

duck's egg. Add: **b.** *to break one's duck's egg*: see *break *v.* 7 d.

c. Freq. *duck-egg blue, duck's egg blue.*

1899 *Daily News* 13 Feb. 8/5 One of the prettiest of the new colours is a soft, duck's-egg blue. 1934 *Times* 18 July 17/6 This year pastel colours are much used, particularly a rather dark duck-egg blue. *a* 1954 F. Brett Young *Wistanslow* (1956) iv. 87 The cool duck-egg blues and greens made fashionable by the Adam brothers. 1967 *Harper's Bazaar* Mar. 8 (Advt.), Sweater...In maple..duck egg blue.

d. *duck('s)-egg china, porcelain* (see quots.).

1897 W. Turner *Ceramics of Swansea* 72 Some of this ware has a greenish hue when held up to a strong light. Hence locally, it is called the 'duck's egg.' *Ibid.* 343/2 'Duck's Egg' China at Swansea. 1904 *Burlington Mag.* July 397/1 Old Swansea, or what is locally called the 'duck-egg' porcelain, has been collected, especially in South Wales, for many years. 1957 Mankowitz & Haggar *Encycl. Eng. Pott. & Porc.* 78/2 *Duck Egg Porcelain*, a much sought after variety of Swansea soft-paste porcelain showing a greenish translucency made about 1816–17.

ducky, *sb.* Add: **2.** (Later examples.) Also **duckie.** Cf. Duck *sb.*[1] 3.

1908 J. Masefield *Capt. Margaret* v. 124 You are afraid. You're in love with Olivia, ducky. D'ye think you're going to fight me? 1913 C. Mackenzie *Sinister Street* I. i. iv. 61, I believe you'll miss your poor old Mrs. Frith, eh, ducky? 1945 G. Millar *Maquis* i. 21 You must wear it tonight, though, duckie. 1958 E. Hyams *Taking it Easy* 54, I must have sounded disagreeable, because Matilda said, 'Don't be narky, ducky.' 1968 *Listener* 19 Sept. 370/2 'Hullo, duckies,' he said, 'guess what I'm doing.'

ducky (dɒ·ki), *a. colloq.* [f. Duck *sb.*[1] + -y[1]; cf. Ducky *sb.*] Sweet, pretty; fine, splendid. Chiefly in affected or familiar use.

1897 *Westm. Gaz.* 5 Apr. 10/1 She remembers making his first dress with its 'ducky buttons.' 1905 M. Barnes-Grundy *Vacil. Hazel* 111 The duckiest, little bronze beetle, ..mounted in gold. 1905 *Punch* 8 Mar. 178/1 Only see how prettily he's scratching his ducky little ear. 1927 *Ibid.* 14 Sept. 285/1 You can wear one of those ducky little lace caps. 1949 F. Brown *Dead Ringer* ii. 16 'Is everything okay, Am?' 'Everything's ducky.' 1958 M. Dickens *Man Overboard* xv. 248, I shall tell the tradesmen that you have no authority to give orders...That's going to make you look just ducky, isn't it?

duct, *sb.* **4.** Delete †*Obs.* and add later examples.

1954 N. Denholm-Young *Handwriting* iv. 32 The method of tracing the strokes, and the resulting general appearance of the script can conveniently be termed the duct or *ductus*. 1957 N. R. Ker *Catal. MSS. containing Anglo-Saxon* p. xxv, The change from Anglo-Saxon minuscule to caroline minuscule..involved the duct of the handwriting of all manuscripts. 1969 M. B. Parkes *Eng. Cursive Book Hands 1250–1500* p. xxvi, The duct of a hand is the distinctive manner in which strokes are traced upon the writing surface: it represents the combination of such factors as the angle at which the pen was held in relation to the way in which it was cut, the degree of pressure applied to it, and the direction in which it was moved.

5. Add to def.: *spec.* = Ink-*trough*; also *attrib.*

1880 [in Dict.]. 1888 *Encycl. Brit.* XXIII. 706/1 A trough, which contains the ink..is fitted with the *duct roller* of cast iron. 1968 *Gloss. Terms Offset Lithogr. Printing* (*B.S.I.*) 38 *Duct*, the trough..which contains the supply of ink, and by means of which the ink is presented to the duct roller.

b. A pipe or tube through which air is conveyed for cooling, ventilation, etc.

1884 J. S. Billings *Princ. Ventilation & Heating* xi. 190 Into this chimney empties a foul-air duct..which receives the air from lateral ducts opening beneath the foot of each bed. 1908 A. G. King *Pract. Steam & Hot Water Heating* ix. 92 It is well to take the hot-air duct from the boxing at the end opposite to that where the cold air enters. 1930 *Engineering* 28 Feb. 279/1 The arrangement of the fans and ducts. 1947 T. N. Adlam *Radiant Heating* xiv. 308 For circulating the air through the various offices..a system of metal ducts has been installed. 1962 *Which?* (Car Suppl.) Oct. 139/1 Demister duct trim screws [were] slack.

c. A conduit for an electric cable or the like.

1893 T. O'C. Sloane *Stand. Electr. Dict.* 193 *Duct*, the tube or compartment in an electric subway for the reception of a cable. 1901 *Westm. Gaz.* 11 Apr. 7/2 The work of laying the cable ducts has practically finished. 1945 '*Electr. Engineer' Ref. Bk.* xxvi. 7 The lightest and most convenient system is probably to run the cables in open ducts. 1962 P. Dunsheath *Hist. Electr. Eng.* xiv. 241 The [telephone] cables were at first drawn into cast-iron pipes and later earthenware single and multiple ducts.

6. a. Also used *attrib.* in such phrases as *duct-cancer, -cyst, -papilloma*, = (cancer, etc.) affecting the epithelium of the ducts of the mammary glands.

1864 T. Holmes *Syst. Surg.* IV. 680 Duct-cysts. Perfectly closed cysts..but having an opening communicating with a duct. 1889 *Lancet* 21 Dec. 1278/1 In duct cancer of the breast he had not observed eczematous appearances. 1910 *Practitioner* Apr. 469 When a duct-papilloma obstructs one of the large ducts near the nipple. 1966 Wright & Symmers *Systemic Path.* I. xxviii. 982/2 Duct papilloma is considerably less frequent than fibroadenoma.

duct (dɒkt), *v.* [f. the sb.] *trans.* To convey through a duct; usu. in the form **ducted** *ppl. a.*, conveyed through a duct; situated or operating in a duct. Cf. *ducting *vbl. sb.*

1936 *Aircraft Engin.* VIII. 218/3 The ideal efficiency of the ducted radiator at high speed is about 50 per cent greater. 1938 *Encycl. Brit. Bk. of Yr.* 20/2 Ducted cooling has been developed sufficiently to recover..more than one half of the radiator losses. 1945 *Jrnl. R. Aeronaut. Soc.* XLIX. 698/1 The ducted fan system of propulsion, as we understand it to-day, consists of a fairly large diameter ducted fan or axial compressor of relatively low compression ratio at the intake to the nacelle. 1945 '*Electr. Engineer' Ref. Bk.* xxvi. 7 Ducted cables are screened by the metal box in which they run. 1958 *Times Rev. Industry* Aug. 39/1 Compressed air is ducted from the compressors.. to..pressure jets. 1965 *Economist* 13 Feb. 670/3 A revolutionary new type of helicopter which uses hot ducted turbine exhaust gases to drive the rotor blades through tip vents.

ducting (dɒ·ktiŋ), *vbl. sb.* [f. *duct *sb.* + -ing 1 g.] A system of ducts; material in the form of a duct or ducts.

1945 E. Molloy *Princ. Heating & Ventilation* xiv. 263 They are available as wall fixtures where no ducting is necessary on the motor side of the fan. 1955 *Archit. Rev.* CXVIII. 340/3 A system of telephone ducting embedded in the floor screen has been provided throughout the building. 1957 *Times* 19 Aug. 2/7 Grain..can be ventilated..in heaps by means of portable ducting. 1958 *New Scientist* 16 Jan. 16/2 A length of ducting in which noise level can be measured. 1969 F. G. Thompson *Electr. Installation* II. iv. 73 Ducting is used to provide a network of cable ducts in concrete floors.

ductule. Delete *rare* and add further examples in *Anat.*

1924 H. E. Jordan *Text-bk. Histol.* (ed. 3) xv. 502 The ductuli aberrantes are blind tubules...The superior ductule opens into the epididymis..; the inferior opens at the globus minor. 1939 E. R. A. Cooper *Human Histol.* xii. 203 Interlobular ducts continue into intralobular ones, from which terminal ductules communicate with collections of secretory cells known as alveoli. 1951 H. A. Cates *Primary Anat.* (ed. 2) viii. 222/2 Each efferent ductule, finer than the lead of a pencil, is highly coiled and, when unraveled, about eight inches long. 1967 G. M. Wyburn et al. *Conc. Anat.* i. 15 Twelve to sixteen small ducts, the efferent ductules, connect the testis to the head of the epidi[d]ymis.

ductus (dɒ·ktɒs). [L. (see Duct).] **1.** *Anat.* In various (mod.) Latin phrases = Duct 6 a.

1699 M. Lister *Journey to Paris* 72 The same happens to the *Ductus Bilaris* in its insertion into the Guts. 1722 W. Cheselden *Anat. Humane Body* (ed. 2) iv. iii. 232 When the Umbelical-Vein is stop'd, the Ductus Venosus soon shrinks and Disappears, having no longer any Blood flowing through it. 1811 *London Dissector* (ed. 3) viii. 178 The aorta..is connected to the pulmonary artery by a ligament, which in the fœtus, was a large canal, the Ductus Arteriosus. 1846 *Dublin Q. Jrnl. Med. Sci.* II. 505 Fungous Growth round the Orifice of the Ductus Choledochus. *Ibid.*, The ductus communis was greatly dilated, but the obstruction to the flow of the bile had never been complete. 1927 Haldane & Huxley *Animal Biol.* iv. 97 Ductus arteriosus, derived from part of the embryonic arterial arch system, and connecting pulmonary artery with..the aorta. 1962 Davson & Eggleton *Princ. Human Phys.* (ed. 13) xiii. 297

As the arterial blood of the umbilical vein passes from the placenta, the greater part of it traverses the ductus venosus and enters the inferior vena cava. *Ibid.* 299 About 78% of the right ventricular output passes into the descending aorta via the ductus arteriosus. *Ibid.* lxiv. 1480 The vasa efferentia lead to a single long duct, the ductus epididymis.

2. = Duct 4; the method of making strokes with the pen, etc.

1922 *New Internat. Encycl.* (ed. 2) X. 11 The Bulgarian ductus of the Glagolitsa is round, while the Croatian is more angular. 1933 *Burlington Mag.* July 23/1 The 'ductus' appears to me to be absolutely that of Dürer's own pen. 1934 Priebsch & Collinson *German Lang.* ii. ix. 404 The script is round Anglo-Saxon, though of a more or less cursive ductus. 1954 [see *duct sb. 4].

|| **ductus litterarum** (dʌ·ktŭs litĕrā·rŭm). [mod.L., f. L. *ductus* (see prec.) + *litterarum*, gen. pl. of *littera* letter.] The process by which errors are introduced into a text at the copying stage, study of which may make possible the restoration of true readings in a corrupt text.

1888 *Athenæum* 7 Jan. 25/1 Of all our literature there is none more carelessly printed than our early drama—none in which conjecture, founded on the *ductus literarum*, comes more legitimately into play for the correction of its errors. 1899 A. W. Ward *Hist. Eng. Dram. Lit.* (ed. 2) II. 175 The extraordinary blunders of a printer, who when he could not read a word substituted another that most readily suited the *ductus literarum* without reference to meaning. 1926 *Daily Colonist* (Victoria, B.C.) 16 July 4/6 It [sc. a hypothesis] may be rejected on various grounds; the ductus litterarum may not satisfy some purists. 1966 *English Studies* XLVII. 272 The standard methods of textual criticism (analysis of ductus litterarum, minimum errors, common types of misprints).

ductwork (dʌ·ktwɜːk). [f. Duct *sb.* + Work *sb.*] A system of ducts for the conveyance of liquids, gases, etc.

1934 E. L. Joselin *Ventilation* v. 70 When considering a length of ductwork whose sizes are not yet determined, it is usual to assume an allowance for bends, etc. 1941 Sheldon & Farquhar *Progress Sci.* 23 Air washers or finned coil banks with separate supply fans are provided, connected to guest rooms through systems of ductwork. 1959 *Times* 5 Jan. 2/3 The flow of liquids or gases through branched ductwork systems and steam power cycles. 1960 R. W. Marks *Dymaxion World of B. Fuller* 34/1 Ventilation was provided by a small fan located under the lavatory; air was drawn from the nearest room and exhausted through ductwork to the outside. 1971 *Engineering* Apr. 127 Automatic viscous filters and ductwork.

dud, *sb.* Add: **4.** A counterfeit thing, as a bad coin, a dishonoured cheque; in the war of 1914–18 applied *spec.* to an explosive shell that failed to explode; hence (cf., however, sense 3 in Dict.) applied contemptuously to any useless or inefficient person or thing. (Cf. next.)

1897 *Daily News* 14 Jan. 2/2 He admitted that he knew that he ought not to have sold the piracies, and that such works were known as 'Duds'. 1908 *Westm. Gaz.* 28 Jan. 4/1 Gambling with 'Duds'...A 'dud' car is a worthless contraption, which..has arrived at a stage when it would be dear at any price. 1908 *Captain* Apr. 23/2 We want talent, not duds. 1915 *Blackw. Mag.* Feb. 141 Our weary hearts rejoice When Silent Susan sends us down a dud! 1915 'Boyd Cable' *Between Lines* 254 One of these [shells] was a dud an' didn't burst. 1920 *Punch* 1 Sept. 168/1 He..has..been irritated by his school-boy son derisively addressing him as an 'old dud'. 1923 *Public Opinion* 30 Nov. 531/1 All the torpedoes they carry are duds. 1928 Galsworthy *Swan Song* iii. xvi. 345 It's when you don't understand that you feel such a dud. 1951 R. Graves *Poems & Satires* 39 An expert on shell-fish, otherwise a dud.

dud (dʌd), *a.* [app. adj. use of Dud 3, *4.] Counterfeit; failing to answer to its description or to perform its function; worn out; useless; unsatisfactory.

1903 *Daily Chron.* 29 July 9/4, I..got him to give me half a crown for a dud ring. 1904 *Ibid.* 13 May 6/3 Wanted comedy and dramatic sketches. Something with life and go in it. No Dud stuff required. 1908 [see prec.]. 1915 H. Rosher *In R.N.A.S.* (1916) 130 As luck would have it, the weather was dud. 1917 *Blackw. Mag.* May 803/2 It was soon afterwards that our engine went dud. 1917 'Contact' *Airman's Outings* 267 They wanted a plan of some new defences on which the Hun had been busy during the spell of dud weather. 1918 W. J. Locke *Rough Road* xviii, 'It's a dud sort of a place, Durdlebury,' said he. 'Dud?' He laughed. 'It never goes off.' 1929 *Star* 21 Aug. 14/4 Hitherto, he has met with rotten luck in Africa. Seemed to strike one dud patch after another. 1948 J. Betjeman *Coll. Poems* (1958) 159 'You going to the Hanks's hop to-night?'..'It's pretty dud though,—only lemonade.' 1958 *Times Lit. Suppl.* 7 Feb. 73/4 The dud violinists rehearsing in the next room.

dude. Add: **2.** A non-westerner or city-dweller who tours or stays in the west of the U.S., esp. one who spends his holidays on a ranch; a tenderfoot; **dude ranch**, a ranch which provides entertainment for paying guests and tourists; so **dude rancher**, one who owns a dude ranch. Also **du·dess**, a female dude. Chiefly *U.S.*

1883 *Prince Albert Times* (Saskatchewan) 4 July 5/1 The dude is one of those creatures which are perfectly harmless and are a necessary evil to civilization. 1885 *Weekly New Mexican Rev.* 28 May 2/5 The dudes and dudesses of Vegas are rehearsing for the opera entitled 'The Doctor of Alcantara'. 1889 'Mark Twain' *Yankee* viii. 80 Reverence for rank and title..had disappeared—at least, to all intents and purposes. The remnant of it was restricted to the dudes and dudesses. 1921 *Scribner's Mag.* Mar. 343/1 'Is this Scott Lawson's dude ranch?' soberly inquired the rider of the pinto. 1924 H. Croy *R.F.D. No. 3* 148 I'm going to put up the finest cattle barn in the state—that is, belonging to a real dirt farmer, not to one of them city dudes. 1940 E. Fergusson *Our Southwest* vi. 108 Every dude rancher saw his profits disappear underground in pumps, pipes, and septic tanks. 1949 'J. Tey' *Brat Farrar* xiii. 116 There was paint galore at the dude ranch, but there was also a tradition of toughness. 1967 *Boston Sunday Herald* 14 May ii. 13/1 It features in abundance the attractions that people who want to live in high, wide and handsome fashion will appreciate in its various seasons—riding, fishing, hunting, golfing, skiing, dude ranches and plush hotels.

dude (diūd), *v. colloq.* (orig. *U.S.*). Also **dood**. [f. the sb.] *intr.* With *up*: to dress oneself as or like a dude; also *refl.* Usu. in phr. *duded up*.

1899 B. Tarkington *Gentleman fr. Indiana* xiii. 240 Why should Cale Parker be wearing a coat, and be otherwise dooded and fixed up beyond any wedding? 1924 D. Marquis *Old Soak's Hist.* xxv. 124 If I gotto be all duded up.. I'm a goanto dude myself up. 1958 *New Statesman* 22 Feb. 243/1 The country cousins duding up to impress less snappy dressers back home. 1960 *Guardian* 8 Nov. 1/6 The two men, shaved and rested and all duded up.

due, *a.* and *adv.* Add: **9. d.** *due to*, as prepositional phr. = *owing to* (Owing *ppl. a.* 3 b).

Described as 'erroneous' by W. A. Craigie in the *Dict. of Amer. Eng.*, and said by H. W. Fowler in *Mod. Eng. Usage* (1926) to be 'often used by the illiterate as though it had passed, like *owing to*, into a mere compound preposition', this use is now widely current though still firmly rejected by many grammarians.

1897 S. T. Clover *Paul Travers' Adv.* 190 [The Koturah] was taxed to her capacity, due to the fact that..she was advertised to go first to Adelaide. 1926 in Fowler *Mod. Eng. Usage* 123/2 The old trade union movement is a dead horse, largely due to the incompetency of the leaders. 1953 *Sat. Even. Post* 5 Dec. 173 So far, due to engineering controls, more precise than any known to industry, this has never happened. 1955 *Times* 25 July, Largely due to the defence efforts of the Western Powers, Europe was in a state of stalemate. 1957 Elizabeth II in *Times* 15 Oct. 10/6 Due to inability to market their grain, prairie farmers have for some time been faced with a serious shortage of funds to meet their immediate needs.

due, *sb.* Add: **4. b.** *pl.* The fee for membership or use of a college, club, etc.

1670 in *Publ. Col. Soc. Mass.* (1925) XV. 52 To require the College dues from him forthwith. 1790 *Let. to V. Knox* 10 The Collegiate and University dues are peculiarly trifling. 1937 *Discovery* June p. xlix/2 Annual dues, 5 dollars. 1966 *Rep. Comm. Inquiry Univ. Oxf.* I. v. 168 There are only small variations in those charges, if dues and establishment charges are added to those made for board and lodging.

c. *fig.* (pl.). Responsibilities or obligations; esp. in phr. *to pay one's dues*, to fulfil obligations, undergo hardships, or gain experience. *U.S. slang.*

1943 'S. G. Wolsey' *Call House Madam* xiv. 403 She was mixed up later in one of the rottenest shooting messes ever staged in Hollywood, but she got away with her end of it and never paid her dues. 1961 N. Hentoff *Jazz Life* (1962) ii. 29 'Paying dues' is the jazz musician's term for the years of learning and searching for an individual sound and style while the pay is small and irregular. 1968 *Crescendo* Apr. 12/3, I guess I've, as they say, paid my dues. Looks like I'm still paying 'em. 1969 *Down Beat* 17 Apr. 19/2 Duke, Thad, Mel and myself, we've paid considerable amounts of dues in trying to get this thing off the ground.

|| **duende** (duɛ·nde). [Sp.] **a.** A ghost, an evil spirit. **b.** *transf.* Inspiration, magic, 'fire'.

1924 *Contemp. Rev.* Jan. 98, I never heard of the existence of any good fairies, but *duendes* or bad ones abound... Often when a man has laid down his axe..a *duende* will seize it. 1960 E. Sitwell in R. Campbell *Coll. Poems* III. 5 Roy Campbell had this '*duende*'. 1967 McCormick & Mascareñas *Compl. Aficionado* v. 167 According to good Mexican and Spanish aficionados, Manolete and Belmonte possessed both angel and *duende*, hence their extraordinary interest. 1967 'La Meri' *Sp. Dancing* (ed. 2) vi. 76 The most important part of gypsy dancing is the *duende*, the emotional motivation. The word *duende* means, literally demons or poltergeists. The dancer, then, must be possessed. 1968 G. Household *Dance of Dwarfs* 64 For Pedro they were real...For Geronimo they are clearly duendes. Yet his treatment of the subject differed from the matter-of-fact way in which he usually describes the various spirits which surround us. 1970 *Daily Tel.* 31 Mar. 12/1 His own dancing was polished but lacking in what Spaniards call *duende*—the demoniacal intensity which sweeps audiences off their feet.

duff, *sb.*² Add: **4.** Something worthless or spurious; counterfeit money; smuggled goods; also, the passing or selling of such things. *slang.*

1781 G. Parker *View of Society* II. 158 Sham leggers. The duff. Whispering dudders. 1879 *Macm. Mag.* XL. 502/1 Men at the duff (passing false jewellery). 1895 H. L. Williams *Love & Lockjaw* 7 Lucky my money won't do him any good. It is duff that I carry for a hold-up. 1935 G. Ingram *Cockney Cavalcade* viii. 120 'That's all duff,' he proceeded. The announcement that so much was rubbish, as it consisted of imitation jewellery [etc.].

duff, *v.*¹ Add: **4.** *to duff up.* **a.** *intr.* To become foggy or hazy (see also quot. 1876). *colloq.* or *dial.*

1876 F. K. Robinson *Gloss. Whitby* 58/1 *Duff up*, to drift like road dust on a hot day. 1942 I. Gleed *Arise to Conquer* x. 94 The weather looks O.K...If it duffs up, Derek, give us a shower of Verey lights to show us the way to come home.

b. *trans.* To beat (someone) up, to thrash. Hence **duffing-up**, a thrashing. *slang.*

1961 *Technology* May 131/1 Jeff and Bill was duffed up on saturday at st Albans drillhall. 1967 *Listener* 31 Aug. 261/3 He may have been taught—either through a fine he couldn't pay or a duffing-up he couldn't sustain—that this age has rid itself of some semantic taboos only to enthrone others. 1968 R. Lait *Chance to Kill* i. 3 They had been duffed up at the police station. 1970 A. Draper *Swansong for Rare Bird* i. 11 Even when I was getting a good duffing up I just kept on going; as a result I never lost a punch-up.

duff, *v.*² [Back-formation f. Duffer *sb.*²] *trans.* and *intr.* In Golf, to perform (a shot) badly (see quot. 1897). Also *transf.*, to make a mess of (something), to muff. Hence **duffed** *ppl. a.*, **duffing** *vbl. sb.* and *ppl. a.*

1897 *Encycl. Sport* I. 469/1 The verb 'to duff' does not mean..to play as a 'duffer' or hopelessly bad player, but simply to hit the ground first, behind the ball, so that the ball is struck with the upper edge of the face, and sent only a short way into the air. *Ibid.* 469/2 Duffing is very frequently caused by the player having his ball too near his right foot. 1906 *Westm. Gaz.* 8 Sept. 3/1 At the first duffed shot. 1909 *Ibid.* 30 Apr. 12/2 Mayo duffed his approach. 1909 J. R. Ware *Passing Eng.* 119/2 'He duffs everything he touches.' 'He is the most duffing duffer that ever duffed.' 1924 *Glasgow Herald* 15 June 11 He made one solitary slip when he duffed his putt on Wee Bogle. 1927 *Sunday Express* 29 May 21/6 The ninth provided Landale's crowning error, for he duffed two mashie shots.

duff (dʌf), *a. colloq.* [f. Duff *sb.*²; cf. Dowf *a.*] Worthless, spurious, false, bad, 'dud'.

1889 Barrère & Leland *Dict. Slang* I. 336/2 *Duff* (thieves), spurious. 1910 *Sessions Paper* CLIV. 24 Nov. 82, I rang it [sc. a coin] on the counter; he said 'Break it up—it is duff.' 1944 G. Netherwood *Desert Squadron* 10 It was said by the erks that he once sold rock on Blackpool sands. This was just 'duff gen'. 1956 R. Robinson *Landscape with Dead Dons* ix. 79 Pity we got off to a duff start. 1965 J. Lymington *Green Drift* i. 8, I went down to the pub because the play was so duff. 1967 *Crescendo* June 24/1 A duff piano player will still sound duff on a Bosendorffer Grand.

duffel, duffle. (The more common form is now **duffle**.) Add: **1. b.** Short for *duffle coat.

1957 C. MacInnes *City of Spades* 108 You'd better buy yourself a duffel. 1957 M. Spark *Comforters* v. 112 Caroline pulled their spare duffle from the back seat and arranged it over her head and shoulders. 1959 J. Verney *Friday's Tunnel* ix. 85 Instead of a duffle, he wore a bright scarlet roll-neck pullover.

3. (Earlier and later examples.) So **duffle bag** orig. *U.S.*, a cylindrical canvas bag for carrying personal belongings; **duffle coat**, a coat made of duffle; *spec.* a short coat with a hood and fastened at the front with toggles.

1684 in *New Jersey Archives* (1880) 1st Ser. I. 459 He..gave them [sc. Minisinck Indians] four Duffle Coats &ca. 1863 Mrs. Gaskell *Sylvia's Lovers* ii. 20 She was going..to buy a bran new duffle cloak all for herself. 1915 'Bartimeus' *Tall Ship* vi. 105 The hood of his duffel-suit fell back. 1917 E. E. Cummings *Let.* 1 Oct. (1969) 37, I had a duffle-bag, chuck full. 1919 W. Lang *Sea-Lawyer's Log* x. 124 The guns' crews, wrapped in thick, blanket-like garments known as duffle coats. 1939 A. Keith *Land below Wind* xx. 312 There were our two brief-cases..and the duffle bag. 1953 P. Frankau *Winged Horse* iii. i. 181 He wore a duffle coat and corduroys. 1957 M. Spark *Comforters* v. 114 Laurence could not see her face, it was behind the duffle-coat. 1961 L. van der Post *Heart of Hunter* i. vii. 114 In the end the suitcase and duffle bag we had given Dabé were nearly bursting with presents. 1968 J. Ironside *Fashion Alphabet* 38 *Duffel coat*, a warm coat of heavy cloth...The front is fastened with toggles. This 'fashion' originated in the Navy when duffel-coats were worn as protection against the elements.

b. *Comb.*, as *duffle-clad*, *-coated*, *-dressed* adjs.

1901 Kipling *Kim* xiii. 330 The sallow, greasy, duffle-clad people. 1957 *Times Lit. Suppl.* 18 Oct. 629/4 Amorous students and duffle-clad professors. 1953 E. Taylor *Sleeping Beauty* iv. 56 Wind-nipped, duffle-coated, they climbed out of what they called their vintage Bentley. 1958 P. Mortimer *Daddy's gone a-Hunting* i. 8 A pillioned girl with hair streaming, duffle-coated arms clasped tightly round his waist. 1776 J. Trumbull *M'Fingal* (1826) iii. 89 Call'd forth each duffil-dress'd curmudgeon.

duffer, *sb.*² Add: **1. b.** *duffer's* (or *duffers'*) *fortnight*, a fortnight of the angling season during which trout are supposed to be caught easily.

1927 *Observer* 19 June 27/4 That period of imbecility the so-called 'duffer's fortnight'. 1928 *Daily Express* 28 May 4/5 This annual festival of the Mayfly inaugurates the 'Duffers' Fortnight'.

3. Also *N.Z.* Also *attrib.*

1863 V. Pyke in *App. to Jrnls. House of Reps. N.Z.* D vi. 3 The place was rushed by about 500 men, who speedily deserted it, and declared the Waitahuna to be a 'duffer'. 1869 R. Waite *Narr. Disc. West Coast Goldfields* 15 Those first arrivals chose to call the expedition a duffer rush. *a* 1928 D. McK. Wright in Currie *Centennial Treasury*

Otago Verse 54 We had sunk a hundred holes that was duffers.

duffer, *v.* Add: Also *N.Z.*

1894 C. J. O'Regan *Voices of Wave & Tree* 10 Life's claim is almost duffered. 1897 D. McK. Wright *Station Ballads* 111 A good bit of ground duffers out.

dufterdar, var. *defterdar.

dug, *ppl. a.* Add: **2.** *dug-in*, entrenched; firmly established in a position; (see also quot. 1948). Cf. Dig *v.* 11.

1919 W. H. Downing *Digger Dial.* 21 *Dug in*, in a safe or comfortable position. 1944 *Times* 6 July 4/6 From a ridge behind the airfield dug-in German tanks have a clear field of fire. 1948 Partridge *Dict. Forces' Slang* 62 *Dug-in job*, a base job; such a job within a unit..as carried certain privileges.

dug-out, *sb.* Add: **2.** (Earlier U.S. examples.) Also *attrib.*

1855 in J. A. Thomson *80 Yrs'. Remin.* (1904) I. 171, I live in a dug-out tent [at Balaklava], which is pretty warm, with a capital fire-place made out of potato tins! 1860 *Jrnl. Discourses* VIII. 293/1 When you have built splendid habitations, be as willing to leave them as you would to leave a dug-out. 1873 J. H. Beadle *Undevel. West* xxxi. 685 The unhappy traveler, if compelled to seek shelter in winter, will find it in a Swedish 'dug-out'.

b. *spec.* Applied to the roofed shelters used in trench warfare. Also *attrib.*

1904 *Westm. Gaz.* 7 Dec. 7/1 The following telegram from General Sakharoff..has been received at St. Petersburg:..Our troops, thanks to their dug-outs, warm clothing, and plentiful food, do not suffer from the cold. 1914 D. O. Barnett *Lett.* (1915) 19 A dug-out in the reserve trenches. *a* 1917 E. A. Mackintosh *War the Liberator* (1918) 23 And I shall see no more The gallant friendly faces Framed in my dug-out door. 1919 G. K. Rose *2/4th Oxf. & Bucks Lt. Infty.* 26 Desire Trench..was a shallow disconnected trough upholstered in mud and possessing four or five unfinished dug-out shafts. *Ibid.* 63 A bombed dug-out is the last word in 'unhealthiness'.

3. A person of out-dated appearance or ideas; *spec.* a superannuated officer, etc., recalled for temporary military service. *slang.*

1912 *Blackw. Mag.* June 805/2 From his turn-out, he was probably a prehistoric 'dug-out', a 'was-bird' of 'weird' early Victorian ideas. 1915 (*title*) Tales of a 'dug-out'. *Ibid.* 5 [Publishers' note], These War Stories are written by an Officer—a 'dug-out', returning to the Service after 20 years' absence. 1916 H. G. Wells *Mr. Britling* ii. ii. 232 A new untried man—usually a dug-out in an advanced state of decay—is stuck into the job. 1918 W. J. Locke *Rough Road* vii. 75 The Colonel was immensely proud of them and sang their praises to any fellow dug-out who would listen to him. 1920 *Q. Rev.* July 139 Retired officers and civilians, the much sneered at 'Dug-outs'..saved the situation. 1939 A. J. Toynbee *Study Hist.* VI. 98 These 'elder statesmen' are the last people to whom a community can safely commit its destinies in an emergency, since..these 'dug-outs' are doubly incapacitated. 1958 P. Kemp *No Colours or Crest* iii. 31 The Assistant Provost Marshal..was a dug-out major of a famous cavalry regiment.

du·g-up, *a.* [See Dig *v.* 14.] Exhumed; unearthed. Usu. *fig.*

1897 *Daily News* 23 Jan. 5/5 When the loud laughter which greeted this dug-up relic had died away. 1900 *Westm. Gaz.* 20 Jan. 4/2 The facts which this 'dug-up' material discloses are all to Thackeray's credit. 1921 Galsworthy *To Let* i. xii. 106 Thought her father had some 'ripping' pictures and some rather 'dug-up'.

|| **duiker**² (dəi·kəɪ, dœ̄ü·kər). *S. Afr.* [Afrikaans, f. Du.: see Duiker in Dict.] Any of several cormorants of the genus *Phalacrocorax*, esp. *P. carbo*.

1838 D. Moodie tr. *Record* i. 13 The yacht returned from Robben Island, bringing about an hundred black birds, called duikers, (cormorants) of a good flavour. 1856 C. J. Andersson *Lake Ngami* ii. 16 The way in which the 'duikers' (cormorants and shags) obtain their food is not uninteresting. 1950 L. G. Green *At Daybreak for Isles* iii. 129 Besides the trek-duiker and white-breasted duiker, there is a species on the islands known as the bank duiker... The reed duiker..also nests on the islands; but this is really a fresh-water bird. *Ibid.* iii. 29 The shy duikers, the cormorants that can only be approached on their nests or in the darkness.

duiweltjie, var. *dubbeltjie.

|| **duka** (dū·kā). [Swahili, = shop, store, business.] In Kenya, a shop, store. Also *attrib.*

1924 *Chambers's Jrnl.* Jan. 12/2 He revels in exploring the *duka's* stock of caps, hats, shirts, cotton vests, and drill trousers. 1942 *E. African Ann.* 1941–2 88/2 Stock..can usually be depended upon to give fair..returns on the capital which it represents, and which is always bringing in something with which to pay the wages and the duka bill. 1963 *Listener* 21 Mar. 485/1 They [sc. the Asians] have played a vital part in Kenya's development, pushing retail trade out into the tribal areas through their little shops, their *dukas*. 1966 D. Bennett *Stranger in Grave* xviii. 148 Simbega Town was comprised of an Indian 'duka', an administrative building..and an African market.

duk-duk (du·kduk). [Native name.] A secret society among the natives of New Britain, Bismarck Archipelago, which executes justice on the rest of the tribe and practises sorcery and mysterious rites; also, a member of this society.

1883 W. Powell *Wand. Wild Country* 61 The Duk-duk..may be spoken of as the administration of law, being judge, policeman, and hangman all in one. 1884 *Encycl. Brit.* XVII. 372/2 Justice is executed, and tabus, feasts, taxes, &c., arranged, by a mysterious disguised figure, the 'duk-duk'. 1896 A. J. Butler tr. *Ratzel's Hist. Mankind* I. 133 The weak chiefs of Melanesia, in order not to be quite powerless, apply the mystic Duk-Duk system to their own purposes. 1908 *Athenæum* 11 Apr. 444/2 The Dukduk of Melanesia. 1929 E. A. Weber (*title*) The Duk-Duks. Primitive and historic types of citizenship.

duke, *sb.* **7.** Add to def.: Usu. *pl.* Also **dook**. (Further examples.)

1874 Hotten *Slang Dict.* 153 'Put up your dooks' is a kind invitation to fight. 1898 J. D. Brayshaw *Slum Silhouettes* i. 3 'E could 'andle 'is dooks, an' no error: the way 'e set abaht Bill was a fair treat. 1952 Partridge *From Sanskrit to Brazil* 4 He can handle his fives or dooks..or hands, i.e. he can box well. 1963 J. Mitford *Amer. Way Death* vi. 191 The funeral men are always ready with dukes up to go to the offensive.

Dukhobor, var. *Doukhobor.

D.U.K.W., dukw: see *duck *sb.*⁴

dulce, *sb.* **2.** (Earlier and later examples.)

1844 G. W. Kendall *Narr. Santa Fé Exped.* II. i. 31 Among the higher order of Mexicans the dinner finishes with fruits, dulces or sweetmeats. 1846 R. Ford *Gatherings from Spain* xi. 130 The sweet hams of the Alpujarras..are called *dulces* or sweet, because scarcely any salt is used in the curing. 1923 J. Hergesheimer *Bright Shawl* 127 Minute fragile dulces, cakes, glazed in green and pink, and ornamental confections of almond paste.

dulcin (dʌ·lsin). *Chem.* Also **dulcine**. [f. L. *dulcis* sweet + -in¹.] **a.** = Dulcite, *dulcitol.

1862 H. Watts tr. *Gmelin's Hand-bk. Chem.* XV. 384 Dulcite $C_{12}H_{14}O_{12}$... Dulcose (Laurent); Dulcine (Jacquelain). 1920 C. T. Kingzett *Pop. Chem. Dict.* 116 Dulcitol or Dulcin or Dulcite ($C_6H_{14}O_6$).

b. A synthetic crystalline derivative of phenetidine which has been used as a sweetening agent; *p*-ethoxyphenylurea, $C_9H_{12}N_2O_2$; *p*-phenetylurea.

1893 J. Attfield *Chem.* (ed. 15) 514 Para-phenetol-carbamide or dulcin is a body having a very powerful sweet taste, and proposed for use, like saccharin, in place of sugar. 1912 *Jrnl. Chem. Soc.* CII. ii. 104 Detection and identification of 'saccharin' and 'dulcin' in beverages, foods, drugs, cosmetics, etc. 1948 Gardner & Cooke *Chem. Synonyms* (ed. 5) 187/1 Dulcine..is 200 times sweeter than cane sugar. 1959 S. Davidson et al. *Human Nutrition & Dietetics* lvii. 788 The use of dulcin as an alternative sweetening agent to saccharin..was found to cause liver tumours in animals.

dulcitol (dʌ·lsitɒl). *Chem.* [f. Dulcit(e + -ol.] = Dulcite.

1884 *Jrnl. Chem. Soc.* XLVI. 1284 Dulcitol is neutral to litmus. 1938 *Thorpe's Dict. Appl. Chem.* (ed. 4) II. 296/1 Haas has detected both dulcitol and sorbitol in a red seaweed. 1970 R. W. McGilvery *Biochem.* xxvi. 639 Some infants have defects of galactose metabolism...If the kinase is missing..the accumulation of galactose leads to formation of cataracts through reduction to the corresponding hexitol, dulcitol.

dulcitone (dʌ·lsitōᵘn). [f. L. *dulcis* sweet + Tone *sb.*] A keyboard instrument in which steel forks are struck by hammers (see quot. 1888).

1888 T. Machell *Brit. Pat.* 4071 1 My invention, which I call 'The Dulcitone', has been designed to meet the want now generally felt of a portable keyboard percussion instrument...In the 'Dulcitone' the sound producers (hereinafter referred to as forks) are bars of steel bent into a U form, the forks being of graduated lengths for the various musical notes. 1909 *Chambers's Jrnl.* Mar. 206/2 The dulcitone is a musical instrument on the lines of the pianoforte. 1923 R. Noble *Shakespeare's Use of Song* 17 The dulcitone, which is portable and is capable of being performed upon by an actor or actress of moderate musical accomplishment. 1970 R. Smith Brindle *Contemp. Percussion* 63 The dulcitone, as its name suggests, is soft and sweet, with a bell-like sound not unlike the celesta.

dull, *a.* Add: **8. dull emitter**, a thermionic valve in which the filament operates at a relatively low temperature, and so does not glow brightly; also, such a filament; also *attrib.* Hence **dull-emitting** adj.

1922 *Wireless World* 30 Sept. 858/2 The set has been designed that either R or 'Dull Emitter' valves can be fitted, the latter permitting of the operation of the set from dry cells. 1923 *Ibid.* 5 May 137/2 The dull-emitting filaments of the valves are made from thoriated tungsten and..if thoriated tungsten had not proved useful in electric lamp manufacture, there would not exist to-day the 'dull-emitter' valve. 1923 *Exper. Wireless* Nov. 97/2 The object of the dull emitter is to obtain the same filament emission with only a fraction of the energy previously required to render the ordinary filament sufficiently incandescent. 1970 D. F. Shaw *Introd. Electronics* (ed. 2) x. 209 Oxide-coated cathodes..glow a dull red and they are therefore called dull emitters, contrasted with tungsten filaments which are bright emitters.

dulosis (diulōᵘ·sis). *Ent.* [mod.L., ad. Gr. δούλωσις, f. δουλοῦν to enslave, f. δοῦλος slave.] The practice, exhibited by certain genera of ants, of enslaving other ants or colonies of ants. So **dulo·tic** *a.*, slave-holding.

1904 *Biol. Bulletin* VI. 257 Dahl is mistaken in supposing that *L. bismarckensis* is a dulotic ant, as a perusal of the above quoted passages from Wroughton's work will suffice to show. 1905 W. M. Wheeler in *Bull. Amer. Mus. Nat. Hist.* XXI. 2 This observation, taken in connection with the close taxonomic affinities of *perpandes* with *sanguineus*, forcibly suggests dulosis. 1925 A. D. Imms *Gen. Textbk. Entom.* 571 From temporary social parasitism the next step is exhibited by dulosis or slavery. 1928 W. M. Wheeler *Social Insects* 189 Yet another Myrmicine genus, *Harpagoxenus*..must be included among the dulotic, or slave-making ants. 1970 Brown & Taylor in *Insects of Australia* (C.S.I.R.O.) xxxvii. 953/1 The special type of permanent parasitism known as dulosis, or slave-making, in which workers of the 'slave' species are constantly replenished by the addition of pupae acquired through mass raids on neighbouring pure nests of the host species, is so far not known to occur in Australia.

duma (dū·mă). Also (in Fr. form) **douma**. [Russ. *dúma*.] In Russia, an elective municipal council. *The Duma*, the elective legislative assembly (*Gosudárstvennaya Duma*), which was established in 1905 by a ukase of Tzar Nicholas II and lasted until the Revolution of 1917. Hence **du·maist**, a member of a duma or the Duma.

1870 *Nation* 1 Dec. 364/2 Municipal Reform in Russia... The city is governed by a city council (*duma*) and a city regency (*uprava*). 1886 *Encycl. Brit.* XXI. 79/2 Since 1870 the municipalities have had institutions like those of the zemstvos... The executive is in the hands of an elective mayor and an *uprava* which consists of several members elected by the duma. 1905 *Daily Chron.* 16 Feb. 5/1 The Duma will consist of delegates of the district councils, each sending five. 1905 *Outlook* (N.Y.) 12 Aug. [illegible] 1905 *Times* 19 Aug. 7/2 The Duma is established for the preliminary study and discussion of legislative propositions which..will go, in due course, to the supreme authority through the Council of the Empire. 1906 G. Meredith *Let.* 5 July (1970) III. 1565 The members of the Duma..have shown a remarkable self-control. 1906 *Daily Chron.* 2 Aug. 1/7 M. Herrenstein, a wealthy Dumaist, has been shot dead. [illegible] 1955 G. B. Carson *Electoral Practices in U.S.S.R.* 2 The first important electoral machinery was that created for elections to the imperial duma.

dumb, *a.* Add: **1. b.** Esp. in phrases *dumb chum*, *dumb friend*, applied to domesticated animals.

1870 *Animal World* July 163/1 My considerable experience..combined with the affection I still feel for the dumb friends of my childhood, induces me to note down a few remembrances of their favourites. [illegible] 1934 —— *Right Ho, Jeeves* i. 13 [illegible] 1940 'N. Blake' *Malice in Wonderland* i. vii. 84 Don't leave your dumb friends at home! Bring them to Wonderland and intern them in our superbly equipped Pets' Corner! 1957 R. Campbell *Portugal* 62 Cassius's dumb-chum, the morse, adored him.

7. b. Foolish, stupid, ignorant (chiefly of persons); *spec.* **dumb blonde**, a conspicuously attractive but stupid blonde woman; **dumb bunny**, a stupid person; **dumb Dora**, a stupid girl; **dumb ox**, a stupid, awkward, or uncommunicative man (cf. Dumb *a.* 1 b in Dict.); *spec.* a nickname for St. Thomas Aquinas. *colloq.*

Possibly reinforced by G. *dumm* or Du. *dom*.

[*c* 1323 P. Calo *Vita* in D. Prümmer *Fontes Vitae S. Thomae Aquinatis* (1912) i. 78 Ceperunt eum Fratres vocare bovem mutum.] [illegible] Feb. 441/1 My, but men are dumb. A woman would have caught on long ago. 1914 R. Brooke *Let.* 15–17 Aug. (1968) 609 This is a badly-written, dumb, letter. 1919 F. Hurst *Humoresque* 259, I been so dumb not to right away see it. 1922 *Dialect Notes* V. 141 *Dumb-bunny* [in College slang], a somewhat stupid person...'My dear, you are the preshest old dumb-bunny.' *Ibid.* 147 *Dumbdora*, a stupid girl. 1930 [see *ball *sb.*¹ 18]. 1932 'A. Bridge' *Peking Picnic* xx. 255 He is that dumb, if you'll pardon the word, madam, that not a bit of sense could I get out of him. 1936 'J. Tey' *Shilling for Candles* vi. 62 A sulky fair girl, who played 'dumb' blondes from year's end to year's end. 1938 E. Bowen *Death of Heart* i. i. 25 One has got to see just how dumb Mr. Quayne was. He had not got a mind that joins one thing and another up. 1945 L. Shelly *Jive Talk Dict.* 24 *Dumb bunny*, stupid person. 1947 *Times Lit. Suppl.* 15 Nov. 594/2 The cult of 'dumb-ox' individualism is certainly unsuitable for New Zealand, where the individual has rarely bulked as large as the State. 1959 'M. Derby' *Tigress* ii. 88 The dumb blonde to whom all instruments and machinery were insoluble riddles. 1959 *Listener* 16 July 110/3 An amiable dumb-ox. 1965 G. McInnes *Road to Gundagai* ix. 145 They [sc. hens] would then wait expectantly, heads

cocked on one side with a sort of dumb-Dora inquisitive chuckle.

8. *dumb ague* (earlier and later examples); also, an irregular form of malarial fever, which lacks the usual chill; also *dumb chill*, *fever*.

1793 *Trans. Amer. Philos. Soc.* III. p. iv, Fever and ague..also visits the borders of limpid streams. The lesser degree of it generally called *dumb ague*, is not rare in the most salubrious places during the months of September and October. 1832 R. Baird *View Valley Mississippi* viii. 73 These maladies are intermitting and remitting bilious fevers..which..have received the names of 'ague', 'dumb ague', and 'chill and fever'. 1859 Bartlett *Dict. Amer.* (ed. 2) 133 *Dumb chill*, or *dumb ague*, an expression common in malaria regions to denote that form of intermittent fever which has no well defined 'chill'. 1893 Dartnell & Goddard *Gloss. Wiltshire* 49 'Tis what 'ee do caal the dumb-agey.

B. *sb.* **3.** *N. Amer.* A foolish or stupid person. Also **dum(b)-dum(b)**. *colloq.*

1928 *Daily Express* 4 Dec. 10/3 A 'dumb' is a stupid person, and if he's dumb enough he'll probably drive you 'cuckoo' or crazy. 1970 L. Sanders *Anderson Tapes* xxxi. 81 You're no dumdum, are you? 1970 *Calgary Herald* 24 Aug. 9/1 Better they should employ some dumb-dumb.

C. *Comb.* **b.** **dumb-play** = Dumb show 2.

1920 *Chambers's Jrnl.* 374/1 A violent dumb-play of smoothing the hair and arranging the coats of pyjamas. 1921 *Glasgow Herald* 10 Feb. 6 Certain M.P.'s did take the part of actors and were duly taken and their dumb-play shown on the screen.

dumb-bell, *sb.* Add: **2. b.** [After *dumb *a.* 7 b.] = *dumbhead. Also *attrib. slang* (orig. *U.S.*).

1920 *Collier's* 3 July 8/1 The gent..stands alone as the Crown Prince of dumb-bells. 1922 S. Lewis *Babbitt* xviii. 227 The poor old dumb-bells that you can't get to dance. *a* 1930 D. H. Lawrence *Etruscan Places* (1932) i. 21 They gave the usual dumb-bell answer. 1936 *Punch* 15 Apr. 430/3 Next came one of those series of Dumb-bell Letters which seem to be very popular, a dumb-bell being the kind of person who writes to the manufacturer asking him to replace a gadget that has been lost, and then adds a postscript telling him not to bother as the missing gadget has just been found.

dumb cluck, du·mb-cluck. *slang* (orig. *U.S.*). [f. *dumb *a.* 7 b + *cluck *sb.* 5.] A dull or stupid person; a fool.

1929, 1931 [see *cluck *sb.*]. 1934 J. M. Cain *Postman always rings Twice* vi. 59 This Greek had had a fracture of the skull, and a thing like that don't happen to a dumb cluck like him every day. 1939 C. Morley *Kitty Foyle* 164 They're just dumb clucks, and they haven't any ambition to go off on their own. 1944 *Coast to Coast* 1943 125 She had to get up and speak, and tell these dumb-cluck women they'd just got to pull their weight. 1945 D. Wheatley *Man who missed War* vi. 62 Do I look the kind of dumb-cluck who would be wastin' her time joy-ridin' in this boat instead of gettin' her a liner? 1958 *Economist* 1 Nov., There emerges an elite of Top People (no longer encumbered with hereditarily privileged dumb-clucks) in the 140-and-up IQ range. 1960 O. Manning *Great Fortune* ii. 15 For the last half-hour I've been telling these dumb clucks to find me a bloke who can speak English.

du·mbhead. *U.S.* and *Sc. slang.* [f. *dumb *a.* 7 b + Head *sb.*, after G. *dummkopf*, Du. *domkop*.] A blockhead.

1887 H. Frederic *Seth's Brother's Wife* 339 We wouldn't elect such a dumb-head to be a hog-reeve. 1895 S. R. Crockett *Men of Moss-Hags* xxxix. 278 What a dumbhead I was, to bide with an empty belly in a place where at least there must be plenty of fish near at hand. 1921 C. E. Mulford *Bar-20 Three* xi. 125 Have I got to do all the thinking for this crowd of dumbheads?

dumb-iron (dʌ·maiəɹn). [See Dumb *a.* 8.] A carriage-spring composed of two half-elliptic springs joined at the ends. In the chassis of a motor car, each of the pair of curved forward ends of the frame-side members to which one end of each front spring is fixed. (See also quot. 1963.)

1907 *Westm. Gaz.* 19 Dec. 4/2 The distinctive frame is of pressed steel, slightly narrowed from the dashboard, tapered, and turned down in front so as to form its own dumb-irons, thus making the connexion for the forward ends of the springs. 1920 *London Mag.* June 334/1 The sharp dumb-irons dug into the bank, and the car reared itself on end. 1928 *Daily Tel.* 16 Oct. 17 Sankey's tool box..which fills the usual blank space between the dumb irons, curved to follow the line of the dumb iron. 1963 R. F. Webb *Motorists' Dict.* 86 *Dumb irons*, the portions of the chassis of a car which carry the shackles for the front springs. Recently the term has also come to mean the iron or steel supports carrying the fenders or bumpers.

dumble (dʌ·mb'l). *dial.* A dumbledore.

1893 in Dartnell & Goddard *Gloss. Wilts.* 50. 1909 Hardy *Time's Laughingstocks* 67 The dumbles thin their humming.

dumbness. Add: **2.** Stupidity; ignorance (cf. *dumb *a.* 7 b). *U.S. colloq.*

1858 Thoreau *Jrnl.* 10 Oct. in *Autumn* (1892) 94, I see dumb-bells in the minister's study, and some of their dumbness gets into his sermons. 1924 *Chicago Tribune* 26 Oct. (Comics) 6 You can tell me how to take some of the dumbness outta the kid.

dumb-waiter. Add: **1.** (Earlier example.)

1749 J. Cleland *Mem. Woman Pleasure* I. 156 A bottle of Burgundy, with the other necessaries, were set on a dumb-waiter.

2. For *U.S.* read *orig. U.S.* and add examples.

1847 Webster 372/2 When the kitchen is in the basement, the dumb-waiter is made to rise and fall by means of pulleys and weights. 1876 J. S. Ingram *Centenn. Exposition* iii. 69 The general kitchen whence the food is distributed throughout the upper stories by means of dumb waiters. 1903 A. H. Lewis *Boss* vi. 69 There's a dumb waiter from the bar to send up beer and smokes. 1960 H. Pinter *Dumb Waiter* 16 Disclosed is a serving-hatch, a 'dumb waiter'. A wide box is held by pulleys.

dum casta (dʌm kæ·stă). *Law.* [L.; short for *dum sola et casta vixerit* as long as she shall live alone and chaste.] A clause in a legal instrument conferring on a woman a benefit which is to cease should she cease to lead a chaste life, or should she marry or remarry.

[1628 E. Coke *First Pt. Inst. Laws Eng.* iii. 234ᵛ Dum, also maketh a limitation, as if a Lease be made, *Dum sola fuerit*, or *Dum sola & casta vixerit*.] 1883 J. M. Lely *Wharton's Law Lex.* (ed. 7) 280/2 This proviso is termed the 'dum casta clause'. 1905 *Law Jrnl. Rep., Prob.* LXXIV. 3/1 The main subject to be regarded is the temptation to which the wife may be exposed; and in this respect it appears a material precaution that the limitation *dum casta* should be inserted. 1906 *Westm. Gaz.* 29 Mar. 7/2 The question..was whether it was a condition of the allowance of £100 that a *dum casta* or chastity clause should be inserted. 1960 *Times* 15 Jan. 15/7 The defendant..forfeited..her right to maintenance..by committing a breach of the *dum sola et casta* clause in the order.

dum-dum (dʌ·mdʌm). Also **Dum Dum, Dum-Dum.** [f. *Dum-Dum*, name of a town and arsenal near Calcutta, where they were first produced.] In full *dum-dum bullet*: a metal-cased bullet with a soft core uncovered at the point, which expands on impact. Hence **du·m-dum** *v. trans.*, to convert into a dum-dum bullet.

1897 *Westm. Gaz.* 14 Dec. 7/3 The piper hero, Findlater, was wounded in the ankle with a Dum Dum bullet. 1898 *Ibid.* 25 Mar. 3/1 That the War Office authorities are not responsible directly for Dum-dumming the Lee-Metford bullets. 1899 *Ibid.* 25 Nov. 6/1 Any man can 'Dum-Dum' his own cartridges in a very few minutes by merely filing the nickel envelope off the nose of the bullet. 1906 *Ibid.* 27 Mar. 9/2 It is impossible that the Kaiser can know or approve of the adoption of the dum-dum. 1968 *Peace News* 12 Apr. 8/3 Dum-dum bullets were outlawed by 22–2 votes. 1970 *Daily Tel.* 9 Jan. 3/7 Use of the dum-dum has been classified as a war crime by a number of international conventions, including the Hague Declaration of 1907.

|| **dumka** (du·mkă, dʌ·mkā). *Mus.* Pl. **dumkas, dumky.** [Czech. = plaintive song, elegy.] An alternately melancholy and gay piece of music, found chiefly in the work of Slavonic composers.

[1886 A. Dvořák (*title*) Dumka a Furiant.] 1895 W. H. Hadow *Stud. Mod. Mus.* 2nd Ser. 221 Even more distinctive is his [sc. Dvořák's] treatment of the Dumka or 'Elegy', a complex form which, like a sonnet-sequence, holds in combination a series of separate poems. 1947 A. Einstein *Mus. Rom. Era* xvii. 302 Dvořák was less regionally limited than Smetana; although he still wrote polkas, dumkas, and furiants, he also wrote waltzes and mazurkas.

dummerer. (Later example.)

1931 J. Buchan *Blanket of Dark* xvi. 304 There was a troop of crowders in the little town, and as many cozeners and dommerers as if it had been an abbey-gate.

|| **dummkopf** (du·mkɒpf). *colloq.* (orig. *U.S.*). Also **dom cop, dum(b)kopf.** [G., see *dumbhead.] = *dumbhead.

1809 'D. Knickerbocker' *Hist. N.Y.* II. v. ii. 16 As great a dom cop, as if he had been educated among that learned people of Thrace, who..could not count beyond the number four. 1834 Carlyle *Sart. Res.* ii. vii. 59/1 Not being born purely a Loghead (*Dummkopf*), thou hadst no other outlook. 1894 G. du Maurier *Trilby* I. ii. 166 Great big she-fool that you are—sheep's head! Dummkopf! Donnerwetter! 1923 M. Watts *L. Nichols* 15 All I got to do is wait on dumkopfs! 1951 M. Lowry *Lett.* (1967) 237 He was..as brilliant a teacher as I was a dummkopf of a pupil. 1968 *Listener* 31 Oct. 567/3 They may turn out, after all, to have been fall guys, dumbkopfs, dupes of their own chicanery.

dummy, dumby, *sb.* (The usual modern form is **dummy**.) Add: **1. b.** A deaf-mute; a tramp or beggar who pretends to be deaf and dumb. *slang.*

1874 in Hotten *Slang Dict.* 1918 'A No. 1' in N. Anderson *Hobo* (1923) vii. 100 *Dummy.* Pretends to be deaf and dumb. 1926 J. Black *You can't Win* vi. 70 What do you do if you bump into a natural dummy when you're D. D.-ing? 1940 C. McCullers *Heart is Lonely Hunter* (1943) i. iv. 47 But a dummy!..'Are there any other deaf-mute people here?' he asked.

2. b. In auction or contract Bridge, the hand of the declarer's (in the old type of Bridge, of the dealer's) partner, which is displayed and is played by the declarer; also, the player whose cards are thus displayed; *double dummy* (see quot. 1964). Also *attrib.*

1886 *Biritch, or Russian Whist* 2 No suggestions as to play may be made by the one standing out (Dummy) to the dealer. 1894 'Boaz' *Pkt. Guide to Bridge* 5 After the first player has played a card, the dealer plays his partner's hand, which, like Dummy, is placed face upwards on the table. 1901 R. F. Foster *Bridge* 17 After laying down his cards..the dummy takes no further part in the play, and is not allowed to make any remarks or suggestions. 1901 W. Dalton *Bridge Abridged* 23 With very bad cards in your own hand, there is a slightly increased possibility of an exceptional hand in your Dummy. *Ibid.*, After the first card is led the Dummy hand is exposed. 1903 [see *bridge *sb.*³ b]. 1904 *Bridge & Progressive Bridge* 18 Dummy Bridge, or Bridge for three players, can be played in several ways. The following is the best system:—The player who cuts the lowest card takes Dummy against the other two players..and the Dummy hand is placed opposite to him. *Ibid.* 22 Double Dummy Bridge, or Bridge for two players, is not at all a bad game. 1910 W. Dalton *'Saturday' Bridge* (9th imp.) 14 After exposing Dummy, the dealer's partner has no part whatever in the game. *Ibid.* xvi. 230, Z leads to the first trick, and A's hand is exposed on the table, and A becomes the dummy for that deal. 1936 E. Culbertson *Contract Bridge Complete* xxxix. 443 South leads a small trump and dummy's Queen is taken by East's King. 1955 *Times* 6 July 4/7 It is not a double-dummy problem to find the best line of play by the declarer. 1964 *Official Encycl. Bridge* 131/1 *Double dummy*, play of a hand that could not be improved upon, as though declarer were looking at all four hands...It can also be used to refer to perfect play by the defenders. Originally, double dummy was a two-handed form of whist in which each player had a dummy. Some players exposed all four hands, thus giving rise to the modern usage. 1969 E. H. Pinto *Treen* 224 Playing card stands..are used in double dummy bridge.

5. a. (See also *tailor's dummy* s.v. Tailor *sb.* 6 b.) **e.** (Later examples.) (See also quot. 1964.) So *in dummy*, in dummy form.

1898 A. J. Balfour in *Hansard Commons* 20 June 768 The Report has been presented in dummy, and is now being proceeded with. 1929 F. W. Carter *Secrets of your Daily Paper* 59 They are supplied with what is known as a 'dummy' or a 'make-up'. 1964 *Gloss. Letterpress Rotary Print. Terms (B.S.I.)* 12 *Dummies*, blank or printed copies usually made up for editorial or production purposes. 1967 *Times Rev. Industry* Aug. 69/1 A mammoth *Encyclopaedia of Education*, the dummy of which looks most impressive even though not all the contents are in it.

f. In full *dummy teat*. An indiarubber teat put into a baby's mouth to soothe it. Also *fig.*

1903 *Science Siftings* 22 Aug. 269/1, I never saw the child but it had a dummy in its mouth. 1906 *Chemist & Druggist* LXIX. 648/2 There has been little progress in the shape of the 'dummy teat'. 1915 D. H. Lawrence *Let.* 12 Feb. (1962) I. 316 He [sc. E. M. Forster] sucks his dummy—you know, those child's comforters—long after his age. *a* 1930 —— *Last Poems* (1932) 273 The British Public..gets bigger and bigger..and its dummy-teat has to be made bigger and bigger and bigger.

g. (Rugby) Football. *to give* (or *sell*) *the* (or *a*) *dummy*: to feign to pass the ball so as to deceive one's opponent. *colloq.*

1907 'Old International' *Rugby Guide* 27 Feinting, 'giving a dummy', or pretending to pass is a useful adjunct to the numerous other qualifications of a good centre. *a* 1914 J. E. Raphael *Mod. Rugby Football* (1918) 125 A little judicious 'dummy' giving might be very effective in securing an opening. 1920 *Times* 8 Nov. 6/3 R. C. Pickles 'sold the dummy' really cleverly to score again for Gloucestershire.

h. The punishment cell in a prison. *N.Z. slang.*

1936 'R. Hyde' *Passport to Hell* iii. 60 It is always dusk in the Dummy, which lies underground. *Ibid.* iii. 65 Twenty-one days on bread and water in the Dummy. 1945 O. Burton *In Prison* vii. 107 The aggressor in this case was promptly led off and incarcerated in the 'dummy'.

i. A figure representing a human being, animal, etc., used by a ventriloquist. Cf. *doll *sb.*¹ 2 b. Also *fig.*

1936 E. Sitwell *Victoria* xix. 229 The miserable, hallucinated ventriloquist-dummy hero of one of the most notorious murder trials. 1950 [see *doll *sb.*¹ 2 b]. 1960 *Economist* 8 Oct. 175/1 Is he no more than the ventriloquist's dummy for any scientific pressure group that is not getting its way with the Treasury?

j. In a double bail in a milking shed, the structure which separates the foreparts of the two cows, keeps them in position for milking, and serves as a stand for parts of the milking machine. *N.Z.*

1950 *N.Z. Jrnl. Agric.* Apr. 378/3 The use of galvanised iron piping or iron piping for erection of internal dummies and partitions. *Ibid.* May 479 (*caption*) A small dairy with suspended dummy and bucket milking machine for one cow.

k. *Computers.* An instruction or a sequence of data that merely occupies space, used either to regularize the position of other items or to allow a later insertion. Usu. *attrib.*

1948 Goldstine & von Neumann in *Coll. Wks. J. von Neumann* (1963) V. 158 The three *a*'s must obtain numbers of the same parity. This may necessitate the insertion of dummy (ineffective, irrelevant) orders in appropriate places. 1951 M. Wilkes et al. *Preparation of Programs for an Electronic Digital Computer* 148 Order 13 is I F during input of punched digits, T F for dummy zeros which make up remainder of 10 digits. *c* 1964 *KDF9 Programming Man.* iv. 22 Any instruction preceded by an asterisk will be compiled as the first instruction in a new word, any redundant spaces in the preceding word being filled with dummy instructions. 1969 Lehman & Bailey *Digital Computing* x. 193/2 The function might be defined as SUM(A,B,C,D) = A+B+C+D. In this case, the dummy variables are A, B, C, and D. When the function reference is executed, these dummy variables take on the values of the variables named in the reference. 1970 F. Stuart *Fortran Programming* (ed. 2) ix. 203 All arguments mentioned as dummies may be duplicated in other subprograms or in the main program, without causing error.

7. (Later examples.)

1971 J. Leasor *Love-All* i. 6 Now for the flat, number 1792...Floor 17, flat two. The nine was just a dummy digit to make the building seem more important...Some people did that with their names and added an extra initial..Harry S. Truman for example. 1971 *Listener* 22 Apr. 521/2 Of 91 human volunteers, 47 received three grams of ascorbic acid daily for nine days, and 44 others took dummy tablets.

b. *Comb.* **dummy-head(ed)** *a.*, applied to a torpedo which is provided with a thin copper head and filled with water for target practice; **dummy run** *colloq.*, orig. *Naval slang* (see quots. 1916 and 1929); hence, a practice attack, exercise, landing, etc.; a 'trial run', a rehearsal.

1906 *Daily Chron.* 8 Sept. 5/3 A large proportion of the dummy-headed torpedoes struck her hull. 1923 *Daily Mail* 22 June 5 The 'planes released six dummy-head aerial torpedoes. 1916 *In Northern Mists* xvi. 60 At certain firing practices the ship steams along a buoyed course, aiming the guns at the target but not actually firing, as a rehearsal or dummy run. 1929 F. Bowen *Sea Slang* 42 *Dummy run*, a practice evolution in the Navy. 1943 C. H. Ward-Jackson *Piece of Cake* 27 *Dummy run*, a bombing raid or target approach during which no bombs are dropped. 1959 *Punch* 6 May 604/1 A man of my known and respected equanimity was considered the ideal vehicle for a dummy run. 1962 *Listener* 31 May 942/1 The pied wagtails made dummy runs over the hawks' heads. 1967 E. Grierson *Crime of one's Own* vii. 57 It was a dummy run...What I saw today was part of the cover story.

dummy, *v.* Add: **2.** *to dummy up*: to refuse to talk or give information; to keep quiet. *U.S. slang.*

1926 J. Black *You can't Win* xix. 282 He dummies up on the natives an' in a couple of days they let him go. 1942 R. Chandler *High Window* (1943) xx. 145 You can't dummy up on a murder case. 1962 K. Orvis *Damned & Destroyed* xii. 84 All right, dummy up, then.

3. *trans.* To prepare a dummy (Dummy *sb.* 5 e) of (a book, document, etc.). Also *intr.* (with *in*) of copy: to fit into a lay-out. *U.S.*

1928 in *Funk's Stand. Dict.* 1952 *Time* 21 Apr., *Dummy* (v.t.), to lay out a sample issue of the magazine, showing where editorial material and ads will be printed. 1967 L. J. Braun *Cat who ate Danish Modern* iv. 33 The copy dummied in perfectly, cut-lines spaced out evenly.

4. *trans.* In Football, to feint (a pass); to deceive (an opponent) by means of a feigned pass, a body-swerve, etc. Also *intr.* So *to dummy one's way*. Cf. *dummy *sb.* 5 g.

1958 *Times* 13 Oct. 15/1 Once Ridd nearly dummied his way through a wall of humanity. 1960 E. S. & W. J. Higham *High Speed Rugby* xviii. 262 The fly-half dummies a pass to him and then slips the ball back inside to the blindside wing. 1960 T. McLean *Kings of Rugby* 148 Jackson made dummying runs which lightened the gathering gloom. 1961 *Times* 8 May 4/6 Smith..dummied King one way, swivelled the other and on the turn cracked a fierce shot to the roof of the Leicester net. 1967 *Times* 9 Oct. 13/8 A typically brilliant try by Pickering, who dummied his way over near the posts.

dump, *sb.*² Add: **b.** Colloq. phr. *not to care a dump*: not to care at all; to regard as unimportant.

1843 *Ainsworth's Mag.* IV. 315 Mrs. Dodger didn't care a 'dump' if she didn't. 1908 G. K. Chesterton *All Things Considered* 70, I do not care a dump whether they know the alphabet.

dump, *sb.*⁴ Add: **1.** *spec* A pile of ore, earth, etc., which accumulates during mining operations; esp. *U.S.* and *S. Afr.* Cf. *mine-dump* (*mine *sb.* 6).

1865 *Harper's Mag.* Feb. 287/1 A number of Mexicans were at work getting out the ore and..I took a seat a little on one side of the 'dump'. 1871 *Daily News* 21 Sept., The dump is being overhauled and the pay ore selected for the company's mill. 1909 *Westm. Gaz.* 11 Feb. 3/3 Such statements as that..a mine dump can contain 40,000 tons, and that such a dump consists of 10 per cent. of pitchblende. 1931 J. Mockford *Khama* xxiii. 155 The grey dumps of the world's greatest gold mines. 1948 A. Paton *Cry, Beloved Country* ii. viii. 174 He..looked out over the veld, out to the great white dumps of the mines, like hills under the sun. 1956 H. G. Dines *Metalliferous Mining Region S.W. Eng.* I. iv. 117 The dumps were being worked over for uranium ores in 1907.

b. The practice of dumping goods (see *dump *v.*¹ 2 c); also, the goods dumped.

1884 *Congress. Rec.* 1 May 3663/1 It is this dump that we want to stop; it is protection against this dump that the protective system seeks to accomplish. 1908 *Westm. Gaz.* 8 Apr. 9/2 The present 'dump', which has assumed such amazing proportions in the eyes of the alarmists.

c. A collection of provisions, ammunition, equipment, etc., deposited in a convenient place for later use; also, the place where such supplies are deposited.

1915 *Daily Mail* 30 Dec. 4/6 The Dump! Just a pile of old clothes, battered helmets, decrepit boots, kit bags, shirts, socks, boots—all the little personal properties of soldiers. 1916 J. Buchan *Battle of Somme* 55 That same day we..took a 'dump' of German stores. 1918 E. M. Roberts *Flying Fighter* 61 One night about six o'clock I received orders to report at an engineers' dump known as Hyde Park Corner. 1919 R. H. Reece *Night Bombing with Bedouins* 6 The gunners may be called upon to fire at certain targets, such as cross-roads or houses used as infantry headquarters or ammunition and stores dumps. 1925 E. F. Norton *Fight for Everest: 1924* 353 A dump for stores..would seem to be essential half-way between Camps II and III. *Ibid.* 361 To provide dumps of reserve cylinders on the mountain. 1926 T. E. Lawrence *Seven Pillars* (1935) iii. xxxi. 178 The collection of the necessary food-dumps for the army. 1937 *N. & Q.* CLXXIII. 19/2 Should the Air Ministry succeed in establishing their ammunition 'dump' at Acorn Bank [etc.]. 1939 [see *arms-dump* s.v. *arm *sb.*² V].

d. An act of defecation.

1942 Berrey & van den Bark *Amer. Thes. Slang* § 124/1 *Defecation*,..call of nature, crap, dump. 1966 Auden *About House* 26 To start the morning With a satisfactory Dump is a good omen All our adult days.

2. b. A place, building, house, etc.: usu. as a pejorative or contemptuous term. *colloq.* (orig. *U.S.*).

1899 'J. Flynt' *Tramping with Tramps* 393 *Dump*, a lodging-house or restaurant; synonymous with 'hang-out'. 1903 *Cincinnati Enquirer* 9 May 13/1 *Dump*, a house; saloon, hang-out for a gang. 1914 Jackson & Hellyer *Vocab. Criminal Slang* 30 *Dump*, a rendezvous; an establishment of any kind; a hangout; a joint; a meeting place. 1919 F. Hurst *Humoresque* 321 You never got in your life to live in a worse dump. 1929 P. Johnson *Four Plays* 26 We'll see what's doin' in this bloody dump. 1932 Wodehouse *Hot Water* i. 27 Do you think if I had any money I'd be living in a dump like the Château Blissac? 1941 K. Tennant *Battlers* xxviii. 192 Commercial travellers..were hurrying through these little 'dumps' of towns. 1942 *Daily Express* 8 Jan. 2/7 A uniformed cop patrolled the bar...I didn't think that mattered much at a dump like this. 1959 J. Burke *Echo of Barbara* ii. 13 We'll go and have an evening in the town here. Not much of a dump, but you'll find quite a good crowd there.

3. b. *Surfing.* = *dumper d.

1935 *Bulletin* (Sydney) 9 Jan. 11/3 It gave the Duke of Gloucester his first experience of a dump, in the Mooloolaba surf. 1963 *Observer* 13 Oct. 15/3 A rider..must be able to..escape the 'dump' by flicking back off the top of the wave or sliding across to some section where the water is too deep for the wave to break. 1967 J. Severson *Great Surfing* Gloss. 153 *Dump*, a wipe-out in surfing.

4. Esp. used *attrib.* (chiefly *U.S.*) to designate vehicles having a body that tilts or opens at the back for unloading materials, as *dump car, cart, truck, wagon*; **dump condenser** (see quot. 1960); **dump tank**, a tank used for receiving a sudden discharge of liquid from a reactor or for storing radioactive liquids while they are dangerous; **dump valve**, a valve which releases the contents of a container quickly.

1912 *Out West* Feb. 133/1 The small dump cars were wont to carry the yellow gravel from the cut. 1868 *Mich. Agric. Rep.* VII. 347 Joram Priest, Detroit, [manufactured the] 2 dump carts. 1958 *Nuclear Power* III. 170 Instrumentation system of a typical gas-cooled reactor..dump condenser. 1960 *Gloss. Atomic Terms* (*H.M.S.O.*) 20 *Dump condenser*, a water-cooled steam condenser which allows the heat output of a power reactor to be 'dumped' into the cooling water system should the turbine system become inoperative. 1959 *Nuclear Energy Engineer* XIII. 337 Reactor for Chalk River... Dump tanks. 1964 C. F. Bonilla in *Reactor Handbk.* (ed. 2) IV. iii. 122/1 Dump tanks in any of the coolant systems may receive hot discharge from the system at any time. 1930 *Water Works & Sewerage* Dec. 24/3 Where to buy..Trucks, Dump. 1936 J. Steinbeck *In Dubious Battle* xv. 295 Up the road from Torgas a huge Mack dump-truck rolled. 1959 *New Scientist* 29 Oct. 801/1 Dump trucks..shifted, on an average, over 500 yards an hour. 1930 *Flight* 4 July 760/1 It is intended to let air into this reservoir quickly when the safety dump valve is used. 1955 *Jrnl. Brit. Interplanetary Soc.* XIV. 16 To give a clean shut-down a dump valve was fitted to the cooling jacket which drained overboard any fuel remaining in it. 1869 *Rep. Comm. Agric. 1868* (U.S.) 357 Wheelbarrows, carts, or dump-wagons will be necessary. 1969 *Islander* (Victoria, B.C.) 9 Nov. 2/1 There were no trucks as today, 120 mules and 50 to 60 dumpwagons on the move from dawn till dark did the trick.

dump, *v.*¹ **2. a.** For Chiefly *U.S.* read orig. *U.S.*, and add earlier and later examples. Also, to make a dump of (*dump *sb.*⁴ 1 c).

1784 J. Hiltzheimer *Diary* 16 Mar. (1893) 62 The Street Commissioners selected sites to dump the dirt from the streets. 1919 A. P. Herbert *Secret Battle* viii. 165 Philpott..accused him hotly of dumping the rations carelessly anywhere. 1919 G. K. Rose *2/4th Oxf. & Bucks Lt. Infty.* 35 A pile of logs dumped in the wrong place. 1925 E. F. Norton *Fight for Everest: 1924* 52 To prepare the camp and dump tents and stores for it. 1961 *New Eng. Bible Acts* xxvii. 38 They lightened the ship by dumping the corn in the sea.

c. *trans.* To export, or throw on the market, in large quantities and at low prices; *spec.* to offer for sale (surplus goods), esp. abroad, at less than the ordinary trade prices. Also *absol.* Often in *ppl. adjs.* and *vbl. sb.*

1868 *Commerc. & Financ. Chron.* VI. 326/1 New stock secretly issued [was] 'dumped' on the market for what it would fetch. 1884 *Congress. Rec.* 1 May 3663/1 The surplus dumped from foreign pauper markets is the great bane of our industries. 1903 *Westm. Gaz.* 8 Sept. 2/2 Those who base their case on (*a*) decreasing exports, and (*b*) the dumped state of iron and steel. 1903 *Daily Chron.* 25 Sept. 3/7 'Dumping' is in our eyes a great sin. 1904 *Treasury* Oct. 8/1 The..capitalists desire this, as it..enables them to dump their surplus production on foreign countries. 1908 *Westm. Gaz.* 30 Mar. 10/2 As for the dumping scare,..there is nothing in it. *Ibid.* 4 May 2/2 You appear to think that he dumps for the sake of dumping. 1916 *Economist* 4 Mar. 458/1 We hear of large stocks of cheap manufactures that will be dumped upon us. 1928 *Britain's Industr. Future* (*Lib. Ind. Inq.*) i. v. 50 They showed that the practice of dumping demoralises the world-market to the ultimate disadvantage of all concerned. 1957 *Act 5 & 6 Eliz. II* c. 18 (*title*) Customs Duties (Dumping and Subsidies) Act. *Ibid.*, For the purposes of this Act imported goods shall be regarded as having been dumped..if the export price from the country in which the goods originated is less than the fair market price of the goods in that country. 1970 *Financial Times* 13 Apr. 12/6 Continental companies could 'dump' steel in the U.K. if there are no tariff barriers.

d. To discard, abandon, get rid of. *colloq.*

1919 *Athenæum* 15 Aug. 759/1 'To dump' a thing that it is a nuisance to carry means to get rid of it. 1944 *Korero* (N.Z.) 9 Oct. 27 The pony will try..to 'dump' its burden [the rider]. 1946 D. Stivens *Courtship of Uncle Henry* 29 You've dumped plenty of fellows before. *Ibid.* 30 Dumping me like this for a couple of dumb sailors.

e. Of a wave: to hurl (a swimmer or surfer) down. Cf. *dumper d. Chiefly *Austral.*

1938 J. Moses *Nine Miles fr. Gundagai* 88 W'en de breakers dumped Me at Curl Curl. 1963 *Observer* 13 Oct. 15/4 The wave traps and dumps the rider, burying him for half a minute or longer and churning him over and over on the ocean floor. 1966 *Surfer* VII. 54 A really good body surfer got banged up pretty bad when a Yokohama wave dumped him right on the reef.

dumper. Delete *U.S.* and add earlier and later examples.

1856 *Trans. Mich. Agric. Soc.* VII. 334 There are on the road..20 gravel dumpers. 1958 *Engineering* 7 Feb. 165/3 Designed for a payload of 28 short tons, the end-tip dumper can travel at 24 m.p.h. 1967 *Guardian* 18 July 3/7 A dumper driver.

c. One who, or a country or community which, dumps goods (*dump *v.*¹ 2 c).

1903 *Daily Chron.* 24 Oct. 5/2 They would not have supported the fiscal policy of Mr. Chamberlain, for they were 'dumpers', who sold shoes to the poor below cost-price. *Ibid.* 21 Nov. 6/3 How long do you think that the dumpers will be content with only dumping unfinished goods? 1919 *Economist* 11 Oct. 568/2 Germany as a 'Dumper'.

d. A large wave which breaks suddenly and hurls a swimmer or surfer down with great force. *Austral.* and *S. Afr.*

1933 *Bulletin* (Sydney) 5 Apr. 11/3 One morning he arrived totally gummy in the lower jaw...'Done them in in a dumper yesterday at the surf,' he told us. 1956 S. Hope *Diggers' Paradise* 166 The dumper is a wave that builds up into a solid mass of water that advances closer inshore without breaking. 1963 A. Ross *Australia 63* ii. 51 On the surf beaches beyond Manly, the dumpers thunder in. 1970 *Studies in English* (*Univ. of Cape Town*) I. 27 The most popular expression in South African surfing idiom would appear to be *dumper*. A 'dumper' occurs most often when there is an abrupt rise of the ocean bottom to the shore. The wave thus breaks in shallow water and, instead of breaking slowly from the top, falls suddenly in an arc.

dumping, *vbl. sb.*¹ Add: (See also *dump *v.*¹ 2 c.) **b. dumping-ground**, a place where refuse, etc., is deposited; also *transf.* and *fig.*

1857, 1885 [in Dict.]. 1899 *Westm. Gaz.* 17 May 3/1 A corrupt young married woman..whom the mother finds a convenient dumping-ground for this too mature daughter. 1936 J. C. Beaglehole *New Zealand* i. 15 England needed a new dumping-ground for convicts. 1969 J. Mander *Static Soc.* viii. 269 The sub-continent is being used as a dumping-ground for obsolescent weaponry.

2. *Path.* The abnormally rapid emptying of the stomach via the bowels such as sometimes occurs after partial gastrectomy. Freq. *attrib.* and as *ppl. a.*, esp. in *dumping syndrome* (see quot. 1970).

1922 C. L. Mix in *Surg. Clinics N. Amer.* II. 617 Patient suffering from severe gastric disturbance following gastro-jejunostomy. Fluoroscopic examination revealed a 'dumping stomach'. 1935 G. B. Eusterman et al. *Stomach & Duodenum* x. 241 In performing gastro-enterostomy, the stoma should be neither too large nor too small...If too large, so-called 'dumping' results with precipitant evacuation of gastric content. *Ibid.* lix. 845 Too rapid emptying of the stomach, 'dumping stomach', following gastro-enterostomy or gastric resection may cause a train of disturbances, chiefly intestinal. 1954 *Brit. Med. Jrnl.* 13 Nov. 1131/2 Serious 'dumping'..happens to have been associated with bilious regurgitation. *Ibid.*, It is customary..to suggest that the 'dumping' syndrome subsides..within a few months of [gastrectomy] operation. 1970 Passmore & Robson *Compan. Med. Stud.* II. xvii. 3/2 After surgical removal of part of the stomach, patients sometimes suffer from attacks of sweating and flushing, weakness and occasionally fainting together with intestinal discomfort and explosive diarrhoea. These symptoms..are known as the dumping syndrome.

dumpoked (dʌ·mpōᵘkt), *a.* *India.* [ad. Hind. *dampukht*, f. Pers. *dam* breath + *pukhteh* cooked.] Applied to a steamed dish of boned and stuffed meat. Hence **du·mpoke**, a steamed dish of this kind, especially a boned and stuffed chicken or duck.

1696 Ovington *Voy. Suratt* 397 A dumpoked Fowl. 1698 J. Fryer *Acc. E. India & P.* 93 These Eat highly of all Flesh Dumpoked, which is Baked with Spice in Butter. *Ibid.* 404 Baked Meat they call Dumpoke which is dressed with sweet Herbs and Butter, with whose Gravy they swallow Rice dry Boiled. 1879 Mrs. A. E. James *Ind. Househ. Managem.* 89 *Dumpoke* is a boned chicken, its form preserved, only all the bones extracted; it is stuffed with a rice forcemeat, and you cut it through in slices. 1954 G. S. Rao *Indian Words in Eng.* iii. 17 Indian dishes figured on their tables: cabob, pilau, dumpoked fowl, curry, kedgeree.

dun, var. *DHOON.

dunam (du·nām). Also **dunum.** [mod. Heb., f. Turkish *dönüm*.] A measure of land, used esp. in Israel, equal to 1,000 sq. metres or about a quarter of an acre.
1920 *Glasgow Herald* 31 Dec. 8 The [Zionist] Commission has assisted in the planting and draining of 1600 duna[m]s of land. 1922 JOYCE *Ulysses* 60 They plant a dunam of land for you. 1930 *Economist* 1 Nov. 796/2 An area of at least 130 'dunums' is required in order to maintain an Arab peasant family. 1946 KOESTLER *Thieves in Night* 23 In Khubeira they paid six pounds for the dunum and another five hundred to the Mukhtar. 1962 *Economist* 28 Apr. 345/2, 2½ million dunams (something over half a million acres) of land have been distributed [in Iraq]. 1971 L. DAVIDSON *Smith's Gazelle* iii. 50 The 5,000 dunams, 1,250 acres, were basically suitable for a mixed sheep/wheat economy.

dunch, *v.* For '*Sc.* and *north. dial.*' read 'chiefly dial.' and add later examples.
1930 W. S. MAUGHAM *Cakes & Ale* ii. 23 Verbs that you only know the meaning of if you live in the right set (like 'dunch'). 1963 *Times* 9 Jan. 4/3 It [*sc.* mud] catches one in the eye when a [golf] shot is 'dunched'.

dunch, *a.* Add: **4.** Stupid, slow of comprehension; dull.
1845 *Ainsworth's Mag.* VII. 368 The boy is either so dunch..or he is so sharp that you can scarcely trust him out of your sight. 1889 A. GISSING *Both of this Parish* I. xv. 324 William Stretch be a trifle dunch in some of his faculties. 1927 M. SADLEIR *Trollope* 169 The *Autobiography* contains a number of judgments on novels, and.. they are 'dunch' and unconvincing.

Dundee (dʊndi·). [Name of a Scottish city on the Firth of Tay.] **1.** Used *attrib.* to designate a variety of rambling rose.
1837 T. RIVERS *Rose Amateur's Guide* 43 The Ayrshire Rose (Rosa Arvensis Hybrida)...Dundee Rambler, is the most double, and one of the best in this division. 1899 G. JEKYLL *Wood & Garden* vii. 79 What a fine thing, among the cluster roses, is the old Dundee Rambler!
2. *Dundee marmalade*: a kind of marmalade manufactured in Dundee (registered as a trade-mark by James Keiller & Son in 1880). Also *ellipt.*
1856 DICKENS in *Household Words* 28 June 555/2 Anchovy Paste, Dundee Marmalade, and the whole stock of luxurious helps to appetite. 1962 *Housewife* Apr. 92/2 Their silver-topped jar of Dundee marmalade. 1963 C. MACKENZIE *My Life & Times* II. 59, I do not remember ever seeing him eat any other preserves than Beech's strawberry jam and Keiller's Dundee marmalade. 1964 WODEHOUSE *Frozen Assets* iii. 51 The marmalade, too, had a tang which even Henry Blake-Somerset's imported Dundee could not have rivalled.
3. Used *attrib.* or *absol.* to designate a kind of rich fruit cake, usually decorated with split almonds.
1892 T. F. GARRETT *Encycl. Pract. Cookery* I. 234/2 Dundee Cake. Put 1 lb. of butter into a basin..loaf sugar.. flour..candied peel..and lastly thirteen eggs. 1920 'K. MANSFIELD' *Jrnl.* Sept. (1954) 218 A paper parcel that held a very large wedge of cake—of the kind known as Dundee. *Ibid.*, 'This is the last of our precious Dundee,' said she,.. cutting it so tenderly that it almost seemed an act of cannibalism. 1928 M. LOWRY *Let.* (1967) 3, I heard the girl opposite me order some more Dundee cake. 1938 E. BOWEN *Death of Heart* II. i. 187 Portia ate doughnuts, shortbread and Dundee cake. 1943 G. GREENE *Ministry of Fear* I. i. 8 He had always liked cakes, especially rich Dundees.

Dundonian (dʊndōᵘ·niăn). [f. *DUNDEE, after *ABERDONIAN *a.* and *sb.*] A native or inhabitant of Dundee.
1898 *Westm. Gaz.* 6 Oct. 10/1 It is common for Dundonians wishing salmon from their own river to send for it to London. 1965 *Guardian* 26 Nov. 9/2 Some Dundonians harbour a continuing misgiving about the implications [etc.].

Dundreary (dʊndriə·ri). [Name of Lord *Dundreary*, a character in T. Taylor's comedy *Our American Cousin* (1858).] In allusive attrib. uses, esp. *Dundreary whiskers*, long side whiskers worn without a beard. Also *absol.*, usu. in *pl.* (See also quot. 1864².)
1862 *Englishwoman's Dom. Mag.* Aug. 183 Bodger.. came to understand (in a Dundreary manner) a little more about Bradshaw. 1864 *Chambers's Jrnl.* 17 Sept. 595/2 It was only a summer scarf, of the sort that is called Dundreary. 1864 HOTTEN *Slang Dict.* 127 *Dundreary*, an empty swell. 1867 'PIPS' *Lyrics & Lays* 141 Full proud is he, I ween, Of his Dundreary whiskers. 1882 'F. ANSTEY' *Vice Versa* xvii, Bushy black whiskers, more like the antiquated 'Dundreary' type than modern fashion permits. 1894 C. G. HARPER *Revolted Woman* ii. 39 This fashion was the 'Piccadilly-weeper' variety of adornment, known at this day—chiefly owing to Sothern's impersonation of a contemporary lisping fop—as the 'Dundreary'. 1906 GALSWORTHY *Man of Property* I. i, His cheeks, thinned by two parallel folds, and a long clean-shaven upper lip, were framed within Dundreary whiskers. 1929 C. H. SMITH *Bridge of Life* ii. 38 The older men wore beards, Dundrearys or side whiskers; the middle-aged, mustaches.

dunducketty (dʊndʊ·kėti, -ĕti), *a. colloq.* or *dial.* Also **-ety, -ity.** [app. f. DUN *a.* + DUCK *sb.*¹] In phr. *dunducketty mud-colour*: (of) a dull, drab colour.
1818 LADY MORGAN *Let.* 26 Aug. in *Autobiogr.* (1859) 29 A colour I have supplied her with (not from the rainbow), dun-ducketty mud colour. [1847 HALLIWELL, *Dunduckitymur*, an indescribable colour, but rather dull. *Suffolk.*] 1897 M. KINGSLEY *Trav. W. Africa* 420 It is better than all white, or dunduckety mud-colour paint.

dune. Add: Used more widely of any mound of drifted sand; also, a similar ridge or mound of clay formed by the action of wind. (Further examples.)
1879 *Encycl. Brit.* X. 266/1 Captain Sturt found vast deserts of sand in the interior of Australia, with long lines of dunes 200 feet high. 1909 *Jrnl. Geol.* XVII. 754 These 'clay dunes' were almost always associated with a lagoon. 1926 T. E. LAWRENCE *Seven Pillars* IV. xliv. 256 The ground was flat and featureless till five o'clock, when we..found ourselves..amid sand-hills coated slenderly with tamarisk. ..The bushes and the dunes broke the wind. 1944 A. HOLMES *Princ. Physical Geol.* xiii. 264 One of the most remarkable features of desert dunes is their apparent power of collecting all the sand in their neighbourhood. 1961 *Jrnl. Sedimentary Petrol.* XXXI. 246/2 Clay dunes are..limited to the shores of clay-floored saline playas and tidal mud flats. 1963 *Ibid.* XXXIII. 766/2 The eolian particles become fixed in the dune when the clay regains moisture from the air or from rain.
b. dune-bedding (see quot. 1940); **dune sand,** sand formed into dunes by the wind; **dune-slack** = SLACK *sb.*¹ 2.
1940 *Chambers's Techn. Dict.* 271/2 *Dune bedding*, that type of current bedding commonly exhibited by sand dunes and interpreted in sandstones as evidence of desert conditions. 1946 L. D. STAMP *Brit. Struct. & Scenery* xxii. 229 The Permian Mauchline Sandstones..are believed to represent desert sand-dunes for they exhibit dune-bedding on a huge scale. 1792 A. YOUNG *Trav. France* 74 Mons. le Brun has an improvement on the Dunes, which he very obligingly shewed me. Between the town and that place are..one or two fields inclosed of most wretched blowing dune sand. 1916 *Jrnl. Geol.* XXIV. 242 Red dune sands are exceptional rather than the rule in the desert regions of today. 1967 *Oceanogr. & Marine Biol.* V. 132 So far as the various environments along the shore are concerned, beach sands and dune sands have been distinguished by several authors. 1938 *Nature* 7 May 817/2 The parallel ridges often rest on a relatively impermeable substratum, and the water, of which the dune ridges are the catchment area, drains into the intervening hollows and sometimes forms what are known as 'dune slacks'...The dune slacks are the especial home of the bog pimpernel. 1964 V. J. CHAPMAN *Coastal Vegetation* vi. 152 Most of the species to be found in dune slacks are common or reasonably common.

dung, *sb.* Add: **5. a.** *dung-cake.*
1901 KIPLING *Kim* iv. 91 For luxury's sake, Kim bought a handful of dung-cakes to build a fire. 1929 [see *CHOOLA]. 1949 M. L. DARLING *At Freedom's Door* 213 There is one feature of village life which is so universal that I have hardly mentioned it—the dung cakes.

dung, *v.* Add: **2. a.** Also of human beings.
1865 H. J. HUNTER *Rep. Med. Off. P.C. 1864* App. VI. 183 in *Parl. Papers* XXVI, The women and children dung into pots.., the men dunging away from home. 1928 E. & C. PAUL tr. *Marx's Capital* II. 761 A house whose 'tenants dunged against the house-side'.

dunga, var. *DOONGA.

dungaree. Now usu. pronounced (dʊŋgărī·). **b.** (Further examples.) Now usually made of blue dungaree or similar material.
1928 *Granta* 2 Nov. 74/2, I myself, as a matter of fact, sweated through nine dungarees. 1940 *Geogr. Jrnl.* XCV. 260 Dressing more and more in..American mass-produced dungarees.

dunk (dʊŋk), *v.* orig. *U.S.* [Pennsylvanian G. *dunke* to dip (cf. G. *tunken*).] *trans.* To dip (bread, cake, etc.) into a beverage or other liquid. Also *absol.*, *transf.*, and *fig.* Hence **du·nker, du·nking** *vbl. sb.*
1919 *Quill* Feb. 12 (*title*) Some notes on dunking. *Ibid.* 16 It should be remembered that the really fastidious dunker never burns his thumb. 1927 *Dialect Notes* V. 474 Them young-uns is allus a-dunkin' their bread in th' bottom-sop. 1931 *Daily Progress* (Charlottesville, Va.) 21 Feb. 4/8 No information is given about dunking. *Ibid.* 10 The pone should be 'dunked' in the likker. *Ibid.* 2 Mar. 10 Governor Long..is champion of the dunkers. 1940 in *Amer. Speech* (1941) XVI. 147/2 Dunk your nylons in rich suds of neutral soap. 1940 *Chambers's Techn. Dict.* 271/2 *Dunking*, a colloquialism for the process of dipping the [cinematographic] film into chemical solutions for any stage of processing. 1941 'R. WEST' *Black Lamb & Grey Falcon* II. 311 Constantine sat dunking a roll in his coffee. 1951 M. MCLUHAN *Mech. Bride* 4/2 The same man would rather dunk himself in the newspaper than have any esthetic or intellectual grasp of its character and meaning. The incorrigible dunker would probably do well to skip the next few pages. 1958 G. MITCHELL *Spotted Hemlock* xvi. 187 'For heaven's sake, don't stop eating,' said Miss Paterson, herself dunking a doughnut. 1959 M. M. KAYE *House of Shade* iii. 36 'He's a one, isn't he? Your gentleman friend,' said the blonde, dunking Dany's head into a basin. 1959 *Manch. Guardian* 26 June 5/3 Nylon sheets..can be 'dunked' along with underclothes or put in a washing machine. 1960 *Times* 30 Nov. 13/6 In naval parlance a 'dunker' never becomes immersed except in dire emergency, and only a part of his equipment, in the form of a sonar buoy, is lowered into the sea. This is the device which enables the helicopter crew to 'listen in' for submarines. *Ibid.*, My interest in dunking arose through my being in the aero engine business. 1971 *Nature* 26 Feb. 597/1 But the possibility that the RNA picks up basic proteins when dunked in the cytoplasm cannot be disregarded.

Dunkirk. Add: **2.** (dʊnkɜ̄·ɹk). The (scene of the) evacuation of the British forces from Dunkirk between 29 May and 3 June 1940; hence used allusively for any similar withdrawal, crisis, etc.; so *to do a Dunkirk*, to make such a withdrawal. Also *attrib.*, esp. in *Dunkirk spirit.*
1941 *Time* 9 June 9/2 In far too many items we are still lagging behind and..there may well be more Dunkirks. 1941 *New Statesman* 20 Sept. 275/1 America can look for no Dunkirk To underline the warnings Presidential. 1943 *Hutchinson's Pict. Hist. of War* 4 Aug.–26 Oct. 21/2 Evacuation of some of the enemy's forces, harassed by allied aircraft and naval units, has already begun, but it remains to be seen how successful this minor 'Dunkirk' will be. 1944 *Amer. Speech* XIX. 278, I have heard 'to do a Dunkirk'. 1948 *Hansard Commons* 28 Mar. 2853 He asked me to bring about what he called a 'Dunkirk evacuation' of Palestine. 1955 *Times* 8 Aug. 3/3 The Navy had saved the situation, and Dunkirk had become 'D' Day. 1956 J. CHRISTOPHER *Death of Grass* vii. 113 If the country only shows the Dunkirk spirit, we can pull through. 1961 *Listener* 19 Oct. 589/1 The Dunkirk spirit of only starting to try hard when it becomes really necessary is deeply ingrained in the British character. 1961 H. WILSON in C. Booker *Neophiliacs* (1969) ix. 249, I myself have always deprecated..appeals to the Dunkirk spirit as an answer to our problems. 1969 *Guardian* 18 July 20/8 The Church of England faces 'an ecclesiastical Dunkirk' unless it recognises the contribution young people can make.

Dunlop (dʊ·nlǫp). Also **Delap, Dulap, Dunlap.** The name of a parish in Ayrshire, Scotland, used (chiefly *attrib.*) to designate an unskimmed-milk cheese originally made there.
1821 J. GALT *Annals of Parish* vi. 66 They have, under the name of Delap-cheese, spread far and wide over the civilized world. 1825 J. C. LOUDON *Encycl. Agric.* III. VII. v. 990 Dunlop cheese, so named from the parish..where it was originally made. *c* 1842 in *Scottish Hist. Soc. Misc.* (1939) VI. 271 Who has not heard..of Dunlop cheese. 1864 J. B. GREENSHIELDS *Annals Lesmahagow* xvii. 278 Sweet milk cheese, known as 'Dunlop', is the staple commodity. 1878 *Chambers's Encycl.* II. 786 Dunlop Cheese..is now manufactured in the dairy districts of Scotland generally. 1905 W. H. SIMMONDS *Pract. Grocer* III. 79 The Dunlop is a rather rich cheese, similar in general quality to Derbyshire or Gouda, but larger in size. 1966 P. V. PRICE *France* 94 Take a pound of Double Gloucester, Lancashire, Dunlop, or a whole baby Cheddar.

dunnage, *sb.* Add: **2. dunnage bag,** a kit-bag.
1904 *Chicago Evening Post* 23 Aug. 7 A dunnage bag..is a sack of heavy brown canvas nearly four feet long...In it the aspiring mountaineer may pack fifty pounds of necessities. 1925 FRASER & GIBBONS *Soldier & Sailor Words* 86 *Dunnage bag*, kit bag (Navy). 1929 M. DE LA ROCHE *Whiteoaks* xxviii. 381 The dogs..trotted without rest from point to point of interest—the dunnage bag, the provisions, the weapons.

dunnamany (dʊ·nāme:ni), colloquial form of (*I*) *don't know how many*. So **du·nnamu:ch,** (I) don't know how much. Cf. next.
1831 R. LOWER *Tom Cladpole's Jurney to Lunnun* 14 So arter dun-a-much more talk, He said he must be gwyn. 1853 W. D. COOPER *Gloss. Sussex* (ed. 2) 43 *Dunnamany* and *Dunnamuch*, corruptions of *I don't know how many* or *much*. 1906 KIPLING *Puck* 267 Justabout tore the gizzards out of I dunnamany. *Ibid.* 273 The Pharisees just about flowed..down the beach to the boat, *I* dunnamany of 'em. 1946 G. HEYER *Reluctant Widow* iii. 41 Your lordship knows I've been with Mr. Hitchin a dunnamany years.

dunno (dʊ·no, dənōᵘ·), also **dunna(w),** etc., colloquial forms of (*I*) *do not* (or *don't*) *know.* See DO *v.* 29 and cf. prec.
1842 G. P. R. PULMAN *Rustic Sketches* 34 There I vlounder'd like a zow, An' ramm'l'd out I dun-no how. 1848 LOWELL *Biglow Papers* 1st Ser. vii, I dunno as it's ushle to print Poscrips. 1867 P. KENNEDY *Banks Boro* xxiii, Now indeed I *dunna* what to do. 1868 [see *BLOW *v.*¹ 12 b]. 1888 B. L. BURNETT *Stable Boy* ii, I dun naw wat ta du! 1902 J. MASEFIELD *Salt-water Ballads* 85 Dunno a heap about the what an' why, Can't say 's I ever knowed. 1926 R. MACAULAY *Crewe Train* I. ii. 10 If you had asked her why, she would have replied, 'Dunno. It's a bother speaking to people when you're out.' 1938 C. MORGAN *Flashing Stream* II. ii. 186 He'll begin to say to himself: 'Well, I dunno. P'raps I'm wrong.' 1946 [see *GET *v.* 38 g]. 1966 P. MOLONEY *Plea for Mersey* 22 A sed 'Wharar thee wack?' 'A dunno,' she said back.

dunny, *a.*¹ (Later examples.)
1819 T. THOMPSON in *Coll. Songs, Comic and Satirical* (1827) 10 Tyneside seem'd clad wiv bonny ha's An furnaces sae dunny. 1951 J. FRAME *Lagoon* 38 The house with..its dunny roses on the trellis-work.

dunny (dʊ·ni), *sb.*² Also **danna, dunikin, dunnakin, dunnee, dunniken, dunnyken** [Origin unknown, but cf. DUNG *sb.* and KEN *sb.*²] **1.** *dial.* and *slang.* An earth closet, (outside) privy (see also quot. 1859). Also *attrib.* (Freq. in *Austral.* and *N.Z. slang.*)
a 1790 H. T. POTTER *New Dict. Cant* (1795) 26 *Dunnakin*, a necessary. 1812 J. H. VAUX *Flash Dict.* 184 *Knapping a jacob from a danna-drag.* This is a curious species of robbery..; it signifies taking away the short ladder from a nightman's cart, while the men are gone into a house, the privy of which they are employed emptying, in order to effect an ascent. 1846 *Swell's Night Guide* 57 'Where's the plant, cully?'..'Fenced, in a dunniken.'..'What? Fenced in a crapping ken?' 1859 HOTTEN *Dict. Slang* 29 *Danna*, excrement. *Danna drag*, a nightman's or dustman's cart.

Ibid. 35 *Dunny-ken*, a watercloset. 1943 *Horizon* VIII. 156 The crazy dunikins, outside w.c.s listing away from the prevailing wind. 1947 [see *BUGGER *v.* 2 b]. 1952 J. BAXTER in Chapman & Bennett *Anthol. N.Z. Verse* (1956) 258 Cigarette stink from a hole in the rushes Dark as a dunny. 1960 N. HILLIARD *Maori Girl* I. iii. 19 She delighted in giving cheek to the boys and taking refuge in the girls' dunny. 1962 C. ROHAN *Delinquents* 137 How is the dunny? ..Does it have a chain? 1965 G. MCINNES *Road to Gundagai* iii. 38 The lavatory, or the 'dunny' as Dad used to call it. 1970 *Private Eye* 16 Jan. 16 It seems a bit crook for old bazza to spend the night in the dunnee!
2. *Sc.* [Perh. a different word.] An underground passage or cellar, common in old tenement buildings.
1918 N. MUNRO *Jaunty Jock* ii. 24 Flat on flat it rose for fourteen stories, poverty in its dunnies (as they called its cellars), poverty in its attics. 1922 J. BUCHAN *Huntingtower* vii. 152 Sleepin' in coal-rees and dunnies and dodgin' the polis. 1947 *Forward* 4 Jan. 3 Broken-down sewage system which periodically overflows into the dunnies and back-courts.

dunt, *sb.*¹ Add: **1. d.** (See quot. 1924.)
1924 *Jrnl. R. Aeronaut. Soc.* Mar. 198 Special temperature conditions are brought about by vertical currents of air. The vertical currents usually consist of a hot stream of air rising or a cold stream falling. In flying into such a current the airship will experience a 'dunt'. The 'dunt' received will depend on two actions—one the dynamic action of the current, and the other the sudden change in lift due to change in temperature. 1928 E. F. SPANNER *Gentlemen prefer Aeroplanes* vi. 43 Moderate clouds—unavoidable temperature 'dunts'—and so on. *Ibid.* x. 71 When the vessel meets 'temperature "dunts"' and other atmospheric irregularities.

duo. Add: **2.** Two people; a couple; esp. a pair of entertainers. Also *fig.* (Cf. quot. 1872 in Dict.)
1887 *Lantern* (New Orleans) 16 July 2/2 The morals of this duo were very much impaired. 1930 *Amer. Speech* VI. 115 (*newspaper headline*) Date set for bank theft trial of duo. 1942 BERREY & VAN DEN BARK *Amer. Thes. Slang* § 583/33 *Pop duo* or *trio*, a popular team of two or three [entertainers]. 1958 *Times* 23 Sept. 3/4 They [*sc.* two Spanish dancers] are a delightfully matched duo. 1958 *Listener* 16 Oct. 595/2 Richards..ejected the poet, leaving a duo of poem and reader. 1969 *Stage* 18 Sept. 23/2 (Advt.), Organ-drums duo with vocals.

duodenectomy (diū:odėne·ktŏmi). *Surg.* [f. DUODEN(UM + *-ECTOMY.] Partial or total excision of the duodenum.
1908 *Practitioner* Sept. 454 Codivilla is the only surgeon, who, operating on the pancreas, has performed a duodenectomy. 1962 *Lancet* 22 Dec. 1308/1 A further partial gastric resection, partial pancreatectomy, and duodenectomy with node dissection.

duodeno- (diū,odī·no), comb. form of DUODENUM, as in **duodeno-jejunal** *a.*, pertaining to the duodenum and the jejunum.
1886 *Buck's Handbk. Med. Sci.* II. 537 Duodeno-chole-cystostomy. 1887 *Ibid.* V. 606 The little pocket which lies behind this (duodeno-jejunal recess). 1890 BILLINGS *Med. Dict.* 417/1 *Duodeno-renal*, relating to duodenum and kidney. 1900 DORLAND *Med. Dict.* 215/1 *Duodenocholecystostomy*, the operation of forming a communication between the gall-bladder and the duodenum. *Duodenocholedochotomy*, surgical incision of the duodenum and bile-duct. *Duodeno-enterostomy*, the artificial creation of a passage from the duodenum to another part of the small intestine. 1906 *Practitioner* Dec. 761 The duodeno-jejunal junction. 1908 *Ibid.* Nov. 712 The duodenopyloric constriction. 1966 *Lancet* 24 Dec. 1391/1 The whole of the intestine from 2 ft. below the duodenojejunal flexure to the mid-transverse colon was grossly ischæmic. 1968 S. TAYLOR et al. *Short Textbk. Surg.* (ed. 2) xxviii. 427 After the stomach contents have been aspirated a duodenoduodenostomy is performed.

duopoly (diu,ǫ·pŏli). [f. DUO- + Gr. πωλ-εῖν to sell, after MONOPOLY.] A condition in which there are only two suppliers of a certain commodity, service, etc.; the domination of a particular market by two firms. Hence **duo·polist,** one member of a duopoly.
1920 A. C. PIGOU *Econ. of Welfare* xii. 232 Cournot decided..that the resources devoted to production under duopoly are a determinate quantity, lying somewhere between the quantity that would have been so devoted under simple competition and under simple monopoly respectively. 1951 J. R. WINTON *Dict. Econ. Terms* 29 When the total supply of a commodity is produced by two firms or individuals, that state of affairs is known as 'duopoly'. 1961 *Guardian* 28 Jan. 1/1 He..believes.. monopoly is..preferable to the present duopoly—the domination of the magazine industry by two powerful groups. 1965 SELDON & PENNANCE *Everyman's Dict. Econ.* 132 If a duopolist moves his price above or below the monopoly price he will be worse off because profits are maximized at the monopoly price.
b. *transf.* Control or domination by two persons or groups.
1959 *Listener* 1 Jan. 4/2 The party duopoly of Britain. 1959 *Economist* 22 Aug. 525/2 As we move from nuclear duopoly to nuclear oligopoly, the range of circumstances in which these weapons could be rationally used becomes narrower.

duotone (diū·ŏtōᵘn), *sb.* [f. DUO- + TONE *sb.* 10.] A half-tone illustration printed in two colours from two plates made from the same original, but using different screen angles; also, the process by which such an illustration is printed. Also *attrib.*
1907 AMSTUTZ & JENKINS *Hand-Bk. Photoengraving* viii. 406 *Duo-tones*, the printing of two superposed impressions from one engraved half-tone plate, in two colors, one a black and the other a tint, placing them slightly out of register, but in alignment; or the printing of two superposed colors from two separate plates. 1913 *Dial* 16 Dec. 530/1 Of the numerous illustrations..some are 'collotype plates',.. and..others..are 'duo-tone illustrations'. 1967 KARCH & BUBER *Offset Processes* v. 174 A true duotone is a two color reproduction made by two halftone negatives at two different screen angles, with the halftone plates printed in complementary colors. 1971 *Publishers' Weekly* 23 Aug. 63 The extensive coverage in this issue of 'The New Gift and Art Books',..including a 16-page duotone picture portfolio of outstanding books.
Passing into general use as *adj.*
1934 WEBSTER, *Duotone*, having or yielding two tones or colours. 1966 'A. HALL' *9th Directive* vi. 61 The duotone Chevrolet in the main parking area was moving away.

duotype (diū·ŏtəip). [f. DUO- + TYPE *sb.*¹] A half-tone illustration printed from two plates made from the same monochrome negative, but etched differently to give two sets of colour-values; also, the process used to print such an illustration.
1913 S. H. HORGAN *Halftone & Photomechanical Processes* 207 *Duotype*, two half-tone plates made from the same negative, but etched differently. 1967 KARCH & BUBER *Offset Processes* v. 176 A duotype is a two color halftone made from one halftone negative.

dupe (diūp), *sb.*² *colloq.* [Abbrev. of DUPLICATE *sb.*] A duplicate; *spec.* in *Cinemat.*, a duplicate negative made from a positive print (see quots.). Also *attrib.* or as *adj.*
1916 E. W. SARGENT *Technique Photoplay* (ed. 3) 361 *Dupe*, a pirated print of a film. Made by passing a genuine positive print through a printer in company with negative film whereby an imperfect duplicate of the negative is obtained from which prints are made. 1937 A. BUCHANAN *Film Making* iii. 51 Frequently a 'dupe' negative is made... A dupe is created by printing a special lavender positive print from the original negative, and from that a second or duplicate negative is made in the laboratory. 1948 [see next]. 1969 C. YOUNG *Todd Dossier* 152 Rosen..sent me down to 326 to pick up a dupe file copy of Todd's surgery report.

dupe (diūp), *v.*² *Cinemat. colloq.* [f. prec.] *trans.* To make a 'dupe' of. Hence **duped** *ppl. a.*, **du·ping** *vbl. sb.* Cf. prec.
1912 F. A. TALBOT *Moving Pictures* xviii. 206 His most successful subjects were seized by American houses and 'duped'—as unauthorized reproduction is called in cinematographic parlance in that country. 1923 *Ibid.* (ed. 2) x. 134 These 'duped' (duplicated) films were highly indifferent. 1933 A. BRUNEL *Filmcraft* 158 A duped negative has until recently always given prints considerably inferior to positives taken from an original negative. 1948 E. LINDGREN *Art of Film* 205 *Duping print*, special soft print..made from an original negative so that a dupe negative can subsequently be made from it. 1963 *Movie* July/Aug. 33/1 Fleapit projectionists..duping and scratching the first sequence of cut-up film to make it look old. 1969 J. ELLIOT *Duel* III. ii. 233 She went to the cutting-rooms to learn the mysteries of..fine grain duping prints.

dupion. Delete '? *Obs.*' and add later example. Also **douppion, duppion.**
1933 T. WELFORD *Textile Student's Man.* vi. 61 *Dupions*, double cocoons.
2. A rough silk fabric woven from threads from double cocoons; such a thread. Now also applied to imitations of such fabric made by other processes.
1922 A. F. BARKER *Textiles* (ed. 2) xv. 329 Not many of the Chinese or Japanese duppions are exported, being reserved for native use. 1938 *Times* 28 Feb. 19/4 For cruising frocks there is a douppion flamella at 7s. 11d. 1961 *Times* 13 Feb. 13/3 Available in silk twills, chiffon and duppions. 1968 J. IRONSIDE *Fashion Alphabet* 226 *Dupion* (*Douppion*)..is a silk thread made from two cocoons which have been so closely 'nested' that a coarse double filament results. The double thread is always woven as one and the resultant silk fabric is irregular and slub-like. The yarn is now simulated in rayon and acetate fibres. 1971 *Scotsman* 20 May 6/4 The carpet is a Wilton-weave twist pile in Grecian olive and the curtains are in turquoise dupion.

duplet. Restrict †*Obs. rare* to sense in Dict. and add: **2.** *Physics.* A pair of electrons with opposite spins, esp. one forming a covalent or co-ordinate bond between two atoms.
1921 I. LANGMUIR in *Science* LIV. 60/2 According to Postulate 1, the first complete layer in any atom consists of two electrons close to the nucleus. Let us call this stable pair of electrons a *duplet* and let us broaden the definition of duplet to include any pair of electrons which is rendered stable by its proximity to one or more positive charges. We may now state the second postulate...Two atoms may be coupled together by one or more duplets held in common by the completed sheaths of the atoms. 1940 GLASSTONE *Textbk. Physical Chem.* i. 8 Sharing of electrons occurs in pairs, each pair or duplet being equivalent to one conventional chemical bond.
3. *Mus.* A group of two notes or beats; *spec.* (see quot. 1938). (Cf. TRIPLET 2 c.)
1922 S. GREW *Art of Player-Piano* v. 33 We are..able to take the duplet-division of the beat. 1938 *Oxf. Compan. Mus.* 273/1 A duplet is a group of two [notes] or its equivalent, to be performed in the time of three.

duplex, *a.* Add: **1. b.** *duplex process*, a process for making steel in which the charge undergoes treatment by two of the standard processes or in two furnaces in succession.
1902 *Jrnl. Iron & Steel Inst.* LXI. 589 The quantities of slag made were also very great..compared with 0·1 ton of slag in the present duplex process. 1926 *Ibid.* CXIII. 571 This is a continuous duplex process carried out in two tilting open-hearth furnaces. 1939 J. DEARDEN *Iron & Steel Today* vii. 101 Duplex processes..are designed to combine the advantages of the Bessemer and open-hearth processes, while avoiding their disadvantages. 1951 G. R. BASHFORTH *Manuf. Iron & Steel* II. xii. 353 Duplex processes usually employ a high percentage of hot metal.
c. Designating paper or board which is formed by uniting two separate layers of paper, or which is coloured differently on either side.
1901 *Brit. Pat.* 22,418 2 In making duplex papers that is papers that may be white on the one side and coloured on the other, or be coloured on one side and coloured with a different colour on the other side, a brass or metal roller D would be used for spreading the colour. 1914 E. A. DAWE *Paper* xvii. 120 Duplex Papers may be made of two layers of differently coloured papers brought together in the wet state and rolled together, or may be coated with different colours, after the paper is made, as duplex art papers. 1962 F. T. DAY *Introd. to Paper* iv. 46 In the case of Duplex tinted boards, different coloured pulps are used to produce a board with a different colour on each side.
d. Of an eye: having pigment on the anterior surface of the iris as well as on the posterior surface, as in eyes that are a colour other than blue.
1908 C. C. HURST in *Proc. R. Soc.* B. LXXX. 86 The eyes in which two kinds of pigment are present; the one, yellow-brown in colour, deposited on the outer or anterior surface of the iris; the other, blue-black in colour, deposited on the inner or posterior surface of the iris. Such eyes I propose to call *duplex*. *Ibid.*, To the duplex type belong the various shades of eyes with both anterior and posterior pigments. 1911 A. D. DARBISHIRE *Breeding & Mendelian Discov.* 276 Duplex eyes are those which have a layer of brown pigment in front of the iris. 1946 R. R. GATES *Human Genetics* I. v. 88 Hurst..classified eyes as duplex or simplex.
e. *Biol.* Of a polyploid organism: having the dominant allele of any particular gene represented twice.
1923 A. F. BLAKESLEE et al. in *Bot. Gaz.* LXXVI. 345 The convenient terms nulliplex, simplex, duplex, triplex, and quadriplex, can be used in regard to the number of chromosomes containing a given gene, without thereby admitting the amount of so-called dominance of the factors in question. 1963 LEWIS & JOHN *Chromosome Marker* iv. iii. 327 Thus, a tetraploid of the type *AAaa* (duplex) can produce gametes of three kinds—*AA*, *Aa* and *aa*.
f. *Biochem.* Applied to a molecular structure, or part of one, in which two polynucleotide strands are linked together side by side, as in a DNA molecule.
1969 *Nature* 22 Nov. 771/1 There should be a unique sequence of complementary material in each strand giving rise to a single, homogeneous set of duplex regions in the heteroduplex molecules. 1970 *Proc. Nat. Acad. Sci.* LXVI. 197 It was predictable that the flow of biological information in the living cell, postulated to start with the transcription of the duplex DNA to yield a series of messenger RNA species, must originate from only one of the two complementary DNA strands. 1971 *Nature* 19 Feb. 530/1 The enzyme shows a strong preference for single stranded DNA, which it degrades about 150 times faster than duplex DNA.

duplex (diū·pleks), *sb.* [f. the adj.] **1.** orig. *U.S.* A house or other building so divided horizontally or vertically that it forms two dwelling-places; also, one of the dwellings in such a building. Also *attrib.* or as *adj.*
1922 *Daily Ardmoreite* (Ardmore, Okla.) 6 Jan. 9/1 For Rent—6 room duplex bungalow. 1931 'D. STIFF' *Milk & Honey Route* iii. 32 You live in a duplex with ten baths, or a cold-water walk-up flat, according to your ability to pay rent. 1936 *Archit. Rev.* LXXX. 101 It consists of an office for the architect-owner on the ground floor, and on the upper floors two 'duplex' (i.e. self-contained two-storey) apartments, one of which is occupied by the architect. 1937 *Sunday Dispatch* 28 Feb. 2/7 A duplex is a house of two storeys, with its own front door on the ground floor. 1938 *Archit. Rev.* LXXXIV. 165 (*caption*) Duplex (two-storey) flats occupy the centre portion. 1959 *Observer* 15 Nov. 18/3 We do not live in a spacious house..but a Chelsea duplex. 1968 *Globe & Mail* (Toronto) 17 Feb. 49/1 (Advt.), 3 bedroom duplex or equivalent. 1969 *Sydney Morning Herald* 24 May 37/1 (Advt.), The owner has had plans drawn for a very impressive duplex. 1971 *Rand Daily Mail* (Home Owner) 27 Mar. 7/2 City dwellers are gravitating towards high density living (flat complexes, town houses, duplex..and so on).
2. *Biochem.* A double-stranded or duplex polynucleotide molecule (see *DUPLEX *a.* 1 f).
1963 *Proc. Nat. Acad. Sci.* L. 910 No evidence for an RNA-replicating duplex, analogous to that of the single-stranded DNA virus ϕX174, has thus far appeared, despite the fact that RNA–RNA duplexes are known to be very stable structures. 1967 I. H. HERSKOWITZ *Basic Princ. Molec. Genetics* iii. 31 Hybrid DNA–RNA duplexes (each composed of one DNA strand and one RNA strand) can be made. 1969 *Prog. Nucleic Acid Res.* IX. 304 The stability of the duplex is affected both by the base composition and by the sequences of bases along the individual strands.

duplexer (diū·pleksəɹ). [f. DUPLEX *a.* + -ER¹.] A device by means of which signals

are alternately transmitted from and received by a single radar or radio aerial.
1952 REINTJES & COATE *Princ. Radar* (ed. 3) i. 10 The use of a single antenna for both transmission and reception in nearly all radar sets is made possible by a *transmit-receive device*..also called a *duplexer* and a *receiver-protective device*. 1954 H. C. MINNETT in E. G. Bowen *Textbk. Radar* (ed. 2) ix. 294 An aerial duplexer or TR/RT switch is an electronically operated switching device, which enables the same aerial to be used for both transmitting and receiving in pulsed radar equipments. 1962 *Engineering* 5 Oct. 434 Discussions on valves..will cover..magnetrons, modulators, duplexers.

du·plexing, *vbl. sb. Metallurgy.* The utilization of a duplex process (see *DUPLEX *a.* 1 b).
1918 *Iron Age* 30 May 1398/1 'Reversed Duplexing'.. consisted of making steel in electric furnaces..at least one of which was run with basic lining, and the rest of the furnaces were to be run with acid lining. 1959 *Economist* 9 May 557/1 One can use hot-metal in them [*sc.* electric furnaces] by 'duplexing', blowing the hot iron in a Bessemer convertor..before feeding it into the arc furnace.

‖ **duplex querela** (diū·pleks kwėrī·lă). [Law Latin, lit. twofold complaint.] (See quot. 1763.)
1713 E. GIBSON *Codex Juris Ecclesiastici Anglicani* XXXIII. viii. 823 He may appeal to the immediate Superiour ..and thereupon obtain an Instrument directed to the Ordinary, and called *Duplex Querela*. 1763 R. BURN *Eccl. Law* I. 113 When the bishop doth..refuse, or unduly delay to admit and institute a clerk to the church to which he is presented, the clerk may have his remedy against the bishop in the ecclesiastical court..complaining to the judge of appeals thereof; the judge is wont to write to the bishop in the form of law, and this writing they call a duplex querela. This duplex querela is to contain a monition to the bishop. 1845 H. W. CRIPPS *Pract. Treat. Laws rel. Ch. & Clergy* v. i. 480 Where a refusal is appealed against, the dean of the arches, or other judge or judges of the court of appeal, sends a letter to the bishop so refusing, which letter or rescript is called *duplex querela*. 1895 J. WILLIAMS *Briefless Ballads* 62 Aid me, Muses!..for *duplex querela* is an uninviting theme. 1931 *Church Times* 20 Feb. 223/1 The Clerk might have commenced proceedings by *Duplex querela* in the Arches Court, but for weighty reasons this was turned down.

duplicate, *a.* and *sb.* Add: **A. 5.** *duplicate bridge, whist*, a type of bridge or whist in which the hands are replayed by different persons.
1891 J. T. MITCHELL (*title*) Duplicate whist. Its rules and methods of play. 1894 R. F. FOSTER *Duplicate Whist* 27 The theory of duplicate whist, or Rejoué,..is that the play of each of the competitors..shall be contrasted with that of the others, by giving to each the same cards, with the same advantages or disadvantages of position at the table, an equal number of times. 1929 M. C. WORK *Bridge Pointers & Tests* 184 *Duplicate Auction Bridge*, a form of the game in which the hands are played more than once (*i.e.*, overplayed). 1959 *Listener* 13 Aug. 262/1 In duplicate bridge 50 points are added for all part-score contracts.
6. *Genetics.* Designating one of two or more non-allelic genes having indistinguishable effects.
1914 G. H. SHULL in *Zeitschrift für Induktive Abstammungs- und Vererbungslehre* XII. 96 (*heading*) Duplicate genes for capsule-form in *Bursa bursa-pastoris*. *Ibid.* 120 By 'duplicate' determiners I understand those which, when separated from each other, produce characters so like that they can not be distinguished from one another. 1949 DARLINGTON & MATHER *Elem. Genetics* 410 Two genes of similar but non-cumulative effect are said to be *duplicate*. 1951 *Cold Spring Harbor Symp. Quant. Biol.* XVI. 162/1 We also assume that the component genes are not identical; i.e., that they do not represent duplicate genes which have yet to diverge in function.
B. 4. *Ellipt.* for *duplicate bridge, whist* (see sense A. 5 above). Also *attrib.*
1894 R. F. FOSTER *Duplicate Whist* 23 'Duplicate' is looked upon as the coming game. 1898 C. E. LEIBOLD *Woman Proposes* xvi. 171, I am primed for a spirited game of duplicate. Come, get ready. 1929 M. C. WORK *Complete Contract Bridge* 237 *Duplicate*, a form of the game in which the hands are played more than once (i.e., overplayed). 1963 *Listener* 17 Jan. 137/1 This point is dealt with in a different manner in the Duplicate Laws.

duplicating, *vbl. sb.* (In Dict. s.v. DUPLICATE *v.*) Add: Also *attrib.*, as *duplicating machine* (cf. *writing engine*), *paper.*
1893 J. T. DAVIS *U.S. Pat.* 494,060 4/2 The duplicating machine writes and spaces simultaneously with the initial machine. 1914 E. A. DAWE *Paper* xvii. 120 *Duplicating papers*, unsized or half-sized papers used for taking copies on cyclostyle, mimeograph and similar duplicating machines. 1962 A. NISBETT *Technique Sound Studio* ii. 40 Use a fairly stiff paper (duplicating paper is quite good). 1969 D. TRIESMAN in Cockburn & Blackburn *Student Power* 153 A minority report which the Executive turned out on some duplicating paper.

duplication. Add: **1. e.** *Genetics.* The existence in a set of chromosomes of two copies of a particular chromosome segment; the process by which this comes about; also, the duplicated segment.
1917 C. B. BRIDGES in *Genetics* II. 454 As evidence that pieces may be lost bodily from chromosomes and that fragments may join together, there may be offered two distinct cases of 'duplication' (unpublished), a phenomenon, the explanation of which seems to be that a section taken from the mid-region of one X has become attached to the end of the other X, its mate. 1945 *Genetics* XXX. 161 The position effect, here, would appear to extend over a distance at least as great as the length of the duplication. 1949 DARLINGTON & MATHER *Elem. Genetics* v. 105 Heterozygous deficiencies and duplications are..for segments what monosomics and trisomics are for whole chromosomes. 1965 PEACOCKE & DRYSDALE *Molec. Basis Heredity* vii. 77 These results are those expected if the interphase chromosome before duplication contains two components each of which remains intact ..during chromosome duplication. 1965 A. M. SRB et al. *Gen. Genetics* (ed. 2) vii. 198 Extra parts of chromosomes are called duplications. Various kinds of duplications have been observed. Some exist attached to the chromosome whose segments are 'repeated'; some are attached to different chromosomes; others may exist as independent fragments.

duplicator. (Earlier and later examples.)
1893 J. T. DAVIS *U.S. Pat.* 494,060 (*title*) Combined type-writing machine and duplicator. 1916 'BOYD CABLE' *Doing their Bit* ii. 36 You know what a duplicator is? Thing for printing copies off a typed stencil sheet. 1939–40 *Army & Navy Stores Catal.* 327/1 A compact duplicator...Will copy Handwriting, Typewriting, Drawings, Plans.

duplicity. Add: Hence **dupli·citous** *a.*, showing duplicity, deceitful.
1961 in WEBSTER. 1966 *New Statesman* 11 Mar. 350/3 Peggy Mount, as the duplicitous washerwoman,..subdues her comic extravagance. 1969 G. LEFF *Hist. & Soc. Theory* ii. 44 Whether John was contrite or merely duplicitous in acceding to the barons' demands in 1215 is irrelevant to the meaning of Magna Carta.

dupondius (diupǫ·ndiǒs). *Roman Antiq.* Earlier also **dipondius.** Pl. **dupondii.** [L.] A bronze or brass coin of the value of two asses.
1601 HOLLAND *Pliny* XXXIII. iii. II. 462 Like as the weight in brasse of two pound, they named Dipondius. 1853 H. N. HUMPHREYS *Coin Collector's Man.* II. 378 The second bronze, which was called the dupondius, or double as, was ..exactly half the sestertius. 1937 *Oxoniensia* II. 67 The large bronze coins, *sestertii*, *dupondii* and *asses*, which are representative of the first two centuries of Roman occupation

duppion, var. *DUPION.

Dupuytren (düpüi·trɐn). The name of Baron Guillaume *Dupuytren* (1777–1835), French surgeon, used in the possessive to designate certain conditions observed or appliances, etc., invented by him. **Dupuytren's contraction** or **contracture,** a painless disorder of the fibrous tissue of the palm that leads to fixed flexion of one or more fingers and affects chiefly old people; **Dupuytren's fracture,** fracture of the fibula just above the malleolus; **Dupuytren's method** (see quot. 1887); **Dupuytren's paste** (see quot. 1886); **Dupuytren's splint,** a splint to prevent eversion in Pott's fracture.
1876 O. W. MADELUNG (*title*) The causes and operative treatment of Dupuytren's finger contraction. 1880 *Brit. Med. Jrnl.* 19 June 919/1 (*title*) Dupuytren's fracture of the fibula. 1883 HOLMES & HULKE *Syst. Surg.* (ed. 3) I. 1093 Dupuytren's splint is often used, and is good when properly applied. 1886 *Buck's Handbk. Med. Sci.* II. 7 Dupuytren's paste was made of from six to ten parts by weight of arsenic, and one hundred of calomel. *Ibid.* III. 159 Dupuytren's Finger-contraction.—This deformity is dependent chiefly upon chronic disease and contraction of the palmar aponeurosis. 1887 *Ibid.* V. 197 Dupuytren's Method. A method of amputating at the shoulder-joint. 1895 *Treves's Syst. Surg.* I. 870 Dupuytren's splint..a 'long splint' in miniature, should be padded with increasing thickness from the knee down to the ankle. 1908 *Practitioner* Feb. 279 Treatment of Dupuytren's contraction by thiosinamine. 1947 E. B. KRUMBHAAR tr. *Castiglioni's Hist. Med.* (ed. 2) xix. 713 Baron Guillaume Dupuytren...His descriptions of varicose aneurysms, of fracture of the lower end of the fibula (Dupuytren's fracture, 1819) and the characteristic flexion of one or more fingers due to contraction of the palmar fascia (Dupuytren's contraction,..) are classic. 1966 WRIGHT & SYMMERS *Systemic Path.* II. xxxviii. 1447/2 Fibromatosis of the plantar fascia of the feet is much less frequent than Dupuytren's contracture.

dura², var. DURRA, DHURRA.
1925 *Blackw. Mag.* Sept. 423/2 Largesse was forthcoming in the shape of increased rations, both meat and dura. 1926 *Spectator* 24 Apr. 757/1 The only agriculture carried on was the cultivation of 'dura'. 1966 C. SWEENEY *Scurrying Bush* xii. 177 A man from the village of Korongo Tabanya was walking along a path..to his field of *dura*.

durable, *a.* Add: **2. b.** *spec.* Designating a class of goods the usefulness of which continues over a period of time, as distinguished from goods produced for immediate consumption. Hence as *sb. pl.* (rarely *sing.*), goods of this kind (cf. *consumer durable* s.v. *CONSUMER 2 c).
1930 *Economist* 31 May 1206/2 It includes an estimate of the value of services rendered to their owners by durable, direct or consumers' goods. 1948 G. CROWTHER *Outl. Money* (rev. ed.) v. 142 Some of them are for immediate consumption, the rest are goods whose value will last beyond the immediate present. These two categories can be called current goods and durable goods. 1951 *N.Y. Times* 14 Oct. F. 1 (*headline*) Production eases on some durables. 1957 *Economist* 7 Sept. 777/2 One can trace the rapid growth of engineering exports in the immediate postwar years, while investment and consumption of metal 'durables' grew more slowly. 1958 *Spectator* 12 Sept. 353/3 The spending on durables'—cars, refrigerators, television and radio sets and household furniture—increased by 64 per cent. 1959 *Listener* 15 Jan. 92/1 Bicycles, sewing machines, radio sets, three-piece suites, and all the various sorts of durable consumption goods that help to make life tolerable for Africans like the rest of us.

durain (diūə·rĕⁱn). [f. L. *dūrus* hard + *-ain* as in FUSAIN (sense *2).] A type of hard, compact coal forming a dull layer in a bituminous seam.
1919 [see *CLARAIN]. 1930 *Engineering* 18 July 92/1 Vitrain and durain belonged to the former [coking coal] and fusain to the latter [non-coking coal] class. 1950 WILSON & WELLS *Coal, Coke, & Coal Chemicals* iii. 59 Exact correlation of the Bureau of Mines and the British terminologies, particularly the clarain and the durain of the latter, is not possible on account of the different bases for the classification. 1961 W. FRANCIS *Coal* (ed. 2) vi. 312 Generally American coals appear to contain much less durain than British coals.

dural (diūə·răl), *sb.* Abbrev. of next.
1937 *Jrnl. R. Aeronaut. Soc.* XLI. 495 Standard dural tanks weigh approximately 0·47 to 0·38 lb. per gallon. *Ibid.* 588 Before these plastic materials can be considered as being on a par with steel, dural, or spruce. 1951 'N. SHUTE' *Round Bend* 145 A seating scheme of long, hard dural benches.

Duralumin (diuræ·liumin). Also **duralumin, duraluminium.** [Perh. f. L. *dūrus* hard + ALUMIN(IUM: but see note below.] The proprietary name of a number of heat-treatable wrought aluminium alloys which contain copper and other elements and are comparable to mild steel in strength and hardness but are much lighter.
The name may have been derived from the German place-name *Düren*. Heat-treatable light aluminium alloys were first produced by the Dürener Metallwerke Aktien Gesellschaft.
1910 *Blackw. Mag.* July 4 In this airship, and in the Barrow ship, another aluminium alloy called duralumin is employed. 1910 *Trade Marks Jrnl.* 7 Dec. 1941 Duralumin ...An aluminium alloy. The Electric and Ordnance Accessories Company, Limited,..Aston, Birmingham. Manufacturers. 1917 A. G. CLARK *Motor Car Engin.* II. 53 A metal which has recently come into prominence on account of remarkable properties that it possesses is duralumin, which is manufactured by Vickers, Limited...This material combines lightness with great strength, has an excellent elongation, and may be forged. 1920 ANDREWS & BENSON *Aeroplane Design* 15 Duralumin is composed of aluminium, copper, manganese, and magnesium. 1921 *Glasgow Herald* 8 July 8/4 The aerohydrotor parts will be made chiefly of duraluminium wire stayed. 1925 E. F. NORTON *Fight for Everest, 1924* 131 The duralumin carriers. 1928 C. F. S. GAMBLE *N. Sea Air Station* 10 This hull was twelve-sided, the framework being built of the alloy duralumin. 1958 *Engineering* 7 Feb. 164/3 A skin of Duralumin..should be adequately resistant to meteoric particles. 1968 E. R. PETTY *Physical Metall. Engin. Materials* viii. 153 The true high strength alloys are the duralumins and the aluminium-copper-magnesium-zinc type.

duration. Add: **1. c.** *Phonetics.* The quantity or length of a sound.
1888 H. SWEET *Hist. Eng. Sounds* 9. § 21 A sound which can form a syllable by itself is called *syllabic*. Syllabicness implies an appreciable duration and force. 1933 BLOOMFIELD *Lang.* vii. 109 *Duration* (or *quantity*) is the relative length of time through which the vocal organs are kept in a position. 1957 S. POTTER *Mod. Ling.* iii. 60 Many degrees of *length*, *duration* or *quantity* may be detected by observation and experiment in most languages. 1962 A. C. GIMSON *Introd. Pronunc. Eng.* 24 We shall..refer later to the 'long' vowels of English such as those of *bean* and *barn*, as compared with the 'short' vowel in *bin*. But, in making such statements, we shall not be referring to absolute duration values.
d. The time during which a war lasts, used first of the 1914–18 war from the term of enlistment 'for four years or the duration of the war'; esp. in phr. *for the duration*, until the end of the war; hence, for a long or an unconscionably long time. Also *attrib.*
1916 *Punch* 12 July 51 'I've got a lot of contracts to finish.' 'How long will they take?' 'Oh, about three years—or the duration of the War.' 1925 FRASER & GIBBONS *Soldier & Sailor Words* 86 *Duration, for the*, a phrase often used colloquially to express weariness and impatience. Men, for instance, whose relief was long overdue might be heard complaining, 'Are we going to stop here for the duration?' 1930 *Times Lit. Suppl.* 17 Apr., Nothing so prosaic as 'doing one's bit' would have kept an Italian heart up 'for the duration'. 1939 *Punch* 1 Nov. 483/1 Miss Dodge's motor-bicycle..has been laid up for the duration. 1940 *New Statesman* 19 Oct. 381 We have received a number of letters from country readers offering week-end hospitality for those who must work in London, or 'duration' hospitality for their children. 1941 M. TREADGOLD *We couldn't leave Dinah* xi. 176 That ought to keep you busy for the duration. 1960 L. DURRELL *Clea* II. iii. 157 The war had intervened... The Brigadier was pinned down..'for the duration'.
3. *Comb.*, as **duration-block** (see quots.).
1890 W. JAMES *Princ. Psychol.* I. xv. 610 The unit of composition of our perception of time is a *duration*, with a bow and a stern, as it were—a rearward- and a forward-looking end. It is only as parts of this *duration-block* that the relation of *succession* of one end to the other is perceived. 1935 *Mind* XLIV. 33 There is no such thing as a simple present; the present, to use William James's expression, is a 'duration-block' made up of 'succeeding' parts.

durational, *a.* (Later examples.)
1930 F. R. BLAKE in *Curme Vol. Ling. Studies* 38 There is only one temporal durational case. 1942 *Mind* LI. 150

Only when they have been redigested and creatively synthesized as metaphysical derivatives of eternal action can their durational causality be discovered.

durationless (diurēⁱ·ʃənlės), *a.* [f. DURATION + -LESS.] Having no duration.
1919 A. N. WHITEHEAD *Enquiry Princ. Nat. Knowledge* I. 2 The ultimate fact embracing all nature is distribution of material throughout all space at a durationless instant of time. 1933 *Mind* XLII. 308 The absurd suggestion of everything being contained in a single durationless instant.

durative (diūə·rătiv), *a.* [f. DURAT(ION + -IVE.] Continuing; not completed; *spec.* in *Gram.* applied to a form which marks action as going on. Also as *sb.* Hence **du·ratively** *adv.*, **durati·vity.**
1889 W. R. MORFILL *Russ. Gram.* 40 Durative verbs on taking a prefix become perfective. 1895 [see PERFECTIVE *a.* 3]. 1904 *Expositor* 6th Ser. X. 360 In οἱ ἀπολλύμενοι, strongly durative though the verb is, we see its perfectivity in the fact that the goal is *ideally* reached. *Ibid.* 361 'Ἀγωνίζεσθαι is only used in the durative present. *Ibid.* 444 The effective aorist κατήντησαν is very different from a durative like ἐπορεύοντο. *Ibid.* 441 Other futural presents..have no lack of durativity about them. 1912 WRIGHT *Compar. Gram. Gk. Lang.* § 424 (2) An action is said to be cursive, durative or imperfective when it denotes continuous action without any reference to its beginning or end, as in English *I am striking* compared with *I strike.* 1921 H. POUTSMA *Char. Eng. Verb* 5 Iterative..distinguished into..duratively iterative. 1924 [see *ASPECT *sb.* 9 b]. 1932 *Jrnl. Eng. & Gmc. Philol.* XXXI. 251 The durative aspect indicates duration: 'He *is working* in the garden.' 1935 [see *EFFECTIVE *sb.* 3]. 1957 *Essays in Crit.* VII. 121 *Je vais* is not quite the same as the durative present 'I am going'.

‖ **durchkomponiert** (du:rχʸkǫmpōᵘniᵉ·ɪt), *a. Mus.* [G., f. *durch* through + *komponiert* composed.] Of songs, etc.: having a different musical setting for each verse, stanza, or strophe; through-composed.
1897 J. S. SHEDLOCK tr. *Riemann's Dict. Mus.* 209/1 *Durchkomponirt* (Ger., 'through-composed'), a term applied to a song, when the different strophes of the poem have each their own melody, and are not, as in the volk-song and simple art-song, sung to one and the same melody. 1957 E. T. CONE in N. Frye *Sound & Poetry* 10 Schubert..made the decision to renounce the simple quatrains of Goethe's ballad in favor of a *durchkomponiert* design following the climactic narrative. 1958 *Listener* 20 Nov. 850/3 Puccini's *Girl of the Golden West*..is more completely *durchkomponiert* than his earlier operas.

‖ **du reste** (dü rɛst), *adv. phr.* [Fr.] Besides, moreover.
1827 M. WILMOT *Jrnl.* 14 Aug. in *More Lett.* (1935) 285 Du reste he only drank one cup of weak tea. 1889 E. DOWSON *Let.* 3 Feb. (1967) 32 The scenery is everything that could be desired & the dresses excellent. *Du reste*, I am still suffering from it. 1919 W. S. MAUGHAM *Moon & Sixpence* xxii. 96 *Du reste*, it has still to be proved that this friend of yours has merit.

Durham (dʊ·rəm). [Name of a town and a county in the north of England.] **a.** Used to designate a breed of shorthorn cattle originating in Durham, now generally called shorthorns.
1810 J. BAILEY *Agric. Durham* xiv. 230 The Durham Ox was bred by Mr Charles Colling, of Ketton, in the year 1796. 1855 *Poultry Chron.* III. 391/2 Amongst them [*sc.* breeds of cattle] were to be seen short horns, Durhams, Devons, and others in great perfection. 1861 GEO. ELIOT *Silas M.* vi. 89 P'rhaps you didn't say the cow was a red Durham. 1902 *Encycl. Brit.* XXV. 190/2 The Shorthorn..is still termed the Durham breed in most parts of the world except the land of its birth.
b. Durham Mustard, ground mustard orig. produced by a Mrs. Clements in Durham in the 18th century.
1799 *Times* 1 June 1/4 Best Durham mustard 2s. 1861 MRS. BEETON *Bk. Househ. Managem.* 216 From the circumstance of Mrs. Clements being a resident at Durham, it obtained the name of Durham mustard. 1911 *Encycl. Brit.* XIX. 98/1 The bright yellow farina..produced under the name of 'Durham mustard' pleased..George I. 1970 SIMON & HOWE *Dict. Gastronomy* 274/1 In 1729 she [*sc.* Mrs. Clements] hit on the idea of grinding mustard seeds... It is said she made a small fortune selling what became known as Durham mustard.

Durkheimian (dɜ̄ɹkhəi·miăn), *a.* and *sb.* [f. the name of Émile *Durkheim* (1858–1917), French sociologist + -IAN.] **A.** *adj.* Of, pertaining to, or characteristic of Émile Durkheim or his sociological theories. **B.** *sb.* A follower or supporter of Durkheim.
1915 C. E. GEHLKE *Durkheim's Contrib. to Sociol. Theory* iii. 68 To admit a 'psychological' explanation of the social phenomenon would be to waive the central requirement of the whole Durkheimian sociology. 1936 B. MALINOWSKI in R. Firth *We, the Tikopia* p. vii, In confirmation of that Durkheimian principle of solidarity personified in *l'Etre Moral*. 1938 R. H. LOWIE *Hist. Ethnol. Theory* xii. 223 Radcliffe-Brown agrees not only with other Durkheimians. 1951 PARSONS & SHILS *Toward Gen. Theory of Action* II. v. 239 A Durkheimian 'theory of social facts'.

durn, *U.S.* var. *DARN *sb.*², *adv.*, *a.*, and *v.*² Cf. DERN, DURNED.
1835 A. B. LONGSTREET *Georgia Scenes* 18 Old Boler's.. broke a dish and two plates all to durn smashes! 1866 C. H. SMITH *Bill Arp* Durn the staff and Joe Brown, too. 1867 G. W. HARRIS *Sut Lovingood* 32, I can't say that es a human shut [shirt] I'd gin a durn fur a dozin ove em. 1888 *Portland Transcript* (Farmer), I'll bet I could make as good-lookin' a burst as any o' these,—an' mebbe a durn sight better. 1898 H. S. CANFIELD *Maid of Frontier* 176 It was as much as a man's life was worth to say 'durn' out loud. 1918 *Sat. Even. Post* 5 Jan. 12 If I'd been as big as you be they wouldn't have cared a durn about my eyes. 1936 W. FAULKNER *Absalom, Absalom!* 44 Boys, this time he stole the whole durn steamboat!

duro, var. *DOURO.

Durrellian (dʊre·liăn), *a.* [See -IAN.] Of or pertaining to the English writer Lawrence *Durrell* (born 1912), or his style. Also **Durrelle·sque** *a.*
1961 *New Statesman* 21 July 92/3 In the background there is that damned baroque sea, going through its daily transformations in a spray of Durrellian metaphors. 1961 *Spectator* 22 Dec. 922 The Durrellesque fantasy of a Coptic-Zionist alliance. 1966 *Economist* 17 Sept. 1152/3 One or two of the portraits, notably that of the old Imam Ahmad of the Yemen, are almost Durrell-ian gems. 1970 *Guardian* 26 Mar. 11/6 The usual Durrellian whirligig of allusions and illusions, quotes and echoes.

durrie, durry, varr. DHURRIE.
1921 E. M. FORSTER *Let.* 17 May in *Hill of Devi* (1953) 80 The gifts were dumped on a durry..where sat H.H., among clerical assistants. 1924 *Chambers's Jrnl.* July 472/2 He returned with a roll of Kashmir *durries*, and one by one each *durrie* became a fatter roll than it had been. 1934 M. L. DARLING *Wisdom & Waste in Punjab Village* i. 11 The durry in my room was impregnated with dirty foot-marks.

durum (dⁱū·rʊm). [f. L. *durum*, neut. of *dūrus* hard, used as the specific epithet of *Triticum durum* (R. L. Desfontaines *Flora Atlantica* (1798) I. 114).] In full *durum wheat.* A species of wheat, *Triticum durum*, or one of its varieties, characterized by hard seeds rich in gluten and yielding a flour used in the manufacture of spaghetti, etc.
1908 P. T. DONDLINGER *Bk. Wheat* i. 8 There are several kinds of the less common wheats, such as Polish wheat, spelt and durum wheat. 1911 *Encycl. Brit.* XIV. 276/2 A wonderful region for growing the *durum* or macaroni wheat. 1921 *Nature* CVII. 251/1 The common wheats, such as Marquis, were susceptible, the durums such as Kubanka, 'commercially resistant'. 1961 *Economist* 26 Aug. 822/1 Durum wheat, an extremely hard wheat normally grown in hot climates and used to make pasta dishes like spaghetti and macaroni, has risen steeply in price.

durwan (dǭɹwā·n). *India.* Also 8 **derwan, dirwan, door-van,** 9 **darwan.** [Hind. (a. Pers.) *darwān*.] A porter or door-keeper.
1773 E. IVES *Voy. to India* 50 Derwan. These are properly porters, who sit at the gate to receive messages, &c. 1781 *Bengal Gaz.* 14 Apr. (Yule), A Door-van is well known to be the alarm of the House. 1784 in W. S. Seton-Karr *Sel. fr. Calcutta Gaz.* (1864) I. 12 A most extraordinary and horrid murder was committed upon the Dirwan of Thomas Martin, Esq. 1874 *Calcutta Rev.* LIX. 207 In the entrance passage..is a raised floor with one or two open cells in which the *darwāns* (or door keepers) sit, lie and sleep. 1926 *Blackw. Mag.* June 726/2 His durwan..refused to awaken a sleeping man.

durzee (dɜ̄·ɹzi). *India.* Also **derzie, -y, dhirzee, dhurzee, dirge, dirgee, dir(d)jee, dirzie, durjee.** [Hind. (a. Pers.) *darzī* (Pers. *darz* sewing, hem).] A tailor.
1812 M. GRAHAM *Jrnl. Resid. India* 30 The *derdjees*, or taylors, in Bombay, are Hindoos of a respectable caste. 1834 A. PRINSEP *Baboo* II. x. 180 The milliners and durzees of the City of Palaces. *a* 1847 M. M. SHERWOOD *Life* (1854) xvi. 300 We took other servants, Dirges, and Dobes. 1848 *Alfred in India* 18 The needlework of the family is done by a *durzie*, or tailor. 1884 D. AUBREY *Lett. fr. Bombay* 204 Their parents have no wish to make 'dirzees' or tailors of them. 1894 M. DYAN *Man's Keeping* i, Manufacturing, with the incompetent aid of her durzee, a gown for the ball. 1922 *Blackw. Mag.* Feb. 183/2 The regimental dhurzee. 1969 *Islander* (Victoria, B.C.) 6 July 2/3 And the names! Ayahs, kansamahs, chuprassis, peons, chowkidars, durzis, dhobis, kitmutgars. Wonderful names and each with a different job.

dusk, *a.* and *sb.* Add: **B. 2. b.** *attrib.*, as *dusk-hour, -light, -time.*
1881 A. B. & S. EVANS *Leics. Words* 143 *Dusk-hour*, late evening twilight. 'Ah shouldn' like to mate his ogly mug upo' *dusk-hour* in a daa'k leane.' 1937 W. DE LA MARE *Poems*, Perhaps 'twas the talk of chance farers..In the dusk-light clear? 1957 J. KEROUAC *On Road* (1958) IV. vi. 300 We saw all of Mexico City stretched out..spewing city smokes and early dusklights. 1890 J. D. ROBERTSON *Gloss. Gloucester* 42 *Dusk time*, evening.

dust, *sb.*¹ Add: **3. f.** *dust and ashes* (in allusion to the legend of the Dead Sea Fruit): used to indicate severe disappointment or disillusionment.
1902 W. JAMES *Var. Relig. Exper.* vi. 143 Trustful self-abandonment to the joys that freely offer has entirely departed from both Epicurean and Stoic; and what each proposes is a way of rescue from the resultant dust-and-ashes state of mind. 1911 BEERBOHM *Zuleika D.* xxi. 310 But there was no spark of triumph now in her eyes; only a deep melancholy; and in her mouth a taste as of dust and ashes. 1930 A. HUXLEY *Vulgarity in Literature* iii. 13 The spirit of the time..demands that we should 'press with strenuous tongue against our palate' not only joy's grape, but every Dead Sea fruit. Even dust and ashes must be relished. 1945 —— *Let.* 2 Apr. (1969) 518 The most wildly romantic adventures all turned into dust and ashes.
4. (See also BITE *v.* 16 in Dict. and Suppl.)
8. a. *dust-haze, -heap* (further examples; also *fig.*), *-screen, -spout, -yard* (examples; also *fig.*). **b.** *dust-catcher*; *dust-catching, -free, -laying, -proof, -raising* adjs. **c.** *dust-clogged, -filled.* **d.** *dust-white* adj.
1939 D. R. G. CRAWFORD *Gas Producer Operator's Handbk.* v. 70 The simple gas-cleaning plant..is for use..where a clean cool gas is required...Where hot raw gas is required the complete cleaning plant is replaced by a static dust-catcher. 1940 *Chambers's Techn. Dict.* 273/1 *Dust catcher*, a chamber in which dust is extracted from furnace gases by causing a sudden change in the direction of the gas stream. 1953 D. J. O. BRANDT *Manuf. Iron & Steel* viii. 64 This will normally lead direct to the dust-catcher. 1902 *Encycl. Brit.* XXVII. 428/1 A dust-catching apparatus has been.. erected at Edinburgh. 1894 H. NISBET *Bush Girl's Romance* xxxi. 293 The bushranger's dust-clogged brow became corrugated. 1908 *Westm. Gaz.* 25 July 3/2 A man with a dust-filled throat. 1925 SHAW & OWENS *Smoke Probl. Gt. Cities* xi. 211 When breathing dust-laden air, it is only after long periods of quiet breathing that the air from the deep parts of the lungs is dust free. 1934 *Discovery* July 184/1 Bringing the surfaces together in a clean and dust-free condition. 1925 SHAW & OWENS *Smoke Probl. Gt. Cities* x. 197 Smoke or dust is normally dispersed in an upward direction ..and it is evident from ordinary observations of dust haze that the upper limit is sometimes very well defined. 1945 *Finito! Po Valley Campaign* 36 In the dust-haze..factories ..collapsed in ruins. 1901 *Westm. Gaz.* 1 May 3/2 The Salvation Army deserves to be helped in its work of sifting the dust-heap of our lowest social strata. 1940 R. G. COLLINGWOOD *Ess. Metaphys.* 120 The distinction between truth and falsehood is part of that antiquated lumber which has at last..been thrown on the dust-heap. 1899 *Westm. Gaz.* 18 July 2/1 Water-carts sprayed the road with grateful dust-laying streams. 1902 *Ibid.* 11 Sept. 7/3 Oil is the latest dust-laying agent. 1869 *Rep. Comm. Agric. 1868* (U.S.) 15 The museum has been partly filled with absolutely dust-proof cases. 1882 *Leisure Hour* 414/2 The fittings are massive and dust-proof. 1898 J. SOUTHWARD *Mod. Printing* I. iv. 26 One of the most useful recent innovations is to make the racks 'dust-proof'. 1903 Dust-proof [see *BENNY¹]. 1934 *Archit. Rev.* LXXV. 142/2 Dust-proof electric light fittings are available. 1903 B. HARRADEN *Kath. Frensham* 272 A long, straggling, dust-raising line of about 50 conveyances. 1908 *Westm. Gaz.* 22 Dec. 4/2 On the mere off-chance of minimising in an infinitesimal degree their dust-raising propensities. 1963 P. DRACKETT *Motor Rallying* i. 10 There were even dust-raising tests and noise checks. 1899 W. H. MAXWELL *Removal of Town Refuse* vii. 175 Dust screens are ineffective. 1918 W. OWEN *Let.* 15 June (1967) 559, I have now a waterproof tent with long grass & buttercups all round to act as dustscreens. 1926 T. E. LAWRENCE *Seven Pillars* (1935) III. xxxv. 205 Two dust-spouts, tight and symmetrical chimneys, advanced. 1937 A. HUXLEY *Let.* 3 June (1969) 422 Western Texas, which we crossed in the midst of a premature heatwave (dust-spouts in a temperature of 105° in the shade). 1923 E. SITWELL *Bucolic Comedies* 48 And in the street dust-white and lean, Two black apes bear her palanquin. 1938 W. DE LA MARE *Memory* 16 Dust-white hedge. *a* 1852 MAYHEW *London Labour* (1861) II. 216/1 The dust-yards must not be confounded with the 'night-yards'. 1854 DICKENS *Hard T.* II. ix. 232 Her father was usually sifting and sifting at his parliamentary cinder-heap in London..and was still hard at it in the national dustyard. 1904 *Daily Chron.* 27 Sept. 8/2 In a few odd corners of London there still exist dustyards in which the refuse of the great city is sifted and sorted.
e. dust-bowl orig. *U.S.*, a region subject to drought where, as a result of the loss or absence of plant cover, the wind has eroded the soil and made the land unproductive; hence, any region that is arid or unproductive; also *attrib.*; **dust-cap,** a cap (CAP *sb.*¹ 12) to protect something from dust; **dust-coat** (earlier and later examples); **dust-core** *Electr.*, a core of magnetic powder in which the insulating properties of the binding agent result in reduced core losses; **dust-counter,** an instrument for counting the dust particles in a known volume of air; **dust-cover, dust cover,** a cover to protect something from dust; *spec.* a detachable paper cover or jacket in which a new book is normally issued and which often contains information about the book or its author; also *fig.*; **dust-devil** (earlier and later examples); **dust-flow,** a stream or landslide of volcanic ashes saturated with water; **dust-jacket, dust jacket** = JACKET *sb.* 2 b (cf. *dust-cover); **dust-pan** (freq. *dustpan*) (earlier and later examples); hence **dustpanful,** as much as a dust-pan will hold; **dust-sheet,** a sheet for covering furniture or the like to keep off dust (cf. quot. 1888 *s.v.* SHEET *sb.*¹ 1 a); hence **dust-sheeted** *a.*, covered or provided with a dust-sheet; **dust-storm** (examples); **dust-trap,** something in or on which dust collects; also *attrib.*; **dust-wind,** a wind bringing dust-storms; **dust-wrapper** = WRAPPER *sb.* 1 b (cf. **dust-cover*).
1936 *Durant* (Okla.) *Daily Democrat* 26 Mar. 1/7 The panhandle 'dustbowl' was outside the path of the wind. 1936 *Dallas Morning News* 26 Dec., They say he nearly defeated himself by urging Landon's election among the

dust bowl farmers. 1937 *Ann. Reg. 1936* 288 Some pastoral areas where over-feeding of live-stock had completely killed the pasturage were called 'dust-bowls'. 1951 B. Russell *New Hopes for Changing World* 33 Will all the arable land be turned into dust-bowls as it has been in large parts of the United States? 1959 *Listener* 20 Aug. 276/1 A more depressing picture, widely supported at the present time, is that Venus is an arid dust-bowl. 1898 *Springtime* Apr. 103/1 There are..five separate pieces in the dust-cap alone. 1930 *Engineering* 16 May 647/1 After screwing down the needle valve, disconnecting the pump and screwing on the dust cap, the strut is ready for use. 1702 C. Fiennes *Journeys* (1947) 261 The wind soone dry'd my dust coate. 1916 H. G. Wells *Mr. Britling* I. iii. 80 A lady in a motoring dust coat. *Ibid.* 86 The dust-coat lady. 1968 J. Ironside *Fashion Alphabet* 39 *Dust-coat*, any lightweight coat worn mainly to protect the clothes and not necessarily for warmth. [1920 *U.S. Pat. 1,523,109* Telephone loading coil cores of the so-called 'dust' type.] 1924 S. R. Roget *Dict. Electr. Terms* 70/1 *Dust core*, an iron core for induction in telephone lines in which eddy current and hysteresis losses are negligible. 1928 *Trans. Amer. Inst. Electr. Engin.* XLVII. 436/2 The commercial use of permalloy-dust-core loading coils..has brought about a number of very important improvements. 1954 E. Molloy *Radio & Telev. Engin. Ref. Bk.* xxx. 11 Cores made of ferrites..are commercially available now, and but for their high cost would have replaced the dust core over the frequency range in which they are advantageous. 1970 D. F. Shaw *Introd. Electronics* (ed. 2) v. 87 These materials [*sc.* ferrites] have many advantages over the earlier 'dust-core' materials. 1892 J. Aitken in *Proc. R. Soc. Edinb.* XVIII. 39 A simple pocket dust-counter. 1936 *Discovery* Nov. 348/1 The Aitken dust-counter has been considered by some to give counts too high for the actual dust particles. 1902 D. Salomons in A. C. Harmsworth et al. *Motors* vi. 95 Every car should have mackintosh rain-covers..; also dust-covers, which are useful on many occasions. 1921 *Sat. Westm. Gaz.* 17 Sept. 14/1 The dust-cover..suggests that the book will be of service not only in schools but also as 'an entertainment for home-reading'. 1923 *Times Lit. Suppl.* 22 Feb. 126/1 The lurid dust cover. 1942 'N. Shute' *Pied Piper* 49 Dozing uneasily in the chair, half-covered by the dust-cover from the bed. 1962 *Which?* (Car Suppl.) Oct. 138/2 Rubber dust covers on front brake cylinders disintegrated. 1968 *Listener* 1 Aug. 144/2 Soon after you came back from your recent honeymoon, your wife was quoted in one newspaper as saying that she must help you to take the dust covers off certain areas of your personality. 1888 Kipling *Plain Tales fr. Hills* 43 The wheeling choking 'dust-devils' in the skirts of the flying storm. 1926 T. E. Lawrence *Seven Pillars* (1935) iii. xxxv. 205 At last I saw that part of the yellow cloud off Serd was coming slowly against the wind in our direction, raising scores of dust devils before its feet. 1955 H. Klein *Winged Courier* vii. 46 The airmen experienced a new African flying hazard in the form of dust devils, some of which rose as high as 8,000 ft. 1904 *Science* 1 July 24/2 Clouds of steam rising from the crater, accompanied from time to time by dust-flows. 1928 *S. J. Looker's Booklover's Catal.* Jan. 5 The Life and Letters of Emily Dickinson..in dust jacket. 1928 *Observer* 24 June 8 The book is sent out by Constable's in a particularly attractive dust-jacket. 1957 *Times* 25 Nov. 11/3 Henry Fielding's *Tom Jones* in a practical transparent dust jacket at 12s. 6d. 1785 F. Hopkinson *Misc. Essays* (1792) II. 158 It was soon after swept out with the common dirt of the room, and carried in a dust pan to the yard. 1857 Dickens *Dorrit* xxv. 125 Ladies would fly out at their doors crying, 'Mr. Baptist—tea-pot!' 'Mr. Baptist—dust-pan!' 1966 J. Betjeman *High & Low* 17 Brooms and plastic dustpans hang from the ceiling. 1882 F. A. Kemble *Rec. Later Life* I. 60 Three and four dustpanfuls a day would be swept away. 1965 M. Echard *I met Murder* (1967) xvii. 137 The maid told me she swept up dead roaches by the dustpanful. 1854 Mrs. Gaskell *Let.* 17 May (1966) 290 Not even a book to beguile the time—five fathoms deep they lie beneath dust-sheets. 1907 *Westm. Gaz.* 17 Aug. 4/3 The big town-house was depressing in its shroud of dust-sheets. 1928 *Daily Mail* 25 July 4/2 Hundreds of dust sheets, 2 yards wide by 2¾ yards long, for covering furniture, are being sold by a West End firm. 1936 W. de la Mare *Wind blows Over* 159 Having muffled the furniture with their sepulchral dust-sheets. 1917 C. S. Lewis *Let.* 18 July (1966) 38 Some of the rooms were all dust-sheeted. 1879 Mrs. A. E. James *Ind. Househ. Managem.* xii. 82 Dust-storms come on often very quickly. 1936 'F. Gerald' *Millionaire in Memories* ii. 42 At Port Pirie a dust-storm swept down upon us. 1905 *Daily Chron.* 17 Apr. 8/2 Fussy, dust-trap trimming near the hem. 1906 *Westm. Gaz.* 15 Jan. 2/1 Hailstones, the slowly falling flakes of snow, drops of rain, are literally dust-traps. 1967 R. Rendell *New Lease of Death* i. 8 The primrose venetian blinds..were dust-traps. 1901 *Geogr. Jrnl.* XVIII. 91 Observations, outline and relief of the region,..temperature in the interior, dust-winds, temperatures of wells and springs [etc.]. 1932 *Book-Collector's Q.* Apr.–June 10 The somewhat more humble 'dust-wrapper' is to be found in the catalogues of the greater and more conscious booksellers. 1934 *Punch* 24 Oct. 476/2 The love-interest which I guessed (from the dust-wrapper) must be contained somewhere.

dust, *v.*¹ Add: **6. b.** *to dust a person's jacket* (further examples).

1807 *Eagle* (Staunton, Va.) 28 Aug. 4/2 Go in peace, or I will dust thy jacket with this horse-whip. 1842 Barham *Ingol. Leg.* 2nd Ser. 52 Old Shylock was making a racket, And threatening how well he'd dust every man's jacket. 1895 'Rosemary' *Under Chilterns* i. 31 Master told me as 'ow 'ee'd dusted 'is jacket for 'im.

7. a. Delete † and add later examples. Now *colloq.* or *dial.*

1884 'Mark Twain' *Huck. Finn* xxxix. 395 So she took and dusted us both with the hickry [*sic*]. 1950 *Time* 30 Jan. 14/2 [Miners] dusted one of [the district leader's] lieutenants with an old shoe for trying to talk them back to work. 1970 H. E. Roberts *Third Ear* 6/2 *Dust* v., beat up.

10. (Further examples.) For 'Now *U.S. slang* or *colloq.*' read: 'Chiefly *U.S. slang* or *colloq.*'

1860 *Mesilla* (Ariz.) *Times* 18 Oct. 1/2 The 'gold seekers' thought prudence the better part of valor and 'got up and dusted'. 1888 'R. Boldrewood' *Robbery under Arms* II. xi. 190 And you're a going to dust out right away, you say? 1909 J. R. Ware *Passing Eng.* 120/2, I quickly got inside, locked the door, and dusted out the back way.

12. to dust off. a. *intr.* (See sense 10 in Dict.) **b.** *trans.* = sense 6 a in Dict. (orig. *U.S.*).

1948 A. Huxley *Ape & Essence* (1949) 127 The charvessels can dust off the tables and wash the floors. 1959 *Economist* 18 Apr. 218/1 The clerics are..dusting off their copies of Fox's *Book of Martyrs*.

c. To bring to ruin, defeat, kill; to discard, get rid of. *slang* (orig. *U.S.*).

1938 H. Asbury *Sucker's Progress* xii. 385 He had been dusted off by Vanderbilt. 1942 Berrey & Van den Bark *Amer. Thes. Slang* § 27/5 Eliminate; discard; get rid of,..*dust off. Ibid.* § 118/3 Kill; murder,..*dust off.* 1960 *Times* 13 Dec. 4/1 They have always been dusted off in the inter-zone matches.

dust-bin, dustbin. Add: Also *fig.* and *attrib.*

1901 M. Beerbohm in *Sat. Rev.* 23 Feb. 234/2 The Drama of the Dustbin. This phrase was not coined by me... Mr. Clement Scott issued it into our currency as a description of 'Mr. & Mrs. Daventry'. 1930 C. Dixon *This Way to Paradise* i. 29 Wasn't it just one of the usual dust-bin experiences? *Ibid.* 30 The whole of existence is an affair of dust-bins. 1931 R. Graves *Poems 1926–1930* 53 He leads the sick words into parliament To rule a dust-bin world with deep-sleep phrases. 1940 *Times* 25 Apr. 8/7 When two..men were summoned..for removing refuse from dustbins without authority..solicitor for the prosecution said, 'This offence is known as "dustbin totting".' 1958 E. H. Carr *Socialism in one Country* I. iv. 148 Trotsky relegated his defeated opponents to the dustbin of history. 1959 *Times* 23 May 5/6 Plainly the *Ulysses* of James Joyce is the great source of the modern dustbin drama.

2. A gun-turret of an aircraft, esp. the one beneath the fuselage. *slang.*

1934 *Flight* 15 Feb. 160/1 Our photographer..took his photographs from the 'dustbin', the retractable and rotatable turret for a Lewis gun beneath the fuselage. 1942 *Gen* 1 Sept. 14/1 The rear gunner's turret is the 'dust bin'. 1967 C. H. Barnes *Shorts Aircraft* 377 Here there had originally been mounted a twin-Browning F.N.25 retractable ventral turret..; this 'dust-bin' turret was deleted from early production Stirlings.

dust-box. Add: **2.** (Austral. examples.)

1896 E. Turner *Little Larrikin* ii. 17 On the footpaths there were dust-boxes twice a week. 1906 E. Dyson *Fact'ry 'Ands* xvi. 219 They've dumped Toucher in ther dust-box for immedjit removal.

3. (See quot. 1893.)

1893 *Funk's Stand. Dict.*, *Dust box*, a box supplied with dust for the use of poultry, to aid them in expelling lice. 1894 *Vermont Agric. Rep.* XIV. 176 Fix a dust box so that biddy can take a bath every day.

duster. 1. Delete † from sense 'a dust-brush' and add later examples. Cf. *feather-duster* (Feather *sb.* 19).

1895 *Montgomery Ward Catal.* 104/2 Duster, 8-inch handle, all dark down, 7×9 inch, double faced brush. 1939–40 *Army & Navy Stores Catal.* 110/3 Household brushes...Turnover Library Duster, White Lily bristles.

4. Now usu. a woman's loose casual full-length coat. In full, *duster coat.*

1959 W. Camp *Ruling Passion* xviii. 154, I adore that duster coat, but it's at least a couple of inches longer than your frock. 1960 *Guardian* 25 June 10/3 The Princess wore a royal blue duster coat over a..flowered frock. 1962 'S. Ransome' *Without Trace* ix. 98 A dark-haired woman..was wearing a duster while hanging the family wash on a line.

5. *Naval slang.* An ensign or flag; *the red duster* (see quot. 1929).

1904 *Eastern Morn. News* 27 June, A regatta..of the duster boats for the championship of the Humber. 1918 *Punch* 27 Mar. 206/1 She's dipped her dingy duster in the spray of all the seas. 1929 F. C. Bowen *Sea Slang* 111 *Red Duster*, the Red Ensign.

dustering (dʌ·stəriŋ). [f. Duster + -ing¹.] Material such as is used for dusters.

1910 H. G. Wells *Mr. Polly* ii. 47 He was now hanging long strips of grey silesia and chilly-coloured linen dustering.

dusting, *vbl. sb.* Add: **1. b.** The sprinkling of powdered insecticide, fertilizer, etc., on crops, usu. from the air. Usu. in form *crop dusting.*

1926 H. H. Arnold *Airman & Aircraft* 212 (*index*) Crop 'dusting' by planes. 1930 *Flight* 17 Oct. 1151/1, 11 engaged on crop dusting. 1958 *Times Rev. Industry* July 76/2 The prototype of the Wirraway war-time trainer, rebuilt for crop-dusting.

c. With *adv.*, as *dusting-on.*

1879 *Telegraphic Jrnl.* 15 Oct. 344/2 The 'dusting-on' process [of phosphorescent photography] consists in coating a plate with a preparation of dextrine, honey, and bichromate of ammonia which..becomes hardened..remaining tacky where it is protected from..light.

3. dusting-brush (example); **dusting-powder** (example), also = *talcum powder.*

1907 *Yesterday's Shopping* (1969) 358/1 Typewriter sundries...Dusting Brushes. *Ibid.* 506/1 Dusting Powder for wounds..tin 0/10½. 1926 *Toilet Goods Economist* 12 June 59 Bath salts, face powders, dusting powders. 1947 E. H. Young *Chatterton Square* xxiii. 158 She had bathed and changed and smelt faintly of dusting powder. 1951 'J. Tey' *Daughter of Time* i. 11 The Midget smelt of lavender dusting powder.

dustless, *a.* (Later examples.)

1904 *Westm. Gaz.* 9 Aug. 1/3 An August morning that..is sunnily cool, and breezily fresh and deliciously dustless. 1970 *Sunday Tel.* 11 Oct. 21/2 The council's automatic dustless loading bins have made the dustmen's lot a happier one. 1970 *Interior Design* Dec. 767/4 Little progress seems to have been made in finding a dry writing/erasing system, acceptable to lecturers, which is dustless.

dust-up (dʌ·stʌp). *colloq.* [f. Dust *sb.*¹ + Up *adv.*¹] A quarrel, fight, disturbance; = Dust *sb.*¹ 5 b.

1897 *Daily News* 6 Mar. 7/3 They turned at the Lasher, and after a dust-up for about a minute in Iffley Reach did a nice piece of paddling back to the raft. 1897 S. L. Hinde *Fall Congo Arabs* 152 An American nigger said..they ain't had such a dust-up in this hole since creation. 1926 D. L. Sayers *Clouds of Witness* i. 31 [He] had thought Wednesday evening's dust-up none of his business. 1930 H. Tomlinson *All our Yesterdays* v. i. 391 Some fellows always thought they'd pulled it off..if they got out of a dust-up on the cheap. 1944 'N. Shute' *Pastoral* v. 123 He had a bit of a dust up with one of his girl friends.

dusty, *a.* Add: **1. b.** Of wine: containing sediment.

1886 J. Noble *Handbk. Cape Gd. Hope* 275 Notwithstanding the large amount of alcohol which they [*sc.* Cape wines] contain, they are not clear and always somewhat 'dusty'.

4. b. (Further examples.) Also *none so dusty* = 'not so bad'. **d.** Also, unsatisfying, inconclusive, poor; esp. in phr. *dusty answer.*

1856 F. E. Smedley *Harry Coverdale* xlii. 312 None so dusty that—eh? for a commoner like me? 1862 G. Meredith *Modern Love* 82 Ah, what a dusty answer gets the soul When hot for certainties in this our life! 1864 Hotten *Slang Dict.* 128 'What do you think of this?' 'Well, it's not so dusty,' *i.e.*, not so bad; sometimes varied to 'none so dusty'. 1926 Galsworthy *Escape* i. i, Out of Germany! Cripes! That was none so dusty! 1927 R. Lehmann (*title*) Dusty answer. 1929 J. B. Priestley *Good Compan.* i. i. 16 'You're a swell tonight all right!'..'Not so dusty, Mar,' said Leonard. 1936 'I. Hay' *Housemaster* xix. 235 The applicants met with what is technically known as a dusty answer.

5. dusty boy *Naval slang*, a naval stores rating; **dusty miller,** (*c*) any of several noctuid moths with speckled markings on their wings (*U.S.*).

[1909 J. R. Ware *Passing Eng.* 120/2 *Dusty*, a ship's steward's assistant.] 1916 'Taffrail' *Pincher Martin* vii. 114 Puts on a lot o' swank fur a bloomin' dusty boy. 1909 G. Stratton-Porter *Girl of Limberlost* xii. 243 Small insects of night gathered, and at last a little dusty miller, but nothing came of any size.

Dutch, *a.*, *sb.* (*adv.*). Add: **A.** *adj.* **2. b.** *S. Afr.* Of, pertaining to, or designating South Africans of Dutch descent; Afrikaans-speaking.

1731 Medley tr. *Kolben's Pres. State Cape of Good Hope* v. 58 The Terror of the *Dutch* Arms was spread through all the Nations about the Cape. 1791 G. Carter *Narr. Loss Grosvenor* xvii. 50 They had got out of the country of the Caffrees, and had reached the northermost of the Dutch settlements. 1852 C. Barter *Dorp & Veld* vi. 52 Dutch families on their way to Maritzburg for the half-yearly 'Nacht maal' or sacrament. 1871 J. Mackenzie *10 Yrs. North of Orange River* i. 18 The 'Nachtmaal', or celebration of the Lord's Supper, by the Dutch Church, takes place several times in a year. 1954 D. D'Ewes *Mydorp* xii. 70 He attended the Dutch Reformed Church regularly every Sunday. 1970 *Cape Times* 28 Oct. 1/5 He had never had much hope in the liberal movement there, but had contacts with Dutch Reformed Church ministers and had never found them completely impervious to suggestion.

3. b. Dutch barn (earlier example); **Dutch cap,** (*a*) a woman's cap of lace or muslin with a triangular piece rolled back at each side; (*b*) a type of contraceptive pessary; = *diaphragm *sb.* 2 b; **Dutch doll,** a jointed wooden doll; **Dutch door** (see quot. 1890); **Dutch elm disease,** a fungous disease of elms, first discovered in Holland, caused by *Ceratocystis ulmi*; **Dutch garden** (see quots.); **Dutch hoe** (see Hoe *sb.*² 1 b); **Dutch interior,** a painting of the interior of a Dutch room or house, esp. by the Dutch painter Pieter de Hooch (1629–83); also *transf.*; **Dutch oven** (see Oven *sb.* 2 a); also *slang*, a person's mouth; **Dutch pink** [Pink *sb.*⁵], a yellow lake pigment; also *slang*, blood; **Dutch roll,** (*a*) a roll in ice-skating, executed by gliding with the feet parallel and pressing alternately on the edges of each foot; (*b*) *Aeronaut.* (see quot. 1960); **Dutch sauce,** a sauce served with fish; = *Hollandaise; **Dutch tile,** a kind of glazed tile frequently painted in colours; **Dutch white,** a pigment consisting of one part of white lead and three parts of barium sulphate; **Dutch wife** (see quots.).

1743 W. Ellis *Mod. Husb.* June x. 76 In order to enjoy his Hay finer than his Neighbours, he built him a *Dutch* Barn, in 1738. 1726 Mrs. Johnson *Let.* Oct. in E. Hamilton *Mordaunts* (1965) vii. 145 Misses coat fitts her very well, the Dutch cap Miss Mordaunt is not at all reconciled to but will wear it if her Cosens doe. 1857 D. H. Strother *Virginia* 68 His head..was surmounted by a tiny Dutch cap. 1922 M. Stopes in *Lancet* 12 Aug. 357/2 Recourse may then be had to the inverted or Dutch cap. 1943 J. H. Peel *Textbk. Gynæcol.* xxiii. 342 The best and most widely applicable is undoubtedly the Dutch cap pessary. 1960 C. Watson *Bump in Night* iv. 44 Don't ever

again describe bridesmaids as wearing Dutch caps. 1962 'H. Lourie' *Question of Abortion* iv. 34 One patient..had proved not to be pregnant..she had come back for a Dutch cap. 1967 M. Drabble *Jerusalem the Golden* vii. 155 Spread before her on the floor was a..dutch cap, an instruction leaflet, and various other accoutrements of contraception. 1797 Lady A. Barnard *Let.* 10 July (1901) ii. 57 What they [*sc.* Dutch ladies] most want is shoulders and manners. I know now what is meant by a 'Dutch doll'; their make is exactly like them. 1824 Scott *Redgauntlet* Let. iii, All thy motions, like those of a great Dutch doll, depending on the pressure of certain springs. 1926 W. Deeping *Sorrell & Son* vii, Her head was as neat as the head of a Dutch doll. 1890 Webster, *Dutch door*, a door divided into two parts, horizontally, so arranged that the lower part can be shut and fastened, while the upper part remains open. 1945 Nelson & Wright *Tomorrow's House* v. 57 The kitchen also opens into the main room by way of..a Dutch door. 1927 *Gardeners' Chron.* 19 Feb. 133/3 (*heading*) The Dutch Elm Disease...The disease was first observed in Holland in September, 1919. 1931 *Science* 30 Oct. 437/1 After the identification of the Dutch elm disease in Ohio, during the summer of 1930. 1968 *New Scientist* 5 Sept. 489/1 The beetles that carry Dutch elm disease are attracted to dead or dying trees for egg laying. *a* 1772 T. Whately *Obs. Mod. Gardening* (new ed.) (1801) iv. 153 To get too, as far as can be, the advantage of natural prospects, the artificial mounts of the flat Dutch gardens should here be introduced. 1872 A. Smee *My Garden* 584 The chief peculiarities of a Dutch garden may be said to consist in its being seen at one glance;..in the utmost symmetry being observed in all its parts..; in its trees being clipped sometimes into curious shapes and figures..; in its having long serpentine or straight walks..; [etc.]. 1899 S. R. Hole *Our Gardens* 277, I asked an old gardener whether he could tell me anything about Dutch Gardens, and he made answer, 'They be bits o' beds with edgings o' box, and gravel walks, and four sloping banks forming a square outside, and they be pretty toys for children, and very snug for varmint.' 1902 H. I. Triggs *Formal Gardens* pl. 58 Holland House, Kensington. The Dutch Garden. 1928 L. Archer-Hind tr. *Gothein's Hist. Garden Art* II. xiii. 218 People were misled by the term 'Dutch garden', as it came to be used derisively in the eighteenth century. *Ibid.* 230 The Dutch garden must be reckoned as of the French school. [1816 M. Bryan *Dict. Painters & Engravers* I. 559 His [*sc.* P. de Hooch's] favourite subjects were the interiors of Dutch apartments..the sun shining through a window.] 1886 Bryan & Graves *Dict. Painters & Engravers* I. 370/1 Dutch Interior [by P. de Hooch]; a Lady playing the lute and singing, whilst a cavalier accompanies her. 1913 E. Wharton *Custom of Country* i. v. 73 The hall, with..the quiet 'Dutch interior' effect of its black and white marble paving. 1940 'F. O'Connor' (*title*) Dutch interior. 1966 M. Catto *Bird on Wing* ix. 131 The lights were dim. It made an intimate picture. Like one of those cosy Dutch interiors: the burgher at home with his wife. 1970 V. C. Clinton-Baddeley *No Case for Police* iii. 63 Like some picture of a 'Dutch Interior', the open door revealed a hall, and, beyond that, another open door...Only the traditional distant figure was lacking. 1769, 1849 Dutch oven [see Oven *sb.* 2 a]. 1922 Joyce *Ulysses* 419 O, cheese it! Shut his blurry Dutch oven with a firm hand. 1968 *Islander* (Victoria, B.C.) 11 Aug. 7/2 Other relics of trail days, which time has not completely erased, are three beehive Dutch ovens built from native stone. 1758 Dutch pink [see Pink *sb.*⁵]. 1835 Dutch pink [see *English pink* s.v. *English *a.* 2 e]. 1853 'C. Bede' *Verdant Green* ii. iv, That'll take the bark from your nozzle, and distil the Dutch pink for you, won't it? 1881 J. Bell *Anal. Foods* i. 22 The leaves were slightly coloured with Dutch pink to impart a bloom. 1893 *Durham Univ. Jrnl.* X. 103 Others have in spite of honourable endeavour been obliged to content themselves with mediocre achievement and Dutch roll. 1939 *Jrnl. R. Aeronaut. Soc.* XLIII. 795 Lateral oscillations, or 'Dutch roll', as they are sometimes called. 1960 *Electronic Engin.* XXXII. 407 A 'dutch roll' is the characteristic short period lateral oscillation of an aircraft, involving yaw, roll and sideslip, which is excited either by rudder application, or a lateral gust. 1573 'C. Hollyband' *French Schoolemaister* fol. 55ᵛ Will you eate of a Pike with a high dutche sauce? 1893 T. F. Garrett *Encycl. Pract. Cookery* II. 387/2 Dutch or Holland Sauce (à la Hollandaise). 1955 *Oxf. Jun. Encycl.* XI. 396/2 *Hollandaise* (or Dutch sauce) is made entirely of butter emulsified with egg-yolks and lemon juice—a sort of butter mayonnaise served warm. 1727–41 Dutch tile [see Tile *sb.*¹ 1 c]. 1753 H. Walpole *Let.* 12 June (1903) III. 168 A cool little hall..hung with paper to imitate Dutch tiles. 1844 Dutch tile [see Tile *sb.*¹ 1 c]. 1862 Rossetti *Let.* 9 Jan. (1965) II. 435, I have had the fireplace covered with real old blue glazed Dutch tiles. 1957 *Granta* 9 Mar. 19/3 He walked over the steeply pitched, loose Dutch tiles of the roofs as if he were on a dance-floor. 1968 L. O'Donnell *Face of Crime* i. 11 The fireplace faced with authentic Dutch tiles in the traditional pale lavender depicting scenes from the Bible. 1886 H. C. Standage *Artists' Man. Pigments* i. 5 White lead (known also as Ceruse, Cremnitz, Dutch, Flemish, Hamburg, Venetian, or Roman White). 1891 Farmer *Slang* II. 349 Dutch-wife, a bolster. 1965 W. Young *Eros Denied* xxvii. 271 We call..a masturbation machine a Dutch husband or wife. 1966 G. Black *You want to die, Johnny?* vi. 114 'What's this great long bolster for?..' 'Colonial invention. For the hated Imperialists. Known as a Dutch wife.' 1967 *Guardian* 19 May 9/6 He will liberate man from dependence on the opposite sex by constructing what seems to be known in Japan as a 'Dutch Wife'; a kind of life-size mechanical doll with built-in electric heating and all the other refinements.

4. *Dutch act* (see sense B. 4 below); *Dutch auction* (earlier example); so *Dutch auctioneer*; *Dutch feast* (earlier example); *Dutch lunch, party, supper, treat* (orig. *U.S.*), one at which each person contributes his or her own share.

1859 Sala *Tw. round Clock* 21 The sale is conducted on the principle of what is termed a 'Dutch auction', purchasers not being allowed to inspect the fish in the doubles before they bid. 1830 *Virginia Lit. Museum* 632 A Dutch auctioneer, whose practice is to set up his wares at the highest price, and thence bid downwards till he meets with a purchaser. *a* 1700 Evelyn *Diary* 25 Nov. an. 1682 (1955) IV. 296, I was exceedingly afraide of Drinking, (it being a Dutch feast). 1904 *Columbus Post-Dispatch* 21 Aug., Dancing was enjoyed by all as was the Dutch lunch which was partaken of at intervals during the evening. 1954 J. Symons *Narrowing Circle* xxv. 107 'Shall we make this a Dutch lunch?'..He was all there when it came to money. 1927 *Observer* 8 May 13/3 Dutch parties are rather more elaborate, in that while the hostess provides the dance floor, music, table, service, and cutlery, her friends bring along the drinks and the viands, raiding their family cellars and larders. 1904 *Dallas Morning News* 10 Sept. 6 Depriving themselves of money they need to buy plug-cut and Dutch suppers with. 1887 *Lippincott's Mag.* Aug. 191 'You'll come along too, won't you?' Lancelot demanded of Ormizon. 'Dutch treat vous savez.' 1937 *Sunday Express* 14 Feb. 25/3 Are you a 'Dutch treat' addict? (In a Dutch treat every one buys his or her own drinks.) 1945 'L. Lewis' *Birthday Murder* (1951) iii. 39 It's Dutch treat; he pays his own way and makes the women pay theirs. 1958 'A. Gilbert' *Death against Clock* 81 We arranged to go to the pictures next night. Dutch treat.

B. *sb.* **2. c.** *S.Afr.* = *Afrikaans *sb.*; in full, **Cape Dutch* or *South African Dutch.*

1731 Medley tr. *Kolben's Pres. State Cape of Good Hope* iii. 26 The People far up the Country, on the Appearance of Strangers, are us'd to say in *Dutch*, *wat Volk*, i.e. *What People?* 1798 Lady A. Barnard *Jrnl.* 11 May in *Lives of Lindsays* (1849) III. 437, I doubt much if my whole stock of Dutch amounts to two dozen of words. 1849 N. J. Merriman *Kaffir, Hottentot & Frontier Farmer* (1854) 51 He knew Dutch well, and between the three tongues we contrived to make ourselves intelligible. 1936 F. R. Thompson *Matabele Thompson* i. 25, I..became proficient in Dutch, and in the various native languages that I came across.

d. *High Dutch* (S.Afr.) [tr. Afrikaans *Hooghollands*], Netherlands (literary) Dutch as distinguished from Cape Dutch or Afrikaans.

1901 Logeman & van Oordt *How to speak Dutch* (ed. 3) i. 31 The main points of difference between so-called 'High Dutch' and Cape-Dutch phonetics and spelling may be enumerated as follows. 1911 H. H. Fyfe *S. Afr. To-Day* viii. 96 High Dutch..is not the language of the Dutch people in South Africa..the 'taal'..is the common speech. 1936 Haarhoff & van den Heever *Achievement of Afrikaans* i. 13 While we tried to write in High Dutch our thoughts were cast in rigid moulds;..and the result was often secondhand rhetoric. 1958 L. van der Post *Lost World Kalahari* iii. 60 In High Dutch I wrote: 'I have decided to-day.'

3. c. (Later examples.) *U.S.*

1906 M. E. W. Freeman *By Light of Soul* xx. 277 Well, you women do beat the Dutch. 1939 *Amer. Speech* XIV. 267 If it is startling news, it 'beats the Jews' or 'beats the Dutch'.

d. *S. Afr.* The Afrikaans-speaking people; the South Africans of Dutch descent.

1731 Medley tr. *Kolben's Pres. State Cape of Good Hope* v. 57 There are at this Time the strictest Alliance and the closest Friendship subsisting between the Dutch and the several Hottentot Nations. 1776 Masson *Phil. Trans. R. Soc.* 305 There is another species of animal called by the Dutch Bles-moll. 1850 J. W. Appleyard *Kafir Lang.* 10/2 The Dutch language as generally spoken..by the Dutch themselves in the country districts, is very different from the Dutch as used in Holland. 1897 [see Hottentot 1 a].

4. Slang phrases (orig. *U.S.*): (*a*) *in Dutch*, in disfavour, disgrace, or trouble; (*b*) *to do a* (or *the*) *Dutch* (*act*), to desert, escape, run away; also, to commit suicide.

1904 H. Hapgood *Autobiogr. of Thief* vi. 112 A week later Dal was found dead in his cell, and I believe he did the Dutch act (suicide). 1909 J. R. Ware *Passing Eng.* 120/2 We did a dutch with everything—even down to the coal-hammer. 1912 A. H. Lewis *Apaches of N.Y.* iv. 70, I don't want to put you in Dutch with your fleet. 1948 M. Allingham *More Work for Undertaker* xxv. 286 He'll be in dutch if there's nothing to show at the end of it. 1953 P. Frankau *Winged Horse* ii. iv. 134 Maybe Baron'll fire me when he knows I'm in Dutch with his family. 1958 M. A. de Ford in J. Macdonald *Lethal Sex* (1962) 115 You can't face it..so you're doing the Dutch and leaving a confession. 1959 E. Fenwick *Long Way Down* xx. 155 Scare the poor kid to death, probably—and get her in dutch with her people, too. 1965 'R. L. Pike' *Police Blotter* (1966) ii. 37 The day Caper Connelly does the dutch, my guess is it'll be against somebody else. 1968 J. Dos Passos *Best Times* ii. 69 While I plodded around..trying to explain my position and getting myself deeper in Dutch every time I opened my face, I saw marvellous scenes.

C. *adv.* Restrict † *Obs.* to sense in Dict. and add: **2.** With each person paying for his own food, drink, etc.; esp. in phr. *to go Dutch* (cf. *Dutch lunch*, etc., under sense A. 4 above). orig. *U.S.*

1914 S. Lewis *Our Mr. Wrenn* v. 63 We'll go Dutch. 1957 *Economist* 5 Oct. 14/1 To suggest a free trade area to any of them in such circumstances looks rather like proposing to a teetotaller that you and he go dutch on daily rounds of drinks. 1962 *Ibid.* 29 Sept. 1213/3 There is 'Dutch auction', 'Dutch uncle', to eat out 'Dutch' with one's friends, and many more.

dutch, *sb.*² *slang.* [Abbrev. of Duchess.] A costermonger's wife; *gen.* a wife; often *old Dutch.*

a 1889 Mitchell *Jimmy Johnson's Holiday* (Barrère & Leland), He made a vow he'd never row With his old Dutch again. 1889 Barrère & Leland *Dict. Slang* I. 341/2 *Dutch* (popular), a wife. 1893 A. Chevalier *My Old Dutch*, There ain't a lady livin' in the land As I'd 'swop' for my dear old Dutch! 1901 R. C. Lehmann *Anni Fugaces* 128, I detected a coster..with some one to act as his Dutch. 1926 *Calgary Daily Herald* 7 June, Joe Brown, Sal Gratton, and the rest of the quaint coster characters of 'My Old Dutch' come to life and live over their romantic story at the Strand theatre.

Dutchman. Add: **1.** (Later U.S. examples.)

1871 E. Eggleston *Hoosier Schoolm.* (1872) vii. 74 The robbery at 'the Dutchman's' (as the only German in the whole region was called). 1931 'D. Stiff' *Milk & Honey Route* iii. 38 Germans of all kinds are 'Dutchmen', 'square-heads' or 'Heines'.

b. A European; a foreigner (see quots.). *colloq.*

1857 J. D. Borthwick *3 Yrs. in California* 311 Europeans..save French, English, and 'Eyetalians' are in California classed under the general denomination of Dutchmen, or more frequently 'd—d Dutchmen', merely for the sake of euphony. 1892 Stevenson & Osbourne *Wrecker* xii. 194 In sea-lingo (Pacific) *Dutchman* includes all Teutons and folk from the basin of the Baltic. 1907 *Daily Chron.* 21 Dec. 6/6 'Only fifty years ago,' he remarked, 'we gave the generic name of Dutchman to all the representatives of Western civilisation in the Far East.' 1910 G. C. Eggleston *Recoll.* 3 To us in the West, at least, all foreigners whose mother tongue was other than English were 'Dutchmen'. 1925 Fraser & Gibbons *Soldier & Sailor Words*, *Dutchman*, the British seafarers' name for sailormen in general, natives of Northern Europe: Dutchmen proper, Danes, Swedes, Russians, Germans. Finns are excepted. 1928 *Daily Express* 20 July 2/7 British sailors refer to foreigners employed on vessels as 'Dutchmen'.

4. Also in other applications (see quots.). Chiefly *U.S.*

1859 Bartlett *Dict. Amer.* 134 *Dutchman*, a flaw in a stone or marble slab, filled up by an insertion. 1905 *Terms Forestry & Logging* 36 *Dutchman*, a short stick placed transversely between the outer logs of a load to divert the load toward the middle and so keep any logs from falling off. 1909 *Cent. Dict.* Suppl., *Dutchman*, a layer of suet fastened with skewers into a roast of lean beef or mutton. 1957 *Brit. Commonwealth Forest Terminol.* 64 *Dutchman*, a prop used in logging for such purposes as preventing the binding of a saw when crosscutting, or for supporting the coupling of an arch while it is being hooked to a tractor. 1960 *New Yorker* 3 Sept. 20/3 He mended the [marble] lion by cutting recesses several inches deep wherever the stone was damaged, and fitting new pieces of stone therein. These pieces are known in the trade as dutchmen.

Dutchy (dʌ·tʃi), *sb. colloq.* (orig. *U.S.*). Also **Dutchee, Dutchie.** [f. Dutch *a.*, *sb.* + -y⁶.] A familiar or contemptuous name for a Dutchman or a German.

1835 C. F. Hoffman *Winter in West* II. 165 Where's Yankee and Dutchee? the bacon and greens are smoking on the table. 1864 J. T. Trowbridge *Cudjo's Cave* (1868) iv. 18 See here, Dutchy! ye hain't been foolin' us, have ye? 1888 M. Grigsby *Smoked Yank* xvi. 139 Then some one behind would yell: 'Go it, Dutchie.' 1901 *Daily Chron.* 12 Aug. 6/1 The captain of the ship insulted him by saying, 'Here's another Dutchy who wants to be an American.' 1949 *Sci. Monthly* July 44 Among Western 'civilized' peoples we hear the constant use of such terms as..'nigger', 'Dutchy', 'flip', 'greaser', [etc.]. 1959 *Amer. Speech* XXXIV. 225 Nicknames in Australian lower-class society...*Dutchy*, a German.

‖ du théâtre (dü teātr'), *adj. phr.* [Fr.] 'Of the theatre'; characteristic of or suited to the theatre; theatrical.

1895 G. B. Shaw *Our Theatres in Nineties* (1932) I. 7 His plays are *du théâtre* when the right people are in the theatre. 1911 M. Beerbohm *Let.* 6 Dec. (1964) 211 You who are so very *du théâtre* and love being behind the scenes among T-lights and properties.

‖ du tout (dü tu), *adv. phr.* [Shortening of Fr. *pas du tout* not at all.] Not at all; by no means.

1824 T. Creevey *Let.* 23 Sept. in *Creevey Papers* (1903) II. iii. 82 'Do you know what happened last night?'—'Du tout,' says I. 1904 H. O. Sturgis *Belchamber* vii. 91 'It is very good of you to take an interest.'..'Oh, *du tout*,' she said suavely. 1936 A. Christie *ABC Murders* ix. 70 'Mr. Poirot..[would] like to see it.' '*Du tout*,' said Poirot quickly. 'You misunderstand me.'

duty. Add: **5. f.** (Earlier example.)

1825 H. Wilson *Mem.* II. 175 Such a thing on his head, doing duty for a hat!

g. *to do* (*one's*) *duty*: euphemism for 'to defecate, urinate'.

1935 A. J. Cronin *Stars look Down* iii. iv. 509 The lamb ran away and stood in the middle of the field doing duties at an adjacent haystack. 1938 I. Goldberg *Wonder of Words* vi. 108 The child..never defecates or urinates; he ..does his 'duty'.

7. *duty call* (examples), *duty-man*; **duty-bound** *a.*, bound by duty; morally or legally obliged (cf. quot. 1591 *s.v.* Bound *ppl. a.*² 7 a); **duty cycle** (see quot. 1924); **duty-pay,** a bonus paid for work done outside the ordinary routine work (see also quot. 1879).

1908 Hardy *Dynasts* III. iii. i. 93, I was duty-bound To let him know. 1957 H. Roosenberg *Walls came tumbling Down* viii. 191 This Dutch officer..would be duty-bound to stop it. 1864 C. M. Yonge *Trial* viii. 142 Forgetting her has not been easy to the payers of duty calls. 1905 *Daily Chron.* 2 Aug. 4/6 When he [*sc.* the Kaiser] made his duty-call to the Danish capital. 1924 S. R. Roget *Dict. Electr. Terms* 70/2 *Duty cycle*, the cycle of operation (starting, running and stopping) which a motor on intermittent duty performs each time it runs. *Duty cycle factor*, the ratio of the equivalent constant current during a duty cycle to the steady running current of the motor. 1962 Simpson & Richards *Junction Transistors* xvi. 419 Such low duty-cycle devices as radio beacons. 1971 *Nature* 19 Mar. 160/2 The higher pulse repetition rate of the radar does not increase the accuracy if the signal-to-noise ratios and overall duty cycles (fraction of time that the transmitter is

on) are equal in the two cases. 1906 *Westm. Gaz.* 15 Sept. 4/1 Prepared to join the suggested training battalion for a further period of six months, as non-commissioned officers and 'Duty-men'. 1942 *Gen* 1 May 42/1 Special dutymen to your stations! 1879 *Cassell's Fam. Mag.* V. 103/2 'Duty pay' (*i.e.* extra pay awarded to men whose work is of a specially onerous or responsible kind). 1879 C. Marvin *Our Public Offices* 67 Most of these extras in the way of nomenclature had something handsome attached to them in the shape of gratification money or 'duty-pay'.

b. *attrib.* or quasi-*adj.* Designating a visit, work, etc., undertaken as a duty (opp. to the same undertaken voluntarily or for pleasure); also applied to a person whom one visits, etc., as a duty. Cf. *duty call, dance* in Dict.

1806 M. Edgeworth *Leonora* II. xlii. 4 If it be duty-work kindness, I would not give thanks for it. 1852 Geo. Eliot *Let.* 5 June (1954) II. 32 Mrs. M[ackay] is disagreeably nervous and wanting in ease. One feels very glad when she has done her duty-talk. 1873 C. M. Yonge *Pillars of House* IV. xxxix. 151 Their grandson never went near them if he could help it, only enduring a duty-visit by the help of shooting. 1939 C. Isherwood *Goodbye to Berlin* 268 It was more probably a duty-party, given once a year, to all the relatives, friends and dependents of the family. 1941 M. Treadgold *We couldn't leave Dinah* iv. 64 Dreary duty-visits to the tall sombre house in Eaton Square. 1946 A. Christie *Hollow* i. 13 We had him to lunch with some other Duty people. 1953 J. Masters *Lotus & Wind* vii. 97 Her father circled sedately from time to time with duty-partners. 1970 P. Moyes *Who saw her Die?* i. 14 I'll go and have a duty dinner with Kitty Prestwether. She's been pestering me ever since we got here.

duty-free, *a.* and *adv.* Add: **b.** *Comb.* **duty-free shop,** a shop at an airport, on a boat, etc., at which duty-free goods can be bought.

1965 *Which?* May 150/1 Not all airports have duty-free shops. 1969 *Ibid.* Mar. 79/1 Some 'duty-free' shop prices are too high. 1970 *New Yorker* 16 May 44/2 Maybe I should get him a gift at the duty-free shop. 1971 P. Purser *Holy Father's Navy* iv. 22 We were already on the second of the two bottles we'd bought at the duty-free shop.

duvet. Add: (Later examples.)

1967 R. Petrie *Foreign Bodies* ii. 22 Marian..turned back the sheets, heaping the plump feather *duvet* over a chair. 1970 *Cabinet Maker & Retail Furnisher* 31 July 169 Continental quilt or duvet? Call it what you will, this is a market with potential. 1971 *Guardian* 3 May 9/5 The cult of the Duvet is undoubtedly spreading even if there are those who find continental quilts..too..hot.

2. A downy growth.

1934 in Webster. 1957 M. Sharp *Eye of Love* x. 88 A slight *duvet* of black hair by comparison softened them [*sc.* her forearms].

Duvetyn (diū·vėtin, dʌ·vtin). Now usu. with lower-case initial. Also **-tine, -tyne.** [f. Fr. *duvet* down.] The trade name of a soft material of worsted and silk with a fine downy nap, used for women's coats and dresses. Also *attrib.*

1913 *Trade Marks Jrnl.* 10 Sept. 1434 Duvetyn..Silk Piece Goods. 1921 *Glasgow Herald* 19 Feb. 5 The upper part, which had a high calyx collar buttoned at the left side, was of dark duvetine. 1922 *Tatler* 18 Oct. Advts. p. w, This Hat can also be copied in Duvetyn, in all shades, at 2½ Gns. *Ibid.* 130/2 A new fabric known by the name of crocodile duvetyn. 1923 G. G. Denny *Fabrics & how to know Them* 41 The so-called silk duvetyn usually has a fine mercerized cotton warp and spun silk filling. 1924 *Scribner's Mag.* Aug. 200/2 She was wearing a new dark blue duvetyne suit. 1939 Joyce *Finnegans Wake* i. 148 Your delighted lips, love, be careful! Mind my duvetyne dress above all! 1960 *Times* 25 July 13/4 Pale sage green duvetyn.

dux. Add: **2.** (Earlier and later examples.) Also, the subject of a canon; the leading voice or instrument in a fugue or canon.

1740 J. Grassineau *Mus. Dict.* 68 *Dux*, in fugues is the first voice or instrument that begins, and serves as a guide to the other parts, which are called *comes*, or followers. 1885 G. B. Shaw in *Mag. Music* Nov. 178/3 Gounod often gives us a few pretty bars in canon, or a theme, with a bold skip or two at the beginning, introduced and answered in the rococo 'dux and comes' style. 1938 *Oxf. Compan. Mus.* 135/1 The voice first entering with the melody in a canon is called *Dux* ('leader') or *Antecedent.*

Duxeen (dʌ·ksīn). [Patented by the *Dux* Chemical Solutions Co., Bromley-by-Bow, London. See *-een¹.] A paper used as a book covering, made in imitation of bookbinders' cloth.

1920 in *Trade Marks Jrnl.* 24 Nov. 2215. 1921 *Caxton Mag.* XXIII. 127 (Advt.), Duxeen is a strong, flexible hard-surfaced fibre. 1927 in *Longman's Class. Catal. Educ. Works* 11. 1956 *Bookman's Conc. Dict.* 90 *Duxeen*, a patented form of bookbinding material, simulating cloth.

duxelles (‖düksel, dʌ·kselz). Also **d'Uxelles.** [f. the name of the Marquis *d'Uxelles*, 17th-c. Fr. nobleman.] A seasoning (see quots.). Also *attrib.*

1877 E. S. Dallas *Kettner's Bk. of Table* 162 Duxelles is the name given to a combination of mushrooms, parsley, and shalots, which are chopped together finely and used for flavouring. *Ibid.* 163 The name of Duxelles..commemorates..La Varenne..[who] was lord of the kitchen of the Marquis d'Uxelles. *Ibid.* 164 Duxelles Sauce is made by adding Duxelles to about six times as much of Brown or Spanish sauce. 1964 L. Joyce-Cowen *Million Menus* 223/1 Duxelles is made of mushrooms..chopped and sautéed with ..onion and parsley. 1964 A. Launay *Caviare & After* 130 *Duxelles*, small cuts of meat garnished with mushrooms pounded into a hash with onions and shallots cooked in a white wine.

dvandva (dva·ndvā). *Philol.* Also **dwandwa.** [Skr. *dvandvá*, the repeated nom. of *dva* pair, couple.] In full *dvandva compound.* A compound word in which the elements are related to each other as if joined by a copula.

1846 Monier Williams *Elem. Gram. Sanscr.* ix. 158 Native grammarians class compound nouns under..*dwandwa*, or those formed by the aggregation into one compound of two or more nouns..which, if uncompounded, they would all be in the same case, connected by a copulative conjunction [etc.]. 1932 W. L. Graff *Lang. & Languages* iii. 119 A dvandva compound is one in which the relationship between the two constituents may be rendered by the conjunction *and*; for example, *prince-consort.* 1946 *Trans. Philol. Soc. 1945* 86 The compound *šilonča*- then came to be misunderstood as a dvandva. 1962 A. Campbell in Davis & Wrenn *Eng. & Medieval Stud.* 18 The lays seem to have preserved the dvandva compound, a type otherwise unknown in Germanic.

dvornik (dvǭ·rnik). [Russ. *dvórnik*, f. *dvor* courtyard.] A house-porter.

1903 *Westm. Gaz.* 17 Jan. 10/1 The duties of the dvorniks, a sort of assistant police. 1919 H. S. Walpole *Secret City* i. xxii. 160, I said good-night to every one. I could hear their laughter as I waited at the bottom of the stairs for the Dvornik to let me out. 1923 *Blackw. Mag.* Feb. 203/2 The 'dvornik' had been with the family for years.

dwa-grass: see *twa-grass.

dwarf, *sb.* Add: **2. b.** *Astr.* One of the class of smaller stars of greater density as distinguished from the larger diffuse stars or 'giants'; without qualification or as *dwarf star* the term usu. denotes a star of the class comprising the majority of main-sequence stars (including the sun), as distinguished from a *white dwarf* (*white *a.*).

1912 [see *giant B. 1 c]. 1913 *Observatory* Aug. 326 One can predict the real brightness of a dwarf star from a knowledge of its spectrum. 1921 *Discovery* Sept. 236/1 After this stage is reached and the star attains a certain density, it falls off rapidly in temperature, and becomes a cooling dwarf. 1956 H. S. Jones in A. Pryce-Jones *New Outl. Mod. Knowl.* ii. 114 Other stars are so small that their diameters are less than one-hundredth of the Sun's. Such stars are called dwarfs. 1968 P. Moore *Sky At Night* II. xxix. 214 It used to be thought that a typical star..would end its career as a feeble Red Dwarf.

dwarfism (dwǭ·ɹfiz'm). [f. Dwarf *sb.* + -ism.] The condition or character of being a dwarf or considerably undersized.

1891 *Brit. Med. Jrnl.* 5 Dec. 1187/2 When found as a racial feature dwarfism appears to be propagated indefinitely. 1919 *Nature* 3 Nov. 304/1 One of these forms of dwarfism is known to medical men as achondroplasia, because the growth of cartilage is particularly affected. 1934 [see *achondroplasia]. 1958 *Times* 1 July (Agric. Suppl.) p. vi/3 The genetics of dwarfism, over-simplified in the beginning, are now being studied thoroughly. The only present advice on how to avoid it is to breed for size. 1962 R. H. Smythe *Anat. Dog Breeding* 19 Toy Pomeranians are examples of true dwarfism.

dwarf-man. [f. Dwarf *sb.* and *a.* + Man *sb.*¹] A very small man; a dwarf.

1877 *Encycl. Brit.* VII. 568/2 The dwarf man..was less attractive than the dwarf boy. 1922 Joyce *Ulysses* 380 It was upheld by four dwarfmen of that country. 1955 Tolkien *Return of King* 202 A sort of small dwarf-man.

dwerg (dwǭɹg). Pseudo-archaic form of Dwarf *sb.* after OE. *dweorg.*

1892 Kipling *Lett. Trav.* (1920) 76 A puff of the warm night wind among the flares set the whole line off again in a crazy dance of dwergs. 1892 —— *Seven Seas* (1896) 124 And the troll and gnome and dwerg, and the Gods of Cliff and Berg Were about me and beneath me and above.

dwile (dwail). *dial.* Also **dwily, dwyle.** [cf. Du. *dweil* mop, f. *dweilen* to mop.] A house-flannel, floor-cloth, or mop (see also quot. *a* 1825).

1823 E. Moor *Suffolk Words* 117 *Dwily*, a towel. *a* 1825 R. Forby *Vocab. E. Anglia* (1830) I. 101 *Dwile.* 1. A refuse lock of wool. 2. A mop made of them. 3. Any coarse rubbing rug. 1887 E. R. Suffling *Land of Broads* (illustr. ed.) xvii. 251 *Dwyle*, a dishcloth. 1962 A. Jobson *Window in Suffolk* vi. 100 She called a house-flannel a *dwile.*

dy (dai). [a. Sw. *dy* mire, ooze (H. von Post 1862, in *K. Sven. Vetenskaps-Akad. Handl.* 1), f. ON. *dý*, perh. f. Gmc. **dunhia*, rel. to **dungia*, the antecedent of OE. *dung* Dung (see Falk and Torp); but the etym. of the ON. word is disputed.] A type of sediment rich in organic matter deposited in unproductive lakes.

1936 S. A. Waksman *Humus* iv. 66 Formation of 'dy', or colloidal humus material, in environments supplying amounts of air that are limited as a result of water cover. 1939 W. H. Twenhofel *Princ. Sedimentation* ii. 88 Black muds..have been differentiated into *gyttja*..and *dy* if the organic matter was brought to the lakes in colloidal form. 1957 G. E. Hutchinson *Treat. Limnol.* I. xvii. 882 The distinction between autochthonous (*gyttja*) and allochthonous (*dy*) organic sediments was early made by the Swedish students of sedimentation.

dy, dyah: see *daye.

dyad. Add: **2. d.** *Math.* An operator *ab* so defined that $F.(ab) = (F.a)b$ for all F, where a and b are vectors and F is any linear vector function.

1884 J. W. Gibbs *Coll. Wks.* (1928) II. ii. iii. 53 An expression of the form αλ or βμ will be called a dyad. 1933 H. B. Phillips *Vector Anal.* x. 217 A linear equation satisfied by dyads will remain valid if each dyad is replaced by the dot or cross product of its two vectors. 1969 Ll. G. Chambers *Course Vector Anal.* vi. 181 Consider now the dyad, or indefinite product of two vectors a and b, ab...This has the properties that (ab). F=a(b. F) and F.(ab)=(F.a)b.

dyadic, *sb.* Add: **2.** *Math.* Any quantity formed by the addition or subtraction (or both) of dyads.

1884 J. W. Gibbs *Coll. Wks.* (1928) II. ii. iii. 53 An expression consisting of any number of dyads united by the signs + or — will be called a *dyadic binomial, trinomial* etc., ..or, more briefly, a *dyadic.* The latter term will be used so as to include the case of a single dyad. 1924 C. E. Weatherburn *Adv. Vector Anal.* v. 81 Then r′=r·ai+r·bj+r·ck. We write this more briefly r′=r·(ai+bj+ck), where the expression in brackets is an operator called a dyadic. 1969 Ll. G. Chambers *Course Vector Anal.* vi. 189 Dyadics may be formed in which the vector operator ∇ is a component part.

Dyak (dai·æk). Also **Daya(c)k, Dayakker.** [f. Mal. *dayak* up-country.] A member of one of the aboriginal peoples inhabiting parts of Borneo and Sarawak, often divided into *Land Dyaks* and *Sea Dyaks*; the language of these peoples. Also *attrib.* or *adj.*

1836 *Penny Cycl.* V. 189/2 The mines are only wrought by the Dayacks. 1840 J. Brooke *Jrnl.* 2 Oct. in E. Hahn *J. Brooke of Sarawak* (1953) iv. 51 The arrival of various Dyak tribes. 1876 *Encycl. Brit.* IV. 58/1 The Dyaks, Dayaks, or Dayakkers are..broken into numerous tribes. 1880 *Ibid.* V. 818/1 Among the other languages which have been reduced to writing and grammatically analysed are the Balinese..the Dayak, and the Macassarese. 1882 *Jrnl. Straits Branch R. Asiatic Soc.* (1883) Dec. 213 (*title*) Sea Dyak religion. *Ibid.*, They may be regarded as the racial gods of the Sea-Dyaks. 1893 Russan & Boyle *Orchid Seekers* xii. 134 Squatting on their 'tail-mats' to the Collector's right hand..were the Dyaks; on the left lay the Malays. *Ibid.* xiii. 151 On every side rose cries of terror in Malay and Dyak. 1924 D. H. Lawrence in *Adelphi* I. 883 How apparently untamed the savage may be, Dyak or Hottentot, you may be sure he is grinding upon his own.. ideas. 1937 *Discovery* Jan. 7/1 The Dyak tribes. *Ibid.*, The inhabitants inland are divided into two distinct races, known as the Sea and Land Dyaks. 1957 W. R. Geddes *Nine Dyak Nights* v. 44 *Orang Kaya*, meaning 'Rich Man' ..is not a traditional Dyak title—the words are not even Dyak, but Malay. 1966 L. W. Jones *Pop. Borneo* iii. 35 A half of the Dyak population..had disappeared in ten years as a result of forced labour, slavery, sickness and the sword. 1968 *Listener* 4 Apr. 427/2 West Borneo['s].. population of one and a half million Dayaks and Chinese.

dyarchal, etc.: see *diarchal *a.*, etc.

dybbuk (di·buk). Also **dibbuk.** Pl. **dybbukim, dybbuks.** [Heb. *dibbūḳ*, f. *dāḇaḳ* to cling, cleave.] In Jewish folk-lore, the malevolent spirit of a dead person that enters and controls the body of a living person until exorcized.

1903 *Jewish Encycl.* IV, *Dibbukim*, transmigrated souls. 'Dibbuk'..is a colloquial equivalent, common among the superstitious Jews in eastern European countries, for a migrant soul. 1926 Alsberg & Katzin tr. 'S. Ansky' (*title*) The dybbuk. 1929 T. Wolfe *Look Homeward, Angel* (1930) xxx. 431 An old Jew who muttered jargon into a rabbi's beard as if saying a spell against Dybbuks. 1959 *Times* 30 Dec. 9/1 *The Tenth Man* is about a Jewish girl who thinks she is possessed by a dybbuk, or demon.

dydle, var. Didle *sb.*

dye, *sb.* Add: **3.** *dye-pot* (examples); **dye-coupled, -coupling** *Photogr.* (see quot. 1958).

1943 C. Duncan *Man. Miniat. Camera* (ed. 2) x. 100 (*heading*) Toning with a dye-coupled developer. 1958 M. L. Hall et al. *Newnes' Compl. Amat. Photogr.* xxxv. 328 *Dye-coupled Developing.* Where a wide range of colours is required probably the best method is provided by dye-coupling developers. The general principle is that the by-products of the developer giving the silver image will couple up with another compound, to produce a coloured substance. 1891 'Ganconagh' *J. Sherman & Dhoya* ii. iii. 57 Some mischievous goblin always runs off with the dye-pot. 1930 E. Pound *XXX Cantos* xvii. 78 Dye-pots in the torch-light.

dye, *v.* Add: **1. c.** (See also Grain *sb.*¹ 10 b; Wool *sb.* 1 g.)

2. b. Delete † before *into.* (Further example.)

1891 C. Graves *Field of Tares* 62, I saw a lovely flush rise in her cheeks and dye her sweet white throat into crimson.

d'ye. Colloq. contraction of *do ye, do you.*

1697 [see How-do-you-do 1]. 1773 [see Do *v.* 26]. *a* 1832 [see Do *v.* A. 2 d]. 1866 J. W. Graves in *People's Eng. Songs* (1903) 9 D'ye ken John Peel..? 1922 Joyce *Ulysses* 420 D'ye ken bare socks?

dyeable (dai·əb'l), *a.* [f. Dye *v.* + -able.] That can be dyed. Hence **dyeabi·lity.**

1934 Webster, Dyeable. 1939 *Nature* 19 Aug. 313/2 It [*sc.* the structure of the rayon fibre] is responsible for the amount of swelling, the dyeability, and the resistance to creasing. 1969 *Sci. Jrnl.* July 78/1 The sheath polymer not engaged in bond formation remains distributed along the fibre where it can still contribute to properties of abrasion resistance, covering power and dyeability. 1970 *Ibid.* Jan. 23/1 New techniques for producing coloured, patterned effects on dyeable fabrics have been developed.

dye-line, dyeline (dai·lain). [f. Dye *sb.* + Line *sb.*] = *diazo, Diazotype; also, a print or copy prepared by this process. Freq. *attrib.*

1951 *O & M Bull.* Oct. 23 The simplicity of the dyeline process and the cheapness of diazo papers..have led to the development of equipment for ordinary office use. *Ibid.* 24 The Mervac Copier..is suitable both for the preparation of the transparent master and as a dyeline printer for the subsequent production of additional copies. 1953 *Jrnl. Photogr. Sci.* I. 78/1 Photographic copies of documents may be of many different types; they may differ in the process used (silver, dye line, or blueprint). 1959 *Economist* 21 Mar. 1087/2 Dyeline, or Diazo processes have been used for many years in drawing offices, but have been introduced in small fast-copying machines only since the war.

dying, *vbl. sb.* Add: **1. b.** *dying-back*: see Die *v.*¹ 14, *die-back.

1921 *Times Lit. Suppl.* 8 Sept. 574/3 Its silviculture is very difficult, more especially the question of the dying-back of its seedlings. 1959 *Jrnl. R. Hort. Soc.* LXXXIV. 483 Many plants..suffer some dying back of their top growth.

dying, *ppl. a.* Add: **1. b.** *dying god* (also with capitals), a god whose death is commemorated annually, typifying the seasonal death of vegetation.

1890 J. G. Frazer *Golden Bough* II. iii. 206 If we ask why a dying god should be selected to take upon himself and carry away the sins and sorrows of the people, it may be suggested that in the practice of using the divinity as a scapegoat we have a combination of two customs which were at one time distinct and independent. *Ibid.* 207 These features become at once intelligible if we suppose that the Death was not merely the dying god of vegetation, but also a public scapegoat. 1911 *Ibid.* (ed. 3) III. (*title*) The Dying God. 1912 *Ibid.* VII. i. 33 In that case..we should have to confess that Greece had what we may call its Good Friday and its Easter Sunday long before the events took place in Judaea which diffused these two annual commemorations of the Dying and Reviving God over a great part of the civilised world. 1947 C. S. Lewis *Miracles* xiv. 138 The records..show us a Person who *enacts* the part of the Dying God, but whose thoughts and words remain quite outside the circle of religious ideas to which the Dying God belongs. 1952 O. R. Gurney *Hittites* vii. 137 He may have been a typical 'dying god' like Adonis, Attis, and Osiris, representing the vital forces of nature which appear to die in winter and revive in the spring.

dyingness. (Later example.)

1955 E. Bowen *World of Love* v. 98 She could not suffer dyingness to usurp.

dymaxion (daimæ·ksiən), *a.* Also **Dymaxion.** [Arbitrary; see note below.] In the philosophies of construction and design of R. Buckminster Fuller: yielding the greatest possible efficiency in terms of the available technology, 'doing the most with the least'. Hence *dymaxion car, house, map,* etc.

In a private communication (July 1969) Mr. Buckminster Fuller said that the word was coined for him in 1929 by his business associates as a 'word-portrait' of him and his work. They were concerned to form a euphonious word of four syllables based on words that occurred in Fuller's own description of his prototype (*Dymaxion*) house, viz. *dy*(*namism*, *max*(*imum*, and *ion.*

1929 R. B. Fuller in *Archit. Forum* July 104 (*caption*) Elevation and Isometric Drawing of the Dymaxion House. 1934 *Sat. Rev.* 2 June 1/1 The author [*sc.* H. G. Wells], during his recent visit to America, with the new Dymaxion car. 1960 R. W. Marks *Dymaxion World of B. Fuller* 9/1 Another dimension of this Weltenschauung is expressed by the term *Dymaxion*, a label Fuller has used to qualify the implication of his various inventions, developments, and projected ideas. *Ibid.* 34/1 The Dymaxion bathroom was designed by Fuller for ultimate production in plastics, when plastics had been developed to an adequate point. *Ibid.* 49/1 The Dymaxion map was not a shadow projection, as are other global maps, but a topological transfer of a high frequency form of Fuller's totally-triangulated systems from the surface of a sphere to the equivalent triangular spaces on the faces of a polyhedron. 1962 R. Davenport in R. B. Fuller *Epic Poem on Industrialization* p. vi, At his own factory in Bridgeport he designed and built the first Dymaxion car.

dynamic, *a.* and *sb.* Add: **A. 8.** *Mus.* Of, pertaining to, or indicating the volume of sound from a musical instrument or in a musical performance. Also *transf.*

1893 J. S. Shedlock tr. *Riemann's Dict. Mus.* 226/1 The natural dynamic shading of a musical phrase is therefore *crescendo*..and *diminuendo*...Dynamic and agogic shadings must be used with economy. 1931 G. Jacob *Orchestral Technique* iii. 29 It is excellent as a bass-strengthener, however, in dynamic ranges of *mf* and upwards. 1941 *B.B.C. Gloss. Broadc. Terms* 10 *Dynamic range*, range of intensities within which the volume of a programme fluctuates. 1948 *Penguin Music Mag.* June 129 Fearless attack, great volume, and subtle dynamic changes.

9. *dynamic braking* = **electric braking*; *dynamic equator* (see quot. 1883); *dynamic* (also *dynamical*) *metamorphism* Geol., metamorphism produced by mechanical forces; *dynamic pressure*, the pressure in a fluid that is due to its motion.

1930 *Engineering* 24 Jan. 105/1 Driven by two 100 h.p. mill-type motors..equipped with solenoid and dynamic braking. 1950 *Ibid.* 22 Dec. 528/2 Slip-ring induction motors with dynamic braking. 1883 *Encycl. Brit.* XVI. 165/1 The line connecting all the points where the magnetic intensity is least is called the dynamic equator. 1889 Dynamical metamorphism [see *dynamometamorphism 2]. 1893 *Jrnl. Geol.* I. 854 The dynamic metamorphism of the eruptive rocks occurring in it. 1965 A. Holmes *Princ. Phys. Geol.* (ed. 2) viii. 170 Since the main process is dynamic, slate is said to be a product of dynamic metamorphism. 1919 A. B. Eason *Flow & Measurem. Air & Gases* ix. 163 Threlfall.. mentions tests made upon a Pitot tube in which..there was a large suction effect, and..the difference between the static and dynamic pressures was of the form $H = k(u)^n$, where *n* was greater than 2. 1949 O. G. Sutton *Sci. of Flight* ii. 36 The dynamic pressure of the flow..is the pressure felt on the nose of a body at the point where the impinging stream is actually brought to rest. 1970 A. C. Kermode *Flight without Formulae* 43 A fluid can exert pressure for two reasons: first because of its movement... The pressure due to movement we will call dynamic pressure.

B. 4. *Mus.* = *dynamics 3.

[1884 H. Riemann (*title*) Musikalische Dynamik und Agogik.] 1959 D. Cooke *Lang. Music* iii. 159 The minor system, soft dynamic, and slow tempo.

dynamically, *adv.* Add: **b.** *Mus.* With regard to dynamics (sense *3).

1939 A. Copland *What to listen for in Music* vii. 95 They [*sc.* percussion instruments] are generally used in one of three ways: to sharpen rhythmic effects, dynamically to heighten the sense of climax, or to add color to the other instruments. 1947 A. Einstein *Mus. Romantic Era* xi. 136 The overture to the *Francs Juges*..is most delicately modeled both orchestrally and dynamically.

dynamicism (dainæ·misiz'm). = *dynamism 2.

1947 E. Bowen *Coll. Impressions* (1950) i. 18 A life of inner dynamicism which covered, geographically, little space. 1948 L. Spitzer *Linguistics & Lit. Hist.* 183 Falstaff, perhaps, comes very close to the French figures, only that his belly does not make him fit for dynamicism.

dynamicist (dainæ·misist). [f. Dynamic(s + -ist 2.] One who studies dynamics.

1956 *Nature* 18 Feb. 321/1 It is particularly important that meteorologists (especially dynamicists) should develop the difficult studies of individual clouds. 1957 G. E. Hutchinson *Treat. Limnol.* I. v. 251 Some empirical work was done in the eighteenth century, largely by engineers, but it seldom conformed to the expectations of the deductive dynamicists. 1971 *Nature* 23 Apr. 491/3 The assumptions that are made by planetary dynamicists regarding the state of the Earth's interior in the geological past are particularly difficult to investigate.

dynamics. Add: **3.** *Mus.* The variation in, or amount of, volume of sound from a musical instrument or in a musical performance. Also *transf.*

1883 in *Encycl. Dict.* 1908 *Jrnl. Folk-Song Soc.* III. iii. 155 He will..find no difficulty in keeping well together with the record, as regards rhythms, twiddles, added syllables, dynamics, etc. 1939 A. Copland *What to listen for in Music* iii. 25 Merely by changing the dynamics, that is, by playing it loudly and bravely or softly and timidly, one can transform the emotional feeling of the very same succession of notes. 1942 E. Blom *Mus. England* vii. 116 The new instrument, capable of producing not only harmony but a range of dynamics undreamt of by harpsichord players. 1962 A. Nisbett *Technique Sound Studio* 252 Graphical representation of the envelope (or dynamics) of a single note may show separate distinctive features.

dynamism. 2. Delete 'In various *nonce-uses*' and add further examples. Now usu. in sense: energizing or dynamic action, energy, 'drive'.

1940 *Economist* 20 July 70/1 Admiral Yonai's Ministry was brought down by the Army simply on the grounds of 'insufficient dynamism'. 1951 R. Firth *Elem. Social Org.* v. 165 The primitive artist had hardly anything like the theory of æsthetic dynamism lying behind the use of the Chinese brush. 1955 P. Heron *Changing Forms of Art* 148 The solid earth itself is far too convulsed to be really convincing in the long run *as* the solid earth. I enjoy these landscapes very much, but their unity is violated by so much 'dynamism'. 1966 G. N. Leech *Eng. in Advertising* vi. 65 Board-room clichés which imply the dynamism, success, and health of the firm's activities.

dynamitard. (Later examples.)

1892 G. B. Shaw *Fabian Soc.* 5 Not that we were dynamitards..how innocent we were of any practical acquaintance with explosives. 1917 T. E. Lawrence *Lett.* (1938) 232 The Ageyl dynamitards were inefficient, and our supply of dynamite small.

dynamite, *sb.* Add: **1. b.** *fig.* Something or someone potentially unsettling, dangerous, or disastrous.

1922 Hart Crane *Let.* 23 Jan. (1965) 78 Another small magazine, full of compressed dynamite. 1926 J. Black *You can't Win* (1927) xiv. 199 I'm dynamite with them old brums in the cribs. 1945 P. Cheyney *I'll say she Does!* v. 128, I call it goddam urgent. It's dynamite. 1961 *Countryman* 441 The 'Church question' is still dynamite in the field of education. 1970 'W. Haggard' *Hardliners* vi. 62 He had reason to be cautious for the proposal..had been naked political dynamite.

c. Heroin or a similar narcotic (see quots.). *slang* (orig. *U.S.*).

1924 G. C. Henderson *Keys to Crookdom* xxiv. 301 So powerful was the 'kick' of heroin that it was called 'dynamite'. 1938 *Amer. Speech* XIII. 183/2 *Dynamite.* 1. Bootleg dope with an unusually high narcotic content... 2. A knock-out dose given to an addict under the guise of narcotics. 1967 M. Culpan *In Deadly Vein* vii. 149 'A little bit of horse? Some dynamite?' Horse was heroin; so was dynamite.

dynamite, *v.* Add: Also *fig.*

1903 *Westm. Gaz.* 30 July 3/1 Mr. Chamberlain..tried to dynamite discussion by declaring that the rejection of the Bill would mean the infringement of treaty obligation. 1938 *Amer. Speech* XIII. 156/1 *Dynamite an order*, to rush an order through as quickly as possible.

dynamo, *sb.* Add: (Later examples.)

1916 *Autocar Handbk.* (ed. 6) xvii. 240 Electric lighting failed to secure adoption until the complete system involving the fitting of a dynamo upon the car came forward. 1936 *Encycl. Sports* 212/2 Many firms make cheap, light, and reliable dynamos driven off the edge of the front or rear tire.

b. *fig.*

1892 G. Meredith *Let.* 4 Feb. (1970) II. 1064 Ask Jim whether his whirring dynamo of a brain has flung to nothing Will's questions. 1904 E. Dowden *Browning* 327 Any stream of moral electricity worked from a dynamo of the will. 1938 *Time* 24 Oct. 2/2 In short, I am to be..the 'alter ego' of a human dynamo. 1963 D. Ogilvy *Confess. Advert. Man* (1964) i. 18, I have to rely on..empirical techniques for spotting creative dynamos.

dy:namometamo·rphism. [f. Dynamo- + Metamorphism.] **1.** (In Dict. s.v. Dynamo-.) **2.** *Geol.* = *dynamic metamorphism* (*dynamic *a.* and *sb.* A. 9). Hence **dy:namometamo·rphic** *a.*

1889 *Geol. Mag.* 243 Rosenbusch..seeking to bring clearly into view the consideration that the special kind of metamorphism referred to is only produced when the mechanical forces effect movement, and thus do work, suggested the term Dynamical metamorphism or dynamometamorphism. 1904 *Nature* 4 Aug. 317/2 The observations of..Barrois in Brittany thus receive confirmation from the stronghold of the dynamometamorphic school. 1909 F. P. Mennell *Introd. Petrol.* xix. 152 It is usual to discriminate between dynamo- and pyrometamorphic types. *Ibid.*, These rocks which owe their immediate characters to dynamometamorphism..are usually grouped together under the name of schists. 1966 *McGraw-Hill Encycl. Sci. & Technol.* VIII. 296a/2 Cataclastic rocks are mechanically sheared and crushed. They represent products of dynamometamorphism, or kinetic metamorphism.

dynamometer. Add: **1. b.** *attrib.*, as **dynamometer car,** a railway vehicle with equipment for continuously measuring and recording the force exerted by the locomotive pulling it and other information about the running of the train.

1879 *Railroad Gaz.* 30 May 298/2 Mr. F. M. Wilder..has during the past year been occupied..in perfecting a dynamometer car. *Ibid.* 298/3 The train consisted of 45 loaded cars, with a caboose and the dynamometer car. 1953 W. W. Hay *Railroad Engin.* I. xi. 153 Many of the test data used in deriving drawbar pull, tonnage ratings, and other locomotive performance data are obtained with a dynamometer car.

dynamometric, *a.* (Earlier example.)

1837 *Ann. Electr., Magn. & Chem.* I. 444 The mechanical effect..was estimated at half the force of a man. I shall apply it hereafter to an exact dynamometric apparatus.

dynamotor (dai·nāmōᵘtǫɹ). [f. Dyna- + Motor *sb.*] A combined electric motor and generator with one field magnet and two armatures, or with one armature upon which are two windings, one which receives current as a motor and the other which generates current as a dynamo.

1908 *Westm. Gaz.* 30 Mar. 5/2 For starting purposes supplementary power is obtained by means of an electric machine, best described as a dynamotor. 1910 *Hawkins's Electr. Dict.* 133/2 Another form of dynamotor is called the continuous alternating transformer. 1942 *Electronic Engin.* June 36/2 The basic power supply is 24v. from the aircraft battery, and this operates a compact dynamotor.

dynatron (dai·nătrǫn). [f. Dyna- + *-tron.] (See quots.)

1918 A. W. Hull in *Proc. Inst. Radio Engin.* VI. 35 A new, hot cathode, three electrode vacuum tube, the dynatron, is described. 1918 *Wireless World* June 148 The dynatron utilises the secondary emission of electrons by a plate upon which the primary electrons fall. 1932 *B.B.C. Techn. Tables & Gloss.* 48/2 *Dynatron*, a thermionic valve, generally with four electrodes, which can be arranged to have negative resistance over a portion of its grid volt anode current characteristic, and used, by virtue of this property, to generate continuous oscillations. **1944** *Electronic Engin.* Mar. 432/3 The dynatron does not work on the feedback principle as in the case of most oscillators.

-dyne (dəin), *suffix*, forming sbs. repr. Gr. δύναμις power, used in the formation of scientific, esp. electrical, terms. Examples: *aerodyne, amplidyne, autodyne, heterodyne.*

Dynel, dynel (dəi·nĕl). orig. *U.S.* [Proprietary term.] A synthetic fibre that is a copolymer of vinyl chloride and acrylonitrile and resembles wool; also, the fabric made from it.

1950 *Trade Marks Jrnl.* 20 Dec. 1094/1 Dynel. Yarns and threads, all of synthetic textile materials. Union Carbide & Carbon Corporation. **1952** *Economist* 2 Feb. 282/2 There is..war..in the textile industry, that between wool and the new synthetic substitutes—dynel, dacron and others. **1962** J. T. MARSH *Self-Smoothing Fabrics* ii. 10 Dynel goods should not be heated above about 120° C for the fibre contains a high proportion of vinyl chloride. **1967** *Daily Tel.* 18 Jan. 13/7 A dress made completely of Dynel hair.

dynode (dəi·nōᵘd). [f. Gr. δύν(αμις power + ELECTR)ODE.] An electrode which emits secondary electrons.

1939 L. M. MYERS *Electron Optics* v. 307 These secondary emitting surfaces are termed the dynodes, and this term is applicable to all other forms of secondary emitting surfaces in electron multipliers. **1952** *Electronic Engin.* May 214/1 Only recently..has it been found possible to manufacture dynode surfaces with an operating life of 1000 hours. **1953** *Ibid.* Feb. 48/2 An electron multiplier having beryllium-oxide coated dynodes. **1965** *New Scientist* 2 Sept. 558/3 The image is formed by directing the secondary electrons to the first dynode or phosphorescent screen of the electron multiplier.

dys-. Add: **dysbasia** (-bē¹·siă) [Gr. βάσις stepping, step], difficulty in walking; **dyschezia** (-ki·ziă) [Gr. χέζ-ειν to defecate + -IA¹], difficult or painful defecation; **dy·schronous** *a.*, not agreeing as to time, separate as to time; spec. in *Ecology*, not overlapping in period of blooming; **dyslexia** [first formed as G. *dyslexie* (R. Berlin 1883, in *Med. Correspondenz-Blatt des Württemberg. ärztl. Landesvereins* LIII. 209)] (earlier and later examples); *spec.* = *word-blindness* (WORD *sb.* 29); hence **dysle·ctic, dysle·xic** *adjs.* and *sbs.*; **dysme·tria** [Gr. μέτρον measure], inability to control the range of movement in a muscular action; **dyspareunia** (-pǎrū·niă) [Gr. πάρευνος (f. παρά beside + εὐνή bed) lying beside or with + -IA¹], difficult or painful sexual intercourse; **dysphemia** (-fi·miă) [Gr. δυσφημία ill language], stammering (see quots.); hence **dysphe·mic**, one who stutters; **dyspho·tic** *a.* *Ecology* [Gr. φῶς, φωτ- light], poorly lighted, used *esp.* of those depths in oceans, lakes, etc., where some light penetrates but brings about little or no photosynthesis; also erron. **disphotic**; **dysrhy·thmia**, an abnormal or disordered rhythm; *spec.*, an abnormal rhythm in the electrical waves shown in an electro-encephalogram; hence **dysrhy·thmic** *a.* and *sb.*; **dysthymia** (-þəi·miă) [Gr. δυσθυμία despondency], despondency or depression; hence **dysthy·mic** *sb.*, a person affected with dysthymia (cf. *dysthymic* adj., in Dict.); **dystrophic** *a.*, (*b*) *Ecology* [ad. G. *dystroph* (A. Thienemann *Binnengewässer Mitteleuropas* (1925) iv. 198, 201)], of a lake: having much dissolved organic matter.

1890 GOULD *New Med. Dict.* 138/1 *Dysbasia*, difficulty of walking. Proposed instead of abasia, since in the affection there is rarely absolute inability to walk. **1962** R. N. DEJONG in A. B. Baker *Clin. Neurol.* (ed. 2) I. i. 52 In the hysterical dysbasia there is often marked swaying from side to side. **1848** DUNGLISON *Dict. Med. Sci.* (ed. 7) Dyschezia. **1915** R. KNOX *Radiography* I. 233 Dyschezia, in which the passage through the colon is normal, but defaecation is inefficiently performed. **1969** M. PAULSON *Gastroenterologic Med.* xlv. 1196/1 Rectal constipation (dyschezia, simple constipation, simple retention) may be defined as a failure of the defecation reflex. **1902** *Science* 4 July 5/1 Consciousness..can make synchronous impressions dyschronous in their effects, and dyschronous impressions synchronous. **1924** *Ecology* V. 393 A species, genus or family which does not overlap [in blooming time] with any of its relatives, or a group which contains species which do not overlap with their relatives, may be called *dyschronous*. **1888** *Arch. Ophthalmol.* XVII. 307 The term 'dyslexia'..signifies an inability to read to one's self or aloud for a short time. **1957** L. E. TRAVIS *Handbk. Speech Path.* (1959) ii. 54 *Dyslexia*, partial inability to read characterized by associative learning difficulty; a form of dysphasia. **1960** *New Scientist* 15 Sept. 738/2 There might be a special category of reading backwardness which could be delimited and termed 'specific dyslexia'—though the deplorable name of 'word blindness' should be avoided. **1961** M. CRITCHLEY in *Trans. Ophthalmological Soc. U.K.* LXXXI. 473 Some dyslexics cannot place in correct serial order days of the week or months of the year. *Ibid.*, Poor muscular co-ordination has often been witnessed in the younger dyslexic children. **1964** L. KAISER in D. Abercrombie et al. *Daniel Jones* 108 Several factors may lead to dyslexia. *Ibid.*, The right hand ..may be undeveloped in the dyslectic child. **1965** *Sunday Times* 10 Oct. 58/4 The most inspiriting contender for the title of the World's Greatest Dyslexic: Hans Christian Andersen. **1966** *Sat. Rev.* 16 Apr. 82/1 (*heading*) Teaching the dyslexic child. **1911** T. L. STEDMAN *Pract. Med. Dict.* 257/2 *Dysmetria*, a form of dysergia in which the subject is unable to arrest a muscular movement at the desired point or, sometimes, to prevent an involuntary reverse movement. **1913** G. HOLMES in White & Jelliffe *Mod. Treatm. Nerv. & Mental Dis.* II. xiv. 630 Cerebellar disease produces a disproportion in the elements of a movement, and there results therefrom what Babinski has termed a dysmetria. **1952** *New Biol.* XIII. 55 The dysmetria characteristic of cerebellar disease consists in an incapacity to stop a movement at the intended place, with resulting overswing. **1970** *Nature* 19 Sept. 1228/2 Dysmetria will result from the malfunctioning of both cerebellar nuclei and cerebellar cortex. **1873** R. BARNES *Clin. Hist. Dis. Women* vi. 61 We want a word to express the condition of difficult or painful performance of the sexual function...I have determined to adopt the word 'dyspareunia'. **1962** *Lancet* 12 May 1011/2 An assortment of pains in various sites, sexual frigidity, dyspareunia, and general nervousness, fears, and depression. **1894** C. F. COXWELL in *Intercolonial Q. Jrnl. Med. & Surg.* I. 3 Dysphemia, or stammering, is regarded by most persons as an annoying trick. *Ibid.*, The dysphemic can speak, as a rule, tolerably well in private. **1933** S. M. STINCHFIELD *Speech Disorders* vi. 116 Dysphemia is defined as intermittent and variable nervous disorders of speech accompanying certain psychoneuroses. **1940** *Q. Jrnl. Speech* Apr. 203 *Dysphemia*, speech difficulty, usually related to a psychoneurotic or nervous condition. **1957** L. E. TRAVIS *Handbk. Speech Path.* (1959) ii. 55 *Dysphemia*, a nervous disorder of speech arising from psychological disturbance; includes stuttering. **1903** W. R. FISHER tr. *Schimper's Plant Geogr.* III. v. iii. 811 The flora of the dysphotic region is composed almost exclusively of microphytes. **1958** P. LAKE *Physical Geogr.* (ed. 4) xviii. 424 The water column can..be divided into.. the disphotic zone with insufficient light for photosynthesis. **1970** B. H. MCCONNAUGHEY *Introd. Marine Biol.* i. 17 Below the euphotic zone are the dysphotic, or mesopelagic, depths extending down to about 1000 metres. **1909** *Cent. Dict.* Suppl., Dysrhythmia. **1939** *Jrnl. Amer. Med. Assoc.* CXIII. 1002/2 [Electro-encephalographic] records were made..of twenty-six patients who had cerebral dysrhythmia. **1962** *Listener* 7 June 994/2 In all susceptible patients bursts of abnormal electric activity of the brain (so-called dysrhythmia) could be induced by flicker from the stroboscope. **1965** W. R. BRAIN *Speech Disorders* (ed. 2) xiii. 150 The dysrhythmias [of speech] include cluttering and stuttering. **1969** W. CLELAND et al. *Med. & Surg. Cardiol.* ii. 43 Absence of the P wave may be due to sinoatrial block or atrial dysrhythmia. **1941** W. G. LENNOX *Science & Seizures* (ed. 3) xi. 90 These 'asymptomatic dysrhythmic' individuals form a reservoir from which persons with seizures are drawn. **1944** *Proc. R. Soc. Med.* XXXVII. 319 Comparison between dysrhythmics in whom fits occur and those in whom they do not occur may be artificial. **1965** W. R. BRAIN *Speech Disorders* (ed. 2) xiii. 150 Truncated dysrhythmic and incoherent utterance. **1844** DUNGLISON *Med. Dict.* 254/2 *Dysthymia*, depression, despondency. **1944** H. J. EYSENCK in *Jrnl. Mental Sci.* XC. 855 The term 'affective disorder' or 'dysthymia' is suggested for the anxiety-depression-obsessional group. **1947** *Brit. Jrnl. Psychol.* May 135 Twenty-five female dysthymics. **1961** *New Scientist* 26 Oct. 225 An anxious neurotic patient, or dysthymic. **1931** R. N. CHAPMAN *Animal Ecology* xvi. 305 The dystrophic lake, or brown-water lake, is found among peat bogs. **1959** J. CLEGG *Freshwater Life Brit. Isles* (ed. 2) ii. 38 The so-called *Dystrophic* lake..which is generally understood to be a lake with a bottom composed of peaty humus and very soft, acid water. **1967** G. E. HUTCHINSON *Treat. Limnol.* II. xxii. 380 The water color involved in the separation of Thienemann's dystrophic type is ordinarily regarded as due to extractives from bog soils and peat.

dyscalculia (diskælkiū·liă). *Med.* [f. DYS- + CALCUL(ATE *v.*¹ + -IA¹.] Severe disturbance of the ability to calculate, resulting from cerebral injury.

1953 M. CRITCHLEY *Parietal Lobes* vii. 208 Dyscalculia is the fourth characteristic feature of the Gerstmann syndrome...The most striking defects are apparent when the patient attempts to make calculations upon paper. **1964** —— *Developmental Dyslexia* vi. 38 There may exist a higher-level dyscalculia, made up partly at any rate by an inability to visualise numbers. **1965** W. R. BRAIN *Speech Disorders* (ed. 2) iii. 27 Dyscalculias of the spatial type indicate a lesion of the right hemisphere. **1965** LUCHSINGER & ARNOLD *Voice-Speech-Language* II. B. i. 596/1 True dyscalculia is present when the processes of calculation are disturbed independently of the reading act and in all forms of mathematical manipulation.

dysfunction (disfʊ·ŋkʃən). [f. DYS- + FUNCTION *sb.*] Any abnormality or impairment of function. Hence **dysfu·nctional** *a.*, **dysfu·nctionally** *adv.*, **dysfu·nctioning** *vbl. sb.*

1916 in *Gould's Practitioner's Med. Dict.* (ed. 3) 307/2. **1927** HENDERSON & GILLESPIE *Text-Bk. Psychiatry* xi. 355 An aplasia or dysfunction of the tissues. **1933** *Mind* XLII. 114 The physiological [point of view] (which subordinates neural to glandular, circulatory, etc., dysfunctioning). **1936** *Brit. Med. Jrnl.* 28 Mar. 630/1 The school of endocrine dysfunction incriminates variously the thyroid, parathyroid, ovarian, and pituitary glands. **1949** R. K. MERTON *Social Theory & Social Struct.* (1951) I. i. 51 Unintended consequences of action are..those which are irrelevant to the system which they affect neither functionally nor dysfunctionally. **1952** T. PARSONS *Social System* ii. 35 They must be counteracted by 'mechanisms of control' unless dysfunctional consequences are to ensue. **1959** G. D. MITCHELL *Sociology* I. ii. 38 That it is functional..for mothers of young children to go out to work does not mean that it is not dysfunctional for the well-being of their children. **1959** B. WOOTTON *Social Sci.* vii. 225 A social judgment is still implied in the decision to rank the thieving tendency together with its bodily concomitants as symptoms of disease or dysfunction. **1969** R. BLACKBURN in Cockburn & Blackburn *Student Power* 186 Thus the 1929 crash was 'dysfunctional' for capitalism and so was the Chinese People's Liberation Army 'dysfunctional' for the social order of Nationalist China in the nineteen thirties and forties. **1971** *Optometry Today* (Amer. Optometric Assoc.) 12 Corrective lenses or visual training techniques to correct or improve the perceptual-motor dysfunction.

dysgenic (disdʒe·nik), *a.* [f. DYS- + *-GENIC.] Exerting a detrimental effect on the race, tending towards racial degeneration, *spec.* opposed to *eugenic*. Hence **dysge·nically** *adv.*

1915 W. R. INGE *Outspoken Ess.* (1919) 41 Its [*sc.* frequent war's] dysgenic effect, by eliminating the strongest and healthiest of the population, while leaving the weaklings at home to be the fathers of the next generation. **1922** *Edin. Rev.* July 46 Encouraging the dysgenic art of fortune-hunting. **1926** *Spectator* 1 May 804/2 Conceptive control has been an almost entirely harmful or dysgenic factor. **1928** G. B. SHAW *Intell. Woman's Guide Socialism* xxxviii. 150 Division of society into classes, with the resultant dysgenic restrictions on marriage. **1934** C. P. BLACKER *Chances of Morbid Inheritance* iv. 122 If our propaganda succeeds another barrier against dysgenic marriages will have gone. **1937** A. HUXLEY *Ends & Means* viii. 79 A residual population, dysgenically selected for its lack of spirit and intellectual gifts. *Ibid.* ix. 90 So far as individuals are concerned, war selects dysgenically. **1971** *Daily Tel.* 5 Oct. 14/2 The 'two child family' or 'Zero Population Growth'..have the merit that they are less dysgenic than is the present irresponsible propagation.

B. *sb. pl.* [Cf. EUGENIC *sb.*] Racial degeneration, or its study.

1920 A. HUXLEY in *London Mercury* June 182 A process which we may be permitted to call dysgenics—the carrying on of the species by the worst members. **1951** *New Biol.* XI. 24 Eugenics or dysgenics..are necessarily in progress all the time. **1958** *Times Lit. Suppl.* 11 July 386/3 Racial dilution, dysgenics, and national and social loss of status. **1970** *New Scientist* 28 May 438/3 'Dysgenics', as Shockley describes it, has to do with 'retrogression evolution'.

dysgraphia (disgræ·fiă). *Med.* [f. DYS- + Gr. γραφία writing.] Inability to write coherently (as a manifestation of brain damage). Hence **dysgra·phic** *a.* and *sb.* Cf. AGRAPHIA.

1934 H. C. WARREN *Dict. Psychol.* 87/1 *Dysgraphia*, a cerebral disorder characterized by inability to write. **1953** M. CRITCHLEY *Parietal Lobes* vii. 208 Agraphia, or better, dysgraphia, forms the third cardinal feature of Gerstmann's syndrome. This may take the form of a disturbance in the execution of letters..or of words. **1962** A. L. BENTON in J. Money *Reading Disability* 84 So-called 'parietal' dyslexia,..is associated with severe dysgraphia, all aspects of writing being disturbed. **1963** B. HAIGH tr. *Luria's Restoration of Function after Brain Injury* v. 183 In less extensive lesions the disintegration of writing takes the form of dysgraphia..and examination of the writing of such patients reveals a disturbance of the clear distinction of the sounds of words, confusion between closely related phonemes, and other signs of a defect of acoustic analysis. **1964** M. CRITCHLEY *Developmental Dyslexia* vi. 36 The true 'dysgraphic' is an adult of at least normal education, who has developed a difficulty in expressing himself on paper as the result of a disease. **1970** —— *Aphasiology* i. 11 Not every instance of poor writing in cases of cerebral disease is dysgraphic.

dyslexia, dyslexic: see DYS- in Dict. and Suppl.

dysphemism (di·sfīmiz'm). [f. DYS- + *-phemism* as in EUPHEMISM; cf. F. *dysphémisme*.] The substitution of an unpleasant or derogatory word or expression for a pleasant or inoffensive one; also, a word or expression so used; opp. EUPHEMISM. Hence **dysphemi·stic** *a.*, of the nature of or containing such an expression.

1884 L. A. TOLLEMACHE *Safe Studies* 142 The great system which Comte, and other assailants, call by the euphemism, or dysphemism, of Catholicism. [**1927** A. CARNOY *La Science du Mot* xxii. 351 Le dysphémisme..est impitoyable, brutal, moqueur. Il est une réaction contre le pédantisme, la raideur et la prétention, mais aussi contre la noblesse et la dignité dans le langage.] **1933** *John o' London's* 7 Oct., Professor Carnoy, the distinguished French philologist, has coined the word 'dysphemism' as a set-off to 'euphemism', to which it is directly opposed...'It consists, above all, in the substitution for dignified or simply normal terms, of expressions borrowed from spheres more vulgar, familiar, and joyous.' **1940** E. PARTRIDGE in *S.P.E. Tract* LV. 181 A minor species of dysphemism is the pejorative suffix, as in 'robustious'. **1955** *Archivum Linguisticum* VII. 159 Mothers addressing their babies as *Sübling* etc., are clearly unaware of the dysphemistic force of the suffix. **1962** *John o' London's* 14 June 583/1 'Robber' may also be one of those political dysphemisms used to discredit a nationalist rebel.

dysphoria (disfôᵊ·riă). [ad. Gr. δυσφορία malaise, discomfort, f. δύσφορος hard to bear, f. δυσ- DYS- + φέρειν to bear.] A state or condition marked by feelings of unease or (mental) discomfort (see quots.). Hence **dysphoric** (disfǫ·rik) *a.*, pertaining to, characteristic of, or characterized by dysphoria.

1842 DUNGLISON *Dict. Med. Sci.* (ed. 3) 248/1 *Dysphoria*, dissatisfaction, restlessness, suffering. **1933** *Jrnl. R. Anthrop. Inst.* LXIII. 511 *Kunta kunta* means more than merely immoral; it is used for an *incestuous* marriage..when a condition of dysphoria results; when the people feel

'shamed'. **1934** WEBSTER, Dysphoric. **1951** PARSONS & SHILS *Toward Gen. Theory of Action* II. ii. 142 Such a dysphoric feeling..may be called *guilt*. **1959** G. D. MITCHELL *Sociology* II. v. 81 The social *euphoria*, or well-being, has been turned into a state of *dysphoria*.

dysplasia (displē¹·ziă). *Path.* [mod.L., f. DYS- + Gr. πλάσις moulding, conformation, f. πλάσσειν to form, mould + -IA¹.] Abnormal development or growth of tissues, cells, etc.

1935 *Edinb. Med. Jrnl.* XLII. 550 Study of this male skull led to the conclusion that the primary dysplasia was of the skull-base. **1962** *Lancet* 6 Jan. 46/1 Dysplasias of the external ears and of the eyes. **1962** R. H. SMYTHE *Anat. Dog Breeding* 159 Hip dysplasia does not seem to be a particularly painful condition.

dysplastic (displæ·stik), *a.* *Path.* [ad. G. *dysplastisch* (E. Kretschmer *Körperbau und Character* (1921) v. 53), f. DYS- + Gr. πλαστός formed, f. πλάσσειν to form, mould: see -IC.] Of, or characterized by, abnormal growth or development (of tissues, cells, etc.). Also as *sb.*, a dysplastic person.

1925 W. J. H. SPROTT tr. *Kretschmer's Physique & Character* v. 65 We understand under 'dysplastic' such forms of growth as vary very markedly from the average and commonest form of the type in question. *Ibid.* 67 The dysgenital groups to which so many of the schizophrene dysplastics bear..resemblances. **1939** *Nature* 9 Sept. 469/2 Properly designed exercises..can be used to correct faults of carriage..and fit the mildly dysplastic to take their place without shame among the euplastic. **1949** M. MEAD *Male & Female* xiv. 286 Girls and boys who are so physically immature and dysplastic. **1964** M. CRITCHLEY *Developmental Dyslexia* v. 17 Most neurologists, however, would be reluctant to visualise in developmental dyslexia any focal brain lesion, dysplastic, traumatic or otherwise.

dysprosium (disprōᵘ·ziə̆m). *Chem.* [mod.L., f. Gr. δυσπρόσιτος difficult of access; named by L. de Boisbaudran 1886, in *Compt. Rend.* CII. 1004, who showed that the 'holmium oxide' of de Cleve was a mixture containing compounds of holmium and of dysprosium (see *HOLMIA).] A paramagnetic metallic element of the lanthanide series, present in yttria-rich minerals such as gadolinite and forming yellowish green salts in which it is trivalent. Symbol Dy; atomic number 66; atomic weight 162·5.

1886 *Jrnl. Chem. Soc.* L. 667 The earths which do not show the bands of holmium but give the bands of dysprosium are relatively rich in terbia. **1906** *Nature* 5 Apr. 551/2 The isolation and some atomic characters of dysprosium. **1962** *New Scientist* 27 Dec. 724 The RCA laser of the solid-state type (using dysprosium in calcium fluoride). **1966** PHILLIPS & WILLIAMS *Inorg. Chem.* II. xix. 18 Dysprosium is strongly ferromagnetic below —168° C. and exhibits a remarkable magnetic anisotropy. **1967** *New Scientist* 20 July 158/3 Accurately machined pole-pieces, such as could be turned out of dysprosium or holmium, would ensure better symmetry than currently used iron-free magnet coils.

dysprosody (disprǫ·sŏdi). *Med.* [f. DYS- + PROSODY.] A speech disorder affecting inflexion, stress, and rhythm, sometimes found in aphasic conditions.

1947 G. H. MONRAD-KROHN in *Brain* LXX. 405 (*heading*) Dysprosody, or altered 'Melody of Language'. **1965** W. R. BRAIN *Speech Disorders* (ed. 2) 114 Disturbances of inflection, stress and rhythm, termed dysprosody by Monrad-Krohn..may make the speech [in Broca's aphasia] sound as though pronounced by a foreigner. **1970** ESPIR & ROSE *Basic Neurol. Speech* v. 27 Disturbances of inflexion, stress and rhythm (dysprosody) are other features which may occur.

dyss (dis). *Archæol.* Also **dysse.** Pl. **dysser.** [ad. Da. *dysse*.] (See quot. 1970.)

1938 A. SERNER *'Dyss' Burial* i. 8 The Younger Stone Age begins with the 'dyss' period. *Ibid.* 27 My research below intends to show that the oldest form of megalithic tombs in South Scandinavia, the rectangular simple dolmen called below 'dyss' by me, is spread in prehistoric times outside South Scandinavia. **1947** V. G. CHILDE *Dawn Europ. Civilization* (ed. 4) x. 179 One classical method of disposal of the dead..which gives its name to the whole period in Denmark, was ceremonial burial in a megalithic dolmen or dyss. In its oldest form a dyss is a small chamber formed by four uprights supporting a single large capstone, and less than 6ft long by 2 ft. wide. **1968** G. JONES *Hist. Vikings* I. i. 18 Builders of dolmen and dysse. **1970** BRAY & TRUMP *Dict. Archaeol.* 78 *Dyss*, a megalithic cist of the final stage of the Danish Early Neolithic...The oldest dysser are slab cists roofed with capstones and containing from one to six skeletons. The burial chamber is covered with a mound which rises to the height of the capstone and has a retaining kerb of stones.

dystopia (distōᵘ·piă). [mod.L., f. DYS- + U)TOPIA.] An imaginary place or condition in which everything is as bad as possible; opp. UTOPIA (cf. CACOTOPIA). So **dysto·pian** *sb.*, one who advocates or describes a dystopia; **dysto·pian** *a.*, of or pertaining to a dystopia; **dysto·pianism**, dystopian quality or characteristics.

1868 J. S. MILL in *Hansard Commons* 12 Mar. 1517/1 It is, perhaps, too complimentary to call them Utopians, they ought rather to be called dys-topians, or caco-topians. What is commonly called Utopian is something too good to be practicable; but what they appear to favour is too bad to be practicable. **1952** NEGLEY & PATRICK *Quest for Utopia* xvii. 298 The *Mundus Alter et Idem* [of Joseph Hall] is..the opposite of *eutopia*, the ideal society: it is a *dystopia*, if it is permissible to coin a word. **1962** C. WALSH *From Utopia to Nightmare* 11 The 'dystopia' or 'inverted utopia'. *Ibid.* 12 Stories..that seemed in their dystopian way to be saying something important. *Ibid.* ii. 27 A strand of utopianism or dystopianism. **1967** *Listener* 5 Jan. 22 The modern classics—Aldous Huxley's *Brave New World* and George Orwell's *Nineteen Eighty Four*—are dystopias. They describe not a world we should like to live in, but one we must be sure to avoid. **1968** *New Scientist* 11 July 96/3 It is a pleasant change to read some hope for our future...I fear that our real future is more likely to be dystopian.

dytiscid, *a.* Add: (Examples.) Also as *sb.* Hence **dytisci·diform** *a.*

1866 E. C. RYE *Brit. Beetles* x. 64 When ponds, etc., are dried up in the summer, the *Dytiscides* take to the wing. **1882** *Sci. Trans. R. Dublin Soc.* 2nd Ser. II. 207 The most important characteristics of the Dytiscid head besides its short, broad form, are first, its great extension transversely behind the eyes..and second, the great extension of the eyes on the under surface. *Ibid.* 228 Its [*sc.* Omophron's] epimeron is Dytiscidiform in shape...With respect to the size of the anterior pieces [of the mesothorax of Pelobius] it is Dytiscid rather than Carabid. **1950** F. BALFOUR-BROWNE *Brit. Water Beetles* II. xii. 250 Sub-aquatic eggs of many insects, including those of Dytiscids, develop quite well on damp cotton-wool. *Ibid.* 252 The Symmetrid or Dytiscid type has rounded and very numerous suckers.

dytiscus (diti·skə̆s). [mod. L. (Linnæus *Systema Naturæ* (ed. 10, 1758) I. 411): see DYTISCID *a.*] A member of the genus of water-beetles so named. Also *attrib.*

1866 E. C. RYE *Brit. Beetles* x. 63 The females of *Dytiscus* are, however, sometimes found without these dorsal furrows. **1895** L. C. MIALL *Nat. Hist. Aquatic Insects* i. 30 No one can hunt long for water-insects without coming across the rapacious Dytiscus. **1902** C. CORNISH *Naturalist on Thames* 26 The great carnivorous water-beetle, the dytiscus..will rise by night from the surface of the Thames. *Ibid.*, A dytiscus beetle. **1952** M. K. WILSON tr. *Lorenz's King Solomon's Ring* iii. 17 A larva of the water-beetle Dytiscus. *Ibid.*, The Dytiscus larva..is a slim, streamlined insect, rather more than two inches long. **1969** R. F. CHAPMAN *Insects* xvi. 308 The fore legs of *Dytiscus* and some other beetles bear suckers.

dzo, var. *DSO.

dzong, var. *JONG.

E

E. Add: **II. 9.** *E-layer, -region*: a stratum in the ionosphere, above the D-layer, that reflects medium frequency radiation; = **Heaviside layer*.

1930, 1935 [see *D. II. 1 c]. **1955** *Sci. Amer.* Sept. 128/2 The E layer (extending from about 60 to 90 miles above the earth's surface). **1959** DAVIES & PALMER *Radio Stud. Universe* ix. 162 Far below the F-region is the E-region which is centred at a height of 110 km.

10. *E-type*, a type of 'Jaguar' sports-car.

1961 *Autocar* 17 Mar. 403/2 Independent rear suspension is still a rarity on British cars, and Jaguar have not introduced it on the E-type without a great deal of investigation. **1964** J. WELCOME *Hard to Handle* v. 56 The note of the 'E' type's exhaust. **1967** L. DEIGHTON *Expensive Place* v. 41 On the corner an E-type was parked.

III. E.A., E/A, enemy aircraft; E.A.M., Ethnikon Apeleutherotikon Metopon (National Liberation Front); E.B.S., emergency bed service; E.C.G., electrocardiogram; E.C.T., electro-convulsive therapy; E.D.C., European Defence Community; E.D.D., English Dialect Dictionary; E.D.P., e.d.p., electronic data processing; E.D.S., English Dialect Society; E.E., Early English (see EARLY *a.* 4 b); E.E.C., European Economic Community; E.E.G., electro-encephalogram (see *ELECTRO-); E.E.T.S., Early English Text Society; E.F.T.A. (see *EFTA as a main entry); *e.g.* (examples); E.H.P., effective (or electrical) horse-power; E.H.T., extra high tension; E.I., East India; E.I.C., East India(n) Company; E.L.A.S., Ethnikos (or Ellinikos) Laikos Apeleutherotikos Stratos (National (or Greek) Popular Liberation Army); E.L.D.O. (see quot. 1962); E.M.F., e.m.f., electromotive force; E.N.I.A.C., also *eniac* (see quots.); E.N.S.A. (see *ENSA as a main entry); E.N.T., ear, nose, and throat; E.P., electro-plate(d); E.P., extended-play (record) (see quot. 1962); E.P.D., excess profits duty; E.P.N.S., electro-plated nickel silver; E.P.T., excess profits tax; E.P.U., European Payments Union; E.S.N., e.s.n., educationally subnormal; E.S.P., extra-sensory perception (see *EXTRA-SENSORY *a.*); E.S.R.O. (see quot. 1961); E.S.T., Eastern Standard Time (in the eastern parts of the U.S and Canada); E.S.T., electro-shock (or electric shock) treatment (see *ELECTRO-); E.S.U., e.s.u., electrostatic unit (see *ELECTROSTATIC *a.*); E.T.A., estimated time of arrival; E.T.D., estimated time of departure; e.V., eV, electron volt; E.V.A., extravehicular activity (activity outside a space-craft).

1916 T. E. LAWRENCE *Lett.* (1938) 214 The feed-block of one of the E.A. German Maxims. **1942** 'B. J. ELLAN' *Spitfire* 42 A faint silhouette even if the E/A was not actually illuminated. **1944** *Time* 16 Oct. 45/3 The Greek resistance forces wore the insignia of E.A.M. (left-wing National Liberation Front) or E.L.A.S. (E.A.M.'s fighting arm). **1945** *Spectator* 21 Sept. 265/1 There were those..who regarded E.A.M. as little more than a group of unscrupulous adventurers who were employing the brigand bands of E.L.A.S. to seize power by violent means. **1966** P. P. ARGENTI *Occupation of Chios* vi. 61 The Communist E.A.M. (*'Εθνικὸν 'Απελευθερωτικὸν Μέτωπον*, i.e. National Freedom Front). **1952** E.C.G. [see *electrocardiogram* s.v. *ELECTRO-]. **1962** *Lancet* 5 May 946/2 The first three cases show that ST and T-wave depression can occur in the left chest leads of E.C.G.s in acute pulmonary embolism...The reasons for these E.C.G. changes are debatable. **1952** *Times* 22 May 5/4 (*heading*) E.D.C. talks end in Paris. **1954** *Encounter* Nov. 31/1 And so [certain elements of the Right] were led to oppose EDC, which was conceived precisely in order to safeguard France against the extreme risks of German rearmament pure and simple. **1960** *Electronic Engin.* Mar. 153/2 In the early days of business e.d.p. it was not unusual for the manufacturer's programmers and analysts to guide, control, and even supplement the efforts of the user's personnel. **1960** *Times Rev. Industry* Dec. 3/1 Electronic data processing (E.D.P.). **1873** *N. & Q.* 4th Ser. XII. 342/1 Walter W. Skeat, Hon. Sec. E.D.S. **1877** G. F. CHAMBERS *Handbk. Sussex* 26 The Ch...is chiefly Perpendicular, with an E.E. tower. **1966** *Listener* 3 Feb. 166/2 Things like that little E.E. job in Palermo, now closed, could not possibly be mistaken for any Sicilian place of worship. **1958** *Times Rev. Industry* July 75/1 Exchanges with the other members of E.E.C. accounted for 27 per cent. of imports. **1961** *Listener* 10 Aug. 196/2 The preferences British goods enjoyed on the Australian market would disappear if Britain joined E.E.C. **1969** *Times* 27 Jan. 10/6 Union thinking..is strongly against exchanging present guaranteed prices for any 'target' prices on the E.E.C. model. **1867** SKEAT *Piers Plowman* p. xii, My tract published for the E.E.T.S. in 1866. **1891** SCHICK *Lydgate's Temple of Glas* p. xii, Dr. Erdmann's forthcoming edition of the *Story of Thebes*, for the E.E.T.S. **1958** *Times Lit. Suppl.* 17 Jan. 35/1 Few E.E.T.S. introductions can show such grace, style and wit. **1682** R. BAXTER *Answer to Mr. Dodwell* 226 What if they hold, e.g. Arrianism, Socinianism, Manichisme, &c... Are they not Heretical? **1818** MOORE *Fudge Fam. Paris* (ed. 5) 106 Like him, Tiberius loved his joke, On matters, too, where few can bear one; *E.g.* a man, cut up, or broke Upon the wheel—a devilish fair one! **1871** B. JOWETT *Dial. Plato* I. 110 In this or that passage—e.g. in the explanation of good as pleasure. **1922** JOYCE *Ulysses* 700 Requisite farming implements, e.g., an end-to-end churn. **1920** *Conquest* June 360/1 An average of about £50 per E.H.P. at power house must be allowed. **1948** *Electronic Engin.* XX. 160 The circuit.. derives its H.T. from the E.H.T. supply which feeds the cathode ray tube. **1814** *Statutes* 54 *Geo. III* Index 2/2 E. I. Co...Cap. xxxiv. The East India Company. **1898** J. J. M. INNES *Sir H. Lawrence* 103 The E. I. Board. **1944, 1945** E.L.A.S. [see *E.A.M. above]. **1945** W. S. CHURCHILL *Victory* (1946) 7, I have been told that I made a mistake in under-rating the power of the Communist-directed E.L.A.S. **1966** P. P. ARGENTI *Occupation of Chios* vi. 67 E.L.A.S. (*'Εθνικὸς Λαϊκὸς 'Απελευθερωτικὸς Στρατός*, i.e. National Popular Liberation Army). **1962** *Daily Tel.* 22 Feb. 24/4 ESRO is independent of the European Launcher Development Organisation (ELDO),..which seeks European support for converting Britain's abandoned Blue Streak defence rocket into a satellite launcher. **1881** HEAVISIDE in *Jrnl. Soc. Telegr. Engin.* X. 271 The phase of the current is behind that of the E.M.F. **1905** N. H. SCHNEIDER *Study Electr.* 76 The e.m.f. is raised. **1946** *N.Y. Times* 15 Feb. 1/7 The Eniac, known more formally as 'the electronic numerical integrator and computer', has not a single moving mechanical part. **1946** *Lancet* 30 Nov. 795/1 Another war secret now disclosed is an electrical calculating machine which has been built in the United States and has been called an 'electronic brain', or more accurately an Electronic Numerical Integrator and Computer—ENIAC. **1947** *Ann. Reg.* *1946* 392 The one [electronic calculating machine] which has gained most publicity is the Electronic Numerical Integrator and Automatic Calculator (E.N.I.A.C.). **1944** *Dorland's Med. Dict.* 503/2 E.N.T. Abbreviation for *ear, nose* and *throat*. **1952** 'R. GORDON' *Doctor in House* xiv. 154 The E.N.T. clinic was busy..for the London atmosphere silted up patients' sinuses and roughened their lungs. *c* **1895** *Catal. R. S. Rowell, Oxford* 43 Best E.P. Entrée Dish 50/–. **1954** *Gramophone* May 491/2 The second batch of EPs to appear. *Ibid.* Sept. 165/2 The other side of the new EP. *Ibid.*, Two of them..now make an EP appearance. **1962** A. NISBETT *Technique Sound Studio* 252 *E.P.* (*extended play*) *record*, a 45 rpm 7-inch fine groove record which may carry more material than a coarse-groove 10 or 12-inch 78. **1921** *Times* 4 Feb. 11/4 The End of E.P.D. **1897** *Illustr. Price List* (Edward Morrison, Parsonstown) 22, E.P.N.S. Preserve Dish with Glass, 9/–. E.P. Butter with Glass, 3/6. **1949** *Canning Handbk. Electro-Plating* (ed. 16) xxi. 338 For best quality hotel plate and E.P.N.S. spoons and forks it is customary to give a deposit of 0.002 inch in thickness. **1970** *Daily Tel.* (Colour Suppl.) 24 Apr. 3/2 If he's after antiques you know he's an amateur if he takes EPNS stuff and reproduction porcelain and ignores the real thing. **1940** *Economist* 6 Jan. 3/1 There is the unknown yield of E.P.T. **1950** *Times* (Suppl.) 18 Sept. p. iv/5 A number of European countries,..might not have felt justified in exposing themselves to the inevitable risks of E.P.U. **1955** *Special Schools Jrnl.* June 13/2, I have a strong conviction that Mary is not E.S.N. at all. On the orthodox tests of intelligence she certainly gains a score which puts her in the upper level of the E.S.N. group. **1960** *Where?* III. 13/2 *Educationally Subnormal* (*ESN*), an expression used to describe children whose I.Q. is between about 55 and about 80. **1961** *Engineering* 21 July 71/1 In the scientific field there would be close cooperation with the proposed European Space Research Organization (E.S.R.O.). **1968** *Times* 19 Nov. 6/8 Britain is not getting value for money from the European Space Research Organization (Esro). **1937** *Printers Ink Monthly* Apr. 53/1 *EST*, a term used by NBC to denote Eastern Standard Time. **1966** *Punch* 4 May 658/2 The National Broadcasting Company found a sponsor willing to put the British General Election up on prime time (9.30 EST to be precise). **1957** S. H. KRAINES *Mental Depressions* xvi. 456 Many names are given to this technique and its modifications: electric shock therapy (E.S.T.), shock therapy (S.T.), electrocoma (E.C.), electric treatments (E.T.), electroconvulsive therapy (E.C.T.), [etc.]. *Ibid.*, In the Manic-Depressive Illness, E.S.T. has proved to be of value in shortening the illness. **1969** L. L'ABATE in C. M. Franks *Behavior Therapy* III. xiv. 480 More concrete mediators in general psychiatric practice would be drugs, EST, and allied ancillary practices. **1939** *Flight* 26 Oct. 335/1 A quickly estimated E.T.A., too, will narrow down the area of search if the next outstanding landmark fails to turn up. **1958** 'N. SHUTE' *Rainbow & Rose* vi. 231 E.T.A. is seven-fifteen. **1939** *Jrnl. R. Aeronaut. Soc.* XLIII. 85 Three hours before the E.T.D. of the aircraft. **1965** *Newsweek* 14 June 30/3 White climbed into a new 'extravehicular activity' (EVA) suit. **1969** *Daily Tel.* 24 Feb. 21/3 In connection with the space-walk, Schweikart who 'will go Eva' (extra-vehicular activity) said 'I am looking forward to it'.

each, *a.* Add: **1. d.** *each way*: a betting term denoting that a horse, etc., has been backed for both a win and a place. Also *attrib.* and *transf.*

1869 *Cassell's Mag.* IV. 170/2 One pound on Blue Gown each way. **1897** E. H. COOPER *Mr. Blake* vi, I'll just go and put a little bit on Highborn for this race..; only fifty pounds or so each way. **1926** [see *BRADBURY]. **1929** *Star* 21 Aug. 16/2, I should think he may be a good each-way bet. **1930** *Daily Mail* 17 June 14/2 Sun Glory..has a useful each-way chance. **1951** E. RICKMAN *Come racing with Me* x. 84 A very small bet 'each way'. **1963** *Times* 15 Feb. 11/4 Why should Britain back the 625 line horse to win when we can place an each-way bet on the dual standard?

eager, *a.* Add: **6. eager beaver** [BEAVER¹ (cf. *1 d)]: a glutton for work; an over-zealous or officious person. Also *attrib.* and *transf.* *colloq.* (orig. *U.S.*).

1943 *Daily Nebraskan* 7 Nov., Eager Beaver, anyone who has the slightest conception of what physics is all about. **1947** *Sat. Even. Post* 16 Aug. 122/1 Before photostating my application, some eager beaver.. helpfully typed the address. **1948** *Time* 5 July 1/1 It would be futile to play that game against Republican Earl Warren, one of the foremost spokesmen of the eager-beaver West. **1958** E. DUNDY *Dud Avocado* I. i. 11 The Eager-Beaver-Culture-Vulture with the list ten yards long, who *just* manages to get it all crossed off before she collapses of aesthetic indigestion each night. **1959** *Observer* 24 May 27/8 The British pack were like a set of eager beavers. **1959** *Spectator* 29 May 753/1 Well-meaning state social workers..and eager-beaver politicos lavish with praise and perquisites, are not helping him.

eagle, *sb.* Add: **1.** Also in U.S. allusive use.

1798 T. G. FESSENDEN *Orig. Poems* (1806) 4 The Eagle of Freedom with rapture behold, Overshadow our land with his plumage of gold! **1865** *Trans. Ill. Agric. Soc.* *1861–2* V. 203 The clang of mighty forges, fashioning out the sinews of war, to guard the eagle of liberty. **1872** 'MARK TWAIN' *Speeches* (1910) 415 You won't mind a body bragging a little about his country on the Fourth of July. It is a fair and legitimate time to fly the eagle. **1904** *N.Y. Even. Post* 14 Sept. 1 Mr. Fassett was toying with those dear but haggard phrases..and generally making the eagle scream. **1942** R. G. LILLARD *Desert Challenge* 194 He introduced the orator of the day..and for a while the eagle screamed.

c. (Later U.S. example.)

1893 *Congress. Rec.* 16 Oct. 2561/2 The grand old man, known as the Bald Eagle of Rhode Island rose to reply.

d. *Golf.* A score of two under bogey or par.

1922 C. LEITCH *Golf* xii. 202 Barnes..holed his putt for a 3, so securing what is known in American golfing parlance as an 'eagle'. **1926** WODEHOUSE *Heart of Goof* iv. 135 When he had got that 'eagle' on the third, he had looked bored. **1927** *Observer* 3 Apr. 29/1 [He] let loose a whole flock of 'birdies' and 'eagles' with which he captured the remainder of the holes. **1928** *Daily Express* 9 July 17/7 He accomplished an 'eagle' two for the 351 yards. **1966** WODEHOUSE *Plum Pie* ii. 66 It was on the fourth green, after he had done an eagle, that he spoke again.

2. d. *U.S.* = *eagle button* (*10).

1789 W. MACLAY *Deb. Senate* (1880) 97 The Cincinnati had seats allotted for themselves—worn their eagles at their button-holes. **1812** *Niles' Weekly Reg.* II. 131/1 Cockades and eagles..have been bought and made.

e. *U.S. Mil.* A device worn on the shoulder, indicating the rank of a colonel.

1861 *Army Regulations* (U.S.) 485 Shoulder-Straps... For a Colonel—..a silver-embroidered spread eagle on the centre of the strap. **1865** L. N. BOUDRYE *Fifth N.Y. Cavalry* 184 Col. Hull..was killed. He was a gallant young officer, who had but recently borne the eagle.

5. (Earlier and later U.S. examples.)

1786 in *Amer. Museum* (1789) II. 182 There shall be two gold coins; one..to be stamped with the impression of the American eagle, and to be called an eagle. **1789** *Gazette of U.S.* (N.Y.) 14 July (Th.), The Eagle containing 10 dollars of 50d. is worth 2 guineas, and rather more. **1809** T. G. FESSENDEN *Pills Poetical* 16 Were eagles, in embargo times, As plentiful as cents and dimes. **1816** *Niles' Weekly Reg.* IX. 370/2 The president tendered several eagles in gold to cover the costs. **1854** M. J. HOLMES *Tempest & Sunshine* xiii. 84 At the same time managing to slip an eagle into the hands of the honest clerk. **1905** T. DIXON *Clansman* 139 A beggar asked him for a night's lodging, and he tossed him a gold eagle.

10. b. eagle button, a button bearing the device of an eagle.

1897 J. L. ALLEN *Choir Invis.* ii. 15 As he joined them, three other young men—Federalists—sauntered past, wearing black cockades, with an eagle button on the left side.

eagle-hawk. Add: **2.** An Australian bird of prey, *Uroaëtus audax*.

1827 in J. Bischoff *Van Diemen's Land* (1832) 177 The eagle-hawk pouncing upon them for his prey. **1848** J. GOULD *Birds of Australia* I. pl. 1 *Aquila Fucosa*.. Wedge-tailed Eagle. Eagle Hawk, Colonists of New South Wales. *a* **1882** H. C. KENDALL *Poems* (1886) 159 The fierce-featured eagle-hawk flies—afraid as a dove is afraid. **1911** C. E. W. BEAN *'Dreadnought' of Darling* xii. 105 A big black eagle-hawk. **1965** H. FRAUCA *Bk. Austral. Wild Life* 93 Some cattlemen..suggested that the local wedgetails be classified as vermin because they were scratching the cattle. One cattleman said that two of his cows had been 'badly scratched by them eagle-hawks'.

ean, *v. intr.* Delete † *Obs.* and add later example.

1965 *Times* 18 Feb. 14/6 The..ewes who were due to ean..were driven down..to the fold.

ear, *sb.*¹ Add: **1. c.** Phrases: *up to the ears* (later examples); *to* (*be willing to*) *give one's ears* (earlier examples); *on one's ear*, drunk; *dry behind* (occas. in the U.S. *back of*) *the ears*, adult, experienced, mature; also, *wet behind the ears*, immature, naïve; *to have* (or *keep*) *one's ear*(*s*) *to the ground*, (fig.) to be on the alert regarding rumours or the trend of public opinion; *out on one's ear*, dismissed, ejected ignominiously.

1889 W. B. YEATS *Let.* Sept. (1954) 136, I am up to the ears in Irish novelists. **1966** *Listener* 6 Jan. 15/1 He.. is up to his ears in work.

1796–1801 T. G. FESSENDEN *Orig. Poems* (1806) 128 Jove tells his peers He'd give his ears For such an hour as this is. **1853** DICKENS *Bleak Ho.* xxxix. 395 There are people in London who would give their ears to be you.

1906 E. DYSON *Fact'ry 'Ands* xiv. 180 It's these cheap 'n' easy shickers rollin' round on their ear what brings discredit on beer. **1921** K. S. PRICHARD *Black Opal* ii. 17 The old chap has 'got on his ear' in Sydney. **1932** J. VAN DRUTEN *Behold, We Live* I. 17, I shall be on my little ear if I don't get some food soon.

1914 *Dialect Notes* IV. 105 *Dry back of the ears*, mature;—of persons. **1931** BROPHY & PARTRIDGE *Songs Brit. Soldier* (ed. 3) 375 Wet behind the ears, a term of reproach imputing ignorance or youth. **1939** *Amer. Speech* XIV. 264 The classic chastening phrase applied to the adolescent who feels that he is grown is that he 'isn't dry behind the ears yet'. **1939** J. STEINBECK *Grapes of Wrath* ix. 109 When you bastards get dry behin' the ears, you'll maybe learn to let an ol' fella sleep. **1962** J. F. STRAKER *Coil of Rope* vii. 71 You're still wet behind the ears, darling. It's time you grew up. **1968** W. J. BURLEY *Three Toed Pussy* iv. 68, I am not an abortionist but neither am I wet behind the ears. I've been around.

1920 *Nat. Rev.* Apr. 142 On the eve of a Presidential Election campaign, when practical politicians have their ears to the ground. **1955** G. GREENE *Quiet American* III. i. 190 What's the gossip of the market, Tom? You fellows certainly do keep your ears to the ground. **1966** 'H. CALVIN' *Italian Gadget* ix. 146 Honestly, the way I've had my ear to the ground, I simply don't see it.

1951 M. KENNEDY *Lucy Carmichael* vi. ii. 294 If I pipe down and give no trouble I'm safe. If not, I'll be out on my ear. **1953** J. TRENCH *Docken Dead* ix. 129 I'm going to throw the sleuth out on his ear.

h. *to get* (a person) *up on his ears*, to make him indignant; so *to be on one's ear*; *to get up* or *go off on one's ear* (U.S.), to rouse or bestir oneself.

1871 L. H. BAGG *4 Years at Yale* 44 A man somewhat offended or indignant is said to be *on his ear*. **1871** *Galveston News* 4 May (De Vere), They..said that I was lightning, when I got up on my ear. **1882** W. D. HOWELLS *Mod. Instance* II. xxix. 144 'I can cut your acquaintance fast enough,..if you're really on your ear!' 'I'm on my ear,' said Ricker. **1889** FARMER *Americanisms* s.v., *To get up* or *go off on one's ear*, to bestir oneself; to rouse oneself to a great effort. **1907** M. C. HARRIS *Tents of Wickedness* III. iv. 255, I only hope Paul Fairfax won't read it and get on his ear!

i. *Oh my ears and whiskers!*, a joc. exclamation.

1865 'L. CARROLL' *Alice in Wonderland* i. 7 Alice..was just in time to hear it [*sc.* the White Rabbit] say..'Oh my ears and whiskers, how late it's getting!' **1956** M. STEWART *Wildfire at Midn.* iii. 33 He gave a little bark of laughter. 'Oh my ears and whiskers!' **1963** 'M. ERSKINE' *House in Belmont Square* i. 20 Oh, my ears and whiskers! Why didn't you say so before?

3. d. *to go in at one ear and out at the other* (later examples); *to believe one's ears*, to credit that one can have heard aright (chiefly with negative expressed or implied); *to bite one's ear*, see *BITE *v.* 16.

1862 MRS. GASKELL *Let.* ? 16 May (1966) 923 You know things so often go in at one ear with him & out at another that it will be as well to be quite sure. **1930** 'BRENT OF BIN BIN' *Ten Creeks Run* v. 52 You might hear a lot that isn't true if you listen to gossip. It's best to let such things go in one ear and out the other.

1807 LAMB *Tales fr. Shakes.* I. 200 Lear at first could not believe his eyes or ears, nor that it was his daughter who spoke so unkindly. *a* **1885** 'H. CONWAY' *Living or Dead* (1886) viii, He..blamed my partner, who could scarcely believe his ears.

5. *by ear* (later examples). Fig. phr. *to play* (*it*, etc.) *by ear*, to proceed step by step according to results.

1839 *Edin. Rev.* July 496 Miss Austen is like one who plays by ear, while Miss Martineau understands the science. **1873** L. TROUBRIDGE *Life amongst Troubridges* (1966) 7 She plays the piano *beautifully* and can play anything by ear. **1961** L. GRIBBLE *Wantons die Hard* xiii. 161 'I'm playing this by ear,' he grunted once when the American queried the devious route he was following. **1961** A. SMITH *East-Enders* xi. 183 'What happens then?' 'I don't know...We're playing it by ear at the moment.' **1962** A. NISBETT *Technique Sound Studio* xi. 195 The relative [sound] levels are judged by ear.

15. a. *ear-scoop, -spoon*.

1895 *Arnold & Sons' Catal. Surg. Instrum.* 83 Ear scoop and Eye spud. **1928** V. G. CHILDE *Most Anc. East* viii. 189 Tweezers..were combined with prickers and earscoops in regular reticules. **1966** *Punch* 7 Sept. 344/2 Diggers in the newly excavated Roman villa in Hampshire have found a lot of ear-scoops. *c* **1325** *Gloss. W. de Bibbesworth* in Wright *Voc.* (1857) 146 Cakenole *gloss* herespon. **1878** *Sunday at Home* 6 July 424/2, I made several purchases, such as..ear spoons. **1921** *Chambers's Jrnl.* 763/2 Toilet implements as toothpick, tweezers, ear-spoon. **1937** *Jrnl. R. Anthrop. Inst.* 322 Inca metal-work includes many small objects such as pins, pincers, earspoons, etc. **1969** E. H. PINTO *Treen* 323 The prevalence of ear spoons in *Atuis*.

16. ear-biter *slang*, †(*a*) *U.S.*, a special agent of the Post Office *obs.*; (*b*) an habitual borrower of money, a cadger; hence **ear-biting** *vbl. sb.*; **ear-bob** (later U.S. example); **ear-bone,** a bone of the ear; **ear-cap** (later example); **ear-clip,** an ear-ring, esp. one that clips on; **ear-defenders** *pl.*, (a pair of) plugs or ear-muffs designed to protect the ear-drums from damage by loud or persistent noise; **ear-flap,** also, a flap of material covering the ear; hence **ear-flapped** *a.*; **ear-fly,** a gad-fly belonging to the genus *Chrysops*, esp. *C. vittatus*, which attacks the ears of horses; **ear-guard:** restrict † to sense in Dict. and add: (*b*) a protection for the ears; (*c*) *slang* (see quot. 1941); **ear-hole** (later examples); **ear-hoop** *U.S.*, an ear-ring; **ear-lappet** = **ear-tab*; **ear-lid:** delete † and add later examples; **ear-lock** (later example); **ear-muff** orig. *U.S.*, a protection for the ears, in cold weather, from noise, etc.; **ear-phone,** (*a*) a device applied to the ears for listening in to radio broadcasting; (*b*) a device to aid defective hearing; (*c*) a woman's hair-style (current in the 1920s) of a shape reminiscent of the listening device; **ear-piece,** an apparatus or a part of one designed to be fitted to the ear, as of a telephone or a radio receiver; **ear-plug,** (*a*) an ornament worn in the lobe of the ear; (*b*) a wad of cotton-wool, wax, or other substance placed in the ear to prevent an inrush of cold air or water, or to exclude excessive noise; **ear-roll,** in a leather helmet, a roll of leather behind the ear; **ear-stud,** a stud, freq. ornamented, worn in a hole pierced through the lobe of the ear; **ear-tab** *U.S.*, a tab, esp. one affixed to either side of a fur cap, to protect the ear in cold weather (= TAB 13); also, = **ear-phone* (a); **ear-training** *vbl. sb.*, the cultivating of the ability to distinguish and reproduce sounds and rhythms in speech or music.

1855 J. HOLBROOK *Ten Yrs. among Mail Bags* 27 The restoration of the 'ear-biters' (as they were then sometimes facetiously called). **1934** *Bulletin* (Sydney) 17 Oct. 21/2 No..earbiters anxious to give you a moral for the lars'. **1940** WODEHOUSE *Eggs, Beans & Crumpets* 19 Two things which rendered Oofy Prosser a difficult proposition for the ear-biter. **1954** —— *Jeeves & Feudal Spirit* i. 10 It is not often that one is confronted with ear-biting on so majestic a scale, a fiver till next Wednesday being about the normal tariff. **1938** M. K. RAWLINGS *Yearling* xii. 123 Now you jest give me them ear-bobs, you pirate. **1681** GREW *Musæum* I. v. i. 82 Part of the Ear-Bone of a Whale. **1901** *Jrnl. R. Microsc. Soc.* 138 Ear-bones of Opossum. **1915** SHIPLEY & MACBRIDE *Zool.* (ed. 3) 644 The evolution of the ear-bones in Mammalia. **1963** *Times* 4 May 6/4 They will admire the earcaps knitted to protect the horses from flies. **1945** A. SELWYN *Retail Jeweller's Handbk.* viii. 100 To the diamond mounters must be credited the idea of a new ornament, the ear-clip (*not* same as clip-on earring); this lies over the lobe and follows the ear up for a short way. **1951** *Catal. of Exhibits, South Bank Exhib., Festival of Britain* 39/1 Gold, sapphire and diamond clip, with ear clips and ring. **1958** *Times* 15 Dec. 11/6 Citrine, diamond and 18 ct. gold Earclips. **1915** *Illustr. London News* 16 Jan. 72 (*caption*) Ear-defenders for British: a Device to prevent deafening by big guns. **1917** W. OWEN *Let.* 9 Jan. (1967) 424 There is a Howitzer just 70 or 80 yards away...I can't tell you how glad I am you got me the ear-defenders. **1934** *Discovery* Dec. 345/2 Ear-defenders, small plugs which limit the intensity applied to the ear drum. **1961** *Flight* LXXX. 933/3 Headsets incorporating ear defender fittings. **1907** *Yesterday's Shopping* (1969) 882/2 Motor caps...Seal fur caps, with earflaps. **1968** W. GARNER *Deep, Deep Freeze* xxi. 198 He let down the ear-flaps of his fur hat. **1894** A. CONAN DOYLE *S. Holmes* 2 Sherlock Holmes, with his sharp, eager face framed in his ear-flapped travelling cap. **1925** E. F. NORTON *Fight for Everest, 1924* 51 Ear-flapped caps. **1806** M. LEWIS *Orig. Jrnls.* (1905) V. 87 Found here [are] the butterflies, common house and blowing flies, the horse flies, except the goald coloured ear fly. **1917** SANDERSON & PEAIRS *School Entom.* 136 The Horse-flies. *Tabanidæ*...Some are called Ear-flies, some Gad-flies. **1939** AUDEN & ISHERWOOD *Journey to War* ii. 50 Chinese in skin caps with ear-guards like airmen's helmets. **1941** BAKER *Dict. Austral. Slang* 27 *Earguards*, small side-whiskers. **1923** MANCHON *Le Slang* s.v., Give 'im a plug in the ear-hole! **1962** *John o' London's* 11 Jan. 27/3 Before you know it you'll be out on your earhole. **1966** *New Statesman* 25 Feb. 265/2 [In the Wallis Islands] a man can be fined two dollars for entering church with a piece of tobacco stuck in his earhole. **1808** *Massachusetts Spy* 18 May (Th.), A large assortment of Earhoops, of different sizes. **1845** S. JUDD *Margaret* I. x. 64 Many wore ear-hoops of pinchbeck, large as a dollar. **1864** B. TAYLOR *H. Thurston* vi. 79 Woodbury recognised, projecting between ear-lappets of fur, the curiously-planted nose..which belonged to the Rev. Mr. Waldo. **1869** Ear-lappet [see LAPPET *sb.* 4]. **1871** T. H. HUXLEY *Anat. Vert.* 250 The tympanic membranes [of the crocodile] are exposed, but a cutaneous valve, or earlid, lies above each, and can be shut down over it. **1930** *Times* 3 July 10/5 Nature..has not provided against assaults upon the hearing, and we have no 'ear-lid' by which we can shut out noise. **1867** J. R. LOWELL *Poems* (1912) 578 His ear-locks gray, striped with a foxy brown, Were braided up to hide a desert crown. **1859** *Rep. Comm. Pat.* *1858* (U.S.) II. 572 William P. Ware..[patented an] Ear, Cheek, and Chin Muff..July 6, 1858. **1889** *Kansas Times & Star* 30 Nov., A young chap got off a train from the South today...Said they wore no flannels, gloves or earmuffs down at Shreveport. **1906** E. DYSON *Fact'ry 'Ands* xvi. 215 Well, y' orter wear ear muffs this bitin' weather. **1940** *Chambers's Techn. Dict.* 276/2 *Ear muffs*. 1. Pads of rubber or similar material which are placed on head telephone receivers to minimise discomfort during long use. 2. Large pads for reducing the effect of noise on factory workers. **1954** A. W. FIELDING *Hide & Seek* ii. 29 Theatrical side-whiskers as thick and furry as ear-muffs. **1924** *Glasgow Herald* 24 Dec. 6 With a slight effort of the imagination we can forget earphones and valves. **1927** *Radio Times* 22 Apr. 145/1 One might even suppose that you had never in your life put on a pair of earphones or listened to a loud speaker. **1928** J. F. BARNHILL *Nose, Throat & Ear* 583 Ear phones..are the most serviceable present means of aiding impaired hearing. **1964** M. DRABBLE *Garrick Year* iv. 57, I spent hours trying to put my hair into earphones. **1965** *Punch* 17 Mar. 404/3 Only two of the sound films..have subtitles, the others being shown with 'earphone commentary'. **1967** *Lebende Sprachen* XII. 136/2 *Earphones, headphones*, a device worn in or on the ear for listening to a sound source. *a* **1843** Ear-pieces [in Dict.]. **1853** W. O. MARKHAM tr. *Skoda's Auscult.* 29 The ear-piece may be convex, concave, or plane, provided the disk forming it be large enough to close the ear completely. **1882** *Leisure Hour* XXXI. 312/1 In front of the coil is placed a thin disc of iron (H), which is fitted into an ear-piece (I) [of the telephone]. **1904** F. LYNDE *Grafters* xxvii. 352 The train-master dropped the ear-piece of the telephone. **1909** *Practitioner* Dec. 862 The ear-pieces should fit closely and accurately into the external auditory meatus. **1916** 'BOYD CABLE' *Action Front* 185 The earpiece receiver strapped tightly over one ear. **1923** A. HUXLEY *Antic Hay* x. 156 For business, tortoiseshell rims and nickel ear-pieces lend incisive poise. **1928** GALSWORTHY *Swan Song* I. iv. 29 Soames..took up the earpieces of the wireless. **1962** *Which?* Feb. 42/2 Table I shows which radios had earpieces, sometimes included in the price. **1904** *Nature* 9 June 138/1 Numerous copper ornaments..such as pendants and ear-plugs. **1904** *Daily Chron.* 5 Sept. 8/3 We should like..to see the use of ear-plugs spread, for many cases of ear trouble are brought into activity by the inrush of water during diving or swimming. **1940** *Nature* 17 Aug. 219/2 The methods of protection of the ears against explosive sounds are very simple. One of these, the earplug, has..always been used extensively by gunners and others in the presence of intensive noise. **1955** E. POUND *Classic Anthol.* II. 140 In the old Capital officers wore ear-plugs fittingly Of sea stones (common jade). **1959** *Daily Tel.* 19 Aug. 10/5 B.O.A.C. stewardesses, ground receptionists and other staff who have to work near jet airliners are to have miniature ear plugs. **1909** *Westm. Gaz.* 23 Sept. 12/3 One of the regulations was that competitors had to wear ear-rolls to facilitate their hearing when other riders wished to overtake them. **1922–3** *Halford Cycle Co. Ltd.* 110 Helmets. Leather, with Ear Rolls. [**1873** *Young Englishwoman* Sept. 443/1 The most recherché..is the small diamond earring, riveted into the ear like a stud.] **1939–40** *Army & Navy Stores Catal.* 463/1 Pair Cultured Pearl Earstuds. **1961** *Times* 6 June 22/6 A pair of circular-cut diamond earstuds. **1966** R. PETRIE *Dead Loss* vi. 44, I could get stud earrings, vulgar-big, from Woolworths.... Big white plastic ear-studs. **1855** *Knickerbocker* Feb. 199 In stable-yards, old-looking black boys, in cat-skin caps, with ear-tabs to them. **1909** *N.Y. Even. Post* 28 Jan. (Th.), With the first really cold weather of the winter, there appeared on Broadway the vendors of ear-tabs. **1922** *Short Stories* Feb. 36/1 An experience much like the one I had had in the wireless station...Sprague gave me one set of ear tabs and took the other for himself. **1899** F. G. SHINN *Elem. Ear-Training* I. p. iii, The author believes that the placing of the subject of direct ear-training in a prominent position in schemes of musical education, will..raise the standard demanded of public and private performances. **1921** H. E. PALMER *Princ. Language-Study* 17 If his ear-training is neglected during the elementary stage, he will replace foreign sounds by native ones. **1929** I. C. WARD *Phonetics Eng.* (1931) iii. 17 Ear-training exercises. **1947** *Penguin Music Mag.* May 49 The lessons should be as varied as possible: a little ear-training, a little sight-reading.., a few finger exercises.

ear, *sb.*² Add: **b.** *ear of corn*, a head of maize. Also *ellipt.* *N. Amer.*

1622 *Mourt's Relation* 21 We..found a fine new Basket full of very faire corne of this yeare, with some 36. goodly eares of corne. *c* **1665** P. E. RADISSON *Voyages* (1885) 78 Each takes an ear of corne and putts in their mouths. **1831** J. M. PECK *Guide for Emigrants* II. 156 The ears [of Virginia corn] are usually five or six feet, and often more from the ground. **1880** WEBSTER Suppl. s.v. *Head*, In the U.S., *ear* is used of Indian corn alone. **1894** *Vermont Agric. Rep.* XIV. 42 The ears (grain and cobs) are of course digestible. **1958** *Edmonton* (Alberta) *Jrnl.* 26 July 39/4 The staff..reported harvesting about 300 ears of Altagold corn.

ear, *v.*² Add: **b.** Of maize. *N. Amer.*

1624 CAPT. J. SMITH *Virginia* VI. 239 The stalke was first set, began to eare ere it came to halfe growth, and the last not like to yeeld any thing at all. **1896** *Vermont Agric. Rep.* XV. 71 Usually it [*sc.* the Red Cob] does not ear.

ear-bash (iᵊ·ɹbæʃ), *v. trans.* and *intr.* *slang* (chiefly *Austral.*). [f. EAR *sb.*¹ + BASH *v.*²] To talk inordinately (to someone). Hence **ear-bashing** *vbl. sb.* and *ppl. a.*; **ear-basher,** a chatterer; a bore.

1944 L. Glassop *We were Rats* xxxvi. 205 'Time for you bastards to do some spine bashing,' said one. 'Are you going to sit there ear bashing all night?' 1944 R. J. Oakes in *Austral. Short Stories* (1951) 374 The little ear-bashing engineer would take him along. 1953 K. Tennant *Joyful Condemned* iii. 22 She was ear-bashing me all over tea. 1955 'N. Shute' *Requiem for Wren* (1956) v. 121 The Commander escaped after a quarter of an hour of ear-bashing. 1956 S. Hope *Diggers' Paradise* xvii. 157 Most golf clubs have a share of 'ear-bashers' as the Aussies call the type who verbally replay their strokes *ad nauseum* [*sic*] at the 'nineteenth'. 1957 I. Cross *God Boy* (1958) x. 81 We went on ear-bashing each other like that for the next hour or so. 1962 *Daily Tel.* 24 Nov. 1/6 (*heading*) Ear-bashing exercise by Mr. Wigg. [*Ibid.*, Mr. Wigg bashes the ear as once he bashed the square.] 1966 'L. Lane' *ABZ of Scouse* II. 32 *Ear-basher*, one who talks too much; a garrulous person.

ear-drop. 1. (Examples.)
1720 Defoe *Capt. Singleton* 211 A few Diamonds in the Ear-drops. 1861 [see *Albert 1]. 1936 L. C. Douglas *White Banners* ix. 201 While finding the jabot, she had come upon a pair of long imitation-jade eardrops.

earful (iə·ɹful). *colloq.* [f. Ear *sb.*[1] + -ful 2.] As much (talk) as one's ears can take in at one time; a large quantity (*of* talk, gossip, etc.); a strong reprimand, a 'piece of one's mind'.
1917 R. W. Lardner *Gullible's Travels* (1926) 27, I and my Missus and Mrs. Hatch clubbed together on the straps and I got a earful o' the real dope. 1922 S. Ford *Trilby May crashes In* x. 159 A parlor Bolshevist..had started to give me an earful about the downtrodden. 1927 A. Clarke *Son of Learning* iii. 56, I have heard an earful of good stories. 1929 W. R. Burnett *Little Caesar* iv. i. 106, I want the boys to get an earful. Go ahead and spill it. *Ibid.* ii. 119 I'll have to go over and give that bird an earful. 1944 W. Stevens *Let.* 13 Mar. (1967) 462 He said that the Germans had not changed since Tacitus described them. Considering that I was only waiting for the shop to open, this was rather an earful. 1945 L. A. G. Strong *Othello's Occupation* 49 She got one of the hottest earfuls I ever heard. 1946 F. Sargeson *That Summer* 56, I tried to get an earful when I heard somebody out on the landing-place. 1964 *Times* 11 Feb. 11/7, I used to put a bottle on the seat and if it rolled off when the pupil let his clutch out, he got an earful.

earlet. Add: **2. d.** = Tragus, esp. when largely developed as in some bats.
1837 *Penny Cycl.* VII. 25/2 The ears are large, insulated and lateral, with an internal earlet. 1845 *Encycl. Metrop.* XXV. 1072/1 Ears large, the tragus or earlet always existing. 1904 *Speaker* 24 Dec. 315/2 The earlet, a curious development of the tragus in insectivorous bats. 1957 *Encycl. Brit.* V. 596/2 The 'earlet', or tragus, is generally large in the insectivorous bats but is never present in the Megachiroptera nor in the insectivorous Rhinolophidae.

early, *a.* Add: **4. b. Early English** (earlier example).
1807 G. Millers *Descr. Cathedral Ch. Ely* 17 Sketch of the Characteristics of English Church Architecture. 1. Age.—Saxon..2. Age.—Norman..3. Age.—Early English; from 1200 to 1300, comprehending the reigns of John, Hen. III. and Edw. I.
7. Special collocations: as, **early American** *a.*, belonging to or characteristic of the colonial and post-Revolutionary period in America; **early-closer,** one whose place of business is closed one afternoon in the week; **early closing,** orig. designating a movement for the reduction of the daily hours of labour in wholesale and retail trades; now, the system by which business premises are closed for the day at the end of the morning on a particular day of the week; also, the closing of public houses earlier at night; also *attrib.* or *adj.*; **early days,** early in time for something (to happen, etc.); **early leaver,** a pupil who leaves school without completing the full course of study; hence *early leaving*; **early Victorian** *a.*, belonging to or characteristic of the early years of Queen Victoria's reign, its literature, fashions, etc.; also as *sb.*; so *early Victorianness*; **early wood,** the less dense part of the annual ring of a tree.
1895 *Critic* 9 Nov. 310/2 (*title*) Early American paintings at the Metropolitan Museum. 1915 W. A. Dyer (*title*) Early American craftsmen. 1922 *Country Life* Sept. 45/1 Furniture from the workshop of Duncan Phyfe holds distinctly a place of its own in the history of early American utilitarian art. 1926 *Ladies' Home Jrnl.* Sept. 98/4 There was first of all a charm created by the primitive simplicity of the antique Early American furniture as seen against nut-brown panelings of old pine. 1937 E. St. V. Millay *Conversation at Midnight* ii. 47 They fill the place so full of Early American furniture that there isn't a comfortable chair in the house. 1945 O. Nash *Many Long Years Ago* (1954) 96 He was too Early-American to write for advice from Emily Post. 1863 *Sat. Rev.* 17 Jan. 78/1 The public would look with a more favourable eye upon the early closers if [etc.]. 1906 *Daily Chron.* 28 July 7/6 Thus giving early-closers and Saturday-afternooners a chance. 1845 *Chambers's Edin. Jrnl.* 25 Aug. 123/1 Those who are striving for the early closing of shops. 1846 *Punch* 11 Apr. 165/1 (*heading*) The Early Closing Movement. 1849 *Hogg's Weekly Instructor* III. 24/2 The Metropolitan Early Closing Society. 1851 *Household Words* 1 Nov. 126/2 That early-closing movement which has fastened the portals of all those magnificent palaces of linen-drapery. 1881 *Ibid.* 12 Nov. 53/2 That Saturday 'early closing'..must be a vast boon to shop-people. 1888 *Boy's Own Paper* Christmas Number 34/1 Wednesday was our early-closing day. 1902 *Encycl. Brit.* XXXII. 685/2 Scottish Early Closing of Public-houses Act. 1534 T. More *Wks.* (1557) 1156/1 She telleth hym then that it is but earely dayes, & he shal com time inough. 1739–40 Richardson *Pamela* (1740) I. xiv. 26 'Tis early Days with Pamela; and she does not yet think of a Husband. 1828 J. Constable *Let.* 25 Apr. (1962) 244 No impediments have yet arisen, and it is early days. 1935 G. Heyer *Death in Stocks* xi. 145 It's early days yet. 1957 *Times* 23 Dec. 11/2 As regards the current year, it is early days to express any considered opinion, but trading conditions are bad. 1951 *Times Educ. Suppl.* 12 Jan. 21/1 That there is a substantial number of early leavers is not to be disputed. *Ibid.* 21/2 To tolerate early leaving is to destroy the structure and purpose of the grammar school. 1878 C. L. Eastlake *Hints Household Taste* (ed. 4) vi. 161 Fifty years hence most of our early 'Victorian' upholstery will have fallen into useless lumber. 1883 *Eng. Illustr. Mag.* Nov. 89/2 The late Georgian or early Victorian age. 1896 *Lady's Realm* Dec. 129/1, I was a youth—in the mahogany age—early Victorian, you know. 1895 F. Harrison (*title*) Studies in early Victorian literature. 1906 *Fortn. Rev.* Nov. 880 We invite the humour of the gods if we look down on the early Victorians. 1908 *Westm. Gaz.* 31 Oct. 15/3 The bonnets..were positively early-Victorian in size, and quite early-Victorian in the dainty pale silk frillings which framed most becomingly the face and the hair. 1959 *N. & Q.* CCIV. 123/1 Mr. G. M. Young's excellent two-volume survey of *Early Victorian England* began in 1830. 1914 Wodehouse *Man Upstairs* 181 It was the gruesome Early Victorianness of it all that took the heart out of him. 1914 S. J. Record *Mech. Prop. Wood* ii. 44 The inner portion [of the annual ring] was formed early in the season, when growth was comparatively rapid and is known as early wood. 1931 G. A. Garratt *Mech. Prop. Wood* ii. 115 That [wood] originally nearest the pith, which was formed in the spring of the year and is consequently known as springwood or early wood, is generally more open textured and of lighter colour than the outer portion of the ring. 1967 F. D. Silvester *Timber* iv. 26 The discoloration tends to be uneven as if a stain..has bled into the early wood portion of the growth rings.
b. In collocations used *attrib.*; *spec.* **early-warning,** used of equipment, bases, etc., designed for the early detection of aerial, etc., attacks; also *fig.*
1897 M. Kingsley *Trav. W. Africa* 65 The pottery has a very early-man look about it. 1900 *Westm. Gaz.* 9 Jan. 2/1 The shock of that early-dawn disaster. 1906 *Daily Chron.* 2 June 9/1 The abolition of early-season events for two-year-olds. 1909 *Westm. Gaz.* 23 Apr. 8/2 Much of the early-morning work is performed entirely alone. 1916 R. Frost *Let.* 8 Aug. (1964) 41 Regarded as an attack of early-marriage nerves, it seems even a little funny. 1946 A. Lee *German Air Force* iv. 44 With no early warning system to help them operate fighters economically.., the Polish Air Force could not be expected to survive. 1949 E. Coxhead *Wind in West* iii. 61 The first recipient of her early-morning tea. 1956 *Ann. Reg. 1955* 135 President Eisenhower's 'early warning' plan..for a mutual exchange of military blueprints and aerial reconnaissance by Russia and America. 1958 *Listener* 17 July 78/1 The early warning missile station is about to be constructed. 1961 *Flight* LXXX. 654/1 The same hardware and techniques used to launch an orbiting scientific capsule can also be used to orbit an early-warning satellite. 1970 *Daily Tel.* 28 Apr. 2/3 An early warning system to alert doctors of the latest drug, Mandrax, being misused has been called for.
B. *sb.* **a.** An early fruit or vegetable. Chiefly *pl.*
1846 *Times* 13 Feb., Regents..kidneys, earlies, Scotch reds. 1883 R. Fremlin *Potato* 4 Second Earlies and Late Sorts. 1900 *Daily News* 10 Sept. 2/3 Hops..A few samples of earlies were offered at market to-day. 1925 *Daily Mail* 25 Apr. (Advt.), Chrysanthemums. A nice collection of well rooted outdoor earlies. 1961 J. Seymour *Fat of Land* iv. 53 In March we start potato planting. We plant a few earlies and later a lot of main crop.
b. *pl.* Early years or days.
1927 'A. Horn' (*title*) The Ivory Coast in the earlies. 1928 *Sunday Express* 24 June 8/5 So much for the old show days in the earlies. 1936 *Times Lit. Suppl.* 22 Feb. 167/2 A sheep-droving expedition on the scale of the one described here was more common in the 'earlies' of Australia than now.

early, *adv.* Add: **5. b.** *early on* [f. *earlier on* after Later on], at an early stage.
1928 D. L. Sayers *Bellona Club* xv. 170 'It might have been *given* him earlier.'..'Well—not too early on, Peter. Suppose he had died a lot too soon.' 1942 A. L. Rowse *Cornish Childhood* iv. 82, I was a very 'forward' child: very early on throwing my bottle out over the cradle. 1953 D. D. C. P. Mould *Ireland of Saints* viii. 103 The difficulties surrounding the relation between philosophy and the Christian revelation and faith had been studied from early on in the Church's history. 1958 *Spectator* 1 Aug. 175/3 Two of the principal characters..are involved early on in a session of the Law.

earlyish (ɜ·ɹli,iʃ), *a.* and *adv.* [f. Early *a.* + -ish[1] 3.] Somewhat early; at an early hour; = *early on.*
1838 Mrs. Gaskell *Let.* 17 July (1966) 18 We set off earlyish 4 o'clock say. 1927 W. Deeping *Kitty* xxviii. 352 It was earlyish, and the big room—empty and noiseless—seemed to envelope old Jermyn with velvet arms. 1950 in K. Jackson *Lang. & Hist. Early Brit.* (1953) ii. xi. 294 Its epigraphy..is earlyish. 1963 *Times* 23 Jan. 12/7 The wheels of their cars, in which they hoped to make an earlyish get-away, were already jammed with ice. 1968 H. R. F. Keating *Inspector Ghote hunts Peacock* ix. 116 'It was early,' he said. 'Quite earlyish. Say about seven..give or take five minutes either side.'

earmark, *v.* Add: **3.** To set aside (money, etc.) for a particular purpose. So **ea·r-marking** *vbl. sb.* (also *concr.*). Also *fig.*
[*a* 1868 W. Stebbing in M. Pattison *Acad. Org.* i. 13 Corporate property..unearmarked..by the donor.] 1890 *Spectator* 28 June, With large sums ear-marked and accumulating for the extinction of licences. 1893 *Times* 28 Apr. 9/4 A perfectly arbitrary earmarking of moneys paid into a common fund. 1928 *Manch. Guardian Weekly* 31 Aug. 179/2 Sundry amounts of sovereigns were 'earmarked' for the Commonwealth Bank of Australia. 1929 S. McKenna *Happy Ending* iii. ii, I need only earmark sufficient time in the summer for certain people whose hospitality I've accepted. 1931 *Economist* 10 Jan. 64/2 The usual year-end earmarkings have cut down the usual gain, so that the year may end with the total under earmark virtually unchanged at around $135,000,000. 1966 *New Statesman* 22 July 123/2 Already the unions give about £270,000 to Transport House without ear-markings.

ea·r-mi:nded, *a. Psychol.* [f. Ear *sb.*[1] + Minded *a.*] Having a marked tendency to carry on mental operations most readily by auditory images; thinking in sounds.
1888 *Pop. Sci. Monthly* XXXIII. 604 Tests for distinguishing between an eye-minded and an ear-minded person..reveal whether the words are better apprehended by the eye or by the ear. 1900 W. B. Secor in *Amer. Jrnl. Psychol.* XI. 228 Notwithstanding the fact that he was decidedly visual, he had a strong tendency to be ear-minded. 1901 E. B. Titchener *Exper. Psychol.* I. i. 196 The purely ear-minded man would recognise persons, things and places by the sounds connected with them. 1925 C. Fox *Educational Psychol.* iv. 76 Classification into ear-minded, eye-minded, etc.
Hence **ea·r-mi:ndedness.**
1893 *Jrnl. Proc. & Addresses Nat. Educ. Assoc. U.S.* XXXII. 780 Some..did show extreme eye-mindedness, and others..showed extreme ear-mindedness. 1902 Baldwin *Dict. Philos.* II. 571 The varying prominence which visual and auditory and muscular processes occupy in different minds—eye-mindedness, ear-mindedness, motor-mindedness.

ea·r-ringed, *a.* [f. Ear-ring + -ed[2].] Wearing ear-rings.
a 1846 B. R. Haydon *Autobiogr.* (1927) iii. xiii. 235 We were saluted as we entered by one of those ear-ringed, red-capped blackguards. 1883 J. Payn *Thicker than Water* xvi, The ear-ringed, ringleted sailors. 1892 Stevenson & Osbourne *Wrecker* viii. 123 Some ear-ringed fisher of the bay. 1920 *Chambers's Jrnl.* 561/1 His black-browed, ear-ringed kind knew him of an older race. 1924 J. Buchan *Three Hostages* xiii. 192 The earringed Jewess.

earth, *sb.*[1] Add: **4.** *to run to earth*: to chase (the quarry) to its earth; *fig.* to capture or find (something sought for) after a long search. Similarly *to go to earth*, said of the quarry; also *fig.*
1857 Kingsley *Two Y. Ago* xxviii, Frightened—beat—run to earth myself, although I talked so bravely of running others to earth just now. 1876 [see Run *v.* 42 *fig.*]. 1888 *Spectator* 7 Jan. 20/2 All the men who helped to run to earth the various members of the Ruthven family..were richly rewarded. 1913 *Punch* 26 Feb. 153/1 Men who used to go to earth behind evening papers on the entrance of a woman now spring to their feet in platoons without a moment's hesitation. 1917 M. Webb (*title*) Gone to Earth. 1950 R. Macaulay *World my Wilderness* xvi. 194 The policeman..turned back to assist his colleagues in flushing Barbary, so mysteriously gone to earth. 1953 'F. O'Connor' *Stories* 63 Eventually he would run her to earth in some snug with a couple of cronies.
6. *Electr.* (Substitute def. and add examples.) Connection of a wire conductor with the earth, either accidental (with the result of leakage of current or dangerous differences of potential) or intentional (as for the purpose of providing a return path for a telegraph current, etc.). (Cf. Ground *sb.* 15 b.)
[1773 H. Cavendish *Jrnl.* 9 Feb. in *Electr. Researches* (1879) 267 It was suspected that this increase of separation of the balls before they closed was owing to the wire designed to carry off the el[ectricity] to earth not conducting fast enough.] 1868 L. Clark *Elem. Treat. Electr. Measurement* vi. 42 The earth connections should therefore be carefully looked to occasionally. If a station have a defective earth, and have two wires leading to it, the evil will generally disclose itself. 1870 [in Dict.]. 1876 Preece & Sivewright *Telegraphy* 225 Upon no account whatever is a leaden gas-pipe to be employed for the purpose of affording earth. *Ibid.* 243 Earths are indicated by an increase in the strength of the current at the sending end, and by a decrease in the strength, or the entire cessation of it, at the other end. *Ibid.* 253 If the earth at B is bad while that at A and at C is good, then a part of A's current, on reaching B, instead of going to earth there, will take the course of the wire to C, working C's apparatus, and go to earth at C. 1901 L. M. Waterhouse *Conduit Wiring* 17 When the cables are pulled through, the braiding (and perhaps the rubber) is torn off and the result is a bad 'earth' at some future time. 1911 *Encycl. Brit.* XXVI. 523/2 The signals received on such sensitive instruments..are liable to be disturbed by the return currents of other systems..and to obviate this it is necessary to form the 'earth' for the cable a few miles out at sea. 1966 *Buying Secondhand* (Consumers' Assoc.) 71 Earth is always green or green/yellow except in German-made appliances where earth is red.

9. c. (Examples illustrating wider use, chiefly in interrogative and negative contexts.) Also, with a superlative, used as an intensive phr.
1774 Goldsm. *Retal.* 103 With no reason on earth to go out of his way, He turned and he varied full ten times a day. 1847 J. Carlyle *Let.* 15 July (1883) I. 389 If I could have done anything on earth but cry. 1859 Princess Royal *Let.* 26 Aug. in R. Fulford *Dearest Child* (1964) 207, I cannot see what on earth he can have of very urgent business here in November. 1873 'Mark Twain' & Warner *Gilded Age* 29 I've got the biggest scheme on earth—and I'll take you in! 1876 R. Broughton *Joan* xiii, You people really have the worst small-beer in Europe. Where on earth did you get it? 1882 Mrs. J. H. Riddell *Daisies & Buttercups* i. iv. 121 What on earth did it all matter to me? 1885 'F. Anstey' *Tinted Venus* 128 Why on earth was she making this dead set at him? 1910 Wodehouse *Psmith in City* xviii. 158 Master Edward Waller..in frocks, looking like a gargoyle;..in sailor suit, looking like nothing on earth.
d. Colloq. phr. *the earth*, used in intensive expressions indicative of great or excessive ambition, cost, expense, etc.; *to cost the earth*: see *Cost *v.* 1 d.
1928 Wodehouse *Money for Nothing* vii. 132 What's the matter with you that you always want the earth? 1952 —— *Barmy in Wonderland* xiv. 137, I pay a director the earth. Where is he? 1958 *Engineering* 4 Apr. 427/2 The customer has a perfect right to ask for the earth, but the supplier, if he is wise, will not necessarily let him have it. 1961 A. Christie *Pale Horse* xii. 129 Would it be terribly expensive?..She'd heard they charged the earth.
B. I. 1. a. *earth-child, -god* (examples), *-line, -lord* (later example), *-magic, -time, -year.* **c.** *earth-colour* (example; hence *earth-coloured* adj.), *-smell.*
1906 *Westm. Gaz.* 2 June 6/2 Earth-child, struggle no more. 1931 Blunden *To Themis* 56 Age cannot wither you, Tiny philosopher, Earth-child, musician. 1935 T. S. Eliot *Murder in Cathedral* i. 12 The labourer bends to his piece of earth, earth-colour, his own colour. 1918 D. H. Lawrence *New Poems* 50 The waste all dry..Stirring with earth-coloured life. 1871 Swinburne *Poems* (1904) II. 124 The earth-god Freedom. 1904 *Folk-Lore* Sept. 312 As an embodiment of the earth-god the king was responsible for the fruits of the earth. 1866 G. M. Hopkins *Jrnl.* 6 May (1959) 135 A charming day, sky pied with clouds, near the earth-line egg-blue. 1907 Kipling *Twenty Poems* (1918) 2 They are concerned with matters hidden—under the earth-line their altars are. 1944 Blunden *Shells by Stream* 13 Something between a castle and a cave..For that earth-lord to pace. 1901 'L. Malet' *Hist. R. Calmady* vi. x. 603 All this, the unity and secrecy of the place..circling them about with something of earth-magic. 1928 C. Day Lewis *Country Comets* 9 For his was the simpleness Born of earth-magic. 1895 K. Grahame *Golden Age* 14 The air was wine, the moist earth-smell wine. 1942 T. S. Eliot *Little Gidding* i. 7 There is no earth smell Or smell of living thing. 1951 A. C. Clarke *Sands of Mars* ii. 15 We keep normal Earth-time—Greenwich Meridian—aboard the [space-]ship. 1951 S. Spender tr. *Rilke's Life of Virgin Mary* 47 Something endured Still, rest of earth-time, canker withered. 1953 E. F. Russell *Somewhere a Voice* (1965) 18 It would take them most of an Earth-year to reach the fortieth parallel.
4. *earth-bound* (later examples); also *transf.* and *fig.*, and indicating motion towards the earth; *earth-long.*
1869 W. James *Coll. Ess. & Rev.* (1920) 1, The 'Sadducees', as our author [*sc.* a spiritualist] loves to call the earth-bound portion of the community. 1931 C. Day Lewis *From Feathers to Iron* xv. 31 Earth's first faint tug at the earthbound soul. 1935 *Discovery* Feb. 43/1 To an earth-bound rocket pressure is more important than velocity. 1950 S. Spender *Sel. Poems Whitman* 11 With all his loftiness and idealism, he is peculiarly earth-bound. 1903 W. S. Blunt *7 Golden Odes* 15 Herds knelt, their necks stretched earth-long. 1935 C. Day Lewis *Time to Dance* 55 Earth-long and heaven-outfacing woes.
II. earth-almond, delete def. and substitute = Chufa (earlier U.S. examples); **earth colour, pigment,** a pigment obtained from native earth, as the ochres and umbers; so **earth white**; **earth-current** (earlier example); **earth-life,** terrestrial existence; **earth-man,** (*a*) a human being (or occas. a mythical creature) whose life and instincts are closely allied with the natural or material (as opposed to the spiritual) world; (*b*) esp. in science fiction, an inhabitant or native of the planet Earth; also **earthsman, earth-woman; Earth-Mother** [tr. G. *erdmutter*], in mythology and folklore, a spirit or being taken as a symbol of the earth; a sensual and maternal woman; also = Mother earth 1; **earth-mover** orig. *U.S.*, a vehicle or machine designed for the excavation or shifting of large quantities of earth; so **earth-moving** *vbl. sb.* and *ppl. a.*; **earth-pig,** transl. Du. *aardvarken* = Aard-vark; **earth-plate** (examples); **earth-return** (*a*) *Electr.*, an earthed return circuit, as distinguished from a metallic return; also *attrib.*; (*b*) *attrib.*, returning to the planet Earth; **earth satellite,** an artificial satellite projected into orbit around the earth; also *attrib.*; **earth-shaking** *ppl. a.*, also *fig.*; **earth-shine** (later examples); **earth-side,** also *attrib.* or as *adj.*, and used adverbially; **earth-soul,** (*a*) *Philos.*, the supposed collective consciousness of the earth, including as its parts the consciousnesses of all earth's inhabitants (cf. *anima mundi); (*b*) the soul of a former earth-dweller; **earth-wax** = Ozocerite; **earth-wire** *Electr.*, wire carried from a conductor into the earth, esp. to prevent contact from the leakage of current from one wire into another; hence **earth-wire** *v.*, **-wired** *ppl. a.*, **-wiring** *vbl. sb.*; **earth-woman** (see *earth-man above).
1856 *Rep. Comm. Pat.: Agric. 1855* (U.S.) p. xiii, The Earth Almond, or *Chufa*, (*Cyperus esculentus*), a small tuberous esculent, from the south of Spain, has naturalised itself to our climate and soil. 1860 Earth-almond [see Chufa]. 1913 N. Heaton *Hurst's Man. Painters' Colours* (ed. 5) v. 155 Iron oxide is also the colouring principle of the group of pigments known as 'earth colours'. 1951 *Oxf. Jun. Encycl.* VII. 325/1 Examples of such earth colours are yellow ochre, siennas (dark yellows), and umbers (browns). 1872 *Phil. Mag.* XLIII. 186 It is almost impossible to have two earth-plates inserted any distance apart without a difference of tension,..This is due in some cases to earth-currents. 1906 W. De Morgan *Joseph Vance* xix. 191 The black Shadow that oppressed me was bidden to..scatter itself over the remainder of my earth-life. 1906 *Daily Chron.* 28 May 3/4 One brief day—as long as seven years of this earth-life. 1922 O. Lodge *Raymond Revised* 47 Humour does not cease with earth-life. 1958 *New Statesman* 15 Mar. 353/2 The spy from Outer Space..will be happy to discover..an authentic smell, that is, of mid-century earth-life in general. 1860 H. B. Tristram *Great Sahara* i. 18 A negro from Timbuctoo engaged to remove the plague, and taught sacrifices to the 'Earth-men', or demons who roam the earth. These are believed to be harmless when once they have obtained a human residence. 1904 G. K. Chesterton *G. F. Watts* 126 He would see giants and the sea..and brown earth-men and red earth-women lying in the heaps of greens and browns and reds. 1905 *Daily Chron.* 16 Mar. 8/2 The sensual earth-man must be killed, beyond all chances of reviving, before the man after the divine pattern and will can live. 1930 A. H. Krappe *Sci. Folk-Lore* i. 20 In at least one type [of fairy tale], the story of the Earthman, the helper, a dwarf, sometimes a witch, has to be overcome by the hero first. 1936 C. S. Lewis *Alleg. Love* vii. 312 Mammon is the gold-hoarding earthman of immemorial tradition, the gnome. 1947 W. K. Richmond *Poetry & People* i. 14 At heart we are still Saxons, and deeper still we are countryfolk, peasants, earth-men. 1949 R. Heinlein *Red Planet* (1963) i. 3 The Mars creature saw an elderly pale Earthman. 1960 *Guardian* 26 Aug. 2/7 If their site did indeed become the set of the space film, the corps members would not be suitable for the parts of earth-men. 1904 *Edin. Rev.* Jan. 38 The Indian women disraimented still enact the ancient ritual of the Rain-Goddess or Earth-Mother. 1906 *Inst. Mag.* Apr. 312 When the great, good Earth-Mother saw this, she called to April and sent her back to gain a victory over her malicious enemy. 1907 *Academy* 31 Aug. 837/1 Soft to his neck earth-mother clings. 1961 S. Lloyd *Art Anc. Near East* iii. 82 His symbolic marriage with Inanna, the 'earth-mother'. 1962 *John o'London's* 31 May 529/2 An earth-mother barmaid. 1939 *Civil Engineering* XXXIV. 228 (*title*) The development of earth moving equipment for highway construction. 1941 *Agricultural Engineering* XXII. 19/1 Earth moving makes up the principal portion of construction work. *Ibid.* 24/1 Profit by experience of earth movers. 1959 *B.S.I. News* June 10/2 Giant earth-mover tyres. 1963 *Times* 24 Jan. 11/7 The earthmover, the caterpillar tractor. 1968 *Daily Tel.* 1 Nov. 19/4 Three or four heavy earth-moving vehicles started levelling the adjacent land. 1785 G. Forster tr. *Sparrman's Voy. Cape Gd. Hope* I. 270 The *aard-varken*, or earth-pig, which, probably, is a species of *manis.* 1962 M. Burton *Syst. Dict. Mammals* 195 (*heading*) Aardvark, earth-pig, ant-bear (*Orycteropus afer*). 1923 L. C. Martin *Colour & Methods of Colour Reprod.* vi. 73 Generally speaking, the 'earth pigments' are the most stable and satisfactory. 1847 Brett & Little *Compendium Improvements Electric Telegraphs* 22 An earth plate..which carries the current back by the conducting powers of the earth. 1872 Earth-plate [see *earth-current]. 1871 *Eng. Mech.* 8 Sept. 627/1 Earth return currents are not practical. 1902 *Encycl. Brit.* XXXIII. 227/2 Steinheil of Munich..discovered the use of the earth return. 1940 *Chambers's Techn. Dict.* 277/2 *Earth return circuit*, a telegraphic current using one transmission wire, the return current passing through the earth and thereby encountering a low resistance. 1962 F. I. Ordway et al. *Basic Astronautics* v. 191 Later, when Earth-return vehicles become feasible, samples can be brought back. 1968 *Times* 23 Dec. 6/4 There is no reason in logic why it should not..eject them into an earth-return orbit. 1949 *Rocket Jet Flying* Spring 6 The 'earth satellite vehicle program'..is the most imagination-firing news we've heard in quite a while. 1950 *Jrnl. Brit. Interplanetary Soc.* IX. 155 As performances improve, so we may expect to see the appearance of the close-orbit Earth satellite vehicle. 1956 *Collier's Year-Bk.* 48/2 Plans to launch an earth satellite were announced in the middle of 1955. 1959 Davies & Palmer *Radio Studies of Universe* x. 180 A most spectacular and ambitious project..has been the launching of earth satellites by Russia and the U.S.A. 1807 J. Barlow *Columbiad* iv. 10 p. 135 Earth-shaking storms and constellated skies. 1948 E. Sitwell *Notebk. on Shakes.* viii. 104 With the exception of two earth-shaking sentences, and one speech of great beauty..Iago never speaks 'above a mortal mouth'. 1966 Ogilvy & Anderson *Excurs. Number Theory* xi. 144 Besides, what if a study is not of earth-shaking importance? 1946 *Nature* 21 Dec. 907/1 The portion of the moon's surface that is lighted up by earthshine. 1963 *Daily Tel.* 20 May 26 (*heading*) Space man slept well...Kept out 'earthshine'. 1949 R. Heinlein *Red Planet* (1963) iv. 53 Many's the time he's told me stories about the school he went to back Earth-side. 1956 *Galaxy Sci. Fiction* XLIII. 11/2 Some Earth-side official of the Interstellar Prison Service. 1851 H. Melville *Moby Dick* 79 It smells like another world, more strangely than the moon would to an Earthsman. 1871 Swinburne *Songs bef. Sunrise* 149 The earth-soul Freedom, that only Lives, and that only is God. 1905 W. James *Ess. Rad. Emp.* (1912) iv. 136 Speculations like Fechner's, of an Earth-soul, of wider spans of consciousness enveloping narrower ones throughout the cosmos, are..philosophically quite in order. 1948 C. Day Lewis *Poems 1943–47* 64 You might well surmise They are earth-souls. 1884 Earth-wax [see Ozocerite]. 1958 W. T. O'Dea *Social Hist. Lighting* 216 Ozokerit, or 'earth-wax', found in the region of the Roumanian oil wells, later proved..superior, at a price, to paraffin wax candles. 1908 F. Maire *Mod. Pigments* iv. 40 *Earth whites* are so named to distinguish that class of pigments which owe their origin to mother earth in contradistinction to those which are derived from a metallic origin. 1868 E. Atkinson tr. *Ganot's Physics* (ed. 3) 650 Into the other hole of the fuse a wire is placed which serves as earth wire. 1876 Preece & Sivewright *Telegraphy* 215 Earth-wiring...The object of the earth-wires is to prevent contact from arising through the leakage of currents from one wire at its point of support into another. *Ibid.* 216 In dry sandy soil or in rock the earth-wiring is therefore to be avoided. *Ibid.* 258 It is always advisable to earth-wire at least the last five supports on each side of every office, as a protection against the effects of lightning. 1911 *Encycl. Brit.* XXVI. 512/1 For protection from lightning each pole has an 'earth wire' running from the top, down to the base. 1966 Earth wire [see *earth *v.* 8]. 1904 Earth-woman [see *earth-man]. 1955 'J. Wyndham' in 'E. Crispin' *Best S.F.* 72 He greeted Lellie [*sc.* a Martian] just as if she were an Earth woman.

earth, *v.* Add: **8.** *Electr.* To connect (a conductor) with the earth. Hence **earthed** *ppl. a.*; **earthing** *vbl. sb.* (also *attrib.*).
1885 *Jrnl. Soc. Telegraph Engineers* XIV. 454, I have myself seen a circuit 'earthed' at an intermediate station in the middle of a message. 1888 *Science* 13 July 18/1 In dry weather they [*sc.* conductors] are not earthed at all well. 1889 *Daily News* 25 Dec. 6/7 A piece of mechanism known as an 'earthing device', the invention of Major Cardew, which infallibly cuts off the current if a condition of danger occurs. 1898 *Ibid.* 3 May 5/3 The swaying to and fro of the earthed line in the field due to terrestrial magnetism. 1902 *Encycl. Brit.* XXV. 773/1 Let a conductor—say, a metallic sphere—be supported by a metal rod of negligible capacity whose other end is earthed. 1906 A. F. Collins *Man. Wireless Telegr.* 212 Earthed terminal. The wire connecting the plate buried in the earth and the aerial wire. 1909 *Install. News* III. 80/1 Mr. Leckie recommended earthing through a resistance. 1966 *Buying Secondhand* (Consumers' Assoc.) 72 If the appliance is intended to be earthed, make sure there is an earth wire fitted. With earthed appliances the continuity of the earth wire ought to be checked.

earthling, *sb.*[2] **1.** (Later examples.)
1949 R. Heinlein *Red Planet* (1963) viii. 125 An Earthling has no good way to estimate the age of a Martian. 1965 J. Blish *Mission to Heart Stars* vi. 77 The Earthlings are now deep into the Heart Stars. 1967 *Ampleforth Jrnl.* Summer 163 To receive the overspill by immigration..the planets might come to the rescue of the Earthlings.

earthly, *a.* Add: **1. c.** Colloq. phr. *no earthly, not an earthly*, (*a*) with *sb.*, no—whatever, (*b*) *ellipt.*, no conceivable (chance, hope, use, etc.).
1838 Dickens *Let.* 25 Jan. (1965) I. 359 The fifty pounds..makes no earthly difference. 1868 [see Dict.]. 1899 *Referee* 22 Oct. (Ware, s.v. *No*), The actors who have not booked their seats via Mr. Henry Dana, are hereby notified that they have now no earthly, as all seats have been allotted. 1907 *Hockey* 13 Dec. 101/2 The poor goal-keeper had not an 'earthly'. 1910 Wodehouse *Psmith in City* xxvii. 239 There's no earthly need to get out. 1911 W. J. Locke *Clementina Wing* xii. 134 Just an extra fiver on Punchinello. He's got no earthly—you know that as well as I do. 1912 Galsworthy *Eldest Son* iii, *Christine.* Let's go and beg him, Ronny! *Keith.* No earthly! The only hope is in the girl. 1922 —— *Loyalties* i. ii, The Inspector's no earthly. 1936 J. Curtis *Gilt Kid* xxix. 279 'Going to get away with it?' 'Not a hope, mate. Not a bleeding earthly.' 1965 *Listener* 11 Nov. 765/3 Received standard, like the Liberals, won't stand an earthly.
5. Comb. *earthly paradise*: see Paradise *sb.* 1.

earthquaky (ɜ·ɹþkwē̆ki), *a.* [f. Earthquake + -y[1].] Resembling the effect of or suggesting the motion of an earthquake.
1837 Dickens *Pickw.* xlv, Legs shaky—head queer—round and round—earthquaky sort of feeling—very. 1846 Thackeray *Cornhill to Cairo* ii. 15 Many churches..had a dry, uncomfortable, earthquaky look, to my idea.

earthwork. Add: **2.** The action or process of excavating (the bed of a canal, line of a railway, or other civil engineering work); the soil so cast up; Embankment 3.
1831–3 *Encycl. Metropol.* (1845) VIII. 247/1 The earth-work for a complete line of communication between Liverpool and Manchester. 1842 Francis *Dict. Arts*, *Earth-work*, a term applied to cuttings, embankments, and all other works where earth is to be removed or collected together. *a* 1854 C. Tomlinson *Cycl. Useful Arts* II. 448/2 The inclination of the earth-works, whether for excavations or embankments, must be determined mainly by their height. 1862 *Chambers's Encycl.* IV. 27/2 In the formation of canals, railways, and other roads, embankment and excavation go hand in hand, and, under the name of earthwork, form..a vast branch of industry. 1899 *Daily News* 8 Mar. 7/1 The wretched earthwork labourers..would be earning from 2½ to 4 piastres per day near their own homes. 1911 *Daily Colonist* (Victoria, B.C.) 30 Apr. 1/7 Considerable heavy rock work here presents itself, as well as a fair amount of medium earthwork. 1955 *Times* 30 June 13/5 It is easy to forget that the gigantic earthworks of the railway age..were performed in the main by spade and barrow, horse and cart.

earthy, *a.* Add: **1. d.** *Electr.* (See quot. 1940.)
1876 Preece & Sivewright *Telegr.* 257 If an underground wire becomes earthy, owing to the insulating covering being partly removed, and the conductor being thus laid bare, [etc.]. 1940 *Chambers's Techn. Dict.* 278/1 *Earthy*, said of (1) circuits when they are connected to earth, either directly..or through a condenser; (2) any point in a communicating system..which is at earth potential, although not actually connected to earth, through zero impedance. 1945 *Electronic Engin.* XVII. 735 The 'earthy' terminal is a shrouded insulated terminal which is connected to the earth terminal of the case. 1949 *Ibid.* XXI. 359 The feedback may be very conveniently taken from the earthy end of the cathode load resistor.
7. *Comb.*
1923 W. Deeping *Secret Sanctuary* vi. 62 A lean, peevish, earthy-faced man in a hard felt hat. 1922 D. H. Lawrence *England, My England* 240 Then he removed her saturated, earthy-smelling clothing.

ease, *v.* Add: **1.** Also with *out of* and (*U.S.*) with *up*.
1916 'Boyd Cable' *Action Front* 148 There is nothing we can do for him..except morphia again, to ease him out of his pain. 1923 R. D. Paine *Comr. Rolling Ocean* v. 83 The doctor is down there easing up the guys that got the hide burned off 'em.
c. Delete † (obs.) and add later example.
1926 T. E. Lawrence *Seven Pillars* (1935) xliv. 256 Gasim explained that he had dismounted to ease nature.
4. c. (Later examples.)
1832 J. P. Kennedy *Swallow Barn* II. i. 4 He became notorious for picking up stragglers..and now and then, for easing a prisoner of his valuables. 1882 *Sydney Slang Dict.* 4/1 *Ease*, to rob. To Ease a Bloak, to rob a gentleman.
d. *to ease in* or *into*: to 'break in' gently (to work, etc.); to move or manœuvre into (a place, appointment, etc.). Similarly, *to ease out.*
1892 *Leisure Hour* May 462/1 He is gradually eased in to his work. 1951 *N.Y. Times* 28 Nov. 1/6 There have been many precedents in the Soviet satellites for easing prominent Communist leaders into jail by incessive steps. 1955 *Times* 10 May 8/7 It is being suggested that the Russians are easing him out of office because of the resistance he is said to have shown towards some of their demands. 1968 *Listener* 1 Aug. 133/3 A fair number..have either been pensioned off or eased into some job which gives them a living.
e. *Dressmaking.* To join two pieces of material whose edges are of unequal length in such a way that the extra fullness of the larger section is distributed evenly along the join. Freq. with *in*.
1932 D. C. Minter *Mod. Needlecraft* 108/2 Allow sufficient width at armhole to allow for easing in. 1967 *Simplicity 7391* (Simplicity Pattern Co. Inc.), To ease top of sleeve, stitch along seam line and ¼" *inside* seam line between notches, using a long machine-stitch. *Ibid.*, Slip-stitch, easing in fulness. 1968 J. Ironside *Fashion Alphabet* 82 *Ease*, v. When one section of a garment is fuller than another to which it is to be joined, the excess material is distributed evenly without gathers or pleats and usually pressed away.
7. b. To move, lift, or shift *down* gradually or gently; or spec. *into* a person's pocket.
1688 S. Sewall *Diary* 31 Oct. in *Mass. Hist. Soc. Coll.* (1878) 5th Ser. V. 234, I help'd to ease the Corps into the Grave. 1850 N. Kingsley *Diary* 154 The road is verry steep..so bad in some places that a rope is necessary to ease them down with. *a* 1875 'Mark Twain' *Works* (1900) XIX. 246 The widow caught the limp form and eased it to the earth. *Ibid.* 252 'Ease this down for a fellow, will you?' I eased the gravestone down till it rested on the ground. 1926 J. Black *You can't Win* iv. 41 The only thing to do is to blow back his money. Either the lawyer or one of the girls eased it into his coat pocket.
8. Delete †*Obs.* and add later example.
1937 *New Statesman* 25 Dec. 1089/1 The fighting in northern China has eased a little since the fall of Nanking.
b. To relax or cease one's efforts, *spec.* in rowing; also with *up*; *to ease all*, to stop rowing (cf. *Easy *v.*).
1863 *Rowing & Sailing* 55 This spurt should be continued till the boat begins to rock, when it is better to 'ease all' than to attempt altering the stroke into a milder one. 1882 *Daily Tel.* 2 Mar. (Cassell), They also row right through to Iffley without easing. 1883 'Mark Twain' *Life on Miss.* 44, I eased up and went slow and cautious. *Ibid.* xii. 115 When the measurements indicate that the yawl is approaching the shoalest part of the reef, the command is given to 'Ease all!' 1889 J. K. Jerome *Three Men in Boat* v. 73 On catching sight of Harris and me.., he eased up and stared. 1915 'Boyd Cable' *Between Lines* 25 'Can't you keep on belting 'em for a bit?' asked the Platoon officer. 'Might make 'em ease up on us.' 1945 E. Waugh *Brideshead Revisited* ii. iv. 262 She's playing up the religious stuff at the moment for all it's worth. I daresay she'll ease up a bit when she's settled.
c. *to ease up* intr. To become less burdensome.
1939 M. Spring Rice *Working-Class Wives* iv. 79 Things will ease up a bit soon when the children get older.
10. ease off. c. *intr.* To fall away with a gentle slope.
1880 'Mark Twain' *Tramp Abroad* xli. 477 The slope eased off,..and Croz and I, dashing away, ran a neck-and-neck race. 1925 E. F. Norton *Fight for Everest, 1924* 79 The slope began to ease off towards the edge of the shelf above.
d. *trans.* To fire off.
1916 'Taffrail' *Carry On* 93 What I do want is a nice little cruiser or a destroyer; something, at any rate, to ease off the guns at. 1962 W. Granville *Dict. Sailors' Slang* 44/1 *Ease off a fish*, fire a torpedo (Submariners).
e. *intr.* To take things easily.
1925 E. F. Norton *Fight for Everest, 1924* 35 It enabled us to..have a welcome bath, and generally ease off after our first taste of high-level marching.
11. *intr.* Of shares: to become easier, fall in value. Hence **ea·sing** *vbl. sb.*
1900 *Daily News* 5 Mar. 10/7 There is no easing in the prices for coal. 1904 *Financial Times* 23 Nov. 1/7 Sewing Cottons eased off to 12s. 3d. 1927 *Daily Tel.* 22 Nov. 2 Marconi eased 3d. and Eastern Telegraph declined 3.

easily, *adv.* Add: **8.** *colloq.* At least (a specified number or time); more than.
1947 *Library Jrnl.* Aug. 1097/2 Easily, a million cards were handled in the process and the difference between that and 100,000 destroyed.

east. B. *sb.* Add: **2. b.** The states of eastern Europe; the Communist powers. Freq. *attrib.* and appositively with *west* (or *West*).
1951 *Ann. Reg. 1950* iii. 188 There were more 'espionage' convictions..and the closing of the Czech Consulate-General in New York. These incidents, significant of the growing estrangement of East and West, were also..evidence that neither party wished to push its claims to the limit. 1959 *Daily Tel.* 10 Mar. 1/5 Russia had been told her proposals could not be accepted because there was no trust between East and West. 1959 *News Chron.* 5 Aug. 4/2 The harsh reality of the cold war, of East–West tension. 1959 *Manch. Guardian* 11 Aug. 7/4 East–West experts agreed..on the feasibility of detecting tests in the outer orbit. 1968 K. Martin *Editor* xv. 318 It would be very difficult to maintain an objective view of East/West relations.
4. *Bridge.* (With capital initial.) The player sitting opposite 'West'.
1926 *Auction Bridge Mag.* Aug. 126/2 West and North passed, and East doubled. *Ibid.*, East's top cards are all sure tricks. *Ibid.*, East is justified in expecting to set South at least one trick. 1929 *Bridge World* Oct. 7/1 The Five Spade bid by East is very good. 1958 *Listener* 2 Oct. 541/2 East had been able to recognise his plan.
D. 1. b. *East Coast fever*, in East and South Africa, a disease of cattle caused by a protozoan organism (*Theileria*) and transmitted by ticks.
1904 A. Theiler in *Rep. S. Afr. Assoc. Advancem. Sci.* 203 The disease was originally called Rhodesian Redwater...An improvement in the nomenclature was the words Rhodesian Tick Fever: a still better name is East Coast Fever. 1955 J. H. Wellington *Southern Africa* II. i. v. 79 *East Coast fever*, or 'African Coast Fever', a very virulent and highly fatal form of piroplasmosis.
2. *east-bound*: also *ellipt.*
1909 R. A. Wason *Happy Hawkins* 222 This time the west-bound had to take a sidin' and wait twenty minutes for the east-bound. 1948 H. Drake-Brockman in B. James *Austral. Short Stories* (1963) 111 The east-bound was on the way towards Wallu.

east-about, *adv. U.S.* [Cf. *west-about*, West *adv.* E. b.] In an easterly direction.
1886 *Sci. Amer.* 2 Jan. 7/1 The cause, whatever it was, gradually spread, moving east-about.

East-end. (Earlier examples of *spec.* sense.)
1846 R. Ford *Gatherings from Spain* xxi. 297 Our East end tailors' pattern-book of the last new fashion. 1867 *Punch* 23 Nov. 216 In the regions of sport and of money, In the circles of East-end and West.

Easter, *sb.*[1] Add: **1. b.** R.C. Ch. *to make* (†*do*) *one's Easter* (see Make *v.* 57 e): to perform one's Easter duties (see below). Hence *Easter* is used for an individual performance of these.
1700 T. Marwood *Diary* 8 Apr. in *Cath. Rec. Soc. Publ.* (1909) VII. 62, I was at S[t] Gomars & Saw ye Quire do their Easter. *Ibid.* 11 Apr., I was at my Easter at S[t] Gomars. 1885 E. H. Dering *Lady of Raven's Combe* I. ii. 20 Mick..is..very happy about everything, when he has made his Easter. 1892 *Month* May 37 Taking in those who have made their Easters at Melior St...we may count the Easters as 2000.
3. b. Easter-dues (examples); **Easter duty** (or duties), the religious duties (viz. of confession and communion) obligatory at Eastertide; **Easter-eggs** (earlier example); now chiefly, egg-shaped forms of confectionery presented at Easter; **Easter lily,** any of several species of (white) lily, or other spring-flowering plant (chiefly *U.S.*); **Easter sitting**(s) = *Easter term* (a); **Easter term,** (*a*) a term in the law-courts formerly movable and falling between Easter and Whitsuntide, now fixed within a certain period each year; (*b*) in the older universities, a term which was kept formerly between Easter and Whitsuntide, but which is now included in the Trinity term; in some universities and schools, the term between Christmas and Easter; **Eastertide, time,** the church season of Easter, either Easter Sunday, or the Sunday with the following days until Ascensiontide; **Easter week,** the week beginning with Easter Sunday.
1720 in *Jrnl. Derbysh. Archæol. Soc.* (1905) XXVII. 215 Easter dues 1. 11. 0. 1848 in J. T. Bunce *Old St. Martin's, B'ham* (1875) 35 Easter dues 4d. for a man and his wife, and 4d. for each single person above the age of 16, and 2d. from each housekeeper. 1723 S. Ll. *Gen. Instruct. Hist. & Tenets Relig.* ii. v. 256 What Punishments hath the Church decreed against those who have not perform'd their Easter Duty? 1809 J. Milner in F. C. Husenbeth *Life* (1862) viii. 166 To be particularly anxious that all should make their Easter duty. 1804 M. Wilmot *Let.* 11 May in *Russian Jrnls.* (1934) i. 97, I must not forget Easter Sunday..the service is the same, and after it is over Easter Eggs are presented painted and carv'd and decorated in a variety of ways. 1894 G. du Maurier *Trilby* I. i. 34 They would..marvel at the beautiful assortment of bonbons..especially, at this particular time of the year, the monstrous Easter-egg, of enchanting hue. 1949 A. Huxley *Let.* 6 Mar. (1969) 594, I am enclosing an Easter egg. 1877 Bartlett *Dict. Amer.* (ed. 4) The Calla is frequently called the Easter Lily. 1894 *Jrnl. Amer. Folk-Lore* VII. 101 *Zephyranthes Atamasco*,..Easter lily. 1896 T. W. Sanders *Encycl. Gardening* (ed. 2) 206 Lilium..Bermuda Easter Lily. 1909 Webster 693/2 *Easter lily*, *a.* The Annunciation lily (*Lilium candidum*). *b.* A large-flowered and early forcing variety ..of the common trumpet lily (*Lilium longiflorum*). 1932 *Discovery* Aug. 256/2 The Easter lily,..the royal lily, and the leopard or tiger lily of California. 1875 *Act* 38 & 39 *Vict.* c. 77. Sched. 1. Order lxi, The Easter sittings shall commence on the Tuesday after Easter week and terminate on the Friday before Whitsunday. 1910 *Law Times* 9 Apr. 509/2 On Tuesday last the Easter Sittings commenced with 145 appeals and 1514 causes awaiting hearing. 1913 Halsbury et al. *Laws Eng.* XXVII. 436 The third, Easter sitting. 1603 J. Stow *Survey of London* (ed. 2), *Of towers & castles*, Easter tearme beginneth not afore xvii. dayes after Easter. 1641 G. Cavendish *Negotiations Cdl. Woolsey* xi. 30 Untill such time as the Cardinall resorted thither to him, where after Easter terme was ended, he kept his feast of Whitsontide. 1672 Cowell *Interpr.* s.v. *Terme, Terminus Paschae*, Easter term, which begins the Wednesday fortnight after Easter-day, and ends the Monday next after Ascension-day. 1728 Chambers *Cycl.* s.v. *Term*, Oxford Terms...Easter-term begins the 10th day after Easter, and ends the Thursday before Whitsunday. Cambridge Terms...Easter-term begins the Wednesday after Easter-week, and ends the week before Whitsunday. 1818 Easter term [in Dict.]. 1905 H. A. Vachell *Hill* viii, Racquets, the chief game in the Easter term. 1930 *Law Times* 26 Apr. 373 The Easter Law Term will commence on..the 29th April, and will end on..the 26th May. *c* 1000 Ælfric *Hom.* I. 312 Nu is his ðrowung and his ærist ure Easter-tid. 14.. Eastertyde [in Dict., 3 a]. 1856 Smyttan *Hymn, 'Forty days'* vi, That with thee we may appear At the eternal Eastertide. 1868 Morris *Earthly Par.* ii. 213 If one chanced to fare Into that place at Easter-tide. 1885 Eastertide [in Dict., 3 a]. *c* 1440 Esterne tyme [in Dict., 3 a]. 1579 L. Vaux *Catech.* f. 77 Euery man and woman..should receiue the blessed Sacrament at Easter time. 1849 M. Arnold *Strayed Reveller* 104 'Twill be Easter-time in the world. *c* 1000 *Ags. Gosp.* John xx. i. *rubric*, Ðys sceal on sæternes dæȝ on þære easter wucan. *c* 1406 in G. R. Owst *Preaching Med. Eng.* (1926) i. 23 At Seint Marie Spitel, in Estir Weke. 1549 *Bk. Com. Prayer* 54 b, Tewesdaye in Easter weke. *a* 1558 G. Cavendish *Wolsey* (1959) 133 The thursday in Ester weke. *a* 1670 S. Collins *Pres. State Russia* (1671) 18 In the Easter week all his Majesties Servants and Nobility kiss the Patriarchs Hand, and receive either gilded, or red Eggs. 1728 [see *Easter term*]. *a* 1773 A. Butler *Moveable Feasts & Fasts Cath. Ch.* (1839) viii. 233 Every day during Easter week. 1884 Addis & Arnold *Cath. Dict.* 285/1 Down to the twelfth century each day in Easter week was a holiday of obligation.

easterliness (ī·stəɹlinės). [f. Easterly *a.*[2] + -ness.] Easterly quality or condition.
1927 J. Adams *Errors in School* 222 Each of them has acquired an absolute quality; Edinburgh of easterliness and Liverpool of westerliness.

easterly, *a.*[2] and *adv.* Add: **C.** *sb.* An easterly wind.
1901 F. T. Bullen *Sack of Shakings* 265 The brave west wind..being succeeded by baffling easterlies. 1955 A. L. Rowse *Expansion Elizabethan England* xi. 416 The English often had to wait for weeks for a favouring easterly.

Eastern, *a.* Add: **1. b.** Situated in, of or pertaining to, the (north-)eastern parts of the United States.
1776 *Jrnl. Cont. Congress* 24 Dec. VI. 1039 That the delegates of the eastern states confer together, and also those of the southern states. 1831 J. M. Peck *Guide for Emigrants* 250 Notice of their plans..was inserted in an Eastern periodical. 1837 *Southern Lit. Messenger* III. 373, I set out to make a tour of the Eastern States. 1885 *Century Mag.* Nov. 33/1 Mr. Tully..admitted willingly that he was an Eastern man—a Down East lumberman and boat builder. 1908 C. E. Mulford *Orphan* xiii. 153 At one time an Eastern woman had tried to live there, but..New York regained and kept its own. 1943 J. D. Hicks *Short Hist. Amer. Democracy* 334 Eastern farmers were unable to meet in full the needs of their new industrial centers for foodstuffs.
c. *the Eastern Shore*: that part of Maryland and Virginia lying between Chesapeake Bay and the ocean. Also *attrib.*
1624 Capt. J. Smith *Virginia* iv. 141 To plant the Secretaries land on the Easterne shore neere Acomack.

1777 *Maryland Jrnl.* 5 Aug. (Th.), [He] now has a family living near Choptank, on the Eastern Shore. 1785 Washington *Diary* 31 Dec. (1925) II. 461 Landed 230 Bushels of Oats to day from an Eastern Shore vessel. 1786 *Ibid.* 8 Apr. III. 39 Afterwards I sowed..the common Oat of the Eastern Shore. 1854 W. G. Simms *Southward Ho!* 179 In approaching the 'Eastern Shore' of Virginia..you find yourself gliding toward..scenes of repose, delicacy, and quiet beauty. 1948 Mencken *Amer. Lang., Suppl.* II. vii. iv. 128 On the Eastern Shore [of Maryland], south of the Choptank river, the dialect shows the influence of Tidewater Virginia...It is often difficult to say of a strange Eastern Shoreman..whether he comes from below the Choptank or above.

Easternism (ī·stəɹniz'm). [f. Eastern + -ism.] Eastern characteristics, practices, etc.; tendency to make Eastern in character.
1916 S. Graham *Through Russian Central Asia* i. 8 This year I go into the depths of the Russian East and..continue my study of Easternism and Westernism in the Tsar's Empire. 1921 *Glasgow Herald* 23 June 4 A new Easternism which shall enable the Eastern races to stand on a level with the other races of mankind. 1921 *Chambers's Jrnl.* 819/1 Cairo can show nothing more oriental than this; its [*sc.* Tetuan's] easternism..is nearly complete.

East Indian, *a.* and *sb.* Add: **A.** *adj.* **1.** (Later examples.)
1951 R. Mayer *Artist's Handbk. of Materials & Techniques* xii. 415 *East Indian Laurel*, a dark, reddish-brown wood with a wavy grain, strong and elastic. 1961 *B.S.I. News* Sept. 13/2 Specifications for oil of East Indian lemongrass.
B. *sb.* (Earlier examples.)
1801 Jane Austen *Let.* 8 Jan. (1952) 106 Mrs. Laurel is going to be married to a Mr. Hinchman, a rich East Indian. 1817 M. Edgeworth *Harrington* I. vii. 147 'Who is she?' 'An East Indian, I should guess, by her dark complexion.'
2. A man descended from one of the indigenous peoples of the Indian sub-continent, esp. if resident in the West Indies.
1931 *Everyman Encycl.* (ed. 2) III. 23/1 In 1928 the pop. [of British Guiana] was 307,784, which includes E. Indian immigrants. 1961 *Outlook* Feb. 214 East Indians is the name given to the descendants of those who came to British Guiana from India as indentured labourers between 1838 and 1917; the name also includes the few who themselves came from India. East Indians make up the largest group in British Guiana's total population of 540,000.

Eastlake (ī·stlēık). The name of Charles Locke *Eastlake* (1836–1906), English designer, used *attrib.* and *ellipt.* to denote furniture associated with his book *Hints on Household Taste* (see quot. 1878).
1878 C. L. Eastlake *Hints on Household Taste in Furniture* (ed. 4) p. viii, I find American tradesmen continually advertising what they are pleased to call 'Eastlake' furniture, with the production of which I have had nothing whatever to do, and for the taste of which I should be very sorry to be considered responsible. 1881 C. C. Harrison *Woman's Handiwork* iii. 141 There is lovely Queen Anne furniture, successor to the more ponderous 'Eastlake'. 1967 *Boston* (Mass.) *Sunday Herald* 9 Apr. (Show Guide) 15/1 Eastlake furniture..was and still is second-rate—a mixture of shoddy workmanship, and a debased Gothic style which was not true to Charles L. Eastlake's ideas.

East Side. *U.S.* [f. East D. 1 b + Side *sb.*[1] ii.] That section of New York City which lies on the east side of Manhattan (to the east of Fifth Avenue). Also *attrib.* Hence **Ea·st-si:der,** one who lives on the East side.
1882 J. D. McCabe *New York* iii. 79 This immense volume of travel is now being absorbed by the East Side Elevated Railroad. *Ibid.* li. 642 The original 'Bowery Girl'..passed at a quick gait peculiar to herself along the Bowery, or through Chatham square—it was the perfection of East Side poetry. 1894 C. D. Warner *Golden House* v, She was so full of sympathy with the East-Side work. 1898 *Westm. Gaz.* 28 Apr. 8/2 The Bowery brigade, recruited from..patrons of the eastside dime hotels and doss-houses. 1899 J. L. Williams *Stolen Story* 4 The busiest bank presidents..opened their mouths to him [*sc.* a reporter] quite as readily as East Side saloon-keepers. 1903 *N.Y. Even. Post* 21 Aug., The Health Commissioner to-day made a tour of the East Side. 1903 *N.Y. Tribune* 25 Oct., 15,000 East Siders attended the final dedication ceremonies at William H. Seward Park. 1957 *Times Lit. Suppl.* 11 Oct. 603/2 He was born to Italian parents, who brought him in infancy to New York, where he had the usual East Side immigrant upbringing.

eastward, *adv.* and *a.* Add: **A.** *adv.* **1. b.** *Comb.*
1911 M. Newbigin *Mod. Geogr.* iv. 90 The long, eastward-stretching, inland sea. 1914 G. A. J. Cole *Growth of Europe* viii. 157 The eastward-running rivers. 1915 E. Pound *Cathay* 14 And evening drives them on the eastward-flowing waters.
B. *adj. eastward position*: the position of the celebrant standing on the west side of the altar (and so facing east) in the Communion Service.
1873 J. B. Dykes *Let.* 5 July (1897) 306 The eastward position of the Celebrant. 1876 J. Harrison *Eastward Position* 9 The phrase 'eastward position' as descriptive of the posture claimed by some for the minister at the Communion Table is inaccurate, misleading, and deceptive. 1890 *Times* 5 Feb. 3/2 The third charge [against the Bishop of Lincoln]—relating to the eastward position—was contained in article 5. 1921 M. Baring *Passing By* 256 She said that the local clergyman was so low—no eastward position.

eastwardly, *a.* Add: **a.** Also, facing the east. **b.** (Earlier Amer. example.)
1703 *Essex Inst. Hist. Coll.* XLII. 360 Thence on a Straigh[t] lien to a heap of Stones on the Eastwardly sied of a hill. 1768 Washington *Diary* 2 Aug. (1925) I. 286 Wind Eastwardly—with appearances of Rain.

east wind. Add: **b.** In the game of mah jong the name given to one of the four tiles called winds, and to a disc representing this tile; hence, the player drawing this disc, who is the first to play.
1922 R. E. Lindsell *Ma-cheuk* 9 Four players make up a table, and seats are usually determined by chance, the four discs ('East', 'South', 'West', and 'North') being placed face-down on the table and each player drawing one in turn. The player who draws 'East' has choice of seats. *Ibid.* 25 In the East round, South has originally one East wind. 1923 *Daily Mail* 3 Mar. 7 Why 'East Wind' should have played his 'Red Dragon'. 1924 *Mah Jong Rules of Queen's Club* 9 'Jong' is always East Wind, the player on his right South. 1960 R. C. Bell *Board & Table Games* vi. 152 The tiles are grouped into..The East, South, West, and North Winds. *Ibid.* 153 The person drawing East Wind takes the seat in front of the tong and drops his disc into it. *Ibid.* 156 The first round is East Wind's, and continues until each player in turn has been East.

easy, *a.* and *adv.* Add: *adj.* **3.** (Later U.S. example.)
1902 O. Wister *Virginian* iv. 44 'She's easier this morning, since the medicine.' This was the engineer, whose sick wife had brought a hush over Medicine Bow's rioting.
8. b. *easy street*: comfortable circumstances, affluence. Esp. preceded by *on*. *colloq.* (orig. *U.S.*).
1901 G. W. Peck *Peck's Red-Headed Boy* iii. 18 This rich old Jew..who has been economical until he has got a million and is residing on easy street, will forget the traditions of two thousand years. 1902 G. V. Hobart *It's up to You* 31 A young man who could walk up and down Easy Street. 1903 A. H. Lewis *Boss* 205 Just as a sport finds himself on easy street. 1923 Wodehouse *Adv. Sally* xiv. 180 Honestly,..it's the chance of a lifetime. It would put you right on easy street. 1932 M. de la Roche *Master of Jalna* v. ix. 1011 'Are you sure you can spare it?' 'Good Lord! I hope so—after the sale! I'm in Easy Street.' 1938 F. Scott Fitzgerald *Let.* 25 Nov. (1964) 44 There will have to be a period of tough sledding before you come to Easy Street. 1957 L. P. Hartley *Hireling* viii. 60 He knew that she lived in Easy Street, but then so did most, if not all, of his customers.
9. c. In phrs. *easy to look at, easy on the eye(s)*, affording pleasure to the beholder; esp. of women: comely, beautiful. *colloq.* (orig. *U.S.*). Also in various combs., as *easy-to-make, easy-to-operate, easy-to-use, easy-to-wear*, etc.
1902 Ade *Girl Proposition* 141 He put his Tag on a blonde Canary 17 Years of Age who spelled Sure with an H and had from 7 to 9 Thoughts every 24 Hours. But she was very Easy to Look at. 1922 Wodehouse *Clicking of Cuthbert* x. 253 Her Highness is the easiest thing to look at these eyes have ever seen. 1937 C. Beaton *Diary* 3 June in *Wandering Years* (1961) 310 A pliable, easy-to-pose subject. 1937 *Punch* 8 Sept. 260/1 The same incorrigibly cheerful creature, very easy to look at, very pleasant to listen to. 1938 *Amer. Speech* XIII. 205 [The dictionary] is a substantial compilation, easy on the eyes, comparatively rich in idiomatic expressions. 1939 *Vogue's Cookery Bk.* i. 15 This is an easy-to-make soup for a hot day. 1943 D. E. Stevenson *Two Mrs. Abbotts* v. 35 Miss Walters was certainly easy on the eye. 1949 *Consumer Reports* Feb. 68/2 Many prospective purchasers want an easy-to-operate device. 1951 in M. McLuhan *Mech. Bride* 153/1 The same easy-to-follow lessons. 1958 *Oxf. Mail* 1 July 6/5 One of the good Westerns, easy on the eye and mind, and pretty tough on the pulses. 1959 *Times* 9 Mar. Suppl. p. x/3 Easy-to-serve packed foods. 1959 Easy-to-use [see *applicator]. 1959 *News Chron.* 10 Aug. 6/5 This hat has the soft easy-to-wear line. 1960 *Farmer & Stockbreeder* 15 Mar. 122/1 This tough, easy-to-use, real wood building panel. 1960 *Guardian* 25 Apr. 4/1 Easy-to-follow recipes.
10. c. Of water, etc.: not rapid, swift, or turbulent.
1875 'Mark Twain' *Old Times Mississippi* iii. 52 Follow along close under the reef—easy water there—not much current.
12. *easy game, mark, meat*: see sense *13 b.
c. Esp. in colloq. phr. *I'm easy*, I'm ready to comply (with whatever is proposed), without having any strong feelings (about the proposal); I don't mind one way or the other.
1941 in Baker *Dict. Austral. Slang* 27. 1945 in C. H. Ward-Jackson *Piece of Cake* (ed. 2) 26. 1948 D. Ballantyne *Cunninghams* (1963) ii. xviii. 80 'How about you, honey?' 'I'm easy.' 1968 'L. Marshall' *Blood on Blotter* xii. 84 You can believe 'em or not. I'm easy. *Ibid.* xxi. 145 You'll have to make up your own mind. I'm easy.
13. *easy money*: money obtained without effort, and, often, illegally. Also with an amount specified (*easy dollar*, etc.). orig. *U.S.*
1882 *Sydney Slang Dict.* 5/1 The money earned by a prostitute is said to be 'honest', as distinguished from that earned by a thief. Probably from the story of the converted burglar who determined to sin no more himself and who lectured against dishonesty, but sent his wife out every night to earn an *honest shilling*. '*Easy shilling*' is synonymous in the foregoing sense. 1896 Ade *Artie* x. 79, I guess it's easy money too from the way he lets go of it. 1909 'O. Henry' *Roads of Destiny* vi. 88 The boarding houses were corralling the easy dollars of the gamesome law-breakers. 1914 G. Atherton *Perch of Devil* i. i. 15 It may be this..spirit that inspires the midnight burglar.., not merely the desire for 'easy money'. 1923 Wodehouse *Adv. Sally* vi. 79 If you want to make a little easy money, you go and bet somebody ten seeds that I'm going to interrupt it again. 1935 *Discovery* Feb. 50/1 It is not to be assumed..that fox farming is a short cut to easy money. 1956 'J. Wyndham' *Seeds of Time* 100 To enable quick-turnover spivs to make easy money out of suckers. 1967 *Observer* 26 Mar. 9 Here the pursuit of happiness fuses with the pursuit of the easy dollar.
b. *easy meat* (colloq.): someone or something overcome, mastered, or persuaded without difficulty; anything compassed with ease. Similarly *easy game, mark.*
1896 W. C. Gore in *Inlander* Jan. 147 *Easy mark*, an easy prey to a joke. *Ibid.*, *Easy meat*, one easily duped. 1899 'Mark Twain' in *Harper's* Dec. 49/1 So I disguised myself and came back and studied you. You were easy game. 1927 T. E. Lawrence *Let.* 8 Dec. (1938) 557 That..made him easy meat for all the politicians. *a* 1928 in C. F. S. Gamble *North Sea Air Station* (1928) ix. 143 Had the Zeppelin been picked up by a search-light, it would..have been easy meat. 1929 Wodehouse *Gent. of Leisure* i. 9 If a man's fool enough to be an easy mark—. *a* 1935 T. E. Lawrence *Mint* (1955) i. xix. 66 The driver is an old sweat, not a rookie's easy meat. 1935 *Evening News* 29 June 3/1 An immense number of names has been invented for the victims [of confidence men]—..suckers, easy marks, wise guys, come-ons. 1943 *Archit. Rev.* XCIII. 23 All are easy meat, because a civilization is disintegrating and has lost its standards. 1944 W. E. Harney *Taboo* (ed. 4) 92 He had a gentle nature, ever out to please. He was the type that is known both to white and black as 'an easy mark'. 1955 *Sci. News Let.* 28 May 342/1 American pipelines are easy game for an enemy. 1958 *Times* 29 Oct. 3/2 The play was chosen, no doubt, on the principle that comedy is easy meat for the inexperienced. 1967 Partridge *Dict. Slang Suppl.* 1105/1 *Easy mark*, a girl easy to persuade into sexual intercourse: since ca. 1920.
14. c. *easy rider* (*U.S. slang*): (*a*) a sexually satisfying lover (see also quot. 1926); (*b*) a guitar.
1912–13 W. C. Handy *Memphis Blues*, Mr. Crump don't 'low no easy riders here. 1926 in R. de Toledano *Frontiers Jazz* (1947) iii. 37 'Rider', 'easy rider', which term means both lover and (not either, or) procurer... Fidelity to his woman is expected of the easy rider. 1927 *Jrnl. Abnormal & Social Psychol.* XXII. 16 'Easy rider'. This apt expression is used to describe a man whose movements in coitus are easy and satisfying. It is frequently met both in Negro folk songs and in formal songs. 'I wonder where my easy rider's gone', is a sort of by-word with Southern negroes. 1949 R. Blesh *Shining Trumpets* vi. 128 In rural Negro parlance..*easy rider* meant the guitar..carried suspended by its cord. In the double meaning of Negro imagery, the femininely formed guitar..typifies also a woman companion. In Negro 'city talk', the term *easy rider* has come to mean either a sexually satisfying woman or a male lover who lives off a woman's earnings. 1958 P. Oliver in P. Gammond *Decca Bk. Jazz* i. 24 For the blues singer, the most valuable instrument was the guitar,..and, as his 'easy rider', could be slung across his back when he wished to travel.
17. (Earlier and later examples.)
1836 D. Webster *Private Corr.* (1856) II. 21 The deposit and distribution bill has become a law, and money is already getting to be much easier, as the phrase is. 1848 Bartlett *Dict. Amer.* s.v., 'The money market is easy'; i.e. loans of money may easily be procured. 1870 J. K. Medbery *Men & Myst. Wall St.* 69 The lender seeks the borrower. Money becomes a drug. Technically it is 'easy' or 'inactive'. 1873 *Money Market* (ed. 3) 4 When the total supply of surplus money, unemployed in a reproductive manner, is abundant, the market is said to be 'easy'; when it is scarce, the market is 'tight'. 1875 *Economist* 2 Jan. 18/2 The prevailing anticipation of an easy money market. 1957 *Ibid.* 7 Dec. 867/2 The objective expert inquiry which Mr Eisenhower sought was swamped in the House when the advocates of easy money climbed aboard.
B. *adv.* **4. a.** Also, *to go easy* (*on* or *with*), to use sparingly; to act cautiously, to proceed with caution; also *absol.* (cf. quot. 1885 in Dict.); *easy does it*, go carefully, take your time.
1891 Farmer *Slang* II. 352/1 *Easy does it!* (popular). An exclamation of encouragement and counsel='Take your time and keep your coat on.' 1900 C. W. Winchester *Victories of Wesley Castle* vii. 143 You will have to go easy on that subject. 1928 J. P. McEvoy *Showgirl* 21 No high pressure stuff, sis. Easy does it with Dick. 1934 R. Macaulay *Going Abroad* xxx. 279 I've more or less gone easy on the powder and lipstick. 1935 'A. Bridge' *Illyrian Spring* xi. 143 Easy does it—to be easy was the thing. 1947 D. M. Davin *For Rest of Lives* xxii. 108 Go easy with that torch. 1955 L. P. Hartley *Perfect Woman* xxx. 272, I won't ask her yet what's the matter he decided. Easy does it. 1965 *Times Lit. Suppl.* 22 Apr. 317/1 A couple of foreign translations of my works should have warned me to go easy.
b. *sb.* (Examples of extended uses.)
1892 G. R. Lowndes *Camping Sk.* v. 161 We took a day's easy at Tyn-y-groes, by the salmon pool. 1922 T. E. Lawrence *Let.* 7 Sept. (1938) 365, I..read it in an easy, as we sat on the stye roof. 1928 *Observer* 19 Feb. 14 In the fourth [movement], being unable to switch off, I took an easy by thinking of something else.
c. *stand easy*: an order in military use

allowing a greater freedom of posture than 'stand at ease'.

1859 *Field Exerc. Infantry* 5 If the command to Stand-at-Ease is followed by the word Stand Easy, the men will be permitted to move their limbs, but without quitting their ground. 1883 *Ibid.* I. 6 On the word *Squad* being given to men standing easy, every soldier will at once assume the position of standing at ease. 1914 *Recruit Training (Infantry)* 5 Stand at Ease. Feet sufficiently apart. Easy position. Dressing maintained. Men perfectly still till 'Stand easy' given. 1920 Galsworthy *Foundations* III. 62 Form fours—by your right—quick march!..Left turn!..Stand easy!

C. c. easy-paced *a.*, *Cricket* and *Golf*, said of the ground or pitch when the ball comes at an easy pace off or along it; **d. easy-care**, used *attrib.* of (the properties of) man-made and other fabrics: convenient, serviceable (implying rapid drying after laundering, and crease-resistance); **easy-clean**, used *attrib.* of a fabric, etc., that is easy to clean.

1960 *Farmer & Stockbreeder* 12 Jan. Suppl. 2/1 The easy-care properties common to all synthetic fibre or man-made fabrics. 1962 J. T. Marsh *Self-Smoothing Fabrics* ii. 11 The production of 'easy-care' goods. 1963 *New Yorker* 8 June 88 Walking shorts, made expressly for us in cool, easy-care cotton-and-acetate seersucker. 1937 *Times* 5 Oct. 6/5 The easy-clean wheels have large hubs—a break from Daimler tradition. 1966 *Daily Tel.* 26 Oct. 13/2 Easy-clean fabrics for chair covers are not yet all available by the yard in the shops. 1905 *Westm. Gaz.* 25 July 8/2 Cotter, though he made the ball bump considerably at times, was scarcely suited by the easy-paced wicket. 1928 *Observer* 1 July 28/1 Batting first on an easy-paced pitch, Leicestershire began well against Warwickshire. 1959 *Times* 12 Sept. 3/4 Easy-paced greens.

easy, *v.* Restrict †*Obs.* to senses in Dict. and add: **c.** *intr.* Of an oarsman or crew: to cease rowing. **d.** *trans.* To give (an oarsman or crew) the order to stop rowing.

1852 J. F. Bateman *Aquatic Notes* iii. 32 The University steerer, supposing he had bumped them there, 'easied all', but his Crew, perceiving their mistake, pulled on again. 1881 *Rowing, Steering & Coaching on Cam* 25 All boats going down are supposed to give way to boats coming up—*i.e.*, to easy and pull in their oars. *Ibid.*, You must always easy for the 'Varsity trials in the October term. *Ibid.* 26 You must take care to easy some distance from where you want to stop. 1882 *Daily Tel.* 2 Mar. (Cassell), They..were not easied until reaching Iffley Lasher. 1890 S. Lane-Poole *Barbary Corsairs* II. xvi. 213 She..'easies' with every blade suspended motionless above the waves. 1959 *Times* 13 Mar. 18/1 King's easied opposite the Doves.

eat, *v.* Add: **1. f.** *colloq. fig.* To receive (esp. a stage performance) with gusto; to acclaim. Also *eat up*. (Cf. Devour *v.* 6.)

1911 L. Merrick *Peggy Harper* iv. 197 They *ate* the piece—it was only Galbraith they were guying. 1917 Wodehouse *Uneasy Money* iv. 23 I'm an English countess, doing barefoot dancing to work off the mortgage on the ancestral castle, and they eat me. 1919 F. Hurst *Humoresque* 195 You wait until you see the way they're going to eat me up in the court scene in 'Saint Elba'. 1928 'Ian Hay' *Poor Gentleman* iii. 58 The highbrow and pacifist reviewers there simply ate it [*sc.* the book] up, and said that if this was war, war ought to be stopped. 1949 N. Mitford *Love in Cold Climate* 261 London society..simply ate Cedric up, occasional echoes of his great success even reaching Oxford. 1958 K. Amis *I like it Here* 158 He held forth instead in a series of essays... The *Sunday Times* would absolutely eat this chap.

2. a. Also, *to eat dinners*.

1850 Thackeray *Pendennis* xxix, In term time, Mr. Pen showed a most praiseworthy regularity in performing one part of the law-student's course of duty, and eating his dinners in Hall. 1856 H. Mayhew *Gt. World of London* I. 72 Lawyerlings 'qualify' for the bar by eating so many dinners. [1867 *Cassell's Mag.* I. 287/2 These dinners he must eat in hall in his own person.] 1879 *Chambers's Jrnl.* 23 Aug. 539/2 No student shall be called to the Bar until he has eaten a certain number of dinners at his Inn. 1929 A. Waugh *Three Score & Ten* 71 The eating of dinners in the Temple, and the attendance of lectures.

g. *to eat dirt*: see Dirt *sb.* 6 c.

h. *to eat one's hat*: see Hat *sb.* 5 c.

3. d. *to eat out of another's hand*: to be completely submissive to a person, to be under another person's control.

1915 Conrad *Victory* II. vii. 139 He's like that—sometimes that familiar you might think he would eat out of your hand, and at others he would snub you sharper than a devil. 1921 H. S. Walpole *Young Enchanted* II. v. 185, I won a glorious victory and Victoria has eaten out of my hand ever since. 1957 J. Masters *Far, far the Mountain Peak* xxiii. 236 This reconnaissance would have him eating out of his hand before it was done. 1968 M. Carroll *Dead Trouble* iii. 53 Shaun won't turn me out now. I've got him eating out of my hand.

7. (Earlier and later U.S. examples.)

1837 *Crockett Almanac* 17 Well, Capting, do you ate us, or do we ate ourselves? 1842 *Spirit of Times* (Philad.) 4 Mar. (Th.), [The *Bay State Democrat* says that Mr. Dickens] has declined the invitation of the Philadelphians to eat him. 1855 M. Thomson *Doesticks* vii. 53, I resolved ..to quit the premises of the Emerald Islander who agreed to 'lodge and eat' us. 1889 Farmer *Americanisms* s.v., A steamer is alleged to be able to eat 400 passengers and sleep about half that number. 1928 S. V. Benét *John Brown's Body* 367 You ought to be et. We'll eat you up to the house when it's mealin' time.

8. d. Colloq. phr. *he won't eat you*, and varr.: he (we, they) will not injure or harm (the person addressed): an exhortation not to be alarmed.

1738 Swift *Polite Conv.* II. 126 Why, we won't eat you, Sir *John*. 1848 Trollope *Kellys & O'Kellys* III. vii. 161 Why, man alive, the ogre can't eat you! 1904 A. Bennett *Great Man* xxiv. 261 Don't be afraid. She won't eat you. 1949 F. Swinnerton *Doctor's Wife comes to Stay* 96 Don't be afraid. We shan't eat her, much as we should like to. You'll have her safe home again, with plenty dollars and the hearts of all our people over there. 1965 J. Symons *Belting Inheritance* ii. 23 You don't want to let Mamma worry you. She won't eat you, though she may look as if she will.

e. To disturb, vex. Also *intr.* in *to eat at*. (Cf. *Bite *v.* 13 b.) orig. *U.S.*

1893 S. Crane *Maggie* (1896) 90 'Well,' he growled, 'what's eatin' yehs?' 1904 W. H. Smith *Promoters* xvi. 238 What's eatin' 'em? Are they trying to hog the whole game? 1910 C. E. Mulford *H. Cassidy* xi. 74 'What's eating him, anyhow?' 'I don't know. I never saw him act that way before.' 1929 S. Anderson in *Mercury Story Bk.* 235 There was something else eating at me. 1934 R. Stout *Fer-de-Lance* xi. 158 You can answer my simple question without running a temperature. Was that what was eating you? 1938 E. Bowen *Death of Heart* II. iii. 232 'Now, what's eating him?' said Mr. Bursley. 1957 I. Cross *God Boy* (1958) xv. 118 'What's eating you?' asked Joe. 'Nothing,' I said.

16. eat off. b. To clear off (a crop) by feeding it to cattle: said also of the cattle. Also *intr.* (for *pass.*) of a crop.

1733 W. Ellis *Chiltern & Vale Farm.* 28 One of our best Farmers..eat off his Turneps early, and chalked his Ground well. 1764 *Museum Rust.* II. xxv. 79 We eat it [*sc.* coleseed] off with sheep..to make them fit for the butcher. 1841 *Jrnl. R. Agric. Soc.* II. I. 126 Nitrate of soda may be sown..on pasture that does not eat off regularly. 1921 W. E. Heitland *Agricola* 266 Stock must be kept on the farm, partly to eat off your own fodder-crops.

17. eat out. c. Also in phrases with *heart*.

1886 G. T. Stokes *Ireland of Celtic Ch.* xi. 212 Image worship..which had largely eaten out the heart of religion among them. 1890 [see Heart *sb.* 47]. 1890 *Review of Reviews* II. 323/1 General Boulanger is not eating out his heart in vain regrets. 1919 M. K. Bradby *Psycho-analysis* 70 Commonsense says that it is better..'to let off steam' than to 'eat your heart out'.

f. *intr.* To take a meal elsewhere than at one's residence.

1933 *Maclean's Mag.* 1 Aug. 35/1 They had planned to 'eat out', but presently Norma's chances of being 'discovered' in some smart café became completely nil. 1945 'L. Lewis' *Birthday Murder* (1951) iv. 47 Why don't we eat out tonight?. 1954 R. Macaulay *Let.* 8 July (1962) 159 Women..stay at home while the men eat out in cafés and restaurants.

18. eat up. g. *fig.* To traverse (a distance, ground) rapidly.

1898 H. S. Canfield *Maid of Frontier* 111 If I don't put my spurs into him an' make him eat up the groun'. 1905 R. T. Sloss *Bk. Automobile* 179 One of the keenest pleasures in possessing a car is being able to annihilate a hill or 'eat it up'. 1919 Wodehouse *Damsel in Distress* x, Ten minutes in the gray car ate up the distance between the links and George's cottage.

eat, *sb.* Delete †*Obs.* and add: **1.** (Later examples.) Now freq., esp. in *pl. colloq.*

1782 Mme. D'Arblay *Diary* 26 Oct. (1842) II. 158, I was too much tired to choose appearing at dinner, and therefore eat my eat upstairs. 1889 *Kansas Times & Star* 7 May, A majority..adjourned to the Coates House for 'eats' and refreshment. 1897 Kipling *Captains Courageous* ix. 213 How shall I take money when I make so easy my eats and smokes? 1912 W. Owen *Let.* 23 Apr. (1967) 130, I suppose I must thank you for the eats too. 1918 'Ian Hay' *Last Million* p. xiii, There is no ice-water, no ice-cream, no soda-fountains, no pie. It is hard to get the old familiar eats in our restaurants. 1955 J. P. Donleavy *Ginger Man* (1962) xxi. 214 On the tables were eats the like of which I'm sure have never been seen on this isle.

2. (Later examples.)

1844 J. T. Hewlett *Parsons & Widows* III. liv. 269 What was he to do 'between the eats'? 1904 *Westm. Gaz.* 20 May 10/1 One Tennessee innkeeper described his establishment as..25 cents a sleep, 25 cents an eat. 1951 J. Frame *Lagoon* 60 Goodbye and thank you for the little eat.

eatage. 2. (Earlier example.)

1843 [see Eddish 2 b].

eater. Add: **2.** A fruit that eats well, or is intended to be eaten uncooked (cf. Cooker 2).

1906 *N. & Q.* V. 156/1 The apples with none, or only small pips, were not so good as eaters or cookers. 1930 *Observer* 11 May 13/2 The blossom is as thick..on the hardy 'cooker' as on the shy and delicate 'eater'. 1953 E. Hyams *Vineyards in England* ix. 78 This [grape] is quite a good eater.

eatery (ī·tĕri). *colloq.* (orig. *U.S.*). [f. Eat + *-ery 2 b.] An eating-house.

1901 'H. McHugh' *Down the Line* 52 Muttheimer's is one of those eateries where the waiters look wise. 1923 Wodehouse *Inimit. Jeeves* i. 11 Why, then, was he lunching the girl at this God-forsaken eatery? 1927 *Daily Express* 30 Nov. 10/5, I think you served me with tea and toast at one of those cheap, colossal eating-houses in the West-End...No doubt you think that that clattering eatery..is 'life' and independence. 1938 M. Allingham *Fashion in Shrouds* xx. 366 Lugg and I have been round every fishy club and suspicious eatery in London. 1959 *Times* 26 May 8/1 His inability to make contact with a really good hunk of beefsteak in the eateries of Germany, Italy and France. 1959 *Vogue* June 124 Pull in at one of the curb-service eateries.

eau. Add: **eau-de-Cologne** (earlier example); **eau de Javelle,** see *Javelle; **eau-de-nil** (erron. **-du-**), **eau-de-Nil** [lit. 'water of (the) Nile'], a pale green colour supposed to resemble that of the Nile; **eau de Portugal,** a perfume comprising an essential oil known as essence of Portugal; **eau-de-vie** (examples); **eau sucrée** (o sükre), water with sugar in it.

1802 C. Wilmot *Let.* 19 June in *Irish Peer* (1920) 72 The Ladies most frequently have their Baths perfum'd with Eau de Cologne, Rose Water, or some perfume of that kind. 1870 *Young Ladies' Jrnl.* VII. 482 A pretty toilette of eau-du-nil. 1890 [see *crevette]. 1891 *Truth* 10 Dec. 1240/2 A row of begonia leaves in eau-de-Nil velvet. 1905 E. F. Benson *Image in Sand* ix, Four greens—*eau-de-nil*, *vert-doré*, aquamarine, and emerald—shone and shimmered together. 1928 *Times* 9 May 11/5 A robe de style of eau-de-nil tulle. 1960 *Times* 30 Jan. 7/4 *Eau-de-nil* and *tilleul* green. 1825 H. Wilson *Mem.* 43 'A little Eau de Portugal would do no harm..' I remarked,..alluding to his dislike of perfumery. 1848 Mrs. Gaskell *Mary Barton* I. vi. 104 Did not I get you that eau de Portugal from town. 1938 L. MacNeice *Zoo* xiv. 229 We poured a bottle of eau-de-Portugal over our heads. 1963 'M. Albrand' *Call from Austria* vii. 64 He smelled of Eau de Portugal, the lotion he used on his hair; he smelled a little like an orange. 1748 Smollett *R. Random* II. xlii. 54 We..were treated at breakfast with chocolate and *l'eau de vie* by our paramours. 1804 M. Wilmot *Let.* 24 Jan. in *Russ. Jrnls.* (1934) I. 78 The desert was in another room, dry'd fruits, Cakes, and *eau de vie*. 1840 Barham *Ingol. Leg.* 1st Ser. 168 Many a flaggon Of double ale..and *eau-de-vie*. 1846 Dickens *Pict. fr. Italy* 86 He..produces a jorum of hot brandy and water; for that bottle..now holds nothing but the purest *eau de vie*. 1911 E. M. Clowes *On Wallaby* vii. 184 Delicately burning his spoonful of eau-de-vie over his coffee. 1825 H. Wilson *Mem.* III. 69 The Frenchman..drank eau sucré, and studied in his dictionary. 1844 Thackeray *Misc. Ess.* (1885) 61 There was *eau sucrée* in the dining-room if the stalwart descendants of Du Guesclin were athirst. *a* 1845 Barham *Ingol. Leg.* 3rd Ser. (1847) 189 Lemonade, *eau sucrée*—and drinkables mild. 1872 E. Braddon *Life in India* viii. 314 Playing dominoes and drinking *eau sucrée*.

eaves. Add: **3.** *Comb.* *eave(s)-shoot, -spout, -trough* (designating various forms of gutter or spout to catch the drip from eaves); also *eaves-troughing*.

1889 'Herring' & 'Ross' *Irish Cousin* II. III. iv. 207 The noisy splashing of the water that fell from a broken eaveshoot on to the gravel. 1899 Somerville & 'Ross' *Exper. Irish R.M.* i, The rain sluiced upon me from a broken eaveshoot. 1846 in N. Eliason *Tarheel Talk* (1956) 270 Put up smart of the eve spout. 1865 H. B. Stowe *House & Home P.* 103 The water-barrel which stood under the eaves-spout. 1889 R. T. Cooke *Steadfast* xxxv. 369 A wild November storm shrieked and wailed in the eave-spout. 1851 H. Melville *Moby Dick* III. xxxv. 211 Same with cocked hats; the cocks form gable-end eave-troughs. 1878 B. F. Taylor *Between Gates* 176 Every day a wooden spout, a great eaves-trough was laid from the top of the steps. 1968 *Globe & Mail* (Toronto) 17 Feb. 53/3 (Advt.), Roofing & Eavestroughing. Chimney, eavestrough, roof repairs, free estimates, guaranteed.

‖**ébauche** (ebof). [Fr., sketch.] **1.** *Painting and Sculpture.* A sketch, an outline drawing; a maquette, a rough-hewn version. Also *fig.*, a first draft or attempt.

1722 J. Richardson *Statues in Italy* 310 The Work is extremely Good; not highly Finish'd, but rather a sort of *Ebauche*. 1841 J. S. Mill *Let.* Aug. in *Wks.* (1963) XIII. 483 Its deficiencies, as is usually the case with an *ébauche*, are chiefly in the introductory part. 1844 —— *Ess. Pol. Econ.* v. 126 As a mere anticipation or *ébauche* of a definition,..we do not quarrel with the received formula. 1873 C. M. Yonge *Pillars of House* III. xxviii. 107 His sketch..the hastily-designed and partly-coloured *ébauche*. 1958 *Times* 1 Feb. 8/5 A Deposition in our Sketch Club ..would not have been out of place among the *ébauches* of..Tintoretto.

2. *Horology.* An unfinished watch movement, consisting of plates, bridges, etc.

1929 G. H. Baillie *Watches* xii. 251 In England, the *movement in grey* consisted of an *ébauche* with train of wheels and pinions, all more or less in the rough. 1965 *Economist* 18 Dec. 1348/1 These [Swiss] firms make their own *ébauches* (the top and bottom plates of the movement and the train of gears and pinions within).

Ebenezer. Add: **1.** (Earlier and later Amer. examples.)

1693 C. Mather *Wond. Invis. World* 14 Many an Ebenezer has been Erected unto the Praise of God by his Poor People here. 1745 *Essex Inst. Hist. Coll.* VI. 181/2, I can..here set up my Ebenezer, & say that hitherto God has preserved me. 1840 *Knickerbocker* XV. 184 These things..were only inducements to him to set up his Ebenezer in this place. 1870 J. H. B. Nowland *Early Reminisc. Indianapolis* 86 At a love-feast, the old man said..he was able to raise his Ebenezer, and that the cloud had passed away.

b. Temper, passion. *U.S. slang.*

Evidently by a misunderstanding of 'raise'.

1836 *Public Ledger* (Philad.) 27 July (Th.), Says I, Deb, ..I'll send you wool enough to make a wig. That ris Deb's ebenezer. 1838 T. C. Haliburton *Clockm.* 2nd Ser. xxii. 333 If you go for to raise your voice at him,..his Ebenezer is up in a minit. 1849 D. Nason *Jrnl.* 14 Our Steward is under the constant necessity of a check-rein upon his ebenezer.

2. (Earlier and later examples.)

1849 Thackeray *Pendennis* I. xv. 133 Clavering New Church prospered...Their efforts had thinned the very Ebenezer hard by them. 1871 [see Zion].

‖**ébéniste** (ebenist). [Fr., a furniture-worker who makes use of ebony, f. *ébène* ebony.] An Ebonist, *spec.* a French cabinet-maker who veneers furniture (orig. with ebony).

1906 R. Glazier *Man. Hist. Ornament* (ed. 2) 133 The famous master *ébénistes* of the Renascence, of whom the best known is André Boule. 1934 *Burlington Mag.* Oct. 157/1 The first maîtres ébénistes in Paris are mentioned in 1638, and from about that date ebony became one of the most appreciated kinds of 'bois des Indes' for furniture in Europe. 1955 *Times* 11 July 3/4 The French furniture.. will include many fine pieces signed by the noted *ébénistes* M. Carlin, N. Petit, R. Lacroix, and C. C. Saunier.

Hence **ébénisterie** (ebenistri), cabinet-making in the manner of the *ébénistes*.

1910 *Encycl. Brit.* IV. 322/1 Boulle..had not only inherited the rather flamboyant Italian traditions of the late Renaissance, but had *ébénisterie* in his blood. 1945 *Burlington Mag.* Oct. 260/2 The bulk of the ébénisterie was carried out by a certain George.

Ebionitism. (Earlier examples.)

1780 N. Lardner *Hist. Heretics* I. 20 Epiphanius's Introduction to his Account of Ebionitism must be allowed to be a remarkable instance of harshness, not to say railing. 1856 Bagehot *Coll. Works* (1965) I. 388 Strange confused beliefs, Millenarianism, Gnosticism, Ebionitism, were accepted.

E-boat. Used, esp. in the war of 1939–45, as an abbreviation for an enemy torpedo-boat.

The view, frequently expressed, that *E* represents G. *Eile* 'speed' is purely speculative. The usual G. word for a speedboat is *Schnellboot*.

1940 *Hutchinson's Pict. Hist. of War* 7 Aug.—1 Oct. 16 One of our motor torpedo-boats, thinking that the enemy ahead was an E-boat, and being too close to take action, rammed the enemy. 1941 *Times Weekly* 30 July 7/3 One ship of the convoy received damage from an E-boat attack. 1958 *Listener* 30 Oct. 691/1 E-boat scares on the convoy routes.

Eboe. Add to def.: The same term as *Ibo* in West Africa, but used with wider application in the W. Indies.

ebony. Add: **4.** *ebony skin* (earlier example).

1823 C. Mathews *Let.* 23 Feb. in *Memoirs* (1839) III. 388 A Hottentot Adonis appeared, with..an effluvia arising from his ebony skin.

b. A Negro. *U.S.*

1851 H. Melville *Moby Dick* II. xxii. 165 The old black..came shambling along from his galley;..this old Ebony floundered along. 1863 'E. Kirke' *My Southern Friends* iv. 69 The scented ebony roared. 1889 Farmer *Americanisms* s.v., An *ebony* is a negro in common parlance.

ebullioscopic (ĭbɒ·lioskǫ·pik), *a.* [f. Ebullioscope + -ic.] Of or pertaining to the ebullioscope or ebullioscopy. So **ebullioscopy** (ĭbɒliǫ·skŏpi), the study and use of the ebullioscope.

1902 *Nature* 4 Sept. 436/2 Molecular weight determinations in liquid nitrogen peroxide by the ebullioscopic method. 1914 *Jrnl. Chem. Soc.* CVI. II. 170 Of the following five formulæ for calculation of the ebullioscopic constant. 1914 *Jrnl. Amer. Chem. Soc.* XXXVI. II. 1411 The Constants of Ebullioscopy and Cryoscopy. 1922 *Jrnl. Chem. Soc.* Abstr. II. 117 Extension of ebullioscopy and its application to binary mixtures. 1948 Glasstone *Physical Chem.* (ed. 2) ix. 636 The proportionality constant K_e is called the molal elevation constant, or the ebullioscopic constant, and..it can be seen to be equivalent to the boiling point elevation for a solution of unit molality. *Ibid.* 637 The experimental procedure is often described as the ebullioscopic method for the determination of molecular weights, and the whole subject is referred to as ebullioscopy. 1956 *Nature* 21 Jan. 128/1 Molecular weight determination (by ebullioscopic method in methanol) gave 445.

ecad (i·kæd). *Ecology.* [f. Gr. οἶκ-ος house + -ad.] An organism modified by its environment.

1905 F. E. Clements *Res. Methods Ecol.* iii. 148 On the basis of their..behavior in the production of new forms, species may be distinguished as variable, mutable, or adaptable. The new form which results from variation is a *variant*; the product of mutation is a *mutant*, and that of adaptation, an *ecad*. 1920 *Nature* CIV. 708/1 The so-called species of the mycologist is comparable with the 'ecad' of the ecologist, and is the resultant of the organism and its environment. 'Ecads' indistinguishable from each other may be produced from two distinct organisms interacting with one and the same environment, or with two different environments. 1964 V. J. Chapman *Coastal Veget.* ii. 47 Another free-living fucoid, *Ascophyllum nodosum* ecad *scorpioides* is to be found..on the salt marshes of the western Scottish lochs. 1970 *Watsonia* VIII. 26 It is concluded that *P. pseudodistans* and *P. fasciculata* are ecads of one species.

écarté. Add: **2.** *Ballet.* (See quot. 1957.)

1922 Beaumont & Idzikowski *Man. Theory & Pract. Class. Theatr. Dancing* I. 29 (*caption*) Position of the body in the direction of the body termed écarté. 1957 G. B. L. Wilson *Dict. Ballet* 109 *Écarté*,..a position of the body, oblique to the audience, in which the arm and the leg, on the ground or raised nearest to the audience, are extended on the same vertical and diagonal plane as the rest of the body. 1960 J. Lawson *Classical Ballet* vi. 148 This *écarté* position for the commencement of a turn is extremely important when the dancer has to circle the stage.

ecca, var. *ekka.

ecchondroma (ekǫndrōu·mă). *Path.* [mod.L., f. Gr. ἐκ out + *chondroma.] A chondroma growing outwards from the surface of a bone or cartilage.

1890 in Billings *Med. Dict.* 1897 *Allbutt's Syst. Med.* IV. 826 Ecchondromas are usually firmly attached, hard, sessile growths. 1929 Wakeley & Buxton *Surg. Path.* xv. 124 Chondroma..may burst through the shaft and become pedunculated (*ecchondroma*). 1954 G. P. Wright *Introd. Path.* (ed. 2) xxix. 537 Chondromas arise most frequently in bony structures, and are termed enchondromas or ecchondromas according to their internal or external situation.

Eccles (e·k'lz). [Name of a town in Lancashire.] *Eccles cake*, a kind of fancy cake.

1872 *Young Englishwoman* Nov. 620/1 Can you tell me how to make Bride Cake and Eccles' Cakes? 1881 E. Skuse *Confect. Hand-Bk.* (ed. 3) 155 Eccles Cake. Roll out a sheet of paste..about a quarter of an inch thick, then roll out another sheet same size,..spread on the first sheet some Banbury meat..then cover it with the second sheet. 1889 *Manch. Sch. Board Cookery Classes* 38 Eccles Cakes. 2 oz. brown sugar, 1 oz. butter, ¼ lb. currants, 1 oz. candied peel, a little grated nutmeg and lemon rind. 1908 J. Kirkland *Mod. Baker* III. 354 Eccles cakes are made in nearly the same manner and with the same ingredients as Banburys, only the usual shape is round. 1957 *Encycl. Brit.* VII. 881/1 Eccles cakes, made of pastry with currants, have a wide reputation.

ecclesial, *a.* Delete †*Obs.* and add later examples.

1961 B. R. Wilson *Sects & Society* III. xii. 250 The significant charismatic element at work at the ecclesial level. 1966 J. D. Crichton in *Studia Patristica* (Internat. Conf. Patr. Stud.) VIII. 208 The pleas for liturgical reform..achieved an ecclesial dimension.

ecdemite, var. *ekdemite.

1882 J. D. Dana *Man. Min.* (ed. 4) 152 Ecdemite, a lead chloro-arsenite. 1951 C. Palache et al. *Dana's Syst. Min.* (ed. 7) II. 1036 Found as crystals and masses associated with heliophyllite in manganoan calcite at Långban, Wermland, Sweden. Ecdemite apparently also constitutes the uniaxial parts of the intergrowths with heliophyllite found..in Wermland.

ecdysiast (ekdi·ziæst). [f. Gr. ἔκδυσις (see Ecdysis) + -αστής, after type ἐνθουσιαστής.] A strip-teaser. Hence **ecdy·siasm**, the activity or occupation of strip-teasing. Also *fig.*

1940 Mencken *Let.* 5 Apr. in *Amer. Lang.* Suppl. I. (1945) VI. vi. 585 It might be a good idea to relate strip-teasing in some way..to the associated zoölogical phenomenon of molting...A resort to the scientific name for molting, which is *ecdysis*, produces both *ecdysist* and *ecdysiast*. 1947 I. Brown *Say the Word* 50 And there we can leave the problems and pride of ecdysiasm. 1957 *Times Lit. Suppl.* 11 Oct. 611/2 As happily sustained a piece of autobiographical ecdysiasm as we remember. 1958 *Time* 7 July 44/1 Is it possible for a nightclub to lose money with famed Ecdysiast Sherry Britton stripping to bugle beads and pearls?

ecdysone (e·kdisōun). Also **ecdyson.** [ad. G. *ecdyson* (P. Karlson 1956, in *Ann. des Sci. naturelles* (*Zool.*) XVIII. 133), f. Gr. ἔκδυσις: see Ecdysis.] A steroid, present as a hormone in the young forms of insects and some other arthropods, that controls moulting.

1956 *Vitamins & Hormones* XIV. 234 The name ecdysone,..pointing to the significance of the active substance for all molting processes, was proposed..for the pure, chemically defined, substance after isolation of the prothoracic gland hormone. 1961 *New Scientist* 2 Nov. 279/3 The moulting hormone (ecdyson) is produced at intervals by the thoracic glands under the influence of another hormone secreted by certain cells in the brain. *Ibid.*, Adults do not normally produce any ecdyson, but if this is provided artificially they can be induced to moult again. 1968 *Nature* 9 Nov. 601/2 The ecdysones control moulting in three diverse classes of arthropods besides insects.

ecesis (eki·sis). [f. Gr. οἴκησις act of inhabiting.] Of plants, the entry into, or establishment in, another habitat.

1904 F. E. Clements in *Bot. Surv. Nebraska* VII. 50 (*Cent. Dict.* Suppl.), In a word, Ecesis is the adjustment of a plant to a new habitat. It comprises the whole process covered more or less incompletely by acclimatisation, naturalisation, accommodation, etc. 1938 Weaver & Clements *Plant Ecology* (ed. 2) i. 3 Migration alone, however, cannot produce vegetation...The migrants must make themselves at home, an idea that is expressed in a single term *ecesis*. *Ibid.* vi. 131 Ecesis is the adjustment of the plant to a new home. 1952 P. W. Richards *Tropical Rain Forest* xvii. 383 The ecesis of secondary species is always effected by seeds or fruits.

Hence **ecesic** (eki·sik) *a.*, of, or belonging to, ecesis.

1916 F. E. Clements *Plant Succession* 63 (*title*) Ecesic Causes.

ecgonine (e·kgǫnin). *Chem.* [ad. G. *ecgonin* (F. Wöhler 1862, in *Ann. d. Chemie u. Pharm.* CXXI. 372), f. Gr. ἔκγονος, f. ἐκ out of, from + γον-, γεν- to produce + -ine[5].] An alkaloid base, $C_9H_{15}NO_3$, obtained by the decomposition of cocaine. Hence **ecgonic** (ekgǫ·nik) *a.*, defining an acid obtained from ecgonine; **e·cgonate**, a salt of this.

1864 Watts tr. *Gmelin's Hand-Bk. Chem.* XVI. 304 Chloroplatinate of Ecgonine. A mixture of hydrochlorate of ecgonine with chloride of platinum and strong alcohol. 1885 *Jrnl. Chem. Soc.* XLVIII. II. 913 It would seem that cocaine, ecgonine, and isotropine are derivatives of ethyl tetrahydropyridine. 1891 *Jrnl. Chem. Soc.* LX. I. 749 The acid $C_7H_{11}NO_3$, obtained by the oxidation of lævo-ecgonine, is termed ecgonic acid. *Ibid.*, Ethyl ecgonate, $C_7H_{10}EtNO_3$, is a colourless, viscid liquid. 1901 *Ibid.* LXXX. I. 291 Ecgonic acid..crystallizes from ethyl acetate or benzene in colourless needles. 1920 [see *diamorphine]. 1957 *Encycl. Brit.* V. 908/2 On hydrolysis with mineral acids or baryta, cocaine breaks up into *ecgonine* (tropine carboxylic acid), benzoic acid and methyl alcohol...All the 'cocaines' are found to be derivatives of a simple base, ecgonine, which contains both a hydroxyl (.OH) and a carboxyl (.COOH) group.

echard (e·kård). *Ecology.* [f. Gr. ἔχειν to hold, with-hold + ἄρδεια irrigation.] Water in the soil which is not available to plant roots.

1905 F. E. Clements *Res. Methods Ecol.* ii. 32 The terms, physiological water-content, and physical water-content, are awkward...It is here proposed to replace them by short words which will refer directly to the availability of the soil water for absorption...The terms suggested..are respectively, *holard*..*chresard*..*echard* (ἔχω, to withhold). *Ibid.* 32 This is the non-available water, or the echard. 1926 Tansley & Chipp *Study of Vegetation* vii. 127 When the plants show signs of permanent wilting..the amount [of water] which the soil then contains is called the *echard* (water held back by the soil from the plant). 1938 Weaver & Clements *Plant Ecology* (ed. 2) viii. 203 The water retained in the soil at the time of permanent wilting is nonavailable for growth or the *echard* (to withhold). 1960 N. Polunin *Introd. Plant Geogr.* x. 301 It is often necessary to distinguish between water which is available to plants.. and that which is so strongly held as to be unavailable (*echard*).

echelon. Add: **1. b.** Each of the subdivisions in the rear of the main supply service for troops in warfare.

1922 *Encycl. Brit.* XXXII. 493/2 One corps..was subdivided at all échelons into telephone units, wireless units, and visual units 1924 L. E. Vining *Held by Bolsheviks* 88 People are coming to me in numbers asking me to get their trains and echelons away. *Ibid.* 104 Each echelon commander is demanding to have his train put on the west-bound track. 1948 A. Baron *From City, from Plough* xvii. 155 There was a clutter of signboards.. with the emblems of battalions and brigades blazing a trail for their rear echelons to follow. 1955 *Times* 7 June 6/5 Their transport echelon, supplying them from the comparative comfort of the reserve, is now known as 'the passive wing'.

c. *N.Z.* (See quot. 1941.)

1941 Baker *N.Z. Slang* iii. 24 The New Zealand application, dating from 1939, of *echelon* to a division of an Expeditionary Force..represents a new meaning applied to an old term. 1947 C. V. Smith in J. Reid *Kiwi Laughs* (1961) 152 They are only a few short of the number that went overseas with the first Echelon.

3. *transf.* A grade or rank in any (esp. civil) administration or profession. orig. *U.S.*

1950 in M. McLuhan *Mech. Bride* (1951) 2/3 'The catalogue will be confined to posts in the higher echelons,' said Mr. Ross. 1951 *Here & Now* (N.Z.) May 20/2 The United States delegation on any one day may, on the other hand, amount to as many as 100 people (with reserve echelons of advisers, experts, consultants and translators in Washington). 1957 P. Frankau *Bridge* 203 Do educationalists get as drunk when they convene as the lower echelons do? 1958 *Times* 12 Jan. 5/3 The higher echelons of the law and politics have become almost an Oxford preserve. 1969 *Daily Tel.* 8 Feb. 15/1 A security leak in the top echelon of the Civil Service.

echeloned, *ppl. a.* Add: Also *transf.*

1924 Skerl tr. *Wegener's Orig. Cont. & Oceans* iv. 65 The Cordilleras..run from south to north along the coast, and terminate in echeloned folds which retreat successively westwards.

echinate (e·kinĕit), *v.* Restrict *nonce-wd.* to sense in Dict., and add: **b.** *trans.* Of a sponge spicule: to project from (the fibre) at an acute angle. So **e·chinating** *ppl. a.*

1882 *Cassell's Nat. Hist.* VI. 328 The skeleton is.. composed of chief spicules lying parallel to form a fibre, which is spined by other (echinating) spicules projecting from it. 1900 E. R. Lankester *Treat. Zool.* II. 140 The spicules so placed are said to 'echinate' the fibre. 1940 L. H. Hyman *Invertebrates* I. vi. 347 Spongin fibers may have both enclosed ('coring') and echinating spicules.

echino-. Add: **echi·nochrome,** a brown or yellowish brown pigment found in some echinoderms; **echi:nococco·sis** *Path.*, disease caused by infection with tapeworms of the genus *Echinococcus*; hydatid disease; **echi:no-plu·teus** *Zool.*, the free-swimming larval form of an echinoid.

1883 C. A. MacMunn in *Proc. Birmingham Philos. Soc.* III. 380 *Echinochrome*.—It is unfortunate that as one meets with new colouring-matters long names have to be invented to distinguish them from other colouring matters. This has been the case with the pigment which I have now to mention, and for which I propose the name of *echinochrome*. 1886 *Jrnl. R. Microsc. Soc.* 48 Dr. C. A. MacMunn describes the spectroscopic or chemical characters of the blood of various worms and molluscs; one of the most interesting pigments which he has detected is that which he calls echinochrome..obtained from the perivisceral cavity of *Strongylocentrotus lividus*. 1912 *Jrnl. Chem. Soc.* CII. I. 520 Echinochrome is probably held in the same way as chlorophyll is held in the plant cell. 1959 *Chambers's Encycl.* III. 763/1 They [*sc.* naphthoquinones]..are apparently limited to sea-urchins. Many such species display conspicuous amounts of naphthoquinoid echinochromes in their thin skin and in the calcareous material of spines and shells, which thus assume red, purple, pink or greenish colours. 1900 *Veterinarian* Dec. 655 The liver was an interesting specimen of echinococcosis. 1961 Chandler & Read *Introd. Parasitol.* (ed. 10) xv. 362 The malignant alveolar form of echinococcosis known from North and Central Europe..is caused by a distinct species, *E. multilocularis*. 1909 *Cent. Dict.* Suppl., *Echinopluteus*, the pluteus, or free-swimming larva, of a sea-urchin. 1913 T. Mortensen in *Jrnl. Marine Biol. Assoc.* X. 16 (*caption*) Part of the skeleton of the *Echinopluteus* of *Spatangus purpureus*. 1932 L. A. Borradaile et al. *Invertebrata* xviii. 556 The 'posterolateral arms'..if they are present in the *Echinopluteus*, are there small. 1955 L. H. Hyman *Invertebrates* IV. xv. 490 The echinopluteus occurs in an infinite variety of shape and structure. 1967 P. A. Meglitsch *Invert. Zool.* xi. 409/1 The echinopluteus develops as the gastrula becomes flattened on the future oral surface.

echitamine (eki·tămin). *Chem.* [ad. G. *echitamin* (O. Hesse 1878, in *Ber. d. Deut. Chem. Ges.* XI. 1548), f. mod.L. *echītēs* generic plant-name + Amine.] An alkaloid found in the bark of *Echites scholaris* and other species of *Echites*. So **echita·midine, echi·teine, echi·tenine, e·chitine** (see quots.).

1876 *Jrnl. Chem. Soc.* XXIX. 277 Echitein crystallizes from hot strong alcohol in light needles. *Ibid.*, Echitin, $C_{32}H_{52}O_2$, forms white scales. 1879 *Ibid.* XXXVI. 71 Dita bark (the bark of *Alstonia scholaris* or *Echites scholaris*) contains two alkaloids, ditamine and echitamine. 1881 *Ibid.* XL. 185 The action of hydrochloric acid on echitamine hydrochloride. 1886 *Buck's Handbk. Med. Sci.* II. 519/2 Three alkaloids, ditamine, echitamine, and echitenine. 1906 *Watts' Dict. Chem.* II. 413/1 Echitin..accompanies echicerin, from which it differs in being less soluble in ligroïn. 1932 *Jrnl. Chem. Soc.* II. 2626 (*title*) Echitamine in Alstonia Barks. Barks of various species of *Alstonia* have been examined in the hope of finding a source of echitamine richer in that alkaloid than is the bark of *A. congensis*. *Ibid.*, A new crystalline alkaloid, $C_{20}H_{24}O_2N_2$, which it is proposed to call *echitamidine*, has been obtained..from the mother-liquors from echitamine hydrochloride.

echium (e·kiəm). [mod.L. (J. P. de Tournefort in C. Linnæus *Genera Plantarum* (1737) 38), f. Gr. ἔχιον the name used by Dioscorides, f. ἔχις viper, in allusion to the spotted stem.] A plant of the genus of shrubs or herbs so called, belonging to the family Boraginaceæ and native to Europe; cf. *viper's bugloss* (Viper 5 b).

1883 W. Robinson *Eng. Flower Garden* 118/2 These five species..are representative of the annual and biennial Echiums. 1924 W. Deeping *Three Rooms* v. 40 Later, the spikes of the echium would make a wonderful and rich blueness. 1963 W. Blunt *Of Flowers & Village* 267 There were some splendid echiums on Teide.

echiuroid (ekiyūə·roid), *a.* and *sb.* [f. mod.L. *Echiurus*, generic name, f. Gr. ἔχις adder + οὐρά tail + -oid.] **A.** *adj.* Of or pertaining to marine worms of the phylum *Echiuroidea* or genus *Echiurus*. **B.** *sb.* A member of this phylum or genus. So **echiu·rian, echiu·rid.**

1886 *Jrnl. R. Microsc. Soc.* Dec. 984 The armed Gephyrea or Echiuroids. *Ibid.* 985 The nerve-trunk has in adult Echiurians lost all trace of segmentation. 1897 *Q. Jrnl. Microsc. Sci.* Dec. 367 A new British Echiuroid Gephyrean. 1902 *Encycl. Brit.* XXVII. 625/2 The female [of *Bonellia*] has the normal Echiuroid structure. 1909 *Jrnl. R. Microsc. Soc.* 50 New Deep-sea Echiurid. 1941 J. Steinbeck *Log from Sea of Cortez* (1958) 273 *Echiuroid*, a worm-like animal related to the sipunculids. 1967 P. A. Meglitsch *Invertebrate Zool.* xv. 646/2 Echiuroid nephridia are much like the nephridia of sipunculids. *Ibid.* 648/1 The sedentary life of the echiuroids is reflected in the simplicity of the neuro-sensory equipment.

echo, *sb.* Add: **1. c.** *Telephone Engineering.* (See quots.) Also *attrib.*

1925 *Electr. Commun.* IV. 40/1 Echo effects are caused by reflections of voice waves which take place whenever electrical irregularities are encountered in telephone circuits. 1959 *Post Office Electr. Engineers' Jrnl.* LII. 69/1 A problem which continually confronts the communication engineer is the avoidance of echoes. *Ibid.*, These echoes arise from reflections occurring at unavoidable electrical discontinuities.

d. In full *radio echo*. A radio wave which has been reflected or otherwise returned with sufficient magnitude and delay to be perceived as a wave distinct from that transmitted (see also quot. 1941). Also *attrib.* and *fig.*

1928 C. Størmer in *Nature* 3 Nov. 681/1 (*heading*) Short wave echoes and the Aurora Borealis. *Ibid.*, On Feb. 29 of this year I received a letter from Engineer Jørgen Hals..in which he says: '..I repeatedly heard signals from the Dutch short-wave transmitter station... At the same time as I heard the telegraph-signals I also heard echoes.' *Ibid.* 17 Nov. 768/1 (*heading*) Radio echoes and magnetic storms. 1930 *Times* 25 Mar. 11/4 Neither..appear to attach very much importance to these 'echo' effects...The effect of these 'echoes' at the receiving station can..be overcome by a directive aerial system which does not accept radiation from the backward direction. Echo signals may..also be produced by radiation leaving the transmitter..and travelling more than once round the world. 1941 *B.B.C. Gloss. Broadc. Terms* 10 Echo. (1) Repetition of a sound after an interval of time, caused for example by reflection of sound waves, or occurring as a result of propagation of electromagnetic waves over more than one path. (2) Reverberation artificially added to the output from a studio or hall. 1965 *New Statesman* 11 June 904/3 The hire-purchase controls just introduced, small enough not to induce substantial 'echo' effects in the future.

e. A radar wave which has been reflected from an object; also, the representation of it on a radar screen.

1944 [see *echolocation]. 1945 *Electronic Engin.* XVII. 719/2 The measurement of interval between pulse and echo on the time base of the C.R. tube. 1947 etc. [see *angel *sb.* 10]. 1953 R. Chisholm *Cover of Darkness* iii. 36 In the air the 'noise' (interference) was usually worse and often the echo was less clear. On the elevation (left-hand) tube there is slightly more of the echo above than below the horizontal trace.

8. Also in *Bridge*, a signal to one's partner, esp. by the playing of a higher card of a suit followed by a lower, indicating how many cards of a suit are held, or requesting a specific lead.

1899 A. Dunn *Bridge* 51 The plain suit echo. 1902 —— *New Ideas on Bridge* 80 A player must have either led or 'called for' trumps before his partner gives the 'three-trump echo'. 1939 N. de V. Hart *Bridge Players' Bedside Bk.* 140 High–low play as a signal to the partner is termed a *peter* if it occurs in a suit contract and an echo at No-Trumps. *Ibid.* 141 There is..no longer any difference between a peter and an echo. 1960 T. Reese *Play Bridge with Reese* iii. 20, I should imagine that West has led from four hearts and the diamonds may well be 3–3 since no one began an echo.

9. *echo-detection, device, mixture, -phrase, -question, -ranging, -reflex, -word*; **echo box** (see quot.); **echo chamber** orig. *U.S.*, a confined space where sound reverberates; also *attrib.* and *fig.*; **echo organ**, a set of pipes in an organ, enclosed in a wooden box to give a distant sound effect; **echo-plummet** [*Echo *sb.* 1 d], a radio sounder for measuring height above ground level or the depth of water in the sea; **echo room** = **echo chamber*.

1950 *Gloss. Terms Radar (B.S.I.)* 7 *Echo box*, a cavity resonator, having small damping, energized by pulses of energy radiated from a nearby aerial or by a probe in a waveguide. 1937 *Variety Radio Directory* 1937/38 341 *Echo chamber*, a reverberant room used to add hollow effects and actual echoes. 1953 *New Yorker* 4 Apr. 6/3 There's free-wheeling music in the echo-chamber grill. 1955 *N.Y. Times* 16 Jan. VI. 26/4 New York, an echo chamber of squeezed apartment buildings, is even more in need of a honking ban than..Paris. 1958 *Punch* 22 Jan. 152/1, I was going to write about echo-chambers, which are currently giving such pleasure to all listeners by piping the broadcast voice into an empty, thirty-foot room and piping it out again with sonorous accretions of overtones. 1947 Crowther & Whiddington *Science at War* iv. 156 The numerous break-downs in service almost caused the use of the echo-detection to be dropped by the Navy. 1933 *Discovery* June 174/1 Use will also be made of an 'echo' device which measures the depth of the sea by the time which the echo takes to return from the bed. 1962 A. Nisbett *Technique Sound Studio* xi. 192 When echo is wanted on only one voice, as..in a conversation between two people, one of whom is at the bottom of a well, control over this can be exercised with the echo-mixture switch. 1855 E. J. Hopkins *Organ* xv. 66 The Echo organ consists of a duplication of the treble portion of some of the stops found on the other manual organs, closed in a wooden box, to render their tone soft and more distant-sounding. 1884 *Encycl. Brit.* XVII. 830/1 The fifth manual, where it occurs, is the echo organ. 1938 *Oxf. Compan. Mus.* 659/1 In certain very large instruments there is a manual devoted to delicate stops, with a far-away sound, and this is called the Echo Organ. 1962 Y. Olsson in F. Behre *Contrib. Eng. Syntax* 90 The echo-phrase 'We have'. 1931 *Flight* 27 Mar. 274/2 The echo-plummet used extensively for maritime navigation seems to encounter certain difficulties in aeroplanes. 1933 *Discovery* Oct. 307/1 The well known acoustic echo plummet which is employed in measuring ocean depths. 1956 *Kenyon Rev.* XVIII. 433 Rising juncture.. signals an 'echo question'. 1961 *New Scientist* 11 May 304 To point the beam out from the side of the ship (echo-ranging) and tilt it down a few degrees below the horizontal. 1957 D. L. Bolinger in *Publ. Amer. Dial. Soc.* XXVIII. 26 The difference between yes–no echo-reflex and how–why echo-reflex. 1933 'R. Stranger' *Dict. Wireless Terms* 59 *Echo Room*, a term used in broadcasting to indicate the place where echoes are produced artificially in order to give a more life-like effect to broadcasting. 1922 O. Jespersen *Lang.* xvi. 313 Fr. *pisser*—an echo-word if ever there was one. 1938 *Michigan Daily* 14 July 1/1 In general, an 'echo-word', Dr. Emeneau elucidated, is formed by the reduplication of a part of the word after an inserted pattern syllable that has no meaning of its own.

echo, *v.* Add: **4. b.** *absol.* Also in *Bridge* (cf. *Echo *sb.* 8), to indicate how many cards of a suit are held, or to request a specific lead.

1885 R. A. Proctor *How to Play Whist* 96 You cannot readily echo, as you can signal, by the discard. 1900 A. Dunn *Bridge* (ed. 3) 53 The 'trump suit echo' is played to inform a partner that the player who 'echoes' originally held four, or more, trumps. *Ibid.* 54 He should take the first opportunity of echoing in a plain suit, which will inform the leader that he has still one trump remaining. 1959 Reese & Dormer *Bridge Player's Dict.* 77 It is possible to echo by playing a three followed by a two. *Ibid.* 78 In the next example a defender echoes with four cards. *Ibid.* 236 If East had held a slightly different hand ..he would not have echoed in trump.

echogram (e·kogræm). [Echo *sb.* 1 + -gram.] The record made by an echo-sounder.

1936 *Discovery* Dec. 388/1 Some of the motor skiffs were fitted out with the necessary apparatus, and directly they came in contact with a shoal of 'invisible' herring the echogram or recorded echo gave evidence of their presence. 1947 *Times* 3 Sept. 5/6 The bottom is so hilly that the record dissolves into three or more intersecting convexities, which appear simultaneously on the echogram, as if the ship were moving over numerous hills or hummocks. 1958 *New Scientist* 2 Oct. 947 Tracings of echograms of some leveed channels found off the California coast.

echograph (e·kogrɑf). [Echo *sb.* 1 + -graph.] A device which automatically records echograms.

1947 *Times* 3 Sept. 5/6 The ship is equipped with a powerful echograph specially constructed for us. 1950 *Ann. Reg.* 1949 428 A continuous record of the bottom profile..registered by an ultrasonic echograph specially constructed in London. 1959 *Oceanogr. & Marine Biol.* I. 84 (*caption*) Echo-graphs and cathode ray tube records.

echoic (ekōu·ik), *a.* [f. Echo *sb.* + -ic.] Of the nature of an echo: a term proposed by J. A. H. Murray and used in this Dictionary to describe formations which echo the sound which they are intended to denote or symbolize.

1880 [in Dict. s.v. Echo *sb.*]. 1886 *N.E.D.* s.v. Botch *v.*[1], App. related to Du. *botsen* to knock, dash, Ger. dial. *butschen, butzen* to strike, knock; according to Franck an onomatopœic word of echoic origin. 1950 Partridge *Here, There & Everywhere* 182 *Grurmstipth*, an omnibus: obviously an echoic term.

echolalia (ekolēi·liă). [mod.L., f. Gr. ἠχώ Echo + λαλιά talk.] **1. a.** *Path.* The meaningless repetition of words and phrases. **b.** *Educational Psychol.* The repetition of words and phrases by a child that is learning to speak.

1885 *Buck's Handbk. Med. Sci.* I. 290/1 Periphrastic forms of speech and the recurring or barrel-organ utterances, constituting what is known as *echolalia*. 1890 W. James *Princ. Psychol.* II. xxiv. 407 His condition is that of 'Echolalia',—instead of answering the question, he simply reiterates it. 1924 R. M. Ogden tr. *Koffka's Growth of Mind* v. 311 At the time when speech is being learned, there begins a period of echolalia in which the child repeats with tireless continuation all the words or sentences it hears. 1924 [see *echopraxia]. 1930 *Times Lit. Suppl.* 27 Mar. 264/2 *Echolalia* or imitation of definite phonetic patterns. 1943 *Mind* LII. 317 The uttering of a tautology may be a natural sign-function which..truly mediates in a particular context many useful inferences, such as, that the person is an unconscious metaphysician, is subject to echolalia, [etc.]. 1948 [see *echopraxia]. 1955 T. Morrow in G. Caplan *Emotional Probl. Early Childhood* III. 343 He displayed, and displays, classical echolalia and delayed echolalia. 1964 B. Rimland *Infantile Autism* (1965) III. 180 There are numerous instances of affirmation by repetition and echolalia in the literature on autism.

2. A depreciatory term for a succession of sounds in poetry which subordinates sense to sound.

1895 *Westm. Gaz.* 3 Dec. 2/1 Our suspect poets, with.. their liking for echolalia. 1897 R. Vallance *William Morris* iii. 31 The 'Eve of Crecy' contains two magnificent examples of that mode of poetic expression, dubbed 'echolalia' by Max Nordau, and as such condemned by him.

Hence **echola·liac, echola·lian, echola·lic** *adjs.*

1927 C. Gray in H. J. Foss *Heritage of Music* 181 Von Bülow..was at least constant in his echolalian obsessions. 1931 E. Wilson *Axel's Castle* vii. 252 Most of us balk at her soporific rigmaroles, her echolaliac incantations. 1938 J. Eisenson *Psychol. Speech* iii. 47 When the infant reaches the echolalic and voluntary utterance levels, he has reached a developmental stage where rather finely differentiated muscular responses can be made. 1959 S. Sitwell *Bridge of Brocade Sash* viii. 170 The priest..who with echolalic persistence reiterated the same name over and over again.

e:choloca·tion. [f. Echo *sb.* 1 + *location *sb.* 7.] The location of objects by means of the echo reflected from them by a sound-signal, as of ultrasonic sounds emitted by bats or by man-made devices. Hence **e·cholocate** *v.*, **e:choloca·ting** *vbl. sb.* and *ppl. a.*

1944 D. R. Griffin in *Science* C. 589 (*title*) Echolocation by Blind Men, Bats and Radar. *Ibid.*, To describe this process of locating obstacles by means of echoes, I suggest the word *echolocation*. *Ibid.* 590/1 A radio signal is sent out from the airplane and electronic devices measure the time required for this signal to be reflected back...Finally we have radar, a refinement of the radio altimeter so beautifully perfected that it can echolocate airplanes at great distances by sending out radio waves and picking up the reflected energy returning from the distant aircraft. 1957 *New Biol.* XXIII. 43 Another theory about moth behaviour, in which sound acts as an aid to navigation, suggests that they use a system of echo-location analogous with that of bats. 1958 *Observer* 3 Aug. 5/5 The Harvard scientists estimate that the bat's echo-location system is incomparably more efficient for its size, weight and power, than the most modern radar system. 1959 *Observer* 14 June 5/4 Asdic, another wartime device used for echolocating underwater objects by ultrasonic vibrations, has given us more knowledge about whales. 1964 *Oceanogr. & Marine Biol.* II. 438 The conversational squeals and echolocating clicks of porpoises.

echometer. Add later example. Also, a type of echo-sounder.

1937 *Nature* 18 Sept. 499/1 An 'echometer' depth sounder. *Ibid.* 25 Sept. 553/2 Another exhibit..shows how one type of echometer can indicate to a trawler the depth of the sea-bed and also that at which a shoal of fish may be passing. 1961 *Times* 2 June 14/7 Dry-paper recording echometer.

echopraxia (ekopræ·ksiă). *Path.* Also **echopraxis.** [mod.L., fr. Gr. ἠχώ Echo + πρᾶξις action.] The meaningless repetition or imitation of the movements of others. Hence **echopra·ctic** *a.*

1904 T. Johnstone tr. *Kraepelin's Lect. Clin. Psychiatry* iii. 25 All his movements showed a certain constraint and want of freedom. His limbs remained for some time in the position in which you placed them. If you raised your arms quickly in front of him, he imitated the movement, and he also clapped his hands when it was done before him. These phenomena, called respectively *flexibilitas cerea*, 'waxen flexibility', or catalepsy and *echopraxis*, are familiar to us from experiments in hypnotism. 1905 A. J. Rosanoff tr. *Rogues de Fursac's Man. Psychiatry* I. iv. 92 Some [patients] repeat exactly the words (*echolalia*) or the gestures (*echopraxia*) of the persons around them. 1924 A. A. Brill tr. *Bleuler's Textbk. Psychiatry* ii. 152 The impulse for such an action can also be given by means of example alone, as in *echopraxia* and *echolalia*. *Ibid.* 153 The echopractic patients imitate whatever strikes them in the actions or words of their surroundings... It is partly a question of hysteria-like mechanism..and partly a matter of an incapacity to get away from a conceived idea, so that instead of giving an answer the question is repeated, or instead of a new action the preceding act is imitated (organic echolalias and echopraxias). 1948 *Brit. Jrnl. Psychol.* Dec. 89 Mutism..echolalia, echopraxia, negativism..were the most common symptoms of this group [of schizophrenics].

e·cho-sounding, *vbl. sb.* [f. Echo *sb.* + Sounding *vbl. sb.*[2]] The action or process of sounding or ascertaining the depth of water or of an object below a ship by means of the echo reflected from a transmitted sound-signal. Also in extended uses, e.g. of the flight of bats. Also *attrib.* or as *adj.* Hence **echo-sounder,** a device used for this purpose; (as back-formations) **echo-sound** *v.* and *sb.*

1923 *Hydrographic Rev.* I. 72 (*title*) Echo Sounding. Test carried out by the U.S.S. 'Stewart' 20th to 29th June 1922. 1927 S. H. Long *Navigational Wireless* xi. 148 Admiralty Type Echo Sounder. A second type of echo depth sounder..developed by the Admiralty Research Department. 1928 *Marine Observer* June 115/2 This vessel is fitted with the Admiralty Echo Sounding machine.., a wonderful aid to navigation, as one is able to get soundings up to 135 fathoms with accuracy. 1933 *Jrnl. R. Aeronaut. Soc.* XXXVII. 343 A very efficient 'height over the ground' indicator is installed in the 'Graf Zeppelin'. This is the 'Behm' echo sounding device. 1935 *Nature* 1 June 898/1 For navigation the echo sounder was employed originally to delineate the 100 fathom line. 1936 *Discovery* Dec. 388/1 The electrical echo-sounder is to be used for the detection of shoals of fish. 1947 *Times* 3 Sept. 5/6 Thanks to echo-sounding, one of the most marvellous inventions of our days, belief in the perfect smoothness of the deep-sea bottom has been much revised, especially in regard to the Atlantic. 1953 J. Y. Cousteau *Silent World* vi. 63 We were passing the same area in the research ship *Élie Monnier* when the echo-sound tape recorded another massive object on the floor. *Ibid.* x. 97 Our echo sounder located the wreck of a French submarine. 1958 *Observer* 3 Aug. 5/4 Scientists at Harvard University have been studying the 'echo-sounding' behaviour of bats. 1959 A. Hardy *Fish & Fisheries* iii. 51 His having previously echo-sounded herring shoals in the same area. 1968 *Times* 18 Oct. 16/7 By a combination of echo-sounding from the air and rapid sampling it was found possible to determine in only 20 minutes the main features of any lake.

ECHO virus (e·ko vəiə·rɒs). *Path.* Also **echovirus.** [f. the initials of *e*nteric *c*yto-pathogenic *h*uman *o*rphan: 'orphan' because when first discovered there was no disease of which they were known to be the cause.] Any of a group of enteroviruses infection with which is usually asymptomatic but which may produce symptoms resembling those of the common cold, mild meningitis, or (rarely) poliomyelitis. So *ECHO group*, etc.

1955 *Science* 16 Dec. 1188/1 These viruses—some of which have been referred to in previous literature as 'orphan viruses' and others as 'human enteric viruses'—are now classified as the 'enteric cytopathogenic human orphan (ECHO) group'. *Ibid.* 1188/2 Kidney cells of different monkey species vary in their susceptibility to the ECHO viruses. 1957 [see *enterovirus]. 1961 M. Hynes *Med. Bacteriol.* (ed. 7) xxv. 378 Three groups of viruses—polioviruses, Coxsackie viruses and ECHO viruses—are primarily invaders of the human intestinal tract. All are small, 25–35 mμ in diameter, and survive well outside the body. *Ibid.* 384 ECHO viruses can often be isolated from the fæces or nasopharynx of healthy individuals, but they can probably cause aseptic meningitis, acute diarrhœa in children, and mild respiratory infections. 1965 *Times* 3 Sept. 14/5 The echoviruses, of which over 30 types are known,..are closely related to the poliovirus, and produce a form of meningitis, but they have also been isolated from children with an upper respiratory infection resembling the common cold.

echoy, *a.* Add: Also, liable to resound with an echo.

1906 J. Joyce *Let.* 7 Aug. (1966) II. 147 When he goes into an 'echoy' place he shouts: in the Colisseum [*sic*] and in S. Peter's. 1967 *Listener* 18 May 652/3 There are three tiers of galleries..it's very echoey.

‖**echt** (eχt), *a.* [G., real, true, genuine.] Authentic, genuine, typical. Also as *adv.*

1916 G. B. Shaw in *New Age* 25 May, Many Englishmen who know Germany, and whose social opinions are *echt* Junker opinions, hail this war as a means of forcing England to adopt the Prussian system. 1917 E. Pound 19 Dec. in *Lett. J. Joyce* (1966) II. 414 The opening is echt Joice [*sic*]. 1934 C. Lambert *Music Ho!* iii. 173 England has never produced an artist so 'echt-English' as Mussorgsky is 'echt-Russian', or Renoir 'echt-French'. 1950 D. Gascoyne *Vagrant* 56 His endlessly varied *echt*-lyrical lute-ditties. 1956 *Essays & Studies* IX. 14 Those passages in his letters which are *echt*-Coleridge do not belong at all to letter-writing. 1962 N. Freeling *Love in Amsterdam* II. 70 'Are you married?' he asked... 'I see your ring, but is that camouflage or *echt*?'

ecize (i·səiz), *v. intr. Ecology.* [f. Gr. οἰκίζειν to colonize; cf. *ecesis.] Of plants: to become established in a new habitat; to Colonize (sense 3).

1916 F. E. Clements *Plant Succession* iv. 69 Ecesis..has proved so definite and convenient in use that it seems desirable to employ a corresponding verb, *ecize*, from οἰκίζω, to make a home, colonize. *Ibid.* 71 The plants which can ecize in such extremes are necessarily restricted in number and specialized in character. 1938 Weaver & Clements *Plant Ecology* (ed. 2) iii. 72 A habitat in which the seeds..may germinate and the seedlings successfully ecize (i.e. make a home).

ecka, var. *ekka.

éclair (ēiklɛə·ɹ). [Fr., lit. lightning.] A small finger-shaped cake made of choux-pastry, and filled with any of various kinds of cream.

1861 *Vanity Fair* 2 Feb. 50/1 A waiter, whereon stood.. a plate of macaroons, éclairs and sponge cake. 1870 *London Society* XVIII. 506/2 You get..meringue à la crême, eclairs. 1889 R. Wells *Pastrycook & Confect. Guide* 103 Eclairs..after baking fill with cream. 1902 *Daily Chron.* 15 Feb. 8/4 Chocolate éclairs are not difficult to make at home. 1925 A. Huxley *Those Barren Leaves* IV. iii. 291 At one of the little tea-tables..Miss Elver.. was eating chocolate éclairs and meringues, messily, with an expression of rapture on her cream-smeared face. 1927 C. Mackenzie *Vestal Fire* I. iii, You knew that the éclairs of the Villa Amabile would effuse authentic cream and not discharge stale custard like those dreadful waistless middle-aged éclairs you had met in some houses. 1956 *C. Spry Cookery Bk.* xxix. 866 Pipe out in finger or éclair shapes on to lightly greased baking-sheets. 1969 H. MacInnes *Salzburg Connection* xi. 154 Have a chocolate éclair.

eclaircise (ĭklɛ̄ə·ɹsəiz), *v.* Delete †*Obs.* and add later U.S. examples.

1872 *Chr. World Pulpit* I. 309/1 Till the enigma is solved; till the mystery is eclaircised. 1884 *Ibid.* XXV. 238/3 If men..listen to preaching that stimulates them, eclaircises their imagination.

eclecticism. (Earlier example.)

1798 Willich *Elem. Crit. Philos.* 1 There arose a sort of *Eclecticism*, which discouraged party-spirit, and recommended philosophical discretion.

eclipse, *sb.* Add: **2. b.** (See quot. 1838.) Hence *eclipse-dress, -feathers, -plumage*.

1838 C. Waterton *Ess. Nat. Hist.* 202 At the close of the breeding season, the drake undergoes a very remarkable change of plumage..and..is..so completely clothed in the raiment of the female, that it requires a keen..eye to distinguish the one from the other...Thus we may say that once every year..the drake goes, as it were, into an eclipse. 1906 C. W. Beebe *Bird* 48 The invisible cloak of his brooding mate is dropped over him for a while—his colours vanish, and by a partial moult..the hues of his plumage change to an inconspicuous mottling of brown, hardly distinguishable from the female... This has been happily termed the 'eclipse' plumage. 1913 *Brit. Birds* VII. 2 The short eclipse-feathers.. differ so little in general tint from the feathers of the winter- and breeding-plumage..that it is difficult to see what advantage the bird derives from the change. *Ibid.* 74 The second eclipse is for the most part similar in colouring to the first eclipse,..but these second eclipse Eiders can always be distinguished..from the first eclipse..birds. 1914 *Bull. Brit. Ornith. Club* XXXIII. 67 The pigment for the coloration of the eclipse-dress was beginning to form. 1930 Kirkman & Jourdain *Brit. Birds* 159 Gadwall..In eclipse much like duck. 1958 Bannerman *Birds Brit. Isles* VII. 1 The male [*sc.* mallard] in eclipse resembles the female closely. *Ibid.* 31 The adult drake [of the teal] begins to assume eclipse plumage in June..by August the eclipse is complete.

eclipsing, *ppl. a.* **1.** Delete † and add later example.

1923 *Discovery* Sept. 245/2 Mr. F. C. Jordan..announces the discovery of a remarkable eclipsing variable star.

3. (Later example.)

1891 Hardy *Tess* I. ii. 25 The name of the eclipsing girl..has not been handed down.

eclosion (ĭklōu·ʒən). [a. F. *éclosion*, n. of action of *éclore*, f. *é-* = Ex- + *clore*:—L. *claudĕre* to shut.] Emergence from concealment; *spec.* in *Ent.*, the emerging of an insect from the pupa case, or of a larva from the egg.

1889 in *Cent. Dict.* [1906 W. James *Let.* 7 Apr. (1920) II. 246 It is queer to be assisting at the *éclosion* of a great new mental epoch, life, religion, and philosophy in one.] 1925 A. D. Imms *Gen. Textbk. Entomol.* 186 As the time for the eclosion of the imago approaches the pupa noticeably darkens in colour. 1928 S. Thompson *Battle of the Horizons* v. vii, Her speculation vaguely pursued the fate of that simper—the man waiting for the savour of its rehearsed eclosion. 1953 H. Miller *Plexus* (1963) vii. 282 What intrigues me to this day, is the glory and the wonder of eclosion.

ecmnesia (ek,mni·ziă, -siă). *Psychol.* [mod.L., f. Gr. ἐκ out + μνῆσις memory.] Loss of memory with regard to the events of a particular period.

1890 Billings *Med. Dict.*, *Ecmnesia*, a form of amnesia in which there is normal memory of occurrences prior to a certain date, with loss of memory of what happened for a certain time after that date. 1967 Brussel & Cantzlaar *Chambers's Dict. Psychiatry* 77 *Ecmnesia*, amnesia with poor memory for recent events, but with relatively intact memory for events in the remote past.

ecoclimate (ī·kokləimĕt). [f. *eco-* as in *ecology + Climate *sb.* 3.] The climate as an ecological factor; the climate of a habitat; cf. *microclimate. So **e:coclimato·logy.**

1931 B. P. Uvarov in *Trans. Entomol. Soc.* LXXIX. I. 128 In the study of insect ecology we must, therefore, pay special attention to the climate of the actual habitat of the insect, or to the *ecoclimate*, as I would suggest that the sum-total of meteorological factors within a habitat should be called. 1937 *Nature* 16 Oct. 664/1 Another weak point is the lack of discussion on ecoclimates of the actual animal habitats, which very often enable an animal to survive in a region the general climate of which is wholly unsuitable for it. 1966 R. L. Smith *Ecology & Field Biology* i. 6 Early investigation of the physical environment of organisms..led to ecoclimatology.

ecological, ecology, now the more usual forms of Œcological, Œcology. So **eco-lo·gic** *a.*, **ecolo·gically** *adv.*, **eco·logist.**

A supposed use of *ecology* in Thoreau's letters represents a misreading of *geology*: see *Science* (1965) 13 Aug. 707 and *Bull. Thoreau Soc.* (1973) No. 123. 6.

1896 *Pop. Sci. Monthly* May 72 These ecologic color adaptations. *Ibid.* Dec. 185 Botany..especially with reference to the physiology and ecology of plants. 1902 *Nature* 17 Apr. 574/1 The ecology of a glacial lake. 1904 C. L. Laurie *Flowering Plants* 6 The study of plants that grow together, forming plant associations, in some respects the most interesting part of Ecology. *Ibid.* 8 Ecological classification of plants. 1909 Webster, Ecologically. 1916 F. E. Clements *Plant Succession* 73 It is one of the most important tasks of ecology to determine the root and shoot relations of communal plants. 1926 *Spectator* 25 Sept. 492/1 Part of the distinctively modern progress in palaeontology has just been this ecological outlook. 1928 R. S. Troup *Silvicultural Syst.* xix. 183 The development of ecology as a science has given a special impetus to the study of the physical and physiological conditions bearing on natural regeneration. 1930 C. Elton *Anim. Ecol. & Evolution* 7 Evolution..is not at all a popular subject among animal ecologists to-day. 1931 H. G. Wells *Work, Wealth & Happiness of Mankind* (1932) i. 29 Economics..is spoken of in the *Science of Life* as a branch of ecology; it is the ecology of the human species. 1933 —— *Shape of Things to Come* I. § 12. 111 Of human ecology he betrays no knowledge. 1935 *Nature* 2 Feb. 168/1 The study of such interactions under the given conditions of habitat is the most important part of what is called plant ecology. 1935 *Times* 6 Mar. 10/3 We know that in some cases ecologically related trees can well gain the freedom of another country. 1936 H. G. Wells *Anat. Frustration* ix. 86, I assume the world community..subject to general ecological laws. 1941 Auden *New Year Letter* I. 23 And grasped in its complexity The Catholic ecology. 1955 M. Gluckman *Custom & Conflict in Africa* i. 7 The ecological needs for this friendship and peace lessen as the distance grows greater. 1959 A. Hardy *Fish & Fisheries* xvi. 302 The work of the marine ecologists is only in its infancy. 1967 *Listener* 6 Apr. 459/3 In different ecologies territorial systems will vary or even be absent altogether.

econometric (ĭkǫnome·trik), *a.* and *sb.* *Econ.* [f. Econo(my + Metric *a.*¹ 2.] **A.** *adj.* Of, or relating to, or characterized by, the application of mathematics to economic data or theories.

1933 (*title*) Econometrica: Journal of the Econometric Society, vol. I. **1956** R. F. Harrod in A. Pryce-Jones *New Outl. Mod. Knowl.* 490 Trade Cycle study appeared an especially good field for the use of econometric methods. **1959** *Economist* 28 Feb. 809/2 The econometric model.. represents the actual workings of the economy by a system of mathematical equations.

B. *sb.* (*pl.*) The branch of economics concerned with the application of mathematical economics to economic data by the use of statistical methods.

1933 *Econometrica* I. 5 (*heading*) The Common Sense of Econometrics. **1951** *Univ. Birmingham Faculty of Commerce & Social Sci.* Regs. 63 Econometrics. deals mainly with the difficulties encountered in applying statistical technique to economic data. **1954** *Times Lit. Suppl.* 26 Mar., Econometrics now vies with mathematical philosophy..as an exercise in formal logic. **1970** *Daily Tel.* 23 Feb. 9/7 American research economists have pioneered the science of 'econometrics', in which factors affecting the economy are written as mathematical symbols.

econometrician (ĭkǫ:n-, ĭkǫ:nometri·ʃăn). *Econ.* [f. prec. + -ian.] A student of, or specialist in, econometrics.

1947 J. Tinbergen in S. E. Harris *New Economics* 219 It seems worth while to consider the many-sided contributions made to economic thought by John Maynard Keynes from the angle of the econometrician. **1956** R. F. Harrod in A. Pryce-Jones *New Outl. Mod. Knowl.* 490 It was the ambition of the econometricians.. to formulate economic laws in terms which could be statistically verified. **1968** *Listener* 18 July 68/2 It is mere playing with words to call management consultants specialists in the same sense as chemical engineers or econometricians.

economic, *a.* and *sb.* Add: Also commonly with pronunc. (ekonǫ·mik).

2. c. Practical or utilitarian in application or use, e.g. *economic botany, geography*, etc.

1861 *Jrnl. Soc. Arts* 22 Mar. 295 (*heading*) The Economic History of Paraffine. **1882** W. Cunningham *Growth Eng. Ind. & Comm.* 5 Economic History is not so much the study of a special class of facts, as the study of all the facts of a nation's history from a special point of view. **1882** B. D. Jackson (*title*) Vegetable technology; a contribution towards a bibliography of economic botany. **1914** J. McFarlane *Economic Geography* 1 Economic Geography may be defined as the study of the influence exerted upon the economic activities of man by his physical environment. **1922** C. K. Leith *Econ. Aspects Geol.* i. 1 The application of geology to practical uses, resulting in the development of the science generally known as *economic geology*. **1959** *N.Z. Timber Jrnl.* Apr. 52/2 Economic Forestry..is directed mainly towards marketing and utilization of forest products.

d. economic man, a convenient abstraction used by some economists for one who manages his private income and expenditure strictly and consistently in accordance with his own material interests. Cf. **economical man*.

1889 G. B. Shaw *Fabian Essays in Socialism* 25 There is no such person as the celebrated 'economic man'. **1890** A. Marshall *Princ. Econ.* I. vi. 78 When the older economists spoke of the 'economic man' as governed by selfish, or self-regarding motives, they did not express their meaning exactly. **1929** A. Huxley *Do what you Will* 217 Historical materialists, who deal..with abstract 'Economic Men'. **1930** *Economist* 19 July 115/2 There is the danger..that the State..may prove to be a mechanism..so forgetful of human personality in its concentration on the barren concept of 'economic man' that ordinary men, hungry for individual well-being, may revolt against it. **1965** Seldon & Pennance *Everyman's Dict. Econ.* 138 *Economic Man*, concerned with the immediate aim of obtaining the largest possible command over resources with the minimum of sacrifice.

e. economic rent, a rent that brings a fair return on capital and current expenditure (see also quots. 1965 and 1966).

1889 G. B. Shaw *Fabian Essays in Socialism* 6 Here is a clear advantage of £500 a year to the first comer. This £500 is economic rent. **1894** —— in *Fortn. Rev.* Apr. 480 The difference between the worst acre and the best (otherwise the 'economic rent') can be reduced finally by equality of cultivation. **1950** G. Brenan *Face of Spain* v. 103 On their present wages, no working-class family can pay an economic rent. **1965** Seldon & Pennance *Everyman's Dict. Econ.* 142 *Economic rent*,.. the earnings of a factor of production in excess of the minimum sum necessary to keep it in existing use. **1966** A. Gilpin *Dict. Econ. Terms* (1967) 60 *Economic rent*, a payment to a unit of a factor of production which is in excess of the minimum amount necessary to keep that unit in its present occupation. A firm may pay a wage sufficient to retain its present staff. In attempting to increase its staff, however, the firm may find it necessary to raise wages and attract workers from other employment. The increase in wage now enjoyed by the original staff is economic rent.

f. economic system, the sum of the economic institutions and arrangements of a society.

1898 A. P. Atterbury tr. *Sombart's Socialism & Social Movement* i. 4 By a 'social class' I understand.. men who are interested in a specific system of production and distribution. We must, in understanding any social class, go back to this economic system. **1909** M. Epstein tr. *Sombart's Socialism & Social Movement* (ed. 6) 1 In using the words 'economic system' I mean a given social order, or an economic condition of things, which is characterized by one or more prominent economic principles. **1929** W. Sombart in *Econ. Hist. Rev.* II. 13 The general conception which I employ in order to distinguish, describe and correlate economic phenomena is that of *the economic system*. *Ibid.* 14 By an economic system I understand a mode of satisfying and making provision for material wants, which can be comprehended as a unit, wherein each constituent element of the economic process displays some given characteristic. **1937** R. L. Hall (*title*) The economic system in a socialist state. **1951** R. Firth *Elem. Soc. Org.* iv. 127 The role of the anthropologist here is rather that of a watch-dog— to see that no one takes away the reality of the economic systems of primitive peoples by default.

g. economic war(fare). the use of economic measures as a means of bringing pressure to bear on another country, or in retaliation for such measures taken against the user.

1916 G. L. Dickinson (*title*) Economic war after the war. **1939** W. S. Churchill in *War Illustr.* 2 Dec. 374/3 Nazi Germany is all the time under the grip of our economic warfare falling back in oil and other essential war supplies. **1949** E. Pound *Pisan Cantos* lxxviii. 64 And the economic war has begun.

h. economic growth, the growth per head of the population in the production of goods and services over a stated period of time; the rate of expansion of the national income. Cf. *growth¹ 1 c.

1940 C. G. Clark *Conditions of Economic Progress* x. 337 (*heading*) The morphology of economic growth. **1948** *Proc. Amer. Philos. Soc.* XCII. 229/2 The economic growth of the United States can thus be defined. **1953** J. Viner *Internat. Trade & Econ. Devel.* vi. 103 It is not necessary to look for other factors..to explain pervasive poverty and slow economic growth. **1965** *Times* 17 Feb. 19/6 Economic growth is no longer regarded as the cure-all for the nation's ills.

economical, *a.* Add: **3. b.** *economical man* = **economic man* (*economic *a.* 2 d).

1858 H. C. Carey *Princ. Social Sci.* I. i. 29 We have the politico-economical man, on one hand influenced solely by the thirst for wealth, and on the other..under the control of the sexual passion. **1880** Bagehot *Econ. Studies* ii. 83 The nature of the 'man' who first occupied new countries did not 'conform' to the standard of economical man; the being of reality was not the being of the hypothesis. The first men..nearly approached in nature to the present savage man... They could not make any of the articles which we now call 'wealth'.

economism (ĭkǫ·nŏmiz'm). [f. Fr. *économisme*, f. *économ(ie* economy + *-isme* -ism.] A belief in the primacy of economic causes or factors (see also quots. 1949, 1967).

1919 W. R. Inge *Outspoken Essays* i. 23 Ruskin..saw ..the danger to which spiritual values were exposed at the hands of the dominant economism. **1940** *Mind* XLIX. 423 Marx's economism—his emphasis on the economic background as being the ultimate basis of any sort of development—is exaggerated. **1948** J. L. Adams tr. *Tillich's Protestant Era* (1951) p. xxxiii, Religious socialism was always interested in human life as a whole and never in its economic basis exclusively. In this it was sharply distinguished from economic materialism, as well as from all forms of 'economism'. **1949** I. Deutscher *Stalin* ii. 31 The first strikes.. stimulated a new trend called 'Economism'. This peculiar label was used by Russian socialists to describe what the French called Syndicalism, that is non-political Trade Unionism. **1967** *Times* 19 Jan. 7/2 The past week has seen the coining [in China] of a new derogatory term, 'economism', which was linked today with the parallel errors of better wages and conditions.

economist. Add: **4. d.** One who practises or advocates economism.

1949 I. Deutscher *Stalin* ii. 31 The 'Economists' wanted to confine their activities to supporting workers' claims for higher wages and better conditions of work, without bothering about politics. **1955** H. Hodgkinson *Doubletalk* 46 An Economist is one who accepts a Marxist analysis of society and believes in the inevitable rise of socialism, but maintains that 'the unaided evolution of inevitable economic changes can bring about the desired revolution without theoretical guidance'.

economy. Add: **4. d.** The cheapest class of air travel; also *attrib.*

1959 *World Air Transport Statistics* iii. 38/1 Scheduled Operations..First..Tourist..Economy. **1963** *Daily Tel.* 13 May 12/2 A five per cent. increase in the cost of 'economy-class' travel across the Atlantic. **1970** *Times* 3 Nov. 8/3 America's domestic airlines..are competing to see which can carry the fewest possible economy class passengers in the greatest possible comfort.

9. economy-size, applied *attrib.*, esp. in *Advertising*, to objects which are sold in a size that is said to be economically advantageous to the customer; also *economy-sized* adj.; also in extended uses. orig. *U.S.*

In practice freq. but not always the largest packet, etc., in a series.

1950 R. P. Bissell *Stretch on River* xix. 185 The only trouble with the economy size tube of shaving cream is that it takes up more room in the valise than a pair of rubber boots. **1953** *Harper's Mag.* Mar. 100 [His] mother..treats him as something between a lapel ornament, a trained marmoset, and an economy-sized male escort. **1955** *N.Y. Herald-Tribune* 12 Nov., Lou Gehrig had just hit a large economy-size home run into the right-field bleachers. *Ibid.* 1 Dec. ii. 1/5 The investigators managed to arrange 'economy sized package deals which included a little bit of everything'. **1957** Wodehouse *Over Seventy* v. 65 It was a bear. And not a small bear either. One of the large economy size. **1958** *Times Lit. Suppl.* 23 May 281/3 Another giant, economy-size book about the hard lot of Scandinavian females in the bad old days! **1962** L. Deighton *Ipcress File* vi. 36 Economy-sized cafés. **1962** *Times* 17 Oct. 13/3 'Economy' sized cars. **1966** H. Yoxall *Fashion of Life* xiv. 134 The small tube [of shaving cream] is called 'large',..we have a medium tube called 'king size', and a large tube called 'economy size'.

ecophene (ī·kofīn). *Ecology.* [f. *eco-* as in *ecology + *phen(otype + *-e*.] (See quots.)

1922 G. Turesson in *Hereditas* III. 346 The reaction-types of the ecotypes called forth by the modificatory influences of extreme habitat factors may appropriately be termed *ecophenes*. *Ibid.* 347 The term *ecophene* is proposed to cover each of the reaction-types of the ecotypes arising through the modificatory influences of the combinations of extreme habitat factors given in nature. **1949** Darlington & Mather *Elem. Genetics* 389 *Ecophene*, the range of phenotypes produced by one genotype within the limits of habitat under which it is found in nature. **1957** *Encycl. Brit.* XXI. 177/2 Ecophenes are illustrated by climatic modifications at the tree line, by shade types of plants that normally range through a variety of light conditions, and by temperature influences on growth at different latitudes. **1968** R. Daubenmire *Plant Communities* i. 33 Whether these variations represent ecotypes or ecophenes, the habitat type is the best indicator of the extent of each type of variance.

‖**écorché** (ekorʃe). *Painting* and *Sculpture.* [Fr., pa. pple. of *écorcher* to flay.] A subject so treated as to expose the muscular system.

1854 Thackeray *Newcomes* lxxviii, If you will have the kindness to look by the *ecorché* there, you will see that little packet which I have left for you. **1862** *Chambers's Encycl.* III. 761/2 It is not uncommon to represent the ecorché in action, in the form of the Fighting Gladiator. **1883** J. W. Mollett *Illustr. Dict. Art & Archæol.* 120/2. **1891** 'L. Malet' *Wages of Sin* iv. v, Try to put the bones into this upper figure and make an *écorché* of the lower one. **1963** *Encycl. World Art* VII. 662 One of the earliest examples is ascribed to Baccio Bandinelli and, because of its peculiar dancing movement, is usually called the 'dancing *écorché*'.

fig. **1891** Hardy *Group of Noble Dames* 85 Lifting her eyes as bidden she regarded this human remnant, this *ecorché*, a second time. But the sight was too much. **1908** —— *Dynasts* iii. ii. iv. 376 The contorted and attenuated écorché of the Continent appearing as in an earlier scene, but now obscure under the summer stars.

ecospecies (ī·kospīʃiz). *Ecology.* [f. *eco-* as in *ecology + Species.] A subdivision of a species of which the individual members are interfertile.

1922 G. Turesson in *Hereditas* III. 102 In the efforts made by the writer to arrive at an understanding of the Linnean species from an ecological point of view—of the ecospecies, as I prefer to say in the following—studies have been made of a number of plant species. *Ibid.* 344 The term ecospecies has been proposed..to cover the Linnean species or genotype compounds as they are realised in nature. *Ibid.* 347 The Linnean species represents an ecological unit..narrowed down to the ecological combination-limit. A genotype compound of this order is here termed an ecospecies. **1957** M. Abercrombie et al. *Dict. Biol.* 76 *Ecospecies*, group of plants comprising one or more ecotypes within a coenospecies whose members can reproduce amongst themselves without loss of fertility in offspring. Approximates to conventional 'species'.

e·cosphere. [f. *eco-* as in *ecology + Sphere *sb.*] The region of space including planets whose conditions are not incompatible with the existence of living things.

1953 H. Strughold *Green & Red Planet* (1954) iv. 36 Only a small zone about 75 million miles wide—out of the 4,300 million that stretch between the sun and Pluto at its farthest point—provides a planetary environment well-suited to the existence of life. We might call this zone the thermal ecosphere of the sun. Other stars may have such ecospheres of their own. **1965** Dole & Asimov *Planets for Man* iv. 109 To be habitable, a planet must be inside the ecosphere.

écossaise. Add: **b.** A dance to such a tune.

1841 Mrs. Gaskell *Let.* (1966) 43 A very fashionable waltz step came up while we were at Heidelberg the ecossaise,—and the little girls with their empty milk pails went dancing it along the road. **1954** *Grove's Dict. Mus.* (ed. 5) II. 881/2 Authorities differ about the question whether the Écossaise and the Schottisch are..one and the same kind of dance... The modern view is that the dance has nothing to do with Scotland, but was.. originally a French dance.

e·cosystem. *Ecology.* [f. *eco-* as in *ecology + System.] (See quots.)

1935 A. G. Tansley in *Ecology* XVI. 299 There is constant interchange..within each system, not only between the organisms but between the organic and the inorganic. These *ecosystems*, as we may call them, are of the most various kinds and sizes. They form one category of the multitudinous physical systems of the universe. *Ibid.* 306 The fundamental concept appropriate to the biome considered together with all the effective inorganic factors of its environment is the *ecosystem*, which is a particular category among the physical systems that make

up the universe. **1939** —— *Brit. Islands* iii. x. 228 A unit of vegetation..includes not only the plants..but the animals habitually associated with them, and also all the physical and chemical components... Such a system may be called an *ecosystem*. **1952** P. W. Richards *Tropical Rain Forest* v. 111 Certain animals..which play an important part in the rain-forest ecosystem. **1963** *New Scientist* 28 Mar. 684/2 The unit of ecology is the ecosystem, which includes the plants and animals occurring together plus that part of their environment over which they have an influence.

ecotone (ī·kotōᵘn). *Ecology.* [f. *eco-* as in *ecology + Gr. τόν-ος tension, Tone.] A transitional area between two ecological communities. Hence **e·cotonal** *a.*

1904 F. E. Clements in *Bot. Surv. Nebraska* VII. 153 Zonation in a habitat... The line that connects the points of accumulated or abrupt change in the symmetry is a stress line or *ecotone*... Ecotones are well marked between formations, particularly where the medium changes: they are less distinct within formations. It is obvious that an ecotone separates two different series of zones in the one case, and merely two distinct zones in the other. **1926** Tansley & Chipp *Study of Vegetation* iv. 53 Transitional belts between well-marked communities are called *ecotones* or 'tension belts'. **1950** *Jrnl. Ecology* XXXVIII. 70 Apart from interdigitation of typical stands, ecotonal communities are to be found. **1952** P. W. Richards *Tropical Rain Forest* xv. 338 Within the Closed forest there is manifestly a transition or ecotone from the.. evergreen forests of the wettest areas to the..deciduous forests bordering the savannas. **1967** *Oceanogr. & Marine Biol.* V. xi. 265 Mixing along the boundaries of two water masses may result in nature in an increase of the diversity of frontier or ecotone populations.

ecotype (ī·kotəip). *Ecology.* [f. *eco-* as in *ecology + Type *sb.*¹] A subdivision of an ecospecies the members of which are the product of genotypical adaptation to a particular habitat.

1922 G. Turesson in *Hereditas* III. 112 The term *ecotype* is proposed here as ecological unit to cover the product arising as a result of the genotypical response of an ecospecies to a particular habitat. The ecotypes are then the ecological sub-units of the ecospecies. **1949** W. C. Allee et al. *Princ. Animal Ecol.* v. xxxiv. 674/2 The interbreeding population would indicate either a hybridization between the river and cave forms, or that the cave form is an ecotype (ecological sub-species) of the river form. **1949** *Q. Jrnl. Forestry* XLIII. 88 The new concept of ecotypes—races of species adapted to various environments—which the ecologists claim to distinguish although taxonomists cannot discern any constant structural differences between them. **1959** E. P. Odum *Fund. Ecol.* (ed. 2) i. iv. 144 If marked ecotypes exist, the occurrence of the same series of taxa in different localities does not necessarily mean that the same conditions exist.

ecphore (e·kfōᵊɹ), *v.* *Psychol.* Also **ek-**. [ad. G. *ekphorieren* (R. Semon *Die Mneme* (ed. 3, 1912) iii. 93), f. Gr. ἐκφορεῖν, f. ἔκφορος (to be) made known (cf. ἐκφέρειν to carry or bring forth, produce, disclose).] *trans.* To evoke or revive (an emotion, a memory, or the like) by means of a stimulus. So **ecphoric** (ekfǫ·rik) *a.*, pertaining to or characterized by ecphory; whence **ecpho·rically** *adv.* Also **ecphorize** (e·kfŏrəiz) *v.*; whence **e·cphorizable** *a.*; **ecphory** (e·kfŏri) [ad. G. *ekphorie* (R. Semon *ibid.* 25)], the evocation of a disposition from a latent to a manifest state.

1917 *Brit. Jrnl. Psychol.* June 429 An 'ekphored' feeling is always a new state of feeling and never the memory image of a previous one. *Ibid.* 453 If we look at..a red rose and perceive it, and after a little while ekphore its memory-image, we note immediately how unlike..this memory-image is to the original perception. *Ibid.* 456 The ekphory of the memory-image of a pain. **1921** L. Simon tr. *Semon's Mneme* 39 Groups of influences may act ecphorically on an engram. *Ibid* 73 The diurnal periodic leaf movements of plants are ecphorised chronogeneously for some time after the cessation of the light-stimulus that normally liberates them. *Ibid.* 138 The ecphory of an engram. *Ibid.*, The ecphoric factor..consists of the partial or entire repetition of that energetic condition which formerly acted engraphically. **1921** B. Russell *Analysis of Mind* iv. 84 The second mnemic principle, or 'Law of Ekphory'. *Ibid.*, When two stimuli occur together, one of them, occurring afterwards, may call out the reaction for the other also. We call this an 'ekphoric influence', and stimuli having this character are called 'ekphoric stimuli'. **1923** B. Duffy tr. *Semon's Mnemic Psychol.* 155 An engram which, when evoked into life (*ecphorized*), will produce a mnemic sensation in consciousness. *Ibid.* 314 Homophonously ecphorizable engrams. **1925** C. Fox *Educat. Psychol.* 10 The process by which future stimuli touch off the engrams is known as ecphory. *Ibid.*, The partial recurrence of the excitation-complex which left behind it a simultaneous engram-complex acts ecphorically on the latter. **1937** G. W. Allport *Personality* (1946) v. xix. 525 One variation of the associational theory is the doctrine of redintegration, or as it is sometimes called, *ecphory*. **1958** H. B. & A. C. English *Dict. Psychol. Terms* 169 Ecphory, the activation of a memory trace, or engram.

‖**écrevisse** (ekrəvis). [Fr.] A freshwater lobster.

1854 Thackeray *Newcomes* I. xxviii. 266 Pass me the écrévisses [*sic*], they are most succulent. **1930** E. Waugh *Labels* viii. 188 There were heaps of rather dangerous-looking lobsters and *écrevisses*. **1966** *Punch* 14 Sept. 384/2 We import frozen fragments of dogfish tail and call them *scampi*, while not bothering to gather the *écrevisses*..in our..chalk-streams.

‖**écrin** (ekræ̃). [Fr.] A casket for jewellery.

1854 Thackeray *Newcomes* II. xxi. 203 The cigar-boxes given over to this friend, the *écrin* of diamonds to that, et cætera. **1910** S. R. Crockett *Dew of their Youth* xxxviii, A hoop of rubies..was placed in a lined box of morocco leather, called an 'ecrin'.

ecstasiate, *v.* Add: (Later examples.) Also [after F. *s'extasier*, f. *extasie*, early form of *extase* Ecstasy.] *intr.* or *refl.*, to go into an ecstasy.

1838 J. Pardoe *River & Desert* I. 10 He extasiated on the Emperor, and shrugged his shoulders at all other crowned heads. **1889** *Sat. Rev.* 7 Dec. 662/2 What we admit that we cannot sincerely extasiate ourselves before or admire is M. Huysmans's idea of style. **1923** A. Huxley *Antic Hay* xv. 213 The holy Teresa's quivering and ecstasiated flank. **1957** *Observer* 8 Dec. 14/3 The appearance of the Chimp Father Christmas, scattering soot, is so ecstatiating [*sic*] that [etc.].

ectad (e·ktæd), *adv.* *Anat.* [f. Gr. ἐκτός outside + *-ad.] On the outward side of, external to. Const. *of.* *rare.*

1882 Wilder & Gage *Anat. Technol.* 27 The dura (mater) may be described as *ectad* of the brain, but *entad* of the cranium. **1900** Dorland *Med. Dict.* 220/1. **1953** *Faber Med. Dict.* 133/1 *Ectad*, on or towards the outer part.

ectal (e·ktăl), *a.* *Anat.* [f. Gr. ἐκτός outside + -al.] External, superficial. *rare.*

1881 B. G. Wilder in *Science* (N.Y.) 19 Mar. 125/1 *Ental*, and *ectal* are here first proposed as substitutes for the more or less ambiguous words *inner* and *outer*, *interior* and *exterior* [etc.]. **1940** *Chambers's Techn. Dict.* 280/1 *Ectal layer*, a thin membrane at the extreme edge of an excipulum.

ectene (e·ktĭni). *Gr. Church.* Also **ektene.** [eccl.Gr., f. ἐκτενής extended.] A litany recited by a deacon and choir.

1850 J. M. Neale *Holy East. Ch.* I. iii. ii. 361 The Ectene for the first Sunday in Lent. **1882** *Encycl. Brit.* XIV. 707/1 A series of short intercessions resembling the Greek 'Ektene', or deacon's litany. **1916** N. F. Robinson *Monasticism Orthodox Ch.* 89 Then followeth the customary Ectene. **1957** *Oxf. Dict. Chr. Ch.* 437/2 *Ectene*..In the E. Church, a prayer constructed like a Litany for use in the Liturgy. It consists of short petitions said by the deacon to which choir or congregation respond with Kyrie Eleison.

ecthesis (e·kþĕsis). [ad. Gr. ἔκθεσις exposition, f. ἐκτιθέναι to put forth.] An edict of the Emperor Heraclius promulgated A.D. 638, maintaining the doctrine that Christ has only one will.

1728 Chambers *Cycl.* s.v., The Ecthesis favour'd the error of the *Monothelites*. **1850** J. Miley *Hist. Papal States* I. i. 225 Meddlers in theology, attempting by *ectheses* and *types*..to dictate what the vicars of Christ were to teach the Church. **1854, 1902** [see Type *sb.*¹ 5 b]. **1902** H. K. Mann *Lives of Popes* I. i. 365 Monothelism and the *Ecthesis* were condemned. **1956** J. Hussey tr. *Ostrogorsky's Hist. Byzantine State* ii. 97 He [*sc.* the Patriarch Sergius] relegated the single energy to the background, and now propounded the doctrine of a single will (θέλημα) in Christ. This new monothelete formula..was promulgated by the Emperor [*sc.* Heraclius] in 638 under the name of the *Ecthesis*.

ecto-. Add: **ectobla·stic** *a.*, of or belonging to an ectoblast; **ectocho·ndral** *a.* *Anat.*, situated or occurring outside cartilage; **e·ctochone** *Zool.*, the outermost structure of a chone; **e:ctodynamomo·rphic** *a.* (see quots.); **ecto-e·thmoid** *Zool.*, the prefrontal bone of the skull; **ectole·cithal** *a.* *Embryol.*, having the food yolk situated outside the formative yolk; **e·ctoloph** *Zool.*, the outer ridge on the crown of a lophodont tooth; **e·ctomere** *Embryol.*, each of the cells produced by the segmentation of the ovum; **ectoneu·ral** *a.* *Zool.*, of or pertaining to nerves situated superficially in an echinoderm; **e·ctophyte**, an external vegetable parasite; hence **ectophy·tic** *a.*; **ectopla·stic** *a.*, pertaining to ectoplasm; **ectorhi·nal** *a.*, situated on the outer side of the nose; **e·ctosome**, the outer layer of the body wall of certain sponges; hence **ectoso·mal** *a.*; **ecto·steal** *a.* *Anat.*, of, pertaining to, or situated on the exterior surface of a bone; hence **ecto·steally** *adv.*

1889 *Cent. Dict.*, Ectoblastic. **1902** *Daily Chron.* 16 Sept. 5/2 A pair of ectoblastic invaginations. **1910** *Practitioner* Jan. 40 The core of the chorionic villus is mesoblastic in origin,..whereas the trophoblast is ectoblastic. **1889** A. Macalister *Text-bk. Hum. Anat.* 38 The mesoblastic tissue..is replaced by bone in either of two ways, which are called respectively ectochondral and entochondral ossification. **1913** *Cunningham's Text-bk. Anat.* (ed. 4) 86 Cartilage bones are those which are preformed in cartilage, and include most of the bones of the skeleton. Their growth is often described as endochondral and ectochondral, the former term implying the deposition of membrane bone in the centre of the cartilage, while the latter signifies a deposit of membrane bone on the surface of the cartilage. **1887** Sollas in *Encycl. Brit.* XXII. 415/1 A transverse muscular sphincter, which defines an outer division or ectochone from an inner or endochone. **1888** Rolleston & Jackson *Forms Anim. Life* 795 The..chonae..are divided..into an outer part, the ectochone, long and cylindrical, and an inner part, the endochone. **1927** C. F. Marbut tr. *Glinka's Great Soil Groups* 34 We propose to designate them as endodynamomorphic soils to distinguish them from the ektodynamomorphic soils in which external factors predominate over the internal. **1930** *Nature* 19 July 89 Ektodynamomorphic soils, that is, those in which climatic and other external factors in soil formation predominate over internal ones. **1888** Rolleston & Jackson *Forms Anim. Life* 91 Anteriorly to the septum is a large bone, the ecto-ethmoid..(= the pre-frontal of Huxley). **1928** C. F. Cooper *Parker & Haswell's Text-Bk. Zool.* (ed. 4) II. 78 The outer walls of the olfactory capsules may be ossified by paired ecto-ethmoids. **1884** Sedgwick & Heathcote tr. *Claus' Zool.* I. 112 The first processes of segmentation in these at first ectolecithal ova are withdrawn from observation, since they take place in the centre of an egg covered by a superficial layer of food yolk. **1905** *Amer. Geol.* Apr. 244 The outer border (ectoloph) rises higher than the cross crests. **1966** A. S. Romer *Vert. Paleont.* (ed. 3) xx. 239 An ectoloph, forming the outer wall of the tooth, connecting paracone and metacone and often..with a W-shaped contour. **1889** *Cent. Dict.*, Ectomere. **1890** Billings *Med. Dict.*, *Ectomere*, the larger of the two primitive cells produced by the segmentation of the mammalian ovum after impregnation. **1924** E. V. Cowdry *Gen. Cytology* 584 The subdivisions of these ectomeres are much more numerous in large eggs than in small ones. **1909** A. Sedgwick *Student's Text-Bk. Zool.* III. iii. 123 The ectodermal part of this plexus, which is continuous with the endodermal and may be called the ectoneural, is especially concentrated in an annular tract round the mouth. **1962** D. Nichols *Echinoderms* ii. 28 The main sensory part lies orally, and has been called the superficial oral or ectoneural system. **1883** D. Macalister tr. *Ziegler's Path. Anat.* i. 354 Ectophyte. **1902** *Encycl. Brit.* XXV. 439/1 There is a reciprocal symbiosis between the Angiosperm and the fungus which is endophytic, only rarely ectophytic. *Ibid.*, More or fewer of their rootlets have their extremities invested by a weft of hyphal mycelium as an ectophytic mycorhiza. **1885** *Encycl. Brit.* XIX. 834/1 Ectoplastic products in which the material produced by the protoplasm is separated from it. **1900** G. C. Bourne *Introd. Comp. Anat. Anim.* 109 The cell-protoplasm..has given rise to substances which..are formed on the outside of the cell as a sort of envelope..in which case they are called ectoplastic products. **1890** *Jrnl. Anat.* XXV. 106 The demarcation between these two parts is due to the presence of a fissure, more or less distinctly defined in different animals, which has been named the rhinal or ectorhinal fissure. **1904** A. Keith *Hum. Embryol.* (ed. 2) 258 Hippocampal and Ectorhinal Fissures. **1888** Sollas in *Challenger Rep.* XXV. p. xvi, This cleavage occurs parallel to the outer surface, dividing the sponge at once into a choanosomal and ectosomal portion. **1887** *Encycl. Brit.* XXII. 415/1 An outer or ectosome and an inner or choanosome. **1959** *Chambers's Encycl.* XIII. 101/2 The ectosome consists of a single-layered epithelium which may become several layers thick. **1868** W. K. Parker *Monogr. Struct. & Devel. Shoulder-Girdle Vertebrata* 66 In all the known Urodela the endosteal deposits are entirely ruled by the much earlier and more potent ectosteal sheath. **1870** W. H. Flower *Introd. Osteol. Mamm.* vii. 74 They ossify, according to Parker, ectosteally, or from without inwards.

ectocrine (e·ktokrəin, -in). [f. Ecto- + *-crine* as in *endocrine.] An external metabolite which on release to the external world influences the vital processes of members of the same or other species. Also *attrib.*

1947 C. E. Lucas in *Biol. Reviews* XXII. 291 Many organisms have adapted themselves to tolerate or take advantage of the external metabolites of their neighbours. ..Ecological relationships appear to have arisen in this way, both stimulating or inhibiting..community integration, competition and succession, and probably symbiosis. The term 'ectocrine' is suggested for a substance mediating such processes. **1961** —— in M. Sears *Oceanography* 508 The production of a very small number of specific algae..may depend on..particular harmful ectocrines not reaching dangerous concentrations. *Ibid.* 509 Ectocrine processes may also be relevant in any attempts which may yet be made to culture 'food' algae on a commercial scale. **1963** Henderson *Dict. Biol. Terms* (ed. 8) 162/1 *Ectocrine*, organic substances or decomposition products in the external medium which inhibit or stimulate plant life. **1967** *Oceanogr. & Marine Biol.* V. iii. 104 The fertility of the English Channel during the following winter much depends on the partition of nutrients and ectocrines which occurs at these several discontinuities.

ectogenesis (ektodʒe·nėsis). *Biol.* [mod.L.: see Ecto- and Genesis.] The production of structures or bodies outside the organism. So **ectogenetic** (ektodʒėne·tik), **ectogenic** (ektodʒe·nik), **ectogenous** (ektǫ·dʒėnəs) *adjs.*, pertaining to ectogenesis, producing or produced from without.

1883 D. Macalister tr. *Ziegler's Path. Anat.* I. 291 Some of the pathogenous bacteria are accustomed to develope and multiply without the body, while others only do so within it. The former kind we may describe as ectogenous, the latter as endogenous. *Ibid.*, Sometimes the ectogenous bacteria proceed to multiply within the body. **1900** B. D. Jackson *Gloss. Bot. Terms* 83/2 Ectogenic. **1909** *Cent. Dict.* Suppl., *Ectogenesis*, the pro-

duction of or the giving rise to structures from without. **1923** Haldane *Daedalus* (1924) 63 It was in 1951 that Dupont and Schwarz produced the first ectogenetic child. *Ibid.* 64 France was the first country to adopt ectogenesis officially, and by 1968 was producing 60,000 children annually by this method. **1926** *Times Lit. Suppl.* 21 Oct. 710/4 Mr. Shaw's affinities..are with the biological school, whose most startling forecast, so far, is Mr. Haldane's ectogenetic baby. **1930** *Ibid.* 24 Apr. 11 By the twenty-first century science will have solved the problem of ectogenesis, will be able, that is to say, to develop a human infant from a fertilized cell by laboratory methods. **1957** I. Asimov *Naked Sun* (1958) xviii. 223 Dr. Delamarre himself was planning a future in which ectogenesis would be possible and marriage unnecessary.

ectomorph (e·ktomǭɹf). *Anthropometry.* [f. Ecto- + Gr. μορφ-ή form.] A person with the lean body-build in which the physical structures developed from the ectodermal layer of the embryo, i.e. the skin and the nervous system, predominate: one of W. H. Sheldon's three constitutional types (cf. *endomorph 2, *mesomorph). Hence **e·ctomorphic** *a.*, **e·ctomorphy.**

1940 W. H. Sheldon *Varieties Human Physique* i. 5 In proportion to his mass, the ectomorph has the greatest surface area and hence relatively the greatest sensory exposure to the outside world. *Ibid.*, Ectomorphy means relative predominance of linearity and fragility. *Ibid.* iii. 45 Hair of ectomorphic people is of nearly the same abundance as in the case of endomorphy. **1954** R. Fuller *Fantasy & Fugue* iv. 75 Now you..unfortunately incline to the cerebrotonic ectomorph—you worry too much, you're too good looking, and you can't abandon yourself happily to booze. **1956** C. P. Snow *Homecomings* li. 367 His profile confronted hers, each of them firm and beautiful in their ectomorphic lines. **1961** *New Scientist* 6 Apr. 854/1 The ectomorph is much more of an introvert and more shrewd and calculating. **1962** *Listener* 24 May 905/2 A sort of parlour game, deciding how many points he scores for endomorphy, mesomorphy, and ectomorphy. **1963** Auden *Dyer's Hand* 345 The chirpy social endomorph will give a different picture of it from that of the melancholic withdrawn ectomorph.

-ectomy, a combining form representing Gr. ἐκτομή excision, used to form words denoting a surgical operation for the removal of a part, as *appendicectomy, colectomy, hysterectomy, lobectomy.*

ectopic (ektǫ·pik), *a.* *Path.* [f. Ectopia + -ic.] Characterized by ectopia: said esp. of pregnancy and gestation. Hence **ecto·pically** *adv.*

1873 R. Barnes *Clin. Hist. Dis. Women* 424 The gestation is ectopic, that is, proceeding in an abnormal locality. **1893** *Brit. Med. Jrnl.* 1 Apr. 50/1 Ectopic gestation with ruptured sac. **1929** G. S. Dodds *Essent. Hum. Embryol.* 78 In ectopic pregnancy the uterus hypertrophies. **1938** *Proc. Prehistoric Soc.* IV. 61 It is unknown outside Wessex, save in one or two obviously ectopic grave-groups. **1961** *Lancet* 26 Aug. 454/2 One or both of the following conditions must exist to produce ectopic calcification. **1962** *Ibid.* 5 May 937/1 Very rarely, a single kidney may be drained by an ectopically opening ureter.

ectoplasm (e·ktoplæz'm). **1.** (See Dict. s.v. Ecto-.) **2.** A viscous substance which is supposed to emanate from the body of a spiritualistic medium, and to develop into a human form or face. So **ectopla·smic, ectopla·stic** *adjs.*; **ectopla·smically** *adv.* Also *fig.*

a **1901** F. W. H. Myers *Human Personality* (1903) II. 548 In describing..imperfectly aggregated ectoplasms we have already touched on the next class, that of quasi-organic detached ectoplasms. *Ibid.* 549 Utterance may be referable to an ectoplastic throat as distinctly as grip to ectoplastic fingers. **1922** *Daily Mail* 2 Dec. 13 Frederick Munnings-Gaulton..prominent as a medium ..appears to have been an adept at producing 'ectoplasm'. **1923** *Ibid.* 5 Mar. 5 Ectoplasm..is described as being to the touch 'a cold and viscous mass comparable to contact with a reptile'. **1923** S. De Brath tr. *Richet's Thirty Years Psychical Research* i. 12 In this book.. will be found two chapters on each variety of phenomena,..whether the matter in hand concerns..the movement of objects (telekinesis), or materializations (ectoplasmic forms). **1926** A. Conan Doyle *Hist. Spiritualism* I. v. 114 The ectoplasm pictures photographed by Madame Bisson and Dr. Schrenck Notzing ..may in their first forms be ascribed to the medium's thoughts or memories taking visible shape in ectoplasm. *Ibid.* xii. 259 The mediumship of the Eddy brothers..has probably never been excelled in the matter of materialization, or, as we may now call them, ectoplasmic forms. **1926** J. Laird *Our Minds & their Bodies* 1 Those hearers look for photographic evidence of 'auras' and 'ectoplasms'. **1926** C. E. M. Joad *Babbitt Warren* 116 Ectoplastic mediums, quack doctors and inspired clairvoyants. **1927** S. De Brath tr. *Geley's Clairvoy. & Material.* 175 It seems to me desirable to substitute for 'materialization' the term 'ectoplasmic form'. **1952** M. McCarthy *Groves of Academe* (1953) v. 84 Mulcahy's rather ectoplasmic voice effused itself into Furness's ear. **1953** A. W. Fielding *Stronghold* ii. i. 105 Each bearded *pallikari*..glared angrily from his frame as though emerging from an ocean of ectoplasm. **1954** Koestler *Invisible Writing* iv. xxxvii. 400 As I read on, I had the impression of meeting the *Doppelgaenger*, the spectral doubles of Rubashov and Ivanov—a ghostly, ectoplastic regurgitation by reality of the characters and events of my imagination. **1960** *Spectator* 29 July 176 He had always evaporated, ectoplasmically, and swirled on.

ectotrophic (e·ktotrǫfik), *a.* *Bot.* [ad. G. *ectotrophisch* (A. B. Frank *Lehrb. d. Pflanzenphysiologie* (1890) II. 135), f. Ecto- + Trophic *a.*] Chiefly of Mycorrhiza: forming a tissue on the surface of roots. Opp. to *endotrophic *a.*

1899 [see *endotrophic *a.*]. [**1903** H. C. Porter tr. *Strasburger's Text-bk. Bot.* (ed. 2) i. ii. 209 The fungal hyphæ are sometimes present..within the cells..In other cases the fungus surrounds the young roots with a dense investment of interwoven hyphæ. The former arrangement is spoken of as endotrophic, the latter as exotrophic [later edd. ectotrophic] mycorhiza.] **1926** Tansley & Chipp *Study of Vegetation* ix. 158 Ectotrophic [fungi].. are found..mainly as a mantle around the rootlets of trees. **1938** Weaver & Clements *Plant Ecology* (ed. 2) xi. 329 If the fungal mycelium occurs on the outside of the root and between its cells..the mycorrhiza is ectotrophic (i.e., nourished outside). **1959** *New Scientist* 24 Sept. 528/2 The well-defined ectotrophic mycorrhizas of certain forest trees, and the more variable endotrophic condition of the majority of plants.

ectrodactyly (ektrodæ·ktili). *Anat.* Also **ectrodactylia.** [ad. mod.L. *ectrodactylia*, f. Gr. ἔκτρω-μα, -τρω-σις, etc. (f. ἐκ- + τρω- to damage) + δάκτυλος finger.] Congenital absence of digits. Also **ectroda·ctylism.**

1848 Dunglison *Med. Lex.* (ed. 7) 296/2 *Ectrodactylia*, a malformation in which one or more fingers or toes are wanting. **1883** *Index-Catal. Libr. Surg.-Gen. U.S.* IV. 51/1 Ectrodactylism. **1893** *Jrnl. Anat.* XXVII. 422 Schæfar gives an account of five cases of ectrodactyly. **1894** W. Bateson *Study of Variation* 358 Absence of Digits (Ectrodactylism). **1899** *Jrnl. Anat.* XXXIII. 524 A case in which ectrodactyly and syndactyly of the right hand co-existed with double ectrodactyly of the feet. **1909** W. Bateson *Mendel's Princ. Heredity* xii. 228 Several pedigrees of ectrodactylism..are recorded in medical literature. **1946** R. R. Gates *Human Genetics* I. xi. 434 Lobster claw appears to be an extreme form of syndactyly (i.e., combined with ectrodactyly).

ectromelia (ektromī·liă). [mod.L., f. Gr. ἔκτρω-σις miscarriage + *-melia* f. Gr. μέλ-ος limb + -ia¹.] **1.** The congenital absence of a limb or limbs.

1908 *Jrnl. Amer. Med. Assoc.* LI. 2002/2 (*title*) Case of Ectromelia. Associated with Other Deformities. G. W. Kosmak..Ectromelia. Kosmak reports, with an illustration, a case showing entire absence of the upper extremities...There is..not a dimple indicating where the arms ought to appear on the shoulder-girdle.

2. A virus disease of mice often causing loss of limbs. Also *attrib.*

1930 J. Marchal in *Jrnl. Path. & Bacteriol.* XXXIII. iii. 713 (*title*) Infectious Ectromelia. A hitherto undescribed virus disease of mice. *Ibid.* 726 As regards the proposed name of this condition, for which I have to thank Mr. Clifford Dobell who proposed it, I hope that any suggestion of a congenital abnormality in the word 'ectromelia' may be dissipated by the adjective 'infectious'. **1946** *Nature* 27 July 119/1 Ectromelia virus in mice.. agglutinated the same limited range of fowl cells as were agglutinated by vaccinia virus preparations. **1963** *New Scientist* 12 Sept. 540 Ectromelia, a killing disease of mice, is produced by a pox virus.

Ecuadorian (ekwădōᵊ·riăn), *a.* and *sb.* Also **-ean.** [f. *Ecuador* name of country in S. America + -ian.] **A.** *adj.* Of, belonging to, or characteristic of Ecuador. **B.** *sb.* A native or inhabitant of Ecuador.

1860 *Jrnl. R. Geogr. Soc.* XXX. 70 The Pailon of San Pedro, or St. Peter's Cauldron, was surveyed by Captain Kellett, of H.M.S. *Herald*. A copy of the Survey was presented to the Equatorian Government. **1861** *Ibid.* XXXI. 181 It is known to every one that gold exists at a short distance,..if the Ecuatoreans would only take the trouble to search for it. **1877** *Encycl. Brit.* VII. 645/1 The only real glacier known to exist in the Ecuadorian Andes. **1886** A. Simson *Trav. Ecuador* ii. 27 The dinner, in an Ecuadorian sense, was capital. **1892** E. Whymper *Trav. Gt. Andes* 179 Ecuadorian Hand-Made Lace. *Ibid.* 421 The head-coverings worn by Ecuadorians shew that the sun's rays are seldom so powerful as to be dangerous. **1927** *Glasgow Herald* 24 Oct. 7 Ecuadorian Hats [commonly called 'Panama' hats]. **1936** *Discovery* Aug. 249/2 The Ecuadorean Government. **1959** *Chambers's Encycl.* IV. 788/1 The territory in dispute between Ecuador and Peru consisted of..a small..area around Túmbez..and practically the whole of the Ecuadorian Oriente.

ecuelle (ēkiue·l). Also **écuelle.** [ad. F. *écuelle*:— pop.L. **scūtella*, altered from *scŭtella*.] **1.** A two-handled porringer used for soup.

1856 J. C. Robinson *Inv. Mus. Ornamental Art* 19 'Ecuelle' or Shell Tazza, ornamented with interlaced arabesque ornament in ruby and yellow lustre. **1872** W. Chaffers *Keramic Gallery* I. p. xi, Ecuelle and Dish, painted with Carnival figures. **1956** G. Taylor *Silver* vi. 129 The écuelle, a similar French vessel with two opposed flat handles, somewhat resembling the Scottish quaich, but with a domed cover, was introduced early in the [eighteenth] century. **1963** *Times* 15 May 8/7 A Meissen chinoiserie ecuelle and cover.

2. The process or apparatus by which oils are extracted from the peel of citrus fruit. Also **ecue·lling.**

1884 *Encycl. Brit.* XVII. 812/1 The process known as *écuelle*, in which the skin of the ripe fruit is scraped against a series of points or ridges arranged upon the surface of a peculiarly-shaped dish or broad funnel. **1891** *Bull. R. Bot. Gardens Kew* May–June 120 From the rind of the fruit, by a process known as 'ecuelling', which consists of gently rubbing the fruit on rounded projections arranged inside a brass basin, a very fine essence of limes is obtained. **1949** J. B. S. Braverman *Citrus Products* v. 184 The écuelle.. can be regarded as the prototype of all machines rasping whole fruit. It is a very old process.

ecumenic, -ical, -icalism, -icality, -ically, -icity, now the more usual forms of Œcumenic, Œcumenical, etc. So also **e·cumenism**, the doctrine, or quality, of universality (esp. of the Christian church); **e·cumenist.**

1840 Ecumenicity [see Œcumenicity]. **1845** *Encycl. Metrop.* XVIII. 387/2 Ecumenical. **1864** Ecumenical [see Œcumenical *a.* 1]. **1941** W. Temple *Citizen & Churchman* v. 87 What is known as the Ecumenical Movement represents a movement of the spirit. **1948** W. A. Visser 't Hooft in *Univ. Church in God's Design* iv. iv. 183 There is still in much of our present ecumenism a strong element of relativism. **1949** *Theology* LII. 15 To endeavour along the lines of a misconceived ecumenicalism to smooth out these differences. **1963** T. Ware *Orthodox Church* xii. 257 The ecumenicity of a council cannot be decided by outward criteria alone. **1964** *Economist* 25 Jan. 315/2 Monsignor Cardinale is a noted ecumenist. **1970** *Daily Tel.* 19 May 18 The cause of ecumenism has received no setback. It might indeed be enhanced, if the Archbishop [of Canterbury] were to receive and accept an invitation to be present in Rome at the canonisation ceremony.

Ed. or **ed.,** abbrev. of *edited* (*by*), *edition*, *editor*.

1843 *Ladies' Companion* (N.Y.) Feb. 167/1 Mr. Poe's article concludes with the following words.—*Eds.* **1878** S. G. C. Middlemore tr. *Burckhardt's Civilisation Renaissance in Italy* II. iv. iv. 38 Roscoe, *Leone X.*, ed. Bossi, viii. 174. **1892** K. E. Digby *Introd. Hist. Law Real Property* (ed. 4) i. ii. 39 Camden, Britannia, ed. 1610, p. 348. **1934** 'N. Bell' *Winding Road* ix. 180 He..said the ed. fellow was an old rascal. **1953** F. Shehab *Progressive Taxation* ii. 11 Adam Smith, The Wealth of Nations (ed. Cannan). *Ibid.*, Sir John Sinclair, The History of Public Revenue of the British Empire (3rd ed., 1803). **1957** S. N. Sen *Eighteen Fifty-Seven* 421 Forrest, G.W., ed. Selections from the Letters, Despatches and other state papers..of the Government of India. **1969** D. Waley *Italian City-Republics* 245 Giovanni Villani, Cronica, ed. F. G. Dragomanni.

Edam (ī·dăm). In full *Edam cheese*: a cheese of a type originally made at Edam, near Amsterdam.

1836 J. Murray *Handbk. Trav. Continent* 52/1 A vast quantity of these cheeses (called here Edam cheeses, but known all over the world as Dutch cheeses). **1885** *Times* 8 Oct. 6/6 Specimens of Gorgonzola, Roquefort, Gruyere, Parmesan, Edam and Gouda cheese. **1890** J. Macdonald *Stephen's Bk. Farm* (ed. 4) iv. 515 The two important cheeses made in Holland, both of which are sold in the English markets in very large quantities, are known as round or Edam, and flat or Gouda Cheese. **1902** *Encycl. Brit.* XXVII. 355/2 The Edam and Gouda are the common cheeses of Holland. **1951** *Good Housek. Home Encycl.* 449/1 To serve Edam cheese, the top is sliced off and used as a cover, and the cheese is scooped out from the centre with a spoon. **1955** J. G. Davis *Dict. Dairying* (ed. 2) 191 The real Edam has at least 40 per cent. fat in the dry matter and the skimmed milk cheese at least 20 per cent.

edaphic (ĭdæ·fik), *a.* *Ecology.* [ad. G. *edaphisch* (A. F. W. Schimper *Pflanzen-Geogr.* (1898) i. i. 5), f. Gr. ἔδαφος floor + -ic.] Pertaining to, produced or influenced by, the soil.

1900 Jackson *Gloss. Bot. Terms* Additions. **1902** *Encycl. Brit.* XXV. 430/1 The varying climatic or environmental conditions to which Angiosperms may be exposed in their wide distribution, including those of the soil, edaphic, those of the atmosphere, epedaphic, and those of water, aquatic. *Ibid.*, Geophytes are..subject to the influence of both edaphic and epedaphic factors of environment. **1906** Chamberlin & Salisbury *Geol.* II. 343 Adaptation to the immediate physical environment, particularly the nature and depth of the sea-bottom (edaphic adaptation). **1930** *Nature* 25 Jan. 120 Wherever domesticated animals come upon the scene the biotic factor undoubtedly ranks with the edaphic and climatic as of supreme importance. **1930** *Forestry* IV. 29 The term edaphic association is used by Tansley and other British ecologists for relatively permanent communities which do not reach the climatic climax owing to soil conditions. **1936** *Geogr. Jrnl.* LXXXVIII. 566 'Social ecology' in relation to the edaphic factor. **1938** *Nature* 7 May 815/1 The sand dune.. can be aptly described as an edaphic desert—a desert that is due, not to climatic but to soil conditions, since the coarse texture and very low organic content render sand a very poor retainer of water. **1949** *New Biology* VI. 54 Thus under semi-arid conditions, the 'natural' vegetation type may be open grassland, without trees of any kind... Such grasslands..may provide an example of an edaphic climax, seral development being arrested because of soil influences.

edaphon (e·dăfǫn). *Ecology.* [G. (R. Francé *Das Edaphon* (1913)), f. Gr. ἔδαφ-ος floor + *-on* as in *plankton*.] The community of microflora and microfauna in the soil.

1927 S. A. Waksman *Princ. Soil Microbiol.* xxiv. 642 We are..justified in speaking of a soil population and may even accept the idea of an Edaphon as suggested by Francé, although his conclusion that the edaphon is an indicator of soil fertility may not be fully justified. **1932** Fuller & Conard tr. *Braun-Blanquet's Plant Sociol.* iv. 76 The quantitative investigation of the edaphon and aeroplankton..the soil and air floras. **1960** N. Polunin *Introd. Plant Geogr.* xi. 316 Mention should also be made..of the 'edaphon', the flora and fauna of the soil.

eddy, *sb.* Add: **4. eddy-chamber,** a chamber in which a current of fluid is compelled to whirl in eddies; **eddy-current,** a current of electricity induced in places where such currents are undesirable and cause waste of energy; also *eddy-current brake, loss.*

1899 *Yearbk. U.S. Dept. Agric.* 153 W. S. Barnard, working under the direction of Riley, invented the admirable eddy chamber, or 'cyclone system' of nozzles. **1887** *Soc. Telegr. Engin. Jrnl.* XVI. 65 The coil is surrounded..by a circular pole piece, cut away at one point to hinder eddy-currents. **1892** W. P. Maycock *Electr. Lighting* i. vi. 158 Most armatures of direct current machines have iron cores, and the revolution of the cores in the magnetic field would..cause currents to be induced in the core...These currents [are] called *foucault* or *eddy currents*. **1930** *Engineering* 6 June 724/3 Three systems were in use, operated, respectively, by hydraulic, pneumatic and electro-magnetic (eddy currents in the wheels) means. **1934** Webster, Eddy-current brake. **1940** *Chambers's Techn. Dict.* 281/1 Eddy-current loss. **1943** *Electronic Engin.* XVI. 194 If a solid conductor is replaced by a number of insulated strands..it appears obvious at first sight that the eddy current losses must be reduced.

edenite (ī·dənəit). *Min.* [ad. G. *edenit* (E. F. Glocker *Grundriss d. Min.* (1839) 410), f. *Eden*ville, Orange County, New York: see -ite¹.] A light-coloured sodian variety of hornblende containing relatively little iron.

1842 L. C. Beck *Mineral. New York* ii. vi. 302 About a mile north of Edenville is another locality, at which the crystals [of hornblende] are of a hair-brown or grey color, associated with..mica and chondrodite, in white limestone. They vary in size from very small to an inch in diameter. From the peculiarity of its colour, it has sometimes been called *Edenite*. **1868** J. D. Dana *Syst. Min.* (ed. 5) 235 Aluminous Magnesia-Lime Amphibole. (*a*) Edenite... (*b*) Smaragdite. **1886** *Jrnl. Chem. Soc.* L. 519 Edenite (Hornblende). Colour green..or greyish-yellow. **1932** W. E. Ford *Dana's Textbk. Mineral.* (ed. 4) 575 The varieties of hornblende here included range from the light-colored edenite, containing but little iron..to the dark-colored or black hornblende.

edestin (ĭde·stin). *Chem.* [f. Gr. ἐδεστός eatable + -in¹.] A globulin occurring in hempseed; also, any of several globulins in wheat, rye, etc.

1894 Osborne & Voorhees in *Rep. Connecticut Agric. Exper. Station 1893* 216 This substance agrees in composition with the vitellin which exists in the seeds of wheat, maize... For this body we adopt the name *Edestin*. **1896** *Jrnl. Chem. Soc.* LXX. i. 400 The proteid which is soluble in saline solutions the author shows to be edestin, or vegetable vitellin. **1926** J. S. Huxley *Essays Pop. Sci.* viii. 88 Many others [*sc.* proteins], such as edestin from hemp-seed. **1947** *New Biol.* III. 121 Edestin (molecular weight 300,000) and other protein particles have also been photographed.

edge, *sb.* Add: **2. a.** (Further examples.) Also, *to give* (a person) *the rough edge of one's tongue*, to abuse, revile.

1915 Wodehouse *Psmith, Journalist* (1923) xiii. 90 It will give our output precisely the edge it requires. **1932** H. Crane *Let.* 31 Mar. (1965) 405 Dangers that give the same edge to life here that the mountains give to the horizon. **1936** W. Holtby *South Riding* iii. iii. 175 That girl gives them the rough edge of her tongue. **1939** W. H. Baumer *Sports as taught and played at West Point* 348 Often talk by the trainer on any other subject than the game will take off the 'edge'. **1939** F. Thompson *Lark Rise* i. 14 It took the edge off the appetite. **1946** *Penguin Music Mag.* Dec. 103 The tone is full and incisive, with a distinct edge to it—perhaps rather too distinct. **1957** *Essays & Studies* X. 42 'Then came still Evening on.' The stately order takes the edge off the homely expression 'Evening came on'.

b. (Later example.)

1868 W. Collins *Moonstone* I. v. 55 'Betteredge, your edge is better than ever...' 'He's a wonderful man for his age.'

d. Of temper (cf. *edgy *a.* 4).

1916 'Boyd Cable' *Action Front* 224 The company commanders found him with rather a sharp edge to his temper.

e. *to have an* (or *the*) *edge on, to get the edge on* (and similar phrases with *over*): to have a grudge against (quot. 1896); to have (or acquire) an advantage over. Also used without a following *on* or *over* in sense 'advantage, superiority'. orig. *U.S.*

1896 *Daily News* 18 Mar. 7/5, I expect that when I am gone, especially from your own people, who always had an edge on me, and for no reason. **1911** C. E. Mulford *Bar-20 Days* (1921) xiv. 147 I'll do anything to get th'edge on that thief. **1914** Joyce *Dubliners* 48 She had always had an edge on her, especially whenever there were people listening. **1929** *Publishers' Weekly* 14 Sept. 1060 Here we have the edge on our rivals, not only because of our superior location, but also because we are reputedly reckless about reducing prices. **1932** *Latimer Co. News-Democrat* (Wilburton, Okla.) 19 Feb. 1/6 Ray Tucker of the Scripps-Howard Newspaper Alliance gives Murray the edge over Roosevelt in the North Dakota primary. **1932** *N.Y. Herald-Tribune* 2 May 1/4 (*heading*) Herriot's party obtains edge and is held sure to win. **1936** M. Mitchell *Gone with Wind* liv. 943 Belle's got the edge on you..because she's a kind-hearted, good-natured soul. **1946** Wodehouse *Money in Bank* xxiv. 214 Jeff..thought he saw now where she had the edge on Myrtle Shoesmith. She was a quicker starter. **1949** M. Mead *Male & Female* xv. 316 He has to compete.. with girls who have an edge in almost all the activities for which reward is given. **1952** *Times* 19 Dec., Scottish schools also had the edge on the English and Welsh in the matter of pullovers and sweaters. **1953** G. S. Coffin *Acol & New Point Count* 11 If you hold an average hand..it has no edge over other hands around the table. **1957** *New Yorker* 12 Jan. 79 The Cincinnati Daily Enquirer is solid No. 1 right down the line. Including a 21,000 edge in City Zone circulation..a healthy lead in Retail Display Linage. **1958** *Daily Express* 15 July 6/6 Nancy Mitford's great edge is that she never went to school. **1959** *Listener* 23 Apr. 738/3 The edge was with Mrs. Durran who goes through to the final. **1959** *Spectator* 8 May 654/2 Goat's milk..makes a yoghourt which, therapeutically speaking, has the edge over any made from cow's milk. **1960** *Times* 14 June 16/1 They still hold the edge, especially with Statham in his present form. **1966** *Listener* 3 Mar. 299/2 Labour's special relationship with the working classes..was supposed to give it an edge over the tories.

f. The state of being drunk. *U.S. colloq.*

1920 F. S. Fitzgerald *This Side Paradise* (1921) i. iii. 117 We'll drink to Fred Sloane, who has a fine, distinguished edge. **1925** E. Hemingway *In our Time* iv. 60 'How do you feel?'..'Swell. I've just got a good edge on.'

4. b. *to be on* (occas. *upon* or *on the*) *edge*: to be excited or irritable (cf. *2 d).

1872 Minto *Man. Eng. Prose Lit.* ii. ii. 261 Whitgift's strenuous hostility and unsparing rigour of argument set his opponent on edge. **1900** *Daily News* 4 Oct. 6/1 For me there was to be no sleep far into that night, for my nerves were upon edge. **1908** R. W. Chambers *Firing Line* v. 56 I'm all on edge over this landscape scheme. **1924** J. Buchan *Three Hostages* xvi. 227 His manner had not the ease it used to have. He seemed on the edge about something. **1932** *Daily Express* 29 June 6/2 Her nerves were plainly on edge. **1938** S. V. Benét *Thirteen o'Clock* 290 He felt fagged and on the edge already. **1951** J. B. Priestley *Festival at Farbridge* iii. i. 394 Laura had in fact worked much too hard, and now she was altogether too fine-drawn and too much on edge.

c. *over the edge*: insane.

1929 E. Bowen *Last September* iii. xvii. 217 He would go over the edge, quite mad.

6. b. See also Razor *sb.* 1 b.

7. c. *edge of regression*: see Regression 3 b.

10. c. Delete †*Obs.* and add later example.

1856 Mrs. Gaskell *Let.* ? 22 July (1966) 397 The white ..& the lavender edge—did you look after black, (imitation,—Cambray, or Maltese) lace.

11. c. *fig.* Often with defining word, as *absolute, outside*: the 'limit', the very extreme. Phr. *over the edge* (see quot. 1945).

1911 'Ian Hay' *Safety Match* i. 8 Cheating again! My word, Nicky, you are the absolute *edge*! **1925** Wodehouse *Carry on, Jeeves* ix. 218, I am still blushing all over at the recollection of something she says in paragraph two... You can take it from me that it's the edge. **1945** Baker *Austral. Lang.* vi. 134 *Over the edge* (or *fence*), unreasonable, beyond the pale of fairness or decency.

12. a. *edge-to-edge*, attrib.; **b.** *edge nailing*; **edge-metal** (see quots.); **edge-runner,** an apparatus for crushing stone, fibrous matter, etc.; **edge-tone** *Mus.*, a sound-wave produced when a stream of air is deflected by an edge; **edge-zone** (see quot 1902.).

1839 Ure *Dict. Arts* 961 The coal-seams thus upheaved [*sc.* at a high angle], are called edge-metals by the miners. **1883** Gresley *Gloss. Coal-m.* 95 *Edge coals, edge metals, edge seams* (Sc.), highly inclined seams of coal. **1876** *Encycl. Brit.* IV. 493/1 For side or edge nailing..clasp-nails..are used. **1871** *Cassell's Techn. Educator* II. 267/1 The clay..is conveyed to the edge-runners or other machinery used to pulverise it. **1883** H. G. Harris in H. J. Powell *Princ. Glass-Making* 46 The pieces of limestone are further crushed under a pair of 'edge-runners'. **1883** *Encycl. Brit.* XVI. 463/2 Edge-runners (Chilian mills). **1891** *Engineer* 9 Jan. 36 Edge runner grinding mill. **1963** R. R. A. Higham *Handbk. Papermaking* ii. 24 Kollergang or edge-runner. This machine consists of two large circular stones, driven by a central shaft, which revolve in a metal or stone pan. **1952** C. W. Cunnington *Eng. Women's Clothing* vii. 259 Lined 'edge-to-edge' coat in wool georgette. **1961** F. H. Burgess *Dict. Sailing* 81 *Edge to edge*, descriptive of the planking of a carvel-built boat. **1963** C. R. Cowell et al. *Inlays, Crowns & Bridges* vii. 70 (*caption*) For an edge-to-edge bite the tip is prepared horizontally. **1931** G. Jacob *Orchestral Technique* ix. 95 The contrast between the 'edge-tone' of the trumpets and the rounder tone of the horns. **1962** A. Nisbett *Technique Sound Studio* iii. 67 The breathy edge tone that we hear with the flute. **1902** *Encycl. Brit.* XXV. 459/2 A fold of soft tissue extending to a greater or less distance over the theca, and containing..a cavity continuous over the lip of the calicle with the cœlenteron. This fold of tissue is known as the edge-zone. **1904** *Ann. & Mag. Nat. Hist.* XIII. 22 The epitheca is that part of the skeleton secreted by the edge-zone.

edge, *v.*¹ Add: **6. c.** To defeat by a small margin. *N. Amer.*

1953 *Springfield* (Mass.) *Sunday Republican* 20 Sept. 8B (*heading*) Bellows Falls edges Windsor eleven, 12–6. **1966** *N.Y. Times* (Internat. ed.) 22 Apr. 12/1 The Los Angeles Dodgers edged the Houston Astros, 3–2. **1968** *Globe & Mail* (Toronto) 17 Feb. 41/8 Malvern edged Parkdale 4–3 in the first game.

7. *Cricket.* To deflect (the ball) with the edge of the bat.

1904 P. F. Warner *How we recovered Ashes* xi. 215 The only blemish in his display was that he once edged a ball from Rhodes between the wicket-keeper and slip. **1906** *Daily Chron.* 28 July 7/4 Seeing his first ball edged by Rhodes over the slips. **1927** *Observer* 7 Aug. 18/2 Being inclined to edge the ball. **1970** *Times* 19 Aug. 6/2 He was fortunate now to edge Wilson only just wide of Fletcher in the gully.

edger. Add: **4.** An operative in various crafts.

1909 *Daily Chron.* 26 June 8/5 Optician's Edger wanted. **1921** *Dict. Occup. Terms* (1927) § 105 *Edger*, a fettler who smoothes edges and joints of clay ware articles with a knife, leather or sponge before firing. *Ibid.* § 429 *Black edger* (gloves);..stains, with a blackened pad, as received from machinist, the white edge left on black kid gloves.

edge-ways, -wise. 2. b. (Example with *get*.)

1854 S. Hale *Lett.* (1918) i. 9 Mrs. Otis herself talked all the time, so there was no room for me to get in a word edgewise.

edgily (e·dʒili), *adv.* [f. Edgy *a.* + -ly².] In an edgy manner; irritably, testily.

1837 H. E. Manning *Let.* 8 Mar. in S. Leslie *Life* (1921) xvi. 269 Newman did not like my joking letter. He answered rather edgily and defensively. **1922** M. Arlen *Piracy* iii. ix. § 3 'You are making a butt of me, Virginia,' he complained edgily.

edginess. Add: (Later examples.) Also, irritability.

1962 *Listener* 7 June 991/1 This edginess in the presence of the actual world characterizes a great deal of transcendent metaphysics. **1963** *Times* 8 June 5/2 Her tone sometimes acquired an edginess out of character. **1965** M. Morse *Unattached* iv. 131 There was a certain amount of edginess..between the groups.

edging, *vbl. sb.* Add: **6.** *attrib.* in various senses.

1841 Mrs. J. W. Loudon *Ladies' Comp. Flower Garden* 117/2 *Gentiana acaulis* is frequently used as an edging plant. *a* **1877** Knight *Dict. Mech.* I. 774/1 *Edging-machine*, a machine for edging boards to a given pattern. An edger. **1880** *G. Turner's Catal. Tool Wks. Sheffield* 24 Common Grass Plot Edging Knives. *a* **1884** Knight *Dict. Mech.* Suppl. 293/1 *Edging and dividing bench*, a circular saw of special adaption for sawing blocks into voussoir shapes. **1907** *Macm. Mag.* 825 Lepidoptera seek out this sunny retreat and suck the edging plants. **1957** *N.Z. Timber Jrnl.* Feb. 45 *Edging strips*, pieces fixed on the edges of flush doors to cover and secure the edges of the plywood.

edgy, *a.* Add: **4.** Having one's nerves on edge; irritable; testy.

1837 [implied in *edgily *adv.*]. **1864** in Webster. **1914** Kipling *Divers. Creatures* (1917) 149 'I say,' he began hurriedly, 'do you mind if I come in here for a little? I'm a bit edgy.' **1915** J. Buchan *39 Steps* i. 33 Then I could see that he began to get edgy again. He listened for little noises. **1924** *Chambers's Jrnl.* 555/1 The horses were very nervous and edgy for the rest of the day. **1929** S. Aumonier *Ups & Downs* 514 That its mood was tense and 'edgy' was evident from the way in which the members spoke and behaved. **1958** *Times Lit. Suppl.* 23 May 280/3 An American family of harassed father, edgy mother and irritated crop-headed boys.

edh (eð). Also **eth.** Name of the Anglo-Saxon and Old Icelandic letter, or the phonetic symbol, ð ('crossed d').

1846 E. J. Vernon *Guide Anglo-Saxon Tongue* i. 4 ð (*eth*) our soft *th*, as in *other*. **1875** *Encycl. Brit.* I. 612/2 In order to express the corresponding sonant (heard in '*that*', and confusedly denoted by the same compound *th*) a stroke was drawn across the simple d (ð), and the new letter was called *edh*. **1965** C. Barber *Flux of Language* vii. 130 The Old English scribes confused matters a little by using two symbols,..'thorn', and..'eth'..indiscriminately. **1969** *English Studies* Suppl. p. ii, Conybeare..made no use of thorn, Thk [*i.e.* Thorkelin] none of eth in transcription, though both sometimes wrote *th* as in modern English.

edifier (e·difəiəɹ). Restrict *rare* to sense a and add later examples of sense b in a wider application.

1832 Carlyle *Crit. & Misc. Ess.* (1872) IV. 117 The true Spiritual Edifier and Soul's-Father of all England. **1909** *Westm. Gaz.* 22 Jan. 3/1 The real edifiers. **1921** *Glasgow Herald* 19 Feb. 6 The professional edifiers of British youth.

edisonite (e·disənəit). *Min.* [f. name of Thomas Alva *Edison*, American inventor (1847–1931) + -ite¹.] A name formerly given to a supposed variety of rutile. *Disused.*

1888 W. E. Hidden in *Amer. Jrnl. Sci.* 3rd Ser. XXXVI. 274, I therefore propose for it the name *Edisonite*, after Mr. Thomas Alva Edison. **1889** *Jrnl. Chem. Soc.* LVI. 354 Edisonite, a fourth form of Titanic Anhydride. **1944** C. Palache et al. *Dana's Syst. Min.* (ed. 7) I. 560 Edisonite, a supposed orthorhombic polymorph of TiO_2.

edit, *v.* Add: **2. d.** To prepare a film for the cinema or recordings for broadcasting, etc. (by eliminating unwanted material, etc.); = *cut *v.* 21 e. Also with *in* and *out*.

1917 *Sci. Amer.* 8 Dec. 441/3 Editing a film is perhaps the most interesting phase of laboratory work. **1933** A. Brunel *Filmcraft* 94, I have myself edited..35 mm. films with just a bench, scissors..winder, film cement, a brush to put it on with, and a bin to unroll my film into. **1938** *Encycl. Brit. Bk. of Yr.* 122/1 Mobile recording vans with a new method of editing records. **1961** *Encounter* XVI. 49 Her [*sc.* Ada Leverson's] frequent use of film scenario devices..'editing in' scenes and themes apparently unrelated to the one that, in realistic terms, she is just then evoking. **1962** A. NISBETT *Technique Sound Studio* 265 *Pot-cut*, editing a short segment of unwanted material out of a programme..by quickly fading out and fading in again. **1969** J. ELLIOT *Duel* III. v. 312 Keep running... We can edit this out.

editing, *vbl. sb.* Add: (Earlier example.)
1840 J. S. MILL *Let.* 16 Apr. in *Wks.* (1963) XIII. 427 There, I think, is a full account of all the world has got by my editing and reviews.
b. = *CUTTING *vbl. sb.* 1 d.
1921 A. C. LESCARBOURA *Cinema Handbk.* (1922) v. 236 With the positive films printed up..a very necessary operation in the preparation of motion picture films is the editing and cutting. **1944** S. COLE (*title*) Film editing. **1944** R. MANVELL *Film* iv. 37 Editing is the art of putting the film together shot by shot from the celluloid strips themselves. **1962** A. NISBETT *Technique Sound Studio* vii. 117 'Editing' can mean various things: cutting and rejoining a tape..; *copy editing*, in which selected sections..are copied..on to a make-up tape; *mixing*, where the combined output of two tapes are fed to a third recorder. **1963** *Movie* Apr. 32/2 A director shouldn't use his camera and editing bench to impose something which he hasn't been able to put into the action.

edition, *sb.* **3. b.** *fig.* (Earlier and later examples.)
1612 DONNE *Sec. Anniv.* 309 All the vertuous Actions they expresse, Are but a new, and worse edition Of her some one thought. **1802** C. WILMOT *Let.* 30 Aug. in *Irish Peer* (1920) 83 Mr. Richard Trench..is..so beautiful that if I was to describe him, you'd fancy it was the Apollo Belvedere in a Second Edition. **1823** CARLYLE *Let.* 22 Oct. in *Early Lett.* (1886) II. 232 A kind of theological braggadocio, an enlarged edition of the Rev. Rowland Hill. **1892** *Daily News* 18 Oct., The exhilarating ballet of 'Don Juan'..seems likely to run into many 'editions', as the phrase goes, before it exhausts its popularity. **1920** H. CRANE *Let.* 24 Sept. (1965) 43 About every other person..are [*sic*] enlarged editions of Lord Douglas. **1943** J. S. HUXLEY *TVA* x. 74 The slightly younger edition of the architect.

édition de luxe: see LUXE 2.

‖ **editio princeps** (ĭdi·ʃio pri·nseps). Pl. editiones principes (ĭdiʃiōᵘ·nīz pri·nsipīz). [mod. L.] The first printed edition of a book.
1802 T. F. DIBDIN *Introd. Knowl. Rare & Val. Ed.* 4 This editio princeps contains but nine comedies. **1815** J. SCOTT *Visit to Paris* xv. 294 The room of the *Editiones principes*, contains every thing to gratify the taste of the bibliographer. **1875** *Encycl. Brit.* III. 656/2 Azzoguidi's *editio princeps* of Ovid. **1885** J. B. LIGHTFOOT *Apostolic Fathers* I. 113 A misprint of the editio princeps. **1930** *Times Lit. Suppl.* 6 Feb. 85/4 Laid down the plan of the *editio princeps*. **1955** *Times* 9 May 13/2 He published..an *editio princeps* of several minor works.

Editola (editōᵘ·lă). Also **editola.** [Proprietary name.] A machine used for editing films (see quots.).
1935 *Trade Marks Jrnl.* 31 July 952/2 *Editola*, apparatus included in Class 8 for reproducing motion picture films and sounds simultaneously. Photographic Electrical Co. Ltd., 80/2 Wardour Street, London, W.1, Manufacturers. **1939** *Times* 18 Nov. 10 The 'editola' on which the picture film and sound track are synchronized and checked. **1967** *Listener* 14 Sept. 326/3 The rattling chipmunk chatter of the Editola.

editor, *sb.* Add: **3. b.** A person in charge of a particular section of a newspaper, e.g. of the financial news (*city editor*: see *CITY 9).
1843 *Knickerbocker* XXII. 494 We cannot permit the young associate-editor of that print..to misrepresent us. **1894** E. L. SHUMAN *Steps into Journalism* 19 On the larger papers the work of the managing editor is divided, giving him an assistant, the managing news editor. **1936** *S.P.E. Tract* XLV. 188 In American newspaper offices, a member of the staff who is in charge of a single feature or department is dignified by the title of *editor*. This practice is being introduced into English journalism, but not without protest. **1954** *Manch. Guardian Weekly* 14 Oct. 7/1 'Say,' said the American photographer..'My editor's hopping mad for pictures.'
c. The literary manager of a publishing house, or head of one of its publishing departments. orig. *U.S.*
1915 *Bookman* XLI. 306/2 You now find erstwhile stonewall editors wax in your hands. **1930** *Publishers' Weekly* 5 Apr. 2096 Far more attention might well be given the West's peculiar needs by Eastern editors. **1958** *Oxford Books* (catal.) 3 *Oxford Progressive English*..was founded by E. C. Parnwell, while oversea editor of the Oxford University Press.
d. *attrib.* (appos.), as *editor-author*, *-manager*, *-proprietor*, *-publisher*; also **editor-in-chief**, the chief editor of a publication, in a publishing-house, etc.
1952 M. LOWRY *Let.* Mar. (1967) 292, I had a right..and the aim..to achieve an editor–author relationship. **1873** J. M. BAILEY *Life in Danbury* 287 Our highest ambition has been to be the editor-in-chief of a large New York daily, and help do up the mail. **1913** E. C. BENTLEY *Trent's Last Case* ii. 19 He was..editor-in-chief of the *Record*. **1955** *Times* 2 July 4/3 The editor-in-chief was Mr. Humphry House, Fellow of Wadham College, Oxford. **1959** *Encounter* July 67/1 He..is now editor-in-chief of Stanford University Press. **1959** *Manch. Guardian* 21 July 6/7 Sir William [Haley] had then been for one year editor-in-chief [at the B.B.C.]. **1961** (*title*) Webster's Third New International Dictionary. Editor in Chief Philip Babcock Gove, Ph.D. **1899** *Westm. Gaz.* 20 May 1/3 Editor-manager of the *Oxford Chronicle*. **1906** *Ibid.* 20 Sept. 10/2 The editor-proprietor of the 'Studio'. **1907** *Ibid.* 8 Nov. 12/1 The staff, from the editor-publisher downward.
5. One who cuts and edits a film.
1917 *Sci. Amer.* 8 Dec. 441/3 The editor, usually the director himself in dramatic productions, directs the cutting and assembling of the various strips of film into the complete production; and just as the editor of a magazine ..makes his corrections, so does the film editor view the assembled film on the screen and make corrections and changes. **1933** A. BRUNEL *Filmcraft* 158 *Editor*, one who cuts, assembles, edits and titles a film. **1937** H. G. WELLS *Brynhild* v. 61 The brightest and best producers, directors, scenarists, cameramen, special effects men, editors.

editor, *v.* Delete quot. 1883 (MS. reads 'edited') and add U.S. example in sense 'to edit'.
1961 *Amer. Speech* XXXVI. 138 The Britannica is now editored by Walter Yust.

editorial, *a.* and *sb.* Add: **A.** *adj.* **b.** *spec.* Written, or ostensibly written, by the editor of a newspaper, as distinct from news items.
1802 *Deb. Congress U.S.* 25 Feb. (1851) 796 The editorial part of the paper..was supposed to come from the pen of Mr. Hobby. **1816** *Niles' Weekly Register* X. Index, Editorial articles. **1849** *Knickerbocker* XXXIV. 9 We shall set forth in this editorial leader, that our friends may know exactly where we are. **1877** *Harper's Mag.* Dec. 109/1 [He] thoroughly worked an idea into an editorial leader. **1920** P. C. BING *Country Weekly* 145 [They minimized] the importance of the editorial page.
c. *editorial we* [WE *pron.* 2], the pronoun *we* used by a single person to denote himself, as in an editorial.
1841 DICKENS *Let.* 24 Aug. (1969) II. 368 Every rotten-hearted pander who..struts it in the Editorial We once a week. **1888** KIPLING *Phantom 'Rickshaw* (1890) 137 Under special patronage of the editorial We. **1964** E. A. NIDA *Toward Sci. Transl.* ix. 204 Such 'editorial we's' must be shifted to 'I' if they are to be intelligible in some languages. **1964** R. H. ROBINS *Gen. Ling.* vii. 287 A different anomaly in relation to number in pronouns is seen in what is often called the 'editorial *we*'.
B. *sb.* (Earlier U.S. example.)
1830 *Collegian* (Cambridge, Mass.) 44 The great green table in the centre groaning under the weight of editorials, and friendly correspondence.

edito·rialist. [f. EDITORIAL *sb.* + -IST 2.] One who writes editorials. Also *transf.*
1901 *N.Y. Even. Post*, The syndicator..talks to the would-be 'editorialist'. **1941** KOESTLER *Scum of Earth* 26 It seemed impossible that the editorialist of the paper had dared to write this off his own bat. **1964** *Spectator* 28 Feb. 270 Its education comes from street gossip and the exhortations of radio editorialists. **1965** H. KAHN *On Escalation* xi. 223 An editorialist or commentator may be invited to express a strong view of the matter.

edito·rialize, *v.* orig. *U.S.* Also **-ise.** [f. EDITORIAL + -IZE.] *intr.* To write editorials; to make editorial comment; to introduce editorial comments or an editorial slant into a factual account, etc. Also *trans.*
1856 G. D. BREWERTON *War in Kansas* 75 As Mrs. Partington feelingly remarked, when Ike tumbled into a barrel of soft soap: 'Isn't it a blessed thing to editorialize for an appreciative public?' **1928** *Nation* (N.Y.) 24 Oct. 416 The Italian press is free..to editorialize with the utmost braggadocio about Italy's plans for world domination. **1960** *New Left Rev.* May–June 3/1 'A clear statement that the party remains committed..'—the *New Statesman* editorialises (5 March 1960). **1961** *Sunday Times* 12 Nov. 48/5 The only 'editorialising' slip that I spotted was Bill Grundy's remark, while anonymous and unseen square-squatters spoke: 'Listen to the voices of progress.' Bernard Braden editorialises almost nonstop in *The Time, The Place and the Camera*. **1963** *Economist* 26 Oct. 353/1 That feeling that led him..to editorialise too much.
Hence **edito·rialized** *ppl. a.*; **edito·rializing** *vbl. sb.*, the action of the verb; also *concr.*, editorial comment.
1917 M. L. SPENCER *News Writing* 87 The necessity of presenting news from an unbiased standpoint,..of avoiding 'editorializing'. **1958** BLESH & JANIS *They all played Ragtime* i. 34 Prissily Victorian as these editorializings are, they were the trivia of a new age. **1961** *Guardian* 30 Mar. 13/3 The newspapers..rushed into print with editorialised reports of President Kennedy's press conference. **1967** R. J. STERLING *President's Plane is Missing* (1968) ii. 20 Stories that were..hard-hitting with no tinge of biased editorializing.

† **edmondsonite** (e·dmənsənait). *Min. Obs.* [f. name of George *Edmondson* (1798–1863), headmaster of Queenwood College, Hampshire + -ITE¹.] = TÆNITE 2.
1882 W. FLIGHT in *Proc. R. Soc.* XXXIII. 344 Lying on the plates of meteoric iron..were found thin metallic plates..of a flexible mineral, which had the composition Fe_2Ni_2... I propose to call this compound Edmondsonite.

edriophtha·lmate, *a.* [See EDRIOPHTHALMIAN.] Of or pertaining to the Edriophthalma. Also **edriophtha·lmatous, -mic** *adjs.*
1889 *Cent. Dict.*, Edriophthalmatous, edriophthalmic. **1904** *Ann. & Mag. Nat. Hist.* Feb. 154 In the Edriophthalmate orders no similar arrangement is known.

educand (e·diŭkænd). [ad. L. *ēducandus*, gerundive of *ēducāre* to EDUCATE.] One who is to be or is being educated.
a **1909** PETTY (Cent. D. Suppl.), We wish therefore that the educands be taught to observe. **1917** J. ADAMS *Student's Guide* 6 We have a case where the teacher is all educator, and the pupil all educand. **1927** —— *Errors in School* 44 Preparing the educand to take his place in his surroundings.

educatable (e·diŭkĕⁱtăb'l), *a.* [f. EDUCATE *v.* + -ABLE.] = EDUCABLE *a.* Hence **educata·bility.**
1868 A. B. ALCOTT *Tablets* 100 Not letters but life chiefly educate if we are educatable. **1894** G. C. STOPES *Brit. Free Women* 153 A constantly expanding view of the educatability even of ordinary girls. **1908** *Glasgow Herald* 21 July 10 Some fishes are 'more intelligent', or, as Dr. Longley puts it, 'have a greater degree of educatability' than others.

educated, *ppl. a.* Add: Phr. *educated guess*, a guess based upon a background of experience of the matter in hand.
1954 J. C. INGRAHAM *Mod. Traffic Control* 58 A lot of 'educated guesses', based on experience coupled with study, must go into answering questions like that. **1958** *Wall St. Jrnl.* 15 Dec. 14/1 It's an educated guess of top cavers, for example, that more than half of the chain's tinned fruit is sold under its private labels [illegible]. **1960** A. BATTERSBY *Guide to Stock Control* iv. 37 'An educated guess,' said General Patton, 'is just as accurate and quicker than compiled error.' **1966** M. CATO *Bird on Wing* vi. 94 'What's happened to her?' 'I wouldn't be sure. I could make an educated guess. I think she's gone home.'

educational, *a.* Add: **2.** (Later examples.) Also in special Combs.: **educational psychologist** (see quot. 1961); so *educational psychology*; **educational sociology**, the study of educational methods or systems from a sociological standpoint.
1917 W. R. SMITH (*title*) Introduction to educational sociology. **1923** D. SNEDDEN *Educational Sociology* I. iii. 33 Educational sociology has as its chief province the scientific determination of educational objectives. It constitutes an application of diverse sciences [illegible]. **1928** W. R. SMITH *Princ. Educational Sociology* i. 6 Educational sociology may be defined as the science which describes and explains the institutions, social groups, and social processes..in relation to the study and practice of education. **1950** *Discovery* May 142/1 The experience of the forces educational broadcasts [illegible] **1961** *Ibid.* 815/1 Further educational broadcasting includes all types of serious broadcasting. **1961** *Ibid.* 816/1 *Educational psychology*..includes observation of human beings in action and inquiry into all the means by which their behaviour is modified through their contact with their teachers, either directly or indirectly. **1960** *Where?* III. 13/2 *Educational psychologist*, a man or woman who studies human behaviour and is mainly concerned in dealing with the problems of children, and, sometimes, their teachers. **1961** WEBSTER, *Educational psychology*, a field of study that deals with the application of objective psychological methods and esp. of standardized tests to such problems as the selection of students for advanced or specialized training, the assessment of a student's progress, and the development of more effective methods of instruction.

educationally, *adv.* Add: **educationally subnormal** *adj. phr.*, applied to children who are mentally backward and cannot be taught in ordinary schools.
1953 C. BURT *Causes & Treatm. Backwardness* (ed. 2) 5 Special arrangements for those children..who appeared to be educationally subnormal. **1960** [see E.S.N. *s.v.* *E III].

Edwardian, *a.* and *sb.* Add: Also with pronunciation (edwǭ·ɹdiăn). **2.** (Earlier example.)
1879 *Dublin Rev.* Jan. 92 He [*sc.* R. Edgeworth] had also denounced..the pulling down of altars and the Edwardian communion service.
3. Belonging to or characteristic of the reign of Edward VII.
The spelling *Edwardian* is occasionally found.
1908 *Westm. Gaz.* 28 Sept. 2/2 That the Edwardian age is more placidly disposed towards such a threat [*sc.* 'your beer will cost you more'] than the times of the King's great-grandfather, George III. **1909** BRABROOK *Yet Again* 199 Let special suburbs be founded for Edwardian buildings. **1931** —— *Zuleika Dobson* ii. 12 Miss Dobson now, in the midst of the Edwardian Era, was the toast of two hemispheres. **1946** *Chambers's Jrnl.* 462/1 He sleeked his Edwardian moustache. **1927** M. SADLEIR *Trollope* 3 To the critical eye of Edwardian and Georgian enlightenment the mid-Victorians have appeared smug and hypocritical and selfish. **1930** *Birmingham Post* 24 Apr. 15/2 Most Edwardian little girls enjoyed a foretaste of the permanent wave—but without the permanence—suffering the infliction of tightly-twisted 'curl rags' at night. **1968** *Woman's Own* 28 Sept. 28/1 Can you tell me what is the difference between a veteran car and a vintage car? A veteran is a car built before 1905 and a vintage car is one manufactured

between the beginning of 1919 and the end of 1930. The period in between is covered by the category 'Edwardian car'.
4. Designating or characteristic of the clothes worn by Edwardians (senses 2 and 3).
1954 D. C. CALTHROP *Eng. Dress* v. 108 Let our pious prayer be that we shall not see our men and women in Edwardian clothes with a few exceptions, and those..are the grey frock coat and top hat and white waistcoat. **1954** 'N. BLAKE' *Whisper in Gloom* i. iv. 55 A..young man, in Norfolk jacket and tight Edwardian trousers. **1959** *Cath. Times* 23 Oct. 12/1 Any boy turning up in 'this exaggerated form of Edwardian dress' will be sent home.
B. *sb.* **1.** An alumnus of a school of King Edward VI's or VII's foundation, of St. Edward's School, Oxford, or of King Edward VII School, Sheffield.
1863 *St. Edward's Sch. Chron.* Apr. 15/2 Then with a kick S. Edward's chief raises the ball on high; And above their adversaries' heads the Edwardians see it fly. **1875** *Ibid.* July 254/2 The captain of the Old Edwardians. **1920** [illegible] [His] unrivalled knowledge and memory of Old Edwardians. **1923** *Birmingham Post* 16 Feb. 6/3 (*heading*) Distinction for an Old Edwardian. **1967** *King Edward VII School (Sheffield) Mag.* Spring 332 The Headmaster opened his report by expressing gratitude to..the Old Edwardians.
2. A person belonging to the period of Edward VII. Also *transf.*
1930 *Times Lit. Suppl.* 25 Sept. 617/2 Beguiled to sleep towards the end of last century..and suddenly awoken in a world of earnest Edwardians. **1929** [illegible] He was an Edwardian of Edwardians..surprisingly gracious, tactful and charming. **1930** V. SACKVILLE-WEST (*title*) The Edwardians. **1960** BIRD & HUTTON-STOTT *Veteran Motor Car* [illegible]
3. A person wearing clothes of Edwardian style; a 'Teddy boy'.
1954 *Newsweek* 17 May 47/1 They are the Edwardians—the bored drug-end kids of London. The Edwardians take their name from the costume that is, with variations, their uniform.. Tightly buttoned, knee-length 'Milord' coat with soft, black-velvet collar; double-breasted, pearl-gray waistcoat; and tight, drainpipe trousers.

Edwardia·na, *sb. pl.* [f. *EDWARDIAN + -ANA *suff.*] **1.** Matter relating to a school of King Edward VI's foundation, St. Edward's School, Oxford, or King Edward VII School, Sheffield.
1902 *S. Edward's Sch. Chron.* June 231/2 A very interesting collection of what I may perhaps term 'Edwardiana', that is, all printed matter designed for the use of the School.
2. Objects made in the reign of Edward VII, or imitations thereof.
1951 N. MARSH *Opening Night* iii. 62 There was a tarnished looking-glass, upon the surface of which someone had..painted a number of water-lilies and leaves. Among this growth, as if drowned in Edwardiana, Jacko's and Martyn's faces were reflected. **1961** *Times* 9 June 15/4 The old wafer-thin genteel affairs..have either gone the way of Edwardiana. **1966** *Guardian* 5 Dec. 6/5 All sorts of modish Victoriana and Edwardiana including parasols with elaborate handles re-covered to make umbrellas.

Edwa·rdianism. [f. *EDWARDIAN *a.* 3 + -ISM 3.] The collective characteristics of the reign of Edward VII; an Edwardian sentiment or expression.
1935 N. MARSH *Enter a Murderer* i. 18 'Has Gardener cut me out?' 'My sweet, what an Edwardianism.' **1936** J. BETJEMAN *Ghastly Good Taste* [illegible]

Edwardine (e·dwǭɹdin), *a.* and *sb.* [f. EDWARD + -INE¹.] **A.** *adj.* **1.** Belonging to the reign of Edward VI.
1859 *Sat. Rev.* 6 Aug. 161/2 [illegible] **1936** M. V. RONAN in *European Civilisation* IV. 589 The proclamation of the Edwardine Liturgy.
2. Belonging to or characteristic of the reign of Edward VII.
1908 *Tablet* 14 May Suppl. 792/1 One event of the Edwardine reign afforded to the Catholics of this country satisfactions that were all their own, the union of the King of Spain and the Princess Ena.
B. *sb.* An adherent of the religious principles of Edward VI.
1949 P. HUGHES *Rome & Counter-Reformation* iv. 67 The French ambassador..was initiated into the anti-Catholic publicity campaign which the Edwardines were preparing.

eel, *sb.* Add: **6. eel-pot** (earlier Amer. and *fig.* examples); **eel-schuit,** an eel-boat.
1630 in T. Hutchinson *Coll. Orig. Papers Massachusetts-Bay* (1769) 51 The yeele potts you sent for are made. **1648** 28 All winter they sell Brooms, Staves, Elepots, Baskets. **1838** DICKENS *Nickleby* xvi. 143 Manchester Buildings is an eel-pot, which has no outlet but its awkward mouth. *a* **1865** SMYTH *Sailor's Word-bk.* (1867) 269 *Dutch eel-skuyt*, a flat-bottomed somewhat cutter-rigged sea-boat. **1899** Eel schuyt [see SCHUIT]. **1905** M. A. WYLLIE *London to the Nore* iv. 42 Gaily painted and varnished Dutch eel-schuits.

eel (īl), *v.* [f. the sb.] *intr.* **a.** To fish for eels. **b.** To move or progress sinuously like an eel. Also (with *way*) *trans.* Cf. *EELING *gerund* and *vbl. sb.*
1922 JOYCE *Ulysses* 226 Five..sandwichmen..eeled themselves. **1953** DYLAN THOMAS *Lett.* (1966) 416, I must..ooze and eel up wheezily. **1962** M. DRABBLE *Summer Bird-Cage* xi. 185, I..use a small red plastic colander, and everything eels into the sink as often as not. **1969** *Daily Tel.* (Colour Suppl.) 17 Jan. 6 (Advt.), The Herald 13/60..simply eels its way through town traffic.

ee·l-grass. Chiefly *U.S.* [See EEL 6.] A plant with long narrow leaves: (*a*) = *grass-wrack* (GRASS *sb.*¹ 13); (*b*) = *tape-grass* (TAPE *sb.*¹ 4); (*c*) *fig.*
1790 S. DEANE *New-England Farmer* 19/1 The farmer ..may cart into [stockyard]..marsh-mud, eel-grass, flats, or even sand and loam [for making manure]. **1806** *Baltimore Even. Post* 19 Feb. 3/3 (Th.), A young man at Sullivan [Maine] saw a Fox go down to some eel-grass, and roll himself up in it. **1838** *Rep. Agric. Mass.* 1837 58 The eel-grass is of little value for manure. [illegible] **1867** *Amer. Naturalist* May 164 Go to the nearest brook, gather..a root of the Eel grass, [etc.]. **1880** C. E. BESSEY *Botany* 473 Order Hydrocharideæ.—This contains the Eel Grass, *Vallisneria spiralis*. **1888** GOODE *Amer. Fishes* 65 Jungles of eel-grass and meadows of [illegible] **1969** [illegible]

eeling (ī·liŋ), *gerund* and *vbl. sb.* [f. the vb.] Fishing for eels.
1780 in *Narragansett Hist. Reg.* (1882) I. 104 Made an eel spear. **1811** *Ibid.* Went eeling. **1843** *Knickerbocker* XXI. 426 No clammin', no eelin', and no pastor to feed your cows onto. **1895** *Outing* (U.S.) XXVI. 406/2 When he couldn't go eeling. **1955** E. S. DUCKETT *St. Dunstan* v. 135 In later years the fowling, eeling, and fisheries of the Fens..were to provide ample maintenance..for the abbeys.

eel-pout. 1. Substitute for def.: A marine fish of the family Zoarcidæ, esp. the viviparous blenny, *Zoarces viviparus*; also formerly applied to the BURBOT. (Add further examples.)
1772 in P. L. Phillips *Notes B. Romans* (1924) 124 There are three species of Eel Pouts [*sic*], but these are in every Sea or River on the South Sea. **1842** *Nat. Hist. N.Y., Zoology* iv. 155 The thick-lipped Eel-pout..is caught on the coast. **1885** [see LAMPER-EEL 2]. **1905** D. S. JORDAN *Guide to Study of Fishes* II. xxix. 518 The Zoarcidæ, or eel-pouts, have the body elongate. **1969** A. WHEELER *Fishes Brit. Isles & N.W. Europe* 448 Eelpouts often lie under stones or buried in soft mud.

ee·l-worm. [f. EEL + WORM *sb.* 4.] A nematode worm, esp. one parasitic on plants.
1888 E. A. ORMEROD *Notes of Observations of Injurious Insects* XI. 8 If the Eelworms are once established in land there is difficulty in clearing them. **1897** *Yearbk. U.S. Dept. Agric.* 568 What appeared to be eel worms, or nematodes, were observed in potato stalks, causing the foliage to turn yellow. **1930** *Dundee Advertiser* 15 Jan. 4/5 'Tulip-root', a disease [of oats] caused by the eel-worm. **1935** *Discovery* Sept. 266/1 Eelworms that live in vinegar barrels and the many species which inhabit the soil and by their effects upon it influence the welfare of man. **1951** *Times Lit. Suppl.* 14 Dec. 882/2 Rust in the eastern tropics, hitherto fortunately never confused with eelworm attack, is due to *Phytophthora*. **1959** *Times* 14 July 10/6 Maris Piper is resistant to [illegible] eelworm.

-een (īn), *suffix*¹, ultimately derived from L. *-īnus*, *-īna* adj. suffix, through F. *-in*, *-ine*, as in † *damaskeen* (1551), *canteen* (1737), *tureen* (1706), *bombazeen* (18th–19th century variant of *bombasine*). *Bombazeen* apparently furnished the model for *velveteen* (1795, *velvatean* 1776), on which followed *sateen* (an alteration of *satin*).

-een (īn), *suffix*², ad. Ir. dim. suffix *-īn*, as in *buckeen* (1793), *colleen* (1828), *dudeen* (1841), *girleen* (1836), *jackeen* (1840), *poteen* (1812), *spalpeen* (1780), *squireen* (1809).

e'enamost (ī·nămōᵘst), *adv. Eng.* and *U.S. dial.* Also **e.** **e'en a most, a'most,** 'a most; **enymost, eeny**(-)**most, enermost.** [f. *e'en* EVEN *adv.* + *a'most* ALMOST *adv.*] Almost; nearly.
a. **1735** *Pegge Alphabet of Kenticisms* (1876), *E'en a most*, almost. **1781** *Pennsylvania Jrnl.* (Philadelphia) 20 June, I once heard a man say, another 'swore terribly; he swore e'en amost like a good while pie'. **1819** *Kansas* (N.Y.) *Republ.* 21 Aug. (Th.), [The whale's head] was e'en a most off. **1843** T. C. HALIBURTON *Attaché* I. xiv. 253 The repudiation of debts has lowered us down e'en a'most to the bottom of the shaft. **1844** 'J. SLICK' *High Life in N.Y.* I. iv. 43 He looked eenamost tuckered out. **1891** H. FREDERIC *Copperhead* (1894) 208 'I could e'en a'most 'a' thought it was Alvy talkin',' was what he said. **1911** in *Dialect Notes* III. 543.
β. **1813** JANE AUSTEN *Let.* 18 Oct. (1932) II. 356 Poor Will Amos said the fleas were so starved that they..enermost eat him up. **1833** C. A. DAVIS *Major Downing* 168 It was made so long ago he has enymost forgot it. **1840** *Crockett Almanac* 2 The sun will be enermost hid. **1845** S. JUDD *Margaret* I. xiv. 113 He..has got the whole Bible eeny most by heart. **1870** STOWE *Oldtown Fireside Stories* (1871) 189 [illegible] eenymost beat out.

eeny (ī·ni). A nonsense word (also *eny*, *eenee*, *eena*) used in the first line of the most popular counting-out rhymes among children in Britain and the U.S.
1855 *N. & Q.* 1st Ser. XI. 113/2 The Schoolboy Formula. ..Eeny, meeny, mony, mite. **1888** W. W. NEWELL *Games & Songs Amer. Children* xiv. 199 Eny, meny, mony, mite. **1923** KIPLING *Land & Sea Tales* 279 Eenee, Meenee, Mainee, Mo! Catch a nigger by the toe! **1944** H. G. WELLS *'42 to '44* 199 However counting began, it is evident from the first it produced a profound effect of mystery and magic upon the human mind. 'Eena, meena, mina, mo'—they counted men out for definite purposes. **1958** A. HACKNEY *Private Life* xii. 123 [illegible]

-eer, *suffix*¹. Add: The spelling *-eer*, replacing the older *-ier*, became frequent in new words in the early 17th century. *Mountaineer* and *waistcoateer* (a prostitute) afford early instances, and are also exceptional examples, of the use of this suffix. A few formations denote inanimate objects, as *gazetteer* (1704), *muffeteer* (1806–7).
In the latter part of the 17th century gerundial and (to a less extent) participial formations on agent-nouns in *-eer* appeared, and increased in the course of the following century, as *auctioneering* (1733), *buccaneering* (1703), *electioneering* (1774), *engineering* (1720), *parliamenteering* (1711), *privateering* (1664), *volunteering* (1691). These, being formed directly on the sbs. in *-eer*, do not necessarily imply the existence of a corresponding infinitive or finite verbal form, though an early example actually appears in *mutineered* (1682). Some of these formations, e.g. *parliamenteering*, were in commoner use than the original sb.; occasionally, as in *revolutioneering*, no sb. exists, *-eering* itself being used as a suffix.

-eer, *suffix*², representing Du. *-eren*, ad. F. infinitive ending *-er*; as in *domineer* (Shakespeare), ad. Du. † *domineren*, ad. F. *dominer* to DOMINATE; *commandeer* (1881), ad. Cape Du. *commanderen*, ad. F. *commander* to COMMAND.

eff (ef), *v.* [Variant of EF, name of the letter F, euphemistically representing *FUCK *v.* 2, 3.] **1.** (Used as an expletive on its own account, as a milder alternative to the full form of the word *fuck*, or else as a euphemistic report of an actual use of the full word.)
1950 HEMINGWAY *Across River & into Trees* ix. 78 'Eff Florence,' the Colonel said. **1958** K. AMIS *I like it Here* xiii. 164 You young people eff off. **1958** B. SANDERSON *Night of Horns* xxv. 165 You sure effed things up. **1963** L. MEYNELL *Virgin Luck* viii. 102 'Eff off,' Johnny told Antonio. Antonio effed off to the other end of the ship. **1967** K. GILES *Death & Mr. Prettyman* [illegible] 'What did he matter?' 'Effed if I know.'
2. To utter the word *fuck* or some equivalent word. Also *to eff and blind*, to use strong expletives, to swear continuously.
1943 M. HARRISON *Reported Safe Arrival* 31 They'd eff and blind till yer ear-'oles started ter frizzle. **1959** A. WESKER *Chicken Soup & Barley* i. ii, He started effing and blinding and threw their books on the floor. **1965** J. GASKELL *Fabulous Heroine* 96 He would argue and eff in an intellectual way at afternoons.
Hence **e·ffer; e·ffing** *vbl. sb.* and *ppl. a.*; also as *adv.*
1944 J. AGATE *Red Letter Nights* 45 (Carrying a sandbag.) 'Arf of effing France,' and 'ad to walk the other 'arf to fetch it.' **1949** *Penguin New Writing* XXV. 186 He's a bastard, but I'll effing well get him. **1959** E. PARTRIDGE *Here, There & Everywhere* 28 The British workman's exclusive preference for *bloody*.. A preference shared with 'effing'. **1961** S. PRICE *Just for Record* vi. 55 You're effing right. **1963** P. WEST *Mod. Novel* ii. i. 142 On it goes, the brief effing-and-blinding. **1964** S. BELLOW *Herzog* 81 What good are those effing eggheads? **1967** C. DRUMMOND *Death at Furlong Post* ii. 16, I had an argument with that effer, [illegible]. **1969** *Observer* 2 Feb. 31/1 Mary felt the situation was 'diabolical' and there was plenty of effing and blinding. **1969** *Private Eye* 24 Oct. 14 The relatives get effing tough. **1970** K. GILES *Death in Church* i. 11 I'm a tolerant effer myself, but blasphemy I can't come at.

effect, *sb.* Add: **1. d.** Any of various phenomena of physical science, e.g. those connected with electric currents, usually named after the first discoverer or describer of the appearance. See also *DOPPLER, *EINSTEIN, *FARADAY, *ZEEMAN.

1881 S. P. THOMPSON *Elem. Less. Electr. & Magn.* 343 This phenomenon of heating (or cooling) by a current, where it crosses the junction of two dissimilar metals (known as the 'Peltier effect', to distinguish it from the ordinary heating of a circuit where it offers a resistance to the current, which is sometimes called the 'Joule effect'). *Ibid.* 346 This effect, known as the Thomson effect from its discoverer Sir W. Thomson, is opposite in iron to what it is in copper or zinc. **1894** *Ibid.* 563 Kerr's Effect. Dr. Kerr showed in 1877 that a ray of polarized light is also rotated when reflected at the surface of a magnet or electromagnet.
e. *Psychol.* (See quots.)
1905 E. L. THORNDIKE *Elem. Psychol.* (1907) x. 166 This law..[of acquired connections] might be stated..as follows. (1) The line of least resistance is,..that resulting in the greatest satisfaction to the animal... We may call (1) the Law of Effect. **1922** R. S. WOODWORTH *Psychol.* 392 The law of effect, stated as objectively as possible, is simply that the successful or unsuccessful outcome or effect of a reaction determines whether it shall become firmly linked with the stimulus, or detached from the stimulus and thus eliminated. **1929** *Encycl. Brit.* XVIII. 683/1 The so-called 'Law of Effect',..which refers..more particularly to the acquisition of skill, and to the formation of habit. **1951** J. C. FLUGEL *Hundred Years Psychol.* (ed. 2) II. ii. 83 Thorndike and others owe to Bain the first clear formulation of what later became known as the 'law of effect' (the 'stamping in' of movements under the influence of pleasure).
3. b. Also of music (see also quot. 1938).
1870 *Porcupine* 19 Mar. 492/3 The other perfectly unique 'Spohr effects' produced during the performance of this cantata. **1938** *Oxf. Compan. Mus.* 285/1, *Effects*, a term used in dance-band parlance..for imitative instruments such as various forms of whistle and anvil, baby cry..whip crack, &c. **1955** KEEPNEWS & GRAUER *Pictorial Hist. Jazz* ii. 19 They went in for cowbells and other dubious 'novelty' effects.
c. (Cf. STAGE-*effect*, *sound-*effect*.) Now usu. in pl.: the various aids and contrivances (appropriate 'noises off', lighting, etc.) used to accompany and vivify the production of plays, films, or broadcasts. Also *attrib.*, as in *effects studio*, *microphone*, etc.
1881 P. FITZGERALD *World behind Scenes* I. 46 Few pieces excited more mysterious interest on its first production than the spectral effect in the drama. *Ibid.* 70 One of the most startling and successful effects was given in a revival of 'Sardanapalus'... In the midst of the banqueting, a thunderbolt descended. **1911** C. N. BENNETT et al. *Handbk. Kinematogr.* ix. 69 We append a table of well-known tinting effects... Moonlight effect,..candle light and lamp light effects,..firelight effect. **1914** E. A. DENCH *Playwriting for Cinema* xviii. 78 Lighting effects often present a difficulty. **1928** *B.B.C. Handbk.* 268/2 *Effects Studio*, a studio in which the noise effects incidental to a transmission are made. **1933** A. BRUNEL *Filmcraft* 217 Salaries..wages..set-building materials..costumes.. properties, laboratory charges, trick effects. **1941** *B.B.C. Gloss. Broadc. Terms* 10 *Effects*, sounds characteristic of a scene or incidental to an event, either produced artificially in order to create illusion (e.g. in a dramatic broadcast) or occurring naturally (e.g. in an outside broadcast); sound effects. Hence *effects microphone*, microphone specially placed to pick up such sounds. **1950** T. S. ELIOT *Cocktail Party* 40 She has simply faded—into some other picture—Like a film effect. **1951** N. MARSH *Opening Night* iv. 98 A stage-hand..carried an effects-gun. This was fired at the appropriate moment. **1961** K. REISZ *Technique Film Editing* (ed. 9) 279 *Effects track*, sound-track of sound effects other than speech and music.
6. *for effect.* (Example.)
1824 J. S. MILL in *Westm. Rev.* II. 347 He who writes for effect..must address himself to the prevalent feeling.
7. *to put into effect*: to accomplish, to realize.
1936 *Discovery* Sept. 296/1 A tireless film director who was forever having ideas and would not rest until they were put into effect successfully.

effect, *v.* **3.** (Later example.)
1817 A. CONSTABLE *Let.* 16 Jan. in *J. Constable's Corr.* (1962) 153, I..hope you will..endeavour to make all right with the Doctor, even tho' as by this time you know I dare say that your first letter did not effect.

effectism. Delete *nonce-wd.* and add later example.
1889 *Harper's Mag.* Nov. 964/1 The vice which has been very graphically called *effectism*, or the itch of awaking at all cost in the reader vivid and violent emotions.

effective, *a.* and *sb.* Add: **A. 3. e.** *effective temperature*: see quots. 1929, 1930, 1957.
1915 *Sci. Amer.* 3 Apr. 328/2 Whereas in the case of terrestrial illuminants the actual temperature can be gaged, we must be satisfied with determining the *effective* temperatures of celestial bodies. Supposing the sun to radiate like a 'black' body,..its 'effective' or 'black' temperature can be determined from the sum total of radiation sent to the earth (solar constant). **1924** H. DINGLE *Mod. Astrophysics* ix. 119 The wavelength of maximum energy, from which the effective temperature is determined. **1929** R. A. WARDLE *Princ. Applied Zool.* 197 There would seem to be for each species of insect a range of temperatures between whose maximum and minimum extremes the insect is active... This range of temperatures may be termed the range of Effective Temperatures. **1930** *Engineering* 28 Nov. 671/1 The effective temperature of a room, that is to say, the temperature at which, in still air, a sizeable black body at standard temperature will lose heat at the same rate as it is being lost in its environment. **1957** *Encycl. Brit.* XVII. 847F/2 The term 'effective temperature' was sometimes used to denote what we now call 'colour temperature'. 'Effective temperature' is now used to denote the temperature of a black body for which the total emitted radiation..is the same as the total radiation.. emitted from the surface of a star.
f. *Philology.* Denoting the completion or result of an action. Cf. *EFFECTIVE *sb.* 3.
1932 *Jrnl. Eng. & Germanic Philol.* XXXI. 251 The latter class may be called the effective aspect: 'His strength *gave out*', i.e., he came to the end of his strength. **1970** *Language* XLVI. 300 The *-t* forms were predominantly selected..for preterits in contexts suggestive of non-durative ('effective') aspect.
g. Optics. *effective aperture*, the diameter of the widest beam of light incident on an optical system that goes to form an image, whether formed at a principal focus or not.
1893 *Proc. R. Soc.* LII. 412 The effective aperture of one or more of the various stops supplied with the lens is found by a well-known method. **1965** M. J. LANGFORD *Basic Photogr.* iii. 54 As the aperture closes the diameter of this incident light beam or 'Effective Aperture' narrows proportionally.
B. *sb.* **3.** An effective verb or an effective aspect or part of a verb. Cf. *EFFECTIVE *a.* 3 f.
1935 CURME *Gram. Eng. Lang.* II. xii. 237 In duratives, ingressives, effectives..the present participle represents the act as incomplete.

effector (efe·ktǫɹ). **1.** As a variant of EFFECTER.
2. *Biol.* In attrib. use or as adj., or sb., applied to an organ which shows the specific effect of a nervous reaction.
1906 C. S. SHERRINGTON *Integrative Action Nerv. Syst.* i. 7 An *effector* organ, e.g., gland cells or muscle cells. *Ibid.* ix. 309 The conductor mediating between receptor and effector. *Ibid.*, At the deep, i.e. effector, end the branching of the conductive stem places it in touch not with one effective cell but with many. **1920** T. P. NUNN *Education: its Data & First Princ.* 166 Connector-axons which make their way to effector-neurones entirely outside the cord. **1927** W. SHUMWAY *Vert. Embryol.* 186 The sense organs (receptors) from the nerves (transmittors) by which stimuli are passed on to the muscles or glands (effectors). **1927** S. W. RANSON *Anat. Nerv. Syst.* (ed. 3) 18 A sensitive mechanism for receiving stimuli and conducting them to the appropriate organs of response. These organs..are known as effectors. **1952** *Brit. Jrnl. Psychol.* Feb. 18 To keep his attention one step ahead of his effector performance, or any decrease of effector accuracy. **1962** *Gray's Anat.* (ed. 33) 943 (*caption*) Effector neurones. Effector organs. *Ibid.*, Specialized receptors are differentiated at the surface of the body for the receipt of the stimulus and specialised effectors, either muscle or secretory cells, are differentiated for the response. **1968** *Lebende Sprachen* XIII. 39/1 Such an adaptive system requires..an effector mechanism for adaptively seeking that goal condition.

effer: see *EFF *v.*

efferve·scingly, *adv.* [f. EFFERVESCING *ppl. a.* + -LY².] In an effervescing manner, sparklingly.
1898 *Westm. Gaz.* 15 June 3/1 He speaks briskly and effervescingly.

efficiency. Add: **2. c.** *spec.* in *Economics*, as *economic*, *marginal*, *technical efficiency* (see quots.).
1906 A. SHADWELL (*title*) Industrial efficiency. A comparative study of industrial life in England, Germany and America. **1916** A. MARSHALL *Princ. Econ.* (ed. 7) IV. v. 193 Health and strength, physical, mental and moral.. are the basis of industrial efficiency, on which the production of material wealth depends. **1926** *Industrial Management* Jan. 4/1 The choice is restricted to a definitely established range of sizes and types, outside of which the designer of the average car would venture with considerable trepidation, and at the sacrifice of manufacturing efficiency. *Ibid.* 34 (Advt.), Sure and efficient power drives make for higher overall plant efficiency. **1936** J. M. KEYNES *Gen. Theory Employment, Interest & Money* IV. ii. 135 The relation between the prospective yield of a capital-asset and its supply price or replacement cost, *i.e.* the relation between the prospective yield of one more unit of that type of capital and the cost of producing that unit, furnishes us with the *marginal efficiency of capital* of that type. **1953** G. HUTTON *We too can Prosper* i. 13 Productivity is the efficiency..of production. **1965** J. L. HANSON *Dict. Econ. & Commerce* 133/1 *Economic efficiency*, the maximum average output per employee. *Ibid.*, Economic efficiency is a much wider term than technical efficiency, which applies only to the efficiency of one factor of production, namely capital. **1965** SELDON & PENNANCE *Everyman's Dict. Econ.* 137 *Economic efficiency*, relates to output per unit cost of the resources employed; contrasted with *technical* efficiency, which measures output of energy per unit of energy applied. **1966** A. GILPIN *Dict. Econ. Terms* 57 *Economic efficiency*, the efficiency with which scarce resources are used and organised to achieve stipulated economic ends. In competitive conditions, the lower the cost per unit of output, without sacrifice of quality, in relation to the value or price of the finished article, the greater the economic efficiency of the productive organisation.
3. *Mech.* and *Physics.* † **a.** The work done by a force in operating a machine or engine; the total energy expended by a machine. *Obs.*
1827 D. GILBERT in *Phil. Trans. R. Soc.* CXVII. 27 Since therefore a machine is efficient in producing duty, or effect, in proportion to the force applied, multiplied into the space through which it acts, I propose to denominate this function ($f \times s$) *efficiency*; retaining the word *duty* for a similar function indicative of the work performed... By a comparison of these two quantities, the *efficiency* expended on any machine, and the *duty* performed by it, an exact measure will be ascertained of its intrinsic worth. *Ibid.* 31 The received efficiency of a water wheel being represented by the pounds of water passing over it × by the fall in feet, less the height due to the velocity with which the periphery moves in its rotation. **1832** W. WHEWELL *First Princ. Mech.* iv. 54 It has been proposed by Mr Davies Gilbert..to call the product $f \times s$ the Efficiency of the force *f*. *Ibid.* 55 The work done by any machine is always equal to the Efficiency of the moving force... But machinery may very much increase the convenience of application of the force. *Ibid.* 64 To find the theoretical Efficiency of a Steam Engine working for one minute.
b. The ratio of useful work performed to the total energy expended or heat taken in.
Radiant or *luminous efficiency*, the ratio of the total luminous flux produced to the total power consumed.
1855 W. J. M. RANKINE in *Proc. Philos. Soc. Glasgow* III. vi. 398 Efficiency of engines. An engine is a contrivance for transforming energy by means of the periodical repetition of a cycle of variations of the accidents of a substance. The *efficiency* of an engine is the proportion which the energy permanently transformed to a useful form by it bears to the whole energy communicated to the working substance. **1858** —— *Man. Appl. Mech.* vi. 610 The object of improvements in machines is to bring their efficiency as near to unity as possible. **1879** G. B. PRESCOTT *Speaking Telephone* xiv. 465 The true comparative measure of the efficiency of dynamo-electric machines. **1887** *Encycl. Brit.* XXII. 496/1 The efficiency of furnace and boiler is the ratio which the amount of heat taken up by the water bears to the whole potential energy of the fuel. In good boilers this efficiency is about 0·7. **1902** *Ibid.* XXXIII. 418/1 The ratio between the power given out by a transformer and the power taken up by it is called its efficiency. **1929** *Ibid.* VIII. 290/1 The 'efficiency' of a lamp is measured in lumens per watt. **1958** *Van Nostrand's Sci. Encycl.* (ed. 3) 561/1 Usually there is a load for which the efficiency is a maximum. **1960** R. W. MARKS *Dymaxion World of B. Fuller* 51/1 In engineering, the term 'efficiency' means the ratio of foot-pounds of work realized to the foot-pounds of energy consumed by a given mechanical system. **1967** CONDON & ODISHAW *Handbk. Physics* (ed. 2) v. 6/1 Of all engines working between two heat reservoirs of temperature τ_1 and τ_2 ($\tau_2 > \tau_1$), the reversible engine is the one of maximum efficiency.
4. An efficiency apartment; a room with limited facilities for washing and cooking. *N. Amer.*
1952 *Times-Picayune* (New Orleans) 11 May v. 8 (Advt.), Modern, furnished efficiency. Living room, bedroom combined, kitchen and bath. **1959** H. GARDNER *So what else is New?* 70 Irving Hoffman..spotted this room-for-rent sign down in Greenwich Village: 'One room efficiency, no bath, suitable for artist.' **1959** *Times Lit. Suppl.* 6 Nov. p. xiii/2 Even in..the motel, you will find..a curiously abrupt sign labelled 'Efficiencies', by which token you will learn that your expensive quarters are furnished with..an up-to-date kitchenette. **1968** *Globe & Mail* (Toronto) 17 Feb. 34 (Advt.), Efficiencies 67.50 per week.
5. *attrib.* **efficiency apartment** *N. Amer.*, one with limited facilities for washing and cooking; **efficiency audit**, an examination of a business organization, etc., for the purpose of establishing the efficiency of its procedures; **efficiency bar**, a restriction of salary to a maximum figure which may be increased only when satisfactory evidence of efficiency has been produced; **efficiency engineer** *U.S.*, = **efficiency expert*; **efficiency expert**, one who examines the efficiency of industrial or commercial organization or production.
1930 *San Antonio* (Texas) *Light* 31 Jan. 14/8 (Advt.), 641 N. Flores..*Efficiency Apts.* **1965** L. H. WHITTEN *Progeny of Adder* (1966) 25 The efficiency apartment he really couldn't afford in southwest Washington. **1952** *Public Administration* XXX. 77/2 Accountancy is developing the technique of standard costs; and the new science of business management is devising techniques of 'efficiency audit'. **1965** J. L. HANSON *Dict. Econ. & Commerce* 133/1 If a nationalised industry is to be run deliberately at a loss 'as a service' it loses the economic test of its efficiency. In such a case the profit test would have to be replaced by some kind of 'efficiency audit'. **1897** *Daily News* 16 Mar. 2/2 (*heading*) An efficiency bar. No officer whose salary has reached 112 *l.* is to be allowed to pass that figure without a satisfactory certificate of the excellence of his conduct and of his ability to perform the highest duties of his class. **1964** *New Statesman* 10 Apr. 580/1 (Advt.), Lecturer £1,250..–£2,150 with an efficiency bar at £1,750. **1913** *Engineering Mag.* Aug. 715/1 Whenever an efficiency engineer is engaged to study a business or a manufactory, he must of necessity deal to a great extent with the rank and file in obtaining the vital data upon which to base his recommendations. **1954** J. A. C. BROWN *Social Psychol. Industry* x. 276 The efficiency engineers Taylor and Gilbreth and the early industrial psychologists. **1913** *Engineering Mag.* Aug. 717/2 So important is the work of this latest class of engineers, the efficiency experts, that no factor should be passed lightly over that has a bearing on their success. **1924** E. SHEPHARD *P. Bunyan* 184 Paul hired a kind of efficiency expert..and Gerber used to walk around..and keep tab on the men and count up how they spent their time. **1959** D. EWEN *Compl. Bk. Amer. Mus. Theater* 295 This musical spoofs an efficiency expert.

effigurate (efi·giŭrĕⁱt), *a. Bot.* [f. EF- (= *ec-*, *ex-*) + FIGURATE.] Having a definite outline.
1872 W. A. LEIGHTON *Lichen-Flora* (ed. 2) p. xxiv,

Effigurate, having a distinct form or fringe. **1886** HOLMES & GRAY *Brit. Fungi, Lichens* 6 When its [*sc.* of the thallus] circumference is radiate or stellate, it is effigurate. **1921** A. L. SMITH *Lichens* iii. 80 A frequent type of squamulose thallus is that termed 'placodioid', or 'effigurate', in which the squamulose character is chiefly apparent at the circumference.

effigy. Add: **3. effigy-mound,** a prehistoric earth mound in the shape of an animal.
1885 *Science* V. 131/2 So few earthworks resembling animals in their shape are known beyond the limits of Wisconsin, that I send you an account..of the most interesting..Minnesota effigy mounds. **1929** *Encycl. Brit.* (ed. 14) XV. 928/2 In Wisconsin the most interesting mounds are the effigy mounds—earthen forms of mammals, birds and reptiles. *Ibid.*, The purpose of these effigy mounds is probably totemic. **1948** A. L. KROEBER *Anthropology* (ed. 2) xviii. 820 Some early Woodland cultures..Proto-Effigy Mound or Old Copper, Wisconsin... Some middle-period cultures..Effigy Mound, Wisconsin.

effing: see *EFF *v.*

‖ **effleurage** (eflörāʒ), *sb. Massage.* [Fr., f. *effleurer* to touch or stroke lightly.] A centripetal stroking movement made with the flat or the heel of the hand. So **effleurage** *v. intr.*, to massage a part or a limb with this movement.
1886 MURRELL *Massage* iii. 10 *Effleurage*..is a stroking movement made with the palm of the hand passing with various degrees of force over the surface centripetally. **1890** A. KELLGREN *Man. Treatment* 9 The effleurage may be very superficial or deep. **1893** A. C. HALE *Massage* 133 Separate the fingers and effleurage throughly over and round the malleoli. **1961** I. FLEMING *Thunderball* iv. 36 Two hands, clad in..fur, were rhythmically passing..up and down the whole length of his body... Presently he said sleepily, 'Is that what they call effleurage?'

effluent, *sb.* Add: **c.** Waste discharged from an industrial works.
1930 *Jrnl. Iron & Steel Inst.* CXL. 159A The effluent is very clear..and is not harmful to human or fish life. **1959** *Listener* 8 Jan. 48/2 The nuisances of smoke and factory effluent. **1969** MORSE *Fur Trade* II. vi. 85 This is the result of industrial effluent from the two paper mills.
d. Radioactive waste discharged from an atomic plant.
1955 *Rep. U.K. Atomic Energy Auth.* I. ix. 24 The Authority are required to analyse samples of all liquid effluent discharged.

effort, *sb.* Add: **2. b.** Often used somewhat trivially for any kind of achievement, artefact, or result of activity.
1871 L. W. M. LOCKHART *Fair to See* xxiii, His first attempt [*sc.* the letter itself] ran thus..and he tried again. ..This effort was also torn up in despair. **1924** A. D. SEDGWICK *Little French Girl* I. vii. 59 The Venus is an effort of Ruth's. **1945** C. H. WARD-JACKSON *Piece of Cake* (ed. 2) 26 *Effort*, applied to anything that has been made, especially mechanical. **1967** *Lebende Sprachen* XII. 161/1 Companies who have extensive development efforts under way. *Ibid.*, The research effort will be carried out by the development of laboratories already in operation. *Ibid.*, The government recognizes the benefits of a broadbased space effort.
c. *Mech.* (See quots. 1842, 1875.)
1842 FRANCIS *Dict. Arts*, *Effort*, the force with which a body in motion tends to produce an effect, whether the effect be really produced or impeded. **1875** RANKINE & BAMBER *Mech. Text-Bk.* (ed. 2) 205 A direct force is.. distinguished, according as it acts with or against the motion of the point..by the name of effort, or of resistance, as the case may be. **1883** *Encycl. Brit.* XV. 764/2 If the component along the direction of motion acts with the motion, it is called an effort. **1959** *Chambers's Encycl.* IX. 200/2 The ratio of the load to the effort is the mechanical advantage of the machine, and the ratio of the distance moved by the effort to the distance moved by the load is the velocity ratio.

effortful (e·fǫɹtfŭl), *a.* [f. EFFORT *sb.* + -FUL.] Exhibiting, full of, or requiring effort.
1900 *Academy* 1 June 473/1 At its most effortful it is strained and excited—a painfully obvious striving beyond the poet's power. **1926** *Brit. Weekly* 4 Nov. 120/4 A life of value is always a strenuous and effortful life. **1927** C. C. MARTINDALE *Christ is King* ii. 34 It is usually more effortful to do right than to do wrong. **1927** *Manch. Guardian Weekly* Sept. 169/2 The effortful conscious striving towards an æsthetic and philosophical ideal. **1964** *Economist* 31 Oct. 469/1 The effortful field of the Government's attempts.

e·ffortlessness. [f. EFFORTLESS *a.* + -NESS.] The character or quality of being effortless.
1889 W. B. CARPENTER *Perm. Elem. Relig.* vii. 253 He passed from the effortless stage, through the stage of pain and effort, and on again to the stage of effortlessness. **1904** *Daily Chron.* 18 Apr. 3/2 And, in spite of its effortlessness, how splendid in technique. **1923** *Public Opinion* 28 Dec. 643/2 He gave an impression of effortlessness in the calm ease of his manner. **1928** *Observer* 8 Jan. 21/4 That effortlessness and comparative noiselessness of the engine.

effusive, *a.* Add: **1. b.** *Geol.* [G. (H. Rosenbusch *Mikrosk. Physiographie d. Massigen Gesteine* (ed. 2, 1887) 6).] Of an igneous rock: poured out on the earth's surface in a state of fusion and afterwards solidified; so *effusive period*, the period in which effusive rocks were formed. Also *sb.*, an effusive rock.
1888 F. H. HATCH in J. J. H. Teall *Brit. Petrogr.* 429 *Effusive*, a term lately used abroad for those rocks which have been poured out at the surface, the word *eruptive* now being generally used for the whole group of massive rocks. **1895** A. HARKER *Petrol.* 128 The two periods of consolidation were styled by Rosenbusch the 'intratelluric' and the 'effusive'. **1897** G. P. MERRILL *Treat. Rocks* II. i. 60 To divide the eruptive rocks into two general groups: first, the intrusive or plutonic rocks; and second, the effusive or volcanic rocks. **1903** *Amer. Jrnl. Sci.* XVI. 121 An origin contemporaneous with that of the Rossland effusives. **1903** A. GEIKIE *Text-bk. Geol.* (ed. 4) 197 The effusive or volcanic rocks (Erguss-gesteine). **1905** J. GEIKIE *Struct. & Field Geol.* 206 Effusive rocks..are of two types, crystalline and fragmental. **1915** W. LINDGREN in W. N. Rice et al. *Probl. Amer. Geol.* v. 273 Basalt is probably the most widely spread of the Tertiary effusives. **1939** A. JOHANNSEN *Descr. Petrogr.* (ed. 2) I. i. 5 The effusive or extrusive rocks (vulcanites of Scheerer) are those which were poured out upon the surface of the earth; for example, our modern lavas... In this group are also included certain other rocks which, strictly speaking, are not extrusives, namely, the pyroclastics.

Efta (e·ftă). Also **E.F.T.A., EFTA.** [f. the initials of *E*uropean *F*ree *T*rade *A*ssociation.] An economic alliance, on free trade principles, comprising the United Kingdom and several other European countries.
1959 *Cmnd.* 906 5 in *Parl. Papers 1959–60* XXIV. 21 The individual freedom of action of E.F.T.A. members in their external tariffs. **1960** *Times Rev. Industry* Dec. 4/2 European Free Trade Association (Efta). **1961** *Economist* 17 June 1252/3 Markets in which a decisive majority felt prospects were improving were EFTA, North America, the Common Market and the Communist block. **1963** *Times* 1 Feb. 13/5 Commercially we can mark our sense of unity with Europe by urging the Efta countries to join with us in inviting the E.E.C. to join the European Free Trade Area.

egalitarian (ĭgælitēᵊ·riăn), *a.* Delete *nonce-wd.* and add later examples. Also *sb.*, one who asserts the equality of mankind. Hence **egalita·rianism**, the doctrine or condition of such equality.
1894 *Daily News* 21 July 5/2 The Scot..is, of course, the most 'egalitarian' of mankind. **1898** J. E. C. BODLEY *France* I. 162 In warlike but egalitarian France. **1909** BELLOC *Marie Antoinette* 377 The violent egalitarian mood which had now for a year and more driven the military fury of the Republic. **1920** *19th Cent.* July 10 A variant of the superman as noxious as the mediocrity or the egalitarian. **1930** W. K. HANCOCK *Australia* iii. 67 The economic and social foundations of the homogeneous egalitarian society. **1931** A. HUXLEY *Music at Night* II. 123 The theatres in which the egalitarians will enjoy the talkies, tasties, smellies, and feelies. **1932** *Nation* (N.Y.) 6 July 16/1 There is apparently a further identification of egalitarianism, communism, and materialism. **1937** *Times* 8 June 22/5 With an eye to the growing egalitarianism of social life..the ceremonial side of the monarchy should be scrutinized afresh. **1955** M. GLUCKMAN *Custom & Conflict in Africa* ii. 31 Not even the first introduction of.. European trade-goods broke up this egalitarianism.

Egeria (īdʒīᵊ·riă). In *Roman Mythology*, the name of a goddess, supposed to be the instructress of Numa Pompilius, and regarded as the giver of life; *transf.* a tutelary divinity; a patroness and adviser.
1621 BURTON *Anat. Mel.* (1624) To Rdr. b, My Mistris Melancholy, my Ægeria, or my *malus Genius*. **1818** LADY MORGAN *Fl. Macarthy* III. ii. 82 Why may not I have my Egeria or my dæmon, as well as another? **1826** DISRAELI *Viv. Grey* III. vi, It is in these moments, that Nature becomes our Egeria. **1890** *Athenæum* 24 May 670/2 With the help and counsel of Beatrice, who turns his Egeria, he wins fame in law and politics. **1932** H. NICOLSON *Diary* 9 Dec. (1966) 124 She has a sort of *Egeria* look which must be put on. **1934** E. POUND *Eleven New Cantos* xxxv. 23 His Wife now acts as his model and the Egeria Has, let us say, married a realtor.

egg, *sb.* Add: **3. b.** In full *egg coal*: see quots. orig. *U.S.*
1855 *Santa Barbara* (Calif.) *Gaz.* 22 Nov. 1/5 The attempt to make omelets out of 'egg' coal has been abandoned. **1880** *Bradstreet's* 2 Oct. 5/4 The sizes used are 'lump', 'steamboat', 'broken', and 'pea'; while for family use the sizes are 'egg', 'stove' and 'nut'. **1900** *Coal & Metal Miners' Pocketbk.* (ed. 6) 434 Egg passes over 2" mesh, and through 2½" mesh. *Ibid.* 585 Egg coal, anthracite coal that will pass through a 2½" square mesh and over a 2" square mesh. **1924** A. T. SHURICK *Coal Industry* 144 The broken coal..is again screened into egg, stove, and nut coal. **1970** F. MCKENNA *Gloss. Railwaymen's Talk* 35 *Eggs*, ovoid briquettes, made of coal dust and cement dust, used during coal shortage.
c. *Cricket.* = DUCK'S EGG b.
1861 *Bell's Life* 25 Aug. (Suppl.) 2/1 Dowson 'laid an egg'; R. D. Walker made 10 in an hour and a quarter. **1898** K. S. RANJITSINHJI *With Stoddart's Team in Australia* x. 195 Gregory..was yorked first ball...Iredale also secured an 'egg'.
d. A bomb, a mine. *slang.*
1917 *War Illustr.* 13 Jan. 524/2 That seaplane..having some explosive 'eggs' to drop. **1918** E. M. ROBERTS *Flying Fighter* 335 *Eggs*, bombs weighing twenty pounds and upward filled with high explosives and 'laid' in Hunland. **1929** F. C. BOWEN *Sea Slang* 44 *Eggs*, submarine mines, a war-time phrase. **1939** *War Illustr.* 9 Dec. 399/1 The Germans are thought to be using relays of U-boats. Even the smallest of these can carry up to a dozen 'eggs'... A fast surface layer can put down more than 200 mines 'at a sitting'. **1947** AUDEN *Age of Anxiety* (1948) i. 18 But we laid our eggs Neatly in their nest.
4. a. *a bad egg* (earlier and later examples); *good egg* (slang): (*a*) an excellent person or object; (*b*) an exclamation of enthusiastic approbation. Also with other preceding adjs., esp. *tough*.
1855 'P. PAXTON' *Capt. Priest* 319 In the language of his class, the Perfect Bird generally turns out to be 'a bad egg'. **1903** KIPLING *Traffics & Discoveries* (1904) 138 'Us'll find they ships!'..'Good egg!' quoth Moorshed. **1910** GALSWORTHY *Justice* I, A real bad egg. **1914** C. MACKENZIE *Sinister St.* II. III. x. 711 It doesn't look a hundred quid to a tanner on his blue. Bad luck. He's a very good egg. *Ibid.* xii. 739 Oxford was divided into Bad Men and Good Eggs. **1915** D. O. BARNETT *Lett.* 56 We are going to do this regularly, and I think it is a very good egg. **1915** WODEHOUSE *Something Fresh* x. § 3 'She isn't going to sue me for breach?' 'She never had any intention of doing so.' The Hon. Frederick sank back on the pillows. 'Good egg!' he said with fervour. **1920** GALSWORTHY *In Chancery* I. xii. 100 He was a rotten egg. **1929** S. AUMONIER *Ups & Downs* 418 Hullo, Pan! Good egg! **1930** E. H. LAVINE *Third Degree* (1931) iii. 30 Occasionally, a really tough egg is trained to be a killer. **1938** WODEHOUSE *Summer Moonshine* i. 18 She's a tough egg.
b. *to have all your eggs in one basket* (examples); also *to put* († *venture*) *all one's eggs in one basket*, etc.
1666 G. TORRIANO *Second Alphabet of Proverbial Phrases* 125/2 To put all ones Eggs in a Paniard, *viz.* to hazard all in one bottom. **1710** S. PALMER *Proverbs* cxxiii. 344 (*heading*) A Mouse that has but one Hole, is soon Catch'd: or, Don't venture all your Eggs in One Basket. **1874** WHYTE MELVILLE *Uncle John* III. xxvii. 140 'Annie, my own darling, may I carry your basket all my life?' 'If you'll put all your eggs in it, yes,' answered Annie boldly. **1925** D. H. LAWRENCE *Refl. Death Porcupine* 179 It is a pity that we have insisted on putting all our eggs in one basket: calling love the basket, and ourselves the eggs. **1955** *Times* 3 May 3/6 An area which had all its eggs in one basket. **1969** *Times* 5 Nov. 23/3 The earl is evidently feeling a little uncomfortable that all his family eggs are in one basket.
c. *old egg*: a familiar form of address = old chap, old fellow, old sport.
1919 *Punch* 5 Mar. 190/2 Cheerio, old egg. **1927** 'A. ARMSTRONG' *Patrick Engaged* ix. § 3 'You'd be arrested.. and spoil the whole show,' replied Patrick tersely. 'Sorry, old egg, sorry!'
6. a. *egg-basket* (examples), *-coloration*, *-mass*, *-season*, *-stage*.
1773 J. WEDGWOOD *Let.* 21 Nov. (1965) 156 Egg Baskets; Egg Cups, with covers and without. **1867** G. W. HARRIS *Sut Lovingood* 132 He wer histin aig-baskets. **1911** J. A. THOMSON *Biol. Seasons* II. 177 A stereotyped kind of egg-coloration. **1889** M. E. BAMFORD *Up & Down Brooks* 45 The bright-yellow egg-mass. **1921** *Brit. Museum Return* 119 An exceedingly fine egg-mass of *Natica sp.* from Scotia Bay, South Orkneys. **1956** *Nature* 10 Mar. 489/2 Fifteen days after sowing, second-generation larvæ were hatching within the egg-masses. **1953** N. TINBERGEN *Herring Gull's World* x. 94 Once I saw the results of repeated raids by a fox in the egg-season. **1883** F. DAY *Indian Fish* 26 They have their enemies in the egg stage..and during their maturity.
b. *egg-collecting*, *-gatherer*, *-hunting*, *-robber*, *-slicer*.
1936 *Discovery* Feb. 56/2 What branch of science can egg collecting now advance? **1855** *Knickerbocker* XLVI. 223 Upon the approach of the egg-gatherers, with little or no dissenting clamor, they rise up in one vast, dangling-legged body. *Ibid.*, As egg-hunting is viewed by our country people as a species of 'picnicking', lovers and their mistresses..are the principal actors in these excursions. **1936** *Discovery* Jan. 12/1 Saddlebags are strapped to horses, and egg-gatherers fill these bags in no time. **1953** N. TINBERGEN *Herring Gull's World* xvi. 133 Potential egg-robbers. **1951** *Catal. Exhibits, South Bank Exhib., Festival of Britain* 63/1 'Skyline' Egg Slicer.
c. *egg-ended*, *-eyed*, *-faced* adjs.
1859 W. J. M. RANKINE *Man. Steam Engine* § 63 The ends of 'egg-ended' cylindrical boilers. **1875** *Plain Needlework* 10 Abel Morrell's 'egg-eyed' needles. **1921** *Glasgow Herald* 26 Sept. 6 He will find the egg-faced man there.
7. egg albumin, albumin obtained from the white of an egg, esp. ovalbumin; also = ALBUMEN 1; **egg-and-spoon race,** a foot race during which the competitors are required to carry an egg in a spoon (see also quot. 1894); **egg-apparatus** *Bot.*, the group of three cells at the micropylar end of the embryo-sac in seed plants, only one of which is fertile; **egg-bread** *U.S.*, bread made of the meal of Indian corn, eggs, etc.; **egg-burster,** a thickened area on an embryonic insect helping it to burst the egg membranes when hatching; **egg-capsule,** a natural envelope containing eggs; **egg-cosy** [COSY *sb.* 2], a cover to keep a boiled egg warm; **egg-eater** *S. Afr.*, a snake of the genus *Dasypeltis*, capable of crushing eggs with internal projections from its vertebrae; **egg-eating snake** = prec.; **egg-flip** (earlier example); **egg-fruit,** the fruit of the egg-plant; **egg-hot** (later examples); **egg-membrane,** a membrane surrounding an egg; = *vitelline*

membrane; **egg powder**, an artificially prepared substitute for eggs in cookery; **egg-purse** = *egg-capsule; **egg-raft, -rope, -string**, a connected series of eggs laid by various insects; **egg-stand**, a stand or frame for holding a set of egg-cups; **egg-timer**, (*a*) a device for timing the cooking of an egg; (*b*) a device for boiling an egg; **egg-tooth**, a small, hard, white protuberance developed in the embryo bird and reptile which is used to crack the egg and is cast off after hatching; **egg-tube**, an oviduct, esp. of an insect; **egg-whip**, an egg-whisk; **egg-whisk** (examples).

1871 *Jrnl. Chem. Soc.* XXIV. 572 Some properties of Egg Albumin. **1919** J. B. COHEN *Class-Bk. Chem.* II. 94 Of the albumins, egg- and serum-albumin are the most important. **1956** *Nature* 18 Feb. 330/1 The cells were smeared on microscopical slides coated with egg-albumin or human serum. **1894** *Daily News* 8 Sept. 5/3 The gentlemen had a turn in the egg-and-spoon race, in which the competitors had to punt with one hand and balance an egg on a spoon with the other. **1936** *Punch* 12 Aug. 187 (*caption*), I would like you to meet Mr. Scooter, the winner of our Parents' Egg-and-Spoon race. **1902** *Encycl. Brit.* XXV. 434/2 Of the three energids of the egg-apparatus, one alone is normally functional as the egg. **1854** M. J. HOLMES *Tempest & Sunshine* vii. 118 Egg-bread which Southern cooks know so well how to make. **1862** in *Southern Hist. Soc. Papers* (1884) XII. 26 The table was spread with rich egg-bread, fried ham, and pure coffee. **1911** R. D. SAUNDERS *Col. Todhunter* ii. 29 The Missouri supper of fried chicken, egg-bread, butter-beans and corn on the ear. **1920** *Brit. Museum Return* 121 The Egg-burster of Encephalous Fly-larvæ. **1953** *New Biology* XIV. 116 There are three larval stages in most species [of fleas], two in a few, very alike except in size and in the presence of a conical egg-burster on the top of the head of those of the first stage. **1959** SOUTHWOOD & LESTON *Land & Water Bugs* 411 *Egg-burster*, a thickened, often sclerotized, usually Y- or T-shaped area on the vertex of the embryonic cuticle; it bears a tooth and its movement is considered to help in rupturing the chorion and egg membranes. **1883** *Encycl. Brit.* XVI. 653/2 When egg-capsules are formed they are often of large size, have tough walls, and in each capsule are several eggs floating in a viscid fluid. **1921** *Brit. Museum Return* 119 A string of egg-capsules containing young examples of *Buscyon perversum*. **1936** RUSSELL & YONGE *Seas* (ed. 2) ii. 50 Common on the under side of rocks are the egg-capsules of the dog-whelk. **1894** SOMERVILLE & 'ROSS' *Real Charlotte* III. xxxv. 23 The egg-cosy that his wife had crocheted for him. **1906** *Westm. Gaz.* 3 Nov. 6/2 The sale of pin-cushions, d'oyleys, and knitted egg-cosies was unprecedented. **1909** *Lady's World* Dec. 284 The egg cosy made in the form of a cock's head and comb. **1967** A. WILSON *No Laughing Matter* III. 352 Journalism and worked up righteous anger, that's all she'd written. Egg cosies and cloaks! **1887** *Encycl. Brit.* XXII. 194/2 A..genus of snakes, *Dasypeltis*...In Cape Colony these snakes are well known under the name of 'eyer-vreter', i.e. 'egg-eaters'. **1911** *East London Dispatch* 1 Sept. 7 (Pettman), The Egg-eater lives almost entirely on eggs, which it eats in a curious fashion. **1969** A. BELLAIRS *Life of Reptiles* I. iv. 183 An egg-eater is able to demolish all the eggs of a clutch in rapid succession. **1897** *Daily News* 19 July 3/1 Natal..now claims to possess an egg-eating snake of phenomenal abilities. **1931** *Times Educ. Suppl.* 27 June p. iv/3 A very young egg-eating snake from Africa. **1965** R. & D. MORRIS *Men & Snakes* viii. 174 The egg-eating snake eats nothing but eggs and it is not surprising to find that many African birds have evolved elaborate anti-snake devices. **1832** J. ROMILLY *Diary* 23 June (1967) 16 Egg-flip & sack in Sen[ate] H[ouse] at 11. **1811** W. J. TITFORD *Sks. Hortus Bot. Amer.* 53 (*heading*) Egg fruit or Mad Apple, *Solanum Melongena*...It is called in India, Branjaw, in Jamaica, Garden Egg and Valanghanna, Brown Jolly or Bolangena. **1817** W. DARBY *Geogr. Descr. Louisiana* (ed. 2) 222 All the solanums (Irish potatoe, peppers, and egg-fruit,) whose leaves are easily killed by the slightest degree of freezing. **1887** *Harper's Mag.* Jan. 310/1 A dozen well-grown plants will supply a large family with egg-fruit. **1850** DICKENS *Dav. Copp.* xi. 119 She..made a little jug of egg-hot afterwards to console us. **1886** *Folk-Lore Jrnl.* IV. 116 At the plentiful supper always provided on this night [sc. Christmas Eve], egg-hot, or eggy-hot, was the principal drink. It was made with eggs, hot beer, sugar, and rum, and was poured from one jug into another until..covered with froth. **1885** *Science* V. 425/1 In the Scombresocidae the entire egg-membrane is covered with strong filaments, which intertwine with those of contiguous eggs. **1893** J. TUCKEY tr. *Hatschek's Amphioxus* 74 The movement of the embryos inside the egg membrane..is a quite peculiar one. **1862** *Englishwoman's Dom. Mag.* Sept 239/2 With the mysteries of making egg-powder we are quite unacquainted. **1864** *Chemist & Druggist* 193 It seems that certain cooks..are in the habit of buying and using egg powders. **1907** H. W. WILEY *Foods & their Adulteration* II. 115 The egg powder..formed is almost devoid of moisture and..may be kept for a long time without deterioration. **1921** *Chambers's Jrnl.* 375/2 The cockroach..is very careful in the selection of a suitable site to place her egg-purse. *Ibid.*, Each egg-purse [of a cockroach] contains sixteen eggs, arranged in two rows, with the ends from which the larvæ will emerge pointing towards the top. **1927** *Observer* 18 Sept. 8 The egg-raft is laid by the many species of that group of mosquitoes, the *Culicine*, represented by our common gnat. **1891** *Nature* 10 Sept. 457 Each egg-rope is moored to the bank by a thread. **1839** *Rep. Constabulary Force Commissioners* 25 in *Parl. Papers* [169] XIX, P— slipped into the parlour and brought out a watch and a silver egg-stand. **1909** BENNETT *Old Wives' Tale* II. i. § 3 The resplendent egg-stand holding twelve silver-gilt egg-cups and twelve chased spoons to match. **1969** E. H. PINTO *Treen* 136 Double-decker egg stands..made to hold 12, 24 or 48 eggs, were to be found in all Victorian and Edwardian larders. **1888** ROLLESTON & JACKSON *Forms Anim. Life* (ed. 2) 222 In *Nephelis* and *Clepsine*..'egg-strings', produced by the continuous division of a cell, lie free in the capsular cavity. **1884** Egg-timer [see EGG *sb.* 6 b]. **1909** *Cent. Dict. Suppl.*, *Egg-timer*, an apparatus for the automatic cooking of eggs. It consists of a vessel containing boiling water and a series of..baskets...When the time has elapsed the basket automatically rises out of the water. **1962** *TV Times* 28 Dec. 6/2 An egg-timer, I repeated with the assurance of a man who knew which way the sand trickled. **1893** A. NEWTON *Dict. Birds* 36 The 'egg-tooth'..is developed in the embryos of all birds as a small whitish protuberance or conglomeration of salts of calcareous matter, deposited in the middle layers of the epidermis of the tip of the upper bill. **1959** *New Biology* XXX. 87 In viviparous reptiles the eggshell is reduced to a thin, soft membrane or lost entirely, and the egg-tooth, which assists the young of oviparous forms to break out of their eggs, is sometimes rudimentary. **1960** M. BURTON *Wild Animals* 145 The young grass snakes..make their way out of the egg by tearing several rents in it with a special egg-tooth projecting from the front of the jaws. **1826** KIRBY & SPENCE *Entomol.* IV. xlii. 148 The ovaries, or egg-tubes as they are sometimes called. **1895** D. SHARP in *Cambr. Nat. Hist.* V. 137 The number of egg-tubes varies greatly in different Insects. **1909** *Cent. Dict. Suppl.*, *Egg-whip*. **1910** *Daily Chron.* 23 Apr. 7/5 Beat with an egg whip until smooth and glossy. **1868** M. JEWRY *Warne's Model Cookery* 36/1 Egg whisk, for beating eggs. **1882** [see WHISK *sb.*¹ 3]. **1924** *Week-end Bk.* 261 A large jug, and an egg-whisk..efficiently replace the [cocktail] shaker.

egg-beater. 1. (See Dict. s.v. EGG *sb.* 6 b.) (Earlier U.S. example.) **2.** = *HELICOPTER (see also quot. 1946). *U.S. slang.*

1828 E. LESLIE *Receipts* 49 Beat the eggs in a broad shallow pan with a wooden-egg-beater or whisk. **1937** *Atlantic Monthly* Jan. 26/1 Pilots of airplanes contemptuously termed autogiros 'egg beaters', and the whole idea appeared to be a colossal flop. **1946** *Amer. Speech* XXI. 310/1 *Egg-beater*, a twin-engine training plane, so termed because of the small size of the engines. **1948** *Canadian Alpine Jrnl.* June 178 One of those strange contraptions known as a helicopter or 'egg-beater'. **1962** L. DEIGHTON *Ipcress File* xxiv. 156 [Amer. loq.] 'Egg beaters.' The two helicopters came in.

egg-box. A box in which eggs are packed. Also *attrib.* and *transf.* (= *EGG-CRATE).

1854 DICKENS *Hard T.* I. iv. 20 That was the cot of my infancy; an old egg-box. **1937** *Discovery* May 134/2 The men [sc. egg-stealers]..were equipped with egg-boxes, coils of rope, climbing irons. **1952** *Archit. Rev.* CXI. 17/2 'Eggbox' screens may be designed across roof openings to obscure views of the sky for all normal angles of vision.

egg-crate. A crate in which eggs are packed; also *transf.* and *attrib.*, esp. in *egg-crate ceiling*, etc., denoting a construction which diffuses light.

1943 *Newsweek* 16 Aug. 54 The rickety 'egg crate' plane was shunned by all but true pioneers and thrill seekers. **1949** *Archit. Rev.* CVI. 307 The upper gallery at this point has an 'egg-crate' ceiling of steel sheeting to preserve a human scale. **1959** *Engineering* 13 Feb. 198/3 To meet the need for improved lamp screening appearance, translucent opal plastics egg-crate louvres are made.

egg-cup (e·gkʌp). A cup-shaped vessel to hold an egg. Also *transf.* Hence **e·ggcupful** (usually **egg-cupful**), as much as will fill an egg-cup.

1773 [see *egg-basket* (*EGG *sb.* 6 a)]. **1837** DICKENS *Pickw.* xvi. 160 There's nothin' so refreshin' as sleep, Sir, as the servant-girl said afore she drank the egg-cup-full o' laudanum. **1848** H. R. FORSTER *Stowe Catal.* 117, 6 egg-cups. **1870** [see EGG *sb.* 6 a]. **1871** *City-Road Mag.* I. 263 We drank all but boiling coffee at half a piastre an egg-cupful. **1905** H. G. WELLS *Kipps* I. v. 88 An egg-cupful of brandy. **1936** G. POLLETT *Song for Sixpence* x. 82 But the oak-trees shower their 'egg-cups' everywhere. **1962** E. O'BRIEN *Lonely Girl* xix. 237 Egg-cupfuls of Joanna's home-made Advocaat. **1969** E. H. PINTO *Treen* 73/2 The early Victorian revolving egg cup stand..is of mahogany, with ebony egg cups.

egger, *sb.*² (Earlier and later examples.)

1834 AUDUBON *Ornith. Biogr.* II. 370 Turtles..deposit their eggs in the burning sand, and clouds of sea-fowl arrive every spring for the same purpose. These are followed by persons called 'Eggers'. **1849** D. J. BROWNE *Amer. Poultry Yd.* (1855) 310 A class of persons called 'eggers', who follow..the avocation of procuring the eggs of wild birds. **1908** *Daily Chron.* 16 Apr. 3/7 Driven out of Cornwall by the 'Eggers' a few pairs for a time managed to breed..in South Wales.

eggery (e·gəri). [f. EGG *sb.* + -ERY.] A collection of eggs; an establishment for producing eggs.

1846 WORCESTER, *Eggery*, a nest of eggs. **1910** *Daily Chron.* 21 Mar. 4/7 A Western Canadian paper declares that next to the discovery of a gold-mine the most profitable investment is the 'importation of a batch of well-disposed hens and the establishment of an up-to-date eggery'. **1938** *Reader's Digest* Mar. 93/1 An Arndt eggery consisting of four sections.

e·gg-head, egghead. *colloq.* (orig. *U.S.*). [f. EGG *sb.* + HEAD *sb.*] An intellectual, a 'highbrow'. Also *attrib.* So **eggheadish** *a.*; **eggheadery, eggheadism.**

1952 *Cleveland* (Ohio) *Plain Dealer* 27 Sept. 10/3 A good many intelligent people..obviously admired Stevenson. 'Sure,' was the reply, 'all the egg heads love Stevenson.' **1952** *N.Y. Times* 19 Oct. E1/7 Writers of letters to editors tend to be in the intellectual or 'egg-head' category where Stevenson sentiment is strong. **1953** R. CHANDLER *Long Good-Bye* xlvii. 289, I told him I didn't think it would do Springer any harm. 'Only with the eggheads.' **1954** *Daily Tel.* 23 Mar., Mr. Stevenson..said..'Eggheads of the world, unite—you have nothing to lose but your yolks.' 'Eggheads' is a derogatory description for intellectuals in America. **1955** *Sci. Amer.* Apr. 2/3, I fear that, while publicly unspoken, anti-intellectualism and suspicion of 'eggheads' may have been a factor. **1956** *N.Y. Times Mag.* 9 Sept. 68/3 People said it was 'eggheadism' or 'moderation'. **1957** *Listener* 2 May 725/2 A former American ambassador to India and one of the Democratic Party's leading egg-heads. **1960** *Ibid.* 13 Oct. 627/1 This autumn six British firms are trying out one special kind of paper-back—the 'egghead' or quality or highbrow paper-back. **1962** *Punch* 30 May 827/1 The theatre (alas for would-be eggheadery) is nowadays scarcely noticed at all. **1963** *Times* 25 Feb. (Canada Suppl.) p. xiv/4 Often excessively eggheadish in their choice of material, these little dramatic islands have still proved [etc.]. **1963** *New Society* 26 Sept. 4/1 There is a reaction against eggheadery in America. **1965** *Spectator* 22 Jan. 98/1 Not all American egg-heads are equally self satisfied about the standard of their own criticism.

e·gg-headed, *a.* [cf. HEADED *a.* 1 c.] With a head shaped like an egg; also *transf.* (see prec.). So **egg-hea·dedness.**

1919 W. DEEPING *Second Youth* iv. 37 A little egg-headed pedant. **1938** *Time* 27 June 31/1 That man was starchy, six-foot-six Sir John Charles Walsham Reith, a dour, egg-headed, ascetical Aberdonian. **1957** H. CROOME *Forgotten Place* xiii. 163 Howard's own Old Intellectual tie had always been so conspicuous and Howard's views..so uncompromisingly egg-headed. **1959** V. PACKARD *Status Seekers* (1960) xv. 216 The four factors thus far cited—money, ancestry, distance of home from center of the city, and eggheadedness. **1965** *Economist* 10 July 154/1 More eggheaded discussion of these matters brings in the concepts of 'social rates of discount' and a 'shadow price' for capital. *Ibid.* 25 Sept. 1226/1 The Department of Economic Affairs was at pains..to underplay economic eggheadedness.

egging, *vbl. sb.*² Add: **2.** The laying or production of eggs, as in *egging season, time.*

1905 *Westm. Gaz.* 4 May 4/1 Among the risks which attend the breeding of game birds, frost at the 'egging' time is not the least serious. **1909** *Ibid.* 14 May 5/1 This is the 'egging' season, and outlying nests of pheasants and partridges may be disturbed.

eggless (e·glės), *a.* [f. EGG *sb.* + -LESS.] Without eggs.

1904 H. G. WELLS *Food of Gods* I. iv. 92 The two surviving hens..spent their remaining years in eggless celebrity. **1909** *Daily Chron.* 2 Feb. 4/7 The recipe for an eggless omelette. **1915** *Evening News* 20 Jan. 7 Eggless and Less Egg Cookery. **1967** A. WILSON *No Laughing Matter* II. 73 Sukey..was trying out a new eggless recipe.

egg-shell. Add: Also **eggshell. c.** Used *attrib.* or as *adj.* as a term of colour or of a paint finish intermediate between flat and glossy, e.g. *eggshell enamel, finish, glaze.* (See also quot. 1929.)

1894 F. B. GARDNER *Painters' Encycl.* 119 *Egg-shell gloss*, a term frequently used by painters. **1898** *Westm. Gaz.* 3 June 8/3 A costume of egg-shell blue cheque silk. **1908** *Daily Chron.* 12 Mar. 3/3 The sun is reflected by its front of dull white terra-cotta stone with egg-shell glaze. **1909** *Westm. Gaz.* 27 Apr. 5/2 People argue of paints, and enamels, and varnishes, and..whether the true 'egg-shell gloss' has been obtained or not. **1909** *Chambers's Jrnl.* 25 Sept. 684/1 That beautiful, characteristic, velvety, egg-shell enamel is highly artistic. **1925** *Arts & Decoration* Mar. 82/2 The egg-shell finish of the freshly laid egg..is the dull finish the decorators now prefer. **1929** A. E. OWEN-JONES *Retail Stationer's Handbk.* 193 *Eggshell Finish*, a finish imparted to notepapers by omitting the calendering; the surface is dull, and is covered with pin-point pores. **1936** C. DAY LEWIS *Friendly Tree* vii. 96 A sky of egg-shell blue. **1943** W. WILCOX in *55 Short Stories from N. Yorker* (1949) 139 She was wearing dark blue..and a neat little eggshell blouse. **1953** *Gloss. Paint Terms* (B.S.I.) 27 The following stages in increasing order of gloss are normally recognized:—Flat (or matt)...Eggshell flat. Eggshell gloss. **1955** E. BOWEN *World of Love* v. 88 Electric candelabra..round all the eggshell walls. **1958** *Woman* 22 Feb 2/1 Gloss, matt, flat, eggshell paints go over themselves or each other.

egg-white. [WHITE *sb.* 1.] The white of an egg, the albumen. Also *attrib.*, esp. in *egg-white injury*, a disease caused by eating an excess of raw egg-white.

1898 *Jrnl. Physiology* XXIII. 130 Fresh egg-white is thoroughly whipped into a froth. **1916** E. POUND *Lustra* 52 Green arsenic smeared on an egg-white cloth. **1936** *Discovery* Dec. 388/2 A preparation from egg-white, which reduces the clotting time of blood, provides new hope for haemophiliacs. **1937** H. T. PARSONS et al. in *Biochemical Jrnl.* XXXI. I. 425 (*heading*) Egg white injury and protective factor. **1959** *Chambers's Encycl.* XI. 260/1 Albumins and globulins..are frequently found together as in..milk and egg white. *Ibid.* XIV. 350/2 Biotin is a..factor..identified with the so-called vitamin H..or anti-egg-white-injury factor.

eggy, *a.* Add: **a.** (Examples.) Also, *spec.*, as in quot. 1901. **b.** (Later example.)

1901 *Farm, Field & Fireside* 13 Dec. 357/3 If it has laid eggs, or is within a short time of laying, it is also detected. These birds are called 'eggy'. **1924** *Blackw. Mag.* Feb. 249/2 Abdul brought his most welcome offering an eggy rice pudding. **1929** E. BOWEN *Last September* III. xix. 238 The sponge cake..was moist and eggy. **1951** J. B. PRIESTLEY *Festival at Farbridge* II. i. 263 You might push that eggy plate away then with your foot.

eglestonite (e·g'lztənait). *Min.* [f. name of Thomas *Egleston*, an American mineralogist (1832–1900) + -ITE¹.] A native oxychloride of mercury, occurring in brownish-yellow isometric crystals.

1903 A. J. MOSES in *Amer. Jrnl. Sci.* 4th Ser. XVI. 258 The eglestonite crystals are usually easily recognized. **1912** *Brit. Museum Return* 196 Eglestonite and calomel from Palo Alto, California. **1933** *Amer. Mineralogist* XVIII. 7 Descending acid solutions altered the cinnabar, producing the secondary quicksilver minerals, metacinnabar, calomel, and eglestonite. **1968** EMBREY & PHEMESTER tr. *Kostov's Mineralogy* 204 Eglestonite is isostructural with garnet.

‖ **églomisé** (eglomi·ze), *a.* and *sb.* [Fr., f. name of *Glomy*, a Parisian picture-framer of the 18th cent. So It. *agglomizzato*, G. *eglomisiert*.] Applied to glass painted on the back, and used by Glomy for frames.

1877 LADY C. SCHREIBER *Jrnls.* (1911) II. 19 His Eglomisé Glass is also lent on view. **1897** A. HARTSHORNE *Old Eng. Glasses* 343 To the last quarter of the eighteenth century belong also those florid painted panels—'églomisés'—inserted in the bottoms. **1912** *Catal. Wks. Art J. E. Taylor sold at Christie's* Lot 85 A portable altar of eglomisé and silver-gilt. The base is further enriched with two eglomisé plaques. **1967** *Times* 21 Feb. 21/7 (Advt.), A Louis XVI gold and verre eglomisé snuff-box by Joseph-Etienne Blerzy, Paris, 1798.

ego. Delete 'The pronunciation (i·go) is now seldom heard in England.' Add: **2.** In speech: I, the speaker. Hence **e·go** *v. trans.*, to say 'ego' when claiming an object, in response to 'quis?'. *Schoolboy slang.*

1913 C. MACKENZIE *Sinister Street* I. I. vii. 103 He was often first with the claimant 'ego', when someone shouted 'quis?' over a broken pocket-knife found. **1959** I. & P. OPIE *Lore & Lang. Schoolch.* viii. 134 In private schools a child who wishes to dispose of something..calling out 'Quis?' and the boy or girl who first replies 'Ego' receives the object and may say (to the horror of the classicist) 'I egoed it'.

3. Self-esteem, egotism, self-importance.

Some examples are influenced by sense 4.

1891 KIPLING *Light that Failed* v. 59 I've made a discovery. Torp, there's too much Ego in my Cosmos. **1907** *Daily Chron.* 13 Feb. 7/4 By 'exaggerated ego',..he meant a disproportionate idea of the importance of oneself and a belief that one was clothed with powers, capacities, and ability far above the normal or above those actually possessed. **1952** S. KAUFFMANN *Philanderer* (1953) vi. 97 The quarrels—which had begun between him and his wife simply because his ego could not possibly be satiated even when fortune was good—increased terribly now. **1962** J. D. SALINGER *Franny & Zooey* 166 You keep talking about ego. My God, it would take Christ himself to decide what's ego and what isn't.

4. *Psychol.* That part of the mind which is most conscious of self; *spec.* in the work of Freud that part which, acted upon by both the id and the super-ego (ego-ideal), mediates with the environment.

1894 *Brain* XVII. 130 By reason of the clouding over of the ego produced in one case by hypnotism, in another by nervous shock, an idea once installed..will further develop and acquire sufficient force for objective realisation. **1910** H. W. CHASE (tr. Freud) in *Amer. Jrnl. Psychol.* XXI. 193 The incompatibility of the idea in question with the 'ego' of the patient was the motive of the repression. **1922** [see *EGO-IDEAL]. **1927** J. RIVIERE tr. *Freud's Ego & Id* v. 83 Like the dweller in a borderland that it is, the ego tries to mediate between the world and the id, to make the id comply with the world's demands and, by means of muscular activity, to accommodate the world to the id's desires. **1943** *Psychol. Rev.* L. 454 In a semi-doze we lose all sense of our egos though we may be conscious enough of impersonal items. **1943** H. READ *Education through Art* vi. 197 The super-ego is the direct representative of the unconscious, of the id, and hence the possibility, indeed, the inevitability of a conflict with the ego, a conflict between what is perceptual and real and what is imaginative and ideal.

5. *attrib.* and *Comb.* (senses 3 and 4), as *ego-attitude, -complex, -consciousness, -instinct, -satisfaction*; *ego-hood* (later example); *ego-bound, -less* adjs.; **ego-trip**, an activity, period of time, etc., devoted entirely to indulging in one's own interests or in self-expression; hence **ego-tripper.**

1937 *Brit. Jrnl. Psychol.* Jan. 265 When the object is encountered in a new environment with a different Ego-attitude, then the communication takes place between object-process and object-trace. **1929** D. H. LAWRENCE *Pansies* 67 As a plant becomes pot-bound Man becomes ego-bound Enclosed in his own limited mental consciousness. **1916** C. E. LONG tr. *Jung's Coll. Papers Analyt. Psychol.* i. 80 They are disturbances which only belong to the superficial, and none reaches so deep as to attack the strong-knit foundation of the ego-complex. **1922** *Brit. Jrnl. Psychol.* Oct. 115 Jung then goes on to describe the 'ego complex' (*Ichkomplex*) which in the normal mind is the dominant psychic instance. **1926** W. McDOUGALL *Introd. Soc. Psychol.* (ed. 20) Suppl. iv. 404 The Freudians have recognised the importance of this rôle [sc. self-regard] in all that they have written of the function of the 'ego-complex', and the 'ego instincts' in inhibiting, controlling, conflicting with, and repressing the sexual tendencies. **1917** GLUECK & LIND tr. *Adler's Neurotic Constitution* (1921) ii. 21 Consciousness of guilt and conscience are fictitious guiding principles of caution, like religiosity and subserve the craving for security. Their object is to prevent a lowering of the ego-consciousness when the irritated aggressiveness impels immoderately to selfish deeds. **1963** AUDEN *Dyer's Hand* 96 An ego-consciousness which paints himself painting himself. **1906** S. S. LAURIE *Synthetica* II. 241 This is his return to God, from Whom his negating Egohood for a time separates him. **1917** C. R. PAYNE tr. *Hirschmann's Freud's Theories of Neuroses* x. 241 Freud is perfectly clear on the point of the decisive rôle played in life and also in the neurosis by the egotistic or ego instincts alongside of the sexual instinct. **1922** C. J. M. HUBBACK tr. *Freud's Beyond Pleasure Principle* iii. 54 Our discussion so far results in the establishing of a sharp antithesis between the 'ego-instincts' and the sexual instincts, the former impelling towards death and the latter towards the preservation of life. **1937** *Sunday Times* 12 Dec. 5/2 Those subconscious ego-less depths of it [sc. the consciousness] from which..only the greatest things in art can come. **1954** J. A. C. BROWN *Soc. Psychol. Industry* vii. 189 Work becomes an avenue for securing ego satisfactions. **1969** *It* 13–28 June 11/2 They're using the music as a vehicle for character and personality building. I don't think they're half as much a musical ego-trip as people imagine. **1970** *Observer* 20 Sept. 26/1 His was no musical ego-trip, no money-grubbing Tin-Pan Alley beano feast. **1970** *Melody Maker* 12 Sept. 29 Ego tripper...That's me, folks!

egocentric (egose·ntrik), *a.* [f. EGO + CENTRE *sb.*, after *geocentric, heliocentric.*] Centred in the ego; in vague or popular use: self-centred, egoistic.

1900 in *Ann. Rep. Bur. Amer. Ethnol.* 1897–8 831 An earlier ethnocentric system born of the primeval egocentric cosmos of inchoate thinking. **1901** J. M. BALDWIN *Dict. Philos. & Psychol.* II. 194/2 For the ego-centric point of view is substituted the homocentric. **1919** in B. H. Streeter *Spirit* iii. 97 The instincts, the radical fault in most of which is their selfish and egocentric character. **1926** *19th Cent.* July 83 The girl of to-day absorbs a freedom-loving and egocentric conception of life. **1943** H. READ *Education through Art* v. 140 When an attempt is made, not merely to express egocentric sensation..but also to represent an external object as it presents itself to the sensations of the artist..then some control is exercised by other than tactual and somatic sensations. **1953** R. NIEBUHR *Christian Realism & Political Problems* (1954) i. 22 It might be noted that, without a sense of the universality of an egocentric corruption, the passion for a universal humanity quickly degenerates. **1964** M. CRITCHLEY *Developmental Dyslexia* xii. 70 The age at which a child..turns from an autistic, egocentric individual, to a societal, ethnocentric being.

b. *Philos.* *egocentric predicament*, the supposed impossibility of knowing anything outside one's own mind.

1910 R. B. PERRY in *Jrnl. Philos.* VII. 5 (*title*) The Ego-Centric Predicament. *Ibid.*, I shall seek to discover whether a certain circumstance, which has never been disputed, does or does not constitute evidence for a theory that has been much disputed. The circumstance I shall call the *ego-centric predicament*, and the theory, *ontological idealism*. **1917** A. S. PRINGLE-PATTISON *Idea of God* x. 192 This is what an American Realist, in a phrase worthy of Kant, in its full-flavoured technicality, has dubbed 'the ego-centric predicament'. The Ego is..the pre-supposition of all its knowledge.

Also as *sb.*, one who is self-centred. So **e:gocentri·city, egoce·ntrism**, the state or quality of being egocentric; self-centredness.

1903 *Amer. Jrnl. Psychol.* July–Oct. 100 Ziehen limits the hysterical constitution to emotional instability, egocentricity, craving for attention. **1918** E. JONES *Psycho-analysis* (ed. 2) 632 Its ruthless and absolute egocentricity. **1926** M. WARDEN tr. *Piaget's Lang. & Thought of Child* v. 238 All ego-centrism is designed by its structure to stand half-way between autistic thought which is 'undirected'..and 'directed' intelligence. Ego-centrism is therefore obedient to the self's good pleasure and not to the dictates of impersonal logic. **1928** *Brit. Weekly* 22 Mar. 593/2 In this book the first personal pronoun occurs with a frequent and monotonous egocentricity. **1932** D. H. LAWRENCE *Last Poems* 179 The hordes of the ego-centric, the robots. **1934** *Mind* XLIII. 87 Egocentrism, in so far as it means confusion of the ego and the external world, and egocentrism in so far as it means lack of co-operation, constitutes one and the same phenomenon. **1943** H. READ *Education through Art* viii. 276 Then by stages the player grows out of his egocentrism, observes other players, gradually begins to co-operate with others, and finally in conjunction with a group arrives at a reciprocal agreement embodying definite rules. **1957** *Times Lit. Suppl.* 25 Oct. 637/1 The..complete ruthlessness and egocentricity of the very young female person. a **1963** L. MACNEICE *Astrol.* (1964) iii. 86 Gemini is the patron not only of intellectuals but of egocentrics. **1970** *Daily Tel.* 13 Mar. 17 Behind their sham chivalry, I contend, lurk some nasty tendencies to sadism, egocentricity, rebellion and neurosis.

ego-ideal (e·go,aidi·ăl). *Psychol.* [f. EGO + IDEAL *sb.*; tr. G. *Ichideal.*] **a.** That part of the mind which, in Freudian theory, is evolved from the ego as it becomes aware of parental and social standards, and imposes upon it concepts of ideal behaviour with which it should conform. **b.** = *SUPEREGO. **c.** In vague or popular use: a conception of oneself as one would like to be.

1922 J. STRACHEY tr. *Freud's Group Psychol.* vii. 69 We have been driven to the hypothesis that some such faculty develops in our ego which may cut itself off from the rest of the ego and come into conflict with it. We have called it the 'ego ideal', and by way of functions we have ascribed to it self-observation, the moral conscience, the censorship of dreams, and the chief influence in repression. **1943** H. READ *Education through Art* vi. 176 A spearhead..passes through the pre-conscious and conscious levels of the ego, and emerges above everything as the ego-ideal or super-ego. **1949** J. I. M. STEWART *Character & Motive in Shakes.* 109 If we regard the Othello-figure on the stage as..Othello's ego-ideal or self-exemplar, then Bradley is really right in point of dramatic feeling. **1951** E. C. TOLMAN in Parsons & Shils *Toward Gen. Theory of Action* III. iii. 311 The ego-ideal consists of acquired positive values and valences for those types of behavior in which one *should* engage. **1952** W. SPROTT *Social Psychol.* ix. 185 The recipients of praise are their [sc. children's] models and bit by bit what has conveniently been called an 'ego-ideal' is elaborated. **1964** GOULD & KOLB *Dict. Soc. Sci.* 593/2 The repressing force derives from cultural standards which have been incorporated by the individual to serve as his 'ego-ideal' or 'super-ego'.

e·go-invo·lvement. *Psychol.* [f. EGO + INVOLVEMENT.] The process or fact of the ego being identified with various aims, attitudes, values, etc., so that behaviour defending and furthering these is strongly reinforced, and affects one's self-esteem. Hence **e·go-invo·lved** *a.*, having one's self-esteem dependent on something external.

1936 M. SHERIF *Psychol. Soc. Norms* ix. 179 As the social values..include positive values..the ego involvement does not appear only as a checking or inhibiting factor, but also as a positive indicator of certain lines of action and striving. **1940** *Jrnl. Abnormal & Soc. Psychol.* XXXV. 500 One might expect different levels of aspiration for a given task in ego-involved subjects who have..experienced different amounts of success or failure with that task. **1955** T. H. PEAR *Eng. Social Differences* i. 15 In studying social stratification, it is necessary to free oneself..from ego-involvement. **1963** *Daily Tel.* 14 Feb. 11/5 Some enjoy it even though its barbs threaten their own ego-involved concepts.

egomania. Delete *humorous* and add later examples.

1895 tr. *Nordau's Degeneration* 241. **1895** *Times* 21 Nov. 11/2 He panders either to the morbid egomania of the sitter or to the snobbism of his readers. **1901** *Academy* 18 May 427 There is not much reasoning with nerves and temperament; and as for 'egomania', was she not shut up with a mirror all her life? **1924** W. B. SELBIE *Psych. Relig.* 140 Instances abound of religious individualism degenerating into egoism and producing even egomania. **1957** *New Yorker* 26 Oct. 50/2 A profession that has more than its share of artists given to displays of egomania and other irritating forms of temperament.

egomaniac (egomēi·niæk). [f. EGOMANIA, after *monomaniac*, etc.] One who suffers from egomania. Also *attrib.* and *transf.*

1890 *Daily News* 8 July 6/1 To the end she is an egomaniac, and her religion is but another expression of the sensuousness that is the active principle of her being. **1895** *Westm. Gaz.* 25 Feb. 2/1 The Ego-maniac—the degenerate who is too feeble of will to control his thoughts by a regard to the welfare of society. **1914** *Daily Express* 17 Nov. 4 The crazy egomaniac who has deluged Europe with blood. **1915** C. S. JONES *Hohenzollern* 199 So ruthless and imperious as to become that most dangerous of lunatics—a confirmed egomaniac! **1920** *Glasgow Herald* 20 Nov. 5 Pretensions that had grown to egomaniac heights. **1921** GALSWORTHY *To Let* II. ii, I know those enthusiastic egomaniac gentry. **1963** *Punch* 30 Oct. 640/2 And cats, egomaniacs.

Egyptian, *a.* and *sb.* Add: **A.** *adj.* **2. Egyptian lily**, the white arum, or trumpet lily, *Zantedeschia æthiopica*; = CALLA 2; **Egyptian millet**, *Pennisetum spicatum*; **Egyptian onion**, a form of the common onion, *Allium cepa aggregatum*; **Egyptian pea**: see PEA¹ 3; **Egyptian privet**, henna.

1847 A. H. LINCOLN *Lect. Bot.* (ed. 2) xxxvi. 191 The elegant exotic, Calla ethiopica, or Egyptian lily. **1959** R. M. CARLETON *Index Common Names Herbaceous Plants* 41/2 *Egyptian-lily*, Agapanthus africanus (A. umbellatus); Zantedeschia (Richardia). **1829** *Virginia Herald* (Fredericksburg, Va.) 25 Apr. 2/4 The leaves are of a beautiful green—long, narrow, dagger shaped, not unlike those of the Egyptian millet. **1864** Egyptian millet [see MILLET 2]. **1907** L. H. BAILEY *Cycl. Amer. Agric.* II. 471/1 Various other names have been applied to this plant, such as penicillaria, cat-tail millet, Egyptian millet. **1880** *Encycl. Brit.* XII. 285/1 The Tree Onion or Egyptian Onion..produces small bulbs instead of flowers. **1959** R. M. CARLETON *Index Common Names Herbaceous Plants* 41/2 *Egyptian onion*, Allium cepa. **1825** LOUDON *Encycl. Agric.* § 1057 The henné or Egyptian privet. **1956** *Colour Index* (ed. 2) I. 1751 The powdered dried leaves of the Egyptian privet.

6. Egyptian binding (see quot. 1728); **Egyptian cotton**, a cotton of fine quality developed in Egypt; **Egyptian pound** (see POUND *sb.*¹ 3 d); **Egyptian wheel** (see quots.).

1728 CHAMBERS *Cycl.* s.v. *Bookbinding*, The Manner of binding Books in Volumes, i.e. of sewing the leaves together, to roll 'em on round Pieces or Cylinders of Wood, appears the most Ancient..which we may call *Egyptian binding*. **1877** *Encycl. Brit.* VI. 485/1 The new description of Egyptian cotton has since been known by the name 'Jumel' in France. **1957** Egyptian cotton [see *BACK-DOOR 2]. **1793** *Descr. Estate of Culross* 9 Sir George Bruce erected machinery, consisting of the Egyptian wheel, commonly called chain and bucket, to drain the Coal to the dip of the old workings. **1867** M. E. HERBERT *Cradle Lands* vii. 196 A well of water worked by a thoroughly Egyptian wheel. **1880** *Encycl. Brit.* XII. 435/1 The Egyptian Wheel or Noria.

7. *Zool.* **Egyptian cobra**: see *COBRA.

B. *sb.* **4.** The Hamitic language of Egypt.

1556 A. VELE in R. Robynson tr. *More's Utopia* (ed. 2) Printer to Reader, It is a tongue to vs muche straunger then the Indian, the Persian,..the Arabicke, the Egyptian, the Macedonian. **1646** J. GREGORY *Notes & Obs.* (1650) To Rdr., This Booke of ours..may be read in..Coptick or Ægyptian. **1842** *Visitor or Monthly Instructor* 409/1 It ain't Greek at all: except, perhaps a few words. What ain't Greek is Egyptian; and what ain't Egyptian is Greek. **1857** S. BIRCH in J. G. Wilkinson *Egyptians* 182 Enlightened rulers prided themselves in speaking foreign tongues...Cleopatra spoke seven, Egyptian among the number. **1875** W. D. WHITNEY *Life Lang.* 254 In this [sc. 'Hamitic'] family, the Egyptian occupies the same commanding position as the Chinese. **1877** *Encycl. Brit.* VII. 721/2 The inscribed and written character of Egyptian was the hieroglyphic. **1960** S. POTTER *Lang. in Mod. World* viii. 112 It is customary to divide the history of Egyptian into three periods; Old or hieroglyphic, Middle or hieratic, and New or demotic.

5. a. A mother-of-pearl shell from Egypt. **b.** An Egyptian cigarette.

1885 *Encycl. Brit.* XVIII. 447/2 The Arabs still obtain from this district [sc. Jiddah and Koseir] a quantity of mother-of-pearl shells, which are shipped from Alexandria, and come into the market as 'Egyptians'. **1892** *Whitehall Rev.* 22 Oct. 8/1 She could tell Russians from Egyptians, and sometimes took a mild Havannah with her B and S in the smoking-room. **1905** *Westm. Gaz.* 28 Oct. 10/2 A box of Egyptians. **1965** P. ROBINSON *Pakistani Agent* viii. 107, I only smoke Egyptians, thank you.

Egyptianize, *v.* Add: **c.** To develop a country or district according to the methods adopted in Egypt; in later use, see quot. 1958; **Egyptianized** *ppl. a.* (later examples).

1899 *Westm. Gaz.* 4 Aug. 9/1 Were we prepared to..begin the work of Egyptianising the Yangtse Valley? **1958** *Britannica Bk. of Year 1957* 519/1 *Egyptianize*, meaning to nationalize foreign holdings in the manner of the Egyptian government in 1956. **1959** *Daily Tel.* 3 Mar. 16/3 The U.A.R. government is paying to the British Government a lump sum of £27,500,000 in compensation for British private property Egyptianised, that is nationalised or compulsorily acquired. **1959** *Times* 24 Mar. 15/4 The 4,500 registered owners of sequestrated or Egyptianized property will wish to go to Egypt to recover their property or assess their loss. **1962** *Listener* 1 Mar. 366/2 The new 'Egyptianized' middle class which the regime has put into power.

Egy:ptianiza·tion. [f. EGYPTIAN *a.* + -IZATION.] **a.** The compulsory acquisition by the Egyptian government of foreigners' property and interests in Egypt. **b.** The rendering Egyptian in character or organization; the placing of a country under Egyptian officials.

1957 *Times* 22 Feb., It is since reported that 'verbal assurance' has been given that the Egyptianization of Greek business interests will not occur 'for at least five years'. **1961** *Daily Tel.* 29 Sept. 1/1 Syria..is seething with discontent over the Egyptianisation of the country. **1962** *Listener* 1 Mar. 365/2 Then there is the gradual, but all-pervasive Egyptianization. Tens of thousands of foreigners have left Egypt. **1967** *Ibid.* 25 May 674/1 The expulsion of thousands of Britons..the removal of all foreign doctors, dentists, accountants..from professional organizations; those were all part of Nasser's Egyptianization process.

Egy·ptianizing, *vbl. sb.* Making Egyptian or like the Egyptians.

1921 G. A. F. KNIGHT *Nile & Jordan* xv. 193 To promote the Egyptianizing of the province, Amenhotep III built at Soleb..a gigantic temple.

Egy·ptianizing, *ppl. a.* = *EGYPTIZING *ppl. a.*

1949 W. F. ALBRIGHT *Archaeol. of Palestine* viii. 185 A small Egyptianizing statue dating about the eighteenth century B.C.

Egypticity (īdʒipti·sĭti). [f. *Egypt* + -ICITY.] The character or quality of being Egyptian (see quot. 1895).

1888 *Expositor* Sept. 219 The Pentateuch—Egypticity and Authenticity. **1895** W. H. TURTON *Truth of Christianity* x. 147 The Egypticity of the narrative. By this is meant that the part of the Pentateuch in which reference is made to Egyptian customs, seasons, and names appears to be written with correct details throughout. **1921** G. A. F. KNIGHT *Nile & Jordan* xxviii. 384 The reference..is thoroughly Egyptian, and testifies to the 'Egypticity' of the book.

Egyptizing (ī·dʒiptaiziŋ), *ppl. a.* [f. *Egypt* + -IZE + -ING².] Becoming Egyptian in character; adopting Egyptian characteristics.

1847 J. LEITCH tr. *Müller's Anc. Art* § 240 The Egyptising image of the god of Heliopolis. **1923** *Weekly Westm. Gaz.* 24 Mar. 16/2 The art of two great and widely different empires knew its Egyptising phases. **1934** *Discovery* Oct. 278/1 The Egyptizing and Phoenicizing character of much of the art of South Eastern Cyprus in the early archaic period must not be allowed to disguise the fact of its Cypriote origin.

Egypto-, repr. stem of Gr. Αἴγυπτος Egypt, in comb. (= Egyptian and...), as *Egypto-Abyssinian, -Arab, -Arabic, -Caucasian, -Semitic, -Syrian, -Turk.*

1831 *Athenæum* 3 Dec. 787/3 All..were obviously of Arabian or Egypto-Caucasian extraction. **1870** P. B. RANDOLPH *Seership!* (1884) 73 Nor do I see any reason why the white women of Western Europe..should not be as..successful..as their Arabian and Egypto-Syriac sisters. **1879** *Chambers's Jrnl.* 174/2 The Egypto-Syrian forces. **1884** *Illustr. Lond. News* 9 Feb. 130/1 Peace had been completely restored on the Egypto-Abyssinian frontier. **1893** F. ADAMS *New Egypt* 16 Even these fanatical Egypto-Turks,..even these degraded Egypto-Arabs. **1909** WEBSTER s.v. *Saracenic*, The square or polygonal minarets in diminishing stories of the Arabic or Egypto-Arabic style. **1937** J. R. FIRTH *Tongues of Men* iv. 52 The diffusion of the Egypto-Semitic word-craft.

eh, *int.* Add: **3.** Used interrogatively, as a request for the repetition or explanation of something that has just been said: = What do you say? *colloq.* or *vulgar.*

1837 T. BACON *Impressions in Hindostan* II. 149 Eh? What's that, Sackville? a **1845** HOOD *Hood's Own* 2nd Ser. (1861) 26 'The mail bags are on board—and it's more than my post is worth to put back.' 'Eh? What? How?' exclaimed the Oddity. **1869** *Chambers's Jrnl.* 595/2 'Eh?' said the old clerk, at last detecting my intrusion. I repeated my business. **1903** C. MARRIOTT *House on Sands* xvi, 'Ass!' muttered Lanyon, with a diabolical grin. 'Jelly-belly!' 'Eh?' said Sir Peter, blankly. 'Your health,' replied Lanyon, gravely, raising his glass. **1905** H. G. WELLS *Kipps* III. i. 321 'I'm going to build a house, Ann.' 'Eh?' said Ann suddenly, as if awake. 'Build a house.' **1969** *Listener* 27 Feb. 280/2 Most of what we think characteristically modern already existed by 1914: aircraft, radio-telegraphy, psycho-analysis, Post-Impressionism, motion-picture palaces, and the Labour Party (eh?).

eicos-, eicosa-, eikos-, varr. ICOS-, etc. (in Dict. and Suppl.).

1909 WEBSTER, *Eikosane*, Chem., a solid hydrocarbon, $C_{20}H_{42}$, of the methane series, of artificial production, and also occurring in petroleum. **1958** *Jrnl. Brit. Interplanetary Soc.* XVI. 442 Values of specific impulse for heptane, octane, decane and eicosane oxidized by WFNA have been calculated. **1964** J. A. LOVERN in *Oceanogr. & Marine Biol.* II. 188 Lipids from fish mitochondria are, like fish depot lipids, low in arachidonic acid but rich in eicosapentaenoic and docosahexaenoic acids. **1968** *New Scientist* 18 July 142/1 This eicosanucleotide (a string of 20 nucleoside beads)..is the longest that needs to be synthesized by purely chemical methods.

eider-down. Also **eiderdown. 2.** Substitute for def.: A quilt filled with eider-down or any similar soft material.

1950 *Times* 26 Apr. 7/7, I ask you..to lend your pen to scotching the unwarrantable..term 'eiderdown' when applied to the ordinary goose-down quilt. **1955** *Oxf. Jun. Encycl.* XI. 20/2 In the 20th century, eiderdowns, stuffed with down or flock and covered in silk or chintz, became popular. **1970** J. PRIOR *Soft Furnishing* vii. 80 An eiderdown is made like a very large cushion, with a design stitched right through the cover to keep the feathers in fixed pockets.

3. A heavily napped wool or cotton or man-made fabric of thick texture, in plain and fancy colours, used for petticoats, cloaks, bath-robes, etc. Freq. *attrib.*

1872 A. MERCIER *Our Mother Church* xvi. 360 For more elaborate texts..letters may be made of cotton wool, of the fluffy calico called eiderdown, [etc.]. **1873** *Young Englishwoman* Nov. 572/2 Could the kind Editor..give directions..for making an eider-down petticoat. **1890** *Amer. Mail Order Catal.* (1961) 11 Handsome Eider-down short cloak, with astrachan shoulder cap. **1905** *Smart Set* Sept. 155/2 A red eiderdown bath-robe.

eidetic (əide·tik), *a.* *Psychol.* [ad. G. *eidetisch* (E. R. Jaensch), f. Gr. εἰδητικός, f. εἶδος form.] Applied to an image that revives an optical impression with hallucinatory clearness, or to the faculty of seeing such images, or to a person having this faculty. Also *sb.*, **a.** one who sees eidetic images; **b.** *pl.* (see quot. 1930). Hence **eide·tically** *adv.*, by an eidetic image.

1924 G. W. ALLPORT in *Brit. Jrnl. Psychol.* Oct. 100 The true eidetic image, in distinction to the visual memory-image, revives the earlier optical impression when the eyes are closed, in a dark room, and sometimes when the eyes are normally open, with hallucinatory clearness. *Ibid.* 101 Children with less eidetic ability frequently need to be instructed regarding the nature of the phenomenon. *Ibid.* 120 Eidetic imagery is a common possession of children. **1925** C. FOX *Educat. Psychol.* 81 The inability to discriminate between hallucinations and normal mental imagery and the mental confusion thereby entailed is responsible for a psychological abortion called the 'eidetic image'. **1929** *Encycl. Brit.* VIII. 112/1 An object may be eidetically seen either immediately after it has been removed from sight or after a considerable period of time..has elapsed. **1930** O. OESER tr. *Jaensch's Eidetic Imagery* 1 Eidetics—the theory of eidetic or perceptual images. **1931** *Brit. Jrnl. Psychol.* July 95 The most vivid and original fantasias were produced by eidetic children. **1939** J. D. S. PENDLEBURY *Archaeol. of Crete* v. 275 Snijder has produced some remarkable parallels between Minoan art and that of a class of people known as 'eidetics'. **1943** H. READ *Education through Art* iv. 83 But generally Jaensch's attempt to force racial distinctions..contradicts the whole genetic theory of eidetics. **1949** KOESTLER *Insight & Outlook* xxvi. 361 There are people endowed with the faculty of eidetic imagery. **1970** *Nature* 17 Jan. 227/1 Many people may wish for a 'photographic memory', an ability which has been referred to by psychologists as eidetic imagery. *Ibid.* 227/2 Only thirty-five of the children were classified as eidetic. *Ibid.*, The eidetics claimed that they were not simply remembering what had been in the picture. *Ibid.* 27 June 1268/1 Can a purely binocular pattern..be stored eidetically?

Eidophusikon (əidǒfiū·zikǫn). Also **-con.** [f. Gr. εἶδο-ς form + φυσικόν natural, f. φύσις nature.] A kind of magic lantern showing a series of pictures illustrative of some occurrence, invented by Philip de Loutherbourg *c* 1780.

a **1800** S. PEGGE *Anecdotes* (1803) 252 How otherwise is the next generation to understand what is meant by the *Lyceum*, the *Eidophusicon*, Sir Ashton Lever's *Holophusicon*, Walker's *Eidouranion*? **1866** *Nature & Art* I. 162/2 In 1782 De Loutherbourg took the theatre for the exhibition of his *Eidophusikon*... He confined the Eidophusikon for the most part to the exhibition of English landscapes under different conditions of light and shadow. **1934** *Burlington Mag.* Feb. p. xv/1 De Loutherbourg's 'Eidophusikon' of movable representations of natural effects which caused the world to wonder when displayed in 1782.

Eidouranion. (Earlier example.)

a **1800** [see *EIDOPHUSIKON].

Eifel (əi·fəl). The name of a plateau in Western Germany, used *attrib.* to designate a subdivision of Middle Devonian in the region so named, and the rocks. Hence **Eifelian** (əifī·liăn) *a.* and *sb.* [ad. F. *eifelien* (A. H. Dumont 1848, in *Mém. de l'Acad. R. des Sciences*, etc., *de Belgique* XXII. 67)].

1853 *Q. Jrnl. Geol. Soc.* IX. 20 The Eifelian limestone and schists of Belgium and the Rhenish provinces. *Ibid.* 23 The *Système eifelien* of M. Dumont is formed of three distinct groups,—the Eifel limestone, Eifel fossiliferous schists, and the red sandstones and conglomerates. **1879** *Encycl. Brit.* X. 341/2 Stringocephalus group, consisting of the great Eifel limestone with underlying crinoidal beds. **1895** J. D. DANA *Man. Geol.* (ed. 4) 626 The Lower, Middle, and Upper divisions are named (1) the Rhénan, (2) the Eifelian, and (3) the Famennian. **1912** A. J. JUKES-BROWNE *Strat. Geol.* (ed. 2) 217 Middle Devonian. This also comprises two divisions, the Eifelian and the Givetian. **1957** *Encycl. Brit.* VII. 289/2 The Eifelian (Couvinian in Belgium) consists of limestones and shales abounding in corals and brachiopods. **1963** D. W. & E. E. HUMPHRIES tr. *Termier's Erosion & Sedimentation* xiii. 290 Most of the others are Eifelian, and the reef barrier seems to have persisted almost to the Late Devonian.

eigen- (əi·gən). G. *eigen* OWN, proper, peculiar, characteristic, used in adoptions or partial translations of G. compounds in *Math.* and *Physics*, as *EIGENTON, *EIGENFUNCTION; hence employed as comb. form in many scientific compounds, as *eigenload, -period, -solution, -state, -vector, -vibration* (see quots.).

1930 Eigenstate [see *EIGENVALUE]. **1940** *Chambers's Techn. Dict.* 283/1 *Eigenperiod, eigenton*, frequencies at which acoustic resonance is experienced in rectangular chambers, because of continued reflections between opposite walls. **1947** *Jrnl. R. Aeronaut. Soc.* LI. 204 A complete theory of the conditions for the existence, and the form of the self-equilibrating orthogonal and load systems (eigenloads)..which vary similarly along all generators, is given. **1951** *Rev. Mod. Physics* I (*title*) Approximate eigen-solutions. **1955** *Electronic Engin.* XXVII. 200 An infinite line of similar quadripoles each with matrix A connected in cascade..must be a vector such that the effect of operating on it with A is to leave it unaltered except for a change in magnitude...Any vector..fulfilling this condition is called an eigen-vector of the operator matrix H. **1955** W. PAULI *Niels Bohr* 39 The transformation..can be interpreted as replacing every emission operator by an absorption operator of the same eigen-vibration *without* interchange of particle and anti-particle. **1955** H. B. G. CASIMIR in W. Pauli *Niels Bohr* 131 The remarkable feature of Bose–Einstein condensation is that a sizable fraction of the particles is forced into the very lowest eigenstate, a state which should have curious and essentially non-classical properties. **1957** L. FOX *Two-point Boundary Probl.* vii. 162 Third, if the vector *y* is an approximation to any eigenvector y_k, so that, in 18, a_k is much greater than any other a_r, then first-order changes in *y* give rise to second-order changes in λ, and the eigenvalue is determined more accurately than its eigenvector. **1968** C. G. KUPER *Introd. Theory Superconductivity* ix. 150 Likewise, $b\phi_n$ is either zero or an eigenstate with energy $E_n - \hbar\omega$.

ei·genfrequency. *Physics.* [tr. G. *eigenfrequenz.*] One of the frequencies that cause a given system to vibrate; a resonant frequency.

1955 W. HEISENBERG in W. Pauli *Niels Bohr* 14 Schrödinger attempted..to resolve quantum theory into a simple classical wave theory. The motive for this attempt was the discovery that the discrete eigenvalues appeared in wave mechanics not as energies, but as the eigenfrequencies of waves.

ei·genfunction. *Physics.* [tr. G. *eigenfunktion.*] A solution of a differential equation possessing solutions only for special values of a parameter.

1926 *Proc. R. Soc.* A. CXII. 661 A set of independent solutions, which may be called eigenfunctions. **1927** [see *EIGENVALUE]. **1938** *Nature* 16 Apr. 668/1 The homopolar or bond eigenfunction method associated with the names of Heitler, London, Slater and Pauling. **1946** E. C. TITCHMARSH (*title*) Eigenfunction expansions associated with second-order differential equations. **1951** *Rev. Mod. Physics* XXIII. 1/2 The eigenfunction of a bound state.

eigenton, -tone (əi·gən,tōᵘn). *Acoustics.* [a. and tr. G. *eigenton.*] A frequency at which acoustic resonance characteristically occurs in an enclosure.

1940 [see *EIGEN-]. **1962** A. NISBETT *Technique Sound Studio* iii. 57 The even harmonics of the eigentones—the basic resonances of the room—will be missing. **1968** CHOMSKY & HALLE *Sound Pattern Eng.* 352 The feature of 'proper pitch' (*Eigenton*)..subsumes the phonetically quite different features of rounding and backness.

ei·genvalue. *Physics.* [tr. G. *eigenwert.*] One of those special values of a parameter in an equation for which the equation has a solution (see quot. 1938).

1927 *Nature* 23 July 117/1 Among those..trying to acquire a general acquaintance with Schrödinger's wave mechanics there must be many who find their mathematical equipment insufficient to follow his first great problem—to determine the eigenvalues and eigenfunctions for the hydrogen atom. **1930** P. A. M. DIRAC *Princ. Quantum Mech.* iii. 35 The physical meaning of an eigenvalue is that there exists a state, namely, the eigenstate belonging to it, such that a measurement of the observable when the system is in this state will certainly give for result just this eigenvalue. **1938** R. W. LAWSON tr. *Hevesy & Paneth's Man. Radioactivity* (ed. 2) viii. 93 A single-valued, finite, and continuous solution of the differential equation is found only for particular values of the parameter *E* which occurs in the equation, and these are the 'proper'- or 'eigen'-values of the differential equation. **1939** A. S. EDDINGTON *Philos. Phys. Sci.* 162 In mathematical language [existence is represented by] a symbol *J* with two eigenvalues, which are most conveniently taken to be 1, standing for existence, and 0, standing for non-existence. **1951** *Rev. Mod. Physics* XXIII. 1/2 The reason why we choose *a* as the known parameter and *b* as the eigen-value. **1955** [see *EIGENFREQUENCY]. **1958** P. A. M. DIRAC *Princ. Quantum Mech.* (ed. 4) ii. 36 The set of eigenvalues of a real dynamical variable are just the possible results of measurements of that dynamical variable and the calculation of eigenvalues is for this reason an important problem.

eight, *a.* and *sb.* **B. 2. b.** For 'at the Universities of Oxford and Cambridge' read 'at the University of Oxford and elsewhere'. (Add further examples.) Also, a boat for eight oarsmen.

1890 *Pall Mall Gaz.* 27 May 6/1 (Oxf. Notes), To-morrow the eights week will come to an end. **1898** S. LE BLANC-SMITH in W. A. Morgan *The 'House' on Sport* 302 It is more difficult to obtain eight such men than to obtain four, added to which the eight being the heavier boat..greater rigidity is necessitated in the thowl-pin. **1908** *Daily Chron.* 21 May 4/6 Oxford has every right to distinguish her annual inter-college races..as 'The Eights'. **1911** BEERBOHM *Zuleika D.* vi. 81 Isn't it a lovely day for the Eights? **1955** *Times* 6 June 7/7, I noticed in your issue of May 30 your Eights Week chart. ..You record a double overbump...I should like to know whether this feat has ever been accomplished before in Eights? *Ibid.* 25 July 3/6 Beaumont College, who won their juniors at Molesey a week ago, followed it up by winning the junior-senior eights. **1962** *Times* 13 Feb. 4/6 They have offered to build a racing eight by the hot-moulded method.

4. Slang phr. *one over the eight*: one alcoholic drink too many. (Cf. *ONE *numeral a.*, etc., 1 d.)

1925 FRASER & GIBBONS *Soldier & Sailor Words* 88 *One over the eight*, one drink too many. Slightly intoxicated, the presumption being that an average 'moderate' man can safely drink eight glasses of beer. **1928** *Daily Express* 3 Aug. 7/4 Luton magistrate: What does he mean by 'one over the eight'? ('A glass too many'?)

C. *eight-manned, -oared, -rowed, -spoked, -wheeled* adjs.; *eight-bore, -ounce*; *eight-bearer, -company, -dog*; *eight-pointer* [POINTER 10], *-wheeler, -yarder*; **eight-coupled** *a.*, having eight coupled wheels; **eight-day clock** (earlier example); **eight-foil** *Her.* (see quot.).

1908 *Westm. Gaz.* 17 Dec. 7/2 An eight-bearer yellow chair. **1874** J. W. LONG *Amer. Wild-Fowl Shooting* 23 For flight-shooting, an 8-bore [gun] is as large as is advantageous, and a 10 is sufficiently small. **1900** *Daily News* 26 June 3/5 An eight-company battalion. **1893** *English Mechanic* 14 Apr. 181/1 (*heading*) Eight-coupled goods engines. **1903** *Westm. Gaz.* 3 Oct. 7/3 One of the standard eight-coupled goods engines. **1904** C. S. LAKE *Locomotive* 60 Goods engines with eight-coupled wheels. **1741** RICHARDSON *Pamela* IV. xiii. 77 Being wound up..once a Week, like a good Eight-day Clock. **1876** *Coursing Calendar* 56 An eight-dog stake was added to the card in the evening. **1847** *Gloss. Terms Brit. Her.*, *Huit-foil, Eight-foil*, or *Double quatrefoil*, an eight leaved flower used as a mark of cadency for the ninth son. **1897** *Daily News* 3 Feb. 6/4 A woman riding one among seven men on an eight-manned wheel. **1874** K. H. DIGBY *Temple of Memory* iii. 40 Their eight-oar'd races. *Ibid.* 41 Their eight-oar'd crew felt quite in Heav'n. **1886** *Outing* (U.S.) VIII. 161/1 On one side an eight-ounce rod, a thread of silk [etc.]. **1909** *Daily Chron.* 28 Sept. 4/5 A good eight-pointer, weighing over 15 stone. **1838** *Mass. Agric. Rep.* 1837 24 The Pickwacket corn, an early eight-rowed variety. **1869** *Rep. Comm. Agric.* 1868 (U.S.) 431 A small variety of eight-rowed corn. **1944** *Burpee's Seeds* 26 If you want the best and sweetest for your table, it's true 8-rowed Golden Bantam. **1884** F. KROHN tr. *Glaser de Cew's Magn. & Dyn.-electr. Mach.* 33 The armature consists of an eight-spoked wheel. **1906** *Daily Chron.* 1 Mar. 7/2 Eight-wheeled first and third-class carriages. **1904** *Westm. Gaz.* 28 Dec. 3/1 A special mail train consisting of twelve eight-wheelers. **1930** *Morning Post* 17 June 14/7 He holed an eight-yarder for a 2 at the sixth.

eight ball, eight-ball, eightball (ē̆i·tbǫl). *U.S.* [f. EIGHT *a.* and *sb.* + BALL *sb.*¹] **a.** The black ball, numbered eight, in the North American variety of pool (POOL *sb.*² 3); this variety of pool. **b.** Phr. *behind the eight ball*, at a disadvantage; 'snookered'. **c.** *slang.* A Negro. **d.** *slang.* A stupid person. **e.** (See quot. 1937.)

1932 J. T. FARRELL in *This Quarter* Dec. 379 Well, after this, you walk in yere own neighbourhood, eight-ball, or else you can be the guest to yere own funeral. **1932** *World's Work* Feb. 26/1 Mr. Ells, the wizard of the cue, in a position he expertly diagnosed as squarely behind the 8-ball. **1934** J. O'HARA *Appointment in Samarra* (1935) 12 You get signing checks for prospects down at the country club, and you wind up behind the eight-ball. **1937** *Printers' Ink Monthly* Apr. 53/1 *Eight ball*, an astatic microphone characterized by non-directional pick-ups and having a 360-degree beam. **1944** *N.Y. Herald* (Bk. Rev.) 24 Sept. 12/4 An attempt to describe what makes the drawings funny lands you behind the eight ball. **1945** L. SHELLY *Jive Talk Dict.* 11/1 *Eightball*, colored man or obstacle. **1964** D. G. WISBESKI *Otter in House* (1965) x. 157 Okee..broke up a pool game by climbing on the table and taking off with the eight ball.

eighteen, *a.* and *sb.* Add: **2. b.** *sb. pl.* A sheet of eighteen pages; a book in 18mo.

1683 MOXON *Mech. Exerc.*, *Printing* II. 55 When a Twelves, Eighteens, etc. is wrought. **1795** *Hull Advertiser* 19 Dec. 4/3 A purposely manufactured wove paper, in Octodecimo or Eighteens. **1808** C. STOWER *Printers' Gram.* 190 A plan for imposing a half sheet of eighteens. **1839** T. C. HANSARD *Print. & Type-Founding* 168 Works done in sixteens, eighteens, twenty-fours, or thirty-twos. **1937** E. J. LABARRE *Dict. Paper* 139 *Eighteenmo*, or *eighteens*, are other terms for *decimo-octavo*.

3. *eighteen-hole(s* (golf-course); *eighteen-tonner.*

1907 *Westm. Gaz.* 21 Sept. 9/1 The opening of Matlock's new eighteen-holes golf course. **1944** *Mod. Lang. Notes* Dec. 515, 18-hole golf course. **1888** (*title*) To Gibraltar and back in an eighteen-tonner.

b. As in *eighteen-twenties*, the years between 1819 and 1830.

1906 *Daily Chron.* 16 Oct. 4/4 A collection of poems by Frances Ridley Havergal, belonging to the eighteen-seventies. **1909** *Westm. Gaz.* 24 Mar. 2/1 She has abundance of 'sensibility', as that word was understood in the eighteen-twenties. **1924** GALSWORTHY *White Monkey* II. ix, The Hotch-potch Club went back to the eighteen-sixties. **1929** S. ERTZ *Galaxy* x, The eighteen-nineties came, with their revival of interest in literature and painting. **1929** (*title*) The eighteen-seventies: essays by Fellows of the Royal Society of Literature. Ed. H. G. Granville-Barker. **1930** (*title*) The eighteen-eighties. Ed. W. de la Mare.

eighth, *a.* Add: **3.** *eighth note* = QUAVER *sb.*¹ *U.S.*

1889 in *Cent. Dict.* **1958** BLESH & JANIS *They all played Ragtime* iv. 77 Unaccented eighth notes alternating with accented quarter notes.

B. *sb.* **2. c.** = *eighth note.*

1956 M. STEARNS *Story of Jazz* (1957) xxi. 273 Iturbi produces the feeling that he is playing straight (not dotted) eighths.

eight hours. A period of time regarded as a fair working-day; freq. *attrib.* (or with possessive apostrophe) and in form *eight-hour* (esp. in phr. *eight-hour day*).

1845 DISRAELI *Sybil* VI. vi. 430 We have a right to four shillings a day wages, eight hours' work and two pots of ale.—A fair day's wage for a fair day's work. **1865** *Nation* (N.Y.) I. 517/2 It is enough to condemn the scheme of this eight-hour labor league to say that..it would diminish production. **1867** in S. & B. Webb *Hist. Trade Un.* (1920) 309 Such a measure of legislative restriction as shall secure a uniform Eight Hours Bill in factories, exclusive of meal-times. **1869** C. L. BRACE *New West* v. 60 The effort to gain a larger share of the profits of capital by means of an Eight-Hour Law. *Ibid.* 61 From all we can hear, the eight hour movement will soon fall to the ground. **1891** S. WEBB & H. COX (*title*) The eight hours day. **1895** in *Encycl. Brit.* (1902) XXXII. 668/1 Higher wages and eight hours for Government workmen—all these things were in the direction of helping the unemployed. **1899** *Daily News* 6 Mar. 7/1 It should never be forgotten that Alfred was the inventor of the eight hour day. **1902** *Encycl. Brit.* XXXII. 668/1 English public opinion was sceptical when the Trades Union Congresses declared themselves powerless to establish an eight hours' working day without the help of the State. **1910** J. F. FRASER *Australia* 202 The eight hours' day is operative throughout the Commonwealth. **1948** J. BETJEMAN *Coll. Poems* (1958) 227 An eight-hour day for all.

eightsome (ē̆i·tsəm), *a.* Add: [See -SOME².] (Further examples.) Also *sb.* = an eightsome reel.

1745 A. MURRAY *Let.* 3 July in J. J. Murray *Chrons. Atholl & Tullibardine Families* II. 481, I am to be in the 'Eightsome Minuet' if I be at the Ball. **1875** W. ALEXANDER *Sks. Life among Ain Folk* 246 They..caper through the 'eightsome' figure with louder 'hooch-hoochs!' than before. **1899** *Daily Mail* 1 June 4/5 After two o'clock reels and eightsomes were indulged in. **1905** *Westm. Gaz.* 15 Sept. 7/2 A host of charming young Highland girls, with the tartans and badges of their clans, danced the eightsomes and the Reel of Tulloch. **1926** *Glasgow Herald* 7 June 8 The dancing of Scottish reels, foursomes, eightsomes, and sometimes even a sixteensome. **1952** 'J. TEY' *Singing Sands* vii. 109 There was room, with some squeezing, for three eightsomes.

eight-square, *a.* (Later U.S. example.)

1907 S. E. WHITE *Arizona Nights* I. i. 4 He..carried across his saddle a heavy 'eight square' rifle.

eighty, *a.* and *sb.* Add: **2. c.** Eighty acres of land. *U.S.*

1842 C. M. KIRKLAND *Forest Life* II. 207 Happy he whose far reaching 'eighties' enclose a sugar-bush. **1872** *Amer. Naturalist* VI. 77 The whole surface of the plains is sere and brown save some 'eighties' or larger tracts that are fenced. **1936** M. H. BRADLEY *Five-Minute Girl* v. 74 So his ax had set to work on the maples and hemlocks in his north eighty.

3. *eighty-ton* (gun).

1874 *Porcupine* 18 Apr. 37/1 The construction of an 80-ton gun at Woolwich is stated to be decided upon. **1880** *Encycl. Brit.* XI. 292/2 In the 80-ton gun powder cubes of 1½ in. edge are used. **1888** A. C. GUNTER *Mr. Potter of Texas* iii, That was the first eighty-ton gun fired in war.

-ein, -eine, variants of -IN¹ and -INE⁵ used as suffixes in the names of certain chemical compounds containing anhydrides, esp. phthalic anhydride. In bases the ending is usually *-eine*, and in non-bases *-ein*.

1871 *Jrnl. Chem. Soc.* XXIV. 834 (*heading*) Fluorescein and Fluorescin. *Ibid.*, By heating together phthalic anhydride and resorcin, Baeyer has obtained a body which he calls fluorescein. **1884**, etc. [see *CYSTEINE]. **1889** *Jrnl. Chem. Soc.* LVI. 1153 Catecholphthalein..is obtained when phthalic anhydride (3 parts) is heated with catechol and zinc chloride. **1932** *Chem. Abstr.* 1388 One of the alleged secondary alkaloids, nicoteine, isolated by Pictet, is shown to have no existence. **1949** *Blakiston's New Gould Med. Dict.*, *Narceine*, an alkaloid contained in morphine.

‖ **einfühlung** (əi·nfūluŋ). *Psychol.* Also **Ein-.** [G., see *EMPATHY.] = *EMPATHY.

1904, etc. [see *EMPATHY]. **1925** I. A. RICHARDS *Princ. Lit. Crit.* xv. 108 The work done by Lipps, Groos and others on *einfühlung*, or empathy. **1929** C. I. LEWIS *Mind & World-Order* ii. 41 'Real duration'..is something which is immediate, in his own case, and is to be apprehended in its other manifestations only by empathy or *einfühlung*. **1936** *Times Lit. Suppl.* 2 May 374/2 If..a doctrine of Einfühlung could be established. **1966** *English Studies* XLVII. 311 Some of the elements which have caused Vernon Lee..to be almost forgotten by a later age, to the point of not even crediting her with the introduction of the idea of *Einfühlung* or empathy.

einkorn (əi·nkǫən). [G., f. *ein* one + *korn* seed.] An inferior species of wheat, *Triticum monococcum*.

[**1884** tr. *De Candolle's Orig. Cultivated Plants* v. 365 The one-grained wheat, or little spelt, *Einkorn* in German, is distinguished..by a single seed in the little ear.] **1904** T. F. HUNT *Cereals in America* iv. 48 Einkorn never..gives rise to a fertile cross with common wheat. **1924** J. A. THOMSON *Science Old & New* xliii. 250 Inferior cultivated wheats like Emmer, Spelt, and Einkorn. **1965** *Jrnl. R. Hort. Soc.* XC. 284 The basic crops of neolithic agriculture were the wheats einkorn..and emmer.

Einstein (əi·nʃtəin, -stəin). The name of Albert *Einstein* (1879–1955), German physicist and mathematician, used *attrib.* or in the possessive to designate certain theories and principles enunciated by him or arising out of his work.

1922 GLAZEBROOK *Dict. Appl. Physics* II. 359/2 According to Einstein's theory all energy possesses mass. *Ibid.* 595/1 Richardson has shown that Einstein's equation may be derived by thermodynamic and statistical methods. **1923** J. M. MURRY *Pencillings* 68 To some the Einstein theory may show the way of reconciliation. **1927** A. S. EDDINGTON *Stars & Atoms* 52 The Einstein effect is proportional to the mass divided by the radius of the star. **1934** WEBSTER, *Einstein shift*, the difference in wave-length..between light emitted by a source of a definite nature in the sun..and that emitted by a like source on the earth. **1958** *Listener* 11 Dec. 973/1 Eddington remained faithful to this idea that the universe evolved from the static but unstable Einstein universe. **1967** *Handbk. Chem. & Physics* (Chem. Rubber Co.) (ed. 48) F-69, *Einstein theory for mass–energy equivalence*, the equivalence of a quantity of mass *m* and a quantity of energy *E* by the formula $E = mc^2$. The conversion factor c^2 is the square of the velocity of light.

Einstein–Bose (əi·nʃtain,bōᵘs, -bǭᵘz, -st-). *Physics.* [The names of Albert *Einstein* (see prec.) and S. N. *Bose* (see *BOSON).] A designation used as an alternative to *BOSE–EINSTEIN, as *Einstein–Bose particle, statistics*, etc.
[**1929** DIRAC in *Proc. Cambr. Philos. Soc.* XXV. 62 The so-called 'statistics' of Einstein–Bose or Fermi applies only to an assembly of actual systems which could interact with each other.] **1931** *Physical Rev.* XXXVII. 333 We.. justify the assumption that the clusters satisfy the Einstein–Bose or Fermi–Dirac statistics according to whether the number of particles in each cluster is even or odd. **1938** *Proc. R. Soc.* A. CLXVI. 127 (*heading*) Quantum theory of Einstein–Bose particles and nuclear interaction. **1955** R. D. EVANS *Atomic Nucleus* iv. 178 Photons and α particles obey Einstein–Bose statistics.

Einsteinian (əinʃtai·niăn, -st-), *a.* [f. the name of Albert *Einstein* + -IAN.] Pertaining to or characteristic of Einstein or his theories.
1925 C. E. M. JOAD *Mind & Matter* 46 In an Einsteinian universe the velocity of light is the greatest velocity possible. **1928** *Observer* 25 Mar. 9/4 Einsteinian physics. **1937** *Mind* XLVI. 101 He rejects the view which would combine the importance of the Einsteinian relativity theory to scientific method. **1941** D. L. SAYERS *Mind of Maker* iii. 33 We may say: that Einsteinian physics has superseded Newtonian physics. **1959** *Encounter* Dec. 64/2, I confused Bergsonian time with Einsteinian time. **1968** E. MCGIRR *Lead-Lined Coffin* iii. 55 Once in every few years one of these thousands of men shows he has a brain of Einsteinian capacity.

einsteinium (əinʃtai·niŏm, -st-). *Chem.* [f. the name of Albert *Einstein* + -IUM.] An artificially produced radioactive element. Symbol Es. Atomic number 99. Atomic mass of isotope 253 (1967) 253·0847.
1955 [see *FERMIUM]. **1958** *Times* 29 Aug. 6/5 The following 10 [transuranic elements] were known.. einsteinium (99), fermium (100). **1967** *New Scientist* 6 July 22/2 Oak Ridge National Laboratory says it has succeeded in separating roughly one ten millionth of an ounce of einsteinium, the transuranic element first discovered in 1954.

eirenic, *a.* Delete *rare* and add later examples. (See also IRENIC *a.* and *sb.*)
1938 *Sunday Times* 13 Feb. 14/3 In view of the eirenic purpose of the Commission, this omission is the more regrettable. **1964** P. F. ANSON *Bishops at Large* i. 37 The Bishop of Iona made good speed on his eirenic journey.

eirenical, var. IRENICAL *a.*
1890 GASQUET & BISHOP *Edw. VI & Bk. Com. Prayer* 28 The choice of Quignon's work for a model had an aspect almost eirenical. **1891** *Tablet* 3 Jan. 11 The whole Pastoral.. is decidedly eirenical. **1958** *Spectator* 30 May 702/3 The scholarly writer and preacher and most eirenical divine, Richard Baxter.

eisegesis (əisīdʒī·sis). [f. Gr. εἰς in, into + *-egesis* of EXEGESIS.] The interpretation of a word or passage (of the Scriptures) by reading into it one's own ideas. Hence **eisege·tical** *a.*
1878 P. SCHAFF *Through Bible Lands* I. v. 53 The eisegetical manner of those allegorical and typological exegetes who make the Scriptures responsible for their own pious thoughts and fancies. **1892** *N.Y. Evangelist* 3 Mar. 4/4 (Funk), Dr. Elliot.. held firmly to the doctrine that exegesis, and not 'eisegesis', is the province of the student of the Scriptures. **1924** H. E. FOSDICK *Mod. Use Bible* iii. 87 The reformers.. could use eisegesis instead of exegesis on many a passage which they thought they were literally interpreting. **1958** *Times Lit. Suppl.* 3 Jan. 10/1 Fully aware of the perils of *eisegesis*, Dr. Grant pronounces quite firmly against any such reverence for philosophy or theology as would empty history of meaning.

Eisteddfodic (əi:steðvǫ·dik), *a.* [f. EISTEDDFOD + -IC.] Of or belonging to the Eisteddfod.
1877 J. RHYS *Lect. Welsh Philol.* vi. 314 *Ofydd*.. is defined to be an Eisteddfodic graduate who is neither bard nor druid, and translated into ovate. **1894** *Daily News* 31 May 5/4 One of the Eisteddfodic conductors. *a* **1922** H. JONES *Old Mem.* (1923) i. 29, I pass over the Eistedfodic prizes we won.

eis wool (əis). [G. *eis* ice.] (See quots. 1882, 1957.)
1882 CAULFEILD & SAWARD *Dict. Needlework*, *Eis wool* (sometimes written 'Ice Wool'), a very fine glossy description of worsted wool, made of two-thread thickness, and employed double for making shawls. **1891** *Queen* 17 Oct. 639/3 Twelve balls of eis wool are required. **1957** M. B. PICKEN *Fashion Dict.* 110/1 *Eis wool*, fine, glossy, wiry woolen yarn. Used for clouds, scarfs, etc.

either. Add: **B. 3. c.** *either-or*, *either/or* [in some examples reflecting Da. *enten-eller* (title of book by the Danish philosopher Kierkegaard, 1843)], used as *sb.*, a necessary or unavoidable choice between alternatives. Also *attrib.* or as *adj.*, black and white, susceptible of only one of two (often extreme) solutions, responses, etc.
1931 *Times Lit. Suppl.* 10 Sept. 672/3 Dr. Harris is a philosopher and a logician with a little too much of the 'either-or' in his mental make-up. **1931** *Church Times* 9 Oct. 388/2 Catholicism.. provides.. a check upon humanism, which any 'either/or' theology cannot give. **1941** AUDEN *New Year Let.* II. 44 The either-ors, the mongrel halves Who find truth in a mirror. **1942** 'G. ORWELL' in *Partisan Rev.* IX. 498 One can predict the future in the form of an 'either-or': either we introduce Socialism, or we lose the war. **1944** W. LOWRIE tr. *Kierkegaard's Either/or* II. 134 Although my life now has to a certain degree its either/or behind it, yet I know well that it may still encounter many a situation where the either/or will have its full significance. **1951** C. S. LEWIS *Let.* 23 Apr. (1966) 228, I have no use for mere *either-or* people. **1953** *Economist* 10 Jan. 58/1 Too much rigid logic of the black-and-white either-or variety. **1958** *Spectator* 30 May 709/1 An 'either-or' attitude, leading to.. the Cawnpore Well and the hanging-parties at Benares. **1965** *Language* XLI. 258 Either-or questions.

-eity [cf. Fr. *-éité*], termination of nouns of quality or condition corresp. to adjs. in -EOUS, on the model of L. *idoneus* IDONEOUS, late L. *idoneitās* IDONEITY (1617), scholastic L. *homogeneus* HOMOGENEOUS, *homogeneitās* HOMOGENEITY (1625). Among other early examples are *spontaneity* (1651), *subterraneity* (1686), *consentaneity* (1798). Two exceptional mod.L. formations *ecceitās* (f. *ecce* lo, behold), *velleitās* (f. *velle* to will) gave ECCEITY (1549), VELLEITY (1618).

ejaculate (ĭdʒæ·kiŭlĕt), *sb.* [f. the vb.] Ejaculated seminal fluid.
1927 *Jrnl. Amer. Med. Assoc.* LXXXVIII. 901/2 Since the diagnostic vasostomy, the patient has not had sperms in his ejaculate. **1938** *Amer. Jrnl. Med. Sci.* CXCVI. 369 A determination of the number of sperm in the total ejaculate is obtained by multiplying the count per cc. by the volume of the entire ejaculate. **1966** *Biol. Abstr.* XLVII. 5708/1 Spermatozoa were observed in coagulated spontaneous ejaculates of isolated male guinea pigs.

ejaculation. Add: **2.** *spec.* The discharging of the male sperm.
1888 E. L. KEYES *Surg. Dis. Genito-Urinary Organs* I. xxvi. 442 Ejaculation of semen may be produced by a variety of causes. **1928** [see next]. **1966** R. D. AMELAR *Infertility in Men* ii. 32 The semen is ejaculated in a liquid form and becomes a gel or coagulum, immediately after ejaculation, only to liquefy again within 5 to 20 minutes.

‖ **ejaculatio præcox** (ĭdʒækiŭlēi·ʃio prī·kǫks). [mod.L.: see EJACULATION, PRÆCOCIAL *a.*] (See quot. 1925.)
1892 C. G. CHADDOCK tr. *Krafft-Ebing's Psychopathia Sexualis* iii. 196 He became neurasthenic, being afflicted with weakness of erection and ejaculatio præcox. **1921** E. J. KEMPF *Psychopath.* vii. 326 During his marriage, he was virtually impotent, and only at times was he able to perform as much as ejaculatio præcox. **1925** STEDMAN *Med. Dict.* (ed. 8) 307/2 *E[jaculatio] præcox*, premature ejaculation, rapid termination of the sexual act on the part of the male. **1928** EISENDRATH & ROLNICK *Text-Bk. Urology* xxviii. 441 The term impotence conveys a relative meaning, varying in degree from premature ejaculation (ejaculatio praecox) to total inability to have erection. **1959** L. P. WERSHUB *Sexual Impotence in Male* viii. 102 Some forms of *ejaculatio praecox* can often be due to *physical* causes.

ejecta (ĭdʒe·ktă), *sb. pl.* [neut. pl. of pa. pple. of L. *eicĕre*, f. *e-* out, forth + *iacĕre* to cast.]
1. The matter ejected from a volcano.
1886 *Amer. Meteor. Jrnl.* III. III. 109 Dust and other ejecta played but a secondary part in the production of the phenomena. **1890** *Nature* 16 Oct. 601/2 It may be thought that any volcanic ejecta would speedily melt the snow upon which they fell. **1902** *Daily Chron.* 11 Sept. 6/2 On the afternoon of the 3rd the ejecta was of the colour of sulphur. **1957** G. E. HUTCHINSON *Treat. Limnol.* I. i. 25 The summit of the volcano usually has a well-defined crater from which ejecta escape to continue the building of the cone.
2. *Path.* That which is ejected from the body.
1890 BILLINGS *Med. Dict.* I. 433/1 *Ejecta*, matters thrown out. **1908** *Practitioner* Oct. 601 Frequent vomiting,.. the ejecta being often blood-stained and sometimes faecal. **1953** *Faber Med. Dict.* 134/2 *Ejecta*, matter thrown out; excretions.

ejection. Add: **1. d.** *Aeronaut.* The mechanically contrived 'baling out' of a pilot from an aeroplane or space-craft. Also *attrib.*, as *ejection seat*, on which this is effected. Cf. *EJECTOR 2.
1945 *Aeroplane* 16 Nov. 569/1 It was the first German aeroplane to employ a pilot-ejection seat... The single-seat cockpit is positioned well forward and the pilot ejection seat is of the explosive cartridge type. **1946** *Aeroplane Spotter* 10 Aug. 182/1 The first automatic high-speed 'baling-out' ejection by Mr. Bernard Lynch.. on.. July 24. **1956** W. A. HEFLIN *U.S. Air Force Dict.* 184/2 Ejection chute, cockpit capsule, seat, seat-trainer. **1967** *New Scientist* 27 Apr. 195/1 Ejection seats were omitted from the multi-man *Voskhod* spacecraft.

ejective. Add: **3.** *Philol.* Of voiceless consonants: articulated by means of non-pulmonic air-pressure, created by closing and raising the glottis. Also as *sb.*
1932 D. JONES *Outl. Eng. Phonetics* (ed. 3) xvii. 141 Consonants.. in which the necessary air pressure is produced by some other means than by the lungs. Sounds in which the air is forced outwards by these means are called *ejective* consonants. **1939** L. H. GRAY *Found. Lang.* iii. 57 Ejectives (plosives with a simultaneous glottal stop). **1964** R. H. ROBINS *Gen. Ling.* iii. 103 Three non-pulmonic types of consonant articulations are used in some languages: ejectives, implosives, and clicks. *Ibid.*, Ejective consonants, which are sometimes also called glottalized consonants, are most commonly found with plosive or affricate release. **1968** CHOMSKY & HALLE *Sound Pattern Eng.* 323 Ejectives have a transition with a somewhat higher termination frequency than the corresponding nonejectives.

ejector. Add: **2.** *ejector gun, rifle*; **ejector seat** *Aeronaut.* = *ejection seat.
1892 W. W. GREENER *Breech-loader* 42 Fine guns and ejector guns cannot be purchased under £35. **1907** *Yesterday's Shopping* (1969) 636 The Society's hammerless ejector rook rifle. **1920** G. BURRARD *Notes on Sporting Rifles* 68 Spare fore sights, extractors in the case of ejector-rifles. **1945** *Flight* 15 Nov. 524/1 Two types of German pilot-ejector seats. **1948** *Flight* 22 July 100/1 There are no known reports of a pilot getting out of the Vampire. To fit an ejector seat now would be a major operation. **1955** *Times* 6 July 10/6 Lieutenant-Commander Rickell used his ejector seat to bale out.

‖ **ejido** (exiðo, ēhī·ðo). [Mexican Sp., f. Sp. *ejido* common land, f. L. *exitus* departure, f. *exire* to go out.] In Mexico, land farmed communally; a co-operative farm; land to which communal title is held. Also *attrib.*
1889 in *Cent. Dict.* **1931** C. BEALS *Mexican Maze* ii. 37 Their *ejidos* (village commons).. were again menaced. **1938** *New Statesman* 12 Feb. 241/1 To facilitate the harvesting of profitable crops, the lands are in some places worked collectively, the members of each neighbourhood or 'ejido' helping each other in co-operative fashion. **1946** H. F. INFIELD *Co-operative Living in Palestine* 4 The Russian Kolkhoz, the Palestinian Kvutza, the Mexican Ejido.. have made effective use of comprehensive co-operation. **1964** *New Statesman* 28 Feb. 327/3 Zapata stressed the return of the lost communal holdings to the rural Indian villages. Thus he proposed the '*ejido*' system in which title to lands was to be vested in the landholding village.

eka- (ī·kă), *pref.* *Chem.* [a. Skr. *eka* one.] Used 1872 in *Ann. d. Chemie u. Pharmacie*, VIII. Supplementband 196 by D. I. Mendeléeff (1834–1907), Russian chemist, to denote a predicted element that should occupy the next lower position to that so qualified in the same group in the periodic system.
1889 *Jrnl. Chem. Soc.* LV. 648 *Gallium*, which proved to correspond to eka-aluminium of the periodic law.. *scandium*, corresponding to eka-boron.. and.. *germanium*, which proved to correspond in all respects to eka-silicum. **1931** *Observer* 18 Oct. 22/4 Eight years ago there were six [atomic] numbers corresponding to unknown elements... Both 43 and 75 were known before their discovery as eka-manganeses, from their anticipated similarity to manganese... 85 should be chemically akin to iodine, and so is provisionally called eka-iodine... On Friday the news came from America that eka-caesium had been discovered. **1938** *Ann. Reg. 1937* 355 Curie & Savitch.. obtained three new series of short period radioactive elements ending possibly in eka-gold, eka-iridium, and eka-rhenium. **1955** *Ann. Reg. 1954* 398 It was announced in February that physicists of the University of California had produced element No. 99 provisionally called eka-holmium. **1969** *Daily Tel.* 5 May 13/7 Russian scientists at Dubna and Leningrad are looking for element 114, eka-lead, in old mirrors made from lead glass.

ekdemite (e·kdĭməit, ekdī·məit). *Min.* Also **ecdemite.** [ad. Sw. *ekdemit* (A. E. Nordenskiöld 1877, in *Geol. För. Förh.* III. 379), f. Gr. ἔκδημ-ος unusual + -ITE[1].] A yellow oxychloride of lead and arsenic, of uncertain formula, found in Sweden and the U.S.A.
1879 *Jrnl. Chem. Soc.* XXXVI. 22 Ekdemite.., a new mineral, coarse-crystalline, foliated, monoaxial, with a distinct basal cleavage. **1882, 1951** [see *ECDEMITE]. **1968** EMBREY & PHEMISTER tr. *Kostov's Mineralogy* 467 Ekdemite is tetragonal, dimorphous with heliophyllite.

eking, *vbl. sb.* **3.** (Earlier example.)
1819 J. ROSS *Voy. Discovery* p. xviii, Hooks and ekeings were placed in the bows above the lower-deck hook.

ekistics (ĭki·stiks). [f. mod.Gr. ἡ οἰκιστική, f. Gr. οἰκιστικός relating to settlement, f. οἰκίζ-ειν to settle (a colony), f. οἶκος house, dwelling.] A name given by C. A. Doxiadis to the study of human settlements and the way they develop and adapt themselves to changing circumstances. Hence **eki·stic** *a.*, relating to ekistics; **ekisti·cian** (-ʃăn), one who studies or is versed in ekistics.
1959 C. A. DOXIADIS in *Rep. Proc. Town & Country Plan. Summer Sch.* (Town Plan. Inst.) 119 Ekistics, as a scientific discipline, can be divided into Ekistic geography, Ekistic economy, social Ekistics, and the other branches which.. make up its whole and.. relate it to the neighbour sciences of geography, economics, sociology, etc. **1968** *New Society* 26 Sept. 456/1 Only the new science of ekistics can embrace the whole study of human settlements and their planning, in all their manifold elements. **1968** C. A. DOXIADIS *Ekistics* ii. 75 Among all the experts who deal with the problems of human settlements, the Ekistician should be the one who has the most inter-disciplinary approach. *Ibid.* 516 *Ekistics*, science of human settlements. Term coined by the author..; first used in his lectures of 1942 at the Athens Technical University.

ekka (e·kā). *Anglo-Ind.* Also **ecca, ecka.** [a. Hindi *ekkā* lit. unit, f. Skr. *eka* one.] A small one-horsed vehicle used in India. Also *attrib.*
1811 F. B. SOLVYNS *Les Hindous* III. xxxi. pl. 4 Ekka.. perhaps the simplest carriage that can be imagined, being nothing more than a chair covered with red cloth, and fixed upon an axle-tree between two small wheels. The Ekka is drawn by one horse. **1848** J. H. STOCQUELER *Oriental Interpreter*, *Ecka*, a light pony gig on two wheels, with crimson cloth cushions on the top, on which the natives of India (who alone use them) sit cross-legged. **1887** KIPLING *Departm. Ditties* (1888) 21 The Waler jumped an *ekka* just above the City Drain. **1888** —— *Plain Tales* iii. 21 The *ekka*-pony.. had gone home. **1892** *Pall Mall Gaz.* 21 Dec. 3/2, I am going to take an ecka. **1895** B. M. CROKER *Village Tales* 148 His wife had real silver tyres to the wheels of her ekka! **1899** *Westm. Gaz.* 25 Oct. 3/1 Ekka bells. **1922** *19th Cent.* Oct. 592 The bullock carts, the loaded camels, the eccas, the elephants, of typical India. **1966** E. NEWBY *Slowly down Ganges* 101 The only vehicle.. was an ekka, a small pony cart with a canopy over it and side curtains.

ekker (e·kəɹ). *University* or *School slang.* [See *-ER[6].] = EXERCISE *sb.* 7.
1891 S. J. DUNCAN *Amer. Girl in London* xxiii. 254 Walking for what in the vulgar tongue might be called exercise, but here [*sc.* in Oxford] was 'ekker'. **1901** *Winchester Coll. Notions* 40 Most houses have Ekker Rolls, the amount of compulsory exercise varying from four to six hours a week. **1970** *Wykehamist* 21 Mar. 447/1 Whatever the supposed range of activities that qualify as ekker, the demands of the major games.. usually over-ride all others.

ekphore, etc.: see *ECPHORE *v.*, etc.

ektene, var. *ECTENE.

ektodynamomorphic, var. *ectodynamomorphic* (*ECTO-).

El, el (el), colloq. U.S. abbrev. of *elevated railroad*; **el train** = *elevated train* (*ELEVATED *ppl. a.* 1).
a **1906** 'O. HENRY' *Trimmed Lamp* (1916) 120 Behold Ikey as he ambles up the street beneath the roaring 'El' between the rows of reeking sweatshops. **1932** E. WILSON *Devil take Hindmost* xii. 126 Irma Meyer lives under the El [in Brooklyn]. **1936** J. DOS PASSOS *Big Money* ii. 9 An el train clattered raspingly through the.. streets. **1959** *Times Lit. Suppl.* 6 Nov. p. xvii/4 John Frederick Nims has a poem 'Penny Arcade', set in Chicago, which begins with a description of 'This pale and dusty palace under the El'.

elaborate, *v.* Add: **3.** *intr.* To become elaborate.
1876 H. SPENCER *Princ. Sociol.* I. § 103 This custom elaborates as social development goes through its earlier stages. **1903** R. LANGBRIDGE *Flame & Flood* xx, These [*sc.* preparations] she discovered had augmented and elaborated to a considerable extent.

elaioplast (ĭləi·ǫplæst). *Cytology.* Also **elæo-.** [a. Du. *elaioplast* (J. H. Wakker 1887, in *Maandblad voor Natuurwetenschappen* XIII. 110), f. Gr. ἔλαιον oil + -PLAST.] A type of plastid, found esp. in some angiosperms, that contains or secretes oil.
1888 *Jrnl. Microsc. Soc.* 443 Elaioplast.—Under this term Herr J. H. Wakker describes nearly globular strongly refringent yellow bodies which he finds.. in epidermal cells of the leaves of *Vanilla planifolia*. *Ibid.*, The elaioplasts are formed gradually in the epidermal cells during the development of the leaf. **1969** BROWN & BERTKE *Textbk. Cytol.* xiii. 244/1 Leukoplasts containing one or more large starch grains are often called amyloplasts; ..or if droplets of oil, they are called elaioplasts.

Elamite (ī·lăməit), *sb.* and *a.* Also 4–6 **Elamyt.** [f. *Elam*, the name of an ancient country in Mesopotamia + -ITE[1].] **A.** *sb.* An inhabitant of Elam; the language of its people. **B.** *adj.* Of or pertaining to Elam, its inhabitants, or their language.
[*c* **1000** *Genesis* (Junius MS., 1931) l. 2081 Fleonde wæron Elamitarna aldorduguðe dome bedrorene.] *c* **1384** WYCLIF *Bible* (1850) IV. 511/1 Party, and Medy, and Elamyte, and thei that dwellen in Mesopotamye. **1526** TINDALE *Acts* ii. 9 Parthians, Medes and Elamyts, and the inhabiters of Mesopotamia. **1614** RALEIGH *Hist. World* (caption of map between pp. 178 and 179), Elam. The eldest sonn of Sem possest the regions of Persia and therfore were those nations first called Elamites. **1874** [see *AKKADIAN *a.* and *sb.*]. **1880** *Encycl. Brit.* XI. 155/1 It [*sc.* Nestorian Christianity] was successfully preached to the Bactrians, the Huns, the Persians, the Indians, the Persarmenians, the Medes, the Elamites. **1894** A. H. SAYCE *'Higher Criticism' & Verdict of Monuments* iii. 164 Chedor-laomer.. would have been written in Elamite Kudur-Lagamar. *Ibid.*, Eri-Aku.. bore a Sumerian and not an Elamite name. *Ibid.* xi. 516 The Elamite kings.. entitle themselves lords 'of the kingdom of Anzan, kings of Shushan'. **1928** C. DAWSON *Age of Gods* iv. 83 Resemblances certainly exist between the Dravidian Brahvi and the ancient Elamite tongue, and it is possible that all three languages are early representatives of the Caucasian linguistic group. **1937** *Discovery* Sept. 287/1 Many of them resemble certain seal-amulets of early Elamite date. **1970** *Ashmolean Mus., Rep. Visitors 1969* 15 Three cylinder seals, Mitannian, Cappadocian and Elamite.

eland. Delete *Boselaphus oreas* and substitute *Taurotragus oryx*. Also found in E. Africa (*T. derbianus*, giant eland, W. Africa and the Sudan).

‖ **élan vital** (elañ vital). [Fr. (H. Bergson *L'Évolution créatrice* (1907) iii. 276); see ÉLAN.] In the philosophy of Henri Bergson (1859–1941), a vital impulse or life force, of which we are aware intuitively; *spec.*, an original impetus of life supposed to have brought about the variations which during the course of evolution produced new species; a creative principle found in all living beings. Hence *gen.*, any mysterious vital principle.
1907 W. JAMES *Let.* to Bergson 13 June (1920) II. 292 The *élan vital*, all contentless and vague as you are obliged to leave it, will be an easy substitute to make fun of. **1912** G. B. SHAW *Let.* 8 Dec. (1952) 61 Your illness.. is a trap of the Life Force—the *Élan Vital*. **1926** J. S. HUXLEY *Ess. Biologist* i. 33 Bergson.. ascribes it [*sc.* biological progress] to his *élan vital*.. But to say that biological progress is explained by the *élan vital* is to say that the movement of a train is 'explained' by an *élan locomotif* of the engine. **1938** L. MACNEICE *Mod. Poetry* i. 18 The *élan vital* of Whitman or of Lawrence. **1959** *Listener* 28 May 937/1 No other theory as to the basic mechanism of evolution (Lamarckism, orthogenesis, *élan vital*, etc.) is consonant with the facts of general biology. **1962** *Ibid.* 20 Dec. 1062/2 The *élan vital*? A complete red herring, Dr. Crick asserted; biologists just didn't ask questions like that any more.

elaphure (e·lăfiuᵊɹ). [ad. mod.L. *Elaphurus* (A. Milne-Edwards 1866, in *Comptes rendus Acad. Sci.* LXII. 1091), f. Gr. ἔλαφος stag + οὐρά tail.] A species of reddish-tawny deer, Père David's deer (*Elaphurus davidianus*), discovered in northern China, and introduced into England in captivity. So **e·laphurine** *a.*
1872 *Trans. Zool. Soc.* VII. 336 The Elaphure is.. very distinct in the form of its horns from every other described species of the genus [*Cervus*]. **1891** FLOWER & LYDEKKER *Mammals* ix. 320 The Elaphurine group is represented.. by *Cervus davidianus*.

elapid (e·lăpid), *a.* and *sb.* *Zool.* Also **elapide, elapine, elapoid.** [ad. mod.L. *Elapidæ* or *Elapinæ*.] **A.** *adj.* Of, pertaining to, or resembling a venomous colubrid snake of the family Elapidæ or subfamily Elapinæ. **B.** *sb.* A member of this family.
1885 H. C. BUMPUS in J. S. Kingsley *Standard Nat. Hist.* III. 379 The *Elapides* are characterized by having the head rounded and depressed. **1895** *Athenæum* 14 Dec. 838/2 *Boulengerina stormsi* an elapoid snake from Lake Tanganyika. **1910** R. DITMARS *Reptiles of World* iv. 196 The Elapine snakes swarm in Australia. **1926** T. BARBOUR *Reptiles & Amphibians* iii. 22 The mambas or tree cobras are the most widespread and prominent Elapine types. *Ibid.* 35 The venom of the Elapid types is much more neurotoxic than is that of the vipers. *Ibid.*, The venom of the Old World Elapids.. causes a paralysis of the phrenic nerve. **1969** A. BELLAIRS *Life of Reptiles* I. iv. 121 The elapid and back-fanged snakes.. tend to hold on to their prey.

elasmobranchian (ĭlæ:zmǫbræ·ŋkiăn), *a.* and *sb.* *Zool.* = ELASMOBRANCH.
1889 in *Cent. Dict.*

elasmosaurus (ĭlæ:zmosǫ·rǫs). *Palæont.* Also anglicized **-saur.** [mod.L. (E. D. Cope 1868, in *Proc. Acad. Nat. Sci. Phila.* 93), f. Gr. ἐλασμός metal plate + σαῦρος lizard.] An extinct marine reptile.
1879 *Encycl. Brit.* X. 359/2 One of the most extraordinary of these reptilian forms was the *Elasmosaurus*—a huge snake-like form 40 feet long, with slim arrow-shaped head on a swan-like neck rising 20 feet out of the water. **1889** *Cent. Dict.*, Elasmosaur. **1924** *Glasgow Herald* 5 July 4 One of them, Elasmosaurus, seems to have been 45 feet in length, half consisting of neck. **1961** *Observer* 28 May 30/5 Loch Ness contains the answer to the sea-serpent problem.. a relict population of Elasmosaurs.

elasmotherium (ĭlæ:zmoþīᵊ·riŏm). *Palæont.* Also **elasmothere.** [mod.L. (G. Fischer de Waldheim 1808, in *Mém. Soc. Imp. Nat. Moscou* II. 250), f. Gr. ἐλασμός metal plate + θηρίον beast.] A large extinct rhinoceros of the genus so named, the remains of which are found in the Pleistocene of Eurasia.
1879 H. A. NICHOLSON *Man. Palæont.* (ed. 2) II. xlii. 329 We may mention here the singular *Elasmotherium* of the Post-Pliocene. **1905** E. R. LANKESTER *Extinct Animals* iii. 144 One great extinct beast (the Elasmotherium), allied to the rhinoceroses, had a great horn. **1927** PEAKE & FLEURE *Hunters & Artists* iii. 32 Another grass-eating rhinoceros from Siberia, usually known as the Elasmothere. **1932** J. S. HUXLEY *Probl. Rel. Growth* v. ii. 151 The extinct Elasmotherium possessed a horn in the shape of a flattened cone. **1966** A. S. ROMER *Vertebrate Paleont.* (ed. 3) xxx. 345/2 Some of the contrasting types, such as the woolly rhinoceros and the giant tundra-dwelling *Elasmotherium* of Eurasia, were cold-climate forms.

elastance (ĭlæ·stăns). *Electr.* [irreg. f. ELASTIC + -ANCE.] The capacity of a dielectric for opposing an electric charge or displacement.
1885 O. HEAVISIDE in *Electrician* 9 Oct. 408/2 (*bb*). If in the final state (*b*), we unite P to Q by a second dielectric wire also of infinite capacity, we make a closed dielectric circuit which has no elastance (or elastic resistance to displacement). **1890** W. P. MAYCOCK *Pract. Electr. Notes* (ed. 2) 48 Elastance is the reciprocal (or reverse) of Permittance; the comparative elastance of a dielectric is called its elastivity. For instance, a dielectric of great permittance has little elastance and *vice versâ*. **1890** *Electrician* 14 Nov. 41 The use of the words decalescence, reluctancy, reluctivity, elastance, elastivity, permittivity, might be well deferred till they are more generally adopted. **1893** O. HEAVISIDE *Electromagn. Theory* I. ii. § 31. 29 The formal relation of reluctance to reluctivity with magnetic force and induction, is the same as that.. of elastance to elastivity with electric force and displacement.

elastase (ĭlæ·stēis, -z). [f. ELAST(IC *a.* + *-ASE.] An enzyme, usu. isolated from the pancreas, which decomposes elastin. Also *Comb.*, as *elastase-inhibitor*; *elastase-inhibiting* ppl. a.
1949 BALÓ & BANGA in *Nature* 17 Sept. 491/1 (*title*) Elastase and Elastase-Inhibitor. *Ibid.*, The extract of fresh pancreas as well as that of acetone-dried pancreas powder contains a specific enzyme which we called 'elastase'. **1950** —— in *Biochem. Jrnl.* XLVI. 387/2 Pancreatic extract contains an enzyme which dissolves the elastic fibres of the aorta wall and which is called elastolytic enzyme or elastase. **1956** *Nature* 10 Mar. 468/1 The elastase-inhibiting effect of the serum of Ehlers Danlos patients was found to be quantitatively increased over that of normal serum... It would not seem too unlikely to suggest that the elastase inhibitor can be an essential factor in the formation of elastic fibre. **1969** *Times* 21 Jan. 5/6 The structure of two other protein splitting enzymes, known as ox chymotrypsin and pig elastase, were also announced. **1970** *Nature* 28 Feb. 802/1 The proteolytic enzyme elastase.

elastic, *a.* and *sb.* Add: **A.** *adj.* **3.** *elastic limit* (later examples). Also *elastic collision* (G. *elastisch* used in this sense, e.g. by Franck and Hertz 1913, in *Verh. d. Deut. Physik. Ges.* XV), a collision between two particles in which the total kinetic energy is conserved; *elastic constant*, a constant that expresses the reaction of a material to stress; *elastic hysteresis* = *HYSTERESIS 2; *elastic modulus* = *modulus of elasticity*; *elastic scattering*, the scattering of particles without loss of kinetic energy; *elastic strain*, a temporary deformation of a material under strain; *elastic wave*, a wave consisting of elastic deformations propagated through a medium.
1848 R. MALLET in *Trans. R. Irish Acad.* XXI. 97 The determination of the time of transit of the elastic wave through the earth's crust. **1886** J. MILNE *Earthquakes* III. 44 An earthquake consists of elastic waves of compression and distortion. **1909** *Cent. Dict.* Suppl., *Elastic hysteresis*, an effect, analogous to magnetic hysteresis in iron, observed in the relation of strain to stress when the stress to which an elastic body is subjected is alternately increased and diminished. **1913** *Proc. R. Soc.* A. LXXXVIII. 299 We shall suppose for the sake of generality that the collision of ion and molecule is not perfectly elastic, an assumption which allows roughly for a possible loss of energy on collision. **1925** *Jrnl. Iron & Steel Inst.* CXI. 589 Ultimately the whole question of the tensile strength of metals and of the elastic limit resolves itself into an investigation of variation with temperature. **1927** H. N. RUSSELL et al. *Astronomy* II. xvii. 551 After colliding with the far more massive atom the electron may be found moving at the same speed as before; such a collision is called *elastic*. **1930** *Engineering* 14 Feb. 231/3 This must not be so high as to cause heating of the piece by elastic hysteresis. **1930** A. W. JUDGE *Engineering Materials* III. i. 12 It will be seen, then, that the elastic strains occurring in engineering work are very small indeed. **1930** RUTHERFORD et al. *Radiations from Radioactive Substances* v. 134 The difficulties.. may be illustrated by the distinction made to-day between an 'elastic', and 'non-elastic' collision of an electron with an atom. **1931** *Discovery* Apr. 123/1 Elastic waves travel through different rocks at different velocities. **1933** *Physical Rev.* XLIII. 112 (*heading*) Elastic scattering of electrons by mercury atoms. **1936** F. FOSTER *Mech. Testing Metals & Alloys* i. 2 If, after the application and removal of the load, the strain disappears completely, the material is said to be perfectly elastic and the strain is then referred to as *elastic strain*. **1940** *Chambers's Techn. Dict.* 284/1 Elastic modulus. **1941** in M. Gowing *Brit. & Atomic Energy 1939–1945* (1964) App. II. 403 The diffraction effect (elastic scattering) is confined mainly to small angles. **1948** *Sci. News* VII. 14 For rubber, firmness therefore depends on the nearly constant ratio of pressure to amount of deformation, which is called an 'elastic modulus'. **1950** *Engineering* CLXX. 97/2 The links.. will break before the elastic limit of.. the guide apparatus is reached. **1955** H. B. G. CASIMIR in W. Pauli *Niels Bohr* 122 The elastic constants in the superconducting state are not appreciably different from those in the normal state. **1955** *Gloss. Terms Radiology* (*B.S.I.*) 14 *Elastic scattering*, in which the scattered radiation has the same quantum energy as the incident radiation. **1958** W. K. MANSFIELD *Elem. Nuclear Phys.* iv. 30 When the neutrons are deflected the collisions are called *elastic*, since the kinetic energy is conserved and the collisions can be treated by the normal dynamics of billiard ball collisions. **1966** *New Scientist* 5 May 296/3 An electromagnetic interaction between the electron and the nucleus can cause the electron to veer off in its path at some angle while the interacting nucleus recoils in a different direction. This phenomenon is called 'elastic scattering'.

4. (Partly attrib. use of the sb.) Applied to fabrics, or articles made of them, containing threads or thin strips of rubber usually covered by a woven material. So *elastic sides*, of elastic boots; also *attrib.*, as *elastic side boots*, and *ellipt.* = such boots; *elastic-sided* adj. Also *elastic band* (BAND *sb.*[2] 2); *elastic stocking*, a remedial stocking made partly of rubber.
1835 DICKENS *Sk. Boz* 1st Ser. (1836) I. ii. 15 Elastic waistcoats, bosom friends, and warm stockings, poured in upon the curate. *c* **1835** in A. Adburgham *Shops & Shopping* (1964) iii. 27 Madame L. begs also to recommend her highly approved *elastic Parisian corsets*. **1839** URE *Dict. Arts* s.v., *Elastic bands*. The manufacture of braces and garters, with threads of caoutchouc.. seems to have originated.. in Vienna. **1846** A. BAIN *Brit. Pat. 11,480* 5 In order to support.. the paper I employ elastic bands of india rubber. **1851** *Great Exhib. Catal.* II. III. xvi. 525/1 Elastic side, dress, and other boots. *Ibid.* xx. 579/2 Corset.. made to fasten in the front with patent spring clasp, and gauze elastic sides... Boots and shoes, with elastic sides, made by sewing pieces of India-rubber to the upper leather and then to the sole. **1855** J. H. NEWMAN *Let.* in M. Trevor *Newman: Light in Winter* (1962) 63 My accident of the autumn not being well I had just been obliged to put on my elastic stocking again. **1856** [in Dict.]. **1858** SIMMONDS *Dict. Trade*, *Elastic-bands*, belts, braces, gaiters, &c., made with threads of caoutchouc, either naked or covered. *a* **1877** KNIGHT *Dict. Mech.* I. 776/2 *Elastic-fabric loom*, one having mechanical devices for stretching the rubber threads or shirrs, and holding them at a positive tension while the fabric is woven. **1880** *Encycl. Brit.* XII. 842/1 The threads used in making elastic webbing are usually cut from spread sheets. **1899** SOMERVILLE & 'ROSS' *Irish R.M.* 295 Narrow feet in elastic-sided cloth boots. **1902** M. BARNES-GRUNDY *Thames Camp* 30 A man with a soft felt hat and elastic-sided boots. **1922** V. WOOLF *Jacob's Room* vii. 136 One pair of elastic stockings for Mrs. Page, widow, aged sixty-three. **1928** GALSWORTHY *Swan Song* II. vi. 157 He's made millions.. out of the elastic band—has some patent for making them last only just long enough. **1937** F. STARK *Diary* 25 Nov. in *Winter in Arabia* (1940) 29 The cook, who, in elastic-sided boots, announces dinner. **1970** *Focus* June 10/2 *Elastic waistband* means ordinary elastic is sewn on at the waist.
c. *elastic fibre*, a fibre of a kind present in certain types of connective tissue and cartilage, characterized by being branched, easily stretched, and giving tissue a yellow colour when present in bulk.
1849 A. H. HASSALL *Microsc. Anat. Human Body* I. xvii. 331 Acetic acid applied to a portion of mixed cellular tissue, at once allows the elastic fibres to be clearly seen. **1968** PASSMORE & ROBSON *Compan. Med. Stud.* I. xxix. 10 The lung's elastic fibres form a system of branching springs.
B. *sb.* (Earlier example.)
1847 in H. Howe *Hist. Collect. Ohio* 48 With the *elastics* supplied by the ladies, for a halter.. the young dog passed from the shores of time.

ela·sticated, *pa. pple.* and *ppl. a.* [f. ELASTIC *a.* + -ATE[3] + -(E)D[1].] Of cloth, etc.: woven or stitched with indiarubber thread and so made stretchable. Also *fig.*, rendered flexible.
1925 *Chambers's Jrnl.* Feb. 130/1 A sense of the joy of power silkened and elasticated. **1931** *News Chron.* 26 Mar., Let the whole matter be considered.. by the League Council so that this instrument may be so elasticated as to start a movement for the general lowering of tariff walls. **1959** *Observer* 5 Apr. 13/3 The fitted slacks are elasticated. **1959** *Manch. Guardian* 3 July 5/6 Elasticated sides and reinforced bra top. **1963** *Times* 10 June 15/6 The laziest boy in the group asked for a swivelling head and elasticated arms so that he could turn around and reach for things without moving. **1969** *Daily Tel.* 17 Nov. 12/6 Trousers which do have an elasticated nylon lining.

elasticity. Add: **2. b.** (Earlier example.)
1801 JANE AUSTEN *Let.* 8 Jan. (1952) 106 Appleshaw, that village of wonderful Elasticity, which stretches itself out for the reception of everybody who does not wish for a house on Speen Hill.
c. *Econ.* Flexibility in supply or demand in response to variations in other factors.
1925 S. E. THOMAS *Elem. Econ.* iv. 37 The Elasticity and Inelasticity of Demand. **1959** *Oxf. Univ. Gaz.* 16 Mar. 797/1 Analysis of the interaction of beef, milk and feed prices and the elasticities in livestock production within the United Kingdom. **1966** A. BATTERSBY *Math. in Management* ix. 232 The quantity 'e' is related to the economists' 'elasticity of demand'.

ela·sticized, *pa. pple.* and *ppl. a.* [f. ELASTIC *a.* + -IZE + -(E)D[1].] = *ELASTICATED *pa. pple.* and *ppl. a.*
1909 in WEBSTER. **1957** *New Yorker* 16 Nov. 164/2 A dog ski suit.. elasticized at all four ankles. **1958** *Times* 1 Dec. 13/6 The washable velvet shorts.. have an elasticized waist. **1970** *Daily Tel.* 6 July 13 A battle-dress jacket, elasticised at the waist.

elastivity (ilæsti·vĭti). *Electr.* [irreg. f. ELASTIC + -IVITY.] The property of a dielectric by virtue of which the flow of current between points having difference of potential is restrained.
1890, 1893 [see *ELASTANCE].

elastomer (ilæ·stŏməɹ). [f. ELAST(IC *a.* + *-omer* as in ISOMER.] Name applied to any of various synthetic rubbers or plastic substances resembling rubber. Hence **elastome·ric** *a.*
1939 *Ind. & Engin. Chem.* (*Ind. ed.*) XXXI. 944/2 The writer [*sc.* H. L. Fisher] also comes into the discussion with new suggestions, the first of which is 'elastomer'. This word is reminiscent of isomer, electromer, metamer, and polymer... The writer presents elastomer to cover the elastic or rubberlike substances. **1940** *Ibid.* (*News ed.*) XVIII. 757 (*title*) Chemigum—a new elastomer. *Ibid.* 757/2 Out of the study of more than 300 elastomers developed on a laboratory scale, the decision to study further.. ten of these was arrived at. **1951** *Engineering* CLXXI. 480/3 Brochure entitled 'Elastomeric Engineering Examples'. **1954** *Sci. News* XXXII. 52 Rubbers, for which the appropriate name 'elastomers' has been coined and is used in the United States, are chain polymers which have been cross-linked to form a widely meshed network. **1958** *New Scientist* 17 Apr. 22/2 Flexible silicone elastomer tape. **1963** *Economist* 16 Nov. 700/3 The new elastomeric synthetic fibres. **1965** *New Scientist* 24 June 857/1 Although very expensive to produce, the elastomer yarns are finer and so the amount used is less than in the case of rubber. **1970** *Guardian* 5 Sept. 4/4 Nylon/elastomeric support tights.

elatement (ĭlēi·tmĕnt). Delete †*Obs.* and add later examples.
1894 *Leisure Hour* Jan. 146/2 Impossible to describe the child's elatement. **1922** C. SIDGWICK *Victorian* xv. 117 A note of elatement in his voice.

elative, *a.* Restrict *rare* to senses in Dict. and add: **c.** *Gram.* Having a superlative or intensive force. **d.** Denoting the case used in some languages to express motion away from. Also as *sb.* **a.** A superlative or intensive. **b.** An elative case.
1860 *Trans. Philol. Soc. 1857* 34 The declension of the personal pronouns [in the Tushi language] is as follows.. Elative.. Comitative.. Terminative. **1906** J. H. MOULTON *Gram. N.T. Greek* I. v. 78 In the NT the obsolescence of the superlative, except in the elative sense, is most marked. *Ibid.* 79 Mt 11[20] may show the elative—'those very numerous mighty works'. *Ibid.* 97 The repetition of an adjective produces an elative. **1915** A. H. MCNEILE *St. Matt.* 59/2 Ἐλάχιστος.. may be elative, 'very small'. **1922** FOAKES-JACKSON & LAKE *Beginnings Christianity* II. 505 This is what the grammarians call an elative superlative. **1951** *Archivum Linguisticum* III. i. 60 An elative-ablative type of case suffix. **1959** J. C. CATFORD in Quirk & Smith *Teaching of English* vi. 186 Roughly the same field is covered by the system of 'local cases' in Finnish.. 'illative'.. 'ablative'.. 'elative'.

elbow, *sb.* **1. b.** Delete †*Obs.* and add later examples.
1908 *Animal Managem.* 330 For the horse, a small sausage-shaped pillow, long enough to surround the pastern.. will be found to prevent the elbow touching the ground when lying down. *Ibid.* [see *elbow-brushing* in *5].
4. g. *more power to one's elbow*: may you (he, etc.) succeed (in a laudable enterprise).
1832 LOVER *Leg. Ireland* 1st Ser. (ed. 2) 133 More power to your honour's elbow. **1860** [see POWER *sb.*[1] 2]. **1928** F. M. FORD *Let.* 5 Mar. (1965) 177 Anyhow more power to your elbow! **1956** A. WILSON *Anglo-Saxon Attitudes* II. i. 196 Some Members of Parliament.. are to ask questions.. about the unfortunate mishandling of Mr. Cressett's market-garden. More power to their elbow, I say.
h. *to lift one's* (or *the*) *elbow*: to drink immoderately. Similarly *to bend the elbow*; *to crook the elbow*: see *CROOK *v.*[1] 1 d.
1823 GROSE *Dict. Vulgar T.* s.v. *lift*, To lift one's hand to one's head; to drink to excess, or to drink drams. To lift or raise one's elbow; the same. **1915** CONRAD *Victory* I. vi. 130, I would watch them lifting their elbows at my expense. **1928** *Daily Express* 7 Mar. 15/4 Many a young man who has been lifting his elbow too frequently in Dublin is packed off to Melleray for a cure. **1938** D. RUNYON *Take it Easy* xiii. 247 A great change comes over Haystack Duggeler. He stops bending his elbow and helps Hattie cook and wash the dishes. **1939** J. B. PRIESTLEY *Let People Sing* iv. 86, I only hope to goodness this couple we're picking up this afternoon don't lift the elbow too much. **1967** *Coast to Coast 1965–6* 145 He's not much chop. Too fond of bending the elbow.
5. elbow-brushing (see quot.); **elbow-length** *a.*, reaching to the elbow; **elbow-pipe,** a pipe having a bend resembling an elbow; cf. ELBOW 2 c; **elbow-sleeve,** a sleeve reaching only to the elbow.
1908 *Animal Managem.* 330 Elbow brushing in the camel is a serious condition resulting from the friction of the elbow pad against the side in animals which are tied in at the elbow and whose toes are turned out. [**1897** *Sears, Roebuck Catal.* 231/2 This length is very stylish, and known as the elbow length.] **1904** *Westm. Gaz.* 26 May 10/1 Fashion launched on us the mandate that all sleeves must be elbow length. **1908** *Ibid.* 9 Jan. 4/1, I find for everyday wear the elbow-length cream 'Viyella' gloves hard to beat. **1932** D. C. MINTER *Mod. Needlecraft* 148/2 An elbow-length sleeve. **1967** E. SHORT *Embroidery & Fabric Collage* ii. 54 Gloves.. elbow length with more elaborate decoration for evening wear. **1861** W. FAIRBAIRN *Iron* viii. 166 An elbow-pipe.. establishes a communication.. between the blast-pipe and the tuyere. **1883** *Specif. Alnwick & Cornhill Rlwy.* 22 Proper elbow pipes and connections to be made with existing drains. **1875** L. TROUBRIDGE *Life amongst Troubridges* (1966) 117 A pretty white muslin with.. elbow sleeves, plissées with lace. **1899** *Daily News* 24 June 4/6 Some of the smart people compromise by adopting elbow sleeves with very long gloves rucked up the whole length of the fore-arm. *Ibid.* 22 July 4/3 Elbow sleeves have come in with the heat. **1908** L. M. MONTGOMERY *Anne of Green Gables* xiii. 130 Diana is having a new dress made with elbow sleeves.

elbow-chair. (Earlier and later examples.)
1655 tr. *Sorel's Comical Hist. Francion* x. 9 He.. permitted that the two men should put him in an Elbow chair. **1887** R. N. CAREY *Uncle Max* xiv. 278, I fetched father's big elbow-chair with a cushion or two.

Elchee. (Earlier example.)
1824 J. MORIER *Adv. Hajji Baba* III. xiii. 148 Ignorant of what is due to the character of *Elchi*, or ambassador.

elder, *a.* and *sb.*[3] Add: **A.** *adj.* **1. c.** *elder statesman*, in Japan, a member of the *GENRO; *transf.*, a person of ripe years and experience whose counsel is therefore sought and valued.
1921 *Contemp. Rev.* July 8 Prince Yamagata.. continues the most influential member of the *Genro*, or Elder Statesmen. **1923** *19th Cent.* Jan. 138 The exclusive powers of the *genro*, or elder statesmen, are passing with the men themselves. **1932** *Fortune* Sept. 100 This choice was exercised by the last of the *genro*, or 'Elder Statesmen', Prince Saionji, aged eighty-three. **1934** WEBSTER, *Elder statesmen*. **a** In Japan, an informal body (*genro*)... **b** Any similar class of persons. **1935** G. GREENE *Bear fell Free*, One should have made some prescient elder-statesman pronouncement. **1937** *John o' London's* 15 Jan. 667 (*heading*) Balfour, the Elder Statesman. **1955** H. SPRING *These Lovers fled Away* ii. 61 When you are twelve, a boy of fifteen.. seems almost an elder statesman. **1958** *Observer* 19 Oct. 22/4 Professor Tawney is the elder statesman of English economic history. **1959** T. S. ELIOT *Elder Statesman* II. 56 The difference between being an elder statesman And posing successfully as an elder statesman Is practically negligible.
B. *sb.* **4. b.** An order of priests in the Catholic Apostolic Church.
1828 E. IRVING *Sermons* I. p. xxiii, These Sermons on the Incarnation.. you received with all acceptation; and the Elders whom God hath set over you made choice of them to stand first in these volumes. **1876** *Encycl. Brit.* V. 238/1 Four-and-twenty priests, divided into the four ministries of 'elders, prophets, evangelists, and pastors'... The understanding is that each elder, with his co-presbyters and deacons, shall have charge of 500 adult communicants in his district.
c. A minister of any denomination. *U.S. local.*
1792 in *Mass. Hist. Soc. Coll.* II. 30 In the year 1673 settlers.. employed one Elder Jones as their preacher. **1851** *Advent Rev. & Sabbath Herald* 21 July 3/3 Elder Jesse Thompson and his companion [wife], (at whose table we are now writing,) were of this number. **1874** B. F. TAYLOR *World on Wheels* I. xix. 140 Take a young fellow from Hamilton or Rochester.. and call him Elder, as his country brethren and sisters always will. **1921** R. M. JONES *Later Periods Quakerism* I. iv. 120 Even now in the rural districts of New England a minister of any denomination is called 'Elder'. **1925** Z. A. TILGHMAN *Dugout* 7, I can remember some of the elder's sermon that day.

elder, *v.* Add: **2.** *intr.* To become older, to begin to show age. So **e·ldering** *ppl. a.* *poet.* and *colloq.*
1876 G. M. HOPKINS *Wr. Deutschland* (1918) st. 18 Never-eldering revel and river of youth. **1885** S. W. MITCHELL *In War Time* xii. 186 Before she went away she was what my nurse used to call 'eldering'. **1949** O. NASH *Versus* 121 In my eldering age.

Eldonian (eldǭᵘ·niăn), *a.* [f. the name of John Scott, Earl of *Eldon* (1751–1838) + -IAN.] Belonging to or characteristic of Lord Eldon, regarded as typical of 'diehard' toryism. Also **Eldonine** (e·ldənəin) *a.*
1855 *National Rev.* Oct. 257 Those years were the commencement of what is called the Eldonine period. **1876** W. CORY *Lett. & Jrnls.* (1897) 419 A fusion of Benthamic legislation with Eldonine evolution. **1898** *Daily News* 8 July 4/7 His Toryism is Eldonian and antediluvian. **1900** *Ibid.* 31 Dec. 4/7 The century, considered from a political point of view, may be divided into the Eldonian, the Peelite, the Palmerstonian, and the Gladstonian eras. **1926** *London Merc.* May 104 An Eldonian Tory still describes a type peculiar to the English temperament, and never, perhaps for that reason, personally unpopular.

elect, *a.* and *sb.*[1] Add: **A.** *adj.* **3.** *bride elect* (example). See also BRIDE *sb.*[1] 1.
1806 C. WILMOT *Let.* 21 Oct. in *Russ. Jrnls.* (1934) II. 239 The Bride elect dissolved in tears.

election. Add: (Also 9 *U.S.* 'lection.)
7. b. The choosing between two rights by a person who derives one of them under an instrument in which a clear intention appears that he should not enjoy both.
1628 COKE *On Litt.* II. xii. ccxix. 145a If a man granteth a Rent.. or a robe to one and to his heires, the Grantor shall haue the election, for he is the first Agent, by payment of the one, or deliuerie of the other. **1891** G. SERRELL *Equitable Doctrine Election* 6 Election is often said to rest on an implied condition. **1959** JOWITT *Dict. Eng. Law* I. 699/2 The equitable doctrine of election is founded on the principle that there is an implied condition that he who accepts a benefit under an instrument must adopt the whole of it, conforming with all its provisions and renouncing every right inconsistent with them.

8. *election dinner, expenses*; **election bun, cake** *U.S.*, varieties of fancy bread; **election commissioner,** one of a body of men appointed to inquire into corrupt practices during an election or (*U.S.*) to take charge of an election; **election district** *U.S.*, a district created for the purposes of elections; **election petition,** a petition brought against a member of parliament for illegal practices during his election campaign.
1860 O. W. HOLMES *Prof. Breakf.-t.* ii. 52 [He] recollects he had a glazed 'lection bun, and sat eating it, and looking down on the Common. **1805** *Pocumtuc Housewife* (1906) 30 Elections Cake. **1832** L. M. CHILD *Amer. Frugal Housewife* 71 Old-fashioned election cake is made of four pounds of flour [etc.]. **1947** R. BEROLZHEIMER et al. *U.S. Regional Cook Bk.* 53 Election Cake was always served on Election Day. **1899** *Kansas City Star* 17 Feb. 1/3 He said that Julius Wurzberger, an election commissioner of St. Louis,.. had spent $10 getting him drunk. **1902** *Encycl. Brit.* XXVI. 369/1 A prosecution for any of the above offences cannot be instituted more than a year after the offence was committed, unless an inquiry by Election Commissioners takes place. **1751** G. STONE *Let.* 18 May in *9th Rep. Hist. MSS. Comm.* App. III. (1884) 40/2 Of the two sorts of champagne that sealed with yellow wax might go off at balls... The red wax is too bad for an election dinner at Dover. **1839** R. PEEL *Let.* 8 Dec. in *Corr. C. Arbuthnot* (1941) 212 May not a half-pay captain stand a contested election—or may not he be present at his brother's election dinner? **1799** in *Deb. Congress U.S.* (1851) 7th Congr. 2nd Sess. App. 1411 In every election district in the country. **1835** *Southern Lit. Messenger* I. 218 One or two.. get a resolution passed for a general caucus of the whole party, in the town, or election district. **1902** E. C. MEYER *Nominating Systems* 19 The township or ward, is included in a number of different election districts, each of which has its own convention. **1859** J. S. MILL *Parl. Reform* 19 Let law and opinion conspire to the end that election expenses be suppressed. **1869** W. F. COLLIER *Let.* 28 Mar. in B. & P. Russell *Amberley Papers* (1937) II. 159 You are fully aware that my views with respect to Election expenses are very nearly identical with your own. **1835** J. A. ROEBUCK *Short Rev. of Long Session* 9/1 Connected with this matter of elections was one peculiar and distinguishing work of this Parliament—I mean Election Petitions. **1839** *Act 2 & 3 Vict.* c. 38 (*title*) An Act to amend the Jurisdiction for the Trial of Election Petitions. **1840** *Penny Cycl.* XVII. 278/2 In 1741, Sir Robert Walpole.. was.. driven from office by a vote upon the Chippenham election petition. **1885** *Encycl. Brit.* XVII. 705/1 By the Act of 1879 the trial of an election petition is conducted before two judges instead of one, as before.

electioneer, *v.* (Later U.S. examples.)
1836 R. P. SMITH *Col. Crockett's Exploits & Adv.* 11 For this reason he came out openly to electioneer against me. **1843** 'R. CARLTON' *New Purchase* xviii. 139 Numbers go to see their neighbours, or to hear the news, and not a few to electioneer. *a* **1922** C. C. ANDREWS *Recoll.* (1928) 185 Soldiers have electioneered to some extent for different candidates.

electioneer (ĭlekʃənīᵊ·ɹ), *sb.* [f. ELECTION + -EER, after *auctioneer*, etc.] = ELECTIONEERER.
1848 J. R. LOWELL *Biglow P.* 1st Ser. vi. 77, I du believe hard coin the stuff Fer 'lectioneers to spout on. **1895** *Pall Mall Gaz.* 26 July 8/1 In the Yorkshire Constituencies. The Champion Electioneer. **1905** *Jrnl. Educ.* Apr. 272/2 To secure that the professorship shall not necessarily go to the most successful electioneer. **1928** *Daily Express* 11 Oct. 11/1 The Liberal electioneers at once denounced it as a 'Tory trick'.

electioneerer. (Earlier example.)
1800 M. EDGEWORTH *Parent's Assistant* (ed. 3) VI. 150 A damn'd bad electioneerer!.. Why, unless he bought a vote, he'd never win one.

electioneering, *vbl. sb.* (Earlier examples.)
1760 E. PYLE *Let.* 22 Nov. (1905) 330 He will never give them any more trouble on the article of electioneering. **1785** G. A. BELLAMY *Apology for Life* III. 89 A solicitor, who was a very bustling man, and well versed in electioneering affairs.

elective, *a.* and *sb.* Add: **A.** *adj.* **3. b.** Of college or high-school studies: subject to the student's choice; optional. So *elective system*. orig. *U.S.*
1847 in *Ann. Rep. Harvard Coll. 1883–4* 14 The elective system is now given up in this department. **1868** C. W. DILKE *Greater Brit.* I. vii. 89 The system of elective studies pursued at Michigan [University] is one to which we are year by year tending in the English universities. **1880** *Harper's Mag.* July 254/1 In the German Universities the studies are all elective and optional; in the colleges of the United States compulsory. **1890** J. G. FITCH *Notes Amer. Schools* 59 In the high schools and universities the practice of prescribing 'elective' subjects is very common. **1957** G. G. READER in R. K. Merton *Student-Physician* 84 Two months of elective work. **1969** *Brit. Jrnl. Med. Educ.* III. 102 There should be an elective period of, say, 10 weeks, in which the student should be able to choose from a wide variety of elective topics.
4. Revived in medical use: optional, not urgent (see quot. 1941).
1941 *Dorland's Med. Dict.* (ed. 19) 476/2 *Elective*, subject to the choice or decision of the patient or physician.., applied to procedures that are only advantageous to the patient but not necessary to save his life. **1963** *Lancet* 19 Jan. 132/1 A self-contained, thirty-bed hospital which catered solely for elective surgical cases. *Ibid.*, The theatre was used twice weekly for elective operating sessions. **1964** G. L. COHEN *What's Wrong with Hospitals?* iv. 65 An elective operation such as tonsillectomy. **1966** *Lancet* 24 Dec. 1394/1 A reversed intestinal segment was inserted as an elective procedure after massive resection for a mesenteric embolus.
8. b. *fig.* (Additional examples.)
1827 CARLYLE *German Romance* IV. 12 In the Romance department, Goethe has written.. *Wilhelm Meister's Apprenticeship*, and *Die Wahlverwandtschaften* (The Elective Affinities). **1860** *Harper's Mag.* July 205/1 We hear much of *passional attraction*, of *elective affinity*, etc. **1926** M. BARING *Daphne Adeane* xvii. 212 These things happen: 'elective affinities', you know.
B. *sb.* **2.** A subject of study specially selected by the student in a college or university; an optional subject or course of study. orig. *U.S.*
1850 *Docs. City Boston* (Mass.) Doc. No. 38, 45 Making some studies electives and giving to the members of the first class some liberty of choice. **1876** J. D. WHITNEY in E. T. Brewster *Life J. D. W.* (1909) 330, I shall have an elective this winter in economical geology. **1901** *Westm. Gaz.* 30 May 12/1 At Yale, where the study of this language had been neglected for many years, more than a hundred students have chosen it as one of their 'electives'. **1902** J. CORBIN *American at Oxford* 167 The idea of grouping electives is the fundamental difference between English and American education. **1926** *Amer. Oxonian* July 100 Oxford is a school for specialists. There are no minors, no electives, nothing but majors. **1930** *Lambeth Conference Rep.* 174 For the Priesthood he [*sc.* a candidate] must pass a further examination in.. The Bible.. Pastoral Care and one of a long list of Electives. **1957** G. G. READER in R. K. Merton *Student-Physician* 84 The student.. would divide his time between pediatric and psychiatric clinics and part-time electives. **1962** B. LENNOX *Rep. Visit to U.S.A. & Canada* ix. 25 Once we have broken ground with the new curriculum [at Glasgow], I think we should next consider the introduction of electives. **1969** *Brit. Jrnl. Med. Educ.* III. 182 An exciting range of electives in community medicine.

electoral, *a.* Add: **1.** *electoral college*: see COLLEGE *sb.* 1 b; *spec.* in *U.S.* (see quot. 1889).
a **1691, 1790** [see COLLEGE *sb.* 1 b]. **1800** *Deb. Congress U.S.* 23 Jan. (1851) 31 If this body of the Electors of all the States had been directed by the Constitution to assemble in one place, instead.. of being formed into different Electoral colleges [etc.]. **1812** W. BOYLAN *Let.* 5 Sept. in J. Steele *Papers* (1924) II. 689 The sense of each district may then be pretty generally represented in the Electoral College. **1845** [in Dict.]. **1889** *Cent. Dict.*, *Electoral College*, a name informally given to the electors of a single State, when met to vote for President and Vice-President of the United States, and sometimes to the whole body of electors. **1903** R. L. ASHLEY *Amer. Govt.* 274 The work of the Electoral College. **1949** *Pacific Discovery* Jan.–Feb. 3/2 Once we even elected a president of the United States through open and brazen corruption of the electoral college. **1969** *Pioneer* (Lucknow) 13 Aug. 1/1 The Congress has a clear majority in the Presidential electoral college.

Electra (ĭle·ktră). Name in Greek tragedy of the daughter of Agamemnon and Clytemnestra, responsible for the murder of the latter, used *attrib.*, esp. in *Electra complex*, a term used by psychoanalysts to denote a daughter's feelings of attraction towards her father and hostility towards her mother.
1913 [see *COMPLEX *sb.* 3]. **1927** B. MALINOWSKI *Sex & Repression in Savage Society* I. viii. 65 This is the configuration of the Electra complex; it is therefore of an entirely different nature from the Oedipus complex. **1963** A. HERON *Towards Quaker View of Sex* 58 This is the Oedipus situation... Its female counterpart is called the Electra situation, but this appears a rather less powerful and significant event in the life of the girl.

electret (ĭle·ktrĕt). [f. ELECTR(ICITY + MAGN)ET.] A permanently polarized piece of dielectric material, the electric counterpart of the permanent magnet.
1885 *Electrician* 7 Aug. 230/1 Solid insulating substances may be made electrets artificially, with a greater or less amount of permanency. **1936** *Nature* 18 July 130/1 An 'electret' is produced by solidifying certain types of wax, for example, carnauba wax, in a strong electric field... It is known that electrets can retain their original charge for a considerable period. **1960** *New Scientist* 7 Apr. 884/3 An electret which tends to depolarize when exposed to light. **1968** *Ibid.* 4 July 30/1 The advantages of using electrets in telephony are economy and performance.

electric, *a.* and *sb.* Add: **A.** *adj.* **2. b.** Additional phraseological combs. and examples: **electric action** *Mus.*, the action of an organ in which the mechanical parts are operated by electricity; **electric arc,** also *attrib.*, as *electric-arc welding* (see also *ARC WELDING); **electric balance** (example); **electric battery** (examples); **electric bell,** a bell operated by electricity; **electric blanket,** an electrically warmed blanket; **electric brake,** (in electrically driven vehicles) a brake operated by the temporary use of the driving motor as a generator, the resulting current being either returned to the supply line or dissipated as heat in a resistance; hence *electric braking*; **electric calamine** *Min.* (*U.S.*) = HEMIMORPHITE; **electric car,** (*a*) *N. Amer.* a trolley car, (*b*) a motor car propelled by electrical power; **electric chair** *U.S.*, an instrument of capital punishment by electrocution; **electric circuit** (earlier examples); **electric clock** (earlier and later examples); **electric column** (examples); **electric convulsive therapy** = **electro-convulsive therapy*; **electric cooker,** an electrically operated cooker; also *electric cooking*; **electric discharge,** also *attrib.*; **electric displacement,** see DISPLACEMENT 2 d; **electric eel** (examples); cf. *electrical eel* (ELECTRICAL *a.* 2); **electric egg** (examples); **electric eye,** = **magic eye*, (*a*) a photo-electric cell on which a beam of light is directed, the interruption of which acts as a trigger for the operation of electric relays; (*b*) a miniature cathode-ray tube used as a tuning indicator in a radio receiver, etc.; **electric fan,** a fan driven by an electric motor; **electric fence,** a fence, often consisting of a single strand of wire, charged with electricity; hence *electric fencing*; also *fig.*; **electric field** (see FIELD *sb.* 17, quot. 1881); **electric fire,** (*b*) an electric heater (mainly for domestic use); **electric flux** (see quots.); **electric furnace** (examples); **electric generator** (see GENERATOR 2, quot. 1879); **electric guitar,** a guitar in which the sound from the plucked strings is picked up electro-magnetically or by means of a form of condenser microphone and then amplified; hence **electric guitarist,** one who plays the electric guitar; **electric hare,** a dummy hare made to run by electricity, used in greyhound racing; also *transf.* (quot. 1941); **electric heater** (examples); **electric hedge** *nonce-wd.*, an invisible barrier charged with electricity (see *N. & Q.* (1959) CCIV. 338); **electric iron,** an iron (sense 5) heated by an electric element; **electric kettle,** a kettle heated by electric power; **electric lamp** (see LAMP *sb.*[1] 1; examples); **electric lobe** (see LOBE 1 d, quot. 1849; additional example); **electric locomotive,** an electrically driven railway locomotive engine, powered either by Diesel-driven motors or from a contact wire or rail; cf. *DIESEL-*electric* adj.; **electric machine** (example); **electric mixer,** an electrically driven machine or mechanical contrivance for mixing food (cf. MIXER 2); **electric motor,** a motor (MOTOR *sb.* 3) for transforming electric energy into mechanical energy; **electric nerve** *Zool.* (see quot. 1940); **electric organ,** (*a*) *Zool.* (later example): see ORGAN *sb.*[1] 5, quot. 1773; also quot. 1851 s.v. ELECTRICAL *a.* 2; (*b*) *Mus.*, an organ [ORGAN *sb.*[1] 2] with an electric action; **electric piano,** a piano, or an instrument resembling this, in which the mechanism is worked or the sounds are produced electronically; **electric potential** [POTENTIAL B. 4], see quots. 1941 and 1962; **electric power** (cf. POWER *sb.*[1] 14); **electric precipitation** = **electrical precipitation*; **electric railway** (examples); **electric range** [RANGE *sb.*[1] 12], an electric cooker; **electric ray** (examples); also, (*b*) = **electric wave*; **electric razor,** a razor operated by electricity; **electric recording** = **electrical recording*; so *electric record*; **electric shaver** = **electric razor*; **electric shock** (earlier and later examples); also *attrib.*, esp. of medical treatment by means of an electric shock (cf. **shock therapy, treatment*; **electro-shock*); **electric sign,** a sign illuminated by electricity; **electric steel** (see quot. 1919); **electric storm** *Meteorol.*, a violent disturbance of the electrical condition of the atmosphere, often causing interference with electrical transmissions; **electric strength,** the stress, usually expressed in kilovolts per millimetre, which an insulating material can withstand without breakdown; dielectric strength; also, the property of withstanding such stress (cf. STRENGTH *sb.* 1 j, esp. quot. 1873); **electric stress** (see quot. 1943); **electric telegraph** [see TELEGRAPH *sb.* 2] (earlier example); **electric toaster,** a toaster operated by electricity;

electric toothbrush, a toothbrush which is controlled by electricity; **electric torch**, (*a*) a gas-lighter operated by electricity (Knight *Dict. Mech.* a 1877); (*b*) a contrivance consisting essentially of an electric lamp enclosed in a portable case containing a battery; **electric traction** (see quot. 1940); **electric train**, a train operated by electricity; **electric tramway**, a tramway operated by electricity; **electric typewriter**, a typewriter operated by electricity (*Funk's Stand. Dict.*, 1893); **electric wave**, an electrical wave (see WAVE *sb.* 5 a, esp. quot. 1889); **electric welding** = **electric-arc welding*; **electric wind** = AURA 3 b.

1889 *Cent. Dict.*, Electric action. 1948 *Penguin Music Mag.* VII. 25 Electric action..and similar delights enable modern organists to gambol through Bach fugues at a hair-raising speed. 1885 *Electrician* 31 July p. iv (Advt.), Electric Arc Lamps. 1893 *Operator & Electrical World* XXII. 14 (*title*) Electric arc welding and metal working. 1936 *Economist* 18 Jan. 150/2 Its business in electrodes, a natural consequence of the development of electric-arc welding. 1887 *Nature* 29 Sept. 522/1 New electric balances. These balances are founded on the mutual forces, discovered by Ampère, between the fixed and movable portions of an electric circuit. 1795 *Gentl. Mag.* LXV. 141/2 Perhaps the electric batteries..may be realized. 1946 *Whittaker's Electr. Engin. Pocket-Bk.* (ed. 7) 859 (*heading*) Commercial electric battery vehicles. 1877 *Telegraphic Jrnl.* V. 7/2 The *Manchester Evening News* says that the electric bell system has been carried out.. at the Manchester City Hall. 1891 A. E. KENNEDY in *Electr. in Daily Life* 239 The first application of electricity to household purposes was presented by the electric bell early in the [19th] century. 1930 *Punch* 16 Apr. 425/1 Had a rotten night. My electric blanket fused and I had to get up to mend it. 1938 S. R. ROGET *Dict. Electr. Terms* (ed. 3) 34/1 *Electric blanket*, a blanket having woven into it resistance wires by which a certain amount of warming effect can be procured by a current. 1959 *B.S.I. News* Feb. 6/2 The Home Secretary was asked recently in the Commons if he was aware of the number of fires and accidents caused by electric blankets. 1885 *Electrician* 3 July 147/2 Arrangements are being made to render this electric brake automatic, so that the main circuit will be broken and the brake circuit with the motors closed automatically. 1910 *Stand. Handbk. Electr. Engin.* (ed. 3) 986 An efficiency of 60 per cent could be obtained during retardation by electric braking. 1836 W. T. BRANDE *Man. Chem.* (ed. 4) II. 860 Silicate of Zinc occurs native under the name of electric calamine. 1946 J. R. PARTINGTON *Gen. & Inorg. Chem.* xv. 383 The hydrated silicate..is electric calamine or hemimorphite. 1888 F. M. A. ROE *Army Lett.* (1909) 366 Most of my outings are on the electric cars. 1899 'MARK TWAIN' in *Cosmopolitan* XXVII. 592 The merry electric car replaced the melancholy bus. 1901 *Daily Colonist* (Victoria, B.C.) 27 Oct. 1/4 An electric car ran into a cow last night near Cedar Cottage. 1947 *Newsweek* 8 Sept. 85/3 Electric cars would do in town. 1966 *Economist* 17 Dec. 1246/3 The motor car companies are working hard to improve the battery—the main block in the way of an electric car. 1889 *Peel City Guardian* 8 June 6/2 The preparations, which are to consist of taking a seat in an electric chair. 1903 *N.Y. Even. Post.* 28 Oct. 12 Two men were sentenced to die in the electric chair. 1948 *Chicago Daily News* 18 Sept. 3/6 A 24-year-old former convict..must die in the electric chair. 1767 J. PRIESTLEY *Hist. Electr.* VIII. 651 The conductors which form an electric circuit. 1782 *Brit. Pat.* 1318 3 Making the part..a portion of that electric circuit. 1845 *Brit. Pat.* 10,838 1 Improvements in Electric Clocks. 1922 JOYCE *Ulysses* 109 They ought to have..an electric clock or a telephone in the coffin. 1814 G. J. SINGER *Elem. Electr.* I. i. 27 The electric column (invented by J. A. De Luc, esq.)..consists of 800 or 1000 small discs, of silver, zinc, and paper. 1840 *Ann. Electr. Magn. & Chem.* IV. 43 The dry electric column..will emit sparks and charge coated glass. 1953 HINSIE & SHATZKY *Psychiatric Dict.* (ed. 2) 767/1 Electric convulsive therapy, or E.C.T., is indicated in mania, depressions, and certain cases of schizophrenia. 1963 H. BURN *Drugs, Med. & Man* (ed. 2) xi. 122 Treatment by taking a certain number of tablets each day is much pleasanter than electric convulsion therapy, but electric convulsion therapy is still retained for those patients in whom the drugs are ineffective. 1907 *Electr. Bull.* Jan. 5/1 Tell your husband it was that wretched gas stove and by and bye he will send home an electric cooker. 1933 *Archit. Rev.* LXXIV. p. xlii, What is the outstanding fact about an ordinary electric cooker? Its capacity for being switched on and off. 1907 *Electr. Bull.* May 9/1 The..most promising field for the introduction of electric cooking methods is the first-class hotel. *Ibid.* 9/2 Electric cooking will not appreciably increase the room temperature. 1935 *Economist* 23 Mar. 690/2 There is no doubt that the electric lighting of streets has made a very big step forward by the use of the electric discharge lamp, which..illuminates roads without any dark patches. 1794 R. J. THORNTON *Med. Extracts* (1796) I. II. xii. 121 Vandelott makes two species of the electric eel, the black and reddish. 1881 *Encycl. Brit.* XII. 650/1 The electric eel is the most powerful of electric fishes. 1954 G. DURRELL *3 Singles to Adv.* ix. 205 The rare specimen turned out to be a baby electric eel, some two feet long. 1868 *Q. Jrnl. Sci.* V. 117 Electric discharge in vacuo, such as..an electric egg. 1877 *Phil. Mag.* III. 176 P. Thiess attempts to derive the action of the electric egg from the fact that the air which surrounds the electrode is charged with the electricity. 1930 *Electronics* Aug. 216/2 'Electric Eye' for Glass Industry.. a photo-electric cell for determining weak spots in glass construction. 1933 *Ibid.* Apr. 103/1 (*heading*) Bar and rod heating controlled by an electric eye. *Ibid.*, Valve-stem heaters are also available with electric-eye control. 1937 *Daily Express* 3 Mar. 5/2 Take this 'electric eye', for example. We're installing it in the hotel driveway to warn people in the street whenever a car's coming out. 1966 *Punch* 23 Nov. 784/1 The period is obviously some time in the future..: electric eyes open the average door. 1883 *Electr. Light* 1 Mar. 173/1 Electric fans. The small electric motors..have been ingeniously applied, under a patent recently taken out, to the working of fans for drawing-room use. 1905 A. BENNETT *Sacred & Profane Love* II. iv. 168 The electric fans revolved ceaselessly. 1901 L. G. BROWN *Among Tuaregs* in D. Ker *Knight of Honour* 46 'It is my electric fence,' he replied. 1944 G. HENDERSON *Farming Ladder* xv. 174 The calves..are confined with a single, thin strand of wire, electrically charged, and which they graze up to, but do not touch. We are great believers in electric fences. 1952 M. MCCARTHY *Groves of Academe* (1953) vi. 98 Her personality was posted with all sorts of No Trespassing signs and criss-crossed with electric fences, which repelled the intruder with a smart shock. 1958 *Times* 26 May 9/7 Electric fences are used everywhere and the grass so saved conserved as silage or hay. 1942 *Farming Handbk.* 41 Considerable success has been reported from several quarters with electric fencing. 1943 *Gloss. Terms Telecomm.* (*B.S.I.*) 4 In contradistinction [to electric circuit], an electric field implies action which can only be specified uniquely in terms of time and two or three dimensions. 1962 CORSON & LORRAIN *Introd. Electromagn. Fields* ii. 29 We define the electric field intensity E to be the force per unit charge exerted on a test charge in a field. 1921 *Spectator* 9 Apr. 455/1 A 'Magicoal' electric fire, a device which..deceives our short-sighted friends into believing that we are burning an unconscionable amount of the best coal. 1933 *Sunday Referee* 2 July 16/2 In the living-room is..a ply-wood built-in book-case, at one end of which is an electric fire. [1904 GOODCHILD & TWENEY *Technol. & Sci. Dict.* 230/2 *Flux* (Elect.), the total number of lines of force which pass through any given space or cross any given area.] 1904 *Electr. Rev.* (N.Y.) 24 Dec. 1040/1 (*title*) A hydrodynamic model of electric flux. 1934 WEBSTER, *Electric flux*, the flux existing in the space between charged bodies. 1885 *Electrician* 9 Oct. 411/2 A dynamo electric machine..will permit of the establishment of the electric furnace on a larger scale. 1937 *Discovery* May 153/2 There is an increasing tonnage of high speed steel being melted in the high frequency electric furnace. 1938 *Oxf. Compan. Mus.* 288/1 Electric Guitars, Mandolins, &c., were by 1936 on the market. 1940 *Sears, Roebuck Spring & Summer Catal.* 561 Give your guitar the volume and tone of an electric guitar. 1959 H. HOBSON *Mission House Murder* ii. 13 A nice, clean, unusual noise from vibraharp, electric guitar, a piano and a..bass-viol. 1957 *Observer* 1 Dec. 14/2 The Negro electric-guitarist Charlie Christian, who died in 1942 [etc.]. 1927 *Punch* 27 July 85/3 A greyhound race at the White City had to be re-run last week because the electric hare ran too fast. 1941 *Hutchinson's Pict. Hist. War* 19 Mar.–13 May 219 An ingenious device now in operation for the training of air gunners is an 'electric hare' plane. Mounted on a truck, it is run at high speed round the track while the gunner works from a mobile turret. a 1877 KNIGHT *Dict. Mech.* I. 780/2 *Electric heater*, a device in which a fine platinum wire heated by a passing electric current is made to communicate sensible heat as a means of warming or burning, as the case may be. 1935 *Discovery* May 145/1 A small electric heater is placed under the table. 1921 G. B. SHAW *Back to Methuselah* IV. I. 169 *The Elderly Gentleman*:..When I tried to step off the pier on to the road, I received a shock, followed by an attack of pins and needles which ceased only when I stepped back on to the stones. *Zoo*: Yes: there is an electric hedge there. It is a very old and very crude method of keeping animals from straying. 1907 *Electr. Bull.* Aug. 32/2 (*heading*) Electric Irons. *Ibid.*, Ironing benches, each equipped with an electric iron. 1920 *Whittaker's Electr. Engin. Pocket-Bk.* (ed. 4) 491 per lb. [1907 *Yesterday's Shopping* (1969) 270 Electric Copper Kettle.] 1910 H. G. WELLS *New Machiavelli* II. iv. 252 Everything was put ready for me to make tea..electric kettle, infuser, biscuits. 1965 M. ERSKINE *Take Dark Journey* vi. 68 The electric kettle was plugged in..but not turned on. 1856 *Brit. Pat.* 2456 1 We, Joseph Lacassagne, Chemist, and Rodolphe Thiers, Manufacturer,..do hereby declare the nature of the said Invention for 'An Improved Electric Lamp', to be as follows [etc.]. 1882 *Electric Light* 1 June 21/1 Incandescent lamps, in which a carbon thread is enclosed in a glass bulb..form the safest kind of electric lamp. 1935 *Discovery* Nov. 327/2 A special brain-centre [in the Electric torpedo fish] known as the electric lobe. The organ as a whole resembles a beehive, for the electrogenic units or elements are grouped in batteries of prismatic form. 1884 *Pop. Sci. Monthly* XXIV. 745 This motor is connected with the driving-wheels by gearing, belting, [etc.]..so that its revolution produces..a consequent progressive motion of the electric locomotive. 1760 WESLEY *Desideratum: Electr. Made Plain* 35 That the Electric Machine may as properly be term'd a Fire-Pump, as Mr. Boyle's Machine, an Air Pump? 1933 *Pop. Sci.* Jan. 50/1 Electric Mixer... Driven by a small motor, this new kitchen utensil..is said to cause no radio interference. 1959 *Which?* June 44/1 Many women who have long ago accepted every other kind of electric domestic gadget seem to be still vehemently divided about the merits of the electric mixer. 1886 *Electrician* 15 Jan 194/1 With electric motors..one has to be contented with a much lower efficiency. 1936 *Discovery* Aug. 238/1 Each of the shutters for the nine lenses is actuated by an individual electric motor. 1935 *Ibid.* Nov. 329/2 (*caption*) Cross section of *Malopterurus electricus*. Nfe, electric nerve; oe, electric organ. 1940 *Chambers's Techn. Dict.* 285/2 *Electric nerve*, a modified motor nerve, serving an electric tissue. 1885 *Electr. Engin.* Mar. 149/1 We are enabled to imagine in our houses..electric pianos and organs. 1886 *Telegraphic Jrnl.* 17 Sept. 294/2 (*caption*) An electric organ. 1936 *Discovery* July 223/2 This cupboard full of valves replaces, in the Coupleux electric organ, the pipe-chest of the ordinary organ. 1944 J. S. HUXLEY *On Living in Rev.* iv. 50 The eel's electric organs. 1885 Electric piano [see **electric organ*]. 1927 A. HUXLEY *Let.* 17 May (1969) 286 Everything happening to the strains of the steam organs and electric pianos of the inevitable Flemish kermesse. 1938 A. E. WIER *Macmillan Encycl. Mus.* 519 *Electric piano*, an instrument to which the principle of electric amplification is applied; the soundboard and tri-chord strings are dispensed with, and a microphone picks up the sound from the strings in groups of five. 1969 *It* 13–28 June 12/2 Eclection were seen for the first time with new member John Palmer, who plays excellent flute and vibes but rather erratic electric piano. 1871 *English Mechanic* 12 May 193/1 With electric potentials of less amount than sufficient to decompose water, they can be charged and discharged like condensers. 1902 Electric potential [see POTENTIAL B. 4]. 1902 *Encycl. Brit.* XXX. 449/1 A transverse difference of electric potential (Hall effect). 1941 S. R. ROGET *Dict. Electr. Terms* (ed. 4) 282/2 *Electric potential*, the quality, analogous to pressure, which tends to cause a flow of electricity from a point where it is higher to a point where it is lower. 1962 S. HANDEL *Dict. Electronics* 112 *Electric potential*, of a point, the potential difference between the point and some equipotential surface, usually that of the earth which is chosen to have zero potential. 1883 *Encycl. Brit.* XX. 249/2 The first attempt to apply electric power for propulsion on railways was made..in 1842. 1922 *Ibid.* XXX. 955/2 A Departmental Committee..on electric power supply was formed. *Ibid.* 958/1 The latest statistics available for Canada..show 795 central electric power stations. 1956 A. H. COMPTON *Atomic Quest* v. 340 Electric power came from curiosity about the relation between electricity and magnetism. 1920 *Whittaker's Electr. Engin. Pocket-Bk.* (ed. 4) 651 The Electric Precipitation of Smoke and Fumes... The precipitation of dust and fume suspended in gases, by means of the electrostatic field..has only been applied to industrial purposes within the last six years. 1882 *Proc. R. Inst. Gt. Brit.* X. 69 Siemens' Lichterfelde Electric Railway, now running at Berlin. 1902 *Encycl. Brit.* XXVI. 228/1 An overhead electric railway runs from the Zoologischer Garten to the Schlesisches Thor. 1921 *Daily Colonist* (Victoria, B.C.) 26 Mar. 3/1 A complete little home, 4-room, modern and almost new..hot water heating system and *very good electric range*. 1922 *Encycl. Brit.* XXX. 957/2 Another activity curtailed by the war was the organized sale of electric ranges. 1774 J. WALSH in *Philos. Trans. R. Soc.* LXIV. 464 The Torpedo, or Electric Ray, frequents the shores of this island, contrary to a received opinion among Naturalists. 1862 COUCH *Brit. Fishes* I. 119. 1880 *Encycl. Brit.* XII. 649/2 The electric rays (*Torpedinidæ*). 1896 *Pearson's Mag.* II. 754/2 A solid wall..is very transparent to the electric ray. 1897 *Strand Mag.* Mar. 273/1 Electric rays or waves.. will penetrate all substances at all distances. 1958 G. DURRELL *Encounters with Animals* II. 88 The electric-ray or torpedo-fish, a curious creature that looks rather like a frying-pan run over by a steam-roller. 1933 *Chambers's Jrnl.* Dec. 1007/1 (*caption*) A Vibrating Electric Razor. 1945 S. LEWIS *Cass Timberlane* (1946) xxi. 126 She made coffee for them, and she lent them an electric razor. 1926 *Glasgow Herald* 16 Dec. 13/6 Columbia introduced Electric Recording into Great Britain... This is the first Christmas that a..programme of Electric Records has been available for the gramophone. 1940 F. & R. LOCKRIDGE *Norths meet Murder* iii. 55 He..had recently bought an electric shaver. 1959 *Listener* 30 July 161/1 Infra-red cookers, electric shavers, stereophonic amplifiers. 1767 J. PRIESTLEY *Hist. Electr.* p. xiii, The electric shock itself..will appear almost as surprising, as any discovery that he [*sc.* Sir Isaac Newton] made. 1779 J. WEDGWOOD *Let.* 8 Nov. (1965) 244 The Doctor gives us great hopes of our poor little girl's limbs being restored, even without the assistance of the electric shocks. 1881 [see COLLOQUIALISM 2]. 1934 H. C. WARREN *Dict. Psychol.* 90/2 *Electric shock method*, a variety of the method of punishment in animal training, whereby a habit is broken or a new habit formed by applying an electric shock at some point in the behavior series. 1947 P. L. HARRIMAN *Dict. Psychol.* 118 *Electric-shock therapy*, treatment of mental disorders by electric shocks to the head of the patient. 1958 *Listener* 13 Nov. 799/1 The inventor of electric shock treatment. 1907 *Electr. Bull.* Jan. 31 (Advt.), Art Fittings, Lamps..Electric Signs and Letters. 1937 *Discovery* July 217/1 Electric signs, posters, magazines, catalogues, all use the appeal of colour to enhance their commercial value. 1909 *Cent. Dict.* Suppl. I. 415/1 Electric steel process. 1919 E. HENDRICK *Chem. Everyday Life* xi. 99 Electric steel is the latest method whereby the steel is heated, and the required ingredients added in an electric furnace. 1872 F. M. A. ROE *Army Lett.* (1909) 52 There was a terrific rain and electric storm last evening. 1906 *Springfield Weekly Republ.* 14 June 16 A severe electric storm, accompanied by a high wind, swept over Springfield. 1944 'G. GRAHAM' *Earth & High Heaven* 277 'Must have been a bad electric storm lately,' he remarked. 1934 WEBSTER, Electric strength. 1886 *Electrician* 15 Jan. 187/1 U_1 is the intensity of the electric stress. 1935 *Discovery* July 213/2 The behaviour of air, solids, liquids and the vacuum respectively as dielectrics when under high electric stress. 1943 *Gloss. Terms Electr. Engin.* (*B.S.I.*) 12 *Electric stress*, the stress occurring in an insulating material when subjected to an electric force. 1840 *Monthly Chron.* I. 383/2 Electric Telegraph.—This extraordinary machine is now being worked on the great western rail-road [in Britain]. 1913 *Technical World Mag.* 152 (Advt.), The economy electric toaster. 1926 *Daily Colonist* (Victoria, B.C.) 3 July 10/2 (Advt.), Upright Electric Toasters. Will toast two pieces at the one time. Heavy nickelplated finish. 1951 *Catal. of Exhibits, S. Bank Exhib., Festival of Britain* 133/2 Automatic electric toaster. 1936 *Punch* 29 Apr. 478/2 The latest invention is an electric toothbrush. 1963 *Which?* Dec. 374/2 An electric toothbrush is not very different in principle from an ordinary toothbrush. You still have a stick with a brush on the end, but instead of moving the brush by hand, the brush is driven by a small electric motor. 1875 *English Mechanic* 9 July 441/3 (*title*) Frictional electric torch. 1902 *Windsor Mag.* June p. xxv/2 (Advt.), The 'Ever-Ready' Portable Electric Torch. 1925 E. F. NORTON *Fight for Everest, 1924* v. 102 Somervell and I had to carry..a light rucksack apiece, with compass, electric torch, [etc.]. 1888 *Encycl. Brit.* XXIII. 495/2 Another method of effecting electric traction is to carry a store of energy on the car or on a special locomotive. 1940 *Chambers's Techn. Dict.* 286/1 *Electric traction*, the operation of a railway or road vehicle by means of electric motors, which obtain their power from an overhead contact wire or from generators or batteries mounted on the vehicle. 1901 *Trans. Amer. Inst. Electr. Engin.* 1900 XVII. 90 In the future there may be a.. tendency to reduce the weight of electric trains. 1908 A. BENNETT *Buried Alive* vi. 139 The roar of an electric train. 1939 'R. CROMPTON' *William & A.R.P.* vii. 185 'Steal that electric train of his he's always swankin' about,' suggested Ginger. 1885 *Electrician* 3 July 146/2 With electric tramways the energy (electricity) is conveyed from the generating station to the rails. 1958 *Oxf. Mail* 30 Sept. 10/5 Apart from the fact that an electric typewriter makes typing easier and faster, many more clear carbon copies can be obtained than on an ordinary typewriter. 1964 T. L. KINSEY *Audio-Typing & Electric Typewriters* iv. 35 It takes twenty-two times more mechanical energy to operate a manual typewriter than it does to operate an IBM electric typewriter. 1871 *Leisure Hour* 18 Nov. 720/2 The electric wave, produced by a few pieces of copper and zinc. 1895 S. P. THOMPSON *Electr. & Magn.* (new ed.) II. xiv. § 516 In the case of true electric waves, portions of the energy of the current..are thrown off. 1902 *Encycl. Brit.* XXVIII. 55/2 Electric waves must be passing through the dielectric surrounding a condenser in the act of discharging. 1890 *Times* 17 Jan. 3/3 Electric Welding..is the invention of Professor Elihu Thomson, of Lynn, United States... The development of electric welding in America has been rapid. 1907 [see WELDING *vbl. sb.*[1] 3]. 1930 *Engineering* 8 Aug. 164/1 The chapter..giving..information on..electric and gas welding. 1899 *Phil. Mag.* XLVIII. 401 (*title*) On the velocity and mass of the ions in the electric wind in air.

4. electric blue (examples); also, a brilliant light blue. Also in extended use and *ellipt.* **electric green** = **electric blue*.

1873 *Young Englishwoman* Feb. 77/1 One model..is of bronze velvet and electric green silk. 1882 *Queen* 23 Dec. (Advt.), Opera hood..Black, Brown, Cream, Pale Blue, and Electric. 1893 *Ladies' Home Jrnl.* Oct. 2/2 Her preference is for subdued tints, her favorite color being a soft shade of electric blue. 1912 *Queen* 4 May 750/2 Lady Mary C. in electric-blue cloth with black and white hat. 1955 *Times* 9 July 3/3 A huge black fish with a brilliant electric blue tail.

B. *sb.* **3.** Short for (*a*) *electric lamp, light*; (*b*) *electric motor car, railway, tramcar*, etc.; (*c*) *electric circuit*; (*d*) *electric company shares*.

1886 *Harper's Mag.* July 314/1 The light of common day ..is preferable to any manner of..alabaster lamps, or even the latest improvement in electrics. 1890 *Boston* (Mass.) *Jrnl.* 3 Mar. 1/7 Do the electrics travel too fast for the public safety? 1892 S. HALE *Lett.* (1919) 269 There are cables, and electrics, but these haven't entirely driven out the horse. 1896 *New Peterson Mag.* Jan. 65/2 [They] would no more enter an 'electric' than they would give up wearing mitts all day. 1897 KIPLING *Capt. Cour.* 215 Harvey..was asleep before his father could shade the electrics. 1905 E. GLYN *Viciss. Evangeline* 166 In the twinkle of an eye we were rolling in the electric to Willis's. 1909 *Westm. Gaz.* 2 Sept. 9/1 Underground electrics. 1916 E. WALLACE *Clue Twisted Candle* (1917) xiii. 149 He felt for the switch of the electric. 1924 S. GLASPELL *Fidelity* iv. 33 She told him..that Mrs. Blair had come for Mrs. Franklin in her 'electric' and they had gone to a tea. 1931 *Economist* 10 Jan. 78/2 Rates, especially for electrics, potash and stores, were far below the level of the beginning of December. 1946 A. PHELPS *I couldn't care Less* xiv. 118 Checking some of the other electrics and finding them working I realized.. there must be a break in that circuit somewhere. 1947 *Jrnl. R. Aeronaut. Soc.* LI. 92/2 Electrics in the small and medium aircraft may absorb only a small fraction of the auxiliary power load. 1959 *Engineering* 16 Jan. 86/2 Booklets on glass fibre, yacht electrics, sail rigs. 1961 *Countryman* LVIII. 435 Last summer I returned, glad to find so much the same, except for 'the electric' which stretches its ugly spider poles to every croft. 1963 *Sunday Times* (Colour Suppl.) 7/1 He tried to lower his flaps part way—forgetting, in the anxiety of the moment, that without electrics his flap position indicator would not work. 1970 *Which?* May 143/1 The electrics are described as 'solid state', which means they use printed circuits instead of wires.

electrical, *a.* Add: **2.** *electrical clock*; *electrical eel, machine* (earlier examples); **electrical engineer**, an engineer devoting himself to electrical aspects of engineering; **electrical engineering**, the work done by, or the profession of, an electrical engineer; **electrical precipitation**, precipitation of a substance, esp. in a gas, by the action of a uni-directional electric field; cf. **electric precipitation*; **electrical recording** (see quot. 1940).

1746 W. WATSON *Sequel to Experiments in Electr.* 32 If the electrical machine is placed upon originally-electrics, the man..gives no sign of being electrified. 1775 *Phil. Trans. R. Soc.* LXV. I. 102 An Account of the *Gymnotus Electricus*, or Electrical Eel. 1842 *Brit. Pat.* 9465 40 For connecting any number of distant clocks together in the manner of what are called electrical clocks, so that they will all move with an exact uniformity. 1883 (*title*) The Electrical Engineer, a Journal of Electrical Engineering. *Ibid.* 1 May 3/1 In view of the larger extension of electrical interests..we have deemed it advisable to alter the title ['Electric Light'] to 'The Electrical Engineer'. 1883 *Pop. Sci. Monthly* XXIV. 255 Electrical engineering..embraces a knowledge of cables, telegraphy, electric lighting, electrical measurement. 1912 *Metall. & Chem. Engin.* X. 686/2 Electrical precipitation of suspended particles... Review of the processes of condensing dust and fume by means of high-tension electric current. These processes are the invention of Dr. F. G. Cottrell. 1925 *'His Master's Voice' Album Ser.* No. 3 The Rose Cavalier..Electrical Recording. 1930 *Engineering* 27 June 833/3 Considerable progressive development could be recorded in the use of electrical precipitation plants. 1935 *Discovery* Apr. 111/2 This new and valuable testing instrument..put into its hands by the electrical engineer. 1940 *Chambers's Techn. Dict.* 286/1 *Electrical recording*, the use of amplified currents from microphones for operating electromagnetic or electrodynamic drives for the cutting stylus in wax recording. 1965 M. MORSE *Unattached* i. 18 A large electrical-engineering firm.

electric light. (Earlier and *attrib.* examples.)

1767 J. PRIESTLEY *Hist. Electr.* I. 9 A much finer appearance of electric light..was observed by Dr. Wall. 1777 T. CAVALLO *Compl. Treat. Electr.* p. v, He [*sc.* Mr. Hawkesbee] first remarked various appearances of the electric light. 1878 *Boston Transcript* 16 Nov. 4 The electric-light patents he [*sc.* Edison] has applied for are already antiquated. 1893 *Outing* (U.S.) XXII. 132/2 The famous Spokane Falls..furnish abundant power for.. electric-light plants. 1936 *Discovery* June 189/2 The natives [in Kano, Nigeria] run their own electric light plant and water services.

electrifiable (ele·ktrifəi,əb'l), *a.* [f. ELECTRIFY + -ABLE.] Capable of being electrified.

1828–32 WEBSTER.

electrification. Add: **1. b.** The action or process of converting (a railway system, factory, etc.) to the use of electricity.

Earlier attempts to provide a term for this were *electrolisation* and *electrilisation*, with a verb *electralise* [sic] = *ELECTRIFY *v.* 1 b.

[1900 *Westm. Gaz.* 15 Feb. 4/1 The electrolisation of the inner circle, with its twenty-six stations. *Ibid.* 15 Aug. 3/1 The complete electrolisation of London tramways. 1901 J. BURNS in *Standard* 4 Mar., If Londoners had only done in 1895 what they did on Saturday, three-fourths of the tramways would have been electralised. 1901 *Westm. Gaz.* 28 Aug. 4/3 Americanised companies which go in for the 'standard' system of electrilisation.] 1901 *Westm. Gaz.* 2 Jan. 8/3 The character of the coal consumed..has altered the position for the worse... The one possible remedy is electrification. *Ibid.* 6 June, District Railway Electrification. 1904 *Daily Chron.* 26 Aug. 7/4 Electrification Perils. Risk Entailed by the Transformation of the 'Underground'. 1921 *Dict. Occup. Terms* (1927) § 561 Electrification ganger. 1963 M. GOLDSTEIN *Dict. Mod. Acronyms* 123 REA, Rural Electrification Administration. 1967 [see *ELECTRIFIER].

3. *fig.* The condition or state of being electrified or excited. (Cf. ELECTRIFY *v.* 2.)

1878 F. A. KEMBLE *Rec. Girlhood* I. iv. 113 The tragedy was ended, and I had electrified the audience, my companions, and, still more, myself; and so, to avert any ill effects from this general electrification, Mrs. Rowden thought it wise and well to say to me [etc.]. 1892 *Leisure Hour* Aug. 657/1 Her electrification by Mr. Belport's proposal.

electrified, *ppl. a.* (Earlier examples.)

1745 W. WATSON in *Phil. Trans.* XLIII. 485 In electrified Bodies you see a perpetual Endeavour to get rid of their Electricity. *Ibid.* 491 The Finger of the electrified Person is brought near thereto.

electrifier. (Examples of literal use.)

1906 *Chambers's Jrnl.* Sept. 703/2 Then it [*sc.* peat] falls into the electrifier. 1967 *Listener* 26 Jan. 122/2 Now that electrification from Euston to Manchester and Liverpool is finished, the British Railways Board have become eager electrifiers again.

electrify, *v.* Add: **1.** (Earlier examples.)

1745 W. WATSON in *Phil. Trans.* XLIII. 490, I procur'd an iron Bar..; this I electrified lying on Cakes of Wax and Resin. *Ibid.* 498, I was desirous to know if I was able to electrify a Drop of cold Water.

b. To introduce electric power into (a system of railways, a farm, factory, etc.).

1900 *Westm. Gaz.* 17 Oct. 9/3 It is not very astonishing that the directors of the District Railway should be in no violent hurry to start upon the electrifying of their line. 1970 *Yankee* (Dublin, New Hampshire) Aug. 140 You can electrify a flat wick oil lamp with a Gyro Queen Converter. 1970 *Railway Mag.* Oct. 574/2 The electric locomotive was supplied to the East Sussex County Council when the hospital railway was electrified in 1903.

3. *intr.* To become charged with electricity.

1745 W. WATSON in *Phil. Trans.* XLIII. 490 Whilst the Bar was at one End electrifying, a Spoon lay upon the other.

electrifying, *vbl. sb.* (Earlier example.)

1747 W. WATSON in *Phil. Trans.* XLV. 58 The Room where the electrifying Machine was placed.

electro-. Add: **ele:ctro-acou·stics**, acoustics investigated by electrical methods; hence **ele:ctro-acou·stic, -acou·stical** *adjs.*; **ele:ctro-ana·lysis** *Chem.*, electrolytic analysis; so **ele:ctro-analy·tical** *a.*; **ele:ctroballi·stic** *a.* (example); **ele:ctro-bra·sser** (see quot. 1921); † **ele·ctrobus** *Obs.*, an omnibus propelled by electricity; also *attrib.*; **ele:ctroca·rdiogram** *Med.* [ad. G. *electrocardiogramm* (W. Einthoven 1894, in *Arch. Ges. Physiol.* LVI. 541)], a record of the electric currents produced in the body by the heart-beats of a patient; **ele:ctroca·rdiograph**, the apparatus used in registering electrocardiograms; hence **ele:ctrocardiogra·phic** *a.*, **ele:ctrocardio·graphy**; **ele:ctro-cau·tery** *Med.*, cautery by means of an electrically heated instrument; **electro-chemical** *a.*, hence *electro-chemical equivalent*, the weight of a substance deposited during electrolysis by the passage of a specified quantity of electricity, usu. expressed in grams per coulomb; *electro-chemical series* (see quot. 1943); **ele:ctro-che·mically** *adv.*, in accordance with the laws of electro-chemistry; **ele:ctro-che·mist**, one who practises electro-chemistry; **electro-chemistry** (earlier and later examples); **ele:ctrochroma·togram**, a chromatogram produced by electrochromatography; **ele:ctro-chromato·graphy** [ad. F. *électrochromatographie* (H. Lecoq 1944, in *Bull. de la Soc. R. des Sciences de Liège* XIII. 21)], chromatography in which the migration and separation of the constituents of a mixture take place under the influence of a constant electric field; zone electrophoresis; so **ele:ctrochromatogra·phic** *a.*; **electro-chronograph** (earlier U.S. example); **ele:ctro-coagula·tion** [a. F. *électro-coagulation* (Doyen)], the hardening and destruction of tissue or the control of bleeding by the direct application of an electrically heated surgical instrument (= **electro-cautery*) or by means of a high-frequency electric current; **ele:ctro-co·ntact**, used *attrib.* of a submarine mine which is exploded by means of electricity; **ele:ctro-convu·lsive** *a.*, of or pertaining to a convulsive response to electric shock treatment, spec. *electro-convulsive therapy* (abbrev. *E.C.T.*), *treatment*, the treatment of mental disease by the application of an electric shock to the central nervous system; **ele:ctro-co·pper** *v. trans.*, to coat with copper by electrolysis; **ele:ctroco·rtin** *Biochem.* [*CORTIN], = *ALDOSTERONE; **ele:ctro-cu·lture**, the use of electricity in the growing of crops; **ele:ctrocy·clic** *a. Chem.*, of a chemical change, characterized either by the formation of a single bond between the two ends of a linear conjugated organic molecule, or by the converse process of breaking such a bond; **ele:ctro-desicca·tion**, the drying and consequent destruction of tissue or the control of bleeding by the local application of a high-frequency electric current; **ele:ctro-dia·lysis**, dialysis accelerated by the passage of an electric current; **ele:ctro-ence·phalogram** (occas. **electrencephalogram**) [ad. G. *elektrenkephalogramm* (H. Berger 1929, in *Arch. f. Psychiatrie u. Nervenkrankheit* LXXXVII. 527 ff.)], the record made by an electro-encephalograph; abbrev. **E.E.G.**; **ele:ctro-ence·phalograph**, an instrument for recording the electrical activity of the brain; hence **ele:ctro-encephalo·grapher, ele:ctro-encephalo·graphist, ele:ctro-encephalo·graphy** *sbs.*, **ele:ctro-encephalogra·phic** *a.*; **ele:ctro-extra·ction** [EXTRACTION 3], the electrolytic recovery of a metal from a solution of its salts (Webster 1934); **ele:ctrofo·rming** *vbl. sb.*, the production of metallic objects by a process of electro-deposition; hence (as a back-formation) **ele·ctroform** *v. trans.*; also **ele·ctroform**, an electroformed metallic object; **ele·ctroformed** *ppl. a.*; **electro-galvanic** *a.* (examples); **ele:ctro-ga·lvanizing** *vbl. sb.* (see quot. 1940); **electrogenic** *a.* (examples); **ele:ctrogravi·tics** (see quot. 1960); so **ele:ctro-gravi·tic** *a.*; **ele:ctro-hydrau·lic** *a.*, (*a*) pertaining to a hydraulic system that is powered or controlled electrically; (*b*) pertaining to electro-hydraulics; hence **ele:ctro-hydrau·lically** *adv.*; **ele:ctro-hydrau·lics**, the use of high-voltage electrical discharges in liquids; **ele·ctrojet**, an intense electric current that occurs in a narrow belt in the lower ionosphere, esp. in regions of strong auroral displays and near the magnetic equator; **ele:ctro-lumine·scence**, electrical luminescence, as created in a gas-filled tube or by the excitation of a layer of phosphor (cf. LUMINESCENCE, quot. 1896); hence **ele:ctrolumine·scent** *a.*; **ele:ctro-mecha·nical** *a.*, of or pertaining to the application of electricity to a mechanical process, device, etc.; **electro-medical** *a.* (examples); **ele:ctromo·tograph**, see MOTOGRAPH; **ele:ctromy·ogram**, a graphical record of the variations in electric potential in a muscle; **ele:ctromy·ograph**, an instrument for making electromyograms; so **ele:ctromyogra·phic** *a.*, **ele:ctromyogra·phically** *adv.*; **ele:ctromyo·graphy**; **ele:ctronarco·sis**, the production of a narcotic state by the passage of an electric current through the brain; hence **ele:ctronarco·tic** *a.*; **ele:ctro-ne·gative** *a.*, (*b*) a term used by Berzelius to describe atoms that go to the anode in electrolysis; hence, having a tendency to form negative ions; so **ele:ctro-ne·gatively** *adv.*; **ele:ctro-negati·vity**, the degree to which atoms or radicals attract electrons; the state of being electro-negative; **ele:ctro-o·culogram**, the record produced by an electro-oculograph; **ele:ctro-o·culograph**, an instrument for studying the eye by registering the varying electric potentials of the skin around it; hence **ele:ctro-oculogra·phic** *a.*, **ele:ctro-oculogra·phically** *adv.*; **ele:ctro-oculo·graphy**, the technique of using the electro-oculograph; **ele:ctro-o·ptic, ele:ctro-o·ptical** *adjs.*, of or pertaining to electro-optics; **ele:ctro-osmo·sis**, osmosis under the influence of an electric field; also called *electro-endosmosis, electrosmosis*; hence **ele:ctro-osmo·tic** *a.*, **ele:ctro-osmo·tically** *adv.*; **ele:ctrophi·lic** *a.*, having an affinity for electrons; **ele:ctro-photo·graphy**, the production of photographic prints by electronic means (see also quot. 1894); so **ele:ctro-photogra·phic** *a.*; **ele:ctro-phre·nic** (-fre·nik, -fri·nik) *a.*, of or pertaining to electrical stimulation of the phrenic nerves; **electro-physiology** (earlier and later examples); *electro-physiological* adj. (examples); also, *ele:ctro-physio·togist*; **ele:ctrople:xy** *Psychiatry* [after APOPLEXY], electro-shock or electro-convulsive therapy; **ele:ctro-pneuma·tic** *a.*, pertaining to a combination of electrical and pneumatic power; **ele:ctro-po·lish** *v. trans.*, to polish the surface of a metal by electrolysis; hence *ele:ctropo·lished* ppl. a.; **ele:ctro-po·sitive** *a.* (earlier example); also, (*b*) a term used by Berzelius to describe atoms that go to the cathode in electrolysis; hence, having a tendency to form positive ions; so **ele:ctro-po·sitively** *adv.*; **ele:ctro-positi·vity**, the degree to which atoms or radicals tend to form positive ions; the state of being electro-positive; **ele:ctro-redu·ction**, the reduction which takes place in an electrolytic cell (cf. REDUCTION 10); **ele:ctro-re·tinogram**, a graphic record of the electrical activity of the retina; hence **ele:ctro-retino·graphy**, the study of, or the technique of obtaining, such records; **ele:ctro-sho·ck**, an electric shock, used esp. *attrib.*, as *electro-shock treatment*, medical treatment by means of an electric shock (cf. **electric shock*, **electro-convulsive therapy*); **ele:ctro-si·lver** *v.*, hence as *sb.*; also *ellipt.* **electro**; **ele·ctro-slag**, a name given to a process for refining or welding metal in which an electric current is passed through a layer of slag in contact with it and into the metal itself; **ele:ctrostimula·tion** *Med.*, electrical treatment sometimes used in the therapy of impaired nerves or muscles; in *Psychiatry*, a form of electro-shock therapy; **ele:ctrostri·ction** [STRICTION] (see quot. 1942); hence **ele:ctrostri·ctive** *a.*; **ele:ctro-sy·nthesis**, chemical synthesis effected by electricity; **ele:ctrota·xis** *Biol.* [TAXIS 6], the response of an organism to an electric current; so **ele:ctrota·ctic** *a.*; **electro-therapeutics** (earlier example); so **ele:ctro-therapeu·tical** *a.*, of or pertaining to electro-therapeutics; **ele:ctro-the·rapist, -therapeu·tist**, one who practises electro-therapy; **ele:ctro-the·rmal, -the·rmic** *adjs.*, relating to heat derived from electricity; **ele:ctro-ti·n** *v. trans.*, hence **ele:ctro-ti·nning**, tinning (sense 1) by electrolysis; so **ele·ctro-tinned** *ppl. a.*, **ele:ctro-ti·npla·te**, electrolytic tinplate; also *electro-tin plating* vbl. sb.; **ele:ctro-tro·pic** *a.*, of, pertaining to, or exhibiting electrotropism; **electro-tropism**, curvature of growth in plants due to slight electric currents; **ele:ctro-va·lence, -va·lency**, a chemical bond in which two oppositely charged ions are linked by the electrostatic attraction between them; an ionic bond; so **ele:ctro-va·lent** *a.*; **ele:ctro-visco·sity**, the property of being electroviscous, the component of the viscosity of a substance that is due to the electroviscous effect; **ele:ctrovi·scous** [ad. G. *elektroviskos* (A. Passynski 1935, in *Kolloid Zeitschr.* LXX. 182/2)], (of a liquid or dispersion) having a viscosity that depends on the applied electric field; of or pertaining to electroviscosity; **ele·ctro-win** *v.*, hence **ele:ctro-wi·nning** *vbl. sb.*, the process of extracting a metal from a solution of its salts or from its ore by electrolysis (cf. WINNING *vbl. sb.*[1] 5).

1927 Electro Acoustics [see *ABSTRACTOR]. 1934 *Amer. Speech* IX. 309/2 A method for measuring small intensities by means of electro-acoustical apparatus. 1935 *Ibid.* X. 72/2 An electro-acoustic determination of the factors of voice quality which distinguishes one voice from another. 1943 *Endeavour* II. 123/2 The term 'electro-acoustics', which not only indicates the combination of electrical and mechanical apparatus but also the methods by which masses, restoring forces represented by springs and the like, and mechanical resistance find their analogies in electrical terms. 1955 *Gloss. Acoust. Terms* (*B.S.I.*) 17 *Electro-acoustical transducer*, a transducer which is actuated by energy from an electrical system and supplies energy to an acoustical system or vice versa. 1956 *Nature* 10 Mar. 464/2 Electro-acoustic measuring techniques. 1959 *Times* 11 Mar. 3/1 Important Position with small American Company open for electro-acoustic technical man. 1970 *Sci. Jrnl.* May 45/2 The obvious first use for the electro-acoustic strip is as a store for signals. 1903 *Electr. World & Engin.* 28 Mar. 530/2 The difference observed in the electro-analysis of mercury from a potassium cyanide solution are due to an attack of the platinum disc serving as cathode and to the solubility of platinum in potassium cyanide. *Ibid.* 21 Nov. 853/2 Electro-analytical methods. 1964 R. JOHNSTON in *Oceanogr. & Marine Biol.* II. 112 The modern electro-analytical techniques. 1890 *Nature* 18 Sept. 502/2 In 1860..I carried on for the Government the first electro-ballistic experiments made in this country. 1921 *Dict. Occup. Terms* (1927) § 278 *Electro-brasser*..effects electrolytic deposition of brass on better types of finished screws. 1960 *Classification of Occupations* (General Register Office) 30/2 Electro-brasser (screws). 1906 *Westm. Gaz.* 20 Apr. 11/1 The London Electrobus Company. *Ibid.* 4 Dec. 6/3 About January.. the first of the electrobuses will be placed in regular running. 1904 *Sci. Amer.* 5 Mar. 197/3 The human electrocardiogram discovered by A. D. Waller. 1927 A. V. HILL *Living Machinery* iii. 117 The second record is a normal 'electro-cardiogram' accompanied by a 'phono-cardiogram'. 1952 *Electronic Engin.* XXIV. 102 The electrical output of the heart, which when recorded is termed the electrocardiogram or E.C.G. 1913 *Q. Jrnl. Med.* VI. 442 The two installations, electro-phonograph and electro-cardiograph, are arranged side by side. 1927 A. V. HILL *Living Machinery* iii. Pl. xii (*caption*), String galvanometer ('electrocardiograph') as used for recording currents produced by the human heart. 1958 *Observer* 9 Nov. 17/5 With the electrocardiograph one records the electrical activity of the hearts of both child and mother, superimposed on one another. 1910 *Canad. Jrnl. Med. & Surg.* XXVII. 276 As a result of electrocardiographic work, we can for the first time be perfectly sure that the nature of excitation..in an extra-systole is wholly different from that of the normal heart beat. 1913 *Lancet* 21 June 1784/1 The advances made in the application of electrical methods to the study of the heart are well illustrated in a catalogue..of electro-cardiographic apparatus. 1961 *Ibid.* 22 July 183/2 Electrocardiographic changes often suggest or confirm the presence of an aneurysm, but they do not always agree..with the radiological..evidence. 1910 *Canad. Jrnl. Med. & Surg.* XXVII. 273 (*title*) Electrocardiography and electro-phonography as aids in clinical diagnosis. 1957 *Electronic Engin.* XXIX. 132 The striking advances which have been made in electrocardiography in recent years. a 1884 KNIGHT *Dict. Mech.* Suppl. 305/2 Electro-cautery. 1931 *Times Lit. Suppl.* 27 Aug. 650/4 His view that electro-cautery..affords permanent relief. 1953 *Jrnl. Brit. Interplanet. Soc.* XII. 235 The behaviour of a normal mouse was contrasted with that of a labyrinthinectomized mouse (i.e., an electrocautery needle had been applied to the horizontal semi-circular canals in the membranous labyrinth of the middle ear). 1848 H. WATTS tr. *Gmelin's Hand-bk. Chem.* I. 155 The elements succeed one another in the electro-chemical series of Berzelius. 1857 [see **absolute unit*]. 1883 E. ATKINSON tr. *Mascart & Joubert's Treat. Electr.* I. iii. 247 The electromotive force of an electrolyte is equal to the mechanical equivalent of the heat of combination of its electro-chemical equivalent. 1922 GLAZEBROOK *Dict. Appl. Physics.* II. 354/1 Since the ratio of the mass to the charge for a hydrogen ion in solution, *i.e.* the electro-chemical equivalent of hydrogen, is $1{\cdot}04 \times 10^{-4}$ e.m.u., the mass of the hydrogen atom is given [etc.]. 1943 *Gloss. Terms Electr. Engin.* (*B.S.I.*) 94 *Electro-chemical series*, a tabular arrangement of the elements in the order of the electrode potential developed when an element is immersed in a solution of normal ionic concentration. 1834 [see ELECTROLYZED *ppl. a.*]. 1881 Electro-chemically [in Dict. s.v. ELECTRO-]. 1946 *Nature* 19 Oct. 548/1 These advances led to the 'ground' mine, laid on the sea-bed, operated either magnetically, electro-chemically or acoustically. 1837 *Ann. Electr., Magnet., & Chem.* I. iii, I shall shortly take an opportunity of calling the attention of electro-chemists to two points. 1885 *Jrnl. Franklin Inst.* Feb. 81 The man skilled in its science [*sc.* electro-metallurgy] and art may appropriately be styled an electro-chemist. 1814 G. J. SINGER (*title*) Elements of electricity and electro-chemistry. 1956 *Nature* 28 Jan. 181/2 (*heading*) Electro-chemistry of stainless steel in sulphuric acid. 1951 *Analytical Chem.* XXIII. 27/2 (*caption*) Arrangement for formation of electrochromatogram in paper strip, *P*, between glass plates. *Ibid.* 30/1 For ions of identical sign, chromatographic sequences are usually identical with electrochromatographic sequences. 1948 *Chem. Abstr.* XLII. 6703 A new analytical technique, electrochromatography, in which the adsorption is effected with the aid of an e.m.f., is described. 1966 *McGraw-Hill Encycl. Sci. & Technol.* III. 97/2 Electrochromatography is especially useful for the examination of ionized substances such as acids, bases, and salts. 1851 C. CIST *Sk. Cincinnati* 302 The invention of the electro-chronograph by Professor Locke of our city. 1913 *Jrnl. Advanced Therapeutics* XXXI. 29 Electro-coagulation is used to destroy the neoplasms of the skin and of the mucous orifices. 1960 *Surg. Forum* X. 439 Positive electrocoagulation is very effective in creating hemostasis in tissues not usually amenable to suture, using currents up to 100 ma. 1888 *Encycl. Brit.* XXIII. 449/2 This is done..in electro-contact mines by a circuit closer in the mine. 1903 *Daily Chron.* 25 June 4/3 The relative merits of electro-contact and observation mines. 1952 *Brit. Jrnl. Psychol.* Feb. 38 A course of electroconvulsive therapy..the weakening or breaking up of certain behaviour patterns promoted by affective disturbances. *Ibid.*, If E.C.T. blots out or breaks up the depressive pattern. 1958 A. WILSON *Middle Age of Mrs. Eliot* II. 251, I hope they don't give your friend E.C.T. If it's a very expensive place they probably won't. But at the hospitals they do. 1965 *Nursing Times* 5 Feb. 188/2 Electroconvulsive therapy (ECT) is used in the treatment of both depressive and manic states. 1968 *Times* 8 Nov. 5/8 A new legitimate opportunity for studying human brain function has recently been afforded by the use of electroconvulsive shock treatment, or ECT as a therapy for depressive illness. 1873 E. SPON *Workshop Rec.* 1st Ser. 212/2 Steel, iron, zinc, lead, and tin which have been previously electro-coppered. 1954 Electrocortin [see *ALDOSTERONE]. 1955 *Sci. News Let.* 5 Feb. 82/1 At the Mayo Clinic, Rochester, Minn., Dr. Philip S. Hench, co-discoverer of cortisone and Nobel Prize winner, and a number of his associates have been trying out two of the new anti-arthritis drugs. These two are aldosterone, also called electrocortin, and fluorohydrocortisone. 1902 WEBSTER Suppl., Electro-culture. 1917 *Electrical Rev.* 6 July 21/1 Though only just beginning to attract public attention, electro-culture is by no means a new idea. 1932 *Times* 13 May 9/3 Even if electroculture in the vegetable garden were to become common, [etc.]. 1965 WOODWARD & HOFFMANN in *Jrnl. Amer. Chem. Soc.* 20 Jan. 395/1 We define as *electrocyclic* transformations the formation of a single bond between the termini of a linear system containing *k* π-electrons.., and the converse process. *Ibid.* 5 May 2045/1 The stereochemical course of an electrocyclic ring closure is determined by the symmetry of the highest occupied molecular orbital of the open-chain reactant. 1919 W. L. CLARK in *Amer. Jrnl. Electro-therapeutics & Radiology* XXXVII. 211 (*title*) The treatment of some lesions of the skin and mucous membranes by electrodesiccation. 1926 *Arch. Dermatol. & Syphilol.* XIII. 344 Electrodesiccation..should be evaluated as an additional therapeutic weapon in the hands of the dermatologist. *Ibid.* 345 Electrodesiccation produces a dehydration of the tissue, with shrinking of the cellular elements. 1962 *Southern Med. Jrnl.* LV. 391/1 Postoperative bleeding was controlled by electrodesiccation under halothane anesthesia. 1921 *Chem. Abstr.* XV. 3983 The Ca is removed by 'electrodialysis'. 1959 *Observer* 7 June 10/6 With brackish water it is cheaper to remove the salt from the water. The most promising method of doing this is electrodialysis, which uses an electric charge across sets of special membranes to separate out the salt. 1965 *New Scientist* 4 Nov. 341/1 Electro-dialysis can reduce the acidity of citrus fruits by at least 50 per cent. 1934 *Arch. Neurol. & Psychiatry* (Chicago) XXXI. 469 Electroencephalogram of the cortex in stimulation of the labyrinth. 1936 W. G. WALTER in *Lancet* 8 Aug. 305 (*title*) Location of cerebral tumours by electroencephalography. *Ibid.*, He [*sc.* Hans Berger] called the record of this activity..the Electrenkephalogram, abbreviated as 'EEG' by analogy with the electrocardiogram—ECG. In English-speaking countries the term has come to be 'electroencephalogram'. *Ibid.* 307/2 This confirmed the electro-encephalographic diagnosis. 1939 *Times* 9 Mar. 11/5 The exhibits included an electrencephalograph, lately brought from the United States by Dr. Denis Williams. The apparatus is an amplifying device for recording on paper the minute electrical waves from the human brain. 1941 BEAUMONT & DODDS *Rec. Adv. Med.* (ed. 10) xiii. 317 Electro-encephalography..is now accepted as a genuine method of studying electrical changes in the cerebral cortex. 1951 *Sci. News Let.* 17 Feb. 106/2 The Eastern Association of Electroencephalographers. 1958 *New Yorker* 5 Apr. 74/3 Here was the electroencephalographist's summary of his findings. 1958 *Observer* 21 Sept. 13/4 Electro-encephalographic recordings of the brain of the unborn child. 1968 *Sunday Tel.* 28 Jan. 4/4 The donor, Miss Denise Darvell, a motor accident victim, was then given electrocardiograph and electroencephalograph tests on heart and brain. 1969 *Times* 15 Apr. 9/3 Recording the pattern of brain waves, or electroencephalogram..emitted. 1934 WEBSTER, Electroform, v., electroforming, n. 1950 *Jrnl. Iron & Steel Inst.* CLXV. 245/1 The bending fatigue properties were determined for electroformed sheets of iron-nickel. 1954 *Electronic Engin.* XXVI. 325/2 Electroforming..is particularly suited to the production of waveguides and micro-wave equipment. *Ibid.* 325/3 Copper is the metal most commonly deposited, but electroforms can be produced with an initial deposit of ·003/·005 in. silver or nickel. 1837 J. S. COYNE *Queer Subject* I. iii, You shall see me make this subject more by the action of the Electro-Galvanic Battery. 1847 BRETT & LITTLE *Impr. Electr. Telegr.* 13 The nature of electro-galvanic action. 1922 *Encycl. Brit.* XXX. 966/1 *Electrogalvanizing*, the electrolytic deposition of a coating of zinc, from sulphate solutions, upon iron articles is now a well-established industry in all the leading manufacturing countries. 1940 *Chambers's Techn. Dict.* 287/1 *Electro-galvanising*, the electro-deposition of a protective coating of zinc on metal objects. 1935 *Discovery* Nov. 327/2 The organ as a whole resembles a beehive, for the electrogenic units or elements are grouped in batteries of prismatic form. 1970 *Nature* 28 Feb. 820/1 Electrogenic explanations of ionic pumps have been proposed. 1956 *Interavia* XI. 373 An electrogravitic field acting on all parts simultaneously. *Ibid.* 374/1 The communication possibilities of electro-gravitics, as the new science is called, confound the imagination. 1956 *Jrnl. Brit. Interplan. Soc.* XV. 387 'Electrogravitics' may revolutionize methods of transport by control of gravity fields. 1960 C. H. GIBBS-SMITH *Aeroplane* 296 *Electrogravitics*, a popular, but increasingly used, term for the general study of gravity and the force it exerts. 1922 J. W. M. SOTHERN *Marine Diesel Oil Engines* xi. 359 In the electro-hydraulic gear the power unit is placed directly beside the steering gear, and consists of a constant speed, non-reversing electric motor E driving a pump unit. 1966 *Science* 7 Oct. 155/1 Electrohydraulics is a new process for converting electrical energy directly to other forms of energy. It depends upon the discharge of a high-voltage arc under the surface of a liquid medium. *Ibid.* 157/2 Electrohydraulic treatment is an extremely effective..way to sterilize water and sewage. 1967 *Jane's Surface Skimmer Systems* 18/1 Similar electro-hydraulic signalling and feed-back systems are used to control propeller pitches. *Ibid.* 117 (*caption*) Power is provided by four 850 hp M-50 diesels,..controlled electro-hydraulically from the forward wheelhouse. 1955 E. BURGESS *Frontier to Space* v. 99 This experiment appears to confirm that the equatorial electrojet does in fact exist..and that it occurs in the E layer. 1956 *Jrnl. Brit. Interplan. Soc.* XV. 276 At 100 km. is the E layer in which electrojets flow and which were detected by early Aerobee rockets penetrating into this region. 1902 MANN & MILLIKAN tr. *Drude's Theory of Optics* III. iii. 540 The sodium light is produced by an electric discharge in a vacuum tube. In this electro-luminescence the temperature is..lower. 1909 *Cent. Dict.* Suppl., Electro-luminescence, electroluminescent. 1936 *Times Lit. Suppl.* 15 Feb. 125/3 Electroluminescence, such as is exhibited in neon signs, perhaps shows the way to greater efficiency. 1956 *Nature* 4 Feb. 240/2 The electro-luminescent properties of some zinc sulphide phosphors have been tentatively ascribed to a partial transformation from hexagonal to cubic. 1956 *Sat. Rev.* 6 Oct. 55/1 This new form of light is called electroluminescence. The glow comes from a powdered phosphor..sprayed on the surface of a glass plate... On the side that emits the light the glass is coated..with a transparent stuff that conducts electricity. 1888 *Encycl. Brit.* XXIII. 450/1 Electro-mechanical mines can be made by placing a voltaic battery inside the mine itself. 1927 *Observer* 22 May 13/1 The national exhibit of hydro-electricity.. consists principally of drawings, graphs..which all give an idea..of the electro-mechanical constructions..in Italy. 1943 *Gloss. Terms Electr. Engin.* (*B.S.I.*) 135 *Electromechanical brake*, a brake consisting of friction shoes applied..and released electrically. 1963 B. FOZARD *Instrumentation Nucl. Reactors* vii. 72 Where an electro-mechanical register is used..care must be taken to ensure that this does not lead to lost counts. 1967 *Jane's Surface Skimmer Systems* 12/1 The craft..is balanced prior to moving off by an electro-mechanical weight shifting arrangement. 1909 R. B. MATTHEWS *Electr. for Everybody* I. ix. 87 (*heading*) Electro-medical apparatus. 1961 *Engineering* 2 June 753 The large number of electro-medical appliances now in use. 1917 *Amer. Jrnl. Physiol.* XLII. 229 The undulations seen in the electromyogram of human muscles in voluntary contraction. 1963 *Lancet* 19 Jan. 152/2 A 'myopathic' pattern in the electromyogram can be obtained from a high proportion of carrier females. 1964 *Amer. Speech* XXXIX. 231 Electromyograms of muscular activity are similar for all positions of [f]. 1958 *IRE Trans. Med. Electronics* XI. 45 (*heading*) A new six-channel electromyograph for studies on muscle. 1923 *Amer. Jrnl. Physiol.* LXV. 234 (*heading*) Electromyographic studies of muscular fatigue in man. 1970 *Language* XLVI. 315 We will present some new electromyographic data obtained from the laryngeal and supralaryngeal muscles during speech. 1928 *Brain* LI. 517 It is possible that the evidence of physiological fatigue could be demonstrated electromyographically. 1926 *Jrnl. Amer. Med. Assoc.* 16 Oct. 1300/1 Work in electromyography is complicated by variations in the resistance of the body in different individuals. 1949 J. F. FULTON *Textbk. Physiol.* (ed. 16) v. 133 Electromyography, as a convenient method for detection and measurement of neuromuscular disorders, can be used only on the basis of deviation from the known normal patterns of muscular activity. 1950 *Electronic Engin.* XXIV. 43/1 The measurement of phenomena associated with the nervous, cardio-vascular and respiratory systems. This includes..electromyography (E.M.G.). 1951 *Sci. News* XXII. 7 Electrical potentials can be recorded from contracting muscles... This technique is called electromyography. *Ibid.*, The great advantage of the electromyographic method..is that the activity of the muscles can be studied in the intact individual, practically without any interference whatsoever with the normal modes of posture or movement. 1956 *Nature* 18 Feb. 340/2 Electromyography of the internal laryngeal muscles has been applied in the diagnosis of laryngeal palsy. 1949 POLATIN & PHILTINE *How Psychiatry Helps* (1951) vi. 168 Electronarcosis..is a new form of shock therapy for mental diseases. *Ibid.*, The term electronarcosis means, literally, coma or sleep induced by electrical current. *Ibid.* 169 Results of Electronarcosis Therapy. 1961 *Times* 17 Mar. 19/6 Electrical anaesthesia, or electronarcosis. 1952 *Brit. Jrnl. Psychol.* May 92 Bernberg showed that lengthy electronarcotic treatment of young rats..impedes their later learning. 1813 BERZELIUS in *Ann. Philos.* II. 453 If..an acid or electro-negative oxide is neutralized by a quantity of base or electro-positive oxide. 1834 FARADAY in *Phil. Trans.* CXXIV. 79 Substances are frequently spoken of as being *electro-negative*, or *electro-positive*, according as they go under the supposed influence of a direct attraction to the positive or negative pole. 1881 Electronegative [in Dict.]. 1968 *New Scientist* 30 May 460/1 Oxygen is one of the most electronegative of elements, having a greater 'acquisitiveness' for electrons than most other elements. 1905 W. SIMON *Man. Chem.* (ed. 8) II. viii. 94 Each element in the table below behaves electro-negatively toward those following, and electro-positively toward those preceding. 1926 *Jrnl. Amer. Chem. Soc.* XLVIII. 3130 A study of the electronegativity of organic radicals is of much importance from the standpoint of the electronic conception of valence. 1947 CARMICHAEL & DEARBORN *Reading & Visual Fatigue* (1948) viii. 266 The term electro-oculogram..is used to apply to the records traced in ink upon the recording paper... These records are made..by an ink-writing oscillograph which is actuated by amplified currents originating in electrodes placed about the orbits of the eyes. 1951 *Arch. Ophthalmology* XLV. 184 Many optimistic predictions or suggestions have been made for the future of electro-oculography in visual research. *Ibid.* 185 There remains much that could be investigated using the electro-oculograph as an instrument of research. 1955 *Brit. Jrnl. Ophthalmology* XXXIX. 398 This recorded electro-

oculographic response (EOG) depends on the angle and speed at which the globe rotates. 1958 *Amer. Jrnl. Ophthalmology* XLV. 158/2 In 25 subjects with unilateral amblyopia there was uncertain fixation which is.. electrooculographically demonstrable. 1879 *Telegr. Jrnl.* VII. 306/2 (*title*) Electro-optic properties of liquids. 1906 *Westm. Gaz.* 20 Sept. 2/1 Lorentz made an electro-optical model which led Zeeman to recognise the action of a magnetic field on the lines of the spectrum. 1963 G. TROUP *Masers & Lasers* (ed. 2) ix. 162 Optical modulation techniques, using electro- and magneto-optical methods. 1906 *Jrnl. Soc. Chem. Industry* 15 Feb. 127/2 (*heading*) Water and juices from mineral, animal and vegetable substances; continuous process for extracting — by electro-osmosis. 1944 S. FIELD *Princ. Electro-deposition* xx. 279 This motion of the dispersion medium is known as electro-osmosis. 1962 *New Scientist* 15 Mar. 630 Electro-osmosis..is the flow of liquid through the pores in a membrane under the influence of an electric field. 1907 *Chem. Abstr.* I. 2206 The solid constituents remain as separate as in mixtures such as sand, clay, and lime, the electrosmotic components separating readily while the non-electrosmotic components remain in the liquid. 1946 *Nature* 10 Aug. 205/1 The variations of speed [of colloidal particles]..are due to electro-osmotic currents. 1969 *Times* 30 July 10/6 The electrical method of damp-proofing is known as electro-osmotic damp-proofing. 1907 *Chem. Abstr.* I. 2651 Any substances that do not wander electro-osmotically remain in the liquid. 1942 S. GLASSTONE *Introd. Electrochem.* xvi. 527 The volume *v* of liquid transported electro-osmotically per second. 1941 E. D. HUGHES in *Trans. Faraday Soc.* XXXVII. 605 The substituting agent uses its electrons to attack the nucleus..or ..its nucleus to combine with the electrons... The latter [type of substituting agent is termed] 'electrophilic', and we name the substitution processes accordingly. 1964 J. W. LINNETT *Electronic Struct. Molecules* vi. 92 The most common type of reagent that attacks an aromatic system is an electrophilic one. 1959 HALAS & MANVELL *Technique Film Animation* 338 *Electro-photographic transfer*, the photographic recording of pictures projected electrically, as through the cathode ray tube of television. 1967 E. CHAMBERS *Photolitho-Offset* xiv. 205 The electro-photographic system..is described. 1894 E. L. WILSON *Cycl. Photogr.* 138/1 Electro-Photography, the production of photographs by means of electricity, or electric light. 1944 *Electronic Engin.* XVII. 145/2 (*title*) Electro-photography..a new system of producing photographic prints by electronic action. 1948 *Science* 29 Oct. 482/1 (*heading*) Electrophrenic Respiration... A new type of artificial respiration has been developed which uses electrical stimulation of one or both phrenic nerves. 1939 *Brit. Jrnl. Psychol.* Jan. 260 Hartline's electrophysiological studies of nerve impulses in the vertebrate eye. 1962 *Oxf. Univ. Gaz.* 9 Mar. 786/1 An electro-physiological study of the way in which the bee discriminates the visual patterns which are known to be of importance in its normal life. 1889 *Cent. Dict.*, Electro-physiologist. 1948 *Mind* LVII. 245 In the book before us [by Adrian] we have renewed proof of the debt owed by the modern student of nerve and brain to the electro-physiologist. 1838 *Ann. Electr., Magn., & Chem.* II. 228 Mr. Leithead's late work on electro-physiology. 1945 *Electronic Engin.* XVII. 564 In electrophysiology it is becoming increasingly important for workers to familiarise themselves with quantitative procedures such as wave analysis. 1950 *Jrnl. Mental Sci.* XCVI. 514 (*title*) 'Regressive' electroplexy in schizophrenics. *Ibid.* 520 In the light of our experiences we have discontinued the use of 'regressive' electroplexy. 1958 *Sunday Times* 6 July 10/5 Before the days of Electroplexy (electrical shock treatment). 1965 J. POLLITT *Depression & its Treatment* iv. 50 Electro-convulsive treatment (E.C.T.) also referred to as electro-shock treatment (E.S.T.) or electroplexy is still the most rapid and effective treatment. 1899 *Daily News* 1 July 4/5 This Westinghouse electro-pneumatic signalling plant. 1933 *Discovery* Feb. 65/2 A novel electro-pneumatic system of synchronizing the changing of the two gears has been evolved in the latest models with outstanding success. 1956 *Electronic Engin.* XXVIII. 484/2 The losses produced by electropolished copper wires left out-of-doors for several months increased by 40 per cent. 1813, 1834 Electro-positive [see **electro-negative*]. 1850 Electro-positive [in Dict.]. 1967 C. W. DAVIES *Electrochemistry* xii. 139 Metals like cadmium, zinc or sodium, which tend to liberate hydrogen from acids, are often called electro-positive... The term 'electro-positive' implies that the metal has a relatively strong tendency to ionize as a positive ion. 1905 Electro-positively [see **electro-negatively*]. 1955 *Chem. Abstr.* 7303 (*heading*) Electropositivities of iodine. 1927 *Chem. Abstr.* 57 (*heading*) The preparation of amino acids by electroreduction of oximino esters. 1946 *Nature* 16 Nov. 706/1 The irreversible electro-reduction of each compound involves two electrons per molecule. 1936 *Amer. Jrnl. Physiology* CXVII. 338 Certain visual phenomena can be recorded not only from the eye-ball itself in the usual electro-retinogram, but from other stations in the optic pathway such as the optic nerve. 1951 *Electronic Engin.* May 170/1 An electro-retinogram is the voltage change generated by the retina in response to a flash of light falling on the eye. *Ibid.*, The amplifier.. has been used..for recording human and animal electro-retinograms. 1964 J. Z. YOUNG *Model of Brain* vii. 118 The electro-retinogram produced by a moving spot of light changes with the direction of movement. 1948 *Brit. Jrnl. Psychol.* Dec. 80 Experiments on the electro-retinography of various animals. 1944 *Dorland's Med. Dict.* 486/2 *Electroshock*, shock treatment by applying electric current to the brain. 1955 *Sci. News Let.* 29 Jan. 73/1 Electroshock treatment of the mentally ill can be 'softened'..by a new drug, succinyl-choline chloride. 1879 F. W. ROBINSON *Coward Conscience* II. II. xxi. 160 Dish-covers..radiant with electro-silver. 1884 R. JEFFERIES *Dewy Morn* I. xxv. 282 For all their silver and electro 'the family' went to London. 1959 *Metalworking Production* 29 May 950 (*heading*) Electro-slag welding comes to UK. *Ibid.* 950/1 A molten pool of highly-ionized and consequently electrically conductive slag which transfers the heat instead of the arc itself—therein lies the secret of the 'electro-slag' welding technique. 1966 *New Scientist* 7 July 6/2 Electro-slag refining ..is what the steelmaker calls a secondary refining process—one, that is, which refines an ingot of steel to higher standard of purity and structural perfection. 1953 *Confinia Neurologica* XIII. 300 (*title*) Physiodynamic differentiation with non-convulsive electrostimulation. *Ibid.* 301 My preference as to order of therapeutic priority based on safety, psychodynamic productivity and economy is as follows: carbon dioxide therapy has first priority unless there are specific contraindications, then non-convulsive electrostimulation (E.S., nc.), then convulsive electrostimulation (E.C.T.). 1969 *Nature* 9 Aug. 645/2 It is difficult to compare precisely the various forces produced by electrostimulation with those obtained when the subject volitionally tried to isolate and drive the same muscle groups. 1909 *Cent. Dict.* Suppl., Electrostriction. 1942 E. B. UVAROV *Dict. Sci.* 66 *Electrostriction*, state caused in a solution of an electrolyte by forces of attraction and repulsion between charged ions, giving rise to decreased mobility. 1956 *Electronic Engin.* XXVIII. 132/3 The distinction between piezostrictive (linear) and electrostrictive (square-law, etc.) response to applied field. 1957 *Sci. News* XLIV. 98 For ultrasonic vibrations magnetostrictive, piezo-electric, and electrostrictive vibration generators are used. These convert electrical energy into mechanical energy and vice versa. 1898 H. W. F. LORENZ tr. W. Löb (*title*) Electrolysis and electrosynthesis of organic compounds. 1965 *New Scientist* 23 Dec. 877/1 The application of large scale potentiostats in..electrosynthesis and electrorefining,—particularly in organic and molten salt solvents. 1900 R. PEARL in *Amer. Jrnl. Physiol.* IV. 96 Nearly all of the work on electrotaxis among the Protozoa has been done either to determine the sense of reaction..or to analyze the effect of the current on protoplasm. *Ibid.* 122 He [*sc.* Carlgren] does not attempt to make the cataphoric action the cause of all electrotactic phenomena. 1968 *Biol. Abstr.* XLIX. 2347/2 The collection of the active zoospore by electrotaxis was not successful. 1908 *Practitioner* Oct. 609 The other sections, namely, Dermatological, Electro-therapeutical, Epidemiological. 1909 *Ibid.* Dec. 870 The Electro-therapeutical Department, West London Hospital. 1868 W. JAMES in *N. Amer. Rev.* CVII. 326 The 'laws' of nerve and muscle electricity..have not..furnished a single hint even towards electro-therapeutics. 1899 *Westm. Gaz.* 19 July 10/1 Oculists, and electro-therapeutists. 1928 *Sunday Dispatch* 23 Dec. 5/6 Even Grosvenor-street having a dressmaker and an electro-therapeutist. 1928 *Daily Express* 16 May 7 An electro-therapist. 1884 W. THOMSON in *Pract. Applications Electr.* 174 Joule's electro-thermal method. 1902 *Encycl. Brit.* XXVIII. 123/2 In 1885 the brothers Cowles patented a process for the electrothermal reduction of oxidized ores. 1961 *Engineering* 10 Nov. 611 A new non-explosive process for breaking rock has been studied by the United States General Electric Company. The process is known as electrothermal forcing. 1899 *Engineering Mag.* June 495/1 Electrothermic methods have been in use for a number of years in connection with the production of aluminium, sodium, [etc.]. 1959 *Jrnl. Iron & Steel Inst.* CXCI. 197/2 Results of electro-thermic smelting tests carried out on these ores. 1889 *Cent. Dict.*, Electrotin v. 1937 *Jrnl. Iron & Steel Inst.* CXXXVI. 27 A, It appeared that in the acid bath, speeds of electro-tinning comparable with those used for hot-tinning are feasible. 1942 *Iron & Steel Engineer* Feb. 90/1 The importance of the matter of electro-tin plating of steel strip stems from the magnitude of the tin plate industry. 1944 *Jrnl. Iron & Steel Inst.* CXLIX. 152 P, It was understood from results of American experience in the use of electro-tinned strip that steel electroplated with 6–8 oz. of tin per basis box..could be made into cans..without difficulty. 1945 HOARE & HEDGES *Tinplate* xii. 256 Electro-tinplate is a new material, the full scope of which is as yet untried. [1898 H. C. PORTER tr. *Strasburger's Text-bk. Bot.* 263 The existence of electropism in plants shows clearly that an irritability may be present, from which no direct benefit is ordinarily derived.] 1907 R. J. H. GIBSON tr. *Jost's Lect. Plant Physiol.* xxxi. 481 Hegler..established negative electrotropism in *Phycomyces*. 1965 BELL & COOMBE tr. *Strasburger's Textbk. Bot.* III. 374 Roots and shoots behave inversely.., roots bending towards the side which becomes positively charged, and shoots away from it (electrotropism). 1921 I. LANGMUIR in *Science* LIV. 61/1 We shall see that the positive and negative valence differ from one another fundamentally only in algebraic sign, so that we shall find it convenient to include both positive and negative valence under the term *electrovalence*, which we may designate by the symbol v_e. *Ibid.* 61/2 The *electrovalence* of an atom in any compound may thus be defined as the number of electrons which the neutral atom must give up in forming that compound. 1923 *Jrnl. Chem. Soc.* CXXIII. 822 Whilst a single bond may be either a covalency or an electrovalency, a double bond in organic chemistry usually reacts as if it contained one covalency and one electrovalency. 1927 N. V. SIDGWICK *Electronic Theory Valency* 84 Covalencies, unlike electrovalencies, are directed forces. *Ibid.* 83 The electronic structure which we assign to an atom in a molecule depends on whether we regard its attachments to other atoms as electrovalent or covalent. 1964 N. G. CLARK *Mod. Org. Chem.* ii. 23 Atoms bound by this type of valency must remain in close proximity..and there is no question of their leading separate existences, as do the ions in electrovalency. 1947 *Nature* 18 Jan. 100/2 The long-chain molecules of which collagen fibres are composed are mainly held together by lateral bondings of an electrovalent character. 1950 *Sci. News* XV. 58 The whole thing is held together by the fact that it is made up of mutually attracting particles..the ionic or electrovalent link, as it is called. 1936 *Chem. Abstr.* XXX. 6621 (*heading*) The theory of electroviscosity. 1941 *Jrnl. Phys. Chem.* XLV. 953 The electroviscosity accounts for 59 per cent of the specific viscosity in a 1 per cent calcium caseinate system. 1935 *Chem. Abstr.* XXIX. 4236 (*heading*) The effect of an electric field on the viscosity of colloidal solutions. I. The external electroviscous effect. 1967 *New Scientist* 30 Mar. 666/1 (*caption*) Apparent viscosity of a typical electroviscous dispersion varies directly with the strength of the applied electric field, and inversely with the shear rate. 1934 WEBSTER, Electrowin, v. 1940 *Chambers's Techn. Dict.* 291/1 Electro-winning. 1961 *Times* 13 Apr. 2/4 The fields of mining..electrowinning and refining. 1967 *New Scientist* 5 Oct. 21/1 If we extend the generally understood meaning of the term 'electro-winning' to include the electrochemical recovery not only of metals, but of all elemental substances from aluminium to chlorine. *Ibid.* 23/1 Aqueous electrowinning processes mostly take place in acid electrolytes.

electrocute (ĭle·ktrŏkiūt), *v.* [f. ELECTRO-, after EXECUTE *v.*] *trans.* To put to death by means of a powerful electric current.

1889 *Voice* (N.Y.) 1 Aug., Kemmler, the murderer sentenced to be 'electrocuted'. 1890 *Boston* (Mass.) *Jrnl.* 3 May 5/6 The important thing to consider is that the State has a large number of murderers which it can neither hang nor 'electrocute'—as the new phrase hath it. 1890 *Congress. Rec.* 8375/1 That the gentleman.. should be 'electrocuted' by the Kemmler process recently adopted in the state of New York. 1920 A. HUXLEY *Limbo* x. 84 It was as though he were about to be electrocuted. 1970 *Capital Times* (Madison, Wis.) 16 Sept. 27/4 In 1890 the court cleared the way for William Kemmler of Buffalo, N.Y. to be the first person to be electrocuted in the United States.

b. *transf.* To kill in any way by electricity.

1909 *Yorkshire Post* 4 Aug. 4/5 [A boy] who was electrocuted on the Mersey Railway last Saturday. 1913 *Daily Mail* 13 Jan. 3/2 The horse..was struck by the wire and instantly electrocuted. 1970 *N.Y. Times* 18 Sept. 24/4 A girl electrocuted herself..when she got into a bath wearing electric hair curlers.

electrocution (ĭlektrŏkiū·ʃən). [f. prec.: see -TION.] Execution by electricity.

1890 *Evening News* 6 May 4/1 The Supreme Court of the United States has refused to grant a writ of habeas corpus in the case of the wife murderer under sentence of electrocution. 1890 *Columbus* (Ohio) *Dispatch* 4 Aug., Buffalo parties invited to witness the Kemmler electrocution. 1902 *Encycl. Brit.* XXVI. 579/1 Sentence of death is executed by hanging by the neck except in New York and Ohio, where it is carried out by 'electrocution', or by passing through the body of the convict a current of electricity of sufficient intensity to cause death and until death is caused.

b. Death caused in any way by electricity.

1940 in *Chambers's Techn. Dict.* 1948 *Nat. Geogr. Mag.* Mar. 395/1 A few whiffs of gas or chloroform are sometimes used, but most ranchers prefer electrocution.

electrode. For def. read: A conductor by means of which electricity enters or leaves an electrolyte, gas, or other medium, or a vacuum. **b.** *spec.* A welding-rod, a filler-rod.

1930 *Engineering* 14 Mar. 353/2 About 75 per cent. of his work was with bare wire electrodes. 1940 *Chambers's Techn. Dict.* 194/1 *Continuous electrode*, a type of carbon electrode used in electric furnaces; the electrode is gradually fed forward as the lower part burns away, and the upper part is renewed by adding fresh material. 1958 *Engineering* 11 Apr. 478/2 Automatic machines utilising..continuous electrodes were employed for the automatic welding of the vessel.

2. *attrib.* and *Comb.*, as *electrode efficiency, holder, material*; **electrode potential,** the difference of potential between an electrode and the electrolyte with which it is in contact (*Gloss. Terms Electr. Engin., B.S.I.*, 1943).

1940 *Chambers's Techn. Dict.* 287/1 *Electrode efficiency*, the ratio of the quantity of metal deposited in an electrolytic cell to the quantity which should theoretically be deposited according to Faraday's laws. 1944 S. FIELD *Princ. Electrodeposition* iii. 26 Another convenient term is 'electrode efficiency', representing the relative efficiencies at anode and cathode. 1909 *Cent. Dict. Suppl.* s.v. *Arc*, (*caption*) Electrode-holder. 1936 *Discovery* July 221/1 A ferruled electrode holder of the clamp type. 1940 *Chambers's Techn. Dict.* 287/1 *Electrode-holder*, in electric arc-welding, a device used for holding the electrode and leading the current to it. 1902 *Encycl. Brit.* XXVIII. 123/2 Ordinarily carbon is used as the electrode material. 1907 *Trans. Faraday Soc.* July 70 Electrode potentials in liquid ammonia. 1931 *Jrnl. Iron & Steel Inst.* CXXIII. 742 The electrode potentials of pearlitic steels of various carbon contents..were also measured. 1944 S. FIELD *Princ. Electrodeposition* xiii. 167 Two of these electrode potentials, or 'single' potentials as they are sometimes called, contribute to the e.m.f. of the Daniell cell.

Hence **ele·ctrodeless** *a.*, destitute of electrodes.

1893 *Athenæum* 13 May 609/2 Mr. E. C. Rimington read a paper 'On Luminous Discharges in Electrodeless Vacuum Tubes'. 1899 *Daily News* 29 June 6/5 The electrode-less or ring discharge in various gases. 1966 *McGraw-Hill Encycl. Sci. & Technol.* XII. 589/1 A simple and extremely useful light source has been developed since 1946...It consists of an electrodeless glass or quartz tube containing a minute amount of volatile metal or compound plus a trace of argon gas excited in alternating electromagnetic fields of ultra-high frequency.

electrodynamic, *a.* Add: (Earlier example.) Also, of a loudspeaker, microphone, etc.: = **moving-coil* (*loudspeaker, microphone*). Also, **electrodynamical** (example); **electrodynamics** (earlier example).

1827 H. DAVY in *Phil. Mag.* I. 197 What may be called the electro-dynamic relations. 1827 J. CUMMING *Man. Electro Dynamics* p. vi, A new science..treating of the effects of electricity in motion through a continued system of conductors, has been termed Electro-dynamics. 1907 *Nature* 11 Apr. 554/1 In spite of the fact that the word 'electrodynamical' has gone out of fashion, and that it is more proper nowadays to say 'electromagnetic', the old word is here retained. 1934 *Discovery* Oct. 301/2 Electro-dynamic sound producer. 1940 *Chambers's Techn. Dict.* 287/1 Electrodynamic loudspeaker.. electrodynamic microphone.

electrolier. (Later examples.)

1923 A. HUXLEY *Antic Hay* xiii. 192 The convex reflections of the electroliers. 1942 *Times* 21 Mar. 2/4 'Look at all these lights!' he exclaimed, with a motion of the arm towards the numerous ornate electroliers with their dazzling brilliance.

electrology. Add: (Example.) Also occasionally used in other senses (see quots. 1896, 1940, and 1970).

1840 W. WHEWELL *Philos. Induct. Sci.* I. p. lxxii, Electricity is in the same condition as heat; having only one word to express the property and the science. M. Le Comte proposes Electrology: for the same reason as before, I should conceive Electrics more agreeable to analogy. 1896 *Daily News* 13 Feb. 2/1 The newly-elected president [*sc.* Capt. Abney] avowed his preference for 'electrology' as against 'photography'. 1940 *Chambers's Techn. Dict.* 290/2 *Electro-therapy*.., the treatment of diseases by electric currents, or by electrically produced radiations. Also called *electrology*. 1970 *Daily Hampshire Gaz.* (Northampton, Mass.) 1 Oct. 2/1 (Advt.), Electrology. Permanent Hair Removal.

electrolysis. Add: **1.** (Earlier example.)

1834 W. WHEWELL *Let.* 5 May (1876) II. 182 If you take anode and cathode, I would propose for the two elements resulting from electrolysis the terms anion and cation.

2. b. (See quots.)

1909 *Daily Mirror* 4 Oct. 10/3 (Advt.), Electrolysis. Superfluous hair permanently removed; ladies only; consultations free. 1934 WEBSTER, *Electrolysis*, surgical destruction, as of hair roots..with an electric current. 1959 *Woman* 5 Sept. 18/1 The only permanent way of getting rid of superfluous hair is by electrolysis.

electrolyte. For def. read: **a.** A substance which dissolves in water or another suitable medium to give a solution capable of conducting an electric current; also, such a solution. **b.** The ionized or ionizable constituents of a biological system. Add later examples.

1950 J. OSBORNE *Dental Mech.* (ed. 3) xxiii. 384 Care must be taken to maintain the electrolyte solution in its correct condition, otherwise satisfactory plating cannot be achieved. 1960 *Times* 11 Apr. 1/4 Neuro-physiology, electrolyte regulation, hormone assay. 1962 *Lancet* 6 Jan. 32/1 Awareness of impaired function of the distal tubule..alerts the surgeon to the possible need for electrolyte replacement in patients who do badly after prostatectomy. 1966 *McGraw-Hill Encycl. Sci. & Technol.* XII. 426/2 The pH of acid soils varies between 3 and 7, the reaction depending on the ion saturation and the soluble electrolyte content. 1968 *Practical Motorist* Nov. 333/1 *Electrolyte*. In a car battery, the liquid filling the cells, in which the plates are immersed. It is a mixture of sulphuric acid and distilled water.

electrolytic, *a.* Add: (Additional examples.) Also as *sb.* = electrolytic copper (shares).

1892 *Daily News* 7 Dec. 5/6 The united production of all the electrolytic copper works in the world is estimated ..at between thirty and forty thousand tons of copper every year. 1897 *Ibid.* 4 Jan. 2/6 The electrolytic refinery at Anaconda has now the capacity of treating the other half of the Company's product. 1909 WEBSTER, *Electrolytic dissociation theory*, the theory that the molecules of electrolytes in solution dissociate into ions. 1912 *Times* 19 Dec. 20/3 The level of prices in this market [is] considered favourable to buyers in comparison with those at present ruling in New York for electrolytic. 1923 U. R. EVANS *Metals & Metallic Cpds.* I. 55 Electrolytic Dissociation or Ionization: a splitting into electrically charged ions, of which the 'positive ion' consists of the most characteristic metal present, bereft of some or all of its valency electrons. 1929 *Bell System Techn. Jrnl.* VIII. 41 The aluminum electrolytic condenser..utilises the dielectric property of the film to provide electrostatic capacity. 1930 *Engineering* 11 Apr. 475/2 Precautions are necessary to ensure that only electrolytic copper is used. 1930 *Economist* 20 Sept. 538/2 Some of the smaller independent producers in America are now offering electrolytic at only 10½ c. per lb. 1931 S. R. ROGET *Dict. Electr. Terms* (ed. 2) 105/2 *Electrolytic rectifier* or *valve*, an apparatus in which alternate half waves of an alternating current are suppressed..so that a unidirectional current is produced. 1937 *Jrnl. Iron & Steel Inst.* CXXXVI. 42 A, Electrolytic polishing—really a type of etching—allows the properties of the crystal faces to be studied. 1940 *Chambers's Techn. Dict.* 287/2 *Electrolytic capacitor*, a capacitor in which the dielectric between the plates is an electrolyte, instead of the more usual solid insulating material. *Ibid.*, *Electrolytic cell*, a vessel in which electrolysis is carried out. *Ibid.*, *Electrolytic refining*, the method of producing pure metals, by making the impure metal the anode in an electrolytic cell and depositing a pure cathode. 1958 *Chambers's Techn. Dict.* Suppl. 976/1 *Electrolytic polishing*, forming an apparently polished surface on metal by making it an anode in an electrolytic bath. 1958 *Newnes Compl. Amat. Photogr.* 113 A 22.5 V. hearing aid battery is used to charge an electrolytic capacitor through the flashbulb.

electrolyzation. (Earlier example.)

1834 FARADAY in *Phil. Trans.* CXXIV. 117 When it has performed its full work of electrolyzation, it has only separated the elements of a single grain of water.

electro-magnetic, *a.* (Earlier and later examples.) Now freq. **electromagnetic** (without hyphen).

1821 FARADAY *Diary* 3 Sept. (1932) I. 49 Electro-magnetic expts. with Hare's Calorimotor. 1837 *Ann. Electr., Magn., & Chem.* I. 387 We need scarcely despair of seeing the electro-magnetic telegraph established for regular communication from one town to another. 1905 *Harmsworth Encycl.* III. 2174/3 It was shown by Clerk Maxwell that a repeated electrical or magnetic disturbance or change would set up and radiate waves of an electro-magnetic nature in the ether, and that, with certain assumptions concerning the ether, these waves would travel with the velocity of light. Upon this is based his electro-magnetic theory of light. 1908 *Chem. Abstr.* 372 Calculation of the resistance of a wire when reflecting electro-magnetic waves. 1929 E. MALLETT *Telegr. & Teleph.* xi. 281 Whenever an electric current changes in a circuit, energy in the form of electromagnetic waves is radiated away from the circuit. 1937 J. E. SHRADER *Physics for Students* lv. 539 (*caption*) The electromagnetic spectrum. 1955 *Gloss. Terms Radiology* (*B.S.I.*) 10 *Electromagnetic radiation*, radiation associated with a periodically varying electric and magnetic field. 1957 *Technology* Mar. 13/1 The common feature of these new electromagnetic pumps is that, instead of using electromagnetic forces to cause solid metal to rotate, they use these forces to drive liquid metal under pressure along a pipe. 1965 M. GARBUNY *Optical Physics* i. 2 Altogether, it is now known that the electromagnetic spectrum.. encompasses radio- and microwaves, infrared, visible, and ultraviolet light, *x* (or Roentgen-) rays, gamma, and cosmic radiation.

electromatic (ĭle:ktrŏmæ·tik), *a.* [f. ELECTRO- + AUTO)MATIC *a.*] Of, pertaining to, or designating electrical equipment operated automatically, particularly road traffic light signals.

Electromatic was registered as a trade mark for a particular kind of telephone by Messrs. Gent & Co. Ltd., Leicester, in 1912; and (in the form *Electro-Matic*) for 'street traffic visual signalling apparatus' by the Automatic Electric Company Ltd., Liverpool, in 1932.

1933 *Boys' Mag.* XLVII. 170/1 The greatest advance [in traffic signals]..has been the introduction of the 'Vehicle Operated' or Electromatic System. 1937 *Times* 13 Apr. p. viii/1 With the wide adoption of electromatic vehicle-actuated traffic lights of recent years, complaints of unnecessary delay at signal-controlled crossings have been much dispelled. 1960 *Times* 13 Jan. 17/7 Electromatic washing machines.

electromeric (ĭle:ktrŏme·rik), *a.* [f. ELECTRO- + *-meric* as in ISOMERIC *a.*] Of, pertaining to, or characterized by a displacement of electrons in a molecule postulated to occur during some chemical reactions. Also **ele·ctromer,** one of two or more hypothetical forms of a compound that differ only in the distribution of the electrons between the constituent atoms.

1923 G. N. LEWIS *Valence* xi. 132 We no longer..see any justification for the assumption that nitrogen trichloride must be either $N^{+++}Cl_3^-$, or $N^{---}Cl_3^+$. Such a pair of hypothetical substances have become known in the modern dualistic theory as electromers. *Ibid.*, It is hardly to be expected that we shall be able to segregate and isolate the several species of molecules so as to obtain electromeric *substances*. 1941 *Nature* 28 June 813/2 The effect depends on the possibility of the electromeric displacement ..in the transition state. 1964 N. G. CLARK *Mod. Org. Chem.* xxv. 529 This temporary electronic displacement is known as the electromeric effect, and it operates in the same sense as the mesomeric effect.

electrometer. (Later examples.) Cf. also *BINANT.

1946 *Electronic Engin.* XVIII. 277/3 The two control grids are both highly insulated in accordance with electrometer valve requirements. 1958 *Chambers's Techn. Dict.* Suppl. 976/1 *Electrometer tube* or *valve*, one with high grid insulation, specially designed for measuring minute currents or voltages.

electrometrical *a.*, **electrometry.** (Examples.)

1779 VISCT. MAHON *Princ. Electr.* I. 8 The electrometrical Balls may thereby be made to divaricate. 1814 G. J. SINGER *Elem. Electr. & Electro-Chem.* III. iv. 312 Some difficulty attends the demonstration of the electrometrical effects. 1868 *Brit. Assoc. Rep.* 1867 489 The first step towards accurate electrometry in every case is to deduce from the scale-readings numbers which shall be in simple proportion to the difference of potentials to be determined. 1956 *Nature* 14 Jan. 57/2 Aviation medicine research depends on a detailed application of electrometrical recording techniques.

electromobile (ĭle:ktrŏmobī·l). (Disused.) [f. ELECTRO- after *automobile*.] A motor vehicle driven by electricity. Also *attrib.*

1899 *Automotor* Nov., Accumulators for Electromobiles. 1900 G. ILES *Flame, Electr. & Camera* 148 An electromobile cab. 1906 (*title*) Steam car and electromobile review. 1907 *Daily Chron.* 12 Nov. 7/4 The new electromobile garage, the largest in the world.

electromotive, *a.* Add: (Earlier example of *electromotive force*.) Also, *electromotive series* (see quot. 1940).

1827 J. CUMMING *Man. Electro Dynamics* i. 1 To the power which certain agents possess, of separating the two electricities, Volta has given the name of electromotive force. 1889 *Cent. Dict.*, Electromotive series. 1940 *Chambers's Techn. Dict.* 288/2 *Electromotive series*, the chemical elements arranged in order of their standard electrode potentials.

electron² (ĭle·ktrǫn). [f. ELECTR(IC *a.* + *-on* as in *anion, cation, ion*.] A stable elementary

particle which has an indivisible charge of negative electricity, is a constituent of all atoms, and is the carrier of electric current in solids; also, the anti-particle of this, having a positive instead of a negative charge (see *POSITRON). Orig. the name of the magnitude of the electronic charge.

1891 G. J. STONEY in *Trans. R. Dublin Soc.* 2nd Ser. IV. 583 A charge of this amount is associated in the chemical atom with each bond...These charges, which it will be convenient to call *electrons*, cannot be removed from the atom; but they become disguised when atoms chemically unite. 1902 *Fortn. Rev.* June 1023 Larmor's hypothesis of electrons, which supposes a kind of an electric atom, a charge not associated with ordinary matter. 1902 *Nature* 18 Sept. 488 The conception..that the ultimate atoms of matter involve positive and negative electrons. 1902 [see *CORPUSCLE 2 c]. 1927 A. S. EDDINGTON *Stars & Atoms* 16 The electron is the lightest thing known, weighing no more than 1/1,840 of the lightest atom. 1933 *Discovery* Mar. 69/2 Experimentally there is now a case for the existence of a positive electron, and.. this particle is apparently one of several pieces which can result from impact of matter in the atmosphere with the cosmic or penetrating radiation. 1955 *Sci. News Let.* 30 Apr. 274/3 Electrons are light-weight, fundamental particles of the atom having a negative charge. 1958 [see *CORPUSCLE 2 c]. 1968 M. LIVINGSTON *Particle Physics* iv. 62 This has the effect of producing, out of energy, a pair of electrons, one with negative and one with positive charge.

b. *fig.*

1904 *Sat. Rev.* 30 Jan. 138/1 Mr. Hardy's puppets are infinitesimal—mere 'electrons', shifted hither and thither, for no reason, by some impalpable agency. 1913 *Empire Rev.* Dec. 1336 The imponderable electrons of sentiment and feeling which allow our far-away peoples and clans to cohere. 1929 D. H. LAWRENCE *Pansies* 26 The swan within vast chaos, within the electron.

2. a. Examples of *attrib.* use.

1903 *Science* 26 June 1001/2 The electron theory.. explains Ampère's idea that magnetism is due to a rotating current of electricity round each atom of iron. 1921 *Discovery* Sept. 226/2 The corresponding electron velocities are so small that it is difficult to measure them. *Ibid.*, We must recognise that wave radiation and electron radiation are in a sense mutually convertible. 1923 J. S. HUXLEY *Essays of Biologist* vii. 252 From this state there evolved one in which the various electron-systems that we call atoms first appeared. 1926 *Atlantic Monthly* Apr. 377/2 The study of electron-emission has given us.. radio broadcasting. 1926 R. W. LAWSON tr. *Hevesy & Paneth's Man. Radioactivity* iii. 42 The electron shells of the atoms. 1938 *Ibid.* (ed. 2) iv. 90 The energy levels of atoms as established by experiments on electron collisions. 1955 *Bull. Atomic Sci.* Mar. 94/3 The basic methods of operation of the synchrocyclotron, the electron synchrotron, the linear accelerator, and the proton synchrotron. 1955 H. B. G. CASIMIR in W. Pauli *Niels Bohr* 120 In many metals the electron bands may well have a complicated structure. 1956 *Nature* 10 Mar. 483/2 Straightforward hydrogen atom transfer, and electron transfer..followed by proton expulsion. 1962 *Listener* 10 May 810/1 The living muscle is made up of atomic nuclei, mostly of carbon, linked together by fixed electron bonds. 1967 *New Scientist* 21 Dec. 707/2 To understand the electron-transfer process in detail has proved to be a stiff test for the theoreticians.

b. *Comb.* **electron beam,** a beam or stream of electrons, a cathode-ray beam; **electron camera** *Television* (see quot. 1940); **electron cloud,** a cloud-like mass of electrons; **electron diffraction** [DIFFRACTION 1], the diffraction of a beam of electrons; **electron gas,** a system of free electrons; **electron gun,** a device in which electrons obtained by thermionic emission from a heated cathode are emitted as a narrow beam; **electron lens** (see quot. 1960); **electron microgram** = next; **electron micrograph,** a micrograph produced by an electron microscope; hence **electron micrography**; also *electron micrographic*; **electron microscope** [ad. G. *elektronenmikroskop* (E. Brüche 1932, in *Naturwissensch.* 15 Jan. 49)], a microscope in which the resolution and magnification of minute objects is obtained by the passage of a stream of electrons through a system of electron lenses (cf. MICROSCOPE *sb.* 1); hence **electron microscopy**; also *electron microscopic, microscopical, microscopist*; **electron multiplier** [MULTIPLIER 4], an instrument used for amplifying the intensity of a current of electrons; **electron optics** [ad. G. *elektronenoptik* (Knoll & Ruska 1932, in *Ann. d. Physik* XII. 607)], a branch of physics concerned with the influence of electric and magnetic fields on the movement of electrons (quot. 1916 represents a different sense); hence **ele:ctron-o·ptical** *a.*; **electron pair,** (*a*) = *DUPLET 2; (*b*) (see quot. 1955¹); **electron spin,** the intrinsic angular momentum of an electron, described by the quantum number *s*; the property of an electron by virtue of which it possesses this momentum; **electron telescope,** a telescope of electron-optical type in which the image is obtained by the use of infra-red rays; **electron tube,** a vacuum tube in which a current of electrons passes between electrodes; **electron volt** (abbrev. eV, e.V.), a unit of energy used in nuclear physics (see quot. 1962); **electron wave,** a (hypothetical) wave associated with the movement of an electron and held to account for its various wave-like properties; a de Broglie wave of an electron.

1927 *Physical Rev.* Dec. 705 Electron beams resulting from diffraction by a nickel crystal. 1956 *Nature* 3 Mar. 419/2 A series of remarkable electron micrographs.. obtained by..using a new fine electron-beam technique. 1958 O. R. FRISCH *Nuclear Handbk.* vii. 5 The electron beam takes the place of the current in the secondary coil. 1937 *Discovery* Nov. 329/1 The electron camera at the transmitter, with a cathode-ray tube at the receiving end. 1940 *Chambers's Techn. Dict.*, *Electron camera*, a generic term for any device which converts an optical image into a corresponding electric current directly by electronic means, without the intervention of mechanical scanning. 1926 R. W. LAWSON tr. *Hevesy & Paneth's Man. Radioactivity* vii. 62 Atomic sphere..at the disposal of the electrons...The dimensions of this electron cloud [etc.]. 1946 *Nature* 16 Nov. 717/1 The electron-cloud in the outer incomplete shells of the atom or the molecule react on the nucleus. 1927 *Physical Rev.* Dec. 705 The spacing factor concerned in electron diffraction by a nickel crystal. *Ibid.* 731 At sufficiently high voltages there is no difference between the occurrence of x-ray and electron diffraction beams. 1939 *Nature* 25 Feb. 372/2 Messrs. W. Edwards and Co. .. have recently published details of their Finch electron diffraction camera intended for industrial research. 1958 *Jrnl. Iron & Steel Inst.* CLXXXIX. 56/2 The marked thinness of the blades and prisms..resulted in electron diffraction patterns being observed. 1955 H. B. G. CASIMIR in W. Pauli *Niels Bohr* 129 There is something rather tempting in the idea that due to some sort of interaction the electron gas 'coagulates' into blocks. 1962 *Listener* 10 May 809/2 A few free electrons, drifting about hither and thither and forming an electron-gas in the metal. 1924 *Brass World* Feb. 52/1 The electron gun is not swung about, but its stream of projectiles is deflected by means of electric or magnetic forces. 1930 *Chem. Abstr.* 2044 The electron gun..consists of a hot filament near the small hole in a diaphragm and on the axis of a hollow cylinder at a pos. potential, which produces a 'visible' beam of electrons. 1953 AMOS & BIRKINSHAW *Television Engin.* I. iv. 52 The electron gun is an electrode assembly designed to produce a narrow beam of electrons. 1957 R. W. G. HUNT *Reprod. Colour* iii. 31 Three electron guns are used in the [mosaic colour television] tube and they fire from the three positions R, G and B. 1931 *Physical Rev.* 1 Aug. 585 (*heading*) Electron lenses. 1960 *Gloss. Terms Telecommunic.* (*B.S.I.*) 61 *Electron lens*, a device producing a system of electric and/or magnetic fields capable of controlling the convergence or divergence of a beam of electrons or ions. 1943 *Jrnl. Amer. Med. Assoc.* 16 Oct. 400/2 Figure 1 shows an electronmicrogram (1 × 14,500) of a sample of purified SK murine virus. 1934 *Sci. Abstr.* A. XXXVII. 203 The electron micrographs, taken with a kathode temperature of 1000° C., showed the crystalline surface structure of the nickel. 1956 *Sci. News* XXXIX. 114 Electron micrographs..showed the partial splitting of nucleic acid and protein [from tobacco mosaic virus]. 1959 *New Biol.* XXX. 61 Electron micrographs can now be made of sections of cells in much the same way as light micrographs were previously made and the result looks more familiar although larger and sharper in detail. 1944 *Jrnl. Biol. Chem.* CLVI. 598 The difference probably was related to the measurement of the images irrespective of orientation on the electron micrographic plates. 1941 *Amer. Jrnl. Path.* XVII. 576 (*heading*) The electron micrography of purified viruses. 1932 *Brit. Chem. Abstr.* A. 209/1 (*title*) Electron microscope. 1941 *Ann. Reg.* 1940 345 An important advance during the year was the construction of an electron microscope with about twenty times the resolving power of the best light microscope with oil immersion. 1955 *Bull. Atomic Sci.* Mar. 75/1 The electron microscope..permitted us to observe objects measured in hundred thousandths of a millimeter. 1968 *Times* 13 Nov. 16/1 Scanning electron microscopes.. have a much greater depth of focus than electron microscopes have. 1933 *Chem. Abstr.* XXVII. 5631 (*heading*) Electron microscopic images from secondary electrons. 1961 *Lancet* 12 Aug. 380/2 Electron-microscopic appearances in pinocytosis. 1945 *Jrnl. Appl. Physics* XVI. 731/2 Thin replicas suitable for electron microscopical studies. 1948 *Ibid.* XIX. 119 (*heading*) The training of technical electron microscopists. 1960 *New Biol.* XXXI. 30 A complicated intracellular system of double membranes..which is known to the electron-microscopists as the 'endoplasmic reticulum'. 1934 *Nature* 16 June 911/2 It seems that the osmium impregnation method can be applied..to electron microscopy. 1946 *Nature* 14 Sept. 363/2 The size and approximate shape of bacteriophage particles has been investigated by electron-microscopy. 1940 *Chambers's Techn. Dict.* 289/1 Electron multiplier. 1953 AMOS & BIRKINSHAW *Television Engin.* I. v. 107 If the return beam is directed into an electron multiplier the tube output can be increased until it is well above the noise level of conventional vision-frequency amplifiers. 1955 *Gloss. Terms Radiology* (*B.S.I.*) 22 *Electron multiplier*, a device, generally in the form of a vacuum tube..in which small electron currents are amplified in several stages by a cascade process employing secondary emission. 1933 *Sci. Abstr.* A. XXXVI. 199 A combined electric and magnetic deflection method which gives electron-optical enlargements of emitting surfaces. 1951 *Engineering* 13 July 40/2 A tube with sufficiently good electron-optical focus for high-definition pictures was made. 1916 *Chem. Abstr.* 299 Electron optics of the hydrogen molecule. 1932 *Sci. Abstr.* A. XXXV. 645 Application of geometric electron optics makes it possible to explain the behaviour of the beam in an electron tube. 1938 *Nature* (Suppl.) 20 Aug. 331/1 A system of 'electron optics' has been elaborated which shows how a beam of cathode rays issuing from a point can be reassembled into an image by passing through a localized electrostatic or magnetic field having axial symmetry. 1934 WEBSTER, *Electron pair*, a group of two electrons, as called for in certain atomic theories. If the pair acts as a bond between atoms, it is called a *sharing pair*, otherwise a *lone pair*. 1940 *Chambers's Techn. Dict.* 289/1 Electron pair (equated with *duplet*). 1955 *Gloss. Terms Radiology* (*B.S.I.*) 15 *Electron pair*, an electron and a positron arising from pair production. 1955 H. B. G. CASIMIR in W. Pauli *Niels Bohr* 131 There does not seem any possibility to reconcile the idea of electron pairs with the thermodynamics of superconductors. 1958 O. R. FRISCH *Nuclear Handbk.* viii. 12 Electron pairs can be created by other processes including: (*a*) Passage of a heavy particle through matter..(*b*) Passage of a fast electron through the field of a nucleus..(*c*) Collision between two electrons. 1964 J. W. LINNETT *Electronic Struct. Molecules* ii. 28 This difficulty of solving the Schrödinger equation for systems containing two or more electrons means that the development of the electron-pair bond has to be based, to a large extent, on empirical reasoning. [1926 *Nature* 20 Feb. 264/2 The dotted lines represent the position of the energy levels to be expected in the absence of the spin of the electron.] 1930 RUARK & UREY *Atoms, Molecules & Quanta* iii. 96 It is not surprising that the discovery of the electron spin was delayed until 1926. 1941 *Mind* L. 279 Although we cannot pretend to describe the nature of electron-spin as an activity or a condition in the external world, we may say this much about it: spins can be differentiated from one another by specifying their components in a set of mutually orthogonal planes. 1952 *Physical Rev.* 15 Nov. 951/1 We have observed conduction electron spin resonance absorption in fine particles of metallic sodium.. at a frequency of 9240 Mc/sec. 1948 *Jrnl. Brit. Interplan. Soc.* Nov. 250 The 'electron telescope' has been used to make visible stars which could not otherwise be detected. 1923 *Chem. Abstr.* 2801 The positive pole and the auxiliary positive plate in an electron tube consist of a hollow tube and a spiral. 1930 *Electronics* Apr. 25/2 The electron-tube control of the stage lighting in the.. Chicago Civic Opera House. 1932 *Iron & Steel Engineer* Mar. 124 (*title*) The application of electron tubes in the steel industry. 1930 RUTHERFORD et al. *Radiations from Radioactive Subst.* xvii. 530 In dealing with the changes of energy in atomic nuclei, it is convenient to express the energy in terms of the kinetic energy gained by an electron..in volts. The energy per atom associated with a change of mass 1 on the atomic mass scale can be shown to be 933 million electron volts. 1938 R. W. LAWSON tr. *Hevesy & Paneth's Man. Radioactivity* (ed. 2) iii. 48 Energies of less than 100 e.V. 1962 *Gloss. Terms Nuclear Sci.* (*B.S.I.*) 41 *Electronvolt*, a unit of energy (abbreviation: eV) equal to the kinetic energy acquired by an electron when accelerated through a potential difference of one volt ($1 \text{ eV} = 1{\cdot}60 \times 10^{-19}$ erg). 1927 *Proc. Nat. Acad. Sci.* XIII. 460 A single plane of atoms reflects a very appreciable fraction of the electron wave, whereas the same plane would reflect only an inappreciable part of an X-ray wave. 1930 G. P. THOMSON *Wave Mechanics of Free Electrons* ii. 36 On de Broglie's theory the space round a nucleus has a refraction index for electron waves which varies with the distance from the nucleus. 1942 J. D. STRANATHAN *Particles Mod. Physics* xiv. 540 These electron waves travel with a velocity greater than that of light.

electronic (ĭlek-, elektrǫ·nik), *a.* [f. *ELECTRON² + -IC.] **1.** Of or pertaining to an electron or electrons.

1902 J. A. FLEMING (*title*) The electronic theory of electricity. 1905 A. M. CLERKE *Mod. Cosmogonies* x. 175 An electronic theory of gravitation. 1906 J. B. BURKE *Orig. Life* xi. 191 Three states of electronic aggregation. 1922 J. MILLS *Within the Atom* xi. 147 Whenever electronic impacts give rise to radiation, the energy associated therewith is always proportional to the frequency. *Ibid.* xii. 159 Only in a highly evacuated tube would there be the possibility of large electronic orbits. 1923 *Times Lit. Suppl.* 15 Nov. 763/2 The earth's atmospheric circulation is..dependent upon the sun's electronic emissions. 1927 N. V. SIDGWICK (*title*) The electronic theory of valency. 1928 *Phil. Mag.* VI. 1254 (*title*) Electronic waves and the electron. 1937 *Discovery* July 226/2 Ideas concerning electronic energy and electron levels in atoms *in vacuo*. 1955 C. G. DARWIN in W. Pauli *Niels Bohr* 4 He [*sc.* Planck]..derived the electronic charge as $4{\cdot}69 \times 10^{-10}$ e.s.u.

2. Of or pertaining to electronics; esp. of something operated by the methods, principles, etc., of electronics. *Spec.* of music: produced electronically, without pipes, strings, etc. Also applied to devices producing such music.

1930 *Electronics* Dec. 435 Electronic Musical Instruments...Examples of such instruments are the electronic organ of M. Coupleaux of Paris. 1931 *Ibid.* July 18/1 All recent developments in electronic music in Europe have used the same principle..that of the beat note produced by the interference of two tube-oscillators. 1939 *Nature* 25 June 1119/2 Electronic organs are making us acquainted with the synthesis of musical sounds. 1951 *Times* 23 Oct. 6/3 The electronic camera claims many advantages over its optical counterpart. 1952 KOESTLER *Arrow in Blue* xxix. 271 The final conclusion was suddenly present in my mind—like the result which appears on the dial of electronic calculators. 1953 *Sci. News* XXX. 70 The problem of electronic computing is basically simple, for the fundamental requirement is electrical manipulation of information. 1959 *Observer* 23 Aug. 7/3 Electronic music..makes use of electrically generated sounds recorded on tape and then rearranged into patterns. 1962 *Gloss. Terms Autom. Data Processing* (*B.S.I.*) 7 Electronic data processing (e.d.p.) is a term widely used in the U.S.A. for data processing largely performed by electronic means. 1962 M. MCLUHAN *Gutenberg Galaxy* 275 As we experience the new electronic and organic age with ever stronger indications of its main outlines, the preceding mechanical age becomes quite intelligible.

3. Special Combs., as **electronic brain** *colloq.*, = **electronic computer* (see also quot. 1945); **electronic computer**, a computer operated electronically; **electronic flash**, a flash produced by an electrical discharge through a gas-filled tube and used for high-speed photography; **electronic heating**, dielectric heating (cf. DIELECTRIC B. 1).

1945 *Aero Products* Jan. 22 An 'electronic brain', which helps pilots test-fly new airplanes, has been invented by flight research engineers of Consolidated Vultee Aircraft Corporation. Technically, the device is known as a 'flight recorder'. 1946 *Aircraft Yearbk.* 165 Electronic 'brains' will guide the counter-missile with precision. 1958 *Listener* 11 Dec. 983/1 Automatic digital calculators, sometimes called 'electronic brains', are just over a decade old. 1946 *Electronics* Aug. 110/1 The servomechanism is part of the computer, and..computers of this type have become known as electronic computers. 1962 *Mod. Lang. Rev.* LVII. 162 An electronic computer [has] been used to index the *Summa Theologiae* of St. Thomas Aquinas. 1947 *Americana Ann.* 561 The Electronic Flash Gun, made by the Abrams Instrument Corporation. 1949 *British Birds* XLII. 240 The photographs shown.. were taken by high-speed electronic flash. 1962 *Listener* 22 Feb. 337/2 The invention of the electronic flash camera has permitted photographs to be taken at much higher speeds than ever before. 1968 *Gloss. Terms Offset Lithogr. Printing* (*B.S.I.*) 12 *Electronic flash lamp*, a discharge lamp giving a high light output for a very brief period. 1946 *Sci. News Let.* 26 Jan. 53 The drying and curing of rubber by electronic heating is six times faster than conventional processes. 1959 *Gloss. Packaging Terms* (*B.S.I.*) 82 *Electronic heating*, heating produced by dielectric loss.

electro·nically, *adv.* [f. *ELECTRONIC *a.*; see -LY *suffix*².] By electronic means.

1923 *Science* 31 Aug. 164/2 A vibrationless, but electronically excited molecule. 1943 *Times* 21 Sept. 4/5 Details of 'one of America's best kept military secrets', an electronically-controlled auto pilot for bombing aeroplanes, were disclosed here to-day. 1954 E. E. CUMMINGS *Let.* 8 Dec. (1969) 238 With a flexible (electronically) instrument I could have roared as softly as a seashell. 1954 *Economist* 13 Mar. 791/1 A machine that calculates electronically.

electronics (ilek-, elektrǫ·niks). [Plural of *ELECTRONIC *a.* used *subst.* and construed as a singular; cf. *dynamics, physics*, etc.] That branch of physics and technology which is concerned with the study and application of phenomena associated with the movement of electrons in a vacuum, a gas, a semi-conductor, etc., as in thermionic valves, X-ray tubes, etc.

1910 *Chem. Abstr.* IV. 2767 Radio activity and electronics. 1945 *Times* 15 Oct. 3/2 The committee's report said..that 'knowledge of electronics promises the ability to detonate atomic bombs at great distances by radio'. 1954 *Economist* 30 Oct. 412/2 The stream-lining of operations that..electronics engineers believe to be the main..attraction of computers. 1958 *Listener* 23 Oct. 632/1 The three industries in the American economy where employment is likely to grow the fastest are education, electronics, and chemistry.

electrophone (ĭle·ktrŏfōᵘn). [f. ELECTRO- + -PHONE.] An instrument for transmitting sounds by means of electric currents. Hence **ele·ctrophone** *v. trans.*, to transmit (a speech, etc.) by this instrument; **electropho·nic** *a.*, of or pertaining to an electrophone; also, (of music, sound) produced electronically; **electropho·nically** *adv.*

1864 *Trans. R. Scottish Soc. Arts* 185 (*title*) On the electrophone and some of its applications. By T. S. Wright. *Ibid.*, The best form for a sonorous condenser or electrophone for the production of audible sounds by.. induction currents. *Ibid.*, The electrophonic relay had a further advantage also, that the most delicate apparatus might be used for the working of the relay. 1868 *Chambers's Encycl.* X. 510/2 The electrophone may be adapted to the telephone. 1898 *Westm. Gaz.* 22 July 5/1 People who could hear Melba, Jean de Reszke, and the rest through the electrophone. 1900 *Ibid.* 13 Feb. 10/2 An electrophonically-heard sermon. 1904 *Ibid.* 20 Jan. 8/2 The rapidity with which the words spoken by Mr. Chamberlain in the Guildhall were electrophoned to Langham-place. 1921 *Glasgow Herald* 30 Mar. 9 By means of wireless and an electrophone device British troops on the Rhine are to be entertained at bi-weekly concerts. 1922 *Ibid.* 22 Dec. 9 The Radio Association anticipates a popularity for wireless telephony far exceeding that of the electrophone. 1958 *Times Lit. Suppl.* 4 July 372/5 The problems of structure, of rhythmic debility and of such future complications as electrophonic glissandi, which have been thrown up by the abandonment of key and the adoption of dodecaphony. 1960 *20th Cent.* Nov. 459 Electrophonic music has already threatened the performer of serious music.

electrophoresis (ĭle:ktro,fōrī·sis). [f. ELECTRO- + Gr. φόρησις being carried.] The migration of colloidal particles suspended in a liquid under the influence of an electric field (cf. *CATAPHORESIS). Hence **ele:ctrophore·tic** *a.*, **ele:ctrophore·tically** *adv.*

1911 *Chem. Abstr.* 641 Electrophoresis of Lampblack. 1925 *Ibid.* 1376 An examn. of the electrophoretic migration velocities. 1938 *Nature* 4 June 1000/2 The study of molecular migrations in the electric field has been facilitated by the development of an improved electrophoresis apparatus by Tiselius in Svedberg's laboratory. 1942 *Jrnl. Gen. Physiol.* XXV. 507 This material was electrophoretically homogeneous, *i.e.* gave a single sharp peak..in the electrophoretic pattern. 1944 [see *CATAPHORESIS]. 1946 *Nature* 28 Sept. 447/1 The clinical properties of the electrophoretically homogeneous fibrinogen obtained by these authors have not yet been examined. 1957 *Sci. News* XLVI. 21 When a charged particle of colloidal dimensions is suspended in a liquid between two charged electrodes, it moves towards the electrode with a charge opposite to its own...This is known as electrophoresis. 1960 *Times* 8 Apr. 2/3 The successful applicant..should have experience in chemical chromatographic and electrophoretic methods of analysis. 1962 *Lancet* 8 Dec. 1185/1 The abnormal haemoglobin was identified by paper electrophoresis. 1962 M. RABAEY in A. Pirie *Lens Metabolism Rel. Cataract* 312 It was not possible to demonstrate α-crystallin electrophoretically in embryos of less than 10 days incubation.

electro-plate, *sb.* (Earlier example.)

1844 *Penny Mag.* XIII. 417 (*title*) A day at an electro-plate factory.

ele·ctro-plated, electroplated, *ppl. a.* [f. the vb.] Coated with a metal by electrolysis.

1851 *Art-Jrnl. Illustr. Catal.* (1970) 38/2 A bridle, contributed by Mr. Penny,..a metal chaser, who has executed the whole of the ornamental work in electro-plated silver. 1861 MRS. BEETON *Bk. Househ. Managem.* 174 Where silver fish-carvers are considered too dear to be bought, good electro-plated ones answer very well. 1875 G. H. LEWES *Problems* 1st Ser. II. 135 On the bleached damask stand the silver teapot and electro-plated toast rack. 1967 *Times Rev. Industry* June 71/2 Significant inroads are being made into this market by the use of electro-plated plastics.

electroscope. (Earlier and later examples.)

1810 J. A. DE LUC in *Jrnl. Nat. Philos.*, etc. XXVII. 83 The pile produces the same effects with so small a quantity of the [electric] fluid, as to be often hardly sufficient to move the gold leaf electroscope. 1844 H. M. NOAD *Lect. Electr.* (ed. 2) 448/1 Condensing Electroscope. 1879 *Encycl. Brit.* VIII. 34/1 Advantage of this principle has been taken in the condensing electroscope of Volta, which is an ordinary gold-leaf apparatus, except that the knob is replaced by a circular disc. 1908 *Practitioner* Sept. 364 By means of the electroscope the passage of the tube can be controlled, if it should meet with any obstruction from pathological displacements of the oesophagus. 1955 *Gloss. Terms Radiology* (*B.S.I.*) 21 *Electroscope*, an instrument for indicating a potential difference by electrostatic means.

electrostatic, *a.* Add: **electrostatical** *a.*, **electrostatics** (earlier examples). Also Special Combs., as *electrostatic field* (cf. FIELD *sb.* 17, quot. 1881), *generator, induction* (cf. INDUCTION 10), *lens* (= *electron lens), *precipitation* (= *electrical precipitation), *voltmeter*; **electrostatic unit,** an electrical unit based on the C.G.S. system (1893 in *Funk's Stand. Dict.*), abbrev. **E.S.U., e.s.u.**

1827 J. CUMMING *Man. Electro Dynamics* 4 M. Ampère has proposed to apply the term electro-statics to all the effects of electricity in its state of tension. 1853 W. THOMSON in *Phil. Mag.* June 403 Results..enable us to determine the numerical relation between electro-statical and electro-magnetic units. *Ibid.*, Resistance measured as above in terms of the electro-statical unit. 1860 —— *Ibid.* Sept. 233 In..June 1853, I described a method for measuring differences of electric potential in absolute electrostatic units. 1863 *Brit. Assoc. Comm. Electr. Standards Rep.* 26 Aug. IV. 34 *Electrostatic System of Units.* This new measurement of quantity forms the foundation of a distinct system or series of units, which may be called the electrostatic units. *Ibid.* 39 Electric resistance in electrostatic units is measured by the reciprocal of an absolute velocity. 1880 *Jrnl. Soc. Telegr. Engin.* IX. 430 These examples are perhaps sufficiently elucidative of the part that electro-static induction plays during the establishment of a current in a wire. 1906 GOODCHILD & TWENEY *Technol. & Sci. Dict.* 829/2 *Electrostatic voltmeters.* These depend on the principle of the Electrometer. 1912 *Metall. & Chem. Engin.* X. 173/2 (*caption*) Cross-section through one of the electrostatic precipitation chambers. 1916 *Chem. Abstr.* 562 At present the differences between the electrostatic units and the electromagnetic units of measurement are the cause of much confusion. 1920 *Whittaker's Electr. Engin. Pocket-Bk.* (ed. 4) 651 The precipitation of dust..by means of the electrostatic field..has only been applied to industrial purposes within the last six years. 1931 S. R. ROGET *Dict. Electr. Terms* (ed. 2) 109/2 *Electrostatic Generator*, a machine which produces charges at high potentials when driven mechanically, by an electrostatic process. 1938 R. W. LAWSON tr. *Hevesy & Paneth's Man. Radioactivity* (ed. 2) I. iv. 7 Since 1 coulomb = 3×10^9 electrostatic units, it follows that the charge borne by a monovalent ion is $1{\cdot}6 \times 10^{-19} \times 3 \times 10^9 = 4{\cdot}8 \times 10^{-10}$ e.s.u. 1953 AMOS & BIRKINSHAW *Television Engin.* I. x. 237 A practical electrostatic lens of the type used in cathode-ray tubes has an optical equivalent containing several convex and concave elements. 1958 O. R. FRISCH *Nuclear Handbk.* xvi. 1. 14 In such a system of electrostatic lenses, the ideal sought is one having fixed optical properties independent of the mass number and initial energy of the ions. 1962 *Gloss. Terms Nuclear Sci.* (*B.S.I.*) 6 The electrostatic generator (first designed by Van de Graaff), in which a conveyor belt is used to charge an insulated electrode to the desired potential.

electrotechnic (ĭlektrote·knik). [f. ELECTRO- + TECHNIC B. 3.] *pl.* The technics of electricity. Hence **ele:ctro-te·chnical** *a.*; **ele:ctrotechni·cian,** one who is versed in electrotechnics.

1884 S. P. THOMPSON (*title*) Dynamo-electric machinery: a manual for students of electrotechnics. 1885 *Electrician* XVI. 3/2 The following synopsis..in electro-technics at the various high schools in Germany. 1896 *Daily News* 1 Dec. 5/5 A commercial as well as an electro-technical success. 1902 *Encycl. Brit.* XXXII. p. xxxii, In close touch with the Engineering Laboratory should be a Laboratory for Electro-Technical Industries. 1927 *Daily Express* 21 Dec. 11/7 Engineers who are to take charge of large machines, locomotive drivers and electro-technicians, where instant and accurate response to any change in conditions is essential, are tested by another machine. 1945 *Archit. Rev.* XCVII. 103 (*title*) Electro-technical industry and Highlands. 1952 *Electronic Engin.* XXIV. 382 M.K.S. units..are now the internationally agreed system for electrotechnics.

elegance. 3. a. (Later example.)

1965 *Encounter* Aug. 57/1 The elegance of the solution and the economy of the thought and work that went into it are qualities scientists give some weight to.

elegant, *a.* Add: **5. a.** (Later examples.)

1952 G. MANLEY *Climate & Brit. Scene* iii. 32 We owe to Sir Geoffrey Taylor of Cambridge one of the most elegant discussions of the factors governing the depth and density of such fogs. 1966 'A. HALL' *9th Directive* viii. 74 This operation..was sensitive and it was elegant. 1966 *Listener* 1 Sept. 304/1 MacNeish's demonstration is particularly elegant because he could trace the whole processes in a small, almost self-contained area.

8. (Earlier and later examples.)

1764 J. BLAIR *Acct. Coll. New-Jersey* 12 The college had..'an elegant hall of genteel workmanship'. 1817 M. BIRKBECK *Notes Journey America* 152 You hear of an *elegant* mill, an *elegant* orchard, an *elegant* tan-yard, &c. and familiarly of *elegant* roads, meaning such as you may pass without extreme peril. The word implies eligibility or usefulness in America, but has nothing to do with taste. 1822 J. WOODS *Engl. Prairie Illinois* 203 This negro said, some very *elegant* potatoes grew on this land last year. 1833 DICKENS *Lett.* (1965) I. 20, I..will not give you the trouble of deciphering any more of my iligant writing. 1931 O. T. BARCK *New York City* iv. 83 The most elegant of these homes were occupied by officers.

9. *elegant-looking* adj.

1799 MALTHUS *Jrnl.* 3 July (1966) 117 We had met with many pretty elegant looking women.

elegante². (Later examples of the form *élégante*.)

1922 E. SITWELL *Façade* 8 And Madame A .. the élégante, With Madame X .. the elephant. 1927 —— *Rustic Elegies* 79 Our élégantes favouring bonnets of blond. 1965 *Harper's Bazaar* Nov. 80 Gamine turns into élégante when the clock strikes six.

elegiast. Delete *rare*⁻¹ and substitute *rare*. (Later example.)

1969 *Daily Tel.* 3 July 20/2 She is the elegiast of bleakness and decay, an original artist of major attainment.

element, *sb.* Add: **4. c.** The resistance wire carrying the current in an electric heater; (also used of) the bar or collection of pieces of asbestos, etc., in an electric or gas stove.

1906 *Nature* 17 May 60/2 The method exemplified is the use of silicated carbon upon a terra-cotta base, forming an 'element'. 1925 *Jrnl. Iron & Steel Inst.* CXI. 535 The dimensions, temperature..distribution of heating elements..and various furnace types are considered. 1926 *Gloss. Terms Electr. Engin.* (*B.S.I.*) 153 *Heating element*, the complete resistor, including the element carrier on which it is wound, as used in ovens, electric fires, radiators, etc. 1952 'N. SHUTE' *Far Country* iii. 75 The girl stared at the hot elements of the fire.

5. c. *Math.* Any of the symbols or quantities which, set out in an array, constitute a determinant or matrix.

[1859 G. SALMON *Lessons Mod. Higher Algebra* i. 1 The coefficients a_1, b_1, &c., which enter into the expression of a determinant, are called the constituents of that determinant, and the products a_1b_1, &c., are called the elements of the determinant.] 1867 C. L. DODGSON *Elem. Treat. Determinants* ii. 6 If *mn* quantities be so placed as to form *m* rows and *n* columns: they are said to form a Block; and the *mn* quantities are called the Elements of such a Block. 1881 [see CONSTITUENT *a.* and *sb.* B. 4]. 1939 A. C. AITKEN *Determinants & Matrices* i. 4 A matrix may possibly consist of a single row, or of a single column, of elements. 1968 P. A. P. MORAN *Introd. Probability Theory* iii. 109 $p(t)$ is a column vector whose elements are $p_1(t), \ldots, p_n(t)$.

d. *Math.* and *Logic.* [cf. G. *element* in same sense (G. Cantor in *Math. Ann.* (1882) XX. 114, (1883) XXI. 587, (1895) XLVI. 481, etc.).] Any of the (real or conceptual) entities of which a set is composed; an entity that satisfies the criterion or criteria used to define a set.

1901 L. E. DICKSON *Linear Groups* i. 5 A set of *s* distinct elements satisfying the above four conditions is said to form a field of order *s*. 1953 A. A. FRAENKEL *Abstract Set Theory* i. 22 In fact '*x* is red' corresponds to the relation '*x* is an element of the set of all red things'. 1965 SHIH-CHÊN HU *Elem. Mod. Algebra* v. 119 The only nilpotent element in an integral domain is the zero element 0.

element, *v.* Restrict †*Obs.* to senses 1 and 3 in Dict. and add later U.S. example of sense 2.

1905 F. GREENSLET *J. R. Lowell* i. 2 When we endeavor to add to our portrait of his personality some analysis of the things that elemented it.

elemental, *a.* Add: **B.** *sb.* An entity or a force which is regarded by occultists as capable of producing physical manifestations.

1877 H. P. BLAVATSKY *Isis* p. xxx, These elementals are the principal agents of disembodied but never visible spirits. 1894 H. NISBET *Bush Girl's Rom.* 235 So that we may not be horrified or shy aside at the sight of the strange beings and elementals that surround us. 1923 *Westm. Gaz.* 7 Apr., Did they suppose that God Almighty would permit a Pharaoh..thousands of years after his own death..to loose what spiritualists called an Elemental? 1965 *Listener* 27 May 792/2 He said gently.. 'there is an Elemental out there.'..I said, 'No, there isn't.' His reaction was..something between disappointment in me and extreme elation that I should have denied the apparition.

elementalism (elĭme·ntăliz'm). Delete *nonce-wd.* and substitute: **1.** A method or theory which divinizes the elemental powers of nature.

1863 [in Dict.]. 1882 OGILVIE Suppl., *Elementalism*, the theory which identifies the divinities of the ancients with the elemental powers. *Gladstone.*

2. A system based upon elemental forces or characters.

1921 *Glasgow Herald* 8 Jan. 4 An incompetent charlatan, performing in public under the guise of 'sublime elementalism' his worthless exercises. 1921 *Times Lit. Suppl.* 17 Feb. 106/4 To reduce two women, one the embodiment of inherited fastidiousness, the other touched with intellectual greatness, to a common level of elementalism.

3. The verbal separation into separate concepts or entities of things which cannot be separated empirically or physically, e.g. space and time, mind and body. So **elementali·stic** *a.*

1933 A. KORZYBSKI *Science & Sanity* i. 24 Meanings, and the meanings of meanings, with their inseparable affective components, give us..the *non-elementalistic* foundation on which all civilization and culture depends... The psychological mechanism is extremely simple and necessitates a breaking away from the older elementalism. But it is usually very difficult for any given individual to break away from this older elementalism, as it involves the established *s.r.*..the *semantic reaction.* 1946 S. A. HAYAKAWA in W. S. Knickerbocker *20th Cent. Eng.* 1. 50 We have been defining man in non-functional and elementalistic terms.

elementary. Add: **6.** *elementary particle*: now used in *Physics* spec. for any of a number of particles smaller than an atom (the leptons, mesons, and baryons) which are not known to be composed of any simpler particles and which are characterized by having a definite mass, a lifetime that is long compared with the interaction time, and well-defined electromagnetic properties, and are capable of an independent existence.

1934 *Chem. Abstr.* 2258 Theory of elementary particles. ..A relativistic wave equation of the Dirac type for a particle of given rest mass will allow only a spin of ½ in the usual units. 1937 F. RASETTI *Elem. Nuclear Physics* 8 If the elementary particles are to be kept together in a stably bound nucleus, we must expect energy to be released in the formation process. 1949 GAMOW & CRITCHFIELD *Theory Atomic Nucleus* i. 8 It is possible to develop the theory of the fundamental nuclear properties and the various nuclear reactions..on the basis of.. interaction-laws between the nucleons from which the composite nuclei are built. Here lies a convenient,.. though not sharply defined, boundary between nuclear physics proper, and the next, as yet rather unexplored, division of the science of matter which can be called tentatively the physics of elementary particles. 1963 K. W. FORD *World of Elementary Particles* i. 2 The elementary particles represent the deepest-lying substructure of matter to which man has been able to probe. *Ibid.* 17 The heaviest known 'elementary' particle, the xi particle..has not yet been explained as a composite of any of the lighter particles. That *all* of these particles are built up from some more primordial material remains ..a strong possibility. 1963 *Oxf. Univ. Gaz.* 9 May 1183/2 The University has established a Professorship of Elementary Particle Physics.

7. Also *elementary education, schoolmaster.*

1870 *Act* 33 & 34 *Vict.* c. 75 § 1 This Act may be cited as 'The Elementary Education Act, 1870'. 1912 W. OWEN *Let.* 2 July (1967) 148 By him sat an Elementary School-master, of Elementary Education. 1944 H. C. DENT *Education Act, 1944* 12 Under Clause 7 of the Act the category 'elementary' disappears from the English educational system, and *ipso facto* local education authorities for elementary education only disappear also.

eleonorite (e:lĭ,ǫnō°·rəit). *Min.* [ad. G. *eleonorit* (A. Nies 1880, in *Ber. d. Oberhessischen Ges. f. Natur- u. Heilkunde* XIX. 111), f. *Eleonore*, woman's name used as the name of a mine near Giessen.] A name formerly given to a supposed variety of beraunite.

1881 *Jrnl. Chem. Soc.* XL. 525 New Minerals from the Eleonora Mine in the Dünsberg near Giessen... 1. Eleonorite...This mineral occurs, together with strengite, in an isolated block of brown hæmatite. 1923 *Nature* 24 Nov. 779/1 A very well-defined specimen of the variety of beraunite known as eleonorite, found at the iron and manganese mine of Roury Glen, Glandore, Co. Cork, consists of a mass of diverging fibrous crystals of a reddish-brown colour, between walls of limonite. The fibres are elongated in the direction of the *b* axis and show very strong pleochroism.

elephant. Add: **1. d.** *to see the elephant* (U.S. slang): to see life, the world, or the sights (as of a large city); to get experience of life, to gain knowledge by experience. Also *to show* or *get a look at the elephant.* (Cf. LION *sb.* 4.)

[1835 A. B. LONGSTREET *Georgia Scenes* 10 (Th.), That's sufficient, as Tom Haynes said when he saw the elephant.] 1844 G. W. KENDALL *Narr. Santa Fé Exped.* I. 108 There is a cant expression, 'I've seen the elephant' in very common use in Texas. *a* 1847 W. T. PORTER *Quarter Race Kentucky* (1847) 87 (Th.), I axed him if he'd ever seen the elephant. 1849 N. KINGSLEY *Diary* (1914) 86 [I] went up town and saw the Elephant, and it almost baffles description. 1849 T. T. JOHNSON *Sights in Gold Region* 324 (Th.), If you think we have not shown you enough of the elephant..please to mount him and take a view for yourself. 1857 *Quinland* II. II. xviii. 126 The 'Fox and Crow' is one of the famous sights in New York. It is never missed by the countryman or the foreigner, who is searching after the 'elephant'. 1878 J. H. BEADLE *Western Wilds* iii. 45 My friend Will Wylie, who had seen the elephant in its entirety, from trunk to tail. 1906 'O. HENRY' *Four Million* 87 He makes his rounds every evening, while you and I see the elephant once a week. 1960 T. V. OLSEN *High Lawless* (1961) iii. 30 Saturdays some of the boys from the three big outfits come in to see the elephant.

e. As the emblem of the Republican party in the United States.

[1860 *Railsplitter* in Sperber & Trittschuh *Amer. Pol. Terms* (1962) 141/2 Woodcut: elephant..announcing a Lincoln demonstration. 1874 *Harper's Weekly* XVIII. 912 (*caption under an elephant*) The Republican Vote.] 1876 *Ibid.* 28 Oct. 868 (*caption*) 'The elephant walks around'—and the 'still hunt' is nearly over. 1904 *Chicago Tribune* 20 June 6 The selection..will..handicap the republican elephant in the coming race. 1952 *Economist* 12 July 89/1 It would now take some sort of a miracle for Mr. Taft to catch the Republican elephant.

f. In full *elephant-colour*: a fashion shade simulating the grey colour of the elephant. Cf. *elephant-grey* below.

1875 *All Year Round* 278/1 So admirably is elephant-colour copied. 1894 *Queen* 6 Oct. 574/1 The shade of cloth used..being known as 'Elephant'. 1923 *Daily Mail* 7 June 6 In Almond Green,..Mole, Elephant, Honey.

2. a. For def. read: A rare albino variety of elephant which is highly venerated in some Asian countries. **b.** Also, an object, scheme, etc., considered to be without use or value. Add earlier and later examples.

1607 TOPSELL *Four-f. Beasts* 208 An Indian, who had brought vp from a Foale a white Elephant. 1851 G. E. JEWSBURY *Let.* 23 July (1892) 414 His services are like so many white elephants, of which nobody can make use, and yet that drain one's gratitude, if indeed one does not feel bankrupt. 1928 GALSWORTHY *Swan Song* II. ii. 113 'You look so well in that hat, Uncle.' Soames took it off again. 'White elephant,' he said. 'Can't think what made Fleur get me the thing!' 1931 'R. CROMPTON' *William's Crowded Hours* viii. 182 Let's have it a sort of second-hand stall—what they call a White Elephant Stall. 1950 G. J. RENIER *History* III. iii. 203 The miser often has a white elephant upon which he lavishes money: a house, or an unworthy friend. 1963 *Times* 4 Mar. 11/4 The original white elephant was presumably an elephant, but he or she might well have been a poor man's mediocre racehorse. 1970 *Daily Mail* 27 Oct. 2/8 The new rail link may prove to be ideologically satisfying, but in financial terms it looks like becoming the biggest white elephant in all Africa.

10. b. *Army slang.* (See quots. 1925 and 1943.)

1917 A. G. EMPEY *From Fire Step* 152 One gun..had the exact range of our 'elephant' dug-out entrance. 1919 *War Terms* in *Athenæum* 15 Aug. 759/1 Elephant, corrugated iron shelter. Baby elephant, small corrugated iron shelter. 1919 G. K. ROSE *2/4th Oxf. & Bucks Lt. Infty.* 10 Battalion Headquarters..were comfortable enough with many 'elephant' dug-outs and half a farm-house for a mess. 1925 FRASER & GIBBONS *Soldier & Sailor Words, Elephant* (*and Baby Elephant*) *Dug-Out*: a dug-out made with semi-circular linings of heavy corrugated iron. The two names refer to the two sizes issued. 1943 HUNT & PRINGLE *Service Slang* 30 *Elephant hut*, a Nissen hut (shaped somewhat like the beast).

11 and **11. b.** *elephant bell, house*; **elephant bird,** a large fossil bird of the genus *Æpyornis*, found in Madagascar; **elephant grass,** now the usual form of *elephant's grass*; for def. substitute: various kinds of grass or grass-like plants, esp. *Pennisetum purpureum*; **elephant-grey** = *1 f; **elephant-path,** a path trodden by elephants; **elephant-rain** (see quot.); **elephant's ear** (earlier example); **elephant's foot,** (*b*) a plant belonging to the genus *Elephantopus*; **elephant('s) trunk** *rhyming slang*, drunk, also *ellipt.*

1923 D. H. LAWRENCE *Birds, Beasts & Flowers* 170 The elephant bells striking slow, tong-tong, tong-tong. 1889 *Cent. Dict.*, Elephant-bird. 1933 A. S. ROMER *Vert. Paleont.* x. 214 In Madagascar lived a number of species of 'elephant-birds'. 1969 *New Scientist* 13 Mar. 20/1 The peculiar and often bizarre forms evolved in isolation: species such as..the elephant-bird in Madagascar. 1832 W. ROXBURGH *Flora Indica* III. 566 Elephant grass... Elephants are fond of it. 1895 MRS. CROKER *Village Tales* 15 We marched two and two,..glancing askance at every bush, at every big tuft of elephant grass. 1906 *Westm. Gaz.* 28 Dec. 2/1 Long grass in Uganda, 'elephant grass', grows from fifteen feet to twenty feet high. 1968 J. W. PURSEGLOVE *Tropical Crops* II. 472 Elephant grass. *Pennisetum purpureum* Schum., is the common mulch. 1896 *Daily News* 12 Sept. 6/2 'Elephant' grey is another favourite. 1906 *Westm. Gaz.* 7 Apr. 18/2 A tailor suit of elephant-grey facecloth. 1921 G. JEKYLL *Colour Schemes Flower Garden* 2 The trunks of the Spanish Chestnuts are elephant-grey. 1922 JOYCE *Ulysses* 153 Molly had that elephantgrey dress with the braided frogs. 1895 C. J. CORNISH *Life at Zoo* 7 The Elephant and Antelope Houses. 1922 JOYCE *Ulysses* 92 A tall blackbearded figure..stumping round the corner of Elvery's elephant house. 1943 HUNT & PRINGLE *Service Slang* 30 *Elephant houses*, old forts at Dunkirk. 1853 *Edin. New Philos. Jrnl.* LV. 79 Passages made formerly by the gigantic elephant, which are well adapted for bridle-paths... The knowledge of these various elephant-paths forms the resource of the marauding Caffre. 1895 KIPLING *2nd Jungle Bk.* 215 A light spring rain—elephant-rain they call it—drove across the Jungle. 1866 LINDLEY & MOORE *Treas. Bot.*, Elephant's-ear, the common name for *Begonia*. 1789 W. AITON *Hortus Kewensis* III. 280 *Elephantopus scaber.* Rough-leav'd Elephant's foot. Nat. of the East Indies. Cult. 1695, in Chelsea Garden. 1845–50 MRS. LINCOLN *Lect. Bot.* 186 The elephant's-foot (*Elephantopus*), a low, hairy-leaved plant, with purple, ligulate florets. 1901 C. T. MOHR *Plant Life Alabama* 759 Carolina Elephant's-Foot..[grows in] open dry woods, borders of fields, pastures. 1966 H. W. RICKETT *Wild Flowers of U.S.* I. 508 The genus *Elephantopus*—elephant's foot—is mainly tropical. 1859 HOTTEN *Dict. Slang* 143 Elephant's trunk, drunk. 1909 J. R. WARE *Passing Eng.* 123/2 *Elephant's trunk*: drunk. The phrase became incomprehensible by the dropping of the rhyming. 'Oh he's elephants' (*i.e.*, intoxicated) will, in time to come, exercise many an etymologist. 1931 *Evening Standard* 19 Aug. 10/1 He came home and he found the artful dodger elephant trunk in the bread and butter (He found the lodger drunk in the gutter).

elephantiasis. Add: **2.** *fig.* A great or undue expansion or enlargement.

1866 G. MEREDITH *Let.* 22 Dec. (1970) I. 349 You have become the victim of a kind of mental elephantiasis—you fancy all things as immensities. 1928 R. CAMPBELL *Wayzgoose* ii. 53 But ah! her Soul's dimensions to report—Elephantiasis comes far too short! 1955 *Bull. Atomic Sci.* Oct. 276/1 It is a tremendous building, saved from elephantiasis by the skill of its architects in blending classical and modern. 1969 *Daily Tel.* (Colour Suppl.) 27 June 5/3 The Monopolies Commission itself no longer acts...Even if it *is* asked to investigate, the managements later have almost complete freedom to develop corporate elephantiasis.

Elers ware (e·ləɪz wɛ̄°ɹ). Also **Elers' ware.** A kind of red stoneware made in Staffordshire about 1690–1700 by the two Elers brothers, David (1656–1742) and John Philip (1664–1738). (See also quot. 1903).

[1691 *Newcastle-under-Lyme Corp. Minutes* in *Evening Sentinel* 2 Feb. 1944, 3/4 At an Assembly in the Gildhall, 18 August, 3 William and Mary, 1691, William Burslem, Mayor..Ordered that a present be made to my Lord Cheife Justice Holt..of some of Mr. David Elers earthen ware to the vallew of three pounds or thereabouts.] 1869 LADY C. SCHREIBER *Jrnl.* (1911) I. 8 Elers ware teapot with effigies of King and Queen. *Ibid.* 32 An Elers ware coffee pot. 1870 W. CHAFFERS *Marks & Monograms on Pottery* 508 The sharply moulded ornaments on Elers' red Staffordshire ware was but a continuance of the moulded enscrollments of the stone ware of Germany. 1903 R. L. HOBSON *Catal. Eng. Pott. in Brit. Mus.* 168 The term 'Elers ware' has passed into general use as the name for all the red stoneware made in Staffordshire during the eighteenth century. 1924 RACKHAM & READ *Eng. Pott.* vi. 95 Lacework ornament of the kind associated with Elers ware.

eleusine (elĭusī·n). *Bot.* [mod.L. (J. Gaertner *De Fructibus et Seminibus Plantarum* (1788) I. 7), f. the name of the Attic town *Eleusis*, site of a temple to the corn-goddess Demeter + -INE⁴.] A member of the genus of annual tropical grasses so named, some of which are grown for grain in Africa and Asia; also called RAGI, finger millet, or birdsfoot millet.

1829 LOUDON *Encycl. Plants* 1089. 1914 H. KRAEMER *Applied & Economic Bot.* v. 467 The fruit is a grain..the seed being always firmly united with the thin pericarp (except in Sporobolus, Eleusine, etc.). 1932 *Antiquity* VI. 337 The usual method of planting [among the Nandi] is to have a little of everything in the same patch—eleusine, millet, maize, beans, and sweet potatoes. 1938 *Nature* 6 Aug. 259/1 The staple food [of the Ciga of Uganda] is eleusine, eked out by peas, beans, corn and wild greens. 1969 F. R. ALLCHIN in Ucko & Dimbleby *Domestication & Exploitation of Plants & Animals* 327 The probability that certain of the millets, notably *Eleusine* and *Pennisetum*, were first domesticated in Africa.

elevate, *v.* Add: **6. b.** *absol.*

1886 'M. GRAY' *Silence of D. Maitland* III. III. vi. 154 It was the kind of sorrow that purifies and elevates. 1967 *Listener* 5 Oct. 448/2 The first step is to brainwash the pop audience of its pirate-induced distrust of the Corporation..then, slyly and imperceptibly, *elevate*.

elevated, *ppl. a.* Add: **1.** Also (U.S.) *elevated highway, railroad, railway* (examples), *road*; so *elevated station, train.* Also *ellipt.* as *sb.* = elevated railway, etc.

1868 *N.Y. Tribune* 1 July 8/2 The remarkable work of constructing an elevated railway..was again on trial yesterday. **1868** *N.Y. Herald* 1 July 8/6 The elevated railroad through Greenwich street has 'gone up' higher than was contemplated in the charter of the company. **1868** *Comm. & Financ. Chron.* VI. 361/1 Three tiers of roads could be constructed; a basement road, a surface road, and an elevated road. **1880** *Harper's Mag.* Sept. 563 The buzzing which vibrates in the cars on the elevated railway. **1881** *Scribner's Mag.* May 159/2 The clatter and roar and grinding wail of the Elevated train. **1881** W. G. Marshall *Through Amer.* 24 The effect of the 'elevated'—the 'L', as New Yorkers generally call it—is, to my mind, anything but beautiful. **1884** *N.Y. Herald* 27 Oct. 2/2 Commodious First Flat; Rent $37; Elevated Station 66th st. **1890** *Century Mag.* Nov. 45 In those days there were no elevated roads. **1901** *Scribner's Mag.* XXIX. 454/1 In the street the Ninth Avenue Elevated train roared by..The cobble-stones on Sixth Avenue were shining under the Elevated. **1906** 'O. Henry' *Four Million* 8 Standing under a gas-light and looking over the elevated road at the moon. **1945** W. Maxwell *Folded Leaf* 257 They rode on the elevated railway. **1947** *Harper's Mag.* May 453/2 The remaining eighty per cent [of the population] cling to the greasy straps of the antiquated, unsanitary, dilapidated, and dangerous Elevated. **1963** *P.M.L.A.* Dec. p. vii/2 (*list*) U.K. flyover: U.S. elevated highway.

6. b. (Earlier and later examples.)

1827 J. Wight *More Mornings at Bow St.* 51 A *leetle* elevated in liquor. **1868** C. M. Yonge *Founded on Paper* xviii. 211 Though the landlord declared him to have been fairly sober, he was what one awe-stricken youth termed 'a little elevated'.

elevation. Add: **1. c.** *Ballet.* A dancer's leap or jump (steps of elevation) off the ground; the point attained in such a leap; in modern dance, an act or the action of tightening the muscles and improving the general lift of the body in a dancer's stance. (Also in Fr. form *élévation.*)

1830 R. Barton tr. *Blasis's Code of Terpsichore* ii. vi. 77 In all your high caperings, develope a manly vigour, and let your *steps of elevation* be agreeably contrasted, by the rapidity of your *terre-à-terre* steps. **1880** G. B. Shaw *London Mus. 1888–89* (1937) 23 The *entrechats, battements, ronds de jambes, arabesques, élévations,* that are the stock-in-trade of the art of theatrical dancing. **1934** A. Haskell *Balletomania* xi. 223 She has gained an elevation that allows her to do thirty-six consecutive *entrechats six!* **1948** *Ballet Ann.* II. 36 A magnificent soaring elevation. *Ibid.* 126 A slender dancer of exceptional elevation. **1949** Shurr & Yocum *Mod. Dance* i. 14 Elevation refers not only to 'inches off the floor'—as in jumping, leaping, and leaping—but also to the body lift. **1961** [see *A TERRE].

elevational (elivēⁱ·ʃənăl), *a.* [f. Elevation + -al.] Of or pertaining to elevation. Cf. Elevation 11.

1928 *Daily Tel.* 13 Nov. 20/6 The use of a model for an intended building was, he considered, far superior to any plan of elevational drawing. **1948** *Archit. Rev.* CIV. 238 (*caption*) The illustration shows an elevational view. **1970** H. Braun *Parish Churches* i. 16 In the same way that elevational architecture had been examined from the pictorial aspect, the plan came to be investigated from the point of view of its accommodation.

elevator. Add: **3. a.** (Earlier examples.)

1787 in *Rep. Comm. Pat.* 1848 (U.S.) (1849) 574 One of which [machines], denominated by the said Oliver Evans an elevator, is calculated by its own motion to hoist the wheat or grain from the lower floor..to the upper loft of such mill. **1791** J. Weld *Trav. N. Amer.* iii. 21 The elevators are inclosed in square wooden tubes.

d. (Earlier, later, and *attrib.* examples.)

1853 *Harper's Mag.* VII. 130/2 The introduction of a steam elevator, by which an indolent, or fatigued, or aristocratic person may be borne up to the third, fourth, or fifth floor. **1872** M. E. Holley *Betsy Bobbet* (1891) 395 She spoke up and says she, 'Here is the elevator, be carried up.' **1879** *Chicago Tribune* 8 May 8/4 As is the custom with elevator-boys—a reprehensible one it is too—the lad in charge of the elevator started it before closing the door. **1885** *Century Mag.* XXX. 579/1 With staircases and elevator-shafts which must remain open, [etc.]. **1890** *Congress. Rec.* 8123/2 On this list there are firemen, watchmen, elevator men. **1902** *Encycl. Brit.* XXVIII. 119/1 A..plunger, which carries the elevator-cage on its upper end. *Ibid.* 119/2 The walls..of the elevator shaft. **1945** *Chicago Tribune* 26 Jan. 27/1 (Advt.), Elevator operator for office building. **1947** Auden *Age of Anxiety* (1948) ii. 47 When elevators raise blondes aloft to bachelor suites.

4. *Aeronautics.* **a.** An elevating screw. **b.** The movable part of the tailplane of an aeroplane used to change the attitude of the aeroplane to its flight-path.

1871 *English Mechanic* 27 Jan. 448/2 The side elevators would not only raise but poise the car. **1910** R. Ferris *How it Flies* v. 83 The large elevator planes in front have been a distinctive feature of the Wright machine. *Ibid.* xx. 460 *Elevator,* a shorter name for the elevating planes or elevating rudder, used for directing the aeroplane upward or downward. **1911** *Reports & Mem.* (Adv. Comm. Aeronaut.) No. 59, 103 The most convenient arrangement of controls of elevator..might be investigated. **1915** G. Bacon *All about Flying* 3 In monoplanes, of course, the elevator has always been in the tail. **1934** J. A. Sinclair *Airships in Peace & War* iv. 84, I had remained inside the control car with only the elevator-man and we both left the ship [*sc.* a zeppelin]. **1969** *Listener* 1 May 596/1 An elevator bracket broke.

eleven, *a.* and *sb.* Add: **A.** *adj.* **2. c.** *eleven o'clock* (dial. and formerly in U.S.A.), *eleven hours* (Sc.), a refreshment or slight repast taken at about eleven o'clock.

1805 A. Scott *Poems* (1808) 120 At length, 'le'en hour's time brought the dame. **1808** Jamieson, *Eleven-hours,* a luncheon; so called from the time that labourers or children get their meridian. **1845** S. Judd *Margaret* ii. i. 214 Men and boys were seen going to the tavern for their eleven o'clock. **1898** 'S. Tytler' *Mrs. Carmichael's Goddesses* xv. 163, I was trying my 'eleven hours' at sawing and hammering and polishing till my 'eleven hours' came. **1900** *Eng. Dial. Dict.* II. 247/2 *Eleven hours,—o'clock* or *o'clocks.*

B. *sb.* **2.** (Earlier example.)

1743 in T. Waghorn *Cricket Scores* (1899) 30 The above match was played in the Artillery Ground between the above eleven.

C. elevenpenny, comb. form of *eleven pence,* as in **elevenpenny** bit formerly, in the U.S., a coin of the value of twelve and a half cents; = Levy *sb.*[3]; **eleven-plus,** the age (between 11 and 12) at which pupils leave primary schools; also, an examination taken at the same stage before entering one of the various types of secondary schools.

1807 C. W. Janson *Stranger in Amer.* xiv. 186 Beggars will stipulate with you as to the sum they expect to be given them—they will name a quarter of a dollar, a quarter-dollar or eleven penny bit. **1826** *New-Harmony Gaz.* 3 May 256/2 (*h.*), There were many poor people who would have made the shirts for three elevenpenny bits apiece. **1859** [see Levy *sb.*[3]]. **1891** Kipling *Light that Failed* ix. 182 The one-and-elevenpenny umbrella. **1937** E. Garnett *Family from One End Street* iii. 27 The age whereat in state educational circles as 'eleven-plus' (that is, fit for secondary examination for all elementary school children with any ambition). **1946** *Lancet* 10 June 823/1 A universal age of entry at 11+ and a leaving age of 16 will go far. **1957** *Listener* 21 Nov. 833/2 The so-called eleven-plus examination. *Ibid.* The eleven-plus examination. *Ibid.* Eleven-plus. **1958** *Economist* 12 Apr. 99/1 Labour..turns to the Rent Act, the block grant, the 'iniquitous eleven-plus'. **1959** I. & P. Opie *Lore & Lang. Schoolch.* xi. 227 They [*sc.* children] are particularly conscientious about bringing charms to the 11-plus examination, the 'scholarship' as they call it, which determines whether they shall go on to a grammar school or to a secondary modern.

elevener. Add: (Earlier example.)

1823 E. Moor *Suffolk Words & Phr.* s.v. *Bever,* the 'levener'—this..is the interstitial snack between the *prime* and the next.

2. One who takes a drink at 11 a.m. *rare*⁻¹. *U.S.*

1807 C. W. Janson *Stranger in Amer.* xxii. 299, I know of no custom more destructive than that which is practised by slingers and eleveners.

elevens. Add: (Earlier example.) In form **ele·venses** *pl.* (orig. *dial.*, now in general use), light refreshment about 11 a.m.

1849 W. & H. Raynbird *Agric. Suffolk* vi. 296 The name 'fourses' and 'elevens', given to these short periods of rest and refreshment, show when taken. **1887** Parish & Shaw *Dict. Kent. Dial.* 51 Elevenses. **1895** C. J. Cornish *Wild Eng. Today* 243 The workmen rest for their 'elevenses' and 'fourses'. **1923** *Daily Mail* 30 July 6 The men at work with bare chests up enjoying their drink in the shade of the hedge at 'elevenses' or 'fourses' according to the hour. **1927** Wodehouse *Meet Mr. Mulliner* ii. 56 At eleven o'clock he had his 'elevenses', consisting of coffee, cream, more bread and more butter. **1930** T. Thompson *Man in Black Hall* vii. 143 Charlotte—at eleven o'clock with a cup of tea. **1933** *Punch* 2 Aug. 124/1, I came upon him drinking his elevenses (these functions usually lasted from eleven-thirty to twelve—the time of his dinner—to five-thirty to ..). **1947** L. Brown *Say the Word* 17 'Elevenses' is middle-class and particularized, vary with coffee and a biscuit at the hour named. **1951** *Lilliput* Sept.–Oct. 111/2 On the desk in front of him was a cup of coffee and some sandwiches. 'Elevenses', he said.

eleventh, *a.* and *sb.* Add: **A. 1.** Also *eleventh-hour* used *attrib.* or as *adj.*

1870 Rossetti *Let.* 17 Mar. (1965) II. 820 But I am getting into that pleasing old-man state..with 11th hour work there is plenty to say. **1897** C. M. Flandrau *Harvard Episodes* 230 So, in response to his eleventh-hour prayers, he did what he could. **1904** *Daily Chron.* 24 Oct. 5/4 An eleventh-hour alteration in the arrangements for the return of Queen Alexandra from Copenhagen. **1966** *Listener* 1 Dec. 810/2 To commit a boy to preparing for *either*..examination..deprives the eleventh-hour developer of a chance of an award.

B. *sb.* **2.** *Mus.* A note eleven diatonic degrees above or below a given note; also (usually) the interval between this and the given note, equivalent to an octave and a fourth; a chord containing two such notes.

1597 [in Dict., sense A. 4]. **1811** T. Busby *Dict. Mus.*, Eleventh, an interval consisting of ten conjunct degrees, or eleven diatonic sounds. **1880** Grove *Dict. Mus.* I. 437 The chord of the dominant eleventh, when complete..is hardly likely to be found unabridged. *Ibid.* 438 [Some] theorists..repudiate the chords of the eleventh and thirteenth. **1934** C. Lambert *Music Ho!* i. 26 The ninths and elevenths and whole-tone chords that form the stock-in-trade of Debussy's early mannered style.

elevon (e·lĭvǫn). *Aeronaut.* [f. Elev(ator (sense *4 b) + *aileron.] The movable section of the trailing edge of a delta wing, so called because it fulfils the function of both aileron and elevator on the wing and tailplane of an aeroplane so equipped.

1945 *Sphere* 22 Sept. 358 (*caption*) Elevons. **1946** *Aircraft Recognition Jrnl.* Dec. 65/2 It is not absolutely necessary to have sweep-back in the flying wing's plan form, but it is useful because elevons (the 'stream-lined' control surfaces) would have such poor leverage about the centre of gravity. **1948** *Jane's Aircraft* 27c/1 Longitudinal and lateral control by 'controllers' or 'elevons' which are hinged on each outer wing to serve as both elevators and ailerons. **1953** 'N. Shute' *In the Wet* ix. 301 He could see the four engines, the elevons, the flaps, the long line of the fuselage.

elf, *sb.*[1] Add: **6.** *elf-craft, -flower, -folk, -friend, -key, -kingdom, -light, -speech, -wing.*

1869 W. de la Mare *Flora* 5 Lovely as elf-craft. **1921** —— *Crossings* 47 A steeple hat bound round with Elf flowers. **1922** —— *Down-adown-derry* 12, I saw from concealment..a company of elf-folk. **1937** Tolkien *Hobbit* iii. 62 The master of the house was an elf-friend. **1924** R. Graves *Mock Beggar Hall* 26 The elf-key at the rainbow's rise. **1954** Tolkien *Fellowship of Ring* i. xi. 203 A story of the Elf-kingdoms. **1913** W. de la Mare *Peacock Pie* 172 Elf-light, bat-light, Touchwood-light and toadlight. **1955** Tolkien *Return of King* App. F 409 Fragments of Elf-speech. **1929** Blunden *Near & Far* 49 Elf-wings set out on visit and patrol.

elf, *sb.*[2] *S. Afr.* Also elft. [Afrikaans, f. Du. *elft* shad.] The spiny-finned sea-fish *Pomatomus saltatrix.*

1731 G. Medley tr. *Kolben's Pres. State Cape G. Hope* II. 190 In the Table-Bay, and in Bay-Falzo, is caught a sort of Fish the Dutch call Elft. The Elft is Three Quarters of a Yard long or more; and is scaled much like a Herring. **1902** *Trans. S. Afr. Philos. Soc.* XI. 215 The Elf (*Temnodon saltator*), which somewhat resembles the Elft (*Alosa alosa*) of Holland, though the two are by no means scientifically related. **1947** K. H. Barnard *Pict. Guide S. Afr. Fishes* 111, 106 The Elf (*Pomatomus saltatrix*) ..known in Natal as the Shad (quite different, of course, from the true Shad of the Herring family), and in other parts of the world as the Bluefish, Skipjack, Tailer, etc. **1949** *Cape Times* 22 Nov. 13/6 Three elft..have been landed by..anglers.

elfin, *a.* and *sb.* Add: **B.** *sb.* **4.** **elfin-tree,** a dwarfed tree found commonly in alpine regions; **elfin-wood,** a wood composed of such trees. Hence *elfin timber, woodland.*

1903 W. R. Fisher tr. *Schimper's Plant-Geogr.* III. ii. 794 We contrast alpine grassland, alpine shrubland, and alpine desert with those of the lowlands, and retain for alpine forest merely the title elfin-wood. *Note,* 'Elfin-wood' and 'elfin-tree' are the terms introduced here as the equivalents of 'Krummholz'. *Ibid.* 705 Elfin-tree..is characterized by a short, gnarled, often oblique or horizontal stem, and long serpentine branches, which are bent in all directions. **1929** Weaver & Clements *Plant Ecol.* xi. 257 The gnarled, sprawling, much-branched, elfin timber grows through the short summer of high altitudes. **1952** P. W. Richards *Trop. Rain Forest* xvi. 346 In many places..the Montane Rain forest consists of dwarf crooked trees smothered with..mosses; this very characteristic type of vegetation is often called..Elfin woodland.

elf-locked, *ppl. a.* (Later example.)

1946 W. de la Mare *Traveller* 23 Plaiting cramped fingers in the elf-locked mane.

Elgarian (elgā·riăn, elgĕə-), *a.* [f. the name of the English composer Sir Edward William *Elgar* (1857–1934) + -ian.] Of, pertaining to, or characteristic of Elgar or his works.

1936 E. Newman *Elgar* i. 8 Those mannerisms of vocabulary and rhythm that so often enable us to mark a tune of his as being typically Elgarian. **1914** *Everyman* 13 Mar. 717/1 Taking Elgarian works at unauthorised tempi. **1942** *Scrutiny* X. 219 The noisy Elgarian climaxes. **1953** C. Day Lewis *Italian Visit* i. 21 A great Elgarian clash and bray of sunshine.

Elgin Marbles (e·lgin mā·b'lz). A collection of sculptures and architectural fragments from Athenian buildings, chiefly from the frieze and pediment of the Parthenon, which were collected, transported to England, and sold in 1816 to the British Government by Thomas Bruce, 7th Earl of Elgin (1766–1841).

1809 B. R. Haydon *Diary* Nov. (1960) I. 95 Mengs said fifty years ago, that perhaps the statues they then possessed were not the works the Ancients accounted their best, and had Mengs been happy enough to see the Elgin Marbles, he would have been convinced of the truth of his conjecture. **1816** *Rep. Sel. Comm. Earl of Elgin's Collection* (*Parl. Papers*) III. 14 With respect to the state of mutilation in which the Elgin marbles are left now, and of the corrosion of much of the surface by the weather reduce their value, it is difficult precisely to ascertain. **1817** Keats (*title*) On seeing the Elgin Marbles for the first time. **1926** S. T. Warner *Lolly Willowes* II. 93 He admitted that there were no wigs in the Elgin Marbles. **1967** W. St.Clair *Lord Elgin & Marbles* xxiv. 272 Ever since Mr. Hugh Hammersley's motion in the House of Commons in 1816 that the Elgin Marbles should be held in trust and returned to Greece, the question of their possible return has been a political issue.

Elian (i·liăn), *a.* Pertaining to or characteristic of the *Essays of Elia* (1823), or their author, Charles Lamb (1775–1834). So **E·lian** *sb.*, an admirer of 'Elia'. Also **Eliaism** (i·liă,iz'm), a characteristically Elian essay.

1854 Patmore *Friends & Acq.* I. 99 Lamb, in his exquisite Eliaism, 'Blakesmoor in H—shire'. **1903** in *Dobell's Catal.* June 31/2 This is a comfort to think that there are so many good and true Elians in the world. **1905** *Athenæum* 3 June 683/1 Many of the new letters are of no literary account; perhaps a score have the true Elian touch. **1923** *Times Lit. Suppl.* 23 Aug. 555/2 Mr. Tillyard renounces all claims to satisfying 'every Elian's' appetite. **1959** *Daily Express* 3 Jan. 8/5 The Elia-man's device for roasting pork.

elidible (flai·dĭb'l), *a.* [f. Elide *v.* + -ible.] That may be elided.

1878 G. Gosnell *Versification* 67 Theoretically, there is no interval between the elidible vowels of separate phrases.

eligibleness (e·lidʒĭb'lnės). [f. Eligible *a.* + -ness.] Eligibility.

1692 Locke *Third Letter for Toleration* i. 45 You leave the Matter very perplex'd, when you defend the Eligibleness of vesting a power in the Magistrates Hands. **1877** G. W. Foote *Philos. Beggin. Christ.* ii. 49 The right of suffrage, and eligibleness to office.

eliminability (fli:minăbi·lĭti). The property or quality of being eliminable.

1947 H. Reichenbach *Elem. Symb. Logic* ii. 13. 63 How is this concept compatible with the eliminability of the rule of replacement? **1957** P. Suppes *Introd. Logic* viii. 254 We may now formulate the concept of eliminability. **1967** S. C. Kleene *Math. Logic* § 39. 219 The application of the eliminability theorem.

eliminant. Add: **B.** *sb.* **2.** *Med.* An agent which eliminates deleterious matter from the system.

1887 C. A. Moloney *Sk. Forestry W. Afr.* 437 Recommended as an eliminant in malignant cholera. **1907** *Practitioner* Oct. 557 Stoeltzner states that the intestinal mucosa acts both as absorbent and eliminant of lime-salts. **1922** *Daily Mail* 15 Nov. 12 (Advt.), It consists of uric acid solvents and eliminants which exert an irresistible solvent action upon uric acid.

eliminating (fli·minēⁱtiŋ), *ppl. a.* [f. Eliminate *v.* + -ing[2].] That eliminates.

1888 *Pall Mall Gaz.* 13 Sept. 4/1 The eliminating influence of the battle-field. **1909** *Westm. Gaz.* 9 June 14/3 To face the ordeal of the eliminating competition.

eliminator. Add: **b.** Any apparatus which eliminates the need for a battery (high or low tension or grid bias) by enabling a wireless set to be worked from any electricity supply.

1928 *Morn. Post* 9 July, The rectifier intended for inclusion in high tension battery eliminators.

elinvar (e·linvăɹ). [f. Fr. *élinvar* (C. E. Guillaume 1920, in *Compt. Rend. Acad. Sci.* CLXXI. 83), f. *élasticité invariable* invariable elasticity.] (See quot. 1923.)

1922 *Raw Material* Nov. 474/1 Elinvar permits of its application in the very good watches and even to chronometers of the non-metallic pendulum. **1923** Glazebrook *Dict. Appl. Physics* V. 394/1 *Elinvar* (short for elasticity *invariable*) is an alloy of iron and nickel, with a considerable admixture of other metals or metalloids, possessing an invariable modulus of elasticity (Young's modulus). **1930** *Daily Tel.* 8 Jan. 8/6 Dr. Charles Edouard Guillaume..is the inventor of three metallic alloys of great importance, invar, elinvar, and platinite. **1934** *Nature* 1 Sept. 318/1 (*heading*) Elinvar hairsprings in watches. **1950** *Electronic Engin.* XXII. 2 Tuning forks made from elinvar are available.

elision. Add: **1. b.** *fig.*

1893 in *Funk's Stand. Dict.* **1936** R. Campbell *Mithraic Emblems* 20 Seven hues in white elision. **1962** *Sunday Times* 28 Jan. 12/2 The elision of pay pause into pay freeze is that at this stage scarcely been attempted. **1966** M. Critchley *Developmental Dyslexia* viii. 52 The process of learning to read entails the elision from the focus of attention of the continuing memory-images of the non-dominant hemisphere.

eli·sionable, *a.* *rare.* [f. Elision + -able.] That may be elided.

1877 G. M. Hopkins *Let.* 3 Apr. (1935) 40, I treat *ng* as elisionable, like a Latin *m.*

élite, *sb.*[2] Add: Also *attrib.* and *Comb.*; **élite stand, tree** (see quot. 1953).

1852 C. M. Yonge *Two Guardians* viii. 143 She did not belong to the very élite circles. **1912** J. H. Moore *Ethics & Educ.* xiv. 111 The fighting instinct persists so strongly, even in élite peoples, that all our great armies..are arranged on the plan of a battle. **1922** Joyce *Ulysses* 644 The *élite* society of oilskin and company. **1936** *Forestry* X. 35 By this process, in every fifth row of trees *élite* trees were selected, roughly every fifth stem when suitable. **1948** T. S. Eliot *Notes towards Def. Culture* ii. 36 An élite-governed society. **1953** *Brit. Commonw. For. Terminol.* I. 128 *Elite stand,* one selected for seed collection on account of its specially good quality. *Ibid.* 139 *Elite tree,* a tree selected for seed collection or for vegetative propagation on account of some specially good quality it may possess [etc.]. **1955** Koestler *Trail of Dinosaur* 209 The proposal aims at the creation of an élite force, within the framework of the Atlantic Pact. **1958** *Listener* 30 Oct. 683/1 Formal schooling which a generation ago was still confined to a small élite group. **1968** *Listener* 4 Apr. 445/2 The *New Statesman* was, and remains, an élite paper, a 'quality' paper.

2. (Normally without accent.) A standard size for letters used in typewriting, measuring horizontally twelve letters to the inch.

1920 H. Etheridge *Dict. Typewr.* 103 Elite Type. **1923** *Amer. Type Founders Co. Specimen Bk.,* Popular Typewriter Faces..10 Point Elite Underwood. **1927** F. Haines *Advanced Typewriting* 51 'Elite' type, which is most suitable for private correspondence, measures twelve (letters or spaces) to the inch. **1928** *Big Six* (*Remington Typewriter Co. Ltd.*), Type Styles..Elite. **1951** J. Steinbeck *Log f. Sea of Cortez* (1958) p. liv, The next page G of small elite type.

élitism (ėli·tiz'm). [f. Élite *sb.*[2] + -ism.] Advocacy of or reliance on the leadership and dominance of an *élite* (in a society, or in any body or class of persons). Hence **éli·tist** *sb.* and *a.*

1950 D. Riesman in *Psychiatry* XIII. 303/1 He [*sc.* Freud] shared with..Nietzsche and Carlyle elements of an élitist position. **1951** —— *Individualism Reconsidered* (1954) 30 In Elton Mayo and in other recent writers, one can find a similar élitism or concern with the formation of group consensus through strong leadership. **1957** C. Hoyt *Guide to Communist Jargon* xlix. 161 His [*sc.* Lenin's] conception of the 'narrow' party consisting of an *élite,* whose more highly developed class consciousness enables it to see further than those among whom it works... But this *élitism* is anti-democratic. **1960** *New Left Rev.* Nov.–Dec. 30/1 The élitist manipulation of the Communist Party. **1961** *Ibid.* Jan.–Feb. 59/2 Thompson doesn't know what to do with..us.. Goodness knows why he 'suspects' that cause us with elitists? **1968** *New Scientist* 30 May 446/1 The theme of his speech was 'Elitism and Excellence', the excellence being that of the mathematically gifted, and the elitism being the varying degrees of regard paid to this talent.

Elizabethan, *a.* and *sb.* Add: **A.** *adj.* (Earlier example.) Also, belonging to the period of Queen Elizabeth II (1952–).

1817 I. D'Israeli *Curiosities of Lit.* (ed. 5) I. 383 The fashions of the Elizabethan age have been chronicled by honest John Stowe. **1957** D. L. Emblen *Apothecary's Shop* 235 The 'New Elizabethan' theme, cocking in the minds of statesmen, leads to nothing more glorious than the devastation of an Arab quarter.

2. (Earlier example.)

1823 Lamb *Elia* 190 What a collegiate aspect has that fine Elizabethan hall.

B. *sb.* (Earlier example.) Also, a person of the period of Queen Elizabeth II.

1817 D. Masson *Life Milton* I. 69 The..literature of England would be represented to Milton, in the year 1624, by that cluster of men..who had been already named 'the Elizabethans'. **1953** F. E. Halliday in R. Carew *Surv. Cornwall* 11 It still remains..one of the best accounts that we possess of life in Elizabethan England, a quality that should appeal to Elizabethans of the new age.

Elizabethanism (ĭlizăbī·păniz'm). [f. Elizabethan *a.* and *sb.*: see -ism.] A manner or style, or a (literary, etc.) work, or a particular feature of these, characteristic or imitative of the style, language, etc., of the reign of Queen Elizabeth I (1558–1603).

1865 S. Butler in H. F. Jones *Mem.* (1919) II. xxxiii. 207 He who would write a translation like those of the Elizabethans must above all things avoid Elizabethanisms. **1897** *Daily News* 10 May 9/1 Complaint is heard of 'affected Elizabethanisms'. **1909** G. K. Chesterton in D. Figgis *Wiz. Life* p. vii, The latest Elizabethanism has differed not only from the actual Elizabethan work, but even from revivals of it. **1928** *Times* 27 Feb. 10/3 The small string orchestra..provided a pleasant Elizabethanism in a consort of recorders.

elk[1]. Add: **1. b.** *pl.* (With capital initial.) In full: the Benevolent and Protective Order of Elks, formed in New York City in 1868, orig. a society of actors and writers, later a social and charitable organization; *sing.* a member of this organization.

1884 (*title*) Constitution, By-Laws..of the Second annual benefit of Chicago Lodge, No. 4, 'D.', Protective Order of Elks, was given yesterday afternoon. **1922** L. Mumford in H. Stearns *Civilis. U.S.* 6 In every American city, small or big, Odd Fellows, Elks,..and other orders which number..found for themselves a prominent place. **1950** W. Stevens *Let.* 27 Feb. (1967) 670 The other hotels are full of Elks. **1957** W. H. Whyte *Organization Man* vii. xxi. 268 He can move upward (from the Elks, say, to the Rotary) only by sanction of the next upper group.

2. b. In Anglo-Indian use, the Sambur.

1884 W. Rice *Indian Game* vi. 95 The deer in the foreground..are the samber stag, beautiful animals one sometimes hears miscalled the 'elk'. **1890** S. Baker *Wild Beasts* xxv. 306 Sambur deer, (miscalled elk in Ceylon).

4. elk-hound, a dog of Scandinavian origin specially adapted for hunting the elk, having a thick and weather-resisting coat of a grey colour with black tips, and a thick tail curled over the back; **elk-wood** (U.S. examples).

[**1835** C. F. Hoffman *Winter in West* II. 12 A very successful experiment has been made here in crossing the greyhound and Newfoundland...If the race be continued, they ought to be dubbed elkhounds.] **1878** *Kennel Club Stud Bk.* 213 Norwegian Elkehound [*sic*]. **1889** *Kennel Gaz.,* Swedish Elk Hound. **1901** E. Jesse *Dogs* xvi. 491 The Elk-hound..may be termed the Scandinavian Pointer, for, as well as for elk and bear-hunting, it is used as a gun-dog for blackcock. **1908** *Kennel Encycl.* II. 588 The true Elk or Bear hound is distinctly by nature a hunting dog, hailing originally from Swedish Lapland or Jemtland... The dogs are designated by their owners 'Svenske Hunder'. **1936** C. L. B. Hubbard *Observer's Bk. Dogs* 63 In 1901..the British Elkhound Society was formed. **1907** F. Pursh *Jrnl. Bot. Excursion* (1923) 27 Acer montanum, very common throughout these woods, called Elkwood. **1834** *Southern Lit. Messenger* I. 97 The underwood is mostly streaked maple or elkwood (the *Acer Striatum* of Michaux). **1880** *Harper's Mag.* July 182/2 Vines and elk-wood cover both sides [of the rock].

ell[1]. Add: **5.** In full **ell coal,** a type of coal normally found in seams one ell or more in thickness. *Sc.*

1794 J. Naismith *Agric. Clydesdale* 36 About 16 or 17 fathoms under this, lies the ell coal, so called, because it was first found of this thickness, but it is frequently from 4 to 5 feet thick. **1845** *New Statistical Acct. Sc.* V. 613 Seven workable coal seams, in the following ascending order, viz. the stone-coal 42 feet; ell coal, 3; splint. **1876** *Geol. Fife* (Geol. Survey) I. 82 The Ell Coal lies 1 to 7 fms. above the Upper Eight Foot... Sometimes it is a seam of 3 to 4 ft... The ell is a steam coal. **1931** *Times* 16 Mar. 19/7 Lanarkshire.—Ell best, 15s. 6d.

ell[3]. Delete def. in Dict. and substitute: *U.S.* = L 2 a. Also *attrib.*

1773 *Rec. Early Hist. Boston* (1887) XVIII. 198 A Major part of the Selectmen of said Town..[have] Established a New Street..beginning at the North Easterly Corner of the Brick Tenement, [or Ell, so called] belonging to [etc.]. **1875** J. G. Holland *Sevenoaks* xii. 157 Can't ye put on an ell when we want it? **1876** 'Mark Twain' *Tom Sawyer* ix. 81 A single minute later he was..creeping along the roof of the 'ell' on all fours. **1891** W. D. Howells *Mercy* xix. 367 Matt could see a light in the ell-chamber. **1894** T. Watson *Bethany* (1920) 9 So it came to pass that two handsome rooms were built in the front of the west end of the old house, forming an 'ell' thereto. **1924** H. Croy *R.F.D. No.* 3 60 Higbee's house..was dirty white in color, with a series of ells and lean-tos fastened to it.

ellagitannin (elăd3ītæ·nin). *Chem.* [f. Ellagic *a.* + Tannin.] Any tannin that on hydrolysis gives ellagic acid and a sugar.

1895 *Naturalist* 25 Ellagotannin $C_{14}H_{10}O_{10}$, which is the most highly oxidised of vegetable astringents. **1898** *Proc. Chem. Soc.* 104 The presence of ellagic acid has also been detected, and thus besides gallotannin, ellagitannin is also present. **1956** Hickinbottom & Garwood in E. H. Rodd *Chem. Carbon Compounds* IIIB. xii. 806 The hydrolysable tannins are usually divided into gallotannins which on hydrolysis give gallic acid, and ellagitannins which give ellagic acid. **1966** E. Haslam *Chem. Vegetable Tannins* iv. 112 The most important sources of ellagitannins are..Valonea, Myrobalans, Divi-divi and Algarobilla.

ellestadite (elĭstæ·dait). *Min.* [f. the name of R. B. *Ellestad* (b. 1900), American analytical chemist + -ite[1].] A halogen-containing mineral of the apatite group in which phosphorus is largely replaced by sulphur and silicon.

1937 D. McConnell in *Amer. Mineralogist* XXII. 977 The name ellestadite is proposed for the end-member of the apatite group which contains SO_4 and SiO_4 groups substituting for PO_4 groups. **1959** P. F. Kerr *Optical Min.* (ed. 3) xii. 230 Fermorite, strontian apatite, and ellestadite are similar to apatite in physical properties.

Ellingtonian (elĭŋtōu·niăn), *a.* and *sb.* [f. the name of the American musician, Edward ('Duke') *Ellington* (b. 1899) + -ian.]

A. *adj.* Of, pertaining to, or characteristic of 'Duke' Ellington, or his musical productions. **B.** *sb.* A devotee or follower of Ellington. So **Ellingto·nia** *sb. pl.,* music written by, or in the style of, Ellington.

1933 *Fortune* (U.S.) Aug. 93/1 By the time the orchestra plays them they are Ellingtonian. **1952** B. Ulanov *Hist. Jazz Amer.* (1958) xv. 180 The shaking of the maracas, the rattling gourds which several other Ellingtonians quickly picked up. **1958** in P. Gammond *Decca Bk. Jazz* viii. 103 From this solid root of Ellingtonian music..grew..mainstream jazz. *Ibid.* xxi. 262 Some of these..were mere adaptations of pop tunes, but..pop tunes and Ellingtonia. **1961** *John o' London's* 6 July 55/1 Duke and his faithful equipe..take two warm-ups from Grieg and turn them into authentic Ellingtonia.

elm, *sb.* Add: **2. Dutch elm:** see Dutch A. *adj.* 3 c.

5. *elm-arched, -bordered, -fringed, -grey*; **elm bark beetle, elm (leaf) beetle, Elm City** (also *City of Elms*) *U.S.* (see quots.); **elm butterfly,** a butterfly whose larva feeds on the leaves of the elm, as the comma-butterfly (*Grapta comma-album*).

1860 *Harper's Mag.* June 4/1, I wandered through the elm-arched streets [of New Haven] in solitude as absolute as though I trod the aisles of a primeval forest. **1868** J. R. Lowell *Al Fresco* (1896) 61 Upon these elm-arched solitudes No hum of neighbor toil intrudes. **1909** *Cent. Dict.* Suppl., *Elm bark-beetle,* a scolytid beetle, *Phlæophthorus liminaris,* which bores the bark of elm-trees. **1936** *Discovery* Feb. 41/2 The Elm-Bark Beetle (*Eccoptogaster scolytus* Fab.) has increased remarkably in recent years and as it carries the fungus of elm disease, its increase is a serious economic problem of the future. **1961** *New Scientist* 16 Mar. 665/1 DDT..is used to control the elm-bark beetle. **1876** *Field & Forest* II. 12 One [insect] found destroying the foliage of the elm, pronounced..to be *Galeruca calmariensis,* the elm beetle. **1902** C. J. Cornish *Naturalist on Thames* xxxvi. 224 The elm-bordered meadows of the Vale of White Horse. **1872** A. S. Packard *Study Insects* Index 692/1 Elm butterfly. **1843** *Yale Lit. Mag.* VIII. 328 Some inconsiderate hard-hearted beauty, that was supposed to reside somewhere in the 'City of Elms'. **1871** Schele de Vere *Americanisms* (1872) 664 New Haven in Connecticut, is known throughout the United States as Elm City, from the number and magnificent size of the elm-trees that adorn the public squares and most of the principal streets. **1909** *Westm. Gaz.* 14 Apr. 5/1 The elm-fringed arm of the Tyburn stream. **1935** E. Bowen *House in Paris* I. v. 70 The elm-grey autumn park. **1881** *Amer. Naturalist* XV. 242 Inquiries about the imported elm leaf-beetle.

Elmenteitan (elmentēⁱ·tăn), *a.* [f. *Elmenteita,* a lake in Central Kenya + -an.] Of, or relating to, the mesolithic culture which belongs to the Makalian wet phase, and which was found in deposits of the East African lake Elmenteita by Dr. L. S. B. Leakey in 1927. Also as *sb.,* a member of this culture; the culture itself.

1929 L. S. B. Leakey in *S. Afr. Jrnl. Sci.* XXVI. 754 The Elmenteitan. This is a localised development from the Upper Kenyan Aurignacian in Makalian times. It has no exact parallel in Europe, but has close affinities with the Magdalenian. **1931** —— *Stone Age Cultures Kenya Colony* viii. 172 The Elmenteitan culture certainly represents a very late and aberrant development of the upper Kenya Aurignacian culture. **1939** C. S. Coon *Races of Europe* iv. 85 The East African Elmenteitans represent..a gerontomorphic or sexually differentiated Mediterranean or Galley Hill form, and in cranial features [are] closer to Galley Hill itself than any other branch. **1964** K. P. Oakley *Framework for dating Fossil Man* II. 268 The Elmenteitan was named after Lake Elmenteita, for the typical occurrence is in Gamble's Cave which is close to the lake.

elocute (e·lŏkiūt), *v.* [Playful back-formation f. Elocution.] To practise elocution; to declaim in an elocutionary manner.

1884 'Mark Twain' *Huck. Finn* 315 They didn't yellocute long till the audience got up. **1896** *Idler* Mar. 183/1 'Elocute' as he might, his reputation was always overshadowed by that of a 'past' boy. **1908** *Dialect Notes* III. iv. 308 *Elocute, v.i.,* to recite in elocutionary style. **1920** S. Lewis *Main Street* x. 121 That was fine. I don't know but what you can elocute just as good as Ella. *Ibid.* iv. 47 Ella is our shark at elocuting. **1963** *Times* 14 May 15/4 Dorothy Reynolds's Oenone, flutingly elocuted.

elocution. Add: Hence **elocu·tional** *a.,* **elocu·tionally** *adv.*

1933 *Times Lit. Suppl.* 30 Mar. 220/3 Sentence-form being in the main elocutional, there may be sentences without locutional sentence-form. **1932** A. H. Gardiner *Theory Speech & Lang.* v. 322 No sentence can be really elocutionally formless, since utterance itself imposes a certain minimum of form.

elodea (elōu·diă, elodī·ă). *Bot.* [mod.L. (A. Michaux *Flora Boreali-Americana* (1803) I. 20), f. Gr. ἑλώδης marshy.] A member of a small genus of aquatic plants, belonging to the family Hydrocharidaceæ and native to temperate America; cultivated in aquaria and pools. (Cf. Water-thyme 2.)

1894 Darwin & Acton *Pract. Physiol. Plants* i. 16 An Elodea leaf is mounted in water. **1937** E. C. B. Wright *Gen. Plant Physiol.* viii. 226 The main study was made of the chloroplasts of the aquatic Elodea. **1955** *Sci. News Let.* 16 July 46/1 Elodea, the moss-like weed in most American goldfish bowls, yielded from 12 to 14 tons green weight to the acre in a Kansas pond. **1960** *Times* 9 July 9/5 Soft weed (such as Elodea, Hornwort).

Elohim. Add: **c.** *attrib.* = Elohimic *a.*

1875 *Encycl. Brit.* III. 637/1 It is still possible to reconstruct at least the Elohim document. **1886** *Ibid.* XX. 30/1 The Elohim psalms..have undergone a common editorial treatment distinguishing them from the rest of the Psalter. **1936** J. E. Carpenter in A. S. Peake *Commentary on Bible* 122/1 He [*sc.* Jean Astruc] noticed that in different narratives the Deity was designated by different names...On this basis he distributed the contents of Genesis into two main documents, an Elohim narrative A and a Yahweh story B. **1963** S. Sandmel *Hebrew Scriptures* xxvi. 330 There was a source additional to the Yahve and Elohim source, which could be called the Deuteronomic code.

Elohism (elōu·hiz'm). [f. Elohim + -ism.] The worship of Elohim.

1888 *Edin. Rev.* Apr. 502 It was the task of the great prophets..to bring Israel back to the primitive Elohism of the patriarchs.

elongation. Add: **6.** *Mech.* In mechanical testing, the amount of extension of a test piece when stressed, usu. expressed as a percentage of the original length; also *attrib.*

1877 *Encycl. Brit.* VII. 818/1 Eight different specimens..bore from 43½ to 46 lb. (average 45·2) just before breaking, with elongations of from 17 per cent to 22 per cent. *a* **1884** Knight *Dict. Mech.* Suppl. 868/2 The index needle also moves two slides which show the maximum of elongation or compression. **1909** *Cent. Dict.* Suppl., *Elongation strain,* the percentage of elongation of a piece of material being tested in a testing-machine. **1922** Glazebrook *Dict. Appl. Physics* I. 144/1 The elongation and contraction of area are required in order to obtain a measure of the ductility of the material. **1930** *Engineering* 3 Jan. 31/2 The percentage elongation observed for arsenical copper was 58 per cent.

eloquent, *a.* Add: **4.** *fig.* Effectively expressive *of.*

1870 Rossetti *Poems* 191 Close kissed and eloquent of still replies. **1873** R. A. Proctor *Expanse of Heaven* xi. 123 To the Almighty every atom in infinite space is eloquent of the universe itself. **1885** A. Forbes *Souvenirs of some Continents* ix. 208 His whole attitude eloquent of discouragement.

Elsan (e·lsæn). [App. f. initials of *E*phraim *L*ouis Jackson, a chemical manufacturer, + San(itation.] The trade name of the Elsan Manufacturing Co. (1924 *Trade Marks Jrnl.* 9 Jan., 47) for a type of lavatory in which the sludge is rendered inoffensive by chemical means.

1939–40 *Army & Navy Stores Catal.* 248/2 Elsan chemical closet...No drains, no water-flush..Elsan model '33'. **1943** Hunt & Pringle *Service Slang* 30 *Elsan gen,* news which cannot be relied upon. Literally, 'news invented in the gentlemen's toilet', Elsan being the trade name of the excellent chemical lavatories with which bombers are equipped. **1959** *Now & Then* No. 103 17, I..loaded all my belongings, the old beds and chairs and tables, the paraffin lamps and stoves, even the Elsan, into a turf lorry. **1961** J. B. Priestley *Saturn over Water* xii. 159 In a recess..were two pails of water and an Elsan kind of lav.

else, *adv.* Add: **4. b.** Also *or else,* with aposiopesis (the alternative to be imagined), as a *colloq.* form of warning or threat.

1833 *Examiner* 6 Jan. 1/2 The *Chronicle* puts a case in strict analogy:—Suppose a landlord were to say to his tenant, 'I have got a cause coming on at Chelmsford Assizes; you must give me your pledge to be there as a witness, and swear to such and such a falsehood, or else..' **1879** J. McCarthy *Hist. our own Times* II. 275 Correspondence..between two Irish chieftains. 'Pay me my tribute,' wrote the one, 'or else!' **1940** Harrisson & Madge *War begins at Home* viii. 205 Air-Raid Wardens.. could knock on your door and tell you that your lights must be properly masked or else. **1958** M. Dickens *Man Overboard* xiii. 196 You've been engaged to make this small-boys' penitentiary show a profit, or else. **1959** J. Verney *Friday's Tunnel* vi. 61 If I do..have children ..they'll jolly well obey me and no argument. Or else.

elsewards (e·ls,wəɹdz, e·ls,wǫɹdz), *adv.* *rare*⁻¹. [f. Else + -wards.] In the direction of, towards some other place.

a **1882** Trollope *Autobiogr.* (1883) II. xviii. 173 These earthly sufferers know that they are making their way heavenwards,—and their oppressors their way elsewards.

elsewhence, *adv.* For †*Obs. rare*⁻¹ read *arch.*, and add later examples.

1855 W. Smith *Lat.-Eng. Dict.* 51 *Āliundĕ,* adv. (ali unde), lit. else-whence. **1892** *Pall Mall Gaz.* 9 Dec. 1/2 He has learnt..not only the lesson of reticence from his Chief, but a certain lesson in vivacity to boot elsewhence.

eluate (e·liu,ĕt). *Chem.* [f. L. *ēlu-ĕre* + -ate[1] c.] The solution resulting from elution. Also *attrib.*

1934 in Webster. **1939** *Ann. Reg. 1938* 371 In the vitamin B complex..factors..accepted by many workers were..Macrae's yeast-filtrate and yeast eluate factors. **1949** H. W. Florey et al. *Antibiotics* I. iv. vii. 283 Foster (1944) who adsorbed the active substance.. eluted it with methanol, and concentrated the eluate *in vacuo.* **1962** Waley & van Heyningen in A. Pirie *Lens Metabolism Rel. Cataract* 343 Fractions of the eluate were assayed..for peptidase.

eluent (e·liu,ĕnt). *Chem.* Also **eluant.** [f. L. *ēluent-em,* pres. pple. of *ēluĕre* to wash out: see -ent, -ant[1].] A fluid used to elute adsorbed material.

1941 Bacharach & Robinson tr. *Zechmeister & Cholnoky's Princ. & Pract. Chromatography* ii. 75 The individual homogeneous portions of the column are coarsely broken up..and then immediately dropped into the eluent..and stirred round. **1942** H. H. Strain *Chromatographic Adsorption Anal.* v. 65 Eluants, as a rule, are chosen with the object of obtaining rapid and complete liberation of adsorbed compounds. **1966** E. Haslam *Chem. Vegetable Tannins* iv. 100 The tannin may then be isolated using acetone or methyl ethyl ketone as eluant. **1969** J. A. Dean *Chem. Separation Methods* iv. 68 Assuming a fixed flow rate for the eluent, the time..at which each component in the sample will emerge from the column depends on the individual partition coefficients.

Elul (i·lʊl). [a. Heb. *ĕlūl.*] The name of one of the Jewish months, being the twelfth of the civil and sixth of the ecclesiastical year.

1535 Coverdale *Neh.* vi. 15 The wall was fynished on the fyue & twentyeth daye of the moneth Elul. **1611** Bible 1 Macc. xiv. 27 The eighteenth day of the moneth Elul, in the hundred threescore and twelft yeere. **1753** in Chambers *Cycl.* Suppl. **1886** *Encycl. Brit.* XXI. 126/2 What we do know from a calendar of the intercalary month Elul II. is that in that month the 7th, 14th, 19th, 21st, and 28th days had a peculiar character. **1907** *Daily Chron.* 10 Aug. 4/6 With the advent of the Jewish month of Ellul, of which tomorrow is the first day, a religious revival annually takes place in Jewry. **1962** C. Roth *Standard Jewish Encycl.* 617 Elul..has become the traditional month of penitence and spiritual preparation for the Day of Judgment, and the *shophar* is sounded after morning service on weekdays throughout Elul.

elute, *v.* Delete †*Obs. rare* and add: [after G. *eluieren,* in same sense.] Now *spec.* to wash (adsorbed matter) away from the substance that has adsorbed it. Hence **elu·ted** *ppl. a.*

[**1925** Fodor & Reifenberg in *Biochem. Jrnl.* XIX. 189 With a view to separating the proteins from other nitrogenous substances, we have tried to adsorb them by suitable adsorption media and afterwards to eluate [*sic*] them. [*Note*] We are adopting the terminology of Willstätter.] **1934** *Nature* 14 Apr. 554/1 From the adsorption on fuller's earth..the pigments were eluted by mixtures of pyridine and water. **1934** *Jrnl. Biol. Chem.* CV. 527 The solution of the eluted carotene was evaporated and the β-carotene recrystallized from *n*-heptane. **1946** *Nature* 20 July 96/1 By varying the nature and concentration of the solution used in eluting the adsorbed cations, it should be possible to isolate them in successive fractions of eluate. **1951** W. G. Berl *Physical Methods in Chem. Anal.* II. 593 It is not necessary in this case to elute until the solutes appear in the effluent. **1966** *Lancet* 31 Dec. 1432/1 Iron-solubility tests were carried out on the eluted fractions of unlabelled gastric juice.

elution. Add: [G. *elution,* in same sense.] In later use, the removal of adsorbed matter. (Cf. *elute *v.*)

1922 *Chem. Abstr.* XVI. 2338 (*heading*) Remarks on the 'elution' of saccharase and maltase from their adsorption compounds. **1934** *Nature* 1 Sept. 307/2 Experiments dealing with the inactivation of enzymes on adsorption and their reactivation by elution. **1966** *Lancet* 31 Dec. 1432/1 When gastric juice of controls was resolved by molecular sieving, a consistent elution pattern was found.

elutriator (fliū·triēⁱtəɹ). [f. Elutriate *v.* + -or.] An apparatus for sorting finely divided material according to mass and size by means of a stream of liquid or gas.

1904 *Econ. Proc. R. Dublin Soc.* I. 278 The residue now left in the elutriator consists of particles from 0·1 to 1 mm. diameter. **1957** R. A. Scott in H. W. Cremer et al. *Chem. Engin. Pract.* III. 352 Air elutriators are commonly employed in certain branches of the food industry..for the separation of husk and of small foreign seeds from grain.

eluvial, *a.* Add: **2.** *Soil Sci.* Pertaining to a layer of soil which has undergone *eluviation, as in *eluvial horizon.*

1924 *Geol. Mag.* LXI. 451 Three main horizons are generally recognized,..the A or eluvial horizon, the B or illuvial horizon, and the C horizon, which consists of the parent material. **1934** *Forestry* VIII. 26 The leached layer above being known as the eluvial horizon. **1952** P. W. Richards *Tropical Rain Forest* ix. 208 All soils of the rain-forest region..are therefore strongly leached and have an eluvial horizon, or leached zone, above the illuvial horizon. **1963** D. W. & E. E. Humphries tr. *Termier's Erosion & Sedimentation* vi. 138 A soil is composed of an impoverished zone, the *eluvial horizon,* and also a zone of enrichment or accumulation, the *illuvial horizon.*

eluviate (iliū·viēⁱt), *v.* *Soil Sci.* To undergo eluviation. Usually in pa. pple. or as **elu·viated** ppl. a.

1926 Tansley & Chipp *Study of Vegetation* vii. 116 With lapse of time not only are calcium carbonate and other soluble salts leached from the surface soil..but the finer *insoluble* particles tend also to be mechanically carried down, so that the coarser particles alone remain in the surface layers, which are then said to be 'eluviated' as well as 'leached'. **1963** S. R. Eyre *Vegetation & Soils* ii. 36 In most soils which are subject to leaching, it is possible to distinguish two quite distinct sets of horizons... The upper set are referred to as '*eluviated horizons*'.

eluviation (fliūviēⁱ·ʃən). *Soil Sci.* [as if ad. L. *ēluviātiōn-em,* noun of action f. *ēluĕre* to wash off.] The lateral or vertical movement of material in solution or suspension through the soil.

1928 in B. D. Jackson *Gloss. Bot. Terms* (ed. 4) Suppl. **1930** L. A. Wolfanger *Major Soil Divisions of U.S.* ii. 45 The extraction and concentration are brought about in part through eluviation (the mechanical transfer of material). **1932** G. W. Robinson *Soils* iii. 53 We may refer to the translocation of material either mechanically or in solution, as eluviation, and two main types of eluviation may be distinguished..mechanical eluviation, in which.. the finer fractions of the mineral portions of the soil are washed down to lower levels, and..chemical eluviation in which decomposition occurs and certain products thus liberated are translocated in true or colloidal solution. **1952** L. M. Thompson *Soils & Soil Fertility* vi. 73 The term eluviation designates the movement or shifting of the materials through the soil body. **1957** *Gloss. Geol.* (Amer.

Geol. Inst.) 94/2 Eluviation may take place downward or sidewise according to the direction of water movement. **1968** R. W. Fairbridge *Encycl. Geomorphol.* 1001/1 Infiltrating precipitation passes through the slope prism, exiting by the lower boundary and carrying down ions, colloids and perhaps even mineral particles of clay and silt grades, a process generally termed *eluviation,* or simple *leaching.*

elven. Restrict †*Obs.* to sense in Dict. and add: **2.** *Comb.* (referring to a kind of imaginary being in the works of J. R. R. Tolkien). **a.** appositive, as *elven-kin*; **b.** attributive, as *elven-king, -tongue*; *elven-wise* adj.

1937 Tolkien *Hobbit* xvii. 285 But the Elvenking said: Long will I tarry. **1954** —— *Fellowship of Ring* I. iii. 88 It was singing in the fair Elven-tongue. *Ibid.* II. i. 248 Words unheard were spoken then of folk of Men and Elven-kin. **1955** —— *Return of King* 341 He was elven-wise.

Elysée (elize). In full *Elysée Palace*: a building in Paris, on the Champs Elysées, the official residence of the French head of state: hence used as a synonym for the head of state and his advisers, or the Government, of the day.

1851 *Observer* 7 Dec. 4/1 The President..had proclamations printed at a private printing press in the Elysée. **1870** E. G. E. Ward *Jrnl.* 7 Oct. in *Outside Paris* (1871) 26, I had seen William of Prussia enter the Château of Francis the First as a conqueror. When will he enter the Tuileries or the Elysée? **1934** A. Thirkell *Wild Strawberries* xiii. 297 Alas for the lilies of France!..The tricolore still flew over the Elysée. **1966** *Guardian* 11 Nov. 13/3 The new direct communication link between the Elysée Palace and the Kremlin. **1967** *Ibid.* 9 Oct. 7/7 The phrase..has been the Elysée's line since Mr Wilson's first advances.

elytro-, comb. form of Gr. ἔλυτρον sheath, used = Vagina in various medical terms (see Elytron 4 and subseq.).

em. Add: (Earlier example.) Also *attrib.*, as *em quad, quadrat,* a square Quadrat (see also quots. 1927 and 1967). Cf. M 2.

[**1683–1824** see Quadrat 2.] **1793** in Stower *Printer's Gram.* (1808) xvii. 419 That em and en quadrats..be included. **1808** Stower *Printer's Gram.* xvii. 419 The ems and ens at the beginnings and ends of the lines not to be reckoned in the width. **1863** W. Blades *Life & Typogr. W. Caxton* II. iv. p. xliii, When a 'white' line was wanted under a chapter head or over a colophon, em quadrats were ranged side by side for the purpose. **1871** Em quad [see *mutton-quad*]. **1927** R. B. McKerrow *Introd. Bibliogr.* I. ii. 13 Quads were no doubt usually square spaces, as broad as the height of a line. From the fact that the letter m (or M) used to be cast on a square body, such a square space is called an em-quad. **1967** E. Chambers *Photolitho-Offset* ii. 12 Quads and spaces are the separation material used between letters and words. The em quad is the square of a given body size; the en quad is half the square width.

emaciate, *v.* **2.** Delete † *Obs. rare* and add later examples.

1873 T. H. Green *Introd. Pathol.* (ed. 2) 21 When a person emaciates, the fat is gradually removed from the cells. **1934** R. Frost *Let.* 23 Feb. (1964) 237, I have been in bed with a bug..etiolating and emaciating.

‖ **email** (emēⁱ·l). [Fr., = enamel.] Used *attrib.* in **email ink,** ink used on glass, porcelain, etc.; **email ombrant** *Pottery,* a process in which the impressions of the design appear as shadows (see quots. *a*1877 and 1957).

a **1877** Knight *Dict. Mech.* I. 796/1 *Email-ombrant,* a process which consists in flooding colored but transparent glasses over designs stamped in the body of earthenware or porcelain. **1880** C. A. Janvier *Pract. Keramics* viii. 95 In these the pattern is..exactly the opposite of the *émaux ombrants.* *a* **1884** Knight *Dict. Mech.* Suppl., *Email ink*...Colored inks—black, white, red, blue—used with a quill on glass, porcelain, ivory, marble, bone, mother of pearl, or metal. **1957** Mankowitz & Haggar *Encycl. Eng. Pott. & Porc.* 83/2 *Email ombrant,* an illusionist style of decoration like lithophane but the intaglio decoration was filled with coloured glaze (usually green) which produced a monochrome picture in a variety of tones. Developed at the Rubenes [*sic* = Rubelles] factory by Baron A. du Tremblay in the 1840's.

emanant (e·mănănt), *sb.* *Math.* [f. the ppl. adj.] The result of operating upon a quantic with the operator $(x'd/dx+y'd/dy+\ldots)$.

1852 J. J. Sylvester in *Camb. & Dublin Math. Jrnl.* VII. 56 Every emanant is..itself a covariant of the function to which it belongs with respect to each of the related classes of variables which enter into it. **1860** A. Cayley *Math. Papers* (1891) IV. 604 The coefficients of the successive terms $\lambda^m, \lambda^{m-1}\mu,$ etc. are said to be the emanants of the quantic $(*)(x,y)^m$.

emanate, *v.* Add: (Earlier example.)

1756 *Crit. Rev.* II. 340 The French author is not much oblig'd..to his English translator. We meet with the words,..misanthropes, tranquilly, emanate, legerity.. and many others.

4. Delete *rare* and add later *fig.* examples.

1929 W. Faulkner *Sartoris* I. ii. 34 Bookcases.. emanating an atmosphere of..meditation. **1939** S. de Madariaga *Columbus* xii. 136 In this land [*sc.* Castille].. the mental-moral soil was broken..by forbidding abysses which emanated all kinds of poisonous gases and murderous flames.

emanation. Add: **1. d.** *Math.* The process of finding successive emanants.

1853 J. J. Sylvester in *Phil. Trans.* CXLIII. 545 The process of emanation is one of incessant occurrence in the theory of invariants. **1856** A. Cayley *Math. Papers* (1889) II. 321 The facients $(X, Y, \ldots)$ may be termed the facients of emanation, or simply the new facients.

2. c. *Chem.* A radioactive gas produced by the radioactive decay of a solid; *spec.* any of the three gases radon, actinon, and thoron produced respectively by radium, actinium, and thorium (so *radium* etc. *emanation*); also used as a name for the element radon, of which these gases are now known to be isotopes.

1900 Rutherford in *Phil. Mag.* XLIX. 1, I have found that thorium compounds continuously emit radio-active particles of some kind, which retain their radio-active powers for several minutes. This 'emanation', as it will be termed for shortness, has the power..of passing through thin layers of metals. **1903, 1907** [see Thorium]. **1906** Rutherford *Radioactive Transformations* iii. 72 A large quantity of emanation was introduced into a glass tube..and the ionization due to the issuing rays was measured. **1927** N. V. Sidgwick *Electronic Theory Valency* iii. 29 The names radon, thoron, and actinon are now accepted for the three isotopic emanations. It is desirable that there should be some name for element no. 86 irrespective of any particular isotope, and I have retained the name emanation (with the symbol Em) for this purpose. **1940** Glasstone *Physical Chem.* ii. 125 The emanations, particularly that from actinium, lost their activity relatively rapidly. **1950** —— *Sourcebk. Atomic Energy* v. 116/2 The production of a gaseous emanation provides a convenient means of separating this radioactive species from those which precede it in the disintegration series. **1957** *Encycl. Brit.* XVII. 882 B/1 There are six elements collectively known as the inert gases...Their atomic numbers and their names are: 2 Helium, 10 Neon, 18 Argon, 36 Krypton, 54 Xenon, 86 Emanation. **1964** J. R. Partington *Hist. Chem.* IV. xxvii. 941 Some doubts about the existence of radium emanation were removed by Rutherford and Soddy, who liquefied it by cooling in liquid air.

emanationism (emănēⁱ·ʃəniz'm). [f. Emanation + -ism.] = Emanatism.

1881 G. S. Hall *Aspects German Culture* 315 It [superstition] settled very thickly again in the first Christian centuries, as cabalism, emanationism, neo-platonism.. with their hierarchies of spirit-hosts. **1951** N. Annan *Leslie Stephen* v. 163 These two conceptions of God, though logically opposed, were united by Plato in the *Timaeus* by the device of emanationism: the Transcendental God, the goal of all desires, is perfect Good and therefore envious of nothing; hence He will wish everything to be as He, and He accordingly overflows into the imperfect and projects Himself into the universe. **1956** E. L. Mascall *Chr. Theol. & Nat. Sci.* iii. 92 Emanationism, according to which the existence of the universe follows with *physical* necessity from the existence of God.

emana·tionist, *a.* [f. as prec. + -ist.] **1.** In sociology, relating to the theory that behaviour is conditioned exclusively by environment.

1937 T. Parsons *Struct. Social Action* xiv. 536 A naive 'emanationist' view of the mechanisms of the influence of nonscientific ideas.

2. Relating to the philosophical theory of the origin of the world by emanation.

1941 G. G. Scholem *Major Trends in Jewish Mysticism* vii. 256 The emanationist doctrine of Neoplatonism.

Emanatist. Delete *rare* and add *attrib.* examples.

1867 H. P. Liddon *Divinity of Lord* vii. 646 These Emanatist doctrines. *Ibid.* 647 The Nicene Fathers.. were able..to vindicate for the word its Catholic sense, unaffected by any Emanatist gloss.

emancipated, *ppl. a.* Add: **2.** Esp. of or relating to women.

1850 G. E. Jewsbury *Let.* 29 Aug. (1892) 365, I don't think he knew what an 'emancipated woman' he was showing off to. **1888** Mrs. H. Ward *R. Elsmere* III. v. xxxi. 7 His wife..was..in all matters of religious or political opinion 'emancipated' to an extreme. **1904** Conrad *Nostromo* II. iii. 129 The tall Antonia..offered him her hand (in her emancipated way). **1970** *Guardian* 16 Sept. 11/5 As emancipated as the Lebanese are educated Palestinian women, who are influential in many Middle East countries.

emancipation. **2. a.** (Earlier U.S. example.)

1785 Jefferson *Let.* 7 Aug. (1853) I. 377 Emancipation is put in such a train that in a few years there will be no slaves northward of Maryland.

emancipist. (Earlier example.)

1827 P. Cunningham *Two Years in N.S.W.* II. xxiv. 118 The grand division..of the free classes..is into that of emigrants, who have come out free from England, and emancipists, who have arrived here as convicts, and have either been pardoned or completed their term of servitude.

embanker (embæ·ŋkəɹ). [f. Embank *v.* + -er[1].] One who makes an embankment.

1852 J. Wiggins *Pract. Embanking Lands* ii. 80 The embanker may..always test the time it will take to freshen and to admit the growth of clover. **1896** *Contemp. Rev.* Aug. 203 Barrowers,..embankers,..and bricklayers.

embarkation. **1.** (Further *attrib.* examples.)

1899 *Westm. Gaz.* 9 Nov. 10/2 As the ship was about to sail, investigations were made, and..the embarkation officer went aboard to see the supply. **1904** *Daily Chron.* 6 Sept. 6/7 The embarkation staff at the port. **1923** Kipling *Irish Guards Gt. War* I. 2 Every one was new to embarkation-duty. **1946** Koestler *Thieves in Night* III. 322 And back home in Suffolk on embarkation leave he had listened to a sermon.

embarras, *sb.* Add: **1.** *embarras de richesse* [Abbé d'Allainval (*title*) L'embarras des richesses (1726)] (earlier and later examples).

1751 Chesterf. *Let.* 29 Mar. (1774) II. 6 Say..you dread *l'embaras des richesses* ever since you have seen what an incumbrance they were to poor Harlequin. **1826** M. Wilmot *Let.* 11 July (1935) 243 A too large Stock in trade, proving now and then an *embarras de richesse.* **1932** C. Williams *Eng. Poetic Mind* iii. 77 Iago..is action incarnate only he cannot find his own reason for it. Othello has wronged him..; Othello has played him false with Emilia; so has Cassio; he loves Desdemona also—*embarras de richesses.*

2. Delete entry in Dict. and substitute: *Canada* and (local) *U.S.* [N. Amer. Fr.] (See quots. 1814.) ? *Obs.*

1793 A. Mackenzie *Voy. Montreal* (1801) 226 In passing over one of the embarras, our dog..fell in. **1814** H. M. Brackenridge *Views Louisiana* 205 At the distance of every mile or two,..there are *embarras,* or rafts, formed by the collection of trees closely matted, and extending from twenty to thirty yards. *Ibid.* 208 Passed an *embarras,* N.E. side, the most difficult since we started.

embed, *v.* Add: **1. b.** Examples (also of derivatives) in *Linguistics.*

1961 C. S. Smith in *Language* XXXVII. 346 When one sentence contains another, the latter sentence will be said to be 'embedded'. *Ibid.* 348 Together the three embedding rules produce all sentences that contain adjectives and exclude ungrammatical sentences. *Ibid.* 360 An embedded compared adjective has the widest possible scope in any given sentence. **1966** G. N. Leech *Eng. in Advertising* ii. 20 The prepositional phrase in 'a pipeful of good tobacco' is an embedded adverbial group. **1968** *Language* XLIV. 32 Any rules that are not concerned with embedded sentences..always treat a constituent sentence as an unanalysable unit.

Also **embe·ddedness.**

1922 Whitehead *Princ. Relativ.* ii. 15 Fact..is not the sum of factors; it is rather the concreteness (or, embeddedness) of factors, and the concreteness of an inexhaustible relatedness among inexhaustible relata. **1937** *Mind* XLVI. 83 When we know how to deal with the erst-while novelty, when we have 'got it taped', it falls into embeddedness and becomes, or engenders, a part of ourselves. **1963** *Economist* 7 Sept. 832/1 Man's social embeddedness.

emblem, *sb.* Add: **6. emblem book,** a book containing drawings with accompanying interpretations of their allegorical meaning; so *emblem poem, writer,* etc.

1870 H. Green *Shakesp. & Emblem Writers* ii. 30 (*heading*) Sketch of emblem-book literature previous to A.D. 1616. *Ibid.* iii. 107 Of the Emblem-books in Spanish, German, Flemish, Dutch, and English, only the last would be available for Shakespeare's benefit. **1888** G. E. Sears (*title*) A collection of emblem books by Andrea Alciati. **1945** D. Bush *Eng. Lit. in Earlier 17th Cent.* 581 The latest and fullest study of Quarles's emblem imagery is by E. James. **1948** R. Freeman *Eng. Emblem Books* App. I. 238, I have taken for my criterion four characteristics commonly agreed to be essential by the emblem writers themselves. These are: 1. An emblem book should be a collection of moral symbols. 2. It should have pictures, or..should postulate the existence of pictures. 3. Attached to each picture should be a motto or brief *sententia*... 4. There should be an explanatory poem or passage of prose in which the picture and motto are interpreted and a moral..is drawn. **1963** *N. & Q.* May 168/1 Some of the illustrated *Arma Christi* verses..are rudimentary emblem poems. *Ibid.,* The medieval poems are not deliberately collected into 'emblem books'.

emblema (emblī·mă). Pl. **-mata.** [L.: see Emblem *sb.*] An ornament in relief, either carved or mounted, on jewellery, vases, etc.

1842 W. Smith *Dict. Gk. & Rom. Antiq.* **1880** C. T. Newton *Ess. Art & Archæol.* vi. 265 Embossed and cut out in outline, like the *emblemata* of later Greek art. **1885** *Encycl. Brit.* XIX. 181/1 A seated figure of Athene—an 'emblema' soldered on, in very high relief. **1957** *Encycl. Brit.* XIX. 404/1 Small mosaic-pictures isolated in geometrical pavements were called *emblemata.*

embolectomy (embole·ktŏmi). *Path.* [f. Embol(us + *-ectomy.] The surgical removal of an embolus.

1923 *Surgery, Gynecology & Obstetrics* XXXVI. 309 (*heading*) Embolectomy in the treatment of circulatory disturbances in the extremities. **1961** *Lancet* 19 Aug. 402/2 Pulmonary embolism not infrequently complicates a variety of..conditions...Until recently Trendelenburg's operation of pulmonary embolectomy afforded the one hope.

embolic, *a.* Add: **2.** *Embryology.* Characterized by emboly.

1877 E. R. Lankester *Notes Embryol.* i. 13 Two degrees of this Epibolic mode of Invagination (so called by Selenka, in distinction from the Embolic mode) are distinguished by Haeckel as Amphiblastic and Discoblastic. **1881** F. M. Balfour *Compar. Embryol.* II. xiii. 276 One half of the blastosphere may be pushed in towards the opposite half, and a gastrula be thus produced...

-ITE¹.] A name formerly given to a supposed variety of boulangerite.

1854 J. D. DANA *Syst. Min.* (ed. 4) II. 81. **1868** [see BOULANGERITE]. **1883** *Encycl. Brit.* XVI. 395/1 Plumbostib or Embrethite [*sic*], from Nertchinsk, is only a variety.

embri·ttle, *v.* [f. EM- + BRITTLE *a.*] *trans.* To render brittle. Hence **embri·ttling** *vbl. sb.* and *ppl. a.*

1902 *Encycl. Brit.* XXIX. 574/2 Sudden cooling hardens and embrittles steel and cast iron. **1903** H. M. HOWE *Iron, Steel & other Alloys* ix. 257 The coarsening and embrittling of low-carbon steel. **1930** *Engineering* 11 Apr. 494/1 Lead pipes embrittled by the prolonged application of water-hammer. **1960** *New Scientist* 20 Oct. 1067/2 The plasticity of..steels and of molybdenum alloys decreases markedly when subjected to neutron irradiation. This embrittling phenomena [*sic*] is important. **1962** J. T. MARSH *Self-Smoothing Fabrics* v. 45 As early as 1919, it was found..that cotton and rayon could be made crease resisting..although the products were greatly embrittled and commercially valueless.

embri·ttlement. [f. prec. + -MENT.] The action of the vb. *EMBRITTLE, or the result of such action; loss of ductility.

1920 *Amer. Inst. Mining & Metall. Engin. Trans.* LXIV. 440 It showed no evidence of embrittlement. **1926** *Fuel Economist* II. 115 (*heading*) Treatment of boiler feed water to prevent caustic embrittlement. **1939** *Nature* 11 Mar. 443/2 The full effect of the temper-hardening treatment..cannot be employed with advantage on account of the accompanying embrittlement. **1962** J. T. MARSH *Self-Smoothing Fabrics* ix. 113 Many investigators appear to have assumed that the embrittlement of cotton in conventional processes was due to the use of an acid catalyst at a high temperature. **1969** *Physics Bull.* Mar. 89/2 The embrittlement of alloy steels containing small amounts of P, As, Sb or Sn. *Ibid.*, Embrittlement occurs after long heating at temperatures between 350 and 550°C, while tempering to remove it is achieved by brief heating to 650°C.

embroidery. Add: **6.** *attrib.* and *Comb.*

1695 [sense 4 in Dict.]. **1841** LADY WILTON *Art Needlework* xxiv. 392 The Empress Josephine..exercised ..her needle and embroidery-frame, with beautiful address. **1880** L. HIGGIN *Handbk. Embroidery* i. 1 'Embroidery needles' for ordinary crewel handwork. *Ibid.* v. 59 We give a recipe for embroidery paste. **1882** CAULFEILD & SAWARD *Dict. Needlework* 197/1 Embroidery needles..for canvas work..are short, thick, and blunt, and the eye is wide and long. *Ibid.*, Embroidery paste is used for two purposes in needlework: one to make two materials adhere together, the other to strengthen and stiffen Embroidery at the back. **1883** [sense 1 in Dict.]. **1895** *Montgomery Ward Catal.* 90/3 Embroidery cotton on spools. *Ibid.* 126/3 Embroidery Hoops, 4, 5, and 6 inches in diameter. *Ibid.* 448/3 Embroidery Scissors, polished laid steel. **1899** *Daily News* 28 June 8/4 A gigantic embroidery-like cloth of gold. **1909** *Englishwoman* Apr. 232 The lace-making and embroidery industry. **1909** *Embroidery* V. 119/1 Amongst the many existing varieties of embroidery stitches some of the prettiest and most attractive are those which bring to mind, more or less directly, natural forms and growth. *Ibid.* 124/1 That book decorations were frequent sources for embroidery designs is beyond doubt. **1938** A. G. I. CHRISTIE *Eng. Medieval Embroidery* 18 With the solitary exception of St. Dunstan..no name of an English medieval embroidery designer has yet been found. **1967** E. SHORT *Embroidery & Fabric Collage* iv. 99 The invention of sewing and embroidery machines during the Victorian era had its effect mainly on dress and furnishings. *Ibid.* 115 Other articles which give great scope to the embroidery designer are kneelers, banners and dossal hangings.

embryo-, combining form of Gr. ἔμβρυον EMBRYON, as in **e:mbryoca·rdia,** a condition of the heart in which its sounds resemble those of the fœtal heart (now *rare*); **e:mbryopla·stic** *a.*, pertaining to or participating in the formation of the embryo (*Cent. Dict.* 1889); **e·mbryoscope,** an instrument for examining embryos; so **e:mbryosco·pic** *a.* (*Cent. Dict.* 1889).

1890 GOULD *New Med. Dict.* 142/2 *Embryocardia*,..an affection of the heart, characterized by a heart-beat like that of a foetus. **1908** OSLER & MCCRAE *Syst. Med.* IV. 275 Tachycardia with embryocardia exhibits phenomena which correspond to a prolonged series of extra-systole. **1959** J. R. CHRISTIAN in A. A. Luisada *Cardiology* II. III. 147 The rapid rate and equal intensity of the sounds in the new-born produce a tic-tac type of rhythm or embryocardia, similar to that heard in the fetus. **1889** GEDDES & THOMSON *Evol. Sex* viii. 103 The minute area of formative protoplasm [that] the observers of to-day look down upon through their embryoscopes.

embryoma (embriōᵘ·mă). *Path.* Pl. **-mata.** [f. EMBRYO + *-OMA.] A tumour composed of tissues resembling those of the fœtus or thought to arise from fœtal tissues or fœtal remnants: applied esp. to some malignant tumours seen in childhood.

1903 J. H. TEACHER in *Jrnl. Obstet. & Gynæcol.* July 54 Either at once (congenital tumours) or after an interval (mixed tumours of later life) this developes into an imperfect organism—a teratoma or embryoma. **1954** L. B. AREY *Developmental Anat.* (ed. 6) x. 180 An immature type [of teratoma] with poorly differentiated tissue (embryoma) is highly proliferative and usually malignant. **1968** B. M. PATTEN *Human Embryol.* (ed. 3) viii. 167 To be classed as an embryoma a tumor must exhibit enough of the fundamental parts characteristically present in a normal embryo to suggest that it arose from an aggregation of cells which under other conditions might have produced a complete individual.

embryonically (embriǫ·nikăli), *adv.* [f. EMBRYONIC *a.*: see -ICALLY.] In (the) embryo.

1883 FOSTER & BALFOUR *Elem. Embryol.* (ed. 2) viii. 255 The atrophy of the dorsal section of the embryonically large canal of the spinal cord. **1899** E. GRIFFITH-JONES *Ascent through Christ* III. ii. 377 In prophecy He was as it were embryonically incarnated.

embryoniferous (embriǫ̆ni·fĕrəs), *a. Bot.* [f. EMBRYO + -(I)FEROUS.] Producing or bearing an embryo.

1819 J. LINDLEY tr. *Richard's Observ. Fruits & Seeds* 29 An examination of the embryoniferous cavity. **1834** R. BROWN *Misc. Bot. Wks.* (1866) I. 570 The remains of the embryoniferous areolæ, from four to six in number, were still visible.

embus (embʊ·s), *v.* [f. EM- 1 a + BUS *sb.*², after *embark.*] **a.** *intr.* To mount a bus or transport vehicle. **b.** *trans.* To transport by bus.

1915 [see *DEBUS *v.*]. **1927** *Daily Tel.* 13 Sept. 10/3 Using the mechanised transport thus released for embussing the headquarters. **1929** D. RORIE *Medico's Luck* vii. 141 On the 24th July we left Clairmarais for Lederzeele, and 'embussed'..for the XVIII Corps. **1940** *Spectator* 16 Aug. 161/2 Embussing happened in the last war; so did debussing...Archeologists embus when they go on their expeditions, at any rate in Wales. The term occurs in Field Service Regulations. **1955** *Daily Tel.* 20 Oct., Troops..had..on arrival 'embussed' for an unknown destination.

‖ **embusqué** (anbüske). [subst. use of pa. pple. of F. *embusquer* to ambush.] One who avoids military service (as by securing a post in a government office or the Civil Service). Also *attrib.* and *transf.*

1916 J. BUCHAN *Greenmantle* i. 4 Not some *embusqué* business in an office, but a thing compared to which your fight at Loos was a Sunday-school picnic. **1920** *Blackw. Mag.* May 586/2 These particular *embusqués*, who made Egypt a byword during the War. **1925** R. GRAVES *Welchman's Hose* 51 Brave fellows, no *embusqués* in that class. **1926** *Spectator* 7 Aug. 207/2 The best scenes are those between the poor *embusqué* and his wife. **1932** 'A. BRIDGE' *Peking Picnic* v. 51 Perhaps the real use of the presence of Europeans in Peking was to afford shelter to *embusqués* Chinese. **1955** *Times* 21 May 9/4 A number of idle and ill-disposed *embusqués*, whose principal aim in life is to blight their husbands' military career.

emcee·. Also (erron.) **emsee.** The names of the letters M and C, used to denote Master of Ceremonies, the introducer of a show or other form of entertainment. *slang* (orig. *U.S.*).

1933 B. J. CHIPMAN *'Hey, Rube'* 194/1 *Emsee*, Master of Ceremonies. **1938** *Sat. Even. Post* 2 Apr. 8/2 The emcee introduces the guest artists. **1940** *Variety* 24 July 34 Emcee announcer. **1955** L. FEATHER *Encycl. Jazz* vii. 227 He spent three years..in vaudeville as singer, dancer and emcee. **1964** *Listener* 27 Aug. 319/3 Emcee Steve Race knows his subject from the inside..and he gives *Jazz 625* an air of mature authority.

Hence **emcee** *v. trans.* and *intr.*

1937 *Hollywood Citizen* (Los Angeles) 14 Sept., Leon Errol..will emcee for the occasion. **1948** *Variety* 25 Aug. 2/1 George Jessel emceed that event.

-eme (īm), *suffix*, in *Linguistics* the termination of many names of significant or distinctive units of structure of some kind in the lexicon, grammar, and phonology of languages, e.g. *grapheme, lexeme, morpheme, phoneme, sememe, toneme.*

1953 W. J. ENTWISTLE *Aspects of Lang.* iii. 79 A by-product of Linguistic Analysis..has been the sudden burgeoning of the *-eme* family. **1962** H. A. GLEASON in Householder & Saporta *Probl. Lexicogr.* 98 Some kind of -emes;..no satisfactory term is at hand, though 'sememe' has been used.

emerald. Add: **5. d. emerald cuckoo,** an African cuckoo, *Chrysococcyx cupreus*, with green and gold plumage.

1876 H. BROOKS *Natal* iv. 136 Amongst the climbers of the coast bush there is one bird known as the emerald cuckoo. **1937** *Nature* 3 July 18/1 Among the birds [found in Arabia], three are new to science, namely, a race of the common magpie, a small Scops owl, and an emerald cuckoo. **1953** R. CAMPBELL *Mamba's Precipice* xi. 115 The four-note whistle of an emerald cuckoo. **1964** A. L. THOMSON *New Dict. Birds* 170/1 The Emerald Cuckoo *Chrysococcyx cupreus* has plumage of brilliant golden-green with bright yellow on the body.

emerge, *v.* Add: **4.** Also said of the production of a type by such a process as evolution.

1913 G. E. SMITH in *Rep. Brit. Assoc. 1912* 582 When the true mammal emerged. **1915** [see next].

emergence. Add: **2.** Also said of the result of an evolutionary process: cf. prec. and EMERGENT B. *sb.* 3.

1911 GEDDES & THOMSON *Evolution* 102 It is undeniably difficult to discover the factors in his emergence and ascent. **1913** G. E. SMITH in *Rep. Brit. Assoc. 1912* 577 The gradual emergence of human traits from the uncouth simian features of our ancestors. **1915** *Scientia* XVIII. 255 The emergence of anything new in the world... If intrinsic structure and external conditions are.. strictly similar, nothing new emerges. But if with like intrinsic structure the conditions are different, or *vice versa*, something new may emerge. And if genuinely emergent (as contrasted with resultant in accordance with G. H. Lewes's distinction) it may be unpredictable. **1920** S. ALEXANDER *Space, Time, & Deity* II. III. ii. 45 The emergence of a new quality from any level of existence means that at that level there comes into being a certain constellation or collocation of the motions belonging to that level, and possessing the quality appropriate to it, and this collocation possesses a new quality distinctive of the higher complex.

emergency. Add: **4. c.** *Cricket*, etc. An emergency man, a substitute. (No longer current.)

1851 *Nottingham Rev.* 5 Sept. 3/4 Emergency Williams, Esq., b. Goodrich. **1862** in W. G. Grace *Cricketing Remin.* (1899) i. 12 With this ball (presented by M.C.C. to E. M. Grace), he got every wicket in 2nd innings, in the match played at Canterbury, August 14, 15, 1862, Gentlemen of Kent *v.* M.C.C. for whom he played as an emergency, and in which, going in first, he scored 192 not out. **1885** J. LILLYWHITE *Cricketers' Compan.* 59 George Alexander ..only played as an emergency.

d. *spec.*, as a political term, to describe a condition approximating to that of war; *occas.* as a synonym or euphemism for war; also *state of emergency*, wherein the normal constitution is suspended.

1893 C. G. LELAND *Memoirs* 251 The rebels..had penetrated into Pennsylvania, and Philadelphia was threatened. This period was called the 'Emergency'. **1958** *Spectator* 17 Jan. 65/2 He has declared a state of emergency to suppress a strike of African railway workers. **1958** *Oxf. Mag.* 13 Mar. 374/2 The unmentionable word 'war' decently euphemised as 'emergency'.

5. Now esp. in general adj. sense 'used, issued, called upon, or arising in an emergency'.

1896 *Daily Chron.* 15 Aug. 11/6 She had been asked by the medical officer to take charge of the emergency brandy. **1898** *Daily News* 13 May 5/2 The emergency ration is never served out for more than five days consecutively. **1900** *Jrnl. Soc. Arts* 21 Sept. 802/2 The fitting of emergency brakes. **1902** *Young Engineer* I. 47/2 Doors are provided both for regular use and as emergency exits. **1903** *Westm. Gaz.* 11 Sept. 8/1 Directly the current between the Bank Station and the stations as far as the British Museum was cut off, an emergency current was turned on. **1904** F. F. MOORE *Original Woman* xxiv. 265 He was a man who had always at hand an emergency exit opening outward by which he could escape from any situation that was getting too hot for him. **1915** *Lit. Digest* (N.Y.) 21 Aug. 348/1 All of the [beaver-]ponds are equipped with 'emergency exits' in the form of holes in the bank. **1920** *Act* 10 *Geo. V* c. 5 (*title*) An Act to continue temporarily certain emergency enactments. **1921** *Dict. Occup. Terms* (1927) § 729 *Emergency man*..a tramway conductor..who is held in readiness to replace anyone of depot cashier's assistants who may be on leave or sick. **1925** W. DEEPING *Sorrell & Son* xxx. 296 He asked you to do an emergency job for him in the [operating] theatre. **1925** A. S. M. HUTCHINSON *One Increasing Purpose* I. xx, 'Have you a girl on your staff called Glade?' 'Not on my staff... Miss Glade is one of my emergency-calls. I get her when I want her, if she is available.' **1929** *Star* 21 Aug. 12/1 Glamorgan's emergency bowlers. **1935** *Archit. Rev.* LXXVII. 206 (*caption*) Looking along the outside wall of the lecture hall, and at one of the lecture hall emergency exits. **1940** *Ann. Reg. 1939* 397 Many of whom [*sc.* lawyers] were chosen as chairmen of tribunals and committees under one or other of the Emergency Acts. **1944** *Living off Land* v. 102 Carry an emergency ration in the shape of a tobacco tin of salt and oatmeal, well mixed. **1944** A. THIRKELL *Headmistress* iii. 64 An emergency card for her rations. **1956** A. H. COMPTON *Atomic Quest* 259 The fliers returned to an emergency landing at Okinawa. **1961** *Lancet* 12 Aug. 338/2 Data for the occurrence of disease were derived from the Emergency Bed Service (E.B.S.) of the King Edward Hospital Fund for London. **1970** *Ibid.* 19 Sept. 604/2 The past decade has seen the establishment of emergency call services to supplement practice rota systems.

emergent. Add: **A.** *adj.* **2. c.** *Science.* That emerges unpredictably as the result of an evolutionary process, *spec.* in *emergent evolution.*

1915 [see *EMERGENCE 2]. **1923** C. L. MORGAN (*title*) Emergent evolution. *Ibid.* i. 1 Under what I here call emergent evolution stress is laid on this incoming of the new. **1928** *Observer* 1 June 5/3 That growing body of thought called 'Emergent Evolution'. **1932** *Discovery* Apr. 108/2 One of the salient features in recent aetiology (i.e. evolution lore) has been the recognition of the more or less open secret expressed in the term 'emergent evolution'... It has become evident that the Ascent of Life has been a succession of 'emergent' steps, novelties that are creative rather than 'additive', such as birds from ancestral reptiles.

3. b. Of a nation: that is newly independent; of a people: that is conscious of its national identity.

1954 *N.Y. Times Mag.* 31 Oct. vi. 13 The ambitions and the unquenchable hope of emergent Africa. **1957** H. THOMAS *World's Game* xv. 191 The chauvinistic young foreign ministers of emergent countries. **1960** *Daily Tel.* 13 Jan. 10/2 Each of the 'emergent' territories in Africa has different problems, to which each must find its own best solution. **1963** *Listener* 7 Feb. 233/1 They [*sc.* the Fijians] just will not be emergent or emancipated.

characteristic of the period of the French Empire (see *5 b (c) and (d)).

1869 Lady C. Schreiber Jrnl. (1911) I. 29 Green cup and saucer, imitating 'Empire' Sèvres. 1870 O. Logan Before Footlights 292 How do you manage to pay $60 for your new but ugly little Empire bonnet? 1879 Encycl. Brit. IX. 849/2 The 'empire' style, a stiff, affected classicalism, prevailed in France during the reign of Napoleon. 1887, 1888 [see Dict., 8]. 1889 R. Brook Elem. Style Furnit. 29 As in all other French styles, 'Empire' was closely imitated in this country. Ibid., It is impossible to have a better authority on 'Empire' Furniture, than the book of designs published in Paris, by the architects, Percier and Fontaine, in 1809. 1901 E. Singleton Furnit. of our Forefathers II. 573 Empire sofa owned by Mrs. William Young, Baltimore, Md. 1904 H. E. Binstead Furnit. Styles x. 116 It is never difficult to determine what is Empire. 1905 A. Hayden Chats on Old Furnit. 208 The wood used for..Empire cabinets is rich mahogany. a 1910 'O. Henry' Sixes & Sevens (1916) xiii. 129 High-collared, baggy, empire-waisted, ample-skirted. 1958 Vogue Apr. 18 Choose this Empire-line charmer in swirling full length. 1967 E. Short Embroidery & Fabric Collage ii. 41 The same design would have been unthinkable on say an Empire dress of seventy years earlier. 1968 Harrods Xmas Catal. 15/2 An Empire line slip in nylon. 1970 Oxford Times 25 Sept. 11/5 She wore a full length white empire line dress with a guipure lace bodice and circular train trimmed with guipure lace.

c. Applied to wines and spirits grown in the British Empire (Commonwealth) and imported into Great Britain.

1937 Auden in Auden & MacNeice Lett. f. Iceland v. 57 Someone may think that Empire wines are nice. 1954 P. Frankau Wreath for Enemy III. v. 210 You would look like an advertisement for Empire Burgundy. 1965 R. Jeffries Dead against Lawyers ix. 94, I can offer you Empire sherry or a beer?

empirical, *a.* Add: **4.** *empirical ego* = *empirical self; *empirical philosophy* = Empiricism 2 b (cf. also Pragmatism 4); *empirical psychologist*, an exponent or adherent of empirical psychology; *empirical psychology*, the science of the mind developed by observation and experiment, rather than by deduction from general principles (opp. *rational psychology*); *empirical self* (see quot. 1890).

1851 H. L. Mansel Proleg. Logica ix. 275 Among modern philosophers, empirical psychology..is frequently classified as metaphysical. 1870 S. H. Hodgson Theory of Practice I. ii. 252 Every feeling and every object in the whole empirical ego stands in some relation to it [sc. the emotion of moral sense]. 1890 W. James Princ. Psychol. I. x. 291 The Empirical Self of each of us is all that he is tempted to call by the name of me. 1892 —— Coll. Ess. & Rev. (1920) 321 We certainly need something more radical than the old division into 'rational' and 'empirical' psychology. 1901 —— in Proc. Soc. Psychical Res. XLII. 21, I can therefore speak ..as a mere empirical psychologist, of Myers's general evolutionary conception. 1902 —— Var. Relig. Exper. xiv. 374 According to the empirical philosophy..all ideals are matters of relation. 1949 Mind LVIII. 124 The empirical psychologist is almost of necessity an epiphenomenalist in practice. 1963 R. P. Wolff Kant's Theory Mental Activ. I. ii. 144 Only the empirical self is knowable. 1967 S. J. Todes in R. P. Wolff Kant 164 The distinguishing feature of the empirical ego is that all empirical knowledge must be in terms of it.

empiricism. 2. (Earlier examples.)

1796 F. A. Nitsch Kant's Princ. concerning Man 218 The comfortless attacks of Fatalism, Scepticism, and Empiricism. 1798 A. F. M. Willich Elem. Crit. Philos. 7 A slavish dependence upon the Empiricism of Locke.

empiricist. Add: (Earlier and later examples.)

1698–1712 Shaftesbury Phil. Regimen (1900) 207 The prescriptions of the vulgar-wise, like those of the Empiricists. They know only the symptom; apply only to the symptom. 1955 Sci. Amer. Aug. 86/1 Now, through Bertrand Russell, almost all British and American empiricists are to some degree disciples of Hume.

B. *adj.* Of, or relating to, or characterized by, philosophical empiricism.

1890 W. James Princ. Psychol. I. vii. 195 Empiricist writers are very fond of emphasizing one great set of delusions which language inflicts on the mind. 1907 —— Pragmatism ii. 51 Pragmatism represents a perfectly familiar attitude in philosophy, the empiricist attitude. 1953 H. H. Price Thinking & Exper. viii. 252 In any Empiricist theory of thinking..there has to be a doctrine of cashability. 1965 N. Chomsky Aspects Theory Syntax 206 There is no justification for the common assumption that there is an asymmetry between rationalist and empiricist views.

empirico- (empi·riko), taken as comb. form of Empirical *a.*, Empiricism, etc., usually in some such sense as 'partly empirical and partly...'. Also **empi·rio-**.

1895 A. C. Armstrong tr. Falckenberg's Hist. Mod. Philos. II. xvi. 622 The preponderance of natural science and the empirico-skeptical tendency of the philosophy of the day conditioned by it. 1897 C. H. Judd tr. Wundt's Outl. Psychol. v. 318 This general principle is known as the principle of psycho-physical parallelism. It has an empirico-psychological significance and is thus totally different from certain metaphysical principles. 1937 C. B. Weinberg (title) Mach's empirio-pragmatism in physical science. 1938 Times Lit. Suppl. 22 Jan. 52/3 A distinction of prime importance between the 'empirico-mathematical' and the 'empirico-schematic' sciences. Ibid. 52/4 The empiriometric and empirio-schematic analysis of observable reals. 1942 Mind LI. 306 Intuitionism..is particularly out of harmony with the more extreme types of empirico-formalism which are so influential at the moment.

empirio-criticism (empi:rio,kri·tisiz'm). *Philos.* [tr. G. *empiriokritizismus* (R. Avenarius 1894, in *Vierteljahrschr. f. Wiss. Philos.* XVIII. 138), f. *empirio-* *Empirico- + *kritizismus* criticism.] The philosophy of Richard Avenarius (1843–96), a form of positivism consisting primarily in the rejection of the dualism of body and mind, and the restriction of knowledge to a basis in experience treated critically and so deprived of all metaphysical elements.

1897 Mind VI. 449 (title) Richard Avenarius and his General Theory of Knowledge, Empiriocriticism. Ibid. 451 'Critical empiricism'..looks for the universal element in experience...Empiriocriticism, on the other hand, takes up the position that everything is experience when it has been stated as experienced by an individual—though it may be that..it is only experience for this one individual in question...This empiriocriticism also approaches experience critically, but it does not determine its concept of experience beforehand. 1933 Mind XLII. 379 All his [sc. Lenin's] philosophical work, the greater part of which appears in the volume Materialism and Empirio-Criticism, reveals his partisanship. 1955 H. B. Acton Illusion of Epoch I. i. 19 Books by Ernst Mach and Richard Avenarius in which, under the name of 'empirio-criticism', a phenomenalist account of matter was advocated.

Hence **empi:rio-cri·tical** *a.*, of, or relating to, or characterized by, empirio-criticism.

1909 in Cent. Dict. Suppl. 1925 F. Znaniecki Laws Soc. Psychol. i. 1 The arguments against the possibility of psychological laws may be reduced to three. The first has been put forward by the so-called empirio-critical school of German philosophers (Avenarius, Petzoldt, and others), and..by the neo-criticists (Natorp, etc.). 1933 Mind XLII. 379 The Russian followers of the empirio-critical philosophy of Mach and Avenarius. 1949 I. Deutscher Stalin iv. 106 Neo-Kantian or empirio-critical philosophy.

emplace, *v.* Delete *rare*, and add examples, chiefly in spec. sense of providing an emplacement for guns.

1900 Daily News 22 Jan. 3/2 This morning the Boers emplaced a five-pounder at the brick fields, and proceeded to drop shells into the market square. 1904 Daily Chron. 28 May 5/4 A series of batteries, strongly emplaced, crowded the crest of the hill. 1915 J. Buchan Nelson's Hist. War V. 26 They had an ordinary range of four to five miles, and this allowed them to be emplaced well to the rear out of any danger from the enemy. 1958 P. Kemp No Colours or Crest xii. 269 A detachment of our escort had emplaced itself behind a low bank.

emplane (emplē·n), *v.* Also **enplane.** [f. Em- 1 a + *Plane *sb.*[5]] *trans.* and *intr.* To take or go on board an aeroplane.

1923 Westm. Gaz. 2 July 3/2 Two Vickers-Victoria machines arrived, and emplaned the distressed garrison. 1930 Air July 292 The Prime Minister and Miss Ishbel MacDonald emplaning for a week-end in Edinburgh. 1941 F. H. Joseph in M. Wheeler Britain at War 95 Group was first enplaned, then those passengers taken on. 1944 Cape Times 25 May, Men of the Fifth Indian Division had no idea where they were going when they emplaned. 1966 New Statesman 14 Jan. 58/3 A popular radio actress..is emplaned to Paris in the early Sixties. 1966 Catholic Standard 21 Oct. 4/6 He [sc. Krushchev] suddenly pulled up stakes and enplaned with all haste for Moscow. 1970 Daily Tel. 25 July 10 The usual Kennedy Airport congestion caused us to wait 2½ hours, after emplaning, for takeoff.

empleomania (e:mplēŏmē·niă). [Sp., f. *empleo* employment + Mania.] A mania for holding public office.

1878 J. R. Lowell in Century Mag. (1898) Nov. 144/1 The empleomania, which is the dry-rot of Spain. 1920 Glasgow Herald 23 Oct. 12 The evil which has sapped the vigour of so many nations—'empleomania'—has made its insidious way into British administrative departments. 1922 Ibid. 4 Jan. 4 This creed, adoption of which begets the insanity of empleomania..has had a stranglehold upon the whole Ibero-American Continent.

employability (emploi,ăbi·liti). [f. Employab(le *a.* + -ility.] The character or quality of being employable.

1926 A. M. Carr-Saunders Eugenics vii. 157 Categories (a) and (b) of employability account for 89·4 per cent. of the men and 88·2 per cent. of the women. 1927 Daily Tel. 28 June 7/2 The scheme..is for the purpose of so improving the general employability of young unskilled men. 1959 B. Wootton Social Sci. & Social Path. vii. 223 Even the test of employability or of working efficiency does not serve to distinguish the sick from the sound.

employee. Now, also with pronunc. (emploi·ī), a more usual form than *employé*. Add earlier and later examples.

1850 L. H. Garrard Wah-To-Yah xii. 172 Horses and mules..were here herded, by their employees. 1891 Pall Mall Gaz. 23 Oct. 2/1 To arrange a forty-eight hour week for the few binders, while retaining the fifty-four hours for the bulk of the employees. 1906 Daily Chron. 9 May 5/5 'I don't like this affectation of "employee",' observed Judge Addison, in the Southwark County Court. 'I prefer English words.' 1909 Ibid. 15 Dec. 1/3 The employee shares in the company are 50,000 of £1 each. 1928 Britain's Industr. Future (Lib. Ind. Inq.) III. 141 The stimulation of employee-ownership under schemes of profit-sharing and investment by employees. 1954 J. A. C. Brown Soc. Psychol. Industry iii. 84 The supervisors of high production groups were those who were..more employee-centred.

¶ In U.S. often written **employe.**

1904 N.Y. Times 26 Mar. 1 Receiver Taft called the employes of the failed firm into his office. 1923 Childs & Cornell Office Administr. 258 The training of a new employe. 1930 Chicago Daily News 25 Aug., The first annual picnic of employes of The Daily News and their families.

employment. Add: **5.** *attrib.* (in sense 2 c) *employment agency, agent, bureau, exchange*, etc., professional intermediaries between applicants for work and employers.

1886 J. A. Porter New Stand. Guide Washington 204 Dondore and Morse, Attorneys..also Employment Bureau. 1888 12th Rep. Ohio Bur. Labour Statistics 263 'Employment agencies'..have very appropriately been characterized as 'a class who trade on the needs of the inexperienced searcher for honest employment'. 1903 Westm. Gaz. 23 Dec. 5/2 All the corporations have resolved to support the bakers, with the object of obtaining the complete suppression of the employment agencies [in Paris]. 1909 Lancet 11 Sept. 830/1 (heading) An employment exchange for university undergraduates. 1921 Dict. Occup. Terms (1927) 771 Employment agent; employment bureau manager. 1945 Archit. Rev. XCVIII. 115/1 Mushroom employment-agencies and dubious correspondence courses play on the soldier's anxiety and issue him promissory notes which will never be honoured. 1961 Technology Feb. 45/2 Youth employment service. Ibid., The company contact the youth employment officer. Ibid., It is probably this kind of situation that evoked from Miss Avent the words 'employment exchange'.

empolder (empōu·ldər), *v.* [f. Em- 1 b + Polder[1].] *trans.* To make a polder of; to reclaim from the sea. Cf. *impolder *v.* So **empo·ldered** *ppl. a.*

1889 Cent. Dict., Empoldered, reclaimed and brought into the condition of a polder; brought under cultivation. 1922 Blackw. Mag. July 14/2 A few acres have been cleared and empoldered with mud-dams. 1951 New Biol. XI. 77 By 1940 flood-fallowing experiments had been begun. An area of 100 acres..was empoldered (i.e. surrounded by earth walls).

‖**empressé** (aṅprɛse), *a.* [Fr., pa. pple. of *empresser*, f. *em-* = *en-* + *presser* Press *v.*[1]] Eager, zealous.

1837 J. F. Cooper Recoll. Eur. II. ii. 42 All the French women were exceedingly empressées in their manner towards the Great Unknown. 1878 L. W. M. Lockhart Mine is Thine II. xvii. 23 Be low, be depressed, but, at the right moment, empressé and earnest. 1906 W. De Morgan Joseph Vance xxxiii. 325 'Was he very empressé in his manner?' I asked. 'Spooney, do you mean?'

empressite (e·mprěsəit). *Min.* [f. the name of the *Empress*-Josephine mine, where it was first found: see -ite[1].] A telluride of silver, of disputed composition (see quots.).

1914 W. M. Bradley in Amer. Jrnl. Sci. 4th Ser. XVIII. 163 Some years ago, what appeared to be a new mineral was found in the Empress-Josephine Mine in the Kerber Creek District of Colorado by Professor R. D. George of the University of Colorado...The mineral gives a ratio of practically Ag:Te::1:1...The name empressite has been given to it by Professor George. 1956 Amer. Mineralogist XLI. 722 Ag_5Te_3 is synthetic empressite. $Ag_{5-x}Te_3$ is the formula deduced, for the mineral, from cell dimensions and density of analyzed crystals. 1964 Ibid. XLIX. 326 Chemical analysis..and x-ray single crystal data on natural material from the type locality show that Bradley's original composition of AgTe was correct, and it is proposed that the name empressite be retained for this material. 1965 Ibid. L. 797 The mineralogy of the silver tellurides is still far from completely understood. Is it not then premature to redefine the term 'empressite' for which a composition near Ag_5Te_3 has long been accepted? 1968 Ibid. LIII. 1513 'Empressite' from the type locality consists of two optically distinct phases with the formulae AgTe and $Ag_{5-x}Te_3$.

empting, *vbl. sb.* Add: (Earlier and later U.S. examples.) Also **emptin.**

1650 Salem (Mass.) Q. Court Recs. in Essex Antiquarian (1903) VII. 28 Elizabeth Pinion,..having come into the house to borrow some emptings, Tobiah took her and threw her [etc.]. 1839 'M. Clavers' New Home xviii. 120 Mrs. Doubleday..keeps her husband's house..in unexceptionable trim. Her emptin's are the envy of the neighbourhood. [For 1884 in Dict. read 1848.] 1889 R. T. Cooke Steadfast vi. 74 Everybody wanted some of her 'empt'ins' to set their own. 1894 M. E. Wilkins Pembroke 46, I—wanted to see—if—Sylvy had any emptins. 1902 Clapin Dict. Amer. 173 To run emptins, to show signs of not holding out well, as for instance a speech or an enterprise of any kind.

empty, *a.* and *sb.* Add: **A.** *adj.* **3. d.** Of a cow or other farm animal: not pregnant. Cf. Full *a.* 1 e.

1900 in J. Wright Eng. Dial. Dict. II. 252/1. 1950 N.Z. Jrnl. Agric. Feb. 100/1 Many cows thought to be empty have proved to be in calf when slaughtered at the works. 1960 Farmer & Stockbreeder 23 Feb. 80/3 Empty gilts shared in the general enthusiasm, averaging £41 15s 2d.

4. spec. *Empty Quarter* [rendering Arab. *Ruba el-Khali*], the great southern desert of Arabia (formerly identified with the Dahna); also *fig.*

[1875 Encycl. Brit. II. 240/1 This greater desert, the 'Roba El Khaliyeh' or 'Empty Space' of geographers—the 'Dahna' or 'Crimson' of modern Arabs.] 1910 Encycl. Brit. II. 260/1 The great desert known as the Dahna or the Rub'a el Khali ('the empty quarter'). 1913 T. Houtsma et al. Encycl. Islam I. 370/2 Central Arabia... The desert has several names..generally it is called al-Dahnā', 'the red country', after the colour of the sand. On the maps it is indicated al-Rub' al-Khālī, i.e. the empty quarter. 1929 T. E. Lawrence Let. 12 July (1938) 663 They can pass over the Ruba el Khali, the so-called 'Empty Quarter' of Arabia. 1932 B. Thomas (title) Arabia felix, across the Empty Quarter of Arabia. 1959 Listener 19 Nov. 893/1 The Empty Quarter stretches for nine hundred miles across southern Arabia. 1963 Guardian 5 July 8/3 One hails him as the Hakluyt of that empty quarter where the parched intellect croaks for a swig of feeling.

B. *sb.* Also, an empty cab or taxi; an empty house or premises; an empty bottle, container, etc. *colloq.*

1879 F. H. Grundy Pictures of Past vi. 166 'Well,' says Leigh Hunt, 'I found him [sc. a cabman] returning from Hammersmith, and he said as an empty he would take me for half-fare.' 1905 Westm. Gaz. 23 Aug. 8/2 Property owners throughout the various suburbs of London are making loud complaint of the steady increase in the proportion of 'empties'. 1914 Evening Post (Wellington, N.Z.) 24 Feb. 4 Get as many 'empties' [= bottles] as possible. 1924 Ladies' Home Jrnl. June 100 He delivers it by the case and calls for the empties. 1938 G. Greene Brighton Rock VII. i. 277 A cupboard stood open full of empties. 1961 L. van der Post Heart of Hunter I. v. 82 Some fuel-drums lay in a neat line..all empties.

C. *attrib.* and *Comb.* **empty-cell** used *attrib.* (see esp. quot. 1946); **empty word** *Linguistics*, a word which has no meaning in itself but serves a grammatical function, e.g. *at, but, for*, etc.

1917 A. J. Wallis-Tayler Preservation of Wood vii. 199 This process [sc. the Rueping or Rüping Process] is on what is known as the empty cell treatment. 1946 Cartwright & Findlay Decay of Timber xiii. 260 It is usual to treat poles and timber for building use by the empty cell process, by means of which the walls of cells of the wood are left coated with a layer of preservative, but the cavities are empty. 1968 W. E. Willis Timber iv. 86 The two empty cell methods are known as the Rueping and Lawry processes. 1892 H. Sweet New Eng. Gram. I. 22. § 58 When a form-word is entirely devoid of meaning, we may call it an empty word, as opposed to full words such as earth and round. 1953 W. J. Entwistle Aspects of Lang. x. 298 Chinese.. makes considerable use of 'empty words', that is of auxiliaries in the broad sense of the term, which includes prepositions as well as subordinate members of the verbal complex. 1968 P. Kratochvíl Chinese Lang. Today iv. 117 Traditional Chinese linguists considered practically all minimal forms, beside what would be called nouns, as 'empty words'.

empyema. The more frequent pronunciation is now (empəi,ī·mă).

em quad, quadrat: see *em.

emu. Add: Now the usual spelling of Emeu.

1936 I. L. Idriess Cattle King v. 46 In a good season emus and kangaroos are fairly plentiful. 1944 Living off Land ii. 16 Grubs, snakes, emus, and kangaroos can be palatable. Ibid. 28 Emu meat, when cooked in the skin, is inclined to be oily. 1968 K. Weatherly Roo Shooter 76 Many people say that the emu is the fastest creature in the bush.

2. **emu-apple**, an Australian tree, *Owenia acidula*, which bears a red fruit; **emu-bush**, either of two Australian shrubs (see quot. 1889).

1898 Morris Austral Eng. 137/2 Emu-apple. 1905 J. H. Maiden Forest Flora N.S.W. II. 90 It [sc. Owenia acidula] is called 'Sour Plum', 'Native Peach or Nectarine', and 'Emu Apple'. 1889 —— Useful Nat. Plants Austral. 132 Heterodendron oleæfolium..Emu Bush.. All the colonies except Tasmania. Ibid. 317 Eremophila longifolia..'Emu Bush', owing to emus feeding on the seeds of this and other species...In all the colonies except Tasmania. 1910 C. E. W. Bean On Wool Track 86 The emu-bush, which droops like the bunch of an emu's tail and is very good to eat—as the rabbits have found. 1931 V. Palmer Separate Lives 211 Country that was a maze of emu-bush and mulga scrubs. 1936 F. Clune Roaming round Darling xvii. 165 The currant or emu bush: needle-leafed, with fruit like blackberries, each one the size of a pea—not caviare to emus.

emulsifier (ĭmv·lsifəi,əɪ). *Chem.* [f. Emulsify *v.* + -er[1].] An agent or apparatus which effects the emulsification of an oil, etc.

1888 Chemist & Druggist XXXII. 28/2 An Emulsifier Wanted. 1902 C. J. S. Thompson Pract. Dispensing (ed. 3) iii. 28 When making an emulsion the oil to be emulsified should be added to the emulsifier. 1928 Manch. Guardian Weekly 26 Oct. 323/4 Milk..is being produced in increasing quantities from an 'emulsifier'. 1954 Thorpe's Dict. Appl. Chem. (ed. 4) XI. 345/2 The emulsifier, which promotes the dispersion of one liquid in another, acts as a bridge between the continuous phase and the dispersed phase.

emulsion. Add: **4.** *Photogr.* A mixture consisting of a silver compound held in suspension in collodion or gelatin, used in coating plates, films, etc. Also *attrib.*

1840 J. F. W. Herschel in Phil. Trans. CXXX. 8 My first attention was directed to the discovery of a liquid, or emulsion, which by a single application, whether by dipping or brushing over, should communicate the desired quality [of sensitiveness]. 1877 English Mechanic 23 Nov. 261/2 Emulsions..consisting of collodion and silver, ready for pouring upon plates. 1878 W. W. Abney Treat. Photogr. xvi. 115 All the different varieties of the emulsion processes. Ibid. xxii. 153 If a glass plate has to be coated with the emulsion, the same directions as those given for coating emulsion plates should be followed. 1880 [see Dict., 3]. 1908 Westm. Gaz. 18 Jan. 14/2 The emulsion side of the film or plate. Ibid. 25 July 14/2 This positive can be made on a silver emulsion plate. 1958 M. L. Hall Newnes Complete Amat. Photogr. 75 Individual semi-rigid sheets of thicker, transparent film, emulsion-coated, designed for darkroom loading into film-holders.

5. **emulsion paint**, a water-thinned paint containing a non-volatile substance, e.g. synthetic resin, as its binding medium.

1939–40 Army & Navy Stores Catal. 238/2 Hall's distemper. Oil-emulsion paint. 1944 G. M. Sutheim in J. J. Mattiello Protect. & Decorat. Coatings IV. xi. 331 Emulsion paint is a surface coating in which the vehicle is an emulsion. Ibid. 333 The main difference between emulsion paint and ordinary or improved water paints is the fact that emulsion paints form continuous hydrophobic films on drying. 1958 Woman 22 Feb. 2/1 Emulsion paint goes over old emulsion. 1960 House & Garden June 72/1 Emulsion paints..need no special undercoat, are speedy to apply, dry in a few hours.

emulsoid (ĭmv·lsoid). *Chem.* [G. *emulsoid* (P. P. von Weimarn 1908, in *Zeitschr. f. Chem. & Ind. d. Kolloide* III. 27/2), f. Emuls(ion + -oid.] A liquid which appears to contain another liquid in colloidal suspension (see also quot. 1956).

1909 Chem. Abstr. III. 394 Classification of Colloids... The physical aggregations..are divided into two classes suspensoids..and emulsoids, in which the particles are minute drops of liquid. 1913 Ibid. VII. 2144 An experimental test for emulsoids lies in their coagulability to two liquid layers. 1925 J. Alexander Colloid Chem. (ed. 2) 27 The reversible colloids are therefore called emulsoids and the irreversible colloids suspensoids. 1938 Science 4 Mar. 212/1 Originally 'emulsoids' included coagula, jellies and emulsions, the term being based on the assumption that jellies are fine emulsions. 1956 J. S. Anderson tr. Remy's Treat. Inorg. Chem. II. xvi. 686 Colloidal emulsions are called emulsoids. [Note] The terms 'emulsoid' and 'suspensoid' are often applied also to the sols of lyophilic and lyophobic colloids, respectively, without regard to the state of aggregation of the disperse phase.

en (en) *sb.* The name of the letter N. In *Printing*, the half square, formerly of the type n, used as a unit for measuring the amount of printed matter in a line, page, etc. Also *attrib.*, esp. in *en quad, quadrat*, a block of metal half the width of the *em quad (see also quot. 1967); *en rule*, a dash of the width of an en. Cf. N I. 1 b.

1793, 1808 [see *em]. 1875 J. Southward Dict. Typogr. (ed. 2) 32 En dash.—A dash..half the length of the em of the fount to which it belongs. 1888 Pall Mall Gaz. 10 Sept. 11/2 The 'justifier'..spaces out the lines with great regularity and in so short a time that 20,000 ens per hour is about the average output. 1891 Printers' Register 7 Dec., 49,000 American ems, equal to 98,000 English ens. 1893 Hart Rules for Compositors 21 En rules are to be used in such cases as 1880–1, and not hyphens or em rules. 1904 Goodchild & Tweney Technol. & Sci. Dict. 204/1 En quads, spaces exactly half of an em. 1921 Glasgow Herald 22 Feb. 10 The first scale paid to linotype operators was 3d. per thousand 'ens'. 1967 En quad [see *em].

‖**en** (aṅ), *prep.* [Fr., = in; as (a).] **en attendant**, in the meantime, while (one is) waiting; **en avant**, forward; **en axe** (see quot.); **en barbette** (see Barbette *sb.*); **en beau**, in a favourable manner; **en brosse**, of hair, cut short, giving a bristly effect; **en cabochon** (see Cabochon and quot. 1940); **en casserole** (see Casserole 2); **en clair**, in ordinary language (not in cipher); **en cœur**, in dress-making, heart-shaped, V-shaped; **en coquille** (see quot. 1882); **en déshabillé**, in undress; also *fig.* (see Dishabille); **en évidence**, in evidence, in the forefront, conspicuous(ly); **en face**, from the front, facing forward, (*Bibliogr.*) opposite; **en famille**, in or with the family, as one of the family, at home; **en fête**, in festival array, keeping holiday; **en garçon**, as a boy, as a bachelor; **en grand seigneur**, like a lord; **en grande tenue**, in full dress; **en gros**, in general, in broad terms; **en l'air**, 'in the air'; (of troops) unsupported; **en noir**, on the black side; **en pantoufles**, lit. 'in slippers', hence, in a free and easy atmosphere; **en pension**, as a boarder in lodgings; **en permanence**, permanently; **en pointe(s)** *Ballet*, on the extremity of the toe(s); **en poste**, of a diplomat, in an official position (at a specified place); **en prince**, like a prince, in a princely manner; **en principe**, in principle; **en prise** *Chess* etc., in a position to be taken; **en rapport** (see Rapport *sb.* 3 and in Dict. as separate entry); **en regard** *Bibliogr.* = *en face*; **en règle**, in due form; **en retraite**, in retirement, on half pay; **en revanche**, in return, as a quid pro quo; **en route** (see Route *sb.* 5); **en secondes noces**, by a second marriage; **en suite** (see Suite 5 and *en suite); **en train**, in progress; concerned, occupied (usu. const. *with*); **en ventre sa mère** [legal Fr.], in the womb, unborn; **en ville**, away from home. See also *en bloc, En masse, *en passant, *en plein, *en tout cas.

1743 H. Walpole Let. 12 Oct. (1903) I. 389, I hope in time to have poems upon him...En attendant, I have sent you some pamphlets to amuse your solitude. 1793 Gibbon Let. 25 Nov. (1956) III. 365 Leave to go abroad to-morrow and to go out of town when I please en attendant the future measures of a radical cure. 1801 M. Edgeworth Belinda xii, En attendant here are your two hundred guineas. 1931 N. & Q. CLXI. 73/2 We are still waiting..for his work to be accessible in English. They [sc. our readers] may, en attendant, like to know of an article..in the current number of the Dublin Review. 1815 J. Simpson Visit to Flanders iii. 122 The man who, when the lives of others were concerned, could only cry, 'en avant, en avant'. 1823 Byron Let. 22 Apr. in Moore Lett. & Jrnls. (1830) II. 641 But never mind—en avant! live while you can. 1836 tr. Bourrienne's Mem. N. Bonaparte xiv. 176 'Very well,' said the chief consul, 'en avant—let us proceed.' 1845 R. Ford Handbk. Spain I. II. 218/2 The easy victories of the French over the Spaniards were mainly owing to their dashing en avant charges. 1901 R. Sturgis Dict. Archit. 178/2 A monument is said to be en axe with a street when the centre line of the street passes through the centre of the monument. Two rooms are said to be en axe, or an opposite window or door, or two opposite doors, are said to be en axe when the axis of the room, pavilion, wing, or whole building passes through them, or even when they are centred upon one another with deliberate care to bring them exactly opposite. 1818 Edin. Rev. XXX. 315 We have every reason certainly to be satisfied with the account she gives of us... We are certainly painted en beau. 1832 Mill Let. 27 Dec. in Wks. (1963) XII. 132 A sketchy kind of book,..intended to throw light upon France, painting it & all it contains en beau. 1875 Contemp. Rev. XXVI. 553 Painted en beau by the Ins or en noir by the Outs. 1901 'L. Malet' Hist. Sir R. Calmady VI. ii. 500 Zimmermann, the colossal Swiss-German courier, with his square, yellow beard and hair en brosse. 1909 H. G. Wells Tono-Bungay II. iii. 192 Ewart had returned with his hair cut en brosse. 1924 M. Kennedy Constant Nymph i. 15 He..wore his dark hair en brosse. 1957 R. Mason World Suzie Wong II. viii. 239 The familiar shape of the head under the en brosse hair. 1889 En cabochon [in Dict., def. of Cat's-eye 2]. 1910 J. Buchan Prester John (1922) xi. 150 There were fifty-five rubies in it...In shape they were oval, cut on both sides en cabochon. 1940 Chambers's Techn. Dict. 294/2 En cabochon, a style of cutting used in the case of certain gemstones, notably garnets...Such stones are not facetted, but a smooth-domed surface is produced, the plan of the stone being circular or oval. 1897 En clair [in use at the Foreign Office]. 1920 'Taffrail' H.M.S. Anonymous ix. 189 The German commandant..made the following wireless signal 'en clair' to the British men-of-war. 1928 Daily Tel. 17 July 10/7 Making 'en clair' wireless signals. 1960 C. Mackenzie Greece in my Life 53 He went back to the Chancery and sent off a rousing telegram to the Foreign Office en clair. 1873 Young Englishwoman Feb. 78/1 A dinner dress..with corsage open en cœur. 1874 Echo 30 Dec. 2/2 These bodies [sc. corsages] may be.. opened en cœur or square. 1882 Caulfeild & Saward Dict. Needlework, En cœur. 1878 Cassell's Fam. Mag. Apr. 295/1 The skirt is..bordered with a flounce en coquille. 1882 Caulfeild & Saward Dict. Needlework, En Coquille, the French term to denote 'shell-shaped'. The ribbon or lace is laid like a succession of scallop-shells, one above or over the other, in groups of threes. 1699 M. Lister Journey to Paris 35 A young Gentleman in a Fur Cap en dishabille. 1808 Edin. Rev. Oct. 133 A powerful mind en deshabille, and free from the fetters of study. 1877 C. Reade Woman Hater I. vii. 172 Let me catch her en déshabille, with her porter on one side, and her lover on the other. 1818 Lady Morgan Fl. Macarthy II. i. 6 Mr. Crawley now placed himself en evidence at his window. 1858 Lytton What will he do with It? II. v. x. 226 George Morley and his fair cousin walked boldly, en évidence, before the prying ghostly windows. 1888 S. Buxton Handbk. Pol. Quest. Day (ed. 7) 153 The anxiety of members to keep themselves en évidence. 1911 E. M. Clowes On Wallaby vi. 162 The lack of courtesy, of patience, and consideration is all fully en évidence. 1768 Sterne Sent. Journ. I. 41, I..eyed him as he walk'd along in profile,—then en face. 1931 Periodical (Suppl.) Dec., The first edition text is printed with a transcript of the manuscript en face. 1936 Burlington Mag. Mar. 117/2 She has turned slightly to the right and is looking at the spectator, her head en face. 1726 A. Pope Let. June (1956) II. 380, I shall be much pleased to see you here en famille, but pray tell me the time..it being else a great uncertainty whether I may not..have strangers with me. 1787 P. Beckford Italy (1834) II. 55 We passed the day quite en famille at Belem with a whole legion of Marialvas. 1843 Thackeray Misc. (1857) IV. 86 It was vary kaind of you to come upon us en famille, and accept a dinner sans cérémonie. 1915 P. Jones War Lett. (1918) 140 Flemish is almost always used by the people en famille. 1926 D. H. Lawrence Plumed Serpent i. 14 How could any man's coat-collar look so home-made, so en famille! 1965 C. D. Eby Siege of Alcázar (1966) iv. 82 A quiet luncheon en famille, and a nap in an olive grove. 1865 'Ouida' Strathmore I. xi. 120 The city was in its spring-tide gaiety, the dense crowds were sweeping down towards the barrières of the west, Paris was en fête. 1888 Lancs. Even. Post 3 Feb. 2/4 Haslingden Liberalism..is to be en fête this evening on the occasion of its annual soirée. 1942 G. Mitchell Laurels are Poison xvii. 185 The whole of the College was prepared to be en fête. 1802 C. Wilmot Let. 13 Mar. in Irish Peer (1920) 51 The little girls, till they are eleven or twelve years old most frequently dress 'en garçon'. 1811 L. M. Hawkins C'tess & Gertr. I. xv. 248 He soon settled himself again, en garçon, in chambers. 1866 Mrs. H. Wood Elster's Folly xvii, He was living quite en garçon, with only one man. 1929 S. McKenna Happy Ending II. i. 129 Mark Tollesbury will be in London en garçon. 1968 'M. Innes' Appleby at Allington I. i. 9 He had invited Appleby to dinner—en garçon, since he was a bachelor and since Lady Appleby would not get back from London until next day. 1805 T. F. Fremantle Let. 24 July in Wynne Diaries (1940) III. 185 If I succeed in my band, I shall indeed be en grand Seignieur [sic]. 1831 Disraeli Young Duke IV. i, [It] prevented him from receiving his friends en grand seigneur in his hereditary castle. 1887 G. B. Shaw Don Giovanni Explains in Wks. (1932) VI. 109 Speaking to her..en grand seigneur, I bade her do instantly whatever she was told. 1834 H. Greville Diary 27 Nov. (1883) I. 41 A farce at which all the Court en grande tenue was obliged to attend. 1872 W. Chaffers Keramic Gallery I. 59 A lady in the costume of Louis XIV., en grande tenue. 1927 Blackw. Mag. June 749/2 He found the Rissaldar-major en grande tenue just back from a wedding. 1953 C. Day Lewis Italian Visit iii. 42 Garish Campagna knocks at the back door, Rome calls en grande tenue. 1727 M. W. Montagu Let. 23 June (1966) II. 77, I was glad to hear Mr. Remond's History from you, thô the newspaper had given it me en gros and my Lady Stafford in detail some time before. 1831 Lady Eliz. Grosvenor Let. (1965) iv. 98 In some few things it might be advantageously altered, but en gros he is well pleased with it. 1904 W. James Coll. Ess. & Rev. (1920) 446 Taking it [sc. a book] en gros, what strikes me most in it is the great sense of concrete reality with which it is filled. 1706 P. Siris tr. Feuillet's Art of Dancing vi. 9 A Bending and Springing Step, with a Foot en l'Air. Ibid. Pl. xv. 39 Forwards en l'Air, and afterwards falling on the same Leg. 1808 Wellington Disp. (1838) IV. 130 Its retreat to the sea should be considered in some degree en l'air. 1853 C. Brontë Villette II. xxvii. 280 What had a person devoted to a serious calling..to do with Counts and Countesses, hotels and châteaux? To him, I seemed altogether 'en l'air'. 1891 'L. Malet' Wages of Sin v. i, I wait on circumstances. My plans are rather distractingly en l'air every way just at present. 1918 Farrow Dict. Mil. Terms 204 En l'air...Said of troops when too far from the main body, either to render assistance or to receive support. 1964 W. B. Pemberton Battles of Boer War vi. 164 The extreme right of the Boers was practically en l'air and deserted. 1860 E. Eden Semi-attached Couple II. iii. 40 Perhaps I see things en noir today. 1875 En noir [see *en beau]. 1905 Mrs. H. Ward Marr. W. Ashe II. xi. 197 She must know that everything she does is seen there en noir. 1912 A. Lang Shakes., Bacon & Gt. Unknown xiii. 282 No wonder that he turned to tragedy..and saw life en noir. 1921 J. Conrad Wks. XVIII. p. viii, The only thing that will not be found amongst those Figures and Things that have passed away will be Conrad en pantoufles. 1930 Time & Tide 2 May, His secretary..had a little work up his sleeve, all ready to be printed and published on the morrow of France's burial, in which he was to be exposed 'en pantoufles'. 1955 Times 12 May 13/4 The authority and erudition he can pack into eight pages on 'The North-East in the Eighteenth Century'.. is yet one more proof that he is never en pantoufles. 1816 En pension [see Pension *sb.* 6 c]. 1889 Athenæum 14 Sept. 351/1 The persistence with which Mrs. Markenfeld ..remains en pension in a gloomy house with such abominable company. 1848 Thackeray Van. Fair liii. 474 There were dirty trays, and wine-coolers en permanence. 1872 E. Braddon Life in India v. 178 His Lares and Penates are erected en permanence. 1934 A. J. Toynbee Study Hist. III. 12 To make themselves at home upon it en permanence. 1955 Times 29 Aug. 10/5 Juliet expresses her character, and especially her emotional agitation, en pointes. 1959 Times 1 Sept. 11/3 Miss Doris Lainé..did a wonderfully neat hop en pointe. 1932 N. Mitford Christmas Pudding ix. 137 He..since being en poste at Cairo, had interested himself mainly in Egyptology. 1962 John o' London's 31 May 517/1 While he was en poste in Paris he gathered much of the material. a 1700 Evelyn Diary 16 Jan. an. 1679 (1955) IV. 162 A french Merchant, who had his house furnish'd en Prince. 1885 'L. Malet' Col. Enderby's Wife I. II. iv. 143, I was assured that you would be treated en prince. 1954 E. Jenkins Tortoise & Hare vii. 69 When her father was alive they did a lot of racegoing...He used to do it en prince. 1928 T. S. Eliot For Lancelot Andrewes v. 88 For Mr. Symons there is, at least en principe, a ritual.. of 'vice' or 'sin'. 1959 Encounter XIII. 87, I agree en principe with most. [1750 Chess made Easy p. ix, When one Piece can take another, that other is said to be in Prise of the first.] 1821 Kaleidoscope 13 Mar. 294/2 [He] had the privilege of taking such of the pieces..as might be en prise. 1902 Encycl. Brit. XXVIII. 758/2 B moves anywhere not en prise. 1960 R. C. Bell Board & Table Games ii. 48 Alquerque...If another piece is then en prise it is taken in the same move by a second short leap. 1909 I. Bywater Aristotle on Art of Poetry p. xxxiii, I give it as rendered into Latin by Dr. Margoliouth..with the Greek of A° en regard for purposes of comparison. 1959 T. S. Eliot tr. St.-J. Perse's Anabasis 13, I have always refused to publish the translation except in this way, en regard with the French text. 1816 T. L. Peacock Headlong Hall vi. 83 You are right, Miss Graziosa: your taste is correct—perfectly en règle. 1833 C. MacFarlane Lives of Banditti & Robbers I. 63 Many of these Calabrians were banditti..and afterwards became robbers en règle. 1880 L. Fagan Panizzi I. 42 Panizzi's passport being perfectly en règle. 1893 G. B. Shaw Widowers' Houses I. 4 On the steamboat a little négligé was quite en règle. 1936 C. S. Lewis Allegory of Love i. 6 A treatise, with rules and examples en règle for the nice conduct of illicit loves. 1848 Thackeray Pendennis i, Specimens of the costume of a military man en retraite. 1860 W. H. Russell Diary in India I. 57, I receive as a reward for my services..a post, with which I am well satisfied. I live in it for years. I use all my energies in it, and for me there is no future except en retraite. 1827 E. Belgrave in G. Huxley Lady Eliz. & Grosvenors (1965) vii. 134 Dinner..is rather a long proceeding. But, 'en revanche', they are a very short time at dessert. 1841 M. G. Blessington Idler in France I. 308 She offers him, en revanche, a cane, buttons, or a pin—in short, some present. 1883 L. Oliphant Altiora Peto vii. 125 She gave a comical look at that lady's waist and elbows, which was evidently en revanche for the well-bred stare to which she had been subjected. 1891 W. James Let. 25 Aug. (1920) I. 316 My Cousin Elly Emmet..is about to marry en secondes noces a Scotchman. 1909 Mrs. H. Ward Daphne vi. 142 The Duchess had been..married en secondes noces..by an epicurean duke. 1961 P. Green tr. Oldenbourg's Massacre at Montségur vi. 192 Petronilla, already married en secondes noces to Nuno Sanche. 1778 H. Walpole Let. 27 Nov. (1904) X. 349 You cannot justly expect him to be very punctual at first, till he is got en train. 1803 E. Wynne Diary 9 Sept. (1940) III. 90 We found the Ball already en train by the time we arrived. 1855 Mrs. Gaskell Let. ? July (1966) 360, I am en train for applying to Miss 'Temple'. 1891 E. Simcox in K. A. McKenzie E. Simcox & G. Eliot (1961) ii. 56 The Miners had another big Congress at Paris, when I was en train with Babylonian weights and measures. 1772 Ruffhead & Morgan Jacob's New Law Dict. s.v. Infant, II. Who are minors, and how far the law regards infants in ventre sa mere...An infant in ventre sa mere, or in the mother's womb, is supposed in law, to be born for many purposes. 1848 Wharton Law Lex. 9/2 A child en ventre sa mère is..a life in being. 1891 Law Times XC. 461/2 Beyond all question, for many purposes, a child en ventre sa mère is considered as being alive. 1860 Once a Week 11 Feb. 152/1 She was so huffy that I told Blot I would dine en ville for a short time. 1884 F. Boyle On Borderland 336 The horrid cookshops which send dinners en ville.

en-, *prefix*[1]. Add: **2, 3.** **endragon** (in example as *endragoned* ppl. adj.); **enwiden**: delete † and add later example.

1876 G. M. Hopkins Wr. Deutschland (1918) 20 In wind's burly and beat of endragonèd seas. 1965 D. J. Millen in Earth & Sky (Cath. Poetry Circle) 20 Man applauding man, enwidens And unfolds the praiser's joy.

-en, *suffix*[6], the ending of the past participle of many strong verbs, as *broken, spoken, sunken.* OE. *-en*, corresp. to OFris. *-en*, OS. *-an*, (M)Du. *-en*, OHG. *-an* (MHG., G. *-en*), ON. *-enn, -inn* (Sw. *-en*), Goth. *-ans*:— prim. Germ. *-inaz, -anaz* (Indo-Eur. *-ĕnos, -ŏnos*), of which some languages generalized one and some the other, the first type being represented by mutated forms in OE., eg. *cymen* (:— *kuminaz), pa. pple. of *cuman* to Come, beside *cumen* (:— *kumanaz-).

enable, *v.* **5.** Delete † *Obs.* and add later examples.

1892 Daily News 2 Feb. 3/3 It will also enable the inclusion of others. 1902 A. Nutt Leg. Holy Grail 27 This has been held by some scholars to enable the dating of the romance. 1963 A. Heron Towards Quaker View of Sex iv. 42 What God asks he enables, provided only and always that we will to do his will.

enablement. Delete † *Obs.* and add later example.

1964 Gould & Kolb Dict. Soc. Sci. 460/2 A natural right may be defined as a liberty or immunity that ought to be protected or a service or enablement that ought to be provided for all men at all times and under all conditions.

enabling, *ppl. a.* Add: *enabling act*, a legislative enactment enabling or empowering a person or corporation to take certain action.

1856 F. Pierce in Messages & Papers of Presidents (1896–9) V. 358 The States of California, Michigan and others were self-organized, and as such were admitted into the Union without a previous enabling act of Congress. 1873 J. H. Beadle Undevel. West xix. 364 Let Congress pass an enabling act for that Territory. 1965 Listener 20 May 745/1 This final power of approving or rejecting Assembly measures was reserved to Parliament in the Enabling Act of 1919.

enamel, *sb.* Add: **1. d.** Also used *attrib.*, as *enamel paint.*

1865 M. W. Brown U.S. Patent 49,708 2/2 In the use of my enamel-paint for the purposes of covering iron, steel, or other solid metallic substances..I apply a heat to the articles so covered or coated..not to exceed 300° Fahrenheit. 1895 Montgomery Ward Catal. 623/2 Enamel paint..for general decorative purposes. 1911 Encycl. Brit. XX. 458/1 The term 'enamel paint' was first given to a compound of zinc white, petrol and resin, which possessed on drying a hard glossy surface. The name is now applied to any coloured paint of this nature. 1946 M. Dickens Happy Prisoner viii. 153 She had managed to buy a whole range of enamel paints.

3. (Earlier example.)

1762 H. Walpole Let. 6 June (1903) III. 95, I have not lost one enamel, nor bronze.

5. *enamel-colour* (earlier example), *-kiln* (earlier example).

1784 S. Jones Let. 8 Mar. in Sel. Lett. J. Wedgwood (1965) 288 A man that can make as good..Enamel coulers as any man in the country. 1774 J. Wedgwood Let. 18 Dec. (1965) 170 Mr Rhodes must fix them with a little Borax &c in his enamel Kiln.

enamel, *v.* Add: **2. b.** (Examples.)

1889 Pall Mall Gaz. 17 Oct. 1/3 The craze for enamelled furniture and enamelled nick-nacks gets worse and worse. Everybody enamels. Bachelors enamel their own furniture and rooms. 1937 Discovery Feb. 57/2 Enamelled kitchen stoves.

d. (Earlier example.)

1804 M. Wilmot Let. 9 Apr. in Russ. Jrnls. (1934) 92 The fashion of their country ordains that every woman shou'd enamel their face.

enamouring, *ppl. a.* (Later example.)

1939 Times 2 June 21/2 The large sinking fund..is not an enamouring feature for many trustees.

enanthem (enæ·nþėm). *Path.* [Anglicized form of next or a. G. *enanthem.*] = next.

1885 C. H. Fagge Princ. & Pract. Med. I. 170 He seems to have thought that an important argument.. could be based upon an analogy which he drew between the specific cutaneous eruption of a contagious fever and the lesions in this disease [sc. enteric fever] which affect Peyer's patches and the solitary follicles. To express this very relation, the word enanthem had been already coined by German writers as a correlative to the term exanthem. 1962 Lancet 12 May 997/1 This is an atypical history [of smallpox]—the rash developing on the first day of fever and illness. There was a well-marked enanthem, and between 5000 and 10,000 lesions on the skin. Ibid. 999/1 The enanthem varied from one or two lesions on the palate to a severe eruption all over the palate and fauces.

enanthema (enænþī·mă). *Path.* [mod.L., ad. Gr. ἐνάνθημα eruption.] A mucosal, as opposed to a cutaneous, eruption.

1842 Dunglison Dict. Med. Sci. (ed. 3) 263/1 Enanthema, a name recently given to certain eruptions of the mucous membrane. 1964 Brit. Med. Jrnl. 25 Apr. 1121/3 How then would he describe the difference between exanthema and enanthema?

enantioblastic (enæ:nti,oblæ·stik), *a.* *Bot.* [f. Gr. ἐναντίος opposite + βλαστός (-blast) + -ic.] Having the radicle turned away from the micropyle. Also **ena:ntiobla·stous** *a.*

1858 A. Gray Introd. Struct. & Syst. Bot. 531/2 Enantioblastous, with the embryo at the end of the (orthotropous) seed diametrically opposite the hilum, as in Tradescantia. 1878 M. T. Masters Henfrey's Elem. Bot. (ed. 3) I. ii. 157 The radicle generally points to the hilum (homoblastic), rarely away from it (enantioblastic). 1887 Bentley Man. Bot. (ed. 5) 347 We have already observed, that the radicle as a general character is turned towards the micropyle...Some apparent exceptions to these relative positions occur..when the radicle is described as enantioblastic.

enantiodromia (enæ:nti,odrōu·miă). [a. Gr. ἐναντιοδρομία running in contrary ways, f. ἐναντίος opposite + δρόμος running.] The process by which something becomes its opposite, and the subsequent interaction of the two: applied esp. to the adoption by an individual or by a community, etc., of a set of beliefs, etc., opposite to those held at an earlier stage. Hence **enantiodro·miacal, enantiodro·mic** *adjs.*, resulting from enantiodromia.

1917 D. Hecht in C. E. Long tr. Jung's Coll. Papers Anal. Psychol. (ed. 2) xiv. 415 Heraclitus..discovered the most wonderful of all psychological laws, namely, the regulating function of antithesis. He termed this 'enantiodromia' (clashing together), by which he meant that at some time everything meets with its opposite. Ibid. 417 Enantiodromia is the being torn asunder into the pairs of opposites. 1923 H. G. Baynes tr. Jung's Psychol. Types xi. 542, I use the term enantiodromia to describe the emergence of the unconscious opposite...A good example of enantiodromia is seen in the psychology of Saul of Tarsus and his conversion to Christianity. 1943 E. L. Mascall He who Is x. 128 Islam is submission...And by that enantiodromia, which is so marked a feature of human activity, Mohammedanism becomes the most militant religion in history, for once the believer has made his submission he..sees himself as an instrument of the divine ruthlessness. 1946 Brit. Jrnl. Med. Psychol. XX. 214/1 Just as the repressions that went with the increasing dogmatism and formalism of the Middle Ages had produced the enantiodromiacal reaction of the Renaissance, so the repressions of the age of materialism and respectability produced its deadly challenge out of itself. 1953 Times Lit. Suppl. 25 Sept. 612/5 We are reminded also of that other—excruciatingly enantiodromic—answer..: 'Nes. Yo.' 1959 R. F. C. Hall tr. Neumann's Archetypal World of Henry Moore 4 The dialectal law of Heraclitus, the law of enantiodromia, according to which any given position is always superseded by its negation.

enantiomorph (enæ·nti,omǭɪf). [a. G. *enantiomorph* (C. F. Naumann *Elemente d.*

theoretischen Krystallogr. (1856) II. ii. 104), f. Gr. ἐναντίος opposite + μορφή form.] A form which is related to another as an object is related to its image in a mirror; a mirror image. Also *adj.* = **enantiomo·rphic, -mo·rphous** *adjs.*; whence **-mo·rphously** *adv.* So **enantiomo·rphism, enantiomo·rphy**, the condition or property of being enantiomorphous, esp. in *Cryst.*

1885 A. C. Brown in Encycl. Brit. XIX. 312/1 Two figures or two portions of matter are said to be enantiomorph to each other when these forms are not superposable, i.e., the one will not fit into a mould which fits the other, but the one is identical in form with the mirror image of the other. Ibid., As an example of enantiomorphism we may take our two hands, which will not fit the same mould or glove, but the one of which resembles in figure the mirror image of the other. Ibid. 313/2 The crystallographic theory of enantiomorph crystals. Ibid. 314/1 We now know a considerable number of cases where ..both enantiomorphs have been discovered, and many where only one has as yet been found. 1892 Science XX. 89/1 Optical isomers..may..be taken as the analogues of enantiomorphous crystals, as of quartz, right-handed and left-handed; the pairs in each case being perfectly equivalent, but not superposable. 1895 Story-Maskelyne Crystallogr. vi. § 150. 169 The configuration of the one tetartohedron will then correspond to that of the other as seen in a mirror. In a word, the two tetarto-symmetrical forms are enantiomorphous. 1895 Bloxam's Chem. (ed. 8) 600 The crystals..of the racemate differ from each other in the position of a certain unsymmetrical face; this is on the right hand in the one kind and on the left hand in the other (enantiomorphous). 1898 Nature 8 Sept. 454/1 Enantiomorphism is possible only in the case of asymmetric solid figures. Ibid. 454/2 The special one-sided asymmetry of the base will modify its mode of combination with the two enantiomorphous acids. Ibid. 455/1 Asymmetric agents can only display selective action in dealing with enantiomorphs. 1898 Jrnl. Chem. Soc. LXXIII. 608 The question of the proportion in which enantiomorphously related crystals are deposited. 1900 A. Hill Introd. Sci. ii. 86 The lævo-rotary crystals look like the dextro-rotary when they are seen in a looking-glass. They are reversed or enantiomorphic in the language of crystallography. 1900 J. Larmor Æther & Matter 209 Enantiomorphy [of a molecule] reverses the signs of all its electrons and perverts their relative position. 1902 H. A. Miers Min. 50 Two supplementary forms which are similar but not identical are said to be 'enantiomorphous'; all forms which have neither a centre nor a plane of symmetry are enantiomorphous to another form. 1929 Times 2 Feb. 8/3 The whole is a perfect enantiomorph (mirror image) of the coast of Holland and the Zuyder Zee. 1951 N. F. M. Henry et al. Interpr. X-Ray Diffraction Photogr. I. 6/1 Enantiomorphous crystals lack all elements of symmetry which would produce a right-handed arrangement from a left-handed one, and vice versa. Ibid., Enantiomorphism is an important property because of its connection with optical activity, and because of its occurrence in substances of great biological and technical importance. 1956 Daily Tel. 6 Feb. 6/2 Guessing the time from enantiomorphic clock faces. 1964 N. G. Clark Mod. Org. Chem. iii. 30 Optical isomerism occurs whenever a molecule is not identical with its image as seen in a plane mirror; the two spatial arrangements ('object' and 'image') thereby give rise to two isomeric forms of the compound. These are known as enantiomorphic forms, enantiomorphs or antimers. 1965 New Scientist 5 Aug. 352/3 The melting point attributed by your correspondent to the D form [of thalidomide] referred to a compound used in the synthesis of this enantiomorph.

enantiotropy (enæ:nti,ǫ·trŏpi). *Physical Chem.* [ad. G. *enantiotropie* (O. Lehmann *Molekularphysik* (1888) I. 119), f. Gr. ἐναντίο-ς opposite + τροπή turning.] The existence of two stable polymorphs of a substance which at a certain transition temperature are inter-convertible. Hence **enantiotro·pic** *a.*

1900 Jrnl. Chem. Soc. LXXVIII. II. 83 Enantiotropy of Tin. 1903 H. C. Jones Princ. Inorg. Chem. 172 Substances which like sulphur exist in two phases of the same state of aggregation, and the two phases can be reciprocally transformed into one another by changing the temperature, are known as enantiotropic. 1904 A. Findlay Phase Rule 42. 1937 Nature 31 July 202/2 Barium carbonate shows an enantiotropic change at 800° from the rhombic to the hexagonal form. 1948 Glasstone Physical Chem. (ed. 2) vi. 469 Two crystalline forms of a substance are said to be enantiotropic, or to exhibit enantiotropy, when each has a definite range of stability, and the change from one form to the other takes place at a definite temperature in either direction.

en attendant, en avant, en axe, en beau: see *en *prep.*

‖**en bloc** (aṅ blok), *adv. phr.* [Fr.] In a block, as a whole. Also *attrib.*

1861 J. S. Mill Repr. Govt. xvii. 312 The amount assigned to each being levied by the local assembly..and paid en bloc into the national treasury. 1878 L. W. M. Lockhart Mine is Thine I. xii. 241 You judge all your old friends, en bloc, simply from your own point of view. 1888 Contemp. Rev. Jan. 81 We are bound to take Nature en bloc, with all her laws and all her cruelties, as well as her beneficences. 1900 Westm. Gaz. 11 July 2/2 We agree, prima facie, that there is every kind of objection to en bloc disfranchisements. 1905 Spectator 28 Jan. 111/1 He was all but ready to..accept the tenets of the elder Church en bloc. 1908 Westm. Gaz. 17 Nov. 5/3 This machine is the new 1909 model, the en bloc engine, four-speed gear-box, and thermo-syphon cooling denoting the only departure from standard practice. 1930 G. R. de Beer Embryol. & Evol. i. 10 It is not the 'stage' which is shifted en bloc, but certain characters which may be peculiar to that stage. 1934 Discovery June 174/1 Nearby..are villages whose inhabitants in summer migrate en bloc,..to the islands off the coast where they spend their time fishing. 1966 English Studies XLVII. 65 A mere en bloc rendering of the phrase obscures an image.

en brosse, en cabochon: see *en *prep.*

encallow (enkæ·lōu), *sb. local.* = Callow *sb.* 3. Hence **encallow** *v.*

1836 Penny Cycl. V. 408/1 The encallow, as it is technically called, or the top-soil. 1850 E. Dobson Bricks & Tiles 21 The first operation [in digging clay] is to remove the mould and top soil...In London the vegetable mould is called the encallow, and the operation of removing it, encallowing. 1884 C. T. Davis Manuf. Bricks 103.

encapsulate, *v. trans.* and *refl.*, **encapsulation.** Now the more usual forms of Incapsulate *v.*, Incapsulation, esp. in *fig.* senses. So **enca·psulated** *ppl. a.*, **enca·psulating** *vbl. sb.* and *ppl. a.*

1861, 1888 Encapsulation [see Incapsulation]. 1876 [see Incapsulate *v.*]. 1913 Chambers's Jrnl. Aug. 553/2 The cell becomes encapsulated and pointed. 1958 N. R. Hanson Patterns of Discovery i. 27 Language can encapsulate scenes and sounds, teeth and growls, smiles and laughs. 1958 Times Rev. Industry May 70/1 The bulk of encapsulating production is pharmaceutical. 1958 Oxford Mag. 20 Nov. 109/1 They are forced to work for two terms for an exam...which encapsulates that dreary anachronism, compulsory Latin. 1959 Oxford Mail 2 Feb. 5/4 Another interesting, though less successful, experiment involves a small nude encapsulated in a large abstract design. 1959 New Scientist 13 Aug. 192/2 One way of encapsulating is akin to blowing bubbles. 1960 Aeroplane XCVIII. 254/2 Encapsulation of electronic equipment is undesirable. 1960 New Left Rev. Nov.–Dec. 4/2 The polite encapsulation of the Liberal Press. 1961 Guardian 22 May 5/1 May one, if need arise, be encapsulated and shot into space, but not, I beg, with members of one's family. 1967 Listener 29 June 863/3 This charming little creation..encapsulates an aspect of history for all time.

encased (enkē·st), *ppl. a.* [f. Encase *v.*] In phr. *encased knot*, a knot of dead wood that has not been connected by living matter with the adjacent wood but is often partly or wholly enclosed in bark.

1917 C. H. Snow Wood & Org. Struct. Materials iv. 42 An encased knot is one which is surrounded wholly or in part by bark or pitch. 1931 G. A. Garratt Mech. Prop. Wood ii. 55 Dead branches result in what are known as encased knots, which are, in effect, nothing more than pegs in a hole. 1968 G. Tsoumis Wood as Raw Material x. 176 The knots resulting from such branches [sc. dead branches enclosed in living stems] and called encased or loose.

encasement. Add: **2.** The act of encasing; the condition of being encased; *spec.* in *Bee-keeping*. Also *concr.* (see quot. 1884).

1825 New Monthly Mag. XVI. 598 This incasement of our feet in icy boxes was very annoying. 1884 J. Hunter Man. Bee-keeping (ed. 3) 15 In the case of a strange Queen's appearance in the hive..the stranger is seized by the Workers, who gather round her and form a closely packed ball..technically called 'an encasement'. 1888 F. R. Cheshire Bees II. 431 During examination..an encasement of the queen in her own hive is commenced.

encashable (enkæ·ʃăb'l), *a.* [f. Encash *v.* + -able.] Capable of being cashed.

1913 J. M. Keynes Indian Currency & Finance iii. 42 If the Government had made its notes encashable at a great variety of centres, it would have been taking on itself the expense and responsibility [etc.]. 1927 Daily Tel. 7 Mar. 2 This loan bears ½ per cent. more interest than the Savings Bonds, but whereas the latter are encashable at any time without loss of capital, the price receivable for the Conversion Loan would depend on the market value when sold.

encephalitogenic (ense:făləitodʒe·nik, enke:f-), *a.* Also occas. **encephalogenic.** [f. Encephalit(is + -o + *-genic *suffix.*] Producing or able to produce encephalitis.

1923 Jrnl. Amer. Med. Assoc. 24 Nov. 1788/2 Levaditi distinguishes the two kinds of active materials or viruses contained in the naso-pharyngeal secretions as 'salivary keratogenic' and 'salivary encephalogenic', according as they resemble..the herpes virus or..the encephalitis virus. 1925 Jrnl. Exp. Med. XLI. 215 (heading) An exotic strain of encephalitogenic virus. 1940 van Rooyen & Rhodes Virus Dis. Man lx. 841 Beans, corn, and other substances were alleged to be acted on by bacteria and in this way an encephalitogenic agent produced. 1958 Immunology I. 111 Encephalitogenic bacillary fractions produce histological reactions at the injection site.

encephalo- (ense·fălo, enke·fălo), comb. form of Gr. ἐγκέφαλος brain, as in **encephalogram** (ense·fălogræm, enke·f-), an X-ray photograph of the brain; **encephalograph** (ense·fălogrɑf, enke·f-), (*a*) = *encephalogram; (*b*) an instrument for recording the electrical activity of the brain, *electro-encephalograph; **encephalo·graphy** [ad. G. *encephalographie* (A. Bingel 1921, in *Fortschr. auf d. Gebiete d. Röntgenstrahlen* XXVIII. 205)], the radiological examination of the brain (see quot. 1955); hence **encephalogra·phic** *a.*; **ence·phalolith**, a concretion in the brain (Billings); **encephalo·logy**, a description of the brain; the science of the brain; **ence:phalomala·cia**, softening of the brain; **ence:phalomeni·ngocele**, protrusion through a fissure in the skull of brain-substance with the attached membranes; **ence:phalomyeli·tis**, inflammation of both the brain and the spinal cord; any of various virus diseases characterized by fever and lack of co-ordination and damage to the central nervous system.

1928 Bull. N.Y. Acad. Med. IV. 828 (title) Case report illustrating the early diagnostic significance of the encephalogram. 1959 B. Wootton Social Sci. & Social Path. x. 306 The possibility that the encephalograms are themselves affected by experience deserves at least to be considered. 1934 Webster, Encephalograph. 1955 Oxf. Jun. Encycl. XI. 307/1 The 'encephalograph', a device which records electrical effects of the activity of the brain. 1922 Jrnl. Amer. Med. Assoc. LXXVIII. 622/2 Bingel discusses his..technic for encephalography or the roentgenographic representation of the brain. Ibid., He reported 100 cases in which encephalographic data had been secured. 1955 Gloss. Terms Radiology (B.S.I.) 21 Encephalography, the radiological examination of the ventricles and subarachnoid space following the injection of air by cisternal or lumbar puncture. 1962 Lancet 8 Dec. 1205/1 The electroencephalogram shows that the cerebral cortex has been extensively destroyed, and this can readily be confirmed by air encephalography. 1824 (title) Encephalology, or a very brief Sketch of Dr. Hirnschadel's Ologies of the Cranion and Phren perfected by the Rationals. 1842 Dunglison Med. Lex. (ed. 3) 263/2 Encephalomalacia, mollities cerebri. 1863 D. Maclachlan Pract. Treat. Dis. Advanced Life viii. 172 (heading) Encephalomalacia, necrencephalus, ramollissement cérébral, or softening of the brain. 1912 Adami & McCrae Path. 487 The result of such embolism is encephalomalacia. 1900 Index Catal. Libr. Surgeon-General's Off., U.S. Army 2nd Ser. V. 1/2 Encephalomeningocele, See Brain (Hernia of). 1905 T. H. Green's Path. (ed. 10) 11 They [sc. pouches in the cranium] may contain brain-substance (encephalocele), or brain-substance and fluid (encephalomeningocele). 1908 Jrnl. Nerv. & Ment. Dis. XXXV. 388 (title) A case of probable encephalomyelitis. 1939 Ann. Reg. 1938 376 Man is susceptible to equine encephalomyelitis. 1963 Greenfield & Norman in W. Blackwood Greenfield's Neuropath. (ed. 2) viii. 475 Encephalomyelitis as a sequel to acute infectious disease, especially smallpox..and measles.., has been known for two centuries. 1966 Wright & Symmers Systemic Path. II. xxxiv. 1197 Rabies is essentially an acute encephalomyelitis.

‖**enchaînement** (anʃɛ̄nmaṅ). *Ballet.* Also **enchainement.** [Fr.] A connected series or sequence of steps.

1830 R. Barton tr. C. Blasis' Code of Terpsichore II. vii. 85 Previous to the commencement of a pirouette, either from the inside or outside, the dancer may pause in any sort of attitude or arabesque in which he pleases to end his enchaînement. 1934 A. Haskell Balletomania iii. 67 Dancing is so subtle that six ballerinas will perform the same enchaînement in an almost identical manner, and yet create an entirely different impression. Ibid. xi. 229 Some Maryinsky teacher, who has a rich repertoire of enchaînements. 1936 —— Prelude to Ballet iii. 19 Any competent dancer can arrange a small dance..by drawing upon various enchaînements (combinations of steps) with which she is familiar. 1948 Ballet Ann. II. 69 The student's..ability to assimilate new steps and new enchaînements.

encharm (entʃā·ɪm), *v.* Delete † *Obs.* and add later examples.

1890 M. W. Hungerford Life's Remorse I. xv. 200 Such a knowledge of horseflesh as should encharm the soul of any Irishman. 1905 W. Holman Hunt Pre-Raphaelitism I. xiii. 358 [Christ]..encharming the simple by His love and lovableness.

‖**enchilada** (entʃilā·dă). [Amer. Sp., fem. of *enchilado*, pa. pple. of *enchilar* to season with chili, f. Sp. *en* in + *chile* chili.] A tortilla served with a sauce seasoned with chili.

1887 F. C. Gooch Face to Face with Mexicans xii. 410 They go to some stand..and eat enchiladas and tamales and drink pulque. 1895 Jrnl. Amer. Folk-Lore Jan.–Apr. 62 Enchiladas are practically corn fritters allowed to simmer for a moment in chile sauce, and then served hot with a sprinkling of grated cheese and onion. 1934 J. M. Cain Postman always rings Twice i. 5 Fried eggs and bacon, enchilada, flapjacks, and coffee. 1968 N. Benchley Welcome to Xanadu viii. 180 He..ordered a bowl of chili and enchiladas.

enchondroma. Add: In later use usu. restricted to mean: a chondroma arising within a bone. (Later example.)

1958 H. L. Jaffe Tumors of Bones xi. 170 An enchondroma beginning within a metaphysis may eventually involve a large portion of the shaft.

enchondrosis (enkǫndrōu·sis). [f. Enchondr(oma + -osis.] A chondroma arising from cartilage.

1871 T. H. Green Introd. Path. xvii. 132 The enchondromata must for the most part be regarded as innocent growths. Those homologous forms which originate from cartilage, and have been called 'Enchondroses', differ in all respects from the heterologous. **1908** *Practitioner* Dec. 763 On the inner aspect of the second toe there is an enchondrosis. **1967** R. A. Willis *Path. Tumours* (ed. 4) xliii. 686 A once prevalent idea that multiple enchondroses and multiple exostoses are only different varieties of one disease has been discarded.

enchylema (enkilī·mă). *Biol.* [G. (J. von Hanstein *Das Protoplasma als Träger d. pflanzlichen & thierischen Lebensverrichtungen* (1880) iv. 39): see En-[2] and Chyle.] In some old theories of the structure of cytoplasm: **a.** A clear liquid supposed to constitute the ground-substance of a cell nucleus and to contain a fibrillar network. **b.** A liquid supposed to exist in the form of 'alveolar spheres' in a continuous, usu. more viscous liquid.

1886 *Science* VIII. 125/1 This basal substance, enchylema, is probably more or less nearly fluid during life. **1888** Rolleston & Jackson *Forms Anim. Life* p. xxi, Protoplasm..appears sometimes to be structureless, but as a rule it is more or less vesicular, consisting of a denser substance (mitome) enclosing droplets of a more fluid character (enchylema, paramitome). **1896** E. B. Wilson *Cell* i. 17 According to the view most widely held, one of its [*sc.* protoplasm's] essential features is the presence of two constituents, one of which, the ground-substance, cytolymph, or enchylema, is more liquid, while the other, the spongioplasm or reticulum, is of a firmer consistency, and forms a sponge-like network. **1940** L. H. Hyman *Invertebrates* I. i. 2 The spumoid or alveolar theory [of protoplasm], developed by Bütschli, and widely accepted, even today, states that protoplasm is composed of spheres, the alveoli, filled with a clear fluid, the hyaloplasm or enchylema, and suspended in a continuous interalveolar substance.

‖ **encierro** (enþīē·ə·ro). [Sp., lit. act of enclosing or locking up.] The driving of bulls through the streets of a Spanish town from a corral to the bull-ring, freq. for the specific purpose of giving amateurs an opportunity to 'play' the bulls.

1845 R. Ford *Handbk. Trav. Spain* I. ii. 179 The *encierro*, the driving them to the arena, is a service of danger. **1901** *Encycl. Sport* I. 153/2 Soon the whole herd of bulls..is rattled along the lanes..in the direction of the bull-ring. The horsemen are assisted in this operation, called the *encierro* ('enclosing') by numerous amateurs. **1959** *Times* 10 July 8/6 A quaking fear seizes every English stomach before the *Encierro* begins. **1966** G. Sims *Sleep no More* xii. 84 He's taken part in the *encierro* at Pamplona—that's a run of about half a mile down zig-zagging boarded-up streets to the bull ring. They let loose six young bulls to chase you.

encirclement (ensɜ̄·ɪk'lmĕnt). [f. Encircle *v.* + -ment.] The act or fact of encircling or of being encircled: used esp. with reference to countries that believe themselves to be surrounded by hostile nations.

1919 G. B. Shaw *Peace Conference Hints* ii. 20 The famous encirclement (*Einkreisung*) which was the master-stroke of the Allied strategy. **1920** H. G. Wells *Outl. Hist.* 456/2 Their [*sc.* the French] government set about the encirclement of the colonies and their subjugation in a terrifyingly systematic manner. **1927** *Observer* 24 July 14/2 A ring of Verey lights..that exaggerated..the encirclement and nearly completed it. **1940** *Ann. Reg. 1939* 63 The Reich Government in official declarations raised the cry of 'encirclement', and accused Great Britain of planning the destruction of German trade. **1957** *Times* 11 May 6/2 The threat of attack by Israel to prevent her encirclement by an Arab federation bent on her destruction. **1966** *Listener* 10 Mar. 339/2 The Chinese now see their encirclement by what they like to call 'imperialist' intrigues. **1967** S. Beckett *Eh Joe* 38 E [camera] now begins a much wider encirclement.

en clair: see *en *prep.*

enclave, *sb.* Now usually pronounced (e·nklēⁱv).

enclisis (e·nklisis). *Gram.* [mod.L., f. Gr. ἔγκλισις, f. ἐν on + κλίνειν to lean.] Pronunciation as an enclitic; the transference of accentuation to a previous word.

1885 *Amer. Jrnl. Philol.* VI. 218 Retaining the convenient terms orthotonesis and enclisis to designate this alternating accent. **1949** *Archivum Linguisticum* I. 169 The expanded or composite form, which results from enclisis with members of the paradigm of the third personal pronoun. **1963** in Brown & Foote *Early Eng. & Norse Studies* 134 An unusual postpositive use of the article, though with no suggestion of enclisis as in Scandinavian.

encloistered, *ppl. a.* Delete †*Obs.* and add later *fig.* example.

1908 Belloc *On Nothing* 72 The smoke also of the train as it skirts the Downs is part and parcel of what has become (thanks to the trains) our encloistered country life.

enclose, inclose, *v.* **3. b.** (Earlier and later examples.)

1707 Addison *Lett.* (1941) 68 My Lord Hartford desiring me to enclose this Letter to the Electoress. **1708** *Ibid.* 109, I this morning receiv'd..your Lordship's letter..with several others enclosed in the same packet. **1835** Dickens *Let.* 8 Aug. (1965) I. 72 My dear Tom.—I inclose you the proof we spoke of, agreeably to your request. **1838** —— *Let.* 28 Dec. 478, I inclose to Mrs. Blanchard an Invitation from my worser half. **1903** J. Joyce *Let.* 9 Mar. (1966) II. 35, I enclose you self-explaining documents.

enclosed, inclosed, *ppl. a.* Add: *spec.* of religious communities who are secluded from relations with the outside world.

1884 Addis & Arnold *Cath. Dict.* (1897) 911/2 The nuns were to be strictly enclosed. **1905** *Athenæum* 30 Sept. 431/1 Catherine de' Ricci belonged to an enclosed community of Dominican Tertiaries. **1957** *Oxf. Dict. Chr. Ch.* 296/2 The enclosure is more strict in women's orders than men's, and most strict of all in the enclosed orders.

b. Delete †*Obs. rare*[−1] and add earlier and later examples. For def. read 'a letter or other enclosure within a letter'.

1618 J. Chamberlain *Let.* 5 May (1939) II. 162 The inclosed stands me in two shillings. **1707** Addison *Lett.* (1941) 72, I was to wait on you..with the enclosed from the Ambassadour Cornara. **1839** Dickens *Let.* 25 Jan. (1965) I. 494 The best way of answering the inclosed which I found lying at home last night.

enclosure. Add: **4. a.** *spec.* on a racecourse (see quot. 1963).

1867 'Ouida' *Under Two Flags* I. iii. 46 The horses were being led into the enclosure for saddling. **1910** *Encycl. Brit.* XIII. 737/2 [France] There are none of the clubs and special enclosures such as at Sandown, Kempton, Hurst, Lingfield, Gatwick, etc., though portions of the stand are set apart for privileged persons. **1963** Bloodgood & Santini *Horseman's Dict.* 72 *Enclosure*, space set apart on the grounds of a racecourse for some specific purpose; the Royal Enclosure, the Jockey Club Enclosure, the Members' Enclosure, etc. **1968** J. Fleming *Kill or Cure* ix. 115 She wasn't allowed into the Royal Enclosure at Ascot.

b. (Early U.S. example.)

1776 *Jrnls. Cont. Congress* IV. 107 Two letters from General Schuyler..with an account of his expedition to Tryon county, with 9 enclosures.

encode (enkōᵘ·d), *v.* [En-[1] + Code *sb.*[1] 3 b.] *trans.* To translate into cipher or code; also *techn.* of computers (see quot. 1955). Also *transf.*

1919 in L. Tissot-Dupont *Dict. Termes de Telegraphie-Telephonie.* **1931** H. O. Yardley *Amer. Black Chamber* xiv. 187 If his cable, even after translation into Japanese and encoded in his code, still retains the proper name, I'm sure I can find it. **1932** D. L. Sayers *Have his Carcase* xxvi. 341 It's the kind of thing that young Alexis could easily learn to encode and decode. **1955** *Gloss. Terms Autom. Digital Computers (B.S.I.)* 10 *Encode*, to express information by means of a *code*. Colloquially, 'to code'. **1959** B. C. Brookes in Quirk & Smith *Teaching of English* v. 155 The lecturer..encodes [his] information in appropriate speech signals. **1959** E. Pulgram *Introd. Spectrogr. Speech* i. 12 The membranes of telephones and microphones are obstacles of this type receptive to air waves The energy that strikes them is translated, or encoded, into electric currents, which are in the receiver electronically decoded again into sound waves.

Hence **enco·ding** *vbl. sb.* (also *attrib.*).

1953 *Encounter* Oct. 21/1 Communism itself [is] only a secondary encoding of the completely unmentioned Soviet Union. **1954** *Electronic Engin.* XXVI. 84 The technique of encoding and programming, the 'training of the robot'. **1956** J. Whatmough *Language* i. 13 A striking utterance is found, on inspection, to disturb commonplace encoding and decoding processes.

encoder (enkōᵘ·dəɹ). [f. the vb.] Someone or something that encodes; *spec.* a part of a computer that converts data under specific conditions.

1944 *Harper's Mag.* Oct. 468/1 The Italian radio operator..did not have the mechanical encoder and had to encode the messages by hand. **1956** *Electronic Engin.* XXVIII. 37 At the rear of the machine is the perforator or 'encoder' which produces the punched tape. **1958** *Ibid.* XXX. 685 The comparator described..compares two voltages at constant intervals of time, a process required for example in voltage encoders. **1959** *Times Rev. Industry* Feb. 86/1 Separation of the directional information from two normal left and right stereo signals in an encoder. **1964** J. Z. Young *Model of Brain* ix. 159 Cells with long spreading dendrites provide the encoders. **1966** *Electronics* 17 Oct. 38 Low-drift gyros are still desirable, but the encoder function and analog-to-digital requirements tied to gimbaling are bypassed.

en cœur: see *en *prep.*

encoffined (enkǫ·find), *ppl. a.* [f. Encoffin *v.* + -ed[1].] Enclosed in a coffin.

1904 M. Corelli *God's Good Man* 59 The encoffined saint whose sarcophagus had been unearthed. **1907** E. Wilder tr. *Werder's Hamlet's Myst.* i. 49 The secret of the encoffined and unprovable crime. **1908** *Daily Chron.* 17 Aug. 5/6 An encoffined body.

‖ **encoignure** (aṅkwanʸür). [Fr., f. *en-* in + *coin* corner.] A piece of furniture, esp. of ornamental design, made with an angle to fit into a corner.

1848 H. R. Forster *Stowe Catal.* 22 A pair of very handsome encoigneures, of rich buhl on tortoiseshell. **1936** *Burlington Mag.* Oct. 187/1 Among the masterpieces of French furniture in Windsor Castle are a commode and a pair of *encoignures*. **1962** *Internat. Art Treasures Exhib.* (*Vict. & Albert Mus.*) 21/2 A Louis XVI suite of lacquer furniture attributed to P. Garnier comprising a Commode and a pair of Encoignures.

encomienda (enkōᵘmie·ndā). *Hist.* [Sp., = commission, charge, sb. corresp. to the vb. *encomendar* to commit, charge; cf. med.L. phr. *in commendam* (see Commendam).] An estate granted to a Spaniard in America, with powers to exact taxation and corvée from the Indian inhabitants; such authority; a system derived from such authority. Also **encomendero**, the holder of an encomienda.

1810 *Eclectic Rev.* VI. II. 1065 The systematic slavery of the *encomiendas* having been annulled by Charles III. **1818** *Docs. Congr. U.S. For. Rel.* (1834) IV. 325 (Stanford), All these regulations were found ineffectual to secure the Indians against the rapacity of the encomenderos, and encomiendas were abolished. **1877** *Encycl. Brit.* VI. 174/1 That system of *repartimientos* or *encomiendas* which was afterwards to work such cruel mischief among the conquered. **1885** *Ibid.* XVIII. 677/2 'Encomiendas', or grants of estates on which the inhabitants were bound to pay tribute and give personal service to the grantee. **1952** Bertrand & Petrie *Hist. Spain* (ed. 2) xxiii. 198 The *encomenderos*, to whom territories and whole populations of Indians were granted on the condition of feeding them and instructing them in the Christian faith—these colonists declared that there was nothing to be done with the Indians. **1964** M. Harris *Patterns of Race in Americas* ii. 18 In the highlands..the dominant form of labour appropriation..was known as the *encomienda*...A man who had performed service..in the conquest of the new territories was rewarded with the privilege of collecting tribute and drafting labor among a stated group of Indians...Cortes..received an *encomienda* consisting of twenty-two townships.

encompass, *v.* Add: **6.** Used for Compass *v.*[1] 2.

1882 P. Robinson *Under Sun* III. v. 201 Whatever the method employed for encompassing his death. **1889** Mrs. H. L. Cameron *Lost Wife* I. iv. 69 What earthly reason could Captain Thistleby have for encompassing my destruction?

encoop (enkū·p), *v. poet.* [f. En-[1] + Coop *sb.*[1]] *trans.* To coop up.

1867 J. B. Rose tr. *Virgil's Æneid* 267 Again besieged, again encooped in hold. **1906** Hardy *Dynasts* II. I. viii. 172 Her fleet at any minute can encoop Yours in the Baltic.

enco·ppicement. [f. En-[1] 2 + Coppice *sb.* + -ment.] The promotion and preservation of coppices.

1935 *Forestry* IX. 11 The object of the encoppicements ..was coppice in which hazel and ash figured largely. **1958** *Wilts. Arch. Mag.* LVII. 65 Although encoppicement was legally sanctioned, the hedges of a coppice are [etc.].

en coquille: see *en *prep.*

encountering, *ppl. a.* Add: In the use of G. M. Hopkins (see quot.).

1878 G. M. Hopkins *Let. to Bridges* 30 May (1935) 53 The rhythm is anacrustic or, as I should call it, 'encountering'.

encroach (enkrōᵘ·tʃ), *sb.* Delete †*Obs. rare* and add later examples.

1920 *Conquest* Nov. 39/3 Rocks are affected by micro-fungi, and may crumble as a result of their encroach into crevices and their subsequent action. **1924** *Chambers's Jrnl.* Nov. 714/1 From a line square of posts, bordering on the marshes, and on the mud-flats' farthest encroach, the one safe track leads shorewards.

encrustation, var. Incrustation.

1920 H. G. Wells *Outl. Hist.* 271/2 There was no effective prohibition of superstitious practices..and supplementary worships. At an early stage [of Buddhism] a process of encrustation began. **1923** *Daily Mail* 24 Jan. 6 The heavy encrustation of swollen buds on every twig. **1957** V. W. Turner *Schism & Continuity in Afr. Soc.* vi. 197 Each core..in a village may have a number of classificatory encrustations.

enculturation (enkɒ·ltiŭrēⁱ·ʃən). *Social Sciences.* [f. as Culturation: see En-[1].] (See quot. 1948.)

1948 M. J. Herskovits *Man & Works* iii. 39 The aspects of the learning experience which mark off man from other creatures, and by means of which..he achieves competence in his culture, may be called enculturation. This is in essence a process of conscious or unconscious conditioning, exercised within the limits sanctioned by a given body of custom. **1955** *Yearbk. Anthropol.* I. 315/1 The training of students involves the usual problems of enculturation. **1968** *Indian Mus. Jrnl.* V. 24 Normally, education is part of the process called by some anthropologists 'enculturation' (the culture inculcating its own values). **1968** N. O. Lurie in J. A. Clifton *Introd. Cultural Anthropol.* 297/1 Our biological evolutionary heritage means that we must learn our cultural adaptations for survival, and for this vital technique to work, a culture must be firmly implanted in the members of the society which shares it. The term 'enculturation' is sometimes used to describe this process.

encyclopædically (ensəiklopī·dikăli), *adv.* [f. Encyclopædic: see -ically.] In an encyclopædic manner; comprehensively.

1855 Rossetti *Let.* 25 Nov. (1965) I. 280, I found his [*sc.* Browning's] knowledge of early Italian art beyond that of anyone I ever met—*encyclopaedically* beyond that of Ruskin himself.

encyst, *v.* Add: Also *absol.* or *intr.*

1896 tr. *Boas' Zool.* 86 A great many forms..have the power of encysting. **1966** *McGraw-Hill Encycl. Sci. & Technol.* VI. 439/2 The ability of protozoans to encyst is at least one of the reasons for their world-wide distribution.

end, *sb.* Add: **1. d.** *the end. fig.* and *colloq.* Of persons and things, a term to express the extreme in disparagement; the 'limit' (cf. *limit *sb.*); the 'last straw'.

[**1929** Wodehouse *Mr. Mulliner Speaking* ii. 55 'This,' he said in a shaking voice, 'is the end. From this moment I go off the stuff.'] **1938** N. Marsh *Death in White Tie* iii. 43 The sort of people who go there are just simply The End..the most unspeakable curiosities. **1944** 'Brahms' & 'Simon' *Titania has Mother* v. 41 But the damage had been done... It was The End. **1953** E. Taylor *Sleeping Beauty* x. 175 You simply *are* the end. **1959** G. Freeman *Jack would be Gentleman* v. 85 Donald, you really are the absolute end.

2. b. An outlying part of a village or small country town, the end of an estate, or an outlying property, usually preceded by a descriptive name. (For place-names and field-names in *-end* see A. H. Smith *Eng. Place-Name Elements* (1956) s.v. *ende*[1].)

1866 Geo. Eliot *F. Holt* I. viii. 196 Mr Goffe, of Rabbit's End, had never had it explained to him that.. land must inevitably be given up when it would not yield a profit. **1906** *Bungalow* Dec. 8/2 It is the typical 'end' so beloved of the novelist, the disreputable quarter which the parsimonious squire neglects. **1910** E. M. Forster (*title*) Howards End. **1925** *Victoria Hist. County Bucks.* III. 465/1 The soil [at Winslow] is of Oxford clay, which has been worked in a pit at Tinker's End. **1954** M. Beresford *Lost Villages* x. 339 Bedfordshire... The form of settlement..gives us the characteristic 'Ends'. Here, farms are scattered.

3. e. *Archery.* (*a*) The place at which a mark is set up. (*b*) The number of arrows shot from one end of a range.

1801 T. Roberts *Eng. Bowman* 288 *End*, the place where a mark is fixed. **1836** in A. E. Hargrove *Archery* (1845) 89 Any member who shall draw an arrow before the *end* has been determined by the Judges, shall forfeit his right to count for such arrow. **1879** M. & W. Thompson *Archery* x. 52 By the rules of the York Round three arrows to each archer constitute an end. **1887** W. Butt *H. Ford's Archery* 289 On July 5, 1877, he made 3 golds in one end at 100 yards.

f. Of a sports pitch or court; also, the half occupied by a team or individual player. Phr. *to change ends*: to play in the opposite direction to that in which the team or individual had previously played; so *change of ends*; *to choose ends*: to indicate the direction in which one prefers to play; so *choice of ends*.

1851 W. Bolland *Cricket Notes* v. 87 The safest plan is to stipulate..that the disputed man shall not be put on at either end—as a doubtful bowler should never be permitted. **1865** J. Pycroft *Cricketana* xiii. 235 He used at Lord's to bowl from the Pavilion end in order that he might have the slope against him. **1867** G. H. Selkirk *Guide to Cricket-Ground* iv. 59 The finest bowler might change ends every over, and bowl continuously. **1881** *Two Codes Football* (ed. 3) 28 Ends shall only be changed at half-time. **1890** C. G. Heathcote in J. M. Heathcote *Tennis*, etc. 253 The winner of the toss should choose the best end, taking note in doing so not only of the position of the sun at the moment, but of what it will be in the course of the match. **1933** D. L. Sayers *Murder must Advertise* xviii. 313 The slogger smote a vigorous ball from the factory end. **1960** J. R. Witty in Fabian & Green *Assoc. Football* I. II. iii. 167 Choice of ends has always been decided by a toss of the coin. **1963** R. B. Stratford *Netball* ix. 70 Unless the weather conditions make the choice of ends desirable, it is better [for the captain who wins the toss] to choose the centre pass. **1970** *Times* 22 Oct. 13/4 (*Soccer*) Five minutes after the change of ends came the body blow.

g. In American and Canadian football, one stationed at the end of a line or team of players; a wing; the position occupied by such a player.

1892 *Outing* (U.S.) Dec. 50/1 Long gains at the tackles, ends, and around the ends, were common. **1899** A. H. Quinn *Pennsylv. Stories* 22 The ends and the backs came together as though drawn by a magnet and the pyramid toppled and fell. **1944** *Greeley* (Colo.) *Daily Tribune* 24 Sept. 2/1 The Wizards team is well spiked with veterans this year having a veteran backfield, two ends and a letterman tackle. **1968** *Globe & Mail* (Toronto) 13 Feb. 29/1 Of their 10 selections, seven were offensive ends or flankers.

h. Phr. *the end of the road* (*transf.* and *fig.*).

1924 Dillon & Lauder *End of Road* 4 Keep right on to the end of the road. Keep right on to the end. **1954** M. Procter *Hell is City* III. vi. 108 It'll be the end of the road for him. He won't care what he does. **1959** 'A. Fraser' *High Tension* x. 99 It isn't any good pretending. We've come to the end of the road. **1968** *Guardian* 8 Oct. 1/2 The end of the road for Mr. Dubcek's Czechoslovakia may not have been reached after all.

5. d. A share or portion; a part or side. orig. *U.S.*

1903 A. H. Lewis *Boss* xiv. 181 That's always th' Tammany end; forty per cent. **1907** in Asher & Heal *Send No Money* (1942) 115 After the genial and affable derelicts in the money changing end received the money.. no stove has arrived. **1926** J. Black *You can't Win* ix. 105 Didn't him and Smiler bring it [*sc.* $200] up here for my end of that chippy gambling house's bankroll? **1928** *Publishers' Weekly* 30 June 2598 To talk to such a person about the editorial end of a publishing business means little or nothing. **1948** 'N. Shute' *No Highway* v. 126 Honey would have to come back to this country to tell us his end of it. **1962** B. Knox *Little Drops of Blood* ii. 39 How about your end of it?

6. a. *Coalmining.* Also phr. *on the end* (see quots.).

1867 W. W. Smyth *Treat. Coal* xii. 140 A far better proportion of round coal will be obtained by working on the end, *i.e.*, in the direction of such cleat. **1892** H. W. Hughes *Text-bk. Coal-min.* vii. 158 If the face is parallel to the cleat, the coal is said to be 'on the end'.

e. *end of steel* (also formerly *end of the steel*): the limit to which tracks have been laid during the construction of a railway; a railway terminus; so *end-of-steel town*, a town at a railway terminus. *Canad.*

[**1884** *Prince Albert Times* (Sask.) 4 July 3/1 A number of leading citizens of Calgary waited on Inspector Steele.. on the eve of his departure to End of Track.] **1909** A. D. Cameron *New North* ii. 21 Edmonton is the end of steel. Three lines converge here. **1912** H. Footner *New Rivers of North* 276 We came to the end of the steel, but there was no construction work going on. **1912** J. B. Bickersteth *Let.* 3 Nov. in *Land of Open Doors* (1914) x. 199 An end-of-steel town is a wicked place. **1933** *Meccano Mag.* Mar. 195/1 Greatly reduced the time required for the journey from the 'end of steel', as the railway terminus is called in Canada. **1933** Thompson & Edgar *Canadian Railway Devel.* 247 An engine drawing a train of ballast trucks..steamed to the end of steel. **1962** R. Slobodin *Band Organiz. Kutchin* 12 The railroad terminus—the 'end of steel'—is at Waterways, Alberta.

f. *big end*: see *big *a.* B. 2.

7. c. in attrib. uses of phrases, as *end-account*, *end-August*, *-year* (also followed by a specified year-date); *end-of-December*, *end-of-season*, *end-of-term* (also *end-of-termy* adj.), *end-of-the-century*, *end-of-the-year*.

1891 *Literary World* 20 Nov. 422/3 Our end-of-the-century civilisation. **1902** *Westm. Gaz.* 12 June 7/2 The end-of-December account. **1903** *Ibid.* 21 Aug. 9/1 The final details of the end-August settlement. **1909** *Ibid.* 9 Dec. 10/1 The end-of-the-year requirements are likely to be satisfied. **1911** H. Walpole *Mr. Perrin* iii. 42 The end-of-termy feelings. **1923** *Manch. Guardian Weekly* 10 Aug. 106/1 Smith Minor has been cruelly flogged by an avaricious head master who had discovered that his end-of-term rose-bowl was only electro plate instead of the solid silver which he had stipulated in his letter to the lad's parents. **1928** *Daily Chron.* 9 Aug. 8/6 The falling off in the average shown in latest developments, accentuated by end-account sales. **1938** *New Statesman* 8 Jan. 39/1 The end-year issue of *The Bookseller*. **1947** Wyndham Lewis *Let.* 21 Dec. (1963) 423, I was sorry to hear of your end-of-the-year feeling of nervous exhaustion. **1954** *Economist* 11 Sept. 1/2 The show [*sc.* Farnborough]..is both a shop-window and an end-of-term report. **1955** *Times* 6 July 16/1 The effects of end-account selling. **1959** *Times* 12 June 15/1 An end-of-season concert. **1960** *Times* 13 Jan. 17/2 There has recently been some movement of funds from London—quite apart from the Swiss end-year transactions. **1966** *Rep. Comm. Inquiry Univ. Oxf.* I. 182 At end-1965 prices. **1966** *Listener* 23 June 914/3 But it was the end-of-term exam that really made him anxious.

8. a. *It isn't* (or *wouldn't be*, etc.) *the end of the world*, it is not a calamity, it is not a matter of great importance. Also *end-of-the-world* used attrib. or as adj.

1907 G. B. Shaw *Major Barbara* III, Nothing's going to happen to you;.. it wouldnt be the end of the world if anything did. **1964** J. Creasey *Guilt of Innocence* xvi. 136, I know what a shock you had... But it isn't the end of the world, you know. **1964** A. Wilson *Late Call* ii. 78 'I don't see the numbers very clearly...' 'Well, it isn't the end of the world if you don't.'

1916 D. H. Lawrence *Let.* 7 Nov. (1932) 375 The book frightens me: it is so end-of-the-world. *a* **1963** L. MacNeice *Astrol.* (1964) ix. 286 A recent 'end-of-the-world' forecast.

c. Now used, somewhat trivially, in the sense 'to be the downfall of, to put an end to (ambition, promotion, etc.)'.

1934 G. B. Shaw *On Rocks* I. 200 He was just ripe for the Cabinet when his wife..made money by journalism. That was the end of him. **1936** J. B. Priestley *They walk in City* iv. 76 She and the bow-legged chap would be pushing a pram up the road..and that..would be the end of them.

17. b. Also *right* or *straight on end*: (*a*) consecutively, uninterruptedly; (*b*) immediately.

[**1778** Strait an end: see Straight *adv.* 2 c.] **1837** T. Hook *Jack Brag* i. 16 The fox going away right on-end across a heavy country. **1867** Smyth *Sailor's Word-bk.* 574 *Right on end*, in a continuous line; as the masts should be. **1883** S. Baring-Gould *John Herring* I. xi. 154, I be going to die right on end, I be.

20. a. *end on.* Recent instances of attrib. use. Also *transf.*, applied to an educational course in which students spend six months at college and six months in industrial work alternately.

1888 *Encycl. Brit.* XXIV. 366/1 Several ships have been built for the 'end-on attack'. **1894** *Pall Mall Gaz.* 23 Oct. 2/1 Ships built mainly for end-on fire. **1898** Kipling in *Morn. Post* 10 Nov. 5/1, I meant my shot for an end-on shot. **1908** *Chambers's Jrnl.* Mar. 256/1 The greatest care should be taken on this road, else an end-on collision is possible. **1910** *Installation News* IV. 59/1 In a Tungsten lamp only 50% of the light is directed below the horizontal and the end-on candle power. **1959** P. F. R. Venables *Sandwich Courses* vi. 101 If half [the students] are away, and half at the firm..training and production [are] facilitated by this Cox and Box arrangement of end-on courses, i.e. two equivalent groups within the year.

(*b*) 'Working a seam of coal, &c., at right angles to the cleat, or natural planes of cleavage' (Gresley, 1883).

b. Esp. in phr. *all ends up*: completely, utterly.

1921 A. W. Myers *Twenty Yrs. Lawn Tennis* 19 Barrett beat him 'all ends up' in an early round. **1932** *Times Educ. Suppl.* 10 Dec. 1/3 Larwood bowled him all ends up. **1954** P. Frankau *Wreath for Enemy* III. iv. 203 In two months' time you'll be cursing me all ends up.

22. f. *to keep one's end up* (also *to keep* or *hold up one's end*): to sustain one's part or bear one's share fully in an undertaking or performance.

1867 *Baily's Monthly Mag.* Aug. 291 Mr. Appleby kept up his end very seasonably. **1878** Mrs. Stowe *Poganuc People* vi. 63 Nobody can say she hain't been a good yoke-fellow; she's kept up her end. **1887** Tourgée *Button's Inn* 131 She'd be worn out..trying to keep up her end [of the work]. **1892** Stevenson & Osbourne *Wrecker* i. 21 Do you think..that a man who can paint a thousand-dollar picture has not grit enough to keep his end up in the stock market? **1899** *Westm. Gaz.* 24 Nov. 5/1 Colonel Baden-Powell and his gallant garrison will have to keep their end up unassisted. *a* **1910** 'O. Henry' *Rolling Stones* (1916) 90 The Diamond-Cross'll hold its end up with a man who'll look after its interests. **1926** P. C. Standing *Anglo-Australian Cricket* xxiii. 103 Ranjitsinhji, who scored 42 and not out 93 and was still 'keeping up his end' when time was called. **1928** *Observer* 18 Mar. 15/3 Not a single woman who appears..in this play is able to keep her end up after the cosmic announcer informs the inhabitants of the world that the world is about to end. **1970** E. McGirr *Death pays Wages* v. 99 Piron had a certain enthusiasm for the Great Apes and was able to hold his end up, thanks to his good memory.

24. *to come to the end of one's tether*: delete definition in Dict. and substitute: see Tether *sb.* 4; *to go (in) off the deep end*, etc.: see *deep *a.* 1 d.

25. *Comb.*, with sense 'placed at the end', 'coming at the end'; as *end-body*, *-bud* (example in *Zool.*), *-process*, *-result*, *-rhyme*, *-situation*, *-spurt*; **end-artery** *Anat.* [ad. G. *endarterie* (J. Cohnheim *Untersuchungen ueber d. embolischen Processe* (1872) i. 18), f. End *sb.* + Artery], an artery which supplies almost all the blood to a part of the body and does not anastomose with itself or with other arteries; **end-board,** (*b*) a Board (sense 4) of a book; **end-fire array,** a radio antenna array in which the direction of reception of electromagnetic waves is in line with the elements of the array; so *end-fire radiation*, etc.; **end game** *Chess*, *attrib.* uses; also *Bridge*; **end-gate** *U.S.* = Tail-board 1; also *attrib.* and *fig.*; **end-hole** (see quot. 1796); **end-leaf,** a (usually blank) leaf inserted at one or other end of a bound book; **end moraine** = *terminal moraine* (Terminal *a.* 3); **end-paper** usu. in *pl.*, the blank leaves placed at the beginning and end of a book; **end-piece,** a piece forming the end of a box, etc.; in watchmaking, the support for the end of a pivot; **end-play** *Bridge*, any of various methods of play which (usually at about the eleventh trick) force an opponent into making a lead which will cost him a trick; hence as *v. trans.*; **end-position** *Philol.*, the position at the end of a clause or phrase; **end-scraper** *Archæol.* = *grattoir; **end table** *U.S.*, a table suitable for placing at the end of a couch or beside another piece of furniture; **end-time,** the end of a period of time; *spec.* the end of the world; **end-use,** the final specific use to which a product is put; so *end-user*; **end zone** *N. Amer.*, (*a*) in football (see quots. 1916 and 1935); (*b*) in ice hockey, either of two sections of the rink which extend to the goal line from the neutral zone.

1880 *Glasgow Med. Jrnl.* Dec. 490 Although not anatomically, the superior mesenteric is *functionally* an 'end artery'. **1883** J. Coats *Man. Path.* 35 There are parts of the body in which the arteries are distributed to a perfectly definite piece of tissue, and have no anastomotic connections. In the case of such arteries, to which Cohnheim gives the name of End Arteries, the results of obstruction are very serious. **1964** S. Duke-Elder *Parsons' Dis. Eye* (ed. 14) i. 11 The retinal arteries are end-arteries and have no anastomoses at the ora serrata. **1936** *Burlington Mag.* May 213/1 Adhering to the inside of the end-boards of a great ninth-century book of the

Gospels. **1875** *Encycl. Brit.* I. 861/2 Nerve fibres at their peripheral extremities terminate in connection with peculiar structures, named end-bodies, terminal bodies. *Ibid.*, The peripheral end-bodies in the skin. **1903** Dorland *Med. Dict.* (ed. 3), *End-body*, that one of the two elements of the serum of an immunized animal which serves actually to kill the bacteria. The other element, the Between-body or Immune-body, simply fastens the end-body to the bacteria. The end-body is also called the complement and addiment. **1900** J. S. Kingsley *Text Bk. Vert. Zool.* 68 Allied to the sense organs of the lateral line are structures known as end buds. **1936** R. S. Glasgow *Princ. Radio Engin.* xiv. 458 If..there is a progressive phase difference between the currents of the adjacent antennas, the radiation will be concentrated along the axis of the array instead of at right angles to it. This type of structure is known as an end-fire array. **1939** *Amat. Radio Handbk.* (*Radio Soc. Gt. Brit.*) x. 147 Aerials do not usually radiate uniformly in all directions. ..The effect of the ground is generally disadvantageous.. but it does help in allowing end-on or 'end-fire' radiation from a horizontal wire. **1943** *Gloss. Terms Telecomm.* (*B.S.I.*) 66 *End-fire array*, a type of directive aerial-array the principal direction of radiation of which is in the direction of the array line. **1949** *Electronic Engin.* XXI. 346 A wealth of information on broadside and end fire arrays. **1897** *Westm. Gaz.* 15 Feb. 4/1 A fine end-game player. **1899** *Ibid.* 1 June 6/3 Skill in end-game play. **1947** T. Reese *Reese on Play* iv. 143 (*heading*) End game. **1952** I. Macleod *Bridge* xv. 190 Bridge writing tends to concentrate on the end game. **1964** V. Nabokov *Defence* viii. 121 We'll simply take the endgame position at the point it was interrupted today. **1873** *Newton Kansan* 15 May 2/2 An iron end gate rod was thrust easily into the excavation. **1905** *Emporia* (Kan.) *Gaz.* 3 Mar., Henry kept right on lamming the end-gate of the band wagon of reform with a poker. **1911** H. Quick *Yellowstone Nights* vi. 165 Two boys..tied to the feed-rack by Allen's hired man and spanked with the end-gate of his wagon. **1796** C. Jones *Hoyle's Games Improved* 296 (*Cribbage*) It is of Advantage to the last Player to keep as close Cards as possible, in Hopes of coming in for Fifteen, a Sequence, or Pair, besides the End Hole, or Thirty-one. **1877** *Encycl. Brit.* VI. 576/1 (*Cribbage*) The go, end hole, or last card is scored by the player who approaches most nearly to thirty-one. **1905** F. H. Collins *Author & Printer* 108/1 *Endleaves*, the blank fly-leaves at the beginning and end of a book. *Ibid.*, *End papers*, see end leaves. **1954** N. R. Ker *Pastedowns in Oxf. Bindings* p. vii, This method of strengthening the binding has been used since the Middle Ages, the only difference being that the endleaves then consisted of parchment instead of paper. **1903** A. Geikie *Text-bk. Geol.* (ed. 4) II. Index 1446/2 Terminal moraines (end-moraines). **1954** W. D. Thornbury *Princ. Geomorphol.* xv. 374 Three types of glacial deposits—end moraine, lateral moraine, and ground moraine—may be distinguished, depending upon whether deposition took place at the end of, at the side of, or beneath an ice stream. **1960** B. W. Sparks *Geomorphol.* xiii. 291 The edges of former ice sheets are often marked by terminal or end moraines, which are ridges of glacial material not usually exceeding 150–200 feet in height in lowlands. **1818** H. Parry *Art Bk.-binding* 31 Common marble paper pasted between the first and second leaf of the end-papers. **1846** Dodd *Brit. Manuf.* VI. 94 The end-papers are..glued to the boards. **1885** W. J. E. Crane *Bookbinding* vii. 59 The 'end papers' are those blank leaves which are found at the beginning and end of every bound book. **1901** *Athenæum* 4 May, The end-papers were a little spotted, but the general condition was splendid. **1960** *Times* 6 Oct. 17/4 There are endpaper maps. **1839** *Penny Cycl.* XIII. 118/1 The end-pieces, when real diamonds are used, are what are called rose-diamonds, and are procured from Holland, where they are cut. **1878** W. W. Abney *Treat. Photogr.* xxx. 216 A movable end-piece through which the plate passes into the holder. **1881** End-piece [in Dict.]. **1931** *Bridge Mag.* VI. 213/1 All end plays are divided into three large groups; negative, positive and neutral. **1936** E. H. Downes (*title*) Squeezes coups and end plays. *Ibid.* 51 The Declarer..knows that the opponent being End-Played will be forced to make a return lead in a suit in which the Declarer holds a tenace position. **1950** G. S. Coffin (*title*) Endplays. *Ibid.* p. ix, Although an endplay situation usually occurs when all four hands are played down to a few cards, endplays sometimes embrace so many cards that a player is 'endplayed' even on the opening lead! **1928** H. Poutsma *Gram. Late Mod. Eng.* (ed. 2) viii. 442 When they [*sc.* adverbial adjuncts] have to be thrown into prominence, they not infrequently have front-position or end-position. **1962** S. Stubelius in F. Behre *Contrib. Eng. Syntax* 198 End-position adverbials. *a* **1942** B. Malinowski *Sci. Theory Culture* (1944) x. 92 The supply of physical material, the conditions in which the digestive processes can be carried out and the sanitary arrangements of the end-processes. **1963** *Times Lit. Suppl.* 1 Mar. 156/5 The end-process of a long evolution. **1912** J. S. Huxley *Individual in Animal Kingdom* ii. 52 These substances—its food—may be varied..and the end-result, its protoplasm, yet be the same. **1958** *Spectator* 1 Aug. 157/1 Mr. Gaitskell can claim to have been surprisingly successful; but the end-result is unappetising. **1966** *Lancet* 31 Dec. 1458/1 The same end-result can be achieved by the subsequent use of a promoting agent. **1855** H. Clarke *New Dict. Eng. Lang.* 332/1 *End-rhyme*, chiming at the end of each verse. **1907** *Camb. Hist. Eng. Lit.* I. 62 The consistent use of end-rime and alliteration in one and the same poem. **1927** E. V. Gordon *Introd. Old Norse* 296 End-rhyme was called *runhending*. **1953** G. Turville-Petre *Orig. Icelandic Lit.* i. 41 Egill was before his time in using end-rime regularly. **1915** W. J. Sollas *Anc. Hunters* (ed. 2) 298 The grattoirs or end scrapers are generally short and rough. **1937** Garrod & Bate *Stone Age Mt. Carmel* I. I. iii. 32 End-scrapers. **1924** R. M. Ogden tr. *Koffka's Growth of Mind* iii. § 8. 103 When the animal has attained his goal, he has arrived at a situation which to him is an end-situation. **1918** C. S. Myers *Present-Day Applic. Psychol.* 13 Thus when the fatigue involved is slight, and the influence of practice or of end-spurt is sufficiently great, a larger instead of a smaller output of work will occur towards the end of the day's work. **1963** J. M. Fraser *Psychol.* (ed. 2) xvii. 222 A very dramatic effect in speeding up the tempo..by..extending the influence shown in the 'end-spurt'. **1851** C. Cist *Cincinnati* 206 Circular, center, card, and end tables. **1917** J. H. McConkey *End of Age* 28 Similar to this will be the conditions in the end-time. **1953** *Scottish Jrnl. Theol.* VI. 162 We believe that..although the full pattern and purpose belongs to the End-Time, reflections of it can be mirrored in the corporate meetings of the Church. **1953** Berg *Dict. New Words* 74/2 End-use. **1958** *Times Rev. Industry* Feb. 11/1 Forecasts..are not made simpler by the ease with which one kind of fuel can be substituted for another in many of the end-uses. **1963** *Economist* 16 Feb. 627/2 Esso policy of not selling chemicals to the end-user. **1916** J. H. Bancroft *Handbk. Athletic Games* 219 End-zone, ten-yard territory between the end line and the goal line. **1935** *Encycl. Sports* 528/1 American rugby...The playing field is..divided into the field of play and end zones. These are, respectively, 100 yds. and 10 yds. each in length. **1961** F. C. Avis *Sportsman's Gloss.* 245/1 *Face-off spot*...in the end zone, they are 15 ft. from goal line. **1963** *Dict. Canadian English* (*Intermediate*) 292/2 *End zone*. 1. in rugby football, the part of the field between each goal line and the corresponding end of the field. 2. in hockey, the ice between each blue line and the corresponding end of the rink. **1968** *Globe & Mail* (Toronto) 11 July 32/5 Then Gabler..threw a perfect one [*sc.* pass] into the end zone to Profit.

end, *v.*[1] Add: **2.** Also with *off*, *up*.

1884 G. Allen *Philistia* III. 250 Capital sentence to end off one's speech with. **1926** *Ladies' Home Jrnl.* Aug. 109 Those things you use to divide off words and end up sentences with.

c. Colloq. phr. *to end it* (*all*), to commit suicide.

1911 *Maclean's Mag.* Sept. 172/1 Sometimes I wonder if it's all worth while; sometimes I'm half inclined to end it. **1925** F. Harris *My Life & Loves* IV. iv. 90 This is the end; the sooner I put a bullet through my head the better... I took up the rifle to end it all when suddenly my eye caught sight of the..tins of sardines.

III. 8. *trans.* To furnish with an end of a particular kind, for protection or ornament.

1889 *Cent. Dict.* s.v., To end a cane with an iron ferrule.

9. To put *up* on end; to up-end. (Examples are U.S. and Sc.)

1889 *Cent. Dict.*, *End*..to set on end; set upright. **1890** *Century Mag.* Aug. 617/1 We ended-up an old plank.. against the twelve-foot brick wall. **1890** J. Service *Thir Notandums* x. 72 There's aye a wheen toom barrels endit up and waiting to be filled.

¶ *to end or mend*: see *to mend or end* (Mend *v.* 12 c).

a **1670** S. Collins *Pres. State Russia* (1671) ii. 9 It is a strange chastisement to kill, seeing the design hereof was never intended to end people, but to mend them. **1820** Scott *Monast.* II. v. 177 My fate calls me elsewhere, to scenes where I shall end it or mend it.

-end, a formative element representing the termination *-endus, -a, -um* of the gerundive of Latin verbs in *-ēre* and *-ĕre*. Examples are *agend*, *dividend*, *minuend*, *repetend*, *reverend*, *subtrahend*. The meaning of these words is passive, thus *agend* 'that which is to be done'. Like *-and *suffix*[2] this element has never been a living suffix, having no separate existence apart from the Latin gerundive form from which it is derived. The gerundial endings are sometimes retained in their Latin (neut.) form (with pl. *-a*), as in *addendum*, *agendum*, *corrigendum*, *reddendum*, *referendum*.

endarch (e·ndaɹk), *a. Bot.* [f. Gr. ἔνδον End(o- + ἀρχή beginning, origin.] Having a primary xylem developing outwards and a central protoxylem.

1900 B. D. Jackson *Gloss. Bot. Terms.* **1902** *Encycl. Brit.* XXV. 413/1 When..there are several protoxylem strands situated at the internal limit of the xylem..the stele is endarch. **1910** J. M. Lowson *Textbk. Bot.* (ed. 5) xiv. 361 The lower part of the stem has a single central protoxylem, i.e. it is endarch and monarch. **1957** H. C. Bold *Morphol. Plants* xix. 370 Development of the xylem is endarch, for the first-formed annular and spiral protoxylem cells arise near the inner limit of each procambium strand.

endarteritis: see Endo-.

endaspidean (endæspi·diăn), *a.* [f. mod.L. *Endaspideæ* (C. J. Sundevall *Methodi Naturalis Avium Disponendarum Tentamen* (1872) I. 55), f. Gr. ἔνδον within + ἀσπίς shield.] Pertaining to passerine birds that have an anterior series of scutella on the inner side of the tarsus.

1889 in *Cent. Dict.* **1902** R. Ridgway *Birds North & Middle Amer.* IV. 328 The several modifications of the tarsal envelope in the present group may be described as follows...II. *Endaspidean.*—The above arrangement (exaspidean) reversed, the acrotarsium extending to and around the tarsus from the *inner* side, the narrow plantar space being thus external instead of internal. **1959** Van Tyne & Berger *Fund. Ornith.* ii. 48 Several types of passerine tarsal scutellation have been described... Endaspidean...Anterior, scutellated segment of tarsal sheath extends across the inner side of the tarsus.

endeavourer. Restrict †*Obs.* to sense in Dict. and add: **b.** In full, *Christian Endeavourer*: a member of the Young People's Society of Christian Endeavour, a religious association which originated in the United States in 1881.

1893 *Rep. Third Nat. Christian Endeavour Convention* 41 Wherever she went she met as many of the Endeavourers as she could, and all gave her one promise..'We mean to try and do all we can for Jesus'. **1896** *Harper's Weekly* 22 Aug. 822/2, I found restless Endeavorers still going into and coming out of their state-rooms. **1897** *Helping Words* Nov. 247 'There are some directions in the Bible that you can't follow.'..'That is a remarkable admission for an Endeavourer to make.' **1900** H. Lawson *On Track* 136 At one end of the table a Christian Endeavourer endeavouring.

endellionite (ende·liǒnəit). *Min. Obs.* exc. *Hist.* [orig. called *endellion*, f. the place-name St. *Endellion*, Cornwall: see -ite[1].] = Bournonite.

1809 J. L. de Bourbon in *Jrnl. Nat. Philos., Chem. & Arts* XXIV. 229, I have given this the name of *endellion*; which avoids the termination in *ite*, so frequent in the nomenclature of mineral substances. **1837** Dana *Syst. Min.* 413 Other localities are..Endellion near Redruth, in Cornwall, where it was first found, and whence it was called Endellionite, by Count Bourbon. **1954** G. W. Himus *Dict. Geol.* 17 *Bournonite*...Also called Wheel-Ore and Endellionite.

endemic, *a.* and *sb.* Add: **A.** *adj.* **a.** [ad. Fr. *endémique* (A. P. de Candolle 1820, in *Dictionnaire des Sciences Naturelles* XVIII. 412).] Now used *spec.* of plants and animals that are indigenous only in a specified area.

1905 F. E. Clements *Res. Methods Ecol.* iv. 227 (*heading*) Endemism... Since its first use by De Candolle, the term endemic has been employed..by phytogeographers with the meaning of 'peculiar to a certain region'. *Ibid.* 228 In its proper sense, endemic refers to distribution, and not to origin. **1937** *Discovery* July 204/1 Of the *Carabidae*, 169 genera are represented, of which 63 are endemic. **1951** *Jrnl. Ecol.* XXXIX. 215 The term 'endemic' is relative in that it means, as now generally used in biology, a taxon or an ecological group limited in range to the geographical area under consideration.

c. *fig.*

1852 [in Dict.]. **1905** *Westm. Gaz.* 15 Nov. 2/1 The same endemic series of murders of a more or less political character as in Macedonia. **1969** *Times* 15 Jan. 7/2 Until a few years ago warfare was endemic. **1970** *Daily Tel.* 14 Apr. 16 That bane of British economic policy-makers—an endemic tendency for imports to rise more rapidly than exports.

B. *sb.* **2.** A plant native to a certain limited area.

1932 Fuller & Conard tr. *Braun-Blanquet's Plant Sociol.* xi. 282 The original dry sod, untouched by fire, is composed exclusively of the old Mascarene Tertiary endemics. **1947** R. Good *Geogr. Flowering Plants* iii. 48 While one part of a large region possesses a high proportion of endemics another and adjacent region may have considerably fewer. **1960** N. Polunin *Introd. Plant Geogr.* vii. 206 Some endemics are confined to very limited areas.

endemism (e·ndĕmiz'm). [f. Endem(ic + -ism.] The character or quality of being endemic; *spec.* of a plant or animal species, the state or condition of being indigenous only in a specified area.

1886 *Encycl. Brit.* XX. 126/1 In their fauna also the Pyrenees present some striking instances of endemism. **1905** F. E. Clements *Res. Methods Ecol.* iv. 228 The primary causes of endemism are two, lack of migration and presence of barrier... Endemism is readily recognized by methods of distributional statistics, applied to areas limited by natural barriers to migration. **1931** *Nature* 7 Mar. 338 The endemism is tremendous, and the fauna has the aspect of great antiquity. **1951** *Jrnl. Ecol.* XXXIX. 88 One of the most striking features of chasmophyte communities is the restricted distribution of their constituents; endemism is the rule.

en déshabillé: see *en *prep.*

e·ndingless, *a.* [f. Ending *vbl. sb.* + -less.] Without ending.

1946 *Language* XXII. II. 160 Here one may question the analysis of endingless forms in attributive positions as uninflected variants of overtly marked forms. **1957** R. W. Zandvoort *Handbk. Eng. Gram.* iv. 188 Adjectives and adverbs of *one syllable* usually take *-er* and *-est*... This also applies to the endingless adverbs discussed in 953: *He could not speak any plainer.*

endi·stance, *v. Theatr.* and *Cinematogr.* [f. En-[1] + Distance *sb.* 3.] *trans.* To produce an effect of 'alienation' (sense *1 d) in an audience; to 'distance' (cf. sense *1 a). So **endi·stancement**; **endi·stancing** *vbl. sb.* (attrib.).

1961 *Times* 23 May 15/1 The experiment in 'endistancement' involved in putting most of the characters in ..masks. **1961** *Times Lit. Suppl.* 30 June 400/2 There are few if any exponents of 'the new drama' who have not at some stage leavened realism with..the exploitation of 'endistancing' techniques..according to the dictates of Brecht's heroic theatre. *Ibid.* 400/3 Here we are 'endistanced' from the action with a vengeance. **1963** *Times* 20 Apr. 9/3 But the idea of screens so big that they

filled the audience's wide field of vision, or as much as possible, and left no possibility of 'endistancement', has continued to haunt the cinema industry.

endite (e·ndəit), *sb. Zool.* [f. Gr. ἔνδ-ον within + -ite[1].] An appendage on the inner side of a limb of a crustacean.

1881 E. R. Lankester in *Q. Jrnl. Microsc. Sci.* XXI. 348 The median portion may be spoken of as the axis or corm, whilst the processes may be called 'phyllites' or 'apophyses', those ranged along the ventral or neural border of the corm being called 'endites', and those given off from the dorsal border being called 'exites'. **1888** Rolleston & Jackson *Forms Anim. Life* 532 The Phyllopod type of appendage,..with its branchia and external respiratory plate, and its series of internal lobes or endites. **1902** *Encycl. Brit.* XXV. 695/2 It is by the specialization of two 'endites' that the endopodite and exopodite of higher Crustacea are formed. *Ibid.* 696/1 The conversion of the Arthropod's limb into a jaw..is effected by the development of an endite near its base into a hard..gnathobase. **1965** Parker & Haswell *Text-bk. Zool.* (ed. 6) I. VIII. 390 The proximal endite..is small, and bears strong spines.

endlichite (e·ndliχəit, e·ndlikəit). *Min.* [f. the name of F. M. *Endlich*, Amer. mining engineer + -ite[1].] An arsenio-vanadate of lead.

1885 Genth & vom Rath in *Proc. Amer. Phil. Soc.* XXII. 367 Endlichite, or Vanadium-Mimetite, a new species...The name has been suggested by Mr. N. H. Muhlenberg in honor of Dr. F. M. Endlich, Superintendent of the Sierra Mines at Lake Valley [New Mexico]. **1944** C. Palache et al. *Dana's Syst. Min.* II. 897 Endlichite...Contains As in substitution for V, with As:V = 1:1 in the material from Hillsboro, New Mexico.

e·nd-man. Also **end man.** [f. End *sb.* + Man *sb.*[1]] A man at an end of a line or row; *U.S.* the man at either extremity of the semi-circle of performers in a nigger-minstrel entertainment, a corner-man.

1865 Sala *Diary* II. 395 He propounded conundrums to his brother 'end man' who played the banjo. **1884** [see End *sb.* 25]. **1886** *Harper's Mag.* Nov. 837/1 Binns.. sang..appearing to Roxy as he sang as delightful a personage as an end man. **1889** *Cent. Dict.* s.v., In the early days of negro minstrelsy each troupe had two end-men... The larger troupes have since had two, and sometimes four, of each class of end-men. **1909** T. C. de Leon *Beaux, Belles & Brains of 60's* xxxi. 356 Emmett was the star of Birch and Backus, as endman. **1956** M. Stearns *Story of Jazz* (1957) xi. 116 The usual half circle of players with the end-men and interlocutor cracking jokes and doing their various specialties.

endo-. Add: **endarteritis,** hence **e:ndarteri·tic** *a.*, relating to or affected with endarteritis (*Cent. Dict.* Suppl., 1909); **e:ndobio·tic** *a. Bot.* (see quot.); **e·ndoblast** *Biol.* = Hypoblast 2; = *entoblast (b); **endobro·nchial** *a.*, situated or occurring within a bronchus; **e:ndoca·nnibalism** [ad. G. *endokannibalismus* (R. S. Steinmetz 1896, in *Wien. Anthrop. Ges. Mitth.* XXVI. 1)], the practice of eating parents and relatives; **e:ndocervici·tis** *Path.*, inflammation of the lining of the canal of the cervix uteri; **endocho·ndral** *a. Anat.*, situated or occurring within the substance of a cartilage; **e·ndochone** *Zool.*, the innermost structure of a chone; **endocli·nal** *a. Geol.*, of the nature of an **e·ndocline,** a fan-fold of anticlinal type; **e:ndocorpu·scular** *a. Path.*, within a corpuscle; **endocra·nial** *a.*, of or pertaining to the endocranium (Cent. Dict., 1889); **endocy·clic** *a.*, *Chem.* situated within the ring; **endocyto·sis** *Biol.*, the taking in of matter by a living cell; **e:ndodynamomo·rphic** *a.* (see quot. 1930); **endoga·stric** *a. Zool.*, situated within the abdomen; **e·ndognath** *Zool.*, the inner branch of the oral appendages of a crustacean; **endometrial** *a.*, substitute for def.: pertaining to the endometrium; add later example; **e:ndometrio·ma** [*endometri(um + *-oma], a tumour containing endometrial tissue; a localized swelling due to an area of endometriosis; **e:ndometrio·sis** *Path.*, a condition resulting from the development of endometrial tissue in an abnormal, esp. extra-uterine, situation; so **e:ndometrio·tic** *a.*; **endome·trium** *Anat.*, the mucous membrane lining the uterus; **e:ndomito·sis** *Biol.* [ad. G. *endomitose* (L. Geitler 1939, in *Chromosoma* I. 7)], division of chromosomes in a nucleus without subsequent division of the nucleus; hence **e:ndomito·tic** *a.*, pertaining to or involving endomitosis; **endope·ptidase** *Biochem.*, any of a group of proteolytic enzymes which split peptide bonds other than the terminal ones; **endophy·tic** *a. Bot.*, of, pertaining to, or of the nature of an endophyte (also **endophytous** *a.*); **endoplasm** *Biol.*, add def.: the inner soft layer of the cytoplasm; hence **endopla·smic** *a.*, of, pertaining to, or consisting of endoplasm (Cent. Dict., 1889); **endoplast** (earlier example); **e·ndopod** = Endopodite; **e:ndopodi·tic** *a.*, of or pertaining to the endopodite; **endopsy·chic** *a.*, that is within the mind; **e:ndora·diosonde,** an encapsulated electronic device which when placed within the body transmits radio signals giving information about conditions inside the body; **endore·ism, endorhe·ism** *Geogr.* [ad. F. *endorrhéisme*, 'ou, si l'on admet une orthographe simplifiée', *endoréisme* (E. de Martonne 1926, in *Compt. Rend.* CLXXXII. 1396)], land drainage not reaching the sea; interior drainage; so **endore·ic, endorhe·ic** *adjs.*; **endosi·phon,** the internal tube of certain cephalopods; Siphuncle 1; hence **endosi·phonal, -nate** *adjs.*; **e:ndosoma·tic** *a. Biol.*, designating or pertaining to a device that an animal uses which is part of its own body; hence **e:ndosoma·tically** *adv.*, in the manner of an endosomatic organ; **e·ndosome** (*a*), the innermost part of a sponge; (*b*) *Cytol.*, a deeply staining mass of chromatin in the middle of a vesicular nucleus in certain protozoans; **endo·steal** *a.* (earlier example); **endo·stracum,** the inner layer of the shell of a crustacean; **endothe·cium** *Bot.* [mod.L. (J. E. Purkinje *De cellulis antherarum Fibrosis* (1830) i. 1], a layer in the wall of the anther; **endotho·rax** *Anat.*, the system of internal processes of the thorax or cephalothorax of arthropods; **endoto·xin,** a toxin contained within a bacterial cell, *spec.* one that is not released until disintegration of the cell; **e:ndotra·cheal** *a.*, within the trachea (Billings, 1890).

1883 *Jrnl. Anat. & Physiol.* XVII. II. 180 (*title*) Obliterative endarteritis, and the inflammatory changes in the coats of the small vessels. *Ibid.* 182 It is now known that the so-called organisation of thrombus is in reality an obliterative endarteritis, in which the thrombus plays merely a passive part. **1900** B. D. Jackson *Gloss. Bot. Terms*, *Endobiotic*, living within as a parasite. **1930** H. M. Fitzpatrick *Lower Fungi: Phycomycetes* iii. 44 When, in parasitic species, the summer sporangia or resting spores are formed within the host cell they are termed endobiotic (endophytic when the host is a plant, endozoic when it is an animal). **1895** Endoblast [see *cœloblast]. **1928** G. H. Carpenter *Biol. Insects* vii. 152 The insinking of a mass of cells ('middle plate') along the axis of the germ-band to form a lower layer (endoblast). **1932** *Jrnl. Thoracic Surg.* I. 435 By means of closed endobronchial anesthesia..it is possible to maintain complete control of anesthetic concentration in a single lung. **1966** *Lancet* 31 Dec. 1449/2 Fundamental principles of physics as applied to anæsthesia..are described in some detail, and so also is endotracheal and endobronchial intubation. **1900** tr. *Deniker's Races of Man* v. 148 Endocannibalism is but the remains of a natural state of primitive man. **1960** R. & C. Needham tr. *Hertz's Death & Right Hand* 44 Endocannibalism..secures for the flesh the most honourable of sepultures. **1879** *St. George's Hosp. Rep.* IX. 443 One case [of retroversion of uterus] was accompanied by severe endocervicitis. **1882** *Quain's Anat.* (ed. 9) II. 113 Since the cartilage grows in every dimension by interstitial expansion, the bone which is invading it (endochondral bone) becomes gradually wider as the ossification advances. **1913** [see *ectochondral adj.]. **1887, 1888** Endochone [see *ectochone]. **1901** *Nature* 19 Sept. 514/1 The Silurian tableland,..its endoclinal and exoclinal structures. **1889** *Geol. Mag.* Feb. 62 We must naturally expect to find the deepest strata in the 'fan structure' (*endocline*) or pseudo-synclinal form and the highest in the folds of the inverted fan structure (*exocline*) or pseudo-anticlinal. **1901** *Practitioner* Mar. 276 The new generation of endocorpuscular parasites. **1924** *Glasgow Herald* 10 May 7 Sir Arthur Keith.. delivered..the annual lecture..his subject being 'Phrenological Studies of the Skull and Endocranial Cast of Sir Thomas Browne'. **1946** *Nature* 6 July 5/1 The study of endocranial casts assumes some importance in palæontology, and particularly in the palæontology of man and the other primates. **1913** Endocyclic [see *exocyclic adj. s.v. *exo-]. **1963** C. de Duve in De Rueck & Cameron *Lysosomes* 412 The new term..simply indicates that materials are coming into the cell (endocytosis) rather than going out (exocytosis). **1969** N. S. Cohn *Elem. Cytol.* (ed. 2) v. 124 Foreign protein or other material could undergo digestion within the cell as a result of such endocytosis (pinocytosis and phagocytosis). **1927** Endodynamomorphic [see *ectodynamomorphic adj. s.v. *ecto-]. **1930** *Nature* 19 July 89 Endodynamomorphic soils are those in which external factors have not yet exerted their full influence and which are therefore immature. **1932** Fuller & Conard tr. *Braun-Blanquet's Plant Sociol.* ix. 256 The endodynamomorphic soils are either limited to desert and mountain regions with sparse vegetation or they are to be considered merely as early stages of development of the ectodynamomorphic climax soils. **1888** Rolleston & Jackson *Forms Anim. Life* 795 The endogastric septa of Haeckel. **1902** *Encycl. Brit.* XXX. 795/2 Ultimately the coil becomes ventral or endogastric. **1899** *Proc. Zool. Soc.* 705 The middle lobe of the endognath (the proximal division of the *lacinia externa* in Boas's nomenclature). **1932** S. Zuckerman *Soc. Life Monkeys & Apes* v. 71 Endometrial growth during the ovarian phase of follicular growth. **1922** W. B. Bell in *Jrnl. Obstet. & Gynæcol.* XXIX. 443 These extra-uterine growths have been described as 'adenomyomata' and 'adenofibromyomata'... I have referred to such tumours as 'endometriomata' or as 'endometriomyomata' and 'endometriofibromyomata'. I hope that gynæcologists will adopt this in the place of the older nomenclature. **1962** J. W. Huffman *Gynecol. & Obstetrics* xlv. 1035 The ovaries are buried in the adherent mass or are converted into endometriotic cysts or endometriomas. **1925** J. A. Sampson in *Amer. Jrnl. Obst. & Gyn.* X. 462 (*title*) Inguinal endometriosis (often reported as endometrial tissue in the groin, adenomyoma in the groin, and adenomyoma of the round ligament). *Ibid.*, Ectopic endometrial tissue in the wall of the uterus.. has arisen from the direct invasion of the myometrium by the mucosa lining the uterine cavity. We might call such a condition a direct or primary endometriosis of the uterus. **1940** *Amer. Jrnl. Obstet.* XL. 554 Peritoneal endometriosis associated with endometrial cysts of the ovary. **1961** *Lancet* 23 Sept. 696/2 It is still debatable whether endometriosis can be permanently cured by hormone therapy. **1942** A. H. Curtis *Textbk. Gynecol.* (ed. 4) xxxiii. 511 Endometriotic tissue lodged in a vein. **1966** Wright & Symmers *Syst. Path.* I. xxvii. 894 Endometriotic lesions of the uterus. **1882** *Syd. Soc. Lex.*, Endometrium. **1907** *Practitioner* Dec. 792 The endometrium was normal. **1912** Adami & McCrae *Text-bk. Pathol.* 647 The histological appearances of the endometrium in certain stages of the menstrual cycle. **1956** *Nature* 14 Jan. 59/1 Mechanical irritation of the endometrium during the luteal phase of the œstrous cycle will provoke the formation of decidual cells. **1942** *Biol. Abstr.* XVI. 1930/2 Chromocenter formation depends on failure of the daughter chromatids to move apart in the endomitoses by which polyploidy occurs. **1943** *Ibid.* XVII. 1344/2 Inner multiplication of chromosomes, i.e., polyploidy arisen through 'endomitosis'. **1959** Southwood & Leston *Land & Water Bugs* xii. 351 The phenomenon of endomitosis has been studied in this bug. **1951** M. J. D. White in G. H. Bourne *Cytol. & Cell Physiol.* (ed. 2) v. 209 Geitler has also described what he calls 'endomitotic' cycles of condensation and decondensation during the formation of endopolyploid nuclei. **1960** L. Picken *Organiz. Cells* iv. 128 This cycle of changes in the chromosomes only becomes linked with the formation of the gigantic quasi-spindle..after the genome population has been built up to a high level by repeated endomitotic divisions. And..the 'anaphase' movement does not separate daughter chromosomes. **1936** Bergmann & Ross in *Jrnl. Biol. Chem.* CXIV. 723 It is therefore possible to classify the two types of peptidases as *exopeptidases*, which are restricted to terminal peptide linkages, and *endopeptidases*, which are not thus restricted. **1962** A. Spector in A. Pirie *Lens Metabolism Rel. Cataract* 330 An endopeptidase called β-protease..will attack lens protein in acid pH. **1887** H. M. Ward tr. *Sachs' Lect. Physiol. Plants* xxiii. 371 (*heading*) Endophytic parasites. **1898** H. C. Porter tr. *Strasburger's Text-bk. Bot.* 308 Some species also are endophytic and inhabit cavities in other plants. **1902** Endophytic [see *ectophytic* adj. s.v. *ecto-]. **1935** F. E. Fritsch *Struct. & Reprod. Algae* I. 155 An endophytic green alga..has been reported in the skin of the carp. **1964** V. J. Chapman in *Oceanogr. & Marine Biol.* II. 210 The asexual generation is represented by a unicellular endophytic plant. *a* **1889** C. V. Riley (Cent. Dict.), The larvæ of the castnians are..endophytous, boring the stems and roots of orchids and other plants. **1934** *Nature* 19 May 761/2 The endoplasmic streaming which is the most striking feature in the pseudopodium of an amœba is not.. present in most other rhizopods. **1955** *Sci. News Let.* 28 May 338/2 The endoplasmic reticulum. **1853** Endoplast [see Periplast]. **1893** T. R. R. Stebbing *Hist. Crustacea* iv. 36 In describing a crustacean appendage he [*sc.* T. H. Huxley] names the first two joints the protopodite, which bears on its extremity on the inner side the endopodite, and on the outer side the exopodite. For these terms the shortened forms exopod and endopod will here be preferred—exopod for exopodite, and endopod for endopodite and protopodite combined. **1959** *Chambers's Encycl.* IV. 278/2 In the walking legs of crabs and lobsters only the endopod remains, but in the corresponding thoracic limbs of some more primitive Malacostra the exopod is presented as a many-pointed flagellum and is used for swimming. **1880** T. H. Huxley *Crayfish* 218 The inner or endopoditic division of the antenna. **1913** *Lancet* 19 Apr. 1116/1 The repressing factor in the waking consciousness is named by Freud the *endopsychic censor*. **1922** J. Riviere tr. *Freud's Introd. Lect. Psycho-anal.* 240 The *why* of the symptom, its tendency, is..always an endo-psychic process. **1927** C. Mackenzie *Vestal Fire* I. i, A *frisson* that no endo-psychic censor is capable of providing. **1932** *Brit. Jrnl. Psychol.* Oct. 160, I shall include in the term anxiety that special form which by reason of its endopsychic conditioning is thought of as guilt. **1957** *Lancet* 15 June 1224/1 The endoradiosonde described here embodies a mechano-chemical transducer that changes its dimensions with pH. **1927** E. de Martonne in *Geogr. Rev.* XVII. 397 In this paper we shall attempt to define the extent of the endoreic, or interior-basin, domain and explain its causes. **1957** G. E. Hutchinson *Treat. Limnol.* I. iv. 226 The Caspian basin, into which the Volga drains, adds greatly to the area of the endorheic regions of the north temperate zone. **1927** E. de Martonne in *Geogr. Rev.* XVII. 397 It is convenient to describe each of these two great regions by a single word: we may term through-flowing, or ocean, drainage 'exoreism' (ex-o-ré-ism, from the Greek ἐξ, out, and ῥεῖν, to flow), interior basin drainage 'endoreism' (en-do-ré-ism, from ἐν, in, and ῥεῖν). **1963** D. W. & E. E. Humphries tr. *Termier's Erosion & Sedimentation* 404 *Endorheism*, inland drainage, drainage toward the center of a land mass. **1883** Hyatt in *Proc. Boston Soc. Nat. Hist.* 261 The 'endosiphon', here spoken of for the first time by that name, is the internal tube long known in Actinoceras, and lately demonstrated in Piloceras. *Ibid.* 273 The endosiphonal tube is narrow and regular. **1883** —— *Proc. Amer. Assoc. Adv. Sci.* XXXII. 328 The endosiphonate..types [of cephalopods]. **1951** *New Biol.* XI. 21 Nearly all the effector and receptor organs of nearly all animals are of the type..called endosomatic. Roughly, this means that they are material parts of the organism and are made of the organism's own stuff. 1957 P. B. MEDAWAR *Uniqueness Individual* vii. 140 Cameras have eye-like and clothes have skin-like functions, and motor-cars the functions endosomatically performed by legs. 1887 *Encycl. Brit.* XXII. 415 A reticulation of ectosome on the one side and of endoderm and mesoderm, i.e., endosome, on the other. 1912 E. A. MINCHIN *Introd. Study Protozoa* vi. 73 In the condition with a single, or one greatly preponderating, mass of chromatin, the nuclear space..presents the appearance of a vesicle containing the chromatin-mass at or near its centre;..the chromatinic mass may be termed..an endosome. *Ibid.* 76 The endosome..in most cases..is composed of a matrix or ground-substance of plastin in which the chromatin is lodged. An endosome of this kind is termed a karyosome. 1926 G. N. CALKINS *Biol. Protozoa* ii. 59 Endosomes may consist entirely of chromatin..or they may be composed of chromatin and plastin in various combinations. 1961 MACKINNON & HAWES *Introd. Study Protozoa* I. 12 The term karyosome (the endosome of some authors) is here used descriptively as the name of any conspicuous, deeply staining body lying in the nuclear sap. 1868 Endosteal [see *ectosteal* adj. s.v. *ECTO-]. 1880 T. H. HUXLEY *Crayfish* iv. 194 This zone [*sc.* the ectostracum] may be distinguished from the endostracum which makes up the rest of the exoskeleton. 1960 I. F. & W. D. HENDERSON *Dict. Sci. Terms* (ed. 7) 160/2 Endostracum...The inner layer of mollusc shell. 1832 LINDLEY *Introd. Bot.* I. ii. 128 The lining of the anther has received particular illustration from M. Purkinje, who calls it *endothecium*. 1959 FOSTER & GIFFORD *Compar. Morphol. Vascular Plants* xix. 495 This well-marked outer layer of the microsporangium wall is usually designated as endothecium. 1878 J. BELL tr. *Gegenbauer's Comp. Anat.* v. 249 These processes..are found chiefly in the head and thorax in many orders of the Insecta..where they form a complicated structure known as the 'endothorax'. 1957 RICHARDS & DAVIES *Imms's Gen. Textbk. Ent.* (ed. 9) I. 60 Under the term endothorax is included the endoskeleton of the thorax. 1905 A. C. ABBOTT *Princ. Bacteriol.* (ed. 7) xxv. 568 We now regard the toxic action of these bacteria to be due to the formation of endotoxins or intracellular toxins. 1907 *Jrnl. Amer. Med. Assoc.* 923 The serum injected into the patient has brought about a local disintegration of the gonococci and a liberation of endotoxins. 1964 M. HYNES *Med. Bacteriol.* (ed. 8) vi. 67 The toxins of bacteria which do not produce exotoxins are termed endotoxins and are thought to be firmly bound to the bacterial proteins in a non-diffusible form. 1910 *Practitioner* June 861 Endotracheal stenosis, due to syphilis. 1962 *Lancet* 28 Apr. 879/2 To protect the patient against obstruction of the airway, endotracheal intubation with a cuffed tube is highly desirable during abdominal surgery.

endocentric (endose·ntrik), *a.* *Linguistics.* [f. ENDO- + CENTRIC *a.*] Applied to a construction or compound having the same grammatical function as one of its constituent parts. Opp. to *EXOCENTRIC *a.* Hence **endoce·ntrically** *adv.*

1933 BLOOMFIELD *Lang.* xii. 194 The forms *John* and *poor John* have, on the whole, the same functions. Accordingly, we say that the English character-substance construction (as in *poor John, fresh milk*, and the like) is an *endocentric* construction. *Ibid.* xiv. 235 Since a *blackbird* is a kind of a *bird*, and a *door-knob* a kind of a *knob*, we may say that these compounds have the same function as their head members; they are endocentric. 1942 BLOCH & TRAGER *Outline Ling. Analysis* v. 76 If a phrase has the same function as one or more of its immediate constituents, it is an endocentric phrase and has an endocentric construction. 1957 S. POTTER *Mod. Ling.* v. 114 Inasmuch as *good men* is a phrase without predication, it is said to be an *endocentric construction*. As a unit *good men* has the same function as one or more of its constituents. 1964 R. H. ROBINS *Gen. Ling.* vi. 238 Only *all* and *both* can be endocentrically preposed to constructions closed in this way (*both my big black dogs*, etc.). 1968 B. M. H. STRANG *Mod. Eng. Structure* (ed. 2) v. 77 Nominal groups (*every conceivable kind of flower, the finest recording I've ever heard*) have heads (*flower, recording*) and these heads are, like the groups they belong to, nominal. Structures of this..kind are *endocentric*.

endocrine (e·ndǫkrəin, -in), *a.* and *sb.* *Physiol.* Also -crin. [f. Gr. ἔνδον ENDO- + κρίνειν to separate. Cf. It. *glandole endocrine* (N. Pende, 1909), *endocrinico* (ibid.), F. *glandes endocrines* (Gottignies, 1912), Sp. *glándulas endocrinas* (R. Mollá, 1912).] Denoting a gland having an internal secretion which is poured into blood or lymph; a ductless gland, as the thyroid, pituitary, and adrenal glands. As *adj.* also = endocrinal.

Stedman's *Med. Dict.*, 1911, has *endocrin* with the erroneous def. 'The internal secretion of a gland'.

1914 E. A. SCHÄFER *Introd. Study Endocrine Glands* 5 Organs..passing such material into the blood or lymph are termed internally secreting or endocrine organs. 1914 *Lancet* 12 Sept. 714/1 The organs of internal secretion, or endocrine glands. 1921 *Amer. Jrnl. Med. Sci.* CLIX. 800 Thyroid and other endocrin disturbances. 1922 *N.Y. Med. Jrnl.* CXV. 393 The diagnostic and therapeutic rôle of the endocrines. 1924 R. MUIR *Text-bk. Pathol.* 721 Several of the endocrine glands exert..a co-ordinated action on the metabolism of carbohydrates. 1925 J. LAIRD *Our Minds & Their Bodies* 52 Secretions from the endocrines. 1925 W. J. H. SPROTT tr. *Kretschmer's Physique & Char.* 84 Whether the thyroid plays here a rôle as primary cause of an endocrine nature..one cannot say for certain. 1927 HALDANE & HUXLEY *Anim. Biol.* ix. 190 The endocrine system or sum total of the ductless glands. 1954 *Sci. News* XXXIII. 17 In spring, the lengthening hours of light or rising temperature cause endocrine changes in many birds and fishes. 1957 *Encycl. Brit.* X. 392/1 The mode of secretion may be toward the outside of the body (exocrine) or toward the blood and lymph vessels (endocrine).

Hence **e·ndocrinal**, **endocrinic** (-kri·nik), **e·ndocrinous** *adjs.*, pertaining to, of the nature of, or relating to endocrine organs; **e:ndocrino·logy**, the physiology or study of the endocrines; whence **e:ndocrino·logist**; **e:ndocrinolo·gical** *a.*

1913 DORLAND *Med. Dict.* (ed. 3), Endocrinology. 1914 *Brit. Med. Jrnl.* 14 Feb. 369/1 Deficiency of endocrinic glandular secretion. 1914 *Lancet* 4 Apr. 952/2 Influence of the Endocrinous Glands upon Uterine Hæmorrhage. 1919 *Nature* CIV. 208/1 'Endocrinology', or physiology of the internally secreting glands. 1923 *Glasgow Herald* 10 Nov. 4 Variations in the endocrinal or regulatory system. 1930 *Times Lit. Suppl.* 25 Dec. 1103/1 For the endocrinologist there are three papers dealing respectively with the chemical properties of the oxytocic principle of the pituitary gland, with oestrin and with a comb-growth-promoting substance obtained from testes and urine. 1949 M. MEAD *Male & Female* ii. 29 Nor do we..find what..endocrinological clues set them on these paths. 1956 E. L. MASCALL *Christian Theol. & Nat. Sci.* vi. 212 Some new discovery of the endocrinologists about the influence of our glandular secretions upon our behaviour.

endocuticle (e·ndokiūtik'l). Also **endocuti·cula**. [f. ENDO- + CUTICLE.] The inner part of a cuticle. **a.** The flexible, laminated, innermost part of the cuticle of an insect or other arthropod.

1929 F. L. CAMPBELL in *Ann. Entomol. Soc. Amer.* XXII. 415 Following the suggestion of Mr. R. E. Snodgrass, of the Bureau of Entomology, the writer calls the extremely thin outermost layer (Grenzlamelle) the epicuticula, the brittle, pigmented outer layer (epidermis, Pigmentschichte) the exocuticula, and the flexible, colorless inner layer (dermis, Hauptlage) the endocuticula. 1935 BORRADAILE & POTTS *Invertebrata* (ed. 2) x. 309 The arthropod cuticle has a thin, impermeable, non-chitinous external layer (epicuticle) and a thick, elastic, permeable, lamellar inner layer (endocuticle), largely composed of chitin. 1967 P. A. MEGLITSCH *Invert. Zool.* xvi. 676 The endocuticle is a continuous layer..and particularly important in the flexible membranes between the skeletal plates.

b. The innermost part of the cuticle that surrounds animal fibres or hairs, etc.

1949 [see *EPICUTICLE b]. 1956 *Nature* 18 Feb. 319/1 [They] described sections of hair in their study of the cuticle layer. They now believe..that the endocuticle is non-keratinous in character. 1962 W. J. ONIONS *Wool* ii. 16 Electron microscopy of silver stained, reduced fibres suggests that the exocuticle contains more sulphur than the endocuticle, which is a layer resistant to alkaline extraction but not to trypsin.

endoderm. Add: **2. c.** *attrib.*

1885 *Encycl. Brit.* XIX. 14/2 The endoderm cells..are almost wholly taken up in the chemical work of digesting and assimilating the food received into the cavity, the lining of which they form. 1914 E. W. MACBRIDE *Text-bk. Embryol.* I. viii. 171 The formation of a cap of small ectoderm cells resting on larger endoderm cells and gradually investing the latter by the process called epibole. 1940 *Chambers's Techn. Dict.* 297/1 *Endoderm disc*, in certain *Malacostraca*, a posterior unpaired thickening on the ventral surface of the blastoderm during early development.

endodontia (endodǫ·ntia). [See next.] = next.

1946 (*title*) Journal of endodontia. 1949 M. COHEN in V. R. Trapozzano *Rev. Dentistry* xiii. 374 (*heading*) Endodontia. What is the objective of root canal therapy? 1969 *Gloss. Terms Dentistry* (*B.S.I.*) 21 *Endodontia*, the part of dentistry concerned with the pulp cavity and periapical tissues.

endodontics (endodǫ·ntiks). [f. Gr. ἔνδον END(O- + ὀδούς, ὀδόντ- tooth + -ICS.] The branch of dentistry that deals with the prevention, diagnosis, and treatment of diseases of the pulp and periapical tissues.

1946 *Dentistry* VI. 771/1 (*index*) Endodontics, penicillin in. 1960 H. J. HEALEY *Endodontics* iii. 85 Endodontics requires..ability in the digital manipulation and use of specially designed instruments.

Hence **endodo·ntal**, **endodo·ntic** *adjs.*, pertaining to endodontics; **endodo·ntically** *adv.*, according to the methods of endodontics; **endodo·ntist**, one who practises or specializes in endodontics.

1946 *Jrnl. Endodontia* I. 1 The 'endodontist' is not any more a rebel against the ruling of 'science' but a recognized specialist of dentistry. *Ibid.* 25 Only exact endodontic technique can save a tooth without in the least endangering the patient. 1960 H. J. HEALEY *Endodontics* i. 15 Odontalgia of this nature is of particular importance to the endodontist. *Ibid.* iii. 71 The selection of cases for endodontic therapy is of utmost importance in the ultimate success or failure in endodontically treated teeth. 1960 A. B. WADE *Basic Periodontology* vi. 274 Endodontal therapy. 1965 L. I. GROSSMAN *Endodontic Practice* (ed. 6) xx. 450 Endodontal treatment..concerns itself with the health of the supporting structures of the teeth.

endogamy. Add: **2.** *Bot.* The fusion or coalescence of two or more female gametes.

1937 GWYNNE-VAUGHAN & BARNES *Struct. & Devel. Fungi* (ed. 2) 5 The power of movement of attached gametangia is very limited, so that, when endogamy is possible, it is likely to occur.

endogenesis (endodʒe·nĕsis). *Biol.* [mod.L., f. Gr. ἔνδον ENDO- + γένεσις origin, production.] The production of structures or bodies within the organism.

1882 *Syd. Soc. Lex.*, *Endogenesis*, the development of one or more cells in the interior of a parent cell.

endogenetic (e:ndodʒĭne·tik), *a.* [f. as prec.: see *GENETIC.] **a.** *Path.* Developed internally. **b.** *Biol.* Produced from within. **c.** *Geol.* (i) (See quot. 1904); (ii) formed by processes originating within the earth.

1874 DUNGLISON *Dict. Med. Sci.* 358/2 *Endogenetic*,..having an origin from internal causes, as *endogenetic diseases*. 1902 *Encycl. Brit.* XXV. 439/2 The inflorescence may be endogenetic, the ovule may consist of nucellus alone, and frequently there is no ovule. 1904 A. W. GRABAU in *Amer. Geologist* Apr. 249 A group which owes its origin chiefly to chemical agents or agents acting from within, or so intimately associated with the forming rock mass that the process of formation may be called *endogenetic*. Endogenetic rocks may also be called non-clastic. 1914 T. CROOK in *Min. Mag.* XVII. 73 *Endogenetic rocks*, formed by processes of internal origin, which processes operate deep-seatedly or from within outwards. 1937 WOOLDRIDGE & MORGAN *Physical Basis Geogr.* v. 54 Internal uplifting, distorting or disrupting forces, which may be grouped as 'endogenetic'. 1953 CZECH & BOSWELL tr. *Penck's Morphol. Analysis of Land Forms* i. 2 The activity of those *endogenetic forces*, originating within the planet, which are responsible for raising individual portions of the crust above sea level.

endogenic (endodʒe·nik), *a.* *Geol.* [f. ENDO- + *-GENIC.] = *ENDOGENETIC *a.* c (ii).

1942 O. D. VON ENGELN *Geomorphology* iv. 56 The relief features..are created by internal, that is, endogenic forces. 1961 L. D. STAMP *Gloss. Geogr. Terms* 176/1 Endogenetic is the usual English form, endogenic the American. 1963 D. W. & E. E. HUMPHRIES tr. *Termier's Erosion & Sedimentation* ii. 202 Deformation will occur simultaneously with diagenesis, giving rise to domes or folds. It has been called 'endogenic folding' by Grabau.

endogenous, *a.* Add: **d.** *Geol.* Formed within a mass of rock, or within the earth's surface; *spec.* applied to intrusive rock changed by contact with surrounding rocks.

1845 A. PRICHARD tr. *von Humboldt's Kosmos* I. 432, I entitled (1832) the plutonic and volcanic eruptive rocks *endogenous* (that which is engendered in the interior,) the sedimentary and fossil rocks *exogenous*, (externally engendered). 1859 D. PAGE *Handbk. Geol. Terms*, *Endogenous*, fossil stems and fragments exhibiting the endogenous structure are so termed. 1878 T. S. HUNT *Chem. & Geol. Ess.* (ed. 2) xi. 196 The endogenous character of this granite is well shown by its banded structure. 1899 *Ibid. Geol. Soc. Amer.* I. 4 The growth of the continent, so far as through marine waters, may be said to have been endogenous. It began to be exogenous on the Atlantic side in the Cretaceous era. 1954 W. D. THORNBURY *Princ. Geomorphology* xix. 488 Volcanism arises from forces which are endogenous in nature and are produced by physical and chemical changes taking place in the earth's interior.

e. *Psychiatry.* Applied to disorders originating within the individual.

1925 G. V. T. HAMILTON *Introd. Objec. Psychopath.* ix. 283 Conceivably acting exogenous and endogenous stimuli ..may lead to seriously morbid habits of indirect reaction. 1962 *Lancet* 29 Dec. 1341/2 Genetic factors are known to be important in the aetiology of so-called endogenous depressions, the predisposition being inherited, and the depressive illness often being precipitated by other factors which may be physical or psychological.

endogeny (endǫ·dʒĭni). [f. ENDO- + -GENY.] = *ENDOGENESIS.

1882 *Syd. Soc. Lex.*, *Endogeny*, the condition of growing or developing from within, as when young or daughter cells are developed in the interior of a mother cell.

endomixis (endomi·ksis). *Biol.* [f. ENDO- + Gr. μίξις mixing, (sexual) intercourse.] A process that occurs in some lower ciliates, in which the old macronucleus disintegrates and a new one is formed from micronuclear material without conjugation taking place.

1914 WOODRUFF & ERDMANN in *Jrnl. Exper. Zool.* XVII. 425 This [rearrangement] involves a more profound intermingling of nuclear and cytoplasmic substances than is possible during the typical vegetative life of the cell. Since this intermingling occurs within a cell we term this reorganization process endomixis. 1936 *Jrnl. Morphology* LIX. 13 The term 'endomixis' has been applied to a variety of nuclear reorganization processes which may not have very much in common. 1961 MACKINNON & HAWES *Introd. Study Protozoa* 293 Many of the books continue to give detailed accounts of endomixis, which almost certainly does not occur in *Paramecium*...Endomixis does occur in some of the lower ciliates. 1965 V. A. DOGIEL *Gen. Protozool.* (ed. 2) vii. 383 In its simplest form endomixis is accomplished in *Epistylus articulata*.

endomorph (e·ndomǭɹf). [f. ENDO- + Gr. μορφ-ή form.] **1.** *Min.* [in Dict. s.v. ENDO-.] **2.** *Anthropometry.* A person with a soft round body-build in which the physical structures developed from the endodermal

layer of the embryo predominate: one of W. H. Sheldon's three constitutional types (cf. *ECTOMORPH, *MESOMORPH).

1940 W. H. SHELDON *Varieties Human Physique* iii. 34 Those whose physiques show a predominant endomorphy we call *endomorphs*. *Ibid.* vii. 251 The endomorph likes soft, overstuffed furniture, luxurious general furnishing, and ceremonial eating equipment. 1951 AUDEN *Nones* (1952) 56 An endomorph with gentle hands. 1954 R. FULLER *Fantasy & Fugue* 75 I've a soft body, big chest, bigger stomach. A typical viscerotonic endomorph—[illegible] jugged hare. 1963 [see *ECTOMORPH].

Hence **e:ndomo·rphic** *a.*, (*a*) *Min.* [ad. F. *endomorphique* (J. Fournet 1858, in *Ann. Sci. Physiques Soc. d'Agric., Lyon* II. Procès-verbaux, p. iii)], of or pertaining to an endomorph; (*b*) *Anthropometry*, characterized by endomorphy. Also **e·ndomorphy** (see quots.).

1888 F. H. HATCH in J. J. H. TEALL *Brit. Petrography* 430 *Endomorphic*, applied, first by Fournet, to contact-metamorphism when produced in the erupted rock. It is used in contradistinction to *exomorphic*. Syn. Endogenous. 1940 W. H. SHELDON *Varieties Human Physique* i. 5 *Endomorphy* means relative predominance of soft roundness throughout the various regions of the body. *Ibid.* iii. 38 The whole endomorphic head is constructed on a spherical plan. 1964 A. HUXLEY *Let.* 19 July (1969) 508 The massive mixtures of endomorphy and mesomorphy [illegible]

endopterygote (e:ndoptĕ·rigōt), *a.* and *sb.* *Ent.* Also -ic, -ous *adjs.* [f. mod.L. *Endopterygota* (D. Sharp 1899, in *Proc. IV Int. Congr. Zool. 1898* 246), f. Gr. ἔνδον within + πτερυγωτός winged.] Belonging to the division Endopterygota of insects, which develop their wings internally until the pupa stage. Also as *sb.* Hence **endopterygo·tism**.

1899 *Proc. Intern. Congress Zool. 1898* 248 The great majority of existing insects are endopterygotic. 1902 *Encycl. Brit.* XXIX. 402/1 Some of the Palaeozoic insects, though we infer them to have been exopterygotous, were really endopterygotous. [illegible] 303/2 The change from exopterygotism to endopterygotism..by an intermediate period of anapterygotism. 1928 G. H. CARPENTER *Biol. Insects* xii. 329 The orders of winged insects—whether exopterygote or endopterygote—are distinguished mainly, as regards the adults, by the characters of their jaws and their wings. 1957 RICHARDS & DAVIES *Imms's Gen. Textbk. Ent.* (ed. 9) iii. 255 There is no doubt that the Megaloptera..include some of the most primitive recent Endopterygotes. 1964 V. B. WIGGLESWORTH *Life of Insects* 320 This [*sc.* Neuroptera] is the first of the 'endopterygote' orders.

endorse, *v.* Add: **1. e.** To make an entry of an offence on (a licence, e.g. of a publican or motorist).

1902 *Hansard Commons* 30 Jan. 1436 The justices will not exercise the power given them of endorsing licences, the effect of which, on the third endorsement, would be to deprive a man of the value of his licence. 1970 *Oxford Times* 18 Sept. 2/1 For driving without insurance, he was jailed for three months..and his licence was endorsed.

2. c. To declare one's approval of, 'crack up' (a person or thing).

1914 *Concise Oxf. Dict.*, *Endorse* (vulgar in advertisements), declare one's belief in. 1914 H. G. WELLS *Englishman looks at World* vi. 61 Politicians comes to endorse me since this was written. 1925 *Publishers' Weekly* 26 Dec. 2013/1, I am told..that [illegible] by prominent stage-folk is now not so eagerly sought after by astute manufacturers. 1926 *Ibid.* 20 Feb. 563 The New York Times, The New York Post,..and many others endorsed the book so highly that we are now setting it up ourselves. *Ibid.* 18 Sept. 960 From the standpoint of the man or woman who endorses the standard college or college education. 1968 *Down Beat* 7 Mar. 38/1 He didn't endorse any breakfast cereals.

endorsement. Add: **1.** *spec.* The entering of an offence on a licence.

1902 [see *ENDORSE *v.* 1 e]. 1903 'IAN HAY' & WODEHOUSE *Baa, Baa, Black Sheep* II. 47 May I see your licence, my dear B. B.? I hope it has no endorsements on it! 1960 *Act* 8 & 9 *Eliz. II* c. 16. 86 (*heading*) Endorsement of licences.

endosmo·tically, *adv.* [f. ENDOSMOTIC *a.*: see -ICALLY.] In an endosmotic manner.

1881 K. SEMPER *Nat. Conditions* vi. 184 They [*sc.* desert-snails] may be capable of absorbing a larger amount of water endosmotically through the skin than the snails in our damp climates. 1884 SEDGWICK & HEATHCOTE tr. *Claus's Zool.* 307 The nutritive fluid passes endosmotically into the body parenchyma.

endothelioma (e:ndoþīliō̆ᵘ·mă). *Path.* Pl. -ata. [ad. G. *endotheliom* (C. O. Block 1875, in *Arch. d. Heilkunde* XVI. 412), f. ENDOTHELI(UM + *-OMA.] A tumour of endothelium. Hence **endothelio·matous** *a.*

1880 H. OSGOOD tr. *von Ziemssen's Cycl. Pract. Med.* IX. 344 Under the name of primary melanotic endothelioma of the liver, Block recently described a case of diffuse or infiltrated pigment cancer. 1886 *Buck's Handbk. Med. Sci.* II. 812/1 *Endothelioma* (psammoma) of the optic nerve consists of alveolar connective tissue, in which cells lie embedded in more or less concentric layers. 1888 J. F. PAYNE *Gen. Path.* 302 Endothelioma is a name sometimes given to this [*sc.* psammoma] and other growths originating in, and composed of, endothelium. 1894 D. J. HAMILTON *Text-bk. Path.* II. II. lxxxv. 738 Cylindromatous endotheliomata are found in various parts of the brain or projecting into the brain from the membranes. 1906 *Practitioner* Nov. 666 The endotheliomatous areas consisted of tubules, columns or sheets of cells. 1964 S. DUKE-ELDER *Parsons' Dis. Eye* (ed. 14) xxv. 369 Some of these may be endotheliomatous growths. *Ibid.* xxxiii. 530 Primary tumours of the optic nerve...A few are endotheliomata or fibromata.

endothermic (endǫþə̄·ɹmik), *a.* *Chem.* [ad. F. *endothermique* (M. Berthelot *Essai de mécanique chimique* (1879) II. iv. ii. 18), f. ENDO- + THERMIC *a.*] Characterized by, or attended with, the absorption of heat.

1884 M. M. P. MUIR *Princ. Chem.* I. iv. 254 To found a system of classification on the difference between exothermic and endothermic changes. 1890 *Bloxam's Chem.* (ed. 7) 141 When C is burnt into CO_2 by 2 N_2O, it evolves 40,400 more units of heat than when burnt in O_2, showing that, contrary to the usual law, heat is evolved in the decomposition of the N_2O, amounting to 20,200 units per molecule. Such a compound is said to be endothermic. 1898 *Nature* 18 Aug. 375 The true chemical equivalent of light energy can only be measured by means of an endothermic irreversible reaction. 1902 *Encycl. Brit.* XXV. 37/1 Such endothermic bodies are nearly always found to show considerable violence in their decomposition. 1927 *Jrnl. Iron & Steel Inst.* CXVI. 479 Practically all clays undergo an endothermic reaction between 600° and 650°C. 1945 *Electronic Engin.* XVII. 670 The individual process of induced transformation is just as likely to be endothermic as exothermic. 1960 *New Biol.* XXXI. 62 The formation is analogous to an endothermic compound or to any molecule in which energy is stored. 1963 HIGHAM *Handbk. Papermaking* v. 109 In the case of the sulphate process the reaction given above is endothermic and, as such, requires considerable heat to achieve completion. 1965 *New Scientist* 4 Nov. 345/1 Beyond Mach 5 the heat to be absorbed will be so great that it will be necessary to use 'endothermic fuels' which undergo chemical reactions which absorb heat.

endothermy (e·ndǫþə̄ɹmi). [f. ENDO- + Gr. θερμός heat + -Y³.] (See quot. 1922.)

1922 *Lancet* 3 June 1109/1 Endothermy is the production of heat in the tissues from within, the active electrode being cold when applied, the process differing therefore from other methods of cauterisation by heat. 1933 *Times* (Electricity Suppl.) No. 5 Dec. p. xxiii/6 Endothermy, a modification of diathermy..which enables him to carry out certain operations that were previously impossible.

endotrophic (-e·ndotrǫfik), *a.* *Bot.* Also (erron.) **endotropic**. [ad. G. *endotropisch* (B. Frank 1887, in *Ber. d. Deut. Bot. Ges.* V. 398).] Chiefly of Mycorrhiza: nourished from within; growing within the cells of plant roots. Opp. *ECTOTROPHIC *a.*

1899 *Ann. Bot.* XIII. 7 The presence of an endotropic mycorhiza in the roots of the *Ophrys* [illegible]. 1903 [see *ECTOTROPHIC *a.*]. 1926 TANSLEY & CHIPP *Study of Vegetation* ix. 158 Endotrophic [fungi] are found mainly within the cells of the roots of herbaceous plants. 1959 [see *ECTOTROPHIC *a.*].

endowment. Add: **5.** *attrib.* and *Comb.*: (sense 2) **endowment assurance, insurance,** a form of life insurance providing for the payment of an endowment or fixed sum to the insured person at a specified date, or (usually) to his representatives on his death, should that take place before the specified date of payment; **endowment (insurance) policy,** a policy which provides for payment according to the above method. So **endowment plan.**

1865 *Nation* (N.Y.) I. 157 Endowment Assurance Policies..are issued to persons desirous of making provisions for advanced life. 1871 *Harper's Mag.* Aug. 477 Benjamin P. Gunn..explaining the endowment plan to his guide. 1880 *Encycl. Brit.* XIII. 168/2 Endowment-assurances. 1889 *Cent. Dict.*, Endowment policy, or, in full, Endowment insurance policy. 1895 E. CARROLL *Princ. Finance* 298 Insurance is divided into Accident and Casualty Insurance, Endowment Insurance, Fire Insurance, Life Insurance, Marine Insurance, etc. 1898 *Westm. Gaz.* 5 Feb. 6/3 The Equity and Law seems to be the only office to which the public should take their endowment assurance business. 1898 *Daily News* 26 Feb. 8/1 The favourite system of insurance is the endowment plan, is it not? 1962 *Which?* June 183/1 In this report, we compare non-profit and with-profits endowment assurance policies.

e·nd-point. [f. END *sb.* + POINT *sb.*¹] The end or latter part of a period, process, etc. **1.** *spec.* in *Chem.*, etc. The point in a titration, or the stage of a process of dilution, at which a definite effect is observable.

1899 HENDERSON & PARKER *Introd. Anal. Chem.* 114 The exact point at which neutralization takes place ('the end point') is determined by means of an indicator. 1919 *Chemist & Druggist* XCI. 243 The solution is very slowly decolorised at first...The end point was taken as the faintest visibly pink tint, permanent for one minute. 1946 *Nature* 19 Oct. 556/2 The concentration of chloride in the blood of agouti and black mice has been determined..not certain. 1956 *Ibid.* 18 Feb. 309/2 Tobacco seedlings show systemic symptoms sooner when inoculated with tobacco sap diluted almost to the infection end-point than when they are inoculated with undiluted sap from infected Prince beans. 1961 *Lancet* 29 July 228/1 In the tube dilution methods the end-point was taken as the last tube showing visible growth after 48 hours incubation at 35–37°C. 1963 *Ibid.* 5 Jan. 34/2 'Total acid' meant titratable acidity to an end-point about pH 8–20 (phenolphthalein).

2. *transf.* and *fig.*

1921 E. SAPIR *Language* (1922) v. 135 The starting-point and end-point of the flow of activity. 1927 J. CLENDINNING *Human Body* iv. 109 (*caption*) An air sac. The end point of a lung branch. 1959 J. C. CATFORD in Quirk & Smith *Teaching of English* vi. 189 The variations in the vertical dimension..represent..different types of end-point, or 'second term', for the relations represented. 1965 W. S. ALLEN *Vox Latina* iii. 60 Scaurus..who comments on the current end-point of the diphthong as *e* rather than *i*.

end-pro·duct. [f. END *sb.* + PRODUCT *sb.*¹ 3.] **1.** *Chem.* A stable, non-radioactive nuclide that is the final product of a radioactive series.

1905 RUTHERFORD in *Phil. Mag.* X. 306 The view that lead is the final or end product of the transformation of radium is supported by the fact that lead is always found in the radioactive minerals in about the amount to be theoretically expected from the content of uranium, when the quantity of helium, present in the mineral, is used to compute its probable age. 1907 B. C. A. WINDLE *Sci. Facts* 10 The element thorium appears to be constantly engaged in generating from itself another solid element which again decays, its end-product being so far unknown. 1922 *Encycl. Brit.* XXXII. 223/2 The atomic weight of the end-product of uranium is close to that of lead. 1958 F. E. ZEUNER *Dating the Past* (ed. 4) 320 Actinium..decays until finally, after the emission of 7 atoms of helium, the inactive end-product is reached, which..is actinium-lead with the atomic weight of 207.

2. A finished article in a manufacturing process.

1939 *Fortune* Apr. 1 An automobile..is the end-product of a long line of manufacturing, mining and farming activities. 1950 *Engineering* 8 Dec. [illegible] The engineering product..may be an end-product of a component. 1958 *Listener* 1 June 93/1 We speak of the end-product in a factory. 1959 *Times Lit. Suppl.* 13 Mar. 140/4 [illegible]..can rely on the interpretative sensibility of craftsmen..whose end-products he describes as..'authentic in their plastic statement'.

3. *transf.* and *fig.*

1923 J. S. HUXLEY *Ess. Biologist* ii. 79 His adult self, the end-product of that development. 1942 J. GLOVER in E. Jones *Social Aspects Psycho-Analysis* 44 The earliest nascent manifestations of these instincts and their most complicated and remote end products. 1931 *Nature* 7 Feb. 214/2 The end-products of soil formation are closely similar from the two contrasted types of parent rock. 1942 M. MCCARTHY *Company she Keeps* (1943) v. 231 The prodigality was merely an end-product of asceticism. 1961 J. WILSON *Reason & Morals* iii. 112 Of course 'I must not fight' must be the end-product of a calculation: perhaps for some pacifists it is.

endragoned, *ppl. a.*: see *EN- *prefix*¹ 3.

endrin (e·ndrin). [Prob. f. EN-³ + -*drin* as in *ALDRIN, *DIELDRIN.] An insecticide that is a stereo-isomer of dieldrin.

1953 *Jrnl. Econ. Ent.* XLVI. 164/1 (*title*) Control of Mosquitoes with DDT, Aldrin, Dieldrin, Endrin and Isodrin. *Ibid.* 199/1 Report of the Committee on Insecticide Terminology. The Committee has approved endrin as a coined name for the insecticidal chemical [etc.]. 1956 *Nature* 25 Feb. 356/1 DDT and some of its analogues, rotenone and endrin. 1964 *New Statesman* 3 Apr. 538/2 Almost coincident with the recommendation to withdraw endrin and dieldrin came a report on the damage done by these two pesticides..in the Mississippi basin.

Endura (endiū·rā). [mod.L., f. O. Prov. *endurar* to fast, endure.] A practice among the Cathari, in which those who had received the consolamentum underwent physical privations, freq. resulting in death, in order to prevent recontamination of the soul.

1888 H. C. LEA *Hist. Inquisition* I. iii. 95 When this heretication occurred on the death-bed, it was commonly followed by the 'Endura' or 'privation'. 1922 H. J. WARNER *Albigensian Heresy* iv. 85 Every inducement was now made to the sick man to end his life by any means other than by direct violence. He was urged to undergo the *Endura*, which took various forms. 1961 P. GREEN tr. *Oldenbourg's Massacre at Montségur* ii. 51 Certain *perfecti* were so passionately addicted to extreme fasting that they incurred the charge of wanting to put an end to their own lives. This is the explanation behind the legend of the *endura*, or voluntary death by starvation. 1970 R. HAUGHTON *Love* iii. 94 The Cathars, whose extreme doctrine could culminate in the ecstatic suicide of fasting of the 'Endura'.

endurably (endiū̆ᵊ·răbli), *adv.* [f. ENDURABLE *a.* + -LY².] In an endurable manner; so as to be endured.

In recent Dicts.

endurance. Add: **1. c.** Of inanimate things: the power of holding out; the capacity (e.g. of steel) of withstanding strain.

1890 *Daily News* 13 Nov. 5/7 Her speed was 18 knots an hour. Her coal endurance is given as 475 tons stowage, and with that stowed she could have steamed 7,000 miles. 1911 *Proc. Inst. Mech. Engin.* IV. 878 Annealed cast-steel showed rather less endurance than annealed mild-steel, in spite of the much greater tenacity of the cast-steel. 1930 *Jrnl. Iron & Steel Inst.* CXXII. 597 The heat-treatment of the rails was found to have markedly increased the endurance of the steel. 1932 *Discovery* May 143/2 The endurance of spiral and laminated springs ..is much less than that of the steel of which they are composed.

4. *attrib.* (esp. in spec. sense of the durability of metals), as *endurance limit, range, test* (= **fatigue limit, range, test*).

1925 *Jrnl. Iron & Steel Inst.* CXI. 563 Increase in the elastic ratio increase, the ratio of endurance limit to tensile and torsional strength. *Ibid.*, With low elastic ratio the full endurance range can be utilised only when the range is between nearly equal and opposite stresses. 1902 *Phil. Trans.* A. CXCIX. 289 The breaking stress for annealed specimens of the type used in the endurance tests was 23·1 tons for Lowmoor iron, and 48 tons for cast-steel. 1930 [see *ATTRITION 2 b *fig.*]. 1930 *Engineering* 3 Jan. 27/3 Endurance tests on the complete springs. 1932 AUDEN *Orators* III. 109 Are you sure of passing the endurance test?

endways, *adv.* **2. b.** (Later U.S. example.)

1871 'MARK TWAIN' *Screamers* 31 He was all ready for the dog too, and knocked him endways with a rock when he came to tear him.

enemy, *sb.* and *a.* Add: **A.** *sb.* **1. b.** *enemy of the people*, a common form of indictment used by popular leaders, esp. Communists, against their political opponents.

1888 E. M. AVELING tr. *An Enemy of Society* IV. in H. Ellis *Plays by Ibsen* 288 He's an enemy of the people! 1904 CONRAD *Nostromo* III. xiii. 477 Comrade Fidanza,.. you have refused all aid from that doctor. Is he really a dangerous enemy of the people? 1938 *Encycl. Brit. Bk. of Yr.* 1938 680/1 The purge against Trotskyists and the hunt for 'enemies of the people and socialism'. 1955 *Treatment Brit. P.O.W.'s in Korea* (*H.M.S.O.*) 13 The battered condition of those who had just returned to the compound as an example of what could happen to anyone who showed himself to be 'an enemy of the people'. 1958 *Economist* 15 Nov. 589/2 The deputy minister of the interior recently admitted that about 700,000 enemies of the people..still exist in Hungary.

e. Phr. (*to be*) *nobody's enemy but one's own*: (to be) responsible only for one's own misfortunes.

[1592 GREENE *Upst. Courtier* sig. H2r, I thinke him an honest man if he would but liue within his compasse, and generally no mannes foe but his owne.] 1639 J. CLARKE *Parœmiologia* 21/1 He is no mans enemy but his owne. 1719 [see Dict., 1]. 1849 LYTTON *Caxtons* II. XIII. iv. 325 Guy Bolding, with all his faults, was one of those excellent creatures who are nobody's enemies but their own. 1850 DICKENS *Dav. Copp.* xxv. 263 He is quite a good fellow—nobody's enemy but his own.

III. *Comb.*: instrumental, as *enemy-controlled, -held, -occupied* adjs.

1918 *Act* 8 & 9 *Geo. V* c. 31 § 8 Any property belonging to a company which is an enemy-controlled corporation. 1937 *Daily Tel.* 19 Oct. 15/2 Enemy-held Channel ports would no more have deprived England of the command of the sea than enemy-held Ostend and Zeebrugge did. 1946 W. S. CHURCHILL *Secret Session Speeches* 45 The enemy-held Atlantic ports and airfields. 1919 J. M. KEYNES *Econ. Conseq. Peace* (1920) 108 To maintain the civilian French population in the enemy-occupied districts. 1944 J. S. HUXLEY *On Living in Rev.* iii. 38 The enemy-occupied countries.

B. 2. Delete '*rare* in modern use' and add examples.

1891 MEREDITH *One of our Conq.* III. vi. 108 The young ..have either emotion or imagination to fold them defensively from an enemy world. 1902 *Westm. Gaz.* 5 Aug. 6/3 Stock requisitioned during the late war from private enemy owners. 1909 *Ibid.* 5 Apr. 1/3 The destination..is presumed to exist if the goods are consigned to enemy authorities, or to a contractor established in the enemy country who..supplies articles of this kind to the enemy. *Ibid.* 2/1 If goods consigned to any trader supplying an enemy population could be seized. 1915 J. H. MORGAN tr. *German War Bk.* 113 Usages of war in regard to enemy territory and its inhabitants. 1915 *N.Y. Tribune* 30 Mar. 8/3 British naval officers in their reports have also referred to 'enemy' ships and fleets. 1942 E. WAUGH *Put out more Flags* III. v. 233 'No, not an enemy agent.' 'Certainly not, sir, but a frivolous, talkative girl.' 1946 'G. ORWELL' *Crit. Ess.* 137 Koestler..was once again thrown into prison as an enemy alien.

eneolithic, var. *AENEOLITHIC *a.*

1911 J. L. MYRES *Dawn Hist.* x. 224 The result was a long chalcolithic (or as the Italians say, eneolithic) phase, in which good cheap stone and bad expensive bronze were in use concurrently. 1932 *Antiquity* VI. 480 Rock-engravings near Lake Gokcha (Sevang) [Armenia].. rather akin to paintings on eneolithic pottery. 1934 A. J. TOYNBEE *Study Hist.* I. I. C. III. 168 Local conditions of life in the 'Eneolithic Age'. 1956 *Antiquity* XXX. 99 Two round eneolithic pits also penetrated the palaeolithic level.

energid (enə̄·ɹdʒid). *Biol.* [ad. G. *energide* (J. Sachs 1892, in *Flora* LXXV. 57), f. *energie* energy: see -ID³.] The nucleus of a cell together with its active cytoplasm regarded as a vital unit.

1897 *Nat. Sci.* Dec. 393 We may introduce that change from the word cell to that of energid (Sachs). *Ibid.*, The distinguishing characteristic of an energid is the living element (protoplasm and nucleus), whilst that of a cell is the membrane. 1900 I. B. BALFOUR tr. *von Goebel's Organ. Plants* I. i. 24 A polyergic plant is either an *energid-colony* or *cœnobium* (cellular or non-cellular) in which..each energid is capable of living for itself; or the energids exhibit a division of labour and..form an *energid-dominion*. 1960 L. PICKEN *Organization of Cells* x. 444 It is by no means the case, that mitosis and cytoplasmic cleavage are inseparable phenomena, or the formation of a polynuclear energid an exceptional event. 1965 BELL & COOMBE tr. *Strasburger's Textbk. Bot.* 14 It can be envisaged that each nucleus has its own volume of protoplasm. The term 'energid' has been suggested for these physiological units lacking morphological delimitation.

energism (e·nəɹdʒiz'm). *Metaphysics.* [ad. G. *energismus* (F. Paulsen *Einleitung in d. Philos.* (1892) 432), f. late L. *energīa* ENERGY: see -ISM.] The theory that the supreme good does not lie in pleasure but in a contented activity of mind (see also quot. 1931). Hence **energi·stic** *a.*

1895 F. THILLY tr. *Paulsen's Introd. Philos.* (1898) 421 This view is opposed by another theory, which does not seek the highest good in subjective feelings, but in an objective content of life, or, since life is activity, in a *specific mode* of life. Permit me to call this view *energism*. 1913 *Hastings's Encycl. Relig.* VI. 511/1 'Energism' is Paulsen's title for his revived Greek position. In this third use of the term it [*sc.* happiness] includes, rather than excludes, perfection as an end. 1931 A. WOLF in W. Rose *Outl. Mod. Knowl.* 544 The view which identifies it [*sc.* the sole attribute of ultimate reality] with energy is called *energism*. *Ibid.* 548 Philosophy which was more or less materialistic, or energistic, or at least positivist.

energize, *v.* Add: **1. b.** (In later technical use.)

1886 P. BENJAMIN *Age of Electricity* vi. 80 An electromagnet is energized or de-energized..by simply establishing or stopping the current in the coil. 1909 *Westm. Gaz.* 12 Jan. 4/3 The Blaisdell energised solid tyre. 1958 *Times* 26 July 4/7 When the fence was energized it was highly dangerous.

2. (Examples illustrating more recent usage.)

1895 *Contemp. Rev.* Jan. 16 That Law would revive and energise the moment our backs were turned. 1908 *Westm. Gaz.* 11 Dec. 2/1 The stray Tories who chanced to go with the Liberal tide in 1906, and who now energise on the Liberal benches. 1920 *Pilgrim* Oct. 105 The act of speaking makes words..the actual vehicle and expression of a concrete personality here and now locally energising in them. 1922 J. Y. SIMPSON *Man & Attainm. Immort.* i. 8 Theology..must in some degree be a knowledge of the world as an expression of God, and of Him as energising in and through it.

energizer. Add: (Later examples.)

1959 *Daily Tel.* 4 June 10/2 We do not want a sponsor of South Sea bubbles, but we do want an energiser of practical plans. 1961 *New Scientist* 6 July 13/2 Of the 120 patients selected, half were given the 'tranquillizer' and half the 'energizer'.

b. *spec.* A carbonate mixed with charcoal to increase carburizing speed.

1938 D. K. BULLENS *Steel & its Heat Treatment* (ed. 4) I. II. x. 264 Energizers. The richness of the CO may be further enhanced by mixing barium or sodium carbonate with the C...These compounds are called 'energizers' because they produce a more energetic carburization. 1948 J. E. GARSIDE in H. W. Baker *Mod. Workshop Technol.* I. vii. 154 The materials consist essentially of a mixture of wood, bone, or leather charcoal, with compounds such as carbonates of barium, calcium, or sodium, which are termed 'energizers'.

energy. Add: **6.** *Physics.* (Later examples.) With preceding adjs., *atomic energy*, see *ATOMIC *a.* 2 d; *radiant energy* (see quots. 1901, 1942).

1901 S. NEWCOMB in J. M. Baldwin *Dict. Philos.* I. 326/1 *Radiant energy or radiance*. Every hot body radiates its heat through the space around it, thus imparting energy to the ether. The energy..is now believed to take a form of waves of electric energy..and its amount is exactly equal to that which the hot body loses. 1936 *Nature* 25 Jan. 135/1 Mass and energy are equivalent, and mass is to be regarded in a sense as a concentrated source of energy. 1942 J. D. STRANATHAN *Particles* ix. 374 Although sufficient quantitative evidence has already been presented..to convince one of the interchangeability of mass and energy..when an electron and a positron combine the two masses disappear and there appears simultaneously a definite amount of radiant energy. 1956 G. THOMSON *Atom* (ed. 5) xvi. 156 It is one of the consequences of the theory of relativity that mass and energy are closely connected, so closely that when one appears the other must disappear. *Ibid.* 161 Uranium..splits into two nuclei..with a diminution of mass and consequent release of energy.

d. (*mental*) *energy* (see quots.).

1901 S. NEWCOMB in J. M. Baldwin *Dict. Philos.* I. 326/2 It is sometimes supposed that thought or mental action..may be a distinct form of energy. ...But the energy itself can be only that of the matter comprising the brain and nervous system. 1944 J. S. HUXLEY *On Living in Rev.* xv. 192, I shall use the term mental energy in the broad popular sense, as denoting the driving forces of the psyche, emotional as well as intellectual.

7. *attrib.* and *Comb.*, as *energy-change, -consumption, -density, -exchange, -level, -producer, -production, -state, unit, value*; *energy-carrying, -giving, -producing, -rich* adjs.

1905 *Westm. Gaz.* 28 Apr. 2/1 The energy-carrying power of a beam of light. 1884 M. M. P. MUIR *Princ. Chem.* II. iii. 453 The energy-changes attending the formation of various compounds. 1909 *Installation News* III. 109/1 The energy consumption should not be below 500 watts. 1958 H. J. GRAY *Dict. Physics* 175/2 *Energy density*, amount of (e.g. radiant) energy in unit volume. 1910 W. JAMES *Let.* 17 June (1920) II. 345 Physically a dinosaur's brain may show as much intensity of energy-exchange as a man's. 1941 J. S. HUXLEY *Uniqueness of Man* III. iv. 96 The fertilizing, energy-giving belt of cyclonic weather. 1951 *Good Housek. Home Encycl.* 374/2 Energy-giving food. 1910 W. JAMES *Let.* 17 June (1920) II. 345 Certain arrangements of matter *on the same energy-level* are, from the point of view of man's appreciation, superior, while others are inferior. 1922 A. D. UDDEN tr. *Bohr's Theory of Spectra* III. iv. 116 The values of the atomic energy corresponding to these states are frequently referred to as the 'energy levels' of the X-ray spectra. 1930 G. THOMSON *Atom* xi. 164 The first outstanding fact in the theory of these extra nuclear electrons is the existence of what are called Energy Levels. The idea is that each electron in the atom requires an amount of energy to remove it completely from the atom, which is characteristic of the particular atom, and the particular electron, though more than one electron may be at the same energy level. 1912 J. S. HUXLEY *Individual in Animal Kingdom* vi. 146 Some of the existing materials must be sacrificed as energy-producers. 1909 *Daily Chron.* 8 July 6/4 An energy-producing food. 1927 HALDANE & HUXLEY *Animal Biol.* iii. 92 Children need a great deal more food per pound of their own weight for energy-production alone, besides their requirements for growth. 1952 *New Biol.* XIII. 125 The limited stock of abiotically formed energy-rich molecules. 1955 W. HEISENBERG in W. Pauli *Niels Bohr* 14 Schrödinger had recognized that the wave functions were the elements of the transformation matrices for the transition from energy states to position states. 1948 *Mind* LVII. 299 We pay for our electric current in kilowatt-hours, which are energy units. 1938 R. W. LAWSON tr. *Hevesy & Paneth's Man. Radioactivity* (ed. 2) xi. 109 For the disruption of aluminium nuclei it is found that there are certain energy values (resonance levels) of the incident α-particles.

en évidence, en face, en famille: see *EN *prep.*

‖ **enfant gâté** (aṅfaṅ gāte). [Fr., = spoilt child.] A person who is accorded excessive adulation or indulgence.

1802 GIFFORD tr. *Juvenal's Satires* p. xlvi, In a word, Horace seems to have been the *enfant gaté* of the palace. 1833 SCOTT *Introd. Novels* I. 479 The Author of the Waverley Novels..might..have been termed *L'Enfant Gâté* of success. 1892 E. DOWSON *Lett.* (1967) 275, I.. think that there is..beneath her double perversity of enfant gâtée and jeune fille coquette a solid foundation of affection. 1937 A. J. CRONIN *Citadel* III. vi. 298 Frances.. had an amused affection for the *enfant gâté*. 1960 *Times* 2 June 12/3 As a person he has survived the period of being the *enfant gâté* of post-war Germany.

enfantillage (aṅfaṅtiyāʒ). [Fr., f. *s'enfantiller* (f. *enfant* child): see -AGE.] A childish action or prank.

1827 [see *BÊTISE]. 1914 E. SIDGWICK *Duke Jones* 384 The talk distracted itself again owing to Charles' enfantillages.

‖ **enfant terrible** (aṅfaṅ tęribl). [Fr., = terrible child.] A child who embarrasses his elders by untimely remarks; *transf.* a person who compromises his associates or his party by unorthodox or ill-considered speech or behaviour; loosely, one who acts unconventionally.

1851 *Fraser's Mag.* Mar. 322/1 He..seemed to tell all these stories just as an *enfant terrible* might, without fully understanding them. 1854 THACKERAY *Newcomes* I. xxi. 194 The *enfant terrible*, young Alfred,..announcing to all the company at dessert, that Ethel was in love with Clive. 1885 'L. MALET' *Col. Enderby's Wife* II. III. v. 27 That *enfant terrible* of Mrs. Farrell's is not coming back, I trust? 1930 T. KARSAVINA *Theatre Street* III. xxiv. 293 A permanent figure of the theatre was Jean Cocteau, the *enfant terrible* of rehearsals. 1939 D. CECIL *Young Melbourne* iii. 87 She..spoke her mind with *enfant terrible* candour. 1940 *Ann. Reg.* 1939 10 The *enfant terrible* of the party, who once more did not see eye to eye with his colleagues.

enfeebler. (Later example.)

1846 *Edin. Rev.* LXXXIV. 362 It is well said by Plato in the *Phaedrus*, that the invention of letters was the great enfeebler of memory.

en fête: see *EN *prep.*

enfe·vered, *ppl. a.* [f. ENFEVER *v.* + -ED¹.] Fever-stricken.

1893 *Pall Mall Mag.* I. 887 His enfevered brain. 1901 *Daily Chron.* 2 July 3/1 Whilst the last chill comes To that enfevered clay and makes it pure.

Enfield. Now a town, not a village. (Earlier and later examples.)

1854 C. A. WINDHAM *Crimean Diary* (1897) 91 The Enfield rifle..should be the arm for the infantry. 1860 [see BROWN BESS]. 1876 [see MINIÉ]. 1959 *Chambers's Encycl.* V. 674/2 The Enfield system of rifling..was adopted. In 1895 the Lee-Enfield rifle was introduced.

enfleurage (aṅflörāʒ). [Fr.] The process of extracting perfumes from flowers by means of fats such as lard and olive oil; INFLOWERING.

1855 G. W. S. PIESSE *Art of Perfumery* i. 15 Absorption, or Enfleurage. 1880 *Encycl. Brit.* XIII. 595/2 The aroma is extracted by the process known as 'enfleurage', *i.e.*, absorption by a fatty body, such as purified lard or olive oil. 1885 [see INFLOWERING]. 1966 *New Scientist* 15 Sept. 624/1 Enfleurage, a very old process in which essential oils from blooms are extracted by absorption into thin films of olive oil or lard.

enforceability (ẹnfōᵊɹsăbi·lĭti). [f. ENFORCEAB(LE + -ILITY.] The character or quality of being enforceable.

1921 B. M. SQUIRES in J. R. Commons *Trade Unionism & Labor Problems* xliv. 786 An admission of illegality must necessarily reflect upon the enforceability of the provisions of the act. 1925 *Glasgow Herald* 9 Sept. 10 The danger of 'a Legislature which has grown accustomed to pass any number of laws without concern for their consequences or their enforceability'. 1928 *Britain's Industr. Future* (*Lib. Ind. Inq.*) III. xvii. 212 The possibility of obtaining legal enforceability for their decisions.

enforcement. Add: **7.** *attrib.* and *Comb.* (sense 6) *enforcement officer, staff, work.*

1946 *How Britain was fed in War Time* (Ministry of Food) iv. 40 At the end of 1944 there were 415 Ministry of Food Statutory Orders in force...The wide new field of offences resulting from the Ministry's Orders made it necessary to appoint a special enforcement staff. 1947 *Evening News* 17 Dec. 1/1 After weeks of secret inquiry by his enforcement staff, Mr. Strachey [Minister of Food] has..been presented with a report. *Ibid.*, All Ministry enforcement officers are now empowered to search premises and confiscate any meat suspected of having been illegally killed. 1948 *Hansard Commons* 11 Mar. 1405 Eighty-four inspectors are employed on all forms of enforcement work. 1959 *Times* 17 July 8/3 Devon County Council agreed to spend £500 on the appointment of four plain clothes litter wardens...The wardens, officially called enforcement officers, will be in action for eight weeks.

engage, *v.* Add: **5. b.** (Earlier Amer. example.)

1760 WASHINGTON *Diary* 7 Jan. (1925) I. 109 Accompanied Mrs. Bassett to Alexandria and engaged a Keg of Butter of Mr. Kirkpatrick, being quite out of that article.

6. e. (Example.)

1766 J. WEDGWOOD *Let.* 15 Sept. (1965) 42, I should expect him to engage for three years at least.

‖ **engagé** (aṅgaʒe), *ppl. a.* [F.] Of (the work of) writers, artists, etc. = *COMMITTED *ppl. a.* b. (See also *ENGAGED *ppl. a.* 4.)

[1948 H. READ *Art Now* (ed. 3) 139 'L'art engagé', art in the service of the revolution.] 1955 G. GREENE *Quiet American* 121, I don't know what I'm talking politics for. They don't interest me and I'm a reporter. I'm not *engagé*. 1956 *Essays in Crit.* VI. 388 His [*sc.* Hazlitt's] temper was active, sympathetic and *engagé*, rather than philosophical. 1959 *Encounter* Oct. 54/1 Surrealism built a functional, utilitarian literature—the first *engagé* art of our time. 1966 *Listener* 17 Mar. 378/1 We hear a lot of talk about the duty of the artist to be 'committed' or '*engagé*'.

engageants, *sb. pl.* (Later example.)

1872 *Young Englishwoman* Nov. 594/1 The sleeves, open to the elbow over print lace engageantes, are edged.. with..velvet.

engaged, *ppl. a.* Add: **1. d.** Hence *engaged ring* (? Obs.) = *engagement ring.

1869 *London Society* XVI. 513 (*title*) The Engaged Ring. 1873 C. M. YONGE *Pillars of House* II. xxii. 252 It is not a proper engaged-ring. You can't wear it.

3. *Teleph.* Of a number or line: in use and therefore unavailable to a second caller; of a person: telephoning. Hence *engaged signal, -test, -tone.*

1891 J. POOLE *Pract. Telephone Handbk.* viii. 151 (*heading*) Engaged test. *Ibid.* 153 If she then hears a click in her receiver, she knows that the number is engaged. *Ibid.* 154 If a subscriber's line is engaged. 1893 PREECE & STUBBS *Man. Telephony* xvi. 271 (*heading*) Switch with distinct 'engaged signal' circuit. 1921 G. B. SHAW *Back to Methuselah* III. 137 Engaged! Who is she calling up now? 1927 *Electr. Communication* VI. I. 14/2 If the busy signal is heard in the receiver..then the called subscriber is engaged. 1935 G. GREENE *England made Me* i. 17, I hoped it was the engaged tone, but I knew it was the ringing tone.

4. A term of literary and artistic criticism: = *COMMITTED *ppl. a.* b. (Cf. *ENGAGÉ *ppl. a.*)

1947 J. HAYWARD *Prose Lit. since 1939* 47 This is not to say that literature must become 'engaged', as one school of continental writers now insists; that it must.. 'take sides' in the social revolution. 1957 *Observer* 27 Oct. 14/3 He [*sc.* François Mauriac] had written a rather wry little note in *L'Express*, suggesting that I had treated him, in his capacity as an 'engaged' writer, like some sole-surviving specimen of an extinct race. 1958 *Sunday Times* 20 July 6/5 The faults usual in 'engaged' literature.

engagement. Add: **2. b.** Also *attrib.*, as *engagement book*.

1850 THACKERAY *Pendennis* II. ii. 15 Anatole, his man..got a sight of her ladyship's engagement-book. 1959 T. S. ELIOT *Elder Statesman* I. 19 But what a time for your engagement book! You know what the doctors said: complete relaxation.

d. (Earlier example.) Also, *engagement ring*.

1811 JANE AUSTEN *Sense & Sens.* II. vii. 114 'If your engagement had been carried on for months and months.. before he chose to put an end to it.'..'Engagement!' cried Marianne, 'there has been no engagement.' 1861 *Cassell's Illustr. Family Paper* 2 Feb. 160/2 The gentleman wears the engagement ring on the third finger of the left hand; the lady on the third finger of the right hand. 1875 MRS. STOWE *We & Neighbors* xxxix. 372 Angie wore on her finger an engagement-ring. 1909 *Daily Chron.* 5 Aug. 7/1 Though they had not yet reached the engagement-ring stage, all of her friends were wondering how soon they could begin to plan for the wedding. 1970 C. KERSH *Aggravations of Minnie Ashe* xiv. 202 He bought an engagement ring (later to be pawned by his widow and never redeemed).

4. c. In *Literary Criticism*, etc.: = *COMMITMENT 6 c. Sometimes with Fr. pronunc. (aṅgaʒmaṅ).

1948 P. MAIRET tr. *Sartre's Existentialism & Humanism* 16..[An] important Sartrean concept—*engagement*—is here translated as 'commitment'...The French word.. conveys a shade of meaning different from that of our word 'engagement'; the existentialism *engagement* is essentially unilateral. 1950 *Theology* LIII. 477 This kind of Christian engagement has been continued. 1952 H. READ *Philos. Mod. Art* iv. 79 Modern protagonists of 'engagement' in art, of socialist realism, of nationalism, etc. 1956 C. WILSON *Outsider* ii. 38 Sartre, whose theory of commitment or 'engagement'..led him to embrace a modified communism. 1960 *Times* 16 Mar. (Canberra Suppl.) p. xv/1 In this age of *engagement*, escapism is deplored.

engagingness. Delete † *Obs.* and add earlier and later examples.

1727 BAILEY vol. II, *Insinuatingness*,..insinuating Nature, Engagingness. 1906 G. SAINTSBURY *Minor Poets of Caroline Period* II. 371 A certain quality of engagingness which it has.

en garçon: see *EN *prep.*

Engelmann (e·ŋgəlmăn). The name of Dr. George *Engelmann* (1809–84) of St. Louis, a botanist, used *attrib.* or in the genitive to designate a spruce (*Picea engelmannii*) native to western North America.

1866 'J. SENILIS' *Pinaceæ* 48 Abies Engelmani [*sic*]: Engleman's [*sic*] Spruce Fir. 1868 J. HOOPES *Bk. Evergreens* 177 A[bies] Engelmanni, *Parry*.—Engelmann's Spruce. 1908 *Bot. Gaz.* May 333 The two trees that dominate the society are the Engelmann spruce (*Picea Engelmanni*) and the subalpine fir (*Abies lasiocarpa*). 1948 A. L. RAND *Mammals E. Rockies* 8 The forests of the east slopes of the mountains are characterized by Engleman spruce, lodgepole pine, alpine fir (near timberline), white-barked pine. 1964 GLEASON & CRONQUIST *Nat. Geogr. Plants* xx. 366 (*caption*) The trees here [*sc.* in Idaho] are alpine fir, Engelmann spruce, and white-bark pine.

engine, *sb.* **5. c.** Delete †*Obs.* and add: Still used for appliances used in the illicit catching of salmon, etc.

1861 *Act* 24 & 25 *Vict.* c. 109 § 11 No fixed Engine of any Description shall be placed or used for catching Salmon in any inland or tidal Waters. 1873 J. W. W. BUND *Law rel. Salmon Fisheries* x. 270 The nets were illegal fixed engines. 1923 *Act* 13 & 14 *Geo. V* c. 16 § 11 No fixed engine of any description shall be..used for taking.. salmon or migratory trout. 1968 *Times* 23 May 17/5 The appellant was convicted for unlawfully using..a fixed engine—a net which he had left unattended, secured by anchors.

11. a. *engine-hose, -house* (example in sense 'railway engine'); **b.** *engine-driver* (earlier example); also **engine-bell,** (*a*) a device for internal communication on a ship; (*b*) a bell rung as a warning on a railway train; **engine-pit,** (*a*) a pit for an engine; (*b*) a trough in the ground or floor constructed to allow a mechanic to work on the underside of a machine (esp. a vehicle); (*c*) (see quot. 1940); **engine-sized (paper)** (example); hence **engine-sizing,** the process of coating paper fibres in a beating machine with size or resin-mixture; the size itself; **engine-turned** (earlier example); **engine-turner** (examples); **engine-turning** (earlier example).

1845 *Knickerbocker* XXV. 59 The short, sharp ring of our engine-bell was heard. 1846–7 THOREAU *Walden* (1957) 145 When the engine bell rings. 1828 F. A. KEMBLE *Let.* Nov. in *Rec. Later Life* (1882) I. 180 Engine drivers, and persons connected with the railroads and coaches. 1838 *Knickerbocker* XII. 373 A small engine-hose ..coiled up like a huge snake on the deck. 1840 F. WHISHAW *Railways of Gt. Brit. & Irel.* 4 The carriage-house and engine-house. 1839 URE *Dict. Arts* II. 970 In every winning of coal, the shape of the engine-pit deserves much consideration. 1853 J. A. BEIL *Technol. Wörterbuch* 365/2 The engine pit (the pit in which the engine works), *der Maschinenschaft, puits pour la machine*. [1883 *Encycl. Brit.* XX. 237/2 Between the rails of each radiating line a pit is constructed to afford access below the engines for inspection.] 1886 *Proc. Inst. Civil Engineers* LXXXVII. 392 The engine-pits are 2 feet 7 inches deep, 3 feet 10 inches wide, and extend the full length of the roads. 1904 GOODCHILD & TWENEY *Technol. & Sci. Dict.* 201/1 *Engine pit*, a depression or pit into which a man can get to examine the lower parts of a locomotive, motor car, etc. 1940 *Chambers's Techn. Dict.* 298/2 *Engine pit*..an engine sump or crank pit; the box-like lower part of the crank-case. 1880 J. DUNBAR *Pract. Papermaker* 29 Engine-sized papers. 1911 *Encycl. Brit.* XX. 732/1 Some form of animal or vegetable size or glue ..mixed with the pulp in the beating engine...Known as 'engine-sizing'..filling up the interstices of the fibres with a chemical precipitate. 1954 J. SOUTHWARD *Mod. Printing* (ed. 7) II. xlii. 443 Engine sizing—a soap consisting of resin and alum, colouring matter, [etc.]. 1765 J. WEDGWOOD *Let.* 6 July (1965) 35, I intend sending two setts of Vases, Creamcolour, engine-turned. 1769 *Ibid.* 19 Nov. 84 We have not one Engine Turner left Here now. 1909 *Westm. Gaz.* 7 Jan. 5/2 The engine-turner who used the wonderful 'rose engine' to engrave the background. 1764 J. WEDGWOOD *Let.* 28 May (1965) 27, I..have sent you a semple of one hobby horse (Engine turning).

engineer, *sb.* Add: **3.** *electric engineer*: now usually **electrical engineer*.

5. b. (Earlier and later U.S. examples.)

1832 *Amer. Railroad Jrnl.* I. 356/2 Engineers and attendants on the Engine. 1946 *Chicago Daily News* 10 July 11/2 The engineer was killed and several passengers shaken up.

6. (With defining word, as *human engineer, spiritual engineer*), one who is claimed to possess specialized knowledge, esp. as regards the treating of human problems by scientific or technical means.

1931 J. C. RICHEY (*title*) The social engineer. 1936 MENCKEN *Amer. Lang.* (ed. 4) VI. vi. 290 *The Engineering News-Record* also discovered..a socio-religious e[ngineer] (an uplifter), a social-e[ngineer] (the same),..a human-e[ngineer] (another variety of psycho-analyst).

engineer, *v.* Add: **2. b.** (Earlier U.S. and later examples.) Also, to manœuvre, (occas.) to 'shepherd'.

1864 *Daily Tel.* 7 July, The lobbying or engineering a bill through the Legislature. 1865 S. S. COX *Eight Yrs. Congress* 99 When he..undertakes to engineer a resolution through this House for the expulsion of a brother member, [etc.]. 1889 *Sat. Rev.* 16 Mar. 299/2 The jealousies and interests of workmen against employers are engineered remorselessly by professional wirepullers. 1890 *Fortn. Rev.* May 716 To 'engineer' a party.. throughout a holiday expedition in a foreign country is an arduous..undertaking. 1959 *Daily Tel.* 28 July 1/1 The late debate had been engineered by the Opposition.

engineering, *vbl. sb.* Add: **1.** *chemical, electrical, human, social engineering*: see the first elements (in Suppl.).

2. *engineering geology, shop, yard.* Also **engineering science**, engineering regarded as a field of study, esp. that part of it which can be treated according to the laws of mathematics and the physical sciences; used esp. as the name of a department in places of higher education.

1865 BRANDE *Dict. Sci.* (ed. 2) I. 779/2 Engineering geology. 1961 J. CHALLINOR *Dict. Geol.* 68/1 *Engineering geology*, geology applied to engineering, including building construction. 1901 *Engineering* 6 Sept. 331/2 The policy of the University [of Glasgow]..is to offer to students a course of study in engineering science, and to leave them to acquire their experience and practice in the office or in the workshop. 1907 *Oxf. Univ. Gaz.* 15 Oct. 68/1 The above Statute proposes to establish a Professorship of Engineering Science...The Professor will give Laboratory—but not Workshop—Instruction. 1911 *Engineering* 7 July 40/3 The classification and analysis of the phenomena presented to the engineer in the practice of his art conveniently called 'Engineering Science', occupies a position intermediate between biology..and physics...It is founded on the laws of motion, of heat, and of electricity, whose action is investigated in the physical laboratory; but it is concerned with the working of these laws under conditions comparable in their complexity with those under which the same laws work in living things. 1930 *Ibid.* 24 Jan. 98/3 His laboratory..was used to enlarge the boundaries of engineering science. 1919 *Brit. Manufacturer* Nov. 24/1 Engineering shops. *Ibid.* 24/2 Engineering yards.

englacial (englē̆i·ʃiăl), *a.* [f. EN-¹ + GLACIAL *a.*] Embedded in or being within a glacier. So **engla·cially** *adv.*

1891 *Amer. Geologist* Dec. 376 The purpose of this paper is to call attention of glacialists to the means of discrimination of the portion of the drift which, at the time of final melting of the ice-sheet, was enclosed within the ice and therefore is called englacial drift by Pres. Chamberlin, and the portion which was subglacial, lying under the ice. 1896 *Jrnl. Geol.* IV. 961 Superglacial and englacial streams might be supposed to make deposits in their channels. 1904 CHAMBERLIN & SALISBURY *Geol.* (1905) I. 268 A surface load..buried by snow and ice..is englacial. 1925 ODELL in E. F. Norton *Fight for Everest: 1924* 315 The medial moraine from the north-east shoulder of Everest, carried englacially. 1933 *Geogr. Jrnl.* LXXXI. 304 The greatest part of the englacial debris. 1935 *Nature* 20 Apr. 605/2 The blanketing effect of the silt brought down by en-glacial and sub-glacial streams.

England. Add: **2.** *Old England*: the 'old country' (as distinguished from *New England*).

1631 T. DUDLEY *Let.* Mar. in *Coll. New-Hampshire Hist. Soc.* (1834) IV. 243 Wee were free enough in Old England, to turne our in sides outwards. 1638 *Essex Inst. Hist. Coll.* (1862) IV. 184/1 George Ropes is to have 20 acres of Land to be Laid out for him at his retorne from old England. *a* 1700 EVELYN *Diary* 26 May an. 1671 (1955) III. 579 To know in what condition *New-England was*; which appearing to be very independent as to their reguard to old England, [etc.]. 1775

ENGLANDER

in *Mass. Hist. Soc. Coll.* (1836) 3rd Ser. V. 87 Thus stands the matters betwixt Old England and America. **1884** *Boston (Mass.) Jrnl.* 30 Dec. 2/4 Our goods are crossing the water to keep alive old England.

3. *England, home, and beauty* a phr. from a poem of *c* 1812 (see quot.) expressing idealistic patriotism, used allusively or ironically.

c **1812** J. Braham *Death of Nelson* 4 In honor's cause I fall at last. For England home and beauty, For England, home and beauty. **1850** Dickens *Dav. Copp.* liii. 536 Let it be..said of me, as of a gallant and eminent naval Hero..that what I have done, I did..'For England, home, and Beauty'. **1874** M. Clarke *His Natural Life* i. vii. 54 'England, home, and beauty!' said Vetch, with a mock-heroic air.

Englander. Add: **2.** *Little Englander*: see Little A. *adj.* 13.

English, *a.* and *sb.* Add: **A.** *adj.* **2. e.** Misc. special Combs.: **English bond**: see Bond *sb.*¹ 13 a (in Dict. and Suppl.); **English breakfast,** a large breakfast including cooked food, opp. *Continental breakfast; **English Canadian,** an English-speaking Canadian, as opp. to a *French Canadian; so **English-Canadian** *a.*, of, or pertaining to, English Canadians or the predominantly English-speaking parts of Canada; **English disease,** (*a*) (in Dict. s.v. 2 c); (*b*) rickets; (*c*) loosely applied to various disorders in the English economy or foreign policy; also **English sickness; English finish,** a relatively smooth machine-finish given to paper, or paper so treated; **English horn**: see Horn *sb.* 13 d; **English Miss** [Miss *sb.*² 4] a somewhat derisive term for an unmarried woman implying primness, or prudishness, etc.; **English pink** [Pink *sb.*⁵] (*a*) = **Dutch pink*, a yellow lake pigment, or shade of yellow; (*b*) [Pink *sb.*⁴] a pink colour (see quot. 1963); **English rose,** a typically attractive light-complexioned English girl; **English setter** [Setter *sb.*¹ 11 a], a breed of long-haired sporting dog, usu. white or white with patches of colour; a dog of this breed; **English sickness** (see *English disease* above); **English-speaker,** a speaker of English; **English springer** [Springer¹ 8 b], a small variety of spaniel; **English Sunday,** Sunday kept as a day of rest and worship, as traditionally in England; opp. **Continental Sunday*; **English toy terrier,** a variety of miniature terrier; cf. Toy *sb.* 11 c.

1807 C. Wilmot *Let.* 5 Aug. in Londonderry & Hyde *Russ. Jrnls.* (1934) ii. 256 Cavanaugh..routed up all his House & prepared an English Breakfast. **1857** C. M. Yonge *Dynevor Terrace* II. v. 71 The English breakfast, which had been established..had quite vanished; each of the family had a cup of chocolate in private. **1964** P. Jones *Month of Pearl* v. 46 Continentals have a name for the kind of breakfast we habitually eat..the 'English Breakfast'. It means an impossible mess of fried eggs, ham, and toast, and the biggest pot in the house full of tea. **1968** P. Loraine *Dead Men of Sestos* x. 134, I used to like what I believe is called an English breakfast—a great deal of bacon and egg and sausage, and then a great deal of toast and butter and that delicious bitter marmalade. *a* **1820** G. Head *Forest Scenes* (1838) 307 There was an old man among them, an English Canadian, called Mr. Weller. **1886** J. G. Bourinot *Intell. Devel. Canadian People* iii. 80 The most widely circulated English and French Canadian papers. **1897** —— *Canada* xxix. 448 The spirit of conciliation and justice..has happily influenced the action of leading English and French Canadian statesmen. **1964** M. Gallant in R. Weaver *Canad. Short Stories* (1968) 2nd Ser. 79 The English Canadians in the room agreed, glancing nervously at the French. **1967** *Times* 28 Feb. (Canada Suppl.) 24 Similarly, English-Canadians would be adversely affected. **1871** C. E. Hackley tr. *T. Billroth's Gen. Surg. Path. & Therapeutics* xvi. 451 It [sc. rachitis] was often called 'English disease'. **1939** Drummond & Wilbraham *Englishman's Food* viii. 182 Rickets..came to be generally known on the Continent as 'The English Disease' (*Die englische Krankheit*). **1969** *Physics Bull.* Apr. 132/1 British industry suffered grievously from the re-establishment of the university-industry barrier in the years that followed the war. It would be very interesting to discover why this happened, for it is, perhaps, one of the symptoms of the socalled 'English disease' in economics. **1970** *Daily Tel.* 9 Feb. 12 There can be no doubt that to a very large degree the improvement of Britain's trade figures during recent months has been due to the spreading of the 'English disease' over other industrial countries. **1970** B. Levin *Pendulum Years* ix. 149 The 'English disease' can be defined in a dozen different ways, and has been, but they have in common that irreducible minimum of amateurism that seems to cling to matters British. **1934** Webster, English finish. **1960** G. A. Glaister *Gloss. Bk.* 124/1 *English finish*, a calendered paper with a smooth, matt surface. **1817** H. C. B. Campbell *Jrnl.* 3 Sept. in *Journey to Florence* (1951) 59 Two ugly English Misses very stiff and prim. **1866** Trollope *Claverings* xviii. 80 [Frenchman *loq.*] Your English meess is so much and so grand. **1879** C. M. Yonge *Magnum Bonum* II. xxvi. 520 Just the insipid English Mees...You should hear what the French think of the ordinary English girl. **1928** R. H. Mottram (*title*) The English Miss. **1960** *Housewife* Oct. 34/2 You are not used to strong drink...The typical little English Miss. **1703** English pink [see Pink *sb.*⁴]. **1835** G. Field *Chromatogr.* ix. 84 Dutch Pink, English and Italian Pinks, are sufficiently absurd names of yellow colours prepared by dyeing, whitening, &c. with vegetal yellow tinctures, in the manner of rose pink, from which they borrow their name. **1901** G. H. Hurst *Dict. Chemicals used in Paints* 147 *English pink*, a name given to yellow pigments prepared from Persian berries; they are similar to Dutch pinks. **1963** *Times* 25 May 11/6 Chrometin pink..became known on the Continent as 'English pink'. **1902** B. Hood *Merrie Eng.* ii. 211 Dan Cupid hath a garden Where women are the flow'rs...And Oh! the sweetest blossom That in the garden grows,—The fairest Queen, it is, I ween, The perfect English rose. **1918** 'D. Lyall' *Eng. Rose* xx. 275 The two girls kissed one another, and the English rose was received in that happy corner of Ireland's garden and found it good. **1967** D. Cilento *Manipulator* i. 13 Rose had..won a seven-year contract with a film company on the strength of her English rose good looks. **1968** H. R. F. Keating *Inspector Ghote hunts Peacock* v. 67 She was the ideal English Rose..her hair was crisply golden...Her complexion was a vigorous pink and white. **1859** 'Stonehenge' *Dog, in Health & Disease* i. iv. 95 The English setter imitated the pointer; but whether it was effected by crossing with that dog is difficult to say. **1910** *Encycl. Brit.* VIII. 378/1 The English setter should have a silky coat with the hair waved but not curly; the legs and toes should be hairy, and the tail should have a bushy fringe. **1950** A. C. Smith *Dogs since 1900* 120 At the beginning of last century English Setters were often described as spaniels by old-fashioned sportsmen. **1963** *Economist* 14 Sept. 886/2 An article [German]..called 'The English Sickness' [=isolationism]. **1969** *Daily Tel.* 18 Apr. 18 The only obvious result of so much effort..has been an abysmal economic growth rate scarcely known elsewhere to post-Keynesian economics. Humiliatingly, the English Sickness has become almost a by-word for economic inefficiency and economic failure. **1970** *Sci. Jrnl.* Apr. 26/1 'The English sickness' is a term widely used in Europe to describe high levels of absenteeism, restrictive practices and wildcat strikes. **1853** Thackeray *Let.* 4 Mar. in *Lett. A. T. Ritchie* (1924) iv. 48 A great hearty nation [sc. the U.S.] of 26 millions of English-speakers. **1894** J. J. Astor *Journey in other Worlds* 38 This has given the English-speakers, especially the United States, a free hand. **1961** *Guardian* 30 June 12/2 English-speakers [of South Africa] opposed to the excesses of nationalism. **1808** English springer [see Springer¹ 8 b]. **1922** R. Leighton *Compl. Bk. Dog* 186 It is only recently that the Kennel Club has officially recognized the variety known as the English Springer. **1879** C. M. Yonge *Magnum Bonum* II. xvii. 333 The afternoon was spent in an easy-going, loitering way, more like a foreign than an English Sunday. **1903** A. Bennett *Leonora* iii. 79 The smooth calm of the English Sunday. **1966** *Guardian* 22 Nov. 1/1 The grim visage of the traditional English Sunday looks like disappearing quite soon. **1907** F. T. Barton *Terriers* xi. facing p. 98 (*caption*) Typical hind-quarters and stern of a white English Toy Terrier. **1962** *Times* 10 Feb. 6/3 English Toy Terriers (Black and Tan).

B. *sb.* **1. b.** (Examples of *Old English* and *Middle English*.)

[**1822** J. Grimm *Deutsche Grammatik* I. i. 506 Mittelenglische buchstaben.] **1836** M. de Larenaudière tr. *T. Wright's Lit. Anglo-saxonne en Angleterre* 38 Que l'on peut appeler l'anglais intermédiaire (middle english). **1839** T. Wright *Lit. & Learning under Anglo-Saxons* 107 The form of our language during the twelfth and the first half of the thirteenth century is generally termed Semi-Saxon; from that period to the time of the Reformation it has received from modern philologists the name of *Middle-English*. **1871** H. Sweet *Alfred's transl. Gregory's Pastoral Care* pref. p. v, I use 'Old English' throughout this work to denote the unmixed, inflectional stage of the English language, commonly known by the barbarous and unmeaning title of 'Anglo-Saxon'. **1879** —— in *Trans. Philol. Soc.* 1878 377 It is a relief to turn to Germany, where Old and Middle English are not regarded as 'fringes'. **1927** *Englische Studien* Nov. 75 This tendency grows stronger in later Middle English. **1933** Bloomfield *Lang.* i. 17 By comparing our records of Old English (say, in the writings of King Alfred) with modern English, we can see how English has changed in the last thousand years.

c. *English English*, English as spoken in England as differentiated from that spoken, e.g., in the United States of America.

1804 M. Wilmot *Let.* 30 May in Londonderry & Hyde *Russ. Jrnls.* (1934) i. 102 If the other side is not English English, it is just the sort of language that might make one blush for what it is. **1943** *Spectator* 5 Feb. 120/1 Of the two hundred million people speaking English nearly seven-tenths live in the United States, and another tenth in the British Dominions are as much influenced by American as by English English. **1958** *Listener* 18 Dec. 1050/1 To get round the difficulty of putting Lorca across full-bloodedly, producers have had recourse to Irish, Cockney, and West Country speech... Lorca's plays suffer in English English because we do not like to grasp his nettles. **1961** H. R. F. Keating *Rush on Ultimate* i. 13 'I never know what's Australian and what's English...Can you call it the gen over here?'..'From what I gather from the boys it's a bit dated now, but it's English English all right.' **1966** S. Harvester *Treacherous Road* i. 9 Most educated Egyptians spoke English English.

2. Delete † *Obs.* and add later examples. Also, the English word or equivalent (*for*).

1824 *Oriental Herald* I. 90 Whose patience is equal to the reading of the 'Bahar Damash' in Dr. Scott's English? **1890** T. Hardy *Melancholy Hussar* in *Three Notable Stories* 170 Phyllis used to say that his English, though not good, was quite intelligible to her. **1926** R. Kearton *Naturalist's Pilgr.* vii. 74 'What are they mining for.., Herr Sonbergh?' 'Ah,' he exclaimed, 'I know quite well, but I cannot remember the English for it.' **1930** *N. & Q.* 11 Oct. 270/1 Throughout the English is apt, for spaces, to be careless and dull.

3. d. English language or literature as a school or university subject or examination.

1889 F. Levander in A. Herbert *Sacrifice of Education to Examination* 42 The school in question was a small one...The subjects were English, Geography,..German, and Drawing. **1911** W. Owen *Let.* 11 Sept. (1967) 79 Have just got home from the 'English'. Am fairly confident of a pass in this. **1926** Kipling *Debits & Credits* 276 Howell, a favourite in 'English' as well as Latin. **1950** A. Wilson *Such Darling Dodos* 107 Mr. Rogers said we should read *Barnaby Rudge* for English, it will be the last book I shall read in class at St. Bertram's. **1968** *Listener* 29 Feb. 275/3 The journalistic addiction of our academic intellectuals..is now the menacing disease of university 'English'. **1970** *Guardian* 29 May 9/1 The UCCA clearing scheme..doesn't hold much hope if he wants to read English.

7. *Billiards.* = Side *sb.*¹ 14 d. Also *transf.* *U.S.*

1869 'Mark Twain' *Innoc. Abr.* 116 You would infallibly put the 'English' on the wrong side of the ball. **1898** R. Hughes *Lakerim Athletic Club* xv. 242 Eaton would slash the ball with a stiffened wrist, an elbow swing, and a quick, hard jump into the air at the same time, to put the 'English' on. **1959** *Sunday Times* 5 Apr. 4/5 The billiard term 'putting on the english', which Atticus states is current parlance in American bowling circles. The story goes that an enterprising gentleman from these shores travelled to the United States during the latter part of the last century and impressed the Americans with a demonstration of the effect of 'side' on pool or billiard balls. His name was English. **1966** H. Nielsen *After Midnight* (1967) iii. 53 When Simon tried to close the door..he encountered difficulties. The officer lent a hand. 'You have to put a little English on it,' he explained. 'There's a defect in the catch.'

C. *Comb.* **a.** *English-reading, -speaking* (earlier and later examples).

1907 W. James *Pragmatism* i. 17 Religious philosophy in our day and generation is, among us English-reading people, of two main types. **1959** *Listener* 24 Dec. 1120/3 The great insufflators of artistic interest in the English-reading world. **1829** Bentham *Justice & Cod. Petit.* iv. 6 Thence was created the necessity of employing these so little trustworthy trustees..as interpreters between the English-speaking parties and the French-speaking judges. **1937** *Discovery* June 185/1 The English-speaking nations. **1940** 'G. Orwell' *Inside Whale* 79 Dickens is scarcely intelligible outside the English-speaking culture. **1970** C. L. Cline *Lett. George Meredith* p. xxix, Meredith stood almost without rival at the head of letters in the English-speaking world.

English, *v.* **3. b.** Delete *nonce-use* and add later examples.

1934 H. G. Wells *Exper. Autobiogr.* II. viii. § 5. 622, I think Conrad owed a very great deal to their early association; Hueffer helped greatly to 'English' him and his idiom. **1965** *Evening Standard* 10 Dec. 6/6 A New York tailor is advertising:..Let us take your Stateside suit and English it up.

Englishism. Add: **2.** An English idiom or form of speech.

1893 in *Funk's Stand. Dict.* **1923** *Blackw. Mag.* Aug. 203/2 We..did the best we might with blunt Englishisms.

englishite (i·ŋgliʃəit). *Min.* [f. the name of George L. English (1864–1944), American mineral dealer + -ite¹.] A hydrated phosphate of potassium, calcium, and aluminium found at Fairfield, Utah County, Utah.

1930 Larsen & Shannon in *Amer. Mineralogist* XV. 328 Layers of a clear, glassy, cleavable mineral that resembles crusts of gordonite..were found on microscopic examination to be a different species...The name englishite is proposed for this mineral. **1951** C. Palache et al. *Dana's Syst. Min.* (ed. 7) II. 957 Englishite... Probably monoclinic. Occurs as layers and aggregates of subparallel plates.

Englishize (i·ŋgliʃəiz), *v.* [f. English *a.* + -ize.] *trans.* To make English.

1858 *Brownson's Q. Rev.* Apr. 190, I want the Church Americanized no more than I want her..Englishized or Gallicized. **1922** *Blackw. Mag.* Sept. 281/1 F...had been accustomed to meet the Englishised Indian in the privacy of his board-room in London. **1928** *Sunday Dispatch* 22 July 22/3 The Englishised sport from the other side of the Atlantic. **1949** *Hansard Commons* 16 May 50/2 The conditions that exist in the North of Ireland, in those Six Counties which have been cut away and 'Englishised'.

Englishman. Add: **2.** An English ship.

1823 J. F. Cooper *Pilot* xxxiii, Griffith saw his own ship borne away from the Englishman. **1885** W. C. Russell *Strange Voyage* I. iv. 36 She was not an Englishman, though I really forget the nationality of the colour she flew at the peak.

Englishry. Add: **1. b.** = Englishness, Englishism.

1894 *Westm. Gaz.* 21 Sept. 2/1 Our Englishry is often a shallow veneer. Scratch the Teuton, and you find an O'Shaughnessy. **1924** *Year's Work in Eng. Stud.* 1923 37 Related..is the fable of the decline and fall of English philology since the mid-nineteenth century, and its loss of Englishry and of appeal to the English. **1940** E. C. Bentley *Those Days* i. 3 Palmerston's death brought out in Brooks the essential Englishry of that time. **1958** *Times Lit. Suppl.* 26 Sept. 543/3 The high romance, the doggedness, the talent for nonsense that are still, to the connoisseur of Englishry, staples of life in this drab, unhygienic island.

2. b. (Later U.S. example.)

a **1910** 'Mark Twain' *Autobiogr.* (1924) I. 120 The Norman Conqueror came over to divert the Englishry.

Englishy, *a.* Delete *nonce-wd.* and add further examples.

1873 W. D. Howells *Chance Acquaintance* vi. 133 The people, too, had such Englishy faces. **1903** *Westm. Gaz.* 4 Feb. 10/1 He spoke in a high Englishy way, which rather jarred on the ears of a Scotch boy. **1912** D. H. Lawrence *Let.* 13 May in F. Lawrence *Not I, but the Wind* (1934) 39 It's a quiet, dead little village [in Germany]..a bit Englishy. **1930** *Times Educ. Suppl.* 26 Apr. 183/3 The English teacher [in Canada] needs much tact, for if she is suspected of being 'Englishy', there will not be much reward of gratitude from the parents. **1967** *Guardian* 1 Feb. 9/6 Englishy pubs..have sprung up in New York.

Eng. Lit. Short for *English Literature*, considered as an academic subject of study.

1850 Mrs. Gaskell *Let.* 25 Jan. (1966) 103 Mr Tom Taylor (the late Professor of Eng. Lit. in University College). **1869** Lady Amberley *Diary* 24 Feb. in B. & P. Russell *Amberley Papers* (1937) II. 260, I went to the two Introductory Lectures given by Prof. Morley on Eng. Lit. and by Prof. Forster on Experimental Physics. **1937** S. Potter *Muse in Chains* 17 Eng. Lit. is an example of the interpretation of the greater by the lesser: of great English writers by anecdotalists, antiquarians, hero-worshippers, pedants, and collectors. **1948** *N. & Q.* 6 Mar. 89/2 Joan Curl's pleasant story has a distinct flavour of Eng. Lit. **1957** *Listener* 19 Dec. 1025/1 The donnish or Eng. Lit. attitude toward the drama. **1958** *Observer* 30 Nov. 19/7 Earnest students of Eng. Lit. in the U.S.S.R. **1967** G. Steiner *Lang. & Silence* 82 It seems to me that the wide gap between the orthodox academic formulation of 'Eng. Lit.'..and the realities of our intellectual and psychological situation may account for the general *malaise* in the field.

englobe, *v.* Add: **2.** *Biol.* To absorb (bacteria, etc.) within a phagocyte or the like. So **englo·bed** *ppl. a.*; **englo·bement,** the process or state of being englobed; **englo·bing** *vbl. sb.*

1900 Gould *Pocket Med. Dict.* (ed. 4) 233 *Englobing*, the taking in of an object by a phagocyte. **1902** *Encycl. Brit.* XXXI. 536/2 Red blood corpuscles are often englobed by this amœba. **1902** *Jrnl. Exper. Med.* VI. 155 The bodies of englobed parasites. *Ibid.*, The englobement of parasites in the liver is more active at certain periods of the cycle.

engram (e·ngræm). [ad. G. *engramm* (R. Semon *Die Mneme* (1904) ii. 20), f. Gr. ἐν En- + γράμμα letter.] A memory-trace; a permanent and heritable physical change in the nerve tissue of the brain, posited to account for the existence of memory. So **engraphy** (e·ngrăfi), the action of exciting an organism in such a way that a permanent change or engram results. So **engra·phic** *a.*, of or pertaining to engraphy; **engra·phically** *adv.* Cf. *ecphore.

1908 *Mind* XVII. 275 Forel..uses the 'engram' theory of Semon. **1921** L. Simon tr. *R. Semon's Mneme* ii. 24 When an organism has been temporarily stimulated and has passed, after the cessation of the stimulus, into the condition of 'secondary indifference', it can be shown that such organism..has been permanently affected. This I call the engraphic action of a stimulus, because a permanent record has been written or engraved on the irritable substance. I use the word engram to denote this permanent change wrought by a stimulus; the sum of such engrams in an organism may be called its 'engram-store'. *Ibid.* 32 Neither can such influences act engraphically. *Ibid.* 274 The engram-association is a result of engraphy and becomes manifest on ecphory. **1923** Ogden & Richards *Meaning of Meaning* iii. 138 Past strikings [of a match] have left, in our organisation, engrams, residual traces which help to determine what the mental process will be. **1925** C. Fox *Educat. Psychol.* 10 When a child simultaneously sees his nurse and receives food, both the optical stimulus and the taste stimuli produce their engraphic effects and the engrams are permanently associated. **1927** Joad *Mind & its Workings* 40 What I am aware of when I appear to remember something is not the past occurrence which, as I say, I remember, but a present state or modification of my body. This present state or modification is called an engram. **1951** G. Humphrey *Thinking* viii. 226 The tendency to substantialize the end product of a part of the activity, and to speak..of engrams, neural imprints, sensation, [etc.]. **1967** *New Scientist* 27 July 206/1 The two most popular chemical contenders for this elusive 'memory trace' or 'engram' have been ribonucleic acid (RNA) and protein.

engrammatic (engrămæ·tik), *a.* [f. prec. + -atic.] Of, or pertaining to, an engram.

1925 *Brit. Jrnl. Med. Psychol.* V. 119 The author considers that the 'engrammatic sediment' of primitive symbolism is part of the reflex pattern of human cerebration. **1944** J. Laird in P. A. Schilpp *Philos. B. Russell* 310 If..mnemic connection is not physiologically engrammatic, there is room for the logical construction of minds which are not material. **1964** *New Scientist* 10 Sept. 643/2 The 'engrammatic' molecules that..store information in the central nervous system.

en grand seigneur, en grande tenue: see *en *prep.*

engraver. Add: **3.** *attrib.* and *Comb.*, as *engraver student*; **engraver beetle,** any beetle of the family Scolytidæ; a bark-beetle.

1896 *Daily News* 11 Dec. 8/6 In the first year..the number of students [at the Royal Academy] was 77..and four were engraver students. **1897** J. H. Comstock *Insect Life* v. 216 These smoothly cut figures are the mines of the engraver-beetle. Many kinds of these engravings can be found, each characteristic of a particular kind of engraver-beetle.

‖ **engrenage** (aŋgrənāʒ). [Fr.] *fig.* Gearing, putting into gear, meshing.

1918 G. B. Shaw *What I wrote about War* (1930) 237 The prison machine would not release the two men caught in its engrenage. **1931** *Times Lit. Suppl.* 10 Sept. 675/2 The new political romanticism of Communism believes that the reality can be dissipated by a new ordering of society; but probably it will succeed only in loosening one *engrenage* by imposing another. **1963** *Economist* 19 Jan. 203/1 'Good Europeans'..sought consolation in the..theory of *engrenage* (getting into gear).

en gros: see *en *prep.*

enhance, *v.* Add: **4. b.** Also, in more recent use, (of property, etc.) to increase in value or price.

1889 *Cent. Dict.* s.v., A debt enhances rapidly by compound interest. [Rare.] **1892** *Pall Mall Gaz.* 15 Nov. 3/2 Until the property enhanced sufficiently to sell. **1904** *N.Y. Even. Post* 31 Mar. 2 With stationary or diminishing incomes, their living expenses are constantly enhancing.

enhanced, *ppl. a.* Add: **b.** *Spectroscopy.* Applied to the lines of a metallic spectrum which are strengthened, or which only appear, under the action of the spark.

1897 J. N. Lockyer in *Proc. R. Soc.* LX. 476 In these stars iron is practically represented by the enhanced lines alone. **1903** *Phil. Trans.* Ser. A. CCI. 211 The majority of the lines are due to metallic vapours, the enhanced lines and the arc lines being of about equal prominence. **1907** N. Lockyer *Spectroscopic Comp. Metals* 24 The enhanced lines of titanium are..considerably weaker in the stellar spectrum than in the sun. **1922** *Encycl. Brit.* XXX. 298/2 The 'enhanced lines' of strontium 4077 and 4215 are relatively strong in stars of high luminosity and weak in those of low luminosity. **1951** L. H. Aller in J. A. Hynek *Astrophysics* ii. 31 The chromosphere lines are enhanced lines.

enhancive (enha·nsiv), *a.* Also *U.S.* **enhansive.** [f. Enhance *v.* + -ive.] That tends to enhance or intensify; *spec.* designating a sentence or word of which the second part is more forcible than the first, or the second part itself.

1853 [see *annexive *a.*]. **1889** *Blackw. Mag.* Nov. 708 What so..enhancive of the glow of the bright firelight..as that delightful ball of warm dark-grey fur coiled up upon the hearth? **1904** A. B. Davidson *Theol. O.T.* iii. 99 *Elohim*—a plural not numerical, but simply enhancive of the idea of *might*.

enharbour, *v.* Delete † *Obs.* and add later example.

a **1907** F. Thompson *Coll. Poetry* (1913) 322 As lovers..do their hearts Enharbour in its continent heart.

enhat (enhæ·t), *v.* [f. En- 1 b + Hat *sb.*] *trans.* To invest with a cardinal's hat.

1925 *Times Lit. Suppl.* 26 Nov. 795/1 The rings ceremonially presented to Cardinals when enhatted. **1925** *Times* 18 Dec. 13/2 Five Cardinals Enhatted.

enhypostasia (enhəipostē̆·ziă). *Theol.* [mod. L., f. Gr. ἐνυπόστατος really existent.] **a.** Substantial or personal existence. **b.** Personality existing not independently but in union with another personality; often describing the human nature of Christ as related to His divine nature as God the Son. So **enhyposta·tic** *a.*

1877 [see *anhypostasia]. **1889** *Cent. Dict.*, Enhypostatic. **1917** H. M. Relton *Study Christology* 78 Hypostasis distinguishes an individual by its characteristic properties, enhypostatic shows that it is not an attribute, which has existence in something else and is not seen in itself. *Ibid.* 226 The doctrine of the Enhypostasia.. secures that the self-consciousness of the God-man is a single-consciousness which is not purely human, not merely human, but truly human. **1946** [see *anhypostasia]. **1954** *Scottish Jrnl. Theol.* VII. 249 By *enhypostasia*..it [sc. classical Christology] asserted that in the *assumptio carnis* the human nature of Christ was given a real and concrete subsistence within the hypostatic union—it was enhypostatic in the Word.

enjambement. Add: (Also aŋʒaŋbmaŋ.) Now applied less restrictedly to the carrying over of a sentence from one line to the next.

1929 *Trans. Amer. Philol. Assoc.* LX. 202, I use the term *enjambement* by itself in its largest sense, that of the running over of the sentence from one line to another. The word is often used by writers on prosody with the narrower force which it originally had, that of the running over of a group of closely joined words.

enjoinder (endʒoi·ndəɹ). [f. Enjoin *v.* after *rejoinder*.] Something imposed upon one, a duty, an obligation.

1894 *Westm. Gaz.* 1 Sept. 3/1 No distinction is made in the analysis of conscience between its objectively rational evolution and its subjectively ultra-rational enjoinders. **1951** I. Compton-Burnett *Darkness & Day* ii. 64 Then act on Alice's unspoken enjoinder.

enjoy, *v.* **4. b.** Delete † *Obs.* and add later example.

1950 A. Clarke *Coll. Plays* (1963) 284 *Brother.* Her present marriage has been consummated. *Abbot.* But lacking true consent, is null and void. *Brother.* Although her husband frequently enjoyed her?

en l'air: see *en *prep.*

enlarge, *v.* Add: **2. d.** *Photogr.* To make a picture larger than (the original negative). Also *absol.*

1866 J. Towler *Negative & Print* xiv. 132 The screen will have to be shoved further off from the lens on the opposite side, and the picture becomes thereby enlarged. **1871** *English Mechanic* 24 Feb. 549/3 Cheap Enlarging Camera. *Ibid.* 5 May 166/2 [A] condenser..for enlarging with a ¼ plate lens. **1878** Abney *Treat. Photogr.* Index, Enlarged photographs. **1903** A. Watkins *Photogr.* (ed. 2) 43 Daylight Enlarging. *Ibid.* 44 In commencing..to calculate enlarging exposures. *Ibid.*, The indicated exposure will be right for a decidedly dense negative without taking into account the increase for the enlarging factor. *Ibid.* 47 Most photographers want to enlarge from a negative of settled size to one size of paper. **1959** *Chambers's Encycl.* X. 688/2 The position of the enlarging lens is adjustable with respect to the negative.

enlargement. Add: **1. c.** *Photogr.* The process of enlarging a picture; a negative or print made of a larger size than the original.

1866 J. Towler *Negative & Print* xiv. 129 (*heading*) The solar camera and solar enlargements. **1871** *English Mechanic* 17 Mar. 621/2 The inner body of large camera for enlargement. **1878** Abney *Treat. Photogr.* xxix. 209 It can also be shown that an enlargement from a small negative is better than a picture of the same size taken direct as regards sharpness of detail. **1884** —— *Instr. Photogr.* (ed. 6) 188 Enlarged negatives can be produced either by making an enlarged transparency, or by enlarging the negative from it in the camera. In all cases of enlargement the camera must be employed. **1903** A. Watkins *Photogr.* (ed. 2) 47 The five separate influences which decide an enlargement exposure. **1919** C. C. Turner *Struggle in Air* xv. 203 Often enlargements of these photographs revealed important changes undetected by the naked eye.

enlarger. Add: **1. b.** *Photogr.* One who, or an apparatus which, enlarges photographs.

1886 in W. D. Welford *Photographer's Indispensable Handbk.* (1887) 334 Bromised paper affords wonderful facilities to the enlarger, whether he be a professional or only a casual. **1894** *Brit. Jrnl. Photogr. Almanac* 1895 1034 (Advt.), The snap shot enlarger. **1911** B. E. Jones *Cassell's Cycl. Photogr.* 221/2 The fixed-focus enlarger, a piece of apparatus in the form of a double box, which allows one degree of enlargement only. **1940** F. J. Mortimer *Wall's Dict. Photogr.* (ed. 15) 245 (*heading*) The vertical enlarger. **1959** *Chambers's Encycl.* X. 688/2 In some types of enlarger focusing is automatically adjusted to match the distance from enlarger lens to baseboard.

enneastyle (e·nĭăstəil), *a.* *Arch.* [f. Gr. ἐννέα nine + στῦλος column.] Having nine columns or pillars.

1875 *Encycl. Brit.* II. 410/2 An enneastyle arrangement.

en noir: see *en *prep.*

Enochian (inǭ·kiăn), *a.* [f. *Enoch* + -ian.] Of, belonging to, or characteristic of Enoch (see Gen. v. 24), or the apocryphal Book of Enoch. Also **Eno·chic** *a.*

1893 R. H. Charles *Bk. Enoch* 313 The term 'the Son of Man'..may have been derived from a current Enochic usage. **1902** *Encycl. Brit.* XXV. 496/1 Slavonic Book of Enoch...This new fragment of the Enochic literature has only recently come to light. **1919** S. C. Carpenter *Christianity acc. Luke* iv. 41 [Our Lord's] actual use of the Enochian conception [of the Messiah]. **1920** V. Burch in Harris & Burch *Testimonies* II. x. 85 Peter's use of the Enochic writings makes in the direction of a close relation between his citations from them and his citations from Isaiah.

enol (ī·nǫl). *Chem.* [f. -en(e + -ol.] Any organic compound that contains the unsaturated alcoholic group ·CH:C(OH)· and is a tautomeric form of a corresponding keto compound. Hence **eno·lic** *a.*, of or pertaining to an enol; **e·nolase,** a crystalline enzyme that acts as a catalyst; **e·nolate,** a metallic derivative of an enol; **enoliza·tion,** conversion into enols or enolic groups.

1894 *Jrnl. Chem. Soc.* LXVI. ii. 433 Compounds containing the group :C:C(OH) are said to be 'enolic'. *Ibid.* 434 No evidence supporting the existence of the tautomeric enolic form sometimes attributed to malonic acid was obtained. **1937** *Nature* 31 July 198/1 An *in vitro* intramolecular shifting of the enol linkage along the carbon chain in hexoses...Such a shifting of enol linkages would make it possible to understand how *l*-lactic acid can derive from *d*-glucose *in vivo*. *Ibid.*, The enolase and dismutase of hexosediphosphate breakdown. *Ibid.*, The occurrence of enolization and dismutation would explain the inhibition of glucolysis proper by fluoride and iodoacetate. **1946** *Ibid.* 21 Dec. 910/1 Furthermore, N-phenylsydnone exhibits its relationship to enols in that it undergoes instantaneous bromination in glacial acetic acid solution. **1962** S. G. Waley in A. Pirie *Lens Metabolism Rel. Cataract* 354 The product (the enolate) has the negative charge on the oxygen atom. **1964** N. G. Clark *Mod. Org. Chem.* xv. 297 It is possible to prepare one or other of the pure tautomers from ordinary ethyl acetoacetate by reason of the greater volatility of the 'enol' form. **1965** Phillips & Williams *Inorg. Chem.* I. xvii. 646 Other interesting carbon carriers are enolates which pick up CO_2 via a carbanion mechanism.

enophthalmus, -mos (enǫfþæ·lmŏs, -ǫs). *Ophthal.* [mod.L., f. Gr. ἐν in + ὀφθαλμός eye.] Abnormal retraction of the eyeball into the orbit.

1883 *Index-Catal. Libr. Surg.-General's Office, U.S. Army* IV. 253 (*heading*) Enophthalmus. **1892** Gradle in *Hare's Syst. Pract. Therap.* 1075 The..receding of the eyeball into the orbit, enophthalmos, is apparently produced by any condition of emaciation or extreme anæmia. **1907** *Practitioner* Nov. 734 The intermittent exophthalmos occurs when the head is depressed, such as occurs in stooping, whereas enophthalmos follows when the head is in the erect position. **1964** S. Duke-Elder *Parsons' Dis. Eye* (ed. 14) x. 109 When all the sympathetic function on one side is lost, resulting in miosis, a narrowed palpebral fissure and slight enophthalmos (due to loss of tone of Müller's muscle) [etc.].

enormousness. 2. (Earlier example.)

1802 C. Wilmot *Let.* 15 Nov. in *Irish Peer* (1920) 114 The Gothic Cathedral [at Milan]..is not yet finish'd from the enormousness of the expense.

‖ **enosis** (enǭ·sis, e·nǫsis). [f. mod. Gr. ἕνωσις, f. ἕν-, εἷς one.] The proposed union of Cyprus and Greece. Also *transf.* Hence **eno·tic** *a.*, of enosis; **e·notist,** one who advocates enosis.

1948 *Times* 13 Aug. 4/4 The constitutional issue has been to some extent obscured by manifestations in favour of Enosis or self-government. **1954** L. Durrell *Let.* (1969) 126, I am grappling with the moribund Information Services of the island, trying to make our case against the united howls of Enotists, British pressmen and fact-finding M.P.s. **1955** *Times* 2 July 6/7 The Archbishop and his *enosis* followers remain uncompromising in their demand for immediate self-determination. **1956** *Ann. Reg.* 1955 125 The Turks..asked for assurances that *enosis* with Greece would not be permitted. **1958** *Punch* 29 Jan. 172/3 He himself [sc. Gladstone] was not ablaze with the enotic flame. **1958** *Economist* 18 Oct. 219/1 Next year Somalia is likely to become independent, whereupon the cry of enosis may go up in British Somaliland.

enough, *a.*, *sb.*, and *adv.* Add: **A.** *adj.* **1.** Also *U.S. dial.* in phr. *enough sight better*, etc. Cf. Sight *sb.*¹ 2 b.

1845 S. Judd *Margaret* i. xiv. 110 Their music is enough sight better than ours. **1856** A. Cary *Married, not Mated* 63 Granmam likes Hal, in fact, enough sight the best. **1887** M. E. Wilkins *Humble Romance* 160 If it's got to be done by anybody I'd enough sight rather 'twoud be done by the town. **1891** —— *New Eng. Nun* 407 They'd keep dusted 'nough sight cleaner. **1911** J. C. Lincoln *Cap'n Warren's Wards* xvi. 251 It was enough sight damper amongst the seats than in those cloth waves.

3. e. *Phr.* *enough is enough*, one must be satisfied with what has been achieved, etc.

1546 J. Heywood *Prov.* ii. xi, Here is enough, I am satisfied (saide he) Sens enough is enough (saide I)..For folke saie, enough is as good as a feast. **1721** J. Kelly *Scot. Prov.* 93 Enough is enough of Bread and Cheese. **1834** Southey *Doctor* I. xx. 199 As for money, enough is enough. **1941** G. B. Shaw *How to become Musical Critic* (1960) 319, I could multiply instances; but enough is enough. **1947** N. Streatfeild *Grass in Piccadilly* 225 It's been very nice, but enough is enough.

B. *adv.* **1. b.** (Later example.)

1878 Hardy *Return of Native* I. ii. ii. 247 Looking at a spot into which she was not climber enough to venture.

3. With comparatives: amply, sufficiently. *U.S. dial.*

a **1852** F. M. Whitcher *Widow Bedott P.* (1883) xvi. 59 It's enough ginteeler 'n them flambergasted blue and yaller things. **1897** R. M. Stuart *In Simpkinsville* 18 You'd see one thet was enough pinker an sweeter 'n the rest to make you climb for it.

en pantoufles: see *en *prep.*

en passant, *adv.* **2.** (Example.)

1818 W. S. Kenny *Pract. Chess Exerc.* 106 You prevent him by pushing immediately your queen's knight's pawn upon his knight, which..obliges the adversary to take your pawn *en passant*.

en pension, en permanence: see *en *prep.*

enplane, var. *emplane *v.*

‖ **en plein** (aŋ plæŋ). [Fr., = in full.] **1.** Of a bet in roulette and other banking games: placed entirely upon one number or side; of a number or side, backed with the whole bet.

1881 Hardy *Laodicean* II. iv. iv. 258 Leaving off backing numbers *en plein*, he laid his venture..*à cheval*. **1933** Mrs. C. S. Peel *Life's Enchanted Cup* xiii. 161 Suddenly I knew that I must risk my last five francs *en plein* on number thirteen. **1966** G. Greene *Comedians* i. ii. 88, I had already won once *en plein*. **1966** C. Robertson *Judas Spies* i. 9 He had been sitting at a roulette table..playing the even chances mostly, occasionally trying his luck *en plein*.

2. *en plein air*, in the open air.

1888 *Practitioner* Oct. 258 Typhus..is..very uncommon amongst the class of people..who..live largely a life *en plein air*. **1962** *Observer* 1 July 19/8 The Greek tradition of performing the classics *en plein air* is not.. continuous since antiquity.

3. In other phrases.

1865 'Ouida' *Strathmore* I. vi. 103 We are so proud of our stolen nuts that we crack them *en plein jour* instead of keeping them to enjoy in the darkness of night. **1896** E. Dowson *Let.* 5 July (1967) 371 Have I not been peacefully rusticating these five months *en pleine campagne*? **1940** J. Joyce *Let.* 11 Feb. (1966) III. 465 He had practically to kidnap the child *en pleine rue*. **1944** A. L. Rowse *Eng. Spirit* viii. 62 The Portuguese Empire was *en pleine déchéance*.

en pointe(s), en poste, en prince, en principe, en prise, en regard: see *en *prep.*

enregister, *v.* For '*rare* in mod. use' substitute 'Revived in recent use as a gallicism', and add later examples.

1896 *McClure's Mag.* VI. 479/2 The works of men too numerous to be enregistered here. **1918** *Times Lit. Suppl.* 21 Mar. 137/2 The three hundred thousand enregistered by the barrister Chenaux. **1928** *New Statesman* 27 Oct. 81 The young spiders..are obeying what is nowadays a racially enregistered tropism to climb.

enregistration (e:nredʒistrē·ʃən). [f. Enregister *v.*: see -ation.] The registering, on the brain, of previous actions, so that performance becomes automatic or instinctive.

1922 J. Y. Simpson *Man & Attainm. Immortality* xi. 241 Increased cerebral development involves in some way a wider and more complex range of enregistration and combination of action and reaction, and so, through the presence of alternatives, of choice. **1927** *Glasgow Herald* 24 Mar. 4 Experiments with rats that quickly master a labyrinth of the Hampton Court maze type point to an enregistration of tactile and muscular sensations. **1930** *John o' London's* 8 Mar. 878/3 Enregistration of the past is characteristic of life.

en règle, en retraite, en revanche: see *en *prep.*

enrich, *v.* Add: **6. b.** To raise (gas) to a required calorific value by the admixture of another gas.

1921 *Chem. Abstr.* XV. 1070 The gas to be enriched enters through a pipe and..passes through an outlet provided with a metal grid. **1958** *Times* 2 June p. iii/2 This gas is then enriched to the declared calorific value by the automatic addition of neat refinery gas.

c. To increase the abundance of a specific isotope in (a material); occas., to increase the abundance of (a specific isotope).

1945 H. D. Smyth *Devel. Methods of using Atomic Energy* 130 Uranium which had already been partially enriched. **1949** *Atomics* Oct. 66 The O^{18} isotope of oxygen is being enriched by a factor of 100 at Harwell by a thermal diffusion plant. **1970** *Daily Tel.* 13 Nov. 5/1 Russia has offered to enrich uranium for Sweden's boiling water reactors.

enriched, *ppl. a.* (Examples in senses of *enrich *v.* 6 b and c.)

1936 *Physical Rev.* XLIX. 404 The enriched sample [of carbon] may be removed for analysis. **1940** *Chambers's Techn. Dict.* 136/2 Carburetted (or enriched) water-gas. **1945** H. D. Smyth *Devel. Methods of using Atomic Energy* 22 Such proposed arrangements are usually called 'enriched piles'. *Ibid.* 23 A chain reaction bomb in pure, or at least enriched, U-235 or plutonium. **1955** *Ann. Reg.* 1954 393 A full scale reactor to generate electricity..was to use slightly enriched uranium as fuel. **1957** *Times* 22 Aug. 4/4 Enriched uranium...Enriched nuclear fuel... Enriched fuel. **1970** *Daily Tel.* 4 Nov. 5 Enriched uranium with a uranium-235 share of more than 20 per cent. is the material used in atomic bombs.

enriching, *ppl. a.* Add: Also in sense 6 b of *enrich *v.*

1958 *Times* 2 June p. iii/1 Gas released as a by-product from a neighbouring petrochemical industry has been used for some years on a small scale as an enriching gas at Partington gas works.

enrichment. Add: **1. a.** *spec.* Of a mixture of isotopes, esp. nuclear fuel. Cf. *enrich *v.* 6 c.

1936 *Physical Rev.* XLIX. 404 (*heading*) Evidence for the enrichment of carbon in the heavier isotope by diffusion. **1945** H. D. Smyth *Devel. Methods of using Atomic Energy* 40 The amount of enrichment to be expected in a single production unit..was very small. **1958** *Engineering* 4 Apr. 423/3 Though it is stated that the new reactor uses enriched fuel, enrichment is normally unnecessary when the moderator is heavy water.

Ensa (e·nsă). Also **E.N.S.A., ENSA.** [f. the initials of *Entertainments National Service Association*.] An organization established in 1939 to arrange entertainments for the services during the war of 1939–45.

1939 *War Illustr.* 11 Nov. 275 During the first two weeks of the war ENSA—Entertainments National Service Association—came into being with its headquarters at Drury Lane Theatre, London, to organize concert parties. **1940** *Times Weekly* 10 Jan. 19 Miss Gracie Fields..gave her third concert (organized by E.N.S.A. on behalf of N.A.A.F.I.) to the men of the Forces. **1941** G. Greene *Nineteen Stories* (1947) 181 Not even an Ensa show? **1943** J. B. Priestley *Daylight on Saturday* xi. 68 General Office rang up to say that there had been a message from ENSA reminding him that they were giving a show in the canteen to-day. *Ibid.*, Proscot had cheered up a little when first he had been reminded that an ENSA show was coming to-day.

en secondes noces: see *en *prep.*

ensemble, *adv.* and *sb.* **A.** *adv.* Delete † *Obs.* and add later examples.

1861 G. Meredith *Let.* 19 Nov. (1970) I. 115 Before dinner we all bathed in Como, ladies and gentlemen ensemble. **1966** Auden *About House* 44 He, she, or both ensemble Emerge from a private cavity to be reborn.

B. *sb.* **1.** (Further examples in the context of dancing.)

1802 M. Berry *Jrnl.* 16 Mar. (1866) II. 137 A long *pas de deux* was performed with such a perfect *ensemble* and precision, that [etc.]. **1915** M. E. Perugini *Art of Ballet* iii. xxxiii. 293 There had hardly been, perhaps, quite that unity and perfection of *ensemble* which the coming of a dancer of superb technique made possible.

b. A woman's dress, hat, etc., as a complete whole.

1927 *Weekly Dispatch* 6 Nov. 16 A simple ensemble..in shades of brown. *Ibid.*, The afternoon ensemble is by no means dead. **1930** *Daily Express* 8 Sept. 5/5 White rabbit, brocade, velvet..add considerable chic to the evening ensemble. **1969** *J. C. Penney Catal.* Fall & Winter 105 A coat and dress *ensemble* made of Orlon acrylic knit.

3. *Mus.* The united performance of all voices or all instruments in a piece of concerted music, or of a chorus and orchestra; also, the manner in which this is done; the musicians comprising such a concert group or orchestra.

In quot. 1951 applied to the united performance of voices in an operatic chorus.

1844 *Musical Examiner* 28 Sept. 809 It was really possible for five principal vocalists to achieve a perfect *ensemble*. **1880** Grove *Dict. Mus.* II. 659/2 A feeling of carelessness..which the conductor must be quick to detect lest the *ensemble* be marred thereby. **1915** *Musical Q.* I. 83 We must put up with *ensemble* when we want to talk of that part of music, which is produced by the co-operation of several performers. *Ibid.*, Much care is required to secure a good ensemble in a vocal piece. **1927** *Observer* 27 Nov. 14/4 The ensemble between pianoforte and violoncello was good. **1929** *Encycl. Brit.* VIII. 616/2 The 'ensemble numbers' of an opera (trio, quartet and so forth). By extension the term is applied to the process of combining in this manner and to the skill with which it is accomplished. Thus in this sense it may be said that the ensemble of a choir or of a quartet was poor. **1934** S. R. Nelson *All about Jazz* iv. 71 His band can rank with the world's finest symphony orchestras in precision..and all those other attributes which go to make the finest musical ensembles. **1946** *Penguin Music Mag.* I. 20 The present management of Covent Garden..will cordially welcome occasional visits from complete foreign ensembles. **1947** A. Einstein *Mus. in Romantic Era* xvi. 242 Here again are ensembles like the quintet at the end of the first half of the last act, even though it is no dramatic Mozartian ensemble. **1951** Auden & Kallman *Rake's Progress* iii. 45 At the end of the ensemble, the voices of Rakewell and Shadow are again heard from the street. **1970** *N.Y. Times* 8 Nov. 84/5 The New York Rock Ensemble.

b. In ballet, musical comedy, or variety, a 'chorus' of dancers; a dance performed by such a group.

1915 M. E. Perugini *Art of Ballet* iii. xxxiv. 303 'Ship Ahoy!' a nautical one-scene *divertissement*...The final *ensemble*, when the lady passengers..danced beneath the soft glow of the swinging lanterns was a particularly novel, pretty and inspiriting picture. **1921** Wodehouse *Jill the Reckless* xi. 161 The ladies of the ensemble were changing their practice-clothes after a particularly strenuous rehearsal. **1922** Beauclerk & Evrenov tr. *Svetlov's Thamar Karsavina* 61 Karsavina's *rôle* is..devoid of incident and overwhelmed by the too brilliant and confusing *ensemble* that surrounds her. **1930** T. Karsavina *Theatre Street* II. xvi. 213 The ensemble lagged; Fokine..suddenly..flew at me. 'How can I blame the *corps de ballet* if the star herself gives a bad example.' **1933** A. G. Macdonell *England, their England* vii. 102 Complete with..hand-made smocks for ye gaffers..and aluminium Eezi-Milk stools for the dairymaids (or Ladies of the Dancing Ensemble). **1936** A. Haskell *Prel. Ballet* x. 50 Instead of one brilliant individual and a mechanical group in the background there must now be an *ensemble* of dance artists, whose function it is to interpret instead of merely being occupied in keeping time and keeping line. **1958** *Times* 11 Sept. 4/5 Lifar's opening *Noir et Blanc* made an impressive stage spectacle as it put the company through its paces in a series of testing solos and ensembles.

4. *Math.* A collection or combination of systems of identical constitution but in possibly differing states.

1902 J. W. Gibbs *Statist. Mech.*, p. xi, We consider especially ensembles of systems in which the index..of probability of phase is a linear function of the energy. *Ibid.* 116 A microcanonical ensemble of systems. *Ibid.* 169 The time-ensemble, or ensemble of phases through which a single system passes in the course of time. *Ibid.* 190 A grand ensemble is therefore composed of a multitude of petit ensembles. **1938** R. C. Tolman *Princ. Statist. Mech.* xiii. 524 The properties of a thermodynamic system..may be studied with the help of the average properties of an appropriately chosen representative ensemble of systems. **1965** D. Middleton *Topics Communication Theory* i. 1 We are concerned not with the transmission and reception of just one particular 'message', but rather with the set, or ensemble, of all possible messages for the purpose at hand.

5. *attrib.*

1929 [see sense *3 a]. **1938** G. Ffrangcon-Davies in R. D. Charques *Footn. Theatre* iv. 241 One gathers from contemporary records that 'ensemble' playing as we know it to-day was non-existent. **1939** W. Hobson *Amer. Jazz Music* 204 *Do Your Duty* has an ensemble introduction after which Bessie sings thirty-two bars with a loose ensemble background. **1958** A. Miller *Coll. Plays* 16, I saw the productions of the Group Theatre..the brilliance of ensemble acting. **1958** *Spectator* 27 June 831/3 The Youth Theatre concerns itself with ensemble playing, not with turning out miniature stars.

ensemble, *v.* Delete † *Obs.* and add later example.

1966 *Observer* 20 Mar. 11/1 Our driver carefully changed his bowler for the modish Cossack tea-cosy, ensembled with string gloves.

ensete (ensī·ti). [Native name.] An African relative of the banana, *Ensete ventricosum* (*Musa ensete*).

1790 J. Bruce *Trav. to Nile* V. 36 The Ensete is a herbacious [*sic*] plant. It is said to be a native of Narea. **1864** J. A. Grant *Walk across Africa* p. xv, Ensete; a species of wild plantain..called m'seegwah and m'tembah by natives, who make necklaces of its seeds. **1961** H. P. Huffnagel *Agric. Ethiopia* xxviii. 292 The main product of the ensete is a fermented starch of high nutritive value.

ensiform, *a.* Add: **B.** *sb.* = *ensiform cartilage.*

1907 *Practitioner* Oct. 467 Even when it travels directly downwards, in the same line as the pulmonary murmur, it [*sc.* an aortic murmur] travels much further, and is usually clearly audible at the ensiform.

ensign, *sb.* Add: **5.** In the Royal Navy pronounced (e·ns'n).

9. ensign-fly, a parasitic hymenopterous insect of the family Evaniidæ.

1894 Comstock *Man. Insects* 628 We have named these insects Ensign-flies, because they carry the abdomen aloft like a flag. **1897** —— *Insect Life* 86.

ensnarl, *v.*¹ Delete † *Obs.* and add later *fig.* examples.

1896 P. Brooks *New Starts in Life* xiv. 239 Shall life be one great deep stream of joy, ever and anon darkening and ensnarling itself in suffering? **1894** W. R. Thayer *Poems New & Old* 19 The soul of Halid is ensnarled in a secret and pardonless crime. **1924** *Public Opinion* 1 Aug. 104/2 Exhibiting an amazing capacity to get at the ensnarled kinks of the mind and soul.

‖ **ens necessarium** (enz nekesā·rium, nesĕsē̆ə·riŏm). *Philos.* [L., necessary being.] A necessarily existent being; God.

1900 J. H. Stirling *What is Thought?* ii. 32 We draw into consideration what concerns..the *constitution* of the ens necessarium. *a* **1910** W. James *Some Probl. Philos.* (1911) iii. 43 God..cannot lack being therefore: He is *Ens necessarium, Ens realissimum*, as well as *Ens perfectissimum*. **1937** A. Huxley *Ends & Means* xiv. 278 The argument..that if there is an *ens necessarium* it must be at the same time an *ens realissimum*. **1949** *Mind* LVIII. 75 If by God we mean an *ens necessarium* or a Being that exists with 'necessity'.

En-Soph (ē̆n sōuf). Also **-Sof.** [See Ensophic *a.*] In Cabbalistic doctrine, the absolute infinite and incomprehensible God.

1791 Brucker & Enfield *Hist. Philos.* II. iv. iii. 217 The chief heads of the Cabbalistic Doctrine are these... The Being from whom all things proceed is a Spirit, uncreated, eternal,..existing by the necessity of its nature, and filling the immensity of space. This spirit is *En-Soph*, the Infinite Deity. *Ibid.* 221 The En-soph, or Deity, contains all things within himself. **1865** C. D. Ginsburg *Kabbalah* ii. 64 No one has seen the *En Soph* at any time. **1902** *Jewish Encycl.* III. 468/1 An expression of the will of the En-Sof is not necessary in the act of emanation.

ensophic, *a.* Delete † *Obs.* and add later example.

1873 C. G. Leland *Egypt. Sketch-Bk.* xvi. 202 The supermundane and ensophic universe.

ensorcell, *v.* Add: Hence **enso·rcellment,** magic, enchantment.

1931 M. Summers *Supernatural Omnibus* 10 Egypt the ancient..is the very womb of wizardry, of ghost lore, of ensorcellment, of scarabed spells and runes.

‖ **ens rationis** (enz ræ̆tiō̆u·nis, ræʃ-). *Philos.* Pl. **entia rationis** (e·ntiă, e·nʃiă). [L., being of the mind: see Ens.] An entity of reason, a being that has no existence outside the mind (opp. *ens reale*). Cf. Entity 3.

1565 T. Harding *Confutation Apol. C. of E.* iii. v. fol. 144ᵛ, Disputations..about *genus* and *species*, and the rest of the universals.., whether they were *entia relia* [*sic*], or *rationis*. **1652** N. Culverwel *Lt. Nature* 201 The soul..will set up Beings of its own, *Entia Rationis*; Reason's creatures. **1707–8** Berkeley *Philos. Commentaries* (1944) No. 535 The distinction betwixt entia Realia & entia rationis may be made as properly now as ever. **1883** F. H. Bradley *Princ. Logic* I. vi. 175 It is a metaphysical *ens rationis*, an abstract universal which can not be real. **1948** *Mind* LVII. 364 Later the view that its [*sc.* the supersensible world's] existence is not even problematic, that it is a mere *ens rationis*, 'a concept not an existent', is considered.

‖ **ens reale** (enz rē̆ā·li, riˌē̆·li). *Philos.* Pl. **entia realia** (e·ntiă rē̆ā·liă, e·nʃiă riˌē̆·liă). [L., real being.] A being that exists independently of any finite mind (opp. *ens rationis*). Hence *superl.* **ens realissimum** (enz rē̆·āli·simum, riˌāli·simŏm), the most real being; God.

1565, 1707–8 [see *ens rationis]. **1847** J. D. Morell *Hist. Philos.* (ed. 2) I. ii. 257 Kant..allows that we inevitably trace all conditions of existence up to the supreme condition, the 'ens realissimum', and thus attain to the idea of a God. **1877** [see Hypostatize *v.*]. *a* **1910, 1937** [see *ens necessarium]. **1958** W. Stark *Sociol. Knowl.* iv. 187 Movement is the *ens realissimum*, and not fact as commonly understood.

enstool (enstū̆·l), *v.* [f. En-¹ + Stool *sb.* 1 e.] *trans.* To place (a west African chief) on his 'stool'. Hence **enstoo·lment.**

1895 *Times* 16 Nov. 5/4 The King of Kokofu..is one of the three chiefs who, by ancient custom, perform the ceremony of 'enstooling' the King of Kumassi as King paramount [of the Ashanti Confederation]. **1895** *Daily News* 25 Nov. 3/5 The enstooling of Prempeh on March 26, 1886. **1925** *Public Opinion* 11 Oct. 396/1 Should a Chief be enstooled. **1959** A. Abbs *Ashanti Boy* v. 191 We had many people in Dwamanoa..because of the white Queen's enstoolment.

ensuant (ensiū̆·ănt), *a.* Restrict † *Obs.* to sense in Dict. and add: **2.** Following or consequent *on*.

1897 *Daily News* 21 Oct. 7/7 His condition, ensuant on a paralytic stroke, left little room for hope. **1900** *Ibid.* 2 July 5/6 The stoppage of trade ensuant on the war.

en suite. Now sometimes written as one word, *ensuite*.

1968 *Globe & Mail* (Toronto) 13 Feb. 31/3 (Advt.), Large and formal dining room, ensuite a roomy bright kitchen. *Ibid.* 17 Feb. 45 Main floor laundry. Ensuite bath to every bedroom.

enswamped, *ppl. a.* (Later U.S. example.)

1821 T. Nuttall *Trav. Arkansa* vi. 108, I was now obliged more deeply to wade through the enswamped forests..which surround the habitable prairie lands.

entad (e·ntæd), *adv.* *Anat.* and *Zool.* *rare.* [f. Gr. ἐντός within + *-ad.] On or towards the inner side or interior *of*; in or into a position nearer to the centre *of*.

1882 [see *ectad *adv.*].

entailment². Add: **2.** *Philos.* [cf. Entail *v.*² 5.] The strict or logically necessary implication of one proposition by another.

1933 L. S. Stebbing *Mod. Introd. Logic* xv. 286 Professor Moore's analysis of the distinction between material implication and entailment makes it possible. **1957** G. H. von Wright *Logical Stud.* 166 (*title*) The Concept of Entailment. **1965** P. Caws *Philos. Sci.* xiv. 106 The weak kind..is called material implication, as distinct from strong implication or entailment.

ental (e·ntăl), *a.* *Anat.* and *Zool.* [f. Gr. ἐντός within + -al.] Inner; internal.

1881 [see *ectal *a.*]. **1889** *Buck's Handbk. Med. Sci.* VIII. 111 The ental surface of the pia.

entamœba (e:ntămī·bă). Also *U.S.* **entameba.** [mod.L. *entamoeba* (Casagrandi and Barbagallo 1897, in *Ann. d'Igiene Sperimentale* VII. 163), f. Ent(o- + Amœba.] A member of a genus of amœbas so called, including several species parasitic on vertebrates.

1914 Fantham & Porter *Some Minute Animal Parasites* vii. 147 Entamœbae are extremely important, as they occur commonly in the intestine of man in tropical countries. *Ibid.* 151 The multiplication of Entamœba occurs usually by simple division into two daughter forms. **1926** C. M. Wenyon *Protozool.* I. ii. 227 Entamœbae are of common occurrence in the intestine of animals. **1962** *Lancet* 8 Dec. 1209/1 The presence of the entamœbae was held to be proven only if the trophozoites could be found in permanent fixed preparations. **1964** M. Hynes *Med. Bacteriol.* (ed. 8) xxviii. 426 Two entamœbæ are found in the fæces of man: *Entamœba coli* a harmless commensal, and *Entamœba histolytica*, the cause of amœbic dysentery and liver abscess.

entamœbiasis (e:ntæmibəi·ăsis). *Path.* Also *U.S.* **entamebiasis.** [f. Ent(o- + Amœb(a + -(i)asis.] A disease caused by entamœbæ, esp. *Entamœba histolytica*.

1929 Thomson & Robertson *Protozool.* xiii. 151 Relapses in entamœbiasis are of common occurrence.

entanglement. Add: **1.** *spec.* A compromising relationship, an unsuitable liaison. (Cf. quot. 1748 in Dict.)

1863 Mrs. Gaskell *Dark Night's Work* v. 60 This foolish entanglement of Ralph's; they would not call it an engagement. **1957** 'D. Rutherford' *Long Echo* i. 24 Maria..dead could not involve him in any entanglement.

3. *Mil.* An extensive barrier arranged so as to impede an enemy's movements; an abatis formed of trees and branches, or an obstruction formed of stakes and barbed wire.

1834 J. S. Macaulay *Field Fortif.* iv. 83 The boughs of the brushwood..interlacing with one another, will thus form a very good obstacle, called an entanglement. **1876, 1879** [see *wire entanglement*]. **1899** *Westm. Gaz.* 17 Nov. 2/1 Where a wood enters into the scheme of defence, an abattis—in this case called an 'entanglement'—forms naturally one of the best resources of the defenders. **1916** 'Boyd Cable' *Action Front* 47 Slowly and cautiously, with the officer leading, they began to wend their way out under their own entanglements. **1917** [see Wire *sb.* 1 e]. **1922** Blunden *Shepherd* 69 They've all died on the entanglements.

‖ **entente** (aŋtaŋt). [Fr.] **a.** An understanding; most freq. used as a shortening of *entente cordiale. **b.** A group of states or powers connected by an *entente cordiale*.

1854 Ld. Bloomfield *Let.* 9 Dec. in G. Bloomfield *Remin.* (1883) II. 47 Just at the moment that the King [of Prussia] was proclaiming the happy *entente* with Austria. **1878** L. W. M. Lockhart *Mine is Thine* II. xxxv. 325 Esmé held out her hand, which he took; and thus the friendly *entente* was sealed! **1907** *Times* 20 Apr. 8/2 But was there not an *entente* which must be the great ideal of every Briton—the *entente cordiale*, the *entente commerciale* between Great and Greater Britain? **1907** *Westm. Gaz.* 16 Aug. 2/2 The reforms introduced in the Macedonian vilayets by the *entente* Powers. **1908** *Daily Chron.* 14 Aug. 1/5 Why, asked Mr. Lloyd George, cannot we have an agreement with Germany, seeing that Great Britain has already concluded ententes with France, Russia, and the United States? **1920** *Glasgow Herald* 5 Aug. 7 The young Austrian republic, whose existence would be endangered unless the Entente intervened. **1923** *Westm. Gaz.* 23 Jan., The Little Entente (Roumania, Czecho-Slovakia, and Jugo-Slavia).

Hence **ente·ntist,** one who favours an entente.

1915 *Times* 30 Aug. 5/7 Oh! you stupid Quadruple Ententists! **1924** *Glasgow Herald* 24 Jan. 7 Carlotti, the Italian Ambassador at Petersburg and an Ententist.

‖ **entente cordiale** (aŋtaŋt kordyal). [Fr.] A friendly understanding, esp. one between two or more political powers or states; *spec.* with reference to the understanding arrived at between England and France in 1904, and between these two countries and Russia in 1908.

1844 T. Raikes *Jrnl.* 3 June (1857) IV. 400 Still less had he [*sc.* the Tsar] the slightest wish or intention to derange or counteract this *entente cordiale* between England and France. **1845** Ford *Hand-bk. Spain* i. 227 But all this breeds bad blood and mars, on the Spaniards' part, the *entente cordiale*. **1847** H. Greville *Diary* (1883) I. 189 If Guizot remains in office Normanby must be recalled, as the only chance of a renewal of the *entente cordiale*. **1860** F. Nightingale *Notes on Nursing* vi. 36 The *entente cordiale*..between the doctor and his head nurse. **1870** J. R. Lowell *Among my Bks.* 320 Something that would break the *entente cordiale* of placid mutual assurance. **1904** *Spectator* 31 Dec. 1070/1 The Cologne Gazette last week discounted M. Jaurès's reply to Count von Bülow, and warned its readers that the *entente cordiale* with England was aimed against Germany. **1908** *Times* 11 Mar. 11/6 The agreements which constitute the *entente cordiale* with France. **1965** A. J. P. Taylor *Eng. Hist. 1914–45* xii. 417 There had always been something strained and artificial in the *entente cordiale* with France. **1970** *New Yorker* 25 July 8/2 The dining is an *entente cordiale* between East and West. Dancing. Closed Mondays.

ententophil, -phile (aŋtă·ntofil), *a.* and *sb.* [f. *entente + -(o)phil, -phile.] (One who is) friendly to a particular *entente*.

1920 *Glasgow Herald* 9 May 9 (Greece) A Government formed with the least possible delay should declare itself ententophile. *Ibid.* 15 June 7 Signor Meda is the most Ententophile, or the least Germanophile, personality of the 'Popular Party'. *Ibid.* 14 Oct. 6 M. Take Jonescu,..the staunchest Ententophile in Eastern Europe. **1932** R. H. B. Lockhart *Mem. Brit. Agent* IV. iii. 231 Trotsky was not to be placated... 'While your fools of spies are trying to prove that I am a German agent, my friends down there.. are calling me an Ententophile.'

entepicondylar (entepikǫ·ndilăɹ), *a.* *Anat.* [f. Gr. ἐντός within + Epicondyle, after *condylar*.] *entepicondylar foramen*: a foramen in the humerus of many vertebrates just above the medial epicondyle.

1893 *Athenæum* 18 Nov. 701/2 An entepicondylar (ulnar) foramen in the humerus. **1949** *Buchanan's Man. Anat.* (ed. 8) vi. 282 The supracondylar process represents a portion of bone that forms an entepicondylar foramen in many vertebrates. **1964** Parker & Haswell *Text-bk. Zool.* (ed. 7) II. 771 In the armadillos..the humerus is short and powerful, with well-developed processes and ridges, and with a foramen (entepicondylar foramen) above the inner condyle.

enter, *v.* Add: **4.** (Example const. *at*.)

1854 *Prospective Rev.* X. xl. 525 He..entered at Oriel College.

20. d. To get (land) recorded in a land-office in one's name as the intending occupier. *U.S.*

1799 *Columbian Centinel* (Boston, Mass.) 1 July 3/1 Real Estates, entered at the Register of Deed's Office. 1835 in H. Howe *Hist. Coll. Ohio* (1847) 387 Much land was entered in the county, and many settlements made. 1843 'R. Carlton' *New Purchase* xiv. 100 Out there, a settlement usually takes its name from the person that first 'enters the land', i.e. buys a tract at the land office. 1871 Schele de Vere *Americanisms* (1872) 173 All other lands can be obtained by entering them. 1884 H. Butterworth *Zigzag Journ. Western States* 217 'What does it usually cost to make a farm on government land?'..'It costs fourteen dollars to enter one hundred and sixty acres of land.'

22. c. To put down or cause to be put down upon the record.

1692, 1809 [see Docket *v.* 2]. 1860 *Act* 23 & 24 *Vict.* c. 38. § 3 No Judgment which has not already been or shall not hereafter be entered or docketed under the several Acts..shall have any Preference against Heirs. 1896 A. Pulling *Law Rep., Five Years' Digest* 679/1 The defendant failed to appear, and judgment was entered. 1930 *Daily Express* 6 Nov. 7/2 No judgment was entered.

enteric, *a.* Add: **B.** *sb.* Enteric fever.

1900 *Westm. Gaz.* 31 Jan. 5/2 The preventive inoculation for enteric. 1926 W. R. Inge *Lay Thoughts* 99 Two of the worst scourges, enteric and tetanus.

entering, *vbl. sb.* Add: **3. entering edge** *Aeronaut.*, that edge of a surface which is the front edge in flight; now called *leading edge*.

1908 H. Maxim *Artif. & Nat. Flight* vii. 100 Our planes must have a certain length of entering edge—that is, the length of the front edge must bear a certain relation to the load lifted.

entero-. Add: **entera·lgia** *Med.* [Gr. ἄλγος pain], pain in the intestines, colic; **e:ntere·ctomy** *Surg.* [*-ectomy], removal of a portion of the intestine; **e:ntero-anastomo·sis** *Surg.*, the joining of two portions of an intestine so as to make a continuous tube; **enterobi·asis** *Path.* [*-iasis], infection of the intestines by pinworms of the genus *Enterobius*; **e:ntero-chlo·rophyll** *Chem.*, a form of chlorophyll present in some animals; **enteroco·ccus** *Bacteriol.* [ad. F. *enterocoque* (Thiercelin 1899, in *Compt. rend. Soc. Biol.* V. 55)], one of several species of streptococcus found in the intestine; **e·nterocœle, -cœl** *Zool.*, (part of) a cœlom or body-cavity that is or has been in communication with the archenteron; hence **enterocœ·lic, -ous** *adjs.*; **e:nterocoli·tis** *Path.*, inflammation of the small intestine and colon; **enterocri·nin, ente·rocrinin** *Biochem.* [Gr. κρίνειν to separate: cf. *endocrine], an intestinal hormone that stimulates digestion; **e:ntero-dy·nia** *Med.* [Gr. ὀδύνη pain] = *enteralgia*; **e:ntero-entero·stomy** *Surg.*, an operation for forming a permanent opening between two non-continuous portions of the intestine; **enteroga·strone** *Biochem.*, a humoral agent from the intestinal mucosa that retards gastric secretion and motor activity; **e:ntero-hepa·tic** *a. Med.*, relating to the intestines and liver; **e:ntero-hepati·tis** *Med.* and *Vet.*, inflammation of the intestines and liver, *spec.* in gallinaceous birds; = *black-head 5; **e:nteroki·nase** *Biochem.*, an enzyme found in the intestinal mucous membrane that brings about the conversion of trypsinogen to trypsin; **entero-ne·phric** *a. Zool.*, designating a nephridial system in which the septal nephridia open into the intestine; **e:nteropto·sis** *Path.* [Gr. πτῶσις falling], abnormally low position of the small intestines; **e:ntero·rrhaphy** *Surg.* [Gr. ῥαφή suture], the sewing up of a wound in the intestines; **e·nterospasm** *Path.*, spasmodic contraction of the intestine; **entero·stomy** *Surg.* [Gr. στόμα mouth], the operation for making a permanent opening into the intestine; **e·nterotome** [Gr. τομίς knife], an instrument for opening the intestinal canal; **e:ntero-toxæ·mia** *Vet.*, toxæmia caused by disease of the small intestine; **enteroto·xin** *Med.*, a toxin originating in the intestine, causing food poisoning; *spec.* a cytotoxin.

1830 J. P. Kay in *N. of England Med. & Surg. Jrnl.* 1 Nov. 220 (*title*) Gastralgia and enteralgia, or morbid sensibility of the stomach and bowels. 1840 A. Tweedie *Syst. Pract. Med.* IV. 152 The treatment of enteralgia may be commenced by insuring the removal of any cause of irritation within the canal. 1897 *Trans. Amer. Pediatric Soc.* IX. 119 These cases are common in the practice of every physician and are commonly diagnosed as 'gastralgia' or 'enteralgia'. 1877 L. A. Stimson tr. *Ziemssen's Cycl. Pract. Med.* VII. 662 In case..the volvulus cannot be withdrawn,..it has been recommended to excise it completely (enterectomy), and then..to insert the upper end through an opening into the cæcum (Hacken), and to ligate the other: entero-anastomosis. 1908 *Practitioner* Mar. 368 To..wait for an improvement in the patient's condition to perform an enterectomy. *Ibid.* Sept. 459 Entero-anastomosis without resection. 1930 *Q. Cumul. Index Medicus* VII. 894/2 (*title*) Diagnosis and epidemiology of enterobiasis. 1970 Passmore & Robson *Compan. Med. Stud.* II. xix. 30 Communal living, infrequent bathing and contaminated clothing combine to produce epidemics of enterobiasis. 1883 C. A. MacMunn in *Proc. R. Soc.* XXXV. 133 It..can be detected in the bile of specimens of *Helix* after a six months' fast; for this colouring-matter, since it is found in the appendages of the enteron, the name enterochlorophyll is proposed. 1888 Rolleston & Jackson *Forms Anim. Life* (ed. 2) 117 The secretion of the liver is acid, and has been found to have a diastatic and a peptic action in *H. pomatia*. It contains enterochlorophyl in *Helix pomatia*. 1908 Park & Williams *Pathogenic Micro-Organisms* (ed. 3) xx. 252 *Enterococcus.*—Thiercelin in 1903 described the enterococcus proteiformis (Fig. 93) as occurring as a coccus. *Ibid.* (*caption*), Fig. 93 Represents the gradation of the Enterococcus (Thiercelin) from the apparent bacillary forms to the coccus without a capsule. 1925 R. E. Buchanan *Gen. Syst. Bacteriol.* III. 301 *Enterococcus*, a casual name given by various writers to cocci found in the intestinal tract. 1955 *Sci. News Let.* 19 Feb. 118/1 A member of a streptococcus germ family is now blamed for causing tooth decay. Studies showing that this strep., called an enterococcus, caused tooth decay in rats [etc.]. 1877 Enterocœle [see Endodermal *a.*]. 1951 G. de Beer *Vertebrate Zool.* (ed. 2) xix. 238 In *Amphioxus*, the cœlomic cavities of the somites, when they arise, are in open communication with the gut, and are hence known as enterocœls. 1952 J. Z. Young *Life of Vertebrates* iii. 48 In earlier chordates..the cœlom is continuous with the archenteron and is said to be an enterocoele. 1888 *Nature* 2 Feb. 334/2 The 'Schlauchförmiger Kanal'..being enterocœlic in origin. 1888 Rolleston & Jackson *Forms Anim. Life* (ed. 2) 592 The anterior enterocoelic pouch. 1964 Parker & Haswell *Text-bk. Zool.* (ed. 7) II. 61 The cavities in these [folds], continuous with the archenteron, are the beginnings of the enterocœlic system. 1875 E. R. Lankester in *Q. Jrnl. Microsc. Sci.* XV. 165 You have..in the former an 'entero-cœlous' condition, to use Professor Huxley's terminology. 1857 Dunglison *Med. Lex.* (ed. 15), Enterocolitis. 1886 *Buck's Handbk. Med. Sci.* II. 435/1 In entero-colitis the inflammation is mostly confined to the lower end of the ileum. 1964 M. Hynes *Med. Bacteriol.* (ed. 8) xi. 145 It [*sc. Staph. aureus*]..can cause a very dangerous enterocolitis when it invades an intestine sterilized by antibiotic therapy. 1938 E. S. Nasset in *Amer. Jrnl. Physiol.* CXXVIII. 481 The name *enterocrinin* (intestinal secretagogue) is proposed for this hormone. 1957 *Encycl. Brit.* VII. 380/2 The existence of a hormone, enterocrinin, has been demonstrated and a highly active, crystalline product has been obtained. Injection of this hormone produces an outpouring of intestinal juice. 1848 Dunglison *Med. Lex.* (ed. 7), Enterodynia. 1870 Garrod & Baxter *Mat. Med.* (ed. 3) 406 Painful affections of the stomach and duodenum, as in gastrodynia, enterodynia. 1896 A. E. Maylard *Treat. Surg. Alim. Canal* lxvi. 543 Entero-enterostomy (short-circuiting).—This operation is performed when it is impossible to remove the diseased portion of the bowel. 1955 R. Maingot *Abdominal Operations* (ed. 3) xxix. 655 Entero-enterostomy is then carried out proximal to these anastomoses in order to deflect the gastric contents into the distal jejunum. 1930 Kosaka & Lim in *Proc. Soc. Exper. Biol. & Med.* XXVII. 891 An inhibitory agent may be formed in the mucosa of both the small and large intestine as the result of contact with fat...The name Entero-gastrone (derived from *entero[n, gastr[on* and *chal[one*) is suggested for the gastric inhibitory agent. 1937 *Amer. Jrnl. Physiol.* CXVIII. 475 A method is described for the preparation of an extract (enterogastrone) from the duodenal mucosa of hogs, which exhibits an inhibitory action on gastric mobility and secretion. 1961 *Lancet* 19 Aug. 395/2 The absorption of large amounts of urobilinogen from the gut into the entero-hepatic circulation which were then re-excreted by the liver. 1895 *U.S. Dept. Agric. Bull.* viii. 7 (*heading*) An infectious disease among turkeys caused by protozoa (infectious entero-hepatitis). 1900 Dorland *Med. Dict.* 233/2 *Enterohepatitis*, inflammation of the bowel and liver. 1906 [see *black-head 5]. 1959 *Times* 7 Dec. (Agric. Suppl.) p. vii/4 Blackhead disease (enterohepatitis). 1902 W. H. Thompson tr. *Pawlow's Work of Digestive Glands* ix. 160 We had, therefore, discovered a ferment, not for this or that constituent of the food, but a ferment of other ferments. I propose to give it the name of *Enterokinase*. 1907 Enterokinase [see Trypsin]. 1923 T. R. Parsons *Fund. Bio-Chem.* iii. 32 This enterokinase is an example of the group of substances termed 'kinases', which have the power of activating the mother substances or precursors of enzymes. 1961 W. J. E. Jessop *Fearon's Introd. Biochem.* (ed. 4) xiv. 223 In acid solution, pH5 to 6, the conversion of trypsinogen to trypsin is catalysed by the intestinal enzyme, enterokinase. 1919 K. N. Bahl in *Q. Jrnl. Microsc. Sci.* LXIV. 1. 101 The elaborate 'enteronephric' type of the nephridial system in Pheretima. 1963 R. P. Dales *Annelids* i. 33 The enteronephric system is apparently related to water conservation. 1897 Enteroptosis [see Viscero-]. 1907 *Practitioner* Dec. 771 Patients with very lax abdominal walls and marked enteroptosis. 1949 H. W. Florey *Antibiotics* I. i. 55 One of the most ingenious uses to which yeast has been put was due to Günzburg (1896), who proposed to treat enteroptosis with it. 1807 Morris & Kendrick *Edin. Med. Dict.*, *Enteroraphia*, the suture of a gut when wounded. 1889 *Buck's Handbk. Med. Sci.* VIII. 299/1 The immediate restoration of the continuity of the gut by circular enterorrhaphy. 1897 H. Illoway *Constipation* viii. 88 Enterospasm occurs most frequently in gastric and intestinal indigestions. 1908 *Practitioner* Aug. 219 A localised enterospasm. 1923 P. L. Mummery *Dis. Rectum & Colon* xvii. 408 In view of..the presence of blood in the stools, it seemed probable that the enterospasm was set up by an ulcer in the colon. 1878 J. Ashhurst *Princ. & Pract. Surg.* (ed. 2) xxxviii. 747 Enterostomy..to provide an opening for the introduction of food into the small intestine, has been suggested by Surmay. 1948 H. & J. Devine *Surg. Colon & Rectum* xi. 143 An enterostomy is a great affliction: the liquid faeces of the small intestine pour on to the abdominal wall, irritate and excoriate it,..and thus make the patient almost a complete invalid. 1842 Dunglison *Med. Lex.* (ed. 3) 267/1 Enterotome. 1882 Wilder & Gage *Anat. Technol.* i. 70 The enterotome supplied in post-mortem cases is a pair of long scissors, one blade of which is enlarged and rounded, and projects beyond the other so as to precede it in opening an intestine. 1948 H. & J. Devine *Surg. Colon & Rectum* xxi. 233 Should the small bowel be included in the enterotome the patient will complain of severe pain. 1932 H. W. Bennetts in *Coun. Sci. & Ind. Res. Bull.* LVII. 10 In order to remove confusion, the name infectious entero-toxaemia is herewith adopted for the Western Australian disease. 1957 *Times* 3 Sept. 15/5 Enterotoxaemia often occurs when the flock is suddenly moved to better pasture. 1936 *Jrnl. Infectious Dis.* LVIII. 321 The ability to produce an enterotoxin may be restored to certain food poisoning organisms which have lost this power by serial transfers on starch agar. 1964 M. Hynes *Med. Bacteriol.* (ed. 8) xi. 145 Enterotoxin-producing strains are a common cause of food poisoning.

enteron (e·ntĕrǫn). *Anat.* Pl. **entera.** [mod. L., a. Gr. ἔντερον an intestine.] The alimentary canal; the intestine or gut.

1842 Dunglison *Med. Lex.* (ed. 3) 267/1 Enteron. 1878 F. J. Bell tr. *Gegenbaur's Comp. Anat.* iii. 165 Separate glands are almost always absent from the mid-gut of the Vermes, but the epithelium is generally found to be different from the epithelia of the other divisions of the enteron. 1880 *Encycl. Brit.* XII. 548/1 By the formation of a mouth to the sac, the enteron acquires the functions of a digestive retort. 1927 Haldane & Huxley *Anim. Biol.* xii. 268 The mouth leads into a cavity called the coelenteron, because it fulfils the functions both of the coelom and of the enteron or gut of higher forms. 1963 Borradaile & Potts *Invertebrata* (ed. 4) v. 155 The enteron or cavity of the colony [of polyps] is continuous and common to all its members.

enterovirus (e:ntĕro,vəiə·rŏs, e·ntĕro,vəiə:rŏs). *Path.* [f. Entero- + Virus 2.] Any of a group of RNA viruses which primarily infect the lymphoid tissue of the alimentary tract but which may also involve the nervous system or other tissues, causing poliomyelitis or other diseases. Hence **enterovi·ral** *a.*

1957 *Amer. Jrnl. Publ. Health* XLVII. 1556/1 The National Foundation for Infantile Paralysis, recognizing that the poliomyelitis, Coxsackie, and ECHO viruses all inhabit the alimentary tract as well as share other properties, has changed the name of its Committee on the ECHO Viruses to the Committee on the Enteroviruses. 1961 *Lancet* 29 July 248/2 The enteroviruses such as poliomyelitis and Coxsackie. *Ibid.* 23 Sept. 684/2 Our methods of stool extraction were suitable for enteroviral isolations.

entertainment. Add: **13.** *attrib.* and *Comb.*

1904 *Westm. Gaz.* 29 Aug. 3/1 A prince among provincial entertainment-mongers of the humbler order. 1922 *Punch* 25 Oct. 385 No entertainment tax. 1937 *Burlington Mag.* Feb. 96/2 Gives the book a high entertainment-value. 1946 G. B. Shaw *Shaw on Theatre* (1958) 268 The critics reported that my plays were not plays, whatever other entertainment value they might possess. 1951 McLuhan *Mech. Bride* 75/1 The reciprocal connections between *chic* and the entertainment industries. 1958 *Listener* 6 Nov. 742/2 She was wonderful entertainment value and a rare spirit. 1959 Halas & Manvell *Technique Film Animation* 10 Potentialities [of animation] which..extend much more widely than the entertainment film.

enthalpy (e·nþălpi, enþæ·lpi). *Physics.* [ad. Gr. ἐνθάλπειν to warm in, f. ἐν in + θάλπειν to heat.] A thermodynamic property of a system: the sum of its internal energy and the product of its volume and the pressure exerted on it; usu. measured relative to an arbitrarily defined zero.

1927 H. Moss tr. *Mollier's Steam Tables* 3 The quantity *E* is known as the internal or intrinsic energy while the quantity *H* is termed the total heat, enthalpy or (sometimes) heat content. 1930 *Engineering* 8 Aug. 162/1 The calculation of the enthalpy, or heat contained in saturated air will show that most of the heat present..is the latent heat in the associated steam. 1950 *Jrnl. R. Aeronaut. Soc.* LIV. 73/2 This is the usual enthalpy-entropy diagram, drawn on the assumption of no internal losses or increase in entropy. *Ibid.* 76/1 Since there can be no loss in energy, the enthalpy levels of each of the points of the cycle diagram prior to the final expansion remain the same as for the ideal cycle. 1962 *Engineering* 23 Feb. 276/2 The total heat content (enthalpies) of the coolant gas at inlet. 1967 Condon & Odishaw *Handbk. Physics* (ed. 2) v. i. 5 In a change occurring at constant pressure..all the heat added to the system goes to increase the enthalpy.

enthuse, *v.* For '*U.S.*' read 'orig. *U.S.*' and add earlier and later examples.

1827 in *Amer. Speech* (1947) XXII. 286/2 My humble exertions will I trust convey and enthuse, and draw attention to the beautifully varied verdure of N.W. America. 1852 *N.Y. Weekly Tribune* 24 Jan. 3/2 We were probably not so much 'enthused' as they were. 1859 *Congress. Globe* 16 Feb. 1058/3 They are what they call in the country 'enthused'—run mad on the subject [of Cuba]. 1909 *Daily Chron.* 27 Oct. 3/3 A little too much of the enthusing forest lover. 1912 W. Owen *Let.* 22 Sept. (1967) 162, I cannot enthuse over the things as Leslie does. 1918 E. Wallace *Down under Donovan* v. 44 A million francs, think of it! Isn't it sufficient to enthuse a man without a profession? 1938 'D. Hume' *Good-Bye to Life* v. 61 The prospect of a stay at Three Gables was nothing to enthuse about. 1957 *Times* 11 Nov. 13/3 She enthused rapturously on the advantages of being on the telephone. 1964 C. Chaplin *Autobiogr.* xxv. 432 *The Great Dictator* opened..to a glamorous audience who were elated and enthused. 1968 *Guardian* 13 July 8/6 Edmund Blunden's sudden resignation, in mid-term, has enthused nobody.

enthusiasm. Add: **3. b.** An object of (freq. temporary) enthusiasm; an action or idea about which one feels enthusiastic, a 'craze'.

1916 Joyce *Portrait of Artist* ii. 102 The letters cut in the stained wood..stared upon him, mocking his bodily weakness and futile enthusiasms. 1931 M. Sadleir *Bulwer* iv. v. 224 Godwin, whose enlightened political philosophy and actual achievements in fiction were among Bulwer's youthful enthusiasms. 1937 H. G. Wells *Brynhild* v. 58 Cummington's enthusiasms always annoyed him and this last enthusiasm just now he felt might annoy him very much.

entire, *a.* Add: **4. b.** (Earlier example.)

1799 Malthus *Jrnl.* 13 July (1966) 146 Entire horses are in general use here.

5. d. *Her.* Of a bearing, e.g. a cross: attached to the sides of the shield.

1688 [see Fixed *ppl. a.* 1 b]. 1825 W. Berry *Encycl. Her.* I. Bb 2/1 *Entire*, or *Through out*, sometimes called *fixed* and *firm*, being attached to the sides of the shield, as a *cross pattée entire*. 1889 [see Firme *a.*]. 1969 Franklyn & Tanner *Encycl. Dict. Heraldry* 96/1 Cross-crosslet ..when extended to the limits of the field..is not.. 'anchored'..but is cross-crosslet 'entire'.

C. *sb.* **5.** *Philately.* (See quot. 1967.)

1897 O. Firth *Postage Stamps* vii. 148 The advisability of collecting 'entires', or 'Ganz Sachen', as the Germans more expressively have it, is a problem that sooner or later presents itself to the ardent collector. 1937 *Daily Tel.* 16 Oct. (Stamp Suppl.) p. ix/1 The air mail has stimulated the collecting of entires. 1967 *Scott's New Handbk. Philatelists* 13 *Entire*, a whole stamped envelope, letter sheet or wrapper, used or unused.

entitative, *a.* **1.** (Later examples.)

1890 *Tablet* 29 Nov. 860 The vexed question as to the entitative simplicity or not of the living principles of mere plants and brute animals. 1907 *Dublin Rev.* July 188 St. Thomas, with all the Scholastics, maintained the absolute entitative distinction of God from creatures. 1909 M. H. Dziewicki *Wyclif's De Ente* 243 The commission of sin as an entitative act.

ento-. Add: **entoblast,** also (*b*), an inner germ-layer of an embryo; = *endoblast, Hypoblast 2; **entocho·ndral** *a. Anat.*, situated or occurring within cartilage; **e·ntocœle** *Zool.*, that portion of the gut-cavity of certain polyps which lies between a pair of mesenteries (see quot. 1885); so **entocœ·lic** *a.*; **entode·rmal, -mic** *adjs.* = Endodermal, -ic *adjs.*; **entomere** *Embryol.*, each of the more granular cells produced by segmentation of the primitive ovum; **entopla·stral** *a.*, pertaining to the entoplastron; **entopla·stron** in turtles (see quot.); **entoscle·rite** *Ent.*, an internal sclerite; **entose·ptum,** in corals, a septum developed interiorly; **entoste·rnite** *Anat.*, an internal fibro-cartilaginous plate giving support to a series of muscles in various arthropods; **entoste·rnum** *Ent.*, an internal process or system of processes of the sternum of an arthropod.

1892 E. L. Mark tr. *Hertwig's Textbk. Embryol.* v. 86 The inner germ-layer (entoblast or entoderm) lines the cœlenteron and provides for nutrition. 1926 Jordan & Kindred *Textbk. Embryol.* ix. 67 We are compelled to postulate an earlier solid stage of the blastula in which ectoblast (epiblast) and entoblast (hypoblast) are prelocalized. 1945 W. J. Hamilton et al. *Human Embryol.* viii. 105 An *entoblast* which gives origin to the epithelial linings of the digestive and respiratory tracts. 1889 Entochondral [see *ectochondral* (*ecto-)]. 1885 G. H. Fowler in *Q. Jrnl. Microsc. Sci.* XXV. 578 For the chambers (Radialtaschen, Loges,) into which the cœlenteron is periaxially divided by the mesenteries, I am compelled to coin new names; to those chambers which lie between a 'pair' of mesenteries the term entocœle is applied..; to those chambers of which one lies between every two pairs of mesenteries the term exocœle. 1902 *Trans. Linnean Soc.* Oct. 304 The appearance of a new mesenterial pair is followed very closely by the outgrowth of a tentacle from its entocœle. 1963 Borradaile & Potts *Invertebrata* (ed. 4) v. 182 In the typical sea-anemone..and in coral polyps..the secondary mesenteries ..are situated in the spaces between two adjacent pairs (exocoeles), never between two members of a pair (entocoeles). 1888 *Q. Jrnl. Microsc. Sci.* XXVIII. 5 The tentacles are probably entocœlic only. 1902 *Trans. Linnean Soc.* Oct. 304 The tentacles are outgrowths of both the entocœlic and exocœlic mesenterial chambers. 1884 Sedgwick & Heathcote tr. *Claus' Zool.* iii. 100 The entodermal lining of the gastro-vascular canals. 1886 *Buck's Handbk. Med. Sci.* III. 172/1 The division of the margin of the ectodermal disk into two parts, one resting directly on the entodermic yoke. 1890 Billings *Med. Dict.* I. 455/1 Entomere. 1896 R. Lydekker *R. Nat. Hist.* V. 77 Owing to the absence of the unpaired entoplastral bone. 1871 T. H. Huxley *Man. Anat. Vert.* v. 202 The entoplastron and the two epiplastra correspond with the median and lateral thoracic plates of the Labyrinthodont *Amphibia*. 1902 *Proc. Zool. Soc.* 17 June 174 From the middle of its area arises a stout, hooked entosclerite, which projects backwards into the cavity of the prosoma. 1885 G. H. Fowler in *Q. Jrnl. Microsc. Sci.* XXV. 578 The septa lying in these two classes of chambers are similarly called exosepta and entosepta. 1903 *Ann. & Mag. Nat. Hist.* Feb. 147 The union of the entoseptum within each pair of the second cycle mesenteries with the adjoining exosepta. 1884 E. R. Lankester in *Q. Jrnl. Microsc. Sci.* XXI. 547 In order to make a close comparison of these Entosternites, it will be necessary to determine exactly the insertions of the muscles to which they give origin. 1888 Rolleston & Jackson *Forms Anim. Life* 526 An entosternite or chitinoid fibro-cellular plate. 1902 *Nature* 25 Sept. 529 The entosternite of Mygale. 1902 *Encycl. Brit.* XXV. 527/2 The affinity between Limulus and the Arachnids, indicated by the presence of a free suspended entosternum or plastron or entosternite in both.

entomo-. Add: **entomophagan** (example); **entomostracan** *sb.* (earlier example).

1840 W. Whewell *Philos. Induct. Sci.* I. p. cx, The words *Carnivores* and *Insectivores* are better..than Greek terms; otherwise we might..speak of *Zoophagans* and *Entomophagans*. 1835 [see Malacostracan *a.* and *sb.*].

‖ **en tout cas** (aṅ tu ka). [Fr., = in any case or emergency.] **1.** A parasol which also serves the purpose of an umbrella.

1874 *Cassell's Fam. Mag.* Dec. 20/1 'En-tout-cas' are larger than parasols and smaller than the usual umbrellas, and are suspended by a chain. 1889 *Chambers's Jrnl.* 28 Dec. 827/1 The parasol and the popular *en tous cas*. 1915 Galsworthy *Little Man* 39 Maud comes running, ..dragging a bull-dog..by the crutch end of her en-tout-cas. 1928 *Daily Express* 17 July 5/2 The useful en-tout-cas that will withstand a summer shower. 1951 E. Paul *Springtime in Paris* vi. 124 A torn *en-tout-cas* (combination umbrella and sunshade) turned inside out.

2. (Properly *En-Tout-Cas.*) A trade name for an all-weather hard lawn-tennis court.

1928 *Trade Marks Jrnl.* 17 Oct. 1644 *En-Tout-Cas*, marking tapes, and pins used..for marking out hard lawn tennis courts...The En-Tout-Cas Company (Syston) Limited..London E.C.2. 1931 *U.S. Lawn Tennis Assoc. Official Program* 23 Mar. 4 'En-Tout-Cas' courts are ready for play thirty minutes after heavy rain. 1953 *Landfall* 7 Feb. 121 But mostly we played..in doubles.. shouting and sweating on the mossy en-tout-cas court. 1955 *Newsweek* 29 Aug. 46/3 Italians..prefer the slow, high bounces of European entout-cas to the speedy skids of a tennis ball on grass.

entr'acte. (Earlier examples.)

1842 J. F. Warner *Univ. Dict. Mus. Terms* p. xxxvi, *Entr' Acte*, the music played between the acts of a French play. 1849 (*title*) The theatrical programme, and entr'acte. 23 July.

entrail, *sb.*[1] **3. a.** Delete '*rare* in mod. use' and add later examples in sense of human organs.

1889 'Mark Twain' *Conn. Yankee* 46 They would have dug his entrails out..to get at that tale and squelch it. 1915 W. S. Maugham *Of Human Bondage* c. 527 Hunger was gnawing at his entrails. 1916 Joyce *Portrait of Artist* iii. 159 He sprang from the bed, the reeking odour pouring down his throat, clogging and revolting his entrails.

entrain, *v.*[1] Add: **2.** *spec.* Of a fluid: to carry (particles) along by its flow; *spec.* of steam which carries along particles of water through a pipe or particles of sugar from an evaporating pan during the manufacture of sugar; also, to incorporate (air-bubbles) in concrete. Hence **entrai·ner,** 'a device for saturating a current of gas or steam with liquid, usually a hollow or pocket for collecting a liquid in such a way that it will be picked up by a passing current of gas or steam' (*Cent. Dict.* Suppl.); **entrai·ning** *vbl. sb.* and *ppl. a.*; **entrai·nment**[1], the action of the verb in these senses.

1892 *Mod. Lang. Notes* Nov. 393 *Entrain, entrainement*—evidently from Fr. *entrainer*. The action of carrying over particles of syrup or sugar by the steam exhausted from vacuum-pans in boiling sugar. 1930 *Engineering* 14 Feb. 213/1 Resulting in further steam being entrained in the jet. 1955 Brown & Dey *India's Mineral Wealth* (ed. 3) ix. 350, 0.63 ton of cement would be used for 100 cubic feet of concrete, without the help of pozzolanas or air entrainment. 1955 *Bull. Atomic Sci.* Jan. 2/1 He wrote down the mathematical equations for the entrainment of air in the balance wheel of the watch. 1956 Entraining [see *Air *sb.*[1] B. I. 1]. 1958 *New Scientist* 7 Aug. 573/1 There are actual droplets of liquid boiler water which have been carried out of the boiler, or entrained, by the steam. 1959 *Oceanogr. & Marine Biol.* I. 157 This figure shows the existence of an entrainment current (that is, water dragged along). 1962 *Engineering* 30 Mar. 430 Manufacturers of air-entraining Portland cement.

entrain, *v.*[2] Add: **2.** *intr.* To go on board a train.

1890 *Daily News* 8 Apr. 3/3 The troops should be back ..in sufficient time to..entrain for London. 1899 *Ibid.* 21 Nov. 5/4 The debarkation and entraining of the troops as they arrive here is being carried out rapidly..The troops entrain at the docks. 1914 R. Brooke *Coll. Poems* (1918) Mem. p. cxxx, We..entrained in the last train left. 1938 Auden & Isherwood *On Frontier* ii. i. 68 The clerk entraining for the office. 1968 G. Jones *Hist. Vikings* ii. i. 65 If he entrains at Malmö..for Stockholm and Uppsala.

‖ **entrain** (aṅtræṅ), *sb.* [Fr.] Enthusiasm, animation.

1859 *Once a Week* 8 Oct. 304/2 The result of all this is to be seen in a greater degree of *entrain* than can perhaps be found in any other congregation of holiday-seeking Britons. 1888 Mrs. H. Ward *R. Elsmere* III. v. xxxi. 12 The instruments dashed into the opening allegro with.. an *entrain* that took the room by storm. 1919 J. C. Snaith *Love Lane* xxiii. 117 Corporal Hollis could not be expected to display the entrain of a sergeant of the Black Watch. 1954 E. Jenkins *Tortoise & Hare* ix. 93 She wanted to put her head out of the window, but that would have shown too much *entrain*.

en train: see *en *prep.*

entrainment[2] (entrē¹·nmĕnt). [f. Entrain *v.*[2] + -ment.] The act or fact of entering a train.

1891 *Pall Mall Gaz.* 8 July 2/2 The early hour of entrainment is..still maintained.

entrance, *sb.* Add: **2. d.** (Later U.S. example.)

1887 C. D. Warner *Their Pilgr.* (1888) xi. 253 Paying their entrance, and passing through the turnstile..they stood in the Congress Spring Park.

7. (Later example.)

?1857 Mrs. Gaskell *Let.* ?Aug. (1966) 462 The contradiction..involved in the 3£ entrance..on the one page, and the 1£ entrance for clothes on the next.

8. *attrib.*, as *entrance-channel, -cue* (Theatr.), *-door, -hall* (earlier and later examples).

1839 De la Beche *Rep. Geol. Cornwall* xv. 520 The entrance-channel would be much better than at present. 1955 Friedman & Weisskopf in W. Pauli *Niels Bohr* 148 In the optical model, compound nucleus formation and absorption..represent the removal of the particle from the entrance channel. 1901 C. Morris *Life on Stage* (1902) ix. 59 The canary breeches were always there, ready to.. break a buckle just at the moment of my entrance-cue. 1961 Bowman & Ball *Theatre Lang.* 124 *Entrance cue*, a cue for an actor to come onstage. *a* 1817 Jane Austen *Persuasion* (1818) IV. viii. 149 As she ceased, the entrance door opened again. 1880 'Mark Twain' *Tramp Abroad* 168 At the entrance-door up-stairs. 1677 in M. M. Verney *Verney Lett.* (1930) I. iv. 55 Is there an Entrance Hall?—No, but a wide passage that serves the purpose. 1970 *N.Y. Times Mag.* 25 Oct. 87 The neo-Gothic entrance hall is furnished with Victorian chairs and bench.

‖ **entrecôte** (aṅtrəkōt). [Fr., lit. = between rib.] In full *entrecôte steak*. A boned steak cut off the sirloin.

1841 Thackeray in *Fraser's Mag.* June 715/1, I dined.. with a friend. We had..two entrecotes aux épinards. *Ibid.* 715/2 Whenever..you go to Paris, call at once for the *entrecote*; the filet in comparison..is a poor *fade* lady's meat. 1877 E. S. Dallas *Kettner's Bk. of Table* 63 The rib-steak or entrecôte, cut from the ribs of beef. 1928 A. Christie *Myst. Blue Train* xix. 150 The Comte de la Roche had just finished..an *entrecôte Béarnaise*. 1962 *Punch* 28 Nov. 774/1 Tartan Plate, consisting of fried egg, entrecôte steak, grilled tomato, roll and butter.

‖ **entredeux** (aṅtrədȫ). *Sewing.* [Fr., lit. = between two.] An insertion of lace, linen, or other material; = Beading *vbl. sb.* 2.

1850 *Harper's Mag.* I. 432 Morning cap trimmed with Valenciennes and gauze ribbons..muslin *guimpe bouillonné*, with embroidered *entre-deux*. 1902 *Daily Chron.* 1 Feb. 8/3 A milk-white gown, most delightfully arranged about the waist in corselet form by means of lace entre-deux. 1904 *Ibid.* 3 May 8/5 A way of using medallions and entredeux of stitched linen on gowns of fine cloth. 1928 *Observer* 12 Feb. 23 Such felts are..simple compared to many another which is complicated by endredeux of straw. 1964 *McCall's Sewing* ii. 28/2 *Entre-deux*, another term for beading or veining.

entrée. Add: Also **entree. 1. b.** (Earlier and later examples.) Also *spec.* the privilege of admission to a Royal Court.

1762 Earl of March *Let.* 20 Oct. in J. H. Jesse *Selwyn & Contemp.* (1843) I. 212 It is a charming house, and as I have..a partiality for the French, I am very glad to have the *entré*. 1822 Scott *Peveril* IV. ix. 210 The.. nobility who had from birth..the privilege of the *entrée*. 1898 *Westm. Gaz.* 25 Feb. 1/1 Several papers have announced that the Queen has granted the privilege of the *entrée* for this afternoon's Drawing Room to..the two daughters of Lord Rosebery. As a matter of fact the wife and unmarried daughters of an ex-Cabinet Minister have the *entrée* for life as a matter of right. One of the *entrée* presentations this afternoon will be Lady Marjorie Carrington. 1966 *Listener* 10 Mar. 354/2 He was portraying a luxurious social life to which he had no *entrée*. 1969 H. Perkin *Key Profession* v. 161 This..gave the A.U.T. the entrée to Government offices.

c. The entrance of performers in any large spectacle or show; the ceremonial procession of circus performers round the arena; the coming of an actor upon the stage; = Entrance *sb.* 1 b.

1824 in M. W. Disher *Greatest Show on Earth* (1937) 101 It [*sc.* the programme] began with a Grand Amazonian Entrée. 1824 K. Decastro *Mem.* 121 Their 'Grand Entree', as it is now called. 1830 N. Ames *Mariner's Sks.* 278 An actor no more comes on the stage, he makes his 'entree'. 1870 O. Logan *Before Footlights* xxviii. 367 In the circus dressing-room they are preparing for the 'grand *entree*'.

2. (Earlier examples.) Also *attrib.*, as *entrée dish*.

1759 W. Verral *Cookery* 46 Roasted ham. For this *entrée* is generally provided a new Westphalia or Bayonne ham. 1773 J. Wedgwood *Let.* 21 Nov. (1965) 156 Poach'd Egg Cups; Entre Dishes; Cover'd Dishes. 1846 Soyer *Gastron. Regenerator* 713 (*heading*) New pagodatique entrée dish. 1901 *Connoisseur* Dec. 275/2 A nice pair of Sheffield plate entrée dishes.

3. (Examples.) Also, a number or divertissement in a ballet or ballet-opera.

1724 *Short Explic. For. Wds. in Musick Bks.* 29 *Entree*, or *Entre*, is a particular Kind of Air so called. 1776 J. Hawkins *Hist. Sci. & Pract. Mus.* IV. ix. 239 The operas of Lully consist of recitatives, short airs,..choruses in counterpoint, with entrées, and splendid dances. 1949 *Ballet Ann.* III. 92 The choreographers..retain the extremely rigid forms of *entrée* and build-up of *enchaînement*, which are found in the old French ballets.

4. Phr. **entrée en matière** [Fr., lit. = entry into the matter], the beginning (of discourse, etc.), the broaching of a subject.

1930 E. F. Thomas *Nevertheless the Duke* xii. 203 'I have known you before!'..As an *entrée en matière* it was, to say the least, time-worn. 1938 *N. & Q.* CLXXIV. 272/1 Mr. Peter Quennell, who makes his *entrée en matière* with the pleasantly disputable assertion that [etc.]. 1960 *20th Cent.* Sept. 279 A most delightful *entrée en matière* for the very full and substantial biography.

entremets. Add: **1. c.** Freq. in *sing.* A sweet dish.

1846 E. Acton *Mod. Cookery* (ed. 5) Index b. § 592 All sweet puddings are served as *entremets*, except when they replace the roasts of the second course. 1846 'A Lady' *Jewish Man.* vi. 110 Fill *vol-au-vents* with fruits richly stewed with sugar..; it forms a very pretty entremêt. 1877 E. S. Dallas *Kettner's Book of Table* 380 There is no sweet entremet more simple..than..plain boiled rice with preserved or stewed fruit and cream. 1935 M. Morphy *Recipes of all Nations* 86 Pancakes have been one of the most popular of French entremets. 1969 R. & D. DeSola *Dict. Cooking* 94/2 *Entremets*, dessert course served in France after the cheese course.

entrench, intrench, *v.* Add: **2.** *spec.* in *Politics*, to safeguard the position of (an individual, a group, etc.) by constitutional provisions.

1649, 1796 [in Dict.]. 1962 *Listener* 12 Apr. 627/1 The most weighty criticism which can be made of all schemes to draw up a list of human rights and to entrench them in a constitution relates to the practical problem of selecting the rights to be protected and of formulating them in legal language.

entrenched, intrenched, *ppl. a.* Add: **c.** *Politics.* Esp. in *entrenched clauses, provisions*, constitutional legislation that may not be repealed except under more than usually stringent conditions.

1950 *Newsweek* 23 Jan. 45/3 It was pretty well agreed that the right to two official languages, freedom of religion, and provincial control over education would be 'entrenched' clauses of the constitution [of Canada], not to be changed without approval of every province. 1952 *Ann. Reg.* 1951 103 The Government were still bound by the 'entrenched clauses' of the South Africa Act. 1957 Elliott & Summerskill *Dict. Politics* 106 *Entrenched Provisions*, those sections of the South Africa Act, 1909, which can be altered or repealed only by a Bill passed by both Houses of Parliament sitting together, and agreed to at the third reading by not less than two thirds of the total number of members of both Houses. 1962 *Listener* 12 Apr. 626/2 A constitutionally entrenched Bill of Rights.

entrenching, intrenching, *vbl. sb.* Add: (Earlier U.S. example of *entrenching tool.*) Also *entrenching clause* = *entrenched clause.

1775 *Let.* 26 June in N. Bouton *Provincial Papers* (1873) VII. 531 We now send what entrenching Tools we have prepared. 1956 *Times* 10 Nov. 7/2 The supposition that the 'entrenching' clause 152 was so fundamental to the constitution that any legislation to make its evasion even indirectly possible was bad law.

entrenchment, intrenchment. Add: **1. b.** Also *fig.*, *spec.* the establishment of constitutional safeguards by entrenched clauses.

1961 *Ann. Reg.* 1960 92 The only entrenchment in the Constitution was that guaranteeing the equal status of the two official languages, English and Afrikaans.

‖ **entre nous** (aṅtrənu), *phr.* [Fr.] Between ourselves; in private.

1689 Shadwell *Bury F.* i. ii. 10 I'll tell the more *entre nous*: But, in the mean time [etc.]. 1705 Cibber *Careless Husband* ii. ii. 17 Why then, *Entre Nous*, there is a certain *Fille de Joye* About the Court here. 1766 Goldsmith *Vicar* I. xii. 112 Entre nous, I protest I like my Lady Blarney vastly. 1819 Keats *Cap & Bells* xxxiv, And master is too partial, *entre nous*. 1833 Dickens *Let.* ?Jan. (1965) I. 14, I suppose the story is to be *entre nous*. 1889 E. Dowson *Let.* 29 June (1967) 87 It's sublimely fatuous that I should mind—but entre nous it has quite spoofed me quite. 1922 Joyce *Ulysses* 613 A religious silence of the strictly *entre nous* variety.

entrepreneur. Add: **a.** (Earlier example.)

1828 J. Ebers *Seven Years King's Theatre* iv. 115 The payment of the benefit expenses by the unfortunate *entrepreneur*. A manager is..an animal whom it is supposed lawful and commendable to bleed at every vein.

b. (Earlier example.)

1838 *Actors by Daylight* I. 230 On the first landing of the Bayaderes at Bordeaux this active *entrepreneur* despatched a messenger *instanter* to that city.

c. *Pol. Econ.* One who undertakes an enterprise; one who owns and manages a business; a person who takes the risk of profit or loss.

1852 Carlyle *Let.* 15 Sept. in J. A. Froude *Life of Carlyle* II. xx. 107 A public set of rooms—*Kursaal* they call such things, finer than some palaces, all supported by gambling, all built by one French gambling *entrepreneur*. 1883 F. A. Walker *Pol. Econ.* i. i. 203 The employer, or entrepreneur, receiving profits. *Ibid.* vi. xi. 432 The state as capitalist is at no small disadvantage; as entrepreneur, that disadvantage is vastly aggravated. 1889 R. T. Ely *Introd. Pol. Econ.* (1891) 170 We have..been obliged to resort to the French language for a word to designate the person who organizes and directs the productive factors, and we call such a one an *entrepreneur*. 1922 F. Lavington *Trade Cycle* iii. 19 In modern times the entrepreneur assumes many forms. He may be a private business man, a partnership, a joint stock company, a co-operative society, a municipality or similar body. 1930 J. M. Keynes *Treat. Money* I. ix. 124 The individuals who perform entrepreneur functions. *Ibid.* xi. 159 Entrepreneurs will sometimes begin to act before the price-changes which are the justification of their action have actually occurred. 1959 J. Braine *Vodi* viii. 123 Tom's father looked every inch the successful *entrepreneur*. 1959 *Listener* 26 Nov. 915/2 Where their predecessors were flanked by engineers and scientists, the new-style entrepreneurs will be buttressed by sales managers and advertising experts.

Hence **entrepreneu·rship.**

1934 in Webster. 1959 *Economist* 14 Feb. 588/1 Entrepreneurship..might be common to all developing economies. 1968 *Sci. Jrnl.* Dec. 79/1 MIT-based entrepreneurship.

entrepreneurial (ãntr'prənö·riăl, -prəniüə·riăl), *a.* [f. Entrepreneur + -ial.] Of or pertaining to an entrepreneur or entrepreneurs.

1922 H. L. Reed *Devel. Federal Reserve Policy* i. 6 Occasionally entrepreneurial activity is injured by the rising scale of prices, as..in the public utility and railroad fields. 1940 *Economist* 24 Feb. 333/1 The mature economy theory is confined to academic..circles; it is rarely encountered in business or financial, and never in entrepreneurial, groups. 1951 C. W. Mills *White Collar* i. ii. 26 The small businessman has been deprived of his old entrepreneurial function. 1964 *Economica* Aug. 332 He was a solid Kent landowner and farmer of the businesslike kind with his risks well spread in land, trade and government contracting...Dr. Coleman's biography is a model of what real entrepreneurial history can do for economic history. 1966 *New Scientist* 17 Nov. 366/2 It is an entreprenurial business venture, designed to take advantage of the rapid development of data processing.

So **entrepreneu·rially** *adv.*

1960 *New Left. Rev.* Sept.–Oct. 28/2 The entrepreneurially-minded who could not move so rapidly in the managerial world of the big corporations.

entropy. Add: **1.** (Later examples.) Also *transf.* and *fig.*

1925 A. & J. Strachey tr. *Freud's Coll. Papers* III. v. 599 In considering the conversion of psychical energy no less than of physical, we must make use of the concept of an entropy, which opposes the undoing of what has already occurred. 1933 W. E. Orchard *From Faith to Faith* xi. 280 The deduction which one of our greatest physicist astronomers draws from the second law of thermodynamics: namely, that since there must be a maximum entropy, there must have been once its maximum opposite. 1955 *Sci. Amer.* May 124/2 Certain combinations of balls yield a greater change in entropy than others. Those combinations in which entropy change reaches maximum value lead to solutions. 1955 *Ibid.* June 64/1 This equilibrium state..is the thermodynamic condition of maximum entropy—the most disordered state, in which the least amount of energy is available for useful work. 1965 *Financial Times* 11 Aug., Moralising by those whose industrial entropy is an accepted fact of life is neither likely to persuade the workers nor assist the trade unions in the task of trying to meet the nation's difficulties.

2. a. *Communication Theory.* A measure of the average information rate of a message or language; *esp.* the quantity $-\Sigma p_i \log p_i$ (where the p_i are the probabilities of occurrence of the symbols of which the message is composed), which represents the average information rate per symbol.

1948 *Bell Syst. Techn. Jrnl.* XXVII. 396 Consider a discrete source of the finite state type...There is an entropy H_i for each state. The entropy of the source will be defined as the average of these H_i weighted in accordance with the probability of occurrence of the states in question. ..This is the entropy of the source per symbol of text. 1953 S. Goldman *Information Theory* 329 The amount of language information (*i.e.*, entropy) in the sequence is.. a measure of the choice that was available when the sequence was selected. 1953 D. A. Bell *Stat. Methods Electr. Engin.* x. 145 Since entropy increases as the arrangement of a system becomes less distinguishable from other possible arrangements, while the value of a pattern for conveying information depends on its uniqueness, the information capacity of a signal is the negative of its entropy. 1960 D. Middleton *Introd. Stat. Communication Theory* vi. 301 H(X) is called the (communication) entropy of the X ensemble, since..it is the direct mathematical analogue of the more familiar entropy measure of statistical mechanics. 1964 *Language* XL. 210 The basic probability concept, 'entropy', and its quantum, the 'bit', are now part of the metalanguage of linguistics.

b. *Math.* In wider use: any quantity having properties analogous to those of the physical quantity; *esp.* the quantity $-\Sigma x_i \log x_i$ of a distribution $\{x_1, x_2, \ldots\}$.

1951 *Jrnl. R. Statistical Soc.* B. XIII. 60 The idea of selective entropy provides us with a new and important concept in the analytical theory of probability. 1961 *Proc. Cambr. Philos. Soc.* (*Math. & Physical Sci.*) LVII. 839 The analogue of Boltzmann's H-theorem is not a statement about the monotonicity of the entropy of the chains..but a statement about the 'entropy' of the frequency distribution $(s_1, .. s_n)$. 1968 P. A. P. Moran *Introd. Probability Theory* i. 50 Since $-x \log x$ is a convex function the entropy of a finite set of events is a maximum when their probabilities are equal.

entry. Add: **1. e.** The beginning of the part of a performer or instrument in a canon or other musical composition; also *attrib.* in *entry sign*.

1879 Grove *Dict. Mus.* I. 390/2 Those who can stand the enormous strain which is implied in the recollection of every *nuance* and the exact entry of every instrument in a long and complicated work. 1893 J. S. Shedlock tr. *Riemann's Dict. Mus.* 220/1 Entry Signs are the marks in a canon (of which only one part is written out) for the entry of the imitating parts...The sign which a conductor gives to a player or singer to come in after a long pause is also called an Entry Sign. 1965 *New Statesman* 9 Apr. 586/2 The Mahler performance..had all the marks of under-rehearsal..and every now and again a fluffed entry.

f. In bridge, (a card providing) an opportunity to transfer the lead to one's partner or to one's dummy. Also *Comb.*, as *entry-creating, entry-killing* ppl. adjs.

1884 'Cavendish' *Whist* (ed. 14) 135 In case his only card of entry in that suit should be an honour, not an ace. 1906 W. Dalton *'Saturday' Bridge* ii. 53 When you hold six or more cards of a black suit, thoroughly established, and one other card of entry, No Trumps should always be declared at the score of love. 1934 E. Culbertson in *Amer. Speech* IX. 10/1 Contract Bridge terms that are almost an indispensable part of every player's vocabulary ..such as..*entry*. 1939 N. de V. Hart *Bridge Players' Bedside Bk.* 149 An entry-killing play which consists in refusing to take a trick in an opponent's suit until the other opponent is exhausted of the suit. 1959 *Listener* 19 Mar. 530/2 The ace of spades still provided an entry to the dummy. *Ibid.* 17 Sept. 462/1 There are countless entry-killing and entry-creating plays.

4. c. The initial training of young hounds (cf. Enter *v.* 18 b); now, more commonly *collect.*, young hounds who are being entered. Also *transf.*, the younger generation (see also quot. 1946).

1845 W. Youatt *Dog* iii. 83 There must always be a little flesh in hand for the sick, for bitches with their whelps, and for the entry of young hounds. 1856 'Stonehenge' *Man. Brit. Sports* ii. iv. 124/2 Your chance of good sport through this season and the next depends more upon your young entry than upon the old draft-hounds. *Ibid.* 125/1 In order to have an opportunity of rating the young ones for speaking to 'riot', while under the fresh recollections of the encouragement which they have received in their entry to their own particular game. 1881 *Encycl. Brit.* XII. 315/2 The young entry are sure to run riot. 1897 *Encycl. Sport* I. 543/2 It is better to keep steadily on, confining hounds as much as possible to covert, or the entry will forget what they have learned. 1899 Somerville & 'Ross' *Irish R.M.* vi. 130 Dr. Jerome Hickey was having a stirring time with the young entry and the rabbit-holes. 1924 J. Buchan *Three Hostages* vii. 101 Thank God that we have a man like him among the young entry. 1946 M. C. Self *Horseman's Encycl.* 453 Hounds are 'entered' when they are first put into the pack during the cubbing season. Young riders are 'entered' by being brought by their fond parents to the covert side. Both are known as 'Young Entry'.

6. Delete † *Obs.* and substitute *Obs.* exc. *U.S.* Add later example.

1907 *Springfield Weekly Republ.* 24 Oct. 1 An act of the Legislature which became operative with the entry of the month.

9. b. (Earlier example.) Also *fig.*

1721 W. Webster *Ess. Bk.-keeping* (ed. 2) 1 Book-keeping is the Art of stating our Accompts,..to which end, the Italian manner of Debtor and Creditor, by double entry, is by experience, found most conducive. 1961 S. Chaplin *Day of Sardine* i. 22 But if my writing's bad my double entry memory is good. It all goes down and sooner or later comes shooting out.

10. *entry-card, -list, -money* (earlier example), *-point, -way*; **entryman** *U.S.*, one who enters upon public land with the intention of settling.

1907 *Strand* Nov. 507/1, B might easily have an entry card in spades or clubs. 1908 *Westm. Gaz.* 13 Aug. 4/1 It was only with the greatest difficulty that the R.A.C. managed to get anything like a representative entry list. 1886 *N. Amer. Rev.* Jan. 59 The entryman, under the timber culture act, is not compelled to plant any trees until the third year from date of entry, when if he likes he may file a relinquishment of his claim, and the land is again open for entry. 1912 *Out West* June 418 His report was heard read with great satisfaction by upwards of a hundred of the entrymen of Los Angeles and vicinity. *c* 1490 *Cely Papers* in *English Studies* (1961) XLII. 143 Coket siluer othirwyse callid entree money. 1926 J. S. Huxley *Ess. Pop. Sci.* 234 The sperm entry-point. 1964 A. Edel in I. L. Horowitz *New Sociology* 218, I should like to..make a start on analyzing the entry-points, if we may so call them, where value issues enter into social science. 1746 *Probate Rec. N. H.* III. 391 It also agreed by us that the said Cellar great Doors and the yard the Entryway Stairs..all be in Common. 1854 M. J. Holmes *Tempest & Sunshine* iii. 44 There was no entry way to the building. 1889 R. T. Cooke *Steadfast* xxii. 236 His study door opened from the left hand of the little entry-way into which they stepped from without. 1893 S. Merrill in M. Philips *Making of Newspaper* 96 Patrolman Blucher had finished his nap in the entry-way leading to Eckstein's cigar factory.

‖ **Entscheidungsproblem** (entʃəi·duŋs,problə:m). *Math.* and *Logic.* [G., f. *entscheidung* decision + *problem* Problem.] = *decision problem* (*decision 5).

[1922 *Mathematische Annalen* LXXXVI. 163 (*title*) Beiträge zur Algebra der Logik, insbesondere zum Entscheidungsproblem.] 1930 *Proc. London Math. Soc.* XXX. 271 The *Entscheidungsproblem* is to find a procedure for determining whether any given formula is valid, or, alternatively, whether any given formula is consistent. 1938 *Mind* XLVII. 445 Gödel's example belongs to the field of investigations of the *Entscheidungsproblem*. This problem is to discover whether the accepted primitive propositions and rules of inference of mathematical logic allow us to conclude either the truth or the falsehood of every propositional formula, and if so to give a general method by which this can be done. 1958 M. Davis *Computability & Unsolvability* viii. 134 Hilbert declared that the decision problem..(often referred to simply as the *Entscheidungsproblem*) was the central problem of mathematical logic.

enubilate, *v.* Delete † *Obs.*⁻⁰ and add examples.

1903 *Sat. Rev.* 10 Jan., Maeterlinck is gradually enubilating himself from those enchanted mists in which first he strayed. 1928 'Brent of Bin Bin' *Up Country* xv. 252 The M'Eacherns..asked Bert to help the mounted troopers in clearing out the Eagle Hawk Gullies crowd, and thus enubilate the atmosphere.

enucleate, *v.* Add: **2. b.** *Surg.* To extract (an eye) from the socket.

1867 G. Lawson *Injuries Eye, Orbit, & Eyelids* vii. 277 The opening in the conjunctiva, through which the eye has been enucleated, may be closed by drawing it together by a fine thread. 1906 A. Bennett *Whom God hath Joined* vii. 260, I had to have an oculist at last. I couldn't sleep, and he enucleated the eye, as the term is. 1968 *May & Worth's Man. Dis. Eye* (ed. 13) iv. 137 When an eyeball containing a malignant growth is enucleated, as much of the optic nerve as possible should be removed.

3. *Biol.* To deprive (a cell) of its nucleus.

1909 in *Cent. Dict.* Suppl. 1968 *Science* 7 June 1115/3 Kinetics of protein synthesis were measured in oocytes of *Rana pipiens* enucleated at various times during and following pituitary-induced maturation.

enucleate (ĭniū·kli,ĕ¹t), *a. Biol.*, etc. [ad. L. *enucleatus*, pa. pple. of *enucleare* to Enucleate.] Without a nucleus.

1889 in *Cent. Dict.* 1910 F. Keeble *Plant-Anim.* iv. 113 The enucleate green cell may be connected by fine processes with another green cell still possessed of nuclear substance. 1921 L. W. Sharp *Introd. Cytology* iv. 69 The mammalian erythrocyte..loses its nucleus at an early stage and may continue to exist in the enucleate state for..30 days. 1968 H. Harris *Nucleus & Cytoplasm* i. 2 (*heading*) Lessons from an enucleate cell.

enu·cleated, *ppl. a.* [f. Enucleate *v.* + -ed[1].] Having the kernel or nucleus extracted. Also *fig.*

1885 R. F. Burton *Arab. Nts.* VII. 14 Arab "Ajwah", enucleated dates pressed together into a solid mass. 1896 E. B. Wilson *Cell* vii. 248 The nucleated fragments quickly heal the wound...On the other hand, enucleated fragments, consisting of cytoplasm only, quickly perish. 1900 *Daily News* 17 May 6/3 The most enucleated invalid might stand the intellectual strain of the 'Southern Cross' and the 'Northern Lights'. 1960 L. Picken *Organization of Cells* iv. vi. 139 The possibility of dividing cells into nucleated and enucleated halves.

enucleation. Add: **2. b.** *Surg.* The removal of an eye from the socket.

1867 H. Power *Dis. Eye* x. 582 Enucleation of the eye. 1949 A. D. Ruedemann in C. Berens *Eye & its Diseases* (ed. 2) lxix. 983 The nerve may be grasped with enucleation forceps to facilitate the excision of a long piece of the optic nerve.

3. *Biol.* The extraction of a nucleus from its cell.

1909 in Webster. 1960 L. Picken *Organization of Cells* iv. vi. 139 The amoeba is unable to spread after enucleation and remains rounded up.

enumer (ĭniū·məɹ), *v. trans.* = Enumerate *v.* 1.

1936 Auden *Look, Stranger!* 36 Among the foes which we enumer You are included. 1936 —— & Isherwood *Ascent of F6* (1937) ii. iii. 103 Ignore the Law, it's a bore, Don't enumer all the rumours of a war.

enumerability (ĭniū:mĕrăbi·līti). [f. next + -ity.] The fact or quality of being enumerable.

1939 *Yale Rev.* Spring 497 Arithmetic abstracts objects of all qualities save their enumerability; the symbol π abstracts in a single Greek letter a complicated relation between the parts of all circles. 1941 *Mind* L. 286 Mr. Ayer might have used the same argument in the case of enumerability. 1941 J. S. Huxley *Uniqueness of Man* i. 28 Arithmetic abstracts objects of all their qualities save their enumerability.

enumerable (ĭniū·mĕrăb'l), *a.* [f. Enumer(ate *v.* + -able.] That can be enumerated; having a definite number; numerable; *spec.* in *Math.* = *denumerable *a.*

1889 in *Cent. Dict.* 1907 E. W. Hobson *Theory of Functions* ii. 66 An aggregate which contains an indefinitely great number of elements is said to be enumerable, or countable (*abzählbar, dénombrable*), when the aggregate is such that a (1, 1) correspondence can be established between the elements and the set of integral numbers 1, 2, 3, [etc.]. *a* 1910 W. James *Some Probl. Philos.* (1911) xi. 171 The terms are not 'enumerable' in that order. 1940 A. J. Ayer *Found. Emp. Knowl.* ii. 124 If the sense-data do not appear to be enumerable, they really are not enumerable. 1968 P. A. P. Moran *Introd. Probability Theory* i. 1 We suppose that the number of ways in which the event can occur is either finite.., or 'enumerable', i.e. that the events can be numbered off, or put into one-to-one correspondence with the integers.

Hence **enu·merably** *adv.*

1968 P. A. P. Moran *Introd. Probability Theory* i. 4 If the number of E_i is enumerably infinite, the number of A_i is non-enumerably infinite.

enunciator. b. (Earlier U.S. example.)

1847 *Rep. Comm. Pat.* 1847 (U.S.) 101 Letters patent have also been granted for an improved enunciator, for use in hotels, &c.

enuresis (eniurī·sis). [mod.L., f. Gr. ἐνουρεῖν to urinate in.] Incontinence of urine, esp. minor incontinence not associated with any organic disease. Hence **enure·tic** *a.* and *sb.*

1800 tr. *W. Cullen's Nosology* 173 (*heading*) Enuresis. Involuntary but not painful flux of urine. 1822 J. M. Good *Study Med.* IV. 492 Paruria incontinens..is the enuresis of most of the nosologists. 1907 *Practitioner* Apr. 542 For a year past he had terrors practically every night, and often..enuresis. 1909 *Lancet* 1 May 1245/2 He had frequently known nocturnal enuresis to persist after very thorough removal of adenoids and tonsils. 1937 C. L. C. Burns in E. Miller *Growing Child* vi. 180 It might be said that the neurotic attitude of an enuretic child is an effect of the enuresis rather than a cause. 1943 *Our Towns* iii. 85 The mother of the enuretic..does not teach her child..control of its natural functions. 1968 *Encycl. Brit.* VIII. 619/2 Enuresis..usually means bed-wetting.

envelope, *sb.* Add: **1. b.** (Later examples.)

1952 *Brit. Jrnl. Psychol.* July 239 The 'envelope' is the late Dr. Plant's name for the processes by which the individual selects from the environment those features with which he can deal and shuts out the rest. 1957 N. Frye *Sound & Poetry* 136 Traditional literary scholarship, chiefly..concerned with the sender and receiver in reaction with the cultural envelope. 1963 *Listener* 28 Feb. 370/2 The Liverpool building..is a palpable *volume*, held in place by a semi-translucent envelope of concrete and glass, to which the ribs are clearly accessory.

5. b. *Electr. Engineering.* A curve formed by joining the successive peaks of a graph of an oscillation, esp. a modulated wave.

1928 *Exper. Wireless & Wireless Engin.* V. 302/1 When the voltages..are applied to the grid of an anode bend detector whose A.C. characteristic is a straight line, the envelopes indicated by curves 1 and 2 (Fig. 8) are rectified. 1929 *Ibid.* VI. 620/1 R, when plotted against time, will give the outline of the envelope representing the succession of carrier wave peaks. 1970 D. F. Shaw *Introd. Electronics* (ed. 2) xi. 256 The audio-frequency signal is represented by the envelope of the radio-frequency wave and..a device is employed which will give an output voltage proportional to the positive (or negative) envelope of the waveform.

6. *Aeronaut.* The gas or air container of a balloon or airship.

1901 *Sci. Amer.* 10 Aug. 89/3 The balloon is inflated with hydrogen, and in order to maintain at all times a tension on the envelope—that is to say, perfect inflation—a compensating balloon filled with air is placed in the interior. 1909 *Rep. Advisory Committee Aeronautics* 15 July 121 There will be some risk of [electrical] discharges between the airship and a charged cloud..while with a non-conducting envelope, if wet, there would be the same risk. 1950 *Gloss. Aeronaut. Terms* (*B.S.I.*) I. 49 Envelope. *a.* The gas-containing unit of a balloon or non-rigid or semi-rigid airship. *b.* The outer cover of an airship in which the gas-containing units are surrounded by a layer of air or inert gas. 1961 R. Higham *Brit. Rigid Airship* vii. 118 This was despite the fact that the engineer had to climb on to the top of the envelope each time it was necessary to switch to a new tank of petrol.

7. The outer containing-vessel of a vacuum tube.

1932 F. E. Terman *Radio Engineering* xi. 395 The 280 tube has two cathodes and two anodes and so is essentially two half-wave rectifiers enclosed in one envelope. 1959 *Listener* 12 Mar. 454/1 Extra high pressure mercury vapour lamps in quartz envelopes.

8. *attrib.* and *Comb.*, as *envelope cementer, cutter, flap, folder*; **envelope conditions** *Aeronaut.* (see quot.); **envelope delay** *Electr. Engineering* (see quot. 1940).

1904 *Daily Chron.* 12 Jan. 10/6 Envelope Cementer wanted...Envelope Folder wanted. 1944 *Jrnl. R. Aeronaut. Soc.* XLVIII. 117 Tests at other heights can then be confined to what are termed 'envelope' conditions; that is, the engine conditions which will give the maximum economy at any given speed. 1901 *Daily Chron.* 16 Apr. 8/6 Envelope Cutter. 1940 *Chambers's Techn. Dict.* 301/1 *Envelope delay*, the time taken for the envelope of a signal to travel through a transmission system, without reference to the time taken by the individual components. 1943 F. E. Terman *Radio Engineers' Handbk.* v. 443 Envelope delay is of particular importance in tuned amplifiers handling television signals, because any variation in the envelope delay with modulating frequency adds to the total delay error of the system. 1956 Amos & Birkinshaw *Television Engin.* II. i. 20 A more sensitive indication of the seriousness of any phase distortion is given by the slope of the curve itself, usually measured by the slope of the tangent to the curve...The slope of the tangent is the *envelope delay* or *group delay*. 1891 Kipling *Light that Failed* xi. 191 A letter with a black M. on the envelope flap. 1904 Envelope folder [see *envelope cementer*].

en ventre sa mère: see *en *prep.*

enviabi·lity. [f. Enviable *a.*: see -ity.] The character or quality of being enviable.

a 1916 H. James *Middle Years* (1917) 74 The enviability of certain complacencies of faith and taste. 1930 *Times Lit. Suppl.* 16 Oct. 839/3 The enviability or ultimate fatuity of his pleasant life.

en ville: see *en *prep.*

environing, *vbl. sb.* (Later example.)

1879 *Athenæum* 11 Oct. 476/2 In the midst of his customary environings the Duc de Richelieu receives a visit from a young lady.

environing, *ppl. a.* (Later examples.)

1951 D. Riesman in A. W. Loos *Relig. Faith & World Culture* 62 We can distinguish conceptually between the needs of society (as a system of social organization) and those of the environing group (as a system of psychological ties and expectations). 1957 G. Clark *Archæol. & Society* (ed. 3) i. 36 Their environing habitats and biomes.

environment. Add: **2.** (Later examples.)

1956 P. S. Sears in W. L. Thomas *Man's Role in changing Face of Earth* ii. 473/1 The situation is clouded by a widespread confidence that this impact of man upon environment can continue indefinitely. 1967 K. Mellanby *Pesticides & Pollution* ii. 31 Perhaps the most obvious way in which man has contaminated his environment is by polluting the air with smoke. 1968 *Biol. Conservation* I. 70/1 EDF is attempting to establish..a body of common law under which the general public can assert its constitutional right to a viable, minimally-degraded, environment.

c. *spec.* in *Phonetics.* (See also quot. 1951.)

1951 Z. S. Harris *Methods Struct. Ling.* ii. 15 The environment or position of an element consists of the neighbourhood, within an utterance, of elements which have been set up on the basis of the same fundamental procedures which were used in setting up the element in question. 1960 *Medium Ævum* XXIX. 27 There was evidently a phonemic distinction between forms which ultimately had the assibilated consonant and those which did not, even in the environment of front vowels. 1963 *Amer. Speech* XXXVIII. 50 Consonant and pause probably made up just about this percentage of environments for all finals. 1966 *Ibid.* XLI. 258 In all other phonetic environments.

3. *attrib.*, as *environment area, control, minister*.

1963 *Daily Tel.* 28 Nov. 16/2 The future pattern of cities should be conceived as a patchwork of 'environment areas' of residence, commerce or industry from which traffic other than that concerned with the area would be excluded. 1968 *Listener* 26 Sept. 393/2 A house to Bucky is an environment control. 1970 *Times* 27 Oct. 2/7 (*heading*) Mr Walker defines role of environment minister.

environmental, *a.* Add: (Later examples.) Also in Special Combs., as *environmental area, resistance, variation*. Hence **environme·ntally** *adv.*, with reference to or by means of (one's or the) environment.

1884 *Mind* July 338 Environmentally-initiated Sensations are classified according to the nature of the agent by which they are aroused. 1891 *Blackw. Mag.* CL. 853/1 Some general environmental cause appears to be necessary for the explanation of the facts. 1902 W. James *Var. Relig. Exper.* xiv. 360 Utopian dreams of social justice.. are, in spite of their impracticability and non-adaptation to present environmental conditions, analogous to the saint's belief in an existent kingdom of heaven. 1917 A. S. Pringle-Pattison *Idea of God* 75 Terms like stimulus, response, behaviour, all imply the notion of selection, the power of adaptation to environmental change. 1918 *Times Lit. Suppl.* 2 May 205/1 A cell may environmentally acquire a new property and keep it. 1928 *Daily Tel.* 21 Aug. 13/1 If you cannot indict a nation, neither can you fully describe a nation, environmentally or spiritually. 1928 R. N. Chapman in *Ecology* IX. 114 It seems evident that we have in nature a system in which the potential rate of reproduction of the animal is pitted against the resistance of the environment, and that the quantity of organisms which may be found is a result of the balance between the biotic potential or the potential rate of reproduction, and the environmental resistance. 1945 Koestler *Yogi & Commissar* iii. ii. 171 The socialist attitude to criminology, based on environmental psychology. 1950 *New Biol.* VIII. 18 Any peculiar character induced in a parent by special environmental conditions does not appear in the offspring, unless the same environmental conditions are reproduced. Non-hereditary variation due to external conditions is known as environmental variation. 1957 *Financial Times Ann. Rev. Brit. Industry* 85/5 This raises a question wider than occupational safety—namely environmental safety. 1958 *Punch* 12 Feb. 238/2 Both..knew that the most important thing in life—it is something we first heard of from the Americans—is to get 'environmentally well-adjusted'. **1962** *Listener* 24 May 903/1 There must be a proper relationship between the traffic capacity of the main distributory system and the capacity of the intervening areas (which for convenience I call the 'environmental areas'). *Ibid.* 30 Aug. 304/1 Allied to the earth sciences are the environmental sciences. Meteorological forecasts will be greatly improved in the future. **1965** *Language* XLI. 277 An environmentally conditioned allophone of terminal fade. **1967** K. Mellanby *Pesticides & Pollution* i. 15 The rather special case of pesticides, which have recently been shown to constitute such an important contribution to environmental pollution. **1970** P. R. & A. H. Ehrlich *Population, Resources, Environment* xi. 275 Environmental consulting firms have begun to appear. **1970** *New Yorker* 15 Aug. 42/1 Under its control is an environmental-health commission.

environme·ntalism. [f. Environmental *a.* + -ism.] A theory of the primary influence of environment on the development of a person or group.

1923 A. A. Goldenweiser *Early Civilization* xiv. 292 Attempts to interpret forms of civilization..by environmental conditions...The staunch environmentalism of Buckle still has its charms for many of his readers. **1931** *Encycl. Social Sci.* V. 561/1 Environmentalism is the tendency to stress the importance of physical, biological, psychological or cultural environment as a factor influencing the structure or behavior of animals, including man. **1957** K. A. Wittfogel *Oriental Despotism* i. 11 Static theories of environmentalism.

environme·ntalist. [f. as prec. + -ist.] One who believes in or promotes the principles or precepts of environmentalism; also, one who is concerned with the preservation of the environment (from pollution, etc.).

1916 *Amer. Jrnl. Sociol.* XXI. 628 The environmentalist will often agree with the anti-environmentalist that certain changes in a culture may be due..to certain cultural features..but, objects the environmentalist, these cultural features were, in their turn, produced by the physical environment. **1925** F. Thomas *Environmental Basis of Society* xi. 288 The environmentalist holds, for example, that the snow house of the Eskimo is determined by the surrounding Arctic environment. **1936** *Nature* 26 Sept. 530/1 Whether our outlook be mainly that of the eugenist or that of the environmentalist, we must not 'cease from mental fight' until we have..'built Jerusalem in England's green and pleasant land'. **1940** R. S. Woodworth *Psychol.* (ed. 12) vii. 225 The age-long dispute between the hereditarians and the environmentalists. **1941** J. S. Huxley *Uniqueness of Man* ii. 39 Sentimental environmentalists, who adhered to the crudest form of Lamarckism. **1959** *Archit. Rev.* CXXV. 304/2 The people are the Environmentalists...Their means of expression are the relationships between buildings and between spaces, and the element in which they work is time. **1970** *New Yorker* 9 May 33/2 Dr. Robert N. Rickles, a thirty-four-year-old chemist and an environmentalist,..took the helm of the city's Department of Air Resources last month. **1970** *Nature* 15 Aug. 655/2 The project to build a supersonic transport has run into renewed complaints from the environmentalists.

Hence as *adj.*, of, pertaining to, or relating to environmentalism.

1934 in Webster. **1952** W. Sprott *Social Psychol.* ii. viii. 147 Blackburn who champions the environmentalist position shows that..it is only among the poorer classes that the negative correlation is significant.

Also **envi:ronmentali·stic** *a.*

1941 *Natural Hist.* Feb. 112 The truths of environmentalism...serve Doctor Peattie well in this revival of the environmentalistic philosophy. **1962** H. J. Eysenck *Know your own I.Q.* 36 This method of investigation..favours the hereditary rather than the environmentalistic point of view.

envision (envi·ʒən), *v.* [f. En-[1] + Vision *sb.*] *trans.* To see or foresee as in a vision; to envisage; to visualize. So **envi·sioning** *vbl. sb.*

1921 L. Strachey *Q. Vict.* vii. 221 His blackest hypochondria had never envisioned quite so miserable a Catastrophe. **1925** *Chambers's Jrnl.* Dec. 800/1 Namitie..resigned himself apparently to envisioning her in the arms of the perfidious cousin. **1927** *Observer* 15 May 6 Karel Capek has..envisioned a world in which atomic energy, having been harnessed, first provides mankind with a new religion and then sets all the world at war. **1933** J. Baillie *And Life Everlasting* (1934) ii. 24 The indictment is..that the glories of heaven have always blinded the eyes of those that envisioned them to the more intimate glories of earth. **1938** *Observer* 18 Dec. 11/2 Mr. Houghton..is addicted to words like 'envisioning' which are not to be found in Webster or in the New English Dictionary. **1952** M. McCarthy *Groves of Academe* (1953) iii. 40 And the more he envisioned this prospect, the more he was of two minds about it. **1953** S. Spender *Creative Element* iii. 67 The development of modern literature today away from such highly individualistic envisionings. **1963** *Economist* 31 Aug. 744/2 The Negro firebrands, who..envisioned the march as an excuse for swarming into the very offices of Congressmen.

enwheel, *v.* For '† *Obs.*' read '*Obs.* exc. in echoes of Shakespeare's use'.

1897 F. Thompson *New Poems* 42 The Presence-hall where Angels Do enwheel their placèd King. **1912** L. A. Harker *Mr. Wycherly's Wards* ix. 129 Enwheel'd around with love on every hand.

Enzed (enze·d). *Austral.* and *N.Z.* [repr. pronunc. of initial letters of *New Zealand*.] A popular written form of 'New Zealand'; also, a New Zealander. Hence **Enze·dder**, a New Zealander.

1918 *Chrons. N.Z.E.F.* 24 May 179/1 Another interesting struggle was the tug-of-war between the 'En-Zeds' and the 'Ossies'. *Ibid.* 21 June 225/1 We christened the Adjutant 'Kiwi'—the symbol of En-Zed. **1934** *Bulletin* (Sydney) 1 Aug. 11/4 During the war years it was an honored title for all Enzeds serving overseas. **1941** Baker *N.Z. Slang* v. 43 *Enzed* and *Enzedder* were also coming into colloquial use by the end of last century. **1944** J. H. Fullarton *Troop Target* iii. 25 Pack my..body back to Enzed in a wheel-chair. **1946** T. E. Haughey *Railway Reminisc.* 6 He was back in En-Zed on a holiday. **1948** R. Finlayson *Tidal Creek* ix. 214 Uncle Ted, like a real Enzedder with no use for this snobby business, has renounced the title. **1956** S. Hope *Diggers' Paradise* 99 Two ace shearers, an Aussie and an Enzedder.

enzyme (e·nzəim). [f. Gr. ἐν in + ζύμη leaven.] **1.** The leavened bread with which the Eucharist is administered in the Greek Church.

1850 J. M. Neale *Eastern Ch.* 1074 'If,' says he [Theorianus, A.D. 1170], 'the Divine virtue changes the oblations into the Body and Blood of Christ, it is superfluous to dispute whether they were of Azymes or Enzymes, or of red or white wine.'

2. *Biochem.* [ad. G. *enzym* (W. Kühne 1877, in *Verhandl. Naturhist.-Medic. Ver. Heidelberg* I. 194), f. mod. Gr. ἔνζυμος leavened.] Any of the proteins produced by cells which catalyse specific biochemical reactions; formerly called (*unorganized*) *ferment*. Also *attrib.* and *Comb.* Hence **enzyma·tic, enzy·mic** *adjs.*; **enzyma·tically, enzy·mically** *advbs.*

1881 W. Roberts in *Proc. R. Soc.* XXXII. 146, I would suggest the desirability of adopting this term [G. *enzym*] into English, with a slight change of orthography, as 'enzymes', and also of coining from this root the cognate words which are requisite for clear and concise description. The action of an enzyme may be designated *enzymosis*, and the nature of the action may be spoken of as *enzymic*. **1890** [see Zymolysis]. **1898** *Ann. Bot.* XII. 491 The alcoholic fermentation of sugar is effected by the activity of an enzyme or soluble ferment. **1899** [see Zymase]. **1900** A. J. Ewart tr. *Pfeffer's Physiol. Plants* I. viii. 503 Even in an actively growing plant an abundant supply of the products of enzymatic action may partly or wholly inhibit the formation of the enzyme in question. **1902** *Jrnl. Chem. Soc.* LXXXI. 373 The phenomenon of fermentation is caused by enzyme action. **1903** *Westm. Gaz.* 20 Oct. 8/2 An enzyme-secreting organism..has been conveyed to the farm. **1908** *Chambers's Jrnl.* July 541/2 The hydrogen peroxide is decomposed by the enzyme catalase present in the milk into water. **1927** Haldane & Huxley *Anim. Biol.* iv. 106 Each digestive enzyme is a definite substance with the property of bringing about, or enormously speeding up, a particular chemical reaction. **1930** *Chem. Abstr.* 1654 The polypeptidase group includes enzymes which require the presence of a free NH_2 and others which require a free COOH in the substrate. Probably these groups are the point of attachment in the formation of enzyme-substrate compd. **1942** *Ann. Reg. 1941* 344 [Enzyme research] suggested strongly that enzymatic activity is vested in the protein molecule itself. **1944** *Ann. Rev. Physiol.* VI. 277 Renin which was released by the ischemic kidney acted enzymatically upon a serum globulin. **1948** *Sci. News* VII. 125 They are the parts of digestive juices which break up and dissolve the fat, starch and meat of food; they are also responsible for fermentation: the word 'enzyme' in fact means merely 'in yeast'. **1949** Miles & Pirie *Nature Bacterial Surface* i. 5 There is a group of mucopolysaccharides whose presence in certain organisms is essential for the stability of the cell membrane, and..these can be destroyed enzymically. **1953** *New Biol.* XV. 112 The enzyme-catalyzed reactions which make up this machinery [of a living organism] are nearly always reversible. **1956** *Nature* 14 Jan. 80/1 Since..cresolase activity is an enzymic..process, molecular oxygen must combine at some stage with enzyme or enzyme-substrate complex. *Ibid.* 24 Mar. 575/1 Even in the presence of enzymatically generated peroxide, catalase did not catalyse the oxidation of *p*-hydroxyphenylpyruvic acid to homogentisic acid. **1962** A. Pirie *Lens Metabolism Rel. Cataract* 429 Lactic acid was estimated colorimetrically..and enzymically. **1968** *Listener* 11 July 49/1 The biochemical enzymic machinery in the muscles. **1968** *Times* 19 Nov. 6/7 Sulphur compounds are known to be important inhibitors of enzymatic processes. **1968** *Physics Bull.* Dec. 433 Several different species of spherical cells were treated enzymatically to remove their outer coats. **1969** *New Scientist* 16 Oct. 122/1 Its great breakthrough with the enzyme wash powders brought troubles at both ends of the product.

enzymo·logy. [f. *enzym(e + -ology.] The division of biochemistry which deals with enzymes. Hence **enzymo·logist**; **enzymolo·gical** *a.*

1900 B. D. Jackson *Gloss. Bot. Terms.* Add., Enzymology. **1939** *Industr. & Engin. Chem.* (*News Ed.*) 10 Oct. 652 Biochemist–Enzymologist, experienced in Bromelin and Papain production and research publications. **1941** *Manch. Guardian Weekly* 17 Jan. 52 Just after I had read about the triumphs of Swedish enzymology, I met the village milkman. *Ibid.*, Meanwhile we may remain careless of 'what the Swede intend' and let the enzymologists go their way. **1956** *Nature* 24 Mar. 559/1 Enzymological studies on polysaccharide syntheses in plants. **1961** *Times* 17 July 3/1 Biochemistry with emphasis on enzymology. **1970** *Nature* 9 May 500/1 The creeping computerization that is spreading through biochemistry is now beginning to permeate the activities of enzymologists.

E O. (Earlier example of *attrib.* use.)

1801 M. Edgeworth *Belinda* III. xxv. 53 He likes the lady's EO table better than the lady.

Eoanthropus (ī,oæ·nþrŏpŏs). [f. Eo- + mod. L. *anthropus*: see Anthropo-.] The name given to a genus or a member of a genus represented by what was formerly believed to be the skull of a prehistoric man. See *Piltdown.

1913 A. S. Woodward in *Q. Jrnl. Geol. Soc.* LXIX. 135 The facial parts of the skull..differed..from those of any typically human skull. I therefore propose that the Piltdown specimen be regarded as the type of a new genus of the family Hominidæ, to be named *Eoanthropus* [*Ibid.* 137 The species of which the skull and mandible have now been described in detail may be named *Eoanthropus dawsoni*, in honour of its discoverer.] **1935** Huxley & Haddon *We Europeans* ii. 59 It is perhaps a million years since Eoanthropus and Sinanthropus roamed the earth.

eobiont (ī,obəi·ǫnt). [f. Eo- + Gr. βιουντ- pr. pple. stem of βιοῦν to live, f. βίος life.] A hypothetical chemical structure, supposed to arise during biopoesis, that has certain characteristics of living matter but is not alive in the fullest sense.

1953 N. W. Pirie in *Discovery* Aug. 242/2 In early systems, which may be called eobionts, functions may have been performed by other materials, inefficiently no doubt but well enough to get things started. [*Note*] Eobiont = 'dawn organism'. **1967** J. D. Bernal *Orig. Life* iii. 27 He [*sc.* Oparin] showed that such colloidal bodies could carry on complex chemical reactions and could gradually form what were afterwards referred to as eobionts or pre-vital masses which could carry on a chemical evolution of their own.

eocene, *a.* **1.** (Earlier example.)

1831 W. Whewell *Let.* (1876) II. 111, I propose for your four terms 1 acene, 2 eocene, 3 miocene, 4 pliocene.

‖**eo ipso** (ī·o i·pso), *advb. phr.* [L., ablative of *id ipsum*, that (thing) itself.] By that very act (or quality); through that alone; thereby. Cf. Ipso facto.

1696 J. Sergeant *Method to Science* i. v. 50 Nothing can be said to be *Divisible*, or capable to be made *more*, but it must be said *eo ipso* to be Actually and truly *One*. **1846** W. Hamilton *Suppl. Diss.* in *Reid's Wks.* (ed. 2) Note A. § 4. 754/2 For when a belief is necessary it is, *eo ipso*, universal. **1890** W. James *Princ. Psychol.* II. xxi. 300 What is lively and interesting stimulates *eo ipso* the will. **1901** A. E. Taylor *Probl. Conduct.* v. 295 A piece of benevolent legislation which compelled the consumer to pay more for the necessaries of life would *eo ipso* diminish the funds he can afford to devote to the pursuits of the higher culture. **1955** J. L. Austin *How to do Things with Words* (1962) viii. 98 To perform a locutionary act is in general..also and *eo ipso* to perform an illocutionary act, as I propose to call it.

E.O.K.A., EOKA, Eoka (ēǒu·kă). [f. initials of mod. Gr. Ἐθνικὴ Ὀργάνωσις Κυπριακοῦ Ἀγῶνος National Organization of Cypriot Struggle (for the furthering of the Greek cause in Cyprus).] The name of a Greek-Cypriot liberation movement.

1955 *Times* 11 July 10/1 These were the first terrorist outrages for a week, and may have been caused by a 'lone wolf' terrorist, as Eoka, the underground movement, has refrained from all activity. **1956** *Ann. Reg. 1955* 124 The extremist societies, the Communists on the one hand and E.O.K.A. (National Organization for the Cyprus Struggle) on the other. **1957** *Observer* 28 July 3/3 Leaflets..spoke of an organization calling itself EOKA..which had decided to begin the 'struggle for liberty'. **1957** *Times* 11 Nov. 10/4 In a public declaration this weekend the [Greek] Government strongly rebuked the Eoka Cypriot underground organization for interfering in Greek foreign affairs.

éolienne, eolienne (i,ōuli̇e·n). Also æ-. [ad. F. *éolienne*, fem. of *éolien*, f. Gr. αἰόλος quick-moving, glittering, sheeny + *-ien* = -ian.] A fine dress fabric of silk and wool. Also *attrib.*

1902 *Westm. Gaz.* 27 Feb. 3/2 Crêpes de Chine and éoliennes and taffetas. **1909** *Ibid.* 28 June 5/3 Éolienne skirts trimmed with lace. **1918** *Home Chat* 4 May 106 A tunic..of..fine coating serge, or wool Eolienne. **1920** *Glasgow Herald* 17 Apr. 4 A navy blue aeolienne frock. **1957** M. B. Picken *Fashion Dict.* 117/2 Eolienne..or aeolian.

eolith (ī·oliþ). *Archæol.* [f. Eo- + Lith, after *neolith*. Cf. F. *éolithe*.] The name given to certain flints which have been found in Tertiary deposits in England, France, and elsewhere, which have been claimed to be the earliest traces of human handiwork, but whose origin is much disputed.

1895 A. H. Keane *Ethnol.* 74 Other modern savages, who are quite incapable of fashioning any of these British *eoliths*, as they are called. **1907** T. R. Holmes *Anc. Brit.* 26 Flints of divers shapes..which have been termed 'eoliths', or stone implements of a dawning age. **1921** R. A. S. Macalister *Text-bk. European Archæol.* I. 148 Certain flint flakes to which the name 'Eoliths' has been given. **1950** G. E. Daniel *Hundred Years Archæol.* vii. 230 Benjamin Harrison, a village shopkeeper of Ightham in Kent..began to collect stone tools in Kent, especially in the plateau gravels. Here he found what are now referred to as Harrisonian eoliths.

eolithic (ī,oli·þik), *a.* *Archæol.* [ad. F. *éolithique* (G. de Mortillet *Le Préhistorique* (1883) i. iv. 18), f. Eo- + Lithic *a.*[1], after *neolithic, palæolithic*.] Pertaining to the earliest age of man that is represented by the use of worked flint implements. Also *fig.* and as *sb.*

1890 T. Wilson in *Rep. Smithsonian Inst. 1888* 604 The first or Eolithic period belongs entirely to the tertiary geologic epoch. **1893** J. A. Brown in *Jrnl. Anthropol. Inst.* XXII. 94, I venture to suggest the following four divisions of the Stone age..1. Eolithic; Roughly hewn pebbles and nodules and naturally broken stones, showing work with thick ochreous patina, found on the plateaux of the chalk. *Ibid.* 97 Flint implements of the Eolithic period (Plateau). **1920** Quiggin & Haddon *Keane's Man Past & Pr.* 10 The tools exhibit deliberate flaking, and mark the transition between eolithic and palaeolithic work. **1926** T. E. Lawrence *Seven Pillars* (1935) li. 294 We.., after killing the sheep in relay, had recourse to stray flints to cut them up...We used them in the eolithic spirit. *a* **1936** Kipling *Something of Myself* (1937) vii. 177 *The Times* leaders on 'motor-cars' were eolithic in outlook. **1948** A. L. Kroeber *Anthropol.* (ed. 2) xvi. 631 Rutot's Mesvinian stage of the Eolithic is recognized as probably a Belgian facies of the oldest Levalloisian or Pre-Mousterian. But his still earlier Eolithic stages..find little acceptance by the prehistorians as containing real artifacts.

Eonism (ī·oniz'm). Also **eonism.** [f. the name of the Chevalier Charles d'Éon (1728–1810), a French adventurer who wore women's clothes: see -ism.] Transvestism, esp. by a man. So **E·onist,** one who wears the clothes of the opposite sex.

1928 H. Ellis *Studies Psychol. Sex* VII. i. 10 It was clearly a typical case of what Hirschfeld later termed 'transvestism' and what I would call 'sexo-aesthetic inversion', or more simply, 'Eonism'. *Ibid.* 12 The Eonist (though sometimes emphatically of the apparent sex) sometimes shows real physical approximations towards the opposite sex. **1970** *Times* 5 Sept. 8/4 Today we can see that the Chevalier was an a-sexual transvestite. From his name Havelock Ellis coined the term *eonism* to describe this minor deviation.

‖**eo nomine** (ī·o nōu·mini), *advb. phr.* [L., ablative of *id nōmen*, that name.] Under that name; that is so called; explicitly.

1627 *Let.* 23 Nov. in Birch *Court & Times Chas. I* (1848) I. 292 The Earl of Bridgewater hath, *eo nomine*, disbursed £10,000. **1822** [see Arminianism]. **1870** S. H. Hodgson *Theory of Practice* II. ii. 69 He who seeks happiness, eo nomine, misses it. **1938** R. G. Collingwood *Princ. Art* iv. 72 The case of religious art *eo nomine*..hardly needs analysis.

eosinophil (ī,osi·nŏfil), *a.* and *sb.* *Med.* Also **-phile.** [G. *eosinophil* (P. Ehrlich 1878–9, in *Farbenanalyt. Untersuch. z. Histol. d. Blutes* (1891) 7), f. Eosin + -o + -phil, -phile.] **A.** *adj.* Having an affinity for eosin; staining readily with eosin. **B.** *sb.* A cell readily stained by eosin. Hence **eosinophi·lic, eosino·philous** *adjs.* = A above; also, pertaining to eosinophilia; **eosinophi·lia,** a condition of the blood marked by the formation and accumulation of an excess of eosinophil cells.

1886 H. M. Biggs tr. *Hueppe's Meth. Bacteriol. Invest.* ii. 68 The elements of the blood..are divided, according to Ehrlich, into—1. Lymphoid elements...2. Myeloid cells (eosinophile). 3. Undetermined (spleen and..marrow). *Ibid.* 69 The *α* or eosinophile granule..can be stained in all the acid aniline-dyes. **1892** Gould *Pocket Med. Dict.* 105 Eosinophilous. **1899** J. R. Green *Soluble Ferments* xxii. 382 An eosinophilous substance diffused out of the nucleus into the cytoplasmic zone. **1900** Dorland *Med. Dict.* 234/2 Eosinophil, eosinophilic, eosinophilous. **1900** W. Myers tr. *Ehrlich & Lazarus' Histol. Blood* 149 By eosinophilia we understand an increase only of the polynuclear eosinophil cells in the blood. Confusion of this form of leucocytosis with leukæmia is quite impossible. **1905** *Medical Annual* 140 Of 158 non-infected persons 91.1 per cent had under 5 per cent of eosinophiles. *Ibid.*, The eosinophilia may persist some time after the disappearance of ova. **1907** J. G. Adami *Inflammation* ii. xi. 82 During the height of the infection the eosinophils were found collected in the blood-vessels, actively migrating into the peritoneal cavity. **1907** *Practitioner* Sept. 455 The Eosinophil Cells. **1910** H. W. Armit tr. *Ehrlich & Lazarus' Anæmia* iii. 167 The Post-infective Form of Eosinophilia...There may even be a distinct eosinophilic leucocytosis. **1912** Adami & McCrae *Text-bk. Path.* 99 All verminous parasites set up eosinophilia, an increase in the number of eosinophile leukocytes in the circulating blood. *Ibid.* 128 The leukocytes that take part [in inflammation] are the polynuclear..cells, the lymphocytes, and the eosinophiles. **1949** H. W. Florey et al. *Antibiotics* II. xlv. 409 Eosinophilia developed in 14 out of 16 patients at some time during 120 days of treatment with 3 g. daily of highly purified streptomycin. In 7 patients the eosinophil cells amounted to 10 per cent. or more of the white blood cells. **1958** *Immunology* I. ii. 116 A homogeneous eosinophilic cytoplasm interspersed with areas of necrosis.

eotechnic (ī,ote·knik), *a.* [f. Eo- + Gr. τεχνικ-ός (see Technic *a.* and *sb.*).] Denoting or belonging to the first stage of industrial development, when technical advances are first applied to the making of machines.

1934 L. Mumford *Technics & Civilization* iii. 109 The demonstration that industrial civilization was not a single whole..was first made by Professor Patrick Geddes...In defining the paleotechnic and neotechnic phases, he however neglected the important period of preparation, when all the key inventions were either invented or foreshadowed. So, following the archeological parallel he called attention to, I shall call the first period the eotechnic phase: the dawn age of modern technics. *Ibid.* 112 At the bottom of the eotechnic economy stands one important fact: the diminished use of human beings as prime movers, and the separation of the production of energy from its application and immediate control. **1954** J. A. C. Brown *Social Psychol. Industry* i. 22 The eotechnic or medieval phase of industry stretches roughly from about A.D. 1000 to 1750.

Eötvös (ö·tvöʃ). The name of R. *Eötvös* (1848–1919), Hungarian physicist, used *attrib.* or in the possessive to designate certain apparatus invented and principles enunciated by him.

1913 *Jrnl. Chem. Soc.* A. CIV. ii. 300 The surface-tension formulæ of Young and Eötvös...Eötvös' formula. **1920** A. S. Eddington *Space Time & Gravit.* vii. 112 But it was shown by experiments with the Eötvös torsion-balance that the ratio of weight to mass for uranium is the same as for all other substances. **1922** *Encycl. Brit.* XXXI. 207/1 The quantities which the Eötvös balance is capable of measuring are almost incredibly small. **1930** N. K. Adam *Physics & Chem. of Surfaces* v. 153 Eötvös' 'law'...[154] Eötvös deduced his equation theoretically from considerations of 'corresponding states' of liquids of similar molecular constitution...The central point of the theory is, however, that surfaces should be compared on the basis of the number of molecules per unit area. *Ibid.* 159 The orientation and shape of the surface molecules.., their mutual attractions and details of their shape, size, and packing..may all affect the value of the Eötvös 'constant'. **1957** *Encycl. Brit.* VII. 846/1 The Eötvös or *Polflucht* force. **1963** Jerrard & McNeill *Dict. Sci. Units* 46 *Eotvos unit* (E), a unit used in geophysical prospecting to indicate the change in the intensity of gravity with change in horizontal distance. It has the dimensions of 10^{-9} gal per horizontal centimetre.

epacris (epæ·kris). [mod.L. (J. R. & J. G. A. Forster *Characteres Generum Plantarum* (1776) 19), f. Gr. ἐπί + ἄκρις summit; see Epacrid.] A plant of the genus of Australian evergreen heath-like shrubs so called.

1805 *Curtis's Bot. Mag.* XXII. 844 (*heading*) Rigid Epacris. **1807** *Ibid.* XXV. 982 This beautiful Epacris was sent us in blossom. **1841** *Florist's Jrnl.* (1846) II. 137 A 'New Subscriber'..wishes to know the reason his Epacrises are losing their foliage. **1884** 'R. Boldrewood' *Old Melbourne Mem.* xi. 77 Special species of Epacris grew there. **1885** H. H. Hayter *Carboona* 7 Of our tribe she is the flower, Lily, epacris and orchid. *a* **1963** B. James in *Austral. Short Stories* (1963) 48 And an Australian garden for..epacris, for mint bush, Geraldton wax and indigofera.

epagomenal, epagomenous (epægǫ·mĕnăl, -əs), *adjs.* = Epagomenic *a.*

1906 *Expositor* Apr. 342 The five epagomenous days of the year. **1924** W. M. F. Petrie *Relig. Life Anc. Egypt* v. 165 On the first of the epagomenal days at the end of the year. **1928** C. Dawson *Age of Gods* vii. 151 The Egyptian solar calendar with its 12 months of 30 days and 5 epagomenal days, which is so marked an improvement on the Babylonian lunar year.

epana-. Add: **epanalepsis** (earlier example); so **epanaleptic** (-le·ptik) *a.* [Gr. ἐπαναληπτικός], characterized by epanalepsis or repetition of a word or phrase; also as *sb.*, such repetition; **epana·phoral** *a.*, characterized by epanaphora.

1584 D. Fenner *Artes Logike & Rethorike* sig. D3ᵛ, Epanalepsis, which signifieth to take backe. **1927** F. J. E. Raby *Hist. Chr. Latin Poetry* iii. 96 Luxorius..shows a vicious taste for the cento and for epanaleptic verses. **1956** Auden *Old Man's Road*, The language they loved was coming to grief, Expiring in preposterous mechanical tricks, epanaleptics, rhopalics, anacyclic acrostics. **1906** *Athenæum* 10 Mar. 303/2 Under cover of all this epanaphoral fury..Mr. Campbell has in more than one instance shifted his ground.

eparterial (epaɹtīə·riăl), *a.* *Anat.* [See Ep-.] Situated above the pulmonary artery (see quots.).

1882 E. A. Schäfer *Quain's Elem. Anat.* (ed. 9) II. 511 The right bronchus..soon gives off a branch..which comes off *above* the place where the right pulmonary artery crosses the bronchus (eparterial branch). **1921** A. Keith *Hum. Embryol.* (ed. 4) xxii. 349 The bronchus of the upper right lobe..commonly lies above its artery—that is to say, it is eparterial. The other bronchi are hyparterial. **1962** *Gray's Anat.* (ed. 33) 1340 After giving off a branch to the upper lobe of the right lung—termed the superior lobe (eparterial) bronchus because it arises above the right pulmonary artery—the right (hyparterial) bronchus passes below the artery.

‖**épatant** (epataṅ), *a.* [Fr., pres. pple. of *épater* (see next).] Shocking (to the conventionally-minded). (Usually with implied reference to the next.) Also *absol.*

1925 G. B. Shaw *Shaw on Theatre* (1958) 171 The old anti-bourgeois weeklies..always Frondeur and *épatant* no matter who was in power, have disappeared. **1945** W. Stevens *Let.* 16 May (1967) 500 No one seems to be more addicted to the *épatant* (but it is not in any meretricious sense). **1961** *Listener* 16 Nov. 825/3 It has to be striking, astonishing, titillating, in a particular way—it has to be *épatant*. **1963** Auden *Dyer's Hand* 502 When Benedick says, 'Well, a horn for my money when all's done,' he is being deliberately *épatant*.

‖**épater** (epate). [Fr., to flabbergast.] Phr. *épater le(s) bourgeois*: to startle or shock the 'man in the street' or the uncritical adherent of traditional (artistic or ethical) theories.

The Fr. phrase 'Je les ai épatés, les bourgeois' is attributed to Alexandre Privat d'Anglemont (d. 1859).

1903 G. B. Shaw *Man & Superman* Epist. Dedic. p. v, You once asked me why I did not write a Don Juan play...You meant me to épater le bourgeois. **1920** O. W. Holmes *Let.* 30 Aug. (1953) I. 277 One is that you refer to things not generally known by your readers without explanation—a thing which inferior people do to '*épater les bourgeois*'. **1938** W. S. Maugham *Summing Up* 233 Sometimes he [*sc.* an author]..to *épater le bourgeois* has paraded the red waistcoat of Théophile Gautier. **1941** L. MacNeice *Poetry of Yeats* v. 95 Moore..is writing like a naughty boy *pour épater les bourgeois*. **1958** B. Nichols *Sweet & Twenties* i. 21 Tallulah's wickedness is really a pose—a mask to *épater les bourgeois*.

‖**épaulement.** Add: **2.** *Ballet.* (epolᵊmaṅ). [Fr.] The placing of the shoulders, one forward and the other drawn back, with the head turned to look over the forward shoulder.

1830 R. Barton tr. *Blasis's Code of Terpsichore* 67 *Epaulement*, opposition of the body. *Ibid.*, Opposition, *épaulement du corps*; half-arm in opposition and legs in the third position. **1952** *Ballet Ann.* VI. 121/1 The shortcomings of the French school—poor *port de bras* and *épaulement*, bent knees and weakness in *pirouettes*. **1961** *Guardian* 8 June 8/4 Sticklike arms and an almost complete lack of *épaulement*.

‖**épée** (epe). [Fr., = sword.] The sharp-pointed sword used in duelling and (blunted) in fencing. Hence **épé(e)ist** [cf. F. *épéiste*, also used], an épée fencer.

1889 W. H. Pollock et al. *Fencing* vii. 112 The fencer who has never handled the practice *épée* may at first find some difficulty when he exchanges the foil for it. **1910** *Encycl. Brit.* IX. 668/1 English épéists have also been coming to the front. *Ibid.* 668/2 Épée fencing can be, and often is, conducted indoors. **1910** *Westm. Gaz.* 11 Apr. 14/1 Mr. Seligman is a very fine épéeist. **1922** G. F. Laking *Rec. Eur. Armour* V. 59 The Court or 'small' sword, the use and rules of which are practically identical with those of the modern *épée*. **1955** *Times* 4 July 4/1 In the men's bouts, fought with the *épée*, London had 13 wins to three. **1958** *Times* 28 Nov. 16/6 There are 14 of the current top 16 amateur épéeistes entered.

epeiric (epəiə·rik), *a.* *Geol.* [f. Gr. ἤπειρος mainland, continent + -ic.] *epeiric sea*: see quot. 1930.

1925 J. Joly *Surface-Hist. Earth* iii App. 65 We are justified in ascribing the epeiric seas, attending the coming of a revolution, to the density-changes arising from the change of state of a basaltic substratum. **1930** *Ibid.* (ed. 2) 208 *Epeiric seas*, shallow marine waters connected with the oceans but wholly within the continental platform. **1964** A. B. Shaw *Time in Stratigr.* ii. 7 Great currents of the type characteristic of today's open oceans probably did not occur in epeiric seas of the past.

Epeirid (epəiə·rid). *Zool.* [ad. mod. L. *Epeiridæ* (see below), ? f. Gr. ἐπί on + εἴρειν to string together.] A member of the family of web-spinning spiders, formerly called Epeiridæ, now known as Argiopidæ. Also *attrib.* or as *adj.*

1881 O. Pickard-Cambridge *Spiders of Dorset* ii. 587 The studding of the lines of their snares, by some Epeirids, with viscid globules intended to entrap their prey. *Ibid.*, The cross-lines of Epeirid snares. **1902** *Trans. S. Afr. Philos. Soc.* XI. p. xlvi, The garden..was tenanted by numerous Epeirid spiders (*Argyope australis*). **1909** A. E. Shipley in Sedgwick *Text-bk. Zool.* III. xv. 408 Perhaps our commonest Epeirid, *Meta segmentata*.

epeirogeny (epəiərǫ·dʒeni). Also **epeiroge·nesis** [-genesis]. [f. as *epeiric *a.* + -geny.] The formation of continents; the deformation of the earth's crust by which continents and ocean basins are produced. So **epeirogene·tic** [-genetic], **epeirogene·tical, epeiroge·nic** [*-genic] *adjs.*

1890 G. K. Gilbert *Lake Bonneville* viii. 340, I shall take the liberty to apply to the broader movements the adjective *epeirogenic*...The process of mountain formation is *orogeny*, the process of continent formation is *epeirogeny*, and the two collectively are diastrophism. **1894** *Geol. Mag.* 449 The gentle but varying amount of epeirogenic deformation. **1898** J. E. Marr *Princ. Stratigr. Geol.* 32 Those wide-spread, fairly uniform movements which are spoken of as epeirogenic or continent-forming. **1903** A. Geikie *Text-bk. Geol.* (ed. 4) II. 1428/1 Epeirogeny or continent-making. **1909** Webster, Epeirogenetic. **1925** J. Joly *Surface-Hist. Earth* x. 162 The relations which exist between epeirogenic movements and glacial phenomena. *Ibid.* 189 Orogenesis and epeirogenesis..are forms of diastrophism.

1937 Wooldridge & Morgan *Physical Basis Geogr.* v. 62 Radial movements, due to forces acting vertically, which have been styled epeirogenic (continent building or plateau building). **1937** *Geogr. Jrnl.* XC. 91 Epeirogenetical changes. **1939** *Nature* 4 Feb. 184/1 So far as the post-Franciscan history of Southern California is concerned, tectonically comparable areas seem to have been affected at all times by orogenic and epeirogenic forces. **1963** D. W. & E. E. Humphries tr. *Termier's Erosion & Sedimentation* i. 13 In fact, a universal epeirogeny occurred during the late Precambrian and the Carboniferous.

epencephalic, *a.* (Earlier example.)

1851 [see Rhinencephalon].

ependyma. Add: So **epe·ndymal** *a.*, pertaining to the ependyma; **ependymi·tis,** inflammation of the ependyma.

1874 A. E. J. Barker tr. *H. Frey's Histology & Histochem. Man* 575 To this [*sc.* the neuroglia of the spinal cord] several names have been given, such as 'central ependymal thread', 'grey central nucleus'. **1889** *Cent. Dict.*, Ependymitis. **1898** Ependymal [see *astrocyte]. **1902** Cunningham *Text-bk. Anat.* 419 The ependymal cells. *Ibid.* 436 The ependymal layer. **1910** *Practitioner* July 65 Definite signs of post-basic meningitis or ependymitis about the fourth ventricle were found. **1962** E. C. Crosby et al. *Correl. Anat. Nerv. Syst.* ix. 577 The ependymal cells of the choroid plexus are usually regarded as secreting the cerebrospinal fluid. **1965** R. P. Morehead *Human Path.* xxxv. 1411 This condition, known as granular ependymitis, may cause obstruction of the ventricular system.

epenthesis. Add: (Later examples.) Now used in a wider sense to account for the presence of an unetymological vowel (cf. *anaptyxis) or consonant.

1954 Pei & Gaynor *Dict. Linguistics* 66 Epenthesis, the interpolation in a word or sound-group of a sound or letter which has no etymological justification for appearing there. **1955** *Sci. Amer.* Aug. 79/2 Sometimes you hear a consonant inserted where the spelling of the word suggests no such sound: fambly for family, chimbley for chimney. The name of this phenomenon, from the Greek, is 'epenthesis'. **1968** *Language* XLIV. 281 A number of phonetic treatments which are shared by Gathic and Younger Avestan but not by either Sogdian or Ossetic: (1) There is no trace of *i-* or *u-*epenthesis.

epe·nthesized, *ppl. a.* [See -ize.] Of a letter or sound: inserted by or resulting from epenthesis.

1880 Wilkins & England tr. *Curtius' Gr. Vb.* xi. 216 The epenthesised ι.

epeolatry (epi,ǫ·lătri). [f. Gr. ἔπος, ἔπεος word: see -latry.] The worship of words.

1860 O. W. Holmes *Professor* v. 104 Time, time only, can gradually wean us from our *Epeolatry*, or word-worship, by spiritualizing our ideas of the thing signified. **1928** *Daily Express* 28 Jan. 8/7 Many writers suffer from this disease of epeolatry, or word-worship. **1968** *Observer* 21 Jan. 27/7 A long farewell to Marshall McLuhan, most treacherous of clerks and a threat to all who cherish epeolatry.

ephebeum (efibī·ǒm). *Antiq.* [L., a. Gr. ἐφηβεῖον, f. ἔφηβος Ephebe.] A court in the palæstra for the young men to exercise themselves.

1697 Potter *Archæol. Græcæ* i. viii. 40 (*caption*) The Ephebeum. **1851-2** *Archit. Publ. Soc., Detached Ess.* (1853) *Baths* 3/1 The ephebæum (or young men's hexedra). **1901** R. Sturgis *Dict. Archit.*, *Ephebeion*; *-eum*. In Greek archæology, a place for the youths (*epheboi*) to exercise; hence, in Greco-Roman archæology, any place for gymnastic exercises, as in connection with Roman thermæ.

ephedra (efe·dră). [mod.L. (Tournefort in Linnæus *Genera Plantarum* (1737) 312), f. Gr. ἐφέδρα sitting upon.] A member of the genus of low evergreen trailing shrubs so called, belonging to the family Gnetaceæ and sometimes called shrubby horsetails. Hence **e·phedroid** *a.*, resembling ephedra.

1914 W. J. Bean *Trees & Shrubs Hardy in Brit. Isles* I. 515 The Ephedras..usually inhabit dry, inhospitable regions. **1933** *Tropical Woods* XXXVI. 7 Ephedroid Perforation Plate.—A plate having a small group of bordered, circular openings (as in *Ephedra*). **1951** *Dict. Gardening* (R. Hort. Soc.) II. 748/2 The hardy Ephedras are best accommodated in well-drained places in the rock-garden. **1953** K. Esau *Plant Anat.* xi. 225 (*caption*) End walls of vessel members showing..different types of perforation plates..foraminate (ephedroid). **1963** V. Nabokov *Gift* ii. 118 Those plants that to the last remain faithful to travellers: stunted ammodendrons, lasiagrostis, and ephedras.

ephedrine (e·fĕdrin). *Med.* Also -in. [S. Nagai 1887, in *Pharmazeut. Zeit.* XXXII. 700, f. *Ephedra* (see prec.) + -ine[5].] An alkaloid, $C_{10}H_{15}NO$, that occurs in certain species of *Ephedra* in four optically active forms and is made synthetically; 2-methylamino-1-phenylpropanol; *spec.* one of the lævorotatory isomers, used in the form of its salts as a sympathomimetic drug.

1889 *Jrnl. Chem. Soc.* LVI. 1020 Nagai, some years ago, isolated an alkaloid, ephedrine, from the *Ephedra vulgaris*, but through lack of material was unable to do more than determine its physiological action. **1900** Dorland *Med. Dict.*, Ephedrin. **1929** *Irish Jrnl. Med. Sci.* Apr. 184 In cases where the blood-pressure falls very low, ephedrin will control the fall. **1940** *Thorpe's Dict. Appl. Chem.* (ed. 4) IV. 317/1 The difference between ephedrine and *ψ*-ephedrine is limited to the carbon atom bearing the hydroxyl group. **1949** T. A. Henry *Plant Alkaloids* (ed. 4) 642 Being more stable to metabolic conditions, ephedrine can be given by mouth, whereas adrenaline has to be used by injection. **1967** G. A. Swan *Introd. Alkaloids* ii. 9 As ephedrine has been used successfully in the treatment of bronchial asthma, hay fever and other allergic conditions, many syntheses of it have been devised.

ephelis (efī·lis). *Med.* Pl. **ephelides** (efi·lidīz). [a. L. *ephēlis*, f. Gr. ἔφηλις (or ἐφηλίς), in pl. rough spots on the face, or perh. freckles.] A freckle; also *collect.* for various kinds of skin discoloration.

1756 J. Greive tr. *Celsus' Of Medicine* vi. v. 327 It is almost a folly to cure vari, lenticulæ and ephelides; but 'tis impossible to prevent women from being nice in what regards their beauty. *Ibid.*, An ephelis is cured by resin. **1813** T. Bateman *Pract. Synopsis Cutaneous Dis.* viii. 317 The term Ephelis denotes not only the freckles..which appear on persons of fair skin, and the larger brown patches, which likewise arise from exposure to the direct rays of the sun, as the name imports; but also those large dusky patches, which are very similar in appearance, but occur on other parts of the surface, which are constantly covered. *Ibid.* 318 Sauvages has improperly classed with Ephelis, the mottled, and dusky red hue of the shins of those, who expose their legs constantly to strong fires in the winter; and also the livid patches of scurvy. **1833** J. Forbes et al. *Cycl. Pract. Med.* II. 70 Ephelis diffusa, an eruption of distinct or confluent large, irregular, round patches, of a tawny, yellow, or brown colour. **1908** *Practitioner* Aug. 348 Some dermatologists include freckles, lentigo, or ephelides under the heading of sexual pigmentation. **1954** Ormsby & Montgomery *Dis. Skin* (ed. 8) v. 703 Ephelides do not occur on the palms or soles. **1961** P. Borrie *Roxburgh's Common Skin Dis.* (ed. 12) xx. 370 When heat is applied frequently to the skin..a dark brown network is formed, the so-called Ephelis ab igne.

ephemeral, *a.* Add: Hence **ephe·merally** *adv.*; **ephe·meralness.**

1848 Mill *Pol. Econ.* II. xiv. § 4. i. 467 The most ephemerally celebrated, and the really greatest poets (Byron and Shelley). **1911** G. H. Mair *Eng. Lit.: Mod.* ii. 48 Pamphlets or text books that have been preserved by accident from the ephemeralness which was the common lot of hundreds of their fellows. **1922** W. J. Locke *Tale of Triona* ix. 101 Let him make good, not ephemerally..but definitely. **1941** Wyndham Lewis *Let.* 10 Aug. (1963) 295 The book-business, in America has been..reduced to a level of ephemeralness, news-value, and mere fact-finding past belief. **1969** *Daily Tel.* 21 Apr. 16/3 Some of the enjoyment of a good performance lies in its very ephemeralness.

ephemeris. Add: **3. b.** **ephemeris time,** a uniform time scale used in astronomy, defined in terms of the orbital motions of the moon and planets and taking as the fundamental unit the **ephemeris second,** equal to a certain fraction of the tropical year 1900 (see quot. 1966).

1950 *Colloques Internat. du C.N.R.S.* XXV. Constantes Fondamentales de l'Astronomie..129 The Conference recommends that, in all cases where the mean solar second is unsatisfactory as a unit of time by reason of its variability, the unit adopted should be the sidereal year at 1900.0; that the time reckoned in these unit [*sic*] be designated Ephemeris Time. **1964** R. H. Baker *Astron.* (ed. 8) iii. 78 The *American Ephemeris and Nautical Almanac* and the British *Astronomical Ephemeris*..tabulate the fundamental positions of the sun, moon, and planets at intervals of ephemeris time. *Ibid.*, In the present century, ephemeris time has been gaining on universal time and in 1960 was ahead by 35 seconds. **1966** Kaye & Laby *Tables of Physical & Chem. Constants* (ed. 13) 8 Since 1956 the ephemeris second, defined as the fraction 1/31 556 925.974 7 of the tropical year for 1900 January 0 at 12ʰ ET, has been adopted as the fundamental invariable unit of time by the International Committee of Weights and Measures.

ephorate. Add: = Ephoralty in both senses.

1841 [in Dict.]. **1897** *Daily News* 26 Apr. 8/6 The two native bodies, the Athenian Ephorate and the Society of Antiquities. **1923** W. W. Tarn in J. B. Bury et al. *Hellenistic Age* 134 Then, having captured the ephorate, they were able to prevent him [*sc.* Agis] carrying out both his proposals together. *Ibid.* 135 He [*sc.* Cleomenes] also abolished the ephorate.

‖**ephymnium** (efi·mniǒm). Also **-ion.** [ad. Gr. ἐφύμνιον the burden of a chorus or hymn.] **a.** *Antiq.* In Poetry, a refrain, a short colon subjoined to a strophe. **b.** In some Eastern Churches, a refrain to a hymn; an antiphonal refrain.

[**1853** H. Burgess *Sel. Metr. Hymns & Homilies of Ephraem Syrus* p. liv, These ἐφύμνια, or supernumerary verses, have a pleasing effect.] **1910** *Encycl. Brit.* XIV. 182/2 When a strophe contained five lines, the fifth was generally an 'ephymnium', detached in sense, and consisting of a prayer, invocation, doxology or the like, to be sung antiphonally, either in full chorus or by a separate part of the choir. **1931** *Ess. & Stud.* XVI. 21 The main Ode, which consists of four pairs of strophes or stanzas in various rhythms, the first three pairs being divided by ephymnions or refrains. **1949** *Oxf. Class. Dict.* 568/1 In an *Ephymnion* (refrain) words, as well as metrical form, are repeated: e.g. Aesch. *Supp.* 117–75.

ephyra (e·firă), **ephyrula** (efi·riŭlă). *Zool.* [mod.L. (F. Peron 1809, in *Ann. du Muséum d'Hist. naturelle* XIV. 354), f. Gr. Ἐφύρα, Ionic -η, L. *Ephyrē*, name of a Nereid and of an Oceanid. *Ephyrula* is a diminutive.] A larval jellyfish, after separation from the scyphistoma.

1861 J. R. Greene *Man. Anim. Kingdom* II. ii. 66 Each Ephyra soon acquires a nutritive system. **1897** Parker & Haswell *Text-bk. Zool.* I. iv. 161 The saucer-like bodies separate from one another, and each, turning upside down, begins to swim about as a small jelly-fish called an *Ephyrula*...The umbrella of the ephyrula is divided into eight long bifid arms. **1898** A. Sedgwick *Student's Text-bk. Zool.* I. iv. 166 In rare cases (*Pelagia*) the development is simplified, and the larva passes directly into the *Ephyra*. **1916** H. S. Pratt *Man. Invertebr. Anim.* 125 Each disc is called an ephyra..and is a young medusa or jellyfish, which on becoming free grows in time to be a sexual animal. **1940** L. H. Hyman *Invertebrates* I. vii. 507 This [*sc.* the scyphistoma] is often followed by a young, free-swimming medusan stage, the *ephyra* (or *ephyrula*). **1967** P. A. Meglitsch *Invert. Zool.* vi. 153 (*caption*) Scyphistomae live for several years..producing larval medusae known as ephyrae by transverse fission.

epi-. Add: **2. epichlorhydrin** (examples).

1857 W. A. Miller *Elem. Chem.* III. 382 Epichlorhydrin ($C_3H_5O_2Cl$)..is a limpid oil. **1910** *Encycl. Brit.* I. 532/2 It [*sc.* acrolein] is also produced by the action of sodium on a mixture of epichlorhydrin and methyl iodide. **1956** *Nature* 21 Jan. 122/1 A polymer of repeating unit..which may be prepared from 4,4'-dihydroxydiphenoxyethane and epichlorhydrin in the presence of sodium hydroxide.

b. [a. G. *epi-* (E. Votoček 1911, in *Ber. d. Deut. Chem. Ges.* XLIV. 360).] Occas. prefixed to the name of a sugar or sugar derivative to indicate that a second compound, bearing the prefix, is an epimer of the first.

1911 [see *epimer]. **1932** H. Pringsheim *Chem. Monosaccharides & Polysaccharides* iii. 38 Mannose might be termed 'epiglucose'. **1963** K. Mayer tr. *Staněk et al. Monosaccharides* xviii. 507 The term epi-glucosamine is today reserved for D-mannosamine (2-amino-2-deoxy-D-mannose).

epialid: see *hepialid *a.* and *sb.*

epibiont (epibəi·ǫnt). [f. Epi- + Gr. βιουντ- pr. pple. stem of βιοῦν to live, f. βίος life.] An organism that lives on the surface of another, esp. one that is not normally parasitic on it.

1949 W. C. Allee et al. *Princ. Animal Ecology* xvii. 244/1 Representatives of many different phyla grow as epicoles (epibionts) on the shells or the skin of others without becoming noticeably parasitic. **1951** *Arch. Néerlandaises de Zool.* X. 38, I found no evidence that any of the epibionts can be considered as true parasites feeding on the living tissues of the oyster itself. **1957** *New Biol.* XXIII. 62 An epibiont is an organism which lives on the surface of another organism, without causing any direct harm. **1961** J. Green *Biol. Crustacea* vii. 104 Sometimes one finds that there is competition for space on the surface of a popular host. The epibionts, which are the organisms living on the surface, of *Daphnia*, show this rather well. **1965** [see *epibiotic *a.* and *sb.* 2].

epibiotic (epibəi,ǫ·tik), *a.* and *sb.* [f. Epi- + Gr. βιωτικ-ός pertaining to life, f. βίος life.]

1. a. *adj.* Designating one of a few isolated plants or animals that are members of an otherwise extinct population of a species. **b.** *sb.* An epibiotic organism.

1930 H. N. Ridley *Dispersal of Plants* p. xviii, A number of..plants of limited area, however, are the relics of an earlier flora which has nearly disappeared from change of climate or environment. These are known as *Epibiotics*, or survivors. *Ibid.*, Many plants..became scarce, then Epibiotic, then perhaps disappeared entirely. **1947** R. F. Daubenmire *Plants & Environment* x. 377 A species may suffer one or more catastrophes which destroy all but a fragment of the total population. The remnants are called relics, epibiotics, or depleted species.

2. *adj.* (See quot. 1960.)

1960 I. F. & W. D. Henderson *Dict. Sci. Terms* (ed. 7) 165/1 *Epibiotic*,..growing on the exterior of living organisms. **1965** V. A. Dogiel *Gen. Protozoology* (ed. 2) x. 574 The great majority of Protozoa of the plankton are free-living. Among them only a few epibionts belonging to the Suctoria lead an attached mode of life...The total number of such epibiotic forms is limited to a few dozen. **1967** *Oceanogr. & Marine Biol.* V. 518 The most characteristic species may be the following..: the sponge *Thenea muricata* (with its epibiotic zoanthid *Parazoanthus marioni*); [etc.].

epiblast. Add: **2.** (Earlier example.) So **epibla·stic** *a.*

1875 [see Hypoblast 2]. **1887** A. C. Haddon *Introd. Study Embryol.* ii. 36 The blastoderm of a newly-laid egg..consists of a definite epiblastic layer and an inferior irregular mass of rounded cells. **1893** A. M. Marshall *Vert. Embryol.* 371 The small epiblastic villi of the lower pole of the vesicle. **1921** A. Keith *Hum. Embryol.* (ed. 4) i. 12 An epiblastic or ectodermal set [of cells].

epiboly (epi·bŏli). *Embryol.* Also **epibole** (-ŏli). [Gr. ἐπιβολή a throwing or laying on.] The inclusion of one set of segmenting cells within another by reason of the more rapid division of the latter. Hence **epibolic** (epibǫ·lik) *a.*

1875 E. R. Lankester in *Q. Jrnl. Microsc. Sci.* XV. 163 Embolé and epibolé are but two extreme forms of one and the same process. *Ibid.* 165 (*heading*) Epibolic invaginate Planulæ. **1877** T. H. Huxley *Man. Anat. Invertebr. Anim.* xii. 683 The process of inclusion of the hypoblast within the epiblast may have the appearance of the growth of the latter over the former, or what is termed epiboly. **1887** A. C. Haddon *Introd. Study Embryol.* ii. 33 The gastrula in the Frog is thus formed partly by invagination (*embolé*), partly by overgrowth (*epibolé*). **1897** Parker & Haswell *Text-bk. Zool.* I. iv. 205 The stage has been produced, not by a process of invagination or tucking-in, but by one of epiboly or overgrowth. *Ibid.* v. 257 The process by which the germinal layers have become formed is..a process of epibolic gastrulation. **1914** [see *endoderm]. **1956** K. F. Lagler *Freshwater Fishery Biol.* (ed. 2) vi. 97 When epiboly is about half completed in the Atlantic salmon, a few pairs of body segments (somites) are already visible. **1966** Epibolic [see *embolic *a.* 2].

epibranchial (epibræ·ŋkiăl), *a.* *Zool.* [See Epi-.] Of or belonging to the segment next below the pharyngobranchial in a branchial arch. As *sb.*, this segment.

1846 [see *pharyngo-branchial* (Pharyngo-)]. **1877** Huxley & Martin *Elem. Biol.* (ed. 4) 119 Two lamellae united by their ventral edges and enclosing a central cavity which opens into a chamber (*epibranchial*) above. **1892** C. S. Minot *Hum. Embryol.* 651 A chain of epibranchial organs. **1893** A. M. Marshall *Vert. Embryol.* 40 Along the mid-dorsal line of the pharynx is a deep epibranchial groove..lined by a single layer of long columnar ciliated cells. **1921** A. Keith *Hum. Embryol.* (ed. 4) 243 At the upper end of each cleft depression there develop remarkable sense-organs, known as the epibranchial placodes. **1929** *Q. Jrnl. Microsc. Sci.* LXXII. 152 Epibranchial flaps eventually grow downward over the gill-slits of both sides to form the belated atrium. **1946** *Nature* 12 Oct. 524/1 They open posteriorly at the basal portion of the vertical tube referred to above, which communicates between the epibranchial chamber and the exhalant chamber.

epic, *a.* and *sb.* Add: **A.** *adj.* **2.** (Earlier example.) *epic theatre*, a play or plays characterized by realism and an absence of theatrical devices.

1731 A. Hill *Advice to Poets* 35 Then, might our great, Third Edward's awful Shade..Pale, from his Tomb, in Epic Strides, advance. **1935** *Life & Letters To-day* Sept. 74 This method of theatrical presentation was dubbed by Brecht the 'epic' as opposed to the 'dramatic' style. **1957** R. Hoggart *Auden* 15 In some of their techniques for presenting social problems and for obtaining a sense of urgent participation from the audience they seemed to have learned something from the early 'epic theatre' of the German Communist playwright Bertolt Brecht.

B. *sb.* **2. b.** (Additional examples.)

1916 A. Huxley *Let.* 19 Mar. (1969) 95, I want very much to see the *Birth of a Nation*, which is said to be a really great film, an epic in pictures. **1940** M. Gorelik *New Theatres for Old* ix. 412 An 'epic' is a large-scale film in which the events, usually historical, take precedence over the 'love interest'. **1957** N. Frye *Anat. of Crit.* iv. 314 *Ulysses*, then, is a complete prose epic. **1965** *Movie* Spring 36/1 The budget was supposed to have guaranteed an action-packed epic.

epicacuana, illiterate var. Ipecacuanha.

epicanthus (epikæ·nþǒs). *Anat.* [f. Epi- + Canthus. Cf. Gr. ἐπικανθίς = Encanthis.] A downward fold of skin which sometimes covers the inner angle or canthus of the eye, esp. in Mongols. Hence **epica·nthic** *a.*

1865 Dunglison *Dict. Med. Sci.* (rev. ed.) 355/1 Epicanthis, epicanthus. **1913** A. Keith *Hum. Embryol.* (ed. 3) 196 The curious epicanthic fold..is represented in all races during foetal life. *Ibid.* 197 (*caption*) The epicanthic or Mongolian fold. **1936** *Jrnl. R. Anthrop. Inst.* LXVI. 37 The presence or absence of an epicanthic fold is not recorded. **1961** *Listener* 31 Aug. 317/1 He gave me a sidelong, rather quizzical glance from his dark epicanthic eyes. **1964** S. Duke-Elder *Parsons' Dis. Eye* (ed. 14) xxxi. 509 Epicanthus is a semilunar fold of skin, situated above and sometimes covering the inner canthus.

epicardium (epikā·ɹdiǒm). Pl. **-ia.** [f. Epi-, after Pericardium.] **1.** *Anat.* The innermost layer of the pericardium, closely investing the heart.

1865 in Dunglison *Dict. Med. Sci.* (rev. ed.). **1894** D. J. Cunningham *Man. Pract. Anat.* II. i. 35 The investing part [of the pericardium] which covers the heart is called the visceral part or the epicardium. **1962** *Gray's Anat.* (ed. 33) 730 The visceral portion, or epicardium, covers the heart and the great vessels, and from the latter it is reflected to form the parietal layer, which lines the fibrous pericardium.

2. *Zool.* In certain ascidians, each of two hollow outgrowths from the pharynx, connected with the process of budding.

1893 *Q. Jrnl. Microsc. Sci.* XXXV. 125 The perivisceral cavity of Ciona corresponds to the epicardium of Clavellina. **1914** E. W. MacBride *Text-bk. Embryol.* I. xvii. 626 The conjoined inner walls of the epicardia form a kind of visceral peritoneum, enwrapping heart, pericardium, and intestine. **1929** *Q. Jrnl. Microsc. Sci.* LXXII. 158 Pharynx produced behind the endostyle into a pair of diverticula ('epicardia'). **1950** J. Z. Young *Life of Vertebrates* iii. 62 The situation is complicated by the presence of a pair of outpushings from the pharynx, the epicardia, or perivisceral sacs.

Hence **epica·rdiac, epica·rdial** *adjs.*

1893 *Q. Jrnl. Microsc. Sci.* XXXV. 123 Two tubes are first formed as outgrowths of the pharynx [in Clavellina], called by van Beneden and Julin [1887] the 'epicardiac tubes'. **1908** *Practitioner* Nov. 638 In the first group of cases there is simple adhesion between the peri- and epicardial layers. **1911** *Encycl. Brit.* XXVII. 390/1 There are..two main types of budding [in compound Ascidians]...There is first the 'stolonial' or 'epicardiac' type. **1914** E. W. MacBride *Text-bk. Embryol.* I. xvii. 626 The epicardial tubes or epicardia. **1962** *Lancet* 29 Dec. 1373/1 The infection was cured without removing the electrode, which was left in the ventricle until the insertion of epicardial wires 10 days later.

epicentre. Add: (Later examples.) Also *fig.*

1905 *Westm. Gaz.* 15 Feb. 2/1 The Ischian shock was chosen because of its relations to volcanic action; that of Charleston because of the discovery of the double epicentre. **1937** Wooldridge & Morgan *Physical Basis Geogr.* v. 58 In the region of the epicentre (the point, or small area, immediately above the focus) actual waves or swells occur, which..are responsible for most of the damage. **1955** *Times* 21 June 10/2 A 'severe' earthquake, with its epicentre probably located either in the Aleutian Islands off Alaska, or in Siberia. **1970** *Times* 10 Apr. 3/1 It has been a good year for the Maltings. Those at Snape, epicentre of the Aldeburgh Festival, have been lovingly rebuilt after last year's fire.

epiclastic (epiklæ·stik), *a.* *Geol.* [f. Epi- + Clastic *a.*] Designating rocks which are formed by the breaking up of pre-existing rocks upon the earth's surface.

1887 J. J. H. Teall in *Geol. Mag.* IV. 493 Some modification in the use of the term clastic is rendered necessary...I venture to suggest that it should be applied to all rocks which consist largely of mineral fragments and that we should distinguish between the three types of clastic rocks..by using the terms epiclastic, cataclastic, and pyroclastic. *Epiclastic*—Rocks formed of fragments resulting from the breaking up of older rocks occurring *upon* the earth's surface.

epiclesis (epikli·sis). *Liturgiology.* Also **epiklesis.** [Gr. ἐπίκλησις, f. ἐπικαλεῖν to call upon, invoke.] A part of the prayer of consecration in which the presence of the Holy Spirit is invoked to bless the eucharistic elements, or the communicants, or both.

[**1832** W. Palmer *Orig. Liturg.* II. 134 The immediate or proper prayer of consecration..may be divided into two particulars: first, the prayer itself, or ἐπίκλησις, in the language of the primitive church; and, secondly, the commemoration of our Lord's deeds and words at the last supper.] **1878** *Encycl. Brit.* VIII. 653/2 In Quæstio 107 it is laid down that immediately on the pronunciation of the *Epiclesis*, transubstantiation takes place. **1912** [see *anamnesis b]. **1925** *Contemp. Rev.* Oct. 426 Many Anglo-Catholics desire the introduction of the *Epiklesis*, or Invocation of the Holy Spirit, at the time of the consecration of the elements. **1959** J. Gill *Council of Florence* viii. 272 The problem as to whether it is the dominical words or the *epiclesis* in the Liturgy that should be deemed to effect the mystery. **1966** *Cath. Dict. Theol.* II. 226/2 There are *epikleses* with a similar drift in old Gallican liturgies.

e:picontine·ntal, *a.* *Geol.* [f. Epi- + Continental *a.* 1.] (See quot. 1905.)

1905 Chamberlin & Salisbury *Geol.* I. 11 Those shallow portions of the sea which lie upon the continental shelf, and those portions which extend into the interior of the continent with like shallow-depths, such as the Baltic Sea and the Hudson Bay, may be called epicontinental seas, for they really lie *upon* the continent, or at least upon the continental platform. **1935** *Geogr. Jrnl.* LXXXVI. 498 An epicontinental embayment of the Pacific. **1959** *New Biol.* XXIX. 8 There is plenty of evidence..that at various times nearly all the continents have been submerged by such 'epicontinental' seas.

epicormic (epikǫ·ɹmik), *a.* *Forestry.* [f. Epi- + Corm[2] + -ic.] Of a shoot or branch: growing from a dormant bud which has been suddenly exposed to the light and air.

1909 P. T. Maw *Pract. Forestry* 149 With some trees, especially Oak and Chestnut, an unduly thick canopy will often cause..the flushing of latent buds along the stem, and epicormic branches will be thrown out. **1921** R. C. Hawley *Silviculture* 135 The isolated position of the standards with full light from all sides..often results in the formation of epicormic branches on trees which have not an adequate crown development. **1953** *Brit. Commonw. For. Terminol.* i. 44 *Epicormic branch*, a branch originating from a dormant or adventitious bud arising on the trunk or an older branch.

epicritic (epikri·tik), *a.* *Psychol.* [f. Gr. ἐπικριτικός adjudicatory.] An epithet used to designate the finer discriminations of the senses, particularly of touch, as contrasted with *protopathic*. In terms of behaviour, sometimes used to refer to emotional and instinctive forms, as opposed to rational.

1905 H. Head et al. in *Brain* XXVIII. 107 To this form of sensibility we propose to give the name 'epicritic', since it is peculiarly associated with the localisation and discrimination of cutaneous stimuli. **1920** W. H. R. Rivers *Instinct & Unconscious* iv. 23 Epicritic sensibility may be only a greater perfection of protopathic sensibility. **1920** *Discovery* Nov. 340/1 From a physiological aspect, the distinction of protopathic from epicritic sensibility is, so far, devoid of neurological basis. **1942** *Brit. Jrnl. Psychol.* July 61 Head..stipulates expressly that the elaboration of spatial schemata depends upon processes occurring only at the level of *epicritic* integration. **1966** B. Haigh tr. *Luria's Higher Cortical Functions* ii. 146 A lesion of these cortical areas causes the most persistent disturbances of complex epicritic sensation.

epicuticle (e·pikiŭtik'l). Also **epicuti·cula.** [f. Epi- + Cuticle.] The outer layer of a cuticle. **a.** The thin outermost waxy covering on an insect or other arthropod.

1929 [see *endocuticle a]. **1934** A. D. Imms *Gen. Textbk. Entomol.* (ed. 3) i. 7 In most insects a very thin stratum or epicuticula (less than 1*μ* in thickness) is present outside the exocuticula. **1957** *Encycl. Brit.* XII. 417 F/1 Outside this [*sc.* the exocuticle of insects] there is an extremely thin layer, the 'epicuticle', which contains no chitin. **1967** P. A. Meglitsch *Invert. Zool.* xvi. 676 Three distinct layers can be recognized in the cuticle. A thin outermost layer of tanned proteins, waxes, and other fatty compounds makes up the epicuticle.

b. A thin outer membrane surrounding animal fibres or hairs, etc.

1949 J. Lindberg et al. in *Textile Research Jrnl.* XIX. 674/1 Proceeding from the outside of the fiber inward..the following have been described: (*a*) a very thin *epicuticular membrane*..forming the external surface of the fiber, (*b*) an *exocuticle* layer, and (*c*) an *endocuticle*...The epicuticle, as seen in the electron microscope, appears as a thin, uniformly thick..film falling easily into folds. **1956** *Nature* 18 Feb. 319/1 An interesting suggestion was that in human hair..the epicuticle seems to be tucked in beneath the scale which it is covering; this feature is less obvious in wool fibres. **1962** W. J. Onions *Wool* ii. 16 The epicuticle, a membrane about 100 Å thick (i.e. about 1/100th of the thickness of the total cuticle in wool) was isolated..by treating wool with a solution of sodium sulphide.

Hence **epicuti·cular** *a.*

1939 V. B. Wigglesworth *Princ. Insect Physiol.* ii. 20 The unicellular glands are lined only by the epicuticular cuticulin. **1945** *Ann. Reg. 1944* 376 The insecticidal action of chemically inert dusts which abrade the epicuticular wax film and so cause the insect's death.

epicyclic, *a.* Add: Applied esp. to the gear of motor vehicles. Also as *sb.*

1902 H. Sturmey in A. C. Harmsworth *Motors* x. 197 A Crypto or epicyclic gear..consists of two gear wheels, a small one A, with external teeth, and a large one B arranged on the outside of it with internal teeth. **1906** *Daily Chron.* 9 June 3/5 Another kind of gear, known as the epicyclic, its principle resembling that of the differential. **1907** *Westm. Gaz.* 13 Nov. 9/1 Three-speed epicyclic gear. **1910** *Ibid.* 1 Feb. 4/2 Many of the old epicyclics failed solely for the reason that their designers were too niggardly with the wearing surfaces. **1937** *Times* 13 Apr. p. xv/4 The manufacture of epicyclic gears is highly specialized work. **1948** *Autocar Handbk.* (ed. 20) vi. 120 The Wilson gear box..uses epicyclic or 'planetary' gears.

epicyte (e·pisəit). *Biol.* [f. Epi- + -cyte.] The investing membrane of a cell.

1900 in Dorland *Med. Dict.* **1903** E. A. Minchin in E. R. Lankester *Zool.* i. ii. 179 The cuticle or epicyte [in Gregarines] is a membrane secreted by the ectoplasm, usually of some thickness.

epideictic, *a.* Add: **b.** *spec.* in *Ornith.*: applied to collective displays and other conventional behaviour regarded by some as having evolved from the need to control the distribution of population.

1959 V. C. Wynne-Edwards in *Ibis* CI. 440 There is a very large class of social phenomena in animals which partake of this character...In birds they usually involve collective displays, singing, aerial acrobatics and the like: the so-called 'crow's parliament' of Rooks and Jackdaws..is a familiar example in this country. They may be designated as *epideictic* demonstrations. **1962** —— *Anim. Dispersion* i. 16 Epideictic displays are especially evolved to provide the necessary feed-back when the balance of population is about to be restored...They generally involve conventions that have evolved away from the direct primitive contest. **1966** D. Lack *Population Stud. Birds* 311 Most of the behaviour considered epideictic by Wynne-Edwards can be satisfactorily interpreted in other ways, and no positive evidence has been presented for an epideictic function.

epidendrum (epide·ndrǒm). Also **-dendron.** [mod.L. (C. Linnæus *Genera Plantarum* (1737) 272), f. Gr. ἐπί upon + δένδρον tree. Cf. Epidendral, -ic *adjs.*] Any of various chiefly epiphytic orchids of the genus so named, native to the western hemisphere (esp. tropical America).

1791 W. Jones *Let.* 18 Oct. (1821) II. 155 The most lovely *epidendrum* that ever was seen..grew on a lofty *amra*. **1842** *Florist's Jrnl.* (1846) III. 212 A fine orchidaceous plant, with the habit of an Epidendron. **1890** Watson & Bean *Orchids* xxxi. 237 The list of useful garden Epidendrums is a fairly long one. **1910** C. H. Curtis *Orchids* 100 All Epidendrums are evergreen. **1951** *Dict. Gardening* (R. Hort. Soc.) II. 749/1 The so-called true Epidendrums have cylindric reed-like, often slender, stems. **1955** E. Pound *Classic Anthol.* I. 30 Feeble as a gourd stalk (epidendrum).

epidermolysis (epidəɹmǫ·lisis). *Path.* [G. (H. Köbner 1886, in *Deut. Med. Wochenschr.* 14 Jan. 21/1, f. Gr. ἐπιδερμίς Epidermis + -o + λύσις a loosening or releasing.] A loosened state of the epidermis, giving rise to large blisters with little or no external cause.
1894 Gould *Illustr. Dict. Med.* 433/2 *Epidermolysis*, Acantolysis bullosa; Dermatitis bullosa; a rare skin-disease in which bullæ form upon the slightest pressure. **1897** *Brit. Jrnl. Dermatol.* IX. 301 (*title*) Epidermolysis bullosa. *Ibid.* 313 In Haemophilia bleeding occurs, in Epidermolysis exudation. **1909** W. Bateson *Mendel's Princ. Heredity* I. xii. 220 In epidermolysis transmission through unaffected persons occurred in some of the strains. **1922** R. C. Punnett *Mendelism* (ed. 6) 202 Epidermolysis bullosa, a disease in which the skin rises up into numerous bursting blisters. **1966** Wright & Symmers *Systemic Path.* II. xxxix. 1463 Epidermolysis bullosa is a rare dermatosis affecting both sexes.

epidiascope (epidəi·äskōup). [f. Epi- + Dia-[1] + -scope.] A magic lantern made to project images of both opaque and transparent objects. Hence **epidiasco·pic** *a.*
1903 *Nature* 19 Feb. 376/1 The epidiascope, a new optical lantern,..is primarily intended for the projection on the screen of opaque bodies, such as insects, coins, fossils, diagrams, &c., in their natural colours. It is equally serviceable for projection of transparent objects. **1907** *Practitioner* Dec. 860 An epidiascopic demonstration. **1927** *Brit. Jrnl. Photogr. Alm.* 345 The Ica Hand Epidiascope. **1940** *Geogr. Jrnl.* XCV. 73 Being printed on one side only the plates are readily adaptable for wall illustration and use in the epidiascope.

epididymectomy (e:pididime·ktŏmi). *Surg.* [f. Epididym(is + *-ectomy.] Excision of the epididymis.
1900 Gould *Pocket Med. Dict.* (ed. 4) 236. **1903** *Jrnl. Trop. Med.* 15 Jan. 25/1 The operation of epididymectomy can have only a very limited application in Egypt, owing to the late stages at which the disease is met with. **1938** *Brit. Jrnl. Urol.* X. 129 *Epididymectomy* is the operation of choice in tuberculous epididymitis.

epidi·dymo-, used as comb. form of Epididymis, = 'pertaining to the epididymis and', as *epididymo-orchitis*.
1897 *Ann. Surg.* XXV. 309 Those adhesions which are found in gonorrhoeal epididymo-orchitis are entirely wanting. **1963** *Lancet* 12 Jan. 95/1 In torsion of the fully descended testis the signs and symptoms resemble those of epididymo-orchitis.

epidosite (epi·dŏsəit). *Petrogr.* [ad. G. *epidosit* (L. Pilla 1845, in *Neues Jahrb. f. Min., Geognosie, Geol.* 64), altered form of Epidote: cf. Gr. ἐπίδοσις a free or additional giving, and see -ite[1].] A rock composed chiefly of granular or fibrous epidote with some quartz.
1863 *Rep. Progress Geol. Survey Canada* xvii. 497 In many parts occur beds which are entirely made up of quartz and epidote...This rock..is the epidosite of some lithologists. **1866** S. H. Lawrence tr. *Cotta's Rocks* 355 Epidosite, or pistacite rock.—Epidote usually combined with some quartz. **1868** J. D. Dana *Syst. Min.* (ed. 5) 284 [Epidote] sometimes forms with quartz an epidote rock, called epidosyte. **1880** —— *Man. Geol.* (ed. 3) 75 Epidosyte.—Pale green to pistachio-green. Consists of epidote mixed with quartz. **1893** A. Geikie *Text-bk. Geol.* (ed. 3) II. vii. 183 Epidote-Rocks.—Epidosite (Pistacite-rock)—an aggregate of bright green epidote with some quartz, occurs with chlorite-schist (Canada), with granite and serpentine (Elba), and with syenite. **1937** A. Johannsen *Descr. Petrogr.* III. 211 With sufficient iron, the final product of alteration may be a mixture of epidote and quartz, producing the rock called epidosite, which has been described..as derived from various rocks—augite-porphyries, diabases, åkerites, etc.

epidotized (e·pidotəizd), *ppl. a.* [f. Epidote + -ize + -ed.] Altered metamorphically into epidote. So **e:pidotiza·tion.**
1888 J. Teall *Brit. Petrography* 150 The epidotisation of the felspar is an alteration process. **1890** [see *chloritization]. **1930** Peach & Horne *Geol. Scotl.* 79 Beds with abundant pebbles of epidote and epidotized felspar. **1941** *Proc. Prehist. Soc.* VII. 58 The rock is identical with the epidotised tuffs. *Ibid.*, It is only however near the rhyolite that they exhibit..epidotization. **1965** G. J. Williams *Econ. Geol. N.Z.* x. 158/2 All rocks are hydrothermally altered, resulting in uralitization and chloritization of pyroxene, saussuritization of felspar, and epidotization of the rock as a whole. *Ibid.* xii. 181/2 Hutton, McCraw (1950) reported chalcopyrite as scattered grains in epidotized breccia in greywacke and argillite in the Takitimu Mountains, Southland.

epidural (epidiūə·răl), *a. Anat.* [f. Epi- + Dura (mater) + -al.] Situated upon or outside, or affecting, the dura mater.
1882 E. A. Schäfer in *Quain's Elem. Anat.* (ed. 9) II. 374 The spaces between the fibrous trabeculæ..can be injected from the epidural space where this exists, and the injecting fluid can be forced along them..into the subdural space. **1895** *Buck's Handbk. Med. Sci.* Suppl. 608/2 There are two layers of dura [in the spine], an ectal (periosteal) and an ental (myelic), the interval between them constitutes an epidural space. **1902** Cunningham *Text-bk. Anat.* 558 The spinal dura mater..is separated from the walls of the spinal canal by an interval, the epidural space, which is occupied by soft fat and a plexus of thin-walled veins. **1968** *Guardian* 12 Sept. 1/1 Epidural analgesia—a procedure which means a mother can get continuous pain-relief throughout her confinement without any loss of clarity or consciousness. A cocaine derivative is passed into the membrane surrounding the fluid round the spinal cord.

epifauna (e·pifǭnă). [a. Da. *epifauna* (C. G. J. Petersen 1913, in *Beretn. f. d. Danske biol. Station* XXI (in *Fiskeri-Beretn. 1912*). 15), f. Epi- + Fauna.] A collective term for the animal life of a region that lives on the surface of a marine deposit or another animal or plant.
1914 C. G. J. Petersen in *Rep. Danish Biol. Station 1913* XXI. 16, I would therefore propose, that..we can distinguish between an In-fauna, which is connected with the ordinary, level, uniform sea-bottom,..and various On-faunas (epifaunas), which occur in connection with such objects on the sea-bottom as stones, piles, shells, rocks or plants; the occurrence of such epifaunas is dependent on the distribution of these objects in the bottom. **1951** *Arch. Néerlandaises de Zool.* X. 38 Many members of the oyster's epifauna find their food on the shell itself. **1959** H. Barnes *Oceanogr. & Marine Biol.* I. 46 For the animals living on the surface or on other animals, the so-called epifauna, the material must be 'scraped' off the bottom and when the animals are relatively sparsely distributed a considerable amount of ground may have to be covered. **1964** *Adv. Marine Biol.* II. 172 Nearly every type of marine deposit..is inhabited by an assemblage of animals, some of which live on the surface (the so-called 'epifauna'), while others (the 'infauna') burrow or lie buried beneath the surface.
Hence **epifau·nal** *a.*, pertaining to, having the characteristics of, epifauna.
1960 *McGraw-Hill Encycl. Sci. & Technol.* I. 340/2 Caprellids commonly occur on epifaunal growths in shallow water. **1969** *Science* 28 Mar. 1419/2 In the past most deep-sea collections were made with coarse-meshed trawls which collected predominately larger epifaunal organisms.

epifocal (epifōu·kăl), *a. Geol.* [See Epi- 1.] Situated above the focus of an earthquake.
1898 J. Milne *Seismology* x. 195 Mallet's well known method..might..outside an epifocal area, be equally well used for the..determination of the maximum slopes of surface waves. **1902** *Encycl. Brit.* XXVII. 608/1 Causes which should produce magnetic effects within an epifocal district. **1910** *Ibid.* VIII. 817/2 (*Earthquake*) That part of the surface of the earth which is vertically above the centre is called the *epicentre*; or, if of considerable area, the epicentral or epifocal tract.

epigamic (epigæ·mik), *a. Zool.* [f. Gr. ἐπί upon + γάμος marriage + -ic. Cf. Gr. ἐπίγαμος marriageable.] Relating to the mating of animals and the characteristics of colour, etc., which serve to attract the opposite sex during courtship.
1890 E. B. Poulton *Colours of Animals* xvii. 338 Epigamic Colours. Colours displayed in courtship. **1926** J. S. Huxley *Ess. Biologist* (ed. 3) iv. 135 The human species..has epigamic characters of both kinds. Some, like voice and moustache, are different in the two sexes, others, such as colour of eyes and lips,..are common to both. **1959** *Chambers's Encycl.* II. 328/2 The principal types of display are epigamic, or display in which a sexual element is involved; aposematic or threat display; and distraction or diversionary display.

epigeal, *a.* Add: **b.** Of cotyledons: borne above ground after germination.
1861 R. Bentley *Man. Bot.* 773 They burst through the coats, and rise out of the ground in the form of green leaves..,in which case they are epigeal. **1884** A. Henfrey *Elem. Bot.* (ed. 4) 652 The seed-leaves..are pushed up above ground, and become epigeal. **1909** W. Bateson *Mendel's Princ. Heredity* 31 In *Phaseolus* hypo-geal cotyledons are dominant to epi-geal. **1964** A. J. Brook *Living Plant* i. 13 Because they are raised above the soil on germination, these cotyledons..are termed *epigeal*.

epigenesis. Add to etymology: The word is used by W. Harvey, *Exercitationes* 1651, p. 148, and in the English *Anatomical Exercitations* 1653, p. 272. It is explained to mean 'partium super-exorientium additamentum', 'the additament of parts budding one out of another'. Also add earlier *fig.* example.
1798 A. F. M. Willich *Elem. Crit. Philos.* 156 Epigenesis of pure Reason has been called the Kantian explanation concerning the coincidence of the pure intellectual conceptions..with the objects of experience.

epigenetic, *a.* Add: **2.** *Phys. Geogr.* and *Geol.* (*a*) Of a natural drainage system, = Superimposed *ppl. a.* 1 b (*Cent. Dict.* Suppl., 1909); (*b*) applied to a deposit of later origin than the rocks among which it occurs (opp. Syngenetic *a.* 2).
[**1886** F. Richthofen *Führer f. Forschungsreisende* vi. 174 Man kann die hier dargestellten Bildungen als *epigenetische Erosionsthäler* bezeichnen.] **1905** J. Geikie *Struct. & Field Geol.* xvi. 233 *Epigenetic Ore-Formations.* The formations included under this head are of later age than the rocks with which they are associated or in which they occur. **1914** *Min. Mag.* XVII. 72 'Epigenetic'..still is used antithetically to the term 'syngenetic' by writers on ore deposits, to describe a deposit the age of which is subsequent to that of the rocks in which it occurs. **1937** Wooldridge & Morgan *Phys. Basis Geogr.* xiv. 205 Such a drainage system is called *superimposed* or *epigenetic*. *Ibid.* xxii. 401 'Epigenetic gorges' through ridges, due to superimposition of drainage. **1965** [see *autoradiograph *sb.*].
B. *sb. pl.* (See quot. 1942.)
1942 C. H. Waddington in *Endeavour* Jan. 18/2 The..more important part of the task is to discover the causal mechanisms at work, and to relate them as far as possible to what experimental embryology has already revealed of the mechanics of development. We might use the name 'epigenetics' for such studies, thus emphasizing their relation to the concepts, so strangely favourable to the classical theory of epigenesis, which have been reached by the experimental embryologists. **1952** —— (*title*) The epigenetics of birds. *Ibid.* (title-p.), 'Epigenetics'. The science concerned with the causal analysis of development.

epigeous, *a.* Add = *epigeal *a.* b.
1847 J. Lindley *Elem. Bot.* p. xxxvi/2 *Epigeous*, growing close upon the earth.

epiglottiditis (e:piglǫtidəi·tis). *Path.* [f. Gr. ἐπιγλωττιδ-, ἐπιγλωττίς epiglottis + -itis.] = next.
1864 G. D. Gibb *Dis. Throat & Windpipe* (ed. 2) i. 63 A severe instance of acute epiglottiditis..occurring in a man..in whom the epiglottis could be felt as a hard solid ball, filling up the lower end of the pharynx. **1901** P. W. Williams *Dis. Upper Respiratory Tract* (ed. 4) viii. 165 Local variations in the character and distribution of the inflammatory changes have been distinguished by the terms *epiglottiditis*, *arytenoiditis*, or *chorditis*.

epiglottitis (e:piglǫtəi·tis). *Path.* [f. Epiglott(is + -itis.] Inflammation of the epiglottis.
1830 *London Med. Gaz.* VI. 314/2 Epiglottitis, though rare in its simple form, may nevertheless be frequently present in combination with inflammation of the neighbouring parts. **1888** *Brit. Med. Jrnl.* 15 Sept. 618/2 Before the invention of the laryngoscope, independent inflammation of the epiglottis was recognised and classified as epiglottitis. **1958** *New Eng. Jrnl. Med.* CCLVIII. 874/2 Acute epiglottitis is a serious and often fatal condition that, although infrequently recognized, lends itself to accurate bedside diagnosis.

epigon (e·pigǫn). More recent form of Epigone[1].
1890 J. H. Stirling *Philos. & Theol.* xii. 231 Any true follower of Hume, any genuine aufgeklärt *epigon* of his. **1890** W. Cory *Lett. & Jrnls.* (1897) 553 Anglo-Catholics, a second crop, the Epigons, started vestments. **1926** *Contemp. Rev.* July 22/1 The epigons of the fighters for freedom.

epigone[1]. Examples of pl. form *epigoni*.
1937 H. Read *Art & Society* 257 That task may perhaps be left to future epigoni. **1964** F. H. Blum in I. L. Horowitz *New Sociology* 162 Unlike the *epigoni* of his time..Mills grasped this problem in its totality.

epigyne (e·pidʒəin). *Zool.* [ad. mod.L. *epigynum*, f. Gr. ἐπί Epi- + γυνή woman, female. Cf. Epigynous *a.*] The ovipositor in arachnids; the external genital plate in spiders. Also in mod.L. form **epigynum** (epidʒəi·nŏm).
1875 *Encycl. Brit.* II. 273/1 These [*sc.* ovaries] have a common exterior orifice (vulva)..and connected with this opening there is frequently an epigyne, or ovipositor. **1892** *Trans. Linnean Soc.* Zool. V. 297 The epigynum or external genital plate..entirely closes the genital aperture. **1900** *Proc. Zool. Soc.* 387 The under surface of the body is whitish except the small yellow-brown epigyne. **1916** H. S. Pratt *Man. Invertebr. Anim.* 415 Between them [*sc.* the lung spiracles] is the genital pore which, in the female, is covered by a plate..called the epigynum. **1951** Locket & Millidge *Brit. Spiders* I. iii. 24 The term 'epigyne' (or 'epigynum') has been used in slightly different ways. In the present work it will be applied to..the general appearance of the female genital area from the outside, when the spider is viewed ventrally.

epiklesis, var. *epiclesis.

epileptically (epile·ptikăli), *adv.* [-ly[2].] As if affected with epilepsy; because of epilepsy.
1883 E. C. Mann *Man. Psychol. Med.* xxv. 483 There are on record many homicides committed by epileptically insane persons. **1912** Kipling *Diversity of Creatures* (1917) 211 The House sang at the tops and at the bottoms of their voices,..epileptically beating with their swelled feet. **1929** D. H. Lawrence *Pansies* 128 Dostoevsky..With his sham christianity Epileptically ruined The last bit of sanity..Of the Russian nobility.

epileptoge·nic, epilepto·genous, *adjs. Path.* [See *-genic and -ous.] Producing epileptic attacks.
1885 *Alienist & Neurologist* (St. Louis) VI. 449 Seppilli..believes..that basilar motor centers..may acquire..the epiloptogenous property by exaggerating their proper function. **1889** *Cent. Dict.* II. 1970/2 Epileptogenic. **1908** *Practitioner* Sept. 466 The fits re-appear after a time, because the scar itself becomes in turn an epileptogenous centre. **1949** H. W. Florey *Antibiotics* II. xxxviii. 1215 Their critical epileptogenic dose. **1961** *Lancet* 22 July 179/2 Electroencephalographic examination..showed multiple epileptogenic areas.

epilimnion (epili·mniǫn). Pl. **epilimnia.** [f. Epi- + Gr. λιμνίον, dim. of λίμνη lake.] The upper, uniformly warm layer of water in a stratified lake: cf. *hypolimnion. Hence **epilimne·tic** [cf. Gr. λιμνήτης living in marshes], **epili·mnial** *adjs.*, situated in or relating to the epilimnion.
1910 E. A. Birge in *Trans. Wisc. Acad. Sci., Arts & Lett.* XVI. II. 1005, I employ two new words in this paper. ..These terms are *epilimnion*, for the upper warm layer of water..and *hypolimnion*, for the lower colder water. **1936** *Nature* 5 Dec. 971/1 The result of this warming is that, in summer, a warm upper layer of less dense water, the epilimnion, comes to lie over a cold deeper water mass, the hypolimnion. **1940** *Ibid.* 24 Aug. 255/1 The epilimnial current..opposes the hypolimnial current, causing turbulence at the thermocline. **1957** G. E. Hutchinson *Treat. Limnol.* I. vii. 447 In the former case, the 15-m. depth was evidently ordinarily epilimnetic, in the latter case metalimnetic. **1959** J. Clegg *Freshwater Life* (ed. 2) ii. 38 The epilimnion..is well lighted, warm and well oxygenated, and..a wealth of minute plants is able to thrive. **1959** *Times* 20 June 9/3 They [*sc.* tench] have been termed 'epilimnion feeders'.

epilobium. (Earlier examples.)
1775 T. Blaikie *Diary Scotch Gardener* (1931) 5 July 49 Found several plants on the channells of the Lake [of Geneva] especially the new sort of *Epellobium*..with a strong root deep in the hand gravell...Most of the other *Epellobiums*..in general has creaping roots. **1809** Smith & Sowerby *Eng. Bot.* XXVIII. 2000 Epilobium alsinifolium...This *Epilobium* was gathered on the Cheviot hills by Mr. Winch, who rightly judges it to be the plant of Ray, always mistaken for *alpinum*.

epimedium (epimi·diŏm). *Bot.* [mod.L. (Linnæus *Systema Naturæ* (1735)), f. L. *epimēdion*, Gr. ἐπιμήδιον.] Any plant of the genus of hardy perennial herbs so named.
1798 Smith & Sowerby *Eng. Bot.* VII. 438 Epimedium alpinum, *Barren-wort*...Dr. Withering has lately introduced the *Epimedium* into his work on the authority of Dr. Richardson, who is said in Blackstone's *Specimen Botanicum* to have found it in Bingley Woods, Yorkshire. **1900** J. M. Abbott in W. D. Drury *Bk. Gard.* ix. 293 Epimediums are useful plants for shady positions. **1901** *Cassell's Dict. Gard.* 321/2 The Epimediums will thrive well in shady places, and even under trees. **1961** *Amat. Gardening* 2 Dec. 19 A plant of epimedium. This is a hardy herbaceous plant which does well in a rock garden.

epimer (e·piməɹ). *Chem.* [a. G. *epimer*, pl. *epimere* (E. Votoček 1911, in *Ber. d. Deut. Chem. Ges.* XLIV. 360), f. Epi- + *-mer.] Either of two isomers differing in the configuration of the atoms about a single asymmetric carbon atom in a compound containing other asymmetric carbon atoms; orig. in a more restricted sense (see quot. 1948[1]). Hence **epime·ric** *a.*, of, having the qualities of, or standing in relation to another compound as, an epimer; **epi·merism** [ad. G. *epimerie* (E. Votoček 1911, *loc. cit.*): see -ism], the fact or condition of having epimers.
1911 *Chem. Abstr.* V. 1606 It is proposed that sugars which are related in the same manner as is glucose..to mannose..should be distinguished by prefixing epi- to the name of 1 member of the pair. The relationship itself should be termed epimerism; the 2 related compds. would then be called epimers. *Ibid.* 1607 Epimeric aldoses give the same osazone. **1948** W. W. Pigman *Chem. Carbohydrates* ii. 26 In the sugar series, substances which differ only in the configuration of the carbon atom immediately adjacent to that carrying the carbonyl or carboxyl group are known as epimers. *Ibid.*, It may be well to extend the definition of epimers to mean any pair of stereoisomers that differ solely in the configuration of a single asymmetric carbon atom. **1962** E. L. Eliel *Stereochem. Carbon Compounds* iv. 40 Diastereoisomers that differ in configuration at only one asymmetric center are sometimes called 'epimers'; thus menthone and isomenthone are epimers. **1965** K. Mislow *Introd. Stereochem.* ii. 91 The diastereomeric ephedrines and pentaric acids are necessarily epimeric.

epimerite (epi·mĕrəit). *Zool.* [f. Gr. ἐπί Epi- + μέρος part + -ite[1].] (See quots.)
1885 *Encycl. Brit.* XIX. 854/1 (*Gregarinidea*) There is frequently if not always present, either in early growth or more persistently, an anterior proboscis-like appendage (the epimerite) growing from the protomerite. **1921** Parker & Haswell *Text-bk. Zool.* (ed. 3) I. ii. 85 Sometimes the protomerite is produced in front into a process ending in a rounded enlargement, the epimerite, which may be provided with radiating spine-like projections. **1961** Mackinnon & Hawes *Introd. Study Protozoa* III. 169 The Cephalina, which are gut parasites in which the body is typically divided into different regions, of which the anterior develops an organ of fixation called the epimerite.

epimerization (epi:mərəizēi·ʃən). *Chem.* [f. *epimer + -ization.] The conversion of one epimer into the other. Hence **epi·merize** *v. trans.*, to convert (one epimer) into the other; **epi·merizing** *vbl. sb.*
1929 *Jrnl. Chem. Soc.* I. 345 The authors have effected the epimerisation of several methylated lactones. **1930** *Ibid.* II. 2623 This is epimerised to tetramethyl glucose by alkali. **1956** L. F. & M. Fieser *Org. Chem.* (ed. 3) xiv. 361 A sugar carrying a methoxyl group at the 2-position is epimerized by alkali at C_2 with deuterium exchange at this position. **1959** *Jrnl. Chem. Soc.* I. 637 Titanium chloride is a powerful epimerising agent. **1965** G. Hallas *Org. Stereochem.* ii. 43 In both cases, epimerization takes place by enolization.

epimeron. Add: Also in Insecta (see quot. 1964).
[**1869** A. S. Packard *Guide to Study of Insects* 9 The episternum is situated between the epimerum and sternum.] **1895** J. H. & A. B. Comstock *Man. Study of Insects* xxi. 502 The first of these [sclerites] is the episternum, the second is the epimeron. **1925** A. D. Imms *Gen. Textbk. Ent.* I. 23 The pleuron consists of an anterior sclerite or episternum and a posterior sclerite or epimeron. **1964** R. M. & J. W. Fox *Introd. Compar. Ent.* ii. 53 The principal pleural sclerites of each segment are the episternum, lying anterior to the legs, and the epimeron, above and behind the leg bases.

epinephrine (epine·frin). *Chem.* Also **-in.** [f. Gr. ἐπί upon + νεφρός kidney + -ine[5].] = *adrenaline.
1899 *Jrnl. Chem. Soc.* LXXVI. I. 395 Epinephrine, the active material of suprarenal capsules. **1900** Dorland *Med. Dict.*, Epinephrin. **1908** *Practitioner* Mar. 422 The active principle from the medulla of the suprarenal capsule is now generally known as adrenalin, though other terms have been applied, such as suprarenin, epinephrin. **1949** H. W. Florey et al. *Antibiotics* II. xxxviii. 1234 The use of propylene glycol with penicillin, epinephrine hydrochloride, and arachis oil. **1962** A. Huxley *Island* xiv. 250 Their pink epinephrin, my visions; their white epinephrin, my crimes. **1963** H. Burn *Drugs, Med. & Man* (ed. 2) ii. 23 Most people have heard of the substance adrenaline, which is called epinephrine in the United States Pharmacopoeia.

epineural, *a.* Add: **2.** *Zool.* Lying over and parallel to a nerve. Said of various processes in echinoderms, etc.
1906 *Cambr. Nat. Hist.* I. xviii. 515 In Echinoids the ambulacral groove has become converted into a canal called the 'epineural canal'. **1914** E. W. MacBride *Text-bk. Embryol.* I. viii. 254 The epineural sinus..derives its name from the circumstance that it lies above the rudiments of the ganglia of the ventral nerve cord. *Ibid.* xvi. 495 Epineural flaps. *Ibid.*, The epineural roof which covers in each radial nerve cord of the adult. *Ibid.* 514 The epineural ridges. *Ibid.*, These epineural canals meet in a central epineural space, roofed over by a membrane termed the epineural veil. **1955** L. H. Hyman *Invertebrates* IV. xv. 140 To the outer side of the radial nerve lies a cavity, the epineural canal or sinus.

epipalæolithic (e:pipæli,ǒli·þik), *a. Archæol.* [ad. Sp. *epipaleolítico* (H. Obermaier *El Hombre Fósil* (1916) x. 314), f. Epi- 1 + Palæolithic, Paleo- *a.* (*sb.*).] Belonging to the period or cultural stage between the end of the palæolithic and the beginning of the mesolithic; formerly = Mesolithic *a.*
1921 R. A. S. Macalister *Textbk. Europ. Archæol.* x. 549 The 'Epipalaeolithic' Azilian. **1924** C. D. Matthew tr. *Obermaier's Fossil Man in Spain* x. 323 The final phase of the Capsian, the Tardenoisian, the Azilian, and the northern Maglemose industries are the posthumous descendants of the Palæolithic, on which account we have adopted for the group the name of 'Epipalæolithic'. **1928** V. G. Childe *Most Anc. East* ii. 33 Others [*sc.* rock-paintings] belong to the immediately succeeding epipalæolithic period. **1936** *Discovery* Aug. 262/2 Some would regard it [*sc.* the period between the Old and New Stone Ages] as a continuation in attenuated form of activities fundamentally developed by palæolithic man. They, therefore, prefer to term it 'epi-palaeolithic'. Others regard it as essentially transitional, and call it 'mesolithic'. **1970** D. A. Roe *Prehistory* iii. 91 Between the fully Upper Palaeolithic Cultures..and the fully Mesolithic ones..there are cultures which are not clearly either one or the other, and the sensible name of 'epi-palaeolithic' is sometimes given to them. *Ibid.* 95 The various epi-palaeolithic cultures may be regarded as ending with the close of Late Glacial conditions and the beginning of the Pre-Boreal phase.

epipharynx (epifæ·riŋks). *Zool.* [f. Epi- 1 + Pharynx b.] In insects, a membranous lining (see quot. 1826); esp. in Diptera, a piercing organ (see quot. 1897); also in Arachnida, a chitinous plate (see quot. 1909).
1826 Kirby & Spence *Introd. Entomol.* III. xxxiii. 359 Epipharynx (the Epipharynx). A small valve under the Labrum, that in many Hymenoptera closes the Pharynx, and is an appendage of its upper margin. **1897** Parker & Haswell *Text-bk. Zool.* I. xi. 592 A projection of the roof of the mouth cavity (epipharynx) is present in some Insects. **1909** *Cambr. Nat. Hist.* IV. xviii. 459 In those free living Mites which have undergone no great modification of the mouth parts two other portions can be distinguished, the upper lip or 'epipharynx', and the 'lingua'. **1957** Richards & Davies *Imms's Textbk. Ent.* (ed. 9) III. 589 (*Diptera*) The labrum..is dorsally well sclerotized but ventrally more membranous, this surface sometimes being termed the epipharynx. *Ibid.* 669 (*Siphonaptera*) The epipharynx is a long slender organ which is ventrally grooved. *Ibid.* 676 (*Hymenoptera*) An epipharynx is well developed and trilobed in the higher forms, the median lobe being pointed and projecting. **1962** *Science Survey* XVII. 273 The flow of air passes between the floor of the pharynx and the little flap-like epipharynx, and is interrupted by movements of the latter.

epiphenomenal (e:pifinǫ·minăl), *a.* [f. Epiphenomenon + -al.] Of the nature of an epiphenomenon; relating to epiphenomena.
1899 J. Ward *Naturalism & Agnosticism* II. xiv. 100 From the standpoint of naturalism a world described in such terms is epiphenomenal. **1924** *Glasgow Herald* 29 Aug. 10 He maintained that mind was not epiphenomenal. **1929** C. S. Myers *Psychol. Conceptions* 22 The old view expressed by Spencer..was that consciousness has arisen as an 'epiphenomenal' product of living matter when physiological processes became too complex to work automatically. **1943** A. M. Farrer *Finite & Infinite* ix. 99 'Beauty' has been regarded as a quality epiphenomenal upon sense-data. **1957** G. E. Hutchinson *Treat. Limnol.* I. i. 3 To the geologist the dam therefore is an epiphenomenal afterthought, while to the limnologist it makes all the difference in the world.

epipheno·menalism. [f. prec. + -ism.] The theory that consciousness is an epiphenomenon, i.e. a secondary result and by-product of the material brain and nerve-system. So **epipheno·menalist**, one who holds this theory; also as *adj.*
1899 J. Ward *Naturalism & Agnosticism* II. xii. 34 Doctrine of conscious automatism or psychical epiphenominalism examined. **1903** A. E. Taylor *Elem. Metaphysics* 317 The epiphenomenalist theory is largely adopted by the workers in the physical sciences. **1915** G. F. Stout *Man. Psychol.* (ed. 3) 98 Perhaps the word which best characterises the general position is epiphenomenalism. Consciousness is regarded as a superfluous apparition or phenomenon which unaccountably crops up at a certain stage in the course of material processes. **1922** W. R. Inge *Outspoken Ess.* 2nd Ser. 6 Not only does epiphenomenalism (as Plotinus said long ago) 'make soul an affection, or disease, of matter'. **1949** *Mind* LVIII. 124 Epiphenomenalism has outstanding merits, heuristic and educational. It teaches the student to take very seriously indeed the search for the neural correlates of his mental processes. *Ibid.*, It is refreshing..to read a clear defence of a frankly epiphenomenalist position.

epiphenomenon. Add: **b.** *spec.* in *Psychol.* Applied to consciousness regarded as a by-product of the material activities of the brain and nerve-system.
1890 W. James *Princ. Psychol.* I. v. 129 But this would be a quite autonomous chain of occurrences, and whatever mind went with it would be there only as an 'epiphenomenon', an inert spectator. **1899** J. Ward *Naturalism & Agnosticism* II. 37 The newly coined phrase *epiphenomenon* (or, as the Germans say, *Begleiterscheinung*). **1913** J. M. Baldwin *Hist. Psychol.* II. iv. 60 This charge [of materialism] is frankly accepted..by those, such as Maudsley, who accept the 'epiphenomenon' theory of consciousness; to them consciousness is merely a by-product, a spark thrown off by the engine, the brain. **1952** W. Sprott *Social Psychol.* 208 Marxists have never taken ideas as mere epi-phenomena. **1965** H. Kuhlenbeck in J. R. Smythies *Brian & Mind* 156 Yet, in this respect, consciousness remains either an 'epiphenomenon' or a parallel, not 'causally' involved phenomenon...The term 'epiphenomenon' stresses the 'vectorial' or one-way, open transformation from public physical space-time into private perceptual space-time.

epiphloem (epiflōu·em). *Bot.* Also **-phlœum.** [f. Gr. ἐπί upon, Epi- + φλόος = φλοιός bark: see Phloem.] The outermost or corky bark.
1839 J. Lindley *Introd. Bot.* (ed. 3) I. 89 When the substance called *cork* is formed, the epiphlœum consists of polyhedral cells. **1866** Lindley & Moore *Treas. Bot.* 1109/2 *Suber*.., cork. The epiphlœum of bark, when it acquires an elastic soft texture, and is preternaturally enlarged. **1900** B. D. Jackson *Gloss. Bot. Terms*, Epiphloem.

epiphysial, *a. Anat.* Also (and now more commonly) in the form **epiphyseal.**
1842 *Penny Cycl.* XXIII. 135/2 Between the posterior extremity of the ilia and the first three caudal vertebræ, a distinct epiphyseal piece of bone is wedged in. **1871** J. Ashhurst *Princ. & Pract. Surg.* xxiii. 418 The epiphyseal cartilages (cartilages of conjunction) become enlarged. **1908** *Practitioner* Oct. 531 The slightest tap on the point of either of these fingers gave her intense pain, referred to a spot corresponding with the epiphyseal line. **1964** L. Martin *Clinical Endocrinol.* (ed. 4) iii. 124 Radiological examination will show the characteristic changes of epiphyseal dysgenesis.

epiphysis. Add: **3.** The pineal body of the brain. In full *epiphysis cerebri*.
1882 A. Thomson in *Quain's Elem. Anat.* (ed. 9) II. 831 (*heading*) Pineal gland. Epiphysis cerebri...This body is formed by an out-folding from the back part of the inter-brain roof, at a place where the opposite sides remain united by nervous matter afterwards giving rise to the pineal peduncles. **1887** A. C. Haddon *Introd. Study Embryol.* 199 Vertebral epiphyses are peculiar to Mammals. **1889** *Buck's Handbk. Med. Sci.* VIII. 175/1 The *epiphysis* (pineal gland) is not regarded as an important neural ingredient of the brain. **1951** O. Larsell *Anat. Nerv. Syst.* (ed. 2) xiv. 349 The pineal body or epiphysis..is considered by many students of the brain as a vestigial organ, related to the parietal eye of lower vertebrates.
4. In sea-urchins, a calcareous piece attached to each half of an alveolus.
1906 *Cambr. Nat. Hist.* I. 526 [Echinus] Each tooth is firmly fixed by a pair of ossicles.—Their upper ends are connected by a pair of ossicles called 'epiphyses'. These two epiphyses meet in an arch above. **1959** *Chambers's Encycl.* I. 592/2 The divergent ends of the limbs of each jaw are bridged by ossicles called epiphyses.

epiphysitis (epifisəi·tis). *Path.* [-itis.] Inflammation of an epiphysis (sense 1) or of the

cartilage which separates it from the main bone.
1874 in Dunglison *Dict. Med. Sci.* (new ed.). **1889** *Buck's Handbk. Med. Sci.* VIII. 346/1 The most common disease of the hip joint is tubercular epiphysitis, occurring in the growing bone of children. **1908** *Practitioner* Apr. 507 The left elbow was swollen, and the condition was thought to be an epiphysitis. **1970** Passmore & Robson *Compan. Med. Stud.* II. xxi. 20 Syphilitic infection of growing bones takes the form of epiphysitis and perichondritis.

epiplankton (epiplæ·ŋktǫn). *Zool.* [Epi- 1.] That portion of the plankton occurring from the surface of the sea to a depth of about one hundred fathoms. Hence **e:piplankto·nic** *a.*
1898 *Proc. Zool. Soc.* 13 Dec. 1024 As the *Collozoum* occurred in 30% of the Epiplankton hauls..I think we are justified in regarding it as essentially epiplanktonic. **1923** W. A. Herdman *Founders Oceanogr.* viii. 231 Epi-, Meso- and Bathy-plankton. **1967** *Oceanogr. & Marine Biol.* V. 320 The animal in question has a larval stage, or by epiplanktonic drift, in which mature stages are carried across the ocean adhering to floating material.

epiplasm (e·piplæz'm). *Biol.* [f. Epi- + Plasm.] Protoplasm remaining over after the formation of spores. Hence **epipla·smic** *a.*
1887 Garnsey & Balfour tr. *A. de Bary's Compar. Morphol. Fungi* iii. 77 After the orientation of the spores the protoplasm of the ascus shows the characters of a substance, for which I formerly proposed the name of epiplasm. **1906** *Cambr. Nat. Hist.* I. 96 (*Sporozoa*) Some of the cytoplasm of the original cells remains over unused, as 'epiplasm', and ultimately degenerates. **1947** *Nature* 11 Jan. 64/1 When the cytoplasm is cut off from the nuclear vacuole by the spore wall, volutin accumulates in the epiplasm.

epipod (e·pipǫd). *Zool.* Shortened form of Epipodite.
1893 T. R. R. Stebbing *Hist. Crustacea* 41 Huxley supposes that this valve may represent the epipod. **1911** *Encycl. Brit.* XVII. 457/1 In the maxillipeds and the trunk-legs it is common to find..a seven-jointed stem, the endopod, from which may spring two branches, the epipod from the first joint, the exopod from the second.

epipteric (epipte·rik), *a. Anat.* [f. Gr. ἐπί upon + πτερόν wing + -ic.] *epipteric bone* or *ossicle*, a small Wormian bone sometimes found between the parietal and the great wing of the sphenoid. Also as *sb.* = epipteric bone.
1879 W. H. Flower *Catal. Osteology* I. p. xi, The small supplementary ossifications at the upper end of the alisphenoid, which may be called epipterics. **1884** *Jrnl. Anat. & Physiol.* 219 Professor Flower has drawn attention to the fact, that..where the frontal and squamosal bones articulate..the union is brought about by means of a Wormian bone..uniting with the frontal or squamosal, this intercalary ossicle he termed the epipteric bone. **1902** Cunningham *Text-bk. Anat.* 127 They [*sc.* Wormian bones] occur commonly about the pterion, and in this situation are called epipteric bones (Flower). **1967** *Gray's Anat.* (ed. 34) 375 One or more [sutural bones], the pterion ossicles or epiteric [*sic*] bones, sometimes exist between the sphenoidal angle of the parietal bone and the greater wing of the sphenoid bone.

epipterygoid (epipte·rigoid). *Anat.* [Epi- 1.] A slender vertical bone which is situated above the pterygoid in the skull of certain living reptiles and of primitive tetrapods. Also *attrib.* or as *adj.*
1897 Parker & Haswell *Text-bk. Zool.* II. 297 Extending nearly vertically downwards from the pro-otic to the pterygoid is a slender rod of bone, the epipterygoid. **1912** J. S. Kingsley *Compar. Anat. Vertebr.* 82 An 'ascending process' which reaches the upper margin of the trabecula, and which, in many reptiles, often ossifies as the epipterygoid bone. **1956** A. S. Romer *Osteol. Reptiles* iii. 64 The epipterygoid extends as a rodlike structure toward the skull roof.

epipubic (epipiū·bik), *a. Anat.* [See next.] Situated on the pubis; of the nature of an epipubis; = Marsupial A. *adj.* 2 b.
1897 Parker & Haswell *Text-bk. Zool.* II. 162 [Elasmo-branchii] In some cases a median epipubic process projects forwards from the pelvic arch. *Ibid.* 489 [Marsupials, Echidna] With the anterior border of the pubes are articulated a pair of large epi-pubic or 'marsupial' bones. **1951** G. R. de Beer *Vert. Zool.* (ed. 2) xli. 390 Nearly all Marsupials have a marsupial pouch in the female and epipubic bones in both sexes.

epipubis (epipiū·bis). *Anat.* [Epi- 1.] A cartilage or bone in front of the pubis in amphibians, marsupials, or reptiles.
1897 Parker & Haswell *Text-bk. Zool.* II. 282 Attached to the anterior border of the pubic region there occurs in many Urodela and in Xenopus, a rod of cartilage, forked in front, the epipubis. **1942** L. H. Hyman *Compar. Vertebr. Anat.* (ed. 2) iv. 57 [In marsupials] this pouch..is supported by a V-like pair of marsupial bones (epipubes) projecting forward from the anterior end of the pelvic girdle.

Epirot. Restrict † *Obs. rare*[−1] to sense in Dict. and add: Also **Epirote** (ĭpəi·rōut). **b.** A native or inhabitant of ancient or modern Epirus in north-western Greece and southern Albania. Also as *adj.*
[**1642** Howell *For. Trav.* (1868) x. 50 You must go amongst the Mountaines and places of fastnesse, as the Epirotiques in Greece, the Heylanders in Scotland, the Brittaines in Wales.] **1835** W. M. Leake *Trav. N. Greece* ii. 68 The Epirotes seem to have been quite prepared to receive the Roman yoke. **1844** G. Finlay *Greece under Romans* iv. 411 Carians, Paionians, Epirots, and Macedonians. *Ibid.* 435 The slave population of Attica and Laconia were replaced by tribes of Epirot or Albanian peasants. **1915** J. Buchan *Thirty-Nine Steps* i. 21 He has a bodyguard of Epirotes that would skin their grandmothers. *Ibid.* iv. 82 Epirote guards. **1932** G. N. Cross *Epirus* i. 5 Theopompus..speaks of fourteen Epirot tribes. *Ibid.* ii. 20 In the days of King Pyrrhus..the Epirots were a united people. **1957** D. M. Nicol *Despotate of Epiros* i. 15 The native Epirotes..were a warlike people. *Ibid.* 17 The possession of most of the Epirote and Albanian coast-line and islands..had been granted to Venice.

epirrhema (epiri·mă). *Antiq.* [ad. Gr. ἐπίρρημα, f. ἐπί upon, after + ῥῆμα word, saying.] In the Attic Old Comedy, a speech addressed by the Coryphæus to the audience after the Parabasis. Hence **epirrhema·tic** *a.*
1835 W. R. Hamilton tr. *Süvern's Ess. Birds Aristoph.* 96 The second epirrhematic parabasis. **1887** *Amer. Jrnl. Philol.* VIII. 183 His [*sc.* Zielinski's] theory of the original 'epirrhematic' composition of a comedy as compared with the 'epeisodic' of a tragedy.

episcopalia (episkǒpēi·liă), *sb. pl.* [neut. pl. of late L. *episcopālis* Episcopal.] Episcopal belongings, e.g. vestments, buildings.
1865 Brande & Cox *Dict. Sci.* (ed. 3) I. 798/1 *Episcopalia* or *Onera Episcopalia*, in ecclesiastical history, synodals or other customary payments from the clergy to their bishop. **1903** *Daily Chron.* 23 Jan., To St. Joseph's College at Mill-hill testator left his episcopal vestments and other episcopalia. **1914** W. E. Beet *Med. Papacy* 13 Symmachus was a busy builder, and in his erection of *episcopalia* over against the vestibule of St. Peter's may be seen the first beginnings of the Vatican.

episcope[1] (e·piskōup). [f. Epi- + -scope.] A magic lantern for projecting images of opaque objects. So **episcopic** (-skǫ·pik) *a.* (see quot. 1911).
1909 in *Cent. Dict.* Suppl. **1911** *Encycl. Brit.* XVI. 186/1 If the light traverses the object, the projection is said to be diascopic, if by reflected light, episcopic. **1922** *Nature* CIX. 56/1 The episcope, shown by Newton and Co., is a marked improvement on forms previously seen, and should come into more general use for the projection on screens of opaque objects. **1937** [see *diascope].

episcope[2] (ĭpi·skŏpi). See *episcopy.

episcopy. Substitute for etym.: [ad. Gr. ἐπισκοπή watching over, f. ἐπί over + σκοπή watch.] Add: **1.** Later examples in form *episcopé*: spec., of pastoral government.
1957 *Relations betw. Anglican & Presb. Churches* 20 Among the functions of the ordained ministry..is..that of exercising oversight or 'episcopé'. Apostolic mission and authority continue to be exercised through this 'episcopé', which includes pastoral care, the continuance of the ordained ministry through ordination, and the guardianship of truth and exclusion of error. **1963** R. S. Louden *True Face of Kirk* iv. 39 In Presbytery..there has been gathered up this apostolic authority to ordain and induct—the oversight of the flock of God (*Episcopé*), regularly continued from Christian generation to generation, is corporately exercised and transmitted through Presbytery, in good church order.

episcotister (episkǫti·stəɹ). [f. Gr. ἐπισκοτίζειν to throw a shadow or darkness over, f. ἐπί upon + σκότος darkness: see -ist and -er[1].] An apparatus for admitting light into a darkened room by means of adjustable discs.
1905 Titchener *Exper. Psychol.* II. 80 The episcotister is placed as close as possible to the screen. **1939** *Brit. Jrnl. Psychol.* Oct. 140 Adjustment of the variable episcotister aperture.

episematic (episimæ·tik), *a. Biol.* [f. Epi- + Sematic *a.*] Designating natural colours, markings, etc., which serve to assist animals of the same species to recognize each other. (Opp. to *aposematic *a.*)
1890 E. B. Poulton *Colours of Animals* xvii. 337 Recognition Markings..assist an individual of the same species, and are termed Episematic. *a* **1908** —— *Ess. Evol.* 357 Episematic Colours are the logical antithesis of Aposematic, their object being to assist in keeping friends together instead of keeping enemies at a distance. **1926** A. S. Pearse *Animal Ecol.* ix. 299 Epigamic..might have been considered the same as episematic colors if they had not been believed to be concerned with..appreciation of beauty. **1934** J. S. Huxley in *Proc. 8th Internat. Ornith. Congr.* (1938) 454 Episematic function can also in many cases among birds be readily deduced for certain colour-patterns. **1959** Van Tyne & Berger *Fund. Ornith.* iii. 102 A means (especially in gregarious birds) of attracting the attention of other individuals of the same species...('Episematic' colors and patterns.).

episiotomy (episiǫ·tŏmi). *Surg.* [f. Gr. ἐπίσιον pubic region + -tomy.] Incision of the vulval orifice to facilitate parturition; an operation of this kind.
1878 A. E. Broomall in *Amer. Jrnl. Obstetrics* XI. 524 In 56 cases, when the perineum was unyielding,..I incised the vaginal orifice, and in nearly all of these cases I demonstrated..that episiotomy saved the perineum. **1953** Carter & Dodds *Dict. Midwifery* 166/1 A midwife may perform an episiotomy if she considers it to be in the interest of her patient. **1954** G. I. M. Swyer *Repr. & Sex* xi. 127 A properly timed and executed episiotomy will nearly always prevent a third degree tear.

episode. Add: **4.** Also in other musical forms (see quot. 1947[1]).
1947 *Penguin Music Mag.* Dec. 29 *Italian Symphony*. A fresh subject (in technical language an 'episode') is introduced. **1947** A. Einstein *Mus. Romantic Era* xvii. 311 It might even be called a rondo with varied theme and alternating episodes.

episome (e·pisōum). *Genetics.* [f. Epi- + *-some[4].] **a.** A small particle which together with a larger 'protosome' was postulated to constitute a gene, and which by its absence was held to cause gene mutation. (No longer current.)
1931 D. H. Thompson in *Genetics* XVI. 268 The gene [is pictured] as consisting of a main particle firmly anchored in the chromosome with varying numbers of one or more kinds of other particles attached. The main particle is called the protosome and the attached particles the episomes. **1966** E. A. Carlson *Gene* xiii. 111 Using the position effect hypothesis with Thompson's protosome-episome chain theory of the gene, Dobzhansky tried to account for the various Bar mutations and their reversions.
b. [ad. F. *épisome* (Jacob and Wollmann 1958, in *Compt. Rend.* CCXLVII. I. 154).] A particle sometimes found in cells, esp. in bacteria, which has a genetic function and which can either exist and multiply as an independent unit or become attached to a chromosome and multiply with it. Hence **episo·mal** *a.*, pertaining to or having the character of an episome; **episo·mally** *adv.*, by means of or as a result of episomes.
1960 *Bacteriol. Rev.* XXIV. 209/1 For genetic factors that are additions to..the genome,..Jacob and Wollman..have proposed the term 'episomes'. **1962** *New Scientist* 6 Sept. 514/3 An episome is a genetically active particle which can exist in two alternative forms. *Ibid.*, These ideas have led Jacob and Wollman to propose the 'episome' hypothesis of extra-chromosomal inheritance. Some examples of episomally controlled inheritance..would be those controlling non-essential cell functions. *Ibid.*, The episomal determinants of bacteria. **1965** *Times* 26 Feb. 15/4 They are known collectively as episomes, a term which conveys merely that they are particles additional to the chromosome. **1968** J. A. Serra *Mod. Genetics* III. xix. 24 Certain properties in bacteria are transmitted by the type of particle called an episome. *Ibid.* xxiii. 643 Variable location of the sex locus is better explained in terms of transposition of an episomal element, as in the dipteran *Megaselia scalaris*.

episperm (e·pispəɹm). *Bot.* [f. Gr. ἐπί upon + σπέρμα seed: see Epi- and Sperm *sb.*] The exterior covering of a seed. Hence **epi-spe·rmic** *a.*
1819 J. Lindley tr. *Richard's Observ. Fruits & Seeds* 17 The covering or integument peculiar to the seed being the most exterior of its constituent parts, I give it the name of Episperm. *Ibid.* 32 If..the entire kernel..is immediately covered over by the episperm, it is called epispermic. **1861** R. Bentley *Man. Bot.* 337 The terms most frequently used, are *testa* or *episperm* for the outer coat.

epispore. Add: **2.** *Zool.* In Sporozoa (see quot.).
1903 E. R. Lankester *Zool.* I. II. 189 The spore-envelope or sporocyst [in Gregarines] consists of two layers, an outer clear and delicate epispore, and an inner refringent and tough endospore.

epistasis (epi·stăsis). [Gr. ἐπίστασις a stopping, stoppage, (also) scum on urine, f. ἐφιστάναι to stop, check.] **1.** *Med.* **a.** The checking of any discharge, as of blood or menses. **b.** A pellicle that forms on the surface of urine after it has stood. *rare*[−0]
[**1849** F. Adams *Genuine Wks. Hippocrates* II. Aphorisms vii. no. 35 When the scum on the surface is fatty and copious, it indicates acute diseases of the kidneys. [*Note*] It appears from Galen, that in some copies he found *epistasis*, and in others *hypostasis*; the one evidently referring to the scum on the surface, and the latter to the sediment: he favours the former reading, and understands it to refer to melting of the fat in the neighbourhood of the kidneys.] **1807** Morris & Kendrick *Edin. Med. & Physical Dict.* II. Gloss., *Epistasis*, see Epischesis [= 'a suppression of excretions']. Also the substance on the surface of the urine. **1848** in Dunglison. **1889** in *Cent. Dict.* **1934** in Webster; and in mod. med. dicts.
2. *Genetics.* [Back-formation from *epistatic *a.*] The expression of one hereditary character to the exclusion of another when the two are controlled by alleles of different genes; also more widely, any interaction of non-allelic genes.
1917 *Genetics* II. 613 (*index*) Epistasis. **1932** Sinnot &

Dunn *Princ. Genetics* (ed. 2) v. 82 Epistasis..is similar to dominance except that it occurs between different factors instead of between the two members of an allelomorphic pair. **1965** J. A. Serra *Mod. Genetics* I. iii. 63 Some of the most striking examples of inhibitor-type epistasis are provided by genes which control pigmentation in animals or plants. **1966** *McGraw-Hill Encycl. Sci. & Technol.* II. 327 Each gene may have favorable effects in some combinations and unfavorable effects in others, such as over-dominance and many kinds of epistasis, so that its average effect may be small.

epistasy (epi·stăsi). *Genetics.* [ad. Gr. ἐπιστασία dominion: see -y[3] and cf. prec.] = *epistasis 2.
1918 R. A. Fisher in *Trans. R. Soc. Edin.* LII. 404 A similar deviation from the addition of superimposed effects may occur between different Mendelian factors. We may use the term Epistacy [*sic*] to describe such deviation. **1949** Darlington & Mather *Elem. Genetics* vii. 157 Epistasy comes in at all stages of development, early and obvious or late and inferential.

epistatic (epistæ·tik), *a. Genetics.* [prob. f. Epi- + Gr. στάσις standing, position, state, after Hypostatic *a.*, Hypostasis; but cf. Gr. ἐπιστάτης one who is set over (see Epistates), ἐπιστασία dominion; a derivation directly from ἐπιστάτικος of government, (also) standing still, or from ἐπίστασις Epistasis is unlikely.] Of, exhibiting, or caused by epistasis (sense 2). Const. *to*, *on*, *over*. (Opp. to *hypostatic*.)
1907 W. Bateson in *Science* 15 Nov. 653/2 Till lately we spoke of the relations between the gray color of the mouse to the black color in terms of dominance. Those terms, strictly speaking, should only be applied to members of the same allelomorphic pair. We can perhaps best express the relation between the gray and the black by the use of the metaphor 'higher and lower', and I therefore suggest the term *epistatic* as applicable to characters which have to be, as it were, lifted off in order to permit the lower or *hypostatic* character to appear. **1909** —— *Mendel's Princ. Heredity* ii. 41 (*Primula*). The magenta shades have a factor epistatic to crimson and pink. *Ibid.* v. 98 The pale ivory is due to a factor epistatic on the bright yellow. **1910** L. Doncaster *Heredity* vi. 75 The presence of a higher member of the series obscures or prevents the development of the lower. This is expressed by saying that grey is 'epistatic' over black and chocolate, and black over chocolate. **1949** Darlington & Mather *Elements of Genetics* vii. 157 A gene removing an organ is..obviously epistatic to one modifying that organ. **1968** M. W. Strickberger *Genetics* xv. 239 Epistatic or interaction effects are also known..in which the phenotypic expression caused by one pair of genes depends upon alleles at another gene pair.

epistemic (episte·mik, -ī·mik), *a. Philos.* [f. Gr. ἐπιστήμη knowledge + -ic.] Of or relating to knowledge or degree of acceptance. Hence **episte·mically** *adv.*
1922 W. E. Johnson *Logic* II. p. xix, The charge of circularity or petitio principii is epistemic. **1933** *Mind* XLII. 82 Proof of the actual existence of a material world epistemically depends on just these elements of confusedness and involuntariness characteristic of our sense-awareness. **1936** *Mind* XLV. 270 The analysis of propositions so as to exhibit their reference to the qualitative continuum is called *epistemic* reduction. **1953** K. Britton *J. S. Mill* vi. 188 The Experience Philosophy arranges judgements in an 'epistemic order'.

epistemological, *a.* (Earlier and later examples.)
1884 *Mind* VIII. 16 Considerations which are plainly extra-logical and can only be called epistemological. **1895** W. James *Coll. Ess. & Rev.* (1920) 399 No conventional restrictions *can* keep metaphysical and so-called epistemological inquiries out of the psychology books. **1955** *Sci. Amer.* Apr. 100/3 This epistemological problem is best solved by following the classical dictum, 'If you can't beat 'em join 'em'.

epistemolo·gically, *adv.* [-ly[2].] In an epistemological manner; with reference or in regard to epistemology.
1882 *Mind* VII. 535 Epistemologically irrelevant. **1906** *Westm. Gaz.* 9 June 3/2 Epistemologically..Von Hartmann is a transcendental realist who ably defends his views. **1919** *Mind* Jan. 4 There is nothing which I believe to be epistemologically more unsound than this identification of the knower's knowledge or experience with the reality of the object he knows. **1963** J. Lyons *Structural Semantics* iii. 43 A psychologically and epistemologically satisfying theory of our understanding of other languages.

epistemologist (e:pistimǫ·lŏdʒist). [-ist.] One who is versed in epistemology.
a **1897** W. Wallace *Lect. Nat. Theol.* (1898) 36 The epistemologist..examines the history of our ordinary physical knowledge. **1902** *Encycl. Brit.* XXXII. 55/2 It is only epistemologists, and notably Kant, who so describe individual experience. **1925** J. E. Turner *Theory Direct Realism* 34 This systematic connection is ignored by Stout, as also by epistemologists in general. **1964** E. A. Nida *Toward Sci. Transl.* iii. 33 A problem with which epistemologists since Descartes have continued to wrestle.

epistom(e. Add: The Latin form *epistoma* is also used. Also applied to similar parts in arachnids, molluscs, etc.
1860 Mayne *Expos. Lex.* Add. 1491/2 *Epistoma, epistomis, epistomium*, names for the space between the buccal cavity and origin of the intermediary *antennæ* in the *Crustaceæ Maxillares*; also for the hood of the *Condylopa*. **1880** T. H. Huxley *Crayfish* 155 In front of the mouth, the sternal region which appertains, in part, to the antennæ, and, in part, to the mandibles, is obvious as a broad plate..termed the *epistoma*. **1897** Parker & Haswell *Text-bk. Zool.* I. 328 (Molluscs), A small lobe—the epistome..—overhangs the mouth and lies between it and the anus. *Ibid.* 499 The head exhibits no segmentation: its sternal region is formed largely by a shield-shaped plate, the *epistoma*, nearly vertical in position. **1909** *Cambr. Nat. Hist.* IV. 291 In *Pterygotus* there is a well-developed epistome..between the mouth and the front margin of the carapace. **1957** Richards & Davies *Imms's Textbk. Ent.* III. 587 In holoptic flies.. The epistoma is the distal border of the face. *Ibid.* 752 Among Curculionoidea the reduced fronto-clypeal region is often termed the epistoma.

episyllogism. Add: (Earlier examples.)
1837–8 W. Hamilton *Logic* (1860) I. 364 That syllogism which contains the consequent of another, is called an *Episyllogism*. **1851** H. L. Mansel *Proleg. Logica* 218 The subsequent consequence is related as an episyllogism.
Hence **episyllogi·stic** *a.*, pertaining to or characterized by an episyllogism.
1886 P. K. Ray *Text-bk. Deductive Logic* (ed. 2) III. vi. 216 The first form is called Synthetic, Progressive, or Episyllogistic. **1896** J. Welton *Man. Logic* (ed. 2) I. 391 The progress of thought has been from prosyllogism to episyllogism. Such a demonstration is called *Progressive, Episyllogistic*, or *Synthetic*.

epitaxy (epitæ·ksi, e·pitæksi). Also in mod.L. form **epita·xis.** [ad. F. *épitaxie* (L. Royer 1928, in *Bull. de la Soc. française de Min.* LI. 8), f. Epi- + Gr. τάξις arrangement: see -y[3].] The growth of crystals on a crystalline substrate that determines their orientation; the orientation of crystals so grown.
[**1929** *Mineral. Abstr.* IV. 175 'Épitaxie' is possible when the reticular distances are approximately identical only in two dimensions and not in three.] **1931** *Ibid.* 566 (*index*) Epitaxy. **1934** *Ibid.* V. 495 For epitaxy there is required both geometrical agreement and similarity of polarity between particles which replace one another at the points of union. **1949** *Proc. Physical Soc.* LXIIA. 344 (*heading*) The structure and epitaxy of lead chloride deposits formed from lead sulphide and sodium chloride. **1951** H. E. Buckley *Crystal Growth* x. 382 The formation of small crystal elements of the impurity growing, on account of dimensional similarity, in parallel position on the plane in the host crystal (epitaxis). **1961** *New Scientist* 26 Oct. 244/1 Epitaxis. This is the latest weapon in the semiconductor engineers' armoury.
Hence **epita·xial** *a.* [-al], grown by, characterized by, or resulting from epitaxy: used of crystals, crystal growth, and crystal orientation; **epita·xially** *adv.*, so that epitaxy occurs; by an epitaxial process.
1949 *Proc. Physical Soc.* LXIIA. 352 When strong epitaxial orientation occurs there are..usually two or more lattice row types which are parallel in the substrate and overgrowth crystals. **1960** *Proc. IRE* XLVIII. 1642/1 Epitaxial Diffused Transistors. This letter describes diffused base transistors, the structures of which incorporate thin semiconductor layers epitaxially deposited on low-resistivity substrates of the same semiconductor. **1960** *New Scientist* 25 Aug. 504/3 Epitaxial crystal growth (in which the growing crystal has the same orientation as the 'seed'). **1962** L. P. Hunter *Handbk. Semiconductor Electronics* (ed. 2) iv. 46 The transistors currently referred to as 'epitaxial' are diffused junction transistors made on epitaxially deposited material. **1966** *New Scientist* 30 June 848/1 It is now also possible to grow epitaxial silicon crystal layers on to a single-crystal silicon substrate.

epithalamus (epiþæ·lămŏs). *Anat.* [f. Epi- 1 + Thalamus (sense 1).] A part of the diencephalon (see quot. 1962).
1902 Cunningham *Text-bk. Anat.* 501 The epithalamus, which comprises the pineal body and the habenular region. **1921** A. Keith *Hum. Embryol.* (ed. 4) 103 In the upper region will be differentiated the optic thalamus, the epithalamus (the pineal body with its ganglia and commissures) and the metathalamus or geniculate bodies. **1962** *Gray's Anat.* (ed. 33) 1033 The epithalamus comprises the trigonum habenulae, the pineal body, and the posterior commissure.

epithallus (epiþæ·lŏs). *Bot.* [f. Epi- 1 + Thallus.] (See quots.)
1879 W. A. Leighton *Lichen-flora Gt. Brit.* (ed. 3) 510 *Epithallus*, cortical layer. **1911** *Encycl. Brit.* XVI. 579/1 (Lichens), The term *epithallus* is sometimes applied to the superficial dense portion of the cortical layer.
So **epithalline** (epiþæ·ləin) *a.*, situated or growing upon the thallus.
1943 Ainsworth & Bisby *Dict. Fungi* 112 *Epi-thalline*, of a falsely thalline apothecial edge in lichens.

epithecium (epiþī·ʃiŏm). *Bot.* [mod.L., f. Epi- 1 + Thecium.] The surface layer of the fruiting body in certain lichens and fungi. So **epithe·cial** *a*
1879 W. A. Leighton *Lichen-flora Gt. Brit.* (ed. 3) p. ix, The *paraphyses* are slender filaments..which, by their close proximity, constitute the disk or *epithecium*. **1921** A. L. Smith *Lichens* iv. 158 In some genera [of lichens] there is a profuse branching of the paraphyses to form a dense protective epithecium. **1952** C. J. Alexopoulos *Introd. Mycology* xii. 290 In some apothecia, the paraphyses..have their tips united above the asci, forming a layer we call the epithecium. **1967** V. Ahmadjian *Lichen Symbiosis* ii. 17 Some lichens have algal cells within their fruiting tissues..either hymenial or epithecial algae, depending on their location within the fruiting body.

epitheliomatous (e:piþiliōu·mătəs), *a. Path.* [-ous.] Pertaining to, or of the nature of, epithelioma.
1881 *Trans Obstetr. Soc. Lond.* XXII. 7 The disease was an ordinary epitheliomatous cervix. **1908** *Practitioner* Sept. 355 A squamous-celled epitheliomatous ulcer.

epitheted, *ppl. a.* Add: **c.** Designated by an epithet too coarse or violent to repeat. *jocular*.
1896 *Punch* 21 Nov. 241/2 Mr. Jones's compliments, Sir, and *when the epitheted substantive* is this noise going to stop? **1905** *Daily Chron.* 1 July 4/4 No London fire engine has such precedence as the 'Varsity boat. If you make it 'easy all' with your epithetted clumsiness in a boat—it is ten-and-sixpence.

epitokous (epi·tokəs), *a.* [ad. G. *epitok* (E. Ehlers *Die Borstenwürmer* (1868) I. 453), f. Gr. ἐπίτοκος fruitful, bearing offspring, f. ἐπί Epi- + τόκος birth.] Of or relating to the **epitoke**, the posterior sexual part of the body of certain polychætous worms.
1896 W. B. Benham in *Cambr. Nat. Hist.* II. x. 276 Polychaeta...The immature forms may become ripe and lay eggs while still retaining the 'Nereid' characteristics, or these immature forms may become 'Heteronereids'...Ehlers employed the term 'epitokous', whilst he called the 'Nereid' phase 'atokous'. **1902** *Boston Soc. Nat. Hist.* XXIX. 399 The Heteronereis (epitokous form of Ehlers) is not uncommon. **1911** J. A. Thomson *Biol. Seasons* III. 261 In some of our British shore-worms..a remarkable change occurs in the body at the breeding season. So striking is the transformation in some instances, that the breeding ('epitokous') phase has been mistaken for a distinct species. **1916** H. S. Pratt *Man. Invertebr. Anim.* 283 In many (Polychæta) the anterior part of the body is sexless and is called the atoke, while the hinder part is sexual and is called the epitoke. **1963** R. P. Dales *Annelids* viii. 165 Durchon has shown that immature worms are prevented from metamorphosing into epitokous individuals by inhibitory hormones produced in the brain.

epitoxoid (epitǫ·ksoid). [Epi- 1.] A toxoid which has less affinity than the toxin for the corresponding antitoxin.
1900 in Dorland *Med. Dict.* **1903** [see Toxoid].

epitrichium (epitri·kiŏm). *Anat.* [mod.L., f. Gr. ἐπί upon + τρίχιον, dim. of θρίξ hair.] A thin cellular membrane which overlies the epidermis and hair during fœtal life, usually disappearing before birth. Hence **epitri·chial** *a.*
1882 *Proc. Boston Soc. Nat. Hist.* XXI. III. 301 The surface is covered by a thin 'epitrichial' layer of large, flat, polygonal cells more or less granular, with inconspicuous nuclei. **1885** *Science* VI. 226/1 A new membrane of the human skin, which he homologizes with the epitrichium of the lauropsida. **1887** A. C. Haddon *Introd. Study Embryol.* 100 The epidermis of Amphioxus permanently remains as a single layer. In all other embryo Vertebrates, the epiblast, from being single, becomes double layered, owing to the primitive epiblast giving rise to a layer of flattened epithelial cells, the epitrichial layer. **1913** A. Keith *Hum. Embryol.* (ed. 3) 451 The epitrichium,..so named because the hairs are developed beneath it, and when they grow out in the sixth month this surface layer of flat epithelium is shed. **1962** *Gray's Anat.* (ed. 33) 138 The ectoderm at first consists of a single stratum of cells, but in the sixth week two strata can be recognised, a superficial, named the epitrichium, consisting of flat cells, the nuclei of which stain readily, and a deep, named the stratum germinativum.

epitrochoidal *a.* (Later examples.)
1959 *Times* 15 Dec. 7/1 The shape of the bore in which the triangular rotor revolves is described as epitrochoidal: the shield-shaped piston moves round the inside of a cavity like the hollow skin of an Edam cheese. **1969** *Observer* (Colour Suppl.) 23 Mar. 27/1 Take the revolutionary Wankel engine. Twin rotors in an epitrochoidal bore. That's two rotors going round and round, instead of four or more pistons going up and down.

epizootic, *a.* **1.** (Earlier example.)
1790 *Gentl. Mag.* June 496/2 Diseases are not so prevalent amongst our cattle (at least *epizo-otic* diseases are not)...Epizo-otic diseases are, in the brute creation, what epidemic diseases are in men.

epoch. Add: **7. epoch-making** *a.* (earlier example); now extended to designate any remarkable or sensational event, publication, etc.; **epoch-marking** *a.*, journalistic alteration of *epoch-making*.
1863 *Athenæum* July 9/1 He has produced what the Germans call an *epoch-making* book. **1895** *Westm. Gaz.* 15 Jan. 2/2 Every author of an epoch-making or epoch-marking book is liable to pass through two stages. **1928** R. Campbell *Wayzgoose* ii. 59 He produced an epoch-making article. **1919** J. L. Garvin *Econ. Found. Peace* 272 Consent by the United States to administer Constantinople and the Straits, Armenia and Palestine, would be an epoch-marking step in itself. **1923** *Daily Mail* 16 Jan. 7 This epoch-marking experiment. **1940** Graves

& Hodge *Long Week-End* vi. 93 Most of these inventions, all described as 'epoch-making', were never heard of again after the first news-thrill. **1963** *Times Lit. Suppl.* 4 Jan. 6/2 The evil of the Fulton speech..was not that it was 'epoch-making' but that it was 'epoch-marking'.

‖ **éponge** (epon̄ʒ). [Fr., = sponge.] Sponge cloth (cf. *RATINE).

1928 *Daily Express* 11 July 5/5 For a bathing-coat try to find flowery éponge or any other bright fabric. **1952** C. W. Cunnington *Eng. Wom. Clothes Pres. Cent.* 294 *Eponge*, a soft, loose fabric of cotton, wool or silk; similar to ratine.

eponychium (eponi·kiə̆m). *Anat.* [mod.L., f. Gr. ἐπί upon + ὄνυξ nail.] **a.** The horny embryonic structure whence the nail is developed. **b.** A film of epidermis covering the dorsal part of the nail; = Quick *sb.*¹ 4.

1885 tr. P. G. Unna in H. Ziemssen *Handbk. Dis. Skin* i. 23 A trace of the horny layer of the first foetal months remains only on the ungual phalanx till a later period, and to this..I have given the name 'eponychium'. **1892** C. S. Minot *Hum. Embryol.* 555 The epitrichial layer over the area has received the special name of *eponychium* from Unna. **1902** Cunningham *Text-bk. Anat.* 734 The part [of the epitrichium] which persists over the root of the nail is termed the eponychium, and covers the proximal part of the lunula. **1968** Bloom & Fawcett *Textbk. Histol.* (ed. 9) xxii. 495 Under the proximal fold, the horny layer spreads onto the free surface of the nail body as the eponychium.

epoophoron (epo,ǫ·fŏrǫn). *Anat.* [mod.L., f. Ep- + *oophoron* ovary (f. Gr. ᾠόν egg + -φόρος bearing, bearer).] = Parovarium.

1883 Foster & Balfour *Elem. Embryol.* (ed. 2) 224 The sexual part becomes in the cock the after-testes or coni vasculosi...In the hen it forms part of the epoophoron of Waldeyer, and is composed of well-developed tubes with yellow pigment. **1887** *Buck's Handbk. Med. Sci.* V. 520/2 The parovarium (Syn., paroöphoron, epoöphoron, corpus pampiniforme (Wrisberg), organ of Rosenmüller) in the human female, is a structure having a vertical diameter of 15 to 25 mm. **1962** *Gray's Anat.* (ed. 33) 1538 The epoöphoron lies in the lateral part of the mesosalpinx between the ovary and the uterine tube.

epoxide (ĭpǫ·ksəid). *Chem.* [f. Epi- + Oxide.] A chemical compound the molecules of which contain an oxygen atom attached to two carbon atoms directly or indirectly bonded to one another; a cyclic ether. Also *attrib.*, as *epoxide resin* = *epoxy resin.*

1930 [see *EPOXY *prefix* and quasi-*adj.*]. **1938** Allen & Blatt in H. Gilman *Org. Chem.* I. vi. 545 Peracetic acid and perbenzoic acid form epoxides by addition of an atom of oxygen to ethylenic linkages. **1950** *Chem. Abstr.* XLIV. 3298 *Epoxide resins*...Products for enamels, paints, adhesives, films, molded articles, etc. **1958** *Engineering* 31 Jan. 7/3 (Advt.), The epoxy or epoxide resins are among the newest and most versatile modern plastics.

Hence **epoxida·tion,** the formation of an epoxide by the addition of an atom of oxygen to two doubly bonded carbon atoms; **epo:xidi·zable** *a.*, capable of undergoing epoxidation; **epo·xidize** *v.*, to convert into an epoxide by means of epoxidation; so **epo·xidized, epo·xidizing** *ppl. adjs.*

1944 *Jrnl. Amer. Chem. Soc.* LXVI. 1927/2 The epoxidation of oleic acid. **1945** *Ibid.* LXVII. 414/2 Epoxidized oils, a new class of chemical reaction products from triglycerides, have been prepared. **1946** *Ibid.* LXVIII. 1507/2 The same olefins were epoxidized with peracetic acid in acetic acid solution. **1958** *Industrial & Engin. Chem.* June 873/1 Use of this versatile peracetic acid as the epoxidizing agent results in greater freedom from undesirable side reactions. **1960** J. G. Wallace in Kirk & Othmer *Encycl. Chem. Technol.* Suppl. II. 327 Epoxides may be produced by decomposing the hemiperoxyacetate in the presence of an epoxidizable olefin. *Ibid.* 329 Epoxides..may also be obtained by reactions not involving epoxidation. **1963** W. R. Moore *Introd. Polymer Chem.* vi. 178 New epoxy resins have recently been produced in which the epoxide groups are obtained by epoxidation of a carbon-carbon double bond.

epoxy (ĭpǫ·ksi), *prefix* and quasi-*adj.* *Chem.* [f. Epi- + Oxy-.] **A.** *prefix.* An element used in the names of chemical compounds that are epoxides. **B.** Hence as quasi-*adj.*: pertaining to or deriving from an epoxide; containing the ring structure characteristic of epoxides; **epoxy resin**, any of various synthetic polyether resins that contain epoxy groups and benzene rings, are usu. made by polymerizing an epoxide, and are convertible into a thermoset form.

1916 *Chem. Abstr.* X. 3400/2 Epoxy, —O— (to different atoms). **1930** *Jrnl. Amer. Chem. Soc.* LII. 1191 The epoxides of oleic and elaidic acids have been obtained in crystalline form... Three directions in which these epoxy acids decompose on heating have been discussed. **1945** *Ibid.* LXVII. 412/1 Peracetic acid functions as an oxidizing agent..to yield epoxy compounds, but..in acetic acid solution the epoxy ring is opened and mixtures of hydroxy-acetoxy compounds are obtained. **1950** S. O. Greenlee *U.S. Pat.* 2,494,295 5 The complex epoxides are in general resinous in character...The complex epoxy resins are soluble, unless too highly polymerized, in solvents such as acetone. **1960** 'A. Burgess' *Doctor is Sick* viii. 58 If only the head could be, perhaps painlessly, severed and then, with some epoxy resin or other, fitted back. **1960** *Handbk. Chem. Soc. Authors* iii. 108 Oxygen linked, in a chain of carbon atoms, to two of these atoms will be denoted by the prefix epoxy in all cases where it would be unprofitable to name the substance as a cyclic compound. Examples: ethylene oxide = epoxyethane. **1966** *McGraw-Hill Encycl. Sci. & Technol.* X. 469/1 The epoxy resins form a relatively new class of cross-linked polyethers characterized by excellent chemical resistance, adhesion to glass and metals, electrical insulating properties, and ease and precision of fabrication. **1969** *Jane's Freight Containers 1968–69* 21 (Advt.), An extensive range of abrasion resistant phenolic, vinyl and epoxy linings with non-toxic, odourless and tasteless characteristics.

‖ **épris** (epri), *a.* fem. **éprise** (-z). [Fr.] Enamoured.

1793 Dr. Burney *Let.* May in *Diary Mme. d'Arblay* (1842) V. 426, I have for some time seen..that you are *éprise*. **1848** Thackeray *Van. Fair* lxi. 555 That Major seems to be particularly *épris*. **1850** C. M. Yonge *Kenneth* viii. 90 You are quite *épris* with those children. **1893** Beerbohm *Let.* 13 Oct. (1964) 75 Nor need you have said that you have been *épris* of little artistes yourself. **1917** A. Huxley *Let.* 22 Nov. (1969) 138 Who is Flora Fo[r]ster? I am so much épris with her poem that I shall write to her. **1949** 'C. Hare' *When Wind Blows* 54 As if anybody else could not see that Mr. Ventry is *épris* of Nicola. **1949** A. Wilson *Wrong Set* 155 He would have felt it necessary to come to my defence, he was so completely *épris*.

epsilon (epsəi·lǫn). [Gr. ἒ ψιλόν, lit. 'bare e', i.e. 'e and nothing else', = short e written ε and not αι.] The fifth letter of the Greek alphabet (Ε, ε); *Astr.* denoting the fifth brightest star in a constellation.

c **1400** Maundev. (1839) iii. 20. **1842** *Penny Cycl.* XXII. 448/2 The following stars, γ Leonis, ε Bootis, ζ Herculis, δ Serpentis, and γ Virginis, were made out to be revolving double stars, by W. Herschel, in 1803. **1930** A. J. Quirke *Forged, Anon. & Suspect Docs.* v. 77 Affectation of letter-forms of unconventional type...A typical example is the use of the Greek 'epsilon' for small *e*. *Ibid.* vii. 100 Examination of small 'e'...If Greek 'epsilon' form is used, [etc.]. **1931** *Times Lit. Suppl.* 1 Jan. 10/3 The epsilon-shaped E [i.e. ε]. **1959** *Listener* 3 Dec. 971/1 An interesting system is that of Epsilon Lyræ.

2. a. An examiner's fifth-class mark.

1907 [see *ALPHA 4].

b. A person of low intelligence.

1932 A. Huxley *Brave New World* i. 15 'In Epsilons,' said Mr. Foster very justly, 'we don't need human intelligence.' *Ibid.* iv. 69 The liftman was a small simian creature, dressed in the black tunic of an Epsilon-Minus Semi-Moron. **1958** *Listener* 11 Sept. 374/2 With the people down below—'the epsilons'—enjoying very little—do you think this kind of situation can long endure?

Epsom. Add: **3.** The racecourse on Epsom Downs in Surrey, where the Derby and the Oaks are run; the principal race-meeting held there.

c **1810** W. Hickey *Mem.* (1960) 326 We all went..to Epsom races. **1827** P. Egan *Anecdotes Turf* 53 A young man had once the ambition of riding over Epsom against Dick. **1838, 1867** [see *Derby 1 b]. **1880** Trollope *Duke's Children* I. xvii. 202 It was impossible that Lord Silverbridge should separate himself from the Major,—at any rate till after the Epsom meeting.

epuration. Add: Usu. in Fr. form *épuration* (epürasioṅ): The removal of politicians or officials from office, or their prosecution (in France); a purge.

1883 *Sat. Rev.* 20 Oct. 489/1 There is..no reason to suppose that any section of the French people..has been offended by..the *épuration* of its magistracy. **1948** W. Stevens *Let.* 7 Jan. (1967) 573 The *épuration* has always been something that I have not been able to follow very attentively...I have always thought that Petain..was quite badly treated.

epyllion (epi·liǫn). Pl. **epyllia** (-iă). [Gr. ἐπύλλιον dim. of ἔπος Epos.] In classical use: see quot. 1949. Also applied to post-classical poems of short or moderate extent resembling an epic in style or matter; a miniature epic.

The word seems to have been revived by German scholars in the mid-19th c.: see *Class. Jrnl.* XLIX (1953–4), 111 ff.

1876 R. Ellis *Commentary Catullus* 228 Unfortunately the contemporaneous epyllia, the Io of Calvus, the Smyrna of Cinna, the Glaucus of Cornificius, have perished. **1890** A. S. Wilkins *Rom. Lit.* 75 The fashion at this time was to substitute for long and tedious epics brief and highly finished pictures of episodes or idylls (*epyllia*). **1895** J. W. Mackail *Latin Lit.* II. i. 104 Pieces in hexameter verse, belonging broadly to the class of the *epyllion*, or 'little epic', which was invented as a convenient term to include short poems in the epic metre that were not definitely pastorals either in subject or treatment. **1931** M. M. Crump (*title*) The epyllion from Theocritus to Ovid. **1949** *Oxf. Class. Dict.* 335/1 *Epyllion*, a literary type popular from Theocritus to Ovid, was a narrative poem of about 100 to 600 hexameters; the subject was usually taken from the life of a mythical hero or heroine, the love motif being prominent in later epyllia. **1952** C. S. Lewis *Hero & Leander* 24 Marlowe's success is most easily seen if we compare him with other sixteenth-century specimens of the erotic epyllion.

equal, *a.* **2.** (Additional examples.)

1834 *N.Y. Evening Post* 4 Apr. 2/2 A war of exclusive privileges against equal rights. **1889** C. M. Yonge in *Monthly Packet* Christmas No., 31 It was of no use to talk to an old wretch like that about social movements and equal rights. **1923** *Management Engin.* IV. 343/2 *Equal Pay*, a principle whereby, given a specific occupation, the rate of pay should be the same for both sexes, output being the same. **1956** *Ann. Reg. 1955* 3 On 25 January Mr. Butler announced that an agreed basis had been found for the introduction of 'equal pay' (for both sexes) in the non-industrial civil service.

equalita·rianism. [-ism.] Belief in the natural equality of mankind.

a **1866** J. Grote *Moral Ideals* (1876) 409 The saying that the happiness of each *man* is to be consulted alike..: we will call it *equalitarianism*. **1920** H. G. Wells *Outl. Hist.* 512/2 The dominant liberal ideas were freedom and a certain vague equalitarianism. **1921** *Q. Rev.* July 140 In fifth-century Athens, where equalitarianism ran to such a ridiculous extreme that gradually all civil offices were distributed among the citizens by lot. **1922** G. M. Trevelyan *Brit. Hist. 19th Cent.* v. 84 Equalitarianism on board ship is proverbially dangerous.

equality. Add: **6. Equality State,** popular name in the U.S. for Wyoming, the first state to introduce women's suffrage.

1891 M. F. Sweetser *King's Handbk. U.S.* 904 The Commonwealth [*sc.* Wyoming] has been called the equality state, because ever since its organization men and women have been accorded equal rights to vote. **1963** R. I. McDavid Jr. *Mencken's Amer. Lang.* x. 700 Wyoming is the *Equality* or *Suffrage State*, so called because its Territorial Legislature made the first grant of the suffrage to women voters in 1869.

equalize, *v.* Add: **4. b.** *absol.* To make the score (in Football, etc.) equal to that of the opponent.

1925 S. J. Southerton in Marshall & Tosswill *Football* (rev. ed.) xviii. 318 Soon after half-time Millington should have equalised from a penalty kick. **1937** *Times* 15 Feb. 7/3 The South were not long in equalizing.

7. c. *Electr. trans.* To correct or modify (a signal or frequency response) by the use of an equalizer. Also *intr.*, to provide compensation (in the shape of an equalizer) *for*.

1928 Sterling & Kruse *Radio Man.* xi. 385 These circuits are good for equalizing frequencies up to 3000 and 5000 cycles respectively. **1943** F. E. Terman *Radio Engineers' Handbk.* iii. 244 For most purposes it is sufficient to equalize only for attenuation. **1958** *Van Nostrand's Sci. Encycl.* 606/1 The equalizer shows a frequency response which is the inverse of the system it is intended to equalize so the result..is to restore the overall response to a flat characteristic. **1962** A. Nisbett *Technique Sound Studio* iv. 80 When the high frequency losses are progressive, measures may be taken to equalize for them.

8. *intr.* To become equal.

1906 *Amer. Naturalist* June 427 The temperature in the outer and inner layers had equalized to a large extent. **1971** M. G. Scroggie *Found. Wireless & Electronics* xix. 324 These pressures tend to equalize.

equalizer. Add: **b.** *Electr.* A passive network designed to modify the frequency response of a circuit (as a transmission line or an amplifier), esp. in order to compensate for frequency-dependent attenuation or phase shifts.

1928 Sterling & Kruse *Radio Man.* vi. 222 The equalizer is employed to correct the frequency characteristic of the telephone line by forming a shunt which affords a variable impedance for different frequencies. **1930** *Bell Syst. Techn. Jrnl.* IX. 587 Since the amplifier has a flat gain characteristic, an attenuation equalizer is called for to correct the distortion introduced by the cable. **1943** F. E. Terman *Radio Engineers' Handbk.* iii. 244 A phase equalizer is an all-pass filter designed to introduce a desired phase shift as a function of frequency in the load current. **1958** *N.Z. Listener* 26 Sept. 9/2 Here also are the variable equalisers that compensate for the frequency losses in the lines to the recording centre from outside points. All the lines come to the central rack and the equalisers there, except the fixed programme lines which go to fixed equalisers in the main equipment room. **1958** [see prec. 7 c].

c. *Football.* A goal that equalizes (cf. *EQUALIZE *v.* 4 b).

1930 *Daily Express* 9 Sept. 12/2 Wednesday were playing very well..and Rimmer headed the equaliser just inside half an hour. **1960** *Times* 25 Jan. 16/5 Jackson scored a scrambling equalizer.

d. A revolver. Also, a cosh; (occas.) some other weapon. *slang* (orig. *U.S.*).

1931 D. Runyon *Guys & Dolls* (1932) iv. 89 He outs with the old equalizer and starts blasting away. **1954** Wodehouse *Jeeves & Feudal Spirit* xvi. 162, I reached a hand into my pocket and got a firm grasp on the old Equalizer. **1958** B. Hamilton *Too Much of Water* xi. 252, I..got..my little equalizer...I gave him a very sound bang, and he just slithered down. **1961** I. Jefferies *It wasn't Me!* xi. 157 He just thought anybody running about with a nasty look and an equalizer was a foreigner.

equatable (ĭkwē̆i·tăb'l), *a.* [f. Equat(e *v.* + -able.] Capable of being equated.

1893 R. H. I. Palgrave *Dict. Pol. Econ.* I. 581/2 Difficulty of attainment experienced by the purchaser is equateable with the difficulty of production. **1935** G. K. Zipf *Psycho-Biol. Lang.* 236 A corresponding degree of crystallization of the configuration, which is equatable with relative frequency of occurrence. **1967** *Technology Week* 23 Jan. 26/1 (Advt.), This is typical of the growing number of space guidance..assignments..in 'controlled software', ..the establishment of equatable simulations for various missions.

equation. Add: Also (ĭkwē̆i·ʒən). **3.** *personal equation* (earlier examples).

1845 *Penny Cycl.* Suppl. I. 535/2 If A and B are severally in the habit of noting events 3-tenths of a second after and 4-tenths of a second before they take place, their personal equations may be described as being +0ˢ.3 and 0ˢ.4. **1865** Brande & Cox *Dict. Sci.* (ed. 3) I. 800/2 The term personal equation has of late been introduced into Astronomy.

b. *human equation*: *transf.* sense from *personal equation*.

1938 *Reader's Digest* Apr. 77/2 The Oakland Bridge suffers from such a simple, unpredictable human equation as the preference of truck drivers to loaf on a ferry. **1964** F. Bowers *Bibliogr. & Textual Criticism* III. ii. 71 We must throw out the human equation as much as we can in our search to find an explanation for seeming aberrancies.

7. equation box = *equational box* (see next); **equation table,** a table showing the time a clock should indicate when the sun is on the meridian.

1893 *Funk's Stand. Dict.* s.v. *Equational*, Equation box. **1731** (*title*) An explanation of the nature of equation of time, and use of the equation table for adjusting watches and clocks to the motion of the sun. **1850** Denison *Clock & Watch-m.* 19 This [sundial]..with the equation-table will give the means of correcting a clock on any fine day.

equational, *a.* Add: **2.** *Mech.* Of gearing, etc.: equalizing, adjusting; *equational box* (see quot.).

a **1877** Knight *Dict. Mech.* I. 807/2 *Equational box*... A differential gearing used in the bobbin and fly machine for the adjustment of different degrees of twist, for different yarns.

3. *Biol.* [tr. G. *aequationstheilung* 'equation division' (A. Weismann *Ueber die Zahl der Richtungskörper* (1887) ii. 42).] Of a division of the chromosomes in a cell: taking place longitudinally and resulting in two equal segments that are incorporated into the daughter nuclei.

1920 W. E. Agar *Cytol.* 50 Now in a meiosis with tetrad formation of the Copepod type it follows that if both the joints represent division planes, one division must be longitudinal (or *equational*, since the resulting daughter chromosomes receive similar sets of chromatin elements), and the other division must be transverse (or *reductional*, since each resulting daughter chromosome receives only one half of the set of chromatin elements). **1949** Darlington & Mather *Elem. Genetics* ii. 52 The distribution of the chiasmata in the pairing segment relative to the centromere and the differential segments then decides whether the first division is reductional for these segments or equational.

4. *Gram.* Of a sentence or phrase (see quots.).

1933 Bloomfield *Lang.* xi. 173 The equational type [of sentence] was less common [in Latin] than in Russian: *beātus ille* 'happy (is) he'. **1961** R. B. Long *Sentence & its Parts* v. 115 The common-voice forms of *be* are reversible when *be* is a true equational verb. The worst thing is the lights. **1964** *Language* XL. 272 'Equational sentence' is a good term for a sentence containing a copula. **1964** R. H. Robins *Gen. Ling.* vi. 234 In some languages a favourite, productive, sentence type is found beside those containing a verb, namely a noun+noun or a noun+adjective structure, with no verb required to complete it. This type, which is sometimes labelled equational, is found in Russian.

equative (ĭkwē̆i·tiv), *a.* [f. Equat(e *v.* + -ive.] Used to denote identity or degree of comparison.

1913 J. M. Jones *Welsh Gram.* 241 The adjective in Welsh has four degrees of comparison, the positive, the equative, the comparative, and the superlative. **1926** *Glasgow Herald* 24 June, The difficulty of equative terms in dealing with matters of religious belief. **1965** *Language* XLI. 72 An equative clause. **1966** G. N. Leech *Eng. in Advertising* ii. 18 Semantically, it [*sc.* apposition] signifies an equative or attributive relation.

equator. Add: **3. b.** *thermal equator*, the imaginary line on the earth's surface which denotes the location of highest mean air temperature for a particular period.

1930 *Meteorol. Gloss.* (ed. 2) 172. **1945** E. Bennett-Bremner *Front-line Airline* x. 65 Making due allowance for south-easterly winds to the inter-tropical front which lies on the thermal equator.

d. *Aeronautics.* 'The line along which the plane of the maximum horizontal section cuts the envelope of an aerostat' (W. B. Faraday *Gloss. Aeronaut. Terms* 1919).

1784 *New Review* July 61, 64 men who kept the machine in by ropes fixed to its equator. **1815** J. Smith *Panorama Sci. & Art* (1826) II. 151 From this equator proceeded ropes, by which was suspended a car in the form of a boat, a few feet below the balloon. **1838** M. Mason *Aeronautica* 268 An opening, however, of about four feet in length which appeared above the equator of the balloon, soon brought it down again. **1900** *Westm. Gaz.* 1 May 2/1 A balloon loses but little of its lifting power if a rent is made below the equator.

e. *Biol.* The plane of division of a cell or nucleus lying midway between the poles and perpendicular to a line joining them.

1888 [see Pole *sb.*² 7]. **1896** [see *EQUATORIAL *a.* 3]. **1908** J. A. Thomson *Heredity* (*caption facing* p. 51) The nucleus spindle has at its equator four groups of tetrads. **1913** J. W. Jenkinson *Vertebr. Embryol.* 95 The structure and symmetry of the egg, its axis, poles, and equator. **1961** M. J. D. White *Chromosomes* (ed. 5) v. 83 If there is a considerable distance between the chiasmata and the centromeres..the centromeres will be attached to the spindle about midway between the equator and the poles.

equatorial, *a.* Add: **1. c.** Pertaining to the equator of a balloon (cf. prec. 3 d).

1838 M. Mason *Aeronautica* 30 In shape it [*sc.* a balloon] somewhat resembles a pear; its upright or polar diameter exceeding the transverse or equatorial by about one-sixth.

3. *Biol.* (Cf. prec. 3 e.) *equatorial plate* [tr. F. *plaque équatoriale* (E. van Beneden 1875, in *Bull. de l'Acad. R. d. Sci. de Belg.* (*Classe des Sciences*) 2me Sér. XL. 733)], in mitosis and meiosis, the position occupied by the chromosomes when arranged in the equatorial plane of the spindle during metaphase.

1887 A. C. Haddon *Introd. Study Embryol.* ii. 19 At the stage when the chromatin is equatorially situated (the 'equatorial plate', which is the equivalent of the wreath and aster stage), the achromatin forms a well-marked spindle-shaped bundle of fibres. **1896** E. B. Wilson *Cell* ii. 49 The chromosomes group themselves in a plane passing through the equator of the spindle, and thus form what is known as the equatorial plate. **1913** J. W. Jenkinson *Vertebr. Embryol.* iii. 40 The plane at right angles to the axis and including the centre of the egg is equatorial. **1952** A. Hughes *Mitotic Cycle* iv. 116 During metaphase the chromosomes move from random positions into the 'equatorial plate'. **1958** C. P. Swanson *Cytol. & Cytogenetics* iii. 52 The spindle..serves to bring the chromosomes onto the metaphase, or equatorial, plate. **1960** L. Picken *Organization of Cells* x. 501 Either before fertilization or at the eight-cell stage, an equatorial separation of the two halves of the egg gives rise to qualitatively different larvae.

equi-. Add: **equice·llular** *a.* *Biol.*, made up of similar cells; **equicohesive temperature** *Metall.* (see quot. 1917); **equiconti·nuous** *a.* *Math.* [tr. It. *egualmente continuo* (C. Arzelà 1895, in *Mem. d. R. Accad. d. Sci. dell'Ist. di Bologna* V. 226)], of a set of functions $f_i(x)$: having the property that for all x_1 and all $\epsilon > 0$ there corresponds a $\delta > 0$ such that, if $|x - x_1| < \delta$, $|f_i(x) - f_i(x_1)| < \epsilon$ for all i; **equimo·lar** *a.* = *equimolecular* (*b*); **equimole·cular** *a.*, (*a*) having an equal number of molecules; (*b*) having an equal number of moles (cf. *MOLE *sb.*⁷); **equisi·gnal** *a.* *Aeronaut.*, of a radio beacon or guidance system of overlapping zones (see quot. 1951).

1890 E. R. Lankester in *Encycl. Brit.* XXIV. 810/2 The unicellular or equicellular Protozoa. **1917** Z. Jeffries in *Jrnl. Amer. Inst. Metals* Dec. 300 (*title*) The Amorphous Metal Hypothesis and Equi-Cohesive Temperatures. *Ibid.* 311 There must be some intermediate temperature in any given metal at which the cohesion of the amorphous and crystalline phases is the same. I will refer to this temperature as the 'Equi-Cohesive Temperature'. **1959** *Jrnl. Iron & Steel Inst.* CXCI. 208/2 The author first describes methods for determining equicohesive temperature and for ascertaining its dependence on grain size, strain and strain rate. **1926** E. W. Hobson *Theory Functions* (ed. 2) II. ii. 168 In case the family of continuous functions is such that, for each value of ε, a single set of sub-intervals or sub-cells can be so determined that, for every function $f(x)$ of the family, the fluctuation in a sub-interval, or a sub-cell, of a set, is less than ε, the family is said to consist of *equi-continuous* functions. **1959** G. & R. C. James *Math. Dict.* 21/1 *Ascoli's theorem.* From any set of uniformly bounded functions equicontinuous on a bounded closed (compact) set (such as a closed interval) it is possible to select an infinite sequence which converges uniformly to a limit function which is also continuous. **1946** *Nature* 14 Dec. 877/1 Carré and Libermann have shown..that it is of great advantage to use equimolar quantities of acid, pyridine, and thionyl chloride. **1961** *Lancet* 2 Sept. 566/2 The solution contained equimolar amounts of α-ketobutyric acid and sodium hydroxide. **1909** J. W. Jenkinson *Exper. Embryol.* iii. 140 Equimolecular solutions of monobasic lithium salts. **1922** F. W. Aston *Isotopes* 19 Solutions of two different compounds of lead in equimolecular proportions. **1963** A. J. Hall *Textile Sci.* ii. 72 An equimolecular mixture of the two monomers. **1931** P. V. H. Weems *Air Navig.* xiv. 273 A radio beam is broadcast by transmitters known as equi-signal beacons. Transmitters employ two cross-loops, radiating a characteristic dot-and-dash signal. **1936** *Jrnl. R. Aeronaut. Soc.* XL. 172 The system of directional equi-signal beacons established to mark the chain of air routes throughout the U.S.A. **1951** *Gloss. Aeronaut. Terms* (*B.S.I.*) iii. 31 *Equi-signal zone*, a zone within which, with an overlapping signal pattern system, indication is given that an aircraft is on a track.

equid (e·kwid). [ad. mod.L. *equidæ*, f. *equus* horse.] A member of the family Equidæ, which embraces horses, asses, mules, zebras, and certain extinct ungulates with undivided hoofs.

1889 in *Cent. Dict.* **1928** V. G. Childe *Most Anc. East* ii. 25 The zebra, and perhaps another equid. **1932** *Antiquity* VI. 254 The calm, conventional figures of equids from Gross Spitzkopje. **1937** *Proc. Prehistoric Soc.* III. 185 A naturalistic engraving of an equid covered by the archaeological deposit. **1968** A. S. Romer *Procession of Life* xvi. 252 In the equids the front foot, like the hind, eventually became monodactyl.

equilenin (ĭkwili·nin). Also -ine. [ad. F. *équilénine* (A. Girard et al. 1932, in *Compt. Rend.* CXCV. 981), f. *EQUILIN + infixed -*en*-.] A sex hormone, $C_{18}H_{18}O_2$, present in the urine of pregnant mares.

1933 *Chem. Abstr.* XXVII. 1929 Besides folliculin, $C_{18}H_{22}O_2$, 2 new hormones have been isolated from pregnant urine. Equilenine $C_{18}H_{18}O_2$..was obtained in a 1·50 g. quantity from 52 tons of urine. **1943** *Endeavour* Jan. 29/1 Analogy for aromatization of the hydrogenated ring system is provided by the oestrus-producing hormone equilenin. **1958** *Sci. News* XLVII. 88 Two other closely related oestrogens, produced by the equine species, are equilin and equilenin.

equilibrator (ĭkwi·librē̆itǫɹ). [f. Equilibrate *v.* + -or.] A device for maintaining or restoring equilibrium, as in an aircraft, or between the primary and secondary currents in a phototelegraphic apparatus.

1908 *Westm. Gaz.* 18 Aug. 4/2 All flying-machines..are liable to be brought more or less suddenly to earth by a wrong movement given to the equilibrator. **1909** *Ibid.* 11 Mar. 4/2 The pilot..might..by skilful manipulation of the equilibrator, succeed in recovering his equilibrium.

equilibrium. Add: **4. equilibrium diagram,** a diagram representing the limits of temperature and composition within which the various phases or constituents of an alloy system are stable; a constitutional diagram; **equilibrium moisture content** (see quot. 1966).

1925 *Jrnl. Iron & Steel Inst.* CXII. 345 In the equilibrium diagram of the iron-carbon system..the A2 line is shown by a dotted line. **1934** Webster, Equilibrium moisture content. **1948** *New Biol.* IV. 91 Wood being a hygroscopic material adjusts itself to the moisture condition of its surroundings by losing or absorbing moisture until an equilibrium moisture content is attained. **1966** A. W. Lewis *Gloss. Woodworking Terms* 27 *Equilibrium moisture content*, normal stable state of timber when its moisture content corresponds with the humidity of the air.

equilin (ĭ·kwilin). Also -ine. [ad. Fr. *équiline* (A. Girard et al. 1932, in *Compt. Rend.* CXCIV. 911), f. L. *equī*(*nus* (see Equine *a.*) + *follicu*)*line*.] A sex hormone, $C_{18}H_{20}O_2$, present in the urine of pregnant mares.

1932 *Chem. Abstr.* XXVI. 4090 The other 2 fractions on being extd. with ether yielded a new cryst. substance with chem. and phys. properties very similar to those of folliculin...The name equilin is suggested for this new hormone. **1958** [see *EQUILENIN].

equine. b. *sb.* Delete 'nonce-use' and add later example. Also, an equid.

1909 *Westm. Gaz.* 25 Mar. 4/2 One of the essential differences between the engineer's horse-power and the equine's effort is that a properly made engine..will give off its maximum effort for prolonged periods; but..this cannot be done by a horse. **1947** J. Stevenson-Hamilton *Wild Life S. Afr.* i. 9 The curious striped equines, known as zebras.

equinely (ĭ·kwəinli), *adv.* [f. Equine *a.* + -ly².] In an equine manner; like a horse.

1899 *Pearson's Mag.* Apr. 395/1 At the end of this course a man is supposed to be entirely at home on his horse, and with it to do everything that is humanly and equinely possible. *a* **1910** 'O. Henry' *Trimmed Lamp* (1916) 86 The chestnut, approving his new rider, danced and pranced, reviling equinely the subdued bays.

equinity. (Later example.)

1899 *Westm. Gaz.* 29 Aug. 3/1 The femininity of other animals in proportion to their caninity, or felinity, or equinity.

equipartition (ī:kwi,pāɹti·ʃən). *Physics.* [f. Equi- + Partition.] In full *equipartition of energy*. A principle of thermodynamics which states that when a system is in thermal equilibrium the kinetic energy is equally divided amongst the degrees of freedom.

Earlier called 'the law of equal partition', 'the law of partition of kinetic energy', etc.

1902 *Phil. Mag.* IV. 385 The object of the present paper is, firstly, to give a proof of Boltzmann's Theorem on the Equipartition of Energy from a somewhat new point of view; and, secondly, to examine what are the precise conditions under which equipartition will take place, and whether these conditions..will occur in an actual gas. **1905** *Nature* 18 May 54/2 Equipartition of energy is supposed to establish itself within a small fraction of a second. **1929** Lowry & Sugden *Physical Chem.* 19 According to Maxwell's principle of equipartition of energy, each degree of freedom calls for an equal increment of internal energy. **1957** *Encycl. Brit.* XIII. 386/2 According to a famous principle known as the equipartition of energy, when a dynamical system is in statistical equilibrium at a given temperature..the same amount of average kinetic energy is associated with each degree of freedom. **1962** W. B. Thompson *Introd. Plasma Physics* iii. 25 At the same time, collisions between the original and secondary electrons result in an equipartition of energy, so that the electrons acquire a roughly Maxwellian distribution at some temperature T₋. **1966** K. Mendelssohn *Quest for Absolute Zero* vi. 114 Another aspect of this picture which will later acquire great significance is the so-called 'law of equipartition'. This says that the energy will be distributed over all degrees of freedom.

equipotential, *a.* Add: **2.** (Earlier examples.)

1865 Brande & Cox *Dict. Sci.* (ed. 3) I. 804/1 Equipotential surface. **1879** *Encycl. Brit.* VIII. 25/2 A surface constructed so that the potential at every point of it has the same value is called an equipotential or level surface.

3. *Biol.* Of a germ or embryo: having equal potentialities (see quot. 1920). Hence **e:quipotentia·lity.**

1908 H. Driesch *Sci. & Philos. Organism* I. 83 Though ectoderm and endoderm have their potencies equally distributed amongst their respective cells, they possess different potencies compared one with the other. And the same relation is found to hold for all cases of what we may call elementary organs: they are 'equipotential', as we may say, in themselves, but of different potencies compared with each other. **1909** J. W. Jenkinson *Exper. Embryol.* 282 This conception, of the absolute equipotentiality of the parts, as we have already had occasion to remark, is erroneous. **1920** L. Doncaster *Introd. Cytol.* 118 It is maintained by Driesch and others that eggs of this type are equipotential systems, that is to say, they are not differentiated before fertilisation into regions of different potentiality, and any part of the egg is capable of giving rise to any part of the embryo. **1964** M. Critchley *Developmental Dyslexia* xiii. 77 The frequent occurrence of cortical equipotentiality.

equiprobability (ĭkwi,prǫbăbi·liti). *Logic.* [f. next + -ity.] The characteristic of being equally probable.

1921 J. M. Keynes *Treat. Probability* iv. 41 The discovery of a rule, by which equiprobability could be established, was, therefore, essential. **1937** *Mind* XLVI. 485 In any actual case the evidence for equi-probability is always empirical. **1958** *Oxf. Mag.* 1 May 408a/2 This conclusion..does not rest..on any gratuitous equiprobability assumption; the only equiprobability postulate is [etc.].

equiprobable (ĭkwi,prǫ·băb'l), *a.* *Logic.* [f. Equi- + Probable *a.*] Equally probable.

1921 J. M. Keynes *Treat. Probability* v. 65 A set of exclusive and exhaustive *equiprobable* alternatives. **1937** *Mind* XLVI. 485 No-one imagines that he can tell *a priori* what are the probabilities of the various alternatives when they cannot be seen to be equi-probable. **1949** W. Kneale *Probability* 151 Even in games of chance we do not know *a priori* that the various alternatives are equiprobable. **1953** J. B. Carroll *Study Lang.* vii. 200 There are thirty-two characters..which can be transmitted; if these characters are regarded as equiprobable, the transmission of one character represents five units of information, five being the logarithm of 32 to the base 2.

equity. Add: **5. c.** (See quot. 1966.) orig. *U.S.*

1904 E. S. Meade in *Pol. Sci. Q.* Mar. 50 Its preferred stock is quoted at..prices which indicate a general conviction that the equity in the company is worth little. **1928** *New Statesman* 28 July, Finance Suppl. p. vi, Out of the combined issued capital of £16,629,000 the public put up 93 per cent. of the cash required, but received only 21.8 per cent. of the equity—that is the balance of profits remaining after the fixed dividends have been paid on the Preferred capital. **1930** *Times* (Financial Rev.) 11 Feb. p. iii/2 It was widely imagined that more money was to be made in high pressure equities than in anæmic mortgages. **1966** A. Gilpin *Dict. Econ. Terms* (1967) 72 *Equities*, the ordinary shares of a limited company. They carry the right to the residue of a company's assets after it has paid all its creditors, and share in the distribution of profits, if any, after interest has been paid to preference share-holders and debenture holders each year. **1969** *Times* (Suppl.) 5 May p. iii/1 The shift in portfolio preferences of institutional investors from bonds to equities..is even more rapid..than the statistics suggest. **1970** *Money Which?* Sept. 143/2 The ordinary shares of companies (also called equities) are bought and sold on a Stock Exchange.

attrib. **1930** *Daily Express* 8 Sept. 10/2 Purchasers of equity securities of the speculative type. **1931** *Ibid.* 16 Oct. 14/5 To pay 5 per cent. on the equity shares and meet the preferential and debenture interests, a trading profit of £111,000 is necessary. **1953** *Economist* 4 Apr. 18/2 Canadian tax practice..has made loan finance more attractive to corporations than equity finance. **1965** *McGraw-Hill Dict. Mod. Econ.* 181 *Equity capital*, the total investment in a business by all its owners.

equivalence, *sb.* Add: **1. d.** *principle of equivalence*: one of the fundamental postulates of the general theory of relativity, which states that at any point of space-time the effects of a gravitational field cannot be experimentally distinguished from those due to an accelerated frame of reference. Also *equivalence principle*.

The principle was proposed by Einstein in *Ann. d. Physik* (1911) XXXV. 898–908, and was first called *Äquivalenzprinzip* by him in *Ibid.* (1912) XXXVIII. 360, 443.

1918 A. S. Eddington *Rep. Relativity Theory Gravitation* ii. 19 The hypothesis that gravitation may be of essentially the same nature as the geometrical forces introduced by the choice of co-ordinates..which was put forward by Einstein, is called the Principle of Equivalence. **1955** O. Klein in W. Pauli *Niels Bohr* 99 In a generalized quantum-relativity theory, comprising also electromagnetism and perhaps meson fields corresponding to the nuclear forces, there would probably be some kind of generalized equivalence principle.

equivalent, *a.* and *sb.* Add: **A.** *adj.* **5. b.** Optics. *equivalent focal length* (see quots.).

1867 Sutton & Dawson *Dict. Photogr.* (ed. 2) 116 Suppose that a lens..when presented towards a distant object, renders the image of it a certain size upon the ground glass. Then the 'equivalent focal length' of that lens..is equal to the principal focal length of a single lens, having a small stop in close contact with its central portion, which gives the same sized image of that object, when taken from the same point of view. **1961** A. L. M. Sowerby *Dict. Photogr.* (ed. 19) 335 The focal length, or equivalent focal length, of a lens is defined as the distance from the node of emission, to the position at which the lens forms a sharp image of a distant object.

c. Electr. *equivalent circuit*: an electric circuit consisting usually of resistance, inductance and capacitance and having characteristics equivalent to those of other electric circuits or apparatus.

1920 in R. E. Neale *Whittaker's Electr. Engin. Pocket-Bk.* 242 When speaking of the constants of a transformer, it is permissible to consider an equivalent circuit, the impedance of which is such that the same current at the same power-factor would flow if the same pressure were applied to it as to the primary of the transformer. **1943** *Electronic Engin.* XVI. 149 It is convenient to consider an equivalent circuit composed of pure reactances and resistances. **1962** Simpson & Richards *Junction Transistors* v. 75 These usually take the form of equivalent circuits which can be used to represent the transistor under different circuit conditions.

B. *sb.* **3. c.** *mechanical equivalent of heat* (earlier example); cf. Joule; *mechanical equivalent of light*: the amount of radiant flux that corresponds to unit luminous flux (or, in quot. 1908, that is emitted by a source of unit luminous intensity), usu. expressed in watts per lumen.

1845 *Rep. Brit. Assoc.* ii. 31 (*heading*) On the Mechanical Equivalent of Heat. **1908** *Proc. R. Soc.* A. LXXX. 19 The labours of workers in Germany and in the United States [have not]..sufficed to permit of definite values being adopted for..the mechanical equivalent of light. *Ibid.* 22 The mechanical equivalent of light

$$M = \frac{4\pi p}{I} = \frac{W_r}{Kf}\left(\frac{d}{D}\right)^2,$$

giving the mechanical equivalent in watts per candle. **1959** W. S. Sharps *Dict. Cinemat.* 110/2 The reciprocal of the luminous efficiency of radiant energy is often termed, the 'mechanical equivalent of light'.

equivalve (ĭ·kwivælv), *a.* *Zool.* [f. Equi- + Valve *sb.*¹ 2.] Of a bivalve mollusc: having both valves alike in shape and size.

[**1856** P. H. Gosse *Man. Marine Zool.* II. 51 Shell oval or oblong, equal-valved.] **1862** J. G. Jeffreys *Brit. Conchol.* I. 28 Shell equivalve, oblong, inequilateral. **1897** Parker & Haswell *Text-bk. Zool.* I. xii. 650 (Pelecypoda), As a general rule the right and left valves are alike, or nearly so, the shell being therefore equivalve. **1968** N. F. McMillan *Brit. Shells* 17 Equivalve. Where the two valves of a bivalve are similar.

equivolu·minal, *a.* *Physics.* [See Equi-.] Having an equal unchanging volume (see quot.).

1899 Kelvin in *Phil. Mag.* XLVII. 480 On the Application of Force within a Limited Space, required to produce Spherical Solitary Waves, or Trains of Periodic Waves, of both Species, Equivoluminal and Irrotational, in an Elastic Solid. *Note*, By 'equivoluminal' I mean every part of the solid keeping its volume unchanged during the motion.

er (ɜ̄ɹ). Used to express the inarticulate sound or murmur made by a hesitant speaker. Also, preceded by *Mr.* (or *Miss*, etc.), used in place of a name of which the speaker is uncertain. Also as *v. intr.*

1862 *St. James's Mag.* Mar. 481 Oh!—er—Well I think it was a—er—little fête at—at—Dear me—where was it? **1878** *Porcupine* XX. 681/2 Those ladies and gentlemen..who have had the advantage of—er—being —er—confined in Newgate prison. **1904** V. L. Whitechurch *Canon in Residence* i. 9 Mr. Smith said, 'Er, can I have a bed?' **1912** R. W. Wright-Henderson *Annabel* i. 9, I think, Mr. Buckle, if you would slightly modulate —er—that is—. *Ibid.* viii. 122 Have you decided yet what—er—. **1924** R. Macaulay *Orphan Island* xxi, Grandmamma's—er—adultery (if you will pardon the crude word). **1931** *Time & Tide* 12 Sept. 1057 He ahs and ers, and hums and hawes his way through an incredibly fatuous pronouncement. **1935** Marsh & Jellett *Nursing-Home Murder* viii. 116 That's what makes it so dreadful, Mr.—er—I never could have believed it. **1940** 'N. Blake' *Malice in Wonderland* ix. 129 You sound quite breathless, Miss Er. **1958** *Aspects of Translation* 37 The really astute Englishman..must feign a certain diffident hesitation, put in a few well-placed —ers. **1967** R. Rendell *Wolf to Slaughter* vi. 56, I happened to see your advert in Grover's window, Mrs. Er..?

-er, *suffix*⁶. Also **-ers.** Introduced from Rugby School into Oxford University slang, orig. at University College, in Michaelmas Term, 1875; used to make jocular formations on sbs., by clipping or curtailing them and adding *-er* to the remaining part, which is sometimes itself distorted. Among the earliest instances are *FOOTER *sb.*¹ 3b (=football) (1863), Rugger² (1893), Socker (1891); Togger (1897), Tosher³ (1889); *BONNER, *BREKKER (1889), *DIVVERS, *EKKER (1891).

Other familiar examples are *BEDDER³, *bed-sitter* = bed-sittingroom, *collekkers* = collections (see Collection 8), *CUPPER², and (formerly) *rudders* = rudiments (of divinity), *stragger* = stranger. Examples of proper names are *Adders* = Addison's Walk, *Bodder* = Bodleian, *Jaggers* = Jesus (College), *Quaggers* = The Queen's College. (Cf. F. Madan *Oxford outside the Guidebooks*, 1923.)

1892 *Isis* 8 June 26/1 At the close of the Lancashire match we heard one man ejaculate..'This is breath-ers'. ..This..is all that remains of the..expression 'breathless excitement'. **1899** *Daily Tel.* 14 Aug. 9/5 The triumph of this jargon was reached when some one christened the Martyrs' Memorial the 'Martyrs' Memugger'. **1903** D. F. T. Coke *Sandford of Merton* ii. 10 'Wagger-pagger-bagger' for the receptacle of torn-up letters and the like. **1904** *Daily Chron.* 25 Mar. 4/7 Mr. Gladstone was 'the Gladder'. An undergraduate left his 'bedder' in the morning to eat his 'brekker' in his 'sitter'; later he attended a 'lecker', and in the afternoon he might run with the 'Toggers' (torpid races) or take some other form of 'ecker'. **1912** *Tatler* 23 Oct., Suppl. 40 The 'Pragger-Wagger', it should be explained, is the new name given to the Prince of Wales. **1914** D. O. Barnett *Lett.* (1915) 13 The P-Wagger came to see us yesterday. I met him coming off parade, and threw a hairy salute. **1914** C. Mackenzie *Sinister St.* II. iii. 517 'They say we shall all have to interview the Warden tomorrow.' 'They say on Sunday afternoon the Wagger makes the same speech to the freshers that he's made for twenty years.' *Ibid.* 543 'You'd better let me put your name down for the Ugger'. 'The what?'..'The Union.' **1963** *Sunday Times* 22 Sept. 18/4 The twenties were the age of the 'er' at Repton. The changing-room was the chagger and a sensation a sensagger. A six at cricket was a criper.

era. Add: **4. c.** *era of good feeling(s*, in U.S. Hist., a period during the presidency of Monroe (1817–24), when there was virtually only one political party. Also *transf.*

1817 *Columbian Centinel* 12 July (Th.) (*heading*) Era of Good Feelings. **1851** H. A. Garland *Life J. Randolph* II. 278 During the 'era of good feelings', and the undisturbed repose of Mr. Monroe's administration, [these ideas] had been widely disseminated. **1904** *N.Y. Herald* 22 Sept. 4 The nomination was made unanimously, and the era of good feeling continued throughout the naming of the rest of the ticket. **1945** *Chicago Daily News* 16 Aug. 10/2 We do not wish to disturb the new 'Era of Good Feeling' in traction. **1948** *Pennsylvania Mag. Hist.* Apr. 113 The era of the 1820's—the so-called but ludicrously named 'Era of Good Feeling'. **1951** G. Dangerfield (*title*) The Era of Good Feelings.

6. era-making *a.* = *epoch-making*.

1894 J. Millar tr. *Weizsäcker's Apostolic Age* I. ii. iii. 188 The era-making significance of the treaty. **1929** R. A. Cram *Cath. Ch. & Art* iv. 53 The great monastic sovereignty..that had grown out of the first and era-making beginnings at the hands of St. Benedict. **1965** *Times Lit. Suppl.* 27 May 438/3 We recognize easily enough in the history of cultures how era-making an invention writing has been.

erase, *v.* Add: **1. b.** *Electr.* To remove the recorded signals from a magnetic tape or medium.

1945 *Sci. News Let.* 9 June 363 Any part of the wire can be erased without harm to the sound recorded on adjacent parts. **1949** Frayne & Wolfe *Elem. Sound Recording* xxix. 592 Magnetic media..that require high magnetizing forces to produce saturation are more difficult to erase. **1962** *Gloss. Terms. Autom. Data Proc.* (*B.S.I.*) 68 *Erase*, in a magnetic store, to obliterate stored data by returning the magnetic state of a cell to a uniform null condition.

erase (ĭrē̆i·z), *sb.* = *ERASING *vbl. sb.* c. Also *attrib.*, as *erase head*, etc.

1948 Wiegand & Zenner in *Trans. Amer. Inst. Electr. Engin.* LXVII. 507 (*title*) A turn-in-gap erase head for magnetic recorders providing intense high-frequency fields. **1949** Frayne & Wolfe *Elem. Sound Recording* xxix. 593 The equipment required for such erase is very simple. **1952** *Gramophone* Aug. 68 The Tape-desk..contains..an erase unit. **1954** *Electronic Engin.* XXVI. 294 The erase head consists of a specially designed permanent magnet which subjects the individual elements of the tape to reversing magnetic fields. **1957** *Encycl. Brit.* X. 618/2 The erase and re-use features of magnetic recordings are important economic advantages in many commercial applications [of tape recorders]. **1962** A. Nisbett *Technique Sound Studio* 273 *Tape recording amplifier* differs from an ordinary amplifier in that it includes an oscillator for bias and erase.

erasing, *vbl. sb.* Add: **c.** *Electr.* The action of the verb *ERASE 1 b. Also *attrib.*

1949 Frayne & Wolfe *Elem. Sound Recording* xxix. 580 The recording material first passes over an erasing magnet or head to eliminate any previous recording.

1950 G. A. Briggs *Sound Reprod.* (ed. 2) xx. 139 A combined head for recording, reproducing and erasing has been designed. *Ibid.*, The operations of erasing and re-recording are likely to be carried out only at infrequent intervals. **1959** W. S. Sharps *Dict. Cinemat.* 94/1 An erasing head is a device for obliterating any previous recordings on magnetic media. **1960** H. Carter *Dict. Electronics* 94 *Erasing*, the process of demagnetizing the tape used in a tape recorder.

erasure. Add: **1. c.** *Electr.* = *ERASING *vbl. sb.* c.

1950 G. A. Briggs *Sound Reproduction* (ed. 2) xx. 139 Thorough erasure of the tape or wire is necessary before it passes over the recording head. **1962** A. Nisbett *Technique Sound Studio* 252 *Erasure*, the removal of recorded signals from a tape so that it is ready to re-use. **1967** *Electronics* 6 Mar. 79/2 (Advt.), Complete, gradual or selective erasure is possible.

erbswurst (ɜ̄·ɹbzwŭ̄ɹst). [G.] Seasoned pease-flour compressed into sausage shape and used for making soup (see also quot. 1885).

1885 J. Nixon *Compl. Story Transvaal* x. 180 Twice a week we had 'erbswurst', or dried beans. **1890** Kipling *Life's Handicap* (1891) 36 Erbswurst, tinned beef..and meat-biscuits may be nourishing but,..Thomas Atkins needs..bulk. **1913** W. Lee White in F. H. Harris *Dartmouth out o' Doors* 117 With erbswurst and beef tablets for a base [of a soup]. **1941** V. Nabokov in *Atlantic Monthly* Nov. 619/2 This churlish, heavy man, who fed mainly on *Erbswurst* and boiled potatoes.

Ercles (ɜ̄·ɹklīz), Bottom's pronunciation of 'Hercules'; 'Ercles vein' is a stock quotation. (See *Mids. N. D.* I. ii. 31, 42.)

1898 G. F. R. Henderson *Stonewall Jackson* II. xv. 100 Even the Northern press made sport of Pope's "Ercles vein", and the Confederates contrasted his noisy declamation with the modesty of Lee and Jackson. **1901** *Standard* 9 Apr. 4/7 Mr. J. B. Glacier, in his opening speech, was in the Ercles vein, and dwelt on perils overcome and triumphs gained. **1922** J. Buchan *Huntingtower* ii. 36 The Ercles vein—'God's in His Heaven, all's right with the world.' No good, Mr McCunn. **1927** T. S. Eliot in T. Newton tr. *Seneca's Trag.* I. p. xxxiii, This is the proper Ercles bombast, ridiculed by Shakespeare, Jonson, and Nashe.

erector. Add: **1. b.** *spec.* An engineer who works at the assembling of engines and other iron and steel structures.

1892 *Daily Chron.* 28 Apr. 9/2 Engineer, Fitter, or Erector. **1905** *Westm. Gaz.* 18 Dec. 7/1 Charles Grove, a foreman erector of steel work.

c. A machine used in erecting iron and steel structures.

1895 *Daily News* 31 Oct. 6/5 These 'erectors' are two hydraulic appliances for lifting up and fixing in position the ponderous segments of the iron ring constituting the exterior of the tunnel. **1909** *Daily Chron.* 19 May 3/4 The tunnel shield, complete, with hydraulic erector.

eremurus (erĭmiū̆ə·rŏs). *Bot.* [mod.L. (F. A. Marschall von Bieberstein *Centuria Plantarum* (1818) 61), f. Gr. ἐρῆμος solitary + οὐρά tail.] A hardy herbaceous perennial plant of the genus of Liliaceæ so named, native to west and central Asia, and cultivated for its dense racemes of white, yellow, or reddish flowers.

1829 Loudon *Encycl. Plants* 278 *Eremurus*...Its long spikes of yellow flowers may be easily imagined to merit such an appellation. **1855** *Curtis's Bot. Mag.* LXXXI. 4870 (*caption*) Showy Eremurus. **1900** W. D. Drury *Bk. Gard.* x. 325 Eremuri are noble plants. **1901** *Cassell's Dict. Gard.* I. 323/1 A good but not too heavy soil suits the Eremuruses. **1962** *Amat. Gardening* 17 Mar. 4/2 One of the most effective spike plants is probably the eremurus, often called Foxtail Lily.

erepsin (ĭre·psin). [a. G. *erepsin* (O. Cohnheim 1901, in *Zeitschr. f. physiol. Chem.* XXXIII. 460), f. L. *eripere* to take away, f. Gr. ἐρείπειν + Pe)psin.] Name given to a mixture of enzymes in the intestinal juice.

1902 *Jrnl. Chem. Soc.* LXXXII. ii. 93 Erepsin has no action on ordinary proteid, but only on peptone, and on a part of the proteoses. **1940** *Thorpe's Dict. Appl. Chem.* (ed. 4) IV. 325/1 The early literature refers to an enzyme 'intestinal erepsin' able to hydrolyse polypeptides, peptones, etc...The term is now abandoned in favour of polypeptidases. **1961** *Brit. Med. Dict.* 513/1 *Erepsin*, a mixture of enzymes, chiefly polypeptidases, present in the intestinal mucosa and capable of converting polypeptides into amino acids.

erethic (ĭre·þik), *a.* [f. Ereth(ism + -ic.] = Erethismic *a.*

1888 *Amer. Jrnl. Psychol.* I. 375 My mental make-up is inherited mostly from the paternal side and is erethic in quality. **1894** [see Spurty *a.*].

Erewhonian (erĭwōᵘ·niăn), *a.* and *sb.* [f. *Erewhon*, title of a book (partial reversal of *Nowhere*) by Samuel Butler, published 1872 and describing a form of utopia: see -ian.] Of, belonging to, or characteristic of the book *Erewhon*, the utopia described in it, or the principles inculcated therein. **b.** *sb.* An inhabitant of Erewhon.

1897 *Daily News* 16 June 5/2 The Erewhonian plan of counting disease as a crime. **1900** *Ibid.* 10 Dec. 9/6 The Erewhonian Professors, Hanky and Panky. **1921** *Edin. Rev.* July 28 The exhortation which persuaded Samuel Butler's Erewhonians to destroy all their machines. *Ibid.* 36 The Erewhonian policy of breaking up the machines is manifestly impossible in this country. **1927** *Observer* 25 Sept. 16/6 The Erewhonian paradox of imprisoning the invalid is not wholly a barren one. *a* **1930** D. H. Lawrence *Collier's Friday Night* (1934) ii. 49 *Beatrice*:..But was it the Gypsy, or the Athletic Girl that does Botany? *Ernest* (*shaking his head*): No. It was an Erewhonian. **1954** P. Capon *Down to Earth* iv. 68 The Antigeosians had an almost Erewhonian attitude to physical disabilities...Illness was not something that could be talked about without shame.

erf². Add: (Earlier and later examples.) Pl. **erfs, erven;** † **erbes.** Also *attrib.*, as *erf-holder*.

1812 A. Plumptre tr. *M. K. H. Lichtenstein's Trav.* I. xxiii. 335 The proposal, that instead of extensive farms, it should be divided into small parcels of land, or *erbes*. **1818** C. I. Latrobe *Jrnl. Visit S. Afr. 1815–16* xvi. 262 His industry put him in possession of this *erf*, a name given to a small lot of ground, not being a complete farm. **1851** J. J. Freeman *Tour S. Afr.* viii. 184 The party came to Buxton on Thursday morning, when burning commenced immediately, and no entreaties of erf-holders, tears of mothers and children, availed. **1902** *Westm. Gaz.* 14 Apr. 10/1 The erven (acres) in question were granted to these Reservists by the Government. **1947** *Cape Argus* 11 Jan. 6/4 Erfs were allocated. **1948** *Cape Times* 26 Mar. 3/2, 21 erfholders who dwell on the mainland but till their land on the island. **1959** *Star* (Johannesburg) 12 Jan. 20/3 We will sell by Public Auction the following:..Five Erven..Certain Erf No. 136 situate on Olive Avenue.

erg¹. Substitute for def.: A unit of work or energy; the amount of work done when a force of one dyne moves its point of application one centimetre in the direction of the force.

‖ **erg²** (ɜ̄ɹg). Pl. **areg, ergs.** [Fr. (E. Fromentin *Un été dans le Sahara* (1857),) f. N. Afr. Arabic *'irj*, of Berber origin.] A type of desert area in the Sahara consisting of shifting sand dunes.

Fromentin's book appeared in serial form in the *Revue de Paris* 1854.

1875 *Encycl. Brit.* I. 250/1 The most truly desert region of the Sahara is an irregular belt of shifting sand dunes, the 'Erg' or 'Areg'. **1937** Wooldridge & Morgan *Physical Basis Geogr.* xx. 301 The sand areas of the Old World deserts ('Erg' of the Sahara, 'Koum' of Asia). *Ibid.* 319 It was at a later date..that the red Kalahari sands extended widely in South Africa. They appear to represent an 'erg phase'. **1943** *Algeria* (Naval Intelligence Div.) I. ii. 27 The sand-dune areas are called *ergs*. *Ibid.* v. 155 Two fairly distinct types [of sand desert] can be distinguished: the *nebkas* or small areas of sand,..and the erg (pl. areg) or large dune system... The erg consists of drinn steppe with a drought-resisting grass as the dominant species. **1950** G. Brenan *Face of Spain* ix. 195 He..had spent his life among the sand *ergs* and *ashab* pastures of the Sahara.

ergastic (ɜ̄ɹgæ·stik), *a.* *Cytol.* [ad. G. *ergastisch* (A. Meyer 1896, in *Bot. Zeitung* I. 212), f. Gr. ἐργαστικός able to work.] Of or pertaining to the non-living products of the cell.

1902 E. B. Wilson *Cell* (ed. 2) i. 30 The passive energid-products (ergastic structures..) are formed as enclosures..or excretions. **1935** *Nature* Suppl. 9 Mar. 378/2 Plastids, Golgi material, chondriosomes and ergastic substances. **1957** *Encycl. Brit.* XVIII. 7/1 All living cells in plants contain variable amounts of non-living materials that are collectively designated as inclusions or ergastic substances...One of the most common examples of ergastic substances is the cell sap.

ergastoplasm (ɜ̄ɹgæ·sto,plæz'm). *Biol.* [ad. F. *ergastoplasme* (C. Garnier 1897, in *Bibliographie Anat.* V. 288), f. as prec. + Plasm.] (See quot. 1925.) Hence **ergastopla·smic** *a.*

1902 E. B. Wilson *Cell* (ed. 2) vi. 322 The 'ergastoplasmic' (Garnier) fibrillæ of gland-cells. *Ibid.*, The conception of a dominating or 'superior' cytoplasm (including 'archoplasm', 'kinoplasm', 'ergastoplasm'). **1903** *Nature* 12 Mar. 455/2 (*title*) On the structure of the tracheal cell of the gad-fly, and on the origin of the ergastoplasmic formations. **1925** E. B. Wilson *Cell* (ed. 3) iv. 264 The term 'ergastoplasm', originally applied by Garnier and Bouin to the basophilic fibrillar formations in certain types of cells..has recently been resuscitated..as a non-committal term applied to strongly basophilic cytoplasmic material. **1964** G. H. Haggis et al. *Introd. Molecular Biol.* i. 12 Two differentiated regions of the cytoplasm can be recognized in the microscope—one being usually called the Golgi region, the other the ergastoplasm. *Ibid.* v. 137 The ergastoplasm or rough-surfaced endoplasmic reticulum.

ergastulum (ɜ̄ɹgæ·stiŭlə̆m). *Rom. Antiq.* Pl. **-ula** (-iŭlă). [L.] A prison-like building on a large estate, used for housing slave workers (in quot. 1891, an urban prison for slaves).

1833 W. Blair *Inquiry into Slavery amongst Romans* v. 100 When Columella wrote, some distinction appears to have existed between the *ergastulum* for ordinary workmen, and that for ill-behaved slaves, as he directs the latter to be built under ground. **1865** Brande & Cox *Dict. Sci.* I. 808/2 *Ergastulum*, a workhouse, or a dwelling place for the slaves. **1885** R. S. Storrs *Div. Orig. Christianity* v. 159 As workers on farms, they [*sc.* slaves]..slept at night in the cells of the ergastula, under-ground. **1891** F. W. Farrar *Darkness & Dawn* I. xxxi. 303 The ergastulum is half-subterranean. Its windows are narrow, and high above our heads.

ergate (ɜ̄·ɹgē̆it). *Ent.* Also in mod.L. form **ergates** (ɜ̄·ɹgătīz). [ad. Gr. ἐργάτης workman.] The worker ant.

1901 W. M. Wheeler in *Amer. Nat.* XXXV. 877 (*title*) The parasitic Origin of Macroërgates among Ants. **1910** —— *Ants* 97 The worker (ergates) is characterized by the complete absence of wings.

So **ergata·ndromorph** [Gr. ἀνήρ, ἀνδρός man, -μορφος -form], an ant that combines worker and male characters; **e·rgatane:r** (-ĭəɹ) [Gr. ἀνήρ man], a male ant resembling a worker; **e·rgatogy:ne** [Gr. γυνή woman], a worker-like ant with female characters; **e·rgatoid** *sb.*, a wingless sexually competent ant, as an ergatogyne; also as *adj.*; **e:rgatomo·rphic** *a.*, worker-like; **e:rgatote·lic** *a.* (see quot.).

1899 *Cambr. Nat. Hist.* VI. 140 The adult, sexually capable, though wingless forms, are called ergatoid, because they are similar to workers. **1901** *Amer. Nat.* XXXV. 885 The *fusca* workers attempt to change worker larvæ of Polyergus into queens but succeed only in producing the wingless ergatoids. **1910** W. M. Wheeler *Ants* vi. 94 The *ergatanēr*, ergatomorphic, or ergatoid male resembles the worker in having no wings and in the structure of the antennæ. *Ibid.* 96 The ergatogyne, ergatomorphic, or ergatoid female, is a worker-like form, with ocelli, large eyes, and a thorax more or less like that of the female, but without wings. *Ibid.* 99 The ergatandromorph..is an anomaly similar to the [gynandromorph] but having worker instead of female characters combined with those of the male. *Ibid.* vii. 120 In these [*sc.* honey-bees] only the secondary instincts are manifested in the queen, while the worker retains the primary series in full vigor and thus more clearly represents the ancestral female of the species. This type may therefore be called ergatotelic. **1915** H. St. J. K. Donisthorpe *Brit. Ants* 114 Wasmann records a mixed ergatandromorph with only the colour of the head like that of the worker. **1927** *Glasgow Herald* 16 July 4 The ergatoids can reproduce ergatoids, besides workers and soldiers. **1970** Brown & Taylor in *Insects of Australia* (C.S.I.R.O.) xxxvii. 952/2 Various wingless intermediates between queens and workers occur in some species. These are the *ergatogynes* (ergatoids).

ergative (ɜ̄·ɹgătiv), *a.* *Gram.* [f. Gr. ἐργάτης workman + -ive.] In full *ergative case*. A term used of languages such as Eskimo, Basque, and some others, where the subject noun of an intransitive verb and the object noun of a transitive verb have the same case, to designate this case.

1943 J. Marouzeau *Lexique Terminol. Linguistique* (ed. 2) 89 Ergatif (Ergativ, Ergative). Cas désignant l'agent dans les langues comme le basque. **1950** *Archivum Linguisticum* II. ii. 100 In syntactically transitional types of language like Georgian and Laz..the ergative and nominative are found side by side. **1964** E. Palmer tr. *Martinet's Elem. Gen. Ling.* vi. 179 This ending..was necessarily the morpheme of an 'ergative' case. **1968** J. Lyons *Introd. Theoret. Ling.* viii. 352 The term..for the syntactic relationship that holds between (1) [*the stone moved*] and (3) [*John moved the stone*] is 'ergative': the subject of an intransitive verb 'becomes' the object of a corresponding transitive verb, and a new ergative subject is introduced as the 'agent' (or 'cause') of the action referred to.

ergatocracy (ɜ̄ɹgătǫ·krăsi). [f. Gr. ἐργάτης workman: see -ocracy.] Government by the workers. So **ergatocra·tic** *a.* (in example *fig.*).

1920 E. & C. Paul (*title*) Creative revolution. A study of Communist ergatocracy. **1920** *Glasgow Herald* 28 Oct. 4 So far from Bolshevism being an experiment in democracy, or even in..ergatocracy, it is a reversion to the crudest oligarchy. **1944** G. Barker *Eros in Dogma* 33 The common growing that gives the grass its ergatocratic green.

ergo, *sb.* **1.** (Later U.S. example.)

1867 J. R. Lowell *Fitz Adam's Story* in *Poems* (1912) 569 His fancy's thrall, he drew all ergos thence, And thought himself the type of common sense.

ergocornine (ɜ̄ɹgŏkǭ·ɹnīn, -in). *Chem.* [ad. G. *ergocornin* (Stoll & Hofmann 1943, in *Helv. Chim. Acta* XXVI. ii. 1570), f. Ergo(t + Cornin + -e.] An alkaloid derived from ergot, $C_{31}H_{39}N_5O_5$.

1944 *Chem. Abstr.* XXXVIII. 2957 Alkaloids of the ergotoxine group—ergocristine, ergocryptine and ergocornine. **1963** *New Scientist* 21 Feb. 389/1 The alkaloid drug, ergocornine,..stops the production of the progesterone hormone.

ergodic (ɜ̄ɹgǫ·dik), *a.* *Math.* [ad. G. *ergoden* (L. Boltzmann 1887, in *Jrnl. f. d. reine und angewandte Math.* C. 208), f. Gr. ἔργον work + ὁδός way + -ic.] Of a trajectory in a confined portion of space: having the property that in the limit all points of the space will be included in the trajectory with equal frequency. Of a stochastic process: having the property that the probability of any state can be estimated from a single sufficiently extensive realization, independently of initial conditions; statistically stationary. Also, of or pertaining to this property.

1928 tr. *E. Schrödinger's Coll. Papers Wave Mech.* 143 The 'ergodic hypothesis' (Boltzmann) is what Maxwell called the 'principle of continuity of path'. **1931** G. D. BIRKHOFF in *Proc. Nat. Acad. Sci.* XVII. 651, I propose.. to establish a general recurrence theorem and thence the 'ergodic theorem'. **1947** COURANT & ROBBINS *What is Mathematics?* (ed. 4) vii. 354 A rectangular box..leads in general to an ergodic path; the ideal billiard ball going on for ever will reach the vicinity of every point, except for certain singular initial positions and directions. **1962** E. PARZEN *Stochastic Processes* iii. 74 In general, a stochastic process is said to be ergodic if it has the property that sample (or time) averages formed from an observed record of the process may be used as an approximation to the corresponding ensemble (or population) averages. **1967** CONDON & ODISHAW *Handbk. Physics* (ed. 2) v. ii. 19 The justification for the use of this ensemble would be a consequence of an ergodic theorem; that is, a theorem which states that almost all systems in the ensemble spend the same amount of time in any regions of the same (nonvanishing) area on the constant energy surface.

Hence **ergodi·city**, the quality or property of being ergodic.

1949 W. FELLER *First Berkeley Symposium on Math. Statistics & Probability* 418 What ergodicity might possibly mean in practice may be illustrated by the much discussed Pareto law of income distribution. **1960** G. HERDAN *Type-Token Math.* xx. 296 Under certain circumstances a system will tend in probability to a limiting form which is independent of the initial position from which it started. This is the Ergodicity Property.

ergogram (ə·ɪgŏgræm). [f. Gr. ἔργον work + -GRAM.] A record made by an ergograph.

1904 G. S. HALL *Adolescence* I. 150 Endurance as measured by ergograms. **1918** C. S. MYERS *Present-Day Applic. Psychol.* 8 After sufficient rest, complete recovery occurs, so that a second ergogram equal to the first is obtainable. **1948** A. G. BILLS in T. G. Andrews *Methods of Psychol.* 469 The pattern of the ergogram, which was long considered typical for the fatigue of all isolated muscle groups.

ergograph (ə·ɪgŏgrɑf). [ad. It. *ergografo* (A. Mosso 1888, in *Atti d. R. Accademia d. Lincei* 4th Ser. V. 411), f. Gr. ἔργον work: see -GRAPH.] An instrument which measures and records the work done by particular groups of muscles. Hence **ergogra·phic** *a.*, **ergo·graphy**.

1892 *Sat. Rev.* 9 Apr. 430/2 The action of the brain on the muscles, as demonstrated by experiments made with the ergograph. **1897** E. W. SCRIPTURE *New Psychol.* 230 The ergograph..consists of a rest in which the arm is fixed so that the middle finger can be moved alone without involving any of the others. **1898** *Daily News* 25 Nov. 6/3 His ergographic curves, and his abdominal respiratory curves, are duly recorded. **1909** C. S. MYERS *Text-bk. Exper. Psychol.* xiv. 184 Ergography.—The work performed by an active muscle..may be best determined by means of graphic records. *Ibid.* 186 An ergographic record. *Ibid.* 383 The ergograph..is especially adapted for the study of simple movements in which very few muscles are involved. **1948** A. G. BILLS in T. G. Andrews *Methods of Psychol.* 468 The most widely used type of ergograph is that invented by Mosso...Its purpose is to force an isolated muscle group..to work against strong resistance repeatedly until a point of apparent exhaustion is reached. **1953** F. BARTLETT in Floyd & Welford *Sympos. Fatigue* i. 2 The model for the experimental study of muscle fatigue in the intact organism: some form of ergographic technique.

ergometrine (əɪgŏme·trin, -in). *Biochem.* [f. ERGO(T + Gr. μήτρ(α- womb + -INE⁵.] An amide of lysergic acid, $C_{19}H_{23}N_3O_2$, which is one of the most powerful alkaloids present in ergot and is used as an oxytocic.

1935 DUDLEY & MOIR in *Brit. Med. Jrnl.* 16 Mar. 520/2 In 1932 one of us (C. M.) drew the conclusion..that the characteristic and traditional effect of ergot was caused..by a substance still unidentified...We are now..reporting the isolation of the substance to which ergot rightly owes its long-established reputation as the '*pulvis parturiens*'. We propose to call it 'ergometrine'. **1937** GWYNNE-VAUGHAN & BARNES *Struct. & Devel. Fungi* (ed. 2) 253 The sclerotia..contain a number of alkaloids one of which, ergometrine, causes rapid contraction of the uterus. **1961** *New Scientist* 2 Nov. 278/2 They isolated ergometrine from ergot, simultaneously with American workers, who called it ergonovine. **1963** *Lancet* 19 Jan. 139/1, 0.5 mg. of ergometrine maleate was given intravenously and intramuscularly, and the drip was discontinued. **1970** J. CROSSLAND *Lewis's Pharmacol.* (ed. 4) xvi. 428 Ergometrine is responsible for the abortions seen in ergotism.

ergonomics (əɪgŏnǫ·miks). [f. Gr. ἔργον work: see ECONOMICS.] The scientific study of the efficiency of man in his working environment. Hence **ergono·mic** *a.*, of or pertaining to ergonomics; **ergono·mically** *adv.* Also **ergo·nomist**, one who is skilled in ergonomics.

1950 *Lancet* 1 Apr. 645/2 In July, 1949, a group of people decided to form a new society for which the name the 'Ergonomics Research Society' has now been adopted. **1952** *Oxf. Mail* 2 July 2/5 He [*sc.* K. F. H. Murrell] found it necessary to invent the word 'ergonomics' to cover the kind of work he and other people are doing in studying work in relation to environment. **1954** W. E. LE GROS CLARK in Floyd & Welford *Sympos. Human Factors in Equipment Design* ii. 5 A man and his machine may be regarded as the functional unit of industry, and the aim of ergonomics is the perfection of this unit so as to promote accuracy and speed of operation, and at the same time to ensure minimum fatigue and thereby maximum efficiency. *Ibid.*, Anatomists have not given so much attention as other groups of workers to the application of their studies to ergonomic problems. **1958** *Engineering* 21 Feb. 226/2 This attempt to promote the study of ergonomics, that is, the study of the interaction of men and their environment (now usually with special reference to the machine environment) deserves every success. **1958** *Archit. Rev.* CXXIV. 42/2 The design of an instrument to be gripped by such a diversity of hands must be an ergonomic compromise, but on the whole the new G.P.O. model achieves a fairly satisfactory balance between function and aesthetics. **1959** *Daily Tel.* 8 Apr. 16/3 (*heading*) Ergonomists seek easier design for working. **1961** *Engineering* 6 Oct. 432/3 The keyboard has been ergonomically designed and is contoured..to make every key more accessible to the typist's fingers. **1969** *Times* 5 May 23/5 A seat designed by the one car-maker with an ergonomics department. **1970** *Sci. Jrnl.* Apr. 28/2 The whole subject [of absenteeism] indeed is of interest..to sociologists and ergonomists (industrial psychologists and work physiologists).

ergonovine (əɪgŏnōᵘ·vīn, -in). *Biochem.* [a. It. *ergonovine* in *Boll. Soc. Ital. Biol. Sper.* (1937) XII. 232–4, f. ERGO(T + L. *nov(us* NEW *a.* + -INE⁵.] = *ERGOMETRINE.

1938 *Chem. Abstr.* XXXII. 303 (*heading*) The determination of ergonovine in ergot preparations...The ergonovine content is detd. by the intensity of coloration. **1955** *Sci. News Let.* 30 July 71/3 The radioactive fungus ergot, and the drugs ergotamine and ergonovine to be extracted from it. **1957** *Encycl. Brit.* VIII. 683/2 Ergonovine is used widely in obstetrical practice in the management of the placental stage in the prevention of excessive blood loss. **1961** [see *ERGOMETRINE].

ergophobia (əɪgŏfōᵘ·biă). *jocular.* [f. Gr. ἔργον work + -PHOBIA.] Fear of work. Hence **ergopho·bic** *sb.*, one who fears work.

1905 W. D. SPANTON in *Brit. Med. Jrnl.* 11 Feb. 300/2 He has discovered that it often pays better to idle and loaf about than to work, and the consequence is that a new disease has been engendered, which I have termed 'ergophobia'. **1905** *Daily Chron.* 26 May 5/7 Ergophobia..means a hatred or terror of work...It is a new disease which a medical paper has recently called attention to. **1921** *19th Cent.* Dec. 986 Ergophobia—a recognised if not a notifiable disease. **1960** *Daily Mail* 24 Nov. 10 (*headline*) Ergophobics? *Ibid.* 10/3 Boys suffering from ergophobia (the dilemma of having to choose between working and stealing).

ergosome (ə·ɪgŏsōᵘm). *Biochem.* [f. Gr. ἔργον work + *RIBO)SOME.] (See quots.)

1963 F. O. WETTSTEIN et al. in *Nature* 2 Feb. 434/2 In this article we have described a novel ribosomal particle which appears to be the functional unit of protein synthesis and which we therefore propose to call 'ergosome'. **1964** *Inst. Biol. Jrnl.* II. III. 94 *Ergosome*, an aggregate of ribosomes which is active in the synthesis of proteins. Same as polysome.

ergosterol (əɪgǫ·stĕrǫl). *Biochem.* [f. ERGO(T + *-STEROL.] A steroid alcohol, $C_{28}H_{44}O$, present esp. in ergot and in many yeasts and fungi, and productive of vitamin D_2 under ultraviolet irradiation. Also earlier **ergosterin** [ad. F. *ergostérine* (C. Tanret 1889, in *Compt. Rend.* CVIII. 98); cf. CHOLESTERIN].

1889 *Jrnl. Chem. Soc.* LVI. 408 Ergosterin..is completely dissolved by sulphuric acid without discoloration. **1906** *Ibid.* XC. II. 202 As obtained by Bömer's method..from the fat of rye, ergosterol contains a small amount of a product which forms small, white flocks melting at 60–61° and is non-saponifiable, but the nature of which is unknown. **1927** *Nature* 24 Sept. 440/2 Irradiated ergosterol possesses extraordinarily potent anti-rachitic activity. **1928** A. B. CALLOW *Food & Health* 51 Ergosterol..was present in minute quantities in what was formerly thought to be pure cholesterol...When ergosterol is irradiated it acquires the property of preventing or curing rickets, that is to say, it becomes vitamin D. **1934** *Nature* 2 June 838/1 Irradiated ergosterol determines the fixation of calcium in the animal organism, and hence induces good ossification...These ergosterol-containing oils readily become rancid. **1963** D. KRITCHEVSKY in Florkin & Stotz *Comprehensive Biochem.* X. i. 14 Ergosterol..was isolated initially from ergot but..also occurs in most fungi, including yeast, in lichens, algae and some vegetable oils.

ergotamine (əɪgǫ·tămin, -in). *Biochem.* [ad. G. *ergotamin* (Spiro & Stoll 1921, in *Actes Soc. Helv. Sci. Nat. 1920* 235), f. ERGOT + AMINE.] An alkaloid, $C_{33}H_{35}O_5N_5$, present in some kinds of ergot and used chiefly in the treatment of migraine. Hence **ergota·minine** [ad. G. *ergotaminin* (Spiro & Stoll 1921, in *Actes Soc. Helv. Sci. Nat. 1920* 236), f. ERGOT + AMIN(E + -INE⁵], a physiologically inactive ergot alkaloid, isomeric with ergotamine.

1921 *Brit. Pat.* 170,302 The principal alkaloid of ergot in the pure crystallised state..has since been designated..by the scientific name of 'ergotamine'. **1922** *Jrnl. Chem. Soc.* A. CXXII. I. 47 Ergotamine, obtained in crystallisable form from ergot, on treatment with methyl alcohol gives a new, less active alkaloid, *ergotaminine*. **1952** L. MARION in Manske & Holmes *Alkaloids* II. 376 Ergotamine and ergotaminine do not seem to occur in rye ergot, but are found in other ergots such as that growing..on marrom grass. **1955** [see *ERGONOVINE]. **1961** *New Scientist* 2 Nov. 278/1 The first pure ergot alkaloids were obtained by Stoll in 1920, namely the pharmacologically-active compound ergotamine and the inactive mirror-image molecule ergotaminine. **1964** S. DUKE-ELDER *Parsons' Dis. Eye* (ed. 14) xxiv. 362 Rest, warmth, and sleep are the best measures to combat the attacks [of migraine]; they can sometimes be warded off or alleviated by ergotamine tartrate. **1968** J. H. BURN *Lect. Notes Pharmacol.* (ed. 9) 75 Ergotamine has now lost much of its importance since the discovery of ergometrine (ergonovine).

ergotinine (see after ERGOTINE). (Additional examples.)

[**1875** C. TANRET in *Compt. Rend. Acad. Sci.* LXXXI. 896 Un alcaloïde nouveau...Je propose de l'appeler *ergotinine*]. **1940** *Thorpe's Dict. Appl. Chem.* (ed. 4) IV. 329/1 The first alkaloid ergotinine isolated by Tanret..was inert.

ergotoxine (əɪgŏtǫ·ksīn, -in). *Biochem.* [f. ERGO(T + TOX(IN + -INE⁵.] A mixture of three similar alkaloids present in ergot and used as an oxytocic, formerly thought to be a single compound; also, any of the alkaloids of this mixture.

1906 BARGER & CARR in *Chem. News* XCIV. 89/2 The crystalline alkaloid ergotinine was obtained from ergot by Tanret more than thirty years ago...We have now obtained the second alkaloid in a state of chemical purity, and suggest for it the name *ergotoxine*. **1937** GWYNNE-VAUGHAN & BARNES *Fungi* (ed. 2) 253 Owing to the presence of ergotoxine the sclerotia, if included in the grain used for bread, give rise to serious disease. **1952** L. MARION in Manske & Holmes *Alkaloids* II. 377 This anomaly..has recently been explained away by the discovery that ergotoxine is not homogeneous, but a complex mixture. **1958** FRUTON & SIMMONDS *Gen. Biochem.* (ed. 2) xxxiii. 862 Ergot contains the pharmacologically active alkaloids ergometrine, ergotamine, and several ergotoxines.

eria. Add: Also **eri.** The silkworm that produces this silk, *Attacus ricini*. Also *eria cocoon, moth, worm*.

1837 *Jrnl. Asiatic Soc. Bengal* VI. 23 The *eria* worm and moth differ from the mulberry worm or moth in every respect. **1866** *Chambers's Encycl.* VIII. 724/1 The 'Eria', or 'Arrindy' silkworm, native of India. **1887** *Encycl. Brit.* XXII. 60/2 The eria or arrindi moth of Bengal and Assam, *Attacus ricini*, which feeds on the castor-oil plant. **1909** H. MAXWELL-LEFROY *Indian Insect Life* 479 Moths reared from true eri cocoons cultivated in Assam proved to be the latter [*sc. Attacus ricini*]. **1923** *Nature* CXI. 411/2 It appears that all the recognised diseases are prevalent, and those of the mulberry, muga, and eri worms are the same. **1944** *Indian Farming* V. 365/1 The rearing of *eri* is much easier than that of any other silk worm. *Ibid.* 365/2 There is a large trade in silk and the *eri* cocoons. **1956** W. J. B. CROTCH *Silkmoth Rearer's Handbk.* (ed. 2) x. 128 Philosamia Grote (Eri Silkmoth).

ericoid (e·rikoid), *a. Bot.* [f. ERIC(A: see -OID.] Resembling or belonging to a plant of the genus *Erica*, the family Ericaceæ, or the subfamily Ericoideæ.

1900 B. D. JACKSON *Gloss. Bot. Terms* 92/2 *Ericoid*, used of leaves which are like those of heaths. **1954** *Proc. Prehist. Soc.* XX. 234 Beaulieu IX..appears to be the oldest, though it actually shows a higher quantity of ericoid pollen than Beaulieu VI. **1965** *Jrnl. Linn. Soc. Bot.* LIX. 119 (*heading*) The Ericoid Group...If we ignore *Calluna*, the floral and fruit characters used by Drude are correlated with the presence of the characteristic 'ericoid' habit.

erigeron (ĕri·dʒərǫn). Delete *Obs.* and add: **b.** [Linnæus *Hortus Cliffortianus* (1737) 407, from a name used by Theophrastus for a similar plant.] A hardy herbaceous annual, biennial, or perennial plant bearing daisy-like flowers and belonging to the large genus so called, of the family Compositæ.

1815 *Bot. Reg.* I. 10 (*caption*) Fordyce's Erigeron. **1900** J. M. ABBOTT in W. D. Drury *Bk. Gard.* viii. 270 Erigerons are summer-flowering composites with beautiful star-shaped flowers. **1901** *Cassell's Dict. Gard.* I. 325/2 Snails and slugs are frequently very destructive to Erigerons in late autumn. **1922** *Glasgow Herald* 16 Dec. 8 Mauve erigerons, scarlet geums, arabis. **1964** M. FISH *Ground Cover Plants* xi. 82 For years I have grown a low, fleshy erigeron..and nothing could make better ground cover for a bank or wall. **1969** HAY & SYNGE *Dict. Garden Plants* 295/1 A wide range of hybrid erigerons have been raised in England and in Germany.

erinaceous, *a.* See also HERINACIOUS.

erineum (ĕri·niǒm). *Bot.* [mod.L. (C. H. Persoon *Mycologia Europaea* (1822) I. 2), f. Gr. ἐρίνεος woolly, f. ἔριον wool.] A pathological growth of the epidermis of plants caused by mites of the family Eriophyidæ. Hence **eri·neal** *a.*

Formerly supposed to be caused by a fungus to which *Erineum* was given as the generic name.

1889 *Cent. Dict.* III. 1994/2 The erinea were formerly considered to constitute a genus of fungi. **1901** H. M. WARD *Disease in Plants* xxiii. 214 In true galls the hypertrophy may consist merely in the enlargement of cells already present..e.g...the hairy outgrowths of the epidermis known as *Erineum*. **1916** H. S. PRATT *Man. Invertebr. Anim.* 438 A fuzzy spot or erineum is a dense mass of twisted hairs, among which the mites live. **1937** F. D. HEALD *Introd. Plant Path.* ii. 16 An abnormal development of hairs or trichomes from the surfaces of leaves giving feltlike patches, first believed to be of fungous origin and named *erineum*, but now known to be caused by parasitic mites and designated *erinose*. **1964** M. S. MANI *Ecol. Plant Galls* iii. 50 Erineal galls with papillate hairs are not uncommon. **1969** N. W. HUSSEY et al. *Pests of Protected Cultivation* vi. 223 *Aceria lycopersici* (Wolffenstein), known as the tomato erineum mite, causes white and densely hairy patches on the stems and petioles.

erinose (e·rinōᵘs, -z). *Bot.* [f. as prec.: see *-OSE.] = *ERINEUM. Freq. *attrib.*

[**1910** G. MASSEE *Dis. Cultivated Plants & Trees* 541 Vine erinosis.—This disease is caused by *Eriophyes vitis*.] **1926** F. D. HEALD *Man. Plant Dis.* xvi. 416 Erinose spots are always convex above. **1937** —— *Introd. Plant Path.* ii. 16 Erinose spots on grapes are white at first. **1951** *Dict. Gardening* (R. Hort. Soc.) II. 772/1 Erinose often follows an attack of Red Spider on Cinerarias, Pelargoniums, and Vines.

eristic, *a.* and *sb.* **B.** *sb.* **1.** (Later example in sense 'a controversialist'.)

1965 *Times Educ. Suppl.* 26 Mar., Arrogance characterized the eristic while humility marked the heuristic.

erk (ɜɪk). *slang.* Also **irk.** [Of obscure origin.] **a.** A naval rating. **b.** An aircraftman, esp. an A.C.2. **c.** *transf.* Used as a term of contempt.

1925 FRASER & GIBBONS *Soldier & Sailor Words* 89 *Erk*, a rating. (Navy). Lower deck colloquialism for any 'rank' not that of an officer. **1928** T. E. LAWRENCE *Let.* 20 Jan. (1938) 570 Cranwell, which was a home from home, for the irks. **1940** *Reader's Digest* May 31/2 The aviators..call their mechanics *erks*, apparently a corruption of A.C., the abbreviation for aircraftsman. **1943** P. BRENNAN et al. *Spitfires over Malta* iii. 65 The erks came running up to tell us that..the 109 had been diving down. **1944** E. PARTRIDGE in *19th Cent.* Apr. 182 An *erk*, now used for an A.C.2..meant an air mechanic. This odd word is, the writer believes, a shortened pronunciation of the italicised letters in *air mechanic* (perhaps in the form of 'air mech')...Some airmen less convincingly maintain that it comes from 'lower-deck hand'. **1959** I. & P. OPIE *Lore & Lang. Schoolch.* x. 175 Somebody they dislike..may be called..erk, gawp, kid.

Erlang (ɜ·ɪlæŋ). The name of a Danish mathematician, Agner Krarup *Erlang* (1878–1929). **1.** Used to designate various formulæ, functions, etc., derived by him or arising out of his work; as *Erlang('s) distribution*, a distribution function of the time required for the occurrence of *n* random events (see quot. 1960¹); also, an Erlang loss formula; *Erlang('s) formula*, a formula giving the probability that a given number of lines in a communication system are occupied; *spec.*, *Erlang('s) congestion* (or *B-* or *first loss*) *formula*, a formula giving the probability of congestion assuming that calls finding all lines occupied are cleared; *Erlang('s) delay* (or *second loss*) *formula*, a formula giving the probability of congestion assuming that such calls are delayed until a line is available. Hence **Erla·ngian** *a.* [-IAN].

1918 *Post Office Electr. Engin. Jrnl.* XI. 151 These durations are very far from complying with either of Mr. Erlang's distribution laws, yet the results agree closely with the theoretical curve. I think, therefore, Mr. Erlang's formula can be looked upon as being of quite general application. **1920** *Ibid.* XIII. 215 Erlang's formula states that if the average number of simultaneous calls is *A*, the probability of exactly *r* switches being engaged if the total number of switches is *x*, is $(A^r/r!)/(1+A+A^2/2!+\ldots+A^r/r!+\ldots+A^x/x!)$. **1948** *Bell Syst. Techn. Jrnl.* XXVII. 425 The..assumption underlying the Erlang B formula is that of no delay. A call which encounters *all trunks busy* is cleared out. **1953** *Ann. Math. Statistics* XXIV. 339 We may also mention an intermediate type of input: E_k (Erlangian). **1954** A. JENSEN *Distribution Model* v. 16 Erlang's distribution corresponds to Poisson's distribution just as that of Pascal corresponds to the binomial distribution. **1960** R. SYSKI *Congestion Theory Teleph. Syst.* iv. 139 One has: $f_n(t)dt = (n\mu t)^{n-1}e^{-\mu nt}\,n\mu dt/(n-1)!$, $t \geq 0$ where μ is a positive constant. This form is known as the Erlang distribution...In several cases, the Erlang distribution is the only one which yields explicit solutions to congestion problems. *Ibid.* v. 252 The famous Erlang's loss formula (his B-formula—denoted usually by $E_N(A)$)..serves as the basis of traffic calculations for most telephone administrations. **1968** P. A. P. MORAN *Introd. Probability Theory* ii. 87 We can also truncate the Poisson distribution on the other tail and obtain a bounded distribution

$$p_n = \ldots = \lambda^n \bigg/ \left\{ n! \sum_{j=0}^{N} \lambda^j (j!)^{-1} \right\},$$

$n = 0, 1, \ldots, N$. This is also known as 'Erlang's distribution' and arises in the theory of telephone traffic.

2. A unit of the intensity of traffic in a communication system (see quot. 1960).

1947 *Post Office Electr. Engin. Jrnl.* XXXIX. 174/2 The proposal by the Swedish Administrations to use the name 'erlang' for the unit of intensity of traffic was adopted [by the 6th and 7th Commissions de Rapporteurs of the C.C.I.F.]. **1948** E. BROCKMEYER et al. *Life & Wks. A. K. Erlang* 20 The name of 'Erlang' has been used in the Scandinavian countries to denominate the unit of traffic since the beginning of 1944. **1951** *Jrnl. R. Statistical Soc.* B. XIII. 153 A parameter of special importance is the traffic-intensity $\rho = b/a$; in a long period of time, *T*, there will be approximately *T/a* calls for service, each of mean length *b*, and so ρ is the expected service demanded per unit of time, in 'erlangs'...(Properly it is not a unit at all, but an indication of how the preceding figure has been calculated; in this respect it resembles the octave, the decibel and the stellar magnitude.) **1960** *Gloss. Terms Telecomm. & Electronics (B.S.I.)* 216 *Erlang*, the unit of traffic flow...An Erlang is..equivalent to the traffic flow in one circuit continuously occupied.

Ernestine (ɜ·ɪnĕstəin), *a. Hist.* [f. *Ernest*, proper name = G. *Ernst*.] Designating the elder or electoral line of the house of Frederick of Saxony, originating from his son Ernest (1441–86).

1841 *Penny Cycl.* XX. 493/2 Albert and Ernest, who were the founders of the Albertine and Ernestine lines. **1866** *Chambers's Encycl.* VIII. 515/2 The Ernestine, which was also the elder or electoral line. **1904** *Westm. Gaz.* 23 Nov. 2/3 The Ernestine line of the House of Saxony to which the late Prince Consort belonged. **1911** *Encycl. Brit.* XXIV. 270/1 This deed transferred the electoral title..from the Ernestine to the Albertine branch of the house. **1943** A. J. GRANT *Hist. Europe 1494–1610* (ed. 3) i. 5 The electoral title went with the Ernestine line. **1962** R. A. G. CARSON *Coins* 366 In the original partition the electorate of Saxony was in the portion of Ernest but was transferred from the Ernestine line to the Albertine by the emperor Charles V in 1547.

Ernie (ɜ·ɪni). [f. *e*lectronic *r*andom *n*umber *i*ndicator *e*quipment.] Name given to the device for drawing the prize-winning numbers of premium bonds.

1956 *Electronic Engin.* XXVIII. 400 It has been decided that the only feasible method of making the draw is by electronic means. For this purpose an Electronic Random Number Indicator Equipment (ERNIE) has been evolved. **1957** *Times* 1 June 4/1 Ernie here is a pleasant enough fellow and..will be a lot more efficient at picking winning premium bonds. **1966** C. BERMANT *Diary of Old Man* 115 He won the premium bonds once, I know that for sure. We were sitting in the library and he was going over the Ernie numbers with a magnifying glass when he gave a sudden loud whoop. **1970** 'A. GILBERT' *Death wears Mask* ix. 137, I might have come from Ernie, to tell him he's won £25,000 in the draw.

erode, *v.* **2.** For *absol.* read *intr.* and add *fig.* examples.

1969 *Daily Tel.* 15 Oct. 16 (*heading*) Are the joys of canal boating to erode away? **1970** *Ibid.* 14 Oct. 17/2 Over the years their power has eroded considerably.

erogenous (erǫ·dʒinəs), *a.* = EROGENIC *a.* Both words are incorrectly formed.

1889 in *Cent. Dict.* **1905** H. ELLIS *Stud. Psychol. Sex* IV. I. i. 9 In some hysterical subjects there are so-called 'erogenous zones' simple pressure on which suffices to evoke the complete orgasm. **1915** C. R. PAYNE tr. *Pfister's Psychol. Method* 155 Those places which are important for the gaining of sexual pleasure, we call erogenous zones. **1925** C. FOX *Educat. Psychol.* 237 Any portion of the skin or mucous membrane which yields a feeling of pleasure when stimulated is described as an erogenous zone. **1949** G. BATESON in M. Fortes *Soc. Structure* 38 All the modes associated with the erogenous zones..define themes for *complementary* relationship. **1967** G. STEINER *Lang. & Silence* 90 Once the maximum number of erogenous zones of the maximum number of participants have been brought in contact..there is not much left to do or imagine.

-eroo, factitious slang suffix as in *BOOZEROO, **brusheroo* (BRUSH *sb.*² 8 b), *FLOPPEROO.

U.S. formations in *-eroo*, *-aroo* (e.g. *BUCKAROO) are discussed in *Amer. Speech* (1942) XVII. 10f, and in T. Pyles *Words & Ways Amer. Eng.* (1952) 199.

1964 *Guardian* 8 July 7/6 Those jerkeroos feel embarrassed.

Eros (ĭə·rǫs, e·rōᵘz). Pl. **Erotes** (erōᵘ·tez); **Eroses** (ĭə·rǫsĕz, e·rōᵘzĕz). [L. *Erōs*, a. Gr. ἔρως.] **1.** Love, the god of love, or a representation of him: = CUPID.

[*c* **1386** CHAUCER *Knt.'s T.* 1374 Nat oonly lik the loueris maladye of Hereos but rather lyk Manye]. **1671** PHILLIPS *New World of Words* (ed. 3), *Eros*, according to the *Ethnic* Poets the God of Love, who in Latin is commonly called *Cupido* also the name of *Mark Anthony's* servant who killed himself..the word in Greek signifying Love. **1775** J. BRYANT *Mythol.* I. 510 Under this characteristic they represented an heavenly personage, and joined her with Eros, or divine Love. **1817** BYRON *Manfred* II. ii. 35 He from out their fountain dwellings raised Eros and Anteros. **1864** TENNYSON *Enoch Arden, etc.* 157 A bevy of Eroses apple-cheek'd. **1870** S. H. HODGSON *Theory of Practice* I. 196 Love..in this sense should be called Eros. ..The resulting Eros will be felt in strength proportional to the combined intensity of these two constituents. **1877** PATMORE (*title*) The Unknown Eros and other Odes. **1888** A. H. SMITH *Catal. Engr. Gems Brit. Mus.* 127 Silenus.. threatening to flog Eros, who is held up by two other Erotes. **1896** L. R. FARNELL *Cults Gk. States* II. xxi. 625 The only ancient centres of Eros-worship were Thespiae and Parion, where he was regarded probably not merely as the personification of human love, but as a physical and elemental force, a divinity of fertility. **1904** BUDGE *3rd & 4th Egypt. Rooms Brit. Mus.* 229 Erotes, or Cupids, holding grapes and thyrsus with wreath. **1928** *Times* 14 Dec. 10/4 Mr. John Murray's suggested new site for the Shaftesbury memorial..would be an excellent one but for the fact that 'Eros' would then be lost to view from the principal approach roads to Piccadilly-circus. **1960** V. NABOKOV *Invitation to Beheading* xiv. 139 We were discussing..the pleasures of life, and had just examined Eros in a general way.

b. *Spec.* in Freudian Psychology: the urge towards self-preservation and sexual pleasure. Also, in recent Christian writings, earthly or sexual love, contrasted with *AGAPE 2.

1922 C. J. M. HUBBACK tr. *Freud's Beyond Pleasure Principle* vi. 64 Thus the Libido of our sexual instincts would coincide with the Eros of poets and philosophers, which holds together all things living. *Ibid.* 67 We are the more compelled now to accentuate the libidinous character of the self-preservative instincts, since we are venturing on the further step of recognising the sexual instinct as the Eros, the all sustaining. **1932** A. G. HEBERT tr. *Nygren's Agape & Eros* vii. 166 Eros is longing and desire; but God can have no unsatisfied needs. *Ibid.* 167 Eros-love is the movement of man to God, whereby man's need seeks to find its satisfaction in the Divine fullness and blessedness. *Ibid.* 175 Eros-religion is always æsthetic. **1940** W. EMPSON *Gathering Storm* 29 The Freudians regard the death-wish as fundamental, Though 'the clamour of life' proceeds from its rival 'Eros'. **1955** [see *AGAPE 2]. **1960** C. S. LEWIS *Four Loves* v. 106 That sexual experience can occur without Eros, without being 'in love', and that Eros includes other things besides sexual activity, I take for granted. **1967** A. EHRENZWEIG *Hidden Order of Art* xiii. 219 Both phenomena have their common source in a more fundamental rhythm that may be associated with the interaction of the life instincts, Eros and Thanatos.

2. *Astr.* An asteroid discovered by Witt at Berlin in 1898.

1900 *Ann. Reg. 1899* II. 102 Dr Witt, exercising a discoverer's right, has named the new planet Eros. **1901** G. C. COMSTOCK *Observations of Eros* 3 The following observations of Eros were made with the 40-cm. (Clark), equatorial telescope of the Washburn Observatory. **1926** H. MACPHERSON *Mod. Astron.* 92 Eros..comes at perihelion within the orbit of the Earth, from which it is then distant but thirteen million miles. **1959** SPITZ & GAYNOR *Dict. Astron.* 29 *Eros*, 20 mi. in diameter comes within 14 million miles, every 44 years.

erosion. Add: **3.** *erosion cycle, territory.*

1946 L. D. STAMP *Brit. Struct. & Scenery* v. 37 It may be said that land-forms depend..thirdly on the phase or stage within the erosion cycle. **1937** *Geogr. Jrnl.* XC. 376 Erosion-territory of the Great Ice sheet.

erosional (ĭrōᵘ·ʒənăl), *a.* [f. EROSION: see -AL.] Caused by erosion.

1903 *Science* 2 Jan. 26/2 The terraced character of the outlet at Horseheads was also described, and the opinion expressed that the broader terrace is an erosional and not a constructional..feature. **1947** *Endeavour* V. 43 That stratified rocks are the sands and muds of bygone ages and have been thus formed out of the erosional debris of pre-existing land masses began to be appreciated at the end of the eighteenth century. **1963** D. W. & E. E. HUMPHRIES tr. *Termier's Erosion & Sedimentation* i. 17 Heavy erosional phases represented by conglomeratic deposits of 'molasse' type. **1969** *Nature* 11 Oct. 164/1 Such streams are usually associated with extensive flood-plains leading out from upland V-shaped erosional valleys.

erotica (erǫ·tikă). [a. Gr. ἐρωτικά, neut. pl. of ἐρωτικός amatory, f. ἐρωτ-, ἔρως love.] Matters of love; erotic literature or art (freq. as a heading in catalogues).

1854 (*title*) Erotica. The Elegies of Propertius. **1913** H. JACKSON *Eighteen Nineties* v. 122 The romance,..in its unexpurgated form, suggests deep knowledge of that literature generally classed under *facetiæ* and *erotica* by the booksellers. **1956** K. CLARK *Nude* viii. 323 Supplied the Court with presentable *erotica*. **1957** N. FRYE *Anat. Criticism* ii. 110 The silly ones [*sc.* studies], which simply project the author's own erotica. **1960** *Spectator* 21 Oct. 595 In 1939, Girodias took over the erotica industry. **1961** *Ibid.* 3 Mar. 296 The artist had drawn the erotica. **1967** G. STEINER *Lang. & Silence* 91 Above the pulp-line..lies the world of erotica, of sexual writing with literary pretensions or genuine claims.

eroticism. Add: Also = EROTISM sense a.

eroticize (erǫ·tisəiz), *v.* [f. EROTIC *a.* and *sb.* + -IZE.] *trans.* To make erotic; to stimulate sexually.

1914 GEDDES & THOMSON *Sex* vi. 122 At the time of sexual maturity, according to Steinach, the brain is greatly influenced by the internal secretions, it is 'eroticised'. **1915** E. POUND *Let. c* 7 Sept. in *Lett. J. Joyce* (1966) II. 366 Bed rooms [*sic*] scenes where the audience can be tittivated, eroticised. **1955** H. MARCUSE *Eros & Civilization* (1956) ii. 39 Their unrepressed development would eroticize the organism. **1970** *Jrnl. Gen. Psychol.* Apr. 154 More women [than men] were embarrassed by eroticized stories and would not attend a sexy show if they could avoid it.

erotize (e·rǒtəiz), *v.* [f. Gr. ἐρωτ-, ἔρως love + -IZE.] *trans.* To transform (an emotion, etc.) into sexual feeling.

1936 *Brit. Jrnl. Psychol.* Jan. 288 The erotizing of aggressive impulses is a remarkably general process which accounts for much of the complexity of life. **1966** *New Statesman* 23 Sept. 441/1 The key could well be the primate and human capacity for erotising experience.

erotogenic (erŏᵘtodʒe·nik), *a.* [f. Gr. ἐρωτο-, ἔρως love + *-GENIC.] = EROGENIC *a.*

1909 in *Cent. Dict.* Suppl. **1922** J. RIVIERE tr. *Freud's Introd. Lect. Psycho-analysis* 264 The gratification obtained can only relate to the region of the mouth and lips; we therefore call these areas of the body *erotogenic zones*. **1924** J. RIVIERE et. al. tr. *Freud's Coll. Papers* II. 39 A certain degree of directly sexual pleasure is produced by the stimulation of various cutaneous areas (erotogenic zones). **1955** H. MARCUSE *Eros & Civilization* (1956) ii. 39 The pleasure of the proximity senses plays on the erotogenic zones of the body. **1968** *Economist* 25 May 25/3 Libido transcends beyond the immediate erotogenic zones.

erotogenous (erŏtǫ·dʒĕnəs), *a.* [f. as prec.: see -*GENOUS.] = *EROTOGENIC *a.*

1928 H. ELLIS *Studies Psychol. Sex* VII. ii. 116, I adopted the term 'erogenous zones' or, as I now prefer, 'erogenic zones'. The English psycho-analysts have sometimes put forward the form 'erotogenous'.

erotoma·niac. [f. EROTOMANIA: see -MANIA.] One affected by erotomania. Also *attrib.* or *adj.*

1858 BUCKNILL & TUKE *Man. Psychol. Med.* 212 The erotomaniac is..the sport of the imagination. **1895** *Contemp. Rev.* Apr. 494 The impudences..of the erotomaniacs. *Ibid.* 496 Erotomaniac literature. **1895** *Church Times* 5 Apr. 398/1 This kind of erotomaniac fiction. **1897** WILDE *Let.* 10 Aug. (1962) 631 He is the most learned erotomaniac in Europe. **1921** *Spectator* 16 Apr. 497/1 Outside an asylum for erotomaniacs, the world is not governed by perpetual storms of unrestrainable animal passion. **1967** *Listener* 30 Nov. 723/1 The erotomaniac gleam at the tip of the phallic cigar [of Groucho Marx].

errand. Add: **5. errand-boy** (earlier example); so *errand-girl*.

1765 *Minute Bk. St. Anne's Sch., Blackfriars* in D. Owen *Eng. Philanthropy* (1965) I. i. 25 [The Mother] suffering her said Boy to go from Place to Place as Errand-boy instead of keeping him at School. **1782** F. BURNEY *Diary* 28 Dec. (1842) II. 220 The green-woman.. sent her own errand-girl. **1860** C. M. YONGE *Hopes & Fears* II. v. 100 Augusta will be ready to take her in—she is pining for an errand girl. **1933** D. L. SAYERS *Murder must Advertise* ii. 31 'And who do you think I am,' I said, 'the errand-girl?'

erraticism (eræ·tisiz'm). [see -ISM.] Erratic tendencies.

1889 E. DOWSON *Let.* 14 July (1967) 92, I discovered the erraticism of the Sunday posts. **1907** *Daily Chron.* 18 Jan. 9/6 Pronounced Erraticism [in golf-play]. **1920** *19th Cent.* July 6 In various ways his erraticism threatened actual danger to the allied cause. **1925** DREISER *Amer. Trag.* II. II. xlvi, (Strange, equivocal, frivolous erraticism of his in this instance) the tripod of a recently purchased camera [etc.]. **1928** T. C. CHAMBERLIN *Two Solar Families* p. xxi, Their courses were all-wayward. And yet their erraticisms are accompanied with spectacular brilliancy.

error. Add: **4. d.** *law of error, random error* (see quots.); *probable error, standard error* (see under the first element in Suppl.).

1875 F. GALTON in *Phil. Mag.* 4th Ser. XLIX. 37 The law of frequency of error says that 'magnitudes differing from the mean value by such and such multiples of the probable error, will occur with such and such degrees of frequency'. **1876** Law of error [see DISPERSION 6]. **1910** *Encycl. Brit.* IX. 754/2 In mathematics, 'error' is the deviation of the observed or calculated quantity from its true value. The calculus of errors leads to the formulation of the 'law of error', which is an analytical expression of the most probably true value of a series of discordant values. **1936** *Jrnl. R. Aeronaut. Soc.* XL. 77 The distribution of the components of the velocity fluctuation at any given point appears to follow the 'random error law'. **1951** M. JAHODA et al. *Res. Methods Social Rel.* I. IV. 100 Random error is due to those transient aspects of the person, of the situation of measurement, of the measurement procedure, etc., which are likely to vary by chance from one measurement to the next. **1959** *Chambers's Encycl.* VIII. 220/1 The component of molecular velocity along any chosen direction is distributed according to the so-called 'error law', i.e. the number of molecules whose component velocity *u* lies between narrow limits *u* and *u+du* is proportional to $e^{-Au^2}du$.

6. *error-analysis*; *error-correcting, -detecting* vbl. sbs. (so *error-correction, -detection*); *error-free* adj.

1963 P. STREVENS *Papers in Lang.* (1965) i. 8 The study of error-analysis. **1968** FOX & MAYERS *Computing Methods for Scientists & Engineers* V. 119 Wilkinson (1963) also gives more details in a reasonably elementary way, of the error analysis for matrix problems. **1962** *Gloss. Terms Automatic Data Processing* 22 *Error correcting code*, an error detecting code which uses additional code elements so that for certain errors the mutilated representation resembles more closely the original than any other valid representation. **1965** *Math. in Biol. & Med.* (*Med. Res. Council*) III. 124 The similarity to, and the difference from, either the 'substantialization' of sign-sequences or error-correcting codes may be noted. **1964** T. W. MCRAE *Impact of Computers on Accounting* vi. 164 These error-correction procedures are very complicated indeed. **1927** J. ADAMS *Errors in School* 236 Applying the parallel to error-detecting. **1962** *Gloss. Terms Automatic Data Processing* 22 *Error detecting code*, a code in which each representation of a character conforms to specific rules of construction, so that for certain errors the mutilated representation corresponds to no valid character. **1927** J. ADAMS *Errors in School* 248 Responsibility of error-detection. *Ibid.* 38 Error-free material. **1964** C. DENT *Quantity Surveying by Computer* iii. 32 Checking devices to ensure error-free tapes.

ersatz (ɜ·ɪzæts, ||e·rzats). [G., = compensation, replacement.] A substitute or imitation (usually, an inferior article instead of the real thing). Also *attrib.* or *adj.*, and *fig.*

1875 *Encycl. Brit.* II. 594/2 (German army), Those who are exempted..are passed into the *Ersatz* reserve. **1892** J. ROYCE *Let.* 17 Oct. in R. B. Perry *Tht. & Char. of W. James* (1935) I. 803 To me he is a great comfort, although..no *Ersatz* for the aforesaid condition of my heart. **1910** *Pedagogical Seminary* 305 When names are forgotten owing to such a disturbance, an *Ersatz* name appears. **1919** *War Terms* in *Athenæum* 1 Aug. 695/1 Another word not seldom met with is 'ersatz'. It is the German 'substitute'. **1927** *Daily Express* 30 Sept. 1/1 It will merely be an imitation Parliament, an 'Ersatz' Parliament, designed to fulfil the immediate needs of the Dictatorship. **1928** T. S. ELIOT *For Lancelot Andrewes* 89 Of course Mr. Shaw and Mr. Wells are also much occupied with religion and *Ersatz-Religion*. **1930** *Observer* 9 Mar. 12 The coffee..will be..tempered with a judicious mixture of 'ersatz'. **1938** *Jrnl. R. Aeronaut. Soc.* XLII. 919 The problem of 'Ersatz' materials is well considered and modern aeronautical materials (like plastics, compressed wood, resin glues, etc.) are described. **1939** *Punch* 22 Nov. p. vii/2 (Advt.), A *real* tobacco. There is no Ersatz in Four Square. **1940** *Nature* 13 Jan. 60/1 Present-day brands of margarine cannot be considered in any way *Ersatz* butter. **1942** L. B. NAMIER *Conflicts* 58 This set a high premium on Hitler's jack-boots and *ersatz* uniforms. **1944** E. H. W. MEYERSTEIN *Let.* 24 Nov. (1959) 297 He [*sc.* Swinburne] went now and then to places where women beat him, which must have been a wretched ersatz for the sort of thing he wanted, e.g. to be flogged by the Duke of Wellington. **1949** F. MACLEAN *Eastern Approaches* III. v. 362 A breakfast of black bread and captured ersatz coffee made from roasted grain. **1950** *Mind* LIX. 20 Propositions regarded as *ersatz* facts or quasi-things. **1952** H. READ *Philos. Mod. Art* I. iii. 69 All we can create in that way is an *ersatz* culture, the synthetic product of those factories we call variously universities, colleges or museums. **1956** C. WILSON *Outsider* v. 133 Kant, with his *ersatz* morality, is a special target.

erstwhile, *adv.* Add: Also *adj.* = former. (Cf. *whilom*.)

1903 'A. MCNEILL' *Egregious English* iii. 31 The erstwhile portly mother of daughters. **1909** *Westm. Gaz.* 21 Aug. 13/1 A tottering pleasure-resort, whose erstwhile patrons look more longingly every year at the pretty and easily reached villages of Normandy and Brittany. **1925** L. HOUSMAN *Odd Pairs* i. 24 Her erstwhile rival.

Ertebølle (ɜ̄ɪtəbö·lə). Also **Ertebölle, Ertebolle.** [Place-name in Jutland, Denmark.] An early Danish culture illustrated by remains found at Ertebølle. Also *attrib.*

1927 PEAKE & FLEURE *Hunter & Artists* vii. 109 This period, known as the Littorina period, is the time of the 'kitchen-middens' or shell-mounds...Two phases of this culture are usually recognised...The second or chief phase..is called the Ertebölle phase. **1933** *Antiquity* VII. 47 Sites of the Ertobolle cultures which are in the Atlantic period and correspond roughly to the Litorina maximum. **1948** A. L. KROEBER *Anthropol.* (ed. 2) xvi. 670 *Ertebølle*, first called Kjøkkenmødding, then kitchen-midden culture. Type site in Denmark, distribution northern. Time, late Mesolithic..5000–3000 B.C. **1954** S. PIGGOTT *Neolithic Cultures* iii. 84 Bone combs..occur in the Ertebølle culture. **1959** *Chambers's Encycl.* V. 453/2 The Ertebölle culture is well known from the extensive middens formed of the shells of the molluscs on which the people largely lived.

e·ructated, *ppl. a.* [f. ERUCTATE *v.* + -ED¹.] Vomited.

1869 C. L. BRACE *New West* i. 13 The reeling state-rooms, smelling of oil, bilgewater and eructated dinner.

erven, pl. of ERF¹.

-ery, *suffix.* Add: **2. b.** In modern, chiefly U.S., use, after *bakery* (= baker's shop or works), and similar words, this suffix has gained considerable currency in denoting 'a place where an indicated article or service may be purchased or procured', as *beanery, bootery, boozery, breadery, cakery, carwashery, drillery, drinkery, eatery, hashery, lunchery, mendery, toggery, wiggery*.

eryngium (ĭri·ndʒiǒm). [mod.L., f. L. *ēryngion*, a. Gr. ἠρύγγιον: see ERYNGO. Adopted by Linnæus (*Systema Naturæ* 1735) and earlier botanists as a genus name.] A hardy herbaceous perennial plant of the genus so called, belonging to the family Umbelliferæ, and bearing blue or white flowers.

[**1548** [see SEA-HOLLY].] **1578** H. LYTE tr. *Dodoen's Herb.* IV. 518 Two kindes of Eryngium, the one called the great Eryngium, or Eryngium of the Sea, and the other is called but Eryngium onely. **1616, 1626, 1861** [see ERYNGO I b]. **1899** G. JEKYLL *Wood & Garden* xi. 138 Sometimes a Poppy or an Eryngium comes up with one thick root, impossible to divide. **1900** W. D. DRURY *Bk. Gard.* 270 Eryngiums are handsome plants with..large branching heads of Thistle-like flowers. **1962** *Amat. Gardening* 17 Mar. 5/3 Echinops and eryngiums..are others which need little attention.

erysipeloid (erisi·pĭloid). *Path.* [a. G. *erysipeloid* (Rosenbach 1887, in *Arch. f. klin. Chir.* XXXVI. 346), f. ERYSIPEL(AS + -OID. J. Daudé (*Traité de l'érysipèle épidémique* (1867) II. 128) had earlier used *érysipéloïde* as the name of some skin diseases that he distinguished from erysipelas and from erythema, but this is not the source of the mod. sense.] Dermatitis of the hands due to infection with the bacillus causing swine erysipelas *Erysipelothrix rhusiopathia*.

[**1882** A. STILLÉ in J. Ashhurst jr. *Internat. Encycl. Surg.* I. 181 The nature of most of these recurrent cases is open to question, and..not a few of them seem to have been instances of erythema rather than of erysipelas, or, as they have been called by Daudé, erysipeloids.] **1888** *Jrnl. Cutaneous & Genito-Urinary Dis.* VI. 13 Rosenbach describes Erysipeloid as being a form of wound infection. **1955** G. O. DAVIES in Gaiger & Davies *Vet. Path. & Bacteriol.* (ed. 4) xv. 319 Swine erysipelas infection sometimes occurs in man from handling infected pigs and is spoken of as erysipeloid. **1961** *Lancet* 2 Sept. 530/1 Does penicillin cure erysipeloid in a few days? **1969** *New Scientist* 13 Feb. 349/1 Infections including..erysipeloid in Man—a wound infection especially common in veterinarians.

erythema. Add: **ery·themal** *a.* [-AL], of, pertaining to, or causing erythema.

1940 *Jrnl. Appl. Physics* June 422 Other emissivities such as an erythemal emissivity might be defined. **1961** *Lancet* 26 Aug. 449/1 Other spectral regions—e.g., 260–320 mμ produced normal erythemal responses. **1962** H. C. WESTON *Sight, Light & Work* (ed. 2) v. 156 The amount of erythemal..ultra-violet radiation emitted by fluorescent tubes..is about the same..as that which accompanies noon sunlight.

erythræmia (eriþrī·miă). *Path.* Also **erythremia.** [mod.L., ad. G. *erythrämie* (W. Türk 1904, in *Wiener klin. Wochenschr.* 18 Feb. 192/1), f. Gr. ἐρυθρός red + αἷμα blood.] An excess of red cells in the blood; any disease so characterized, esp. polycythæmia vera.

[**1860** MAYNE *Expos. Lex.* Add. 1493/1 *Erythræmia*, term for the full preparing of arterial blood in the lungs.] **1908** W. OSLER in *Lancet* 18 Jan. 145/1 For the recognition of the disease a blood count is necessary, not simply a blood examination, as in cases of leukæmia. The essential feature, the polycythæmia, the erythræmia, can be determined only by counting the number of red blood corpuscles in a cubic millimetre. **1968** PASSMORE & ROBSON *Compan. Med. Stud.* I. xxvi. 3 The red cell mass is naturally low in anaemia; when it is raised the patient is said to have erythraemia.

Hence **erythræ·mic** *a.*, characterized by or characteristic of erythræmia.

1938 HARROP & WINTROBE in H. Downey *Handbk. Hematol.* IV. xxxiv. 2378 Many members of such families may possess the 'erythremic constitution'. **1962** *Lancet* 1 Dec. 1140/2 Acute erythræmic myelosis, or acute immature-cell erythræmia, has become well known since its original description by Di Guglielmo (1928).

erythrism (e·riþriz'm). [f. Gr. ἐρυθρός red + -ISM.] Abnormal or excessive redness (in the plumage of birds or hair of mammals); a red variety (of some recognized species).

1886 *Proc. Zool. Soc. Lond.* 77 Erythrism is particularly common among the Mungooses. **1893** *Athenæum* 28 Jan. 124/3 Mr. Hose thought that this species might possibly be only an erythrism of *S*[*emnopithecus*] *chrysomelas*. **1908** *Zoologist* Apr. 136 A good example of the same erythrism was that by Mr. McLean in the summer of 1906. **1930** E. W. HENDY *Wild Exmoor* 297 Erythrism, or excess of red colouration, may be produced..by feeding birds upon red pepper. **1962** *New Scientist* 9 Aug. 284/2 All examples of albinism (whitening),..xanthism (yellowing) or erythrism (reddening) are throwing light on genetic divergence.

Hence **erythri·stic** *a.* exhibiting erythrism.

1910 *Encycl. Brit.* V. 489/1 A third colour-phase, the 'erythristic' or red, is represented by the sandy cat. **1923** *Glasgow Herald* 17 Mar. 6 The variety, technically called 'erythristic', occurs among ferrets as well as among polecats. **1936** *Discovery* Feb. 57/1 Another collector kicked guillemots' eggs into the sea in the hope that subsequent layings might produce the erythristic eggs for which he was looking. **1953** BANNERMAN *Birds Brit. Isles* II. 11 (Calandra lark), Erythristic eggs occur.

erythritol (eri·þritǫl). *Chem.* [f. ERYTHRITE + -OL.] A tetrahydric alcohol, 1,2,3,4-butanetetraol, occurring chiefly in lichens and algæ, and obtainable by reduction of erythrose.

1891 *Jrnl. Chem. Soc.* LX. 657 (*heading*) Alkaline derivatives of erythritol. *Ibid.*, Cold concentrated solutions of erythritol and sodium hydroxide were mixed..and the solution was allowed to evaporate. **1900** *Jrnl. Chem. Soc.* LXXVIII. I. 579 The algæ *Trentepholia Jolithus*..when extracted with ether, yields erythritol. **1923** J. J. SUDBOROUGH *Bernthsen's Org. Chem.* (new ed.) viii. 210 Erythritol (Butane-tetrol) occurs in the free state in *Protococcus vulgaris*. **1968** *New Scientist* 30 May 445/1 Certain tissues in the foetus, but not in the adult, contain erythritol, which stimulates the growth of *Brucella abortus*.

erythro-. Add: **ery·throblast** [G. (M. Löwit 1886, in *Sitzungsber. d. Akad. d. Wissenschaften in Wien* 3rd Ser. XCII. 56)], any of the normal series of nucleated cells recognizable as precursors of erythrocytes; hence **erythrobla·stic** *a.*; **ery·throcyte**, a red blood corpuscle; also *attrib.*; hence **erythro-**

cy·tic *a.*; **ery:throcythæ·mia** = *ERYTHRÆMIA; **ery:throcyto·sis**, erythræmia, esp. when a secondary condition resulting from some other disturbance; **ery:thromela·lgia**, dilatation of the arteries of the extremities, esp. the feet; **ery:thropho·bia**, (*a*) fear of blushing; (*b*) hypersensitivity to the colour red; **ery:thropla·stid** = **erythrocyte*; **erythro·psia** [Gr. -οψια seeing], a form of chromatopsia in which all objects appear red.

1890 BILLINGS *Med. Dict.*, Erythroblasts. **1898** *Allbutt's Syst. Med.* V. 651 None of the leucocytes of the blood becomes transformed into red corpuscles, these being formed from special cells—'erythroblasts'—in the bone-marrow. **1908** OSLER & MCCRAE *Syst. Med.* IV. 600 A considerable number of the immature cells of the adult marrow, including many of the erythroblasts, possess a considerable degree of affinity for basic coloring matters. **1968** H. HARRIS *Nucleus & Cytoplasm* iv. 86 An electron microscopic study of nuclear elimination from the late erythroblast. **1908** *Practitioner* Feb. 239 The erythroblastic tissues of the marrow. **1908** OSLER & MCCRAE *Syst. Med.* IV. 679 The disease is due to a primary hyperplasia of the erythroblastic bone-marrow. **1894** G. M. GOULD *Illustr. Dict. Med.* 441/1 *Erythrocyte*, a red blood-corpuscle. **1908** OSLER & MCCRAE *Syst. Med.* IV. 600 We speak of the..blue-stained erythrocyte as polychromatophilic. **1946** *Nature* 30 Nov. 793/1 Regular erythrocyte counts and hæmoglobin determinations were done in a drop of blood. **1961** *Lancet* 2 Sept. 522/2 The erythrocyte-sedimentation rate was 35 mm. in the 1st hour. **1905** OSLER *Princ. Med.* (ed. 6) 748 An increase in the number of the red blood-corpuscles—erythrocythæmia. **1908** *Practitioner* Feb. 234 According to the degree in which the parent cell has developed along the leucocytic or erythrocytic route. **1946** *Nature* 16 Nov. 707/2 Therapeutic potency..is apparent not only against the erythrocytic but also against the exo-erythrocytic forms of the malaria parasite. **1908** *Practitioner* Feb. 238 It would almost seem better to bring the nomenclature of this new disease into line with that adopted to describe an increase of the various forms of leucocytes, and to call it polycythæmia rubra, erythrocytosis, or, better still, erythrocythæmia. **1966** WRIGHT & SYMMERS *Systemic Path.* I. iv. 175 The number of red cells in the blood is seldom as large in erythrocytosis as in polycythæmia vera. **1878** S. W. MITCHELL in *Amer. Jrnl. Med. Sci.* July 17 The foot and hand disorder I am about to describe may be conveniently labelled Erythromelalgia. **1956** *New Gould Med. Dict.* (ed. 2) 419/2 *Erythromelalgia*...Also called acromelalgia, Mitchell's disease. **1894** G. M. GOULD *Illustr. Dict. Med.* 441/2 *Erythrophobia*, morbid intolerance of red colors: sometimes observed after operations for cataract. **1947** P. L. HARRIMAN *Dict. Psychol.* 126 *Erythrophobia*, morbid fear of blushing; also, hypersensitivity to reds, a symptom reported as occurring in some patients who have had cataracts removed. **1936** *Mind* XLV. 389 They are then 'neurotic' symptoms.. (neurotic vomiting..blushing and erythrophobia, [etc.]). **1960** KOESTLER *Lotus & Robot* II. viii. 213 In this is included fear of blushing when appearing before a person, or erythrophobia. **1921** A. KEITH *Hum. Embryol.* (ed. 4) xxi. 336 At every period of life the red blood corpuscles (erythroplastids) arise from erythroblasts. **1885** *Boston Med. & Surg. Jrnl.* 24 Dec. 615/2 The patient reported the appearance of the red vision in four months after having had the cataract extracted...It came on at bedtime...The erythropsia had gone on the following morning. **1964** S. DUKE-ELDER *Parsons' Dis. Eye* (ed. 14) xxiv. 363 *Erythropsia* (red vision) occurs particularly after cataract extraction if the eyes are exposed to bright light.

erythroblastosis (eri:þroblæstōᵘ·sis). *Path.* [ad. G. *erythroblastose* (H. Rautmann 1912, in *Beitr. z. Path. Anat.* LIV. 347), f. **erythroblast*: see -OSIS.] The presence of erythroblasts in the blood; *esp.* hæmolytic disease of the newborn, erythroblastosis fœtalis, a severe anæmia that results when the mother develops antibodies against the blood of the fœtus. Hence **ery:throblasto·tic** *a.*

1931 *Amer. Jrnl. Path.* VII. 293 Erythroblastosis may occur without edema. **1938** *Nature* 29 Jan. 202/2 Researches aiming at investigating the nature of the virus of aviary erythroblastosis. **1957** *Encycl. Brit.* I. 863/1 Once a woman has borne an erythroblastotic infant, she is likely to bear others. **1964** M. HYNES *Med. Bacteriol.* (ed 8) xxv. 404 Erythroblastosis from rhesus incompatibility.

erythrogen. Add: Hence **ery:throge·nic** *a.*, (*a*) producing a sensation of redness; (*b*) pertaining to the production of erythrocytes; (*c*) causing redness of the skin.

1902 C. L. FRANKLIN in Baldwin *Dict. Philos.* II. 780/2 *Colour* should be used in its subjective sense,..and *erythrogenic radiations* should be substituted for the physical cause of colour. **1951** WHITBY & HYNES *Med. Bacteriol.* (ed. 5) xi. 158 Only streptococci of Lancefield's Group A produce the erythrogenic toxin. **1951** [see *ERYTHROPOIESIS].

erythroid, *a.* Add: **2.** Of or pertaining to erythrocytes.

1927 SABIN & DOAN in *Proc. Soc. Exper. Biol. & Med.* XXV. II. 122 In the new-born rabbit the erythroid-myeloid ratio of the bone marrow is nearly the reverse of that of the adult. **1961** *Lancet* 26 Aug. 491/1 Moderate to severe folic-acid deficiency existed in the absence of megaloblastic characteristics in the erythroid or myeloid series of the blood and marrow.

erythrol (e·riþrǫl). *Chem.* [f. ERYTHR(ITE + -OL.] **1.** = *ERYTHRITOL.

1879 *Jrnl. Chem. Soc.* XXXVI. 664 *Erythrol* (erythrite) mixed with hay-water yielded a trace of alcohol. **1910** *Practitioner* Feb. 277 Erythrol tetranitrate is said to have less effect than liquor trinitrini. **1940** A. J. CLARK *Appl. Pharmacol.* (ed. 7) xxi. 402 Organic nitrates, such as nitro-glycerine and erythrol tetranitrate, act like nitrites.

2. (See quot. 1908.) (Disused.)

1900 in DORLAND *Med. Dict.* **1908** *Practitioner* Feb. 279 For butyric fermentation, is used erythrol, a red powder insoluble in water, which is a double iodide of bismuth and cinchonidine.

erythromycin (eri:þromai·sin). *Med.* [f. ERYTHRO-: see *-MYCIN.] An antibiotic isolated from *Streptomyces erythreus*.

1952 *Antibiotics & Chemotherapy* II. 279 The announcement..of the discovery of erythromycin (Ilotycin), a new broad spectrum antibiotic..emphasizes again the extensive efforts necessary to isolate an antibiotic. **1960** M. E. FLOREY *Clinical Appl. Antibiotics* IV. ii. 36 Like erythromycin, the antibiotic carbomycin..was isolated in 1952 and has similar antibacterial properties. **1962** *Lancet* 24 Nov. 1096/1 Erythromycin has an antibacterial range somewhat similar to that of penicillin. **1967** *New Scientist* 5 Oct. 35/2 The extremely useful macrolide group of antibiotics, which includes erythromycin and oleandomycin.

erythronium (eriþrōᵘ·niǒm). [mod.L. (Linnæus *Systema Naturæ* 1735), ad. Gr. σατύριον ἐρυθρόνιον, Dioscorides' name for the red SATYRION, referring to the colour of the flowers of the European species.] A member of the genus of ornamental bulbous plants so called, native to North America, Europe, Russia, and Japan, and belonging to the family Liliaceæ.

1797 *Encycl. Brit.* VI. 722/1 *Erythronium*, dog's-tooth violet. **1868** *Curtis's Bot. Mag.* XCIV. 5714 It would appear that the species or forms of West American *Erythronia* retain their colours over considerable areas. **1900** W. D. DRURY *Bk. Gard.* x. 326 Erythroniums range between 6 in. and 9 in. in height. **1961** *Amateur Gardening* 30 Dec. 5 Erythroniums like a good loam to which the peat and leaf mould has been added.

erythropoiesis (eri:þropoi,ī·sis). *Physiol.* [f. ERYTHRO- + Gr. ποίησις creation.] The production of red blood cells (erythrocytes).

1918 *Amer. Jrnl. Anat.* XXIV. 24 In the liver..and the omentum, erythropoiesis has been repeatedly described as taking place extravascularly. **1941** *Lancet* 23 Aug. 208/2 A scheme of normal and pathological erythropoiesis is proposed which shows that disease may cause changes in the development of the normoblastic group of cells or maturation may be deflected to a megaloblastic group. **1951** G. H. BOURNE *Cytol. & Cell Physiol.* (ed. 2) x. 432 Perhaps the most favourable material for observing the details of erythropoiesis in living tissue is provided by small explants of the presumptive erythrogenic mesoderm of an early chick blastoderm...The erythrogenic cells rapidly disintegrate into small, almost spherical primitive blood cells.

ery:thropoie·tic, *a.* [f. ERYTHRO- + POIETIC *a.*] Of or pertaining to the production of erythrocytes.

1908 *Practitioner* Apr. 457 The bone-marrow of dogs which have developed a polycythaemia from being kept at high altitudes, has been..found to give abundant evidence of increased erythropoietic activity. **1949** FLOREY & JENNINGS in H. W. Florey *Antibiotics* II. xxxvii. 1209 Harmlessness to the erythropoietic tissues was an important desideratum for penicillin. **1966** *Lancet* 24 Dec. 1416/2 Iron utilisation is a marker of bone-marrow erythropoietic activity.

ery:thropoi·etin. [f. *ERYTHROPOIET(IC *a.* + -IN¹.] A protein formed in the kidney that stimulates the formation of red blood cells (erythropoiesis).

1948 *Acta Physiol. Scand.* XVI. 152 Our present knowledge of the formation of erythropoietins and their role in anoxic conditions is limited. **1958** *Nature* 31 May 1537 Recently the presence of a substance capable of stimulating erythropoiesis (erythropoietin) in the peripheral blood of man and anæmic animals has been demonstrated repeatedly by a variety of assay methods. **1963** *Lancet* 12 Jan. 89/2 He supports the suggestion that there is a single erythropoietin produced or activated by the kidney.

erythrose (eri·þrōᵘz, e·riþrōᵘz, -s). *Chem.* [f. ERYTHR(ITE + -OSE².] A tetrose sugar, $CHO\cdot[CH(OH)]_2\cdot CH_2OH$, isolated as a liquid and existing in two molecular configurations; it differs from threose in having the hydroxyl groups on the second and third carbon atoms on the same side of the carbon chain.

1900 *Jrnl. Chem. Soc.* LXXVIII. I. 139 *d*-Erythrose.. shows mutarotation, and has an approximate maximum specific rotation $[\alpha]_D - 14.5°$ at 20°. **1963** E. PERCIVAL in Florkin & Stotz *Comprehensive Biochem.* V. i. 11 The simplest sugars are the tetroses D- and L-erythrose..and D- and L-threose.

erythrosin (eri·þrǫsin). *Chem.* Also **-ine.** [irreg. f. Gr. ἐρυθρός red + -IN¹.] A red colouring matter obtained from fluorescein by the action of iodine, used in photography and as a histological stain.

1884 *Photogr. Jrnl.* Dec. 52 Each of the following colouring matters was found to exercise a sensitizing action..for the extreme red rays of the solar spectrum. Iodine green,..erythrosin, pyrosin. **1891** *Jrnl. Camera Club* V. 123–7. **1892** *Photogr. Ann.* II. 72 Plates sensitised with erythrosin and erythrosin silver. **1908** *Jrnl. Chem. Soc.* XCIV. I. 669 When silver nitrate is added to an excess of a dilute solution of erythrosin, a much more intensely colloidal solution of the silver salt of erythrosin is obtained. **1926** *Encycl. Brit.* Suppl. III. 133/2 Erythrosine, that is the tetra-iodo derivative of fluorescein, was found by Eder to be the most suitable dye for sensitising in the green and yellow. **1968** *Brit. Corrosion Jrnl.* III. 31/1 The fruit salad contained a few cherries dyed with erythrosin.

Esau (ī·sǭ). Name of the elder of the twin sons of the patriarch Isaac, who sold his birthright to his brother Jacob for 'a mess of pottage' (Gen. xxv. 25 ff.), used symbolically for: one who prefers present advantage to permanent rights or interests.

1662 T. BROOKES *Crown & Glory of Christianity* 82 He that is a Cato without, and a Nero within; a Jacob without, and an Esau within..is ripened for the worst of torments. **1822** *Nat. Intelligencer* (Washington) 14 Feb. 2/3 The elder brethren of the confederacy—the political Esaus of our tribe. **1845** BROWNING *Flight of Duchess* xvii, Smooth Jacob still robs homely Esau. **1859** KINGSLEY *Misc.* II. 240 Esau is a dumb soul, especially here in England; but he has as deep a heart in him as Jacob. **1859** THACKERAY *Virginians* II. xxxvi. 297, I turned from Esau, and I clung to Jacob. And now I have my reward, I have my reward! **1888** STEVENSON *Master of Ballantrae* iv. 105 'Ah! Jacob', says the Master. 'So here is Esau back.' **1923** D. L. SAYERS *Whose Body?* ii. 26 It's no good talking as if you were announcing dinner—you're spilling the brandy. The voice is Jacob's voice, but the hands are the hands of Esau.

escalate (e·skălĕⁱt), *v.* [Back-formation f. *ESCALATOR.] **1. a.** *trans.* To climb or reach by means of an escalator. **b.** *intr.* To travel on an escalator.

1922 *Granta* 10 Nov. 93/2, I dreamt I saw a Proctor 'escalating', Rushing up a quickly moving stair. **1927** *Atlantic Monthly* Jan. 48/1 With almost incredible ingenuity we ward off the bumps, plane the sharp corners, 'escalate' the heights. **1959** *Observer* 21 June 14/6 Why does other people's luggage always look so neat and articulate as it comes escalating along at the airport?

2. *fig.* (*trans.* and *intr.*). To increase or develop by successive stages; *spec.* to develop from 'conventional' warfare into nuclear warfare.

1959 *Manch. Guardian* 12 Nov. 1/1 The possibility of local wars 'escalating into all-out atomic wars'. **1961** *Economist* 28 Oct. 317/2 Using tactical nuclear weapons which would be likely to escalate hostilities into a global nuclear war. **1962** *Listener* 29 Mar. 543/2 To keep the Communists from escalating the war to a stage in which they are capable of..taking over the reins of local government themselves. *Ibid.* 549/1 Something short of a deliberate Soviet attack, some sort of accident, some sort of border incident, will escalate so rapidly into nuclear war that we shall be unable to stop it. **1963** *Horizon* (U.S.) July 37/1 The wish of the author to magnify or escalate (favorite new word in Washington) the importance of a trivial utterance by grandiloquent terminology. **1963** *Oxf. Mail* 26 July 6/9 Peace as well as war, can escalate. **1966** *Punch* 28 Sept. 482/2 The animal book boom is terrific and still escalating. **1967** *Listener* 3 Aug. 131/3 Only a tiny percentage of cannabis-smokers escalate to heroin. **1970** *Guardian* 4 June 12/1 The National Council has before it a demand to 'escalate Palestinian action in all fields'.

escalation (e:skălĕⁱ·ʃən). [f. *ESCALATE *v.* + -TION.] **a.** The act or process of increasing armaments, prices, wages, etc. **b.** An increase or development by successive stages; *spec.* the development of 'conventional' warfare into nuclear warfare, or the use of successively more powerful types of weapons in war.

1938 *Kansas City Star* 29 Mar. 2/4 *Escalation* means the building of bigger battleships when other nations do so. **1947** *Times* 14 Feb. 8/3 Mr. Dalton said that the report [of the War Damage Commission] did advise a certain escalation. **1949** *N.Y. Times* 14 Aug. III. 1/1 Sales contracts permit upward adjustment through 'escalation' as a result of wage increases granted by the vendors. **1959** W. YOUNG *Strategy for Survival* iii. 32 The main argument against this policy is the danger of what strategists call *escalation*, the danger that the size of the weapons used would mount up and up in retaliation until civilization is destroyed. **1959** *New Statesman* 5 Dec. 784/2 The risk of starting the process of escalation towards total war. **1960** *Guardian* 14 Dec. 8/1 The 'escalation' of a minor conventional war into a major nuclear one. **1962** H. KAHN *Thinking about Unthinkable* vi. 185 There is a tendency for each side to counter the other pressure with a somewhat stronger one of its own. This increasing pressure step by step is called 'escalation'. **1964** *Ann. Reg. 1963* 136 In addition both leaders..were anxious to check the increasingly painful escalation of defence costs. **1964** *Financial Times* 23 Mar. 14/3 Shell Oil Company and the Moonie Oil Group have agreed on a price of $2.83 a barrel for Moonie oil without an escalation clause. **1965** *Spectator* 12 Feb. 187/3 A conviction that President Johnson would not risk escalation. **1967** *Ibid.* 21 July 87/1 The case for liberalising the law re marihuana seems almost proven...The doubts remaining are..that it may produce escalation to hard drugs.

escalator (e·skălĕⁱtəɪ). orig. *U.S.* [f. ESCALADE *v.*, after ELEVATOR.] **1.** (Orig. a trade-name.) A moving staircase made on the endless-chain principle, so that the steps ascend or descend continuously, for carrying passengers up or down.

1900 *N.Y. Jrnl.* 25 Nov. 59/2 The escalator..is a movable stairway built by the Otis Elevator Company for the use of passengers of the Manhattan Elevated Railway. **1904** *Daily Chron.* 21 Nov. 5/5 [New York] There are to be four elevator shafts, besides stairways and 'escalators'. *a* **1910** 'O. HENRY' *Trimmed Lamp* (1916) 131 You will perceive that the Bee-Hive was not a fashionable department store, with escalators and pompadours. **1910** *Daily News* 2 July 10 In the course of the hearing counsel referred to a proposed moving staircase as an 'escalator'. **1911** *Engineer* 6 Oct. 356/3 The escalator or moving stairway connecting the 'Piccadilly' and District Railway. **1923** *Spectator* 29 Sept. 411/2 Three escalators will serve the Bakerloo Tube. **1968** *Listener* 20 June 817/1 A counterblast to the prurience of *Playboy* and the escalator ads.

b. *fig.*

1927 *British Weekly* 14 Apr. 38/2 The mechanism of a great social 'escalator' whereby the 'down and outs' of Manchester have restored to them the loving ministry of Christian service. **1927** L. MACNEICE in *Oxf. Poetry* 18 And watching the Before become Beyond Down the escalator shifting of the Past. **1943** J. D. HICKS *Short Hist. Amer. Democracy* xxxv. 729 He climbed aboard the political escalator in 1899 when he became a councilman. **1950** *Economist* 9 Dec. 1003/1 Prices and wages are fellow-travellers on the same upward escalator. **1967** D. BOULTON *Objection Overruled* i. 20 The Serbian assassin's shot..put the world on an irreversible escalator to multiple war.

2. *attrib.*, esp. designating a clause, contract, agreement, or the like, that provides for an increase (occas., a decrease) in prices, wages, armaments, etc., to meet specified contingencies.

1930 *N.Y. Times* 18 Apr. 2/1 The so-called safeguarding, or contingency, clause, sometimes referred to as the escalator provision. **1930** *Times* 1 May 16/3 The existence and implications of the 'contingent' or 'escalator' clause of the Treaty are naturally of importance from the American point of view. **1932** *Daily Express* 1 July 9/1 France..wants an 'escalator' clause inserted in the agreement which would link up the war debts to America with this proposed fund. **1948** *Time* 7 June 5/3 Labor leaders have never liked cost-of-living 'escalator' contracts, on the grounds that they tie the worker to a fixed standard of living. **1960** *Farmer & Stockbreeder* 23 Feb. 7/2 Cattle prices are subject to escalator adjustment. **1963** *Economist* 5 Jan. 50/1 'Escalator' mortgages, which deliberately bunch repayment towards the end of the life of the mortgage.

escalope (eˈskălōᵘp). Also **escalop, escallope.** [OF. *escalope* shell: see ESCALLOP.] Thin slices of boneless meat (occas. of fish), prepared in various ways; *esp.* a special cut of veal taken from the leg.

1828 LYTTON *Pelham* III. viii. 128 You have neither applauded my jokes, nor tasted my *escallopes*. **1845** BREGION & MILLER *Pract. Cook* 41 *Escalopes*, small pieces of meat cut in the form of some kind of coin. **1877** E. S. DALLAS *Kettner's Book of Table* 179 Many cooks here..begin to speak of escalopes of veal and of other meat. **1894** C. RANHOFER *Epicurean* 690 Escalops of tenderloin of beef with truffles. **1900** S. BEATY-POWNALL *'Queen' Cookery Bk.* IV. ii. 16 These [entrées]..are also termed *escalopes*,..*médaillons*, &c...These latter terms are nowadays more frequently applied to beef or veal, or even fish, than to mutton. **1956** R. FROST in J. C. Trewin *Plays of Year 1955* XIII. 154 It says here that this is an escalope of veal Viennoise. In the appalling bilingual jargon of small hotels that means to say that it's a Wiener Schnitzel. **1959** R. POSTGATE *Good Food Guide* 97 One member notes for special praise the escalope Holstein at 6/-. **1963** *Listener* 24 Jan. 187/2 The veal escallops are cooked in cider for half an hour.

escape, *sb.*¹ Add: **1. b.** *fig.* Mental or emotional distraction, esp. by way of literature or music, from the realities of life.

1853 DICKENS in *Househ. Words*, extra Xmas no. 35/2 Labouring people..in need of mental refreshment and recreation...Come! Amuse me harmlessly, show me something, give me an escape! **1919** *English Rev.* Mar. 183 It has been said by Mr. Arthur Symons that 'all art is a mode of escape'. **1923** *Time* 14 Apr. 14/3 For the cities are saturated with the literature of escape. **1933** *Amer. Speech* VIII. III. 34/1 Those who use drugs merely as an escape-mechanism while incarcerated. **1934** *Punch* 18 Apr. 440/1 *Libel* makes no pretence to belong to any other class than it is now the contemptuous fashion to call the literature of escape. **1942** *Sphere* 25 July 120/1 Music always enjoys a boom in wartime. It acts as a magic carpet, as an escape. **1954** M. F. RODELL *Mystery Fiction* (ed. 2) ii. 4 Mysteries belong to the vast category of escape fiction. **1969** *Islander* (Victoria, B.C.) 16 Nov. 16/3 But far from being escape reading, these fantasties often make more penetrating comments on the 'real' world than realistic fiction does.

c. The action of escaping from the region where the gravitational force of a planet is predominant. Usu. *attrib.* (See 8 below.)

1947 *Jrnl. Aeronaut. Sci.* XIV. 471 (*heading*) The problem of escape from the earth by rocket.

2. c. A bird that has escaped from captivity.

1937 *Brit. Birds* XXXI. 27 There is no reason to think this bird an escape as the place is very remote. **1953** D. A. BANNERMAN *Birds Brit. Isles* I. 108 There is a doubt as to whether they were genuine wild birds or merely 'escapes' from captivity.

4. c. An outlet for a liquid; *esp.* a sluice or gate for the outlet of surplus water in a canal. Also *attrib.*, as *escape drain, head, water*.

1889 W. WILLCOCKS *Egypt. Irrig.* ii. 80 The Ibrahimia Canal regulator lock works except when silt deposit interferes. The escape headlock does not work. **1891** *Daily News* 31 Oct. 6/4 The Koshesha Escape, or Barrage...The escape is constructed to discharge 2,000 million cubic metres in twenty days in good Nile years. **1893** *Pall Mall Gaz.* 30 Jan. 3/1 The revolution of these caps whips the skim-milk through a broad escape. **1907** W. G. BLIGH *Irrig. Wks.* x. 274 In places on a canal where an Escape or a Branch takes off, a regulating bridge across both works is generally necessary. *Ibid.* 280 An Escape Head should, if practicable, be combined with a fall. **1911** W. WILLCOCKS *Irrig. Mesopot.* 27 Every canal system needs its escapes. **1913** WILLCOCKS & CRAIG *Egypt. Irrig.* (ed. 3) II. viii. 465 The escape waters of the canals and larger water-courses should be allowed to enter the escape-drain by means of regulators at the tails of the canals.

8. escape clause, a clause in a formal or legal document that specifies the conditions under which a contracting party is free from obligations; **escape hatch,** an emergency exit in a ship, submarine, or aircraft; also *fig.*; **escape rocket** (see quots.); **escape velocity** (of space vehicles, etc.), a speed sufficient to overcome the gravitational force of a planet; also *escape speed*.

1945 *Time* 5 Nov. 29 These prohibitive terms were eased a bit..by an escape clause under which interest could be postponed in years when Britain had an unfavorable balance of trade. **1948** *Ann. Reg. 1947* 72 There were so many necessary 'escape clauses' that the importance of a permanent organisation to supervise the working of them was apparent. **1966** M. R. D. FOOT *SOE in France* xi. 363 He was given a long list of instructions to carry out in France, but a pair of escape clauses let him omit any that were inconvenient. **1971** *Where* Sept. 279/1 Pretty well all universities and colleges have some escape clause in their entrance regulations by which they can avoid having to turn away somebody they wish to admit merely because of a lack of formal qualifications. **1925** *Sci. Amer.* Dec. 373 The submarine..carried four torpedo tubes at the forward end of the torpedo room, which was provided with an escape hatch. **1940** *Flight* 17 Oct. 332/1 He could have followed the rear gunner through the bottom escape hatch. **1948** R. DE KERCHOVE *Internat. Maritime Dict.* 244/1 *Escape hatch*, 1. A small hatch..in a 'tween-deck or weather deck for the safe exit of men engaged in trimming or stowing coal or grain. **1955** *Sci. News Let.* 6 Aug. 86/1 The continuous reconnaissance of people and traffic, preferably from the air for maximum scanning of 'escape hatches'. **1962** W. NOWOTTNY *Lang. Poets Use* iv. 84 These crucial words..may serve as escape-hatches from conceptual terms. **1951** *Flight* 3 Aug. 141/2 The propellant thus assembled, being stored at a high fraction of escape velocity, will represent a substantial energy potential and its transfer into a single 'escape rocket' will mean that this vehicle supports itself only from the time it leaves the terminal orbit. **1963** M. CAIDIN *Man-in-Space Dict.* 76/2 *Escape rocket*, a rocket used in a system to ensure the escape of an astronaut from a booster in the event of malfunctioning. **1951** *Tuscaloosa News* 16 Oct. 4/6 The problem of reaching the moon is basically that of getting rockets to fly fast enough to attain an 'escape' speed—an escape speed being about 23,000 miles an hour—sufficient to escape from the pull of the earth's gravity. **1934** *Space* (Cleveland) I. IV. 1 The equation for escape velocity is derived from the equation for the orbital velocity of a particle about a mass. **1945** H. S. ZIM *Rockets & Jets* 207 Calculations based on the law of falling bodies indicate that a speed of about seven miles per second or 25,000 per hour is essential before 'escape velocity' is attained. **1954** Escape velocity [see *ASTRONAUT 2]. **1959** *Listener* 22 Jan. 161/1 The Moon has a diameter of only 2,160 miles, and an escape velocity of one and a half miles per second.

escape, *v.* Add: **1. d.** *fig.* To avoid or retreat from the realities of life. (Cf. *ESCAPE *sb.*¹ 1 b.)

1939 *Punch* 20 Dec. 675/2 Many a publisher has had the good idea of advising you to escape really thoroughly by way of an eight hundred-and-fifty page novel about family life in the Victorian era.

f. Of a space-craft: to attain sufficient velocity to enable it to overcome the gravitational force of a planet.

1947 *Jrnl. Aeronaut. Sci.* XIV. 471/1 The problem of imparting to a..pay load the..velocity to escape from the earth is re-examined. **1949** *Jrnl. Brit. Interplanetary Soc.* VIII. 79 A spaceship will escape from Earth at 11·2 kilometres a second.

escapee. Now also with pronunc. (eskēˈpi). Add to def.: Also an escaped military or political prisoner.

1875–6 [see Dict.]. **1902** *Westm. Gaz.* 21 Aug. 3/1 The military watched him in his home in Middelburg as a cat watches a mouse, and..traps in the way of sham rebels and escapees were laid for him. **1945** *News Chron.* 20 Apr. 4/4 A large number of the escapees were members of air crews who had baled out over France. **1954** KOESTLER *Invisible Writing* IV. xxxix. 421 The five of us, plus some fifty other escapees.

escapement. Add: **3.** In a pianoforte (see quot. 1896).

1896 A. J. HIPKINS *Pianoforte* 114 *Escapement*, a space that is left between the hammer at its full rise and the strings, necessary for the strings to vibrate and to prevent jarring. **1961** C. CLUTTON in A. Baines *Mus. Instruments* 89 The essentials of the modern pianoforte action:..double action, a check, an escapement, and an *una corda* mechanism.

escapism (eskēˈpizˈm). [f. *ESCAP(E *sb.*¹ 1 b + -ISM.] The tendency to seek, or the practice of seeking, distraction from what normally has to be endured.

1933 *Encycl. Social Sci.* IX. 533/1 The bibulous, aphrodisiac lyrics strummed out by Anacreon of Teos at the banquets of Polycrates, tyrant of Samos, are an example of escapism, comparable to the songs of Alcaeus and Sappho in strife ridden Mytilene. **1935** L. MACNEICE *Poems* 26 This escapism of yours is blasphemy. **1937** H. READ *Art & Society* 260 Many of the critics of abstract art..dismiss it as the most evident byzantinism, escapism, absolutism. **1940** L. D. WEATHERHEAD *This is Victory* ix. 188 Religion that was mere escapism. **1946** J. CARY *Moonlight* 16 Was she, after all, an escapist? Amanda had a great contempt for escapism. **1954** *Essays in Criticism* IV. 50 He is not entering a plea for mere 'escapism' in literature.

escapist (eskēˈpist). [f. ESCAP(E *sb.*¹ + -IST.] **1.** One who escapes, or who tries to escape, from captivity, prison, etc. Cf. ESCAPEE (in Dict. and Suppl.).

1934 in WEBSTER. **1938** *Punch* 22 June 700/2 Escapists..The efforts of these nefarious ruffians to dodge the police. **1943** *Hutchinson's Pict. Hist. War* May–Aug. 43 Pantellaria was the destination of these rubber dinghy escapists, but they were rounded up by the Royal Navy. **1958** *Times* 22 Mar. 7/4 He rushed out of court before he could be stopped, whereupon the judge..fined the escapist £20.

2. *fig.* Esp., one who seeks distraction from reality or from routine activities. Cf. *ESCAPISM.

1933 C. S. LEWIS *Pilgrim's Regress* VI. iii. 125 'And you have never heard Mr. Halfways either.' 'Never. And I never will. Do you take me for an escapist?' **1937** *Essays & Studies* XXII. 148 The satirist is a kind of escapist. **1942** E. WAUGH *Put out more Flags* 252 Turning their backs on the world of effort and action. Fortunate islanders [*sc.* the Irish],..happy, drab escapists. **1946** [see *ESCAPISM]. **1958** *Spectator* 6 June 738/1 Both Hawthorne and Henry James were censured as escapists.

Hence *attrib.* or as *adj.*, that provides escape from reality or routine; pertaining to escapism.

1930 J. C. RANSOM *God without Thunder* 178 It is much more likely that they betoken a defeated and escapist people. **1933** *Archit. Rev.* LXXIII. 206/2 Those great pioneers of the past..took up the escapist attitude of William Morris. **1940** W. EMPSON *Gathering Storm* 62 A critic like Dr. Leavis can speak with the same tone of moral outrage about an Escapist (sentimental) novel as a customs official would about *Lady Chatterley's Lover*, say. **1959** *Manch. Guardian* 15 Aug. 3/6 The escapist peace of a nudist camp.

escapologist (eskăpǫˈlŏdʒist, eskēˈp-). [f. ESCAP(E *sb.*¹ + -OLOGIST.] A performer skilled in extricating himself from knots, handcuffs, confinement in a box, etc. Also *transf.* and *fig.* (= *ESCAPIST 1, 2).

1926 *Glasgow Herald* 1 Oct. 6 An Australian 'Escapologist'..gave a demonstration of his powers to a private audience at the London Coliseum yesterday. **1936** C. BELFRAGE (*title*) Away from it all. An escapologist's notebook. **1943** A. R. D. FAIRBURN *Strange Rendezvous* (1952) 16 Philosophers, sportsmen, and men who are very busy about their own and other people's business are three kinds of escapologists. **1946** G. MILLAR *Horned Pigeon* vii. 81 Cram was an escapologist. He had already escaped once, from a transit camp in Palermo. **1955** J. THOMAS *No Banners* x. 85 On Christmas night the [Madrid] Embassy threw a party for the 'escapologists', mostly R.A.F. and B.E.F. personnel. **1958** *Times* 22 Oct. 6/2 Randi, the Canadian escapologist who recreates a number of famous Houdini tricks. **1970** *Daily Tel.* 22 Sept. 12/4 Yesterday's edition included an escapologist who freed himself from a mail bag and manacles inside a locked chest.

So **escapoˈlogy,** the methods and technique of escaping, esp. from captivity or danger; the calling of an escapologist.

1939 T. KOEVES *Timetable for Tramps* 22 At the higher escapology, the escape from reality by the use of the imagination aided by fashionable dopes, I had found myself to be a duffer. **1959** *Times* 29 Dec. 6/6 This was P—'s seventeenth escape from prisons and approved schools..At his trial..he was described as 'a past master in the art of escapology'. **1965** *Spectator* 12 Mar. 341/3, I carefully read all the books in the library about disasters by sea, fire, earthquake..and marked those passages which seemed to contain some lesson in escapology. **1966** *Daily Tel.* 19 Dec. 14 (*caption*) Escapology adviser to the musical, 'Man of Magic'.

‖ **escargot** (ęskargo). [Fr.] An edible snail.

1892 in T. F. GARRETT *Encycl. Pract. Cookery* I. viii. 575/1. **1926** *Glasgow Herald* 26 Mar. 10 I've savoured an escargot. **1960** *Times* 30 Jan. 9/6 The increase in appreciation of food, that has called forth *escargots* in Inverness.

-escence, *suffix*, forming sbs. corresponding to adjs. in -ESCENT, as *effervescence, iridescence*.

eschatocol (ęˈskätokǫl). [G. *eschatokoll* (It. *escatocollo*, F. *eschatocole*), f. Gr. ἔσχατος last + κόλλα glue. Cf. πρωτόκολλον PROTOCOL. Martial (II. vi. 3) has ἐσχατοκόλλιον end of a papyrus roll.] The concluding section of a charter, containing the attestation, date, etc.; a concluding clause or formula.

1897 F. W. MAITLAND *Domesday Bk.* 247 Such words as a charter has about 'consent'..may occur in the

eschatocol, the clause which deals with the execution and attestation of the instrument. **1904** *Eng. Hist. Rev.* Jan. 152 In regard to the dating we find originally a separation, the year going into the protocol, the month and regnal year into the eschatocol. But by the middle of the twelfth century the whole date is put into the eschatocol.

eschatoloˈgically, *adv.* [f. ESCHATOLOGICAL *a.* + -LY².] In relation to eschatology.

1910 W. MONTGOMERY tr. *Schweitzer's Quest Historical Jesus* xix. 349 The eschatological school itself..represents Jesus as thinking and speaking eschatologically in some of the most important passages of His teaching. **1922** A. E. GARVIE *Beloved Disciple* ix. 199 Whether we interpret these functions spiritually or eschatologically. **1934** *Theology* XXVIII. 153 The Holy Spirit is present in us eschatologically. **1963** AUDEN *Dyer's Hand* 78 Today, Christian and Atheist alike are eschatologically minded.

eschatoˈlogize (-dʒəiz), *v.* [See -IZE.] *trans.* To give an eschatological character to.

1919 S. C. CARPENTER *Christianity acc. S. Luke* vii. 101 The first Evangelist has 'eschatologized' an originally non-eschatological saying.

eschatology. Add: **b.** In recent theological writing, esp. as 'realized eschatology' (see quot. 1957), the sense of this word has been modified to connote the present 'realization' and significance of the 'last things' in the Christian life.

1950 *Scottish Jrnl. Theol.* III. 90 Correspondingly for Christians..eschatology is being anticipated in the here and now, and the glory of the parousia throws its light backwards into the present life of the Church. **1957** *Oxf. Dict. Chr. Ch.* 462/1 In modern Protestant theology eschatology has been given a new meaning, esp. through the studies of A. Schweitzer...Acc. to him, primitive Christianity was exclusively an eschatological preaching of judgement and salvation. Eschatological considerations are also a dominant factor of the teaching of K. Barth and his school..which sees the life of the individual Christian and of the Church as a series of 'decisions' invested with an eschatological character...Recently many attempts have been made to draw out the present and abiding significance of future eschatological happenings, e.g. by C. H. Dodd in his conception of 'realized eschatology'.

‖ **eschaton** (eˈskătǫn). *Theol.* [a. Gr. ἔσχατον, neut. of ἔσχατος last.] (See quots.)

1935 C. H. DODD *Parables of Kingdom* vi. 193 The *eschaton*, the divinely ordained climax of history, is here. **1936** —— *Apostolic Preaching* 199 In prophecy and apocalypse alike, the divine event, the *eschaton*, is always 'round the corner'. The prophet never conceives himself as standing midway in the course of history. *Ibid.* 210 For the New Testament writers in general, the *eschaton* has entered history. **1950** *Scottish Jrnl. Theol.* III. 301 Behind the parable, or rather beside it, there is the kingdom which has broken into the midst, the *eschaton* which is here and now.

escheaˈtment. [f. ESCHEAT *v.* + -MENT.] Forfeiture or lapsing by escheat.

1869 BLACKMORE *Lorna D.* I. xxiv. 283 On pain of a heavy fine or escheatment. **1919** E. MACNEILL *Phases Irish Hist.* xi. 312 The new grants were not preceded..by..any escheatment or invalidation of the existing contract.

eschscholtzia. Add: Also **eschscholzia.**

esclandre. (Earlier example.)

1832 GREVILLE *Mem.* 1 June (1874) II. xviii. 306 Threatening to make an *esclandre* and leave the château.

escort, *sb.* Add: **1. b.** *Naut.* A warship, or group of warships, accompanying merchant ships or other vessels for protection. Also used of aircraft in a similar role. Freq. *attrib.*, as *escort carrier, destroyer, vessel*; so *escort duty, work*. Also with defining word, as *convoy escort, surface escort*.

1914 *Sphere* 31 Oct. 121/2 Numerous letters have now been received from men in the torpedo boat destroyers. One writes: 'We are doing patrolling duties...' Another says: 'After a period as escorts for transports we are now having a change.' **1918** *Ibid.* 9 Mar. 223 (*caption*) Some British battleships at sea with their escort of protecting destroyers. *Ibid.* 30 Mar. 273 (*caption*) A convoy escort returning to port. *Ibid.* 29 June 234/2 Two British destroyers..which formed the anti-submarine escort, at once engaged the enemy vessels. **1919** L. FREEMAN *Sea-Hounds* v. 129 As I heard one of the men put it, it was the 'bruisiest' bit of escort-work they had ever been..called upon to face. **1923** W. S. CHURCHILL *World Crisis* I. xxi. 496 The First Cruiser Squadron..had been..employed on escort duties at sea. **1938** *Jane's Fighting Ships* p. v, The heavily armed escort vessel *Egret* obviously designed to protect convoys against air attack. **1944** *Jane's All World's Aircraft 1943–44* 25a/2 The new class of Escort Carriers..adapted for flying operations by installing a 'flat top' deck. **1948** *Jane's Fighting Ships 1947–48* 222 New Construction Programme, 1948–50. Six escort destroyers to be laid down. **1954** S. W. ROSKILL *War at Sea 1939–1945* I. xvii. 358 The tactical employment of air escorts was now being improved steadily. *Ibid.*, Aircraft employed as convoy escorts or on anti-submarine patrols. *Ibid.*, Surface escorts..were being formed into Escort Groups in the Western Approaches Command. *Ibid.* xxi. 464 It was now possible to provide convoys with an average strength of five anti-submarine escorts.

c. *Aeronaut.* A fighter aircraft used to accompany bombers; usu. *attrib.*, as *escort fighter*.

1946 A. LEE *German Air Force* i. 11 Messerschmitt 110..As a long-range day escort fighter it was to fail. **1965** *New Statesman* 30 Apr. 690/3 The rapid conversion of the Mustang into a long-range escort.

2. b. A person (usu. a man) accompanying a woman to a dance, party, etc.

1936 *Lit. Digest* 12 Dec. 29/1 The escort merely announces himself downstairs at the client's address and presently she appears. **1946** K. TENNANT *Lost Haven* (1947) xvii. 267 Miss O'Shea was drinking ginger-beer and her escort had a shandy. **1955** T. STERLING *Evil of Day* viii. 95 The forget-me-not corsage..she had bought for herself, explaining to her escort that gardenias gave her a headache. **1955** J. P. DONLEAVY *Ginger Man* (1956) xi. 95 I've just walked into a bar, and I was frightened to death that the barman would tell me that women without escorts couldn't come in.

escort, *v.* Add: **c.** To 'keep company' or 'walk out' with (a woman). *U.S.*

1890 *Harper's Mag.* Oct. 716/2 A whisper also went the rounds that Dick Jones was escorting Miss Turner.

escortage (eskǭˈɹtėdʒ). *U.S.* [f. ESCORT *v.* + -AGE.] The action of escorting.

1894 P. L. FORD *Hon. Peter Stirling* xxvii. 153 She likes my company, and finds my escortage very convenient. **1911** H. S. HARRISON *Queed* xiii. 146 At nine, as it chanced, she was to go out under the escortage of Charles Gardiner West.

escrow. Add: **b.** A deposit held in trust or as security; *in escrow*, phr. used of money, etc., so held. *N. Amer.*

1888 *Boston* (Mass.) *Jrnl.* 17 Oct. 1/7, $800,000 is held in escrow to pay off unmatured bonds. **1932** *Amer. Mercury* May 44/2 Ralston now had in his possession diamonds believed to be worth at least $1,500,000, ample security for the $300,000 which had been placed in escrow, and which he immediately paid over to the prospectors. **1957** *Economist* 30 Nov. 777/2 About $100 million in fees from the disputed area off Louisiana have been put in escrow pending the court's decision. **1967** WODEHOUSE *Company for Henry* i. 27 'I am temporarily not in possession of a watch.' 'Pawned it?' 'It is for the moment in escrow.'

escudo (eskiūˈdo). [Sp., Pg.:—L. *scūtum* shield. Cf. ÉCU, SCUDO.] A Portuguese silver coin, originally of the value of a crown. Also applied to other former coins, gold or silver, in Spain, Portugal, and Spanish America.

1821 P. KELLY *Universal Cambist* (ed. 2) II. 281 *Escudo*,..a gold coin of Portugal and Spain. **1914** E. C. HARGROVE *Progressive Portugal* 262 In 1913, the Portuguese Authorities altered the denomination of the coinage from milreis and reis to escudos and centavos. **1920** *Chambers's Jrnl.* 404/2 The escudo, which normally..is of the value of nearly four shillings and sixpence. **1925** *Glasgow Herald* 14 Aug. 6 The paper escudos nominally equivalent to a French franc. **1957** *New Statesman* 2 Nov. 554/1 Scruffy black-marketeers..offered me wads of *escudos* for my ticket.

escutcheon. Add: **3. g.** = *milk escutcheon* (MILK *sb.* 10).

1847 tr. *F. Guenon's Treat. Milk Cows* 33 Cows of each class have their print or escutcheon marked by the hair, which starting first from the middle of the four teats, as the center, extends under the belly...It may be said in general, that the cows whose escutcheon is formed of the finest hair, are the best. **1881** [see *milk-escutcheon*]. **1912** F. T. BARTON *Cattle, Sheep & Pigs* i. 18 The escutcheon may extend over the whole of the hind quarters and the udder.

-ese. Add: On the model of derivatives from authors' names were formed *Americanese, cablese, headlinese, journalese, newspaperese, novelese, officialese*, etc.

1898 F. HARRISON in *19th Cent.* June 941 As Mat Arnold said to me..'Flee Carlylese as the very devil!' Yes! flee Carlylese, Ruskinese, Meredithese and every other *ese*. **1899** *Golf Illustr.* 14 July 134 American 'golfese'. **1906** *Daily Chron.* 2 Aug. 3/2 Deplorable guide-bookese. **1935** E. E. CUMMINGS *Let.* 11 Mar. (1969) 140 Am fighting..to retranslate 71 poems out of typewriter language into linotype-ese. **1951** *Amer. Speech* XXVI. 172 (*heading*) Washingtonese.

esker. Add: No longer restricted to Ireland or to Irish usage.

1917 *Nature* 2 Aug. 441/2 He considers that the ice-barrier tongues from the Antarctic glaciers are afloat and do not rest on esker-like embankments built of their moraines and subglacial gravels. **1955** *Sci. Amer.* Sept. 85 Winding across the center [of the Tasnuna Valley, Alaska] is an esker, a narrow ridge of gravel left by a stream that ran through the glacier. **1966** *Times* 28 Feb. (Canada Suppl.) p. xvi/2 Frozen lakes, eskers and beaches permit the landing of heavily loaded machines for six months of the year.

Eskimo (eˈskimōᵘ), *sb.* and *a.* Also **Esquimaux** (6–8 **Eskemo, Esquimaw, Esquimo,** etc.). Pl. **-oes, -os** (ōᵘz), **-aux** (ōᵘ). [a. Da. *Eskimo* (Sw. *Eskimå*), ad. F. *Esquimaux* pl., from some Algonquian Indian language; cf. Proto-Algonquian **ašk-* raw, **-imo* eat, Abnaki *askimo* (pl. *askimoak*), Eskimo, eater of raw flesh.] **A.** *sb.* **1.** A member of a widely spread people inhabiting the Arctic from Greenland to Eastern Siberia. (Their own name for themselves is *INNUIT.) Used as sing. and pl.

1584 HAKLUYT *Disc. Western Planting* (1877) xiii. The more northerly partes of the lande amonge the Esquimawes of the Grande Bay. *c* **1685** P. E. RADISSON *Voyages* (1885) 260 Wee put into Harbour..to take fresh Water..at the Coast of the Indians called Esquimos. **1689** H. KELSEY *Jrnl.* 5 July in *Kelsey Papers* (1929) 27 Not suffering me to speak aloud in pretence ye Eskemoes would hear us. **1792** G. CARTWRIGHT *Labrador* 7 The Esquimaux from Ice and snow now free In Shallops and in Whale-boats go to Sea. **18**.. *Eskimaux & Eng. Vocab.* p. vii, If the extent of country be considered, the Innuit or Eskimaux is one of the most widely spread nations on the globe. **1850** R. G. LATHAM *Nat. Hist. Var. Man* I. F. 290 The Eskimo is essentially a Mongol in physiognomy. **1895** KIPLING *2nd Jungle Bk.* 145 Kadlu was an Inuit—what you call an Esquimau. **1915** *Lit. Digest* 16 Jan. 130/1 A month in the land of Gold, Glaciers, Totems, and Esquimos. **1936** *Discovery* July 200/2 According to his view the eastern or Caribou Eskimo are neither original nor degenerate, but represent a fusion of Eskimo and Indian cultures. **1965** R. F. SPENCER et al. *Native Americans* 119 Even though the territorial expanse covered by the Eskimo is tremendous, there is a remarkable unity to Eskimo culture.

2. Any of the several languages of this people, of which one set of dialects or languages, also called Inupik, is spread from Norton Sound, Alaska, to Greenland, and another set, also called Yupik, is in southwest Alaska and the eastern tip of Siberia. These languages, together with those of the Aleut, form the *Eskimo-Aleut, -Aleutian* family. Also applied to a jargon used in intercourse between Eskimos and persons of European origin, consisting mainly of Eskimo words.

1819 J. ROSS *Voy. Discovery* v. 80 A reply was again made in the Eskimaux. **1850** R. G. LATHAM *Nat. Hist. Var. Man* I. F. 290 From the grammar of Fabricius, the Eskimo was soon known to be a language of long compound words. **1916** E. W. HAWKES *Labrador Eskimo*.. Lieut. Lucas, who had acquired a knowledge of Eskimo. **1921** E. SAPIR in *Science* LIV. 408 (*heading*) Eskimo-Aleut. **1932** W. L. GRAFF *Lang. & Langs.* xi. 405 A common Uralian language with which also Eskimo-Aleutian appears to be related. **1939** L. H. GRAY *Found. Lang.* xiii. 374 Eskimo-Aleut has been regarded by highly competent authorities as ultimately cognate with Altaic. **1958** C. F. HOCKETT *Course in Mod. Ling.* 260 In Eskimo..ordinary nouns have a case system of the ergative type. **1964** M. SWADESH in Jennings & Norbeck *Prehist. Man in New World* 531 The last extant language to enter the New World was apparently the forerunner of Eskimo-Aleutian.

3. = **Eskimo dog*.

1856 KANE *Arct. Expl.* I. xiv. 163 My dogs,..the nine splendid Newfoundlanders and thirty-five Esquimaux..had perished. **1872** W. F. BUTLER *Great Lone Land* (ed. 2) xxii. 339 The other two [dogs] were a curious compound of Esquimaux and Athabascan. **1909** W. T. GRENFELL *Adrift on Ice-Pan* (1910) 3 'Sue', a large, dark Eskimo, the image of a great black wolf, with her sharp pointed and perpendicular ears. **1948** C. L. B. HUBBARD *Dogs in Britain* xvii. 178 The Husky Proper..has long been officially recognised..as a show dog, under the name of Eskimo.

4. = **Eskimo pie*.

1922 *Glasgow Herald* 11 Oct. 8 Chocolate eggs filled with ice-cream and known as 'Eskimos'.

B. *adj.* **1.** Of or pertaining to this people or their language.

1744 A. DOBBS *Hudson Bay* 49 The East Main...Here the Nodway or Eskimaux Indians live. *Ibid.* 203 A vocabulary of English and Eskima words. **1770** G. CARTWRIGHT *Jrnl.* 6 Dec. (1792) I. 66 A very imperfect vocabulary of the Esquimaux language. **1791** J. LONG (*title*) Voyages and travels of an Indian interpreter..To which is added..a list of words in the Iroquois, Mohegan, Shawanee, and Esquimeaux Tongues. **1845** YOUATT *Dog* iii. 56 The Esquimaux Indian goes in pursuit of the seal. **1850** R. G. LATHAM *Nat. Hist. Var. Man* I. F. 289 The difficulties presented by the Eskimo language. **1875** H. RINK *Tales & Trads. Eskimo* p. vii, Words..peculiar to Eskimo life. **1918** [see *ESKIMOID *a.*]. **1965** *North* (Ottawa) Nov.–Dec. 12 Letters from the children written in English and Eskimo script.

2. *Eskimo dog*, a large, powerful dog used in the Arctic to draw sledges and for hunting.

1774 B. LA TROBE *Acc. Mission Esquimaux Indians* 29 The great number of Esquimaux dogs..tear out and devour the fish. **1845** YOUATT *Dog* iii. 55 The Esquimaux Dog is a beast of burden and of draught. **1909** W. T. GRENFELL *Adrift on Ice-Pan* (1910) 38, I determined..to kill a big Eskimo dog.

3. *Eskimo curlew* (see quots.). *U.S.*

1813 A. WILSON *Amer. Ornith.* VII. 22 The Esquimaux Curlew..is called by our gunners on the sea-coast, the Short-billed Curlew. **1870** *Amer. Naturalist* III. 8.. Esquimaux Curlew (*Numenius borealis*) breeds near Fort Benton. **1921** *Outing* (U.S.) May 65/1 The passenger pigeon, the great auk.., the Eskimo curlew are no more. **1966** *New Statesman* 11 Nov. 710/3 The Eskimo curlew..winters in Patagonia..and then flies the whole length of the two continents to breed inside the Arctic Circle.

4. *Eskimo pie* (proprietary name), a bar of ice cream coated with chocolate. Chiefly *U.S.*

1928 TURNBOW & RAFFETTO *Ice Cream* 57 Chocolate-coated ice cream bars were introduced in 1921 as 'Eskimo Pies'. **1937** L. BROMFIELD *Rains Came* I. i. 13

The cries of vendors of eskimo pies and *pan*. **1970** D. M. DAVIN *Not Here, Not Now* IV. i. 219 'Let's bring her back some Eskimo pies,' Martin said. 'I haven't had one since I was at school.'

5. *Eskimo roll*: in Canoeing (see quots.).

1946 P. W. BLANDFORD *Canoeing To-Day* 4 In them [*sc.* folding canoes] experts are able to perform the 'Eskimo roll'—a complete rollover, under the water and return to upright at the opposite side. **1965** *Clarendonian* XIX. III. 116 There are one or two gymnastic exercises that canoeists like to perform. The most famous is the so-called 'Eskimo roll' for which there are at least three possible paddle strokes. **1968** W. WARWICK *Surfriding in N.Z.* 10/2 When you have to paddle out through larger waves, over about 8 ft., it is safest to use the Eskimo roll, as a means of getting past a broken wave.

Eskimoid (eˈskimoid), *a.* [f. *ESKIMO *sb.* + -OID.] Eskimo-like; similar in racial type to the Eskimo; resembling an Eskimo. Also as *sb.*

1918 E. A. HOOTON in *Amer. Jrnl. Phys. Anthropol.* I. 53 (*title*) On certain Eskimoid characters in Icelandic skulls. *Ibid.* 54, I have ventured to call these 'Eskimoid characters', because they occur in association predominantly in Eskimo skeletons. **1939** *Peking Nat. Hist. Bull.* XIII. 172 An Eskimoid skull very similar to that of the Upper Cave [in China] is, for instance, represented among the remains of the pre-Columbian Indians from South Western Texas. **1946** F. E. ZEUNER *Dating the Past* viii. 272 Three different racial elements, best to be classified as primitive Mongoloid, Melanesoid and Eskimoid. *Ibid.* ix. 300 Another [racial] type...that of Chancelade, supposed to have eskimoid affinities.

esophoria (esofǭˈriă). *Ophthalmology.* [f. Eso- + *-phoria* ad. Gr. φέρειν to bear.] A latent convergent strabismus of the eyes.

1886 [see *exophoria* s.v. *EXO-]. **1891** R. J. MCKAY in *Trans. Amer. Ophthalmol. Soc.* VI. 136, I put a dark red glass in front of one eye..and see whether there is diplopia, hyperphoria, esophoria, or exophoria. **1906** S. THEOBALD *Prevalent Dis. Eye* xii. 473 In the causation of esophoria hypermetropia and hypermetropic astigmatism play even a more important rôle than myopia does in that of exophoria. **1964** S. DUKE-ELDER *Parsons' Dis. Eye* (ed. 14) xxx. 472 If the latent deviation is one of convergence the condition is called *esophoria*, of divergence *exophoria*, if vertical, *hyperphoria*.

esoterica (esoteˈrikă, iso-), *sb. pl.* [mod.L., f. Gr. ἐσωτερικ-ός (see ESOTERIC *sb.*).] Esoteric objects or products; esoteric details; = ESOTERIC *sb.* 1; pornography.

c **1929** O. NASH *Coll. Verse* (1961) 107 The postal authorities of the United States of America Frown on Curiosa, Erotica and Esoterica. **1950** A. LOMAX *Mr. Jelly Roll* (1952) 78 He..can bore you to death with the esoterica of forgotten bands. **1958** *Spectator* 13 June 770/2 In Soho it is usually possible to get esoterica such as passion fruit, mangoes, [etc.]. **1958** *Times Lit. Suppl.* 31 Oct. 629/4 A collection of oddments..on Jungian esoterica, on sex, on dreams and nightmares.

esotericism (esoteˈrisizˈm, iso-). Add: (Further examples.) Also, a tendency toward esoteric thought or language, obscurity; an example of such thought or language. So **esoteˈricist,** one who holds esoteric doctrines. Also **eˈsoterist** (*Funk's Stand. Dict.* 1893).

1885 MRS. H. WARD tr. *Amiel's Jrnl.* II. 315 Yes, but still a certain esotericism [*orig.* ésotérisme] is inevitable, since critical, scientific, and philosophical culture is only attainable by a minority. **1891** *Review of Rev.* 15 June 522/2 She [*sc.* Madame Blavatsky] made all her 'esotericists', as she called them, send her their photographs. **1892** *Ch. Times* 1 Apr. 337/1 Further on the esotericist is identified with the Rationalist. **1923** J. M. MURRY *Pencillings* 1 One of the most peculiar features of literature during the last thirty or forty years has been the steady movement of a part of it towards esotericism. **1954** W. Y. EVANS-WENTZ *Tibetan Bk. Great Liberation* 33 The Christian exotericists, derived largely from uncultured slave populations, inaugurated a religious revolution against the Christian esotericists, the cultured and well-born followers of the Gnosis. **1955** *Essays in Criticism* V. 421 Yeats's intellectual esotericisms (such as the phrase 'perne in a gyre' in *Sailing to Byzantium*). **1963** *Times* 14 Mar. 16/2 It is still possible to write..without code words or esotericism.

E.S.P.: see *EXTRA-SENSORY *a.*

espacement. Add: **2.** *Forestry* and *Agriculture.* The distance at which trees or crops are set apart when planted.

1934 *Forestry* VIII. 21 Corsican pine is usually planted in this country at an espacement of 4 feet 6 inches or 5 feet. **1950** *Q. Jrnl. Forestry* XLIV. 13 In order to economize in the number of trees per acre, espacements of less than 5 ft. have seldom been adopted.

‖ **espada** (espāˈda). Also erron. **espardo.** [Sp., = It. *spada*, Fr. *épée* sword.] = MATADOR 1.

1882 [see MATADOR 1]. **1893** [see *CUADRILLA]. **1902** *Encycl. Brit.* XXVI. 460/1 Two *matadors* or *espardos* are engaged in each day's fight. **1923** LD. CURZON *Tales Trav.* 215 Just as the *espada* lunged with the sword..the bull..was upon him. **1926** *Blackw. Mag.* Sept. 290/2 Bull-fighting requires great courage in the bull-fighters from the espada to the mozo.

espadrille (espadriˈl). Also **-illo.** [Fr., ad. Prov. *espardillo*, f. *espart* ESPARTO.] A canvas shoe with soles of twisted rope, originally worn in the Pyrenees.

[**1865** M. EYRE *Lady's Walks* xxiv. 263 The Spaniards..wear..the *espartilla*, or hempen sandal.]. **1892** *Daily News* 16 June 5/4 'Espadrilles', or shoes with woven rope soles, are likely to come into favour with running men. **1905** W. J. LOCKE *Morals M. Ordeyne* I. xii. 146 The bathing hour, when she..divests her of *peignoir* and *espadrilles*. **1907** BARING-GOULD *Bk. Pyrenees* vii. 123 Mauléon..has become the centre of manufacture of the espadrillos. **1923** *Blackw. Mag.* Aug. 162/1, I crept through the thickets in my rope-soled espadrilles. **1945** E. BOWEN *Coll. Impressions* (1950) 230 *Their* sons and daughters wanted something more *plage-y*: shorts, espadrilles, the drink after the swim. **1941** A. CHRISTIE *Evil under Sun* iv. 61 Over it [*sc.* bathing dress] she flung on a bath-robe and laced espadrilles on her feet.

‖ **espagnole** (espanʸol). [Fr., lit. = Spanish.] In full *espagnole sauce*, a simple brown sauce.

1845 E. ACTON *Mod. Cookery* iii. 88 (*heading*) Espagnole (Spanish sauce.) (A highly flavoured Gravy.) **1846** 'A LADY' *Jewish Man.* iv. 58 A little Espagnole sauce heightens the flavour. **1951** *Good Housek. Home Encycl.* (1956) 397/2 A tablespoonful of chopped mushrooms..may be added to the Espagnole sauce. **1961** *Harper's Bazaar* Feb. 72/2 The basic brown sauce—*espagnole*.

Esperanto (espęræˈnto). [Orig. the pen-name (Dr. *Esperanto* = Dr. Hoping-one) used by the inventor on the title-page of his book *Langue internationale; préface et manuel complet*, 1887.] An artificial language invented for universal use by Dr. Ludovik Lazarus Zamenhof, a Polish physician. Its vocabulary consists of roots common to the chief European languages, with endings normalized. Also *attrib.*, *fig.*, and *transf.*

[**1888** 'J. St.' (title) Dr. Esperanto's International Tongue...Edited for Englishmen.]. **1892** *Science* XIX. 32/1 Esperanto's grammatic rules are few in number, for they are all gathered upon four pages only. **1898** R. H. GEOGHEGAN (*title*) The International Language 'Esperanto'. **1903** *Daily Chron.* 13 Jan. 5/2 Esperanto is, in fact, a vocal international shorthand. **1905** *Westm. Gaz.* 22 Sept. 2/1 Latin, the Esperanto of the Middle Ages and Renaissance. **1907** *Daily Chron.* 23 Oct. 4/6 If this Esperanto sort of spirit could be extended to the coinage, how easy would holiday-making become! **1922** C. R. BUXTON *In Russian Village* xvi. 70 Petrov..called himself by the Esperanto name, 'Pečnego'. **1936** [see *ESTABLISHMENT 8 b]. **1958** *Times Lit. Suppl.* 17 Jan. 29/2 The promise not to use 'technical jargon or the special esperanto of *ciné-club* experts'. **1960** KOESTLER *Lotus & Robot* I. i. 37 It is true that there were a few huge concrete blocks in the usual Esperanto architecture. **1968** *Listener* 19 Dec. 814/2 The way in which we are less and less able to understand what is said or done to us unless we can readily translate it all into the Esperanto of common pragmatism.

Hence **Esperaˈntic** *a.*, resembling Esperanto; **Esperaˈntist,** one who is versed in Esperanto; an advocate of the spread of Esperanto as a world-language; also *attrib.*

1905 *Westm. Gaz.* 4 May 10/2 The Congress of Esperantists. **1909** M. L. JONES *Advantages of Esperanto* 2 (*dedication*) To my Esperantist friends of the Yorkshire Federation. **1909** 'IAN HAY' *Man's Man* vii, He first of all abused them with all the resources of an almost Esperantic vocabulary. **1916** H. G. WELLS *Mr. Britling* I. v. 164 'Just Esperantists,' said Teddy. **1922** C. R. BUXTON *In Russian Village* xvi. 70 He was a fluent Esperantist. **1955** *Times* 13 Aug. 5/5 Economists, students, or esperantists.

espresso (espreˈso). Also **expresso.** [It. *caffè espresso*, lit. 'pressed-out coffee'.] Coffee made under steam pressure; the apparatus used for making this; a coffee-bar where it is sold. Also *attrib.*

1945 'A. BOUCHER' in M. & G. Gordon *Pride of Felons* (1964) 80, I was drinking a caffè espresso, a strong, bitter, steamed coffee. **1955** *N. Y. Times* 13 Mar. x. 11/5 Also new are the numerous small Coffee Expresso Snack Restaurants off Dublin's Grafton Street. **1955** *20th Cent.* June 549 The brittle conversation of the Espresso Cafés. **1957** A. WILSON *Bit off Map* 137 The even more degrading swamps of espresso bar rebellion. **1957** *Observer* 3 Nov. 18/2 The piece on lunch-time in an espresso. **1958** M. SPARK *Robinson* xii. 186 When I am walking down the King's Road or sipping my espresso in the morning. **1958** *Spectator* 22 Aug. 248/3 And what about changes in fashionable pronunciation, from table to espresso usage. *Ibid.* 7 Nov. 611/1 It was not easy to accept a character like the expresso poet with his impeccable Oxford accent, grovelling in dirt. *Ibid.* 21 Nov. 700/2 The rise of the expresso with its skiffle groups. **1958** *Times* 23 Oct. 16/1 An espresso machine. **1967** 'J. CROSS' *To Hell for Half-a-Crown* iii. 50 The waiter came in with a pot of espresso and a bottle of fifteen-year-old Meukouw.

esprit. Add: **2. a. esprit de corps** (earlier and later examples.) Also erron. **esprit du corps.**

1780 H. WALPOLE *Let.* 27 Sept. (1904) XI. 285 How *l'esprit du corps* absorbs all feelings! **1790** W. MACLAY *Jrnl.* 12 Feb. (1890) 195 This I considered as something in the taste of *esprit de corps*, for he is a lawyer. **1814** JANE AUSTEN *Mansf. Park* I. v. 94 Well done, sister! I honour your *esprit du corps*. **1955** *Times* 15 Aug. 11/3 All observers expressed high praise for the esprit de corps and élan of the infantrymen.

c. esprit de l'escalier (ęspri də leskalʸe), **esprit d'escalier** [Diderot, *Paradoxe sur le Comédien*: F. *escalier* staircase], a retort or remark that occurs to a person after the opportunity to make it has passed.

1906 H. W. & F. G. FOWLER *King's Eng.* i. 32 No one will know what *spirit of the staircase* is who is not already familiar with *esprit d'escalier*. **1911** BEERBOHM *Zuleika D.* xvi. 248 He prayed, as he followed the victorious young woman downstairs, that *l'esprit de l'escalier* might befall him. **1936** A. HUXLEY *Eyeless in Gaza* xxi. 305 One might be allowed a bit of poetic licence—make the *esprit d'escalier* happen at the same time as the romantic affair. **1959** *Listener* 15 Jan. 121/3 The ripostes which come so readily to their lips..smell slightly of *l'esprit de l'escalier*. **1964** S. BELLOW *Herzog* (1965) 325 This mixture of clairvoyance and spleen, *esprit de l'escalier*, noble inspirations, poetry and nonsense.

-esque, *suffix.* Add: Other formations separately noticed include **Audenesque*, **Bramantesque*, **Browningesque*, **Caravagg(i)esque*, *Carlylesque*, **Chaplinesque*, **Dantonesque*, *Dickensesque*, **Disneyesque*, *Macaulayesque*, *Turneresque*.

esraj (esrāˈdʒ). [Bengali.] An Indian musical instrument with three or four strings, and extra sympathetic strings.

1921 H. A. POPLEY *Music of India* 109 The *Esrāj* is the Bengal variety of the sārangī. It is a little smaller than the latter, and uses all wire strings instead of gut...This is the common instrument that one finds to-day in the houses of cultured people in Bengal. It is played with a bow like the sārangī. **1952** A. DANIELOU *Catal. Indian Mus.* 32 The Esrāj. A long and narrow bowed string instrument used mainly in Bengal to accompany singing. **1957** O. GOSVAMI *Story Indian Mus.* xxvii. 305 The difference between Esraj and Dilruba is that the latter is a little bigger. **1966** *Daily Tel.* 13 Aug. 10/4 Four leading folk singers of Bengal and the Punjab..will be travelling from India for the Festival and will be accompanied by stringed instruments like the esraj.

essayiˈstic, *a.* [f. ESSAY *sb.* + -ISTIC.] In the style of a literary essay, or of an exposition; discursive, informal; essay-writing.

1862 *Independent* (N.Y.) 29 May 3/6 Good specimens of De Quincey's writings—autobiographical, narrative, critical, and essayistic. **1895** *Thinker* VIII. 404 The young pastor thinks himself obliged by the weight of custom to press much that is humanistic and essayistic into the ministrations of the truth. **1935** *Times Lit. Suppl.* 9 Nov. 768/4 It is as difficult to predict for him an essayistic future as it is for Mr. James Laver. **1936** WIRTH & SHILS tr. *Mannheim's Ideology & Utopia* i. 47 This essayistic-experimental attitude in thought also explains why here and there repetitions have not been eliminated. **1960** K. AMIS *New Maps of Hell* (1961) iv. 88 Even at his most essayistic, Wylie shows none of that complacent indifference to his reader.

Ess Bouquet (es bukēˈ). [Short for F. *essence de bouquet*.] The trade name of a perfume. Also *ellipt.* as *Ess*.

1855 G. W. S. PIESSE *Art of Perfumery* vi. 118 'Ess' Bouquet..is but a mere contraction of 'essence' of bouquet. **1875** BROWNING *Inn Album* II, With nose in want of snuff Rather than Ess or Psidium. **1892** G. W. ASKINSON tr. *Furst's Perfumes* 175 The title Ess. Bouquet is an abbreviation of the full name given above [*i.e.* Essence des Bouquets]. **1905** *Westm. Gaz.* 8 Aug. 10/1 Queen Alexandra..has recently abandoned her favourite Ess Bouquet to some extent and taken to Sweet Pea. **1920** A. H. J. KEANE tr. *Deite's Man. Toilet Soap-Making* (ed. 2) 222 'Ess-Bouquet' Soap. **1926–7** *Army & Navy Stores Catal.* 487/1 Bayley's Ess Bouquet—bot. 4/6.

esse. Add: **2.** Revived in recent use. (Later examples.)

1920 *Life of Faith* 23 June 619/2 The great missionary meeting on the Saturday morning..is not the *esse* of the movement. **1929** I. M. CLARK *Hist. Ch. Discip. in Scot.* 208 Some form of law will be necessary to regulate her [*sc.* the Church's] life and protect that distinctive character which is her *esse*.

b. Esp. in contrast with *bene esse*. Cf. WELL-BEING *vbl. sb.*

1621 BURTON *Anat. Mel.* I. ii. Mem. 3 Subs. 15 p. 170 How many poore Schollers haue lost their wits..neglecting all worldly affaires, and their owne health, wealth, *esse*, and *benè esse* to gaine knowledge? **1899** *Daily News* 21 Oct. 7/1 So long as your clergy believe that Episcopacy is essential not only to the bene esse, but to the esse of the Church. **1958** *Listener* 12 June 977/1 No Anglican could possibly dream of claiming that Establishment..is of the *esse* of the Church. Our question is simply whether it is of its *bene esse*.

essence, *sb.* Add: **8.** *of the essence* (*of*): indispensable (to). (Cf. F. *de l'essence de*.)

1843 MILL *Logic* I. vii. § 5. I. 172 It is said that genus and species must be of the essence of the thing. **1873** *Act* 36 & 37 *Vict.* c. 66 § 25 (7) Stipulations..as to time or otherwise, which would not before the passing of this Act have been deemed to be or to have become of the essence of such contracts in a Court of Equity. **1931** *Morn. Post* 21 Aug. 8/2 While time is of the essence of the contract to retrieve the situation, discussion still proceeds. **1936** WODEHOUSE *Laughing Gas* iii. 37 It seemed to me that pomposity was of the essence...You can't tick a bloke off properly unless you come over a bit mid-Victorian. **1958** *Listener* 30 Oct. 677/2 The only way for the workers to defend themselves in such a case is by immediate action. Time is of the essence.

9. c. Name of a variety of Tokay wine.

1773 [see TOKAY¹]. **1862** C. TOVEY *Wine & Wine Countries* v. 208 Until recently, the only wine known in England as the produce of Hungary, was the *Imperial Tokay*, or Tokay essence. **1872** THUDICHUM & DUPRÉ

Treat. Orig. Wine xix. 629 *Tokay*.—1. *Essence*: very sweet. **1911** *Encycl. Brit.* XXVIII. 728/1 The most precious variety of Tokay is the so-called *essence*.

11. essence-peddler *U.S.*, (*a*) a pedlar of medicines; (*b*) *transf.*, a skunk.

1838 HAWTHORNE *Note-Bks.* (1868) I. 119 He was not exclusively an essence-peddler. **1849** J. R. LOWELL *Let.* 9 Mar. (1894) I. 170 A skunk was shot in our back-kitchen this morning. There were two of these '*essence-peddlers*', as the Yankees call them, gambolling there the night before. **1860** *Knickerbocker* Apr. 361 It is a vulgar mistake that the porcupine has the faculty of darting his quills to a distance, as the essence-peddler has of scattering his aromatic wares. **1862** LOWELL *Biglow P.* 2nd Ser. II. 77 With means about 'em (Like essence-peddlers) thet'll make folks long to be without 'em. **1890** E. B. CUSTER *Following Guidon* xiv. 200 As soon as that essence-peddler saw fit to move on, the major-general commanding would issue his order to march. **1944** E. A. HOLTON *Yankees were like This* 179 Who later established some of the best known shops in the county; the tin peddler; the essence peddler.

‖ **essenhout** (eˈsənhōᵘt). *S. Afr.* [Afrikaans, ad. MDu. *eschenhout*, f. *esch* ash + *hout* wood.] = *ESSENWOOD.

[**1789** W. PATERSON *Narr.* 80 A tree very useful in making waggons, which the Dutch call Essen, or Ash.] **1819** *Account Colony Cape Good Hope* 71 Essenhout. **1950** *Cape Argus* (Mag. Section) 18 Mar. 7/7 The Cape ebony, white stinkwood, flatcrown, essenhout and umzimbiti, trees that yield beautiful timber for furniture-making, grow in profusion in every kloof. **1961** PALMER & PITMAN *Trees S. Afr.* 277 The essenhout or dog plum is a medium sized tree found in most of the forests of the country.

Essenism. a. (Earlier example.)

1852 GEO. ELIOT *Let.* 21 Jan. in J. W. Cross *Life* (1884) I. v. 218 He..traces Essenism back to Egypt, and thence to India.

essentialism (ėseˈnʃăliz'm). [f. ESSENTIAL *a.* and *sb.* + -ISM.] **1.** *Educ.* A theory advocating the teaching, on traditional lines and to everyone, of certain ideas and methods supposed to be essential to the prevalent culture (opp. to *PROGRESSIVISM).

1939 J. S. BRUBACHER *Mod. Philos. Educ.* xiv. 336 Over against the philosophy of progressive education..stands that of essentialism or traditionalism. **1950** T. BRAMELD *Patterns Educ. Philos.* vii. 211 What then is the central outlook of essentialism?..It views the established beliefs and institutions of our modern heritage as not only real but true, and not only true but good.

2. *Philos.* The belief in real essences of things, esp. the view that the task of science and philosophy is to discover these and express them in definitions.

1945 K. R. POPPER *Open Society* I. iii. 26 Methodological essentialism, i.e. the theory that it is the aim of science to reveal essences and to describe them by means of definitions. **1948** *Mind* LVII. 302 Essentialism is not only deep-rooted in men's thought habits—or linguistic habits; as it penetrates different departments of human thought it works on these. **1957** K. R. POPPER *Poverty of Historicism* i. 27 This 'realist' theory has also been called 'idealist'. I therefore propose to rename this anti-nominalist theory 'essentialism'.

b. [tr. F. *essentialisme*.] The doctrine that essence is prior to existence (opp. to *EXISTENTIALISM).

1949 E. L. MASCALL *Existence & Analogy* i. 11 However much vestigial Platonic essentialism there may have been in the thought of the great Greek father [*sc.* St. John of Damascus], St. Thomas interprets him in the most existential manner possible.

essentialist. Restrict † *Obs.* to sense in Dict. and add: **2.** One who follows or advocates essentialism.

1945 K. R. POPPER *Open Society* I. iii. 25 These methodological essentialists also agreed with Plato.

B. *adj.* Of or relating to essentialism (in its various senses).

1938 *Newsweek* 14 Mar. 26 (*heading*) Essentialist group urges pupils be coddled less and taught more. **1943** P. WYNDHAM LEWIS *Let.* 17 Aug. (1963) 365, I have been hearing a lot of talk about what appears to me a most promising movement in American education: namely the socalled 'essentialist', as opposed to the 'progressive' principle. **1945** K. R. POPPER *Open Society* I. iii. 27 The problems of the social sciences..are still..treated by essentialist methods. **1948** *Mind* LVII. 302 Nineteenth century Idealist theories..are vitiated through and through by the 'essentialist fallacy'. **1949** E. L. MASCALL *Existence & Analogy* ii. 19 A typical Thomist statement..tells us that in God essence and existence..are identical. Understood from the essentialist standpoint, this statement simply affirms that existence is analytically contained in the notion of God's essence.

essentialize, *v.* Restrict † to sense in Dict. and add: **b.** To formulate in essential form, to express the essential form of.

1913 *Westm. Gaz.* 22 Feb. 5/1 Its difficulties and restrictions, instead of essentialising his thought, have frustrated it. **1922** *Times Lit. Suppl.* 5 Jan. 10/1 A poet [*sc.* Dante] in whom the manifold passions and cultural movements of his time were essentialized and ennobled into the highest poetical utterance. **1923** *Glasgow Herald* 19 Apr. 4 The essentialising flame of Rembrandt's genius. **1923** A. HUXLEY *On Margin* 153 That consolation and strength..essentialized and distilled in the form of art. **1930** N. SHEPHERD *Weatherhouse* x. 181 Garry went drunk to bed, but not with whiskey. Again he had seen life essentialised.

essenwood (eˈsənwud). [Partial transl. of Afrikaans *essenhout*: see *ESSENHOUT.] (*a*) The South African ash, *Ekebergia capensis*; (*b*) Cape mahogany, *Trichilia roka* (*emetica*).

1891 R. M. SMITH *Great Gold Lands S. Afr.* xx. 179 Essen-wood, the South African ash (*Eckebergia capensis*). **1910** J. BUCHAN *Prester John* vi, We took a path up the Berg among groves of stinkwood and essenwood.

Essex. Add: *Essex board* (see quot. 1933); *Essex calf* (examples); *Essex pig* (also *black Essex* and *absol.*), a pig of a kind bred originally in Essex.

1933 C. W. GLOVER *Pract. Acoustics for Constructor* 429 Essex Board...Composition—compressed fibre in plies, bound together by strong, fire-resisting cement. **1954** I. MURDOCH *Under Net* xii. 170 It's only made of plastic and Essex board. **1573** G. HARVEY *Let.-bk.* (1884) 135 Will you haue all in a worde and a halfe, Foes must be frende, quoth an Essex kalfe. *a* **1661** FULLER *Worthies* (1840) I. 497 A Cumberland cow may be bought for the price of an Essex calf in the beginning of the year. **1838** *Penny Cycl.* X. 19/2 Essex has been long noted for a superior breed of pigs...The common Essex pigs have long ears. **1861** MRS. BEETON *Bk. Househ. Managem.* 365 Varieties of the Domesticated Hog..*Native*—Berkshire, Essex, York, and Cumberland. **1877** F. D. COBURN *Swine Husbandry* 83 The Essex are so very similar to the Suffolk, in nearly every respect except color...Their skin, from its color, affords them one advantage over the Suffolks. **1897** S. SPENCER *Pigs* i. 15 The Small Black or Suffolk, or again, as it is styled in the States, the Essex,..and a few other breeds brought the Small Black pig within measurable distance of perfection. **1906** J. LONG *Bk. Pig* (ed. 2) i. 6 (*caption*) Neapolitan..a black race of pigs..crossed with the old Black Essex with excellent results. *Ibid.* 7 The Essex and the Dorset..although still recognized by farmers and breeders in their particular districts, are not recognized by experts and the agricultural societies as special breeds. **1911** *Encycl. Brit.* XXI. 595/1 The *Small Black* or *Black Suffolk* was produced from the old Essex pig by crossing with the Neapolitan. **1960** *Farmer & Stockbreeder* 22 Mar. 85/1 Demand was keen for all four breeds—Large Whites, Essex, Wessex Saddlebacks and Landrace.

essive (eˈsiv). [ad. Finn. *essiivi*, f. L. ESSE, with termination ad. L. *-īvus*: see -IVE.] The designation of one of the fifteen cases of Finnish, a declension expressing a continuous state of being, existence in a specified state or capacity.

1890 [see *ADESSIVE *a.*]. **1905** JESPERSEN in *Englische Studien* XXXV. 9 Essive, indicating the state in which anybody or anything is. **1924** —— *Philos. Gram.* xiii. 183 In Finnish the predicative is..in the essive to denote the state in which the subject is at a given time.

establish, *v.* Add: **5. d.** *Cinemat.*, etc. To introduce and secure the identity or position of (a character, set, etc.).

1948, etc. [see *ESTABLISHING *ppl. a.*]. **1960** O. SKILBECK *Film & TV Working Terms* 48 Artists, or items in sets, are said to be established once they have been photographed. From then on, their positions must be held. **1962** A. NISBETT *Technique Sound Studio* 252 *Establishing an effect.* Allowing it sufficient time (and volume, which may be greater than that subsequently used) for it to register in the listener's mind. **1966** *Listener* 22 Sept. 426/3 Lillian Hellman's screenplay..deserves credit for the swiftly efficient manner in which the numerous main characters are established.

establishable, *a.* Delete † *Obs.* and add later examples.

1918 E. HOLLOWAY in W. P. Trent et al. *Hist. Amer. Lit.* II. III. i. 259 The facts of his biography which are well established or establishable. **1940** W. FAULKNER *Hamlet* I. iii. 48 A thick squat soft man of no establishable age between twenty and thirty. **1964** R. H. ROBINS *Gen. Linguistics* v. 194 They are equally well establishable in unwritten languages.

establishing, *ppl. a.* Delete † *Obs.* and add later *Cinemat.* examples, *spec.* in *establishing shot*. (Cf. *ESTABLISH *v.* 5 d.)

1948 E. LINDGREN *Art of Film* 205 *Establishing shot*, long shot introduced at the beginning of a scene to establish the inter-relationship of details to be shown subsequently in nearer shots. **1953** K. REISZ *Technique Film Editing* 280 *Establishing shot*, shot (usually long shot) used near the beginning of a scene to establish the inter-relationship of details to be shown subsequently in nearer shots. **1959** W. S. SHARPS *Dict. Cinemat.* 94/1 *Establishing shot.* This is usually a long shot for establishing a mood or location and so tying together the separate parts of the scene or sequence. **1961** J. MCCABE *Laurel & Hardy* (1962) viii. 159 Some idiotic fool comes along and cuts the film up...Continuity, establishing shots—most of them gone. The pictures just don't make sense on television. **1962** *Listener* 13 Dec. 1025/1 Most of the blame must attach to the director.., whose long establishing tours..managed to create an edgy tedium without ever really conveying the curious characteristic atmosphere.

establishment. Add: **8. b.** Esp. as *The Establishment*: a social group exercising power generally, or within a given field or institution, by virtue of its traditional superiority, and by the use esp. of tacit understandings and often a common mode of speech, and having as a general interest the maintenance of the *status quo*. Also *attrib.* Hence *establishment-miˈnded* adj., *-miˈndedness*. Cf. *anti-Establishment* (*ANTI-¹ 4).

Quot. 1955 is the *locus classicus* for this modern sense, though occasional earlier uses are recorded.

1923 R. MACAULAY *Told by an Idiot* II. xiv. 117 The moderns of one day become the safe establishments of the next. **1936** H. PEARSON *Labby* x. 260 They spoke the common language, the Esperanto, of the Establishment. **1945** D. GOLDRING *Nineteen Twenties* I. viii. 110 It was a head-on collision between two acknowledged leaders of the literary avant-garde and the powerful forces of what Ford Madox Ford used to call the Establishment. **1955** H. FAIRLIE in *Spectator* 23 Sept. 380/1 By the 'Establishment' I do not mean only the centres of official power—though they are certainly part of it—but rather the whole matrix of official and social relations within which power is exercised. **1957** LD. ALTRINCHAM in *National & Eng. Rev.* Sept. 108/2 He delivered his well-known attack on 'the Establishment', a term generally taken to denote those elements in society and politics which are self-satisfied and opposed to all radical change. **1958** *Times Lit. Suppl.* 17 Jan. 26/3 Sir Maurice Bowra, in his dexterous résumé of what might be called the Establishment view of the ancient Greek world. **1958** *Listener* 6 Nov. 716/1 In Russia, Mr. Khrushchev has been trying with some success to shake up an ossified Communist Establishment. **1958** C. P. SNOW *Conscience of Rich* xxxiv. 254 'That gang' meant the people who had the real power, the rulers, the establishment. **1959** *Encounter* Dec. 57/1 The charge of being Establishment-minded. **1959** C. HOLLIS in H. Thomas *Establishment* 181 The power of the Establishment..comes..from the fact that there is in all of us a degree of establishment-mindedness—that we feel it right that the opinions of such persons should have attention paid to them. **1962** *Listener* 8 Feb. 269/1 Always ready for a dig at the musical establishment. **1969** *Mind* LXXVIII. 26 Where scepticism is thorough and pervasive it is usually directed against some entrenched intellectual Establishment. **1969** *Oz* May 19/1 The Establishment is IBM, Xerox, the Kennedys, the London and New York Times, Harvard University, LSE, the Courts.

estamin. Add: Also spelt **estamene,** as the name of a woollen cloth for dresses.

1862 *Illustr. Catal. Internat. Exhib.* II. xxi. No. 4019 (*heading*) Elworthy, W. & T., Wellington, Somersetshire.—Serges, blankets, yarn, &c. The following goods, of which samples are exhibited, are manufactured by this firm, viz. Serges, estamenes, long ells for China, and swanskins for Newfoundland, &c. **1897** *Daily News* 8 Apr. 8/5 Estamene serges with their diagonal ribs or plain woolly surface.

estaminet. Add: Now, any small establishment selling alcoholic liquor. (Earlier and later examples.)

1814 G. W. BRIDGES *Alpine Sks.* ii. 47 A delightful shade of oak and elm, beneath which are numerous *estaminets* and *restaurateurs*. **1915** W. H. L. WATSON *Adv. Despatch Rider* ii. 19 We..had some coffee at a little *estaminet*. **1955** J. THOMAS *No Banners* i. 15 The brothers went into the *estaminet*, hoping to find a bottle of wine or something to eat.

estampage (estæˈmpėdʒ, ‖ ęstańpāʒ). *Archæol.* [Fr., f. *estamper* to stamp.] A squeeze or impression on paper of an inscription.

1887 *Academy* 24 Dec. 427/3, I made complete estampages..of the Asoka inscription at Kālsi. **1888** *Epigraphia Indica* I. 1 Methods of taking direct inked impressions from stone, also, have been devised; and *estampages* can thus be taken, having the advantages of presenting a direct reading of the inscription on the inked face and a mould from it on the back, so that the slightest scratch on the stone is retained on a carefully-taken impression. **1904** M. DE Z. WICKREMASINGHE *Epigraphia Zeylanica* I. p. iii, From these estampages..it was arranged that I should edit the texts for publication. As to the scrupulous care taken in the preparation of ink-estampages there need not be the slightest doubt.

estate, *sb.* Add: **7. c.** *the fifth estate*: in various applications (see quots.).

1932 *Times Educ. Suppl.* 7 May 157/2 Small wonder that radio has been called 'the fifth estate'. **1955** *Times* 14 July 8/1 Unions were now the fifth estate of the realm, Mr. Tiffin continued, and when they wanted a shorter week they would go to the employers and tell them to give it to them.

13. *spec.* A property on which a crop, as rubber, tea, etc., is cultivated; also, a vineyard. Freq. preceded by a defining word.

1851 C. REDDING *Hist. Mod. Wines* (ed. 3) vi. 158 The value of wine estates is very considerable in this department [*sc.* the Gironde]. **1862** in C. TOVEY *Wine & Wine Countries* iv. 142 That wine is..possessed of *finesse*, *delicatesse*, and *bouquet*, which are quite peculiar to the estate. **1878** E. MONEY *Cultiv. & Manuf. Tea* (ed. 3) i. 2 Making..1,000 acres the outside area..that should ever have been purchased for any one estate. *Ibid.* ii. 11 Coolies are well treated on Tea estates. **1911** *Encycl. Brit.* XXVI. 480/1 The finest teas are produced at high elevations in Darjeeling and Ceylon..but the quality from individual estates varies much from season to season. *Ibid.* XXVIII. 721/2 The yield of the principal estates of the Médoc are kept distinct and reach the consumer as the products of a particular growth and of a particular year. **1911** STEVENS & BEADLE *Rubber* ii. 13 Sometimes the rubber plants are set amongst matured coffee or tea, with the idea of..gradually transforming a tea or coffee estate into a rubber estate. **1916** P. SCHIDROWITZ *Rubber* (ed. 2) iv. 26 The *working expenses* of an estate containing say 10,000 to 15,000 trees. **1921** A. L. SIMON *Wine & Wine Trade* vii. 86 All Clarets which have a claim to a more or less high degree of excellence are too proud of their birthright not to go into the world under their own name—the name of the estate or château whence they came. **1965** T. Eden *Tea* (ed. 2) i. 3 From the progeny of the plants thus raised the first estate was planted in 1891. *Ibid.*, In 1924 a commercial estate was established near Tukuyu. *Ibid.* vii. 69 The ideal conditions on a tea estate would be to have no bare soil but a moderate growth of weeds. **1968** Krug & De Poerck *World Coffee Survey* v. 374 The Coffee Board annually collects data from a representative set of estates.

b. = **housing estate.*

[**1906** W. A. Harvey *Model Village* 13 Nearly all the old trees and woodland on the Estate have been preserved. *Ibid.* 68 The planning and working out of the Bournville Estate.] **1915** G. Cadbury *Town Planning* iv. 68 In laying out estates it is advisable to plan the roads so that they do not form through routes for traffic. *Ibid.*, Other road problems are practically confined to estate development. **1923** E. Bowen *Encounters* 142 Their house was among the first two or three on a new estate. **1939, 1958** [see *council (housing) estate* s.v. *council 17]. **1960** C. S. Lewis *Studies in Words* i. 13 When I was a boy *estate* had as its dominant meaning 'land belonging to a large landowner', but the meaning 'land covered with small houses' is dominant now. **1970** *Woman* 14 Nov. 76/4 Till my Dad died we lived in a council house, but then we moved into a larger estate house and my mother became very snobbish.

14. *attrib.* and *Comb.*: *estate-owner*; **estate agent**, one who acts as steward or manager of a landed estate; one who conducts business in the sale of houses and land; hence *estate agency*; **estate-bottled** *a.*, (of a wine) bottled at the vineyard of its growth; **estate car**, a light saloon motor car spec. constructed or adapted to carry both passengers and goods; also *ellipt.*; **estate duty**, a graduated charge levied by the State on real or personal property at the death of the owner; **estate management**, the art of administering an estate (sense 13 or *13 b); hence *estate-manager*; **estate wagon** = *estate car.*

1912 *Estate Agents, Architects & Surveyors* (*Pitman's Shorthand*) 5 The Estate Agency profession. **1880** *Harper's Mag.* Sept. 565 'Estate agent?' he next asked. **1884** *The 'Estate Agent'* 3 House and Estate Agents. **1886** S. W. Mitchell *R. Blake* xx. 188 It seemed to her natural that an unknown Yankee estate-agent should wish to marry a woman of assured social place. **1940** H. J. Grossman *Guide to Wines* 378/2 *Estate-bottled*, wine bottled by the vineyard owner or producer. **1959** *Times* 21 Sept. 13/2 High prices for estate-bottled wines. **1950** *Motor Industry* July 118/2 (*heading*) Latest car price list. ..Standard...Vanguard saloon...Estate car. **1958** *Times* 1 July 6/6 The Hillman Minx..has its own estate car version. **1961** *Times* 5 Oct. 11/4 Enter the swish new Anglia Estate. **1966** *Guardian* 19 Oct. 9/3 Both the saloon and the estate carry 5 in comfort. **1889** *Act 52 Vict.* c. 7 § 5, Estate Duty on personal property passing by will or on intestacy. **1896** *Act 59 & 60 Vict.* c. 28 § 16 The estate duty payable in respect of any annuity. **1915** G. Cadbury *Town Planning* v. 89 Town Planning is ..estate management on a large scale. **1921** in Jeffery & Neville *House Property* 92 The London University now grants a Degree in Estate Management. **1937** *Discovery* Sept. 268/1 Industrial management needs sound traditions just as much as estate management or the professions. **1962** H. R. Loyn *Anglo-Saxon England* iv. 146 A little treatise on eleventh-century estate management. *Ibid.* ix. 370 Estate-managers to look after the lands of the church. **1937** *Discovery* May 164/1 Rich estate owners in Brazil, Peru and Nicaragua. **1962** H. R. Loyn *Anglo-Saxon England* i. 16 Their eponymous Gallo-Roman estate-owners. **1959** M. Steen *Tower* i. vii. 91 There was an estate wagon and a couple of big private cars.

esteemed, *ppl. a.* Add: Much in vogue as a complimentary epithet in commercial correspondence.

1903 *Daily Chron.* 25 Feb. 7/2 Sir, yours of even date to hand. If you can make it convenient, we should be glad if you could come on Wednesday morning to try on your esteemed favour. **1905** H. G. Wells *Kipps* 1. ii. 37 He [*sc.* a draper] never acknowledged an order that was not an esteemed favour.

ester (e·stəɹ). *Chem.* [G. (L. Gmelin *Handb. d. Chemie* (1848) IV. 161), prob. f. *essig* vinegar + *äther* ether.] A derivative of an acid in which one or more acidic hydrogen atoms are replaced by an alkyl, aryl, or similar group; **acrylic ester** (see quot. 1951). Also *attrib.* and *Comb.*; **ester gum** (see quot. 1940).

[**1852** H. Watts tr. *Gmelin's Hand-bk. Chem.* VII. 190 *Ethers du troisième genre.* I formerly distinguished these compounds by the name of Naphthas produced by oxygen-acids (*Naphthen durch Sauerstoffsäuren erzeugt*); but I now propose for them the term *Ester*.] *Ibid.* 215 Compound Ethers formed by Oxygen-acids (Ester). **1889** Muir & Morley *Watts's Dict. Chem.* II. 466 A compound ether (or ester) is a hydrogen salt in which the typical hydrogen has been displaced by an alkyl. **1899** E. F. Smith tr. *von Richter's Org. Chem.* I. 139 Just as salts result from the union of metallic hydroxides with acids, so esters are formed by the combination of alcohols with acids. **1904** [see Thion-]. **1906** *Practitioner* Nov. 593 The rapidly fermentable fruit and malt sugars, esters, and higher alcohols. **1907** *Jrnl. Chem. Soc.* XCII. i. 383 Ester-Acids of Sulphur-substituted Carbonic Acids with Aliphatic Hydroxy-Acids. **1921** *Jrnl. Soc. Dyers & Colourists* XXXVII. 288/1 Acetyl silk, the new ester-silk. **1930** *Flight* 17 Jan. 134/2 Dope is a solution of cellulose esters in various solvents and diluents. **1940** *Chambers's Techn. Dict.* 309/2 *Ester gums*, rosin or gums which have been esterified with glycerine; raw material for varnishes. **1946** *Nature* 28 Dec. 930/2 It may conveniently be obtained by direct esterification of the glycol or by catalysed ester-interchange between the glycol and dimethyl terephthalate. **1951** *Gloss. Terms Plastics* (*B.S.I.*) 9 *Acrylic ester*, an ester of acrylic acid, or of a structural derivative of acrylic acid. **1958** *Engineering* 28 Feb. 281/3 The production of cross links of the ester type formed by the introduction of the COOH and OH groups. **1965** *Nomencl. Org. Chem.* (I.U.P.A.C.) C. 126 Neutral esters of carboxylic acids, *etc.*, are named in the same way as their neutral salts except that (*a*) the name of the alkyl or aryl, *etc.*, radical replaces the name of the cation and (*b*) a periphrase such as '(alkyl or aryl) ester' replaces '(metal) salt'.

Hence **este:rifica·tion**, the process of forming an ester; the conversion of an acid into its ester; **este·rify** *v.* [-fy], to convert into an ester.

1898 *Nature* 6 Jan. 239/1 The authors have made experiments on the esterification of many cinnamic acids. **1903** *Rep. Brit. Assoc.* (1902) 586 The rate of esterification of methyl hydrogen succinate. **1907** *Practitioner* June 864 Arhovin..is a product of diphenylamin and the esterified thymol-benzoic acid. **1920** Cross & Bevan *Paper-Making* (ed. 5) 27 Esterifying reagents. **1940** Esterify [see above]. **1946** Esterification [see above]. **1956** *Nature* 24 Mar. 575/2 Vitamin A is absorbed through the lymphatic system mostly in the esterified form and is stored in the liver predominantly as ester. **1966** *Lancet* 24 Dec. 1421/2 The galactose is esterified with the sulphate at position 6.

esterase (e·stərēɪs, -z). *Biochem.* [f. *ester + *-ase.] An enzyme which brings about the hydrolysis of an ester into the corresponding acid and alcohol.

1916 A. E. Porter in *Biochem. Jrnl.* X. 523 (*title*) The distribution of esterases in the animal body. *Ibid.*, Esterases such as cholesteryl esterases and wax-splitting ferments, that is ferments which are able to split the esters of higher alcohols. **1949** H. W. Florey et al. *Antibiotics* II. xxxi. 1076 The blood of the mouse, rat and guinea-pig was shown to contain an esterase which hydrolysed the methyl and benzyl esters. **1956** *Nature* 28 Jan. 185/2 The esterase activity was expressed as the amount of carbon dioxide evolved from the bicarbonate buffer by the acid formed during the enzymatic hydrolysis of acetylthiocholine.

Esth (esþ, est), *a.* and *sb.* [G. *Esth, Ehst.*] = next. Now *rare.*

1868 *Fraser's Mag.* LXXVIII. 535/1 The Esths..speak a language closely allied to the Finnish. *Ibid.* 536/1 The poem was published in Esth, with a translation into German. *Ibid.* 542/2 Turja is the Esth name for Norway. **1910** *Encycl. Brit.* IX. 798/1 The Esths, Ehsts or Esthonians. **1920** *Glasgow Herald* 28 Dec. 4 Warsaw expects an attack at any moment, the Esths and Letts in February. **1935** Huxley & Haddon *We Europeans* vii. 221 The Esths have reddish-flaxen hair, blue eyes and a cephalic index of 79.

Estonian (estǒu·niăn), *a.* and *sb.* Also **Esthonian.** [f. mod.L. *Est(h)oni(a*, f. *Esth *a.* and *sb.*, after *Saxonia*, etc. + -an.] **A.** *adj.* Of or belonging to Estonia (native name *Eesti*), since 1940 a constituent republic of the U.S.S.R., stretching along the south coast of the Gulf of Finland. **B.** *sb.* **1.** A native or inhabitant of Estonia. **2.** The Finno-Ugrian language of Estonia.

1795 in W. Tooke *Varieties of Literature* I. 23 Esthonian poetry. *Ibid.*, The Esthonians..have an extremely soft, delicate, and tender articulation. **1841** Lady Eastlake *Resid. Shores Baltic* II. xvi. 62 Nothing can exceed the hospitality of the Esthonians. **1863** R. G. Latham *Nationalities of Europe* I. xii. 129 *Ma*, in Estonian, means *land*. *Ibid.* 132 The Estonian instrument is the harp. **1874** A. H. Sayce *Princ. Compar. Philol.* 322 *Wanna Isa* in Esthonian means 'the old father'. **1894** W. F. Kirby *Hero of Esthonia* I. p. xvi, The Finns, the Esthonians, and the Lapps..speak very similar languages. **1925** O. Rutter *New Baltic States* xi. 199 Estonian folklore owes something to the Finnish. **1941** J. H. Jackson *Estonia* i. 20 The Soviets..were asking for naval bases on Estonian territory. *Ibid.* iii. 58 One of the eight chairs on the original foundation was a professorship in the Estonian language. **1948** A. Oras *Baltic Eclipse* i. 20 The Estonian shipyards, railway works, plywood factories. **1952** E. Uustalu *Hist. Estonian People* v. 82 These tax and work registers, or 'Vakus-books', as they were called in Estonian.

estrada (estrā·da). [Pg.:—L. *strāta* (see Street *sb.*).] In the Brazilian rubber trade, a winding path or road connecting a series of trees. Also in Fr. form **estrade.**

1906 *Westm. Gaz.* 12 July 9/3 The estradas (or roads) which are being worked produce at least 250 kilos of rubber per harvest. **1913** R. H. Lock *Rubber* 20 To each seringueiro or collector are assigned from 100 to 150 trees, which are connected by a winding path, or *estrade*, cut through the undergrowth...Beginning at sunrise the seringueiro makes two rounds of his estrade. **1921** *Glasgow Herald* 3 May 10 From one estrada in six months about one ton of rubber may be obtained.

estradiol, var. *œstradiol.

estray, *a.* (Earlier and later U.S. examples.) Also *transf.*

1789 *Kentucky Gaz.* (Lexington, Ky.) 28 Mar. 1/3 All persons shall have access to the Estray-book, without paying any fee therefor. **1876** *Rep. Vermont Board Agric.* III. 426 Many..of these new varieties of grain are not new...They are old varieties estray..from remote quarters..of the globe. **1889** *Harper's Mag.* June 158/2 A farmer living near a middle Georgia town, one day found an estray cow in his pasture.

estrin, estriol, varr. *œstrin, *œstriol.

estro. Delete *Obs.* and add later examples.

1817 Byron *Let.* 2 Jan. (1830) II. 70, I have not done a stitch of poetry since I left Switzerland, and have not at present the *estro* upon me. **1849** Rossetti *Let.* 18 Oct. (1965) I. 81 The verse gets its measure of *estro* accordingly.

estrogen, estrogenic, estrone, estrous, varr. *œstrogen, etc.

estufa. *U.S.* (Earlier example.)

1844 J. Gregg *Commerce of Prairies* I. 271 The famous *estufas*, or subterranean vaults.

-et, *suffix²*, forming neuter verbal and denominative sbs., represents OE. *-et(t* (= Goth. *-iti*, OHG. *-izzi*), as in *thicket.*

-et, *suffix³*, representing the 3rd pers. sing. pres. tense ending of Latin verbs of the second conjugation, as *scilicet, tenet.*

eta¹ (ī·tă, ē·tă). [a. Gr. ἦτα (see Etacism).] The name of the seventh letter (*H*, *η*) of the Greek alphabet, corresponding to English *e.*

c **1400** in Maundeville *Trav.* (1839) iii. 20. **1642** [see Upsilon 1]. **1883** I. Taylor *Alphabet* II. vii. 85 The Latin H and the Greek eta are identical in form and in alphabetic position, though not in value.

eta² (ī·tă). In full, *eta palm.* = Ita-palm.

1769 E. Bancroft *Ess. Nat. Hist. Guiana* 61 The Eta Tree is of the same species with the foregoing [*sc.* cabbage tree], but smaller. **1866** G. W. Bennett *Illustr. Hist. Brit. Guiana* 228 The Eta Palm (*Mauritia flexuosa*) is a beautiful tree and is adored by the Indians for the numerous uses to which it can be applied. **1922** *Blackw. Mag.* July 6/1 The road used to end..before a great eta-palm swamp.

Eta, eta³ (ē·tă). [Jap.] A member of a class of Japanese outcasts. Also *collect.* and *attrib.*

1897 A. M. Knapp *Feudal & Modern Japan* I. v. 173 The cause of the intense repulsion and contempt with which the Japanese have regarded the *eta* class is unknown. *Ibid.*, No pariah class of any nation has ever been under a greater ban of disdain and contempt than have the *eta* of Japan. *Ibid.* 174 The spot where an *eta* has been standing must be sprinkled with salt if a Japanese would tread there without contamination. **1906** *Westm. Gaz.* 28 July 12/2 Hoshiko, eta though she be, full of the supreme devotion. **1955** *Times* 14 July 11/4 He visited the Ainu...the Etas (those mysterious outcastes), and strip-tease shows. **1958** *Times* 9 Sept. 9/7 The pitiable existence of many people of *eta* origin..has hampered their emancipation. **1960** Koestler *Lotus & Robot* ii. vi. 176 The eta—the sweepers, scavengers and tanners—ceased to be untouchables. **1969** *Daily Tel.* 6 Mar. 16/3 The continuation of a practice in Japan..which permits 3,000,000 people known as 'eta' or 'the untouchables' to live in 6,000 outcast hamlets isolated from the people by centuries-old discrimination.

etærio (etiə·rio). *Bot.* Also rarely Hetærio. [ad. F. *étairion, etairium* (C.F.B. de Mirbel 1813, in *Nouveau Bulletin des Sciences, par la Société Philomatique* III. 317), f. Gr. ἑταιρεία association.] An aggregate fruit, like that of the buttercup, strawberry, or raspberry.

The form *hetærio* (in Dict. s.v.) never attained wide currency.

1832 Lindley *Introd. Bot.* I. ii. 176 The parts of an Etærio are Achenia. **1861** R. Bentley *Man. Bot.* iv. 316 When the achænia borne by a single flower are so numerous that they form more than a single whorl or series, they constitute collectively an *etærio.* **1866** [see Hetærio]. **1896** G. Henslow *Wild Flowers* 102 The cluster of drupels is called an etaerio. **1956** *Dict. Gardening* (R. Hort. Soc.) Suppl. 204/1 Etaerio or Aetaerio. An aggregate fruit made up of drupes or achenes as in the Strawberry or Blackberry.

‖ **et al.** (et æl), *phr.* [Abbrev. of L. *et alii* (masc.), *et aliæ* (fem.) or *et alia* (neut.), 'and others'.] And others. (Used esp. to avoid giving a list of authors, etc., in full.)

1883 in S. Fallows *Handbk. Abbrev.* 10/2. **1907** W. James *Let.* 13 Sept. in R. B. Perry *Tht. & Char. of W. J.* (1935) II. 482 As overcomer of Baumgarten, Crusius *et al.*, he [*sc.* Kant] was no doubt great. **1959** *Listener* 13 Aug. 256/2 He relegates Hamilton *et al.* to a prologue and epilogue. **1962** G. K. Hunter *John Lyly* ii. 39 Lyly no doubt completed his school career..having read, construed and explicated innumerable passages of Ovid, Cicero, Virgil, *et al.*

‖ **étalage** (etalā·ʒ). [Fr., f. *étaler* to display.] A display.

1900 *Daily News* 4 Sept. 4/7 A constant étalage of jewellery. **1925** *Blackw. Mag.* Oct. 493/2 The *étalage* in the window betrayed a small grocer's shop. **1953** B. Campbell *Finding Nests* ii. 22 Artifact arrangements of the vegetation, for example the étalage of leaves at the entrance to many robins' nests,..cannot easily be seen from above.

étalon, etalon (‖etalon, e·tălǫn). [F. *étalon* (Fabry & Pérot 1902, in *Ann. de Chimie et Physique* XXV. 107), lit. 'a standard (of weights, measures, etc.)':—OF. *estalon, estelon* prob. of Gmc. origin: see Stallion.] A device used to produce interfering beams of light, consisting essentially of two plane parallel reflecting plates of fixed separation and (in some kinds) adjustable orientation.

1905 E. C. C. Baly *Spectroscopy* ix. 302 Fabry and Perot, profiting by their experience with their interferometer, have adopted fixed air layers as interference apparatus, the widths of which are determined once and for all. They call these apparatus standards (*étalons*). **1914** *Astrophysical Jrnl.* XXXIX. 185 Interferometer, echelon grating, parallel interference-plates, etalon, etc. **1934** *Discovery* July 184/1 The interferometer..is of the type known as the Fabry-Perot étalon, and in this particular form it consists of a metal tube having extremely flat and parallel end surfaces, to which are contacted polished flat plates of glass or natural quartz. **1963** R. W. Ditchburn *Light* (ed. 2) ix. 358 A determination of the length of the shortest etalon in terms of the wavelength of the cadmium red line.

etamine. Add: (Further examples.) Also in Fr. form *étamine.*

1759 W. Verral *Cookery* p. xxviii, An etamine or two for the straining your thick soups, cullies or creams. **1794** A. Young *Trav.* (ed. 2) I. xix. 552 Here are etamines, linen, stockings, bleach-grounds, &c. **1966** *Harper's Bazaar* Sept. 50 Evening dress..in bougainvillia pink étamine.

‖ **étang** (etaṅ). [Fr., ad. OF. *estanc*: see Stank *sb.*] A shallow pool or small lake, esp. one of the type found along the French coast and resulting from the blocking of streams by sand dunes.

c **1845** C. Brontë *Professor* (1857) II. xxi. 109 Near it [*sc.* a place of public resort in the outskirts of Brussels] were several of those lakelets called *étangs*; and there was one *étang*, larger than the rest, where..people were accustomed to amuse themselves by rowing round it. **1879** *Encycl. Brit.* X. 265/2 The dunes travel inland in parallel, irregular, and often confluent ridges, between which rain-water is sometimes arrested to form pools (*étangs* of the French coasts), where formations of peat occasionally take place. **1940** M. Sadleir *Fanny by Gaslight* 10 One of the several solitary and reed-grown *étangs* which occur..in the forest above Les Yvelines. **1960** W. W. Williams *Coastal Changes* ii. 38 Probably the most important effect of eustatic and isostatic movements..is the part that they play in the formation of many coastal features, notably..bars like those which lie between the *étangs* and the sea in the South of France. **1969** *Daily Tel.* (Colour Suppl.) 10 Jan. 32/1 There is a bit of French Provincial between Nantes and Bordeaux, which soon fades into the bleak pine barren of Landes and the sinister *étangs* of the sandy wastes to the south.

eta patch. *Aeronaut.* [f. Gr. ἦτα, the name of the Greek letter *η*; it was the designation of the first airship, built in 1913 at the Royal Aircraft Factory, Farnborough, to have this means of attachment.] (See quot. 1919.)

1918 W. E. Dommett *Dict. Aircraft* 25 *Goosefoot*, a method of attaching the rigging on the envelope, often called an 'Eta Patch'. **1919** W. B. Faraday *Gloss. Aeronaut. Terms* 57 *Eta patch*, a means of attaching rigging to the envelope consisting of a large fan-shaped patch made up of several thicknesses of fabric and reinforced with webbing bands and terminating in a metal ring. **1955** *Aeroplane* LXXXIX. 51/1 Because her suspension was distorting the envelope, 'Eta' patches were introduced, and cured the trouble.

‖ **étatisme** (etatizm). Also **etatism.** [Fr., f. *état* State *sb.* + *-isme* -ism.] Extreme development of the power of the State over the individual citizen.

1923 *Contemp. Rev.* Aug. 195 What one might almost call the fetishism of the State, an extreme form of *étatisme.* **1926** *Spectator* 2 July 6/2 That hideous thing which later the French called *Etatisme* produced a social and political atmosphere in which a free man could hardly breathe. *Ibid.* 24 July 122/2 It must not be thought that we should like to see the British spirit of individual liberty subjected to any such trial of extreme *étatisme.* **1952** V. A. Demant *Relig. & Decline of Capitalism* ii. 57 Absolute individualism or absolute *étatisme.* **1960** *Encounter* Mar. 63/2 He denounces the over-mighty Communist states because he is opposed to all *étatisme.* **1960** *Spectator* 2 Dec. 889 The excessive economic control and 'Etatism' of the Populist regime.

‖ **état-major** (eta,maʒor). *Obs.* [Fr., f. *état* state + *major*, major, adjutant; cf. Sp. *estado mayor.*] The staff of an army, regiment, etc.; = Staff *sb.*¹ 21. Also *transf.*

1805 *Wynne Diaries* 1 Dec. (1940) III. vii. 220 Fremantle had still on board Villeneuve's état major (who he says divert him greatly with their gasconnades). **1830** Hazlitt *Life Nap. Buonaparte* IV. xlviii. 29 Lorence, general-in-chief of the *état-major*, ordered several Jews to be brought before him. **1848** Thackeray *Van. Fair* lxv. 591 The second floor apartments occupied by the *état major* of the gambling firm. **1890** [see Typicality].

et cetera, etcetera. Add: **1.** (Examples of *Yours etc.* used as an ending in letters.)

1825–6 Dickens *Let.* (1965) I. 1 Cheaper in comparison than a Leg. Yours &c C Dickens. **1920** F. W. Crofts *Cask* xi. 124 An unexpected call to England prevented me ordering this in person. Yours, etc., *Leon Felix.* **1970** *Times* 13 Nov. 11/4 He came..from Italy. Yours, &c., *F. Bennett.*

2. a. (Further *attrib.* examples.)

1898 *Daily News* 15 July 6/1 Colonial affairs..formed a sort of etcetera department of the War Office. **1900** *Ibid.* 23 May 6/3 The nickname, Etcetera Department, sometimes given to the English Home Office.

b. (Later examples.) *spec.* (*pl.*), trousers.

1794 Scott *Let.* 5 Sept. (1932) 36 A pair of new boots & Buckskin &c's in which the soldier is to be equipt. **1843** J. R. Planché *Fortunio* i. i. 8 Hush! hush! for up I go, To put a light Silk pair of tight Etceteras on below. **1937** E. Pound *Fifth Decad Cantos* xlii. 7 And how this people can in this the fifth et cetera year of the war, leave that old etcetera up There on that monument! **1942** J. Cary *To be a Pilgrim* xix. 36 'All the things you silly geese have muddled up till you don't know your etc. from an etc.' A very coarse comparison, such as never before..had I used to a woman.

etch (etʃ), *sb.*² [f. Etch *v.*²] The action or process of etching. Freq. *attrib.* and *Comb.* (see Etch *v.*² 4).

1896 *Amer. Bookmaker* Mar. 81/1 After the first etch the plate should be perfectly cleaned. **1967** E. Chambers *Photolitho-Offset* x. 142 After 30 sec. immersion in the etch-bleach bath, the room light (white light) is switched on permanently. **1967** *Times* 23 Nov. 4/7 (*heading*) Improved etch for integrated circuits. *Ibid.*, The new development is called an anisotropic etch, which ensures that the slot does not get wider as it gets deeper.

etch, *v.*² Add: **4. etch figure** [tr. G. *aetzfigur* (H. A. Baumhauer 1869, in *Ann. d. Physik und Chem.* CXXXVIII. 563)], a depression on the face of a crystal caused by the action of a solvent; also *etch pit.*

1879 *Jrnl. Chem. Soc.* XXXVI. 439 'Etch-figures' on quartz crystals. *Ibid.* 440 Different 'etch-figures' are produced by acids upon the octohedral faces from those produced by the action of water upon those faces. **1922** A. E. H. Tutton *Crystallography* (ed. 2) II. iii. iii. 1200 These etch-figures in every case are symmetrical to the single plane of symmetry. **1923** Glazebrook *Dict. Appl. Physics* V. iii. 376/2 The etched facets on crystal surfaces..appear as more or less isolated markings on the crystal surface. In the latter case they exhibit the typical geometrical characteristics of the crystal, and are known as 'etch figures'. **1950** *Sci. News* XV. 67 The etch pits are seen to go down in a series of terraces like Dante's Inferno. **1959** *Jrnl. Inst. Metals* LXXXVII. 376/1 A knowledge of the factors that influence the formation and distribution of etch pits is of primary importance, since these factors govern the validity of the etch-figure technique as a method of determining the distribution of dislocations in a material.

etchant (e·tʃănt). [f. Etch *v.*² + -ant¹.] A corrosive agent used for etching. Also *attrib.*

1930 *Engineering* 26 Sept. 388/2 Dr Voce had used the term 'etchant' in his paper, instead of writing the words 'etched by', and he would like to protest against the use of such an unpleasant word. **1950** *Jrnl. Iron & Steel Inst.* CLXVI. 270/1 A list of ten etchants for revealing austenitic grain size. **1960** *New Scientist* 11 Aug. 391/1 The tests will obviously have to be extended to different compositions, etchant concentrations and temperature regimes. **1968** *Ibid.* 1 Aug. 241/3 The clue that led to the development of the new technique which may eventually replace acid etchants.

etched, *ppl. a.* Add: *etched figure* = *etch figure* (s.v. *etch *v.*² 4).

1876 *Jrnl. Chem. Soc.* II. 273 (*heading*) Etched figures on cubes of rock-salt. **1902** H. A. Miers *Min.* i. v. 112 Etched figures are of the greatest possible value in indicating the true symmetry of a mineral. *Ibid.* 113 A closer study of the etched figures upon various minerals shows that they are of two sorts; some faces of the crystal become covered with minute depressions or pits, others with minute elevations.

etching, *vbl. sb.* Add: **3. etching figure, pit** = *etch figure* (s.v. *etch *v.*² 4).

1877 E. S. Dana *Textbk. Min.* ii. i. 118 Analogous to the etching-figures are the figures produced on the faces of some crystals by the loss of water. **1925** *Jrnl. Iron & Steel Inst.* CXI. 588 Deep etching with picric or nitric acid fails to develop etching pits. **1935** A. Sauveur *Metallogr. & Heat Treatment* (ed. 4) ii. 17 The figures thus outlined, in reality small cavities, are often called 'etching pits' or 'etching figures'.

Eteocretan (ī:tio,krī·tăn, e:tio-), *a.* and *sb.* [f. Gr. Ἐτεόκρης lit. 'true Cretan', f. ἐτεό-ς true + Κρητ-, Κρής Cretan, + -an.] **A.** *adj.* Of, pertaining to, or designating a pre-Greek people of Crete or their language. **B.** *sb.* **1.** A member of this people or linguistic group. **2.** The language of this people.

1615 Chapman tr. *Homer's Odysses* xix. 245 There [*sc.* in Crete] Greekes suruiue, There the great-minded Eteocretans liue. **1895** A. J. Evans *Cretan Pictographs* 85 The language here found represents that of the Eteocretans of whom, as we know, Praesos was a principal stronghold. *Ibid.* 86 It is possible that in the early period during which the indigenous Cretan script, both pictographic and linear, seems to have taken its origin the sole or preponderating element in the island may have been the 'Eteocretan'. **1904** *Ann. Brit. School at Athens 1901–2* VIII. 155 Messapian and Illyrian are clearly related to Venetic, a language to which we have found certain resemblances in Eteocretan. **1939** J. D. S. Pendlebury *Archæol. of Crete* v. 270 Of the famous post-Minoan 'Eteocretan' inscriptions from Praisos it is again unsafe to say anything with certainty. **1962** *Daily Tel.* 4 Apr. 15/3 Prof. Cyrus H. Gordon..claimed..'Linear A' ..and also 'Eteo-Cretan', long taken to be the pre-Greek speech of Crete, are both actually Phoenician. The Eteo-Cretan is written with the Greek alphabet. **1962** *Listener* 10 May 810/2 The Eteocretans of Homer.

-eteria: see *-teria.

eternal, *a.* Add: **4. a.** *Eternal city* (later example).

1869 'Mark Twain' *Innoc. Abr.* (1870) xxvii. 218 Michael Angelo..designed the Eternal City.

d. *eternal triangle*: see *triangle *sb.*

5. Esp. in phr. *eternal truths* or *verities*, necessary truths, as being valid at all times.

1690 Locke *Hum. Und.* iv. xi. 323 Propositions that are once true, must needs be eternal Verities. **1694** *Ibid.* (ed. 2) iv. xi. 366 Propositions are therefore call'd *Eternal Truths*, not because they are Eternal Propositions actually formed...But because being once made..they will, whenever they can be supposed to be made again at any time..always actually be true. *a* **1700** [in Dict.]. **1883** F. H. Bradley *Logic* II. ii. ii. 589 No vehicle conveys the eternal verities half so well as does the labyrinth of a fantastic genealogy. **1900** B. Russell *Philos. of Leibniz* ii. 18 The eternal truths, he says, are all hypothetical. **1960** K. Amis *Take a Girl like You* xxvi. 301 'Do you think it's right to give up your principles for somebody you're in love with?' 'I say, we are getting down to the eternal verities, aren't we?'

eternality (ītəɹnæ·lĭti). Delete † *Obs.* and add later example.

1925 A. N. Whitehead *Sci. & Mod. World* (1926) 121 Every scheme for the analysis of nature has to face these two facts, change and endurance. There is yet a third fact to be placed by it, *eternality*, I will call it.

eternally, *adv.* **2. b.** (Later U.S. example.)

1850 W. R. Ryan *Upper & Lower California* I. 310 If you'll sell it me..I'll be eternally obliged to you.

eternity. Add: **7.** *attrib.*, as eternity ring (see quots.).

1939–40 *Army & Navy Stores Catal.* 449/1 Diamond and Platinum Eternity Ring from £10 10 0. **1950** 'P. Wentworth' *Eternity Ring* iv. 30 'You know the kind of ring they call an eternity ring?' She smiled. 'An old fashion which has come back—a circle of small stones set continuously.' **1960** H. Hayward *Antique Coll.* 112/1 *Eternity ring*, a ring in the form of a plain circle symbolising eternity, often set with a single row of stones.

eth, var. *edh.

ethanol (e·þănǫl). *Chem.* = *ethyl alcohol.*

1900 *Jrnl. Chem. Soc.* A. LXXVIII. i. 618 Ethanol mercury salts in alkaline solution yield no precipitate with potassium sulphide. **1946** *Nature* 23 Nov. 744/2 A solution of the substance in 10 per cent aqueous ethanol gave no coloration with ferric chloride. **1964** N. G. Clark *Mod. Org. Chem.* viii. 125 The first three alkanols are methanol, ethanol, propanol.

ethanolamine (eþănǒu·lămīn). *Chem.* [f. prec. + Amine.] Any of three compounds in which one or more of the hydrogen atoms in ammonia is replaced by a group CH_2CH_2OH, which are viscous, high-boiling liquids usu. manufactured from oxirane (ethylene oxide) and ammonia with extensive industrial uses; *spec.* 2-aminoethanol, $H_2NCH_2CH_2OH$; also, any other amino alcohol structurally related to ethyl alcohol.

1897 *Jrnl. Chem. Soc.* LXXII. i. 313 (*heading*) Amidoethylic alcohol (1:2-ethanolamine). **1950** Kirk & Othmer *Encycl. Chem. Technol.* V. 855 The ethanolamines, combined with fatty acids, yield soaps that are extensively used in industry as emulsifiers and detergents. **1951** C. R. Noller *Chem. Org. Compounds* xxxiv. 694 Ethanolamine and its trimethylammonium salts, the cholines.., constitute a portion of an important class of biological substances known as the phospholipids. **1959** *Times* 1 Apr. 7 (Advt.), Even the photograph was processed with the help of *Shell* ethanolamines.

ethenoid (e·þĭnoid), *a. Chem.* [f. Ethen(e + -oid.] Resembling ethylene in structure or properties; ethylenic.

1907 *Jrnl. Chem. Soc.* XCII. i. 307 The interaction of ethenoid compounds and bromine in solution is of the second order. **1932** *Canad. Jrnl. Research* VII. 475 The bromine absorption of the polymers is very small but..measurable. This fact might be taken to prove the existence of an open-chain structure with an ethenoid linkage. **1951** *Gloss. Terms Plastics* (*B.S.I.*) 7 *Ethenoid plastics*, a class generally understood to comprise the acrylic, vinyl and styrene groups of plastics.

ether. Add: **5. a.** Formerly also thought to be the medium through which radio waves and electromagnetic radiations generally are propagated. (Later examples.)

Later views of the ether were that it provided a frame of reference for the universe, with respect to which Maxwell's equations of electromagnetism or other field equations are valid, but that it possibly lacked any material properties. Following the special theory of relativity, which assumed that there is no absolute frame of reference and which explained why the Michelson-Morley experiment failed to detect an 'ether wind', the concept was gradually abandoned by most physicists, and has received little support since the 1920s.

1924 H. L. Brose tr. *Born's Einstein's Theory Relativity* iv. 75 So far as this conception of the ether is still used nowadays it is taken to mean nothing else than empty space associated with certain physical states or 'fields'. **1925** O. Lodge *Ether & Reality* ii. 43 Fearfully rapid tremors or ether vibrations which can be excited electrically in a form which we know as X-rays. **1926** B. Russell *ABC of Relativity* iii. 31 If there is an ether wind, it is clear that, relatively to an observer on the earth, light-signals will seem to travel faster with the wind than across it. **1935** A. S. Eddington *New Pathways in Sci.* iv. 39 Some distinguished physicists maintain that modern theories no longer require an ether... I think all they mean is that, since we never have to do with space and aether separately, we can make one word serve for both; and the word they prefer is 'space'. **1951** E. T. Whittaker *Hist. Theories Aether & Electricity* (ed. 2) I. p. v, It seems absurd to retain the name 'vacuum' for an entity so rich in physical properties, and the historical word 'aether' may fitly be retained. **1961** M. Čapek *Philos. Impact Contemp. Physics* xiv. 251 There are even now physicists who are looking hopefully toward aether as a clue to the explanation of micro-physical phenomena. **1965** J. D. North *Measure of Universe* iii. 50 It was not long before attempts were being made to design an aether suitable for the representation of [Newtonian] gravitation. **1967** Condon & Odishaw *Handbk. Physics* (ed. 2) vi. 158 It is not profitable to identify the mechanical ether of the eighteenth and nineteenth centuries with the 'polarizable vacuum' of modern quantum field theories.

Hence *colloq.*, wireless, 'the radio'.

1899 *Daily News* 26 May 5/1 An age of ether-wave telegraphy and the Röntgen rays. **1923** *Daily Mail* 11 Jan. 7 The climax of an entrancing week of ether-borne opera...She possesses the real 'wireless voice'. **1930** J. S. Huxley *Bird-Watching* p. vii, A request I made over the ether for more information. **1933** *Discovery* Sept. 280/1 The Babel of a thousand languages will soon assail him on the 'ether'. **1966** G. Millerson *T.V. Production* ii. 20 This interlacing method of scanning effectively saves 'etherspace', allowing more stations to work in any band. **1968** *Listener* 4 July 26/1 The authentic atmosphere, the enthusiasm, [illegible] can be felt over the ether, but it's even better on TV.

6. a. For 'sulphuric and other acids' read 'sulphuric acid'. (Cf. next sense.) The modern systematic name is *diethyl ether*, usu. given as *ethyl ether*; *ethoxyethane* is occas. used.

† **a*.** Any of various compounds analogous to ethyl ether, formed by the reaction of ethyl alcohol and either an acid (other than sulphuric acid) or a salt and sulphuric acid. *Obs.*

The compounds were named according to the acid or salt from which they were derived: e.g. nitrous acid gave *nitrous ether* (now called ethyl nitrite C_2H_5ONO). The extension of meaning of the word *ether* thus took place in two stages, the first including other acids in the preparation, the second other alcohols (giving sense 6 b).

1796 R. Heron tr. *Fourcroy's Elem. Chem. & Nat. Hist.* III. 424 The nitric ether obtained by these processes is a yellow fluid, equally volatile and susceptible of evaporation with sulphuric ether. *Ibid.* 427 M. de la Planche, apothecary, has proposed the preparing of muriatic ether, by pouring..sulphuric acid and alcohol upon decrepitated muriate of soda. *a* **1797** *Encycl. Brit.* IV. 475/2 With the nitrous acid and spirit of wine, may also be made an exceedingly volatile liquor, called nitrous ether, to distinguish it from the vitriolic above mentioned. **1833** *Penny Cycl.* I. 156 The acids which occasion the formation of the ethers already described do not enter at all into their composition. **1867** C. L. Bloxam *Chem.* 521 Acetic ether..has a share in the perfume of cider, perry, vinegar, and of many wines. **1962** M. P. Crosland *Hist. Stud. Lang. Chem.* v. i. 269 The authors [of the *Méthode de nomenclature chimique* (1787)] had decided that each of the etherial compounds obtained by the reaction of alcohol with various acids should be known as 'ether' qualified by an adjective according to the acid used to prepare it, e.g. *éther acétique* (i.e. ethyl acetate). Most of these 'ethers' were what we should now call esters.

b. (Additional examples.) In mod. use, *ether* is restricted to compounds of the type R_2O (*simple ethers*) and ROR′ (*mixed ethers*), where R, R′ are organic groups (other than acyl groups and their analogues) linked directly to the oxygen by a carbon atom; *compound ethers* are now known as *esters.

1852 *Rep. Brit. Assoc. 1851* i. 129 Chemists..seem to consider it sufficient to place all those bodies, the basis of which is an ether, under the head of compound ethers. *Ibid.* 130 If the term, compound ether, be retained at all, it should be restricted to bodies..in which a simple ether is united with an ether radical, as the oxide of ethyl with methyl or with amyl. **1950** L. F. & M. Fieser *Org. Chem.* (ed. 2) vi. 134 Ethers are regarded more appropriately as derivatives of alcohols, from which they are usually prepared. **1964** N. G. Clark *Mod. Org. Chem.* viii. 142 The ethers are all colourless, neutral, inflammable compounds, sparingly soluble in and less dense than water. **1966** *McGraw-Hill Encycl. Sci. & Technol.* V. 85 The names of the ethers correspond to the hydrocarbon groups present. Thus, $CH_3—O—CH_3$ is methyl ether, rarely dimethyl ether, and $C_6H_5—O—CH_3$ is phenyl methyl ether.

etherical (iþe·rikăl), *a.* Delete † *Obs.* and add later example.

1920 *Conquest* May 317/2 The plant..responds to different rays of the vast etherical spectrum, from the extreme ultra-violet to the longest wireless electrical waves.

etherism. Add: **2.** Addiction to taking ether; cf. next, sense 2.

1888 N. Kerr *Inebriety* vii. 108 Etherism is the antipodes of opiumism.

etherist (ī·þĕrist). [-ist.]

1. One who administers ether to a patient.

1884 H. Thompson *Tumours of Bladder* 25 If the etherist permits the patient [illegible] the abdominal muscles, the effort is hopeless.

2. One who is addicted to taking ether as a stimulant or intoxicant.

1888 N. Kerr *Inebriety* vii. 109 At first he took chloral and opium, then he devoted himself to ether, and has been an etherist for some four years.

etheromania (ī:þĕrǒmēɪ·niă). *Path.* [f. Ether + -o + -mania.] A morbid addiction to the consumption of ether as a stimulant or intoxicant. So **etheroma·niac**, an ether addict.

1889 N. Kerr *Inebriety* (ed. 2) vii. 122 Etheromania has been treated at asylums in the North of Ireland. **1894** *Ibid.* (ed. 3) vii. 138 A gentleman aged 35, had been an etheromaniac for 3 years. **1909** *Westm. Gaz.* 23 Feb. 14/3 Etheromania is largely on the increase in South-west Russia.

ethical, *a.* Add: **4.** *Med.* Of a medicine or drug: advertised only in the professional press, not to the general public, and often available only on a doctor's prescription (see quots.). Hence as *sb.*, such a medicine.

1935 *Lancet* 24 Aug. 463/2 Ethical Pharmaceuticals Association (E.P.A.). A number of firms representing manufacturers of ethical pharmaceutical preparations—i.e., preparations not advertised to the public—have formed themselves into an association. **1952** Mercer & Whale in Kirk & Othmer *Encycl. Chem. Technol.* X. 229 [illegible] **1963** *Times Rev. Industry* July [illegible] **1968** D. E. Allen *Brit. Tastes* v. 114 Sales of medicines only to the medical profession ('ethicals') [illegible]

ethicalism (e·þikăliz'm). [f. Ethical *a.* + -ism.] Devotion to ethical ideals.

1899 A. B. Bruce *Apologetics* ii. ii. 192 The ethical monotheism of Hebrew prophecy has for one of its necessary presuppositions the intense ethicalism of the prophets themselves. **1955** *Theology* LVI. 209 The all-too-human ethicalism of a progressive movement asserting a brotherhood of men.

ethicality (eþikæ·lĭti). [f. Ethical *a.* + -ity.] Ethical quality, behaviour, or principles.

1890 J. H. Stirling *Philos. & Theol.* v. 89 Ethicality is the right of rights, the right which is the right of the community, while morality is right doing according to the conscience of the individual. **1934** *Punch* 19 Sept. 314/3 [illegible] **1965** M. Bradbury *Stepping Westward* vii. 333 Froelich..was a man beyond outmoded ethicalities.

ethicism (e·þisiz'm). [-ism.] Devotion to ethics or ethical ideals; tendency to moralize.

1865 *Harper's Mag.* Nov. 807/2 Their art..was marred by the intense ethicism that pervaded the New England mind for two hundred years. **1906** *Daily Chron.* 12 Nov. 3/4 A scientific theory of social duty..ethicism founded on a solid and comprehensive science of man. **1908** *Mod. Philol.* V. 643 This means that he..rejected ethicism, declamation, narration, and description.

ethicist. Add: (Examples.) Also, one who supports ethics or morality in opposition to religion.

1891 *Monist* I. 556 A scientific ethicist has to proceed like any other naturalist. **1905** *Daily Chron.* 29 Nov. 3/3 Able to make their understanding 'issue in act', as the Oxford ethicists used to say. **1907** H. Begbie *Vigil* xix. 307 He would play the part of ethicist and point people to the highest example of human perfection. **1908** *Lit. Guide* 1 Aug. 121/1 A Rationalist or an Ethicist can seldom get credit for what he does, however good his intentions. **1964** P. Meadows in I. L. Horowitz *New Sociology* 446 Among the ethicists..the new technological age was seen as a function of an immanent axiological model.

ethico-. Add: **e:thico-so·cial.**

1905 *Spectator* Lit. Suppl. 28 Jan. 119/1 Modern ethico-social guides, such as Stevenson, Whitman, and Maeterlinck. **1920** *Red Triangle* May 352/2 Every Church is trying to overcome every danger which is trying to destroy its ethico-social being. **1924** W. B. Selbie *Psychol. Relig.* 102 The social element in these early forms of ritual..points to a tribal consciousness which has in it at least the beginnings of an ethico-social outlook.

ethinylœstradiol (e·þinil:strādi·ǫl). Also ethinyl estradiol, oestradiol. [f. Ethine + -yl + *œstradiol.] A synthetic œstrogen, $C_{20}H_{24}O_2$.

1939 *Chem. Abstr.* XXXIII. 3465 (*title*) The action of ethinylestradiol, a derivative of follicular hormone, given by mouth. [illegible] Drug Administration as an oral contraceptive [illegible] containing 9.85 milligrams of norethynodrel, and 0.15 milligram of an oestrogen ethinyl oestradiol 3-methyl ether. **1966** J. Martin *Clinical Endocrinol.* (ed. 4) vii. 244 Ethinyl oestradiol is a modification of the natural hormone. [illegible]

ethionine (eþai·ǒnin, -in). *Chem.* [f. Eth- + *-ionine* as in *methionine.] A crystalline amino-acid that is the ethyl homologue of methionine and inhibits growth.

1938 H. M. Dyer in *Jrnl. Biol. Chem.* CXXIV. 519 We have now synthesized the S-ethylhomocysteine (ethionine). **1955** *Jrnl. Exper. Med.* CII. 151 (*title*) Effect of ethionine-induced pancreatic damage on liver phospholipid. *Ibid.*, Rats were fed diets containing *dl*-ethionine (α-amino-γ-(ethylmercapto)-butyric acid) in quantities sufficient to damage the exocrine cells of the pancreas. **1966** *New Scientist* 17 Mar. 708/3 A medium containing puromycin or ethionine, both of which cause almost complete inhibition of protein synthesis in these cells.

Ethiopian, *a.* and *sb.* Add: **A. 1. c.** *Ethiopian sour gourd* = Baobab.

1640 etc. [see *sour gourd* s.v. Sour *a.* 10].

2. c. Of or pertaining to Ethiopianism.

1904 H. O. Dwight et al. *Encycl. Missions* (ed. 2) 8/1 The seceders from [the Wesleyan Methodist Society churches in the Transvaal] formed a new body which adopted the name 'Ethiopian Church'. The movement was..a protest against a color line in the churches of Africa. **1911** *Encycl. Brit.* XVIII. 593/1 [illegible] **1915** J. Hastings *Encycl. Relig. & Ethics* VIII. 728/2 S. Africa...The racial factor is especially in evidence in the 'Ethiopian Movement', composed of congregations of natives who in 1892 formally seceded from their missionary connexions.

B. a. (Later examples.)

1884 A. H. Sayce *Fresh Light from Anc. Monuments* vi. 155 Tirhakah the Ethiopian [illegible] **1935** *Times* 17 Oct. 10/4 [illegible] **1946** *Times Lit. Suppl.* 15 June 278/4 [illegible] of the Ethiopians.

c. An advocate or supporter of Ethiopianism.

1911 *Encycl. Brit.* XVIII. 593/1 [illegible] **1961** *Listener* 30 Nov. 918/1 The term 'Ethiopian' was first employed by the African secessionists themselves, who took it from the Authorized Version of the Bible, in which Africa and black men are usually called vaguely 'Ethiopia' and 'Ethiopians'.

Ethiopianism (īþiǒu·piăniz'm). [-ism.] An African nationalist movement in South Africa (see quot. 1961).

1906 *Daily Chron.* 13 Feb. 5/4 The rising in Natal is now officially declared..to be the result of the teaching of Ethiopianism, namely 'South Africa for the Black races'. **1910** J. Buchan *Prester John* viii. 131 It is what they call 'Ethiopianism', and American negroes are the chief apostles. [illegible] So-called Ethiopianism appears to have been started in South Africa by various disgruntled native ministers of the Gospel. **1961** *Listener* 30 Nov. 918/1 The term 'Ethiopianism', then, is properly applied to early religio-political manifestations of African nationalism in South Africa.

ethisterone (eþi·stĕrōun). [f. Ethi(ne + *(testo)sterone.] A synthetic hormone similar to progesterone (see quots.).

1947 *Brit. Chem. Abstr.* A. iii. 224 (*heading*) Biological properties of ethinyl testosterone (ethisterone). *Ibid.*, Ethisterone has the following testosterone-like effects: it causes growth of the sex organs in the normal or hypophysectomised castrated immature male rat. **1948** *Brit. Pharmacopœia* 44 Ethisterone. *Synonyms.* Pregneninolone: Anhydrohydroxyprogesterone: Ethinyltestosterone...Ethisterone is 17-ethinyl-Δ^4-androsten-17-ol-3-one. **1962** *Lancet* 25 Aug. 406/2 [illegible] 100 mg. ethisterone (17-α-ethinyl testosterone) and 0.01 mg. [illegible] **1966** J. Martin *Clinical Endocrinol.* (ed. 4) vii. 246 Ethisterone, a synthetic compound very similar to progesterone, is active by oral administration but, owing to its feeble progesterone-like action, comparatively large doses must be given.

ethnic, *a.* and *sb.* Add: **A.** *adj.* **2.** Also, pertaining to or having common racial, cultural, religious, or linguistic characteristics, esp. designating a racial or other group within a larger system; hence (*U.S. colloq.*), foreign, exotic.

1935 Huxley & Haddon *We Europeans* iv. 136 Nowhere does a human group now exist which corresponds closely to a systematic sub-species in animals, since various original sub-species have crossed repeatedly and constantly. For existing populations, the non-committal term *ethnic group* should be used. *Ibid.* vi. 181 The special type of ethnic grouping of which the Jews form the best-known example. **1936** *Discovery* June 167 [In Africa] linguistic divisions are a very fair indication of ethnic groups. **1939** C. S. Coon *Races of Europe* xi. 444 The Jews are an ethnic unit, although one which has little regard for spatial considerations. Like other ethnic units, the Jews have their own standard racial character. **1964** *Listener* 6 Feb. 233/2 There are many groupings of people, ethnic units, population aggregates —call them what you will—that may be distinguished from each other. **1965** *Sun* 6 Dec. 7/6 Ethnic..has come to mean foreign, or un-American or plain quaint. **1969** *New Yorker* 30 Aug. 76/2 Its hopelessly reactionary nature is best exemplified not..even by the ethnic comedians. **1970** *Daily Tel.* 16 Apr. 18 The situation is fast becoming greatly complicated by the presence in Cambodia of large numbers, put at 400,000 to 500,000, of 'ethnic' Vietnamese.

B. *sb.* **2.** *Greek Antiq.* An epithet denoting nationality, derived from or corresponding to the name of a people or city [= ἐθνικόν (Steph. Byz.)]. Also *gen.*

1828 J. A. Cramer *Anc. Greece* III. Index p. i, The Greek ethnic of each town or place has been subjoined where there was authority for it. **1902** D. G. Hogarth *Nearer East* 194 Where the 'Arab' (to use the ethnic widely) lives under conditions similar to the Greek, he resembles him. **1921** C. T. Seltman *Temple Coins Olympia* 103 The dies..upon which the full ethnic ΦΑΛΕΙΩΝ appears. **1921** *Brit. Mus. Return* 79 The ethnics of Damastium and Pelagia. **1959** A. G. Woodhead *Study Gk. Inscriptions* 44 Sometimes the single name, without further elaboration, sometimes with patronymic and demotic or ethnic, or with one of the two.

3. A member of an ethnic group or minority. orig. *U.S.*

1945 Warner & Srole *Social Syst. Amer. Ethnic Groups* (Yankee City Ser. III) v. 68 The Irish..had their origins largely in the peasant stratum...The Jews were of the burgher class...These differences in the ethnics' social-class backgrounds will be seen later to have important bearing on their adaptation. *Ibid.* 93 The ethnics have conspicuously succeeded in 'getting ahead' in the Yankee City social hierarchy. **1961** *Times Lit. Suppl.* 17 Nov. 828/4 The former 'ethnics', a polite term for Jews, Italians, and other lesser breeds just inside the law. **1963** T. & P. Morris *Pentonville* iii. 62 It is the general view of the prison staff that the majority of 'coloureds' and 'ethnics' are West Indians. **1964** S. M. Miller in I. L. Horowitz *New Sociology* 297 As the white ethnics—first the Irish, later the Jews, and still more recently the Italians..gained strength.

ethnicity. Restrict †*Obs. rare* to sense in Dict. and add: **2.** Ethnic character or peculiarity.

1953 D. Riesman in *Amer. Scholar* XXIII. i. 15 The groups who, by reason of rural or small-town location, ethnicity, or other parochialism, feel threatened by the better educated upper-middle-class people. **1964** P. Worsley in I. L. Horowitz *New Sociology* 384 Existing barriers of ethnicity imported into office could thus be removed. **1970** *Oxf. Univ. Gaz.* C. Suppl. vi. 14 In Hilary Term Dr. Leslie Palmier..gave a series of lectures entitled 'Ethnicity in Indonesia'.

ethnobotany (eþnobǫ·tăni). orig. *U.S.* [f. Gr. ἔθνο-ς nation + Botany.] The traditional knowledge and customs of a people concerning plants; the scientific study or description of such knowledge and customs. Hence **e:thnobota·nical** *a.*, **ethnobo·tanist.**

1896 J. W. Harshberger in *Bot. Gaz.* XXI. 146 The study of ethno-botany aids in elucidating the cultural position of the tribes who used the plants for food, shelter or clothing. **1899** *Smithsonian Rep.* 65 Dr. Walter Hough was detailed to carry on ethno-botanical researches in Mexico. **1934** *N. & Q.* CLXVII. 129/1 The importance of ethno-botany has long been understood by systematic anthropologists. *Ibid.*, An ethno-botanical note on *Datura Metel.* **1934** Webster, Ethnobotanist. **1955** *Times* 3 Aug. 7/4 Its distinguished American ethnobotanist.

ethnocentric (eþnose·ntrik), *a.* [f. Gr. ἔθνο-ς nation + Centric *a.*] Regarding one's own race or ethnic group as of supreme importance. So **ethnoce·ntrism, e:thnocentri·city.**

1900 W. J. McGee in *Ann. Rep. Bur. Amer. Ethnol. 1897–98* 831 In primitive culture the epocentric and ethnocentric views are ever-present and always-dominant factors of both mentation and action. **1907** W. G. Sumner *Folkways* i. 13 *Ethnocentrism* is the technical name for this view of things in which one's own group is the center of everything, and all others are scaled and rated with reference to it. *Ibid.* 15 The state..became the object of that group vanity and antagonism which had been ethnocentric. **1951** E. E. Evans-Pritchard *Social Anthrop.* vi. 127 This ethnocentric attitude has to be abandoned if we are to appreciate the rich variety of human culture and social life. **1957** W. S. Allen *Ling. Stud. Lang.* 8 Mar. 7 A familiarity with many languages.. may do much to reduce the ethnocentrism with which, as native speakers of a language, we are inevitably burdened. **1959** *Listener* 1 Jan. 27/2 Is it really necessary to resuscitate the white man's burden at the very time when so much depends upon our getting rid of this form of ethnocentricity? **1964** I. L. Horowitz *New Sociology* 34 The present ethnocentricity reflects a fascination with machines at the expense of minds. **1964** M. Critchley *Developmental Dyslexia* xii. 70 The age at which a child normally begins to read with facility is also the age at which..he turns from an autistic, egocentric individual, to a societal, ethnocentric being.

ethnohistory (eþnohi·stǒri). [f. Gr. ἔθνο-ς nation + History *sb.*] The study of the history of races or cultures, esp. non-Western races or cultures. So **e:thnohisto·rian**, a student or expert in ethnohistory; **e:thnohisto·rical** *a.*, of or pertaining to ethnohistory.

1936 *Times Lit. Suppl.* 2 May 378/2 Each of us must have within reach five hundred ethnohistorians. **1936** L. P. Weaver tr. *Székely's Cosmos, Man & Soc.* 12 Hitherto ethnohistorical research has depended on three sources: archæological documents, linguistic documents and authentic writing. **1952** *Amer. Anthropologist* LIV. 331 Ethnohistory is blessed with an abundance of source materials. **1965** *Language* XLI. 203 Evidence that the ethnohistorian must use. **1966** *New Statesman* 13 May 710/2 (Advt.), The study..of internal change within the non-Western societies (i.e. ethnohistory).

ethnolinguistic (e:þnoliŋgwi·stik), *a.* [f. Gr. ἔθνο-ς nation + Linguistic *a.*] Of or pertaining to ethnolinguistics.

1950 D. L. Olmsted *Ethnolinguistics So Far* iv. 10 The frequency of morphemes in a given ethnolinguistic situation is of importance in determining the relations between talk and action. **1964** E. A. Nida *Toward Sci. Transl.* vii. 148 One important element in this ethnolinguistic model of translation is the nature of the response implied.

ethnolinguistics (e:þnoliŋgwi·stiks). [f. as prec.] The study of the relations between linguistic and cultural behaviour.

1950 D. L. Olmsted *Ethnolinguistics So Far* 1 The word *ethnolinguistics* has been used in a number of contexts by social scientists. **1964** R. H. Robins *Gen. Linguistics* ix. 352 The specific study of the inter-disciplinary links..between anthropology and linguistics has been called *ethnolinguistics.*

ethnomusicology (e:þnomiūzikǫ·lŏdʒi). [f. Gr. ἔθνο-ς nation + *musicology.] The scientific study of the music of a culture or sub-culture, considered either as a combination of sounds or as an aspect of socio-cultural behaviour; also, the comparative study of the music of more than one culture. Hence **e:thnomusicolo·gical** *a.*, pertaining to this study; **e:thnomusico·logist**, a student of or expert in ethnomusicology.

1950 J. Kunst *Musicologica* Pref., It is intended as a general introduction to ethno-musicology, before going on to the study of the forms of separate music-cultures. **1951** *Africa* Jan. 319 This work fulfilled a long-standing need by ethnomusicologists. **1958** *Listener* 15 May 824/3 The book reflects the present state of ethnomusicological studies. **1959** *Times Lit. Suppl.* 5 June 332/1 The new subject of ethnomusicology has emerged in no more than a generation to put system into a mass of authentic material gathered from all parts of the globe. **1967** W. P. Malm *Music Cultures of Pacific* i. 9 An ethnomusicologist should be a man of many viewpoints if he is to appreciate fully the meaning of music in a given culture. **1970** *Daily Tel.* 30 Sept. 10/5 Dr Frank Harrison, Reader in the History of Music at Oxford University.., has been appointed to the Chair of Ethnomusicology at Amsterdam University.

ethnoscience (eþnosai·ĕns). [f. Gr. ἔθνο-ς nation + Science.] The study of races or cultures; ethnography.

1964 E. A. Nida *Toward Sci. Transl.* iii. 38 B. N. Colby in an unpublished paper, entitled 'Eidos, Semantics and Ethnoscience', has dealt extensively with a number of these approaches to studies of semantic fields. **1964** *Language* XL. 230 The terminology which is now appearing in anthropological literature devoted to ethnoscience.

et hoc genus omne: see *hoc genus omne.

ethological, *a.* Add: **3.** Of or pertaining to *ethology 4.

1908 *Westm. Gaz.* 1 Aug. 15/2 Animal psychology in France is no longer the quoting of instances of the super-natural intellectual qualities of the tame dog...The ethological method has sprung into existence. **1959** *Ibis* No. 101, p. 357 (*heading*) The ethological analysis of incubation behaviour. **1959** *New Biol.* XXVIII. 71 The reproductive segregation which it brings about is commonly called ethological isolation.

ethologist. Add: **3.** One who is engaged or versed in the study of *ethology 4.

1953 *New Biol.* XIV. 7 Contemporary study of vertebrate behaviour is increasingly influenced by the school of ethologists, led by Nils [*sic*] Tinbergen and Konrad Lorenz. **1964** *Language* XL. 215 Ethologists find little novelty in the proposal that there is a qualitative difference between human and nonhuman cognitive faculties. **1966** R. Ardrey *Territorial Imperative* (1967) i. 27 Ethologists have assumed that there must be a neurological foundation in the central nervous system providing an anatomical switchboard for handling messages.

ethology. Add: **4.** *Zool.* [Gr. ἦθος nature or disposition of animals; in pl., customs, haunts.] That branch of Natural History which deals with the actions and habits of animals, and their reaction to their environment; esp. the study of instinctive animal behaviour.

[**1859** I. Geoffroy Saint-Hilaire *Histoire Naturelle Gen. des Règnes Organiques* II. 285 C'est à l'éthologie.. qu'appartient l'étude des relations des êtres organisés.. dans *l'agrégat* et *la communauté.*] **1897** Parker & Haswell *Zool.* I. 9 The whole question of the relation of the organism to its environment gives us a final and most important branch of Natural History which has been called Ethology or Bionomics. **1910** W. M. Wheeler *Ants* viii. 124 Their ethology, that is, their functional aspect (physiology and psychology). **1956** O. L. Zangwill in A. Pryce-Jones *New Outl. Mod. Knowl.* 170 A new type of behaviour study which endeavours to combine the rigours of scientific materialism with genuine understanding of the ways of animals..has been christened *ethology.* **1962** *Listener* 9 Aug. 207/1 Ethology is a method developed for studying automatic behaviour in animals, by looking at their movements and postures, but it can also be applied to man.

ethyl. Add: Also commonly with pronunc. (ĭ·þəil). For etym. and def. read: [a. G. *ethyl* (J. Liebig 1834, in *Ann. d. Pharm.* IX. 18), f. Eth(er: see -yl.] The univalent hydrocarbon radical C_2H_5—, either as a group in an organic compound or as a short-lived free radical. (Add earlier example and further selected examples of its use in the names of compounds.) *ethyl acetate* $CH_3COOC_2H_5$, *ethyl acetoacetate* $CH_3COCH_2COOC_2H_5$, *ethyl ether* = Ether 6 a, *ethyl formate* $HCOOC_2H_5$, *ethyl iodoacetate* $CH_2I.COOC_2H_5$ (a kind of tear-gas), *ethyl nitrite* $C_2H_5NO_2$.

1838 *Penny Cycl.* X. 49/2 Ethereum (as it is termed by Dr. Kane, and *ethule* by Berzelius). **1874** *Index to Jrnl. Chem. Soc. 1848–72* 144/2 Ethyl acetate, action of sodium and isopropylic iodide on. **1940** *Chambers's Techn. Dict.* 310/1 *Ethyl acetate*,..colourless liquid of fruity odour, used as a lacquer solvent and in medicine. **1874** *Jrnl. Chem. Soc.* XXVII. 883 The complicated changes involved in all previous processes are at once simplified if ethyl acetoacetate, $CH_3.CO.CH_2.CO.OC_2H_5$, be taken as the starting point. **1964** B. G. Clark *Mod. Org. Chem.* xv. 292 Ethyl acetoacetate, or ethyl β-ketobutyrate, is the most important of the β-keto-esters. The free acid occurs in the urine of uncontrolled sugar diabetics. **1878** *Jrnl. Chem. Soc.* XXXIV. 1065/1 Ethyl ether, brominated. **1885** I. Remsen *Introd. Study Compounds of Carbon* iv. 42 Ethyl ether, $C_4H_{10}O = (C_2H_5)_2O$.— This is the substance commonly known simply as *ether*, or *sulphuric ether.* **1868** *Jrnl. Chem. Soc.* XXI. 528/1 Ethyl formate. **1960** A. E. Bender *Dict. Nutrition* 46/2 *Ethyl formate*, $H.COOC_2H_5$. Fumigant used against raisin moth..etc...Chemical intermediate—in synthesis of B_1, sulphadiazine, etc. **1905** *Jrnl. Chem. Soc.* LXXXVII. 483 Ethyl iodoacetate, ethyl bromoacetate, and methyl bromoacetate interact at 25° with one equivalent of sodium thiosulphate. **1932** A. Huxley *Brave New World* iii. 54 Phosgene, chloropicrin, ethyl iodoacetate. **1870** *Jrnl. Chem. Soc.* XXIII. 455/1 Ethyl nitrite, action of sulphurous acid on. **1961** *Brit. Med. Dict.* 523/2 *Ethyl nitrite*, $C_2H_5.O.NO$, a compound which, dissolved in alcohol.., has been used in angina pectoris with limited effect.

Hence **e·thylate** *v. trans.*, to convert into an ethyl compound, to introduce ethyl into; also *e·thylated* ppl. adj., *e·thylating* vbl. sb.; **e:thylbe·nzene**, a hydrocarbon, C_8H_{10}, usually made from ethylene and benzene; **e:thylhydrocu·prein(e**, a derivative of quinine having various bactericidal uses; **ethylmo·rphine**, a toxic homologue of morphine, $C_{19}H_{23}NO_3$.

1850 *Phil. Trans. R. Soc.* CXL. 110, I have not attempted to ethylate this compound any farther. *Ibid.*, In..the other ethylated bases, the properties of the mother-compound are only slightly modified. **1861** G. Fownes *Man. Elem. Chem.* (ed. 8) 607 (*heading*) Methylated and ethylated derivatives of natural bases. **1920** *Jrnl. Soc. Chem. Industry* XXXIX. T1/1 Although English manufacturers of ethylated products are persistently demanding ethyl sulphate and other ethylating agents, none of these can compete with ethyl chloride. *Ibid.* 3/1 As to the method of ethylating amino compounds with ethyl chloride, it is..impossible to generalize. **1920** *Chemical Age* III. 557/2, I have to learn, for example, how to ethylate *alpha*-naphthylamine [*printed* naphthalymine] very expeditiously at 100°C. on a commercial scale by means of ethyl chloride. **1873** *Jrnl. Chem. Soc.* XXVI. 1028 (*title*) Action of bromine on boiling ethylbenzene. **1934** *Jrnl. R. Aeronaut. Soc.* XXXVIII. 315 In addition to benzene and toluol derived from coal tar, higher aromatic hydrocarbons such as ethyl-benzene, pseudo-cumene and mesitylene have been synthesised and used..as fuel for aviation engines. **1964** N. G. Clark *Mod. Org. Chem.* xix. 376 The first homologue is methylbenzene, $C_6H_5.CH_3$, known as 'toluene', which is followed by ethylbenzene, $C_6H_5.C_2H_5$, and higher alkylbenzenes. **1913** Thron & Freund *U.S. Pat. 1,062,203* 2/2 We claim ..the process of producing ethyl-hydrocupreine which comprises treating hydrocupreine with an ethylating agent. **1914** *Chem. Abstr.* 2756 In human syphilis marked effect was obtained with..injection of ethyl hydrocuprein. **1925** C. H. Browning *Bacteriology* xi. 252 But, so far, the only success obtained in generalised infections due to the ordinary bacteria is the cure of pneumococcal infection in mice by means of Morgenroth's 'optoquine' (ethylhydrocuprein, a derivative of quinine). 1912 *Jrnl. Chem. Soc.* CII. I. 797 (*heading*) Ethylmorphine and ethylmorphine hydrochloride (Dionine). 1964 S. Duke-Elder *Parsons' Dis. Eye* (ed. 14) 564 Therapeutic preparations. Eye-drops. Ethylmorphine (Dionene).

ethylene. *Chem.* [f. Ethyl + -ene.] **1.** (In Dict. s.v. Ethyl; earlier example.)

1852 H. Watts tr. *Gmelin's Hand-bk. Chem.* VII. 32 Ethylene (olefiant gas) = C^4H^4.

2. The bivalent hydrocarbon radical $—CH_2CH_2—$. So **e:thylenedi·amine,** a viscous, strongly alkaline liquid, $C_2H_4(NH_2)_2$ that acts as a chelating agent in the formation of complexes and is used industrially; **ethylenediamine tetra-acetic acid,** $[—CH_2N(CH_2COOH)_2]_2$, a compound widely used, esp. in the form of its salts, as a chelating and sequestering agent in industry, biology, and medicine; abbrev. *EDTA*; **ethylene glycol,** a sweet-tasting liquid, $HOCH_2CH_2OH$, used chiefly in anti-freezes; **ethylene oxide,** the simplest epoxide, $(CH_2)_2O$, a colourless gas used chiefly as a chemical intermediate and fumigant.

1966 *Nomencl. Org. Chem.* (*A & B*) (I.U.P.A.C.) (ed. 2) 16 The names of bivalent radicals derived from normal alkanes by removal of a hydrogen atom from each of the two terminal carbon atoms of the chain are ethylene, trimethylene, tetramethylene, etc. 1861 G. Fownes *Man. Elem. Chem.* (ed. 8) 606 (*heading*) Ethylene-diamine and diethylene-diamine. 1967 *Martindale's Extra Pharmacopoeia* (ed. 25) 252 Ethylenediamine hydrate is used in the manufacture of aminophylline and in the preparation of aminophylline injections. 1942 *Brit. Chem. Abstr.* A1. 334 (*heading*) Copper, nickel, and uranyl compounds of ethylenediaminotetra-acetic acid. 1959 *Sci. News* LII. 7 The degree of hardness of water..could be measured precisely by an elegant titration with an organic compound then mainly used as a textile auxiliary, ethylene diamine tetra-acetic acid (usually abbreviated to EDTA). 1964 S. Duke-Elder *Parsons' Dis. Eye* (ed. 14) xxvi. 378 Irrigation with..a solution of the sodium salt of ethylene-diamine tetra-acetic acid..diminishes scarring in lime burns. 1964 L. Martin *Clinical Endocrinol.* (ed. 4) iv. 158 Edetic acid (ethylenediamine-tetra-acetic acid, EDTA) is a chelating agent which reduces the amount of ionized calcium in the blood. 1901 *Jrnl. Chem. Soc.* LXXX. I. 307 The boiling points of ethylene glycol under varying pressures have been determined. 1959 Ethylene-glycol [see *anti-freeze *a.* and *sb.*]. 1898 *Jrnl. Chem. Soc.* LXXIV. I. 399 Ethylene oxide reacts vigorously with ethylamine. 1933 *Discovery* Aug. 250/2 Control [of the Cacao moth] can also be exercised by vacuum fumigation with ethylene oxide. 1962 *Listener* 20 Sept. 430/1 Another method of sterilization [of a space capsule] is to use the gas known as ethylene oxide.

Hence **ethylenic** (-ī·nik) *a.*, containing, derived from, or characteristic of ethylene; used spec. of a double bond between two carbon atoms.

1880 *Jrnl. Chem. Soc.* XXXVIII. 604 (*heading*) Heats of combustion of glycerol and of ethylenic glycol. 1946 *Nature* 14 Dec. 876/2 The lachrymatory activity of some ethylenic compounds. 1964 N. G. Clark *Mod. Org. Chem.* ii. 14 The pair of valency bonds..which is the essential feature of the olefins and upon which their chemical properties depend, is referred to as an ethylenic or olefinic bond.

ethylidene (eþi·lidīn). *Chem.* Also 9 -den. [f. Ethyl + -id(e + -ene.] The bivalent hydrocarbon radical $CH_3CH=$.

1859 H. Watts tr. *Gmelin's Hand-bk. Chem.* XIII. 451 Bromide of ethylidene. $C^4H^4Br^2$. *Ibid.* 453 The term *ethylidene* is applied to a radical (C^4H^4), supposed to exist in a series of compounds derived from aldehyde, and isomeric with the ethylene-compounds. 1952 Morton & Hoggarth in E. H. Rodd *Chem. Carbon Compounds* IB. xiv. 903 Ethylidenediurethane, $CH_3CH(NH·CO_2Et)_2$, m.p. 126°, from urethane and acetaldehyde.

etic (e·tik), *a.* [f. Phon)etic *a.* Coined by K. L. Pike.] Describing a generalized, non-structural approach to the description of language and behaviour. Cf. *emic *a.*

1954 K. L. Pike *Language in Rel. Human Behavior* I. ii. 8/1 In..the *etic* approach to the data, an author is primarily concerned with generalized statements about the data. *Ibid.* 8/2 The etic approach is comprised of a complex of goals and procedures. 1965 *Language* XLI. 95 His concern is primarily with the etic–emic dichotomy. 1969 *English Studies* L. 586 One person will shake hands with you by lifting your hand up to about shoulder height and then drop it, another will move your hand less high and then down again, a third will 'pump' it up and down two or three times; in Western culture these may be called *etic* differences and can be viewed as various realizations of the one *emic* element: 'shaking hands'.

-etin, *suffix. Chem.* Forming the names of aglycones, esp. those not chemically characterized when discovered and named, usually by replacing *-in* in the name of the glycoside from which the aglycone is obtained; so Fraxetin, Phloretin, Quercetin.

etiologically, var. Ætiologically *adv.*

1905 *Daily Chron.* 23 Aug. 3/1 It is an effort to show etiologically, with cold-blooded impartiality and freedom from partisanship, how [etc.]. 1956 *Nature* 14 Jan. 90/1 Both a virus and pleuro-pneumonia-like organisms have been etiologically implicated.

etiquette. Add: **4.** *attrib.* and *Comb.*, as *etiquette-book*; *etiquette-bound* adj.

1881 'Mark Twain' *Lett. to Publishers* (1967) 133 Send me a collection of etiquette books. 1952 *N.Y. Hist. Soc. Q.* XXXVI. 226 Etiquette books of various years yield information concerning mourning cards and notices. 1954 *Neuphilologische Mitteilungen* LV. 26 The rules laid down in the etiquette books need not always be strictly observed. 1902 *Westm. Gaz.* 28 May 2/1 It is many weary months before the small English lady—etiquette-bound as the most world-worn *mondaine*—is at home in foreign 'Society'.

etiquettical (etike·tikăl), *a.* [-ical.] Pertaining to etiquette.

1838 J. Pardoe *River & Desart* II. iv. 33 Death..has resumed its etiquettical observances. 1887 Mrs. Custer *Tenting on Plains* (1889) ii. 66 It was too much for his etiquettical instincts. 1900 G. Swift *Somerley* 107 We shouted directions at one another, and, in the interest of the moment, forgot the 'Miss' and 'Mr.' and other etiquettical matters.

Eton (ī·t'n). The name of a college, the largest of the ancient English public schools, founded by Henry VI on the Thames opposite Windsor, used *attrib.* or *Comb.* **a.** *Eton boy*, a pupil at Eton College.

1842 T. Morton (*title*) The Eton boy. 1883 [see Point *sb.*¹ 13 b]. 1929 D. H. Lawrence *Assorted Articles* (1930) 40 What could be more uncanny than the present pattern of the Eton-boy girl with flower-like artificial complexion?

b. *Eton jacket*, a short black broadcloth jacket, with an open front and broad lapels, pointed at the back and cut square at the hips, formerly worn by the younger boys at Eton (and elsewhere); also a garment of similar cut worn by women; hence *Eton-jacketed* adj. So *Eton suit*, a suit consisting of an Eton jacket with trousers and waistcoat; also, a woman's suit with a similar jacket and a skirt. Often ellipt. *Eton* (and in pl.).

1881 *Punch* 15 Oct. 170/3 The coat having no tails, could serve as an Eton jacket to a growing youth. 1882 'F. Anstey' *Vice Versâ* ii, Two small boys..in Eton jackets and broad white collars. 1883 *Graphic* 14 Apr. (Advt.), Eton suit, 32/6, 37/6. 1892 [see Toreador c]. 1892 *Queen* 28 May 888 'With a stout heart', published in 1874,..delighted an Eton-jacketed public. 1894 *Million* V. 377/2 Among the ladies riding was Lady ——, in an Eton suit and blue sailor hat. 1899 *Westm. Gaz.* 6 Jan. 3/2 The natty little Eton that in the South so often replaces the coat in the *costume tailleur*. 1900 *Daily News* 8 Sept. 6/3 Two-thirds of the smart dresses are now made with an Eton or bolero over a more or less ornamental front. 1900 *Captain* III. 405/2 The spectacle of Jim in Fourth Form Etons would have been too entrancing a sight. 1913 C. Mackenzie *Sinister Street* I. I. viii. 124 A new Eton suit well became him. 1918 —— *Early Life Sylvia Scarlett* I. ii. 68 Sylvia was wearing Etons at Monkley's suggestion. 1925 *Blackw. Mag.* July 80/2 The Eton jacket (or 'tum-coat') was reserved for Sundays.

c. *Eton blue*, a light blue adopted as the school colour.

1883 J. Brinsley-Richards *Seven Yrs. at Eton* 306 A birch tied with Eton blue ribbon and rosettes. 1899 T. M. Ellis *Three Cat's-eye Rings* 93 Looking at her wonderingly from his wide Eton-blue eyes. 1899 *Captain* II. 14/2 A very gorgeous equipage, with much Eton blue about it. 1963 *Harper's Bazaar* Feb. 74/2 Eton-blue water.

d. *Eton collar*: a broad stiff white collar formerly worn outside the jacket by Eton boys (and others); also, a similar collar as part of a woman's dress or jacket. Also *ellipt.*

1887 *Army & Navy Stores Catal.* 1095 Boys' Collars and Cuffs. The Eton. 1895 *Ibid.* 1160 Boys' front with Eton collar. 1925 *Queen* 22 Apr. 11 Its prim Eton collar, its daintiness of finish with its jabot and cuffs of pleated lawn. 1955 C. S. Lewis *Magician's Nephew* i. 7 In those days, if you were a boy you had to wear a stiff Eton collar every day.

e. In the phrases *Eton-and-Harrow, Eton-and-Oxford*, used (chiefly *attrib.*) to designate a person who has been at Eton or Harrow, or Eton and Oxford, or the characteristic manner, speech, behaviour, etc., of such a person.

[1824 Southey *Let.* 7 Oct. (1965) II. 268 He is an excellent scholar, thoroughly bred at Eton and at Oxford.] 1894 Mrs. H. Ward *Marcella* I. II. iii. 271 The young Eton and Oxford athlete, just home for his Christmas vacation. 1915 J. Buchan *Thirty-nine Steps* i. 18 An elegant young man who talks Eton-and-Harrow English. 1931 Wodehouse *If I were You* iii. 38 He may be a bit Eton and Oxford, but he's working in your best interests. 1952 R. Knox *Hidden Stream* xii. 110 It is possible to regard the Crusades as a kind of vast Eton-and-Harrow, in which both sides rallied to their respective ties by instinct, rather than by reasoned preference. 1962 'A. Gilbert' *No Dust in Attic* xiv. 186 Every man for himself, none of the old Eton-and-Harrow spirit.

f. *Eton crop*, a style of cutting women's hair close to the head all over. Hence *Eton cropped* ppl. adj.

1926 E. Wallace *Square Emerald* 9 The masculinity of the powerful face was emphasized by the grey hair cut close in an Eton crop. 1927 A. Monkhouse *Alfred the Great* 248 Maud was a trim, competent schoolgirl, long-legged, Eton-cropped. 1930 *Punch* 29 Jan. 126/1 A young thing whose essential plainness is emphasised rather than concealed by an Eton crop. 1958 B. Nichols *Sweet & Twenties* 137 She is an..Eton-cropped..mannish young woman.

Etonian (ītōᵘ·niăn), *a.* [f. as the sb.] Of, pertaining to, or characteristic of Eton College.

1749 Fielding *Tom Jones* V. xiii. i. 6 O Learning..in thy favourite Fields, where the..Thames washes thy Etonian Banks. 1841 Lady Stanley *Let.* in N. Mitford *Ladies of Alderley* (1938) 11 Etonian ways have no charm for him. 1883 J. Brinsley-Richards *Seven Yrs. at Eton* 321 It held a quart of beer, and the ceremony of drinking out of it constituted an initiation into the higher circle of Etonian swelldom. 1931 T. H. Pear *Voice & Personality* 75 There is an Etonian voice, the soft cultured drawl.

‖ **étourderie** (eturdəri). [Fr.] Thoughtlessness, carelessness, blundering.

1763 D. Hume *Let.* 24 Feb. in Boswell *London Jrnl.* (1950) 206 By this *etourderie*, to give it the lightest name, you were capable of making a quarrel between me and that irascible little man. 1814 Jane Austen *Mansf. Park* III. xv. 276 Henry is blameless, and in spite of a moment's *etourderie* thinks of nobody but you. 1886 T. Hardy *Mayor Casterbr.* I. xxii. 279, I ought to endeavour to disperse the shade which my *étourderie* flung over my name, by asking you to carry out your promise to me. 1958 I. Murdoch *Bell* vii. 101 His love affairs appeared as the *étourderies* of a much younger man.

‖ **étourdi** (eturdi), *a.* Also (fem.) **étourdie.** [Fr.] Thoughtless, irresponsible, flighty. Also as *sb.*, a thoughtless irresponsible person.

1688 T. Brown *Reasons of Mr. Bays* 11 The first that I begin with shall be that *etourdy bete*, that humble admirer of Jest and Quibble, the Melancholy Clergyman. 1750 Chesterfield *Lett.* 26 Apr. (1774) II. 3 All those French young fellows are excessively *étourdis*, be upon your guard against scrapes and quarrels. 1794 in G. C. Paget *One-Leg* (1961) 45, I must begin this letter by owning that I am the greatest *Etourdi* that ever lived yet that I am always lucky enough to get well out of every Scrape. 1828 C. H. Phipps *English in France* II. 347 Had she but the animal spirits to be *étourdie*, she would be so. 1848 Thackeray *Van. Fair* vi. 45 'I beg a thousand pardons..,' said the young *étourdi*, blushing. 1905 Mrs. H. Ward *Marriage W. Ashe* ii. 29 What a little *étourdie*! How lacking in the reserves, the natural instincts and shrinkings of the well-bred English girl!

‖ **et patati et patata** (epatati,epatata), *phr.* [Fr.] And so on and so forth; = Et cetera 1.

1904 M. Beerbohm in *Sat. Rev.* 14 May 619/2 Vanity of actor-managers, intolerable attempts to stifle honest criticism, shall such things be? et patati, et patata. 1920 F. M. Ford *Let.* 19 Sept. (1965) 128 We have pulled this house to pieces a good deal and successfully—et patati et patata! 1929 R. Aldington *Death of Hero* I. i. 38 'We have brought you up to be a God-fearing Christian man..' *et patata et patati.*

‖ **étrenne** (etrɛn). [Fr., older *estreine*:— L. *strena.*] A New Year's gift; a Christmas box, gift.

1802 C. Wilmot *Let.* 3 Jan. in *Irish Peer* (1920) 31 A hundred whimsical little 'bon bons', I have been given by 'Monsieur et Madames', as 'Étrennes' or New Year's gifts. 1834 K. H. Digby *Mores Cath.* v. vii. 234 The faithful were forbidden to give etrennes. 1880 Disraeli in Buckle *Life* (1920) VI. 505 Osborne has sent me, as an *étrenne* a most beautiful book. 1883 *Sat. Rev.* 24 Mar. 377/1 Mme. de Witt's magnificent volume belongs to the class of *étrennes*. 1928 *Observer* 29 Jan. 12/3 The concierges..demand..the abolition of their étrennes.

‖ **étrier** (etri,e). *Mountaineering.* Also **etrier(s.** [a. Fr. *étrier* stirrup.] A short rope ladder with two or three wooden or metal rungs (see quots.). Also *attrib.*

1955 J. E. B. Wright *Technique Mountaineering* iv. 64 Étriers are rope steps usually of one or two rungs. *Ibid.* v. 82 *Etrier steps.* Etrier-stirrups are generally used in pairs and each consists of one to four light alloy rungs placed a foot or so apart and fastened to two short pieces of rope. The result is a short portable ladder which can be hung from pitons on rock or ice pitches. 1956 C. Evans *On Climbing* iii. 53 Étriers..are small two- or three-rung rope-ladders, clipped to pitons. 1970 *Daily Tel.* (Colour Suppl.) 27 Mar. 42/1 He had left his etriers (a little three-rung rope ladder) hooked on to a piton a few feet below.

Etruscan, *a.* and *sb.* Add: **A.** *adj.* **2.** Designating encaustic pottery made by Josiah Wedgwood and his followers in imitation of ancient pottery discovered in Etruria. Also *ellipt.* as *sb.*, a vase of this kind.

1768 J. Wedgwood *Let.* 21 Nov. (1965) 69 Mr Cox is as mad as a march Hare for Etruscan Vases. 1769 *Ibid.* 6 Feb. 70 Three vases, one of them the large blue.., the other two Etruscans at a Guinea each. 1875 E. Meteyard *Wedgwood Handbk.* 264 The body used for the larger portion of the Etruscan painted vases was basaltes. 1957 Mankowitz & Haggar *Eng. Pottery & Porcelain* 233/2 The first day's production consisted of six Etruscan ware vases.

3. Designating a style of decorative bookbinding (see quots.).

1835 J. A. Arnett' *Bibliopegia* II. 98 *Etruscan.* This style is, where, instead of covering with gold, the book is ornamented with gothic or arabesque compartments, or imitations of Greek borders and Etruscan vases, in their proper colours. *Ibid.* 210 Etruscan style of binding. 1893 S. T. Prideaux *Hist. Sketch Bkbinding* i. 134 John Whitaker initiated the style termed Etruscan, in which designs from the decoration of Etruscan vases were copied in colours. 1907 C. Davenport *Book* viii. 201 Such books are known as 'Etruscan', because many of the designs are of classical feeling.

Etruscology (ītrɒskǫ·lŏdʒi). [f. L. *Etrusc(us* Etruscan + -ology.] The study of Etruscan history and antiquities. Hence **Etrusco·logist,** one versed in Etruscology.

1889 *Archæol. Rev.* III. 376 The melancholy career, not progress, of Etruscology during the last 150 years. *Ibid.* 377 Two other Etruscologists remain to be noticed. 1928 *Times* 7 June 12/3 The International Congress recently held at Florence demonstrated the great interest taken in the science of Etruscology. 1963 *Punch* 22 May 755/1 Etruscologists..have been overcome by gloomy feelings..and..projected those feelings on·to the Etruscans.

-ette, *suffix.* Add: **2.** Used to denote a female, as in *majorette, Suffragette, *usherette.

1921 H. L. Mencken *Amer. Lang.* (ed. 2) vi. 187 The wide use of the suffix *-ette* in such terms as *farmerette, conductorette,..usherette* and *huskerette*, is due to the same effort to make one word do the work of two. 1939 *New Yorker* 11 Nov., The girls employed to annoy visitors to some kind of Chamber of Commerce festival in Southern California will be called welcomettes. 1942 in *Amer. Speech* (1943) XVIII. 147 Roosevelt Signs 'Sailorette' Bill... Usherette in the..movie theater... Chicago tries 'Copettes'. 1970 *Women Speaking* Apr. 5/2 Female teams are called Rockettes, Mercurettes, Atomettes.

‖ **étude** (etüd). *Mus.* [Fr.] = Study *sb.* 11.

1837 J. A. Hamilton *Dict. Mus. Terms* (ed. 4) 31 *Etude* (French), a study. 1855 Chopin (*title*) Grand Study, extracted from the 24 Études for the Pianoforte. 1882 *Pall Mall Gaz.* 22 Dec. 20/1 A fantasia on opera airs or an impromptu or an étude. 1947 A. Einstein *Mus. in Romantic Era* xv. 223 Alkan..wrote..an étude 'The Railroad'. 1962 A. Harman et al. *Man & his Music* 810 The justly celebrated melody of the E major Etude..has the external features of Bellinian *bel canto*, while being unmistakably Chopin in that its rise and fall reflects the sensuously fluctuating tensions of the harmony.

-etum (ī·tɒ̆m), *suffix*, from Latin *-ētum*, neut. of *-ētus, -tus* (see -ed²), is appended to names of trees or other plants, (*a*) to designate a collection or plantation of various species of a single genus or group of plants, as in L. *arboretum* and *pinetum* (see Arboretum, Pinetum); (*b*) in *Ecology*, to designate an association dominated by the species or genus named, as in *characetum, ericetum, salicornietum.*

1688 Evelyn *Diary* 28 Mar. (1955) IV. 576 We went to Kew, to Visite Sir Hen: Capels, whose Orangerie & Myrtetum, are most beautifull. 1905 F. E. Clements *Res. Methods Ecology* iv. 299 The suffix -etum is used to designate a consocies of a formation, e.g. *Picetum, Caricetum*, etc. 1939 *Geogr. Jrnl.* XCIII. 403 The termination -etum in ecological studies denotes an association. The suffix is added to the root of the name of the dominant species. 1964 V. J. Chapman *Coastal Veg.* i. 11 It is convenient to recognize such communities [*sc.* associations] by use of the termination -etum. Thus for a community dominated by species of *Salicornia* one refers to the Salicornietum.

eubacterium (yubæktiə·riɒ̆m). *Bacteriol.* [mod.L. (A. Janke 1930, in *Zent. f. Bakt.* II Abt. LXXX. 490; A.-R. Prévot 1938, in *Ann. Inst. Pasteur* LX. 294), f. Eu- + Bacterium.] A genus of bacteria of the order Eubacteriales, comprising Gram-positive anaerobic bacilli found in the intestines of vertebrates and in soil and water; formerly a subgenus (see quot. 1939).

1939 D. H. Bergey et al. *Man. Determ. Bacteriol.* (ed. 5) 29 Janke..1930..proposes an outline which is slightly modified...Two new generic terms are used, *Anaerobacillus* and *Eubacterium*. *Ibid.* 30 Genus I. *Bacterium* Sub-genus..b. *Eubacterium*. 1946 [see *bacteriolytic *a.*]. 1957 G. E. Hutchinson *Treat. Limnol.* I. xi. 715 In extreme cases, another iron eubacterium, *Ochrobium tectum* occurred at slightly lower levels.

eucaine (yukēi·n). *Pharm.* [f. Eu-, after Cocaine.] A synthetic compound, $C_{15}H_{21}NO_2$, used as a local anæsthetic.

1896 *Chemist & Druggist* XLVIII. 597 Eucaine is employed like cocaine as a hypodermic solution of the hydrochloride. 1897 *Jrnl. Chem. Soc.* LXXII. I. 499 Eucaine,..whose chloride is used in commerce as a substitute for cocaine. 1907 *Practitioner* Apr. 473 Under eucaine, combined with a very little general anaesthetic, the abdomen was quickly opened. 1950 T. Longstaff *This my Voyage* vi. 117 A hypodermic injection of eucaine accomplished the magic.

eucalypt. Delete *rare* and add earlier and later examples.

1877 F. von Müller *Botanic Teachings* i. 7 The vernacular name of Gum-trees for the Eucalypts. 1880 T. W. Nutt *Melbourne Palace Ind.* 11 Stems of the soaring Eucalypts that rise Four hundred friendly feet to glad blue skies. 1893 D. C. Murray *Time's Revenges* vii, A forest of eucalypt closed in the band. 1898 *Queensland Year Bk.* 96 Great eucalypts spread their huge arms overhead. 1953 A. Upfield *Murder must Wait* xx. 179 Bark shed by the eucalypt instead of its leaves. 1970 *Daily Tel.* 10 Jan. 8/6 Dense forests of the ubiquitous eucalypt.

attrib. and *Comb.* 1956 *Nature* 3 Mar. 417/1 Sylvicultural research in Cyprus..on eucalypt forests. *a* 1963 J. Fountain in B. James *Austral. Short Stories* (1963) 268 A forest of spindly eucalypt-suckers pressing in darkly from either side. 1965 G. McInnes *Road to Gundagai* iii. 40 Lonely eucalypt-clad valleys. *Ibid.* x. 172 The eucalypt-scented darkness.

eucalyptian (yūkăli·ptiăn), *a.* (*sb.*) [See -ian.] Belonging to the genus *Eucalyptus*. Also as *sb.*, a tree of this genus. Also **eucaly·ptic** *a.*

1870 A. L. Gordon *Bush Ballads* Ded. 8 The gnarl'd knotted trunks Eucalyptian. 1873 J. B. Stephens *Black Gin* 6 This eucalyptic cloisterdom. 1901 *Harper's Mag.* CII. 708/1 The huge, white-armed eucalyptians overhead.

eucalyptus. Add: **b.** Used *ellipt.* for *eucalyptus oil*, any of many essential oils distilled from the leaves of eucalypts and used medicinally and industrially.

1885 *Buck's Handbk. Med. Sci.* I. 262/2 Odor and expense serve to tell against eucalyptus preparations. 1895 *Lancet* 3 Aug. 268 The soap is pleasantly medicated with 'sanitas' and eucalyptus. 1970 J. Crossland *Lewis's Pharmacology* (ed. 4) xxviii. 832 Cough stimulants are many in number...They include tolu, menthol, eucalyptus, guaiacol, terpin hydrate, benzoin and oil of turpentine.

euchromatic (yūkromæ·tik), *a. Cytology.* [f. next + -ic.] Of chromosome material: staining normally throughout the nuclear cycle.

1936 *Biol. Abstr.* X. 1620/2 Chromosome I (X) is heterochromatic from the proximal end to almost the exact centre; from here to the distal end euchromatic. 1942, 1951 [see *heterochromatic *a.* 2]. 1968 H. Harris *Nucleus & Cytoplasm* iv. 75 It is probable that other highly condensed heterochromatic or heteropyknotic regions in interphase nuclei also synthesize very much less RNA than the euchromatic regions.

euchromatin (yukrōᵘ·mătin). *Cytology.* [a. G. *euchromatin* (E. Heitz 1928, in *Jahrb. f. wissenschaftliche Botanik* LXIX. 764), f. Eu- + Chromatin.] Euchromatic chromosome material.

1932 *Biol. Abstr.* VI. 347/1 Chromosomes are composed of sections of heterochromatin and of euchromatin attached end to end. 1942 *Nature* 17 Jan. 67/1 In the euchromatin the regions between the chromomeres contain globulin-type proteins. 1951 [see *heterochromatin]. 1960 Darlington & La Cour *Handling of Chromosomes* (ed. 3) 67 Heterochromatin is a part of the chromosome which is out of step with the major bulk of the euchromatin. 1961 *Lancet* 26 Aug. 463/1 The synthesis of desoxyribose nucleic acid in heterochromatin and in euchromatin.

euchromocentre (yukrōᵘ·mŏsentəɹ). *Biol.* [a. F. *euchromocentre* (V. Grégoire 1931, in *Académie Royale de Belgique: Bulletins de la Classe des Sciences* 5th Ser. XVII. 1435), f. Eu- + Chromocentre.] (See quot. 1968.)

1934 L. W. Sharp *Introd. Cytol.* (ed. 3) iii. 57 In many nuclei there are at certain stages one or more conspicuous masses of karyotin at several points in the reticulum. Of the many terms applied to these the most suitable seems to be *chromocenters*... In some cases it has been shown that they represent definite chromosomal regions which remain condensed and highly chromatic... These *euchromocenters* appear to correspond in part to the 'prochromosomes'. 1937 *Nature* 22 May 889/2 During prophase, in addition to the euchromocentres, the nucleolus and nuclear sap also stain faintly with Feulgen. 1968 R. Rieger *Gloss. Genetics & Cytogenetics* 152 *Euchromocenter*, in interphase nuclei, a chromocenter which consists of the heterochromatic segments..located on both sides of the centromere.

euchromosome (yukrōᵘ·mŏsōᵘm). *Biol.* [f. Eu- + *chromosome.] = *autosome.

1914 C. E. McClung in *Jrnl. Morphology* XXV. 697 The word 'euchromosome' I shall employ to distinguish those without any marked peculiarities of form or behavior...'Heterochromosome' might advantageously be used as an antonym. 1925 E. B. Wilson *Cell* (ed. 3) xi. 839 The autosomes or euchromosomes..include those [*sc.* chromosomes] of the more normal type.

eucomis (yukōᵘ·mis). *Bot.* [mod.L. (C. L. L'Héritier *Sertum Anglicum* (1788) 17), ad. Gr. εὐκόμης fair-haired, beautiful-headed, f. εὐ- (see Eu-) + κόμη hair.] A plant of the liliaceous genus so named, comprising certain bulbous plants native to southern and tropical Africa, having a crown of large leaves on the flower-spike.

1804 H. Andrews *Botanist's Repos.* VI. Pl. 369 Purple-stalked Eucomis... Eucomis, with a clavated scape, leaves pointing many ways expanded orbicular-spatula-shaped. 1884 [see *pine-apple flower* s.v. Pine-apple 3 b]. 1966 *House & Garden* Dec. 61/1 (*caption*) The delicate Eucomis, from South Africa, raises its splendid head of flower. 1966 E. Palmer *Plains of Camdeboo* xvii. 283 The other green or green and white flowers of the veld, such as the Eucomis..which bears a great tuft of leaves above its greenish inflorescence.

eucone (yū·kōᵘn), *a. Ent.* [a. G. *eucone* (H. Grenacker 1877, in *Klin. Monatsblätter f. Augenheilkunde* Suppl. XV. 18), f. Eu- + Cone *sb.*¹] Of the eyes of certain insects: having a well-developed or true cone (see quots.).

1885 S. J. Hickson in *Q. Jrnl. Microsc. Sci.* XXV. 230 The crystalline cone of the 'eucone eyes'. 1888 Rolleston & Jackson *Forms Anim. Life* 502 The vitreous cells..are reduced to feeble remnants inclosing a solid crystalline cone composed of 2–5 parts, the eucone eyes of other Insecta...In pseudocone and eucone eyes the seven cells are grouped round a central axis. 1957 A. D. Imms *Gen. Textbk. Ent.* (ed. 9) 104 *Eucone eyes.* In eyes of this type each ommatidium contains a true crystalline cone, which is a hard refractive body formed as an intra-cellular product of the cone cells.

eucyclic (yusəi·klik), *a. Bot.* [ad. G. *eucyclisch* (A. Braun 1858, in *Jahrb. f. wissenschaftliche Botanik* I. 367), f. Eu- + Cyclic *a.*] Cyclic with alternate isomerous whorls (see quots.).

1875 Bennett & Dyer tr. *Sachs's Text-Bk. Bot.* II. v. 524 Braun calls those flowers *eucyclic* in which the members of all the whorls are equal in number and alternate. It also happens however that members of the same kind arise subsequently between those of a whorl already formed; as..in many eucyclic flowers with ten stamens. 1878 Henfrey & Masters *Elem. Bot.* (ed. 3) I. ii. 91 A flower thus presenting all the whorls is called complete or eucyclic. 1910 *Encycl. Brit.* X. 560/1 When a flower consists of parts arranged in whorls it is said to be *cyclic*, and if all the whorls have an equal number of parts and are alternate it is *eucyclic*.

eugenically (yudʒe·nikăli), *adv.* [f. Eugenic: see -ically.] In regard to eugenics; from a eugenic point of view.

1912 *Q. Rev.* July 65 The propagation of the eugenically fit. 1922 *Edin. Rev.* July 46 It is eugenically bad, making early marriage impossible, or encouraging the dysgenic art of fortune hunting. 1928 *Daily Express* 1 June 10/4 Wondering why he has not gone in for the great adventure...Is he eugenically unfit?

eugenist (yū·dʒĭnist). [f. Eugenic: see -ist.] A student or advocate of eugenics. Also *attrib.* or as *adj.*

1908 C. W. Saleeby *Health, Strength & Happiness* xxiv. 420 There is thus..a quasi-fatalistic element with which the hygienist, as well as the *eugenist*—to coin a word now necessary—must reckon. 1921 W. R. Inge *Lay Thoughts* (1926) 247 Prudent eugenists are in no hurry to advocate legislation. 1921 W. McDougall *Nat. Welfare* p. vii, This great gap in the eugenist argument. 1928 *Daily Express* 16 Jan. 8/7 Candidates for marriage..may be faced..with a searching Eugenist standard of physical fitness. 1959 B. Wootton *Soc. Sci.* ii. 55 This multiplicity of their social irregularities which first evoked the interest of the eugenists in these families.

eugeogenous (yūdʒi,ǫ·dʒĕnəs), *a. Geol.* [ad. F. *eugéogène* (J. Thurmann *Essai de Phytostatique* (1849) iv. 95), f. Eu- + Geo- + *-genous.] Of rock: that readily decomposes into detritus and is therefore able to form good soil. Opp. to *dysgeogenous*.

1850 *Phytologist* III. 918 The subjacent rocks, with reference to their mode of disintegration, their power of absorption, and their permeability, are divisible essentially into *eugeogenous* and *dysgeogenous*. The eugeogenous rocks produce an abundant detritus. When it is of a clayey (*pélique*) nature, it makes humid stations that are often inundated...The dysgeogenous rocks produce a scanty detritus, sometimes sandy, usually clayey, and always making stations more dry than those of the eugeogenous rocks. 1863 J. G. Baker *North Yorkshire* 152 They are due south of the hills of the eugeogenous range and based upon their slope. 1903 *Geogr. Jrnl.* Aug. 151 Eugeogenous..rocks..yield a plentiful detritus, and the overlying soils are cool and moist.

euglenoid (yuglī·noid), *a.* and *sb. Biol.* [f. mod.L. *Euglena* (C. G. Ehrenberg 1830, in *Ann. d. Physik und Chem.* XCIV. 507), f. Eu- + Gr. γλήνη pupil of the eye + -oid.] **A.** *adj.* Resembling *Euglena*, a member of the genus of single-celled aquatic flagellates so called, which are treated as either protozoa or algæ. **B.** *sb.* An organism of this kind.

1885 *Encycl. Brit.* XIX. 852/2 The movements now become neither vibratile nor amœboid but definitely restrained, and are best described as 'euglenoid'. *Ibid.* 857/1 Euglenoidea...Fam. 6. Astasina. Colourless, metabolic, or stiff Euglenoids. 1888 Rolleston & Jackson *Forms Anim. Life* 843 Chromatophores..are numerous, small, round, or oval in Euglenoids. 1906 *Cambr. Nat. Hist.* I. 124 *Euglena*..shows a peculiar wriggling motion, waves of transverse constriction passing along the body from end to end... Such motions are termed 'euglenoid'. 1955 *New Biol.* XIX. 114 Organisms..of euglenoid affinities. *Ibid.* 119 In euglenoids without the stigma, this phototactic response does not occur.

euglobulin (yuglǫ·biŭlin). *Biochem.* [f. Eu- + Globulin (F. Hofmeister 1899, in *Beitr. z. chem. Physiol.* (1901) I. 361).] The name of one of the fractions into which serum globulin is divided, so called because of its true globulin properties; it is insoluble in pure water but soluble in saline solutions and is precipitated by half-saturation with ammonium sulphate. Also *attrib.*

1905 W. B. Hardy in *Jrnl. Physiol.* XXXIII. 333 Englobulin [*sic*] itself may be merely a step in a series of proteid compounds. 1907 *Jrnl. Biol. Chem.* III. 233 The two serum globulin fractions of the Hofmeister classification..the lower or euglobulin fraction (2·9 to 3·4 saturated) comprises diphtheria and tetanus antitoxin. *Ibid.* 235 The repeatedly precipitated euglobulin..contained..10 per cent of the antitoxin. 1962 *Lancet* 22 Dec. 1302/1 The euglobulin lysis-time..will probably have to serve..as a measure of fibrinolytic activity.

euhedral (yuhi·drăl, -he·drăl), *a.* [f. Eu- + Hedral *a.*] = Idiomorphic *a.*

1906 J. P. Iddings *Rock Minerals* ii. 64 *Euhedral*, well faced, completely bounded by crystal planes, automorphic, idiomorphic. 1913, etc. [see *anhedral *a.*].

eulachon, var. Oolakan.

1866 J. K. Lord *Naturalist in Vancouver Isl.* I. 88 A fish..called by the natives *Eulachon* or Candle-fish. 1888 Goode *Amer. Fishes* 476 They [*sc.* trout] are taken in Frazer River at the time of the eulachon run, but they probably then ascend the river to feed upon the eulachon, and not for spawning purposes. 1955 *Sci. News Let.* 8 Jan. 32/1 The candlefish, or eulachon, of America's north Pacific coast is prized by Indians as food and for its oil.

Euler (oi·ləɹ). The name of Leonard Euler (see Eulerian *a.*) used *attrib.*, in *Comb.*, or in the possessive to designate principles, effects, etc., discovered by him or arising out of his work.

1847 *Phil. Mag.* 3rd Ser. XXX. 424 Recent researches.., in reference to the new analytical theory of imaginary quantities, have revived attention to Euler's theorem, that the sum of four squares multiplied by the sum of four squares produces the sum of four squares. 1889 *Cent. Dict.*, *Euler's numbers*, the numbers E_2, E_4, etc., which occur in the development of sec x by Maclaurin's theorem: namely, sec $x = 1 + E_2x^2/2! + E_4x^4/4! +$ etc. *Ibid.*, *Euler's solution*, a solution of a biquadratic after the second term has been got rid of. 1940 *Jrnl. R. Aeronaut. Soc.* XLIV. 43 The term quasi-Euler is used to distinguish the failure which occurs by buckling over the greatest wave length which the supports allow from the corrugated failure fixed by the minimum condition. Although the skin fails in an Euler curve it is not a true Euler effect. 1947 Courant & Robbins *What is Math.?* (ed. 4) v. 240 On the basis of Euler's formula it is easy to show that there are no more than five regular polyhedra. 1953 W. Rudin *Princ. Math. Anal.* viii. 163, $1 + \frac{1}{2} + \ldots + (1/n) - \log n$..converges. (The limit, often denoted by γ, is called Euler's constant. Its value is 0·5772...) 1961 *New Scientist* 16 Mar. 698/3 The result $V + F - E = 2$ was originally derived for polyhedra: this is the *Euler-Descartes relation*, known to Descartes but first explicitly proved by Euler.

eunomia (yunōᵘ·miă). = Eunomy.

1861 Mill *Repr. Govt.* x. 195 The ballot..conduced to the Eunomia by which Athens was distinguished among the ancient commonwealths. 1952 A. R. Radcliffe-Brown *Struct. & Function Prim. Soc.* ix. 182 They conceived..the eunomia of a society as being in each instance a condition of the harmonious working together of its parts.

Eunomian (yunōᵘ·miăn), *sb.* (*a.*) *Ch. Hist.* [See -an.] A follower of Eunomius, bishop of Cyzicus in the 4th century A.D., who developed Arianism to an extreme form; an Anomœan. Also as *adj.* Hence **Euno·mianism.**

c 1449 Pecock *Repr.* v. iii. 499 The sect of Acyanys and of Ennomynyarys [*sic*]. 1559 Abp. Heath *Speech* 30 Jan. in *Cobbett's Parl. Hist.* (1806) I. 645 Damase, then bishop of Rome.., did give sentence against the heretics, Macedonians, Sabellians and Eunomians. 1577 [see Macedonian *a.*² and *sb.*²]. 1607 T. Rogers 39 *Art.* i. 6 Some thinke there be three Gods..not distinguished onely, but divided also, as did the Eunomeans, and Tretheites. 1780 A. Butler *Lives Saints* (ed. 2) VII. 109 The sixth Volume [*sc.* of St. Ephrem's writings] gives us ninety other polemical *Discourses* against the Arian and Eunomian heretics. 1788 Gibbon *Decl. & F.* IV. xli. 205 Theodosius had been educated in the Eunomian heresy. 1880 E. Venables in W. Smith *Dict. Chr. Biogr.* II. 288 Eunomianism, as a cold, logical system, wanted the elements of vitality. 1893 W. M. Ramsay *Church in Empire* xviii. 448 Did the Eunomian differ from the Catholic only in point of doctrine? 1959 *Chambers's Encycl.* V. 433/2 An extreme 'Eunomian' party of the Arians. *Ibid.*, After his death the Eunomians..broke completely with the orthodox church.

eunomic (yunǫ·mik), *a.* [f. Eunom(y + -ic.] Law-abiding; (socially) well adjusted or ordered.

1952 A. R. Radcliffe-Brown *Struct. & Function Prim. Soc.* ix. 182 Whether a given society at a given time is normal or pathological, eunomic or dysnomic. 1957 R. K. Merton *Social Theory* (ed. 2) II. v. 166 The behavior of 'anomic' and 'eunomic' *individuals* within groups having a designated degree of objective anomie could be systematically compared.

eunuch, *sb.* Add: **1. a.** Also *fig.* (freq. preceded by a descriptive adj.).

1593 Shaks. *2 Hen. VI* iv. ii. 175 Lord Say hath gelded the Commonwealth, and made it an Eunuch. 1819 Byron *Don Juan* Ded. in *Wks.* (1833) XV. 105 The intellectual eunuch Castlereagh. 1819 Shelley *Peter Bell 3rd* iv. xi, But from the first 'twas Peter's drift To be a kind of moral eunuch. 1963 *Times* 18 Jan. 9/5 It seems that in Rhodesia one cannot remain neutral, that after all one is provoked to react, or become a political eunuch.

2. eunuch flute, a type of mirliton (see quot. 1928).

[1635 M. Mersenne *Harmonicorum Libri* II. ii. 79 De Instrumentis Harmonicis... *Fleuste Eunuque.* Fistula monofora.] 1928 E. W. Naylor *Poets & Music* v. 97 The 'eunuch' flute..is not a flute at all, but a tube covered at the small end with a thin membrane. The performer sings, or rather hums, through a hole in the side, near the membrane. 1961 A. Baines *Mus. Instruments* 19 The Eunuch Flute, and the Bumbass.

eunuchoid (yū·nɒ̆koid), *a.* (*sb.*) [See -oid.] Resembling, or characteristic of, a eunuch. Also as *sb.* Hence **eu·nuchoidism.**

1906 *Rep. Brit. Assoc.* 703 (*heading*) Observations made on an 'Eunuchoid' Subject in the Cambridge Anatomy School. 1925 W. J. H. Sprott tr. *Kretschmer's Physique & Char.* 26 A growth in length which indicates eunuchoidism. *Ibid.* 73 Disposition of fat according to the eunuchoid plan. *Ibid.* 86 Schizophrenes and eunuchoids. 1928 *Cowdry's Spec. Cytology* 1231 Eunuchoid conditions. 1959 *Times* 3 Apr. 14/6 Farinelli..looks down ..slightly eunuchoid in shape under his smart cerise frockcoat. 1970 *New Yorker* 3 Oct. 36/1 The eunuchoid subtleties of a well-trained choirboy.

euonymin (yu,ǫ·nimin). *Pharm.* Also -ine. [f. Euonym(us + -in¹.] A crude extract of the bark of *Euonymus atropurpureus* (see quot. 1912).

1862 W. T. Wenzell in *Amer. Jrnl. Pharmacy* XXXIV. 390 The bitter taste of the principle Euonymin. 1886 *Jrnl. Chem. Soc.* L. 72 The bark of *Euonymus europæus* contains no euonymin. 1912 *Jrnl. Chem. Soc.* CI. 1041 A crude, resinous product, as well as an alcoholic extract of the drug, has met with somewhat extended use under the name of 'Euonymin', and this appellation has also been given to various other substances obtained from the bark, all of which, however, appear to have been of a very indefinite character. 1967 T. E. Wallis *Textbk. Pharmacognosy* (ed. 5) vii. 84 Euonymin (the powdered extract) is an hepatic stimulant, direct cholagogue, and mild cathartic.

Eupad (yū·pæd). *Pharm.* [f. the initials of Edinburgh University Pathological Department (where the mixture was invented) with jocular reference to Eu- and Pad *sb.*³, quasi 'good pad'.] A mixture of chlorinated lime and boric acid, used as an antiseptic dry dressing.

1915 J. L. Smith et al. in *Brit. Med. Jrnl.* July 129/1 For convenience we have given the name 'Eupad' to a powder consisting of equal weights of finely ground bleaching powder and powdered boric acid intimately mixed. 1920 Martindale & Westcott *Extra Pharmacop.* (ed. 17) I. 56 A large boil on the leg..was opened and a drain of gauze powdered with Eupad inserted. 1967 *Martindale's Extra Pharmacopoeia* (ed. 25) 313/2 Eupad... A mixture of equal parts of finely ground chlorinated lime (dry) and boric acid, containing about 15% of available chlorine.

eupeptically (yupe·ptikăli), *adv.* [f. Eupeptic *a.* + -al + -ly².] In a eupeptic manner; so as to assist digestion.

1939 *Scrutiny* Mar. 403 Eupeptically gay composers like Virgil Thomson. 1966 *New Statesman* 7 Oct. 522/3 A eupeptically progressive consideration of human affairs.

euphausiacean (yu:fǫzi,ēi·ʃən). [f. mod.L. *Euphausiacea*, f. Gr. εὐ- Eu- + φα- root of φαίνειν to show + οὐσία substance + -acea.] A member of the order Euphausiacea of shrimp-like planktonic crustaceans. So **eu·phau·sian,** a member of the genus *Euphausia* of crustaceans; loosely, a euphausiacean; **eu·phau·sid, -idian, -iid,** a member of the family Euphausiidæ. Also **euphausid** *a.*

1885 G. O. Sars in *Challenger Reps.* (*Zool.*) XIII. 63 In regard to the capture of Euphausidians..scarcely a single haul failing to yield some of the forms comprised in this group. 1893 *Funk's Stand. Dict.*, Euphausiid. 1893 T. R. R. Stebbing *Hist. Crustacea* xviii. 265 The Euphausid larva. 1909 W. T. Calman in R. Lankester *Treat. Zool.* VII. 245 Order *Euphausiacea*...The phosphorescence of certain Euphausians was first observed by J. V. Thompson. 1924 W. M. Tattersall in *Brit. Antarctic* (*Terra Nova*) *Exped.* (*Zool.*) VIII. 1 The Euphausians in this collection were provided entirely by the tow nettings and other plankton gatherings. 1928 Russell & Yonge *Seas* v. 117 Euphausiids..are about an inch and a half in length, but are so abundant that they form a large part of the food of many of the northern fishes. 1934 *Geogr. Jrnl.* LXXXIV. 126 Thousands of the Euphausid, *Thyanoessa inermis*. 1950 J. S. Colman *Sea & its Mysteries* v. 106 The animals that make up the krill belong to the group of crustacea known as Euphausians. 1956 A. C. Hardy *Open Sea* ix. 171 The Euphausiaceans..have..a wonderful set of luminous organs. 1959 H. Barnes *Oceanogr. & Marine Biol.* 93 Planktonic organisms, such as the shrimp-like Euphausids, are responsible for this deep scattering layer. 1965 *New Scientist* 1 July 41/1 The euphausiaceans..are characterized by possessing well developed light-producing organs. 1968 *Ibid.* 24 Oct. 205/1 Whales feed on krill, composed largely of planktonic, relatively large, shrimp-like creatures known as euphausids.

euphemizer (yū·fĭməizəɹ). [f. Euphemize *v.* + -er¹.] One who speaks euphemistically.

1890 C. Martyn *Wendell Phillips* 186 He was the one outspoken man in a nation of euphemizers.

euphenics (yufe·niks). *Biol.* [f. Gr. εὐ- Eu- + φαίν-ειν to show, appear + -ics.] A practice analogous to eugenics in which physical improvement of man is sought through the modification and control of his development (as by chemical or surgical means).

1963 J. Lederberg in *Nature* 4 May 428/2, I propose the term 'euphenics' as the counterpart of 'eugenics', in the same sense that 'phenotype' is opposed to 'genotype'. 1965 *New Scientist* 9 Dec. 714/2 By 'cultural euphenics'..one could make revolutionary changes in a generation. 1967 *Technology Week* 23 Jan. 50/3 Euphenics [is] the modification of his [*sc.* man's] biological development.

euphonion, an early form of Euphonium.

1862 *Catal. Internat. Exhib., Brit.* II. No. 3377, An Euphonion or solo bass in B flat.

euphoria: see Euphory in Dict. and Suppl.

euphoriant (yufǫ·riănt), *a.* and *sb. Med.* [f. Euphori(a + -ant¹.] **A.** *adj.* Of drugs: inducing euphoria or an exaggerated feeling of well-being. **B.** *sb.* A drug with this property.

1947 *Brit. Med. Jrnl.* 28 June 919/1 The ideal euphoriant for clinical use..must induce a high degree of euphoria, must be stable and fully active by the oral route. *Ibid.*, There is a common and widespread belief that all euphoriant drugs are necessarily habit-forming, whereas this is in fact not the case. 1957 H. McIlwain *Chemotherapy & Central Nervous System* x. 243 Synhexyl proved euphoriant, producing drowsiness, lightheadedness, and feelings of strangeness and of floating. 1965 J. Pollitt *Depression & its Treatment* iv. 48 Treatment with drugs may be indicated in certain cases, either in the form of sedatives, tranquillisers or euphoriants. 1968 A. Goldstein et al. *Princ. Drug Action* ix. 592 Although ..the narcotics produce similar actions in all animal species, the euphoriant actions can be observed directly only in man.

euphoric, *a.* Add: (Further examples.) Now gen. in extended sense (cf. *euphory 2): characterized by a feeling of well-being, cheerful; also, producing or causing cheerfulness. Hence **eupho·rically** *adv.*, cheerfully.

1925 W. McDougall *Introd. Soc. Psychol.* (ed. 2) 394 When we have laughed, we commonly experience..its pleasing euphoric effects. 1947 Auden *Age of Anxiety* (1948) ii. 53 Come, peregrine nymph, display your warm Euphoric flanks in their full glory Of liberal life. 1951 R. Hoggart *Auden* vi. 196 Behind the slightly glittering euphoric exterior, the misery of a whole society. 1957 *Times Lit. Suppl.* 12 July 426/4 The euphoric pipe-smoking President. 1958 *Observer* 8 June 15/6 The picaresque Sgt. Bilko has been crackling on ingeniously and euphorically now for week after week. 1958 A. Wilson *Middle Age of Mrs Eliot* II. 271 It seemed impossible that she could truly be as euphoric as she seemed. 1959 *Listener* 20 Aug. 296/2 Two criminal alcoholics, one desperate, the other euphoric. 1963 *Times* 15 Jan. 11/2 The euphorically minded husband.

eupho·ric, *sb.* [f. the adj.] A drug, etc., which produces a sense of euphoria; = *euphoriant *sb.*

1934 in Webster. 1951 A. Burger *Medicinal Chem.* ii. 9 The South American Indians..chewed coca leaves as a stimulant and euphoric. 1958 A. Huxley *Brave New World Revisited* (1959) viii. 102 The classical euphoric, alcohol..in excessive doses, causes illness and addiction.

euphory. Add: **2.** (Further examples.) The spelling **eupho·ria** is now usual, freq. in non-technical contexts with the sense 'a state of cheerfulness or well-being, esp. one based on over-confidence or over-optimism'.

1882 W. James *Let.* 2 Nov. (1920) I. 211 Having taken the plunge, the cutaneous glow and 'euphoria' (*vide* dictionary) succeeded. 1922 R. S. Woodworth *Psychol.* vii. 120 The warmed-up person *feels* ready for business, full of 'ginger' or 'pep'—in short, full of life. The name 'euphoria', which means about the same as 'feeling good', is given to this condition. 1927 F. Brett Young *Portrait of Clare* 607 In this detached euphoria she began to approve of Dr. Boyd. 1939 A. Huxley *After many a Summer* I. iii. 37 The delightful condition of euphoria into which those poor kids and Clancy's good news had plunged him. 1954 A. W. Fielding *Hide & Seek* 215 In this abnormal silence, which only intensified our state of euphoria, we settled down to sleep. 1960 Auden *Homage to Clio* 89 Good Queen Victoria, In a fit of euphoria, Commanded Disraeli To blow up the Old Bailey. 1964 *Ann. Reg.* 1963 9 But this euphoria was not to last. 1971 *Physics Bull.* Apr. 216/1 In the post war euphory, it was easy to obtain support for fundamental research.

euphotic (yufōᵘ·tik), *a.* [f. Eu- + *photic *a.* b.] Pertaining to or designating the upper layer of the photic zone in sea-water where sufficient light penetrates to permit photosynthesis.

1909 Groom & Balfour tr. *Warming's Oecol. Plants* IV. xxxvi. 150 *Euphotic* vegetation, which receives an abundance of light. 1959 *New Scientist* 1 Oct. 586/1 Only a relatively thin surface layer of the ocean, the 'euphotic layer', is illuminated, and so only in this layer can photosynthesis and primary production of organic matter take place. 1964 *Oceanogr. & Marine Biol.* II. 360 The observation at 70m represents the luminescent activity of euphotic zone inhabitants plus that of scattering layer organisms. 1970 *Nature* 25 July 375/1 The flux of organic carbon from the euphotic zone (0–100 m) into the deep sea and into the bottom sediments.

Euphratean (yufrēi·tiăn), *a.* [f. *Euphrates* + -an.] Of, pertaining to, or bordering on the river Euphrates.

1877 J. W. Dawson *Origin of World* xii. 253 This primitive Cushite empire called Ethiopia..occupied the Euphratean valley. 1902 D. G. Hogarth *Nearer East* 220 Trade between east and west has deserted the painful Euphratean routes. *Ibid.* 222 The route taken by the Persian *haj*, after visiting the Euphratean shrines. 1934 A. Toynbee *Study Hist.* II. 385 The Nilotic and Euphratean jungle-swamps.

euploid (yū·ploid), *a. Biol.* [a. G. *euploid* (G. Täckholm 1922, in *Acta Horti Berg.* VII. 234), f. Eu- + *-ploid.] Of a cell, an organism, or tissue: having each of the different chromosomes of the set in equal numbers; having an exact multiple of the haploid chromosome number. So **eu·ploidy,** the condition or state of being euploid.

1926 L. W. Sharp *Introd. Cytol.* (ed. 2) xvii. 385 When the total number [of chromosomes] is..an exact multiple of the fundamental haploid number the term *euploid*..may be employed. *Ibid.* 387 (*caption*) The diploid chromosome complements of various species of *Rosa*, illustrating euploidy. 1933 *Times Lit. Suppl.* 9 Nov. 779/1 Euploidy (regular multiplication of the basic chromosome number) is characteristic of cultivated apples. 1956 *Nature* 25 Feb. 384/2 By developing more constant euploid (or polyploid) cells.

Eurafrican (yuəræ·frikăn), *a.* and *sb.* [f. Eur(opean + *African *sb.* and *a.*] **1.** *Anthrop.* Designation of a dark-skinned race which inhabited regions on both sides of the Mediterranean.

1890 D. G. Brinton *Races & Peoples* iv. 105 The white race is..an *African* race. I have calculated..the area of its control of the three continents...These figures vindicate for the race the title I have given it—Eurafrican. 1899 A. H. Keane *Man: Past & Present* 444 The right of citizenship is to be withdrawn from such time-honoured names as 'Hamitic', 'Semitic'..in favour of 'Mediterranean', 'Eurafrican', and other upstarts. 1910 *Encycl. Brit.* XIV. 217/1 Whether this type is more conveniently designated by the word *Iberian*, or by some other name ('Eur-african', 'Mediterranean', &c.). 1928 V. G. Childe *Most Anc. East* ii. 39 The so-called Eur-african, Mediterranean, or Brown race.

2. Pertaining to or involving both Europe and Africa, or their peoples together.

1909 in *Cent. Dict.* Suppl. 1957 *Time* 18 Feb. 18/2 Pineau said..'[France] would like to promote the formation of a Eurafrican whole.' 1958 *Economist* 25 Jan. 286/2 A Eurafrican bargain [i.e. between France and Algeria]. 1962 *Ibid.* 23 June 1192/1 The 'Eurafrican' negotiations at Brussels.

3. Designation of the 'Coloured People' of S. Africa, descendants of the native peoples with white admixture.

1922 S. G. Millin *Adam's Rest* II. v. 156 Frances no longer looked as if she might be a beautiful Spaniard or Italian. She was obviously a Eurafrican. 1927 W. M. Macmillan *Cape Colour Question* 288 All recent restrictive legislation, designed for the 'segregation' of the Natives, classes the 'Eurafricans' with the Europeans. 1927 *Times* (weekly ed.) 24 Nov. 586/1 The coloured or Eurafrican workers.

Eurasiatic, *a.* Add: Also as *sb.*, a person of Eurasiatic origin.

1935 [see *alpine *sb.* 2]. 1944 J. S. Huxley *On Living in Revol.* 175 The French claimed that the Aryan language and..civilization came into Europe with the Alpines (Eurasiatics).

Eureka. Add: **3.** The proprietary name of an alloy of copper and nickel used for electrical filament and resistance wire. Also *attrib.*

1914 *Trade Marks Jrnl.* 21 Jan. 85 *Eureka.* Electric Conduits and Union Joints and Fittings therefor... John Birch & Sons..Manufacturers. 1915 *Phil. Mag.* XXIX. 368 Of the following metals and alloys all..gave a positive emission..platinoid, eureka, tinned copper, and 'galvanized' iron. 1930 *Engineering* 28 Nov. 672/1 Winding, for which 47-gauge Eureka wire is used. 1939 *Trade Marks Jrnl.* 2 Aug. 1070 *Eureka*...Uncovered copper-nickel electrical resistance wire. The London Electric Wire Company and Smiths Limited..Manufacturers. 1943 *Electronic Engin.* XVI. 78 The heater wire is generally made of eureka or a copper nickel alloy.

eurhythmic, *a.* Add: **B.** *sb. pl.* A system of rhythmical bodily movements, *esp.* dancing exercises, with musical accompaniment, freq. used for educational purposes. Hence **eu·rhy·thmic(al** *adjs.*, **eurhythmi·cian, eurhy·thmist.**

1912 *Standard* 27 Nov., Eurythmics [*sic*] is no longer a mysterious art—it is the new craze. Eurythmics is a word which Professor Jacques-Dalcroze has invented to describe his 'rhythmic gymnastics'. 1920 *Challenge* 21 May 44/3 In the Eurythmics of M. Jacques Dalcroze..the dancer expresses by improvized postures the emotions aroused by music. 1921 H. F. Rubinstein tr. *Jaques-Dalcroze's Rhythm, Music & Educ.* 195 The art of the Eurhythmist is self-sufficient. *Ibid.* 196 Eurhythmic exercises enable the individual to feel and express music corporally. *Ibid.* 206 Eurhythmicians watching exercises performed by fellow-students. 1923 *Daily Mail* 17 Mar. 10 An eurhythmical display under the command of M. Jaques-Dalcroze. 1924 J. J. Findlay in V. Davis *Modern Teaching* (1928) 101 The eurhythmist does not neglect or despise the body; he accepts the counsel of the trainer, but he leads his pupils to fix their attention on rhythmic values. 1925 C. Fox *Educat. Psychol.* 84 Dancing and eurhythmics in school.

Euripidean (yuri:pidī·ăn), *a.* [f. L. *Euripidēus*, ad. Gr. Εὐριπίδειος, f. Εὐριπίδης (see def.).] Of, pertaining to, or characteristic of Euripides, the Athenian tragic poet, or his works, style, etc.

1821 *Q. Rev.* XXIV. 369 The prefatory narrative, which forms..an Euripidean prologue to Miss Austin's novels. 1871 Browning *Balaustion's Adv.* 168 So might..no Euripidean pathos plague Too much my critic-friend of Syracuse. 1924 E. B. Osborn *Our Debt to Greece & Rome* vi. 93 In the Euripidean drama..the chorus merely provides charming interludes of song and dance.

euro. Now the usual spelling of Uroo.

1898 Morris *Austral Eng.* 140/1 *Euro*, one of the aboriginal names for a Kangaroo; spelt also *Yuro*. 1931 I. L. Idriess *Lasseter's Last Ride* (1933) iii. 22 Wallaby, euro, and dingo tracks showed how popular this cool rock-hole was. *Ibid.* iv. 32 They found the hills alive with euros and wallabies. 1934 *Bulletin* (Sydney) 17 Jan. 20/4 Winnie, our pet euro (Westralian rock kangaroo), had a liking for softgoods. 1944 *Living off Land* ii. 27 A number of men are required to capture the wallaby or the euro (hill kangaroo). 1950 'N. Shute' *Town like Alice* 81 You'll find little diggings all over in the sand, where the kangaroos and euros have dug for water.

Euro- (yūᵊro), combining form of *Europe*, in **Euro-African** *a.*, = *Eurafrican 2, 3; **Euro-American** *a.*, pertaining to both Europe and America; = Western *a.* 4 a; also as *sb.* = Western *sb.* 1; **Euro-Asiatic** *a.*, = Eurasiatic *a.*; **Eurocrat** [-crat] (see quot. 1962); **Euro-dollar** (see quot. 1965); **Eurovision** [after *television], television of European range (see quot. 1951).

1952 B. Davidson *Rep. S. Afr.* I. iv. 53 The Euro-African, or Coloured, rate of increase. 1958 *Optima* Mar. 25/2 A Euro-African market, which would, as its name implies, favour trade and exchanges between the two continents. 1963 *Guardian* 7 Jan. 5/7 The school's Provisional Committee of nine Africans and three Euro-Africans. 1928 *Critical Crown* I. 28 Economists lose even the shade of the 'economic man' as they contemplate cultures other than the Euro-American. 1949 M. Mead *Male & Female* ii. 27 Hygienic fears with which most Euro-Americans surround themselves in native villages. 1961 *New Left Rev.* July–Aug. 10/2 A vast..Euro-American entity. 1935 *Discovery* Oct. 293/1 The Euro-Asiatic origin of Man. 1961 *Economist* 29 July 449/1 These new 'Eurocrats' are worth watching. 1962 *Britannica Bk. of Year 1961* 546/1 *Eurocrats*, administrative staff of the various European organizations such as the European Economic Community. 1968 *New Scientist* 14 Nov. 386/3 Eurocrats and au pair girls, bankers from Basle..all are examined. 1960 *Times* 24 Oct. (Financ. Rev.) p. viii/3 London became..the most active dealer in the so-called 'Euro-dollar'. 1965 *Economist* 19 June p. xiii/1 Euro-dollars are deposits of United States dollars with banks outside the United States. The overseas banks need not be in Europe. 1969 *Guardian* 12 Feb. 3/4 (Advt.), When you want Euro-dollars, or any other Euro-currency, a Chaseman can get them for you in minutes rather than hours. 1951 *Evening Standard* 5 Nov., What is Eurovision? A system of collaboration among the West European countries including Britain by which television programmes will become interchangeable. 1954 *N.Y. Times* 6 June 85/3 Europe's leading television engineers will twirl dials with crossed fingers tomorrow when they tune in the first eighteen-nation 'Eurovision' hook-up through Lille. 1958 *Listener* 13 Nov. 796/1 Millions more saw the coronation..on television, through the Eurovision link. 1965 *BBC Handbk.* 39 Twenty-three television organizations in eighteen Western European countries..make up 'Eurovision'. 1968 *Guardian* 22 Aug. 6/5 Czech television screened the invasion...Austrian television was able to pick up the pictures and send them to Britain via Eurovision.

European, *a.* and *sb.* Add: **A.** *adj.* **1. c.** Used in the names of certain economic and defence organizations or unions of western European countries.

1952 *Ann. Reg. 1951* 164 Signature on 18 April in Paris of the treaty establishing a European Coal and Steel Community among the six participating countries. *Ibid.*, The..project for a European Defence Community was the subject of intensive negotiations. 1955 *Times* 30 July 5/3 The council of the Organization for European Economic Cooperation to-day approved the plans that had been submitted to it for renewing for a year the European Payments Union. 1957 *Times* 8 Oct. 11/1 The European Common Market now being organized by the 'Six'—France, Germany, Italy, and Benelux. 1958 *Spectator* 7 Feb. 184/3 The Common Market Treaty, laying down the conditions under which the European Economic Community..is to be established, came into force on 1st January 1958. 1961 *Listener* 17 Aug. 252/1 Britain's application to join the European Community. 1969 *Times* 6 Jan. 7/7 It may be rather less pressing in Britain than in the European Economic Community, where the cost of financing the mounting butter surplus has threatened the whole structure with collapse.

2. *European-minded, -style* adjs.; **European corn-borer** *U.S.*, the larva of the moth *Pyrausta nubilalis*, which attacks maize and other crops; **European plan** *U.S.*, the method or practice at a hotel of charging for lodging and service without inclusion of meals (opp. **American plan*).

1920 G. W. Herrick *Insects of Economic Importance* 122 The European Corn Borer..has lately been discovered in the United States. 1955 *Sci. News Let.* 10 Sept. 164/1 European corn borers and boll weevils appear to be causing the nation's farmers the greatest losses at present. 1932 Hemingway *Death in Afternoon* xix. 268 European-minded politicians. 1961 A. Wilson *Old Men at Zoo* vii. 312 A literary critic of that older, more European-minded school. 1834 *Sun* (N.Y.) 26 May 2/1 The hard times did not warrant his taking board on the 'European Plan'. 1889 *Harper's Mag.* Aug. 484/3 Is this hotel on the European plan? 1914 [see *American plan* s.v. *American *a.* 3]. 1907 *Daily Chron.* 24 July 3/3 A prosperous European-style city. 1957 P. Worsley *Trumpet shall Sound* vi. 115 He..had built himself a European-style house.

B. *sb.* **2.** A person of European extraction who lives outside Europe; hence, a white person, esp. in a country with a predominantly non-white population.

1696 J. Ovington *Voy. to Suratt* 79 She takes her liberty with Subject or Foreigner, African or European at her will. *Ibid.* 87 The more wary Europeans, who traffick with these People. 1791 G. Carter *Narr. Loss of Grosvenor* xli. 154 This party or detachment consisted of one hundred Europeans, and three hundred Hottentots. 1832 A. Earle *Narr. Residence N.Z.* 73 Several Europeans who had accompanied various tribes to battle. 1857 Livingstone *Missionary Trav. S. Afr.* v. 104 By Griquas is meant any mixed race sprung from natives and Europeans. 1905 W. B. *Where White Man Treads* 286 In fact, that no 'native'..may bequeath or divert his land to a European without valuable consideration. 1913 W. W. Thompson *Sea Fisheries Cape Col.* ii. 42 At Somerset West Strand the industry is carried on by Malays and coloured men with a few Europeans; at Hermanus and East London nearly all those engaged in boat fishing are Europeans. 1954 R. St. John *Through Malan's Africa* I. i. 12 'European' in the South African lexicon, means anyone whose skin appears to be white, regardless of where he may have been born or brought up and what his parents may have been. 1959 *Cape Argus* 23 Jan. 4/3 More and more the landladies of London are talking of 'Europeans' when they mean people of white skin. 1961 *Observer* 19 Mar. 23/3, I, a 'European' from Canada, listened to an African Sunday orator in Trafalgar Square.

3. A member(-country) of the European Economic Community; a person who advocates membership thereof.

1959 *Listener* 31 Dec. 1142/2 By 'convinced Europeans' I mean those who are consciously constructing a new European entity...'Europeans'..feel that the Community of the Six is still very fragile. 1962 *Ibid.* 8 Feb. 241/2, I am also worried that perhaps the Europeans do not value the adhesion of Britain sufficiently. 1963 *Guardian* 21 Jan. 9/2 'Europeans' of the five countries feel it is of the utmost importance to make this combined stand..against the French boycott of further meetings with the British. 1970 *Daily Tel.* (Colour Suppl.) 20 Feb. 7/1 As a dedicated European of long standing I must now face the plain fact that the Common Market is in a mess.

Europe·anly, *adv.* [-ly².] In a European way or style.

1902 S. Lane-Poole *Story of Cairo* i. 27 Mean and uneven offices and tenements, neither Europeanly regular nor Orientally picturesque. 1935 E. Weekley *Something about Words* 41 We are told nowadays to think internationally, or at any rate Europeanly. 1958 *Times Lit. Suppl.* 24 Jan. 48/2 The Europeanly aesthetic meaningfulness of certain pictures by Van Eyck, Brueghel and Rembrandt.

europium (yūᵊrōᵘ·piɒ̆m). *Chem.* [mod.L. *europium* (E. Demarçay 1901, in *Compt. Rend. Acad. Sci.* CXXXII. 1485), f. *Europe* + -ium.] A metallic element of the rare-earth group, symbol Eu, atomic number 63.

1901 *Jrnl. Chem. Soc.* LXXX. II. 511 The oxide of the new element *europium*..has been obtained so pure that it does not give any indication of the samarium spectra. 1938 *Nature* 24 Sept. 579/2 Comparison of the lattice constants with those of strontium and barium fluorides shows that in this respect europium compounds approximate more closely to strontium than to barium compounds. 1967 [see *bastnäsite].

Eurovision: see *Euro-.

eurybathic (yūᵊribæ·þik), *a. Biol.* [f. Gr. εὐρύ-ς wide + βάθος depth.] Of aquatic life: able to live at varying depths.

1902 *Encycl. Brit.* XXXIII. 934/1 Similarly, in regard to depth, species have been classed as eurybathic and stenobathic. 1903 [see *EURYTHERMAL a.]. 1942 H. U. Sverdrup et al. *Oceans* xvii. 806 Many species are eurybathic, that is, they endure great ranges of depth. 1967 *Oceanogr. & Marine Biol.* V. 467 It is possible to suppose that eurybathic species may go down beyond the lower limit of the bathyal zone.

eurygnathism (yūəri·gnǎþiz'm). [See -ISM.] Eurygnathous character.

1890 H. Ellis *Criminal* iii. 52 Microcephaly of the frontal region, ..eurygnathism [*sic*], etc. 3 (1901) 50 eurygnathism). 1911 *Encycl. Brit.* XXII. 424/2 Eurygnathism, in the lateral projection of jawbones so characteristic of the Mongolic races.

euryhaline (yūərihē·lin, -in), *a.* [a. G. *euryhalin* (K. Möbius 1871, in *Jahresbericht der Commission zur wissenschaftlichen Untersuchung der deutschen Meere in Kiel* (1873) I. 139), f. Gr. εὐρύ-ς wide + ἅλινος of salt.] Able to tolerate a wide range of salinity.

1888 *Challenger Rep.: Zool.* XXIX. 421 All Baltic marine animals can live in water of varying saltness, and are therefore said to be *euryhaline*. 1964 J. A. Lovern in *Oceanogr. & Marine Biol.* II. 169 Migratory, euryhaline or estuarine species are treated in full where appropriate.

euryphagous (yūəri·fǎgəs), *a. Zool.* [f. Gr. εὐρύ-ς wide + -PHAGOUS.] (See quots.)

1926 A. S. Pearse *Animal Ecol.* iii. 72 Animals that..eat a wide variety [of foods] are *euryphagous*. 1953 *New Biol.* XV. 23 Euryphagous predators (i.e. those which eat a large number of different species), such as insect-eating birds.

euryscope (yūə·riskōᵘp). [f. Gr. εὐρύ-ς wide + -SCOPE.] A rapid rectilinear photographic lens of wide aperture. (*Disused.*)

1882 *Photogr. Ann.* II. p. v, The Rapidity of this series is equal to that of the Rapid Euryscopes. 1902 *Encycl. Brit.* XXXI. 694/1 Voigtländer's 'Euryscopes'..are still largely in use. 1906 R. C. Bayley *Compl. Photographer* vi. 72 Many [rapid rectilinear lenses] have been made which could be used at F/6; these are known as 'Euryscopes'.

eurythermal (yūəriþə·ɹmǎl), *a. Biol.* [ad. G. *eurytherm* (K. Möbius 1871, in *Jahresbericht der Commission zur wissenschaftlichen Untersuchung der deutschen Meere in Kiel* (1873) I. 139), f. Gr. εὐρύ-ς wide + THERMAL *a.*] Able to tolerate a wide range of temperatures. So **eu·rytherm, eurythe·rmic, eurythe·rmous** *adjs.*

1881 [see *stenothermal* s.v. STENO-]. 1888 *Challenger Rep.: Zool.* XXIX. 421 Such animals are distinguished as *eurytherm*, in opposition to *stenotherm* animals, which can live only in warm or only in cold water. 1903 *Nature* 5 Nov. 23/2 Twelve [species] were eurythermic and eurybathic, ranging from the surface to 700 fathoms in both areas. 1940 *Chambers's Techn. Dict.* 311/2 Eurythermous. 1964 V. J. Chapman *Coastal Veg.* iii. 73 *F*[*ucus*] *serratus*..can be regarded as a eurythermal species.

Eusol (yū·sǫl). *Pharm.* Also **eusol.** [f. initial letters of Edinburgh University + SOL(UTION *sb.*: cf. *EUPAD.] A solution of chlorinated lime and boric acid that is used as a general antiseptic.

1915 J. L. Smith et al. in *Brit. Med. Jrnl.* July 129/2 The solution of free hypochlorous acid prepared in this way we have named 'Eusol'. *Ibid.* 134/2 The eusol treatment was kept up. 1920 Martindale & Westcott *Extra Pharmacop.* (ed. 17) I. 52 Eusol has been most extensively used to wounds. 1962 *Pharm. Jrnl.* 31 Mar. 270/2 Its instability in comparison with the more alkaline hypochlorite preparations has always been a serious disadvantage to the use of eusol.

eustasy (yū·stǎsi). *Physical Geogr.* [anglicized back-formation from next (see -Y²), after mod.L. *-stasis* sb. corresponding to *-static* adj.] A uniform change of sea-level throughout the world.

1946 F. E. Zeuner *Dating the Past* iii. 47 Such movements of the sea-level are called eustatic, and the phenomenon, glacial eustasy. 1962 *New Scientist* 2 Aug. 243/1 If the whole level of the ocean has risen or fallen (we call it eustasy), then the total quantity of water in the ocean has been changed from time to time.

eustatic (yustæ·tik), *a. Physical Geogr.* [ad. G. *eustatisch* (E. Suess *Das Antlitz der Erde* (1888) II. xiv. 680); see EU- and STATIC *a.*] Of, pertaining to, or caused by eustasy. Hence **eustatically** *adv.*; **eu·statism** = *EUSTASY.

1906 H. B. C. Sollas tr. *Suess's Face of Earth* II. iii. xiv. 538 We must commence by separating from the various other changes which affect the level of the strand, those which take place at an approximately equal height, whether in a positive or negative direction, over the whole globe; this group we will distinguish as *eustatic movements*. The formation of the sea basins produces spasmodic eustatic negative movements. 1934 R. A. Daly *Changing World Ice Age* iv. 226 Assume that in the last Inter-glacial stage, when the sea level rose on the bank eustatically, a main atoll reef..grew up. 1935 H. Baulig *Changing Sea Level* 4 We have to consider ..eustatic movements, resulting from changes in the capacity of the oceanic basins, a kind of eustatism which might be called deformational. 1935 *Nature* 28 Sept. 492/2 But the more recent the eustatic emergence, the lower are the dangers of mistaking the traces of the corresponding strand-lines on the rocks of continents and islands. 1939 *Proc. Prehist. Soc.* V. 264 Sea-level..rose and fell eustatically in response to the melting and growth of ice-sheets. 1946 [see prec.]. 1954 W. D. Thornbury *Princ. Geomorphol.* vi. 142 *Diastrophic eustatism* is change of sea level resulting from variation in capacity of the ocean basins, whereas *glacio-eustatism* refers to changes in sea level produced by withdrawal or return of water to the oceans. 1970 R. J. Small *Study of Landforms* xii. 423 Successive eustatic falls, independent of those attributable to glacial eustatism, occurred either as the ocean basins were deepened and/or the continental land areas raised *en masse*.

Euston Road (yū·stǝn rōᵘd). The name of a road in London, site of a short-lived School of Drawing and Painting (1938–39), used *attrib.* or *absol.* to designate a group of English post-Impressionist realistic painters of the late 1930s or their type of art. Hence **Eusto·nian, Eu·ston Roa·der**, a member or follower of this group.

1941 *Horizon* May 349 The painters of the Euston Road school resemble each other in that they all paint direct from nature. 1945 F. Nash *Let.* 24 June in A. Bertram *Nash* (1955) ix. 264 The Euston Road boys, and their ruddy realism. 1945 C. Bell *V. Pasmore* 14/1 The Euston Road School was founding. In that school Pasmore was to be the principal teacher; yet he was and is far from being a typical Eustonian. Euston Road was a call to order and an antidote to the sensationalism and amateurism of the School of Paris. 1959 *Listener* 3 Dec. 981/3 The Euston Roaders owed a little to Tonks.

eutectic, *a.* and *sb.* Substitute for def.: **A.** *adj.* That is eutectic; of or pertaining to a eutectic or its liquefaction or solidification; *eutectic point*, the melting-point of a eutectic, or the point representing it in a constitutional diagram.

1884 F. Guthrie in *Phil. Mag.* 5th Ser. XVII. 462 The main argument..hinges upon the existence of compound bodies, whose chief characteristic is the lowness of their temperatures of fusion. This property..may be called Eutexia, the bodies possessing it eutectic bodies or eutectics (εὔ τήκειν)... It will, however, perhaps be better to make the term more useful by limiting its application. I shall use it..for bodies made up of two or more constituents, in such proportion to one another as to give to the resultant compound body..a lower temperature of liquefaction than that given by any other proportions. 1885 *Nature* 5 Nov. 21/1 When metals do unite in atomic ratios the alloy produced is never eutectic, i.e., having a minimum solidifying-point. Thus pure cast-iron is..an eutectic alloy of carbon and iron. 1902 [see B. 16. below]. 1920 *Encycl. Brit.* I. 705/2 The two sloping lines cutting at the eutectic point are the freezing-point curves of alloys. 1949 [see Dict.]. 1947 *Jrnl. Inst. Metals* 330/1 This mixture, which is known as the eutectic mixture, has the lowest melting-point of any which can be formed from these minerals. 1950 *Engineering* 16 June 672/2 Fusible plugs of eutectic alloys. 1967 A. H. Cottrell *Introd. Metallurgy* xvi. 253 This is the ternary eutectic point at which the liquid is in equilibrium with all three solids *A, B* and *C*.

B. *sb.* **1.** A mixture which is distinguished from other mixtures of the same constituents in different proportions by having a single temperature at which it melts and freezes, this temperature being lower than the freezing-point of any of the constituents or of any other mixture of them. Also *fig.*

1884 [see A. adj. above]. 1902 *Encycl. Brit.* XXVIII. 569/2 The eutectic F[reezing] P[oint] is of longer duration, but still at the same temperature. For an alloy of the composition of the eutectic itself there is no arrest until the eutectic temperature is reached. 1923 Glazebrook *Dict. Applied Physics* V. 11. 454/2 This most fusible alloy of a system is termed the 'eutectic' and its microscopic appearance is often characterized by fine laminations. 1926 Alden in *Oxford Poetry* 2 Love mutual has reached its first eutectic. 1948 Glasstone *Physical Chem.* (ed. 2) x. 750 Only a pure substance, or a mixture having the composition of the eutectic, melts sharply at a definite temperature. 1954 *Proc. Prehistoric Soc.* XX. 76 It is probably true that speculum is not very far removed from the eutectic of copper and tin. 1968 *Van Nostrand's Sci. Encycl.* (ed. 3), Eutectic... By a similar usage in metallurgy, a eutectic is a discrete mixture of two or more phases, in definite proportions, which have simultaneously crystallized from the liquid solution, with the mutual solution of their components.

2. A eutectic point.

1940 Glasstone *Physical Chem.* x. 740 The temperature at which the first minute drops of liquid appear..is the eutectic for the given system. 1967 A. H. Cottrell *Introd. Metallurgy* xiv. 210 Such a singular point of phase equilibrium is known as a eutectic or peritectic.

eutecticferous (yūtekti·fĕrǝs), *a.* [f. EUTECT(IC + -IFEROUS.] Bearing or producing eutectics; of eutectic form or kind.

1925 P. Hidnert *Thermal Expansion Aluminum* 715 Previous observers found that these alloys form a eutectiferous series with no chemical compounds. 1930 *Engineering* 8 Aug. 185/1 Cast irons are frequently described as eutectiferous alloys. 1953 *Sci. News* XXVIII. 111 In a simple 'eutectiferous' system the liquid breaks down into the elementary metals.

eutectoid (yute·ktoid), *a.* and *sb. Metallurgy.* [f. EUTECT(IC *a.* and *sb.* + -OID.] **A.** *adj.* That is a eutectoid; of or pertaining to a eutectoid or its transformation; *eutectoid point*, the transformation temperature of a eutectoid, or the point representing it in a constitutional diagram. **B.** *sb.* A solid analogous to a eutectic, in which the high-temperature phase is a solid solution instead of a liquid and the constituents separate out simultaneously when it is cooled to the transformation temperature.

1903 H. M. Howe *Iron, Steel, & Other Alloys* 440 The word 'eutectoid' occurred to me (June 7, 1903)... It suggests having the shape or other important properties of the 'eutectic', while the two words differ from each other enough to indicate that they refer to really distinct things. 1910 *Encycl. Brit.* XIV. 805/1 Excess of iron over this eutectoid ratio. *Ibid.*, Far below the freezing-point, transformations may take place in the solid metal, and follow a course quite parallel with that of freezing, though with no suggestion of liquidity. A 'eutectoid' is to such a transformation in solid metal what a eutectic is to freezing proper. 1925 *Jrnl. Iron & Steel Inst.* CXI. 584 The eutectoid occurring at 0·89 per cent. carbon theoretically should be regarded as a eutectoid of iron and a solid solution containing about 2·0 per cent of carbon. 1946 *Thorpe's Dict. Appl. Chem.* (ed. 4) VII. 596/2 In this case..it is the solid solution beta breaking down into the two solutions alpha and gamma, and the point P is known as a eutectoid point. 1950 *Engineering* 15 Dec. 492/3 Austenite in a plain eutectoid steel could decompose isothermally. 1961 J. N. Anderson *Appl. Dent. Mat.* (ed. 2) vi. 56 One such reaction..is known as a eutectoid reaction. This involves the breakdown as a solid solution at a definite temperature to produce an intimate mechanical mixture of two solids.

eutelegenesis (yū:telĭdʒe·nĕsis). [mod.L., f. EU- + TELE- + -GENESIS.] (See quots.)

1935 H. Brewer in *Eugenics Rev.* XXVII. 123 For the process of reproduction from the germ cells of individuals between whom is no bodily contact, I propose the name telegenesis. The possible application of the process to the eugenic breeding of man may be termed eutelegenesis. 1959 *Nature* (Suppl.) 12 Sept. 44/2 Recent successful experiments with mammals in parthenogenesis and eutelegenesis. 1962 *Daily Tel.* 7 June 19/7 What is known as eutelegenesis, namely insemination by sperm from some admired donor to 'father' their children.

euthenics (yuþe·niks), *sb. pl.* [f. Gr. εὐθην-εῖν to thrive, flourish + -ICS.] The science and art of improving the well-being of man by the betterment of the conditions of life.

1905 Mrs. E. H. Richards *Cost of Shelter* i. 12 The student of social ethics—Euthenics, or the science of *better living*—may well ask..Are the people growing more healthy..stronger, happier? 1910 —— (*title*) Euthenics, the science of controllable environment; a plea for better living conditions as a first step toward higher human efficiency. 1926 *Daily Colonist* (Victoria, B.C.) 25 July 24/1 Euthenics as a word is comparatively young. Various definitions are given for it, the simplest being efficient living. 1967 *Technology Week* 23 Jan. 50/3 Euthenics [is] the amelioration of his [*sc.* man's] environmental opportunity, i.e., his education.

eutherian (yuþiə·riǎn), *sb.* and *a. Zool.* [f. mod.L. *Eutheria* (T. N. Gill 1872, in *Smithsonian Misc. Coll.* XI. p. iii), f. Gr. εὖ- EU- + θηρία beasts.] **A.** *sb.* A member of the Eutheria, an infraclass which comprises the placental mammals. **B.** *adj.* Of or pertaining to this group.

1880 T. H. Huxley in *Proc. Zool. Soc.* 657 An undifferentiated Eutherian. *Ibid.* 658 Eutherian forms with deciduate placentation. 1894–5 [see Metatheria sb. pl.]. 1950 J. Z. Young *Life of Vertebrates* xxi. 545 All the eutherians (placentals) have been derived from small, perhaps nocturnal, insectivorous or omnivorous animals. 1969 L. H. Matthews *Life of Mammals* I. ix. 197 The introduction of European eutherian predators..have [*sic*] played havoc with the metatherian fauna of Australia.

euthyroid (yupəi·roid), *a. Med.* [f. EU- + THYROID *a.*] Having a thyroid gland with normal function. So **euthy·roidism.**

1924 J. Snowman tr. *F. de Quervain's Goitre* v. 43 The two opposing conditions, euthyroidism and athyroidism require no comment. 1932 C. A. Joll *Dis. Thyroid Gland* xii. 233 There are..certain cases on the borderline..between diminished functional activity and the normal or euthyroid state. 1961 *Lancet* 2 Sept. 550/2 The patient was euthyroid. 1964 L. Martin *Clinical Endocrinol.* (ed. 4) iii. 110 Relapse does not take place after radioiodine treatment provided that euthyroidism has been attained.

eutrape·lia. [f. as EUTRAPELY.] Wit, repartee; liveliness; urbanity; = EUTRAPELY.

1956 *Essays & Studies* IX. 65 The satire extends through sex, religion, politics: to the Statesman who is an 'easy' man (the overtones of that word are impossible to define; it is often something like *eutrapelia*). 1965 Battershaw & Quinn tr. Rahner (*title*) Man at play, or did you ever practise eutrapelia? *Ibid.* v. 95 The object of eutrapelia is play for the sake of seriousness.

eutrophic, *a.* and *sb.* Add: **A.** *adj.* **2.** Of a lake, swamp, etc.: (over-)rich in organic or mineral nutrients and having as a result an excessive growth of algæ and other plants, with depletion of oxygen and consequent extinction of animal life; *spec.* (see quot. 1931). Hence **eutrophica·tion**, the process of becoming eutrophic.

1931 R. N. Chapman *Animal Ecol.* xvi. 305 The eutrophic type of lake is characterized by the paucity or absence of oxygen in the bottom waters. 1947 A. D. Hasler in *Ecology* XXVIII. 383/1 Enrichment of water, be it intentional or unintentional, is called eutrophication.... In this paper eutrophication will be interpreted in the broadest sense; namely, lake and reservoir owing to any and all nutritive substances. *Ibid.* 390/1 Eleven oligotrophic and eutrophic lakes have taken on eutrophic characteristics since comparatively recent times. Most rapid eutrophication has taken place in regions where there are deposits of postglacial clay;..the process has been accelerated by 'cultural influences' such as effluents from flax retting plants, barnyards, fertile soil, urban settlements. 1950 J. Clegg *Freshwater Life* (ed. 2) Pl. 2 (*caption*) An evolved lake in which eutrophic conditions occur. *Ibid.* xix. 317 Large reservoirs..exhibit all the characteristics of a typical eutrophic lake. 1967 *Technology Week* 23 Jan. 70/3 Lake Erie's eutrophication. 1968 *Observer* 17 Nov. 9/4 Decaying plant matter consumes all the oxygen in the water, which becomes lifeless.. This process of 'eutrophication' (or over-fertilisation) has overtaken at least 40 lakes in Britain and the US. 1970 *Nature* 11 Apr. 101/1 The characteristic of a eutrophic lake is the way in which the bottom layers are depleted of oxygen during the summer as organic matter sinks to the bottom and decays. 1970 *Motor Boat & Yachting* 16 Oct. 25/1 Eutrophication (dense growth of weeds) tends to occur in enclosed areas of water.

eutrophy. Add: Also, the state of being eutrophic.

1947 A. D. Hasler in *Ecology* XXVIII. 392/1 Erosion has a two-fold effect..increasing sediment volume as well as contributing a higher flow of nutrients which promote eutrophy. 1967 *Oceanogr. & Marine Biol.* V. 284 Such ordination of communities, along a succession is similar to the delineation of freshwater communities along the line, eutrophy—oligotrophy.

euxanthic (yuzæ·nþik), *a. Chem.* [tr. G. *euxanthinsäure* euxanthic acid (O. L. Erdmann 1844, in *Jrnl. f. prakt. Chem.* XXXIII. 192), f. EU- + Gr. ξανθ-ός yellow + -IC.] *euxanthic acid*: a glycoside, $C_{19}H_{16}O_{10}$, yielding glucuronic acid and euxanthone on hydrolysis; purreic acid.

1852 [see Purree]. 1943 *Thorpe's Dict. Appl. Chem.* (ed. 4) VI. 424/2 Euxanthic acid is isolated by digesting puri with dilute hydrochloric acid. 1959 E. H. Rodd *Chem. Carbon Compounds* IV. viii. 966 Xanthen derivatives have long been known and include euxanthic acid and the related euxanthone.

euxanthone (yuzæ·nþōᵘn). *Chem.* [ad. G. *euxanthon* (O. L. Erdmann 1844, in *Jrnl. f. Prakt. Chem.* XXXIII. 205), f. prec. + -ONE.] A yellow derivative of xanthone, $C_{13}H_8O_4$; purrenone; 1:7-dihydroxyxanthone.

1852, 1857 [see s.v. Purree]. 1952 K. Venkataraman *Chem. Synthetic Dyes* II. xxiv. 740 Euxanthone, found in the urine of cows fed on mango leaves, is 1,7-dihydroxyxanthone. 1959 [see prec.].

evacuate, *v.* Add: **8. b.** *spec.* To remove (inhabitants of an area liable to aerial bombing or other hazards) to safer surroundings. Also *transf.*

1938 *Times* 28 Oct. 15/2 Authorities of our large towns will wonder whether or not to evacuate more than children. 1940 *Ann. Reg.* 1939 87 Mr. Greenwood and others strongly appealed to the Prime Minister to evacuate children at once. 1942 *Essays & Studies* XXVI. 61 This statement it has proved impossible to check, as the Burney newspapers are evacuated and the journal is not at Bodley.

evacuated, *ppl. a.* Add: (Examples in sense of *EVACUATE *v.* 8 b.)

1939 *Times* 28 Oct. 7/2 Parents of evacuated children should contribute to their maintenance. 1942 E. Waugh *Put out more Flags* ii. 144, I am the district billeting officer. I'm looking for a suitable home for three evacuated children.

evacuation. 3. (Further *attrib.* examples.)

a 1854 G. Furman *Antiquities Long Island* (1875) 269 The Evacuation day..has been observed as a species of holiday on the west end of Long Island. 1939 *Times* 2 Nov. 8/7 Parents should remember that such educational facilities as are available in the reception areas are not likely to be nearly so good as those available in the reception areas. 1940 *Archit. Rev.* LXXXVII. 101 (*title*) Evacuation camp. 1940 E. T. Seton *Trail of Artist-Naturalist* 141 Monday, which, being Evacuation Day, was a public holiday.

evacuee (ĭvækiu,ī·). Also ǁ **évacué(e).** [ad. F. *évacué.*] A person who has been evacuated (see *EVACUATE *v.* 8 b). Also *attrib.*

1934 Webster, *Évacué* n. masc., *évacuée* fem. 1938 *Times* 24 Nov. 10/4 The Minister..will determine the number and type of evacuees and of residents in each area to be cared for. 1939 *Times Weekly* 1 Nov. 12 The embarcation of these, mostly unwilling, evacuees is illustrated. 1939 in *English Studies* (1940) XXII. 29 Some evacuee mothers have complained...Parents visited their evacuee children in the reception areas. 1950 E. Hyams *From Waste Land* 11 There sprang up a gaudy and elaborate legend concerning the evacuees.

evaluative (ĭvæ·liu,ētiv), *a.* [f. EVALUATE *v.* + -IVE.] Of, pertaining to, or effecting evaluation; appraisive, estimative.

1927 L. V. Koos *Amer. Secondary School* vii. 221 Compilation and subsequent evaluative scrutiny of these claims. 1934 *Philos. Rev.* XLIII. 127 The expression of such feeling of life, and of our evaluative reactions, is..admitted to be a legitimate and worth-while activity. 1939 *Mind* XLVIII. 469 It is only confusing to talk of a 'valuational logic' whose basic rule is that an evaluative conclusion cannot be deduced from non-evaluative premises. *Ibid.* 319 The existential functional interpretation is not the only one. 1952 R. M. Hare *Lang. Morals* ii. vi. 111 Two sorts of things that we can say..about strawberries; the first sort is usually called *descriptive*, the second sort *evaluative*. Examples of the second sort of remark are 'This is a good strawberry' and 'This strawberry is just as strawberries ought to be'. 1964 E. A. Nida *Toward Sci. Transl.* iii. 44 The intention may be described as evaluative or appraisive, that is to say, the source not only designates some referent, but also provides its own evaluation of it.

Evangeliar (ĭvændʒe·liaɹ). Also **Evangeliary** (-iǎri) and in L. form. [ad.eccl.L. *evangeliarium*. Cf. OF. *evangelier*, mod.F. *évangéliaire*.] = EVANGELISTARY I.

1846 W. Maskell *Monumenta Ritualia Ecclesiæ Anglicanæ* I. p. cv, [In] the 'Evangelistarium', 'Evangelium', or 'Evangeliarium', is not involved in so great difficulty. 1863 F. C. Conybeare in *Expositor* Oct. 244 The titles of Matthew', 'of Mark',..in this Evangeliar at the heads of their respective Gospels. 1900 *Jrnl. Theol. Stud.* Apr. 453 As Mr. Kenyon points out, at least one Byzantine Evangeliarium was written in France, in uncials and with rubrics. 1952 F. Wormald *Mod. Lang. Rev.* XLVII..the Evangeliar of Saint...has no significance for the purpose in hand. 1953 P. D. Record *Summary Catal. Western MSS. in Bodl. Libr.* VII. 160/1 Evangeliary. Grk. (xii), 18406.

evaporate (ĭvæ·pŏrèit), *sb. rare.* [f. the vb.] = *EVAPORITE.

1920 A. W. Grabau *Textbk. Geol.* I. x. 215 Such salts are called evaporation products, or briefly, evaporates. 1924 [see *EVAPORITE].

evaporative, *a.* Add: **b.** *evaporative cooling*: the cooling (of an engine) in which the heat is removed by the evaporation of a liquid coolant.

1931 A. H. R. Fedden in *Handbk. Aeronaut.* (R. Aeronaut. Soc.) vi. 450 The commonest methods of maintaining the cylinder temperatures within reasonable limits are by water cooling, air cooling, evaporative cooling and liquid cooling. 1934 *Aircraft Engineering* Nov. 290/3 In a system working on the evaporative cooling principle the water is introduced into the engine jackets at..boiling point.

evaporite (ĭvæ·pǒrəit). *Geol.* [f. *EVAPOR(ATE *sb.*: see -ITE¹.] A deposit of sodium chloride or other salts resulting from the evaporation of a body of water.

1924 C. P. Berkey in *Bull. N.Y. State Mus.* CCLI. 116 The genetic idea is carried better if the resulting products are called evaporates, as Doctor Grabau does in his work on Salt Deposits, and the product of reaction by precipitation may be called a *reactionite*. Perhaps one could write *evaporite* and *reactionite*. 1940 *Mineral. Mag.* XXVIII. 621 (*heading*) The lower evaporite bed. 1962 A. Holmes *Princ. Physical Geol.* (ed. 2) iv. 78 Salt deposits (evaporites), such as are left behind when salt lakes dry up, or when enclosed bodies of sea water are strongly evaporated. 1967 *Oceanogr. & Marine Biol.* V. 131 Evaporite sediments of Mississippian age have caused similar uplift in nearby Nova Scotia.

Hence **evapori·tic** *a.*, pertaining to, characteristic of, an evaporite.

1951 *Jrnl. Sedimentary Petrol.* XXI. 75/1 In the upper Gulf Coast area, evaporites occur in the Ferry Lake, with showings of gypsum and evaporitic dolomite on the outcrop near Austin. 1969 *Daily Tel.* 13 Jan. 19/7 An evaporitic area is an arid region where evaporation prevails.

evapotranspiration (ĭvæ:potrænspirē·ʃən). [f. EVAPO(RATION + TRANSPIRATION 2.] The collective conversion of water to vapour, from the soil by evaporation and from plants by transpiration; the amount of water so lost.

1948 *Geogr. Rev.* XXXVIII. 55 The combined evaporation from the soil surface and transpiration from plants, called 'evapotranspiration', represents the transport of water from the earth back to the atmosphere, the reverse of precipitation. 1958 *New Biol.* XXV. 52 The hydrologic cycle of an area of land receiving all its water as precipitation can be expressed by the equation P=E+R+S where P=precipitation, E=evapotranspiration (interception by vegetation, evaporation from the soil and transpiration by plants), R=run-off and S=changes in water storage.

evasive, *a.* Add: **2.** Esp. in phr. (*to take*) *evasive action*: of a pilot, (to take) action designed to avoid having his aircraft attacked; hence, in general use, (to take) action to avoid any kind of trouble or unpleasantness.

1940 *War Illustr.* 10 Feb. 124/1 The formation opens out and we do a bit of dodging. 'Evasive Action' they call it in the official reports. 1941 M. W. Monks in Michie & Graebner *Lights of Freedom* iv. 58 Fighter pilots' combat reports include 'I took evasive action', and the W.A.A.F.s adopted it in describing their adventures on dates. 1961 J. Heller *Catch-22* (1962) v. 49 There was no established procedure for evasive action. All you needed was fear.

eve, *sb.*¹ Add: **4. eve-of-poll** *a.*, of, pertaining to, or occurring in the period immediately preceding the polling in an election.

1960 D. Potter *Glittering Coffin* Postscript, p. iii, The most savagely controversial eve-of-poll issue. 1960 *Guardian* 26 Feb. 6/3 A parliamentary candidate dying..while actually making his eve-of-poll speech. 1970 *Ibid.* 5 Nov. 2/5 Arguments..forcefully delivered by Senator Muskie, on an eve-of-poll television broadcast.

even, *a.* Add: **10. d.** *to get even* (*with*): to take one's revenge (on), to retaliate (against). orig. *U.S.*

1846 S. F. Smith *Theatr. Apprent.* 148, I took my seat with the hope of getting even. 1880 'Mark Twain' *Tramp Abroad* I. xxv. 250 One should always 'get even' in some way, else the sore place will go on hurting. 1889 Barrère & Leland *Dict. Slang* I. 403/1 Those who think this country [*sc.* America] fails to get even with France for her unjust discrimination against American pork. 1906 F. H. Burnett *Shuttle* xxxiii. 330 There exists for people of a certain type a pleasure full-fed by the mere sense of having 'got even' with an opponent. Through-out his life he had made a point of 'getting even' with those who had irritatingly crossed his path. 1910 *Granta* 11 June 9, I feel that I can never get quite even with him again. 1923 L. J. Vance *Baroque* 40 Crooks..blow the works to get even.

e. (*on*) *even terms* (see quots.). *Austral.* and *N.Z.*

1931 L. Acland in *Press* (Christchurch) 23 Sept. 13/7 *Cadet*, a young man working on a station to learn sheep-farming..often worked 'on even terms' but is now usually paid a low wage. 1941 Baker *Dict. Austral. Slang* 27 *Even terms*, working for one's food.

14. a. *even break*: see *BREAK *sb.*¹ 18; *even chance*: an equal chance that something will or will not happen; *even money*: odds in betting that offer the gambler the chance of winning as much as he has staked; also sb. pl. *evens.*

1816 Jane Austen *Emma* II. viii. 149 It was an even chance that Mrs. Churchill were not in health or spirits for going. 1891 N. Gould *Double Event* xxviii, The bookmakers were roaring themselves hoarse. 'Even money Perfection, 3 to 1 Captain Cook.' 1907 *Westm. Gaz.* 25 June 10/2 Two even-money chances—Troutbeck and the White Knight—won. 1964 C. Wyken *Gambling* ii. 223 The Biarritz system is more for players who find the even-chance systems too slow and too tame. *Ibid.* 225 The *noir* is an even-money bet.

d. *even Stephen* (or *Steven*): a colloq. rhyming phrase used as an intensive for 'even' in various senses; *spec.* = *FIFTY-FIFTY *adv.* and *a.*

[1711 Swift *Jrnl. to Stella* 20 Jan. (1948) I. xiv. 171 Now we are even, quoth Stephen, when he gave his wife six blows for one.] 1866 C. H. Smith *Bill Arp* 64 Dick says you allowed the members to exchange two hundred dollars for two hundred dollars of their money even steven. 1928 Wodehouse *Sam the Sudden* xiii. 96 Do you mean to say..that if Soapy was sitting in with the Archbishop of Canterbury on a plan for skinning a sucker the archbishop wouldn't split Even Stephen? 1930 C. Mokey *Kitty Foyle* xxxi. 320 Some of his stories were a bit corny, but a few of Party Beyman's old Racquet Club favorites were even-Steven with his. 1955 R. Bradbury *October Country* (1956) 80 It's a buy-off... Even Stephen. 1958 A. Frome *Cats Again* (1964) xi. 95 It's even-Steven we'll catch him at it. 1970 A. Draper *Swansong for Rare Bird* i. 10 It's even stevens I'll con my way out of it with no more than a big whipping or probation. *Ibid.* vi. 84 We were going to break even stephen by helping himself from her purse.

17. b. *in evens*: of a hundred-yard run, done in ten seconds.

1900 *Baily's Mag.* May 398 He ran the hundred in 'evens' in 1904. 1955 *Times* 23 July 2/7 The Mill Hill schoolboy..rarely, if ever, fails to do the 100 yards in evens.

even-. Add: **1. even-aged** *a.*, of a forest: composed of trees that are of approximately the same age; **even-even** *a.*, of a nucleus: having an even number of both protons and neutrons; **even-odd** *a.*, of a nucleus: having an even number of protons and an odd number of neutrons.

1905 *Terms Forestry* (*U.S. Dept. Agric.*) 9 Even-aged forest. 1928 R. S. Troup *Silvicultural Systems* ii. 23 Even-aged crops..they are more susceptible to damage by wind and snow than uneven-aged crops. 1962 *Times* 1 Jan. 6/4 Trees in even-aged woods had gone down in swathes. 1940 *Physical Rev.* LVIII. 1041 States with higher angular momenta of the core alone (even-even nucleus) are known in many cases to be very near the normal state. 1949 Gamow & Critchfield *Theory Atomic Nucleus* iv. 93 An even-even nucleus can transform into another even-even nucleus of the same A by the simultaneous emission of two β-particles. 1955 J. A. Wheeler in W. Pauli *Niels Bohr* 167 Heavy even-even nuclei always have zero spin. 1955 *U.N. Proc. Peaceful Uses Atomic Energy* 82 Even-odd nucleus. 1966 Phillips & Williams *Inorg. Chem.* II. xxxv. 627 Two hundred and one beta-stable nuclides contain an even number of protons and an even number of neutrons... Sixty-nine are even-odd, i.e. contain an even number of protons and an odd number of neutrons.

even, *v.* **1. c.** Delete † and add later examples; also, to make even or level.

1931 *Economist* 18 July 127/2 Company practice may rightly go beyond the mere creation of secret reserves, and cover their employment to 'even out' fluctuations in earning power. 1962 A. Nisbett *Technique Sound Studio* v. 102 Volumes have to be controlled; they have to be evened out and held back.

4. a. Delete † *Obs.* and add later example.

1947 T. Maynard *Humanist as Hero* xiii. 158 He now had a wonderful chance to even old scores.

b. (Further examples.) Also, to make even or equal; to balance.

1896 *Westm. Gaz.* 28 Feb. 7/1 So this morning a big collapse was provided to even things up. *Ibid.* 2 Nov. 9/1 When they return to-morrow it is quite possible that those who sold yesterday in order to even up their books may be again purchasers. 1908 *Daily Chron.* 9 Mar. 3/3 But all things are evened up in every age. 1963 K. Williams *Culture & Society* iii. 307 The result of the new educational provision was in part..an evening-up between the fortunate places and the unfortunate.

9. *to even out*: to become even or normal.

1920 A. L. Rowse *England of Elizabeth* v. 158 Things were beginning to even out by then.

evener. c. (Earlier U.S. example.)

1850 *Rep. Comm. Patents 1849* (U.S.) 371, I claim..the exclusive use of said spring rests and 'evener'.

evening, *sb.*¹ Add: **2. c.** Afternoon. *dial.* and *U.S. local.*

1788 in C. F. Jackson *Shropshire Word-Bk.* (1879) 136 The meeting held on Monday evening last was adjourned to be holden to-morrow Evening at three of the Clock. 1790 *Pennsylv. Packet* 5 Jan. 3/1 It was..dark from about two o'clock until about half after four in the evening. 1806 M. Lewis in *Lewis & Clark Exped.* (1905) IV. 319 It was one in the evening before he returned. 1876 'Mark Twain' *Tom Sawyer* ii. 3 He'll play hookey this evening. 1880 G. W. Cable *Grandissimes* xiv. 94 This evening (the Creoles never say afternoon) about a quarter to five. 1882 Mrs. Chamberlain *West Worcs. Words* 10 A woman lately wished me 'good marnin'' at 1.30 p.m., then, having passed, turned back to apologize: 'Good evenin' ma'am, I should 'a' said.' 1886 C. E. Craddock' *Broomsedge Cove* x. 177 Air yer obligated ter stay thar in the middle o' this narrer bridge all evening? 1966 G. W. Turner *Eng. Lang. in Austral. & N.Z.* viii. 164 In Queensland evening may be used to refer to any time after midday.

d. *evenings*, in the evening; of an evening. (Cf. Nights *adv.*) *colloq.* or *dial.* (chiefly *U.S.*).

1652, c 1740 [see Morning *sb.* 3 c]. 1862 O. W. Norton *Army Lett.* (1903) 40 We have rather dull times, but evenings we write letters or sing. 1885 *Century Mag.* XXXI. 35/1 We had some real good talks evenings down on the rocks. 1906 B. Ruck *Her Pirate Partner* xvi. 209 So, for all they keep you so close, you can go out—evenings. 1962 *Globe & Mail* (Toronto) 17 Feb. 32 (Advt.), Evenings, there's dancing to do, nightclubs to visit.

e. *Ellipt.* for 'good evening'. *colloq.*

1912 Masefield *Widow in Bye St.* iv. 17, 'Evening,' she said. 'Good evening.' 1969 I. Fleming *Man with Golden Gun* v. 70 The eyes flirted. 'Evenin'.' 'Good evening. Could I have a Red Stripe?'

f. *Ellipt.* for *evening paper* (see *5 b below). *colloq.*

1901 'B. Wells' *Day Earth caught Fire* vii. 109 The Covent Garden blaze had turned out to be a natural for the evenings, but even more so for the mornings. 1967 C. Douglas *Death and Hunting* v. 36 We've missed the final edition of the local evening now.

5. b. *evening meal*, *school*; **evening dress** (earlier and later examples); hence **evening-dressed** adj.; **evening paper**, a newspaper published later than a morning paper, usually so as to be on sale from about midday onward; **evening party** (earlier example); **evening primrose** (earlier example); **evening suit**, a suit of formal clothes as prescribed by fashion to be worn in the evening.

1797 N. Heideloff *Gallery of Fashion* Nov. in Jane Austen *Novels* (1966) I. 387 Evening dresses. 1825 H. Wilson *Memoirs* I. 91 Ponsonby..had put on an evening dress. 1865 Mrs. Gaskell in *All Year Round* IX. 74/1 There stood Mrs. Forbes in her handsome evening dress. 1936 E. A. Baker *Hist. Eng. Novel* VII. iii. 186 Evening dress is of any colour that suited the fancy of the wearer was discarded for the black coat. 1970 *Times* 6 Nov. 2/9 News readers at B.B.C. Lewis have been told to wear evening dress at night. 1896 *Westm. Gaz.* 15 Feb. 3/1 An evening-dressed audience. 1933 P. Godfrey *Back-Stage* xiv. 179 An evening-dressed gentleman. *Ibid.*, Comedy sketches which allowed of being evening-dressed were booked in preference to those which did not. 1860 Evening meal [see Meal *sb.*² 2 a]. 1954 A. S. C. Ross in *Neuphilologische Mitteilungen* LV. 43 U-speakers eat..*dinner* in the evening;.. *Evening meal* is non-U. 1958 *Listener* 10 July 68/2 A time when many of us are either cooking or eating our evening meal. 1727–41 Evening paper [see Paper *sb.* 8]. 1857 Dickens *Dorrit* II. xii. 219 The evening paper was full of Mr. Merdle. 1940 L. MacNeice *Last Ditch* 24 Divided by the morning tea By the evening paper. 1816 Jane Austen *Emma* II. xvi. 299 Dinner-parties and evening-parties were made for him. 1806 Smyth & Sowerby *Eng. Bot.* XXII. 1534 *Œnothera biennis*. Common Evening-Primrose. 1822 *Missionary Herald* (Boston, Mass.) XVIII. 51 Avails of an evening school. 1823 *Chambers's Edin. Jrnl.* I. 365/2 On the discovery of his literary taste, Mr. Laidlaw put him to an evening school. 1937 *Discovery* Sept. p. lxxxi, Evening Schools of History and Geography are specially arranged to meet the needs of Adult Students. 1862 G. Meredith *Lett.* (1970) I. 166 If you want me to dine, know that I can only do so if you are absolutely *alone*, having no evening suit. 1912 J. Joyce *Let.* 11 Sept. (1966) II. 319 My evening suit is in the lower drawer of the wardrobe. 1915 D. H. Lawrence *Let.* 4 Mar. (1962) I. 328 Will you tell me if I must bring evening suit.

event, *sb.* Add: **1. a.** (Later examples.) Also (*U.S.*) without *of*.

1832 Dickens *Let.* 29 Oct. (1965) I. 84, I have instructed the Bearer to wait for an answer, in the event of your being at home. 1922 Joyce *Ulysses* 404 The juridical and theological dilemma in the event of one Siamese twin predeceasing the other. 1966 *Publ. Amer. Dial. Soc.* 1964 XLII. 8 *Roll bar*, a metal tubular structure over the cockpit which protects the driver in the event of car overturning.

2. a. Also in various *spec.* uses (see quots.). Also *attrib.*, as **event-particle**, one of the abstract minimal elements into which, according to A. N. Whitehead, space-time can be analysed.

1919 A. N. Whitehead *Enquiry Princ. Nat. Knowledge* II. vi. 72 The 'constants of externality' are those characteristics of a perceptual experience which it possesses when we apprehend it..A fact which possesses these constants of externality, that is, what we call an 'event'. *Ibid.* III. ix. 121 An event-particle is an instantaneous point viewed in the guise of an atomic event. The point which an event-particle covers gives it an absolute position in the instantaneous space of that moment. 1920 —— *Concept of Nature* viii. 172 Thus we finally reach the ideal of an event so restricted in its extension as to be without extension in space or extension in time. Such an event is a mere spatial point-flash of instantaneous duration. I call such an ideal event an 'event-particle'. 1920 A. S. Eddington *Space Time & Gravit.* iii. 45 A point in this space-time, that is to say a given instant at a given place, is called an 'event'. An event in its customary meaning would be the physical happening which occurs at and identifies a particular place and time. However, we shall use the word in both senses. 1924 R. M. Ogden tr. *Koffka's Growth of Mind* v. § 10. 329 Here a process of uniting two event-configurations into one is indicated. The transference from a one-word to a many-word sentence is carried out. 1933 *Mind* XLII. 161 The instantaneous punctiform 'events-particles' of certain modern theories. 1959 *Times* 13 Mar. 10/6 The site of a suspected 'event' (the technical term for a nuclear explosion). 1961 *Operations Research* VIII. 649 An 'event', depicted by circled numbers in Fig. 1, is defined as a distinguishable, unambiguous point in time that coincides with the beginning and/or end of a specific task or activity in the Research and (D)evelopment process. 1963 J. Lyons *Structural Semantics* vi. 118 If the antecedent is an 'event' 'state' verb, then the sort of the antecedent expresses the 'event' which issues in the 'state' expressed by the imperfective of the consequent. It follows..that we can analyse the 'event'..verb of the event as a 'perfect' the consequent of an 'event' 'state' verb or merely a 'state' verb. 1964 K. G. Lockyer *Introd. Critical Path Analysis* ii. 9 An arrow diagram is made up of only two basic elements—(i) *An event*, which must be a definable recognizable nature and..must be a point in time'... (ii) *An activity*, which is the 'work' or 'job' which is needed to move on to an event. *Ibid.* x. 100 This 'event-orientation' is quite common in PERT systems. 1968 *Gloss. Terms Project Network Anal.* (*B.S.I.*) 6 *Event*, a point in the progress of a project after the completion of all preceding activities but before the start of any succeeding activity. 1968 P. A. P. Moran *Introd. Probability Theory* iv. 186 The e-field and its corresponding measure is known as a 'probability space', and the sets, (A), are 'events'.

eventide. Add: **eventide home**, a home for elderly persons (orig. one maintained by the Salvation Army).

1918 H. Barnett *Canon Barnett* II. 317 An eventide-home for those who are near the end of their pilgrimage. 1959 J. Fleming *Miss Bones* viii. 99 She was in touch with dozens of good causes, eventide homes, hostels for distressed gentlefolk, etc. 1960 —— *When I grow Rich* x. 120 It looks like an Eventide Home for old liners.

eventuality. 1. (Earlier example.)

1789 *Monthly Rev.* XX. 448 Our Historian [*sc.* W. Harte] uses..*eventualities*, where *contingencies*, or *contingent events*, would be as expressive.

ever, *adv.* Add: **1. c.** *Yours ever*: a phrase used in the subscription of a letter (cf. Yours *poss. pron.* 1 c). Also ellipt. *ever.*

1601 R. Lytton *Let.* 19 Sept. in S. Williams *John Chamberlain's Lett.* (1861) 116 All, as ever..Yours, Rowland Lytton. 1611 T. Heywood *Golden Age* sig. A2, As this is received, so you shall find the rest. 1672 Sir H. Ryalls *Lett.* 12 July, My good friend..so I rest. 1763 Boswell *Let.* 28 July (1924) I. 38 My kind service to Bob. Yours ever, with most sincere affection, James Boswell. 1807 Byron *Let.* 6 Mar. (1898) I. 122 Poetic fame is by no means the 'acme' of my wishes.—Adieu. Yours ever, Byron. 1914 T. E. Lawrence *Lett.* (1938) 169 Yours ever, T. E. Lawrence. 1934 *Ibid.* 809 It's history. Ever, T. E. S. 1954 A. S. C. Ross in *Neuphilologische Mitteilungen* LV. 28 People who know each other really well..sign *Yours ever*.

5. b. Also hyperbolically, (for) *a very long time*; *for ever and a day*: delete 'now only humorous, but formerly in serious use' and add later examples.

1823 Carlyle *Let.* 28 Sept. (1886) II. 225 One youth is to go to Germany, the other to Oxford, and I to take my leave I supposed for ever and a day. 1934 J. O'Hara *Appointment in Samarra* (1935) i. 13 It was only a little after three o'clock, but the party had been going on for ever. 1966 J. Aiken *Trouble with Product X* vii. 138 They [*sc.* traffic lights] always take for ever to change. 1967 *Listener* 18 May 656/3 How else can one explain why the Second Reform Bill of 1867 did not sweep the Conservatives from power for ever and a day?

7. d. In the (orig. intensive) colloq. phrase *as ever is* or *was*.

1708 *Essex Inst. Hist. Coll.* X. 78 Bad riding as ever was. 1842 *Knickerbocker* XX. 96, I am seventy-two as ever was this very spring. 1850 Dickens *Dav. Copp.* xxi. 216 She has been as good a girl as ever was. 1890 S. Hale *Lett.* (1919) 248 Louis Church..a dear as ever was, aged twenty-one. 1929 F. M. Ford *Let.* 2 Sept. (1965) 98 If you..we'd care to adventure either next Week End as ever is or next but one, a train leaves Victoria at 3:40 p.m.

e. Colloq. phr. *did you ever?* = did you ever see or hear the like?, would you believe it? Similarly *did anybody ever?*, *if ever I*, *was there ever?*, *who'd a ever?* (Cf. *well I never*, and quot. 1817 for sense 7 a in Dict.)

1825 J. Neal *Bro. Jonathan* I. 150 'My stars!—well, if ever'—wiping her fat hands very carefully. 1828 T. Creevey *Let.* 6 Dec. in Gore *Creevey's Life & Times* (1934) xiii. 294 Lady Louisa..believed Papa would not have come...if I had not been coming. Was there ever? 1840 *Spirit of Times* 21 Nov. 447/3 What a fib! did you ever? well, I never did hear the beat of that. 1844 G. E. Jewsbury *Let.* 24 June (1892) 136 He..found that the surgery-boy had..given calomel instead of ipecacuanha! Did you ever? 1854 J. E. Cooke *Virginia Comedians* I. xiv. 77 'Did anybody ever!' said Miss Aletha. 1861 Reade *Cloister & H.* xliv, So then, if they take us to task we can say, alack we knew nought; we thought no ill; now, who'd ever? and so forth. 1892 *Peel City Guardian* 23 Jan. 3/3 'And where is she now?' 'In a studio.'.. 'Did you ever?' said Mrs. Fansham. 1909 Masefield *Tragedy of Nan* iii. 61 Fifty pou-und. Fifty pou-und. Did you ever.

f. Qualifying a superlative (usu. an adjective) = ever known, experienced, etc.; 'on record'. orig. *U.S.*

1906 'O. Henry' *Four Million* (1916) 71 Anna and Maggie worked side by side in the factory, and were the greatest chums ever. 1924 *Winston. Gaz.* 12 Aug., Mr. Coolidge is expected to reach the largest audience ever in his acceptance address as Republican candidate. 1955 *Times* 12 May 4/3 The last amateur side from the States..had proved in 1951 the first-ever to beat the A.B.A. in one of their own rings. 1958 *Listener* 17 July 94/3 Cambodia's first-ever seaport. 1960 *Sunday Express* 11 Sept. 1/4 Mr. Matthews had not expected to be able to see his son become the youngest-ever champion. 1969 *Guardian* 6 Sept. 9/1 The biggest ever postbag of telegrams.

g. Colloq. phr. *if ever there* (or *if there ever*) *was* (*one*): an assertion that the person or thing referred to is a perfect or undoubted example of its kind.

1822 Carlyle *Let.* 12 Jan. (1886) II. 30 In composition with 'fence' to be sure, and not governing 'shrubs', but still a preposition if there ever was one. 1872 Hardy *Under Greenw. Tree* I. ii. v. 189 Geoffrey Day is a clever man if ever there was one. 1912 T. Dreiser *Financier* xix. 213 A great young man, if there ever was one. 1929 G. Goodwin *Conversations with G. Moore* xxvi. 161 Richard Wormald was..a large-hearted millionaire if ever there was one. 1966 *Guardian* 21 May 10/4 The Playboy Club—an example of how 'architecture can help to solve the social, visual, technical, and economic problems of the twentieth century' if ever there was one.

8. f. Used as a colloquial intensive, usu. to emphasize an exclamation. (Cf. sense 8 d in Dict.) Chiefly *N. Amer.*

1934 F. Pickersgill *Let.* 29 May (1948) ii. 9 Was I ever glad to get off the beastly boat! 1945 S. Lewis *Cass Timberlane* (1946) vii. 46, I thought they were all so nice, and oh boy! are they ever learned! *Ibid.* xvi. 91 You're damn tootin'... Are we ever! Oh boy, I'll say we are! 1949 *Canadian Forum* Aug. 113/2 Do you remember it George?..Do I ever!

9. b. (Earlier examples of *were it ever so*, etc.)

1854 A. E. Baker *Gloss. Northants Words* 215 'I wouldn't do it if it were ever so,' i.e. nothing should induce me to do it. 1858 C. M. Yonge *Christmas Mummers* viii. 115, I couldn't lead the worship if it was ever so!

10. a. (Further examples.)

1742 F. Blyth *Sermons* II. 281 The immense Sea of God's ever-flowing Mercy. 1822 Carlyle *Let.* 2 June (1886) 218 The minute and ever-present knowledge of his duties in every emergency. 1846 W. Hamilton in *Wks. of T. Reid* 798/2 No answer could be afforded to the ever-recurring questions. 1848 Mill *Pol. Econ.* II. III. xxiv. § 5.215 The ever growing attributions of the government. 1855 A. L. Phillipps *Mahometanism* vi. 249 Anarchical and revolutionary outbreaks, that..have, with ever-increasing force, shaken..the civilized world. 1901 W. James *Let.* 17 May (1920) II. 146 Plane rising behind plane of flat dark relieved against flat light in ever-receding gradation. 1909 E. Pound *Exultations* 39 This ever-flowing monotony. 1933 *Mind* XLII. 146 Immortality is of no value unless it is a ground for the experience of eternal life: 'the continuous enjoyment of an ever-present good'. 1937 R. H. Lowie *Hist. Ethnol. Theory* viii. 109 An ever-recurring principle. 1939 M. Spring Rice *Working-class Wives* iii. 50 Years of ever-increasing toil. 1957 P. Worsley *Trumpet shall Sound* ii. 39 These new wants..spurred the ever-growing army of recruits. 1969 *Jane's Freight Containers 1968–69* 188/3 To speed the clearance and handling of the ever-increasing number of containers.

b. (Further examples.) Cf. Never *adv.* 6 c.

1778 F. Burney *Evelina* III. ii. 22 He wrote the ever-to-be-regretted letter. *Ibid.* xvii. 200 My ever-to-be-regretted mother. 1815 Wordsworth *Poems* I. 366 If Names are more acceptable than images, where is the ever-to-be-honoured Chaucer?

11. ever-bearer, a plant which bears flowers and fruits (sometimes simultaneously) for a long time; hence *ever-bearing* adj.; **ever-bloomer** = *ever-bearer* (orig. applied to a rose); hence *ever-blooming* adj.; **ever-loving** *a.*, that loves for ever; usu. as a stock colloq. epithet for a wife; hence as *sb.*, one's wife; **ever-ready** *a.*, always accessible or prepared; hence as *sb.*, a person or thing that is always ready; *spec.* (usu. with capital initials) a member of a territorial army or similar force that is liable to be mobilized at any time.

1929 Weaver & Clements *Plant Ecology* xiii. 326 While..most plants have a comparatively short period of flowering and fruiting each year..reproductive activity in some continues through several months. The latter are known as ever-bloomers or ever-bearers. *Ibid.*, Several species of plant..show a tendency to exhibit the 'ever-blooming' or 'ever-bearing' habit, *i.e.* the ability to continue both vegetative and reproductive activities more or less successfully together. 1921 *Daily Colonist* (Victoria, B.C.) 13 Oct. 15/1 Evergreen blackberries and everbearing strawberries are coming in but in small quantities. 1887 *N.Y. Semi-Weekly Tribune* 3 May (Cent. Dict.), We have grown over sixty named ever-bloomers or tea-roses. 1931 D. Runyon *Guys & Dolls* (1932) i. 26 Maud Milligan, who is well known to one and all as the ever-loving doll of Big Nip. 1938 —— *Furthermore* v. 87 His ever-loving wife is passing a tambourine around. 1968 'P. Barrington' *Accessory to Murder* ii. 34 The old fool was devoted to his ever-loving. I reckon she was damn lucky. 1828 Mill in *Westm. Rev.* IX. 257 A summary appeal to that ever-ready standard of comparison, English practice. 1842 Ever-ready [see sense 10 a in Dict.]. 1917 'Contact' *Airman's Outings* iii. 60 That vindictive exponent of frightfulness, Archibald the Ever-Ready [*i.e.* an anti-aircraft gun]. 1936 C. Day Lewis *Friendly Tree* 15 An ever-ready, over-sweet smile. 1962 *Times* 21 Mar. 7/4 This [*sc.* the revival of National Service] is undesirable for a number of reasons. The purpose of the Ever-Readies is to avoid such a measure. 1966 *Guardian* 19 Aug. 1/8 The searchers..included 190 'ever-readies' of the Metropolitan Police's Special Group.

everdamp (e·vəɹdæmp). [f. Ever *adv.* + Damp *a.*] (See quot. 1968.)

1933 N. Montague in W. Atkins *Art & Practice Painting* III. vi. 43 'Everdamp' Transfer Paper. This paper, which, in addition to the composition, contains hygroscopic compounds (these keeping the paper moist) and requires no damping book, may be run on a dry plate. ..'Everdamp' is an extremely useful paper. 1967 E. Chambers *Photolitho-Offset* xvii. 258 Yellow everdamp can be purchased either smooth or slightly rough surfaced and embodies a thin non-dry coating. *Ibid.*, It is usual for a good conditioned everdamp paper to require no damping. 1968 *Gloss. Terms Offset Lithogr. Printing* (*B.S.I.*) 24 *Everdamp*, a type of transfer paper which remains limp by having a hygroscopic content in its coating.

Everest (e·vərest). [The name of the highest mountain in the world, on the frontier of Nepal and Tibet, f. the name of Sir George *Everest* (1790–1866), Surveyor-General of India.] Used *transf.* to designate the greatest conceivable achievement, spiritual height, difficulty, etc.

1929 R. Aldington *Death of Hero* I. ii. 52 Isabel..displayed signs of that..talent for violent invective she afterwards developed to such Everest peaks of unpleasantness. 1956 A. H. Compton *Atomic Quest* 346, I see him [*sc.* Jesus] as the Everest among the world's many high mountains. 1958 Duke of Edinburgh in M. Holdgate *Mountains in Sea* p. v, Every man has to face his Everest some time during his life.

everlasting, *a.* and *sb.* Add: **A. 5. b.** (Further examples.)

1832 S. Smith *Major Downing* 95, I had rather fight forty New Orleans battles than to govern this everlasting great country one year. 1848 Lowell *Lett.* I. 136 It's everlasting hot to-day. 1903 K. D. Wiggin *Rebecca* 262 She'd kick the ladder from out under her, everlastin' quick.

B. 3. b. (Earlier examples.)

1738 J. Munn *Observ. Brit. Wool & Manuf.* 5 Damasks, Russets, Everlasting, Cantiloons, Worsted Plush..with many other sorts of Plain and Figur'd Stuffs. 1754 *South Carolina Gaz.* 1 Jan. 4/2 Check mantuas and Irish stuffs, everlastings, bombazine. 1763 *Essex Inst. Hist. Coll.* XLIX. 142 Had on..either a pair of black Everlasting Breeches, or cloth colour'd Leather ones.

Everton (e·vəɹtǫn). The name of a town in Lancashire, used *attrib.* in **Everton toffee**, (*a*) (see quot. 1951); (*b*) rhyming slang for 'coffee'.

1857 'Ducange Anglicus' *Vulgar Tongue* 7 *Everton toffee*, coffee. 1861 Mrs. Beeton *Bk. Househ. Managem.* 800 (*heading*) To make Everton toffee. 1869 Trollope *Phineas Finn* I. 93 His pockets were crammed full of gingerbread and Everton toffy. [1907 *Vict. Hist. Count. Engl., Lancaster* III. 20/2 Molly Bushell's original manufactory of the sweet to which Everton has given a name was in Village Street... She was living in 1759.] 1951 *Good Housek. Home Encycl.* 456/2 *Everton toffee*, a brittle sweetmeat similar to butterscotch, but containing cream or evaporated milk. 1960 J. Franklyn *Dict. Rhyming Slang* 62/2 *Everton toffee*, coffee... Now obsolescent.

evertor (ĭvə̄·rtǫɹ). *Anat.* [f. Evert *v.* + -or.] Any muscle which turns or rotates a part outward.

1903 *Lancet* 4 July 56/2 Either the evertor or invertor is out of use. 1967 *Gray's Anat.* (ed. 34) 717 When the foot is off the ground, it [*sc.* the peroneus longus] is an evertor.

every, *a.* Add: **1. e.** *every time*, on all occasions, without fail or exception, certainly; freq. used as an affirmative exclamation. *colloq.* (orig. *U.S.*).

1865 *Trans. Ill. Agric. Soc.* V. 318, I advise every body to plant it,..but always charge them to also plant Delaware and Catawba without fail, 'every time'. 1925 *Times Lit. Suppl.* 16 July 469/2 The Americans, if only because they have twice the population, are bound to win every time. 1927 A. B. Cox *Mr. Priestley's Problem* ii. 30 'It could be done...What do you say, Doyle?'.. What he did say, tersely, was: 'Every time! Let's!' 1928 S. Lewis *Man who knew Coolidge* I. 41 If he'd study his mother and me a little more..he'd be a lot better off! You bet! Every time! 1939 G. B. Shaw *Good King Charles* I. 50 Thats what I have done, and you havnt. And that puts me ahead of you with the British people every time.

f. *every which way*: for U.S. read orig. and chiefly *U.S.*; also *ever*(*y*) *which a way*(*s*); (add earlier and later examples); so *every which*: whichever, every; *every last*: absolutely every (*colloq.*, orig. *U.S.*); *every so often* (or *oft*): from time to time, at intervals.

1824 'A. Singleton' *Lett.* 82 Children [in Virginia] learn from the slaves some odd phrases; as, every which way. 1844 'J. Slick' *High Life N.Y.* I. iv. 39 He kept a flinging his arms about every which way. 1887 A. W. Tourgée *Button's Inn* 80 You've done..more'n any other man would have done, working and contriving every which way. 1888 'Mark Twain' in *Century Mag.* Jan. 460/2, I know it is with me—every last sentence of it. 1896 —— *Tom Sawyer, Detective* (1897) xi. 117 He got that big di'mond out..and let it flash and blaze and squirt sunlight everwhichaway. 1896 J. K. Snowden *Web of Old Weaver* x. 118 Every so oft I could hear him say a word or two. 1905 *Eng. Dial. Dict.* Suppl. 99/1 'Er goes charing at Mrs. Long's every so often. 1928 'M. Chapman' *Happy Mountain* i. 8 [His hair] had a habit of sticking out every-which-a-ways. 1928 R. Frost *Let.* 11 Oct. (1964) 191 Thought on in these bookish ways our lives go everywhichway. 1931 D. Runyon *Guys & Dolls* (1932) iv. 89 Other coppers are coming from every which direction. *Ibid.* xi. 237 People running every which way. 1949 R. Harvey *Curtain Time* xiii. 128 Sleepily, monotonously, every so often, a drop would form, slide down the umbrella and fall with a soft plop on the floor. 1953 R. Graves *Poems* 29 From every-which-a-way, hot as a two-buck pistol. 1958 *Times Lit. Suppl.* (Suppl.) 15 Aug. p. x/2 Stacked-up copies of every-which volume published in Great Britain since the prosperous thirties. 1959 *Listener* 15 Jan. 115/1 The bullocks bellowed as one, and surged away in a great sea of cattle, every last 900 head of them. 1962 L. Deighton *Ipcress File* v. 30 Hemmed in every-which-way was Ross. 1965 *Listener* 3 June 836/3 As for driving a motor-car while looking every whichway and singing a duet, I hope that the Road Safety Council has already protested. 1968 J. Wainwright *Edge of Extinction* 200 Clear 'em out—every last one of 'em. 1970 *Sunday Times* 22 Nov. 35/5 Every so often we would hear the tender melancholy note of the horn.

11. b. every-nighter, one who attends every performance of a play, series of concerts, etc.

1905 G. B. Shaw in *Grand Mag.* Feb. 126 He specially appeals to those who have seen King Henry VI. five times..to discontinue their visits... The every-nighters..have no excuse for their selfishness. 1947 *Penguin Music Mag.* II. 35 The orchestral player will admit the strain..and so will the most fanatical every-nighter after a few seasons.

everydayness. Delete *rare* and add further examples.

1862 *Temple Bar* V. 263 The every-dayness, the common-placeness of life oppressed me. 1892 *Sat. Rev.* 26 Mar. 364/1 The everydayness of this nineteenth century. 1904 M. E. Durham *Through Lands of Serb* 289 Their dull 'everydayness'. 1954 L. MacNeice *Autumn Sequel* 19 The need for everydayness.

Everyman (e·v'rimæn). [Name of the leading character in a 15th-c. morality play.] The ordinary or typical human being. Also *attrib.*

1906 (*title*) Everyman's Library. 1914 *Scotsman* 22 Oct. 1/2 The 'Everyman' Belgian Relief Fund. 1929 R. Bridges *Test. Beauty* IV. 1337 Nor might he escape the fall of Ev'ryman. 1930 *Ess. by Divers Hands* IX. 49 Everyman's rhyme—'Good friend, for Jesus sake forbear'—does well enough for Philip-sober. 1930 D. L. Morse-Boycott (*title*) God and Everyman. 1935 *Discovery* Mar. 74/2 A concession to the stumbling steps of Everyman, who knows nothing of biology. 1964 *English Studies* XLV. 354 He is not presenting us with an Everyman figure.

everything, *pron.* Add: **1. c.** Colloq. phr. *to have everything*: to possess every kind of attraction, advantage, requirement, etc.

The early examples lack the colloq. flavour of the current phrase.

1845 Wellington *Let.* 25 Feb. (1890) 141, I was very sensible of your kind offer of Assistance— But he had everything. 1852 Mrs. Stowe *Uncle Tom's C.* x. 63/1 Mas'r George, you has everything,—l'arnin', privileges, readin', writin'. 1903 *Westm. Gaz.* 27 May 2/2 The little boy who, when told that if he did a certain thing he would not go to heaven, retorted that you couldn't have everything. 1921 R. Macaulay *Dangerous Ages* v. 95 She was to tell stories of Nan, who had everything. 1925 *New Yorker* 6 June 9/1 In the language of baseball, the fellow has everything. 1928 T. Wolfe *Let.* 21 May (1956) 135 I've got a *new* book in mind..I don't see how this one can fail—it has everything: rich people, swank, a poor but beautiful girl, romance, adventure, [etc.]. 1958 Wodehouse *Cocktail Time* xviii. 156 The Beetle and Wedge home-brew..touched the spot. It had everything. It ran like fire through Albert Peasemarch's veins and made a new man of him. 1969 *New Yorker* 20 Dec. 70/2 The latest expensive, worthless gift for the person who has everything.

everywhere, *adv.* Add to def.: Also, loosely: in many places; of frequent occurrence.

1906 Mrs. Beeton *Bk. Househ. Managem.* lxix. 1811 Patent Carpet Sweepers..are in use everywhere. 1912 H. H. Munro *Unbearable Bassington* vii. 122 'Real country scenery; apple blossom everywhere.' 'Surely only on the apple trees,' said Lady Caroline. 1960 *Times Chron.* 19 July 3/1 The Italian fashions are glitter, glitter, all the way...Lamé is everywhere.

3. As *sb.* All places or directions; *the everywhere*, the infinite; omnipresent space.

a 1631 Donne *The Good Morrow* 11 in *Poems* (1633) 195 For love, all love of other sights controules, And makes one little roome, an every where. 1893 G. Macdonald *Poems* II. 158 Out of the everywhere into here. 1922 D. H. Lawrence *England, my England* 207 Everywhere seemed silent, but for the rattle of trains in the crossing.

everywheres, U.S. colloq. and dial. var. Everywhere *adv.*

1865 'Mark Twain' *Sketches* (1872) 18 Fellers that had travelled and been every-wheres. 1876 —— *Tom Sawyer* vi. 66 He's been nearly to Coonville and most everywheres. 1957 B. & C. Evans *Dict. Contemp. Amer. Usage* 164/2 In the United States *everywheres* is often heard in the speech of well educated people, but it does not appear in print.

Évian (evyañ). Also **Evian.** The name of a town in the department of Haute-Savoie in eastern France (in full, *Évian-les-Bains*), used *attrib.* and *ellipt.* to designate a mineral water obtained from springs there.

1857 R. M. Glover *Mineral Waters* III. xv. 308 Évian (Savoy). This is a water quite of a Malvern type; is one of the most charming sites in the world. 1865 E. Lee *Princ. Baths Switzerland & Savoy* 29 The Evian waters have great analogy with those of Contrexeville, though they are less exciting, and are consequently more applicable..where there exists much irritability of the urinary passages. 1886 *Encycl. Brit.* XXI. 332/2 Mineral waters of various kinds abound (Amphion and Evian, chalybeate; ..La Caille, hot, sulphurous). 1908 *Practitioner* Sept. 468 Milk diet and water, alternating the glasses of milk with glasses of Evian-water. 1933 E. A. Robertson *Ordinary Families* x. 230 Evian water, revita biscuits and cheese were laid out for him. 1936 'R. West' *Thinking Reed* i. 12 André..poured out a glass of Evian and sat back on the duchesse sofa. 1945 J. B. Priestley *Three Men in New Suits* v. 81 Markinch appeared to be dining on Evian, white tablets and cigarettes. 1967 *Time* 30 Nov. 40/2 Serving the teetotalers Vichy water instead of the first wine, Evian water instead of the second and ginger ale instead of champagne.

Evipan (e·vĭpæn). *Pharm.* Also **evipan.** A proprietary name of *HEXOBARBITONE.

1932 *Trade Marks Jrnl.* 8 June 719/1 *Evipan*, a medicine for human use as a hypnotic. Bayer Products, Limited..Manufacturers. 1933 *Lancet* 1 July 18/2 Evipan-sodium is the sodium salt of N-methyl-C-C-cyclo-hexenyl-methyl barbituric acid. 1936 A. Christie *Cards on Table* xxx. 278 N-methyl-cyclo-hexenyl-methyl-malonyl urea... Known more simply as Evipan. Used as an anaesthetic for short operations. 1960 'G. Sava' *Ring at Door* ix. 133, I took out a hypodermic syringe from my bag and filled it with a dose of Evipan. 1968 J. H. Burn *Lect. Notes Pharmacol.* (ed. 9) 42 As anaesthetics thiopentone..and hexobarbitone (Evipan) are used.

evocation. Add: **7.** *Biochem.* and *Embryol.* The action of an *EVOCATOR.

1934 [see *evocator b]. 1940 C. H. Waddington *Organisers & Genes* iv. 24 Evocation can be produced by compounds of several radically different kinds. 1963 J. Cohen *Living Embryos* 52 The chemical stimulation [resulting in induction] is called evocation.

evocator. Add: **b.** *Biochem.* and *Embryol.* A chemical substance in part of an embryo that stimulates the development of another part.

1934 J. Needham et al. in *Proc. R. Soc.* B. CXIV. 409 We suggest that the first type of determination be spoken of as Evocation, since it consists in the evocation of an embryonic axis from the competent ectoderm... The organiser, or the evocator, as it might now be called, is soluble in ether and petroleum ether. 1936 *Nature* 15 Feb. 251/2 This stimulus, which the author [*sc.* C. H. Waddington] calls the evocator, is a chemical substance and is not specific for a particular kind of tissue. But the tissue acted upon must be in a 'competent state' for the evocator to work. 1958 B. M. Patten *Found. Embryol.* vi. 134 The chemical substance it [*sc.* the tissue] gives off is known as the organizing substance or evocator.

ǁ **évolué** (evolü,e), *sb.* (and *a.*). [Fr., pa. pple. of *évoluer* to evolve.] An African (from a part of Africa formerly Belgian or French) who

has been educated on European principles; an African who has adopted European modes of thought. Hence as *adj.*, designating, pertaining to, or characteristic of such a person; Europeanized.
1953 J. Packer *Apes & Ivory* vii. 80 'Exactly what is an *evolué*?' 'A Native of some education—a clerk or an office worker.' **1956** K. Hulme *Nun's Story* xi. 183 A first generation to work with whites..that queer lonely society of the *évolués* which was neither black nor white. **1958** *Listener* 9 Oct. 549/1 The man of colour, tribal as well as *évolué*, in every French territory had to say 'yes' or 'no' to a continued relationship within the French community. **1961** P. Mason *Common Sense about Race* iv. i. 113 The man who had become entirely 'évolué' or assimilated to Western ways. **1970** D. Caute *Fanon* i. 12 The black French *évolué* (the relatively educated, Europeanized and privileged native).

evolute (ɪ·vǒliūt), *v.* orig. *U.S.* [Back-formation from Evolution.] **1.** *intr.* To develop by evolution; = Evolve *v.* 8.
1884 *Cambridge* (Mass.) *Tribune* 15 Aug., If those miserable vagrants could only evolute into respectable people there would be converts to evolution at once. **1891** *Daily News* 11 Dec. 6/3 No one had started lower than himself, but now he had evoluted. **1907** *Daily Chron.* 21 Oct. 4/4 This movement, which started so promisingly, and ought by now to have evoluted into honourable well-paid work. **1926** W. J. Locke *Old Bridge* iv. xv, You must let me evolute my own way, *carissima*. **1966** *New Statesman* 22 Apr. 575/3 Teenagers who had evoluted in their own style like the fauna of Australia.
2. *trans.* To evolve, develop. *Journalese.*
1885 *Rep. Indian Affairs* (U.S.) 33 The changed mode of life..will eventually 'evolute' 'Poor Lo' to a higher sphere. **1896** *Daily News* 29 Feb. 6/2 It was to be an attempt to 'evolute' Mr. Tom Hughes's 'Tom Brown' in various directions, to glorify him and bring him up to date. **1899** *Ibid.* 28 Dec. 6/2 The book plate of a millionaire who yesterday was a barman..may in the course of a few generations be 'evoluted' into a family emblem fit to take rank with the arms of any aristocratic Briton. **1930** *Publishers' Weekly* 4 Jan., Many more individual factors which are evoluted from knowledge gained by years of experience as well as teaching.

evolution. Add: **9.** *social evolution*, the development of human societies.
1853 H. Martineau tr. *Comte's Positive Philos.* II. vi. 156 The elements of our social evolution are connected, and always acting on each other. **1865** Geo. Eliot in *Fortn. Rev.* I. 54 The high complexity of the causes at work in social evolution. **1878** [see Dict.]. **1907** J. London *Iron Heel* viii. 113 You fellows have studied business..but you have not studied social evolution at all. **1958** A. R. Radcliffe-Brown *Method in Social Anthropol.* ii. v. 179 In social evolution societies with more complex structure or organisation have been progressively developed from less complex forms.

evolu·tionally, *adv.* [-ly[2].] In an evolutional way.
1922 O. Lodge *Raymond Revised* 207 They would not be apparent to us now, with our particular evolutionally-derived sense organs. **1969** *Sci. Jrnl.* Jan. 48/2 High fertility is evolutionally profitable only under favourable conditions.

evolu·tionarily, *adv.* [f. Evolutionary *a.* + -ly.[2]] In an evolutionary way; from an evolutionary point of view.
1945 *Scrutiny* XIII. iii. 163 The development of a balanced, critical prose style is of course a compensating (and evolutionarily necessary) virtue, but one that itself is perhaps evidence of a failing creativity. **1969** R. B. Fuller *Operating Man. Spaceship Earth* viii. 117 We cannot have mass production unless we have mass consumption. This was effected evolutionarily by the great social struggles of labor to increase wages and spread the benefits. **1970** *Nature* 5 Sept. 1006/2 The transport mechanism, probably universal in cells, is primitive and conservative evolutionarily.

evulse (ɪvʌ·ls), *v.* [f. L. *ēvuls-, ēvellĕre* to pluck out.] *trans.* To pluck or pull out, tear away. Cf. Evulsed *ppl. a.*
1827 Lamb *Let.* 18 Sept. (1935) III. 131 Twas with some pain we were evuls'd from Colebrook. **1910** *Practitioner* June 786 Polypi (myomatous or mucous) may be evulsed or scraped away. **1926** *Daily Tel.* 6 Aug. 5/7 Until the tooth is loosened and finally evulsed.

Evzone, evzone (e·vzōᵘn). [ad. mod.Gr. εὔζωνος *sb.*, f. Gr. εὔζωνος dressed for exercise, f. εὐ- Eu- + ζώνη girdle (see Zone *sb.* 3).] A member of a select infantry regiment in the Greek army, originally recruited from the Greek highlands and conspicuous for their uniform which includes a fustanella. Also *attrib.*
1897 W. K. Rose *With Greeks in Thessaly* iii. 36 The Colonel placed at my disposal a guard of half-a-dozen Evzones. **1927** *Times* (weekly ed.) 10 Mar. 276/3 The massive upturned pompon clogs worn by the Evzone soldiers. **1941** 'R. West' *Black Lamb* II. 529 The petticoated Evzones who dealt with the Italians were predominantly Albanians.

Ewe (ē·wē), *a.* and *sb.*[3] [Native name.] **A.** *adj.* Of or pertaining to a Negro people of West Africa or their language. **B.** *sb.* **a.** A member of this people; also *collect.* **b.** The language of this people.
1861 (*title*) Nya nyuie h'akpāle ene le wegbe me. The Four Gospels in the Ewe language. **1884** *Encycl. Brit.* XVII. 319/1 *Ewe Group*: Acra (Ga), Fantee, Ashantee, [etc.]. **1890** A. B. Ellis *Ewe-speaking Peoples of Slave Coast* p. v, Tshi, Ga, Ewe, and Yoruba are now four distinct languages. **1902** *Encycl. Brit.* XXI. 115/1 *Upper Guinea family*..Ewe (Awuna, Agbosimi, Togo, Dahoman, Krepe, Fra, Appi). **1936** *Discovery* June 168/2 The Twi-, Ewe-, and Yoruba-speaking peoples succeed each other from west to east. **1950** D. Jones *Phoneme* 100 The Ewe word ma (not) is indistinguishable from mā (to divide). **1962** [see *Ashanti].

‖ **Ewigkeit, ewigkeit** (e·viχᵞkait). [G., = eternity.] Eternity, infinity; *spec.* in jocular phr. *in(to) the ewigkeit*, 'into thin air'.
[**1857** C. G. Leland *Hans Breitmann's Party* (1869) 15 Hans Breitmann gife a barty—Vhere ish dat barty now?..All goned afay mit de lager beer—Afay in de ewigkeit!] **1877** E. S. Dallas *Kettner's Book of Table* 75 There is more gluttony now in the world..and none the less because we are quiet over it—and pretend to think a great deal more of the *ewigkeit*. **1900** W. Tuckwell *Reminisc. Oxford* 255 The old monastic Oxford has evaporated into the *Ewigkeit*. **1924** 'L. Brock' *Deductions of Col. Gore* viii, The thumb of one of his hands, which had been rubbing the pad of his second finger thoughtfully, flicked the chances of any other supposition's being the right one into the *ewigkeit*. **1958** *Observer* 18 May 14/2 The sage of Stuttgart [*sc.* Hegel] should have been viewing last week, selectively, from his antifireside in the *Ewigkeit*. **1969** D. Cory *Night Hawk* 61 We leaped together parachute-less out into the *Ewigkeit*.

ex, *prep.* Add: **2.** (Further examples of common commercial phrases.)
1874 *Porcupine* XV. 775/2 It is no unusual thing for brokers and merchants to sell their goods 'ex quay'. **1878** E. C. Maddison *Specul. Stock Exch.* ii. 13 After prices have been quoted 'ex dividend'. **1882** R. Bithell *Counting-house Dict.* 122 *Ex drawing*. Since the prices of stocks and shares quoted in the official list carry with them the right to claim all accruing advantages in respect of those stocks or shares; and since the 'drawings' for the Sinking Fund or amortization are among those advantages, it is usual to state, about the time when drawings take place, whether the prices carry with them the right to the drawing, or whether that right has ceased. This is done by inserting, after the price, the phrase 'ex drawing' or 'cum drawing'. **1885** [see X, 7]. **1893** R. Bithell *Counting-house Dict.* (ed. 2) 118 *Ex all* (x all.). When these words are added to the quotation of the price of any stock, they signify that the coupon or dividend just due on such stock, and any preference claim to new stock, bonus, or other privilege arising from the possession of the stock sold, is retained by the seller. **1903** *Pitman's Business Man's Guide*, *Ex coupon*,..without the interest coupon. **1906** in Poley & Gould *Stock Exch.* (1907) 292 When Securities on which Options are open are quoted 'Ex Rights' an official price will..be fixed for the Rights. **1928** *Daily Mail* 25 July 19/3 Ex rights. Ex all. Ex bonus. Ex return capital. Ex drawing. Ex cash bonus. **1955** *Times* 28 July 12/2 It was on an ex works basis. **1962** H. O. Beecheno *Introd. Bus. Stud.* vi. 59 The terms 'xr' (ex rights) or 'xc' (ex capital)..seen against the price. This means that the special rights extended to existing shareholders do not apply. **1962** *Times* 13 Feb. 4/6 The cost, ex yard, is the same as for a boat built in this country. **1965** J. L. Hanson *Dict. Econ.* 164/2 *Ex-quay*, goods sold on this condition must be taken charge of by the purchaser after they have been landed from the ship. **1967** K. Giles *Death in Diamonds* ix. 170 All the billing is done ex London.

ex (eks), *a. colloq.* [Ex-[1] 3.] Former, quondam; outdated, passé.
1823 [see Ex-[1] 3 a]. **1892** Kipling *Lett. of Travel* (1920) 91 Nothing looks so hopelessly 'ex' as a President 'returned to stores'. **1952** Wodehouse *Barmy in Wonderland* xiii. 126 'I allude to my fiancée. Or, rather, my ex-fiancée.' 'Is she ex?' 'Ex to the last drop. You never saw anything Ex-er.' **1955** *Times* 16 June 12/2 'Is she a member, or an ex-member, of the Communist Party?'—'She is an ex-member'. 'When did she go 'ex'?'

ex (eks), *sb.*[1] *colloq.* Pl. **exes, ex's, exs.** [Ex-[1] 3.] One who formerly occupied the position or office denoted by the context; *spec.* a former husband or wife.
1827 Moore *Cash, Corn & Catholics* 125 'But *don't* you perceive, dear, the Church have found out That you're one of the people call'd *Ex's* at present?' 'Ah, true—you have hit it..(his Lordship replies), And, with tears, I confess—God forgive me the pun!—We X's have proved ourselves *not* to be Y's.' **1915** *Boston Pilot* 13 Mar. 4/4 The various kinds of Ex's [*sc.* ex-Catholics] are allowed to advertise their nasty anti-Catholic talks. **1929** E. Wilson *I thought of Daisy* ii. 96 'Phil was your first husband, was he?' 'Yes: he's my ex,' she said. **1930** Byrne *Golden Goat* ix. 68 Here was an ex-king, one of the first exes. **1944** A. Christie *Towards Zero* 25 Leonard's new wife and his Ex were the best of friends. **1971** *Ladies' Home Jrnl.* Feb. 30/3 His 'ex' also got away with every stick of furniture and household equipment.

ex (eks), *sb.*[2] Colloq. abbrev. of Expense 3 c. Always in pl. forms *exes, ex's, exs.* (Cf. quot. 1894 s.v. X 7.)
1864 Hotten *Slang Dict.* 129 *Exes*, expenses; written thus—EXS. **1874** *Ibid.* 156 Just enough to clear our exes. **1878** *Porcupine* 10 Aug. 295/2 You are..pretty sure to cover your 'ex's'. **1883** *Referee* 18 Mar. 3/3 The piece was ready, but the 'pieces' were not, and without the exes Morton would not allow..the curtain to go up. **1908** A. Bennett *Old Wives' Tale* iv. i. 436 The mater will fork out all my exes. **1915** F. M. Hueffer *Good Soldier* iii. v. 203, I have offered to pay her ex's myself. **1929** J. B. Priestley *Good Companions* iii. i. 482 I'll fix that too—stand all the exes. **1970** K. Giles *Murder Pluperfect* 85 Their ten thousand bucks per year plus exes.

ex (eks), *v.* [f. the pronunciation of the letter X.] *trans.* To cross *out*; to delete with an x.
[*a* **1849** Poe *Tales* (1896) 109, 'I *shell* have to *x* this ere paragrab,' said he... So *x* it he did, unflinchingly and to press it went *x-ed*.] **1935** R. Stout *League of Frightened Men* xiii. 157 He wasn't so good at the typewriter; he had exed out a good deal. **1958** D. Niland *Call Me when Cross turns Over* iv. 126 Ex the lot out and give you the whole blank page to yourself. **1965** K. Giles *Some Beasts no More* vi. 163 'Was he a good typist?' 'Quite fast, sir, but a lot of exing out, not expert.'

ex-[1]. Add: **3. a.** *ex-wife.*
1876 *Gentl. Mag.* XVII. 147 His ex-wife..is dragging out slow years. **1962** *Oxford Mail* 19 Feb. 6/5 His daughter is in his ex-wife's custody.
b. *ex-service(s)* adj., having formerly belonged to one of the fighting services; of or pertaining to former servicemen.
1907 *Daily Chron.* 5 Dec. 6/7 March of the unemployed ex-service men through the West-end of London to Hyde Park. **1910** *Vanity Fair* 13 Jan. 55/1 Employment for ex-Service men is always a pressing question. **1940** J. Betjeman *Old Lights for New Chancels* 25, I have my ex-Service man and Mamie's done a lino-cut. **1941** *Times Weekly* 23 Apr. 12/3 (Advt.), Please send your donation to..Ex-Services Welfare Society. **1945** *Ann. Reg. 1944* 156, Maori ex-servicemen and ex-servicewomen can participate fully in all benefits.

exact, *a.* Add: **7. b.** Calculus. *exact differential*, a linear differential function of two or more variables that is equal to the total differential of some function of those variables. Also *exact* (*differential*) *equation* (see quot. 1959).
1825 D. Lardner *Elem. Treat. Differential & Integral Calculus* ii. xvii. 284 As there are many differentials of two variables which are not exact differentials, so also there are many differential equations which are not the immediate differentials of any primitive equation. **1877** B. Williamson *Differential Calculus* (ed. 3) vi. 143 Condition that $Pdx+Qdy$ shall be a total Differential.—This implies that $Pdx+Qdy$ should be the exact differential of some function of x and y. **1920** H. T. H. Piaggio *Elem. Treat. Differential Equations* ii. 12 The expression $ydx+xdy$ is an exact differential. Thus the equation $ydx+xdy=0$.. is called an exact equation. **1958** *Van Nostrand's Sci. Encycl.* (ed. 3) 511/1 Suppose that an integrating factor, v exists so that $vL(u)dx$ is an exact differential. **1959** G. & R. C. James *Math. Dict.* (ed. 2) 116/1 *Exact differential equation*, a differential equation which is obtained by setting the total differential of some function equal to zero.

exacta (egzæ·ktă). *N. Amer.* [f. Exact *a.*] (See quot. 1964.)
1964 A. Wykes *Gambling* viii. 193 The bettor tries to pick the first and second horse of a race... (With fewer than eight horses, the bettor must predict their order; in New York state, this bet is called an 'exacta'.) **1968** *Globe & Mail* (Toronto) 13 Feb. 29/9 Exacta (7-3) paid $78.80.

exactly, *adv.* Add: **5. e.** *not exactly*: used ironically for 'not at all; by no means'. *colloq.*
1893 Yonge & Coleridge *Strolling Players* xxxii. 292, I wasn't exactly going to send in my checks this time. **1905** A. Lang *Adv. among Books* 222 Though some of our modern novelists think it coarse and degrading, Hawthorne did not think so, and they are not exactly better artists than Hawthorne. **1949** M. Muggeridge *Affairs of Heart* vii. 125 She was not exactly pretty even then. In fact, she was not pretty at all—which is what the use of the word 'exactly' usually signifies. **1964** E. Ambler *Kind of Anger* vi. 169 'And you're not laughing?' 'I'm not exactly bursting my sides, no.' **1968** *Win* 15 Oct. 16/1 Yeah, I'm scared out of my mind. The thought of prison doesn't exactly excite me.

‖ **exalté** (egzalte), *a.* Also fem. **exaltée.** [Fr., pa. pple. of *exalter* to Exalt.] Excited, elated, abnormally happy. Hence as *sb.*, a person in a state of elation; a fanatic.
1831 J. S. Mill *Let.* 20–22 Oct. (1910) I. i. 7 The agricultural people are as determined as the manufacturers. The West is as *exalté* as the North. **1842** *Ibid.* Nov. iii. 123 Capable of almost any degree of *exalté* feeling from poetry. **1922** J. Agate *Red Let. Nights* (1944) 76 It [*sc.* Ibsen's *Hedda Gabler*] contains nothing of awe or terror, nothing to make us fearful lest our *exaltées* should take to pistol-practice in the garden. **1937** *New Statesman* 30 Oct. 688/1 Costals detests cerebral women, and *exaltées* are not his type at all. **1941** Auden *New Year Let.* i. 17 Singing or sighing as they go: Exalté, piano, or in doubt. **1946** M. Dickens *Happy Prisoner* ix. 200 She was *exaltée*, beside herself. **1967** *Punch* 1 Mar. 321/1 Gerold Frank has enough *exaltés* on his hands to set Dickens and Dostoevsky up in business for a dozen lifetimes.

examining, *ppl. a.* (Earlier and later examples.)
a **1608** Donne *Sat.* iv. in *Poems* (1633) 338 One, to whom, the examining Justice sure would cry. **1928** A. Christie *Mystery of Blue Train* xxvii. 218 You will, perhaps, accompany us immediately to the office of the

Examining Magistrate. **1937** *Discovery* June 191/2 Higher School Certificate, Open Scholarships and other examining bodies. **1966** *Listener* 9 June 827/1 Their formal title was bestowed on the examining justices by statute as long ago as 1848. Their role is now mainly ceremonial.

exarch (e·ksāɹk), *a. Bot.* [f. Ex-[2] + Gr. ἀρχή beginning, origin.] Having the protoxylem adjacent to the pericycle.
1891 H. E. F. Garnsey tr. *Solms-Laubach's Fossil Bot.* xi. 257 Had he called..the leaf-strand of Cycadeae mesarch, and that of Isoëtes exarch.., we might..have been spared this misconception. **1900** B. D. Jackson *Gloss. Bot. Terms*, *Exarch*, used of vascular bundles in which the whole primary wood is centripetal, almost the same as perixylic. **1902** *Encycl. Brit.* XXV. 413/1 When the protoxylem strands are situated at the periphery of the stele, abutting on the pericycle, as in all roots, and many of the more primitive Pteridophyte stems, the stele is said to be *exarch*. **1959** Foster & Gifford *Compar. Morphol. Vascular Plants* iii. 46 Xylem differentiation which occurs *centripetally*..is termed exarch.

Exarchist (e·ksaɹkist, eksā·ɹkist). [f. Exarch + -ist.] A supporter of the Exarch of Bulgaria against the Patriarch of Constantinople. Also *attrib.*
1903 *Daily Record & Mail* 10 Apr. 5 The Greeks.. declare that they will kill two exarchists in the towns for every patriarchist killed in the country. **1903** *Westm. Gaz.* 28 Sept. 2/2 Although the Christians are divided among themselves, Patriarchists and Exarchists being at daggers drawn, the Turkish soldiers and Bashi-bazouks are treating all alike. **1907** A. Fortescue *Orthod. Eastern Ch.* iv. x. 321 In 1890..the Sultan gave his firman for the erection of two more Exarchist sees (Ochrida and Skopia). **1927** *Contemp. Rev.* Apr. 734 Hostility between Exarchists and Patriarchists..long poisoned the life of Macedonia. **1937** *Times Lit. Suppl.* 3 Apr. 246/3 An Exarchist Bulgarophone Hellene.

exaspidean (eksæspi·diăn), *a. Zool.* [f. mod. L. *Exaspideæ* (C. J. Sundevall *Methodi Naturalis Avium Disponendarum Tentamen* (1872) I. 57), f. Gr. ἔξω outside + ἀσπίς shield.] Pertaining to passerine birds that have an anterior series of scutella on the outer side of the tarsus; cf. *endaspidean *a.*
1889 in *Cent. Dict.* **1907** R. Ridgway *Birds N. & Middle Amer.* IV. 328 The several modifications of the tarsal envelope in the present group may be described as follows: 1. *Exaspidean*.—The anterior envelope (acrotarsium) extends entirely across the *outer* side of the tarsus and around the posterior side, sometimes meeting the starting point on the posterior portion of the inner side, the two edges usually separated by a narrow strip or groove of smooth or nonscutellate membrane. **1959** Van Tyne & Berger *Fund. Ornith.* ii. 48 Several types of passerine tarsal scutellation have been described... Exaspidean: With anterior, scutellated segment of tarsal sheath extending across the external side of tarsus.

excalate (e·kskălēit), *v.* [f. Ex-[1] 2 + *-calate* of *intercalate*.] *trans.* To remove from a series: opposed to Intercalate.
1900 *Phil. Trans. R. Soc.* B. CXCII. 342 There remains the assumption that vertebræ have been excalated in front of the pelvis. *Ibid.*, Six vertebræ must have been excalated in front of the pelvis.
So **excala·tion**, the omission, absence, or elimination of a part from the middle of a series; *spec.*, in a race of organisms, the absence of any part, such as one of the middle digits or one of the vertebræ.
1898 *Nature* 22 Dec. 171/2 Kükenthal's discovery of excalation of fingers in the Cetacea. **1900** *Phil. Trans. R. Soc.* B. CXCII. 343 Hence the supposition of excalation of vertebræ in front of the girdle [of *Mustelus vulgaris*] leads also to the necessary corollary that a vast amount of both inter- and excalation must go on at another spot.

excavator. 2. a. (Earlier U.S. examples.)
1843 *Niles' Weekly Reg.* 25 Nov. 200/1 With this excavator he is levelling hills. **1848** *Rep. Comm. Patents 1847* (U.S.) 72 Two patents have been granted for excavators.

excelsin (ekse·lsin). *Biochem.* [f. L. *excels-a* (fem. of *excelsus* high), specific epithet of *Bertholletia excelsa*, Latin name of the Brazil nut + -in[1].] A crystalline globulin contained in the Brazil nut.
1896 Osborne & Campbell in *Jrnl. Amer. Chem. Soc.* XVIII. 621 The globulin of the Brazil-nut...This proteid, being evidently different from all others hitherto examined, deserves a distinct name, and we accordingly propose to designate it *Excelsin*. **1907** *Amer. Jrnl. Physiol.* XIX. 53 The greater part of the protein substance of the Brazil-nut (*Bertholletia excelsa*) consists of the globulin excelsin. **1963** E. G. Young in Florkin & Stotz *Comprehensive Biochem.* VII. i. 6 Some well recognized globulins in plants are..excelsin of the Brazil nut.. and amandin of almonds.

excelsior. Add: **1. b.** (Further examples.) Also as *sb.* and *attrib.*
1858 Trollope *Three Clerks* I. xi. 244 His motto might well have been 'excelsior!' if only he could have taught himself to look to heights that were really high. **1863** *Miss Jemima's Swiss Jrnl.* 30 June (1963) ii. 28 Our guide of Excelsior disposition, who would push on more quickly. **1874** G. M. Hopkins *Let.* 29 Apr. (1956) 124 The garden is all heights, terraces, Excelsiors, misty mountain tops..flights of steps. **1929** D. H. Lawrence *Pansies* 143 Up he goes! Up like a bloomin' little Excelsior In his Sunday clothes!
3. For *U.S.* read orig. *U.S.* and add further examples.
1928 Kipling *Limits & Renewals* (1932) 47 Mr. Wilham's fashionable West End pet-shop, where dogs lived in excelsior-floored cubicles. **1934** J. M. Cain *Postman always rings Twice* xv. 175 She lined it [*sc.* a carton] with excelsior, and on top of that put some woollen clothes.
4. *Printing.* (Usu. with capital initial.) The name of a very small size of type (see quots.). Chiefly *U.S.*
1902 T. L. De Vinne *Practice Typogr.* ii. 68 Excelsior, or 3-point (the half of nonpareil), is a body used in America for music, piece-fractions, and borders only. It seems to be the same body as the English 'minikin'. **1936** D. Thomas *Type for Print* II. 83 'News' faces: Ionic..; Excelsior..; Ideal News...Faces such as these, which are designed for the body-matter of newspapers, are rarely cut larger than 9 pt. **1963** Kenneison & Spilman *Dict. Printing* 62 *Excelsior*, the name of a former standard size of type, approximately 4-point.

excentric, *a.* Add: The preferred spelling for senses 2 and 3 of Eccentric *a.* in *Botany*, and in contexts where a writer wishes to avoid the associations of sense 6.
1866 Lindley & Moore *Treas. Bot.* 483/1 *Excentric*, out of the centre. **1895** W. Schlich *Man. Forestry* III. i. v. 73 The trees grow generally more or less excentric. **1896** *Ibid.* V. i. i. 11 The pith of a tree is frequently excentric. **1948** L. Spitzer *Linguistics & Lit. Hist.* i. 19 By then he will know whether he is really permanently installed in the center, or whether he finds himself in an 'excentric' or peripheric position. **1962** F. White *Forest Flora N. Rhodesia* 116 B[*rachystegia*] *wangermeeana*..midrib v. excentric. **1965** F. Sargeson *Memoirs of Peon* vii. 231 The individual isolation which I had hitherto always associated with any ex-centric standpoint.

excepted, *ppl. a.* and *prep.* Add: **A. 2.** *excepted district* (see quot. 1945).
1944 Beattie & Taylor *New Law of Education* 13 The council of an excepted district shall then make, after consulting the Local Education Authority, their own scheme of divisional administration. **1945** *Educational System* (H.M.S.O.) 58 *Excepted District*, borough or urban district excepted from the scheme of divisional administration prepared by a county L.E.A. and granted special status as a divisional executive. **1958** *Times* 19 Dec. 2/3 (Advt.), Lancashire Education Committee. Widnes Excepted District... Headmaster required.

‖ **exceptis excipiendis** (ekse·ptis eksipie·ndis), *advb. phr.* [late Lat., f. L. *exceptis*, *excipiendis*, ablative pl. respectively of pa. pple. and gerundive of *excipĕre* to Except.] Excepting whatever is to be excepted, with proper exceptions. Cf. *mutatis mutandis.
1877 Geo. Eliot *Let.* 30 Jan. (1956) VI. 335 The cheap edition of my books—which, exceptis excipiendis, is a beautiful edition. **1887** *Athenæum* 9 July 50/2 Nothing is more notorious..in the spiritual régime of the Republic than the extension of toleration to all *exceptis excipiendis*. **1932** *S.P.E. Tract* xxxvii. 546 He is naturally impatient of the common delusion that we speak (*exceptis excipiendis*) much as we spell.

excerptor (eksɜ·ɹptǫɹ). Delete †*Obs.* and add later example.
1892 *Graphic* 27 Aug. 254/1 What the monographer does for the classic, that the excerptor does for his works.

excess, *sb.* Add: **6. b.** *excess luggage* (examples); *excess profits* (see quot. 1915[2]); also *attrib.*
1911 A. Bennett *Card* xi. 261 How much did you pay for the excess luggage? **1915** W. Owen *Let.* 1 Aug. (1967) 350 The said friend, having no baggage, will lighten my excess-luggage-charge. **1915** *Act* 5 & 6 *Geo. V* c. 89 § 38 Excess Profits Duty...There shall be charged, levied, and paid on the amount by which the profits arising from any trade or business to which this Part of this Act applies, in any accounting period which ended after the fourth day of August nineteen hundred and fourteen, and before the first day of July nineteen hundred and fifteen, exceeded, by more than two hundred pounds, the pre-war standard of profits as defined for the purposes of this Part of this Act, a duty (in this Act referred to as 'excess profits duty') of an amount equal to fifty per cent. of that excess. **1915** *Chemist & Druggist* LXXXVII. 521/1 As regards the excess profits tax, the special appeal tribunal will be competent [etc.]. **1940** G. Crowther *Paying for War* 22 It is very right and proper that there should be an Excess Profits Tax and that it should have been imposed in the first month of the war. **1955** *Times* 2 May 20/1 At least we were freed from the ill-conceived Excess Profits Levy.

excess, *a.* Restrict †*Obs.* to sense b and add later examples for sense a. Cf. Excess *sb.* 6 b in Dict. and Suppl.
1902 *Encycl. Brit.* XXV. 646/1 There are also a large number of the 'reserve' who are not required to fill up the vacancies in the battalion going out. These become what are known as 'excess numbers'. **1909** Webster s.v., Excess baggage on a railroad. **1920** A. E. Housman *Let.* 15 Aug. (1971) 176 A charge for 'Passengers' Excess Baggage'. **1940** *Chambers's Techn. Dict.* 605/1 Damage caused by an excess voltage, i.e. a voltage above normal. **1958** G. Bellairs *Corpse at Carnival* iv. 48 The charge for excess baggage was disgusting. I shall complain. **1965** *New Statesman* 29 Oct. 632/3 Only the 'excess demand' (extra long order-books) for construction activity has been snipped away. **1970** *Daily Tel.* 19 Oct. 8/6 Grotowsky's productions strip away all the excess baggage of drama—conventional stage, scenery, make-up, props.

exchange, *sb.* Add: **7. b.** (Earlier examples.) Also *attrib.*
1798 *Deb. Congress U.S.* 27 Mar. II. 1318/2 The great number of exchange papers which pass between the printers of newspapers. **1837** in *Canadian Hist. Rev.* (1938) XIX. 15 Nineteen-twentieths of our exchanges are printed on a fairer quality [paper]. **1848** *Yale Lit. Mag.* XIV. 47 (Th.), Our exchanges.
10. c. = *telephone exchange* (Telephone *sb.* 3).
1887 J. M. W. Yerrington *Trial H. K. Goodwin* 11 Mr. Swan..became engaged in the telephone business.. having charge of the Lawrence Exchange. **1888** *Encycl. Brit.* XXIII. 133/2 In a large town it is neither practicable nor desirable to connect each subscriber directly with all the other subscribers, hence a system of 'exchanges' has been adopted. **1938** D. Du Maurier *Rebecca* xxv, Something was buzzing in the telephone. I heard Beatrice shouting, 'Hullo, hullo, don't cut us off, exchange,' and then there was a click, and silence.
12. exchange clearing, a method of bilateral payment through central banks; **exchange control**, governmental control of purchases of foreign currency and of transfer of currency to foreign countries; **exchange force** *Physics*, an assumed force between atomic particles responsible for the exchange of charges or other co-ordinates; **exchange paper** (see sense *7 b above); **exchange rate** = Rate *sb.*[1] 5 b (cf. *rate of exchange* under Exchange *sb.* 4 a in Dict.); **exchange transfusion**, the removal of some of a person's blood with the simultaneous transfusion of other (normal) blood.
1934 P. Einzig *Exchange Control* xiii. 138 It is difficult even to get an adequate answer to the question as to what Exchange Clearing really is. Some people simply class it with exchange restrictions. Technically, they are right, for the act of compelling importers to pay the purchase price to the Central Bank instead of..to their creditors undoubtedly interferes with..exchange operations. **1931** *Times* 5 Dec. 16/3 The Financial Committee and the Economic Committee of the League of Nations have placed the question of foreign exchange control on their agenda. **1932** *Times* 1 Apr. 12/4 Concern is expressed..over the consequences of the exchange control measures and counter-measures in and around Austria. **1934** L. Robbins *Gt. Depression* viii. 179 A régime of this sort must necessitate the most extensive measures of exchange control. **1934** *Nature* 30 June 981/1 The possibility of deducing the exchange forces between neutrons and protons. **1962** *Gloss. Terms Nuclear Sci.* (*B.S.I.*) 44 *Exchange force*, a type of force, acting between two particles, the mathematical expression of which involves an interchange of their coordinates. Such forces are thought to act between nucleons. **1896** R. Barclay *Disturb. Stand. Value* (ed. 2) v. 134 The silver equivalents of the gold prices..would fall in view of the difference in exchange rates. **1928** *Britain's Industrial Future* (*Lib. Ind. Inq.*) v. xxviii. 413 Stable foreign exchange-rates are not enough. **1946** *Jrnl. Laboratory & Clinical Med.* XXXI. 1017 The main obstacle to the successful performance of an exchange transfusion in infants is not the injection but the process of withdrawal, which is hampered by the coagulation of the infant's blood. **1963** *Lancet* 12 Jan. 86/1 For some years the umbilical vein in the newborn has been used for exchange transfusions. **1970** *Glasgow Herald* 19 Nov. 1/7 Because of the deterioration of the sugar in the blood it was decided, after consultation, to carry out an exchange blood transfusion.
b. *spec.* used *attrib.* to denote a reciprocal arrangement whereby two teachers, students, etc., occupy each other's position for a limited period of time; also designating one of the two parties in such an arrangement.
1912 *Nation* 6 June 555 The system of 'exchange professors' between France and the United States. **1934** H. Miller *Tropic of Cancer* 265, I had been offered a trivial post as exchange professor of English, one of those Franco-American amity arrangements. **1936** L. C. Douglas *White Banners* viii. 164 Springer of Cambridge.. had just arrived as an exchange lecturer. **1953** D. Parry *Going Up—Going Down* ii. 51, I might begin with an exchange-fellowship at Yale or Harvard. **1959** *Listener* 26 Feb. 364/1 Nowadays, more emphasis is placed on teaching foreign languages phonetically and on trying to tie lessons in with exchange visits of pupils abroad. **1960** *Encounter* Mar. 77/1 Young exchange students. **1966** J. Caird *Perturbing Spirit* xxiii. 253 She'd got an exchange teaching job for a year to the U.S.A. **1967** *Guardian* 29 May 8/5 British exchange teachers in the United States, and vice versa, are tax exempt.

excise, *sb.* Add: **5. excise law**, a law relating to excise; spec. *U.S.*, the licensing or liquor law.
1765 [in Dict.]. **1792** *Steele Papers* I. 82 Repealing the excise law since that vote was taken Seems to be out of the question. **1903** *N.Y. Even. Post* 3 Dec. 3 His proposed amendment of the Excise Law to allow the saloons to keep open during certain hours on Sunday.

excitation. Add: **5.** *Physics.* **a.** The action or process of causing the emission of a characteristic spectrum of radiation by a substance.

1914 O. W. Richardson *Electron Theory of Matter* xx. 532 When the green mercury line is used for excitation it is found that..the fluorescent lines are made up of fine lines having a structure similar to the..absorption lines covered by the exciting spectrum. **1922** A. D. Udden tr. *Bohr's Theory of Spectra* ii. ii. 34 Experiments on the excitation of spectral lines and production of ionization by electron bombardment. **1963** R. W. Ditchburn *Light* (ed. 2) xvii. 657 (*heading*) Excitation of spectra by slow electrons.
b. The action or process of raising an atom, etc. to a state of higher energy. Freq. *attrib.*
1921 *Chem. Abstr.* XV. 1854 The excitation of an atom by electron impact consists in the removal of an electron from a stationary orbit to one with a higher quantum number. **1923** *Ibid.* XVII. 683 Their excitation potentials were not accurately det[ermine]d. **1931** *Nature* 30 May 838/1 An accurate experimental determination of excitation energy by electron impact in helium. **1951** L. H. Snyder in *Frontiers in Medicine* (*N.Y. Acad. Med.*) 52 Excitation is the raising of an electron in a molecule or atom to a state of higher energy. **1958** W. K. Mansfield *Elem. Nucl. Physics* i. 7 An excited state is one in which energy is given to the atom, either by collisions, or by absorption of electromagnetic radiant energy, permitting the electron to revolve in an empty outer orbit. The energy required to produce this change is known as the *excitation energy*. **1968** M. S. Livingston *Particle Physics* x. 170 One type is the excitation function which is a plot of the yield of products from a particle interaction as a function of bombarding energy.

excite, *v.* Add: **5.** Also *intr.* and *absol.*
1821 P. Egan *Life in London* i. vi. 85 If some of the plates should appear rather *warm*, the purchasers of 'Life in London' may feel assured, that nothing is added to them tending to excite. **1968** *Listener* 22 Feb. 252/3 Last week's legitimate television drama failed to excite.
7. *Physics.* **a.** To induce a condition in (a substance) in which it emits a characteristic spectrum of radiation; to bring about the emission of (a spectrum). **b.** Hence, to render (an atom, etc.) excited (see *excited *ppl. a.* 2 e).
a. 1913 *Proc. R. Soc.* A. LXXXVIII. 24 Elements.. which emit secondary fluorescent X-radiation when excited by a suitable beam of Röntgen rays. **1926** [see *exciting *ppl. a.* b]. **1959** *Chambers's Encycl.* XIII. 70/1 The electric arc is most suitable for exciting the line spectra of elements. **1966** *McGraw-Hill Encycl. Sci. & Technol.* XII. 581/1 Sources of radiation for spectrography are incandescent or electrically excited.
b. 1921 *Chem. Abstr.* XV. 1854 (*heading*) Observations on atoms excited by electron impact. **1934** H. E. White *Introd. Atomic Spectra* vi. 92 If in collision the energy exchange between..an electron and atom is all energy of translation, the atom is not excited and the collision is said to be elastic. **1953** P. Morrison in E. Segrè *Exper. Nucl. Physics* II. vi. xi. 150 One nucleon enters the nucleus..while four more are excited but 'captured', leaving the residual nucleus excited by 66 Mev. **1969** *Times* 11 July 14/2 Radio signals emitted by helium atoms excited by the high temperatures in the interstellar gas clouds.

excited, *ppl. a.* Add: **2. d.** *excited radioactivity*: Rutherford's name for artificial or induced radioactivity, as opposed to that which occurs naturally.
1900 Rutherford in *Phil. Mag.* XLIX. 161 Thorium compounds under certain conditions possess the property of producing temporary radioactivity in all solid substances in their neighbourhood... Attention was first drawn to this phenomenon of what may be termed 'excited radioactivity' by the apparent failure of good insulators..to continue to insulate. **1913** —— *Radio-active Substances* x. 391 The activity thus produced on inactive substances was..known as 'induced' or 'excited' radio-activity.
e. *Physics.* Of an atom or an orbiting electron, or a nucleus, molecule, etc.: not in its ground state; able to lose energy by emitting electromagnetic radiation or a particle; *excited state*, a state of a quantized system having more energy than the ground state.
1921 *Chem. Abstr.* XV. 3931 Recent researches..lead to the establishment of the existence of excited atoms in a metastable state. **1927** T. Verschoyle tr. *Haas's Atomic Theory* 176 His observations..lead to an approximate value of 10^{-8} sec. for the mean duration of the excited state. **1938** R. W. Lawson tr. *Hevesy & Paneth's Man. Radioactivity* (ed. 2) ix. 99 Excited states have only a limited life, and..the energy surplus of the excited nuclei will gradually be emitted in the form of γ-rays. **1953** P. Morrison in E. Segrè *Exper. Nucl. Physics* II. vi. v. 63 The outgoing particles will in general take away only a small part of the excitation energy, leaving behind an excited residual nucleus. **1968** M. S. Livingston *Particle Physics* x. 170 We might expect the mesonic cloud surrounding a nucleon core to be capable of existing in excited states of higher energy..and to 'radiate' mesons.

exciter. Add: **3.** *Electr.* An apparatus to produce excitation; a machine, as a small auxiliary dynamo, used to energize the field magnets of a dynamo; a device to charge the plates of an electrostatic generator; a sparking device to generate electric waves. Also *attrib.*
1885 W. Garnett *Physicists* 318 The idea which occurred to Siemens, Varley, and Wheatstone was to use the whole, or a part, of the current produced by the armature to excite its own electro-magnet, and thus to dispense with the magneto-electric machine which served as the separate exciter. **1902** *Encycl. Brit.* XXVII. 592/2 A small auxiliary continuous-current dynamo, called an exciter. **1930** *Engineering* 24 Jan. 97/2 The generator is of the standard vertical-shaft type with direct connected exciter. **1940** *Chambers's Techn. Dict.* 314/1 *Exciter lamp*, the electric lamp for providing the light to be modulated for recording sound photographically on a sound-track, or the light-source for modulation by the sound-track in the sound-head of a projector. **1959** *Times* 16 Oct. 15/5 The exciters are gear-driven from the generator shaft.

exciting, *ppl. a.* Add: **b.** That excites an electric current, a magnetic field, an atom, etc., a spectrum, or radioactivity.
1884 F. Krohn tr. *Glaser de Cew's Magn.- & Dyn.-Electr. Mach.* 218 When the thickness of the exciting coil is equal to the diameter of the iron core. **1914** [see *excitation 5 a]. **1926** R. W. Lawson tr. *Hevesy & Paneth's Man. Radioactivity* v. 51 The initial velocity of the β-rays excited in matter by γ-rays is independent of the intensity of the γ-rays... On the other hand, the initial velocity is dependent on the hardness of the exciting γ-radiation. **1964** N. N. Hancock *Matrix Analysis of Electr. Machinery* viii. 113 All the exciting functions (i.e. applied voltages and currents). **1967** Condon & Odishaw *Handbk. Physics* (ed. 2) vi. vii. 145/1 Absorption of the exciting radiation..produces only excited Hg^{198} atoms.

excitingness (eksəi·tiŋnės). [f. Exciting *ppl. a.* + -ness.] Exciting character or quality.
a **1910** W. James *Some Probl. Philos.* (1911) iv. 49 The perceptual flux..shows duration, intensity, complexity or simplicity, interestingness, excitingness, pleasantness or their opposites. **1930** A. Huxley *Brief Candles* 7 You must leave the discovery of the excitingness to the artists. **1965** *Listener* 28 Oct. 677/2 A wonderful hybrid of Kafka and Sapper, combining the paranoiac horror of the one with the sheer low-brow excitingness of the other.

exciton (e·ksitǫn, eksəi·tǫn). *Physics.* [f. *excit(ation 5 + *-on.] In a crystalline semi-conductor or insulator, an excited electron and an associated hole which together form a concentration of energy with certain properties characteristic of a particle.
1936 J. Frenkel in *Phys. Zeitschr. der Sowjetunion* IX. 159 Just as the positive hole can be pictured as a collectivized positron, the excited state can be pictured as a kind of particle which we shall call a (collectivized) exciton. **1938** *Nature* 7 May 839/1 An exciton can move a certain distance through the crystal by transferring its energy from atom to atom, before the energy is all dissipated. **1967** Condon & Odishaw *Handbk. Physics* (ed. 2) vi. vii. 155/2 An exciton in a crystal such as KCl is considered to result from the transfer of an electron from Cl^- to K^+, forming an excited and neutral KCl combination... Excitons can migrate through the crystal by exchange of condition with lattice ions. **1968** *New Scientist* 14 Nov. 384/1 Excitons are quasiparticles which arise in semiconductor crystals.
Hence **excito·nic** *a.*
1958 *Physical Rev. Lett.* I. 452/1 For these materials the dissociation energies of the 'excitonic ion' $++-$ and the 'excitonic molecule' $++--$ are an appreciable fraction of the binding energy of the exciton itself. **1966** *New Scientist* 27 Oct. 175/1 Known as the excitonic molecule, this stable entity is built up of two electrons and two 'holes'. **1969** *Sci. Jrnl.* Mar. 15/1 Fenton's arguments do not favour a resistanceless flow of electrons in DNA, but rather that the molecule might be what he calls an excitonic insulator.

excitron (e·ksitrǫn). [f. *exci(tation 5 + *-tron.] A kind of mercury-arc rectifier having a pool cathode and a single anode, and capable of handling large currents; it differs from an ignitron in having a cathode hot spot maintained continuously and a grid to control the output.
1940 O. K. Marti in *Trans. A.I.E.E.* LIX. 928/2 Tests encouraged the Allis-Chalmers Manufacturing Company to build a six anode rectifier, consisting of six single tanks using an ignition-excitation system for the main arc with a continuous cathode spot...This type of rectifier has been called the Excitron rectifier. *Ibid.* 929/1 (*caption*) A 600-volt Excitron with control cabinet. **1950** P. G. Andres *Survey Mod. Electronics* vi. 248 An important improvement in the field of mercury-arc rectification consisted in the development of a single-anode rectifier..equipped with auxiliary electrodes for establishing and controlling the arc current. A group of such individual tanks are assembled to produce a polyphase rectifier termed an *Excitron*. **1958** *Engineering* 14 Mar. 341/3 The excitron uses the conventional continuous excitation (supplied by an auxiliary d.c. source) throughout every cycle.

exclamation. 4. c. For (U.S.) read (orig. *U.S.*) and add further examples. Also *fig.*
1824 [see Point *sb.*[1] 3 a]. **1841** W. Savage *Dict. Art Printing* 668 The Exclamation point, ! *Ibid.* 669 The Interrogation and exclamation points are indeterminate as to their quantity or time. **1896** S. R. Crockett *Cleg Kelly* xiii. 93 The egg..made a long yolky mark of exclamation on the ground. **1926** Fowler *Mod. Eng. Usage* 569/2 Excessive use of exclamation marks is.. one of the things that betray the uneducated or unpractised writer. **1959** *Times* 29 May 18/2 Characterization [of a play] is in exclamation points. **1963** V. Nabokov *Gift* iii. 153 She was slowly mixing a white exclamation mark of sour cream into her borshch.

exclamative, *a.* Add later example. Also as *sb.*
1938 in *Language Sciences* (1968) Oct. 10/1 Such exclamative constructions as 'Oh Charley!' 'What a man!' **1964** E. A. Nida *Toward Sci. Transl.* ix. 209 There are..certain minor types in English, e.g..., exclamatives (*Fire! Police! Ouch!*).

exclave (e·ksklēiv). [f. Ex-[1] + En)clave *sb.*] A portion of territory separated from the country to which it politically belongs and entirely surrounded by alien dominions: seen from the viewpoint of the 'home' country (as opp. to an *enclave*, the same portion of territory as viewed by the surrounding dominions). Also *transf.* and *fig.*
1888 *Encycl. Brit.* XXIII. 331/2 Besides these, the term Thuringia..includes the various 'exclaves' of Prussia, Saxony, Bavaria, and Bohemia which lie embedded among them. **1939** S. Van Valkenburg *Elem. Pol. Geogr.* vii. 111 Small disconnected sections in a foreign territory are enclaves of the surrounding area and exclaves of the country to which they politically belong. **1951** E. W. Zimmermann *World Resources* (ed. 2) 129 An economic exclave may be defined as a splinter of one economy lying inside another economy. **1956** N. Pevsner *Englishness of English Art* 179 New Towns, which are towns and not garden suburbs with odd shopping centres as urban exclaves and a trading estate along the railway.

exclosure (ekslōᵘ·ʒiŭɹ). [f. Ex-[1] + Closure *sb.*, after Enclosure.] An area from which unwanted animals, etc., are excluded.
1920 F. E. Clements *Plant Indicators* 313 Each pasture contains an exclosure termed an isolation transect. **1938** Weaver & Clements *Plant Ecology* (ed. 2) ii. 37 Fenced areas of varying size and shape..the exclosure keeping out one or more species of animals and the enclosure restricting them to a definite area. **1968** A. G. Hart et al. *Small Animals & Birds* (U.S. Forest Service) 2 Exclosure to exclude birds but admit small mammals.

exclusion. Add: **5. exclusion principle** *Physics*, the hypothesis that no two particles of the same kind can exist in states designated by the same quantum numbers, found to be true for the particles known as fermions.
1928 *Chem. Abstr.* XXII. 4351 (*heading*) The Pauli exclusion principle. **1930** *Physical Rev.* XXXV. 580 Pauli's exclusion principle..was stated in the form that no two electrons in an atom can have the same set of quantum numbers. **1948** *Mind* LVII. 539 Broadly speaking, the Exclusion Principle comprises two distinct features concerning the individuality of electrons and their occurrence in atomic systems: (a) electrons are regarded as intrinsically indistinguishable, and (b) in a given atom, no two electrons can occupy the same 'energy level'.

exclusive, *a.* and *sb.* Add: **A.** *adj.* **6. a.** Esp. of journalistic news or other published matter.
1841 *Punch* 17 July p. iii, An experienced nobleman.. who..is frequently in a position to supply exclusive reports. *Ibid.* 28 Aug. 81/2 (*heading*) Further particulars. (Particularly exclusive.) *Ibid.* 13 Nov. 205/1 Our positive tone on the occasion serves to show the exclusive nature of all our intelligence. **1845** *Douglas Jerrold's Shilling Mag.* I. 262/1 What you get from me will be exclusive—from your 'own' correspondent. **1847** *Sporting Life* 18 Sept. 52/2 It paid for extensive and exclusive reports. **1885** [in Dict.]. **1928** D. L. Sayers *Unpleasantness at Bellona Club* xix. 247 Hardy, with Fleet Street's delicate reticence towards the man with an exclusive story, did not press the question. **1966** *New Statesman* 13 May 680/3 The rival's story is no longer exclusive—it can now be decently lifted: i.e. hastily rewritten.
b. Delete *rare* and add further example.
1959 *Times* 12 Jan. 11/3 Anti-Flam: Permanent flame-resistant finish exclusive to Horrockses Crewdson.
c. Of clothing, furniture, etc.: of a pattern or model exclusively belonging to or claimed by a particular establishment or firm.
1901 *Tatler* 18 Oct. p. iv (Advt.), Some very charming artistic novelties in exclusive and original designs are now ready for inspection. **1919** *Ibid.* 2 July (Advt.), Charming and exclusive designs. **1924** *Queen* 2 July p. xiv (Advt.), The absurdly low prices of the most exclusive gowns in London. *Ibid.* 9 July p. ix (Advt.), Practical designs for golfing, country and travelling wear. Exclusive but inexpensive. **1959** [see sense *6 b].
9. Hence high-class, expensive; highbrow.
1942 *London Calling* 15 Jan. 10/1 Claridges, probably the most exclusive hotel in the world. **1942** *John o' London's* 10 Apr. 6/2 The short story of today is roughly one of two kinds—what is called the Magazine Story; and the newer kind which derives from Tchekov and is cultivated by the more exclusive periodicals.
B. *sb.* **4.** An article, news-item, etc., contributed exclusively to, or published exclusively by, a particular newspaper or periodical. Also *transf.*
1901 *Westm. Gaz.* 28 Aug. 2/2 When he goes beyond this to supply his paper's demand for 'exclusives', he has to rely on second-hand versions of events. **1903** *Daily Chron.* 9 Dec. 5/2 [He] had the experience this week of reading his own obituary notice, which the 'Fermanagh Times' had as an 'exclusive'. **1904** F. Lynde *Grafters* xxviii. 367 You'll have all the exciting details for an 'exclusive', to say nothing of the batch of affidavits in the oil scandal. **1917** C. N. Bennett *Kinematography* 122 A good exclusive will have a 'life' of six months at least. **1936** [see *break *v.* 39 b]. **1959** F. Usher *Death in Error* x. 156 He had a good story and

was..hoping to clean up with a world-wide exclusive when it broke. **1967** *Punch* 8 Feb. 191/1 The *Sun*..has had a number of useful exclusives.

exclusivity (eksklūsi·vĭti). [ad. F. *exclusivité*, f. Exclusive *a.* + -ity.] = Exclusiveness, Exclusivism.
1926 *Science* 2 July 5 From the reactions of oxidation we deduce the principle of the exclusivity of oxidation. **1930** *Economist* 20 Dec. 1166/2 Inducing his Party not to accompany their acceptance of his request for their co-operation with any declaration of 'exclusivity' against M. Tardieu or anybody else. **1934** E. Pound *ABC of Reading* 183 The plenum of letters is not bounded by primaeval exclusivity. **1960** 'M. Caine' *S Man* 107 There is nothing society clings to so determinedly as the notion of exclusivity. **1967** 'La Meri' *Sp. Dancing* (ed. 2) xi. 130 He took an exclusivity contract and presently got me an engagement in Mexico City.

ex-co·njugant. Also **ex-conjugate.** [f. Ex-[1] + L. *conjugānt-em* pr. pple. of *conjugāre* (see Conjugate *v.*); *conjugate* f. the vb.] One of a pair of protozoa, bacteria, etc., that have recently been in conjugation. Also *attrib.* or as *adj.*
1902 *Proc. R. Soc. Edin.* XXIII. 420 He succeeded in getting exconjugates to conjugate within that small number of divisions. **1903** *Archiv für Entwicklungsmechanik der Organismen* XV. 174 After separation, the two ex-conjugants have been isolated and their divisions carefully recorded. **1953** R. P. Hall *Protozool.* ii. 95 Exconjugant and non-conjugant lines of *Paramecium caudatum* and *Pleurotricha lanceolata*. **1960** *New Biol.* XXXI. 109 The two paramecia, when they separate after conjugation (they are then known as ex-conjugants) have the same nuclear (genic) material. **1964** W. Hayes *Genetics of Bacteria & their Viruses* xxii. 557 A different approach is to carry out a pedigree analysis of single, exconjugant, recipient bacteria and their progeny with the aid of a micromanipulator.

ex-convict. [Ex-[1].] A former convict. Also *attrib.* Also in shortened slang form *ex-con.*
1867 *Cassell's Mag.* II. 471/2 The clutches of some vulgar ex-convict mistress. **1887** *Chambers's Jrnl.* 1 Oct. 637/2 An appointment is made with the ex-convict. **1907** J. London *Road* (1914) 146 An 'ex-con' who had done five years in Sing Sing. **1955** *Publ. Amer. Dial. Soc.* xxiv. 34 Better than fifty per cent of Pinkertons are ex-cons. *Ibid.*, This estimated percentage of ex-convicts hired by Pinkerton is much too high. **1965** *Austral. Encycl.* III. 38/1 An ex-convict wanting to return had either to work his passage or pay his fare from his savings.

excretal (ekskrī·tăl), *a.* [f. Excret(a + -al.] Of or pertaining to excreta.
1864 *Jrnl. R. Agric. Soc.* XXV. 496 Human excretal matters. **1964** M. Hynes *Med. Bacteriol.* (ed. 8) xxxi. 478 A lower ratio suggests excretal contamination.

excursion, *sb.* **3.** Add after *Obs.*: except in phrase *alar(u)ms and excursions* (see *alarm *sb.* 4 and 11).
5. (Further examples.)
1860 Mrs. Gaskell *Let.* ?26 May (1966) 621 Fare *by excursion* to London, is 12-6, second class...Now second class *excursion* to London wd not be agreeable I think. **1911** *Daily Colonist* (Victoria, B.C.) 2 Apr. 14/6 (Railway Advt.), Low excursion rates to eastern points. **1920** Galsworthy *Foundations* ii, 1 tuk 'ee therr [*sc.* to Margate] by excursion when yu was six months. **1966** J. Betjeman *High & Low* 29 By tramway excursion.. In search of diversion The millworkers come.

excusal. For 'Now *rare*' read: In later use chiefly of local rates.
1898 *Daily News* 4 Jan. 2/2 The justices had been in the habit of signing excusal lists at the rate collector's office and elsewhere, but not in petty sessions, as the law required. **1899** *Punch* 19 Apr. 190/3 When they press him to tax them excusal he begs. **1906** *Edin. Rev.* Apr. 507 The excusal of rates being one of the most abused forms of out-door relief.

excuse, *v.* Add: **6. b.** Also used as a polite form in addressing a stranger, or in interrupting the speech of another. Hence as *sb.* (in full, *excuse-me dance*), a dance in which one may 'cut in' (see *cut *v.* 54 d).
1814 Jane Austen *Mansf. Park* II. vi, Excuse me, your Ladyship must not see your cards. **1831** J. Banim *Smuggler* xii, Excuse me, sir; but though you have been my guest..this is the first time we have met. **1894** *Home Chimes* XVI. 353 'Excuse me, sir.' Mr. Gill stopped, and addressed a passer-by. **1901** R. S. W. Bell *Tales of Greyhouse* 255 Excuse me—you are an old boy? **1902** Kipling *Just So Stories* 66 ''Scuse me,' said the Elephant's Child most politely, 'but have you seen such a thing as a Crocodile in these promiscuous parts?' **1924** W. Deeping *Three Rooms* iii, Excuse me, but would you care to make up a four? **1940** Harrisson & Madge *War begins at Home* ix. 229 Whereas pre-war excuse-me's were almost always quicksteps, now fox-trots and waltzes are frequently included. **1942** A. P. Jephcott *Girls growing Up* v. 121 'Excuse me' and 'Buzz off' dances give the girl an opportunity to go up to any couple and abstract the boy for her partner. **1953** E. Simon *Past Masters* ii. vi. 113 Jonquil..had emerged as mistress of ceremonies, decreeing now an excuse-me-dance, now a Paul Jones. **1965** J. P. Carstairs *Concrete Kimono* xv. 125 'You actually danced with Reba?' 'Is that so crazy?..It was an *Excuse Me*, anyway.'
7. b. *to excuse oneself*: to ask permission or apologize before leaving.
1926 T. E. Lawrence *Seven Pillars* (1935) vii. lxxxix. 496 After an hour he excused himself, because he had just married a Shobek wife. **1964** B. Malamud *Idiots First* 69 The waiter brought drinks and when Mary Lou had finished hers she excused herself, went to the ladies' room.
c. *to be excused*: to be allowed to leave a room, esp. a schoolroom; *spec.* to go to the lavatory. *colloq.* (chiefly in school language).
1954 *Harrap's Stand. Fr. & Eng. Dict.* Suppl. (ed. 2) 55/2 *Sortir*..*Est-ce que je peux sortir?*, May I be excused? **1963** S. Marshall *Exper. in Education* ii. 60 He could not tie up his own shoe, put on his own coat, or even 'be excused' without help. **1968** 'P. Hobson' *Titty's Dead* xiii. 138 'Please may I be excused?' she said.

excusing, *ppl. a.* Add: **2.** = Excepting *prep.* and *conj.* *U.S.* (chiefly *dial.*).
1887 T. N. Page *In Ole Virginia* 199 Torm..said 'he was the grettest gent'man in the country skusin him [*sc.* Torm] and the Colonel'. **1938** M. K. Rawlings *Yearling* iv. 42, I ain't done much today, excusin' fret and worry, and mess with the sausage. *Ibid.* ix. 85 You kin tame arything, son, excusin' the human tongue. **1964** F. O'Rourke *Mule for Marquesa* (1967) i. 12 He had not slept in two days, excusing catnaps during the train ride.

excusive (ėkskiū·siv), *a.* Delete †*Obs. rare*[-1] and add further examples.
1641 *Arminian Nunnery* 4 The Priestlike Prolocutor did not want a premeditated excusive Iustification. **1855** T. T. Lynch *Lett. to Scattered* (1872) ix. 118 Say whatever may be found excusive on behalf of the very bad. **1903** Crichton-Browne *New Lett. Jane W. Carlyle* I. p. lxix, Hurried, excusive scribbles. **1929** E. Linklater *Poet's Pub* iii. 52 Mr. Wesson smiled all round, an excusive, propitiating smile.

ex-dire·ctory, *a.* [f. Ex *prep.* + Directory *sb.*] Denoting a telephone-number that is not listed in the telephone-directory; also of a person who has such a telephone-number.
1936 *P.O. Teleph. Sales Bull.* Mar. 41/2 It was possible..to allow the separate exchange lines for the house exchange system to be ex-directory...This..provides.. an isolated example where ex-directory facilities may be used as a selling argument. **1963** 'R. East' *Pin Men* ii. 32 Virginia's home number was not ex-directory. **1964** C. E. Maine *Never let Up* xv. 151 'He's ex-directory.' 'But *you* must know his number.' **1966** 'W. Haggard' *Power House* xi. 116 He'd been woken..by a call on an ex-directory number.

exec, colloq. abbrev. of Executive *sb.* (in Dict. and Suppl.).
1896 G. B. Shaw *Let.* 20 Mar. (1965) 614 The Execs will be safe, I should think, to sanction the expenditure. *Ibid.*, We could appeal to the exec. to grant us the price of the report. **1933** B. J. Chipman *Hey Rube!* 194/1 *Execs*, executives. **1958** 'P. Bryant' *Two Hours to Doom* 16 The exec..did not enjoy staff work and administration. **1966** *Punch* 30 Mar. 464/1 The execs must have fallen into transports of delight.

executionary (eksėkiū·ʃənări), *a.* [f. Execution + -ary[1].] Of or pertaining to execution.
1909 *New Schaff-Herzog Encycl. Relig. Knowl.* III. 313/1 The executionary order caused the *titulus* to be fastened at the upper end of the cross. **1920** *Chambers's Jrnl.* Dec. 849/2 The Prince..was considerably relieved to find that he was not on the visitor's executionary list. **1928** *Observer* 8 Jan. 11/5 Dr. Guillotin, who persuaded the Assembly to accept the principle of executionary equality.

executive, *a.* and *sb.* Add: **B.** *sb.* **3.** A person holding an executive position in a business organization; a person skilled in executive or administrative work; a business man. Also *attrib.* orig. *U.S.*
1902 G. H. Lorimer *Lett. Merchant* 22 They will never climb over the railing that separates the clerks from the executives. **1927** Huntington & Whitney *Builders Amer.* 15 All over the United States business men deplore the scarcity of good executives. **1930** B. Colby *Close of Wilson's Admin.* 17 Decisions were.. reached..with a rapid ease that marked the born executive. **1930** *Monotype Recorder* Special No. Mar.–Apr. 5 Chosen as the chief executive of one of our most important industries. **1936** J. B. Priestley *They Walk in City* i. 6 Mr. Welkinghurst, who knew his business, spotted it at once. So did all the executives, of course. **1951** M. McLuhan *Mech. Bride* 35/1 Do you have a personality? Our executive clinic will get rid of it for you. **1958** *Spectator* 7 Feb. 173/1 The type of men who used to go into union politics..are now becoming technicians, administrators and executives. **1958** *Manch. Guardian* 20 Mar. 10/3 In New York..over two thousand business banquets were held last year, involving three to four million executive man-hours. **1959** *Times Lit. Suppl.* 6 Feb. 74/3 He is away from the sophisticated anxieties of his executive suite, and back in Aunt Mabel's brownstone house on 62nd Street. **1962** *Guardian* 5 Oct. 15/2 The comforts of executive-class prosperity. **1968** 'R. Simons' *Death on Display* xv. 77 Most of the cars parked in the driveways were large, new and expensive, and it was the kind of area that estate agents describe gladly as 'executive class'. **1971** *Times* 8 Jan. 8/2 An executive toy can mean—and has meant—anything from worry beads to a laughing box.

exemplarism (egze·mplăriz'm). [f. Exemplar *sb.* + -ism.] **1.** The doctrine that divine ideas are the source of finite realities.
1893 in *Funk's Stand. Dict.* **1908** *Cath. Encycl.* IV. 271 Exemplarism, or the doctrine of archetypic ideas and the supposed knowledge of things in these divine ideas. **1958** G. Leff *Medieval Thought* vii. 200 Now the whole of St Bonaventure's outlook is governed by what has been called exemplarism. That is to say, all creation was a sign of God..a creature was to be regarded only insofar as it provided a trace of God.
2. The doctrine that Christ became incarnate as an exemplar to mankind.
1926 K. E. Kirk in E. G. Selwyn *Ess. Cath. & Crit.* viii. 275 One final criticism of exemplarism may be mentioned. The theory fails to show any grounds for belief in the *necessity* of the death of Christ.

exemplarist (egze·mplărist). [f. prec. + -ist.] One who believes in or advocates exemplarism. Also as *adj.*
1926 K. E. Kirk in E. G. Selwyn *Ess. Cath. & Crit.* viii. 256 If we must have a name for theories of this type, let us..call them 'exemplarist'—because their emphasis is upon the moral value of Christ's example. *Ibid.* 275 A somewhat confused criticism of the appeal to gratitude in Exemplarist theories, based on the fact that *some* Exemplarists had forgotten to emphasise the objective value of an example of loyalty. **1936** *Times Lit. Suppl.* 22 Feb. 166/3 But his book shows him to be very far from the 'exemplarist' type of thought about the Atonement which dominated the nineteenth century. **1958** *Times* 19 Apr. 8/4 To help the 'exemplarists' to enter into the necessity of Christ being a sacrifice as well as an example.

‖ **exempli gratia** (egze·mpli grēⁱ·ʃiă), *phr.* Also **exempli causa.** [L.] For the sake of example; for instance. Freq. abbreviated to *e.g.* (occas. *ex. gr.*).
For further examples of *e.g.* see *E III.
1569 Abp. Parker *Let.* 1 July (1853) 352 *Exempli causa*, I urge the Injunction upon all ministers. **1654** Gayton *Pleas. Notes* 102 The intrinsecall radicall moysture must be supplied, recruited, and replenished with the extrinsecall liquids, that is, *exempli gratia*, in the morning with a sphericall Tost in a pot of Ale of good capacity. **1765** Gibbon *Let.* 31 Oct. in *Misc. Wks.* (1796) I. 437 *Exempli gratiâ*, your letter is dated Vienna, October 12th, 1765; it made its appearance at Beriton,.. October 29th. **1798** Coleridge *Lett.* (1884) 61 There are some ludicrous blunders—exemp. gratiâ—this erudite philosopher mistakes Moses's autograph for the publication of the law. **1802** *Ibid.* 85 In English now, *exem. causâ*, we might say, [etc.]. **1868** G. V. Cox *Recoll. Oxford* iv. 45 His subtle and ingenious narration of Aesop's Fable of the Dog and the Wolf, and his running comments on it, Ex. gr. 'Just see how fat and sleek I am,' said the Dog. **1892** J. Wright *Primer Gothic Lang.* viii. 42 In the former case they are said to be voiced (e.g. the mediæ), and in the latter voiceless (e.g. the tenues). **1933** *Nature* 7 Oct. 533/1 The concept and the word are freely scattered through Wheatstone's Bakerian lecture of 1843, where we find, *ex gr.*, 'in two circuits..when the same resistance is introduced, the strength of the two currents may be weakened'. **1970** J. McN. Dodgson *Place-Names of Cheshire* p. xlv, Sources whose dates cannot be fixed to a particular year are dated by century (e.g. 12, 13 etc.), by regnal year (e.g. E1, H2 etc.) or a range of years (e.g. 1189–1217).

‖ **exemplum** (egze·mplǝm). Pl. **exempla.** [L., see Example *sb.*] An example; *spec.* a moralizing tale or parable; an illustrative story.
1890 T. F. Crane *J. de Vitry's Exempla* p. lxxx, The *exempla* which we have thus far examined have been illustrative stories intended for insertion in sermons. The moral lesson to be drawn from them was left..to.. the preacher. In other words, the *exemplum* was a story which had no independent value, and..was..to be expanded at the preacher's will. **1910** *Encycl. Brit.* IX. 610/2 Homilies..abounding in *exempla* or illustrative stories. **1954** B. White in *Year's Work Eng. Studies 1952* 63 Considered as a piece of 'practical advice for comfortable living in a world not operated altogether under reason and justice', an exemplum told to illustrate and enforce a thesis, the tale is artistically and dramatically satisfying. **1956** *Essays & Studies* New Ser. IX. 57 Shakespeare, Blake and Dante: all three offering.. *exempla* of the energy to be derived from sexual experience. **1968** *Neophilologus* 171 In order to make the sermon more convincing, the preacher..here introduces an *exemplum*—that of the thief on the cross who, by sincere penitence, entered through the Gates of Paradise with his Lord.

exemption. Add: **5.** *attrib.* (sense 2).
1898 *Daily News* 5 July 2/5 Under the exemption clauses of the Acts. **1902** *Westm. Gaz.* 16 Aug. 2/3 Exemption certificates. **1909** *Englishwoman* Apr. 286 The Home Secretary's withdrawal of the Exemption Order.

exercise, *sb.* Add: **8. g.** *pl.* Formal acts or ceremonies on some special occasion. *U.S.*
1841 J. S. Buckingham *America* II. 47 The First Reformed Dutch Church, where the 'exercises', as all proceedings of public meetings are here called, were to take place. **1863** 'E. Kirke' *Southern Friends* xxvi. 262 A few minutes before the 'exercises' [*sc.* an auction sale] commenced, the negroes were marched upon the lawn. **1891** in M. A. Jackson *Memoirs* (1895) 640 The exercises [*sc.* unveiling of a monument] were held upon the campus of Washington and Lee University. **1911** *Springfield Weekly Republ.* 16 Nov. 9 The chief feature of the inauguration exercises [of the Lord Mayor of London]

was a pageant and tableaux. 1926 *Publishers' Weekly* 1 May 1474 Dedicatory exercises.

h. *the object of the exercise*: the (whole) point or purpose of (something stated in the context).

1958 *Spectator* 24 Jan. 102/1 Its report is a living document which..will gradually influence public opinion. That was the object of the exercise. 1959 *Times* 20 June 4/4 The main object of the exercise is to merge Theatre Tickets and Messengers' turnover with ours. 1970 J. Sangster *Touchfeather* i. 2 If we knew what it looked like..we would simply bribe a member of the crew..to take it from his cabin. But that is not the object of the exercise.

11. (sense 7) *exercise-ground, -time, -yard; -loving* adj.; **exercise bone** (see quot.); **exercise book** (examples); also, a book containing set exercises.

1890 Billings *Med. Dict.*, *Exercise bone*, bony deposit produced in or over a tendon by continued and repeated use or pressure. 1838 Mrs. Gaskell *Let.* 17 July (1966) 17 We are 'here today, & gone tomorrow', as the fat scullion maid said in some extract in Holland's Exercise book. 1813 M. Edgeworth *Let.* 1 May (1971) 33 Saw Edward 6th's famous little manuscript exercise book. 1873 *Young Englishwoman* Apr. 207/3 Can the Editor mention a good musical exercise book? I believe the German books are the best. 1932 L. Golding *Magnolia St.* ii. xiv. 461 The little shiny blue-backed exercise-book he used as a diary. 1788 W. Dyott *Diary* Sept. (1907) I. 54 His Royal Highness went on shore to see a most extensive display of fireworks on the exercise ground. 1906 J. Joyce *Let.* 4 Oct. (1966) II. 171 The interspaces being used as military exercise-grounds. 1930 Blunden *Leigh Hunt* viii. 103 It became an exercise-ground in which, edition by edition, its author tried fresh evolutions. 1897 *Daily News* 30 Aug. 5/7 Exercise-loving England. 1897 P. Warung *Tales Old Régime* 209 To be deprived of their exercise-time added fresh pangs to the punishment of the virtuous. 1901 *Wide World Mag.* VIII. 170/1 He was to put up a solid palisade round the outer edge of my exercise-yard. 1966 *Listener* 18 Aug. 237/1 There was no bath..and there was no big exercise yard as there had been in Pretoria.

exercise, *v.* 6. d. Delete †*Obs.* and add later examples.

1877 H. James *Amer.* i, If it was necessary to walk to a remote spot, he walked, but he had never known himself to exercise. 1897 *Daily News* 1 Dec. 5/3 The other prisoners exercised as usual yesterday.

exes: see *ex *sb.*¹ and *sb.*²

Exeter. Add: **Exeter Hall**, a building in the Strand, London, erected in 1830–31, used chiefly for religious and philanthropic assemblies till 1907; often used *allusively* to denote a type of evangelicalism.

1835 Moore *Fudge Fam. England* i. 78 'Tis rumour'd our Manager means to bespeak The Church tumblers from Exeter Hall for next week. 1849 Carlyle *Latter-d. Pamph., Nigger Q.* (1858) 3 Exeter Hall, my philanthropic friends, has had its way in this matter. *Ibid.*, A state of matters..which has earned us not only the praise of Exeter Hall,..but lasting favour (it is hoped) from the Heavenly Powers themselves. *Ibid.* 5 Exeter-Hall Philanthropy. *Ibid.* 15 We must be patient, and let the Exeter-Hallery and other tragic Tomfoolery rave itself out. 1860 W. H. Russell *Diary India* I. 356 Our Christian character in Europe, our Christian zeal in Exeter Hall, will not atone for usurpation and annexation in Hindostan. 1888 *Contemp. Rev.* Apr. 531 Thither [sc. Africa] Manchester turns her longing eyes, thither the heart of Exeter Hall is yearning. 1907 *Daily Chron.* 4 Mar. 6/6 The vanishing of Exeter Hall from the world of Evangelicalism.

exflagellation (e:ksflædʒĕlēi·ʃən). *Zool.* [f. Ex-¹ + *flagellation 2.] The formation or shedding of flagella; the development of microgametes resembling flagella. So **exfla·gellate** *v. intr.*

1909 in Webster. 1912 E. A. Minchin *Introd. Study Protozoa* xv. 357 The male sporonts..form filamentous male gametes resembling flagella, and are consequently said to 'exflagellate'. *Ibid.* 364 The 'exflagellation', or formation of microgametes, which takes place, under normal circumstances, in the stomach of the mosquito, can be seen also in blood freshly drawn and examined on a slide, if ripe sporonts are present. 1926 C. M. Wenyon *Protozool.* I. i. 85 The microgametocyte gives rise in the course of a few minutes to six or ten microgametes by a violent process known as exflagellation. 1964 M. Hynes *Med. Bacteriol.* (ed. 8) xxviii. 441 From the male cell, flagella-like gametes develop, a process known as ex-flagellation.

ex. gr.: see *exempli gratia.

‖ **ex gratia** (eks grēi·ʃiă), *adj.* and *advb. phr.* [L.] Of or by favour; done or given as a favour and not under compulsion; *spec.* implying the absence of any legal obligation.

1769 W. Blackstone *Commentaries Laws Eng.* IV. xxx. 385 Writs of error to reverse attainders in capital cases are only allowed *ex gratia*; and not without express warrant under the king's sign manual. 1920 *Act* 10 & 11 *Geo. V* c. 32 Sch. B. xvi, Ex gratia grants in respect of losses and injuries. 1928 *Daily Chron.* 9 Aug. 5/6 An ex-gratia payment in consequence of his wrongful..imprisonment for nearly 19 years. 1955 *Times* 30 May 11/5 In addition to our existing ex-gratia pensioners there is a sizable body of employees who by reason of their age are not eligible to join the scheme.

exhaust, *sb.* Add: 1. a. Also, the expulsion of combustion products from the cylinder of an internal-combustion engine, the products so expelled, or the valve or pipe by which they escape.

1896 B. Donkin *Text-bk. Gas, Oil, & Air Engines* (ed. 2) ii. xxvii. 391 The engine, oil tanks, and exhaust are arranged in the same way as in the Capitaine launches. 1902 *Daily Chron.* 4 Sept. 7/4 The exhausts crackling like quick firers. 1904 A. B. F. Young *Compl. Motorist* iii. 53 The problem of silencing has been to reduce the sound of the exhaust to a minimum and to retain the maximum of power given off by the engine. 1906 *Macm. Mag.* Nov. 60 Offensive exhaust is the Committee's polite name for what..we must dignify with its proper title, an intolerable stink. 1912 *Motor Man.* (ed. 14) 235 There is no mistaking the somewhat pungent odour of an over-rich mixture exhaust, whilst an excessively over-rich mixture produces a 'black' exhaust. 1915 D. O. Barnett *Lett.* 176 Suddenly he blew a cloud of smoke out of his exhaust, and up went his tail, and he began going down in spirals. 1957 *Encycl. Brit.* VII. 348/2 The four-stroke cycle engine operates as follows: (1) intake..; (2) compression..; (3) power..; (4) exhaust. 1961 L. Mumford *City in History* x. 296 The reek of gasoline exhaust.

3. *exhaust-box, -cylinder, -gas, -lap, -manifold, -pipe* (examples), *-stroke, -valve* (further examples).

1903 *Motoring Ann.* 141 Few things are more annoying than an intermittent, loud report from the exhaust box of a petrol motor. 1912 *Motor Man.* (ed. 14) 235 A car that is addicted to exhaust-box explosions. 1892 *Daily News* 4 Oct. 3/3 The proceeds of combustion pass from an exhaust cylinder in form of a gas that cannot be seen. 1902 A. C. Harmsworth et al. *Motors* vii. 118 The exhaust pipe from the engine which conducts off the exhaust gases after they have done their work in the cylinder. 1904 Goodchild & Tweney *Technol. & Sci. Dict.* 212/1 In gas and oil engines the exhaust gases consist of the products of combustion, together with any unburnt gases remaining after the explosion. a 1930 D. H. Lawrence *Last Poems* (1932) 32 The weather in town is always benzine, or else petrol fumes Lubricating oil, exhaust gas. 1888 Lockwood *Dict. Mech. Engin.* 130 *Exhaust lap*, the reduction or narrowing of the inner faces of a slide valve to less than that distance which would correspond with a length measured between the inner edges of the steam ports, by which difference the ports are closed earlier than they would be if their edges coincided exactly with those of the arch of the valve. 1919 *Gloss. Aeronaut. Terms* (R. Aeronaut. Soc.) 48 The exhaust pipe extends from the exhaust manifold to the silencer. 1889 *Cent. Dict.*, *Exhaust-pipe*, in a steam-engine, the pipe that conveys waste steam from the cylinder to the condenser, or through which it escapes to the atmosphere. 1902 Exhaust pipe [see *exhaust-gas* above]. 1894 B. Donkin *Text-bk. Gas, Oil, & Air Engines* i. i. 16 There are always two strokes, the forward or motor stroke, and the return or exhaust stroke. 1913 W. E. Dommett *Motor Car Mech.* 8 On the next up stroke, and the exhaust valve being open, the burnt gases are forced out of the cylinder, the piston performing the exhaust stroke. 1899 *Motor-Car World* I. 54/2 To the casual observer the greatest failing of the Bollée is the noise, but to the owner the exhaust valve will probably be the most troublesome part. 1919 H. Shaw *Text-bk. Aeronaut.* x. 118 When the piston is nearing its lowest position, the exhaust valve is opened by a cam.

exhibit, *sb.* Add: 1. c. *exhibit A*: the first document or object produced in court as evidence; hence *transf.* and *fig.*, an object or person considered as a piece of evidence, esp. the most important evidence.

[1902 J. M. Lely *Wharton's Law-lex.* (ed. 10) 303 *Exhibit*, a document..referred to in, but not annexed to, an affidavit... Usually the deponent merely refers to it in the affidavit as 'the exhibit hereto annexed marked A', or as the case may be.] 1906 E. Dyson *Fact'ry 'Ands* iv. 49 John was..so limp that the policeman had to hold him up, like exhibit A. 1932 W. Faulkner *Light in August* (1933) xv. 325 To preach to them [sc. Negroes] humility before all skins lighter than theirs, preaching the superiority of the white race, himself his own exhibit A. 1948 C. Day Lewis *Otterbury Incident* iv. 47 The button—let us call it Exhibit A—was found by me. 1963 K. Eble *F. Scott Fitzgerald* x. 158 His short stories will be the supporting evidence—*The Great Gatsby* is Exhibit A—of his lasting claim to attention. 1970 J. Porter *Rather Common Sort of Crime* xiv. 163 Pimp..tossed a crumpled looking paper bag over to the Hon. Con. 'Exhibit A,' said Jack the John.

exhibition. 5. e. Substitute for def.: to behave in such an ostentatious or conspicuous manner as to appear contemptible or laughable.

1853 Dickens *Child's Hist. Eng.* (1854) III. xxxii. 162 His Sowship was making such an exhibition of himself..as is not often seen in any sty. 1914 G. B. Shaw *Misalliance* 83, I know Ive made a silly exhibition of myself here. 1935 I. Compton-Burnett *House & its Head* i. 10 Am I to believe it was you, who chose to make an exhibition of yourself with a maid-servant behind the house?

7. A public examination or display of the attainments of students. *U.S.*

1786 P. Freneau *Poems* 352 Lines, intended for Mr. Peele's Exhibition, Philadelphia, May 10, 1784. 1829 *Regul. Boston School Comm.* 11 There shall be two general visitations of the schools annually, for the purpose of exhibition. *Ibid.*, These exhibitions. 1887 J. Kirkland *Zury* 186 It was customary to have on February 22nd a school 'Exhibition' with speeches, dialogues, and so forth. 1899 E. E. Hale *Lowell* 29 They were within the number of twenty-four students [at Harvard] who had had honors at the several exhibitions. 1943 A. G. Powell *I can go Home Again* 51 At the close of the school next spring, we had the 'Exhibition', with charades, dialogues, recitations, and so forth.

exhibitionism (eksibi·ʃəniz'm). [f. Exhibition + -ism.] 1. Indecent exposure of the sexual organs, esp. as a manifestation of sexual perversion. Also *fig.* and *gen.*, a tendency towards display; indulgence in extravagant behaviour.

1893 C. G. Chaddock tr. *Krafft-Ebing's Psychopathia Sexualis* v. 394 In his dreams he exhibited himself to young, voluptuous women... In this hereditary and degenerate impulsive exhibitionism, it is interesting to note how the perverse sexual impulse is awakened from its latency by..alcohol. 1908 *Practitioner* Jan. 10 A hyperaesthesia sexualis, with an insufficient power to satisfy his impulses, will lead to strained marital relations, immorality, and promiscuous intercourse, exhibitionism, and even sexual perversions. 1919 M. K. Bradby *Psycho-analysis* 136 An expression of sexuality on a level with childish 'exhibitionism'. 1921 *19th Cent.* Mar. 476 Its [sc. love's] regression and narcissism, exhibitionism, masochism, [etc.]. 1924 C. Gray *Surv. Contemp. Music* 15 Others for whom art is an opportunity for spiritual exhibitionism and literature a confessional. 1926 T. Beer *Mauve Decade* v. 180 A species of exhibitionism that makes one grin. 1928 *Sunday Dispatch* 5 Aug. 11/2 Heroin is a drug that promotes exhibitionism. The user..is prompted to false bravery, extravagant behaviour, showiness. 1964 A. Storr *Sexual Deviation* ix. 91 Of these deviations, by far the commonest is exhibitionism, or, as the law terms it, indecent exposure.

2. The mania for exhibitions (sense 6). *rare.*

1920 W. R. Lethaby *Form in Civiliz.* (1922) 182 This exhibitionism is one of the many symptoms of 'the sickness of acquisitive society'.

exhibitionist. Restrict *rare*⁻¹ to sense in Dict. and add: 2. One who indulges in exhibitionism. Also *attrib.*

1893 C. G. Chaddock tr. *Krafft-Ebing's Psychopathia Sexualis* v. 383 There are various categories of exhibitionists. 1928 *Daily Express* 15 Aug. 15 Two 'exhibitionist' bathers at Deauville. 1949 M. Mead *Male & Female* vii. 156 Some societies do not..forbid his playing sexual games, exhibitionist games with other boys. 1958 *Times* 18 Feb. 5/2 A 'hard core of exhibitionists' wanting to get on television. 1963 A. Heron *Towards Quaker View of Sex* 67 Male exhibitionists in particular are seeking through their sexually aggressive act to assert their impoverished sexuality. 1965 F. Sargeson *Memoirs of Peon* vii. 206 Apart from my exhibitionist drive, I had some warrant for my desire.

exhibitionistic (eksibi:ʃəni·stik), *a.* [f. prec. + -ic.] Pertaining or given to exhibitionism. So **exhibitioni·stically** *adv.*

1928 A. Huxley *Let.* 23 May (1969) 297 I'd do a little lying out with my belly exhibitionistically exposed to the natural sun. 1930 *Psychol. Abstr.* Dec. 539/1 Descriptions of exhibitionistic acts which manifest atypical phases are here presented. 1931 *Nation* (N.Y.) 23 Dec. 702 Harris, the exhibitionistic satyr, finally worried the prudish Shaw into giving him a statement concerning the latter's sex life. 1932 S. Zuckerman *Soc. Life Monkeys & Apes* xix. 310 It is difficult to say what is the significance of such behaviour. There is no obvious justification for the view expressed by Tinklepaugh that it is exhibitionistic. 1937 *Scrutiny* Mar. 439 The 'good student', whose fountain-pen slides and skates exhibitionistically over the page. 1949 M. Mead *Male & Female* vii. 156 The lusty, exhibitionistic self-confidence of the five-year-old. *Ibid.* xviii. 369 He is proudly, exhibitionistically sure of his masculinity. 1970 *Daily Tel.* 15 Jan. 16/2 Technically testing yet not unduly exhibitionistic, both cadenzas avoided..excessive length and complexity.

exhorter. 2. (Later U.S. examples.)

1845 A. Wiley in *Indiana Mag. Hist.* (1927) XXIII. 202 He joined the church..and was made class-leader, then exhorter, and then local preacher. 1901 W. N. Harben *Westerfell* 253, I got hugged by a whole string of exhorters.

‖ **ex hypothesi** (eks haipǫ·þisai), *advb. phr.* [mod.L.] From or according to the hypothesis; as a result of the assumptions made; supposedly, hypothetically.

1603 C. Heydon *Jud. Astrol.* 211 The Spring and neape tides, the foure seasons of the yere, with infinite like, they are phisically necessarie, they are ineuitable *ex hypothesi*. 1694 J. Howe *Enq. conc. Trinity* 49 Yet that necessity not being absolute, but *ex hypothesi* only. a 1734 North *Exam.* (1740) i. iii. 215 The Jury, *ex Hypothesi*..inclined on the Plot Side. 1829 *Edin. Rev.* Oct. 219 The universe, *ex hypothesi*, is only an effect. 1887 *Athenæum* 1 Oct. 430/2 A people to whom it was *ex hypothesi* unknown. 1941 W. Temple *Citizen & Churchman* ii. 33 Inasmuch as *ex hypothesi* there is no God for the State to serve, the State becomes an absolute. 1945 R. Knox *God & Atom* vi. 82 To give priority to Something which, *ex hypothesi*, claims absolute priority.

Exilarch (e·ksilāɹk). *Jewish Hist.* [f. Exile *sb.*² + Gr. ἀρχός ruler, translating Aramaic *rēsh galuta* 'chief of the captives'.] One of a line of Jewish princes or rulers in Babylon who exercised authority over, and received tribute from, Jews in all countries from about the third century to the tenth century A.D. Hence **Exila·rchate**, the period during which there were exilarchs; also, the people over whom the exilarch had power.

1893 *Q. Rev.* Jan. 111 Under a succession of Exilarchs,..they found themselves in another Holy Land.

existence. Add: 5. *attrib.*, as **existence-proposition**, a proposition asserting existence, an existential proposition; **existence-theorem** (see quot. 1903).

1937 *Mind* XLVI. 53 This is certainly intended to be or to contain an empirical existence-proposition. 1939 *Ibid.* XLVIII. 144 Plato was at this time deeply concerned with the logic of existence-propositions. 1944 M. Weitz in P. A. Schilpp *Philos. B. Russell* 86 An existence proposition is the traditional 'I' or 'O' proposition. For Russell it is a proposition which asserts the truth of at least one value of a propositional function; e.g., 'Some men are brutal'. 1900 J. Royce *World & Indiv.* I. v. 213 Modern Analysis, and the Theory of Functions, contain very many propositions of the class that are sometimes called 'Existence-Theorems'. 1903 B. Russell *Princ. Math.* lix. 497 The existence-theorems of mathematics—*i.e.* the proofs that the various classes defined are not null—are almost all obtained from Arithmetic.

existential, *a.* Add: 2. (Earlier example.)

1819 Coleridge *Philos. Lect.* (1949) ix. 276 This necessarily led men..to doubt whether a logical truth was necessarily an existencial one, i.e. whether because a thing was logically consistent it must be necessarily existent.

b. *existential import*: significance concerning the existence of something, usually of items denoted by the subject-term of a proposition; the implication that something exists.

1887 J. N. Keynes *Formal Logic* (ed. 2) ii. viii. 138. § 101 (*title*) Formal logic and the existential import of propositions. 1898 A. N. Whitehead *Treat. Univ. Algebra* p. vii, A conventional mathematical definition has no existential import. 1946 *Mind* LV. 109 The *Ethics* [of Spinoza] is a prolonged effort to deduce truths of existential import out of a combination of definitions. 1962 W. & M. Kneale *Devel. Logic* 412 Those Aristotelian inferences which depend on existential import.

3. *Philos.* Concerned with or relating to existence (freq. as distinct from 'essence'), esp. human existence as seen from the point of view of existentialism; *existential philosophy* = *existentialism.

[1846 Kierkegaard *Afsluttende Uvidenskabelig Efterskrift* in *Wks.* (1902) VII. ii. iv. 374 Den existentielle Pathos er Handling, eller Existentsens Omdannelse.] a 1937 M. Geiger in *Philos. & Phenomenol. Res.* (1943) III. 256 'Existential Philosophy' is a collective term for many problems, many methods of thinking, many points of view. *Ibid.* 257 The distinguishing feature of all existential philosophy is the fact that its basic category is *existential significance*. 1939 V. A. Demant *Religious Prospect* vii. 186 The 'existential' theology' declares that human life can gain religious meaning only by a bare act of faith which has no relation to and is even falsified by attempts to give the object of faith, God, any content in man's consciousness. 1940 H. Read *Annals of Innocence* ii. vii. 182 That philosophy is an existential one: it is the expression of all my faculties, of the whole consciousness of a living organism. 1941 *Philosophy* XVI. 267 Its influence permeates the whole school both of existential philosophy and dialectical theology. 1941 Swenson & Lowrie tr. *Kierkegaard's Concl. Unscientific Postscript* ii. ii. 74 The existing subjective thinker is in his existential relation to the truth as negative as he is positive. *Ibid.* iv. 386 Existential pathos is action, the reconstruction of the individual's mode of existence. 1944 [see *existentialism]. 1956 C. Wilson *Outsider* ix. 273 Kierkegaard's attitude is so Existential that his Christianity is a religion that regards God as the intermediary between himself and his fellow human beings, and cannot even accept their existence without first accepting the existence of God. 1962 R. G. Olson *Introd. Existentialism* iv. 129 Existential psychoanalysis, the method for discovering an individual's fundamental project of being. 1964 *Amer. Philos. Q.* I. 122/1 Existential psychiatrists seem to be publishing much more.

existentialism (egziste·nʃăliz'm). *Philos.* [ad. G. *existentialismus* (see below), f. Existential *a.* + -ism.] A doctrine that concentrates on the existence of the individual, who, being free and responsible, is held to be what he makes himself by the self-development of his essence through acts of the will (which, in the Christian form of the theory, leads to God).

The existentialist movement was mainly originated by the Danish writer Søren Kierkegaard (1813–55), who frequently used the term *Existents-forhold* 'condition of existence, existential relation'. It was developed in the 20th c. chiefly in continental Europe by Jaspers, Sartre, and others, and the Eng. word *existentialism* answers to G. *existentialismus*, which is first recorded in 1919 (see below).

[1846 Kierkegaard *Afsluttende Uvidenskabelig Efterskrift* in *Wks.* (1902) VII. ii. ii. 62 Den existerende subjective Tænker er i sit Existents-Forhold til Sandheden ligesaa negativ som positiv. 1849 R. Nielsen *J. Climacus og H. Martensens Christelige Dogmatik* 126 Thi det skal vist Ingen negte, at der i denne Dogmatik forefindes en Mængde Grundsætninger 'Speculationen' og det religiøse 'Existentialforhold' betræffende. 1919 W. Moog *Logik, Psychologie und Psychologismus* ii. 195 Der Relativismus..ist in der Regel auch ein Existentialismus, das heisst, er behauptet die Abhängigkeit des Logischen vom Existierenden. *Ibid.* 231 Die Fiktion ist nach Husserl..'das Lebenselement der Phänomenologie'..des Existentialismus. *Ibid.* 195 Wie der Fiktionalismus auf existentialistischen Voraussetzungen beruht.] 1941 J. Kraft in *Philos. & Phenomenol. Res.* I. 345 Kierkegaard, Nietzsche, and pragmatism are examples of real or possible starting points of existentialism, capable of being multiplied by further examples. *Ibid.* 357 The philosophy of existence is philosophizing only what the practice of the time realizes: the frenzy connected with the conjuring up of highest meanings. The authoritarian thoughts of existentialism are only a dissonant accompanying music to the authoritarian acts under which man in the twentieth century groans. 1944 *Ibid.* V. 127 Existentialism or existential philosophy..is the extreme attempt to find a meaning in the existence of modern man whose philosophy and science originated in the will to make himself, through knowledge, the master of the world and whose philosophy and science have ended in a complete nihilism. 1945 *Harper's Mag.* Sept. 270/2 Sartre's existentialism is profoundly pessimistic, but in the colloquial sense that the human lot is an absurd and unhappy one, rather than in the formal sense that man is 'by nature' evil...In his view..no 'Human Nature' transcends the actual, which is composed of infinitely diverse existences. 1948 *Mind* LVII. 258 What is Existentialism? Historically speaking it is a philosophy of will that stems from that unhappy Danish genius, Soren Kirkegaard. 1957 H. Read *Tenth Muse* xix. 168 That is existentialist dogma, but it is also the romantic dogma, and existentialism is nowadays the philosophical aspect of romanticism. 1957 *Observer* 8 Sept. 6/4 We asked Miss Greco what Existentialism implied. Apparently its essence is summed up by 'whatever you do, you become'. 1958 W. Barrett *Irrational Man* i. 7 Existentialism was a literary movement as well. 1961 N. Mailer *Advts. for Myself* 91 Willy-nilly I had had existentialism forced upon me. 1965 *Eng. Studies* XLVI. 390 If we define Existentialism as the belief that man is born into an alien..universe..then..Melville must be recognized as America's first existentialist writer. 1967 A. MacIntyre in *Encycl. Philos.* III. 147/2 Existentialism may perhaps be considered most fruitfully as a historical movement in which connections of dependence and influence can be traced from one writer to another...The key themes are the individual and systems; intentionality; being and absurdity; the nature and significance of choice; the role of extreme experiences; and the nature of communication.

existentialist (egziste·nʃălist), *sb.* and *a.* [f. as prec. + -ist.] **A.** *sb.* An exponent or adherent of existentialism.

1945 A. J. Ayer in *Horizon* July 12 Philosophically, he [sc. Sartre] is usually described as an Existentialist. The use of this label is justified in so far as he clearly owes much to Husserl, from whom the group of contemporary German philosophers who are commonly known as Existentialists has chiefly drawn its inspiration. 1950 A. Huxley *Themes & Variations* 115 Biran was an existentialist, who insisted that man must always be considered as he really is, an incarnate spirit or mind-body. 1952 M. McCarthy *Groves of Academe* (1953) vii. 142 He wore a long dark back-belted coat of a cheap shaggy material much affected by priests and young existentialists. 1961 *Daily Tel.* 24 Feb. 19/5 The long, greasy hair, the scraggy beards, the short, tight jeans and grubby fingernails now dubbed 'Beat-nik' were immediately after the war the uniform in Paris of young people who called themselves Existentialists.

B. *adj.* Of or pertaining to existentialism or existentialists.

1946 *Time* 23 Dec. 40/2 Jean-Paul Sartre, the leader of the French Existentialist movement, vigorously, often brilliantly, drags a shady topic into the light. 1948 *Mind* LVII. 370 Existentialist reaction against the abstract ideas of personality characteristic of Kantian and post-Kantian idealism. 1956 A. J. Ayer *Probl. Knowl.* i. 23 Existentialist philosophers have gone so far as to deny the law of identity. 1957 *Times* 28 Dec. 7/6 The decline of the existentialist movement has left French intellectual life dispersed and fluid. 1957, 1965 [see prec.].

existentially, *adv.* Add: (Further examples.) Also, in an existential manner; according to existentialism.

1895 W. James *Coll. Ess. & Rev.* (1920) 400 The various [mental] 'contents'..are integers, existentially, and their parts only live as long as they live. 1938 M. R. Adamson tr. *Maritain's True Humanism* i. 13 A perfect natural wisdom, of which man considered existentially was supposed to be in fact capable. 1941 *Philos. & Phenomenol. Res.* I. 349 According to the fundamental principle of the philosophy of existence..even the fundamental analysis is in the last analysis 'existentially rooted' and is in agreement with the system's phenomenological claims. 1948 E. Waugh *Loved One* 7 Sir Ambrose had a more adventurous past but he lived existentially. He thought of himself as he was at that moment, brooded fondly on each several excellence and rejoiced. 1959 *Times Lit. Suppl.* 20 Mar. p. x/5 Interpreting myth 'existentially', which is what the New Testament is trying to do.

exit, *sb.* Add: 2. b. *Cards* (esp. *Bridge*). A means of deliberately relinquishing the lead; also, the action of doing this. Freq. *attrib.*

1935 *Bridge Mag.* Oct. 231/1 Possibly the term 'exit play' is not the best name for the manœuvre of which I speak. It is called an end play, or an elimination, or a throw-in. 1959 *Listener* 24 Sept. 506/2 If South has a doubleton club he will have the safe exit which was denied him. 1960 T. Reese *Play Bridge with Reese* iii. 20, I have got to eliminate their cards of exit.

4. Esp. a door affording exit from a public building. Also *attrib.*

1881 D'Oyly Carte in W. Hamilton *Aesthetic Movement* (1882) 39 There are exits and entrances on all four sides, giving two exits from every part of the house. 1890 A. Lansdowne *Life's Remin. Scotland Yard* xiii, To leave the Monument Station by the exit staircase. 1911 *Encycl. Brit.* XXV. 766/1 Exit doors must open outwards. 1939 Joyce *Finnegans Wake* (1964) 127 Endurses his doom at chapel exit.

6. **exit cue**, the cue for an actor to make his exit; **exit line**, the line spoken by an actor before making his exit; also *transf.*; **exit permit, visa**, a permit or visa authorizing a person to leave a particular country.

1860 G. Vandenhoff *Dramatic Rem.* xiii. 200 If he perceived any eggs or harder missiles flying, not to wait, but to take the first shot for his *exit-cue*. 1919 A. Platt *Pract. Hints Playwriting* viii. 108 An exit line should be so written that it can be spoken at the door. 1933 F. Baldwin *Innocent Bystander* xvi. 306 That's always a good exit line. 1941 Koestler *Scum of Earth* 232 Both Breitscheid and Hilferding had visas for the United States, but their French exit permits had been refused. 1955 G. Greene *Quiet American* i. i. 9 The police..could..refuse me an exit permit. 1949 H. Read *Gauguin 1848–1903* 3/1 A world where innocence and naïveté no longer exist, where currency restrictions and exit visas effectively deprive us of even Gauguin's illusion of liberty. 1968 *Listener* 23 May 657/1 An exit visa from Vietnam now costs a fortune in bribes.

exit, *v.* Add: b. *Cards* (esp. *Bridge*). To lose the lead deliberately; to use an exit (*exit *sb.* 2 b).

1958 *Listener* 2 Oct. 541/2 East..exited with a further Club, defeating the contract. 1959 *Ibid.* 12 Nov. 850/2 South won with the King and exited with the three.

exite (e·ksəit). *Zool.* [f. Gr. ἔξω outside + -ite¹ 3.] In Crustacea, a process on the outer side of the phyllopodium.

1881 [see *endite *sb.*]. 1893 A. E. Shipley *Zool. Invert.* 260 The abdominal appendages [of *Astacus*]..present an unjointed axis which bears on its inner edge six processes termed endites, which bear numerous setae. The axis ends in a sub-apical lobe, and carries on its outer sides two exites. 1967 G. A. Kerkut *Borradaile & Potts's Invertebrata* (ed. 4) xi. 351 There is in this limb [sc. the phyllopodium] an axial portion or corm which bears..on the outer side one or more lobes known as exites.

exlex (e·ks,leks), *a.* Also **ex-lex.** [f. L. *ex* outside + *lex* law.] Beyond the law, outside the law. Also as *sb.*

1909 G. Drage *Austria-Hungary* 560 The *ex lex* years had shown that the relations between the two partners in the monarchy rested upon shifting sand. 1909 *Westm. Gaz.* 21 Aug. 4/3 The correct date for the beginning of ex-lex in Hungary. 1917 Denney *Chr. Doctr. Reconcil.* 7 The only alternative is to pronounce God *exlex*—without law—which is as good as to abandon thinking altogether.

ex-librist. Delete *rare*⁻¹ and add further examples.

1892 E. Castle *Eng. Book-Plates* 12 The ex-librist of advancing centuries. 1893 *Sat. Rev.* 11 Feb. 157/2 The modern American can be a very keen Ex librist.

So **ex-librism** (eks,ləi·briz'm), the collecting and study of ex-libris.

1893 *Sat. Rev.* 11 Feb. 158/1 Among the public that is curious of ex-librism Mr. Hamilton is widely known.

ex-meridian (eksmĕri·diăn), *a. Navigation.* [f. L. *ex* outside + Meridian.] Of an observation of the sun or other heavenly body: not taken on the meridian, but sufficiently close to it to be reducible to a meridian altitude. Also, of, pertaining to, or calculated from such an observation.

1849 J. T. Towson *Tables Reduct. Ex-Meridian Altitudes* 3 Where cloudy weather prevails, the reduction of ex-meridian observations, for the determination of the latitude, is only second in importance to the observed meridian altitude. 1881 S. T. S. Lecky *'Wrinkles' in Pract. Navig.* iv. 147 If the sun's meridian altitude be about 80°, the time from noon of an Ex-meridian observation should be less than ten minutes. 1904 J. Gill *Text-bk. Navig.* (new ed.) xiv. 145 On account of the proximity of the Pole Star to the North Pole..it is..a very suitable object for determining latitude by ex-meridian method at any time when visible. 1953 C. H. Cotter *Elem. Navig.* xxxvi. 397 The ex-meridian tables in Norie seem to be the most popular with officers of the Merchant Navy. 1961 C. H. McDowell *Dict. Math.* (new ed.) 48 *Ex-meridian altitude*, observations of the sun a few minutes before or after noon (when the sun is obscured at noon).

Exmoor (e·ksmūəɹ, -mōəɹ). The name of a district in Somerset and Devon used *attrib.* or *absol.* to designate the particular breeds of ponies and sheep which it produces.

1808 C. Vancouver *Gen. View Agric. Devon* 338 A cross was some years since made at Chittlehampton, of the old Leicester upon the Exmoor. *Ibid.* 345 A Dartmoor or Exmoor ewe..which in November would cost about 18s. will by the June following produce a lamb worth 20s. 1831, 1902 [see *Dartmoor]. 1831 W. Youatt *Horse* iv. 58 The Exmoor Ponies, although generally ugly enough, are hardy and useful. 1837 —— *Sheep* vii. 255 The sheep..bore considerable resemblance to the Exmoor sheep, and to the Dorsets. 1862 *Chambers's Encycl.* IV. 195/1 Throughout this tract there is a native breed of ponies, known as Exmoor ponies. 1902 *Encycl. Brit.* XXV. 194/2 The Exmoors are delicately formed about the head and neck, and they have a close, fine fleece of short wool. 1937 Hull & Whitlock *Far-Distant Oxus* i. 24 A dark-brown Exmoor with a trailing tail and a shaggy mane. 1960 *Farmer & Stockbreeder* 8 Mar. 15/1 Sheep... Hill types include..Rough Fell,..Welsh Mountain and Exmoor Horn. 1969 *Times* 13 Jan. 11/3 A Suffolk..ewe would not be regarded as outstanding unless she produced four lambs at a time, but three would be remarkable for an Exmoor Horn.

‖ **ex nihilo** (eks nai·hilo), *phr.* [L. *ex* out of + *nihilō*, ablative of *nihil* nothing.] Out of nothing.

1573–80 G. Harvey *Letter-Bk.* (1884) 132 And then, in a fantasticall fitt, I cried owte, Ex nihilo nihil fitt. 1656 T. Reck *Annotations upon Relig. Med.* in T. Browne *Relig. Med.* (ed. 4) 209 They had defined creation to be the production of a thing *ex nihilo*. 1681 T. Goodwin *Wks.* I. i. xxvi. 341 The work of Grace is a work of Creation; and why a Creation? Because 'tis *ex nihilo*. 1855 *Punch* 17 Feb. 69/2 (*heading*) Ex Nihilo nihil fit—or unfit. 1920 M. Beer *Hist. Brit. Socialism* II. iv. xiv. 277 No man creates *ex nihilo*, but out of the materials supplied to him by his predecessors, his contemporaries, and his own experience. 1963 Auden *Dyer's Hand* i. 23 Even the language of *Finnegans Wake* was not created by Joyce *ex nihilo*. 1965 J. D. North *Meas. of Universe* xviii. 399 The difficulty of accepting the idea of a creation *ex nihilo* was the source of much medieval casuistry.

exo-. Add: **exoatmosphe·ric** *a.*, occurring or working outside the atmosphere; **exobio·logy** (see quot. 1960); hence **exobio·logist**; **exoca·nnibalism**, the custom of eating the flesh of persons belonging to another tribe; exophagy; **exoce·llular** *a.*, outside the cell; **exocli·nal** *a. Geol.*, characteristic of or pertaining to an exocline; **e·xocline**, an inverted fan-fold (see quot.); **exocœle** (e·ksosīl) *Zool.*, the space which lies between different pairs of mesenteries of a zoantharian polyp; also **exocœl**; so **exocœ·lic** *a.*; **exocy·clic** *adj.*, (*a*) *Zool.*, of an irregular sea-urchin, having the anus displaced from the apical position in which it is found in the regular forms; (*b*) *Chem.*, situated outside the ring; **exocyto·sis** *Biol.*, the expulsion of matter by a living cell; **exo-erythrocy·tic** *a.*, existing outside the red blood-corpuscles; **exomo·rphic** *a. Geol.*, designating contact metamorphic changes in the surrounding rocks by the intrusion of igneous matter; so **exomo·rphism**, the process of causing or undergoing exomorphic changes; **exope·ptidase** *Biochem.*, any of a group of proteolytic enzymes which split terminal peptide bonds only; **exopho·ria** *Med.*, a tendency of the visual axes to diverge outwards from parallelism; latent divergent squint; hence **exopho·ric** *a.* and *sb.*; **exopod** (e·ksopǫd) *Zool.*, an exopodite; **exor(h)e·ic** *a. Geogr.*, characterized by exoreism; **exor(h)e·ism** *Geogr.* [ad. F. *exorrhéisme*, 'ou, si l'on admet une orthographe simplifiée', *exoréisme* (E. de Martonne 1926, in *Compt. Rend.* CLXXXII. 1396)], land drainage reaching the sea; **exose·ptum** *Zool.*, each of the calcareous septa appearing in the exocœle of a coral polyp; **exosoma·tic** *a. Biol.*, designating or pertaining to a device that an animal uses which is not one of its own organs; **e·xosphere**, the layer of the atmosphere farthest from the earth; hence **exosphe·ric** *a.*; **exote·ntacle** *Zool.*, a tentacle arising from an exocœle in certain polyps; **exoto·xin**, a toxin liberated by a living bacterium or other micro-organism into the medium in which it grows; **exotro·pia**, divergent strabismus; hence **exotro·pic** *a.*; **exo·tropism** (see quot.).

1966 *Economist* 5 Mar. 898/1 A 'new, long-range exo-atmospheric interceptor' to which the short-range Sprint missile would be a supplement. 1967 *Listener* 9 Feb. 185/2 The resulting pulse of radiation should make almost everything in range boil; this would happen beyond the atmosphere and the principle is called exoatmospheric interception. 1964 *New Scientist* 26 Nov. 573/2 One of the greatest difficulties the exobiologists are up against is that of making sure that their equipment does not carry any extraneous terrestrial bacteria. 1969 *New Yorker* 12 Apr. 85/1 Exobiologists will be looking for traces of life on the moon. 1960 *Daily Tel.* 14 Jan. 11/1 Dr. Lederberg is building up a team for work on exobiology, as this branch of the study of life on other planets is called. 1960 *Space Research* I. 1153 The problems of exobiology have important applications for the development of theoretical biology and the understanding of the mechanism of the evolution of life. 1969 *Times* 11 Jan. 15/6 The novel feature of the research carried out by staff of the exobiology division of the Nasa Ames Research Centre in California is that the amino-acids discovered in these ancient rocks have the asymmetry characteristic of living things. 1900 tr. *Deniker's Races of Man* 148 'Exocannibalism', that is to say the habit of eating the flesh of strangers. 1946 *Nature* 23 Nov. 745/1 The excellular enzymes previously elaborated by the growing myxococci. 1901 Exoclinal [see *endoclinal* s.v. *endo-]. 1889 Exocline [see *endocline* s.v. *endo-]. 1885, 1963 Exocœle [see *entocœle* s.v. *ento-]. 1940 L. H. Hyman *Invertebrates* I. vii. 568 In the case of paired septa, the space between the members is named endocoel, between pairs, exocoel. 1967 P. A. Meglitsch *Invertebrate Zool.* vi. 159 (*caption*) The retractor muscles face the exocoel in directive septa. 1902 Exocœlic [see *entocœlic* s.v. *ento-]. 1888 Rolleston & Jackson *Forms Anim. Life* 562 Anus exocyclic. 1913 J. B. Cohen *Org. Chem. Adv. Stud.* II. iv. 298 Dispersivity..is increased by the double bond in the side-chain (exocyclic) as compared with that in the nucleus (endocyclic). 1965 Phillips & Williams *Inorg. Chem.* I. xvi. 594 The exocyclic atoms are markedly electronegative. 1966 *Treat. Invertebrate Palaeontol.* (Amer. Geol. Soc.) U. I. U289/2 This exocyclic tendency was not very successful in the Diadematacea. 1963 C. De Duve in De Rueck & Cameron *CIBA Symp. Lysosomes* 126 The process sometimes referred to as reverse pinocytosis and believed to be involved in secretory mechanisms could be called exocytosis. 1970 *Nature* 8 Aug. 620/2 He proposed that chromaffin cells released their contents after attaching to the cell membrane, through 'exocytosis'. 1946 *Nature* 16 Nov. 707/2 Therapeutic potency is..apparent not only against the erythrocytic but also against the exo-erythrocytic forms of the malaria parasite. 1968 J. H. Burn *Lect. Notes Pharmacol.* (ed. 9) 115 The exo-erythrocytic parasites are very important in the chemotherapy of malaria because they are much more resistant to treatment than the red cell forms. 1888 F. H. Hatch in J. J. H. Teall *Brit. Petrogr.* 430 *Exomorphic*, applied by Fournet to contact-metamorphism when produced in the rock through which the molten mass is erupted. Used in contradistinction to endomorphic. Syn. Exogenous. 1903 *Amer. Jrnl. Sci.* CLXV. 280 The density increment due to exomorphic changes. 1931 F. H. Lahee *Field Geol.* (ed. 3) vi. 133 The contact metamorphic zone (exomorphic zone)..varies in thickness from a fraction of an inch to many hundred yards. 1909 *Cent. Dict. Suppl.*, *Exomorphism*, that variety of contact-metamorphism which is developed, in the surrounding walls, by an intruded mass of eruptive rock. 1966 *McGraw-Hill Encycl. Sci. & Technol.* X. 84/1 The crystallization of the larger intrusives may result in profound alterations in the adjacent wall rocks (exomorphism). 1936 Exopeptidase [see *endopeptidase* s.v. *endo-]. 1962 A. Spector in A. Pirie *Lens Metabolism Rel. Cataract* 334 In neither investigation were sufficient substrates studied to indicate that the activity..was associated with a true exopeptidase. 1886 G. T. Stevens in *N.Y. Med. Jrnl.* 4 Dec. 626/2 The different relations of the visual lines..may be defined and arranged as follows..I. *Generic Terms.—Orthophoria*: A tending of the visual lines in parallelism. *Heterophoria*: A tending of these lines in some other way. II. *Specific Terms.*—Heterophoria may be divided into: 1. *Esophoria*: A tending of the visual lines inward. 2. *Exophoria*: A tending of the lines outward. 3. *Hyperphoria*: A tending of the right or left visual line in a direction above its fellow. 1964 Exophoria [see *esophoria]. 1906 *Jrnl. Amer. Med. Assoc.* 1 Sept. 674/1 Practically two-thirds of the whole number of exophorics were what might be called near workers. 1909 *Cent. Dict. Suppl.*, Exophoric, *adj.* 1962 H. C. Weston *Sight, Light & Work* (ed. 2) ii. 41 People who are exophoric for distance are usually progressively more so as the point of fixation approaches the eyes. 1893, 1959 Exopod [see *endopod* s.v. *endo-]. 1927 *Geogr. Rev.* XVII. 399 The boundary between the endoreic and the exoreic regions can be drawn without hesitation. 1957 G. E. Hutchinson *Treat. Limnol.* I. iv. 226 The exorheic regions contain the main lake districts of the world. 1927 Exorheism [see *endoreism* s.v. *endo-]. 1963 D. W. & E. E. Humphries tr. *Termier's Erosion & Sedimentation* 404 *Exorheism*, drainage towards the oceans surrounding a land mass. 1885, 1903 Exoseptum [see *entoseptum* s.v. *ento-]. 1904 *Biol. Bulletin* July 82 The dorsal and middle pairs of exosepta arose bilaterally in advance of the two ventral pairs. 1945 A. J. Lotka in *Human Biol.* Sept. 188 In place of slow adaptation of anatomical structure and physiological function in successive generations by selective survival, increased adaptation has been achieved by the incomparably more rapid development of 'artificial' aids to our native receptor-effector apparatus, in a process that might be termed exosomatic evolution. 1951 *New Biol.* XI. 21 Exosomatic adaptations. *Ibid.*, Most exosomatic organs are instruments or tools..like spectacles, radio sets,..hammers, scalpels, motor cars and guns. 1951 *Jrnl. Brit. Interplan. Soc.* X. 18 The atmosphere can conveniently be divided into four parts, the troposphere, the stratosphere, the ionosphere and the exosphere. 1955 E. Burgess *Frontier to Space* 150 The density and extent of the exosphere. *Ibid.*, The exospheric gases. 1969 *Sci. Jrnl.* May 67/1 The lunar atmosphere is a true exosphere in direct communication with space, to which its molecules can escape freely unless inhibited by gravity. 1904 *Biol. Bulletin* July 88 The exotentacles in *Siderastrea radians* have been found to appear throughout in advance of the entotentacles. 1920 H. Zinsser in *Jrnl. Immunology* V. 286 There was much unclearness..regarding the so-called exotoxins of bacteria. 1964 M. Hynes *Med. Bacteriol.* (ed. 8) vi. 67 The reaction between endotoxins and their antibodies does not follow the law of multiple proportions which is so characteristic of exotoxin–antitoxin reactions. 1897 G. T. Stevens in Norris & Oliver *Syst. Dis. Eye* II. viii. 172 The class Heterotropia..may be divided into two sub-classes... The specific divisions of the sub-class *a* are—1. Esotropia, a deviation of the visual lines inward. 2. Exotropia, a deviation of the visual lines outward. 3. Hypertropia (right or left), a deviation of one visual line above the other. 4. Hyper-esotropia and Hyperexotropia are the compound deviations. 1961 *Lancet* 7 Oct. 812/1 Increase of the inter-pupillary distance with exotropia. 1906 G. T. Stevens *Motor Apparatus of Eyes* iii. xlviii. 384 Exotropic persons are, much more than esophoric subjects, accustomed to close one eye when reading or working. 1898 H. C. Porter tr. *Strasburger's Text-bk. Bot.* 258 A torsion must..occur when a geotropic organ, which has become curved over toward its parent axis, turns itself about so as to face outwards (exotropism).

exocentric (eksose·ntrik), *a. Linguistics.* [f. Exo- + Centric *a.*] Of compounds or constructions: having a different grammatical function from that of the constituent parts (see quots.). Opp. *endocentric *a.*

1914 Bloomfield *Introd. Study Lang.* v. 161 The so-called 'exocentric' compounds..denote an object *having* the thing named in the compound, as *long-nose, short-horn, swallow-tail*. 1933 —— *Lang.* xii. 194 *John ran* is neither a nominative expression (like *John*) nor a finite verb expression (like *ran*). Therefore we say that the English actor–action construction is exocentric: the resultant phrase belongs to the form-class of no immediate constituent. 1968 J. Lyons *Introd. Theoret. Ling.* vi. 232 *In Vancouver* is exocentric, since its distribution is different from that of either the preposition *in* or the noun *Vancouver*.

exocrine (e·ksokrəin, -in), *a. Physiol.* [f. Gr. ἔξω outside + κρίνειν to separate, cf. *endocrine *a.* and *sb.*] Having external secretion; designating or pertaining to the discharge of a secretion through a duct or the secretion itself.

Quot. 1911 is an erroneous definition.

1911 Stedman *Med. Dict.* 296/2 *Exocrin*, external secretion of a gland. 1937 *Ann. Internal Med.* X. 1848 (*heading*) The exocrine functions of the pancreas. 1946 *Nature* 21 Sept. 419/1 Spaces..permit the passage of small quantities of the exocrine secretion of the gland into contact with special thin-walled capillaries. 1964 A. R. Muir in G. H. Haggis *Introd. Molecular Biol.* v. 138 The exocrine cells of the guinea-pig pancreas fulfil these criteria, for they commence secretion if a meal follows a period of starvation. 1967 *New Scientist* 2 Nov. 295/1 The exocrine pancreas cell, which secretes an enzyme that digests protein. 1970 T. S. & C. R. Leeson *Histol.* (ed. 2) v. 87/1 A gland of the exocrine type passes its secretion to a duct system and thus to a body surface.

exocuticle (e·ksokiūtik'l). Also **exocuti·cula.** [f. Exo- + Cuticle.] The middle layer of a cuticle. **a.** The hard, chitinous outer part of the procuticle of an insect or other arthropod.

1929 [see *endocuticle a]. 1951 A. G. Richards *Integument of Arthropods* xvi. 149 In calcified cuticles, at least of decapod crustacea, the exocuticle and outer portion of the endocuticle become calcified. 1967 P. A. Meglitsch *Invert. Zool.* xvi. 676 Beneath it [sc. the epicuticle] is the procuticle, made up of two layers, an outer exocuticle, often pigmented and impregnated with the substances found in the epicuticle, and an inner endocuticle. It is the exocuticle that is thickened to form the skeletal plates or sclerites.

b. The middle part of the cuticle that surrounds animal fibres or hairs, etc.

1949 [see *epicuticle b]. 1962 W. J. Onions *Wool* ii. 16 The exocuticle, which constitutes about half of the scale, is a protein resistant to digestion by trypsin.

exodontia (eksodǫ·nʃiă). [mod.L., f. Ex-¹ + Gr. ὀδούς, ὀδοντ- tooth.] The extraction of teeth. Hence **exodo·ntist**, a specialist in exodontia.

1913 G. B. Winter (*title*) Exodontia; a practical treatise on the technic of extraction of teeth. *Ibid.* iv. 69 Dental anatomy, from the standpoint of the exodontist, should be studied in the light of mechanics. 1927 M. H. Feldman *Man. Exodontia* i. 17 The development of skill in inducing anesthesia..has made it possible to perform the ofttimes intricate phases of exodontia in such a manner as to grant the patient..freedom from the sensation of pain. 1936 Dunning & Davenport *Dict. Dental Sci. & Art* 203/1 Exodontist. 1960 *Times* 30 Mar. 3/5 (Advt.), Oral surgery including exodontia.

ex officio, ex-officio, *advb. phr.* (and *adj.*). Add to def.: *spec.* designating a particular type of oath or legal suit (see quots.).

Discussed further in *N. & Q.* (1960) Oct. 365 f.

1532 *Division betw. Spiritualtie & Temporaltie* fol. 16ᵛ Diuers sutes..haue ben taken in the spiritual courtes of office, that is called in latyn, ex officio: so that the partyes haue not knowen who hath accused them. 1637 E. Rossingham *Let.* 4 Jan. in *Court & Times Chas. I* (1848) II. 260 In one of my last letters I wrote, that Mr. Burton had refused to take the oath *ex officio*. I hear since that he hath..been suspended from..his ministry. 1661 *Act* 13 *Car. II* st. 1, c. 12 The oath usually called the oath *ex officio*, or any other oath whereby such person..may be charged or compelled to confess or accuse, or to purge him or herself of any criminal matter or thing, whereby he or she may be liable to any censure or punishment. 1729 G. Jacob *Law Dict.* s.v. *Information*, When the Attorney General exhibits an *Information*, he does it *ex Officio*. 1769 Blackstone *Commentaries Laws Eng.* IV. xxiii. 304 Informations, that are exhibited in the name of the king alone, are..of two kinds: first those which are truly and properly his own suits, and filed *ex officio* by his own immediate officer, the attorney general. *Ibid.* [Index], *Ex officio* informations. 1797 Jacob & Tomlins *Law-Dict.* I, *Ex Officio Informations*, informations at the suit of the King, filed by the Attorney-General, as by *virtue of his office*. 1886 *Athenæum* 24 July 111/2 It was enacted that any one in Great Britain or Wales 'who is or shall be a Popish Recusant or Papist, or is or shall be educated in the Popish Religion, or whose parent or parents shall be a Papist or Papists', shall on or before January 20th, 1716, take the usual *ex officio* oath. 1936 M. H. Maguire in *Essays in Hist. & Pol. Theory in Honor of C. H. McIlwain* 200 Otho's emphasis fell on the point of answering truly every question that might be asked...It is this specialized form of the oath of calumny which becomes practically equivalent to the oath *ex officio*. 1938 *Act* 1 & 2 *Geo. VI* c. xliii. § 12 Criminal informations, other than informations filed ex officio by His Majesty's Attorney-General, are hereby abolished. 1970 M. Stocks *My Commonplace Book* 106 A Proctor has ex-officio the right to move resolutions in the Hebdomadal Council. *Ibid.* 203 And as Principal of a constituent college of London University I was *ex-officio* a member of its Senate and of its Collegiate Council, as well as of many *ad hoc* appointments committees.

B. As *sb.*, a person serving *ex officio*; an *ex-officio* officer.

1817 *Black Dwarf* I. 329 Their cabinet deliberations On soldiers' caps, or wars of nations, On ex-officios, new suspensions. 1886 *Lett. from Donegal* 13 The ex-officios go for laying out most on the great..arterial roads. 1893 *Westm. Gaz.* 22 Dec. 2/2 The first [principle] is that the Local Government shall appoint ex-officios, the second is that the Guardians themselves shall co-opt ex-officios.

exogenesis (eksodʒe·nĕsis). *Med.* [mod.L., f. Gr. ἔξω Exo- + γένεσις Genesis.] Origin (of a disease) from external causes.

1903 *Jrnl. Trop. Med.* VI. 227, I am satisfied that the exogenesis of cancer is clinically and logically proved.

exogenetic (e:ksodʒĭne·tik), *a.* [f. as prec.: see -genetic.] Having an external cause or origin. **1.** *Med.* Of a disease: having a cause outside the body.

1874 Dunglison *Dict. Med. Sci.* (new ed.) 385/2 *Exogenetic*, having an origin from external causes, as exogenetic diseases.

2. *Geol.* **a.** Of rock: produced by the action of forces external to the material from which it was formed.

1904 A. W. Grabau in *Amer. Geologist* Apr. 229 Contrasted with this group is that of the clastic rocks...These owe their origin chiefly to agents acting from without, and hence may be termed exogenetic. 1957 F. J. Pettijohn *Sedimentary Rocks* (ed. 2) i. 4 It is desirable to discuss the textures of the exogenetic (clastic) rocks, the endogenetic (chemical) rocks, and the epigenetic (diagenetic) rocks separately.

b. Of rock or processes affecting rocks and landforms: formed or occurring at the surface of the earth.

1914 T. Crook in *Mineral. Mag.* XVII. 72 The terms 'endogene' and 'exogene' have long been used by German and French geologists, chiefly as the equivalents of 'eruptive' and 'sedimentary'...It seems permissible to use them as here required, in the forms *endogenetic* and *exogenetic*. *Ibid.* 73 Exogenetic processes are best subdivided..with reference to..weathering and denudation...Exogenetic rocks..are formed at ordinary or comparatively low temperatures. 1937 Wooldridge & Morgan *Physical Basis Geogr.* v. 54 Exogenetic forces co-operate in all cases with the forces of upheaval. 1961 L. D. Stamp *Gloss. Geogr. Terms* 184/1 Exogenetic is the usual English form, Exogenic..the American. 1970 H. H. Read *Rutley's Elem. Mineral.* (ed. 26) i. v. 212 Exogenetic deposits are due to surface-processes such as deposition in a salt-lake or sorting by river-action.

exogenic (eksodʒe·nik), *a.* [f. Exo- + *-genic.] Exogenetic. **1.** *Med.* Of a disease.

1900 Gould *Pocket Med. Dict.* (ed. 4) 246 *Exogenic, exogenous.* See Exogenetic. 1961 *Lancet* 23 Sept. 706/1 During the epidemic stage the study of an exogenic factor would be complicated by various confusing factors—social and immunological as well as virological.

2. *Geol.* = *exogenetic *a.* 2 b. Chiefly *U.S.*

1942 O. D. von Engeln *Geomorphology* iv. 56 Some of the relief features..result from external, that is, exogenic, forces and processes. 1954 W. D. Thornbury *Princ. Geomorphol.* ii. 19 In general, the endogenic processes tend to build up or restore areas which have been worn down by the exogenic processes. 1961 [see *exogenetic *a.* 2 b]. 1968 R. W. Fairbridge *Encycl. Geomorphol.* 309/1 In geomorphology, one may distinguish structural land forms (endogenic) from denudational land forms (exogenic).

exogenous, *a.* Add: **b.** (Later examples in *Psychiatry*.)

1925, 1962 [see *endogenous *a.* e].

d. *Geol.* Formed or occurring outside some structure or mass of rock (specified or understood); *spec.* = *exogenetic *a.* 2 b.

1845, 1890 [see *endogenous *a.* d]. 1888 [see *exomorphic* s.v. *exo-]. 1905 J. Geikie *Struct. & Field Geol.* xiv. 204 Exogenous or intrusive veins.—These are simply protrusions proceeding from a mass of granite into the contiguous rocks. 1933 R. A. Daly *Igneous Rocks & Depths of Earth* viii. 150 Originating within the crater, all of these young bodies may be described as endogenous domes and are thus distinguished from the vastly bigger exogenous domes of the Mauna Loa or Icelandic type. 1954 W. D. Thornbury *Princ. Geomorphol.* iii. 34 The agencies thus far mentioned and the processes performed by them originate outside the earth's crust, and..have been designated by Lawson as epigene and by Penck as exogenous. 1968 R. W. Fairbridge *Encycl. Geomorphol.* 209/1 Rittmann (1962) refers to the volcanoes that build up from simple surface flows and cinders as exogenous cones and those that freeze over readily (the viscous ones) and push up from the inside as endogenous domes.

Exonian (eksōu·niăn). [f. L. *Exonia* Exeter: see -ian.] A native or inhabitant of Exeter.

1871 The English Boy Himself Again or the 'Dame's' Wigging Denied, by An Exonian. 1891 *Where to buy at Exeter* 7/2 In the wars between Stephen and Matilda the Exonians upheld the interests of the Queen, and were subject to a severe siege. 1903 *Daily Chron.* 16 July 5/1 In Durham, the Cathedral is affectionately called 'The Abbey' while that of Exeter is named 'St. Peter's', perhaps more often still shortened by Exonians to 'Peter'. 1913 F. B. Rose-Troup *Western Reb.* xviii. 282 Reference to the 'entrie' and other events occur in certain depositions of Exonians a few years later.

exotericism (exote·risiz'm). [f. Exoteric *a.* and *sb.* + -ism.] Exoteric doctrines, or belief in these. Hence **exote·ricist**, one who holds such doctrines.

1886 Webster Exotericism. 1954 P. Townsend tr. *Schuon's Transcendent Unity Religions* iii. 53 We must now answer more explicitly the question as to the truths that exotericism must of necessity ignore. 1954 Exotericist [see *esotericism]. 1955 *Jrnl. Theol. Stud.* VI. 339 Exotericism as such is according to Schuon not blameworthy, but necessary.

exotherm (e·ksoþəɹm). *Chem.* [f. Gr. ἔξω Exo- + θέρμη heat; cf. next.] A compound which liberates heat during its formation and which absorbs heat or energy during its decomposition. So **exothe·rmal** *a.* = *exothermic *a.*; **exothe·rmally** *adv.*

1895 C. S. Palmer tr. *Nernst's Theoret. Chem.* IV. v. 584 The direction of the chemical change does not necessarily coincide with the direction in which the reaction progresses exothermally. 1906 A. Smith *Introd. Gen. Inorg. Chem.* ii. 27 Exothermal actions, as those which produce heat are called, are usually of the spontaneous order. 1909 *Cent. Dict. Suppl.*, Exotherm. 1954 *Electronic Engin.* XXVI. 139 It is possible to control the exotherm by maintaining the resin and the hardener at a low temperature. 1959 *Jrnl. Iron & Steel Inst.* CXCII. 389/3 Compositions and physical properties of insulating firebricks and fireclay bricks are compared, and compositions and exothermal qualities..are given. 1965 Phillips & Williams *Inorg. Chem.* I. xiii. 506 Water reacts exothermally with very many oxides.

exothermic (eksoþə·ɹmik), *a. Chem.* [ad. F. *exothermique* (M. Berthelot *Essai de mécanique chimique* (1879) II. iv. ii. 18), f. as prec. + -ic.] Characterized by, or attended by, the development of heat; *spec.* designating an exotherm. Cf. *endothermic *a.* Hence **exothe·rmically** *adv.*

1884, 1945 [see *endothermic *a.*]. 1958 *Engineering* 4 Apr. 431/2 The neutral hydrogen mesic atoms drift around and can exchange their muons with deuterium atoms. The exothermic nature of this exchange makes it highly probable. 1962 W. B. Thompson *Introd. Plasma Physics* iv. 58 It is known that many exothermic nuclear reactions with high intrinsic cross sections are inhibited at low energies by the Coulomb repulsion between the like charges on the nuclei. 1964 N. G. Clark *Mod. Org. Chem.* vi. 108 Ammonia and amines react readily and exothermically with ethylene oxide. 1966 Phillips & Williams *Inorg. Chem.* II. xxii. 142 They are formed exothermically from the elements. *Ibid.*, It is a refractory substance, strongly exothermic.

exotic, *a.* and *sb.* **A. 2. b.** Delete † (*obs.*) and add later examples. Also, having the attraction of the strange or foreign, glamorous.

1919 F. Hurst *Humoresque* 85 She was about as exotic as a flowering weed which can spring so strongly and so fibrously from slack. 1933 F. Baldwin *Innocent Bystander* (1935) ix. 172 She hadn't bothered to be exotic and provocative with him. 1933 *Punch* 15 Nov. 546, I want a perfume thuitable to an ecthotic blonde with innothent green eythe and a lithp. 1958 *Listener* 7 Aug. 202/3 An enormous hangar fitted with exotic devices.

c. *spec.* Of or pertaining to strip-tease or a strip-teaser. orig. *U.S.*

1954 *Newsweek* 8 Nov. 96/2 One mushrooming by-product of burlesque,..is what the 1954 billboards call the exotic dancer. *Ibid.* 97/1 Night-club owners who want to bolster attendance hire an exotic dancer. 1955 *Evening Sun* (Baltimore) 19 July 1 The League of Exotic Dancers..a group within the American Guild of Variety Artists, through which all the girl entertainers are booked. 1965 R. Hardwick *Plotters* (1966) xiii. 114 'Exotic dancer—' 'A euphemism for stripper.' 1967 E. Ambler *Dirty Story* i. ii. 20 My wife..is an exotic dancer.

d. Designating any of various high-energy fuels, esp. non-hydrocarbon ones, developed orig. for rocket engines; also applied to certain metals not previously used to a significant extent in technology. orig. *U.S.*

1957 in *Amer. Speech* (1960) XXXV. 285 Fuels that combine hydrogen with boron or lithium are considered to be exotic fuels. 1958 [see *beef *v.* 2 b]. 1958 C. C. Adams et al. *Space Flight* iii. 83 During recent months the terms 'exotic propellants', 'Zip-fuels', and 'Hi-Cals' have come into the pages of the trade and technical literature of rocketry. These generally embrace efforts outside strictly conventional lines to develop high-energy power sources but do, on occasion, include merely high-energy types of liquids and solids. 1961 *Technology* Nov. 282 Alloys of columbium (niobium), one of the new 'exotic' metals noted for its strength and corrosion resistance. 1963 *Machinery* 20 Mar. 631/2 Quite recently it has become increasingly fashionable to describe certain metals as exotic. 1970 M. Smith *Aviation Fuels* iv. 414 In the early stages of the development programme on these exotic fuels they were prohibitively expensive.

B. 1. c. An exotic dancer (see sense *A. 2 c, above), a stripper. orig. *U.S.*

1954 *Daily Oklahoman* 24 Jan. 29 (*heading*) New exotics at the Derby [Club]. 1965 L. Whitten *Progeny of Adder* (1966) 30 A still-famous tenor shared billing with the immense breasts of an 'exotic' named Telly Stahr.

‖ **exotica** (egzǫ·tikă), *sb. pl.* [L., neut. pl. of *exōticus* Exotic *a.*] Exotic objects or items.

1876 W. E. A. Axon (*title*) Exotica...Translations in verse. 1957 A. Graf (*title*) Exotica. Pictorial cyclopedia of indoor plants. 1958 Blesh & Janis *They all played Ragtime* iv. 73 The brasses included exotica like flügelhorn, euphonium, French horn, and Sousaphone. 1959 *Listener* 2 Apr. 592/2 Glass-sided office buildings, banks, cafeterias, department stores—all have become filled with *exotica*. 1970 *Sunday Times* 6 Dec. 43/5 Mrs. Boston, whose small shop..will contain a wide spectrum of exotica.

exotically, *adv.* Delete † and add later example.

1947 'W. Targ' *Carrousel for Bibliophiles* 398 Arthur Morrison's exotically titled *The Green Eye of Goona*..was transmuted by the American publisher—and one wonders why—into *The Green Diamond*.

exotospore (eksōu·tospōəɹ). *Zool.* [f. Gr. ἐξωτικός Exotic + -o + Spore.] = *sporozoite* (s.v. Sporo-).

1902 E. R. Lankester in *Nature* 27 Mar. 500/1 The malaria-germ which is brought by the stab of the Anopheles into the human blood-vessels..is needle-like in shape,..but the most important fact about it for description and comparison is that it has been formed outside the human body, and is introduced as a strange element into the human blood by the agency of the mosquito. I..call it the Exotospore.

expandable (ekspæ·ndăb'l), *a.* [f. Expand *v.* + -able.] That can be expanded; expansible.

1926 H. W. Fowler *Mod. Eng. Usage* 3/1 For the italicized words..it is recommended to substitute the accompanying form in *-able*..*expansible*, expandable. 1965 *Language* XLI. 239 P (preposition) and C (conjunction) are expandable only to the extent that PP..and CC sometimes functions as P and C respectively. 1969 *Jane's Freight Containers 1968–69* 113/3 Pole trailers (expandable to 55 ft).

expanded, *ppl. a.* Add: **1. b.** *expanded metal, steel*, sheet metal slit and stretched into a lattice, used for making screens and lockers, and for reinforcing concrete, etc.

1890 *Builder* 28 June 477/3 A strip of steel..is fed into a powerful automatic machine, which..produces a sheet of expanded metal. 1891 [see Lathing *vbl. sb.*¹ 2]. 1913 *Lockwood's Dict. Mech. Engin.* (ed. 4) 434 *Expanded steel*, steel sheet which is pierced and opened out in a machine, yielding lozenge-shaped spaces, which are filled, and the whole embedded in concrete. 1937 *Discovery* January 20/2 A concrete jacket reinforced by expanded metal and pieces of scrap iron. 1958 *Observer* 3 Aug. 7/3 The first expanded metal..was to replace timber laths in plaster work. 1968 *Gloss. Terms Mechanized & Hand Sheet Metal Work* (*B.S.I.*) 10 *Expanded metal*, a mesh consisting of a lattice-work of undulating strands.

2. b. *Gram.* Designating the tense in which a form of the verb *to be* is used with a present participle (see quots.).

1931 O. Jespersen in *S.P.E. Tract* xxxvi. 524 By the side of the simple tenses we have in English expanded tenses, e.g.: *simple*: he works, he worked, he has worked; *expanded*: he is working, he was working, he has been working. *Ibid.* 527 The expanded infinitive often serves to emphasize the present moment, as in *What can he be doing?* as against the more general *What can he do?* 1932 *Jrnl. Eng. & Germanic Philol.* XXXI. 469 The expanded form must always be employed where the simple form, if used, might be construed as having general force. 1963 F. T. Visser *Hist. Syntax* I. iii. 189 When..*to be* occurs in the 'expanded' or 'progressive' form, as in 'You are being naughty'.

expanding, *ppl. a.* Add: **1.** *expanding universe*, the universe regarded as continually expanding, so that the galaxies are steadily receding from one another.

1931 *Times* 7 Nov. 7/1 'The Expanding Universe' was the subject of Sir Arthur Eddington's presidential address to the Physical Society at their meeting..last evening. 1932 A. S. Eddington (*title*) The expanding universe. 1942 *Time* 1 June 62/2 Chandrasekhar's inspiration was the 'expanding universe' theory of Belgium's Canon Georges Lemaître, who eleven years ago suggested that all the matter in the universe was once condensed into a single primordial atom. 1958 *Listener* 4 Dec. 934/1 The modern concept of the expanding universe.

2. *expanding bullet* = *dum-dum bullet.

1859 *National Mag.* VI. 56/1 Rifles employing expanding or upsetting bullets. 1873 J. Whitworth *Guns & Steel* 21 With all expanding bullets, a quick burning powder must be employed. The expansion depends on the sudden action of the powder upon a bullet possessing inertia. 1909 *Westm. Gaz.* 29 Dec. 7/4 The ammunition from Europe consists of expanding bullets.

expansion. Add: **1. c.** *Naval Arch.* The mathematical enlargement of a ship's lines from a drawing or model to the full size of building. Also *attrib.*

1869 E. J. Reed *Shipbuilding* 186 Either a model of one side of the ship or an expansion drawing is prepared, on which to set off the edges and butts of the plates. *Ibid.* 439 An expansion batten is applied to the line on the floor representing the moulding edge of the frame. 1877 Thearle *Theor. Naval Archit.* I. 163 When an expansion drawing is made, the several strakes of plating can be shown upon it, also their thicknesses...It is obviously impossible to calculate the position of the centre of gravity from an expansion.

4. d. Extension of the territorial rule or sway of a country.

1882 J. R. Seeley in *Macm. Mag.* XLVI. 456 (*title*) The Expansion of England in the Eighteenth Century. **1903** *Sun* (N.Y.) 1 Dec. 2 When he indorsed the doctrine of expansion the cheers were pronounced.

8. expansion apparatus, chamber = *cloud chamber* (see *cloud *sb.* 12); see also quot. 1968; **expansion box,** a chamber fitted to a pipe to allow for the expansion of the liquid, gas, etc., which flows through the pipe.

1897 C. T. R. Wilson in *Phil. Trans. R. Soc.* CLXXXIX. A. 267 (*heading*) Conditions to be satisfied by the expansion apparatus. **1838** *Civil Eng. & Arch. Jrnl.* I. 237/2 As the water warms, it rises through the pipe into the expansion box...To guard against the danger of exceeding the proper degree of heat, the expansion box is furnished with a pipe. **1939** *Jrnl. R. Aeronaut. Soc.* XLIII. 748 The exhaust gases from a group of cylinders are discharged through a number of expansion boxes, the expansion boxes being shaped as a flattened streamline member lying in the airstream. **1933** *Discovery* Apr. 107/2 This recoiling nucleus..is an efficient ionizer and so can be detected either by its electrical effect or its track can be photographed in an expansion chamber. **1968** *Practical Motorist* Nov. 333/1 *Expansion tank* (*chamber*), a small tank connected to the overflow of the radiator on some modern cars which collects water and steam forced through the overflow pipe. The water is drawn back into the main system when the engine cools.

expansional (ekspæ·nʃənăl), *a.* [f. Expansion + -al.] Of or pertaining to expansion.

1925 J. Joly *Surface-Hist. Earth* vii. 115 The sum of the fluctuating changes of volume of the one sign (either expansional or contractional).

expansionary (ekspæ·nʃăn'ri), *a.* [f. Expansion + -ary[1].] Of, tending to, or directed towards expansion.

1936 *Yale Rev.* Sept. 25 Labor..insisted..on an expansionary policy. **1958** *Listener* 31 July 151/2, I imagine the British Government, at any rate, will soon start to turn towards positively expansionary policies. **1960** *Times* 22 Jan. 18/2 Are we so certain that the economy will run into bottlenecks, heavy over-stocking, and all the signs of overstrain quite so early in an expansionary phase? **1966** *New Statesman* 30 Sept. 468/2 The central agency can only function in an expansionary situation. **1969** *Daily Tel.* 30 Aug. 18/5 This will tend to mitigate the inflationary and expansionary effect of the increase in bank lending.

expansionism (ekspæ·nʃəniz'm). [f. Expansion + -ism.] Advocacy of, or furtherance of, a policy of expansion, esp. of territorial expansion.

1900 *Daily News* 2 May 3/5 By Imperialism British Liberals ought not to understand militarism or even expansionism. **1959** *Economist* 28 Feb. 761/1 The expansionism which marks the common market's view of the future. **1960** *Spectator* 7 Oct. 512 We are afraid of Russian expansionism.

expansionist. Add: (Further examples.) Also esp. an advocate of territorial expansion (see *expansion 4 d).

1898 *Daily News* 3 Jan. 6/4 Truly a tempting country for an Expansionist son of Japhet to fix eye upon. *Ibid.* 14 Mar. 4/5 The editor of 'The Free Press' has been expelled from Siam for..advocating a French expansionist policy. **1898** *Westm. Gaz.* 11 Nov. 1/2 Expansionist Republicans. **1909** *Ibid.* 17 Apr. 2/1 It seems.. doubtful if those reasons will commend themselves either to 'Irredentists' or to 'Expansionists'. **1940** *Economist* 10 Aug. 190/1 The banks, much more than the insurance companies, have the compensatory advantage that expansionist war finance means a progressive increase in earning assets. **1961** *Daily Tel.* 29 Nov. 20/4 The Ethiopians are traditionally wary of a Somalia 'expansionist' policy.

expat, colloq. abbrev. of Expatriate *sb.*

1962 D. J. Enright in *Times Lit. Suppl.* 10 Aug. 582 (*title of poem*) Expat. **1968** *Listener* 25 Jan. 102/3 The 'expats', as the expatriate British refer to themselves, are understandably fond of Ghana.

expatriate, *sb.* Add to def.: In modern usage, a person who lives in a foreign country.

1902 *Daily Chron.* 26 Feb. 3/5 'The Expatriates' is a novel by Miss Lilian Bell...Its principal characters are rich Americans and titled Parisians, and the action takes place largely in Paris. **1961** *Economist* 25 Mar. 1193/1 In Dar-es-Salaam all the talk is about 'expatriates', the technical name for the Europeans who run the country alongside or behind the African ministers. **1969** *Age* (Melbourne) 24 May 2/4 Mr. Barnes said the Gordon estate sub-division would be one of several new area developments which would enable closer integration of Papuans and New Guineans and expatriates.

expect, *v.* Add: **4. a.** *expect me when you see me*: a colloq. phr. implying that the speaker is uncertain as to when he will return.

1876 R. D. Blackmore *Cripps* II. xi. 179 If I do not ring then, send everybody to bed. And do not expect me until you see me. **1923** J. Buchan *Midwinter* xii. 191, I..may be absent for days. Expect me back when you see me. **1963** 'B. Graeme' *Almost without Murder* xi. 124 Sometimes I travel at an hour's notice. My brothers expect me when they see me. **1967** B. Whitaker *Of Mice & Murder* vi. 73, I might go off for the day somewhere. All right if you expect me when you see me? **1968** 'P. Hobson' *Titty's Dead* v. 54, I must be off. Expect me when you see me.

c. (*b*) Substitute for def.: To be pregnant; usu. in phr. *to be expecting. colloq.* (Add earlier and later examples.)

1817 Jane Austen *Let.* 23 Mar. (1952) 489 She expects much about this day three weeks, & is generally very exact. **1906** Galsworthy *Man of Property* I. vi. 82 Whether you expect Rogers's wife was really—expecting. **1939** M. Spring Rice *Working-class Wives* iv. 97 She is 32 and has three children, very well spaced and is 'expecting again'. **1957** R. Longrigg *Switchboard* 208 'Make him do a Charleston.' 'Have a heart,' said Sue. 'I'm expecting.'

expectant, *ppl. a.* Add: **1. b.** *expectant mother*, a pregnant woman; also *transf.* So *expectant father*.

1861 M. Reid *Hero in spite of Himself* I. 12 Whenever a Zapoteque woman is about to add one to the number of their community, the expectant father of the child assembles all his relations in his cabin. **1878** Trollope *Is he Popenjoy?* III. vi. 56 There was a brutality about this which for a time made the expectant father almost mad. **1882** R. K. Douglas *China* iv. 87 The expectant mother's next desire is to discover of what sex her child will be. **1928** *Act* 18 & 19 *Geo. V* c. 29 § 1 Arrangements..for attending to the health of expectant mothers and nursing mothers. **1960** M. Burton *Wild Animals Brit. Isles* 97 In fox-hunting countries artificial burrows are constructed of which the expectant-mother vixen will avail herself. **1963** 'G. Bagby' *Murder's Little Helper* (1964) ii. 14 'What kind of a burglar is it that swipes maternity clothes?' 'Expectant father?'

‖ **ex pede Herculem** (eks pedi hɜ·rkiulem), *phr.* [L., 'Hercules from his foot'.] Inferring the whole of something from some small or insignificant part of it, as Pythagoras is said to have calculated the height of Hercules from an estimate of the size of his foot.

1665 R. Head *Eng. Rogue* xiii. 134 Ex pede, I perceived she was endued with ingenuity, by the quaintness of her expressions, *Ex pede Herculem*. *a* **1734** R. North *Examen* (1740) I. ii. 102 But, *Ex pede Herculem*. They that let so much be seen, have not failed of... **1881** W. D. Hay *300 Years Hence* p. vii, From it he may glean some knowledge of Professor Mensier's character. *ex pede Herculem*. **1936** *Times Lit. Suppl.* 10 Oct. 805/2 Ex pede Herculem; these changes may not seem serious except to connoisseurs of English prose rhythm. **1953** W. V. Quine *From Logical Point of View* iii. 63 The finished lexicon is a case, *ex pede Herculem*... In projecting Hercules from the foot we risk error.

expellee. Delete *nonce-wd.* and add to def.: esp. from his own country. (Add further examples.)

1947 *Current History* Apr. 323/1 The Trades Union Congress finally agreed..that 500,000 D.P.'s and expellees in Germany might be admitted. **1961** R. Kee *Refugee World* iii. 21 For the Germans, left to deal with the expellees almost entirely by themselves, there was no sense in complaining. **1967** *Guardian* 15 Feb. 6/2 The neo-Nazi NPD party has probably rendered a service in drawing off the hard-core expellees.

expendable, *a.* Also **expendible.** Delete *rare* and add to def.: considered as not worth preserving or salvaging; normally consumed in use; *spec.* of military personnel: that may be allowed to be sacrificed to achieve a military objective. Hence as *sb.*, an expendable person or object.

1942 W. L. White *They were Expendable* 7 In a war anything can be expendable—money or gasoline or equipment or most usually men. **1942** *Reader's Digest* Oct. 40/1 They would be considered in part as expendable ammunition much as the Navy considers its PT boats. **1942** *Topeka Jrnl.* 9 Nov. 4/4 When an army is retreating, a small force is left behind to cover the retreat and be sacrificed to the enemy. They are 'expendables'. **1956** A. Toynbee *Historian's Approach to Religion* xix. 266 The true purpose of the expendable is simply to serve as a means for promoting the welfare of human beings. In truth it is not sacrosanct but is 'expendable'. **1966** D. Holbrook *Flesh Wounds* 81 We're expendable, see, so you want to watch out. **1966** *Aviation Week & Space Technol.* 5 Dec. 22/3 With five years of supplies and all the expendables, including a crew.

expenditor (ekspe·nditǝr). Delete †*Obs.* and add later *gen.* examples.

1899 *Westm. Gaz.* 23 Sept. 4/2 Expenditor of the level of Romney Marsh. **1901** *N. & Q.* 9th Ser. III. 303/1 General Expenditor to East Kent Sewers.

expense. Add: **5. expense account, sheet,** an account of expenses or expenditure; *spec.* an account of the expenses incurred by an employee in the course of his work and payable by his employer(s); also, the money so paid; freq. *attrib.*

1872 Expense account [in Dict.]. **1922** S. Lewis *Babbitt* x. 141 It went down on my expense-account—gosh, if I'd been paying it instead of the firm, I'd 'a' tramped the streets all night. **1933** J. B. Priestley *Wonder Hero* iii. 68 What about a quick drink and then some lunch?..It'll all go down on my expense account to the paper. **1937** W. S. Maugham *Theatre* ix. 75 When he directed a play he put down on the expense account the fee that a director of the second rank would have received. **1953** 'S. Ransome' *Drag Dark* (1954) xi. 119 They went off together...To lunch. Nice, expense-account place. **1959** R. Postgate *Good Food Guide 1959–1960* 87 Offering a good, solid, 'expense-account' type menu. *Ibid.* 222 Designed in menu and price for business executives with expense accounts. *Ibid.* 367 The eaters were given precisely the same attentive service as the expense-account-wallahs. **1960** *New Yorker* 23 Apr. 66/2 Pills for expense-account indigestion. **1962** G. Greene *Brit. Dramatists* 18 The scenes of Tamburlaine, outlined by pasteboard sets and an expense-sheet, shifted boldly all over Asia. **1965** J. B. Priestley *Three Men in New Suits* v. 91 We'll pay you thirty-five pounds a week and give you a reasonable expense sheet.

expensive, *a.* Add: **2. c.** *Comb.*, as *expensive-looking* adj.

1925 T. Dreiser *Amer. Trag.* I. ii. v. 184 Such expensive-looking and apparently smart displays. **1969** F. Sargeson *Memoirs of Peon* vi. 147 A fancy and somewhat expensive-looking purse.

experience, *sb.* Add: **9.** *attrib.*, as **experience philosophy,** experiential philosophy; **experience school,** the school of empiricism; **experience table,** a table of mortality computed from the experience of one or more life-assurance companies.

1865 J. S. Mill *Lett.* (1910) I. 215 The experience philosophy and the association psychology are getting up again. **1909** W. James *Pluralistic Universe* 9 ... experience-philosophy ... **1882** A. Bain *J. S. Mill* iii. 60 Mr. Ward has continued to uphold his peculiar tenets against the Experience-school. **1936** *Mind* XLV. 242 He holds (as some members of the 'experience-school' do) that feeling...is also a cognitive function. **1879** W. S. B. Champness *Insur. Dict.* (1883) 105 The Experience Table is based on that of Seventeen British Life Offices.

experience, *v.* **2. d.** (Earlier example.)

1827 *Knickerbocker* IX. 356, I have 'experienced religion', as it is called...

experienceable (ekspɪǝ·riɛnsăb'l), *a.* [f. Experience *v.* + -able.] Capable of being experienced.

1907 W. James *Pragmatism* vi. 224 So far as reality means experienceable reality, both it and the truths men gain about it are everlastingly in process of mutation. **1927** J. W. Dunne *Experiment with Time* xxiii. 170 Like all other sensations, its [sc. pain's] range of experienceable intensity must be limited. **1959** *Times Lit. Suppl.* 4 Jan. 3/4 The former is experienceable, can be lived from within as a style of life; the latter are stereotypes used or abused from without.

experient, *a.* and *sb.* Restrict †*Obs.* to B. *sb.* in Dict. and add: **A.** *adj.* (Later examples.)

1926 A. N. Whitehead *Sc. & Mod. World* ix. 211 The ego-object, as consciousness here-now, is conscious of its experient essence as constituted by its internal relations to the world of realities, and to the world of ideas. **1933** — *Adventures of Ideas* xiv. 216 In this transmutation the percipient occasion in question belongs to the personal succession of occasions which is the soul of the animal.

B. *sb.* **2.** One who experiences something; one who undergoes an experience.

1899 J. Ward *Natural. & Agnost.* II. 181 ... **1932** ... *Mind* ... Experience implies an experient. **1965** ... Religious Experience...is personal and incommunicable; however abundant the assurance of Reality it brings to the experient, he cannot share that certainty with others as he can the normal experience of the senses. **1925** J. E. Turner *Theory of Direct Realism* 16 Sense-contents..depend in part for their own existence and character directly upon the existence and activity of the percipient observer or experient. **1938** R. G. Collingwood *Princ. Art* 252 The emotions it expresses are not only the emotions of a merely conscious experient, they include the emotions of a thinker.

experiment, *sb.* Add: **4. b.** *concr.* Experimental apparatus.

1967 *Lebende Sprachen* XII. 161/1 *Experiment* wissenschaftliche Apparatur(en), Messgerät(e). The experiment will measure the distribution of oxygen in the atmosphere. **1969** *New Yorker* 12 Apr. 104/3 The three pots are the passive seismic experiment, the solar-wind experiment, and the suprathermal-ion-detector. *Ibid.* 110/3 A small stool...for the experiment to sit on.

7. *attrib.*, as **experiment farm, station,** an institution provided with means for carrying out scientific research into methods of agriculture, etc.

1893 J. Auld *Picturesque Burlington* 128 State Experiment Farm, including farm house, barn, creamery [etc.]. **1905** *Daily Chron.* 28 Aug. 4/4, I have seen a crop of 100 bushels of oats growing on one of the government farms in Canada. **1874** *Connecticut Bd. Agric. Rep.* 66 The establishment of experiment stations. **1892** *Pall Mall Gaz.* 4 Apr. 3/1 The most careful investigations at many of the experiment stations show that not only are all traces of the poisons removed before the fruit ripens, but [etc.]. **1964** *Economist* 19 Dec. 1352/1 Their [*sc.* the Americans'] research departments or 'experiment stations'.

experiment, *v.* **3.** Delete †*Obs.* and add later examples (? a gallicism).

1899 *Westm. Gaz.* 18 May 1/3 They are about to experiment another system, on a six months' probation. **1900** *Ibid.* 5 Jan. 10/1 It was he who first experimented the new battle formation of the French Army at the famous manœuvres of 1882–3.

experimental, *a.* and *sb.* Add: **A.** *adj.* **3. b.** (Further examples.) Also *experimental physicist, psychologist, psychology.*

1833 J. S. Mill *Lett.* (1910) I. 59 Detrosier...has some prospect of picking up a living as a lecturer on experimental physics. **1878** S. H. Hodgson *Philos. Reflection* I. 227 Fortunately we possess a genuine *a posteriori* experimental psychology, a true science. **1935** *Discovery* June 183/2 The gestalt-theorists, the behaviourists, and the experimental psychologists... **1944** J. A. Passmore in Stout *God & Nature* p. xxxvii, To Stout...experimental psychology was a science of common sense, not a substitute for thinking. **1957** G. Ryle in C. A. Mace *Brit. Philos. in Mid-Century* 258 A not very ancient Oxford Chair of Physics still retains its old label, the Chair of Experimental Philosophy. **1958** J. Cleugh tr. *Jungk's Brighter than Thousand Suns* i. 17 It was theoretical, not the experimental, physicists who were to blame for the confusion.

6. (Further examples of *experimental farm*.)

1894 *Trans. R. Soc. Canada* XII. iv. 143 Six years ago the testing of trees and shrubs..was begun at the experimental farms at Brandon, Manitoba. **1906** *Daily Colonist* (Victoria, B.C.) 4 Jan. 10/3 The Cigarmakers' union..urged that the experimental farms for the growing of tobacco be established in Canada. **1911** C. E. W. Bean *'Dreadnought' of Darling* xxv. 319 When Australian subeditors want to make the open page of their papers especially attractive, they cast on to it the great plans that are industriously afoot...about new experimental farms to be started in the far back country.

b. *spec.* of a theatre, play, etc.

1929 S. W. Cheney *Theatre* xxiii. 520 Jacques Copeau ...administered one of the...most fruitful of twentieth century experimental theatres. **1931** W. Rothenstein *Men & Memories* II. xxiv. 263 Experimental theatres are expensive things. **1961** Bowman & Ball *Theatre Lang.* 127 *Experimental*, said of drama, staging, a type of theatre, etc., which seeks freshness in the writing and production of plays rather than the traditional formulas for commercial or conventional success. **1969** Gish & Pinchot *Lillian Gish* xiv. 194 An experimental theater was also hit.

experimentee. For *rare*⁻¹ read *rare*, and add further example.

1904 *Daily Chron.* 14 May 4/6 After a year of this 'starvation diet' the 'experimentees' were stronger.

‖ **experimentum crucis** (eksperime·ntǝm krū·sis). [mod.L., 'crucial experiment': see Crucial *a.* 2.] A decisive test that shows which one of several hypotheses is correct; a crucial experiment.

1672 [see Crucial *a.* 2]. **1751** Hume *Enq. Princ. Mor.* v. ii. 84 The *Experimentum crucis*, or that Experiment, which points out the Way we should follow, in any Doubt or Ambiguity. **1814** *Phil. Trans. R. Soc.* CIV. 212 We may therefore consider the preceding result as an *experimentum crucis*, which finally decides the question as to the structure of the agate. **1856** *Spiritual Herald* Mar. 37 The injustice of our opponents in imposing any conditions of their own at an *experimentum crucis* of this kind. **1880** W. James *Coll. Ess. & Rev.* (1920) 169 The partisans of feelings of effort innervation request them [sc. illusions of a certain kind] as *experimenta crucis*. **1949** W. Kneale *Probability* § 22. 110 If..they cannot devise an *experimentum crucis*, they begin to suspect that the two rival hypotheses differ only in their symbolic formulation.

expertise (ekspǝɹti·z). [Fr.] **a.** Expert opinion or knowledge, often obtained through the action of submitting a matter to, and its consideration by, experts; an expert's appraisal, valuation, or report. **b.** The quality or state of being expert; skill or expertness in a particular branch of study or sport.

1868 Reade *Foul Play* III. x. 120, I have distanced my competitors in expertise. **1897** *Daily News* 19 Nov. 5/1 This looks as though Mr. Sellar has been uniformly unfortunate in his expertise. **1898** *Ibid.* 10 Feb. 5/4 He was suffered to attack the 'expertise', which the first court-martial accepted. **1898** *2nd Rep. Sci. & Art Mus.* 28 in *Parl. Papers* XI. 9 Sir Edward Maunde Thompson ..pointed out that expertise is not necessarily gained by having a large staff. **1907** A. Bennett *Grim Smile of Five Towns* 183 How could I be expected..to judge delicate points of expertise in earthenware? **1922** J. D. Beresford *Prisoners of Hartling* ii. 27 Arthur would have liked to give a ready diagnosis of this abnormal condition, but his expertise was not equal to the task. **1924** R. H. Mottram *Spanish Farm* 50 He was measuring fields for an 'expertise', a professional assessment. **1959** *Discovery* June 251/2 The expert is only capable of giving opinions within the special field of his own expertise. **1950** B. Wootton *Testament Social Sci.* p. v, Unfortunately, those who have sufficient expertise are few, and greatly occupied with their own specialisms. **1953** 'M. Innes' *Christmas at Candleshoe* ix. 124 'We'll have to get a fee?' 'For making an expertise.' That, it seems, is the technical term.' **1965** W. Girvan *Flying Saucers* v. 63 One might just as reasonably deny the existence of the stars on which he claims such expertise. **1970** *Insight* Nov. 1/2 A team of people with highly specialised expertise: buyers, economists, journalists, researchers with the knowledge and training to deal with consumer problems.

expertism (e·kspǝɹtiz'm). Delete *nonce-wd.* and add later example.

1958 *Cape Law Jrnl.* VII. 109 Medical Expertism considered from its legal standpoint.

expertize, *v.* Add: (Further examples.) Freq. as **expertizing** *vbl. sb.*

1900 *Spectator* LXXXV. 525/1 (*title*) The morality of 'expertising'. **1922** *Readers' Guide to Periodical Lit.* 1919–21 V. 525/1 Expertising in art. **1956** R. Stout *Before Midnight* 158 A mob of experts was expertizing in every direction, and Fritz was seated on Wolfe's chair behind his desk, watching them.

2. *trans.* (Also **expertise.**) To give an expert opinion on; to evaluate, assess.

1949 *Times Lit. Suppl.* 24 July 613/4 By applying the methods he used in 'expertising' old paintings and furniture John Ballinger, art expert, discovered the murderer. **1932** *Brooklyn Mus. Q.* XIX. 74/1 Professor Jean Capart..has been..expertizing the Wilbour collections.

‖ **experto crede** (ekspǝ·to kri·di), *phr.* [ad. L. *experto credite* (Virgil *Æneid* xi. 283), f. *experto* dat. sing. of *expertus*, pa. pple. of *experiri* to experience + *crede*, *credite* imp. sing. and pl. of *credere* to believe.] Believe one who has experienced or tried; believe me, I know.

1579 Gosson *Sch. Abuse* f. 17ᵛ, Experto crede, I have seene somewhat, and therefore I thinke I may say the more. **1732** Swift *Let.* 4 May (1963) IV. 24 You may beckon a blackguard boy under a gate, near your visiting place (*experto crede*) Save eleven pence; and get half a crown's worth of health. **1872** Geo. Eliot *Middlem.* III. v. xlv. 49 Take care—*experto crede*—take care not to get hampered about money matters. **1906** 'Q' *Mayor of Troy* ix. 124 Sir Felix, whose temper—*experto crede*—is seldom at its best in the small hours. *a* **1936** Kipling *Something of Myself* (1937) viii. 208 The night-sister in the Brush and the Ink.. *Experto crede*. **1937** J. M. Murry *Necessity of Pacifism* 131 Such a discipline...I have come to believe is necessary to the spiritual health of such a community as I here envisage. *Experto crede*.

expiree. Add to def.: Chiefly *Austral.* Also *attrib.*

1847 J. D. Lang *Cooksland* ix. 271 Very many of their servants, being old hands or Expiree convicts from New South Wales and Van Diemen's Land, are thoroughly incorporated into them... **1883** E. M. Curr *Recoll. Squatting in Victoria* v. 40 Wages were high, and labourers (almost all old gaol-birds and expiree convicts) exceedingly independent and rowdy.

e·xpert, *v.*² Chiefly *N. Amer. colloq.* [f. Expert *sb.*] *trans.* To examine as an expert; to have (books, etc.) examined by an expert. So **experting** *vbl. sb.*, expert examination or its results.

1889 in *Cent. Dict.* **1901** *Daily Colonist* (Victoria, B.C.) 26 Oct. 6/6 Colonel Lindley with a party of men has been in the coal region for the past couple of weeks experting the coal properties of the company. **1946** G. H. Johnston *Death takes Small Bites* i. 13 'But what about Japan or the Philippines? Wouldn't you find more there to write about?' 'No. Both played out. They've been experted all the way from page one to the obituaries.' **1946** G. Bowers *Word Assay?* (ed. 2) vi. 74 The next phase of experting a new comb is to scrape or rub the tips on wood. *Ibid.* 95, I trust that the experting as set out in this text will be a practicable guide. **1969** 'H. Pentecost' *Girl Watcher's Funeral* (1970) iii. ii. 148 We've done some experting on the trajectory of the body's fall.

explain, *v.* Add: **3. a.** (Further *absol.* examples.) (In some examples, sense 3 a is indistinguishable from 3 b.)

1856 Reade *Never too Late* II. v. 83 'Then what were you in the black hole for?' 'For obeying orders.' 'Nonsense! hum! Explain.' **1886** 'M. Gray' *Silence of Dean Maitland* xi. 36, He took a card from his pocket...'That will explain to Dr. Everard,' he said. ... *King Washington* 122, I cannot explain about the robbery to your father's garden without disloyalty to my duty. **1922** Joyce *Ulysses* 449 Gentlemen of the jury, let me explain. ... (1970) 28 Horace..lost no time in going on to explain.

7. c. With a direct statement as obj.: to utter in explanation.

1856 Reade *Never too Late* II. v. 94 'Yes sir clamined and no mistake.' 'North-country word for starved' explained Mr. Eden. **1903** R. Langbridge *Flame & Flood* x, 'Tears came into Susette's eyes...' 'I have bitten my tongue,' Susette explained. **1968** ... *Sat.* June 93/1 He slapped down a yellow envelope upon the desk. 'Telegram,' he explained tersely.

explanandum (eksplǝnæ·ndǝm). *Philos.* Pl. **explananda.** [L., neut. of gerundive of *explānāre*: see Explain *v.*] The thing to be explained. Cf. *Explicandum.

1892 *Academy* 6 Aug. 114/2 Whenever a psychical explanandum is run to earth in some feature of nerve structure or function, the pursuit is over. **1948, 1962** [see *Explanans]. **1964** I. Scheffler *Anat. Inquiry* 29 This account construes every explanation as consisting of an explanandum..and an explanans.

explanans (eksplǝ·nænz). *Philos.* [L., pres. pple. of *explānāre*: see Explain *v.*] The explaining element in an explanation; the explanatory premisses. (Opp. *Explanandum.)

1948 Hempel & Oppenheim in *Philos. of Sci.* XV. 136 We divide an explanation into two major constituents, the explanandum and the explanans. **1962** L. J. Cohen *Diversity of Meaning* xi. 133 An explanation of naming that is futile and circular because the explanans cannot be conceived except through the explanandum.

explant, *v.* Restrict †*Obs.* to sense in Dict. and add: **2.** *Biol.* To transfer (living tissue) from its site in the body to some other place, usu. a nutrient medium in which a culture of the tissue is initiated. Hence **expla·nted** *ppl. a.*

1915 *Amer. Jrnl. Anat.* XVII. 342 Around the piece of explanted tissue the new growth forms a more or less radiating reticulum. **1922** *Jrnl. Exper. Med.* XXXVI. 393 If part of an organ, for instance a toe of a chick embryo, was explanted, an uncontrolled proliferation of cells began from the injured parts. **1937** *Nature* 6 Mar. 413/1 A typical maturation division figure was observed in an explanted ovary from a four-day (post-embryonic) rat after 9 days' cultivation. **1966** *Sci. Jrnl.* Nov. 58/3 If the embryos are explanted during these three days, cell multiplication may continue.

explant (e·kspla:nt), *sb. Biol.* [f. the vb.] A piece of tissue, an organ, etc., that has been explanted.

1917 *Amer. Jrnl. Anat.* XXII. 171 The explants consist of small pieces of muscle..taken from the muscles of the back, wing or leg of chick embryos. **1926** *Proc. R. Soc.* B. XCIX. 353 During the first days of life *in vitro* most of the explants showed 'emigration' growth from the cut surface. **1957** *New Biol.* XXIII. 115 Transplanting pieces to abnormal positions on the embryo, or...growing such pieces in isolation as 'explants', away from any influence of other tissues. **1970** Passmore & Robson *Compan. Med. Stud.* II. xviii. 1 It is sometimes advantageous to use a tissue explant to study the growth of a virus.

explantation. Restrict †*Obs. rare* to sense in Dict. and add: **2.** *Biol.* The action of explanting tissue.

1915 *Amer. Jrnl. Anat.* XVII. 341 The trouble lies either in..poor chick material, or some manipulation during the process of explantation. **1937** *Nature* 6 Mar. 413/1 Only spermatogonia were present at the time of explantation. **1970** H. E. Wolff in J. A. Thomas's *Organ Culture* i. 3 Explantation is a term consisting of grafting a fragment of tissue or organ into the body of an adult or embryonic host.

explicand (eksplikæ·nd). *Philos.* [ad. L. *explicandus*, gerundive of *explicāre* to Explicate.] = next.

1882 W. James in *Mind* VII. 195 Such a system of explaining nature by spectrum, of making the explicans borrow credentials from the *explicand*, ..is a strange candidate for the honour of being a complete rationaliser of the world.

explicandum (eksplikæ·ndǝm). *Philos.* Pl. **explicanda.** [L., neut. of gerundive of *explicāre*: see Explicate *v.*] The fact, thing, or expression to be explained or explicated. Cf. *Explanandum.

1867 J. S. Mill *Let.* (1920) II. 92 In the explicandum it is he leaves everything unexplained, or smuggles the explicandum into the predicate. **1948** H. Joachim *Log. Stud.* i. 20 The explicandum..since it is a whole must in some sense consist of parts. **1953** C. S. Barell *Linguistic Form* 106 The explanation is not based on any facts known independently of the explicandum. **1969** R. B. Braithwaite *Sci. Explan.* x. 319 The primary explicanda for science are particular empirical facts. **1966** J. J. Katz *Philos. Lang.* iii. 54 Pragmatical concepts which could serve as explicanda.

explicans (eksplikænz). *Philos.* [L., pres. pple. of *explicāre*: see Explicate *v.*] The explanatory part of an explanation; in the analysis or explication of a concept or expression, the part that gives the meaning. (Opp. *Explicandum.)

1882 [see *Explicand]. **1953** R. B. Braithwaite *Sci. Explan.* x. 321 The explicans in such an explanation is an event.

explicate, *v.* **6.** Delete 'Now *rare*' and add later examples.

1959 *Times Lit. Suppl.* 20 Mar. p. x/5 The *kerygma*... attempts to explicate it by saying what it all happens 'existentially' in human experience. **1966** J. J. Katz *Philos. Lang.* ii. 17 Ordinary language philosophers tried to explicate the standards of usage underlying the linguistic behaviour... **1967** C. L. Wrenn *Word & Symbol* 2 Literary texts..are now being explicated in terms of 'symbolic realism', cultural anthropology, folk-myth and allegory.

‖ **explication de texte** (eksplikasyon dǝ tekst). [Fr.] A detailed examination of the text of a literary work.

1935 G. N. Henning *Representative Fr. Lyrics* (rev. ed.) p. vii, Following the Notes is a new feature in American textbooks, *Explications de textes*, whose minute analyses of *l'Expiation* and *Midi* may serve for a more reasoned appreciation of these and other poems. **1950** W. Stevens *Let.* 13 Oct. (1967) 694 It is not only a skillful example of explication de texte, but it reveals yourself. **1962** N. Nowottny *Lang. Poets Use* i. 19 The tradition of *explication de texte* offers methods of examining verifiable features of poems. **1962** R. Quirk *Use of English* xiv. 244 It requires no pedestrian *explication de texte* to see here the tie between 'a passive bucket' and the grammatical passive 'be pumped into'.

explicit (e·ksplisit). Transfer *Obs.* to sense a. **b.** Substitute for def.: As *sb.*, an instance of the use of this indication; hence, the last words or lines of a volume or section of a book; also *fig.*, conclusion, finis. (Cf. Incipit.)

a **1658** [in Dict.]. **1885** H. A. Dobson *At Sign of Lyre* 45 Tired the hand and tired the wit Ere the final *Explicit*! **1897** [see Incipit]. **1932** Blunden *Face of England* 81 The last man..came..to provide the *explicit*. **1963** *Times Lit. Suppl.* 22 Feb. 143/4 Appendices listing all the *explicits* and *incipits*.

explicit, *a.* Add: **5.** *Math.* Of a function: having the dependent variable defined directly in terms of the independent variable or variables. Cf. *implicit function* (Implicit *a.* 2).

1814 P. Barlow *New Math. & Philos. Dict.* sig. X7 (recto) Having given the methods..of obtaining the derived functions, of functions of one or more quantities, whether those functions be explicit or implicit,..we will now show how this theory may be applied. **1830** *Encycl. Metrop.* (1845) I. 796 If z were expressed by an explicit function of x and y, to determine dz/dx we should consider y as a constant in that function, and then differentiate it as a function of x alone. **1886** J. Edwards *Diff. Calc.* i. 4 If the equation connecting the variables be solved for the dependent variable, that variable is reduced from being an implicit to being an explicit function of the remaining variable or variables. **1955** D. A. Quadling *Math. Analysis* xiii. 223 There is no hard and fast distinction between implicit and explicit functions. Any function defined explicitly by an equation $y=f(x)$ could equally well be defined implicitly by the equation $g(x, y)=0$, where $g(x, y)\equiv y-f(x)$.

explode, *v.* Add: **5. d.** Of population: to increase suddenly or rapidly. Cf. *explosion 4 b.

1959 *N.Y. Times* 15 Nov. 25/2 The population of the Bandung area has exploded from 167,000 in 1930 to 1,200,000 this year. **1962** *Listener* 15 Nov. 795/2 The relentlessly exploding population. **1965** *New Statesman* 30 Apr. 672/2 Population has 'exploded' only in the last three centuries because of our new-found science, technology and humanity.

exploded, *ppl. a.* Add: **4. b.** Of a model or illustrated technical drawing: showing all the separate components as if 'exploded' from the complete unit but retaining their relative positions.

1949 *Jrnl. R. Aeronaut. Soc.* LIII. 60/2 They had to study the exploded drawing of the components. **1951** *Engineering* 13 July 33/2 An 'exploded' model shows the layout of a pit head. **1962** Corson & Lorrain *Introd. Electromagn. Fields* ii. 42 The exploded view shows one of the volume elements in detail.

exploding, *ppl. a.* Add: **c.** *exploding wire*, a wire subjected to a sudden and very high current so that it explodes violently. Also *attrib.*

1920 *Astrophysical Jrnl.* LI. 40 The larger part of the energy of the condenser is expended in the exploding wire. **1961** *Aeroplane* 3 Feb. 110/2 U.S. Air Force is investigating the use of 'exploding wires' made of magnetic materials which can be accelerated and controlled by magnetic fields for the electrical propulsion of space vehicles. **1963** *New Scientist* 16 May 386/1 The term 'exploding wire' has..been applied to almost any situation where a conductor..disintegrated with a loud report. *Ibid.* 386/2 The objective in an exploding wire experiment is to bring a large amount of energy into the wire very quickly.

exploitabi·lity. [f. Exploitable *a.*: see -ity.] Capability of being exploited.

1960 *Commentary* June 469/2 Enemies of capitalism.. never predicted..its bounteous exploitability. **1969** *New Scientist* 20 Feb. 395/2 The 'exploitability' criterion means in effect that the sea-bed and subsoil underneath all the oceans of the world have already been partitioned between the coastal states.

exploitative, *a.* Delete *rare* and add to def.: Also, of or pertaining to the exploitation of people.

1908 *Westm. Gaz.* 11 Jan. 1/3 The Tariff Reformers now attack Lord Cromer for what is called his 'exploitative Imperialism'. **1947** E. Fromm *Man for Himself* 64 The exploitative orientation, like the receptive, has as its basic premise the feeling that the source of all good is outside, that whatever one wants to get must be sought there, and that one cannot produce anything oneself. **1949** M. Mead *Male & Female* ix. 200 Protections for all the young..against any exploitative or inhuman treatment. **1951** *Mind* LX. 287 The receptive, exploitative, hoarding, and marketing orientations.

exploitee (eksploiti·). [f. Exploit *v.* 4 b + -ee[1].] One who is exploited.

1941 *New Republic* 17 Feb. 216, I should feel easier in my mind..if I knew and liked my exploitees. **1951** S. Spender *World within World* 118 This meant I was a potential exploitee, because I could never feel within myself the rightness of a social situation which would rebuke the wrongness of others. **1960** *Spectator* 29 July 192 Exploiters and exploitees were gathered together.

exploitive (eksploi·tiv), *a.* [f. Exploit *v.* + -ive.] Tending to exploit; exploitative.

1921 J. R. Commons *Trade Unionism* 574 The ownership of homes, free from the grasp of exploitive and speculative interests. **1937** *Geogr. Jrnl.* XC. 368 An exploitive and wasteful system of agriculture has ruined the soil. **1959** M. Mead *Male & Female* xvii. 354 Affairs that do not lead to marriage are seen as exploitive.

explorational (eksplorēi·ʃənăl), *a.* [f. Exploration + -al.] Pertaining to, connected with, or involving exploration.

1923 E. F. Wyatt *Invis. Gods* II. iii. 57 The explorational tokens of the past in the Government Building. **1927** *Daily Tel.* 31 May 9/5 A blue-book which has had no superior as a contribution to explorational geography.

explore, *v.* Add: **3. a.** Fig. phr. *to explore every avenue* (or *to explore avenues*), to investigate every possibility.

1922 G. Edwards *From Crow-scaring to Westminster* xii. 142, I intended to explore every avenue during the next few days before the final crash came to secure peace. **1926, 1927** [see *avenue *sb.* 2]. **1940** N. Mitford *Pigeon Pie* xv. 232 In war-time we are bound to explore every avenue, whether it is likely to be productive of results, or not.

b. (Further example.)

1919 M. K. Bradby *Psycho-analysis* 165 Those striking characteristics are on the surface. We do not explore for others less obvious, because those that we see satisfy systems of repressed or under-expressed emotion.

c. To make an excursion; to go on an exploration (*to*).

1816 Jane Austen *Emma* II. xiv. 263 While they are with us, we shall explore a great deal. *Ibid.* 264 We explored to King's-Weston twice last summer. **1924** R. Macaulay *Orphan Island* ii. 29 Wouldn't the Royal Geographical Society finance the expedition? It ought to, as it's to explore to an undiscovered island.

exploring, *vbl. sb.* Add: **exploring coil** *Electr.*, a small flat coil of insulated wire connected with a galvanometer, used for determining the strength of a magnetic field from the current induced in the coil when it is quickly turned over or withdrawn from the field; a search coil; **exploring conductor** (see quot.); **exploring wire** (see quots.).

1913 *Year-bk. Wireless Telegr. & Telephony* 316 The moveable exploring coil is attached to a handle and index. **1879** *Noad's Text-bk. Electr.* 75 The Exploring Conductor at Kew Observatory.—This is a conical tube of thin copper raised 16 feet above the dome of the building, carrying at the top a small lantern or collecting lamp, provided with a little cowl. *Ibid.* 83 The late Mr. Crosse and the late Mr. Weekes examined the electrical condition of the lower regions of the atmosphere..by means of exploring wires insulated on appropriate supports. **1892** Mrs. A. Crosse *Redlet. Days* I. 178, I have an apparatus arranged for testing the electricity of the atmosphere. It is connected with exploring wires carried on high poles for more than a mile round the woods.

explosibility (eksplōᵘzibi·lĭti). [f. Explosible: see -ity.] Liability to explode.

1879 *Mind* IV. 335 All the other peculiarities both of choke-damp and of suffocation, such as convulsions and agony on the one hand, density and explosibility on the other. **1909** C. A. Keane *Mod. Org. Chem.* 315 This condition of strain in acetylene and its derivatives is a factor of importance in regard to their explosibility. **1922** *Daily News* 28 Jan. 5/6 The range of explosibility of water-gas-air mixtures is considerably wider than that of coal-gas-air mixtures.

explosion. Add: **3. d.** *Golf.* An explosive (sense *3 b) shot. Also *attrib.*

1924 C. J. H. Tolley *Mod. Golfer* x. 149 If the ball is lying well, you can either play an 'explosion' or take the ball cleanly. **1926** Wodehouse *Heart of Goof* viii. 259 An explosion-shot out of the bunker on the fourteenth. **1957** L. T. Stanley *Fontana Golf Bk.* 106 *Explosion Shot.* The most reliable recovery shot of all if played firmly.

4. b. A rapid or sudden marked increase or development; esp. in *population explosion*.

1953 *Time* 19 Oct. 28/1 Latin America is in the midst of a 'population explosion'. Its people are multiplying 2½ times as fast as the populations in the rest of the world. **1961** 'R. Macdonald' *Wycherly Woman* (1962) ii. 20 Those sudden institutions of learning that had been springing up..to handle the products of the wartime copulation explosion. **1961** *New Yorker* 28 Oct. 43/2 After the population explosion came the home-run explosion, the rental explosion,..a piety explosion, a culture explosion, [etc.]. **1962** *P.M.L.A.* LXXVII. ii. 4/2 We have all heard of 'explosions', of which the 'population explosion' is only one. Well, here is yet another type of explosion.., 'the sky-rocketing volume of knowledge'. **1965** *Listener* 4 Nov. 714/2 We must achieve..the contraception explosion before the population explosion. **1970** *Daily Tel.* 17 Mar. 7/8 British Rail is being forced to join the current 'wages explosion' and concede new substantial pay increases.

5. *attrib.*, as *explosion point*; **explosion chamber,** a chamber at the end of the cylinder of an internal combustion engine in which the charge is exploded; **explosion machine** (see quot.); **explosion pipette,** a pipette in which an explosive mixture of gases may be fired by an electric spark; so *explosion-tube*.

1888 *Lockwood's Dict. Mech. Engin.*, *Explosion chamber*, the hinder extension of a cylinder of a gas engine in which the charge is exploded. **1903** *Motoring Ann.* 282 It reduces the space of the explosion chamber. *a* **1884** Knight *Dict. Mech.* Suppl., *Explosion machine*, a motor which depends for its force upon the explosion of substances generating a gas which is used under pressure in an engine or apparatus. **1901** M. W. Travers *Exper. Study Gases* 136 In the earlier experiments the tap of the explosion pipette was lubricated with a hydrocarbon grease. **1890** W. James *Princ. Psychol.* II. xviii. 75 The level of tension in the cells does not rise to the higher explosion-point. **1938** W. S. Churchill *Into Battle* (1941) 17 The war-lust of Dictator Powers had reached its culminating explosion point. **1893** *Phil. Trans. R. Soc.* A. CLXXXIV. 550 By opening tap 8 quantities of the mixed gases are drawn over into the explosion-tube.

explosive, *a.* and *sb.* Add: **A. 2. c.** *explosive bolt*, a bolt that can be released by being blown out of position by an explosive charge; *explosive rivet*, a rivet containing an explosive charge by means of which it is fixed in place.

1948 *Aeroplane* 3 Sept. 288/1 The four launching rockets..were then jettisoned, by the shearing of explosive bolts. Such components—a development of the explosive rivets of Heinkel—were frequently applied to German guided missiles. **1962** S. Carpenter in *Into Orbit* 54 The escape tower..is attached to the capsule by explosive bolts which blow apart and separate the two components when the tower is no longer needed. **1966** *Times* 4 June 1/1 The shroud [of an unmanned satellite] is in two sections..and explosive bolts and built-in springs were supposed to thrust it off.

3. b. *Golf.* Causing a ball to jump out of a bunker as if an explosion had taken place beneath it.

1924 C. J. H. Tolley *Mod. Golfer* x. 149 If you are lying badly..the ball must be dug out, and the method employed is called an explosive shot.

B. *sb.* **2.** *high explosive*, an explosive compound, such as dynamite, guncotton, etc., which is more rapid and powerful than gunpowder. Also *attrib.*

1877 *Van Nostrand's Eclectic Engin. Mag.* XVII. 300 (*heading*) Transportation of high explosives. *Ibid.* 300/2 Nitroglycerin, the basis of what are generally known as the high explosive powders, is a light colored, oily liquid. **1890** G. S. Clarke *Fortification* ix. 113 High explosives produce great local destructive effect against masonry. **1892** Hake & Macnab tr. *Berthelot's Explosives* 2 Generally speaking, we mean by 'high' explosives, those in which the chemical transformation is very rapid, and which exert a crushing or shattering effect. *Ibid.*. The more common 'high' explosives are bodies containing a large amount of oxygen, and possessing a definite chemical composition. **1899** J. W. Mackail *Morris* II. 237 High-explosive bombs. **1917** *Nature* C. 101/2 High-explosive and armour-piercing shell. **1941** *Ann. Reg. 1940* 95 The bulk of the damage had been caused not by high explosive ..bombs. **1941** Koestler *Scum of Earth* 284 A man might be happier amongst a shower of high-explosives and incendiaries than under the *Pax Swasticana*.

Expo (e·kspo). [Abbrev. of Exposition 3 b.] A large international exhibition; *spec.* the world fair held at Montreal in 1967.

1963 *Montreal Star* 31 Dec. 3/2 The request of the Expo to study the selection of a new symbol. *Ibid.*, They have been asked by the Expo authorities to meet. **1965** *Economist* 10 July p. xxxviii/1 On April 28, 1967, the world will know whether the Montreal Expo is a work of creative imagination or an overblown exercise. **1968** *Guardian* 16 July 11/1 Ever since the first World Exposition was held in London in 1851, every Expo has been held in the Western Hemisphere. **1969** *N.Z. News* 11 June 3/4 Executive chef for the New Zealand restaurants at Expo 70 in Japan. **1970** *Daily Tel.* (Colour Suppl.) 13 Mar. 17/2 International power-game politics are never far away from the realities of an Expo.

‖ **expolitio** (ekspoli·ʃio). [a. L. *expolitio* adorning, embellishing.] = Expolition b.

[*c* **82** B.C. *Rhetorica ad Herennium* IV. xlii. § 54 Expolitio est cum in eodem loco manemus et aliud atque aliud dicere videmur.] **1589** Puttenham *Eng. Poesie* III. xx. 206 For the glorious lustre it setteth vpon our speech and language, the Greeks call it (*Exargasia*) the Latine (*Expolitio*). **1943** J. W. H. Atkins *Eng. Lit. Crit., Mediaeval Phase* 203 *Expolitio*, enlarging on a topic in different ways. **1964** C. S. Lewis *Discarded Image* vii. 192 One kind of *mora* is *Expolitio*. Its formula is 'Let the same thing be disguised by a variety of form; be different, yet the same.'

exponential, *a.* and *sb.* Add: **A.** *adj.* **2. b.** *exponential horn*, a loudspeaker horn in which the diameter and the transverse cross-sectional area increase exponentially with the distance from the diaphragm. Hence *exponential-horned* adj.

1927 *Wireless World* 16 Nov. 664 Whereas a conical horn increases in diameter by a constant additive factor per unit of length, the exponential horn increases its diameter by a *constant multiple* per unit of length. **1954** K. Amis *Lucky Jim* viii. 87 The rondo of some boring piano concerto Welch had once insisted on playing him on his complicated exponential-horned gramophone. **1962** A. Nisbett *Technique Sound Studio* 269 A flared (exponential) horn is provided to improve coupling.

c. *exponential pile* or *reactor*, a small, subcritical section of a reactor lattice with a neutron source at one end, used to obtain information about the full-sized critical reactor.

1945 H. D. Smyth *Gen. Acct. Devel. Atomic Energy Mil. Purposes* vi. 58 By this time [*sc.* 1942] it was known that the neutron density decreased exponentially with increasing distance from the neutron source (hence the name often used for experiments of this type, 'exponential pile'). **1963** F. A. Valente *Man. Exper. Reactor Physics* viii. 262 In principle, exponential piles have the same structure and composition as the potential or proposed reactor of which they are supposed to be a copy and for which they are supposed to obtain design information by experiment.

exponentially (ekspone·nʃăli), *adv.* [f. Exponential *a.* + -ly[2].] By exponentials; in an exponential manner.

1909 in *Cent. Dict.* Suppl. **1938** R. W. Lawson tr. *Hevesy & Paneth's Man. Radioactivity* (ed. 2) xi. 127 The β-radiation of the iron salt does not remain constant, but decreases exponentially. **1949** G. Bateson in M. Fortes *Social Structure* 49 A vista of unknowable factors whose number increases (probably exponentially) into the future. **1968** H. Harris *Nucleus & Cytoplasm* iii. 54 In cells growing exponentially, ribosomal RNA appears to be essentially stable. **1970** *Nature* 22 Aug. 865/2 Population genetics, like any other branch of science, is growing exponentially.

exponentiation (e:kspŏnenʃiēi·ʃən). [irreg. f. Exponent *sb.* + -ation.] The mathematical operation of raising one quantity (the base) to the power of another (the exponent).

1903 B. Russell *Princ. Math.* xii. 120 Moreover exponentiation unavoidably introduces ordinal notions, since a^b is not in general equal to b^a. **1940** *Mind* XLIX. 246 Positive integers compounded by successive exponentiation. **1953** Birkhoff & MacLane *Surv. Mod. Algebra* (ed. 2) xii. 367 For arbitrary cardinal numbers α, β, and γ the following laws on exponentiation hold: (i) $\alpha^\beta\alpha^\gamma=\alpha^{\beta+\gamma}$; [etc.]. **1968** Corlett & Tinsley *Pract. Programming* ii. 15 The operator ↑ is the sign of exponentiation.

export, *sb.* Add: **3. export drive** (see *drive *sb.* 1 g); **export reject,** an imperfect article withdrawn from export and sold on the home market; also *transf.* and *fig.*; **export surplus,** the amount by which exports exceed imports (see also quot. 1965).

1952 O. Lancaster in *Daily Express* 2 Feb. 1/3 The Americans regard Paul Robeson as an export reject. **1962** M. Urquhart *Frail on North Circular* xxiii. 126, I got a new suit... Off the peg. Export reject. **1967** J. Porter *Chinks in Curtain* xii. 121 Don't you start bandying words with me, laddie, or you'll finish up with a mouthful of export reject teeth. **1931** *Economist* 24 Jan. 165/1 It will be noticed that two of our rivals have maintained an export surplus during the past year. **1964** *Ann. Reg. 1963* 75 The world boom in sugar prices enabled Queensland to sell its record export surplus. **1965** J. L. Hanson *Dict. Econ. & Comm.* 163/1 *Export surplus*, a rather ambiguous term, it may be taken to mean an overall surplus in the balance of payments, the doctrine of the mercantilist system, or a surplus of visible exports over visible imports. *Ibid.* 163/2 The aim has been to secure an export surplus in order to build up investment abroad...Immature economic systems tend to have export surpluses of goods in order to pay the interest on foreign loans.

b. *attrib.* or quasi-*adj.* Designating an article of a quality suitable for exporting, usually of a better quality than that made for home consumption. Also *ellipt.*, beer of export quality; a glass of this beer.

c **1898** in A. Davis *Package & Print* (1967) Plate 40, Export pilsener beer. **1936** Bentley & Allen *Trent's Own Case* ix. 105 M. Dupont proceeded..to order two glasses of export-cassis. **1951** A. Christie *They came to Baghdad* iii. 29 The materials for a suit were considered. 'Fortunately, I can give you our own export quality...' **1959** 'H. Carmichael' *Stranglehold* i. 11 I'll have a small whisky..followed by a bottle of your export special. **1962** *Economist* 5 May 465/2 The..beers..are mainly *Pils*..and *Export* (with a greater proportion of malt). **1964** W. J. Gaston *Drifting Death* ii. 18 Cluny ordered a round of exports. **1969** *Guardian* 16 July 18/2 The work that was specially prepared for the Western market which is now either known as Export Porcelain or *Chine de commande*. **1970** G. Lord *Marshmallow Pie* iv. 32 Get in a couple of crates of Export.

exportation. Add: **4.** *Logic.* The inference that if two propositions together imply a third, then the first of them on its own implies that the second implies the third.

1903 B. Russell *Princ. Math.* ii. 16 If pq implies r, then p implies that q implies r. This is the converse of the preceding principle, and is called exportation. **1954** I. M. Copi *Symbolic Logic* iii. 43 Exportation (Exp.): $[(p.q)\supset r]\equiv[p\supset(q\supset r)]$. **1965** Hughes & Londey *Elem. of Formal Logic* xviii. 125 This is known as the Law of Exportation..since its effect is to 'export' one proposition from the antecedent to the consequent.

expose, *v.* Add: **5. c.** *Photogr.* To submit (a sensitized surface) to the action of actinic rays. Often *absol.*

1839 *Visitor* Dec. 479/1 In one specimen which had been exposed only thirty seconds, the plate was still intensely black, excepting in the sky. **1848** *Chambers's Jrnl.* 24 June 403/2 On another plate, exposed for twenty minutes, a long black stripe was produced. **1878** W. Abney *Treat. Photogr.* xxxi. 246 As regards the exposure to be given to a picture there is one golden rule to follow: 'Expose for the shadows and let the lights take care of themselves'. **1903** A. Watkins *Photogr.* (ed. 2) 9 In the all-important question, 'how long to expose?'..the size of the stop has a most important influence. **1905** *Westm. Gaz.* 2 Sept. 14/2 A simple lesson in exposing a plate. **1959** F. Luther *Microfilm* xii. 116 Dagron exposed his microfilms in the camera as conventional glass plates.

expose (ekspōᵘ·z), *sb.* Chiefly *U.S.* = Exposé.

1715 J. Chappelow *Right Way to be Rich* (1717) 137 Mercy..should, with such an Acclamation of exulting Joy, make a visible and glorious Expose of the Blessed Jesus, to Divine Justice. **1813** *Niles' Reg.* IV. 185/1 (*title*) Expose of the French Empire. **1818** *Massachusetts Spy* 14 May (Th.), The expose of the situation of the interior [of the French empire] exceeds the style of modern romance. **1830** *Ibid.* 12 May, Chilton's expose. [Heading of an article on the shortcomings of the Jackson administration.] **1841** *Congress. Globe* 25 Jan. App. 152/3 And how was this honest expose met? **1888** *St. Louis Globe Democrat* 30 Mar. (Farmer), The expose of the Coal Hill convict camp horror. **1902** E. Banks *Newspaper Girl* 106 'Don't go into it with the idea of an "expose",' said the editor. **1967** *Boston Sunday Herald* 2–8 Apr. 67/1, I am..all for more exposes of the calibre of a recent one on British TV, 'Cathy Come Home'. **1968** *Globe & Mail* (Toronto) 5 Feb. 13/4 Some expose likely in the political arena.

exposé. 2. (Earlier and later examples.) Also *attrib.*

1809 P. Coxe (*title*) The Exposé; or, Napoleone Buonaparte unmasked, in a condensed statement of his career and atrocities. **1828** Lytton *Pelham* I. xxii. 171 Such an *exposé* might be attended with a loss the good woman valued more than reputation, viz., lodgers. **1965** *Times Lit. Suppl.* 25 Nov. 1061/2 A startling series of exposé articles. **1966** *Listener* 17 Nov. 749/1 This talk.. dealt mainly with exposé journalism. *Ibid.*, No newspaper embarked on an exposé without giving really serious thought to the matter.

exposit (ekspǫ·zit), *v.* [f. L. *expositus*, pa. pple. of *expōnere* to set forth.] *trans.* To reveal, exhibit, show clearly; to expound.

1882 C. Van Norden *Outermost Rim* viii. 104 The law of retribution is fully exposited only in the history of a great nation. **1949** Wellek & Warren *Theory Lit.* (1961) ix. 99 Examples could be multiplied indefinitely. One can assemble and exposit the 'world' of each, the part each gives to love and marriage, to business, to the professions. *Ibid.* xvii. 244 That case we have partly put in expositing the principle of aesthetic purity.

exposita (ekspǫ·zită), *sb. Logic.* [L., fem. sing. (cf. Maxim) of pa. pple. *exposit-us* of *expōn-ĕre*: see Expone.] = Convertend.

1827 [see Converse *sb.*[2] 3]. **1851** H. L. Mansel *Proleg. Logica* 202 Either the converted proposition is a new judgment distinct from the *exposita*, or it is merely the same judgment expressed in different language. **1857** W. Spalding in *Encycl. Brit.* XIII. 605/1 The antecedent is called the Convertend, or Exposita (the proposition set forth to be converted).

exposition. Add: **3. b.** (Earlier examples.)

1851 *Illustr. London News* 29 Mar. 259/1 This Exposition usually takes place annually in the Louvre. **1851** Geo. Eliot *Let.* 5 Aug. (1954) I. 358 She seems to prefer London...A glance or two at the Exposition she thinks would do her no harm.

4. c. Substitute for def.: (See quots.) Add further examples.

1889 Grove *Dict. Mus.* IV. 630/2 In forms of the harmonic order the term Exposition is commonly used of the first half of a movement in Binary form, because that part contains the statement of the two principal subjects. **1947** A. Einstein *Mus. in Romantic Era* vii. 67 The exposition of the first is a perfect example of demoniac explosion, while, in the first movement of the C-major Symphony the exposition is a piece of inexorable consummation. **1959** Westrup & Harrison *Collins Mus. Encycl.* 230/1 *Exposition*, the statement of the musical material on which a movement is based...In a movement in sonata form it..consists in the presentation of the principal thematic material partly in the tonic key and partly in a subsidiary key or keys.

ex post facto. a. (Earlier and later examples.)

1632 *Reply to Remonstrance of Netherlands E. India Co. in Justification of Proceedings at Amboyna* (E. India Co.) 11 For first for the notoriousnesse of the pretended conspiracie; although now *ex post facto*, they cry it out for notorious. **1679** J. Goodman *Penit. Pard.* I. iii. 77 Whatsoever sin..is not repented of when it is come to our knowledge, is by that means become a voluntary transgression, increasing its guilt *ex post facto*. **1835** *Tomlins's Law-dict.* (ed. 4) *s.v.*, An act done..may be made good by matter *ex post facto*, that was not so at first. **1966** *Listener* 10 Nov. 705/2 A piano suite by Cornelius Cardew eccentrically titled (one supposes *ex postfacto*) *Three Winter Potatoes*.

expostulant (ekspǫ·stiŭlănt), *a.* [ad. L. *expostulant-em*, pr. pple. of *expostulāre* to Expostulate.] Expostulating.

1880 Ruskin *Elem. Eng. Prosody* 23 The weighty and appellant or expostulant use of the Iambic monometre. **1898** I. Zangwill *Dreamers of Ghetto* vii. 295 He would wander tipsily through the sleeping streets..arguing metaphysics with expostulant watchmen. **1922** L. Housman *Dethronements* 78 Ex-Pres. [Wilson]. You can't reproach me with it, Tumulty. Tumulty (*expostulant*). I'm not doing that, Governor!

exposure. Add: **1. e.** *Photogr.* The exposing of a sensitized surface to the action of actinic rays (see *expose *v.* 5 c); also, the time occupied by this action. Also *attrib.*, as **exposure meter,** a device that indicates the correct time to allow a film, etc., to be exposed.

1839 *Visitor* 479/1 Living objects, if they remain motionless during the short periods of exposure, are given with perfect fidelity. **1847** *Phil. Trans. R. Soc.* CXXXVII. 256 After ten seconds of exposure I put the prepared plate in the mercury box. **1878** W. Abney *Treat. Photogr.* xxxi. 246 As regards the exposure to be given to a picture. **1891** *Anthony's Photogr. Bull.* XXII. 285/2 He would like to know whether these exposure meters would give the correct exposures. **1892** [see *background *sb.* 1 b]. **1905** *Westm. Gaz.* 2 Sept. 14/2 It is little use for the amateur to use exposure tables to guide him as to the correct exposures to be given under certain conditions. **1918** *Photo-Miniature* Mar. (Glossary), *Exposure indicator*, a device attached to plate-holders to show that the shutter has been withdrawn and re-inserted, i.e., exposure of plate. **1919** *Brit. Jrnl. Photogr. Alm.* 244 The taking of a photograph is known as an exposure, *e.g.*, 'I made 6 exposures'; and the term is also used for the time occupied, *e.g.*, '6 seconds exposure'. **1926** [see *actinometer 2]. **1940** A. L. M. Sowerby *Wall's Dict. Photogr.* (ed. 15) 281 Exposure calculators are identical in principle with exposure tables...The calculator-mechanism is simply a substitute for the more usual addition of a series of numbers. **1958** *Oxf. Mail* 19 May 6/4 An exposure meter..shows how much light is falling on the subject you want to photograph, and the exposure required.

f. *indecent exposure*, the action of publicly exposing one's body in an indecent manner. Also *fig.*

1851 *Act* 14 & 15 *Vict.* c. 100 § 29 Any public and indecent Exposure of the Person...It shall be lawful for the Court to sentence the Offender to be imprisoned. **1872** Wharton *Law Lex.* (ed. 5) 470/2 *Indecent exposure*, an indictable offence at common law. **1928** G. B. Shaw *Platform & Pulpit* (1962) 188 Disgraceful conduct in the parks, indecent exposure. **1958** J. Cannan *And be a Villain* iii. 49 He took himself off to deal with a case of indecent exposure on the tow-path. **1959** N. Marsh *Singing in Shrouds* iv. 67 Have you witnessed his weekly exhibitions of indecent exposure on the television? **1970** *Sunday Times* (Colour Suppl.) 6 Dec. 24/2 One hard-working executive would have become the only male in an office containing 100 women, if Ian Withers had not discovered that the man had convictions for indecent exposure.

g. *Mountaineering.* (See quots.)

1935 D. Pilley *Climbing Days* i. 17 You really do feel suspended over the valley with nothing but air below. 'Exposure', in other words, to use the climbing technicality, is continuous. **1940** F. Smythe *Adv. Mountaineer* vii. 102 Exposure..is simply the height and nature of the drop beneath the climber. **1957** R. G. Collomb *Dict. Mountaineering* 66 *Exposure*, a climber's awareness or feeling of height; his appreciation of the open position on a steep cliff or mountain...The effect of exposure is a temporary set-back or nervousness.

express, *a.*, *adv.*, and *sb.*[1] Add: **A.** *adj.* **4. e.** = *high-speed *a.*; *express boiler*, a boiler capable of getting steam up with great rapidity; *express highway*, etc.: see *expressway; *express lift*, a lift which does not stop at every floor. Cf. 4 b in Dict.

1897 Kipling *Day's Work* (1898) 217 Express freight's what pays. **1902** *Encycl. Brit.* XXVI. 285/2 The types, sometimes called 'Express' boilers, which are largely used in torpedo-boats,..where the most important requirement is very high power with a very small weight of boiler. **1909** *Cent. Dict.* Suppl., *Express-pump*, a high-speed pump; one that makes a high number of strokes per minute. **1909** *Westm. Gaz.* 9 June 11/4 Six electric passenger lifts, two of which are known as 'express' lifts. **1938** E. Bowen *Death of Heart* I. ix. 160 Matchett is sending Anna's white velvet dress to the express cleaners. **1967** W. Pine *Protectors* xi. 91 The two men rode up from the basement to the twenty-first floor by express lift. **1971** *Daily Tel.* 4 Jan. 7/2 Those express lifts (sorry elevators) which serve New York's highest buildings.

B. *adv.* **3. b.** Without a stop.

1892 *Harper's Mag.* Feb. 426/1 The managers of certain tall buildings now arrange them [*sc.* elevators] so that some run 'express' to the seventh story.

C. *sb.* **3. a.** For '*U.S.*' read 'Chiefly *U.S.*' (Earlier and later examples.) Also, goods or other articles conveyed by this method. Cf. *pony-express* (Pony *sb.* 6 b).

1794 D. M'Gillivray *Jrnl.* (1929) 40 No goods are sent him by this express. *Ibid.*, We expect an express from him very soon. *Ibid.*, The Canoe containing the express. **1839** *Boston Transcript* 27 Feb. 4/1 Boston and New York Express Car. *Ibid.* 21 Mar. 2/2 Harnden's Express, between Boston and New York, has been running since the 4th of March. **1840** *Boston Daily Advertiser* 7 Feb. 4/3 Mr. Harnden will for the present send his Express via Worcester. **1862** C. R. Thatcher

Dunedin Songster No. 1, 9 There was a little express waggon waiting to convey these Melbourne notables to Dunedin. **1869** *Trans. Ill. Agric. Soc.* VII. 522 They can be handled as roughly, almost, as these express agents handle boxes. **1879** F. R. Stockton *Rudder Grange* vii. 76 My package was wheeled to the express-car. **1882** W. D. Hay *Brighter Britain* I. ii. 32 Omnibuses..express-carts, waggons. **1922** O. Jespersen *Lang.* v. 108 Hilary M...called an express-cart a *press-cart*.

express, *v.*¹ Add: **1. a.** *spec.* To press or squeeze out (milk or other secretion) from the breast.

1932 R. C. Jewesbury *Mothercraft* iii. 37 Milk can be expressed manually, but this often causes pain. **1953** Carter & Dodds *Dict. Midwifery* 82/1 The patient should be taught how to express secretion from the nipple so that the milk may be able to flow freely later on. **1958** B. Spock *Baby & Child Care* 33 If you can learn the knack of expressing a little milk at the same time, it will lure him on.

11. *Genetics.* To display or make manifest in a phenotype (a character or effect attributed to a particular gene); to cause (a gene) to produce its associated character in a phenotype. Chiefly *pass.* or (occas.) *refl.*

1918 Babcock & Clausen *Genetics* v. 69 The condition of dominance..is determined by the fact that in the hybrid that character is expressed to the exclusion of its contrasted character. **1927** *Genetics* XII. 153 In these cultures the radius incompletus character is usually expressed more strongly in those flies which show the character Alae divergentes. **1946** R. R. Gates *Human Genetics* I. ii. 5 Many cases are now known in which the same abnormality is strictly dominant in some pedigrees and sometimes skips a generation in others.. The gene is present in the germplasm, as shown by its transmission to the next generation, but for some reason it has completely failed to express itself. **1965** H. E. Sutton *Introd. Human Genetics* ix. 109 A dominant gene may be defined as one that is expressed when present in a single copy. **1968** *New Scientist* 7 Nov. 313/1 As every cell in a multicellular organism contains an identical set of genes directing the synthesis of its proteins, the problem of development comes down to a question of controlling the expression of genes [a gene is said to be 'expressed' when the protein for which it codes is actually being made in the cell]. Thus in a liver cell, different genes are expressed than in, say, a brain cell. **1970** *Sci. Jrnl.* June 42/3 In parthenogenesis..the lethal gene is dominant and consequently its lethal effect is expressed.

express (ekspre·s), *v.*² orig. *U.S.* [f. Express *a.*, *adv.*, and *sb.*¹] *trans.* (See after Express *sb.*¹) Also, to send (letters, goods, etc.) by a special delivery or by express (Express *sb.*¹ 3 a); to send by express delivery.

1776 *Jrnl. Ho. Rep. Mass.* I. 81 Isaac Winslow Esq., brought down..several letters that had been expressed to his Honour the Lieut. Governour from the Eastward. **1847** J. S. Robb *Streaks of Squatter Life* 122 News, now-a-days,..is not news unless expressed, and..the President's message, received in the old fashioned snail-way, you-get-it manner, would not be read with interest. **a 1866**, etc. [see after Express *sb.*¹]. **1893** Stevenson & Osbourne *Wrecker* x. 161 Did he express his baggage, ma'am? **1898** L. Merrick *Actor-Manager* xv. 210 Perhaps if the news were 'expressed' to the office at once. **1899** *Post Office Guide* Jan. 18 4 Letter may be posted in any Letter Box, from which it will be collected by a Postman and be expressed on reaching the proper office. **1930** *Daily Express* 6 Nov. 1/1 I telephoned to my mother to express my passport to me at Croydon. **1936** J. G. Cozzens *Men & Brethren* i. 128 After expressing their luggage, they had come over the mountains from Bologna on foot.

Hence **expre·ssed** *ppl. a.*², sent by express.

1902 A. Bennett *Grand Babylon Hotel* vi. 72 It was..'expressed' luggage despatched in advance. **1909** [see *curly *a.* 3].

expressage. (See after Express *sb.*¹ 3 b; earlier and later examples.)

1857 C. E. in *Long Jrnl.* 12 May in *Q. Calif. Hist. Soc.* (1930) IX. 143 Paid Farley $5 for expressage. **1936** M. H. Bradley *Five Minute Girl* x. 197 The expressage will entail some expense.

expression. Add: **6*.** *Genetics.* The appearance in a phenotype of a character or effect attributed to a particular gene; also, the process by which possession of a gene leads to the appearance in the phenotype of the corresponding character. Cf. *express *v.*¹ 11. (in quot. 1927) = *expressivity b.)

1927 N. W. Timoféeff-Ressovsky in *Genetics* XII. 131, I use the word 'manifestation' to signify the presence of the given character in the phenotype of the fly, and the word 'expression' to signify the degree and form of the expression of the character. *Ibid.* 137 Some cases of strong hereditary variations of the expression of radius incompletus..enable us to suppose that a great hereditary variability of the phenotypic expression of this gene could be obtained with the aid of a more precise accounting. **1946** R. R. Gates *Human Genetics* I. xi. 422 Just as families can differ greatly in the degree of expression of polydactyly, so they can differ in the degree of dominance of the character. **1957** C. H. Waddington *Strategy of Genes* iii. 91 Geneticists commonly use the word 'expressivity' to indicate the degree of expression of some phenotypic character. **1957** *Jrnl. General Microbiol.* XVI. 115 Expression of the character Az[r] begins at the time of dilution..of the zygotes and then rises exponentially to become complete just before the recombinants which inherit start to divide. **1968** [see *express *v.*¹ 11]. **1969** *Nature* 10 May 541/1 The expression of certain genes is controlled by the products of specific regulator genes which act as repressors.

expressionable (ekspre·ʃənăb'l), *a.* [f. Expression + -able.] Capable of showing expression.

1902 *Blackw. Mag.* CLI. 43/2 Mozart..discovered the soul-reaching and expressionable capacity of instrumental music. **1908** *Daily Chron.* 15 Jan. 8/4 The mouth is an electro's mouth; clear cut, expressionable.

expressionism (ekspre·ʃəniz'm). Freq. with capital initial. [f. Expression + -ism; cf. next.] The methods, style, or attitude of expressionists; *spec.* a style of painting in which the artist seeks to express emotional experience rather than impressions of the physical world; hence, a similar style or movement in literature, drama, music, etc. Cf. *abstract expressionism* (*abstract *a.* 6).

1908 *Edin. Rev.* Oct. 466 The appearance of these later and more extreme forms of expressionism...has aided the understanding of beauty partly through a deeper probing of the sensuous elements in aesthetic experience. **1920** *Arts & Decoration* XIII. 88 (*title*) Mid-European Expressionism. *Ibid.*, The sole aim of expressionism was to reflect the impression of nature upon the artist's eye; expressionism, on the contrary, proceeds from the subjectivity of the artist and seeks to form a picture according to the laws of an arbitrary, inner world. **1921** Galsworthy *To Let* I. i. 13 Expression! Ah, they were all Expressionists now, he had heard, on the Continent... He wondered where this—this Expressionism—had been hatched. The thing was a regular disease! **1923** MacGowan & Jones *Continental Stagecraft* iii. 27 A Symbolism that is far on the way towards Expressionism. *Ibid.* 31 Expressionism..is a violent storm of emotion beating up from the unconscious mind. **1924** A. Dukes *Youngest Drama* 136 Realism and expressionism represent contrary attitudes of mind and uses of theatrical machinery. **1934** A. D. Sedgwick *Little French Girl* II. iv. 132 Giles..felt that he could not adequately defend his theories, which rested upon an objection to the use of the body as a means of primitive expressionism. **1925** H. Carter *New Spirit European Theatre* 220 Expressionism is simply expression taking the form of a new technique for the purpose of giving the most intense effect..to the species of drama that expressed pre-war and war time insurrectionary tendencies. **1926** [see *abstractionism 2]. **1926** F. V. P. Rutter *Evol. Mod. Art* v. 112 The kaleidoscopic paintings of the Italian Futurists were matched at Munich by the Expressionism of the Russo-Polish artist Wassily Kandinsky. **1929** P. Sickert *Listener's Hist. Mus.* III. 142 Critics writing of plays (but it applies all round) have spoken of Expressionism as 'the objectivisation of the subjective'. *Ibid.* 144 The Schönbergians and their followers in modern music. **1937** *English Studies* XIX. 241 (*title*) The birth of expressionism in the work of D. H. Lawrence. *Ibid.*, Expressionism, a term used here in its widest sense to cover all anti-realist as well as anti-idealist movements..reaching from Strindberg in the north to Marinetti in the south. **1938** *Oxf. Compan. Mus.* 303/2 Expressionism in music, judged by its products, is but an explained phase of Romanticism in music. **1958** *Listener* 27 Nov. 897/3 Expressionism implies an uncompromisingly violent utterance, coupled too often with a self-pitying sensibility. **1970** *Oxf. Compan. Art* 396/1 The most important forerunner of Expressionism in this special sense was Van Gogh.

‖ **expressionismus** (ekspre:ʃəni·zmʊs). [G.] = prec.

1924 A. Huxley *Those Barren Leaves* iv. v. 307 German romanticism, a little *détraqué*, turns..into expressionismus. **1926** [see *abstractionism 2].

expressionist. Add: (Further examples.) Freq. with capital initial. *spec.* An artist, composer, writer, etc., who exhibits the style or technique of *expressionism. Also as *adj.* Cf. *abstract expressionist* (*abstract *a.* 6).

1914 W. Lewis in *Blast* June 143 Of all the tags going, 'Futurist'..serves as well as any for the active painters of to-day...We may hope before long to find a new word. If Kandinsky had found a better word than 'Expressionist' he might have supplied a useful alternative. *Ibid.* 144 Balla..is a rather violent and geometric sort of Expressionist. His paintings are purely abstract. **1915** *Ibid.* July 39/1 There have grown up three distinct groups of artists in Europe... The third group is formed by the expressionist movement, that is Kandinsky. **1915** *Observer* 1 Aug. 5/6 The reviled Post-Impressionists,.. Cubists, Futurists, Expressionists, Vorticists of to-day may be the honoured masters of to-morrow. **1920** *Arts & Decoration* XIII. 88/3 Arnold Schönberg, the musical expressionist—who also painted. **1921** [see *expressionism]. **1924** *New Statesman* 2 Aug. 494/2 Elmer Rice's play, *The Adding Machine*..has two or three expressionist scenes, but is for the most part not in the expressionist manner. **1927** *Observer* 11 Sept. 6 A cocaine romance written by one of the most gifted of those men who joined the band of expressionists early in the movement. **1938** *Oxf. Compan. Mus.* 303/2 The music of the composers who call themselves Expressionists..is..simply 'ultra-emotional'. **1962** *Listener* 20 Sept. 453/1 The peculiar tension of the characteristic Expressionist instrumental form lies in the aphoristic compression of a dreamlike state.

expressionistic (ekspreʃəni·stik), *a.* [f. Expressionist + -ic.] Of, pertaining to, or produced by expressionists; characterized by expressionism. Hence **expressioni·stically** *adv.*

1921 Galsworthy *To Let* III. x, She had begun to exchange her Empire for her Expressionistic furniture. **1924** *Sunday Times* 23 Mar. 6/3 The fourth scene shows his feelings expressionistically. **1924** *Glasgow Herald* 17 July 4 An example of the expressionistic type of drama. **1927** *Ibid.* 6 Oct. 6 Careless, expressionistic young men. **1928** *Observer* 29 Jan. 15/4 His expressionistic ingenuity achieved many stimulating effects. **1966** *Listener* 6 Oct. 513/2 Jakov Lind's *Landscape In Concrete* is a German miasma-novel, expressionistic, nightmarish, and confusedly allegorical. **1970** *Daily Tel.* 17 June 16/5 A Bach chorale (expressionistically anticipated at the beginning of the second movement).

expressivity (ekspresi·vĭti). [f. Expressiv(e *a.* + -ity.] **a.** The quality of being expressive.

1940 *N.Y. Herald Tribune* 28 Mar. 16/1 This equalized expressivity ends by making Beethoven sound sometimes a little meretricious as a composer. **1948** L. Spitzer *Linguistics & Lit. Hist.* iv. 155 The self-destruction brought about by excessive expressivity was seen by Diderot as a danger to which any artistic nature is exposed. **1961** L. F. Brosnahan *Sounds of Language* i. 16 It is not at all clear why the relative infrequency of the latter in French should have led to its selection on the grounds of expressivity.

b. *spec. Genetics* [ad. G. *expressivität* (N. W. Timoféeff-Ressovsky 1925)]. The kind or degree of phenotypic expression of a gene.

1934 N. W. Timoféeff-Ressovsky in *Biol. Reviews & Biol. Proc. Cambr. Phil. Soc.* IX. 433 The phenotypic manifestation of the mutant allelomorphs may be full and constant, or it can be variable and even dependent upon other factors (low penetrance, variable expressivity and specificity of the genes). **1946** R. R. Gates *Human Genetics* I. ii. 15 Related to penetrance is *expressivity*, which means the degree to which a dominant gene expresses itself. **1957** [see *expression 6*]. **1962** Henderson & Gillespie *Text-bk. Psychiatry* (ed. 9) iv. 36 The genes have a range of variability in their manifestation which is called their expressivity.

expresso (ekspre·so), var. *espresso.

expressway (ekspre·swei). orig. *U.S.* Also **express way.** [f. Express *a.* + Way *sb.*¹] A wide road for fast motor-traffic (see quots.); an urban motorway. Also *express highway, road, route.*

c 1938 L. Mumford *Report on Honolulu* in *City Development* (1946) vi. 94 The municipality..withholding assent from ill-advised express highways. **1944** *Public Roads* XXIV. vi. 178/1 An analysis was made to determine the traffic which would use an expressway in Tulsa, Okla. **1945** *Britannica Bk. Year* 244/1 A significant event of.. 1944 was submission of the report of the President's National Interregional Highway committee recommending postwar construction...An important feature of the plan is the construction of express routes through cities. **1952** *Newsweek* 15 Dec. 84/3 Another 'solution' is the expressway—the live, dual-lane, high-speed thoroughfare which either by-passes or cuts through cities. Entrances and exits are at selected places; ramps and overpasses are arranged so that vehicles do not cross in front of each other. **1956** *Planning* XXII. 178 The average estimated cost for the 123 miles of new expressways (urban motorways)..proposed for the Detroit region..is more than £5,500,000 a mile. **1959** *Mod. Lang. Notes* LXXIV. 304 'Express ways' is the general term including both toll roads and free roads, for cars and trucks. **1961** L. Mumford *City in History* xii. 215 Just as our expressways are not articulated with the local street system, so the great sewers of Rome were not connected with water-closets above the first floor. **1966** *Guardian* 23 Dec. 4/2 The Right Bank express road through the city. **1967** *Listener* 6 July 10/2 Old people's homes, urban expressways, under poor layouts.

expropriating (ekspro·pri,eitiŋ), *ppl. a.* [See -ing².] Dispossessing, depriving of property.

1908 H. G. Wells *First & Last Things* 104 A class conflict between the expropriated Many and the expropriating Few. **1910** *Daily Chron.* 24 Mar. 4/3 The expropriating authority is in justice bound to appropriate such land for its use.

expulsatory (ekspʌ·lsătəri), *a.* [f. L. *expulsāt-, -āre* (see Expulse *v.*) + -ory².] Of or pertaining to expulsion; that expels.

1594 Chapman *Shadow of Night* sig. C iv verso, Expulsatorie Balme that serues To quench lusts fire. **1768** Ld. Mansfield in J. Wesley *Let.* Dec. (1931) V. 378 The expulsatory power of the House. **1910** *Practitioner* Mar. 461 When this [sc. the uterus'] muscular fabric is engaged in expulsatory efforts..pain of extreme intensity is endured.

expulsion. Add: **a.** Also *attrib.*, as *expulsion order.*

1905 *Act* 5 *Edw. VII* c. 13 § 3 The Secretary of State may..make an order (in this Act referred to as an expulsion order) requiring an alien to leave the United Kingdom within a time fixed by the order, and thereafter to remain out of the United Kingdom.

expunct (ekspʌ·ŋkt), *v.* Delete †*Obs.* and add later examples. Cf. next.

1847 Madden *Layamons Brut* I. 127 Written at first *adadet*, and the *e* subsequently expuncted. *Ibid.* III. 460 The ancient scribes, who often in such cases expunct the superfluous letter. **1868** Skeat *Havelok* 21 *note*, The MS. has 'is', but the *is* expuncted. **1908** — in *N. & Q.* 10th Ser. IX. 116/2 The scribe first of all copied 'Atherodes' as 'Atheredys', and after that he.. divided it as 'At heredys', and then, by expuncting *h*, produced 'At eredys'.

expunction. Add: **1. b.** *Palæogr.* The indication of an erasure to be made in a manuscript by means of dots placed beneath the relevant letter or letters.

1954 N. Denholm-Young *Handwriting in Eng. & Wales* viii. 78 The point in the Middle Ages was used.. beneath letters to denote *expunction*, i.e. they are to be regarded as not written. **1960** N. R. Ker *Eng. MSS. after Norman Conquest* 51 Small errors were repaired by expunction or erasure, the correct reading being added above the line.

expunctuation (ekspʊŋktiu,ēⁱ·ʃən). *Palæogr.* [Blend of Expunction and Punctuation.] = *expunction 1 b.

1957 N. R. Ker *Catal. MSS. containing Anglo-Saxon* 325 Letters are cancelled by expunctuation. **1968** *Medium Ævum* XXXVII. 31 Corrections made by a later corrector..with expunctuation or heavy deletion. **1969** M. B. Parkes *Eng. Cursive Bk. Hands 1250–1500* p. xxviii, [] enclose words and letters which have been deleted by the scribe by means of crossing out, erasure, or expunctuation.

exquisite, *a.* and *sb.* Also pronounced (eks-kwi·zit).

‖ **ex silentio** (e:ks sile·nʃio), *advb. phr.* [L.] 'From silence': a phrase used to designate an argument or conclusion based on lack of contrary evidence. Freq. in phr. *argumentum ex* (or *e*) *silentio* (see *argument 3 c).

1909 L. R. Farnell *Cults Gk. States* V. iv. 88 On this point the argument is not merely *ex silentio*. **1945** *Scrutiny* XIII. ii. 93 Electra..shows them that the dead have no objection to her dancing and argues *ex silentio* that they approve. **1953** W. R. Trask tr. *Curtius's Europ. Lit.* iii. 50 The discussion is very summary, and hence forbids conclusions *ex silentio*. **1961** *Times* 2 Oct. 13/2 The only supportable concentrations of power or wealth are those that are at the direct disposal of the state (an exception in the case of trade unions being implied *ex silentio*).

extend, *v.* **1. b.** Add: Esp. *pass.* and *refl.* of a horse: to exert itself to the full; to go 'all out'; so, of a runner, oarsman, etc.; hence *gen.*, to use all one's efforts; to try one's utmost; to be at full stretch.

1856 'Stonehenge' *Man. Brit. Sports* 332/1 The horse is made to extend himself. **1915** Wodehouse *Something Fresh* v, The Blandings *chef* had extended himself in honour of the house-party, and had produced a succession of dishes which, in happier days, Mr. Peters would have devoured eagerly. **1921** *Baily's Mag.* Jan. 3/2 The flying Pharmacie, who has not only won all the eight races in which she has taken part, but has never been really extended for a single stride. **1923** Wodehouse *Inimitable Jeeves* xiii, He delivered an address of twenty-six minutes by Claude's stop-watch. At a village wedding, mark you! What'll he do when he really extends himself! **1931** *Morn. Post* 25 Feb. 16/3 Corpus held their place at the Head of the River without being extended. **1955** *Times* 11 May 5/1 The main interest will be to see how he extends himself on the race-course. **1964** E. Waugh *Little Learning* v. 137 An indolent, humorous clergyman, who we did not think was extending himself fully in coaching us.

extended, *ppl. a.* **1. a.** Add to def.: Of a horse's gait (see quots.); opp. *collected.* Cf. Extend *v.* 1 b.

1778 [in Dict.]. **1938** H. Wynmalen *Equitation* xi. 52, I start teaching him the extended trot...I begin by varying the pace of his trot a little...I gradually increase the margin between the slowest and the fastest pace. **1952** R. S. Summerhays *Elem. Riding* xxvi. 148 *Extended walk*, The horse should cover as much ground as quickly as possible without haste or breaking the regularity of the beats. **1953** G. Brooke *Introd. Riding* vii. 77 To change from an extended to a collected pace.

b. Of a corpse: buried at full length; *extended burial*, burial in which the corpse is laid at full length. (Some of the examples are participial rather than adjectival.)

1875 *Encycl. Brit.* III. 398/2 At the chief's head lay the skeleton of a female..extended upon a sheet of pure gold. **1939** V. G. Childe *Dawn Europ. Civilization* (ed. 3) xvi. 277 The dead were generally buried, contracted or extended, in the settlements. **1952** — *New Light on Most Anc. East* (ed. 4) ix. 187 At Harappa these are represented only by extended and flexed burials in Cemetery H. **1960** K. M. Kenyon *Archæol. in Holy Land* ix. 227 The burials appear to have been extended, lying on their backs.

2. c. In *Insurance* (see quots.).

1889 C. E. Willard *ABC of Life Insurance* 42 If the death of the insured occurs during the term of the 'extended' insurance, the full amount of the policy is paid. **1909** Webster, *Extended insurance*, insurance for the full face of a policy on which payments are stopped, granted for an additional period in consideration of retaining part or all of the cash surrender value of the policy. **1913** *Jrnl. Chartered Insurance Inst.* XVI. 146 He can discontinue premiums as before, and remain assured for the full amount for a limited period. This, you will perceive, is a Paid-up Term policy, and is known as Extended Assurance. **1925** *Act* 15 & 16 *Geo. V* c. 69 § 4 A further period thereafter ending on such date as the Minister may by order prescribe, not being a date later than the first day of the insurance year commencing next after the end of the aforesaid deficiency period (the aggregate of which two periods is in this section referred to as 'the extended period'). **1928** *Britain's Industrial Future* (*Lib. Ind. Inq.*) iv. xx. 278 Claimants to 'extended' benefit.

d. *Bibliography.* (See quots.)

1952 J. Carter *ABC for Book-collectors* 78 *Extended.* When used of individual leaves, this means that the inner margin has been renewed...Occasionally, however, if a book has had to be made-up from a narrower copy, the alien leaves may be extended so that their outer margins range with those of their neighbours. **1960** H. Hayward *Antique Coll.* 236/2 When one or more of the three outer margins of the leaf of a book has been restored, it is said to be re-margined. If it is the inner margin only, the proper term is extended.

e. *extended-play* adj., used of a gramophone record, tape, etc., which contains a longer recording than a record, tape, etc., of a designated 'normal' capacity; *spec.* denoting a record seven inches in diameter, each side playing for about six minutes at 45 revolutions per minute. Abbrev. *E.P.* (*E. III).

1954 *Gramophone* Apr. p. ix (Advt.), Extended play records. *Ibid.* May 491/2 (*heading*) Extended play. **1960** *Times Lit. Suppl.* 2 Sept. 553/4 The sentimental Extended Play record. **1962** A. Nisbett *Technique Sound Studio* iv. 83 The thinner backing of extended and double-play tape produce a higher print level.

3. c. *Sociology.* Of a family, etc.: that comprises not only parents and children but also consanguine and conjugal relatives living in proximity.

a 1942 B. Malinowski *Sci. Theory of Culture* (1944) 168 The function of the extended family I would define in terms of a more effective exploitation of communal resources. **1951** R. Firth *Elem. Social Organiz.* iv. 142 The help is given as part of the reciprocities of kinship relations—to a second cousin, or uncle, or other member of an extended kin group. **1960** *New Left Rev.* Jan./Feb. 27/2 Among working-people you had extended families, often overlapping. **1966** D. Jenkins *Educated Society* iii. 98 The extended, as distinct from the nuclear family.

extender. Add: **1. b.** A substance added to paint or ink to increase its quantity or bulk, or to dilute its colour; also, a similar substance added to glue, etc.

1920 F. H. Jennison *Manuf. Lake Pigments* (ed. 2) 53 There is a considerable difference between the base of a lake, and the filler or extender that is used in the paint trade to let down strong colours. **1957** *Brit. Commonw. Forest Terminol.* ii. 67 *Extender*, a substance added by the user to synthetic resin glue to provide body...Some extenders, such as grain, clay or wood flours, may be relatively inert. Others, such as soluble dried blood can be used..both as an extender and as a supplementary glue. **1960** *World's Press News* 16 Dec. 40 The relative ratio of pigment to extender in the solid portion of the ink is governed by several considerations. **1967** *Gloss. Paper/Ink Terms Letterpress Printing* (*B.S.I.*) 9 *Extenders*, materials used by the ink maker or printer to adjust the colour strength and working properties of an ink.

4. *University Extender*, a University Extension lecturer (see Extension 9 g).

1893 *National Observer* 11 Nov. 652/1 The University Extender is a familiar type. **1894** *19th Cent.* XXXVI. 207 The University Extender exults in the hapless ones who attend his ministrations.

extension. Add: **1. e.** The utmost lengthening of a horse's stride at a particular pace.

1951 G. A. Bennett *Let's All enjoy Horse Show* xii. 107 These exercises are primarily intended to test the horse's extensions. **1953** G. Brooke *Introd. Riding* vii. 72 Extension signifies that a horse is taking its fullest stride at whatever pace it may be moving.

2. b. The action of adding up a horizontal line of figures or computing a sub-total, as on an invoice; the figure thus obtained. Cf. Extend *v.* 2 c.

1861 F. H. Carter *Book-keeping* iii. 40 The book.. should be ruled with the debtor and creditor columns on the same page,..no necessity existing for an extension of periodical balances. **1891** G. Van de Linde *Bookkeeping* iii. 88 An extension money column for the total Interest charged.

c. *Ballet.* The stretching of the leg at an angle from the body.

1934 Webster, *Extension*, in fancy dancing, a posture with one leg extended. **1952** Kersley & Sinclair *Dict. Ballet Terms* 64 An 'extended' position can look beautiful with dancers of the *jarreté* type who are loose enough to sustain such a position without straining too noticeably and are therefore said to have a good 'extension'. **1957** *Ballet Ann.* XI. 130/1 She has some lovely high, sustained extensions.

9. a. *spec.* In a camera, the distance by which the front part carrying the lens can be drawn away from the back part carrying the photo-sensitive surface. Also *attrib.*

1893 *Jrnl. Soc. Arts* XLI. 3 Mar. 381/1 With this form of lens system there can be only one definite extension of camera..in which the entire system is aplanatic. **1958** M. L. Hall *Newnes' Compl. Amat. Photogr.* xxiii. 199 Miniature cameras..use extension tubes..between the lens and the camera body.

b. Also (orig. *U.S.*), an addition to (esp. the rear of) a house or other building, usually not so high as the main building; an annex.

1852 *Congress. Globe* 24 Mar. 845/2 A statement of the materials to be used in the construction..of the proposed extension [of the Capitol]. **1867** [see *extension-room* in 10]. **1889** *Century Mag.* Mar. 781/1 They were making beds together in the extension. **1889** *Century Dict.* s.v., A dining-room extension. **1903** R. Hall *Pine Grove House* 12 The tin roof blew off the extension one windy night.

d. *spec.* Permission for the sale of alcoholic drinks until a later time than is usual at a particular place. Also *attrib.*, as *extension night.*

1923 Wodehouse *Inimitable Jeeves* xvi. 213 We will toddle along to Ciro's after dinner. It's an extension night, isn't it? **1930** A. Bennett *Imperial Palace* lxiv. 522 I've seen enough alcohol drunk to-night to last me a lifetime... **1940** 'G. Orwell' *Diary* 15 Mar. in *Coll. Ess.* (1968) II. 200 About 200 people there, all busily tucking into beer and sandwiches, though it was only 4.30 pm—they had got an extension for the day. **1965** *New Statesman* 14 May 753/1 To mark the sense of public joy,..the nightclubs in Bonn have got a late extension until 11 p.m. **1967** *Listener* 7 Sept. 297/2 At Bath,..where the pubs normally shut at 10.30, they all have an extension till 11.

g. Also *attrib.* (freq. *ellipt.*), as (*university*) *extension course, lecture, lecturer, student.* Also *transf.* and *fig.*

1890 Mackinder & Sadler *University Extension* i. 37 Cambridge Extension Lecturers do much of the work at University College, Nottingham. *Ibid.* ii. 77 Lectures to wait until the public mind had caught the desire for University Extension courses. *Ibid.* 79 Four Extension students. *Ibid.* iii. 92 The towns which find it most difficult to raise funds for Extension Lectures are generally those for which an expensive course is most needed. **1892** E. J. James in *Proc. First Ann. Mtg. Nat. Conf. on Univ. Extension* (Philad.) 29–31 Dec. 101 The University Extension Lecturer...There is no greater opportunity of showing his special skill to the Extension lecturer than is open to him in the conducting of a class. **1934** C. Day Lewis *Hope for Poetry* vi. 29 Life for the average child of the twentieth century becomes an endless series of extension-lectures on everything under the sun. **1936** C. S. Lewis *Allegory of Love* iii. iv. 142 It was the tendency of the age to make every lengthy poem something of an encyclopedia...It may be thought that this extension-lecture function of the medieval poets has nothing to do with their poetry. *Ibid.* 144 The roles of poet and extension lecturer are no longer habitually docked. **1959** R. Nichols *Fifth Caller* xii. 178 His name was on the roster of an extension course in anthropology.

h. A subsidiary telephone, loudspeaker, etc., connected to, but placed at a distance from, the main instrument; also, the number of such a telephone. Freq. *attrib.*

1906 *Ann. Rep. Amer. Teleph. & Teleg. Co.* 14 There have been developed for a substantial extent new classes of service..such as..extension sets. **1908** *Sears, Roebuck Catal.* 204/1 Telephone Parts...60-ohm Telephone Extension Bell. **1914** A. B. Smith & W. L. Campbell *Automatic Telephony* iii. 82 Arrangements for connecting up a wall telephone with an extension. **1924** S. R. Roget *Dict. Electr. Terms* 85 *Extension bell*, an additional bell arranged to ring at the same time as the ordinary telephone or other bell. *Ibid.*, *Extension lines*, lines radiating over a telephone installation from a private switchboard to the various extension instruments. **1928** *Delineator* July 78 The extension telephones are important in many other ways. **1929** A. Lejeune *Crowded & Dangerous* v. 93 You ring the War Office and ask for this extension. **1930** G. Freeman *Jack would be a Gentleman* xv. 66 She enjoyed owning the extension more than the actual telephone. **1966** N. Freeling *Dresden Green* i. 16 There was a record-player with an extension speaker. **1970** A. Price *Labyrinth Makers* ix. 125 Audley..[made] a reversed call to the department... Extension 28 eventually brought him Stocker.

10. c. *extension bag U.S.*, a bag that can be extended; *extension lens*, a lens that may be used in a combination to increase its focal length (see quot.).

1897 *Outing* (U.S.) XXX. 386/1 Utensils and food for two days in an extension bag. **1904** *Delineator* Oct. 547 With the genial season arrived every kind of drummer. They came with extension bags filled with samples. **1902** *Nature* 17 July 280/1 The replacement of the back component by a lens of greater focal length, increasing the focal length of the objective by about 50 per cent. we referred to some time ago; the alternative back lens being known as an 'extension lens'.

extensional, *a.* Add: **2.** *Logic.* Of, or relating to, logical extension (cf. Extension 8 b); *esp.*, concerned with the objects denoted rather than with the predicates applied.

1852 H. L. Mansel *Aldrich* (ed. 2) App. 221 In this respect it [sc. Aristotelian Induction] is synthetical, the parts and whole being viewed in their logical or extensional relation. **1903** B. Russell *Princ. Math.* i. § 68 Thus Brown and Jones are a class, and Brown singly is a class. This is the extensional genesis of classes. **1926** F. P. Ramsey in *Proc. London Math. Soc.* XXV. 348 In calling mathematics extensional we mean that it deals not with predicates but with classes, not with relations in the ordinary sense but with possible correlations, or 'relations in extension'.

extensionality (ekstenʃənæ·lĭti). *Philos.* [f. Extensional *a.* + -ity.] The state or fact of being extensional (see prec.). So *thesis of extensionality* (see quot. 1956).

1926 F. P. Ramsey in *Proc. London Math. Soc.* XXV. 348 The difficulties of the tautology theory..spring from a fundamental characteristic of modern analysis which.. may be called *extensionality*, and the difficulties may be explained as those which confront us if we try to reduce a calculus of extensions to a calculus of truth-functions.

1937 A. Smeaton tr. *Carnap's Logical Syntax Lang.* iv. § 67. 245 The Thesis of Extensionality. Wittgenstein..put forward the thesis that every sentence is 'a truth-function of the elementary sentences' and therefore (in our terminology) extensional. **1944** K. Gödel in P. A. Schilpp *Philos. B. Russell* 137 The axiom of extensionality (i.e., that no two different properties belong to exactly the same things). **1956** J. O. Urmson *Philos. Analysis* i. ii. 12 This thesis that language is throughout truth-functional was often called the thesis of extensionality. Every statement anyone ever makes must, according to this thesis,..be either a logically simple statement or else a truth-function of such statements.

extensionally (ekste·nʃənăli), *adv.* [See -ly².] By way of extension.

1903 *Nature* 3 Sept. 411/1 A class may be defined either extensionally, by an enumeration of its terms, or intensionally, by the concept which denotes its terms. **1935** *Mind* XLIV. 152 An abstract deductive structure capable of interpretation not only extensionally, for example as the traditional two-valued calculus of truth-values, but also intensionally, as a calculus of unasserted propositional meanings. **1951** Z. S. Harris *Methods in Struct. Ling.* 372 The element X is thus associated with an extensionally defined class consisting of so many features in so many of the speech occurrences in his [sc. the linguist's] corpus. **1964** E. Bach *Introd. Transformational Gram.* vii. 155 Like classes, relations can be considered intensionally or extensionally.

extensionless (ekste·nʃənlĕs), *a.* [-less.] Without extension.

1919 A. N. Whitehead *Enq. Princ. Nat. Knowledge* ii. 7 An extensionless moment of time. **1922** tr. *Wittgenstein's Tractatus* 153 The I in solipsism shrinks to an extensionless point.

extensive, *a.* Add: **3. d.** *Econ.* Applied to methods of cultivation in which a relatively small crop is obtained from a large area with a minimum of attention and expense; opp. Intensive *a.* 5.

1832 [see Intensive *a.* 5]. **1949** P. E. James *Geogr. of Man* 60 Where the expenditure of labor and capital per unit of land is relatively small, the economy is said to be extensive.

extensometer (ekstensǫ·mĭtəɹ). [f. L. *extensus*, pa. pple. of *extendere* to Extend + -ometer.] An instrument for measuring the deformation of metal under stress, or an instrument in which such deformation is used to register the elastic strains borne by other materials (e.g. concrete); (see also quot. 1958).

1887 *Proc. Physical Soc.* VIII. 183 Screw-micrometer Extensometer. **1902** *Encycl. Brit.* XXXIII. 10/1 Extensometers, or apparatus for observing the small deformation which a test-piece in tension or compression undergoes before its limit of elasticity is reached. **1925** E. S. Andrews *Strength of Materials* (ed. 2) 317 Extensometers are instruments for measuring the elastic strains of materials in tension or compression. **1950** *Engineering* 25 Aug. 172/3 Twin extensometers were mounted in a manner similar to that employed in tests on reinforced-concrete beams using the Lamb extensometer. **1958** *New Scientist* 15 May 38/3 Earth strains in the Andes are to be studied by..seismologists working with extensometers... The extensometer or strain-seismograph measures the differential motion between two nearby points, those at the ends of the instrument. The change of length..between the two ends—which may either stretch or squeeze—divided by the original length gives the change in strain. **1966** *Ibid.* 23 June 796/2 A hydraulic extensometer sensitive to dimensional changes in the order of one part in 10000.

extent, *sb.* Add: **7.** *Campanology.* (See quot. 1901.)

1901 H. E. Bulwer *Gloss. Techn. Terms Bells* 14 *Extent.* (a) The full number of distinct 'changes', of which a 'method' admits, is the 'extent' of the 'method'. (b) The full number of times a bell, or some definite combination of bells, can occupy a given position, without repetition of any one of the 'changes' containing such bell, or bells, in that position. **1968** *Daily Tel.* 27 Nov. 19/7 Three 'extents' [i.e. 720 changes each] upon the six large bells there, in three different methods, 2,160 changes in all.

exterior, *a.* and *sb.* Add: **B.** *sb.* **1. b.** An outdoor scene represented on the stage or in a film or television programme; a film, or sequence of a film, photographed outdoors.

1872 *Daily Tel.* 11 Mar. 3/4 The cottage garden exteriors, by Mr. Hawes Craven..are as pretty and effective stage pictures as have ever been seen. **1918** H. Croy *How Motion Pictures are Made* 74 The few pictures that had been made were made outdoors in the open—'exteriors', as they are called. **1949** *Here & Now* (N.Z.) Nov. 27/2 The exteriors, shot unglamorously in a dead daylight, have a harsh gritty quality.

exteriorization. (Later examples.)

1917 C. R. Payne tr. *Pfister's Psychoanalytical Method* xi. 206 The exteriorization by which the patient thinks he can perceive his separate organs in the reality. **1961** R. Crookall *Supreme Adventure* ii. i. 38 The latter also applies to exteriorizations (of both mediumistic and non-mediumistic types) that are enforced by anæsthetics, etc. **1962** *Listener* 17 May 854/2 The final exteriorization of a work was frequently difficult for him [sc. Debussy].

exteriorize, *v.* Add: (Later examples.) Also *absol.* Hence **exte·riorized** *ppl. a.*

1929 *Pall Mall Gaz.* 13 Dec. 3/1 A woman went off into hypnotic sleep close by, and on the girl's exteriorized sensitiveness coming in contact with her she drew away. **1959** *Times Lit. Suppl.* 9 Oct. 576/4 A poem must be set at a distance from its reader..but to exteriorize too soon is to refuse to suffer. **1961** R. Crookall *Supreme Adventure* ii. i. 59 His exteriorized double is often seen by mortals at about this time. *Ibid.* 62 The whole of vitality takes longer to exteriorize than the Soul Body. **1964** *Lancet* 19 May 1055/1 Outside the forearm the cannulas are connected by a shunt..which forms an exteriorized semipermanent arteriovenous fistula.

extermination. Add: **2. b.** *attrib.*, as **extermination camp**, a concentration camp for the mass murder of human beings, applied esp. to the camps set up by Nazi Germany in the war of 1939–45.

1945 *Atlantic Monthly* July 55 Among the workers a threat of being sent to an extermination camp has its uses. **1952** C. P. Blacker *Eugenics* 146 We have witnessed the inhumanities of concentration and extermination camps. **1957** H. Roosenburg *Walls came tumbling Down* 10 The Jews..were sent to the extermination camps.

external, *a.* and *sb.* Add: **A.** *adj.* **6.** In education, of a student: that does not attend a university but takes its examinations; of an examiner: that tests students of a college, school, etc., of which he is not himself a member; hence of such an examination or the degree so obtained.

1888 *Cal. Univ. Coll., Bristol, 1888–9* 21 An Honour Certificate shall be granted, after examination by an External Examiner, in conjunction with the Professor or Lecturer of the subject. **1898** [see Internal *a.* 2 c]. **1900** *Univ. Birmingham Cal. 1900–1901* 85 The Examiners of the University shall be the Professors of the University with such Lecturers of the University as the Council from time to time appoint and such External Examiners not being Professors Lecturers or Teachers in the University as may from time to time appointed by the Council. **1912** *University of London Regs. for External Students* 25 An External Student B.A. Pass Examination... The External Final Examination. **1959** *Listener* 19 Mar. 515/1 What is the deadly effect on education of the device of 'external' marking of papers?

exteroceptor (e:kstĕrose·ptǫɹ). *Physiol.* [irreg. f. L. *externus* exterior + *-ceptor* of Receptor.] A sense organ which receives external stimuli, as those of touch, etc. Hence **exterocep·tive** *a.* Cf. *interoceptor, *proprioceptor.

1906 C. S. Sherrington *Integr. Action Nerv. Syst.* 130 There exist..two primary distributions of the receptor organs, each a field in certain respects fundamentally different from the other. The surface field lies freely open to the numberless vicissitudes of the environment. This field, extero-ceptive as it may be called, is rich in the number and variety of receptors which adaptation has evolved in it. *Ibid.*, The reaction of the animal to stimulation of one of its extero-ceptors excites certain tissues, and the activity thus produced in these latter tissues excites in them their receptors, which are proprioceptors. **1927** Haldane & Huxley *Animal Biol.* i. 24 The exteroceptor organs are the windows of the animal onto the outside world. [illegible] **1932** S. Zuckerman *Soc. Life Monkeys & Apes* xvi. 266 [illegible] **1964** J. Z. Young *Model of Brain* xiv. 232 [illegible]

extinct, *pa. pple.* and *a.* Add: **B.** *adj.* **1. a.** *extinct volcano*; also *fig.*, *spec.* a person who has lost the considerable energy, etc., he once possessed.

1865 H. C. Robinson *Diary* 21 Jan. (1967) 314 Allsop, whose name has been long forgotten... An extinct volcano. **1869** J. A. Symonds *Let.* 20 Aug. (1967) I. 566 My illusions & conceits are Extinct volcanoes. **1914** F. M. Ford *Let.* June (1965) 59 You talk of extinct and semi-extinct volcanoes...is..only window dressing.

extinction. Add: **1. c.** *Physics.* Reduction in the intensity of a beam of light as a result of absorption and scattering as it passes through a medium.

1794 [see sense 1 in Dict.]. **1827–8** J. F. W. Herschel in *Encycl. Metrop.* (1845) IV. 431 The extinction of a beam of homogeneous light in passing through a homogeneous medium. **1957** H. C. van de Hulst *Light Scattering by Small Particles* xiii. 388 Extinction can be measured by observing the..decrease of a beam of light..as it goes through a container with scattering particles. **1963** J. M. Stone *Radiation & Optics* xiv. 323 Extinction is accounted for partly as scattering and partly as absorption. Extinction may..consist entirely of absorption (for example, in an opaque screen of negligible reflectance) or almost entirely of scattering (for example, from..a very thick cloud in the sky).

(ii) The state or condition of darkness in a crystal placed in polarized light between crossed nicols. Hence *angle of extinction*, the angle between the direction of extinction and a principal crystallographic direction. Also *attrib.*

1873 J. Tyndall *Six Lectures on Light* iv. 133 This brings us at once to the part played by the analyser, the sole function of which is to recompound the two vibrations emergent from the gypsum. It reduces them to a single plane, where, if one of them be retarded by the proper amount, extinction will occur. **1912** R. W. Clark tr. *Weinschenk's Petrogr. Meth.* 67 Maximum brightness is obtained when the vibration directions of the crystal are at 45° to those of the nicols. It diminishes upon further rotation and passes gradually over into complete darkness when these directions are respectively parallel. This latter position is also called the position of extinction, and the vibration directions in the crystal, the extinction directions. *Ibid.* 70 Extinction Curve for Diopside. *Ibid.*, In triclinic minerals the determination of the extinction angles is of value only when the orientation of the face upon which they are observed is accurately known. **1921** A. Holmes *Petrogr. Meth.* 126 When a transparent mineral fragment is rotated between crossed nicols various phenomena may be observed... If the object remains dark, and is therefore isotropic, the total extinction indicates an amorphous or cubic (isometric) mineral in an untwinned condition, or a basal section of a uniaxial mineral (tetragonal, hexagonal, or trigonal). **1937** *Discovery* Sept. 283/2 A complete record of the colour variations and the angles of extinction can be permanently obtained.

(iii) A reduction in the intensity of X-rays diffracted by a crystal owing to interference between the incident beam and beams multiply diffracted by the lattice planes (*primary extinction*) and to the progressive weakening of the incident beam in its passage through the crystal as energy is diverted into the diffracted beams (*secondary extinction*).

1914 C. G. Darwin in *Phil. Mag.* 6th Ser. XXVII. 681 The extinction is complete long before the [X-]rays going in a slightly different direction are appreciably absorbed. This fact is important in explaining the reflexion from an ordinary imperfect crystal. **1922** — *Ibid.* XLIII. 804 [illegible] **1934** W. P. Davey *Study Crystal Struct.* x. 308 [illegible] **1948** K. Lonsdale *Crystals & X-Rays* vi. 149 [illegible] **1966** D. G. Brandon *Mod. Techniques Metallogr.* 83 Crystal imperfections will prevent double diffraction and hence prevent extinction effects, producing an increased intensity in regions of lattice strain.

6. *attrib.*, as **extinction coefficient**, any coefficient that gives a measure of the total degree of extinction of radiation by a medium; **extinction frequency** (see quot.); **extinction meter**, a photographic exposure meter which gives readings in terms of visual comparisons.

1902 *Encycl. Brit.* XXX. 237/1 [illegible] **1953** Amos & Birkinshaw *Television Engin.* i. 227 [illegible] extinction frequency. **1931** J. H. Reyner *Ciné-Photogr. for Amateurs* v. 58 These meters are often known as extinction meters, because the image, viewed through them is gradually reduced or worked up from an extinction point. **1951** G. H. Sewell *Amateur Film-Making* (ed. 3) ii. 19 An early type [of exposure meter] used for cinematography was the extinction meter, and this consisted of..looking through a small telescope in which was a density wedge [illegible]

extinguisher. Add: **1.** *spec.* = *fire-extinguisher* (Fire *sb.* B. 2).

1814 [in Dict.]. **1908** *Westm. Gaz.* 14 May 12/2 The extinguishers..[illegible]. **1930** *Engineering* 10 Oct. 458/2 It was necessary to have a sufficient supply of chemical extinguishers on board.

3. *extinguisher-shaped* adj.; **extinguisher-moss**, a moss whose peristome closes inwards when touched by water.

1821 S. F. Gray *Nat. Arr. Brit. Plants* I. 725 Encalypta..Extinguisher-moss. **1876** *Encycl. Brit.* IV. 161/2 The Extinguisher-Moss (*Encalypta*). **1840** Barham *Ingol. Leg.* 23 A shocking bad hat, Extinguisher-shaped. **1903** R. Gower *Rec. & Remin.* 16 The towers with their extinguisher-shaped roofs. **1909** A. Morrison *Green Ginger* 17 The wide fireplace and its blazing embers, stuck with black extinguisher-shaped beer-warmers.

extorsion (ekstǭ·ɹʃən). [f. L. *extorsiōn-em* (see Extortion).] Outward rotation of a part of the body, esp. of the eye.

1899 A. Duane tr. *Fuchs's Text-bk. Ophthalmol.* (ed. 2) xiv. 576 The *inferior rectus*..rotates the eye in such a

way that the upper extremity of the vertical meridian is depressed to the outside (extorsion). **1964** S. Duke-Elder *Parsons' Dis. Eye* (ed. 14) xxviii. 446 An involuntary movement of torsion: intorsion when the upper pole of the cornea rotates nasally, extorsion when temporally.

extra, *a.*, *adv.*, and *sb.* Add: **A.** *adj.* **e.** *extra cover* (*point*), a fieldsman in cricket whose position is between cover-point and mid-off, but more distant than either from the batsman's wicket; also, his position in the field. Also *ellipt.*

1867 G. H. Selkirk *Guide Cricket Ground* ii. 26 Extra cover. When the batsman hits much to cover-point an extra man is placed near to aid in fielding the balls. **1888** A. G. Steel in Steel & Lyttelton *Cricket* iii. 112 If the ground is inclined to be slow, he [sc. third man] may be brought forward to extra cover-point, between cover-point and mid-off. **1897** K. S. Ranjitsinhji *Jubilee Bk. Cricket* ii. 50 There is no need to treat extra-cover separately. The position is a cross between cover and mid-off, and its duties are a mixture of the duties required in those two places. **1921** P. F. Warner *My Cricketing Life* xiv. 268 Skeet..is very quick on his feet at cover or extra cover. **1955** I. Peebles *Ashes* xv. 149 May hit the next ball splendidly past extra.

B. *adv.* **a.** *extra sec* (or *extra dry*), of champagne: very slightly sweetened. Also *ellipt.* as *sb.*

1891 in C. Ray *Compleat Imbiber* (1967) IX. 122 Champagnes. Cuvée de Réserve. Extra Sec, Sec, and Brut 84/- per Doz. Bottles. **1907** *Yesterday's Shopping* (1969) 98 *Champagne*...Ayala, extra quality, extra dry. **1916** A. Bennett *Lion's Share* xxxix. 303 He would always command a half-bottle of the extra dry for himself. **1930** — *Imperial Palace* xxxvi. 247 Their mania for *extra sec.* **1951** R. Postgate *Plain Man's Guide to Wine* v. 95 'Brut' ..should be really dry..'Extra Sec', 'Extra Dry'.. moderately dry. **1965** A. Sichel *Penguin Bk. Wines* iii. 151 The words 'Brut' or 'Extra Dry' or 'Extra Sec' represent the driest of all—a dosage of 1 per cent or 2 per cent of liqueur.

C. *sb.* (Further examples.)

1793 *Kentucky Gazette Extra* 23 Mar. **1796** *Herald Extra* II. 30 Mar. **1832** *Amer. Railroad Jrnl.* I. 409/1 A newsboy thrust a 'Cholera extra' between us. **1838** Dickens *Nickleby* I. iii. 19 Tell your uncle, my dear, how far you went in French and extras. **1840** — *Old C. Shop* xxxiv. 287 'What's the rent?'..'One pound per week...The boots and clothes are extras.' **1858** *Bell's Life* 20 June 8/4 They..accomplished an innings of 161... It will be seen there were 26 extras! **1895** Kipling *Day's Work* (1898) 171 The newspaper extra—a slip printed on one side only, and damp from the press. **1912** A. Brazil *New Girl at St. Chad's* v. 86 The riding course was a special feature of the summer term...It was an 'extra', not part of the ordinary school curriculum. **1955** *Times* 26 May 9/7 It is just the important 'extra' given to the listener by this new two-sound track system which gives true relaxation and ease of listening.

b. A person engaged for a minor part, or to be present during a crowd scene, in a play, film, etc.; = Supernumerary *sb.* e. In full †*extra lady, gentleman.* Also *transf.*

c 1777–8 in Hodgkinson & Pogson *Early Manch. Theatre* (1960) 82 Or more Extras [to be employed] than cannot be avoided—the Bill for them to be signed by the Prompter. **1880** D. K. Ranous *Diary of Daly Débutante* 16 May (1910) 185 Who should come to call on the Daly young ladies but one of the Philadelphia 'extras' at the theatre. **1888** G. Grossmith *Society Clown* vi. 114 The Marines [in 'H.M.S. Pinafore'] were what is theatrically known as 'extra-gentlemen'. They are not engaged to sing, and therefore do not hold such a good position as the chorus. **1890** Barrère & Leland *Dict. Slang* II. 206/2 *Scene rats* (theatrical), extras engaged in ballets or pantomimes. **1895** G. B. Shaw *Our Theatres in Nineties* (1932) I. 284 He..marched off..leaving the Republic and its army looking like the merest crowd of 'extras'. **1896** *Ibid.* II. 179 'Extra' ladies and gentlemen (formerly called 'supers'). **1916** 'R. M. Bower' *Phantom Herd* 125 Extras may be depended upon for carrying gossip from one studio to another. **1916** R. E. Welsh *A-B-C of Motion Pictures* iv. 42 If the story has scenes calling for the use of hundreds of characters,..he will call for 'extras', who are engaged by the day. **1938** Wodehouse *Summer Moonshine* xx. 176 After that [I was] a movie extra. **1958** *Times Lit. Suppl.* 12 Dec. 723/2 Is it that Mr. Crossman is so charmed by the performance of politics that he ignores the subject of the play, so busy in this particular book with the stars that he forgets the extras? **1964** C. Chaplin *Autobiogr.* xvi. 272 One can.. march twenty thousand extras into the Red Sea. *Ibid.* xviii. 317 I'd do anything—get a job—do extra work in films.

c. (See quot. 1838.) Chiefly *U.S.*

1827 Mrs. B. Hall *Let.* 23 Oct. in *Aristocratic Journey* (1931) 106 We wished to get an 'Extra'..but..the Stage Proprietor had the want of conscience to ask thirty-six for an 'Extra'. **1838** J. F. Cooper *Home as Found* ii. 35 Mr. Howel informed him that an extra in America meant a supernumerary coach, to carry any excess of the ordinary number of passengers. **1842** Dickens *Amer. Notes* II. vi. 161 There being no stage-coach next day.. I hired 'an extra', at a reasonable charge. **1846** R. M. Ballantyne in E. Quayle *Ballantyne the Brave* (1967) iv. 67 The stage..was full..but..an 'extra' (or separate sleigh of smaller dimensions than the stage) had been provided for us. **1944** E. M. Kahn *Cable Car Days* 103 After six hours of waiting an 'extra' arrived and we resumed our journey.

d. At a ball, a dance additional to those on the dance-programmes.

1885 C. M. Yonge *Nuttie's Father* I. xiii. 151 'Oh dear! I'm engaged all through [*i.e.* for all the dances]'.. 'Give him one of the extras!' **1900** E. Glyn *Visits Eliz.* 276 When we got to the ball-room an extra was on. **1913** J. Vaizey *College Girl* xxvii. 369 Not until the three programmes were filled to the last extra did he..think of his own pleasure.

extra-. Add: **1. extra-acade·mic,** without or external to a university; **extrabra·nchial,** outside the branchial arches; also as *sb.*, an extrabranchial cartilage; **extraco·nscious,** outside or above what is conscious; hence **extra-co·nsciousness** *sb.*; **extra-contine·ntal,** beyond or outside a continent; *U.S.*, outside the western hemisphere; **extra-curri·cular,** outside the normal curriculum; also *transf.* and *fig.*; **extra-dio·cesan,** outside the diocese; **extradu·ral,** outside the dura mater; **extra-embryo·nic,** outside the embryo; **extra-Europe·an,** also, beyond the boundaries of Europe; **extra-experie·ntial,** outside experience; **extra-fami·lial,** outside the family; **extra-fo·veal,** lying or occurring outside the fovea of the retina; **extrahepa·tic,** outside the liver; **extra-li·ngual,** outside the bounds of language; **extra-lingui·stic,** outside the field of linguistics; = **extra-lingual*; **extra-ma·rginal,** outside the field of consciousness; **extra-ma·rital,** of sexual relationships outside marriage; hence **extra-ma·ritally** *adv.*; **extra-me·ntal,** beyond the mind; independent of mental apprehension; **extramorai·nic,** situated beyond, or not associated with, a moraine; **extra-mu·sical,** outside the field of music; not an intrinsic part of music; **extra-orga·nic, -organi·smal, -organi·smic,** outside the organism; **extra-paro·chial,** outside the parish; **extra-pe·rsonal,** situated or coming from outside a person; **extra-pyra·midal,** outside the pyramidal tracts (see Pyramidal *a.* 3 a); **extrare·nal,** outside the kidneys; **extra-so·cial,** outside society; **extraso·lar,** outside the solar system; **extra-spa·tial,** not in, or not forming part of, space; **extra-syste·mic,** outside the system; not according to a system; **extra-terrestrial** (later examples); hence as *sb.*, a being living or originating beyond the earth; **extra-vehi·cular,** outside a vehicle; *spec.* denoting activity outside a space-vehicle while it is in flight; **extra-ve·rbal,** not employing or describable in words; **extra-vi·sual,** not employing or connected with the faculty of sight; outside the normal range of sight.

1932 T. S. Eliot *Sel. Ess.* 452 Not aspiring..to academic or extra-academic honours. **1877** Parker & Bettany *Morphol. Skull* 39 Four extrabranchial cartilages.., parallel to the ceratobranchials. *Ibid.* 343 The extrabranchials of the Dogfish are..superficial cartilages related to the branchial arches. **1912** J. S. Kingsley *Compar. Anat. Vertebr.* 65 In the branchial region of the elasmobranchs a variable number of extrabranchial cartilages may occur. **1865** A. C. Fraser in *N. Brit. Rev.* XLIII. 29 We have no practical need for the extra-conscious existence of anything that we apply language to. **1900** A. T. Ormond *Found. Knowledge* iii. 90 An extra-conscious sphere which exists as yet only as implicate or postulate. **1949** Koestler *Insight & Outlook* xxiv. 334 Biologists..are..apt to find the solutions of their problems in sleep and extraconscious processes. **1897** W. James *Will to Believe* 311 The 'extra-consciousness', as one may call it, can be kept on tap, as it were, by the method of automatic writing. **1938** *Tablet* 8 Oct. 469/1 England being an extra-continental Power, it is to France that the small nations would most readily turn. **1960** *Organization of American States, Declaration* 28 June in *Ann. Reg. 1960* (1961) 532 The Seventh Meeting.. Condemns emphatically intervention..from an extra-continental Power in the affairs of the American republics. **1925** H. Cushing *Life Osler* I. 491 Lest one lose track of Osler in his daily rounds in the hospital wards during the recital of all these extracurricular matters. **1930** *Times Educ. Suppl.* 30 July 329/3 Is it not possible to incorporate our extra-curricular activities in our curriculum? **1949** L. Charteris in *Queen's Awards* (1951) IV. 294 He's dumb enough to think that Lucy won't catch on to the extracurricular functions of that busty secretary. **1967** N. Freeling *Strike Out* 140 Covert rendezvous, flirtations, extracurricular carryings on. **1789** W. Scott in J. Haggard *Rep. Consist. Cases* (1822) I. 34 Supposing the jury..had been of opinion it was extra-parochial,—that the place may be so, and not extra-diocesan, is not to be denied. **1897** *Westm. Gaz.* 14 July 1/3 The island [*sc.* Lundy] is extra-diocesan and extra-parochial. **1900** Dorland *Med. Dict.* 249/2 Extradural. **1907** *Practitioner* Nov. 731 The causes of exophthalmos.. are..tumours of the optic nerve, which are either intradural or extra-dural. **1962** *Lancet* 27 Jan. 172/1 The 'lucid interval' syndrome, classically associated with extradural hæmorrhage. **1913** Dorland *Med. Dict.* (ed. 7), *Extra-embryonic*, not a part of the embryo proper: applied to that portion of the embryo outside of the umbilical stalk. **1958** *New Biol.* XXVI. 24 The extra-embryonic part of the blastoderm soon becomes divided into chorion and yolk sac. **1905** *Westm. Gaz.* 4 May 9/3 The Morocco incident shows that arrangements ought to be made by France and Germany with regard to their extra-European interests. **1942** L. B. Namier *Conflicts* 1 European interests and entanglements have defeated the extra-European expansion of the Continental nations. **1904–5** W. James *Meaning of Truth* (1909) iii. 69 There may or may not be an extra-experiential 'ding an sich' that keeps the ball rolling. **1951** *Mind* LX. 18 Extra-experiential knowledge. **1952** C. P. Blacker *Eugenics* xi. 312 Their nutrition is surprisingly average—doubtless due to extra-familial feeding in schools. **1904** *Westm. Gaz.* 15 June 1/3 A plate of platinum heated until it is dull and glowing is first seen by the extra-foveal portions of the retina. **1947** *Nature* 4 Jan. 25 Measurements were also attempted for extra-foveal vision. **1961** *Lancet* 12 Aug. 341/2 Extrahepatic biliary exploration had shown no abnormality, but cholecystectomy was performed. **1961** Y. Olsson *Syntax Eng. Verb* ii. 16 The problem of dealing with language as related to extra-lingual phenomena. **1963** J. Lyons *Structural Semantics* iv. 56 Extralingual features (e.g. situations). **1927** *Mod. Philol.* Nov. 212 Extra-linguistic group habits. **1935** *Mind* XLIV. 505 The danger of using the material mode is that it misleads us into thinking that pseudo-object sentences are concerned with extra-linguistic objects such as numbers, things, properties, experiences, space, time, and so on. **1964** Extra-linguistic [see *extra-somatic *a.*]. **1898** W. James *Coll. Ess. & Rev.* (1920) 440 The humbugging and masquerading extra-marginal self is as great a paradox for psychology as the comatose spirits are for pneumatology. **1902** — *Var. Rel. Exper.* 233 Feelings which are extra-marginal and outside of the primary consciousness altogether. **1929** B. Russell *Marriage & Morals* v. 51 We, however, wish to appeal to reason, and we must therefore employ dull neutral phrases, such as 'extra-marital sexual relations'. **1963** in A. Heron *Towards Quaker View of Sex* i. 6 The incidence of extra-marital intercourse is great. **1967** J. Potter *Foul Play* vii. 91 Sandwiched between the pair of them: one said to be queer and the other extra-maritally active. **1853** H. L. Mansel *Lett.* (1873) 84 This is a criterion..of the mental conceivability, not of the extra-mental existence of an object. **1886** J. McCosh *Psychol., Cognitive Powers* i. 27 All knowledge obtained through the senses is discerned as extra-mental, that is, as out of and beyond the perceiving mind. **1903** C. A. Strong *Why Mind has Body* 214 Extra-mental realities. **1958** W. Stark *Sociology of Knowledge* 140 But it is *not* claimed by the sociology of knowledge that ideas are the intra-mental *effects* of extra-mental *causes.* **1891** *Amer. Geol.* VIII. 239 The thickness of this extra-morainic till ranges from 30 to 70 feet. **1968** R. W. Fairbridge *Encycl. Geomorphol.* 460/2 These features are best termed 'proglacial lakes', though the term 'extra-morainic lakes' has also been used. **1925** I. A. Richards *Princ. Lit. Crit.* xviii. 156 These physical laws are, as it were, an extra-musical piece of knowledge. **1968** *Daily Tel.* 12 Nov. 19/1 It would be difficult..to say just how much of the impact which the work made on the large audience was due to extra-musical reasons. **1866** J. McCosh *Exam. of J. S. Mill's Philos.* 121 The world beyond the body..I call the extra-organic world. **1935** *Mind* XLIV. 549 It will be necessary, no doubt, 'to recognise that bodily factors, and not only extra-organic things, may, in certain situations, function as objects'. **1923** *Glasgow Herald* 16 Nov. 4/2 There is obviously considerable change from generation to generation, but most of this is extra-organismal and only repercusses indirectly, if at all, on the flesh and blood constitution. **1955** F. Lounsbury in *Georgetown Univ. Inst. Lang. & Linguistics Monogr. Ser.* VIII. 162 The 'meanings' dichotomized by this criterion may be referred to as *intraorganismic* and *extraorganismic* meanings respectively. **1789, 1897** Extra-parochial [see *extra-diocesan*, above]. **1923** J. S. Huxley *Ess. Biologist* iv. 152 Some great extra-personal flood of soul, into the meagre stream of everyday life. **1964** M. Critchley *Developmental Dyslexia* ix. 60 The result of a failure of lateral orientation with reference to corporeal awareness, and the organisation of self within extra-personal space. **1912** *Brain* XXXIV. 296 This affection, where it occurs in an uncomplicated form, is an extra-pyramidal motor disease. **1965** *Nursing Times* 5 Feb. 188/1 Parkinsonism may be seen fairly frequently in patients who have had chlorpromazine for a considerable period, owing to its effect on the extra-pyramidal system. **1885** C. H. Ralfe *Pract. Treat. Dis. Kidneys* iii. 159 Extra-Renal [albuminuria] in which the albumin is mainly derived from the pus formed in the genito-urinary tract. **1964** O. Kinne in *Oceanogr. & Marine Biol.* II. 302 Both renal and extrarenal active ion transport require oxidative energy. **1934** Webster, Extra-social. **1958** *Times Lit. Suppl.* 17 Jan. 26/4 The extra-social, inner-directed group-life as practised by Epicurus and his friends. **1889** *Cent. Dict.*, Extrasolar. **1953** J. Blish *Case of Conscience* (1958) i. i. 13 Don't forget that Lithia is my first extrasolar planet. **1962** F. I. Ordway et al. *Basic Astronautics* vi. 285 The existence of extrasolar planets would have to be proved by astronomical means. **1931** S. Beckett *Proust* 72 Swann who..spatialises what is extraspatial. **1943** *Mind* LII. 343 The individual consciousness..does not and cannot regard itself as being literally timeless and extra-spatial. **1935** *Mind* XLIV. 509 We are forced to ask whether these purposes can themselves lie wholly within the system constructed, or whether they are extra-systemic and thus extra-linguistic. **1964** *Language* XL. 20 An extrasystemic abbreviating device. **1945** *Wireless World* Oct. 305 (*title*) Extra-terrestrial relays. Can rocket stations give world-wide radio coverage? **1963** M. Caidin *Man-in-Space Dict.* 81/1 *Extraterrestrial*, anything or any being beyond the earth, or from some place other than the earth. **1966** *New Statesman* 8 July 58/3 Contact with extra-terrestrials will..come suddenly. **1967** *New Scientist* 5 Oct. 9/1 MacDonald himself, having started to study UFOs 18 months ago, feels that the idea that the UFOs are extraterrestrial must be given serious attention. **1969** *Times* 28 Apr. 10/8 Scientists have looked for signs of extraterrestrial life in meteorites. **1965** *New Scientist* 18 Mar. 700/3 This 'extravehicular activity', as it is termed, will be the first exposure of man in this hostile environment without the protection of a heavy metal coat. **1966** *Ibid.* 16 June 703/2 Cernan..had to take repeated rests during his extravehicular activities. **1932** A. Gardiner *Theory Speech & Lang.* i. i. 24 'Things' to be spoken about are not simply illusory and..are extra-verbal (i.e. outside the words). **1961** *Lancet* 5 Aug. 308/2 The recording also serves to illustrate the various levels of verbal and 'extraverbal' communication. **1900** Dorland *Med. Dict.* 769/1 The extravisual zone just outside of it is practically incapable of accurately focussing light. **1959** H. Read *Conc. Hist. Mod. Painting* i. 14 Before Cézanne..the artist brought in extra-visual faculties—it might be his imagination, which enabled him to transform the objects of the visible world..or it might be his intellect.

extract, *sb.* Add: **2. c.** (Later example.) Also in fuller form *extract wool*.

1888 *Encycl. Brit.* XXIV. 661/1 Extract wool is that which is recovered from rags of various cloths in which cotton and wool are variously woven together. **1963** A. J. Hall *Textile Sci.* ii. 38 The recovered wool passes under various names such as mungo, shoddy, alpaca and extract (this latter contains cotton fibres also since it is obtained from waste mixture goods).

extracted (ekstræ·ktėd), *ppl. a.* [f. Extract *v.* + -ed¹.] Derived, drawn out, in senses of the vb.; *spec.* in *Biol.*, derived by controlled breeding, selected.

1694 W. Salmon tr. *Bate's Dispens.* 237/2 An extracted and digested Tincture of Mars. **1903** *Proc. Zool. Soc. Lond.* II. 76 It is not stated that the 'extracted' albinos were tested, but there is little doubt that..they would have produced nothing but albinos. **1909** W. Bateson *Mendel's Princ. Heredity* v. 92 It was possible and indeed more usual to find whites exclusively produced by the cross of two extracted F_2 whites. **1910** L. Doncaster *Heredity* vii. 86 The 'extracted' pure individuals in the F_2 generation do not differ recognisably from the original parents in the characters considered. *Ibid.* 136 When two heterozygous individuals are mated together, their homozygous offspring are spoken of as 'extracted' homozygotes.

b. *extracted honey*: honey separated from the uncrushed comb by centrifugal force or by gravity; so *extracted comb*.

1881 T. W. Cowan *Brit. Bee-Keeper's Guide Bk.* 88 Pure extracted honey will usually granulate if kept at a low temperature. **1897** Bartrum & McClelland *Bees in Bar-Frame Hive* 11 Extract at a distance..from the hives, in a room into which the bees cannot penetrate; return extracted comb at night. **1905** P. N. Hasluck *Beehives & Bee Keepers' Appliances* 116 A number of hives are worked for extracted honey.

extractive, *a.* and *sb.* **A.** *adj.* **2.** Substitute for def.: Concerned with extraction; tending to extract or remove resources or products; spec. *extractive industry*, an industry concerned with obtaining natural products, esp. non-replaceable raw materials such as coal, metallic ore, etc. (Add further examples.)

1907 H. W. Macrosty *Trust Movement* iv. 106 The extractive industries dealing with stone and similar products. **1942** *Rep. Comm. Land Utilization* ix. 61 in *Parl. Papers 1941–42* (Cmd. 6378) IV. 421 *Extractive Industries*. These comprise all the mining and quarrying industries. **1962** *Listener* 1 Feb. 205/2 The interest the West has shown towards Latin America has been a commercial, essentially an extractive, interest.

extractor. Add: **3.** Also, an instrument for extracting honey from the combs. Cf. *honey-extractor* s.v. Honey *sb.* 7.

1875 *Encycl. Brit.* III. 501/1 To a German apiarian we are indebted for the invention of a machine called the honey-extractor...There are various patterns of the machine, but the principle of all may be said to be the same, that of centrifugal force. **1885** [in Dict.]. **1886** F. G. Jenyns *Bk. about Bees* 150 When the bee-keeper wishes to obtain the greatest possible quantity of honey he..uses to a great extent the machine called an extractor. **1938–9** *Army & Navy Stores Catal.* 991 All types of Hives, Extractors, etc.

5. *attrib.*, as **extractor fan**, a ventilation fan in a window, outside wall, etc., which replaces stale air indoors with fresh air.

c **1945** *Design at Home* (C.E.M.A.) 13/1 Extractor fan to remove cooking odours and steam. **1958** *Listener* 23 Oct. 671/1 You can fit extractor fans in your windows. **1967** 'M. Hunter' *Cambridgeshire Disaster* iv. 26 He..noted the meal half prepared and the whir of the extractor fan.

‖**extra dictionem** (ekstrā diktiō^u·nĕm), *phr. Logic.* [see Extradictionary *a.*; L. tr. of ἔξω τῆς λέξεως (Aristotle, *Sophistical Refutations*, ch. 4).] Of fallacies: unconnected with the linguistic expression used; not due to ambiguity, etc.; Extradictionary. (Opp. *in dictione.)

1826, 1847 [see *in dictione]. **1852** H. L. Mansel *Aldrich* (ed. 2) 136 Purely logical fallacies belong, not to the *in dictione*, but to the *extra dictionem*. **1906** H. W. B. Joseph *Introd. Logic* xxvii. 534 Later writers..called the fallacies *extra dictionem* fallacies *in re*, or material fallacies. **1970** C. L. Hamblin *Fallacies* i. 13 The traditional Latin terms are *in dictione* and *extra dictionem*.

extradite, *v.* **1. a.** Substitute for def.: To surrender to another country, state, or power a person accused or convicted of a crime committed there. (Add later example.)

1967 *Listener* 6 July 6/3 Congolese Government asks Algeria to extradite Mr Moise Tshombe.

extra-i·llustrated, *a.* [f. Extra- + Illustrated *ppl. a.*] = Grangerized *ppl. a.* So **extra-illustra·tion**, Grangerizing; **extra-i·llustrator**, a Grangerizer.

1889 [see Grangerize *v.*]. **1910** *Encycl. Brit.* XII. 351/2 'Grangerizing' became a term for such an extra-illustration of any work, especially with cuts taken from other books. **1952** J. Carter *ABC for Book-Collectors* 79 Grangerised, or extra-illustrated books as they are now more commonly called, are copies which have had added to them..engraved portraits, prints, etc. **1960** *Times* 1 Mar. 12/5 £180 for an extra-illustrated copy of the first edition of Boswell's Johnson.

extrality (ekstræ·lĭti). Syncopated form of *extraterritoriality.

1925 *Springfield Republican* 29 June, The question of extraterritoriality—or 'extrality', as some writers are beginning to spell a long word that may fill a good deal of space from now on. **1926** *Spectator* 9 Jan. 38/1 The connecting link between the anti-British campaign and the question of abolishing 'Extrality'—otherwise extraterritoriality—is the Russian policy. **1927** *Glasgow Herald* 5 Mar. 11 Such questions as 'extral[ilty' and 'concessions' were not strange words in Japanese ears. **1929** *N.Y. Times* 6 Sept. 24/3 That China has suffered grievously as a result of 'extrality' (as it is now commonly called).

extramural, *a.* Add: *spec.* in *Education*, of institutions or teaching organized by a university or college for persons other than its resident students.

1884 [in Dict.]. **1901** *Daily Chron.* 1 June 3/4 All extramural colleges, high-grade schools, art and technical institutions. **1962** *Lancet* 15 Dec. 1261/1 In the middle 'forties the end of extramural teaching in Glasgow..left the Faculty at a low ebb. **1969** I. & P. Opie *Children's Games* p. xv, The Department of Extramural Studies, Keele University.

b. *transf.* and *fig.*

1955 *Times* 10 June 6/2 The opening of Parliament yesterday was shorn of all such extramural spectacle as the Queen's drive in state. **1964** G. L. Cohen *What's Wrong with Hospitals?* iii. 57 Let the almoner deal with him—a consultant has no time for extra-mural comforting. **1966** J. Wainwright *Crystallised Carbon Pig* xiii. 59 The nightly rendezvous for the city's sugar daddies, extra-mural secretaries, bald-headed Lochinvars.

Hence **extramu·rally** *adv.*, in an extramural way.

1927 *Observer* 5 June 7/2 The University College of the South-West is the youngest of our University institutions; but during the short period of its career it has developed considerably both intra- and extra-murally.

extraneous, *a.* Add: **2. d.** *Mus.* (See quots.)

1801 T. Busby *Dict. Mus.*, *Extraneous*, an epithet applied to those sharps and flats, and those chords and modulations, which, forsaking the natural course of the diatonic intervals, digress into abstruse and chromatic evolutions of melody and harmony. **1839** [see Modulation 4 c]. **1876** Stainer & Barrett *Dict. Mus. Terms* 162/1 *Extraneous modulation*, a modulation to an extreme or unrelated key. *Ibid.* 295/1 When a remote key is reached by relative keys, the modulation is by some said to be *extraneous*. **1938** *Oxf. Compan. Mus.* 590/1 Beethoven.., on occasion, leapt, with little notice to the listener, into unrelated keys (*Extraneous Modulation*).

extrapolate (ekstræ·pŏlē^i^t, e·kstrăpŏlē^i^t), *v.* [f. Extra- + -*polate* of Interpolate *v.*, or back-formation f. Extrapolation.] †**1.** *trans.* To remove (a passage) from written matter. (*Nonce-use* by analogy with Interpolate *v.* 3.)

1831 Gladstone *Let.* in C. Wordsworth *Ann. Early Life* (1891) 91 They inserted the letter.., but *extrapolated* or *metabolised* a part where I had mentioned Canning.

2. a. In mathematical or scientific calculations, to estimate the values of (a function or series) outside a range in which some of its values are known, on the assumption that the trends followed inside the range continue outside it; to continue (a curve) on the basis of points already plotted on the graph; freq. *absol.* Also *intr.* (const. *to*), to reach (a specified value) when extrapolated.

1874 W. S. Jevons *Princ. Sci.* II. xxii. 120 If we wish to assign by reasoning results lying beyond the limits of experiment, we may be said, using an expression of Sir George Airy, to *extrapolate*. **1904** *Biometrika* III. 99 The proportionality of stress and strain is only true within narrow limits, yet the early investigators extrapolated from this linearity all across the mysteries of set, yield-point, and stricture, up to rupture! **1925** *Proc. Nat. Acad. Sci.* XI. 735 This scheme extrapolates to H simply by reducing the shielding to zero. **1933** J. K. Roberts *Heat & Thermodynamics* (ed. 2) iv. 100 The liquid and vapour densities..are determined as near to the critical point as is practicable, and..the results are extrapolated to the point itself. **1955** J. A. Wheeler in W. Pauli *Niels Bohr* 183 For the same energy the electron concentration factor, *a*, extrapolates to a value between $\log_{10} a = 1$ and $\log_{10} a = 2$, depending on the value of *Z*. **1957** G. E. Hutchinson *Treat. Limnol.* I. ix. 579 This series of observations was extrapolated to 0°C. by Whipple and Parker.

b. *transf.* To apply (a theory, etc.) to unknown situations on the basis of its relevance to known situations; to infer (conclusions) from known facts or observed tendencies. Also *absol.* or *intr.*

1905 W. James *Meaning of Truth* (1909) v. 129 The philosopher here stands for the stage of thought that goes beyond the stage of common sense; and the difference is simply that he 'interpolates' and 'extrapolates', where common sense does not. **1907** *Outlook* 17 Aug. 206/1 History, geology, astronomy, are merely these extrapolated, and only demonstrate their relationship with the whole. **1935** *Mind* XLIV. 393 Most scientific theories..are 'extrapolated beyond the possibility of verification'. **1953** *Times Lit. Suppl.* 28 Aug. 542/4 His documents..are..comments on a particular colony at a particular time; and it is hard work to extrapolate from them any general view about all the colonies. **1956** A. J. Ayer *Probl. Knowl.* 153 It may very well be that one does not first form a concept of the relation of temporal precedence, and then extrapolate it to events which are beyond the range of this immediate experience.

Hence **extra·polated** *ppl. a.*, that has been extrapolated; obtained by extrapolation.

1931 Rutherford *Coll. Papers* (1965) III. 255 The effective straggling coefficient for any range was calculated from this extrapolated curve. **1938** R. W. Lawson tr. *Hevesy & Paneth's Man. Radioactivity* (ed. 2) ii. 22 The ranges given in the table are the 'practical' or extrapolated ranges. **1967** J. R. Wolberg *Prediction Anal.* v. 147 The uncertainty for the extrapolated value of *y* is obtained by analysis of all the experimental data (i.e., 62 separate data points).

extrapolation. Add: (Earlier and later examples.) Also *transf.*, the drawing of a conclusion about some future or hypothetical situation based on observed tendencies; the inference resulting from such a process. (Cf. prec., sense 2 b.) Hence **extra·polative** *a.*, characterized by or employing extrapolation.

1872 W. M. Watts *Index of Spectra* p. ix, If the line..lies near to the two reference lines, but not between them, the interpolation formula..must be replaced by one of the two following extrapolation formulæ. **1874** W. S. Jevons *Princ. Sci.* II. xxii. 120 It is a matter of great scientific importance to apprehend precisely how far we can interpolate or extend experimental results by extrapolation. **1889** *Cent. Dict.* s.v., The calculation of the population of the United States in 1900, from the population in 1870, 1880, and 1890, would be an *extrapolation*. **1903** A. M. Clerke *Probl. Astrophysics* 9 The necessity for having recourse to the risky expedient of 'extrapolation'—that is, of applying unrestrictedly to the unknown, rules gathered from observation over a comparatively narrow area. **1922** J. Y. Simpson *Man & Attainm. Immortality* i. 16 By extrapolation of the curve of our knowledge we can reconstruct within the range of conceivability, if not of probability, the course of process. **1926** H. C. Macpherson *Mod. Astr.* 135 Dr. Hubble strongly combats the contention that all non-galactic nebulae are spirals, which he characterizes as a 'daring extrapolation', 'not justified by our present knowledge of nebular forms'. **1929** *Physical Rev.* 2nd Ser. XXXIV. 34 The fact that levels of the same name are approaching the same grouping..might naturally have been expected from our extrapolative way of choosing the names. **1940** *Mind* XLIX. 168 The extrapolative inference from the same set of data to an enormous number of independent propositions. **1965** J. D. North *Measure of Universe* xiv. 297 One of the chief concerns of all scientific theories..is to sanction extrapolation..and the expression of the distant and unknown in terms of what is familiar. **1968** *Sci. Jrnl.* Nov. 87/1 Extrapolative planning.

e:xtraposi·tion. *Gram.* [f. Extra- + Position *sb.*] (See quot. 1933.) Hence **e:xtraposi·tional** *a.*

1927 O. Jespersen *Mod. Eng. Gram.* III. i. 12 In many cases the *to*-infinitive as a primary is placed in extraposition while *it* represents it in the ordinary position. **1933** —— *Ess. Eng. Gram.* (1939) ix. 95 A word or group of words is often placed by itself, outside the sentence proper, in which it is represented by a pronoun; we then speak of 'extraposition': For the raine, it raineth every day. **1961** Y. Olsson *Syntax Eng. Verb* v. 88 The extra-positional, but intra-sectional, units.

extrapunitive (ekstrăpiū·nitiv), *a. Psychol.* Also **extropunitive.** [f. Extra- + Punitive *a.*] Reacting aggressively to frustration; characterized by blaming other people or events unreasonably. Contrasted with *impunitive *a.* and *intropunitive *a.*

1938 S. Rosenzweig in H. A. Murray *Explorations in Personality* vi. 587 He may manifest the emotion of anger and condemn the outer world..for his frustration, adopting an attitude of hostility toward his environment. This type of reaction may be termed 'extrapunitive'. **1945** *Brit. Jrnl. Psychol.* XXXVI. 1. 29 *Extrapunitive*, when the individual tends to blame the external world, reacts with anger and hostility, and psychologically defends himself. **1954** G. W. Allport *Nature of Prejudice* xxiv. 383 There is something exhilarating about extropunitive indignation. To be good and angry at someone else, or even at fate, is like being on a spree. **1969** *Daily Tel.* 2 Oct. 25/5 Is his attitude to life 'intropunitive' (tending to blame himself) or is it 'extrapunitive'—i.e. does he attempt to blame others?

e:xtra-se·nsory, *a.* Also **extrasensory.** [f. Extra- + Sensory *a.*] Of perception: made by other means than those of the known sense-organs, *e.g.* by telepathy, clairvoyance, etc.; hence, received by such means; of or pertaining to the study of such perception. (*Extra-sensory perception* is often abbreviated to *E.S.P., ESP.*)

1934 J. B. Rhine *Extra-Sensory Perception* (1935) p. xxx, Let us merely say..'perception by means that are outside of the recognized senses', and indicate this meaning by 'Extra-Sensory Perception' or E.S.P. **1937** A. Woollcott *Let.* 31 Oct. (1946) 157 It looks like an instance of clairvoyance which might be filed for reference with the extrasensory boys at Duke University. **1941**

J. S. Huxley *Uniqueness of Man* i. 30 Extra-sensory guessing. **1944** J. F. Hendry in H. Treece *Herbert Read* 114 He ignores such phenomena as extra-sensory perception or telesthaesia. **1946** A. Huxley *Perennial Philos.* ii. 36 When tests for ESP can be repeated under standardized conditions, the subject..achieves..a measure of scientific respectability. **1946** J. S. Huxley *Unesco* ii. 37 Extra-sensory knowledge. **1947** J. B. Rhine *Reach of Mind* (1948) iv. 53 Distance..has not shown any effect upon ESP performance. **1958** *Times Lit. Suppl.* 6 June 317/1 Examples of telepathy, of clairvoyance, of precognition (all three grouped together under the heading of extra-sensory perception or ESP). **1968** *Times* 4 Oct. 9/1 A double-blind guessing experiment..is interpreted by its authors to provide evidence for extra-sensory perception (E.S.P.). E.S.P., which covers such terms as telepathy and precognition, is the alleged faculty of perception by means which do not depend on the known sense organs.

extrasomatic (ekstrăsomæ·tik), *a.* [f. Extra- + Somatic *a.*] Deriving from or referring to events external to a person considered as an individual or as a member of society.

1938 C. D. Broad *Exam. McTaggart's Philos.* II. I. xxvii. 59, I begin by dividing ostensible sense-perception into 'extra-somatic' and 'intra-somatic'. In the former the percipient seems to himself to be perceiving foreign bodies or events. *Ibid.*, There are at least three important forms of extra-somatic sense-perception, viz., hearing, sight, and touch. **1947** *Amer. Sociological Rev.* XII. 687/2 Human behavior is, therefore, always and everywhere, made up of these two ingredients: the dynamic organization of nerves, glands, muscles and sense organs that is man, and the extra-somatic cultural tradition. *Ibid.* 693/2 Language has an extra-somatic, non-biological, non-psychological character. It had an existence prior to the birth of any individual speaking it; it comes to each person from the outside. **1964** E. A. Nida *Toward Sci. Transl.* v. 70 While referential meanings are extralinguistic, extrasomatic, and situational.., emotive meanings are extralinguistic, somatic and behavioral.

e:xtra-spe·cial, *a.* (and *sb.*). [See Extra B. a. note and Special *a.* 3 c]. Applied to a special extra edition of a newspaper, etc.; also as *sb.*, such an edition. Also *transf.* and *fig.*, very special; exceptionally good or fine.

1889 E. Dowson *Let.* 24 Mar. (1967) 54 Special Sunday Edition...This [*sc.* letter] appears to be an *extra* special: it is overunning [*sic*] all limits. **1891** *Literary World* 2 Jan. 11/3 An extra-special edition is to appear on the date named in honour of the birthday of the Proprietor-Editor. **1896** G. B. Shaw *Let.* 8 Sept. in *Shaw on Shakes.* (1962) 44 It is this that makes her pay him the extra special compliment of offering to take the chest into her own bedroom. **1897** *Truth* (Christmas Number) 25 Dec. 13/2 Strange forms came out upon her,..And offered 'Hextry speshuls'. **1901** *Punch* 31 July 86/1 The magazines, the newspapers, the extra-specials of the twentieth century. **1902** *Captain* VII. 8/1 I'm playing owl.., for one or two extra-special reasons. **1960** *Times* 19 Feb. 5/6 An even shorter centenary study of Mahler—it is not intended to be much more than an extra-special programme note to a recital.

extraspective (ekstrăspe·ktiv), *a.* [f. Extra- + -*spective* of Introspective *a.*]

†**1.** Facing or turning outwards. *Obs. rare*⁻¹.

1819 J. G. Lockhart *Peter's Lett.* (ed. 2) III. lvi. 25 [Hats without] those lawless curves and twists, prospective, retrospective, introspective and extraspective, from under which the unkempt tresses..may at times be seen 'streaming like meteors to the troubled air'.

2. = *extrospective *a.*

1925 C. D. Broad *Mind & its Place* vii. 328 Those situations in which we seem to be in direct cognitive contact with other minds and their states...For want of a better word, let us call them 'extraspective situations'. **1926** *Public Opinion* 6 Aug. 132/2 He is of the intro-spective rather than the extra-spective order.

extrasystole (ekstrăsi·stŏli). *Path.* [a. G. *extrasystole* (K. F. Wenckebach 1899, in *Zeitschr. f. klin. Med.* XXXVI. 183), f. Extra- + Systole.] A heart-beat outside the normal rhythm.

1900 in Dorland *Med. Dict.* **1907** *Practitioner* Apr. 454 Experimentally, an extra-systole can be easily produced by mechanical, thermal, chemical, or electrical stimulation. **1958** *Spectator* 11 July 60/2 Then the heart 'thumps'. This is a common effect of stress: in medical terms, extra-systoles.

extraterritoriality. Add to def.: Extended later to denote the right of jurisdiction of a country over its nationals abroad, or the status of persons living in a foreign country but not subject to its laws. Cf. *extrality.

1869 *Daily News* 8 Jan., In Rome, at one time, this extra-territoriality was made to extend to the inhabitants of the quarter in which the residence of an Ambassador was situated. **1901** *Westm. Gaz.* 28 Jan. 1/2 To do in China what we have just done in Japan and are anxious that others should do in Egypt—abolish the principle of extra-territoriality, and submit ourselves one and all..to the laws and institutions of China. **1925, 1926** [see *extrality]. **1928** G. W. Keeton (*title*) The development of extraterritoriality in China. **1954** *Ann. Reg. 1953* 299 The form of extraterritoriality represented by the U.S. bases in Japan. **1957** K. A. Wittfogel *Oriental Despotism* 434 The Capitulations, which gave privileged foreigners judicial and economic extraterritoriality, were particularly apparent in Constantinople.

extravaganza. 2. (Earlier example.)

1754 *Connoisseur* 7 Mar. 31 Thalia..was..with difficulty restrained from falling into ridiculous drolleries, and what our author calls *extravaganzas* in her manner.

extravagation (ekstrævăgē^i·ʃən). Delete †*Obs.* and add later examples.

1849 A. D. Fillan *Stories of Rebellions* 39 To check all froward desire of extravagation on the part of Major Drummond. **1899** C. K. Paul *Mem.* 304 M. Renan never could understand how it was people were so little tolerant of his own extravagations.

extraversion. Restrict †*Obs.* to sense in Dict. and add: **2.** *Psychol.* = *extroversion 3. (Cf. quot. 1692–1732 in Dict.) So **extrave·rsive** *a.* = *extroversive *a.*; **e·xtravert** *sb.* = *extrovert *sb.*; **e·xtraverted** *ppl. a.* = *extroverted *ppl. a.*

1915 Jung in *Jrnl. Abnormal Psychol.* IX. 396, I called the hysterical type the *extraversion* type and the psychasthenic type the *introversion* type. *Ibid.* 397 An extraverted individual can hardly understand the necessity that forces the introverted to accomplish his adaptation by first formulating a general conception. **1916** C. E. Long tr. *Jung's Coll. Papers Anal. Psychol.* 288, I propose to use the terms 'Extraversion' and 'Introversion' to describe these two opposite directions of the libido. *Ibid.*, I will call 'regressive extraversion' the phenomenon which Freud calls 'transference' (Übertragung), by which the hysteric projects into the objective world the illusions, or subjective values of his feelings. *Ibid.*, We say that he is extraverted when he gives his fundamental interest to the outer or objective world, and attributes an all-important and essential value to it. *Ibid.* 348 The extraverted type has his libido to a certain extent externally. *Ibid.* 349 An Extravert can hardly conceive the necessity which compels the Introvert to conquer the world by means of a system. **1924** A. G. Ikin in *Brit. Jrnl. Med. Psychol.* IV. 207 In altroversion psychic energy is not consistently directed inwards or outwards as in introversion and extraversion, but can flow freely either way. *Ibid.* 214, I have suggested the use of the term altroversion for the socialisation of either the introverted or the extraverted types..with balance between the self and the environment. *Ibid.*, The personality which thus combines introvert and extravert reactions..can be..called an 'altrovert'..[as that] resulting from a one-sided synthesis of interest and libido, with over emphasis on ego and object respectively, is called an introvert or an extravert. **1926** G. Coster *Psycho-Analysis* 38 The extravert goes out to people and things, enjoying contacts and shrinking from solitude and meditation. **1934** W. Brown *Psychol. & Psychotherapy* (ed. 3) ix. 112 Re-education must include stimulation of their emotional, ethical, and intellectual powers, particularly in the direction of extraversion. **1937** *Scrutiny* Dec. 288 They [*sc.* Wiéner's melodies] have a typically French sexual exuberance, are completely extraversive. **1944** *Mind* LIII. 286 Hoernlé's social conscience was persistent, active and extraversive. **1966** *Illustr. London News* 30 July 34/2 Civilisations as advanced as the ones they had left behind—though less aggressive and extraverted.

extrema, pl. of *extremum.

extremal (ekstrī·măl), *a.* [f. Extreme *sb.* + -al; in sense 2 re-formed on *extrem(um, perh. influenced by next.] †**1.** Farthest from the middle of a line or area; outermost. *Obs.*

1432–50 tr. *Higden* (Rolls) III. 211 And if the wire be distreynede in to thre equalites, and the seide instrumente be putte under the oon extremalle diuision other departenge, the longer parte of the wyre ytowchede yeldethe diapente. **1447** Bokenham *Hooly Wummen* (1938) l. 6903 The extremal marchys of hys regyoun.

2. *Math.* Of or pertaining to extreme qualities or configurations, or highest or lowest values.

1939 *Nature* 4 Mar. 358/1 There seems to be no objection to extremal laws of the local type; but those of the integral type make our modern mind uneasy. **1957** Kendall & Buckland *Dict. Statistical Terms* 104 *Extremal quotient*, the ratio of the absolute value of the largest observation to the smallest observation in a sample. **1966** *Mathematical Rev.* XXXI. 17/1 Turán's original theorem on extremal problems in graph theory.

extremal (ekstrī·măl), *sb. Math.* [ad. G. *extremale* (A. Kneser *Lehrb. der Variationsrechnung* (1900) ii. 24), prob. f. G. *extremum* *extremum.] A function $y(x)$ or its graphical representation that is a solution of the Euler–Lagrange equation and so makes an integral $\int f(x, y, dy/dx)dx$ along an arc of the curve a maximum or a minimum; also applied to a surface the integral over which is a maximum or a minimum. Also *attrib.* or as *adj.*

1901 *Ann. Mathematics* 2nd Ser. II. 108 Any function *y* of *x*..which, at all interior points of an interval (x', x'') throughout which it is considered satisfies Lagrange's equation..is called an *extremal*. **1904** O. Bolza *Lect. Calculus Variations* i. 27 Every solution of Euler's equation (curve as well as function) is called, according to Kneser, an *extremal*. **1950** C. Fox *Introd. Calculus Variations* iii. 69 Surfaces for which I is stationary will be referred to as extremals. *Ibid.*, An extremal surface. **1962** L. A. Pars *Introd. Calculus Variations* i. 18 In the early history of the subject it was often assumed that if there is a unique extremal through A and B the problem is solved. *Ibid.* ii. 43 An extremal arc.

extremist. Add to def.: A member of a party advocating extreme measures. Also *attrib.* or as *adj.* Hence **extre·mism**, the views or actions of extremists; **extremi·stic** *a.*, of or pertaining to extremists or extremism.

1907 *Daily Chron.* 28 Aug. 5/2 Bepin Pal, the Extremist leader. **1920** H. V. Lovett *Hist. Ind. Nationalist Movem.* 69 The reception committee was broken up by a gang of Extremists. *Ibid.* 240 This doctrine is ever the result of Extremism. **1921** *Moplah Reb.* 50 in *Parl. Papers* (Cmd. 1552) XXVI. 237 Certain Extremist Muhammadan agitators..have been..working up the people over the Khilafat. **1921** *19th Cent.* July 148 The Fascismo was born in the provinces, where the extremistic menace was stronger.

extremum (ekstrī·mŏm). *Math.* Pl. **extrema, extremums.** [a. L. *extrēmum*, neut. of *extrēmus* (see Extreme *a.*, *adv.*, and *sb.*). First used as a mathematical term (in German) by P. du Bois-Reymond 1879, in *Math. Ann.* XV. 564.] A value of a function that is a maximum or a minimum (either relative or absolute).

1904 O. Bolza *Lect. Calculus Variations* i. 10 The word 'extremum' will be used for maximum and minimum alike, when it is not necessary to distinguish between them. **1947** Courant & Robbins *What is Math.?* (ed. 4) vii. 343 A point where the derivative vanishes, whether it is an extremum or not, is called a stationary point. **1962** A. H. Frink tr. *Akhiezer's Calculus of Variations* i. 5 An extremum in the whole collection M is called an absolute extremum. We shall also consider relative extrema; to define them we must introduce the notion of neighbourhood. **1968** M. J. Forray *Variational Calculus in Sci. & Engin.* i. 6 Are $f(\frac{4}{5}) = 4^4/5^5$ and $f(0) = 0$ the absolute maximum and minimum for $f(x)$ [$= x^4 - x^5$] in $-2 \leq x \leq 2$? We know that these extremums exist for a continuous function defined over a closed interval.

extrinsic, *a.* Add: **3. c.** *extrinsic factor* (or *element*), vitamin B_{12}: so called because, before its identity with vitamin B_{12} was established, an anti-anæmia factor was known which could be supplied extrinsically, i.e. in the diet, but which was ineffective in the absence of an 'intrinsic factor' secreted by the stomach.

1930 *Amer. Jrnl. Med. Sci.* CLXXX. 306 The process freely permits an interaction between a factor present in the normal gastric juice, which may thus be termed intrinsic, and a factor contained in the beef muscle, which is thus an extrinsic element. **1944** *Science* 28 July 82/1 Extrinsic factor can be partially or completely removed from crude casein by repeated precipitation or by extraction with dilute acid or with alcohol. **1970** W. S. Hoffman *Biochem. Clin. Med.* (ed. 4) xi. 508 The extrinsic factor..is now regarded as identical with vitamin B_{12}.

extrinsicism (ekstri·nsisiz'm). Also **extrinsecism.** [f. Extrinsic *a.* + -ism.] A state or embodiment of extrinsicality. So **extri·nsicist** *a.*

1934 *Downside Rev.* LII. 189 To defend the traditional conception of theological science against the accusations of 'extrinsecism' and 'theologism'. **1935** O. F. Clarke tr. *Berdyaev's Freedom & Spirit* i. 29 The empirical personality..finds itself condemned to a divided existence. Thus spirit is always in such circumstances an 'extrinsicism'. **1967** C. Davis *Question of Conscience* III. iv. 210 The modern historical approach to truth can best be seen in contrast to its opposite extreme. This is extrinsicism. *Ibid.*, The extrinsicist view of truth holds that objective truth..exists already out there, outside history in some unchanging realm.

extrospective (ekstrŏspe·ktiv), *a.* [f. Extro- + -*spective* as in Introspective *a.*] Not introspective; regarding external objects rather than one's own thoughts and feelings. Cf. *extraspective *a.* 2.

1909 *Mind* XVIII. 317 Description or appreciation is based upon..external indications (extrospective psychology). **1925** J. E. Turner *Theory Direct Realism* v. 51 It [*sc.* the inherence of existence] is offered as an accurate description of consciousness, confirmed both by psychological observation, intro- and extro-spective, and the logical exclusion of any alternative.

extroversion. Add: **3.** *Psychol.* The fact or tendency of having one's interests directed exclusively or predominantly towards things outside the self; the turning outwards of the libido; opp. *introversion. Cf. *extraversion 2. Hence **extrove·rsive** *a.*, characterized by or given to extroversion.

1920 A. G. Tansley *New Psychol.* iv. viii. 88 Extroversion is the thrusting out of the mind on to life, the use of the mind in practical affairs, the pouring out of the libido on external objects. **1923** *Westm. Gaz.* 21 Mar. 12/3 Every individual possesses both introversive and extroversive mechanisms. **1932** *Brit. Jrnl. Psychol.* Apr. 300 Rorschach..distinguishes two types, the 'introversive' and the 'extroversive'. They are to all intents and purposes the same as Jung's types. **1951** M. McLuhan *Mech. Bride* 90/1 Mussolini..was all for this Marinetti extroversion of the self and fusion with the activity of the machine. **1959** *Times Lit. Suppl.* 16 Jan. 29/2 A dissatisfaction with life which is notable in some other panoramas of the time which maintain the virtues of extroversion.

extrovert (e·kstrŏvɜːt), *sb.* (and *a.*). *Psychol.* [var. *extravert *sb.* after *introvert: see Extro-.] A person given to or characterized by extroversion; a sociable or unreserved person; also *transf.* Also *attrib.* or as *adj.* So **e·xtroverted** *ppl. a.*, **e·xtrovertish** *a.* = *extroversive *a.*

1918 P. Blanchard in *Amer. Jrnl. Psychol.* Apr. 163 Jung's hypothesis of the two psychological types, the introvert and extrovert,—the thinking type and the feeling type. **1920** *Times Lit. Suppl.* 1 Apr. 205/4 The external always throws him [*sc.* George Herbert] back into himself, and then his thoughts turn outwards for confirmation... He is, in the language of modern psychology, both introvert and extrovert, yet never an egotist. **1920** *Challenge* 21 May 44/2 An extrovert soldier faced with the problem of escape from war conditions. **1923** *Westm. Gaz.* 21 Mar., Any one of these will display either the introverted or the extroverted attitude. **1925** C. Fox *Educat. Psychol.* 254 The first is called the extroverted type, because in the main he goes outside himself to the object. **1926** W. McDougall *Outl. Abnormal Psychol.* 440 The characteristic neurosis of the extrovert is hysteria, while that of the introvert is neurasthenia or psychasthenia. **1946** W. S. Knickerbocker *20th Cent. Eng.* 388 Prufrock is the antithetical brother of Eliot's Sweeney, whose extrovertish animalism functions equally without benefit of an ethic. **1957** *Times Lit. Suppl.* 20 Dec. 774/1 Practical and shrewd, irascible and extrovert, he dominated the college from 1792 to 1834. **1958** P. Tanner in P. Gammond *Decca Bk. Jazz* xi. 138 It is also a happy extroverted music, making up in warmth what it may often lack in subtlety. **1968** *Autocar* 25 Jan. 8/1 Certainly not a budget car, this..is already gaining a name as a noisy, successful little extrovert.

extrude, *v.* Add: **1. d.** To shape (metals, plastics, etc.) by forcing them through dies. Hence **extru·ding** *vbl. sb.*

1913 *Lockwood's Dict. Mech. Engin.* (ed. 4) 435 *Extruded metal*, malleable alloys of copper and other metals which, when heated are forced through dies into various sectional forms. **1923** Glazebrook *Dict. Appl. Physics* V. 427/1 It is..possible to extrude alloys which it is extremely difficult..to roll. **1930** *Engineering* 21 Feb. 252/2 In such operations as embossing, extruding. **1957** *Encycl. Brit.* XIX. 610/1 Tubing machines, or extruding machines, are devices for forcing continuous strips of rubber from a die. **1958** *Engineering* 31 Jan. 133/1 It is made of five-ply with an extruded aluminium framework.

extruder (ekstrū·dəɹ). [f. Extrude *v.* + -er¹.]

1. A machine that extrudes (see prec.).

1959 *Times* (Suppl. Rubber Industry) 27 Apr. p. iv/6 Extruders can be of all sizes up to 15 in. in diameter. **1960** H. Powell *Beginner's Bk. Pottery* II. 63 *Extruder*, a pressing machine for making clay coils. These may be made also by fitting an extruder on to a pug-mill. **1965** *New Scientist* 12 Aug. 385/1 The engineer has improved his extruders (the machines which make the pipe).

2. *Typogr.* (See quot. 1960.)

1938 D. Megaw in *Typography* VII. 28/2 Normal stem projection (*i.e.*, the 'extrusion' of ascender and descenders) is a little more than half the body...Recognizability is aided by the slight differentiation caused by the recurrent shapes of a narrower set (this is where 'extruders' come in). **1960** G. A. Glaister *Gloss. Bk.* 128/2 *Extruders*, the collective term for ascenders and descenders. **1964** P. A. D. MacCarthy in D. Abercrombie et al. *Daniel Jones* 169 English in present roman has a proportion..of about 40% extruders.

extrusion. Add: **1. c.** The pressing of metals, plastics, etc., into the required shape by extruding them through dies; also, the article so extruded (see *extrude *v.* 1 d). Also *attrib.*, esp. in *extrusion press*.

1921 *Chem. Abstr.* 3967 Details as to the app. used and the method of extrusion are given. In complete extrusions the rod produced was sound from end to end. **1937** *Archit. Rev.* LXXXI. 267/3 Aluminium alloy extrusions are used for casement sections, door frames, etc. **1940** *New Statesman* 16 Mar. 360/1 Extrusion of metal is the same, in principle, as squirting tooth-paste or spaghetti, or, better still, macaroni. *Ibid.*, An extrusion press can make bars with curved cross-sections and bars containing longitudinal holes. **1962** M. G. Say *Newnes Conc. Encycl. Electr. Engin.* 104/1 The sheath is applied [to an electric cable] through a point and die by means of a ram or continuous extrusion lead-sheathing press. **1964** M. Gowing *Britain & Atomic Energy 1939–1945* xii. 338 New premises..must be provided for the extrusion work on the rods.

extry, colloq. and dial. (chiefly *U.S.*) var. Extra *a.*, *adv.*, and *sb.* (Cf. quot. 1897 s.v. *extra-special *a.*)

1861 O. W. Holmes *Elsie V.* viii. 100 When we git crowded jest at the end of a term, or when there is an extry number of p'oopils. **1894** J. W. Riley *Armazindy* 10 Extry waste o' sympathy. **1895** H. W. Nevinson *Neighbours of Ours* iii. 51, I was sailin' for a good firm and was makin' from thirty shillin' to two pound reg'lar, and at whiles five bob extry. **1911** W. Owen *Let.* 12 Dec. (1967) 102 Take something 'extry' (as Mrs. Lott would say) in the way of nourishment. **1934** C. Carmer *Stars fell on Alabama* i. 45 On account of havin' had a *little* more experience and playin' *extry* good, Old Man Ventress gets the five-dollar gold piece.

exudate, *sb.* Delete *rare*⁻¹ and add examples.

1903 [see *autolysis]. **1907** M. H. Gordon tr. *Abel's Labor. Handbk. Bacteriol.* 164 Pus and various pathological exudates. **1965** V. J. Chapman *Coastal Veget.* iii. 68 Found mainly in coastal waters and derived from algal exudates. **1964** M. Hynes *Med. Bacteriol.* (ed. 8) xxviii. 430 The 'bacillary exudate' consists largely of polymorphs, with a few large endothelial cells...The 'amœbic exudate', on the other hand, contains few pus or tissue cells.

exudation. Add: **1. c.** *Metallurgy.* (See quot. 1958). Also *attrib.*

1945 *Jrnl. Iron & Steel Inst.* CLI. 283 A reliable method of determining the extent of lead segregation in a sample of leaded steel is by means of the 'exudation test'. *Ibid.*, If segregated lead is present, this treatment results in the exudation of the lead on the surface in the form of globules. **1958** A. D. Merriman *Dict. Metallurgy* 87/2 *Exudation*, the phenomenon in which the liquid produced by partial or complete melting of a solid is liberated and escapes. It is commonly applied to the liberation of liquid metal from the solid, as in the case of the production of molten lead from a leaded brass on heating at temperatures in excess of the melting point of lead.

Exultet. Add: *Exultet Roll* (see quot. 1957).

1882 *Catal. Add. MSS. Brit. Mus. 1876–1881* 70 Exultet-Roll...Written in Lombardic or Beneventine characters of the xii^th^ century, with pneums. **1937** *Burlington Mag.* Nov. 241/2 Perhaps the strangest of all liturgical MSS. are the rolls containing the text of the formulary for the blessing of the Paschal Candle on the evening of Holy Saturday, and usually known from the first words 'Exultet joui [*sic*] angelica turba', as Exultet Rolls. **1957** *Oxf. Dict. Chr. Ch.* 486/2 In S. Italy, from the early 10th cent. to the 13th cent., it was customary to write out this prose [*sc.* the Paschal Proclamation], with appropriate musical directions, on rolls known as 'Exultet Rolls'.

exumbrella (eksɒmbre·lă). *Zool.* [mod.L., f. L. *ex-* out + Umbrella 7 b.] The aboral or outer surface of the umbrella of a jellyfish. Hence **exumbre·llar** *a.*, of or pertaining to the exumbrella.

1886 A. W. Greely *Arctic Service* II. 400 The genus *Nauphanta* is a characteristic one, and is remarkable in the peculiar sculpturing of the exumbrella, the division of the umbrella on the exumbral side into a central and coronal or peripheral zone. **1888** Rolleston & Jackson *Forms Anim. Life* 745 The mesogloea of the aboral aspect of the bell is much thickened to form the umbrella *s.* exumbrella. *Ibid.* 783 If the *Ephyra* be regarded from the dorsal, i.e. exumbrellar aspect. *Ibid.* 784 The sense-bodies are usually protected by a dorsal, i.e. exumbrellar covering-piece or hood. **1897** Parker & Haswell *Text-bk. Zool.* I. 124 The convex outer surface of the bell or umbrella..by which the zooid was originally attached to the blastostyle is distinguished as the ex-umbrella, the concave inner surface as the sub-umbrella. **1907** *Gentl. Mag.* July 97/2 The tentacles..imbedded in ridges of jelly of the exumbrella. **1940** L. H. Hyman *Invertebrates* I. vii. 373 The medusa form resembles a deep to shallow bowl of gelatin, termed the bell or umbrella; the convex aboral surface is called the exumbrella, the concave oral surface, the subumbrella. **1967** P. A. Meglitsch *Invertebrate Zool.* vi. 151/1 Small particles and plankton are caught in slime secreted on the exumbrella.

exurb (e·ksɜːɹb). orig. *U.S.* [f. L. *ex* out of + *urbs* city, or back-formation f. next.] A district outside a city or town; *spec.* a prosperous area situated beyond the suburbs of a city.

1955 [see *exurbia]. **1957** *Economist* 16 Nov. 596/2 They have scattered to the four corners of the country and into the suburbs and 'exurbs'. **1966** L. J. Braun *Cat who could read Backwards* (1967) ii. 16 To reach the fashionable exurb fifteen miles beyond the city limits, Qwilleran drove through complacent suburbs. **1967** *Observer* 1 Oct. 27/6 Levittown..is not any arcadian upper middle-class exurb.

exurban (eksɜ·ɹbăn), *a.* and *sb.* [f. Ex-¹ + Urban *a.* after Suburban *a.*] **A.** *adj.* Of or belonging to a district outside a city or town; suburban; *spec.* pertaining to, or characteristic of, an exurb.

1901 *Westm. Gaz.* 18 Dec. 4/2 Pilgrimages of an ex-urban character, visits to King's Langley.., to Berkhampstead. **1955** *Archit. Rev.* CXVIII. 351/1 The most important architectural consequence of motorized America's post-war flight from the cities has been the emergence of the exurban shopping centre as a differentiated building-type. **1955** [see *exurbia]. **1961** L. Mumford *City in History* xvi. 485 It..anticipates the present 'exurban' emphasis on informal clothing. **1963** *New Statesman* 8 Feb. 188/1 The arc of educated, prosperous suburbia running from near-urban Brookline..to ex-urban Weston and Wayland.

B. *sb.* A resident in an exurban district (see quot.). *rare.*

1957 *Britannica Bk. of Year* 512/1 An exurbanite—or exurban—was one who kept up his city ways and habits in the country [in the U.S.A.].

Hence **exu·rbanite** *sb.*, a resident in an exurban district or in exurbia; *a.* = *exurban *a.* orig. *U.S.*

1955 A. C. Spectorsky (*title*) Exurbanites. **1957** [see *exurban *sb.*]. **1959** *Encounter* Sept. 50/2 A sub-urbanite corner..which..dispenses exurbanite pabulum. **1960** *Spectator* 8 Jan. 47/3 The stock types who go through the moves in this upper exurbanite gavotte. **1960** *Observer* 13 Nov. 3/2 The day-old chicks are farmed out to growers, mostly City exurbanites with a place in Sussex.

exurbia (eksɜ·ɹbiă). orig. *U.S.* [f. Ex-¹ + -*urbia* after Suburbia.] (A quasi proper name for) the region outside the suburbs of a city; exurbs collectively. Hence **exu·rbian** *a.*

1955 A. C. Spectorsky *Exurbanites* ii. 14 Spreading outward from New York City, roughly the first twenty-five miles is solid Suburbia; thereafter, for a belt extending another twenty-five miles, come the exurbs. Exurbia being a recent phenomenon, and a unique one, it may be interesting to explore..how the exurban way of life came to be..superimposed on..earlier cultural patterns in certain areas. **1956** *Archit. Rev.* CXIX. 81/3 (*heading*) Exurbia not necessarily subtopia. *Ibid.*, Some of the brightest American designers are entangled with Exurbian development, and particularly its most characteristic building type, the super-market. **1957** *Technology* Oct. 287/1 First-rate teachers, who are not unduly hard to attract to a new school in a prosperous and pleasant area even in the distant fringes of exurbia. **1968** D. E. Allen *Brit. Tastes* ii. 43 That wide belt of smaller towns and villages ('exurbia') to which so many of those who work in the Metropolis return to sleep.

ex-voto. Add: Also *attrib.*

1823 R. Bakewell *Trav. Tarentaise* I. 354 *Ex voto* inscriptions. **1904** Conrad *Nostromo* II. iii. 131 Anzani's great emporium of boots,..hearts (for ex-voto offerings), rosaries.

eye, *sb.*¹ Add: **2. b.** (Earlier and later examples.) See also Mote *sb.*¹ 1 a.

c **1000** [see Beam *sb.*¹ 3 a]. **1910** Galsworthy *Sheaf* (1916) 120 The old theory, 'an eye for an eye' condemned to death over nineteen hundred years ago, but still dying very hard in this Christian country. **1938** G. Greene *Brighton Rock* I. iii. 58 That's what I always say—an eye for an eye and a tooth for a tooth. **1942** E. Paul *Narrow St.* iii. 27 Thérèse's code was 'an eye for an eye', and the result of her interference was salutary in the extreme.

c. *eyes out* = *all out *advb. phr.* 4; esp. in phr. *to go eyes out*. *Austral.* and *N.Z. colloq.*

1863 E. R. Chudleigh *Diary* 8 Jan. (1950) iii. 63 My horse turning to quickly while I was going eyes out, fell and rowled oaver. **1895** J. Roberts *Diary* 28/1 You weren't travelling 'eyes out' were you? **1907** Mrs. Hawdon *N.Z.ers & Boer War* ix. 185 We went 'eyes out' to catch up. **1945** J. Pascoe in *N.Z. Geographer* Apr. 24 Musterers go 'eyes out' to keep the sheds fed with sheep.

d. (See also Wet *v.* 7 d.)

e. *up to the* (also *one's*) *eyes* (earlier and later examples); also, very much; completely, to the limit; *painted* (*up*) *to the eyes*: heavily made-up with cosmetics.

1778 A. Murphy *Know your own Mind* I. 10 Up to his eyes Sir Richard was in love with her. **1786** E. Sheridan *Let.* 2 July (1960) 91 Miss or Mrs McCartney who was sitting with her poor palsied head dress'd with flowers and painted up to the eyes. **1848** E. Ruskin *Let.* in W. James *Order of Release* (1948) v. 114 Lady Morgan who is..painted up to the eyes. **1866** Trollope *Claverings* (1867) I. viii. 97 All the Burtons are full up to their eyes with good sense. **1885** A. Dobson *At Sign of Lyre* 4 The ladies of St. James's! They're painted to the eyes. **1949** A. Wilson *Wrong Set* 89 Daisy's up to her eyes at the minute trying to jog the local party into action.

f. *to close one's eyes to* (something): to ignore, refuse to recognize or consider; *to shut one's eyes to, against, on*: see Shut *v.* 4 a.

1923 J. S. Huxley *Ess. Biologist* p. x, Most of mankind..close their eyes to this possibility.

g. (See also Wipe *v.* 9 d.)

h. *my eye* (earlier and later examples); also used as an expression of emphatic denial; hence as *sb.*, nonsense. See also *Betty Martin.

1826 T. Creevey *Let.* 11 Aug. in J. Gore *Creevey's Life* (1902) x. 226 My eye, what a spot for a '*walky, walky*'. **1920** D. H. Lawrence *Let.* 25 Jan. (1962) I. 617 One becomes indifferent to all political fates—Fiumes, Jugo-Slavakias and such like my-eye. **1929** W. Faulkner *Sound & Fury* 138 'How about Bigelow's Mill..that's a factory.' 'Factory my eye.'

i. *damn one's eyes*: see *damn *v.* 5.

j. *to give one's eyes*: (hyperbolically) to make a great sacrifice; to be willing to give up anything (*to* be able to do a specified thing). Cf. Give *v.* 9 c.

1857 Trollope *Barchester T.* II. xiv. 273 Bertie would give his eyes to go with you. **1875** L. Troubridge *Life amongst Troubridges* (1966) 120, I gave up directly with a very good grace, considering that I would have given my eyes to go.

k. *to do* (someone) *in the eye*: to defraud, injure, humiliate.

1891 J. M. Dixon *Dict. Idiomatic Phrases* 92 The jockey did your friend in the eye over that horse. **1908** *Punch* 20 May 367/1 Done in the eye again. What on earth do you expect? **1922** F. M. Ford *Let.* 12 Feb. (1965) 138, I have just caught a publisher out, doing me in the eye flagrantly over concealed profits. **1922** F. Hamilton *P. J.: Secret Service Boy* i. 38 That woman means to do you in the eye, and to score off you. **1930** J. B. Priestley *Angel Pavement* v. 218 He'd invented the job five minutes before, just to do mother in the eye.

3. b. *eyes and no eyes*: used to express the difference between an observant and an unobservant person; so, said of or to a person who fails to observe; hence used as the title of a book or series of books dealing with the observation of natural objects; *to keep one's*

eye(s) peeled or *skinned*: see the pples.; *to have eyes to see*: to be observant or discerning; *to keep one's eyes open*: to be watchful or observant.

1794 Aikin & Barbauld *Evenings at Home* IV. 93 (*heading*) Eyes, and no eyes; or, The Art of seeing. **1865** C. M. Yonge *Clever Woman Fam.* iii, 'There is a wonderful charm in a circumscribed view, because one is obliged to look well into it all.' 'Yes; eyes and no eyes apply there,' said Rachel. **1867** (*title*) Eyes and no eyes. A magazine of meteorology and natural history. **1873** Geo. Eliot *Let.* 17 Nov. (1955) V. 461 Now we are keeping eyes and ears open for any hint of another little country place. **1901** (*title*) Cassell's 'Eyes and No Eyes' series. **1912** G. L. Strachey *Landmarks Fr. Lit.* vi. 228 To him who had eyes to see, there might be significance in a ready-made suit of clothes, and passion in the furniture of a boarding-house. **1939** E. M. Forster *What I Believe* 18 With this type of person knocking about, and constantly crossing one's path if one has eyes to see or hands to feel, the experiment of earthly life cannot be dismissed as a failure. **1965** *Listener* 22 July 113/1 During the three weeks I stayed in Dar-es-Salaam, keeping my eyes open and renewing contacts, I was hard put to find evidence of Chinese influence.

d. *spec.* A detective agency or a detective, esp. a private one (see **private eye*); a look-out man; see also quot. 1914. *slang* (orig. *U.S.*).

1914 Jackson & Hellyer *Vocab. Criminal Slang* 31 *Eye* (*the*),..The Pinkerton Detective Agency; an operative of the Pinkerton Detective Agency. Example: 'Blow this joint; it's protected by the Eye.' **1936** J. G. Brandon *Pawnshop Murder* x. 90 As the existence of this watcher had been known for some considerable time to Inspector McCarthy..that astute young gentleman gave the 'eye' no chance to weigh upon him. **1955** *Publ. Amer. Dial. Soc.* xxiv. 141 The [Pinkerton Detective] agency is called *the eye*, from its trademark, the all-seeing eye.

f. Applied in local names to a prominent natural object, such as a hill or island.

1837 *Penny Cycl.* IX. 165/2 Ireland's Eye, a rocky picturesque island of thirty acres. **1891** Dixon *Dict. Idiomatic Eng. Phrases* s.v. *Eye*, The eye of the Baltic—Gothland, or Gottland, an island in the Baltic. **1904** *Daily Chron.* 14 Sept. 5/1 A low rugged hill, nicknamed 'Kuropatkin's eye'.

g. A mechanical or electrical device resembling an eye in its function or appearance; cf. **electric eye*, **magic eye*.

1955 *Sci. News Let.* 15 Oct. 243/1 Humans are still needed to direct the plane until the 15-mile limit, when its radar 'eyes' spot the attacking bomber. **1959** *Listener* 19 Feb. 327/1 The instrument used is called a proton magnetometer. It consists of two parts: the sensing element, or 'eye', which is a half-pint bottle of water with a 1,000-turn coil of wire wound tightly around it [etc.]. **1962** *Daily Tel.* 24 Apr. 20/6 The Ranger's 'eyes', or light-sensitive diodes, were to..keep the space-craft on course.

4. a. *to meet the eye*: see Meet *v.* 2 e and f; *to look* (someone) *in* (occas. *at*) *the eye(s)*: to look directly at; to look in the face (see Face *sb.* 2 b); *to collect eyes*: see *collect *v.* 1 g.

1880, 1896 [see Look *v.* 1 e]. **1931** E. A. Guest *Friendly Way* 23, I want to be able as days go by Always to look myself straight in the eye. **1933** H. L. Ickes *Secret Diary* (1953) I. 97, I looked those mayors in the eye and I told them what the exact truth was. **1965** *Listener* 1 July 4/1 To be modern enough to look the great industrial powers in the eye on a basis of full equality.

c. (Later *fig.* example.)

1900 J. K. Jerome *Three Men on Bummel* i. 17 Mr. Pertwee asked me if I had a skipper in my eye.

d. (See also Mind *sb.*¹ 17 d.)

e. (See quots.)

1933 L. G. D. Acland in *Press* (Christchurch, N.Z.) 21 Oct. 15/7 *Force*..is different from *eye*, the dog's control of sheep by staring them in the face. **1938** J. H. McCulloch *Sheep Dogs* ii. 11 The most striking characteristic of the Border Collie is the one which shepherds refer to as 'The Eye', or the power of the dog to control sheep with its eyes.

5. a. *to see eye to eye*: the sense 'to be of one mind, think alike' is the accepted modern usage (usu. in negative contexts); *to cut eyes* or *one's eye*: to cast a glance or glances (*U.S. slang*) (see also quot. 1961); *eyes on stalks*: eyes protruding with amazement, fear, inquisitiveness, etc.

1827 L. Dow *Jrnl.* (1850) 177/2 Went to New York, took steamboat to New Brunswick thence stage No. 7, strangers crossed words and cut eyes. **1837** *Southern Lit. Messenger* III. 233 'Why, we thought about here' said he 'that you were cutting your eye at Miss Gatty.' **1885** 'C. E. Craddock' *Prophet Gt. Smoky Mts.* xv. 288 Ter see him cut his blazin' eye aroun' at ye, ye'd low ez he'd never hearn o' grace. **1935** C. Isherwood *Mr. Norris changes Trains* v. 84 I'm afraid Schmidt and I don't quite see eye to eye on the subject just at present. **1935** Lady Fortescue *Perfume from Provence* 178, I found myself hugging the edge of a positive precipice...With eyes on stalks I drove on. **1938** M. K. Rawlings *Yearling* xi. 102 Look at him cut his eyes. **1955** *Times* 10 May 9/4 The two Governments do not see eye to eye. **1958** M. Stewart *Nine Coaches Waiting* ix. 130 What *they* call a small private party'd make your eyes stand out on stalks, as the saying is. **1961** F. G. Cassidy *Jamaica Talk* vii. 137 A *cut-eye* is the action of 'cutting' the eye at someone by way of insult—that is, catching the person's eye, then deliberately turning one's own away.

d. *the glad eye*: see *glad *a.* 4 d.

e. *to turn a blind eye*: to refuse to take any notice of a situation, state of affairs, etc.

[**1809** Clarke & M'Arthur *Life of Nelson* II. III. iii. 270 Putting the glass to this blind eye, he [*sc.* Nelson] exclaimed, *I really do not see the signal*.] **1823** M. Wilmot *Let.* 1 Oct. (1935) 197, I turn a blind eye and a deaf ear every now and then, and we get on marvellously well. **1923** J. M. Murry *Pencillings* 92 We turn a blind eye to the signals which a writer hoists against our expectation. **1925** B. N. Odell in E. F. Norton *Fight for Everest, 1924* 290 The Tibetans appear to turn a blind eye to the wholesale slaughter involved in the collection..of over 10,000 specimens by our ardent Natural Historian. **1927** G. K. Chesterton *Coll. Poems* 108 Nelson turned his blindest eye On Naples and on liberty. **1963** *Times* 7 Mar. 16/6 The police turn a blind eye to this problem because they are only too glad to get lorries from parking on the main roads.

f. *eyes down*: with one's eyes looking down, *spec.* of bingo-players (see *bingo²) looking at their cards; hence as *sb.*, the start of a game of bingo.

1962 *Daily Tel.* 25 June 11 The players kept their 'eyes down' for nearly three hours competing for 10 prizes. **1966** P. Moloney *Plea for Mersey* 50 And into the Bingo hall I flew. Eyes down—click click—the game is on. **1969** *Oxf. Mail* 17 Jan. 2 (Advt.), Plaza Bingo Club... Eyes Down 7.45. Doors Open 6.0 p.m.

g. *a gleam* (*glint, twinkle*) *in one's eye*, a barely formed idea; *spec.* a child who has not yet been conceived.

1964 C. Driver *Disarmers* x. 242 As for mass revolutionary civil disobedience, this was never really much more than a gleam in Ralph Schoenman's eye. **1965** E. Lacy *Double Trouble* x. 134 You were on the force when I was a gleam in my pop's sexy eye. **1965** B. Sweet-Escott *Baker Street Irregular* ii. 50 At the time of his arrival.., all this was hardly more than a glint in Colin's eyes. **1966** *Guardian* 10 Nov. 3/7 The proposal remains but a twinkle in the Home Secretary's eye. **1967** *Ibid.* 2 Jan. 8/1 This 'walkers' route is now no more than a gleam in the eye—but so was the Pennine Way when Tom Stephenson first thought of it.

6. a. *to keep, have an* (or *one's*) *eye on*: to keep watch upon; to observe carefully or be wary of; hence, to desire or intend to obtain; to approve of.

c **1430** [in Dict.]. **1605** Chapman *All Fooles* III. i. sig. F1 *Rin.* What would he be, If you should not restrayne him by good connsell? *Gost.* He haue an eye on him, I warrant thee. **1851** W. Clark in W. Bolland *Cricket Notes* 141 There he was sure to be with his eye on every one to see if all was right. **1874** *N. & Q.* 5th Ser. I. 361/2 *Have one's eye on*; *i.e.*, to approve of. **1877** *Independent* 23 Aug. 20/2 The Devil already controls Chicago, and we have heard it intimated that he has his eye on New York. **1895** W. S. Gilbert *Foggerty's Fairy* in *Original Plays* 3rd Ser. 29 I'll keep my eye on that young man. **1900** H. James *Notebooks* (1947) 398 Chad has meanwhile continued to deny..that he has his eye on Mlle de Vionnet, that her mother has..hers on him. **1945** E. Bowen *Demon Lover* 92, I needn't exactly hurry. I just ought to keep an eye on the time. **1956** A. Wilson *Anglo-Saxon Att.* I. iv. 156 'And now you are going to leave us, Mr Middleton, we shall miss you very much.' It was usually said by the mothers who had their eye on rich young bachelors. *Ibid.* II. ii. 319 You asked me to keep an eye on that Larrie Rourke.

c. *to have eyes for*: to pay attention to; to be interested in or attracted to (freq. in contexts excluding all but one person or thing); to desire. (Hence *eyes* is used in *U.S. slang* to mean 'a desire or inclination'.)

1810 J. Porter *Scottish Chiefs* IV. xii. 357 Helen had eyes for none but Wallace. **1923** J. S. Huxley *Ess. Biologist* i. 56 To be so horrifiedly fascinated by it as not to have eyes for anything else. **1934** A. Dubin (*song title*) I only have eyes for you. **1948** *New Yorker* 3 July 28 Have you eyes for a sandwich? **1951** W. Sansom *Face of Innocence* xiii. 189 There's a gaz-and-pneu baron from Bormes has only eyes for her. **1955** L. Feather *Encycl. Jazz* x. 346 *Eyes*, desire, ambition. ('No eyes'—'I'm not interested'.)

d. *to keep, have an eye* (or *a sharp eye*) *out*: to be very alert or watchful.

1889 'Mark Twain' *Yankee* ii. 17, I moved away,..keeping an eye out for any chance passenger in his right mind. **1942** *Horizon* July 57 She's got a sharp eye out, Mrs Pike has.

7. *to have, get, one's eye* (*well*) *in* (earlier and later examples); *a straight eye*: see Straight *a.* 7.

1865 J. Pycroft *Cricketana* xi. 216 As to his guess hits..we can only suppose..that he reserves them till his 'eye is well in', and he has observed the uniform break or rise of the ball. **1882** *Bell's Life in London* 1 July 4/6 Bannerman..though he must have fairly 'got his eye in', scarcely ever attempted to hit. **1912** A. Brazil *New Girl at St. Chad's* vii. 115 When you're in doubt, watch each ball carefully, till you get your eye in.

12. c. An eye-like spot in the wing of an insect; an ocellus. (Cf. *eyelet *sb.* 3 a.)

1658 J. Rowland tr. *Moufet's Theat. Ins.* I. xiv. 959 She hath four great wings, every one of them having eyes of divers colours. **1720** E. Albin *Nat. Hist. Eng. Insects* Tab. IV, On the 6th of July came Forth a beautiful Butter-fly with Eyes in his Wings. **1860** W. S. Coleman *Brit. Butterflies* vi. 72 The 'eyes' are velvety black. *Ibid.* 73 Especially to be admired are the double-ringed 'eyes'. **1876** *Encycl. Brit.* IV. 595/2 The Peacock Butterfly..conspicuous from the 'eyes' on the upper surface of its wings. *Ibid.* 596/2 *Tropæa luna*,..with wings of a lemon colour, each with a 'transparent eye'. **1959** L. H. Newman *Looking at Butterflies* 78 The centre of the eye is black and wine-red...The hind-wings also carry large eye-spots.

d. Also applied to the dark spot in hens' eggs.

1895 *Pearson's Weekly* 18 May 712 The yolk of one average-sized hen's egg (from which the 'eye' has been removed).

e. *Geol.* [tr. G. *auge* eye: cf. *augen.] A lens-shaped inclusion in a rock, esp. gneiss, of a different texture from the groundmass of the rock.

1906 E. H. Adye *Cent. Atlas Microsc. Petrogr.* 52 Sporadic crystals of iron pyrite also occur and often take on the so-called 'eye-structure'. **1920** A. Holmes *Nomencl. Petrol.* 39 *Augen-gneiss*, a general term for gneissose rocks, independently of their origin, containing 'eyes', i.e. phacoidal or lenticular crystals, or aggregates, which simulate the porphyritic crystals of igneous rocks. **1930** Peach & Horne *Geol. Scotl.* iv. 117 The pegmatites show fluxion structure with felspar 'eyes'. **1959** W. W. Moorhouse *Study Rocks in Thin Section* xxv. 409 Augen gneiss is a sheared granulated gneiss containing porphyroclasts or 'eyes' (augen), usually of feldspar.

14. Also *eyes of the ship*, and (simply) *eyes*.

1890 W. C. Russell *Ocean Trag.* II. xix. 134 Sleeping as he did, right in the 'eyes', he got the very full of the motion. **1908** *Westm. Gaz.* 29 Apr. 4/1 There was also a man in the look-out—at what was called the eyes of the ship.

16. b. A mass of ore left in a mine to be worked when other ore is becoming scarce or inaccessible; hence *fig.*, a 'plum', a tit-bit left to the last; (*Austral.* and *N.Z.*) the choicest portion of a piece of land; esp. in phr. *to pick* (or *take*) *the eyes out of* (or *from*).

1839 De la Beche *Rep. Geol. Cornwall* 561 The ores thus left in various places are often termed the eyes of the mine; and when it may be necessary, in abandoning the mine,..to remove them, it is termed, picking out the eyes of the mine. **1865** *Ararat Advertiser* (Vict.) 13 June, Sections were taken up and the 'eye picked from the area'. **1865** *Australasian* 23 June, The great prizes—the allotments which were the eyes of the runs. **1891** R. Wallace *Rural Econ. Austral. & N.Z.* i. 24 The original settlers..had in colonial phraseology 'picked the eyes out of the country' in making their selection. **1895** G. Chamier *South-Sea Siren* xi. 165, I took 'the eyes out of it', as we say—spotted all the best patches, secured all the waterholes. **1905** *Westm. Gaz.* 9 Mar. 9/2 We do not want anybody to come in and 'pick the eyes' out of our districts—to take away the profitable load and leave the unprofitable one. **1945** Baker *Austral. Lang.* iii. 56 The word *eye* became the epitome of all that was choice in land.

c. The centre of a target; = Bull's-eye 7.

a **1877** in Knight *Dict. Mech.*

d. The bright red spot observed through the mica- or glass-covered sight hole of a blast furnace.

1884 W. H. Greenwood *Steel & Iron* vii. 126 A small slide containing a glass or mica plate, through which the state of the furnace may be observed; the bright spot thus seen is known as the 'eye of the furnace'. **1888** *Lockwood's Dict. Mech. Engin.* s.v., The eye of a furnace is that spot or area embraced or commanded by the sight holes.

e. The main mass of lean meat in a rasher of bacon, cutlet, etc.

1934 in Webster. **1951** S. Bull *Meat for Table* vii. 77 The eye is more tender than the remainder of the bottom round and may be fried. **1959** *Times* 30 Mar. 10/7 The eye of lean on the all important..back rasher was good in both breeds. **1960** *Farmer & Stockbreeder* Suppl. 16 Feb. 14 Judges assess the number of marks to award for 'size and shape of the eye of lean'. **1966** *Guardian* 22 July 10/5–6 The noisettes of lamb are..the small circular eye of meat in the cutlet.

20. d. Delete *Obs.* exc. *dial.* and add later examples.

1955 J. G. Davis *Dict. Dairying* (ed. 2) 192 The holes or 'eyes'..are the result of the propionic acid fermentation in the cheese. **1961** *Which?* Sept. 234/2 At times the so-called eyes do not form in Swiss cheese, and the resulting cheese is known as blind cheese.

24. c. *Advertising.* (See quot.)

1924 J. McKechnie *Rational Book-Keeping* viii. 111 In advertising, the line at the top is called the 'eye' of an advertisement.

27. a. *eye-colour, -movement, -slit, -stripe, -trouble, -wear*; **b.** *eye-filling* adj.; *eye-gouging* vbl. sb.

1889 F. Galton *Nat. Inheritance* viii. 138 Stature and Eye-colour..are more contrasted in hereditary behaviour than perhaps any other common qualities. **1922** R. C. Punnett *Mendelism* (ed. 6) 204 It was natural that eye-colour should be early selected as a subject of investigation. **1925** C. Fox *Educat. Psychol.* 26 The physical characters were such things as stature, length of arm, cephalic index, eye-colour, etc. **1900** *Daily News* 11 June 10/3 Mr. Panmure Gordon's eye-filling bay gelding Forrester was third. **1961** *Guardian* 16 Nov. 8/3 'Ben Hur' is both eye-filling and a serious work. **1950** J. Dempsey *Championship Fighting* v. 20 There's no one to prevent low blows, butting, kicking, eye-gouging, biting and strangling. **1924** R. M. Ogden tr. *Koffka's Growth of Mind* iii. 71 Do eye-movements..belong among the inherited reflexes, or are they acquired? **1922** Joyce *Ulysses* 55 He watched the dark eyeslits narrowing with greed. **1933** *Brit. Birds* XXVII. 134 The cock bird had the head and ear coverts slate blue, and the eye-stripe is described..as pale and less conspicuous than might have been expected. **1938** *Ibid.* XXXI. 380 It had a whitish eyestripe and the legs were dark grey. **1896** *Westm. Gaz.* 24 Sept. 1/3 Mr. Gladstone's eye-trouble.

a **1963** J. Lusby in B. James *Austral. Short Stories* (1963) 233 Affecting eye-trouble, Rafe jumped in. **1926** *Glasgow Herald* 13 Apr. 11/2 Cameras, telescopes,.. eyewear, and lenses. **1936** R. Campbell *Mithraic Emblems* 161 You buy dark specs to stare at castles, But I collect such eye-wear free.

28. eye appeal orig. *U.S.*, visual appeal or attractiveness; **eye-area** (see quot.); **eye bank** orig. *U.S.*, a reserve store of human corneas kept for treatment of the blind; **eye-bath,** a cup-shaped vessel designed to fit the orbit of the eye, used to apply a lotion to the eye; **eye-black,** black eye-shadow; mascara; **eye-bugging** *a.* (*U.S.*), having or characterized by bulging eyes (cf. *Bug *v.*^{2}); **eye-catcher,** a person or object that draws the eye; so **eye-catching** *a.*, attractive to the eye; striking; prominent; **eye-clip** *v. trans. N.Z.* (see quot. 1933); so **eye-clipping** *vbl. sb.*; **eye-dialect,** unusual spelling intended to represent dialectal or colloquial idiosyncrasies of speech (see quots.); **eye dog** *N.Z.* (see quot. 1951); **eye-drop,** (*b*) *pl.*, a liquid administered to the eye in drops; **eye-dropper,** a device for administering eye-drops; **eye-fly,** a minute fly which in summer-time in the East is troublesome to the eyes of men and beasts; **eye-ground,** the fundus or back of the eye; **eye-level,** the level of the eyes; also *attrib.*; **eye-liner, eyeliner,** a cosmetic applied in a line around the eye; a brush or pencil for applying this; also *attrib.*; hence *eye-lining* vbl. sb.; **eye lotion,** (*a*) lotion for the eyes; (*b*) *slang* (see quot. 1943); **eye-minded** *a. Psychol.*, tending to a frequent use of visual imagery; having a mental constitution chiefly or exclusively visual, so that thoughts and memories take the form of visual images; thinking in terms of the printed or written word rather than of the spoken word; so **eye-mindedness,** the condition of being eye-minded; **eye muscle,** (*a*) a muscle that moves the eye or one of its components; (*b*) a muscle, the longissimus thoracis, which runs alongside the spine and in an animal gives rise to an eye in a piece of meat (*Eye *sb.*^{1} 16 e); **eye-opener,** (*a*) (earlier and later examples); (*c*) (earlier U.S. example); (*d*) an attractive woman; (*e*) a person who reveals facts or clarifies a situation to others; **eye-pencil,** a pencil (Pencil *sb.* 2 c) for drawing cosmetic lines around the eyes; **eye-plate,** (*a*) a chitinous sclerite in which the eyes of *Acarina* are placed; (*b*) (see quot. 1948); **eye-rhyme, -rime** (see quot. 1936); **eye-ring,** a circular space within which the eye of the user of an optical instrument must be placed in order to obtain the full field of view; **eye-shade,** (*a*) (earlier example); (*c*) = **eye-shadow*; **eye-shadow** *Cosmetics*, colouring applied to the eyelids or around the eyes; hence *eye-shadowed* adj.; **eye stitch** (see quots.); **eye-strain,** weariness or strained condition of the eyes resulting from excessive or improper use of the eyes, or uncorrected defects of vision; so *eye-straining* sb. and adj.; **eye-taking** *a.*, that attracts attention; **eye-veil,** a veil which reaches down as far as the eyes; **eye-ward,** a ward for eye patients in a hospital; **eye-wig** *v. trans.* (*N.Z.*) = **eye-clip* vb.; **eye-wire,** wire forming the metal frames of spectacles; **eye-worker,** one whose work needs special use of the eyes.

1926 *National Provisioner* 19 June 25 Through improved wraps, a manufacturer may give his product 'eye appeal'. **1951** M. McLuhan *Mech. Bride* 63/1 Color tempted him to accept the appetizing eye appeal of the food ads. **1960** *Guardian* 29 Nov. 5/1 Prewar cars lacked the eye appeal of modern cars. **1895** A. C. Haddon *Evol. Art* 36 The six rays are but a symmetrical coalescence of two pairs of eye-areas. *Note*, I have adopted the term 'eye-area' to denote the eye device which includes the eye, the eye-lashes, and often the cheek-fold of that side. **1944** *Amer. N. & Q.* IV. 24/2 Eye Bank: founded at the New York Hospital..May 8; for collecting and preserving 'healthy human corneas'. **1959** Eye bank [see *Bank *sb.*^{3} 7 f]. **1830** *J. & S. Maw's Catal.*, Eye Baths. **1935, 1943** Eye bath [see **eye lotion* below]. **1927** *Sunday Express* 20 Mar. 1 The police found little besides some lipstick and eyeblack. **1963** I. Fleming *On H.M. Secret Service* xvi. 273 Now you've ruined my eye-black. **1927** E. Hemingway *Men without Women* (1928) 102 On the walls of the houses were stencilled eye-bugging portraits of Mussolini. **1937** L. C. Douglas *Forgive our Trespasses* xi. 218 With many long, cheek-distending, eye-bugging exhalations. **1923** K. G. Karsten *Charts & Graphs* 689 The best kind of eye-catcher for calling attention to numerical data. **1946** *Vogue* Aug. 31 Let his wife..try.. to emulate the appearance of those eye-catchers, and she's pretty soon put in her place. **1969** B. Turner *Circle of Squares* viii. 55 She was wearing lacy black stockings, the sort of eye-catchers that used to be considered very sexy. **1933** *Aeroplane* 6 Sept. 418/2 Another worth-while dodge is the provision of labels of a most eye-catching design. **1952** D. Dodge *To catch Thief* (1953) ii. 40 She might be really eye-catching if she would make an effort to be. **1930** L. G. D. Acland *Early Canterbury Runs* 1st Ser. vi. 128 Merino sheep..were not handled so often for eye-clipping and so on. **1933** —— in *Press* (Christchurch, N.Z.) 14 Oct. 15/7 *Eye-clip*, to cut the wool away from round a sheep's eyes. If this is not done, the wool, especially on merinos, is apt to grow over the eyes and make the sheep wool-blind. Some people speak of eye-clipping as *Winking*. **1953** B. Stronach *Musterer on Molesworth* viii. 54 Our next job was to eye clip the lambs, for many of them were wool-blind. **1925** G. P. Krapp *Eng. Lang. in Amer.* I. iv. 228 The impression of popular speech..is often assisted by what may be termed 'eye dialect', in which the convention violated is one of the eye, not of the ear. Thus a dialect writer often spells a word like *front* as *frunt*, or *face* as *fase*, or *picture* as *pictsher*, not because he intends to indicate here a genuine difference of pronunciation, but the spelling is merely a friendly nudge to the reader. **1965** *Amer. Speech* XL. 230 Which of the most fastidious elocutionists could object to the vocalized result of *enuff*, probably the oftenest repeated of Capp's eye-dialect usages? **1951** L. G. D. Acland *Early Canterbury Runs* 376 *Eye dog*, dog that commands sheep by his eye. **1963** R. Casey *As Short Spring* 196 What's more, he keeps all eye dogs—hardly ever bark in their lives. They could work sheep with barely a sound on a clear night. **1938–9** *Army & Navy Stores Catal.* 393/2 Eye Drop Bottle... Eye Drop Tubes. **1961** *Harper's Bazaar* May 95 Improve the whites [of eyes] with..French eye drops, deep blue in colour. **1938** J. Steinbeck *Long Valley* 75 He took an eyedropper from a drawer. **1962** J. Braine *Life at Top* ii. 33, I..poured her out her usual medicinal dose of brandy...Sometimes I made jokes about using an eye-dropper next time. **1815** Kirby & Spence *Ent.* I. App. 8 A very minute black fly.. which, because it flies in swarms into the eyes..is called the eye-fly. **1951** E. A. Mittelholzer *Shadows move among Them* I. ix. 68 A tickling on his eye-lashes..it was an eye-fly. **1900** *Jrnl. Exper. Med.* 25 Oct. 196 The eye grounds..were normal. **1910** *Practitioner* July 97 Mental and moral deterioration in the one.., normal eye-grounds and active pupils in the other. **1909** *Westm. Gaz.* 16 Oct. 14/2 The eye-level is the true position for the hand-camera. **1926** *Army & Navy Stores Catal.* 291/2 The 'New-World' eye-level cooker. **1932** E. Bowen *To North* xxiv. 256 Then she discovered them just at eye-level, ranged in a row. **1951** *Good Housek. Home Encycl.* 75/2 *'Eye-level' cookers*.., electric cookers in which the oven is raised to eye-level and the..grill chamber situated alongside. **1956** *Ibid.* 76/1 Eye-level grills. **1958** M. L. Hall *Newnes Complete Amat. Photogr.* 61 Slower in action than an eye-level camera. *Ibid.*, Many cameras do include eye-level viewfinding arrangements. **1961** *Radio Times* 18 May 6/1 This up-to-the-minute Jackson 'extra' fits all the Highline eye-level grills. **1960** *Harper's Bazaar* July 67/1 Slim eye-liner brush for detailed shading. *Ibid.*, A liquid eyeliner complete with brush. *Ibid.*, A soft eye-liner pencil. **1968** *New Scientist* 24 Oct. 178/2 On being sent out by his daughter to buy her a 5s 5d tube of eye-liner, he was amazed at the minute quantity she got for that sum. **1960** *Sunday Express* 6 Nov. 14/4 Gold eye-lining pencils. **1886** in L. de Vries *Vict. Advertisements* (1968) 34/2 Away with eye-glasses and eye lotions. **1935** R. Macaulay *Personal Pleasures* 154 But you have..left behind you..a toothbrush, and a bottle of eye lotion with eye bath. **1943** Hunt & Pringle *Service Slang* 30 *Eye lotion*, wines (only enough of them to provide an eye bath). **1888** J. Jastrow in *Pop. Sci. Monthly* XXXIII. 597 (*title*) Eye-mindedness and ear-mindedness. *Ibid.* 603 An eye-minded person should read, should reduce everything to visual terms. **1897** *Psychol. Rev.* (Monogr. Suppl.) II. 1. 18 Some persons are ear-minded—they think most readily in auditory ('phonographic') images; others are eye-minded, thinking in visual ('photographic') images. *Ibid.* 21 Eye-mindedness is of course quite different from quickness of visual perception. **1925** Eye-minded [see *Ear-minded *a.*]. **1881** *Index Med.* III. 183/2 (*title*) Spasm of the intra-ocular eye muscles. **1924** R. M. Ogden tr. *Koffka's Growth of Mind* iii. 77 The innervations which the eye-muscles undergo in movements of fixation are determined..by the pre-existing position of the eyes. **1959** *Times* 30 Mar. 10/6 Points for quality.. are given for..size and shape of the eye muscle in the back [of a pig]. **1818** H. B. Fearon *Sk. Amer.* 252 At table there is neither conversation nor yet drinking; the latter is effected by individuals taking their solitary 'eye openers', 'toddy', and 'phlegm dispersers'. **1846** D. Corcoran *Pickings* 75 A 'pig and whistle' is the only reg'lar eye opener. **1863** *Rio Abajo Press* (Albuquerque, N.M.) 3 Feb. 2 Quite a catalogue of similar examples of injustice and meanness..might be made..but we merely allude to them as an 'eye-opener' to the public. **1919** F. Hurst *Humoresque* 72 He 'ain't seen her since a child, and all of a sudden he comes West and finds in front of him an eye-opener. **1928** *Manch. Guardian Weekly* 31 Aug. 174/4 He felt his mission to be that of an agitator, of an eye-opener, of a merciless yet undogmatic critic. **1948** E. N. Dick *Dixie Frontier* 189 Ministers regularly took their morning eye-opener and their nightcap in the evening. **1902** *Westm. Gaz.* 20 Nov. 6/3 Sticks of grease-paint, eye-pencils, lip salve. **1903** *Ann. & Mag. Nat. Hist.* Nov. 505 The comparative structure of the hard chitinous parts of the body, especially of the eye-plates, mouth-organs, and palps. **1923** *Man. Seamanship* II. iv. 93 On its [*sc.* the planing shoe's] upper surface are three eye-plates, to which are shackled the three legs of a chain sling. **1932** *Times Lit. Suppl.* 14 Jan. 30/1 Thrice Bill hung head-downwards over the eyeplate below the shackles of the mizzen rigging. **1948** R. de Kerchove *Internat. Maritime Dict.* 248/2 *Eye plate*, a plate or casting with an eye normal to its surface and formed solid with the plate. **1871** Ellis *E. E. Pronunc.* III. viii. 864/2 The spelling seems to have been changed to make an eye-rhyme. **1873–4** G. M. Hopkins *Note-bks.* (1937) 246 Unlawful rhymes—We may notice (i) mere eye-rhymes; [etc.]. **1909** O. Jespersen *Mod. Eng. Gram.* I. 5 But eye-rimes are of comparatively recent growth, many of them owing their origin to words of formerly identical or similar sound having now become differentiated, thus *war* and *far*. **1913** Ogilvie & Albert *Pract. Course Secondary Eng.* II. 186 Some poets, particularly Spenser in his *Faerie Queene*, resort to what is called eye-rhyme, that is, rhyme between words of similar spelling, though of different sound. **1936** Treble & Vallins *A.B.C. of Eng. Usage* 157 Especially to be noted are 'eye rhymes', i.e. those which exist only to the eye and not the ear, like *quay—day*. **1902** Mann & Millikan tr. *Drude's Theory of Optics* 77 The exit-pupil is often called the eye-ring, and its centre is called the position of the eye. **1845** J. T. Smith *Bk. for Rainy Day* 11 The ladies this year [*sc.* 1768] wore half a flat hat as an eye-shade. **1937** L. MacNeice in Auden & MacNeice *Lett. from Iceland* x. 133 Compacts, lipstick, eyeshade, and coiffures. **1930** H. Rubinstein *Art of Feminine Beauty* iv. 67 The most successful way of lending depth and mystery to the eyes is with the eye shadow. **1950** J. Emerald *Photographic Make-Up* i. 27 Eye-shadow is applied to the eyelids with a flat-top sable brush. **1959** *Daily Sketch* 7 Jan. 9 False eyelashes..give the effect of eye-shadowed lids. **1932** D. C. Minter *Mod. Needlecraft* 53/1 Eye stitch..is formed by sixteen stitches going into a single hole..but spread into a square on the outside edge. *Ibid.* 53/2 Algerian eye stitch..is worked over a square having four threads on each side. **1934** M. Thomas *Dict. Embroidery Stitches* 84 *Eye stitch*, the stitch consists of 16 Satin Stitches all taken into the same central hole but with their outer ends arranged over a square of eight threads, with two threads between each stitch. **1874** *Med. & Surg. Reporter* (Philad.) XXXI. 67 (*heading*) Headaches..from eye strain. **1909** *Practitioner* Dec. 779 Of all the causes of eyestrain the most frequent is the presence of an error of refraction. **1949** H. C. Weston *Sight, Light & Efficiency* iii. 57 The eye-strain-risk associated with occupations. **1871** *English Mechanic* 380/3 Eye-straining. **1923** Kipling *Land & Sea Tales* 185 Cold, nose-running, eye-straining work. **1868** G. M. Hopkins *Jrnls. & Papers* 18 July (1959) 177 An ash rose with eye-taking sky-clusters. **1960** *Harper's Bazaar* July 38 She likes..eye-taking colours. **1928** *Daily Express* 4 June 5/3 The eye-veil fashion..is good for the races. **1879** *St. George's Hosp. Rep.* IX. 465 The average stay of a patient in the eye-wards..was 25·84 days. **1950** *N.Z. Jrnl. Agric.* Apr. 387/3 Each ewe..is eye-wigged if necessary. **1881** *Instr. Census Clerks* (1885) 96 Eye Wire Maker. **1962** L. S. Sasieni *Optical Dispensing* i. 25 This pattern is..produced..by stamping or rolling the eye-wire through dies. **1898** G. M. Gould *Biogr. Clinics* (1905) III. 500 A seamstress or..any hard-pushed eye-worker.

eye, *v.* Add: **7.** *intr.* Of eggs: to form eyes (see Eye *sb.*^{1} 12 c).

1904 *Daily Chron.* 25 Mar. 8/3 The eggs take from six weeks to three months to 'eye', as it is called.

eye-ball, *sb.* Also **eyeball.** Add: **2.** Advb. and attrib. phr. *eye-ball to eye-ball*, confronting closely; with neither party yielding. *colloq.* (orig. *U.S.*).

1962 D. Rusk in *Sat. Even. Post* 8 Dec. 16/1 We're eyeball to eyeball, and I think the other fellow just blinked. **1965** H. Kahn *On Escalation* ii. 43 If there is a direct ('eyeball to eyeball') confrontation.., all participants and observers will take an intense interest in the proceedings. **1966** *Guardian* 16 June 1/3 Mr Wilson said we ought not to 'contract out and leave it to the Americans and Chinese eyeball to eyeball'. **1969** 'E. Lathen' *When in Greece* vii. 68 Ken had not bargained for an eyeball-to-eyeball confrontation. **1970** M. Kelly *Spinifex* iii. 65 Two extremely ambitious dogs are eyeball to eyeball over the same bone.

eye-ball, eyeball (əi·bǭl), *v. U.S. slang.* [f. the sb.] *trans.* and *intr.* To look or stare (at).

1901 *Harper's Monthly* Feb. 443/1 'God!' burst from the lips of the man as he eyeballed his attendant. **1942** *Amer. Mercury* July 85 He would eye-ball the idol-breaker. **1968** *Listener* 22 Aug. 229/2 This movie is so richly risible that I advise all, in John Wayne's phrase, to go down to the Warner and eyeball it. **1970** A. Cameron et al. *Computers & O.E. Concordances* 60 Errors..will be exposed very quickly because you simply 'eyeball' down the centre of the page.

eyebrow. Add: **1. c.** *to raise one's eyebrow*(*s*) (or *an eyebrow*): to show surprise or dubiousness (*at* something).

1918 L. Strachey *Emin. Victorians* 28 The most steady-going churchman hardly raises an eyebrow at it now. **1956** G. H. Vallins *Pattern of Eng.* v. 132 Brown, though he raises his eyebrows a little at the usage, by no means condemns it outright. **1956** A. Wilson *Anglo-Saxon Att.* 407 They will perhaps raise their eyebrows at the old reiterated parrot-cry that the triumph of Roman Christianity brought 'civilization' to our island.

d. *up to the* (or *one's*) *eyebrows*, up to the eyes (see Eye *sb.*^{1} 2 e).

1925 Wodehouse *Carry On, Jeeves!* ii. 53 There was a fellow, one would have said, clear up to the eyebrows in the soup. To all appearances he had got it right in the neck. **1954** N. Coward *Future Indefinite* II. iii. 92 To many of my acquaintances..it was a foregone conclusion that I was involved in espionage up to the eyebrows. **1961** B. Fergusson *Watery Maze* x. 252 The commanders concerned were up to their eyebrows in the Sicilian campaign. **1968** A. Diment *Gt. Spy Race* iii. 36 He's.. tanked up to the eyebrows with Acid.

4. eyebrow pencil (see Pencil *sb.* 2 c); **eyebrow tweezers,** tweezers for extracting unwanted hairs from the eyebrows.

1881 [see *Blanc de perle]. **1936** D. Powell *Turn, Magic Wheel* ii. 135 His wife's eyebrow pencil. **1938** [see *Eye-lash c]. **1933** D. L. Sayers *Hangman's Holiday* 123 A pair of eyebrow tweezers. **1966** P. V. Price *France* 113 Eyebrow tweezers (can be invaluable for far more first-aid operations than just plucking eyebrows).

eyeful, *sb.* Add: **c.** Also **eye-full.** A 'good look' at something; an exhilarating or remarkable sight; *spec.* a strikingly attractive woman; esp. in phr. *to get an eyeful* (*of*), to have a good look (at); to see something remarkable, beautiful, etc. (See also quots.) *colloq.*

1899 S. MacManus *In Chimney Corners* 248 She took an eyeful out of Jack, an' right well plaised she was with his appearance. **1914** Jackson & Hellyer *Vocab. Criminal Slang* 31 *Eye full*, the object of scrutiny or of attentive observation...'Nix crackin'. The mark on your left is getting an eye full.' **1919** W. H. Downing *Digger Dial.* 22 *Eye-full*, a complete view. *Ibid.* 26 *Get an eye-full*, see. **1922** Joyce *Ulysses* 765 Ill put on my shift and drawers let him have a good eyeful out of that. **1929** H. V. Morton *In Search of Scotland* ii. 46 'You have seen the Crown Jewels in London?' 'I'll say so! They're an eyeful.' **1933** E. Caldwell *God's Little Acre* xii. 172 When you've got an eyeful of Griselda, there, you know durn well you've been missing a heap thinking such foolishness all your life. **1934** Wodehouse *Thank You, Jeeves* iv. 47 Unquestionably an eyeful, Pauline Stoker had the grave defect of being one of those girls who want you to come and swim a mile before breakfast. **1947** N. Balchin *Lord, I was Afraid* 52 He thought to himself this is a bit of all right and started right in to get an eye-full, see? **1960** 'A. Garve' *Golden Deed* vi. 39 You're both quite an eyeful.

eyeglass, *v.* For *rare*$^{-1}$ read *rare* and add later examples. Also, to see through an eye-glass; *intr.*, to use an eye-glass. Also *fig.*

1841 S. Bamford *Passages in Life of Radical* (ed. 2) I. v. 26 Some of the members..were eye-glassing across the house [of Commons]. **1893** *Daily News* 23 Feb. 6/4 That fever heat, insurrection, and great emergency which ..Mr. Chamberlain eye-glassed in a not distant future.

eye-glassy, eyeglassy (əi·glɑsi), *a. colloq.* [f. Eye-glass *sb.* + -y^1.] Pertaining to or characteristic of one who wears an eye-glass; *allusively*, haughtily superior or contemptuous.

1871 G. Meredith *H. Richmond* xxi, The interior of the Casino seemed more hostile. I remarked it to him. 'A trifle more eyeglassy' he murmured. **1906** *Cassell's Mag.* Apr. 553/1 Mr. Rodder only laughed—the pleasant laugh that wasn't eye-glassy.

eye-lash. Add: **c.** *attrib.* and *Comb.*

1907 *Yesterday's Shopping* (1969) 538/3 Eyelash Cream —box 1/5. **1926–7** *Army & Navy Stores Catal.* 494/2 Eyelash brushes. **1938** *Eve's Jrnl.* Jan., Make-up... Eyebrow pencil, eyelash brushes, eyelash comb. **1960** *Harper's Bazaar* July 67/1 Eyelash curlers in simulated gold.

eyelet, *sb.* Add: **3. a.** *spec.* In a butterfly's wing: an ocellus; = *Eye *sb.*^{1} 12 c.

1832 J. Rennie *Butterfl. & M.* 23 The Eyed Hawk Wings..; second pair rosy at the base, the tips much paler, with a large blue eyelet near the posterior angle. **1832** T. Brown *Bk. Butterfl. & M.* (1834) I. 120 The wings are angular,..with large compound eyelets, reddish in the centre. **1848** [in Dict.].

4. *spec.* Designating stitching or embroidery composed of eyelets giving an open-work effect (see also quot. 1909).

1909 *Cent. Dict.* Suppl., *Eyelet-stitch*, in sewing-machine work, the method of placing the stitches in radial lines round an eyelet-hole or over a metal reinforcing ring or eyelet. **1911** *Daily Colonist* (Victoria, B.C.) 30 Apr. 3/4 (Advt.), Eyelet embroidery is an effective feature. **1932** D. C. Minter *Mod. Needlecraft* 52/1 Eyelet stitch..may be used as a background stitch. *Ibid.* 253/1 Eyelet or English embroidery as decoration. **1960** B. Snook *English Hist. Embroidery* 53 Other shirts..have four-sided stitch and eyelet stitch on the front opening.

eyer. Add: **b.** A maker of eyes in needles.

1881 *Instr. Census Clerks* (1885) 45 Needle Maker... Eyer.

eyesome, *a.* Delete † *Obs.* and substitute: *Obs.* exc. *poet.*

1922 T. Hardy *Late Lyrics & Earlier* 266, I thought not I should meet an eyesome maiden.

eyespot. Add: **3.** A name given to two fungal diseases: **a.** A disease caused by *Helminthosporium sacchari* that affects sugar-cane. **b.** A disease caused by *Cercosporella herpotrichoides* that affects wheat, other cereals, and some grasses.

1906 *Mem. Dept. Agric. India, Bot. Ser.* I. iii. 43 Several other leaf spotting fungi..are known to occur on the sugar-cane...The following are the chief known:—(1.) The 'Eye-spot' of Java canes, produced by *Cercospora Sacchari*. **1929** *Rev. Appl. Mycol.* VIII. 467 The control of eye spot disease of sugar-cane. **1930** *Ibid.* IX. 639 On Bon Fermier wheat affected by foot rot the author noted eye spot lesions of the type usually attributed to *Leptosphaeria herpotrichoides* but associated with an undetermined fungus whose cultural characteristics were quite different. **1942** *Ann. Appl. Biol.* XXIX. 259 Wheat and barley are susceptible to eyespot but oats show considerable resistance. **1968** *Times* 16 Dec. 7/2 The soil-borne diseases, take-all and eyespot. **1970** Liebscher & Koehler tr. *Fröhlich & Rodewald's Pests & Dis. Tropical Crops* 235 Sugar yield losses due to eye spot disease can be very high.

Eyetalian (əitæ·liăn), *a.* and *sb.* Also **Eye-talian.** Spelling used to represent a non-standard or jocular pronunciation of Italian *a.* and *sb.* with initial (əi) sound.

1840 *Spirit of Times* 21 Nov. 447/3 Boston galls may boast of their spinetts, and their gytars, and their eye-talian airs, and their *ears for music*. **1919** F. Hurst *Humoresque* 189 The Eyetalians maybe didn't know no better. **1930** E. Wallace *Lady of Ascot* i. 14 It's foreign—Eye-talian. **1930** E. Raymond *Jesting Army* I. vii. 102 The Eye-talians'll have landed at Smyrna. **1953** A. Baron *Human Kind* xxiii. 171 You know what I seen them Eyetalians do? **1961** P. White *Riders in Chariot* xi. 402 An Eyetalian boy called Fiddle Paganini was finishing singing a number.

Eyetie (əi·təi), *sb.* and *a.* Also **Eyeto, Eyety, Eyetye, Eytie.** [f. prec.] = prec., used with disparaging overtones.

1925 Fraser & Gibbons *Soldier & Sailor Words* 90 *Eyeties*, Italians. **1941** Baker *Dict. Austral. Slang* 27 *Eyeto* (pronounced 'eye-toe'), an Italian. **1943** B. J. Hurren *Eastern Med* viii. 100 Such tricks..made our troops utterly despise the Eyeties (as the soldiers called them). **1943** C. S. Forester *Ship* iv. 30 There was a buzz that the Eyety navy was out. **1958** E. H. Clements *Uncommon Cold* i. 16 The Yugoslavians, the two Eyetyes, some West Germans. **1967** K. Giles *Death in Diamonds* vii. 127 They speak Eyetie?

eye-tooth. Add **b.** (*one*) *would give one's eye-teeth*: (one) is very eager, or ready to make the greatest sacrifices (*to* do something). Cf. *Eye *sb.*^{1} 2 j.

1930 W. S. Maugham *Cakes & Ale* i. 13 He'd give his eye-teeth to have written a book half as good. **1965** F. Sargeson *Memoirs of Peon* iv. 55, I would give away my eye-teeth to take that girl into the park.

eye view, eye-view. [f. the phr. *bird's-eye view* (Bird's-eye 3).] A view; usu. suffixed to a possessive substantive to denote what is seen from the view-point of the person or thing specified; freq. in phrases *bird's eye view, worm's eye view*.

1762–71, etc. [see Bird's-eye 3]. **1908** *Punch* 10 June 429/2 We fear the population will develop balloon-necks through trying to get a worm's-eye view of the gas-bags in the haze. **1920** A. Huxley *Limbo* 137 He prided himself on being able to see the thing as a whole, on taking an historical, God's-eye view of it all. **1952** H. W. Tilman *Nepal Himalaya* xi. 131 A man looking at a big mountain from a valley has naturally only a worm's-eye view. **1959** *Sunday Times* 4 Oct. 17/3 Here is a bookworm's-eye view of a hazardous trade. **1962** A. Huxley *Island* xiii. 218 How should one look at other people?..Should one take the Freud's-eye view or the Cézanne's-eye view? The Proust's-eye view or the Buddha's-eye view? **1962** J. Robertson (*title*) Hospitals and children: a parent's-eye view. A review of letters from parents to the *Observer* and the BBC. **1963** *Listener* 17 Jan. 129/2 From his own eye-view Sir Compton has no lack of items to add to the history of his time.

eye-wash, eyewash (əi·wǫʃ). [f. Eye *sb.*^{1} + Wash *sb.* 3.] **1.** A wash or lotion for the eye.

1866 [see Eye *sb.*^{1} 28].

2. *colloq.* Something that is intended to obscure or conceal actual facts or motives; humbug, blarney; nonsense; something said or done merely for appearance or effect; *spec.* in *Mil. slang* = *Bull *sb.*^{4} 4.

1884, 1889 [see Eye *sb.*^{1} 28]. **1913** *Aeroplane* 20 Mar. 327/2 Well as this may do as 'eye wash', it is not the real thing. **1916** C. E. W. Bean *Lett. from France* (1917) 197 The ignorance which..flies to the conclusion that everything written and spoken about the horrors of this war is humbug, and what the Army calls 'eyewash'. **1919** 'I. Hay' *Last Million* ii. 24 The greater the fuss a regiment made about its appearance—'eye-wash', we called it—the better its work in the field. **1919** *War Slang* in *Athenæum* 8 Aug. 728/1 Anything complimentary is termed 'eyewash'. **1920** *Blackw. Mag.* July 91/2 Kemp went ashore to pay a polite visit to the local Sheikh, as 'eye-wash' against our real activities. **1930** W. S. Maugham *Cakes & Ale* xi. 124 What the critics wrote about Edward Driffield was eye-wash. **1939** J. Cary *Mr. Johnson* 208 'More eyewash about sanitation,' he says to Celia, throwing a blue circular on the breakfast table. **1957** *Economist* 28 Sept. 1001/1 This does not mean that the proposals..are so much eyewash.

Hence **eye·washer,** one who uses 'eye-wash'; **eye·-washing** *vbl. sb.* and *ppl. a.*

1920 *Nat. Rev.* Apr. 142 The President..refused to.. be humbugged by 'eyewashers' who besought him to accept amendments that destroyed his handiwork. **1937** L. Housman *Unexpected Years* 159 The professions of those who offered them, with 'art for art's sake' as an eye-washing accompaniment. **1939** J. Cary *Mr. Johnson* 177 Nobody..calls Tring an eyewasher, because he is in fact a highly competent officer... But he..can always catch the right eye at the right time.

eye-witness. Add: **2.** Also *attrib.*

1921 *Times Lit. Suppl.* 10 Nov. 722/3 The aim of these volumes is to present eyewitness accounts of some of the most important incidents. **1922** C. E. Montague *Disenchantment* vii. 95 Eked out with 'Eye-Witness' stuff—official 'word-painting' by some Regular Officer. **1945** W. S. Churchill *Victory* (1946) 13 The following is an eye-witness report by another British officer.

Hence **eye·-witness** *v. trans.*, to be an eye-witness of (an event); **eye·-witnessed, eye·-witnessing** *ppl. adjs.*

1844 Kinglake *Eothen* viii. 145 Lamartine's eye-witnessing account of the horse saddled by the hands of his Maker. **1905** 'Mark Twain' *King Leopold's Soliloquy* (1907) 17 They seemed to be always..eye-witnessing the happenings. **1912** W. Owen *Let.* 7 Nov. (1967) 167, I was so glad to hear your own eyewitnessing account of a Flight. **1923** *Glasgow Herald* 25 Oct. 8 An eye-witnessed incident which took place recently in this neighbourhood. **1963** 'N. Blake' *Deadly Joker* xiii. 215 A hundred people had eye-witnessed a murder.

eyra (əi·rā). [ad. Tupi *eirara, irara.*] In full *eyra cat*. A wild cat, *Felis yagouaroundi*, in its red phase, found in an area from Argentina and Paraguay to southern Texas.

1860 *Proc. Zool. Soc.* XXVIII. 415 Eyra Cat, *Felis eyra*. **1873** *Ibid.* XLI. 2 A specimen of the Eyra Cat.. brought from Maranham, Brazil. **1906** E. Ingersoll *Life Anim.: Mammals* 105 It is a question whether the eyra and yaguarundi are not varieties of the same species. **1955** *Sci. News Let.* 30 July 78/2 The southwestern states, bordering on Mexico, have four exotic big cats—ocelots, jaguars, margays, and jaguarundis or eyras. **1965** D. Morris *Mammals* 315 It [*sc.* the Jaguarondi] exists in two colour phases: red and grey. The red phase used to be thought of as a distinct species and was then called the Eyra Cat (*Felis eyra*), but this is no longer accepted.

eyrie. Now the commonest spelling of Aerie.

1667, 1861 [see Aerie 1]. **1902** W. de la Mare *Songs of Childhood* 14 A hawk from his eyrie Swooped down like an arrow. **1922** Joyce *Ulysses* 582 Birds of prey.. swooping from eyries. **1937** J. R. R. Tolkien *Hobbit* vi. 115 You ought not to be rude to an eagle, when you are only the size of a hobbit, and are up in his eyrie at night! **1955** W. G. Hoskins *Making Eng. Landsc.* iii. 62 The limestone crags above Littondale provided eyries for these noble birds. **1965** E. Gowers *Fowler's Mod. Eng. Usage* 13/1 *Aery, aerie, eyry, eyrie*. The victory of the last form over the other three seems to have been undeserved.

F

F. Add: **II. 3.** *F-layer, -region*: the highest and most strongly ionized layer in the ionosphere; the Appleton layer.

1928 E. V. Appleton in *Papers Internat. Union Sci. Radio Telegr.* 1927 I. 2/1 'Reflection'..takes place at an upper layer (F layer) which is richer in ionization. **1930** —— in *Proc. R. Soc.* A. CXXVI. 550 The normally polarised components of primary, secondary and tertiary downcoming rays from F region would be F_1, F_2 and F_3, etc. **1930** Appleton & Green *Ibid.* CXXVIII. 170 During the (*a*) series [of day-time tests]..out of eight runs there were three on which E layer only was recorded, three on which F layer only was recorded, and two days on which both were recorded simultaneously. **1942** *Electronic Engin.* XIV. 632 The F layer..is composed of two parts..the lower part, at about 130 miles' height.. and the upper, at 200 miles. **1968** G. M. B. Dobson *Explor. Atmos.* (ed. 2) viii. 153 During the night..the ionization of the *E* region is much reduced, and that of the F_1 region is reduced enough to make it indistinguishable from the F_2 region and the two become a single *F* region.

III. 1. i. *Biol.* F_1, F_2, etc. = first, second, etc., filial generation (see quot. 1949).

1902 W. Bateson et al. *Rep. Evol. Comm. R. Soc.* I. 160 The offspring of the first cross are the first filial generation F. Subsequent filial generations may be denoted by F_2, F_3, &c. **1949** Darlington & Mather *Elem. Genetics* 391 F_1, the first generation of the cross between two individuals homozygous for the particular genes which distinguished them. F_2, the second filial generation obtained by self-fertilizing or crossing *inter se* individuals of an F_1. F_3, progeny obtained by self-fertilization of an F_2 individual. **1968** M. W. Strickberger *Genetics* vi. 99 The smooth F_1 seeds..produced upon self-fertilization an F_2 of 5474 smooth seeds and 1850 wrinkled.

j. f or F (*Physics* and *Chem.*): originally used to designate one of the four main series of lines in atomic spectra, but now more frequently applied to electronic orbitals, states, etc., possessing three units of angular momentum.

1910 W. M. Hicks in *Phil. Trans. R. Soc.* CCX. 97 We recognise four typical sequences, the P, S, D and the new F type.., the fundamental or primitive type. **1965** Phillips & Williams *Inorg. Chem.* I. ii. 46 There occurs a group of 14 elements, the lanthanides or rare-earths, in which the 4*f* sub-shell is filled.

k. In *Physics* F is the symbol of the hyperfine quantum number, expressing the resultant of the nuclear spin I and the electronic angular momentum J.

1925 Ruark & Chenault in *Phil. Mag.* 6th Ser. L. 947 We call these integers fine quantum numbers, and shall designate them by the letter *f*. **1937** J. W. T. Spinks tr. *Herzberg's Atomic Spectra & Atomic Structure* v. 187 The total angular momentum *F* of the whole atom, including nuclear spin. **1967** Condon & Odishaw *Handbk. Physics* (ed. 2) vii. iv. 68/1 At low magnetic fields *I* and *J* are tightly coupled to form a resultant angular momentum F=I+J, whose quantum number *F* at low fields is a good quantum number.

l. *Bacteriol.* [abbrev. of *fertility*.] *F, F agent, F factor*, etc., a particulate factor which determines the mating type of certain bacteria (esp. *Escherichia coli*); hence F^+, designating cells possessing the factor but not having it integrated into the chromosome; F^-, designating cells lacking the factor; *F′* (or *F prime*), designating cells in which a chromosomal fragment is attached to the factor. Also *F duction*, the conveyance of genes from F′ cells into F^- cells.

1952 J. Lederberg et al. in *Genetics* XXXVII. 721 At the time, W-1607 was not analyzed further, but in view of later work will be labelled F^-, in contrast to the type F^+. **1953** *Jrnl. Gen. Microbiol.* VIII. 92 The majority ..of F agents must remain bound to the cells. **1959** Adelberg & Burns in *Rec. Genetic Soc. Amer.* XXVIII. 57 The wild type sex factor (*F′*) of strain K-12 is characterized by its low affinity for the chromosome. **1960** F. Jacob et al. in *Symp. Soc. Gen. Microbiol.* X. 77 The similarities observed between this process, which might be called *F*-duction or sex-duction, and transduction with phage are extremely striking. **1967** F. E. Swatek *Textbk. Microbiol.* x. 291 In $F^+ \times F^-$ matings the F factor is transferred to F^- bacteria.

3. F.A., = *Fanny Adams 2; F.A., Football Association; F.A.F., Fresh Air Fund; F.A.N.Y., (a member of the) First Aid Nursing Yeomanry; also *Fany*; F.A.O., Food and Agriculture Organization (of the United Nations); f.a.q., fair average quality; F.A.Q. (f.a.q.), free alongside quay; F.A.S. (f.a.s.), free alongside ship; F.B.I., Federal Bureau of Investigation (*U.S.*); F.B.I., Federation of British Industries; F.C., football club; F.C.C., Federal Communications Commission (*U.S.*); F.D.C., *fleur de coin*; F.F.(V.), the first family or families (of Virginia); a member or descendant of one of these (*U.S.*); F.G., fine grain; F.H.B., family hold back (a colloq. intimation to the members of a family that their guests have first claim on the course or helping about to be served); F.I.D.O. (see *FIDO); F.I.T., f.i.t., free in truck; F.L.N., (Fr.) Front de la Libération Nationale, an Algerian nationalist organization; F.M., field-magnet; F.M., f.m., frequency modulation; F.O., Foreign Office; F.O.B., f.o.b., free on board; F.O.B.S., fractional orbital bombardment system; F.O.R., f.o.r., free on rail; F.O.T., f.o.t., free on truck; F.P.S., f.p.s., feet per second; F.P.S., f.p.s., foot-pound-second (cf. *Foot *sb.* 35); F.S., f.s., feet per second; F.W.D., f.w.d., front- (or four-) wheel drive.

1919, 1930 F.A. [see *Fanny Adams 2]. **1944** *Penguin New Writing* XX. 128 Bread—that's about what we got as kids. Bread, and sweet F.A. **1967** J. Gardner *Madrigal* ix. 251 The small industrial organisation whose own security officers know sweet FA. **1892** in Gibson & Pickford *Assoc. Football* (1905) I. 112 The meetings of the F.A. are carried out under rules of procedure. **1930** *Morning Post* 18 Nov. 15/4 The first round proper of the F.A. Cup. **1907** *Westm. Gaz.* 22 July 3/2 Without distinction of creed or race, the F.A.F. takes away the children of the slums from the festering alleys..of London. **1918** E. S. Farrow *Dict. Mil. Terms* 218 *Fanny*, in the parlance of the British soldier, the name given to the women of the F.A.N.Y., or First Aid Nursing Yeomanry. **1940** *War Illustr.* 12 Jan. 599 Two volunteers of the First Aid Nursing Yeomanry, known as 'Fanys', are cleaning their ambulance. **1958** P. Kemp *No Colours or Crest* xi. 243 The Major-General had a suite of offices managed by four or five very tall, very beautiful F.A.N.Y.s known as the Potsdam Grenadiers. **1966** M. R. D. Foot *SOE in France* viii. 200 Mary Herbert a FANY courier reached the Riviera by felucca. **1946** *New Statesman* 14 Sept. 183 (*heading*) F.A.O. at work. **1963** *Listener* 21 Mar. 508/3 The F.A.O. has estimated that about half the population of the world suffers from hunger or malnutrition. **1957** *N.Z. Timber Jrnl.* Feb. 45 *f.a.q.*, abbreviation for fair average quality. **1959** *Times* 3 Jan. 10/2 No doubt Sir Allen Brown would be inclined (in the fashion of his countrymen) to dub himself an 'f.a.q. Australian', but in fact 'fair, average quality' hardly does justice to the man. **1928** *Daily Mail* 25 July 19/3 Oats: Plate f.a.q. afloat 27s. 3d. **1888** *Lockwood's Dict. Mech. Engin.*, *F.A.S.*, free alongside ship. Engages to deliver goods on the wharf without extra charge. **1936** *Lit. Digest* 21 Nov. 36 The outlaws shot their way out killing one FBI man. **1952** *Manch. Guardian Weekly* 30 Oct. 3 The FBI and the Central Intelligence Agency. **1921** *Times* 22 Nov. 14/3 (*heading*) Meeting of the F.B.I. to-morrow. **1958** *Spectator* 4 July 33/2 The FBI has just published the results of its second industrial survey. **1908** A. Bennett *Old Wives' Tale* II. vii. 254 The Bursley F.C. had 'tied' with the Knype F.C. on the Knype ground. **1937** *Printers' Ink Monthly* Apr. 53/1 FCC, Federal Communications Commission. **1955** *Times* 27 July 9/7 The F.C.C. rules require that if it is extended to one political candidate his opponents must be accorded equal accommodation. **1892** *Spink's Numismatic Circ.* Dec. 14/2 F.D.C. = Fleur de coin. **1966** P. A. Rayner *Coin Collecting for Amateurs* i. 5 The following are the main degrees of condition used in this country: F.D.C. (*Fleur-de-Coin*): literally—flower of the coin, a French term used to denote perfect mint state with absolutely no wear or damage, [etc.]. **1813** M. L. Weems *Drunkard's Looking-Glass* (ed. 2) 11 With hearts full charged with F.F. Slings, they came upon the ground. **1850** 'M. Tensas' *Louisiana Swamp Doctor* 178 Major Smith..[was] the first one of the race [of Virginians] to acknowledge that he was not an F.F. **1917** *Birds of Amer.* III 64/2 Better yet, he [*sc.* the cardinal] is an FF of America. **1847** *Knickerbocker* XXIX. 495 We could make out..a set of letters that looked like a disorderly F.F.V...A Virginia scion insisted that they were an abbreviation he had seen used in the navy to represent 'First Family in Virginia'. **1861** R. H. Newell *Orpheus C. Kerr* (1863) I. 253 That piece of American intelleck..worthy of an F.F.V. **1931** W. Faulkner *Sanctuary* iii. 22 An F.F.V., or just an unfortunate sojourner there? **1871** *English Mechanic* 17 Mar. 615/3 Medium rifle powder..is still used..under the name of 'powder shell F.G.' (fine grain). **1911** 'I. Hay' *Safety Match* v. 82 She murmured to such of the family as were within earshot the mystic formula, 'F.H.B.!' **1888** *Lockwood's Dict. Mech. Engin.*, *F.I.T.*, free in truck. Engages to load goods in railway trucks without extra charge. **1956** *Times* 7 Apr. 6/4 An office belonging to the other Algerian Nationalist Movement, the National Liberation Front (F.L.N.), was also raided today. **1958** *Times Lit. Suppl.* 14 Feb. 82/5 The French will inevitably have to negotiate with the F.L.N. **1893** W. P. Maycock *Electr. Lighting* II. viii. § 134 The F.M.s of an alternator must be separately excited. **1942** *Electronic Engin.* XIV. 640 A special f.m. amplifier. **1962** FM [see *Amplitude 6 d]. **1859** Queen Victoria *Let.* 14 June in R. Fulford *Dearest Child* (1964) 194 Lord Palmerston..could not insist on Lord John's not having the F.O. if he insisted on having it! **1959** E. Pound *Thrones* civ. 90 Resigned from the F.O. and 'went into the City'. **1890** *Economist* 22 Feb. (Suppl.) 12/1 Bleach opened firm in January at £7 16s per ton for hard wood, f.o.b., and £7 5s per ton for soft wood f.o.r. **1959** *Ann. Reg. 1958* 490 A total value of about $20 million f.o.b. for the year. **1967** *Economist* 11 Nov. 622/2 FOBS puts a vehicle into orbit round the earth, in either direction, about a hundred miles up; it then launches a missile, when told to do so, probably before completing a full circuit. **1969** *U.S. News & World Rep.* 10 Mar. 31/3 *FOBS*, fractional-orbit bombardment system. A missile that orbits the earth and can deliver its warheads from space. A Soviet development that has no U.S. counterpart. **1888** *Lockwood's Dict. Mech. Engin.*, *F.O.R.*, free on rail. Signifies the placing of goods on the railway without extra charge. **1890** f.o.r. [see *f.o.b.* above]. **1888** *Lockwood's Dict. Mech. Engin.*, *F.O.T.*, free on truck. The same as F.I.T. **1945** T. A. Dickinson *Aeronaut. Dict.* 433/2 *F.P.S.*, feet per second. **1962** J. Glenn in *Into Orbit* 200 It was 25,730 fps instead of 25,738 fps. **1892** G. F. Barker *Physics* I. ii. 10 In applied physics another absolute system is sometimes used, based upon the foot, the pound, and the second... This system is called the foot-pound-second system, or more briefly the F.P.S. system. **1962** A. R. W. Hayes *Revision Physics* § 1. 4 The *foot-poundal* is the f.p.s. unit of *energy* or *work*. **1880** *Encycl. Brit.* XI. 304/2 The bullets themselves should retain a velocity of about 500 f.s. on striking. **1905** *Kynoch Jrnl.* July–Sept. 156 Accuracy does not demand a muzzle velocity higher than 2,300 f.s. **1958** *News Chron.* 25 Nov. 7/5 (Advt.), The Austin commercial range includes all-purpose vehicles.. and the FWD Gipsy with rubber torsion suspension. **1959** *Motor* 11 Feb. 62/1 Nowadays f.w.d. is invariably combined with independent front wheel springing.

b. *Photogr.* F, f = Focal length. Used in combination with numbers to indicate the ratio between the focal length of the lens and the diameter of the stop.

1892 C. H. Bothamley *Ilford Man. Photogr.* v. 34 For example, *f*/8 and *f*/16 mean that the diameter of the stop is, in the first case, one-eighth, and, in the second case, one-sixteenth of the focal length of the lens. **1903** A. Watkins *Photogr.* (ed. 2) 42 Instead of having to work out afresh the value of the diaphragm, it is much more convenient to regard the lengthening of the camera as a separate influence which I have named the F factor. *Ibid.* 43 The F factor to allow for the lengthened focus of the lens, and S the value of the subject. *Ibid.* 45 The F figure (distance of image to lens in terms of focus of lens) is always the same as the degree of enlargement plus 1. **1918** *Photo-Miniature* XV. Mar. (Gloss.), *F*(*f*/·) Numbers—denote the 'speed' of a lens. About the most rapid lens is *f*/3 to *f*/4·5...The *f*/ number is the number of times the diameter of the stop will divide into the focal length of the lens. **1959** *News Chron.* 20 June 6/7 The apertures or stops, as they are sometimes called, are marked in f/numbers. If a lens is described as an f/4 lens, it means that is its maximum or widest aperture. **1968** A. Diment *Gt. Spy Race* vii. 107, I started photographing right then, the light meter said I was okay on F2·8 at a thirtieth. **1970** *Amateur Photographer* 13 May 66/1 The actual diameter of the hole was marked on it, so that f/11 showed that the hole was one eleventh of the focal length.

faam, var. *Faham.

fab, colloq. abbrev. of *Fabulous *a.* 5 b. Also as *sb.*

1961 in Partridge *Dict. Slang* Suppl. 1082/1. **1963** *Times* 26 Jan. 10/6 She stretched her stockinged toes towards the blazing logs. 'Daddy, this fire's simply fab.' **1963** *New Statesman* 18 Oct. 536/1 Those girls in leather gear—the height of fab. **1963** *Meet the Beatles* 22 Most of the Merseyside groups produce sounds which are pretty fab. **1971** *It* 2–16 June 11/4 Seale called the fab philosopher 'a moral coward'.

Fabergé (fabɛrʒe). The name of Peter Carl *Fabergé* (1846–1920), a Russian jeweller famed for his small, intricate ornaments, used *attrib.* to designate pieces of his workmanship. Also *transf.* and *fig.*

[**1902** *Encycl. Brit.* XXIX. 748/1 As a representative of Russia, Fabergé of St. Petersburg deserves to be mentioned.] **1930** V. Sackville-West *Edwardians* iv. 191 The jade ash-trays, the Fabergé cigarette-boxes. **1936** G. Greene *Journey without Maps* II. iii. 162 A beautiful little green snake..poisonous with gentility, a Fabergé jewel. **1952** 'J. Tey' *Singing Sands* xiii. 216 It [*sc.* France seen from an aeroplane] was a small jewelled pattern set in a lapis-lazuli sea; a Fabergé creation. **1959** *Encounter* Oct. 8/1 A sort of Fabergé Russianism clings around parts of Nice.

Fabian, *a.* and *sb.* Add: **A.** *adj.* **1.** (Earlier example.)

1740 J. Dyer *Ruins of Rome* 23 Who has not heard the Fabian Heroes sung?

2. a. (Earlier and later examples.)

1777 *Publ. Colonial Soc. Mass.* (1906) VIII. 277, I know the comments that some people will make on our Fabian conduct. **1965** *New Statesman* 9 Apr. 579/1 His [*sc.* Lincoln's] fabian tactics.

b. (Examples.)

1884 E. Nesbit *Let.* Mar. (1933) v. 63 The Fabian Society takes up a good deal of my thoughts just now. **1908** G. B. Shaw *Fabian Essays in Socialism* p. iii, Since 1889 the Socialist movement has been completely transformed..and the result..may fairly be described as Fabian Socialism. **1915** A. Huxley *Let.* Oct. (1969) 79

A sort of magnified Fabian society state, organised even further than at present. **1959** B. & R. North tr. *Duverger's Pol. Parties* (ed. 2) p. xxxi, The activity of the Fabian Society in the creation of the Labour party illustrates the influence of philosophical societies.

B. *sb.* **2.** (Earlier example.)

1884 E. Nesbit *Let.* ?Feb. (1933) v. 63, I do think the Fabians are quite the nicest set of people I ever knew.

Fabianism (fēi·biăniz'm). [f. Fabian *a.* + -ism.] The doctrines and principles of the Fabian Society. Hence **Fa·bianist** *sb.* and *a.*; **Fa·bianistic** *a.*

1890 G. B. Shaw *Let.* 28 Feb. (1965) 244 A blasphemy against the Fabianism which made it famous. **1909** *Cent. Dict.* Suppl. 457/1 Fabianist, n. **1918** H. Dalton in E. Marsh *R. Brooke* 26 During our years at Cambridge, Fabianism was at its high tide. **1924** *Blackw. Mag.* Sept. 393/1 Fabianism touched her during her last year at school. **1955** Koestler *Trail of Dinosaur* 188 British Socialism which is Christian, non-Marxist, Fabianist, and heaven knows what. **1960** *New Left Rev.* Sept.–Oct. 60/1 An intensified Fabianistic process.

fable, *sb.* Add: **6. a.** *fable-poem, -poetry, -type*; **b.** *fable-writer*; *fable-making, -weaving* vbl. sbs.

1905 *Westm. Gaz.* 25 Feb. 10/3 A fable-writer and dramatic poet. **1911** S. S. Colvin *Learning Process* viii. 117 The tendency to fable-making causes him [*sc.* a child] to make statements..at variance with the truth. **1941** L. MacNeice *Poetry of Yeats* vi. 111 A new kind of fable-poetry which avoids becoming allegory. *Ibid.*, The fable-poems..are dry, unromantic pieces. *Ibid.* ix. 192 In the plays..fable-types, who correspond to the beggars and fools and hermits of the lyrics. **1946** Blunden *Shelley* 24 In these fable-weavings something of the desire to be an author is seen.

fabric, *sb.* Add: **7. b.** Also *spec.* the basic structure (walls, floor, roof) of a building.

1866 [in Dict.]. **1934** Batsford & Fry *Cathedrals of England* 4 The fabrics..form in the majority of cases a remarkable patchwork of building periods. **1955** *Times* 3 May 6/4 The skilled staff needed not only to maintain the fabric of the buildings but to care for the precious pictures, furniture, and gardens. *Ibid.* 9 May 5/2 Much has been done..since a public appeal was opened on behalf of the fabric of Norwich Cathedral, to make this noble building safe.

IV. 9. *attrib.* and *Comb.*, as *fabric glove, hat* (sense 4); *fabric-faired* (sense 3), *fabric-printing* (sense 4), adjs.

1897 *Sears, Roebuck Catal.* 231/1 Ladies' fabric gloves. **1906** *Westm. Gaz.* 22 Sept. 13/2 They offer great opportunities for colour schemes, those fabric hats. **1916** *To-Day* 25 Nov. 92/2 Even his warm winter gloves are made of cotton..'fabric' gloves, such as most men wear in the summer. **1934** *Jrnl. R. Aeronaut. Soc.* XXXVIII. 670 Fabric-faired girder designs are diminishing. **1946** *Nature* 2 Nov. 614/2 A rotational viscometer for fabric-printing thickeners. **1971** 'A. Gilbert' *Tenant for Tomb* i. 8 She.. carried a big old-fashioned tapestry bag and wore fabric gloves.

fabric, *v.* Delete † *Obs.* and add later examples.

1921 *Public Opinion* 14 Oct. 375/1 All these were dreamed and fabricked out for immediate material benefit. **1924** W. A. White *Woodrow Wilson* xvi. 352 This high dream of peace, that he fabricked upon the anvil of a three years' debate.

fabricate, *v.* Add **1. e.** To form (semi-finished metal stock or other manufacturing material) into the shape required for a finished product; also with the product as *obj.*

1926 R. J. Anderson *Metall. Aluminium* xix. 853 Aluminium, when annealed, is readily fabricated by these operations [*sc.* drawing, pressing, and spinning], but the light alloys are not so readily worked. **1935** G. E. Doan *Princ. Physical Metall.* 1 A hundred men may be engaged in fabricating the metals which have been extracted, forming them into countless shapes, such as rails, I-beams, boiler plate, [etc.]. **1951** G. Sachs *Sheet-Metal Fabricating* i. 9 Generally, large parts in very thin metal are difficult to fabricate accurately. **1963** Simonds & Church *Conc. Guide Plastics* (ed. 2) vii. 187 Both thermo-setting and thermoplastic materials can be fabricated into a variety of shapes and forms. **1971** *Helicopter* July 147/1 The rotor blades are fabricated from aluminium alloy sheet bonded together by means of a film adhesive.

fabrication. Add: **1.** (Later *concr.* examples.)

1670 Moxon *Tutor to Astron. & Geogr.* (ed. 2) App. 201 A Spheår is complycated only of Lath-like Circles to represent each Orb, and is not an intire Orb as a Globe is, so that you may see the several Fabrications that are made within it. **1893** *Whitby Gaz.* 8 Dec. 2/7, I thought all our masts, funnel, and the whole fabrication, boats and all, would surely go.

b. The process of fabricating in the manufacture of finished products (see *Fabricate *v.* 1 e).

1926 R. J. Anderson *Metall. Aluminium* xix. 852 The fabrication of various articles from aluminium sheet makes up an important branch of the aluminium industry. **1940** J. D. Jevons *Metall. Deep Drawing & Pressing* p. xi, The fabrication of all kinds of metal articles by the severe plastic deformation of metal sheet is a craft as ancient as that of steel-making itself. **1951** G. Sachs *Sheet-Metal Fabricating* i. 7 Frequently, a slight alteration in the geometry of a part may overcome difficulties in fabrication. **1963** Simonds & Church *Conc. Guide Plastics* (ed. 2) vii. 188 After fabrication or molding, most plastics articles require some form of finishing. **1967** E. Slade *Metals in Mod. World* viii. 189 Traditional metallurgical methods..are used increasingly for the fabrication of other materials, particularly the plastics.

fabricator. Add: **3.** *Archæol.* A rod-shaped flint implement, perh. used in the manufacture of other flint tools. (See also quot. 1954.)

1872 J. Evans *Anc. Stone Implem.* 367 A characteristic specimen of the tool to which I have provisionally assigned the name of 'flaking-tool' or fabricator. **1877** W. Greenwell *Brit. Barrows* 35 Those [*sc.* articles found in barrows] of flint, include hatchets; scrapers;.. fabricators or flaking tools. **1930** *Antiquity* IV. 27 'Fabricators' are rare, borers very rare, and calcined flints ('potboilers') exceedingly scarce. **1954** S. Piggott *Neolithic Cultures* x. 286 There is good evidence for regarding a large proportion of 'fabricators' as strike-a-lights used with pyrites to produce a spark.

fabricless (fæ·briklės), *a.* [f. Fabric *sb.* + -less.] Without substance, baseless.

1905 E. F. Benson *Image in Sand* xiv. 315 Thought with him was no vague and dreamy musing..nor..any fabricless cloud-building.

Fabry-Pérot (fabri,pero). *Physics.* The names of C. *Fabry* (1867–1945) and A. *Pérot* (1863–1925), French physicists, used *attrib.* or jointly to designate the étalon invented by them (see *Étalon), and also an interferometer designed on the same principle but having one of the glass plates mounted so that its distance from the fixed plate may be varied.

1908 *Sci. Abstr.* A. XI. 387 A modified Fabry and Perot apparatus. **1914** *Phil. Mag.* 6th Ser. XXVII. 678 We thus have a problem very similar to that of the Fabry-Perrot [*sic*] étalon, only with an infinite number of parallel equidistant plates. **1923** Glazebrook *Dict. Appl. Physics* IV. 887 The red light from a cadmium lamp and the light from an iron arc are sent simultaneously through a Fabry and Perot interferometer. **1956** *Nature* 3 Mar. 434/2 A sharp fringe pattern has been obtained by substituting a small Fabry-Perot étalon for the 'Polaroid'-calcite unit. **1970** G. K. Woodgate *Elem. Atomic Struct.* ix. 173 He studied the hyperfine splitting of the line..with a Fabry-Perot interferometer.

fabulation. Delete † *Obs.*$^{-0}$ and add examples.

1945 D. Devlin in *Briarcliff Q.* Oct. 66 At last we have stored the windows away, The fabulation of my Lord's glory. **1958** *Partisan Rev. Anthol.* (1962) 474 'Fabulation'..is a process no less singular.

fabulous, *a.* Add: **5. b.** Now freq. in trivial use, esp. = 'marvellous', 'terrific'. Cf. *Fab.

1959 *Cambr. Rev.* 30 May 571/2 Miss Mitchell, looking, one must admit, fabulous, played down her frenzy. **1962** *Radio Times* 4 Jan. 23/2, I think it's [*sc.* Salford] a fabulous place. **1963** A. Ross *Australia 63* vii. 129 Trueman puffed at a cigarette and said he felt fabulous.

façade. Add: **1.** Also *attrib.* or as *adj.*, of an architectural design concerned with elegance, etc., in the façade of a building alone. Hence **faça·dism,** such a practice or principle.

1933 *Archit. Rev.* LXXIV. 111 Façadism with a purpose, unlike the façadism of the new Railway Station at Milan. **1936** *Times Lit. Suppl.* 16 May 421/1 He.. refuses to accept the recent revaluation of Nash and is more than tainted with 'façadism'. **1962** *Listener* 15 Nov. 806/2 A warning against façade architecture.

face, *sb.* Add: **1. e.** A slang term of address.

1923 Wodehouse *Inimitable Jeeves* xvii. 232, I ran into young Bingo Little...'Hello, face,' I said. 'Cheerio, ugly,' said young Bingo. **1938** D. Smith *Dear Octopus* I. 38 Come on, face—don't get mopey.

f. Used with varying contextual degrees of contempt, admiration, etc.: a person. *slang.*

1945 L. Shelly *Jive Talk Dict.* 11/1 *Face*, a white man. **1964** *New Statesman* 10 Apr. 555/2 A 'face' is a person, usually someone worthy of admiration. **1964** *Observer* 24 May 12/5 Fashions are usually started by Faces—the name for trend-pushing Mods. **1967** J. Morgan *Involved* 80 Now this face was the ideal man for me to have a deal with.

2. a. *to open one's face* (*U.S. slang*): to open the mouth, to speak; *to shut one's face* (*slang*): to keep quiet (esp. *imp.*); *to laugh on the other side of one's face*: see Laugh *v.* 1 b.

1893 S. Crane *Maggie* (1896) ix. 73 Shet yer face, an' come home, yeh old fool! **1896** Ade *Artie* iii. 26 If you open your face to this lady again tonight, I'll separate you from your breath. **1899** C. Rook *Hooligan Nights* ii. 29 'That's awright,' muttered young Alf. 'You shut yer face.' **1913** D. H. Lawrence *Love Poems* 53 An' stop thy scraightin', childt. Do shut thy face. **1917** U. Sinclair *King Coal* II. 186 The marshal bade him 'shut his face'. **1939** *Best Short Stories* 45 'Shut your daft grinning face,' growled Arthur. **1959** I. & P. Opie *Lore & Lang. Schoolch.* x. 194 'Shut your chops', 'shut yer face'.

d. (Earlier *attrib.* and further examples.) Also **face-to-face, face-to-faceness** *sbs.*, **face-to-facedly** *adv.*

1833 J. Nyren *Young Cricketer's Tutor* 59 He was a good face-to-face, unflinching, uncompromising, independent man. **1874** W. Wallace *Logic of Hegel* xviii. p. cxxxiv, In both there is the same statement of immediacy or face-to-faceness. **1876** *Gentl. Mag.* Feb. 212 The applause thus..publicly and face-to-facedly showered upon it. *a* **1916** H. James *Sense of Past* (1917) 335 His disconcerted and practically defeated face-to-faceness with the way in which they do take it [*sc.* his modernism]. **1949** M. Mead in M. Fortes *Social Structure* 23 A regiment of soldiers, all from..the occident, come into face-to-face relations with non-occidental people. **1952** W. Sprott *Social Psychol.* 257 Improved social relations in face-to-face groups were found salutary in dealing with mental illness. **1956** J. Whatmough *Language* 12 There is a greater time-lag in a printed book, as in a recording, than there is in face-to-face talk. **1960** *Economist* 31 Dec. 1395/2 This biographical 'face-to-face'. **1971** *Times* 8 Jan. 2/4 A combination of radio, correspondence courses and face-to-face teaching.

g. *to grind the faces of*: see Grind *v.*^{1} 2 b.

3. b. Make-up, cosmetics; esp. in phrs. *to do* or *to put on one's* (or *a*) *face*: (of a woman) to apply cosmetics. *colloq.* (Occas. used of a bag containing cosmetic materials.)

1923 W. S. Maugham *Our Betters* I. 45 A wonderful woman who comes every morning to do my face. **1946** G. Millar *Horned Pigeon* xx. 327 She uncovered a dressing-case of..obvious costliness...'You don't intend to leave that thing lying around?' 'Why not?..It's my face.' **1957** J. Frame *Owls do Cry* I. xiv. 62 How dare you watch me put on my face. **1961** *Guardian* 25 Sept. 3/2 My hair is wrecked and my 'face' totally removed. **1963** A. Lubbock *Austral. Roundabout* 10 My 'face', a small bag with make-up and washing things. **1966** 'C. E. Maine' *B.E.A.S.T.* v. 59 You've caught me with my face off. Give me three minutes. **1969** R. Lockridge *Murder in False Face* xiv. 183 I'll wash my hands and put on another face.

4. e. In ice-hockey and lacrosse: the act of facing off (see *Face *v.* 4 c). Also *face-off.*

1896 *Times* (Niagara, Ont.) 20 Feb. 1/4 In the face-off Bishop lost. **1898** H. E. Byers in W. A. Morgan *'House' on Sport* 211 The 'face' is a feature of the game [*sc.* lacrosse]. **1900** *Daily News* 29 Mar. 7/5 Kent opened proceedings with a goal by Jones immediately after the 'face'. **1900** J. C. Isard in J. H. C. Fegan et al. *Football, Hockey, & Lacrosse* 178 When the game is opening, with the face-off at the centre. **1936** *Times* 25 Jan. 5/7 Then Smith equalized from the face-off [in ice-hockey] with a long shot. **1968** *Globe & Mail* (Toronto) 15 Jan. 19/6, I lined up for a faceoff.

7. *to have the face to* (do something): further examples.

1839 Dickens *Let.* ? Dec. (1965) I. 624 As a family man, I really have not the face to dine out again to-day. **1848** T. Arnold *Let.* 16 June (1966) 53, I dismissed my guide with..two shillings; though the little dog had the face to ask 'sicca-penny more'.

b. *to run one's face* (earlier example). Also *to travel on* or *upon one's face*. *U.S.*

1839 *Spirit of Times* 5 Oct. 368/3 The Picayune says there is a chap in New Orleans who has 'run his face so often for drinks, that it is completely worn off'. **1856** *Knickerbocker* XLVIII. 504 [I] must travel on my face after this, when I want to go through the College. **1859** *Yale Lit. Mag.* XXV. 60 (Th.), If you have not a ready tongue, and cannot travel upon your face, you had better [etc.].

10. b. *to save one's face*: see Save *v.* 8 f; also *to save face*; *to lose face* [tr. Chinese *tiu lien*]: to be humiliated, lose one's credit, good name, or reputation; similarly, *loss of face*. Hence *face* = reputation, good name.

1876 R. Hart *These from Land of Sinim* (1901) App. ii. 225 Arrangements by which China has lost face. **1899** *Harmsworth Mag.* June 400 That will save my face in the City. **1915** W. S. Maugham *Of Human Bondage* cvii. 563 To save his face he began making suggestions for altering it, but Mrs Hodges..advised him to show it to Miss Antonia as it was. **1915** J. London *Lost Face* 33 He had lost face before all his people. **1928** Galsworthy *Swan Song* i. 5 They've got to save face. Saving face is the strongest motive in the world. **1929** *Times* 3 Aug. 11/3 Each wishes to concede only what can be conceded without loss of 'face'. **1945** E. Waugh *Brideshead Revisited* 8 They..sidled away at the approach of an officer for fear that, by saluting, they would lose face with their new mistresses. **1957** L. Durrell *Bitter Lemons* 22 To lapse into Greek with anyone who was not a peasant would involve a loss of face. **1957** [see *Fairy-tale]. **1958** *Times* 5 July 12/1 The importance of 'face', whether individually or collectively, in Japan. **1968** G. Jones *Hist. Vikings* II. iv. 129 Harald had lost face in his dealings with Norway.

12. a. An open slope or hillside. *N.Z.* Also *Golf*, the slope or cliff of a bunker.

1857 *St Leonard's Station Diary* 11 July in L. R. C. MacFarlane *Amuri* (1946) III. 126 Sheep seem all right as all on steep sunny faces which have partly cleared. **1947** P. Newton *Wayleggo* (1949) iii. 37 It was country of great open shingle faces. *Ibid.* 153 *Face*: open hillside. **1949** —— *High Country Days* xiii. 137 The stag up in the face gave tongue again. **1881** R. Forgan *Golfer's Handbk.* 33 *Face*,..the sandy slope of a bunker. **1897** *Encycl. Sport* I. 466/1 Shall I play it backwards, or sideways on to the grass, or try to get it over the face of the bunker nearer the hole? **1910** B. Darwin *Golf Courses Brit. Isles* v. 105 We may be just short with our second..and we shall be battering the bunker's unyielding face till our card is shattered and wrecked.

13. c. *on the face of it* (fig.): on a merely superficial view, *prima facie*; obviously, plainly.

1882 *Knowledge* II. 70 The whole theory was absurd on the face of it. **1923** 'J. J. Connington' *Nordenholt's Million* vi. 71 It certainly seemed on the face of it to be a very useless accomplishment. **1960** R. Davies *Voice from Attic* ii. 74 There seems, on the face of it, to be little harm in wanting to be solvent and comfortable.

15. b. The striking surface of a golf-club, cricket-bat, hockey-stick, or tennis-racket; the inside of the net of a lacrosse-stick.

FACE

1881 R. Forgan *Golfer's Handbk.* 8 The head [of a full-sized Driver] weighs 7 oz. or 8 oz., and is distinguished from those of the 'Spoon' family by its 'face' being straight and almost perpendicular. **1887** W. G. Grace in G. A. Hutchison *Cricket* v. 49 It is much better to hold the face of the bat towards the umpire—nothing is gained by showing him the edge. **1891** H. G. Hutchinson *Hints on Golf* (ed. 6) 15 The maker's name gives you a fine guide to the centre of the face, which is the intended point of impact. **1897** *Encycl. Sport* I. 469/1 (*Golf*) The lofted face enables the player to start the ball straight. *Ibid.* 606/2 (*Lacrosse*) The leading string is then joined to the face of the gut by means of other pieces of gut, and the crosse is complete. *Ibid.* 613/2 (*Lawn tennis*) Avoid lop-sided or small-faced rackets, and see that the grain in the frame runs equally round the face. **1900** H. F. P. Battersby in J. H. C. Fegan et al. *Football, Hockey, & Lacrosse* 85 Sticks are made up with the grain running, broadly speaking, in one of two ways: either parallel to the plane of the face, or at right angles to it. **1909** P. A. Vaile *Mod. Golf* i. 13 It is a mistake in the driver or brassie to have the face too shallow.

26. a. (sense 1) *face-bleach, -glove, -hair, -mask, -massage, -paint, -powder, -screen, -tissue, -towel, -veil, -wash.* **b.** (sense 6 b) *face-pulling* ppl. a. and vbl. sb. **d.** instrumental, as *face-forward, -down, -up* (all adj. and adv.). **e.** similative, as *face-high* adj. and adv.

1907 *Daily News* 3 Sept. 2/7 A face-bleach for removing all discolorations from the skin. **1935** E. Bowen *House in Paris* I. v. 75 Leopold..stared at the face-down cards. **1962** *Gloss. Terms Automatic Data Processing* (*B.S.I.*) 90 Various terms are used to indicate the attitude in which a card is placed in the hopper...Face-up feed and face-down feed. **1907** *Westm. Gaz.* 11 Nov. 6/3 Accommodation for..two on large 'face-forward' seats. **1938** R. Graves *Coll. Poems* 152 Leaping face-forward from their high roofs. **1924** W. Deeping *Three Rooms* ii. 11 She was..using the face-glove with angry vigour. **1927** Peake & Fleure *Peasants & Potters* ix. 122 Scanty face-hair save for a chin tuft in the male. **1905** *Westm. Gaz.* 15 June 7/1 The ball went, face-high, just within reach of Gregory. **1906** *Ibid.* 26 Mar. 2/3 It is unnecessary for the operator [of oxygen-supplying apparatus] to wear any helmet or face-mask. **1936** *Discovery* July 206/2 A liquid compound of ether..given..from a face mask through a drop bottle. **1896** 'Iota' *Quaker Grandmother* xxii. 259 She..would have her usual face massage done then. **1900** *Daily News* 31 May 7/4 Departments for manicure, pedicure, and face-massage. **1915** *Illustr. Lond. News* 23 Jan. 120/1 Egyptian women..used..malachite, as the ingredient of a face-paint. **1858** *Illustr. News of World* 17 Apr. 175/4 Saunders's Face Powder..beautifies the complexion. **1879** *Chemist & Druggist* XXI. 481/1, 60 grammes of..Face Powder—a mixture of talc with a small proportion of white magnesia, coloured faintly red with cochineal. **1920** *Punch* 15 Sept. 219/1 A lady with a Russian name, no back to her gown and green face-powder. **1898** *Westm. Gaz.* 23 June 3/2 His face-pulling man whose countenance is as clay in the hands of a sculptor. **1906** *Ibid.* 7 Sept. 3/2 By dint of much face-pulling I managed to persuade my poor muscles to put me in an upright position. **1941** Koestler *Scum of Earth* xiii. 134 He had performed his task, especially the face-pulling, with so much success that he eventually became second clown. **1818** Keats *Let.* 16 Dec. (1958) II. 6, I brought from her a present of facescreens. **1890** W. J. Gordon *Foundry* 155 The men around in face-screens and leg-guards. **1937** *Woman's Fair* Oct. 50/3 Try creaming it [*sc.* your face] a second time and see if the second face-tissue comes off clean. **1951** E. Taylor *Game of Hide-and-Seek* I. iii. 57 Harriet..dropped screwed-up face-tissues on the floor. **1922** S. Lewis *Babbitt* i. 6 His own face-towel. **1925** R. A. Knox *Viaduct Murder* vi. 59 The face-towel was distinctly mentioned in the washing-list. **1926** A. Christie *Murder of R. Ackroyd* x. 127 He wouldn't even buy new face towels, though I told him the old ones were in holes. **1962** Face-up [see *face-down, above]. **1930** A. Clarke *Coll. Plays* (1963) 67 They wear the face-veil. **1870** R. Tomes *Bazar Bk. Decorum* 29 Now we advise them to overturn into the fire all their face-washes..and to betake themselves to soap. **1907** *Daily News* 3 Sept. 2/7 A face-wash for improving the complexion. **1911** E. Ferber *Dawn O'Hara* iii. 34 A motherly hug..enveloped me in an atmosphere of liquid face-wash, strong perfumery and fried lard.

27. face-ache (earlier example); also (*b*) a mournful-looking person (also as a term of address) *slang*; **face-bone**, delete †*Obs.* and add later example; **face bow** *Dentistry*, a calliper-like frame with adjustable attachments that is fixed round the front and sides of the head in order to take measurements of the mouth and jaws prior to making a denture; **face-brick** *U.S.*, a special brick used for facing buildings, etc.; a facing-brick; also, face-work of brick; **face-centred** *a. Cryst.*, applied to a space lattice in which the lattice-points occur at the centres of the faces and the corners of unit cells; hence also to a crystal, substance, etc., having a structure based on such a lattice; **face-cloth** (earlier example in sense 'a cloth for protecting the face of a baby'); (*b*) a woollen cloth with a smooth napped surface; (*c*) a cloth for washing the face; **face cream**, a cream for the complexion; cold cream; **face cut** *Forestry* (see quot. 1957); **face-cutting** *ppl. a.*, making a plane surface; **face decoration**, decoration (of pottery) with a face or faces; so *face-decorated* adj.; **face flannel** = **face-cloth* (*c*), *flannel *sb.* 1 e; **face-fungus** *colloq.*, the hairy growth on a man's face, esp. a beard; **face-glass**, the glass window of a diver's helmet; **face-harden** *v. trans.*, to harden the surface of (metal) by case-hardening, chill casting, or other process; so **face-hardened** *ppl. a.*, (in quot.) hard-faced; **face-lifting** *vbl. sb.*, a method of restoring a more youthful appearance by a surgical operation in which the skin is tightened and the wrinkles smoothed out; also *fig.* and *transf.*, e.g. the refacing of a building; hence (as a back-formation) **face-lift** *v. trans.*; also **face-lift** *sb.*, the operation of face-lifting (*lit.* and *fig.*); so **face-lifter**; **face-line**, (*a*) the alignment of the face of a structure, etc.; (*b*) *pl.* the lines or wrinkles of the face; **face-making**, (*b*) *Obs. slang*, the begetting of children; (*c*) the pulling of faces; **face-man**, a miner who works at the face; **face-pack**, a preparation beneficial to the complexion, spread over the face and removed when dry; **face-piece**, (*b*) = **face-glass*; (*c*) a mask covering nose and mouth (as used in anæsthetics) or nose and eyes (as used in diving); (*d*) a decorative appendage on a horse's bridle; **face-plate**, (*b*) a plate protecting some piece of machinery; (*c*) the area corresponding to the visor in protective head-gear (of a diving- or space-suit); **face-play**, facial movement in acting, etc.; **face-saver**, something that 'saves one's face' (see sense *10 b above); also **face-saving** *ppl. a.* and *vbl. sb.*; **face-symbol** *Cryst.*, the symbol designating the face or plane of a crystal; **face-urn**, an urn decorated with a face or faces; **face-work**, the exterior of masonry, the material forming the outside of a wall or the like; **face-worker** = **face-man*.

1808 Jane Austen *Let.* 26 June (1952) 203 Henry sends us the welcome information of his having had no face-ache since I left them. **1937** Partridge *Slang Dict.* 261/2 *Face-ache*, a C.20 jocular term of address. **1961** Simpson & Galton *Hancock* 35/2 On a train..a carriageful of the most miserable-looking bunch of face-aches. **1922** Joyce *Ulysses* 44 Raw facebones under his peep of day boy's hat. **1940** J. Osborne *Dental Mechanics* vi. 53 A face bow is also used to register accurately the relationship between the upper jaw and the condyle. **1807** *Independent Chron.* (Boston, Mass.) 21 Sept. 3/2 The Subscriber has been at considerable expense for several years past in the improvement of Face Bricks..superior to the Philadelphia Bricks. **1878** *Congress. Rec.* 25 Jan. 548/1 In consequence of the limit placed upon the cost of the building, it was found necessary to adopt face-brick. **1901** R. Sturgis *Dict. Archit.* s.v. *Brick, Face Brick*, one of a superior quality used for the face of a wall. **1913** *Proc. R. Soc.* A. LXXXIX. 257 The assumption of diffraction by the face-centred space lattice does not completely account for the pattern. **1921** *Physical Rev.* XVII. 587 Indium was found to consist of a face-centered tetragonal lattice of atoms. **1959** *Jrnl. Iron & Steel Inst.* CXCIII. 325/2 For face-centred cubic metals, notch ductility was found to be a linear function of the strain-hardening exponent. **1967** A. H. Cottrell *Introd. Metall.* xiii. 158 There are two fully close-packed crystal structures, known as face-centred cubic (F.C.C.) and close-packed hexagonal (C.P. Hex.). **1602** *Accounts of Lord High Treasurer of Scotland* 97 (D.O.S.T.), Ane ribbenit face claithe. **1685** *Inventory* in *Proc. Soc. Ant. Scotland* (1924) LVIII. 359 A parchment box wherein is 3 flanan face cloaths. **1898** *Westm. Gaz.* 29 Sept. 3/2 Dark plaid skirts and face-cloth coats. **1928** *Daily Express* 27 Dec. 5/2 Smooth face cloth and fox. **1930** *Chemist & Druggist* CXII. 663/1 A novelty in the way of face cloths. **1951** J. Frame *Lagoon* 33 She would buy Nurse Harper a cake of soap and a face-cloth wrapped in cellophane. **1953** A. W. Fielding *Stronghold* I. v. 81 Baggy breeches of royal blue face-cloth. **1968** J. Ironside *Fashion Alphabet* 227 *Face-cloth.* This is a term loosely used for a face-finished, sleek, glossy, luxurious looking woollen cloth. **1906** E. Aria *Costume* i. 1 Queen Victoria Eugénie, whose name has been snatched to honour a face-cream. **1913** E. Wharton *Custom of Country* III. xxi. 315 The little boy was encouraged to scatter the grimy carpet with face-creams and bunches of clippings. **1935** A. Christie *Death in Clouds* xix. 182 Jars of face cream. **1874** W. M. Baines *Narr. E. Crewe* viii. 180 After taking off a 'face cut'. **1950** H. Wilson *My First 80 Yrs.* vii. 105 So far we had asked for 'face-cuts', the first slices off the logs with the bark adhering. **1957** *Brit. Commonw. Forest Terminol.* II. 69 *Face cut*, the waney..or thin piece of timber removed when a straight face is cut along a log or flitch on a saw-bench. **1963** C. R. Cowell et al. *Inlays, Crowns & Bridges* ix. 99 The undercut formed by the bulge of the cingulum is removed, using a face-cutting inverted cone or tapered fissure diamond. **1928** Peake & Fleure *Steppe & Sown* 97 Its curious pottery with face decorations reminds us of the face-decorated handles of pots from Cemetery A at Kish in Mesopotamia. **1939** 'J. Struther' *Mrs. Miniver* 47 All the accumulations on the debit side of parenthood: ..the plasticine on the door-handles, the face-flannels in the bathroom, [etc.]. **1961** A. Wilson *Old Men at Zoo* ii. 101 Scrambled eggs should be scrambled, not cooked like a face flannel. **1907** F. Richardson *Bunkum* 316 He [*sc.* Juan de Castro] offered to pledge one of his whiskers...He had grossly over-capitalised his face-fungus. **1959** *Listener* 22 Jan. 183/2 Svengali..with his face-fungus and rolling eyes. **1896** *Strand Mag.* XII. 356/1 It is useless to butt the face-glass or wildly knock your head against the inside of the helmet. *a* **1941** R. Bedford in *Austral. Short Stories* (1951) 98 The tender screwed the face-glass into the helmet. **1874** *Porcupine* 21 Feb. 742/3 They were seen to join in the revelry and devilry as drunkenly as the most face-hardened of their companions. **1896** *Westm. Gaz.* 28 Feb. 6/3 All thick armour is face-hardened on the Harvey principle. **1934** R. Macaulay *Going Abroad* xxvii. 226 What I needed..was a face-lift... I should have a new, young, tight face. **1939** *Times* 28 Mar. 9/5 Many solidly built Victorian houses..which it should be possible to 'face lift'. **1957** *Times* 20 June 11/5 The restorers who will be face-lifting the exteriors of these historical buildings. **1957** *Times Lit. Suppl.* 11 Oct. 609/2 A special fund for preventing theatres and concert halls from becoming shabby and unattractive, or for giving them a face-lift when they are. **1942** N. Marsh *Death & Dancing Footman* i. 25 What's he like, this face-lifter? **1952** *Time* 30 June 28 (Advt.), That's not a 'face-lift'. That's Forerunner Styling—the years-ahead design that makes other 'new' cars seem out of date. **1956** *Sunday Times* 8 Jan. 7/7 Buyers abroad complain that our manufacturers seem content merely to give their cars a face-lift instead of tackling fundamental features. **1971** *Daily Tel.* 1 Feb. 3 Rail travellers are demanding a facelift for scruffy stations with their peeling paint..and squalid lavatories. **1922** F. Courtenay *Physical Beauty* ix. 57 The 'face-raising' or 'face-lifting' process which does away with wrinkles, mouth and eyelines and sagging cheeks by literally 'lifting' off part of the old face and replacing it. **1928** *Sunday Dispatch* 16 Dec. 9/4 A youthful appearance is considered an advantage, and face-lifting is a common thing among men. **1883** Face-line [see Face *sb.* 26]. **1906** *Daily Chron.* 13 July 3/3 They are the real Rembrandt. There is a deep furrow in the brow; the face-lines are bitten as by acid. **1909** *Westm. Gaz.* 6 Mar. 16/1 The work of erecting the new intermediate ribs between the old bridge and the new face-line is now in progress on all the spans. **1963** *Gloss. Mining Terms* (*B.S.I.*) 1. 10 *Subsidiary survey*, an underground survey made to determine the position of a face line or goaf line or some other specific feature. **1785** Grose *Dict. Vulgar T.* I. 2 *Face making*, begetting children. **1799** Jane Austen *Let.* 8 Jan. (1952) 51 My sweet little George! I am delighted to hear that he has such an inventive genius as to face-making. **1827** P. Egan *Anecdotes of Turf* 177 If ever she committed any more sins in the *face-making* line —quod,..should be her portion. **1921** *Glasgow Herald* 21 Apr. 5 The effect of the new offer on the wages of 'face-men'. **1926** *Amer. Mercury* Feb. 168 The profits of the beauty shop are dependent mainly upon the sale of..face packs and similar preparations. **1944** M. Laski *Love on Supertax* x. 97 Women leaping from drying machines..and one from a face-pack. **1960** *Woman's Own* 19 Mar. 25/2 Face-packs twice a week improve the texture and..brighten the skin. **1908** *Westm. Gaz.* 21 Sept. 5/3 A big round-topped helmet that contains a small glass face-piece through which the wearer can see. **1955** J. Sweeney *Skin Diving* vii. 90 The only area of unequal pressure that can occur in a swimmer's equipment is that of his face-piece or sea mask. **1960** H. Hayward *Antique Coll.* 113/2 Face-piece horse brass. **1961** *Lancet* 22 July 214/1 Immediately after the shock the face-piece was re-applied. *a* **1884** Knight *Dict. Mech.* Suppl. 324/1 *Face-plate*, a covering plate for an object to receive wear or shock. **1962** S. Carpenter in *Into Orbit* 75 We had to open up the faceplate on our helmets and squirt the food into our mouths. **1966** 'L. Holton' *Out of Depths* xvi. 158 He put the faceplate over his nose and eyes and the regulator in his mouth and took a suck of the air. **1970** F. McKenna *Gloss. Railwaymen's Talk* p. x, If you faced the fire-box you were subjected to radiant heat from the fire-box and the faceplate fittings. **1908** T. Hardy *Sel. Poems W. Barnes* p. viii, Each word of theirs [*sc.* of husbandmen] is accompanied by the qualifying face-play which no construing can express. **1958** *Listener* 5 June 953/2 Neither text nor production was helpful: everything was bustled along with voices too loud and face-play too obvious. **1941** E. Snow *Scorched Earth* ii. 19 As a face-saver, however, Doihara was given enough support, from the Kwantung Army in Manchuria [etc.]. **1959** *Times* 4 May 3/1 An equalizer in the dying seconds and yet another face-saver in the extra time. **1922** A. Bennett *Lilian* I. iii. 19 She had been trapped beyond any chance of a face-saving lie. **1931** *Economist* 4 Apr. 716/1 After all, 'face-saving' is common form in the politics of every country. **1957** *Observer* 10 Nov. 1/5 This satellite launching is an isolated face-saving move in a sense. **1899** W. J. Lewis *Treat. Crystallogr.* iv. 27 (*heading*) Face symbol. **1903** *Athenæum* 17 Jan. 86/3 How to convert the face-symbols of Naumann into those of Miller. **1944** C. Palache et al. *Dana's Syst. Min.* (ed. 7) I. 36 Any face symbol (*hkl*) in the old orientation may be transformed to the new equivalent (*h′k′l′*). **1959** C. S. Hurlbut *Dana's Man. Min.* (ed. 17) ii. 32 The zone symbols are inclosed in brackets, as [*uvw*], to distinguish them from face and form symbols. **1927** Peake & Fleure *Priests & Kings* 147 The face-urns..do not seem to belong to the invading people who founded Hissarlik II. **1957** V. G. Childe *Dawn Europ. Civilization* (ed. 6) iii. 43 Anthropomorphic lids and jars ('face-urns'). **1838** *Civil Eng. & Arch. Jrnl.* I. 330/1 What is called Flemish bond consists in the disposition of the bricks on the outside, or face work. **1909** *Daily Chron.* 5 Aug. 1/3 The face-work of the old parapet having been re-erected. **1926** *Rep. R. Comm. Coal Ind.* 118 in *Parl. Papers* (Cmd. 2600) XIV, Persons working at the face..'face' workers. In 1905 there were 58 face workers to 42 off-hand workers. **1957** *Economist* 23 Nov. 706/2 There are as many faceworkers in the mines as there were then.

28. With adverbs forming combs. expressing the action of the corresponding verbal combinations, as **face-off** (see sense *4 e above), **face-up**, a facing-up.

1936 *Discovery* Oct. 330/1 Here, at last, is a face-up to the relations between science and society.

face, *v.* Add: **3.** *to face out of countenance*: to confront and disconcert (cf. Countenance *sb.* 6 b); *to face up*: *intr.* to show a bold face; with *to*, to oppose (a person), confront (a problem, etc.). (See also quot. 1954 for use in *Boxing*.)

1854 *Punch* 27 May 222/1 Is he to be faced out of countenance by a young whipper-snapper? **1920** W. Raleigh *Let.* 4 Sept. (1928) II. 519 He faced up to the paradox of man. **1925** M. A. Lowndes *Some Men & Women* 155 She was a woman who always faced up to the realities of life. **1935** *Punch* 4 Sept. 260/2 It was our duty as guardians of the children to face up to the situation. **1939** [see *appease *v.* 2 c]. **1954** F. C. Avis *Boxing Ref. Dict.*, *Face up*, to square up in the approved way to an opponent. **1958** *Times* 8 Sept. 9/4 No longer does a man face a problem; he faces up to it.

4. a. *to face the music*: see Music *sb.* 11.

c. *Ice Hockey* and *Lacrosse*. (Esp. with *off*). *intr.* To start or restart play by (the referee or linesman) dropping the puck, or ball, between the sticks of two opposing players. Also *trans.*, to place the puck, or ball, in this way.

1867 *Laws of La Crosse* 6 Should the ball lodge in any spot inaccessible to the 'crosse' it may be taken out by the hand and immediately placed on the 'crosse', but should an opponent be checking and cry 'face', it must be faced for. **1882** *Rules of Lacrosse* 15 Should the ball lodge in any place inaccessible to the crosse, it may be taken out by the hand; and the party picking it up must 'face' with his nearest opponent. **1897** *Medicine Hat* (Alta.) *News* 25 Feb. 1/6 Ben Niblock scored the first goal for Medicine Hat a few seconds after the puck was faced. **1910** [see *facing *vbl. sb.* 2 b]. **1958** *Vancouver Province* 12 May 11/5 Reeve John Stolberg *faced off* the ball.

5. b. Esp. *pass.* To be confronted *with*.

1920 *Challenge* 21 May 44/2 An extrovert soldier faced with the problem of escape from war conditions.

6. (Later example.) Also, *to face the facts*; (colloq.) *let us* (or *let's*) *face it*: let us not shrink from recognizing some unwelcome fact(s).

1911 G. B. Shaw *Getting Married* 249 You'll want half a dozen different sorts of contract. *The Bishop*: Well, if so, let us draw them all up. Let us face it. **1913** Galsworthy *Fugitive* 1, We've got to face the facts. **1922** N. Angell *Press & Organiz. Society* ii. 43 Let us face the truth. **1928** Wodehouse *Money for Nothing* v. 83 John—she had to face it—was a jellyfish. **1932** —— *Louder & Funnier* 46 One has got to face facts. **1937** W. S. Maugham *Theatre* xxi. 203 But let's face it, I've never been in love with you any more than you've been in love with me. **1958** I. Murdoch *Bell* i. 11 To the pain of Paul and his friends the expression 'let's face it'..was still frequently on her lips.

8. e. With *off*. To turn aside (*spec.* the current of a stream).

1887 Ruskin *Præterita* II. 384 A little logwork to face off the stream at its angles.

14. Also with *up*.

1883 *Specif. Alnwick & Cornhill Rlwy.* 10 All bolt-holes to have bosses cast on them, which are to be faced up. **1889** P. Hasluck *Model Engin. Handybk.* x. 113 If all the flange joints are faced up absolutely true. **1958** *People* (Broadway, N.S.W.) 19 Mar. 54 Primitive grinding or cutting machines to 'face up' their specimens. **1970** H. Braun *Parish Churches* vii. 83 From this technique develops the type of walling known as 'random rubble' which, properly faced-up.., is frequently met with in pre-medieval days.

faced, *ppl. a.*[2] Add: **1. b.** Of a golf-club, tennis-racket, etc. (see *face *sb.* 15 b), as *long-faced, short-faced, small-faced, straight-faced* adjs.

1893 H. G. Hutchinson *Golfing* 21 These short-faced clubs. **1897** *Encycl. Sport* I. 467/2 A straight-faced club. *Ibid.* 613/2 [see *face *sb.* 15 b]. **1909** P. A. Vaile *Mod. Golf* viii. 120 Don't choose a big-headed club, and avoid a long-faced one.

faceless, *a.* Add: Without a face (examples). Hence anonymous, characterless, without identity. Also *transf.* and *fig.*

1865 G. Meredith *R. Fleming* I. xi. 176 Dahlia..lay..with her handkerchief across her eyes...She lay faceless. **1936** A. Huxley *Eyeless in Gaza* xix. 250 Those voluptuous and faceless bodies created by the stylists had actually come to assume Joan's features. **1936** M. Mitchell *Gone with Wind* xix. 331 Always, faceless soldiers stood on the dark porch. **1953** *N.Y. Times* 4 Oct. 15/1 The 'faceless men'—about 15,000 anti-Communist Chinese and 7,800 anti-Communist North Koreans...In all their wars the Communists have regarded prisoners not as human beings but as hostages and pawns. **1959** *Observer* 28 June 15/1 The Institute of Directors was attacked as a pressure group for industry's 'faceless men'. **1960** *Times* 28 Sept. 15/4 One of those firms that has grown from a compact community into a sprawling, faceless industrial monster. **1967** *New Statesman* 29 Dec. 902/1 We shall go on being lectured and told to draw in our belts by a faceless and endlessly replaceable body of men whom [*sic*] we know quite well are not about to do anything very important with the money we save for them. **1970** *Times* 4 May 8/5 In Brussels..as in Bonn, where there are a surprising number of faceless men, his warmth, genuine jollity and readiness to learn were refreshing.

So **fa·celessness**.

1950 *Sat. Rev. Lit.* 21 Oct. 14 Joe Saul appears in the hospital..as only a voice and a white facelessness. **1961** *Times* 20 Dec. 7/1 It [*sc.* a production of *Macbeth*] claps a mask of facelessness on a tragedy.

facellite (fǣk-, fǣse·ləit). *Min.* Also **phacel(l)ite, phakellite.** [a. It. *facellite* (E. Scacchi 1888, in *Rend. dell'Accad. delle Sci. fis. e matemat.* (Napoli) XXVII. 486), f. Gr. φάκελ(λ)-ος bundle: see -ite[1].] = Kaliophilite.

1891 *Jrnl. Chem. Soc.* LX. I. 22 Phakellite, a new mineral. **1931** *Min. Mag.* XXII. 599 The original labels with these specimens bear the name 'Facellite'..but..the chemical optical properties and the densities of fresh material from both specimens are those for kaliophilite.

facet, *sb.*[2] Add: **1.** (Further examples.)

1909 W. M. Davis *Geogr. Ess.* xxvi. 746 A triangular facet on the block front. **1937** Wooldridge & Morgan *Physical Basis Geogr.* xvii. 257 (*heading*) The facets of relief in a landscape of multi-cycle character. *Ibid.* 258 It is the variously inclined facets of intersecting surfaces which must form the units of detailed geographical study. **1944** A. Holmes *Princ. Physical Geogr.* xiii. 258 Isolated pebbles or rock fragments strewn on the desert surface are bevelled on the windward side until a smooth face is cut...Two or more facets may be cut.

fig. (Further example: see also quot. 1951.)

1931 *Times Lit. Suppl.* 20 Aug. 625/3 Many facets of Tyrolese life. **1951** Palmer & Wells *Fund. Libr. Class.* ii. 31 The sum total of the divisions of each aspect we shall call a facet.

faceted, *ppl. a.* **1.** (Further *transf.* and *fig.* examples.)

1898 *Daily News* 30 Nov. 7/6 Not an hour..when the uneasy faceted surface [of the sea] is not throwing back reflections of the sky. **1930** E. Pound *XXX Cantos* xx. 91 In the sunlight, gate cut by the shadow; And then the faceted air: Floating. **1944** A. Holmes *Princ. Physical Geogr.* xiii. 258 Wind-faceted pebbles. **1959** J. D. Clark *Prehist. S. Afr.* vi. 143 This is known as *faceted platform technique* on account of the care that was usually devoted to the preparation of the platform where the core was struck to remove the flake.

facetiæ. Add: **2.** *Bookselling.* Pornography.

1851 Mayhew *Lond. Labour* I. 293/2 He puts to the end of his catalogue..two pages that he calls 'Facetiæ'...indecent books, indeed. **1913** [see *erotica]. **1926** Fowler *Mod. Eng. Usage* 164/1 *Facetiae*, in booksellers' catalogues, is a euphemism for obscenities. **1947** E. Partridge *Usage & Abusage* 115/1 *Facetiae* is a booksellers' euphemism for 'pornography' or 'a book with a certain amount of sexual interest' and should be avoided by anyone who is not a bookseller.

‖ **Fach, fach** (fāχ). [G., compartment, partition, division, shelf, *fig.* circumscribed branch of knowledge.] A line of work or business, a department of activity; (one's) métier. Also in *Comb.*, 'professional'.

1842 J. S. Mill *Let.* Nov. (1910) I. 121 The thing is not in my *fach*. **1887** W. James *Let.* 6 Feb. (1920) I. 263 Wundt..isn't a genius, he is a *professor*—a being whose duty is to know everything, and have his own opinion about everything, connected with his *Fach*. **1905** *Ibid.* 30 Apr. II. 227 These men..are none of them *Fach-philosophers*, and few of them teachers at all. **1920** G. Santayana *Char. & Opin. U.S.* v. 143 Very professional in tone and conscious of his *Fach*. **1923** *Contemp. Rev.* June 757 The Eldorado of story-tellers whose *Fach* is the portrayal of luxurious vice.

facia. Add: (Examples.) Now usu. (fē̆i·-).

1881 *Instr. Census Clerks* (1885) 52 Sign and Facia writer. **1911** *Tariff Reform League: Rep. Labour & Social Cond. Germany* III. 74 A butcher's shop with this sign on the facia board. **1931** *N. & Q.* 14 Nov. 354/1 A double-fronted shop..and the name 'Deck' in the middle of the facia.

2. The instrument panel or dashboard of a motor vehicle. Also *attrib.*, as *facia board, panel*. Cf. *fascia 5.

1924 H. J. Butler *Motor Bodywork* xviii. 279 A front facia board of 1″ stuff. **1937** *Times* 9 Oct. 9/2 The facia board is clearly planned so that the instruments are easily read by the driver. **1955** *Times* 19 July 11/6 The full width of the facia. **1959** *Motor Man.* 93 A push-pull knob on the facia panel.

facial, *a.* Add: **2. d.** *facial eczema*: a disease of sheep in New Zealand.

1910 *N.Z. Jrnl. Agric.* 15 Nov. 464 The prevalence of so-called facial eczema among sheep last year. **1950** *Ibid.* Oct. 317/1 Adverse weather may temporarily change pasture to a toxic state capable of producing facial eczema in stock.

e. *facial tissue*: a soft, absorbent kind of tissue paper; a **face-tissue*.

1930 H. Rubinstein *Art Feminine Beauty* xvi. 265 Spread cream over the face, and wipe it off with a bit of clean linen or fine facial tissues. **1953** 'S. Ransome' *Drag Dark* (1954) i. 15 Box of face powder unspilled, facial tissues neatly folded.

4. b. *sb.* Beauty treatment for the face. Also *transf.* orig. *U.S.*

1914 G. Atherton *Perch of Devil* I. 84 I've got fourteen heads to dress..and most of them want a facial, too. **1932** *Kansas City Times* 26 Apr. 16 The north side of the courthouse clock is peeling...The newspaper suggests a facial. **1934** [see *beautician]. **1937** *Time & Tide* 12 June, Why don't they dress better and have a facial occasionally? **1959** *Observer* 15 Mar. 13/5, I don't have facials now: I keep that and massage in reserve for when I get older. **1961** *Time* (Atlantic ed.) 6 Jan. 7 The Capitol dome, which has just had a dazzling, million-dollar facial, beamed down on the city.

faciation (fē̆isi,ē̆i·ʃən). *Ecol.* [f. L. *faciēs* + -ation.] A community containing more than one of the dominant species in an association (often related to climatic or geographic differences in the area of the association).

1920 F. E. Clements *Plant Indicators* 276 The grouping of consociations within the association is typical of all climaxes, and seems to warrant a special term...It has seemed desirable to definitize the term *facies* for seral groupings and to make a new word, *faciation*, for climax groupings...The two terms conform to the mutual relation seen in associes and association, consocies and consociation. **1938, 1960** [see *facies 2 c]. **1952** P. W. Richards *Tropical Rain Forest* xv. 319 Four 'faciations' of Evergreen Seasonal forest are recognized; more or less the same species occur in all of them but in different proportions.

‖ **faciendum** (fækie·ndŏm, fē̆iʃ-). *Philos.* Pl. **facienda.** [a. L. *faciendum*, 'thing to be done', neut. of gerundive of *facĕre* to make, do.] A thing that should be done.

a **1832** Bentham *Ess. Logic* in *Wks.* (1841) VIII. xv. 241/1 Different assemblages of these *noscenda* and *facienda*; of these subjects or objects of *disciplines*, have received names. *a* **1866** J. Grote *Exam. Util. Philos.* (1870) iv. 67 This is the idea of the *summum jus*, the *faciendum*, the notion of *duty*. **1958** W. Stark *Sociology of Knowledge* iii. 139 'Imaginative' thought..is neither concerned with *data*, like 'existential' thought, nor yet with *facienda*, like normative ideas.

facies. Add: (Pl. **facies.**) **1. b.** *Med.* The appearance or expression of the face.

[**1684** S. Blancard *Physical Dict.* 128 *Facies Hyppocratica* is when the Nostrils are sharp, the Eyes hollow, the Temples low,..the Complexion pale, livid, of a leaden Colour, or Black. **1807** Morris & Kendrick *Edin. Med. & Physical Dict.* II, *Facies hippocratica*, that particular disposition of the features which immediately precedes the stroke of death.] **1883** J. J. G. Brown *Med. Diagnosis* (ed. 2) i. 14 There are one or two facies which stand out more prominently than the rest, and which deserve special attention. **1907** *Practitioner* Apr. 520, I can only recall one facies which at all suggested the condition of *bien-être* associated with the acquired disease. **1908** *Ibid.* Feb. 292 The facies gray. **1932** Jewell & Kauntze *Handbk. Tropical Fevers* xx. 321 During the pyrexial period the face is flushed, the eyes injected (giving a typical 'ferrety' appearance to the facies). **1965** H. T. Hyman *Differential Diagnosis* 166/1 In contrast with changes in facies, abnormalities of facial contour are relatively fixed.

2. b. *Geol.* The character of any part of a formation as displayed by the fossils it contains, the composition, texture, etc., of the constituent rocks, or other differentiating qualities; also, a part of a formation having a particular character. In *Petrol.*, a mass of igneous rock that differs in some way from the main body of rock of which it is part.

1849 [in Dict. sense 2]. **1882** A. Geikie *Text-bk. Geol.* 615 Every well-marked formation is characterized..by a general assemblage or *facies* of organic forms. **1910** Lake & Rastall *Text-bk. Geol.* xvi. 285 When a geological series or system is in one district composed chiefly of limestone and in another of clays and shales, it is usual to speak of these different types of deposit as different 'facies'. **1913** J. P. Iddings *Ign. Rocks* II. I. i. 5 In one region the composition of a nearly homogeneous rock mass..may assume a certain local petrographical significance, while in another region it may appear only as a facies of a rock mass. **1913** Hatch & Rastall *Petrol. Sedim. Rocks* i. 7 Each of the formations..exhibits two very different facies, controlled for the most part by the conditions of formation. **1938** A. K. Wells *Outl. Hist. Geol.* iii. 20 If rocks of a certain age exhibit a lateral change of lithological facies,..a change in conditions of formation is implied. **1951** J. Gilluly et al. *Princ. Geol.* xix. 517 The evidence is not only geometric. The facies of the rocks composing the thrust masses demand great travel also. **1965** G. J. Williams *Econ. Geol. N.Z.* iv. 38/1 The host-rock is a splintery siliceous bluish-grey slate with some micaceous, talcose and felsic facies. **1970** R. C. Selley *Anc. Sedim. Environm.* i. 4 The geometry of a sedimentary facies may be relatively easy to determine where it crops out at the surface.

c. *spec.* in *Ecol.* (See quots.)

1905 F. E. Clements *Res. Methods Ecol.* iv. 238 It is seldom..that the facies and invaders are so equally matched in height and other qualities that they remain in equilibrium. *Ibid.* 317 *Facies*, a dominant species of a formation: a distinct area controlled by it is a consocies. **1920** [see *faciation]. **1932** Fuller & Conard tr. *Braun-Blanquet's Plant Sociol.* ii. 25 The facies is distinguished wholly by differences in the quantity or distribution of species. **1938** Weaver & Clements *Plant Ecol.* (ed. 2) iv. 107 The facies..is the developmental unit of the associes characterized, as is the faciation of climax vegetation, by the grouping of dominants. **1960** N. Polunin *Introd. Plant Geogr.* xi. 334 The seral counterpart of the faciation is the *facies*.

3. *attrib.* and *Comb.*, as *facies-change*; **facies fossil**, a fossil that is characteristic of a facies as a result of the restricted kind of environment tolerated by the original organism; so **facies fauna**, the (fossil) fauna characteristic of a facies.

1923 L. D. Stamp *Introd. Stratigr.* ix. 139 It seems that the Pendleside fauna is a 'facies-fauna' developed only under special conditions. *Ibid.* xii. 194 In Germany the 'facies-fossils' of the Muschelkalk do reappear at a much higher horizon. **1954** J. F. Kirkaldy *Gen. Princ. Geol.* xvii. 254 They [*sc.* fossils] not only date the beds, but in the case of facies faunas, give us valuable evidence as to the conditions under which the beds were formed. **1969** *Proc. Geol. Soc. Lond.* Aug. 146 Once the marker-points are adequately defined, subsequent demonstrations of facies-changes, diachronous boundaries, lacunae or similar phenomena do not affect their validity.

‖ **facile princeps** (fæ·sili pri·nseps), *phr.* and *sb.* [L.] (A person who is) easily first; the acknowledged leader or chief.

1834 Greville *Mem.* (1874) III. xxii. 64 In the prime of life,..*facile princeps* in the House of Commons. **1858** Trollope *Three Clerks* I. ii. 33 He..soon became *facile princeps* in the list of habitual idlers. **1861** *Two Cosmos* I. vi, The *facile princeps* of the Whig Attorneys. **1863** [see Gipsiologist]. *a* **1878** G. G. Scott *Lect. Med. Archit.* (1879) II. xvi. 253 [St. Sophia's] is *facile princeps* among structures on the pendentive domical principle. **1877** *Times* 20 Dec. 7/3 In that [*sc.* political tergiversation] he is *facile princeps*, and has left all competitors behind. **1931** M. Summers *Supernatural Omnibus* 31 Dr. M. R. James..is, with the exception of Vernon Lee, of all writers of ghost stories to-day *facile princeps*.

‖ **facilis descensus Averni** (fæ·silis dese·nsŏs āve·rnəi), *phr.* [L. (Virgil *Æn.* VI. 126, where most witnesses read *Auerno*), *lit.* the descent of (or to) Avernus (is) easy. *Avernus* was the name of a deep lake near Puteoli, the reputed entrance to the underworld.] It is easy to slip into evil ways. Also *facilis descensus*, used as *sb. phr.*

1618 T. Adams *Happines of Church* II. 76 And downe a hill, for hell is a bottome. *Facilis descensus Auerni.* **1885** 'L. Malet' *Col. Enderby's Wife* VII. v. 296 *Facilus* [*sic*] *descensus Averni*. That remark was made a long time ago; but it holds good still. **1895** G. B. Shaw *Let.* 22 Apr. (1965) 525 Moral ruin followed slowly but surely—facilis descensus Averni. **1939** C. S. Lewis *Let.* 8 May (1966) 165 *Romola* is a most purgative work on the *facilis descensus* because the final state of the character is so different from his original state and yet all the transitions are so dreadfully natural. **1946** 'D. Yates' *Red in Morning* i. 27, I took a first class in Honour Moderations. Unfortunately, shortly afterwards, I took twenty-seven pounds from the monies entrusted to my care. *Facilis descensus averni*. And I have never looked back.

facilitate, *v.* Add: **3.** *Physiol.* To increase the likelihood of, strengthen (a response); to bring about the transmission of (an impulse). (Cf. *facilitation 3.)

1906 C. S. Sherrington *Integr. Action Nerv. Syst.* v. 175 Minimal electrical stimuli applied near together in point of time to the fore-limb region of the rabbit's cortex and to the skin of the crossed foot, exerted a facilitating influence, 'bahnung', on each other. **1934** E. S. Robinson in C. Murchison *Handbk. Gen. Exper. Psychol.* xii. 632 Extraneous, or unessential, stimuli can facilitate a given reaction as well as interfere with it. **1951** M. A. B. Brazier *Electr. Activity Nerv. Syst.* ix. 90 A long persisting effect set up by an initial stimulus would serve to facilitate impulses coming much later.

facilitation. Add: **3.** *Physiol.* The increased excitability of a post-ganglionic neuron, resulting in an increased response to a physical stimulus, that is produced by conditioning of the ganglion by impulses due to other (similar or different) stimuli.

1895 E. B. Titchener in *Amer. Jrnl. Psychol.* VII. 80 *Bahnung*, facilitation. **1901** J. M. Baldwin *Dict. Philos.* I. 367/2 *Facilitation*...The term has been given technical meaning as a translation of the German Bahnung (in neurology), which latter carries the idea of the preferential determination of a function..as to its pathway of discharge. **1906** C. S. Sherrington *Integr. Action Nerv. Syst.* v. 176 He [*sc.* S. Exner] argues..that the seat of production of the facilitation lies in the spinal cord. **1943** J. F. Fulton *Physiol. Nerv. Syst.* (ed. 2) xx. 379 A series of inadequate stimuli may become adequate merely through repetition, and thus additional neurons become recruited through local facilitation. **1961** H. H. Jasper in D. E. Sheer *Electr. Stimul. Brain* xxxviii. 561/1 Critical areas of facilitation are in neuronal systems where there is convergence of impulses from many sources. **1968** M. Monnier *Funct. Nerv. Syst.* I. iii. 112 A third, very prolonged facilitation effect begins 300–500 msec after the first conditioning impulse and lasts for several seconds.

facility. Add: **2. b.** (Further examples.) Also, the physical means for doing something; freq. with qualifying word, e.g. *educational, postal, retail facilities*. Also in *sing.* of a specified amenity, service, etc. orig. *U.S.*

1872 R. G. McClellan *Golden State* (1874) 373 There..was but little need of postal facilities. **1920** *Travel* July 42 To secure better roads, better transportation facilities. **1937** *Discovery* Nov. p. ci/1 [Film] projection facilities for approved bookings. *Ibid.*, Lunch and tea facilities. **1958** *Listener* 3 July 10/1 This 'facility', or 'rest room', as it is also variously called [in the U.S.]. **1962** *Gloss. Terms Autom. Data Proc.* (*B.S.I.*) 99 *Hold facility*,..a means of interrupting the computing action and keeping all variables at the value they had. **1962** F. I. Ordway et al. *Basic Astronautics* xiii. 554 (*caption*) Multiaxis spin test inertial facility used by..astronauts. **1967** *Times Rev. Industry* Aug. 74/2 No one could have predicted..that the completed facility would have the bad luck to run head on into..a metals depression. **1969** *Lebende Sprachen* XIV. 65/2 Many other ports offer container facilities. *Ibid.* 97/1 The opening of Britain's National Giro..makes an entirely new banking facility available to the general public. **1971** *Oxf. Univ. Gaz.* 25 Feb. 708/1 The following decree adds Junior Members to the Committee for Sports Facilities.

facing, *vbl. sb.* Add: **2. b.** *Lacrosse.* (See *face *v.* 4 c.)

1910 *Westm. Gaz.* 22 Apr. 14/3 Facing is no longer to be regarded as a penalty.

10. *facing-bar, -board, -pavior.*

1904 Goodchild & Tweney *Technol. & Sci. Dict.* 214/1 *Facing bar* or *work bar*, a smooth iron plate about 4 in. wide and the full width of the lace machine, over which the web of lace travels on its way to the roller. *Ibid.*, *Facing paviors*, hard burnt malms of good colour and shape. **1839** Ure *Dict. Arts* 843 In friable and shivery rocks there is put behind these beams..*facing boards*, which are planks placed horizontally..,set so close together as to leave no interval.

facio- (fē̆i·ʃio), used as combining form (see -o) of L. *facies* face, as in **facioplegic** *a.*, pertaining to paralysis of the face; etc. (see quots.).

1900 Dorland *Med. Dict.* 250/2 *Faciobrachial*, pertaining to the face and arm. *Faciocervical*, affecting the face and neck. *Faciolingual*, pertaining to the face and tongue. *Ibid.* 252/1 *Facioscapulo-humeral*, pertaining to the face, scapula, and arm. **1910** Osler & McCrae *Syst. Med.* VII. 124 The facio-scapulo-humeral type [of progressive muscular dystrophy]. **1922** F. W. Price *Textbk. Pract. Med.* 1577 When the wasting begins in the face (facio-scapulo-humeral type). **1922** A. A. Stevens *Pract. Med.* 916 In a few instances recurrent facial paralysis seems to have been an accompaniment of migraine (facioplegic migraine). **1946** *Nature* 2 Nov. 612/1 Special hospital units for thoracic, facio-maxillary and head injuries.

‖ **façon de parler** (fasondəparle). [Fr.] A way or manner of speaking; a mere phrase or formula.

1804 W. Wilberforce *Let.* 29 Sept. in J. Aitken *Eng. Lett. XIX Cent.* (1946) 28 B.'s first letter would convey the idea of his having had in his mind from the very first the..plan..but I believe that is only a *façon de parler*. **1806** *Edin. Rev.* Jan. 494 Philostratus..introduces the high priest of Serapis saying, (in the time of Vitellius) 'and who can alter the sacred institutions of Egypt?' Was this a mere *façon de parler*? **1813** Wellington *Disp.* (1838) X. 161, I hope that this mode of considering an omission which can easily be rectified, is only a *façon de parler*. **1886** H. G. Hutchinson *Hints Golf* 56 A mere golfing *façon de parler*. **1887** [see *analysis 9 c]. **1907** W. De Morgan *Alice-for-Short* xliii. 439 Which was palpably a lie, taken literally; but was a *façon-de-parler* that passed muster, taken leniently. **1953** C. E. Bazell *Linguistic Form* i. 8 It is clearly a mere *façon de parler* when *sang* is split up into the segments i) *sing* and ii) *i>a*.

‖ **façonné** (fasone), *a.* Also **faconne, faconné(e.** [Fr., pa. pple. of *façonner* to fashion.] Designating a material into which a design has been woven. Also as *sb.*, the material itself.

1895 *Montgomery Ward Catal.* 13/1 Pongee Faconne; width 21 inches. **1939–40** *Army & Navy Stores Catal.* 637/2 Rich black silk and velvet faconné. **1957** M. B. Picken *Fashion Dict.* 121/1 *Façonné*, French word meaning figured; used of fabrics, particularly those with scattered motifs woven in. **1971** *Daily Tel.* 19 Apr. 12/6 Lots are really battle jackets in *faconée* silk, and could be happily worn over a sleeveless dress.

facsimile. Add: **2. a.** *spec.* A radio, telegraphic, or other system that scans written, printed, or photographic material and transmits signals used to produce a likeness of the original; also, matter so reproduced. Freq. *attrib.*, as *facsimile telegraph, transmission*.

1815 R. Wedgwood in *Commercial Mag.* (1846) I. 259 Fac-similes of a dispatch, written..in London, may with facility be written also in Plymouth, Dover..by the same person, and by one and the same act. **1877** *Pract. Mag.* VII. 10/1 (*heading*) The Fac-simile Telegraph...An instrument which transmits by telegraph the weather maps of the Signal Service. **1927** *Marconi Short Wave Beam Syst.* 19 It [*sc.* the short wave beam system] can be used for facsimile transmission over any distance. **1935** *Sci. Amer.* Mar. 122/1 The home radio set will produce a copy of the printed material that was fed into the broadcasting machine, with picture and text reproduced in facsimile...That is what facsimile radio has in store. **1948** *New Yorker* 28 Feb. 21/3 The facsimile newspaper..travels through the air. **1959** A. G. Cooley in K. Henney *Radio Engin. Handbk.* (ed. 5) xxiii. 1 In a facsimile system the subject copy for transmission is scanned by a light beam, a line at a time. **1969** *Daily Tel.* (Colour Suppl.) 31 Oct. 25/1 Distribution..is mastered..with a centralised point in Tokyo, transmitting facsimile pages by microwave to be printed by offset all over the country.

fact, *sb.* Add: **4. b.** In apposition to a following noun clause: *the fact that..* = the circumstance that.

1803 G. Moore *Diary* 15 Jan. in R. J. Mackintosh *Mem. Sir J. Mackintosh* (1835) I. iv. 175, I would not agree to the fact that ennui prevailed more in England than in France. **1834** *Edin. Rev.* Oct. 73 The only difference between Crabbe and himself is the fact, that the one was raised from the ranks, while the other is still remaining in them with at least equal independence. **1851** Queen Victoria *Let.* 13 Oct. in *Queen Victoria's Early Lett.* (1963) 183 The Queen of Spain ought to be made aware of the fact that among the reigning Sovereigns, the Emperors of Austria and Brazil..etc., etc., have not got the Garter. **1957** D. J. Enright *Apothecary's Shop* 209 The fact—not a new one—that Eliot doesn't pull his punches.

d. *the fact is*: const. with following noun clause introduced by *that* or with *that* understood.

1836 R. Owen *Bk. New Moral World* xii. 79 The simple fact is that the institutions of society have been formed..to oppose one part of human nature to another. **1868** Trollope *He knew he was Right* (1869) II. li. 15 You can remain a few minutes longer. The fact is, I've got something I want to say to you. **1875** [in Dict., sense 6 c]. **1959** *N.Z. Listener* 12 June 4/3 The fact is that Lord Bledisloe..did not realise that everything he said was not entirely worth two or three columns in a busy metropolitan daily.

e. *facts and figures*: an alliterative phrase used in the sense 'precise information'.

1845 Dickens *Chimes* iv. 48 Facts and figures! Put 'em down! **1855** *Westm. Rev.* VIII. 438 The honorable gentleman on one side of the House is liable to have his facts and figures shown up by his honorable friend on the opposite side. **1903** H. James *Ambassadors* IV. ix. 80 Strether..put him in full possession of facts and figures. **1957** J. Braine *Room at Top* xii. 122 The hard materialists, the men of facts and figures.

f. *fact of life*: a (stark) reality of existence; a brute fact; freq. *the facts of life*, spec. as a colloq. euphemism for 'knowledge of human sexual functions'.

1854 Thoreau *Walden* ii. 98, I went to the woods because I wished..to front only the essential facts of life. **1885** Rider Haggard *K. Solomon's Mines* ii. 21 Elephant hunters are a rough set of men, and don't trouble themselves with much beyond the facts of life and the ways of Kafirs. **1893** 'S. Grand' *Heavenly Twins* I. i. 6 He snubbed Evadne promptly..when she mentioned a fact of life...'Only confusion comes of women thinking for themselves on social subjects,' he said. **1895** Hardy *Jude* VI. ii. 426, I told him I was going to have another child...I couldn't bear deceiving him as to the facts of life. **1902** W. D. Howells *Lit. & Life* 305, I have not touched upon these facts of life without the purpose of finding some way out of the coil. **1908** K. Grahame *Wind in Willows* ix. 196 To-day, the unseen was everything, the unknown the only real fact of life. **1908** Mrs. H. Ward *Diana Mallory* III. xx. 435 The first withdrawal of the veil which hides..the more brutal facts of life. **1913** *MacLean's Mag.* June 123/1 The tone of all the speeches at the meeting [of the Eugenics Education Society] emphasized the need of teaching boys and girls the essential facts of life, so as to equip them for the momentous time when they choose life partners. **1930** W. S. Maugham *Breadwinner* I. 14, I shall never forget when I was leaving my prep school, and Dorothy told Alfred he must tell me what she called 'the facts of life'. **1932** Wodehouse *Hot Water* xiii. 225 Didn't your mother ever teach you the facts of life, Mrs. G.? Because one of them is never to be too friendly to people you meet on boats. **1953** L. P. Hartley *Go-Between* x. 124 The facts of life were a mystery to me. **1957** G. Faber *Jowett* v. 84 It must certainly have been at St. Paul's that he became disgustedly aware of the 'facts of life'. **1959** *News Chron.* 22 Oct. 5/4 Telling a child of the facts of life and marriage was..one of the tasks that the family..had abdicated to the schools. **1967** *Boston Sunday Globe* 23 Apr. 20/3 The conservationists' warning that we will bury ourselves in our own trash will become a fact of life. **1969** Auden *City without Walls* 71 Mother, tongue-tied with shyness, Struggling to tell him The Facts of Life he dared not Tell her he knew already.

6. c. *the fact of the matter* (example).

1852 C. M. Yonge *Two Guardians* vi. 101 This is the fact of the matter, as Mrs. Cornthwayte would say.

d. Other phrases of assertion or rejoinder: (*and*) *that's a fact* (orig. *U.S.*): an emphatic addition to a statement stressing its truth; also *this is a fact, that's the fact; is that a fact?*: is that so? (used esp. as a rejoinder (expecting no answer) to a statement).

1779 F. Burney *Diary & Lett.* Jan. (1842) I. 163, I know of nobody else that calls me so. This is a fact, Susy. **1834** *Boston* (Mass.) *Post* 5 Aug. 7/2 To this statement of facts he replied—'I was groggy, I *know*; but what I *did*, I *don't* know, that's a fact.' **1844** Dickens *Mart. Chuz.* xxi. 259 'It may not be so easy to do it.' 'And that's a fact,' said a voice..close in his ear. **1851** Mayhew *London Lab.* I. 417/1 Of the *maimed beggars*, some are really deserving objects, as without begging they must starve to death; that's a fact, sir. **1860** *Cornhill Mag.* II. 254 There is old Dr. Squaretoso (he certainly was very rude to me, and that's the fact). **1899** E. W. Hornung *Amat. Cracksman* 202 'My name is Raffles, and we met at Milchester last year.' 'Is that a fact?' cried the Scotchman. **1909** 'I. Hay' *Man's Man* viii. 135 'We're thinking of staying here.'..'Is that a fact?..Weel, I'll bide too.' **1914** Wodehouse *Man Upstairs* 133 A man I met to-day told me you were engaged. Is that a fact? **1923** R. Macaulay *Told by Idiot* II. xxi. 140 Bicycle bolts are a back number, and that's a fact. **1933** M. Lowry *Ultramarine* III. 162 Can't bloody move and that's a fact. **1960** D. Lessing *In Pursuit of English* VI. 215 There is a lot of money to be made out of the libel law. That's a fact. **1962** N. Marsh *Hand in Glove* iv. 110 'It doesn't explain,' Alleyn said, 'why the wick in the lantern's been turned hard off, does it?' 'Is that a fact!' Raikes remarked, primly.

e. Sometimes with exclamation mark: used as an emphatic assertion of the truth of a statement.

1819 Byron *Don Juan* 115, Note 4, page 25 *They only add them all in an appendix.* Fact. There is, or was, such an edition, with all the obnoxious epigrams of Martial placed by themselves at the end. **1848** J. R. Lowell *Biglow P.* 1st Ser. 4 Fact! it takes a sight o' cotton To stuff out a soger's chest. **1899** R. Whiteing *No. 5 John St.* xxi. 213 'Garn!' 'Fack. It was like this.' **1907** Wodehouse *White Feather* viii. 88 We are, really. Fact. **1964** J. Drummond *Welcome, Proud Lady* xvii. 81 'You astonish me.' 'Fact.'

8. *attrib.* and *Comb.*, as *fact-fetishism, -fetishist* sbs.; *fact-bound, -crammed* adjs.; *fact-collecting, -cramming* vbl. sbs.; *fact-gathering* vbl. sb. and ppl. adj.; **fact-finding** *ppl. a.*, that finds out facts; esp. descriptive of a committee, commission, etc., set up to discover and establish the facts of any matter; also as *vbl. sb.*, the work involved in such a process; hence (as a back-formation) **fact-find** *v. intr.*; also *fact-finder*; **fact-proof** *a.*, impervious to facts; **fact-sheet**, a paper on which facts relevant to a particular issue are set out briefly and clearly.

1959 *Encounter* Sept. 14/2 Their determination to stay precise and fact-bound at all costs. **1937** V. Woolf *Years* 380 I'm good, she thought, at fact-collecting. **1894** *Westm. Gaz.* 4 Apr. 3/1 It was a clever fact-crammed speech. **1907** *Daily Chron.* 16 Jan. 3/2 A fact-crammed encyclopædia. **1933** Dylan Thomas *Let.* Sept. (1966) 25 You've got more in your little finger than they have in the whole of their fact-crammed brains. **1876** C. M. Yonge *Womankind* vi. 49 The contact with a really powerful thinking mind..ought not to be sacrificed to mere fact-cramming. **1964** K. Winetrout in I. L. Horowitz *New Sociology* 149 We wind up with fact-fetishism, with a 'social science of the narrow focus, the trivial detail, the abstracted almighty unimportant fact'. **1960** *Spectator* 7 Oct. 527 The book meets the demands of the most hardbitten fact-fetishist. **1969** J. Mander *Static Soc.* ix. 319 Fact-fetishists as they are, both authors belong to the..spirit of the Enlightenment. **1953** *N.Y. Times* 8 Feb. E1/4 Ostensibly their mission was to fact-find on the problem of speeding the rearmament of Europe. **1927** P. H. de Kruif *Microbe Hunters* ii. 67 The spirit of the searcher, the fact finder, flashed out of his eye. **1909** G. B. Shaw *John Bull's Other Island* p. x, The actual distinction between the idolatrous Englishman and the fact-finding Irishman. **1926** B. Webb *My Apprenticeship* vi. 287 My super-enthusiasm for fact-finding. **1927** *N.Y. Times* 18 July 41/1 A fact-finding committee under the leadership of William P. MacCracken Jr., Assistant Secretary for Aeronautics. **1958** *Punch* 16 July 72/2 Sociologically, of course, we are up against the fact-finders. Fact-finding is to estimating as photography is to painting in oils. **1959** *Times* 24 Sept. 15/1 Account of a fact-finding tour of eighteen months spent in New China. **1969** *New Yorker* 31 May 34/2 A well-known congressman..had gone there..on a fact-finding trip. **1943** *Amer. Antiquity* IX. 208 American archaeology is still in its intense historical, fact-gathering stage. **1958** T. Landau *Encycl. Librarianship* 118/1 His fact-gathering and his several publications about libraries. **1961** J. Wilson *Reason & Morals* iii. 185 A subject [*sc.* philosophy] which is really a kind of conceptual psychoanalysis, and not a fact-gathering subject. **1909** G. B. Shaw *John Bull's Other Island* p. ix, He is never quite the hysterical, nonsense-crammed, fact-proof, truth-terrified, unballasted sport of all the bogy panics..that now calls itself 'God's Englishman'. **1959** *Times Lit. Suppl.* 2 Jan. 11/4 This is partly what the Americans call a fact-sheet on the military strengths of America and Russia. **1969** *Guardian* 23 Oct. 3/2 The distribution in Congress last night of a 'fact sheet' outlining the..steps taken by the Nixon Administration.

factable (fæ·ktăb'l). *Archit.* Also **fa·ctabling.** [Alteration of Fractable *sb.*] = Coping *sb.* 1, Fractable *sb.*

Delete from 'List of Spurious Words'.

1819 P. Nicholson *Archit. Dict.* 290/2 Coping upon the gable end of a house, is called *factabling* in Liverpool. **1842** Gwilt *Archit.* Gloss. 971 *Factabling*, the same as Coping. **1904** W. E. R. Windsor *Let.* (to Clarendon Press), Factable. **1940** *Chambers's Techn. Dict.* 320/1 *Factable*, a coping.

factful, *a.* (Later examples.)

1910 *Daily Chron.* 15 Jan. 3/1 Mr Hyatt has given us the most factful and clear-visioned record of the founding of Rhodesia. **1930** H. Williamson *Village Bk.* 38 Most of them are naturally factful where their money is not concerned.

fa·cthood. *rare.* [f. Fact + -hood.] The summation of all that has really occurred or is actually the case.

1945 C. S. Lewis *Great Divorce* v. 42, I will bring you to Eternal Fact, the Father of all other facthood. **1947** —— *Miracles* xi. 107 He is the opaque centre of all existences, the thing that simply and entirely *is*, the fountain of facthood.

factice (fæ·ktis). Formerly also **factis.** [ad. G. *factis, faktis*, f. L. *factic-ius* artificial (f. *facere* to make).] Any of several friable rubber-like substances made by vulcanizing unsaturated vegetable oils and used chiefly as compounding ingredients with rubber. Also called (*rubber*) *substitute*, and distinguished as *brown* or *white*.

1896 *Jrnl. Chem. Soc.* LXX. I. 204 Brown '*factis*' was next tried. A sample containing 15·48 per cent of sulphur, gave..acids containing 14·14 per cent of sulphur. **1904** W. T. Brannt tr. *Bersch's Cellulose & Artificial Rubber* xiv. 322 Under the name *factis*, masses have for some time been..used as a direct substitute for rubber... Factis masses vary very much in appearance. **1912** *Chem. Abstr.* 937 White factice (substitute) is made from rape-seed oil, linseed oil, or cottonseed oil. **1954** Dinsmore & Juve in G. S. Whitby *Synthetic Rubber* xi. 412 The factices..decrease plasticity but somewhat improve extrusion behavior.

facti·city. [f. Fact *sb.* + -icity.] The quality or condition of being a fact; factuality.

1945 W. Ebenstein *Pure Theory of Law* iii. 114 What is the relation between..the law's normativity..and its 'facticity', that is, the efficacy of the idea of the law? **1956** *Scott. Jrnl. Theol.* IX. 310 [Man's] fall is..from a knowledge of himself as a child, not of chance and facticity, but of God; all which is to say, his fall is into *sin* which is a term not in the existentialist vocabulary. **1960** H. Read *Forms of Things Unknown* IV. xi. 183 Suddenly in the excitement of the battle the soldier loses his self, his 'facticity', his consciousness of himself as a human being, and becomes an anonymous machine. **1970** J. Donceel tr. *Rahner's Trinity* III. 35 It accepts the incarnation and the descent of the Spirit as two facticities connected by a rather extrinsic bond. *Ibid.*, Such mere facticity makes the incarnation smack of mythology.

factionalism (fæ·kʃənăliz'm). [f. Factional *a.* + -ism.] A condition characterized by faction; tendency to factional differences; the factional spirit.

1904 *Nation* (N.Y.) 31 Mar. 244/1 The natural remedy for..such fierce factionalism as the Republicans are now displaying in office, is to turn them out. **1922** *Glasgow Herald* 21 Feb. 8 Italy..suffers in her political life from the ineradicable vice of factionalism. **1926** *Contemp. Rev.* Nov. 594 The provinces would be less subject to civil factionalism if the incentive to obtain control of Peking were absent. **1955** *Times* 23 Aug. 7/5 Whether Signor Fanfani's pleas for less factionalism will succeed in reuniting the party will not be clear for some weeks. **1970** *Daily Tel.* 23 July 6 An anthology refreshingly free both of personal prejudice and that factionalism with which Scottish life and letters abounds.

b. *spec.* In Communist usage.

1935 E. Burns *Handbk. Marxism* 835 The Party is synonymous with unity of will, which leaves no room for any factionalism or division of authority in the Party. **1954** Koestler *Invisible Writing* ii. 25 'Factionalism'—the formation of groups with a policy of their own—was a capital crime in the Party.

fa·ctionalist, *a.* [f. Factional *a.* + -ist.] = Factional *a.*

1958 *Listener* 13 Feb. 266/2 Narrow factionalist issues which concentrate on caste, religion, and education. *Ibid.* 267/1 Instead of nationalist enthusiasm..one sees only factionalist quarrels.

factionally (fæ·kʃənăli), *adv.* [f. Factional *a.* + -ly[2].] By means of faction.

1893 *Columbus Dispatch* 16 Sept., It will be their purpose to not factionally prolong the session. **1953** *N.Y. Times* 22 Feb. IV. E 3 Many Democrats who long have been factionally embroiled.

factis, var. *factice.

factitive, *sb.* [f. the adj.] A factitive verb.

1889 in *Cent. Dict.* **1934** Priebsch & Collinson *German Lang.* II. iii. 230 It [sc. *ver*] further forms factitives from adjectives, e.g. *verbittern, versüssen*.

factor, *sb.* Add:

6. *Math.* Used also of groups, rings, and other sets that can be produced from a given group, etc., by an analogue of division. So **factor group*, etc. (see *9).

1906 W. H. & G. C. Young *Theory Sets of Points* iv. 37 If the sets G_1, G_2,..are all equivalent, and γ be their number,..or their potency in the general case, the equation $g = \gamma g_1$ is substituted for the preceding equation [sc. $g = g_1 + g_2 + \ldots$]; g is then called the product of the factors γ and g_1. **1965** J. J. Rotman *Theory of Groups* iv. 58 If $G = H \times K$, H is called a direct factor of G; in additive notation, H is a direct summand of G.

7. *factor cost* (see quot. 1941).

1941 *Economist* 12 Apr. 489/2 There are two methods of valuation: the first at *factor cost* (that is, at what it costs to produce this output), the second at *market prices* (that is, at cost plus the indirect taxes and rates that have to be added in before the output can be sold). **1956** F. Lafitte in A. Pryce-Jones *New Outl. Mod. Knowl.* 579 Factor-cost measurement gives a true indication of the amount of real economic resources appropriated by State 'consumption' services. **1971** *E. Afr. Jrnl.* Mar. 17/1 In 1963, agriculture, forestry and fishing contributed 8·2% of gross domestic product at factor cost.

b. *Genetics.* A gene or other agent that is transmitted from parent to offspring and influences or determines a hereditary character; now *esp.* an agent that has not been shown to have the precise attributes usu. implied by the word 'gene'.

The sense 'gene' continues esp. in contexts that are elementary or have allusion to the historical development of genetics. For a discussion of the uses of G. *faktor*, *factor* by Gaertner 1849, Mendel 1865, and De Vries 1889, see R. C. Olby *Origins of Mendelism* (1966); the use in English prob. arose naturally out of sense 7 rather than as a consequence of the German.

1901 K. Pearson in *Phil. Trans. R. Soc.* A. CXCVII. 287 Differentiation, whether due to function, position on the individual, season of production, &c., is usually connected with one or two well-marked dominating factors. **1905** Bateson & Gregory in *Proc. R. Soc.* B. LXXVI. 581 The mode of appearance of a new type [of *Primula*] by the recombination of the factors brought in by the pure parental type. **1906** W. Bateson et al. in *Rep. Evol. Comm. R. Soc.* III. 3 The appearance of colour is due to the association, in one zygote, of two factors belonging to independent allelomorphic pairs. **1907** R. C. Punnett *Mendelism* (ed. 2) 23 Every zygotic cell, being formed by the fusion of two gametes, is a double structure, and contains two factors belonging to any given pair of characters. **1910** L. Doncaster *Heredity* 136 In Mendelian inheritance, the hereditary determinant of a particular character is spoken of as the factor for that character. **1920** *Amer. Naturalist* LIV. 97 (*heading*) Are the factors of heredity arranged in a line? **1927** Haldane & Huxley *Anim. Biol.* ii. 62 These units are called factors of heredity, or sometimes still more shortly the *genes*. **1949** Darlington & Mather *Elem. Genetics* ii. 40 We must then say that the two alternative determinants or factors of each kind in a zygote may be alike or different. **1958** Brocklehurst & Ward *Gen. School Biol.* xxii. 303 The alternative characters, round or wrinkled cotyledons and yellow or green cotyledons, are caused to develop by factors of two allelomorphic pairs. **1961** U. Mittwoch tr. *Hadorn's Developmental Genetics* i. 1 Not all hereditary factors which obey Mendelian laws should be described as genes, and this applies particularly to lethal factors. A large number of these are due to a loss of more or less extensive pieces of chromosome.., and such deficiencies are also inherited in a Mendelian manner. **1964** D. Wilkie *Cytoplasm in Heredity* i. 1 Non-mendelian patterns of inheritance are often interpreted as demonstrating genetic control by cytoplasmic factors.

c. *Photogr.* In full *filter factor*. (See quot. 1921.)

1909 C. E. K. Mees *Photogr. Coloured Objects* iii. 19 If we wish to find the multiplying factor of this filter, we must consider it in relation to the sensitiveness curve of the plate. **1912** E. J. Wall *Dict. Photogr.* (ed. 9) 338 It is very important to know the number of times the exposure is to be increased; it is not a constant factor..but varies according to the light. **1921** *Photogr. Jrnl.* Apr. 168 A filter factor is the ratio of the exposure required with a given filter to the exposure without the filter. **1961** G. Millerson *Technique Telev. Production* iii. 40 Any filter introduces some light loss, and the working aperture has to be increased, or lighting augmented to compensate. This light attenuation is known as the filter's *factor*. A 'times four' (×4) filter..requires four times the normal exposure.

8. *factor of safety*: the ratio between the load which a structure or material is capable of supporting and the load which it is required to support, or between the stress which causes it to break and the stress which it is required to stand; also, the ratio between the elastic limit of the structure and the usual load. Cf. Safety 6.

1858 [see Safety 6]. **1866** W. J. M. Rankine *Usef. Rules* 205 Factors of safety for perfect materials and workmanship...Dead Load 2. Live Load 4. **1876** *Encycl. Brit.* IV. 296/2 In designing a girder the load which it will have to carry is multiplied by a number called the factor of safety, varying from 3 to 6. **1877** W. H. White *Man. Naval Archit.* 377 Supposing a factor of safety of 8 to be taken instead of 10, the safe working load..for timber subject to compressive strains would be about three-eighths of a ton per square inch. **1891** [see Safety 6]. **1910** *Hawkins' Electr. Dict.*, *Factor of safety*. 1. A term expressive of a determined limit to which materials or machines shall be subjected; the safety limit. 2. In wiring, the ratio between the breaking stress of a wire and the maximum tension to which it is subjected in overhead suspension. **1962** *B.S.I. News* Aug. 20/1 The factor of safety is really a factor of ignorance. The factor of safety becomes less as we learn more about properties and behaviour of materials. **1970** H. Braun *Parish Churches* x. 132 Their buildings were massive and employed a wide factor of safety to achieve moderate heights.

b. *factor of merit* (Electr. Engin.): a measure of the sensitivity of a reflecting galvanometer (see quot. 1922). Cf. *figure of merit* (*figure *sb.* 20 c).

1890 W. E. Ayrton et al. in *Phil. Mag.* XXX. 84 An important..question to decide is whether the factor of merit of a galvanometer should be stated for a constant periodic time of oscillation..or for a constant controlling-field. **1922** Glazebrook *Dict. Appl. Physics* II. 375/1 The accepted period being 10 seconds, it follows that for any galvanometer the Factor of Merit is 100 $D/T^2 (R_G)^{1/2}$ or 100 $D/T^2 (R_G)^{1/2}$, T being the period in seconds, R_G the resistance of the galvanometer, and D the deflection in millimetres at 1 metre.

9. *Comb.* **factor analysis** *Statistics*, a mathematical technique for calculating the relative importance of each of a set of factors that together are assumed to influence some observed set of values or properties; **factor group** *Math.* [tr. G. *factorgruppe* (O. Hölder, 1889)], a group G/H the elements of which are the cosets in a given group G of a normal subgroup H of G; **factor law** = **factor theorem*; **factor theorem** *Algebra*, the theorem that if $f(x)$ is a polynomial and $f(a) = 0$, then $(x - a)$ is a factor of $f(x)$, and conversely.

1931 L. L. Thurstone in *Psychol. Rev.* XXXVIII. 406 It is the purpose of this paper to describe a more generally applicable method of factor analysis which has no restrictions as regards group factors and which does not restrict the number of general factors that are operative in producing the correlations. **1938** *Mind* XLVII. 102 The results of a factor-analysis of the abilities involved, are..to be investigated. **1946** H. Cramér *Math. Methods Statistics* xxxvii. 555 In the psychological factor analysis of human ability, the variables $x_1, .., x_m$ represent the measurements of m given different variables of a person, while $u_1, .., u_n$ are more or less 'general' factors of intelligence, and $v_1, .., v_m$ are 'specific' factors, each associated with a particular ability. **1969** *Computers & Humanities* III. 145 While there are several varieties of factor analysis, each of which is designed to handle a special type of problem, they all involve rather elaborate mathematical operations. **1897** W. Burnside *Theory Groups Finite Order* vii. 119 If a group has two different composition-series, the number of terms in them is the same and the factor-groups derived from them are identical except as regards the sequence in which they occur. **1965** Patterson & Rutherford *Elem. Abstract Algebra* ii. 49 If *G* is a finite group of order *m* and *H* is a normal subgroup of order *n*, then..the factor group *G*/*H* is a finite group of order *m*/*n*. **1901** W. J. Milne *Acad. Algebra* 104 Factor Law.—If a rational integral expression containing *x* reduces to zero when *a* is substituted for *x*, it is exactly divisible by *x*−*a*. **1908** W. J. Milne *Stand. Algebra* 106 (*heading*) Factor Theorem. **1928** Palmer & Miser *Coll. Algebra* ii. 31 The factor theorem assists in factoring by giving a quick way for testing a possible factor. **1966** J. V. Robison *Mod. Algebra & Trig.* xi. 242 (*heading*) Remainder and factor theorems.

factor, *v.* Restrict *rare* to senses in Dict. and add: **2.** *Math.* To resolve (a quantity) *into* factors; to express as a product of factors; = Factorize *v.* 2. orig. *U.S.*
1848 J. Ray *Algebra* I. ii. 73 The principal use of factoring, is to shorten the work, and simplify the results of algebraic operations. **1859** C. Davies *New Elem. Algebra* iii. 79 Factor $30a^4b^2c-6a^2b^3d^2+18a^2b^2c^2$. *Ans.* $6a^2b^2(5a^2c-bd^2+3c^2)$. **1901** W. J. Milne *Acad. Algebra* 93 From any expression that is to be factored, the monomial factors should usually first be removed. **1941** Birkhoff & MacLane *Surv. Mod. Algebra* iv. 88 The polynomial x^4+1..cannot be factored into polynomials with rational coefficients. **1947** F. S. Nowlan *Coll. Algebra* iii. 31 We factor 6 and 72 when we write $6 = 2\times 3$ and $72 = 2^3\times 3^2$. **1966** J. V. Robison *Mod. Algebra & Trig.* v. 109 To factor a polynomial, we express the polynomial as a product of other polynomials, called factors of the given polynomial.
b. *transf.* To resolve or decompose *into* components.
1958 G. A. Miller in *Jrnl. Exper. Psychol.* LVI. 486/2 It is possible to show that such a finite state generator can be decomposed, or 'factored', into a set of simpler generators.

factorable (fæ·ktŏrăb'l), *a. Math.* [f. prec. + -able.] Capable of being factorized; expressible as a product of factors.
1931 L. J. Rouse *Coll. Algebra* iii. 37 If the trinomial is factorable we must find two numbers whose product is *ac* and whose sum is *b*. **1966** Ogilvy & Anderson *Excurs. Number Theory* 146, 2 and 47 are examples of primes; 20 is composite, being factorable into 2×2×5.

factorial, *a.*[1] Add: **3.** Of or pertaining to a factor (sense 7); *factorial analysis* = *factor analysis*.
Quot. 1845 refers to analysis by means of factorials.
[**1845** T. Tate (*title*) A treatise on factorial analysis with the summation of series; containing various new developments of functions, &c.] **1935** *Human Factor* IX. 182 With this infinite number of possible factorial analyses of any set of correlations,..a psychologist can find one which agrees with any preconceived notions he may have about the linkages between tests. **1935** R. A. Fisher *Design Exper.* vi. 111 With factorial experiments designed to make a large number of comparisons there will..usually be an ample number of degrees of freedom for the estimation of error. **1949** Darlington & Mather *Elem. Genetics* App. I. 391 *Factorial Experiment*, one in which all the treatments or agents under investigation are varied simultaneously, and combined in such a way that any derived effect of one or a group of them may be isolated and separately evaluated. **1950** Cochran & Cox *Exper. Designs* v. 124 If all factors are independent in their effects, the factorial approach will result in a considerable saving of the time and material devoted to the experiments. **1971** *Jrnl. Gen. Psychol.* LXXXII. 53 A factorial analysis of covariance was performed on the data for delayed recall.
4. *Photogr.* Of or pertaining to the determination of the interval of time allowed after the appearance of the image and before the completion of the development of a film.
1910 *Westm. Gaz.* 29 June 16/2 If one be using a factorial system of development it is well to reduce the factor a little, using, say, seven where nine would represent the ordinary time of development.

factorially (fæktōə·riăli), *adv.* [f. Factorial *a.*[1] + -ly[2].] By reference to factors.
1913 W. Bateson *Mendel's Princ. Heredity* (ed. 2) iv. 84 The genetic properties of the pied types can be represented factorially by regarding the pattern or distribution of the colour as due to a distinct factor or to its absence. **1935** R. A. Fisher *Design Exper.* vi. 104 The second advantage ..in a factorially arranged experiment, is that, in addition to measuring the effects of the four single ingredients..it measures also the 11 possible interactions between these ingredients. **1935** L. L. Thurstone *Vectors of Mind* ii. 73 The problem of describing factorially the variables whose experimental intercorrelations are given.

factorist (fæ·ktŏrist). *Psychol.* [f. Factor *sb.* + -ist.] One who seeks to explain intelligence as a measurable phenomenon in terms of the relations of general and specific factors in a person.
1930 C. Spearman in C. Murchison *Psychologies of 1930* xviii. 339 Instead of such negations the factorists find good in everything, even in the other 'ists'. **1938** *Brit. Jrnl. Psychol.* Oct. 190 The doctrine urged by the Samplists turns out to be no other than that which has always constituted the 'main' theory of the Factorists themselves.

factorization (fæ:ktŏrəizēⁱ·ʃən). *Math.* [f. Factorize *v.* + -ation.] The operation of resolving a quantity into factors; also, the product of factors so obtained.
1886 G. Chrystal *Algebra* I. vii. 122 Every known identity resulting from the distribution of a product of such factors, when read backwards, gives a factorisation. **1926** Milne & Robertson *Algebra for Schools* xiv. 89 The pupil is strongly advised to check his answer to any question on factorisation by multiplying the factors..to see whether their product gives the original expression. **1967** W. P. Eames *Elem. Theory Numbers* iv. 117 Now we will show that the factorization of *f* is unique.

factorize, *v.* **2.** (Examples.)
1886 G. Chrystal *Algebra* I. vii. 133 The decomposition of

$$x^2+y^2+xy = (x+y+\sqrt{xy})(x+y-\sqrt{xy}),$$

which is often given by beginners when they are asked to factorise x^2+y^2+xy..is no solution of the problem of factorisation in the ordinary sense, inasmuch as the two factors contain $\sqrt{xy}$, and are therefore not rational integral functions of *x* and *y*. **1923** E. H. Chapman *Elem. Algebra* xiii. 103 When we attempt to factorize an algebraic expression, we regard it as a product, and try to undo the multiplication. **1967** W. P. Eames *Elem. Theory Numbers* iv. 117 Since *f* is a product of polynomials which can be factorized, *f* can also be factorized.
b. *intr.* To be capable of being factorized.
1914 A. G. Cracknell *School Algebra* xv. 199 Thus $3y(2x+a)-b(2y+c)$ does not factorise. **1939** Godfrey & Tait *School Algebra* viii. 100 Quadratic equations will not always factorize.

factory. Add: **5. c.** A prison; a police station. *slang.*
1832 in *Penguin Bk. Austral. Ballads* (1964) 33 But the lass I adore, the lass for me, Is a lass in the Female Factory. **1874** M. Clarke *His Natural Life* II. iii. 91 In the factory—a prison for females—the vilest abuses were committed. **1891** F. W. Carew *No. 747* xxxvi. 426 A stranger..whom a plain-clothes D. from the 'Factory' would most assuredly have catalogued as suspicious. **1938** F. D. Sharpe *Sharpe of Flying Squad* 330 *Factory*, the police-station.
6. *factory bill, boy, butter* (U.S.), *child, chimney, -girl* (earlier and later examples), *hooter, lad, owner* (so *factory-owning* ppl. adj.), *system, whistle, worker, world*; *factory-made* pple. (and as *adj.* and *sb.*); **factory cloth** = *factory-cotton*; **factory farm** orig. *U.S.*, a farm organized on industrial lines; hence **factory farmer, -farming; factory floor,** the work-place of, or a forum for, industrial workers; also *attrib.*; **factory ship,** the base ship of a whaling fleet; **factory trawler** (see quot. 1962).
1832 *Rep. Comm. Lab. Children* in *Parl. Papers 1831–2* XV. 20 But since this Factory Bill has been agitated,..the children have gathered round me..and have said..'Will you get the Ten Hour Bill?' **1833** *Blackw. Edin. Mag.* XXXIII. 443/1 Mr. John Wood..of whom the Rev. G. S. Bull..thus spoke a few days ago at a great Factory Bill meeting. *Ibid.* 444/2 He is perhaps a better judge of fat cattle..than of lean Factory boys and girls. **1840** F. Trollope (*title*) Life and adventures of Michael Armstrong, the factory boy. **1888** *Vermont Agric. Rep.* X. 14 Factory butter secures an average higher price than its patrons could secure if they made it at home. **1832** *Rep. Comm. Lab. Children* in *Parl. Papers 1831–2* XV. 640 Education of factory children in England far behind Scotland. **1833** *Blackw. Edin. Mag.* XXXIII. 427 A Factory child..must be at her work..at four o'clock of a snowy winter-morning. **1848** Mrs. Gaskell *Mary Barton* II. ix. 131 She..looked towards the factory-chimneys, and the cloud of smoke which hovers over Manchester. **1933** V. Sackville-West *Let.* 31 Jan. in *Lett. H. Nicolson* (1966) 136 Factory-chimneys line the cliff. **1873** M. Holley *My Opinions* 113 Merrymac calico and factory cloth. **1890** A. Marshall *Princ. Econ.* I. IV. xi. 351 Our knowledge..would be much increased..if some private persons,..or co-operative associations, would make a few careful experiments of what have been called 'Factory farms'. **1926** *19th Cent.* June 825 Factory-farms..can be multiplied or spread widely enough to affect the whole of British agriculture. **1952** *Economist* 7 June 657/1 The operators of the huge western factory farms..resist the rule that no more than 160 acres of a single owner's land can be supplied with federally financed water. *Ibid.*, This so-called '160-acre limitation'..is still applied when new lands are irrigated, but factory farmers, particularly in Texas and California, have been trying.. to get Congress to repeal it. **1964** *New Statesman* 30 Oct. 649/1 Boycott factory farm food?..Boycott factory farmers?..The essential thing is to amend the Protection of Animals Act (1911) to cover factory-farming techniques. **1968** *Ibid.* 5 Jan. 10/3 Under conditions of intensive 'factory' farming, a lot of animals *did* suffer from true infections. **1968** M. Pyke *Food & Society* v. 63 It is fashionable to sneer at intensive methods of livestock production; they are called 'factory farming'. **1968** *Listener* 30 May 692/2 The militant workers on the factory floor have been voting to continue the strike until their leaders have wrung much more substantial concessions from the regime. **1971** P. Worsthorne *Socialist Myth* 252 Factory-floor agitation..is one of the few active pleasures which an industrial society provides for its less well-endowed members. **1832** D. Jerrold in *N. & Q.* (1963) Mar. 105/2 (*title*) The Factory Girl. **1833** S. Breck *Recoll.* (1877) App. 275 Factory-girls were introduced. **1940** 'G. Orwell' *Inside Whale* 122 The idea is to give the bored factory-girl or worn-out mother of five a dream-life. **1930** D. H. Lawrence *Triumph of Machine*, They will hear far, far away the last factory hooter. **1875** E. Towers in *N. & Q.* (1963) Apr. 137/1 (*title*) Poor Joe the Factory Lad. **1904** *Daily Chron.* 23 Aug. 5/4 The over-decorated factory-made furniture of Michigan and Ohio. *Ibid.* 28 Dec. 4/5 Cheapest line of factory-mades. **1905** *Ibid.* 13 June 7/2 Practically all shoes are factory-made in the United States. **1939** S. Spender tr. *Toller's Pastor Hall* III. 102, I wear factory-made shoes nowadays. Not so beautiful and elegant as yours, but cheaper. **1851** *Baird's Cotton Spinner* 31 It is based upon an error, though a rather common one amongst Factory owners. **1934** E. Pound *Eleven New Cantos* xxxiii. 13 That no factory-owner shall sit as a magistrate in cases concerning the spinning of cotton. **1949** Wyndham Lewis *Let.* 6 Aug. (1963) 502 The U.S. factory-owning class is very raw, arrogant, and ignorant. **1930** *N.Y. Zool. Soc. Bull.* Jan.–Feb. 7 (*caption*) The factory ship *Lansing*. **1958** *Times* 12 Nov. 11/6 The cost of building a factory ship today at about £4m. is almost prohibitive for any private whaling company. **1969** N. W. Pirie *Food Resources* v. 135 Larger vessels, such as the 43,000 ton *Vostok* which the Russians are building as a factory ship. **1832** *Rep. Comm. Lab. Children* in *Parl. Papers 1831–2* XV. 413 Will you state how you account for the melancholy result which you attribute to the factory system as at present conducted? **1843** *Ainsworth's Mag.* V. 42 The factory system, the most deplorable ignorance.., have all excited attention in their turns. **1953** *World Fishing* II. 10/1 (*caption*) United States factory trawler 'Ocean Life'. **1962** J. Tunstall *Fishermen* ii. 45 The factory trawler quick freezes its catch at sea. *Ibid.*, The first British factory trawler..started fishing in 1947. **1920** H. Crane *Let.* 9 Nov. (1965) 45, I am sure I should not miss factory whistles in Pisa or Morocco, but I frankly did miss them in Washington. **1848** Mrs. Gaskell *Mary Barton* I. p. vi, The poor uneducated factory-workers of Manchester. **1937** *Discovery* Aug. 234/2 An unskilled factory worker. **1906** *Daily Chron.* 28 May 3/4 Tired minds in a factory-world. **1930** D. H. Lawrence *Nettles* 26 Hooked fishes of the factory world.

factotum. Add: **1.** Also *fig.*
1940 J. H. Jagger *English in Future* ii. 31 *Have* is another factotum, but *gets* off more lightly than *get*.
2. Delete †*Obs.* and add later example.
1914 R. B. McKerrow in *Bibliogr. Soc. Trans. 1911–13* XII. 238 An ornamental block having a space in the centre for the insertion of a capital letter of an ordinary fount of type is called a 'factotum initial', or more properly a 'factotum'.

factualism (fæ·ktiuăliz'm). *Philos.* [f. Factual *a.* + -ism.] A predominant concern with facts or natural consequences, esp. in moral matters; any theory that treats facts as being of prime importance, or maintains that moral conclusions can be drawn from factual data alone.
1946 R. Wellek in W. S. Knickerbocker *XX Cent. Eng.* 68 A false and pernicious 'historicism' is frequently connected with 'factualism'. **1961** J. Wilson *Reason & Morals* ii. 42 The contrast between Moralism and Factualism seems to take the form of a conflict.

factualist (fæ·ktiuălist), *a.* and *sb. Philos.* [f. as prec. + -ist.] **A.** *adj.* Of or relating to factualism; naturalistic. **B.** *sb.* One who is affected by, or adheres to, factualism.
1935 P. Wheelwright *Crit. Introd. Ethics* iii. 91 The 'factualist' fallacy, of which the 'proof' of hedonism is a prominent example. **1961** J. Wilson *Reason & Morals* ii. 39 Moralists and Factualists seem to defend their beliefs in different ways...A Factualist would perhaps say 'Easy divorce does harm to the children'. *Ibid.* 44 The Factualist metaphysic..need not grind any such axe.

factuality (fæktiuæ·lĭti). [f. as prec. + -ity.] The quality of being factual, factualness; (of representation) realism; (of facts) truth.
1887 *Christian Union* 3 Mar. 18/2 When we find these among the facts, it makes us doubt the factuality of the facts. **1922** A. N. Whitehead *Princ. Relativ.* ii. 15 If for one moment I may use the inadmissible word 'Factuality', it is in some ways better either than 'fact' or 'totality' for the expression of my meaning. **1960** *Times* 15 Jan. 16/1 *The Kitchen*..begins as a study in a frightening factuality.

factualness (fæ·ktiuălnės). [f. Factual *a.* + -ness.] The state of being factual; factuality.
1906 F. von Hügel *Let.* 29 Sept. in Briggs & von Hügel *Papal Comm. & Pentat.* II. 29 The massive *factualness*, the serried, interdependent ranks of fact upon fact. **1925** *Public Opinion* 14 Aug. 153/3 Even the compassionate pity they arouse does not lift the weight of their seriousness and factualness from the reader's breast. **1953** *Ess. in Crit.* III. 31 A kind of Janus word, gathering from one direction its whole factualness. **1971** *Daily Tel.* 18 June 14/3 The play element..was handled with an innocence and factualness which are the exact opposite of 'charm', let alone archness.

facture. Restrict 'Now *rare*' to senses in Dict. and add: **4.** *Painting.* The quality of the execution of a picture, esp. of its surface. (A Gallicism.)
1887 *Atlantic Monthly* LX. 510/2 He was acquiring in the Louvre his laborious and rude facture of successive *impasto*. **1936** H. Read *Surrealism* 63 Dali's neat, tight Vermeerish *facture* has its aesthetic as well as Picasso's bold, plangent, viscous brushwork. **1961** *Times* 21 Mar. 16/5 He knows exactly what makes the *facture* of a French picture interesting. **1971** *Listener* 25 Feb. 254/3 Warhol's method depends..on misregistration and over-inking and (art facture-wise) smudging, lopping, blotching.

facty, *a.* Add: **b.** Of persons: freq. implying 'deficient in emotion or imagination'.
1901 *Westm. Gaz.* 18 June 10/1 Mr. Page, most interesting because most real and 'facty' of Horace's editors. **1930** H. G. Wells *Autocracy of Mr. Parham* II. i. 80 He was facty and explicit after the manner of his type. **1934** —— *Exper. Autobiogr.* II. viii. 540 To him [*sc.* G. B. Shaw].. I have always appeared heavily and sometimes formidably facty and close-set; to me his judgements, arrived at by feeling and expression, have always had a flimsiness.

facultative, *a.* Add **1. c.** *Biol.* Not restricted to the particular function, mode of life, set of conditions, etc., implied in the context; opp. Obligate *ppl. a.*
1887 H. E. F. Garnsey tr. *De Bary's Fungi* vii. 356 Saprophytes..which have also the power of going through their course of development wholly or in part as parasites..may be called with Van Tieghem facultative parasites. **1900** B. D. Jackson *Gloss. Bot. Terms* 98/1 *Facultative*, occasional, incidental, as opposed to *obligate*; ~ *Anaërobes*, organisms which can exist without the presence of free oxygen or air. ~ *Symbiont*, an organism which can either exist and reach maturity independently or in symbiosis with another. **1927** Gwynne-Vaughan & Barnes *Struct. & Devel. Fungi* 13 A species which is usually saprophytic but capable of parasitic existence on occasion is described as a hemi-saprophyte or facultative parasite, and a form which is usually parasitic but sometimes saprophytic as a hemi-parasite or facultative saprophyte. **1935** *Discovery* Jan. 27/2 The parasitic bacteria are a large and important group in which every stage of facultative and obligate parasitism can be found. **1940** *Chambers's Techn. Dict.* 320/2 *Facultative gamete*, a zoospore which can function as a gamete. **1950** J. G. Davis *Dict. Dairying* 13 The following table suggests terms for describing the reaction of micro-organisms to oxygen... Facultative anaerobe, Aerophilic..Facultative aerobe, Aerophobic. **1961** *Encounter* XVI. 77 The 'facultative homosexuals'..are those men who consider themselves heterosexually 'normal' but engage on occasion in some form of homosexual intercourse.

facultatively, *adv.* Delete *rare* and add later examples.
1900 [see *aerobe]. **1914** A. Harrison *Kaiser's War* 6 It [*sc.* military monarchism] has been set forth again and again by the German professors, politicians, and spokesmen, militarily, politically, and (as they say) facultatively. **1965** *Canad. Jrnl. Linguistics* Spring 172 The *qi*?- allomorph of the future marker occurs facultatively before the *yi*- allomorph.

faculty. **5.** Delete †*Obs.* and add later examples.
1889 *Cent. Dict.*, *Faculty*...6. In the law of divorce (commonly in the plural), the pecuniary ability of the husband, in view of both his property and his capacity to earn money, with reference to which the amount of the wife's alimony is fixed. **1894, 1965** [see *faculty theory*, sense 12 below].
9. b. The whole teaching staff of a college, university, or school. orig. and chiefly *U.S.*
1767 in J. Maclean *Hist. College N. Jersey* (1877) I. 292 Concurring with the Trustees of this College in the establishment and support of a Faculty. **1780** in *Docs. Revolutionary Hist. New Jersey* (1901–17) IV. 223 The trustees and Faculty are now exerting themselves with great diligence for the improvement of the seminary. **1843** *Yale Lit. Mag.* IX. 66 That was all I could ever get from him on the subject—'that the Faculty were funny fellows, very—had sent him off for laughing'. **1893** W. K. Post *Harvard Stories* 79 There are many classes and individuals..as firmly established..as the Faculty. **1902** 'G. M. Martin' *Emmy Lou* 264 Once one would have said with 'the teachers', but in the High School one knew them as the Faculty. **1967** *Boston Sunday Globe* 23 Apr. 3/1 Catholic University officials worked behind the scenes Saturday to find quick settlement of a student-faculty strike that virtually closed the school Thursday. **1969** *Listener* 12 June 833/1 Students and some faculty are compelling the American universities to painful self-scrutiny.
12. (sense *9 b) *faculty business, -full* (sb.), *list, meeting, member*; **faculty doctrine** = *faculty psychology*; **faculty psychology,** a term for those systems of psychology in which certain mental faculties were held to be the forces and powers accountable for the phenomena of mind; so *faculty psychologist*; **faculty tax** (example of survival in U.S.); **faculty theory,** the theory of taxation according to which every man should help to bear public burdens according to his ability.
1877 E. S. Ward *Story of Avis* v. 51 Some pressing Faculty business took him..to Professor Dobell's house. **1905** W. James *Mem. & Stud.* (1911) v. 83 In faculty-business he might not run well in harness, but..his influence on the students would be priceless. **1912** W. McDougall *Social Psychol.* (ed. 5) 378 All these doctrines ..are forms of the 'faculty doctrine' whose fallacies have so often been exposed. **1905** W. James *Mem. & Stud.* (1911) v. 84 In a university..a few undisciplinables.. may be infinitely more precious than a faculty-full of orderly routinists. **1911** H. S. Harrison *Queed* xviii. 218 The president sat up late going over his faculty list. **1839** H. Caswall *America* xii. 200 [The professors] form a body denominated the Faculty, and conduct the government of the institution by regulations and laws established by themselves in 'Faculty meetings' from time to time. **1911** H. S. Harrison *Queed* xviii. 218 There was one man on the staff that West objected to from the first faculty meeting. **1903** W. James *Mem. & Stud.* (1911) xiv. 331 The College had always gloried in a list of faculty members who bore the doctor's title. **1886** *Encycl. Brit.* XX. 41/1 To free us from the mythology and verbiage of the 'faculty-psychologists'. **1909** *Cent. Dict.* Suppl. s.v. *Psychology*, C. von Wolff (1679–1754) is regarded as the typical faculty-psychologist. **1890** W. James *Princ. Psychol.* I. ii. 27 Gall..took the faculty-psychology as his ultimatum on the mental side, and he made no farther psychological analysis. **1897** C. H. Judd tr. *Wundt's Outl. Psychol.* 11 The faculty-psychology considered these class-concepts as psychical forces or faculties, and referred psychical processes to their alternating or united activity. **1951** J. C. Flugel *Hundred Years Psychol.* (ed. 2) ii. 34 Beneke equally avoids the dangers of the faculty psychology, inasmuch as the powers or 'faculties' with which he deals are not broad characteristics or abilities,..but quite specific forms of apprehension, feeling or behaviour. **1911** E. R. A. Seligman *Income Tax* 398 The only other state in which the faculty tax lasted during the nineteenth century is South Carolina. **1894** —— *Progress. Taxation* iii. 127 The faculty theory of taxation is very old. That a man should contribute to the public burdens in proportion to his ability or faculty is a principle which dates back to the middle ages. **1896** C. C. Plehn *Introd. Public Finance* II. ii. 84 Each citizen should contribute as he is able. They claim that it is easier to measure ability than it is to measure benefit. This theory is called the faculty theory, the term 'faculty' having been found in this sense in early tax laws. **1965** J. L. Hanson *Dict. Econ. & Commerce* 169/1 *Faculty theory of taxation*, an alternative term for the 'Ability-to-pay' theory of taxation.

facundity (fākɒ·ndĭti). Delete †*Obs.* and add later example.
1921 *Times Lit. Suppl.* 17 Feb. 110/4 Two love sequences..have a fine poetic facundity, but that is all.

faddishly (fæ·diʃli), *adv.* [f. Faddish *a.* + -ly[2].] In a faddish or finical manner.
1928 *Collier's* 10 Nov. 35/2 The sufferer generally eats poorly and faddishly.

fade, *sb.*[1] Delete †*Obs.* and add later examples. Phr. *to do* (or *take*) *a fade* (slang), to disappear.
In quot. 1969 *nix out on the fade* in Surfers' slang = don't go away.
1918 A. Stringer *House of Intrigue* ii. 27, I got so I could face a tight fade without a quaver, and do my gay-cat part in sloughing our make as easily as falling off a log. **1949** A. Miller *Death of Salesman* I. 57 If I'm going to take a fade the boss can call any number where I'm supposed to be and they'll swear to him that I just left. **1962** K. Orvis *Damned & Destroyed* xiv. 92 Then pal, we'll both do a fade. **1965** *Language* XLI. 277 Stockwell suggests that the lesser fade at the juncture between the two 'main' conjuncts can be described as an environmentally conditioned allophone of terminal fade. **1969** *Times* 25 July 5/2 Nix out on the fade with it stashed on the moke. **1970** H. Waugh *Finish me Off* (1971) 86 If this is a sample of her business acumen, that beauty salon will do a fast fade.
2. *Cinematogr.* and *Broadcasting.* The action or an act of 'fading' (see *fade *v.*[1] 9); also, the gradual decrease or (freq. *fade-in*) increase in the brightness or definition of a picture or the loudness of a sound. See also *fade-out.
1918 H. Croy *How Motion Pictures are Made* vii. 175 It was in such experiments that the principle of fade was discovered, by means of which a scene could be made gradually to grow plainer until the full details were before the audience. This in photographic parlance came to be known as the fade-in. *Ibid.*, The second means of accomplishing a fade picture is by means of the dissolving shutter. *Ibid.*, The diaphragm fade is open to the objection that with an iris that never closes completely it is impossible to make a complete fade. **1922** L. C. Macbean *Kinemat. Studio Technique* ix. 71 The number of turns in which a 'mix' or 'fade' should be accomplished. **1937** *Printers Ink Monthly* Apr. 53/1 *Fade*, a diminishing of program volume. **1937** *Amer. Speech* XII. 101 To fade is usually the engineer's duty, reducing volume of an orchestra..while an announcer speaks into another microphone, hence a good or bad fade-in. **1960** D. Wilson *Television Playwright* 15 By means of 'mixes' and 'fades' short or long time-lapses can be established. **1962** A. Nisbett *Technique Sound Studio* ix. 153 When we talk about the fade, the first thing that springs to mind is the use to which it is put in dramatic productions: the convention is that each scene starts with a fade in, and ends with a fade out.
3. *Theatr.* The gradual brightening or dimming of the stage lighting; usu. *fade-in* or *fade-out.
1959 W. C. Lounsbury *Backstage from A to Z* 39 *Fade in*, gradual dim up of lights or sound. **1962** *Listener* 5 July 26/3 Pointless use of fade-in and fade-out lights on a stage confusingly and ineptly split into four parts.
4. The reduction in effectiveness of the braking system of a motor vehicle, e.g. as a result of the generation of heat.
1949 G. Grant *Modern Motorcars* vi. 47 On the average car the efficiency of the linings is not greatly impaired until the brake-drum temperature reaches a critical point, where there is a noticeable falling-off of braking power. This is known as 'brake-fade', and can be a very real problem on ultra-fast cars. **1959** *Motor* 4 Mar. 163/1 Unusual design of..brake drum..produces fade only after severe provocation. **1962** *Which? Car Suppl.* Oct. 141/1 Its brakes..were very susceptible to fade.
5. *Comb.*, as **fade-proof** *a.*, resistant to fading.
1909 *Public Ledger* (Philadelphia) 26 June 7/6 Pure worsted serge suits—guaranteed fade-proof. **1959** *Listener* 5 Mar. 435/1 They [*sc.* the fabrics] are rot-proof and fade-proof. **1961** *Ibid.* 30 Nov. 933/1 His jokes are appreciated for their topicality, which cannot be fade-proof.

fade, *v.*[1] Add: **6. b.** Also, to disappear from the scene; to depart; to faint. Freq. const. *away, out.*
1900 Ade *More Fables* (1902) 10 The Bookie told him to Back Up and Fade and do a Disappearing Specialty. **1910** 'I. Hay' in *Granta* 11 June 12 Next moment everybody seemed to fade away. **1924** A. J. Small *Frozen Gold* i. 39 When that gang fades out of one camp, I hear all about it inside twenty-four hours. **1929** J. B. Priestley *Good Companions* I. ii. 63 'My wife,' Mr. Rathbury muttered, fading out. **1950** *Times* 20 Sept. 2/6 Miller, Lindwall, and Johnson have faded out of the picture. **1954** 'N. Blake' *Whisper in Gloom* I. iv. 60 Copper and Foxy faded.
7. *intr.* Of sound: to die *away* or *out*. Also, with *in, up*, to increase gradually in loudness from a low or inaudible level.
1879 G. B. Prescott *Speaking Telephone* ix. 287 The voice increased or faded out in proportion as the telephone was directed toward or receded from the pole of the dipping needle. **1924** *Wireless Ann.* 21 Without a word of warning, the signal 'fades' to nothing. **1932** F. E. Terman *Radio Engin.* xiii. 488 As the signals fade in or out, the sensitivity of the receiver is varied. **1966** J. L. Bernstein *Audio Syst.* v. 122 Moving the control..causes the signal from one channel to fade out and the other to fade in. **1969** A. Glyn *Dragon Variation* ix. 282 'That's a sort of *bon voyage* present from Miami,' she said, when the orchestra finally faded out. **1971** *Daily Tel.* 30 July 13/3 'Welcome to your Sunday bumper edition of Radio Northwick Park.'..The music fades up and Paul closes the mike for an aside to fellow programme announcer Phil.
8. *slang.* (See quots. 1942, 1964.)
1890 *Dialect Notes* I. 61 *To fade*, to bet against the player shooting. [**1912** J. W. Johnson *Autobiogr. Ex-colored Man* vi. 92, I was soon 'fated'. I threw the dice—seven—I had won.] **1931** D. Runyon *Guys & Dolls* (1932) xiii. 273 Nathan Detroit's crap game is what is called..a fading game, because the guys bet against each other rather than against the bank, or house. **1934** J. T. Farrell *Young Manhood Studs Lonigan* (1936) xvii. 368 Weary frowned at the guy and faded ten of the fifteen. **1942** Berrey & Van den Bark *Amer. Thes. Slang* § 750/6 *Fade*, to cover the bet of the thrower [at dice]. **1957** W. C. Handy *Father of Blues* xvii. 233 Lovejoy faded him for twenty-five cents. **1964** A. Wykes *Gambling* vi. 141 To 'fade' is to accept a bet.
9. *Cinematogr. trans.* To cause (a picture, etc.) to pass gradually *in* or *out*, i.e. to appear or disappear on the screen.
1918 H. Croy *How Motion Pictures are Made* vii. 177 The fourth method of fading a picture is by means of a chemical process. *Ibid.*, This shutter device for fading out a picture may be operated automatically or by hand. **1918** V. O. Freeburg *Art of Photoplay Making* 122 The caption 'The Depths of Shame' is faded in. **1922** L. C. Macbean *Kinemat. Studio Technique* ix. 82 On occasions..it is necessary..to *fade* or *mix* titles into a scene to which they relate. **1958** R. H. Bomback *Handbk. Amat. Cinematogr.* II. 93 For some shots that have to be faded in or out you may be able to increase the camera speed so that each individual frame will receive less exposure. **1969** P. Petzold *All-in-One Cine Bk.* 27 If you can fade a scene this is the ideal answer to show the passing of time.
b. *transf.* Of radio and television transmission. Also with *up, down*.
1927 *Observer* 11 Sept. 21 Why can't we have the crowd noises faded in? If it is properly handled it won't interfere with the man who is reading the race. *Ibid.* 30 Oct. 26 It was faded out in the middle of Mr. Chesterton's speech. **1937** [see *fade *sb.*[1] 2]. **1937** *Discovery* Nov. 330/2 The producer of the programme making his selection of view by fading down the signal from one camera and fading up the signal from another before passing the video signal to the vision control engineer. **1958** G. Barker *Two Plays* 22 Music..fade in Peter's voice. **1958** *Spectator* 20 June 819/2 She had to be hurriedly faded out. **1971** *Radio Times* 7 Oct. 70/2 At the end of *Choral Evensong* (Radio 3) the organist is nearly always 'faded out'.
10. *intr.* Of the brakes of a motor vehicle (see *fade *sb.*[1] 4).
1940 *Jrnl. Research Nat. Bureau of Standards* May 543 A lining is said to fade when the coefficient of friction decreases rapidly to a low value. **1962** *Which? Car Suppl.* Oct. 120/2 The brakes still faded very badly. **1970** *New Yorker* 12 Sept. 113/2 (Advt.), Those..front disc brakes..right in the car's airstream to help them cool. So you can keep cool and not worry about them fading.
Hence **fa·der**, an apparatus for controlling (*a*) the volume of sound in a cinematographic film, (*b*) the signal in sound or television broadcasting.
1931 B. Brown *Talking Pictures* v. 102 The simplest type of fader.., where a variable resistance or potentiometer is placed across the output leads to the main amplifier. **1949** *Electronic Engin.* XXI. 354 The equipment consists of the recording machine and amplifier, microphones and fader unit. **1957** *Encycl. Brit.* IV. 218/2 In..sound broadcasting, the amplified current from each microphone is connected to a fader. **1969** W. Rutherford *Gallows Set* vi. 77 Terry said, 'Fade grams. And—Take One.'..The girl at the vision mixer pulled the fader down.

fade-away. *U.S.* [Fade *v.*[1] 6.] **1.** *Baseball.* (See quot. 1961.)
1909 *Amer. Mag.* May 31/1 Matty dropped his famous 'fade away' over the plate, and Tinker drove a long, high, line fly to left center. **1931** D. Runyon *Guys & Dolls* (1932) x. 224, I..cut loose with the old fadeaway. **1961** J. S. Salak *Dict. Amer. Sports* 150 *Fadeaway*, a pitched ball that tends to slow up and drop as it reaches the plate.

2. An act of disappearance.
1911 H. S. Harrison *Queed* v. 56 She had only pretended to die in order to make a fade-away with the gate receipts. **1912** C. Mathewson *Pitching in a Pinch* i. 11 Many persons have asked me why I do not use my 'fade-away' oftener when it is so effective.

Fade-Ometer (fēⁱdǫ·mitəɹ). Also **Fadeometer, Fadometer.** [Trade name, f. Fade *sb.*[1] or *v.*[1]: see -ometer.] (See quot. 1925.)
1925 *Trade Marks Jrnl.* 5 Aug. 1700 *Color Fade-Ometer*, ..apparatus..for testing the fastness of colours. Atlas Electric Devices Company..Chicago, Illinois. **1934** *Chem. Abstr.* 7021/1 Comparison of the fading produced by the Fade-ometer..and by sunlight... For short periods of time the Fade-ometer is a satisfactory substitute for sunlight in fading tests. **1940** *Chambers's Techn. Dict.* 320/2 *Fadometer*, an instrument used to determine the resistance of a dye or pigment to fading. **1951** R. Mayer *Artist's Handbk.* ii. 90 Machines such as the Fadeometer and the Weatherometer have been designed to give the equivalent of years of exposure to direct sunlight and dampness in a few hours.

fa·de-out. [*Fade *v.*[1] 6 b, 7, 9.] An act or instance of fading away or disappearing; *spec.*: **1.** *Cinematogr.* and *Television*. The gradual blacking out or disappearance of a picture.
1918 H. Croy *How Motion Pictures are Made* vii. 176 Slowly the details of the picture are faded in until not only the girl's features are plain, but also the surrounding setting. The reverse of this—the gradual elimination of the scene—came to be known as fade-out. **1921** Wodehouse *Jill the Reckless* xviii. 276 Marriage isn't a motion-picture close-up with slow fade-out on the embrace. **1934** *Electronics* June 173/1 It is possible not only to change the focus easily so that close ups may be made, but 'zooms' and fade-outs are easily possible. **1959** *Punch* 2 Dec. 543/1 Her protector..sadly lets her go, in time for the fadeout. **1962** *Movie* Oct. 6/2 Hitchcock makes much use of the power of the fade-out for fixing impressions in the mind of the audience.
2. *transf.* and *fig.* Disappearance; death; see also quot. 1967.
1924 A. J. Small *Frozen Gold* i. 39 It isn't natural for a whole tribe of stick-at-nothings to be able to do a fade-out like that. **1928** *Punch* 21 Mar. 318/1 The veriest front-row flapper knows that marriage is the 'fade-out' of love. **1930** *London Mercury* Feb. 324 Personally I don't want a sticky fade-out yet. **1937** J. R. Firth *Tongues of Men* xi. 132 T. S. Eliot may perhaps be one of the 'fade-out' voices of a disappearing world. **1967** *Gloss. Paper/Ink Terms for Letterpress Printing (B.S.I.)* 10 *Fade-out*, a gradual reduction in the strength of successive images, caused by the rate of removal of ink exceeding the rate of supply.
3. A temporary interruption of radio communication, caused esp. by ionospheric disturbance due to solar flares; fading.
1937 *Nature* 9 Jan. 61/2 (*heading*) Solar eruptions and radio fade-outs. *Ibid.*, Three radio fade-outs were reported later as having occurred during the morning of December 30 on short-wave..wireless transmission. **1941** K. Henney *Radio Engin. Handbk.* (ed. 3) xv. 527 Ionospheric storms..resulting in fade outs. **1959** *Daily Tel.* 16 May 11/1 The commercial possibilities of long-distance radio communication, with freedom from ionospheric fade-out and other disturbances. **1962** F. I. Ordway et al. *Basic Astronautics* iv. 145 (*caption*) Recorded in D layer as radio fade-out.
4. *Theatr.* The dimming of the stage lighting. (Cf. *fade *sb.*[1] 3.)
1936 I. Shaw *Bury Dead* 62 Send out the call... Women! (*Fadeout*). **1962** [see *fade *sb.*[1] 3].

fader: see *fade *v.*[1]

fadge, *sb.*[1] (Austral. and N.Z. examples.)
1933 L. G. D. Acland in *Press* (Christchurch) 30 Dec. 13/7 *Wool pack*. Becomes a *bale of wool* when filled and pressed and a *fadge* when loosely filled. **1953** S. J. Baker *Australia Speaks* iii. 60 *Fadge*, a butt of a bale or two bags sewn together, usual weight 60 lb. to 150 lb.

fading, *vbl. sb.* Add: **b.** *Cinematogr.* (See *fade *v.*[1] 9.) Also *ppl. a.*
1918 H. Croy *How Motion Pictures are Made* vii. 176 Fading in or out is accomplished by four different methods. **1922** L. C. Macbean *Kinemat. Studio Technique* viii. 65 The operations of 'fading in' and 'fading out'..are.. effected with the diaphragm. *Ibid.* ix. 78 'Mixing' from a 'Fading-in' Title. *Ibid.* 82 The film is then wound back to the beginning of the 'fading', with the lens closed. **1935** D. C. Ottley *Cine-Amateur's Workshop* vi. 26 The author advises a supplementary iris if fading and like effects are required.
c. Irregular fluctuations in the intensity of received radio signals as a result of varying atmospheric conditions.
1912 Lee de Forest *Let.* 13 Sept. in *Proc. Inst. Radio Engin.* (1930) XVIII. 1600, I think you have not considered with sufficient care what a variety of imaginable atmospheric conditions might be conceived to explain the phenomena of 'fading'. **1923** *Wireless Rev.* 23 June 128/2 Manchester and Newcastle both appear to be suffering very badly from fading. **1927** *Glasgow Herald* 1 Jan. 9 A site where interference and what is technically known as 'fading' would be reduced to a minimum. **1931** Moyer & Wostrel *Radio Handbk.* iv. 178 Fading is probably due to the interference between radio waves which arrive at the receiver along different paths. **1953** F. Langford-Smith *Radio Designer's Handbk.* (ed. 4) xxii. 901 Where night-time fading of signals is encountered, straight vertical aerials should be used. **1968** B. P. Lathi *Communication Syst.* iii. 195 The fading is also sensitive to frequencies...Thus the carrier and each of the sidebands undergo different amounts of fading.

fadingness. (Later example.)
1909 *Times Lit. Suppl.* 3 June 202/2 The quick fadingness of grass.

‖**fado** (fā·du). [Pg., lit. 'fate'.] A Portuguese folk-song of melancholy type; also, the name given to various kinds of dance (music) and song popular in Portugal.
1902 *Jrnl. Amer. Folk-Lore* July–Sept. 165 The nearest we can get to the original significance of the word is to call the '*fado*' the laborer's song of fate; which is more than we can do with the present form, for the Portuguese indiscriminately call '*fados*' what we designate as serenades, ballads, jigs, and sailor's hornpipes. **1936** *Discovery* Dec. 396/2 The *fado*, that remarkable modern product of Portuguese popular genius. **1957** *Times Lit. Suppl.* 6 Dec. 736/3 A brilliant run-through of Portuguese poetry and prose, the inevitable chapter on the *fado*. **1969** *Times* 1 Nov. p. viii/5 The most typical way to spend an evening in Lisbon is to listen to *fados*, the traditional song of the Portuguese town worker; but make sure that you go to a genuine *fado* restaurant.

‖**faena** (fa͵e·nă). *Bull-fighting.* [Sp., lit. 'task'.] (See quot. 1957.)
1927 E. Hemingway *Men without Women* (1928) 43 He looked at the bull, planning his faena, his work with the red cloth that was..to make him manageable. **1957** *Times* 12 Oct. 7/7 Getting by in the bull-ring means giving the bull the least possible chance of making contact—by..cutting short a whole *faena* (the part when the matador plays the bull alone with the red cloth). **1961** H. Thomas *Spanish Civil War* xx. 174 The amputation of the ear of a bull in honour of a matador, following a very successful *faena*.

Faenza (fa͵e·nza). Name of a city in the province of Emilia-Romagna, Italy, used, gen. *attrib.*, to designate the pottery made there in the sixteenth century. Cf. Faience.
1863 W. Chaffers *Marks Pott. & Porc.* 105 There are two plates of the XVIIIth Century in the Sèvres Museum; one in imitation of Faenza, the other an allegory of Luther. **1885** *Encycl. Brit.* XIX. 627/2 Faenza plate (tondino), with border in deep ultramarine blues, and central coat of arms in rich orange and red. **1904** Goodchild & Tweney *Technol. & Sci. Dict.* 214/1 *Faenza Ware*, a distinctive term given to a fine sort of pottery originally made at Faenza near Bologna, in Italy. **1962** *Victoria & Albert Mus. Internat. Art Treasures Exhib.* 76/1 A Faenza cylindrical drug vase with a band in the centre.

Færoese: see *Faroese *a.* and *sb.*

faff (fæf), *v. dial.* and *colloq.* [cf. Faffle *v.*] *intr.* To fuss, to dither. Often with *about*. Also as *sb.*, fuss, 'flap'.
1874 S. Baring-Gould *Yorks. Oddities* I. 179 T' clock-maker..fizzled an' faff'd aboot her, but nivver did her a farthing's worth o' good. **1888** J. Dalby *Mayroyd* III. vi. 99 A flay-crow wench, aw feathers an' faff. **1954** N. Coward *Future Indefinite* v. 319 The Welfare Officers appeared,..faffed about, used either too much initiative or too little, and retired in due course. **1959** 'O. Mills' *Stairway to Murder* x. 117 It's a bad time for the villagers anyway, faffing around after the ewes. **1960** M. Cecil *Something in Common* xx. 228 Dithering about in a perpetual faff.

faffle, *v.* Delete †*Obs.* and add: **c.** (Later example.) **d.** = *faff *v.*
1951 Auden *Nones* (1952) 21 Impassive, cloaked and great on Horseback under his faffling flag. **1965** M. Frayn *Tin Men* xxi. 113 Faffling away at whatever it was she faffled away at.

fag, *sb.*[1] Add: **2.** (Earlier *transf.* examples.)
1813 P. Hawker *Diary* 7 Jan. (1893) I. 66 Mr. Macintosh..a good fag. **1827** *Sporting Mag.* May 12/2 Though by no means good fags, a French chasseur will be in the field before daylight.
b. *Cricket.* A fieldsman. *Obs.*
1840 *Sporting Mag.* Aug. 332 Great Jupiter [shall be] long fag, because he can shy The ball like a thunder-bolt straight through the sky. **1897** W. J. Ford in K. S. Ranjitsinhji *Jubilee Bk. Cricket* ix. 341 The 'fags' in the out-field have their hands full indeed.

fag (fæg), *sb.*[4] *slang.* [Abbreviation of Fag-end. (Cf. Fag *sb.*[2] 2.)] **a.** The fag-end of a cigarette. **b.** A cheap cigarette. **c.** Any cigarette (the current use). Also *attrib.*, as **fag card**, a cigarette card; **fag hag** (see quot. 1945).
1888 *Sat. Rev.* 30 June 786/2 They..burn their throats with the abominable 'fag', with its acrid paper and vile tobacco. **1893** *Pick-me-up* 14 Oct. 45/2 Stimulants he calls 'booze' and a cigarette a 'fag'. **1898** *Eng. Dial. Dict.* s.v., Here [*sc.* at Redruth] we are often asked by youngsters to 'chuck' them 'a fag'—and whole cheap cigarettes are also often called fags. **1908** *Church Times* 7 Feb. 173/1 He gathered into a leather pouch the remains of his cigarettes, and left the room. 'What does he do with all those fags?' asked Conway. **1921** Galsworthy *To Let* III. viii. 284 The fag of Fleur's cigarette..fell on the grass. **1922** Joyce *Ulysses* 70 Smoking a chewed fagbutt. **1928** Galsworthy *Swan Song* iv. 24 Cinemas, fags, and football matches—there would be no real revolution while they were on hand. **1942** C. Barrett *On Wallaby* v. 97 Cobbers of the men in detention had hit upon an ingenious method of smuggling fags to them. **1945** L. Shelly *Jive Talk Dict.* 24/1 *Fag hag*, girl chain smoker. **1959** W. Golding *Free Fall* ii. 49 There was the business of the fagcards. We all collected them.

fag (fæg), *sb.*[5] *U.S. slang.* [Abbreviation of *faggot *sb.* 6 b.] = *faggot *sb.* 6 b.
1923 N. Anderson *Hobo* vii. 103 Fairies or Fags are men or boys who exploit sex for profit. **1931** 'D. Stiff' *Milk & Honey Route* 205 *Fagot* or *fag*, a *road kid* with homosexual tendencies. **1932** Hemingway *Death in Afternoon* 298 *Maricón*, a sodomite, nance, queen, fairy, fag, etc... Interested parties..are continually proving that Leonardo da Vinci, Shakespeare, etc., were fags. **1939** R. Chandler *Big Sleep* xii. 90 A stealthy nastiness, like a fag party. **1957** J. Kerouac *On Road* (1958) III. v. 206 The car belonged to a tall, thin fag. **1964** 'L. Egan' *My Name is Death* v. 49 You can't tell the fags from outside looks.

fag, *v.*[1] Add: **6. b.** To smoke; to supply with a cigarette. Cf. *fag *sb.*[4] *U.S. slang*.
1926 Maines & Grant *Wise-Crack Dict.* 8/1 *Fag me*, give me a cigarette. **1940** *Amer. Speech* XV. 335/2 To smoke is..to fag. **1954** W. Faulkner *Fable* (1955) 324 'Fag me again.' The corporal gave him another cigarette.

fagdom (fæ·gdəm). [f. Fag *sb.*[1] 2 + -dom.] The condition of being a fag.
1902 J. Buchan *Watcher by Threshold* II. ii. 118 From the anomalous insignificance of fagdom Colin climbed up the School. **1914** C. Mackenzie *Sinister St.* II. III. xii. 734 The photographic souvenirs of Lonsdale's fagdom.

fag-end. Add: Hence **fag-e·nder.**
1828 *Examiner* 21 Sept. 618/1 Clinging to the old rubbish of the worn-out Cabinet of former days—the fag-enders of a hated party. **1924** J. M. Murry *Voyage* vii. 121 'There are some religious people left.' 'I don't think she'd deny that. She was only speaking of people like us—the fag-enders,' said Doherty.

fagging, *vbl. sb.*[2] (See Fag *v.*[2]) (Earlier and later examples of *fagging-hook*.)
1837 *Outl. Flemish Husb.* in *Brit. Husbandry* (1840) III. II. iii. 17 Those who are accustomed to the method of fagging in use in Middlesex, Surrey,..will readily see that this scythe is only an improved fagging hook, allowing the reaper to stand upright at his work. **1942** *Archit. Rev.* XCI. 51/1 Scythes, slash-hooks, fagging-hooks, hay-knives or turnip-knives. **1958** *Times* 16 Aug. 7/6 Fagging-hook, frostbeck, hack-hook..are some of the variations on a single agricultural theme.

fagging, *ppl. a.* (Earlier example.)
1795 W. B. Stevens *Jrnl.* 8 Aug. (1965) III. 279 He is the most fagging Student I ever knew and this to the exclusion of all other enjoyments.

faggot, fagot, *sb.* **3. a.** Delete † and add earlier and later examples.
1723 J. Nott *Cook's & Confectioner's Dict.* BA § 27 A Faggot of sweet Herbs. **1906** Mrs. Beeton *Bk. Househ. Managem.* lxi. 1638 The little bunch or fagot of herbs..usually consists of two or three sprigs of parsley, a sprig of thyme, and a bayleaf. **1960** *Harper's Bazaar* Oct. 153/1 The traditional faggot of mixed herbs.
6. (Further examples.) Also in extended uses.
1882 Mrs. Chamberlain *W. Worc. Words* 11 *Faggit*, a term of reproach used to children. **1900** *Eng. Dial. Dict.* II. 278/2 To a stray cow: 'Come out o' that, ye old faggot.' **1922** Joyce *Ulysses* 723 That old faggot Mrs Riordan. **1925** D. H. Lawrence *Refl. Death Porcupine* 176 To me she [*sc.* a cow] is fractious, tiresome, and a faggot. Yet the subtle desirableness is in her, for me. **1969** *Sunday Mirror* 9 Feb. 35 'Urry up wi' that glass o' beer, you lazy faggot!
b. A (male) homosexual. *slang* (orig. and chiefly *U.S.*).
1914 Jackson & Hellyer *Vocab. Criminal Slang* 30 s.v. Drag, Example: 'All the fagots (sissies) will be dressed in drag at the ball tonight.' **1926** Wood & Goddard *Dict. Amer. Slang* 16 *Fagot*, a chorus man; an effeminate man. **1936** J. Dos Passos *Big Money* (1937) 273 The first thing Marge thought was how on earth she could ever have liked that fagot. **1962** H. Kane *Killer's Kiss* xxvii. 207 Duffy was no queen, no platinum-dyed freak, no screaming faggot. **1966** *New Statesman* 18 Mar. 392/3 The American word 'faggot' is making advances here over our own more humane 'queer'. **1970** *New Yorker* 28 Nov. 21/2 A gathering of homosexuals..a Jew, a Negro, a butch faggot, a nellie faggot.
10. faggot filling-stitch = *faggot-stitch*; **faggot-iron**, iron, in the form of bars or masses, made from welding together a faggot or pile of iron bars; **faggot-stitch**, needlework in which two pieces of material are joined together by stitching resembling the faggoting of drawn-thread work; so *faggot-stitching*.
1934 M. Thomas *Dict. Embroidery Stitches* 182 *Sheaf filling stitch*, also known as faggot filling stitch. **1957** M. B. Picken *Fashion Dict.* 121/1 *Fagot filling-stitch*, group of parallel stitches tied together at center to form a bundle. **1870** E. Röhrig *Technol. Dict.* (ed. 2) II. 219/2 *Fagot-iron, faggotted iron*, das Paketeisen, das Schroteisen. Ferraille. **1903** *Daily Chron.* 13 June 8/4 A lace and canvas epaulette pelerine, the seams of which are separated by faggot-stitch. **1928** *Sunday Express* 3 June 8 The frills, which were not more than half an inch wide, were set in with faggot stitching.

faggot, *v.* Add: **1. d.** *Embroidery.* To ornament (needlework) by Faggoting; also, to join (two pieces of material) by faggot-stitch.
1900 M. E. Wilkins *Parson Lord* 117 She has hem-stitched and fagoted everything. **1904** *Daily Chron.* 3 May 5/4 They are joined with an insertion, or are faggoted together to make kimonas for morning wear.

fag(g)oted (fæ·gǫtėd), *ppl. a.* [f. Faggot *sb.* or *v.*] Made into faggots. Spec. *faggoted iron* = *faggot-iron*.
1841 W. Templeton *Locomotive Steam Engine* 29 The best wrought iron, fagoted and welded together, generally termed fagoted iron. **1870** [see *faggot-iron*]. *a* **1877** Knight *Dict. Mech.* I. 823/2 Fagoted iron furnace. **1960** Horner & Abbey *Dict. Mech. Engin.* (ed. 8) II. 134 *Faggoted scrap*, the ends, cuttings, or remnants of wrought-iron bar and plate of the smithy and boiler shop, piled into fagots and welded under the steam hammer.

faggoting, *vbl. sb.* **1.** (Example.)
1837 W. B. Adams *Eng. Pleas. Carriages* vii. 107 The best axle-trees are formed of several flat bars of iron welded together in a mass, which process is technically called 'faggoting'.

faggoty, *a.* Add: **2.** Of, pertaining to, or suggestive of homosexuality; homosexual. Cf. *faggot *sb.* 6 b.
1928 C. M'Kay *Home to Harlem* I. iv. 36 And there is two things in Harlem I don't understan'. It is a bulldyking woman and a faggoty man. **1964** S. Bellow *Herzog* 238 There was a certain faggotty prissiness in his speech. **1967** *Listener* 7 Dec. 740/3 It is..irredeemably vulgar, a vision of 'only faggoty desire'. **1968** A. Binkley *What shall I Cry?* 183 Albie in his faggoty silk pajamas.

Fagin (fēⁱ·gin). The name of a character in Dickens's 'Oliver Twist', a receiver who trained children to be thieves and pick-pockets; allusively used for a thief, a trainer of thieves, or a receiver.
1847 *Punch* 2 Oct. 125 The Fagin of France after condemnation. (*Slightly altered from* 'Oliver Twist'.) **1905** *Daily Chron.* 14 Apr. 6/6 Thieves' kitchen kept by two modern Fagins...This school of crime bore outwardly the innocent semblance of a greengrocery and ice-cream shop. ..The two Fagins who conducted it were..both Italians. [**1907** *Ibid.* 17 May 3/4 The Fagin-like person who has hitherto been King of the Nile.] **1965** *Listener* 26 Aug. 292/2 The young unemployed.., the bastards.., the artful dodgers..thus have 500 Fagins to instruct them in everything from petty thievery to drug addiction. **1970** C. Drummond *Stab in Back* iv. 82 In the world of receivers of stolen property a kind of system of titles prevails, starting with 'Cousin Joe', progressing upwards through 'the Uncle' to 'Father'—very important indeed, but dwarfed by..a gentleman known as 'the Fagin'.

fagine (fēⁱ·dʒin). *Chem.* Also -in. [ad. G. *fagin* (Buchner and Herberger 1832, in J. A. Buchner's *Rep. f. d. Pharmacie* XL. 157), f. L. *fag-us* beech + -ine[5].] An alkaloid reported to have been obtained from the nuts of the common beech *Fagus sylvatica* (but see quot. 1908).
1838 R. D. Thomson in *Brit. Annual* 337 Fagin. **1846** Loudon *Veget. Kingd.* 291 The husks of the common Beech-tree yield a narcotic extractive, called Fagine. **1860** *Chambers's Encycl.* I. 804/2 A volatile, narcotic, poisonous principle, called Fagine, is..found in it [*sc.* beechmast]. **1885** *Jrnl. Chem. Soc.* XLVIII. II. 676 The author [*sc.* J. Habermann]..believes that the existence of fagine as a true alkaloid cannot be doubted. **1908** Hall & Defren tr. *Abderhalden's Text-bk. Physiol. Chem.* vi. 114 Other bases have been isolated from various plants, which in part have been given characteristic names; e.g. amanitine, from toad-stools; fagine, from buchu seeds, etc. They are, however, all identical with choline.

fagmaster (fæ·gmɑstəɹ). [f. Fag *sb.*[1] 2 + Master *sb.*[1]] A boy who has a junior boy as his fag; a fagger.
1876 Ld. Albemarle *50 Yrs. of My Life* I. v. 293 One day I had to take a pair of my fagmaster's shoes to 'Cobbler Foots' to be mended. **1895** *Q. Rev.* Oct. 283 Probably he would have been..cobbed by his fagmaster as an incorrigible toast-burner. **1920** *Blackw. Mag.* Oct. 483/2 One small boy, Harker minor,..was told by his fagmaster that he would be licked for 'fighting' as soon as he was well. **1964** *Punch* 13 May 724/2 Handling his fagmaster's amours.

fagopyrism (fæ:gopəi·riz'm). *Vet.* Also as mod.L. **fagopyri·smus.** [ad. mod.L. *Fagopyr(um* (see Buckwheat), f. L. *fāg-us* beech-tree + -o + Gr. πυρός wheat + -ism.] A skin disease affecting certain white-skinned animals, caused by poisoning by buckwheat and subsequent exposure to sunlight.
1909 in Webster. **1909** *Johns Hopkins Hospital Bull.* XX. 153/2 Dr. Smith said fagopyrismus was derived from Latin *fagus*, beech-tree, and Greek *pyros*, wheat, and meant poisoning from the buckwheat. The condition is not uncommon in pigs, swine, horses..which eat the grains, plant, straw, or stubble of the buckwheat. White or spotted animals are especially susceptible, and the disease is moderated by shelter or a cloudy sky. **1917** E. A. Bruce in *Jrnl. Amer. Vet. Med. Assoc.* III. 189 (*heading*) Fagopyrismus(buckwheat poisoning)and similar affections. **1930** W. A. Billings *Livestock & Poultry Dis.* xiv. 317 Fagopyrism, caused by exposure to sunlight after animals have been feeding on buckwheat. The symptoms closely resemble those of big-head. **1961** R. Seiden *Livestock Health Encycl.* (ed. 2) 373/1 *Photosensitization*,.. also called white-skin disease..fagopyrism.

faham (fā·hăm). *Bot.* [f. the native name for the plant.] An orchid, *Jumellea* (*Angræcum*) *fragrans*, from Réunion and Mauritius, whose leaves are used, especially in France, as a substitute for tea; also, the leaves so used.
1850 *Gardeners' Chron.* 599/3 The leaves of Faham, known also by the name of Fahon or Fahum leaves, are imported from the Mauritius. *Ibid.* 600/1 Like many of the exotic orchids, the Faham is parasitic. **1858** Simmonds *Dict. Trade*, Faham-tea. **1861** [see *Bourbon 3 d]. **1891** *Curtis's Bot. Mag.* CXVII. 7161 The popular name is variously spelled Fahame, Faham, Fahan, Fahon, Fahum, and Faam, of the origin of which I have no information.

faience. Add: Hence **faie·nced** *ppl. a.*, covered with faience.
1954 Koestler *Invisible Writing* xii. 137 The mosaic floors and faienced walls convey even to-day an image of perfection.

fail, *v.* Add: **4. b.** *to fail safe*: of a mechanical or electrical device or machine, aircraft, etc., to revert, in the event of failure or breakdown, to a condition involving no danger. Also **fail-safe** *a.*
1948 *Jrnl. R. Aeronaut. Soc.* LII. 198/1 Automatic power plant control including automatic 'fail safe' provision against mechanical trouble or power failure. **1949** *Ibid.* LIII. 179/1 What was really required was a propeller and c.s.u. combination which would fail safe. **1958** 'P. Bryant' *Two Hours to Doom* 15 The *Failsafe* procedure, the system SAC had dreamed up to prevent any accidental attack sparking off a third world war. **1958** *Times* 20 Aug. 9/7 It is becoming accepted practice that they [*sc.* jet aircraft] must 'fail safe'. **1958** *Listener* 16 Oct. 593/1 Designers [of aircraft] learnt how to design against fatigue, and the 'fail-safe' form of design was developed. **1959** *New Statesman* 790/3 Just as atomic bombers have a technical fail-safe device which recalls them from a mission if anything goes wrong, so it is inevitable that every country will have to devise its own political fail-safe policy, designed to insulate itself from the dangers due to other countries' actions. **1962** F. I. Ordway et al. *Basic Astronautics* vii. 317 The component..is as fail-safe as the one tested. **1962** *Want to run a Railway?* 16 The system is designed throughout on 'fail-safe' principles—if any part of it is not working properly then signals automatically go to danger. **1971** *Daily Tel.* (Colour Suppl.) 8 Jan. 15/4 Because of the possibility of human error and total reliance on communications between pilots and controller the system will 'fail-dangerous' rather than 'fail-safe'.
10. c. With *inf.* as object, of a thing, circumstance, situation, influence: not to have the effect of, not to result in (doing something) (usu. in negative const.).
1920 *Eng. Hist. Rev.* Jan. 143 Turenne..was a protestant, a circumstance which could not fail to prejudice Louis.
14. c. *trans.* To be unsuccessful in (an examination).
1906 *Westm. Gaz.* 22 Apr. 3/1 Such [*sc.* Indian] men applying for positions write, 'I am a failed B.A.'. **1909** H. G. Wells *Tono-Bungay* III. ii. 244 He failed some dreadful examination and had to go into the militia. **1925** E. Wallace *Mind of Mr. J. G. Reeder* v. 168 He..had a sketchy knowledge of the law (on his visiting cards was the inscription 'Failed LL.B.'). **1971** *Daily Tel.* 7 Oct. 2/2 The RIBA said that students in the five schools were failing examinations because of low teaching standards.

fain (fēⁱn), *v.*[2] Chiefly *school slang*, orig. *dial.* = Fen *v.*[2] Used in the expression *fains* or *fain(s I, fain it, fainit(e)s*: see quots.
1870 *N. & Q.* 4th Ser. VI. 415/2 'Fains', or 'Fain it'—A term demanding a 'truce' during the progress of any game, which is always granted by the opposing party. *Ibid.* 517/1 A boy who had 'killed' another at marbles, that is hit his marble, would call out 'Fain it', meaning 'You mustn't shoot at me in return'; or if a boy was going to shoot, and some inequality of surface was in his way, which he would have cleared away, his antagonist would prevent him by calling out 'Fain clears'. *Ibid.* 517/2 If a prefect wants anything fetched for him and does not say by whom, those who wish to get off going say 'Fain I'. **1889** Barrère & Leland *Dict. Slang*, *Faints* [*sic*], in vogue among schoolboys to express a wish temporarily to withdraw from participation in the particular sport or game being played. **1891** Farmer *Slang*, Fains! Fainits! Fain it! **1913** C. Mackenzie *Sinister St.* I. I. vii. 103 He could shout 'fain I' to be rid of an obligation and 'bags I' to secure an advantage. **1927** W. E. Collinson *Contemp. English* 14 The custom of putting oneself out of the game altogether by crossing the fingers and saying pax! or faynights! [feinaits] or both together. **1948** J. Betjeman *Coll. Poems* (1958) 150 'I'd rather not.' 'Fains I.' 'It's up to you.' **1960** *Guardian* 1 July 9/7 The Englishman..could remain absolutely pax and fainites. **1969** I. & P. Opie *Children's Games* i. 18 This rule is so embedded in children's minds that their immediate response to the proposal of a game is to cry out..'Me fains first'. *Ibid.*, He must safeguard himself by saying in one gulp, 'Let's-play-Tig-fains-I-be-on-it'.

‖**fainéantism(e** (fɛ̨ne͵ɑ̃ti·z'm). [f. Fainéant + -ism.] = Faineantise.
1873 *Bookseller* 2 July 549/1 He had little tolerance for anything like fainéantism or shirking in those about him. **1898** *Star* 3 Jan. 1/7 Old Russia, with its fainéantism, its childish superstition, its base and effeminate pleasures. **1958** *Spectator* 31 Jan. 131/1 Its unflattering implications of *fainéantisme* and unemployability.

faint, *sb.* **3.** *faint-fit*. Delete † and add later example.
1892 [see Dwalm *sb.*].
5. d. Used in superl., with ellipsis of *idea, notion*.
1961 Partridge *Slang Dict.* Suppl. 1083/1. **1962** I. Murdoch *Unofficial Rose* ii. 27 'Where's Penny?' said Ann. 'Haven't the faintest,' said Miranda.
7. b. Of food: not fresh; tainted. *local.*
1848 Dickens *Dombey* lvii. 573 A man..chewing a faint apple. **1902** *Westm. Gaz.* 25 Oct. 2/3 This meat hadn't been trimmed. I admit it was 'faint'.

fainter. (Later example.)
1935 W. de la Mare *Early One Morning* 268 If he is not by nature a fainter he may pore over the blood.

fair, *sb.*[1] Add: **1. a.** Now *spec.* (*a*) an exhibition, esp. one designed to publicize a particular product or the products of one industry, country, etc.; freq. with defining word prefixed; (*b*) = *fun fair*. (Add further examples.)
1763 [see Roundabout *sb.* 4 b]. **1805** *National Intelligencer* (Washington, D.C.) 24 Apr. 3/2 Washington Fairs. It will be perceived..that..the premiums offered, may be an inducement to graziers and farmers to bring their cattle and other effects to the city. **1844** *Farmers' Cabinet* 15 Oct. 73/1 New York State Fair and Cattle Show at Poughkeepsie. **1850** *New Eng. Farmer* II. 413 The State Board of Agriculture are making up a collection of samples of Indian Corn for the World's Fair [*i.e.* the International Exhibition of 1851 in London.] **1856** *Spirit of Times* 11 Oct. 95/1 The Ladies' Riding Match at the Monroe County Fair. **1863** [see *book-fair* s.v. Book *sb.* 17 a]. **1864** P. Paterson *Glimpses Real Life* xvii. 164 The penny-showman fights his battle of life, industriously wandering from one fair to another... At these places are usually congregated a multifarious crowd of exhibitions, swings, merry-go-rounds, Punch-and-Judys, and living skeletons. **1903** *Daily Chron.* 7 Dec. 5/1 In America the word 'fair' has been preserved to describe meetings at which prizes are awarded for farm live stock. **1950** *Oxf. Jun. Encycl.* IX. 209/1 All the paraphernalia of a modern fair is carried in vans and on trailers from fairground to fairground. **1952** *Sat. Rev.* 25 Oct. 67/1 This year's Audio Fair will show a tremendous improvement in [loud-]speakers designed in the conventional way. **1968** D. Braithwaite *Fairground Archit.* i. 13 The atmosphere of today's pleasure fair is far removed from the vigorous barter of primitive trading fairs. **1969** A. Cade *Turn up Stone* i. 10 Michael's previous forays overseas had been limited to..an annual pilgrimage to the Frankfurt Book Fair.
d. *fun of the fair*: colloq. phr., in extended use.
1852 Dickens *Bleak Ho.* xxiv. 250 You're half the fun of the fair, in the Court of Chancery. **1899** R. Whiteing *No. 5 John St.* xi. 104 The balance, 11s. 6¼d., must serve for everything—washing, food, odds and ends, household and personal. 'The fun of the fair.' **1940** 'N. Blake' *Malice in Wonderland* I. ii. 25, £3 10. 0. inclusive of everything—all the fun of the fair, as you might say. **1954** L. MacNeice *Autumn Sequel* x. 69 The last Fun of the fair will snap with a snap of elastic.
2. *fair-day* (later example), *-goer, -ground* (earlier and *attrib.* examples); *fair-like* adj. (later example); *fair-wards* adv.
1929 A. Clarke *Later Poems* (1961) 19 When tide had filled the boat-rings, Few dealers could be tempted Who drank upon the fair-day. **1836** Dickens *Sk. Boz* I. 321 The regular fair-goers. **1966** *Times* 28 Feb. (Canada Suppl.) p. x/3 A determined fair-goer could travel through Ontario, Quebec..and find a fair every day. **1741** Lady Pomfret *Let.* 21 June (1805) III. 247 The fair-ground; which is a square enclosure, with..shops of all sorts on each hand. **1953** *Encounter* Oct. 40/1 A blue-green, rococo building..that reminds me of a fair-ground pavilion, and is a temple. **1898** *Daily News* 28 Mar. 4/3 The shooting galleries, cocoanut 'shies' and other fair-like attractions. **1926** W. J. Locke *Stories Near & Far* 298 The gradual traffic going fair-wards passed him by.

fair, *a.* and *sb.*[2] Add: **A.** *adj.* **1. b.** (Earlier example of *the fair sex*.) Also in comparative (cf. quot. 1665 in Dict.).
1690 Locke *Hum. Und.* III. x. § 34 Eloquence, like the fair sex, has too prevailing beauties in it to suffer itself ever to be spoken against. **1878** J. H. Beadle *Western Wilds* xxix. 451 The fairer section of our party are startled at the crowds of men in the streets.
d. Also *arch.*
1820 Scott *Abbot* xi, 'So much for your lineage, fair sir,' replied his companion. **1889** 'Mark Twain' *Connecticut Yankee* 230 Even so, fair my lord.
10. a. *fair game* (earlier and later examples), *wage(s)* (attrib.).
1801 M. Edgeworth *Belinda* I. vii. 209 Quiz the doctor ..he's an author—so fair game. **1825, 1852** [see Game *sb.* 10 b]. **1894** S. & B. Webb *Hist. Trade Unionism* 386 A hundred and fifty local authorities have now (1894) adopted some kind of 'Fair Wages' resolution. **1902** *Encycl. Brit.* XXXII. 692/1 The extent to which a 'fair wages clause', in some form or another, has been inserted in public contracts. **1907** *Westm. Gaz.* 3 Sept. 7/1 The committee appointed to consider the fair-wages resolution. **1909** *Daily Chron.* 11 Mar. 1/7 The Labour party's

FAIR

motion for a stricter fair-wage clause in Government contracts.

b. *a fair field* (earlier U.S. example).

1873 'Mark Twain' & Warner *Gilded Age* 355 For the first time in his life his talents had a fair field.

d. *spec.* in games.

1856 *Spirit of Times* (N.Y.) 6 Dec. 229/1 A player must make his first base after striking a fair ball. 1867 *Routledge's Handbk. Football* 41 Knocking on and throwing forward are disallowed: in case of this rule being broken a catch from such a knock or throw shall be equivalent to a fair catch. *Ibid.* 47 A Fair Catch is a catch from a kick, or a knock on from the hand but not from the arm of the opposite side, or a throw on, when the catcher makes a mark with his heel provided no one else on his side touch the ball. 1896 Knowles & Morton *Baseball* 23 If the ball falls exactly on the foul line, it is a fair hit, unless it rolls into foul territory. 1935 *Encycl. Sports* 518/1 A fair catch can be made in a player's own In-goal. 1960 E. S. & W. J. Higham *High Speed Rugby* xiii. 183 The method of making a fair-catch is to make a mark on the ground with the heel as the ball is caught, and to call: 'Mark!'

e. fair do's: see *do *sb.*^{1} 2 c.

f. *fair enough*, colloq. phr. implying acquiescence: 'that's reasonable'.

[1866 Trollope *Belton Est.* III. iii. 57 This was fair enough, and she..promised him that she would do her best.] 1926 Wodehouse *Heart of Goof* ix. 304 'I am going to..tear him into little bits..and make him swallow himself.' 'Fair enough,' said Mr. Brown. 1934 A. Christie *Parker Pyne Investigates* 29 Wilbraham considered. 'Fair enough,' he said at last. 'I agree.' 1953 J. B. Priestley in *Best One-Act Plays 1952–53* 118 'The mistress must not press any demands upon a man who, whatever their relations may have been, is not the man she really wants, as she herself has said.' 'Fair enough.' 1958 G. Barker *Two Plays* 18 'Why won't he come?' 'Simply because he don't want to.' 'Fair enough. Fair enough. Let's get the dinghy.' 1959 *Cambr. Rev.* 30 May 559/2 If one's employer makes plastic hot water bottles then, as far as humour goes, plastic hot water bottles will be his blind spot. Which, I came to suppose, was fair enough.

g. *fair go*: see *go *sb.* 4 d.

11. c. Unquestionable, absolute, complete, thorough (*dial.* or *slang*); also *Austral.* and *N.Z.* See also *dink *sb.*^{4} and *a.*^{2}, *dinkum B. *adj.*

1872 E. J. Irving *Fireside Lays* 232 The sichts an' the soun's that we witnessed, Amaist made me greet for fair shame. 1889, etc. Fair cop [see *cop *sb.*^{2} 1]. 1902 W. Satchell *Land of Lost* iv. 26 While it [*sc.* others' money] held out it would be a fair pour. 1916, etc. [see *cow *sb.*^{1} 4 c]. 1937 N. Marsh *Vintage Murder* xxi. 174 'A fair nark' or, more emphatically, 'a fair cow' is anything inexpressibly tedious or baffling. 1944 N. Coward *Middle East Diary* 10 They can ramp about among obscure English essayists and have a fair beano.

d. *a fair treat*: something or someone highly enjoyable or satisfactory; also as *advb. phr. colloq.*

1898 J. D. Brayshaw *Slum Silhouettes* 3 The way 'e set abaht Bill was a fair treat: why, Bill couldn't git a look-in nohow. 1899 [see Treat *sb.*^{1} 5 b]. 1902 E. Nesbit *Five Children & It* iv. 119 As Cyril said, and I can't think where he got hold of such a strange expression, 'It does you a fair treat!' 1905 H. G. Wells *Kipps* I. iv. 80 These little Folkestone hills are a Fair Treat. 1936 'N. Blake' *Thou Shell of Death* iii. 52 Miss Cavendish. You're looking a fair treat.

18. a. *fair-ankled* (earlier examples).

c 1624 Chapman tr. *Homer's Batrachomyomachia* 148 Her faire-anckl'd Mother. 1851 Buckley *Iliad* xiv. 261 Fair-ankled Danaë.

b. fair-face *a.*, of brickwork, = *fair-faced *a.* 3; **Fair Hebe jug,** a jug inscribed with the words 'Fair Hebe' (see quots.); **fair-skin,** a fair-skinned person; **fair-to-middling** *a.* (also *adv.*), slightly above average.

1948 *Archit. Rev.* CIV. 186 The north wall is fair-face brick. [1903 R. L. Hobson *Catal. Eng. Pott. Brit. Mus.* 288 Jug..in the form of a tree stump, round the base of which are figures..a man offering a nest full of eggs to a girl seated on a log, between them a scroll inscribed *Fair Hebe*..signed I. Voyez 1788.] 1912 W. Chaffers *Marks Pott. & Porc.* (ed. 13) 698 Some 'Fair Hebe' jugs also bear the mark of Voyez. 1929 H. Read *Staffs. Pott. Figures* 16 We do not know when he [*sc.* Voyez] returned to Staffordshire and have no evidence of his connection with Ralph Wood until 1788, which is the date of the famous 'Fair Hebe' jug... Several copies of this jug are known..they always bear Voyez's signature. 1901 Kipling *Kim* x. 254 Besides, a Pathan is a fair-skin. 1865 'A. Ward' *A. Ward: His Travels* 41 The men are fair to middling. 1884 A. Doherty *N. Barlow* xi. 66, I guess my wallet's fair-to-middling fat. 1900 H. Garland *Eagle's Heart* 223 'How'd they feed ye back there?' ..'Oh, fair to middlin'.' 1961 *Listener* 5 Oct. 527/2 A fair-to-middling detective story.

B. *sb.*^{2} **1. c.** *for fair*: completely, altogether. *U.S. slang.*

1900 Flynt & Walton *Powers that Prey* 180 They're goin' to railroad him for fair. The *World* says the police found the weapon on him. 1903 *N.Y. Times* 5 Dec. 5 'I seem to be putting my foot in it for fair,' said the green marine. 1916 H. L. Wilson *Somewhere in Red Gap* viii. 352 Pete must of [= have] been crazy for fair about that time. 1957 J. Kerouac *On Road* (1958) I. ix. 53 Then we danced and started on the beer for fair.

d. *fair's fair*: (reciprocal) fairness is called for. *colloq.*

1898 S. Weyman *Castle Inn* xxiii. 231 No, but fair's fair, and if I am not in this, it is low. 1938 C. S. Forester *Ship of Line* ix. 118 Fair's fair..I'll spin a coin for it. 1963 C. Mackenzie *My Life & Times* II. 161 'No, no,' said Mr Green, with a sigh of noble resignation. 'Fair's fair. They're your plums now.' 1969 *Private Eye* 25 Apr. 12 Fair's fair Eric, for cripes sake!

fair, *adv.* **9. c.** Delete *Obs.* exc. *dial.* and substitute: Now *dial.*, *U.S.*, *Austral.*, and *N.Z.*

1897 D. McK. Wright *Station Ballads* 55, I feel fair sick. 1904 'G. B. Lancaster' *Sons o' Men* 88 It fair gets me down to see the poor brutes dying. 1928 J. Devanny *Dawn Beloved* xix. 190, I get fair sick of it. 1935 H. W. Horwill *Dict. Mod. Amer. Usage* 125/2 *Fair*, in the sense of completely, fully, or clearly, distinctly, this word is now obs. in Eng. exc. in certain dial., but is still current in Am. 1945 *Coast to Coast 1944* 200 It fair gets my goat. 1966 J. Hackston *Father clears Out* 181 A cow..will come home, and, in a most flagrant manner, deposit a pat fair bang in the bail.

10. b. *fair-seeming* (later example), *-thinking*.

1920 D. H. Lawrence *Lost Girl* xvi. 359 Your fair-seeming face covered the schemes and vice of your true nature. 1915 F. M. Hueffer *Good Soldier* II. i. 108 The upright..fair-thinking, public character.

fair, *v.* **1. b.** (Additional examples.) Also *dial.* and *U.S.* with *away*, *off*, or *up*.

1836 E. L. Willson *Diary* 29 May (1929), It was still rainy; towards night it faired away. *a* 1859 *Western Tales* (Bartlett), He..moved to the North, and whenever he see a fog risin', took to his bed, and kept it till it fair'd off. *a* 1859 *Spirit of Times* (N.Y.) (ibid.), There's going to be a nasty fog to-night,..call me if it fairs up. 1859 W. White *Northumb. & Border* 448 The squall lasted for nearly two hours. When it 'faired up' the son said [etc.]. 1867 G. M. Hopkins *Jrnl.* 29 Aug. (1959) 153 Dull, fairing in afternoon. 1878 Stevenson *Inland Voy.* 160 It faired as the night went on, and the moon came out of the clouds. *Ibid.* 177 The afternoon faired up. 1899 T. D. Price *MS. Diary* 28 Apr., Rain in morning but faired up and good day. 1903 A. Adams *Log of Cowboy* viii. 110 The weather in the meantime had faired off. 1923 W. Stevens *Let.* 30 Jan. (1967) 233 Later in the morning it faired off, as they say. 1951 L. Craig *Singing Hills* 68 He spoke of how it had weathered up considerable the day before, but it had faired up without a rain to do any good. 1960 V. Williams *Walk Egypt* (1961) II. vii. 140 He tipped his head back and scanned the sky. 'Looks like it's fairing off some.'

3. Also of an aeroplane or motor car (cf. *fairing *vbl. sb.*^{2}). Also with *in*, and *transf.* Hence **faired** *ppl. a.*

1922 *Encycl. Brit.* XXX. 21/1 To reduce resistance, exposed parts may be 'faired'. 1930 *Times Lit. Suppl.* 30 Mar. 231/2 The perfectly faired curves. 1933 *Jrnl. R. Aeronaut. Soc.* XXXVII. 77 Faired lines and a retractable landing carriage improved the cruising speed. 1934 *Flight* 25 Jan. 78 e/1 Now that we are familiar with metal monocoque fuselages..it is worth the trouble to fair the lines properly. 1959 *Motor* 2 Sept. 97/2 Chassis members have been faired in. 1959 B. G. D. Salt in Halas & Manvell *Technique Film Animation* 319 Movements in animation are generally 'faired'. This means that the movement in question is smoothly blended with its neighbours, no abrupt change taking place... The term 'to fair' or 'fairing'—an old craft word commonly used in the aircraft industry—is probably the most accurate available. 1963 *Times* 15 Jan. 12/2 The faired-in headlamps.

fair and square, *a.* and *adv.* Add: **B.** *adv.* Also: completely, utterly.

1931 *Daily Express* 16 Oct. 7 Thirty years ago, American boot manufacturers opened shops right and left in this country, but British pluck and skill beat them fair and square. 1957 *New Yorker* 16 Feb. 125 The financial burden ought to be borne fair and square by the home government.

fair-copy, *v.* (Earlier example.)

1819 R. Woodhouse *Let.* 19 Sept. in H. E. Rollins *Lett. Keats* (1958) II. 164 He then read to me Lamia, which he has half fair Copied.

Fair Deal. *U.S. Pol.* [f. Fair *a.* 10 + Deal *sb.*^{2} (cf. sense *4 c, quots. 1928, 1958).] Name given to the policy of social improvement advocated by H. S. Truman, President of the U.S. 1945–53.

1949 *Time* 6 June, The Fair Deal had all but lost steerage-way.

Hence **Fair Dealer,** a proponent of this programme.

1954 *Manch. Guardian Weekly* 14 Jan. 2/3 The false premise that the Fair Dealers were ever in control of the Democratic party.

fairess. (Later confirmatory example.)

1875 G. M. Hopkins *Jrnl.* 7 Feb. in *Note-bks. & Pap.* (1937) 217, I asked Miss Jones in my Welsh lesson the Welsh for *fairy*..(she says *fairess*..for a she-fairy).

fair-faced, *a.* Add: **3.** Of brickwork or stonework: not plastered.

1948 *Archit. Rev.* CIV. 132 (*caption*) The walls are of fair-faced brickwork, distempered. 1958 *Ibid.* CXXIII. 131 The offices are of r.c. construction with brick infill, with columns slate-faced and the edge beams fairfaced. 1970 *Interior Design* Dec. 775/1 An industrial standard of finish has been adopted with fair-faced brick partitions, exposed concrete, and granolithic floor. 1971 *Country Life* 30 Sept. 848/1 In the Tweed valley the walls are beautifully built in 'fair-faced' hand-split blue whinstone.

fair-haired, *a.* Add: **c.** *fig.* Darling, favourite. (Cf. White *a.* 9, White boy.)

1909 *Sat. Even. Post* 24 Apr. 26/2 The old crowd of Fair-haired Correspondent Boys who hung to the ear of President Roosevelt with viselike grip dissolved. 1918 J. W. Gerard *Face to face with Kaiserism* vii. 76 The [German] Government published a certificate in the *Official Gazette* to the effect that I was their fair-haired boy, etc. 1949 *Time* 14 Mar. 30/1 Vishinsky was Stalin's newest fair-haired boy.

fairily (fɛ̄ᵊ·rili), *adv.* [f. Fairy *a.* + -ly^2.] In a fairy-like manner.

1864 Tennyson *Islet* 18 Fairily-delicate palaces shine.

fairing (fɛ̄ᵊ·riŋ), *vbl. sb.*^{2} [f. Fair *v.* 3 + -ing^1.] The action or result of making the lines of a vessel, aircraft, or motor vehicle suitable for its easy passage through water or air; the line or curvature so made, or the structure added for this purpose.

a 1865 Smyth *Sailor's Work-bk.* (1867) 286 *Fairing*, sheering a ship in construction. Also, the draught of a ship. 1914 *Aeronaut. Jrnl.* Oct. 316 *Fairing*, a piece added to any structure to reduce its head resistance or drag. 1916 H. Barber *Aeroplane Speaks* 137 *Fairing*, usually made of thin sheet aluminium, wood, or a light construction of wood and fabric; and bent round detrimental surface in order to give it a 'fair' or 'stream-like' shape. 1927 *Daily Tel.* 6 Sept. 11/3 The fairing of the wings into the fuselage. 1930 *Daily Express* 23 May 11/5 The fairing piece at the extreme end of the tail, which performs no actual function in the operation of the airship, was slightly damaged. 1934 *Flight* 25 Jan. 78 e/1 The section on lines will..apply equally well to a set of 'fairing lines' for an ordinary fairing drawing. 1936 *Times* 19 Oct. 8/4 The car, which has graceful curves and fairings to reduce wind-noise, costs £368. 1962 J. Glenn et al. in *Into Orbit* 244 *Fairing*, part of the structure of an aircraft or spacecraft which provides streamlining in order to reduce aerodynamic resistance.

b. *transf.* (see quot. 1922); also used of the movement of animated figures in cinematography.

1922 *Encycl. Brit.* XXX. 21/1 In other countries cable has continued to be used, frequently duplicated, the cables lying one behind the other with a wood 'fairing' between them. 1959 [see *fair *v.* 3].

Fair Isle (fɛ̄ᵊ·rəil). The name of one of the Shetland islands, used attrib. to designate woollen articles knitted in certain designs characteristic of the island.

1851 *Illustr. Catal. Gt. Exhib.* III. III. 585/2 Fair Isle socks, gloves, vest piece, comforter, and cap. 1923 *Queen* 25 Oct. p. i/3 (Advt.), Allover Fair Isle Jumpers. *Ibid.*, Fair Isle Knitting Yarn, in 2-ply yellow, green, blue and red. 1924 *Tourist* Winter Sports No. 20 Shetland wool with Fair Isle border. 1925 *Queen* 8 Apr. 19 Attired in her Fair Isle jersey. 1955 *Oxf. Jun. Encycl.* XI. 277/2 The well-known Fair Isle patterns in many colours are modern interpretations of..national designs.

b. Fair Isle field mouse, a kind of mouse, an alleged (sub-)species (*Apodemus sylvaticus fridariensis*) of the field-mouse (Field *sb.* 19), found in the Shetlands, of larger size than the ordinary field-mouse.

[1906 N. B. Kinnear in *Ann. Scot. Nat. Hist.* Apr. 68, I think the Fair Isle mice are quite entitled to be ranked as a sub-species, and therefore I propose the name of Mus sylvaticus Fridariensis.] 1915 G. E. H. Barrett-Hamilton *Hist. Brit. Mammals* xvii. 542 (*heading*) The Fair Isle Field Mouse. *Apodemus Fridariensis*. *Ibid.*, Distribution: Fair Isle and Shetland Islands, where it occurs on Yell, and possibly on Mainland. 1960 M. Burton *Wild Animals Brit. Isles* 78 The Fair Isle or Shetland field mouse..., like the yellow-necked but without the collar.

fair-lead. Add: **c.** *transf.*

1937 *Times* 13 Apr. xvi/2 Roller fairleads at the rear end of the vehicle allow the cable to be used immediately to the rear or at any angle to the side. 1957 *N.Z. Timber Jrnl.* Feb. 45 *Fairlead*, a device of 3 or 4 rollers in the form of a U through which a cable can be carried out or reeled in from any direction for hauling logs.

Fairlight Clay. [f. *Fairlight*, name of a town in Sussex + Clay *sb.*] A band of shales and clays of the Wealden series extending through Kent and East Sussex.

1875 W. Topley *Geol. Weald* I. i. 5 In the neighbourhood of Hastings and Fairlight, clays have been separately mapped..and were termed 'Ashburnham Beds'... They are now called *Fairlight Clays*. 1929 P. G. H. Boswell in Evans & Stubblefield *Handbk. Geol. Gt. Brit.* 388 Near Hastings and Eastbourne the lower part is replaced by a lenticular mass of variegated and mottled clays, the *Fairlight Clays* (350 ft. +) with seams of lignite, sandstone and spherulitic ironstone. 1960 L. D. Stamp *Britain's Struct.* (ed. 5) xvi. 189 Fairlight Clays, grey and variegated shales and clays.

fairly and squarely, *adv.* (see under Fair and square). (Earlier examples.)

1638 W. Chillingworth *Relig. Prot.* IV. § 19. 201 Me thinkes he hath little reason to complain, that he hath not been fairely, and squarely dealt with. 1862 *Congress. Globe* 27 Mar. 1402/2, I..doubt..the ability of these guns to remain in their position if..struck fairly and squarely by shot from the enemy.

fair-top boot. *U.S.* [Fair *a.* 1 f.] A boot topped with light-coloured leather.

1799 in C. Cist *Cincinnati* (1841) 159 As an illustration of fashions..fair-top boots. 1826 *Rhode Isl. American* Let. 2 Mar. (Th.), His usual dress in the Senate is a blue coat,..drab small-clothes, and fair-top boots. 1853 J. G. Baldwin *Flush Times Alabama* 22 He dressed in the old-fashioned fair-top boots and shorts. 1860 S. Mordecai *Virginia* iii. 48 His tall and burly form arrayed in fair-top boots [etc.].

fair-trader. Add: (*c*) a smuggler (cf. Fair-trade 1 b, Free-trader 2).

1824 Scott *Redgauntlet* II. xiii. 313 Ay, Mr Alan Fairford—a good name for a fair trader. 1884 D. Pae *Eustace* 217, I am what is called a fair trader—in other words a smuggler.

fai·rwater. [f. Fair *v.* 3 + Water *sb.*] A structure making the lines of a vessel suitable for its easy passage through water.

1910 D. W. Taylor *Speed & Power of Ships* I. ii. 125 About all that can be done for the propeller hub is to fit a conical fair-water behind it. Model experiments show that a long fair-water, say of length about twice the diameter of the propeller hub, offers materially less resistance than a short fair-water of length say about one-half the diameter of the propeller hub. 1927 G. Bradford *Gloss. Sea Terms* 62/1 *Fair water cone*, the conical cap covering the after side of the hole of a propeller which takes the end of the tail shaft. 1948 R. de Kerchove *Internat. Maritime Dict.* 250/2 *Fairwater*, 1. A conical-shaped cap fitted over the locknut at the afterend of the propeller shaft to prevent an abrupt change in the streamlines. Also called fairwater cone, propeller cap, fairwater cap. 2. Any casting or plate affixed to the hull with a view to maintaining smooth streamline flow of water. Plating fitted about the ends of shaft tubes or struts to prevent a sharp change in the direction of streamlines. 1956 *Jane's Fighting Ships 1956–7* 226 Transferred to Italy by U.S.A. after conversion to guppy-schnorkel [submarine]. Modified structure and fairwater.

fairway. Add: **b.** *Golf.* That part of a golf-course between a tee and putting-green which consists of short grass.

1910 B. Darwin *Golf Courses Brit. Isles* i. 18 Hillocks have risen as if by magic in the middle of the fairway. 1923 *Daily Mail* 18 Jan. 9 The maintenance staff..are busy on the fairways and teeing grounds. 1952 *Golf* ('Know the Game' series) 24/2 The player must loft the ball out, thus losing considerable distance in relation to an opponent lying on the fairway.

c. *transf.* and *fig.*

1927 *Daily Tel.* 30 Aug. 10/5 (Lawn tennis) The court..was soft and yielding near the base-lines, even if true enough in the fairway. 1931 H. G. Wells *Work, Wealth & Happiness of Mankind* (1932) xii. 574 Every member must leave his seat in the chamber and walk in queue through the fair way of one or other of two lobbies, the 'Ayes' or the 'Noes'. 1959 *Manch. Guardian* 12 Aug. 5/7, I had walked..through the forest..following the long fairways used to check fires.

fairy, *sb.* and *a.* Add: **A.** *sb.* **5. c.** A male homosexual. *slang.*

1895 *Amer. Jrnl. Psychol.* VII. 216 This coincides with what is known of the peculiar societies of inverts. Coffee-clatches, where the members dress themselves with aprons, etc., and knit, gossip and crotchet; balls, where men adopt the ladies' evening dress, are well known in Europe. 'The Fairies' of New York are said to be a similar secret organization. 1923 [see *fag *sb.*^{5}]. 1925 F. Lonsdale *Spring Cleaning* II. 75 *Mona.* I say, what's the fairy's name? *Richard.* I have happily forgotten it for the moment! 1929 R. Hughes *High Wind in Jamaica* iv, 'Who are they?' Emily asked the Captain... 'Who are who?' he murmured absently... 'Oh, those? Fairies.' 'Hey! Yey! Yey!' cried the mate. 1945 E. Waugh *Brideshead Revisited* I. v. 102 Two girls stopped near our table and looked at us curiously. 'Come on,' said one to the other, 'we're wasting our time. They're only fairies.'

C. *attrib.* and *Comb.* **1. d.** adj. *fairy-fine.*

1904 *Westm. Gaz.* 20 May 2/4 A wonder, a green miracle, More fairy-fine than words can tell. 1925 C. Day Lewis *Beechen Vigil* 21 Give me your hands so fairy-fine.

2. *fairies' bath*: also **fairy bath**; **fairy bells,** a kind of musical instrument; **fairy cake,** a small individual sponge cake, usu. iced and decorated; **fairy-cycle,** a small-wheeled low bicycle for children; **fairy-fly,** a minute insect which deposits its eggs in the eggs of other insects; a chalcid of the family Mymaridæ; **fairy godmother,** a fairy who acts as god-mother or protector to a mortal child; also *transf.*, a benefactress; so *fairy godfather*; also (*nonce-wds.*) *fairy-godmother* v. trans., *-godmotherly* adj.; **fairy lamp,** a lamp containing a candle; **fairy light** (usu. in pl.), (*a*) a small coloured light used in illuminations and often hung among trees; (*b*) a kind of night-light; (*c*) perversion of 'Very light'; hence **fairy-lighted, -lit** adjs.; **fairy moss,** tiny, free-floating, aquatic ferns of the genus *Azolla*; **fairy penguin,** the little or little blue penguin, *Eudyptula minor*, found on the southern coasts of Australasia; **fairy prince,** a prince of the fairies; *transf.*, an idealized person, the ideal husband-to-be; **fairy prion,** a prion, *Pachyptila turtur*, of Australasian and sub-antarctic coasts, with a bluish bill and bluish feet; **fairy queen,** also, the player who takes the part of the fairy queen in a pantomime; such a part; **fairy rose,** *Rosa chinensis* var. *minima*; **fairy tern,** (*a*) *Austral.* and *N.Z.*, a small black-crowned tern, *Sterna nereis*; (*b*) a tropical white tern, *Anous* (*Gygis*) *albus*.

1856 C. M. Yonge *Daisy Chain* I. viii. 72 Flora..sought in the hedge-sides for some crimson 'fairy baths' to carry home. 1895 *Army & Navy Co-op. Soc. Price List* 15 Sept. 1664 Fairy Bells. 1916 E. Crutchfield in G. B. Shaw *Sham Educ.* in *Doctors' Delusions* (1931) 361 A clown was then introduced. He played the 'fairy bells' in imitation of church bells. 1895 *Army & Navy Co-op. Soc. Price List* 4 Fairy Cakes... Price per lb. 0/7. 1967 R. Rendell *New Lease of Death* v. 47 A tea trolley..laden with home-baked pastries, strawberries in glass dishes, fairy cakes in paper cases. 1971 *Guardian* 9 Jan. 11/1 The occasional fairy cake and cuppa from Mrs Purdie the tea-lady. 1926–7 *Army & Navy Stores Catal.* 888/2 'Fairy cycle'. With adjustable saddle. 1927 *Times* 6 July 13 A girl of six..riding her fairycycle. 1928 *Daily Express* 2 June 7/3 The boy was pushing his fairy cycle near his home when a collision occurred between two motor-cars, one of which killed him. 1903 W. F. Kirby *Butterflies & Moths of Europe* p. xxxvi/1 The smallest of all known insects are some parasitic *Hymenoptera* belonging to the family *Mymaridæ*, which lay their eggs in those of other insects. They have elegant battledore-shaped wings, with a fringe of very long hairs... They are popularly known as 'Fairy Flies'. 1949 *Oxf. Jun. Encycl.* II. 82/2 The Fairy Flies, which include the smallest known insects, are Chalcids that spend their larval life inside the eggs of other insects, feeding on the contents. 1935 *Mademoiselle* Aug. 59/2 Acquaintance? He's my fairy godfather. 1959 M. Steen *Tower* I. iv. 65 The little old fairy godfather was nicely covering the sherry he offered his visitors. 1851 M. Gatty (*title*) The fairy godmothers and other tales. 1855 C. M. Yonge *Hist. Sir Thomas Thumb* p. vii, Other fairies..were generally women, enchantresses..and probably all the fairy god-mothers we hear of, are meant for these. 1883 Fairy godmother [see Fairy *sb.* and *a.* C. 1]. 1909 G. K. Chesterton *Orthodoxy* iv. 98 Fairy godmothers seem at least as strict as other godmothers. 1919 W. Deeping *Second Youth* xxxi. 260 Miles's name had been placed upon the list some weeks ago, with Kate's name recording her sponsorship, for each accredited visitor had to be 'fairy' god-mothered' by a W.M.C. member. 1959 *Listener* 18 June 1058/2 Any country which embarks on this programme will be the world's fairy godmother. 1962 *Ibid.* 26 July 146/3 Mrs Eileen J. Garrett, President of the Parapsychology Foundation of New York and fairy godmother of paranormal study. 1900 E. T. Fowler *Farringdons* 272 'I have found your mislaid grandfather; be a mother to him for the rest of your life!' It would give one the most delicious, fairy-godmotherly sort of satisfaction! 1891 Fairy lamp [see Fairy *sb.* and *a.* C. 1]. *a* 1941 V. Woolf *Captain's Death Bed* (1950) 205 In the Amusement Compound..they light a horse-shoe of fairy-lamps above the Jack and Jill. 1871 *English Mechanic* 24 Nov. 244/2 A long crescent of fairy-lights, glimmering on the coast-line. 1886 *Colon. & Ind. Exhib. Official Catal.* 166 Advt., New Patent 'Fairy' Lamps and 'Fairy' Lights. 1891 *Strand Mag.* Aug. (Advt.), The Queen of Lights. 'Fairy' Light. With Double Wicks. 1895 *Army & Navy Co-op. Soc. Price List* 15 Sept. 20 Night Lights... Fairy Lights, to burn in the 'Fairy Lamps'. 1925 Fraser & Gibbons *Soldier & Sailor Words* 91 *Fairy Light*, Very Light, named after its inventor. Very Lights were flares or fire balls fired from a pistol, employed everywhere in the trenches and by airmen. 1927 W. E. Collinson *Contemp. Eng.* 98 Fairy lights (Verey lights to send up the S.O.S. or distress signals). 1955 M. Hastings *Cork & Serpent* vii. 86 A gay little inn with a string of fairy lights illuminating the banks of bottles behind the bar. 1968 *Listener* 12 Dec. 807/3 Even electric fairy lights are not entirely free of fire risk. 1938 *Times* 10 Jan. 10/4 Fairy-lighted Christmas trees. 1953 C. Day Lewis *Italian Visit* I. 22 Let fairy-lit streets run wine through the veins like a ride on a scenic Railway. [1908 T. W. Sanders *Encycl. Gardening* 163 Fairy Floating Moss (Azolla caroliniana).] 1938 F. Perry *Water Gardening* ix. 134 The Azollas are so dainty in appearance that the English name of Fairy Moss is no misnomer. 1968 D. Bartrum *Water in Garden* viii. 116 *Azolla caroliniana*..is closely allied to the Ferns... The popular name of the plant is Fairy Moss. 1848 J. Gould *Birds of Australia* VII. 85 (*heading*) *Spheniscus undina*, Gould. Fairy Penguin. 1901 A. J. Campbell *Nests & Eggs Austral. Birds* II. 1012 The Fairy Penguin is the smallest of its singular tribe. 1966 G. Durrell *Two in Bush* ii. 79 Fairy Penguins appeared in small groups and hopped their way up the rocks towards their nest burrows. 1840 *Fraser's Mag.* July 99/2 Caroline..had set him down for her divinity, her wondrous fairy prince. 1911 'I. Hay' *Safety Match* xii. 194 A purely hypothetical fairy prince, composed of equal parts of Peer of the Realm, Lifeguardsman, Mr Sandow, Lord Byron, and the Bishop of London, whom she had cherished in..her heart. 1966 J. Gloag *Sentence of Life* xlix. 415 It was Mr Maddox this, Mr Maddox that. You'd have thought he was the fairy prince the way she carried on. 1912 G. M. Mathews *Birds of Australia* II. II. 217 Pseudoprion Turtur Turtur. Australian Fairy-Prion. 1930 W. R. B. Oliver *New Zealand Birds* 114 Fairy Prion... *Pachyptila turtur*. The Fairy Prion was first discovered in Bass Strait... The Fairy Prion in all its forms is to be distinguished from the other members of the genus by the form of the bill. 1966 G. Durrell *Two in Bush* ii. 79 Then the Fairy Prions—delicate little swallow-like petrels—started to arrive. 1899 G. B. Shaw *Let.* 12 Oct. in J. Dunbar *Mrs. G.B.S.* (1963) xi. 187 Poor Moony Silly grinned like a fairy queen in a fifth rate pantomime. 1904 *Sat. Rev.* 14 May 620/2 In the seventies pantomime was flourishing still. Demon King and Fairy Queen..were familiar and popular things. 1848 W. Paul *Rose Garden* II. ii. 131 Rosa... The Lawrenceana, or Fairy Rose... In dry soils the Fairy Roses may be planted in masses. 1955 G. S. Thomas *Old Shrub Roses* ix. 78 In 1805, at Colville's nursery in England, Parsons's Pink China gave rise to the Dwarf Pink China, a miniature Rose known in England as the Fairy Rose. 1925 P. Moncrieff *New Zealand Birds* 24 Little White Tern (Fairy Tern). 1926 W. Beebe *Arcturus Adv.* ix. 228 Scores of sea-birds—frigate birds, boobies, and pure white fairy terns. 1954 Fisher & Lockley *Sea-Birds* i. 29 *Gygis alba*, the tropical, white, almost 'transparent' fairy tern breeds north in the Atlantic. 1966 R. A. Falla et al. *Field Guide Birds N.Z.* 163 Fairy Terns reappear at breeding grounds in September.

fairyland. Add: Also *attrib.*

1832 A. Earle *Narr. Res. N.Z.* (1966) 70 A beautiful rainbow..tinged the scene with a fairy-land effect. 1968 *Times* 15 Oct. 13/5 Wedgwood fairyland lustre, a twentieth-century ware especially sought after in America.

fairyology (fɛ̄ᵊriǫ·lŏdʒi). [f. Fairy + -ology.] The study of fairies. Hence **fairyo·logist,** one who studies fairies.

1859 M. A. Denham (*title*) A few fragments of fairyology. 1908 *Daily Chron.* 11 Dec. 4/4 Mr. Dion Clayton Calthrop, the 'fairyologist' of all others—to quote Mr. Tree's quaint phrase.

fairy story (fɛ̄ᵊ·ri stōᵊ:ri). [f. Fairy *sb.* and *a.* + Story *sb.*^{1}] = next.

1850 Dickens *Dav. Copp.* xix. 193 Life was more like a great fairy story, which I was just about to begin to read, than anything else. 1859 Thackeray *Virginians* II. xxv. 199 Our favourite fairy story-books. 1925 G. H. Mullin *Adv. Scholar Tramp* vi. 85, I told her an abominable pack of lies—a fairy-story, as we call such fabrications on the Road. 1927 R. Lehmann *Dusty Answer* I. i. 19 He reminded me of something fabulous—a Chinese fairy-story. 1932 E. Hemingway *Death in Afternoon* Gloss. 299 s.v. Maricón, There are many very, very funny Spanish fairy stories. 1956 'B. Buckingham' *Three Bad Nights* xxii. 202 Please go on with your fairy story if it amuses you. 1966 S. Jackman *Davidson Affair* viii. 100 You'll never persuade them to accept your fairy-story ending.

fairy-tale (fɛ̄ᵊ·ri tēᵊl). [f. Fairy *sb.* and *a.* + Tale *sb.*; rendering F. *conte de fées*.] **a.** A tale about fairies. Also *gen.*, fairy legend, faerie. **b.** An unreal or incredible story. **c.** A falsehood. Also *attrib.* Hence **fai·ry-tale-ish** *a.*

1749 H. Walpole *Let.* 3 May (1903) II. 369 It..was by far the..prettiest spectacle I ever saw: nothing in a fairy tale ever surpassed it. 1750 H. Brooke (*title*) A new collection of fairy tales. 1797 J. Moser (*title*) Moral Tales; consisting of the Reconciliation,..a Fairy Tale, in the modern style. Clementia and Malitia; a Fairy Tale, in the ancient style. 1819 Keats *Let.* 21 Sept. (1958) II. 173 Dont forget to tell Reynolds of the fairy tale Undine. 1864 [see Fairy C. 1]. 1887 Ruskin *Præterita* II. 244 [Shakespeare's] Henry V. is only a king of fairy tale. 1896 Ade *Artie* xiii. 117 Guess what they train on... Cocoanut pie. That ain't no fairy tale, neither. 1900 *Daily Tel.* 4 July 10/2 Mr. Kruger..has told a Chicago journalist one of his pretty little fairy tales, the only truth in which is that some burghers are again taking up arms. 1900 E. T. Fowler *Farringdons* xiv. 272 If I were a man I should belong to the Herald's Office. It would be such fun to be called a 'Red Bonnet' or a 'Green Griffin', or some other nice fairy-tale-ish name. 1914 *Scotsman* 16 Sept. 6/4 The Russians have neither 'landed in nor passed Great Britain on their way to France or Belgium', nor are they now on French or Belgian soil. The whole story has been a gigantic fairy tale. 1920 'O. Douglas' *Penny Plain* ix, It would be very nice and fairy-tale-ish! 1923 [see *bee^1 5 b]. 1957 M. Kennedy *Heroes of Clone* IV. vi. 269 Why should I back up this fairy-tale of yours, just to save your face? 1963 Auden *Dyer's Hand* 212 There is only one fairy-tale motif, to my knowledge, that contains an element of inner conflict. 1971 *Daily Tel.* 16 Jan. 20/3 A full moon then added its light to the fairy-tale scene.

‖ **faisandé** (fəzande), *a.* [Fr.: pa. pple. of *faisander* to hang (game) up until it is high.] Affected, artificial, theatrical; 'spicy'.

1912 C. Mackenzie *Carnival* xxvii. 306 The more imaginative observer would perceive in the group something unhealthy, something *faisandé*, an air of too deliberate enjoyment that seemed to imply a perfect knowledge of the limitations of human pleasure. 1930 *Times Lit. Suppl.* 21 Aug. 664/4 There is nothing *faisandé* about Irène de Benauge; her mind is..hard, clear, clean, beneficent and unimaginative. 1958 *Observer* 1 June 15/1 He plays the part in a *faisandé* Cockney accent straight out of Bruce Bairnsfather's Old Bill cartoons.

‖ **fait accompli** (fetakonpli, fēᵊ·takǫ·mpli). [Fr.] An accomplished fact; an action which is completed (and irreversible) before affected parties learn of its having been undertaken.

1845 R. Ford *Handbk. Spain* I. 337 This is now a *fait accompli*. 1872 E. Braddon *Life in India* v. 179 After the overland route was a *fait accompli*, the rules of the services..restricted the servants of the East India Company from enjoying much of their leave westward of the Cape of Good Hope. 1895 *Bow Bells* 16 Aug. 177/2 It will be very interesting to see..whether various other suggestions..will also become fait accompli. 1905 Mrs. H. Ward *Marriage of W. Ashe* II. viii. 139 Elizabeth knew that his appearance in the conversation invariably meant a *fait accompli* of some sort. 1906 W. De Morgan *Joseph Vance* xxxiv. 337 And Janey knew she was herself, so that no explanation seemed necessary. A *fait accompli* has leisure to wait for an official *raison d'être*. 1955 *Times* 6 Aug. 5/2 He did not propose to proceed by surprise or stealth, or to bind his hands and renounce the prerogatives of the executive. He could give an assurance to the parties represented in the majority that the Government would never be presented with a fait accompli.

‖ **faites vos jeux** (fetvoʒö), *int.* [Fr.] Place your bets (an instruction given by croupiers at a roulette game).

1867 'Ouida' *Under Two Flags* I. ix. 183 Brains..let themselves be lulled with the monotone of '*Faites votre jeu*' [sic]. 1904 A. Bennett *Great Man* xxv. 273 '*Faites vos jeux, messieurs*,' said the chief croupier... Yes, he was at Monte Carlo. 1922 Joyce *Ulysses* 496 Sieurs et dames, faites vos jeux!..(Tiny roulette planets fly from his hands.) Les jeux sont faits! 1963 G. Greene *Sense of Reality* 113 'Faites vos jeux, messieurs,' the croupiers cried. 1964 A. Wykes *Gambling* ix. 217 At the beginning of each round the croupier invites players to place their bets with the phrase *faites vos jeux*.

faith, *sb.* Add: **14.** *faith-philosophy, -state, -value*; *faith-shaking, -sown, -starved, -straining* adjs.; **faith-ladder** (see quot. *a* 1910).

1909 W. James *Pluralistic Universe* viii. 328 In some of my lectures at Harvard I have spoken of what I call the 'faith-ladder'. *a* 1910 —— *Some Probl. Philos.* (1911) App. 224 The following steps may be called the 'faith ladder': 1. There is nothing absurd in a certain view of the world being true, nothing self-contradictory; 2. It *might* have been true under certain conditions; 3. It *may* be true, even now; 4. It is *fit* to be true; 5. It *ought* to be true; 6. It *must* be true; 7. It *shall* be true, at any rate true for *me*. Obviously this is no intellectual chain of inferences, like the *sorites* of the text-books. 1846 J. D. Morell *Hist. View Philos.* II. vii. 311 Haumann had..attempted to found a system of faith-philosophy..but it was Jacobi who first brought the faith-philosophy into repute. 1896 *Westm. Gaz.* 21 Dec. 2/3 Could anything be more faith-shaking than this halt of several weeks in the negotiations? 1844 J. G. Whittier *Wks.* (1898) 197/2 Faith-sown seeds Which ripen in the soil of love. 1946 R. Campbell *Talking Bronco* 69 Where faith-starved multitudes may quarry As in a mountain, and be fed. 1896 *Amer. Jrnl. Psychol.* Apr. 315 The state of confidence, trust,..which we have found..in every conversion considered is the Faith-state. 1924 W. B. Selbie *Psychol. Relig.* 158 To induce what psychologists call the faith state may be a very great and wonderful thing if the object of faith is worthy, i.e. God or Christ. 1897 'Mark Twain' *Following Equator* 172 Here are some faith-straining figures. 1903 G. Tyrrell *Lex Orandi* xxiii. 191 Mistakings of faith-values for fact-values are to be ascribed to that almost ineradicable materialism of the human mind which makes us view the visible world as the only solid reality.

faithful, *a.* Add: **3. b.** (Later example.)

a 1817 Jane Austen *Northanger Abbey* (1818) II. ix. 180 'Isabella promised so faithfully to write directly.' 'Promised so faithfully!—A faithful promise!—That puzzles me.—I have heard of a faithful performance. But a faithful promise—the fidelity of promising!'

8. *Comb.*, as *faithful-hearted*.

1925 J. Gregory *Bab of Backwoods* xix. 244 Faithful-hearted retainer.

faithfully, *adv.* **4. b.** (Later examples.)

a 1817 [see *faithful *a.* 3 b]. 1922 Joyce *Ulysses* 746 And I promised him yes faithfully Id let him block me.

Faithist (fēᵊ·þist). [f. Faith *sb.* + -ist.] A member of a sect whose religion is based on revelations contained in the 'Kosmon Bible' or 'Oahspe' and on angelic communications. Also *attrib.*

1885 *Santa Fé Weekly New Mexican* 23 July 4/4 Tanner has joined the new community or sect of faithists near Las Cruces, N.M. 1907 *Westm. Gaz.* 25 June 10/1 The Faithist Community, which has established a modest footing in Balham, and whose comprehensive gospel ranges from the creation of man to the 'glory and labours of the gods and goddesses of the Etherian Heavens'. 1921 *Glasgow Herald* 21 Sept. 10 Mr. F. T. A. Davies, founder of the Faithist movement in England. 1928 *Sunday Dispatch* 29 July 9/5 The Faithists originated in America. There are only two or three hundred in Great Britain—most of them in London and the Home Counties.

faithless, *a.* **1. b.** absol. *the faithless.* (Later example.)

1944 A. Clarke *Coll. Plays* (1963) 217 The terrible assumption of that woman..is a warning to the faithless.

‖ **faja** (fa·ha). [Sp.] A sash, girdle.

1841 G. Borrow *Zincali* I. II. v. 304 A faja, or girdle of crimson silk, surrounds the waist. 1897 'H. S. Merriman' *In Kedar's Tents* xix. 213 About his person in the form of handkerchief and faja were..touches of bright colour. 1931 *Discovery* Dec. 392/2 (Mexico) The 'faja' (sash)..may be worn by either men or women according to style and design.

fake, *sb.*^{2} Add: **1.** Passing from *slang* to *colloq.* in the sense of 'a counterfeit person or thing'.

1888 *N.Y. Mercury* (Farmer), Both ladies then came to the conclusion that the fortune-teller was a fake, and they decided to notify the police. 1899 R. Whiteing *No. 5 John St.* xxiii. 229 She went so far as to affirm her conviction that I was a 'fake'. 1927 F. Lonsdale *The Fake* III. ii. 59 If you stood at this moment to be judged by these..people, you would have earned from them that which I can only ever feel for you—everlasting contempt. ..you Fake! 1945 Auden *Coll. Poetry* 35 To think of love as a subjective fake.

3. *attrib.* or as *adj.* Spurious, counterfeit; *spec.* used in *Jazz* of a book of music containing the basic chord-sequences of tunes.

1775 W. Howe *Let.* 3 Dec. in *Canadian Archives* (1904) App. I. 355 So many artifices have been practiced upon Strangers under the appearance of Friendship, fake Pilots &c., that those coming out with Stores..cannot be put too much on their guard. 1890 *Stock Grower & Farmer* 15 Mar. 7/1 The farce of surveying for their fake ditch. 1892 [see Dict., sense 1]. 1903 *Daily Chron.* 16 June 3/4 [Amer. loq.] They were all men of position and character, who would not be connected with a 'fake' institution. 1909 *Westm. Gaz.* 16 Nov. 2/3 Swindled by a 'fake' paper, the supposed verdict being simply a prophecy of what it was likely to be. 1920 *Glasgow Herald* 17 Nov. 9 Fake whisky..the symptoms following consumption are similar to those of gastric poisoning. 1927 *Melody Maker* May 495 By blowing the instrument as though you intended to play one long note, but during this long note, changing the fingering alternately from 'legitimate' (or recognised) to fake fingering. 1933 *Mind* XLII. 34 Anyone can write a fake dictionary. 1953 X. Fielding *Stronghold* III. i. 182 These fake Americans usually appear so scornful of what..they call 'the home-town'. 1958 *Amer. Speech* XXXIII. 225 *Bible* (the chord book, or 'fake book', which guides most small combos through this weary world). 1964 *McCall's Sewing* iv. 57/1 *Fake fur*, woven or knitted fabrics made of cotton or synthetic fibres to simulate the fur of animals. 1965 *New Yorker* 2 Jan. 46/2 Bring that fake book, please, in case they ask me to play something I recorded forty years ago. Everybody but me remembers those tunes.

fake, *v.*^{2} Add: **1.** (Further examples.) Also *absol.*

1896 *People* 6 Sept. 10/3 Complainant..denied that..he was accused by a lady of picking her pocket..he did not say he had 'never faked a poke' in his life. 1908 *Smart Set* Sept. 39, I knew that..it had no curative power and I didn't want to be caught faking. 1921 G. B. Shaw *Back to Methuselah* IV. III. 203 What else could the poor old chap do but fake up an answer fit for publication?

b. *spec.* To conceal the defects of (an animal) by colouring hair or feathers.

1874 [see Dict., sense 1]. 1883 W. G. Stables *Our Friend the Dog* vii. 60 *Faking*, dyeing, staining, clipping, or otherwise interfering with the dog's coat or appearance, to hide defects and deceive the judge or public.

c. To feign or simulate.

1941 *London Opinion* Apr. 42/1 Faking an interest in the goods displayed. 1943 E. O'Neill *Moon for Misbegotten* III. 148, I..hid my face in my hands and faked some sobs.

3. *intr.* Of jazz musicians: to improvise. *colloq.*

1926 *Melody Maker* Jan. 20 In those days..the dance band was not studied by the orchestrator as it is now, and one had to 'fake' saxophone and banjo parts from those of such other instruments as were catered for in the score. 1933 *Fortune* Aug. 92/3 It is no exaggeration to say that his band of fourteen can *fake* (improvise) as adroitly as the early five-piece combinations. 1944 *Spotlight* Jan. 18 There was enough good music 'faked' in those days.

faker (s.v. Fake *v.*^{2} in Dict.). (Earlier and later examples.)

1846 R. L. Snowden *Magistrate's Assistant* 345 Umbrella menders..mushroom fakers. 1898 *Westm. Gaz.* 25 June 2/3 The profession of the faker has become quite a recognised one, and happy the man who..shows intelligent anticipation of some important item of war news. 1934 S. R. Nelson *All about Jazz* i. 23 Chas. Washington..was the father of the fakers, as he was unable to read a note of music.

fakes: see Faikes.

fakir. Add: **1. b.** *erron.* for Faker, pronounced (fēᵊ·kəɪ). *U.S.*

1882 in S. Poe *Buckboard Days* (1936) 99 Thieves, Thugs, Fakirs and Bunkco-Steerers. 1902 A. D. McFaul *Ike Glidden* xvii. 127 Each day brought its new characters, fakirs, peddlers, schemers and promoters. 1903 *N.Y. Even. Post* 31 Oct. 5 One may see at almost any of the downtown corners a street fakir selling shoestrings. 1932 E. Wilson *Devil take Hindmost* ix. 87 Some listen to a patent-medicine fakir.

Falange (fala·nhe, falā·nŋ, -læ·nŋ). Also Ph-. [Sp., Phalanx.] A Spanish political party, founded in 1933 as a Fascist movement by J. A. Primo de Rivera and merged in 1937 with traditional right-wing elements to form the ruling party, the Falange Española Tradicionalista, under General Franco. Hence **Fala·ngism**; **Fala·ngist** *sb.* and *a.*

1937 *Ann. Reg. 1936* 247 The employment by the Fascist *Falange Española* of gunmen and thugs to destroy 'Bolshevism'. (The *Falange* was a comparatively insignificant group which had come to model its tactics as well as some of its ideology on those which had proved eminently successful in Italy and Germany.) 1938 *Times* 30 Sept. 13/6 They [*sc.* the Carlists] find little inspiration in the obscurities with which the more wayward Falange writers overload their meditations. 1938 G. T. Garratt *Shadow Swastika* vii. 192 The improved prospects of the Phalangists. 1939 R. Campbell *Flowering Rifle* III. 94 Braided with scarlet arrows by the Sun, A 'Falangist' himself, if there is one! 1941 *Economist* 25 Jan. 97/1 If undiluted Falangism is to be Spain's future, foreign policy must finally fall into line. *Ibid.*, The Falange is trying to force the pace. *Ibid.*, Greater Falangist control in foreign policy. 1955 *Ann. Reg. 1954* 215 The Falangist newspaper Arriba. 1957 P. Kemp *Mine were of Trouble* i. 15 Blue, tasselled *Falange* forage caps. 1961 H. Thomas *Spanish Civil War* l. 415 Franco took the opportunity to decree the establishment of his new party, uniting the Falange and Carlists. *Ibid.*, The Generalissimo, of course, would be the head of the new party though he had never been either a Falangist or a Carlist.

Falasha (falā·ʃā). [Amharic *falasha* exile, immigrant.] One of a group of people in Ethiopia holding the Jewish faith. Also as *collect. sing.*

1710 tr. *Tellez's Trav. Jesuits in Ethiopia* I. viii. 38 Betwixt the Emperor's Dominions and the Cafres..there are still many of these Jews, whom they there call Falaxas, which signifies, Strangers. 1790 J. Bruce *Trav.* I. 404 The Falasha..are a people of Abyssinia, having a particular language of their own;..they are now, and ever were, Jews. 1869 S. P. Goodhart tr. J. M. Flad (*title*) The Falashas (Jews) of Abyssinia. 1908 *Daily Chron.* 6 Mar. 3/5 Proposed School for the 'Falashas' of Abyssinia. 1908 J. Hastings *Encycl. Relig.* I. 165/2 The Falashas..are frequently called the Jews of Abyssinia. But they are certainly not Jews by descent... Possibly their Jewish faith is the survival of some earlier diffusion of Judaism through Abyssinia before the introduction of Christianity, as there is no record of their conversion. 1964 *Guardian* 26 Mar. 16/4 The Falasha, the so-called Black Jews of Ethiopia. 1970 Roth & Wigoder *New Standard Jewish Encycl.* 663 *Falashas*, Abyssinian tribes observing a form of Judaism, living N. of Lake Tana.

falciparum (fælsi·pārə̆m). *Path.* [mod.L., f. L. *falc*(*i*)-, *falx* sickle + *-parum* (see -parous).] The specific epithet of a protozoon of the genus *Plasmodium*, used *attrib.*, to designate a severe type of malaria caused by this protozoon, in which paroxysms recur at approximately two-day intervals. Also called *æstivoautumnal*, *malignant*, and *subtertian malaria*.

1930 M. F. Boyd *Introd. Malariology* ii. 49 The extent of gametocyte production appears to vary directly with the intensity of injection, at least in cases of *vivax* and *falciparum* infections. 1940 *Amer. Jrnl. Public Health* Apr. 410/1 New York City is an endemic area for the occurrence of falciparum malaria. 1952 P. F. Russell *Malaria* 33 Acute falciparum malaria sometimes imitates other abdominal conditions such as pancreatitis..and acute appendicitis. 1965 *Listener* 18 Mar. 406/1 People who inherit a sickle cell gene..possess a greater resistance to falciparum malaria.

falconet. **2.** For def. read: A pygmy falcon, especially one belonging to the genus *Microhierax*.

1883 E. W. Oates *Birds Brit. Burmah* II. 212 *Microhierax fringillarius*, the Black-legged Falconet. *Ibid.* 213 The Black-legged Falconet occurs in Tenasserim. 1931 *Discovery* Nov. 348/1 The black-legged falconet (which breeds in holes). 1961 O. L. Austin *Birds of World* 83/2 The falconets, or pygmy falcons, of the subfamily Poliohieracinae occur in South America, Malaya, the Philippines, and Africa. 1964 Fisher & Peterson *World of Birds* 31/2 The Falconidae include..the tiny insect-hawking falconets, not much bigger than sparrows.

falconish, *a.* For †*Obs. rare*$^{-1}$ read *rare*. (Later example.)

1833 Carlyle *Crit. & Misc. Ess.* in *Wks.* (1869) IV. 351 The vulturous and falconish character of our Isle.

‖ **fale** (fa·le). [Samoan.] A house of the type built by the Samoans (see quot. 1959); also *transf.* (see quot. 1966).

1902 L. P. Churchill *Samoa 'Uma* iii. 36 Her [*sc.* a Samoan woman's] house ceases to be *fale*, but becomes *maota*, a mansion. 1959 A. McLintock *Descr. Atlas N.Z.* 81 The Samoan *fale* with its compounded floor, open sides, and leaf-thatch roof is admirably suited to the climate [of Samoa]. 1961 C. C. Marsack *Samoan Medley* i. 18 The houses are almost without exception the Samoan type of *fale*, a thatched roof standing on posts, open on all sides and with no interior partitions. 1966 *Courier Mail* (Brisbane) 21 Jan. 11 No grass hut here, but a luxurious six-bed holiday home... This is the first of 10 'fales'—Samoan houses—being built on..Fraser Island, off the Queensland coast... The fale is fitted with hot and cold water, [etc.].

Faliscan (fāli·skăn), *sb.* and *a.* Also -ian. [f. L. *Falisc-us* + -an.] **A.** *sb.* **a.** A native or inhabitant of the ancient Etrurian city of Falerii. **b.** The dialect or language of Falerii and its environs, or its alphabet, of Etruscan origin. **B.** *adj.* Of or pertaining to Falerii, its inhabitants, dialect, or produce.

1600 Holland *Livy* v. 197 It was the manner amongst the Faliscians, to entertaine..one that should both teach and instruct them [*sc.* their children] in schoole. 1686 tr. *Roman Hist. written in Latine by Titus Livius* VII. xvii. 183 The whole Nation of the Tuscans took Arms, and led on by the..Faliscans, were advanced as far as Salinae. *Ibid.* 'Table', Faliscan Priests with flaming Torches. 1875 *Encycl. Brit.* I. 610/2 The second [family of Italian alphabets contains] the Latin and Faliscan... These three languages are all written from right to left, in which the Faliscan agrees with them. 1883 I. Taylor *Alphabet* II. 127 The Faliscan and the Latin [alphabets], wedged in between the Etruscan and the Oscan. 1932 *Times Lit. Suppl.* 21 July 530/4 The last five plates figure Italian fabrics, Faliscan, Apulian and Campanian. 1965 W. S. Allen *Vox Latina* iii. 64 A Faliscan inscription.

fall, *sb.*^{1} Add: **1.** *to ride for a fall*: see Ride *v.* 1 d.

2. *fall of the leaf*: delete † and add later example. For 'In U.S. the ordinary name for autumn' read 'In N. Amer. the ordinary name for autumn' and add Canadian examples.

1697 H. Kelsey *Papers* (1929) 95 Thos. Bullears boy..died of ye rivers mouth last fall. 1767 *Quebec Gaz.* 5 Jan. 3/1 A few barrels of pickeled *cod fish*, taken..last Fall. 1849 J. E. Alexander *L'Acadie* II. i. 10 A week at Loughborough in the 'fall' is delightful. 1922 C. E. Montague *Disenchantment* ix. 132 The fall of the leaf had brought, too, a sad shortage of heroes. 1936 D. McCowan *Anim. Canad. Rockies* xi. 99 In the Rockies they stay out until quite late in the fall.

12. *spec.* A descent to the floor in the technique of modern dancing.

1949 Shurr & Yocom *Mod. Dance* v. 174 This fall requires a strong contraction of hip and abdominal muscles. *Ibid.* 190 The use of the floor as a momentary base in a total movement sequence of fall and recovery.

c. spec. of a wicket in Cricket.

1871 'Thomsonby' *Cricketers in Council* 30 Swiping..or rushing far out of your ground to meet the ball..are equally certain to result in the ultimate fall of your wicket. 1966 J. Arlott in B. Johnston *Armchair Cricket 1966* 9 Every run and fall of a wicket.

d. *Tennis.* (See quot.)

1900 G. E. A. Ross in A. E. T. Watson *Young Sportsman* 609 The second contact of the ball with the floor, called the *fall* of the ball, at any point where chase-lines are painted.

18. *spec.* in *Criminals' slang.* **b.** An arrest.

1893 [see *fall money* s.v. sense *29]. 1894 *Reminisc. Chief-Insp. Littlechild* xx. 204 This man..is now in prison on the Continent. The story of his last 'fall' is interesting. 1935 [see sense *18 c].

c. A period of imprisonment.

1934 *Amer. Speech* IX. 26/2 *Fall*, v. to go to prison. n. a term in prison. 1935 N. Ersine *Underworld & Prison Slang* 35 *Fall*. 1. An arrest. 2. An imprisonment. 'That last fall got him the book.'

23. c. (See quot. 1933.) *Austral.* and *N.Z.*

a 1921 G. H. Gibson in *Penguin Bk. Austral. Ballads* (1964) 206 He could..Chop his name with a green-hide fall on the flank of a flyin' steer. 1933 L. G. D. Acland in *Press* (Christchurch) 14 Oct. 15/7 *Fall*, a plain strip of leather at the end of the lash of a whip to which the cracker is attached.

24. b. The long hair hanging down the faces of certain terriers.

1908 J. Maxtee *English & Welsh Terriers* 86 When the coat is sufficiently long to justify its being done, the 'fall' (as the long hair covering the face is termed by fanciers) should be plaited and tied up over the head with ribbon. 1948 B. Vesey-Fitzgerald *Bk. Dog* 996 *Fall*, the long fringe of hair overhanging the face of the Yorkshire, Skye and Clydesdale terriers.

28. a (Further examples.)

1906 *Daily Colonist* (Victoria, B.C.) 17 Jan. 1/5 The condition of the fall wheat of Alberta today is declared to be A1. 1926 *Ibid.* 18 July 2/5 Officials have been working indefatigably in their efforts to this year promote the most successful Fall Fair that this community has ever experienced. 1928 M. de la Roche *Whiteoaks* (1929) ii. 27 He had a new fall overcoat of expensive English melton.

29. fall-block, either of the two lower blocks of a boat's falls; **fall-board** (earlier example); **fall-breaker** (Break *v.* 28 b), that which reduces the impact of a fall or the speed of falling; **fall-front** *sb.* and *attrib.* = *drop-front* (s.v. *drop- a); **fall guy** *slang* (orig. *U.S.*), one who is easily tricked, an easy victim; one who 'takes the rap' for others, a scape-goat; **fall-leaf** *U.S.*, a table-leaf which can be let down; also *attrib.*; **fall(-)line,** (*a*) a line in the eastern United States marked by waterfalls and rapids at points where rivers cross it and forming the boundary between the Atlantic coastal plain and the Piedmont; hence in *Physical Geogr.*, any line joining points of sudden descent on rivers and representing a similar geological discontinuity (see quot. 1968); (*b*) *Ski-ing* [ad. G. *fallinie*] (see quot. 1961); **fall money** *Criminals' slang*, money set aside by a criminal for use if he should be arrested; **fall-pipe,** (*a*) the sloping inlet pipe of a water-ram; (*b*) the pipe conveying rainwater from a roof, etc., to the ground; **fall-rise** *Phonetics* (see quot. 1932); **fall-rope,** a rope used for lifting; **fall zone** = *fall-line (*a*).

a 1877 Knight *Dict. Mech.* 678/1 s.v. *Davit*. When the boat is lowered the hooks of the fall-blocks are cast off simultaneously. 1883 *Man. Seamanship for Boys* 136 Fall Blocks, for Top-Tackle Pendants. 1743 W. Ellis *Mod. Husbandm.* (1744) III. i. July 86 The Entrance [being] afterwards stopped with a Fall-board, so that no Pole-cat..can in the least molest them. 1927 *Daily Express* 31 Oct. 13/2 The competitors climb like cats to the top of a tall tower... They jump and dive eighty-five feet into fall-breakers. 1946 W. F. Burbidge *From Balloon to Bomber* iii. 35 Leonardo da Vinci..in his manuscript *Codex Atlanticus* (*circa* 1514) described a device, aptly called a 'fall-breaker', consisting of 'a tent of chalked linen' which would enable a man to 'fall from any great height without danger to himself'. 1902 W. H. Hackett *Decorative Furnit.* xiii. 160 A satinwood upright secretaire with fall-front. 1907 *Yesterday's Shopping* (1969) 372/1 Fall front spirit stands..2-pint half cut bottles..34/0. 1929 *Times* 2 Nov. 7/5 A Louis XV. kingwood marqueterie escritoire, with..the fall-front drawer, and pigeon holes. 1955 R. Fastnedge *Eng. Furnit. Styles* iii. 87 Small walnut and marquetry fall-front writing cabinets on chests of drawers or stands. 1906 H. Green *At Actors' Boarding House* 226, I never thought I'd be the fall guy for such raw work as this. 1912 G. Bronson-Howard *Enemy to Society* ix. 293 We ain't goin' to be th' 'fall guys' for Steve... If we've got to do time, so has he! 1929 'C. G. Gordon' *Crooks Underworld* 118 The grafter becomes the 'fall guy', and the solicitors ensure immunity from scandal by falling back on the dignity of their profession. 1935 *Amer. Speech* X. 15/1 *Fall guy*, [formerly] one who bungles his work and usually gets caught; a clumsy thief. This word now means one who is made the goat, one who involuntarily *takes the rap* for the rest of the gang. 1946 'P. Quentin' *Puzzle for Fiends* (1947) xxi. 149 Selena..would deliver me as a fall guy to the police without batting an eye. 1956 S. Bellow *Seize Day* (1957) i. 34 Perhaps he was foolish and unlucky, a fall guy, a dupe, a sucker. 1959 'M. Ainsworth' *Murder is Catching* vi. 77 There's never been a political murder yet..without there being some sort of a fall guy, prepared in advance. 1963 *Spectator* 6 Sept. 295 Ward began to hear from friends that he was being cast for the part of fall guy (I know of no equivalent expression here) by Profumo's friends. 1854 B. F. Taylor *Jan. & June* 204 It was the old table with the fall leaves. 1882 C. Waite *Adv. Far West* 189 [A] fall-leaf table. 1882 *Nation* (N.Y.) 13 July 33/1 It is here, at the 'fall-line', that the most available water-powers are to be found. 1902 Ld. Avebury *Scenery of Eng.* xvi. 481 A similar line along the junction of the uplands with the sea-plain is known in the United States as the 'fall-line'. 1955 *Sci. Amer.* Mar. 82/2 In the Atlantic terrace the thin edge of the sedimentary wedge reaches inland to the 'fall line' (a zone of waterfalls) on the eastern side of the Appalachian Mountains. 1961 J. S. Salak *Dict. Amer. Sports* 152 *Fall line*, the natural way down a slope—the course a rolling ball would take. On a perfect inclined plane, it would be the shortest straight line from any point on the slope to the bottom. 1968 R. W. Fairbridge *Encycl. Geomorphol.* 344/1 The term 'fall line' or 'fall zone'..has been adopted in geomorphology as a general term to identify the boundary between a more resistant crystalline upland or plateau province and a coastal plain of weak rocks. 1970 *Times* 13 Feb. 13/5 Some series of gates are ranged down the fall line—the straight line down the mountain. 1893 L. W. Moore *His Own Story* xii. 197 If any accident happened to us, Hall was to stand his part of the 'fall' money. 1929 'C. G. Gordon' *Crooks Underworld* 215 We had often discussed the matter of 'fall money'. 1888 *Lockwood's Dict. Mech. Engin.* 134 *Fall pipe*, a drive pipe. 1929 *Daily Tel.* 22 Jan. 12/4 When cornered he slid down a fallpipe. 1932 D. Jones *Outl. Eng. Phonetics* (ed. 3) xxxi. 282 If there are several syllables following the emphatic fall, the terminal rise is spread over these. But if there is only one unstressed syllable following, the terminal rise is compressed into it. If there is no following unstressed syllable, the terminal rise is compressed into the same syllable as the emphatic fall; the emphatic syllable is therefore said in this case with a fall-rise. 1962 S. Stubelius in F. Behre *Contrib. Eng. Syntax* 200 Rise-endings and fall-rise endings are by no means rare in statements. 1889 *Cent. Dict.*, Fall-rope. 1891 Kipling *Barrack-r. Ballads* (1892) 205 And the fall-rope whines through the sheave. 1933 *Times Lit. Suppl.* 18 May 350/3 Someone had..thrown the body overboard, but a fall-rope caught it by the neck. 1929 H. S. Sharp in *Science* LXIX. 545/1 The erosional surface to which it is proposed to apply the term Fall Zone peneplane is of considerable importance. 1937 Wooldridge & Morgan *Physical Basis Geogr.* xv. 220 The Fall Line, or, more accurately, the Fall Zone of North America, where falls and rapids mark the descent of the streams from the 'old-land' to the coastal plain. 1968 Fall zone [see *fall-line, above].

30. fall-in *Mil.* [Fall *v.* 87 g] (see quot. 1852); **fall-off** (earlier and later examples).

1852 R. Burn *Naval & Mil. Techn. Dict.* (ed. 2) II. 92/1 *Fall-in*, beat of drum or trumpeter's call. 1880 G. M. Hopkins *Let.* 5 Sept. (1935) 109, I think the Lethe mythology of the last stanza is a fall-off and unrealises the whole. 1915 'I. Hay' *First Hundred Thousand* vi. 42 Here the 'fall-in' sounded. 1958 M. L. Hall et al. *Newnes Compl. Amat. Photogr.* x. 116 It [*sc.* bounce flash] produces soft, even lighting and reduces the effect of background fall-off. 1969 *Listener* 5 June 786/1 There is going to be a fall-off in orders.

fall, *v.* Add: **1. g.** *Tennis.* (See quots.) (Cf. *fall *sb.*^{1} 12 d.)

1900 G. E. A. Ross in A. E. T. Watson *Young Sportsman* 610 Y returns the first service so that X cannot reach the ball. The marker, seeing that it 'falls' (*i.e.*, comes into contact with the floor for the second time) on the line 2, calls out 'chase 2'. 1957 *Encycl. Brit.* XIII. 793/1 A chase..is made, the goodness or the badness of which depends upon the spot on the floor which the ball touches next after its first bound... Thus if a ball fell on the line marked 4, he [*sc.* the marker] would call 'chase 4'.

19. b. *to fall flat*: see Flat *a.* 7 c.

d. (Earlier example.) Also, of the batsman, with *to*: to be out.

1773 J. Duncombe in R. Freeman *Kentish Poets* (1821) II. 365 Full fast the Kentish wickets fell. 1882 *Australians in Eng.* 147 Grace fell to a catch at long-on. 1888 R. H. Lyttelton in Steel & Lyttelton *Cricket* 344 Ridley again fell to Patterson, with the total at 16.

22. b. To become pregnant.

1722 *Session Bk. Penninghame* (1933) I. 479 The said..Jannet..confessed that she fell with child and parted with it in May last. 1891 Farmer *Slang* II. 370/2 *Fall* (venery), to conceive. 1957 Young & Willmott in 'C. H. Rolph' *Human Sum* vii. 129 The expression a woman uses when she is pregnant. She says she has 'fallen'. 'We had been married eight months before I fell.'

23. f. *Criminals' slang.* To be arrested. Cf. *fall *sb.*^{1} 18 b.

1879 *Macm. Mag.* XL. 502/2, I fell (was taken up) again at St. Mary Cray for being found at the back of a house. 1938 F. D. Sharpe *Sharpe of Flying Squad* xiv. 158 'What, has he fell *again*!' Getting caught [by the police] is always referred to as 'falling'.

g. *Criminals' slang.* To be convicted; to be sent to prison. Cf. *fall *sb.*^{1} 18 c.

1893 L. W. Moore *His Own Story* xxxiv. 447, I want you to follow my instructions when the case is tried, and if I fall I will find no fault with you. 1926 J. Black *You can't Win* ix. 108 If you do fall, the government don't hang a lot of prior convictions on you. 1932 'S. Wood' *Shades Prison House* ix. 140 A young Jew at Parkhurst who fell for three years at the game simply because he failed..of ascertaining the actual *locale* of the university of which he claimed to be an alumnus. 1934 [see *fall *sb.*^{1} 18 c].

32. b. (Earlier *Sc.* and later northern examples.)

1845 *New Statist. Acct. Scotl.* XV. (Caithness) 16 Eminent men. Under this head falls to be noticed the late Rev. David Mackay. 1892 H. R. Mill *Realm of Nature* i. 3 They alone fall to be considered here.

51. c. (Still current in Australia and N.Z.)

1860 A. F. Ridgway *Voices from Auckland* 67 Large quantities of bush have this season been 'fallen'. *Ibid.* 67 When to fall his bush? When to burn off? 1869 J. May *Guide to Farming in N.Z.* 18 We had a dread of commencing to fall bush. 1882 W. D. Hay *Brighter Britain* I. vii. 186 In felling bush, or 'falling' it, as we say here, advantage is taken of the lay of the land. 1891 R. Wallace *Rural Econ. Austral. & N.Z.* xv. 231 In the case of bush country the trees require to be 'fallen', and the under-scrub cut. 1941 Alley & Hall *Farmer in N.Z.* iv. 99 The bush was then felled, or 'falled'.

59*. fall for ——. To be captivated or carried away by; to yield to the attractions of; (in a bad sense) to be taken in by. *slang* (orig. *U.S.*).

1903 R. L. McCardell *Conversat. Chorus Girl* 28 The mayor fell for it. 1911 L. J. Vance *Cynthia* xi. 179 There's only one sensible thing... And I think I see you falling for it. 1914 *Sat. Even. Post* 23 May 62/3, I fell for her the first time I seen her. 1916 H. L. Wilson *Somewhere in Red Gap* v. 186 Even Mis' Judge Ballard fell for it, though hers were made of severe black with a long coat. 1917 A. G. Empey *From Fire Step* xxii. 152 When firing my gun, I was to play my tune, and Fritz, no doubt, would fall for it, try to imitate me as an added insult. 1924 R. Macaulay *Orphan Island* xix. 250 European critics, for ever falling for new things, would fall most certainly for this. 1926 Galsworthy *Silver Spoon* II. viii, 'He's fallen for Marjorie Ferrar.' '"Fallen for her"?' said Soames. 'What an expression!' 'Yes, dear; it's American.' 1929 J. J. Farjeon *Underground* xxi. 139, I held out my pocket-case, and said I'd found it on the floor of the hotel. 'Is it yours?' I asked. To my surprise, he fell for it beautifully. 1952 L. A. G. Strong *Darling Tom* xvii. 138 So he had fallen for it, had he, the big sap? 1958 *Times Lit. Suppl.* 12 Sept. 509/4 He meets and falls for a twenty-year-old dipsomaniac.

60. fall from ——. c. Delete † and add later example.

1872 C. E. Maurice *Stephen Langton* iv. 254 The followers of Louis were falling from him.

62. fall into ——. f. Delete † and add later example.

1967 E. Short *Embroidery & Fabric Collage* iii. 81 Curtains similarly fall into two main categories, namely opaque and transparent.

64. fall on ——. h. (Earlier example.)

1857 Trollope *Barchester T.* III. xvii. 298 It is well known that the family of the Slopes never starve: they always fall on their feet, like cats.

81. fall back on. b. (Earlier example with reference to money.)

1864 C. M. Yonge *Trial* II. v. 103 Her own fortune had always appeared to her something to fall back on in case of want of success.

84. fall down. f. To 'come to grief', collapse, fail. Freq. with *on*. *slang* (orig. *U.S.*).

1873 J. H. Beadle *Undevel. West* xxxii. 704 We'll reach Sioux City by 5 o'clock if we don't fall down... But we did 'fall down' just at noon, running hard aground. 1899 J. L. Williams *Stolen Story* 223 It amazed him every time a new reporter..fell down on a story. 1928 *Publishers' Weekly* 26 May 2175 We know of many cases where we fell down on buying books written by authors that had had successful books before. 1928 *Daily Tel.* 14 Aug. 10/4 If we fall down on the job of absorbing these 10,000 men of good stock and stamina. 1953 R. Lehmann *Echoing Grove* 300, I felt I couldn't cope: I saw myself falling down on the responsibilities. 1959 *Spectator* 6 Feb. 181/2 The Congress party is falling down on the job of rallying public confidence in the present policies of the Government of India. 1961 A. Christie *Pale Horse* x. 112 Oh dear, I seem to have fallen down badly. I was so very sure.

87. fall in. g. Also, usu. as a command, *get fell in*.

1930 E. Raymond *Jesting Army* III. v. 327, I was standing with the crah'd watchin' the West Essex getting fell in. 1945 C. H. Ward-Jackson *Piece of Cake* (ed. 2) 32 '*Get fell in!*' Fall in! Common way for senior N.C.O.s to give the command. 1959 *N.Z. Listener* 10 July 4/4 It was perfectly dreadful for instance to be fallen in—or as the Navy says, to 'Get fell in'.

90. to fall in with. a. (Later example.)

1955 *Times* 11 Aug. 9/5 Senhor Alberto de Lacerda is a Portuguese writer in his late twenties, who has had the good fortune to fall in with Mr. Arthur Waley.

91. fall off. f. Also of persons.

1914 W. OWEN *Let.* 14 Oct. (1967) 287 If pupils fall off I shall be obliged to pack off sooner!

93. fall out. b. Also *trans.*

1923 KIPLING *Irish Guards in Great War* I. 24 A newly appointed platoon-officer..fell them out. **1942** E. WAUGH *Put out More Flags* ii. 120 Companies paraded at quarter past eight; immediately after inspection men were fallen out for the company commanders' orderly room.

95. fall over. d. *to fall over oneself*: (*a*) see quot. 1895; (*b*) to show great eagerness (freq. to do something), to be desperately anxious to. *colloq.*

1895 W. C. GORE in *Inlander* Dec. 113 *Fall all over one's self*, to get confused. **1904** *Brooklyn Standard Union* 2 Aug. 6 The bonafide independent element is not falling over itself to come to Parker's assistance. **1923** R. D. PAINE *Comr. Rolling Ocean* 130 The thirsty outlaws fall over themselves to hand you ten or twelve dollars a quart for it. **1935** *Forres Gaz.* 30 Oct. 4/7 In your haste to 'fix up' a piece of business, have you ever 'fallen over yourself'? **1947** K. TENNANT *Lost Haven* iii. 48 Why was it that..these rich coots with tons of money nearly fell over themselves grabbing at anything they could get free? **1957** *Ess. in Criticism* VII. 159 [A review] which, though it falls over itself at the end to be as nice as possible, [etc.].

e. *to fall over backwards* = *to bend over backwards* (*BACKWARDS *adv.* A).

1958 B. NICHOLS *Sweet & Twenties* xi. 135 Even nice old ladies..fell over backwards in their efforts to stop the menace of the encroaching scissors. **1958** *Listener* 20 Nov. 848/1 There was a certain complacency, a falling over backwards to be fair. **1966** *Ibid.* 11 Aug. 190/1 In the system of trial is it correct to say, as many people say, that we fall over backwards to protect the guilty?

100. fall together. Restrict †*Obs.* to senses in Dict. and add: **d.** *Phonetics.* To become identical.

1905 J. WRIGHT *Eng. Dial. Gram.* p. iv, Many vowel-sounds, which are generally supposed to have fallen together in Middle English, were in reality kept apart. **1959** A. CAMPBELL *Old Eng. Gram.* 153, æ, e, and i fell together in a sound written e in unaccented syllables.

fallacy. Add: **3.** In certain phrases in the formal terminology of Logic, as *fallacy of accident* (see quots.); *fallacy of composition* (see COMPOSITION 4 b); *fallacy of division*, the fallacy that whatever is true of a whole must be true of any part or member of that whole.

[**1552** R. ASCHAM *Let.* 12 July in H. Ellis *Orig. Lett. Lit. Men* (1843) 12 Lest the *fallax* of composicion and division..inverte the sentence. **1599** BLUNDEVIL *Logike* 166 *Fallacia Accidentis*..: which may bee englished thus: the fallax of the Accident.] **1685** tr. *Arnauld & Nicole's Logic* III. xix. 108 To judge of a thing which only agrees with it by accident. This Sophism is call'd in Schools Fallacia accidentis, the Fallacy of the Accident. *Ibid.* 109 To pass from sence divided to sence compos'd..is call'd Fallacia Compositionis, Fallacy of Composition. *Ibid.*, To pass..from sence compos'd to sence divided..is call'd..Fallacy of division. **1870** W. S. JEVONS *Elem. Lessons Logic* xxi. 176 The fallacy of *accident* consists in arguing erroneously *from a general rule to a special case*, where a certain accidental circumstance renders the rule inapplicable. **1961** J. G. BRENNAN *Handbk. Logic* (ed. 2) x. 213 The Gambler's Fallacy may be construed as an example of the fallacy of division. **1967** *Philosophy* XLII. 7 An avoidance of the fallacy of accident or *secundum quid*.

b. Also in extended use (cf. 4, 5): *fallacy of misplaced argument* (see quot. 1942); *fallacy of misplaced concreteness* (see quots.); *fallacy of simple location*, acc. to A. N. Whitehead, an assumption that underlies the whole of science since the 17th century, viz. a form of materialism which holds that the space-time location of a material object is not dependent on reference to other space-time regions; *fallacy of the inversion of parts* (see quot. 1867); *fallacy of the perfect dictionary* (see quot. 1938).

1867 J. S. MILL *Exam. Hamilton's Philos.* (ed. 3) vii. 127 There is a mode of controversy which I do not remember to have seen in any enumeration of Fallacies, but which will some day find a place there, under some such name as the Inversion of Parts. It consists in indignantly vindicating as against your adversary the very principle which he is asserting against yourself. **1925** A. N. WHITEHEAD *Sci. & Mod. World* (1926) iii. 64 The accidental error of mistaking the abstract for the concrete..is an example of what I will call the 'Fallacy of Misplaced Concreteness'. [*Ibid.* iv. 72 There is no element whatever which possesses this character of simple location.] **1927** B. RUSSELL *Analysis of Matter* xxxii. 340 Dr Whitehead..is not open to such a charge; his 'fallacy of simple location', when avoided, leads to a world-structure quite different from that of common sense and early science. **1929** A. N. WHITEHEAD *Process & Reality* I. i. 9 The 'fallacy of misplaced concreteness'..consists in neglecting the degree of abstraction involved when an actual entity is considered merely so far as it exemplifies certain categories of thought. **1938** —— *Modes of Thought* 235 The very natural belief, that mankind has consciously entertained all the fundamental ideas which are applicable to its experience. Further it is held that human language, in single words or in phrases, explicitly expresses these ideas. I will term this presupposition, the Fallacy of the Perfect Dictionary. **1942** R. G. COLLINGWOOD *New Leviathan* iv. 73 The fallacy of arguing about questions like this is what I call the *Fallacy of Misplaced Argument*; which may be defined as the fallacy of arguing about any object immediately given to consciousness.

fal-lal, *sb.* and *a.* Add: **A.** *sb.* **4.** *fig.*

1957 *Technology* Mar. 3/4 Some people may find it rather odd, in a journal professedly devoted to training and education for industry, to come across film reviews, music notices and such fal-lals.

B. *adj.* **a.** (Later examples.)

1830 H. LEE *Mem. Manager* II. vi. 9 Are no lacqueys kept to administer to idle caprices? To run on fal lal messages, or carry complimentary cards, parrots or lapdogs? **1847** A. & H. MAYHEW *Greatest Plague* vii. 87 Lo and behold! my neat, unpretending chrysalis had changed into a flaunting fal-lal butterfly.

fal-lal, *v.* [f. FAL-LAL *sb.*] *intr.* To behave or dress in an affected or finicking manner; to idle, dally, procrastinate.

1854 A. E. BARKER *Gloss. Northants.* I. 219 *Fallals*, flaunting finery. It is also used participially to a female dressed out beyond her station. 'See how she's *fallal'd* out.' **1857** TROLLOPE *Three Clerks* II. iv. 72 She can't go on fal-lalling with you, and then nothing to come of it. **1864** —— *Small Ho. at Allington* II. xvi. 161 If I was to be fal-lalling with that married woman, you wouldn't like it. **1866** —— *Last Chron. Barset* I. xxvi. 233 My wife can do very well without you fal-lalling here all day.

fall-away, *sb.* Restrict †*Obs.* to sense in Dict. and add: **2.** A falling off.

1878 B. F. TAYLOR *Between Gates* 232 On midsummer the Yo Semite is less of a fall than a fall-away. **1879** G. M. HOPKINS *Let.* 14 Aug. (1935) 86 A fall-away or diminuendo.

fall-back, *sb.* Add: **c.** *concr.* Fall-back pay.

1961 *Observer* 7 May 40/6 Men who failed to find a day's work should get a guaranteed minimum payment (the 'fall-back'). **1969** *Daily Tel.* 11 Apr. 29/3 The men are paid on the basis of the £17 'fall-back' plus piece-work, which gives them average earnings of about £30 a week.

fall-back, *a.* **1.** Of a chaise, etc.: having a back which can be let down. *U.S.*

1767 *Boston Gaz.* 12 Oct. (Th.), A Fall-back Chaise for sale. **1768** *Essex Inst. Hist. Coll.* XI. 38 He also has five fall-back chaises, one fall-back curricle. **1832** *Coll. New H. Hist. Soc.* III. 37 He was the proprietor of a fall back chaise.

2. That may be used in an emergency; *esp.* of a minimum wage paid when work is not available.

1930 *Economist* 6 Dec. 1057/2 The operatives..laid stress on the absence of any provision for a fall-back wage, and a minimum wage to provide for the contingency of under-employment. **1942** W. ROSE *Good Neighbours* iii. 23 The work lasted through the summer, and provided a fall-back job when weather conditions were unsuitable for hay work. **1959** *Times* 8 Dec. 13/6 By the time the tug strike was a few days old 5,000 dockers were without work and receiving 'fall-back' pay.

fallen, *ppl. a.* Add: **3.** *fallen arch*: see *ARCH *sb.* 4 b; so *fallen instep*.

1926 M. D. POST *Man Hunters* 292 (*caption*) Flat foot or a fallen arch. **1948** A. BARON *From City from Plough* 11 All the other ranks 'll have fallen arches or double ruptures. **1954** L. MACNEICE *Autumn Sequel* IX. 59 A life of fallen insteps and surplus fat.

4. *spec.*, of those who have died in battle.

1765 H. BLAIR *Crit. Diss.* (ed. 2) 70 He professedly delights in strife and blood. He insults over the fallen. **1819** [in Dict.]. **1825** G. R. GLEIG *Subaltern* vii. 109 Desiring..a fellow..to keep guard near one of his fallen comrades. 'I don't care for living men,' said the soldier; 'but..don't keep me beside *him*.' **1871** [in Dict.]. **1914** *New Republic* 5 Dec. 14/2 Mr. Shaw referred to the miserable pensions that are being paid to the widows of fallen soldiers. **1914** W. D. NEWTON *War* xi. 119 The corpses of the fallen were trodden and ground beneath stumbling heedless soles. **1917** L. BINYON (*title*) For the Fallen and other poems. **1919** *Times* 12 Nov. 16/5 The majority of the citizens willingly consented to pay a tribute to the memory of the fallen.

6. Comb. *fallen-in* (cf. FALL *v.* 87).

1900 *Westm. Gaz.* 14 Feb. 6/3 Inrush of snow and sleet through the fallen-in roof. **1938** E. BOWEN *Death of Heart* I. ii. 47 Thomas sits so fallen-in, waiting for Anna, that his clock makes the only sound in his room.

fallenness (fǭ·l'n,nės). [f. FALLEN *ppl. a.* + -NESS.] The state of being fallen; *esp.* degeneracy consequent upon the Fall.

1871 W. GRAHAM *Lect. Ephesians* vi. 364 It teaches the fallenness of our whole nature. **1912** W. H. MOBERLY in *Foundations* 283 What is important is not an historical Fall, but the fact of 'fallenness'. **1923** *Expos. Times* July 439/2 It fails to express the fallenness of man's original animalism and degradation. **1952** V. A. DEMANT *Religion & Decl. Capitalism* iv. 86 The fallenness of man consists in a certain net of evil in which mankind is caught because of his refusal to live obediently.

faller. Add: **1.** *faller out* (later example). Also *faller away*.

1835 COBBETT *Legacy Parsons* ii. 54 The disrespectful name of *Dissenters*; that is to say, fallers away from the true faith. **1964** *Cath. Gaz.* June 161/2 Most of the fallers out..tended to be..Catholics who were not deeply involved.

2. Delete 'Only *dial.*'

1905 *Terms Forestry & Logging* 37 *Faller*, one who fells trees. **1908** M. A. GRAINGER *Woodsmen of West* xiv. 84 The 'fallers' had worked along the slope. **1926** K. S. PRICHARD *Working Bullocks* ii. 23 The faller, Bob Carew, who was giving a lift on the rake. **1941** *Time* 11 Aug. 16/2 The faller (who chops and saws the tree). **1966** *Daily Colonist* (Victoria, B.C.) 4 Feb. 17/6 Mr. Peterson is employed as a faller by Butler Brothers. **1971** *Timber Trades Jrnl.* 14 Aug. 72 (Advt.), A gang of experienced fallers required for a parcel of beech and oak near Portsmouth.

5. b. A kind of stamp (see quots.).

a **1884** KNIGHT *Dict. Mech.* Suppl. 326/1 *Faller*, a vertical stamp in a fulling, milling, or stamping machine; usually lifted by cams and allowed to drop vertically and endwise. *Ibid.*, *Faller Machine*, a machine with vertical stamps or fallers, used in milling goods.

fallibilism (fæ·libiliz'm). [f. FALLIBLE *a.* + -ISM.] The principle that propositions concerning empirical knowledge cannot be proved.

c **1897** C. S. PEIRCE *Coll. Papers* (1931) I. I. iii. 61 Fallibilism..only says that people cannot attain absolute certainty concerning questions of fact. *Ibid.* 70 Fallibilism is the doctrine that our knowledge is never absolute but always swims, as it were, in a continuum of uncertainty and of indeterminacy. **1941** *Mind* L. 81 Fallibilism which denies intuitive or certain knowledge even of common-sense propositions.

fa·llibilist, *a.* [f. as prec. + -IST.] Maintaining or accepting the principle of fallibilism; pertaining to or resembling fallibilism.

c **1897** C. S. PEIRCE *Coll. Papers* (1931) I. I. iii. 63 Let us look then at two or three of the grandest results of science and see whether they appear any different from a fallibilist standpoint from what they would to an infallibilist. **1941** *Mind* L. 82 That all propositions must be either logically certain or equally corrigible is a rash bifurcation of which fallibilist philosophers should surely beware.

falling, *vbl. sb.* Add: **6. c.** Also *falling in* (*with*): the action of the vb. *fall in* (FALL *v.* 87, 90); opp. *falling out*.

1838 DICKENS *Let.* 16 Jan. (1965) I. 355 It almost seems as if we had had a mortal falling out. I hope we shall have a lasting falling in again, soon. **1919** H. CRANE *Let.* 13 Dec. (1965) 26 He has had a falling out with Amy Lowell, but a falling in with T. S. Eliot by way of compensation.

d. *falling-short*, the action of the vbl. phr. *fall short of* (FALL *v.* 97): failure, insufficiency; inability to reach.

1857 F. L. OLMSTED *Journey through Texas* p. viii, Certain fallings-short from the standard of comfort and of character in older communities are inevitable. **1916** A. BENNETT *These Twain* xi. 221 His life seemed to be a life of half-measures, a continual falling-short. **1936** *Discovery* Oct. 308/2 'My predecessors,' he said, 'have spoken of the shortcomings of the active world—to me they are but the fallings short of science.'

7. = FELLING *vbl. sb.* 1. Also *attrib.*, in *falling axe, rope, saw, wedge*. Now *N. Amer.*, *Austral.*, and *N.Z.*

1388 WYCLIF *Ps.* lxxiii. 6 Thei castiden down it with an ax, and a brood fallinge ax. **1580** [see Dict. 3]. **1678** *New Castle* (Delaware) *Court Rec.* (1904) 362, 3 falling axses. **1866** J. MURRAY *Descr. Province Southland* III. 29 For the falling, cross-cutting and splitting of his posts and rails, he will find it of advantage again to get a mate. **1875** G. C. DAVIES *Rambles School Field-Club* viii. 67 A 'falling rope'..that men attach to the top of a tree when they wish to cut it down. **1905** *Terms Forestry & Logging* 37 *Falling ax*, an ax with a long helve and a long, narrow bit, designed especially for felling trees. *Falling wedge*, a wedge used to throw a tree in the desired direction, by driving it into the saw kerf. **1943** R. E. SWANSON *Rhymes Lumberjack* 12 So huge they were (both cedar and fir) that days were spent in their falling. **1946** F. DAVISON *Dusty* xvii. 200 [He] came to know as much as a distant observer could of the business of falling, hauling and sawing pine logs. **1960** *Citizen* (N. Vancouver) 10 Mar. 10/1 Swinging balance and cut of keen falling-saw Bite of falling-axe, tap of falling-wedge.

8. *falling-in-love* [FALL *v.* 38 b, LOVE *sb.* 7 f], the state of becoming enamoured.

1859 *Nat. Rev.* Oct. 375 In the first place, a fighting period; and in the next place, a falling-in-love period. **1923** J. S. HUXLEY *Ess. Biologist* vii. 271 The commonest example is 'falling in love', where the simple sex-instinct becomes intertwined with other instincts. **1963** AUDEN *Dyer's Hand* IV. iv. 222 The traditional symbol in Western Literature for this kind of personal choice is the phenomenon of falling-in-love.

falling, *ppl. a.* Add: **2.** *falling diphthong*: see DIPHTHONG *sb.* note. Also *Comb.*, as *falling-rising* (tone).

1924 H. E. PALMER *Gram. Spoken Eng.* I. 13 Falling-Rising Nucleus-tone. **1964** R. H. ROBINS *Gen. Ling.* iii. 111 Falling-rising tones.

5. Applied to weather in which rain, snow, or hail falls or may be expected. *dial.* and *U.S.*

1733 FRANKLIN *Poor Richard's Almanac* 6 Windy and falling weather. **1760** WASHINGTON *Diaries* I. 112 The morning..promised much rain or other falling weather. **1780** in *Coll. New H. Hist. Soc.* IX. 176 Cloudy but no falling weather. **1838** C. GILMAN *Recoll. Southern Matron* xxv. 172 It looks like falling weather, and my old drab will come in well to-day. **1859** BARTLETT *Dict. Amer.* (ed. 2) s.v. *Weather*, 'We are going to have falling weather', means that we are going to have rain, snow or hail. **1884–** in dial. glossaries (Gloucester, Hereford, Warwick, Worcester). **1919** H. L. WILSON *Ma Pettengill* v. 165 Will you look at that mess of clouds? I bet it's falling weather over in Surprise Valley.

b. falling collar, a wide collar which lies flat; **falling leaf**, an aerobatic manœuvre in which an aeroplane is stalled and sideslipped while losing height; also *attrib.* and *transf.*

1862 *Revised Regs. Army U.S.* 478 A sack coat..made loose, without sleeve or body lining, falling collar, inside pocket on the left side. **1866** MRS. GASKELL *Wives & Daughters* I. vi. 62 Two boys, in the most youthful kind of jackets and trousers, and falling collars. **1918** H. BARBER *Aerobatics* 51 (*caption*) Falling Leaf. **1935** H. G. WELLS *Things to Come* IX. 82 Aeroplane looping the loop—then the falling leaf trick. **1957** BANNERMAN *Birds Brit. Isles* VI. 314 When performed at the lower altitudes the technique is often the individual 'falling-leaf', but at higher altitudes the birds rush downwards together and the formation is not broken. **1970** H. KRIER *Mod. Acrobatics & Precision Flying* ii. 56 The falling leaf is a series of checked spins, in which the airplane is allowed to fall off first to the right and then to the left, or vice versa.

fall-less (fǭ·l,lės), *a.* [f. FALL *sb.*^{1} + -LESS.] Having no fall (in various senses).

1897 F. THOMPSON *New Poems* 49 Then leaf, and flower, and falless [*sic*] fruit Shall hang together on the unyellowing bough. **1910** *Daily Chron.* 2 Feb. 8/2 The Indian..was nearly always the aggressor in the eighty minutes' fall-less wrestling which took place.

Fallot (fæ·lo). [Name of Étienne Louis Arthur *Fallot* (1850–1911), French physician.] *Fallot's tetralogy*, a form of congenital heart disease in which four abnormalities occur together, freq. accompanied by cyanosis.

The condition had been previously described by a number of writers before Fallot published his description in 1888.

1927 *Amer. Heart Jrnl.* III. 85 A case of pulmonary stenosis, dextroposition of the aorta, and septal defects (not strictly Fallot's tetralogy) in a man who lived to be fifty-two. **1970** *New Scientist* 26 Mar. 625/1 Techniques and results for the treatment of Fallot's tetralogy—a particularly difficult type of blue baby with which to deal.

fall-out, *sb.* [f. vbl. phr. *fall out*: see FALL *v.* 93.] Radioactive refuse of a nuclear bomb explosion; the process of deposition of such refuse. Also *attrib.*, *Comb.*, *transf.*, and *fig.*

[**1946** *Effects of Atomic Bombs at Hiroshima* (*Home Office*) vi. 15 The fall of a small fraction of the radioactive fission products.] **1950** *Effects of Atomic Weapons* 35 When the violence of the disturbance due to the bomb has subsided, the contaminated dirt particles gradually fall back to earth, giving rise to the phenomenon known as the *fall-out*. **1952** *N.Y. Times* 17 Aug. E9/6 Nevertheless, a good deal of radioactive stuff is picked up and carried by the wind and deposited all over the country.. So far there have been no dangerous concentrations of radioactive 'fall-out', as it is called, that is outside of the proving grounds in Nevada. **1954** *Time* 20 Dec. 66/3 The most recent H-bomb test (by the Russians) was made in Siberia about three months ago, but the fall-out of fear and worry that the H-bomb tests have caused has by no means died away. **1955** *Times* 19 July 8/5 An hour after the explosion, radio-active 'fall-out' material would be dropping 20 miles away. **1957** *Economist* 5 Oct. 16/2 A temporary cessation within the Labour party of personal bomb explosions and consequent deadly fall-out. **1961** *Guardian* 16 Sept. 7/1 Fallout-laden clouds. **1961** *John o'London's* 28 Sept. 345/1 The make-it-yourself fallout shelter. **1964** *Listener* 3 Dec. 865/2 The wasteful method of technical fall-out, of just hoping for ideas to seep through. **1965** *Sunday Times* (Colour Suppl.) 31 Oct. 35/1 After the A.D. 79 eruption this fallout [of ashes, etc.] buried Pompeii in a matter of hours. **1969** *Times* 2 Sept. 10/3 Two Japanese fishermen died after fall-out had reached their vessel.

false, *a.* Add: **2. b.** *false note*, also *fig.*

1876 *Academy* 9 Sept. 253/1 The minutest..examination [of an actress's part] will hardly discover a false note or a dropped stitch. **1917** T. S. ELIOT *Prufrock* 19 Inside my brain a dull tom-tom begins..That is at least one definite 'false note'. **1932** *Week-end Rev.* 13 Aug. 188/2 This episode, with its rather sentimental sequel, seems to strike a false note in the story.

d. or by any restrictive action or influence.

1768 BLACKSTONE *Comm.* III. viii. 127 To constitute the injury of false imprisonment there are two points requisite: 1. The detention of the person; and, 2. The unlawfulness of such detention. **1880** T. E. HOLLAND *Jurispr.* xi. 111 A man has a right to go where he pleases, ..and anyone who prevents him from so doing, whether by constraint actually applied, or by such show of authority or force as has an effect on the will equivalent to actual constraint, is said in English law to be guilty of 'false imprisonment'.

h. *false point* (POINT *sb.*^{1} C. 6): a mistaken act of pointing by a game-dog. Hence *false-pointing* vbl. sb., *false-point* vb.

1954 BANNERMAN *Birds Brit. Isles* III. 214 Ring-ouzels..possess an exceptionally strong scent. They are a frequent cause of 'false point' in gun-dogs..a Border terrier often winding the sitting bird at a greater distance than he would a snipe or woodcock. **1960** *Times* 24 Sept. 9/1 On scenting game they [*sc.* setters] halt..close enough..to avoid 'false-pointing'. **1965** P. WAYRE *Wind in Reeds* iii. 38 The best of dogs can false-point.

6. *false start.* (Examples.)

1815 *Racing Calendar* 102 Industry..was distanced by having been pulled up soon after starting, in consequence of a repeated cry of 'false start'. **1850**, etc. [see START *sb.*^{2} 5]. **1854** J. R. PLANCHÉ *Camp at Olympic* 13 Because I fling your follies in your face, And call back all the false starts of your race. **1957** G. RYLE in M. Black *Importance of Lang.* (1962) 149 It was a tragically false start.

8. b. Law. *false issue*, an issue introduced by counsel in order to conceal the real issue; *false pretences*, misrepresentations made to convey a false impression. Also *false representation*.

1757 *Act 30 Geo. II* c. 24 § 1 All persons who knowingly and designedly, by false pretence or pretences, shall obtain..money, goods, wares or merchandizes, with intent to..defraud any person..of the same. **1778** P. THICKNESSE *Year's Journey* (ed. 2) I. iii. 32 He may indeed say..that obtaining money by false pretences is better than forging. **1831** J. BANIM *Smuggler* I. xi. 253 Made prisoners in it, upon false pretences. **1836** *Mag. Dom. Econ.* I. 92 Indictable for obtaining money under false pretences. **1858** False representation [see REPRESENTATION 4]. **1861** *Act 24 & 25 Vict.* c. 96 § 88 It shall be sufficient in any Indictment for obtaining or attempting to obtain any such Property by false Pretences to allege that the Party accused did the Act with Intent to defraud, without alleging an Intent to defraud any particular Person. **1903** *Westm. Gaz.* 12 June 10/1 Mr. Robson: Then your lordship does not specify any false issues that you say I raised? His Lordship: The false issue you raised was whether or not the uncle was justified in going to the detective to find out the wife's residence instead of going to the father, and there were hours in the Divorce Court occupied upon that point. **1907** *Daily Chron.* 28 Aug., When charged he pleaded guilty to three charges of obtaining goods on false pretences. **1959** JOWITT *Dict. Eng. Law* I. 781/2 It is now practically immaterial whether a prisoner is indicted for false pretences or for larceny.

13. d. *false teeth* (earlier examples). Also *false eyelashes, nose*. Also in more general sense.

1795 E. WYNNE *Diary* 31 Dec. (1937) II. 59 The poor Man has a dozen false teeth in his mouth. **1836** DICKENS *Sk. Boz* I. 329 Gentlemen..with..false noses. **1873** *Young Englishwoman* Oct. 518/2 Will the Editor kindly say if there is anything revolting in wearing false teeth? **1902** *Encycl. Brit.* XXXII. 605/1 Where pheasants exist in any number, a 'false covert' of spruce and fir loppings should be made at the point to which it is desirable to force the birds. **1939** L. MACNEICE *Autumn Jrnl.* xv. 57 False eyelashes and finger-nails of carmine. **1954** —— *Autumn Sequel* xxiii. 141 That Christmas should be white Is something we go on with, like false noses.

e. Also, (*b*) a deceiver, a hypocrite.

1893 STEVENSON *Catriona* xxv. 104 That false-face, Prestongrange; I think shame to own to you that I was ever trusting to a lawyer.

i. Photogr. *false image*: an extra image made on the plate by a defective lens at the same time as the image proper.

1892 *Photogr. Ann.* II. 39 Another troublesome fault is what is called a ghost, or false image. **1918** *Photo-Miniature* XV. Mar. (Gloss.), *False Image*, an extra image, usually unsharp and often inverted, which a defective (doublet) lens will give on the plate at the same time as the image proper.

14. c. *false alarm*: an alarm without foundation, given either purposely to deceive or under misapprehension of danger. Now often *transf.* or *gen.*

1579 GOSSON *Sch. Abuse* 32 That is a vain brag, & a false alarme. **1594** [see ALARM *sb.* 8]. **1770** JOHNSON *False Alarm* 3 One of the chief advantages derived by the present generation from the improvement and diffusion of Philosophy, is..exemption from false alarms. **1902** C. JAMES *Milit. Dict.* s.v. *Alarm*, False-Alarms, are strategems of war, frequently made use of to harrass an enemy, by keeping them perpetually under arms. **1834** tr. *Jacquemont's Journ. India* I. 214 My pistols and watch were almost in his way; but, disturbed..by some noise or false alarm, he had not time to choose. **1847** A. BRONTË *Agnes Grey* xvii, There was 'no need to be in such agitation about the matter—it might prove a false alarm after all'. **1873** *Gentl. Mag.* X. 587 The false alarm..was made the occasion of a discussion.., which ended in the Lord Mayor advising his civic brethren to wait and see. **1900** *Daily News* 7 Dec. 3/2 Any day the giving of a false alarm might cause the deaths of persons endangered by fire, and whom the Brigade did not reach in time. **1931** H. MUTSCHMANN *Gloss. Americanisms* 22/1 *False alarm*, a divorced woman.

15. In various technical uses: see quots.

1854 *Poultry Chron.* I. 488 On inspecting the windpipe from the root of the tongue, I found congestion and swelling of the glottis and rima glottidis, but no false membrane such as I should have found in a child. **1873** T. H. GREEN *Introd. Pathol.* (ed. 2) 265 In croup the exuded materials coagulate principally upon the surface of the membrane, where, together with the newly-formed cellular elements, they form the false membrane. **1880** *Encycl. Brit.* XII. 180/2 Immediately within the opening of the nostril [of a horse], the respiratory canal sends off on its upper and outer side a diverticulum or blind pouch (called 'false nostril') of a conical form. **1882** CAULFEILD & SAWARD *Dict. Needlew.*, *False hem*, this is applied to a fold-over at the extreme edge of any portion of dress or other article..which has the appearance of a hem..but is not one. **1883** *Man. Seamanship for Boys* 97 Cardinal Points...False Points. So called because they borrow their names from the two points between which they come. **1883** *Encycl. Brit.* XV. 235/2 Such a point is called a false north pole, and we see that the existence of two true north poles necessitates the existence of a false north pole. **1886** *Ibid.* XXI. 404/2 These portions go on growing in a direction at an angle with the previous one, but still in contact, and thus produce the 'false-branching' to which the *Cladothrix* owes its name. **1888** *Lockwood's Dict. Mech. Engin.* s.v. *False Water*, When steam is generated very rapidly in a boiler, the immediate effect is a rapid rising of the water level in the gauge cocks, due to the increase in volume caused by admixture of steam. This sudden increase is termed false water. **1902** *Encycl. Brit.* XXXIII. 99/2 *False Station*.—When the theodolite cannot for any reason be placed over the centre of a station, if the distance be measured, and the theodolite reading of it be noted, the observed angles may be reduced to what they would be at the centre of the station. False stations have frequently to be made in practice. **1932** D. C. MINTER *Mod. Needlecraft* 132 False hem machined on. **1948** *New Biol.* V. 14 Colonies such as those of *Gloeotrichia* and *Rivularia* originate by a process known as false branching, in which one part grows out at a break in the filament. **1960** B. SNOOK *Eng. Hist. Embroidery* 106 False quilting, worked through two thicknesses of fabric without padding, was also used.

b. (Further examples.)

1854 THOREAU *Walden* (1937) 182 The *celtis occidentalis*, or false elm. **1940** E. STEP *Wayside &c. Trees* 23 The Sycamore, Great Maple, or False Plane (*Acer pseudoplatanus*).

d. *false conception* (earlier example).

1611 DONNE *Anat. World* sig. B4v, And false-conceptions fill the generall wombs.

e. *false grain*: a fresh crop of small sugar crystals formed during the process of sugar manufacture when syrup is introduced into the crystallizing pans in order to increase the size of the crystals already formed.

1900 S. P. SADTLER *Hand-bk. Industr. Org. Chem.* (ed. 3) 132 The process of admitting successive portions of fresh syrup after the 'grain' has once formed is used in the development of large crystals. It must be used with judgment though, or the new syrup starts a new set of minute crystals, making what is called 'false grain'. **1959** *Chambers's Encycl.* V. 200/2 If more sugar is held in the solution than that given by the super-solubility curve, a fresh crop of nuclei or a 'false grain' will result.

f. *false killer*, a name of the whale, *Pseudorca crassidens*, which partly resembles the KILLER (sense 2); *false scorpion* = *pseudo-scorpion* (PSEUDO- 2).

1875 *Encycl. Brit.* II. 281/2 The impression inevitably created on a comparison of the true and false scorpions is that the latter are little scorpions without tails. **1937** NORMAN & FRASER *Giant Fishes, Whales & Dolphins* II. xiii. 294 In the genus *Pseudorca*..there is only one species admitted, *P. crassidens*, the False Killer Whale. It is nearly related to the true Killer, but many external differences distinguish the two forms from each other. **1949** *Oxf. Jun. Encycl.* II. 140/1 False Scorpions owe their name to the possession of a pair of large, claw-like 'pedipalpi' or specialized feelers, like those of the Scorpions; but they lack the Scorpion's tail and sting. **1959** A. HARDY *Fish & Fisheries* xv. 288 The false killer, *Pseudorca crassidens*, is entirely black and might perhaps be mistaken for the pilot whale.

16. c. *false dawn, morning, sunrise* [tr. Arab. *ṣubḥ kāḏib*]: a transient light which precedes the true dawn by about an hour, a phenomenon common in the East. Also *fig.*

1832 J. MORIER *Zohrab* i. 6 Do tell me..whether that be the dawn or the false dawn? **1868** E. FITZGERALD *Omar* (ed. 2) ii. 1 Before the phantom of False morning died. **1879** E. ARNOLD *Lt. Asia* v. 112 Then slept he.. But rose e'er the False-Dawn. **1924** E. WHARTON (*title*) Old New York. False Dawn. **1928** *Daily Express* 23 Aug. 8/3 A flickering false-sunrise. **1946** K. TENNANT *Lost Haven* (1947) i. 29 When the false dawn came, it was a pale thing compared to the moon-light. **1963** *Times* 16 Jan. 3/3 We are thus back to where we were before the excitement of Melbourne and the optimism of the week that followed. That seems now like a false dawn.

17. a. *false bottom*: also, a partition built close to the bottom, as in a box or trunk; *false core* (see CORE *sb.*^{1} 8; examples).

1800 M. EDGEWORTH *Parent's Assist.* (ed. 3) III. 132 This box has a false bottom—it holds only three quarters as much as it ought to do. **1843** C. & J. HOLTZAPFFEL *Turning* I. 338 The term false-core is employed by the brass founder to express the same thing as the drawback of the iron founder. The former calls every loose piece of the mould not intended for holes, a false core. **1875** [see CORE *sb.*^{1} 8]. **1907** MCWILLIAM & LONGMUIR *Gen. Foundry Practice* xii. 95 In light work these removable parts of a mould are termed 'false cores', and in heavy work 'drawbacks'. **1933** MCLACHLAN & OTTO in W. J. Kearton *Engin. Educator* (ed. 2) II. 606/2 Part of a mould forming an internal shape, whether loose or not, is usually referred to as a 'cod', while a loose part of a mould forming some external part of it, is generally referred to as a 'false-core' or 'drawback'.

e. *false ceiling*, a dummy ceiling fixed below the genuine one (e.g. to accommodate wires, conduits, etc.).

1870 BREWER *Dict. Phr. & Fable* 285/2 *False ceiling*, the space between the garret ceiling and the roof. **1937** *Archit. Rev.* LXXXI. 63 (*caption*) The entrance hall from the living room, showing again the curved screen wall with its false ceiling.

D. *Comb.* **1.** Also (not with agent-noun): *false-innocence* (also *false-innocent* adj.). Cf. *FAUX-NAÏF *sb.* and *a.*

1938 R. GRAVES *Coll. Poems* 187 Their false-innocence assaulting her, Breaching her hard heart. **1928** D. H. LAWRENCE *Woman who rode Away* 272 Her eyes especially were warm and naive and false-innocent.

b. *false-faced* (later example), *-tongued*.

a **1959** E. MUIR *Coll. Poems* (1960) 291 But why was our old friend Everyman Among this false-faced company? **1910** W. DE LA MARE *Three Mulla-Mulgars* 41 Oh, these false-tongued Mulgars!

2. c. *false-brood, -feed, -preen* (whence *false-brooding*, etc., vbl. sbs.).

1948 *Brit. Birds* XLI. 237 He 'false-brooded' by crouching in any convenient depression nearby. *Ibid.* 243 In the same way false-brooding occurs as a substitute activity when uncertainty arises within the bird's mental framework. **1961** D. NETHERSOLE-THOMPSON in Bannerman *Birds Brit. Isles* X. 313 After oystercatchers have copulated, they may preen or 'false preen', feed rapidly or 'false feed', peck at the ground, curtsey. **1948** *Brit. Birds* XLI. 12 She then inserts her bill into that of the male and symbolic 'false-feeding' follows in which food is not actually passed. **1949** *Ibid.* XLII. 7 On one or two occasions the false-preening by the male, mentioned by Makkink, was observed at this stage.

3. false ebony (see quots.); **false nest** (see quot.); **false twist** *Textiles* [TWIST *sb.*^{1} 4] (see quot. 1960), also *attrib.*; whence *false-twist* v. intr. and *false-twisting* ppl. a.

1892 C. M. YONGE *Old Woman's Outlook* v. 96 The inside heart wood [of the laburnum] is so black as to be called false-ebony. **1911** *Encycl. Brit.* XVI. 32/1 The heart wood of the laburnum is of a dark reddish-brown colour, hard and durable, and takes a good polish...The laburnum has been called false ebony from this character of its wood. **1954** BANNERMAN *Birds Brit. Isles* III. 354 This species [wren] builds 'false' nests in the vicinity of the nest which is eventually to hold the eggs..some naturalists considering that they are for the purpose of roosting in, but this is denied by others who consider they are merely the result of the male's craze for building. **1960** *Textile Terms & Defs.* (ed. 4) 62 *False twist*, turns inserted in opposite directions and in equal numbers in adjacent elements of yarn..characterized by its temporary nature. *Ibid.*, The false-twisting element, through which the yarn or sliver passes. **1965** *Guardian* 31 Mar. 17/1 False-twist version of acetate yarn. *Ibid.*, The idea of false-twisting acetate yarns.

false, *v.* **4. a.** Restrict †*Obs.* to other senses in Dict. and add later *arch.* example.

1923 HARDY *Coll. Poems* (ed. 2) 596 Past regretting Loves who have falsed their vow.

false-card, *v. intr.* To play a false card (FALSE *a.* 2 g); esp. in *Bridge*, to play a card other than the normal one, so as to mislead an opponent. So **false-carding** *vbl. sb.*

1589 W. WARNER *Alb. Eng.* VI. xxx. 132 Nay, be it that he should espie false carding, what of it? **1902** J. B. ELWELL *Bridge* 82 The dealer false-cards so that the adversaries will not know that he holds the queen. **1923** *Daily Mail* 23 June 6 A simple case of false-carding is where declarer holding ace, king of a suit led by an opponent takes the first trick with the ace in order to convey the idea that he does not possess the king... Many players false-card without rhyme or reason. **1939** N. DE V. HART *Bridge Players' Bedside Bk.* xxii. 79 The declarer executed his simple piece of false-carding on the edge of an abyss. **1960** T. REESE *Play Bridge with Reese* xxxii. 112 Unless both opponents were false-carding in spades they are 4–4.

falsework (fǭ·lswɒ̄ɹk). [f. FALSE *a.* 17 + WORK *sb.* 10.] A temporary framework or support put up while the main structure is being built and afterwards removed.

1874 [see FALSE *a.* 17 d]. **1930** *Engineering* 18 Apr. 503/3 All substructure and false work were removed. **1954** *Archit. Rev.* CXVI. 336 It was decided to..avoid covering the whole interior with falsework. **1958** *Times* 19 June 10/6 The falsework supporting the unfinished sixth span of the bridge collapsed.

falsies (fǭ·lsiz), *sb. pl. colloq.* (orig. *U.S.*). [f. FALSE *a.* + -IE.] A padded brassière; breast-pads. Also *transf.*

1943 *N.Y. Post* 6 Oct. 42/3 'Falsies'..the term for the pads that convert them [*sc.* nightclub chorus girls] from 32s to 34s. **1946** *Life* 14 Oct. 35 'Falsies', padded rubber or cloth brassières worn to augment movie actresses' bosoms. **1957** R. HOGGART *Uses of Literacy* viii. 210 'Falsies', the most popular new feature in all the sex-joke magazines. **1958** M. DICKENS *Man Overboard* vii. 105 The secretary slouched in..her falsies pushing out her sweater like cardboard cones. **1964** J. S. HUXLEY *Ess. Humanist* 58 My American friends tell me that falsies are now obtainable for the other side of the female anatomy—false bottoms, in fact.

falsifiability (fǭ:lsifəi,ăbi·lĭti). *Philos.* [f. FALSIFIABLE *a.* + -ITY.] The quality or fact of being falsifiable. Also *attrib.*

1937 *Philos. Sci.* IV. 26 We may take Popper's principle of falsifiability as an example. **1945** *Mind* LIV. 119 The concepts of absolute and of relative verifiability (and falsifiability) are of an entirely different character. **1959** *Cambr. Rev.* 30 May 567/2 Whether or not the verifiability criterion is a criterion of the meaningfulness of a statement or of the scientific character of a statement, the defects which it has in either case can be reproduced *quid pro quo* for the falsifiability criterion. **1965** J. D. NORTH *Measure of Universe* xiv. 295 Not all propositions can have an inherent degree of falsifiability.

falsifiable, *a.* (Later examples.)

1959 *Cambr. Rev.* 30 May 568/2 Suppose now that a statement was meaningful if and only if it was falsifiable. **1964** W. P. ALSTON *Philos. Lang.* iv. 70 We should also note the specially wide sense in which the positivists use 'verifiability'. In this use, it is really equivalent to the disjunction 'verifiable or falsifiable', that is, 'capable of being established as true or false'.

falsism. (Earlier example.)

1835 J. S. MILL *Diss. & Disc.* (1859) I. 109 This *dictum* ..is, as Coleridge would say, a falsism.

‖ **falso bordone** (fa·lso bǫɹdo·ni). *Mus.* [It. *falso* false + *bordone* BOURDON^2.] = FA-BURDEN.

1740 GRASSINEAU *Mus. Dict.* **1801** T. BUSBY *Dict. Mus.* sig. G4 *Falso bordone*, a term applied in the early days of descant to such counterpoint as had either a drone bass, or some part constantly moving in the same interval with it. **1876** [see *basso ostinato]. **1961** *Listener* 5 Oct. 535/3 The solemn falsobordoni of Monteverdi's *Dixit Dominus*.

faltboat (fæ·lt,bōᵘt). Also in G. form **faltboot.** [partial tr. of G. *faltboot* collapsible boat, f. *falten* to fold + *boot* boat.] A folding boat (see FOLDING *ppl. a.* 2). Hence **fa·ltbooting** *vbl. sb.*

1926 *Blackw. Mag.* Feb. 241/1 Crossing the English Channel is one of many adventurous voyages in 'Faltboots'. *Ibid.* 254/1 'Faltbooting'..is a fine weather employment. **1936** *Times Lit. Suppl.* 20 June 525/3 The still growing popularity of the *faltboat*, besides revolutionizing canoeing itself, has produced a small library of books on the sport.

falutin, *sb.* and *a.* = HIGHFALUTIN *sb.* and *a.*

1921 T. R. GLOVER *Pilgrim* viii. 118 He does not use of the Church the splendid language of Paul, still less the falutin of some second century Christians. **1928** *Observer* 12 Feb. 17/5 The Swarajists..would have..sought to flout Parliament with falutin phrases.

falx. Add: †**2.** *Zool. Obs.* Each of the paired chelicerae of spiders; sometimes used to mean only the basal section of a chelicera (see quot. 1889) and in this sense = *PATURON.

1866 E. F. STAVELEY *Brit. Spiders* ii. 12 Below the eyes and above the mouth is a pair of jaw-like organs, known as the falces. **1885** H. O. FORBES *Nat. Wand. East. Archipelago* III. vi. 216 With the rapidest motion..it [*sc.* a spider] flashed its falces into my flesh. **1889** H. C. MCCOOK *Amer. Spiders* I. i. 22 Each mandible is composed of two parts, the base or falx and the fang...The falx is covered with hair, especially near the base, and at the apex is formed into a toothed groove or sheath into which the fang is folded down when at rest. **1904** W. H. HUDSON *Green Mansions* xx. 288 He [*sc.* a spider] might come stealthily back at night to plunge his long, crooked falces into my throat.

fame, *sb.*^{1} Add: **5. a.** *fame-destroying, -getting* adjs.; **b.** *fame-favoured* adj.; **c. fame-flower** (see quot. 1891).

1899 R. ST. BARBE *In Mod. Spain* 34 The fiddler with his fame-destroying *mañana* theme. **1893** E. SULLIVAN tr. *Dante's Comedy* II. 4 Rescue me from her, fame-favoured bard. **1879** T. MEEHAN *Native Flowers & Ferns U.S.* II. 55 Terete talinum...Thus Rafinesque proposed to call it *Phemeranthes* or 'Fame-flower'. **1891** *Cent. Dict.* s.v. *Talinum*, *T*(*alinum*) *teretifolium*, a native of the United States from Pennsylvania to Colorado and southward, a low tuberous-rooted perennial, growing on rocks and exceptional in its cylindrical leaves, has been called *fame-flower* from the transitoriness of its elegant purple petals. **1941** R. S. WALKER *Lookout* 47 It appears on the sandstones at Rock City..and travels all the way to the southern extremity...This is fame-flower. **1820** SHELLEY *Let.* 20 Apr. (1964) II. 185 Atoms of the fame-getting..whirlwind.

Famennian (fāme·niăn), *a.* *Geol.* Also **Fammenian,** †**Famennien.** [ad. F. *Famennien* (J. Gosselet 1879, in *Ann. Soc. géol. du Nord* VI. 396), f. *Famenne*, name of a district of western Belgium.] Name of the upper of the two stages constituting the Upper Devonian in Europe; of or pertaining to this stage or the geological period during which it was deposited. Also *absol.*

1882 [see *FRASNIAN *a.*]. **1888** J. PRESTWICH *Geol.* II. 581 Famennian system. **1893** A. GEIKIE *Text-bk. Geol.* (ed. 3) 787 In Brittany also, Devonian strata are found, including representatives of the Famennian groups with Cypridinas and Goniatites. **1894** J. D. DANA *Man. Geol.* (ed. 4) 627 In the Eifel, the three divisions, the Rhénan, Eifelian and Famennian are well developed. **1909** H. B. C. SOLLAS tr. *Suess's Face of Earth* IV. 61 Five species of Avicula (Leptodesma) which the Chemung stage possesses in common with the Famennian of Belgium. **1931** GREGORY & BARRETT *Gen. Stratigr.* vi. 96 In Spain and Portugal the Devonian is widely developed and includes a full succession from Siegenian to Famennian. **1963** D. W. & E. E. HUMPHRIES tr. *Termier's Erosion & Sedimentation* x. 208 At Ben-Zireg near Colomb-Béchar.. banks of Devonian rocks are enclosed in the Visean... They form a *Wildflysch*, resting on Fammenian limestones.

‖ **familia** (fămi·liă). Pl. **familiæ.** [See FAMILY *sb.*] (See quots.)

1708 *Cowel's Law Dict.*, *Familia*, is sometimes taken by our Writers for a Hide, sometimes called a Manse, sometimes Caracuta, or a Plough-land, containing as much as one Plough and Oxen can till in one year. **1729** G. JACOB *New Law-Dict.*, *Familia*, signifies all the Servants belonging to a particular Master; but in another Sense, it is taken for a Portion of Land, sufficient to maintain one Family. **1883** *Encycl. Brit.* XX. 712/2 With the Christian emperors the last traces disappeared of the old conception of the *familia* as an aggregate of persons and estate subject absolutely to the power and dominion of its head. **1912** J. A. WILLIS tr. *Marucci's Christ. Epigr.* 6 Nearly all the cognomens of the most illustrious familiae, as the Scipiones,..the Cicerones, etc., have been already explained by ancient writers. **1933** E. K. CHAMBERS *Eng. Folk-Play* 226, I do not think that the relation of father and sons indicates anything beyond the relation of the leader of the revel to his *familia* or troop. **1965** D. WHITELOCK in Bessinger & Creed *Medieval & Linguistic Stud.* 223 It is..very likely that when Aldred, bishop of Worcester, became archbishop of York in 1060, some members of his familia went with him.

familial (fămi·liăl), *a.* [a. F. *familial*, f. L. *familia* family + -AL.] **a.** *Med.* Occurring among members of a family, hereditary.

1900 GOULD *Pocket Med. Dict.* (ed. 4) 249 *Familial*, pertaining to a family. **1903** *Amer. Jrnl. Psychol.* July–Oct. 100 The chronic constitutional type, said to be to a large extent familial. **1910** *Practitioner* June 809 It is a familial defect rather than an acquired disease. **1925** H. CUSHING *Osler* I. 432 Instances in which the disease [*sc.* Angina Pectoris] has shown hereditary or familial tendencies. **1964** S. DUKE-ELDER *Parsons' Dis. Eye* (ed. 14) xviii. 259 A somewhat similar macular degeneration beginning between the ages of twelve and fourteen is seen as a rare familial disease (Stargardt's Disease). **1971** *Nature* 14 May 97/1 Such particles were readily detected in the milk of 60% of American women with a familial history of breast cancer.

b. Of, pertaining to, or characteristic of a family.

1903 E. C. PARSONS tr. *Tarde's Laws of Imitation* vii. 252 The essentially familial character which this people [*sc.* the Chinese] has retained. **1907** *Daily Chron.* 6 June 4/4 The parental or familial attitude towards education. **1919** CONRAD *Arrow of Gold* IV. ii. 163 A remote ideal which yet may belong to his familial tradition. **1929** M. JOSEPHSON *Zola* pl. facing 492 A Familial Scene in the Garden of Zola's English Retreat. **1953** J. B. CARROLL *Study of Lang.* ii. 62 A technique for inferring, from statistical comparisons of language systems, estimates..of their probable familial relationships. **1957** V. W. TURNER *Schism & Continuity in Afr. Society* viii. 238 The kinship basis of a village.. represents a compromise between familial and lineal principles of organization.

Hence **fami·lially** *adv.*, in a familial manner.

1953 J. B. CARROLL *Study of Lang.* ii. 25 Features of languages can be studied comparatively, whether these languages are familially related or not.

‖ **familistère** (familistę̄r). [Fr.] = FAMILISTERY.

1880 *Encycl. Brit.* XI. 265/2 A palatial *familistère* with accommodation for 400 families. **1886** *Sci. Amer. Suppl.* 10 July 8761/3 In 1859 Godin put up a large building, called the '*familistere*', for the accommodation of 300 families, adding a theater, school-house, etc. **1899** *Daily News* 15 Sept. 3/2 The famillistère [*sic*] founded by M. Goudin at Guise.

‖ **famille** (famiy). [Fr. = family.] **a.** Phr. *famille de robe*: a family founded by a lawyer; a legal family (in pre-Revolutionary France). (See ROBE *sb.* 4 b.)

1857 J. W. CROKER *Ess. Fr. Rev.* iv. 161 Pierre Louis Rœderer, born about 1756 of a respectable *famille de robe*. **1881** A. I. THACKERAY *Mme de Sévigné* ii. 10 A.. gentle heiress belonging to a respectable *famille de robe*.

b. *famille jaune*, Chinese enamelled porcelain of which the predominant colour is yellow; so also *famille noire, rose, verte*, in which the predominant colour is, respectively, black, red, and green.

1872 LADY C. SCHREIBER *Jrnl.* (1911) I. 145 C. S. espied a bowl and cover of the 'famille verte' like the jars we acquired at Cadiz. **1876** *Ibid.* 456 Went up to Le Roy's, to look after some Famille Verte beakers. **1898** W. G. GULLAND *Chinese Porc.* I. 174 *Famille verte*. So called from the decoration being chiefly in green enamel. *Ibid.* 192 *Famille rose*. So named from a rose tint in greater or lesser quantity being employed in the decoration. **1900** F. LITCHFIELD *Pott. & Porc.* 114 The high value placed upon a really fine specimen of the '*famille noire*'. **1903** *Daily Chron.* 7 July 2/7 A large famille-rose cistern. **1904** E. DILLON *Porcelain* 106 We may probably associate with the beginning of this reign [*viz.* of Kien-lung (1735–1795)]..the first use of the *rouge d'or* which has given its name to a well-known class of porcelain—the *famille rose*. **1945** W. B. HONEY *Ceramic Art China* ii. 149 Enamelled Wares of the K'ang Hsi Period (1662–1722)...Big vases with green, black, and yellow grounds, are often of noble form...These coloured grounds of the *famille noire, famille jaune*, etc., are much sought after by European and American collectors. **1962** *Oxf. Univ. Gaz.* 30 May 1280/1 Eight pieces of porcelain, export ware, *famille rose*, Chinese, 19th century. **1969** *Guardian* 16 July 18/1 Rarer examples of famille noire or jaune which were made under the Ch'ing Dynasty.

family, *sb.* Add: **6. a.** (Earlier example in *Philology* and later example in *Physics.*)

1762 J. PRIESTLEY *Theory Lang.* viii. 117 The Eastern tongues of that genus, or family as we may call it. **1950** F. GAYNOR *Encycl. Atomic Energy* 157 A radioactive series, also called radioactive family, is a group of radioactive isotopes, each of which is the product of the radioactive decay of the preceding one.

b. (In Botany 'family' is now used, as in Zoology, for a division of an order, and has therefore superseded the term 'natural order'; e.g. order Rosales, family Rosaceæ. Also *spec.* in Ecology.)

1916 B. D. JACKSON *Gloss. Bot. Terms* (ed. 3) 143/2 *Family*..a group of genera, formerly styled *Order*. **1916** F. E. CLEMENTS *Plant Succession* vii. 139 A family is a group of individuals belonging to one species. It often springs from a single parent plant. **1960** N. POLUNIN *Introd. Plant Geogr.* xi. 334 The seral equivalent is the *family*, derived from the multiplication and gregarious growth of a single immigrant. **1951** G. H. M. LAWRENCE *Taxon. Vascular Plants* iv. 46 An order of plants is composed of one or more families...The family usually represents a more natural unit than any of the higher categories.

9. a. (Further examples.) Also, of unexceptionable nature, suitable for all the members of a family.

1801 M. EDGEWORTH *Belinda* I. xii. 370 She was in hopes that these terrible family quarrels might be made up. **1807** R. SOUTHEY *Lett. from Engl.* vi. 47 To paint the family group is out of my power. **1807** (*title*) The Family Shakespeare. **1817** *Blackw. Mag.* I. 63/2 She.. assented to his proposal of having the usual family-worship in her bedroom. **1845** *Edin. Rev.* LXXXII. 407 The development of domestic feelings and family life. **1853** MRS. GASKELL *Let.* ? Feb. (1966) 223 About Ruth one of your London librarians..has had to withdraw it from circulation on account of 'its being unfit for family reading'. **1859** F. C. L. WRAXALL tr. *Robert-Houdin's Memoirs* I. viii. 154 The family tickets gave admission to four persons at half price. **1874** (*title*) Cassell's Family Magazine. **1889** J. K. JEROME *Three Men in Boat* i. 6 If I was a co-operative stores and family hotel combined, I might be able to oblige you. **1902** H. JAMES *Wings of Dove* i. 15 Show family feeling by seeing what I'm good for. **1926** A. MØLLER tr. *Pedersen's Israel* I. II. 376 A dissolution of the family-unity. **1933** *Discovery* Dec. 381/2 The..trend of policies since the war has been to break these [*sc.* estates of eastern Europe] up into single family units. **1937** M. BORDEN *Black Virgin* x. 188 She was new to the house, if not to family rows. **1940** *Illustr. Lond. News* CXCVI. 608/3 'Family' cars that have been laid up indefinitely. **1960** J. B. PRIESTLEY *Lit. & Western Man* vii. 76 *The School for Scandal* might be described as a Restoration comedy.., without the impudent indecencies, and so entirely suitable for family entertainment. **1960** *Guardian* 22 July 4/6 Much of the output..[is] unsuitable for 'family viewing'.

c. (Further examples.)

1699 M. LISTER *Journey to Paris* 101 He shewed me the Catalogue of Authors..alphabetically disposed by Family Names. **1771** SMOLLETT *Humph. Cl.* III. 182, I was.. engaged in a family partie. **1775** G. WHITE *Let.* 2 Oct. in *Selborne* (1789) 195 The name of their [gypsy] clan is *Curleople*..may not this family-name..be the very name they brought..from the Levant? **1780** *Mirror* 25 Apr. 401 The little family-history I am going to relate. **1795** tr. *C. P. Moritz's Travels* 116, I saw..sundry little family parties, walking arm in arm along the banks of the Thames. **1813** JANE AUSTEN *Pride & Prejudice* III. xviii. 314 The comfort and elegance of their family party at Pemberley. **1829** H. FOOTE *Compan. to Theatres* 36 Private boxes,..family ——. **1846** C. BRONTË *Professor* (1857) II. xxv. 242 He..would dwell on the past times of his house, on his family history. **1901** C. MORRIS *Life on Stage* xv. 203 A short time after that, she sat one evening in Mr. Ellster's family box. **1925** O. JESPERSEN *Mankind, Nation & Individual* ix. 172 The Araukans carefully conceal their personal-name from strangers: in their presence they are called by their family-name. **1967** N. FITZGERALD *Affairs of Death* i. 6 You have the family face at its best. **1969** M. LASKI *Jane Austen & her World* 9 Sir Thomas Leigh..had sheltered Charles I at Stoneleigh Abbey in Warwickshire, the family seat of the younger, ennobled branch of the family.

10. b. Also *to put in the family way*: to make pregnant.

1898 *Sessions Paper* Feb. 266, I did not put his wife in the family way. *a* **1935** T. E. LAWRENCE *Mint* (1955) II. ix, There's only one thing..that the Air Force can't do: put us in the family way.

11. *family portrait*; *family-oriented* adj.; **family allowance**, an allowance paid by the state to parents who have a specified number of children; also any similar allowance paid by an employer to employees with families; in N.Z. called *family benefit*; **family Bible** (earlier and *attrib.* examples); **family coach** (earlier example); **family compact,** (*b*) the name applied to the governing class, esp. the officialdom, of Upper Canada in the first part of the 19th cent.; **family council** (examples); also *transf.*; **family doctor,** a general practitioner (traditionally regarded as a friend and adviser to the family on other than medical matters), hence **family-doctor** *v. trans.*; **family-likeness** (earlier example); **family planning,** the use of various methods of birth control to limit the size of a family; freq. used *attrib.* of an association, centre, etc., from which medical advice and information about contraception are obtainable; **family practitioner** = **family doctor*; **family resemblance** = *family-likeness*; **family room** *N. Amer.*, a living- or recreation room; **family skeleton** = SKELETON *sb.* 1 b; **family-tree** (earlier example); also a diagrammatic representation of the relationship of specified languages.

1924 E. F. RATHBONE *Disinherited Family* v. 193 Family allowances in an extremely rudimentary form were started in France in 1890 by the Railway Companies. **1928** *Britain's Industr. Future* (*Lib. Ind. Inq.*) III. Argt. 139 The minimum wage should be fixed for each industry..; the introduction of Family Allowances may be found desirable by industries to which they are suited. **1928** *Manch. Guardian Weekly* 6 Jan. 5/2 A living wage laid down by the State, to pay which industry must be reorganised by State action, family allowances, [etc.]. **1958** *New Statesman* 11 Jan. 29/1 The most vital and cherished social services—family allowances, school meals, welfare food and welfare milk. **1970** *Times* 4 May 9/6 It is for the Government to make an immediate increase in family allowances. **1955** D. O. W. HALL *Portrait of N.Z.* ix. 179 A family benefit of 10/- weekly is paid the mother for every child below 16 years (up to 18 years if still at school) without any means test. **1740** RICHARDSON *Pamela* II. 359 Good Books, such as a Family-Bible, a Common-Prayer. **1908** *Daily Chron.* 20 June 3/1 A regimental record of the family-bible kind. **1709** Family coach [see Dict., sense 9 a]. **1828** *Toronto Public Library MSS.* B104 153 The measures to be adopted to relieve this province from the evils which a family compact have brought upon it. **1899** J. P. TAYLOR *Cardinal Facts Canad. Hist.* 119 About this time [1820] the 'Family Compact' is said to have been formed in Upper Canada. **1965** *Kingston* (Ont.) *Whig-Standard* 9 Jan. 9/8 He wrested the leadership of the old Family Compact from the Anglican Tories of muddy York and made it into the Conservative party. **1853** MRS. GASKELL *Cranford* iii. 51 The state of the remainder wine was examined into in a family council. **1902** *Little Folks* 251/1 Ted called a family council in his carpentering shed. **1965** *Guardian* 25 Aug. 5/1 The under-16s..will be dealt with by *Family Councils*—appointed by local authorities and composed of social workers, teachers, doctors, and others experienced in dealing with children. **1846** R. FORD *Gatherings from Spain* xvii. 228 Most Spaniards who can afford it have their family or bolster doctor, the *Medico de Cabecera*. **1943** A. CHRISTIE *Moving Finger* viii. 92 He spoke in a comfortable family-doctor kind of way. **1952** 'C. BRAND' *London Particular* v. 56 Fat old Tedward, who had family doctored her since she was born. **1766** GOLDSM. *Vicar* I. i. 8 A family likeness prevailed through all..they had but one character. **1949** M. MEAD *Male & Female* 460 A multi-disciplinary approach to family-oriented treatment of illness. **1965** M. BRADBURY *Stepping Westward* v. 265, I thought England was a family-oriented society. **1939** *Family Planning Assoc., 9th Ann. Rep.* 1 Objects...To advocate and promote the provision of facilities for scientific contraception so that married people may space or limit their families. **1945** *Lancet* 3 Mar. 294/1 Growing numbers of women attending 'family planning' centres. **1957** *Listener* 26 Sept. 467/2 Family planning..will not spread in the villages until someone invents a contraceptive which is easy, very cheap, and does not require privacy. **1970** *Times* 30 Apr. 12/8 It was members of this committee [*sc.* Birth Control Investigation Committee] who persuaded Lady Denman to form what is now the Family Planning Association to coordinate the work. **1814** JANE AUSTEN *Mansf. Park* I. ix. 174 Of pictures there were abundance, and some few good, but the large part were family portraits. **1902** A. CONAN DOYLE *Hound of Baskervilles* xiii. 297 A study of family portraits is enough to convert a man to the doctrine of reincarnation. **1931** *N. & Q.* 20 June 447/2 Is not the fifteenth century too early for family portraits? **1846** DICKENS *Dombey* i. 4 'Mr. Pilkins here, who from his position of medical adviser in this family—no one better qualified to fill that position, I am sure.' 'Oh!' murmured the family practitioner. **1858** *Sat. Rev.* 29 May 566/1 A peculiar and remarkable writer, whose style showed little or no family resemblances with that of any living author. **1901** A. E. TAYLOR *Probl. Conduct* iii. 151 The family resemblance which all systems of judgments of approbation exhibit. **1953** G. E. M. ANSCOMBE tr. *Wittgenstein's Philos. Investigations* § 67, I can think of no better expression to characterize these similarities than 'family resemblances'. **1853** *Harper's Mag.* VI. 443 The first night of my arrival I was honored with a spare family room. **1884** W. D. HOWELLS *S. Lapham* xix. 356 Indicating the family-room, he added, 'She's in there.' *Ibid.* xxv. 461 He heard talking in the family room. **1968** *Globe & Mail* (Toronto) 3 Feb. 11/5 (Advt.), Every home includes a log burning fireplace and a separate family room. **1850** THACKERAY *Pendennis* II. xxi. 208 That ugly closet.. in which, according to the proverb, the family skeleton is locked up. **1881** Family skeleton [see SKELETON *sb.* 1 b]. **1936** C. DAY LEWIS *Friendly Tree* v. 74 Evelyn and Richard glanced at each other with embarrassment. There was some family skeleton peeping out here. **1961** *Guardian* 23 Mar. 9/2 Family skeletons are being brought out of the German cupboard. **1807** R. SOUTHEY *Lett. from Engl.* lx. 368 An English Esquire would as soon walk abroad in his grandfather's wedding suit, as suffer the family Tree to be seen in his hall. **1933** BLOOMFIELD *Lang.* xviii. 311 The comparative method thus shows us the ancestry of languages in the form of a family tree. *Ibid.*, The earlier students of Indo-European did not realize that the family-tree diagram was merely a statement of their method. *Ibid.* 318 The older family-tree theory of linguistic relationship.

famine, *sb.* Add: **5. a.** *famine-prices* (earlier example), *relief* (cf. RELIEF^2 4 a).

1847 *Builder* 29 May 256/2 The famine prices of provisions. **1876** *Correspondence on Famine in W. & S. India* I. 13 in *Parl. Papers 1877* (C. 1707) LXV. 1 In my letter of 25th August last..I pointed out the necessity of starting famine relief works. **1963** *Listener* 17 Jan. 108/2 Such young people seek their own expression in such positive forms of political action as work projects and famine relief.

Fammenian, var. *FAMENNIAN.

famous, *a.* Add: **1. c.** Phr. *famous last words* (cf. WORD *sb.* 25 b): a remark or prediction likely to be proved wrong by events.

1948 PARTRIDGE *Dict. Forces' Slang* 67 *Famous last words*, a catch-phrase rejoinder to such fatuous statements as 'Flak's' [*sc.* Anti-aircraft fire is] not really dangerous'. **1948** *Shell Aviation News* CXVII. 9/1 Leopoldville is easy to find because you cannot miss the Congo River. (Famous last words!) **1962** J. CANNAN *All is Discovered* iv. 89 'Why...It's nothing.' 'Famous last words.'

5. (Earlier and additional examples.)

a **1700** EVELYN *Diary* 13 Nov. an. 1695 (1955) V. 223 Famous & very chargeable Fireworkes. **1800** M. EDGEWORTH *Parent's Assistant* (ed. 3) V. 147 'There's to be the most *famous* doings, that ever were heard of upon the Downs here,..Lady Diana Sweepstakes, you know, is a *famous* rider, and archer.' **1966** *Times* 9 Nov. 1/1 Both parties..were ready to claim a famous victory in the early hours of tomorrow.

fan, *sb.*^{1} Add: **5. d.** *Physical Geogr.* A fan-shaped or conical alluvial deposit formed by a stream or river where its bed becomes less steep (e.g. at the edge of a plain); *esp.* a deposit of little height and gentle slope (cf. *CONE *sb.*^{1} 1 d). Also *attrib.*, as *fan-delta, -terrace.*

1864 J. VON HAAST *Rep. Form. Canterbury Plains* 19 Thinking that giving such a name [*sc.* 'delta'] to the alluvial accumulations of the rivers in this island, showing some peculiarities, would impart an erroneous impression, Dr. Hector and myself, in drawing up a synopsis of the geological formations of New Zealand, have adopted for the formation of those subaerial accumulations the expression 'Fan', for those of regular water-courses; and of 'Half-cone', for those of intermittent mountain torrents, and we shall for the future use these two expressions. **1873** *Q. Jrnl. Geol. Soc.* XXIX. 446 At the mouth of each of these [gorges] are alluvial fans, which project out into the flat of the river-alluvium. *Ibid.* 447 The fan is properly a flat cone, having its apex at the mouth of the ravine. **1883** *Proc. R. Physical Soc. Edin.* VII. 290 In this paper *lateral* delta terraces will be referred to as fan terraces. **1884** [see Dict., sense 5]. **1890** GILBERT in *U.S. Geol. Survey Monogr.* I. 81 The 'alluvial fan' of Drew is the 'alluvial cone' of American Geologists, and there would be some reason for preferring 'fan' to 'cone' if it were necessary to employ a single term only. It is convenient to use them as synonyms, employing 'cone' when the angle of slope is high and 'fan' when it is low. **1902** in Ld. Avebury *Scenery of Eng.* 482 The vale of Neath contains a series of such cones or 'fans' of gravel. **1920** *Nature* 13 May 322/2 The wind-worn grains of magnetite brought northward from the great fan-deltas of the Adour system. **1948** *Proc. Prehist. Soc.* XIV. 34 These alluvial fans extend right down to the sea. **1965** A. HOLMES *Princ. Physical Geol.* (ed. 2) xviii. 551 There are all gradations from wide fans 10–100 miles across that are usually nearly flat.., through fans of moderate width and inclination.., to relatively small steep-sided cones.. built of the coarser debris brought down by short torrential streams.

6. f. In a motor vehicle, an apparatus for sending a current of cold air over the radiator. Also *attrib.*, as **fan belt,** the belt which transmits torque to the fan from the engine.

1900 *Motor-Car World* May 178/2 The motor..is cooled by means of a fan driven from the periphery of the fly-wheel. **1904** GOODCHILD & TWENEY *Technol. & Sci. Dict.* 215/1 *Fan cooling*, a small fan, worked by the motor, is sometimes used to drive a current of cool air on to the outside of the cylinder to keep it cool. **1909** *Westm. Gaz.* 27 Apr. 4/1 The cylinders had kept so cool by the fan-blower that I could almost keep my hand on the combustion head. **1912** *Motor Man.* (ed. 14) iii. 64 An alternative method which avoids the use of a separate fan, is provided by fan-vaned arms in the fly-wheel. **1921** *Daily Colonist* (Victoria, B.C.) 19 Oct. 6/1 (Advt.), We recommend that you carry a spare fan belt of genuine leather. **1939** J. HARRISON *Motor-Cars To-day* vi. 82 Under conditions where slow speed is combined with heavy pulling, such as when climbing a hill, there would be a risk of over-heating if it were not for the fan. **1956** 'T. B. D. SERVICE' *Ford Cars* v. 114 It may be necessary, after extended mileage, to replace a fan belt. **1963** BIRD & HUTTON-STOTT *Veteran Motor Car* 175 The Vivinus.. voiturette had a fan-assisted air-cooled engine. **1963** R. F. WEBB *Motorists' Dict.* 95 *Fan cooling*. This is necessary for any air-cooled engine that is placed in an enclosed engine compartment.

10. a. *fan-blower* (earlier U.S. example). **b.** *fan-leaf*. **d.** *fan-pleated* (earlier example).

1847 *Rep. Comm. Patents 1846* (U.S.) 84 The fan blowers now used in steamboats for blowing the fires in the furnaces are generally made from two to three feet in diameter. **1869** A. R. WALLACE *Malay Archip.* xvii. 422 It is probably the Livistona rotundifolia of botanists, and is the most complete and beautiful fan-leaf I have ever seen. **1879** LADY C. SCHREIBER *Jrnl.* 1 Dec. (1911) II. 248 He had come to England and brought some fan leaves. **1873** *Young Englishwoman* June 286/1 A fan-pleated flounce.

e. In various *attrib.* uses relating to the ventilation of a mine by means of a fan.

1875 Fan-shaft [see Dict., sense 10 a]. **1883** W. S. GRESLEY *Gloss. Coal-mining* 101 *Fan drift*, a short tunnel leading from a short distance from the top of the upcast shaft to the fan chamber or casing in which the fan runs, along which the whole of the return air is drawn by the fan... *Fan shaft*. 1. A shallow pit-shaft sunk beneath a fan connecting it with the fan drift. 2. The upcast shaft where a fan is in use. **1906** *Daily Colonist* (Victoria, B.C.) 14 Jan. 10/3 Firemen, engineers, fanmen and pumpmen.. had their wages increased on November 1. **1908** *Westm. Gaz.* 9 Mar. 7/4 The fan-drift connecting the present downcast pit with the existing fan has at last been completed. **1921** *Dict. Occup. Terms* (1927) § 047 *Fan boy, fan man*,.. turns handle of small ventilating fan to ventilate heading not served by main air current. **1927** *Daily Tel.* 7 June 12/5 The dismissal of three fanmen who refused to do certain work. **1963** *Gloss. Mining Terms* (*B.S.I.*) II. 10 *Fan drift*, an airway leading from a mine shaft, or airway, to a fan. *Ibid.*, *Fan shaft*. 1. The ventilating shaft to

FAN

which a mine fan is connected. **2.** The spindle on which a fan impeller is mounted.

f. *Geol.* Used *attrib.* and in *Comb.* to denote an arrangement of strata in a series of folds which incline outwards from the central fold, the axes of the folds being likened to the diverging lines of a fan.

1882 A. GEIKIE *Text-bk. Geol.* VII. 917 The inward dip and consequent inversion..lead up to the fan-shaped structure, where the oldest rocks of a series occupy the centre and overlie the younger masses. **1902** *Encycl. Brit.* XXVIII. 652/2 The peculiar arrangement in mountains known as fan-structure may be produced by the continued compression of a simple anticline. **1937** WOOLDRIDGE & MORGAN *Physical Basis Geogr.* v. 68 Before the recognition of recumbent folds or nappes..the Alps were usually interpreted as showing 'fan-folding'.

11. fan belt, see *6 f; **fan consonant,** a consonant pronounced with the edges of the tongue more extended than is usual in making analogous sounds, as in the Arabic 'emphatic' consonants; **fan cooling,** see *6 f; **fan dance,** a solo dance in which the performer uses a fan or fans, esp. to conceal her nudity; hence *fan dancer*; **fan-delta,** see *5 d; **fan draught,** a system of supplying air in boiler furnaces by means of mechanically driven fans; **fan flat,** the flat [FLAT *sb.*^{3} 10 b] in which the fans for ventilating the boiler room of a ship are situated; **fan heater,** a heater containing an electric fan that forces air over an electrically-heated element into a room or other place; **fan-jet (engine),** a jet engine in which additional thrust is provided by cold air drawn by a fan through a duct surrounding the rest of the engine, which is used to drive the fan; also (as *fan-jet*), an aeroplane having such engines; = *TURBO-FAN; **fan-lift** *a. Aeronaut.*, fitted with fans to assist the vertical take-off of an aircraft; **fan marker** *Aeronaut.*, a radio marker beacon that transmits a fan-shaped beam; **fan worm,** any of various annelids of the families Sabellidæ and Serpulidæ.

1902 H. SWEET *Primer Phonetics* (ed. 2) 36 Fan (spread) consonants..are modifications of point and blade consonants. [**1908** —— *Sounds Eng.* 45, *tl, dl* occur in Irish English as substitutes for þ, ð respectively; in them the fan modification is supplemented by a slight raising of the back of the tongue.] **1879** Fan dance [see DANCE *sb.* 2]. **1951** T. RATTIGAN *Who is Sylvia?* II. 248 I'd better get Babs to do her fan dance—if she's still vertical. **1936** R. E. SHERWOOD *Idiot's Delight* I. 22 Shirley is the principal, a frank, knowing fan dancer. **1894** W. H. WHITE *Man. Naval Archit.* (ed. 3) xiv. 561 Fan draught is also of great value under unfavourable conditions, such as hot weather, calms, or following winds. **1909** *Westm. Gaz.* 15 May 2/2 A monstrous wave..poured into the fan-flat. **1923** *Man. Seamanship* II. xvii. 285 The fans are situated on an enclosed fan flat from which they draw their air. *Ibid.*, Access to the boiler rooms is arranged through the fan flats. **1961** I. MURDOCH *Severed Head* vi. 48 He dangled his long broad-nailed hand in front of his new fan heater. **1970** *Bodl. Libr. Rec.* VIII. 173 Cold air is deflected from the entrance by three 3kw fan heaters. **1963** *Sat. Rev.* 20 July 14 (Advt.), In 1961, American Airlines introduced a new engine called the fan-jet—with 30% more power than ordinary jets. **1963** *N.Y. Times* 15 Sept. p. xx7/3 The fan-jet engines..have turbine blades spaced for a minimum of sound. **1967** N. E. BORDEN *Jet-Engine Fund.* 47 Fanjets, as they are called by some of the commercial airlines, and turbofans are one and the same thing. **1968** *Daily Colonist* (Victoria, B.C.) 10 Nov. 9/6 The nose gear of an Alaska Airlines 727 fanjet was severed Saturday afternoon when the plane collided with a moose as the plane came in for a landing. **1961** *Flight* LXXX. 504/2 Jet and fan lift aircraft appear to offer good range-speed-payload performance where substantial range is required and where hovering requirements are at a minimum. **1961** *Aeroplane & Astronaut.* CI. 791/1 Two VTOL test-bed aircraft using the G.E. J85-5 fan-lift engine. The complete programme, which covers two years and will cost $10.5 million, includes the construction of the two fan-lift airframes by Ryan at San Diego. **1948** *Shell Aviation News* CXXIV. 8/2 Written examinations are required on radio facilities in the New York area, including radio ranges, homing facilities, fan markers and let down procedures on the heavily congested La Guardia airport. **1851** Fan-worm [see SABELLA]. **1963** R. P. DALES *Annelids* 15 The most specialized tube-dwellers are the sabellid and serpulid fan-worms.

fan, *sb.*^{2} Delete †*Obs.* and add: Re-formed in mod.E. (orig. *U.S.*) to denote: a keen and regular spectator of a (professional) sport, orig. used of baseball; a regular supporter of a (professional) sports team; hence, a keen follower of a specified hobby or amusement, and *gen.* an enthusiast for a particular person or thing.

1889 *Kansas Times & Star* 26 Mar., Kansas City baseball fans are glad they're through with Dave Rowe as a ball club manager. **1896** ADE *Artie* xvii. 158 I'm goin' to be the worst fan in the whole bunch. **1901** *Dialect Notes* II. 139 *Fan*, a base ball enthusiast; common among reporters. **1914** *Daily Express* 3 Oct. 3 First League football 'fans' in London can have a joyous time to-day. **1915** *Film Flashes* 13 Nov. 1 It is quite usual for a picture 'fan' to come out of one theatre and immediately cross the road to another. **1919** W. T. GRENFELL *Labrador Doctor* (1920) iv. 56 Among my acquaintances there were not a few theatre fans. **1921** A. W. MYERS *20 Yrs. Lawn Tennis* 142 This was sheer spectacular tennis, dear to the hearts of the American 'fans'. **1925** H. V. MORTON *Heart of London* 93 The fight fans howling like a pack of hungry wolves. **1928** S. VINES *Humours Unreconciled* xiii. 168 What about..your League of Nations and disarmament fans? **1950** *Manch. Guardian Weekly* 4 May 15/4 The Water Department.. had received..'good-natured' complaints from base-ball fans about the washing-out of two days' play.

2. *Comb.* **fan club,** a group formed by the devotees of some hero, 'star', etc.; **fan letter,** a letter from an admirer to a celebrity; **fan magazine,** a journal specializing in some common object of devotion or in well-known personalities; **fan mail,** the letters sent to a celebrity by his or her followers.

1941 V. FAULKNER in *Sat. Even. Post* 6 Sept. 37/3 Sleepy Hollow, the name Desire had selected for her residence from a contest held by her fan club. **1959** 'O. MILLS' *Stairway to Murder* v. 51, I believe she's the President of your Fan Club. **1932** WODEHOUSE *Louder & Funnier* 41 How many fan-letters did you get last week? **1937** W. S. MAUGHAM *Theatre* x. 82 She was naturally polite and it was, besides, a principle with her to answer all fan letters. **1928** *Amer. Speech* III. 364 It was picked up from a 'fan-magazine'. **1951** *Life* 21 May 130/2 'Fanzines', or fan magazines, which are usually small mimeographed publications devoted to amateur STF, criticism and gossip. **1924** *Motion Pict. Mag.* June 43 (*title*) The business of fan mail. **1937** AUDEN & MACNEICE *Lett. fr. Iceland* 17 A poet's fan-mail will be nothing new. **1955** R. BANNISTER *First Four Minutes* xiv. 194 It was the beginning of a fan mail and of invitations to open bazaars that have continued ever since.

Fan (fæn), *sb.*^{3} and *a.* Also **Fang.** [Fr. *Fan*, presumably ad. Fan *Pangwe*.] **A.** *sb.* **a.** A member of an African people in the Ogowe basin in western equatorial Africa. **b.** The Bantu language of this people. **B.** *adj.* Of or pertaining to this people or their language.

1861 P. B. DU CHAILLU *Expl. & Adv. Afr.* vii. 65 He.. set off to a Fan village. *Ibid.* 67 Great crowds of Fan.. came to see me. **1865** *Trans. Ethnol. Soc. Lond.* III. 37 It is proposed thus to write the very nasal nasals of the Fan language. **1879** *Encycl. Brit.* X. 3/2 The Fan, whose name appears under the various forms of Fanwe, Panwe, Phaouin, and Paouen, are newcomers to the Gaboon district. **1883** R. N. CUST *Sk. Mod. Lang. Afr.* II. xii. 413 As we advance into the Interior, we find only two leading Languages, the Fan, spoken by the invading Oshiba, and the Benga. **1897** M. KINGSLEY *Trav. W. Afr.* xiv. 319 A young Fan man has to fend for himself. *Ibid.* 322 Fan pottery, although rough and sunbaked, is artistic in form. **1911** J. FRAZER *Golden Bough* (ed. 3) I. vi. 349 Thus in the Fan tribe the strict distinction between chief and medicine-man does not exist. The chief is also a medicine-man and a smith to boot; for the Fans esteem the smith's craft sacred, and none but chiefs may meddle with it. **1936** *Discovery* June 172/1 The area of the Western Bantu includes..in the west [the home of] such renowned cannibals as the Fang; it also includes the territory of considerable and highly organised kingdoms, such as the medieval kingdoms of the Kongo and the Balunda, and the later Bushongo Empire. **1962** *Listener* 12 Apr. 630/1 The Fan villages.

fan, *v.* Add: **1. b.** (Later U.S. examples.)

1879 A. W. TOURGÉE *Fool's Err.* xxvii. 261 Whoever got 'fanned out'—it was always our own folk that did it. **1908** A. RUHL *Other Americans* x. 151 One dreams of..a Broadway policeman marching down upon them leisurely with a nightstick and fanning them away.

2. b. Also, to move smartly; to clear *out*. *U.S. slang.*

1902 O. WISTER *Virginian* xv. 168 This hyeh train?.. Why, it's been fanning it a right smart little while. **1904** *Leslie's Monthly Mag.* Aug. 421/2 He saw I was drunk, and fanned out, me shootin' at him with every jump. **1927** W. R. JAMES *Cow Country* viii. 230 Todd.. stuck to his seat and fanned his pony on out to the open.

3. Also, to stir (water) in this way; hence *intr.* or *absol.*

1886 *Outing* (U.S.) VIII. 161/1 The trout..is balancing himself on the hard sandy bottom, his fins slowly fanning the water. **1898** *Daily News* 15 June 4/7 It is commonly in steady weather that you see the trout with their heads a quarter of an inch beneath the surface, and with their tails fanning expectantly.

6. b. Further *intr.* examples in sense 'to spread out in the shape of a fan; to assume a fan-like shape'. Also *fig.*

1926 *Hutchinson's Best Story Mag.* Nov. 59/1 His black tangle of beard, fanning over his knees ruffled in the wind. **1943** F. S. HERMAN *Dynamite Cargo* xi. 95 The warship screen fanned out and closed in. **1962** *Times* 25 Apr. 16/6 There fan out from the incident..not changes of attitude..but a series of comments on the nature of priestly vocation. **1965** *Electronics Weekly* 1 Dec. 17/5 Generating stations..interconnected by.. transmission feeders fanning out to reach principal load centres.

7. a. (Later U.S. examples.)

1916 E. TITUS *I Conquered* vi. 84 His quirt fell...He fanned his pony again, and the beast grunted in his struggles for increased speed in the climbing. **1918** C. E. MULFORD *Man fr. Bar-20* ix. 100 Cussed if I wouldn't 'a' give six pesos, U.S. to 'a' seen that cougar a-fannin' you! **1931** L. STEFFENS *Autobiogr.* I. II. v. 213 You wonder why we fan these damned bums, crooks, and strikers with the stick.

b. Also = *FRISK *v.* 4 a.

1927 E. WALLACE *Feathered Serpent* xvii. 216 Legally no policeman has the right to 'fan' a prisoner until he gets into the police station. **1946** J. IRVING *Royal Navalese* 74 *To fan*, to search a person quickly for symptoms of concealed contraband articles, firearms, etc.

8. *N. Amer.* Of a pitcher in baseball: to cause (a batter) to strike out.

1909 in WEBSTER. **1912** C. MATHEWSON *Pitching in a Pinch* v. 101 He fanned the next two men. **1970** *Globe & Mail* (Toronto) 28 Sept. 19/4 The Indian lefthander..has fanned more than 300 hitters.

b. *intr.* Of a batter: to strike out.

1886 *Outing* (U.S.) July 477/2 The man who..'fans out' or 'pops one up'. **1945** *This Week Mag.* 21 Apr. 10 He fanned in a pinch and the opposition booed.

‖ **fanā** (fānā·). [a. Arab. *fanā'*.] The Sufi doctrine of annihilation (see quots.).

1867 E. H. PALMER *Oriental Mysticism* v. v. 67 Further than this he cannot go, but pursues his habit of self-denial and contemplation until his death, which is, however, merely looked upon as a total re-absorption into the Deity, forming the consummation of his Journey, the last stage, designated *Faná*, Extinction. **1911** *Encycl. Brit.* XXVI. 31/2 The step to pantheism was first decisively taken by the great Persian Ṣūfī, Abū Yazīd (Bāyezīd)..(d. A.D. 874), who introduced the doctrine of annihilation (fanā), *i.e.* the passing away of individual consciousness in the will of God. **1956** E. WOOD *Yoga Dict.* 153/1 In the end the constant thought or remembrance..of God, will lead to union..with Him, in which there will be passing away (*fana*) of all human qualities or nature.

Fanagalo (fā·nagalō̤). Also **Fanakolo, -kalo.** [f. the common phr. *fana ga lo* = 'like this' in the *lingua franca* of southern African mines (see below), f. Zulu *fana* be like + *ka* poss. prefix Class 1a + *lo* demonstr. Classes 1 and 3.] A *lingua franca* of southern Africa, made up of elements of Zulu/Xhosa, English, and Afrikaans. Formerly called *Kitchen Kaffir* or *Mine Kaffir*.

The word *kaffir* being regarded as derogatory by Africans, the new name *Fanagalo* came to be adopted by mining authorities and other employers.

1947 J. D. BOLD *Dict. & Phr.-Bk. Fanagalo* 6 The appelation *Fanagalo* probably derives from *kuluma fana ga lo* meaning 'to speak like this'. **1951** *Cape Argus* 19 Jan. 8 'Fanakolo', the *lingua franca* of the mine compounds of South Africa. **1952** L. MARQUARD *Peoples & Policies S. Afr.* ii. 35 A more or less regularized version of this ['Kitchen Kaffir'] called Fanakolo, is used as the *lingua franca* on the gold-mines. **1956** A. G. MCRAE *Hill called Grazing* xii. 126 He spoke the lingua franca of all South African natives, the so-called Fana-ka-lo.

Fanar, Fanariot(e, varr. PHANAR, PHANARIOT.

1819 T. HOPE *Anastasius* I. ii. 42 He..plunged headlong into all the intrigues of the Fanar. *Ibid.* iv. 74 The persons of the Fanariote grandees were of a piece with their habitations. *Ibid.* 82, I had my share of the second hand insolence, which the Fanariotes take very quietly from the Turks. **1838** *Penny Cycl.* X. 194/2 A crowd of Fanariotes always followed the new Hospodars. *Ibid.*, The bankers of the Fanar. **1856** tr. *F. Perthes' Mem.* I. xxxiii. 411 The cruel exactions of the Greek princes, and insatiable Fanariotes. **1878** DISRAELI in G. E. Buckle *Life* (1920) VI. 320 A perfect Greek of the Fanar. **1886** *Encycl. Brit.* XX. 20/1 The Fanar quarter of Constantinople.

fanatic, *a.* **1. b.** Delete † and add later examples.

1926 W. J. LOCKE *Stories Near & Far* 225 A bearded, fanatic-eyed..figure. **1932** W. FAULKNER *Light in August* xx. 447 Fanaticfaced country preachers.

‖ **fanchon** (fañʃoñ). [Fr., dim. of name *Françoise*.] A kerchief. Also *fanchon bonnet, cap* (see quots.).

1872 *Young Englishwoman* Nov. 611/1 This pretty light fanchon is knitted with white and red Shetland wool. **1928** 'BRENT OF BIN BIN' *Up Country* xvi. 275 They had elegant bonnets richly ornamented, fançon-shaped behind. **1957** M. B. PICKEN *Fashion Dict.* 122/1 *Fanchon (kerchief) bonnet*, bonnet resembling a diagonally folded kerchief. Popular in Victorian period. **1960** CUNNINGTON & BEARD *Dict. Eng. Costume* 76/1 *Fanchon*, 1830's on. A small kerchief for the head, the term being chiefly used for the lace trimming falling about the ears of a day cap or outdoor bonnet. *Ibid.*, *Fanchon cap*, 1840's to 1860's. A lace or tulle cap with side pieces covering the ears, or sloping down to them.

‖ **fanchonnette** (fañʃonet). [f. as prec. + -ETTE.] (See quots.)

1845 E. ACTON *Mod. Cookery* xvi. 432 *Fanchonnettes*.. Roll out..puff paste..cover it..with peach or apricot jam..roll a second bit of paste..and lay it carefully over the other. *Ibid.* 433 This is not the form of pastry called by the French *fanchonnettes*. **1861** MRS. BEETON *Bk. Househ. Managem.* xxvii. 660 Fanchonnettes, or Custard Tartlets. **1958** W. BICKEL tr. *Hering's Dict. Cookery* 682 *Fanchonnettes*, tartlet moulds lined with puff paste, filled with almond cream, half baked, covered with meringue mixture.

fancify, *v.* Add: Hence **fancifica·tion.**

1937 *New Republic* 24 Feb. 74 The constant elaboration, figures of speech, conceits, fancifications, involutions, not seldom characteristic of Elizabethan writing.

1960 V. PACKARD *Waste Makers* (1961) xi. 124 The trend in color went right back to white, with some venturing into fancification with tracery effects. **1962** *Spectator* 14 Dec. 941 Their elusive and often fragmented fancifications.

fancy, *sb.* and *a.* Add: **A.** *sb.* **12. a.** (Earlier example.)

1735 J. MOORE *Columbarium* 40 These Pigeons by their Flight afford an admirable Satisfaction, to those Gentlemen of the Fancy that have time to attend them.

b. (Earlier example.)

1807 SOUTHEY *Lett. fr. Eng.* (1951) lxxi. 451 The Amateurs of Boxing, who call themselves *the Fancy*.

14. Often = **fancy cake* (below, C. 1).

1891 *Confectioners' Union* 15 Oct. p. iii (Advt.), Fondant, Jellies, Gelatine goods,..and other Fancies. *Ibid.* 15 Nov. 633/2 The room where Christmas fancies are being packed by a small army of girls. **1894** E. SKUSE *Compl. Confect.* 110 There are a great number of fancies made from grain sugars, sold about Christmas time. **1968** M. BRAGG *Without City Wall* I. xix. 191 Jars of jam being given by the ladies, a box of teacakes and fancies by Agnes.

B. *attrib.* and *Comb.* **1. b.** *fancy-spinning* vbl. sb.; *fancy-taking* ppl. adj. **c.** *fancy-topped, -waistcoated* ppl. adjs.

1962 W. NOWOTTNY *Lang. Poets Use* iv. 91 Gaunt tries to comfort him with something like the fancy-spinning of Richard. **1855** *National Rev.* July 57 The great features..which make society..remarkable, fancy-taking. **1875** *Atlantic Monthly* Jan. 70 She has two tall fancy-topped chimneys. **1909** *Daily Chron.* 7 Aug. 4/7 He of the well-dressed, fancy-waistcoated order.

2. fancy Dan *slang* (orig. *U.S.*), a dandy; a showy but ineffective worker or sportsman; **fancy-girl** *colloq.* = *fancy-woman*; also **fancy-piece; fancy-woman** (earlier example).

1943 *Amer. Speech* XVIII. 107 Fancy Dan (a pitcher good in practice but sour in a game; also a dressy player). **1950** J. DEMPSEY *Championship Fighting* ii. 12 The amateur and professional ranks today are cluttered with..'fancy Dans'. **1950** A. LOMAX *Mr. Jelly Roll* (1952) 49 Then you could observe the fancy Dans, dressed fit to kill. **1960** T. MCLEAN *Kings of Rugby* xi. 168 Wellington's strong and experienced men..knew too much about defensive play to be gulled by Fancy-Dan stuff. **1892** P. H. EMERSON *Son of Fens* xv. 131 We allust call our scythes arter our fancy-gels or our wives. **1930** A. P. HERBERT *Water Gipsies* xxii, Let's hear the rest now—out with it! You been his fancy-girl? **1969** 'M. INNES' *Family Affair* ix. 101, I thought this fellow might paint a man's fancy girls—see? **1821** P. EGAN *Life in London* I. iii. 47 Or even *smiled* with indifference at the rolls of *soft* which his most captivating *fancy-piece* drew from him repeatedly. **1812** J. H. VAUX *Flash Dict.*, A woman who is the particular favourite of any man, is termed his *fancy woman*, and *vice versa*.

C. *adj.* **1.** (Earlier example.) Also *fancy cake*, etc.

1751 H. GLASSE *Art Cookery* (ed. 4) App. 333 A few Astertion Flowers stuck here and there looks pretty..but, Lemon, and all those Things are entirely Fancy. **1893** E. HÉRISSÉ *Art of Pastry Making* ii. 8 (*heading*) Fondant icing..for icing fancy pastry-cakes. *Ibid.* xiii. 97 (*heading*) Fancy ornamental meringues. (Meringues décorées.) **1912** —— *Ibid.* (new ed.) ii. 20 Dip in the fancy cakes or pastry with the point of a penknife or fork. **1960** *Good Housek. Cookery Bk.* (ed. 5) 387/1 Small iced fancy cakes. ..Cut the cake into fancy shapes before icing.

d. *fancy fair* (earlier examples). Also *fancy shop*.

1835 DICKENS *Sk. Boz.* (1836) I. 136 Flaming accounts of some 'fancy fair in high life'. **1840** H. COCKTON *Life Valentine Vox* xxv. 178 A placard..which announced that a Fancy Fair and a Fete Champetre were about to take place. **1844** J. COWELL *30 Yrs. among Players* (1845) I. vi. 15/2 His address..was at a handsome fancy-shop in George-street. **1911** A. BENNETT *Hilda Lessways* I. iii. 24 She went straight into Dayson's little fancy shop.

e. *fancy ball* (earlier example). Also *fancy dance*.

1809 M. EDGEWORTH *Parent's Assistant* (1848) III. 285 How fine we are!..one might fancy one's-self..at a fancy ball. **1827** J. CONSTABLE *Let.* 10 Oct. (1962) 237 To day..a Garden Promenade—and fancy dance in a temporary pavilion.

f. *fancy goods* (cf. 1 d).

1792 *Columbian Centinel* (Boston, Mass.) 12 May, A large and extensive assortment of staple and fancy goods. *a* **1828** D. WORDSWORTH *Tour Continent* in *Jrnls.* (1941) II. 73 Stalls and shops of jewelry, millinery and fancy goods of all sorts. **1860** DICKENS in *All Year Round* 28 Jan. 321/1, I..have rather a large connexion in the fancy goods way. **1933** *Archit. Rev.* LXXIV. 16/3 A manufacturer of fancy goods and handbags remarks that, 'we study the public taste while we try to originate'.

3. d. *fancy franchise* (earlier example).

1859 *Hansard Commons* 28 Feb. 1025 [John Bright *loq.*] I say, all these fancy franchises are absurd.

e. *fancy religion* (see quot.).

1925 FRASER & GIBBONS *Soldier & Sailor Words*, *Fancy religion*, a very old Service colloquial term in both Navy and Army for a creed or denomination not Church of England, Roman Catholic, or Presbyterian, before the War the three authorized creeds.

5. Applied to a person who is fancied or who fancies himself: mostly synonymous (but in some expressions with slang overtones) with FANCY MAN a, or *fancy-woman*. Cf. also *fancy-girl* (sense B. 2 above).

1891 KIPLING *Light that Failed* v. 73 'E brought some most remarkable fancy young gentlemen up 'ere. **1905** H. ELLIS *Stud. Psychol. Sex* IV. 235, I..told her again I had no money. She..said, 'That does not matter... I want you to be my fancy boy.' **1934** T. S. ELIOT *Rock* I. 12 One o' them fancy lads—a good soldier and fond o' the ladies—but a great one for 'is church. **1938** E. BOWEN *Death of Heart* III. iv. 381 Eddie had been warned ..that one could not go to all lengths as Mrs. Quayne's fancy boy. **1942** BERREY & VAN DEN BARK *Amer. Thes. Slang* § 452.3 *Dandy*,..fancy pants. **1967** 'M. HUNTER' *Cambridgeshire Disaster* viii. 52 Some puffed-up fancy pants..said something which made the barmaid laugh.

fancy, *v.* Add: **1. e.** (Earlier and additional examples.)

1813 JANE AUSTEN *Let.* 6 Nov. (1932) II. 90 Very snug, in my own room, lovely morn[g], excellent fire, fancy me. **1850** E. RUSKIN *Let.* 9 Feb. in M. Lutyens *Effie in Venice* (1965) I. 139 But, only fancy, the Thousands and Thousands of wax lights. *a* **1943** R. G. COLLINGWOOD *Idea Hist.* (1946) v. 270 Whom is she shielding? Either her father or her young man. Is it her father? No: fancy the rector! **1971** E. MCGIRR *No Better Fiend* 83, I did the 'fancy that!' line of patter.

8. c. To view (a horse) favourably as a likely winner of a race.

1922 JOYCE *Ulysses* 391 A racing-horse he fancied. **1923** WODEHOUSE *Inimitable Jeeves* xi. 116 Ocean Breeze is fancied, as I am told the expression is, for a race which will take place..at Goodwood.

fancy bread. [FANCY *sb.* and *a.* C. 1.] Bread not of the ordinary texture, size, and weight of the standard 'household' and 'cottage' loaves.

1801 *Times* 9 Mar., Germans, who make what they call French or fancy bread, particularly to please the appetites of foreigners. **1841** *Guide to Trade, Baker* 65 Fancy bread, ginger-bread, buns, rolls, muffins and crumpets, etc. **1853** [see FANCY *a.* 1]. **1904** *Westm. Gaz.* 23 Aug. 4/1 Fancy bread is for the future to be defined as that which is 'made up into separate rolls, twists, or other shapes, each of which is less than one pound in weight'. **1908** J. KIRKLAND *Mod. Baker* II. xxvii. 165 The greatest diversity of opinion prevails amongst bakers as to what is fancy bread. The rough interpretation of the term as recognized by the Bread Laws is: Bread that cannot readily be mistaken for plain bread. The distinguishing mark.. is some difference in shape or glaze: but the baker.. makes it include all sorts which involve more labour in manufacture, or entail greater cost for materials.

fancy dress, *sb.* Add: **1.** *fancy dress ball* (examples). Also further *attrib.* and *fig.* examples.

1808 *Monthly Pantheon* I. 76 Bending branches press In many folds blithe nature's fancy dress. **1844** G. W. KENDALL *Narr. Santa Fé Exped.* II. 51 Such variety of costume..would put to the blush..any..fancy-dress procession ever invented. **1882** *Let.* 8 Aug. in A. Nathan *Costumes by Nathan* (1960) iv. 60 They would be required for a Fancy Dress Ball the week of the 4th September. **1915** A. D. GILLESPIE *Let.* 8 June (1916) 187 A fancy dress dance given there by the Germans on Christmas Eve. **1919** D. ASHFORD *Young Visiters* vi. 52 Is it a fancy dress party he asked. **1922** C. E. MONTAGUE *Disenchantment* ii. 26 London..was grotesque with a kind of fancy-dress ball of non-combatant khaki. **1938** L. MACNEICE *Mod. Poetry* 39, I realize that this piece [*sc.* The Lady of Shalott], though dressy, is fancy-dress at its best.

fancy man. a. and **c.** (Earlier and later examples.)

1811 *Lex. Balatron.*, *Fancy man*, a man kept by a lady for secret services. **1818** 'A. BURTON' *Adv. Johnny Newcome* iii. 154 The Sweepers e'en, were *fancy men!* **1891** T. HARDY *Tess* III. v. xli. 26 Though your fancy-man was so up about it. **1922** JOYCE *Ulysses* 48 Her fancyman is treating two Royal Dublins. **1960** D. LESSING *In Pursuit of English* vi. 217 My mother's married that fancy man and he's already started to treat her bad. **1966** B. NAUGHTON *Alfie* i. 12 You won't get one husband in ten feels any thanks to the wife's fancy man for the happiness he brings to the marriage.

fancy work. Add: (Earlier and *attrib.* examples.) Also *fig.*

1810 F. CUMING *Sk. Tour Western Country* xxvi. 163 Boarders instructed in..embroidery..and any other fashionable fancy-work. **1816** JANE AUSTEN *Emma* I. iii. 39 His..claim on her to leave her neat parlour hung round with fancy-work whenever she could. **1864** C. M. YONGE *Trial* II. xvi. 308, I admired the pattern, and Blanche inspected it..hoping to convert me to a fancy-work-woman. **1896** E. TURNER *Little Larrikin* xviii. 206 A pair of slippers bought at a fancy-work shop. **1907** *Times Lit. Suppl.* 12 July 217/3 The importance and the novelty of their matter seem to prevent any excessive rhetoric; they cannot afford the time for fancy work. **1929** *Daily Express* 7 Nov. 19/1 They cut out all fancy work and went straight ahead for goal. **1954** F. C. AVIS *Boxing Ref. Dict.* 38 *Fancy work*, 'clever' movements of the hands and body intended to impress the onlookers, but often merely a waste of energy.

fandangle. (Earlier U.S. example.)

1835 *Southern Lit. Messenger* I. 361 What is the use of all these fandangles of lace?

fandango. Add: **1. c.** *fig.*

1841 *Congress. Globe* App. 25 Jan. 153/2 All the fool Federal fandangoes that disgraced the country. **1894** *Monthly Packet* Feb. 152 The hippopotamus does not indulge in these fandangoes. **1928** BLUNDEN *Undertones of War* xix. 197 The usual free-verse fandango of brick mounds and water-holes.

4. *fandango-dancer.*

1809 *Tales of Other Realms* I. 94 We quitted the *fandango dancers* in disgust.

fa·ndom. orig. *U.S.* [f. *FAN *sb.*^{2} + -DOM.] The world of enthusiasts for some amusement or for some artist; also in extended use.

1903 *Cincinnati Enquirer* 2 Jan. 3/1 (*heading*) Fandom puzzled over Johnsonian statements. **1928** *Publishers' Weekly* 30 June, Ty Cobb, the idol of baseball fandom. **1958** *Times* 13 Sept. 7/6 The same editor calculates that at least half his British writers have been recruited from 'fandom'. **1963** *Philos. Rev.* LXXII. 520 Morality has or ought to have its fandom.

fanfare. Add: **c.** A style of bookbinding decoration developed in Paris in the 16th century in which a continuous interlaced ribbon, bounded by a double line on one side and a single on the other, divides the whole surface on both covers into symmetrical compartments of varying shapes and sizes.

1895 J. W. ZAEHNSDORF *Short Hist. Bookbinding* 22 A development of the 'fanfare' sprays of foliage. *Ibid.*, The graceful ornamentation known as 'fanfare' is attributed to the Eves...The name of 'fanfare' was given to this style of work in the last century, when Charles Nodier had a volume entitled 'Les Fanfares et Corvées Abbadesques' bound for him by Thouvenin. **1936** *Times Lit. Suppl.* 13 June 493/4 The popular attribution of *fanfare* bindings to Nicolas and Clovis Eve is..shown to be..untenable. **1959** *Chambers's Encycl.* V. 489/1 All bindings in the so-called 'fanfare' style..have an all-over pattern of ribbons, bounded on one side by a double and on the other by a single line. **1969** *Times* 25 Feb. 16/1 (Advt.), A fine Parisian fanfare binding and other decorated book bindings.

fanfoot (fæ·nfut). Pl. **fanfeet, fanfoots.** [f. FAN *sb.*^{1} + FOOT *sb.*] **a.** A collectors' name for a moth of the family Hyperinæ. **b.** A gecko of the genus *Ptyodactylus*, having fan-shaped toes; also called *fan-footed gecko*.

1832 J. RENNIE *Butterfl. & Moths* 146 Polypogon..The Common Fan-foot..The Clay Fan-foot. **1863** W. SMITH *Dict. Bible* II. s.v. *Lizard*, The Fan-Foot Lizard (*Ptyodactylus Gecko*). **1933** *Proc. Zool. Soc.* II. 764 The Fan-foot is abundant and distributed widely in Palestine. *Ibid.* 765 A Fan-footed Gecko brought in from the desert was turned loose in my room. **1961** R. SOUTH *Moths Brit. Isles* (ed. 2) I. 393 The fan-foot (*Zanclognatha tarsipennalis* Treit.). *Ibid.* 394 The small fan-foot (*Zanclognatha nemoralis* Fab.). *Ibid.* 395 The dotted fan-foot (*Zanclognatha cribrumalis* Hübn.). The clay fan-foot (*Paracolax derivalis* Hübn.). *Ibid.* 396 The common fan-foot (*Herminia barbalis* Clerck).

fang, *sb.* Add: **4. e.** *colloq.* A human tooth. Also *Comb.* and *fig.*

1840 [see Dict., sense 4]. **1891** FARMER *Slang* II. 374/1 *Fang-faker*, a dentist. **1919** W. H. DOWNING *Digger Dial.* 22 'To put in the fangs'—to demand money, etc. **1936** WODEHOUSE *Laughing Gas* v. 57 Possibly because they were old dental college chums,..these two fang-wrenchers shared a common waiting-room. **1943** HUNT & PRINGLE *Service Slang* 31 *Fang farrier*, dentist. **1957** 'N. CULOTTA' *They're Weird Mob* (1958) viii. 109 Jimmy got himself some bread and butter and an open tin of jam. 'Yer good on the fang, mate,' said Joe.

IV. 9. *attrib.* and *Comb.*: **fang-bolt,** a bolt having a spiked nut or washer, used for attaching iron to wood.

1876 J. W. BARRY *Railway Appliances* ii. 73 Fang-bolts consist of bolts long enough to pass through the sleepers, with a screw cut on the lower end to fit a wide flat nut, having on it fangs or short spikes. **1915** C. J. ALLEN *Mod. Brit. Perm. Way* 60 Whereas this type of fang-bolt has in all three separate parts—bolt, nut, and washer—it will be noticed that the Great Southern and Western and Great Eastern fang-bolts..consist of the bolt and a fanged nut only.

Fang, var. *FAN *sb.*^{3} and *a.*

fang, *v.*^{1} **1. c.** (Later example.)

1922 G. BLAIR *Haunted Dominie* 21 O what shall then betide me, when Death shall fang my shoulders.

fanglomerate (fæŋglǫ·mĕrĕt). *Geol.* [f. *FAN *sb.*^{1} 5 d + CON)GLOMERATE *sb.*] A rock consisting of comparatively erosion-resistant fragments of various sizes deposited in an alluvial fan and consolidated into a solid mass.

1912 A. C. LAWSON in *Bull. Geol. Soc. Amer.* XXIII. 72 (*heading*) Fanglomerate, a detrital rock at Battle Mountain, Nevada. **1933** *Geogr. Jrnl.* LXXXI. 156 The gravel was then consolidated into conglomerates and fanglomerates. **1957** F. J. PETTIJOHN *Sedimentary Rocks* (ed. 2) xiii. 630 This facies appears to consist of fanglomerates deposited near the boundary fault scarps and of channels, floodplains, and swamp deposits in the more remote areas.

‖ **fango** (fa·ŋgo). [It., mud, dirt: cf. FANC.] A kind of mud obtained from the thermal springs at Battaglia in the Veneto in Italy, used in the treatment of gout, rheumatism, and other ailments. So **fangothe·rapy.**

1900 GOULD *Pocket Med. Dict.* (ed. 4) 249 Fango. **1903** DORLAND *Med. Dict.* (ed. 3) 264/1 *Fangotherapy*, treatment by the application of fango. **1904** E. RUTHERFORD *Radio-Activity* 363 'Fango'—a fine mud obtained from hot springs in Battaglia, Northern Italy. **1905** *Jrnl. Balneology* Jan. 5 The annual visits of several patients to Italy for the sole purpose of taking a course of fango-packs. **1906** *Christian World* 22 Mar., I have just returned from a trial of 'Fango' at the Royal Hotel and Baths, Matlock Bath. **1906** *Punch* 2 Feb. 161/1 The Italians are rightly proud of their [spa] muds (do they not still send us Fango?). **1913** J. LINDSAY *Gout* xxxi. 204 Peat, mud, fango, and similar semi-solid baths are of therapeutic value on account of their thermal action. **1940** B. I. COMROE *Arthritis* xxvii. 332 In some spas, fango is used extensively. *Ibid.* 333 Fango therapy produces marked hyperemia of the skin.

fank, *v.* (Later example.)

1923 *Glasgow Herald* 19 Nov. 7 Days..set apart for fanking the sheep.

fanny^2 (fæ·ni). *Naut.* [? The female name.] A tin for holding anything to be drunk; a mess-kettle.

1904 *Daily Chron.* 11 Aug. 3/2 Many total abstainers drawing their grog and leaving it in the 'fanny' for the benefit of the mess. **1923** *Man. Seamanship* (*H.M.S.O.*) II. 35 A full set of mess utensils, consisting of, mess kettle..fanny (metal tin 1½ gallons). **1925** FRASER & GIBBONS *Soldier & Sailor Words*, *Fanny*, a name for the receptacle holding the bluejackets' 'tot' of rum. **1926** *Blackw. Mag.* Dec. 823/2 On board a British man-of-war the same..vessel is called a 'fanny'...Tea made in a billy or fanny is the best to be had. **1943** C. S. FORESTER *Ship* i. 13 The 'mess-traps' about which he had worried, the 'fannies' of soup. **1952** G. HACKFORTH-JONES *Dangerous Trade* xxv. 175 Send a fanny full of hot tea while you are about it.

Fanny^3. The word formed by the initials of *F*irst *A*id *N*ursing *Y*eomanry accommodated to the form of the name *Fanny*.

1918 [see F.A.N.Y. (*F III. 3)].

fanny^4. *slang* (orig. and chiefly *U.S.*). [Orig. unknown.] = BACKSIDE 3.

1928 HECHT & MACARTHUR *Front Page* II. 115 Parking her fanny in here. **1930** N. COWARD *Private Lives* I, You'd fallen on your fanny a few moments before. **1937** T. RATTIGAN *French without Tears* II. i. 44 That's it. Progress. *Kit.* Progress my fanny. **1946** R. CAMPBELL *Talking Bronco* 29 Ere you came back to serenade the sentry, Who thanks you with this bayonet in your fanny! **1949** E. POUND *Pisan Cantos* lxxx. 95 And three small boys on three bicycles Smacked her young fanny in passing. **1953** 'R. GORDON' *Doctor at Sea* I. 16 Move over, Second, and let the Doctor park his fanny. **1959** M. STEEN *Woman in Back Seat* II. vii. 284 Classy, isn't it [*sc.* a cardigan]?—that little roll round the fanny. **1960** 'N. SHUTE' *Trustee from Toolroom* iv. 82 I'd never be able to think of John and Jo again if we just sat tight on our fannies and did nothing.

fanny (fæ·ni), *v. slang.* [Orig. unknown.] *trans.* To deceive or persuade by glib talk. Also as *sb.*, glib talk, a tall story.

1933 G. INGRAM *'Stir'* v. 65 'Spin a right fanny to the "Croaker",' advised Smith. **1936** 'J. CURTIS' *Gilt Kid* xiii. 134 He would spin her a fanny about the marriage laws. **1942** G. KERSH *Nine Lives Bill Nelson* ii. 12 A Guardsman comes to Bill with some Fanny about needing some cash. **1949** *John o' London's* 4 Mar. 123/3 'Fannied' a 'Pitch', that is, got together a larger crowd and told them of the excellence of the play to be seen inside. **1965** A. PRIOR *Interrogators* viii. 147 They could not fanny Norris into thinking they believed he might have been out to a woman.

Fanny Adams (fæ·ni æ·demz). [The name of a young woman who was murdered *c* 1867.] **1.** *Naut. slang.* **a.** Tinned meat. **b.** Stew.

1889 BARRÈRE & LELAND *Dict. Slang*, *Fanny Adams* (naval), tinned mutton. **1927** *Blackw. Mag.* Feb. 259/2 'Fanny Adams' (or preserved mutton) brought from the ship. **1962** W. GRANVILLE *Dict. Sailors' Slang* 46/1 *Fanny Adams*, general nautical slang for stew or hash.

2. *slang.* Freq. prec. by *sweet*: nothing at all. Cf. *F.A.* (*F III. 3).

Sometimes interpreted as a euphemism for 'sweet fuck all' in the same sense.

1919 W. H. DOWNING *Digger Dial.* 22 *F.A.*, 'Fanny Adams', or 'Sweet Fanny Adams'—nothing; vacuity. **1930** BROPHY & PARTRIDGE *Songs & Slang 1914–18* 123 F.A. Sometimes lengthened into Sweet F.A. or bowdlerized into Sweet Fanny Adams. Used to mean 'nothing' where something was expected. **1949** J. R. COLE *It was so Late* 61 What do they do? Sweet Fanny Adams!

fanship (fæ·nʃip). [f. *FAN *sb.*^{2} + -SHIP.] The state or quality of being a 'fan'.

1928 *Weekly Disp.* 20 May 17, I, who left fanship.. when I was seventeen..started out on the goose chase for Menjou. **1959** P. BULL *I Know Face* xi. 197, I showed him my fanship was genuine by the comprehensive collection I had of his records.

fantabulous (fæntæ·biŭləs), *a. slang.* [Blend of FANT(ASTIC *a.* + F)ABULOUS *a.*] Of almost incredible excellence.

1959 *N.Z. Listener* 9 Jan. 4/2 Air travel—it's fantabulous. **1961** *She* Sept. 59/2 There's BMC's fantabulous service depot, where owners and cars alike get real kid-glove/red-carpet treatment. **1968** *Punch* 6 Mar. 352/3 A cat involved in a court of law May fairly be called fantabulous. **1971** *Sunday Express* (Johannesburg) 28 Mar. (Home Jrnl.) 4/2 Since the bust up of the fantabulous group, it's been George who's been doing most of the slogging.

fan-tail, *sb.* Add: **3.** Also in New Zealand.

1860 A. S. ATKINSON *Jrnl.* 27 Oct. in *Richmond-Atkinson Papers* (1960) I. 645 Found a fantail nest (young ones) & a riroriro's with two young ones & an egg. **1882** W. L. BULLER *Man. Birds N.Z.* 26 The Pied Fantail, ever flitting about with broadly-expanded tail.., is one of the most pleasing and attractive objects in the New Zealand forest. **1935** J. GUTHRIE *Little Country* xxiii. 356 A fantail perched on a twig near them and flirted its feathers.

4. b. = FAN *sb.*^{1} 6 c.

1934 *Archit. Rev.* LXXVI. 166/3 The elaborate invention of an ingenious Scot, by name Meikle, who, in 1750, conceived a method of turning the mill automatically so that the 'sails', 'sweep', 'arms'—whatever pet name you care to employ—always faced the wind. It—the invention—was known as a 'fantail', and consisted of a fan of six to ten blades...Mounted on a spindle, and so arranged that if the wind should veer slightly, it strikes the 'fan-tail' which turns the mill through appropriate gearing until such time as the sails face the wind again. **1949** K. S. WOODS *Rural Crafts Eng.* II. iv. 74 By these three inventions—automatic sails, fan-tail, and governor—the wind itself has been used to counteract its own fickle and varying moods in man's service.

c. *Naut.* 'The projecting part of the stern of a yacht or other small vessel when it extends unusually far over the water abaft the stern post' (*Cent. Dict.* 1889); so extended to larger craft. (Cf. OVERHANG *sb.*)

1882 *Harper's Mag.* LXIV. 174/2 The stalactites of ice..at the start lent the wheel and 'fan-tail' a novel beauty. **1943** J. STEINBECK *Once there was War* (1959) 206 The boat..threw out a curving V of wake and boiled the water a little under the fantail. **1948** R. DE KERCHOVE *Internat. Mar. Dict.* 252/1 *Fantail stern*, a type of yacht stern in which the shell planking or plating sweeps up to a sharp point to join the deck planking...This type of stern has been almost abandoned in recent years. **1955** C. S. FORESTER *Good Shepherd* 70 Krause strode out on the wing of the bridge as the..gun at the fantail went off. **1957** *Jane's Fighting Ships 1957–58* 424 The aircraft crane on the fantail has not the capacity to handle heavy boats.

fantasia. Add: **1. b.** *transf.* and *fig.*

1818 E. BLAQUIERE tr. *Pananti's Narr. Res. Algiers* xix. 367 These excesses called *fantasias*, or paroxysms of passions. **1880** J. H. SHORTHOUSE *J. Inglesant* xxvi. 372 Wandering amid this brilliant fantasia of life, Inglesant's heart smote him. **1896** 'J. O. HOBBES' (*title*) The Herb-Moon. A Fantasia. **1919** G. B. SHAW *Heartbreak House* p. xlix, Heartbreak House: a Fantasia in the Russian manner on English themes. **1921** D. H. LAWRENCE *Sea & Sardinia* 41 Every wretched bit of would-be extra chic is called a fantasia. **1922** —— (*title*) Fantasia of the Unconscious. **1935** H. G. WELLS *Things to Come* v. 33 Long lines of steel-helmeted men. Lorries full of men. Lorries full of shells. Great dumps of shells. A fantasia of war material in motion. **1957** T. S. ELIOT *On Poetry & Poets* 83 A suspension of the action in order to enjoy a poetic fantasia: these passages are really less related to the action than are the choruses in *Murder in the Cathedral*.

fantasied, *ppl. a.* Delete *arch.* and add recent *Psychol.* examples of sense a.

1960 I. BENNETT *Del. & Neur. Childr.* iii. 69 Lying to support fantasied exploits. **1963** *New Society* 21 Nov. 11/1 Fantasied wishes for power. **1969** *Sunday Times* (Colour Suppl.) 9 Feb. 9/3 A double image of marriage as stability, and unfulfilled romantic passion as the chivalric myth.. kept socially chaste and fantasied longings in their proper place.

fantasist (fæ·ntăsist). [f. FANTASY + -IST.] One who 'weaves' fantasies.

1923 *Glasgow Herald* 10 May 6 Wilde, a chartered fantasist. **1927** E. M. FORSTER *Aspects of Novel* vi. 141 The other novelists say 'Here is something that might occur in your lives', the fantasist 'Here is something that could not occur'. **1928** *Observer* 1 Jan. 5/2 The Fantasists or Special Geniuses! *Ibid.* 2 Sept. 5 There is, I believe, no agreement among statisticians as to the number of H. G. Wells's. There are the sociologist and the fantasist and the novelist and the prophet. **1934** E. BOWEN *Cat Jumps* 190 Mrs. Letherton-Channing could countenance a diluted reality but could not suffer a fellow fantasist. **1960** *Guardian* 21 Sept. 7/1 Pictures that vary in mood between the poetry of a fantasist and the illusionism of a surrealist.

fantasize (fæ·ntăsəiz), *v.* [f. FANTASY + -IZE.] **a.** *intr.* = FANTASY *v.* 1 c.

1926 *Spectator* 13 Mar. 455/2 Certainly one may fantasize to one's own taste. **1960** *Guardian* 28 Sept. 11/2 Only when they listened to each other could one fantasize about their mood.

b. *trans.* To visualize fancifully, to represent in the fancy.

1950 LONDON & CAPRIO *Sexual Deviations* xiii. 593, I used to put my penis between my legs so that nothing but the hair of the pubic region showed and would thus fantasize how my sister looked when I did the peeping. **1961** E. HUNTER *Mothers & Daughters* I. 59 The men were..fantasizing the entire crew in dress uniform. **1968** *Sunday Sun* (Baltimore) 3 Nov. D1/6, I always fantasized myself as a James Dean type.

Hence **fa·ntasized** *ppl. a.*

1964 *Spectator* 8 May 632/2 The manner and matter of these various charges seem to belong to his own fantasised world.

fantastic, *sb.* Add: **5.** That which is fantastic, strange, eccentric or odd.

1908 *Daily Chron.* 17 Mar. 3/3 The tradition of the fantastic which has clung to the memory of Buckingham.

fantasticate, *v.* Add: **c.** *trans.* To render fantastic. So **fanta·sticated** *ppl. a.*

1936 P. FLEMING *News from Tartary* 19 Half a dozen pierrettes and apaches..suitably fantasticated our unserious departure. **1960** *Guardian* 16 June 8/3 Highly fantasticated, and sustained with brilliantly executed tricks. **1962** *Listener* 5 July 4/1 We had a terrific opportunity in the nineteenth century to elaborate and fantasticate our social structure. **1964** W. GOLDING *Spire* vi. 124 The fantasticated, eighty foot fall of the lights. **1966** *Punch* 16 Mar. 393/2 The novel..gives an amusing account, duly fantasticated, of his arrival in London. **1971** *Daily Tel.* 15 Apr. 6/8 Mr West has a racy, impressionistic, fantasticated style that is most engaging.

fantastication (see under FANTASTICATE *v.*). Add to def.: Also = FANTASTRY a. (Additional examples.)

1929 C. MORGAN *Portrait in Mirror* iv. 211 Richard.. would laugh, too, at my folly and be unable to consider it as anything but a boy's fantastication of life. **1931** *Punch* 28 Oct. 471/2 Mr Alan Napier had a well-written part in the *Venetian Ambassador* and brought the man to life, his habitual fantastications being reasonably in the picture. **1955** *Times* 15 July 4/4 His gay fantastication of nature in delicate pinks and greens of a truly theatrical effect plays a great part in the preposterous idyll of the lovers' day in the country. This fantastication includes all sorts of recklessly idiotic birds and snails and kite-flying cows. **1969** *Daily Tel.* 23 July 19/1 This scene comes off well. There is a regrettable lack of high elegance, of aristocratic fantastication, in the proceedings.

fantasy, phantasy, *sb.* Add: **3. b.** A day-dream arising from conscious or unconscious wishes or attitudes.

1926 G. COSTER *Psycho-Analysis* ii. 35 The term *phantasy* is much used in analytical psychology, and the fact that its technical meaning differs subtly from its colloquial one leads to some confusion. A phantasy is a day-dream in which desire, unfulfilled in the world of reality, finds an imaginary fulfilment or satisfaction. **1957** P. LAFITTE *Person in Psychol.* ix. 120 The Rorschach test invites him to enact his very vaguest fantasies, as when he sees pictures in the fire or on the wall.

4. f. A genre of literary compositions.

1949 (*title*) The Magazine of Fantasy and Science Fiction. **1954** M. F. RODELL *Mystery Fiction* ii. 4 Mysteries belong to the vast category of escape fiction. Westerns, 'romances', historical novels, and fantasies (other than satires) all belong in the same category. **1955** F. BROWN *Angels & Spaceships* 9 Fantasy deals with things that are not and cannot be. Science fiction deals with things that can be, that some day may be.

8. *attrib.* and *Comb.*, as *fantasy-building* vbl. sb. and ppl. a.; *fantasy-life, -world.*

1938 *Tablet* 1 Jan. 20/2 Now it is very strange that these judgments of her contemporaries should fit in so exactly with the present day estimate of Teresa as a fantasy-building neurotic..., unable at times to distinguish between imaginative and objective reality. **1959** H. READ *Conc. Hist. Mod. Painting* vii. 287 An immense effort to rid the mind of that corruption which, whether it has taken the form of fantasy-building or repression..constitutes a false witness to sensation or experience. **1937** 'M. INNES' *Hamlet, Revenge!* I. 12 They have their tenure in remaining—remote, jewelled and magical—a focus for the fantasy-life of thousands. **1960** C. DAY LEWIS *Buried Day* 22 It is said that an only child develops a particularly vivid fantasy-life. **1920** T. P. NUNN *Education* vii. 83 Prefers his fantasy-world to reality. **1955** A. C. SMITH *Speaking Eye* vi. 63 The other chairs were littered with film magazines...The fantasy world of sex.

fantasy, *v.* **1. a.** Delete 'Now *arch.* with the sense: To imagine in a visionary manner' and add later examples.

1949 M. MEAD *Male & Female* xii. 262 Behind that schoolgirl complexion..could be phantasied many sorts of conditions. **1960** *20th Cent.* Dec. 519 It is ludicrous to fantasy 'disinventing' the hydrogen bomb. **1970** *New Society* 31 Dec. 1157/1 The long-termers only occasionally fantasy or talk with each other about various styles of 'making-out' on release.

fantasying, *vbl. sb.* (see after FANTASY *v.*). (Later examples.)

1932 O. D. LEAVIS *Fiction & Reading Public* I. iii. 54 A habit of fantasying will lead to maladjustment in actual life. **1960** I. BENNETT *Del. & Neur. Childr.* iii. 69 'Bouts' of lying and fantasying to get other boys into trouble.

Fanti (fæ·nti), *sb.* and *a.* Also **Fante, Fantee.** [Native name.] The name of a Negro people inhabiting Ghana, and their language; a member of this people. Also *attrib.*

1819 T. E. BOWDICH *Mission to Ashantee* 344, I have heard about half a dozen words in the Fantee, which might be said to be not unlike the same nouns in the Welsh language. *Ibid.* 345 The Fantee word *umpa* (*true, indeed*). [**1868** CARR & BROWN (*title*) Mfantsi grammar.] **1875** [see *ASHANTI]. **1879** *Encycl. Brit.* X. 756/1 Fante dialects are spoken, not only in Fanti proper, but in Afutu or the country round Cape Coast. *Ibid.*, A Fanti..can converse without much difficulty with a native of Aquapem or Ashantee. *Ibid.*, The Aquapem, which is based on the Akem but has imbibed Fanti influences. **1923** [see *AKAN *a.* and *sb.*]. **1955** P. STREVENS *Papers in Lang.* (1965) ix. 115 The Kwa Larger Unit, comprising such languages as Twi, Ga, Fante, Ewe, Yoruba and Ibo. **1962** [see *ASHANTI].

b. Phr. *to go fantee*: to join the natives of a district and conform to their habits.

1885 KIPLING *Departmental Ditties* (1886) 59 'Went Fantee'—joined the people of the land. Turned three parts Mussalman and one Hindu. **1887** —— *Plain Tales fr. Hills* (1888) 23 He was perpetually 'going Fantee' among natives. **1906** *Daily Chron.* 9 Aug. 8/5 Caluna, too, had 'gone fanti', and was raiding and pillaging. **1930** CHESTERTON *Four Felons* 190 He was a white man, or whitish man, who had gone fantee, and wore nothing but a pair of spectacles.

fantod, var. FANTAD.

1839 C. F. BRIGGS *Adv. H. Franco* I. 249 You have got strong symptoms of the fantods. **1867** [see FANTAD]. **1884** 'MARK TWAIN' *Huck. Finn* xvii, These was all nice pictures,..but I didn't somehow seem to take to them, because..they always give me the fan-tods. **1910** *Sat. Westm. Gaz.* 1 Jan. 6/1 Sundays inside of a house gives you the fan-tods. **1920** GALSWORTHY *In Chancery* I. v, You mustn't get into a fantod, it'll never do. **1935** J. MASEFIELD *Box of Delights* viii. 220 'I say,' Kay said, 'what a place!' 'It gives me the fantods,' Peter answered. 'I don't like the place.'

fanzine (fæ·nzin). orig. *U.S.* [f. *FAN *sb.*^{2} + MAGA)ZINE.] A magazine for fans, esp. those of science fiction.

1949 *New Republic* 17 Jan. 16 *Fantasy Commentator*, perhaps the best of the fanzines, once ran a history of fan magazines. **1950** *N.Y. Times* 7 May VII. 26/4 The fantasy writers..now in California..are busily forming societies, printing and circulating special fan magazines—'fanzines' is the term. **1951** [see *fan magazine* s.v. *FAN *sb.*^{2} 2]. **1957** P. MOORE *Science & Fiction* 90 So much for the official science-fiction magazines...There remain the amateur publications, known as 'fanzines'.

fapesmo (fāpe·zmo). *Logic.* A mnemonic word for that supposed indirect mood of the first figure of syllogisms in which the major premiss is universal and affirmative, the minor universal and negative, and the conclusion particular and negative. (Later seen to be, by changing the order of the premisses, the fourth-figure mood FESAPO.)

1599 BLUNDEVIL *Logike* 121 Barbara: Celarent: Darii: Ferio: Baralipton Celantes: Dabitis: Fapesmo: Frisesomorum. **1685, 1849** [see *DABITIS]. **1884** J. N. KEYNES *Formal Logic* III. iv. 199 Similarly Fapesmo and Frisesomorum (the Fesapo and Fresison of Figure 4) have no corresponding direct moods.

far, *adv.* Add: **2. a.** *far and wide*: see WIDE *adv.* 1 b.

3. b. *far gone*: see also GO *v.* 48 f in Dict. and Suppl.

4. c. (Earlier examples of *far and away*.)

1852 *Democratic Rev.* 11 Far and away the greatest. **1880** TROLLOPE *Duke's Children* I. xxi. 253 He was far-and-away the cleverest of his party.

6. b. Also, *as far as that goes* (used to express disagreement) = on the contrary; *as far as, so far as* [So. 35], in so far as it concerns (me, etc.); as for.

1905 A. LANG *Adv. among Bks.* 231 As far as that goes ..most of you were highly favoured. **1926** H. W. FOWLER *Mod. Eng. Usage* 170/1 *As* or *so far as x* cannot be used as short for *as far as x goes* or *so far as concerns x*. **1939** H. S. CANBY *Thoreau* xiv. 217 The cabin..was in perfect condition so far as frame and covering until 1868. **1960** J. F. KENNEDY in *U.S. News & World Report* 26 Sept. 76/1 As far as whether I could attend this sort of a function in your church..then I could attend.

c. *as far as* (*I*, etc.) *see him*: normally in contexts implying distrust.

1606 J. DAY *Ile Guls* C2 Wil you trust him? Yes as farre as I see him. **1848** TROLLOPE *Kellys & O'Kellys* I. i. 15 He knows what he's about, and isn't the man to thrust a Protesthant half as far as he'd see him. **1969** 'A. GILBERT' *Missing from Home* xiii. 205 'She doesn't trust us as far as she can see us,' Charlie remarked.

8. a. *far-flying, -go·ng, -ranging, -travelled.*

1852 J. A. ROEBUCK *Hist. Whig Min.* II. 32 The view was exceedingly offensive to the far-going reformers. **1854** *Excelsior* II. 128 His venerable and far-travelled friend. **1870** tr. *F. Marion's Wonderful Balloon Ascents* I. i. 5 The far-flying comets. **1905** *Westm. Gaz.* 9 Sept. 2/3 Far-travelled tourists. **1923** KIPLING *Irish Guards in Great War* I. 5 One single far-ranging rifle-bullet. **1939** *War Illustr.* 4 Nov. 283/3 Far-flying squadrons over home waters and foreign seas had splendidly maintained our cause. **1959** E. H. CARR *Socialism in one Country* II. xix. 201 The intellectuals whose far-ranging thought had provided the inspiration of the revolution. **1969** *Jane's Freight Containers 1968–69* 240/3 Far-going mechanisation and cost-reduction is rendered possible.

c. far-apart *a.*, at a great distance (from); **far-darting** *a.*, esp. as epithet of Apollo, the *far-darter*; **far-distant** *a.*, at a great distance; **far-eastern** *a.* (example); **far-eyed** *a.*, = FAR-SIGHTED *a.*; **far-seeing** (earlier and later examples); **far-thinking** *a.*, = FAR-SIGHTED *a.* 1.

1865 *Punch* 27 May 215/1 What, Sir, is the object of a railway? To do away with distance, and bring far-apart scenes within the easy reach of all. **1955** E. BOWEN *World of Love* xi. 222 The sky..let fall far-apart tepid drops. **1849** THOREAU *Week on Concord* 199 And near at hand the far-darting glances of the god. **1851** BUCKLEY *Iliad* 4 The wrath of Apollo, the far-darting king. **1871** WHITMAN *Passage to India* (1872) 6 The far-darting beams of the spirit! **1793** J. BARLOW in *Amer. Poems* I. 86 Far-distant land. **1936** *Mind* XLV. 549 Modern astronomical researches into far-distant stars. **1913** A. FORTESCUE *Lesser Eastern Churches* ii. 36 Edessa certainly was the chief see of far-eastern Christendom. *a* **1882** EMERSON *Wks.* (1883) IX. 258 The height of Fancy's far-eyed steep. **1903** HARDY *Dynasts* I. VI. i. 109 The wariness That marks your usual far-eyed policy. **1837** LONGF. *Voices Night* (1843) 42 The Poet..far-seeing. **1943** W. S. CHURCHILL *End of Beginning* 240 The wise, far-seeing appeals of the American President. **1937** B. H. L. HART *Europe in Arms* xv. 211 They can hardly fulfil such a far-thinking role.

farad. Delete the clause beginning 'a term' and add: **1. a.** Suggested (but apparently not used) as a name for a unit of charge, equal to approximately $9{\cdot}5 \times 10^{-10}$ coulomb.

The unit was defined as the charge on either plate of a capacitor that has plates one square metre in area spaced one millimetre apart and charged to the voltage of a Daniell cell, there being dry air between the plates.

1861 CLARK & BRIGHT in *Electrician* 9 Nov. 4/2 The farad, or unit of quantity, is as small a one as is likely to be required in practice...One hundred farad, if allowed to pass through a very sensitive galvanometer, produce a visible motion of the needle.

b. A unit of charge (equal to the present microcoulomb) and also of capacitance (equal to the present microfarad). *Disused.*

1868 *Rep. Brit. Assoc. 1867* 488, 10^5 EMF, acting on a circuit of 10^{13}, will pass in one second 10^{-8} absolute units of quantity; and similarly, 10^5 EMF will charge a condenser of absolute capacity equal to 10^{-13} absolute units with 10^{-8} absolute units of quantity...Mr. Clark calls the unit of quantity thus defined (10^{-8}) one Farad, and similarly says that the unit of capacity has a capacity of one Farad, it being understood that this is the capacity when charged with unit electromotive force (10^5). **1868** L. CLARK *Elem. Treat. Electr. Measurement* vii. 44 The farad is that quantity of electricity which, with an electromotive force of one volt, would flow through a resistance of one megohm in one second. **1875** J. T. SPRAGUE *Electricity* v. 156 Microfarad = Microveber/Volt $= 10^{-8}/10^5 = 10^{-13}$. The microfarad here represents what is usually called the farad.

2. The practical unit of capacitance (now incorporated into the International System of Units), being the capacitance of a capacitor in which a charge of one coulomb raises the potential difference between the plates by one volt. (Orig. also used as a unit of charge, equal to the present coulomb.)

1873 J. C. MAXWELL *Electr. & Magn.* II. IV. x. 244 The quantity of electricity which flows through one Ohm under the electromotive force of one Volt during one second, is equal to the charge produced in a condenser whose capacity is one Farad by an electromotive force of one Volt. **1874** *Phil. Trans. R. Soc.* CLXIV. 1 Certain multiples have been adopted in practice and have received names, and are now in almost universal use among electricians. These units are: ..Capacity.—The Farad, equal to 10^{-9} absolute electromagnetic units. *Ibid.* 2 The measure of quantity is the same as that of electrostatic capacity, and in practice generally receives the same name, although it has been sometimes called the 'Weber'; the weber or farad quantity is equal to 10^{-1} absolute units. Electrical currents are defined as currents of so many farads per second. **1881** *Electrician* 24 Sept. 297/2 At the meeting [of the Paris Electrical Congress] the report of the First Section on Electrical Units was received, and the following resolutions adopted:—..7. The name farad will be given to the capacity defined by the condition that a coulomb in a farad gives a volt. **1881, 1892** [see Dict.]. **1882** *Electrician* 15 July 205/2 The unit of static quantity is a *farad*, or that quantity which, if discharged through one ohm in one second, would give a veber current. **1894** *Rep. Brit. Assoc. 1894* 122 Governments represented by the delegates of this International Congress of Electricians.. are..recommended to formally adopt as legal units of electrical measure the following:...As a unit of capacity the international farad, which is the capacity of a condenser charged to a potential of one international volt by one international coulomb of electricity. **1925** BRYANT & CORRELL *Alternating-Current Circuits* ii. 51 The farad is too large a unit for most considerations, and the microfarad is more common, there being 1,000,000 microfarads in 1 farad. **1947** *Jrnl. Inst. Electr. Engineers* XCIV. I. 342/1 From the 1st January, 1948, the units employed will be those derived from the centimetre, gramme and second, i.e. the so-called 'absolute' units... One international farad = 0·99951 'absolute' farad. **1958** W. C. MICHELS in Condon & Odishaw *Handbk. Physics* IV. v. 55/1 Precision standards are readily available in decade values from 10^{-3} ohm to 10^6 ohms, from 10^{-10} farad to 10^{-6} farad, and from 10^{-4} henry to 10 henrys.

Faraday (fæ·rădĕⁱ). The name of Michael *Faraday* (1791–1867), English scientist, used: **1.** *attrib.* or in the possessive to designate certain phenomena observed, apparatus invented, and principles enunciated by him.

Faraday cage, an earthed metal screen surrounding a piece of equipment to protect it from external electrostatic interference; **Faraday's constant** = *FARADAY 2; **Faraday('s) dark space,** in a discharge tube the dark space observed between the positive column and the negative glow when the pressure is moderately low; also called the *second dark space*; **Faraday('s) disc,** a metal disc in which an e.m.f. is induced when it is made to rotate in a magnetic field parallel to the axis of rotation; **Faraday effect,** the rotation of the plane of polarization of light or other electromagnetic waves when transmitted through certain substances in a magnetic field that has a component parallel to the direction of transmission; **Faraday('s) ice-pail experiment,** an experiment used to demonstrate certain principles of electrostatic induction; **Faraday's law,** any of two or three laws: (*a*) when the magnetic flux linking a circuit changes, an e.m.f. is induced in the circuit proportional to the rate of change of the flux linkage (the *law of induction*); (*b*) the amount of any substance deposited or liberated during electrolysis is proportional to (i) the quantity of charge passed and (ii) the equivalent weight of the substance (the *law*(*s*) *of electrolysis*); (quot. 1850 refers to a different phenomenon); **Faraday's line,** a line of force of a magnetic field; **Faraday tube,** a tube of force of an electrostatic field, defined so that one tube arises from a unit charge.

1916 G. KAPP *Princ. Electr. Engin.* I. vii. 103 All the transforming apparatus is in a building which is a huge Faraday cage...If the building should be struck by lightning this would momentarily acquire a high potential, but nothing inside it would be damaged. **1955** A. HUXLEY *Let.* 27 Aug. (1969) 761 Harry, the Dutch sculptor,..goes into trances in the Faraday cages. **1971** *Physics Bull.* Jan. 46/2 Ions leave the oscillator through a rectangular slot along the cathode and parallel to the wires and are collected in a Faraday cage. **1931** PAGE & ADAMS *Princ. Electr.* vi. 201 As the atomic weight of hydrogen is 1·008 and its valence is unity, *F*, called Faraday's constant, is the number of coulombs required to liberate 1·008 grams of hydrogen. **1893** T. O'C. SLOANE *Stand. Electr. Dict.* 249 *Faraday's Dark Space*, a non-luminous space between the negative and positive glows, produced in an incompletely exhausted tube through which a static discharge..is produced. **1958** C. G. WILSON *Electr. & Magn.* xii. 365 At a pressure of 10^{-1} mm. Hg. or less, the Faraday Dark Space and Negative Glow disappear and the Crooke's [*sic*] Dark Space almost fills the tube. **1886** J. A. FLEMING *Short Lect. Electr. Artisans* ii. 36 (*caption*) Faraday's Disk Induction Machine. **1962** CORSON & LORRAIN *Introd. Electromagn. Fields* 530 Consider the Faraday disk or, as it is also called, the homopolar generator. **1889** O. LODGE *Mod. Views Electr.* xv. 278 The only substance in which the Faraday effect is large, is iron. **1966** *McGraw-Hill Encycl. Sci. & Technol.* V. 181/1 The Faraday effect is particularly simple in substances having sharp absorption lines, that is, in gases and in certain crystals. **1888** A. GRAY *Absol. Measurem. Electr. & Magn.* I. i. 24 The results of Faraday's ice-pail experiments..are direct consequences of the following general proposition regarding closed conductors. **1953** E. R. PECK *Electr. & Magn.* i. 35 In the Faraday ice-pail experiment, a charged metal ball is suspended by an insulating string inside a closed metal pail which is initially uncharged and is insulated except for a wire connecting it to an electroscope indicating the potential of the pail. **1850** *Phil. Mag.* 3rd Ser. XXXVII. 245 Faraday's law..may be illustrated by some very curious although extremely simple experiments. **1881** *Jrnl. Chem. Soc.* XXXIX. 286 The more experimental methods were refined, the more completely were the exactness and generality of Faraday's law [of electrolysis] confirmed. **1886** J. A. FLEMING *Short Lect. Electr. Artisans* ii. 31 We are able to group under one law all the effects so far described, and the expression of this is called Faraday's law of Induction. **1904** R. A. LEHFELDT *Electro-chem.* I. i. 3 Faraday's two laws may be conveniently summed up in one statement..96600 coulombs are required for the deposition of one gram equivalent of any substance. **1954** W. E. ROGERS *Introd. Electr. Fields* xi. 307 The interrelations of a circuit with its magnetic flux are investigated most easily by relating the induced emf to the currents which produce the flux, rather than by using Faraday's law directly. **1857** *Rep. Brit. Assoc. 1856* 12 (*heading*) On a method of drawing the theoretical forms of Faraday's lines of force without calculation. **1911** *Encycl. Brit.* XVII. 323/1 Faraday's lines not only show the direction of the magnetic force, but also serve to indicate its magnitude or strength in different parts of the field. **1893** J. J. THOMSON *Recent Res. Electr. & Magn.* 3 The Faraday tubes either form closed circuits or else begin and end on atoms. **1959** *Chambers's Encycl.* XIV. 8/1 The number of Faraday tubes in unit area was called the displacement by Maxwell.

2. As a name for the quantity of electric charge required to flow to deposit or liberate one gramme-equivalent of any element during electrolysis, viz. approximately 96,490 coulombs. (Usu. with lower-case initial letter.)

1904 R. A. LEHFELDT *Electro-chem.* I. i. 3 This fundamental quantity of electricity, which occurs constantly in all writings on electro-chemistry, is called by the Germans a 'faraday', a term which we in England may very well adopt. **1940** S. GLASSTONE *Physical Chem.* xii. 872 One faraday of electricity liberates 16·8 liters of gas at S.T.P. **1958** HAMER & WOOD in Condon & Odishaw *Handbk. Physics* IV. ix. 140/2 The measurements of the faraday by electrochemical methods involve the measurement of the absolute current, the time, and the mass of material reacted.

farandole. Delete def. in Dict. and substitute: ‖ **farandole** (farandol). Also **farandola.** [Fr. *farandole*, ad. mod. Pr. *farandoulo* in same sense; It. *farandola*. Cf. Sp. *farándula* troop of travelling comedians.] A Provençal dance, generally in 6/8 time; the music which accompanies this dance, or any music written to its peculiar rhythm. Also *fig.* Add further examples.

1876 STAINER & BARRETT *Dict. Mus. Terms* 164/2 *Farandola*, a dance popular among the peasants of the South of France and the neighbouring part of Italy. It is performed by men and women taking hands, and forming a long line, and winding in and out with a waving motion. 1890 A. J. C. HARE *S.E. France* viii. 341 Here the peasants still dance the farandola. 1904 *Athenæum* 2 Apr. 426/3 Mr. Chesterton's farandole of farce may easily be wrongly praised.

‖ **farang** (fara·ŋ). [Thai *fa⁴-rang⁴* white race of people, ad. FRANK *sb.*¹; cf. FERINGHEE.] The Thai term for a foreigner, esp. a European.

[1852 F. A. NEALE *Narr. Res. Capital of Kingdom of Siam* vii. 109 'What!' said he, 'do you Franks dare to break the laws of this country, and set my authority in defiance?'] *a* 1861 H. MOUHOT *Trav. Indo-China* (1864) I. iii. 126 The priests were much surprised to see a 'farang' (foreigner) in their pagoda, but some trifling gifts soon established me in their good graces. 1873 F. VINCENT *Land of White Elephant* xv. 184 Notice was sent to His Excellency of the arrival of the 'farangs' (foreigners). 1967 *Listener* 10 Aug. 170/3 'In those days,' said one of the old waiters, 'we had many, many *farangs*.'

far-away, *a.*, *adv.*, and *sb.* Add: **A.** *adj.* **2. b.** Of a voice: sounding faint as if from a distance.

1897 M. KINGSLEY *Trav. W. Africa* x. 216 A quaint, falsetto, far-away sort of voice. 1900 H. LAWSON *On Track* 83 Then he commenced to speak—..to talk in that strange, absent, far-away tone that awes one. 1926 *Glasgow Herald* 30 Sept. 5, I heard his voice sounding in a far-away and curious tone.

C. *sb.* **2.** Part of a cinema film taken at a distance, as distinguished from a 'close-up'. ? *Obs.*

1926 B. MAINE *Receive It So* 29 Seeing that for nine-tenths of the house these things appeared as 'far-aways', the whole point of them was lost, because the essence of a 'close-up' is that you should see a tear drop or a stamp licked, and for the moment nothing else.

fa·r-back, *a.* and *sb.* [f. FAR *adv.* + BACK *a.*]
A. *adj.* **1.** Ancient.

1869 *Atlantic Monthly* XXIII. 200 As of the far-back days the poets tell. 1890 [see FAR *adv.* 8 c].

2. Remote in space; inaccessible.

1900 *Daily News* 4 Oct. 6/1 The jackals and other denizens of the far-back forests. 1934 A. RUSSELL *Tramp-Royal in Wild Australia* xx. 131 Station men in the far-back parts of Australia.

B. *sb.* **1.** (See quot.) *slang.*

1889 BARRÈRE & LELAND *Dict. Slang* I. 354/2 *Far back* (tailors), an indifferent workman or an ignorant person.

2. The most remote districts or back settlements.

1926 *Spectator* 11 Sept. 370/1 It [*sc.* Australia] is no longer a pioneering country, except in the far-back.

farce, *sb.*¹ Delete †*Obs.* and add later example.

1904 *Daily Chron.* 4 May 10/5 The hollow should be filled with a rich veal farce.

farce, *sb.*² Add: **3.** *farce-writer.*

1710 C. GILDON *Life T. Betterton* 174 Nay, after these our very Farce Writers deserve more Esteem. 1859 H. MORLEY *Jrnl.* 5 Nov. (1866) 240 Reasons which a farce-writer is entitled to regard as sufficient.

farce, *v.*¹ Add: **7.** = FARSE *v.*

1857 [see FARSE *v.*]. 1894 *Athenæum* 28 July 128/2 A peculiar feature in the recitation is the 'farcings' of each psalm, *i.e.*, the introduction of sentences, generally giving some application of the psalm to Christ. 1895 *Liturgy of Holy Apostles Adai & Mari* 1 There is no farcing at feasts, but a Hallelujah is said. 1907 *Mod. Philol.* IV. 585 None of these gradual variations caused legend or farced epistle to become anything other than legend and farced epistle.

farceur. (Earlier example.)

1781 G. SELWYN *Let.* 27 Dec. in *15th Rep. Hist. MSS. Comm.* (1897) App. VI. 553 Such *farceurs* as are in opposition.

b. An actor or writer of farces.

1889 in *Cent. Dict.* 1957 *Listener* 10 Oct. 581/2 His company are accomplished farceurs.

c. *attrib.*, as *farceur pianist.*

1910 V. TREE *Let.* 30 Sept. in *Castles in Air* (1926) I. 34 He is a splendid *farceur* pianist.

fardel, *sb.*¹ Add: **1. b.** The omasum, or third stomach, of ruminants. Also *fardel-bag.*

1862 *Chambers's Encycl.* IV. 245/1 Fardel-bound, a disease of cattle and sheep, consists of impaction of the fardel bag. 1908 *Animal Managem.* 16 The..'manyplies' or 'fardel' (omasum).

4. fardel-bound *a.*, costive; so as *sb.*, a condition in cattle and sheep in which food is retained in the folds of the omasum.

1825 LOUDON *Encycl. Agric.* 976 Costiveness also brings on a colic in them, called clue bound, fardel bound, etc. 1844 H. STEPHENS *Bk. Farm* II. 155 The fardlebound of cattle and sheep is nothing more than a modification of the disease in horses called stomach-staggers, which is caused by an enormous distention of the stomach. 1862 [see *1 b].

fa·r-down, *sb.* and *a.* **A.** *sb.* (Also **far-downer.**) An Irish-American belonging to a family which emigrated from the north of Ireland. *U.S.*

1834 *Amer. Railroad Jrnl.* III. 384/1 The parties arrayed against each other are known as the *Fardowns* and the *Corkonians*. *a* 1837 R. J. BRECKINRIDGE *Memoranda* (1845) I. 29 This city [*sc.* Cork] gives name to one of those bloody factions, which under the appellations of *Corkonians* and *Fardowns* divide the lowest classes of Irish Catholics in that distant land [*sc.* U.S.A.]. 1857 *Spirit of Times* 21 Feb. 405/2 They formed into two hostile factions, called Corkonians and Far-Downers. 1899 *Echo* 9 Mar. 1/2 Down in that quarter of Chicago you can hear all the 'various accents of Ireland, from the awkward brogue of the "far-downer" to the mild and aisy Elizabethan English of the Southern Irishman'.

B. *adj.* Situated or existing far below.

1911 FLETCHER & KIPLING *School Hist. Eng.* vii. 134 The far-down shark Shoots glimmering on his ways. 1949 BLUNDEN *After Bombing* 49 In the far-down street That Statue watched the laughing markets meet.

fare, *sb.*¹ **5. b.** (Earlier and later examples of U.S. sense.)

1707 in *Essex Inst. Hist. Coll.* XLII. 165 Loosing their last faires of fish. 1792 J. BELKNAP *Hist. New Hampsh.* III. 214 The fish of the summer and fall fares is divided into two sorts. 1831 A. SHERBURNE *Mem.* (ed. 2) x. 217 The old gentleman had just arrived from the ocean with a fare of fish. 1875 J. G. HOLLAND *Sevenoaks* xi. 139 The whalers had returned with scantier fares year after year. 1886 *Harper's Mag.* Dec. 105/1 He come by with his fare o' fish, an' hove to to see what I was gittin. 1904 *N.Y. Even. Post* 18 June, The prices brought by the 'fares', which are..cargoes of fresh or salted fish.

9. *attrib.* and *Comb.* *fare-board, -card, -stage; fare-paying* adj.

1897 *Daily News* 10 July 9/2 Labels on the outside [of buses] announced that the fares were sixpence or one shilling for any distance, and in each case the painted fare-board was replaced by a printed paper. *Ibid.*, If omnibus proprietors wished to charge special fares they must go to the expense of having a special fare card painted. 1928 *Daily Tel.* 16 Oct. 12/4 The first company of farepaying passengers to cross the Atlantic by airship. 1926 *Year's Work Eng. Stud.* 1924 31 We have seen it [*sc.* a book] devoured on the top of a tram to the oblivion of fare-stages.

Far East. [FAR *a.* 1 a, EAST *sb.*] The extreme eastern regions of the Old World, esp. China and Japan.

[1616 T. ROE *Let.* 17 Jan. (1899) I. 113 Load-stones heere are none. They are in the farre East Countries.] 1852 J. H. NEWMAN *Second Spring* 29 The great St. Francis opened the way to the far East. 1894 G. N. CURZON *Problems of Far East* i. 7 No introduction is needed in presenting the Far East to an English audience. 1898 *Westm. Gaz.* 1 Jan. 2/3 The Emperor's Far-East policy. 1906 *Q. Rev.* Jan. 287 The main interest of Great Britain in the Far East. 1911 *Ibid.* July 245 The great conflict which for many months bathed the Far East in blood weakened Russia in Europe. 1931 *Sat. Rev.* 21 Nov. 645 Apparently they cannot leave the Far East alone. 1950 *Sci. News* XV. 126 The rhesus negative characteristic is almost unknown in the Far East.

So **Far-Eastern** *a.*, of or belonging to the Far East.

1888 *Peel City Guardian* 14 July 7/4 One firm has the monopoly of this Far-Eastern *pari mutuel*. 1900 *Black & White Budget* 25 Aug. 649/1 The Far Eastern question is a new one to American diplomacy.

farewell, *sb.* **4.** (Later examples.)

1932 E. HEMINGWAY *Death in Afternoon* xix. 249 He is..giving a final series of farewell performances. 1938 M. ALLINGHAM *Fashion in Shrouds* viii. 120, I wanted to make sure that he was coming down on Saturday to Ramillies' farewell party. 1959 N. MAILER *Advts. for Myself* (1961) 267 To announce a farewell appearance then be on the scene again the following week is to ooze all the ebbing charm..of the desperate old actor. 1971 R. BUSBY *Deadlock* xv. 224 You could say it was his farewell performance.

farewell, *v.* Add: **a.** spec. *Austral.* and *N.Z.*: to honour a departing or retiring person or persons at a ceremonial occasion.

1902 J. H. M. ABBOTT *Tommy Cornstalk* 235 The North Terrace and the railway station had farewelled us with more delightful fervour than even if we had been an Australian Eleven driving round in drags. 1930 *Tribune* (Melbourne) 15 May 2/3 A popular priest, Fr. Hyland farewelled. 1931 *Auckland Star* 22 Mar., Opportunity was taken to farewell the Rev. E. Drake and Mrs. Drake.

fare-you-well. *U.S. colloq.* [f. phr. *fare you well*, FARE *v.*¹ 9.] *to a fare-you-well*: to the last point; to the utmost degree; completely.

1884 G. W. CABLE *Dr. Sevier* II. liv. 173 And then it means a house.., and milk, anyhow, till you can't rest, and buttermilk to fare-you-well. 1910 W. M. RAINE *B. O'Connor* 77 The little cuss has got me bluffed to a fare-you-well. 1911 R. D. SAUNDERS *Col. Todhunter* i. 3 The fight's begun, and we've got to rally around old Bill Strickland to a fare-you-well. 1924 W. M. RAINE *Troubled Waters* x. 101 'Who is boss of the round-up this year?'..'Rowan is, and believe me he worked us to a fare-you-well. He's some driver, Mac is.' 1943 S. LEWIS *Gideon Planish* x. 95 You can lambast Brother Rood to a fare-you-well.

‖ **farfel** (fā·ɹfĕl). Also **farfal, ferfel.** [Yiddish *farfal, farfil, ferfel* sb. pl., ad. MHG. *varveln* sb. pl. noodles, noodle soup.] Ground or granulated noodle dough.

1892 I. ZANGWILL *Childr. Ghetto* (1893) I. i. 49 In Jewish cookery..*Ferfel*..are *Lockschen* [like vermicelli] in an atomic state. 1925 *Jewish Encycl.* IV. 256/1 Lokshen consists of flour and eggs made into dough.. then cut into long strips...Cut into small squares, these strips are called 'farfil'. *Ibid.* 257/2 Teigachz, or pudding ..is usually made from rice, noodles, 'farfel' (dough crumbs). 1960 A. E. BENDER *Dict. Nutrition* 5 Farfals are ground, granulated or shredded [alimentary pastes].

far-fetch, *v.* (Later example.)

1929 D. H. LAWRENCE *Pornogr. & So On* (1936) 16 Poetry more and more tends to far-fetch its word-meanings.

far-flung, *a.* [FAR *adv.* 8 a.] 'Flung', 'cast', or extended far or to a great distance.

1895 *Times* 14 Oct. 8/1 To the far-flung fenceless prairie. 1897 KIPLING *Recessional* i, Lord of our far-flung battle-line. 1902 *Daily Chron.* 31 Jan 3/5 The tangled, far-flung story of the once semi-Royal house of Douglas. 1925 B. N. ODELL in E. F. Norton *Fight for Everest, 1924* 292 A far-flung head tributary of the Dzakar Chu. 1927 H. E. FOSDICK *Pilgr. Palestine* (1928) i. 23 This far-flung vista of the land he loved. 1933 A. G. MACDONELL *England, their England* xi. 197 He was in favour of..the immediate launching of twenty new cruisers to protect the trade routes of our far-flung Empire. 1960 *Times* 20 Apr. 14/7 The efficiency and self-sufficiency of this far-flung French outpost.

farinaceous, *a.* Add: **5.** Characterized by flour: *farinaceous city, colony*, playful names for Adelaide and South Australia, from the large export of wheat.

1873 TROLLOPE *Austral. & N.Z.* II. 184 [Adelaide] has also been nicknamed the Farinaceous City. 1902 *Daily Chron.* 7 May 7/1 Before emigrating to the 'farinaceous colony'.

fa·rkleberry. *U.S.* = SPARKLEBERRY.

1765 J. BARTRAM *Diary* 11 July in *Trans. Amer. Philos. Soc.* (1942) XXXIII. 14/1 Trees which naturaly [*sic*] grows there is..very fine long-leaved pine pitch pine yapon fartle [*sic*] berry chinkapin. 1829 A. EATON *Man. Bot.* (ed. 5) 344 *Vaccinium arboreum*, farkleberry. 1856 A. GRAY *Man. Bot.* (ed. 2) 248 *Vaccinium arboreum*, the Farkleberry, a tall species of this section, with evergreen leaves, probably extends northward into Virginia. 1942 L. R. TEHON *Fieldbk. Nat. Illinois Shrubs* 232 The Farkleberry ranges in dry sandy soils in open woods from Virginia to Florida and westward into Texas. 1951 *Dict. Gardening* (R. Hort. Soc.) IV. 2186/2 Subgenus Batodendron has only one species, *V. arboreum*, the Farkleberry or Sparkleberry, which forms a large bush or small tree with campanulate flowers, included anthers, and dry hard berries.

farm, *sb.*² Add: **5. b.** Extended to tracts of water devoted to the breeding or rearing of some animals, *gen.* with qualification, as *fish-farm, oyster-farm, terrapin-farm*, etc. (see first terms in Dict.).

1866 *Chambers's Jrnl.* 22 Sept. 601/1, I saw no farm of mussels. 1962 *Daily Tel.* 4 Apr. 12/3 The cultivation of an underwater 'farm' on a one-by-three mile area of sea.

c. Extended to storage installations.

1955 *N.Y. Times* 20 Feb. 20/4 Each farm..consists of several storage tanks with a pump house. 1958 *Globe & Mail* (Toronto) 10 Oct. 20/4 Tankers have piped ashore 29½ million gallons of gasoline and oil products into the tanks of three storage farms.

8. (Examples.)

1879 'TICKET-OF-LEAVE MAN' *Convict Life* vii. 167 After his conviction..he can 'fetch the farm', which is thieves' language for obtaining admission to the infirmary. *a* 1889 *Evening News* (Barrère & Leland), The dodges which would take place to 'fetch the farm'. 1900 *Pall Mall Gaz.* Oct. 202 The 'farm'..in ordinary prison vernacular, the synonym for the hospital. *Ibid.* 203 To 'fetch the farm' was, and still is, the current slang equivalent for getting round the doctor.

9. a. *farm account, -cart, -gate* (examples), *implement, -kitchen, -labour* (earlier example), *-life, -woman, -work* (earlier example).

1813 J. SINCLAIR *Syst. Husbandry Scotl.* (ed. 2) II. App. 173 Gentlemen farmers, who..are not aware, of the real state of their farm accounts. 1816 *Farmer's Mag.* XVII. 477 Other labourers..dependent upon farm-work ..are without employment. 1825 LOUDON *Encycl. Agric.* 737 In the adjustment of farm labor, the great art is to divide it as equally as possible throughout the year. 1831 W. HOWITT *Bk. Seasons* 105 In farm-kitchens..hear a chirping of chickens. 1834 D. LOW *Elem. Pract. Agric.* 630 Farm gates have sometimes been made wholly of hammered iron. 1844 H. STEPHENS *Bk. Farm* I. 18, I have confidence of giving such an *exposé* of farm implements as will surpass every other work of the kind. *Ibid.* III. 1349 The next, and one of the most important books, is the Farm-account book. 1863 D. G. MITCHELL *My Farm Edgewood* 17 The economies of a quiet farm-life. 1864 J. C. ATKINSON *Stanton Grange* xxvi. 286 So much the better, if he could..get the inmates to put him into one of their farm-carts on some straw. 1891 HARDY *Tess* II. xxi. 12 A farm-woman would be the only sensible kind of wife for him. 1905 *Westm. Gaz.* 6 May 10/2 We see the children playing in the farm-kitchen. 1915 D. H. LAWRENCE *Rainbow* i. 2 The heated, blind intercourse of farm-life. 1960 *Farmer & Stockbreeder* 1 Mar. 58/1 The total January farm-gate deliveries of 148¾m gallons.

b. farm-hand (U.S. examples).

1843 *Cultivator* X. 85 In unpropitious weather for out door employment your farm hands can go to threshing out grain. 1878 W. WHITMAN *Specimen Days* (1883) 108 He was about the best specimen of a young country farm-hand I ever knew.

farm, *v.*² Add: **3. b.** With *out.* To send (a university student) to a tutor outside his own college.

1959 *Camb. Rev.* 7 Mar. 399/1 Politics is often taught by history tutors or by research students; there is a great deal of 'farming out' which naturally undermines the purpose of the tutorial system. 1961 *New Statesman* 28 Apr. 669/3 One term 'farmed out' with a man tutor taught me more both of my subject and of its relation to other subjects and to life than I learned from several terms with women tutors.

6. *trans.* and *intr.* In *Cricket*, of a batsman: to contrive to receive the majority of the balls bowled.

1955 I. PEEBLES *Ashes* xii. 125 Maddocks was run out..when trying to farm the strike. 1963 *Times* 23 May 4/3 Richardson..went belligerently on..'farming' skilfully.

farman, var. FIRMAN.

1932 *Times Lit. Suppl.* 10 Nov. 842/2 This volume contains photographic reproductions of seventeen farmans granted to the head of the Vallabha sect of Vaishnavas. 1962 *Economist* 13 Jan. 115/1 No such *farman* or decree has actually been issued by HM the Shah personally.

farmer². Add: **7. c.** An ignorant rustic; a stupid or gauche person.

1864 HOTTEN *Slang Dict.* (ed. 3) 131 *Farmer*..this term..In London..is used derisively of a countryman, and denotes a farm-labourer, clodpole. 1903 A. H. LEWIS *Boss* 263 Me fadder aint such a farmer as to go leavin' his address wit' no one.

d. (See quots.) *Naut. slang.*

1886 R. BROWN *Spunyarn & Spindrift* vi. 89 I'm a farmer to-night, and means to have a quiet and peaceful night's rest. *Ibid.*, A sailor calls himself a farmer when he has neither wheel nor look-out during a night. 1933 P. A. EADDY *Hull Down* viii. 179, I was a 'farmer' that night,..not having any wheel or look-out.

8. a. *farmer-boy.*

1854 G. W. CURTIS *Lit. & Soc. Ess.* (1894) 5 The farmer-boy—sweeping with flashing scythe through the river meadows. 1957 J. KEROUAC *On Road* (1958) iv. 24 The standard North Dakota farmer-boy hat.

c. *farmer's* (or *farmers'*) *lung*, a pulmonary disease resulting from sensitization to fungal spores from mouldy hay, etc.

1945 *Brit. Med. Jrnl.* II. 716/1 *Farmer's lung*..is clinically and radiologically a fairly well characterized disease..commonly believed to be a kind of pulmonary mycosis. 1962 *Lancet* 26 May 1108/2 The test is..often the most convenient way of distinguishing sarcoidosis from..carcinomatosis, pneumoconiosis, or 'farmer's lung'. 1970 *Daily Tel.* 8 Oct. 9/8 Farmers' lung, a rare disease caused by mouldy hay, killed Mr Joseph Bradley.

farmerette (fāɹməre·t). orig. and chiefly *U.S.* [f. FARMER² + -ETTE.] A woman or girl who farms land; a farmeress.

1918 *Independent* (N.Y.) 14 Sept., The farmerettes are producing food which creates the bodies and minds of mankind. 1923 *Daily Mail* 15 Aug. 2/4 Girl Farmers' Popularity...The organisers of the fair heard that the 'farmerettes' were coming. 1963 M. MCCARTHY *Group* iv. 72 Gag up a sketch about the milk strike with some farmerettes in straw hats.

farm-house. Add: **b.** *attrib.* and *Comb.*

1835 W. IRVING *Tour on Prairies* xxv. 267 Regaling themselves with savoury anticipations of farm-house luxuries. 1845 E. ACTON *Mod. Cookery* (ed. 2) xi. 245 Leg of Suffolk farm-house pork. *Ibid.* 261 Common farm-house sausages are made with nearly equal parts of fat and lean pork. 1851 MRS. GASKELL *Let.* May (1966) 836 Mary says she enjoyed the farm house bread. 1879 [in Dict.]. 1903 LD. R. GOWER *Rec. & Reminisc.* xv. 138 All the furniture of the simplest and most farmhouse-like kind. 1955 *Times* 10 May 12/3 Farmhouse cheese probably represents in the popular mind the difference between 'home made' and mass-produced cheese. 1960 *Farmer & Stockbreeder* 8 Mar. 57/1 Farmhouse-gate sales are considerable. 1969 J. FREDMAN *Fourth Agency* vii. 55 We'd all break for a farmhouse tea of hot buttered scones at four o'clock.

farmlet. Delete *rare* and add further examples.

1936 F. CLUNE *Roaming round Darling* i. 3 A general store, a row of workmen's cottages, and a few farmlets chucked in, constitute the altered town. 1952 J. STEINBECK *East of Eden* ii. 9 The owned and fought-over farmlets of Europe. 1963 *Punch* 13 Nov. 718/3 The Minister..giving extra support..to his own farmlets.

farmost, *a.* Delete †*Obs.* and add later examples.

1885 W. TOWERS *Poems* 15 (E.D.D.), He sought the barn's farmost end. 1920 H. D. SHAWCROSS *Road-Wanderer* x. 120 The farmost end is seen crowned by a vision of distant hills.

fa·rm-wife. Also **farmwife.** [f. FARM *sb.*² + WIFE *sb.*] A farmer's wife.

1880 W. D. HOWELLS *Undiscovered Country* 261 The growth of the bargaining passion in the wary farm-wives. 1907 *Daily Chron.* 21 Oct. 8/1 The farm-wife from the Ghyll stepped softly in. 1954 W. FAULKNER *Fable* (1955) 331 Above them stood the farmwife.

farm-yardy (fā·ɹmyaɹdi), *a.* [f. FARM-YARD + -Y¹.] Of, pertaining to, or resembling a farm-yard.

1928 GALSWORTHY *Swan Song* I. xii. 93 Things are much better..than they were—not nearly so stuffy and farmyardy.

farnesol (fā·ɹnĕsǫl, -sōᵘl). *Chem.* [a. G. *farnesol* (Haarmann & Reimer 1902, in (German) *Patentschrift* 149603), f. mod.L. *farnes-iana*, epithet of a species of Acacia, f. the name of Odoardo *Farnese* (1573–1626), Italian cardinal + -OL 1.] A terpenoid alcohol, $C_{15}H_{25}OH$, that occurs in various essential oils and is used in the preparation of scents.

1904 *Jrnl. Chem. Soc.* LXXXVI. I. 514 *Farnesol*, $C_{15}H_{26}O$, a new sesquiterpene alcohol, is obtained from various acacia oils, from musk oil, and from lime-tree blossom oil. 1949 T. F. WEST et al. *Synthetic Perfumes* iii. 29 Pure farnesol appears to the casual observer to be an almost odourless oil. 1961 *New Scientist* 2 Nov. 279/2 A relatively simple organic compound, farnesol (an open-chain terpene alcohol), has the same effects on immature insects as extracts of natural juvenile hormone. 1970 V. B. WIGGLESWORTH *Insect Hormones* iv. 68 Farnesol is..an intermediate in the biosynthesis of cholesterol and of carotenoids; although it had rather low juvenile hormone activity in insects..the natural hormone might well be a related isoprenoid compound.

‖ **far niente** (fār nie·nte). [It., lit. 'to do nothing.'] Idleness. (Usu. in phr. DOLCE FAR NIENTE.)

1819 T. HOPE *Anastasius* II. vii. 132, I determined.. to indulge in the supreme pleasure of the Italians—the *far niente.* 1857 TROLLOPE *Barchester T.* I. ix. 118 The *far niente* of her Italian life had..brought her to regard a state of inactivity as the only earthly good. 1894 'M. O'RELL' *John Bull & Co.* xxxii. 271 Allow him to pass his life in the softest of *far nientes.*

faro¹. Add: **2. faro dealer,** the dealer in a game of faro.

1856 *Harper's Mag.* Dec. 68/2 A grim, imperturbable faro-dealer. 1952 S. KAUFFMANN *Philanderer* (1953) iv. 64 Cameron..had once been a faro dealer in a gambling house.

faro². (Earlier and later examples.)

1848 THACKERAY *Van. Fair* xxxi. 266 A mug of Faro at the bench of a beer-house on the road to Laeken. 1960 *Spectator* 25 Nov. 831 This Belgian beer..[is] called *lambic* or *faro*, and the books all say it's light and rather bitter.

Faroeman (fēᵊ·romæn, fæ·romæn). = *FAROESE *sb.* a.

1898 *Blackw. Mag.* Aug. 246/1 The ancestors of the Faroemen were Norwegians. 1905 T. N. ANNANDALE *Faroes* 2 The taxes are collected by sheriffs, who are always native Faroemen. *Ibid.* 3 The Faroeman is by nature peaceable.

Faroese (fēᵊro,ī·z, færo,ī·z), *a.* and *sb.* Also **Faröese, Færoese, Feroese.** [f. *Faröe* + -ESE, after Sw. *Färöarna*, Da. *Færøerne* Faröe islands = Icel. *Færeyskr* Faroese (adj.), f. *Færeyjar* Faroe islands, f. *fær* sheep + *ey* island.] **A.** *adj.* Of or pertaining to the Faröe Islands, an autonomous dependency of the kingdom of Denmark situated between Great Britain and Iceland, their inhabitants, or the language spoken by them. **B.** *sb.* **a.** A native or inhabitant of the Faroe Islands. **b.** The language of the Faroe Islands, which is a variety of Norse.

1851 *Illustr. Catal. Gt. Exhib.* III. III. 552/1 Faroese and Danish St. Luke. 1855 *Narr. Cruise Yacht Maria* i. 15 Immediately after we anchored, a Feroese gentleman came on board. *Ibid.* ii. 25 The Feroese are very long-lived. 1898 J. RUSSELL-JEAFFRESON *Faröe Isl.* i. 15 The religion of the Faröese..is Lutheran. *Ibid.* v. 99 The child cried out in Faröese, 'Nüg me up.' *Ibid.* 103, I collected..as many of the old Faröese proverbs as I came across. 1908 W. G. COLLINGWOOD *Scand. Brit.* 259 The 'Norn' is shown to be fairly pure Norse, with a very slight sprinkling of Danish, Færoese, Frisian and English words. *Ibid.* 260 Phonetic changes like those in Icelandic and Færoese. 1953 *New Biol.* XV. 57 A Faeroese shearwater, with a nestling. 1969 *Guardian* 13 Aug. 4/8 The enterprise and success of their Faroese neighbours is an irksome example to the Shetlanders. 1969 *Language* XLV. 570 In languages like Faeroese or Danish..sonorants may be completely voiceless after tenues. 1969 J. WOOD *Three Blind Mice* iii. 42 Taking on two belligerent Faroese. 1970 *Sruth* (Inverness) 10 Dec. 6/1 Frederick Petersen, author of the Faroese national anthem.

far-off, *a.* Add: **1. d.** Other *fig.* uses, e.g. of thoughts, looks, etc.

1849 *Hogg's Weekly Instructor* III. 313/1 The far-off thoughts of earthly love and beauty. 1870 W. MORRIS *Earthly Par.* III. 462 Scarce happy 'neath his far-off moody gaze. 1876 *Mr. Gray & his Neighbours* II. 41 Alice Gray was very pale, and with that far-off look in her eyes, which those who are to die young have more than others. 1922 C. E. MONTAGUE *Disenchantment* i. 8 The far-off, longed-for ideal of smartness. 1948 L. MACNEICE *Holes in Sky* 21 The Painter's little daughter, far-off-eyed. 1959 T. S. ELIOT tr. *St. J. Perse's Anabasis* 37 My soul engaged in far-off matters.

2. *absol.* (Earlier and later examples.)

1866 C. M. YONGE *Dove in Eagle's Nest* I. xii. 230 There will be freedom in rushing at last into the great far-off! 1932 V. MCNABB in *God's Way of Mercy* (1937) xxi. 186 Faith brings the far-off very near, and makes the very little large.

Faroish (fēᵊ·ro,iʃ, fæ·ro,iʃ), *a.* Now *rare.* [f. *Faröe* + -ISH¹.] = *FAROESE *a.* Also **Farish** (fēᵊ·riʃ, fæ·riʃ) *a.*

a 1889 CHILD *Ballads* I. 315 (Cent. Dict.), The Swedish ..Danish, and Faroish ballads. 1898 *Blackw. Mag.* Aug. 245 A collection of Farish ballads. 1905 T. N. ANNANDALE *Faroes* 14 It is only some fifty or sixty years since Farish began to be written. *Ibid.* 131 A Farish cottage is generally clean, an Icelandic farmhouse is almost as often airless, filthy, and verminous.

‖ **farol** (fārōᵘ·l). [Sp., lit. lantern.] In *Bullfighting*, a movement in which the bull-fighter draws the bull by passing the cloak back rapidly over his own head.

1932 E. HEMINGWAY *Death in Afternoon* i. 18 In bullfighting..it is the picturesqueness..of farols and molinets. *Ibid.* xv. 169 The different ways of using the cape, the gaonera, the mariposa, the farol. 1957 A. MACNAB *Bulls of Iberia* vi. 61 In the *farol*, he flings both arms upwards and the cape swirls up above his head, with the bull sometimes rearing up spectacularly in an effort to reach it.

farouchely (farū·ʃli), *adv.* [f. FAROUCHE *a.* + -LY².] In a farouche manner.

1931 A. POWELL *Afternoon Men* v. 52 The boy, an ill-conditioned youth..stood farouchely clasping and unclasping his hands. 1967 *Listener* 31 Aug. 267/2 Gillespie..farouchely ambles across the platform.

far-out, *a.* [f. FAR *adv.* + OUT *adv.*] **a.** Remote, distant.

1887 J. SERVICE *Life Dr. Duguid* 84 A faur oot freen of John Paiks' father. 1888 'R. BOLDREWOOD' *Robb. under Arms* I. xvii. 232 These far-out back-of-beyond places. *Ibid.* III. xv. 215 The far-out squatters that were stocking up new country in Queensland. 1929 D. H. LAWRENCE *Pansies* 100 Yet even waste, grey foreshores, sand, and..far-out clay Are sea-bed still.

b. Of jazz: of the latest or most progressive kind. More generally, avant-garde, far-fetched; excellent, splendid. orig. *U.S.*

1954 *Time* 8 Nov. 70 Jazz lingo becomes obsolescent almost as fast as it reaches the public ear...A daring performance was 'hot', then 'cool', and now is 'far out'. 1956 M. STEARNS *Story of Jazz* (1957) xviii. 232 There were too many choices in 'far-out' harmony. 1959 *Listener* 1 Oct. 542/3 Cute, gone, far-out little girls. 1960 *Melody Maker* 31 Dec. 6/3 Oliver Cool is the pseudonym of singer Larry Ellis, who found he won more success when he adopted the 'far-out' monniker. 1962 *Listener* 29 Nov. 911/2 In the spread of serialism, we are faced with another and greater neo-classical craze, the most far-out of all. 1963 K. AMIS *One Fat Englishman* iii. 36 She was..several times more attractive than her with-it off-beat far-out co-religionist deserved. 1970 *Sci. Jrnl.* May 47/2 Talking with computers, so much a far-out idea when this journal discussed IBM's work on it four years ago, now seems quite straightforward.

farrash, var. FERASH.

1873 H. BLOCHMANN tr. *Abu al Fazl ibn Mubarak's Ain I Akbari* I. xvi. 47 Besides, there are employed a thousand Farrâshes, natives of Îrân, Tûrân, and Hindustan. 1894 *Safar Nameh Persian Pict.* 162 It belonged to the Shah's farrash. 1965 B. SWEET-ESCOTT *Baker St. Irregular* iii. 80 Nobody seemed surprised at Ankara when the bags were handed over to the embassy's *farrash* at the station by me and not by my friend at Tripoli.

far-reaching (stress variable), *a.* [f. FAR *adv.* 8 + REACHING *ppl. a.*] That reaches far; extensive (*lit.* and *fig.*); exerting an influence or producing an effect which extends far in space or time.

1824 NEWMAN *Poems, To H. E. N.* xii, The dusky heath far-reaching. 1849 J. R. LOWELL *To W. L. Garrison* 18 in *Poet. Wks.* (1891) 116 Rome's far-reaching bolts. 1860 RUSKIN in *Cornhill Mag.* Sept. 284 That which seems to be wealth may be only the gilded index of far-reaching ruin. 1874 H. SIDGWICK *Meth. Ethics* 245 A fundamental conflict of ideas, which appears more profound and far-reaching in its consequences the more we examine it. 1894 *Pop. Sci. Monthly* XLIV. 572 Geology formed a subject of far-reaching importance. 1900 *Daily News* 1 Mar. 2/1 A number of far-reaching caves. 1913 J. G. FRAZER *Psyche's Task* (ed. 2) 168 Such mental variations, with all their far-reaching train of social consequences. 1950 L. S. THORNTON *Revelation & Mod. World* v. 149 This far-reaching continuity of revelation. 1971 *Guardian Weekly* 24 Apr. 5/1 The implications of the new controversy are far-reaching.

Hence **far-rea·chingly** *adv.*, **far-rea·chingness.**

c 1850 BAGEHOT *To R. C. Ch.* iii. in *Wks.* (1915) I. 12 With hand of power and thoughts of light, As Britain seas, far-reachingly. 1893 *Harper's Mag.* May 821 The most far-reachingly beneficial expression of this fostering policy. 1901 S. BUTLER in H. F. Jones *Mem.* (1919) I. 264, I was oppressed and scared by the far-reachingness and daring of what I had done.

‖ **farruca, Farruca** (fæ̆rū·kă). [Sp., fem. of *farruco* Galician or Asturian, f. *Farruco* nickname of *Francisco* Francis.] A local Spanish dance.

[1914 T. & M. W. KINNEY *Dance* vi. 127 *La Farruca, el Tango*, and *el Garrotin*, the most popular *Flamenco* dances at present, preserve to admiration the Gipsy qualities. *Ibid.* 130 La Farruca probably exploits more completely than any of its fellows the varied resources of the *Flamenco*.] 1931 C. RICE *Dancing in Spain* i. 28 The three dances which constitute the nucleus of the modern flamenco school are the Tango, the Garrotín, and the Farruca. *Ibid.* ii. 52 Of all the dances of Spain it is the Farruca which provides a man with the fullest opportunity of displaying his powers. 1958 *Times* 8 Oct. 6/3 This [*sc.* a suite of songs and dances] was rather too long, but it included a Farruca from Mr. Paco Ruiz. 1967 'LA MERI' *Sp. Dancing* (ed. 2) ix. 121 Forty years ago I danced farruca and bulerias with the gitanas and was taken for one of them.

far-sight. Add: Also *fig.*

1887 *Christian Union* 12 May 22/1 With keen far-sight, with indomitable energy. 1927 V. MCNABB *Cath. Ch. & Philos.* iii. 117 Anyone who would deny that reason had such insight and far-sight as to discern God.

fart, *sb.* Add: **1. c.** A contemptible person.

1937 in PARTRIDGE *Dict. Slang* 267/1. 1968 J. SANGSTER *Touchfeather* xi. 121 I'm supposed to be non-operational. What does the silly old fart want? 1971 *Ink* 24 July 5/1 Marty Feldman said to the judge as he left the witness stand 'I don't think he even knew I was here, the boring old fart.'

fart, *v.* Add: **3.** *intr.* To fool *about* or *around*; to waste time.

1900 in *Eng. Dial. Dict.*, Go bon tha! thoo's allus farten aboot, thoo's warse ner a hen wi' egg. 1937 PARTRIDGE *Dict. Slang* 267/1 *Fart about*, to dawdle; to waste time; play about. 1964 P. LARKIN *Jill* (ed. 2) 223 They scuffled for a few moments, knocking over a pile of books and papers. 'Don't fart about,' said Patrick wearily. 1968 H. C. RAE *Few Small Bones* II. v. 115 Now don't fart about, Birnie, or I'll get angry. 1969 J. WAINWRIGHT *Big Tickle* 173 Look! It's important. Stop farting around.

farther, *a.* Add: **3. b.** *Farther East* = *FAR EAST.

1876 *Fraser's Mag.* Jan. 1/2 There came another thunderclap tending once more to direct men's eyes from the near to the farther East. 1902 A. T. MAHAN *Retrospect & Prospect* 160 It is..the principal sea route between Europe and the Farther East.

farthing. Add: **5. a.** Also as quasi-*adj.*, trivial, almost valueless, unworthy of respect or notice.

1893 STEVENSON *Catriona* v. 58 They each made me a little farthing compliment. 1896 BELLOC *Verses & Sonnets* 25 His farthing honours.

farthingale. Add: farthingale chair, a seventeenth-century chair with a wide seat, a low straight back, and no arms.

1904 P. MACQUOID *Hist. Eng. Furnit.* vii. 179 The chair..is an early example of what was termed a farthingale chair. 1955 *Oxf. Jun. Encycl.* XI. 51/1 The 'farthingale chairs' designed for ladies wearing the wide skirts.. no longer had solid backs, but upholstered rectangular backs raised above the seats on turned supports.

‖ **fartlek** (fā·ɹtlek). [Sw., f. *fart* speed + *lek* play.] A method of training for middle- and long-distance running, in which the athlete runs over country, mixing fast with slow work.

1952 *Scholastic Coach* XXI. VII. 20 The answer that finally emerged was 'fartlek'—the Swedish system of training. 1957 DUNCAN & BONE *Oxf. Pkt. Bk. Athletic Training* (ed. 2) iii. 23 Fartlek is a pleasant..form of wind sprints across country. *Ibid.*, The essence of a successful fartlek session is enjoyment in hard but varying work. 1958 *Times* 18 Aug. 3/1 He was out training..and was using the Fartlek or run-as-you-please training which originated from Sweden in the early 1940s. 1968 B. TULLOH *Tulloh on Running* v. 46 *Fartlek*, meaning 'speed play'..is a continuous run in which patches of fast striding are interspersed with jogging.

Far West. *N. Amer.* [f. FAR *a.* 1 a + WEST *sb.*¹] A term applied to regions in the Rocky Mountains and along the Pacific coast. Formerly applied to areas lying west of the earliest settlements, i.e. to what is now the *Middle West.*

1830 *Deb. Congress U.S.* 8 May 920/2 Sir, I am from the West, although not from the 'far, far West'. 1832 *Ibid.* Feb. 290 Some of the descendants of illustrious families have gone to the far West. 1835 *Southern Lit. Messenger* I. 615, I do not believe that the prairies of the 'far West' can exhibit more luxuriant fields of corn. 1836 *Bytown* (Ottawa) *Gaz.* 13 Oct. 3/2 Being in the 'far West' at the time. 1843 'R. CARLTON' *New Purchase* I. 2 In a part of what was, at the time of this journey, the Far West. 1845 *Globe* (Toronto) 14 Oct. 1/3 The.. 'Far-West'..what was formerly known as such is not now even thought of in connexion with that term. 1880 A. A. HAYES *New Colorado* (1881) vii. 108 Surely this is at variance with the traditions of the Far West. 1890 [see FAR *a.* 1 a]. 1948 *N.Y. Star* 30 June 18/1 The cities of the Midwest, Far West and East. 1968 J. W. WATSON *N. Amer.* (ed. 2) xxi. 683 The Far West..gained from the battle which the Mid-West had fought for the growth of domestic industry.

Hence **Far-Wester, Far Westerner,** a settler in, or inhabitant of, the Far West; **Far-Western** *a.*, of or belonging to the Far West.

1843 'R. CARLTON' *New Purchase* I. 112 True honest-hearted far westers unadulterated by foreign or domestic scum. 1844 [see FAR *adv.* 8 c]. 1845 *Knickerbocker* XXVI. 283 There is a county in one of our far-western states which numbers just four whigs. 1862 G. C. STRONG *Cadet Life at West Point* 125 The far-Westerner.. declared his intention of being a lifelong bushfighter against the Indians. 1874 B. F. TAYLOR *World on Wheels*, etc. I. vi. 50 Such touches of border-life give a Far Western train a character of its own. 1887 C. B. GEORGE *40 Yrs. on Rail* xi. 234 The hearty 'Wall stranger', with which a Far Wester greets his neighbor on the train, immediately does away with reserve. 1948 *Pacific Discovery* May–June 15/1 The cloud that Far Westerners near the coast see most frequently is the advection fog. 1952 A. HUXLEY *Let.* 14 Aug. (1969) 651 A series of soliloquies having Far Western places as their source and excuse.

‖ **Fasching** (fæ·ʃiŋ). [G.] (See quot. 1963.) Also *transf.*

1911 R. BROOKE *Let.* Mar. (1968) 287 München in Carnival..these smooth, baggy, tired faces in the streets with the watery obstinacy 'It is *Fasching*. We are Enjoying ourselves' in the eye. 1928 *Daily Express* 3 Jan. 8/5 The 'Fasching', or carnival season. 1963 *Guardian* 21 Feb. 9/5 In Munich we are in the middle of Fasching, which is the South German word for what the Rhineland calls carnival. It began on Epiphany (January 6) and lasts until Shrove Tuesday. 1965 *New Statesman* 19 Mar. 465/1 Cardiff, more than ever, staged a secular *Fasching*. International rugby matches in Wales have always been the occasion for carnival. 1970 *Sat. Rev.* 12 Sept. 64 Fasching starts soon after New Year's Day and holds Munich in its thrall until Ash Wednesday.

Fasci (fa·ʃi), *sb. pl.* [It., pl. of *fascio* bundle, burden, assemblage, group :- pop.L. *fascium* for L. *fascis* bundle: see FASCES.] Groups of men organized politically, such as those (*fasci dei lavoratori*) in Sicily *c* 1895, and those of the Fascisti (e.g. the *fascio interventista* of 1915).

1902 *Encycl. Brit.* XXIX. 649/1 To produce in Sicily a discontent of which Socialist agitators took advantage to organize the workmen of the towns and the peasants of the country into groups known as *fasci*. 1921 *Public Opinion* 20 May 464/3 The first Fasci (composed of ex-soldiers) began to show signs of resistance and opposition to the Communists. 1922 *Q. Rev.* Jan. 144 A considerable proportion of the poet's legionaries in Fiume was drawn from the Fasci in different Italian towns. 1959 E. J. HOBSBAWM *Primitive Rebels* iii. 42 The great peasant rising of 1894—the *Fasci Siciliani*—saw it [*sc.* the Mafia] on the side of reaction, or at best neutral...Even then it was observed that the rise of the Fasci had diminished the hold of *Mafia* on the peasants.

fascia. Add: **2.** (Examples in the sense of FACIA.)

1909 H. G. WELLS *Tono-Bungay* I. iii. 100 The bright little shop still saying 'Ponderevo' with all the emphasis of its fascia. 1963 *Gloss. Build. Terms* (*B.S.I.*) 23 *Fascia*, a long and relatively narrow upright face at eaves or cornice or over a shop front. 1971 *Guardian* 21 July 11/5 In worn-out Upper Street..gaudy fascias brighten up derelict façades.

3. a. (Further examples.)

1807 MORRIS & KENDRICK *Edinb. Med. & Physical Dict.* II, *Fascia lata*, a strong tendinous expansion..to surround the muscles of the thigh. 1961 *Lancet* 5 Aug. 318/1 Connective tissue (fascia lata). 1967 G. M. WYBURN et al. *Conc. Anat.* vi. 156 The fascia lata is attached above to the pelvis and inguinal ligament and below to the condyles round the knee joint.

5. *attrib.* (in sense *2 and in the sense of *FACIA 2), as *fascia-board, -panel.*

1924 H. J. BUTLER *Motor Bodywork* 482/2 Fascia boards. 1925 W. DEEPING *Sorrell & Son* i. 9 Painted in white letters on a black fascia-board 'John Verity—Dealer in Antiques'. 1957 'N. SHUTE' *On Beach* i. 21 He could hardly remember what the fascia panel looked like or on which side the seat adjustment lever lay.

fasciculation. Add: **c.** *Physiol.* [f. FASCICUL(US.] Uncoordinated twitching of a muscle, esp. that involving the simultaneous contraction of whole bundles or fasciculi of muscle fibres.

1938 *Brain* LXI. 321 This phenomenon, commonly called 'fibrillation', but more properly called 'fasciculation', may therefore represent the intermittent involuntary discharge of single nerve impulses by an abnormal motor neurone. *Ibid.* 322 In the few seconds following the movement fasciculation is more rapid in that particular unit. 1959 H. GRUNDFEST in J. Field *Handbk. Physiol.: Neurophysiol.* I. v. 164/1 The disorganized contraction of muscles, frequently but incorrectly termed 'fasciculation', is due to the initial excitatory effect of the synaptic depolarization. 1963 *Lancet* 19 Jan. 130/1 With decamethonium muscle fasciculation was only slight, whereas with suxamethonium it was severe and long lasting.

fasciitis (fæʃi-, fæsi,ǝi·tis). *Path.* Also **fascitis** (fæ̆ʃ-, fæ̆sǝi·tis). [f. FASCI(A + -ITIS.] Inflammation of a fascia.

1893 DUNGLISON *Dict. Med. Sci.* (ed. 21) 424/1 Fascitis. 1940 B. I. COMROE *Arthritis* xlii. 611 The main therapeutic attack upon plantar fascitis includes..use of a rubber pad to protect the heel, procaine injections into the plantar fascia..and removal of focal infection. 1961 *Lancet* 30 Sept. 740/1 The term plantar fasciitis was used to describe a condition characterised by painful feet. 1966 R. W. EVANS *Histol. Appearances of Tumours* (ed. 2) i. 5 Nodular fascitis is not a tumour but a fibroblastic proliferation.

fascinatedly (fæ·sinēⁱtėdli), *adv.* [f. FASCINATED *ppl. a.* + -LY².] In a fascinated manner.

1894 M. DYAN *All in Man's Keeping* I. vi. 103 The lady stared fascinatedly on. 1949 T. BONNET *Mudlark* (1950) vi. 65 Mr. Brown performed again,.. watched fascinatedly by Wheeler. *a* 1953 E. O'NEILL *Touch of Poet* (1957) II. 69 She stares at him fascinatedly—then blurts out with impulsive admiration.

fascinator. Add: **c.** A head shawl worn by women, either crocheted or made of a soft material. orig. *U.S.*

1878 K. D. WIGGIN *Let.* 20 Dec. in N. A. Smith *K. D. Wiggin* (1925) 35 Mother crocheting a fascinator. 1897 *Sears, Roebuck Catal.* 305/1 Ladies' Fascinator, made of good quality Shetland yarn..colors, pink and white. 1908 *Scribner's Mag.* Feb. 177 A scant worsted 'enchanter' or 'fascinator', I think she called it, thrown over her head and shoulders. 1909 *Daily Chron.* 8 Dec. 9/4 A lovely shawl of eider, Shetland, or any fine wool..can be utilised as a fascinator, or shoulder wrap. 1949 R. HARVEY *Curtain Time* 110 The ladies in their furred evening cloaks..with snowflakes caught in their lacy fascinators. 1969 *New Yorker* 29 Nov. 59/2 Doris, in a pink cloak trimmed with swansdown and a candy-striped fascinator over her dark head, stepped out.

fasciola (fæsiō·lă). *Anat.* [L., small bandage.] In full *fasciola cinerea.* A thin layer of grey matter that forms the posterior continuation of the dentate gyrus in the hippocampal formation of the brain; the *gyrus fasciolaris.* (See also quot. 1882.) Hence **fasci·olar** *a.*, of or pertaining to the fasciola.

1848 QUAIN *Elem. Anat.* (ed. 5) II. 724 The grey matter in the floor of the fourth ventricle has been named fasciolæ cinereæ. 1882 WILDER & GAGE *Anat. Technol.* 478 Fasciola...The somewhat thickened margin of cinerea along the fimbria. In man the ventral portion has commonly been called *fascia dentata*; in the cat, however, there is no denticulation, and the name fascia is certainly misleading; hence the senior author proposed to employ *fasciola* for the whole. 1889 *Cent. Dict.* III, *Fasciolar*, pertaining to the fasciola, or fascia dentata of the brain. 1954 T. L. PEELE *Neuroanat. Basis Clin. Neurol.* xxiii. 502 Posteriorly, the dentate gyrus is continuous through the fasciolar gyrus below and behind the splenium with the supracallosal gyrus. 1962 E. C. CROSBY et al. *Correl. Anat. Nerv. Syst.* vii. 422 Although the main mass of the hippocampus terminates in the region central to the splenium of the corpus callosum, strands of hippocampal tissue (usually gyrus dentatus) extend, as fasciola cinerea, around the splenium to continue into the induseum griseum.

fasciole (fæ·siōᵘl). *Zool.* [f. prec.] One of the bands of minute tubercles, bearing modified spines, in spatangoid sea-urchins.

1850 E. FORBES *Brit. Organic Rem.* III. plate x. 2 A fasciole of tertiary spines. 1888 ROLLESTON & JACKSON *Forms Anim. Life* 558 Clavulæ, found only in *Spatangidæ*, are minute spines,..attached to minute tubercles which form regular bands known as *fascioles* or *semitæ*. 1962 D. NICHOLS *Echinoderms* viii. 100 Clavulae..occur in special bands called *fascioles*, diagnostic of the order Spatangoida.

fascioliasis (fæ:siŏlǝi·ăsis). *Path.* [mod.L., f. *Fasciola*, name of a genus of trematode worms, f. L. *fasciola* small bandage + *-IASIS.] Infection with the liver-fluke, *Fasciola* (*Distoma*) *hepatica*; liver-rot.

1890 J. H. STEEL *Treat. Dis. Sheep* v. 140 Hence to all intents and purposes we may deal with rot as fascioliasis, i.e. the pathological state which results from fluke invasion. 1968 G. LAPAGE *Vet. Parasitol.* (ed. 2) xvii. 338 About 200 cercariae will produce acute fascioliasis in a rabbit, but much depends on the nutritional state of the host. 1968 E. J. L. SOULSBY *Helminths, Arthropods & Protozoa Domesticated Animals* 23 The fluke is cosmopolitan in its distribution, and is the cause of fascioliasis (liver fluke disease, liver rot), especially in sheep and cattle.

Fascism (fæ·ʃiz'm, occas. fæ·siz'm). Also in It. form **Fascismo** (faʃi·zmo), and with small initial. [ad. It. *fascismo*, f. *fascio* bundle, group: see *FASCI and -ISM.] The principles and organization of Fascists. Also, *loosely*, any form of right-wing authoritarianism.

1921 *19th Cent.* July 148 The Fascismo was born in the provinces, where the extremistic menace was stronger. 1922 *Q. Rev.* Jan. 148 A section of the Press..now veered completely round to the cause of Fascism. The Fascist terror increased in intensity. 1923 *Contemp. Rev.* Jan. 44 We do not want Fascismo in this country. *Ibid.* Nov. 557 Fascism in Germany will never be more than one of several factors. 1925 *Weekly Westm. Gaz.* 10 Jan. 320/2 The outrages which have been associated with Fascism have gradually alienated much of the support which it won two years ago. 1934 tr. *K. Heiden's National Socialism* xvii. 354 The electoral victories all over Europe with which the Labour Parties have replied to German Fascism. 1936 *Discovery* Dec. 378/1, I have strongly criticised modern education and the methods of handling youth generally as inculcating excessive respect for authority and thereby conducing to the growth of

Fascism. 1939 A. COBBAN *Dictatorship* v. 124 In March 1919..Fascism was still..a revolutionary and socialist movement, hostile to the monarchy, to finance, and to parliamentary government, demanding social reform and workers' control, but separated from the other branches of the socialist movement by its intense nationalism. 1965 L. VENNEWITZ tr. *Nolte's Three Faces Fascism* II. ii. 87 Maurras was the first man in Europe who as a thinker and a politician drove conservatism beyond the limits dividing it from incipient fascism. 1971 *Tablet* 26 June 616/2 (*title*) The ghost of Fascism.

Fascist (fæ·ʃist, occas. fæ·sist), *sb.* and *a.* Also in It. form **Fascista** (faʃi·sta) and with small initial. Pl. **Fascists, Fascisti** (faʃi·sti). [ad. It. *Fascista*, formed as prec.: see -IST.] One of a body of Italian nationalists, which was organized in 1919 to oppose communism in Italy, and, as the *partito nazionale fascista*, under the leadership of Benito Mussolini (1883–1945), controlled that country from 1922 to 1943; also *transf.* applied to the members of similar organizations in other countries. Also, a person having Fascist sympathies or convictions; (*loosely*) a person of right-wing authoritarian views. Hence as *adj.*, of, pertaining to, or characteristic of Fascism or Fascists.

1921 *Times* 1 Jan. 9/6 The *Fascisti* are certainly paying back..the Socialists in matters of violence. 1921 *Glasgow Herald* 30 Mar. 9 A party of Fascisti from Perugia visiting Citta di Castello burned the Labour Bureau. 1921 *Public Opinion* 20 May 464/1 For the moment the Fascisti are acting as a sort of Government bodyguard for the elections and Fascist candidates find a place..on the Government lists. 1922 *Daily Mail* 17 Nov. 7 Signor Mussolini, the Fascist leader, to-day made his first speech in the Chamber. 1923 [see *BLACKSHIRT]. 1928 *Outlook* 26 May 645/2 Indeed, I cannot help wondering whether the suggestion does not originally emanate from the British Fascists. 1929 H. S. WALPOLE *Hans Frost* ii. 146 At Venice..a Fascist official at the railway station had been abominably vulgar. 1931 J. S. HUXLEY *What dare I Think?* vii. 231 The Fascist dictatorship. 1936 M. PLOWMAN *Faith called Pacifism* 74 Fascists in all countries are obliged openly to acknowledge their reliance upon war as a means of policy. 1936 *S.P.E. Tract* XLVI. 204 Does not their enthusiasm [for the study of stylistic perfection], when compared with other forms of fascist and fanatical activities, seem almost innocuous? 1938 *Ann. Reg.* 1937 320 The foreign policy of Fascist Italy. 1940 AUDEN in *I Believe* 30, I cannot see how a Socialist country could tolerate the existence of a Fascist party any more than a Fascist country could tolerate the existence of a Socialist party. 1940 N. MITFORD *Pigeon Pie* iii. 39 Luke was an awful old fascist. 1960 S. M. LIPSET *Political Man* v. 133 Fascist ideology, though antiliberal in its glorification of the state, has been similar to liberalism in its opposition to big business, trade-unions, and the socialist state. 1961 H. THOMAS *Spanish Civil War* viii. 70 José Antonio Primo de Rivera ..gradually emerged as the Leader of all the Spanish Young Fascists. *Ibid.* 71 The Socialists..were described by [Communist] party jargon as 'social fascists'. 1963 *Times* 27 Mar. 10/2 As the main body of demonstrators began to move away,..screams of 'Fascist pigs' and 'Gestapoism' continued. 1969 *Times* 17 Nov. 10/4 Taunts of 'Sieg Heil', 'Fascists', and the occasional smoke bomb from youthful demonstrators were bound to invite trouble. 1971 *E. Afr. Jrnl.* Mar. 28/1 The international fascist alliance had already found 'an adventure'. 1971 *Tablet* 26 June 616/2 In remoter mountain villages there are still a few..slogans which..survived..the collapse of Fascist rule.

Hence **fa:sciza·tion, fas:cistiza·tion,** the action or process of making Fascist. Also **fa·scistize** *v. trans.*

1925 *Glasgow Herald* 20 May 10 The complete 'fascistisation' of Italy. 1937 A. HUXLEY *Ends & Means* v. 36 Belief in our ideal postulates has acted as a brake on fascization. 1940 *Ann. Reg.* 1939 196 This Charter [the Fascist School Charter] has for its object to 'fascistise' entirely the three classes of instruction. To 'fascistise'..is 'to give the school a social and political content' of a certain type. 1955 H. HODGKINSON *Doubletalk* 50 This is part of the process of 'fascistisation' or move from bourgeois democracy to full fascism. 1965 L. VENNEWITZ tr. *Nolte's Three Faces Fascism* I. i. 7 Mussolini's theory of the imminent fascistization of the world undoubtedly seems prejudiced and vague.

Fascistic (fæʃi·stik, fæsi·stik), *a.* Also **fascistic.** [f. prec. + -IC.] Of or pertaining to Fascism or Fascists; having Fascist ideals. So **Fasci·stically** *adv.*

1928 *Daily Express* 2 Oct. 9 Mussolini is taking every possible measure to ensure that the children..grow up feeling 'Fascistically'—the word is good current Italian. 1935 *Mind* XLIV. 208 Why should not some Fascistic professor of physiology succeed in discovering some drug or mode of life that would extend indefinitely the organism's power to repair itself, and so to stave off death? 1947 *Ibid.* LVI. 187 One need not wonder that the same man should be a Socialist and also fascistically inclined. 1951 E. MITTELHOLZER *Shadows move among Them* III. vii. 306 Why should you consider it Fascistic to get rid of a man who has proved himself a hopeless crime-addict? 1962 R. B. FULLER *Epic Poem on Industrialization* 58 Witness its panicky Fascistic plotting.

fascitis, var. *FASCIITIS.

fashion, *sb.* Add: **9. c. high fashion** (chiefly *attrib.*), haute couture; also *transf.*

1958 *Sunday Times* 19 Oct. 25/1 'Line for line' copies of French models are the current high fashion rage in New York. 1959 *Observer* 22 Mar. 3/8 Now that high fashion has overtaken the circular, manufacturers have two kinds of worry. 1959 *News Chron.* 10 Aug. 6/5 The high-fashion hats seen in Paris. 1963 *Listener* 28 Feb. 370/1 The Education Board of the R.I.B.A. now finds its stodgy, nineteenth-century type syllabus at least partly in tune with the high fashion of the day.

13. a. and **d.** *fashion-artist, -designer, -journal, -letter, magazine, -model* (so *-modelling* vbl. sb.), *show; fashion-bound, -conscious, -driven, -favoured, -minded, -ridden* adjs.

1903 A. BENNETT *Truth about Author* x. 120, I could instruct a fashion-artist. 1949 KOESTLER *Insight & Outlook* xxi. 297 Implicit, sophisticated, or fashion-bound forms of artistic expression. 1951 M. MCLUHAN *Mech. Bride* 75/2 Mallarmé and Joyce refused to be distracted by the fashion-conscious sirens of contents and subject matter. 1960 *Guardian* 15 July 8/2 The fashion-conscious visitor to Paris. 1909 *Westm. Gaz.* 15 Mar. 5/1 The very red about which fashion-designers are decided. 1891 KIPLING *Light that Failed* iv. 69 The easy applause of a fashion-driven public. 1896 *Westm. Gaz.* 2 July 4/3 Fashion-favoured stuffs. 1905 E. WHARTON *House of Mirth* II. ix. 432 The newspapers, the fashion-journals. 1894 E. L. SHUMAN *Steps into Journalism* 154 Jennie June..who is known all over the country for her syndicate fashion letters. 1903 A. BENNETT *Truth about Author* x. 120 A Paris fashion-letter. 1879 M. E. BRADDON *Vixen* II. xiv. 256 Your *trousseau* should be..described in the fashion magazines. 1966 H. YOXALL *Fashion of Life* viii. 75 The majority of middle-aged women still have distressingly little sense of style...For them the fashion magazines might just as well not exist. 1938 *Chatelaine* Oct. 2/1 The upward sweep has become of intense importance to the fashion-minded. 1940 *Manch. Guardian Weekly* 30 Aug. 142 British manufacturers are continually producing new samples to meet the requirements of the most fashion-minded overseas wearers. 1962 *John o' London's* 22 Mar. 291/1 An ageing fashion-model. 1970 P. MOYES *Who saw her Die?* i. 12 She was thirty-three years old, with the figure of a fashion model. 1964 C. BARBER *Present-Day Eng.* v. 118 Fashion-modelling has been built up..as the ideal career. 1897 G. B. SHAW *Let.* 28 May (1965) 770 The silly visionary fashion-ridden theatres. 1938 *Collier's* 8 Jan. 13 This is, believe it or not, a fashion show.

14. fashion-book, a book describing and illustrating new fashions in dress; **fashion house,** a business establishment for the display and sale of high-quality clothes; **fashion-paper** (further examples); esp. a journal specializing in current fashions in dress; **fashion plate** (earlier and later examples); also applied to other kinds of fashionable display.

1840 *Fiddle Faddle Fashion Book* 8/1 To depart from *Fashion* in a *Fashion-Book* is so shocking a thing that we tremble at the very idea of it. 1853 MRS. GASKELL *Cranford* vii. 134 Three or four handsomely bound fashion-books ten or twelve years old. 1907 B. M. CROKER *Company's Servant* viii. 74, I have brought back ..a fashion-book, ribbons, gloves, and lots of sweets. 1958 *London Mag.* Dec. 33 A woman of character and elegance who ran a fashion house on one of the streets near Central Park. 1970 D. WHEATLEY *Gateway to Hell* iv. 44 She..became the top model in a leading fashion house. 1901 F. H. BURNETT *Making of Marchioness* I. i. 4 You might get on to one of the second-class fashion-papers to answer ridiculous questions about house-keeping or wall-papers or freckles. 1913 KIPLING *Lett. of Trav.* (1920) 246 She showed me what looked like a fashion-paper print of a dress-stuff. 1925 L. P. SMITH *Words & Idioms* iii. 127 Even in fashion-papers we read of 'creations' in millinery. 1851 A. O. HALL *Manhattaner* 21 She unrolls the fashion plate. 1859 L. WILMER *Press Gang* 332 Fashion-plate magazines. 1891 *Stamp Collector* Dec. 48 The latest philatelic 'fashion plates' tell us that the new idea of collecting postal cards is to collect them direct from the countries issuing them. 1908 *Westm. Gaz.* 26 May 4/2 Altering the outline of each new season's model, with the result that the preceding ones are.. rendered old-fashioned...We do not want the fashion-plate methods to become general amongst motor-constructors. 1967 D. YARWOOD *Eng. Costume* (ed. 3) viii. 211 We now have innumerable fashion plates and magazines.

fashioned, *ppl. a.*¹ Add: (Further examples.) Used esp. of stockings that are shaped to fit the contours of the leg. Cf. **full-fashioned*, **fully-fashioned* adjs.

1887 C. A. MOLONEY *Sk. Forestry W. Afr.* 26 Fashioned wood. 1895 *Montgomery Ward Catal.* 300/2 Ladies' lisle finished hose, patented seams, fashioned ankles. 1909 *Daily Chron.* 13 Apr. 5/2 Fashioned hosiery. 1919 *Ladies' Home Jrnl.* May 56 Burson Fashioned Hose..without seams.

fasola (fā·sǫlā, -sōᵘl-). *Mus.* [f. FA *sb.* + SOL *sb.*² + LA *sb.*] (See quot. 1964.)

1933 G. P. JACKSON (*title*) White spirituals...The story of the fasola folk. *Ibid.* 3 Ask any real country person..about fasola singers, and he will very likely be able to direct you to one of them. 1933 *Times Lit. Suppl.* 13 Apr. 257/1 Conventions held..for 'fasola' singing. 1956 M. STEARNS *Story of Jazz* (1957) viii. 126 Yankee singing masters invaded rural areas with their 'fasola' and 'shape-note' systems of reading music. 1964 *Conc. Oxf. Dict. Mus.* 320/1 *Lancashire Sol-fa.* A modern name for a system of sight-singing more properly called 'Old English Sol-fa', since it was universally used in England from at least the early 17th c. and its latest textbook appeared in 1879. It is a method of solmization applied to the normal staff notation; the first 3 notes of every major scale are called *fa-sol-la*, and so are the second 3 notes, the remaining note being called *mi*... In the Amer. Colonies (and later the U.S.A.) it was called *Fasola* or, sometimes (from the special notation there used), *Patent Notes.*

fast, *a.* Add: **1. f.** (Later examples.) Also *fast-colour* attrib.

1939 H. G. WELLS *Holy Terror* I. iii. 72 There's that fool Lord Horatio Bohun and his purple shirts—mauve shirts really they are, for he never had the wit to get his shirts fast colour. *Ibid.* III. iii. 317 We are the Revolution from the West and all fast-colour Revolutions come out of the West. 1962 *B.S.I. News* June 10/1 There is a good deal of loose use of expressions such as 'guaranteed fast colours' and 'fast colours', without any indication of what the colours are fast to—and without even any recognition that a colour that is fast to light is not necessarily fast to washing, and *vice versa.* 1962 J. T. MARSH *Self-Smoothing Fabrics* xviii. 298 Since 1952–53 there has been an increasing use of vat dyes with thermo-setting resins on cotton, for fast-colour work has become a particularly important factor in the U.S.A.

h. Of an organism: resistant to the stain-removing or toxic action of a (specified) agent. Used chiefly in *Comb.* preceded by the name of the agent. (Cf. FAST *a.* 1 f.)

1904 [see *acid-fast adj.]. 1907 *Jrnl. R. Inst. Public Health* XV. 453 The atoxyl-fast strain is also resistant to a number of substances related to atoxyl. 1949 E. J. PULASKI in S. A. Waksman *Streptomycin* xxxiv. 491 Wide variation in susceptibility is observed and initially streptomycin-fast organisms are found. 1951 A. GROLLMAN *Pharmacol. & Therapeutics* i. 26 By subjecting susceptible organisms to inadequate doses of a drug it is possible to render them 'drug fast'. 1964 M. HYNES *Med. Bacteriol.* (ed. 8) v. 48 An 'acid-fast' organism is one which when stained resists decolourization with 25 per cent sulphuric acid. *Ibid.*, The smegma bacillus..is acid-fast, but not alcohol-fast.

2. c. Frozen. *N. Amer.* ? *Obs.*

1706 *Boston News-Letter* 21 Jan. 2/2 New York Jan. 7th..Hudsons River was froze over and continued fast several days. 1743 J. ISHAM *Obs. Hudsons Bay* (1949) XII. 179 The Lakes and Rivers near the ocean some affirm, are fast all the Summer. 1796 E. DRINKER *Jrnl.* 23 Dec. (1889) 296 Clear and very cold. Wind N.W. The river fast to day. 1854 in *Beaver* (1924) Dec. 31 We have had the pleasure of seeing the river fast this morning.

8. *fast and furious*: see FURIOUS *a.* 1 d.

d. *Photogr.* Needing only brief exposure (emulsion, film); shortening the necessary exposure time (lens); providing a brief exposure time (shutter).

1902 A. WATKINS *Photogr.* 115 The folded focussing cloth can be held in front of the lens before drawing the slide. Probably half the subjects usually taken with a fast shutter would be better done in this way. 1923 *Brit. Jrnl. Photogr. Alm.* 1924 165 (Advt.), Ilford Roll Films... Exceptionally fast. 1939 HENNEY & DUDLEY *Handbk. Photogr.* viii. 214 In all the common methods of specifying film speed, the larger numerical units indicate the faster ..emulsions. 1958 *Amat. Photographer* 31 Dec. 914/2 The benefits of a fast lens are greatest with a 35-mm camera. 1968 A. DIMENT *Gt. Spy Race* vii. 107 It was a good, fast film and I worked the re-wind lever like the bolt on a rifle.

e. Used in nuclear physics to denote various processes, devices, etc., that involve **fast neutrons* (see sense 11), as *fast breeder (reactor)*, a breeder reactor that is a fast reactor; *fast fission*, the fission of a nucleus by a fast neutron; *fast pile, reactor*, a nuclear reactor in which fission is caused primarily by fast (unmoderated) neutrons.

1945 H. D. SMYTH *Gen. Acct. Devel. Atomic Energy Mil. Purposes* viii. 79 The enhancement may be taken into account by multiplying the original number of neutrons N by a factor ε which is called the fast-fission effect or the fast-multiplication factor. *Ibid.* 91 The original fast neutrons are slightly increased in number by fast fission. 1947 *Sci. News Let.* 6 Sept. 147 'Fast reactor' uses man-made plutonium. 1949 *Nucleonics* Dec. 40 The 'fast' pile, in which the neutrons produce fission at practically the same energy at which they themselves are released by fission. 1954 *Wall St. Jrnl.* 11 Oct. 3/4 There are two more experimental reactor types to be built under the A.E.C.'s present plans—a so-called homogeneous reactor and a 'fast breeder' reactor. 1954 *Sci. Amer.* Dec. 39/1 The homogeneous reactor and the fast breeder represent the boldest extrapolations among the five reactor projects. 1957 *Technology* July 184 Remoteness was considered essential for the site of an experimental fast-breeder reactor. In this respect, Dounreay was ideal. 1964 A. SALMON *Nuclear Reactor* iii. 31 The fast reactor cannot use natural, or near natural, uranium because inelastic scattering by the ²³⁸U reduces the mean neutron energy into a region where fast fissions are not sufficient to overcome the losses due to leakage of neutrons from the reactor and to capture in the ²³⁸U. 1968 *New Scientist* 1 Feb. 230/2 Most industrialized countries are growing enthusiastic.. about fast reactors, which they foresee as the next successful breed of nuclear power generators.

11. fast back, fastback, (*a*) in *Bookbinding*, a back that adheres to the sheets as distinguished from a loose back or spring back; (*b*) orig. *U.S.*, on a motor car: a back that slopes in one continuous line from the top of the car to the rear bumper; hence, such a car; freq. *attrib.*; **fast buck** *U.S.* [*BUCK *sb.*⁸], a quickly earned dollar; **fast-footed** *a.*, (*a*) *Cricket*, denoting the action or method of keeping the back foot immobile in batting; (*b*) fast-moving,

FAST

speedy of foot; **fast-goer** (earlier example); **fast ice,** ice covering sea-water but usually attached to land (cf. FAST *sb.*^{3} and *FAST *a.* 2 c); **fast lane,** a traffic lane, usu. that farthest from the outer edge of a motorway or dual carriageway, intended for drivers who wish to overtake slower cars; **fast neutron,** a neutron with kinetic energy greater than some arbitrary value, esp. one that has not been slowed down by the action of a moderator after being produced by the fission of a nucleus; **fast one** *slang* (orig. *U.S.*), esp. in phrases, as *to pull* (or *put over*) *a fast one*, to take unfair advantage by (rapid) action of some sort; **fast store** *Computers* (see quot. 1962); also called *fast-*, *quick-access storage*; **fast worker** *colloq.*, one who makes rapid progress, used esp. in amatory contexts.

1912 A. J. PHILIP *Business of Bookbinding* xv. 181 Books which are to have leather backs are now individually examined and tested to find out which should have fast backs and which open backs. 1965 *Economist* 7 Aug. 552/2 The 1600 [Volkswagen car] is given a 'fast' (or sharply sloping) back. 1968 *Autocar* 14 Mar. 27/1 German fastback fitted with new fully automatic transmission. 1969 *Guardian* 7 Oct. 5/4 A fastback family saloon. 1949 *New Yorker* 5 Nov. 82 Tryin' to hustle me a fast buck. 1969 *Observer* 23 Nov. 7/1 A friend who made a fast buck out of selling dirty postcards. 1971 *Publishers' Weekly* 22 Mar. 33 In recent years, the Norman Rockwell kind of vision has been sullied by cynical, fast-buck door-to-door operators. 1897 *Encycl. Sport* I. 228/2 The 'fast-footed' style of hitting. 1907 *Westm. Gaz.* 21 Aug. 7/1 To play right over it in attempting a fast-footed drive. 1960 T. MCLEAN *Kings of Rugby* xi. 143 Watson, who had not been conspicuously fast-footed as Faull made his run, restored his reputation [etc.]. 1837 DICKENS *Sk. Boz* 2nd Ser. 42 [His] great aim..was to be considered as a 'knowing card', a 'fast-goer'. 1932 *Geogr. Rev.* XXII. 81 The relative lightness and flatness of the sea ice over the entire length of our flight classified it as belonging to the pack-ice zone intermediate between the polar cap ice of the central basin and the fast ice of the coastal shelves. 1956 *Nature* 31 Mar. 599/2 About five thousand Emperor penguins were found on the fast-ice in an adjoining bay... A few days after arrival the sea ice on which the penguins had been nesting broke up. 1966 T. WISDOM *High-Performance Driving* xi. 111 One is frustrated on a motorway by the driver ahead in the 'fast' lane (if only he appreciated that it is the *overtaking* lane). 1971 K. ROYCE *Concrete Boot* x. 120 I'd been batting away on the fast lane, peering ahead..through the cars in front of me. 1933 *Physical Rev.* XLIV. 236 (*heading*) Disintegration of neon nuclei by fast neutrons. 1945 H. D. SMYTH *Gen. Acct. Devel. Atomic Energy Mil. Purposes* vi. 53 The program would provide the theoretical and experimental data required for the design of a fast-neutron chain-reacting bomb. 1956 A. H. COMPTON *Atomic Quest* i. 55 It was the fast neutrons coming directly from the atom's fission that would take part in an atomic explosion; in the controlled reaction where a moderator was used, it was the slow neutrons that were most important. 1968 *New Scientist* 8 Feb. 305/1 Experiments in fast-neutron therapy have so far concentrated mainly on animals. 1923 H. C. WITWER in *Cosmopolitan* Nov. 98/1 He's trying to put over a fast one! 1932 J. SAYRE *Rackety Rax* xiii. 103 Brick pulled a fast one in the St. Mary's game. 1932 WODEHOUSE *Hot Water* xvii. 282 The thought that a girl capable of thinking up a fast one like that should be madly throwing herself away on Blair Eggleston..was infinitely saddening. 1933 J. G. COZZENS *Cure of Flesh* II. 157 You never know when they may pull a fast one on you. 1937 R. STOUT *Red Box* ix. 127 There was a chance you were putting over a fast one. 1943 H. BOLITHO *Combat Report* 107, I said..that they must not try to pull a fast one on me. 1943 HUNT & PRINGLE *Service Slang* 31 *A Fast One*: (a) any remark giving rise to Thought. (b) a trick, especially one calculated to shift the onus. 'Pulling a fast one' on your associates is very much the same as 'putting one over' on them. 1953 R. LEHMANN *Echoing Grove* 250 [Amer. loq.] The last thing I meant was to try and pull a fast one, but I guess that's what I'm doing. 1958 'A. GILBERT' *Death against Clock* 96 Mad to think they can pull a fast one.. over the whole community. 1955 *Scope Electronic Computers in Office* (Office Managem. Assoc.) 36 The fast store has a capacity of 390 forty binary digit words. 1962 *Gloss. Terms Autom. Data Proc.* (*B.S.I.*) 64 *Fast store*, an imprecise term referring to a store whose access time is relatively short. 1964 T. W. MCRAE *Impact of Computers on Accounting* i. 8 Apart from these historical or backing stores we need a fast access store for handling that part of the data which is being currently processed by the computer. 1921 S. FORD *Inez & Trilby May* xii. 212 The dark stranger is getting a bit free. He is patting Inez on the arm... 'One of these fast workers, I take it,' says I. 1929 J. B. PRIESTLEY *Good Companions* II. v. 373 He had had two while some fellows..had not been able to secure a single drink... He was undoubtedly a fast worker. 1961 I. MURDOCH *Severed Head* xxiii. 195 'I'm going to get married.'..'Admit you're a fast worker!' 1965 'R. FOLEY' *Suffer a Witch* (1966) v. 95 And you had dinner with him practically the next evening?..He seems to be a fast worker.

fast, *adv.* Add: **8. c.** *fast-locked, -shut* (later example).
1907 *Academy* 27 July 717/1 Some fast-locked gate. 1931 R. CAMPBELL *Georgiad* ii. 46 Nor glide a ghost around each fast-shut door.
d. *fast-breeding*, (sense 10 of adj.) *-going, -growing* (later examples), *-moving, -running, -swimming* ppl. adjs.
1866 TROLLOPE *Belton Est.* I. iii. 81 She was a fast-going girl. 1901 *Daily News* 23 Feb. 6/7 The awkwardness inseparable from fast-growing young creatures who have not yet attained perfect command of their limbs. 1910 *Westm. Gaz.* 24 June 12/2 The river has been very thick and fast-running. 1935 B. RUSSELL *Relig. & Sci.* vi. 159 It *might* happen that..all the fast-moving particles got to one side. 1952 C. DAY LEWIS tr. *Virgil's Aeneid* v. 115 The fast-swimming Tritons. 1957 *Times Lit. Suppl.* 25 Oct. 646/2 *De Luxe Tour* careers along, where the other novels meander. It is fast-moving, slick and professional. 1960 *Farmer & Stockbreeder* 16 Feb. Suppl. 23/1 A completely healthy fast-growing herd of pigs. 1964 G. H. HAGGIS et al. *Introd. Molecular Biol.* vii. 203 Ultra-fine genetic analyses using fast-breeding microorganisms. 1971 *Engineering* Apr. 31/1 With such a fast-growing market there is always a danger of people with little knowledge of vending producing machines which, in the long term, could damage the image.

fast, *v.*^{2} Add: **2. d.** *trans.* To cause to fast or be without food.
1854 *Poultry Chron.* I. 15 Before they are killed, they should be fasted at least fourteen hours. 1970 *Nature* 24 Oct. 383/1 The animals were then fasted, one group for 4–6 days, another for 13 days. 1971 *Ibid.* 5 Feb. 420/2 They [*sc.* rats]..were fasted about 48 h before use.

fasten, *v.* Add: **7. c.** *intr.* To admit of being fixed or fastened.
1850 F. E. SMEDLEY *F. Fairlegh* iv, The macintosh.. fastening round the neck with a hook and eye. 1924 A. D. SEDGWICK *Little French Girl* II. viii, A dark silk dress..fastening at the breast with a great old clasp of wrought gold.
8. Also with *up*.
1908 J. S. FLETCHER *Paradise Crt.* v. ii, Was he..to be fastened up there like a rat in a trap for—how long?
b. *intr.* To admit of being closed with fastenings.
1829 SCOTT *Old Mort.* Note x, The iron hasps [of the window]..fastened in the inside. 1862 G. MACDONALD *D. Elginbrod* II. xxiv, He could find no fastening upon it [*sc.* a door]. 'No doubt,' thought he, 'it does fasten, in some secret way or other.'
12. d. Also with *to*.
1881 C. E. L. RIDDELL *Sen. Partner* xxxv, One of the nephews..insisted on fastening himself to Mr. Snow.
e. = FIX *v.* 6 c.
1881 C. E. L. RIDDELL *Sen. Partner* xvi, Fastening her kinsman with a cold steely eye.

fasting, *vbl. sb.*^{2} Add: **3. fasting blood sugar,** the concentration of sugar in the blood after a period of fasting.
1927 *Practitioner* Feb. 116 If the fasting blood sugar is above 0·14 per cent., especially if it is in the neighbourhood of 0·2 per cent. or over, then diabetes is clearly established. 1968 J. ANDERSON in W. G. Oakley et al. *Clinical Diabetes* xxvi. 689 In uncomplicated hypothyroidism the fasting blood sugar is usually normal but may occasionally be low.

fa·st-talk, *v. colloq.* Chiefly *U.S.* [f. FAST *adv.* + TALK *v.*] *trans.* To persuade by eloquent or deceitful talk.
1946 *Richmond* (Virginia) *News Leader* 29 Mar. 1/3 A marine private who fast-talked a Japanese bank president out of 587,800 yen. 1950 J. S. REDDING *Stranger & Alone* xxii. 225 He wasn't going to let some cracker lawyer.. cheat and fast-talk Nan out of it. 1959 C. WILLIAMS *Man in Motion* x. 126 Purcell and Stedman deliberately fast-talked the liquor store man into an identification. 1960 *Sunday Express* 18 Dec. 17/6 Gantry hits the road again, to fast-talk himself into whatever good bet lies round the next bend. 1969 J. BURMEISTER *Hot & Copper Sky* iii. 64, I fast-talked my way into a bottle after..the party broke up.

fat, *a.* and *sb.*^{2} Add: **A.** *adj.* **2. e.** Of larger size than is usual; large in comparison with others of the same species.
1877 *Encycl. Brit.* VII. 368/2 The Fat Dormouse (*Myoxus glis*) is larger than the British species. 1877 A. B. HORTON in C. A. Moloney *Sk. Forestry W. Afr.* (1887) 38 Nuts well supplied with flesh, or what is technically called 'fat nuts'.
f. *fat herring* = MATIE.
1863 [see MATIE]. 1883 F. M. WALLEM *Fish Supply Norway* 15 The Norwegian fat herring is considered to be the very best herring in the world. 1967 *Oceanogr. & Marine Biol.* V. 436 The 'fat' herring fishery, exploiting one-to-four-year-old, adolescent fish.
4. c. Delete † *Obs.* and add later examples. Substitute for def.: Of an actor's part, offering abundant opportunity for skill, display, etc.; of speeches, etc., impressive, effective.
1901 C. MORRIS *Life on Stage* xix. 139 What is vulgarly known to-day as a 'fat' part, 'fat' meaning lines sure to provoke applause. 1929 WODEHOUSE *Gentleman of Leisure* xv. 119 True acting part, the biggest in the piece, full of fat lines.
10. c. *a fat lot*: a large amount, a great deal: always *ironical* and implying 'very little, hardly anything'. Similarly *a fat chance*, implying 'hardly any opportunity (or possibility)'; also (N.Z.) *a fat show*.
1892 I. ZANGWILL *Childr. Ghetto* (1893) I. ii. 24 'I von't sell you no more tickets,' said Sugarman..'A fat lot I care,' said Becky, tossing her curls. 1899 C. J. C. HYNE *Further Adv. Capt. Kettle* i. 5 Shows what a fat lot of influence this État du Congo has got. 1906 F. METCALFE *Side Show Studies* 99 The Proprietor..suggested sending the announcement..to the papers. 'A fat chance I'd stand of having it printed,' he grumbled. 1908 'I. HAY' *Right Stuff* vi. 83 Rot! Fat lot you know about it, Dilly. 1914 D. H. LAWRENCE *Widowing Mrs. Holroyd* III. 72 Some women could have lived with him happy enough. An' a fat lot you'd have thanked me for my telling. 1916 J. S. CLOUSTON *Two's Two* ix. 69 'And a fat lot of good they'll be!' scoffed Archibald. 1922 S. LEWIS *Babbitt* v. 63 Fat chance! 1933 W. S. MAUGHAM *Sheppey* III. 75 Fat chance I've got of going to France now. 1936 WODEHOUSE *Laughing Gas* viii. 88 A fat chance, of course. I should have known his psychology better. 1947 A. VOGT in D. M. Davin *N.Z. Short Stories* (1953) 374 The men from the mill..had a fat chance of getting in [to the pub] before that [6 p.m.]. 1948 D. BALLANTYNE *Cunninghams* (1963) I. xii. 63 It would be corker if he could go outside with Carole Plowman...Fat show! 1951 E. COXHEAD *One Green Bottle* i. 23 Idiot, thought Cathy. Fat lot you know. 1962 *Coast to Coast 1961–62* 87 You're a fat lot of help to me, moping about alone all the time. 1967 B. WRIGHT tr. *Queneau's Between Blue & Blue* xix. 208 Fat lot of use it was me getting my posterior frozen for a whole night to do my host a favour.
13. *fat-arsed, -hipped*.
1891 J. S. FARMER *Slang* II. 377/1 *Fat-arsed*, broad in the breech. 1923 D. H. LAWRENCE *Kangaroo* xvi. 351 Slaving to keep this marvellous Empire going, with its.. fat-arsed hypocritical upper classes. 1921 —— *Let.* 26 Mar.? (1962) II. 647 The fat-hipped soft fellow we saw at Anticoli last year.
14. fat-back *U.S. local*, a strip of fat from the back of a pig; **fat cat** *slang* (orig. and chiefly *U.S.*), a political backer; also *transf.*; **fat crab** *U.S.*, a crab ready to shed its shell; **fat-headedness,** stupidity; **fat lamb** *N.Z.*, a lamb bred to be exported as refrigerated meat; **fatstock, fat-stock,** marketable live-stock; **fat-tail,** a fat-tailed sheep; **fat-wood** *U.S. local* = LIGHTWOOD^2 a.
1903 *Sears, Roebuck Catal.* (ed. 113) 18/1 Clear back pork. This pork is made from the fat backs of prime hogs, is free from lean and bone. 1932 E. CALDWELL *Tobacco Road* i. 8 Salty soup Ada had made by boiling several fat-back rinds in a pan of water. 1941 A. F. RAPER *Sharecroppers All* 24 The great majority bring for lunch cold baked 'taters' and sandwiches of soda biscuits and fat back. 1928 F. R. KENT *Pol. Behavior* vi. 59 These capitalists have what the organization needs—money to finance the campaign. Such men are known in political circles as 'Fat Cats'. 1949 *Sat. Rev. Lit.* 16 Apr. 4 Hollywood celebrities, literary fat cats. 1960 *Economist* 8 Oct. 157/2 Methods of exhorting [*sic*] money which.. harry so-called 'fat-cats'—rich supporters—to an extent which looks like cruelty to animals. 1966 *New Statesman* 21 Jan. 78/3 The kind of balance-sheet fat cats who characterise the worst side of ITV. 1971 *Flying* Apr. 34/1 Those who view the business jet as a smoke-belching, profit-eating chariot of the fatcat. 1905 W. P. HAY in *Rep. Bureau Fisheries 1904* (U.S.) 411 The individual bearing it [*sc.* a narrow white line] is classed as a 'fat crab', or more vulgarly as a 'snot'. 1840 A. BUNN *Stage* I. 272 Such is the fat-headedness of *John Bull*. 1891 KIPLING *Light that Failed* viii, The fat-headedness of deliberately trying to do work that will live. 1915 WODEHOUSE *Something Fresh* iii, If you want any further proof of your young man's fat-headedness, mark that. 1891 R. WALLACE *Rural Econ. Austral. & N.Z.* xxvii. 369 Any of the various breeds or crosses now in the country [*sc.* N.Z.] will supply suitable mothers for fat lambs, provided they are good milkers. 1923 *N.Z. Jrnl. Agric.* 20 Jan. 50 Which breed of sheep is best for fat-lamb production? 1928 R. G. STAPLEDON *Tour Austral. & N.Z.* viii. 58 Fat lambs are also largely produced on a Romney basis. 1956 G. BOWEN *Wool Away!* (ed. 2) xii. 147 On the flat country..fat-lamb and mutton production predominate. 1880 *Trans. Illinois Dept. Agric.* X. 65 The success of the Fat Stock Show. 1906 *Daily Colonist* (Victoria, B.C.) 30 Jan. 5/4 It was decided to hold a fat stock sale in this city in the middle of March. 1907 R. WALLACE *Farm Live Stock Gt. Brit.* (ed. 4) 281 Steers slaughtered at the Smithfield Fat-Stock Show. 1912 *Jrnl. Bath & West & Southern Counties Soc.* 5th Ser. VI. v. 175 For many years the Middle Whites carried off the highest awards at the fat stock shows. 1955 *Times* 13 July 12/5 The fatstock marketing scheme had been working successfully for the past 12 months. 1888 *Castle Line Handbk. S.A.* 55 (Pettman), The fat-tails held their own for many years. 1909 WEBSTER, *Fatwood*, any pine wood full of pitch. 1938 M. K. RAWLINGS *Yearling* vii. 67 She unbanked the fire on the hearth and threw on fat-wood. 1963 *Times* 19 Mar. 10/7 All kinds of wood, including fat-wood.
B. *sb.*^{2} **1. b.** *pl.* or (in *attrib.* use only) *collect. sing.* Fat cattle or sheep. *Austral.* and *N.Z.*
1886 C. SCOTT *Sheep-farming* 137 Sheep intended for the fat market. 1894 A. ROBERTSON *Nuggets* 132 What say ye to him goin' to Melbourne to see the flock o' fats sold at the Flemington Yards? 1907 W. H. KOEBEL *Return of Joe* 20 The only different thing you'll see, I hope, will be a larger mob of fats. 1921 H. GUTHRIE-SMITH *Tutira* xxxvi. 362, I remember..watching the well-known dealer, Andrew Grant, picking 'fats'. 1936 I. L. IDRIESS *Cattle King* ix. 77 Drovers with their plants were coming in increasing numbers to drove to distant markets the growing herds of 'fats'. 1954 B. MILES *Stars my Blanket* xxiii. 199 Their four-year-olds..will be trucked to the factories as 'fats'. 1969 *Sunday Mail Mag.* (Brisbane) 5 Oct. 2/1 We were mustering fats along the channels of Cooper's Creek.
2. d. Phr. *a bit of fat*: (an instance of) good fortune; (an opportunity for) profit. *colloq.*
1923 BREWER *Dict.* 419/2 *A bit of fat*, an unexpected stroke of luck; also, the best part of anything. 1936 J. CURTIS *Gilt Kid* ii. 19 'Where did you graft in Wandsworth?' 'Cleaner.' 'Blimey, that was a bit of fat for you, wasn't it?' 'Yeah, but you couldn't pinch no grub.' 1951 P. BRANCH *Wooden Overcoat* i. 6 He had been acquitted... Cor! What a bit o' fat! I got away with it!
3. c. *to chew the fat*: see *CHEW *v.* 3 g.

5. d. (Earlier and later examples.)
1812 *Dramatic Censor 1811* 440 Mr Dowton..did not exhibit any propensity to give us too much of the *fat.* 1933 P. GODFREY *Back-Stage* iii. 36 An expression which is likely to puzzle the uninitiated is the term 'fat'. Theatrical 'fat' is determined not by the size of the part, but by its effectiveness.
6. a. *fat-mass.* **b.** *fat-formation, -former, -forming, -splitting* adjs.
1909 *Daily Chron.* 8 July 6/4 Fat-formation in the body..is not to be regarded as a mere accretion or addition of the fat we consume to the tissues of the frame. 1886 C. SCOTT *Sheep-farming* 41 A ton of good linseed cake contains of fat-formers 1508 lbs.; of flesh-formers, 582 lbs. 1893 F. F. MOORE *I forbid Banns* xxxiv, She knew the flesh-formers from the fat-formers, and partook of both. 1848 *Rep. Comm. Patents 1847* (U.S.) 116 With respect to the fat-forming principle..the proportions of the different parts of the grain stand thus. 1872 G. M. HUMPHRY *Obs. Myol.* 5 In addition to the four fat-masses disposed with the longitudinal series of vessels are four deep fat-masses lying beneath the four lateral muscles. 1890 BILLINGS *Med. Dict.*, *Fat-splitting ferment*, steapsin. 1907 *Practitioner* Aug. 320 The steapsin, or fat-splitting ferment of the pancreatic juice. 1955 J. G. DAVIS *Dict. Dairying* (ed. 2) 181 Tributyrin is extremely bitter, and certain fat-splitting bacteria may cause bitterness by producing this substance.
c. fat-body, in arthropods, esp. insects, and amphibians: a mass consisting of connective tissue and globules of fat, forming a reserve store of nutritious matter; **fat boy** (see quots.); **fat-cake** (see quots. 1865, 1898); **fat-liquor,** a mixture of soap and oil used in leather manufacture for softening hides; also as *vb.*; so **fat-liquored** *ppl. a.*, **-liquoring** *vbl. sb.*; **fat-soluble** *a.*, soluble in fats or oils.
1869 A. S. PACKARD *Study of Insects* 37 In the larva of insects is found the *corpus adiposum*, or fat-body, in the form of large lobes of fat-cells. 1927 HALDANE & HUXLEY *Anim. Biol.* iv. 113 In the frog fat is stored in special fat-bodies. 1964 V. B. WIGGLESWORTH *Life of Insects* ix. 136 It may be the 'fat body' or the blood that is coloured. 1940 *Chambers's Techn. Dict.* 324/1 *Fat boys* or *fatters*, boys or men employed to lubricate axles of tubs or wagons and rollers of haulage gear. 1955 *Evening News* 28 Sept. 3/8 *Fat Boy*, a boy who greases wagon-axles and does odd jobs in a quarry. 1839 'T. TREDDLEHOYLE' *Sum Thowts abaght Nan Bunt's Chresmas Tea-Party* 19 A great pile a muffins an fat-cake browt up at table. 1865 W. HOWITT *Hist. Discov. Austral.* II. 15 Fat-cakes, made of flour, and fried in their pan, a bush dainty. 1898 MORRIS *Austral Eng.*, *Fat-cake*, a ridiculous name sometimes applied to *Eucalyptus leucoxylon*. 1953 P. ABRAHAMS *Return to Goli* vi. 200 After this we had fat-cakes and coffee. 1903 L. A. FLEMMING *Pract. Tanning* 22 The leather is then ready to be stained or colored, fat-liquored, put out on the grain and treated in the finishing operations. *Ibid.* 127 The leather treated with it [*sc.* palmetto] takes even cooling and carries the fat-liquor well. 1931 WILSON & MERRILL *Anal. Leather* xi. 395 In both fat liquoring and finishing, most of the fatty constituents of the soap are absorbed by the leather. 1945 MCLAUGHLIN & THEIS *Chem. Leather Manuf.* xxiv. 723 Soap emulsions are used in some special cases for fatliquoring leather. *Ibid.* 739 It [*sc.* egg yolk] clarifies the fatliquor and..gives a drier, or a less oily and greasy feel, to the surface of the leather. *Ibid.* 742 A sample of fatliquored, chrome-tanned calf skin. 1922 *Encycl. Brit.* XXXII. 932/1 These [vitamins] are (1) the anti-scorbutic factor; (2) the water-soluble B.; (3) the fat-soluble A. 1926 J. S. HUXLEY *Ess. Pop. Sci.* viii. 88 Fat-soluble vitamin A. 1956 *Nature* 21 Jan. 123/2 The chlorinated compounds which are fat-soluble and chemically stable tend to accumulate in the fatty tissue.

‖ **fat** (fat, fa), *sb.*^{3} [Fr., a stupid, conceited person, f. Pr. *fat* stupid, ignorant, f. L. *fatuus*: see FATUOUS *a.*] A presumptuous, conceited dandy; a fop.
1832 F. TROLLOPE *Dom. Manners* I. xx. 333 'Monsieur,' said my young *fat*, with an indescribable grimace. 1853 C. BRONTË *Villette* I. xiv. 265, I took my revenge on this 'fat', by making him as fatuitous as I possibly could. 1876 GEO. ELIOT *Dan. Der.* III. v. xxxv. 316 How that *fat* Deronda can bear looking at her.

fat, *v.* Add: **3. d.** *trans.* In the manufacture of leather, to smear over with fat-liquor.
1903 L. A. FLEMMING *Pract. Tanning* 166 The leather is now treated as usual, and fatted or oiled.

fatality. Add: **4. b.** *attrib.*, as *fatality figure, rate.*
1897 *Daily News* 8 Jan. 6/2 What is called the fatality rate, that is to say, the proportion of deaths to cases, varies considerably in the case of diphtheria. 1912 *Maclean's Mag.* Feb. 433 The 'fatality' figures in Toronto..are as follows. 1966 *Lancet* 24 Dec. 1372/2 The daily fatality-rate..shows a peak on the second day in all the groups. 1971 *Brit. Med. Bull.* XXVII. 27/1 The most likely explanation seemed to be a change in the fatality-rate.

fate, *sb.* Add: **3.** *as sure as fate*: see SURE *adv.* 4 a.
4. *a fate worse than death*: see *DEATH *sb.* 17 b.
5. d. fate-line *Palmistry*, a line in the palm of the hand supposed to indicate a person's fate in worldly affairs.
1889 K. ST. HILL *Gram. Palmistry* II. 43 When the Fate Line is tortuous..it is a sign of misfortune, or bad character. 1945 E. BOWEN *Demon Lover* 103 Dirt engraved the fate-lines in Mary's palms.

Fatha, Fatheh, varr. *FATIHAH.

father, *sb.* Add: **1. f.** Also *father-in-church.*
1871 MRS. H. WOOD *Dene Hollow* vi, 'I shall want you to stand father-in-church to this young lady,' said Geoffrey to the clerk.
g. Used colloquially by a wife addressing or referring to her husband.
1855 DICKENS *Dorrit* (1857) I. ii. 12 'Never mind, Father, never mind!' said Mrs. Meagles. 1917 D. CANFIELD *Understood Betsy* (1922) iii. 61 'It's gathering,' said Aunt Abigail... 'Father'll churn it a little more till it really comes.' 1960 C. WATSON *Bump in Night* iv. 40 If anyone was to blame it was Mr Biggadyke. He nearly lost father his job over that business.
h. *the father and mother of a*: used colloquially to indicate extreme severity, exceptionally large size, etc. Also (esp. *Austral.* and *N.Z.*), *the father of a*.
1892 KIPLING *Many Invent.* (1893) 45 It would ha' bin my duty..to give you the father an' mother av a beltin. 1892 —— *Lett. of Travel* (1920) 41 The father and mother of all weed-spuds. 1908 SOMERVILLE & 'ROSS' *Further Exper. Irish R.M.* viii. 197 There's been the father and mother of a row down there between old Sir Thomas and Hackett. 1930 *Bulletin* (Sydney) 3 Dec. 21/3 ''Twas the father of a lickin' I intended for ye, Skinny, for the way ye hoofed it awhile ago,' he said. 1938 N. MARSH *Artists in Crime* xiii. 202 You've led Bathgate into the father and mother of a row for talking out of school. 1947 'A. P. GASKELL' *Big Game* 119 Harry threatened to give him the father of a hiding. 1948 D. W. BALLANTYNE *Cunninghams* II. xi. 177 The local side got the father of a hiding. 1954 G. SMITH *Flaw in Crystal* 139 They've allowed me to arrange the father and mother of a credit overdraft with the banks. 1958 L. A. G. STRONG *Treason in Egg* vii. 143 There's the father and mother of a message on the paper. I can't understand it. 1960 *Punch* 13 July 47/2 The stage is set for the father and mother of a row.
i. In conjunctive phrases: *father-child, -daughter, -son.*
1949 M. MEAD *Male & Female* xvi. 326 A father-daughter household is not as disapproved [etc.]. 1958 *Listener* 21 Aug. 263/2 The father-son relationship is necessarily permanent. 1960 C. DAY LEWIS *Buried Day* vii. 142 Pathetic attempts to revive the old father-child relationship.
4. d. *like a father*: in a paternal, authoritative, or severe manner.
1830 J. K. PAULDING *Chron. Gotham* 64 If she wont listen to reason, I will talk to her like a father. 1922 JOYCE *Ulysses* 316 Talking to him like a father, trying to sell him a secondhand coffin. 1926 F. W. CROFTS *Insp. French & Cheyne Myst.* vi. 74 James talked to him like a father and he seemed to swallow it all down.
8. b. *Father Christmas*: the personification of Christmas as a benevolent old man with a flowing white beard, wearing a red sleeved gown and hood trimmed with white fur, and carrying a sack of Christmas presents. *Father Time*: see TIME *sb.* 25.
1658 J. KING (*title*) Examination and Tryall of Old Father Christmas. *a* 1800 in J. Brand *Pop. Antiq.* (1813) I. 373 Lordings, in these realms of pleasure Father Christmas yearly dwells. 1860 *Christmas Tree* 189 'Tis now, when once more from his lair Old Father Christmas issues forth. *Ibid.* 190 Hail, Father Christmas! Come, and bring Thine ancient merriment and glee. 1864 *Chambers's Bk. of Days* II. 740/2 Old Father Christmas, bearing, as emblematic devices, the holly bough, wassail bowl, etc. 1919 *Punch* 24 Dec. 538 Uncle James (who after hours of making up rather fancies himself as Father Christmas). 1940 *Economist* 20 Jan. 93/1 A Father Christmas State,..which weakly allowed its citizens to believe that wealth came from him not them. 1968 C. CHAUNDLER *Everyman's Bk. Anc. Customs* II. 180 December the sixth is the feast day of St. Nicholas,.. in Britain we usually call him Santa Claus, or Father Christmas.
9. c. (U.S. examples of *Father of waters*, etc.; in later use *spec.* the Mississippi.)
1763 tr. *Le Page du Pratz' Hist. Louisiana* I. II. i. 202 By some savages of the North it is called Meact-Chassipi, which literally denotes, the ancient Father of Rivers, of which the French have, by corruption, formed Missisipi. 1808 T. ASHE *Travels* III. xxxviii. 173 Throughout this great water, this Father of Floods, as the Indians calls [*sic*] it, in some places, islands are seen sinking into annihilation. 1813 *Niles' Weekly Reg.* V. Suppl. 176/2 The Mississippi is the Nile of America. The aborigines who resided on its banks, called it Mechaseba, or Father of waters. 1818 H. B. FEARON *Sk. Amer.* 257 The facilities of export afforded by those 'fathers of waters', the Ohio, Mississippi, and Missouri. 1836 J. HALL *Statistics of West* iii. 46 The traveler is struck with the magnitude..of the stream which has been so appropriately called, the Father of waters. 1878 B. F. TAYLOR *Between Gates* 23 Fox river, Rock river, Mississippi, the old Father of them all. 1917 J. F. DALY *Life A. Daly* 64 A voyage down the Father of Waters in war times. 1949 *New Orleans Times-Picayune Mag.* 27 Mar. 7 With summer approaching, the unique craft will start slipping down Father of Waters.
11. a. *father-symbol.* **b.** *father-sentiment.*
1920 T. P. NUNN *Education* xii. 146 The mother-sentiment appears, to be followed..by the father-sentiment. 1927 B. MALINOWSKI *Sex & Repress. Savage Soc.* 43 Soon the typical father-sentiment is formed, full of contradictory emotions. 1922 *Internat. Jrnl. Psychoanal.* III. 206 A totem is, of course, as we know from Freud, a father-symbol.
12. father-complex (see *COMPLEX *sb.* 3); **father-figure,** one who is regarded as having some of the characteristics of a father; also **father-image, -imago** (see *IMAGO); **father-fixation** [*FIXATION 3 b], a fixation on one's father; hence *father-fixated* adj.; **father-right** [G. *vaterrecht*], the supremacy of the father in a family in which descent follows the male line; **father-rule,** the rule of the father of a family as distinguished from the rule of the male relatives of the mother where descent follows the female line; patriarchy; **Father's Day** *U.S.*, 'A day for recognition of the respect and gratitude felt by children toward their fathers, commonly observed on the third Sunday in June. Father's Day was originated by Mrs. John Bruce Dodd of Spokane and proclaimed by the governor of Washington in 1910 but was not widely observed until twenty-five years later' (Webster 1950); more recently observed also in Britain; **father-substitute, -surrogate,** a person to whom the attachment of a child is directed in place of the father.
1916 C. E. LONG tr. *Jung's Coll. Papers Analyt. Psychol.* iii. 174 As a pious and obedient daughter..Sarah has brought about the usual sublimation and cleavage of the father-complex and on the one side has elevated her childish love to the adoration of God, on the other has turned the obsessive force of her father's attraction into the persecuting demon Asmodi. 1922 D. H. LAWRENCE *England, my England* 20 Let the psychoanalysts talk about father complex. It is just a word invented. 1950 J. STRACHEY tr. *Freud's Totem & Taboo* iv. 143 The same contradictory feelings which we can see at work in the ambivalent father-complexes of our children and of our neurotic patients. 1934 M. BODKIN *Archetypal Patterns in Poetry*, Index 338/2 Father-figure, -imago. 1955 D. HUDSON in *Forgotten King* (1960) 200, I was conscious of William Ewart Gladstone as a father-figure of looming.. import. 1957 *Ann. Reg. 1956* 168 President Eisenhower.. was also something of a 'father figure' whose personality inspired such confidence as often to serve as a substitute for a policy. 1958 *Sunday Times* 1 June 6/4 Samuel Beckett (a father-figure: his 'Murphy' (1938) constantly echoes in today's fiction). 1970 *New Society* 5 Mar. 396/3 Later this was changed to grandfather so that a lost father-figure could be hauled in by the back hairs as an excuse. 1961 *Times* 20 Sept. 16/5 The father-fixated daughter. 1930 *Amer. Jrnl. Psychiatry* 864 The reverse of course is equally true of the woman with a so-called father-fixation who marries not so much a husband as a father symbol. 1939 G. GREENE *Confid. Agent* I. ii. 86 I'm not romantic. This is what's called a father-fixation. 1937 *Harper's Monthly Mag.* Nov. 569/2 Mr. Martin questions whether the labours of the Shamans and witch doctors in creating the perfect 'father image' have not been a little overdone. 1956 E. L. MASCALL *Christ. Theol. & Nat. Sci.* vi. 217 As soon as he [*sc.* Freud] had convinced himself that the idea of God was the projection of a father-image in a wish-fulfilment. 1916 B. M. HINKLE tr. *Jung's Psychol. of Unconscious* I. iii. 55 Difficulties develop in the capacity for erotic expression, which may be reduced analytically to disturbances through a repressive attempt at resuscitation of the father image, or the 'Father-Imago'. 1923 E. JONES *Ess. Appl. Psycho-Analysis* 79 The figure of Polonius may be thus regarded as resulting from 'decomposition' of the paternal archetype... He is but a substitute for the step-father, i.e. a father *imago*. 1943 *Horizon* Oct. 252 Dramas of the soul which aspired either to destroy or reach beyond the traditional father-imago. 1899 F. W. MOORE tr. *Gumplowicz's Outl. Sociol.* I. 53 He [*sc.* J. Lippert] goes on to show the rising 'father right'. 1907 *Folk-Lore* June 245 The passage from motherright to fatherright. 1955 M. GLUCKMAN *Custom & Conflict in Afr.* iii. 73 The two contrasting types of kinship system, extreme father-right and extreme matriliny, are built up on the same principle: the link of mother to child. 1899 F. W. MOORE tr. *Gumplowicz's Outl. Sociol.* III. 112 It is recognized that mother-rule everywhere gave place to father-rule. 1943 *Greeley* (Colo.) *Daily Tribune* 5 June 5/5 Governor Vivian proclaimed Sunday, June 30, as Father's Day. 1956 A. HUXLEY *Adonis & Alphabet* 169 Mother's Day and, despite the growing absurdity of poor Poppa, Father's Day were instituted. 1963 *New Yorker* 15 June 82 And, what an inspired idea for Father's Day! 1938 F. L. LUCAS *Delights of Dictatorship* ii. 46 The hysteria that can find in the Head of that State [*sc.* the U.S.S.R.] a dream-husband, a father-substitute,..a God. 1949 E. COXHEAD *Wind in West* vi. 156 He had almost detached himself from Mother... Transference to a father-substitute was plainly the next stage. 1950 J. STRACHEY tr. *Freud's Totem & Taboo* iv. 148 The totem may be the *first* form of father-surrogate. 1957 N. FRYE *Anat. Criticism* 165 The fury with which these characters are baited..shows that they are father-surrogates.

fatherhood. Add: **1. e.** The position of being the 'father' or oldest member of a society, esp. of the House of Commons.
1899 *Daily News* 24 Apr. 7/3 The House decided informally that the Fatherhood rested with whoever had been longest in the House. 1907 *Westm. Gaz.* 23 May 2/3 The 'Fatherhood' of Sir Henry Campbell-Bannerman.

fathering, *vbl. sb.* Delete † *Obs.* and add later examples.
1894 'G. EGERTON' *Discords* 60 The want-wit inconsistency that forgives the man that begat the brat and treats with pitiless scorn the helpless result of his fathering. 1894 *Athenæum* 24 Feb. 238/3 The fathering of bairns. 1894 G. DU MAURIER *Trilby* III. VII. 19 Taffy.. was equal to any burden or responsibility all this instinctive young fathering might involve. 1903 *Daily Chron.* 16 Oct. 3/4 There was too much fathering. The settlers were fathered by their priests, fathered by their King. 1920 H. G. WELLS *Outl. Hist.* 122 The divine fathering and birth of Amenophis III. 1969 J. ROWE *Yours by Choice* (rev. ed.) vii. 54 Some people..look upon the fathering of an out-of-wedlock child as the 'wild oats' normally sown by young men about town.

fa·therlandless, *a.* [tr. G. *vaterlandslos.*] Unpatriotic.
1898 *Daily News* 11 July 3/4 A Conservative Deputy at Dirschau lately called five Liberal electors who voted for the Polish candidate 'Vaterlandlose gesellen' (unpatriotic 'Fatherlandless' fellows). 1915 T. F. A. SMITH *Soul of Germany* 274 Every Socialist, in my opinion, means an enemy to Empire and Fatherland. They are the Fatherlandless enemies of the divine order of things.

fathership. Add: **b.** = *FATHERHOOD 1 e.
1899 *Westm. Gaz.* 28 Dec. 3/3 The 'Fathership' falls upon Lord Templemore. 1901 *Ibid.* 22 Apr. 1/1 The successor of the late Mr. Villiers in the 'fathership' of the House of Commons. 1910 *Ibid.* 16 Feb. 2/3 Mr. Balfour.. lost his chance of the 'Fathership' by a like accident.

fathogram (fæ·ðogræm). [f. *FATHO(METER + -GRAM.] A tracing, made by an echo-sounder, representing the varying depth of water beneath a moving vessel.
1950 *Internat. Hydrogr. Rev.* XXVII. 56 If both vessels operate their echo sounders..the depth record may be almost impossible to interpret. Diagonal streaks across the fathogram are an indication of interference. 1953 J. Y. COUSTEAU *Silent World* xii. 124 My mind leaped back to the recording of the deep scattering layer. The coincidence of the whale's lunch and the lines drawn on the fathogram may have been entirely fortuitous. 1966 SHEPARD & DILL *Submarine Canyons* ii. 19 The inverted V that often appears on a fathogram below the crossing of the canyon profile may indicate the true depth of the canyon floor.

fathom, *sb.* Add: **3. b.** (See quot.)
1968 *Guardian* 26 Apr. 1/1 The fathom..is to disappear from British Admiralty charts. In future sea depths will be marked in metres.
6. fathom-fish *western N. Amer.* = OOLAKAN.
1849 A. ROSS *Adv. Oregon River* (1904) vi. 109 To prepare them [*sc.* the ulichans] for a distant market, they are laid side by side, head and tail alternately, and then a thread run through both extremities links them together, in which state they are dried, smoked, and sold by the fathom, hence they have obtained the name of fathom-fish. 1897 E. COUES *New Light Greater Northwest* II. 787 Another name of these 'smelts' was fathom-fish, given because they were strung on strings and sold by the fathom.

fathomable, *a.* **b.** *fig.* (Later example.)
1896 *Daily News* 27 May 6/6 No doubt, if the mystery of these rays is fathomable, he will fathom it.

Fathometer (fæðǫ·mɪtəɹ). Also **fathometer.** [f. FATHO(M + -METER.] An instrument for the determination of the depth of the sea by measuring the time taken by a sound-wave to reach the bottom and return.
1925 *Hydrographic Rev.* II. II. 140 (*heading*) The 'Fathometer' of the Submarine Signal Co. of Boston. This apparatus, also based on the same principle as that of Fessenden, differs..by the particular feature only, *viz.* the angle to be measured is observed by means of a lighted index..which is illuminated when it receives the reflected acoustic signal. 1927 S. H. LONG *Navigational Wireless* xi. 149 Another method of determining the sounding of a ship consists of a piece of apparatus..called the 'Fathometer'. This instrument is based on the principle of emitting a sound wave from a sound transmitter, causing this wave to be reflected from the sea-bed, and receiving the reflected sound wave on a receiver. 1937 *Geogr. Jrnl.* XC. 543 Her echo-sounding device (known in the U.S.A. as the Fathometer). 1971 *Islander* (Victoria, B.C.) 14 Mar. 10/3 The speedometer and log was helpful in establishing position and the fathometer very useful when entering an unfamiliar anchorage.

fatigability (fætigăbi·lĭti). Also **fatiguability** (fătī·g-). [f. FATIGABLE: see -ILITY.] Susceptibility to fatigue.
1908 W. MCDOUGALL *Introd. Soc. Psychol.* iv. 119 Differences in respect to fatigability and rapidity of recuperation. 1919 J. A. HADFIELD in B. H. Streeter *Spirit* III. 79 A man's fatiguability is tested by tying a weight to his finger [etc.]. 1964 L. MARTIN *Clinical Endocrinol.* (ed. 4) iii. 112 Some degree of muscular weakness and increased fatiguability exists in most cases of thyrotoxicosis.

fatigue, *sb.* **1. b.** Substitute for def.: The condition of weakness in metals or other solid substances caused by cyclic variations in stress. Now esp. as *metal fatigue*. (Examples of *attrib.* and *Comb.* uses.)
1905 Fatigue test [see *4]. 1913 *Sci. Amer.* Suppl. 13 Dec. 372 (*heading*) A recently installed fatigue testing machine. 1937 *Discovery* July 194/2 The fatigue resistance of some metals. 1949 *Jrnl. R. Aeronaut. Soc.* LIII. 788/2 A ray of hope that non-ferrous alloys could be developed which would show similar features in lack of 'fatigue-memory' to the ferritic steels. 1955 *Times* 8 July 5/1 Her Royal Highness drove across the airfield to see a Bristol Britannia airliner undergoing metal fatigue tests in a water tank. 1966 *McGraw-Hill Encycl. Sci. & Technol.* II. 284/2 The only examples of a fatigue fracture in bone are those seen usually in military training, ..in which a soldier not previously used to repetitive exercise is forced to march 30 or 40 miles in a single day.
c. *Physiol.* A condition of muscles, organs, or cells characterized by a temporary reduction in power or sensitivity following a period of prolonged activity or stimulation.
1872 J. H. BENNETT *Text-bk. Physiol.* ii. 83 (*heading*) Muscular fatigue. 1876 *Kirkes' Hand-bk. Physiol.* (ed. 9) xix. 586 *Fatigue of muscle.*—A muscle becomes rapidly exhausted from repeated stimulation. 1888 M. FOSTER *Text Bk. Physiol.* (ed. 5) I. 149 The sense of fatigue of which, after prolonged or unusual exertion, we are conscious in our own bodies..cannot be taken as an adequate measure of the actual fatigue of the muscles. 1891 A. D. WALLER *Introd. Human Physiol.* xii. 448 Negative after-images are the most obvious sign of retinal—or properly speaking—of retino-cerebral fatigue. 1932 S. DUKE-ELDER *Text-bk. Ophthalmol.* I. xxiv. 969 The lowered visual acuity which is readily demonstrable on continued observation..is due to fatigue in mental or even general bodily processes...Of retinal fatigue there seems to be no evidence. 1949 E. G. WEVER *Theory of Hearing* xii. 322 The specificity of the fatigue effect..is studied by exposing the ear to one tone and then testing with various other tones above and below. 1966 *Lancet* 12 Mar. 585/2 True muscular fatigue and what, for want of a better term, must be called tiredness are plainly different.
†**d.** *Mech.* *fatigue of elasticity*: a decrease in the elasticity of a material after a long period or repeated applications of stress, followed by a gradual recovery after the stress is removed; also *elastic fatigue*. *Obs.*
1877 W. THOMSON in *Encycl. Brit.* VII. 802/1 Experimental exercises..brought to light some very remarkable and interesting results,..showing also a very remarkable fatigue of elasticity, according to which a wire which had been kept vibrating for several hours or days through a certain range came to rest much quicker when left to itself than when set in vibration after it had been at rest for several days and then immediately left to itself. 1895 *Proc. R. Soc.* LVIII. 136 In every case, overstrain has produced a like fatigue of elasticity, and elastic recovery has followed during an interval of some days or weeks of rest. 1899 *Nature* 6 July 239/2 Experiments on an iron wire..showed distinct fatigue of elasticity. 1899 W. WATSON *Text-bk. Physics* xix. 205 This phenomenon is referred to as elastic fatigue. 1920 A. E. H. LOVE *Treat. Math. Theory Elasticity* (ed. 3) iv. 117 An analogous property of bodies is that to which Lord Kelvin has called attention under the name 'fatigue of elasticity'. 1922 *Proc. Physical Soc.* XXXV. 165 In his own experience of electrometers elastic fatigue appeared to be due mainly to the attachment at the two ends of the suspension wire.
e. The reduction of the efficiency of an (esp. luminescent) material during use or exposure.
1904 *Philos. Trans. R. Soc.* A. CII. 443 At pressures below, say, a tenth of an atmosphere, zinc showed little or no signs of fatigue when illuminated by ultra-violet light. 1949 G. F. J. GARLICK *Luminescent Materials* vii. 187 The saturation of the screen material by electrons at low voltages..results in a fatigue of the material.
3. b. (Earlier and later examples in sense 'fatigue-dress'.)
1836 J. HILDRETH *Dragoon Campaigns Rocky Mts.* I. vii. 51 We have not yet received our uniforms..but even in our 'fatigues', we make an imposing appearance when mounted. 1966 D. F. GALOUYE *Lost Perception* viii. 85 Clad in ill-fitting and occasionally torn fatigues, the soldiers were a vivid contrast to the flawlessly uniformed Guardsmen. 1967 *Boston* (Mass.) *Herald* 5 Apr. 59/3 The generals..in their 'pressed fatigues and shining stars'. 1970 *Daily Tel.* 6 Oct. 4/4 A Palestinian major, dressed in jungle green fatigues.
4. *fatigue-cap* (examples), *frock, man, trousers*; **fatigue limit,** the (presumed) value of the fatigue strength for an infinite number of cycles of stress; also called *endurance limit*; **fatigue products, stuff, substances** [tr. G. *ermüdender Stoff, ermüdende Substanz* (J. Ranke *Tetanus* (1865) xiv. 329, xviii. 450)], an accumulation of toxic material formerly held to result from excessive muscular activity and to be the cause of fatigue; **fatigue range,** the maximum variation in stress that a material can be subjected to for a large or infinite number of cycles of stress without its failing from fatigue; **fatigue strength,** half the value of the fatigue range of a material for a given (large) number of cycles of stress, esp. when the mean stress is zero; also called *endurance strength*; **fatigue test,** a test carried out on a material or part to determine its behaviour under cyclic variations of stress and esp. its fatigue strength; also as *vb.*; so *fatigue testing* vbl. sb.; **fatigue testing machine,** a machine for performing fatigue tests; **fatigue-work** (earlier example).
1861 R. A. WISE *Let.* 9 Jan. in *Wm. & Mary Coll. Q.* 2nd Ser. XVIII. 197 Their uniform is to be home spun pantaloons and a red flannel shirt and a fatigue cap. 1914 E. A. POWELL *Fighting in Flanders* vi. 139 Small, round, visorless fatigue-caps. 1803 M. LEWIS *Jrnl.* (1905) VIII. 233, 20 Fatigue Frocks or hunting shirts. 1847 in H. Howe *Hist. Coll. Ohio* 490 It was at night..and they [*sc.* the soldiers] were dressed in fatigue frocks. 1911 *Proc. Inst. Mech. Engin.* Oct. 884 As far as he could ascertain, there was no relationship between fatigue limits and the ordinary elastic limits. 1959 *Chambers's Encycl.* IX. 326/2 When a metal is subjected to many millions of cycles of alternating or vibrating stress, below the elastic limit, but above the so-called 'fatigue' limit, it eventually breaks by 'fatigue'. 1774 J. ANDREWS *Let.* 28 Sept. (1886) 57 A number of *fatigue* men with hand barrows. 1871 L. W. M. LOCKHART *Fair to See* III. xxxi. 49 *Laissez-aller* hospital orderlies and sluggish 'fatigue-men'. 1957 M. K. JOSEPH *I'll soldier no More* (1958) xiii. 255 Tino and a couple of fatigue-men have sweated in the cookhouse. 1909 *Westm. Gaz.* 8 June 9/2 To eliminate from the muscles what pathologists know as 'fatigue products'. 1935 WINTON & BAYLISS *Human Physiol.* (ed. 2) i. 17 It was at one time supposed that a specific physiological mechanism was at the root of the state of fatigue, which was attributed to the action of products of activity, 'fatigue products', circulating in the blood. 1922 GLAZEBROOK *Dict. Appl. Physics* I. 179/2 A determination of the fatigue range occupied a very long time. 1954 L. E. BENSON in W. R. Osgood *Residual Stresses in Metals* 81 With mild steel..the upper limit of the fatigue range of stress (with zero mean stress) is approximately equal to the yield point. 1912 *Min. Proc. Inst. Civil Engin.* CLXXXVIII. I. 10 There only remained the investigation of the effects of the high speed on the fatigue strength of the specimens. 1959 CRANDALL & DAHL *Introd. Mech. Solids* v. 221 The fatigue strength of 2024-T4 aluminium alloy, when tested as unnotched bars, is about 30,000 psi for 10^6 cycles and 24,000 psi for 10^7 cycles. 1971 *Engineering* Apr. 38/1 These bronzes..have a very good fatigue strength. 1885 W. STIRLING tr. *Landois' Text-bk. Human Physiol.* II. x. 667 The cause of fatigue is probably the accumulation of decomposition products, 'fatigue stuffs' in the muscular tissue. 1905 *Jrnl. Iron & Steel Inst.* LXVII. 488 It is not easy to definitely compare the tensile and fatigue tests in this case. 1955 Fatigue test [see *1 b]. 1959 SINES & WAISMAN *Metal Fatigue* vii. 167 Any part design which differs greatly from standard practice should be carefully fatigue-tested before it is adopted for service. 1905 *Jrnl. Iron & Steel Inst.* LXVII. 486 The fatigue testing-machines were of Wöhler's cantilever type. 1908 *Ibid.* LXXVI. 70 The chief object of fatigue-testing was to discover the existence or non-existence of a relation between the elastic properties of a material and its resistance to fatigue. 1813 *Niles' Weekly Reg.* III. 295/2 Fatigue trowsers. 1846 *Aide Mémoire Mil. Sci.* I. 521 Fascine pickets are..usually considered fatigue-work.

fatigue, *v.* Add: **1. b.** To induce a condition of fatigue in (a muscle, organ, etc.): see *FATIGUE *sb.* 1 c.
1872 J. H. BENNETT *Text-bk. Physiol.* ii. 83 The muscles of a strong man are not so easily fatigued as those of a weak one. 1885 W. STIRLING tr. *Landois' Text-bk. Human Physiol.* II. x. 668 Loads may be suspended to perfectly passive muscles without fatiguing them. 1896 W. P. LOMBARD in W. H. Howell *Amer. Text-bk. Physiol.* ii. 97 It is doubtful whether nerves are fatigued by the process of conduction. 1954 WEVER & LAWRENCE *Physiol. Acoustics* ix. 157 The ear is strongly fatigued with a tone of a certain frequency..and then is stimulated with two primary tones of adjacent frequencies.
2. b. *trans.* To weaken by the application of a periodically varying stress; to induce fatigue in (*FATIGUE *sb.* 1 b, d).
1899 J. A. EWING *Strength of Materials* iii. 56 A piece which has been fatigued by many variations of stress. 1905 *Jrnl. Iron & Steel Inst.* LXVII. 491 A broad face of a specimen of oblong section is polished, and the specimen is then fatigued. 1925 TIMOSHENKO & LESSELS *Appl. Elasticity* xvi. 464 The material, after being fatigued, showed an increase in angular strain when frequency of alternation was decreased. 1962 A. J. KENNEDY *Processes Creep & Fatigue in Metals* v. 298 The fracture hardness was lower for copper specimens fatigued at 20×10^3 p.s.i. than for those fatigued at 16×10^3 p.s.i.

fatigued, *ppl. a.* Add: **c.** Worn, shabby. Also *fig.*
1774 F. BURNEY *Diary* 20 Feb. (1907) I. 285 These two subjects he wore thread-bare; though indeed they were pretty much fatigued, before he attacked them. 1894 *Daily News* 22 Jan. 5/1 Preferring a fatigued brown calf binding to a new suit of leather for 'Tom Jones'. 1897 *Ibid.* 20 Nov. 8/3 The renovation of a dress that has seen some wear, and consequently has that fatigued appearance.
d. Affected with, or characterized by, a condition of fatigue (*FATIGUE *sb.* 1 b, c, d).
1853 W. BRINTON tr. *Valentin's Text Bk. Physiol.* xv. 381 It is probable that, even during life, a fatigued muscle is more yielding than when fresh and vigorous. 1877 M. FOSTER *Text Bk. Physiol.* 65 A muscle, even within the body, after prolonged action is fatigued. 1885 W. STIRLING tr. *Landois' Text-bk. Human Physiol.* II. x. 667 In order to obtain the same amount of work from a fatigued muscle, a much more powerful stimulus must be applied to it. 1899 J. A. EWING *Strength of Materials* iii. 56 A period of rest in a 'fatigued' piece tends to restore elasticity. 1931 E. H. SALMON *Materials & Structures* I. xviii. 540 It was concluded that, as a result of mere repetition of loading, a material became less capable of resistance, or, as it was termed, *fatigued*. 1949 E. G. WEVER *Theory of Hearing* xii. 324 The fatigued ear resembles an ear that is suffering from nerve deafness. 1967 J. Y. MANN *Fatigue of Materials* iii. 13 The relative areas of the fatigued zone and the zone of final static failure..can provide information relating to the general stress level on the plane of fracture.

‖ **Fatiha(h)** (fā·tiha). Also **Fatha, Fatheh, Fattah, Fattha.** [Arab. *fātiḥa, fatḥa* opening, f. *fataḥa* to open.] The short first sura of the Koran, used by Moslems as a prayer.
1821 G. F. LYON *Narr. Trav. N. Afr.* ii. 74 Some one of Mukni's men cries out 'The Fattha!' (or first chapter of the Koran), every one joining in that prayer. *Ibid.* iii. 94 We sometimes met people on our road, who invariably at parting recited the Fatha. 1824 J. MORIER *Hajji Baba* III. i. 20 The *Fatheh* (the first chapter of the Koran) was repeated by all present. 1863 *Chambers's Encycl.* V. 818/1 Immediately after the introductory fattah or exordium, follows the longest chapter. 1883 E. ARNOLD *Pearls of Faith* xlv, 'Gabriel! why stay'st thou me?'..'Since at this hour the Fâtihah should be read.' 1916 J. R. HARRIS *Testimonies* I. v. 41 He begins his discourse with an imitation of the Fatha, or opening chapter of the Koran. 1959 *Chambers's Encycl.* VIII. 255/2 The *suras* do not stand in historical order. The first, the Fatiha, is a short prayer, much used by the Moslems.

Fatimite (fæ·timəit), *a.* and *sb.* Also **Fat(h)emite, Fathimite, Fatimid(e.** [f. Arab. *Fāṭima* (see below) + -ITE^1.] **A.** *adj.* Descended from Fatima, the daughter of Mohammed by his first wife, Khadija. **B.** *sb.* A descendant of Fatima and her husband, Ali; a member of the Arabian dynasty which ruled in parts of northern Africa from A.D. 908 to 1171, and during some of that period in Egypt and Syria. Also *attrib.*
1727–38 CHAMBERS *Cycl.*, Fathimites, or Fathemites. 1797 *Encycl. Brit.* III. 4/2 He [*sc.* Obeidallah] pretended to be descended in a right line from Fatema..for which reason..the Arabs called him and his descendants Fatemites. *Ibid.*, Abul Kasem, son to the Fatemite khalif Al Mohdi. *Ibid.* VII. 177/1 Fathemites, Fatemites, or Fathimites. 1838 *Penny Cycl.* X. 207/1 Obeidallah, the first Fatimide caliph. 1872 H. A. RAWES *God in His Works* iii. 55 The kingdoms of the Fatimites. 1883 *Encycl. Brit.* XVI. 587/1 With the reign of Moktadir is connected one of the greatest events in the history of the Caliphate, the foundation of the Fâtimite dynasty. 1897 *Daily News* 8 Apr. 8/1 In 1072 the Holy City was possessed by the Fatimite Caliph of Egypt. 1902 *Daily Chron.* 20 June 4/2 The Fatimid Caliphate. 1940 *Burlington Mag.* Nov. 152/1 There seems to be no connection between Fatimid art and the miniature painting of the Mamluks. 1951 S. RUNCIMAN *Hist. Crusades* I. I. ii. 31 He realized the Fatimites were more dangerous enemies than their Abbasid rivals. *Ibid.* v. i. 267 The coast line from the middle Lebanon to the Fatimid frontier.

fatiscent, *a.* (Later examples.)
1916 B. D. JACKSON *Gloss. Bot. Terms* (ed. 4) 144/1 *Fatiscent*, cracked, or gaping open. 1957 SNELL & DICK *Gloss. Mycology* 57/2 *Fatiscent*, disappearing; breaking up.

fatling, *sb.* Add: Also *transf.*, of a person.
a 1861 T. WINTHROP *John Brent* (1883) vii. 58 'Well, boys!' said the unpleasant fatling, approaching again.

fatness. 1. c. Delete '†*Obs.* exc. in Biblical phraseology', and substitute: *Obs.* or *arch.* exc. *U.S.* (Later examples.)
1814 J. TAYLOR *Arator* (ed. 2) 172 This [*sc.* using bottom lands for tillage and grazing] is owing to the extreme fatness of the soil. 1849 E. BRYANT *California* ii. 23, I never saw a soil indicative of a higher degree of fatness. 1887 MORRIS *Odyss.* XI. 93 Then about his vineyard's fatness where the mother of wine doth abound. 1891 HARDY *Tess* II. III. xxiv. 34 The oozing fatness and warm ferments of the Var Vale.

Fatso (fæ·tso). *slang.* [presumably f. pl. of *fat* sb., fat person + -O.] Used humorously or derisively as a nickname for a fat person.
1944 R. D. BOYER in P. Gammond *Duke Ellington* (1958) I. 28 The other card players looked over and in succession asked, 'How's it going, Pops?' 'Getting anywhere, Sandhead?' and 'You sending 'em, Fatso?' 1962 L. DEIGHTON *Ipcress File* v. 32, I began to envy Fatso his sausage sandwiches. 1964 E. LACY *Pity Honest* iv. 97 Fatso glanced at me with disgust.

Fattah, Fattha, varr. *FATIHAH.

fatty, *a.* Add: **6.** *fatty series*: the open-chain group of carbon compounds; the aliphatic series.
1872 *Jrnl. Chem. Soc.* XXV. 682 A method of forming the nitro-compounds of the fatty series. 1926 H. G. RULE tr. *Schmidt's Textbk. Org. Chem.* I. 69 The numerous compounds classified under this title [*sc.* aliphatic or fatty compounds] may be regarded as derived from the hydrocarbon methane... Since the common animal and vegetable fats similarly fall under this heading, the whole series is frequently known as the aliphatic or fatty series.
B. *sb.* A fat person: esp. as a nickname.
1797 B. WYNNE *Diary* 3 Mar. (1937) II. x. 170 His wife is a poor miserable thing, the daughter a good looking fatty. 1882 R. RICHARDSON *Trav. Medit.* II. xix. 266 Well-built fatties, with double mouldings in the neck and chin. 1891 FARMER *Slang*, *Fatty*, a jocular epithet for a fat man; a comic endearment for a fat woman. 1919 *Holiday Ann. 1920* 120/1 Fatty Wynn was busy... Arthur Augustus..wondered whether Fatty, engrossed by the cake, would think of looking out on the platform. 1960 V. NABOKOV *Invitation to Beheading* x. 101 Cincinnatus..followed every step of the nimble fatty. 1971 *Petticoat* 17 July 32/1 Success stories connected with slimming are few and far between, so any fatties who might be reading this—take note of this tale!

fatuoid (fæ·tiuoid). [ad. mod.L. *fatua* (H. Nilsson-Ehle 1921, in *Hereditas* II. 401), f. L. *fatuus* (see FATUOUS) + -OID.] A variant form of cultivated oat, partially resembling the wild oat (*Avena fatua*).
1922 C. V. B. MARQUAND in *Bull. Welsh Plant Breeding Station* C. II. 4 (*heading*) The 'wild base' type, or 'Fatuoid-mutation' of Cultivated Varieties. 1930 *Times Lit. Suppl.* 30 Oct. 894/4 Fatuoids are forms resembling the wild oat that appear occasionally in cultivated varieties of oat. 1957 *Encycl. Brit.* XVI. 665/1 Fatuoids or false wild oats have been observed in nearly all cultivated varieties. Fatuoid kernels also drop to the ground as in the case of wild oats.

faucalize (fǭ·kăləiz), *v.* [f. FAUCAL *a.* + -IZE.] *trans.* To add a faucal element to (a sound).
1919 H. H. JOHNSTON *Comp. Stud. Bantu & Semi-B. Lang.* I. ii. 36 The slurring of *g* or the faucalizing of *w.* 1920 *Man* XX. 14 How do you faucalise *w*?

Faulknerian (fǭknīə·riăn), *a.* [-IAN.] Of or pertaining to (the writings of) William Faulkner (1897–1962), American novelist.
1951 *Yale Rev.* XLI. II. p. viii/2 Another book with Faulknerian echoes. 1960 *Encounter* Aug. 74/1 A Faulknerian family disaster. 1961 *Times* 12 Jan. 15/5 Mr. Faulkner's style is as Faulknerian as ever. 1966 S. MARCUS *Other Victorians* i. 17 The prevailing tone of this book..is resonant of danger, doom, and disaster (to strike a Faulknerian note).

fault, *sb.* Add: **1. c.** Delete †*Obs.* and substitute: *arch.* Also without *for* (after F. *faute de*). (Later *poet.* examples.)
1867 EMERSON *Poems* (1914) 252 And, fault of novel germs, Mature the unfallen fruit. 1874 SWINBURNE *Bothwell* (1882) II. xiv. 195 Though it be evil made for fault of time.
3. d. (Later example.)
1894 *Times* (Weekly ed.) 22 June p. iii/4 A first folio Shakespeare, imperfect, and sold with all faults, realized £255.
8. c. *at fault*: now the usual form, rather than *in fault*. (Later examples.)
1935 *Discovery* July 206/1 Sometimes it was the glaze.. which was at fault. 1947 MULGAN & DAVIN *Introd. Eng. Lit.* xiii. 167 The natural presumption is that it is not his already proved skill that is at fault but rather the nature of the theories.
9. Also, a flaw or dislocation in ice.
1860 [in Dict.]. 1925 N. E. ODELL in E. F. Norton *Fight for Everest, 1924* 314 That the ice above the Trough was undergoing severe stress was apparent not only from its foliation, but also from the faults—'flaws' or 'tear-faults'—that had been extensively developed in this area.
11. *fault-fissure, -plane, zone*; **fault-block,** a mass of displaced rock (sometimes of extensive area) bounded by or between two faults; **fault breccia,** breccia resulting from movement along a fault, a *crush-breccia or *fault-rock*; **fault-line scarp,** a scarp produced secondarily along a fault-line by erosion, etc.; **fault-line valley,** a valley along a fault-line; **fault-scarp,** a scarp directly produced by throw at a fault; **fault-vein,** a mineral vein filling a fault; **fault-vent,** a volcanic vent occurring at a fault.
1897 W. B. SCOTT *Introd. Geol.* xiii. 248 If two parallel dislocations hade toward each other, they form a trough fault and include a wedge-shaped fault block. 1925 N. E. ODELL in E. F. Norton *Fight for Everest, 1924* 292 At one point in the gorge a fault block had been turned through 90°. 1968 W. A. PRICE in R. W. Fairbridge *Encycl. Geomorphol.* 794 Fault-block origin for the Old Crow lakes is supported by the large square drainage patterns with which the lakes are associated. 1891 *Q. Jrnl. Geol. Soc. Lond.* XLVII. 322 An interesting feature in the Packhouse Hill section is a metamorphosed fault-breccia, which intervenes between the Lower and Middle Coldwell beds. 1969 C. R. LONGWELL et al. *Physical Geol.* xvii. 413 Rocks adjacent to a fault can be crushed into irregular pieces, forming fault breccia. 1855 J. R. LEIFCHILD *Cornwall Mines* 87 In some instances..the fault-fissures are open. 1903 A. GEIKIE *Text-bk. Geol.* (ed. 4) 372 Another remarkable instance of the..fault-fissure type was furnished by the great Japanese earthquake of 28th October 1891. 1911 *Encycl. Brit.* XXVII. 622/1 The removal of the weak strata along one side of the fault line, leaving the harder strata on the other side in relief; such scarps are known as 'fault-line scarps'. 1935 *Geogr. Jrnl.* LXXXVI. 362 They may mistake fault-line scarps for fault scarps. 1968 R. W. FAIRBRIDGE *Encycl. Geomorphol.* 349/2 A fault-line scarp..is a cliff that is subparallel to the fault trace, but is so modified by erosion that it does not reflect the fault plane in any way. 1913 W. M. DAVIS in *Bull. Geol. Soc. Amer.* XXIV. 207 They should be called by some special name, such as fault-line valleys, as here suggested. 1963 A. N. STRAHLER *Earth Sciences* xxxiv. 598 Streams occupying fault lines are classed as subsequent streams... Their valleys may be termed subsequent valleys, or, more specifically, fault-line valleys. 1889 O. FISHER *Physics of Earth's Crust* (ed. 2) xx. 263 A 'fault-plane' or 'thrust-plane'. 1900 *Geogr. Jrnl.* XVI. 466 Inclined fault-planes with down-throw towards one trough. 1913 *Bull. Geol. Soc. Amer.* XXIV. 166 A fault surface is the surface of fracture; it is rarely plane, but where it is without notable curvature over any area it may be called a fault plane for that area. 1939 *Proc. Prehist. Soc.* V. 107 A spring issuing from the Chalk at the fault-plane. 1897 W. B. SCOTT *Introd. Geol.* xiii. 248 It is comparatively seldom that the upthrow side of a fault is left standing as a line of cliffs; when such is the case, the cliffs form a fault scarp. 1960 L. D. STAMP *Britain's Struct.* (ed. 5) xxiii. 229 One of the finest examples of a fault-scarp to be found anywhere—the great wall of the Highlands overlooking a valley excavated in Old Red Sandstone rocks. 1886 J. PRESTWICH *Geol.* I. xviii. 309 Mineral veins may be divided into fissure-veins, fault-veins, and..quartz-veins. 1903 *Nature* 3 Sept. 413/2 During the geological periods when the fault-vent continued intermittently active. 1931 C. M. NEVIN *Princ. Struct. Geol.* iv. 78 Frequently, the displacement caused by faulting may be spread over a number of closely spaced adjacent fractures, instead of being confined to one surface, in which case a fault zone is formed. 1944 A. HOLMES *Princ. Physical Geol.* vi. 79 Sometimes, instead of a single fracture, there are two or more, forming a strip consisting of a sheet of crushed rock of variable thickness. This is distinguished as a fault zone, and the shattered material within is called a fault breccia. 1970 *Encycl. Brit.* VII. 861/1 The bottom of the well was located in a fault zone.

fault, *v.* **7.** *trans.* Delete 'Somewhat *rare*' and add later examples.
1957 *Listener* 9 May 736/1 One can fault the English for their indifference over long periods to Scottish susceptibilities. 1965 *New Statesman* 30 Apr. 670/2 The Hall Green Conservatives can be faulted for selecting him. 1971 *Wall Street Jrnl.* (Eastern ed.) 19 July 1/3 The.. report...faults Agriculture Secretary Clifford Hardin for failing to take over state meat-inspection services that have failed to satisfy federal requirements.
b. Delete *rare* and add further examples.
1958 *Spectator* 6 June 740/3 His final conclusion at least cannot be faulted. 1961 *Times* 11 May 17/2 Perhaps some scholar will fault a detail.
9. Also *intr.*
1927 *Daily Express* 5 July 11/4 A geological inexactitude which appears to have faulted or slipped from peaceful churchyard surroundings to a boisterous life in the ocean waves.

faultage (fǭ·ltėdʒ). *Geol.* [f. FAULT *sb.* 9 + -AGE.] Faults considered collectively, faulting.
1899 *Geogr. Jrnl.* XIII. 272 Well-marked lines of faultage going down to the bases of the mountains.

faun. Add: **2.** *attrib.* and *Comb.*, as *faun-breeze, -face; faun-faced, -like, -loved, -twinkling* adjs.
1925 E. SITWELL *Troy Park* 44 The elegant faun-breeze. *a* 1930 D. H. LAWRENCE *Etruscan Places* (1932) 16 His face is a faun-face, not deadened by morals. *Ibid.* 17 They can't survive, the faun-faced men. 1902 W. B. YEATS *Celtic Twilight* 75 The faun-like feet. 1962 I. MURDOCH *Unofficial Rose* xxxiv. 325 That particular faun-like grace which fades later. 1881 WILDE *Poems* 69 Faun-loved Heliconian glades. 1929 BLUNDEN *Near & Far* 64 Sunshine of vital gold, faun-twinkling groves.

faunally (fǭ·năli), *adv.* [f. FAUNAL *a.*^{2} + -LY^2.] As regards the fauna of a district or epoch.
1872 *Amer. Naturalist* VI. 342 These two regions are as diverse faunally as they are in physical features. 1904 *Science* 5 Feb. 235/1 Faunally the same species characterize the lower and upper members of the Portage. 1949 *New Yorker* 17 Sept. 24 For the last hundred years, faunally and florally speaking, all has been silence.

faunizone (fǭ·nizōᵘn). *Geol.* [f. FAUN(A + -I- + ZONE *sb.*] A zone (cf. ZONE *sb.* 7) characterized by a particular assemblage of fossils.
1902 S. S. BUCKMAN in *Geol. Mag.* IX. 557 Faunizones are the successive faunal facies exhibited in strata.

Fauntleroy (fǭ·ntlėroi). The name of the hero of Frances Hodgson Burnett's novel *Little Lord Fauntleroy* (1885), used *attrib.* to designate the style of dress or hair (for boys) which the book popularized; also applied, often ironically, to a child of the gentle nature of the book's hero. (The phr. *Little Lord Fauntleroy* is also used.)
1911 L. STONE *Jonah* II. i. 149 A Fauntleroy velvet suit. 1913 F. H. BURNETT *T. Tembarom* xiii. 153 Who's this Fauntleroy in the lace collar? 1923 [see *AFTER-LIFE 2]. 1929 T. WOLFE *Look Homeward, Angel* (1930) viii. 93 Fat Fauntleroy curls. 1933 *N. & Q.* 18 Mar. 193/1 Lace collars were sometimes worn with velvet suits before..1886. Mrs. Hodgson Burnett is understood to have invented the Fauntleroy costume by adding a sash and lace collars and cuffs to these suits. 1933 DYLAN THOMAS *Let.* Oct. (1966) 35 Myself aged seven—thick-lipped, Fauntleroy-haired. 1942 D. POWELL *Time to be Born* (1943) viii. 181 Some little Lord Fauntleroy who had just found out there were rotters in the world. 1950 A. WILSON *Such Darling Dodos* 34 The creature would be dressed in a velvet suit and fauntleroy collar. 1959 *Sunday Times* 9 Aug. 3/7 Brixtonites could not imagine why so many little local boys were walking about in 'Little Lord Fauntleroy' suits.

faunule (fǭ·niul). *Ecol.* Also **faunula.** [mod.L., dim. of FAUNA.] The fauna of a specified small habitat, *esp.* a group of fossils from one small area.
1909 in *Cent. Dict.* Suppl. 1928 *Amer. Midland Naturalist* Jan. 21 A zonule is the stratum or strata which contain a faunule. 1936 *Proc. Prehist. Soc.* II. 166 This faunule indicates very damp conditions and it probably represents a willow-alder marsh. 1938 CARPENTER *Ecol. Gloss.* 308 *Faunula*, the entire animal population of a small unit area. 1953 *Breviora* XII (*title*) A cave faunule from western Puerto Rico.

Faustian (fau·stiăn), *a.* [-IAN.] Of or pertaining to Johann *Faust* (in Latinized form

Johannes Faustus), a wandering astrologer and necromancer who lived in Germany *c* 1488–1541 and was reputed to have sold his soul to the Devil: later, the hero of dramas by Marlowe and Goethe. Also *absol.*
1876 *Gentl. Mag.* XVI. 211 The sombre Faustian grandeur of this piece. 1931 *Times Lit. Suppl.* 21 May 404/4 The Faustian idea of eternal striving. 1941 *Amer. Speech* XVI. 178 Man is seen to characterize himself as the Faustian, with his soul constantly aspiring upward, the higher the better. 1943 *Scrutiny* XI. 270 The combination of the Faustian urge and faith in progress with outstanding intellectual ability explains the respect..felt for Bacon. 1958 J. Carew *Wild Coast* vii. 100 The world of..the written word, the Faustian conflict. 1964 S. Bellow *Herzog* 234, I would never grasp the Christian and Faustian world idea.

‖ **faute de mieux** (fot də myǿ). [Fr.] For want of something better; *attrib.*, used for lack of an alternative.
1766 Chesterfield *Let.* 13 June (1774) II. 501 The seals..were offered first to Lord Egmont, then to Lord Hardwicke; ..but, after their going a begging for some time, the Duke of —— begged them, and has them *faute de mieux*. 1828 Lytton *Pelham* xxxii, At seven o'clock, up came a *cotelette pannée—faute de mieux*, I swallowed the composition. 1865 'Ouida' *Strathmore* ii, Still they were better than nothing, and were peppered *faute de mieux* that day. 1885 W. James *Let.* 5 Feb. (1920) I. 238 This position..is a *parti pris* and a *pis aller,—faute de mieux*,—to continue the Gallic idiom. 1941 Auden *New Year Let.* III. 50 And finds romantic, faute de mieux, The sad nostalgie des adieux. 1949 Koestler *Insight & Outlook* xii. 172 This *faute de mieux* procedure may seem to be a confirmation of the Freudian thesis.

fauteuil. Add: **b.** A seat in a theatre, an omnibus, etc., designed to resemble an arm-chair; *spec.* = Stall *sb.*[1] 5 c. Popularly pronounced (fōu·tīl).
1859 Sala *Twice round Clock* 253 The stalls...Those comfortable *fauteuils*. 1901 *Playgoer* 15 Oct. 38/1 Fine *fauteils* for sixpence; or stage box seats for ninepence. 1925 A. P. Herbert *Laughing Ann* 37 Oh, I likes a bit of enjoyment on a Friday, I like to sit in the fautles an' be grand. 1931 *Morning Post* 20 Aug. 8/3 With their covered-in tops, resilient springs, and voluptuous *fauteuils*, the latest omnibuses have become so attractive that one often feels..disinclined to get off. 1970 *Wykehamist* 11 July 498/2 The *real* critic..settles down in his luxurious fauteuil on the lower slopes of the Theatre Workshop Parnassus to have a good laugh.
c. The seat of a member of the French Academy; hence, membership of the Academy.
1883 *Standard* 23 Aug. 5/2 The number of Academic *fauteuils* would be fixed. 1920 *Glasgow Herald* 28 Apr. 9 The creation..of the class of Senior Academicians, whose fauteuils automatically become vacant.

fauve (fōuv). Also with capital initial. [Fr., lit. 'wild beast'.] A member of a movement in painting, chiefly associated with Henri Matisse (1869–1954), which flourished in Paris from 1905, and which is mainly characterized by a vivid use of colour. Also *attrib.* or as *adj.*, and *transf.* Hence (as adj.), of a bright or vivid colour. So **fau·vism** [F. *fauvisme*], the practice of this style of painting; **fau·vist(e)** [F. *fauviste*], an adherent of fauvism; also as *adj.*
The name was coined by the French art critic Louis Vauxcelles at the Autumn Salon of 1905; coming across a quattrocento-like statue in the midst of works by Matisse and his associates, he remarked, 'Donatello au milieu des fauves!'
1915 Wyndham Lewis in *Blast* July 23/1 May the mortality amongst Cubists, Carnivorists, Fauvists and Vorticists at the front be excessive. *Ibid.* 77/2 Mr. Adeney, in pallid and solidified landscapes, brings us back to the 'Fauves'. 1922 *Encycl. Brit.* XXXII. 6/1 Symbolism and Fauvism. *Ibid.* 6/2 The application of the *fauviste* recipe to the painting of fashionable Parisian society. 1932 *Times* 12 Mar., He was one of the group of Scottish painters who threw in their lot with the 'Fauves'..and..the result was a rather interesting blend of fauviste and Glasgow School characteristics. 1942 *Burlington Mag.* Feb. 51/2 The quasi-primitiveness of Gauguin or the Fauvistes. 1953 *Ann. Reg.* 1952 380 'The Pool of London', a colourful painting by André Derain from his *fauve* period. 1958 *Observer* 15 June 15/2 Ann Jellicoe and N. F. Simpson, the most disturbing of *fauve* humorists. 1959 *Listener* 8 Oct. 565/1 The truly decisive break to France, to Matisse and liberation by the Fauves. 1959 H. Read *Conc. Hist. Mod. Painting* ii. 34 In Paris the painters who reacted against Impressionism were known as 'les fauves' (the wild beasts), a name first used as a witticism by the critic Louis Vauxcelles at the time of the Autumn Salon of 1905. The name was apt because the means used by these painters were decidedly violent. 1960 *Guardian* 14 Nov. 7/3 The fauvist intensity of her colour—the red and yellow..conflicting hotly with the virulent greens. 1960 *New Left Rev.* Nov.–Dec. 59/1 It seems probable that [Picasso's] *Les Demoiselles* was designed..as a counterblast to recent trends in fauvism. 1967 *Vogue* June 101 Striking fauve flowered coat.

‖ **faux bonhomme** (fo bonom). [F. *faux* false + *bonhomme* simple good-natured man.] A sly, shifty person who assumes an open and good-natured manner.
1916 W. S. Maugham *Writer's Notebk.* (1949) 113 He looks at you benignly through gold-rimmed spectacles. The *faux bon homme*. 1932 *Sunday Times* 13 Mar. 6/2 The monster [Iago] in the guise of *faux bonhomme*. 1934 G. B. Shaw *On Rocks* I. 204 *The Oxford youth*..Stow all this fo bunnum business, Chavender...*Sir Arthur.* Oh, faux bonhomme, of course. 1958 *Observer* 25 May 14/1 Smiling the perpetual sweet smile of the *faux-bonhomme*.

‖ **faux-bourdon** (fo,būərdoṅ). [Fr.] = Faburden.
1879 in Grove *Dict. Mus.* I. 509/1. 1937 C. Day Lewis *Starting Point* 66 One reed-like liquid tenor..danced and swayed into a faux-bourdon. 1942 *Scrutiny* XI. 11 The enormous influence exerted by English methods on the Continental voice-leading is evident in the wide-spread adoption of the faux bourdon style. 1967 *Times* 23 Feb. 10/6 Part-writings, suggested medieval organum or sometimes faux-bourdon techniques.

‖ **faux-naïf** (fo,naif), *sb.* and *a.* [F. *faux* false + Naïf *a.*] **A.** *sb.* A person who pretends to be simple or unaffected and adopts a childish or naïve manner. **B.** *adj.* **a.** Of a work of art: self-consciously or meretriciously simple and artless. **b.** Of a person: affectedly simple or naïve; pretendedly ingenuous.
1941 *Fortune* July 111 Gibbs..is a *faux-naïf*: he pretends to have no social graces. 1952 W. Plomer *Museum Pieces* 28 Mother loves to pretend to be extremely simple. She's a *fausse naïve*. 1958 *Listener* 3 July 31/3 As music, 'Catulli Carmina' uses the same naive technical procedures as the earlier 'Carmina Burana', but here they have degenerated into *faux-naïf* mannerism. 1958 *Times Lit. Suppl.* 21 Nov. p. ii/4 *Faux-naïf* pictures..the kind calculated to raise a snigger among grown-ups and fury in the young. 1961 *Times* 13 May 5/2 That terrible affliction of the folk-singer, the faux-naïf. 1964 G. Sims *Terrible Door* xxvi. 140 The *fauxnaif* would have been dumped..in Poole Harbour. 1970 *Times* 18 Apr. p. iv/2 The language is faultlessly faux-naif—a recreation of the idioms of a child's thinking.

‖ **favela** (fāve·lā). [Pg.] In Brazil, a shack, shanty; a slum: usu. in pl. *favelas*, a collection of improvised huts, a shanty town. Hence **favela·do**, a person dwelling in a favela.
1961 G. Mikes *Tango* 18 In the midst of all this beauty and elegance, you discover the *favelas*...The *favela* is a wretched, ramshackle, filthy hut run up out of sticks, rotting planks, dirty rags and cardboard, as a rule in less than twenty-four hours...The *favelas* have no electricity (unless, as frequently happens, an enterprising favelado manages to tap an electric cable). 1962 *Guardian* 5 Oct. 13/4 A Sao Paulo *favela*, one of those slum shanty towns. 1964 'M. E. Chaber' *Six who Ran* (1965) iv. 57 We have a little house in the *favelas*. 1964 W. McCord in I. L. Horowitz *New Sociol.* 427 They hastily retreated to the city's healthier quarters, leaving the suffering *favelados* unchanged. 1969 J. Mander *Static Soc.* ii. 81 We cannot admire Rio's skyline for the squalid *favelas* nestling between her sugar-loaf hills.

faveolus (făvi·ōlǿs). Pl. **-i.** [mod.L., dim. of *favus* honeycomb.] A small depression, like a cell of a honeycomb.
1882 *Encycl. Brit.* XIV. 562/1 The apothecia of several calcicole lichens..have the power..of forming minute faveoli in the rocks. 1963 J. H. Kenneth *Henderson's Dict. Biol. Terms* (ed. 8) 199/2 *Faveolus*...A small depression or pit; alveola.

Faverolle(s) (fæ·vǝrǫl, fævǝrōu·l). [Prob. f. the place-name *Faverolles*, in the department of Seine-et-Oise, France.] One of a breed of domestic fowls originated in France by crossing light Brahmas or Dorkings with Houdans.
1902 L. Wright *New Bk. Poultry* 457/1 Faverolles have for some time been common in the northern part of France, where they were regarded as simply useful fowls. *Ibid.* 457/2 The colour of the Salmon Faverolles cock is quite different from that of the hen. 1957 *Encycl. Brit.* XVIII. 378/2 The Faverolle is a French breed with feathered shanks; in England the Salmon Faverolle is crossed extensively with the Light Sussex for roaster production.

favism (fēi·viz'm). *Path.* Also **fabism**, **fabi·smus.** [ad. It. *favism-o*, f. *fava* broad bean (f. L. *faba* bean) + -ism.] A hereditary form of anæmia manifested only after contact with broad beans.
1903 Dorland *Med. Dict.* (ed. 3) 263/1 *Fabism*, a disease of Italy caused by eating of the bean *Vicia faba*. 1933 *Jrnl. Amer. Med. Assoc.* 28 Oct. 1389/1 Favism (fabism) is a syndrome caused by inhalation from bean plants when in blossom or by ingestion of the beans. 1935 Whitby & Britton *Disorders of Blood* xiii. 241 Fabismus is a hæmolytic anæmia which occurs in Italy from eating beans. 1962 *Lancet* 15 Dec. 1274/1 This enzyme deficiency is commonest among Mediterranean peoples, but recently favism has been reported in 3 patients of British stock, and biochemical evidence of the defect was found in 12 of their relatives.

‖ **favori** (favori). [Fr.] = Favourite *sb.* 3. Usu. in pl.
1801 C. Wilmot *Jrnl.* 3 Dec. (1920) 9 Their noses.. peeping out of a bush of black Beard, Moustaches, Favori and ringlets. 1831 Disraeli *Young Duke* II. xix. 248 His beard, his mustachios, his whiskers, his *favoris*. 1864 G. A. Sala *Quite Alone* I. i. 4 Fawn-coloured *favoris*.

favour, *sb.* Add: **2. b.** (Examples of ironic *colloq.* usage.)
1959 [see *drop *v.* 4 c]. 1962 N. Marsh *Hand in Glove* v. 149 'Look,' Leonard drawled, 'do me a favour and get the hell out of this, will you?' 1963 J. T. Story *Something for Nothing* i. 20 'Shouldn't you get an accountant to sort it out?' 'Do me a favour,' Albert said. 'It was getting an accountant that got me into all this.' 1969 *Guardian* 14 May 1/2 Was she hoping to get engaged during the year of the tour? 'Good God, no, do us a favour.'

favour, *v.* Add: **1. b.** In film-making and broadcasting (see quots.).
1960 O. Skilbeck *Film & T.V. A.B.C.* 52 *Favour*, deliberately to turn the mike or camera to face one artist more than another. 1960 D. Davis *Gram. Telev. Prod.* 60 *Favour*, to see more of one person's face than that of another person in the same shot. 1970 *Amateur Photographer* 22 Apr. 84/2 Because of its strong directional properties a cardioid microphone will have to follow the actors around and be pointed at them when they speak. This is called 'favouring' and must be done gently without transmitting vibrations along the boom pole.

favourable, *a.* Add: **5. b.** Of a patient's condition, progress, etc.: satisfactory; favouring recovery.
1903 *Westm. Gaz.* 8 July 7/1 Seeing that his condition was so favourable. 1970 J. L. Gibbons in *Med. Ann.* 278 Favourable prognostic signs were the presence of mild symptoms..and a stable premorbid personality.
6. (Additional example.)
1930 *Economist* 20 Dec. 1158/2 The official returns indicate that the balance of 'current items', visible and invisible, is much less 'favourable' than before the war.

favourably, *adv.* **3.** (Additional example.)
1952 C. P. Blacker *Eugenics* ix. 219 A modest claim.. is made..that wastage among favourably-assessed parachutists proved smaller than among those unfavourably assessed.

favoured, *ppl. a.*[1] Add examples illustrating the *attrib.* use of 'the most favoured nation'.
1902 *Encycl. Brit.* XXVII. 164/1 By extending to a third nation privileges granted to particular countries, the most favoured nation article began to be framed. 1909 *Westm. Gaz.* 17 Aug. 2/1 Having secured a most-favoured-nation clause in her commercial treaty with China in 1896. 1928 *Manch. Guardian Weekly* 26 Oct. 328/2 In the agreement will be two main clauses dealing with most-favoured-nation treatment.

favourite, *sb.* and *a.* **B.** *favorite son*: delete def. and substitute: *U.S.* (*a*) 'a commendatory title given to George Washington' (D.A.E.); (*b*) 'a man who has endeared himself to a particular country or state, used esp. of a candidate for high office favored by the constituency or political leaders of the state from which he comes' (D.A.E.). Also *transf.*
1788 R. Platt in F. Moss *Amer. Metropolis* (1897) I. ii. 266 Motto: Freedom's favorite son. 1789 *N.Y. Daily Gaz.* 1 May 426/1 Washington, the favourite son of liberty, and deliverer of his country. 1806 J. Milner in C. L. Lewis *Romantic Decatur* (1937) viii. 85 That country welcomes your return to her bosom. She hails you as one of her favorite sons. 1835 D. Crockett *Life Van Buren* 198 The views and wishes of your 'favourite son'. 1962 *Guardian* 27 Nov. 9/1 The rumbustious campaigning of Bavaria's 'favourite son', the Federal Minister of Defence, Herr Strauss. 1963 *Ibid.* 22 Jan. 8/6 The Leeds constituency parties are..open-minded about their candidates, not reserving places for some trade union's favourite son.
2. *favourite sentence* attrib. (see quot. 1964).
1933 Bloomfield *Lang.* xi. 171 Certain forms [of sentence-type] are favorite sentence-forms. *Ibid.* 172 In English we have two favorite sentence-forms. One consists of actor-action phrases—*John ran away*... The other consists of a command..*Be good!* 1942 Bloch & Trager *Outl. Ling. Analysis* v. 74 Let us look briefly at the English actor-action construction. The most striking thing about it is its universality: it is the favorite-sentence type of English. 1964 R. H. Robins *Gen. Ling.* vi. 232 The patterns common to large numbers of the sentences of a language may be called its favourite sentence types.

favrile (favri·l). [Formed as a trade name by L. C. Tiffany (1848–1933), the inventor of the glass, in 1894, from Fabrile *a.*] A kind of glass characterized by rich colouring, iridescence, and enamelling.
1902 *Encycl. Brit.* XXIX. 8/1 The so-called 'Favrile' glass of Messrs. Tiffany of New York owes its effect entirely to surface colour and lustre. 1920 E. Bok *Autobiogr.* (1921) xxii. 226 Just as the ancient Egyptians and Persians had used glazed brick and tile, set in cement, as their form of wall decoration, so Mr. Tiffany had used favrile glass, set in cement.

favus. Add: Hence **fa·vic** *a.*, of, pertaining to, or characteristic of this disease.
1892 G. T. Jackson *Dis. Skin* 226 The favic fungus was found implanted upon the mucous membrane of the

stomach. 1958 *New Biol.* XXVII. 55 *Trichophyton schoenleinii*..gives rise to favic lesions in man.

fawn, *sb.*[1] Add: **3.** (Earlier and later examples.)
1881 C. C. Harrison *Woman's Handiwork* III. 165 Among colors in which most of the stuffs mentioned may be had, are seal-brown, nut-brown and fawn. 1914 [see *beaver[1] 2 c]. 1971 P. D. James *Shroud for Nightingale* v. 170 The thick winter coat in fawn was of good quality. *Ibid.* vii. 220 His fawn polo-neck sweater.
4. fawn foot (see quots.); *fawn-eyed, -pink* adjs.
1895 J. R. Lowell *Last Poems* 34 Seraph strong to soar, or fawn-eyed elf. 1938 S. Spender *Trial of Judge* 16 The gross-lipped fawn-eyed nigger-skinned Hook-nosed intellectual Petra. 1950 H. L. Edlin in *Wood* XV. 373/1 *Fawn foot*, swelling at the end of an axe helve, to give a better grip, shaped like a fawn's foot. 1953 A. Jobson *Household & Country Crafts* xvii. 166 Some characteristic bill-hooks with the funniest little kick of a handle, like the spurs of a fighting-cock (I believe they are technically known as fawnfeet, and assuredly they are taken out of the hedgerow). 1909 *Daily Chron.* 29 Apr. 7/3 Tourterelle is a subdued fawn-pink. 1927 D. H. Lawrence *Mornings in Mexico* 30 It [*sc.* the village] lies..tilted on the fawn-pink slope.

fawn, *v.*[2] **1.** (Later examples.)
1942 J. M. Dowsett *Romance Eng. Forests* ix. 243 No hogs were allowed in the chase during the fence month when the does were fawning. 1951 D. M. Stenton *Eng. Soc. Early Middle Ages* iii. 107 The forest was closed for hunting because the beasts were supposed to be fawning.

fawnish (fǭ·niʃ), *a.* [f. Fawn *sb.*[1] 3 + -ish[1].] Somewhat fawn, resembling fawn in colour.
1895 *Daily News* 31 May 2/1 The rest of the dress being a negation in fawnish cream colour. 1909 M. B. Saunders *Litany Lane* I. i, Looming out of the fawnish mist shone the great golden glory of the world.

fay, *sb.*[2] Add: Also *attrib.* and *Comb.*
1904 *Westm. Gaz.* 14 Sept. 2/3 Watching wild swans by some fay-haunted pool. 1962 *Listener* 12 Apr. 647/1 When she made formal use of figures in her landscapes, they were somewhat mannered, almost fay children.

fay (fēi), *sb.*[4] Abbrev. of *ofay.
1927 *Amer. Mercury* Aug. 393 'What a lot of 'fays!' I thought, as I noticed the number of white guests. 1946 Mezzrow & Wolfe *Really Blues* (1957) 62 He was the first fay boy I ever heard who mastered this vital foundation of jazz music. 1966 *Publ. Amer. Dial. Soc.* 1964 XLII. 44 *Fay* is commonly used by Negroes to designate Caucasians.

fay, *v.*[1] **2. b.** (Earlier and later U.S. examples.)
1847 D. P. Thompson *Locke Amsden* vii. 138, I have no notion of spoiling sense to make it fay in with book rules. 1906 P. Lowell *Mars & its Canals* 347 The explanation of the canals as threads of vegetation fays in with the one which has been found to meet the requirements of the blue-green areas.

Fayum (fēi·yūm). The name of a province in upper Egypt, used *attrib.* to designate articles discovered there. Also **Fayumic** (fēi·yūmik), a dialect of Coptic.
1902 *Encycl. Brit.* XXV. 497/2 *Fayum Gospel Fragment.*—This small fragment contains two sayings of Christ and one of Peter. *Ibid.* XXVII. 727/2 The main divisions of Christian Coptic as recognized and named at present are: Sahidic...Fayumic; [spoken] in the Fayum (formerly named wrongly 'Bashmuric'..) [etc.]. 1937 *Burlington Mag.* July 53/2 Comparison of Karanis and other Fayum glasses with those from the other areas of marketing. 1948 R. H. Pfeiffer *Introd. to O.T.* (1952) 115 Portions of the Scriptures are also preserved..in Fayumic. 1961 M. Levy *Studio Dict. Art Terms* 49 *Fayum portrait*, intensely realistic portraits of deceased persons painted by the artists of Fayum..usually on mummy cases. The portraits belong to the first four centuries of the Christian era.

faze, *v.* For *U.S.* read orig. *U.S.*, and add earlier and later examples.
1830 *Western Monthly Rev.* III. 357 They were too well up to these things to be fazed by a little cold lead. 1843 'R. Carlton' *New Purchase* I. xix. 174 This didn't *faze* me, only I steps back for my old camlit cloak. 1859 *Harper's Weekly* 16 July (Th.), Such a stomach that even a dram of nitric acid would not faze it. 1953 F. Robb *Sea Hunters* ii. 17 Although Cope might be only a fishing skipper no gilded plutocrat was going to faze him. 1961 *Coast to Coast 1959–60* 89 Perrot became an anodized schoolmaster, a disciplinarian no boy could faze. 1963 'V. Lucas' *Bell Jar* ii. 19 Walking has never fazed me. 1971 *Listener* 8 Apr. 435/2 It never really fazed me that much. I don't know why it didn't.

fear, *sb.* Add: **3. e.** Phrases: *to put* (occas. *rub*) *the fear of God into*, to terrify (into submission); *without fear or favour*, impartially.
1890 *Macm. Mag.* Jan. 230/2 Rub the fear of God into the people. 1905 H. G. Wells *Let.* 13 Aug. in D. L. Moore *E. Nesbit* (1933) xii. 214 Bland might like to come over when Clodd is here, and help me put the fear of God into him. 1906 *Independent Rev.* X. 110 He [*sc.* Dr. Johnson] judged authors as if they were criminals in the dock, answerable for every infraction of the rules and regulations laid down by the laws of art, which it was his business to administer without fear or favour. 1930 A. Bennett *Imperial Palace* 479 When she's my wife I'll put the fear of God into her. 1956 A. Wilson *Anglo-Saxon Att.* I. i. 4 If you have grievances..send your problems to John Middleton. He will investigate your case without fear or favour. 1959 *Punch* 16 Dec. 614/3 If I may say so without fear or favour. 1960 'J. Wyndham' *Trouble with Lichen* III. ix. 119 We put the fear of God into the girls over that kind of thing.
5. a. The usual sense of *no fear* is now 'not likely', 'certainly not'. (Add further examples.)
1880 [see Pho, Phoh *int.*]. 1817 A. Constable *Let.* 25 Dec. in *J. Constable's Corr.* (1962) 167 Being very clean, and good temper'd, she hop'd there was no fear of her suiting you. 1930 A. Bennett *Imperial Palace* xii. 73, I invite him to dinner! And in his own hotel! No fear! 1966 J. B. Priestley *Salt is Leaving* xii. 172 No fear! I wouldn't be found dead in Birkden.
6. a. *fear-inspiring* (later example); **b.** *fear-driven, -dulled, -palsied* (earlier example), *-stirred, -stricken*; **fear instinct**, the instinctive tendency to be afraid; **fear paroxysm**, a paroxysm caused by fear.
1901 'L. Malet' *Hist. R. Calmady* v. ix. 460 Some fear-driven hurrying ghost. 1938 W. de la Mare *Memory* 9 The fear-dulled eyes in the pallid face. 1904 W. M. Gallichan *Fishing Spain* xii. 118 Tragic dreams of this strange, fear-inspiring flood. 1908 W. McDougall *Introd. Soc. Psychol.* iv. 94 The fear-instinct has..a special perceptual inlet that renders it excitable by the sound of the cry of fear. 1923 D. H. Lawrence *Kangaroo* xvi. 335 The herd instinct, for example, is of many sorts. It has two main divisions, the fear-instinct, and the aggressive instinct. 1811 Shelley *Let.* 11 Jan. (1964) I. 38 Wild horror the fear-palsied Earth is astounding. 1890 W. James *Princ. Psychol.* II. xxiv. 419 A certain amount of timidity obviously adapts us to the world we live in, but the *fear-paroxysm* is surely altogether harmful to him who is its prey. 1921 W. de la Mare *Veil* 87 To the fear-stirred heart And the ancient dread Of man. 1906 B. von Hutten *What became of Pam* x. 176 The girl, fear-stricken..knelt down.

fear, *sb.*[2] = Fiar *sb.*

fearfully, *adv.* **2.** (Earlier example.)
1835 Dickens *Let.* 16 Dec. (1965) I. 106 This place is fearfully dull, and I have seen neither a newspaper or a book since I came down here.

fearless, *a.* Add: **3.** *Comb.* Parasynthetic, as *fearless-eyed, -looking, -seeming.*
1925 J. Gregory *Bab of Backwoods* x. 127 Bright-eyed, quick-eyed, fearless-eyed. 1864 J. A. Grant *Walk across Africa* vi. 91 There are also wild-cats..very bold, fearless-looking creatures. 1915 D. H. Lawrence *Rainbow* xii. 314 A fearless-seeming, clean type of modern girl.

feasibility. Add: **3.** *attrib.*, as *feasibility study.*
1959 *Times* 14 Jan. 2/5 Feasibility studies of new reactor systems. 1969 *Daily Tel.* 14 Feb. 23/2 Initially, the consortium is to make a two-year 'feasibility' study. 1971 *Guardian* 8 May 3/7 Britain is helping with a feasibility study for rice growing.

feasible, *a.* **2.** (Later example of persons.)
1889 E. C. Dowson *Let.* 3 Feb. (1967) 33, I have suggested Swinburne & Pater, Lang, Oscar Wilde & Lady Dilke as the most feasible.

feather, *sb.* Add: **1. b.** *you could* (or *might*) *have knocked me* (etc.) *down with a feather* (and similar expressions): used hyperbolically = I was much surprised, astounded, etc.; *to make the feathers fly*: to cause a disturbance, to 'stir up' someone.
1741 Richardson *Pamela* I. vii. 13, I was so confounded at these Words, you might have beat me down with a Feather. 1825 J. Neal *Bro. Jonathan* I. iv. 94 If my New York master..had hold o' him; he'd make the feathers fly. *Ibid.* II. xxiii. 339 When they are at home.. how the feathers fly! 1843 [in Dict., *sb.* 10 a]. 1853 Lytton *My Novel* III. ix. xiii. 69 You might have knocked them down with a feather! 1924 C. Mackenzie *Heavenly Ladder* v. 75 At present you could knock me down on a feather, as old Mrs. Geary used to say, I'm that overgone by it all. 1930 W. S. Maugham *Cakes & Ale* vii. 89 When I..saw Rosie standing there, you could 'ave knocked me down with a feather. 1970 *Globe & Mail* (Toronto) 26 Sept. 5/5 You could have knocked me over with a feather.
2. b. *in high feather* (earlier example); *in full feather*: see quots.
1781 G. Selwyn *Let.* in *15th Rep. Hist. MSS. Comm.* App. VI. (1897) 505 He [*sc.* C. J. Fox] is in high feather. 1834 W. N. Glascock *Naval Sketch-Bk.* 2nd Ser. I. 149 A lady in full feather approached the sentinel on the sward. 1855 J. E. Cooke *Ellie* 476 No words can describe the serene effulgence of the Heartsease appearance, when in full feather, and high spirits. 1860 Hotten *Slang Dict.* 138 *In full feather*, rich. *a* 1865 Smyth *Sailor's Word-Bk.* (1867) 327 *Full feather*, attired in best dress or uniform. 1886 *Graphic* 30 Jan. 130/2 Watty, by reason of his office [of butler], was of course always 'in full feather'.
6. (Later example.)
1781 S. C. King in *Med. Obs. & Inq.* (1784) VI. xxii. 231, I send you a feather or pen..extracted from the œsophagus of a man.
8. b. *a feather in the cap*: examples of *ellipt.* use.
1825 M. Constable *Let.* 2 May in *J. Constable's Corr.* (1962) 220, I expect your *Barge Horse* will do wonders for you—only think of the Medal, what a *feather* it is. 1958 *Times* 22 Oct. 14/1 Had Slater collected these two feathers to go with yesterday's collection he would have had a memorable match. 1968 J. F. Straker *SIN & Johnny Inch* 211 Strip off if you feel like it; I reckon this has been quite a feather.
11. b. Also on other animals.
1878 C. Hallock *Hallock's Amer. Club List & Sportsman's Gloss.* p. iv/2 *Feather*, the long, flowing hairs which adorn the legs of a dog. 1884 W. G. Stables *Our Friend the Dog* vii. 60 *Feather*, the long hair on chest, legs, and tail, &c., of some breeds. The hair generally. 1897 *Outing* (U.S.) XXX. 413/1 The coat [of the Russian wolfhound] is long.., forming a liberal 'feather' upon the legs, chest, belly and hindquarters. 1914 *N.Z. Farmer* Apr., Although the General Trend of the hair [*sc.* on the part between the hock and the vulva of a cow] is in an ascending direction there are variations on certain points where changes in the direction of the growth makes the hair stand up on ridges, sometimes forming figures. These variations are called feathers. There are in all seven feathers. 1 The Oval feather. 2 The Buttock feather. 3 The Babine feather. 4 The Vulvous feather. 5 The Bastard feather. 6 The Thigh feather. 7 The Dart feather. 1942 'M. Innes' *Daffodil Affair* I. 30 'A nice dog,' Appleby said...Mr Gee swung round. 'Dish-faced,' he said...'Cow-hocked. No feather. Apple-headed. Pily. 1954 E. Megargee *Dog Dict.* 47 There is abundant feather on the chest, belly, hindquarters, and legs [of a field spaniel].
c. The foamy crest of a wave. Cf. Cut *v.* 37 b.
1838 *Civil Eng. & Arch. Jrnl.* I. 272/1 The feather only of each wave would be seen. 1896 *Daily News* 10 July 3/6 White feathers from their bows.
d. (See quot. 1928.)
1928 *Papers Mich. Acad. Sci. Arts & Lett.* X. 292 *Feather*, the wake or ripple left by a protruding periscope. The 'feather' was often visible when the periscope was not. 1943 'T. Dudley-Gordon' *Coastal Command* 28 You may pass by the tell-tale feather of a periscope.
16. e. (Example.)
1908 *Westm. Gaz.* 28 May 4/2 The gear-wheels are mounted on shafts, each having four 'feathers'.
18. a. *feather boa, club.* **d.** *feather-soft, -witted* adjs.
1895 *Army & Navy Co-op. Soc. Price List* 15 Sept. 1041 Black and Coloured Feather Boas. 1902 *Queen* 15 Feb. Advt. 2nd sheet, Lace Berthes and Feather Boas. 1899 R. Whiteing *No. 5 John St.* x. 101 A Feather Club is an institution for supplying our younger womankind with.. ostrich plumes for their hats. 1923 E. Sitwell *Bucolic Comedies* 38 The goose-girl smoothed down her feather-soft Breast. 1930 W. de la Mare *On Edge* 295 That black-eyed, painted-up, feather-witted little Italian Countess.
19. feather ball, a golf-ball stuffed with feathers (cf. *feathery *a.* 4); **feather-brain**, also a light or weak brain; **feather-curler**, one employed in curling feathers; **feather-poke** (*b*) (earlier example); **feather-top** (earlier example in sense of 'feather-topped wig').
1893 H. G. Hutchinson *Golfing* 4 All this went on before the days of gutta-percha. Men played with 'feather balls'—that is, balls of leather stuffed..tightly with feathers. 1888 W. S. Gilbert *Yeom. of Guard* (1889) I. 7 Thou hast a feather-brain. 1941 A. L. Rowse *Tudor Cornwall* xi. 269 Perhaps..it was his feather-brains which saved his head. 1881 *Instr. Census Clerks* (1885) 79 Feather..Curler. 1895 *Daily News* 4 Apr. 5/6 Earning fifteenpence a day as a feather-curler. 1831 W. Howitt *Bk. Seasons* Apr. 91 The little willow-wren builds an oval nest after the fashion of the common wren and the feather-poke. 1775 D. Garrick *Bon Ton* Prol., The Tyburn scratch, thick club, and Temple tye, The parson's feather-top, frizz'd broad and high!
b. feather-wood (see quot. 1884).
1884 A. Nilson *Timber Trees N.S.W.* (*Index*) 135 Featherwood, Polyosma Cunninghamii. Saxifrageæ. 1927 *Blackw. Mag.* Oct. 464/1 The delicate green tracery of the graceful featherwoods and celery-top pines. 1947 R. H. Anderson *Trees N.S.W.* (ed. 2) 218 In addition to the Opossum Wood and the Featherwood there are four related species belonging to the same family (*Escalloniaceae*).

feather, *v.* Add: **11. b.** To fix (paddle-boards) so as to offer the least resistance while entering and leaving the water. (Cf. Feathering *ppl. a.* b.)
1848 *Rep. Comm. Patents 1847* (U.S.) 71 Improvements in the mode of feathering the floats of paddle wheels.
c. *trans.* To rotate (the blades of a propeller) so that they are parallel to the direction of motion and offer the least resistance to the water or air when not driven. Hence, on a rotating-wing aircraft (or helicopter): to vary the angle of incidence of (the blades).
1883 A. E. Seaton *Man. Marine Engin.* xv. 298 Yachts and ships which are required to sail as well as steam, cannot well do the former when the screw is stopped, unless some means be adopted of feathering the blades. 1909 F. T. Jane *All World's Air-Ships* 52/1 The blades can be feathered, and either propeller feathered independently of the other. 1921 S. W. Barnaby *Marine Propellers* (ed. 6) iii. 58 In a twin-screw ship screws that can be feathered fore and aft by internal mechanism will be found advantageous. 1940 *Flight* 15 Aug. e/2 Should a power unit in a multi-engined aircraft suddenly fail.. it is a great advantage to be able to feather the airscrew. 1947 *Times* 2 Sept. 2/5 After gaining height the piston engines were switched off and their airscrews feathered. 1955 Liptrot & Woods *Rotorcraft* v. 44 Displacement (or lag) therefore causes any movement about the flapping hinge..to 'feather' the blade, that is change its incidence. 1967 C. H. Barnes *Shorts Aircraft since 1900* 358 For the first time a fully loaded Sunderland could

be flown safely with two airscrews feathered on the same side.
d. *intr.* Of a propeller: to assume a 'feathered' position (see *feathered *ppl. a.* 8 b).
1933 *Jane's Fighting Ships* 228 Variable pitch propeller on inner shaft which 'feathers' when Turbines only are used. 1951 *Engineering* 9 Mar. 301/2 Propellers of aircraft..so built that they could 'feather'.

feather-bed. Add: **1.** *fig.* (Later example.)
1952 *Economist* 16 Aug. 386 Feather beds for farmers have been a feature of post-war Europe.
2. b. *pl.* (*sing.* constr.). A stonewort (genus *Chara*); a bed of stoneworts.
1807 F. Pursh *Jrnl. Bot. Excursion* 4 Aug. (1923) 50 The Creek was coverd at its bottom with Chara, which the boatmen call Feather beds. 1851 A. Wood *Class-Bk. Bot.* (ed. 23) II. 637 C[*hara*] *vulgaris*...Feather-beds...It appears in dense tufts, like a soft bed, undulating with the motion of the water.
3. a. *featherbed job, landing, practice, rule, treatment.*
1938 *San Francisco Examiner* 3 June 25/1 But he does feel that 'featherbed jobs'..should be abolished. 1942 A. M. Low *Parachutes* xi. 201 The most comfortable landing is said to be in a young plantation of pines... The firemen..call these 'featherbed landings'. 1922 L. F. Loree *Railroad Freight Transportation* vii. 625 The annual cost of these so-called 'feather-bed' practices. 1934 *Railway Age* XCVII. 858/2 Restrictive rules, sometimes called 'featherbed rules'. 1958 *Times Rev. Industry* Sept. 98/1 Strange to give some industries 'feather-bed' treatment.

feather-bedding, *vbl. sb.* [f. Feather-bed + -ing.] (The action of) making comfortable by favourable, esp. economic or financial, treatment; the state of being so treated; *spec.* the employment of superfluous staff. Hence, as a back-formation, **featherbed** *v. trans.*, to provide with, esp. economic, advantages; so **featherbedded** *ppl. a.*; also **feather-bedder**, one devoted to physical or intellectual comfort.
1921 *Bull. Bureau Business Res., Harrow* 25 Feather bedding, getting pay for work not done. 1928 G. Cooke *Theory Mus.* iii. 33 This movement has certainly produced a harshening of our music which should even satisfy those who most object to the feather-bedding of the Victorians. 1943 *Reader's Digest* Mar. 26/2 For the unions, featherbedding has become an established business procedure. 1949 G. Freedley in Andrews & Trilling *Internat. Theatre* 41 This has brought about a tight departmentalising which has 'feather-bedded' union practices. 1950 *Hansard Commons* 7 Mar. 183 Examples of gross feather bedding both in Government service.. and in private industry. *Ibid.* 20 Apr. 426 He can rely on this party..not to feather bed..the farmers. 1952 M. McCarthy *Groves of Academe* (1953) iv. 74 A certain number of seasoned nonconformists and dissenters, sexual deviants, feather-bedders, alcoholics, impostors. 1954 *Daily Tel.* 6 Feb. 4/7 'Featherbedded farmer' is surely one of the most ludicrous phrases ever coined. 1958 *Economist* 11 Jan. 131/1 The unions have been able to impose feather-bedding and under-employment. 1962 J. Cannan *All is Discovered* vi. 149 Those feather-bedded farmers with their sprays and their fertilisers. 1971 *Daily Colonist* (Victoria, B.C.) 17 Feb. 5/5 Why does he..call us the 'social elite featherbedded on the tax-payers' money'?

feathered, *ppl. a.* Add: **1.** Phr. *feathered friend*: a bird (used sentimentally or ironically).
1933 E. A. Robertson *Ordinary Families* xiv. 294, I did not write one letter about 'our feathered friends' which was not published. 1953 E. Simon *Past Masters* I. iii. 35 We keep chickens..and so far our feathered friends are letting us down badly. 1967 *Guardian* 4 Feb. 14/8 The British farmers'..war against our furred and feathered friends might be an enormous mistake.
8. Also of a paddle.
1866 F. W. Brearey *Remarks upon Aërial Navigation* in *Rep. Brit. Assoc. 1865* II. 17 The downward blow of paddles made of bamboo and silk, if returning feathered for each succeeding stroke, would enable him to effect this.
b. Of the propeller of an aircraft: having the blades turned about their axes so as to be in line with the air-flow. Cf. *feather *v.* 11 c.
1935 *Jrnl. R. Aeronaut. Soc.* XXXIX. 1026 The feathered propeller has only about 3 per cent. of the drag of the normal airplane. 1940 *Flight* 15 Aug. e/2 In the 'fully feathered' position the airscrew blades are turned round until the average torque exerted by the windmilling blades becomes nearly zero. 1970 H. A. Taylor *Airspeed Aircraft since 1931* 117 The prototype was controllable, with the propeller of the critical engine feathered.., down to speeds as low as 110kt.

feathering, *vbl. sb.* Add: **1. c.** The action of feathering the floats of a paddle-wheel or the blades of a propeller (see *feather *v.* 11 b, c); also, an angular motion of a revolving rotor blade about its axis, producing a continual variation in its pitch. Also *attrib.*
1850 *Rep. Comm. Patents 1849* (U.S.) 47 Many applications have been made based upon alleged novelties in the feathering of the vertical float paddle wheel. 1909 A. E. Seaton *Screw Propeller* xi. 162 Flat blades..have been tried on propellers where complete feathering..is required for efficient sailing. 1955 J. Shapiro *Princ. Helicopter Engin.* iii. 173 In most rotors the blades are constrained in pitch by the control linkage and the problem of evaluating free feathering motion does not arise. 1962 *Gloss. Aeronaut. Terms* (*B.S.I.*) v. 16 *Feathering hinge*, a blade pivot which allows the blade pitch angle to be varied. 1970 A. C. Kermode *Flight without Formulae* 202 Feathering means turning the blades so that, when the propeller is stopped, they offer the least resistance.
d. Of cream: see Feather *v.* 8 b.
1928 *Jrnl. Dairy Sci.* XI. 249 The formation of a flocculent coagulum, known as feathering, when sweet cream is added to hot coffee, is a problem of considerable commercial importance. *Ibid.*, The acidity of the product is an important factor in the feathering phenomenon. 1955 J. G. Davis *Dict. Dairying* (ed. 2) 465 *Feathering of cream*, a term applied to the flocculation of fat, etc., on the surface of tea or coffee, particularly with homogenised milk.
e. The spreading of ink on paper, resulting in printed or written matter having a blurred appearance.
1942 J. Grant *Lab. Handbk. Pulp & Paper Manuf.* 120 To prevent the spreading (or 'feathering') of writing ink on paper. 1959 *New Scientist* 23 Apr. 930/1 This [*sc.* an ink containing an additive] could be used in concentration and still did not cause feathering. 1963 Kenneison & Spilman *Dict. Printing* 63 *Feathering*...may be caused by using an unsuitable paper or ink.

feathering, *ppl. a.* Add: **c.** Of the propeller of a ship or an aeroplane: see *feather *v.* 11 c.
1885 A. E. Seaton *Man. Marine Engin.* (ed. 4) xv. 298 A feathering screw. 1909 *Westm. Gaz.* 18 Mar. 4/2 The feathering propeller..for airship work will no doubt appeal..as a means whereby the pitch can be altered. 1909 A. Berget *Conquest of Air* II. ii. 141 We have obtained the longitudinal stability of the aeroplane by the use of the 'feathering tail'. 1939 *Flight* 28 Sept. p. *h*, Blade root cuffs are being fitted by the Curtiss concern to their electrically controlled, fully feathering airscrew. 1968 A. J. Jackson *Blackburn Aircraft* 465 Four Hercules engines installed as self-contained power plants driving 14-ft diameter four-bladed Rotol constant-speed, feathering and reversible-pitch airscrews.

feather-stitch. (Earlier example.)
1835 *Court Mag.* VI. xiii/1 Indian muslin pelerine, embroidered in feather-stitch.

feather-weight. Add: **1.** (Later examples.)
1915 F. M. Hueffer *Good Soldier* I. vi. 88 Leonora lifted her up—she was the merest feather-weight—and laid her on the bed. 1932 D. C. Minter *Mod. Needlecraft* 161/1 Now..tweeds and cloths are manufactured in featherweights.
3. Also *attrib.*
1928 *Daily Express* 2 Oct. 17/7 The world's featherweight boxing title. 1933 Wodehouse *Heavy Weather* ix. 143 Getting a featherweight Boxing blue at Cambridge.
4. A light make of bicycle. (*Disused.*)
1901 *Morning Leader* 18 Dec. 6/2 Cycling Gossip. The topic at the moment undoubtedly is the new featherweight fashion. 1902 *Daily Chron.* 17 May 8/6, I had these featherweights also in mind when I said that 'makers have not catered especially for light riders'. 1908 *Ibid.* 3 Aug. 6/6 Green's mount was a featherweight Raleigh.
5. *Paper.* (See quots.)
1905 *Daily Chron.* 28 June 3/2 A 'feather-weight' laid paper of a durable kind is employed. *a* 1912 *Paper Terminol.* (*Spalding & Hodge*) ii. 9 *Featherweight*, antique laid or wove light handling book papers. They are manufactured mainly from esparto, are very loosely woven, and of extreme bulk. The bulkiest are entirely unloaded. 1957 *N.Z. Timber Jrnl.* Feb. 45/2 *Feather weight paper*, a very light antique book paper of woven esparto. 1966 H. Williamson *Methods Bk. Design* (ed. 2) xviii. 290 Thick, fluffy, feeble, antique featherweight papers..are still widely used to give books a deceptive bulk.
So **fea·therweight** *a.* = *feather-weighted* ppl. adj.; weak.
1885 A. Edwards *Girton Girl* II. x. 147 It would do your cousin a vast deal of good to run away from that feather-weight husband of hers. 1958 *Times* 5 Nov. 13/1 He [*sc.* André Masson] remains..one who can just rely enough on Parisian elegance and flair to be shamelessly pretty and featherweight for much of the time.

feathery, *a.* Add: **4.** Of golf-balls: stuffed with feathers. Also as *sb.*
1891 R. Forgan *Golfer's Handbk.* 34 The so-called 'feathery' variety of golf balls exclusively held the field. These 'featheries' were made of leather, stuffed with feathers.

feature, *sb.* Add: **4. a.** In various specific applications: (*a*) Applied to a person, esp. a professional entertainer.
1801 T. Fremantle *Let.* 1 May in A. Fremantle *Wynne Diaries* (1940) III. 34 Living on such very friendly terms with all the leading features in the Squadron. 1838 *Actors by Daylight* I. 123 Covent Garden was advertised to open, and Miss Fanny Kemble announced as the *feature*.
(*b*) A distinctive or prominent article or item in a newspaper, magazine, etc.
1855 Geo. Eliot in *Westm. Rev.* VIII. iv. 438 The Press has no band of critics who..are on the watch for a slip or defect in the preacher, to make a 'feature' in their article. 1933 H. G. Wells *Shape of Things to Come* I. § 9 93 A daily paper..would..have found itself vigorously outdone by more richly endowed competitors, able because of their wealth to buy up all the most attractive features. 1959 *Times* 5 May 13/3 (Advt.), Nowadays it's not so much news as features; features or comment.
(*c*) A feature film. So *feature-length* attrib. Also in *Broadcasting*, a feature programme.
1913 [see *4 c]. 1921 A. C. Lescarboura *Cinema Handbk.* (1922) i. 23 *Feature*, a pictured story, a plurality of reels in length. *Ibid.* x. 351 Paralleling all this is the filming of short features for the news and magazine films. 1930 J. B. Priestley *Angel Pavement* iv. 177 Then came the film of the evening, the star feature. 1932 *N.Y. Times* 10 Apr. Mag. 22/3 In England there is more chamber music, more serious drama,..fewer 'educational features'. 1938 I. Barry in Bardèche & Brasillach *Hist. Film* 388 Since the advent of sound, feature-length travel and documentary pictures produced commercially have been less evident than formerly. 1941 *B.B.C. Gloss. Broadc. Terms* 20 *News Feature*, programme in which news is presented in radio-dramatic form. 1954 *Economist* 11 Sept. Suppl. 11/1 A BBC feature depicting Folland Aircraft and the Gnat brought a host of small buying orders for Folland shares into the market. 1956 *Ann. Reg.* 1955 385 The first feature-length cartoon film, *Animal Farm*. 1965 *Listener* 20 May 746/3 There is a Robert Mulligan going the circuits as a second feature to *Good Neighbour Sam*.
(*d*) *Philol.* (See quots. 1942 and 1957.)
1926 *Language* II. 157 Different morphemes may be alike or partly alike as to vocal features. 1942 *Ibid.* XVIII. 7 It turns out that some things..are constantly associated in groups of two or more, one appearing wherever the others do. Where this is true, the two or more elements clearly constitute a single unit, one of the minimum units into which the utterances of the range may be analyzed. These units are termed *features*. 1948 *Ibid.* XXIV. 33 Some distinctively different segments contain a distinctive combination, or one or more distinctive components in common. The distinctive aspects or components of a segment constitute one or more *features* of the segment. We use the term 'feature' for what is commonly called a distinctive feature of a sound. Features are said to occur in segments; segments are said to contain features. 1952 R. Jakobson et al. (*title*) Preliminaries to speech analysis: the distinctive features and their correlates. *Ibid.* i. 3 The listener is obliged to choose either between two polar qualities in the same category.. or between the presence and absence of a certain quality. ..The choice between the two opposites may be termed distinctive features. 1957 *Trans. Philol. Soc. 1957* 122 Referring to the *two ways of discriminating* elements in the flow of speech—the one, which distinguishes elements as successive (segmentation), and the other, which distinguishes elements as successive or simultaneous (substitution): we establish *two fundamental types of element*—(i) such as *can* be isolated by segmentation, to be called '*segments*', and (ii) such as *cannot* be, to be called '*features*'. 1968 Chomsky & Halle *Sound Pattern Eng.* 66 Our tentative assumption is that the segmental features and the boundary features fall into distinct sets.
(*e*) *Mil.* A topographically noteworthy point or area.
1944 *Daily Tel.* 15 May, They have also taken five more hill features and three towns. 1956 J. Masters *Bugles & Tiger* xviii. 229 M. L. sent my company..up the hill and spread the others out..on smaller features.
c. *attrib.* and *Comb.* (*a*) **feature film**, the principal film in a programme; so *feature picture*; **feature programme**, a programme based on one central theme; **feature writer**, one who writes a special feature (sense *4 a (*b*) above) in a newspaper, magazine, etc. So *feature editor, story*, etc. The pl. form *features* is also used.
1911 D. S. Hulfish *Cycl. Motion-Pict. Work* II. 200 (*caption*) Feature Films. 1912 Harrington & Frankenberg *Essentials Journalism* vi. 62 It may not be amiss.. to speak briefly of the work of trained 'feature' writers. *Ibid.* 294 A feature story is one in which the news element is made subordinate. 1913 *Moving Picture Ann. 1912* 15 The rapid growth of what has come to be known as the 'feature film'..characterizes the year 1912...Features rather than 'first runs' became the popular cry. 1913 *Technical World* XIX. 464 Sunday feature pictures, sports, and advertising are all made in this dark room. 1913 *Writer's Bull.* Oct. 101/2 Here is a little suggestion to those who write feature articles. 1927 *Melody Maker* May 523/1 Fox-trots..occur frequently in feature pictures to accompany dancing. 1928 *B.B.C. Handbk.* 1929 60 Particular attention..is being paid to what have been called 'feature programmes'—composite programmes in which the aid of music and the spoken word is asked to evoke a mood, or to present an idea. 1929 F. A. Pottle *Stretchers* (1930) 318 The editor-in-chief of the *Flare*, Cpl. Herbert M. Davidson..was, until recently, feature editor of the Chicago *Daily News*. 1938 *Ann. Reg. 1937* 371 Feature films, that is films of 3,000 feet or more, showed only a trifling increase. 1938 *Encycl. Brit. Bk. Yr.* 122/1 During 1937, dramatic productions in the studios showed a considerable development of 'feature programmes', dealing with such subjects as the history of journalism,..Hadrian's Wall, the anti-slavery movement..and the Duchy of Cornwall. 1938 E. Waugh *Scoop* I. v. 85 The other agencies are sending feature men. 1939 'N. Blake' *Smiler with Knife* v. 78 One of the *Daily Post* feature-writers. 1956 Features-producer [see *B.B.C.]. 1957 C. Brooke-Rose *Langs. Love* v. 48 The literary types..with families to support on a regular output of criticism, non-fiction books..and long feature programmes for the B.B.C. 1957 *Times Lit. Suppl.* 8 Nov. 670/4 She persuades the great Arthur Brisbane, of the New York *World*..to hire her, and becomes a first-class journalist, feature-writer and columnist herself. 1958 *Punch* 1 Jan. 80/2 At the cinema trailers usually precede feature films by at least one week. 1958 *New Statesman* 25 Jan. 107/1 Success tempts the American novelist to become a kind of Public Relations Officer ad-libbing into feature journalism. 1969 *Listener* 8 May 660/2 The death last week of Joe Burroughs removes one more of the small corps of features pioneers whose common link was..a

belief that all radio is one art. *Ibid.* 5 June 811/3 (Advt.), Features Editor—to be responsible for the quality of the words of the features pages.
(*b*) Relating to a linguistic feature (sense *4 a (*d*) above).
1953 C. E. Bazell *Linguistic Form* i. 2 Sufficient evidence is also provided by the analogical extension of a feature-alternation to phonemes for which it was previously unknown. *Ibid.* 3 It is misleading to describe feature-analysis as 'analysis in depth'. 1964 *Language* XL. 221 Particularly important for an understanding of the decoding process would be experimentation with 'feature opposition'. 1967 *Word* XXIII. 218 Feature analysis in the area of morphological categories is indeed very old, probably at least as ancient as Aristotle.

feature, *v.* Add: **4.** To exhibit as a 'feature'; to make a special feature or display of, make a special attraction of. orig. *U.S.*
1888 *St. Louis Globe-Democrat* 29 Apr. (Farmer), The biggest thing I saw at the wedding was a lot of glassware and block tin knives and forks, which were featured in one of the rooms. 1906 *Westm. Gaz.* 24 Sept. 9/3 The way in which Miss Clifford had been 'featured' and 'billed' in preference to herself. 1907 *Times* (weekly ed.) 28 June 402 Every day for weeks past it has 'featured' articles..on the Japanese question. 1928 *Publishers' Weekly* 12 May 1932 You will know then why so many booksellers feature Macaulay books. 1928 G. B. Shaw *Intell. Woman's Guide Socialism* 444 The Government posters 'feature' precisely the same epithets. 1929 *Times* 1 Aug., The Louvre, Oxford-circus, are featuring coats and skirts and top coats for Scotland in new designs. 1938 E. Waugh *Scoop* I. v. 86 [Newspaperman loq.] They're featuring me as a special service.
b. *spec.* To exhibit as a prominent feature in a play, film, etc.
1897 *Metropolitan Mag.* (N.Y.) Nov. 383/2 A company that includes in addition to the 'featured' members, [etc.]. 1919 J. Buchan *Mr. Standfast* xiii. 245 Were you ever a cinema actor, Dick?..'Featuring Mary Lamington.' How does the jargon go? 1923 *Westm. Gaz.* 22 Jan., Balzac's 'The Eternal Flame', featuring Miss Norma Talmadge. 1927 *Daily Chron.* 29 Mar. 6 She said she had been 'featured' in 22 or 23 different pieces in America. 1928 *Manch. Guardian Weekly* 7 Dec. Suppl. p. ix/1 A handsome middle-aged gentleman featuring himself through five acts. 1936 *Amer. Speech* XI. 220 Is the actor to be *starred* or merely *featured*? If *starred*, the program, the signs outside the theater, and all advertising will read, 'X in Hamlet'. If *featured*, it will be, 'Hamlet with X'.

featurette (fī·tiŭrĕt, fītiŭre·t). [f. *feature *sb.* 4 a (*c*) + -ette.] A short feature film. Also *transf.*
1942 *Variety* 28 Jan. 8 This Roach featurette is not quite up to par of other issues in the series. 1957 M. Kennedy *Heroes of Clone* I. i. 16 I've sold a little screen play. Nothing important. A featurette. 1959 Halas & Manvell *Technique Film Animation* xvii. 147 Disney's *Man in Space*—a featurette which was an excellent piece of documentary science fiction. 1960 *Guardian* 21 Nov. 7/2 Correspondents who write reports in the older tradition, rather than featurettes. 1970 *Oxford Times* 25 Dec. 17 (Advt.), A Walt Disney cartoon featurette. Wind in the Willows.

featureliness. (Later example.)
1867 Bagehot *Eng. Constit.* v. 124 The picturesqueness, the featureliness, of society diminishes.

feaze, *v.*[3] *U.S.* var. Faze *v.* (Cf. Feeze *v.*[1] 2.)
1906 *Springfield Weekly Republ.* 27 Dec. 1 The gentlemen at the head of the Standard oil trust will not be feazed or troubled a bit by these revelations. 1907 C. E. Mulford *Bar-20* ix. 96 I've bit on so many of them rumours that they don't feaze me no more. 1916 'B. M. Bower' *Phantom Herd* ix. 143 That..rather feazed the Happy Family for a few minutes. 1938 [see *battle-axe 4].

febricule (fe·brikiul). Anglicized form of Febricula.
1887 R. L. Stevenson in *Longman's Mag.* I. 679 A febricule when I digest.

Febronian, *a.* Add: Also *sb.*, a follower of Febronius.
1884 Addis & Arnold *Cath. Dict.* 346/2 The Pope's power was to be reduced to that which Febronians supposed him to have exercised in the first three centuries. 1895 *19th Cent.* Nov. 797 Protestants of the sixteenth and seventeenth centuries were succeeded by Febronians and Jansenists in the eighteenth.

Fechner (fe·χʸnəɹ). The name of Gustav Theodor *Fechner* (1801–87), the founder of German experimental psychology, used in the possessive to designate laws, etc., formulated by him in attempting to establish a quantitative relationship between degrees of physical stimulation and the resulting sensation.
1874 J. Sully *Sensation & Intuition* 62 Fechner's law accurately defines this relation. 1876 *Mind* I. 455 It is thus quite possible to give a physical interpretation to Fechner's law without implying anything 'inconceivable'. 1887 G. T. Ladd *Elem. Physiol. Psychol.* v. 366 The significant addition which Fechner has made to Weber's law consists in the assumption that all just observable differences are equally great. It is therefore also called 'Fechner's law'. 1892 van Lien & Beyer tr. *Ziehen's Physiol. Psychol.* 54 The logarithmic formula of Fechner's Law. *Ibid.* 55 The undoubted advantage of being able to explain or account for Fechner's Formula entirely in accordance with the spirit of the natural sciences. 1909 C. S. Myers *Text-bk. Exper. Psychol.* 86 When a white sector upon a black ground is very slowly turned on the colour wheel, a series of black bands in the form of radii may be observed on that part of the white surface which first stimulates the eye...With somewhat more rapid rotation, especially under bright illumination, various colours, called Fechner's colours, may be visible on the white surface. 1931 *Brit. Jrnl. Psychol.* July 19 *Phenomenal Regression and Fechner's Difference Law.* ..This law..is that if a−b=b−c, then A/B=B/C..if a−b represents the sense distance corresponding to the presentation of the stimulus A with B, and b−c the sense distance similarly resulting from the stimuli B and C. 1951 S. H. Bartley in S.S. Stevens *Handbk. Exper. Psychol.* 981/2 Fechner's paradox is the sensory result when two eyes are independently exposed to *unequal* amounts of light. 1970 C. W. Savage *Measurement of Sensation* viii. 298 Until it is shown that the nature of sensations and the nature of measurement make such a discovery impossible, we cannot conclude that Fechner's law is meaningless.

Fechnerian (feχʸnīə·riăn), *a.* [See -ian.] Relating to G. T. *Fechner* (see prec.).
1890 W. James *Princ. Psychol.* I. xiii. 549 The Fechnerian *Maasformel* and the conception of it as an ultimate 'psychophysic law' will remain an 'idol of the den', if ever there was one. 1909 *Westm. Gaz.* 15 May 4/2 These Fechnerian ideas. 1942 *Psychol. Rev.* XLIX. 140 The Fechnerian logarithmic relationship is not appreciably distorted, in cases of intensity transmission, by being reflexly transformed into motor response of an involuntary character.

feck (fek), *v. slang.* [Origin unknown.] *trans.* To steal; see also quot. 1809.
1809 G. Andrewes *Dict. Slang & Cant*, *To feck*—to look out, to discover the best means of obtaining stolen goods. 1880 Jamieson II. 200/2 *To feck*, to attain by dishonourable means, Loth[ian]. 1916 Joyce *Portrait of Artist* (1964) 40 They had fecked cash out of the rector's room. 1922 —— *Ulysses* 268 Fecking matches from counters. 1962 E. O'Brien *Lonely Girl* i. 9 'Feck any samples?'..'How could I take samples with him sitting there in the car?'

fecula: see Fæcula.

Feculose (fe·kiŭlōuz, -s). *Chem.* Also **feculose.** [f. Fæcul(a + -ose[2].] Proprietary name for an acetylated starch made by heating dry starch with glacial acetic acid and formerly used as a substitute for gelatine and vegetable gums.
1903 *Trade Marks Jrnl.* 29 July 820 *Feculose*. Chemical Substances used in Manufactures. The Firm of William Wotherspoon, Glenfield Starch Works. 1909 *Jrnl. Soc. Chem. Industry* 31 Mar. 291/1 Feculose is also employed in making ju-jubes and other sweets. 1920 Cross & Bevan *Text-bk. Paper-Making* (ed. 5) 309 'Feculose'..is being employed in increasing quantities as a substitute for gelatine. 1966 H. Williamson *Methods Bk. Design* (ed. 2) xviii. 298 Less expensive papers are sometimes tub-sized with feculose. 1967 Kruger & Rutenberg in Whistler & Paschall *Starch* II. xv. 375 The dry-heating procedure for 'Feculose' probably could not compete today with other methods of preparing low D.S. acetates.

fecund, fecundity. Distinguished from *fertile, fertility* (see quots.).
1866 J. M. Duncan *Fecundity* 3 By fertility or productiveness I mean the amount of births as distinguished from the capability to bear...By fecundity I mean the demonstrated capability to bear children; it implies the conditions necessary for conception in the women of whom its variations are predicated...In short, fertility implies fecundity, and also introduces the idea of number of progeny; while fecundity simply indicates the quality without any superadded notion of quantity. 1904 *Brit. Med. Jrnl.* 23 Apr. 769 A difference is drawn between fecundity and fertility. Thus women of Irish birth in Australia are less fecund than women born either in New South Wales or in Scotland, but they are more fertile. In other words fewer Irish women have children, but to those who are fecund more children are born. 1922 A. M. Carr-Saunders *Pop. Problem* ii. 51 Fecundity is measured by the number of ripe ova produced. 1936 *Discovery* Sept. 297/2 The ratio of fertility (actual production of children) to fecundity (child-bearing capacity). 1964 Gould & Kolb *Dict. Soc. Sci.* 265/2 Fertility in modern demographic usage relates to the actual frequency of births and carries no overtones of ability to have children. For the latter concept fecundity or reproductive capacity are now used.

Fed, fed, *sb.* (Earlier and later examples.)
1788 [see *anti-fed]. 1801 [see *anti *sb.*]. 1848 *Campaign Flag* (Maysville, Ky.) 24 Mar. 1/1 The feds have got to polling the passengers on steam-boats again. 1947 *West. Pennsylvania Hist. Mag.* Mar.–June 22 There were few marriageable men left in the narrow confines to which the aristocratic 'Feds' had limited themselves.
2. An official of the federal government; *spec.* a member of the U.S. Federal Bureau of Investigation. *slang.*
1916 A. Stringer *Door of Dread* iv. 53 Seein' Kestner and yuh'd told me the Feds had ev'rything fixt, I give him the glassy eye. 1935 *Lit. Digest* 22 June 38 'G Men' have also been called 'feds'. 1955 *Publ. Amer. Dial. Soc.* XXIV. 47 Anyway, the Feds got the letter where I sent him $400.

fed, *pa. pple.* Slang phr. *to be fed up*: to be surfeited or disgusted (*with*), bored or tired to breaking-point. So **fed-up** *a.* Also with emphasizing expressions, as *fed to the* (*back*) *teeth, fed to death*. Also without *up*.
[1882 F. Arnold *Three-Cornered Essays* (1886) 4 Fed up to the eyelids, it is no care to him that there are other people all otherwise than so well off.]
1900 *Westm. Gaz.* 13 Nov. 2/3 It may be quite true that, to use an expression often heard in South Africa just now, the men are 'fed up' with the war. 1906 *Daily Chron.* 22 Mar. 9/5, I am about 'fed-up' over this motor-car. 1914 *Even. News* 19 Sept. 4/1 We have also seen hundreds of German prisoners, mostly looking 'fed up'. 1916 'Boyd Cable' *Action Front* 32 'I'm just about fed up with him,' said Gunner Donovan bitterly. 1921 W. S. Maugham *Circle* III. 100, I should be fed to the teeth with you sometimes. 1921 Wodehouse *Indiscretions of Archie* I. 11 I've never done anything much in England, and I fancy the family were getting a bit fed. 1922 A. S. M. Hutchinson *This Freedom* IV. iv. 293 Oh, those sickening scarves and things, they were eternally knitting, that wasn't war work. It was fun at first. They were fed to death with doing them now. 1924 A. J. Small *Frozen Gold* iii. 78 He had the air of a man who was utterly and finally fed-up with the eternal effort of keeping the upper hand of the worst. 1928 *Strand Mag.* July 4 Her aspect was that of a girl who is fed to the teeth. 1929 D. G. Mackail *How Amusing* 342 It made me feel pretty fed with the old boy. 1946 P. Larkin *Jill* 30 I'm getting definitely fed..with them all. 1955 *Times* 15 June 9/2 A major relief to a frankly fed-up and frustrated nation. 1958 M. Allingham *Hide my Eyes* x. 99 She's been getting fed to the teeth. Edna has been pestering her. 1961 'G. Hythe' *Death of Scapegoat* I. iii. 87 The skipper's gettin' fed to the back teeth. 1963 I. Fleming *On H.M. Secret Service* ii. 21 He was fed to the teeth with chasing the ghost of Blofeld.
Hence **fed-upness** (stress variable).
1928 D. L. Sayers *Unpleasantness at Bellona Club* iii. 22 Job gone—health gone—no money..general fedupness. 1961 *Times* 4 May 13/4 A state of acute 'browned-offness' or 'fed-upness'.

‖ **Fedai** (fe·dā,ī). Pl. **fedai, fedais.** The Pers. forms also occur. [Pers. *fidā'ī* (pl. *fidā'iyān*) devotee, zealot, one who risks his life in a cause.] = Assassin 1.
1875 *Encycl. Brit.* II. 723/1 When the sheikh required the services of any of them, the selected *fedais* were intoxicated with *hashish*. 1909 *Westm. Gaz.* 15 Feb. 2/1 Thousands of them became 'fedai', men ready to sacrifice themselves for the common good. 1921 *Glasgow Herald* 13 Sept. 7 Seventeen Fedai, otherwise reckless desperadoes, are stated to have been specially brought to Constantinople for the purpose of committing murders. 1958 J. A. Boyle tr. *Juvaini's Hist. World-Conqueror* II. 739 *Fidā'ī*, the term applied to the emissaries of the Isma'ilis, the Assassins. 1969 A. J. Arberry *Relig. Middle East* II. 626 Manfred Halpern speaks of the Muslim Brethren in Egypt, Syria and elsewhere, together with movements like *Fidā'iyān-i-Islām* in Persia.

‖ **fedayeen** (fedayī·n), *sb. pl.* Also **-yin.** [Colloq. Arab. *fidā'iyīn*, pl. of Class. Arab. *fidā'ī* one who undertakes perilous adventures.] Arab guerrillas operating against the Israelis.
1955 *Time* 12 Sept. 34/2 Small groups of Arab raiders carried the fight deep into Israel. Known as *Al Fedayeen* (Self-Sacrificers), the sneaker-shod guerrillas are recruited from Palestinian Arab refugees. 1956 *Time* 23 Apr. 33/3 It was the fedayeen (self-sacrificers), members of specially trained Arab assassin squads, who had crept north from the Egyptian-held Gaza strip. 1956 *Economist* 17 Nov., The only answer to fedayeen tactics is reprisals. 1960 *Spectator* 4 Mar. 311 Several rightly critical references are made to Egyptian fedayeen raids on Israel. 1969 *Times* 20 Nov. p. iv/5 Everywhere in Jordan the tourist will see *fedayin* bristling with arms. 1970 *Daily Tel.* 22 Jan. 18 Once again it was activity by the *fedayeen*, not by regular Jordanian forces, in attacking Israeli settlements that triggered the reprisal.

federal, *a.* and *sb.* Add: **A.** *adj.* **4. b.** Applied to a non-political association of quasi-autonomous units.
1911 *Encycl. Brit.* XXII. 293/2 A loose union, called the 'Federal Council of the Reformed Churches in America', was formed in 1894. 1954 *Times Lit. Suppl.* 5 Feb. 93/1 The geography and diverse character of Wales made a federal university of scattered colleges a necessity.
B. *sb.* (Earlier example.)
1861 *Daily Dispatch* (Richmond, Va.) 24 July 3/5 The federals advanced gradually among the masked batteries.

federator (fe·dĕrēitǫɹ). [f. Federate *v.* + -or.] One who or that which makes a federation; one who takes part in a federation.
1879 J. Martineau *Ess.* (1891) IV. 271 It's [*sc.* religion's] power..as a federator of nations..is freely admitted. 1901 *Wide World Mag.* VI. 472/2 Professor Bickerton himself paying on the same scale as any other Federator.

fedora (fĕdōə·ră). orig. *U.S.* [f. *Fédora*, title of a drama (1882) by Victorien Sardou (1831–1908).] A low soft felt hat with a curled brim and the crown creased lengthways. Also *attrib.*
1895 *Montgomery Ward Catal.* 274/3 For ease, comfort and style, the soft Fedora hats lead all others. 1897 *Sears, Roebuck Catal.* 232/3 Our Young Men's Special Nutria Fur Fedora...Men's Full Shaped Fedora Hat. 1910 *N.Y. Even. Post.* 21 Apr., The fedora dented at wrong angles. 1916 S. Leacock *Moonbeams* ii. 23 There was no use in my having bought a sage-green fedora in Broadway. 1925 L. Bromfield *Possession* ix. 46 He took

off..his brown fedora hat. *Ibid.* xvii. 100 In place of a warm skating cap he wore a Fedora hat pulled over his ears. **1926** *Daily Colonist* (Victoria, B.C.) 17 July 1/7 Men's Straw Fedora Shape Hats...A comfortable every-day hat for the hot Summer days; made in Fedora styles with black ribbon. **1930** *Punch* 26 Feb. 236 He carried a cheap black fedora in his hand. **1937** *Times* 11 Aug. 15/5 Soft hats of tweed and jersey material have stitched brims and fedora-shaped crowns. **1952** Wodehouse *Pigs have Wings* ii. 40 The investigations he has been making into the bumping off of the man in the green fedora.

fee, *sb.*^{2} Add: **11. b.** *fee-payer*; *fee-charging* adj.

1897 *Daily News* 2 Mar. 2/6 They did not know the number of fee-charging schools..as distinguished from the general elementary system of the country. **1959** I. & P. Opie *Lore & Lang. Schoolch.* xiii. 298 The private fee-charging establishments. **1931** *Times Educ. Suppl.* 15 Aug. 321/2 The entrance examination for fee-payers.

feeb (fīb). *U.S. slang.* [Short for Feeble-*minded.*] A feeble-minded person; a 'dumb' person.

1914 *Bookman* XXXIX. 432/1 Feeb? Oh, that's feeble-minded. I thought you knew. We're all feebs in here. **1931** W. Faulkner *Sanctuary* (1932) i. 8, I cook. I cook for crimps and spungs and feebs. Yes, I cook. **1968** E. Fenwick *Disturbance on Berry Hill* vii. 51 'He really is the sweetest guy,' Georgia said wistfully. 'Sometimes I envy that pretty little feeb of his. Suppose if I played dumber I could get one too.'

feeble, *a.* and *sb.* Add: **A.** *adj.* **9.** *feeble-minded* (examples of technical definitions).

1908 *Rep. R. Comm. Feeble Minded* VIII. 324 in *Parl. Pap.* (Cd. 4202) XXXIX. 159 'Feeble-minded', *i.e.*, persons who may be capable of earning a living under favourable circumstances, but are incapable from mental defect existing from birth or from an early age: (*a*) of competing on equal terms with their normal fellows; or (*b*) of managing themselves and their affairs with ordinary prudence. **1963** Barnard & Lauwerys *Handbk. Brit. Educ. Terms* 93 *Feeble-minded*, a term usually applied to those whose I.Q. is between 55 and 69.

B. *sb.* **1.** Delete †*Obs.* and add later examples.

1833 J. S. Mill *Let.* 10 July (1963) XII. 166 The consequence is they must take the *feebles.* **1844, 1896** [see Forcible feeble].

feeble, *v.* **1.** Delete †*Obs.* and substitute *arch.* (Later example.)

1889 'Mark Twain' *Connecticut Yankee* 181 Sir Gawaine's strength feebled and waxed passing faint.

feebling (fī·bliŋ). [f. Feeble *a.*: see -ling^1.] A weakling, a feeble person.

1887 H. A. Beers *Outl. Sk. Amer. Lit.* vii. 273 [Henry James] has exhibited types of the American girl, the American business-man, the æsthetic feebling from Boston. **1896** *Westm. Gaz.* 13 June 2/1 To shriek with horror at them, the sign of a feebling. **1914** S. Baring-Gould *Ch. Revival* 151 The Church was supplied with feeblings as candidates for Orders.

feed, *sb.* Add: **3. c.** Also, food, fare (for human beings). *U.S. colloq.*

1818 H. B. Fearon *Sk. Amer.* 194, I guess whiskey is all the feed we have on sale. **1835** *Knickerbocker* V. 304 A John Smith lives next door, to whom half my choice rounds and sirloins, selected personally in the market,—for I love good feed,—are sent without distinction. **1867** S. Hale *Lett.* (1919) 27 The cook is French and feed delicious. **1898** E. N. Westcott *David Harum* 283 You want a change o' feed once in a while, or you may git the colic.

5. c. (Additional example.)

1957 *Encycl. Brit.* IX. 561/2 The feed for controlling the movement of the writing fluid [in a fountain pen] is made of plastic or of hard rubber.

d. *Theatrical slang.* = Feeder 11. Also in extended and *attrib.* uses.

1929 J. B. Priestley *Good Companions* ii. i, Joe over there..is as good a feed as you could wish for. *Ibid.*, You couldn't want a better dancer...The only thing is, he won't feed. I never struck a worse feed. **1936** N. Coward *To-night at 8.30* I. 94, I don't count—I'm only a feed. **1958** Osborne & Creighton *Epitaph G. Dillon* ii. 61, I suppose I give you what is known as the 'feed' line now. **1958** *Listener* 7 Aug. 189/1 He [*sc.* the interviewer] can lure his victim into a false sense of confidence with the easy-paced, perfectly straightforward, guileless 'feed' question. *Ibid.* 28 Aug. 311/1 English voices.. are nearly always used as 'feeds' for knowledgeable Americans [in broadcast discussions]. **1961** *John o' London's* 14 Sept. 307/2 Cantinflas, together with his feed-man Charamusca.

6. a. (sense 3 and 3 b) *feed-bag* (earlier and *transf.* (*joc.*) examples), *-bin, consumption, -house, -lot, -stable, -station, -stuff, -yard*; (sense 5) *feed-bar, cistern, -cock* (earlier example), *-dog* [Dog *sb.* 7], *-hole* (earlier example). **b.** (sense 3) *feed-chopper.*

1840 C. Mathews *Politicians* I. ii. 16 You prefer to be mystified more after the manner of a cartman's horse with his head in a feed bag. **1906** E. Dyson *Fact'ry 'Ands* v. 54 The bloke that comes canoodlin' here gets that in his feed-bag! **1962** *Punch* 4 July 30/1 The notion that viewers would like to see celebrities at their feed-bags. **1873** *Young Englishwoman* Mar. 131/2 The feed bar and stitch mechanism are inclosed in a box. **1876** J. S. Ingram *Centenn. Exposition* ix. 298 The radiating arms..act against the feed-bar. **1898** M. Deland *Old Chester Tales* 250 The open space between the stalls and the feed-bins should be the stage. **1903** *Cincinnati Enquirer* 9 May 14/1 He got a motor for the corn sheller and feed chopper. **1916** H. L. Wilson *Somewhere in Red Gap* vii. 304 Rex II [beagle] didn't get in till next day and looked like he'd come through a feed chopper. **1838** *Civil Engin. & Arch. Jrnl.* I. 237/2 The whole circle of pipes and vessels described, is supplied with water from the feed cistern. **1967** *Gloss. Sanitation Terms* (*B.S.I.*) 14 *Feed cistern*, a cistern for supplying cold water to a hot water system. **1833** B. Silliman *Man. Sugar Cane* 56 It is drawn up into the vacuum pans..by the feed cocks. **1909** *Westm. Gaz.* 27 Aug. 2/3 Barley..or other meal, which..form so large a factor in the feed consumption of this country. **1961** *Observer* 28 May 33/2 Its [*sc.* a sewing-machine's] feed-dog tended to cockle nylon. **1961** *Which?* Nov. 285/1 Normally the feed dog feeds material directly towards or away from you. **1874** J. W. Long *Amer. Wild-Fowl Shooting* 37 A quick loading flask, i.e., one having a large feed-hole to the charger, should also be used. **1961** C. H. Douglas-Todd *Pop. Whippet* iii. 37 It [*sc.* the kennel-building] must incorporate the feed-house..and the room..where all the 'paper work'..may be carried out. **1889** *Las Cruces* (New Mex.) *News* 16 Nov., [The new lower rate] allows Kansas feeders to ship from this territory or Arizona to their feed lots. **1911** H. Quick *Yellowstone Nights* vi. 165 Two boys..had met in Allen's feed-lot to fight a duel. **1964** *Punch* 25 Nov. 807/1 They're..unlikely to..nest in electric fences, feedlots and barbed wire. **1877** H. C. Hodge *Arizona* 154 Tucson has..four feed and livery stables. **1902** O. Wister *Virginian* ii. 13, I took its dimensions, twenty-nine buildings in all,—..one feed stable, and..others. **1910** W. M. Raine *B. O'Connor* 106 Jay Hardman's place, a tumble-down feed-station on the edge of town. **1856** *Porter's Spirit of Times* 4 Oct. 74/3 But a few years since our whole supplies of bread and feed-stuffs [in Calif.] were drawn from abroad. **1909** 'O. Henry' *Roads of Destiny* ix. 157 There was a feedstuff famine in Mexico. **1957** *Times* 2 Dec. p. ii/3 The post-war period..until 1953 when feedstuffs were derationed. **1969** *Daily Tel.* 18 Dec. 18/1 The animal feedstuffs market. **1879** *Chicago Tribune* 14 May 7/4 The feed-yards in Chicago are extensive. **1911** H. Quick *Yellowstone Nights* iv. 109 One, two, three farmsteads we passed, with its white house hidden in trees, low hog-houses, its feed yards.

7. feed-block (see quot. 1902); **feed-board,** a board on a printing machine to hold sheets of paper fed to the machine; **feed check valve** (also *feed check* ellipt.), a valve placed between the feed-pipe and the boiler, to prevent return of feed-water; **feed collector** (see quot.); **feed-door** (earlier U.S. example); **feed-floor** *U.S.*, a floor off which cattle, etc., can eat food; **feed-fo·rward** [after *feedback], (*a*) the use of calculated or presumed future states of a process to provide criteria for its adjustment or control; anticipatory control; (*b*) the modification of the output signal of a circuit by a part of the input signal that has not passed through the circuit; **feed gear,** the mechanism of feed-motion; **feed-heater** = *feed-water-heater*; **feed horn,** a horn-shaped section of waveguide mounted in front of an aerial and used to direct a signal into it for transmission or to receive a detected signal from it; **feed-mouth** = *feed-door*; **feed-room** *U.S.* and *N.Z.*, a room in which food for animals is stored; **feedstock** orig. *U.S.*, 'raw material supplied to a machine or processing plant (as pulpwood to a paper mill)' (Webster 1961); also *transf.*; **feed-strip,** a cartridge belt for a machine gun; **feed-table,** a table indicating the food values of fodders; **feed-tank,** also, a tank containing drinking-water; **feed-trough,** also *U.S.* a trough in which food for animals is placed; **feed-tub,** the supply vessel of an evaporator; **feed-wire** = Feeder 10.

1895 H. T. Lukin *Maxim Machine Gun* 11 What keeps the extractor at its highest until the cartridge is drawn from the feed block? *Ibid.* 21 When loading for rapid fire, the crank handle is turned over to buffer spring twice and the belt is pulled through the feed block. **1902** *Encycl. Brit.* XXX. 401/2 The feed-block through which the belt of cartridges is fed to the gun. **1928** C. F. S. Gamble *N. Sea Air Station* xvi. 295 A left-hand feed-block, which was instantly interchangeable with the right-hand feed-block. **1899** J. Southward *Mod. Printing* iii. ix. 108 The feed board rises, the lays drop, and the grippers take the sheet. **1917** R. A. Peddie *Outl. Hist. Printing* 33 These machines print single sheets on both sides during their traverse from the feed-board to the delivery board. **1967** Karch & Buber *Offset Processes* ix. 372 The back paper-feed control..lever.. permits the paper to feed on to the feedboard. **1895** *Daily News* 17 Dec. 3/5 To repair feed check valve. **1908** *Westm. Gaz.* 25 Apr. 2/3 Off they went into the stokehole, where the Third put two of them to mind the feed-checks. **1902** *Encycl. Brit.* XXVI. 284/1 A horizontal cross-tube of square section, called a feed collector, which extends the whole width of the [marine water-tube] boiler. **1872** *Rep. Vermont Board Agric.* I. 635 The ore..is landed at the feed door of the furnace. **1868** *Rep. Iowa Agric. Soc. 1867* 104 Put them [*sc.* hogs] in small yards with a good plank feed-floor. **1952** I. A. Richards in *Cybernetics: Trans. of Eighth Conf., 1951* 54 You have no doubt fed forward enough to see that what I am going to talk about from now on is feed-forward. I am going to try to suggest its importance in describing how language works. **1961** E. J. Baghdady *Lect. Commun. Syst. Theory* xix. 505 The feedforward operation..can be extended so that the amplifier bridges two or more cascaded narrow-band limiters. **1963** *Engineering* 6 Dec. 726/3 Anticipatory (feed-forward) control predicts the effect of input variables on output variables by solving transfer equations. **1888** *Lockwood's Dict. Mech. Engin.* 136 *Feed*, the amount or distance of the transverse of a lathe or other machine cutter taken transversely to the depth of the cut. The term is also applied to the feed gear itself. **1895** *Daily News* 27 July 3/1 The torpedo boat destroyer Ferret, which broke down..owing to the defective working of her automatic feed gear. **1864** Webster, Feed-heater. **1885** *Marine Engineer* 1 Apr. 14/2 Although with the same area exposed, the feed-heater must be heavier..yet area for area the feed-heater must be much more efficient. **1952** *Electronics* May 126/3 The main beam from the scanner did not vary with feed-horn rotation when the horn dimension was equal to three and to four-channel widths. **1962** F. I. Ordway et al. *Basic Astronautics* vi. 293 Radio energy from the antenna enters the feed horns. **1833** B. Silliman *Man. Sugar Cane* 34 An opening into the furnace, called the feed-mouth, for the supply of fuel and the regulation of the fire, is left. **1887** A. W. Tourgée *Button's Inn* 142 A great towel..hung inside the feed-room door. **1923** 'K. Mansfield' *Doves' Nest* i, There it stayed..beside the feed-room door. **1939** —— *Scrapbook* 2 Every morning I went across to the feed-room where he cleaned father's boots. **1932** Boehtlingk et al. tr. *Sachanen & Tilicheyev's Chem. & Tech. of Cracking* viii. 330 The hot oil pump circulates the clean feed stock from the base of the bubble tower through the heater. **1958** *Times* 15 July 4/6 The erection of the first plant producing gas for town use from oil feedstocks. **1963** *Economist* 28 Sept. 1140/2 These..techniques..extended to use petroleum feedstocks. **1969** *New Scientist* 30 Jan. 233/3 Para-xylene, an essential feedstock in the manufacture of polyester fibres. **1969** *Sci. Jrnl.* Mar. 25/2 A gas centrifuge designed to separate the lighter uranium-235 isotope from the bulk uranium-238 (in the form of gaseous uranium hexafluoride feedstock). **1969** *Times* 1 July 5/8 Molten rocks might occur and thus provide feedstock for volcanoes. **1902** *Encycl. Brit.* XXX. 404/2 The breech-block..driving a cartridge in front of it out of the feed-strip. **1886** C. Scott *Sheep-farming* 33 He can only examine the feed tables..and guess at the quality of his own fodders. **1886** *Marine Engineer* 1 Feb. 283 Storage-tanks..being connected by pipes to the small feed-tanks situated above the boiler. **1923** *Man. Seamanship* (H.M.S.O.) II. 26 *Fresh Water Service.* The shipwright officer is responsible for this service (except supplies to the feed tanks). **1845** A. Wiley in *Indiana Mag. Hist.* (1927) XXIII. 212 To come with their wagons, and grain, and hay, and feed troughs and watering buckets. **1854** J. R. Bartlett *Pers. Narr. Explor. Texas* II. 12 All [the wagons] had to be provided with feed-troughs. **1867** *Rep. Iowa Agric. Soc.* (1868) 148 Our market is in the feed-troughs. **1878** *Rep. Vermont Board Agric.* 109 The feed tub will be high enough for the sap to run from that to the evaporator. **1903** *Work* 1 Aug. 412/3 A feed wire..will connect with the wires of the netting.

feed, *v.* Add: **4.** *to feed the fishes*: see also Fish *sb.*^{1} 1 c.

6. *to feed up.* (Later and *intr.* examples.)

1913 W. Owen *Let.* 23 Apr. (1967) 185 Congratulations on the success of your feeding me up. —— *Let.* 6 Nov. 207 Began to 'feed up' today, but the difficulty is tremendous; i.e. of getting foods. *a* **1951** J. K. Ewers in *Austral. Short Stories* (1951) 335 I'll just help Larry feed up.

c. Also, *trans.* and *intr.*, to supply another character with cues. (Cf. Feeder 11, *feed *sb.* 5 d.)

1921 N. B. Tarkington *Let.* 21 Apr. (1959) 55 I've built it..craftily..so that the part isn't ever visibly 'fed'. **1929** [see *feed *sb.* 5 d]. **1929** D. G. Mackail *How Amusing* 66 I've had laughs I've earned, and laughs the other fellow's earned by my feeding him. **1968** *Listener* 11 July 59/3 The rest of the cast loll around, feeding the soloist with helpful questions.

g. *Football*, etc. To give a pass to.

1889 Barrère & Leland *Dict. Slang* s.v., To *feed*, to support. **1897** *Encycl. Sport* I. 417/2 The way in which he can aid his side in attacking is by passing to, or, as it is called, feeding his forwards.

h. *slang. trans.* and *intr.* To tire or bore (cf. *fed *pa. pple.*).

1933 G. Heyer *Why shoot a Butler?* iii. 47 Anyone can have the super motor boat as far as I'm concerned. Joan, too. She bars it completely, which feeds Brother Basil stiff. **1940** M. Marples *Pub. Sch. Slang* 77 'It's *feeding*, isn't it?' (i.e. calculated to make one *fed-up*).

i. To accompany (a musician, esp. a jazz musician); *spec.* to play accompanying chords for (a jazz soloist); also, to provide (an accompaniment). *colloq.* (orig. *U.S.*).

1949 L. Feather *Inside Be-Bop* ii. 59 The guitar..is employed like the piano to 'feed' or 'bop' the soloists by 'comping' with irregularly accented chords. **1955** in M. Stearns *Story of Jazz* (1957) xviii. 237 Lennie feeds his chords off the beat by use of irregular accents. **1958** P. Gammond *Decca Bk. Jazz* xii. 140 The pianist fed a progression of relevant chords in a manner which gave the soloist the greatest impetus.

7. b. (Additional examples.)

1879 Geo. Eliot *Impressions of Theophrastus Such* xvii. 302 Each new machine needs a new appliance of human skill to construct it, new devices to feed it with material. **1961** [see *damp *a.* 4]. **1969** *Guardian* 18 Nov. 13/1 It is still cheaper to feed the meters (illegally) all day than pay the high price of off-street parking.

c. To relay or supply electrical signals or power to, esp. as part of a larger network or system.

1894 A. T. Snell *Electric Motor Power* iv. 137 They.. feed the distributing network with current at a pressure of about 100 volts. **1952** E. A. Laport *Radio Antenna*

Engin. ii. 132 It is necessary to feed each line with the potential that will transmit the proper amount of power down each line. **1966** *McGraw-Hill Encycl. Sci. & Technol.* XI. 251/2 Along the route there are Bell system offices, which feed local stations.

8. b. (Earlier and later examples.) Also with *out*, and with ellipsis of indirect object. Also, to supply (food) *to* (a person, etc.).

1818 in *Trans. Ill. State Hist. Soc. 1910* 158 They either have to feed out their corn or their cattle get very poor. **1852** *Trans. Mich. Agric. Soc.* III. 145, I feed almost every thing, hay, oats, straw, [etc.]. **1868** *Rep. Iowa Agric. Soc. 1867* 148 Corn is husked and cribbed and fed out to stock. **1904** *Grand Rapids Even. Press* 2 June 3 The professor..fed snake sandwiches to his college class at a party. **1911** R. W. Chambers *Common Law* viii. 237 Rita..fed them bits of cassava and crumbs of cake. **1946** *Harper's Mag.* Oct. 311/1 Now and then my father would have to borrow money to 'feed-out' the steers. **1961** *New Eng. Bible Matt.* vii. 6 Do not feed your pearls to pigs.

e. *to feed back* (*Electr.* and *Cybernetics*): (i) to return (a fraction of an output signal) to an input of the same or a preceding stage of the circuit, device, process, etc., that produced it. Also *transf.* Chiefly in *pass.* (Cf. *feedback.)

1921 *Wireless World* 10 Dec. 571/2 The magnified oscillations are fed back again into the grid circuit. **1940** *Nature* 7 Sept. 321/2 The pencil draws a curve and after an interval the information expressed by this curve is fed back to the machine. **1952** *New Biol.* XIII. 54 Information about this error must be 'fed-back' into the machine and cause it to alter its condition in such a direction as will diminish the error. **1960** O. Skilbeck *Film & TV Terms* 52 Part of an amplifier's own output may be deliberately fed back to reduce inherent sound distortion.

(ii) In *transf.* sense also used *intr.* of a result or effect of a process: to return as feedback; to affect or modify the process that brought it about.

1945 *Jrnl. Franklin Inst.* CCXL. 266 The *y*-shaft is driven by the output of integrator II and feeds back to drive the input of the function unit. **1960** R. W. Marks *Dymaxion World of B. Fuller* 23/2 Since the benefits keep feeding back into the system, such techno-economic patterns are infinitely regenerative. **1964** M. A. K. Halliday et al. *Linguistic Sci.* x. 275 The experience from the teaching of English to foreign learners is feeding back..to the teaching profession in Britain. **1966** *Rep. Comm. Inquiry Univ. Oxf.* I. 56 There are also advances in social studies, at postgraduate level, that are likely to feed back into undergraduate work.

fee·dback, feed-back, *sb.* [f. Feed *v.* + Back *adv.*] **a.** *Electr.* The return of a fraction of the output signal from one stage of a circuit, amplifier, etc., to the input of the same or a preceding stage; *positive, negative feedback*, tending to increase, decrease, the amplification, etc.; also, a signal so returned.

1920 *Wireless Age* VIII. 27/1 An inductive feed-back in relation to the secondary system generates local oscillations. **1923** *Daily Mail* 19 May 4 The feed-back circuit. **1946** *Nature* 9 Nov. 670/1 The signal to noise ratio was increased by the application of positive feed-back to the tuned circuit. **1949** *Electronic Engin.* XXI. 358/3 To obtain the maximum stabilising effect it is desirable to take the feedback from the output and apply it at the input in the correct phase for degeneration. **1962** M. G. Say *Newnes Conc. Encycl. Electr. Engin.* 289/1 The feedback principle is applied in electronic amplifiers, process controls and accurate servo-mechanisms.

b. *transf.* The modification, adjustment, or control of a process or system (as a social situation or a biological mechanism) by a result or effect of the process, esp. by a difference between a desired and an actual result; information about the result of a process, experiment, etc.; a response.

1943 *Philos. Sci.* X. 19 Purposeful active behavior may be subdivided into two classes: 'feed-back' (or 'teleological') and 'non-feed-back' (or 'non-teleological'). **1945** *Jrnl. Franklin Inst.* CCXL. 265 The shaft motion produced as the result of a sequence of operations is connected back into the sequence, a process called 'feed-back'. **1955** *Times* 31 Aug. 9/4 Only by constant 'feed-back' from the receptor can one ascertain how far a message has been understood rightly. **1959** J. L. M. Trim in Quirk & Smith *Teaching of English* iii. 84 In..a lecture ..the live speaker has a reaction, a 'feed-back' from the listeners, and..he can adjust his speech accordingly. **1962** *Listener* 30 Aug. 303/2 The feed-back from technology..has resulted in the production of more powerful and discriminating instruments. **1968** *Times* 6 Dec. 19/2 The production of both hormones seems to be regulated on a feedback mechanism by the level of calcium in the blood. **1970** *Sci. Amer.* Mar. 68/3 The brain has to receive feedback from the muscles and joints to correct the program of impulses directed to the motor apparatus. **1971** *Frendz* 5 Aug. 20/1 We began to get a fairly good feedback from most people who know about it and it looked as though the concerts would be good scenes.

feed-box (fī·dbǫks). [f. Feed *sb.* + Box *sb.*^{2}] **a.** A box for containing fodder. **b.** A box containing the feeding apparatus of a machine.

1836 E. L. Willson *Diary* 21 May (1929), I..stood behind the wagon and held to the feed box. **1886** C. Scott *Sheep-Farming* 80 A number of small feed boxes, suitable for feeding the ewes in the pens. **1895** *Pall Mall Gaz.* 26 June 2/3 As if he had his nose in a feed-box and was looking for something nice. **1902** *Encycl. Brit.* XXX. 404/1 The platform on the top of the feed-box [of a machine gun] through which the teeth of the smaller feed-wheel project. **1947** *Time* 3 Feb. 20/3 Each tried to butt the other out of the feedbox.

feeder. Add: **3. a.** (Later examples of sense 'a child's bib'.)

1895 *Montgomery Ward Catal.* 80/1 Children's Feeders ..fancy printed designs...Size 10×12. **1932** D. C. Minter *Mod. Needlecraft* 246/1 *Feeder.* Oblong 9″×12″ with neck hollowed out..plaited cord sewn on for tying. **1966** *Price List* (Olney Amsden & Sons) 1 Bibs and Feeders. Plastic..Feeders..12/- dozen.

6. d. A branch road, railway line, air service, etc., linking outlying districts with the main lines of communication. Freq. *attrib.* (see *12).

1855 *Chicago Times* 25 Jan. 2/3 This road has many lateral roads as feeders to it. **1858** Simmonds *Dict. Trade*, *Feeder*,..a branch railway, running into the main-trunk line. **1882** [see Dict. 6, *transf.*]. **1902** *Encycl. Brit.* XXIX. 445/1 The construction..of branch [railway] lines..forming feeders. **1904** Belloc *Old Road* 105 It was a branch track..one of the many 'feeders' which confuse the record of the Old Road. **1961** *Guardian* 29 Apr. 12/1 The airport [at Birmingham] will act as a feeder for Manchester and London.

8. c. (Additional examples.)

a **1877** Knight *Dict. Mech.* I. 829/1 *Feeder*,..a device with fingers which take the top sheet from a pile and lead it into the press where it is printed, [etc.]. Also a device by which blanks are taken successively from a pile and carried into an envelope-machine, [etc.]. **1940** *Chambers's Techn. Dict.* 326/1 *Feeder*, a mechanical appliance for supplying broken rock or crushed ore.. to some form of crusher or concentrator.

9. a. (Additional examples.)

1928 W. Rawlinson *Mod. Foundry Op. & Equip.* xviii. 255 For heavy castings, a feeder (or feeders) of suitable size and shape..is employed, this feeder being located to connect to the top of the casting or to the part of thickest section. **1928** *Jrnl. Iron & Steel Inst.* CXVII. 404 A refractory-lined feeder head is invariably used. **1958** *Ibid.* CLXXXIX. 263/2 It is generally held that the feeder head is useful because products of flotation from the ingot body rise into that part, which is eventually scrapped. **1960** R. Lister *Decorative Cast Ironwork* ii. 24 Another reason for the presence of the sprue and pouring basin is to provide a reservoir to ensure that there is sufficient metal to fill out every part of the mould cavity. For the same reason a riser, feeder, feeding head, or shrink head may be used.

10. Substitute for def.: **a.** A heavy untapped main for carrying electrical energy to a distribution point or system. (Earlier and later examples.)

1886 G. Kapp *Electric Transm. Energy* viii. 213 Another kind of junction..is the so-called 'junction safety catch box', designed for connecting so-called 'feeders' with certain points in the network of mains. **1892** [in Dict.]. **1930** H. P. Seelye *Electrical Distribution Engin.* vi. 82 In the general distribution in the built-up districts of a large town or city, it is usually advantageous to divide the territory served into distinct feeder areas. **1968** 'E. McBain' *Fuzz* xii. 188 Into this substation ran high-voltage supply cables ('They're called feeders,' Ahmad said) from a switching station.

b. An electrical connection between an aerial and a transmitter or receiver of electro-magnetic waves.

1928 Sterling & Kruse *Radio Man.* xiv. 529 The single wire transmission line type of feeder requires that the feeder be connected to the antenna at a voltage loop. **1954** E. Molloy *Radio & Telev. Engin. Ref. Bk.* xxi. 24 Co-axial and twin feeders in general communication use for television reception will reduce the signal by about half (6 db) if 200 ft. is used. **1962** *B.S.I. News* Sept. 21 Two-pin connectors for balanced television feeders.

11. (Earlier and later examples.)

1866 W. Davidge *Footlight Flashes* xvii. 220 Not a scene that wasn't a complete feeder from beginning to end. **1957** *Oxf. Compan. Theatre* (ed. 2) 106/1 The comedian..was assisted also..by..the straight man or 'feeder', who was dressed in perfect evening attire.

12. *attrib.* and *Comb.* (sense *6 d) *feeder airport, line, liner, plane, railway, road, service*; **feeder ear,** a metal fitting attached to a tramway contact wire. Also, **feeder-head** (see sense 9 a), **-stream** (sense 6).

1962 *Aeroplane* CIV. 13/1 Colchester industrialist, Mr. Geoffrey Woods has purchased the disused R.A.F. airfield at Wormingford and is to hold it in trust for the town for future development as a 'feeder' airport. **1970** *Daily Tel.* 10 July 12 The Plymouth City Council is proposing to construct a 'feeder' airport in a beautiful situation only a mile from the edge of the town. **1924** Roget *Dict. Electr. Terms* 87/2 *Feeder ear*, a special type of trolley wire ear to which a feeder is connected. **1895** *Westm. Gaz.* 6 June 6/2 The policy of building what are known as feeder lines is one that is much advocated by politicians who take an active interest in the future of India. **1903** Earl of Cromer in *Daily Chron.* 30 Jan. 5/5, I want feeder lines to Kassala. **1922** *Flight* XIV. 198/1 Once the airship service was started, branch or feeder lines operated with aeroplanes or seaplanes would be established almost at once. **1969** *Jane's Freight Containers 1968–69* 452/1 Lighter feeder-line transport in production and world-wide service. **1946** *Aeroplane Spotter* 16 Nov. 266/1 Twenty de Havilland Doves have been ordered by the Argentine Government; these will be used as 'feeder liners', operating in conjunction with the Vickers Vikings operating the main routes. **1949** *Flight* 24 Nov. 683 The C.A.S.A. group in Spain has developed the Type 201 Alcotan, as a feederliner and crew trainer. **1971** J. Stroud *World's Airliners* xiii. 109 Switzerland has produced a successful feederliner. **1934** *Flight* 15 Feb. 148/2 The company will operate without a Government subsidy and will employ..two feeder planes. **1959** *Observer* 8 Nov. 5/1 You start at a roundabout..on a strange feeder road..only two lanes in width. **1961** *Architect & Building News* 21 June 813 It is possible to drain traffic smoothly away into great feeder roads and motorways. **1933** *Meccano Mag.* Feb. 111/1 'Feeder' air services are now being operated in conjunction with both the Indian and African Empire air mail services. **1934** *Jrnl. R. Aeronaut. Soc.* XXXVIII. 508 Small in total prospect, but first in present availability, comes airport feeder service. Vertical flight would enable air travellers to take off from the centre of the city for transfer to air liners at the outlying airports. **1958** *Times* 21 Apr. 14/5 Special road feeder services were provided on numerous occasions for long distance rail excursions. **1971** *Nature* 19 Feb. 518/2 The best arrangement yet for using supersonic transports—a shuttle service between Newfoundland or Maine and Ireland, with subsonic feeder services from the terminals.

feeding, *vbl. sb.* Add: **1.** (Examples of later technical senses.)

1897 *Encycl. Sport* I. 404/2 Many aver that..Alan Rotherham was the first to reduce the art of feeding to a science. **1929** J. B. Priestley *Good Companions* ii. i. 253 This feeding I'm talking about..is a name in the profession for working up to gags.

4. a. *feeding-habit, -hole, -place* (later U.S. example), *-room, -trough.*

1933 *Brit. Birds* XXVII. 4 Until something exact is known of its distribution and numbers no correct interpretation of its feeding-habits can be made. **1959** E. F. Linssen *Beetles Brit. Is.* I. 55 The effect of feeding-habits—environment and type of food—on larvae. **1868** F. H. Joynson *Metals* 71 In this country the feeding hole of the furnace is generally closed. **1901** *Feilden's Mag.* IV. 436/2 With a steeper inclination..to a 'feeding-hole' for the admission of refuse from the platform above. **1907** *North Amer. Rev.* CLXXXVI. 18 Another cheap feeding-place—the 'Miners' Restaurant'. **1910** *Blackw. Mag.* Mar. 614/1, I was wet, miserable, and tired, so I passed into the common feeding-room. **1825** J. Nicholson *Operat. Mechanic* 403 The flax is fed or supplied in handsfull on the feeding-trough. **1867** 'T. Lackland' *Homespun* ii. 170 It will pay you well to feed your beast.. in the feeding-trough under the shed. **1942** W. Faulkner *Go down, Moses* 227 Corrals and feeding-troughs.

b. feeding board = *feed-board* (*feed *sb.* 7); **feeding-time,** (*a*) (earlier example); (*c*) the time at which animals in captivity are fed (recent examples of sense (*a*) usually have joc. allusion to (*c*)).

1892 J. Southward *Princ. & Progress Printing Machinery* v. 29 The feeding board, at first, consisted of merely a board like a desk, attached to the upper part of the machine. **1919** C. T. Jacobi *Printing* xxix. 301 To perform this correctly the second side in printing should be laid to the reverse end of the feeding-board. **1951** S. Jennett *Making of Books* (1964) vii. 99 At the front of the machine is the feeding board, a large wooden surface sloping gently down towards the base of the cylinder. **1832** *Chambers's Edinb. Jrnl.* I. 77/1 He had some reason to believe the Lion of the North..was best worth seeing at feeding time. **1844** S. Bamford *Pass. Life Radical* v. 27 A howl as wild and remorseless as that from a kennel of hounds at feeding time. **1844** Lady Stanley *Let.* 8 Dec. in N. Mitford *Ladies of Alderley* (1938) 105 It is the very large family altogether & especially feeding time that is oppressive. **1961** A. Wilson *Old Men at Zoo* i. 10, I had become quite deaf to the customary loud orchestra of a whole Zoo...I registered the feeding times only subliminally.

feeding, *ppl. a.* **2.** (Earlier example.)

1824 H. Davy *Diary* July (1836) v. 213 The Danes and Holsteiners appear to be rather *fat headed*, and a feeding and smoking people.

feeing (fī·iŋ), *vbl. sb. dial.* and *Sc.* [f. Fee *v.*^{1}] The hiring of servants for a fee. *feeing market*, hiring market; a general semi-annual gathering of farm hands for the purpose of hiring themselves out for the next six months.

1865 W. H. L. Tester *Poems* 127 I'll tell ye a' About the Friday's feein'. **1875** W. Alexander *Sks. Life among my Ain Folk* 220 On the feeing market night he had taken Baubie home to Briggies'. **1891** A. Gordon *Folks o' Carglen* iii. 66 In the little town of Kail..there were three great hiring markets for farm-labourers in the course of the year. These were the 'feein' Friday',.. 'hairst Monday'...By ten or eleven o'clock a huge concourse of people anxious to be 'feed'..had assembled. **1897** *Westm. Gaz.* 29 May 10/1 The Inverness Feeing Market the other day..was like a military pageant...The recruiting sergeants were alert and busy. **1930** *Times Lit. Suppl.* 18 Dec. 1086/3 Hawked at fairs and feeing-markets.

feel, *sb.* Add: **4.** (Further examples.) (For quot. 1966, cf. Feeling *vbl. sb.* 7 c.)

1891 F. W. Bain *Antichrist* v. 216 The skilful batter, in cricket, plays every ball as it comes to him in a slightly different way: and he cannot tell another man how to do it because it is not in his *idea*, but in his *feel.* **1939** I. Baird *Waste Heritage* i. 3 It sounded hysterical, as though it had picked up the feel of what was going on around it. **1962** A. Nisbett *Technique Sound Studio* viii. 140 It is worth while practising with this and getting

the feel of the timing. **1966** *Observer* 10 July 6/3 He can't, he [*sc.* a stockbroker] says, 'get the feel of the market'.

5. (Further examples.)

1931 *Economist* 4 Apr. 720/2 The coxswain on duty at the moment of emergency had not had the opportunity to learn the 'feel' of the ship. **1958** *Which?* I. ii. 21/2 The car's steering..was slightly 'numb', with undesirably little road feel. **1962** *Times* 4 Jan. 3/5 Angus played a neat, economical game, appearing to have the feel of the court.

7. *slang.* An instance of 'feeling up' (cf. *feel *v.* 1 f). Also *Comb.*

1932 J. T. Farrell *Young Lonigan* vi. 228 Wouldn't it have been nice to have had her there..and have a feel-day. **1934** J. O'Hara *Appointment in Samarra* (1935) vi. 168 And that husband of hers, that Harvey. Trying to give me a feel under the table. **1939** Joyce *Finnegans Wake* 95 Fine feelplay we had of it mid the kissabetts frisking in the kool kurkle dusk of the lushiness. **1969** 'J. Fraser' *Cock-pit of Roses* xviii. 143 She let me have a quick feel. **1970** 'Zeno' *Grab* iv. 35, I gave her a feel, and she pulled away.

feel, *v.* Add: **1. d.** Also, to search *out*, to ascertain, by feeling or testing.

1835 A. B. Longstreet *Georgia Scenes* 242 These [chills] I had to bear for the space of a minute or two before I could *feel out* my hat. **1928** *Publishers' Weekly* 9 June 2376, I cannot stress enough..the necessity of feeling out the possibilities of a book before giving a large order.

f. *slang.* To caress the genital parts of (a person). Usu. const. *up.* Cf. *feel *sb.* 7.

1930 J. Dos Passos *42nd Parallel* i. 75 She's awful hot. Jez I thought she was going to feel me up. **1957** J. Kerouac *On Road* (1958) ii. xi. 175 We used to get next to pretty young daughters and feel them up in the kitchen. **1968** M. Richler *Cocksure* xvi. 92 He literally bumped into Ziggy feeling up the prettiest girl at the party in a dark damp corner. **1970** S. Ellin *Man from Nowhere* i. 9 We're supposed to be married. He might wonder about it if I let him feel you up in front of me without showing I didn't like it.

2. b. (Further example.)

1904 P. F. Warner *How we recovered Ashes* xiii. 253 He again got Clem Hill feeling for the ball.

4. Delete †*Obs.* and add later example.

1938 N. Coward *Operette* i. ix. 87 Play slowly at first until you feel the house.

6. b. Slang phr. *to feel the draught*: to be seriously incommoded; to feel insecure, esp. financially.

1925 *Westm. Gaz.* 27 Apr., When the wind changed it might be the Conservative Party which would be feeling the draught. **1941** N. Marsh *Surfeit of Lampreys* vii. 103 'Did he go bust?'..'I don't think so, Curtis. Must have felt the draught a bit.' **1966** *Listener* 9 June 831/2 With only so much national advertising to go round.., the oldest commercial stations are feeling the draught as well. **1970** *Financial Times* 13 Apr. 13/3 If the BSC or the bigger firms in the private sector felt the draught and turned their attention to smaller orders, the lesser firms could suffer badly—to the point of extinction.

d. Also *to feel one's feet.*

1889 *Farmer's Mag.* Dec. 217 Trees put in now cannot be expected to succeed as well as those that are already beginning to 'feel their feet'.

f. *to feel into* (see quots.).

1912 [see *empathy]. **1919** M. K. Bradby *Psycho-Analysis* xviii. 239 He feels himself into the mind of the man or woman..he is studying. **1933** H. Read *Art Now* i. 50 We do not necessarily humanise the rising column or the graceful vase which we contemplate: we feel into their shape..and react to..its rhythmic convolution; and so we invent the word empathy.

g. *to feel no pain*: to be insensibly drunk. *slang.*

1947 C. Morley *Ironing Board* (1950) 116 At the table with three men who were feeling no pain. **1965** B. Sweet-Escott *Baker St. Irreg.* iii. 94 There were a great many Anglo-Russian parties, a vast quantity of vodka was drunk, and twice I saw senior Russian officers being carried out of the room evidently feeling no pain.

9. e. *to feel like* (earlier U.S. example and later examples).

1829 *Virginia Lit. Museum* 30 Dec. 458 *Like*..is also used, as follows, in the south: 'I do not feel like eating'. **1944** M. Laski *Love on Supertax* viii. 82 I'll go straight home. I don't feel exactly like a street-corner meeting. **1970** W. J. Burley *To kill a Cat* ii. 45 They had a drink together, then Helen said that she felt like bed.

feeler. Add: **5.** A thin metal strip made to a stated thickness, used to measure narrow gaps or clearances and usu. forming one of a set. More fully *feeler gauge.*

1931 D. Garnett *Grasshoppers Come* 6 While Wright was tracing out the cause of an intermittent miss on the starboard magneto of a Gipsy engine and was toying with a pair of feeler gauges. **1935** T. E. Lawrence *Lett.* (1938) 856 A set of Power feeler gauges, as supplied in tool kits. **1967** E. Chambers *Photolitho-Offset* xv. 231 The lock nuts and adjusting screws are released in order that, for example, a 0·006 in. feeler can just be inserted between the bearers on both sides of the cylinders.

b. A device which moves a control lever in the weaving of artificial silk.

1927 T. Woodhouse *Artif. Silk* 131 When the weft is nearly exhausted in the shuttle, an electric circuit is completed; this causes a single feeler under the circular magazine to move a control lever which ensures that the next change of weft will place in the shuttle a bobbin with the same kind of weft as that just finished.

feeling, *vbl. sb.* Add: **9. e.** Const. *in, into, out.*

1926 H. J. Laski *Let.* 2 May (1953) II. 838 The acute pleasure that the feeling-out of other minds gives one. **1933** H. Read *Art Now* i. 49 No mere fellow-feeling, or feeling *with*, but rather a form of imaginative identification of the self with the object, a feeling *into.* **1949** Koestler *Insight & Outlook* xxvi. 359 This feeling-in process, or empathy, is..based on the projection of part of our own personality into the shell of the other.

11. *attrib.* and *Comb.*: (sense 9), esp. after G. compounds of *gefühl*, e.g. *gefühlston* (Wundt).

1892 Feeling-quality [see Dict., 9]. **1892** *Rep. 2nd Internat. Congr. Exper. Psychol.* 34 The feeling-tone of belief appeared..completely neutral, except so far as the subject-matter of belief caused it to have a pleasureable or painful quality. **1894** D. Irons in *Mind* III. 93 Under feeling should be included on the one hand pleasure and pain, and on the other that feeling towards the object, which for the present we may call feeling attitude. The term is not unexceptionable, but is used for convenience instead of some such formula as 'feeling in relation to'. **1901** E. B. Titchener *Exper. Psychol.* I. i. ii. 54 Likeness may mean 'likeness of feeling-effect'. **1906** S. S. Laurie *Synthetica* I. 136 This entitative feeling-potency does not contain the 'modes' of sense. **1909** W. M. Urban *Valuation* ii. 37 Feeling or feeling-disposition is always presupposed by desire. **1912** [see *ambivalence, -ency]. **1919** M. K. Bradby *Psycho-Analysis* 58 Certain unconscious factors..give the complex its peculiar feeling-tone. *Ibid.* 150 A woman who heard the first maroon announcing the Armistice on November 11th, 1918, responded with an instantaneous feeling-thought. **1920** T. P. Nunn *Education* 125 Feeling-spread is almost wholly biological imitation. *Ibid.* 134 Pugnacity is an example of the instincts in which the feeling-element is a definite emotion. **1937** *Discovery* July 216/2 The feeling-values of purple and violet are 'uncertain'. **1960** C. Day Lewis *Buried Day* iii. 56, I all unwittingly responded with the feeling-tone of a woman.

feelingful (fī·liŋfŭl), *a.* [f. Feeling *vbl. sb.* + -ful.] Full of feeling.

1907 R. Boughton *Bach* i. 6 Composers of Passions—human and feelingful if crude and naïve. **1943** H. Read *Educ. through Art* v. xvi. 147 Decoration..implies a certain compromise between means and end, the 'feelingful' adaptation of colour and design to a given form or space.

feelingness (fī·liŋnės). [f. Feeling *ppl. a.* + -ness.] Emotional quality or character.

1870 G. Meredith *Lett.* (1912) I. 216 The feelingness of your letter to Mickleham was much felt there. **1901** F. H. Burnett *Making of Marchioness* ii. vi, 'I do *love* him so,' she whispered hysterically. 'I do so *love* him, and I shall so *miss* him!' with the italicised feelingness of old.

feelthy (fī·lþi), *a. slang.* [Jocular imitation of foreign pronunc. of Filthy *a.*] Obscene.

1933 E. A. Robertson *Ordinary Families* vi. 112 Buy a feelthy peecture, lady? **1935** G. Gorer *Africa Dances* iii. i. 193 Little bronze figures of an obscenity which resembles feelthy postcards. **1962** *Spectator* 11 May 622/2 Tiny ant-like creatures go about their feelthy business. **1963** B. S. Johnson *Travelling People* iv. 68 Maurie has a great collection of feelthy books down here—including a first edition of Cleland's *Fanny Hill, or the Memoirs of a Woman of Pleasure.*

feely (fī·li). Usu. in *pl.* **feelies.** [f. Feel *v.* + -y.] A (hypothetical) talking film augmented by tactual effects.

1931 [see *egalitarian *a.*]. **1932** A. Huxley *Brave New World* iii. 39 Going to the Feelies this evening, Henry?..I hear the new one at the Alhambra is first-rate...The most amazing tactual effects. *Ibid.* xi. 197 Take hold of those metal knobs on the arms of your chair. ..Otherwise you won't get any of the feely effects. **1964** *Discovery* Oct. 22/1 We shall certainly need to be most cautious if techniques develop—to colour, three dimensions, or even to the 'smellies' and 'feelies'.

féerie (fē·ĕri). Also **féery, feërie.** [F. *féerie*, f. *fée* fairy; see Fairy *sb.*] A spectacular theatrical production involving the representation of fairy scenes and characters.

1878 *Lloyd's Weekly* 19 May 7/2 A magnificent *féerie*, in which five Nubian lions are announced as about to make their *début.* **1886** *Athenæum* 24 July 116/3 M. Victorien Sardou is at work on a *féerie*, or rather a piece for children, intended for the Porte-Saint-Martin. **1895** *Daily News* 5 Feb. 5/4 It is not strictly an opera, but is described as a dramatic feërie, entitled 'The Royal Infants'.

Fehling (fē·liŋ). The name of Hermann von *Fehling* (1812–1885), German chemist, used *attrib.* or in the possessive to designate an alkaline solution of cupric sulphate and Rochelle salt used in sugar analysis; also *ellipt.* Hence *Fehling('s) reaction, test.*

1873 *Jrnl. Chem. Soc.* XI. 265 The crystals, which are permanent in the air, neither hygroscopic nor efflorescent, are recommended by the author for the titration of Fehling's solution. **1887** *Encycl. Brit.* XXII. 624/1 Every c.c. of Fehling solution oxidises about 5 milligrams..of dextrose. **1909** *Cent. Dict.* Suppl. 464/3 Fehling's reagent, test. **1946** *Nature* 31 Aug. 295/2 The distillate slightly reduced Fehling's solution. **1954** *Thorpe's Dict. Appl. Chem.* (ed. 4) XI. 203/2 Soxhlet's Fehling solution in excess is boiled with sugar solution. *Ibid.* 205/2 Into a flask of 300–400 ml. capacity, measure 10 ml. or 25 ml. of Fehling. **1960** A. E. Bender *Dict. Nutrition* 48/2 *Fehling's test*, for reducing substances, mostly used to distinguish reducing from non-reducing sugars.

fei (fē·i). [Tahitian vernacular name of the plant and fruit.] A type of plantain, *Musa fehi*, native to Tahiti and New Caledonia.

1829 W. Ellis *Polynesian Researches* I. xiii. 374 In several of these islands, the *fei* is the principal support of the inhabitants. The plantain is a fruit that is always acceptable. **1894** Stevenson & Osbourne *Ebb-Tide* i. ii, Five Kanakas..were squatted round a basin of fried feis. **1926** *Univ. Calif. Pub. Bot.* XII. 156 The Fe'i are of the greatest interest ethnobotanically. **1959** Tindale & Lindsay *Rangatira* i. 13 The big red-gold fei plantains grew sparsely on the slopes of the central mountain.

feijoa (fēi͡dʒōu·ā, fī-, feyōu·ā), [mod.L. (O. Berg 1858 in *Linnæa* XXIX. 258), f. the name of J. da Silva *Feijo*, 19th-century Spanish naturalist.] An evergreen shrub or small tree of the genus so named, which is native to South America and belongs to the family Myrtaceæ; also, the fruit of this tree.

1898 *Curtis's Bot. Mag.* CXXIV. Tab. 7620 Feijoa was discovered by the late Fr. Sellow of Potsdam, who, in 1819, accompanied Prince Neuwied in his journey to Brasil...Over and above the beauty of the foliage and flower of the plant, it is remarkable for the rich aromatic odour and flavour of its guava-like fruit. **1920** W. Popenoe *Man. Tropical & Subtropical Fruits* ix. 294 In its native country the feijoa is scarcely known as a cultivated plant. **1958** *N.Z. Listener* 18 Apr. 23/1 Feijoas or Salad Plant...To eat feijoas, scoop out the centre with a spoon similar to the passion fruit method.

Feinne (fēi·nʸĕ), *sb. pl.* Also **Fein.** [Ir. *feinne*, pl. of *Fian(n.] The soldiers of the ancient Irish militia. Cf. Fenian and *Fian(n.

1782 *Gentl. Mag.* LII. 570/2 The songs relating to the Feinne, and their Chieftain, Fion-mac-Coul. *Ibid.* 571/1 The whole host of the Feinne, or Fingalians. **1872** J. F. Campbell *Leabhar na Feinne* 33 Fionn and the Feinne were the successors of Cumhal and Cuchullin. **1891** *Youthful Exploits of Fionn* 96 A company of the old Feinne. **1891** A. Macdougall *Folk & Hero Tales* 35 The Fein looked at each other.

feint, *a.* Restrict 'now *rare*' to sense in Dict. and add: **2.** In commercial use, the usual spelling of Faint *a.* 5 c; freq. quasi-*adv.*

1859 *Stationers' Hand-bk.* (ed. 2) 72 Feint only, the term for a book having merely feint blue lines across the page from left to right. **1895** *Army & Navy Co-op. Soc. Price List* 15 Sept. 525 Foolscap Paper—Ruled with Money Columns and Feint Lines. **1930** *Publishers' Circ.* 13 Sept. 321/2 The actual book itself should be of foolscap size, ruled feint.

feis (feʃ, fēiʃ). Pl. **feiseanna.** Also **fes(s.** [a. Ir. *feis, fess* meeting, assembly.]

1. An assembly of kings, chiefs, etc., being a kind of early Celtic parliament.

1792 *Encycl. Brit.* IX. 313/1 Ollam Fodla..erected a grand seminary of learning, and instituted the *Fes*, or triennial convention of provincial kings, priests, and poets, at Feamor or Tarah in Meath. **1880** *Ibid.* XIII. 250/2 He summoned a convention (*feis* or *fess*)..to assemble at Druimceta. **1898** J. Heron *Celtic Ch.* i. 19 The *Feis* was an assembly of all the kings, chiefs, ollamhs, and other leading men, and lasted for seven days. **1905** *Daily Chron.* 15 June 3/1 The whole scheme of [old Irish] national life turned on central feiseanna—social, legislative, and literary functions.

2. An Irish or Scottish festival of the arts, resembling the Welsh Eisteddfod.

1896 *Daily News* 2 Oct. 5/2 The 'Feis' extending over four days. **1896** *Westm. Gaz.* 3 Oct. 8/1 The *Feis* is announced to take place next May. **1910** *Daily Chron.* 3 Jan. 3/4 In her incidental reference to latter-day Feiseanna. **1916** *N. & Q.* 12th Ser. II. 71 The feis portions of local shows. **1938** L. MacNeice *I crossed Minch* xii. 162 The school-kids recited her songs at the local Feis. **1955** *Times* 25 July 6/4 A similar ban was imposed last year and prevented the holding of a procession in the town [*sc.* Newtown Butler] in connexion with an annual Roman Catholic feis (festival).

feist, var. Fist *sb.*^{2}

feisty (fəi·sti), *a. U.S. slang* (orig. *dial.*). [f. *feist* = Fist *sb.*^{2} 3 + -y^1.] Aggressive, excitable, touchy.

1896, etc. in Wentworth *Amer. Dial. Dict.* s.v. Ficety, etc. **1913** H. Kephart *Our Southern Highlanders* 94 Feisty means when a feller's allers wigglin' about, wantin' ever'body to see him, like a kid when the preacher comes. **1926** E. M. Roberts *Time of Man* 152 That-there feisty bay mare jumped straight upwards and broke the tongue outen the plow. **1965** 'D. Shannon' *Death-bringers* (1966) xiii. 162 Luther gets a little feisty after a few drinks, and he began to argue with him. **1968** J. Potts *Trash Stealer* xiii. 148 He couldn't shake her loose—she hung on to his arm, feisty as a terrier.

felafel (felā·fĕl). Also **falafel, filafil.** [ad. Arab. *falāfil.*] (See quot. 1951.)

1951 *Commentary* (N.Y.) XI. 269 *Falafel*: sharp peppers and fried dried pea balls sandwiched in a flat roll called a *pitah.* *Falafel* is a standard meal around some urban and most interurban bus stops, where one spends a good part of one's life. **1955** P. Smouha *Middle Eastern Cooking* 74

Filafil..Broad beans..garlic..onion..oil...Form mixture..into balls..and fry them. **1962** A. Ramati *Israel Today* 299 *Falafel*, spiced fried balls of chick peas and hot pepper, soaked in a relish and chilisauce. Usually served in bread. **1963** J. Comay *Introd. Israel* xiv. 247 *Felafel.* (Bought off a stand in a street) is made of a highly spiced mixture of deep-fried balls of ground chickpea, small peppers, and pickled cucumbers, served hot between halves of peeta. It can be ordered hot (pepper hot) or mild. **1968** C. Roden *Bk. Middle Eastern Food* 38 Falafel ..is one of Egypt's national dishes. **1968** P. Durst *Badge of Infamy* i. 4 A Christian Arab sold cold drinks and *felafel* in an envelope of bread. **1970** *Guardian* 8 Oct. 13/2 You can now eat a bag of felafel—hot bean fritters—as you wander round the streets of Amman.

‖ **feldgrau** (fe·ltgrau). [G.] Field-grey; hence, a German soldier in a uniform of this colour.

1934 in Webster. **1945** R. Hargreaves *Enemy at Gate* 296 The *feldgraus* crept into the labyrinth of streets. **1946** H. Nicolson *Diary* 30 Apr. (1968) III. 58 Keitel, in *feldgrau* uniform stripped of all badges..looks trim, tough. **1958** H. Williamson *Love & Loveless* xvii. 294 Although the Germans in *feld-grau* were men, too, there was no feeling from them. **1960** C. Becker tr. *Schwarz-Bart's Last of Just* (1961) viii. 388 The Feldgrau buses deposited fifteen hundred orphans in the..snow-covered roll-call ground.

‖ **feldscher** (fe·ldʃəɹ). Also **feldschar, feldsher.** [Russ. *fél'dsher*, ad. G. *feldscher* field surgeon.] In Russia, a person with practical training in medicine and surgery, but without professional medical qualifications; a physician's or surgeon's assistant; a local medical auxiliary.

1877 D. M. Wallace *Russia* I. v. 104 There is not exactly a doctor, but there is a Feldsher in the village. **1916** H. S. Walpole *Dark Forest* i. iv, Like an old feldschar in my village who hates our village Pope. **1925** *Contemp. Rev.* June 752 In Russia the place of doctor or nurse is often filled by a 'feldscher' or half-trained doctor only. **1937** H. E. Sigerist *Socialized Med. in Soviet Union* iii. 139 All European armies used to have surgeons who had not been trained in universities. ..During the nineteenth century most..replaced the feldshers with army surgeons of academic standing.. The Russian army preserved the feldsher. He practised not only in the army but also among civilians...Many rural medical stations were headed by feldshers. **1957** H. Bower *Short Guide Soviet Life* 58 There are also 600 schools for training nurses, *feldshers* (auxiliary nurses) and midwives.

feldspathoid, felspathoid (fe·ldspăþoid, fe·lspăþoid). *Min.* [f. *fel(d)spath* (see Feldspar, Felspar) + -oid.] Any of a group of minerals chemically similar to the feldspars, but containing less silica. Hence **fe:l(d)-spathoi·dal** *a.*

1896 J. F. Kemp *Handbk. Rocks* i. 6 (*heading*) *Feldspathoids.* With the feldspars are placed two other important and closely related minerals, nepheline and leucite. *Ibid.*, The feldspars, together with the feldspathoids nepheline and leucite, are the most important of the rock-making minerals in their relations to the classification of rocks. **1910** *Bull. Geol. Soc. Amer.* XXI. 87 Monzonite may not be rich in alkalies nor carry any feldspathoid. **1912** [see *felsic *a.*]. **1930** *Geol. Mag.* LXVII. 416 The hypothesis of the origin of felspathoidal rocks by reaction between magma and limestone. **1959** C. S. Hurlbut *Dana's Man. Mineral.* (ed. 17) 500 The feldspathoids are chemically similar to the feldspars, in that they are aluminosilicates of chiefly potassium, sodium, and calcium but with minor amounts of other ions. **1968** *Mineral. Abstr.* XIX. 234 Carbonate-rich feldspathoidal rocks from Nsala vent.

‖ **Félibre** (felibr). [F. *félibre*, a. Prov. *felibre* any one of the teachers in the temple to whom Jesus put questions as a child (Luke 2: 46), prob. (Bloch & von Wartburg) f. Low Latin *fellibris*, var. of *fellebris* nursling (of the Muses), f. L. *fellāre* to suck.] A word used by F. Mistral (1830–1914) to designate a member of the brotherhood which was founded in 1854 by seven Provençal writers for the maintenance and purification of Provençal as a literary language, and also for the promotion of the artistic interests of the South of France. Also as *adj.*

1876 G. Meredith *Let.* 6 Oct. (1970) I. 527 As to the song, my wife worked at it Trojanly and I, as it were a drum accompaniment, thumped out the Félibre lingo. **1892** *Ibid.* 13 Dec. II. 1112 Wyse was an enthusiastic félibre, and published Provençal verses. **1902** *Encycl. Brit.* XXXII. 44/1 Greatest of them all, the true and acknowledged forerunner of the *félibres*, Jacques Jasmin (1798–1864), the hairdresser of Agen. *Ibid.* 45/1 The *félibres* are in no sense of the word the direct successors of the troubadours. **1904** *Westm. Gaz.* 27 May 12/2 The Félibres, who have just been celebrating their jubilee. **1964** *Archivum Linguisticum* 38 Roumanille, however, finally succeeded in persuading his fellow *félibres* to adopt the phonetic system.

Hence **Féli·brian** *a.*, relating to the Félibres, or to the Provençal literature produced by them; **Félibrism** (fē·libriz'm), the movement instituted by the Félibres; the principles underlying that movement.

1908 *Daily Chron.* 16 Jan. 4/4 He plunged into the centre of things Félibrian. **1911** *Daily News* 11 Mar. 6/4 The literary output of Félibrism has been mainly poetic.

‖ **Félibrige** (felibrīʒ). [F. *Félibrige*, ad. Prov. *Felibrige*, f. *felibre*: see prec.] The name of the literary school formed by the Félibres.

1902 *Encycl. Brit.* XXXII. 44/2 The most widely read of the *Félibrige* publications is the *Armana Prouvençau*, which..contains much of the best work of the school. **1907** *Daily Chron.* 14 Nov. 3/3 Two other pioneers of the 'Félibrige' movement, Roumanille and Mathieu. **1927** *Observer* 18 Sept. 12 The Félibrige is the association.. whose object is to keep alive the traditions and the language of Provence. **1964** *Archivum Linguisticum* 38 The Félibrige, therefore, outside Provence, has no common literary language.

fell, *sb.*^{2} Add: **3.** *fell-farm, -walker, -walking*; **fell-field** (see quot. 1916); **fell hound,** a variety of foxhound bred for hunting in hill-country.

1908 W. G. Collingwood *Scand. Brit.* ii. ii. 180 In Cumberland..on fell farms. **1951** M. Lloyd (*title*) Fell Farm Holiday. **1916** B. D. Jackson *Gloss. Bot. Terms* (ed. 3) 144 *Fell-fields*, districts of dwarf, scattered plants, chiefly Cryptogams. **1959** A. H. McLintock *Descr. Atlas N.Z.* 28 Subalpine scrub and fellfield in more elevated situations. **1920** R. Clapham *Foxhunting on Lakeland Fells* iii. 51 A fell hound should stand under, rather than over, 22½ inches. **1948** C. E. Lloyd in B. Vesey-Fitzgerald *Bk. Dog* ii. 452 The Royal Fell Hound Show has done a lot not only to improve the breed but also to promote interest and goodwill. **1957** Clark & Pyatt *Mountaineering in Brit.* i. 24 Jackson..was a persistent fell-walker and scrambler. **1956** C. Evans *On Climbing* i. 13 It is as common to start by being taken up a climb as by fell-walking.

Fell (fel). Name of John *Fell* (1625–86), dean of Christ Church and bishop of Oxford, designating the founts of type and matrices procured by him for the Oxford University Press, the use of which has been revived in recent years.

1900 H. Hart *Cent. Oxf. Typogr.* p. ix, All doubts and conjectures as to where most of the Fell types were purchased may..now be regarded as disposed of. **1922** D. B. Updike *Printing Types* II. 199 Caslon and Fell revivals. **1960** G. A. Glaister *Gloss. Bk.* 134/2 *Fell types*, the type-face purchased by Doctor John Fell for the Oxford University Press, *c.* 1672. They were cut by Dirck and Bartholomew Voskens of Amsterdam and were a source of inspiration to English type-designers of the time. They are in use today at the Clarendon Press, having been revived in 1876 by C. H. O. Daniel. **1966** H. Williamson *Methods Bk. Design* (ed. 2) xi. 152 The larger Fell founts..have a bold irregularity which makes most founts now in use look prim.

fella, fellah, representing an affected or vulgar pronunciation of Fellow *sb.* 9: see *feller^2.

1864 J. S. Le Fanu *Uncle Silas* II. xiii. 203, I did see two horses yoked to a shay, and a fellah a pullin' a box up o' top. **1909** A. Huxley *Let.* 2 Apr. (1969) 30 There is to be an huge auction of all the appurtenances of the leaving fellahs. **1912** A. Conan Doyle *Lost World* vi. 84 Young fellah my lad. **1920** 'Sapper' *Bull-Dog Drummond* i. 38 'An engaging fellah,' said Hugh. 'What particular form of crime does he favour?' **1933** *Times Lit. Suppl.* 9 Nov. 777/2 We feel that, stout fellah though Furness.. is, he would look rather nice at the end of a rope. **1934** W. Holtby *Truth is not Sober* 76 Among the things a Fella does, correct grammar is not necessarily included. **1949** E. Hyams *Not in our Stars* 73 Well, young fella-me-lad. **1952** E. Wilson *Equations of Love* 177 Seemed a nice fella. **1968** R. Clapperton *No News on Monday* viii. 99 The fellah took that opportunity.

‖ **fellagha** (fellā·gā). [ad. colloq. Arab. *fəllāga*, pl. of *fəllāg* bandit, robber.] A guerrilla soldier of a nationalist movement in Algeria and Tunisia when the two countries were under French control. Also as *pl.* and *attrib.*

1954 *Time* 5 July 24 The *fellagha*, rising up in Tunisia's rich but savage hinterland, so far number about 400 fighters. **1955** *Ann. Reg. 1954* 202 The French and Tunisian Governments appealed to the *fellaghas*, promising an amnesty. **1955** *Times* 19 May 14/4 The *fellagha* activities last autumn [in Tunisia]. **1958** *Listener* 18 Sept. 433/2 When the *fellagha* say independence, they mean the end of bad times [in Algeria].

fellah. Add: *Pl.* **fellahin** (now the usual form). Also *fig.*

1856 [in Dict.]. **1912** [see *Bedu]. **1932** *Times Lit. Suppl.* 11 Feb. 86/3 Cities and civilization dying of inanition after eating up the best of the rural population and leaving only the fellahin type. **1947** S. Bellow *Victim* i. 3 The people.., barbaric fellahin among the stupendous monuments of their mystery.

fellatio (felēi·ʃio, felā·tio). Also **fellation.** [mod.L., f. *fellātus*, pa. pple. of L. *fellāre* to suck.] A sexual act in which the partner's penis is sucked or licked. Hence **fella·tor,** *fem.* **fella·trix,** the partner who performs such an act. Also **fella·tory** *a.*, of or relating to fellatio; (as a back-formation) **fellate** (felēi·t) *v.*, to practise fellatio on.

1887 L. C. Smithers tr. *Forberg's Man. Classical Erotology* iii. 72 The verge, introduced into the mouth, wants to be tickled either by the lips or the tongue, and sucked; the party who does this service to the penis is a fellator or sucker. *Ibid.* 73 This fellatrix seems to have borne the name of Labda. *Ibid.* 97 With regard to the subject of fellation, we must not pass over in silence the raven, whom our constant authority (Martial xiv. 74) calls a fellator. **1888** tr. *Priapeia* 142 The patient (fellator or sucker) provokes the orgasm by the manipulation of his (or her) lips and tongue on the agent's member. *Ibid.* 159 Martial directs many epigrams against fellators. *Ibid.* (index) 232 Fellation. **1897** H. Ellis *Stud. Psychol. Sex* I. v. 118 Taking the twenty-four inverted men..I find that three..have never had any physical relationship with their own sex...In two or three cases *fellatio* is the form preferred. **1901** A. Allinson tr. *J. Rosenbaum's Plague of Lust* II. xxii. 29 *Fellator* and *fellatrix*..were liable to suffer from ulcers of the throat. *Ibid.* 31 Yet another consequence of fellation was pain in the mouth. **1916** A. A. Brill tr. *Freud's Leonardo da Vinci* ii. 39 The situation contained in the phantasy..corresponds to the idea of fellatio, a sexual act in which the member is placed into the mouth of the other person. **1955** *Psychiatr. Q.* XXIX. 222 She had cunnilinctual and fellatory relationships with him. **1964** S. Bellow *Herzog* 229 Someone must have told him that fellatio was the path to truth and honour. **1968** *New Amer. Rev.* ii. 159 She makes various attempts through the film to rouse him (including the clearest suggestion of fellation that I have seen in an American picture). **1968** J. Updike *Couples* (1969) v. 482 Lazily she fellated him while he combed her lovely hair. **1969** G. Legman *Rationale of Dirty Joke* viii. 552 An old roué marries an innocent virgin, but being impotent he teaches her to fellate him. **1969** *New Society* 1 May 664/1 The man had just submitted to fellatio on the part of a second man.

felled, *ppl. a.*^{1} **1.** (Earlier example.)

1820 D. Wordsworth *Tour Continent* 11 Aug. in *Jrnls.* (1941) II. 126 The first signs of occupancy and labour we had noticed in our descent were *felled* trees.

felled, *ppl. a.*^{2} (Example not in *Comb.*)

1952 Dylan Thomas *Coll. Poems* p. ix, O kingdom of neighbours, finned Felled and quilled, flash to my patch.

feller^1. Add: **4.** One who sews in various trades.

1894 *N. Brit. Daily Mail* 5 Sept. 4 The wages of the fellers, whether on day work or on piece work, ranged from 1s 6d to 2s 6d. **1921** *Dict. Occup. Terms* (1927) § 404 *Finisher*, tailor's finisher; feller; sews by hand, sleeve linings, vest arm holes. *Ibid.* § 411 *Feller*, sews seams of gloves with a particular kind of felling stitch on treadle machine.

feller^2 (fe·ləɹ). Vulgar or affected (cf. *fella) form of Fellow *sb.* 9. Also in jocular phr. *young feller-me-* (or *my-*)*lad*, orig. used vocatively with an implication of disapproval or reproof; hence as compound sb. designating a young man of a frivolous or irresponsible character; also (occas.) of women.

1825 J. Neal *Bro. Jonathan* I. i. 12 A lawyer..put Peters down for 'a confounded smart feller'. **1849** Thackeray *Pendennis* I. xxvi. 252 Mr. Pynsent, you're a good feller. **1880** *Punch* 20 Nov. 234 Why, there was an Actor, by Jingo! and a scientific chap, and an artist feller. **1897** Kipling *Capt. Cour.* 17 You've nigh slep' the clock around, young feller. **1909** Wodehouse *Swoop!* vi. 86 Not so much of your eight hundred and seventy-five, young feller me lad. **1913** L. A. Harker *Ffolliots of Redmarley* xxiv. 321, I beheld Miss Bax seemingly in difficulties with two young feller-me-lads, who evidently had no intention of going on. **1930** D. H. Lawrence *Nettles* 23 She's a marv'lous, delicious, high-spirited feller. **1939** J. Cary *Mr. Johnson* 154 All right, you remember that, young fellermelad. **1946** K. Tennant *Lost Haven* (1947) i. 17 A feller that's just risen from his bed of 'flu. **1971** *Petticoat* 17 July 2/2 If we did walk into a pub alone and not one feller blinked an eyelid we'd probably think there was something wrong with us.

felling, *vbl. sb.* **3.** (Earlier example.)

1841 Lady Wilton *Art of Needlework* (ed. 3) xx. 317 There are..hemming—felling—and basting.

fellmonger. Add: In modern use restricted to an operative who works skins.

1745, 1834 [see Dict.]. **1845** Dodd *Brit. Manuf.* V. 195 There are in Bermondsey several manufacturers called fellmongers, whose business it is to bring sheep-skins into a certain state of preparation before the leather-dresser commences his operations thereon. **1921** *Dict. Occup. Terms* (1927) § 338 *Fellmonger*, general term for any person employed in fellmongery trade, e.g. washing sheep skins, painting pelt side of skins with chemicals to facilitate subsequent pulling [etc.]. **1960** *New Scientist* 9 June 1496/1 (Advt.), A machine for fellmongers who prepare woolled sheepskins.

Hence **fe·llmongered** *a.*, **fe·llmongering** *vbl. sb.*

1895 *Australas. Pastoralists' Rev.* 15 Aug. 284 The very large quantity of fellmongered wool sold in Sydney. **1897** *Daily News* 7 Oct. 2/2 Fellmongering and leather dressing premises.

fellow, *sb.* Add: **3. a.** *spec.* of a woman.

1840 Dickens *Old C. Shop* xxxvi. 298 Nay, he [*sc.* Mr. Swiveller] would sometimes reward her [*sc.* Miss Brass] with a hearty slap on the back, and protest that she was a devilish good fellow. **1876** A. Trollope *Prime Minister* II. xvii. 281 There was a feeling abroad that 'Glencora' was a 'good sort of fellow' and ought to be supported. **1928** E. A. Robertson *Cullum* x. 185 'My darling fellow,' started one [letter], 'That's the only title I can find with which to pay homage to your male companionableness...You are the only woman I've ever known who will argue..on an abstract question.' **1942** P. Abrahams *Dark*

Testament II. iii. 111 He's lucky and you're a fine fellow. You girls were right to call each other 'fellow'.

4. a. Delete †*Obs.* and add later *local* example.

1966 F. Shaw et al. *Lern Yerself Scouse* 39 It's fer my feller's carryin-out. It is for my husband's packed lunch.

7. a. Now extended to women holding such positions.

1899 *Westm. Gaz.* 24 Nov. 10/1 A lady 'research' Fellow already exists in Wales. **1921** *Oxf. Univ. Cal.*, Somerville College..Lady Carlisle Fellow. **1971** *Ibid.* 275 A supplemental charter and new statutes were granted by which the Principal and Fellows [of St. Hilda's College] became the Governing Body.

9. a. *stout fellow* [Stout *a.* and *adv.* A. *adj.* 3].

1915 Kipling *Debits & Credits* (1927) 31, I lay behind this stout fellow and saw him well into the open. **1919** J. Buchan *Mr Standfast* ix. 170 You're going to be a stout fellow and start in two hours' time. And you're going to take me with you. **1922** —— *Huntingtower* iv. 76 'I got inside the House.' 'Stout fellow,' said Heritage. **1970** *Guardian* 6 Apr. 11/1 The stout fella tradition of never striking against the public interest.

b. *young fellow-me-*(or *my-*)*lad*: see *Feller[2] and *Fella.

1926 E. F. Spanner *Navigators* 36 This young fellow-me-lad seems to have spent ten minutes or so diving in and out among the wreckage. **1929** W. Deeping *Roper's Row* iv. § 1 There were young fellow-my-lads who began to take notice.

†**10. d.** A Negro. *U.S. Obs.*

1753 in *New Jersey Archives* (1897) 1st Ser. XIX. 270 Run away..a Mulatto Fellow named Anthony...Whoever takes up said Fellow..shall have Three Pounds Reward. **1860** Bartlett *Dict. Amer.* (ed. 3) 144 *Fellow* or *Black Fellow*, a black man. Southern.

11. b. *fellow-conspirator, -worker* (later examples).

1899 'Mark Twain' *Man corrupted Hadl.* (1900) 160 He shall deceive his fellow-conspirators. **1936** *Discovery* Oct. 321/2 All the members of this pact wore a black pin as a sign to fellow-conspirators. **1951** R. Firth *Elem. Social Organiz.* i. 23 According to his fears or his politics, he may interpret this as a symbol of anger or of solidarity among fellow-workers. **1961** New English Bible *Philemon* 1 From Paul..to Philemon our dear friend and fellow-worker.

fellowess. Delete †*Obs.* and add later example.

1935 S. Desmond *Afr. Log* l. 256 Good fellows and fellowesses all.

fellowlike, *a.* and *adv.* **A.** *adj.* **a.** (Later *poet.* example.)

1928 T. Hardy *Coll. Poems* 5 Change and Chancefulness..wrought us fellow-like.

fellowly, *a.* and *adv.* Delete †*Obs.* and add later examples illustrating revival in *poet.* and *rhet.* use.

1883 G. Meredith *Joy of Earth* 11 Love it [*sc.* the light] so you could accost Fellowly a livid ghost. **1898** T. Hardy *Wessex Poems* 1 Change and chancefulness in my flowering youthtime, Set me sun by sun near to one unchosen; Wrought us fellowly, and despite divergence, Friends interblent us. **1903** *Trawl* May 3 The fellowly enfolding of the night. **1918** W. J. Jupp *Wayfarings* viii. 93 The revealings of a spirit fellowly and accordant with his own.

fe:llow-tra·veller. [Fellow *sb.* 11 b.] **1.** One who travels along with another.

1611 T. Coryat *Crudities* (Epist. Dedic.), Often disswaded by some of my fellow trauellers from gathering any Obseruations at all till I came into Italie. **1665–1829** [see Fellow *sb.* 11 b]. **1814** Wordsworth *Excursion* ii. 55 My Fellow Traveller said with earnest voice, As if the thought were but a moment old, That I must yield myself without reserve. **1850** —— *Prelude* vi. 161 The brook and road Were fellow-travellers in this gloomy strait.

2. *transf.* One who sympathizes with the Communist movement without actually being a party member. Also in extended uses.

The equiv. Russ. *popútchik* (Trotsky) was used of non-communist writers sympathizing with the Revolution.

1936 *Nation* (N.Y.) 24 Oct. 471/1 The new phenomenon is the fellow-traveler. The term has a Russian background and means someone who does not accept all your aims but has enough in common with you to accompany you in a comradely fashion part of the way. In this campaign both Mr. Landon and Mr. Roosevelt have acquired fellow-travelers. **1941** Auden *New Year Let.* ii. 39 A liberal fellow traveller ran With sansculotte and Jacobin Nor guessed what circles he was in. **1942** E. Waugh *Put out More Flags* ii. 131 'I was never a party member.' 'Party?' 'Communist party. I was what they call in their horrible jargon, a fellow traveller.' **1946** [see *Crypto]. **1952** A. Wilson *Hemlock & After* 147 Bernard, if not a fellow-traveller, was certainly the perfect material for Communist propaganda. **1957** S. Jameson *Cup of Tea for Mr. Thorgill* vi. 48 He was also, quite openly, a fellow-traveller, going faithfully through all the rites: adoration of Russia, reverence for that fabled Beast, *the proletariat*. **1964** C. Chaplin *Autobiogr.* xxix. 491 They were carrying signs that read: 'Chaplin's a fellow traveller.' **1969** L. Woolf *Journey not Arrival Matters* iii. 139 He was distinctly a Fellow Traveller, and may, for all I knew, have been a member of the Communist Party.

Hence (as a back-formation) **fellow-travel** *v. intr.*, to be a fellow-traveller; *trans.*, to support (the Communist movement) as a fellow-traveller; esp. **fellow-travelling** *vbl. sb.* and *ppl. a.*

1858 (*title*) Fellow travelling; or, The experience of life. **1941** *Time* 31 Mar. 42 The Communist fellow-traveling American Peace Mobilization. *Ibid.* 30 June 42 His fellow-traveling had done him no good with the brass hats. **1948** *Ann. Reg.* 1947 240 Those [of the Left] who..would have no taint of 'fellow-travelling', broke with Mr. Wallace. **1949** *Life* 4 Apr. 39/1 Its host was the U.S.'s own National Council of Arts, Sciences and Professions, dominated by intellectuals who fellow-travel the Communist line. **1963** *Observer* 18 Aug. 20/8 The Germans who fellow-travelled with Hitler in the 1930s were guilty of a gross dereliction of national duty. **1964** 'W. Haggard' *Antagonists* iii. 29 The high rich Left..[would] talk and they'd fellow-travel but they'd never know the passion.

felon, *sb.*[1] Add: **5. felon-setting** (later examples: see also quot. 1970).

1922 Joyce *Ulysses* 626 Refuse to have anything to do with them..and their felonsetting. **1970** *Guardian* 6 June 11/7 Felon setting is a peculiarly Irish term, which recalls a deep and ancient hatred of the informer, a central figure in Irish history. It is felon setting to whisper to the Guards or the Special Branch about what the boys are up to. To some, apparently, it was felon setting for Jack Lynch to pass the documents in the arms affair to the Attorney-General.

felony, *sb.*[1] **4. b.** No longer differentiated from misdemeanour (see quot.).

1967 *Act Eliz. II* c. 58 Criminal Law Act. An Act to amend the law of England and Wales by abolishing the division of crimes into felonies and misdemeanours.

‖ **felsenmeer** (fe·lzĕnmiəɹ). Pl. **felsenmeere.** *Physical Geogr.* [G., lit. rock-sea.] An expanse of angular frost-riven rocks which may develop on flat terrain in arctic and alpine climates; a boulder field.

1905 *Jrnl. Geol.* XIII. 121 The lack of a commonly used English equivalent for the German word *Felsenmeer*. *Ibid.*, The development of the *Felsenmeer* means a vast increase of rock surface on which frost, changes of temperature, and all the other chief methods of weathering..can act. **1954** W. D. Thornbury *Princ. Geomorphol.* v. 69 It is common to find at high altitudes accumulations of riven rocks known as *felsenmeere*. These boulder fields attest to the rapidity of the weathering processes at high altitudes. **1963** A. N. Strahler *Earth Sciences* xxv. 455 The ground may be covered with large angular blocks of fresh rock in an accumulation known as a felsenmeer (literally, a rock sea), or boulder field.

felsic (fe·lsik), *a.* *Min.* [f. Fe(ldspar + *lenad* (= *Feldspathoid) + s(ilica + -ic.] Of or pertaining to such light-coloured minerals as the feldspars, the feldspathoids, quartz, and muscovite, or to rocks containing a high proportion of these or similar minerals.

1912 W. Cross et al. in *Jrnl. Geol.* XX. 561 We suggest the term *felsic* for the group of modal feldspars, feldspathoids, and quartz. **1965** G. J. Williams *Econ. Geol. N.Z.* iv. 38/1 The host-rock is a splintery siliceous bluish-grey slate with some micaceous, talcose and felsic facies.

felt, *sb.*[1] Add: **2. d.** A piece of woven cloth with a felted nap used in paper-making.

1752 Chambers *Cycl.* II. s.v. Paper, The coucher, who couches it upon a felt laid on a plank, and lays another felt on it. **1839** Ure *Dict. Arts* 927 Felts and paper are alternately stratified, till a heap of six or eight quires is formed. **1957** *Encycl. Brit.* XVII. 236/1 Some machines are provided with an endless felt which presses against the wire to pick off the paper and carry it through the first press.

4. b. felt(-tip, -tipped) pen, a pen with a felt point, used for labelling, etc.; **felt-side,** the upper surface of a sheet of machine-made paper (see sense *2 d, quot. 1957).

1957 *Stationery Trade Ref. Bk.* 33/2 Cushman & Denison Co. Ltd...Manufacturers of felt tip pens and inks. **1964** *Trademark Register* 1963 (U.S.) 268 *Yankee-Doodle*. Zip-Mark Corp...felt-tipped pens. **1965** *Ibid.* 1964 230 *Felt-riter*. Time Saving Specialities..felt marking pens. **1966** *Guardian* 3 Jan. 2/5 The secret bards [*sc.* graffiti-writers] have taken to felt-tipped pens. **1966** D. Francis *Flying Finish* x. 127 Mike..was already writing names on disposable cups with a red felt pen. **1967** T. Baird *Finding Out* x. 87 He charted a course with a felt-tipped marking pen. **1969** *Soviet Weekly* 6 Sept. 2 Anyhow, I got the job of buying all the things that did not require his personal attendance—notebooks, ball-points, feltpens, [etc.]. **1959** R. Hostettler et al. *Techn. Terms Printing Ind.* (ed. 3) 75 (*caption*) Felt side; top side.

felting, *vbl. sb.* Add: **1. b.** Of the hair (see quot. 1848).

1848 Dunglison *Dict. Med. Sci.* (ed. 7), *Felting*,..a term applied to the hair when inextricably interlaced, as occurs occasionally in women from inattention. **1864** W. T. Fox *Class. Skin Dis.* 75 Canities, Felting. **1966** J. S. Cox *Illustr. Dict. Hairdressing* 57/1 *Felting*, the matting or entangling by movement and pressure of hair into a hair mouse.

c. Of woollen fibres and paper: matting and shrinking.

1884 W. S. B. MacLaren *Spinning* iv. 58 By this rough arrangement of the fibres they..lap round each other more firmly in the felting; because as wool shrinks in the process, a fibre which is wrapped round several others will get a firmer grip. **1905** Goodchild & Tweney *Technol. & Sci. Dict.* 496/2 The wet web then passes through press rolls to ensure perfect felting and even thickness. **1920** Cross & Bevan *Paper Making* (ed. 5) ii. 86 The degree in which felting takes place will depend..upon the form or microscopic peculiarities of the fibres. **1960** *Which?* Apr. 91/1 Rubbing the material by hand, in a washing machine, or in a tumbler drier when wet (or even just damp) may cause felting. **1970** *Nature* 14 Feb. 634/2 Felting is the progressive fibre entanglement in wool resulting from mechanical agitation, particularly in the presence of water.

feltness. (Earlier example.)

1884 *Mind* IX. 1 The immediate feltness of a mental state.

female, *a.* Add: **2. a.** *spec.* **female fern** = Lady-fern.

1597 Gerarde *Herball* II. cccclxix. 969 *Filix fœmina.* Female Ferne or brakes. *Ibid.* 970 In English Brake, common Ferne, and Female Ferne. **1728** R. Bradley *Bot. Dict.* I. s.v. *Filix*. The sharp pointed Female Fern hath the mains Stalks about a Foot long. **1879** [in Dict.]. **1908** E. Step *Wayside & Woodland Ferns* 45 The ancients had their Male and Female-ferns, their *Filix-mas* and *Filix-fœmina*.

10. b. Mus. *female cadence, -close* (see quot. 1954). Cf. Feminine *a.* 6 b, quot. 1844.

1928 E. Blom *Limit. Mus.* 85 Mendelssohn uses female closes to excess. **1954** *Grove's Dict. Mus.* (ed. 5) III. 60/2 *Female, feminine cadence*, a cadence in which the concluding chord falls on the weak beat of the bar.

B. *sb.* **2. b.** Freq. in phr. *the female of the species.*

1911 Kipling in *Morning Post* 20 Oct. 7/3 (*title*) The female of the species. **1922** Wodehouse *Clicking of Cuthbert* ix. 220 The Bingley–Perkins combination, owing to some inspired work by the female of the species, managed to keep their lead. **1940** H. G. Wells *Babes in Darkling Wood* ii. i. 144 The female of the species..by the age of fifteen has a clearer sense of reality in these things than most men have to the doddering end of their days. **1961** J. MacLaren-Ross *Doomsday Book* ii. iii. 128 The female of the species first—take hold of her arm, George.

3. (Delete 'in certain nonce-words'.) *female-determiner, -determining* adj.; **female circumcision** = *Clitoridectomy; **female impersonation,** the personating of a female by a male on the stage; hence **female (im)personator**; **female pill,** any preparation intended as an abortifacient.

[**1837** *Penny Cycl.* VII. 197/1 An account of what he calls the circumcision of females..by some of the African tribes is given by Bosman in his 'Description of the Coast of Guinea'.] **1875** *Porcupine* 2 Jan. 635/3 Eugene, still *facile princeps* in female impersonation and falsetto. **1894** Hardy in *Harper's Mag.* XXIX. Dec. 74/1 Physician Vilbert's golden ointment, life-drops, and female pills. **1897** *Sears, Roebuck Catal.* 27/1 *Female Pills*.. If the very complete directions..will be followed closely, all will be well...With useful information and instructions to ladies concerning their troubles. **1905** *Lancet* 18 Nov. 1497/1 Madame Hipolite's world-renowned Female Pills, the safest and surest. **1909** J. R. Ware *Passing Eng.* 129/1 *Female personator*..the performer is a male who impersonates female appearance, singing, and dancing. **1931** J. S. Huxley *What dare I Think* i. 13 Sex is determined at conception by..male-determining and female-determining [cells], of which the female determiners are a little the bigger. **1931** E. Waugh *Remote People* 209 The rite of female circumcision..is one of the battlegrounds between missionaries and anthropologists. **1956** H. E. Lambert *Kikuyu Soc. & Polit. Inst.* vii. 143 An order limiting the operation of 'female circumcision' to an incision of sufficient depth and extent for the removal of the glans clitoridis only. (Passed in 1932.) **1957** *New Biol.* XXIII. 23 The barred horizontal line represents the threshold between male-determining and female-determining concentration zones. **1958** M. Kelly *Christmas Egg* i. 9 A shop..advertising that a long-dead proprietor had been 'Agent for Female Pills, by the King's Letters Patent 1743'. **1963** A. Heron *Towards Quaker View of Sex* 68 Some [transvestites] go on the stage as female impersonators. **1967** *Listener* 5 Jan. 35/2 Leo McKern's unexpected female impersonation as the Duchess. **1970** *Guardian* 16 Sept. 11/6 Female circumcision..is the crude severing or scarring of the clitoral area in girls approaching puberty.

femalely, *adv.* Delete *nonce-wd.* and add later example in sense 'in a female way'.

1966 B. Glemser *Dear Hungarian Friend* ii. 29 She was emphatically a girl, most femalely female.

femaleness. Delete *nonce-wd.* and add later example.

1955 E. Hyams *Slaughterhouse Informer* 6 She gave out..a kind of essence of lusty femaleness.

2. *Biol.* The quality of being female.

1889 Geddes & Thomson *Evol. Sex* iii. 92 After maleness and femaleness have been partly established. **1894** H. Drummond *Ascent of Man* vii. 315 That this division into maleness and femaleness should run between almost every two of every plant and every animal in existence, must have implications of a quite exceptional kind. *Ibid.* 323 What exactly maleness is, and what femaleness, has been one of the problems of the world. **1902** *Encycl. Brit.* XXXII. 210/1 That maleness and

femaleness may be regarded as expressing metabolic alternatives open to the germ-cell in its development. **1930** G. R. de Beer *Embryol. & Evol.* iii. 22 The various organs and parts of the body do not all switch over from maleness to femaleness together. **1949** M. Mead *Male & Female* i. 3 How are men and women to think about their maleness and their femaleness in this twentieth century. **1963** A. Heron *Towards Quaker View of Sex* iii. 32 A friendship..between two women in whom femaleness predominates.

femic (fe·mik), *a.* *Petrogr.* [See quot. 1902.] Of or pertaining to one of the two primary categories erected by Cross, Iddings, Pirsson, and Washington to classify the igneous rocks, and broadly including the non-aluminous ferromagnesian rocks (see also quots.).

1902 W. Cross et al. in *Jrnl. Geol.* X. 573 To express concisely the two groups of standard minerals and their chemical characters in part, the words *sal* and *fem* have been adopted...*Fem* indicates Group II, since its minerals are dominantly *ferromagnesian*. As adjectives to express these ideas the words *salic* and *femic* will be used. **1912** *Ibid.* XX. 560 Some petrographers have fancied the terms salic and femic as short words, which they wish to apply to modal quartz, feldspar, and feldspathoid minerals in one case, and to all modal ferromagnesian minerals in the other. These terms have also been applied to major rock groups. Such applications are not proper uses of these terms. **1920** A. H. Fay *Gloss. Mining & Mineral Ind.* 264/2 *Femic*...Often but incorrectly used in place of Mafic or Subsilicic. **1932** F. F. Grout *Petrography* ii. 46 The terms 'salic' and 'femic' refer to calculated *molecules*, not actual minerals, and should be used only in connection with chemical composition. **1939** A. Johannsen *Descr. Petrogr.* (ed. 2) I. viii. 86 *Femic* does not mean 'ferromagnesian'. **1966** *Mineral. Abstr.* XVII. 704 The problems of the determination of the amounts of femic minerals are discussed.

femina (fe·mină). *S. Afr.* Usu. in pl. **feminas, femina.** The long feathers from the wing-tips of a female ostrich.

1881 A. Douglass *Ostrich Farm. S. Afr.* xiii. 84 To show all the whites together, and then the feminas. **1896** R. Wallace *Farm. Industr. Cape Col.* xi. 235 White and Light Femina were very firm. Dark Femina, 5s. and 10s. per lb. higher.

feminal (fe·minăl), *a.* Restrict †*Obs.* to sense in Dict. and add: **2.** = Feminine *a.* 4.

1907 *Ladies' Field* 10 Aug. 357/1 Combine the perfection of physical strength with the highest type of feminal beauty. **1922** H. M. Barcynska *Webs* xxv. 207 In her emotional tumult, sheerly feminal, she believed every word she said.

feminine, *a.* and *sb.* Add: **A.** *adj.* **6. b.** (Earlier example.) *feminine ending* (further and *Mus.* examples).

1603 S. Daniel *Defence Rhyme* sig. H3[v], Two feminine nūbers (or Trochees, if so you wil call them). **1893** E. Prout *Mus. Form* ii. 9 What is termed by prosodists a 'feminine ending'—that is, the ending of a verse (in musical language, of a 'phrase') on an unaccented note following the accented note on which the actual cadence mostly occurs. **1955** J. A. Westrup in H. van Thal *Fanfare for E. Newman* xiii. 188 The feminine endings in the melody are similar to those found in early eighteenth-century instrumental music.

B. *sb.* **1. b.** *the eternal feminine* as a literal rendering of G. *das ewig-weibliche* (Goethe).

1892 [in Dict.]. **1898** *Daily News* 5 Sept. 6/2 The Eternal Feminine played a larger part in Oxford social life in the earlier part of the eighteenth century than she has since done. **1912** W. J. Locke *Aristide Pujol* ix, His quest being little Jean and not the eternal feminine.

feminineness. (Earlier U.S. example.)

1849 Thoreau *Week on Concord* 392 It was such a feminineness..as is rarest to find in woman.

feminism. Delete *rare* and add: **2.** [After F. *féminisme*.] Advocacy of the rights of women (based on the theory of equality of the sexes). (Cf. Womanism.)

1895 *Athenæum* 27 Apr. 533/2 Her intellectual evolution and her coquettings with the doctrines of 'feminism' are traced with real humour. **1908** *Daily Chron.* 7 May 4/7 In Germany feminism is openly Socialistic. **1909** *Ibid.* 29 May 4/4 Suffragists, suffragettes, and all the other phases in the crescendo of feminism.

3. *Path.* The development of female secondary sexual characteristics in a male.

1882 *Syd. Soc. Lex.* II, *Feminism*, the qualities of a female. Also Lorain's term for the arrest of development of the male towards the age of puberty, which gives to it somewhat of the attributes of the female. **1945** H. Burrows *Biol. Actions Sex Hormones* xxiii. 453 The symptoms of adrenal virilism and feminism are caused by an excessive production of androgen or oestrogen by the adrenal.

feminist (fe·minist), *a.* and *sb.* Also **fe·mininist.** [ad. F. *féministe*, f. L. *fēmina* woman: see -ist.] **A.** *adj.* Of or pertaining to feminism, or to women. **B.** *sb.* An advocate of feminism.

1894 *Daily News* 12 Oct. 5/5 What our Paris Correspondent describes as a 'Feminist' group is being formed in the French Chamber of Deputies. **1895** *Critic* 2 Feb. 90/2 The writer depicts Ford as the deepest 'femininist' in the Shakespearian constellation. **1898** *Daily Chron.* 15 Oct. 5/1 The lady Parliamentary reporter is the latest development of the feminist movement in New Zealand. **1904** *Athenæum* 26 Nov. 730/2 There have been feminists who claimed George Eliot as the rival of Thackeray. **1920** W. J. Locke *House of Baltazar* v. 56 We're out of this feminist hurly-burly. **1930** *Manch. Guardian* 15 Sept. 7/7 Feminists are rare birds in Russia.

Hence **femini·stic, feminini·stic** *adjs.*

1902 M. Beerbohm *Around Theatres* (1924) I. 365 Ibsen's femininistic propaganda. **1908** *Westm. Gaz.* 11 Sept. 6/3 Some thinkers in Hungary anticipate feministic developments even in Turkey. **1912** *Englishwoman* Mar. 261 This society is only feministic in so far as it strives to give women better opportunities.

feminization. Add: **1. a.** (Further example.)

1960 *20th Cent.* May 401 The process of softening in the worker. I should..call it his feminization.

2. *Biol.* The assumption of female sexual characteristics (by a male animal or plant); the occurrence of female sexual characteristics in a person who is genetically male.

1922 F. H. A. Marshall *Physiology of Reproduction* (ed. 2) xv. 697 (*caption*) Feminisation of guinea-pig... The animals are arranged in the following order:—Normal male, two feminised brothers, castrated brother. **1957** *New Biol.* XXIII. 16 The time of transition to female flower formation provides a convenient index of the 'feminization' of the plant. **1959** *Lancet* 17 Oct. 592/1 Apparent females with testicular feminisation are sex-reversed males. **1970** *Sci. Jrnl.* June 50/1 However in rare cases, known as testicular feminization, the male organs do not develop although the foetus has testes and a male genetic constitution.

feminize, *v.* Add: **2.** *trans. Biol.* To induce female sexual characteristics in. Hence **fe·minizing** *ppl. a.*

1922 [see *Feminization 2]. **1924** G. R. de Beer *Growth* ix. 62 By grafting a male gland into a female the latter can be masculinized, and a male can be feminized by the converse graft. **1962** D. J. B. Ashley *Human Intersex* iv. 60 Parabiosis experiments on primitive vertebrates showed inversion of sex in either direction, males were feminized and females masculinized. *Ibid.* v. 75 Tumours of the adrenal cortex which secreted feminizing hormones.

femino- (fe·mino), used as combining form (see -o) of L. *fēmina* woman, in sense 'female'. **feminophobia,** fear of women (see quot. 1960); *Nonce-wd.*

1884 [see Masculo-]. **1960** Koestler *Lotus & Robot* ii. viii. 213 One would have to coin the term feminophobia—the behaviour of shy young men in the presence of members of the opposite sex.

feminoid (fe·minoid), *a.* [f. L. *fēmin-a* woman + -oid.] Feminine (but not female); of female form or appearance.

1923 J. S. Huxley *Ess. Biologist* iv. 145 The 'feminoid' man shows the reverse tendency. **1932** *Times Lit. Suppl.* 18 Aug. 583/2 Man after infancy has a 'feminoid' phase. **1964** L. Martin *Clinical Endocrinol.* (ed. 4) vi. 215 Feminoid external genitalia.

‖ **femme** (fam). [Fr.] **1.** Woman, wife. *rare.* In the *U.S. colloq.* (with pronunc. (fem)), girl. Also, a lesbian who adopts a passive, feminine role (opp. *Butch *sb.*[1]).

1814 Byron *Jrnl.* 16 Jan. (1875) 361 Divorce ruins the poor *femme*. **1928** J. P. McEvoy *Showgirl* iv. 52 Eight femmes and a pair of male hoofers take up the burden when she is off. **1936** *Nat. Geogr. Mag.* LXIX. 778/2 West Point slang...A young lady is a 'femme' or 'fem'. **1944** [see *Brush *v.*[2] 5 b]. **1961** W. Brown *Bedeviled* 71 A step upward on the social ladder are the female transvestites and their '*femmes*' who congregate in the 'gay' bars of Greenwich Village. **1966** [see *Butch *sb.*[1]].

2. In Fr. combinations. **femme de ménage** (fam də menāʒ), a charwoman, or domestic help; **femme du monde** (fam dü mońd), a woman of the world; **femme fatale** (fam fatal), a dangerously attractive woman; **femme incomprise** (fam ańkońpriz), a woman who is misunderstood or not truly appreciated.

1849 Thackeray *Pendennis* I. xxiii. 217 Miss Amory is a *femme incomprise*. **1870** Mrs. Ogier Ward *Jrnl.* 16 Nov. in *Outside Paris* (1871) 41 To cook for them..I am obliged to hire a femme de ménage. **1885** A. Edwards *Girton Girl* II. iii. 35 Poor Linda must practise many a humiliating economy in her lot of femme incomprise. **1894** W. James *Let.* 24 Jan. in R. B. Perry *Tht. & Char. W. James* (1935) II. 188 His wife was a *femme du monde*, however, and fully made up for his lack of conversation. *c* **1903** Wyndham Lewis *Let.* (1963) 10 At this studio there is a femme de menage that does your room. **1912** G. B. Shaw *Let.* 19 Aug. (1952) 39 Here I saw a *Femme Fatale* who was a fine figure of a woman. **1913** D. H. Lawrence *Sons & Lovers* xii. 318 The dormant woman was the *femme incomprise*. **1919** C. Mackenzie *Sylvia & Michael* iii. 71 If she did not know what Eliane was, she might easily have mistaken her for a *femme du monde* like herself. **1924** Femme fatale [see *Blah *sb.*]. **1936** E. H. W. Meyerstein *Let.* 7 Jan. (1959) 180 As if these gay twitterers, like the eternal femme incomprise, refused to impart the secrets of their being to any! **1938** *Times Lit. Suppl.* 28 May 371/1 A pretty *femme fatale* who lost her head. **1944** W. S. Maugham *Razor's Edge* i. 36 There's no better education for a young man than to become the lover of a woman of a certain age and..if she's..a *femme du monde*..it would immediately give him a situation in Paris. **1954** M. F. Rodell *Mystery Fiction* ix. 56 Whirling in a high-powered car from Monte Carlo to Cap d'Antibes on the trail of an exotically beautiful *femme fatale*. **1955** E. Pound *Section: Rock-Drill* lxxxvi. 26 'Le prussien c'est un chic homme.' Said the aged femme de ménage with four teeth out.

femme de chambre. 2. (Earlier example.)

a **1828** D. Wordsworth *Tour Continent* in *Jrnls.* (1941) II. 10 Pretty *Femme de chambre*..her keys by her side.

femoro- (fe·mŏro), used as combining form (see -o) of Femur = pertaining to the femur; also = Femoral *a.*

1839 Hooper *Med. Dict.* (ed. 7) 624/1 Femorocele. **1872** G. M. Humphry *Obs. Myol.* 22 This femoro-fibular muscle appears to represent the short or femoral origin of the biceps. **1897** *Brit. Jrnl. Dermatol.* IX. 146 In the groins, and about the pubes, penis, and femoro-scrotal folds the scar is seen to be breaking down. **1898** [see *Autotomy]. **1900** Dorland *Med. Dict.* 254/1 *Femorocele*, femoral hernia. **1962** *Lancet* 8 Dec. 1195/1 Femoro-popliteal arterial bypass was advised.

femto- (fe·mto), *prefix.* [ad. Da., Norw. *femt-en* fifteen + -o.] Prefixed to the names of units in metric systems to form the names of units 10^{15} times smaller.

1961 *Symbols, Units & Nomencl. Physics* (Internat. Union Pure & Appl. Physics) 5 (*heading*) The following prefixes should be used to indicate decimal fractions or multiples of a unit...Femto. **1963** *Nature* 16 Mar. 1056/1 The International Committee [on Weights and Measures]..adopted two new prefixes for denoting submultiples of units. They are: femto, 10^{-15}, symbol 'f'.

fen, *sb.*[1] Add: **5. fen-reeve** (earlier and later examples); **fen wainscot** (moth), a moth, *Arenostola phragmitidis*, found in marshy places.

1654 in *East Anglian* (1871) IV. 14 They have chosen John Kent to be ffenn Reeve for the parish of Gillingham All S[ts]. **1910** H. M. Doughty *Chron. Theberton* xv. 209 Fen reeves had been elected every year by 'town meeting'. **1860** H. N. Humphreys *Genera Brit. Moths* I. 67 The Fen Wainscot..has the anterior wings rather bluntly lanceolate. **1951** Colyer & Hammond *Flies Brit. Isles* xx. 251 One [*sc.* a larva of *Cnemopogon apicalis*] has been recorded from a caterpillar of the Fen Wainscot Moth.

fence, *sb.* Add: **5. c.** (*to stand* or *sit*) *on* or *upon the fence* (earlier U.S., and later examples); *on the other side of the fence* (examples); *to mend* (or *look after*) *one's fences*, of a member of Congress: to renew contact with the electors; also, by extension, to make one's peace (with a person); *to rush one's fences*, to act precipitately; *over the fence* (Austral. and N.Z. colloq.), see quot. 1941.

1828 *Richmond Whig* 13 Aug. 1/5 (Th.), There are certain Administration Editors, Editors for a long time on the fence, who occasionally undertake..to sit as censors upon their fatigued and dusty brethren. **1829** R. C. Sands *Writings* (1834) II. 160 Mr. Spratt..was 'on the fence'; where, like a wise man, he determined to sit, until he had made up his mind on which side to get off. **1830** *Annals of Cleveland* No. 316, Now all would-but-dare-not-be-politicians who insist in sitting on the fence, will be amerced a penalty for the same. **1868** J. T. Trowbridge *Three Scouts* in *Beeton's Boys' Ann.* Suppl. i. 4, I judge your sympathies are more on t'other side of the secession fence than on ours. **1888** *Congress. Rec.* 16 Aug. 7646/1 [They] are at home seeking renomination or looking after their fences. **1889** Barrère & Leland *Dict. Slang*, *Mend fences, to* (American), to mend or repair fences for a man is to attend to his interests. **1906** *Forum* Apr. 444 An early adjournment of the session is deemed essential in order that the members may go home to mend their fences, as the saying is. **1917** A. Huxley *Let.* 3 Aug. (1969) 132 As Dean Inge said early in this war, 'God is sitting on the fence and it is perfectly uncertain on which side He will come down.' **1918** *Chrons. N.Z.E.F.* 19 July 276/1 'It's over the blooming fence,' he announced...'These..bounds. First they put Paris Plage out of bounds.' **1922** G. Frankau *Love Story A. Brunton* i. 16 Aliette was not the type of woman who liked rushing her fences, either mentally or on horseback. **1925** A. Huxley *Let.* 16 Sept. (1969) 253 We shall be looking at things mostly from the Indian side of the fence. **1927** *Daily Express* 8 Sept. 2/4 The French delegation, which has hitherto been sitting on the fence, has suddenly become exceedingly active. **1937** 'G. Orwell' *Road to Wigan Pier* xii. 243 The fence on which the literary gent sits..is now pinching his bottom intolerably. **1941** Baker *Dict. Austral. Slang* 28 *Over the fence*, unreasonable, beyond the pale of common-sense or justice. **1949** N. Balchin *Sort of Traitors* iv. 81 You fight all your life for something and then before you know where you are you're on the other side of the fence fighting against it. **1959** *Spectator* 21 Aug. 213/3 It is a pity we have not mended our fences with Colonel Nasser. **1966** *Times* 11 May 13/1 There are still fences to be mended and fingers to be kept in pies. *Ibid.* 13 May 1/7 Britain would not let slip any opportunity in this matter, but there was no question of precipitate action or rushing fences. **1970** 'W. Haggard' *Hardliners* viii. 85 It was a thousand pities that Bull's predecessor had been so stupid...There were fences to be mended here and Albert Bull intended to mend them.

10. b. (sense 5) *fence-post* (earlier U.S. examples). **c.** objective (sense 5) *fence-breaker, -breaking.*

1878 E. S. Elwell *Boy Colonists* 218 He knew [where] Geddes' old horse, our old friend the fence-breaker, was feeding. **1900** *Westm. Gaz.* 8 June 4/3, I hope the fence-breaking will be omitted from the programme. **1792**

Trans. Soc. Promotion Agric., Arts & Manuf. (U.S.) I. 26 Ship-trunnels, fence-posts, mill-cogs and fire-wood. **1853** B. F. Taylor *Jan. & June* (1871) 13 Life..laid away in 'Patent Burial Cases' and fastened to rails and fence-posts.

11. fence-arbour, a piece in a combination lock which connects the spindle and the tumblers; **fence corner** *U.S.*, (*a*) one of the four corners of a fenced enclosure; (*b*) one of the many angles made by a zig-zag rail fence; also *attrib.*; **fence-line,** (*a*) = Fence *sb.* 5; (*b*) the straight strip of land on which a fence is to be erected; **fence-man** *U.S.*, one who practises 'sitting on the fence', who avoids taking a side in an issue; **fence-rail** *U.S.*, a long, rough rail for fencing, split from a small log; **fe·nce-ri:der** *U.S.* (see quot. 1920); also *fig.* = **fence-man*; **fence-riding** *U.S.*, 'sitting on the fence'; avoidance of committing oneself to one or other of two contrary policies; **fence-row** *U.S.*, a fence with the row of shrubs and other vegetation which frequently grows up under its protection; **fence-sitter,** one who 'sits on the fence'; **fence-sitting,** the action of 'sitting on the fence'; also *attrib.*; **fence-viewer** (earlier and *fig.* examples).

1902 *Encycl. Brit.* XXXII. 360/1 A balanced fence arbour. **1832** J. P. Kennedy *Swallow Barn* I. xi. 153 He slowly went to the fence corner, and untied his horse. **1846** J. J. Hooper *Adv. Simon Suggs* i. 14 Simon and Bill were in a fence corner very earnestly engaged at 'seven up'. **1855** *Knickerbocker* XLV. 197 Posting himself at night in a fence-corner, he saw her at one end of a hollow log. **1874** E. Eggleston *Circuit Rider* xxviii. 272 Patty climbed upon a fence-corner. **1876** [in Dict., sense 10 b]. **1901** S. E. White *Claim Jumpers* iv. 64 It was..not as large as a good-sized rat, quite smaller than our own fence-corner chipmunks of the East. **1858** J. A Warder *Hedges & Evergreens* ii. 38 Its cheapness..demands its..adoption where fence-lines are to be permanent. **1950** *N.Z. Jrnl. Agric.* Apr. 336/3 To aid adoption..[of orphan lambs] the [foster-parent] ewes have to be tied to a fenceline. **1961** B. Crump *Hang On a Minute* 120 They still had to clear the fence line, lay out the material and erect nearly four miles of boundary fence. **1963** N. Hilliard *Piece of Land* 106 He'd just as soon sleep in a tent along the fence-line. **1828** *Ohio State Jrnl.* 30 Jan. 3/5 It would be well perhaps for him to inform the public as to their politics. How many neutrals, fencemen &c. **1848** *N.Y. Herald* 14 Oct. (Bartlett 1859), All the fence-men, all the doubters, all the seekers after majorities, will now bustle up. **1889** Farmer *Americanisms* s.v. *Fence*, The possessors of highly developed bumps of caution are called *fence men*; they run with the hare and hunt with the hounds, an operation which receives the equally descriptive name of *fence-riding*. **1733** W. Byrd *Journey to Eden* 27 Sept. in *Westover MSS.* (1841) 110 We found the land..very thin of trees, and those that were standing fit for little but fuel and fence-rails. **1814** J. Taylor *Arator* (ed. 2) 177 Small common fence rails..make folds with less labour..than any I have ever tried. **1853** J. G. Baldwin *Flush Times Alabama* 110 Buck Jones..seized a fence rail, grasped it in both hands, and..hurled the same. **1945** *Reader's Digest* Jan. 53/2 He hewed out 300 fence rails, ten feet long. **1834** 'J. Downing' *Life A. Jackson* 91 The fence riders now took courage and jumpt clean off. **1909** R. A. Wason *Happy Hawkins* 207, I met the foreman o' the E.Z. outfit ridin' into town to see if he couldn't pick up a fence-rider. **1920** J. M. Hunter *Trail Drivers of Texas* 298 The fence rider, also called the 'line rider', is employed to ride fences and repair them. *a* **1859** *N.Y. Mirror* (Bartlett), The dividing line..admits of no fence-riding; the candidate must be on one side or the other. **1868** *Congress. Globe* 17 July (De Vere), This question is one of clear right and wrong, and there can be no fence-riding, when the rights of four millions of men are at stake. **1889** Fence-riding [see **fence-man*]. **1842** *Amer. Pioneer* I. 43, I was alone, clearing out a fence row, about a quarter of a mile from the house. **1855** *Trans. Mich. Agric. Soc.* VI. 144 The fence rows were free from weeds and bushes. **1901** N. L. Britton *Man. Flora N. States* 952 Along fence-rows in partial shade. **1948** *Country Gentleman* May 175/2 He had cut the fence rows from the board fence. **1905** *Westm. Gaz.* 8 July 1/2 Fence-sitters..are to be shaken out. **1949** J. R. Cole *It was so Late* 12 Why don't you make up your mind—you fence sitter! **1960** *Spectator* 7 Oct. 507 One of the Labour Party's most accomplished fence-sitters. **1904** *Westm. Gaz.* 11 Feb. 1/3 This situation will..be prolonged until the fence-sitting Unionists..come down on one side or the other. **1905** *Ibid.* 5 Mar. 1/2 It is very well to..denounce Mr. Balfour for timidity and fence-sitting. **1956** *Essays in Criticism* VI. 95 One would have welcomed..less academic fence-sitting. **1959** *Times Lit. Suppl.* 29 May 322/4 The seminar-paper tends to provide..a fence-sitting indecision. **1661** in *Rec. Early Hist. Boston* (1880) IV. 109 Fence viewers: for the necke of land, Richard Withington, [etc.]. **1880** *Scribner's Monthly* Feb. 504 A hare..or..a brown wood wren...These are the fence viewers of the wood lot.

fenced, *ppl. a.* *fenced in* (examples).

1957 *New Yorker* 29 June 64/3 An enormous, fenced-in storage area piled high with crates and packages. **1960** *Farmer & Stockbreeder* 1 Mar. Suppl. 1 He..pushed the pair of us into a..fenced-in yard.

fencer. Add: **2.** Not restricted to Australia. (Further examples.)

1881 *Instr. Census Clerks* (1885) 98 s.v. *General labourer*, Fencer. **1921** H. Guthrie-Smith *Tutira* xxxviii. 382 Owners and employees had worked shoulder to shoulder as..butchers, fencers, bullock-punchers. **1950** *N.Z. Jrnl. Agric.* Oct. 356/2 Two fencers are kept in steady work at Glenaray [sheep station, Southland].

fenchane (fe·ntʃēin). *Chem.* [f. *Fench(one + -ane.] A liquid that is chemically a saturated bicyclic hydrocarbon of the terpene series, $C_{10}H_{18}$, and may be considered the parent compound of a series including the fenchenes and fenchone; also applied to some related isomers of this (see quots. 1949).

1907 *Jrnl. Chem. Soc.* XCII. i. 24 The close relationship of camphane and fenchane. **1949** Simonsen & Owen *Terpenes* (ed. 2) II. vi. 528 By the catalytic hydrogenation of α-fenchene..Zelinski [1904] prepared a saturated hydrocarbon, $C_{10}H_{18}$, to which he gave the name fenchane...In 1911 Wolff and Kishner prepared from fenchonehydrazone an isomeric hydrocarbon having completely different properties...This hydrocarbon was designated fenchane, whilst for Zelinski's hydrocarbon the name *iso*bornylane was adopted. [*Note*] *iso*Bornylane is also termed α-fenchane. *Ibid.* 530 On reduction..it [*sc.* *iso*bornylol] does not give *iso*bornylane, but a hydrocarbon..termed β-fenchane. **1953** Eastman & Noller in H. Gilman *Org. Chem.* IV. vii. 656 The most important naturally occurring terpene having the fenchane carbon skeleton is fenchone.

fenchene (fe·ntʃīn). *Chem.* [ad. G. *fenchen* (O. Wallach 1891, in *Ann. der Chem.* CCLXIII. 151), f. *fenchel* fennel: see -ene.] Any of several isomeric unsaturated terpenes, $C_{10}H_{16}$, that are related to fenchane and are isolated as oily liquids.

1891 *Jrnl. Chem. Soc.* LX. ii. 1083 Fenchone, on reduction, yields fenchyl alcohol, and from this one obtains fenchyl chloride and fenchene. **1941** *Thorpe's Dict. Appl. Chem.* (ed. 4) V. 5/2 *cyclo*Fenchene..is present in the mixture of fenchenes prepared by any of the general methods.

fenchone (fe·ntʃōun). *Chem.* [ad. G. *fenchon* (O. Wallach 1891, in *Ann. der Chem.* CCLXIII. 129), f. G. *fenchel* fennel: see -one.] An oily liquid with a camphor-like smell which occurs naturally in two enantiomorphic forms in various essential oils (as fennel oil and thuja oil) and is chemically a bicyclic ketone, $C_{10}H_{16}O$, of the terpene series.

1891 *Jrnl. Chem. Soc.* LX. ii. 1082 In many fennel oils a liquid, camphor-like substance occurs. This compound is called fenchone, and is isomeric with camphor. **1932** J. L. Simonsen *Terpenes* II. i. vi. 569 Fenchone is distinguished from other dicyclic ketones by its low boiling-point. **1949** E. Guenther *Essential Oils* II. 420 Fenchone serves mainly as an odor adjunct in room sprays and bath preparations of pine character. **1953** [see *Fenchane]. **1955** *Jrnl. Indian Chem. Soc.* XXXII. 228 The essential oil of *Brunella vulgaris*, Linn has been found to consist mainly of *d*-camphor and *d*-fenchone.

fencing, *vbl. sb.* Add: **7.** (sense 1) *fencing-room*; (sense 3) *fencing-gear, post, -wire* (examples).

a **1950** L. Skuthorpe in *Austral. Short Stories* (1951) 132 Get the fencing-gear, lads, and put that fence up again. **1877** H. C. Hodge *Arizona* 59 The forests of juniper will furnish large quantities of the most durable railroad ties, fencing posts. **1950** *N.Z. Jrnl. Agric.* Mar. 233/1 Pre-cast [concrete] products such as blocks and fencing posts. **1897** *Outing* (U.S.) XXX. 200/1 The contest was held March 20th in the fencing-room of the New York Athletic Club. **1878** *Rep. Indian Affairs* (U.S.) 35, I suggested that fencing-wire be supplied. **1908** *Westm. Gaz.* 11 June 2/1 The fencing-wire trade. **1936** A. Russell *Gone Nomad* vi. 45, I had unloosed the wrong tie rope, thereby dropping a coil of fencing wire on Brown's toes.

Fendant (fańdań). [Swiss-Fr., the name of a grape.] A dry white wine produced in south-western Switzerland.

1911 *Encycl. Brit.* XXVI. 242/1 Among the best Swiss wines are those of..Muscat, Fendant and Vin du Glacier (all in the Valais). [**1926** P. M. Shand *Bk. Wine* vii. 213 The generic name of the Vallais wines is that of their informing grape, Fendant de Sion or Fendant de Vallais.] **1962** *Economist* 29 Dec. 1283/1 Fendant served in tumblers in the lesser taverns can be quite sharp and unrewarding. **1970** P. Moyes *Who saw her Die?* xvii. 218 Their tiredness fell away miraculously under the benign influence of..a flask of Fendant.

fender, *sb.* Add: **2. f.** A bumper of a motor-car. *N. Amer.*

1919 S. Lewis *Free Air* 103 Claire..had enjoyed the sight of their duffle-bags stuck up between the sleek fenders and the hood. **1928** *Punch* 25 Apr. p. xxx/3 (Advt.), Fender guards to match for rear. **1932** E. Wilson *Devil take Hindmost* viii. 47 A thousand-ton electric shear..reduces chassis, springs, wheels, fenders and all to a junk fodder of iron spines. **1960** *Times* 14 Sept. 12/6 When we hired a car in California we found that a car..bristles with surprises. You scrape the fender. **1963** H. Garner in R. Weaver *Canad. Short Stories* 2nd Ser. (1968) 41 There was no sign of a flat tire, but its left front fender was loose.

7. fender wall = Fender 3 c.

1894 J. P. Allen *Pract. Building Constr.* ii. 28 Fender walls are those built round fireplaces to carry the hearth-stones and take the ends of the joists, which would otherwise have to be trimmed. **1924** W. M. F. Petrie *Relig. Life Anc. Egypt* 124 Successive coatings of wall, and fender walls before the doorway.

fendered, *a.* [f. Fender *sb.* + -ed[2].] Provided with a fender or fenders.

1795 R. Dodd *Rep. Improv. Hartlepool* 8 This pier..well fendered, piled, &c. **1927** *Blackw. Mag.* Sept. 368/2 The tug thrust her fendered nose against the timbers.

fenestra. Add: **4.** *Surg.* **a.** A perforation in a surgical instrument other than in the handle.

In quot. 1963 *fenestra* has apparently been taken as a plural form.

1876 Dunglison *Dict. Med. Sci.* (rev. ed.) 411/2 The term *fenestra* is also applied to the open space in the blades of a forceps. **1899** C. Truax *Mechanics of Surgery* xxiii. 479 The sac wall, when held in this forceps, will protrude through the fenestræ of the blades, whose sharply serrated edges insure a firm grasp. **1963** Mitchell-Heggs & Drew *Instruments of Surgery* i. 46 Charnley's dissecting forceps is designed in addition to grip suture material firmly when tying knots...The suture is trapped in the fenestrum by the block projection on the opposing jaw.

b. An opening in a dressing, plaster, etc., for access or the relief of pressure.

1876 C. H. Leonard *Man. Bandaging* x. 116 It would be well to coat the margins of the fenestrae..with parrafin [*sic*], so as to prevent the absorbtion [*sic*] of the fluids by the dressing. **1914** E. L. Eliason *Pract. Bandaging* v. 109 The gauze dressing over the wound should be the size and shape of the desired fenestra or window.

c. A hole cut by a surgeon in any structure of the body.

1941 *Surg. Gynec. & Obstet.* LXXII. 472 This surgically made fenestra remains open. **1958** F. B. Korkis *Rec. Adv. Oto-Laryngol.* (ed. 3) iv. 59 Closure of the newly made fenestra..is not as frequent as it was in the early days of the operation.

fenestrated, *ppl. a.* Add: **2. b.** Of a surgical instrument: having one or more fenestræ.

1881 *Trans. Obstet. Soc. Lond.* XXII. 46 These forceps, fenestrated longitudinally. **1915** A. MacLennan *Surgical Materials* vi. 210 (*caption*) Common type of clamp, with milled and fenestrated blades. **1963** Mitchell-Heggs & Drew *Instruments of Surgery* i. 55 Everett's pile forceps..have a pair of fenestrated, grooved blades to accommodate the tissue of the haemorrhoid or 'pile'.

fenestration. Add: **3.** *Surg.* An opening made surgically to provide a passage through an anatomical structure; *esp.* the operation of cutting such an opening into the labyrinth of the ear to restore hearing in cases of otosclerosis.

1935 R. S. Stevenson *Rec. Adv. Laryngol. & Otol.* v. 80 (*caption*) Fenestration of thyroid cartilage: Ledoux's method. **1937** *Bull. N.Y. Acad. Med.* XIII. 675 (*heading*) Operations based on fenestration of the labyrinth. **1949** G. M. Coates in F. Christopher *Textbk. Surg.* (ed. 5) xviii. 771 The theoretical basis of the fenestration operation is that by producing a new window into the internal ear..the impeded fenestra ovale is by-passed and sound impulses are again projected to the fluids of the internal ear and the cochlea. **1958** F. B. Korkis *Rec. Adv. Oto-Laryngol.* (ed. 3) iv. 59 The operation of fenestration itself is being challenged..on account of the introduction of a less radical procedure—mobilization of the stapes.

‖ **feng-shui** (fʊ·ŋʃū:i). Also 8 **fong-choui.** [Chinese, f. *feng* wind + *shui* water.] In Chinese mythology, a system of spirit influences, good and evil, which inhabit the natural features of landscapes; hence, a kind of geomancy for dealing with these influences in determining sites for houses and graves. Also in extended use.

1797 *Encycl. Brit.* IV. 679/1 The greater part of the Chinese are of the opinion that all the happiness and misfortunes of life depend upon the *fong-choui*. **1883** *Ibid.* XV. 204/1 The *feng-shui*, or 'wind-and-water' magic, is a system the practitioners of which regulate the building of houses and tombs by their local aspects. **1906** W. De Morgan *Joseph Vance* xxvii. 252, I think what the Chinese call the Feng-Shui of the sofa-back had a good deal to answer for. *a* **1936** Kipling *Something of Myself* (1937) vii. 178 We entered and felt her Spirit—her Feng Shui—to be good. **1967** 'A. Cordell' *Bright Cantonese* xi. 123, I climbed to the grave of my mother, which was fine for *Feng Shui* with its wind and water.

Fengu: see *Fingo.

Fenn (fen). The name of Charles *Fenn* (1840–1925), English entomologist, used in *Fenn's wainscot*, a moth, *Arenostola brevilinea*, first described by him in 1864.

[**1869** E. Newman *Illustr. Nat. Hist. Brit. Moths* 271/2 Fenn's Nonagria.] **1907** R. South *Moths Brit. Isl.* I. 308 Fenn's Wainscot (*Leucania brevilinea*). **1955** E. B. Ford *Moths* x. 136 The Flame Wainscot, *Senta flammea* Curtis, and Fenn's Wainscot, *Arenostola brevilinea* Fenn, would often be searched for in vain.

fenner. Delete *rare* and add later examples.

1900 *Daily Express* 26 June 4/5 When we poor fenners skate the ice. **1940** F. Kitchen *Brother to Ox* vi. 98 You could pick out the Lincolnshire 'fenners' by their fancy for bright blue cords.

Fennoscandia (fenoskæ·ndiă). *Geol.* Also **Fenno-Scandia, Fennoskandia.** [ad. G. *fennoskandisch* adj. (W. Ramsay 1904, in *Fennia* XXI. vii. 44), f. L. *Fenn-i* the Finns + -o: see Scandian *a.*] An ancient land mass in north-western Europe comprising most of Scandinavia and Finland and parts of the Soviet Union west of the White Sea. Also in extended use, Scandinavia and Finland as a political unit. Hence **Fennosca·ndian, Fe:nnoscandina·vian** *adjs.*

1907 tr. J. J. Sederholm in *Bull. Commission Géol. de Finlande* V. xxiii. 91 The earnest discussion among Fenno-Scandian geologists,..concerning the origin of the gneissose rocks. *Ibid.* 92 Pre-Cambrian sedimentary formations have been traced through the whole eastern part of Fenno-Scandia. **1910** *Encycl. Brit.* XII. 59/2 Fennoscandia (the name for Scandinavia with Finland as a single geological region). **1944** *Proc. Geol. Assoc.* LV. 61 (*heading*) The Fennoskandian contributions. **1956** L. U. de Sitter *Struct. Geol.* xxv. 367 Pre-Cambrian rocks of Fennoskandia and Greenland. **1962** E. W. Spencer *Basic Concepts Physical Geol.* xii. 212 In such areas as the Canadian Shield and the Fennoscandinavian Shield the surface is rising at measurable rates. **1966** *Times* 3 Feb. 13/2 President Kekkonen's proposals for a neutral 'Fenno-Scandia', to include not only Finland and Sweden but also Norway and perhaps Denmark if they were prepared to leave Nato.

fenster (fe·nstəɹ). *Geol.* [G., lit. window.] An opening or 'window' eroded through an older stratum in a region of overfolding or overthrusting, exposing a younger stratum beneath.

1925 *Bull. Virginia Geol. Surv.* xxv. 45 Occurrences, where the erosion of the older, overlying rocks has exposed the younger or underlying rocks, have generally been called fensters. **1933** W. H. Bucher *Deform. Earth's Crust* vii. 176 This is beautifully shown on the map by the 'embayments' and 'fensters' of Palaeozoic formations. **1939** A. K. Lobeck *Geomorphol.* xvii. 605 So called windows or fensters are produced by the erosion of plateau areas, so that the youngest strata of an underlying nappe are exposed at the bottom of the valley thus formed. **1954** W. D. Thornbury *Princ. Geomorphol.* x. 273 In the early stages of dissection of a nappe, erosion may penetrate the overthrust mass and locally expose patches of the younger rocks below it. Such exposures are called fensters or windows.

fent, *sb.* **4.** (Earlier example.)

1844 S. Bamford *Pass. Life Radical* ix. 60 Her outer garments were of very homely material, being seemingly cotton fents dyed blue.

fera, var. *Ferra.

‖ **feræ naturæ** (fīə·rī natʃu·rī, fe·rai natiu·rai). [L., = belonging to the wild part of nature.] Animals living in a wild state, undomesticated animals. Also as quasi-*adj.* and *fig.*

a **1661** Fuller *Worthies* (1662) ii. 21 Such who object that *heads of Stagges*, had been more proper for her, the *Goddesse of the Game*, may first satisfie us, Whether any Creatures *feræ Naturæ* (as which they could not certainly compass at all seasons) were usually offered for Sacrifices. **1668** Dryden *Even. Love* (1671) iv. sig. H3, Women are not compris'd in our Laws of friendship: they are *feræ naturæ*. **1873** C. M. Yonge *Pillars of House* IV. xxxv. 5 He evidently viewed himself as the Underwood who alone could do his duty by the *feræ naturæ* of the estate. **1897** E. A. Bartlett *Battlefields of Thessaly* xiii. 302 The weapons had been fairly purchased by us from the Arnauts, who had found them abandoned, in fact as *feræ naturæ*. **1959** Jowitt *Dict. Eng. Law* I. 120 At common law animals *feræ naturæ* which are fit for human food are the subject of larceny if they either are in confinement or have actually been tamed; but at common law there can be no larceny of such animals as a lion or a gorilla, even though they are tame or kept in confinement.

Feraghan (fe·răgān). [Pers., f. *Fergana*, *Ferghana*, name of a region in Soviet central Asia.] A costly rug made in Persia, usu. of cotton.

1929 *Encycl. Brit.* XIX. 624 Feraghan. A type much prized in the East. The best, usually small, are finely woven with a short pile. They have the Ghiordes knot, a cotton pile, and two lines of weft. The colouring is soft, and evenly toned patterns like the herati are common. **1931** A. U. Dilley *Oriental Rugs & Carpets* iv. 124 Except for a small number woven in medallion pattern upon a plain field, the Feraghans were produced in all-over design. **1960** H. Hayward *Antique Coll.* 117/1 *Feraghan carpets*, Persian carpets woven with a Sehna knot, which..vary slightly in the treatment of the traditional Herat design.

Ferangi: see *Feringhee.

ferbam (fə̄·ɹbæm). *Chem.* [f. *fer*ric dimethyldithiocar*bam*ate, the systematic name.] A black powder used as a fungicide, esp. in the control of rust diseases of plants; $[(CH_3)_2N{\cdot}CS{\cdot}S{\cdot}]_3Fe$.

1950 *Phytopathology* XL. 118 The Subcommittee on Fungicide Nomenclature of The American Phytopathological Society, cooperating with the Interdepartmental Committee on Pest Control, has selected common names for five commercially-available fungicidal chemicals which are useful in the control of various destructive plant diseases...The coined common names and designations are: *Ferbam* for the fungicidal chemical, ferric dimethyl dithiocarbamate. **1956** *Dict. Gardening* (*R.H.S.*) Suppl. 216/1 Ferbam is commonly used in the U.S.A. to control apple scab and apple rust. **1960** *Times Rev. Industry* Sept. 7/2 One group of fungicides, with names suggestive of Old Testament prophets—nabam, thiram, ferbam, ziram and zineb—can all impart a foreign flavour into the fruits to which they are applied. **1967** S. E. A. McCallan in D. C. Torgeson *Fungicides* I. i. 14 In comparison with the older recommendations of 12–16 lb of sulfur per 100 gal of spray..the use of ferbam at 1½ lb/100 gal seemed strikingly effective.

ferganite (fəɹgā·nait, fə·ɹgānait). *Min.* Also **ferghanite.** [ad. Russ. *ferganit* (I. A. Antipov 1908, in *Górnȳi Zhurnál* LXXXIV. IV. 259), f. *Ferg(h)an-a* (see *Feraghan): see -ite[1].] A hydrated vanadate of uranium.

1910 *Mineral. Mag.* XV. 421 Ferganite. **1925** *Ibid.* XX. 294 It has already been suggested by K. A. Nenadkevich that ferganite is identical with tyuyamunite...On the other hand, ferganite may represent a leached or weathered product of tyuyamunite. **1958** E. W. Heinrich *Mineral. & Geol. Radioact. Raw Materials* ii. 106 Ferghanite. Supposedly $U_3(VO_4)_4{\cdot}6H_2O$.

feria. Add: **2.** [Sp.] = Fair *sb.*[1] 1.

1844 J. Gregg *Commerce Prairies* II. 85 At certain seasons of the year, there are held regular *ferias*, at which people assemble in great numbers, as well of sellers as of purchasers. [**1846** R. Ford *Gatherings from Spain* v. 43 The Spanish term *Feria* signifies at once a religious function, a holiday, and a fair.] **1932** E. Hemingway *Death in Afternoon* iv. 41 The local fairs or ferias, which usually commence on the Saint's day of the town. **1939** Spender & Gili tr. *Lorca's Poems* 123 How dazzling in the feria! **1957** A. MacNab *Bulls of Iberia* iv. 43 In provincial *ferias*, the unboxing of the bulls..is an event worth seeing.

Feringhee. Add: Also **Ferangi.**

1919 W. H. Downing *Digger Dial.* 57 *Ferangi*, European. **1936** F. Stark *S. Gates Arabia* xv. 162 No Ferangi goes without a servant.

ferly, *sb.* (General, i.e. not *Sc.* or *dial.*, examples.)

1935 C. Day Lewis *Time to Dance* 23 Those ferlies you'll not behold Till the guardians of that valley have crossed Your hand with fairy gold. **1962** Davis & Wrenn *Eng. & Medieval Studies* 225 But for anthropology 'we' should have taken the ferlies in medieval romance like trivial excitements in a boy's blood. *Ibid.* 227 The idea that ferlies are Jungian outcroppings from the collective unconscious.

Fermat (fę̄·rma). The name of Pierre de *Fermat* (1601–65), French mathematician, used attrib. or in the possessive to designate certain results and concepts introduced by him, as *Fermat's last theorem*, a famous unproved theorem (of which Fermat said he had 'a truly wonderful proof'), viz. that if n is an integer greater than 2, $x^n + y^n = z^n$ has no positive integral solutions; *Fermat's law* = **Fermat's principle*; *Fermat('s) number*, a number of the form $2^{2^n}+1$, where n is a positive integer; *Fermat's principle*, the principle that the path taken by a ray of light between any two points is such that the integral along it of the refractive index of the medium has a stationary value; *Fermat's theorem*, (*a*) that if p is a prime number and a an integer not divisible by p, then $a^{p-1} - 1$ is divisible by p; (*b*) = **Fermat's last theorem*. Also **Fermatian** (fəɹmē·ʃən), *a.* and *sb.* (*rare*).

1811 P. Barlow *Elem. Invest. Theory of Numbers* ii. 48 [This] leads us at once to the demonstration of one of Fermat's theorems, that he considered as one of his principal numerical propositions. **1839** *Phil. Mag.* 3rd Ser. XIV. 48 Horner's extension of Fermat's theorem suggested this extension of Sir John Wilson's to me. **1845** *Ibid.* XXVII. 286 (*heading*) Proof of Fermat's Undemonstrated Theorem, that $x^n + y^n = z^n$ is only possible in whole numbers when $n = 1$ or 2. **1865** Brande & Cox *Dict. Sci., Lit. & Art* I. 879/1 Another theorem, distinguished as Fermat's last Theorem, has obtained great celebrity on account of the numerous attempts that have been made to demonstrate it. **1884** A. Daniell *Text-bk. Physics* v. 125 Fermat's Law. **1887** J. J. Sylvester in *Nature* 15 Dec. 153, I have found it useful to denote $p^i - 1$ when p and i are left general as the Fermatian function, and when p and i have specific values as the ith Fermatian of p. **1888** *Encycl. Brit.* XXIV. 424/1 It follows that the course of a ray is that for which the time..is a minimum. This is Fermat's principle of least time. **1906** *Bull. Amer. Math. Soc.* XII. 449 Fermat's numbers..are known to be prime for $n = 0, 1, 2, 3, 4$, and composite for $n = 5, 6, 7, 9, 11, 12, 18, 23, 36, 38$. **1948** O. Ore *Number Theory* viii. 204 This is the famous Fermat's theorem, sometimes called Fermat's last theorem, on which the most prominent mathematicians have tried their skill ever since its announcement three hundred years ago. **1959** Born & Wolf *Princ. Optics* 737 The laws of geometrical optics may be derived from Fermat's principle. **1966** Ogilvy & Anderson *Excurs. Number Theory* iii. 36 The higher Fermat numbers have been the subject of prolonged study.

‖ **fermata** (fəɹmā·tă). *Mus.* Pl. **fermatas, fermate.** [It.] A pause of unspecified length; the sign indicating such a pause.

[**1842** J. F. Warner *Univ. Dict. Mus. Terms* p. xxxviii/2 *Fermata* is the Italian name for what we call *a hold*.] **1876** Stainer & Barrett *Dict. Mus. Terms* 165/1 Fermata (*It.*), a pause (from *fermare*, to stay, or stop). **1889** *Cent. Dict.*, *Fermata*,..A pause in the accompaniment to give room for an extended cadenza by the soloist. **1955** Auden *Shield of Achilles* ii. 43 In timeless fermatas of awe and delight. **1962** *Times* 6 Apr. 17/6 Tempo-changes were ungainly and dramatic fermate ill-judged. **1969** *Daily Tel.* 18 Jan. 17/5 At the same time the fermatas, traditionally signposts inviting improvised insertions, grow scarcer.

ferment, *sb.* Add: **1. a.** In the sense 'unorganized ferment' the word has been replaced in scientific use by *enzyme*. (Further examples.)

1890 *Jrnl. Chem. Soc.* LVII. 530 The expression *ferment* is more frequently than not employed as the equivalent of *unorganised ferment*...Several words have been coined in place of unorganised ferment, notably *zymase* and *enzyme*. **1945** E. B. Ford *Butterflies* iii. 58 It [*sc.* melanin] is produced by the oxidation of..'tyrosin' (an amino-acid), through the action of a ferment, 'tyrosinase'.

4. *attrib.* and *Comb.*

1863 H. Watts *Dict. Chem.* II. 634 Ferment-oils..are volatile oils, produced by the fermentation of various plants, not originally contained therein, and essentially different from the oils which are extracted from unfermented plants by distillation with water. **1886** *Encycl. Brit.* XXI. 400/2 In 1870 Pasteur had proved that a disease of silkworms was due to a ferment-organism of the nature of a Schizomycete. **1898** H. Manders (*title*) The ferment treatment of cancer and tuberculosis.

fermentability (see after Fermentable *a.*). (Later example.)

1912 A. C. Chapman *Brewing* iv. 55 A wort having just the degree of fermentability required.

ferme·ntal, *a.* Delete †*Obs.* and add later examples.

1753 Chambers *Cycl. Suppl.* s.v. *Zymology*, The fermental principles of acid and sulphur connatural to that oil [*sc.* oil of vitriol]. **1922** *Glasgow Herald* 21 Aug. 7 Sugar or any other kind of fermental food.

fermenter, fermentor (fəɹme·ntəɹ). [f. Ferment *v.* + -er[1].] **a.** A vessel in which fermentation occurs. **b.** An organism that causes fermentation.

1918 *Jrnl. Med. Res.* XXXIX. 34 The presence of these slow lactose fermenters in the material examined presented a serious problem. **1925** *Jrnl. Biol. Chem.* LXII. 790 This solution..and sufficient water to make 30 liters, were placed in a cylindrical stoneware crock which served as a 'fermenter'. **1944** *Chem. & Engin. News* 25 Apr. 592/1 The production of penicillin in vat fermenters. **1947** *Jrnl. Bacteriol.* LIV. 689 This fermentor has an operating capacity of 500 to 2,000 ml. **1949** E. Chain in H. W. Florey et al. *Antibiotics* xvii. 709 The fermenters used in penicillin-manufacturing plants are made from good-quality mild steel or stainless steel. **1959** *Appl. Microbiol.* VII. 238/1 Four fermentations..were made in 5-gal agitated fermentors. **1967** *Biol. Abstr.* XLVIII. 7290/2 B-alanine in conjunction with KNO_3 incorporated into a primary isolation medium, is a vital ingredient in differentiating the genus *Proteus* from other non-lactose fermenters.

‖ **fermentum** (fəɹme·ntŭm). [med.L. use of L. *fermentum* yeast, Ferment *sb.*] In the church in Rome from at least the fifth century, a portion of the eucharistic oblation sent from the Papal Mass to a presbyter about to celebrate the eucharist in one of the neighbouring churches.

1719 J. Bingham *Origines Ecclesiasticæ* VI. xv. ii. 610 The Ancients..many Times spake of leavened Bread, and sometimes the Eucharist is called Fermentum, Leaven, upon that Account. *Ibid.* 611 As appears from..a Letter of Pope Innocent, where he says, it was the Custom at Rome to consecrate the Fermentum, that is, the Eucharist, in the Mother Church, and send it thence on the Lord's Day to the Presbyters in the Tituli or lesser Churches. **1884** Addis & Arnold *Cath. Dict.* s.v. *Reservation*, A supposed decretal of Pope Innocent to Decentius proves that the Bishop of Rome sent the *fermentum* or consecrated host 'per titulos'—*i.e.* to the chief churches of the city. **1937** Fortescue & Thurston *Mass* ix. 368 It is clear really that the fermentum was the Holy Eucharist...As the Sancta were a symbol of the identity of the sacrifice from one Mass to another, so was the fermentum a sign of union between the bishop and his clergy. **1957** *Oxf. Dict. Chr. Church* 500/1 *Fermentum*, in Rome (5th cent.), the fragments of the Bread of the Eucharist sent on Sundays from the Papal Mass to the presbyters in the parish churches (*tituli*) to typify the unity of the faithful in Christ.

Fermi (fə̄·ɹmi). [The name of Enrico *Fermi* (1901–54), Italian-born physicist.] **1.** Used attrib. to designate certain principles and concepts arising out of Fermi's work, as *Fermi (coupling) constant, effect, energy, gas, interaction, level, liquid, sea*; also *Fermi distribution, statistics* (see *Fermi-Dirac); **Fermi surface,** a surface in momentum space representing the maximum energy, at absolute zero, of the electrons in a crystal with respect to their direction of motion.

1927 *Sci. Abstr.* A. XXX. 741 The Fermi statistics were developed for the electronic gas. 1933 Slater & Frank *Introd. Theoret. Physics* xli. 540 In Fig. 83..we plot the potential in which the electrons may be assumed to move. This is constant..throughout the metal, and the horizontal lines symbolize the Fermi levels, filled with electrons at the absolute zero. 1934 *Chem. Abstr.* XXVIII. 7148 (*heading*) The Fermi effect in aluminium. 1936 Mott & Sneddon *Theory of Properties of Metals & Alloys* ii. 53 The *mean* kinetic energy of the electrons will be called the 'mean Fermi energy'. 1937 *Chem. Abstr.* XXXI. 3777/1 (*heading*) Theoretical significance of the Fermi constant. 1951 *Nuovo Cimento* VIII. 749 A universal Fermi interaction (that is, a direct interaction with a unique coupling constant among any four of the five particles of spin ½ħ, [etc.]). 1952 *Proc. R. Soc.* A. CCXV. 491 For the group I metals, copper, silver and gold, distortion of the Fermi surface should be small. 1953 E. Segrè *Exper. Nucl. Physics* II. vi. 144 The simplest of the nuclear models, that of the Fermi gas of non-interacting nucleons contained in a well of assigned diameter and depth. 1955 H. B. G. Casimir in W. Pauli *Niels Bohr* 119 If the behaviour of the electrons can be described by a Fermi distribution of single electron wave functions, then at the absolute zero all levels up to a certain energy ε_0 will be occupied. 1955 *Rev. Mod. Physics* July 252/2 The introduction of the Fermi sea of electrons to rescue the relativistic theory of the electron. 1966 E. J. Konopinski *Theory Beta Radioactivity* iv. 103 A 'Fermi coupling constant', *g*, has been introduced here...It will measure the strength of the β-interaction. 1966 *New Scientist* 20 Jan. 155/1 An attempt to draw an analogy between the behaviour of liquid helium-3 at low temperatures and electrons in metals—which constitute what is called a Fermi liquid. 1970 *McGraw-Hill Yearbk. Sci. & Technol.* 187/1 The electronic properties of metals and degenerate semiconductors are largely determined by the behavior of electrons at and near the Fermi surface.

2. (With lower-case initial.) A unit of length used in nuclear physics, equal to 10^{-15} m.

1956 R. Hofstadter in *Rev. Mod. Physics* XXVIII. 214/2 Henceforth, we shall measure all distances in terms of 10^{-13} cm as a unit and shall call this unit the fermi. For example, ..sphere of gold at a distance of 8·45 fermis from the center of the nucleus. 1965 C. M. H. Smith *Textbk. Nucl. Physics* ix. 263 Nuclear radii are approximately given by the formula $r = 1{\cdot}20A^{1/3}$ fermi. 1971 *Sci. Amer.* July 101/3 The characteristic distance for a rho-nucleon interaction to occur in nuclear matter..is about three fermis.

Fermi-Dirac (fɜ·ɹmi,diræ·k). The names of Enrico *Fermi* (see prec.) and P. A. M. *Dirac* (b. 1902), English physicist, used to designate certain results and concepts in physics arising out of their work, as **Fermi-Dirac distribution** (**function**), a distribution function of the number of particles in a system of fermions that have a given energy; **Fermi-Dirac statistics**, a type of quantum statistics used to describe systems of identical particles that have wave-functions that are antisymmetric with respect to an interchange of co-ordinates of any two particles.

[1927 *Proc. R. Soc.* A. CXIII. 433 It has..seemed worth while to reopen the discussion by examining..a quite general form of statistical mechanics of which the classical form and Einstein's and Fermi-Dirac's are special cases.] 1928 *Proc. Physical Soc.* XL. 329 Degeneracy in the Fermi-Dirac statistics occurs when λ is large and negative. 1931 [see *EINSTEIN-BOSE]. 1933 Harnwell & Livingood *Exper. Atomic Physics* vi. 192 The distribution function..which is known as the Fermi-Dirac distribution. 1957 *Encycl. Brit.* VIII. 218/2 The electrons in the metal obey the so-called 'Fermi-Dirac statistics', which means that they have only a very small specific heat. 1968 C. G. Kuper *Introd. Theory Superconductivity* ii. 33 Then the distribution function for quasiparticles will be the normal Fermi-Dirac distribution $f(E) = \{1 + \exp(\beta E)\}^{-1}$, where $\beta = 1/kT$.

fermion (fɜ·ɹmiǫn). [f. the name of E. Fermi (see *FERMI) + *-ON.] A particle that obeys the Fermi-Dirac statistics. Cf. *BOSON.

1947 P. A. M. Dirac *Princ. Quantum Mech.* (ed. 3) ix. 210 It leads to a special statistics, which was first studied by Fermi, so we shall call particles for which only antisymmetrical states occur in nature fermions. 1953, 1955 [see *BOSON]. 1963 *New Scientist* 1 Aug. 255/1 Only one fermion can exist in a particular state at any given time. 1968 M. S. Livingston *Particle Physics* iii. 59 It is found experimentally that electrons, protons, and neutrons are all fermions and that they have intrinsic spin $s = \frac{1}{2}(h/2\pi)$.

fermium (fɜ·ɹmiǝm). *Chem.* [f. the name of E. *Fermi* (see *FERMI) + -IUM.] An artificially produced radioactive element, atomic number 100. Symbol Fm.

1955 A. Ghiorso et al. in *Physical Rev.* 1 Aug. 1049/1 We suggest for the element with the atomic number 99 the name einsteinium (symbol E) after Albert Einstein, and for the element with atomic number 100 the name fermium (symbol Fm), after Enrico Fermi. 1957 Katz & Seaborg *Chem. Actinide Elements* x. 403 In all respects fermium in aqueous solutions behaves as a tripositive ion. 1966 *New Scientist* 23 June 774/1 Fermium-257, hitherto the heaviest material ever created by man.

fermorite (fɜ·ɹmǒrəit). *Min.* [f. name of Sir Lewis *Fermor* (1880–1954), of the Geological Survey of India: see -ITE¹.] An arsenate and phosphate of calcium and strontium in the apatite group.

1910 Prior & Smith in *Nature* 23 June 513/1 The name fermorite, after Dr. L. L. Fermor, of the Geological Survey of India,..is proposed for this analogue. 1938 *Amer. Mineralogist* XXIII. 11 Fermorite is an example of the substitution of Sr for Ca but this mineral also shows an appreciable substitution of As for P. 1959 [see *ELLESTADITE].

fern, *sb.*¹ Add: **2. a.** *fern-bug*, *-leaf* (later example), *-moth* (also ellipt.), *-weevil*; *fern-lover*.

1959 Southwood & Leston *Land & Water Bugs* viii. 203 *Bryocoris pteridis*,..fernbug. Widely distributed in the British Isles this bug feeds on ferns. 1937 *Burlington Mag.* Aug. 89/2 Fragments of these Fern Leaf Jars. 1865 N. Bellairs *Hardy Ferns* x. 112, I did what I advise other Fern-lovers to do. 1909 *Daily Chron.* 11 Feb. 3/3 Fern lovers throughout the kingdom are waging a war with the fern grubber. 1908 R. South *Moths Brit. Is.* II. 257 (*heading*) The Fern (*Phibalapteryx* (*Coenocalpe*) *tersata*). 1958 W. J. Stokoe *Caterp. Brit. Moths* (rev. ed.) II. 155 (*heading*) The Fern Moth. *Horisme tersata*. 1959 E. F. Linssen *Beetles Brit. Is.* II. 206 The Australian Fern Weevils..one of these, *Syagrius intrudens*..has been introduced into our islands.

b. fern-bird *N.Z.*, a bird of the genus *Bowdleria*; **fern-crushing** *N.Z.* (see quot. 1947); so **fern-crusher**; also *fern-crushed* adj.; **fern-grinding** *N.Z.*, = *fern-crushing*; **fern-house**, a conservatory in which ferns are grown; **fern-land**, (*a*) land covered with fern *N.Z.*; (*b*) a name applied, esp. by Australians, to New Zealand; **Fernleaf** (see quots.).

1882 W. L. Buller *Man. Birds N.Z.* 17 *Sphenocacus punctatus*. Quoy and Gaim. Fern-Bird. Utick. Matata. 1888 Fern-bird [see *swamp sparrow* (b)]. 1963 *Weekly News* (Auckland) 26 June 31 Fern-birds habitually nest among the raupo reeds. 1891 R. Wallace *Rural Econ. Austral. & N.Z.* xv. 231 In the New Zealand climate fern-crushed pasture-land of good and medium quality will support one to three sheep per acre. 1916 *N.Z. Jrnl. Agric.* 20 June 435 Cattle are often looked upon as more efficient fern-crushers than are sheep on land suitable for cattle. 1891 R. Wallace *Rural Econ. Austral. & N.Z.* xv. 230 Fern crushing. Of the fern country, that upon a limestone formation is the best and the least expensive to break in. 1947 A. Clark in H. Belshaw *N.Z.* ii. 40 Hence beef-cattle are used for 'fern-crushing', that is, for keeping in check the undesirable second growth of bracken fern over large areas of the North Island hills. 1921 H. Guthrie-Smith *Tutira* xvii. 134 The operation known..as 'fern-crushing' or 'fern-grinding'. 1847 J. Paget *Mem. & Lett.* 15 Sept. (1901) 157 You who occasionally see green fields cannot imagine the refreshment of my fern-house. 1851 *Gardeners' Chron.* 27 Dec. 823/2 We entertain no doubt about cold Fern houses becoming in time as common as greenhouses. 1891 L. T. Meade *Sweet Girl Graduate* x. 89 I'll take you into our fern-house...We have got such exquisite maidenhairs. 1943 K. Tennant *Ride on Stranger* (1968) viii. 82 He had plenty of time to..tend the ferns in a dirty little fernhouse. 1843 *N.Z. Company Rep.* (1844) 1 Dec. XIV. 121 The facilities, however, which fern land offers, compared with bush or flax land, has induced a large proportion of the settlers to prefer it for their first operation. 1933 *Bulletin* (Sydney) 13 Sept. 10/2 Willeby has written nothing with the slightest flavour of Fernland. 1916 *Chrons. N.Z.E.F.* 15 Nov. 127/1 Call them 'Overseas soldiers' or 'Down-under' men,..Call them 'Cornstalks' or 'Fernleaves'—all out for a fight—But don't call them Anzacs, for that isn't right. 1925 Fraser & Gibbons *Soldier & Sailor Words* 93 *Fernleaves*, a familiar term for the New Zealanders. (From the New Zealand badge.)

fernery. (Later examples.)

1865 N. Bellairs *Hardy Ferns* v. 48, I had a Fernery made for my spoils. *Ibid.* v. 58 Nothing but Ferns should be planted in a Fernery. 1895 E. J. Lowe *Fern Growing* 159 For bog Ferns, a low part of the Fernery well flooded twice a day will be a good substitution for a marsh. 1969 *Sydney Morning Herald* 24 May 48/10 (Advt.), Large fibro/tiled Home, 3 bedrooms,..toolshed, fernery, garage.

Feroese, var. *FAROESE.

ferozone (fe·rozōᵘn). Also **ferro-**. [Formerly a proprietary name, app. arbitrarily formed.] A mineral substance containing ferrous sulphate and other metal salts formerly used as a precipitant in the treatment of sewage.

1888 *Brit. Pat. 11,663* 8 Treatment of Sewage...Use corresponding sized pieces of basic-cinder or slag the refuse of steel works or ferozone or gypsum..or other porous and water purifying material. 1898 [see POLARITE]. 1910 Maxwell & Brown *Encycl. Municip. & Sanit. Engin.* 164/1 The object of 'ferozone' is to act as a precipitant and to assist in the disinfection and deodorisation of the sewage and sludge. 1927 T. H. P. Veal *Disposal of Sewage* v. 66 Alumino-ferric and ferrozone are more expensive than lime but are more easily applied.

‖ **ferra** (fe·rā). Also **fera**, and in Fr. form **féra**. [Local It.] A species of whitefish, *Coregonus fera*, of the family Salmonidæ, found in the Swiss lakes and Lake Como.

1807 [see POLLAN]. 1879 *Encycl. Brit.* X. 152/1 The Lake of Geneva is not so rich in fish as many of the smaller lakes of Switzerland...The 'fera' (*Coregonus fera*) is economically the most important species. 1925 *Glasgow Herald* 5 Aug. 8 The 'ferra' of Lake Geneva. 1930 A. Bennett *Imperial Palace* xl. 286 My Feras have arrived. 1966 P. V. Price *France* 307 The mountain lakes and streams yield fine fish, of which three..are unlikely to be found elsewhere—the *féra*, *lavaret* and *omble chevalier*; all are types of salmon, but the last looks more like a trout.

Ferranti (feræ·nti). The name of S. Z. de *Ferranti* (1864–1930), English electrical engineer, used *attrib.* to designate certain apparatus invented by him, and a phenomenon first observed in connection with the high-voltage cables designed by him.

1893 T. O'C. Sloane *Stand. Electr. Dict.* 251 *Ferranti effect*, an effect as yet not definitely explained, observed in the mains of the Deptford, Eng., alternating current plant. 1893 M. H. Kilgour et al. *Electr. Distrib.* xi. 387 The Ferranti mains are made with a very large margin of safety. 1902 *Encycl. Brit.* XXXIII. 424 The Ferranti rectifier is much employed for rectifying alternating current for arc lighting purposes. 1940 *Chambers's Techn. Dict.* 328/1 *Ferranti effect*, the rise in voltage which takes place at the end of a long transmission line when the load is thrown off. 1942 Meares & Neale *Electr. Engin. Pract.* (ed. 5) II. xvii. 94 The Ferranti rectifier consists of a synchronous motor driving a commutator, alternate segments of which are connected to slip rings on the same shaft. 1946 P. J. Selgin *Electr. Transmission* v. 107 Any value higher than this will result in the Ferranti effect, the line acting in this respect as a step-up transformer.

Ferrarese (ferārī·z), *sb.* and *a.* [a. It. *ferrarese*, f. *Ferrara* the name of a city in Italy + -ESE.] **A.** *sb.* (*pl.*) The Ferrarese people. (Formerly also pl. *Fereareses*.) **B.** *adj.* Of or pertaining to Ferrara.

1573 Gascoigne *Supposes* II. ii. 19 These Ferareses be as craftie as the deuill of hell. *Ibid.*, They are Fereareses all. 1881 *Encycl. Brit.* XII. 535/2 Messrs J. and H. Gwynne constructed some pumps for draining the Ferrarese Marshes. 1930 E. Pound *XXX Cantos* xxiv. 111 To breed in Ferrara among thin-legged Ferrarese. 1932 *Times Lit. Suppl.* 25 Feb. 127/1 The Ferrarese school was itself derivative. 1949 E. Pound *Pisan Cantos* lxxvii. 61 In flat Ferrarese country.

ferratin (fe·rātin). *Biochem.* [f. L. *ferrāt-us* containing iron + -IN¹.] A substance formerly supposed to be an iron-containing protein that occurs esp. in the liver (see quot. 1946).

1893 O. Schmiedeberg in *Practitioner* Dec. 427, I have now..succeeded..in isolating it in an unmixed condition from pig's liver. It contains, on the average, 6 per cent. of iron, and may briefly be referred to as Ferratin. *Ibid.*, Ferratin is the only combination of iron which is ingested with food, and which is found stored up in the tissues as a reserve-material for the formation of blood. 1909 *Chem. Abstr.* III. 2008 Schmiedeberg prepared 'ferratin' and called it a 'ferrialbuminic acid' of 6% Fe content...The compd. is not a 'ferrialbuminic acid' but is an Fe-containing nucleo-protein. 1930 *Proc. Soc. Exper. Biol. & Med.* XXVIII. 294 The serum of a rabbit treated with sheep ferratin gave precipitin reactions with beef, hog and sheep ferratins..but not with any of the corresponding sera. 1946 *Chem. Rev.* XXXVIII. 381 According to Schmiedeberg..th iron was attached to a protein and this compound, which he called Ferratin, was present in spleen, liver, and marrow...(Today we know that this compound was a mixture of denatured ferritin and denatured proteins.)

ferrazite (fe·rāzəit). *Min.* [f. name of J. B. de Araujo *Ferraz*, of the Geological Survey of Brazil: see -ITE¹.] A hydrated phosphate of lead, barium, and aluminium.

1919 Lee & de Moraes in *Amer. Jrnl. Sci.* XLVIII. 354 We feel justified in claiming novelty for the mineral, and suggest for it the name of Ferrazite. 1951 C. Palache et al. *Dana's Syst. Min.* (ed. 7) II. 832 Ferrazite...Found as discoidal pebbles (*favas*) of granular structure in the diamond sands of the Diamantina district, Minas Geraes, Brazil.

ferredoxin (ferēdǫ·ksin). *Biochem.* [f. *fer* (repr. L. *ferrum* iron) + *REDOX + -IN¹.] Any of certain iron-containing proteins of low redox potential which participate in intracellular electron-transfer processes.

1962 L. E. Mortenson et al. in *Biochem. & Biophys. Res. Comm.* VII. 448 An electron-transferring protein that links hydrogenase with a variety of electron donors or acceptors has been obtained from *Clostridium pasteurianum*...The new factor, which at the present stage contains iron but no detectable heme or flavin, has been named 'ferredoxin'. 1966 *Structure & Bonding* I. 112 Ferredoxin is found in representatives of all major groups of organisms except animals. *Ibid.* 121 The bacterial ferredoxins, in general, are similar to each other in amino acid content, but differ in detail. 1970 *New Scientist* 9 July 65/2 The class of molecules known as ferredoxins..may have up to ten iron atoms per molecule.

ferret, *sb.*¹ Add: **1. b.** (Further examples.)

1946 Brickhill & Norton *Escape to Danger* xv. 140 Night and day..German security guards patrolled and snooped...These guards were known by us as 'ferrets'. 1960 *Times* 2 Dec. 17/2 A more recent approach starts from a device known as a 'ferret' which operates in the mains themselves. Its ordinary use is in cleaning out mains and it consists of an arrangement of water-propelled cleaning brushes.

3. *ferret-like* (earlier example).

1801 *Monthly Mirror* June 421 The contour of the face is what is called *ferret-like*.

ferrian (fe·riăn), *a.* *Min.* [f. L. *ferr-um* iron + *-IAN.] Of a mineral: having a (small) proportion of a constituent element replaced by ferric iron.

1930 [see *FERROAN *a.*]. 1954 *Thorpe's Dict. Appl. Chem.* (ed. 4) XI. 883/1 Ferrian wad contains an increasing proportion of ferric oxide, and grades into goethite or limonite. 1959 W. H. Dennen *Princ. Mineral.* vii. 271 Chemical varieties [of cassiterite]: Ferrian, tantalian (ainalite).

ferricrete (fe·rikrīt). *Geol.* [f. FERRI- + CON)CRETE *sb.*, after *CALCRETE.] Any breccia, conglomerate, or soil zone consisting of soil or other fragmentary material cemented by iron compounds.

1902 G. W. Lamplugh in *Geol. Mag.* 4th Ser. IX. 575, I have the hardihood to suggest..'ferricrete' when the binding substance is an iron-oxide. 1949 *Proc. Prehist. Soc.* XV. 48 Single beads of ferricrete grit, bone and kankar were also found. 1955 *Antiquity* XXIX. 206 The site is transversed by ferricrete ridges, marking a preceding wetter period, and this ferricrete postdates the 20–25 ft. beach. 1959 [see *CALCRETE]. 1968 R. W. Fairbridge *Encycl. Geomorphol.* 554 Ferruginous sandstone tends to develop a very remarkable ferricrete induration along its normal joint planes.

ferrimagnetism (ferimæ·gnĕtiz'm). [ad. F. *ferrimagnétism-e* (L. Néel 1948, in *Ann. de Physique* III. 146), f. FERRI- (taken as an altered form of FERRO-) + MAGNETISM.] A type of magnetism which is macroscopically similar to ordinary ferromagnetism but which is attributed to the non-parallel (esp. antiparallel) alignment of neighbouring atoms or ions having unequal magnetic moments.

1948 *Physics Abstr.* LI. 336/1 In some cases (ferrimagnetism) the substance behaves at a low temperature like a ferromagnetic, but..the saturation magnetization at 0°K is much less than that corresponding to complete parallelism of all the spins. 1966 [see *FERROMAGNETISM].

Hence **ferrima·gnet**, a ferrimagnetic solid; **ferrimagne·tic** *a.* and *sb.*, (a substance) possessing the properties of ferrimagnetism.

1950 *Physics Abstr.* LIII. 756/2 It is shown that the ferrites (ferrimagnetics) are a special case of antiferromagnetism. 1951 L. F. Bates *Mod. Magnetism* (ed. 3) viii. 322 The ferrites are ferrimagnetic substances which obey the general chemical formula Fe_2O_3MO, where M is a bivalent metal. 1955 *Amer. Jrnl. Physics* XXIII. 357/2 In a ferrimagnetic material..the complete magnetic structure contains more than one kind of magnetic ion, or more than one kind of crystallographic site. 1963 J. K. Stanley *Electr. & Magn. Prop. Metals* v. 218 Ferrimagnets are much weaker than ferromagnets. 1966 Phillips & Williams *Inorg. Chem.* II. xxviii. 421 Many of these compounds, e.g. $MgFe_2O_4$, show some magnetism despite the aligning of Fe^{3+} spins and they have been called ferrimagnetics. 1968 *Jrnl. Applied Physics* XXXIX. 903/1 If a ferrimagnet is submitted to an external increasing magnetic field individual sublattice rotations start to occur at some threshold field.

Ferris (fe·ris). The name of G. W. G. *Ferris* (1859–96), American engineer, used *attrib.* to designate a fairground machine, the *Ferris wheel* (see quot. 1897), invented and first erected by him. Also *fig.*

1893 *Sci. Amer.* 9 Sept. 169/1 The World's Columbian Exposition—A view from the Ferris wheel. 1897 R. Johnson *Hist. World's Columb. Expos.* I. 77 The feature at the Exposition that corresponded in its character with the Eiffel Tower at the Paris Exposition of 1889 was the Ferris Wheel—an enormous wheel two hundred and fifty feet in diameter, projected into the air, hung upon supports of steel framework by an axle thirty-two inches in diameter, forty-five feet long, and weighing fifty-six tons...On the periphery of the wheel were hung thirty-six passenger cars, each with a seating capacity of forty to sixty persons. 1927 *Blackw. Mag.* Sept. 358/2 Two ferris wheels squeaked like pigs in agony. 1947 Auden *Age of Anxiety* vi. 136 Fortune's Ferris-wheel. a 1963 L. MacNeice *Astrol.* (1964) iv. 120 It was Ptolemy who invented what Arthur Koestler has called 'the ferris-wheel universe'. 1969 Auden *City without Walls* 91 As passive objects, packed tightly together On Roller-Coaster or Ferris-Wheel, mortals Taste in their solid flesh the volitional joys of a seraph.

ferrite. Add: **2.** Substitute for def.: Any compound that can be regarded as being formed from ferric oxide and a basic oxide or from ferric hydroxide and a base; *esp.* any of the compounds with the formula MFe_2O_4 (where M is a bivalent metal ion or a stoicheiometrically equivalent mixture of such ions), many of which have magnetic properties and a low electrical conductivity that make them suitable for use in high-frequency electrical components. (Examples.)

1851 H. Watts tr. *Gmelin's Hand-bk. Chem.* V. 271 Ferrite of Soda.—78 pts. (1 At.) of ferric oxide mixed with excess of ignited carbonate of soda..form a difficultly fusible mixture. 1897 Roscoe & Schorlemmer *Treat. Chem.* (ed. 3) II. 998 Ferric oxide combines with several basic monoxides to form compounds known as ferrites. 1921 J. R. Partington *Text-bk. Inorg. Chem.* xlviii. 987 With very strong bases unstable ferrites, *e.g.*, $Na_2Fe_2O_4$, are formed. 1936 *Physica* III. 463 The 'ferrites'..can be made by thoroughly mixing the finely divided oxide components in powder form, pressing it into a bar and heating the latter for a long time. 1950 *Proc. Inst. Electr. Engineers* XCVII. II. 238/1 Not all the ferrites are markedly magnetic, zinc and cadmium ferrites being notable exceptions. 1950 N. V. Sidgwick *Chem. Elements* II. 1353 A variety of more complicated ferrites, with more than one Fe_2O_3 to each M'_2O or (more often) $M''O$, are known. 1955 *Sci. Amer.* Jan. 68/1 The ferrites and the pressed-powder magnets..are now attracting a great deal of attention. 1958 N. Cusack *Electr. & Magn. Prop. Solids* xv. 350 The natural mineral ferrous ferrite is called magnetite and is the original lodestone.

3. *Metallurgy.* A soft allotrope of pure iron which has a body-centred cubic crystal structure and is stable below 910°C (*alpha iron*) and above 1403°C (*delta ferrite*, *delta iron*); also, a solid solution based on this allotrope, occurring in low-carbon iron alloys as an almost carbon-free constituent.

1888 H. M. Howe *Metallurgy of Steel* in *Engin. & Mining Jrnl.* 18 Aug. 132 Minerals which compose iron. Name suggested here. Ferrite. Nearly pure iron. 1889, 1900 [see PEARLITE]. 1923 Glazebrook *Dict. Appl. Physics* V. 328/2 In commercial iron and steel the ferrite is not pure, but contains dissolved substances, of which the chief are silicon, phosphorus, and manganese. 1957 [see *ALPHA 3 d]. 1958 *Jrnl. Iron & Steel Inst.* CXC. 92/2 The ferrites studied..were seven of the Mn–Zn..and two of the Co–Zn type with low Curie points.

4. *attrib.* and *Comb.* (in sense *2), esp. **ferrite core** (CORE *sb.*¹ 10, *10 b).

1949 *RCA Rev.* Sept. 392 It is easy to obtain a Q of the order of 50 when using a partially saturated ferrite core. *Ibid.* 387 The use of ferrite-cored coils permits the use of exciter frequencies up to a few megacycles. 1954 E. Molloy *Radio & Telev. Engin. Ref. Bk.* xiv. 19 The small size of the ferrite-rod aerial makes it suitable for very small receivers. 1964 *Ann. N.Y. Acad. Sci.* CXV. 654 The LINC is a small stored-program digital computer which uses transistor circuitry and a random-access ferrite-core memory. 1965 *New Scientist* 6 May 368/1 The main interest in the wide range of synthetic ferrite crystalline compounds, fashioned by techniques familiar to the ceramics industry, has lain in their widespread use as small components in the telecommunications and electronics industries. 1970 *Nature* 12 Dec. 1039/1 The access time for the present ferrite core store is about 0·5 μs.

ferritic (fĕri·tik), *a.* [f. FERRIT(E + -IC.] Containing, composed of, or characteristic of ferrite (*FERRITE 3).

1927 *Amer. Inst. Mining & Metall. Engineers Techn. Bull.* No. 14, 19 Farther away from the pipe the martensitic structure..is broken up by grains of distinctly ferritic appearance. 1934 *Jrnl. Iron & Steel Inst.* CXXX. II. 192 The hardness diminishes as the structure of these irons becomes more ferritic. 1958 *Times Rev. Industry* Feb. 40/1 Small additions of about 0·03 per cent. of boron have also been made to some of the heat-resisting ferritic stainless steels. 1961 *Metals Handbk.* (Amer. Soc. Metals) (ed. 8) I. 408/2 The terms martensitic, ferritic, austenitic, and precipitation hardening are used only in an unprecise qualitative sense, to differentiate classes of steels on the basis of their behavior during heat treatment.

ferritin (fe·ritin). *Biochem.* [a. Czech *ferritin* (V. Laufberger 1934, in *Biologické Listy* XIX. 77), f. FERRI- after *FERRATIN.] A water-soluble crystalline protein containing ferric iron that occurs in many animals, esp. in the liver and spleen, and is involved in the storage of iron by the body.

1937 *Chem. Abstr.* XXXI. 3979 Ferritin obtained from horse spleen contained 21% of trivalent Fe, combined with albumin. 1951 A. Grollman *Pharmacol. & Therapeutics* xxxi. 716 The iron..taken up from the gastro-intestinal tract is combined with protein (as ferritin) in the mucosa and released in response to the demands of the body. 1957 *New Biol.* XXIII. 80 The rapid synthesis of the iron-containing protein, ferritin, in the livers of rats following the injection of large doses of 'soluble' iron. 1962 *New Scientist* 21 June 658/2 Biologically active iron compounds such as transferrin, ferritin, and haemosiderin.

ferro-. Add: **1. a.** **fe:rromagne·sian** *a.* [-AN], containing iron and magnesium.

1902 H. A. Miers *Mineralogy* 471 Biotite (Ferro-magnesian Mica). 1944 A. Holmes *Princ. Physical Geol.* iv. 42 Rocks in which olivine is the most abundant mineral (generally in association with other ferromagnesian minerals) are called peridotite. 1970 *Nature* 6 June 927/1 The lunar rocks..contain large amounts of ferromagnesian minerals.

b. *Metallurgy.* In names of ferro-alloys with the meaning 'iron and', as *ferro-aluminium*, *ferrochrome* (= *ferrochromium*), *ferrochromium*, *ferromanganese*, *ferromolybdenum*, *ferrosilicon*, *ferrotitanium*, *ferrotungsten*, *ferrovanadium*. Also **ferro-alloy**, a crude alloy of iron with one or more other elements, esp. metals, used as a means of introducing them into steel, etc.

1905 *Electr. Rev.* 31 Mar. 517/2 The production of special steels and ferro-alloys. 1928 Kingzett *Chem. Encycl.* (ed. 4) 395 Ferro-alloys are used in the steel industry to remove oxygen and nitrogen from molten steel, or to introduce into the steel a small proportion of the metal. 1960 *Times* 31 May (S. Afr. Suppl.) p. iv/4 The ferro-alloys used by the Union's iron and steel industry. 1888 *Chem. News* 6 July 11/2, I have applied this method to all the grades of ferro-aluminium that we manufacture, varying from 1 to 19 per cent Al. 1953 Case & van Horn *Aluminium in Iron & Steel* i. 14 Ferroaluminium is an alloy of Fe and 40–50% Al. 1878 *Jrnl. Chem. Soc.* XXXIV. 772 The discovery of chrome-steel and ferrochrome by Berthier in 1821. 1958 *Times Rev. Industry* Mar. 47/3 Special steelmakers are in a dilemma about carbon-free ferro-chrome, one of the basic raw materials of special alloy steels. 1894 *Jrnl. Chem. Soc.* LXVI. II. 452 The crystals from chrome steel are smaller than those from ferrochromium. 1966 *McGraw-Hill Encycl. Sci. & Technol.* III. 104/1 Chromium is produced in the form of an iron alloy, ferrochromium, by the reduction of chromite ores..in the electric furnace. 1864 O. E. Prieger *Brit. Pat. 1366* 2 June 3 My Invention consists in the production and manufacture of ferro-manganese, that is to say, the combination of metallic manganese with iron in various proportions. *Ibid.* 5 Ferro-manganese is a perfectly metallic substance, hitherto unknown, and containing solely manganese metal and iron with traces of carbon. 1881 Ferromanganese [see FERRO- 1]. 1967 A. H. Cottrell *Introd. Metallurgy* xi. 132 The problem was solved in 1857 by Robert Mushet, who deoxidized the steel by adding a little manganese to it, in the form of ferro-manganese or spiegeleisen..before casting. 1902 *Jrnl. Soc. Chem. Industry* 30 June 832/1 (*heading*) The volumetric determination of molybdenum in molybdenum steel and ferro-molybdenum. 1963 W. H. Dennis *Metallurgy Ferrous Metals* viii. 133 Ferro-molybdenum is added to steel chiefly to improve hardness and increase toughness. 1882 *Jrnl. Chem. Soc.* XLII. 118 (*heading*) Ferro-silicon. 1965 G. J. Williams *Econ. Geol. N.Z.* xvi. 262/2 A source of pure silica is of course fundamental for the manufacture of ferro-silicon. 1893 *Jrnl. Chem. Soc.* LXIV. II. 97 Ferrotitanium can only be dissolved by fusion with sodium hydrogen sulphate and treatment with cold water. 1953 Aitchison & Pumphrey *Engineering Steels* xii. 503 If the steel is to be cast into green-sand moulds..special deoxidisers, such as aluminium or ferro-titanium, should always be added to the ladle. 1881 Ferrotungsten [see FERRO- 1]. 1955 Kirk & Othmer *Encycl. Chem. Technol.* XIV. 359 In recent years..there has been a general trend toward the use of high-purity tungsten concentrates..instead of the ferroalloy by the steel industry due to the disadvantages of high-melting ferrotungsten. 1904 *Jrnl. Chem. Soc.* LXXXVI. II. 824 Ferrovanadium, containing 33 per cent. of vanadium, is made by igniting in the electric furnace the precipitate obtained by mixing sodium vanadate, iron sulphate, and sodium carbonate. 1956 W. D. Hargreaves in D. L. Linton *Sheffield* 280 The city therefore uses quite large tonnages of pig-iron and of alloying metals, either in the pure form or as alloys such as ferro-chrome, ferro-vanadium, [etc.].

c. Used in a general sense to signify some connection with iron, as **fe:rrobacte·rium** = *iron bacterium*; **ferro-ceme·nt** [ad. It. *ferrocemento* (P. L. Nervi 1951, in *Ingegnere* XXV. 17)], a construction material made from cement mortar in the form of thin slabs reinforced with a meshwork of steel rods or wires; **ferro·meter** [a. G. *ferrometer*], a type of instrument for determining the magnetic properties of a ferromagnetic material; **ferrore·sonance** *Electr.*, a type of resonance in which periodic saturation of an iron-cored inductance produces non-sinusoidal oscillations in a circuit; so **ferro-re·sonant** *a.*; **ferrospine·l** [*ferro-* in *FERROMAGNETIC *a.* and *sb.*], a ferromagnetic substance having a spinel structure (as some synthetic ferrites). Also *FERRO-CONCRETE.

1900 B. D. Jackson *Gloss. Bot. Terms* 100/1 *Ferrobacteria*, bacteria which oxidize ferrous to ferric salts. 1963 D. W. & E. E. Humphries tr. *Termier's Erosion & Sedimentation* vi. 143 It is known that the ferrobacteria can precipitate hydrated iron oxide, accompanied by alumina. 1956 tr. *P. L. Nervi* in *C & CA Library Translation No. 60* 1 It may be of interest to give an account..of a new material which, from its composition and method of construction, I propose calling 'ferro-cement'. 1963 *Engineering* 8 Feb. 232/1 Several ships were built during and after the Second World War in Italy from ferro-cement, as distinct from others built elsewhere from reinforced concrete. 1965 W. H. Taylor *Concrete Technol. & Pract.* xxxv. 553 Many spatial frames erected with ferro-cement include the 312-ft span corrugated vault of the Turin exhibition hall in Italy. 1935 *Sci. Abstr.* B. XXXVIII. 272 A description of the extension of the use of the ferrometer to the delineation of hysteresis loops. 1951 R. M. Bozorth *Ferromagnetism* xix. 854 (*caption*) Basic circuit of the ferrometer, for determining current-voltage characteristics of a material, using a phase-shifter and rectifier. 1924 S. R. Roget *Dict. Electr. Terms* 88 *Ferro-resonance*, the peculiar condition of resonance which can be produced in armoured cables when the iron sheathing causes a variation of the inductance according to the current. 1956 T. E. Ivall *Electronic Computers* viii. 103 Ferro-resonance is the condition which exists when a saturable inductor and a capacitor resonate so that the increased current at resonance keeps the inductor core saturated. 1949 *Electronic Engin.* XXI. 138 (*caption*) 5 watt ferro-resonant converter..which reduces the 50 c/s mains frequency to 16⅔ c/s. 1950 R. L. Harvey et al. in *RCA Rev.* Sept. 321 Ferrospinels are unique crystalline materials of spinel structure which are formed at high temperatures by solid-phase reaction of iron oxide and one or more of certain other metal oxides. [*Note*] The coined word *ferrospinel* is used to denote a *ferro*magnetic spinel...

Ferrospinels are sometimes called 'ferrites'. 1953 F. Langford-Smith *Radio Designer's Handbk.* (ed. 4) xi. 459 Ferrospinels are being used increasingly in electronic equipment operating in the frequency range of 10 to 5000 Kc/s.

2. *ferroprussiate* (delete †: see *3).

3. *Photogr.* **ferro-ga·llic** *a.*, designating a photocopying process in which paper is sensitized with a solution containing ferric salts and developed in gallic acid; also, the paper so sensitized; **ferro-pru·ssiate**, used *attrib.* to designate a process for making blueprints in which paper is sensitized with a solution of potassium ferricyanide and ferric ammonium citrate and developed in water, and also the paper so sensitized.

1902 *Encycl. Brit.* XXXI. 703/2 For architects and engineers, cyanotype and ferro-gallic papers are prepared in rolls of considerable width for the direct reproduction of tracings and drawings, as blue or black prints by these methods. 1918 *Photo-Miniature* Mar. Gloss., *Ferro-gallic process*, for copying plans giving black lines on white ground. 1957 Southworth & Bentley *Photogr. Chemicals & Chemistry* (ed. 3) 91 Gallic acid...Its formation of a blue-black ink-like compound with ferric salts is made use of in the ferro-gallic black-line modification of the blue-print process, gallic acid being employed as developer. 1887 S. K. Burton *Pract. Guide Photogr. & Photo-Mech. Printing* II. 103 (*heading*) Marion's Ferro-Prussiate, or Blue Process. *Ibid.*, Of all the various methods for reproducing Drawing, Marion's Ferro-Prussiate Process is still the simplest and most practical. *Ibid.* 105 In the course of printing, the Ferro-Prussiate Paper assumes various tints. 1911 *Encycl. Brit.* XXVI. 93/2 The earliest discovered process [of photo copying], the ferroprussiate, is still the one most largely used. 1953 Kirk & Othmer *Encycl. Chem. Technol.* XI. 146 The common negative blueprint or ferroprussiate process was discovered by Sir John Herschel in 1842, and has undergone only slight modification.

ferroan (fe·rōᵘăn, ferōᵘ·ăn), *a.* *Min.* [f. L. *ferr-um* iron + *-OAN.] Of a mineral: having a (small) proportion of a constituent element replaced by ferrous iron.

1930 W. T. Schaller in *Amer. Mineralogist* XV. 571 Iron—ferroan, ferrian. 1944 C. Palache et al. *Dana's Syst. Min.* (ed. 7) I. 637 Some fibrous brucite is reported to be highly ferroan, but this may be owing to admixed magnetite. 1965 G. J. Williams *Econ. Geol. N.Z.* xi. 167/2 The occurrence of camptonites, vogesites and monchiquite with abundant ferroan dolomite.

ferrocene (fe·rosīn). *Chem.* [f. FERRO- + cyclopentadiene.] An orange crystalline compound, $Fe(C_5H_5)_2$, a molecule of which consists of an iron atom symmetrically situated between two parallel, superimposed hydrocarbon rings of aromatic character.

1952 R. B. Woodward et al. in *Jrnl. Amer. Chem. Soc.* LXXIV. 3458 We now wish to record experiments which demonstrate typically aromatic properties of this unique iron compound, for which we propose the name *ferrocene*. 1957 *Jrnl. Chem. Educ.* XXXIV. 271/1 Ferrocene has demonstrated significant antiknock properties in fuels for spark ignition engines. 1963 J. H. Harwood *Industr. Applic. Organometallic Compounds* 183 Derivatives of ferrocene are effective ultra-violet absorbers for spaceship coatings. 1965 *New Scientist* 2 Dec. 658 (*caption*) Dicyclopentadienyl iron, or ferrocene, the first member of a new class of organometallic compounds.

ferro-concrete (fe:ro,kǫ·nkrīt). [FERRO-.] = *armoured* *CONCRETE or *REINFORCED CONCRETE. Also as *adj.*, composed or constructed of ferro-concrete.

1900 G. L. Mouchel *Brit. Pat. 8831*, This invention relates to building blocks, caissons, piers,..and analogous structures, and it has for its object to employ in the manufacture, construction, or erection of such structures suitable forms of 'ferro-concrete' i.e. concrete having embedded in it a strengthening metal skeleton or framework. 1907 [see *CONCRETE *sb.* 3]. 1914 *Scotsman* 12 Oct. 8/7 The tall white ferro-concrete telegraph posts lining many of the main roads. 1928 *Oxf. Poetry* 42 Fierce ferro-concrete blares, On Woolworth, on! 1957 *London Mag.* Feb. 73 As a result their poems are integrated constructions that look like towers of ferro-concrete.

ferroelectric (fe:ro,ĭle·ktrik), *a.* and *sb.* [f. ELECTRIC *a.* and *sb.*, after *ferromagnetic*.] **A.** *adj.* **a.** Of a body or substance: having the property that a permanent electric polarization can be induced or reversed by the application of an electric field; exhibiting hysteresis between displacement and an applied electric field. **b.** Of or pertaining to ferroelectricity.

1935 H. Mueller in *Physical Rev.* 2nd Ser. XLVII. 180 The dielectric properties of Rochelle salt..are analogous to the ferromagnetic properties near the Curie point. This fact justifies the use of Eq. (7) for the 'ferro-electric' temperature range. 1940 *Ibid.* LVII. 829/2 It neglects the large piezoelectric effects which occur in all ferroelectric crystals. 1955 *Sci. Amer.* June 96/2 The material used in the experimental ferroelectric memory systems is barium titanate.

B. *sb.* A ferroelectric body or substance.

1946 *Nature* 15 June 808/1 As distinct from other known ferro-electrics, barium titanate does not contain hydrogen. 1958 *Electronic Engin.* XXX. 678 The application of a polarizing field will switch the ferro-electric to one or other of its stable states. 1970 J. Grindlay *Introd. Phenomenol. Theory Ferroelectr.* i. 1 Many ferroelectrics possess a spontaneous dipole moment only in certain temperature ranges.

ferroele·ctrically, *adv.* [f. prec. + -AL + -LY².] In a ferroelectric way.

1954 C. Zwikker *Physical Prop. Solid Materials* xii. 216 Rochelle salt..also has a lower Curie point (−20°C) below which the material behaves non-ferroelectrically. 1971 *Nature* 25 June 542/2 Lattice vibrations in ionic crystals which order ferroelectrically..are reviewed.

ferroelectricity (fe:ro,ĭlektri·sĭti). [f. as prec. + -ITY.] The quality of being ferroelectric; the phenomena exhibited by ferroelectric substances.

1946 *Industr. & Engin. Chem.* XXXVIII. 1108/2 This distinguishes barium titanate from potassium dihydrogen phosphate.., where the dielectric anomalies and dielectric saturation (ferroelectricity) are accompanied by pronounced λ-points. 1950 *Physical Rev.* 2nd Ser. LXXVIII. 748 Even a relatively small ionic polarizability for the Ti ions will be enough to lead to ferroelectricity. 1967 B. T. Matthias in E. F. Weller *Ferroelectricity* 182 Ferroelectricity..really has very little or nothing in common with iron and magnetic forces.

ferromagnet (feromæ·gnĕt). [Back-formation from next, after *magnet* and *magnetic*.] A ferromagnetic body or substance.

1941 *Physical Rev.* 2nd Ser. LX. 252 (*heading*) Statistics of the two-dimensional ferromagnet. *Ibid.* 252/2 The present paper is an attempt to gain sound statistical information about some model of a ferromagnet. 1949 *Ibid.* LXXVI. 1195/2 In a single-domain ferromagnet, the bound surface charge creates a depolarizing field. 1968 R. A. Levy *Princ. Solid State Physics* v. 237 Macroscopically a ferromagnet often appears not to have any spontaneous magnetization, because the domain magnetizations are randomly oriented and thereby cancel one another.

ferromagnetic (fe:romægne·tik), *a.* and *sb.* [f. FERRO- + MAGNETIC *a.* and *sb.*] **A.** *adj.* †**1.** Used synonymously with PARAMAGNETIC *a.*: having a positive magnetic susceptibility. *Obs.*

1850 W. Thomson in *Phil. Mag.* 3rd Ser. XXXVII. 250 It appears to me very probable that this assumption is correct for all known diamagnetic substances, and for homogeneous feebly ferromagnetic substances. *Ibid.* 253 A homogeneous feebly ferromagnetic substance, containing no iron. 1872–5, 1881 [see FERRO- 1]. 1880 J. E. H. Gordon *Physical Treat. Electr. & Magn.* II. xxxi. 14 Iron and similar bodies which are attracted by the magnet are called Ferro-magnetic, or sometimes Para-magnetic bodies. Substances which are repelled are called Diamagnetic. 1886 *Phil. Trans. R. Soc.* CLXXVI. 468 Most bodies are either very slightly ferromagnetic or very slightly diamagnetic. On the other hand iron, nickel, and cobalt are enormously magnetic.

2. a. Of a body or substance: having a large, variable magnetic permeability and exhibiting hysteresis. **b.** Characteristic of or pertaining to ferromagnetism.

The distinction between ferromagnetic and paramagnetic substances was first emphasized by P. Curie (*Ann. de Chim. et de Phys.* (1895) 7me Sér. V. 289).

1896 E. Atkinson tr. *H. du Bois's Magn. Circuit* i. 9 By many authors the terms 'ferromagnetic' and 'paramagnetic' are used pretty indiscriminately. For the present it may be as well to keep the two groups separate. 1929 W. Peddie *Molec. Magn.* iv. 57 Ferromagnetic substances become paramagnetic at sufficiently high temperature. 1951 L. F. Bates *Mod. Magn.* (ed. 3) viii. 322 The ferrites are ferrimagnetic substances...They are all ferromagnetic at room temperatures, except zinc ferrite, which is paramagnetic, and cadmium ferrite which is sometimes paramagnetic and sometimes ferromagnetic. 1962 K. J. Standley *Oxide Magn. Materials* i. 2 One may regard each sub-lattice [of a ferrimagnetic material] as ferromagnetic. 1965 A. H. Morrish *Physical Princ. Magn.* i. 4 If the atomic moments are aligned parallel, the substance is said to be ferromagnetic.

B. *sb.* A ferromagnetic body or substance.

1850 *Phil. Mag.* 3rd Ser. XXXVII. 253 A small ball of this body would, when acted upon by a feeble magnetizing force, become on the whole magnetized like a ferromagnetic, and would be urged from places of weaker towards places of stronger force. 1914 Poynting & Thomson *Text-bk. Physics* xvi. 207 Though Faraday only used the two classes, paramagnetics and diamagnetics, and to one or other of these referred all bodies, it is usual now to separate out the three metals iron, nickel, and cobalt, and to these may probably be added manganese, and to class them as Ferromagnetics. 1962 *Times* 12 Oct. 2/6 Studies of paramagnetics and ferromagnetics.

ferromagnetism (feromæ·gnĕtiz'm). [f. FERRO- + MAGNETISM.

Gregory's use of *ferromagnetism* was independent of Thomson's use of *ferromagnetic*; in the preface to his *Lett. Anim. Magnetism* (1851) Gregory states that it resulted from his mishearing the (then new) word *paramagnetism*.]

The quality of being ferromagnetic; the phenomena exhibited by ferromagnetic materials. (Broadly including ferrimagnetism but in some contexts contrasted with it.)

1851 [in Dict. s.v. FERRO- 1, wrongly dated 1850]. 1930 E. C. Stoner *Magnetism* iv. 74 Ferromagnetism is to be attributed to the spin moments of the electrons in the atoms and ions, and not to the spin moments of free electrons, which normally should give rise only to a small paramagnetism. 1957 *Encycl. Brit.* IX. 186/1 Ferromagnetism is the kind of magnetism associated with iron, cobalt and nickel and some alloys. 1966 *McGraw-Hill Encycl. Sci. & Technol.* V. 229/1 Some materials exhibit a special form of ferromagnetism below the Curie temperature called ferrimagnetism.

‖ **ferronerie, ferronnerie** (ferǫnəri). [Fr. *ferronnerie* ironwork, wrought iron.] (See quot. 1960.)

1925 B. Rackham tr. *Hannover's Pottery* I. iv. 281 Thus certain large basins and plateaux (often in 'broken' shapes), with scenes of pastoral gallantry within borders of rococo ornament or *décor ferronerie*, are painted in colours of the Guillibaud class. 1934 *Burlington Mag.* Apr. 170/2 To the Italian velvet designs imitated by Turkish weavers can be added one of the 'ferronerie' type. 1944 *Ibid.* Nov. 284/1 A remarkable polychrome type with *ferronerie* decoration done under Netherlandish influence. 1960 R. G. Haggar *Conc. Encycl. Cont. Pott. & Porc.* 172/1 *A ferronnerie*, ornament consisting of arabesques, scrolls and volutes resembling or imitating ironwork, characteristic of the Antwerp maiolica painters of the sixteenth century and originating with them.

‖ **ferronière, ferronnière** (ferǫniēᵊɹ). [Fr. *ferronnière*, a frontlet; a coronet worn on the forehead: after Leonardo da Vinci's portrait *La Belle Ferronnière*.] (See quot. 1960.)

1840 Thackeray in *Fraser's Mag.* June 681/2 The sisters..with pink scarfs..and brass *ferronières*..were voted very charming. 1908 H. C. Smith *Jewellery* xx. 172 This head ornament is known as the *ferronière*. 1960 H. Hayward *Antique Coll.* 117/1 *Ferronière*, a chain worn as an ornament encircling the head with a jewel in the centre.

ferrous, *a.* Add: **1.** Also, characteristic of ferrous compounds; of or pertaining to iron in its bivalent state; also applied to iron in that state.

1878 W. Abney *Treat. Photogr.* iv. 30 The action of light on all ferric salts, under certain conditions, is to reduce them to the ferrous state. 1940 R. L. Haden *Princ. Hematol.* (ed. 2) iv. 53 The iron attached to the hematin is in the ferrous form. 1950 *Science News* XV. 100 Its absorption bands are visible only when the iron which is in its molecule is in the reduced or ferrous state. 1962 P. J. & B. Durrant *Introd. Adv. Inorg. Chem.* xxiv. 1023 The oxides and sulphides of iron are giant molecule compounds in which either the ferrous or the ferric ion is present. *Ibid.* 1025 Compounds of ferrous iron.

2. Containing iron in significant quantities; of or pertaining to iron.

1887 *Proc. R. Soc.* XLII. 202 There was but little space, outside the section of the iron, enclosed by the coil; and the small amount of magnetic induction in this non-ferrous space was allowed for. 1915 Ibbotson & Aitchison *Anal. Non-Ferrous Alloys* p. v, The chemistry of the ferrous alloys. 1954 J. L. Bray *Ferrous Process Metallurgy* i. 1 For the manufacture of ferrous alloys, large quantities of iron ore..are required.

ferrozone: see *FEROZONE.

ferrule, *sb.* Add: Also 9 **ferrol**. **2.** (Additional examples.)

1833 Loudon *Encycl. Archit.* 750 Ferrol, in plumbing, is a brass tube soldered to the lead pipe at one end, and then driven into the main water-pipe. 1904 Goodchild & Tweney *Technol. & Sci. Dict.* 219/2 *Ferrule*, a small pulley used for rotating drills, or pieces of work that have to be turned or otherwise operated on. 1940 *Chambers's Techn. Dict.* 328/2 *Ferrule*, a slotted metal tube into the ends of which the conductors of a joint are inserted.

b. (See quot.)

1899 C. K. Paul *Memories* viii. 249 'Ferrule of a book', means the corner. 'Do you mean that one with the yellow ferrule.'

ferry, *sb.*¹ **3. b.** Delete †*Obs.* and add later examples.

1951 *Oxf. Jun. Encycl.* IV. 147 (*caption*) Dover-Dunkirk train-ferry in Dover docks. The opening through which the trains run on to the ferry can be seen on the left. 1961 F. H. Burgess *Dict. Sailing* 85 *Ferry*,..any boat employed in ferrying goods and passengers.

c. Used allusively with reference to Charon's ferry (see CHARON); *to take the ferry*: to die.

1895 G. Meredith *Amaz. Marr.* xxviii, The Lethean ferry-boatload. 1928 Galsworthy *Swan Song* III. vi. 266 What are you going to do with your pictures when you take the ferry? Leave them to the nation?

d. Used jocularly to describe the passage between England and New York, Australia, or New Zealand.

1902 C. Hyne *Mr. Horrocks Purser* 57 They see funny things on the Atlantic ferry which rich young men get mixed up in. 1907 *Daily Chron.* 11 June 10/4 Nearly all the Australian States are now offering assisted passages to selected immigrants...Why should not the Imperial Government take a hand in this, and in certain cases contribute the other £6, and thus realise the 'free ferry'? 1930 C. E. Lee (*title*) The Blue Riband: the romance of the Atlantic ferry.

e. Of aircraft: a flight for carrying goods or persons from one place to another, esp. one of a series of flights on a regular course; also applied to the aircraft itself. Of spacecraft: a module that can be separated from the main craft, e.g. for the final descent to the surface of the moon; so *ferry rocket*. Freq. *attrib.* and *Comb.*

1917 *Flying* 28 Nov. 296/1 Should you chance to live in eastern or south eastern counties you may see them going over..flown by ferry pilots. 1933 *Meccano Mag.* Nov. 845/3 If a machine is intended purely for joy riding and short ferry services, arrangement can be made for an eight-seater to be provided. 1939 *Flying* 1 July 4/1 Applications are being invited from those wishing to take part in these long-distance ferry flights. 1939 *Flight* 9 Nov. 373/1 The chances that an enemy machine might masquerade as a ferry are very small. *Ibid.* 373/2 The senior pilots of a ferry unit must remember the layout, characteristics, and fuel systems of as many as ten or a dozen machines. 1940 *Ibid.* 28 Mar. 282/1 Groups were attached to the two Service ferry pools and these groups worked happily. *Ibid.*, Those experts in all-type ferry flying, the R.A.F. pilots. 1942 *Times Weekly* 9 Sept. 9/1 Filling petrol tanks of United States ferry planes at airports newly laid out in jungles or deserts. 1943 *Jane's All World's Aircraft, 1942* 29 a/2 In June, 1941, the R.A.F. formed the Ferry Command to take charge of all trans-Atlantic deliveries by air. *Ibid.* 30 a/1 The West African ferry route to the Middle East is also used by the communications aircraft of the United Nations. 1943 *Times Weekly* 19 May 8/2 A pool of ferry pilots..flew aircraft of all types right across Africa to Khartoum and Cairo. 1951 A. C. Clarke *Exploration Space* 78 At the end of these manœuvres, which would occupy only a few hours, it would be back in a stable, circular orbit waiting to be refuelled and serviced, and the crew could be taken down to Earth by one of the winged 'ferry' rockets. 1960 F. Gaynor *Dict. Aerospace* 90 *Ferry rocket*, the final step of the planned piloted space vehicle, designed for transporting personnel between earth and the terminal orbit. 1969 *Times* 3 June Suppl. p. iii/8 The two-stage module—the disposable landing ferry that takes the astronauts from their circling spacecraft to the lunar surface—is at the heart of the American plans to land two men on the moon this summer.

ferry, *v.* **1.** Delete †*Obs.* and add later example.

1970 H. Waugh *Finish me Off* (1971) 163 Then we warn the doormen and the super on her building that it wouldn't look good for them to ferry customers to her apartment.

2. d. To fly (an aircraft), or to transport by air, from one place to another, esp. on some regular route, as from a factory to an airfield. Hence **ferrying** *vbl. sb.*

1921 *Flight* XIII. 620/2 Both the Aircraft Disposal Company, and the Bristol people, are having difficulty in finding pilots to ferry these machines across to Spain. *Ibid.* 662/2 He is..'ferrying' machines from Liverpool to Croydon. 1932 R. Mahachek *Airplane Pilot's Man.* 353 *Ferrying*, delivering an airplane to another point by flying it. 1940 *Flight* 28 Mar. 282/1 Women pilots..were..attached to a certain factory for the ferrying of trainers. 1943 *Times Weekly* 19 May 8/2 Gradually the ferrying route came into operation. 1958 *Daily Mail* 18 July 1/1 Israel's permission to ferry troop-carriers through her air-space was not at first fully cleared.

ferryage, ferryer: see FERRIAGE, FERRIER.

fertile, *a.* Add: **1. c.** *Fertile Crescent* (see quot. 1914).

1914 J. H. Breasted in Breasted & Robinson *Outl. Europ. Hist.* I. iii. 56 This fertile crescent is approximately a semicircle..with one wing stretching along the eastern shores of the Mediterranean, the other reaching out to the Persian Gulf. *Ibid.* 57 This great semicircle, for lack of a name, may be called the fertile crescent. 1945 Auden *Coll. Poetry* 343 The earth debates: the Fertile Crescent argues. 1950 H. L. Lorimer *Homer & Monuments* ii. 83 The tip of the 'fertile crescent', whence a main trade route leads into the interior. 1958 *Spectator* 7 Feb. 159/2 The rulers of Syria..had no wish so soon afterwards to be engulfed within the Fertile Crescent scheme then being mooted. 1970 *Guardian* 10 Sept. 15/2 There's an old fable told in the lands of the Fertile Crescent.

d. *Nuclear Physics.* Capable of being transformed into a fissile isotope by the capture of a neutron.

1949 *Nucleonics* Dec. 41/2 The concentration of the new fissionable material builds up within the fertile material. 1963 Peterson & Wymer *Chem. Nuclear Technol.* i. 3 This type of reactor contains fertile material, either U^{238} or thorium, which is converted to fuel by neutrons in excess of those needed to maintain the fission process.

4. fertile-brained *a.* = *fertile-headed* (b).

1894 H. Nisbet *Bush Girl's Romance* 275 It seemed to be all plain enough sailing to this fertile-brained..sybarite.

fertility. Add: **d.** Distinguished from *fecundity* (see *FECUND, FECUNDITY).

1866–1936 [see *FECUNDITY]. 1938 E. Charles in L. Hogben *Political Arith.* xi. 73 (*heading*) The effect of present trends in fertility and mortality upon the future population of Great Britain and upon its age composition. 1964 [see *FECUNDITY].

e. *attrib.* and *Comb.*

1913 E. N. Fallaize in J. Hastings *Encycl. Relig. & Ethics* VI. 523 (*heading*) Fertility charms. 1932 R. Knox *Broadcast Minds* vi. 141 Presenting..the Resurrection as one among a series of fertility-legends. 1933 E. K. Chambers *Eng. Folk-Play* 223 He [*sc.* Dionysus] remains primarily a fertility-god, with the bull, and perhaps the goat, and the *phallus* as his attributes. 1933 R. Tuve *Seasons & Months* i. 40 Whatever remnants of fertility cult lie behind such English folk customs as..'Rogation Day'. 1942 *Burlington Mag.* Apr. 98/2 Being used to ecstatic fertility-rites, the more dogmatic and controlled calendar-rites could not catch their emotions. 1952 Gerth & Martindale tr. *Weber's Anc. Judaism* xi. 279 The prophets declaimed against the rural orgiasticism of the fertility cults. 1952 P. Hughes *Witchcraft* x. 128 The crescendo of the fertility dance. 1956 D. H. Willson *King James VI. & I.* vii. 104 The witch cult in Scotland was derived in part from ancient heathen practice..in which the ritual consisted largely of fertility rites. 1962 [see *AMORAL *a.*]. 1964 *Punch* 11 Mar. 394/3 Most large hospitals are now equipped with fertility clinics. 1965 *Ibid.* 25 Aug. 270/2 First the pill, now fertility drugs—what's that but a return to *stop go*? 1968 *Daily Tel.* 12 Nov. 1/6 Mrs. Pennington..has been taking the fertility drug gonadotrophin, the same drug taken by Mrs. Sheila Thorns who gave birth to sextuplets.

fertilization. Add: **1. c.** *fig.*

1922 H. Crane *Let.* 7 Nov. (1965) 104 This wholesale 'fertilization' of America by such half-baked people as the Algonquin gang. 1933 E. K. Chambers *Eng. Folk-Play* 11 An extreme case of fertilization from a distance is at Icomb in Gloucestershire, where a second act must have been added from a Scottish source. 1957 E. Sitwell *Coll. Poems* 423 The fertilisation of his singing breath.

2. *attrib.* **fertilization tube** *Mycology*, a tube-like outgrowth of the antheridium in certain phycomycetous fungi which penetrates the oogonium and forms a passage for a male nucleus to reach an egg.

1887 H. E. F. Garnsey tr. *De Bary's Fungi* v. 134 When the oosphere has been formed in the oogonium, the antheridium sends a delicate cylindrical or conical tube-like process, the fertilisation-tube, from the part of its surface which is in contact with the oogonium into the interior of the latter. 1930 H. M. Fitzpatrick *Lower Fungi* vii. 153 As soon as the oospheres are differentiated the antheridia form delicate fertilization tubes. 1970 J. Webster *Introd. Fungi* 72 Fertilisation is accomplished by the penetration of fertilisation tubes into the eggs.

fertilizin (fɜ̄ɹti·lizin, fɜ̄ɹtiləi·zin). *Zool.* Also **fertilisin**. [f. FERTILIZ(E *v.* + -IN¹.] Any of a class of substances released into the surrounding water by the eggs of some aquatic animals and agglutinating sperm, usually only that of the same species.

1919 F. R. Lillie *Probl. Fertilization* i. 27 The fertilizable condition of the egg depends upon the presence of a specific substance...This substance may therefore be called the fertilizing substance, or fertilizin...It is..a colloidal substance,..exhibiting some of the properties of a ferment. 1940 *Chambers's Techn. Dict.* 328/2 *Fertilisin*, a substance which is present in the cortex of an ovum and assists in the activation of the ovum. 1960 L. Picken *Organization of Cells* viii. 339 Even among echinoids, however, agglutination is not universal, though there is reason to suppose that a fertilizin is none the less secreted.

fervorous. Delete †*Obs.* and add later example.

1920 H. Begbie *Life Wm. Booth* I. ii. 31 A young lover, after parting from his sweetheart late one night, was in so fervorous a mood of happiness that..he threw his stick into the air.

fescue, *sb.* Add: **4.** (Earlier example.) Also *Chewings fescue* [f. the name of Charles Chewings (1859–1937)], a New Zealand pasture and lawn grass, *Festuca rubra* subsp. *commutata*.

1762 W. Hudson *Flor. Angl.* Index, Fescue-grass, hard, meadow, sheep's, tall. 1917 S. F. Armstrong *Brit. Grasses* viii. 152 Seeds of Chewing's Fescue should have a purity of 90 to 95 per cent. 1925 *N.Z. Jrnl. Agric.* XXXI. 357 Mr. Chewings harvested the fescue-field and marketed the seed, presumably as 'Chewings's fescue'. This would be about the year 1885. 1954 C. E. Hubbard *Grasses* 115 The name 'Chewings Fescue' is from a Mr. Chewings who first sold its seed in New Zealand.

fess, 'fess, *v.* orig. *U.S.* Aphetic for CONFESS *v.* Freq. with *up*.

1840 *Knickerbocker* XVI. 112 It would be a sad thing to die here all alone in the woods with a lie in your mouth; so 'fess clean. 1868 L. M. Alcott *Little Women* (1871) ix. 114, I shall tell them myself, all about it, and ''fess' to mother how silly I've been. 1876 C. M. Yonge *Three Brides* ii. 26 'Come 'fess, Julius,' said she...''Fess and make it up.' 1930 *Randolph* (W. Va.) *Enterprise* 20 Nov. 1/1 The joke is on him and he may as well 'Fess up' to it. 1941 R. A. Heinlein *Menace from Earth* (1966) 75 If I had told you that you were going back to meet yourself face to face, would you have believed me? Come now, 'fess up. 1947 N. Marsh *Final Curtain* xiv. 214 I've come to 'fess up', like a good boy.

fest (fest). *U.S.* [a. G. *fest* festival.] With qualifying word, as *GABFEST, *hen fest*, *talk fest*, etc., denoting a festival or special occasion.

1865 *Harper's Weekly* 5 Aug. 490/2 Arrangements were made for the Saengerfest. 1889 *Kansas Times & Star* 24 June, Bob Ricketts won the gold medal at the shooting fest of the Kansas City Gun Club Saturday. 1910 *Chicago Daily Maroon* 10 June 1/2 After the roll call a 'talk fest' was indulged in by some of the old timers. 1924 [see *BULL *sb.*⁴ 3 b]. 1945 A. J. Liebling in *Best Amer. Short Stories* (1946) 275 He explained that a rat fest was a 'rat race, but all bollixed up'. 1952 M. Steen *Phoenix Rising* iv. 86 Some kind of a liquor-fest in an off-colour bar. 1963 M. McCarthy *Group* i. 23 She..loved ..a good hen fest. 1970 *Guardian* 26 Sept. 11/2 'Cinema City', the Round House's filmfest, ended with an open forum.

‖ **festina lente** (fe·stina le·ntė), *phr.* [L. (Suet. *Aug.* 25) *festinā*, imper. of *festināre* to hasten + *lentē* slowly.] Make haste slowly. Also as quasi-*sb.* (Cf. *HASTE *sb.*)

1590 Lodge *Rosalynde* (1592) sig. O2ʳ, *Festina lenter* [*sic*], especially in loue. 1633 T. Adams *Exp. 2 Pet.* II. 1272 *Festina lentè*: which is the golden meane betweene those two extremes, of Sluggishnesse, and Precipitancy. 1749 Fielding *Tom Jones* III. viii. iv. 169 *Festina lentè* is a Proverb which I learnt long before I ever touched a Razor. 1864 R. Browning *Dramatis Personae* 167 He stops me—'*Festina lentè!*' 1893 C. M. Yonge *Girl's Little Book* 43 Be not too quick to do things well. 'Festina Lente.' 1927 W. Cather *Death comes for Archbishop* I. iii. 38 There is enough to do here. *Festina lente.* I have made a resolve not to go more than three days' journey.. for one year. 1932 Spender & Asquith *Life H. H. Asquith* II. xxxvii. 101 The *festina lente* of his method, which sometimes laid him open to reproach.

festinately, *adv.* (Later examples.)

1886 G. B. Shaw *Let.* 11 Sept. (1965) 160 Festinately yrs GBS. 1923 *Glasgow Herald* 13 Dec. 4 Others have said, also too festinately, that Dr Whyte owed his power to his interest in literature.

festival, *a.* and *sb.* Add: **B.** *sb.* **1. b.** (Examples.) Also applied to a series of films, theatrical performances, etc.

1857 Mrs. Gaskell *Let.* Aug. (1966) 466 M. & Mme Mohl..come here, after the Worcester Festival. 1864 *Chambers's Jrnl.* Shaks. Tercentenary No. 2/1 At the present moment, when a grand Tercentenary Festival of the birth of Shakespeare is about to be celebrated at his native place. 1877 G. B. Shaw *How to become Mus. Critic* (1960) 20 The forthcoming Wagner Festival at the Albert Hall. *Ibid.* 32 The aims of the Festival-givers necessarily are, firstly, commercial; secondly, phenomenal; and, lastly, artistic. 1909 *Englishwoman* Apr. 298, I am convinced that there is enough appreciation of Wagner in England to build a modest festival theatre. 1926 *Publishers' Weekly* 1 May 1474/2 The annual Shakespeare festival at Stratford-on-Avon. 1930 *Times Lit. Suppl.* 551/3 A festival production of *The Pretenders*. 1951 [see *film festival* s.v. *FILM *sb.* 7 c]. 1959 *Times* 28 Aug. 11/4 The returned festival-goer can..compare his notes with Burney's. 1970 G. Spanier *And now it's Sables* 133 Every summer in Paris we have an International Theatre Festival.

3. = FAIR *sb.*¹ 1 c. *U.S.*

1869 'Mark Twain' *Innoc. Abr.* 602 Ladies' Festivals where they were importuned to buy by bevies of lovely young ladies. 1944 *Greeley* (Colo.) *Daily Tribune* 21 Sept. 3/3 The annual mission festival, conducted by the Zion Evangelical Lutheran church Sunday.

festoon, *sb.* Add: **5. festoon-lighting** (see quot.).

1931 Roget *Dict. Electr. Terms* (ed. 2) 120/2 *Festoon lighting*, lighting by incandescent lamps arranged at regular intervals in special holders, along a flexible cable.

festooning, *ppl. a.* (U.S. example.)

1876 'Mark Twain' *Tom Sawyer* xiii. 118 Foliage and festooning vines.

‖ **Festschrift, festschrift** (fe·st,ʃrift). Pl. -en, -s. [G., lit. 'festival-writing'.] A collection of writings forming a volume presented to a scholar or savant on the occasion of his attaining a certain age or period in his career.

1901 *Eng. Misc. Furnivall* 491 An English Miscellany or 'Festschrift' in Dr. Furnivall's honour. 1931 *Periodical* June 100 Dr. Cowley..contributed to the Sir Arthur Evans and the Reginald Lane Poole *festschrifts*. 1931 *Times Lit. Suppl.* 1 Oct. 757/3 Lectures, collected papers and *Festschriften*. 1943 *Mind* LII. 370 This volume, a *Festschrift* presented to M. Maritain on his sixtieth birthday, comprises some twenty papers written by American admirers of his writings. 1961 *Times* 5 Oct. 17/3 The 22..historians..who pay tribute to Dr. Gooch in this bumper *Festschrift*. 1967 [see *birthday book* s.v. *BIRTHDAY 3]. 1969 *Nature* 1 Nov. 400/1 The mythology of molecular biology and the festschrift..celebrating Delbrück's sixtieth birthday in 1966, abound with stories of Delbrück's insistence on rigorous evidence for any claim. 1971 *Ibid.* 11 June 399/1 *Festschriften* are for men who have put their stamp on some branch of enquiry during a certain period of time.

festual, *a.* (Later *arch.* example.)

1864 J. A. Symonds *Let.* 23 Jan. (1967) I. 437 A sound of tragic pipe & festual sonorous verse & rapt oration in the porches of the gods.

fetal, var. FŒTAL *a.*

fetch, *sb.*¹ Add: **1.** (Additional examples.)

1938 J. H. McCulloch *Sheep Dogs* iv. 35 (*heading*) *Course*: [for qualifying trials]..Gathering—400 yards. In outrun, dog may be directed on either side. Straight fetch through gate set midway. 1946 F. D. Davison *Dusty*

xi. 117 The trial had four phases..the fetch, when he [*sc.* the dog] brought them [*sc.* the sheep] down the length of the course to where his owner waited [etc.].

3. b. More generally, the expanse of water over which the wind blows before it reaches the point of observation (on the water or at the water's edge); the distance that waves can travel continuously without obstruction. (Later examples.)

1934 *Geogr. Jrnl.* LXXXIV. 272 The fetch must of course be sufficient; a moderate gale would develop on Lake Superior a wave of only half the height that it would have if in the open Atlantic Ocean. **1950** P. H. Kuenen *Marine Geol.* i. 76 In the Mediterranean, where the length of fetch is restricted, the highest waves reported are 4 to 5 m. **1965** H. J. McLellan *Elem. Physical Oceanogr.* xiv. 99/1 Some quite sophisticated methods have been developed for the prediction of wind wave conditions based on the variables Fetch, Duration, and Wind Speed.

fetch, *v.* Add: **1. d.** *fetch-and-carry*, also as *sb.*, the action of fetching and carrying; one who fetches and carries, a subservient person.

1787 W. Cowper *Let.* 24 Dec. (1904) III. 198, I gave him [*sc.* a dog] a lesson in the science of fetch and carry. **1838** J. P. Kennedy *Rob of Bowl* I. 43 Come and go as you list—none of your fetch and carry. **1905** D. G. Phillips *Plum Tree* 284 He is the fetch-and-carry of an impudent and cowardly crowd in Wall Street. *c* **1926** 'Mixer' *Transport Workers' Song Bk.* 35 He's a..sort of fetch-and-carry For the comic 'Welfare League'.

4. (Later examples.)

1931 A. Huxley *Let.* 25 Sept. (1969) 355 Another sign of his cleverness was the exploiting of the psycho-analytical rigmarole, which will fetch 100's of earnest imbeciles. **1957** R. Hoggart *Auden* 36 Jazz enthusiasts say that a solo performer 'fetches' them.

5. c. *Cricket.* To score (a certain number of runs). ? *Obs.*

1735 in H. T. Waghorn *Cricket Scores* (1899) 11 The Londoners went in first and fetched 95. *c* **1806** in *Daily Chron.* (1906) 12 July 4/7 'On Thursday' (a hundred years ago)..'Hambledon fetched 144 and Winchester 107 runs.' **1906** A. E. Knight *Compl. Cricketer* 345 *Fetch* is an almost obsolete word for 'score'.

8. b. Delete †*Obs.* and add later U.S. examples.

1869 'Mark Twain' *Innoc. Abr.* 37, I could throw a rock here without hitting a captain..You'd fetch the captain of the watch, maybe. **1870** —— *Curious Republic Gondour* (1919) 45 And fetch me with the butt-end of the gun.

9. (Later U.S. example.)

a **1910** 'Mark Twain' *Autobiogr.* (1924) I. 237 A brook that never goes straight for a minute..sometimes fetching a horse-shoe three quarters of a mile around.

b. *to fetch the farm*: see Farm *sb.*[2] 8 (in Dict. and Suppl.).

10. a. *transf.* and *fig.* (Later examples.)

1898 H. S. Canfield *Maid of Frontier* 178 We fetched Chisolm's house just a little before sundown. **1930** T. E. Lawrence *Let.* 25 Feb. (1938) 682 Now-a-days I'm lucky to fetch London once in three months.

18*. fetch round (or **around**). *intr.* To recover, to regain consciousness. Also *trans.*

1870 G. Meredith *Let.* ? Sept. (1970) I. 426 This salt-water fetches me round, Tuck. It's the next best to mountain air. **1874** Hardy *Far fr. Madding Crowd* I. ii. 21, I do hope Daisy [*sc.* a cow] will fetch round again now. **1889** 'Mark Twain' *Connecticut Yankee* 135 You feel strange, like somebody that has been struck by lightning and hasn't quite fetched around yet, and can't just get his bearings.

18. fetch through.** *intr.* To win through.

1912 R. F. Scott *Jrnl.* 16–17 Mar. in *Last Exped.* (1913) I. 593 Though we constantly talk of fetching through, I don't think any one of us believes it in his heart.

19. fetch up. i. (Earlier and later examples.) Also *trans.*

1838 J. C. Neal *Charcoal Sks.* 96, I was soon fetch'd up in the victualling line—and I busted for the benefit of my creditors. **1848** Bartlett *Dict. Amer.*, *To fetch up*, to stop suddenly...We often hear the phrase 'He fetched up all standing' that is, he made a sudden halt. It is a nautical vulgarism. **1859** 'N. Hogg' *Poet. Lett.* (1865) 44 Wul tha nex thing thay dood was ta holler out 'Dress!'.. Wat thay main'd wis ta vetch up a little bit zmurt. **1883** W. Whitman *Specimen Days* 139, I made quite a western journey, fetching up at Denver, Colorado. **1890** G. W. Perrie *Buckskin Mose* ii. 25 Waukegan was designated as the place where he might probably fetch up. **1898** E. N. Westcott *David Harum* 175, I..walked alongside the el'phant..till they fetched up inside the tent. **1902** *Chambers's Jrnl.* Mar. 186/2 Before he could more than grab at the rein lying loosely on the pommel, the filly 'fetched up' against a dead box-tree, hard as cast-iron. **1906** *Springfield Weekly Republ.* 11 Oct. 1 If he tries often enough he may fetch up in Congress. **1956** A. L. Rowse *Early Churchills* xviii. 391 Some of her business correspondence..has fetched up on the other side of the Atlantic. **1971** *Listener* 8 Apr. 449/3, I grew used to bummelling around the Bond Street dealers and fetching up for tea at the National Gallery or the Tate.

j. To bring up, rear, or train (children.) *U.S.*

1841 *Knickerbocker* XVII. 156 Harry Cott says he was 'fetched up' on Long-Island. **1869** Mrs. Stowe *Oldtown Folks* xx. 237, I was fetchin' on her up to work for her livin' as I was fetched up. **1890** S. O. Jewett *Strangers & Wayfarers* 205 We've gone and fetched ye up the best we could.

fe·tch-up, *sb.* [f. vbl. phr. *fetch up*, Fetch *v.* 19 i.] A coming to a stand-still; stopping.

Fibonacci('s) sequence, series, the series of Fibonacci numbers, or any similar series in which each term is an integer equal to the sum of the two preceding terms.

1891 *Cent. Dict.* VII. 5509/2 *Fibonacci's series*, the phyllotactic succession of numbers: 0, 1, 1, 2, 3, 5,..etc. **1901** *Ann. Bot.* XV. 481 This series of fractional expressions, which involves the utilization of the Fibonacci ratio series 2, 3, 5, 8, 13, &c., has thus proved for over sixty years the ground-work of all theories of phyllotaxis. **1914** T. A. Cook *Curves of Life* v. 88 The fact that these Fibonacci numbers dominated leaf-arrangements in the case of higher plants was first established by the German botanists, Schimper and Braun (1830). **1929** *Mind* XXXVIII. 54 The side of the decagon..may be read off at once to any required degree of accuracy from our table of the Sectio Divina or Golden Mean, or in other words from our Fibonacci series. **1938** Hardy & Wright *Introd. Theory of Numbers* x. 147 The series (U_n) or 1, 1, 2, 3, 5, [etc.]..is usually called Fibonacci's series. **1939** W. W. R. Ball *Math. Recreations & Ess.* (ed. 11) ii. 57 The ratios of alternate Fibonacci numbers are said to measure the fraction of a turn between successive leaves on the stalk of a plant: ½ for grasses, ⅓ for sedges, ⅖ for the apple, cherry, etc., ⅜ for the common plantain,.. and so on. **1961** H. S. M. Coxeter *Introd. Geom.* xi. 172 Fibonacci numbers as large as $f_{10}=55$, $f_{11}=89$, $f_{12}=144$ arise as the numbers of visible spirals in certain varieties of sunflower. **1964** *Fibonacci Q.* Feb. 34 The only Fibonacci sequence having all primes as divisors [of] one or other of its terms is the one Fibonacci sequence with a zero element, namely: 0, 1, 1, 2, 3, 5, 8, 13,... **1970** *Daily Tel.* 27 June 7, I do not for a moment believe that the musical sense can be traced back to the employment..of numbers from the Fibonacci series (1, 1, 2, 3, 5, 8, 13 and so on), with their ever closer approximation to the golden section.

fibre, *sb.* Add: **2. b.** Esp. an elongated cell that lacks protoplasm, has thick walls and tapering ends, and serves to strengthen plant tissue. (Additional examples.)

1857 A. Henfrey *Elem. Bot.* 513 Prosenchyma is composed of cells elongated greatly in one direction, and attenuated to a more or less acute point at each end, forming what is called a fibre. **1919** F. O. Bower *Bot. Living Plant* ii. 28 The old name *parenchyma* is kept for a tissue of roughly spherical or oblong cells with square ends, while long thick-walled cells with pointed ends are called *fibres*. **1965** Bell & Coombe tr. *Strasburger's Textbk. Bot.* (new ed.) 107 Within the limit of elasticity, the load-bearing capacity of sclerenchyma fibres taken from the living plant is, generally speaking, equal to that of the best wrought iron.

4. c. A structure characteristic of wrought metal in which there is a directional alignment or elongation of crystals or inclusions.

1855 W. Truran *Iron Manuf. Gt. Brit.* xviii. 161 If the several layers [of metal] are thin and cross each other, the tendency of one layer to develope fibre is neutralized by the opposite tendency of the other. **1928** Williams & Homerberg *Princ. Metallogr.* (ed. 2) vi. 200 A study of fiber in steel is of great importance..in the manufacture of such articles as crankshafts, gears, and other forgings. **1966** *McGraw-Hill Encycl. Sci. & Technol.* VIII. 286/2 Hot-working processes, such as forging, align the inclusions (fiber structure).

8. *fibre-dresser, -form, -mute, needle*; *fibre-dressing, -forming, -yielding* adjs.; **fibre-board, fibreboard,** (a piece of) board made from compressed cellulosic fibrous material (as wood pulp); **fibre-faced** *a.*, (*a*) *U.S.* (paper) having a surface composed of visible fibres; (*b*) having a facing or coat of fibre; **fibre optics,** the study and application of the transmission of images by means of total internal reflection through fibres of glass or other transparent solids; so *fibre-optic* adj.; **fibre plant,** any plant that produces a fibre of commercial value; **fibre-saturation point,** in the drying of timber, the point at which the cell cavities have lost all their moisture but none has been removed from the cell walls, corresponding in most woods to a moisture content of about 30 per cent of the dry weight; **fibre-stitch** (see quots.); **fibre stress,** the local longitudinal stress in a body at a point in, or along a line through, a cross-section over which the stress is not uniform; **fibre-tracheid** *Bot.*, a fibre-like tracheid that typically has a thick wall, tapering ends, and bordered pits with slit-like apertures; **fibre tract** [Tract *sb.*[3] 3 b (*a*)], a bundle of nerve-fibres, esp. one in which the fibres have a common origin, termination, and function.

1897 *Sears, Roebuck Catal.* 616/2 Lunch boxes...Well made of compressed fibre board. **1916** *Bull. U.S. Dept. Agric. No. 322* 17 It would seem that tow would be a more desirable as well as a more profitable raw material than straw for the fiber-board manufacturer. **1950** C. C. Handisyde *Building Materials* xix. 263 Fibreboards should be unpacked and stacked on a clean dry surface. **1970** *Nature* 9 May 489/1 The chief problem with making fibre board from bagasse is that vast amounts of dust are created when the material is pulped. **1904** *Daily Chron.* 9 Dec. 7/3 A younger brother, who was employed at a horse-hair and fibre dressers. **1919** *Brit. Manufacturer* Nov. 25/1 Passing on to more specialised agricultural appliances, we find great attention paid to fibre-dressing machinery, especially in the Yorkshire and Lancashire districts. *a* **1884** Knight *Dict. Mech.* Suppl. 332/2 *Fiber-faced paper*, a means of security against the restoration of the surface of check or draft-paper after it has been tampered with. It consists in imbedding in the pulp..a layer of fibers, the outer ends of which are then raised in the form of a nap, [etc.]. **1922** *Times* 20 June 8/5 At right angles to this disc is a fibre-faced wheel which is mounted on a castellated shaft, along which it can be slid by means of the 'gear' lever. **1946** *Nature* 19 Oct. 553/2 Alginic acid yarn can be acetylated without loss of fibre-form. *Ibid.* 28 Dec. 930/1 A new fibre-forming polymer..to which the name 'Terylene' has been provisionally assigned. **1946** A. L. Bacharach *Brit. Mus.* vii. 107 The trumpets should use fibre mutes (now in general use, but then confined to jazz bands) instead of the old pear-shaped brass ones. **1913** *Talking Machine News & Jrnl. Amusements* Feb. 147/2, I should advise 'Troubled' to try the Fibre needles advertised in this paper. **1929** Wilson & Webb *Mod. Gramophones* 153 A fibre needle, in a sound-box specially made for it. **1961** *Flight* LXXIX. 728/1 Windows in the passenger cabin..may be augmented by closed-circuit television or 'fibre optic' devices. **1965** *Amer. Jrnl. Cardiol.* XV. 672/1 The fiberoptic system depends upon the efficient transmission of light lengthwise along glass fibers, each of which has a core of high refractive index and a sheath of low refractive index. **1956** *Time* 3 Dec. 59/1 If one beam of light can be transmitted along a glass tube, why not transmit detailed images along the same path?..Dr. Narinder Singh Kapany, 30, has succeeded by applying a technique he refers to as 'fiber optics'. *Ibid.*, Fiber optics derives its name from its use of hair-thin strands of optical glass as light carriers. **1957** N. S. Kapany in *Jrnl. Optical Soc. Amer.* XLVII. 117 (*heading*) An introduction to fiber optics. **1961** *Engineering* 4 Aug. 134/1 Fibre optics grew out of a need to look round corners, e.g. into complex castings or inside the kidney of a living person. **1970** *Times Lit. Suppl.* 23 July 817/1 The action of the vocal cords during speech can be observed..through a flexible fiber-optics tube which is inserted through the nose so that its end hangs above the vocal cords. **1887** C. A. Moloney *Sk. Forestry W. Afr.* 277 'Buaze' fibre plant of the Zambesi (*Securidaca longipedunculata*). **1957** *Encycl. Brit.* VII. 939/1 The fibre plant *Carludovica palmata* (not a palm), used for making Panama hats. **1930** *Forestry* IV. 34 The loss of the free moisture marks the first stage in drying at any point in the wood, and it is not until this moisture has all gone and the stage known as 'the fibre saturation point' is reached, that the cell-walls begin to dry. **1882** Caulfeild & Saward *Dict. Needlework*, *Fibre stitch*, a stitch used in Honiton and other Pillow Laces to make open leaves, with a fibre running down their centres. **1957** M. B. Picken *Fashion Dict.* 129 *Fiber-stitch*, stitch used in bobbin lace. **1905** A. H. Heller *Stresses in Structures* vii. 91 The maximum unit fiber stress (working stress). **1931** G. A. Garratt *Mech. Prop. Wood* iv. 200 In large beams the weight should be taken into account in calculating the fiber stress. **1952** *Special Rep. Iron & Steel Inst.* XLIII. iv. 171/2 The maximum fibre stress in bending may be distinctly less than that based on an elastic state of distribution. **1898** H. C. Porter tr. *Strasburger's Text-bk. Bot.* 129 In Oaks, Beeches, and in the *Rosiflorae* wood fibres are absent, and the necessary rigidity is provided for by fibre tracheids. **1953** K. Esau *Plant Anat.* x. 204 Wall thickness and particularly the nature of pitting are used to differentiate between the two main categories of wood fibers, fiber-tracheids and libriform fibers. **1904** *Brit. Jrnl. Psychol.* I. 53 Fibre-tracts from corresponding halves of the retinae both go to the occipital region of one and the same hemisphere. **1959** *Chambers's Encycl.* IX. 771/2 The medulla oblongata contains nerve cells and fibre tracts associated with certain of the cranial nerves. **1908** R. W. Sindall *Manuf. Paper* ii. 40 A large and important genus of fibre-yielding plants.

fi·breglass, fi·bre(-)glass. Also **Fiberglas** (U.S. proprietary term). Any material consisting of very thin glass filaments made into a textile or paper, or embedded in plastic or other substances for use as a construction or insulating material; also, glass in the form of filaments suitable for such uses. Freq. *attrib.*; also *Comb.*

1937 *Amer. Dyestuff Reporter* XXVI. P182 (*heading*) 'Fiberglas' products. *Ibid.*, There are many types of Fiberglas fibers. They vary in diameter according to their use. **1938** *Ceramic Age* XXXI. 9/1 Owens-Illinois Fiberglas, the trade-mark of the material discussed in this paper, already has become a modern, commercial raw material..in the textile industry. **1941** *Flight* 6 Feb. 113/2 A fabric made of 'fibre glass' for wing covering. **1951** *Good Housek. Home Encycl.* 135/1 A quilt of fibre glass between the bedroom ceiling and the roof gives protection against cold. **1956** 'N. Shute' *Beyond Black Stump* i. 14 The new styles in outboard motors and fibreglass boats. **1960** *Spectator* 22 July 133/1 As light and fast and manœuvrable as a fibre-glass racing car. **1964** G. Slayter in O. A. Battista *Synthetic Fibers in Papermaking* v. 103 When impregnated with suitable oils or resins, Fiberglas papers have excellent dielectric strength. **1969** *Jane's Freight Containers 1968–69* 464/1 A fibreglass-resin patch is all that is needed to plug puncture holes.

fibrescope (fəɪ·bəɪskōᵘp). Also **fiberscope.** [f. Fibre *sb.* + -scope.] A fibre-optic device used to view inaccessible internal structure, esp. in medicine to examine passages in the body.

1954 Hopkins & Kapany in *Nature* 2 Jan. 39/2 An optical unit has been devised which will convey optical images along a flexible axis. The unit comprises a bundle of fibres of glass, or other transparent material, and it therefore appears appropriate to introduce the term 'fibrescope' to denote it. **1961** *Lancet* 20 May 1074/1 The fiberscope superficially resembles a conventional gastroscope...The shaft of the instrument contains the glass fibre bundle which transmits the image to the eye-piece and is completely flexible throughout its entire length. **1961** *New Scientist* 1 June 494 A new device for viewing the lower part of the gullet, stomach, duodenum and a foot of the small intestine..known as the fiberscope.

fibrid (fəi·brid). [f. Fibr(e + -id.] A synthetic polymer material, used principally for bonding synthetic fibres, esp. in *Paper-making*.

1960 *New Scientist* 10 Nov. 1234/3 The fibrid.. developed by the Du Pont organization in the United States. Fibrids..may be derived from polyamides (nylon), polyesters (like Terylene) or acrylic compounds such as Orlon. **1964** D. G. Bannermann in O. A. Battista *Synthetic Fibers in Papermaking* iv. 92 Papers of 100% Dacron polyester fiber bonded with polyester fibrid 201 retain 91% of their dry strength when wet at room temperature.

fibril. Add: **1. a.** Also, in vegetable fibres and man-made fibres. **b.** Any thread-like molecular formation such as occurs in some colloidal systems and proteins.

1931 S. S. Kistler in *Colloid Symposium Monograph* 63 The greater strength of the gel after it has been converted to an aerogel is doubtless due to dehydration of the fibrils. **1944** *Jrnl. Amer. Chem. Soc.* LXVI. 663/1 Electron micrographs of this material [*sc.* insulin hydrochloride] reveal the presence of uniform fibrils several microns in length, having uniform widths of approximately 200 A (an asymmetry well above 100). **1944** E. Heuser *Chem. Cellulose* ii. 16 Seifriz and Hock separated wood pulp fibers into fibrils, 1·4 μ thick. **1948** Schmidt & Marlies *Princ. High-Polymer Theory & Pract.* xi. 468 Rayons having a high order of crystallite orientation separate into fibrils under the action of appropriate chemical agents. **1959** [see *fibrillar *a.*]. **1968** A. White et al. *Princ. Biochem.* (ed. 4) xxxviii. 874 Hydrogen and electrostatic bonds are responsible for the end-to-end alignment which results in fibrils many times longer than the tropocollagen molecule.

fibrillar, *a.* (Later examples.)

1935 *Chem. Abstr.* XXIX. 2049/2 In certain circumstances the separation of fibrils can be observed; it is probable, therefore, that the 'amorphous' coagula are also fibrillar. **1948** Glasstone *Physical Chem.* (ed. 2) xiv. 1263 A modification of the fibrillar theory is that the gel consists of a three-dimensional network, formed by linking and cross-linking of molecular chains. **1959** *New Scientist* 5 Mar. 511/3 The crystal structure.. continues..in fibrillar form, although the particular molecules of which a fibril is composed are continually changing.

fibrillary, *a.* Add: **2.** Exhibiting fibrillation (sense 1 b).

1875 [in Dict.]. **1911** *Jrnl. Physiol.* XLII. p. vii, The heart passed into a state of fibrillary contraction. **1968** Luisada & Sainani *Primer of Cardiac Diagnosis* xiii. 91 The fibrillary contractions of the atrium are small and not co-ordinated.

fibrillate, *v.* Add: **2.** *intr.* Of muscle: to undergo fibrillation (sense 1 b).

1910 T. Lewis in *Heart* I. 335 These variations are generated in the auricle and..are only present when the auricle fibrillates. **1960** *Brit. Med. Jrnl.* 7 May 1381/1 Atria which were fibrillating at 37°C. ceased to fibrillate when the temperature was reduced to 27°C.

3. a. *trans.* To break down or split into fibrils or filaments (see *fibrillation 2).

1929 Clapperton & Henderson *Mod. Paper-Making* vi. 84 If our beater bars are too sharp we have great difficulty in fibrillating the fibres; instead we may cut them. **1958** *Times Rev. Industry* May 17/2 The suspended pulp is then refined in two stages, chiefly to remove lumps and to fibrillate the individual cellulose fibres. **1969** *Sci. Jrnl.* Nov. 23/3 When a 0·05 mm thick film of polyethylene was fibrillated it was found that string could be produced at nearly 30 metres/minute.

b. *intr.* Of a fibre: to split up into fibrils.

1942 J. Grant *Lab. Handbk. Pulp & Paper Manuf.* v. 104 Esparto will not hydrate or fibrillate to any significant extent, although..it may be cut. **1952** J. P. Casey *Pulp & Paper* I. vi. 339 Softwood fibers tend to fibrillate to a greater extent than hardwood fibers.

fibrillated, *ppl. a.* Add: **b.** Having been split up into fibrils by fibrillation.

1929 Clapperton & Henderson *Mod. Paper-Making* vi. 83 The strength is caused by the fibres being 'fibrillated', and this, we think, should be the proper word to describe the condition. *Ibid.* 85 All the extra surfaces of the fibrillated fibres retain some surface water. **1951** A. B. Green in J. N. Stephenson *Pulp & Paper Manuf.* II. iii. 189 When fibrils become loosened from the original fiber, the fiber appears under the microscope as broomed, or partially fibrillated.

fibrillation. Add: **1. b.** Esp. of the muscles of the heart. (Later examples.)

1906 Cushny & Edmunds in W. Bulloch *Stud. Path.* 95 (*title*) Paroxysmal irregularity of the heart and auricular fibrillation. **1930** *Q. Jrnl. Med.* XXIII. 309 Certain obvious consequences of fibrillation of the auricle, such as the lowered efficiency of the ventricle. **1965** W. I. Gefter et al. *Synopsis Cardiol.* xv. 395 Ventricular fibrillation is usually a terminal event.

2. The process of splitting into fibrils or thin filaments; esp. in *Paper-making*, the beating of vegetable fibres so that they are

(partially) separated into their component fibrils.

1929 Clapperton & Henderson *Mod. Paper-Making* vi. 76 In beating such stock as esparto and wood, very little fibrillation is necessary. **1952** L. G. Cottrall *Introd. Stuff Prep.* ix. 72 By means of wide blunt bars made of suitable material we can effect the fibrillation of the fibres with a minimum of cutting. **1969** *Sci. Jrnl.* Nov. 23/3 Fibrillation technique for making string from plastics films has recently become very popular.

fibrin. Add: **1. a.** In modern use, an insoluble protein, formed from fibrinogen during blood clotting, which polymerizes to give the network of the clot. (Additional examples.)

1804 W. Nicholson tr. *A. F. Fourcroy's Gen. Syst. Chem. Knowl.* IX. 214 (*heading*) Of the fibrous Part of the Blood, or of the Fibrine. **1845** *London Med. Gaz.* 3rd Ser. I. 618/2 The washed clot is the substance which is usually, but very erroneously, named the *fibrin of the blood*. **1927** R. J. S. McDowall *Clin. Physiol.* xv. 147 Fibrinogen, which is converted into the fibrin of the clot. **1961** R. G. Macfarlane in Macfarlane & Robb-Smith *Functions of Blood* vii. 322 During the clotting of fibrinogen, the products are fibrin, which is a polymer formed by union of major residues of the fibrinogen molecules, and fibrinopeptide. **1970** *Nature* 14 Nov. 669 The conversion of fibrinogen to fibrin..is the last stage of the complex process of blood coagulation.

2. † **fibrin-ferment** [a. G. *fibrinferment* (A. Schmidt 1872, in *Arch f. ges. Physiol.* VI. 447)], = Thrombin *Obs.*; **fibrin film,** a thin sheet of fibrin mixed with a plasticizing agent and used mainly in neurosurgery and to treat burns; **fibrin foam,** a spongy preparation of fibrin used as a hæmostatic in surgery.

1876 *Jrnl. Chem. Soc.* I. 945 The author [*sc.* Alex Schmidt] proceeds to show..that fibrin-ferment is not a body pre-existing in the blood, which was originally in the red blood corpuscles, and only after its exit from the body passed over into the plasma. **1898, 1900** Fibrin ferment [see Thrombin]. **1923** A. W. Fuller *Anæmia* i. 8 The blood-plasma is the fluid medium containing.. fibrin-ferment, for coagulation purposes. **1951** A. Grollman *Pharmacol. & Therapeutics* xxviii. 633 Fibrin foam.. is an effective hemostatic agent and may be left *in situ*. It has replaced the fibrin ferments and thromboplastic substances used previously for this purpose. **1944** Ferry & Morrison in *Jrnl. Clin. Invest.* XXIII. 566 We shall describe these clots, as well as two kinds of derived products—the fibrin films and fibrinogen plastics —which we have developed from the same proteins. **1968** M. Gerendás in K. Laki *Fibrinogen* xiii. 294 In essence, fibrin film is a condensed fibrin clot in which the strands are more or less oriented. The film can be stretched and is elastic. **1944** *Jrnl. Neurosurg.* I. 23 A spongy substance, which has been designated as 'fibrin foam'. **1963** *Brit. Pharm. Codex* 955 Human Fibrin Foam is a dry artificial sponge of human fibrin. It is prepared by clotting with human thrombin a foam of a solution of human fibrinogen...Human fibrin foam is used..as a hæmostatic agent in surgery.

fibrino-. Add: **fibrinogen,** substitute for def.: a soluble protein occurring in blood plasma, the precursor of fibrin; add further examples; **fi:brinoly·sin, fibrino·lysin** [*lysin], any enzyme causing fibrinolysis; **fibrino·lysis** [ad. Fr. *fibrinolyse* (A. Dastre 1893, in *Arch. de Physiol. norm. et path.* 5th Ser. V. 661): see *lysis], the enzymic conversion of fibrin to soluble products; hence **fi:brinoly·tic** *a.*

1927 Fibrinogen [see *fibrin 1a]. **1968** Passmore & Robson *Compan. Med. Stud.* I. xxvi. 16 Recent work.. has shown that the conversion of fibrinogen to fibrin is not as simple as previously represented. Thrombin is now believed to split fibrinogen into one major component, fibrin monomer, and several minor polypeptide components. **1915** Fleishner & Loeb in *Jrnl. Biol. Chem.* XXI. 497 The above observations make it appear probable that there exist in various species of animals certain tissue substances which have the power to dissolve blood coagulum when it is in direct contact with the tissues. These substances may be called fibrinolysins. **1958** *Observer* 14 Dec. 4/3 Fibrinolysin is believed to fulfil an important role in removing small clots that form in the blood-stream. **1962** *New Scientist* 16 Aug. 344/3 Fibrinolysin, or plasmin, which is available commercially. **1907** *Jrnl. Physiol.* XXXV. p. xiii, If the carefully washed fibrin is placed into chloroform instead of chloroform-water..fibrinolysis occurs all the same. **1949** H. W. Florey et al. *Antibiotics* I. ix. 383 Neter.. reported that actinomycin inhibited fibrinolysis. **1911** *Chem. Abstr.* V. 3842 If fresh fibrin is first heated to 90°, then treated with H_2O_2, the catalase and the fibrinolytic action is lessened. **1961** *Lancet* 16 Sept. 645/2 The introduction of purified proteolytic and fibrinolytic enzymes made possible a more direct attack on the clot.

fibrinoid (fəi·brinoid). *Path.* [a. G. *fibrinoid* (E. Neumann 1880, in *Arch. f. mikrosk. Anat.* XVIII. 137), f. Fibrin + -oid.] Any of one or more poorly characterized substances staining like fibrin, found normally in the placenta as well as in various diseased tissues. Also *attrib.*, as *fibrinoid degeneration, necrosis*.

1910 *Practitioner* Jan. 43 The muscle so invaded stains feebly, and the presence of these cells in them has brought about a fibrinoid or necrotic change. **1958** *Amer. Jrnl. Med. Sci.* CCXXXVI. 373 Most authorities now agree that the term fibrinoid should be reserved for an intensely acidophilic (eosinophilic), homogeneous, dense and refractile substance with staining characteristics similar to those of fibrin. In the long interval since Neumann's publication much investigation..has failed to resolve the conflicting opinions regarding the origin and nature of fibrinoid. **1964** S. Duke-Elder *Parsons' Diseases of Eye* (ed. 14) xvii. 226 A proliferative infiltration by chronic inflammatory cells surrounding a central area of fibrinoid necrosis. **1967** C. W. M. Adams *Vasc. Histochem.* vii. 265 More severe stenosis..results in rapidly developing hypertension and 'fibrinoid' degeneration. **1970** Passmore & Robson *Compan. Med. Stud.* II. xxx. 3/2 The term fibrinoid belongs to an age of pathology before the cryostat, the integrating microdensitometer and the technique of immunofluorescence... In spite of this, the term fibrinoid will probably continue to be used to describe a characteristic appearance under light microscopy in a variety of diseases.

Fibro, fibro (fəi·bro). **1.** The proprietary name of a viscose rayon staple.

1926 *Trade Marks Jrnl.* 11 Aug. 1880 *Fibro*,..filaments..supplied in untwisted form for all spinning purposes. Courtaulds, Limited. **1935** *Economist* 23 Nov. 1002/2 The chemist has perfected yet another synthetic textile. The new fibre, technically known as staple fibre and popularised as 'fibro', bids fair, with rayon, to oust the older materials from a wide range of uses. **1941** *Daily Sketch* 22 Apr., The blue fibro dress. **1968** *Woollen Carding* (W.I.R.A.) xv. 143/2 For the Fibro/mungo tests two blends were made of 50 per cent. by weight mungo and 50 per cent. Fibro.

2. = **fibro-cement*; also, a house constructed chiefly with fibro-cement; also as *adj. Austral.*

1953 K. Tennant *Joyful Condemned* viii. 63 Front balconies glassed in or filled with stray bits of fibro. **1957** *Times* 11 May 7/7 The cheap fibro shacks are giving way to clean-lined pastel-painted homes, most of them pleasantly designed, a few the inevitable excrescences of too much money and too little taste. **1963** A. Lubbock *Austral. Roundabout* 66 The new fibro-walled, tin-roofed townships. **1969** *Sydney Morning Herald* 24 May 39/6 (Advt.), Sans Souci old type fibro with iron roof.

fibro-. Add: **fi:bro-adeno·ma,** an adenoma containing much fibrous tissue; hence **fi:bro-adeno·matous** *a.*; **fi:brobla·stic** *a.*, of, involving, or relating to fibroblasts; **fibro-ceme·nt** = **asbestos cement*; **fi·brocyte** [-cyte], an inactive fibroblast; **fi:broela·stic** *a.*, consisting of fibrous and elastic tissue; **fi:broelasto·sis,** name given to a congenital heart disease (*congenital* or *endocardial fibroelastosis*) characterized by proliferation of fibroelastic tissue in the endocardium; **fi:bro-lipo·ma,** a lipoma containing much fibrous tissue; hence **fi:bro-lipo·matous** *a.*; **fibropla·sia** [Gr. πλάσις formation], the proliferation of fibrous tissue, as in the healing of wounds.

1892 *Brit. Med. Jrnl.* 23 Apr. 862/2 A large fibro-adenoma removed from the left breast of a girl. **1961** G. M. Bonser et al. *Human & Experimental Breast Cancer* x. 212 Fibroadenomata are circumscribed, benign tumours. **1894** W. R. Williams *Dis. Breast* xviii. 472 The fibro-adenomatous nature of the disease. **1931** Cheatle & Cutler *Tumours of Breast* xi. 471 Whole sections of these tumours, however, reveal its epithelial contents and establish the true fibroadenomatous nature of the growth. **1884** D. MacAlister tr. *Ziegler's Textbk. Path. Anat.* II. 363/1 (*index*) Fibroblastic cells in organisation. **1960** W. W. Washburn in W. W. Nowinski *Fund. Aspects Norm. & Malig. Growth* viii. 693 The period of active fibroblastic proliferation is followed by one of relatively slow cell division during which collagen is formed. **1927** *Census Commonw. Austral. 1921* II. (Stat. rep.) 297 The highest average number of inmates for South Australia was shared by dwellings of stone and of fibro-cement. **1938** *Ann. Reg. 1937* 175 Production of Italian minerals and pyrites, and the replacing of iron by fibro-cement. **1947** *Archit. Rev.* CI. 84/2 The mobile panels, which only occur on the west façade, are of fibro-cement and are painted. **1911** Stedman *Med. Dict.* 312/2 Fibrocyte. **1927** *Biol. Abstr.* I. 944/1 Results were.. especially pronounced upon the fibrocytes and migrating cells in subcutaneous tissue. **1967** K. Hashimoto in A. S. Zelickson *Ultrastruct. Norm. & Abnorm. Skin* xi. 229 In the normal skin, fibroblasts are few in number and the majority reside in a quiescent form, *i.e.*, the fibrocyte. The fibrocyte has scanty cytoplasm, inconspicuous ergastoplasm, flattened Golgi complex, and a relatively large nucleus. **1891** Foster *Med. Dict.* III. 1584/1 Fibroelastic. **1941** *Arch. Path.* XXXI. 169 A simple noninflammatory hyperplasia of fibroelastic tissue. **1965** R. P. Morehead *Human Path.* xxiii. 642/2 A thin layer of fibroelastic tissue covered by endothelium. **1943** Weinberg & Himelfarb in *Bull. Johns Hopkins Hosp.* LXXII. 299 An analysis of the literature..failed to reveal any unrefutable evidence for..fetal endocarditis... The purpose of this presentation is..to report the occurrence of two cases of endocardial fibroelastosis (a much better term in view of the lack of inflammatory stigmata). **1956** *Amer. Heart Jrnl.* LII. 138 Fibroelastosis (fetal endocarditis, endocardial fibrosis, congenital idiopathic hypertrophy of the heart, and congenital fibroelastosis) is a recognized pathologic entity in infants. **1957** *Brit. Heart Jrnl.* XIX. 186 (*heading*) Fibro-elastosis of the heart in adolescence. **1966** Wright & Symmers *Systemic Path.* I. i. 46/1 Nothing is yet known of the aetiology or pathogenesis of fibroelastosis of the endocardium. **1882** *Syd. Soc. Lex.* II, Fibro-lipoma. **1889** *Brit. Med. Jrnl.* 11 May 1062/1 Fibro-lipoma of Labia Majora. *Ibid.*, A fibro-lipomatous growth. **1965** R. P. Morehead *Human Path.* xxv. 759/2 As is the case with mesodermal growths in general, they appear in the literature under a wide variety of names: lipoma, fibroma, fibrolipoma, leiomyoma, rhabdomyxoma, etc. **1929** *Jrnl. Amer. Med. Assoc.* XCII. 44 The phase of fibroplasia is equally important, for it is during this time that the strength of the wound is developed up to a maximal point. **1933** *Arch. Surg.* XXVII. 846 (*title*) Effect of complete and partial starvation on the rate of fibroplasia in the healing wound. **1969** S. M. Levenson in Dunphy & Van Winkle *Repair & Regen.* xxi. 323 Preparations of cartilage powder instilled locally may accelerate fibroplasia.

fibroid, *a.* and *sb.* Add: **A.** *adj.* Also, composed of, or characterized by, fibre or fibrous tissue. (Additional examples.)

1874 Fagge in *Trans. Path. Soc. London* XXV. 64 A series of cases of fibroid disease of the heart. *Ibid.* 67 Towards the apex of the left ventricle..there was advanced fibroid growth. **1918** *Act* 8 & 9 *Geo. V* c. 14 (*title*) The disease known as fibroid phthisis or silicosis of the lungs. **1967** *Cancer* XX. 134 (*heading*) Neurogenic nature of so-called inflammatory fibroid polyps of the stomach.

Fibrolane (fəi·brolēin). The proprietary name of any of a series of wool-like synthetic fibres made from protein.

1946 *Trade Marks Jrnl.* 7 Aug. 416/1 *Fibrolane*. Raw fibrous textile materials. Courtaulds Limited..London. **1954** R. W. Moncrieff *Artificial Fibres* (ed. 2) xv. 204 The original Fibrolane, known as Fibrolane A, was very similar to Lanital. A newer product, Fibrolane BX,.. is the 'Fibrolane' that is mostly used today. **1957** *Economist* 12 Oct. 158/2 But there is no sign that Courtaulds intends to..cease production of its milk-based protein fibre, 'Fibrolane'.

fibrose (fəi·brōᵘz), *v. Path.* [f. Fibrose *a.*] *intr.* To form fibrous tissue. Hence **fi·brosed** *ppl. a.*; **fi·brosing** *vbl. sb.* and *ppl. a.*

1879 *St. George's Hosp. Rep.* IX. 133 Lungs (some fibrosing or caseous). *Ibid.* 156 Indolent fibrosing tubercle at the apices of the lungs. **1897** *Brit. Med. Jrnl.* 11 Dec. 1728/1 The lower lobe of each lung was completely fibrosed. **1909** *Practitioner* Feb. 210 On examination of the fibroid, it is found shrunken and fibrosed throughout. **1956** G. Lapage *Vet. Parasitol.* x. 275 The bile ducts become fibrosed and thickened.

fibrosis (fəibrōᵘ·sis). *Path.* [mod.L., f. L. *fibra* Fibre: see -osis.] The development in an organ of excessive fibrous tissue; fibroid degeneration.

1873 *Brit. Med. Jrnl.* 4 Jan. 4 Arterio-capillary fibrosis. **1886** *Buck's Handbk. Med. Sci.* II. 174/2 Cirrhosis. Synonyms.—Sclerosis,..Chronic Fibrosis. **1893** *Brit. Med. Jrnl.* 2 Dec. 1214/2 Ordinary fibrosis of the kidney. **1906** *Practitioner* Dec. 730 The fibrosis steadily invades the lung parenchyma. **1920** R. Stockman *Rheumatism & Arthritis* ii. 43 These local fibroses are due to small colonies of microbes invading the tissues. **1968** *Cancer* XXII. 1254 Pathologic examination revealed myocardial fibrosis.

Hence **fibro·tic** *a.*, of, pertaining to, or characterized by fibrosis.

1893 *Brit. Med. Jrnl.* 2 Dec. 1214/2 The difficult problems presented by fibrotic kidneys. **1910** *Practitioner* June 781 Between 45 and 50 the liability to fibrotic metritis, endometritis, and mucous polypi diminishes. **1956** *Nature* 3 Mar. 437/1 In a few specimens the lung tissue..was cloudy in appearance, possibly fibrotic.

fibrositis (fəibrosəi·tis). *Path.* [f. Fibros(e *a.* + -itis.] Any rheumatic disorder of the white fibrous tissue that is of unknown or uncertain cause and is characterized chiefly by pain; inflammation of white fibrous tissue. Hence **fibrosi·tic** *a.*, of, pertaining to, or suffering from fibrositis.

1904 W. R. Gowers in *Brit. Med. Jrnl.* 16 Jan. 118/2, I think we need a designation for inflammation of the fibrous tissue which has not such results [*sc.* the production of induration or suppuration]...We may conveniently follow the analogy of 'cellulitis' and term it 'fibrositis'. **1910** *Lancet* 12 Mar. 713/2 Muscular rheumatism.— This affection is always a fibrositis. *Ibid.*, Muscular fibrositis of the shoulder. **1915** Llewellyn & Jones *Fibrositis* i. 14 Herein..lies the most cogent argument that can be adduced in favour of displacement of the term 'chronic rheumatism' by that of 'fibrositis'. **1926** Thomson & Gordon *Chronic Rheumatic Dis.* iii. 17 The fibrositic change is associated with a certain amount of myxœdema. **1963** *Lancet* 5 Jan. 56/2 Much of the pain in rheumatoid arthritis..arises..from anatomically well-defined areas of secondary 'fibrositis' which I have called the 'critical zones'. **1969** W. S. C. Copeman *Textbk. Rheumatic Dis.* (ed. 4) xviii. 507 'Weather sensitivity' in the fibrositic patient.

fibrous, *a.* Add: **1. d.** Of protein: having an elongated molecular structure showing little folding.

1931 *Nature* 2 May 663/2 X-ray photographs of fibrous proteins. **1954** J. C. Kendrew in Neurath & Bailey *Proteins* IIb. xxiii. 846 Structure protein is often macroscopically fibrous; hence the term 'fibrous protein' is more or less synonymous with structure protein...It has, in fact, become conventional to divide the whole range of proteins into two main classes: the fibrous and the globular. **1968** A. C. T. North in W. G. Crewther *Symposium on Fibrous Proteins* 13 In comparing the relationships between primary sequence and secondary structure found in fibrous and globular proteins, account must be taken of the repetitive nature of fibrous protein sequences.

fice (fəis). *U.S.* = Fist *sb.*² 3.
1805 L. Dow *Jrnl.* (1814) 265 Bob Sample..like a little fice (or cur dog) would rail behind my back. 1843 *Missouri Reporter* (St. Louis) 29 June (Th.), Did you ever see a pack composed of five or six little fice dogs, barking furiously? 1854 R. Glisan *Jrnl. Army Life* (1874) xi. 138 Its resemblance is between that of a small fice and grey squirrel. 1872 [see Fise 2]. 1874 E. Eggleston *Circuit Rider* ix. 89 Dogs set up a vociferous barking, ranging in key all the way from the contemptible treble of an ill-natured 'fice' to the deep baying of a bull dog. 1929 W. Faulkner *Sound & Fury* 354 Beside him Luster looked like a fice dog.

ficelle. Add: **1.** Also *ficelle-coloured* adj.
1900 *Westm. Gaz.* 5 Feb. 2/1 Hat of ficelle-coloured straw. 1927 *Observer* 12 June 11/4 Ficelle coloured lace.
2. (Not in Comb.) A trick, artifice, (stage) device.
1890 E. Dowson *Let.* 8 June (1967) 152 There is more psychological motive in it and less of 'ficelle' which Bouthors objected to so in the 'Diary'. 1894 W. Archer *Theatr. 'World' 1893* xviii. 112, I..did not quite..believe in it, taking it rather as a mere *ficelle*. a 1916 H. James *Art of Novel* (1934) 322 Half the dramatist's art, as we well know, is in the use of *ficelles*. *Ibid.* 323 The 'ficelle' character of the subordinate party is..artfully dissimulated. 1920 G. Murray in H. G. Wells *Outl. Hist.* iii. xvii. 107/1 True, raids on women were a real cause of war, but they were also a very favourite *ficelle* of fiction. 1968 *Listener* 26 Sept. 412/3 No Pucelle, Candy's a ficelle; and..her fate is to be briefly grabbed by a series of stereotypes.

‖ **fiche** (fiʃ). [Fr., slip of paper; *fiche policière*, *fiche de voyageur*, registration form.] **1.** The form filled in by foreign guests in French hotels.
1949 M. Laski *Little Boy Lost* ii. ii. 36 The receptionist who so politely gave me my *fiche* to fill in. 1962 C. Forsyte *Diving Death* xiii. 100 My passport is at the hotel. The proprietor still has it, for filling up my *fiche*.
2. See *microfiche.

Fichtean (fi·χ'tiăn), *a.* [f. the name of the German philosopher Johann Gottlieb *Fichte* (1762–1814).] Of, pertaining to, or connected with Fichte or his philosophy.
1817 Coleridge *Biog. Lit.* I. ix. 148 The following burlesque on the Fichtean Egoismus may, perhaps, be amusing to the few who have studied the system. 1859 J. S. Mill *Liberty* ii. 63 The intellectual fermentation of Germany during the Goethian and Fichtean period. 1890 W. James *Princ. Psychol.* I. x. 365 Kant deemed it of next to no importance at all. It was reserved for his Fichtean and Hegelian successors to call it the first Principle of Philosophy. 1910 —— *Mem. & Stud.* (1911) xv. 373 The author's maiden adventure..begins with dialectic reasoning, of an extremely Fichtean and Hegelian type.

ficin (fəi·sin). *Biochem.* [f. L. *fic-us* fig, fig-tree + -in¹.] A proteolytic enzyme obtained from the sap of certain species of fig and used as an anthelmintic.
1930 B. H. Robbins in *Jrnl. Biol. Chem.* LXXXVII. 252 The name ficin..has been given by us to this purified protein powder containing the active anthelmintic principle. 1949 H. W. Florey et al. *Antibiotics* II. xxxiii. 1096 The following enzyme preparations caused no destruction of penicillin: pepsin, trypsin, pancreatin, ..papain,.ficin, emulsin, urease, invertase, and hurain.

Fick (fik). The name of A. E. *Fick* (1829–1901), German physiologist, used attrib. or in the possessive to designate a method of determining cardiac output introduced by him, as *Fick method*, *Fick('s) principle*, and also a law of diffusion enunciated by him and the equation expressing it, as *Fick('s) equation*, *law*. Hence **Fi·ckian** *a.*
1895 C. S. Palmer tr. *Nernst's Theoret. Chem.* i. v. 144 A complete mathematical description of the process of diffusion..was..restated by Fick, who subjected it to a thorough theoretical and experimental proof...Fick's law has nothing to say concerning the nature of the mechanical force: it is simply the 'nature of the case'. 1919 *Jrnl. Physiol.* LIII. 420 The circulation rate determinations after Fick's principle should be improved. 1957 G. E. Hutchinson *Treat. Limnol.* I. v. 254 Equations of the form of equations 11 and 12 are comparable to that expressing passage of a diffusing solute in an undisturbed liquid according to Fick's law; such equations are therefore commonly called Fickian. 1966 Schultze & Heremans *Molec. Biol. Human Proteins* I. iv. i. 620 It is easy to grasp the reason why the diffusion coefficient occurs in Fick's equation. 1968 Kitchin & Julian in Passmore & Robson *Compan. Med. Studies* I. xxviii. 28 The disadvantages of the Fick method. *Ibid.* 29 The advantage of the Fick principle is its simplicity. It is essentially a dilution method, but has none of the problems of recirculation involved in most dilution methods using exogenous indicators.

fickly, *adv.* (Later examples.)
1891 F. Thompson in *Merry England* Aug. 168, O frankly fickle, and fickly true, Do you know what the days will do to you? 1938 W. de la Mare *Memory* 29 Fickly the sunset glimmered through the rain.

fiction. Add: **4. b.** (Later example.)
1939 'G. Orwell' *Let.* 9 Apr. (1968) I. 394 By contract he's supposed to publish my next three fictions.
5. a. (Earlier example of 'legal fiction'.)
1842 S. Lover *Handy Andy* xl. 312 The gold leaf has its representative in 'legal fiction'.
6. *fiction-character*, *-monger* (earlier examples), *-writing*.
1835 J. P. Kennedy *Horse-Shoe Robinson* ii. 32 If any one, hereafter, should tell your story, he will be accounted a fiction-monger. 1850 Kingsley *A. Locke* II. viii. 111 Trials have become lately quite hackneyed subjects, stock properties for the fiction-mongers. 1865 Geo. Eliot *Jrnl.* 20 July (1884) 333, I am anxious to begin my fiction writing. 1909 *Daily Chron.* 12 Mar. 3/4 A second helping of a fiction character..cannot quite be like the first. 1937 Auden & MacNeice *Lett. fr. Iceland* ii. 28 The originals of the fiction-characters are generally well-known. 1966 *Times* 28 Feb. (Canada Suppl.) p. ix/1 Painting, play and fiction-writing.

fiction, *v.* Substitute for def.: *trans.* and *intr.* To feign; to fictionize; to admit of being fictionized. *rare.* (Examples.)
1849 A. Smith *Pottleton Legacy* xiii. 110 The mistress of the house was dimly fictioned as being the only person who had ever read them. 1961 *Amer. Speech* XXXVI. 138 You can see for your self it doesn't fiction. 1966 *Punch* 12 Jan. 64/2 Yes, yes, yes, but why fiction it? Particularly because the fiction is weak.

fictional, *a.* Add: Hence **fi:ctionaliza·tion**; **fi·ctionalize** *v. trans.* = Fictionize *v.*; **fi·ctionalized** *ppl. a.*
1925 T. Dreiser *Amer. Trag.* (1926) II. iii. xvii. 213 Russell, the illegitimate son of Esta..most reservedly fictionalized by his grandparents as an orphan whom they had adopted. 1947 W. Ley *Rockets & Space Travel* (1948) 105 It was a novel with the title *Outside of the Earth*, a fictionalized account of a journey away from the earth. 1954 *Publishers' Weekly* 16 Jan. 248 A fictionalization of the early years of Lucrezia Borgia's life. 1956 C. Wilson *Outsider* vi. 164 [Raskolnikov's] reaction to it all is a fictionalized version of Dostoevsky's fee[l]ing about it. 1960 *Spectator* 25 Nov. 862 An interesting case of this sort of fictionalisation. 1965 N. E. Eliason in Bessinger & Creed *Medieval & Linguistic Stud.* 133 This is briefly recounted in..much later works, both English and Scandinavian, where it is greatly expanded and fictionalized.

fictionalism (fi·kʃənăliz'm). [f. Fictional *a.*: see -ism.] The 'as if' philosophy of H. Vaihinger (see *as if *phr.*). Hence **fi·ctionalist** *a.*
1924 C. K. Ogden tr. *Vaihinger's Philosophy of 'As if'* p. viii, The principle of Fictionalism..or rather the outcome of Fictionalism, is as follows: 'An idea whose theoretical untruth or incorrectness, and therewith its falsity, is admitted, is not for that reason practically valueless and useless.' 1935 *Mind* XLIV. 532 It..disposes of 'fictionalism' naively understood. *Ibid.*, I must admit that my knowledge of the 'fictionalist' school does not extend beyond Vaihinger's 'Philosophy of the "As If"'. 1964 *Language* XL. 136 The fictionalist view of scientific theories.

fictioneer (fikʃəniə·ɹ). [f. Fiction + -eer.] A writer or inventor of fiction (senses 3 and 4). Hence **fictionee·ring** *vbl. sb.* and *ppl. a.*
1923 D. A. Wilson *Carlyle till Marriage* iii. xvi. 213 Carlyle had as little patience as Irving with sentimental fictioneering. *Ibid.* xxvii. 246 Fictioneering biographers. *Ibid.* iv. iii. 265 A fictioneer would say the story has no plot. 1927 *Scots Observer* 30 July 1/3 He does his job of fictioneering gaily. 1956 A. Huxley *Adonis & Alphabet* 137 Moralists and political reformers, satirists and science fictioneers—all have contributed their quota to the stock of imaginary worlds.

fictioner (fi·kʃənəɹ). [f. Fiction + -er¹.] **a.** A believer in fictionalism. **b.** A writer of fiction; a novelist. Hence **fi·ctionism**, fictionalism.
1916 L. P. Jacks *Philos. in Trouble* 116 Divine Fictionism..Actor Camelius will address the Divine Fictioners today in the Temple of Light. 1924 G. Overton *Cargoes for Crusoes* 215 There are present in this novel the two requisites of Mr. Lucas's art as a fictioner. 1931 A. Wolf in W. Rose *Outl. Mod. Knowl.* xiii. 546 Fictionism, as compared with pragmatism, admits..that even avowedly false views may have practical importance.

fictionize, *v.* Add: **a.** (Additional example.) **c.** *intr.* To give a fictional version (of actual happenings). Hence **fi·ctionizing** *vbl. sb.*
1934 A. Merritt *Burn Witch Burn!* ix. 125 'You think she was fictionizing?' 'Not fictionizing exactly. Observing a series of ordinary occurrences through the glamour of an active imagination.' 1938 *Scrutiny* VII. iii. 262 And her work was no mere historical fictionizing. 1959 *Chambers's Encycl.* IX. 246/1 In his next two works Melville settled back into his routine of fictionizing his experiences.

fictious, *a.* **1.** (Later example.)
1886 C. M. Yonge *Chantry House* xiv. 127 Chapman never 'gave heed to them fictious tales', he said.

ficus. Add: **2.** *Bot.* [Adopted by Linnæus in his *Species Plantarum* (1753) II. 1059 as the name of a genus.] A member of a large genus of trees and shrubs so named, belonging to the family Moraceæ and widely distributed in warm regions; it includes the fig, *Ficus carica*, and the rubber plant, *F. elastica*, a common house-plant.
[1597 J. Gerard *Herball* cxxvii. 1328 The Fig tree is called..in Latine Ficus. 1640 J. Parkinson *Theat. Bot.* lxviii. 1494 The tree is called..in Latin *Ficus* both tree and fruite. 1728 R. Bradley *Bot. Dict.* I., Figg-tree, is Ficus.] 1864 J. A. Grant *Walk across Africa* v. 60 The tree, a ficus, whose bark affords the Waganda their clothing. 1959 *Listener* 17 Dec. 1094/3 Ficus or rubber plants can be acclimatised to light or shade. 1969 T. H. Everett *Living Trees of World* 135/2 Several cultivated types of this *Ficus* are grown.

fid, *sb.* Add: **5. b.** A heap; *pl.* 'heaps', 'crowds'. Also as an exclamation = Great!
1898 Kipling *In Ambush* in *Stalky & Co.* (1899) 13 Fids! Fids! Oh, Fids! I gloat! Hear me gloat! 1920 *Blackw. Mag.* Oct. 473/1 Look at the dirty blighters on that hill there! Fids of 'em! 1926 *Ibid.* Mar. 353/2 Little fids of snow. 1935 H. Nicolson *Let.* 26 Feb. (1966) I. 200 There was..a fat fid of letters from home. 1949 —— *Let.* 15 June (1968) III. 171, I would welcome a fid or two of *Anchusa*..among these white and silver objects.
8. fid-hook (see quot. 1905).
1851 J. S. Springer *Forest Life* 108 (Th.), He examines above all the 'fid-hook' and the 'dog-hook', the former that it does not work out, the latter that it loose not its grappling hold upon the tree. 1905 *Terms in Forestry & Logging* 37 *Fid hook*, a slender, flat hook used to keep another hook from slipping on a chain. 1958 W. F. McCulloch *Woods Words* 61 Fid hook—a. A flat hook with a narrow slot and a movable tongue... b. A flat hook without the tongue.

fid, *v.* Add: Also with *out*, and in *ppl. a.* and *vbl. sb.*
a 1865 Smyth *Sailor's Word-Bk.* (1867) 293 Fidded. 1883 *Man. Seamanship for Boys* 30 Holes in the heel of topmasts, for the top tackle pendants to reef through for housing, striking, or fidding. *Ibid.* 220 The cringle is..fidded out. 1901 W. C. Russell *Ship's Adventure* 342 A full-rigged ship must have fidded topmast and fidded top-gallant-mast. 1930 *Sea Breezes* 86 Does anybody know if the *Prince Oscar* had fidded royal masts when she first came out? 1948 R. de Kerchove *Internat. Maritime Dict.* 255/2 Fidded topmast.

-fid (fid), terminal element representing L. *-fidus* cleft, divided, related to *findere* to cleave, as in *bifid*, *palmatifid*, *pinnatifid*, *trifid*, which are ad. L. *bĭfidus*, mod.L. *palmātifidus*, etc.

fidate (fəi·dē¹t), *v. Chess.* [f. med.L. *fidāt-*, ppl. stem of *fidare*; cf. L. *fīdere* to trust, confide.] *trans.* To give (a piece) immunity from capture: used esp. in chess problems. So **fida·tion.**
1910 H. J. R. Murray in *Brit. Chess Mag.* 230 This left the white pieces unprotected on QR 7 and QKt 7 and in typical mediaeval fashion they were fidated, i.e. made immune from capture. *Ibid.*, The necessity of preserving the original solution..made the Rook's fidation a restricted one: the King could not take it, but the Knight could still do so. 1913 —— *Hist. Chess* 570 In one problem, men are fidated (*atreguado*) and their capture is prohibited. *Ibid.* 679, BS 13 prevents this by fidating Bd 4 from the King and renders the fidation of the P unnecessary.

fiddle, *sb.* Add: **1. b.** *as fit as a fiddle* (earlier examples); *as fine as a fiddle* = as fit as a fiddle; *to hang up one's fiddle* (earlier U.S. examples); *to hang up one's fiddle when one comes home* (earlier U.S. example); *to play second fiddle* (earlier example); *to play third fiddle*, to be the third party.
1603 Dekker tr. *La Sale's Batchelars Banquet* iii, Then comes downe mistresse Nurse as fine as a farthing fiddle, in her petticoate and kertle. 1616 W. Haughton *English-men for my Money* sig. G1ᵛ, This is excellent ynfayth, as fit as a fiddle. *Ibid.* sig. K3ᵛ, This is excellent, this is as fine as a Fiddle. 1647 Beaumont & Fletcher *Women Pleased* iv. iii, *Bart.* Am I come fit *Penurio*? *Pen.* As fit as a fiddle. 1809 B. H. Malkin tr. *Lesage's Gil Blas* (1866) x. xi. 378, I am quite at your service to play second fiddle in all your laudable enterprises. 1811 *Massachusetts Spy* 20 Mar. 4/1 But pleasures are brittle as glass, Although as a fiddle they're fine. 1815 D. Humphreys *Yankey in Eng.* 37, I am as fine as a fiddle. 1827 J. K. Paulding *St. Nicholas* (1836) 78 Pleasure sleighs, which, at that period it was the fashion among farmers to have as fine as fiddles. 1833 S. Smith *Life & Writings J. Downing* 90 (Weingarten), You'll have to hang up your fiddle till another year. 1836 W. Dunlap *Mem. Water Drinker* II. 6 He does not hang his fiddle up behind the street-door when he comes home. 1862 *Dialect of Leeds* 407 As fine as a fiddle. 1866 'Mark Twain' *Lett. Sandwich Islands* 9 America..is out in the cold now, and does not even play third fiddle to this European element. 1870 Mrs. Stowe *Oldtown Fireside Stories* (1871) 185 Wal, you see, from the time that Bill Elderkin come and took the academy, I could see plain enough that it was time for me to hang up my fiddle. 1889 'E. Lyall' *Derrick Vaughan* ii, I took care to drop behind, having no taste for the third-fiddle business. 1931 'L. Thayer' *Last Shot* iv. 40 There you are, Stanley. Looking as fine as a fiddle.
4. f. A swindle. orig. *U.S.*
1874 Hotten *Slang Dict.*, *Fiddle*...In America, a swindle or an imposture. 1947 *People* 22 June 4/2

Says Bevin: 'I want peace..and we shan't get it unless we deal with one another as friends. I will be a party to no fiddles.' 1958 G. Mitchell *Spotted Hemlock* xi. 117 Tony and I can do something about it on our own. Not a fiddle, I don't mean. 1959 *Spectator* 4 Sept. 297/2, I know you'll think this is one of my fiddles. At my last parish we raffled a horse and trap,..a clothes horse and a mousetrap.
5. (Later example.)
1839 Dickens *Nickleby* i. 4 'If one *should* lose it, we shall no longer be able to live, my dear.' 'Fiddle,' said Mrs. Nickleby.
8. fiddle-back, (*b*) = *fiddler beetle; (*c*) a chasuble with the front section shaped like a fiddle; (*d*) a grain found in wood used for violin-making; **fiddle-drill,** a drill rotated by a string and bow, a bow-drill; **fiddle idol** (see quot. 1961); **fiddle-pattern** (examples).
1898 Morris *Austral English* 144/1 *Fiddle-back*, name given in Australia to the beetle, *Schizorrhina australasiæ*. 1899 P. Dearmer *Parson's Handbk.* iii. 91 There is no need in an English vestment for the pieces of ribbon without which it seems impossible to keep a 'fiddle-back' in position. 1908 P. Macquoid *Hist. Eng. Furnit.* IV. ii. 62 Harewood or hairwood is the same cutting of sycamore as that used in the manufacture of violins, and consequently termed fiddleback. 1938 H. E. Desch *Timber* iv. 42 Wavy grain..gives rise to a series of..variations in the reflection of light from the surface of the fibres: this is called fiddle-back figure. 1948 T. Corkhill *Gloss. Wood* 185 Fiddleback...Fine wavy grain common to sycamore and maple and used for the backs of violins. *Ibid.* 327 Highly figured wood is called Bird's Eye, Blister, Curly, and Fiddleback Maple. 1956 F. W. Jane *Struct. Wood* x. 227 Because it has become almost a convention to use wood so figured for the backs of violins, it is commonly known as fiddle-back figure. 1960 P. F. Anson *Fash. Ch. Furn.* xxx. 319 Skirts ending at the knees went with fiddle-back chasubles of about the same length in the more extreme Anglo-Catholic churches. *Ibid.* 363 Fiddle-back vestments of Baroque or Rococo style were taboo. 1969 E. H. Pinto *Treen* 439 Fiddle-back..occurs in some logs of sycamore, maple, mahogany, satinwood, etc. 1888 *Lockwood's Dict. Mech. Engin.*, *Bow drill*, a fiddle drill. 1964 W. L. Goodman *Hist. Woodworking Tools* 163 A full range of bow and fiddle drills. 1939 V. G. Childe *Dawn Europ. Civilisation* (ed. 3) iv. 51 The cists..contain several skeletons together with vases and 'fiddle idols'. 1961 *Observer* 24 Sept. 35/3 Neolithic figurines..known as fiddle idols carved by some of the islanders of the Aegean. These are flattish marble plaques with no features beyond the swelling outline of breasts and hips. 1865 *Reading Indust. Exhib. Catal.* in *Reading Mercury* (1968) 17 Aug. 11 R. Bracher..advertises his fiddle-pattern spoons and forks at 7s. 4d. per oz. 1926–7 *Army & Navy Stores Catal.* 184/2 Spoons, forks, etc., nickel silver, fiddle pattern.

fiddle, *v.* Add: **4.** *trans.* and *intr.* Also, to 'wangle', intrigue.
a 1889 *St. Louis Chron.* in Barrère & Leland *Dict. Slang* (1889) I. 360/1 Bob is the man who fiddled himself into Congress. 1938 F. D. Sharpe *Sharpe of Flying Squad* xv. 169 They fiddled into this job. 1955 *Times* 12 Aug. 5/4 William Alfred Powell, in evidence, said he approached Heard about getting a letter 'fiddled out' for him. 1958 S. Spender *Fool & Princess* 172 His own power for 'fiddling' through...His capacity for making deals.

fiddlededee. B. *sb.* (Examples.)
1843 *Ainsworth's Mag.* IV. 176 A large bundle of the latest fiddlededees of ladies' rattletraps hot from London. c 1848 Planché & Dance *Blue Beard* i. 6 Don't believe him, it's all fiddlededee.

fiddler. Add: **2. b.** *slang.* A swindler, cheat; one who 'wangles'.
1857 'Ducange Anglicus' *Vulgar Tongue* 8 *Fiddler*, a sharper, or cheat. Thieves. 1932 'Jock' *Dartmoor fr. Within* v. 107 B—..was..what is termed in prison 'a clever fiddler'. 1958 *Times Lit. Suppl.* 15 Aug. p. iv/1 He is a 'spiv' or a 'fiddler',..for he has offended against the principles of 'fairness' which is..the virtue nearest to the Englishman's heart. 1962 D. Warner *Death of Bogey* i. ii. 10 Spivs and pimps, fiddlers and tweedlers.
c. A meddlesome or interfering person.
1952 A. Grimble *Pattern of Islands* vii. 143 Interfering with the customs of simple peoples..can end by leaving them bereft of their national will to live. The fiddler is a killer on a grand scale.
3. (Earlier examples.)
1846 *Swell's Night Guide* 119/1 *Fiddler*, a sixpence. 1848 *Sinks of London laid Open* 106 *Fiddler*, a sixpence.
5. fiddler beetle, an Australian beetle, *Eupœcila australasiæ*, belonging to the family Scarabæidæ.
[1907 W. W. Froggatt *Austral. Insects* 161 The Fiddler, *Eupoecila australasiae*,..is black and reddish brown, marked upon the thorax and elytra with green stripes.] 1917 *Austral. Naturalist* Apr. 184 At Tuggerah Lakes in September, 1916, a cluster of cocoons of the Fiddler beetle were found at the base of a hollow stump of a Eucalyptus. 1926 R. J. Tillyard *Insects Austral. & N.Z.* xx. 231 The Fiddler Beetle, *Eupoecila australasiae* Don..is common in Eastern Australia in November. 1945 K. C. McKeown *Austral. Insects* 145 One of the commonest species is *Eupoecila australasiae* Don., the Fiddler Beetle; it is black with bright green hieroglyphics on the elytra.

fiddley-did (fi·d'li,did). *Austral. slang.* Also **fiddl(e)y.** [Rhyming slang for Quid *sb.*²] One pound; a pound note.
1941 Baker *Dict. Austral. Slang* 28 Fiddley. 1957 'N. Culotta' *They're Weird Mob* (1958) iii. 42 Here's a couple and a 'alf fiddley dids. 1966 *Sunday Truth* (Brisbane) 23 Dec. 22/2 And for a sovereign, a quid, a pound, a jimmy-o-goblin or a fiddly. 1967 K. Giles *Death in Diamonds* vii. 132, I said I'd give him four Irish fiddlies and take the chance...I gave him the four quid, sir.

fiddling, *vbl. sb.* Add: **3.** Cheating, swindling.
1884 J. Greenwood *Little Ragamuffins* xxxii. 300 So sure as a boy of mine takes to fiddling, I'd manoeuvre him into quod before he sleeps that night. 1959 *New Statesman* 11 July 39/1 The Ministry people would not agree because they thought it would lead to [taxi] drivers taking strangers all round the place and doing all sorts of fiddling.

fiddly (fi·dli), *a. colloq.* Also **fiddley.** [f. Fiddle *v.* + -y¹.] Requiring time or dexterity; pernickety.
1926 *Blackw. Mag.* Sept. 403/1 A fiddly sort of way of translating into action the Political's broad exhortation. 1960 *Times* 29 July 4/6 'Fiddly things' should be done by automatic machines. 1964 *Which?* Apr. 43/2 The Commer's [roof]..needed rather a push to get it started and had fiddley catches. 1970 *Amateur Photographer* 11 Mar. 28/3 Working in total darkness to line up those fiddly bits of film.

fideism (fəi·dĭ,iz'm). Also **F-.** [f. L. *fides* faith + -ism.] Any doctrine according to which all (or some) knowledge depends upon faith or revelation, and reason or the intellect is to be disregarded, as **a.** = Traditionalism; **b.** a Roman Catholic theory developed from Kantian idealism; **c.** in Protestant usage, also derived from Kant, with reference to justification by faith. Hence **fi·deist**, **fidei·stic** *a.*
1885 [see Traditionalism]. 1895 *Dublin Rev.* Apr. 313 As to Fideism, see Dr. Hettinger's interesting classification of its four stages, as corresponding to the four stages of Rationalism, in his 'Fundamental Theologie', 1879, vol. ii. pp. 348–9. 1903 *Hibbert Jrnl.* I. 556 'Fideism' denotes the material principle—the nature and condition of salvation through Christ. 1908 *Programme of Modernism* 142 Such scepticism destroys the certitude of the fact of revelation and ends in blind fideism. 1909 *Cath. Encycl.* VI. 68/2 Fideism owes its origin to distrust in human reason, and the logical sequence of such an attitude is scepticism. *Ibid.*, For some fideists, human reason cannot of itself reach certitude in regard to any truth whatever. *Ibid.*, It is also a fideistic attitude which is the occasion of agnosticism..and other modern forms of anti-intellectualism. 1912 F. von Hügel *Eternal Life* ii. xii. 344 Rome is finely free from all Fideism or Pietism. 1957 *Oxf. Dict. Chr. Ch.* 503/1 Scholastic theologians regularly charged the Modernists with 'Fideism'. 1966 *Cath. Dict. Theol.* II. 296/1 Fideistic tendencies also appeared during the Middle Ages. *Ibid.*, Fideists like the Abbé Louis Bautain...The doctrine of the last-named is the fullest..expression of fideism and indeed has come to be almost identified with fideism. 1967 *Philos.* XLII. 191 (*title*) Wittgensteinian Fideism.

Fidelism (fide·liz'm). [ad. Sp. *Fidelismo* (also used), f. the name of *Fidel* Castro Ruz (b. 1927), Cuban politician + -ism.] The methods and policies of Fidel Castro's political administration in Cuba; = *Castroism. Hence **Fide·list** *a.*, of or pertaining to Fidel Castro or to Fidelism; ‖ **Fidelista** (fideli·stā), an adherent of Fidel Castro; also *attrib.*
1959 *Washington Post* 24 Nov. A 14/1 Fidelismo has become revolution for revolution's sake. 1960 *Business Week* 3 Dec. 87 'Fidelism', or 'Fidelismo', as the Latin Americans call it,..is the Castro-style revolution that's followed by a left-wing, Communist-influenced, perhaps Communist-controlled, government. *Ibid.*, The Fidelistas and Communists. 1961 *Ann. Reg. 1960* 211 This was fertile ground for 'Fidelismo'. 1961 *Economist* 25 Nov. 723/2 An attempted return to dictatorship in the Dominican Republic, and the chance of a fidelist uprising that might have followed it. 1962 *Listener* 1 Feb. 206/1 *Fidelista* cells began to be formed in the cities, in the universities, in the sugar mills. 1968 *Guardian* 1 Oct. 8/4 That generation is now trying out various kinds of Trotskyism, Maoism, and Fidelism. 1970 'J. Morris' *Candywine Devel.* xviii. 212 It may be..the local Fidelistas. 1971 *N.Y. Rev. Bks.* 28 Jan. 6/4 'El Che' was not Castro's brain, but he was the Fidelista conscience.

fidelity. Add: **1. e.** *Ecol.* The degree of association of a species with a plant community.
1932 Fuller & Conard tr. *Braun-Blanquet's Plant Sociol.* iv. 52 The more or less rigid ties by which the species are bound to certain communities—their fidelity. 1955 *Jrnl. Ecol.* XLIII. 232 Species which are of high fidelity regionally become less faithful in other regions and other different climatic conditions. 1967 M. E. Hale *Biol. Lichens* iv. 61 A large group of lichens..have high fidelity for limestone and other basic substrates.
2. c. The degree to which a sound or picture reproduced or transmitted by any device resembles the original; esp. in *high fidelity. Cf. *hi-fi, *lo-fi.
1878 G. B. Prescott *Speaking Telephone* i. 12 The apparatus..reproduces the original vibrations with perfect fidelity. 1896 Marconi *Brit. Pat.* 12,039 6 The space cannot..be excessively shortened without injuring the fidelity of the transmission. 1918 H. Seymour *Reprod. Sound* 269, I introduced baked carbon sheet for the purpose, with remarkable results as to strength and fidelity of tone. 1923 *Sci. Amer.* July 67/2 For..radio programs ..you will..have an extremely sensitive sound converter ..which is of an entirely different order of fidelity from that..of [other] microphonic devices. 1931 *Electronics* Oct. 137/1 We find that the same word—Fidelity—can be used both in connection with the excellence of sound reproduction and picture reproduction. 1935 *Discovery* Sept. 278/1 Fidelity of both picture and sound reproduction is good. 1958 *Economist* 26 July 270 The German industry also put its first complete fidelity recorder on to its domestic market in that year [*sc.* 1948].
3. fidelity insurance, insurance taken out by an employer to indemnify him against losses incurred through the dishonesty or non-performance of an employee. So *fidelity bond*, *guarantee*, such a bond or contract.
1880 *Encycl. Brit.* XIII. 161/2 *Fidelity Guarantee.* The guarantee of employers against the fraud or insolvency of their servants. 1930 *Economist* 26 Apr. 942/1 Its activities are confined to the transaction of fidelity guarantee insurance. 1970 *Daily Tel.* 28 Sept. 3 The insurance companies..will investigate the staff, checking with previous employers and so on. If their character is good they are given what is known as a fidelity bond.

fidget, *sb.* Add: **1.** (Earlier example of 'on the fidget'.)
1809 B. H. Malkin tr. *Le Sage's Gil Blas* (1866) vii. vii. 245 Her tongue was cruelly on the fidget to be let loose.
2. (Earlier example.)
1816 M. E. Bicknell *Let.* 19 Mar. in *Constable's Corr.* (1964) II. 182 You know what a sad *fidget* I am.

FIDO, Fido (fəi·do). [f. the initials of the words *F*og *I*nvestigation *D*ispersal *O*peration.] A system of dispersing fog over aerodromes by heat from petrol-burners.
1945 *Newsweek* 11 June 53/2 Fido stands for 'Fog Investigation Dispersal Operation'. It was suggested in 1942 by Prime Minister Churchill to Geoffrey Lloyd. 1946 D. Banks *Flame over Britain* vii. 157 We saw the glow of Fido in the night sky. Making a circuit at 1,500 feet above the flames the thick rime on his wind-screen cleared and the runway then became clearly visible through the fog. 1959 *Times* 21 Feb. 7/7 [Passengers] would probably find their first FIDO-aided landing an eerie experience. 1971 *Daily Tel.* 6 Jan. 2/3 With the FIDO (Fog Intensive Dispersal Operation) system used during the war, petrol is burned in troughs on either side of the runway to raise air temperature.

fiduciary, *a.* Add: **4.** The amount of the fiduciary issue has varied in modern times.
1930 M. Clark *Home Trade* v. xxx. 240 The £260,000,000 of note issue is known as the fiduciary issue. 1965 Seldon & Pennance *Everyman's Dict. Econ.* 172 In the early 1960's the fiduciary issue stood at over £2,200 million.

‖ **fidus Achates** (fəi·dŏs akē¹·tīz). [L., = faithful Achates (Virgil *Aen.* vi. 158, etc.).] A devoted follower, henchman.
1603 C. Heydon *Def. Judic. Astrol.* xx. 411 Yet I haue tied my selfe to be *fidus Achates* to him. 1771 Smollett *Humph. Cl.* I. 125 Dr. Lewis, who is the *fidus Achates* of my uncle. 1839 Dickens *Let.* 13 May (1965) I. 550 Mr. Macrone and his fidus Achates Mr. Hansard. 1934 *Times Lit. Suppl.* 24 May 377/4 In Portsmouth they arrest the *fidus Achates*.

fiedlerite (fi·dlərəit). *Min.* [ad. G. *fiedlerit* (G. vom Rath 1887, in *Sitzungsber. d. niederrhein. Ges. f. Natur- und Heilkunde zu Bonn* 157), f. the name of K. G. *Fiedler* (1791–1853), German traveller: see -ite¹.] A hydroxychloride of lead found at Lávrion, Greece, and produced by the action of sea-water on lead-slags.
1892 Dana *Syst. Min.* (ed. 6) iv. 172 *Fiedlerite*...Luster adamantine. Colorless. Transparent...Named after the Saxon Commissioner of Mines, Fiedler, director of the Grecian exploration. 1899 *Jrnl. Chem. Soc.* LXXVI. ii. 433 The new mineral, paralaurionite, may at a cursory glance be easily mistaken for laurionite or fiedlerite. 1934 *Mineral. Mag.* XXIII. 583 The crystals of fiedlerite are poor and the angles for any given form show considerable variation.

field, *sb.* Add: **4. e.** = *airfield.
1912 *Sci. Amer.* 24 Aug. 156/1 The field was so narrow that the aviators were compelled to start always toward the north regardless of the direction of the wind. 1930 *Techn. News Bull., Bureau of Standards* (U.S.) June 61/1 The angle..of the high-frequency landing beam has been adjusted so that an airplane may be guided along the proper gliding path to the field. 1958 'N. Shute' *Rainbow & Rose* i. 8 It's not a licensed field.
9. a. (Earlier example.)
c 1742 J. Love *Cricket* (1770) III. 18 To such impetuous Might compell'd to yield The Bail, and mangled Stumps bestrew the Field.
10. a. (Earlier *fig.* example.)
1742 Richardson *Pamela* III. xxxiii. 315 An hundred Guineas to one against the Field.
c. (Earlier example.)
1744 *Norwich Mercury* 15 Sept., And spread the field at distance wait, To break the striker's force.
d. *to play the field*: to avoid an exclusive commitment (to any one person, etc.); to devote oneself to many causes, persons, etc. *U.S. colloq.*

1936 L. Lefko *Public Relations* ii. 18 He hasn't any steady. He plays the field—blonde, brunette, or what have you. 1948 G. Vidal *City & Pillar* (1949) ii. ix. 199 Well, I've come to the conclusion that the only real pleasure is in playing the field. 1966 *New Republic* 3 Mar. 19 Japan Plays the Field. Peace and Trade with Everyone.
11. (Earlier example.)
1825 'B. Blackmantle' *Eng. Spy* I. i. 32 He was accounted..an active *field*.
15. c. Used *attrib.* to denote an investigation, study, etc., carried out in the natural environment of a given material, language, animal, etc., and not in the laboratory, study, or office; also, to denote a person taking part in such an activity, as *field archæologist*, etc. Also in phr. *in the field*.
1789 Field-naturalist [in Dict., 21]. 1898 W. James *Coll. Ess. & Rev.* (1920) 426 In the '50's and '60's Captain Mayne Reid..was forever extolling the hunters and field-observers of living animals' habits, and keeping up a fire of invective against the 'closet-naturalists', as he called them, the collectors and classifiers, and handlers of skeletons and skins. 1904 Field naturalist [see *aviculturist]. 1906 *Nature* 1 Mar. 411/1 A handbook in a paper wrapper is hardly fit for use in the field. 1915 J. P. Williams-Freeman *Introd. Field Archæol.* p. viii, A general introduction to Field Archaeology. *Ibid.* p. xix, Hampshire is..a happy hunting-ground for the Field Archaeologist. 1930 E. E. Hunt *Audit of Amer.* 8 We have as yet no method for summing up individual incomes into a family income, exc. that of direct field studies. 1932 H. C. Brearley *Homicide in U.S.* 10 There is need of a thorough field survey to determine how accurate are the homicide reports now being made to state and federal bureaus of vital statistics. 1933 *Brit. Birds* XXVI. 363 The elementary co-operative services..a national field centre which can collaborate with other national field centres overseas,—still remain to be provided. 1935 *Discovery* Nov. 346/2 Dr. Margaret Mead, by her studies of certain aspects of primitive societies in New Guinea and Samoa, has already established a reputation as a field anthropologist. 1936 *Brit. Birds* XXIX. 264 (*heading*) Interpretation of field-observations. 1936 *Discovery* Dec. 382/2 It was only a matter of organisation to set up a field laboratory. 1937 R. H. Lowie *Hist. Ethnol. Theory* ix. 151 They lead to still another extension of traditional field research. *Ibid.* xiii. 231 In contrast to Radcliffe-Brown, Malinowski is first of all a field investigator. *Ibid.*, His field technique conforms to Boas' standards. 1937 J. Orr tr. *Iordan's Introd. Romance Ling.* 5 The Romance tongues are still spoken to-day..and can thus be observed and studied directly, 'in the field', to use a term borrowed from the natural sciences. 1946 *Lancet* 15 June 910/2 The Council for the Promotion of Field Studies have..established..their first residential centre. 1949 M. Fortes *Social Structure* p. vi, Much as the theory owes to Durkheim, its emphasis on synchronic study and its rejection of conjectural history are the results of field experience. 1951 *Archit. Rev.* CIX. 51/2 An extensive 'field-survey' of the monuments and a prodigious accumulation of new or hitherto scattered documentary material. 1956 O. L. Zangwill in A. Pryce-Jones *New Outl. Mod. Knowl.* 170 Ethology is concerned with field-observations of behaviour and their systematic interpretation. 1957 *N. & Q.* CCII. 233/1 Shrimpton [is] an extraordinary prodigy, perhaps our first field archaeologist. 1959 *Listener* 5 Mar. 402/2 A university must employ a variety of means far wider than those of the lecture room. Seminars, practical work, field trips, a library with access to original material..are all essential to this. 1960 S. Kauffman *If it be Love* ii. 24 It'd be a help if I had some of the field reports. 1961 *Amer. Speech* XXXVI. 201 Not having ready access to the field records, this reviewer must perforce rely for comparisons on his own published investigations. 1962 H. E. Beecheno *Introd. Bus. Stud.* ix. 80 It is at this point that 'field research' must be undertaken. 1968 *Times* 14 Oct. 15/1 The fossils were seen by an undergraduate from Rutgers University, New Jersey, during a field trip to study techniques in palaeontology. 1969 *Times* 13 Jan. 11/2 When this virus is put into sheep experimentally it produces a disease identical with that from which it can be recovered in the field. 1970 E. Leach *Lévi-Strauss* i. 8 By Malinowski standards 'Lévi-Strauss' field research is of only moderate quality.
d. *Math.* An algebraic system with two operations that satisfy certain axioms analogous to those for the multiplication and addition of real numbers; technically, a commutative ring that contains a unit element for multiplication and an inverse for each non-zero element.
1893 E. H. Moore in *Bull. N.Y. Math. Soc.* III. 75 Suppose that we have a system of symbols or marks,..in number *s*, and suppose that these *s* marks may be combined by the four fundamental operations of algebra..and that when the marks are so combined the results of these operations are in every case uniquely determined and belong to the system of marks. Such a system of *s* marks we call a field of order *s*. 1905 *Trans. Amer. Math. Soc.* VI. 183 Closely connected with the theory of groups is the theory of fields, suggested by Galois, and due, in concrete form, to Dedekind in 1871. The word *field* is the English equivalent for Dedekind's term *Körper*. 1941 Birkhoff & MacLane *Surv. Mod. Algebra* ii. 40 A field *F* is an integral domain which contains for each element $a \neq 0$ an 'inverse' element a^{-1} satisfying the equation $a^{-1}a = 1$. 1958 K. S. Miller *Elem. Mod. Abstr. Algebra* ii. 65 A field is an integral domain...In general, an integral domain is not a field. 1963 E. Weiss *Algebraic Number Theory* v. 185 By a global field we mean a field which is either (1) an algebraic number field, or (2) an algebraic function field in one variable over a finite constant field. 1968 E. D. Macdonald *Theory of Groups* xi. 222 Familiar examples of fields are the real numbers, the rational numbers, and the residue classes modulo a prime.
e. *Logic.* The class comprising the domain and the range or converse domain of a relation.
1903 [see *domain *sb.* 4 f]. 1952 F. B. Fitch *Symbolic Logic* vi. 179 The field of a relation has the same members as the field of the converse of that relation.
f. *Computers.* A set of one or more characters in a record, or a group of columns on a punched card, which together represent a single item of information; an item of information that can be so represented.
1946 P. S. Dwyer in *Proc. Research Forum* (I.B.M.) 21/1 By an 'elimination' field I mean a field which eliminates or causes the accounting machine to throw out every variate in any specified accumulation field for which the corresponding variate is missing in the elimination field. 1953 *Proc. Inst. Radio Engin.* Oct. 1337/1 The card columns are grouped in Fields. For example, a very simple invoicing problem would have a Field for Quantity, another for Unit Price,..and maybe a Field for discounts which apply. 1959 *Jrnl. Assoc. Computing Machinery* VI. 7 Figure 3 illustrates a file of payroll data as it would be recorded on magnetic tape. Each record in this file is partitioned into fields. 1966 Abrams & Corvine *Basic Data Processing* ii. 14 A field is one or more characters of data that are meaningful as a unit. The number of children in a family is a field; so is a person's name.
16. b. *field of consciousness, perception; field of vision* (later examples).
1865 S. Hodgson *Time & Space* vi. 295 It has been shown that the whole field of consciousness is occupied by perception and spontaneous redintegration. 1905 W. James *Ess. Radic. Empir.* (1912) vi. 170 The world experienced (otherwise called the 'field of consciousness') comes at all times with our body as its centre, centre of vision, centre of action, centre of interest. 1933 H. Read *Art Now* ii. 78 Take any field of vision—a landscape, the scene before you now, if you lift your eyes. 1942 R. Hillary *Last Enemy* iii. 72, I noticed how small was my field of vision [from an aeroplane]. 1957 H. Read *Tenth Muse* xiii. 109 As the mind perceives, it automatically selects and organizes the field of perception.
c. Photogr. *depth of field*: the distance between the nearest and the farthest objects that give an image judged to be in focus.
1911 B. E. Jones *Cassell's Cycl. Photogr.* 167 'Depth of field' is sometimes used as synonymous with 'depth of focus' and 'depth of definition', the third expression more correctly indicating what is meant. 1920 C. W. Piper *J. E. Fearn's Mod. Photogr.* (ed. 6) xiv. 102 The distance between the nearest and farthest objects in focus with any particular stop is called the depth of field for that stop. 1939 Henney & Dudley *Handbk. Photogr.* iv. 90 A useful accessory included on most miniature cameras and on many larger cameras is a depth-of-field, or depth-of-focus, table. 1951 G. H. Sewell *Amat. Film-Making* (ed. 2) iii. 28 Objects in front of and behind the main object will throw images of sufficient sharpness, and the magnitude of the distance between these front and back objects is known as 'depth of field'. 1965 *Movie* Spring 34/1, I used a telefoto lens a great deal so as not to have any depth of field.
d. *Television.* A set of equally spaced scanning lines extending over the whole picture area and produced by a single passage of the spot.
1943 D. G. Fink *Telev. Stand. & Pract.* iv. 81 There is some question whether the scanning period should be taken as the field period or the frame period in the case of interlaced scanning. *Ibid.* 76 What is the lowest field frequency that will result in naturalness, smoothness, and nonjumpiness..in the picture? 1950 *Sci. Amer.* Dec. 14/2 The CBS [colour television] system..requires six fields to present a single complete picture (i.e., two interlaced fields in each of the three primary colors). 1967 Wharton & Howorth *Princ. Telev. Reception* iii. 38 Two types of field, 'odd' and 'even', are generated and interlaced to form a complete picture. *Ibid.*, Each complete scan is called a field, and two successive scans which provide the complete picture information are called a picture.
17. Also, a state or situation in which a force is exerted on any objects of a particular kind (e.g. electric charges) that are present; the action of such a force; the value (or direction) at any point of the force on an object defined as having unit magnitude, or the set of the vectors that represent this force at each point in a region. So *electric*, *gravitational*, etc., *field*; also *field of force*. (Earlier and additional examples.)
The word is freq. used as if it denoted an identifiable causal entity.
1845 Faraday *Diary* 10 Nov. (1933) IV. 331 Wrought with bodies between the great poles, i.e. in the magnetic field, as to their motions under the influence of magnetic force. 1850 *Ibid.* 23 July (1934) V. 325 When the opposed bodies are on opposite sides of the axis (Magnetic), then the figured forms would give fields of force in which the lines of magnetic power would vary. 1850 W. Thomson in *Phil. Mag.* 3rd Ser. XXXVII. 251 The 'field of force' [of a magnet] occupied by the mercury and watch-glass. 1865 J. C. Maxwell in *Phil. Trans. R. Soc.* CLV. 460 The theory I propose may..be called a theory of the Electromagnetic Field, because it has to do with the space in the neighbourhood of the electric or magnetic bodies. *Ibid.* 493 The intrinsic energy of the field of gravitation. 1897 J. J. Thomson in *Phil. Mag.* 5th Ser. XLIV. 311 If these corpuscles are..projected from the cathode by the electric field, they would behave exactly like cathode rays. 1903 Rutherford *Ibid.* 6th Ser. V. 179 The magnetic field was applied perpendicular to the plane of the paper and parallel to the plane of the slits. *Ibid.* 184 The rate of discharge..with the electric field off and on. 1928 A. S. Eddington *Nature Physical World* vii. 153 It is usually considered that when we use these [*sc.* magnets, electroscopes, etc.] we are exploring not space, but a field in space. 1928 Newman & Searle *Gen. Prop. Matter* iii. 75 According to Newton's Law the nature of the attracting masses is unimportant, and it is the magnitude of the mass..which determines the gravitational field. 1946 *Ann. Reg. 1945* 355 The stability of ordinary atomic nuclei could be explained only by the existence of a meson field analogous to the electro-magnetic field. 1952 Fitzgerald & Kingsley *Electr. Machinery* i. 6 Voltages are induced..by mechanically rotating a magnetic field past the windings. 1956 E. H. Hutten *Lang. Mod. Physics* iii. 93 Energy is thus contained in the volume of space in which a field exists. 1959 B. I. & B. Bleaney *Electr. & Magn.* i. 4 From Eq. (1.2) we find that $E = q_1/(4\pi\epsilon_0 r^2)r$ is the electric field due to the charge q_1. 1962 J. Dougall tr. *Born's Atomic Physics* (ed. 7) iii. 59 According to Maxwell's electromagnetic theory, light..consists of a periodically variable, electromagnetic alternating field. 1962 Corson & Lorrain *Introd. Electromagn. Fields* i. 1 The force between two electric charges will be considered as being due to an interaction between one of the charges and the field of the other. 1962 V. H. Booth *Physical Sci.* xxi. 300/1 The strength of the gravitational field is extremely weak compared to electric and magnetic fields. *fig.* 1943 H. Read *Educ. through Art* vi. 181 They [*sc.* Jungian archetypes] are centres of influence, or fields of force within the unconscious. 1959 *Times Lit. Suppl.* 23 Jan. 45/3 The present Byronic field of force.
b. That part of an electric generator or motor in which the magnetic flux is produced; a field magnet. (An *absol.* use of *field* used *attrib.* in the preceding sense.)
1893 G. Kapp *Dynamos* x. 226 We can..alter the design altogether so as to obtain enough cooling surface without increasing the weight of the field. 1952 Fitzgerald & Kingsley *Electr. Machinery* i. 11 (*caption*) Part of Boulder Dam hydroelectric station showing wound revolving field (suspended from two cranes) of 82,500-kva.. water-wheel generator.
c. *Embryol.* A region of an embryo capable of developing into a particular organ; a supposed system of influences regarded as collectively causing the differentiation of tissue.
1927 G. R. de Beer in *Biol. Rev.* Mar. 189 The only hypothesis which appears tenable is that of a Gradient System or 'field'..for the place of determination of the rudiments of other organs. 1934 Huxley & de Beer *Elem. Exper. Embryol.* viii. 276 The term *field* implies a region throughout which some agency is at work in a co-ordinated way, resulting in the establishment of an equilibrium within the area of the field. 1957 *Encycl. Brit.* VIII. 976/2 Each area or morphogenetic field..is a gradient field in the sense that the capacity for differentiation is highest in the centre and diminishes gradually toward the periphery. 1969 *Jrnl. Theoret. Biol.* XXV. i. 41 It has been a great surprise and of considerable importance to find that most embryonic fields seem to involve distances of less than 100 cells, and often less than 50.
d. *Psychol.* An environment or situation regarded as a system of psychological forces with which an individual interacts.
1934 *Philos. Sci.* I. 328 Since the genesis of ego and superego out of id depend on experience, one might suppose that the activity of these parts of the ego was determined by the structure of the psycho-biological field. 1935 Adams & Zener tr. *Lewin's Dynamic Theory Personality* iii. 79 To understand or predict the psychological behavior..one has to determine for every kind of psychological event..the momentary whole situation, that is, the momentary structure and the state of the person..and of the psychobiological environment... Every fact that exists psychobiologically must have a position in this field. 1950 *Mind* LIX. 569 A stimulus in an environmental field acts upon an organism in such a way as to evoke a 'directional' response, which leads to adaptation to the field—certain 'directional' responses being attitudes, sentiments, and so on. 1952 W. J. H. Sprott *Social Psychol.* i. ii. 30 The central concept in his [*sc.* Kurt Lewin's] scheme is the 'field'. The 'field' can be analysed into 'subjective' and 'objective' elements. 1964 Gould & Kolb *Dict. Soc. Sci.* 270/1 The field concept as specifically used by gestalt psychologists refers to organized conscious behaviour accompanied by a physiological energy pattern and forming part of a larger physiological system.
18. a. (sense 2) *field-character*, *-mark* (of a bird); (sense 4) *field capacity*, *-drain*, *-measure*, *-name*, *-noise*, *-trial* (further examples); (sense 7) *field ambulance*, *-craft*, *kitchen*, *post card*, *punishment*, *-service* (attrib.), *-telephone*; (sense 17) *field coil*, *law*, *physics*, *winding*. **d.** limitative, as (sense 17) *field-free* adj.
1916 F. M. Ford *Let.* 23 Aug. (1965) 69, I have had..a week in Field Ambulance. 1938 Weaver & Clements *Plant Ecol.* (ed. 2) viii. 201 The total amount of water that is held against the downward pull of capillarity and the force of gravity and does not drain through the soil is termed the water-retaining capacity or field capacity. 1937 *Brit. Birds* XXXI. 84 Field characters, habitat, and 'habits' generally. 1953 Bannerman *Birds Brit. Isles* II. 212 Both sexes have the excellent field-character of a long tail. 1892 Crocker & Wheeler *Pract. Managem. Dynamos & Motors* i. 4 Field Magnet.—This consists of one or more iron cores on which are wound the field coils. 1933 B. H. L. Hart *Future of Infantry* App. ii. 80 The following exercises are of use in developing the individual's field-craft. 1967 *Observer* 14 May 2/5 American troops are poor at fieldcraft. 1933 B. W. Adkin *Land Drainage in Brit.* xx. 348 Open drains are never cut as

deep as the ordinary field drains. 1945 'G. Orwell' *Animal Farm* v. 36 He talked learnedly about field-drains..and basic slag. 1955 O. Klein in W. Pauli *Niels Bohr* 116 The field-free Dirac equation. 1962 *Science Survey* IV. 64 In contrast with the microwave maser, no magnetic field is required because energy levels existing in the field-free atom are used. 1970 G. K. Woodgate *Elem. Atomic Struct.* iv. 70 The beam travels a further distance *l'* in a field-free region to a detector. 1915 'I. Hay' *First Hundred Thousand* xiv. 191 We might..let the grooms and drivers go with the..field kitchens. 1942 *Penguin New Writing* XIV. 9 The oppressor's soldiers have offered soup from their field-kitchens to those among the inhabitants who are utterly destitute. 1928 A. S. Eddington *Nature Physical World* xi. 236 The field laws—conservation of energy, mass, momentum and of electric change..are not controlling laws. They are truisms. 1954 Bannerman *Birds Brit. Isles* III. 20 The white wing-bar..was an excellent field-mark and could be distinguished..at a range of thirty to forty yards. 1857 *Phil. Trans. R. Soc.* CXLVII. 627 The Ordnance Survey Office have used, as standards of reference for field-measures, two 10-feet iron bars. 1924 E. Ekwall in Mawer & Stenton *Introd. Surv. Eng. Place-Names* i. iv. 89 A full investigation of field-names..will probably often tell of a strong Scandinavian influence. 1960 P. H. Reaney *Orig. Eng. Place-Names* x. 207 If we are to be accurate in our use of terms, 'field-names' should be used only of the enclosed areas of arable and pasture land which are never found on the Ordnance Survey maps. *Ibid.* 210 Although not strictly accurate, the term 'field-name' is also used of the innumerable minor place-names found in documents of all periods. 1904 W. Stevens *Let.* 6 Aug. (1967) 78 And it sounds, perfectly, like a field-noise at harvest. 1928 A. S. Eddington *Nature Physical World* xi. 236 The material, gravitational and electromagnetic fields are all included...The quantities enumerated above..obey the great law of field-physics. 1932 *Discovery* Oct. 338/1 The whole conception of what is called field physics springs from Maxwell and Faraday. 1916 T. E. Lawrence *Let.* 14 Dec. (1954) 332 This is not a letter: only a substitute for a field post card. 1917 W. Owen *Let.* 1 Jan. (1967) 421 If on my Field Post Card I cross out 'I am being sent down to the base' with a double line..then I shall actually be at the Front. 1907 *Act 7 Edw. VII* c. 2 § 10 Amendments of Army Act...The words 'field punishment' shall be substituted for the words 'summary punishment' wherever those words occur in provisoes (9) and (10). 1834 J. S. Macaulay *Treat. Field Fortif.* xi. 225 [Tools required] 1 Field-service level. 1 Six-feet rod [etc.]. 1915 F. H. Lawrence in *Home Lett. of T. E. Lawrence* (1954) 660, I am sending a field service post card at the same time as this letter. 1940 *N. & Q.* 28 Sept. 217/2 Field Service Cap. 1908 E. J. Stevens (*title*) Field telephones for army use. 1939 Auden & Isherwood *Journey to War* ix. 222 Soldiers installed a field-telephone. 1895 *Westm. Gaz.* 12 Dec. 7/2 That great field-trial authority [on dogs]. 1970 *Encycl. Brit.* IX. 250/1 Field trials are competitions among individual sporting dogs, or gun dogs, as they are called in Great Britain, of the same general type under conditions that approximate or simulate those found in the hunting field. 1893 D. C. Jackson *Electro-Magn.* v. 123 To know the number of leakage lines is therefore a matter of moment..in calculating the field windings.
19. *field-bee*; **field-lark** (examples); *U.S.* = Meadow *lark* (*b*).
1835 *Chambers's Edin. Jrnl.* 7 Mar. 45/1 Seven species of the genus *apis*; the most remarkable of which are the small field-bee, [etc.]. 1918 D. H. Lawrence *New Poems* 10 As a field-bee, black and amber, breaks from the winter-cell. 1678 Ray *Willughby's Ornith.* ii. 207 Mr. Jessop suspects that there is yet another different sort of this bird, which may be called the lesser field-lark. 1768 Pennant *Brit. Zool.* II. 238 The Lesser Field Lark. a 1883 G. W. Bagby *Old Virginia Gentleman* (1910) 126 We hear his gun go off, and he comes back presently bringing a field-lark in his hand.
20. b. In designations of crops covering a large area, as *field-hay*; **field-pea,** *Pisum sativum* var. *arvense*.
1895 C. J. Cornish *Wild Eng.* 242 'Field-hay', as the produce of the rye-grass, sainfoin, clover, and trefoil is called, is a new feature in the country. 1709 J. Lawson *New Voy. Carolina* 77 All the sorts of English Pease.. thrive..in Carolina. Particularly..the common Field-Pease. 1858 Field pea [see Pea¹ 2 b]. 1892 W. Fream *Elem. Agric.* (ed. 2) 240 The field-pea, of which there are several sorts, distinguished by bearing a blue blossom, should be sown as early as possible in the spring. 1949 G. H. Ahlgren *Forage Crops* xiii. 131 The field pea was brought to America by the early colonists, being reported first in Virginia in 1636.
21. field-battery (earlier examples); **field block, blocking** *Surg.*, a technique in which an anæsthetic solution is injected so as to create a zone of anæsthesia around the operative field and block the passage of nerve impulses; **field boot,** a knee-high military boot, usu. laced from the foot to mid-calf; **field-effect transistor,** a semiconductor device in which the majority carriers flow along a channel whose effective resistance is controlled by a transverse electric field produced by a reverse bias applied to a gate region surrounding the channel, the gate and the channel being of opposite conductivity types; **field emission,** the emission of electrons from the surface of a conductor under the influence of a strong electrostatic field as a result of the tunnel effect; hence **field emission microscope,** a device which utilizes this effect to produce an enlarged image of the emitting surface on a fluorescent screen; **field equation** [Field *sb.* 17], one of a series of equations established by Maxwell (1865) and Einstein (1950) which describe conditions existing within electromagnetic and gravitational fields respectively; **field events,** certain athletic events (see quots.), as distinguished from events on the running-track; **field-folk,** agricultural workers; **field goal** esp. *N. Amer.* and *Austral.*, (*a*) in Football, a goal scored from the field of play; (*b*) in Basketball, a goal scored while the ball is in play; (*c*) in other games; **field-grey** [tr. G. *feldgrau*], the regulation colour of the uniform of a German infantryman; **field-gun** (examples); **field-hand,** (*a*) and (*b*) (earlier U.S. examples); **field ion microscope,** a device which works on the same principles as the field emission microscope but which uses ions produced at the surface of the emitter or in the gas near by in place of electrons; **field labourer** = *field-hand* (*b*); **field magnet** (earlier example); **field-monument** *Archæol.* (see quot. 1954); **field-notes** (earlier and later examples); **field pattern,** the way in which the sensitivity or emitted power of a detector or emitter of waves (e.g. a microphone or an aerial) varies with direction; **field-regulator, -rheostat** (see quots.); **field-sequential system** *Television*, a system of colour television in which each of the successive fields composing the picture is of a single primary colour; **field spaniel,** a spaniel trained to retrieve; a variety of spaniel closely allied to but larger than the cocker; **field strength,** the intensity of an electric, magnetic, or other field; **field system** (see quot. 1915); cf. Open field; **field-test** *v. trans.*, to test in the field (sense *15 c); **field tile,** a tile used in the construction of a field-drain.
1825 J. Neal *Bro. Jonathan* III. 42 A loose broken irregular line of ditch and parapet-field batteries. 1834 J. S. Macaulay *Treat. Field Fortif.* vi. 113 A field battery consists of a parapet pierced with embrasures, with epaulments on the flanks, and traverses..to cover the guns from enfilade fire. *Ibid.*, The artillery accompanying an army in the field are divided into batteries, also called field-batteries. 1932 C. L. Hewer *Rec. Adv. Anæsthesia* x. 102 *Infiltration analgesia*, or field block, where the nerve endings are blocked at or near the site of operation. 1937 *Ibid.* (ed. 2) xii. 134 Field blocking consists in creating walls of analgesia encircling the operative field. 1963 D. E. Hale *Anesthesiol.* (ed. 2) xiv. 431/1 Some procedures may be done under either local or field block, and in practice, they are often combined. 1905 E. J. C. Swaysland *Boot & Shoe Design & Manuf.* vii. 120 The field boot..may be fitted by several methods; they have usually a short strap down the front, reaching to the top edge of the tongue, and a back strip covering the back seam. 1926 T. E. Lawrence *Seven Pillars* viii. xciii. 493 Dawnay..was our best card, with his proved military reputation, exquisite field-boots, and air of well-dressed science. 1938 Auden & Isherwood *On Frontier* 41 'If you're foolish enough,' they declare, 'to resist, You shall feel the full weight of fieldboot and fist.' 1969 R. T. Wilcox *Dict. Costume* 37 (*caption*) Leather field boot—English—1940's. 1953 Dacey & Ross in *Proc. Inst. Radio Engin.* XLI. 970 (*heading*) Unipolar 'field-effect' transistor. *Ibid.*, The 'field-effect' transistor was first proposed by W. Shockley. 1955 —— in *Bell Syst. Techn. Jrnl.* XXXIV. 1150 In essence, a field-effect transistor can be regarded as a structure containing a semi-conducting current path, the conductivity of which is modulated by the application of a transverse electric field. 1970 J. Earl *Tuners & Amplifiers* ii. 33 Trend is towards field effect transistors in the f.m. front-end. 1928 *Proc. R. Soc.* A. CXIX. 173 As the higher temperatures, at which ordinary thermionic emission begins, are approached, the strong field emission does become sensitive to temperature and finally blends into the thermionic. 1951 *Bell Syst. Techn. Jrnl.* Oct. 910 (*heading*) Description of field emission microscope tube. 1965 *New Scientist* 17 June 781/1 Field-emission diodes could prove very useful at microwave frequencies if they were placed in a superconducting cavity. 1965 Phillips & Williams *Inorg. Chem.* I. x. 374 In recent years the use of the Field Emission microscope and the Field Ion microscope has shown that diffusion at the surface of metals is considerable at temperatures as low as $0{\cdot}2T_m$. 1922 E. P. Adams tr. *Einstein's Meaning Relativity* 52 Repeated application of the field equations. 1923 J. Rice *Relativity* iv. 95 So strong, however, was the impulse towards making the field equations as alike as possible for all observers ..that these values for the moving observer were not regarded with any great favour. 1924 H. L. Brose tr. *M. Born's Einstein's Theory Relativity* v. 154 Maxwell's 'field equations', as they are called, constitute a true theory of contiguous action or action by contact, for.. they give a finite velocity of propagation for electromagnetic forces. 1955 W. Pauli *Niels Bohr* 34 We assume ..for the sake of simplicity the *local* character of the field equation, which means that all field quantities are spinors or tensors of finite rank. 1959 *Listener* 27 Aug. 320/3 The steady-state cosmological theory..modifies Einstein's field equations. 1899 *Windsor Mag.* IX. 243/1 Irishmen.. have established a monopoly in what are described as field events of late years. 1912 E. H. Ryle *Athletics* 19 'Field events' (*i.e.*, long-, high- and pole-jumping, weight-putting, hammer- and discus-throwing, and hurdling). 1950 *Oxf. Jun. Encycl.* IX. 35/1 *Field Events*...This department of athletic sports includes the Long and High Jumps, the Hop, Step, and Jump, Pole Vaulting, Putting the Weight, and Throwing the Discus, the Javelin, and the Hammer. 1955 R. Bannister *First Four Minutes* v. 60 To raise the standard of British athletic achievement, particularly in field events. 1886 W. Morris tr. *Æneids* vii. 204 Calling to aid the hardy hearts of field-folk thereabout. 1891 Hardy *Tess* II. xxxii. 149 The fatalistic convictions common to field-folk and those who associate more extensively with natural phenomena than with their fellow-creatures. 1902 *Chicago Record-Herald* 28 Sept. 2/3 A try for a field goal was made, but..the kick was easily blocked. 1906 *Off. Bk. Rules Govt. Game of Basketball 1906–1907* 13 Flint..has tallied the most field goals. 1947 *Redbook* Oct. 56/3 No man with a bum leg could kick a field goal from the 37-yard line with the wind against him. 1961 J. S. Salak *Dict. Amer. Sports* 158 *Field goal* (basketball), a basket scored from the floor; has a value of two points. *Field goal* (bowling), on a two split, to send the ball between the pins hitting neither of them. *Field goal* (football), placekicking or dropkicking ball between uprights and over opponent's cross-bar from scrimmage without touching ground or teammate. Three points. 1969 *Eugene* (Oreg.) *Register-Guard* 3 Dec. 1D/6 Shooting: Field goals, 45.3%. 1969 *Sun-Herald* (Sydney) 13 July 36/2 Williams was set up for a field goal on the fourth tackle but his kick was wide and the ball went dead. 1970 *Globe & Mail* (Toronto) 28 Sept. 18/2 Ivan MacMillan converted both Oldham touchdowns and added a 12-yard single on a wide field goal attempt. 1970 *Washington Post* 30 Sept. D4/3 Shugars..has directed the team to just three touchdowns and a field goal. 1915 *Sphere* 17 Apr. 59/2 The men in field grey tramped past him towards the Aisne. 1929 W. F. Morris (*title*) Bretherton, khaki or field-grey? 1828 J. M. Spearman *Brit. Gunner* 179 The detachments for the service of heavy ordnance are told off and numbered in precisely the same manner as for field-guns. 1907 *Westm. Gaz.* 16 May 7/2 The men of the 'Victory'..give field-gun displays. 1826 *Deb. Congress U.S.* 25 June (1832) 3758/2 The price of labor, fields [*sic*] hands, from eighty to one hundred and twenty dollars per annum and found. 1835 J. H. Ingraham *South-West* II. 254 The third and lowest class consists of those slaves who are termed 'field-hands'. 1845 F. Douglass *Life* (1846) 58, I was now for the first time in my life a field hand. 1952 *Sci. Abstr.* A. Mar. 216/1 (*heading*) The field ion microscope. 1967 *New Scientist* 20 July 134/1 The importance of the field-ion microscope lies in its ability to depict the positions and arrangement of the individual atoms of a solid surface. 1853 Mrs. Stowe *Key to Uncle Tom's C.* 17/1 We ask.. whether this does not show that this poor field-labourer had..a true mother's heart? 1860 J. S. C. Abbott *South & North* 279 A little handful of slaveholders may be exempted from paying wages to..their field-laborers. 1883 *Phil. Mag.* XVI. 144 The intensity of a powerful magnetic field, such as that in the space in which the armature-coils of a dynamo move between the poles of the field-magnets. 1937 *Proc. Prehist. Soc.* III. 266 The care of field-monuments in the Free State is entrusted to the Office of Public Works. 1954 S. Piggott *Neolithic Cultures* ii. 17 The field monuments of the culture fall into three classes,..earthwork enclosures..flint mines.. long barrows. 1786 Washington *Diaries* (1925) III. 84 His assistant who had his field notes..had not returned. 1806 *Deb. 9th Congr. U.S.* 2nd Sess. 1002 He was retained as a necessary assistant to the principal surveyor in copying field notes. 1841 C. Cist *Cincinnati* 152 From these field-notes, the plats, or maps..are prepared. 1849 *President's Message Congress* ii. 572 United States geological survey of public lands in Michigan—Field notes. 1958 Bannerman *Birds Brit. Isles* VII. 210 There have been no recent field-notes published on this species. 1961 *Amer. Speech* XXXVI. 201 Numerous allophonic variations, the accurate transmission of which from informant to fieldworker to field notes..produces certain problems in communication. 1936 *Jrnl. R. Aeronaut. Soc.* XL. 191 The radiated field pattern should therefore remain constant from day to day. There was no evidence whatsoever that atmospheric conditions had any effect on the field patterns. 1962 A. Nisbett *Technique Sound Studio* 265 *Polar characteristic, polar diagram* (Am.: field pattern), the response of a microphone, loudspeaker, etc., showing sensitivity (or volume of sound) in relation to direction. 1919 W. H. Marchant *Wireless Telegr.* (ed. 2) 297 Field Regulator. A variable resistance forming part of the field circuit of a motor or a dynamo. 1924 *Harmsworth's Wireless Encycl.* 918/2 A field regulator is a device for varying the strength of the field magnets in a dynamo or electric motor. *Ibid.* 919/1 From the wireless point of view, the chief use of field regulators is to provide a convenient means for regulating the charging rate of a dynamo used to recharge accumulators. 1910 *Hawkins' Electr. Dict.*, *Field Rheostat*, an adjustable resistance used to vary the strength of the magnetic field of a shunt wound dynamo or motor. 1943 *Gloss. Terms Electr. Engin.* (B.S.I.) 55 *Field rheostat* (*field regulator*), a rheostat arranged for varying, at will, the current in the field winding of a machine. 1950 *Sci. Amer.* Dec. 14/2 The 'field-sequential' system developed by the engineers of the Columbia Broadcasting System is perhaps the simplest of the three. 1957 *Encycl. Brit.* XXI. 912N/2 Presenting the colours in such rapid succession to the eye that persistence of vision merges them into a single sense impression. The last method is employed in the field-sequential system. 1867 'Stonehenge' *Dogs Brit. Isl.* 36 Field Spaniels...The heavy, large-eared, well-feathered, short-legged 'field-spaniels', have been known for years as 'springers'. 1897 *Encycl. Sport* I. 319/2 There are four varieties of field spaniels,..the Clumber, the Sussex, the Black, and the any-other-colour. 1960 *Times* 2 Jan. 9/3 Two varieties now much in a minority are the Sussex and field spaniels. 1896 D. C. & J. P. Jackson *Alternating Currents* xiv. 574 If the field strength of a motor is so adjusted that the values of the impressed and counter pressures are equal..then when the motor is switched on..it will fall back in phase with respect to the impressed pressure, sufficiently to permit the proper load current to pass through the armature. 1946 *Nature* 7 Sept. 332/2 If the field-strength inside the dielectric exceeds a critical value, the insulation breaks down. 1962 F. I. Ordway et al. *Basic Astronautics* iv. 164 Sunspot field strengths up to several thousand gauss have been observed. 1915 H. L. Gray *Eng. Field Syst.* 3 The term 'field system' signifies the manner in which the inhabitants of a township subdivided and tilled their arable, meadow, and

pasture land. **1935** *Proc. Prehist. Soc.* I. 10 The archaeological investigation of ancient field-systems. **1962** H. R. Loyn *Anglo-Saxon England* i. 18 The agrarian inheritance in the shape of methods and field-systems was far from negligible from Roman villa to Saxon village. **1950** in *Amer. Speech* (1956) XXXI. 210 The results were field tested in the Yukon. **1961** *Times* 15 Aug. 13/5 At present N.C.R. are field-testing their prototype scanner at a chain store. **1970** *Computers & Humanities* IV. 323 To field-test and perfect the DOVACK Model for effectiveness, adaptability, and economic feasibility. **1958** J. S. Scott *Dict. Civil Engin.* 138 Field tile. **1965** G. J. Williams *Econ. Geol. N.Z.* xx. 364/2 Similar deposits [of clay].. are used for making bricks, field-tiles and refractories at Kamo.

field, *v.* Add: **5. a, b.** (Earlier examples.)
1823 *Lady's Mag.* July 390/2 How well we fielded! **1833** J. Nyren *Young Cricketer's Tutor* 48 The fieldsman ..should not wait and let the ball come to him, but dash in to meet it, fielding it with his right hand.
c. *to field out*: to be or remain in the field as a fieldsman or as the fielding side.
1888 R. H. Lyttelton in Steel & Lyttelton *Cricket* vi. 280 An eleven that is really A1 in fielding very rarely has to field out for 300 runs. **1921** P. F. Warner *My Cricketing Life* xii. 225 He hated fielding, and had no wish to field out the whole summer! **1944** Blunden *Cricket Country* 11 Someone was bowling, someone batting, the rest fielding out.
d. *fig.* To deal with (a succession of items), 'catch', 'pick up'.
1902 *Daily Chron.* 2 Sept. 3/1, I would get an agile and hard-skinned man to field the novels as they come. **1908** *Ibid.* 20 Apr. 4/6 From Good Friday to the following Tuesday, if you stay in London, you have to field splashes of paint and skirt ladders. **1909** *Ibid.* 18 Nov. 4/6 The Correctors of the Press are demanding the proper consideration of men who field the mistakes of careless writers. **1969** *Morning Star* 11 Oct. 5/7 A man who has just emerged from two years in solitary confinement cannot be expected to field rapid fire questions from the Press.
6. *trans. Games.* To select (a team or an individual) to play; to put into the field.
1922 *Daily Mail* 1 Dec. 11 The F.A. played four professionals in the defence, but fielded an amateur forward line. *Ibid.* 6 Dec. 12 North Midlands hope to field a powerful fifteen in to-day's match v. Warwickshire. **1925** *Times* 12 June 7/1 It would have been rather futile to field the remnants of the M.C.C.'s Australian team when its leading batsmen..were not available. **1927** *Morn. Post* 24 Oct. 13/3 The Oxford side fielded against the United Services was a more workmanlike lot. **1948** *Evening Standard* 28 Apr., The Australians are fielding their strongest team. **1955** *Times* 3 Aug. 3/7 Even more significantly, the British side fielded an unfit hooker. **1962** *Listener* 11 Oct. 586/1 The Swedes fielded a new pair in the first half [of a bridge championship].

field-day. Add: **1. b.** (Earlier and additional examples.) Also, a time of great opportunity or success.
1827 Creevey *Let.* 26 Mar. (1934) II. xiii. 236 Saturday was a considerable field day in Arlington Street,..and a very merry jolly dinner and evening we had. **1855** Bagehot *Coll. Works* (1965) I. 313 A 'field-day' controversy is a fine thing. **1925** E. F. Norton *Fight for Everest: 1924* ii. 45 The two experts, who had for days been working every afternoon, and often late into the night, put in a regular field-day. **1953** A. Huxley *Let.* 8 Dec. (1969) 689 Industrial agriculture is having a field day in the million acres of barren plain now irrigated. **1969** *New Yorker* 12 Apr. 98/2 The human-factors men have been having a field day with it.
2. b. (Example.)
a **1878** G. G. Scott *Recoll.* (1879) viii. 354 We had a delightful field-day in the abbey.

fielded, *ppl. a.* Add: **3.** *Furniture.* Of a panel (see quot. 1940).
1900 in *Eng. Dial. Dict.* **1940** *Chambers's Techn. Dict.* 331/1 *Fielded panel*, a panel which is moulded, sunk, or raised, or is divided into smaller panels. **1952** J. Gloag *Short Dict. Furnit.* 254 *Fielded panel*, a cabinet-making term that describes a panel with the central space raised so that it projects slightly beyond the surface of its frame. **1961** *Times* 13 Feb. 14/5 The 14 doors have fielded up leather covered panels with enrichments in gilt.

fielder. Add: **2.** (Earlier and later examples.)
1844 *Spirit of Times* 6 Apr. 67 The reliance of the 'fielders' was undoubtedly Norma. **1867** 'Ouida' *Under Two Flags* I. iii. 46 Taking long odds with the fielders. **1969** *Australian* 24 May 35/3 It appears that the fielders consider the Bernborough Handicap will be dominated by this trio.
3. (Examples in *Cricket*.)
1832 P. Egan *Bk. Sports* 346/1 A bowler and fielder of very great use. **1912** A. A. Lilley *Twenty-four Years Cricket* iv. 45 A magnificent fielder in the slips.

field-glass. Add: **1.** Now usu. in *pl.*
1888 *Century Mag.* XXXVI. 211/1 A minute examination, with the field-glasses, of all the neighboring mountain. **1911** M. Corelli *Life Everlasting* vii. 142 Mr. Harland was..searching for his field-glasses. **1932** *Discovery* Oct. 330/2 The meter reading is recorded by observations carried out with powerful field glasses. **1963** A. Smith *Throw out Two Hands* xv. 155 Through field-glasses one looks for shapes, and shapes are therefore recognized.

fielding, *vbl. sb.* Add: **1. c.** (Earlier and later examples.)
1823 *Lady's Mag.* July 391/1 John Strong did very well; his length told in fielding. **1955** *Times* 9 May 15/1 The South Africans have been hailed as a great fielding combination.
d. The action of a fielder (sense 2).
1873 Hotten *Slang Dict.* 161 s.v. *Field*, Laying against favourites is called fielding, and bookmakers are often known as fielders. **1895** *Daily News* 20 May 3/6 There was a lot of fielding against yesterday's winner.

Fieldingesque (fīˑldiŋeˑsk), *a.* [f. the name of the novelist Henry *Fielding* (1707–54) + -esque.] Of or pertaining to Fielding, his writings, or his style.
1931 A. Huxley *Music at Night* 1. 12 A few Fieldingesque irrelevancies would destroy it [*sc.* 'Othello']. **1933** *Times Lit. Suppl.* 5 Oct. 657/2 The Fieldingesque adventures related by Cibber's own daughter. **1934** *Ibid.* 17 May 356/2 Novels in the Fieldingesque tradition.

field-land. Restrict † *Obs.* to senses in Dict. and add: **c.** Land suitable for cultivation. *U.S.*
1851 A. O. Hall *Manhattaner* 129, I have seen a million dollars worth of property..plantations; field lands; sugar-houses.

fieldless (fīˑldlės), *a.* [f. Field *sb.* 4 + -less.] Without fields.
1890 O. Crawfurd *Round Calendar in Portugal* 2 A great stretch of hedgeless, fieldless land. **1964** *Economist* 14 Mar. 1016/1 The new concrete, fieldless farms.

fieldsman. Add: **a.** (Earlier example.)
1767 R. Cotton *Cricket Song* ix. in F. S. Ashley-Cooper *Hambledon Cricket Chron.* (1924) 184 Ye Fieldsmen look sharp.
c. A person responsible for the management of various agricultural matters (see quots.).
1750 *Court Baron Rules & Orders* 12 Mar. in *Purefoy Lett.* (1931) II. 435 For every Cow put or turned into the Common or Cow pasture shall be paid yearly to the Fieldsmen..6d. **1914** W. H. Hutton *Highways & Byways in Shakespeare's Country* viii. 144 Weston-sub-Edge... Though the lord of the manor seems for all practical purposes to have died out there at the Reformation, the old open field system of husbandry continued. What was practically a corporation of fieldsmen (who have been truly described as the aristocracy of the village and 'did not represent more than one-sixth of the population') managed the agricultural arrangements of the land... The 'fieldsmen' were gradually dying out before the climax of 1852, when the..enclosures were carried out. **1950** H. J. Massingham *Curious Traveller* xi. 237 The Fieldsmen, appointed annually by the village, determined the crops of the year, fixed and paid wages, imposed fines for trespass and strayed livestock, [etc.].

fieˑldstone. [f. Field *sb.* + Stone *sb.*] Stone as found in the field, used esp. for building.
1896 *Folk-Lore* Mar. 47 Cybele, the Magna Mater, whose image in the shape of a rough field-stone had been given by the Phrygian priests. **1959** *Times* 19 June (Queen in Canada Suppl.) p. iv/4 Grey Quebec fieldstone. **1968** *Globe & Mail* (Toronto) 13 Jan. 28/4 The handsome panelled and fieldstone walls. **1970** H. Braun *Parish Churches* vii. 94 Some parts of England..have as their field-stone flint, lying in great nodules.

field theory. Any theory about fields (see Field *sb.* 15 and 17) or in which the idea of a field is the dominant concept. So **fieːld-theoreˑtical** *a.*; **fieˑld-theorist.**
1901 L. E. Dickson (*title*) Linear groups with an exposition of the Galois field theory. **1922** Jeffery & Perrett tr. *Einstein's Sidelights Relativity* 23 We ought not..to reject the possibility that the facts comprised in the quantum theory may set bounds to the field theory. **1929** Einstein in *Nature* 2 Feb. 175/1 The purpose of my work is..to reduce to one formula the explanation of the field of gravity and of the field of electromagnetism. For this reason I call it a contribution to 'a unified field theory'. **1934** J. F. Brown in *Philos. Sci.* I. 325 The central idea of a 'field' theory is that the behavior of an object within the field is determined by the field structure or spatial-temporal configuration of the energy within the field. *Ibid.*, By 'class' theory I mean essentially what Lewin has called Aristotelian theory, and by 'field' theory what Lewin has called Galilean theory. *Ibid.* 334 Aristotle was trying to answer the question: 'Why do bodies move?' Galileo, the first thinker in the field-theoretical tradition, attempted the more modest question: 'How do bodies move?' **1961** M. B. Hesse *Forces & Fields* x. 274 Field-theorists who, since Maxwell, have interpreted matter as itself a manifestation of a physically prior field. **1961** Powell & Crasemann *Quantum Mech.* iv. 87 Quantum field theory is the basic discipline in terms of which modern ideas in fundamental-particle physics are expressed. **1964** R. H. Robins *Gen. Ling.* ii. 70 The theory..of the linguistic field, or the field theory of meaning, is concerned to show that the lexical content of a language..is not a mere conglomeration or aggregation of independent items. **1967** L. Rédei *Algebra* I. ii. 36 The most important chapters of algebra are group theory, ring theory and the theory of skew fields (in particular field theory).

fieldward, -wards, *adv.* (Earlier example.)
1805 Wordsworth *Prelude* (1926) 228 The Dame, That field-ward takes her walk in decency.

field-work. Add: **1.** Also *spec.* in *Surveying.*
1767 R. Gibson *Treat. Surveying* (ed. 2) 173 If the End of the last Station falls exactly in the Point you begin at, the Field-Work and Protraction are truly taken. **1901** *Daily Colonist* (Victoria, B.C.) 16 Oct. 5/3 Mr. Carry, who conducted the survey from the summit westward, returned from the Mainland on Monday night, having completed the field work. **1956** P. Kissam *Surveying Instrum. & Methods* (ed. 2) xi. 192 The procedure (the location survey) is planned to require a minimum of field work.
3. A comprehensive name to describe the practical side of research in archæology, linguistics, the social sciences, etc., carried out in the areas concerned, as distinguished from theoretical or laboratory investigation. Cf. *Field *sb.* 15 c.
1922 B. Malinowski *Argonauts West. Pacific* 4 This exactly describes my first initiation into field work on the south coast of New Guinea. **1930** *Economist* 24 May 1159/2 He not only played his part in shaping the organisation, but he also did what might be termed important field work for it. **1933** Leavis & Thompson *Culture & Environment* 18 As 'field-work', pupils might.. note the effects of advertising on themselves and their friends. **1936** J. C. Brown (*title*) Field work with public welfare agencies. **1937** *Oxoniensia* II. 75 The field-work and excavation undertaken in 1935 and 1936 have proved beyond a doubt that this north Oxfordshire Dyke, so far from being a continuous circumvallation, was made up of numerous unconnected sectors. **1940** *Mind* XLIX. 64 He does not sit all day in his arm-chair, he spends at least part of his time in observational field-work. **1959** *Listener* 14 May 868/2 Rehabilitation, Boarding Out, Adoption, Supervision, and other Field work associated with deprived children. **1962** *Amer. Speech* XXXVII. 164 The Banks were not included in the *Linguistic Atlas* field work. **1971** *Observer* 1 Aug. 26/8 In the 1930s Jacques Soustelle did some very fine anthropological and archaeological field-work in Mexico.

fieˑld-worker. [f. Field *sb.* + Worker.] One who performs field-work (senses 1 and *3).
1832 *Chambers's Edin. Jrnl.* I. 115/3 Being then left by him without any provision, she sunk into the condition of a field-worker. **1915** J. Webster *Dear Enemy* (1916) 26 We need a field worker to travel about the country and pick up all the hereditary statistics she can about our chicks. **1922** B. Malinowski *Argonauts West. Pacific* 9 The field worker relies entirely upon inspiration from theory. **1935** *Amer. Speech* X. 4/1 The personal character of the informant..is of great importance. Unless he is hospitable, the fieldworker has little chance of putting the case before him. Unless he is honest, his answers to the fieldworker's questions lose most of their value. **1950** *Oxoniensia* XV. 13 The thin but widespread scatter of Romano-British potsherds which can be picked up from the surface and lynchets of Celtic fields is a phenomenon which the field-worker soon learns to accept as a matter of course in Celtic field areas. **1961** *Evening Standard* 2 Aug. 12/3 (Advt.), Market research part-time field workers..for retailer research.

fiend. Add: **4. c.** (Examples with qualifying word or phr.)
1889 Farmer *Americanisms* s.v., The free lunch fiend.. is one who makes a meal off what is really provided as a snack. He pays for a drink, but shamefacedly manages in this way to get something more than his money's worth. **1896**, etc. Dope fiend [see *dope *sb.* 5]. **1904** *Philadelphia Even. Tel.* 25 July 6 The camera fiend is after him, hot foot. **1909** Webster 812/1 *Fiend*... An opium fiend...he is a fiend in mathematics; a botany fiend. **1925**, etc. Drug-fiend [see *drug *sb.*[1] 1 b]. **1927**, etc. Fresh-air fiend [see *fresh *a.*[1] 6]. **1929** R. Graves *Good-bye to All That* xxviii. 377 He had been upset that morning by a letter from an autograph-fiend. **1956** M. Swan *Paradise Garden* I. i. 7 A neighbour a small-boat fiend. **1962** N. Marsh *Hand in Glove* v. 196 I'm a bit of a camera-fiend myself.

fiendish, *a.* Add: Also as *adv.*, excessively, horribly.
1861 W. James *Let.* Sept. (1920) I. 38, I have noticed fleeting shades of expression on her face..*unhuman*, ghoul-like, fiendish-cunning! **1891** S. Hale *Lett.* (1919) 258 As all the Continent is fiendish cold, we did wisely.

fiendishly, *adv.* Examples of trivial use.
1879 E. Nesbit *Let.* in D. L. Moore *E. Nesbit* (1933) iii. 47 Remember I am *fiendishly* busy sometimes. **1951** *Mind* LX. 428 A correlation of observations, fiendishly difficult in practice.

fierasfer (fəiᵊræˑsfəɹ). [mod.L. (L. Oken 1817, in *Isis* CXLVIII. 1182/1), f. Pr. *fierasfèr*, *fielat-fèr*, f. *fieras*, f. OPr. *filat* thread, net (f. L. *filum* File *sb.*[2]) + *fèr* fierce, wild (f. L. *ferus* Ferous *a.*).] A small parasitic fish of the family Carapidæ, which includes a genus once called *Fierasfer*, found in the Mediterranean and in British waters; also called *pearlfish*.
1843 *Proc. Zool. Soc.* XI. 92 The occurrence in Madeira of a single individual, imposes the necessity of proposing, in lieu of the barbarous vernacular appellation *Fierasfer*, a name founded on an obvious character. **1876** tr. *P. J. van Beneden's Animal Parasites* ii. 5 We see Fierasfers,..accompanied by Palæmons and Pinnotheres in the same animal. **1936** J. T. Jenkins *Fishes Brit. Isles* (ed. 2) 168 There are two representatives [of *Ophidiidæ*] which wander occasionally into British waters. These are the Bearded Ophidium..and the Fierasfer (*Fierasfer dentatus*). **1969** A. Wheeler *Fishes Brit. Isles & N.W. Europe* 455/1 The pearlfishes or fierasfers are well known for their habit of living within the body of a holothurian (sea cucumber) host.

fierce, *a.* Add: **1. c.** Of things: forceful; acting strongly or violently.
1912 *Motor Manual* 166 Complaints are occasionally made of what is called a 'fierce' clutch. In other words, the clutch will not slide or slip in, but permits the engine to take hold suddenly, and almost takes the starting control from the driver's hands. **1961** *Listener* 7 Dec. 1007/1 If your oven is inclined to be 'fierce', you may find it best to tie a band of folded brown paper round the outside of the tin. **1971** 'D. Halliday' *Dolly & Doctor Bird* viii. 113 The brake was fiercer than I expected, but the thing was stable enough.

fiery, *a.* Add: **4. d.** *Cricket.* Causing the ball to fly up after pitching. (Cf. *fire *sb.* A. 15.)
1877 C. Box *Eng. Game Cricket* xxvi. 450 *Fiery*, one of the ungenerous appellations a ground receives when it is hard, and probably not so verdant as a lawn or smooth as a billiard table. **1882** *Austral. in Eng.* 181 The wicket was fiery and the outfielding rough. **1893** *Baily's Mag.* Oct. 255/1 Fiery wickets are not at all desirable, since they introduce an element of danger into the game which is customarily absent. **1909** *Westm. Gaz.* 7 Aug. 11/2 Mr. Carr is indispensable to an England eleven on any wicket—fast, slow, crumbly, fiery.

fiery cross. Add: Also, a burning cross used by the Ku Klux Klan to intimidate people. *U.S.*
1926 *New Masses* July 7/2 Three hundred klansmen are burning a fiery cross on a hillside. **1936** M. Mitchell *Gone with Wind* lviii. 982 We convinced the hot-heads that watching, waiting and working would get us further than nightshirts and fiery crosses. **1948** *Chicago Tribune* 2 July 1/7 A 10 foot fiery cross was set ablaze in his front yard last night.

fiesta (fieˑstä). [Sp., feast.] In Spain or Spanish America, a religious festival; also, any festivity or holiday. Also *transf.*
1844 J. Gregg *Commerce Prairies* I. 208 These *carretas* ..[are] the 'pleasure-carriages' of the rancheros, whose families are conveyed in them to the towns..or to *fiestas*. **1845** R. Ford *Hand-bk. Trav. Spain* I. iii. 332/2 The *Fiestas* are of the highest order. **1910** 'O. Henry' *Whirligigs* v. 71 There are bathing and fiestas and bull-fights and scandal. **1923** *Blackw. Mag.* Nov. 697/1 The failure of the great June *fiestas* owing to continued rains. **1934** J. B. Priestley *Eng. Journey* x. 341 An agricultural fiesta. **1966** *Illustr. London News* 26 Feb. 2 Seek out the island of sun, gaiety, carnivals, fiestas, a casino.

Fi.fa, abbrev. of Fieri-facias.

fife, *sb.* Add: **1. c.** *fife and drum*: taken as typical instruments of martial music; often *attrib.* in lit. sense, and fig. = martial, militant.
1674 [see Fife *sb.* 1]. **1900** *Westm. Gaz.* 14 Feb. 3/2 The 'Captains Courageous' of the House were by no means unanimous in his favour. The Under-Secretary for War had not many fife-and-drum supporters in their ranks. **1923** B. Whitlock *J. Hardin & Son* I. v. 69 In the line there was a fife and drum corps. **1958** *Times* 29 Dec. 9/4 As a curate at St. Giles-in-the-Fields he started a fife-and-drum band for boys.

fife, *v.* **a.** (Earlier example.)
1598 Florio *Worlde of Wordes* 462/2 *Zuffolare*, to whistle, to pipe, to fife, to blow hard.

Fifer[2] (fəiˑfəɹ). [f. *Fife* + -er[1].] A native or inhabitant of Fife, a county of Scotland.
1887 P. McNeill *Blawearie* ix. 73 He'll be awfu' cunning, for a' the Fifers are burstin' fu' o' that commodity. **1897** *Daily News* 23 Nov. 3/6 He knew that the outside and envious critic was in the habit of telling Fifers that they worked the idea of a kingdom of Fife for a good deal more than it was worth. **1901** *Daily Chron.* 14 Oct. 5/2 The London 'Fifers' do not forget in exile the engaging qualities of their native county. **1908** *Westm. Gaz.* 12 May 7/2 We Scots outside 'the Kingdom' know it takes 'a long spoon to sup wi' a Fifer'. **1927** *Glasgow Herald* 14 May 9 The average Fifer..has more of Gaelic blood in him than the average Lewisman. **1967** I. Finlay *Lowlands* i. 20 The man of Lothian is far different from the man of Galloway, the Fifer from the Borderer.

Fifie, fifie (fəiˑfi). *Sc.* [f. *Fife* in Scotland + -ie.] 'A type of herring fishing boat with vertical stem and stern-posts, common in the second half of the 19th and early 20th cents. on the E. coast, prob. so called from having been first built and used on the Fife coast' (*Scot. Nat. Dict.*).
1905 *Banffshire Jrnl.* 28 Mar., A place for biggin heerin' boats Zulus, Fifies—best afloat. **1914** *Chambers's Jrnl.* Apr. 264/1 The craft at this time were of the 'scaff' or 'fifie' build. **1927** *Glasgow Herald* 3 May 7 Scaffie boats gave way to Zulu and Fifie types. **1934** *Geogr. Jrnl.* LXXXIV. 438 The Scottish craft, the fifies and zulus. **1950** P. F. Anson *Scots Fisherfolk* viii. 104 The most common type of Scottish fishing vessel until about the introduction of steam was the '*Fifie*'.

fifteen. Add: **B.** *sb.* **2. a.** (Earlier and *attrib.* examples.)
1878 *Cliftonian* Apr. 232 No O.C. has ever complained of difficulty in accommodating himself to the fifteen game. *Ibid.* 233 There will..be some competition to get into the fifteen.

fifth. Add: **A.** *adj.* **1. c.** (Earlier and later examples.)
1631 Dekker *Match me in London* I. 14 Thou tyest but wings to a swift gray Hounds heele, And add'st to a running Charriot a fift wheele. **1902** H. James *Wings of Dove* xvii. 255 'Respect', in their game, seemed somehow ..a fifth wheel to the coach. **1960** J. Bayley *Characters of Love* iv. 242 Her distress is partly finding herself the fifth wheel of the coach.
5. *Fifth Estate*, any institution considered as comparable to the four estates (Estate *sb.* 7).
1966 *New Statesman* 15 Apr. 526/2 The achievements of the trade union movement,..the pride the government takes in an historic, responsible and independent Fifth Estate.
B. *sb.* **4.** A fifth part of a gallon of liquor; a bottle containing such a part. *U.S. colloq.*
1938 D. Baker *Young Man with Horn* (1939) i. 69 Davis pulled the glass stopper out of the square bottle.. —Shorty's fifths never came sealed; they never came fifths for that matter. **1967** E. McGirr *Here lies my Wife* ii. 63 Bring a fifth—if you drink—and I'll provide ice.
C. *Comb.* **fifth (cranial) nerve** *Anat.*, either of the trigeminal nerves (see Trifacial, Trigeminal *adjs.*), together called the *fifth pair of nerves*; **fifth-wheel** (further examples).
1685 S. Collins *Syst. Anat.* II. 1047 The Fifth pair of Nerves, called by the Antients the Third, are seated below the other pair of Nerves. **1836–9** *Todd's Cycl. Anat.* II. 270 The fifth nerve is attached to the surface of the brain on either side of the pons Varolii, at a distance of three-fourths of an inch from its middle line. **1875** *Encycl. Brit.* I. 881/2 The Trifacial or fifth is the largest cranial nerve. **1931** *Daily Tel.* 19 Jan. 11/3 The Archbishop is suffering from severe fifth nerve neuralgia. **1968** W. H. Hollinshead *Anat. for Surgeons* (ed. 2) I. 70 The trigeminal, or fifth cranial, nerve is..the largest of all cranial nerves except the optic. **1881** Fifth wheel [see *wheel-plate* (Wheel *sb.* 18)]. **1962** *Engineering* 30 Mar. 418/2 The turntable on a tractor is known as the fifthwheel.
c. In Quaker use; *fifth day* (i.e. of the week), Thursday; similarly *fifth month*, May.
1698 S. Sewall *Let.-Bk.* in *Mass. Hist. Soc. Coll.* (1886) I. 203 Am going to keep Court at Springfield, next Fifth day. **1799** E. Fry in S. Corder *Life* (1853) 63 Fifth Month, 1st. **1821** *Ibid.* 318 My beloved daughter, Rachel, was married last Fifth-day, the 23rd, at Runcton. **1868** G. G. Channing *Early Recoll. Newport, R.I.* 231, I went frequently to hear him at the fifth-day meetings.

fifth column. [tr. Sp. *quinta columna*; cf. Column *sb.* 10.] Orig. the column of supporters which General Mola declared himself to have in Madrid, when he was besieging it in the Spanish Civil War, in addition to the four columns of his army outside the city (see 1936 quots.); hence, allusively, a body of one's supporters in an attacked or occupied foreign country, or the enemy's supporters in one's own country. Also *attrib.* and *transf.* Hence **fifth-columnist**, freq. loosely, a traitor, a spy. Also *fifth columnism*.
[**1936** *N.Y. Times* 16 Oct. 2/2 Police last night began a house-to-house search for Rebels in Madrid... Orders for these raids..apparently were instigated by a recent broadcast over the Rebel radio station by General Emilio Mola. He stated he was counting on four columns of troops outside Madrid and another column of persons hiding within the city who would join the invaders as soon as they entered the capital.] *Ibid.* 17 Oct. 9/4 Prudence counsels the government to forestall as far as possible the activities of this 'fifth column'. **1938** E. Hemingway (*title*) The Fifth Column. **1939** *War Illustr.* 21 Oct. p. ii/3 This looks to me like the Nazis' 'fifth column' in Belgium ready for the invasion. **1940** W. S. Churchill *Into Battle* (1941) 222 Parliament has given us the powers to put down Fifth Column activities with a strong hand. **1940** *Economist* 20 Apr. 723/1 Now that the news is beginning to reach the outside world of the..extent and..efficiency of the Nazi plot by which the key positions in Norway's cities and harbours were sacrificed to the invader..the other neutrals are attempting..to cope with the problem of their own 'Fifth Columns'. **1940** G. B. Shaw *Platform & Pulpit* (1962) 291 If you call Stalin a bloodstained monster you must be shot as the most dangerous of Fifth Columnists. **1941** E. Snow *Battle for Asia* 354 Factions..sabotaged the Eighth Route and New Fourth Armies in ways which elsewhere would be called Fifth Columnism. **1941** A. Huxley *Grey Eminence* vii. 146 His enemies..accused him of using his missionaries as French agents and anti-Hapsburg fifth-columnists. **1942** *Mind* LI. 166 Hence Prof. Collingwood arrives at the startling conclusion that psychologists are a group of fifth column conspirators, whose real aim is to destroy western civilisation from within. **1958** *New Statesman* 26 Apr. 538/3 After denouncing the Free World's readiness to let Communist fifth columns flourish undisturbed in its midst, he rails [etc.]. **1958** *Times Lit. Suppl.* 15 Aug. p. xxviii/4 Sensational elements began to infiltrate like fifth-columnists into the work of serious novelists. **1967** W. Soyinka *Kongi's Harvest* 66 I've put him through the standard tests. He's no fifth columnist. **1971** *Daily Tel.* 6 Apr. 1 Most of them had confessed to entering East Pakistan as fifth columnists.

fifty-fifty, *adv.* and *a. colloq.* (orig. *U.S.*). [f. Fifty *sb.* 1.] **A.** *adv.* On a basis of fifty per cent. (or one half) each; half-and-half, equally. **B.** *adj.* Equal, shared equally; half-and-half.
1913 Wodehouse *Little Nugget* vi. 121 Say, Sam, don't be a hawg. Let's go fifty-fifty in dis deal. *Ibid.* xii. 209 Would a fifty-fifty offer tempt you? *Ibid.* xiv. 248 'Fifteen per cent. is our offer,' he said. 'And to think it was once fifty-fifty!' **1916** H. L. Wilson *Somewhere in Red Gap* vi. 263 And she glared at Cousin Egbert with rage and distrust splitting fifty-fifty in her fevered eyes. **1922** E. O'Neill *Anna Christie* I. 19 Good girl?.. Well, yuh treated me square yuhself. So it's fifty-fifty. **1924** *Daily Mail* 28 Nov. 10/5 [He] did not take a fifty-fifty chance that the bedroom door he would enter by was the right one. It was pointed out to him the day before. **1937** *Times* 13 Apr. p. xxix/4 In many craft the power of the engine is such that one could hardly say into which division she fits; there is, as a fact, a type of motor-sailer which is spoken of as the 'fifty-fifty'. **1949** H. Wadman *Life Sentence* 47 It will take much of the sting out of the opposition if the ownership is fifty-fifty.

fig, *sb.*[1] Add: **1. c.** (Later West Indian examples referring to the banana.)
1871 Kingsley *At Last* ii. 49 At St. Thomas's we had been introduced to bananas (figs, as they are miscalled in the West Indies). **1961** F. G. Cassidy *Jamaica Talk* xvi. 351 Bananas named for their..flavour and other qualities..apple or honey (the smallest), fig (very small), plum (small). **1970** *Country Life* 17–24 Dec. 1221/3 In the West Indies a fig is a small banana.
10. fig-blue (examples).
1786 *N.Y. Directory* 26 Coller, Christopher, fig-blue manufacturer. **1861** Mrs. Beeton *Bk. Househ. Managem.* 1010 The linen..should again be rinsed..in abundance of cold water slightly tinged with fig-blue.

fig, *v.*[4] **2.** (Earlier example.)
1825 M. Wilmot *Let.* 26 Sept. (1935) 223, I figg'd up the petticoat into a broad sash for Wm's waist.

Figaro (fiˑgăro). *slang.* [The name of the hero in *Le Barbier de Séville* and *Le Mariage de Figaro* by Beaumarchais (1732–99), and later in operas by various composers.] A barber.
[**1831** *Figaro in London* 10 Dec. 1/1 As the wood engraving, which distinguishes this page, will appear at the head of every separate publication of the Figaro,.. we beg most distinctly to state that..the sketch of the lively *barber* is *not* intended for Sir Edward Sugden.] **1864** Hotten *Slang Dict.* 133 *Figaro*, a barber. **1886** *Globe* 18 Mar. 3/2 (Farmer), There is wailing and weeping among a certain section of the army, the Figaros, which has been despoiled at one fell swoop [*viz.* by an order of the French War Minister permitting soldiers to wear their beards]. **1922** *Contemp. Rev.* Mar. 334 [He] one day asked his Figaro who he thought was the richest man in the town.

fight, *sb.* Add: **2. d.** *fight-off*, a contest to decide a tie in a fencing match.
1930 *Morning Post* 14 July 15 In the fight-off, Armstrong worried at his enemy's arm with an incessant attack. **1961** *Times* 9 June 5/3 He then only tied for top place, and in the fight-off Howard..won 5–0.
e. *fight-back*, a retaliation, rally, or recovery (see also quot. 1961). *colloq.*
1953 *Quick* 9 Mar. 16 Butter producers mapped a 'fight back' against substitutes which have made inroads into the butter market. **1960** J. Fingleton *Four Chukkas to Australia* v. 153 The great fight-back..the Englishmen made. **1961** *New Scientist* 17 Aug. 397/1 'Fight-back',.. referring to the way in which the cheese pushes back against your thumb when the pressure is released.

fight, *v.* Add: **2. a.** Of an animal: to struggle for freedom or mastery. Also *trans.*, to strive with (a horse, etc.) for mastery. *U.S.*
1850 'H. Hieover' *Pract. Horsemanship* 179 If you find he at all fights against you..stand now on no ceremony with him. **1875** [in Dict.]. **1908** C. E. Mulford *Orphan* i. 13 He mounted and fought the animal for a few minutes, just as he always had to fight it. **1920** J. M. Hunter *Trail Drivers of Texas* 231, I 'fought' cattle for nine years almost night and day.
4. b. *to fight fire.* (Cf. *fire *sb.* B. 2.)
[**1824** W. Owen *Diary* (1906) 83 Then they fight it [*sc.* a fire]..endeavouring to overcome it by striking it with clap-boards.] **1835** J. Abbott *New Eng. & Institutions* 21 For days and nights together, all the physical force of the village has been arrayed in 'fighting the fire'. **1860** *Leisure Hour* 1 Nov. 690/2 Fight fire, fight water, fight Farmer Jackson, wagoner and mate. **1865** *Chambers's Jrnl.* 29 July 470/1 They took away from the local firemen their apparatus, and proceeded in their own way to 'fight fire'. **1944** J. S. Huxley *On Living in Rev.* ii. 22 There are today thousands who, though they may sometimes grumble, at heart have enjoyed fighting fires or acting as wardens.

fighter. Add: **3.** *Aeronaut.* A high-speed military aircraft designed for aerial combat.
1917 [see *bomber 2]. **1936** *Economist* 8 Feb. 294/2 There will still be..two squadrons of bombing machines.. to one squadron of fighters. **1960** C. H. Gibbs-Smith *Aeroplane* I. xvi. 139 (*caption*) The turbojet Gloster Javelin fighter. **1971** H. F. King *World's Fighters* xi. 116 Several different armament combinations are possible on the Lightning fighter.
4. *attrib.* and *Comb.*, as *fighter aerodrome, cover, duty, escort, machine, patrol, pilot, plane, screen, squadron, strip, umbrella*. Also **fighter-bomber**, an aircraft that combines the functions of a fighter and a bomber; **Fighter**

Command, the headquarters controlling the operation of a fleet of fighters.
1941 *Hutchinson's Pict. Hist. War* 19 Mar.–13 May 43 He sent large forces to deal with fighter aerodromes in the south and south-east of England. **1936** *Air Stories* Dec. 544/1 The R.A.F.'s latest fighter-bomber is as fast as any fighter yet in service anywhere in the world. **1959** *Observer* 14 June 16/6 American fighter-bombers equipped with nuclear weapons. **1941** *Aeronautics* Dec. 39/1 'Fighter Command' did not pass into Everyman's vocabulary until well after the beginning of the war. **1942** *Jane's All World's Aircraft 1941* 19a/1 Fighter Command's contribution to the bombing offensive opened on November 1, 1941, when the new Hawker 'Hurricane II' fighter-bomber..went into action. **1941** D. Garnett *War in Air* 107 The number of Fighters with the Air Component of the B.E.F. was increased, and Fighter cover was given on the Western flank. **1934** *Times* 28 July 9/5 It is a tribute to the success of the Territorial experiment..that these squadrons should have been chosen for fighter duty. **1939** *War Weekly* 24 Nov. 139/4 In the hours of darkness fighter-planes will rarely, if ever, attack bombers, and, therefore, raids carried out entirely in darkness have no need of fighter escort. **1919** N. Flower *Hist. Great War* XII. 121/2 The German aviation service was in extreme need of fighter machines..and aerial machine-gunners. **1917** *Flying* 3 Oct. 161/2 A fighter patrol met a large formation of Albatros scouts. **1936** *Air Stories* Dec. 573/2 The likelihood of securing a hit is in the proportion of 9 to 1 in favour of the fighter pilot. **1939** *Flight* 19 Oct. 309 Fighter pilots in crews' quarters on an aerodrome. **1971** *Sunday Times* 1 Aug. 3/6 The fighter pilot..parachuted to safety. **1939** *Economist* 4 Feb. 220/1 Unless the raiding enemy can be located he is immune..from the defending fighter planes. **1941** *Battle of Britain, Aug.–Oct. 1940* 21 The covering fighter screen flew at very great height. **1938** *19th Cent.* Feb. 202 [illegible] fighter squadrons. **1954** *Birmingham (Ala.) News* 25 Apr. 41/1 Australian Royal Air Force engineers worked at night under floodlights..to repair the bomber and fighter strips. **1942** *Flight* 27 Aug. 218/2 Everything..depended on the British fighter umbrella.

fighting, *vbl. sb.* Add: **3. a.** *fighting weight.*
1884 *Boy's Own Paper* 1 Feb. 275/1 Twelve stone two was his fighting weight. **1898** F. H. G. Strong *Shake Hands* iv. 43 Willard's height was six feet five, and his fighting weight in the neighbourhood of seventeen stone.
b. fighting chair *U.S.*, a fixed chair on a launch, for use when catching large fish; fighting-chance, an opportunity of succeeding by great effort; fighting drunk, -tight *adjs.*, *colloq.*, drunk to a state of quarrelsomeness; fighting-fit *a.*, fit to fight; fit enough to take part in a fight; hence *fighting-fitness*; fighting fund, a sum of money raised to finance a cause or campaign; fighting mad *a. colloq.* (orig. *U.S.*), furiously angry (cf. Mad *a.* 5); fighting-top *Naut.*, a circular platform placed at an elevation on the mast of a warship, on which guns and armed men can be stationed.
1950 Gabrielson & La Monte *Fisherman's Encycl.* [illegible] or 'fighting chair' as they are sometimes called. **1967** L. James *Chameleon File* (1968) ix. 110 He walked over to a revolving chair bolted to the deck..'This is the throne from which we catch the marlin.' [illegible] **1889** *Kansas Times & Star* 20 Feb., With a somewhat divided party, but having a fighting chance of success. [illegible] the sailor so as to give him a fighting chance for his life in the hospital. **1894** *Congress. Rec.* [illegible] He will have a fighting chance. **1927** [illegible] a fightin' chance. **1968** *Daily Chron.* [illegible] acting like hooligans or who are 'fighting' drunk. **1909** *Westm. Gaz.* 1 Oct. 3/3 Jim's Sarah she come 'ome fighting drunk the other night. **1891** Kipling *Life's Handicap* 313 He did not feel fighting-fit that morning. **1965** *Lancet* [illegible] 'fighting fit' soldiers. **1894** H. Drummond *Ascent of Man* 167 Fitness in the stormy days of the world's animal youth was necessarily fighting-fitness. **1940** *Economist* 9 Mar. 411/2 The additional proposal that each industry should raise a 'fighting fund' to assist its exporters. **1940** N. Marsh *Death at Bar* ii. 31 Another ten bob for the fighting fund. **1946** W. James [illegible] fighting-mad? **1955** A. Grimble *Pattern of Islands* 86 [illegible] fighting-mad. *a* **1889** *Chicago Tribune* (Barrère & Leland), [illegible] fighting tight. **1896** *Naval Annual* i. 32 The foremast has two fighting-tops. The mainmast has only one fighting-top. **1915** *Nature* XCVI. 182/1 On board our battleships a range-finder of this kind is placed in one of the fighting-tops on the masts. **1958** O. Warner *Portrait Ld. Nelson* xii. 352 Fired from above, from a fighting top in the *Redoutable*, it [*sc.* a cannon-ball] had penetrated deep into Nelson's chest.

fighting, *ppl. a.* Add: **1. c.** Of words or speeches. Also *transf. colloq.* (orig. *U.S.*).
1876 'Mark Twain' *Tom Sawyer* i. 9 You're a fighting liar, and darn't take it up. **1917** R. W. Lardner *Gullible's Travels* 209 You know they's lot's o' words that's called fightin' words. [illegible] **1930** *Economist* 23 Aug. 374/2 [illegible] 'fighting' speeches..should possibly be interpreted with..due reference to this fact. **1959** *Listener* 12 Feb. 302/3 Tom Fallon..came out with fighting if rather catchpenny words.
d. Fighting French, a name given to the Free French armed forces during the German occupation of France in the 1939–45 war.
1943 *New Statesman* 20 Nov. 327/1 Between them, the people of the Lebanon and the Fighting French have made an ugly problem for each other and for us. **1957** *Encycl. Brit.* XXIII. [illegible] Outside France, Gen. Charles de Gaulle had started his 'Free French' (later 'Fighting French') movement as early as June 18, 1940.

fighty, *a.* Delete † *Obs.*⁻¹ and add later examples in sense: pugnacious.
1888 W. B. Churchward *'Blackbirding' in South Pacific* vi. 108 They annoy me, and then I get bad and fighty. **1941** J. Cary *House of Children* xxxviii. 168 He's so fighty all at once—he sounds absolutely ferocious.

fig-leaf. **2.** (Examples of use in *sing.*)
c **1856** Emerson *Jrnl.* in *Sel. Writ.* (1965) 168 Whipple said of the author of 'Leaves of Grass' that he had every leaf but the figleaf. **1897** *Daily News* 20 Oct. 5/6 Court and country in Spain would rejoice if they succeeded in getting rid of Cuba in a decent..manner. They seek the fig-leaf. **1960** C. W. Mills *Castro's Cuba* iii. 69 If they had some kind of puppet regime there, they could 'recognize' it and arm it. That would be their fig leaf.

figment. Add: **2. b.** Phr. *a figment of* (*the* or *one's*) *imagination.*
1847 C. Brontë *J. Eyre* II. xxv. 277 The long dishevelled hair, the swelled black face, the exaggerated stature, were figments of imagination. **1955** *Bull. Atomic Sci.* Mar. 79/1 This condition is almost certainly a figment of the imagination. **1971** *Nature* 2 Apr. 299/3 Another attempt..to read into prehistoric monuments..patterns and explanations which are simply figments of the observer's imagination.

‖ **figura** (figiūᵊˑrä). [mod.L., ad. L. *figūra* Figure *sb.*] **1.** *Theol.* = Type *sb.*[1] Cf. Figure *sb.* 12.
1959 Manheim & Garvin tr. *Auerbach's Sc. fr. Drama Europ. Lit.* 34 Moses is no less historical and real because he is an *umbra* or *figura* of Christ.
2. *Literary.* A person who represents some higher or supervening reality.
1959 Manheim & Garvin tr. *Auerbach's Sc. fr. Drama Europ. Lit.* 71 Virgil in the *Divine Comedy* is the historical Virgil himself, but then again he is not; for the historical Virgil is only a *figura* of the fulfilled truth that the poem reveals. **1968** *Eng. Studies* XLIX. 404 The poet.. wants us to connect the Troy legend with the theme of his poem and to see in Aeneas a *figura* of his own hero.
b. An act or deed that is representative or symbolic.
1959 Manheim & Garvin tr. *Auerbach's Sc. fr. Drama Europ. Lit.* 66 Cato's voluntary choice of death rather than political servitude is here introduced as a *figura* for the eternal freedom of the children of God. **1964** *Eng. Studies* XLV. 111 His [*sc.* Orsino's] sudden dismissal of the music..is a *figura* that sets the pattern of his behaviour.

figural, *a.* **1.** Delete † *Obs.* and add later examples. Cf. prec.
1953 W. R. Trask tr. *Auerbach's Mimesis* 195 A figural schema permits both its poles—the figure and its fulfillment—to obtain the characteristics of concrete historical reality, in contradistinction to what obtains with symbolic or allegorical personifications. *Ibid.* 196 Far more prevalent in the Christian life of the High Middle Ages is the figural realism which can be observed in full bloom in sermons, the plastic arts, and mystery plays. **1955** R. Jakobson in Saporta & Bastian *Psycholinguistics* (1961) 424/2 The two varieties of figural speech—metaphor..and metonymy. **1959** *Encounter* Nov. 78/2 A 'figural' interpretation, as if every particular event signified precisely 'something other'... Modern 'realism' lacks this 'figural' quality. **1968** E. Salter in *Proc. Brit. Acad.* LIV. 85 The essential messages of *Piers Plowman* are conveyed at their greatest intensity by figural or typological means, as are those of the Bible.
3. c. (In present technical use.)
1952 D. T. Rice *Eng. Art 871–1100* 83 Sculptures of a monumental figural character are quite different from those of a decorative or ornamental nature. **1958** *Times Lit. Suppl.* 3 Jan. 4/4 Similar not to the Book of Lindisfarne or the Book of Durrow, but to the figural sculpture on the Ruthwell Cross. **1970** M. Swanton *Dream of Rood* 12 The broad principal faces of the shaft are carved with figural subjects surrounded by identifying Latin inscriptions.
4. (Examples.)
1938 *Oxf. Compan. Mus.* 316/2 Often the word 'florid' is a good translation of 'Figural'. **1959** *Listener* 17 Dec. 1093/2 As well as this simple liturgical form (*Choral-passion*, implying unaccompanied monody), composers cultivated the *Figural*-passion.

figuration. Add: **2. b.** The figurative style of painting. Cf. *figurative *a.* 2 b.
1962 *Listener* 19 July 93/2 Some painters who persisted with figuration during the nineteen-fifties when it was least in favour.
6. A member of the Figures form (see *Figure *sb.* 22 b).
1904 *Ushaw Mag.* XIV. 200 Whilst the Grammarians scored 16 the High Figurations scored 7.

figurative, *a.* Add: **2. b.** Of a style of the visual arts; esp. applied to painting in which the forms are recognizably derived from objective sources without necessarily being clearly representational: *figurative painter*: one who paints in this style.
1960 *Guardian* 2 Feb. 7/4 'Figurative' is a comparatively new word in the critical vocabulary of contemporary art. It implies a kind of painting that is not abstract and..not necessarily representational. **1962** *Listener* 19 July 93/2 There is a new interest in figurative painting today. *Ibid.* 94/1 The work of three figurative painters who have recently shown in London: Francis Bacon, Sidney Nolan, and Arthur Boyd. *Ibid.* 94/2 In their use of chance in the act of painting figurative painters today undoubtedly owe much to abstract expressionism.

figure, *sb.* Add: **1. c.** *spec.* *figure of the earth*: see Geoid and quot. 1931.
1795 [see Oblate *a.*]. **1931** *Bull. Nat. Res. Counc.* [illegible] Figure of the Earth.—The defining elements of the mathematical surface which approximates the geoidal surface. The figure of the earth has been proved to be approximately an oblate spheroid.
6. *father-figure*: see *Father *sb.* 12.
15. b. *spec.* in wood (see quots.).
1875 *Trans. N.Z. Inst.* VII. 185 [illegible] cabinet-makers call 'figure'. **1940** P. Macquoid *Hist. Eng. Furnit.* [illegible] **1953** H. L. Edlin *Forester's Handbk.* [illegible]
17. (Earlier example, with jocular allusion to sense 7.)
1854 J. R. Planché *Camp at Olympic* i. 18 Like a bold wench, resolved at any price To cut a figure, though it's figure on ice.
19. In Cricket, also a bowler's average.
1905 *Times* 9 May 15/1 His figures..were rather battered about by Perks, who hit him far and wide. [illegible]
b. *figure-of-eight shield.*
1960 [illegible] **1965** L. Cottrell *Anvil of Civilisation* [illegible]
c. (Earlier and later examples.) Also *figure four trap.*
1743 J. Isham *Observ. Hudson's Bay* (1949) 162 [illegible] a figure of 4 trap. [illegible]
20. b. *to do things on the big figure*, to go (*or come*) *the big figure*: see *Big *a.* B. 2; *to go the whole figure* *U.S.*, to go the whole way; to act in a thoroughgoing fashion.
1834 *Sun* (N.Y.) 25 Mar. 2/3 (*heading*) Going the whole figure. [illegible]
c. *figure of merit*: a general term for a numerical expression taken as representing the performance or efficiency of a device or material (see also quot. 1865). Cf. *Factor *sb.* 8 b).
1865 Brande & Cox *Dict. Sci., Lit., & Art* I. 889/1 *Figure of merit*, in rifle-shooting, the figure adopted for denoting the individual success of any rifle. **1881** H. R. Kempe *Handbk. Electr. Testing* (ed. 2) [illegible] The degree of sensitiveness of any galvanometer, or its 'figure of merit', is determined by the amount of current which will produce one division of deflection. **1911** *Encycl. Brit.* XXVII. 173/2 The hysteresis loss should not exceed 3·0 watts per kilogram of iron measured at a frequency of 50~ and a flux-density of 10,000 lines per square centimetre. This is now called the 'figure of merit' of the iron. [illegible] **1952** Drysdale & Jolley *Electr. Measuring Instrum.* (ed. 2) [illegible] Some 'figure of merit' or means of judgment therefore appears to be necessary. **1958** Condon & Odishaw *Handbk. Physics* iv. 109/1 Frequently

the inverse of the loss tangent, the quality factor Q of the dielectric,..serves as the figure of merit, especially in waveguide problems.
22. b. *pl.* The name of the first form in certain Jesuit schools and colleges, divided into High (or †Great) and Low (or †Little) Figures: corresponding to the Rudiments or Accidence of other places.
1629 Wadsworth *Pilgr.* iii. 12, I was promoted to the first forme called the Figures. **1713** in B. Ward *Hist. St. Edmund's Coll.* (1893) [illegible] **1893** B. Ward *Hist. St. Edmund's Coll.* [illegible] **1913** *Ushaw Mag.* [illegible]
25. a. (sense 10) *figure-artist, -composition, -piece* (earlier example). **b.** *figure-artist, -painter.*
1903 *Burlington Mag.* Sept.–Oct. 3/1 [illegible] **1857** 'C. Bede' *Verdant Green* [illegible] **1935** *Burlington Mag.* Nov. [illegible] **1770** J. Wedgwood *Let.* [illegible] **1816** Jane Austen *Emma* [illegible]
26. figure-maker, (*a*) (example).
1769 J. Wedgwood *Let.* 25 June (1965) 76 If we get these painters, and the figure makers, we shall do pretty well in those branches.

figure, *v.* Add: **12. b.** (*a*) *trans.*: also with *down.* To reckon, calculate, understand, ascertain. Also with obj. clause, and *absol.*, esp. in colloq. phr. *it* (*or that*) *figures*, it is reasonable, likely, or understandable; it makes sense. orig. and chiefly *U.S.*
1865 *Congress. Globe* 9 Feb. 671/3, I have not figured the number of square miles that there will be. [illegible]
(*b*) *to figure on* or *upon* (fig.): to think over, consider; to count on, anticipate, expect. *U.S.*
[illegible]
15. figure out. c. Also, more widely, to estimate or calculate; hence, to work out, make out. Chiefly *U.S.*
[illegible] *Out* 87, I figured it out that he was staring at the top of the old gumtree opposite.

figured, *ppl. a.* **7. b.** *figured bass* (examples).
1801 T. Busby *Dict. Mus.* s.v. *Figured*, A bass, accompanied with numerical characters, denoting the harmony formed by the upper or superior parts of the composition, and directing the chords to be played by the organ, harpsichord, or piano-forte, is called a figured bass. **1948** G. B. Shaw *How to become Mus. Critic* (1960) [illegible]

Fijian (fīdʒiˑăn), *sb.* and *a.* Also 9 Feegeean, Feejeean, Fejean. [f. *Fiji*, native name of the principal island of the Fiji archipelago + -an.] **A.** *sb.* **1.** A native or inhabitant of the Fiji archipelago. **2.** The language of the Fijian people. **B.** *adj.* Of or pertaining to the Fiji archipelago, the Fijians, or their language.
1809 J. Davies *Jrnl. Missionaries* 16 Nov. [illegible]

fike, var. Fyke.
1871 *Game Laws N.Y.* in *Fur, Fin & Feather* (1872) 48 It shall not be lawful for any person to take eels in fikes or pots.

fikiness (fiˑkinės). Chiefly *Sc.* Also **feikieness, fykiness.** [f. Fikie *a.* + -ness.] Restlessness, agitation; the action of taking much trouble.
1889 Barrie *Window in Thrums* xiv, Her feikieness ended in her surrender. **1892** N. Dickson *Auld Sc. Min.* 128 I'm sure ye ken as weel as me that love's just an unco fykiness o' the mind.

filament. Add: **1. b.** Also, a similar conductor in a thermionic valve that serves as a heater or as a directly heated cathode. (In light bulbs and valves the filament is now commonly made of tungsten.) (Additional examples.)
1885 *Proc. R. Soc.* XXXVIII. 219 [illegible]
c. *Zool.* The shaft of a down feather.
1869 [see Barbule 2]. **1959** Van Tyne & Berger *Fund. Ornith.* [illegible] few feather filaments (cryptoptiles) scattered between them.
d. *Textiles.* A thread of man-made fibre.
1927 T. Woodhouse *Artificial Silk* [illegible]
2. b. *Astr.* A narrow thread-like streamer from the sun's chromosphere or in its corona.
a **1869** J. Nasmyth in E. Dunkin *Midnight Sky* (1869) 224 The filaments in question are seen..at the edges of the luminous surface. **1871** *English Mechanic* 24 Nov. 243/1 The chromosphere is surrounded by filaments like brilliant hairs. [illegible]
5. *attrib.* and *Comb.*, as **filament bulb**, **-cathode**, **current**; **filament battery**, a battery that provides the current for the filament of a valve; **filament lamp**, an electric light bulb in which the light is produced by an incandescent filament.
1919 *Radio Rev.* Oct. 47 The filament battery is used to supply the anode circuit voltage, as well as for lighting the filament. [illegible]

fiˑlamentless, *a.* Without a filament.
1934 *Amer. Year Bk. 1933* 617/2 The demonstration of 'filamentless' tubes, not commercially important as yet. [illegible]

filaria (filēᵊˑriă, filāᵊˑriă). Pl. **filariae** (occas. **filarias**). [mod.L. (O. F. Müller 1787, in *Naturforscher* XXII. 64), f. L. *filum* thread.] A parasitic nematode worm belonging to the genus *Filaria* or to genera closely related to it. Also *attrib.*
1834 H. McMurtrie tr. *Cuvier's Animal Kingdom* 473 The Filariæ in their external appearance are very similar to the Gordii. [illegible]

filariasis (filēᵊrī,ăˑsis, filărī,əˑsis). *Path.* Also filariosis. [f. *Filari(a + -iasis.] Disease resulting from infection with filariæ.
1879 *Veterinarian* LII. 401 (*heading*) Facts in Filariasis. [illegible]

filarid (filēᵊrid, filăˑrid), *a.* [f. *Filari(a + -id[3].] Of or pertaining to a family of parasitic nematode worms, the Filariidæ.
1929 H. A. Baylis *Man. Helminthol.* ii. 205 The name *Filaria*, however, continues to be largely used, in a loose sense, for filarid worms. **1946** *Nature* 20 July 98/2 *Litomosoides carinii* is a filarial parasite of the cotton rat, *Sigmodon hispidus*. **1956** T. W. M. Cameron *Parasites & Parasitism* 158 Some of the filarid worms live in similar habitats.

filarioid (filăᵊˑrɔid, filăˑrɔid), *a.* [f. *Filari(a + -oid.] Of or pertaining to a superfamily of parasitic nematode worms, the Filarioidea.

1930 E. C. Faust *Human Helminthol.* xxviii. 438 Filarioid species are classified under two families. 1956 G. Lapage *Vet. Parasitol.* viii. 216 The usual habitat of filarioid roundworms is the body cavity, the lymphatic system or the connective tissue of their hosts.

filasse (fila·s). [Fr., = Tow *sb.*[1] Cf. *fillis.] Vegetable fibre prepared for manufacture.

1858 in Simmonds *Dict. Trade.* 1895 *Daily News* 3 Dec. 2/3 Ramie, whose fibre produces a substance called filasse, similar to flax.

filbert. Add: **1. b.** *slang.* The head (cf. Nut *sb.*[1] 7).

1886 H. Baumann *Londinismen* 54/1 *Cracked in the filbert,*..dotty. 1936 J. Curtis *Gilt Kid* xx. 198 Get that into your old filbert.

file, *sb.*[1] Add: **5. file snake,** a non-poisonous colubrid snake of the genus *Mehelya,* found in South Africa.

[1908 *Ann. Transvaal Museum* I. 23 *Simocephalus capensis* (Smith). Three-cornered Snake. Vijlslang.] 1912 F. W. Fitzsimons *Snakes S. Afr.* iv. 96 Three-cornered or File Snakes..although found in most parts of South Africa, seem to be rather rare everywhere. 1931 *Discovery* Mar. 74/2 A file snake captured a frog. 1962 R. M. Isemonger *Snakes Afr.* 98 The file snake is generally feared by Africans, who associate its entry to their house with the death of a relative or friend.

file, *sb.*[2] Add: **4. b.** *Computers.* A collection of related records stored for use by a computer and able to be processed by it. Also *attrib.* and *Comb.*

1954 *Jrnl. Assoc. Comput. Mach.* I. 8/2 A 'master' tape ..contains the file of unit records as at the last date of processing. 1967 Cox & Grose *Organiz. Bibl. Rec. by Computer* 19 A file sequence may contain only one file, and this is generally the case when updating procedures are being carried out. 1969 *Computers & Humanities* III. 132 This search (once through the file), whether for a single interrogation or for several, is called a file-pass.

7. d. An individual soldier.

1903 *Med. Record* 7 Feb. 227 (Cent. Dict. Suppl.), The poor file who has to carry it, as well as his gun and various other accoutrements. 1916 'Boyd Cable' *Action Front* 105 'Dusty Miller', the next file on his left,..spoke to him.

11. (sense 4) *file card, copy, cover, signal; file-closer* (earlier U.S. example).

1966 *English Studies* XLVII. 200 It is as though a scholar..were to decide to re-arrange a mountainous stack of file-cards. 1836 J. Hildreth *Dragoon Campaigns Rocky Mts.* I. vi. 48 Next in order of inspection came the adjutant, and, commencing at the right of the line, 'told off' the battalion by equal troops, the subalterns taking their stations in the rank of file-closers. 1899 *Daily News* 23 Jan. 4/7 Prudent swains might find a file copy extremely useful in general emergencies. 1909 *Daily Chron.* 2 Sept. 4/7 The file copy of 'The Daily Chronicle' of the same date. 1968 V. C. C. Baddeley *My Foe Outstretch'd* ii. 27, I don't know why Miss Cragg couldn't give you a tape...I suppose..she's only got a file copy. 1925 J. G. Bruce in E. F. Norton *Fight for Everest, 1924* vi. 349 A few file covers are useful to keep the correspondence on various subjects separate. 1961 *Lebende Sprachen* VI. 69/2 Office furniture, machines and supplies. File signal.

file, *v.*[3] Add: **1. b.** (Additional examples.)

1835 T. P. Granger *Tomlins' Law-Dict.* (ed. 4) I. s.v. *Bankrupt* II, *By what Acts..a Trader may become a Bankrupt*...If any..trader shall file..a declaration..that he is insolvent..the..secretary of bankrupts..shall sign a memorandum that such declaration hath been filed. 1957 *Encycl. Brit.* III. 66/2 An act of bankruptcy is committed ..if he [*sc.* the debtor] files in court a declaration of inability to pay his debts. 1966 *Times* 12 Aug. 14/4 About 80 of the world's larger foreign companies whose shares are traded in the United States have filed public financial information about their operations.

c. *N. Amer.* To file a claim *on* or *upon,* assert a title to, apply for (a piece of land or a mining claim); also *absol.*

1871 *Scribner's Monthly* II. 254 The half-breed who had 'filed on' the claim alongside Lindley's. 1879 [see File *v.*[3] 4 c]. 1893 *Congress. Rec.* 11 Feb. 1470/1 Many persons filed upon these lands. 1910 W. A. Fraser *Red Meekins* (1921) 194 First thing in the mornin' we'll hike to the outside an' file the claim. 1911 J. F. Wilson *Land Claimers* 2 The relinquishment has been made in Portland where you filed. 1932 T. G. Springer *Sagebrush Buckaroo* (1933) xxii. 252 You thought to stake out all the locations, file on them with dummies, and..get the cream! 1948 *Sat. Even. Post* 4 Dec. 70/2 He..had assays made and was on the way to file his claim, but I got him drunk and filed myself. 1968 R. M. Patterson *Finlay's River* 29 F. H. Davis was the first man alert enough to stake the fraction and get in first to the mining recorder's office to file.

d. Of a newspaper reporter: to transmit (a story, information, etc.) to his newspaper.

1954 D. Dodge *Lights of Skaro* ii. 57 A reporter could file one, and only one, story the Party didn't like before they cancelled his visa. 1964 *New Statesman* 13 Mar. 387/2 They agreed not to file their stories for that morning's British papers.

file-fish. b. (Earlier and later examples.)

1815 *Trans. Lit. & Philos. Soc. N.Y.* I. 467 Tut-mouthed file-fish. *Balistes broccus.* 1962 K. F. Lagler et al. *Ichthyology* vi. 192 Both the dorsal and the anal fins are moderately long-based in the triggerfishes and filefishes (Balistidae). 1969 A. Wheeler *Fishes Brit. Isles & N.-W. Europe* 565 Trigger-fish (File-fish).

filet (fi·lėt). Also **fillet.** [a. F. *filet* thread.] **1.** A kind of net or lace having a square mesh. Also *attrib.,* as *filet lace, net, veiling.*

1881 C. C. Harrison *Woman's Handiwork* I. 82 Drawn-work, with darned filet and cut-work. 1904 Goodchild & Tweney *Technol. & Sci. Dict.* 221/1 *Fillet ground,* a net with absolutely square holes, similar to canvas, but more defined and 'lacey' in appearance. 1907 *Westm. Gaz.* 9 Mar. 13/2 The lace might be of cream silk filet. 1907 *Daily Chron.* 25 June 8/3 The new filet designs. 1907 *Westm. Gaz.* 24 Aug. 13/1 A nice tea-gown ..with filet laces. 1908 *Ibid.* 8 Aug. 13/2 The square spotted filet veiling. *Ibid.* 28 Dec. 5/1 A brown filet net trimmed with small rosettes of velvet. 1931 *Evening Standard* 29 Jan. 5/3 Fadeless artificial silk filet net. 1953 M. Powys *Lace & Lace-Making* iv. 33 Lacis or filet, 16th century. A rare pointed pattern made for collars or to border a cover. The groundwork is the knotted filet mesh made like a fish net and with the same implements.

2. *Cookery.* = Fillet *sb.* 6 a, c. Also in phr. *filet de bœuf.*

1841 [see Fillet *sb.* 6 a]. 1861 Mrs. Beeton *Bk. Househ. Managem.* 918 Entrees, Filet de Bœuf and Spanish Sauce. 1864 G. Meredith *Let.* 24 Feb. (1970) I. 245 Smooth flow the Sauces!—may the filets tender be! 1951 E. David *French Country Cooking* 114 Your beef.. should be a fine piece of *filet.* 1962 *Harper's Bazaar* Aug. 69/1 Tournedos steaks cut from the tail end of a *filet de bœuf.*

b. *filet mignon*: a slice cut from the small end of the tenderloin of beef.

1906 'O. Henry' *Four Million* (1916) 67 'We'll have oysters to-night.' 'And filet mignon with champignons.' 1960 L. Cooper *Accomplices* III. iv. 176 A delightful meal—caviare, an iced consommé, a filet mignon. 1961 J. Heller *Catch-22* (1962) xxii. 233 Yossarian and Orr..went..to eat shrimp cocktails and filet mignon in a very fine restaurant.

‖ **fili** (fi·li). Also **fileadh, filid(h.** [Ir. *fili, fileadha, filidh,* pl. forms of *file* poet.] The name given in Irish to an ancient order of poets.

1876 *Encycl. Brit.* V. 303/1 The *Fili* truly represent the Οὐάτεις, or vates, who formed one of the orders of Druids, mentioned by Strabo. 1880 *Ibid.* XIII. 250/2 The *filid,* whom we shall conventionally call bards..were part of the transformed Druidic order...An ollam fili, the highest grade of the order, was entitled to a large retinue of pupils. 1898 J. Heron *Celtic Ch.* 30 Entrance into the ranks of the Ollamhs and the Fileadh (or poets). 1933 *Times Lit. Suppl.* 26 Jan. 53/1 The *filid,* the poet-seers who have descendants in the land. 1957 *Ibid.* 25 Oct. 636/2 The bards were reciters, employed by the *fili,* or professional poets.

filial, *a.* Add: **2. a.** *filial generation* Biol., the offspring of a cross, the *first filial* (or F_1) *generation* being the immediate offspring of the organisms selected for crossing, the *second filial* (or F_2) *generation* being produced usually by self-fertilisation or intercrossing of F_1 individuals, and so on.

1902 [see *F. III. 1 i]. 1909 W. Bateson *Mendel's Princ. Heredity* i. 8 Mendel took a pair of varieties of which one was tall..and the other was dwarf...These two were then crossed together...In our modern terminology such a cross-bred, the first filial generation, is called F_1. 1938 E. B. Ford *Study of Heredity* ii. 30 If we cross a black bird with a white, all the offspring are grey...These are the 'first filial' generation of our cross. *Ibid.,* If we mate the grey F_1 birds with one another, we obtain a 'second filial generation' (F_2). 1969 *Dict. Gardening Suppl.* (R. Hort. Soc.) 454/2 The plants obtained in this first Hybrid Filial (F_1) generation are all tall like one of the parents.

filibuster, *sb.* Add: **2. c.** (Example of spelling *filibustier*; here a gallicism.)

1896 E. Dowson *Let.* 5 July (1967) 372 Yet they have always their Austin, & his praise of *filibustiers.*

4. An act of obstruction in a legislative assembly. Chiefly *U.S.*

1890 *Congress. Rec.* 11 Feb. 1217/2 A filibuster was indulged in which lasted..for nine continuous calendar days. 1915 *Morn. Post* 13 Feb. 8/3 It has been decided.. to suspend the filibuster in order to attend to important appropriations. 1917 *Daily Chron.* 5 Mar. 1/7 The bill.. was talked to death. Its last hours were spent in a filibuster against Senator Lafollette. 1923 E. F. Wyatt *Invis. Gods* III. iii. 121 There was a three hours' filibuster against the civil service clause. 1965 *Daily Tel.* 5 Apr. 23/3 Need we tolerate the use of 'filibuster' as meaning 'obstruction' instead of 'obstructionist'?

filibuster, *v.* Add: **1. b.** (Additional example.)

1955 C. E. Carrington *Rudyard Kipling* ix. 228 Dr. Jameson's attempt to filibuster the Transvaal Republic into the British Empire was defeated.

2. (Earlier example.)

1853 *Congress. Globe* 4 Jan. 194/1, I saw my friend.. filibustering, as I thought, against the United States.

filicetum: see *filix.

filicic (fili·sik), *a. Chem.* Also **filixic.** [ad. G. *filixsäure* filicic acid (E. Luck 1851, in *Jahrb. f. prakt. Pharm.* XXII. 130), f. L. *filic-, filix* fern + -ic.] *filicic acid,* a mixture of phloroglucinol derivatives obtained from the rhizome of various ferns (chiefly the common male fern, *Dryopteris filix-mas*), and one of the anthelmintic agents in the drug Male Fern (*filix mas*) (the name was orig. given to the crude extract of the fern, now called *filicin*). Similarly **fi·licin** [a. G. *filicin* (E. Poulsson 1891, in *Arch. f. exper. Path. & Pharm.* XXIX. 9)], (*a*) the crude extract of anthelmintic principles obtained from male fern; (*b*) the therapeutically inactive lactone of filicic acid; also loosely applied to the acid itself.

1865 Brande & Cox *Dict. Sci., Lit. & Art* (ed. 4) I. 890/1 *Filicic acid,* an acid obtained from fern root. 1874 Flückiger & Hanbury *Pharmacographia* 669 The medicinal ethereal extract, of which the rhizome yields about 8 per cent., deposits a colourless, granular, crystalline substance, noticed by Peschier as early as 1826, and subsequently designated by Luck, Filicic Acid. 1892 *Jrnl. Chem. Soc.* LXII. I. 380 The poisonous properties of fern extract are due to an amorphous substance which is the true filicic acid. The acid is readily converted into its lactone, by simply boiling its ethereal solution. The lactone is named filicin. 1941 *Thorpe's Dict. Appl. Chem.* (ed. 4) V. 181/2 Filicin (filicic acid)..consists of three.. ketone nuclei. 1958 *Progress in Org. Chem.* IV. 120 The ultra-violet absorption spectrum of filixic acid indicates a structure in which the three phenolic nuclei are separated by a methylene group. 1959 *Chem. Abstr.* LIII. 14324 Filicin prepd. with $Ba(OH)_2$ showed lower toxicity for warm-blooded animals and higher toxicity for cold-blooded animals. 1963 *Brit. Pharm. Codex* 449 Filicin is the name given to the ether-soluble product obtained in the assay of the drug [*sc.* Male Fern] and the extract. 1963 *Acta Chem. Scand.* XVII. 191 Naturally occurring filixic acid, the characteristic phloroglucinol derivative from *Dryopteris filix mas,* has been shown to be a mixture of six homologues, which differ from each other in regard to the acyl side chains of the filicinic acid nuclei. 1968 J. H. Burn *Lect. Notes Pharmacol.* (ed. 9) 123 Castor oil..aids the absorption of the filicic acid in the extract of male fern.

filigree, *sb.* Add: **2. filigree paper, paper work:** see quot. 1960.

1796 M. Edgeworth *Parent's Assist.* (ed. 2) II. II. 7 The shop where the filigree paper was to be bought. 1960 H. Hayward *Antique Coll.* 117/2 *Filigree paper work.* Throughout the 18th cent. filigree was a popular amateur ornament for small cabinets...This was achieved solely with tiny rolls of paper.

filing, *vbl. sb.*[3] Add: **c.** *attrib.,* as *filing cabinet, clerk, system.*

1907 G. B. Shaw *J. Bull's Other Island* I. 3 (*stage direction*) A filing cabinet, with a cupboard on it. 1911 H. S. Harrison *Queed* xxv. 329 West went to a filing cabinet in the corner of the room. 1930 *Daily Express* 6 Nov. 3/2 There are scores of new type-writers..and filing cabinets, telephone instruments and..telephone boxes for the use of the ninety delegates and their staffs. 1964 *English Studies* XLV (Suppl.). 26, I have selected a number of details that might..have remained in my filing cabinet. 1922 S. Lewis *Babbitt* ii. 16 Verona had for six months been filing-clerk at the Gruensberg Leather Company offices. 1956 A. H. Compton *Atomic Quest* 338 From filing clerks to administrators. 1926 *Daily Colonist* (Victoria, B.C.) 10 Jan. 6/3 A central filing system is being contemplated. 1936 *Discovery* Nov. 346/2 A New Filing System. 1958 Filing system [see *brain-child* s.v. *brain *sb.* 6].

filio-pietistic (fi:li,opəi,ėti·stik), *a.* [f. L. *filius* son + Pietistic.] Marked by excess of filial piety. (*contemptuous.*)

1893 C. F. Adams *Massachusetts* 49 The historians of the Massachusetts filio-pietistic school. 1897 *Q. Rev.* Apr. 532 The 'filio-pietistic' school [of American historians]. 1902 *Daily Chron.* 19 June 4/1 The 'filio-pietistic' spirit of the New England writers in American history. 1962 *Times Lit. Suppl.* 6 July 485/3 The age of Jared Sparks with his highly selective filiopietistic approach to biography and history.

Filipinize (fi·lipinəiz), *v.* Also **-ise.** [f. *Filipino + -ize.] *trans.* To convert to operation by Filipinos. Hence **Filipiniza·tion.**

1923 *Glasgow Herald* 28 Apr. 8 The administrative services have been almost completely Filipinised. 1924 *Life & Work* Feb. 30/1 A Dual Government for eight years marked by a sweeping Filipinisation of the services. *Ibid.* 31/1 A Reserve Fund of 41 million dollars was handed over by the Americans to the Filipinised Government.

Filipino (filipī·no), *sb.* and *a.* Also **Filipina** (-ī·nă) fem. [Sp., f. (*las Islas*) *Filipinas* the Philippine islands.] **A.** *sb.* A native or inhabitant of the Philippine islands, especially one of Spanish or mixed blood. **B.** *adj.* Of or pertaining to Filipinos or the Philippine islands.

1898 *Daily News* 16 Dec. 4/6 Though there may be no guarantee of American citizenship for the Filippinos, the islands will become a part of the Union. 1899 R. R. Lala *Philippine Islands* 86 (*caption*) A high-born Filipina. 1900 H. W. Lawton in Forbes *Philippine Islands* (1945) III. 61 If I am shot by a Filipino bullet, it might as well come from one of my own men. 1902 *Encycl. Brit.* XXXI. 668/2 Filipinos had for generations been ordained in the priesthood. *Ibid.,* Ever ready to join issue against any Filipino movement. 1909 'O. Henry' *Roads of Destiny* xxi. 352 A Southerner is as good as a Filipino any day. 1918 E. S. Farrow *Dict. Mil. Terms, Filipino ration,* the ration for use of Filipino scouts, consisting of 70

per cent of fresh beef or canned meat, 20 per cent of bacon, and 10 per cent of fish. 1936 G. A. Malcolm *Commonwealth of Philippines* ii. 36 The fair Filipina of today dresses smartly, whether in dainty native costume or the latest Paris mode. 1944 J. S. Huxley *On Living in Revol.* i. 13 The United States..in its encouragement of the Filipino's development towards independence. *Ibid.* v. 66 Certain Filipino tribes. 1959 *Chambers's Encycl.* X. 657/2 Filipinos, ethnically related to Malays.. but christianized by long association with Spaniards, form 92% of the inhabitants.

filix (fəi·liks, fi·liks). *Bot.* Usu. in pl. **filices.** [a. L. *filix* fern.] The group of plants in the division Pteridophyta that includes the ferns. Hence **filice·tum,** a collection of living ferns.

1731 P. Miller *Gardeners Dict., Filix*: Fern. There are great Varieties of this Plant in the different Parts of the World. 1789 G. White *Selborne* I. v. 11 These.. scenes..delight the naturalist..with their curious *filices* with which they abound. 1832 J. Lindley *Introd. Bot.* iii. 195 Filices, or ferns, are plants consisting of a number of leaves, or fronds, as they are called, attached to a stem. 1856 *Athenæum* 19 Jan. 73/1 Every garden its Filicetum for the cultivation of those species which will bear exposure in our climate. 1869 S. Hibberd *Fern Garden* i. 1 The ferns or *filices* are the most noble of all. 1891 G. Massee *Evol. Plant Life* vii. 213 Filices (*Ferns*).. includes some of the largest of Cryptogams living at the present day. 1965 Bell & Coombe tr. *Strasburger's Textbk. Bot.* III. 572 Some of the Filices have distinct long shoots.

filixic, var. *filicic *a.*

filix mas (fəi·liks mæs, fi·liks mæs). [f. the Latin name of the male fern, *Dryopteris filix-mas.*] An anthelmintic drug prepared from the rhizome of *Dryopteris filix-mas,* the male fern.

1789 W. Cullen *Treat. Materia Medica* II. 41 Filex [*sic*] mas..has long been celebrated as an anthelmintic. 1880 Bentley & Trimen *Medicinal Plants* IV. 300 Filix Mas; the dried rhizome with the bases of the foot-stalks and portions of the root fibres. 1907 J. H. Parsons *Dis. Eye* xvii. 384 Tobacco, alcohol, quinine, filix mas, carbon disulphide, and other poisons sometimes produce defective vision. 1940 J. H. Gaddum *Pharmacol.* xx. 320 The tapeworms in his [*sc.* man's] intestine are removed by Filix mas, pelleterine, and cusso.

fill, *sb.*[1] Add: **2. b.** (Earlier U.S. examples.)

1850 *Congress. Globe* Apr., App. 531/1 It was like.. making deep cuts and large fills with a view to construct a railroad. 1873 J. H. Beadle *Undevel. West* xxiv. 507 There is not..a difficult 'cut' or 'fill' for over two hundred miles.

c. In Poker: the act of filling one's hand. Cf. Fill *v.* 1 c.

1866 *Wilkes' Spirit of Times* 10 Mar. 28/3 In a game of draw-poker on the draw or fill, a party..discovers that he has only four cards. 1887 S. Cumberland *Queen's Highway* 276 If I drew for a 'fill' I 'filled', it is true, only to find that some one at the table had drawn a 'full hand' of a higher denomination.

d. Auxiliary or secondary material inserted to ensure continuity in music, literature, etc.

1934 S. R. Nelson *All about Jazz* ii. 46 The pianist used to fill up, with 'breaks'...In fact, the excellence of the player was determined by the nature of these 'fills'. 1962 C. Watson *Hopjoy was Here* xvi. 176 He knew.. what would tickle a sub-editor's fancy and help meet the insatiable demand for short 'fills'.

e. *Archæol.* The body of material found in a pit, ditch, etc., in excavation, of a later period than the digging of the feature itself.

1952 *Antiquity* XXVI. 118 The levels beneath it and associated with it are pre-pottery, while two deep later fills against its face are still pre-pottery. 1955 *Times* 24 Aug. 9/6 In the fills behind each of the walls were found rich deposits of pottery and votive offerings.

II. With adverbs forming combs. expressing the action of the corresponding verbal combinations (Fill *v.* V).

5. fill-in. a. Something put in as a substitute or to fill a vacancy.

1918 in C. W. Cunnington *Eng. Women's Clothing* (1952) iv. 141 Fill-in or vest or waistcoat is an absolute necessity. 1928 *Sunday Express* 15 Apr. 11/5, I can hardly say my interest was aroused in the 'new art', except as a fill-in for the stage. 1936 N. Coward *To-night at 8.30* I. 101 This date's only a fill-in for us. 1944 R. Lehmann *Ballad & Source* 141 'Well no, not really,' I murmured, realising that she had forgotten she was talking to me, but feeling that some non-committal fill-in was required. 1958 *Listener* 6 Nov. 722/2 At twelve there was a fill-in term at a rather arty school. 1968 [see *Bircher].

b. A briefing.

1946 *Sat. Rev. Lit.* 20 July 24/3 George Holmes..had given the President a fill-in on Stimson's literary background. 1953 Pohl & Kornbluth *Space Merchants* (1955) ii. 18, I gave him a fill-in on what Schocken Associates was up to.

c. = sense *2 d. Also *attrib.*

1952 B. Ulanov *Hist. Jazz in Amer.* (1958) iv. 29 You may get two-bar phrases with two-bar fill-ins. 1955 L. Feather *Encycl. Jazz* ii. 55 The French horns join with the xylophone to provide a 'fill-in' phrase before the three themes resume in Bar 5. 1965 *Crescendo* Oct. 10/1 Hasaan's piano runs at times are extremely dazzling...Roach's explosive fill-ins seem to spark him along to experiment further.

d. *Photogr.* (See quot. 1955.)

1955 M. Reifer *Dict. New Words* 81/2 *Fill-in,* a supplementary light to illuminate shadows. 1957 T. L. J. Bentley *Man. Miniat. Camera* (ed. 5) v. 78 A smaller bulb having one half this guide number will be about right for the fill-in flash. 1958 *Amat. Photographer* 31 Dec. 897/2 The strength of the main light in relation to the strength of the fill-in light is also worthy of much consideration.

6. fill-out = Fill *sb.*[1] 1 a.

1838 Dickens *O. Twist* II. xxv. 87 Produce the sustainance, and let's have a quiet fill-out for the first time these three days!

fill, *sb.*[2] **1.** (Later U.S. examples.)

1845 D. Webster *Priv. Corr.* (1856) II. 202 He could not follow them any more than a dray-horse can jump out of the fills. 1904 F. Crissey *Tattlings Retired Politician* 247 Father's old clay-bank mare..[had never] laid down in his stall, in the fills, or in double harness.

fill, *v.* Add: **1. f.** In Poker: to complete (a 'full house', flush, straight, etc.) by drawing the necessary cards; also, to improve (one's hand) by drawing complementary cards; *intr.* or *absol.,* to make a flush, etc.; also, (of the flush, etc.) to become complete.

1865 'Mark Twain' *Sk. New & Old* (1875) 74 His last acts was to go his pile on 'kings-*and*' (calklatin' to fill, but which he didn't fill,) when there was a 'flush' out agin him. 1882 *Poker* 31 Scott drew to 'fill a straight and a flush both'. 1885 [in Dict., sense 1 c]. 1887 [see *fill *sb.*[1] 2 c]. 1889 R. Guerndale *Poker Bk.* 25 To fill your hand, to improve it by the draw. 1895 'Templar' *Poker Manual* 43 Sometimes the ante-man or the straddler will come in, if there has been no raise, on three to a straight or flush, drawing two cards. Such hands rarely fill. 1901 D. Curtis *Sci. Draw Poker* 56 If the three cards held be the Queen, Jack and nine it is evident that either the ten and eight, or the King and ten, would fill. *Ibid.* 78 Theoretically, the Flush should be filled oftener than the Straight. *Ibid.* 79 If B then fills he is getting 11 to 1. 1913 'A. B. Lougher' *Poker* 13 The next process is that of drawing to fill the hands. 1928 *Amer. Mercury* Oct. 136/2 I'd made maybe a straight flush, [h]a[ve] filled somehow anyway, and cleaned him. 1957 *Encycl. Brit.* XVIII. 129/1 The odds are..39 to 8 against 'filling a bob-tail straight'.

3. c. *intr.* Of the eyes: to fill with tears.

1871 *Two Little Bruces* ix, 'I'm..tired..,' said Clemmie, with filling eyes. 1905 *Westm. Gaz.* 14 Jan. 2/3 The blue-grey eyes filled as the girl got quietly into bed.

7. c. *to fill the bill*: see *bill *sb.*[3] 8 d.

12. a. (Later U.S. example.)

1836 *Congress. Globe* Jan., App. 50/2 From age to age, they [*sc.* the Negroes] have filled this saying ['Cursed be Canaan'].

c. (Earlier and later examples.)

1860 *Richmond Enquirer* 2 Nov. 1/7 The Executive of the State is making the most strenuous efforts to fill the orders for arms that come to him from all parts of the State. 1926 D. L. Colvin *Prohibition in U.S.* 171 During the campaign he filled one hundred and twenty-five speaking engagements. *a* 1940 F. Scott Fitzgerald *Last Tycoon* (1947) v. 104 If you were in a drug store..having a prescription filled. 1968 *Globe & Mail* (Toronto) 17 Feb. 12 (Advt.), Phone orders filled for city and suburban delivery only.

14. b. *Dentistry.* = Stop *v.* 4 d.

1848 *Lit. Amer.* 29 July 64/2 Teeth filled with gold (if preferred), extracted and cleaned. 1859 J. Tomes *Syst. Dental Surg.* 334 In the treatment of caries, filling must ever be regarded as the great remedy. *Ibid.,* There is perhaps no other operation..which is attended with the same unqualified success as that of filling teeth. 1945 L. Baker *Party Line* 92 Dick not only filled and pulled teeth, but he went modern and practiced orthodontia. 1963 C. R. Cowell et al. *Inlays, Crowns & Bridges* iii. 17 Any caries remaining is excavated and these localized excavations are filled with cement.

c. To put or throw *into* (a receptacle) by way of filling it.

1884 *Encycl. Brit.* XVII. 742 Measured quantities..of [oil-seed] meal are filled into woollen bags. 1906 *Springfield Weekly Republ.* 12 Jan. 13 The New Englander curses gold mining. Billions of good, hard New England cash have been filled into those little black holes.

15. fill in. b. (Further examples.)

1962 *Rep. Comm. Broadc.* 224 in *Parl. Papers 1961–2* (Cmnd. 1753) IX. 259 The material to fill-in the programmes when local items were not being broadcast would..be obtained by switching in to one or other of the three national sound services. *Ibid.* 231 The form of organisation envisaged by the BBC—that is, of local stations providing their own programmes, and filling in with centrally provided material.

d. *to fill in* (*the*) *time*: to occupy oneself during a period of inaction. Also *transf.*

1905 W. Bodie *Bodie Bk.* 135 It is simply to fill in the time while I am deciding how to act. 1939 *War Illustr.* 201 Nurses usefully filling in time while they are off duty. 1966 *Listener* 15 Dec. 898/2 The film sags, filling in time tediously.

e. *to fill* (someone) *in on*: to make (a person) conversant with. Also *intr.* and without *on.* orig. and chiefly *U.S.*

1945 E. Newhouse *Many are Called* (1952) 192 Can you fill me in on them? 1951 *Time* 19 Mar. 46/3 He felt chipper enough to spend an hour..with General MacArthur 'filling in' on U.S. affairs. 1951 L. Hobson *Celebrity* (1953) vi. 78 He had called Roy Tribble and asked to be filled in on Jim Hathaway. 1955 *Bull. Atomic Sci.* Sept. 248/3 It would be incorrect and unfair to say that the AEC has altogether failed to 'fill in' reporters who have called upon its information division for guidance. 1960 'R. East' *Kingston Black* xix. 178, I filled in for Kitty. 1962 H. Burnett *Nothing Sacred* (*caption*), You've been candid about my faults, so I'd be glad to fill you in about your own!

f. (See quot. 1948.) *slang.*

1948 W. Granville in Partridge *Dict. Forces' Slang* 69 *Fill in, to,* to give someone a good hiding...(Lower-deck.) 1959 *Times* 3 Mar. 3/4 A naval rating accused of murdering..an antique dealer..was alleged to have..said: 'I filled in a chap and took his money.'

16. fill out. b. (Additional example.)

1937 V. Woolf *Years* 334 How nice to see you—you've filled out.

d. = *fill in* (see 15 b), *fill up.* (Later examples.)

1903 W. E. Curtis *True Abraham Lincoln* 59 Mr. Lanman forwarded to him a blank to be filled out with facts and dates. 1941 *Amer. Speech* XVI. 310/1 The borrower fills out a call card. 1963 H. Garner in R. Weaver *Canad. Short Stories* (1968) 2nd Ser. 44, I filled out a 'first off' tag and attached it to the inspected part. 1967 *Boston Sunday Globe* 23 Apr. (*Parade*) 27/4 (Advt.), So fill out the coupon and send it to us. We'll send you all the details about Canada '67.

filled, *ppl. a.* Add: **1.** (U.S. example in sense: stuffed.)

1843 'R. Carlton' *New Purchase* viii. 51 Why should we trespass on patience with the account of..steaks, filled chickens, plum puddings, and the curious dish of what-nots?

b. (Additional examples.)

1934 Webster, Filled milk. 1959 *Observer* 8 Feb. 4/6 'Filled milk'—that is, milk in which the natural animal fats are replaced by special vegetable fats.

2. (Additional examples.)

1899 *T. Eaton & Co. Catal.* Spring & Summer 218 Fine English China tea sets..with filled-in colors. 1905 *Westm. Gaz.* 15 July 12/2 That..filled-in look [of a bodice]. 1951 Auden *Nones* (1952) 42 Halting at each of the now filled-in shafts. 1968 H. Harmar *Chihuahua Guide* 237 *Filled-up face,* one in which the cheek muscles are well developed, such as in the Staffordshire.

‖ **fille de chambre** (fiy də ʃānbr). [Fr.] A chambermaid; a lady's personal maid.

1675 H. Woolley *Gentlew. Comp.* 79 Their *Filles de Chambre*..attending their beauties. 1768 Sterne *Sent. Journ.* I. 141 Madame de L***..rang up her *fille de chambre* to ask about it. 1809 [see *agacerie]. *a* 1846 B. R. Haydon *Autobiogr.* (1927) III. xiii. 231 The *fille de chambre,* Rose Armande, was a very interesting girl.

‖ **fille de joie** (fiy də ʒwa). [Fr.] A prostitute.

1705 [see *entre nous]. 1755 C. Charke *Life* 135 At this Time, my Lord kept in the House with him a *Fille de Joye.* 1867 'Ouida' *Under Two Flags* II. ii. 31 The grisette charms of little tobacconists, milliners,..bonbon-sellers and filles de joie. 1901 A. E. Taylor *Probl. Conduct* iv. 212 The mere purposelessness and joylessness of the routine life led by the professional 'fille de joie'. 1964 C. Mackenzie *My Life & Times* III. iii. 92 That production in which the four *filles de joie* were played by Jean MacKinlay, [etc.]. 1970 *Daily Tel.* 8 Oct. 16/2 A string of odious street pick-ups when she embarks finally and inevitably on the career of a fille de joie.

filler[1]. Add: **1. a.** In *Mining*: see quots.

1883 Gresley *Gloss. Coal-m., Filler,* one who fills at a working place or in a stall. 1921 *Dict. Occup. Terms* (1927) § 042 *Filler*.., shovels coal into trams or tubs or upon conveyor at coal face. 1928 *Observer* 12 Feb. 21/4 The full effect of the reduction will be felt only by the hewers, fillers, and a few others. 1955 *Times* 3 May 10/1 The unofficial miners' strike in the Doncaster area arises out of dissatisfaction with the pay of fillers (the men who load coal on the conveyor belts).

c. A filling machine or apparatus. Also *attrib.,* as *filler cap,* the cap closing the pipe leading to the petrol tank of a motor vehicle; *filler hose,* the hose of a petrol pump at a garage.

1895 *Army & Navy Co-op. Soc. Price List* 15 Sept. 252/2 Boiler Fillers, galvanized, 2 gall. *Ibid.* 309 Oil Fillers, 1 pint. 1901 *Daily Chron.* 10 Sept. 9/1 Mineral Waters.—Experienced respectable young man, accustomed Riley's filler. 1927 Filler-cap [see *air-lock b]. 1962 *Which? Car Suppl.* Oct. 139/2 Petrol filler hoses had damaged the paint finish..and the paintwork was discoloured next to the filler cap.

2. (Additional examples.)

1874 *Congress. Rec.* 10 Jan. 558/1 They will not give beyond so much for a certain class of fillers and a certain class of wrapper. 1895 *Street Railway Jrnl.* Apr. 252 'Filler' wire strand, consisting of 19 wires of the same size with the small wires inserted so as to make the outer wires of the strand tangent to the circumscribed circle. 1904 Goodchild & Tweney *Technol. & Sci. Dict.* 221/1 *Filling up,* the process of bringing a surface to a level before painting or between the application of the coats... A good filler for a plastered surface is made by mixing fine plaster of Paris, whiting, and warm size...A paste filler..is used to fill the grain of the wood. 1904 *Electr. Rev.* 11 Sept. 450 (Cent. Dict. Suppl.), The floor framing consists of six sills, the four centre ones being six-inch 'I' beams with wood fillers extending the entire length of car. 1913 *Writer's Mag.* Dec. 247/1 Fillers of a few hundred words in this field..are desirable in this department. 1916 J. W. Toumey *Seeding & Planting* 70 The original cost of the seeding or planting can be materially reduced by the use of so-called fillers, i.e. inexpensive species which serve to occupy a portion of the area, but which are removed in the early thinnings. 1930 *Engineering* 25 Apr. 545/3 The effects of trass as a pozzuolanic material and as a merely inert filler. 1935 *Economist* 7 Dec. 1142/2 Plastics..may be divided into three main groups: Phenolic and fillers, [etc.]. 1941 E. Culbertson *Contract Bridge Complete* (ed. 3) iii. 48 Minor intermediate

values, or 'fillers'. These fillers consist of isolated jacks, tens and even nines. 1942 H. A. Maddox *Dict. Stationery* (ed. 2) 40 *Filler,* (also known as stuffer), a small inset which is used nowadays as a popular advertising device to enclose in letters or with other literature. A cheap and effective medium of publicity with a saving of postal cost. 1946 *N. & Q.* CXCI. 238/2 While he was editor..there appeared a series of brief papers (obviously fillers) called 'Omniana'. 1951 R. Mayer *Artist's Handbk.* ii. 89 The inert fillers or extenders are..white or nearly white pigments which have low refractive indices and, therefore, when ground in oil in the manner of the usual artist's colour, have little or no opacity or tinctorial effect. 1954 *Paper Terminol.* (*Spalding & Hodge*) 36 *Loading,* mineral material added to paper pulp..imparting solidity and a good printing surface to the paper...Also called Filler. 1958 *Listener* 18 Dec. 1055/2 Primer, filler, undercoat, and finish. 1959 F. W. Householder in Saporta & Bastian *Psycholinguistics* (1961) 25/1 The 'filler' part [of a tagmeme] is either another formula..or a list. 1961 *John o' London's* 13 Apr. 417/3 The American is given the job as a filler while he waits for the post of judge to fall vacant. 1964 M. A. K. Halliday et al. *Linguistic Sci.* 193 Exclamations and 'fillers' such as *oh, ah, quoi!*

‖ **filler**[3] (filē·əɪ). Also **fillér.** [Hungarian *fillér.*] A Hungarian coin, the hundredth part of a forint.

1904 *Statesman's Year-Bk.* 416 Nickel:—The twenty-heller (20-fillér) piece. 1927 *Observer* 9 Oct. 11/3 Buyers of the fifty-filler 'Trianon' brooch. 1928 *Ibid.* 15 Apr. 12 Nádosy earned four *filler* a day. 1947 *Statesman's Year-Bk.* 1001 A decree of 26 July, 1946, issued by the Hungarian Government..instituted a new monetary unit, the 'forint', subdivided into 100 'fillér'.

fillet, *sb.*[1] Add: **11. f.** *Aeronaut.* (See quot. 1950.)

1935 *Jrnl. R. Aeronaut. Soc.* XXXIX. 143 Graphical curves of lift, drag and moment as a function of incidence show comparisons of a low wing and a dropped wing without fillet, with three unsuccessful fillets and with a so-called optimum fillet. 1950 *Gloss. Aeronaut. Terms* (*B.S.I.*) I. 14 *Fillet,* a fairing at the junction of two surfaces to improve the air-flow. 1966 D. Stinton *Anat. Aeroplane* xi. 206 The remaining shape of an aeroplane is largely non-structural, in that it consists of fairings, cowlings and fillets.

12. fillet steak, a steak cut from the fillet (sense 6 a in Dict.); **fillet weld** (see quot. 1965).

1877 E. S. Dallas *Kettner's Bk. Table* 63 Perhaps the most prized of all..the fillet-steak. 1963 N. Marsh *Dead Water* (1964) 27 We never order fillet steak. 1929 *Engineering* 22 Nov. 688/3 The plate is welded..with the final ⅜-in. fillet weld. 1965 *Welding Terms & Symbols* (*B.S.I.*) I. 51 *Fillet weld,* a fusion weld, other than a butt, edge or fusion spot weld, which is approximately triangular in transverse cross-section.

filleter (see after Fillet *v.*). Add: **b.** One who fillets (sense 2).

1961 *Guardian* 18 Jan. 9/4 Filleters will tell you that the bobbers..who unload the trawlers have the best job.

filleting, *vbl. sb.* Add: **1.** (Additional example.)

1962 J. Tunstall *Fishermen* iii. 62 Hand filleting is not a simple job.

3. *Aeronaut.* = *fillet *sb.*[1] 11 f.

1935 *Jrnl. R. Aeronaut. Soc.* XXXIX. 835 Eleven photographs show..four types of glider with filleting between wings and body.

‖ **fillette** (fiyet). [Fr.] A young girl.

1847 C. Brontë *J. Eyre* I. xv. 288 Unluckily the Varens..had given me this fillette Adèle. 1888 Mrs. H. Ward *R. Elsmere* I. I. iii. 77 A *fillette,* unformed, inexperienced. 1928 C. A. Nicholson *Hell & Duchess* I. v. 94 As for your *fillette,* she is quite charming.

filling, *vbl. sb.* Add: **1.** (Additional examples.)

1958 *Spectator* 4 July 13/2 As 'ribbon development' is now a dirty phrase, the new horror is called 'filling in'. 1959 *Gloss. Packaging Terms* (*B.S.I.*) 35 *Filling-in,* a process for applying paint or other materials to a recessed design to obtain a contrasting colour effect. 1967 *Gloss. Paper/Ink Terms for Letterpress Printing* (*B.S.I.*) 5 *Filling in,* the spreading of the printed image, to the point where the small white spaces are obliterated.

2. (Earlier and later examples in *Dentistry.*)

1848 *Lit. Amer.* 29 July 64/2 Indestructible filling for the teeth. 1943 Wyndham Lewis *Let.* 5 Dec. (1963) 371 My wife was greatly impressed by the number of your 'fillings'.

3. (Earlier U.S. example.) Also, in cigar-making and mining: see quots.

1812 *Niles' Weekly Reg.* II. 9/1 Much of it [*sc.* wool].. may be wrought into..worsted chain or warp for woolen weft or fillings. 1839 J. R. Lowell *Let.* Sept. (1894) I. 52 The filling of cigars now belies the wrapper. 1864 *Congress. Globe* 3 June 2706/3 Cigars are made of Connecticut grown wrappers and the best qualities filled with Cuba filling. 1883 Gresley *Gloss. Coal-m.* 106 *Filling,* the places where trams are loaded in the workings. 1898 *Westm. Gaz.* 15 Dec. 2/1 'Fillins! What may they be?' 'Why, they're the inside of noo cigars, of course.' 1901 *Chambers's Jrnl.* May 302/2 Each leaf will give on an average two 'wrappers' or outside covers for cigars and when used for such the remainder of the leaf is used for 'filling'. 1940 *Chambers's Techn. Dict.* 332/1 *Filling,* the loading of tubs or trucks with coal, ore, or waste.

4. filling-station orig. *U.S.,* a depot for the supply of petrol, oil, etc. to motorists; a petrol station.

1921 *Outing* (U.S.) May 66/2 He should not attempt the trip without a small reserve can of gasoline..enough to carry him to a filling station in case of leakage. 1931 Earl of Mayo et al. *Regional Planning Rep. on Oxfordshire* v. 73 No Filling Station..shall be permitted by the occupier thereof to be visible unless [etc.]. 1935 *Amer. Speech* X. 5/2 The garage mechanic.., or the filling-station attendant, usually knows the town as well as anybody. 1951 J. B. Priestley *Festival at Farbridge* ii. 272 Roadhouses..and filling stations were brave in new paint.

fillis (fi·lis). Also **phillis.** [Var. *filasse.] A kind of loosely-twisted string, made of hemp (*hemp fillis*) or of jute (*jute fillis*), used by horticulturists as a tying material.

1900 *Oxford rope-dealer's price-list,* Fillis, Hemp, and Jute. 1954 A. G. L. Hellyer *Encycl. Garden Work* 91/1 Fillis. Soft string used for tying. 1954 E. R. Janes *Vegetable Garden* 242 Soft fillis string is the best material for securing tomatoes to their stakes. 1961 *Catal. E. J. Woodman & Sons, Pinner* 62/2 Fillis, plain, soft, balls, 1 lb. 4/3. 1969 E. B. Le Grice *Rose Growing* vi. 79 Sisal or fillis should be weather-proofed.

fill-up, *sb.* (Earlier and later examples.)

1853 E. Twisleton *Let.* 6 Feb. (1928) iv. 69 The three first [men] are cousins..and the latter is a fill-up [at a dinner-table]. 1899 *Sketch* 1 Nov. 48/1 For years the flower would figure in the 'fill-up par.'. 1908 *Chambers's Jrnl.* Oct. 767/1 Compressing them [*sc.* exhibits in a museum] within the allotted space in accordance with what may be..described as a 'fill-up' policy. 1941 *B.B.C. Gloss. Broadc. Terms* 12 *Fill-up,* stop-gap material broadcast in an interval between the end of one programme and the beginning of the next. 1948 *Scrutiny* XV. 336 The French pianist Monique Haas has made a record of Bach's Italian Concerto with Rameau's *Les Cyclopes* as fill-up. 1960 *John o' London's* 7 Apr. 414/4 As the fill-up is a rather laboured..performance of Wolf's delicious Italian *Serenade,* I cannot recommend this record. 1970 J. Fleming *Young Man, I think you're Dying* xi. 141 He had breakfast and a fill-up with petrol and oil in a motorway restaurant.

film, *sb.* Add: **3. b.** Now *esp.* a thin, flexible, transparent material consisting essentially of a plastic base or support (formerly of celluloid, now commonly of cellulose acetate) coated on one side with one or more layers of emulsion and sold as a rolled strip and as separate sheets; also, a single roll of this material, allowing a small number of exposures for use in still photography or a large number for use in cinematography. (Further examples.)

1895 *Montgomery Ward Catal.* 217/1, 1 Roll Film, for 25 exposures. 1897 *Sears, Roebuck Catal.* 473/2 The Hand Camera is the simplest form of photographic camera... Some are constructed for the use of films only, others for both films and dry plates... A roll of film weighs but a few ounces. 1897 [see *cinematograph *sb.* b]. 1902 *Encycl. Brit.* XXXI. 687/2 Instead of glass plates, flexible films of celluloid and other materials are available in single sheets like plates, or in rolls enclosed in opaque paper. 1903 A. Watkins *Photogr.* (ed. 2) 50 The great convenience of daylight changing with rollable films. 1910 *Punch* 15 June 433/1 Messrs. Pathé, who have conceived the idea of showing the news of the week on films at the cinematograph theatres. 1918 *Photo-Miniature* Mar. (Gloss.) s.v., Roll film, flat or cut film. 1928 *Daily Express* 12 Apr. 8/6 A length of film, comprising a number of small photographs, is placed in the transmitter. 1946 W. Clark *Photogr. by Infrared* ii. 12 Plates and film for general infrared photography require no greater care in storage than do panchromatic materials. 1950 J. Hunt *Introd. Med. Photogr.* x. 188 Filmstrips either of X-ray or photographic films can be made in a similar manner. 1954 C. Wallace *Enjoy your Photogr.* viii. 101 Loading by touch is quite simple...The photographer should try it with an old film. 1961 K. Reisz *Technique Film Editing* (ed. 9) i. 58 Editing of visually telling strips of film. 1966 LaCour & Lathrop *Photo Technol.* vii. 77/1 Photography as we know it today, would not have achieved its present popularity, were it not for the introduction of roll film. 1969 M. J. Langford *Adv. Photogr.* xv. 317 The top, blue sensitive layer of a multi-layer colour film is also sensitive to near ultra-violet radiation.

c. A cinematographic representation of a story, drama, episode, event, etc.; a cinema performance; *pl.* the cinema, the 'pictures', the movies. (See also *7 c.)

1905 *Westm. Gaz.* 21 Jan. 3/2 A firm who took cinematograph films of his operations...The films once obtained have been sold and even exhibited at country fairs. 1911 *Times* 22 Sept. 6/2 The great majority of heroic and patriotic films shown here make United States sailors and roughriders the heroes. 1912 [see *cinema c]. 1913 *Punch* 14 May 388/2 When half-a-dozen persons in the same film write letters they all do it in the same hand-writing. 1923 F. A. Talbot *Moving Pictures* 72 The films emanating from the European studios. 1927 *Times* 23 Dec. 11/4 The American film *The King of Kings.* 1954 I. Murdoch *Under Net* ii. 33 Anna never tried to get into films.

d. Film-making considered as an art-form.

1920 *Q. Rev.* July 185 The Film as Caricaturist opens up a new and slightly more encouraging vista—though it is hard to resist the impression that many of the more serious creations of the studio are really intended for caricatures even now. 1929 [see *art-form* s.v. *art *sb.* V]. 1962 *Listener* 12 Apr. 645/2 Is film any more fundamentally mechanical than, say, architecture?

7. film-colour *Psychol.,* an expanse of colour that has a filmy appearance, neither being transparent nor seen as being on the surface of an object or at a definite distance (opp. **surface-colour*); **film-cooling** (see quot.).

1922 *Psychol. Bull.* XIX. 574 'Uncertainty'..may quite well be a property of visual phenomena, as Katz.. maintained for the 'distance' of his so-called 'film'-colors (*Flächenfarben*). 1951 R. H. Thouless *Gen. & Soc. Psychol.* (ed. 3) xii. 190 Katz would call the colour.. seen in the spectroscope a film colour. 1962 M. D. Vernon *Psychol. of Perception* v. 83 It will appear as 'film' colour and not as surface colour. The colour cannot be definitely localized as belonging to the surfaces of objects, but seems to be soft and hazy and to lie over them like a film. 1964 *Amer. Philos. Q.* I. 7/1 The term 'film-color' will do as an example. To get the reader to understand this technical term of visual phenomenology, we instruct him to look at some fairly distant object through a paper tube. 1950 *Sci. News* XV. 82 Film cooling [in a space rocket]...Small holes are drilled through the wall of the combustion chamber so that fuel can leak through from the jacket...It boils and forms a protective film of comparatively cool vapour between these gases and the metal wall.

b. In sense 3 b, as *film-base, camera, -carrier, -holder, magazine, -maker, -punch, roll, side, speed, -wind, -winding*; **film badge, ring,** a device containing photographic film which registers the amount of the wearer's exposure to radiation; **film cement** (see quots.); **film pack,** an assemblage of cut films fitted in a case or holder; **film recording** (see quot. 1941); **film stock,** unexposed film; **film strip,** a length of film bearing a sequence of still frames, used esp. as a teaching aid.

1945 H. D. Smyth *Gen. Acct. Devel. Atomic Energy Mil. Purposes* viii. 90 The Health Division later introduced 'film badges', small pieces of film worn in the identification badge. 1955 *Gloss. Terms Radiol.* (*B.S.I.*) 67 *Film badge,* a photographic film used as a radiation monitor. It is often partially shielded to differentiate between types and qualities of ionizing radiation. *Film ring,* a film badge worn as a ring to measure the dose of radiation, usually beta radiation, received by the fingers. 1968 *Brit. Med. Bull.* XXIV. 260/2 Counting of breaks and dicentrics is a surer index of damage than film-badge monitoring. 1923 F. A. Talbot *Moving Pictures* 62 It was necessary to discover..a new film-base. Cellulose could not be avoided as the basic constituent. 1897 *Sears, Roebuck Catal.* 473/2 The film cameras are most convenient for the traveller. 1920 *Edin. Rev.* July 73 Charlie Chaplin must practise his own funniments before the film camera. 1890 *Anthony's Photogr. Bull.* III. 312 We have to use 'film carriers' which up to date have been complicated, unsafe. 1897 C. M. Hepworth *Animated Photogr.* xi. 87 The 'solution' or 'film cement' with which the joints of film are effected is either a solution of celluloid and, possibly, one or two other substances, in a solvent such as acetone or amyl, acetate, or else the latter chemical used alone. 1959 W. S. Sharps *Dict. Cinematogr.* 95/1 *Film cement,* a solution that will partially dissolve film base and so enable two pieces of film to be joined together in a splice. 1918 Film-holder [see **film pack,* below]. 1967 E. Chambers *Photolitho-Offset* viii. 106 Vacuum filmholders permit the use of contact screens. 1931 B. Brown *Talking Pictures* x. 247 The film magazines..were found to accentuate vibration and noise from the camera mechanism. 1859 *National Mag.* VI. 332/1 Solution rapidly takes place, and the photographic film-maker is constituted. 1903 *Brit. Jrnl. Photogr.* 19 June 490/2 The Film Pack consists of a light tight sheath made of black press board, with an opening in one side corresponding in size to the exposure, into which are folded thirteen strips of black, chemically pure paper..; each of twelve of these..has fastened to the inner side a sheet of film. 1918 *Photo-Miniature* Mar. (Gloss.) 20 *Film pack,* a card-board or metal holder containing, usually, 12 cut films so placed in it that, by pulling in succession projecting paper tabs, the films are brought into position...The device is..a magazine film-holder. 1961 A. L. M. Sowerby *Dict. Photogr.* (ed. 19) 308 Film packs are no longer made in Great Britain, but imported packs are from time to time available. 1923 F. A. Talbot *Moving Pictures* 88 The film-punch for indicating exposure changes upon the edge of the film in a camera. 1931 B. Brown *Talking Pictures* x. 216 Thus film recording..'starts right', and.. has no real limitation whatsoever, since it is easy to increase film speed, and thus raise the maximum frequency. 1941 *B.B.C. Gloss. Broadc. Terms* 12 Film Recording: (1) Process of registering sound by electro-mechanical means for reproduction by photo-electric means, the recording medium being transparent ribbon coated with an opaque substance. (2) (Also Film.) Ribbon on which sound has been so recorded. (3) (By extension.) Programme material so recorded. 1890 *Anthony's Photogr. Bull.* III. 22 Where a film roll is used, one is tempted to make a great many utterly useless pictures. 1889 *Ibid.* II. 361 The film side of the plate. 1915 B. E. Jones *Cinematograph Bk.* iv. 32 Round the inner movable dial [of an exposure meter] are scales relating to film speed and stop. 1911 C. N. Bennett et al. *Handbk. Kinematography* I. App. 108 Long strips of photographically-coated celluloid, known as 'film stock'. 1920 *Stage Year Bk.* 51 Film stock in particular became scarce, for there was no plant for manufacturing this in Britain at the time. 1930 P. Rotha *Film till Now* 341 On the film strip, a frame is three-quarters of an inch high by one inch wide. 1962 *Engineering* 16 Nov. 649 Three new filmstrips, intended for training student electronic engineers. 1962 *Unesco Bull. Libr.* XVI. 10 The 35 mm. filmstrip, which carries a sequence of pictures to be projected. The size of the frames on the strip is usually 18×24 mm. Filmstrips are mainly used for teaching purposes. 1958 M. L. Hall *Newnes Compl. Amat. Photogr.* 55 All 35 mm. cameras have film-wind mechanisms. *Ibid.* 61 In one camera all the necessary actions—film wind, exposure counter, shutter tensioning..are combined. 1936 *Discovery* Aug. 238/2 Automatic Film-Winding.

c. In sense *3 c, as *film actress, clip, company, composer, critic, cue, -editing, editor, fan, festival, hero, magazine, -maker, -making* ppl. adj. and vbl. sb., *music, producer, production, rights, script, set, society, -struck* adj., *studio, super, trade, unit*; **film-goer,** a frequenter of the cinema; hence **film-going** *ppl. a.*; **film star,** a star actor or actress for the cinema.

1921 *Punch* 12 Jan. 21/1 Two leading film actresses have made runaway marriages. 1958 Film clip [see *clip *sb.*[2] 2 c]. 1962 *Sunday Times* 26 Aug. 25/5 The sort of film clips which only political propagandists would dream of using. 1918 H. Croy *How Motion Pictures are Made* ii. 35 Charles Pathé, of Paris,..later came to head the world's premier film company. 1948 *Penguin Music Mag.* Feb. 67 Vaughan Williams is, as a film composer, best at moments such as these. 1931 *Punch* 8 Apr. 384/1 The film-critic of *The Daily Telegraph.* 1918 H. Croy *How Motion Pictures are Made* vi. 134 The director..gives him what might be called his film cues. 1922 *Moving Picture Stories* 23 June 25/1 Film editors and splicers are working as they haven't done in many months. 1953 T. Dickinson in K. Reisz *Technique Film Editing* 7 The pivotal contribution of the film editor has never been analysed objectively. Film editing has only been dealt with in the personal theories of Eisenstein, Pudovkin and others. 1918 R. Wagner *Film Folk* i. 12 But that was before the film fans became oversophisticated. 1951 N. Mitford *Blessing* II. i. 162 She discouraged Hughie from following her to Venice, saying vaguely that she would be taken up with the film festival. 1919 *Fortn. Rev.* CVI. 455 Some owners of cinema theatres are doing their best to attract a better intellectual class of film-goer. 1927 *Sunday Express* 26 June 4 Every film-goer likes to feel that he is up-to-date in cinema affairs. 1929 *Evening News* 18 Nov. 12/4 They should realise that the film-going public have brains. 1925 W. Deeping *Sorrell & Son* xv. § 1, He hated crowds, he—the crowd's film-hero. 1939 C. Day Lewis *Child of Misfortune* III. ii. 271 Alice, looking up from her film magazine. 1919 *Fortn. Rev.* CVI. 456 Many film-makers..have set their faces against this appeal. 1960 *Guardian* 27 Oct. 8/2 Certain products of Hollywood—about which French film-makers were traditionally contemptuous. 1938 *Observer* 26 June 12/6 A first-class film-making nation. 1944 J. S. Huxley *Living in Rev.* iii. 37 Wartime film-making, especially perhaps in Canada, is making the movie a comprehensive and faithful mirror for the war-time life and purpose of peoples. 1927 *Melody Maker* Sept. 938/2 The art of cue film-music compilers in the fitting of films with appropriate music. 1914 R. Grau *Theatre of Science* iv. 85 He will prove a greater asset to the film producer a year from now. 1920 *Q. Rev.* July 182 The only kind of ending the film-producers can think of for this or any other type of drama. 1927 P. D. Hugon *Morrow's Word Finder* 205 Terms used in film production. 1935 H. G. Wells *Things to Come* 13 The incorporation of original music in film production is still..an unsolved problem. 1913 *Writer's Mag.* Dec. 263/1 Contracts with.. today's most famous authors for the film rights to their.. stories. 1933 A. Brunel *Filmcraft* 194 Selling the film rights at an immensely exaggerated figure. 1970 L. Meynell *Curious Crime of Miss Julia Blossom* xv. 180 Serial rights. American rights, and I saw a para in last night's *Standard* saying somebody was after the film rights too. 1948 Dylan Thomas *Let.* 17 Nov. (1966) 323 My fee for my next unspecified filmscript. 1952 Eliot & Hoellering *Film of Murder in Cathedral* 7 That is in itself a justification for publishing this film script. 1933 C. Winchester *World Film Encycl.* 475/2 The first film sets were called 'flats'. 1931 *Week-end Review* 14 Feb. 212/2 The Film Society. 1968 *Guardian* 27 Dec. 4/8 Film societies..have increased in membership. 1914 R. Grau *Theatre of Science* iii. 50 The greatest film stars in the world. 1923 *Chambers's Jrnl.* Mar. 180/1 A number of very beautiful women have earned untold riches as 'film-stars'. 1963 *Times Lit. Suppl.* 26 Apr. 313/5 The film-star treatment of the Royal Family. 1923 *Weekly Disp.* 28 Jan. 1 She had become absolutely 'film-struck'. 1913 *Writer's Bulletin* July 11 The great photoplays of the future will be those that are created within the confines of the film studio. 1935 *Discovery* Oct. 309/1 In film studios the necessity for absolute silence demands an air-tight enclosure. 1924 Galsworthy *White Monkey* II. ii, The lurid professions—film super, or mannequin. 1914 *Even. News* 8 Oct. 2/1 It is 'up to' the American film trade to see that the evidence of the camera gets a fair reception. 1951 'M. Innes' *Operation Pax* III. vii. 113 Film units..descend upon the place.

Hence **fi·lmcraft,** the technique of making films; **fi·lmdom** [-dom], the realm of cinematographic production, producers, and stars, the cinema world; **fi·lmland,** (*a*) = *filmdom*; (*b*) Hollywood, or a specific locality associated with the film industry.

1928 *Observer* 15 Apr. 5/4 Filmcraft is little understood. 1963 *Guardian* 9 Feb. 5/5 Its sixty minutes contain no remarkable feat of filmcraft. 1914 *Writer's Bulletin* Feb. 35/1 Even in filmdom..there are a dozen who hold the art of the silent drama in reverence. 1927 *Sunday Express* 12 June 18/3 One of filmdom's finest mansions is Pickfair. 1930 *Daily Express* 9 Sept. 8/3 A strange trail of thought in filmdom. 1914 R. Grau *Theatre of Science* p. vii, There are not a few celebrities in film-land averse from publicity. 1928 *Punch* 21 Nov. 578/3 How Film-land Regards a Royal Commission. 'English Royal Group...'—*Los Angeles Paper.*

film, *v.* Add: **3.** To photograph for use in a cinema or cinematographic device; to exhibit as a cinematographic production; to put on 'the films' or 'the screen'.

Quot. 1899 refers to a 'mutoscope'.

1899 *Westm. Gaz.* 21 Sept. 4/1 Professors of medicine are 'filming' their patients' muscles. 1912 F. A. Talbot *Moving Pictures* xv. 174 He succeeded in persuading Sir Herbert Beerbohm Tree to let him film the Shakespearean production of Henry VIII. 1915 A. S. Neill *Dominie's Log* viii, I have seen *Hamlet* filmed. 1915 W. Irwin *Men, Women & War* 15 We filmed a Belgian troop of cavalry going into action. 1920 *Q. Rev.* July 183 Hallo, here's a big success, let's film it! 1920 *Glasgow Herald* 11 Sept. 7 Churchmen are invited to write thrilling Biblical scenarios to be filmed for exhibition at afternoon services. 1934 W. Saroyan *Daring Young Man* (1935) 26, I am not expecting Paramount Pictures to film this work. 1971 *Daily Tel.* 20 Oct. 12/8 Cameras were allowed for the first time to film their lordships inside their red-carpeted sanctum.

4. *intr.* To be (well or ill) suited for film-acting or for reproduction on film.

1920 *Punch* 31 Mar. 246/3 This incident should 'film' well. 1927 E. Hemingway *Fiesta* I. vi. 60 He's leaving me. He's decided I don't film well. 1928 *Observer* 26 Feb. 20/2 [He] 'films' very well indeed. 1934 W. Saroyan *Daring Young Man* (1935) 204, I always knew I had the sort of face that would film well and look good on the screen.

Hence **fi·lming** *ppl. a.* and *vbl. sb.*

1889 *Anthony's Photogr. Bull.* II. 292, I drew a plan of the filming instrument. 1912 F. A. Talbot *Moving Pictures* xi. 120 The procession drew within the field of the camera and the filming commenced. 1918 H. Croy *How Motion Pictures are Made* 110 The filming of this is put last to give the carpenters time to complete it. 1933 *Discovery* Dec. 359/2 The hunting, filming and capture of wild beasts.

filmable (fi·lmăb'l), *a.* [f. *film *v.* 3 + -able.] Capable of being filmed or adapted to the cinema; well suited to reproduction on film.

1920 V Steer *Secrets Cinema* ii. 27 Even if one possesses..filmable features, the beginner will discover that the work is..exceedingly precarious. 1921 *Glasgow Herald* 25 Mar. 7 Full of 'filmable' incident. 1926 *Spectator* 17 July 88/2 A reason, incidentally, why Dickens is not really filmable. 1961 *Guardian* 14 Sept. 10/5 Gestures which are extremely filmable.

filmic (fi·lmik), *a.* [f. *film *sb.* 3 c + -ic.] Of or pertaining to cinematography; suggestive of the cinema; filmable. Hence **fi·lmically** *adv.*

1929 I. Montagu tr. *Pudovkin's On Film Technique* ii. 19 The problem was the expression, filmically, of his joy. 1930 *Time & Tide* 23 Aug. 1080 They do seem to have an eye for the filmic possibilities of subjects. 1930 *Times Lit. Suppl.* 18 Sept. 728/2 Debating the merits of various types of 'filmic' art. 1936 *Archit. Rev.* LXXIX. 137 But while reading this book one frequently feels that Mr. Rotha has more interest in his social message than in his filmic means. 1948 'E. Crispin' *Buried for Pleasure* iii. 22 His blonde visitant's filmic radiance. 1953 *New Statesman* 14 Nov. 599/1 The most gorgeous of them [*sc.* Shakespeare films], and after fearful struggles, the most filmic..was *Henry V.* 1958 *Times Lit. Suppl.* 15 Aug. p. xxviii/2 Arthur Calder-Marshall (whose early novel *About Levy* was a perfect example of filmic-construction). 1962 *Listener* 23 Aug. 273/2 Most people felt that filmically this was an untidy work.

filmlet (fi·lmlėt). [f. *film *sb.* 3 c + -let.] A short film for cinema or television.

1931 *Punch* 18 May, In very advanced cases the 'trailer'..is a specially made filmlet. 1958 *Bookseller* 8 Nov. 1905 (*caption*) Scene from the Book Tokens 'filmlet' which will be televised during the coming weeks. 1959 *Times Lit. Suppl.* 13 Nov. 662/3 One is reminded of those filmlets and television commercials which so often win prizes.

‖ **film noir** (film nwār). [Fr., lit. 'black film'.] A cinematographic film of a gloomy or fatalistic character.

1958 *Spectator* 25 July 134/2 It [*sc.* a play] tries to be a parody of a *film noir.* 1960 *Times* 9 May 16/4 A schoolboy adventure story which turns suddenly and surprisingly into a bleak film noir.

filmography (filmǫ·grăfi). [f. *film *sb.* 3 c, after Bibliography.] **a.** A list of the films of a particular director, producer, actor, etc., or of those dealing with any particular theme. **b.** The systematic description of, and information about, films.

1962 *Movie* Sept. 30/3 We hope to include also a complete filmography of Preminger, with details of the availability of the films. 1963 *Amer. N. & Q.* II. 10/1 *Film Composers in America*..will contain biographies, filmographies, discographies, and bibliographies. 1968 P. Graham *Dict. Cinema* 5 The filmographies that follow each director, actor, scriptwriter, etc., listed in the dictionary are complete.

fi·lmset, *v. Printing.* [f. Film *sb.* + Set *v.* 72.] *trans.* To compose (matter) by projecting images on to photographic film in a photo-composing machine. So **fi·lmset** *ppl. a.,* **fi·lmsetting** *ppl. a.* and *vbl. sb.*

1954 *Printing Rev.* XVIII. lxv. 41/2 The Keybar frames..can be used without alteration for film-setting on a 'Monophoto'. 1956 *Print in Britain* IV. 234/1 Protype provides the..printer with the opportunity of installing a simple film-setting system. 1958 *Times Lit. Suppl.* 18 Apr. 209/3 In 1954 an Italian novel was filmset in Europe, on the Westover Rotofoto. In the United States, two (not one) books were filmset on the Intertype Fotosetter before the end of 1951. 1959 *Penrose Ann.* p. xviii, It is to be hoped that examples of filmset books.. will be seen in the National Book League's exhibitions of book design. 1963 D. Bland in O. Simon *Introd. Typogr.* (ed. 2) p. v, The increasing use of slug-setting and film-setting for bookwork. 1963 *Printing & Mind of Man* 113 This [*sc.* the Rotophoto], the first British filmsetting machine to be placed on the market, was the invention of..George Westover, and was first shown in 1948. 1966 *Print* (Wynkyn de Worde Soc.) ii. 18 Type is still largely the basis of non-pictorial printing, whether the job be done from the type itself..or from a filmset or 'cold-type' basis.

‖ **fils** (fis). [Fr., son.] The son, junior: appended to a name to distinguish between a father and son of the same name.

1886 *Athenæum* 30 Oct. 565/1 Athanase Coquerel *fils* does not seem to have had any great distinction..of style. 1889 E. Dowson *Let.* 24 Mar. (1967) 54 Dumas fils easily first. 1936 Ogden & Richards *Meaning of Meaning* (ed. 4) v. 89 Mill *fils* in his editorial notes on this passage holds [etc.]. 1965 *House & Garden* Mar. 42 Russell *père* had been in banking... Russell *fils*..thought this a bit-too-cramping project.

filter, *sb.* Add: **3. b.** *Photogr.* A screen to cut out rays which interfere with correct colour-rendering; = *colour-filter* (see *colour *sb.*[1] 18).

1900, 1902 [see *colour *sb.*[1] 18]. 1912 E. J. Wall *Dict. Photogr.* (ed. 9) 335 Most makers of orthochromatic ..plates also supply filters specially adapted for use with their plates. 1948 [see *colour *sb.*[1] 18]. 1958 M. L. Hall *Newnes' Compl. Amat. Photogr.* viii. 97 Light filters are transparent, coloured discs which are attached to the camera lens for the purpose of modifying the colour quality of the light transmitted but which do not otherwise affect the image formed by the lens.

c. In a cigarette: a pad of absorbent material fitted at the unlit end to purify the smoke. Also, a cigarette so fitted.

1908 *Lancet* 21 Mar. 907/2 A pad of cotton wool is inserted into the paper casing of these cigarettes, which effectually prevents dust and shreds of tobacco entering the mouth, besides which a large proportion of the objectionable oils formed during smoking is arrested in the wool filter. 1958 *Amer. Speech* XXXIII. 147 A new cellulose fiber for..cigarette filters. 1963 N. Freeling *Gun before Butter* II. 96 Yes, I do smoke but only filters.

d. *Radiol.* and *Nuclear Sci.* A sheet or block of material inserted in the path of a beam of X-radiation or elementary particles in order to reduce the intensity of radiation of certain wavelengths or energies.

1917 Arthur & Muir *Man. Pract. X-Ray Work* (ed. 2) x. 313 For skin treatment it is well to use a weak filter—such as thick paper—which will absorb the softest rays which..might otherwise produce dangerous over-effects. 1957 *Encycl. Brit.* XVIII. 898/1 In some of the work for which X-rays are used, the quality or penetrating ability of the radiation is expressed..in terms of its absorbability in filters of standard composition and thickness. 1969 G. E. Bacon *Neutron Physics* x. 127 We can..use a sufficiently thick block of graphite as a filter which..removes all neutrons with a wavelength greater than 0·67 nm.

e. *Electronics.* A passive circuit that attenuates all signals except those within one or more frequency bands.

1920 *Radio Rev.* July 505 By a combination of the two types of filter only currents between the two frequency limits are transmitted. 1922, etc. [see *band *sb.*[2] 14]. 1938 F. E. Terman *Fund. Radio* x. 267 The pulsating voltage delivered by the rectifier output can be smoothed into a steady direct-current voltage..by being passed through an electrical network called a filter, which ordinarily consists of series inductances and shunt condensers. 1955 G. M. Glasford *Fund. Telev. Engin.* xvii. 581 There must be employed in the video section a 4·5-Mc rejection filter to remove the audio carrier and its sidebands from the video signal. 1959 *Engineering* 30 Jan. 153/3 A conventional voltage doubler serves as the high voltage rectifier, with sufficient capacitance employed in the filter to reduce the r.m.s. ripple to 0.5 per cent.

5. filter aid, any substance added to a liquid or to a filtering medium in order to improve filtration by preventing the formation of an impervious filter cake; **filter-arrow,** a device forming part of a traffic signal (cf. *filter *v.* 3 b); **filter cake,** the insoluble residue deposited on a filter; so *filter-caked* adj.; **filter circuit** *Electronics* = 3 e above; **filter condenser** *Electronics,* a capacitor forming one of the chief elements in a filter circuit; **filter factor**: see *factor *sb.* 7 c; **filter-feeder,** an animal that obtains its nourishment by means of filter-feeding; **filter-feeding** *vbl. sb.,* the filtering and ingestion of nutrient matter suspended in water; also as *adj.*; **filter-paper** (examples); **filter-passer,** a filter-passing 'organism', esp. a virus; **filter-passing** *a.,* not retained by a filter, esp. by one that does retain bacteria; **filter reactor** *Electronics,* a reactor forming one of the chief elements in a filter circuit; **filter shot** *Photogr.,* a photograph taken through a colour-filter; **filter tip** = 3 c above; hence **filter-tipped** *a.,* **filter-tipping** *vbl. sb.*

1946 Dickey & Bryden *Theory & Pract. Filtration* iii. 47 A precoat and/or a filter aid are often necessary, to prevent the deposited particles from being carried.. into the interstices of the filter medium, or filter cake. 1951 Kirk & Othmer *Encycl. Chem. Technol.* VI. 510

Materials such as kieselguhr or diatomite, asbestos fibers,..and sawdust flour are examples of materials used as filter aids. **1963** *Times* 24 May p. xiv/1 (Advt.), Filteraids and chemicals to purify most pipe-borne liquids. **1963** Priestley & Wisdom *Good Driving* xi. 77 A filter-arrow allowing left-turning vehicles to proceed. **1912** *Mining & Engin. World* 20 Apr. 863/3 In forming a filter cake with any material there is a certain point where the rate of filtration begins to drop off very rapidly due to the resistance offered by the cake. **1967** *Filtration & Separation* IV. 471 Most filter cakes show some degree of compressibility. **1956** L. Klein et al. *Disposal of Sewage* xii. 202 The filter-caked sludge may be dried finally by heat and sold as fertilizer. **1920** *Radio Rev.* July 595 A number of special filter circuits designed with the object of allowing certain frequencies to pass and cutting out others. **1964** A. Nisbett *Technique Sound Studio* iv. 82 Correction of a wide range of faults, such as noise, distortion, and unsatisfactory frequency response, is often attempted by means of filter circuits. **1922** A. F. Collins *Book of Wireless Telegraph & Telephone* iii. iv. 197 Connect the filter reactor and the filter condensers to the mid-taps of the filament secondaries. **1951** S. Deutsch *Theory & Design Telev. Receivers* xii. 406 The size of the various filter condensers is determined by consideration of tolerable ripple. **1921** Filter factor [see *factor sb. 7 c]. **1958** M. L. Hall *Newnes' Compl. Amat. Photogr.* vii. 98 Filters reduce the light transmitted to form the image and require a compensating exposure increase, with certain very pale filters..this filter factor is..small enough to be ignored..The increase is specified as a multiplying factor. **1928** *Trans. R. Soc. Edin.* LV. i. 235 Besides being a filter feeder, Hemimysis also feeds from large solid masses of food. **1959** A. Hardy *Fish & Fisheries* v. 109 The most elaborately developed suspension-feeders are the bivalve (lamellibranch) molluscs. They are filter-feeders. Their gills have been enormously enlarged..as sieving devices to collect the fine particles of food from the water which is propelled through them by cilia. **1931** *Trans. R. Soc. Edin.* LVI. iii. 537 A study of the feeding mechanisms and physiology of digestion in Polychaetes has..been begun, choosing a filter-feeding form as a starting point. **1963** R. P. Dales *Annelids* [illegible] With loss of mobility we find more selective methods of feeding, this evolution culminating in the filter-feeding fan or feather-duster worms. **1849** D. Campbell *Inorg. Chem.* 18 A few grains of recently-heated spongy platinum, in a small piece of filter paper. **1882** *Photogr. Ann.* II. 109 Dried on pure filter paper. **1933** J. McFadyean in *XVIIth Internat. Congress Med.* IV. 3 The term 'filter passers' has been adopted..as the preferable name for the micro-organisms which..were called the ultra-visible or invisible viruses. **1919** *Nature* 6 Nov. 210/2 The infective agents of the common exanthemata..are at some period of their life-history so small as to be included amongst the 'filter-passers'. **1946** *Nature* 21 Sept. 398/1 Bacteria, filter-passers, viruses, protozoa and metazoan parasites or fungi. **1915** *Lancet* 4 Dec. 1241/2 Attempts were made to demonstrate the presence of our filter-passing virus. **1930** *Morning Post* 24 Nov. 3 'Ultra-microscopic' or 'filter passing' virus. **1947** Auden *Age of Anxiety* (1948) v. 107 No filter-passing Virus invade. **1957** G. E. Hutchinson *Treat. Limnol.* I. vi. 385 They give the percentage effect of suspended or retained and dissolved or filter-passing material on the absorption. **1922** A. F. Collins *Book of Wireless Telegraph & Telephone* II. 194 The purpose of a filter reactor is to smooth out the pulsating direct current after it is produced by the rectifier tube. **1951** J. B. Priestley *Festival at Farbridge* I. iii. 126 We got some lovely filter and scrim shots. **1934** *Daily Mail* 12 Oct. 8/4 (Advt.), The Filter Tip that holds throat irritants in check. **1963** *Wall St. Jrnl.* 1 Oct. 1/4 The work on a filter-tip cigaret began early in 1951. The company was seeking a non-mineral material to filter the smoke and finally decided upon a highly purified alpha cellulose tip. **1957** *New Yorker* 23 Nov. 100/2 When I ran out of American filter-tips, I chose the nearest equivalent I could find—a Bulgarian make. **1957** *Times* 18 Dec. 4/2 A substantial increase in the sales of filter tip cigarettes was a common feature of tobacco sales in Britain and the United States in 1957. **1954** *Newsweek* 16 Aug. 71/3 The American Tobacco Co... brought out a filter-tipped version of Herbert Tareyton, its king-size and cork-tipped brand. **1954** *Wall St. Jrnl.* 18 Nov. 10/2 American Machine Foundry Co. has developed a new method of applying filter tips to cigarets which it claims 'renders obsolete all other filter tipping devices currently used in the tobacco industry'.

filter, *v.* Add: **1. d.** *Electronics.* To pass (an electrical signal, etc.) through a filter (*filter *sb.* 3 e) as a means of removing or attenuating components of undesired frequencies or undue prominence. Also with *out*: to select or remove (a component of a signal) by means of a filter.

1921 B. Leggett *Wireless Telegr.* xiii. 399 Zenneck has proposed..to filter out and utilise the harmonic by means of a tuned circuit. **1937** W. G. Dow *Fund. Engin. Electronics* xxi. 487 The greater the number of anodes, the smaller is the ripple to be filtered out. **1959** Zimmermann & Mason *Electronic Circuit Theory* iv. 130 The effects of a poorly filtered power supply on an audio system can be heard directly in the form of an objectionable hum.

3. *transf.* and *fig.* (Additional examples.)

1944 *Times* 29 Apr. 3/3 Here the enemy succeeded in temporarily filtering through our lines. **1958** Hayward & Harari tr. *Pasternak's Dr. Zhivago* II. viii. 231 Although the station area was cordoned off, passengers for the local trains had managed, in some unaccountable way, to 'filter through' (as we would say now). **1969** *Listener* 27 Feb. 281/2 Meanwhile foreign literature (Ibsen and Chekhov notably) and painting (the Post-Impressionist Exhibition) were slowly filtering in. **1971** *Daily Tel.* 16 June 11/5 Hints..of an upsurge in Government spending on new by-passes..have filtered out with unusually effective timing.

b. *spec.* Of road vehicles: to join another line of traffic at a road junction, usu. by deviating from the main stream which is held up by traffic lights.

1928 *Traffic Signals to be used by Police & Drivers of Vehicles* (*H.M.S.O.*) ii. 10 At road junctions, when one stream of traffic has been halted, constables should, so far as may be safe or practicable, permit drivers who wish to do so to turn to the left and so filter into the cross stream of moving traffic. **1937** V. Woolf *Years* 334 He filtered slowly round the corner. **1971** *A.A. Continental Handbk.* 396 Some crossroads have traffic lights with a flashing yellow arrow which allows drivers to filter off in the direction of the arrow.

4. b. *to filter out* (trans.): to separate or prevent the passage of by, or as by, filtering.

1917 G. W. C. Kaye *X Rays* (ed. 2) ix. 116 By filtering out the soft rays from the primary beam by the use of a suitable screen, the polarisation can be doubled. **1921, 1937** [see 1 d above]. **1927** Davis & Kaye *Acoustics of Buildings* i. 7 In suitable circumstances it is possible to filter out undesired harmonics. **1936** *Discovery* July 224/1 These [stops] change the tone-colour..the original note produced being deliberately made very rich in overtones, which in turn are filtered out as required. **1951** Kirk & Othmer *Encycl. Chem. Technol.* VI. 516 The sand acts as the filter medium and filters out suspended solids. **1961** G. Millerson *Technique Telev. Prod.* iii. 43 Sound reproduction begins to lose crispness as higher audio notes are filtered out. **1965** N. Chomsky *Aspects of Theory of Syntax* 224 An extraneous factor..filters out certain latent interpretations provided by the deep structures. **1967** M. Frayn *Towards End of Morning* viii. 151 He seemed unaware of the noise Damian was making. Bob's head ached..at the effort of filtering the adult conversation out from it. **1971** *Daily Tel.* 4 Mar. 3/3 Continued flights..could reduce the ozone layer that filters out cancer-producing ultra-violet rays.

filterability (fi:ltərăbi·lĭti). Also **filtrability.** [f. next: see -ity.] The quality or state of being filterable.

1912 *Jrnl. Med. Res.* XXVII. 1 The filterability of viruses. **1928** T. M. Rivers *Filterable Viruses* i. 10 Others, however, do not seem to be so small and concerning the filterability of these there is much discussion. **1952** *Genetics* XXXVII. 724 The infective transmission of F^{+} at once calls to mind the latent bacterial viruses...However, the filtrability of bacteriophage at once differentiates it from the F^{+} agent.

filterable (fi·ltərăb'l), *a.* Also **filtrable.** [f. Filter *v.* + -able.] Able to pass through the pores of a filter, esp. one that retains bacteria; *filterable virus* (see *virus).

1908 M. H. Hayes tr. *Friedberger & Fröhner's Vet. Path.* (ed. 6) II. 425 The causative agents in fowl-pox belong to the group of so-called filtrable agents of infection. **1912** *Jrnl. Med. Res.* XXVII. 1 The diseases caused by filterable viruses include many of the most important of man and animals. *Ibid.* 3 There is also another protozoan which has a filterable stage. **1922** *Glasgow Herald* 21 Feb. 10 A number of animal diseases besides foot-and-mouth disease are caused by filterable viruses. **1928** *Daily Tel.* 10 July 9/4 Dr. Gye's theory of the causation of cancer by a filtrable virus. **1928** T. M. Rivers *Filterable Viruses* i. 3 To designate the active agents discussed in the book the term 'filterable viruses' was chosen from a long list of unsatisfactory names. For many reasons the term 'viruses' is preferable to 'filterable viruses'. **1952** G. H. Bourne *Cytol. & Cell Physiol.* (ed. 2) ix. 402 A filterable tumour, on crushing to a pulp and mixing with saline, yields by Berkefeld-filtration a cell-free filtrate. **1955** *Sci. Amer.* Mar. 65/1 They [*sc.* viruses] were first distinguished from bacteria by the fact that they were filtrable, whereas the pores of the ordinary porcelain filter blocked the passage of bacteria. **1965** T. M. Bell *Introd. Gen. Virol.* i. 2 Today The Viruses are exclusively the 'filterable viruses' and are considered apart from The Microbes.

filth, *v.* Delete †*Obs.* and add mod. *poet.* example.

1955 W. de la Mare *Beginning* 211 You merely filthed and made vile the most sacred misfortune.

filthy, *a.* Add: **1. c.** Of weather: extremely unpleasant.

1875 L. Troubridge *Life amongst Troubridges* (1966) xi. 127 It was a most filthy day, pouring hard. **1895** 'G. Mortimer' *Tales fr. Western Moors* 290 You'm only half a notion o' weather till you'm atop of Darteymoor when 'tis filthy. **1955** *Times* 3 May 5/2 In filthy weather.

4. b. (Later examples.)

1855 Trollope *Warden* xi. 175 No man lived less addicted to filthy lucre than the warden. **1928** A. Huxley *Let.* 19 July (1969) 301 A considerable sale..and a corresponding quantity of filthy lucre. **1931** Wodehouse *If I were You* ii. 27 Just trying to make a bit of the filthy.

6. (Later examples.)

1843 *Knickerbocker* XXI. 122 His trousers [were] 'filthy dirty' and pulled up above the tops of his boots. **1940** M. Banning *Enough to live On* ix. 168 He's filthy rich and didn't earn a cent of it. **1954** J. B. Priestley *Magicians* ii. 28, I organise these parties for her—she's filthy rich. **1970** *Globe & Mail* (Toronto) 28 Sept. 1/3 He told me he was filthy rich.

7. (Later example.)

1946 R. Lehmann *Gipsy's Baby* 125 Filthy tempered woman.

filthy (fi·lþi), *v.* [f. the adj.] *trans.* To make filthy.

1917 J. Masefield *Old Front Line* v. 67 All this mess of heaps and hillocks is strung and filthied over with broken bodies and ruined gear. **1952** C. Day Lewis tr. *Virgil's Aeneid* iii. 60 They tear the banquet to pieces, filthying all with their bestial Touch. **1967** 'La Meri' *Sp. Dancing* (ed. 2) xi. 137 Cigarette butts and spilled liquids filthied the floor.

filtrability, filtrable, varr. *filterability, *filterable *a.*

filtrate, *sb.* Add: **b.** *attrib.*, as **filtrate factor,** any soluble unidentified substance in a filtrate, esp. such a substance affecting the growth of an organism.

1937 Edgar & Macrae in *Biochem. Jrnl.* XXXI. 893 The second constituent was present in the filtrate after fuller's earth treatment of the yeast extract; this we have called the yeast fuller's earth filtrate factor. **1965** *Mycologia* LVII. 273 For *M[orchella] crassipes* the beneficial effect is accounted for mainly..by the minerals supplied by the natural product, especially Mn and Ca. These two elements make up the 'filtrate factor' previously reported for this fungus.

filtration. Add: **1.** (Additional examples.)

1943 *Gloss. Terms Electr. Engin.* (*B.S.I.*) 142 *Filtration*, of X-rays, the absorption of the longer wave-lengths to a greater extent than the shorter wavelengths by placing an absorbing medium in the path of the beam. **1970** D. F. Shaw *Introd. Electronics* (ed. 2) x. 220 If a current of more than one milliamp is to be delivered by the rectifier, *L–C* filtration must be used.

3. *transf.* (Cf. *filter *v.* 3 b.)

1932 *Evening Standard* 21 Jan. 4/4 'Filtration' takes place whenever the lights give the 'All Clear' at a crossing; many vehicles in the stream which is let loose turn left or right despite the traffic light that in the road into which they are turning the traffic lights are then red.

|| **filtre** (filtr). [Fr.] A filtering appliance, fitted over a pot or cup, for making coffee by the passage of boiling water through ground coffee. Also, the coffee so prepared (see *café 3).

1940 M. Innes' *Secret Vanguard* xx. 217 Orchard was..measuring coffee into a filtre. **1959** *Guardian* 9 Nov. 3/2, I have sat in a French café watching a 'filtre' deliver the coffee drop by drop into my cup. **1961** P. V. Price *France* 35 Cona coffee will cost slightly more than a filtre.

fin, *sb.*¹ Add: **2. d.** *pl.* Rubber flippers for the feet, to assist underwater swimming.

1960 *News Chron.* 9 July 6/7 You'd need a pair of fins (£1), a mask (£1) and a breathing tube. **1967** J. Severson *Great Surfing* Gloss., *Fins*, rubber flippers used as aids in swimming or body surfing; called swim fins.

3. d. In aircraft and rockets, esp. as a stabilizer (see quots.).

1843 *Nautical Mag.* IV. 612 An internal balloon is fitted for the purpose of ascending and descending at will, and the whole is intended to be propelled by fins, paddles, or wings. **1852** *Illustr. London News* XX. 468/1 By means of the handles at the outer sides, the apparatus contained in the case can be turned round at pleasure, and the position of the fins altered, thus enabling the operator to steer the balloon. **1910** C. C. Turner *Aerial Navig.* iv. 54 Lepping's Balloon, 1852. It was to be driven by a fin propeller, worked by hand. **1921** *Chambers's Jrnl.* 57/2 An aeroplane..supported largely by the pressure against its body, its wings reduced to mere fins, serving to guide its motion. **1920** *Discovery* Mar. 78/1 An elliptically-shaped gas-bag provided with a conical-shaped tail fitted with air, to which a rudder and stabilising fins are attached. **1935** [see *anti-icer]. **1935** *Jrnl. Brit. Interplanetary Soc.* Oct. 13 Fin—wing-like device for stabilising flight. **1950** *Gloss. Aeronaut. Terms* (*B.S.I.*) 43 *Fin*, A fixed surface outside the envelope or outer cover of a lighter-than-air aircraft providing aerodynamic stability. Those parts of the stabilizers of a kite balloon providing stability in pitch.

e. A fin-like appendage to a ship's bottom; a fin-keel; also, a centreboard.

1885 *Marine Engineer* 1 Apr. 8/1 But it is possible to fit fins, flappers, or shutters,..to the run of a ship so as to be flush with the plates. **1893** *Outing* (U.S.) XXII. 147/1 The fin is of Tobin bronze, one-quarter inch thick, six feet long on upper edge. [illegible] The hulls..are enormously strong, so as to restrain the heavy weight of the fin. 36 Fin stabilisers will be fitted to both ships.

f. On a motor vehicle. Hence **fin-tailed** *adj.*

1959 *Listener* 4 June 983/1 Even family cars are now made impressive with the fins which were once the distinguishing mark of sharks. **1966** *Guardian* 11 Oct. 8/4 The reluctant mums..are induced to step gingerly out of the fintailed cars. **1968** P. M. Postal *Aspects Phonol. Theory* xiii. 283 There is no more reason for languages to change than there is for automobiles to add fins one year and remove them the next.

fin (fin), *sb.*² *slang.* = Finnip (see also quot. 1925). Also *U.S.*, a five-dollar note; the sum of five dollars.

1868 *Temple Bar* XXIV. 538 'What are "fins"?' 'Five-pound notes, or flash notes.' **1925** *Flynn's* 24 Jan. 119/2 Fin, five dollars. **1949** *Penguin New Writing* XXXVI. 97 We slipped them a fin apiece. **1951** W. R. Burnett *Vanity Row* xv. 101 Costs a fin just to check your hat.

fin, *v.* Add: **4.** *intr.* To swim, as a fish; hence used of underwater swimmers. Also *trans.*, *to fin it, to fin a* (or *one's*) *passage, way.*

1807 J. Barlow *Columb.* viii. 285 Renascent swarms by nature's care supplied, Repeople still the shoals and fin the fruitful tide. *a* **1861** T. Winthrop *Canoe & Saddle* (1862) 134 In midsummer salmon fin it along the

reaches of Whulge. **1894** *Outing* (U.S.) XXIV. 140/1 For this [stump] the rascal [trout] steers, as fast as he can fin his way. **1928** T. Hardy *Coll. Poems* 591 Fishes might seem to fin a passage. **1933** W. de la Mare *Lord Fish* 36 He grew hungrier and hungrier as he finned softly on. **1964** B. Gaston *Drifting Death* ii. 24, I jack-knifed down to beyond twenty and finned hard for the yacht. **1964** *Guardian* 20 May 7/6 You fin slowly down, 15 to 20 feet now.

5. *trans.* To provide with fins (sense 3). Hence **fi·nning** *vbl. sb.*

1933 *Meccano Mag.* Mar. 193/1 The finning of the head has also received special attention. **1962** *Times* 8 May 16/5 Pininfarina's finning is sufficient to assist stability and guide the driver when parking.

finagle (finē¹·g'l), *v.* *colloq.* (orig. *U.S.*). Also **fin(n)agel, finaygle, phe-.** [ad. Eng. dial. *fainaigue* to cheat. See *E.D.D.* and cf. Renegue *v.*] *intr.* To use dishonest or devious methods to bring something about; to fiddle. Also *trans.*, to 'wangle', to scheme, to get (something) by trickery. So **fina·gling** *vbl. sb.*

1926 in H. Wentworth *Amer. Dial. Dict.* (1944) 216/1 Finagle, U.S. political cant. **1926** Anderson & Stallings *What Price Glory* iii, I'm a weary man, and I don't want any finnagelling from you. **1933** P. Cain *Fast One* (1936) vii. 250 Turn her over to me in the open and without any finaygling. **1936** *Writer's Digest* Oct. 4 Discounting any possible editorial finageling..the solid fact remains that opposing politically minded people do cancel subscriptions. **1941** *Time* 9 June 19/3 Months of diplomatic finagling necessary to interview key men. **1941** *Word Study* Nov. 1/2 *Variety*..insists on describing Bergen's high art [*sc.* ventriloquism] as 'tonsilar phenagling'. **1954** J. B. Priestley *Magicians* vi. 120 All the time trying it on, fiddling and finagling, selling anybody out for fourpence. **1955** W. W. Denlinger *Compl. Boston* 173 Any attempt to fudge or finagle or to get ahead of the other fellow will be recognized by the judge for what it is. **1955** *Publ. Amer. Dialect Soc.* xxiv. 164 Big business is trying..to phenagle us all out of a dollar. **1959** *Spectator* 11 Sept. 320/1 The Western members [of the Security Council] managed to finagle their resolution..out of reach of the Soviet Union's veto. **1966** *Punch* 28 Dec. 954/2, I soon finagled my way past the innumerable porters.

final, *a.* and *sb.* Add: **A.** *adj.* **1. d.** *final drive* (see quot. 1968).

1907 R. B. Whitman *Motor-car Princ.* ix. 138 From the change-speed mechanism the power is passed to the driving wheels by the final drive. **1912** J. Armstrong *Motor* ix. 173 The final drive..is achieved mostly through the medium of bevel-gear, that..is very nearly as wear-resisting as a pair of constantly mated spur-wheels. **1968** *Practical Motorist* Nov. 333/1 *Final drive*, the last part of the transmission, consisting of the crown wheel and pinion, differential gears and axle drive shafts.

3. b. *final solution* [tr. G. *endlösung*]: name given to the German policy, from 1941, of exterminating Jewish people in Europe; the process of massacring these victims.

1947 *Trial German Major War Criminals* (H.M.S.O.) xi. p. ix, Final solution of the Jewish question. **1949** D. Macardle *Children of Europe* vi. 110 As the 'final solution', camps fitted with gas-chambers, electrocution plants and huge crematoria were erected in Poland. **1953** G. Reitlinger (*title*) The final solution. **1961** *Observer* 17 Dec. 10/3 The formal adoption of the Final Solution, ordaining the extermination of European Jewry,..on July 31, 1941.

B. *sb.* **2. a.** (Later examples.) **e.** The last edition of an issue of a daily newspaper.

1931 *Sat. Rev.* 4 July 17/1 The difference between the English play called 'Inquest' and the American play called 'Late Night Final' is a fundamental one. **1933** Bloomfield *Lang.* viii. 132 In a form like *test* or *text* we call the [-t] a main final. **1938** F. Thayer *Newspaper Managem.* 82 The *Daily News* published nine regular editions as follows...Final..3:25 P.M. **1963** [see *environment 2 c].

finale. 3. (*Transf.* example of *grand finale.*)

1874 L. Troubridge *Life amongst Troubridges* (1966) ix. 92 All the 'homeites' rush downstairs and the grand finale takes place.

finalism (fəi·năliz'm). [f. Final *a.* and *sb.* + -ism.] **1.** (In Dict. s.v. Final *a.* and *sb.*)

2. The doctrine that natural processes (e.g. evolutionary changes) are directed towards some end or goal.

1909 W. R. Sorley *Interpr. Evol.* 24 'The organization of nature', says Kant, 'has in it nothing analogous to any causality we know'. It is not mechanism; nor, again, is it finalism. **1917** A. S. Pringle-Pattison *Idea of God* xix. 370 He develops his own account of 'creative' evolution in contrast with the two rival theories of mechanism and finalism. **1952** G. Sarton *Hist. Sci.* I. xx. 515 The purpose of a being is revealed by the study of its genesis and evolution. We are falling back upon the theory of finalism (or teleology). **1966** *Philos.* XLI. 129 It only leaves the quite obvious streak of finalism in Hegel harder to understand.

finalist. [f. Final *a.* and *sb.* + -ist.] **1.** (In Dict. s.v. Final *a.* and *sb.*)

2. Any of the competitors that are left in for the final contest; also, a candidate in the last of a series of examinations (see Final *sb.* 2 c and d).

1898 *Bury Times* 2 Nov. 2/3 Two of the finalists,..just before the race started, publicly appealed to the Baths Committee to withdraw the objectionable clause. **1902** *Westm. Gaz.* 1 July 4/3 The finalists in the Gentlemen's Doubles Championships. **1906** *Daily Chron.* 25 May 1/7 The finalists in the Amateur Golf Championship. **1930** *Daily Express* 6 Sept. 2/3 Those aspidistras which seem possible finalists. **1959** *Times* 4 Feb. 2/3 Law Graduate or C.I.S. Finalist..preferred. **1970** *Globe & Mail* (Toronto) 28 Sept. 29/3 (Advt.), Blue chip Toronto based mining firm requires an accounting graduate or finalist.

finalistic (fəinăli·stik), *a.* [f. Finalist + -ic.] Of or pertaining to finalism (sense *2).

1927 *Mod. Philology* Nov. 225 But the fact that we are not omniscient..is no warrant for a finalistic hypothesis. **1958** *New Scientist* 16 Oct. 1076/3 The 'mechanistic' and the 'finalistic' descriptions of biological phenomena.

finalize (fəi·nǎləiz), *v.* [f. Final *a.* + -ize.] *trans.* To complete, bring to an end, put in final form; to approve the final form of. Also occas. *intr.*, to bring something to completion, to conclude. Hence **fi:naliza·tion,** the action or process of finalizing; an instance of finalizing.

1922 *Private Let. Standard Publishing Co., Melbourne* (G. & C. Merriam Co. files) 4 May, Our Mr. S. Jaboor writing from New Zealand advises that he was in communication with you again in March last, with a view to our acting as your representatives in Australasia, and says that you are still finalizing matters with Mr. Forbes. **1926** J. Doone *Timely Tips for New Australians* 7 The established usage of the word 'finalise' (to complete) is..illustrative of the Australian variations of the English language...To 'finalise a deal' is an expression in daily use throughout the island-continent. **1927** *U.S. Naval Inst. Proc.* May 606 Communication tests are being continued, and negotiations are proceeding in relation to finalised traffic arrangements. **1930** *John o' London's* 25 Jan. 656/1 The plaintiff..complained that her lover had promised to marry her but failed to 'finalize' matters. **1931** *Times* 26 Mar. 14 (Austral. correspondent) It is expected that these motions will be carried, thus finalizing the split between the New South Wales and the Federal organizations. **1936** F. Clune *Roaming round Darling* x. 90 Banjo..finalizes: And on misty moonlight evenings, while the dingoes howl around, [etc.]. **1936** P. Moncrieff *N.Z. Birds* (ed. 2) p. xii, Finalization of our New Zealand species has by no means been reached. **1937** I. L. Idriess *Forty Fathoms Deep* (1941) 51 If we don't finalize tonight, those two coons will get suspicious and sell to someone else. **1945** *Devel. & Welfare in West Indies 1943–1944* (Colonial No. 189. 1945), The plans for British Guiana and British Honduras have not yet been finalized, but those of the former colony are well advanced. **1952** *N.Y. Times* 17 Feb. 29 Other efforts which will lead to the finalization of a sound doctrine set forth in service manuals. **1953** N. Birkett *Magic of Words* 8 When I hear of..things being adumbrated, or visualized, or finalized..I think of that other aim of this [English] Association, 'To uphold the standards of English writing or speech.' **1955** *Times* 15 July 15/2 Our claims for refund of taxation..were finalized with the Inland Revenue, resulting in a net repayment of £76,347. **1960** *New Left Review* Nov.–Dec. 12/2, I personally would not agree to any finalisation of those matters with two powers...Only this assembly should finalise them. **1966** *New Statesman* 18 Feb. 215/3 If Russia does finalise her break with China soon, [etc.]. **1971** *Financial Mail* (Johannesburg) 26 Feb. 719/1 Subject to finalisation of certain conditions they had agreed to acquire the trading assets.

finance, *sb.*¹ Add: **8. finance bill,** a legislative bill containing financial provisions; **finance company, house,** a company that is primarily concerned with the financing of hire-purchase transactions.

1901 L. H. Courtney *Working Const. U.K.* 24 What used to be called the Customs and Inland Revenue Bill, received in 1894 the name of the Finance Bill. **1959** *Chambers's Encycl.* X. 451/1 Even finance bills are often declared not to be money bills. **1971** *Money Which?* Mar. 4/1 The tax changes the Chancellor proposes in his speech are set out in detail in the Finance Bill—published about a fortnight later. **1926** *Economist* 9 Oct. 576/2 This credit is passed on by the sellers of the goods either to banks with which they have established relations or to one of the many 'finance' companies which have been established solely to cater for this sort of business. **1968** *Listener* 6 June 725/2 Large companies..in turn are often owned by finance companies much larger than themselves. **1929** *Economist* 27 July 181/2 He declared that the hostility at one time evoked by the idea of hire-purchase or instalment selling was passing..and that the long-term credit bank or finance house had an essential contribution to make to the national welfare. **1958** *Spectator* 22 Aug. 262/3 A Birmingham hire-purchase finance house.

financial, *a.* Add: **3.** In possession of money. *Austral.* and *N.Z. slang.*

1926 J. Doone *Timely Tips for New Australians* 11 In Australia, to describe a man as being 'financial' is to describe him as being possessed of means. **1936** M. E. C. Scott *Barbara & N.Z. Backblocks* 96 Paddy, who had been harvesting for neighbours, was magnificently financial. **1940** W. S. Gilkison *Peaks, Packs & Mountain Tracks* 104 In my 'varsity days I dragged my old 'omnibus' over many a trail where, had I been more financial, I would surely have saved myself much trouble and toil by employing motor car or pack-horse. **1961** P. White *Riders in Chariot* xi. 410 'Shall I tell you, Alf,' he called, 'how us girls got to be financial?'

|| **finca** (fi·ŋkă, fi-). [Sp., f. *fincar*: see Ficche *v.*] In Spain and Spanish America: landed property, (country) estate; a ranch.

1909 in *Cent. Dict.* Suppl. **1922** *Glasgow Herald* 18 Apr. 3 Ministerial 'conclaves' at Andalucian fincas. **1934** A. Huxley *Beyond Mexique Bay* 147 A few years ago..the coffee *fincas* were in full production. **1955** *Sci. News Let.* 8 Jan. 26/2 From these mountain coffee 'fincas' come 15,000 to 20,000 tons of coffee a year. **1963** W. McGivern *Choice of Assassins* (1964) iii. 31 The few wealthy men of the village had come down from their *fincas*..for the Sunday bullfights in Málaga.

finch. Add: **1. c.** *S. Afr.* = Weaver¹ 4. Cf. *fink *sb.*¹

1801 J. Barrow *Trav. S. Afr.* I. iv. 247 Different species of small birds, chiefly sparrows, finches, and grossbeaks. **1833** S. Kay *Trav. Caffraria* iii. 85 The long-tailed finches..were here seen flying about. **1944** *Cape Times* 14 Oct. 6/7 Cape canaries, finches (their woven nests are hanging from the trees this month) and bush doves. **1963** Mackworth-Praed & Grant *Birds S. Third Afr.* II. 673 Southern Africa is very well supplied with Finches, particularly in the south.

find, *sb.* Add: **2. b.** A person who is 'discovered' or brought to public notice; a valuable discovery. orig. *U.S.*

1890 *Sporting Life* (Phila.) 8 Jan. 7/1 As to Secretary Rogers' two 'finds', Day and Anderson, little can be said in their favor. **1914** R. Grau *Theatre of Science* 139 Miss Snow..must be set down as a Thanhouser 'find'. **1915** Galsworthy *Punch & Go*, Miss Hellgrove's a find, I think. **1917** W. Owen *Let.* 14 Oct. (1967) 499 Graves was mightily impressed, and considers me a kind of *Find* !!

4. find-place = *find-spot*; **find-spot** (later examples).

1939 *Burlington Mag.* Oct. p. vi/1 A map showing find-places and centres of manufacture. **1962** *N. & Q.* Dec. 450/2 Drake was unable to learn the exact find-place. **1971** E. Okasha *Hand-list of Anglo-Saxon Non-Runic Inscr.* 43 Find-place and date, where known, with brief details of the find. **1938** *Oxoniensia* III. 49 Mortarium. Hard white ware, colour-washed. Find-spot unrecorded. Early second century. **1960** K. M. Kenyon *Archæol. in Holy Land* i. 33 The Samaria ostraca..provide limits for the reigns of the king to which they refer, but their find-spot was not accurately recorded with reference to the structural phases of the town.

find, *v.* Add: **2. b.** *euphem.* To steal. *slang.*

1865 T. Archer *Pauper, Thief & Convict* ii. 22 Too little moral restraint to go back to their miserable lodging with an empty stomach if they can '*find*' anything that will procure them a meal. **1884** J. Greenwood *Little Ragamuffins* xvi. 139 Pinchin', findin', gleanin', some coves calls it. **1936** *Punch* 2 Dec. 640/1 Certain portions have been 'found' in the literal Army sense of the word, which is a polite way of describing petty theft. **1945** G. Millar *Maquis* viii. 155 He left in a rush and flurry on a brand-new woman's bicycle that he had 'found' in Besançon.

9. f. Of a letter, etc.: to reach (a person); of an address: to be adequate to enable correspondence to reach (a person). Also †*to find out* (obs.).

1605 T. Matthew *Let.* 29 May in Mathew & Calthrop *Life Sir T.M.* (1907) iii. 45, I hope you will acquaint me with your mind, by letter, wᶜʰ wᵗʰ direction at Signore Thomaso Yonge,..will find me out. **1922** W. B. Yeats *Let.* 7 July (1954) v. 687, 82 Merrion Square will always find us.

12. b. *to find religion* (and similar phrases): to experience religious conversion.

1877 *Independent* 4 Jan. 11/3 He said that at least 200 boys had found Jesus in the Tabernacle. **1932** H. J. Laski *Let.* 21 Aug. in *Holmes–Laski Lett.* (1953) II. 1402 He..asked me in a loud voice..if I had found Christ. I said that I was..exempt from religious experience. **1957** J. S. Huxley *Relig. without Revelation* i. 22 If..we have in any true sense of the phrase, 'found religion', it means that we shall have so organised our minds that, for flashes at least, we attain to a sense of interpenetration with the reality around us.

18. Also simply *found* = all found. Chiefly *U.S.*

1826 [see *field-hand]. **1830** S. Smith *Major Downing* 13, I get 12 dollars a month and found. **1839** Dickens *Nickleby* xii. 107 An annual salary of five pounds..and 'found' in food and lodging. **1853** B. F. Taylor *Jan. & June* (1871) 273 A story..he wouldn't have whispered for twelve dollars a month 'and found'. **1923** 'B. M. Bower' *Parowan Bonanza* iv. 48, I got him cheap for yuh. Three dollars and found.

19. d. In Harrow School phraseology (see quot. 1905). Also as *sb.* (see quot. 1881).

1881 M. J. Rendall in C. E. Pascoe *Everyday Life in our Public Schools* 210 In a large House there are usually four Sixth Form 'finds' (a Harrow term signifying a mess of three or four senior boys who take tea and breakfast in a room separate from the hall). **1889** Barrère & Leland *Dict. Slang* I. 362/2 *To find*, to mess together. **1905** H. A. Vachell *Hill* x. 218 You will find together. Of course Scaife can find with you, if you wish. *Footnote.* 'Finding' is the privilege, accorded to the Sixth Form, of having breakfast and tea served in their own rooms instead of in Hall. *Ibid.* xi. 244 They no longer shared No. 7..but they still 'found' together.

20. find out. e. *absol.* or *intr.* To make a discovery; to discover a fact, the truth, etc.; also with prep. *about.*

1862 G. Macdonald *D. Elginbrod* ii. ii, 'I don't like the pigs—I don't know where they are.' 'Well, we must

find out.' **1881** C. E. L. Riddell *Senior Partner* xxxi, 'Who might that one be?' 'I am thinking ye'll have to find out for yourself.' **1893** M. E. Mann *In Summer Shade* xix, 'He has found out about Mrs. Le Grice's bill,' said Lally to herself. **1894** B. Harraden *Varying Moods* 132 Perhaps death brings peace. I shall soon find out about that. **1913** W. M. Gallichan *Like Stars that Fall* xi, Supposing your husband found out? There might be a devil of a row. **1927** M. Sinclair *Anthony Waring* xviii, Tony determined to find out. Aunt or no aunt, he would go to her.

finder. Add: **1. d.** Colloq. phr. *finders keepers,* whoever finds something is entitled to keep it (cf. *findings* (*are*) *keepings,* *finding *vbl. sb.* 6).

[**1825** J. T. Brockett *Gloss. N. Country Words* 89 No halfers—findee keepee, lossee seekee. **1828** D. M. Moir *Mansie W.* xi. 98 According to the auld Scotch proverb of 'he that finds keeps, And he that loses seeks'.] **1856** C. Reade *Never too Late* iii. xiii. 127 We have a proverb—'Losers seekers finders keepers'. **1923** G. E. Kelly (*title*) Finders-Keepers. **1961** B. E. Wallace *Death packs Suitcase* vi. 60 In this game it's going to be finders keepers. **1969** *Daily Express* 17 Mar. 9/3 Where I come from it's finders keepers, losers weepers.

3. d. (Further examples.)

1892 in L. de Vries *Victorian Advts.* (1968) 76/1 The 'artist', ¼-plate camera Containing Euryscope lens working at F/6, large Finder with Hood. **1894**, etc. [see *brilliant *a.* 1 b]. **1915** B. E. Jones *Cinematograph Bk.* iii. 15 A finder is provided for focusing purposes, consisting of a long narrow tube..running right across the camera and having a cap..outside. **1951** G. H. Sewell *Amateur Film-Making* (ed. 2) iii. 33 Generally the finder looks along a line different from the axis of the taking-lens, but parallel to it.

fin de siècle. Add: (Further examples.) Also, decadent. Hence **fin-de-siècl(e)ism,** the state or quality of being characteristically *fin de siècle.*

1893 [see *Argonaut 1 b]. **1896** W. Caldwell *Schopenhauer's Syst.* 523 What is called *Fin-de-Siècle-ism.* **1898** *Westm. Gaz.* 4 July 1/2 These days, which pessimists call *fin-de-siècle* with some sort of idea that the phrase indicates a weariness and weakening of purpose. **1908** *Daily Chron.* 22 Apr. 4/6 He has recreated for us the Versailles of Louis XIV. in that fin-de-siècle epoch when the glories of the Grand Monarque were fading. **1930** *Punch* 30 Apr. 498 This charming old comedy can be trusted to delight us on its own merits. *Fin-de-siècle* trimmings are alien to its spirit. **1948** H. Acton *Mem. Aesthete* vi. 119, I had started sweeping away *fin-de-siècle* cobwebs with a paper called *The Oxford Broom.* **1965** *Listener* 25 Nov. 863/1 He..ignores the aspects of Proust that do not interest him personally; the realist, the chronicler of *fin de siècle* society, and the student of inversion are hardly mentioned. **1966** *New Statesman* 27 May 786/1 [The description] conveys none of his fin-de-siècleism. **1967** R. Croft-Cooke *Feasting with Panthers* viii. 166 Even French phrases carried sinister meanings and when people called the literary pseudo-movement of the Nineties *fin-de-siècle* it added an ominous purple hue to the earlier writings of Wilde and of those..who imitated him.

finding, *vbl. sb.* Add: **4. c.** (*b*) (See quot. 1939.) *U.S.*

1896 *Godey's Mag.* Feb. 222/2 The cost of findings for a waist. **1939** M. B. Picken *Lang. Fashion* 57/2 *Findings*, threads, tapes, buttons, bindings, hooks and eyes, slide fasteners, Featherbone, belting, braids, and other sewing essentials used in garment making; carried in notion departments. **1971** *Lebende Sprachen* XVI. 11/1 US findings-BE/US sewing things.

6. Proverbial phr. *findings* (*are*) *keepings.* Also in *sing.* form. Cf. *finder 1 d.

1863 J. H. Speke *Discov. Sources of Nile* v. 121 The scoundrels said, 'Findings are keepings, by the laws of our country; and as we found your cows, so we will keep them.' **1904** *Daily Chron.* 27 Sept. 1/7 Harsh sentences of imprisonment for 'findings-keepings' offences. **1959** I. & P. Opie *Lore & Lang. Schoolch.* viii. 134 Articles may be acquired..by 'finding-keeping'.

7. *Comb.*, **finding-list** (see quot. 1961).

1889 in *Cent. Dict.* **1930** *Library* XI. iii. 260, I had prepared a skeleton finding-list only. **1961** T. Landau *Encycl. Librarianship* (ed. 2) 142/2 *Finding list*, a list of books, often of a special kind or by a particular writer, represented in a library or in the libraries within a given area.

findrinny (fi·ndrini). [ad. Ir. *findruine* (Roy. Irish Acad. *Dict. Irish Lang.*), f. OIr. *findbruine*, f. OIr. *find* (mod.Ir. *fionn*) white + *bruine of unknown meaning: ultimate origin obscure.] White bronze.

[**1873** E. O'Curry *Manners & Customs Anc. Ir.* I. p. ccccl xv, In Norse shields the rim was of iron, but in the descriptions of shields to be found in Irish poems and prose tales, the material of the rim is almost always *Findruini.* *Ibid.*, *Findruini* was probably bronze coated with tin or some white alloy. *Ibid.* III. 100 She had presented him at the outset with..a bracelet of *Findruine* or carved white metal (silver bronze) to fit his left wrist.] **1895** W. B. Yeats *Poems* 6 On a horse with bridle of findrinny. *Ibid.* 22 With hoofs of the ruddy findrinny. **1934** J. Joyce *Let.* 16 Oct. (1957) 348 A 30-year wedding should be called a 'findrinny' one. Findrinny is a kind of white gold mixed with silver.

fine, *a.* Add: **4. c.** In good health, well. (Cf. *fiddle *sb.* 1 b.)

1791–1897 in *Eng. Dial. Dict.* (chiefly northern and Sc.). **1959** *Listener* 9 July 59/1 The sleepy subjects felt fine during the complex interesting tests. *a* **1968** H. Hood in R. Weaver *Canad. Short Stories* 2nd Ser. (1968) 220 There's nothing wrong with my heart. I've just had my checkup and I'm fine, just fine.

7. f. With advb. force in *Cricket*: behind the wicket and near the line of flight of the ball (opp. *square*).

1867 J. *Lillywhite's Cricketers' Compan.* 9 His [*sc.* the fieldsman's] own judgement will tell him..to play fine at one time, and square at another. **1888** Steel & Lyttelton *Cricket* iii. 147 It is a golden rule for every bowler..on these wickets to have a short-slip 'finer' than on ordinary occasions. **1891** W. G. Grace *Cricket* x. 263 Short-leg..must be quick to detect..whether he means to play it fine, square, or more in the direction of mid-on. **1961** F. C. Avis *Sportsman's Gloss.* 119/2 *Fine leg*, the position of a fieldsman on the leg side but behind the wicket. **1963** A. Ross *Australia 63* vii. 129 Harvey swung across the line of the next ball, and it went straight in and out of Sheppard's hands at short fine leg.

g. In colloq. expressions such as *to cut fine, to run fine,* with advb. force: with a very slight margin (of time, space, etc.).

1871 *Chambers's Jrnl.* 9 Sept. 563/1 For purposes of concealment we sometimes ran it very fine. **1884** *Boy's Own Paper* Xmas no. 40/2 In endeavouring to round a headland the Crystal cut matters too fine, got on to rough ice. **1890, 1892** [see Run *v.* 36 d]. **1896** *Daily News* 24 July 8/6 The making of clothes was, he added, 'cut very fine'. **1899** *Chambers's Jrnl.* 30 Sept. 694/1 If by..chance I was cutting it fine. **1908** *Pearson's Weekly* 5 Mar. 606/2 Running it fine. **1930** W. S. Churchill *My Early Life* vii. 107 This was running things very fine, but it was not until my journey was half completed that I realised that I should be almost certainly late for dinner. **1950** 'W. Cooper' *Scenes Provincial Life* i. v. 72 Robert had ordered himself to come last. Tom argued with Robert about cutting it fine.

h. Of a ship's bow: slender.

1927 G. Bradford *Gloss. Sea Terms* 64/1 *Fine*, knife-like forward, where a vessel enters the water. The more a vessel departs from the shape of a box, the finer her lines.

i. *fine print*: applied to those detailed clauses of a contract, warranty form, etc., that are printed in small type and specify the limits, qualifications, etc., of the agreement. Also *fig.*

1960 J. Carswell *S. Sea Bubble* i. 19 He produced the 'fine print' which it is peril to leave unread. **1966** *Guardian* 28 Nov. 1/1 A careful reading of the fine print of Mr Smith's latest proposals is likely. **1968** D. Hitchens *Cabin of Fear* i. 13 He was a fanatic about the fine-print..and he didn't get where he was by being careless.

j. *fine chemicals*: see *chemical *a.* 6 b.

8. *fig.* See also Point *sb.*¹ B. 1 d.

12. In phr. *fine and dandy*: see *dandy *a.* 2.

d. Of printing: intended for display as much as for reading; of the condition of a book: see quot. 1948; of an edition of a book: printed on better-quality paper, esp. in *fine-paper copy.*

1867 [see Press-work]. **1874** *Art of Paper-Making* iii. 23 In a fine-paper mill, the increase of manual labour..is more than compensated by the greater purity of the resulting product. **1912** A. W. Pollard (*title*) Fine books. **1948** J. Carter *Taste & Technique in Book-Collecting* xii. 172 In its most absolute sense..the term 'fine', applied to any book of any period, could be said to mean no more (if no less) than that all its leaves were present, clean, whole and amply margined; that it was sound and undisturbed in its binding; and that that binding, whatever its material, was fresh and unblemished and, if of leather, of a sufficient elegance to raise it above the modest category of 'neat'. **1952** E. J. Labarre *Dict. Paper* (ed. 2) 104/1 *Fine paper*,..indicates the printing of a book on better paper than the bulk of the edition. **1957** R. L. Collison *Book Collecting* 197 Formerly very popular, this process [*sc.* marbling] is now mostly confined to account books and ledgers, though since the war it has also been used in fine printing. **1964** F. Bowers *Bibliogr. & Textual Crit.* iii. vi. 84 One perhaps aberrant fine-paper copy. **1970** *Sotheby's Catal. Sale Children's Books* 16 Mar. 29 The Golliwogg Books, a complete set of 13 vol., the last eight first editions, coloured illustrations, original cloth-backed boards, pictorial upper covers, seven covers slightly spotted, otherwise a very fine set.

15. b. Also *some fine day*, one day in the future.

1846 G. E. Jewsbury *Sel. Lett. to Mrs. Carlyle* (1892) 189 Some fine day she will let you inside that moral mystery, and will hate you for it afterwards. **1890** [see Money *sb.* 3 a].

B. *sb.* **e.** *pl.* Very small particles: used in various technical senses. (See also quot. 1880).

1880 J. Dunbar *Pract. Papermaker* 14 Fines consist of fine white cottons [*sc.* rags]. **1904** *Westm. Gaz.* 28 July 3/2 Breaking up much waste rock into fines. **1908** *Ibid.* 22 Oct. 13/1 All fines [*sc.* fine ores] are discharged into the tank over the head of the funnel with great force. **1930** *Engineering* 4 Apr. 449/2 The fines are entirely withdrawn from the coal. **1950** *Ibid.* 13 Jan. 53/2 It was..found possible to stabilise a powder by mixing it with dry stearic acid or zinc stearate in small amounts...The chief objection to the process was that the flowing properties of the treated powder were poor, the deterioration being more marked than would have been expected from the slight increase in the proportion of 'fines' during the milling. **1951** *Gloss. Terms Plastics* (*B.S.I.*) 21 *Fines*, a qualitative description of the extremely small particles which may be present in a granular powder. The presence of this very fine material in substantial amount may in some instances be undesirable. **1957** *Brit. Commonw. Forest Terminol.* ii. 73 *Fines*, in pulping, the very short or fragmented fibres. **1965** G. J. Williams *Econ. Geol. N.Z.* xvi. 266/2 Many of the quartz sand deposits contain too much fines (i.e. passing 100 mesh BSS) for glass manufacture. **1969** *Times* 2 May (Suppl.) p. iv/6 All companies with the exception of the Savage River group produce iron ore lump and fines. **1970** *Nature* 24 Jan. 325/2 The word fines is accepted in lunar research as describing all those solid particles with a size less than 1 cm. *Ibid.* 326/1 Several others also reported the gradual loss of adhesion on exposure of fines material to a gaseous atmosphere.

f. See *fine champagne.

C. *adv.* **b.** (Additional examples.)

1932 T. S. Eliot *Sweeney Agonistes* 18 We like London fine. **1947** Partridge *Usage & Abusage* 118/2 *Fine* as an adverb ('He's doing fine') is dialectal and colloquial.

D. *Comb.* **1. a.** *fine-looking* (examples).

1799 Malthus *Jrnl.* 2 June (1966) 44 We saw a great number of healthy fine looking children. **1844** J. Cowell *30 Yrs. among Players* (1845) i. x. 25/2 A fine-looking young man and a beautiful girl. **1922** Joyce *Ulysses* 734 Killing any finelooking men there were.

b. *fine-boned, -leaved, -textured, -woolled* (earlier example).

1927 Joyce *Pomes Penyeach*, I wrap him warm And touch his trembling fineboned shoulder. **1959** J. Wain *Travelling Woman* iii. 22 The chin..at the same time fine-boned and intellectual. **1885** T. Baines (*title*) Greenhouse & stove plants, flowering and fine-leaved, palms, [etc.]. **1952** A. G. L. Hellyer *Sanders's Encycl. Gardening* (ed. 22) 174 [*Edraianthus*] *tenuifolius*, blue fine-leaved. **1901** Fine-textured [see Textured *a.*]. **1958** *Yearbk. Agric.* 1957 (U.S. Dept. Agric.) 758/1 *Fine-textured soil*, roughly, clayey soil containing 35 percent or more of clay. **1797** Fine-woolled [see Woolled *a.* 2].

3. fine-axe *v. trans.*, to face (stone) to a smooth surface by tapping with a mason's axe; **fine entrance, entry,** a slender ship's bow; **fine-groove,** the groove on a long-playing gramophone record; usu. *attrib.*, as *fine-groove record*; **fine-toothed** *a.*, (*a*) also of a saw.

1886 H. C. Seddon *Builder's Wk.* 82 Fine axed is a more careful description of single axed work. **1895** *Daily News* 18 June 6/4 All kinds of Scotch granites, polished and fine-axed. **1869** Fine entrance [see Entrance *sb.* 6]. **1961** F. H. Burgess *Dict. Sailing* 86 *Fine entry, fine entrance*, a lean bow. A quality possessed by any vessel with long tapered bows to offer less resistance. **1956** (*title*) Reproducing equipment for fine-groove records. (B.B.C.) **1957** *B.B.C. Handbk.* 52 A further development during 1956 was the introduction of 'fine groove' long-playing disk recording equipment. **1958** *Chambers's Techn. Dict.* 967/1 (s.v. *coarse groove*). Fine groove, or microgroove, used for vinylite long-playing discs. **1960** *Farmer & Stockbreeder* 8 Mar. (Suppl.) 5/1 Cut very carefully using a fine-toothed saw.

fine, *v.*³ **1.** Delete '*Obs.* exc. with reference to beer' and add later examples with reference to wine.

1843 Ure *Dict. Arts* (ed. 3) 1302 In the month of January the clear wine is racked off, and is fined by a small quantity of isinglass. **1906** Mrs. Beeton *Bk. Househ. Managem.* lxix. 1790 There are various methods of fining wine; eggs, isinglass, gelatine and gum Arabic are all used. *Ibid.*, A cask of clarified wine will fine thirteen dozen bottles of port or sherry.

6. c. *to fine down* (earlier example).

1873 C. M. Yonge *Pillars of House* xlvii. 332 Matilda was..better looking at two-and-thirty than at two-and-twenty, for she had somewhat fined down.

7. In Australia const. *up.*

1966 Baker *Austral. Lang.* (ed. 2) xvi. 350 *Bogaduck weather* for heavy rain and *to fine up*, (of weather) to become fine, are other Australianisms.

fine art. 2. (Examples.)

1855 Geo. Eliot *Let.* 16 July (1954) II. 210 Literature is Fine Art, and the man who writes mere literature with insolent slovenliness is..inexcusable. **1856** *Art-Jrnl.* 57/3 Glasgow Fine-Art Association...The paintings..were exhibited in the St. Vincent Street Fine-Art Gallery. **1863** [see *art gallery s.v. *art *sb.* V]. **1942** Blunden *Romantic Poetry & Fine Arts* 17 Mrs. Hemans must have kept a sharp look-out for fine-arts literature. **1955** R. Blesh *Shining Trumpets* (ed. 3) i. 6 A hybridized popular music rather than a fine-art form. **1966** *Penrose Ann.* XIX. 164 One has met fine art students..aware of their awe-inspiring heritage.

|| **fine champagne** (fin ʃampan^y). [Fr., = eau-de-vie *fine* de la *Champagne* 'fine brandy of the Grande Champagne and Petite Champagne vineyards' (of the department of Charente).] Old liqueur brandy (from the vineyards mentioned above). Also abbreviated *fine* (*de la*) *maison*, old brandy 'of the house'; so *fine*, (French) brandy.

1868 G. A. Sala *Notes & Sk. Paris Exhib.* 167 We shall require sixty bottles of cognac, *fine champagne.* **1875** H. Vizetelly *Wines of World* 19 The Enrageat,..Folle Blanche, or crazy vine, which prevails so extensively in the Deux-Charentes, and from whose fruit *fine champagne* cognac is distilled. **1923** E. E. Cummings *Let.* 15 Sept. (1969) 103, I prefer fine to calvados. **1927** E. Hemingway *Fiesta* i. 12 We had several *fines* after the coffee. **1928** A. E. W. Mason *Prisoner in Opal* xxiii, So! Some coffee and some *fine de la maison*? **1928** *Observer* 11 Mar. 12/2 The wine waiter has had the habit of insidiously proposing a 'fine maison' to the customer who wants an old brandy. **1951** N. Balchin *Way through Wood* xii. 174, I ordered two fines. **1964** E. Kelen *Peace in Their Time* xiii. 350 He was afraid that if he bought somebody a *fine* he would one day die of starvation.

fine-cut, *a.* and *sb.* [Fine *adv.*] **A.** *adj.* (See Fine *a.* D.3.)

1837 *Knickerbocker* IX. 268 He ejected a quid from his mouth, that would have shamed in size the largest paper of Lorillard's fine-cut chewing tobacco. **1878** *Congress. Rec.* 5 June 4164/2 Gentlemen of fortune..can very well afford..to chew their fine-cut tobacco.

B. *sb.* **a.** A kind of fine-cut tobacco.

1844 *Knickerbocker* XXIII. 288, I thrust a ball of 'Mrs. Miller's fine cut'..between the sub-maxillary bone and its carnal casement. **1864** *Congress. Globe* 3 June 2707/3 On tobacco, cavendish, plug, twist, fine cut, and manufactured of all descriptions. **1909** 'O. Henry' *Roads of Destiny* xxi. 355 Haven't got a chew of fine-cut on you, have you? **1939–40** *Army & Navy Stores Catal.* 84/3 Smoking mixtures..(coarse and fine cuts).

b. (See quot.)

1959 W. S. Sharps *Dict. Cinematogr.* 96/1 *Fine cut*, in editing,..the term used for the workprint after the rough cut has been further edited.

fine gentleman. Add: (Earlier examples.) Also *attrib.* or as *adj.* So **fine-ge·ntlemanism, fine-ge·ntlemanship.**

1583 C. Hollyband *Campo di Fior* 207 You fine gentelman, do you leane on the table? **1628** Earle *Microcosm.* xxxvi, *A Blunt Man*..is a great enemy to the fine Gentleman. **1728** Gay *Beggar's Opera* i. iv. 4 How many fine Gentlemen have we in Newgate every Year? **1866** *Athenæum* 24 Nov. 676/2 The evanescent air of fine gentlemanism. **1874** L. Carr *Judith Gwynne* xvii, That boredom is one of the necessary traits of fine gentlemanism. **1833** *Contemp. Rev.* June 899 The spirit of fine gentlemanship. **1928** *Smart Set* Sept. 49/1 Your fine-gentleman airs sit ridiculously on you.

fine-grain, *a.* [f. Fine *a.* 7 + Grain *sb.*¹] Having a fine grain (Grain *sb.*¹ 12–15); consisting of particles that are very small (Grain *sb.*¹ 7); *spec.* in *Photogr.*, containing or composed of particles smaller than normal, so that considerable enlargement of the photograph is possible without any graininess becoming apparent; suitable for producing such photographs.

1927 *Brit. Jrnl. Photogr. Alm.* 1928 758 (*Index*) Fine-grain images, metol developer, Eastman formula. **1928** *Ibid.* 1929 247 *Fine-grain developers*,..the following metol developer is advocated..for giving greater contrast than the fine-grain developers usually quoted. **1934** J. L. Baring tr. *Heering's Rolleiflex-Bk.* 61 Miniature photography has compelled the industry..to produce fine-grain film. **1937** *Discovery* Jan. 22/2 Users of the popular miniature cameras may remember that some years ago their choice lay between a fine-grain emulsion with a slow speed, necessitating a longer exposure, and a coarse-grained one that did not enlarge so well but had a faster speed. **1963** C. R. Cowell et al. *Inlays, Crowns & Bridges* ii. 6 It is important, therefore, to use a fine-grain cement. **1966** *Lancet* 31 Dec. 1429/1 X-rayed separately with an aluminium step-wedge on fine-grain industrial film.

Also **fine-grained** *a.* (also *fig.*).

1538, 1634, 1799 [see Grained *ppl. a.*² 2]. **1838** J. M. Spearman *Brit. Gunner* (ed. 2) 252 The proof of fine-grained, or small-arm powder, is with a charge of two drachms from a carbine barrel. **1859** [see Fine *a.* D. 1 b]. **1894** W. James *Coll. Ess. & Rev.* (1920) 342 In this sense he is original, for no one has yet attained to writing up the subject in as fine-grained a way as this. **1906** *Jrnl. Geol.* XIV. 695 Phanerocrystalline rocks have been described as coarse, medium, or fine grained, with little effort to fix absolute values for these terms. **1911** B. E. Jones *Cassell's Cycl. Photogr.* 280 Process dry plates are relatively slow in order to secure a fine-grained emulsion. **1939** A. Johannsen *Descr. Petrogr.* (ed. 2) I. iii. 31 If the grains are less than a cubic millimeter in size, but recognizable as individuals, the rock is said to be fine-grained. **1949** J. E. Garside *Process & Physical Metall.* xvii. 288 Resulting from variation in grain size, it is possible to distinguish between fine- and coarse-grained steels. **1965** *Language* XLI. 202 A single fine-grained quantization of acoustic space.

fine hair, *sb.* [Fine *a.* 7.] (See quot. 1901.)

1897 C. T. Davis *Manuf. Leather* (ed. 2) 109 Besides the hair proper, the animal hide is covered with a short down, similar to what is found on the arms and legs of the human body. This down is known to the tanners as fine-hair. **1901** F. T. Addyman tr. *Villon's Leather Industry* 359 *Fine hair*, long, fine but coarser than wool, as the rabbit. **1962** Stubbings & Fleisch in F. O'Flaherty et al. *Chem. & Technol. Leather* III. xxxviii. 245 The shorter the unhairing process, usually the more fine hair is left on the sides.

Hence **fine-hair** *v. trans.*, to remove the fine hair or down from (a skin).

1897 C. T. Davis *Manuf. Leather* (ed. 2) 158 The best time for fine-hairing is after the drenching, as the grain of the skins is then cleaner and the fine hairs more easily seen. **1903** H. R. Procter *Princ. Leather Manuf.* 180 After bating, the hides are usually 'worked' ('scudded', 'fine-haired') on the beam, to remove dirt and grease.

fine-hand, *a.* [Fine *a.* 7 d.] Written in a fine or delicate hand. Also *fig.*

1845 C. M. Kirkland *Western Clearings* 154, I want a fine-hand copy. **1846** *Knickerbocker* XXVII. 279 How many recollections of early school-days,..of coarse and fine-hand pens. **1878** B. F. Taylor *Between Gates* 228 From this rock the horse-trail climbs to the right for Nevada Fall and a fine-hand affair, a foot-trail, trends up to the left of Vernal Fall.

fine lady. (Earlier examples.)

1707 M. Lovett *Let.* 26 Mar. in M. M. Verney *Verney Lett. 18th Cent.* (1930) I. xii. 200 There was all the Beaux and fine Ladys in town. **1728** Gay *Beggar's Opera* i. iv. 6 She loves to imitate the fine Ladies.

So **fine-la·dyship,** the condition of being a fine lady.

1848 Thackeray *Van. Fair* lxiv. 588 She resumed her fine-ladyship, and tried to look and feel as if she was in May Fair once more.

finely, *adv.* **8.** Add: *finely-axed*, fine-axed (*fine *a.* D. 3); *finely-cut.*

1902 *Daily Chron.* 13 Oct. 7/1 The pedestal is of finely-axed Kemnay granite. **1835** *Court Mag.* VI. 7/1 Her finely-cut features. **1952** A. G. L. Hellyer *Sanders' Encycl. Gardening* (ed. 22) 5 [*Acer Hersii*] *laciniatum*, finely-cut leaves.

fineness. Add: **2. a.** Now *spec.* the number of parts per thousand of gold or silver in an alloy.

1843 Ure *Dict. Arts* (ed. 3) 74 In estimating or expressing the fineness of gold, the whole mass spoken of is supposed to weigh 24 carats. **1862** G. H. Makins *Man. Metallurgy* viii. 144 From numerous examinations of the quality of the fine silver..the author finds that the average fineness is about 998 parts in the thousand. **1957** *Encycl. Brit.* II. 555/2 The gold present is reported as so many parts of gold in 1,000 parts of alloy. This is called 'Fineness', and British gold coinage is said to be '916·6 fine'.

4. b. fineness ratio *Naut.* and *Aeronaut.* (see quot. 1915).

1911 *Reports & Mem.* (Adv. Comm. Aeronaut.) 58 With a view to deciding upon the best fineness ratio for a non-rigid airship, the same calculation has been made for the case of constant *net* lift. **1915** W. E. Dommett *Submarine Vessels* ii. 10 Another matter which has been experimented upon is the value of the ratio between the length divided by the diameter (or width). This ratio is known as the 'fineness ratio'. **1930** *Engineering* 8 Aug. 169/1 A stream-line body of 5·45 fineness ratio.

finery¹. Add: **4.** *Comb.*: **finery folks,** people stylishly dressed; **finery-ironer, -machinist** (see quots.).

1828 W. Dyott *Jrnl.* Apr. (1907) II. 22 Dick and I..became lookers-on at the assembled crowds and the finery folks going to Sᵗ James's. **1895** *Westm. Gaz.* 16 Apr. 3/2 Of the ironers: four made incomplete weeks; one worked 72,..and one (the finery ironer) 75½. **1908** *Daily Chron.* 12 June 9/6 Laundry.—Wanted finery ironers. **1921** *Dict. Occup. Terms* (1927) § 918 *Finery machinist*, general term for any person ironing delicate articles by machine in a laundry.

|| **fines herbes** (finzɛrb'), *sb. pl.* [Fr., lit. 'fine herbs'.] A mixture of herbs used in cooking. Freq. in adj. phr. with *aux*, as *omelette aux fines herbes*, a savoury omelette flavoured with herbs.

1846 A. Soyer *Gastronomic Regenerator* 474 (*heading*) Omelette aux fines herbes. **1861** Mrs. Beeton *Bk. Househ. Managem.* 169 (*heading*) Whiting aux fines herbes. **1908** G. B. Shaw *Let.* 27 Aug. (1956) 137 What a lunch! *Sweet* omelette (fines herbes, indeed!) and crab apples. **1936** Lucas & Hume *Au Petit Cordon Bleu* 129 For a 'fines herbes' add 1 large tablespoon of chopped herbs with a little onion. **1965** *House & Garden* Dec. 84/3 *Fines herbes.* This means a mixture of fresh, chopped parsley, tarragon, chives and chervil.

finesse, *sb.* Add: **4. b.** Also in Bridge (see quot. 1959).

1902 *Encycl. Brit.* XXVI. 371/1 (s.v. *Bridge*), Deep finesses should be made when there is no other way of stealing a trick. **1958** *Listener* 18 Dec. 1053/2 West could look for twelve top tricks and the club finesse at worst. **1959** Reese & Dormer *Bridge Player's Dict.* 89 A finesse is an attempt to profit from a favorable lie of the cards. A player tries to win or establish a trick with a card that is not the highest held by his side.

finesse, *v.* Add: **2. a.** Also in Bridge.

1902 *Encycl. Brit.* XXVI. 371/1 (s.v. *Bridge*), It is in general unsound to finesse against a card that must be unguarded. **1960** J. Betjeman *Summoned by Bells* i. 5 If you'd finessed my heart And played your diamond.

fine structure. [tr. G. *feinstruktur* (A. Sommerfeld 1916, in *Ann. d. Physik* LI. 2).]

1. *Physics.* **a.** (The presence of) multiplets of closely spaced lines in atomic spectra that appear as single lines under low resolution; usu. applied to multiplets that result from the splitting of energy levels by the electron spin-orbit coupling, in contrast with *hyperfine structure.*

1918 *Phil. Mag.* Oct. 345 Here a^2 is the small quantity $(2\pi e^2/hc)^2$, (where e is the electronic charge, h Planck's constant, and c the velocity of light) occurring again as a universal constant in the discussion of the fine structure of lines. **1936** *Discovery* Nov. 363/1 The fine-structure of spectral lines frequently arises through the nuclei of the atoms, to which the spectrum is due, having a spin. **1942** J. D. Stranathan *'Particles' of Mod. Physics* vi. 224 The details of fine structure differ from one element to another, both as regards the number of lines in the multiplet and the spacing of these lines. **1962** H. D. Bush *Atomic & Nuclear Physics* iii. 58 If the fine structure is examined with the aid of very high resolution instruments, each component of the fine structure is resolved into further components with separations of the order of $1\ cm^{-1}$ (*hyperfine structure*).

b. Similar groups of lines or energies in other spectra.

1930 *Proc. R. Soc.* A. 3 Sept. 229 To examine the fine structure of the α-rays, Rosenblum has employed the focussing method so much used in the examination of β-ray spectra. **1942** J. D. Stranathan *'Particles' of Mod. Physics* viii. 344 The previously supposed homogeneous α-rays emitted from ThC actually fall into five distinct groups. The rays are said to have a fine structure. **1955** R. D. Evans *Atomic Nucleus* xiii. 424 Energy levels in the product nucleus determine the 'fine structure', or energy groups, of the reaction products.

2. *gen.* Small-scale variation in structure, texture, appearance, etc.

1933 *Nature* 17 June (*heading*) Fine-structure of the ionosphere. **1935** *Discovery* Apr. 108/1 This [X-ray] technique yields information on the fine-structure of bodies, of a sub-microscopic order, while the radiographic method reveals the gross structure. **1946** *Nature* 12 Oct. 519/1 An account was given of the examination of..a cleavage face of a selenite crystal...Only low magnifications..were used, yet it was observed that the fringes were extremely ragged, indicative of a complex surface 'fine-structure'. **1967** R. Michard in J. N. Xanthakis *Solar Physics* v. 116 One can see that these flocculi are organized in a large cellular network..visible mainly on average-quality pictures where the fine structure is smoothed out. **1968** *Brit. Med. Bull.* XXIV. 236/1 The fine structure of proteins has also been subjected to taxonomic analysis to estimate the relationships between the constituent amino acids. **1971** *Sci. Amer.* July 19/1 Even more striking than the differences among the states was the shifting of populations at the county level—the changes in what might be called the fine structure of the population distribution.

3. *attrib.*, as *fine-structure constant.*

1923 H. L. Brose tr. *Sommerfeld's Atomic Struct. & Spectral Lines* viii. 497 Our representation of the L-doublet is fully determined and expressed rationally by the fine-structure constant..by the Rydberg frequency..and by the nuclear charge. **1930** Ruark & Urey *Atoms, Molecules & Quanta* v. 134 Introducing the dimensionless fine-structure constant, $a = 2\pi e^2/hc = (7{\cdot}284 \pm 0{\cdot}006)\cdot 10^{-3}$, we get [etc.]. **1935** A. S. Eddington *New Pathways in Sci.* xi. 234 The fine-structure constant..has reference to the structure of spectral lines, but the constant occurs more widely; just as the velocity of light occurs in many problems unconnected with optics. The fine-structure constant is really the ratio of two natural units or atoms of action. **1935** S. Tolansky *Fine Structure in Line Spectra and Nuclear Spin* ii. 5 In general the fine structure components of a line lie very closely together. **1970** *Sci. Amer.* Oct. 63/1 The fine-structure constant is the basic constant of quantum electrodynamics.

fine-tooth, *a.* [Fine *a.* 7 d.] Of a comb: having fine and closely-set teeth. Also in *fig.* phrases. So **fine-tooth-comb** *v. trans.*, to comb (the hair of someone) with such an article; also *fig.* Hence **fine-tooth-combing** *vbl. sb.*

1839 H. Malcom *Trav.* I. ii. 37 Friends who wish to make little presents to the Karen Christians, might send fine-tooth combs. **1849** *Rep. Comm. Patents* (U.S.) (1850) 267 Improvement in making Ivory fine-tooth Combs. **1878** B. F. Taylor *Between Gates* 246 As slender in the distance as a fine-tooth comb. **1889** G. B. Shaw *London Music in 1888–89* (1937) 197 That lady benevolently clothed it [*sc.* a child], fine-tooth-combed it. **1891** *Century Mag.* Feb. 595 I'll go through this town like a fine-tooth comb but what I'll find him. **1935** 'N. Blake' *Question of Proof* xi. 221 We've been over the whole ruddy caboodle with a fine tooth comb. **1949** 'N. R. Nash' *Young & Fair* ii. ii. 71 But I'm going to fine-tooth-comb this school. **1957** *Times Lit. Suppl.* 22 Nov. 710/2 The general reader may..skip the fine-tooth combing of evidence. **1966** A. Firth *Tall, Balding, Thirty-Five* xvii. 216 We've gone through the remains of the helicopter with a fine-tooth comb, but there wasn't much left.

finetop. *U.S.* [Fine *a.* 7.] The meadow and pasture grass of eastern North America (*Agrostis tenuis*), also known as *herds-grass*, etc.

1856 *Rep. Mass. Board Agric.* I. 26 Redtop, Finetop, Burden's Grass, Dew Grass, Herds Grass of Pennsylvania and Southern States. **1889** G. Vasey *Agric. Grasses* (ed. 2) 46 *Agrostis vulgaris* (Redtop, Finetop, Herds Grass, Bent Grass, etc.).

finger, *sb.* Add: **3. a.** *to keep* (or *to have*) *one's fingers crossed,* to crook one finger over another to bring good luck; hence *fig.*, to hope for success or good fortune; also *with fingers crossed* (and similar phrases); *to lift, throw,* or *turn up the little finger,* to drink heavily; (*to point the*) *finger of scorn*: see Point *v.*¹ 12 and Scorn *sb.* 1 c; *to pull* (or *take*) *one's finger out* (slang), to get a move on, get 'cracking'; freq. in *imp.*; *to put the finger on* (slang, orig. *U.S.*), to inform against, to denounce to the police; to identify (a victim) to an assassin; also *transf.*; *to stir a finger*: also (*not*) *to lift* (or *move*) *a finger.*

1824 Scott *Redgauntlet* II. xiii. 311 Nanty likes the turning up of his little finger unco weel. **1833** A. Fonblanque *England under Seven Administr.* (1837) II. 321 We don't believe that he would have moved a finger to procure the return of the only man for the chair. **1845** J. S. Le Fanu *Cock & Anchor* iii. xii. 143 Has he been throwing up the little finger, my dear?..He used to be rayther partial to brandy. **1863** Gladstone *Let.* 12 Nov.

in P. Guedalla *Palmerston Papers* (1928) 269, I believe he has not himself lifted a finger in the matter. **1890** F. M. Peacock *Soldier & Maid* i. 9 The best of fellows,.. but liquors a bit,.. lifts his little finger. **1924** *Ladies' Home Jrnl.* Jan. 24/1 This is the year to keep your fingers crossed and announce yourself from Missouri. **1926** J. Black *You can't Win* vii. 84 If I'm grabbed with this junk I'll rot in jail before I put the finger on you. **1929** D. Hammett *Red Harvest* xxiv. 236 You think I killed them, don't you, Dick?..Going to put the finger on me? **1930** in *Amer. Speech* (1931) VI. 155/2 *Put a finger on*, mark for killing. **1938** 'N. Shute' *Kinsond City* xii. 246 I've got my fingers crossed. I keep them that way all the time. **1940** F. Scott Fitzgerald *Let.* 24 Aug. (1964) 122, I have my fingers crossed but.. I think my stock out here is better. **1941** Baker *Dict. Austral. Slang* 28 *Pull out your finger!* Hurry up! **1942** *Observer* 4 Oct. 7/3 We stooged about a bit above our target.. and then we pulled our fingers out, and pranged it. **1944** J. H. Fullarton *Troop Target* i. 16 For Christ's sake pull your finger out, Bill. **1945** *Penguin New Writing* XXIII. 16 We'll.. duck when we hear a mortar, and keep our fingers crossed. **1951** Partridge *Dict. Slang Suppl.* 1047 *Put the finger on*, to point (a wanted man) out to the police. **1955** D. Garnett *Flowers of Forest* ii. 14 Could anyone honestly say that we should have allowed Paris to be occupied and France defeated without lifting a finger? **1958** T. Sutherland *Green Kiwi* i. 22 'They're tigers for toil,' he went on. 'The bloke that takes a job with them wants to be able to pull his finger out.' **1959** *Times Lit. Suppl.* 25 Dec. 753/2 In one episode Brett puts the finger on a 'loaded' script depicting juvenile crime. **1961** *Times* 18 Oct. 8/2 (Duke of Edinburgh) I think it is about time we pulled our fingers out. **1962** *Spectator* 2 Nov. 680 If the publicity.. would 'take its finger out' and show railways.. as.. if they were now a petty pass when even royalty tells us to pull our fingers out. **1966** D. Shannon *With a Vengeance* (1968) x. 141 I'll go see your Papa. With fingers crossed. **1971** *Daily Tel.* 2 July 15/1, I have not heard of anyone who wants to put the finger on me.

5. c. For '*U.S. slang*' read '*slang* (orig. *U.S.*)'. (Add earlier U.S. and later examples.)
1856 *Porter's Spirit of Times* 4 Oct. 73/1 We each took a first mate's drink—i.e. three fingers. **1940** L. MacNeice *Plant & Phantom* (1941) 53 Three fingers of Scotch and a cube of ice. **1959** 'J. Welcome' *Stop at Nothing* xiii. 185, I poured out another four fingers and handed the tumbler to him.

8. *spec.* a banana.
1894 [in Dict.]. **1895** *Daily News* 26 Aug. 5/2 This.. is a shorter and stouter plant than the tropical banana, and often bears from 150 to 250 'fingers' in a bunch. **1961** F. G. Cassidy *Jamaica Talk* v. 99 A single banana or plantain (a 'finger' of the hand) is often called a *seed*.

c. A long narrow pier, etc. (e.g. a feature of airport architecture).
1951 *Progressive Archit.* Jan. 49 The pair of two-level 'finger' concourses allows passage to plane-loading points almost wholly under cover. **1954** *Civil Engin.* (U.S.) Sept. 56 Work on the terminal building and its fingers. **1958** *Times* 30 May 7/6 Aircraft can taxi to the.. 300 ft. long glazed pier, or 'finger', which stretches out from the terminal to provide completely enclosed passenger access. **1962** *Economist* 15 Dec. 1131/2 Narrow, cluttered 'finger' piers such as those in New York. **1966** *New Statesman* 20 Aug. 261/1 Half-a-dozen well-detailed finger-plan airports. **1970** *Times* 18 Mar. (Liberia Suppl.) p. v/5 An additional 750 ft. 'finger' pier was built at Monrovia to accommodate two further ore-carrying vessels.

10. b. (*a*) A policeman or detective; (*b*) an informer; (*c*) a contemptible or eccentric person; (*d*) a pickpocket; (*e*) one who supplies information or indicates victims to criminals. *slang.*
1899 'J. Flynt' *Tramping with Tramps* 393 *Finger*, a policeman. **1914** Jackson & Hellyer *Vocab. Criminal Slang* 33 *Finger*,.. An informer; an investigator for officers. **1916** C. J. Dennis *Songs Sentimental Bloke* 122 *Finger*, an eccentric or amusing person. **1925** *Flynn's* 24 Jan. 119/1 *Finger*, a pickpocket. **1926** *Ibid.* 16 Jan. 638/1 If th' stunt was pulled right an' th' finger did his job, you know th' getaway is in th' stunt. **1932** *Ibid.* 28 May 36/2 The café owner had taken me for a professional finger, one of the scouts of the underworld. **1933** G. Ingram '*Stir*' 8 *Finger*, a term of contempt for man or woman. **1938** F. D. Sharpe *Sharpe of Flying Squad* xiv. 156 They [sc. pickpockets] work in pairs: one is 'the finger', the other obstructs and jostles. **1960** K. Hopkins *Dead against my Principles* xix. 129 He's a finger, works in Fulham mostly. Small profits, quick returns. **1962** *John o' London's* 25 Jan. 82/3 A man who identifies a suspect at an identification parade.. is called a finger.

14. a. *finger-gesture*, *-sign*; **c.** *finger-wagging*.
1930 R. Paget *Babel* 49 Hand- and finger-gestures. **1938** R. G. Collingwood *Princ. Art* xi. 242 Italians.. have a long tradition of controlled finger-gesture, going back to the ancient game of *micare digitis*. **1953** J. S. Huxley *Evol. in Action* iv. 110 She [sc. Helen Keller].. realized that this particular combination of finger-signs 'meant' the wonderful cool something that was there feeling. **1931** *Times Lit. Suppl.* 6 Aug. 607/3 The inconsequent finger-wagging information, the all too 'noble' priggishness. **1966** G. McInnes *Road to Gundagai* vi. 107 We were the object of finger-wagging lectures.

15. finger-bar (earlier U.S. example); **finger-board**, (*c*) *U.S.* = Finger-post; (*d*) a gradient indicator; **finger-glass** (earlier example); (*b*) *pl.* = Harmonica 1 a; **finger-grass** (U.S. examples); **finger-hold**, something by which the fingers can hold; also *transf.* and *fig.*; **finger-impression**, **-mark**, terms formerly used for **finger-print*; **finger lake** [f. *Finger Lakes*, name of a group of lakes in New York State], any long narrow lake in a glaciated valley; **finger-language** (earlier example); **finger-light** *a.*, (of motor-vehicle steering) susceptible to pressure by the fingers alone; **finger man** *U.S. slang* = *Finger *sb.* 10 b (*e*); **finger-paint** *v. intr.*, to apply paint (to a canvas or other surface) with the finger; also *trans.*; so **finger-painting** *vbl. sb.*; **finger paint**, a jelly-like paint used chiefly by children; **finger-play**, (*a*) a game involving an intricate use of the fingers (see also quot. 1945); (*b*) the use of the fingers for caressing or sexual stimulation, *spec.* of the genital parts; **finger-print**, (earlier U.S. example); now with specific reference to the recording by the police of impressions taken from the finger-tips of criminals and suspects; also further *fig.* examples; hence **finger-print** *v. trans.*, to take the finger-prints of; **finger-printing** *vbl. sb.*; **finger-spin**, spin imparted to a ball by the fingers (as in Cricket and Table Tennis in bowling and serving); hence **finger-spinner**; **finger-style** *a.* and *adv.*, using the fingers alone (not a plectrum) in playing certain stringed instruments; **finger-tight** *adv.*, as tightly as possible by hand; **finger-wave** *Hairdressing*, a wave set in wet hair with the fingers; so **finger-waving** *vbl. sb.*; **finger-work**, (*a*) work executed with the fingers; (*b*) the play of, or manipulation by, the fingers.
1865–6 *Trans. Ill. Agric. Soc.* VI. 52 Patent Cold Rolled Reaper and Mower Finger Bars. **1864** *Frederick-town* (Md.) *Herald* 5 Feb. 3/1 Holding the finger board in a wrong direction for Carter's Mountain. **1845** in D. Drake *Pioneer Life Kentucky* (1870) x. 233 At their.. forks there were no finger boards, and not many living fingers to point the true way. **1883** F. S. Williams *Our Iron Roads* 95 Gradients.. vary considerably as we may see by the finger-boards placed on the lines for the guidance of the engine-driver. **1814** Jane Austen *Mansf. Park* III. xv. 296 She was meditating much upon silver forks, napkins, and finger glasses. **1868** G. M. Hopkins *Jrnl.* 20 June (1959) 167 Madlle. Vogt [played] the finger-glasses [Mattauphone], and certainly that instrument is chromatically more perfect than the other. **1821** *Mass. Hist. Soc. Coll.* 2nd Ser. IX. 149 Plants, which are indigenous in the town of Middlebury, [Vermont, include] Finger grass. **1896** *Rep. Mass. Board Agric.* I. 85 The stems of the finger grass are from one to two feet high. **1943** R. Peattie *Great Smokies* 168 In the first year after abandonment.. bore weed and finger grass become predominant. **1909** *Daily Chron.* 21 Aug. 7/4 Wherever it [sc. samphire] can get finger-hold in the rough face of the cliff, it shows in abundance. **1923** A. Dean *His Children's Children* xviii. 219 The relief of him who, having plunged part way down the face of a precipice, has caught a finger-hold in a crevice. **1923** H. L. Wilson *Oh, Doctor!* xii. 135 'Yes,' assented Seaver; 'one of these movie actors that can bite nails in two and throw a horse over a fence and climb up the side of a church with only a finger hold.' **1974** M. Morse *Un-Natural Hist.* v. 51 There seemed nothing I could do about this without endangering the fingerhold I had gained. **1903** *Westm. Gaz.* 18 Nov. 8/1 Witness told him that his finger-impression had been found on a pane of glass on the roof of the building. **1908** *Daily Chron.* 24 July 1/7 Two men who are so much alike that they can be distinguished only by their finger impressions. **1931** C. Carter *Land-forms & Life* xix. 122 Ten or more.. radial streams and valleys may be counted and named from the map; most of these expand into finger lakes. **1939** A. K. Lobeck *Geomorphol.* viii. 273 The long finger lakes of central and northern Sweden on the eastern side of the Scandinavian Highland correspond to the fiords of Norway on the western side. **1968** R. W. Fairbridge *Encycl. Geomorphol.* 435/2 Where the ice sheet passed over an escarpment, it left extra-deep finger lakes. **1842** Dickens *Amer. Notes* I. iii. 93 She often soliloquizes in the finger language. **1927** *Observer* 15 May 21 But no special skill is needed in directing the car, the steering of which is finger-light. **1958** *Times* 22 Oct. 5 Finger-light power assisted steering enables perfect control. **1930** *Flynn's* 15 Mar. 624/1 Martin was knocked off... It was thought Butch Swang had brought in the finger men to put Martin on the spot. **1951** J. Shaw *Troubled Air* viii. 117 They're using you. You're their finger man... You're their respectable front. **1968** E. McGirr *Lead-Lined Coffin* ii. 80 He was right to concentrate on Fall. There could have been the fingerman watching him. [**1880** Faulds in *Nature* 25 Nov. 76/2, I have been taking sign-manuals by means of finger-marks for now more than twenty years.] **1893** Galton in *Proc. R. Soc.* 28 May 540 Methods of indexing Finger-Marks. **1950** Webster *Add.*, Finger paint (n.). **1951** C. Hayes *Ape in House* (1952) xxiv. 210 She learned.. to fingerpaint on the rugs with butter and orange juice. **1966** 'E. V. Cunningham' *Helen* (1967) xii. 176, I dreamed that I was a little boy using finger paints. **1971** *Guardian* 3 July 7/5 Teachers learning to finger paint at the York Teachers' Centre. **1950** Webster *Add.*, Finger painting. **1959** J. Verney *Friday's Tunnel* iii. 34 Splodgy finger paintings in paste and powder colour. **1945** C. V. Good *Dict. Educ.* 171/2 *Finger play*, a dramatization of a verse or poem with appropriate finger movements, used as an attention-getting device with young children. **1963** *Times* 3 June 11/6 A book of fingerplays (great for idle moments),.. and some books of riddles and puzzles. **1965** P. & E. Kronhausen *Sexual Response in Women* vi. 137 His finger-play on my clitoris helps. **1966** Bereiter & Engelmann *Teaching Disadvantaged Children* i. 14 Some time is usually devoted to group games, finger-plays, and the like. **1970** *Sunday Times* 15 Nov. 32/3 Pepys was highly susceptible to any available charmer, but contented himself as a rule with mutual fingerplay. **1859** *North Amer. Rev.* LXXXVIII. 492 The chapel of St. Verena, where the finger-prints of the young maiden still remain in the rock. **1891** *19th Cent.* Aug. 304 Finger-prints have been proposed.. as a means of identification. **1903** *Daily Chron.* 9 Feb. 8/7 The finger-print system of identification. **1905** *Ibid.* 13 Sept. 7/4 A prisoner.. seldom objects to being finger-printed. **1923** *Glasgow Herald* 2 Nov. 6 Demanding that every new-born child be finger-printed when registered. **1937** *Oxoniensia* II. 6 The large vessels.. with finger-print decoration on the shoulder. **1955** *Wall St. Jrnl.* 1 Nov. 9/3 These methods of 'finger printing' chemical compounds often are considered better evidence of the existence of large molecules like those of natural rubber than physical evaluations. **1956** V. M. Ingram in *Nature* 13 Oct. 793/2 It was decided to call the resulting chromatogram the 'finger print' of the protein. **1961** J. Singh *Ideas Mod. Cosmol.* vii. 96 The measurement.. enables us to compute the curvature. In other words, space curvature leaves its fingerprint on the pattern of galactic distribution in depth over the sky. **1969** *Nature* 23 Aug. 832/2 Fig. 1 shows a 'fingerprint' (peptide chromatogram) of the tryptic peptides of normal myoglobin. **1970** *Daily Tel.* 27 Feb. 4/8 The technique uses the radioactive 'fingerprint' of elements to determine their age and origin or to compare one substance with another. *Ibid.* 4 Mar. 17/8 When I realised it might be used in evidence against me I asked for it to be finger-printed. **1970** *Nature* 21 Mar. 1131/2 Fingerprints of the T_1 ribonuclease digestion products of 16S and 23S RNAs are very different, thus indicating dissimilar primary structures. **1922** *Glasgow Herald* 2 Nov. 6 The big Danish prisons, where fingerprinting is an important factor. **1958** *Sci. Amer.* Jan. 19/3 The electrophoretic 'fingerprinting' technique.. to distinguish the hemoglobin of men, horses and whales. **1905** *Westm. Gaz.* 1 July 2/3 Bowlers with plenty of finger-spin are most likely to take wickets on the mat. **1906** *Ibid.* 8 May 2/1 Finger-spin bowlers like Schwarz, Faulkner, and Vogler. **1927** *Observer* 17 Apr. 17/4 [He] only needs more power of finger-spin to be a match-winning bowler. **1938** *Encycl. Brit. Bk. Yr.* 623/1 It is now no longer legal [in Table Tennis] to use the finger-spin in serving. **1958** *Listener* 16 Oct. 604/2 The wrist-spinner.. is inevitably less accurate than the finger-spinner. **1934** S. R. Nelson *All about Jazz* ii. 53 When the banjo was first introduced into the dance band, the players were finger-style performers. **1955** L. Feather *Encycl. Jazz* (1956) vii. 77 Plays unamplified Spanish concert guitar, finger style. **1940** 'N. Shute' *Old Captivity* v. 148 Bolts were put on finger-tight... Finally.. the bolts were tightened down and locked. **1961** *Lancet* 19 Aug. 410/2 Retention is ensured by screwing the female end-piece fingertight over the male. **1934** Webster, Finger wave. **1936** J. Steinbeck *In Dubious Battle* 299 Her hair was damp.. and little, uneven finger-waves were pressed into it. **1950** J. B. Lohle in Foan & Wolters *Art & Craft Hairdressing* (ed. 3) v. 267 If the natural wave is rather weak.. the older type of blow dryer may be used to give a finger-wave. **1963** M. McCarthy *Group* xii. 286 A finger wave in her majestic coiffure. **1926** *Amer. Hairdresser* May 6 The subject of 'Finger Waving'. **1932** L. Golding *Magnolia St.* iii. iii. 508 They were masters of the latest methods of permanent marcel, water and finger waving. **1849** Finger-work [in Dict., sense 14]. **1906** *Practitioner* Dec. 826 The great drawback of this treatment is that it.. must be minutely carried out, demanding.. a great deal of finger-work. **1906** *Westm. Gaz.* 10 Aug. 3/1 There was never before so clear an exposition of balance, of swing, and of finger-work in bowling. **1927** *Observer* 29 May 14/5 [His] first recital showed promise... Mozart does not suit his style: it needs better finger-work and not so much arm and body and pedal. **1967** *Antiquaries Jrnl.* XLVII. 208/1 Simpler finger-work,.. companion thus to cord-impression.

finger, *v.* Add: **2.** (Further *fig.* examples.)
1885 W. B. Yeats in *Dublin Univ. Rev.* Apr., For the dawn the foliage fingereth. **1929** D. H. Lawrence *Pansies* 61 My body need not be Fingered by the mind. **1956** *Kenyon Rev.* XVIII. 417 Often they [sc. skilful poets] 'finger' the speech flow.

4. c. To indicate (a victim) or supply (information) to criminals; to inform on or identify (a criminal) to the police. *U.S. criminals' slang.*
1930 *Flynn's* 13 Dec. 194/1 Frank Lee.. had 'fingered' many, many dealers to the Feds. **1936** *Ibid.* 12 Dec. 8/2 Matt Prance has got me fingered... They're going to kill me. **1949** R. Chandler *Little Sister* xxxiii. 242 She's on her way back.. with.. the pocket money she got.. for fingering her brother. **1954** H. Roth *Content Assignment* xxii. 139 She.. couldn't finger them for any crime.

6. b. *intr.* for *pass.* To be manipulated with the fingers.
1927 W. W. Bishop *Pract. Handbk. Mod. Libr. Catal.* (ed. 2) 33 The tests of a card are the cleanness and speed with which it 'fingers' in consultation, and the fastness of its color.

fingerful (fi·ŋgəɹful). [f. Finger *sb.* + -ful.] As much as can be held on one finger or between two fingers at one time; hence, a small quantity (*of* something).
1889 H. Johnston *Chrons. Glenbuckie* ii. 19 Just put a fingerfu' o' poother i' the pan. **1923** U. L. Silberrad *Lett. J. Armiter* vi. 138 At first I could get them to do nothing more than suck a finger; and it is quite hopeless to try to get enough milk into them by finger-fulls. **1958** *Listener* 13 Nov. 768/1 They put the final fingerful of colour on to our faces. **1968** H. C. Rae *Few Small Bones* ii. v. 114 He peeped out of the side window through a fingerful of curtain.

fingerless, *a.* (Further example.)
1895 *Chambers's Jrnl.* XII. 628/1 They showed their blotched and swollen bodies, fingerless hands, and toeless feet.

finger-post, *v.* [f. the sb.] *trans.* To indicate by means of a finger-post. Also *fig.*
1908 *Councils' Jrnl.* 17 The Parish Council of Orrell-with-Ford.. is properly proud of having 'name-plated all the roads, finger-posted all the footpaths,' [etc.]. **1926** T. E. Lawrence *Seven Pillars* (1935) 7 The contents seem to me to be adequately finger-posted by this synopsis. **1944** *Ess. & Stud.* XXIX. 52 In the drama the playwright may be 'explicit' in finger-posting parts of plot, situation, or character.

fi·nger-tip. The tip of a finger; used *attrib.* to indicate length (of a garment) or sensitivity (of an object) to pressure, and *spec.* in *Archæol.* designating ornamentation made with the tips of the fingers; hence **finger-tipping** *vbl. sb.* Phr. *to the* (or *one's*) *finger-tips* = through one's whole body, 'from top to toe', throughout; *at one's finger-tips* (see Finger *sb.* 3 e).
1842, **1883** [see Finger *sb.* 14]. **1886** Kipling *Departmental Ditties* 45, I redden to my finger-tips. **1905** *Strand Mag.* June 702/1 He has at his finger-tips every stroke in the game. **1907** *Daily Chron.* 21 Jan. 3/5 Alive with science to the finger tips. **1928** Crawford & Keiller *Wessex fr. Air* 10 If it was constructed, as he concludes, by the finger-tip people. *Ibid.* 200 The fragment of a typical finger-tip urn. **1929** F. Hackett *Henry VIII* viii. 417 He was a magnate to his finger-tips. **1929** V. G. Childe *Danube in Prehist.* 364 Biconical or open urns with finger-tip ornament. **1935** *Times* 28 Nov. 19/5 Finger-tip coat and frock. **1937** *Times* 13 Apr. p. xxii/4 A 'finger-tip' adjustment for setting the brake shoes to the correct clearance. **1937** *Oxoniensia* II. 28 Rough grey-brown smoky ware with finger-tipping on the shoulder. **1956** A. H. Compton *Atomic Quest* 54 He had at his fingertips the most recent experimental values of the constants. **1959** *Observer* 7 June 31/1 A great bout of finger-tip handling right across the Australian goalmouth. **1960** *News Chron.* 19 Feb. 8/7 The fingertip light switch. **1961** *Antiquaries Jrnl.* XLI. 36 Two shoulder sherds.. with stick impression or impressed finger-tip decoration. **1961** J. McCabe *Mr. Laurel & Mr. Hardy* (1962) iii. 75 He was an artist to his fingertips once he got going.

fingle-fangle. Delete †*Obs.* and add later examples.
1900 *Daily News* 26 Nov. 4/1 A lot of fingle-fangle, like piano-playing. **1963** *Brewer's Dict. Phr. & Fable* 359/2 *Fingle-fangle*.. is not heard nowadays, except as an archaism.

Fingo (fi·ŋgo). Also **Fengu** (fe·ŋgu). [f. Bantu *amaMfengu* destitute people in search of work, f. *fenguza* to seek service.] A Negro people of South Africa, now resident in the eastern Cape Province, representing a coalescence of various groups of fugitives driven from Natal in the 1820's.
1829 C. Rose *4 Years S. Afr.* viii. 193, I saw many of the fugitives, who are called *Fingos*, wanderers. **1836** *Grahamstown Jrnl.* 109 (Pettman), The Fingoes residing on the missionary station, displayed.. considerable firmness and courage. **1850** J. W. Appleyard *Kafir Lang.* 41 The term *Amafengu* is a conventional national epithet, first applied to the *Fingoes* by the Kafirs, but now in general use amongst themselves. **1883** *Encycl. Brit.* XVI. 517/1 The South African races, Hottentots and Kaffres, Fingoes and Bechuanas, Basutos and Zulus. **1902** *Ibid.* XXX. 3/1 The formerly degraded but now respected and civilized Fingos or Fengus, who give their name to the district of Fingoland. **1955** M. Gluckman *Custom & Conflict in Afr.* iii. 76 A similar attitude was shown by some Fingo elders.

finick, *sb.* **1.** Delete †*Obs.* and add later examples.
1861 G. Meredith *Evan Harrington* iii. 29 The ladies.. signalized him as a 'finick'. **1949** S. Gibbons *Matchmaker* xxx. 370 No one would have credited such a finnick as Mrs Hoadley with being a first-class cook.

finick, *v.* Add to def.: To behave in a fastidious or finicky manner. (Add examples.)
1857 G. Meredith *Farina* 153 The Demon.. pointed his feet, and finicked a few steps away. *a* **1916** E. P. Oppenheim *Golden Beast* (1926) I. xxiv, She's all brain and fancies and temperament. That sort of person finicks with the marriage question. **1931** V. Woolf *Waves* 76, I do not finick about fearing what people think of 'my father a banker at Brisbane' like Louis.

finickiness (fi·nikinès). Also **finn-**. [f. Finicky *a.* + -ness.] The quality or condition of being finicky; excessive concern over trifles, over-fastidiousness.
1833 J. Nyren *Young Cricketer's Tutor* (1902) 114 It was thought a piece of finnikinness by those simple and homely yeomen. **1920** *Glasgow Herald* 23 Dec. 4 Sensitive to the point of finnickiness. **1930** *John o' London's* 18 Jan. 618/2, I received a letter.. which supplies me with a fine example of this grammatical 'finickiness'.

finicky, *a.* For '*dial.* and *U.S.*' read 'orig. *dial.* and *U.S.*' (Add earlier U.S. and later examples.)
1839 Poe *Wks.* (1865) II. 388 He was really the most finnicky little personage that had ever been seen in Vondervotteimittiss. **1859** *Trans. Ill. Agric. Soc. 1857–8* III. 473 We desire to be neither coarse or finicky. **1873** E. Eggleston *Myst. Metrop.* xxxv. 303 You a'n't so finicky 'bout vittles as you was. **1895** *Rev. Reviews* Oct. 355/1 The fine or finicky ear of the critic in the study. **1934** *Brit. Jrnl. Psychol.* July 59 A prolix writer, finicky I am told, about literary style.

finify, *v.* Delete †*Obs.* and substitute *Obs.* exc. *U.S.* and *dial.* (Later examples.)
1847 J. S. Robb *Streaks of Squatter Life* 73 (Bartlett), If this new judge is the slicked up, finefied sort on a character they pictur' him, I don't want to see him. **1891** *Harper's Mag.* Jan. 222/1 They [sc. women] air obleiged ter set out a table all tricked up an' finified off. **1895** Rye *E. Anglian Gloss.*, *Finify*, to be over-nice in doing anything.

Finiglacial (finiglēi·ʃiăl), *a. Geol.* Also **Fini-glacial, finiglacial.** [f. L. *fini-s* end + Glacial *a.*] Epithet of the last of the three divisions or 'sub-epochs' of the Late Glacial epoch in north-western Europe, when the ice-sheet left Finland and retreated across the Gulf of Bothnia to central Sweden.
1910 G. De Geer in *Geol. Forening. Stockholm Förhandl.* XXXII. 1146 The last part of the late glacial epoch, when the ice receded from the great moraines to eastern Jämtland at the point of the first bipartition of the vanishing ice-remnant, may be called the *finiglacial* age. **1946** L. D. Stamp *Brit. Struct.* xiv. 170 The Fini-Glacial lasted from about 8500 B.C. to 6500 B.C. and was the period of the retreat of the ice from Finland. **1957** J. K. Charlesworth *Quaternary Era* II. xlii. 1158 The Daniglacial may have equalled the Gotiglacial and Finiglacial in length. **1965** F. J. Monkhouse *Dict. Geogr.* 122/2 *Finiglacial stage*, the third stage in the retreat of the continental ice-sheet from Scandinavia.

finish, *sb.* Add: **2.** (Earlier example in application to persons.)
1779 J. Wedgwood *Let.* 23 Nov. (1965) 245 We have formed a.. school, which I have.. some notion of.. continuing instead of sending them [sc. children] from home again, unless by way of finish.

3. (Further examples.)
1888 Cross & Bevan *Text-bk. Paper-Making* x. 159 The degree of smoothness or 'finish' that can be given to the paper by the calenders, depends.. upon the degree of moisture which it contains. **1962** F. T. Day *Introd. to Paper* 115/1 *Finish*. This is the degree of smoothness applied to a paper. Finishes may vary: rough antique, smooth, mill finish, super calender, highly glazed or gloss finish.

4. (Earlier examples.)
1796 Grose *Dict. Vulg. T.* (ed. 3) *The finish*, a small coffee-house, in Covent-Garden market,.. open very early in the morning. **1840** *Fraser's Mag.* Oct. 399/1 The pleasures of the metropolis—the theatres, the Cider-cellars, the Finish.

6. *finish line*, a real or imaginary line which marks the end of a race.
1899 *Daily News* 4 Oct. 3/4 Both yachts being then about three miles from the finish line. **1959** M. Shadbolt *New Zealanders* 86 The leaders in the junior race flashed across the finish-line. **1968** *Guardian* 28 June 1/5 Williams had been declared the winner, both for crossing the finish line first and on a corrected time basis.

finish, *v.* Add: **5.** Also with *up*.
1878 *Harper's Mag.* Nov. 892/2 Shall we finish up and have done with it, with a song at the end? **1902** R. Machray *Night Side London* i. 21 Others vanish into the night.. finishing up.. at some night-club, or in some other den.

b. (*a*) Delete (*obs.*) and add later example.
1939 G. B. Shaw *Good King Charles* ii. 102 How often have you told me that you are finished with all women!

d. Of a competitor in a sporting contest: to come to the end of a course or race (in a certain condition or 'place').
1881 [in Dict., sense 5]. **1894** G. Moore *Esther Waters* ii, He told me.. to bring the colt along and finish up close by where he would be standing. **1929** *Star* 21 Aug. 16/2 Every horse he has backed has finished out of the first three. **1971** *Daily Tel.* 29 Oct. 29/1 Lithiot finished threequarters of a length behind Gold Rod to be third in this big mile race. *Ibid.* 31/5 Randle finished third in the 350 cc Italian Grand Prix at Monza.

finishable, *a.* (Later examples.)
1933 G. Stein *Autobiogr. Alice B. Toklas* ii. 27 Picasso thereupon brought out a smaller picture.. very unfinished and not finishable. **1933** *Punch* 8 Nov. 524/1, I cannot pretend that if I had a daughter of finishable age I should care to send her to Geneva to *Madame Domini's*.

finisher. Add: **1.** (Additional example.)
1963 P. Drackett *Motor Rallying* 90 To be classed as a finisher a car must arrive at the final control.. within half an hour or such other period of time as may be specified.

2. c. (Earlier example.)
1771 C. Burney *Pres. State Mus. France & Italy* I. 27 But the last chorus was a *finisher* with a vengeance! it surpassed, in clamour, all the noises I had ever heard in my life.

finishing, *vbl. sb.* Add: **1.** Also with *off*.
1940 *Chambers's Techn. Dict.* 167/2 *Cleaning-up*, the operation of preparing the finished surface of joinery work. Also called *finishing-off*. **1942** W. S. Churchill *Secret Session Speeches* 23 Apr. (1946) 74 The defeat of Hitler left the finishing-off of Japan merely a matter of time and trouble.

3. finishing-post, the post or place which marks the finishing point of a race.
1895 *Daily Tel.* 26 Sept. 3 The referee was sent on ahead to the Bricklayers' Arms, the finishing-post. **1931** *Times Lit. Suppl.* 29 Jan. 83/3 Unable to mark exactly where the finishing-posts are.

finishment. (Later U.S. examples.)
1872 E. Eggleston *End of World* viii. 59 This 'rithmetic preacher can't make a finishment of this sublunary *speer* by addition. **1873** C. G. Leland *Egypt. Sketch-Bk.* 103, I should say, in finishment of this chapter, that one leaves the dervish performance with mixed emotions.

finitary (fəi·nitări), *a. Math.* [f. Finite *a.* after *unitary*, translating G. *finit* (Hilbert & Bernays *Grundl. d. Math.* (1934) I. 32).] Of methods, proofs, etc: involving only a finite number of steps, a finite number of well-defined objects, and so on; capable of being completed within the concrete domain.
1952 S. C. Kleene *Introd. Metamath.* iii. 63 Methods, called finitary by the formalists, which employ only intuitively conceivable objects and performable processes. (We translate the German 'finit' as 'finitary', since the English 'finite' is used for the German 'endlich'.) **1963** G. T. Kneebone *Math. Logic* vii. 205 We can now formulate in finitary terms the usual arguments. **1968** A. S. & E. H. Luchins *Log. Found. Math.* ix. 145 Hilbert did not offer a precise specification of what procedures were regarded by him as finitary.

finite, *a.* and *sb.* Add: **A.** *adj.* **3.** *finite difference*: a difference between two quantities that is finite; *spec.*, any of the differences between the successive values of a function when its independent variable takes on the values of an arithmetical progression; *finite state*: used *attrib.* in communication theory of any system limited to a finite number of alternative states, or of a type that, even if it contains an infinite number of parts, is producible by a generative device so limited.
1807 Finite difference [see Difference *sb.* 2]. **1860** G. Boole *Calculus of Finite Differences* viii. 137 It is not true that every result of the Calculus of Finite Differences merges when the increments are indefinitely diminished into a result of the Differential Calculus. **1910** *Encycl. Brit.* XIV. 545/1 The theorem [sc. Taylor's theorem] was first given by Brook Taylor in his *Methodus Incrementorum* (1717) as a corollary to a theorem concerning finite differences. **1949** Shannon & Weaver *Math. Theory of Communication* 27 The output of a finite state transducer driven by a finite state statistical source is a finite state statistical source. **1957** N. Chomsky *Syntactic Structures* iii. 18 Suppose that we have a machine that can be in any one of a finite number of different internal states... Any language that can be produced by a machine of this sort we call a finite state language; and we can call the machine itself a finite state grammar. **1957** L. Fox *Two-point Boundary Probl.* ii. 21 The rigorous theory of finite differences attaches a remainder term to every finite-difference formula, and sometimes.. the formula diverges and gives no useful result. **1969** *Listener* 10 July 44/1 Finite-state rules are the simplest kind of grammar rule; they can be as simple as 'after "the" produce any noun at all'. **1971** *Mathematical Rev.* Feb. 278/1 The authors treat some elementary aspects of the theory of finite differences.

B. *sb.* **3.** *Gram.* A finite verb or verb-form.
1924 H. E. Palmer *Gram. Spoken Eng.* ii. 91 A Finite is that part of a sentence which predicates, exclusive however of complements, objects or modifiers. **1963** F. T. Visser *Hist. Syntax* I. iv. 497 The verbs *shall, can, may*.. called anomalous finites or auxiliary verbs. **1965** *Language* XLI. 208 If we consider the range of 'finites' that can operate at X in the sequence *He X (to) come every day*.

finitely, *adv.* (Later examples.)
1965 J. J. Rotman *Theory of Groups* xii. 257 A group is finitely presented in case it has a presentation with only a finite number of generators and a finite number of relations. **1968** P. A. P. Moran *Introd. Probability Theory* iv. 197 We have defined ϕ_c as a finitely additive set function on the field of all cylinders in Ω_0.

finitism (fəi·nəitiz'm). [f. Finite *a.* + -ism.] **a.** *Philos.* and *Theol.* The belief that the world, or some realm, or God, is finite. **b.** *Math.* The doctrine that only methods involving a finite number of steps should be used in mathematics; the rejection of actual infinities.
1922 W. R. Inge *Outspoken Ess.* II. 19 Finitism and Infinitism are equally demonstrable and equally refutable. **1935** A. Ambrose in *Mind* XLIV. 186 (*title*) Finitism in mathematics. **1956** G. E. M. Anscombe tr. *Wittgenstein's Remarks Found. Math.* ii. 63 Finitism and behaviourism are quite similar trends... Both deny the existence of something, both with a view to escaping from a confusion. **1964** J. A. Benardete *Infinity* i. 25 With the return of finitism, the infinite series is no longer viewed as *actually* infinite.

finitist (fəi·nəitist), *a.* and *sb. Philos.* and *Math.* [f. as prec. + -ist.] **A.** *adj.* Characterized by, or relating to, finitism.
1904 W. James *Let.* 12 June (1920) II. 203 My philosophy.. is finitist; but it does not attribute to the question of the Infinite the great methodological importance which you and Renouvier attribute to it. **1914** B. Russell *Our Knowl. External World* vi. 155 There is no longer any reason to struggle after a finitist explanation of the world. **1931** F. P. Ramsey *Found. Math.* 243 Provable.. means provable in any number of steps, and on finitist principles the number must in some way be limited, e.g. to the humanly possible. **1935** *Mind* XLIV. 186 These present to the finitist school of mathematicians difficulties. **1940** *Mind* XLIX. 243 He will use only 'finitist' methods. **1966** *Mathematical Rev.* Jan. 9/2 F is complete with respect to some distinguished notion of proof, such as finitist or predicative.

B. *sb.* An exponent or adherent of finitism.
1897 B. Russell in *Mind* VI. 118 When the finitist objects that an infinite collection can never be.. really a whole, the infinitist replies that.. number itself must be contradictory. **1900** J. Royce *World & Indiv.* I. 558 Couturat, in his dialectical discussion between the 'finitist' and the 'infinitist',.. gives full room to.. these arguments. **1935** *Mind* XLIV. 186 Denials by the finitists which the formalist and logistic schools contest. **1960** S. Körner *Philos. of Math.* v. 111 Finitists such as Aristotle, Gauss, and the older and the newer intuitionists, deny all 'real' content or even 'intelligibility' to such mathematical notions as are not characteristic either of finite aggregates or at most of potentially infinite.

Hence **finiti·stic** *a.*
1937 *Mind* XLVI. 58 Gödel.. seemed to have destroyed for ever Hilbert's hope that he would carry out his programme of establishing with finitistic means the non-contradictoriness of mathematics. **1959** E. W. Beth *Found. Math.* iii. 73 Complete induction ranked high among the forms of argument admitted in finitistic meta-mathematics.

finitize (fəi·nəitəiz), *v.* [f. Finite *a.* + -ize.] *trans.* To make finite. Hence **fi·nitizing** *vbl. sb.*
1906 S. S. Laurie *Synthetica* II. 85 The Unconditional has been under a necessity to finitise Itself. **1917** A. S. Pringle-Pattison *Idea of God* xvi. 321 Professor Howison.. seems to use the idea of self-consciousness entirely as a principle of separation and exclusion, which finitizes even what he calls 'the Supreme Instance', the 'absolutely perfect mind, or God'. **1930** *Monument to St. Augustine* 257 In his account of Kant, Hegel reproaches him with a shallow finitizing, whereas Spinoza represents for him an impracticable infinitizing.

finity. Add later example; also semi-*concr.*, an instance of finiteness.
1899 H. Wright *Depopulation* 57 There seemed no finity to the vista of country the eye followed. **1903** B. Harraden *Kath. Frensham* ii. x. 199 Infinities which one ..could see were finities blending with each other imperceptibly.

fink (fiŋk), *sb.*[1] *S. Afr.* [ad. Afrikaans *vink* Finch.] = Weaver[1] 4. Cf. *finch 1 c.
1834 A. Smith *Diary* 9 Dec. (1939) I. 168 Red vink and Caffer fink common along the streams. **1853** *Edin. New Philos. Jrnl.* LV. 82 The yellow and green *finks* may be seen disporting in multitudes.. and entering every now and then into their grass-woven nests. **1896** E. Clairmonte *Africander* 2 A flock of long-tailed mousebirds, called *finks*, would dash past to settle in a *rooikout* tree. **1897** A. Page *Afternoon Ride* 62 The Kafir fink swaying on the grass. **1908** Haagner & Ivy *Sk. S.Afr. Bird-Life* 70 The Red Bishopbird or Kaffir-fink. **1931** [see **bishop-bird*]. **1936** R. Campbell *Mithraic Emblems* 34 The scarlet fink, the chook, the sprew, That seem to call me by my name.

fink (fiŋk), *sb.*[2] *U.S. slang.* [Origin unknown.] A pejorative term of wide application, esp. **a.** An unpleasant or contemptible person. **b.** An informer; a detective. **c.** A strike-breaker.
1903 Ade *People you Know* 60 Anyone who goes against the Faculty single-handed is a Fink. **1914** Jackson & Hellyer *Vocab. Criminal Slang* 33 *Fink*,.. an unreliable confederate or incompetent sympathizer. **1925** *Amer. Speech* I. 151/2 'Dick' and 'bull' and 'John Law' have become established as names for the police, while 'fink' and 'stool' and 'fly-dick' denote the plain-clothes men. **1926** *Amer. Mercury* Jan. 63/1 Dating from the famous Homestead strike of 1892 is the odious *fink*. [It] according to one version was originally *Pink*, a contraction of Pinkerton, and referred to the army of strikebreakers recruited by the detective agency. **1929** E. Booth *Stealing through Life* xi. 259 'That guy.. is a rat'.. 'So's the fink with him.' **1940** R. Chandler *Farewell, My Lovely* v. 25 Now he's looking for the fink that turned him up eight years ago. *a* **1940** F. Scott Fitzgerald *Last Tycoon* (1949) vi. 143 A fink? That's a strike-breaker or a company tec. **1959** C. Williams *Man in Motion* viii. 106 Except for being a rat, a fink, a scab, a thug, and a goon, he's one of the sweetest guys you'll ever meet. **1970** *New Yorker* 15 Aug. 66/2 The film, in sudden want of a through-and-through-fink type, forgets that it has established two.. hippies as loyal.

fink (fiŋk), *v.*[1] Colloq. corruption of Think *v.*[2]
1888 Kipling *Wee Willie Winkie* (1890) 12, I don't fink I'll ever want to kiss big girls. **1913** B. Moore in E. Pertwee *Reciter's Second Treasury Verse* 226, I really fink, 'tween me and you, I'd raver be a little girl. **1941** H. G. Wells *You can't be too Careful* iii. ii. 115 That boy is still as pure as the driven snow. (*I don't fink.* I saw his face.) **1962** N. Marsh *Hand in Glove* vi. 202 Makes you fink, don't it?

fink (fiŋk), *v.*[2] *U.S. slang.* [f. *fink *sb.*[2]] *intr.* To inform *on*.
1925 *Flynn's* 24 Jan. 119/1 *Fink*, to squeal; to inform on. **1953** W. R. Burnett *Vanity Row* v. 43 Only a rat co-operated with the police; only a rat finked! **1953** R. Chandler *Long Good-Bye* ix. 57 Suppose you had to hire a private eye... Would you want one that finked on his friends? **1969** *Rolling Stone* 28 June 4/2 The gang tries to sell their smack to a black hippie pusher who finks on them.

Finlander (fi·nlændəɹ). Now *rare.* [f. *Finland*, name of a country in N.E. Europe + -er[1].] A native or inhabitant of Finland.
The usual form now is *Finn*.
1727 P. Kinck in *Phil. Trans. R. Soc.* XXXV. 357 A Letter to James Theobald, Esq., giving an Account of the Norwegian Finns, or Finlanders. **1808** in W. Monteith *Narr. Conq. Finland* (1854) 53 The Finlander regiments entering the service of His Imperial Majesty shall not be employed against Sweden.. during the present war. **1838** *Penny Cycl.* X. 274/1 The Laplander is of the same extraction as the Finlander, and calls himself a 'Samelada or Same'. **1854** [see Finn]. **1920** *Contemp. Rev.* Dec. 792 Finnish-speaking Finlanders.

Finlay process. *Photogr.* [f. the name of C. L. *Finlay*, a British photographer.] (See quot. 1935.)
[**1911** *Encycl. Brit.* XXI. 519/1 In C. L. Finlay's 'Thames' colour plate (1908) the tricolour screen is formed by rows of circular dots coloured alternately orange-red and green and the intermediate spaces blue.] **1931** F. R. Newens *Technique Colour Photogr.* iv. 92 The Finlay process is a duplicating process... One of the fundamental necessities for success in the Finlay process, is the closest possible contact between the squares of the taking screen and the emulsion of the negative plate. **1935** *Discovery* July 188/2 All photographers are familiar with.. the Finlay screen plate process... The ruled colour screen is on a separate plate which is exposed in contact with the photographic plate; from the negative so exposed a transparency print in black and white is made, and this is finally viewed bound in register with a ruled colour screen plate similar to that which was used when the negative was exposed. **1940** A. L. M. Sowerby *Wall's Dict. Photogr.* (ed. 15) 142 Shorter exposures are required by the Finlay and Dufaycolor processes than with the earlier screen-plates.

finned. (Additional examples.)
1938 *Archit. Rev.* LXXXIII. 32 'Finned' pipes have been carried under the clerestorey behind a splayed capping on top of the upper stacks as a precaution against down draught. **1958** *Times Rev. Industry* Feb. 68/2 The liquified ammonia.. enters another system of finned tubes. **1970** *Cryogenics* X. 239 (*heading*) A new form of finned-tube heat exchanger.

‖ **finnesko, fin(n)sko** (fi·nsko). [ad. Norw. *finnsko*, f. *Finn* Finn + *sko* shoe.] A boot made of birch-tanned reindeer skin with the hair left on the outside.
1890 tr. *Nansen's 1st Crossing Greenland* I. 47 The two Lapps had two pair of 'finnesko' each. **1911** R. F. Scott *Diary* 20 Jan. in *Last Exped.* (1913) iv. 132 We have.. served out two pairs of finnesko (fur boots) to each traveller. **1928** *Daily Tel.* 27 Nov. 10/6 Leather boots to replace the finsko which the men were wearing.

finney (fi·ni). Local name of finnan haddock. Also **finnie haddie.**
1851 Mayhew *Lond. Labour* I. 77/1 Smoked or dried haddocks (often called 'finnie haddies'). **1906** *Daily Chron.* 10 May 1/4 Finneys, kippers, bloaters, 2s. 6d. per box... Fish Docks, Grimsby. **1966** 'L. Lane' *ABZ of Scouse* 36 Got a face like a stale finny-addy.

finnip. Add: Also **fin(n)if(f), finnup, finuf.** (Earlier and later examples.) Also *U.S.*, five dollars.
1839 W. A. Miles *Poverty, Mendicity & Crime* 112 If he finds any 'finnips' (*5l.* notes).. he gives them to Nelson to fence. **1859** Hotten *Slang Dict.* 39 *Finuf*, a five-pound note. **1859** G. W. Matsell *Vocabulum* 32 *Finniff*, five dollars. **1889** Barrère & Leland *Dict. Slang* I. 343/1 *Finnup ready*, a five-pound note. **1931** D. Runyon *Guys & Dolls* (1932) vii. 141 He asks me if I happen to have a finnif on me, but of course I am not giving finnifs to guys like Feet Samuels, and finally he offers to compromise with me for a deuce, so I can see things must be very bad with Feet for him to come down from five dollars to two. **1966** R. Stout *Death of Doxy* (1967) v. 47, I.. got out my wallet and extracted a finif.

Finnish, *a.* Add: In adjectives of the type *Finnish-Ugrian* = Finnish and —. Also *Finnish Spitz*, a small, stocky breed of dog, with a coarse, reddish-brown coat. Also *Finnish-speaking* adj.
1911 Frazer *Golden Bough* (ed. 3) II. ix. 10 The tribes of the Finnish-Ugrian stock. **1920** [see *Finlander]. **1930** E. C. Ash *Pract. Dog Bk.* ii. 22 The Finnish Spitz... They have been known on the Continent for many years, but have only recently been introduced into this country. **1933** Bloomfield *Lang.* iv. 68 We find, to the east of the Indo-European languages.., the great Finno-Ugrian family... The first [branch] is the Finnish-Lapponic. **1945** C. L. B. Hubbard *Observer's Bk. Dogs* 68 The Finnish Spitz.. was introduced into England in 1927. **1947** Wyndham Lewis *Let.* 15 Nov. (1963) 419 With your Finnish-Irish ancestry and American birth. **1965** B. Vesey-Fitzgerald *Dog Owner's Encycl.* 88 The Finnish Spitz characteristics are eagerness to hunt, courage and fidelity.

Finno- (fi·no), used as comb. form of Finn or Finnic to designate race combinations and language groups of Finns or Finnish with other elements.
1864 *Smithsonian Inst. Ann. Rep. 1863* 110 The Scythian family.. is divided into five principal branches: 1. The Ugrian, or Finno-Hungarian, [etc.]. **1877** *Illustr. London News* 23 May 2/2 If the surrounding Finno-Swedish atmosphere has had any influence on him, it has been to develop his inborn Russian patriotism. **1879** *Encycl. Brit.* IX. 219/1 The term Finns.. being, with its adjective Finnic or Finno-Ugric or Ugro-Finnic, the collective name of the westernmost branch of the great Uralo-Altaic family. **1880** A. H. Sayce *Introd. Sci. Lang.* II. vi. 9 The roots of the Finno-Ugrian dialects exhibit the same vowel-harmony. **1885** *Encycl. Brit.* XVIII. 779/2 The nearer relation of the Samoyed is with the Finno-Hungarian. **1888** *Ibid.* XXIV. 1/1 The Ural-Altaic, Finno-Tatar, or 'Turanian' languages. **1895** A. H. Keane *Ethnol.* 200 We have in Europe.. mixed Finno-Slavs, Slavo-Teutons, Kelto-Teutons, but no Finno-Slav, Slavo-Teutonic or Kelto-Teutonic tongues. **1933** [see *Finnish *a.*]. **1955** *Proc. Prehist. Soc.* XXI. 91 The kinship system belongs to the Finno-Ugric system, characterized by bilateral descent and division into age groups. **1961** L. F. Brosnahan *Sounds of Language* v. 92 The zone of slight affrication.. Greek, Romance, Germanic, and Finno-Ugrian.

fino (fi·no). [Sp. *fino* Fine *a.*] A type of dry sherry; a glass of such sherry.
1846 R. Ford *Gatherings from Spain* xiv. 155 Those [sherries] only which exhibit great delicacy, body, and flavour are called *finos* or fine. **1888** *Encycl. Brit.* XXIV. 607/1 The Amontillado class may be.. subdivided into Fino and Oloroso. **1959** W. James *Word-Bk. Wine* 78 The finos which the Spaniards.. drink are considerably drier than those which they export. **1961** [see *amontillado]. **1962** *Times* 7 May 13/4 Old fino sherry. **1964** G. Sims *Terrible Door* xxvii. 144, I had two double finos in Martinez.

Finsen (fi·nsən). The name of Niels R. *Finsen* (1860–1904), Danish physician, used *attrib.* and in the possessive to designate the method introduced by him of treating skin diseases by ultra-violet light, and also the apparatus he devised for the purpose.
1899 *Brit. Med. Jrnl.* 30 Sept. 825/1 (*heading*) Remarks on Finsen's phototherapy. **1902** *Westm. Gaz.* 12 Mar. 9/3 A Finsen lamp for the new treatment of lupus. **1902** *Brit. Med. Jrnl.* 31 May 1324/2 A small ulcerating surface, which resisted treatment by Finsen's light. **1903** *Ibid.* 3 Jan. 24/1 The Finsen treatment in lupus vulgaris. **1907** *Practitioner* Nov. 721 The Finsen-light treatment of lupus. **1911** *Allbutt's Syst. Med.* IX. 480 Intra-nasal lupus can be rarely reached with Finsen's apparatus. **1934** Dore & Franklin *Dis. Skin* xxiv. 343 For small growths ultra-violet light applied with either the Finsen or Kromayer compression lamp is sometimes successful. **1961** *Brit. Med. Dict.* 561/1 *Finsen apparatus*, or *lamp*, a source of ultra-violet-light radiation designed for the local treatment of lupus. It consists of a carbon-arc lamp of high amperage. *Ibid.*, *Finsen treatment*, the treatment of lupus vulgaris by local ultraviolet irradiation with the Finsen lamp; this is now seldom employed.

fip. (Earlier examples.)
1822 *Philad. Freeman's Jrnl.* 5 Sept. (Th.), A dispute now commenced between two persons respecting some cents and a 'fip'..: one asserting that there were two 'fips', and the other that there was but one. **1833** in *Indiana Mag. Hist.* (1919) XV. 244 He would not let us pass through the fense without a 'fip', that is 4½ cents. **1837–47** [see Levy *sb.*[2]].

fipenny. Add: **fipenny bit** *U.S.* = Fip. (*Disused.*)
1802 *Port Folio* (Phila.) 27 Feb. 6/2 We shall gather *fipenny-bits* from every parsley bed. **1856** Mrs. Stowe *Dred* I. 235 What do we want to send our girls there, to get fipenny-bit ideas?

fipple, *sb.* **1.** Delete †*Obs. rare*⁻¹ and add later example; also *attrib.*, as **fipple flute** (see quot. 1956).
1911 *Encycl. Brit.* XXII. 966/2 (*heading*) Recorder, Fipple Flute or English Flute. *Ibid.*, This channel is so constructed within the mouthpiece that the stream of air impinges with force against the sharp edge of a lip or fipple. **1956** *Oxf. Compan. Music* (ed. 9) 865/1 Flutes of the end-blown variety are.. known as fipple flutes.

fir. Add: **4. fir balsam,** the silver fir of Canada, *Abies balsamea*; **fir-bob** (earlier example).
1810 Michaux *Arbres* I. 18 Sylvir fir,.. Fir balsam,.. [ou] Balsam of Gilead tree. **1899** S. O. Jewett *Queen's Twin* 16 Fir balsams comin' up over the hill all green an' hearty. **1854** *Househ. Words* IX. 327/2 The fire was always in that state when somebody had.. to unload his pockets of fir-bobs, which.. threw beautiful dancing lights.

Firbankian (fəɹbæ·ŋkiăn), *a.* [f. the name of Ronald *Firbank* (1886–1926), English author.] Of, pertaining to, or characteristic of Ronald Firbank or his works and opinions.
1931 C. Beaton *Diary* Feb. (1961) 226 Firbankian intrigues. **1958** *Times Lit. Suppl.* 19 Dec. 733/4 A delightfully absurd, almost Firbankian day-dream. **1963** *Listener* 10 Jan. 92/1 The Firbankian name of Mirabeau Buonaparte Lamar.

Firbolg (fiᵊɹbǫ·lg, in Irish firbǫ·ləg). [Ir., ulterior etym. obscure: perh. f. OIr. *fir* pl. of *fear* man + *bolg* gen.pl. of *bolg* bag, belly, thus 'men of the bags (or bellies)'; or possibly (Pokorny) cogn. w. Gaulish *Belgae*:— IE. **bhel-gh-*/**bhol-gh-*.] A name given in Irish legend to an early colonizing people of Ireland. Also *transf.* Hence **Firbo·lgian** *a.*, of or pertaining to this people.
1797 *Encycl. Brit.* IX. 313/1 The Firbolgs were.. expelled or totally subdued.. by the Tuath de Dannans. **1863** *Chambers's Encycl.* V. 626/1 According to ancient native legends, I[reland] was in remote times peopled by tribes styled Firbolgs and Danauns, eventually subdued by Milesians or Gaels. **1916** Joyce *Portrait of Artist* v. 210 The rude Firbolg mind of his listener had drawn his mind towards it and flung it back again. **1936** M. Franklin *All that Swagger* i. 8 Families of the reverberating cognomens and Firbolgian descent. **1938** L. MacNeice *I crossed Minch* ii. xvi. 240 Flora.. answered.. that Firbolgs were prehistoric and anyway ought to be called Ciuthachs. **1968** *Times* 15 Oct. 13/5 Production of fairyland lustre began in November 1915: gnomes, pixies,.. fey creations, including the 'firbolgs', a peculiarly unappealing class of little men with big eyes. **1971** *It* 2–16 June 24/1 The.. Formorians, Fir-Bolgs and de-Danaans are all remembered as being of semi-human stock.

fire, *sb.* Add: **A. 3. a.** Also *ellipt.* for *gas fire, electric fire*, etc.
1895 *Army & Navy Co-op. Soc. Price List* 334 Ball Fire without Fender and Trivet, 15/9. **1926–7** *Ibid.* 344/2 A small 1-unit fire... One bar (1,000 watts). **1939–40** *Ibid.* 271 All fires are wired complete. *Ibid.*, Well designed 2 kw. fire with hammered copper reflector. **1968** 'J. Le Carré' *Small Town in Germany* v. 70 'Very gadget-minded, Mr Harting is... He had a *tea* machine.'.. 'What else?' 'A fire. The new fan type with the two bars over.'

b. Phr. *fire in the* (or *one's*) *belly*: ambition, driving force, initiative.
1951 N. Annan *L. Stephen* ix. 275 There is no fire in the belly, no sense of urgency. **1960** *20th Cent.* Nov. 390 If they rebel.. then they are crackbrained, want a pretext for a rave, or have fire in their belly. **1970** *Guardian* 10 Aug. 9/8 Her success is due, she says, 'to the fact that I have fire in my belly'.

g. *to play with fire*: to trifle with dangerous matters, esp. at the risk of moral disaster or emotional distress.
1887 J. Grant (*title*) Playing with fire. **1887** Kipling *Plain Tales fr. Hills* (1888) 139 There is no sort of use in playing with fire, even for fun. **1907** *Daily Chron.* 9 Oct. 4/6, I should like to sound a note of warning, for, as one who plays with fire, he can only expect to get burnt. **1925** A. S. M. Hutchinson *One Increasing Purpose* iii. xvi, Looking back upon herself in those fire-playing days. **1928** Galsworthy *Swan Song* v. 33 If—on second thoughts, she wanted to play with fire. **1957** L. P. Hartley *Hireling* xix. 149 She led me on, she played with fire, but she wouldn't have me.

5. e. *a brand from* (or *out of*) *the fire*, see Brand *sb.* 3 b; *to save out of the fire*, to preserve as a remnant from a disaster or catastrophe; *to pull* or *snatch out of the fire* [after Jude 23], to rescue from disaster or ruin.
1855 Trollope *Warden* xix, To be sure Puddingdale is only four hundred, but that would be saving something out of the fire. **1893** H. Hutchinson *Golfing* 85 It is wonderful what matches these strong souls now and again pull out of the fire. **1924** G. Lambton *Men & Horses* 120 Well, I pulled that race out of the fire. **1927** *Daily Tel.* 11 Oct. 5/6 The winner pulling the match out of the fire after being down 2 sets to 1. **1928** *Observer* 10 June 27/4 The best snatcher of a match out of the fire that we ever had. **1932** C. Morgan *Fountain* i. i. 9 A man who.. pulls a business out of the fire isn't an incompetent.

f. *to pull the chestnuts out of the fire*: used (with reference to the fable of the monkey using the cat's paw to extract roasting chestnuts from the fire) of the employment of another to undertake the dangerous part of an enterprise. (See *N. & Q.* 6th Ser., 1883, VII. 286, VIII. 34.)
[**1586** G. Whitney *Emblems* 58 The ape, did reache for Chestnuttes in the fire,.. he with a whelpe did close, And thruste his foote, into the Embers quick, And made him, pull the Chestnuttes out perforce.] **1657** [see Cat's paw 2]. [**1754** Richardson *Grandison* III. 358 He makes her fight his battles for him; and become herself the cat's paw to help him to the ready-roasted chestnuts.] **1886** [see Unionist A. 1. c]. **1930** J. C. Snaith *Unforeseen* ii. xxi, He was the guy who pulled the financial chestnuts out of the fire. **1957** *Economist* 28 Sept., The Germans cannot be expected to pull our chestnuts out of the fire.

g. *where's the fire?*: jocular phrase said to a person who is in a hurry.
1924 P. Creswick *Beaten Path* xxxvi. 195 A husky voice enquired of me: 'Where's the fire?' **1963** J. F. Straker *Final Witness* xvi. 174 'Where's the fire, dear boy?' he drawled. 'Do we really have to run for it?' **1971** 'H. Calvin' *Poison Chasers* ix. 124 'Where's the fire?' he snarled.

12. *St. Anthony's fire.* Read: a name for one or more inflammatory or gangrenous diseases of the skin, variously identified with erysipelas, ergotism, etc.
1527, etc. [see Anthony (St.)].

14. b. *under* (*a*) *fire*: also *fig.*, criticized.
1886 'Mark Twain' *Lett. to Publishers* 19 Mar. (1967) 196 It isn't good policy for anybody connected with our publishing firm to be under a fire of newspaper criticism this year. **1965** *New Statesman* 19 Mar. 430/3 Their chief ..has recently been under fire because it appeared he had joined the Nazi Party in his student days. **1968** *Time* 17 May 66 University ties with IDA have also come under fire at Columbia.

15. *Cricket.* Tendency (of a ball) to fly up erratically or (of the ground) to cause the ball to fly up.
1888 A. G. Steel in Steel & Lyttelton *Cricket* iii. 161 A good long run.. gets way and 'fire' on to the ball. **1897** K. S. Ranjitsinhji *Jubilee Bk. Cricket* 70 Coming from a high elevation, a ball.. has more fire or 'devil' in it. *Ibid.* 80 There is something to be done with the ball when.. the wicket has fire in it.

B. 1. a. *fire-folk, -glow, hazard, -mist, season.*
1877 G. M. Hopkins *Poems* (1918) 26 O look at all the fire-folk sitting in the air! **1907** *Westm. Gaz.* 18 Dec. 2/3 What's amiss with the fire-glow, what is awry with the light? **1926** *Daily Colonist* (Victoria, B.C.) 8 July 4/1 Logging methods such as will tend to reduce the fire hazard that now prevails. **1935** *Discovery* Nov. 316/2 The film used will be of the safety type, which presents somewhat less fire hazard than ordinary newsprint. **1875** G. H. Lewes *Problems* II. 393 We begin an inquiry which may lead back to the primeval fire-mist. **1892** *Work* IV. 58/1 The mantleshelf in the 'fire season' is the dustiest place in the room. **1878** W. James *Coll. Ess. & Rev.* (1920) 47 Back of him [sc. the polyp] lay the not-yet-polyp, and, back of all, the universal mother, fire-mist. **1901** *Wide World Mag.* VIII. 194/2 The long grass was as inflammable as tinder; but this not being the 'fire' season we anticipated no danger.

d. *fire-door* (earlier example).
1838 *Civil Engin. & Archit. Jrnl.* Sept. 326/2 It produced more steam than was required, and with the fire-door kept open.

f. *fire-cracker* (later example).
1958 *Punch* 22 Jan. 150/3 Nine weeks ago firecrackers were thrown into the gangway and among neighbouring seats.

g. *fire-apparatus, -call, -chief.*
1901 *Daily Colonist* (Victoria, B.C.) 6 Oct. 2/1 Loss about $225,000, with 25 per cent. insurance. There was no fire apparatus. **1897** *Daily News* 30 Nov. 5/1 Our fire-call system in London. **1906** *Westm. Gaz.* 26 Apr. 7/1 Within a very few minutes of the fire-call being rung. **1889** *Kansas Times & Star* 21 Sept., Our gallant fire chief was elected first vice-president.

2. (sense 5) *fire-extinguisher* (later example), *-fighter*; *fire-fighting* adj. and vbl. sb.; *fire-resistant* adj.; *fire-retardant* adj. and sb.
1928 Galsworthy *Swan Song* iii. xiii. 319 He.. seized a fire extinguisher... He knew vaguely that you dashed the knob on the floor and sprayed the flames. **1903** *Westm. Gaz.* 16 June 4/3 The 'invention' and arrangement of the display.. is the work of a practical fire-fighter of great experience.. who has been for a long time the chief officer of the Hampton Fire Brigade. **1904** *Forum* Oct. 274 The most modern equipment should be at the command of the fire fighters. **1945** S. Spender *Citizens in War* 26 The work of fire-fighters depended greatly on the prompt reporting of fires. **1970** *Globe & Mail* (Toronto) 25 Sept. B7/2 (Advt.), A well equipped Fire Department with a present strength of over 80 Fire Fighters. **1897** L. J. Gage in *Open Court* XI. 213 The fire-fighting force is divided into 109 companies. **1897** *Westm. Gaz.* 22 Nov. 9/1 Chemical engines, water-towers, and other time-saving and fire-fighting apparatus. **1903** *Daily Chron.* 5 May 7/1 The present week is an interesting one in the history of fire-fighting. **1959** *Times Rev. Industry* Sept. 50/1 The standard of fire-fighting equipment has improved. **1971** *Guardian* 7 July 1/8 Fire fighting equipment.. must be available. **1902** *Science* 12 Sept. 424 The treatment of wood with a view of making it fire-resistant is not a matter of recent years. **1946** *Nature* 19 Oct. 462/2 The treatment of the boards to render them fire-resistant. **1915** *Proc. Nat. Fire Protection Assoc., 19th Ann. Meeting* 108 There is nothing new in the idea of treating wood to make it fire retardant. **1952** *Dict. Fire Technol.* (Inst. Fire Engin.) (Advt.), 'Faspos' fire retardants for fireproofing wood. **1961** *Engineering* 6 Oct. 447/2 The properties of bentonite as a fire retardant. **1962** *Punch* 17 Oct. 560/3 A wise pilot dips his overalls in fire-retardant liquid.

3. a. *fire-scorched, -shot, -swept, -tinged, -wrought* adjs.
a **1892** Whittier *Poet. Works* (1898) 309/1 The fire-scorched stones. **1927** *Observer* 18 Dec. 19/3 Mr. Richard Goolden's fire-shot devil. **1898** G. B. Shaw *Man of Destiny* in *Plays Pleasant & Unpleasant* ii. 155 Napoleon Bonaparte.. rushed the fireswept bridge. **1930** Blunden *Poems* 40 The naked fire-swept windows. **1907** *Daily Chron.* 19 Sept. 4/4 The dark pines.. dripped fire-tinged dews. *a* **1892** Whittier *Poet. Works* (1898) 324/2 Its fire-wrought language. **1905** *Daily Chron.* 9 Dec. 8/1 Mr. Jones's fire-wrought temperament.

4. *fire-bright, -hearted, -tongued* adjs.
1916 A. Huxley *Burning Wheel* 15 My fire-bright window-pane. **1902** *Westm. Gaz.* 24 Feb. 2/3 Each minute Fire-hearted followed. **1931** V. Woolf *Waves* 319 The watery fire-hearted jewel. *a* **1892** Whittier *Poet. Works* (1898) 486/2 The fire-tongued miracle.

5. fire-alarm, (*b*) *N. Amer.* a warning of fire; **fire-altar,** 'an altar upon which burnt sacrifices were offered, as distinguished from one used for incense only' (*Cent. Dict.* Suppl. 1909); **fire assay** *U.S.*, an assay in which intense heat is applied to a material; **fire-back,** (*a*) (earlier U.S. example); **fire-bay** *Mil.*, a trench with a concealed parapet from which firing takes place; **fire-blight,** also (chiefly *U.S.*), a bacterial disease affecting (esp. pear and apple) trees, producing a scorched appearance of the leaves; **fire-blitz,** an air attack with incendiary bombs; so *fire-blitzed* ppl. a.; **fire-board,** (*a*) (earlier example); **fire-bomb,** an incendiary bomb; hence as *v. trans.*, to attack or destroy with these bombs; also **fire-bombing** *vbl. sb.*; **fire-boss,** also *Canad.*; **fire-break** (earlier and later examples); for def. read: an obstacle to the spread of (grass or forest) fires, as cleared or ploughed land; also *fig.*; **fire-brigade,** also *fig.*; **fire-bug** (U.K. examples); **fire-chamber** (earlier U.S. example); **fire-company** (U.S. examples); **fire-control,** the regulation or control of the fire of guns (e.g. in batteries); also, the station from which a commanding or gunnery officer may exercise such control; **fire-crack,** a crack formed by heat, *spec.* in metal when it is being reheated or annealed; hence **fire-cracking** *vbl. sb.*; **fire-crome** = Fire-hook a; **fire-cure** *v. trans.*, to cure (tobacco or leather) over a fire; hence *fire-cured* ppl. a., *fire-curing* vbl. sb.; **fire-curtain,** (*a*) a fire-proof curtain in a theatre, etc.; (*b*) *Mil.* = *curtain *sb.*[1] 3 b; **fire-discipline** *Mil.*, the training of men to fire exactly as directed by the commander, so that the work of a unit is co-ordinated; **fire-dog** (earlier U.S. example); **fire-drill,** (*b*) the rehearsal of action to be taken in the event of fire; **fire dropper,** a man who removes the fire from the fire-box of a locomotive at the end of a journey; **fire-fight** *Mil.*, the struggle to establish fire superiority over the enemy; **fire-guard,** (*b*) *N. Amer.* a member of a fire-brigade or fire-watching group; (*c*) *N. Amer.* = **fire-break*; **fire-guard** *v. trans.*, to surround (land) with a fire-guard (*c*); **fire hall** *N. Amer.*, a fire station; **fire-hat** *U.S.*, a hat for a fireman; **fire-lighter,** (*b*) (examples); **fire-line** = **fire-break*; **fire-piston,** a device for making fire (see quot. 1951); **fire-plough** (see quots.); **fire position** *Mil.*, a position from which fire is opened by an attacking force during an advance; **fire-power** *Mil.*, the total effectiveness of the fire of guns, missiles, etc., of a military force; **fire-raid,** an air-raid with incendiary bombs; **fire-ranger** *N. Amer.*, an official who keeps watch against the occurrence of forest fires; **fire-risk,** (*a*) the risk of loss by fire; (*b*) the obligation of a fire-insurance company to make good loss by fire; (*c*) property insured against fire; **fire-room** (earlier example); (*b*) a furnace-room of a building or stokehold of a ship; **fire-step** = **firing-step*; **fire-stick,** (*c*) an implement used for stirring up a fire; **fire-stop,** a device designed to stop the spread of fire; *spec.* incombustible material placed in the open parts of a structure; **fire-storm,** (*a*) *poet.*, a storm of fire; (*b*) *spec.* a high wind or storm following a conflagration caused by incendiary or nuclear bombs; **fire-swallower,** one who entertains by swallowing or pretending to swallow fire; **fire-trace** = **fire-break*; **fire-trench** *Mil.*, a deep and narrow trench from which firing takes place; **fire-walk,** the ceremony of walking barefoot over hot stones, performed as a religious rite by the Fijians and others, and formerly as an ordeal in European countries; so **fire-walker, -walking**; **fire-wall** orig. *U.S.*, a fireproof wall to prevent the spreading of fire; **fire warden** *N. Amer.*, an official concerned with the prevention or extinction of fires; **fire-watcher,** (*a*) one who tends a fire; (*b*) one who is engaged in fire-watching; **fire-watching** *vbl. sb.*, keeping on the alert, esp. at night, to watch for and report the occurrence or spread of fires, esp. those caused by aerial bombardment; hence **fire-watch** *v. intr.*; **fire-wind** = **fire-storm* (*b*); **fire-woman,** a woman member of a fire-brigade; **fire-zone,** an area swept by gunfire.
1876 *Wide Awake* (Boston, Mass.) Apr. 262/1 The three churchbells that give the fire alarm. **1921** *Daily Colonist* (Victoria, B.C.) 7 Apr. 9/4 Fire Chief Parkin reported that the department had answered four fire alarms during the month of March. **1926** 'M. Leinster' *Dew on Leaf* 30 Fire-altar wreathed in clouds of incense. **1935** *Burlington Mag.* Jan. 30/1 The fire-altar on the coinage of the later Kushan princelings. **1869** C. L. Brace *New West* 167 Mills.. guarantee 75 per cent. returns on the fire assay. **1847** *Rep. Comm. Patents 1846* (U.S.) 249 The fire-backs of fireplaces have been made separate. **1917** *Chambers's Jrnl.* Mar. 174/1 They look for the firebays, which are, of course, on the German side. **1940** J. Brophy *Home Guard* 64 In a fire-trench, to be manned with rifles and machine-guns, the sections facing the expected advance of the enemy are known as fire-bays. **1817** W. Coxe

View Cult. Fruit Trees 175 The fire blight frequently destroys [pear] trees. **1869** *Trans. Ill. Agric. Soc.* VII. 503 There are several distinct diseases, all grouped together under the general name of 'Fire Blight'. **1958** *New Scientist* 20 Nov. 1301/1 The recent outbreak of Fire Blight in this country is alarming...It is caused by a bacterium, *Erwinia amylovora*, that usually attacks the fruit trees in the spring. **1941** A. O. POLLARD *Bombers over Reich* vi. 64 This total of fire-bombs seemed to be the mark, in Nazi eyes, of a super 'fire-blitz'. **1940** *Daily Express* 31 Dec., The fireblitzed area. *a* **1828** D. WORDSWORTH *Tour Continent* in *Jrnls.* (1941) II. 316 She replied 'Oh Madame! you will soon be warm here!' taking down a fire-board. **1895** G. GRIFFITH *Outlaws of Air* xxiii. 207 The conflagration that his fire-bombs had started. **1940** *Flight* 24 Oct. 339/2 Many fire-bombs were dropped, as well as high explosives. **1970** *Guardian* 14 May 2/4 More than twenty buildings have been fire-bombed. **1966** *Economist* 30 July 442/3 The disorders were mostly confined to the hit-and-run fire-bombing..and to isolated shooting incidents. **1969** K. VONNEGUT *Slaughterhouse-Five* v. 87 The greatest massacre in European history..was the fire-bombing of Dresden. **1898** 'R. CONNOR' *Black Rock* 87 Mr. Shaw, fire-boss of the mines. **1955** J. S. GOWLAND *Smoke over Sikanaska* x. 180 There's a Greek who used to be a fire boss in the mine. **1841** *Newfoundland Ho. Assembly Jrnl.* App. 178 The lane opposite Messrs. Codner & Jennings's firebreak. **1930** BILLIS & KENYON *Pastures New* viii. 120 Large areas of grass were often saved by burning fire-breaks..round extensive sections of good grass land. **1934** *Bulletin* (Sydney) 17 Oct. 20/3 Doesn't 'T. T.' know that a firebreak is always burnt against the wind? **1935** *Discovery* Mar. 64/1 Fire-breaks are..left open in the new plantations. **1959** *Economist* 31 Jan. 391/1 Independence behind a firebreak of Belgian ..paternalism. **1965** H. A. KISSINGER *Probl. Nat. Strategy* vi. 120 His [*sc.* A. C. Enthoven's] basic argument is that the most reliable 'fire break' is between conventional and nuclear weapons and that once that threshold is crossed, rapid escalation is probable. **1959** *Daily Tel.* 10 Apr. 19/2 S.E.A.T.O. will probably get the 'fire brigade' that Asian members have so long been calling for. **1931** *Times Lit. Suppl.* 24 Sept. 719/2 An even more elusive foe of the fire office than petrol is 'the cold-blooded fire-bug' who trades in arson. **1962** *Punch* 14 Mar. 439/1 Robert Morley as a dedicated firebug is notable. **1847** *Rep. Comm. Patents 1846* (U.S.) 211, I also claim the air passage below the fire chamber. **1744** *Penna. Gaz.* 14 Dec. 4/2 The Union Fire-Company of Philadelphia. **1792** *Fayetteville* (N.C.) *Gaz.* 16 Oct., Proposals from the Maryland Insurance Fire-company. **1835** *Southern Lit. Messenger* I. 259 The remaining expenses are on account of the public markets, fire companies, salaries of officers, [etc.]. **1880** *Harper's Mag.* July 208/2 Several of the members belonged to the volunteer fire-companies, then in the height of their glory. **1915** *Cincinnati Ann. Rep. 1914* 343 Temporary quarters have been provided for Fire Company No. 54. **1886** J. H. A. MACDONALD *Common Sense on Parade* 118 What is wanted is the conviction in the mind of every instructor..that his men should never leave a parade without having gained something in fire discipline,—that is, that fire control drill be one of the main points in view as a necessary part of the work to be performed on every occasion when men are being drilled, [etc.]. **1907** *Westm. Gaz.* 3 May 7/1 Stand on the deck amidships, look up at the vast tripod which supports the fire-control. *a* **1928** in C. F. S. Gamble *N. Sea Air Station* (1928) xv. 241, I had a very pleasant minute or so wondering which I was going to hit—the wireless masts or the monitor's fire-control top. **1935** *Jrnl. R. Aeronaut. Soc.* XXXIX. 875 Thirty diagrams illustrate problems of fire control arising out of attacks by fighter formations on light bombing formations. **1862** *Catal. Internat. Exhib.* II. x. 12 Cliff's patent enamelled clay retort is..adapted for the use of gas works, by its..freedom from fire-cracks. **1898** *Engineering Mag.* XVI. 147/2 The Influence of Bismuth on Brass, and its Relation to Fire-Cracks. **1959** *Jrnl. Iron & Steel Inst.* CXCIII. 388/3 It is shown that thermal stresses were responsible for the appearance of fire-cracks. **1923** GLAZEBROOK *Dict. Appl. Physics* V. 424/2 This 'fire-cracking' can be entirely prevented by a previous heat treatment at a much lower temperature. **1960** D. J. O. BRANDT *Manuf. Iron & Steel* (ed. 2) IV. xxix. 221 *Fire-cracking*, that is, the development of a series of little crazy cracks on the surface caused by alternate heating and cooling. **1846** Fire-croom [see CROME, CROMB *sb.*]. **1899** *Athenæum* 2 Sept. 329/2 The head of a great fire-hook or fire-crome was noted. **1900** *Rep. U.S. Dept. Agric.* LXV. 34 Tobaccos that have been fire-cured, as the plug tobaccos. **1899** *Atlantic Monthly* LXXXIII. 758/2 The pliant sandals of fire-cured skin. **1962** *Times* 12 Oct. (Standard Bank Suppl.) p. vii/3 Dark fire-cured tobacco. **1848** *Rep. Comm. Patents 1847* (U.S.) 170 Fire-curing is not necessary. **1912** P. MCKEON *Fire Prevention* 153 The proscenium arch has a fire curtain. **1932** G. B. SHAW *Emperor & Little Girl* in *Wks.* VI. 15 These curtains..were really shells, showers of bombshells... They were called fire curtains. **1940** *War Illustr.* 19 Jan. 620/1 Two great steel fire-curtains..can be dropped in a few seconds to subdivide each of the hangars where the machines are housed thus localizing any fire outbreak. **1886** Fire discipline [see *fire-control*]. **1897** *Cavalry Tactics* xvii. 121 Fire discipline must be strictly enforced, both to ensure accurate shooting at the indicated object and to control the ammunition expenditure. **1923** KIPLING *Irish Guards in Great War* I. 11 Our fire-discipline, as we fell back, gave them the impression that the forest was filled with machine-guns. **1792** *Massachusetts Spy* 1 Mar. 3/1 [He] caught a fire-dog, which he threw with such force that he knocked down one of the ruffians. **1911** *Daily Colonist* (Victoria, B.C.) 13 Apr. 20/1 Chief Davis submitted a report on the time occupied by the children in fire drill in clearing the school houses. **1913** J. VAIZEY *College Girl* xviii. 264 [It's] the fire drill! They've had an alarm, and she's told to shut off draughts. **1959** C. V. GOOD *Dict. Educ.* 230/1 *Fire drill*, practice in the systematic, safe evacuation of the children and teachers..upon the sounding of the fire alarm. **1898** *L'pool Weekly Courier* 9 Apr. 2/7 He was a fire dropper—drawing the fires from locomotives. **1899** *Westm. Gaz.* 23 Nov. 2/1 The excitement of a fire-fight at short ranges. **1968** *Listener* 26 Dec. 871/2 One savage, roaring, banging, howling 'fire-fight'. **1833** *Niles' Reg.* XLIV. 259/2 There were twenty-three engine and hose companies,..and four divisions of fire guards. **1874** J. C. MCCOY *Hist. Sk.* 217 An impassable barrier would be created between the unburned grass within the encircled tract, and that upon the outside of the 'fire-guard'. *Ibid.*, A large adjacent tract of land..will be 'fire-guarded', in order to secure a winter range from the ravages of prairie fires. **1912** Fire-guard [see *BACK-FIRE *v.* 1]. **1926** *Daily Colonist* (Victoria, B.C.) 15 July 7/1 It was stated that the fire on the south-west side of the river was practically under control, while a fire guard had been placed between the fire zone and the city watershed. **1963** R. D. SYMONS *Many Trails* 40 The smoky air..choked us as we feverishly plowed fire-guards. **1881** A. BEGG *Gt. Canad. North West* 103 A handsome fire hall centrally located. **1966** *Kingston* (Ont.) *Whig-Standard* 26 Jan. 6/3 Council decided to increase the insurance on the fire hall. **1851** C. CIST *Cincinnati* 213 George E. Minister..makes..fire hats. **1853** GEO. ELIOT *Let.* 5 Nov. (1954) II. 123 Will you..send me the address of the London agent for the Fire-lighters? **1897** *Daily News* 4 June 8/2 Firelighters, specially prepared,..to light all the bonfires throughout the country. **1957** *Times* 9 Sept. 11/5 New-fangled, oil-soaked, 'fire-lighters' and suchlike meretricious aids are nothing but barefaced cheating. **1902** A. C. LAUT *Story Trapper* (1912) 147 The game..must be moving away from the fire line. **1905** *Terms Forestry & Logging* 10 *Fire line*, a strip kept clear of inflammable material as a protection against the spread of forest fire. **1927** W. H. TODD *Tiger, Tiger!* 108, I walked down a jungle ride, or fire line. **1953** *Brit. Commonw. Forest Terminol.* I. 53 *Fire line*, (*a*) the portion of a control line or fire trace from which inflammable materials have been removed... (*b*) A clear felled permanent firebreak..[Ind]. (*c*) A line round an active fire..to..control it [Can]. **1920** *Brit. Mus. Return* 64 A fire-piston of buffalo horn from the Kachin, Upper Burma. **1951** *N. & Q. Anthropol.* (ed. 6) III. 240 The *fire-piston* depends for its working on the development of heat by the sudden compression of air in a confined space. **1893** *Funk's Stand. Dict.*, Fire-plow. **1934** E. EYRE *European Civilisation* I. i. 39 The methods of fire-making in use are: the fire-plough (the rubbing up and down of a piece of hard wood in a groove of soft timber), which may well be the oldest device of its kind. **1955** *Antiquity* XXIX. 132 In the fire-plough a pointed stick is rubbed vigorously along a groove or depression in a piece of wood laid flat on the ground. **1905** *Westm. Gaz.* 15 May 4/1 Cavalry could do next to nothing, and artillery could seldom find really effective fire-positions. *Ibid.* 13 June 3/1 The most favourable spot for your individual fire-position. **1913** R. MEINERTZHAGEN *Army Diary* 6 Oct. (1960) 55 What I should have liked to see is more automatic fire-power in the hands of both the battalion commander and the company commander. **1928** *Daily Tel.* 11 Sept. 12/4 The greatest possible strength of fire-power—from guns and machine-guns—should be concentrated rapidly on one part of the front only. **1955** *Bull. Atomic Sci.* Mar. 94/1 The authors readily admit that the tactical A-bomb has grave limitations, yet they feel that it grants us superior firepower to counter the conventional strength of the Soviets. **1967** *New Scientist* 17 Aug. 328/1 A lieutenant colonel of the Viet-cong..confessed himself unimpressed by the white men's mobility in jungle conditions, but he had no doubts about the effectiveness of their firepower. **1945** *Daily Tel.* 20 July, Over 600 of the giant bombers visited four cities, including Hitachi, with fire raids. **1960** KOESTLER *Lotus & Robot* II. vi. 177 The two conventional fire-raids on Tokyo, earlier in the same year, had claimed another 100,000 victims. **1909** A. D. CAMERON *New North* 82 The Fire-Ranger of the district..has his barley and rice spread out on sheeting. **1920** *Outing* June 137/2 If you discover a fire, put it out if possible. If you cannot, get word to the nearest U.S. Fire Ranger or State Fire Warden. **1928** *Daily Express* 29 Aug. 2/5 The British Columbia fire ranger, whose lookout station is on Mount Cartier, 9,000 feet above sea level. **1865** *Atlantic Monthly* XV. 86 Joint-stock companies for the underwriting of fire-risks. **1867** C. J. BUNYON *Law of Fire Insurance* iii. 40 Gasworks are esteemed a fire risk of special hazard. **1907** *Install. News* Mar. 7/1 The Gunpowder Manufactory at Minden, where Simplex Screwed Conduits have been used because of the fire risk. **1966** *Lancet* 24 Dec. 1405/2 Little is known of the fire-risks to man in oxygen-rich environments. **1632** *Inventory* 9 Apr. in M. W. Barley *Eng. Farmhouse & Cottage* (1961) App. 278 Itim in the Chamber over the fire roume a trundle bedstedle. **1836** *Southern Lit. Messenger* II. 734 Passing the fire-room, where they were just firing up, I stopped. **1880** *Harper's Mag.* Oct. 793 Yet even on the *Seawanhaka* it appears that the fireroom..was unprotected. **1903** A. H. LEWIS *Boss* i. 8 As for a bed, if it should be summer time, what should be finer than the docks? Or if winter, than the fire-rooms of the tugs? **1916** *War Illustrated* 23 Dec. 449/1 Bill got upon the fire-step, placed his rifle against the wall, [etc.]. **1954** W. FAULKNER *Fable* (1955) 54 Two British privates were resting on the firestep of a frontline trench. **1896** J. C. HARRIS *Sister Jane* 100 Sister Jane, armed with a fire-stick (a heavy piece of metal weighing four or five pounds)..was..making an effort to get to the door. **1900** *Ann. Rep. Smithsonian Inst.* (Nat. Mus.) 180 In the same plate are included a pair of wooden fire sticks or tongs [of the Tulare Indians]. **1897** F. C. MOORE *How to build Home* 7 The ceiling should be plastered on metallic lathing. This will be an effectual fire-stop in case of fire starting in the cellar. **1926** C. E. MULFORD *Cassidy's Protégé* iv. 41 The farms he had dreamed of were now no more than memories, their boundary furrows and fire-stops rank with triumphant bunchgrass. **1581** Fire storme [see Dict., B. 1 a]. **1905** *Westm. Gaz.* 13 May 6/2 Glowing embers as it were of the fire-storm [*i.e.*, a grass fire]. **1945** in *Amer. Speech* (1951) XXVI. 236/2 These storms soon combined in an immense 'fire storm' (high winds blowing inwards toward the center of a large conflagration) similar to those caused by ordinary mass incendiary raids. **1959** *Times* 26 June 11/2 Nearly half of these could be expected to die instantly, killed by blast or incinerated in the fire-storms caused by the explosions. **1924** A. HUXLEY *Little Mexican* 57 Fat women, fire-swallowers, elastic men. **1932** 'R. CROMPTON' *William—the Pirate* vi. 134 'My hubby's just gone to have a drink with the fireswallower,' continued Mrs. Tom Thumb. **1891** W. SCHLICH *Man. Forestry* II. iii. 192 Protection is afforded by removing all inflammable matter, or clearing fire-traces around the area. **1953** *Brit. Commonw. Forest Terminol.* I. 55 *Fire trace*, a cleared (often burnt) line used as a base from which to counterfire; also frequently as a syn[onym] for *firebreak*. **1909** *Strand Mag.* Apr. 365/2 He dived down into the fire-trenches. **1916** 'BOYD CABLE' *Action Front* 103 The deep ditch with a narrow platform along its front that was the forward fire trench. **1928** BLUNDEN *Undertones of War* xviii. 193 Men..asleep in narrow chilly firetrenches. **1898** *Westm. Gaz.* 26 July 3/2 The Fijian fire-walk. **1900** *Proc. Soc. Psychical Res.* Feb. 11 Colonel Haggard saw the fire-walk done in Tokio, on April 9th, 1899. **1895** *Folk-Lore* Sept. 246 A photograph of the fire-walkers. **1898** *Westm. Gaz.* 28 June 4/3 The fire-walkers then approached..and..walked leisurely across and around the oven. **1938** *Times Lit. Suppl.* 1 Jan. 5/4 He is inclined to attribute the immunity of the fire-walkers to the power of faith. **1899** *Daily News* 25 Jan. 4/6 There was a fire-walking ceremony at Benares the other day. **1904** *Athenæum* 13 Feb. 216/1 An interesting lecture on 'The Fijians and their Fire-Walking'. **1963** *Times* 2 Mar. 10/6 A Chinese fire-walking ceremony. **1851** C. CIST *Cincinnati* 230 Three smoke-houses..are separated by twelve inch walls..with fire-walls on the roof. **1947** *Aircraft Engin.* Jan. 15/1 The firewall is a stout bulkhead. **1724** *New-Eng. Courant* 10–17 Aug. 2/2 Leave first obtain'd from the Firewardens. **1817** *Upper Canada Gaz.* (York, Ont.) 12 June 95/5 The said Fire Warden shall..carry about with him on occasion of Fires, a staff or some other visible distinguishing badge of office. **1875** *Chicago Tribune* 2 July 3/4 The seven Fire Wardens were..abolished. **1968** R. M. PATTERSON *Finlay's River* 41 Truly this butte was..well sited for the fire warden's lookout. **1941** R. GREENWOOD *Mr. Bunting at War* xix. 269 I've got to fire-watch. **1910** W. DE LA MARE *Three Mulla-Mulgars* xvi. 217 The fire-watcher raised his sad countenance. **1941** *Ann. Reg. 1940* 95 All factories employing not less than thirty persons must have fire watchers. **1944** *Times* 31 Jan. 2/3 In the cities and towns the Home Guard give a good deal of help in the Civil Defence services, as fire-watchers now. **1941** *Manch. Guardian Weekly* 10 Jan. 18 Everyone now realises the great importance of fire-watching and fire prevention. **1952** O. R. GURNEY *Hittites* vii. 150 There were even regulations about fire-watching and nightly patrols. **1945** in *Amer. Speech* (1951) XXVI. 236/2 The wind velocity in the city had been less than 5 miles per hour before the bombing, but the fire-wind attained a velocity of 30–40 miles per hour. **1899** *Westm. Gaz.* 15 Apr. 8/1 A brigade of fire-women should be attached to the staff of such sisterhoods. **1964** *Guardian* 29 May 2/4 The Fire Brigades' Union executive today promised to examine firewomen's conditions of service. **1916** *Brit. Dominions Year Bk. 1917* 248 The pilot..can only secure these valuable pictures by making a dash into the fire-zone. **1923** KIPLING *Irish Guards in Great War* I. 125 All four companies worked their way..out of the fire-zone.

c. fire-beetle = CUCUY; **fire-brat**, a BRISTLE-TAIL (*Thermobia domestica*), found in hot parts of buildings; **fire-bug**, (*a*) *U.S.*, the fire-fly; (*b*) an insect of the family Pyrrhocoridæ; **fire-crest**: substitute for def.: a warbler (*Regulus ignicapillus*), closely resembling the gold-crest; add earlier example; **fire-finch**, a name given to several African weaver birds from the red plumage of the male in the breeding season; **fire-worm** *U.S.*, the larva of any of several moths belonging to the superfamily Tortricoidea.

1842 T. W. HARRIS *Insects Injur. Veget.* 49 *Elater* (*Pyrophorus*) *noctilucus*, the night-shining Elater, is the celebrated *cucuio* or fire-beetle of the West Indies. **1855** K. BUCHELE *Land Nord-Amerika* 32 Den Feuerkäfer (*fire beetle*), 1 Zoll lang, in Louisiana und Texas. **1895** *Cambr. Nat. Hist.* V. 186 The bakers call these Insects [*sc. Thermobia furnorum*] fire-brats, apparently considering them to be fond of heat. **1955** *Sci. News Let.* 22 Jan. 62/3 The firebrat..preferring to live in the vicinity of a fireplace, furnace or other hot spot. **1789** J. MORSE *Amer. Geogr.* 62 Fire Fly or Bug. **1877** J. M. BAILEY *Folks in Danbury* 40 (Th.), [The lamp] don't give more light than a fire-bug. **1921** E. STEP *Brit. Insect Life* 169 A striking representative of this family is a Bug that is known in France as the Beadle and the Soldier (*Pyrrhocoris apterus*), and mentioned sometimes as Fire-bug. **1959** SOUTHWOOD & LESTON *Land & Water Bugs* iv. 72 The firebug occurs throughout much of Europe and is..found in gregarious masses on various mallows and limes. **1843** J. D. HOY in W. Yarrell *Hist. Brit. Birds* I. 324 By the early part of November you will rarely find the Fire Crest. **1889** *Cent. Dict.*, Fire-finch. **1905** G. E. SHELLEY *Birds Afr.* IV. 1. 260 These Fire-finches are represented in the British Museum by a full plumaged male. **1966** C. SWEENEY *Scurrying Bush* x. 140 Little birds, such as cordon-bleus and firefinches, used to come to the crocodiles' pool to drink. **1869** A. S. PACKARD *Guide Study Insects* 339 These larvæ, called the Cranberry-vine worms, hatch about the first of June... Before reaching their full size they, if very numerous, almost wholly destroy the leaves and tender shoots, giving the whole bog a dark dry appearance as though a fire had been over it. This is why they are in some places known as 'fire-worms'. **1954** BORROR & DELONG *Introd. Study Insects* xxvi. 531 The cranberry black-headed fire-worm, *Rhopobota naevana* Hübner, is a serious pest of cranberry plantings in the Eastern states.

fire, *v.*[1] Add: **4. c.** Said of a cylinder of an internal-combustion engine when the fuel inside it is ignited.

1894 W. J. LINEHAM *Textbk. Mech. Engin.* x. 699 The first practical gas engine..was double-acting, charging with air and gas during a half stroke, firing during the remaining half. **1902** J. E. HUTTON in A. C. Harmsworth et al. *Motors* viii. 139 Two cylinders out of the four fire in each revolution. **1959** *'Motor' Manual* (ed. 36) iii. 61 As soon as the engine fires, the left-hand disc valve is

drawn to the right. *Ibid.* ii. 38 When No. 1 is firing or on its power stroke, No. 2 is taking in its fresh charge.

11. b. (Earlier examples.) Also *trans.*

1836 *Southern Lit. Messenger* II. 734 Passing the fire-room, where they were just firing up, I stopped. **1857** E. STONE *Life Howland* xii. 267 The time required to 'fire up', and set the engine again in motion, delayed the arrival. **1867** *Trans. Ill. Agric. Soc.* VI. 320 It can be attached..in less time than is required to fire up a steam fire engine. **1903** *N.Y. Even. Post* 13 Oct. 1 Then the two statesmen fired up their cigars.

15. d. *Photogr.* To release (the shutter). Also *intr.* in *to fire away* or *off*, to take a photograph.

1892 *Photogr. Ann.* II. 51 What is the object of firing off at a street view..with the shutter set at $\frac{1}{10}$ second when $\frac{1}{25}$ would be fast enough? *Ibid.* 349 The shutter release..appears to lift the mirror and fire the shutter much more efficiently. **1900** *Daily News* 21 June 5/1, I..fired away at some Tommies trying to catch a pig. **1958** M. L. HALL *Newnes Complete Amat. Photogr.* iv. 60 There is no doubt that the 35mm. user, with 36 frames on his roll, is much more apt to fire away indiscriminately.

fireable, *a.* Delete †*Obs.*[−1] and add later examples.

1898 *Westm. Gaz.* 26 May 4/2 The *Don Juan de Austria* had only two fireable guns. **1900** *Times* (weekly ed.) 3 Aug. Suppl. p. iii/4 American coal, in consequence of being very firable, is not screened out.

fire-ball. Add: **1. b.** *spec.* The ball of flame resulting from the explosion of a nuclear device.

1955 *Bull. Atomic Sci.* Feb. 45/2 The exploding super-bomb produces a 3 to 4 mile diameter fireball. **1957** *Times* 18 May 6/4 The base of the fireball was well above the surface of the sea.

fired, *ppl. a.* Add: **6.** Of the case of a cartridge: used, having had the cartridge discharged.

1892 W. W. GREENER *Breech-loader* 65 Such processes as cocking the locks and taking out the fired cases. **1902** *Encycl. Brit.* XXX. 402/2 The extractor holds a cartridge and a fired case ready to be pushed into the empty breech and ejector-tube respectively...The lock is withdrawn taking with it the fresh cartridge from the belt and the now fired case.

fire-fang, *v.* (Later U.S. example.)

1896 *Vermont Agric. Rep.* XV. 72 There is danger if ensilage contains more than 25 per cent of dry matter that it will fire-fang in the silo.

fire-fly. Add: **2. b.** *fig.*

[**1897** *Sears, Roebuck Catal.* 162/2 The 'Fire-Fly' Single Wheel Hoe.] **1948** A. TOYNBEE *Civilization on Trial* i. 9 Those pages teeming with firefly flashes of historical insight. **1965** *Sunday Mail Mag.* (Brisbane) 15 Aug. 5/2 The nearest firefly flash of expression came into the generalissimo's face.

fi·re-hunting, *vbl. sb. U.S.* [FIRE *sb.* B. 3 b.] **a.** The firing of timber to drive out game. **b.** Hunting by night in which lights are used to attract or reveal the game. Hence (in sense b) **fi·re-hunt** *sb.*, **fi·re-hunt** *v. trans.* and *intr.*, **fi·re-hunter.**

1705 R. BEVERLEY *Hist. Virginia* II. 39 They [*sc.* Indians] had a better Way of killing the..greater Game, by a method which we call Fire-Hunting: that is,..they wou'd Fire the Woods. **1788** M. DEWEES *Jrnl.* (MS.) 17[v] Had Several Gentlemen to dine..expecting a fire hunt of some deer. **1814** [see FIRE *sb.* B. 3 b]. **1833** *Amer. Turf Reg.* Feb. 305 Here we took our first drive; but..a '*fire hunter*' had anticipated us the previous night. **1852** [see FIRE *sb.* B. 3 b]. **1875** E. KING *Southern States N. Amer.* 417 The sportsman..may 'fire-hunt' the deer in the forests. **1885** [see FIRE *sb.* B. 3 b]. **1891** *Outing* Dec. 236/1 No sooner had he eaten his supper each day than he began to importune the younger men of the party to join him in a 'fire hunt'...Beverly carried the gun, his chief the torch, consisting of 'lightwood' knots blazing in the bowl of a long-handled frying pan. **1948** E. N. DICK *Dixie Frontier* 35 Fire hunting..was most effective on a dark night.

fireless, *a.* Add: **2. c.** *fireless cooker*, an insulated chamber brought to a heat suitable for cooking and capable of maintaining sufficient heat to cook with; also, = *hay-box* (*HAY *sb.*[1]).

1908 E. A. HUNTINGTON (*title*) The fireless cooker. **1945** R. MOLLOY *Pride's Way* 291, I believe Mrs. Wilson approves of a fireless cooker. **1965** G. MCINNES *Road to Gundagai* v. 83 The other gadget was the Fireless Cooker ..a wooden box filled with straw.

fireman. Add: **2.** (Earlier example with reference to a steam-engine.)

1849 F. B. HEAD *Stokers & Pokers* vii. 62 It [*sc.* a locomotive] proceeds to the fire-pit, over which it stops. The fireman here opens the door of his furnace.

fire-proof, *a.* Add: *spec.* of paper.

1922 *Pulp & Paper Mag. Canada* 15 June 494/3 Paper and board may be rendered partially fireproof by treatment with magnesium-sulphate solution. **1962** F. T. DAY *Introd. to Paper* 115/1 *Fireproof papers*...Most are made of asbestos or are suitably treated with standard solutions and recipes.

fire-red, *a.* Delete †*Obs.* and add later examples. Also, reddened by fire.

1896 'MARK TWAIN' in *Harper's Mag.* Aug. 356 Cotton-velvet westcot, fire-red and yaller squares. **1931** V. WOOLF *Waves* 98 The fire-red windows.

B. *sb.* A commercial name for a brilliant red monoazo pigment characterized by strong resistance to oil and light.

1955 H. A. LUBS *Chem. Synthetic Dyes & Pigments* xi. 633 Aktien Gesellschaft für Anilinfabrikation (AGFA) disclosed that replacement of *p*-nitroaniline by 2-chloro-4-nitroaniline..gave the product (12), Fire Red. **1967** KARCH & BUBER *Offset Processes* vii. 269 *Para reds*, and *Fire reds* are quite 'dirty' in color, are semi-transparent and used mostly in poster and label inks.

fireside. Add: **2. b.** (Later example.)

1839 F. REYNOLDS *Let.* 13 May in A. Bunn *Stage* (1840) III. vi. 218 For so many years he formed a part of my fireside.

3. (Further examples.) Also applied *spec.* (orig. in the U.S.) to informal broadcast talks to the nation.

1885 H. JAMES *Little Tour in France* (1900) vii. 70 The sixty years that preceded the [French] Revolution were the golden age of fireside talk. **1926** R. MACAULAY *Crewe Train* II. x. 196 Her fireside chats about the characters and intimate histories of her friends and relations were famed in London. **1935** *Amer. Mercury* July 325/2 Now and again..one may find a half-hearted Hoosier who turns from the lamp to listen to a 'Fireside Chat'. **1939** *Ann. Reg. 1938* 301 In a 'fireside' radio talk he [*sc.* the President] announced that the Administration would initiate a 4 million dollars 'spending and lending' programme. **1963** AUDEN *Dyer's Hand* 344 The fireside-chat I'm-a-plain-fellow manner is still irritating. **1966** *Observer* 7 Aug. 28/4 Fireside chats from Number 10 Downing Street.

fire-work, firework. Add: **4. c.** *pl.* (See quots.) *Services' colloq.*

1925 FRASER & GIBBONS *Soldier & Sailor Words* 94 *Fireworks*, the familiar name for the display of search-lights, star-shells, rockets, etc., over the front-line and No Man's Land, on the occasion of night bombardments. **1943** HUNT & PRINGLE *Service Slang* 32 *Fireworks*. Bomber crews give this name to intensive anti-aircraft fire. It is also used for flare dropping.

firing, *vbl. sb.* Add: **1. c.** The ignition of the fuel in a cylinder of an internal-combustion engine.

1888 *Lockwood's Dict. Mech. Terms* 140 *Firing*, the ignition of a charge in a gas engine. **1902** J. E. HUTTON in A. C. Harmsworth et al. *Motors* viii. 153 When the moment of firing arrives the striker..is caused to drop smartly. **1934** A. D. MERRIMAN *Romance of Engineering* vii. 108 Methods for..compressing the charge before firing were greatly improved upon when Degrand in 1858 suggested compression in the cylinder by the motor piston. **1966** JACKSON & MORTON *Reed's Gen. Engin. Knowledge for Marine Engineers* ii. 41 The time interval between fuel injection and firing, called ignition delay, must not be too long.

8. firing-bay = *fire-bay; **firing-glass**, a table-glass with an unusually thick base; **firing order**, the order in which the cylinders in a motor engine operate; **firing point**, (*b*) in target shooting: the position from which the firing is done; **firing squad**, (*a*) = *firing party* (see Dict.); (*b*) a squad of soldiers detailed to shoot a condemned man; **firing-step** *Mil.*, a board or ledge in a trench, upon which soldiers stand when firing.

1923 KIPLING *Irish Guards in Great War* I. 222 The wreck of a firing-bay. **1905** P. BATE *Eng. Table Glass* 72 'Firing' glasses..having a thick and massive base with which to knock on the table when applause was to be given. **1923** *Weekly Disp.* 30 Sept. 7/4 Short Jacobite firing-glasses. **1910** A. WILLIAMS *Engin. Wonders of World* III. 33/1 The firing order of the cylinders of a seven-cylinder engine is 1, 3, 5, 7, 2, 4, 6, 1, [etc.]. **1959** *'Motor' Manual* (ed. 36) ii. 38 The firing order usually is 1, 5, 3, 6, 2, 4. **1900** *Times* 12 July, The wind has not blown straight from the firing point to the targets. **1958** J. A. BARLOW *Elem. Rifle Shooting* (ed. 5) iii. 38 The register keeper at the firing-point then shouts out the firer's name and the value of his shot. **1904** CONRAD *Nostromo* II. i. 114 The irregular report of the firing squad would be heard. **1917** A. G. EMPEY *Over Top* 291 *Firing squad*. Twelve men picked to shoot a soldier who has been sentenced to death by courts-martial. **1959** T. S. ELIOT *Elder Statesman* I. 27 The ones who don't get out in time Find themselves in gaol..Or before a firing squad. **1916** 'BOYD CABLE' *Action Front* 254 He..leaped on the firing step, and hurled himself over after him.

firm, *sb.*[1] Add: **2. c.** Also used *gen.* of a group of persons working together.

1909 *Westm. Gaz.* 16 Aug. 7/3 The general surgical and medical work is in charge of a medical and surgical 'firm', each consisting of two visiting and one house physician or surgeon. **1933** J. HILTON *Lost Horizon* x. 218 Moderation, y'know—the motto of the firm. **1952** E. F. DAVIES *Illyrian Venture* ii. 35 Our organisation [*sc.* SOE], which was known as 'The Firm', had equipped a hut with every item of kit we were likely to want. **1964** G. L. COHEN *What's Wrong with Hospitals?* vi. 125 Each consultant heads a firm usually comprising registrar and houseman, who get their own beds allocated *en bloc*. **1964** G. LYALL *Most Dangerous Game* xxv. 218 He's one of the Firm...He's not going to talk. **1969** *Daily Tel.* 8 Feb. 13/6 The Kray 'firm' were like income-tax inspectors in assessing what 'pension' or protection money a club should pay them. **1971** R. BUSBY *Deadlock* x. 159 Put in for a transfer to the Met, we could use a good man in my firm.

firm, *a.* and *adv.* Add: **A.** *adj.* **7. a.** (Earlier examples.)

1846 *Times* 19 Nov. 6/2 The trade is very firm, and the prices of Friday are fully maintained. *Ibid.* 30 Nov. 3/3 The wool trade is flat, and few sales can be effected, although prices are firm.

C. 1. b. *firm-bodied*, *-faced*, *-featured*, *-lipped*, *-minded*, *-natured*, *-souled*; **firm-footed**, *spec.* in Cricket = *fast-footed (a).

1915 D. H. LAWRENCE *Rainbow* 315 She was firm-bodied as Diana. **1918** E. POUND *Pavannes & Divisions* 39 One of those firm-faced inspecting women. **1916** JOYCE *Portrait of Artist* v. 273 Cranly's firm featured suffering face. **1907** *Westm. Gaz.* 20 Aug. 3/2 A hitter of the firm-footed school. **1928** *Daily Tel.* 15 May 19/1 He attempts, firm-footed, to drive the ball which is pitched a couple of feet wide of the off-stump. **1961** *Times* 22 May 3/3 He lost Teague, bowled firm-footed. **1921** W. DE LA MARE *Crossings* 27 Firm-lipped, round-browed, keen-eyed. **1913** J. MASEFIELD *Daffodil Fields* 60 He should not have a woman sick with ghosts, But one firm-minded to be his. **1858** BAGEHOT *Coll. Works* (1965) I. 319 They [*sc.* Whigs] are emphatically pure-natured and firm-natured. **1918** W. B. YEATS *Per Amica Silentia Lunae* xvi. 79 Some firm-souled man.

firm, *v.* Add: **7. b.** *Comm.* Of prices that vary with a market, as share prices, rates of exchange, etc.: to become firm (FIRM *a.* 7 a); to rise (slightly), esp. after being weak. Usu. with *up*. Hence **fi·rming-up** *vbl. sb.*

1899 *Daily News* 30 Oct. 2/6 A gradual firming-up of prices. **1907** *Daily Chron.* 9 Dec. 1/7 One or two Argentine Rails firmed up. **1929** *Daily Express* 7 Nov. 2/7 Then all firmed up on trade and Cuban buying. **1938** *Times* 13 Apr. 20/3 Shortly before the close of business they firmed up considerably. **1952** *Sat. Rev.* 20 Sept. 9/3 The firming up of prices resulting from the impact effect of the deficit. **1971** *Daily Tel.* 18 June 23/1 Eurodollar rates firmed up yesterday. **1971** *Timber Trades Jrnl.* 21 Aug. 34/1 The Scandinavian market, which has been fairly weak over recent months, is now showing faint signs of firming.

firmament. **5.** *Comb.* (Later U.S. example.)

1883 'MARK TWAIN' *Life on Mississippi* 244 A firmament-obliterating irruption of profanity.

firmish (fɜ·ɹmɪʃ), *a.* [f. FIRM *a.* + -ISH.] Somewhat firm.

1839 *Fraser's Mag.* XX. 718 Let's see..if the cits stand firmish. **1924** J. MASEFIELD *Sard Harker* III. 124 The hop took him to firmish tussock.

firn. Add: Also *attrib.* and *Comb.*

1934 *Geogr. Jrnl.* LXXXIV. 283 The firn-field, as we know it in the Alps. **1937** *Discovery* Feb. 35/2 The 'firn stoss' or 'ice-quake'. **1958** *Jrnl. Glaciology* III. 265 The firn line..is rather lower than the equilibrium line of the smallest glaciers.

Hence **firnifica·tion**, the conversion of snow into firn.

1923 AHLMANN & TVETEN in *Geogr. Annaler* V. 55 We imagine the process in the firnification of the snow, and its further glaciation to consist mainly in a sublimation, caused by the various vapour pressures on different forms and different sizes of crystals. **1937** *Nature* 30 Jan. 187/1 *Firnification*, or the changes of the snowflake after lying on the ground.

first, *a.* (*sb.*) and *adv.* **A.** *adj.* **1. f.** *the first thing*: the elements or rudiments, esp. in phr. *not to know the first thing about*.

1935 *Punch* 24 July 97/3 But I couldn't start up the car. I don't know the first thing about car-engines. **1938** 'E. CAMBRIDGE' *Spring always Comes* II. xii. 267, I don't know the first thing about painting.

2. b. (Later U.S. examples.) (See also quot. 1944.)

1922 H. TITUS *Timber* xxii. 201 Humphrey was going down state the first of the week to hunt an investor. **1926** *Publishers' Weekly* 16 Jan. 147 When the first of the year had passed. **1944** H. W. HORWILL *Dict. Mod. Amer. Usage* (ed. 2) 132/2 In Eng. one may speak of *the first* of the month, meaning the first day of it. In Am. one may speak also of *the first* of the year, or of the week; not, however, meaning the first day of this period but the first part of it.

4. Also in phr. *of the first order* [F. *du premier ordre*]; (*to put* or *do*) *first things first*: (to give) first place to the most important things.

1794 [see ORDER *sb.* 4]. **1894** G. JACKSON (*title*) First things first; addresses to young men. **1895** *Bookman* Oct. 33/1 A diplomatist of the first order. **1920** W. RILEY *Yorkshire Suburb* 136 The dear lady was..incapable..of putting first things first. **1922** 'SAPPER' *Black Gang* xiv. 217 The whole thing is a boss shot of the first order. **1931** *News Chron.* 28 Apr. 3/3 'First things first' must be the guiding principle. **1961** L. VAN DER POST *Heart of Hunter* xviii. 245 The lion's sense of first-things-first decides the issue: it will go to the water before it eats. **1971** *Scope* (S. Afr.) 19 Mar. 124/4 Let's do first things first.

b. *first Lord* (examples of *ellipt.* use).

1814 JANE AUSTEN *Mansf. Park* II. vii. 155, I would rather find him private secretary to the first Lord than any thing else. **1963** *Times* 26 Apr. 19/4 There is no truth in the allegation that the presence of another spy

inside the Admiralty was known to the First Lord and his Service chiefs for 18 months before Vassall's arrest.

7. g. A first edition; a first-class railway carriage or compartment; the first known or discovered example or specimen of a thing; a first instance or occurrence.

1922 M. SADLEIR *Excurs. Vict. Bibliogr.* 6, I have contrived..to keep myself fairly supplied with 'reading firsts'. **1924** *G. McLeish's Catal.* 13 Essential to the collection of 'modern firsts'. **1925** B. TRAVERS *Mischief* iv, The train arrived and Eleanor..was easily spotted, being alone in the firsts. **1931** BUCK & ANTHONY *Bring 'em back Alive* 215 There is no bigger thrill in the game than bringing back rare 'firsts' [i.e. an animal that has not been seen alive before in the country to which it is brought]. **1955** *Times* 20 June 8/7 Another important French 'first' was the arrival of the Caravelle, the country's first jet airliner. **1960** *Analog Science Fact/Fiction* Dec. 83 Echo I, the sateloon that went up last August, was a true First in space flight. **1960** J. BETJEMAN *Summoned by Bells* vi. 58 Perhaps one day I'll find a 'first' of Keats. **1963** J. F. STRAKER *Final Witness* iv. 48 'I'm sorry about that, Paul.' That made another first: they had not used Christian names before. **1970** *Guardian* 27 Feb. 8/5 Those were days of high adventure: every programme was a first..breaking new ground. **1971** L. BLACK *Death has Green Fingers* ix. 100 I'm in a first, next carriage along the train.

h. The first or lowest gear on a motor vehicle or bicycle.

1909 J. S. V. BICKFORD *Faults & how to find Them* § 1173 If, however, the change from first to second (not through the gate we are assuming) has to be done quickly, [etc.]. **1935** G. HEYER *Death in Stocks* iv. 40 Antonia slammed the gear-lever into first and started with a jerk. **1968** P. DICKINSON *Weathermonger* vi. 66 He wrenched the gear into first and banged through, misjudging it slightly.

B. *adv.* **1. b.** Also as *attrib. phr.*

1960 R. B. GREGG *Power of Nonviolence* (ed. 2) i. 38 Passengers were seated on a first-come, first-served basis. **1964** *Sun* (Brisbane) 23 July 9/1 The Administration Bill would substitute the quota system with one based on a first-come, first-served basis.

f. *first off*: at the first blush, in the first place, to begin with. *U.S. colloq.*

1880 'MARK TWAIN' *Tramp Abroad* xx. 193 First-off, I thought it would certainly give me the botts. **1897** W. D. HOWELLS *Landlord at Lion's Head* 445 First off, you know, I thought I'd sell to the other feller. **1910** W. M. RAINE *B. O'Connor* 39 Four's right. First off Neil, then the fellow I took to be the Wolf. **1915** *Nation* (N.Y.) 10 June 646/1 Men of science..no longer admit first off what simple good sense shows to us.

4. ellipt. for *first class* (in travelling by railway, etc.).

1889 J. K. JEROME *Three Men in Boat* iv. 50 They would..pay the difference and go first. **1919** D. ASHFORD *Young Visiters* ii. 23 They traveled 2nd class in the train and Ethel was longing to go first but thought perhaps least said soonest mended. **1922** M. ARLEN *Piracy* II. xix. 288 Travelling 'first' with a 'third' ticket. **1967** 'J. MUNRO' *Money that Money can't Buy* vi. 75 They..went aboard their Trident...Loomis had insisted on travelling first.

C. *Comb.* **2. first aid** (examples); also *fig.*; **first-aider**, one who is skilled in first aid; **first ascent**, the ascent of a particular mountain for the first time; **first base** (see *BASE *sb.*[1] 15 d); **first blood**, in Pugilism, the first drawing of blood; also *fig.*; **first cross**, the crossing of two pure breeds; the offspring of such a cross; **first-day cover**, an envelope bearing, or designed to bear, stamps post-marked on the day when they were first issued (see *COVER *sb.*[1] 2 d); **first derivative** *Calculus*, the derivative of a function obtained by differentiating it once; **first-ever** *a.* (see *EVER *adv.* 7 f); **first feature**, the main feature in a cinema programme; **first-fleeter** *Austral. slang* (see quots.); **first-footer**, one who goes first-footing; **first gear**, the lowest gear (cf. *GEAR *sb.* 7 b); **first lady**, the most important lady; in the U.S., spec. the wife of the President; in full *the first lady in the land*: **first light** orig. *Mil.*, the time when light first appears in the morning (see also quot. 1952); **first line**, those fighting forces that are most advanced or of the highest quality; **first love**, (*a*) the first time one falls in love; the emotion felt at such a time; (*b*) one's favourite occupation, pastime, possession, etc.; **first night**, **-nighter** (earlier examples); **first novel**, the first novel written by a particular author; **first offender** (see OFFENDER *note*); **first pointed** *a. Arch.*, = Early English (see EARLY *a.* 4 b); **first polar** *Math.*, (of a point $(a_0, a_1, .. a_n)$ with respect to a curve described by the homogeneous function $f(x_0, x_1, ..x_n) = 0$) the curve represented by the equation $\Sigma_i a_i \partial f / \partial x_i = 0$; **first position**, (*a*) the lowest possible position of the left hand on the fingerboard of a stringed instrument; (*b*) *Dancing* (see quot. 1884); **first refusal**, an agreement with a prospective customer that goods, etc., will not be on sale to others until he has refused them (cf. REFUSAL 3); also *transf.* and *fig.*; **first sergeant** *U.S.*, the highest ranking non-commissioned officer in a company or equivalent unit; **first-spear**, in pig-sticking, the first thrust which draws blood; also, the man who makes it; **first-strike**, a first aggressive attack with nuclear weapons; freq. *attrib.*; **First World War**, now the more usual term for the *Great War (b); **first youth**, early youth; esp. in phr. *in* or *past one's first youth*.

1882 PRINCESS CHRISTIAN tr. Esmarch's (*title*) First aid to the injured. **1892** *Medical Annual* 602 Dressing (First Aid). **1926** W. DEEPING *Sorrell & Son* i. § 2 The porter applied first aid to their piece of luggage. **1944** *Living off Land* v. 101 The First Aider should keep cool. **1956** J. A. C. BROWN *Soc. Psychol. Industry* 142 The knowledgeable first-aider. **1935** D. PILLEY *Climbing Days* xvi. 321 Recent chronicles of the French Alpine Club are filled annually with notes of his First Ascents and other exploits. **1809** First blood [see KNOCK-DOWN B. *sb.* 2]. **1899** R. KIPLING in *McClure's Mag.* May 11/2 First blood to me. You flushed, Ansell. You wriggled. **1964** C. WILLOCK *Enormous Zoo* iii. 52 Before first blood was drawn it was decided to see what could be done to the ruined grazing of certain areas by simply keeping the hippo out. **1862** W. YOUATT *Sheep* vii. 260 The first cross evidently detracted much from the beauty of the English sheep. **1910** T. ALLEN *Profitable Pig Breeding* ii. 13 A first cross of pure-bred stock, judiciously selected, is hard to beat. **1937** C. FORMAN *Pig Breeding* iii. 27 Other first crosses might be made with equal success. **1947** J. STEVENSON-HAMILTON *Wild Life S. Afr.* xxi. 165 The two best dogs ..I ever possessed..were a first-cross between a pointer and a bull-terrier, and a first-cross between Great Dane and Irish terrier. **1962** W. E. BOWDEN *Beef Breeding* ii. 64 Milk yields from the first cross would be less than one would expect. **1938** First-day cover [see *COVER *sb.*[1] 2 d]. **1951** J. L. GRUMBRIDGE *Introd. Stamps* ix. 103 This quite intelligible demand for first-day covers as items of philatelic interest. **1971** D. POTTER *Brit. Eliz. Stamps* vi. 66 To accompany the new stamps, a very simple first-day cover envelope was prepared by the Post Office. **1834** *Rep. Brit. Assoc. 1833* 338 If $f(x) = 0$, or $X = 0$ be the equation, $f'(x)$, $f''(x)$, or X', X'' its first and second derivatives. **1852** T. G. HALL *Differential & Integral Calculus* (ed. 5) iii. 31 This process is called successive differentiation, and du/dx, dp/dx, dq/dx, &c. are called the first, second, third, &c. differential coefficients, or the first, second, third, &c. derivatives of u. **1881** First derivative [see DERIVATIVE B. 3]. **1953** P. FRANKLIN *Differential & Integral Calculus* iii. 50 If this function is differentiable, its derivative..will be a new function of x. Suppose that this new function is differentiable. Then the derivative of the *first* derivative is called the *second* derivative of the original function. **1936** *World Film News* May 28/2 To sell a film as a first-feature, a producer must have a personality..well enough known to induce the public to go to the..theatres. **1958** *Observer* 9 Mar. 4/6 This film—a first feature on its first release round a big circuit—was badly presented. **1830** D. MCLEOD in *Hist. Rec. Australia* (1922) 1st Ser. XV. 371 One of the first Fleeters died a few Years ago. **1848** H. W. HAYGARTH *Recoll. Bush Life Austral.* viii. 93 A man who, by his own account, is of so long standing in the neighbourhood as to have been what is called in the colony a 'first fleeter'. **1850** DICKENS in *Househ. Words* I. 418/2 He was a 'first fleeter', that is, came over with Governor Phillips in the first fleet. **1897** *Daily News* 23 Sept. 2/6 Scottish 'first-footers' in the fifteenth century awoke the night echoes with a mild and 'soughing' Tra-la-la. **1926** *Glasgow Herald* 30 Dec. 6 The red herring which your Dundee 'first-footer' carries with him as a lucky emblem. **1909** R. W. A. BREWER *Motor Car* xii. 120 On changing down to the first gear the power of the engine due to its increased rate of revolution rises to 20 B.H.P. **1965** *Motoring Which?* Oct. 136/2 The effect of changing to low ratio is to make all gears lower, so that, for example, in the Land-Rover you then have three gears lower than the ordinary first gear. **1853** LYTTON *My Novel* I. IV. ix. 306 She was so beautiful..and not proud she! though she looked like the first lady in the land. **1861** W. H. RUSSELL *Diary* 3 Nov. (1863) II. 393 Some charming little pieces of gossip about 'the first Lady in the Land'. **1948** *Chicago Tribune* 26 June 1. 1/4 Mrs. Thomas E. Dewey disclosed today what kind of first lady she will be. **1950** 'E. CRISPIN' *Freq. Hearses* i. 27 Madge is one of the First Ladies of British Films. **1965** *Sunday Times* 7 Nov. Suppl. 13/1 [Mrs. Kennedy] disapproved of the term 'First Lady', which had come into semi-official usage in the previous administration. **1970** *Washington Post* 30 Sept. B2/1 The First Lady of the United States was questioned by Italian women reporters today. **1946** *Amer. Speech* XXI. 209 Orders [in the Eighth Army] for an early start never mentioned dawn but instead used *first light*...There might be some squabble about just what time constituted dawn. There couldn't be any at all about *first light*, at least not in the desert. **1952** *Brewer's Dict. Phr. & Fable* 364/1 *First light*, roughly, dawn. Used in World War II to signify the earliest time at which infantry can see to make their way forward. **1969** *Daily Colonist* (Victoria, B.C.) 6 July 1/6 A search-light equipped helicopter was to press the search overnight, with the full armada out in force again at first light today. **1914** H. M. BUIST *Aircraft in German War* 78 They are accounting for such large numbers of the enemy's first line of airmen. **1741** RICHARDSON *Pamela* III. xiv. 63 It was a first Love on both Sides; and so he could not appear to her as a practised Deceiver. **1825** H. WILSON *Mem.* II. 91 First love is all powerful in the head and heart of such an ardent character as Worcester's. **1954** E. MCLEOD tr. *Colette's Vagabond* II. viii. 129 You couldn't mistake it; it was love indeed, *first love*. That was what it was and never again will be. **1971** *Radio Times* 27 May 5/1 Despite a recent ITV series his first love remains radio. **1971** *Daily Tel.* 21 June 19/2 However, she is put off by the apparently poor prospects of a career in biology, her first love. **1711** First Night [see NIGHT *sb.* 4 b]. **1882** First-nighter [see NIGHTER]. **1876** *Gentl. Mag.* Oct. 412 *Daniel Deronda*..is as much a first novel, from a fresh hand and mind, as if no scene of clerical life had ever been penned. **1906** *Daily Chron.* 18 Oct. 3/3 One of the best first-novels of recent years. **1942** E. WAUGH *Put out More Flags* iii. 213 'It won't do,' he always said whenever Mr Bentley produced a new author, 'no one ever reads first novels.' **1849** *Chambers's Edin. Jrnl.* 3 Nov. 283/2 In the assize-yard there was a considerable number of what are called first-offenders. **1935** *Economist* 18 June 1353/1 Not at Borstal only, but in at least one of the prisons for first offenders, the humanising influence is at work. **1875** *Encycl. Brit.* II. 427/1 The First Pointed, Lancet, or Early English style to which the transitional work thus led up. **1886** F. G. LEE *Edw. VI.* i. 58 A first-pointed chapel and charnel-house. **1877** *Encycl. Brit.* VI. 720/2 If the given curve has a node, the first polar passes through this node,..and..if the curve has a cusp, the first polar passes through the cusp, touching the curve there. **1949** SEMPLE & ROTH *Introd. Algebraic Geom.* i. 11 The first polar of an arbitrary point Q with respect to a primal V^n_{r-1} is a primal of order $n-1$, which meets V^n_{r-1} in the locus of points of contact of tangent lines from Q. **1880** H. E. KREHBIEL tr. *Courvoisier's Technics of Violin Playing* (1881) iii. 14 (*caption*) Correct position of the thumb and fingers in the first position. **1884** D. ANDERSON *Compl. Ball-Room Guide* 10 *First position*, heels together, and toes turned out at a right angle. **1957** G. B. L. WILSON *Dict. Ballet* 124 In the First Position the legs are together and the heels touching but not overlapping, with the feet fully turned out in a straight line. **1966** tr. *Flesch's Violin Fingering* I. 9 The basic position of the hand at the lower end of the fingerboard, with the first finger naturally poised to produce a second above the open string, constitutes the 'first position'. **1871** G. H. LEWES *Let.* 5 Apr. (1955) V. 139 We want £100 for the early sheets—the poem to appear in America in June—and of course we give you the first refusal. **1887** First refusal [see REFUSAL 3]. **1966** *Listener* 9 June 833/1 Sometimes I am asked why we have a month's first refusal agreement with ABC. Could we not sell more in an open field? **1970** M. KELLY *Spinifex* i. 12, I made damn sure this department got the first refusal of your services. **1874** J. K. POLK in *Amer. Speech* (1957) XXXII. 209 First Sergeant W. Peat, of this company, who acted with extraordinary daring, lost one of his legs. **1969** I. KEMP *Brit. G.I. in Vietnam* vi. 117 First Sergeant Bryant, the senior N.C.O. in 'A' Force. **1898** *Encycl. Sport* II. 92 The object of the run is to kill the pig, and not entirely for getting the honour of 'first spear'. *Ibid.*, First spear... Transferred frequently to the man who makes it. **1920** *Blackw. Mag.* Jan. 112/2 Much of the supreme joy in pig-sticking lies in the successful struggle for first-spear and in outgeneraling and outriding one's friends. *Ibid.*, A first-spear which merely pricks the pig in the buttock, is a matter for shame. **1962** *Listener* 5 Apr. 605/2 A Russian first strike. **1931** S. JAMESON *Richer Dust* xix. 552 The salvage of what a dear dead and let us piously hope well-damned colonel preferred to call the *First* World War. **1947** *Partisan Rev.* XIV. 237 The mystique of the working class has faded somewhat since the First World War. **1958** *Listener* 19 June 1020/2 The outbreak of the first world war in the present century. **1390** Ferste youthe [see YOUTH 3]. **1789** H. MORE *Let.* (1925) 126 She is past her first youth..and subject to caution. **1878** H. JAMES *Europeans* I. i. 7 She was not pretty...Neither was she in her first youth.

b. In attrib. uses of syntactical collocations; see also FIRST CLASS, FIRST-CLASS B, FIRST-DAY, etc.; *GENERATION 4 b.

1741 RICHARDSON *Pamela* IV. lxiii. 419 Has extraordinary Notions of a First-sight Love. **1840** [see *first cost* s.v. FIRST C. 2]. **1859** [see *first sight* s.v. FIRST C. 2]. **1861** E. COWELL *Diary* (1934) 244 The small audience was one of the usual 'first night' 'critical' ones. **1892** W. W. GREENER *Breech-loader* 174 Loaded with first quality powder. **1894** [see *first night* s.v. FIRST C. 2]. **1897** *Daily News* 1 June 3/2 To establish themselves once more in their former first line positions. **1904** CONRAD *Nostromo* III. xiii. 476 A knot of night prowlers..hung about the door of the first-aid hospital. **1904** *Westm. Gaz.* 28 May 2/3 First-category Cossacks of the Chitinsk, Argunsk, and Veckhni Udinsk regiments. **1905** *Ibid.* 14 Mar. 12/1 Any privileges which go with a first-cabin ticket. *Ibid.* 9 June 10/1 A first edition copy of 'Richard III'. *Ibid.* 13 Nov. 12/1 The only first-magnitude star in the track of the moon. **1906** *Daily Chron.* 21 July 1/7 The third best first-wicket stand on record. **1909** *Ibid.* 23 July 6/7 Suitable employment, treatment &c., for first-stage patients. *Ibid.* 9 Aug. 6/4 At least nineteen first-order stations for agricultural investigations. **1912** J. F. HODGES *Opening & operating Motion Picture Theatre* 13 A film being exhibited for the first time is called 'first run' film. **1912** L. J. VANCE *Destroying Angel* xvi, A light-house—probably a first-order light—with its characteristic flash, not duplicated anywhere along this section of the Atlantic coast. **1926** *Daily Colonist* (Victoria, B.C.) 18 July 9/5 First Aid Kits. For the camp or your car. In metal box, up from 75¢. **1930** *Flight* 24 Jan. p. ii, First-line machines, ready for war, over 1,000; reserves, about 800; total, 1,800. **1933** *Mind* XLII. 312 It is certain that P_u will have an eternal first-order Secondary Part. **1934** F. SCOTT FITZGERALD *Let.* 10 May (1964) 307 It did not have as large a proportion of first-flight stories. **1936** *World Film News* May 28/2 And so we reach the figure of £30,000 as the normal minimum cost of a first-feature film. **1936** *Mind* XLV. 171 The difference between a first-order proposition and a second-order proposition (or proposition of second intention)..is that the former says something about the realm of facts, while the latter says something about 'propositions'. **1940** 'G. ORWELL' *Inside Whale* 56 The two first-person novels are also good stories. **1950** *N.Y. Times* 20 Apr. 1/2 Dr. Sander, accused by the state of first degree murder. **1957** L. FOX *Two-point Boundary Probl.* v. 106 First-order equations are generally amenable only to step-by-step methods. **1960** *New Left Rev.* July–Aug. 2/1 First-strike nuclear weapons. **1961** C. C. T. BAKER *Dict. Math.* 131 *First-order equation*, a differential equation containing only the first differential coefficient dy/dx. **1969** *Daily Tel.* 17 Nov. 14 The most sinister aspect of Russia's nuclear

expansion is her concentration on a variety of giant 'first-strike' weapons, designed to knock out American launchers and control centres. **1970** *Nature* 12 Dec. 1031/2 The reaction follows first-order kinetics. **1971** B. DE FERRANTI *Living with Computer* 87 *First generation computer*, an early machine built in the 1950s or early 1960s using valves. **1971** *It* 9–23 Sept. 21/2 The First New York Erotic Film Festival..will be held in first-run Manhattan cinemas.

first class, first-class. Add: **A.** (Earlier example.)

1789 A. YOUNG *Jrnl.* 17 Aug. in *Travels* (1792) I. 165 In the first class of French families..they undoubtedly are.

b. *ellipt.* = first-class accommodation or fare.

1851 H. W. DULCKEN tr. *Pfeiffer's Visit Holy Land* i. 23 The first-class to Constantinople costs 120 florins. **1930** E. POUND *XXX Cantos* xviii. 82 And Mr. Oige was very choleric in a first-class From Nice to Paris.

B. 1. (Earlier example.)

1837–8 [see SECOND-CLASS *a.* 1].

2. (Additional examples.)

1878 *Rep. Postmaster-Gen.* (U.S.) 33 The efficiency and security of the registry system of first-class mail matter suggested the propriety of extending its provisions to valuable matter of the third class. **1887** *Postal Laws U.S.* 138 A 'drop letter'..is first-class matter and should be returned to the writer, if unclaimed. **1924** H. A. BLACKMAN *Business Mail* 15 In this connection it is well to remember that first-class postage is never an odd number of cents. **1968** *Times* 19 Mar. 2/1 A two-tier system providing a first class letter service for 4d.

3. a. (Examples.)

1847 *Illustr. London News* 22 May 328 (*caption*) Epsom races—1847. The railway—first class. **1971** A. PRICE *Alamut Ambush* x. 124 He borrowed a fivepenny stamp off me..because he wanted it [*sc.* a letter] to go first class.

first-handedness (fɜ̄ɹsthæ·ndèdnès). Also **firsthandness**. [f. FIRST HAND + -ED + -NESS.] The quality or condition of being first-hand or of an original character.

1905 *Daily Chron.* 3 May 8/5 It may breed distrust as to the first-handedness of some brilliant toilettes whose origin we could never account for. **1926** M. BARING *Daphne Adeane* x, That first-handedness; that habit..of taking nothing for granted. **1950** *Scrutiny* XVII. I. 39 For what looks like carelessness..is actually precision and vivid firsthandness.

first name. A person's first or Christian name. Also as *attrib. phr.* Hence as *v. trans.*, to address (a person) by the first name; *first-naming* vbl. sb.

c **1250** [see NAME *sb.* 1]. **1598** FLORIO *Worlde of Wordes* 291/2 Prenome, the first name, the name given at the font. **1839** C. F. BRIGGS *Adv. H. Franco* I. 74 'My first name is Harry,' I said. **1914** 'E. BRAMAH' *Max Carrados* 66 This, Parkinson,..is a photograph of a Mr. —. What first name, by the way? **1935** *Punch* 7 Aug. 142/2 He called her by her first name for the first time. **1940** S. LEWIS *Bethel Merriday* xxv. 267 Zed and..Bill Mattocks..were already first-naming each other as though they had been intimates for years. **1952** P. BONNER *SPQR* (1953) viii. 71 We are not even on a first-name basis. **1959** A. HARRINGTON *Life Crystal Palace* (1960) vii. 109 The company's democratic social organization..permits him to first-name..those several ranks higher. **1968** MRS. L. B. JOHNSON *White House Diary* 23 May (1970) 677 Nearly everybody was on a first-name basis.

firstness (fɜ·ɹstnès). Delete † *Obs. rare* and add earlier and later examples.

c **1380** WYCLIF *Sel. Wks.* (1869) I. 172 Oo firstnesse of love shulde we have to us silf, and to oure fadir and oure modir. **1895** MRS. W. K. CLIFFORD *Flash of Summer* vii, Daffodils..—the bell-like single ones: and their tender firstness appealed to her. **1924** *Glasgow Herald* 4 Sept. 4 A first novel,..with a freshness and spontaneity not always synonymous with 'first-ness' in novel-writing. **1927** *Blackw. Mag.* June 828/2 An odd kind of distinction should halo the person who is the very first to see something or to do something which hitherto humanity has not seen or done...The value of these 'firstnesses' is by no means equal.

fiscal, *a.* Add: **1. c.** *fiscal agent*, 'a bank or trust company acting as the financial representative of a corporation or service organization' (Webster 1961); also *fig.*

1841 in X. D. MacLeod *Biogr. F. Wood* (1856) 75 For forty years this Government has tried a national bank as its fiscal agent. **1856** *Househ. Words* 1 Mar. 153/1 There are..several buffets in Paris where you can be supplied with the cool and cunning drink known..as a Fiscal Agent. **1877** BARTLETT *Dict. Amer.* (ed. 4) 357 The various compounds or mixtures of spirituous liquors and wines served in the United States...Fiscal Agent. **1931** *Times Trade & Engin. Suppl.* 5 Sept. 534/2 Inquiries concerning the heavy fall in the securities of this group have led to the publication of a statement by the British fiscal agents.

B. *sb.* **3.** (Earlier example.)

1801 J. BARROW *Trav. S. Afr.* I. i. 29 Turtle doves, a thrush called the Sprew, and the fiscal bird, the *Lanius Collaris*, frequent the gardens near the town.

fiscalism (fi·skăliz'm). [f. FISCAL + -ISM.] Fiscal principles or policy, esp. in relation to the questions of free trade and protection.

1892 A. C. MORANT tr. *Schäffle's Imposs. Soc. Democr.* 193 The old-fashioned fiscalism. **1902** *Daily Chron.* 15 July 8/7 Two out of every three members spoken to declared that they will fight regardless of fiscalism for the House of Representatives having effective control of the purse. **1906** *Ibid.* 27 Mar. 6/7 Seeing that they are Free Traders, and most of us are Protectionists, their talking fiscalism would not be very helpful.

fiscality. Add: Also, fiscal policy; *pl.* fiscal matters.

1904 W. D. GAINSFORD *What is Fiscal Question?* 7 Such is Cobden Club Fiscality. **1904** *Westm. Gaz.* 13 Jan. 9/1 The rascalities of the Education Act would be set aside by the fiscalities of the food-tax. **1904** *Fabian News* XIV. 4/1 Fiscalities occupy much space. **1905** *Daily Chron.* 14 Dec. 4/6 One eye on Fiscalities and the other on Mr. John Burns's costume.

Fischer-Tropsch (fi·ʃəɪ trɒpʃ). The names of F. J. E. *Fischer* (1877–1948), German chemist, and H. *Tropsch* (1889–1935), German chemist born in Czechoslovakia, used *attrib.* to denote a process introduced by them for the production of hydrocarbons by the catalytic hydrogenation of carbon monoxide.

1933 *Brit. Chem. Abstr.* B. 49/2 The saturated hydrocarbons of a benzine prepared by the Fischer-Tropsch process have been shown to consist of about 33% of hydrocarbons containing *tert.*-bound C atoms and 67% of hydrocarbons unattacked by $SbCl_5$ [etc.]. **1951** KIRK & OTHMER *Encycl. Chem. Technol.* VI. 965 The first commercial Fischer-Tropsch plant at Oberhausen-Holten. **1964** N. G. CLARK *Mod. Org. Chem.* v. 83 The preparation of straight-chain paraffins and olefins by the Fischer-Tropsch process, originating in Germany in 1925.

fisetin (fisi·tin, fi·sĭtin). *Chem.* [ad. G. *fisettin*, f. *fisett(holz)* young fustic + -IN[1].] A yellow crystalline compound, $C_{15}H_{10}O_6$, formerly obtained from young fustic (*Rhus cotinus*) and various other trees for use as a dye.

1865 *Chem. News* 17 Mar. 126/1 M. Chevreul, who has been almost alone in his chemical researches on fustic, found in it—1. A yellow colouring matter, presenting itself when pure in the form of small crystalline needles, and which he called fissetine [*rendering* fisettin], or fisetic acid; 2. A red substance. **1959** N. CAMPBELL in E. H. Rodd *Chem. Carbon Compounds* IVB. viii. 907 A number of naturally occurring hydroxyflavones such as quercetin, luteolin, fisetin, etc. were used as yellow dyestuffs for many centuries.

fish, *sb.*[1] Add: **1. c.** *to drink like a fish* (earlier and later examples).

1744 GRAY *Let.* 26 Apr. in *Wks.* (1807) II. 35 Mr. Trollope and I are in a course of Tar-water;..I drink like a Fish. **1821** J. G. LOCKHART *Let.* 13 July, He..drinks like a fish. **1879** [see DRINK *v.*[1] 11]. **1937** A. J. CRONIN *Citadel* I. ii. 19, I better anticipate the gay tidings —I drink like a fish.

f. Applied to the turtle.

1898 *Westm. Gaz.* 9 Nov. 3/1 The sea round about the West Indies is the happiest hunting-ground for green turtle. The fish (the dealers describe them as fish) are usually taken in the manner described. **1908** *Daily Chron.* 6 Nov. 7/3 The 'fish', as they are called in the trade, are probably as tenacious of life as any animal.

g. *U.S. slang.* A dollar.

1920 *Collier's* 5 June 44/4, I..shoved my way Through the howlin' mob on the en route to the box office To collect our four hundred fish. **1934** WODEHOUSE *Thank you, Jeeves* xii. 168 She was heiress to a sum amounting to more than fifty million fish. **1949** N. ALGREN *Man with Golden Arm* 11 Used to get fifteen fish for an exhibition of six-no-count.

h. *Naut. slang.* (In full *tin fish*.) A torpedo; also, a submarine.

1925 FRASER & GIBBONS *Soldier & Sailor Words* 282 *A tin fish*, a torpedo. **1928** *Papers Mich. Acad. Sci. & Arts* X. 293 *Fish*, torpedo;..submarine. **1943** *Penguin New Writing* XVI. 19 The air seemed full of falling bombs, and tinfish like carelessly dropped cigarettes splashed among the crowded ships. **1946** R. HARLING *Steep Atlantick Stream* ii. 29 They do say the old QM's had a tin-fish under her tail. **1967** B. KNOX *Blacklight* i. 16 The Navy didn't like losing a torpedo...Each 'fish' represented some £3,000 in cash.

3. b. (Later examples.)

1871 J. H. BANKA *State Prison Life* iv. 60 'Fresh fish' is the name applied to all newcomers. **1920** F. SCOTT FITZGERALD *This Side of Paradise* I. i. 36 I'm tired of being nice to every poor fish in school. **1930** A. CHRISTIE *Murder at Vicarage* x. 79 Well—of all the poor fish! If I'd committed a murder, I wouldn't go straight off and give myself up. **1958** *Listener* 9 Oct. 568/1 The old man is revealed as having been a very cold fish.

4. c. *to cry stinking fish*: see CRY *v.* 5 b.

6. a. (sense 1) *fish-bowl*; (sense 4) *fish-dinner* (earlier example), *tea*.

1906 S. W. BUSHELL *Chinese Art* II. viii. 32 A large garden fish-bowl..is decorated in the usual style with enamel colours. **1964** *Listener* 23 Apr. 682/1 The Chinese made their fishbowls out of porcelain. *a* **1680** EVELYN *Diary* *c* Feb. an. 1646 (1955) II. 474 The famous Anna Rencha, whom we invited to a Fishdinner, after 4 daies in Lent. **1930** *Daily Express* 16 Aug. 8/7 If there is any Guards officer who is forced to indulge in fish teas, [etc.].

b. *fish-bait*, *-bar*, *-boat* (delete †), *-craft* (delete †), *-frail*, *-line*, *-net*, *-spear* (later examples).

1870 'F. FERN' *Ginger-Snaps* 275 Won't the laundress rub the skin off her knuckles when she tries to get the fish-bait off your ruffled skirt. **1887** *Mod. London* 195/1 A fish bar where those tempting little fish luncheons popularly denominated 'snacks' may be had at all hours. **1919** J. MASEFIELD *Reynard* 117 Like a fish-boat beached. **1866** *Game Laws Conn.* in *Fur, Fin & Feather* (1872) 34 Shall forfeit the value of the seine and fish-craft used for said purpose. **1953** DYLAN THOMAS *Under Milk Wood* (1954) 33 In you reeled, my boy, as drunk as a deacon with..a fish-frail full of stout. **1639** in *Archives Maryland* IV. 79, 2. fish lines. **1864** J. T. TROWBRIDGE *Cudjo's Cave* xliii. 451 Arms and head hung down, causing him to resemble..a frog hooked on for bait at the end of a fish-line. **1913** J. LONDON *Valley of Moon* III. i, Hall had sent out fish-lines and a swimming suit. **1955** E. POUND *Classic Anthol.* I. 20 Goose to get in a fish-net set! **1837** R. M. BIRD *Nick of Woods* II. xi. 144 Upon this pillar..were laid or suspended sundry Indian utensils of the kitchen and the field,..wooden bowls..fish spears [etc.]. **1908** E. J. BANFIELD *Confessions of Beachcomber* I. iv. 150 To enjoy it to perfection, extricate the creature from his lurking place far down in the blue crevice of the coral, with a fish-spear.

c. *fish-ball* (earlier examples), *-cake* (earlier example), *-chowder* (examples), *-soup* (earlier example), *-stock*.

1854 B. P. SHILLABER *Mrs. Partington* 100 The breakfast was waiting for him, the fishballs were getting cold. **1872** E. A. HART *Runaway* iii. 67 The viands that it might be possible to carry out to Olga. Soup?..Fish-balls? **1854** *Harper's Mag.* May 802/1 The favorite comestible was the *piroga*, a very unctuous kind of fish-cake. **1838** E. C. WINES *Trip to Boston* 79 We had 'clam chowder' and 'fish chowder'. **1897** KIPLING *Capt. Cour.* 84 Harvey stuffed himself to the brim on fish-chowder and fried pies. **1723** J. NOTT *Cook's & Confect. Dict.* No. 136 B, Strain it through a Sieve..and use it to simmer Fish-soops. **1787** J. FARLEY *Lond. Art of Cookery* (ed. 4) 166 Take what quantity may be wanted of fish-stock. **1883** A. THOMAS *Mod. Housewife* 51 The receipt for a fish-stock which is as strengthening and succulent as can well be desired.

d. *fish-breeder* (earlier U.S. example); *fish-breeding* (U.S. example), *fish-producing* adj.

1860 *Rep. U.S. Comm. Patents: Agric. 1859* 232 Practical hints to fish breeders. *Ibid.*, The best methods of securing success in artificial fish breeding. **1890** *Westm. Gaz.* 30 Sept. 3/1 The fish-producing lakes and rivers.

e. *fish-eyed*, *-faced*, *-haunted* adjs.

1937 L. C. DOUGLAS *Forgive our Trespasses* x. 199 Its fish-eyed challenge that dared any man to speed its heart. *a* **1940** W. J. TURNER *Sea Music* in P. M. Jones *Mod. Verse* (1940) 138 Thro' the fish-eyed meadows Flows the herd-pasturing ocean. **1963** WODEHOUSE *Stiff Upper Lip, Jeeves* xviii. 143 He's no worse than that fishfaced blighter. **1887** MORRIS tr. *Homer's Odyssey* x. 458, I myself, I know, How on the sea fish-haunted ye bore a weight of woe. **1952** C. DAY LEWIS tr. *Virgil's Aeneid* XI. 248 The fish-haunted river Padusa.

7. fish and chips, a dish consisting of fried fish and fried chipped potatoes; also *ellipt.* for a shop at which this dish may be bought ready-cooked; also *attrib.*, esp. in form *fish-and-chip*; **fish-basket**, (*a*) a basket used for carrying fish (see 6 b); (*b*) *U.S.*, a creel for catching fish; **fish-bed** (U.S. example); **fish belly**, (*a*) see quot. 1878; (*b*) in Dict.; (*c*) used *attrib.* of a degree of whiteness; **fish-blooded** *a.*, cold-blooded; **fishbone-stitch** (see quot. 1957); **fish-carver** (example); **fish-crow** (earlier examples); **fish dive** (see quot. 1952); **fish-eye**, (*a*) (also *fish's eye*) a variety of moonstone; a diamond or imitation diamond; (*b*) (see quot. 1958); (*c*) used *attrib.* or as *adj.* of a wide-angle lens with a curved front resembling a fish's eye; also *ellipt.*; **fish-face**: delete † and add later example; **fish-farmer** (example); **fish-farming** (examples); **fish-finder**, a device for locating fish; **fish finger**, a small finger-shaped or rectangular section of fish coated in batter or breadcrumbs; **fish-flour** (later examples); **fish-fly** *U.S.*, any of various small insects belonging to the order Megaloptera and family Corydalidæ, especially those included in the genus *Chauliodes*; **fish food**, (*a*) = FISH *sb.*[1] 4; (*b*) the food eaten by fishes; **fish-fry**, (*a*) *U.S.*, a picnic where fish is fried and eaten; (*b*) = FRY *sb.*[1] 3; **fish geranium** *U.S.*, a garden variety of geranium, *Pelargonium hortorum*; **fish guano** (earlier U.S. example); **fish-hatchery**, a place for the rearing of fish by artificial means; so **fish-hatching** *vbl. sb.*; **fish-horn**, a tin horn used on fishing-boats or by sellers of fish; **fish insect** = SILVER-FISH 2; **fish-ladder** (earlier U.S. example); **fish-leaves**, the pondweed *Potamogeton natans*, the flat leaves of which were formerly supposed to give shelter to fish; **fish-manure** (U.S. examples); **fish-meal** (further example); **fish-moth** = SILVER-FISH 2; **fish-net**, used *attrib.* of an open-meshed fabric or garment; **fish paper**, (*a*) paper on which cooked fish is laid; (*b*) (see quot. 1924); **fish paste** = PASTE *sb.* 1 d; **fish-plate**, the perforated draining plate of a fish-kettle; **fish-poison**, a name given to various plants which have an intoxicating effect upon fish, causing them to float helplessly on the surface of the water; **fish pole** *U.S.*, a pole used as a fishing-rod; **fish-sauce** (earlier example); **fish sausage,** a sausage made with fish; **fish-scrap** (U.S. example); **fish-slice** (earlier examples); **fish stick,** (*b*) *N. Amer.* = *fish finger*; **fish-story** (earlier examples); **fish-way** (earlier U.S. example); **fish-worm** *U.S.* = EARTHWORM 1; cf. WORM 8 b.

1876 in *Listener* (1965) 3 June 826/3 Fish and chip shops were a considerable source of nuisance. 1913 ROWNTREE & KENDALL *How Labourer Lives* iii. 169 Fish and chips..3[d.]. 1928 D. SAYERS *Unpleasantness at Bellona Club* x. 122 We'll go and see 'George Barnwell' at the Elephant and have a fish-and-chips supper afterwards. 1940 *Economist* 26 Oct. 517/2 The fish and chips sellers have been blessed by the Ministry of Food. 1948 J. BETJEMAN *Coll. Poems* (1958) 219 Ask at the fish and chips in the Market Square. 1961 E. A. POWDRILL *Vocab. Land Planning* iii. 44 A study of social rank will normally involve an examination of population,..poverty shops (fish and chips, secondhand dealers, pawnbrokers, etc.). 1970 *Which?* Mar. 68/2 If it is classified as a dwelling house, you can't use it as a fish and chip shop without first getting planning permission. 1814 H. M. BRACKENRIDGE *Views Louisiana* 179 They unite, and thus form a semicircle like a fish basket. 1844 S. S. HALDEMAN in Schele de Vere *Americanisms* (1872) 351 Various species are abundantly caught..in fish-baskets, made of lathwork, with diverging walls of stone. 1867 *Game Laws Penn.* in *Fur, Fin & Feather* (1872) 100 It shall not be lawful to take, catch, or kill..any fish, by means of any fish-basket. 1955 E. POUND *Classic Anthol.* I. 16 So come not near my dam and weir, Let my fish-basket be. 1869 *Rep. Comm. Agric. 1868* (U.S.) 75 Bones of marine animals are so abundant as to have induced Professor L. Agassiz, twenty years ago, to call it the 'fish bed' of the Charleston Basin. 1878 W. DICKINSON *Cumbld. Gloss.* 33/1 *Fish belly*, the *Cnicus heterophyllus* plant. The underside of the leaf is white, and turns up in the wind. 1884 'MARK TWAIN' *Huck. Finn* 31 A tree-toad white, a fish-belly white. 1927 A. CONAN DOYLE *Case Bk. S. Holmes* 63 Fish-belly whiteness. 1898 *Westm. Gaz.* 14 Jan. 3/1 Not that the historian is fish-blooded and without predisposition. 1923 *Daily Mail* 27 Feb. 8 The Ministerial policy of fish-blooded neutrality. 1932 D. C. MINTER *Mod. Needlecraft* 182/2 Note fish-bone stitch on frayed edges. 1957 M. B. PICKEN *Fashion Dict.* 131/1 *Fishbone-stitch*, series of diagonal single-purl stitches zigzagged across an unmarked line. 1855 E. ACTON *Mod. Cookery* (rev. ed.) p. xxxvii, (*caption*) Fish Carvers. 1812 A. WILSON *Amer. Ornith.* V. 27 Fish-crow: *Corvus ossifragus*. 1870 *Amer. Naturalist* III. 287 The Fish-crows..are running over the wet sands. 1943 K. AMBROSE *Ballet-Lover's Pocket Bk.* iii. 40 A lift from the climax of the Aurora *pas de deux*; known to dancers as the 'fish dive'. 1952 KERSLEY & SINCLAIR *Dict. Ballet Terms* 78 *Poisson*, a position of the body in which the dancer arches her back, lifts her head, and bends back her legs with the feet crossed. This pose may be sustained while jumping..or in double work when the girl is supported in this position and the term *pas poisson* or *fish dive* is used. 1959 *Times* 26 Jan. 6/5 Her dancing betrayed signs of unsteadiness only in the exacting 'fish-dives' in the final *pas-de-deux*. 1882 E. W. STREETER *Prec. Stones* (ed. 3) 96 The 'Fish's Eye' becomes red by transmitted light, undergoing the same changes as a bead of Schmetze paste. 1914 JACKSON & HELLYER *Vocab. Criminal Slang* 33 *Fish eye*,..a diamond. 1916 F. B. WADE *Diamonds* vi. 139 The stone that is too thin will have a ring of brilliancy around a black and empty centre producing the so-called 'fish eye' effect. 1942 *Metal Progress* XLII. 201 (*heading*) Fish-eyes in steel welds caused by hydrogen. *Ibid.* 203/2 'Coarsely crystalline fracture' surrounded by normal fibrous metal in tensile and impact specimens showing 'flakes', 'snowflakes', 'fish-eyes', and such variously named seats of hydrogen embrittlement..can be immediately recognized by that outstanding characteristic whereby the affected zones stand out brilliantly against the darker fibrous background. 1958 A. D. MERRIMAN *Dict. Metallurgy* 94/1 *Fish eyes*, a term used in reference to microfissures occurring in steel. 1961 R. SCHREYER et al. *Dict. Photogr.* 192/2 Fish-eye lens. 1969 *Amat. Photogr.* 28 May 73/1 The range of lenses is immense. The widest angle is given by fish-eye types which reduce the image scale at the edges more than in the centre. 1971 *Pop. Photogr.* Aug. 60 (Advt.), A fisheye conversion lens. *Ibid.* 90/1 The fisheye could be used to obtain an even broader coverage. 1914 C. MACKENZIE *Sinister St.* II. III. ix. 687 O, shut up, fish-face. 1876 *All Year Round* 29 Apr. 162/1 Broad-leaved aquatic plants are a real blessing to the fish farmer. 1869 *Rep. Comm. Agric. 1868* (U.S.) 330 Area for fish-farming. 1969 *Guardian* 20 Nov. 7/3 Flat fish..spawn only once a year, a productivity level unacceptable to those interested in fish farming. 1961 *New Scientist* 9 Nov. 362/1 In ordinary trawling the fish-finder apparatus has its transducer set in the hull of the ship. 1962 *Ibid.* 2 Aug. 251/1 An ingenious use of a small magnetic memory drum is made in a new type of fish-finder for trawlers. 1962 *Listener* 22 Mar. 510/1 Cornflakes and frozen fish-fingers, oven-ready chickens, and wrapped, sliced bread. 1970 *Which?* Apr. 105/1 Fish fingers are white fish fillets, coated with crumbs. *Ibid.* 106/1 All the fish fingers are a valuable source of protein. Three fish fingers will give an 8-year-old child a fifth of his or her daily protein requirement. 1956 *Nature* 17 Mar. 512/2 Much attention is being given [in S. Africa] to the production of fish flour in a form suitable for enriching bread. 1963 *Spectator* 15 Feb. 191 Fish flour, which has been developed in the US, would seem to be everything that a cheap, protein-rich food should be. 1968 M. PYKE *Food & Society* ii. 19 An argument about fish 'flour'. This is a product composed of fat-extracted, dried and powdered fish. 1969 N. W. PIRIE *Food Resources* v. 140 Initially the product was called 'fish flour'; now however, because of protests from flour millers, it is called 'fish protein concentrate' or FPC. 1866 *Prairie Farmer* 16 June 412/1 (*heading*) Large Fish Fly. 1902 L. O. HOWARD *Insect Bk.* 211 The so-called comb-horned fish-fly..is the commonest form throughout the United States. 1955 *Sci. News Let.* 14 May 313/2 The fishfly, which begins its slow, nocturnal flights about this time of the year, is among the earliest insects with complete metamorphosis, fossil records show. 1883 A. SHEA *Newfoundland Fisheries* 12 Their excellence would give them a high place in the fish-food market. 1909 *Westm. Gaz.* 14 Aug. 2/2 Fish-food from the sea. 1926–7 *Army & Navy Stores Catal.* 1149/3 Zoological department...Fish Food—per pkt.—11½. 1936 *Discovery* Feb. 43/1 On northern streams the Stone Fly and Alder Fly are more numerous, among aquatic insects forming fish food, than the May Fly. 1967 V. CANNING *Python Project* ix. 175 Bags of hound meal, fish and bird food were stacked on the floor. 1824 'A. SINGLETON' *Lett.* 66 Fish-fries are held about once in a fortnight. 1936 M. MITCHELL *Gone with Wind* ii. 25 For two years he had squired her about the County, to balls, fish fries, picnics and court days. 1951 W. DE LA MARE *Winged Chariot* 10 Tiniest fish-fry in a rock-bound pool. 1865 *Trans. Ill. Agric. Soc.* V. 581, I remarked..on viewing some fish geraniums..how much their scent was like that emitted from the scales of a fresh fish. 1901 L. H. BAILEY *Cycl. Amer. Hort.* III. 1261/2 Fish or Bedding Geraniums. 1946 M. FREE *All about House Plants* xvii. 163 House Geraniums (*P[elargonium] hortorum*), also known as Bedding, Horseshoe, Fish and Zonal Geraniums. 1856 *Rep. Maine Board Agric.* 81 The manufacture of 'fish guano', as recently attempted. 1885 Fish hatchery [see HATCHERY]. 1862 *London Society* Aug. 134 All that can be at present done by the press is to show the ease with which a fish-hatching apparatus can be established. 1869 *Rep. Comm. Agric. 1868* (U.S.) 319 The Chinese..have practised fish-hatching successfully for centuries. 1883 G. B. GOODE *Fish. Indust. U.S.* 59 This being needed for fish-hatching purposes, another larger steamer..has just been built. 1856 F. S. COZZENS *Sparrowgr. Papers* iii. 38 Mrs. Sparrowgrass asked me who that was 'blowing a fish-horn'. 1860 J. G. HOLLAND *Miss Gilbert's Career* xix. 351 The instrument leaping out into various angular flourishes, as if a fish-horn had got above its business and were ambitious of the reputation of a key-bugle. 1913 W. W. THOMPSON *Sea Fisheries Cape Col.* iv. 82 The tuneless reverberations of the archaic fish-horn. 1947 L. G. GREEN *Tavern of Seas* (1952) i. 7 The fish horn is music only in the ears of those who love Cape Town. 1905 *Daily Chron.* 21 July 2/7 A common fish insect, which had been injuring photographs and photographic material. 1948 'P. WOODRUFF' *Whatever Dies* 170 There were old copies of the..*Field*, half eaten away by fish-insects. 1865 *Mich. Gen. Statutes* (1882) I. 574 Sufficient and permanent shutes or fish ladders to admit of the free and uninterrupted passage of fish over such dam or dams. 1886 BRITTEN & HOLLAND *Dict. Eng. Plant-Names* 184 Fish leaves. 1788 WASHINGTON *Diaries* (1925) III. 330 The effect of the fish Manure w[hi]ch was put into the Corn hills in May last was visible with the Wheat. 1868 *Rep. Mass. Board Agric.* I. 105 Fish manures, the product of the oil-fisheries on our coast..sell at about forty-five dollars per ton. 1967 *Times* 12 Apr. 28/4 Previously fishmeal was used mainly as a fertilizer, but now it is employed as an indispensable ingredient of animal feed. 1859 R. J. MANN *Natal* viii. 171 It is one of nature's beneficent compensations that the fish-moth is devoid of wings. 1942 E. O. ESSIG *College Ent.* v. 70 The silver fish moth, *Lepisma saccharina* Linnaeus, is now almost cosmopolitan in distribution, being known in North America, Europe, China, Japan, and the Hawaiian Islands. 1881 C. C. HARRISON *Woman's Handiwork* I. 57 Fish-net fringe..can be made of *écru* cord..in ordinary fish-net stitch, in large meshes. 1897 *Sears, Roebuck Catal.* 238/2 Hot Weather Specialities...Men's Fish Net Undershirts..with short sleeves. 1933 D. L. SAYERS *Murder must Advertise* x. 182 My dear! and got up regardless..fish-net stockings and all. 1868 M. JEWRY *Warne's Model Cookery* 98/2 Slide it [*sc.* the turbot] gently on to a hot dish, on which a folded damask napkin or ornamental fish paper has been placed. 1924 C. J. WEST *Class. & Def. Paper* 38 Fish paper, a term applied to a chemically treated board; used..as insulation in building dynamos and motors. 1928 *Daily Mail* 9 Aug. 3/5 Always serve fried fish as hot as possible, dish it on a fish paper, [etc.]. 1920 *Peace Handbks.* (Foreign Office) LXII. 55 Sturgeon..are made into fish-pastes for local consumption. 1939 T. S. ELIOT *Old Possum's Pract. Cats* 31 A spoon and a bit of fish-paste. 1960 A. E. BENDER *Dict. Nutrition* 50/1 *Fish paste* legally must contain not less than 70% fish. 1747 H. GLASSE *Art of Cookery* 88 When the Water boils, lay the Turbott on a Fish-plate. 1802 J. DRAYTON *View South-Carolina* 67 Fish poison, horse chesnut, or buck's eye. (*Æsculus Pavia.*) 1846 LINDLEY *Veget. Kingd.* 384 Serjania triternata is also employed as a fish poison. 1866 LINDLEY & MOORE *Treas. Bot.*, Fish-poison, *Lepidium Piscidium*. Jamaica, *Piscidia Erythrina*. 1884 W. MILLER *Dict. Eng. Names Plants* 46/2 Fish-poison-plant. 1834 *Visit to Texas* ix. 88 We touched [a flame] to a few of the tall canes, at this season as dry as fish poles. 1841 J. F. COOPER *Deerslayer* xiv. 99 Deerslayer..played with the end of a fish-pole in the water. 1957 R. RUARK *Old Man & Boy* 261 'Just the cast net and some fish poles, Lottie,' the old man said sweet as pie. 1728 E. SMITH *Compl. Housewife* 70 English katchop..is good to put into Fish Sauce. 1937 *N.Y. Times* 16 Mar. 15/2 In Frankfort, the hot dog's home town,..fish sausage had been introduced on the market. 1965 *Punch* 12 May 682/1 Technology Minister Frank Cousins gave the Commons an interim report on current progress towards the development of the fish sausage. 1969 N. W. PIRIE *Food Resources* vii. 163 Also in Japan, 150,000 tons of fish sausage are eaten annually though it was almost unknown a few years ago. 1881 *N.Y. Times* in G. B. Goode *Amer. Fishes* (1888) 112 These smacks are engaged..for the oil-rendering and fish-scrap works on Barren Island. 1748 H. GLASSE *Art of Cookery* (ed. 2) v. 118 Have your Fish-Slice ready. 1850 DICKENS *Dav. Copp.* lxi. 602 We pick out the spoons and forks, fish-slices..and sugar-tongs. 1953 *Time* 12 Oct. 103 Birds Eye brought out fish sticks (fresh fish coated with a special batter, breaded, fried, packed and then frozen). 1955 *Kingston* (Ont.) *Whig-Standard* 6 Apr. 25/2 The newly developed frozen fish fillets known as fish sticks. 1819 *St. Louis Enquirer* 8 Dec. (Th.), A fish story!..In consequence of the shoals of white-fish which occupied and choaked the channel between Bois Blanc Island and Amherstburgh, the steamboat could not pass. 1823 *Missouri Intelligencer* 28 Jan. (Th.), That's 'a fish story', but mine's a true one. 1867 *Harper's Mag.* July 183/1 A friend who does not tell fish stories, says he has seen them [*sc.* herring] in such schools that he could not row his boat through them. 1845 *Mass. Acts & Resolves 1843–45* 434 Whenever a fishway shall be constructed..all former laws relating to fishways at said dam shall be repealed. 1854 THOREAU *Walden* 223, I catch shiners with fishworms. 1933 *Amer. Speech* VIII. IV. 15/2 The term *fishworm* has a wide distribution throughout New England. 1971 *Daily Hampshire Gaz.* (Northampton, Mass.) 27 Aug. 1 Billy Middleton..was planning on going fishing when he had the chance to dangle a long fat fishworm in front of Sandra Dobbs.

fish, *sb.*[2] **3. fish-bar** (earlier U.S. example).

1872 W. S. HUNTINGTON *Road-Master's Assistant* (ed. 2) 27 Expansion..is supposed to have been provided for at the rolling-mill, by elongating the bolt-hole both in the rail and fish-bar.

fish, *sb.*[4] Add: **3. fish wire,** a stiff wire, usu. looped at the end, used for pulling or 'fishing' wires through conduits, etc.; also **fish tape.**

1907 W. S. IBBETSON *Electr. Wiring* x. 186 If the fish wires are not put through the whole of the tubing, it is certainly better to put them at the difficult parts round bends, etc. 1958 *Van Nostrand's Sci. Encycl.* (ed. 3) 1820/2 A fish tape or wire, a tempered steel wire of rectangular cross-section, is pushed through the conduit until its end appears at the farther end. A draw line is then attached to it and..the line is drawn through the conduit.

fish, *v.*[1] Add: **3.** *to fish for a compliment* (examples). Also *absol.*

1803 *Lett. Miss Riversdale* I. 264, I feared he would think I was fishing for a compliment. 1814 JANE AUSTEN *Mansf. Park* II. xi. 252, I am not fishing; don't compliment me. 1961 W. BUCHAN *Helen All Alone* 183 'Don't fish!' Helen said childishly...'But please—what is "fish"?'..'Fish for compliments—make people say nice things about you.'

4. b. To use as a bait in fishing.

1922 *Times Lit. Suppl.* 27 Apr. 273/3 Fishing the floating fly with a very fine cast. 1927 *Observer* 24 July 26/3 Mackerel skin, on a single hook, cast and fished like a fly.

5. (Additional examples.)

1653 D. OSBORNE *Let.* 22 Jan. (1903) 36 Where have you fished him out, for I think he is..little known in the world. 1953 H. MILLER *Plexus* (1963) vi. 212, I fished out the money..and handed it to Sadie's brother.

d. To pull (a wire) through a conduit or between floors or walls by means of a stiff looped wire or other device pushed in from the nearer end.

1896 R. ROBB *Electr. Wiring* v. 118 Wires are said to be 'fished' when they are started in at one end of a concealed space and then, so that they may be pulled through, are felt for, or 'fished' for, from the other end, with a hooked wire or other contrivance. *Ibid.*, Wire in a flexible conduit may be fished just as the wire alone would be fished. 1914 H. PENDER *Amer. Handbk. Electr. Engineers* 1957 Flexible conduit possesses the advantage over rigid conduit in that it..may be fished between partitions or floors. 1930 MOYER & WOSTREL *Indust. Electr. & Wiring* vi. 151 In small houses..the boards can be taken up through the doorways, and the wires fished to the ceiling outlets and switches.

9. *intr.* Of water: to provide (good or bad) sport for anglers.

1898 *Daily News* 4 Oct. 9/3 The Arun continues to fish badly. 1904 *Daily Chron.* 13 Jan. 5/2 It is a loch that fishes best in the early part of the year. 1910 *Westm. Gaz.* 26 Feb. 16/4 The Avon was fishing well for roach.

fisher[1]. Add: **6. b.** *fisher-wife.*

1868 M. E. BRADDON *Dead-Sea Fruit* xxxix, The rough fishermen and brawny fisherwives. 1887 RUSKIN *Præterita* II. vi. 200 A fisher-wife doll from Calais. 1895 *Month* Sept. 52 All the fisher-wives we dressed alike in short petticoats of blue stuff.

Fisher[2] (fi·ʃər). *colloq.* [f. the name of Sir Warren *Fisher*, Permanent Secretary to the Treasury 1919–39.] Temporary name for a currency note (esp. of £1). Cf. *BRADBURY.

1922 *Daily Mail* 18 Dec. 7 A 'Bradbury' or a 'Fisher' are terms that have practically passed into the language. 1923 *Motor Cycling* 14 Nov. 35/2 The Bench mulcted him of a couple of Fishers and warned him as to his future behaviour.

fisherman. Add: **2.** (Later example.)

1963 B. VESEY-FITZGERALD *Cat Owner's Encycl.* 63 Some [cats] are expert fishermen.

3. (Later example.)

1927 G. BRADFORD *Gloss. Sea Terms* 64/2 *Fisherman*, a fishing vessel, particularly applied to a schooner.

4. fisherman's daughter, rhyming slang for water; **fisherman's knit,** a type of thick ribbed knitting; **fisherman's knot** (see quots.).

1880 D. W. BARRETT *Life & Work among Navvies* II. ii. 41 If he wants water, he makes a demand for 'the fisherman's daughter'. 1960 *News Chron.* 19 July 6/4 Classic shaped sweaters in heavy fisherman's knit silk. 1961 S. PRICE *Just for Record* v. 35 I'd..slide into my jeans and fisherman's knit sweater. 1876 G. E. VOYLE *Mil. Dict.* 139/2 *Fisherman's knot*, a knot used in pontooning to fasten the cables to the rings of the anchors.

1961 F. H. BURGESS *Dict. Sailing* 88 *Fisherman's knot*, used to bend two small lines together. It is made by forming an overhand knot in the end of each line, in such a way that each contains the other line; both knots are then pulled close to each other, forming one knot.

fish-gig. (Later U.S. example.)

1866 *Atlantic Monthly* Mar. 278/2 They were prevented from ascending by what appears to have been an ordinary fish gig.

fish-hook. Add: **3. fish-hook cactus** *U.S.*, any of several cacti with hooked spines.

1846 *30th Congress 1 Sess. H.R. Ex. Doc.* No. 41, 612 The fish-hook cactus is found here [*sc.* in California]. 1947 *Southern Sierran* (Los Angeles) May 4/2 Rare and very beautiful, the Mohave Fishhook or Pineapple Cactus, *Echinocactus Polyancistrus*, with clustered iridescent magenta-pink blossoms.

fish-in (fi·ʃ,in). [See *-IN *suff.*[3]] A form of protest by American Indians against the loss of fishing rights, characterized by fishing in prohibited waters.

1964 *Post-Herald* (Birmingham, Ala.) 3 Mar. 2/3 They took part in an Indian protest 'fish-in' on the Puyallup River. 1966 *New Statesman* 15 Apr. 530/1 In resisting the injunction, the Indians have staged pathetic 'fish-ins'. 1968 *Daily Tel.* 4 June 22/3 Dick Gregory, the Negro comedian and Civil Rights leader..was sentenced for taking part in a 'fish-in' along the Nisqually river to support local Indians protesting against a State ban on net fishing.

fishiness. Add: **2.** *fig.* 'Shadiness', questionableness. *colloq.*

1919 WODEHOUSE *Damsel in Distress* xiv. 163 The extreme fishiness of Albert's sudden production of a cousin from America.

fishing, *vbl. sb.*[1] Add: **5. a.** *fishing expedition, -gear,* (earlier example), *-light, -limit, -party, -port, rights, -spear, trip, -worm.*

1961 E. S. GARDNER *Case Bigamous Spouse* (1967) xv. 203, I am not going to permit counsel to go on a fishing expedition. 1966 A. SACHS *Jail Diary* iii. 38 They..seemed to be on a fishing expedition. 1839 *Knickerbocker* XIII. 406 To throw by my fishing gear, and sit upon rocks. 1832 T. W. MAGRATH *Let.* Jan. in T. Radcliff *Lett. f. Upper Canada* (1833) xiii. 215 At night the shore was brilliant with the fishing lights in the canoes. 1845 C. M. KIRKLAND *Western Clearings* 120 Nobody broke his windows or pulled the shingles off his roof to make fishing-lights or quail-traps. 1963 *Times* 14 Jan. 6/7 The Anglo-Danish fishing-limit compromise. 1765 *Jrnl. French Traveller* 20 June in *Amer. Hist. Rev.* (1922) XXVII. 72 We went to a fishing party out in the Bay. 1838 C. GILMAN *Recoll. Southern Matron* xxvi. 176 Fishing parties, and the chase soon occupied his leisure moments. 1849 F. PARKMAN *Calif. & Oregon Trail* 35 She sat down and entertained us..with anecdotes of fishing-parties. 1897 KIPLING *Capt. Cour.* 253 Some sort of poem about a fishing-port called Brixham. 1930 *Daily Express* 8 Sept. 9/5 He reached Porthleven, a fishing port in Mount's Bay, at ten o'clock on Tuesday morning. 1936 *Discovery* May 139/2 The rivers and the fishing rights belong to the riparian owners. 1840 C. F. HOFFMAN *Greyslaer* I. xi. 122 The torches..enable them..to approach sufficiently near to destroy him with their fishing spears. 1909 'O. HENRY' *Roads of Destiny* 36 Mr. Robert mentioned his intended fishing-trip. 1943 J. S. HUXLEY *TVA* vii. 51 About a million and a half individual fishing trips..were made on the new TVA lakes. 1884 'MARK TWAIN' *Huck. Finn* xxxvii. 376 Curled him up like a fishing-worm.

b. fishing admiral: see ADMIRAL *sb.* 4; **fishing-pole,** a pole used as a fishing-rod; **fishing-room** (earlier examples); **fishing tool** (see quot. 1963).

1791 T. B. HAZARD *Diary* 12 Sept. (1930) 127/1, I broke my Fishing Pole Cought 4 Pickrel 3 Eeels and 6 Trout. 1873 T. B. ALDRICH *Marj. Daw* 75 He rigged himself up a fishing-pole. 1931 J. BUCHAN *Blanket of Dark* 52 He was..able to spend long days..on the meres with his fishing pole. 1946 H. S. PEARSON *Countryman's Cookbook* 78 Our hooks and lines were ready, and we had made several fishing poles from slender saplings during the winter. *a* 1728 in J. Reeves *Hist. Newfoundland* (1793) II. 76 In such case should the ships fishing rooms of that harbour be taken up before he arrives. 1732 E. FALKINGHAM *Let.* 4 Oct. in *Calendar State Pap. Amer. & W. Indies* (1939) 223 And by that means stake out the very best of the antient fishing room and by that pretence claim a right. 1886 *Century Mag.* July 330/1 The 'fishing tools'..come into important service when the drilling apparatus or the rope breaks in the well. 1963 *Gloss. Mining Terms* (*B.S.I.*) III. 10 *Fishing tools* or *tackle*, tools used to recover objects lost or stuck down a borehole.

fishing, *ppl. a.* Add: **1.** *fishing eagle, hawk, owl* = *fish-eagle, -hawk, -owl* (FISH *sb.*[1] 7).

1688 Fishing hauk [in Dict.]. 1785 T. JEFFERSON *Notes Virginia* (1787) 72 *Accipiter piscatorius*..Fishing hawk. 1795 S. HEARNE *Journey from Prince of Wale's Fort* 398 Eagles of several sorts are found in the country bordering on Hudson's Bay during the Summer; but none, except the common brown Fishing Eagle, ever frequent the Northern parts. 1835 R. M. BIRD *Hawks of Hawk-hollow* I. vi. 85 They are no true fishing-hawks, after all. *Ibid.* vii. 94 The Solitude..[was] broken by..the plaintive scream of the fishing eagles. 1934 *Jrnl. R. Aeronaut. Soc.* XXXVIII. 837 This is the fishing owl of tropical Asia, which goes by the Latin name of Ketupa Flovipes. 1953 G. DURRELL *Overloaded Ark* xiii. 220 The Fishing-Eagles, their black-and-white livery standing out..and their canary-yellow beaks and feet bright in the sun.

fishing-rod. Add: **2.** *attrib.* and *Comb.*

1892 *Vermont Agric. Rep.* XII. 135 At Thetford a large fishing rod manufactory has been quite successful. 1896 *Daily News* 17 June 7/5 The car is to be driven by electricity conveyed by overhead wires, the current being caught by a fishing-rod projection from the car. 1904 *Westm. Gaz.* 1 July 4/1 There is no novelty..in the fishing-rod [golf] clubs. 1953 J. S. HUXLEY *Evol. in Action* iv. 101 Fitting two sticks together by a fishing-rod joint.

fishling. (Earlier U.S. example.)

a 1861 T. WINTHROP *Life in Open Air* (1863) 33 Only petty fishlings, weighing ounces.

fishpond. Add: **2.** For (see FISH *sb.*[2]) read (see FISH *sb.*[3]).

3. A game in which participants try to catch toy fish; also, a receptacle at a fair, etc., from which small parcels can be extracted by means of a fishing-rod.

1892–3 *T. Eaton & Co. Catal.* Fall & Winter 67/2 Games...Fish ponds. 1904 E. NESBIT *Phoenix & Carpet* iv. 81 Anthea and Jane went off..to fish in the fishpond, and dive into the bran-pie. 1921 *Daily Colonist* (Victoria, B.C.) 29 Mar. 6/3 Plain and fancy articles will be on sale, also home cooking and a fish pond. 1939–40 *Army & Navy Stores Catal.* 825/3 Magnetic fishponds..with pond, fishing rods and everything complete.

fish-scale. Add: (Further examples.)

1873 LADY C. SCHREIBER *Jrnl.* (1911) I. 186 Found a pretty German Enamel étui—pink fish-scale ground. 1930 E. POUND *XXX Cantos* iv. 16 Like a fish-scale roof. 1940 *Archit. Rev.* LXXXVII. 12 Fish-scale slates are shown used in the upper part of a shop in the High Street and in a detail of their setting-out. 1961 S. SITWELL *Golden Wall & Mirador* vi. 124 The pair of towers have their columns pitted with a fish-scale.

2. *Metallurgy.* (See quot. 1958.)

1932 *Jrnl. Iron & Steel Inst.* CXXV. 587 Fish-scale in high-speed steel is the result of abnormal grain growth in the austenite due to critical strains. 1958 A. D. MERRIMAN *Dict. Metallurgy* 94/1 *Fish scale*, a defect sometimes encountered in vitreous enamel coatings. It..is often associated with flaking of minute areas of enamel.

fish-skin. 1. (Further example.)

1909 *Westm. Gaz.* 15 May 15/1 The last thing in leather is pearl-tinted fish-skin.

fish-tail, *sb.* (Further examples.)

1931 [see *FISH-TAIL *v.*]. 1939 M. B. PICKEN *Lang. Fashion* 57/3 *Fish-tail*, shaped like tail of fish; having cut-out V in end.—*f[ish-tail] drapery*, train, as on a formal gown, shaped like fish-tail. 1963 *Gloss. Mining Terms* (*B.S.I.*) III. 8 *Drag bit* (fishtail bit, pilot bit), a rotary bit which has two or more cutting blades or wings with hard-faced cutting edges. (Various types are the two-wing, three-wing, fishtail and pilot bits.)

fi·sh-tail, *v.* [f. the sb.] *intr.* To cause the tail of an aircraft or the back of a motor-car to swing from side to side (in aircraft in order to reduce the landing speed; also said of the aircraft or car itself). Hence **fi·sh-tailing** *vbl. sb.*

1927 *N.Y. World* 23 July 2/6 She looped,..fish-tailed and then swooped some. 1931 *Vanity Fair* Nov. 78/2 *Fishtailing*, which is swishing the machine from side to side to reduce the forward speed, explains a *fishtail landing* as zig-zagging after touching ground, in order to slow down when coming into a small field. 1942 J. ALDRIDGE *Signed with their Honour* 14 He put his flaps down and fish-tailed in. 1957 *Life* 29 Apr. 132/1 Causing his rear wheels to spin or the rear end to 'fish-tail', that is, swing back and forth. 1967 L. FORRESTER *Girl called Fathom* xi. 129 She..sat there, stiffly swaying as the car fish-tailed, straightened out. 1971 *Daily Colonist* (Victoria, B.C.) 28 Jan. 31/6 One of the cars appeared to be fishtailing.

fishy, *a.* **6.** (Earlier U.S. example.)

1840 J. P. KENNEDY *Quodlibet* 75 Jesse being..a little amphibious in his politics, or, in Mr. Fog's expressive language, *rather fishy*.

fisk, *v.* Add later *arch.* example.

1906 KIPLING *Puck of Pook's Hill* 120 Take *me* from Pevensey to fisk and flyte through fern and forest.

fissibility. (Later example.)

1908 W. SCHLICH *Man. Forestry* (ed. 2) V. 86 The property of wood owing to which it may be split..is termed fissibility, and depends chiefly on the direction in which the force acts.

fissile, *a.* Add: Now pronounced (fi·səil). **2.** *spec.* in *Nuclear Physics.* Capable of undergoing nuclear fission; sometimes used specifically of materials capable of fission upon absorption of a slow (as opposed to a fast) neutron.

1945 *Nature* 29 Dec. 768/1 The first bomb had not been dropped but..large production plants were rapidly accumulating the fissile material which it was planned to use. 1950 [see *FISSIONABLE *a.*]. 1950 J. COCKCROFT in Crammer & Peierls *Atomic Energy* iv. 75 Fast reactors require considerable quantities of scarce fissile material such as U 235. 1953 *Economist* 14 Nov. 508/1 These ['breeder' reactor] plants, which create fissile fuel out of normally non-fissile materials at a faster rate than they are used up, are in the early stage of development. 1957 *Observer* 7 July 11/8 The 'gaseous diffusion' process used during the war to separate fissile uranium-235 for atomic bombs from natural uranium was based on this work [of Professor S. Chapman]. 1958 *Listener* 19 June 1005/1 There are fissile materials (such as plutonium, and uranium-235) which can undergo nuclear chain-reactions. 1965 [see *FISSION *v.* 2].

fissility. (Later example.)

1955 J. A. WHEELER in W. Pauli *Niels Bohr* 175 The slow neutron fissility of 28 nuclei.

fission. Add: **2.** *transf.* (Further example.)

1950 R. A. KNOX *Enthusiasm* 68 Yet the schism went on; and, as is the way of schisms, propagated itself by fission.

3. *spec.* in *Astr.* The breaking up of one star into two others, as postulated in one theory of the origin of binary stars.

1919 J. H. JEANS *Probl. Cosmogony & Stellar Dynamics* xi. 246 (*heading*) The evolution of binary and multiple stars: the process of fission. 1924 H. DINGLE *Mod. Astrophysics* xv. 236 A binary star may originate in the division, or fission, of a single star which, by its contraction, acquires such a high rotational velocity as to become unstable. 1957 A. W. TITHERLEY *Orig. Solar Syst.* ii. 9 As Algol and most other eclipsing binaries have very short periods..it cannot be doubted that before fission the dense central core had previously become elongated. 1964 R. H. BAKER *Astron.* (ed. 8) xiv. 421 The fission theory..was favored in former times, and some recent efforts have been made to revive it in amended form.

4. *spec.* in *Nuclear Physics.* The splitting, either spontaneously or under the impact of another particle, of a heavy nucleus into two (very rarely three or more) approximately equal parts, with resulting release of large amounts of energy.

1939 MEITNER & FRISCH in *Nature* 11 Feb. 239/2 By bombarding thorium with neutrons, activities are obtained which have been ascribed to radium and actinium isotopes. Some of these periods are approximately equal to periods of barium and lanthanum isotopes resulting from the bombardment of uranium. We should..like to suggest that these periods are due to a 'fission' of thorium. *Ibid.* 240/1 'Fission' of the nucleus. 1942 J. D. STRANATHAN *'Particles' of Mod. Physics* xi. 443 Neutron bombardment of uranium causes the uranium nucleus to split into two parts not far different in mass. This approximately equal splitting is called fission. 1947 J. HAYWARD *Prose Lit. since 1939* 17 It remains to be seen whether man's discovery and immediate abuse of the cataclysmic energy released by atomic fission will fortify or weaken his transcendental aspirations and noumenal gropings. 1957 *Encycl. Brit.* II. 648/2 The fission of one pound of uranium yields about 10,000,000 kw. hr...To produce this same amount of energy would require the combustion of 3,000,000 lb. of coal. 1965 J. S. STRETTAN *Ionizing Radiations* vii. 123 If the neutrons produced by one fission are slowed down sufficiently before they meet the next U^{235} atom, fission will continue and the chain reaction will be established. 1967 ROY & NIGAM *Nuclear Physics* v. 162 Most heavy nuclides undergo spontaneous fission in competition with the emission of an α-particle.

5. *attrib.* and *Comb.*, as (sense *4) *fission bomb, chain reaction, fragment, -fusion-fission, neutron, process, product, reactor, spectrum, yield.*

1941 in H. D. Smyth *Gen. Acct. Devel. Atomic Energy Mil. Purposes* (1945) iv. 42 A fission bomb of superlatively destructive power will result from bringing quickly together a sufficient mass of element U-235. 1957 *Encycl. Brit.* II. 651/1 In a fission bomb the uranium or plutonium explode simultaneously if a sufficient amount of material is assembled. 1950 F. GAYNOR *Encycl. Atomic Energy* 37 The fission chain reaction occurring in the atomic bomb, in which a neutron is captured by a uranium nucleus. 1950 *Amer. Speech* XXV. 26 The process of fission has..created such phrases as..fission spectrum, fission yield, fission fragments. 1965 *Gloss. Atomic Terms* (*H.M.S.O.*) 25 *Fission fragments*, freshly produced fission products before they have been slowed down by nearby matter. 1955 *Sci. Amer.* Aug. 46/3 Speculation about the 'fission-fusion-fission' bomb..was indirectly confirmed last month by Atomic Energy Commissioner Willard F. Libby. This weapon is said to consist of a fission bomb which triggers a thermonuclear reaction which in turn causes ordinary uranium (238) to fission. In a speech at the University of Chicago, Libby mentioned 'a nuclear explosion releasing 10 megatons of fission energy'. 1966 *Observer* 15 May 2/5 'Fission-fusion-fission' means a three-stage bomb with a uranium or fission trigger, a hydrogen or fusion intermediate stage..and an outer case of 'natural' uranium 238. 1956 A. H. COMPTON *Atomic Quest* 26 A means of slowing down the fission neutrons. 1939 MEITNER & FRISCH in *Nature* 11 Feb. 239/2 It seems..possible that the uranium nucleus has only small stability of form, and may, after neutron capture, divide itself into two nuclei of roughly equal size...The whole 'fission' process can..be described in an essentially classical way. 1939 *Nature* 13 May Suppl. 794 (*heading*) Decay curves of uranium and thorium fission products. 1955 *Sci. News Let.* 5 Mar. 147/1 Fission products are atoms of chemical elements newly formed, during the explosion, out of the fragments of uranium and plutonium atoms. 1948 *Physics Abstr.* 488 Nuclear fission reactors. 1955 *Newsweek* 1 Aug. 11/2 A new type reactor seems to be under development in which fusion—the H-bomb principle—supplements the energy provided by the usual fission reactor. 1960 *Gloss. Atomic Terms* (*H.M.S.O.*) 25 *Fission spectrum*, the wide range of elements and isotopes formed in fission is called the fission spectrum. Up to 35 elements

may be formed [etc.]. *Ibid.*, *Fission yield*, the fraction of fissions giving rise to one particular group of fission products all having the same mass number.

fission (fi·ʃən), *v.* [f. the sb.] **1.** *intr.* To undergo fission; to split or divide into a small number of parts comparable in size.

1929 J. H. JEANS *Universe around Us* 226 The development of the hypothetical chaos has now been traced through five generations of astronomical bodies, chaos—nebulae—stars—binary systems—sub-systems, to which a sixth generation must be added if the stars of the sub-system happen to fission further. 1971 *Sci. Amer.* June 87/1 The bubble is pinched at the waist and finally fissions into two bubbles.

2. *trans.* and *intr.* In *Nuclear Physics*: to split or break up into fission products; to cause (a nucleus) to undergo fission.

1947 *Sci. News Let.* 20 Sept. 181 (*heading*) Lighter elements fissioned. 1949 *Nucleonics* May 6/1 The isotope uranium-235..is the only naturally occurring substance that can be fissioned with most of the neutrons available within a pile. 1950 A. C. CLARKE *Interplanetary Flight* 84 This is necessary to maintain the chain reaction in normal uranium, which is fissioned only by slow-moving ('thermal') neutrons. 1955 *Wall St. Jrnl.* 24 Feb. 5/2 Uranium atoms would fission in the graphite core of the proposed reactor. 1965 G. R. KEEPIN *Physics Nucl. Kinetics* i. 3 We shall adopt the convention recommended by the American Nuclear Society. Thus 'fissile' will herein refer to those heavy nuclides which can be fissioned by thermal neutrons. 1967 W. E. MEYERHOF *Elem. Nucl. Physics* v. 219 Another process can start when U^{237} can be created with sufficient excitation energy to fission.

Hence **fi·ssioning** *vbl. sb.* and *ppl. a.*

1955 *Sci. News Let.* 23 July 51/3 Fissioning of uranium 235 atoms in solution provides the gamma rays and neutrons for medical treatment and other research. 1965 G. R. KEEPIN *Physics Nucl. Kinetics* ii. 13 We find that the mass of a fissioning U^{236} nucleus is greater than the total mass of its..fission products.

fissionable (fi·ʃənăb'l), *a.* [f. *FISSION *v.* + -ABLE.] Capable of undergoing nuclear fission. Also *fig.* Hence **fi:ssionabi·lity.**

1945 H. D. SMYTH *Gen. Acct. Devel. Atomic Energy Mil. Purposes* i. 17 Plutonium 239 is the nucleus rightly guessed to be fissionable by thermal neutrons. 1946 *Physical Rev.* 2nd Ser. LXX. 108 Determining the thermal fissionability of several nuclear species. 1948 *Times* 9 Mar. 3/3 The new fissionable element is uranium of the atomic weight of 233. 1950 *Amer. Speech* XXV. 24 Reactors may be named according to fuel (uranium reactor)..addition of fissionable fuel (enriched reactor)..and so on. *Ibid.* 26 *Fissile* and *fissionable*..are sometimes distinguished, with fissile assigned to material that will undergo fission upon absorption of a neutron and *fissionable* reserved for material that fissions spontaneously. 1953 *Ann. Rev. Nuclear Sci.* II. 399 For convenience, this first section on the fissionability of nuclides has been divided into two parts: the first part deals with heavy elements from thorium on up, while the second part deals with medium elements from bismuth on down. 1958 *Spectator* 15 Aug. 211/1 Mr. John Wardle and the Yorkshire Cricket Club continued to throw fissionable material at one another, heedless of the fall-out. 1959 *Times* 22 Jan. 11/2 Parties with an ideology are always more 'fissionable' than those without one. 1968 F. KERTESZ *Lang. Nuclear Sci.* (Oak Ridge Nat. Lab. TM 2367) 18 Fissionable materials were shipped in containers held in a birdcage.

fissiparous. Add: So **fissi·parousness** = *fissiparity* (in examples *fig.*).

1931 A. L. ROWSE *Politics & Younger Generation* x. 295 In England and France the [Communist] party has been reduced to a negligible quantity by its own fissiparousness. 1961 *Times* 14 Jan. 7/6 India, haunted as it is by the fear of 'fissiparousness'.

fissure, *sb.* Add: **4. fissure-eruption** (see quot.).

1882 A. GEIKIE *Text-Bk. Geol.* III. 198 In many parts of the earth..there have been periods..when the crust was rent into innumerable fissures over areas thousands of square miles in extent, and when the molten rock..welled out from the vents, and flooded enormous tracts of country...Of these 'fissure-eruptions',..no examples have occurred within the times of human history, unless some of the lava-floods of Iceland can be so regarded.

fist, *sb.*[1] Add: **1. c.** *to make a* (*good, poor,* etc.) *fist* (earlier U.S. and dial., and later, examples). Also with *of.*

1833 A. GREENE *Dod. Duckworth* ii. 8 You hadn't ought to tax any thing..seeing you've made such a fist of it. 1838 C. GILMAN *Recoll. Southern Matron* v. 46 He reckoned he should make a better fist at farming than edicating. 1841 W. G. SIMMS *Kinsmen* II. 24 (Th.), You made a poor fist of this business. 1869 A. C. GIBSON *Folk-sp. Cumb.* 177 Thoo *hes* mead a fist on't. 1876 HARDY *Ethelberta* xlvi, 'Tis a poor fist I can make at hearing anything. 1920 GALSWORTHY *In Chancery* iii, He made a poor fist of sleeping. 1950 E. A. MCCOURT *Home is Stranger* (1951) ii. 31 He doesn't make much of a fist at farming. 1965 *Listener* 2 Dec. 934/2 You could make a much better fist of it than the experts.

4. fist-fight (later U.S. examples); so *fist-fighter, -fighting*; **fist-note,** in *Printing,* matter of particular importance signalled by a symbol in the shape of a hand with the index finger extended.

1873 J. H. BEADLE *Undevel. West* iv. 88 Dances, drunks and fist-fights met with a sudden interruption. 1961 A. MILLER *Misfits* xi. 128 A blasted look is on Gay's face, as though he had been beaten in a fistfight in a cause he only half-believed. 1950 J. DEMPSEY *Championship Fighting* i. 9 Exploding body-weight is the most important weapon in fist-fighting or in boxing. *Ibid.* xvi. 89 They [*sc.* shovels] are particularly valuable for the fist-fighter. 1934 WEBSTER, Fist-note. 1967 *Dict. Canadianisms* xix/2 Within many entries are short notes (identified by this symbol ☞ and called 'fist-notes').

fist, *v.*[1] Add: **2. b.** spec. in *Football.* Also *absol.*

1909 *Daily Chron.* 1 Feb. 8/5 The latter unobserved by the referee fisted the ball into the net..and a goal was allowed. 1937 *Times* 15 Feb. 6 Bigot first hooked the ball in when Hankey had fisted out a troublesome free-kick by Payen. 1970 *Times* 30 Sept. 15/4 Ferencvaros continued to press—Clemence fisted clear.

fistula, *sb.* Add: **1. c.** Also applied to certain passages in the body made surgically.

1892 KEEN & WHITE *Text-bk. Surg.* II. III. vi. 691 An intentional gastric fistula may require surgical treatment for its closure if the object for which the fistula was made has been accomplished, as when it has been undertaken with a view of dilating a cicatricial stricture of the cardiac orifice. 1963 W. V. MCDERMOTT in R. Warren *Surgery* xxvi. 825/1 In 1877 Nikolai Eck conducted a series of experiments on dogs in which he diverted the portal blood into the inferior vena cava...This type of shunt has commonly been referred to as the Eck fistula.

fit, *sb.*[2] Add: **3. d.** Further examples of hyperbolical phrases.

1861 J. BLACKWOOD *Let.* 23 Dec. in Geo. Eliot *Lett.* (1954) III. 474 Your account of Caliban's exploit..has put me into fits. Archie is in an extasy with it. 1877 S. O. JEWETT *Deephaven* iii. 53 She used to take a notion to set in the dark...I should have forty fits, if I undertook it. 1906 J. LONDON *Let.* 20 Oct. (1966) 213 Bailey Millard is throwing fits all around the shop..because of the way you worded your announcement. 1924 C. BEATON *Diary* 12 Apr. (1961) II. 44, I had fits at the back of the car because Papa kept shouting.

fit, *sb.*[4] **4.** (Earlier example.)

1830 G. COLMAN *Random Rec.* II. vii. 214 A complete *Fit-Out*, on my return to London, at my father's cost.

fit, *a.* Add: **5. b.** (Later examples.) Also, *fit to be tied* (slang), extremely angry, hopping mad.

1856 *Knickerbocker* Oct. 433 Then she laughs fit to kill. 1866 'OLD STAGER' *Stage Reminisc.* vii. 92 The smoke and fumes..came up through the chinks of the stage, fit to choke a dozen Macbeths. 1894 SOMERVILLE & 'ROSS' *Real Charlotte* II. xxviii. 177 The old devil was fit to be tied. 1916 E. F. BENSON *David Blaize* v. 101 Cruikshank's awfully pi: fit to burst. 1922 JOYCE *Ulysses* 754, I was fit to be tied. 1938 R. FINLAYSON *Brown Man's Burden* 30 The roundabout with its hurdy-gurdy shrieking fit to raise the dead. 1953 'E. FERRARS' *Murder in Time* xxi. 189 Mad at you. Fit to be tied. 1956 C. SIMAK *Strangers in Universe* (1958) 193 It threw the place into a tizzy...The boss is fit to be tied. When he gets hold of you...

fit, *v.*[1] Add: **8.** *fig.* Freq. *refl.* of persons, and const. *to, into.*

1919 M. K. BRADBY *Psycho-Analysis* 52 His consequent difficulty in fitting himself in to life. *Ibid.* 247 He cannot fit himself in happily to his surroundings.

c. *to be fitted*: *absol.*, to have a garment, etc., fitted on one.

1926 N. COWARD *Queen in Parlour* I. i. 18, I shall have to go and be fitted. 1958 B. NICHOLS *Sweet & Twenties* 128 Far too busy to come and be fitted in person.

11. d. (Earlier example with *with.*)

1823 J. CONSTABLE *Let.* 29 Aug. (1964) II. 285 Any beautifull Gothick building magnificently fitted up with crimson & gold.

fitch (fitʃ), *sb.*[3] Also **fetch.** [Origin unknown.] In basket-making, a kind of plait in which two canes or osiers are twisted together in the same direction so as to enclose a crossing weft at each half-turn. Also = *fitch-rod*, one of the canes so used. Hence **fitch** *v.*

1907 *Jrnl. Society Arts* 11 Jan. 190/2 The strokes chiefly used are termed: a slew when two or more rods are woven in together.., a fitch when two are woven alternately one under the other. *Ibid.*, Some fitched basket chairs..are nearly 3,000 years old. 1910 *Encycl. Brit.* III. 482/2 The 'fitch'..employed for skeleton work such as cages and waste-paper baskets. 1912 *Rep. Cambr. Anthropol. Exped. Torres Straits* IV. 64. 1953 A. G. KNOCK *Willow Basket-Work* 21 When a round of fitching is completed, the remainder of the two rods may be worked out as a pair. *Ibid.*, The size of the fitch rods is generally a little smaller than that of the stakes.

fitness. Add: **1.** *spec.* the quality or state of being physically fit. Often *attrib.* and *Comb.*

a 1935 T. E. LAWRENCE *Mint* (1950) II. xxii. 158 So I dodge the last weeks of depot training and the orgy of fitness-tests with which it closes. 1939 *Ann. Reg.* 1938 100 The 'fitness' campaign which had recently been launched.

fitted, *ppl. a.* Add: **c.** *fitted carpet,* a carpet cut to fit exactly the floor of a room; **fitted sheet,** a bed-sheet with box ends which fit closely round the mattress.

1898 D. C. PEEL *New Home* iii. 24 In no room would I allow a fitted carpet. 1950 *Vogue* Aug. 98/2 Intellectuals can't afford a fitted carpet. 1963 *Good Housek. Home Encycl.* (ed. 5) 30 *Fitted Tailored Sheets,* those with boxed ends simplify bed-making and are a boon to the restless sleeper. 1969 *Which?* Nov. 372/1 You can get nylon fitted sheets for less than this. 1970 EDWARDES & SMEDLEY *Good Housek. Running Home* 72 Are you going to have a fitted carpet—or not?

fitten, *a.* Add: Later *U.S. dial.* examples.

1873 'MARK TWAIN' & WARNER *Gilded Age* 32 It was fitten to make a body's heart break. 1876 'MARK TWAIN' *Tom Sawyer* 120, I reckon I ain't dressed fitten for a pirate. 1932 W. FAULKNER *Light in August* vi. 98 He would see fitten to reveal it. 1938 M. K. RAWLINGS *Yearling* ii. 19 The scrub's a fitten place for the game to raise, and all the wild things.

fitter, *sb.*[1] Add: **2.** (Earlier and later examples.)

1851 C. CIST *Cincinnati* (Advt.), Brass and iron founders; fitters of wrought iron welded pipe, for steam, gas, etc. 1905 *Terms Forestry & Logging* 37 *Fitter*, 1. one who notches the tree for felling and after it is felled marks the log lengths into which it is to be cut; 2. one who cuts limbs from felled trees and rings and slits the bark preparatory to peeling tanbark. 1948 L. LEVY *Music for Movies* iii. 22 Films being given a preview to the conductor (or as he was sometimes ingloriously known, the 'fitter'), with the object of making some attempt at fitting the music to the mood of the film.

b. One who is engaged to supervise the cutting out and making of garments, or the alteration of ready-made garments to suit customers' requirements.

1858, 1885 [in Dict., sense 2]. 1896 *Daily News* 10 Dec. 7/2 She will find that the fitter who receives her presides..at the table where her orders are being carried out. 1906 *Daily Chron.* 17 July 5/2 She sent the defendants notice that unless they sent a fitter up to make the dress right on the Monday it would be of no use to her. 1921 *Dict. Occup. Terms* (1927) § 405 *Fitter, dressmaker's fitter*, in charge of workrooms in large establishment..; fits garments requiring alterations and takes customers' measurements for length of skirt, etc. 1931 *Times Lit. Suppl.* 16 Apr. 805/2 The women in the workrooms, the fitters, the mannequins.

fitting, *vbl. sb.* Add: **1. b.** The action or an act of fitting on a garment in tailoring and dressmaking.

1900 *Daily Express* 22 June 5/3 Messrs. —— gave her eight fittings. 1930 *London Mercury* Feb. 320 They had had to cancel two fittings and three dress shows. 1942 D. POWELL *Time to be Born* (1943) xi. 274 It was an exciting adventure to lie about a 'fitting'..and then make her dignified dash.

2. (Earlier example.)

a 1817 JANE AUSTEN *Persuasion* (1818) III. xi. 233 The varieties in the fitting-up of the rooms.

5. *fitting-out basin, fitting-room.*

1909 *Daily Chron.* 23 Feb. 1/9 The vessel was..towed to one of the fitting-out basins. 1961 F. H. BURGESS *Dict. Sailing* 88 New vessels..launched and 'taken in hand' for completion, usually in a 'fitting-out' basin. 1909 in A. Adburgham *Shops & Shopping* (1964) xxi. 249 The Costume Department (to which luxurious fitting-rooms are attached). 1962 L. S. SASIENI *Optical Dispensing* ix. 205 Such faults can be corrected in the fitting-room. 1969 R. JAFFE *Fame Game* xiv. 196 He went into a fitting room.

fit-up (fi·tʌp). *Theatr. slang.* [f. verbal phr. *to fit up* (see FIT *v.*[1] 11 d).] A stage or other theatrical accessory that can be fitted up for the occasion. Hence (in full *fit-up company*), a travelling theatrical company which carries makeshift scenery and properties that can be fitted up for the occasion.

1864 P. PATERSON *Glimpses Real Life* xxxiv. 333 The theatre was what is called 'a fit-up', erected in the large room at..a small hotel of the town. 1883 *Referee* 22 July 7/3 The 'fit-up towns'..are the towns which do not boast a theatre, [etc.]. 1885 J. K. JEROME *On Stage* v, He had been a member of a fit-up company that travelled with a complete Shaksperian *répertoire* and four set scenes. *Ibid.* xvii, However, I never did join a booth...The nearest I approached to anything of the kind was this fit-up. 1901 *Westm. Gaz.* 4 Nov. 10/1 His early, struggling, happy days with his father when they took the road with a little fit-up show. 'Fit-up, indeed!' he used to say with a laugh; 'why, our whole fit-up went into a couple of brown-paper parcels!' 1934 S. R. NELSON *All about Jazz* vi. 135 He spent a long time playing in fit-up shows, with a speaking part thrown in gratis. 1957 J. OSBORNE *Entertainer* x. 75, I was..on the Irish fit-ups. 1970 *Daily Tel.* 14 Feb. 12 Today there are some 40 off-Broadway houses. You might add another 40 off-off-Broadway clubs and fit-ups.

FitzGerald (fitsdʒe·răld). [The name of George Francis *FitzGerald* (1851–1901), Irish physicist.] *FitzGerald contraction* or *effect*, the contraction or foreshortening of a moving body in a direction parallel to its direction of motion, small except at speeds comparable to that of light, that was postulated independently by G. F. FitzGerald and H. A. Lorentz in 1892. (The effect is freq. named after Lorentz or after the two men jointly.)

1905 *Phil. Mag.* IX. 680 (*heading*) Report of an Experiment to detect the FitzGerald–Lorentz Effect. 1920 A. S. EDDINGTON *Space, Time & Gravit.* i. 22 Let the reader suppose that he is travelling through the aether at 161,000 miles a second vertically upwards...For this speed the FitzGerald contraction is just ½, so that every object contracts to half its original length when turned into the vertical position. 1967 J. L. ANDERSON *Princ. Relativity Physics* vi. 133 Several attempts were made to explain the negative results of the Michelson–Morley experiment. One of these was the Fitzgerald–Lorentz contraction hypothesis, which postulated that material bodies were contracted as they moved through absolute space.

Fitzrovia (fitsrōu·viă). [f. *Fitzrovius*, Latinized form of the name *Fitzroy* + -IA.] A Bohemian area of London around Fitzroy Square, west of Tottenham Court Road (see quots.).

1958 [see *DOWN AND OUT adj. phr.]. 1971 J. MANTON *Sister Dora* xix. 328 The Medical and Surgical Home, 15 Fitzroy Square...Here among the Italians and artists of Fitzrovia the two friends had ohened one of London's first private nursing homes. 1971 *Guardian* 3 Feb. 8/1 That pub-crawling area of London that Tambi called Fitzrovia, from Fitzroy Square and, more important, the Fitzroy Tavern.

five, *a.* and *sb.* Add: **C. 1. a.** *five-act* (earlier example), *-power.*

1773 J. HOADLY *Let.* in *Corr. Garrick* (1831) I. 506, I have not yet had a sight of Dr. G—'s five-act farce. 1945 W. S. CHURCHILL *Victory* (1946) 151 A five-Power conference bringing in Russia and China.

2. five-and-ten (cent store) *N. Amer.*, a store where all the articles were orig. priced at either five or ten cents; also *colloq.*, *five-and-dime store*; **five corner(s** (earlier example); **five-day week,** a working week of five days; **five-eighth** *Rugby Union* and *Rugby League Football* (*N.Z.* and *Austral.*), one of two players stationed between the scrum-half and the centre three-quarter; **five-figure** *a.*, (*a*) evaluated to or containing tabular entries of five significant figures or five decimal places; (*b*) worth ten thousand pounds or more; **five-finger exercise** (example); also *transf.* and *fig.*, something very easy; **five hundred** orig. *U.S.*, a variety of euchre in which five hundred points constitute a game (Webster Addenda 1918); **Five Nations** [NATION *sb.*[1] 5 d], five confederated tribes of North American Indians collectively known as the *IROQUOIS; later joined by the Tuscaroras to become the Six Nations (SIX *a.* 1 d); **five-nine,** a type of shell used by the German forces in the 1914–18 war; a gun that fires such a shell; **five o'clock** (earlier and later examples); **five o'clock shadow** (see quot. 1937); **five percenter** *U.S.*, an agent who, in return for a percentage fee, assists businessmen to obtain Government contracts, etc.; **five-shooter,** a five-chambered revolver; so **five-shot** *a.*, having five cartridge chambers; **five-spot** *U.S. slang*, (*a*) a five-dollar note; (*b*) a playing-card having five pips; **five-star** *adj.*, (*a*) applied to a hotel classified as first-class and designated by five asterisks; also *transf.* and *fig.*; (*b*) *U.S.*, applied to a general or admiral whose badge of rank includes five stars; **five-to-two** *Rhyming slang*, a Jew; **five-year plan,** a scheme for the economic development of the U.S.S.R. over a period of five years, inaugurated in 1928; later of other countries and repeated in the U.S.S.R.

1880 in *Sat. Even. Post* (1940) 10 Feb. 23/3 Woolworth Bros. 5 & 10 Cent Store. 1907 'O. HENRY' *Trimmed Lamp* 115 Did you ever notice me..peering in the window of the five-and-ten? 1922 L. MUMFORD in H. Stearns *Civilization in U.S.* 9 In the five and ten cent store it is possible for the circumscribed factory operative to obtain the illusion of unmoderated expenditure. 1924 *Antiques* May 229/2 Of course, we are all familiar with the ordinary pink, white and baby blue celluloid thimble of the present-day five-and-ten-cent store! 1934 J. T. FARRELL *Young Manhood* in *Studs Lonigan* (1936) i. 160 A small American flag which he'd copped from the nearby five-and-dime store. 1945 S. LEWIS *Cass Timberlane* (1946) xxv. 166 The lone goldfish in a bowl which she sheepishly brought home from the Five and Ten. 1826 J. ATKINSON *Agric. & Grazing N.S.W.* 19 The native cherry, five corners, jibbong, and others, are merely tasteless berries. 1926 *Sat. Rev.* 2 Oct. 367/2 Mr. Ford has inaugurated a five-day working week...The five-day week cannot be isolated from the whole Ford policy. 1930 *N.Y. Times* 5 Dec. 3/2 The universal adoption of the five-day week by American industry would have the..effect of returning to employment all..now out of work. 1935 *Discovery* Aug. 221/1 The five-day week has already arrived, the four-day week will come and there will be increasing leisure to be employed. 1905 *Westm. Gaz.* 5 Oct. 10/2 Their three three-quarters and two 'five-eighths' gave them..a three-quarter line of five men. 1906 GALLAHER & STEAD *Complete Rugby Footballer* 76 During very many matches of the British tour, one of the present writers (Stead) played five-eighth to Roberts. 1927 *Daily Tel.* 27 Sept. 14/1, I place no value at all on the fact that they indulged in five-eighths, and multiplied the number of potential attackers immediately in rear of the scrummage. 1958 *Times* 22 Oct. 14/3 He has played for Australia against New Zealand in the stead of no less a five-eighth than A. Summons. 1969 *Australian* 24 May 36/4 The class of Balmain five-eighth, Dave Bolton, should be the deciding factor in the backs. 1969 *Sydney Morning Herald* 24 May 28/2 The move of captain Terry Rigney from full-back to five-eighth several weeks ago was one of the trumpets that heralded Gordon's renewed vigour. 1842 *Penny Cycl.* XXIII. 499/1 Lalande, 'Tables de Logarithmes'...Five-figure tables throughout, and no mistake has ever been found in them. *Ibid.* 499/2 'Tables for Nautical Men.' Contains five-figure logarithms, neatly printed. 1861 KNIGHT *Eng. Cycl., Arts & Sci.* VII. 1006 The necessity of dispensing with printed differences is alone almost fatal to the attempt at giving five-figure logarithms on one sheet. *Ibid.* 1007 The first British five-figure table..which gives Gauss's table. 1929 HALL & RIDEAL (*title*) Cambridge Five-Figure Tables. 1948 C. ATTWOOD *Pract. Five-Figure Math. Tables* p. iv, Mean proportional parts have been provided for five-figure values of cubes or areas of circles. 1971 E. LEMARCHAND *Death on Doomsday* vi. 85 They've got some pictures...One or two in the five-figure class at present values. 1903 *Daily Chron.* 20 July 3/3 Teaching small pupils the five-finger exercises. 1936 T. S. ELIOT *Coll. Poems 1909–35* 145 (*title of series of short poems*) Five-Finger Exercises. 1959 W. GOLDING *Free Fall* xii. 234 Leave happiness to the others, Sammy. It's a five-finger exercise. 1963 B. W. ALDISS *Airs of Earth* 64 Early news bulletins had spoken of rioting and arrests here and there, but these were mere five-finger exercises for what was to come. 1920 S. LEWIS *Main Street* 195 What do you say we go down to Jack Elder's and have a game of five hundred? 1936 'R. HYDE' *Passport to Hell* 118 'Five hundred' and louse-catching were the major sports of Gallipoli. 1959 *Sunday Times* 22 Feb. 16/4 *Five Hundred*..never had much success in this country. 1684 L. A. LAHONTAN *Let.* 18 June in *New Voy. N. Amer.* (1703) I. 58 [It was a] pipe of the five nations. 1688 *Pennsylv. Arch.* (1852) 1st Ser. I. 104 The five Nations or Cantons of Indians. 1713 *Col. Rec. N. Carolina* II. 24 To fly to the Indians and hire them to be mediators of a peace..would be derogatory to Her Majesty's honor, and might encourage the five nations to set Indians against Her Majesty's subjects. 1789 *Deb. Congr. U.S.* 25 May (1834) 41 *Resolved.* That the Superintendent of Indian Affairs..inform the Five Nations, the Hurons, and other Indian nations,..that Congress..received their said representation. 1852 J. REYNOLDS *Pioneer Hist. Illinois* 35 The bitter hostility of the Iroquois or Five Nations to the French. 1884 *Encycl. Brit.* XVII. 454/1 (New York) Forming permanent settlements about the headwaters of the streams which gave them passage to the heart of the country, they organized the political league or confederacy known as the Five Nations. 1959 *Chambers's Encycl.* VII. 754/1 *Iroquois* was the name of a great confederation of North American tribes, formed about 1570...At first called the 'Five Nations' (Cayuga, Mohawk, Oneida, Onondaga and Seneca), they became the 'Six Nations' of the 'Long House' when the Tuscarora joined them in 1715. *a* 1918 W. OWEN *Poems* (1963) 55 Deaf even to the hoots Of tired, outstripped Five-Nines that dropped behind. 1930 BLUNDEN *Summer's Fancy* 49 The five-nines drop like hail. 1960 J. BETJEMAN *Summoned by Bells* iii. 27 When Poperinghe and Zillebeke and Mons Boomed with five-nines. 1869 L. M. ALCOTT *Little Women* II. x. 143 Mrs. Kirk asked me if I wouldn't go down to the five-o'clock dinner. 1937 *Time* 11 Oct. 33/1 (Advt.), '5 o'clock Shadow'—that unsightly beard growth which appears prematurely at about 5 P.M., looks bad. 1960 O. MANNING *Great Fortune* II. x. 115 The café..was crowded for the 'five-o'clock', which here [*sc.* in Bucharest] meant coffee or chocolate, and cakes. 1969 *New Yorker* 6 Sept. 115/1 Mr. Nixon, however, was given a deep five-o'clock shadow by the Rumanian artist. 1949 *Life* 24 Oct. 97/2 Be unswervingly loyal to all his party friends, even if they turn into five-percenters. 1848 *Gem of Prairie* (Chicago) 30 Sept. 6/2 These desperadoes are..well armed with the formidable 'five-shooter'. *c* 1900 R. L. HALE *Log of Forty-Niner* 92, I had seen in his belt a bowie, and a five shooter. 1897 *Westm. Gaz.* 30 Nov. 4/3 The Mouzin five-shot magazine rifle of 3 in. 1903 R. L. MCCARDELL *Conversat. Chorus Girl* 42 Della's had a bunch of Broadway blooms that must have cost him a five spot. 1913 C. E. MULFORD *Coming of Cassidy* vii. 118 As the Queen slid off a five-spot showed. 1936 J. DOS PASSOS *Big Money* 78 They went into a hotel..to have a little supper and that just about cleaned his last fivespot. 1913 *Autocar Handbk.* (ed. 5) xxiii. 297 The Association has classified on a 'five star' basis a very large number of British hotels. 1949 *Manch. Guardian Weekly* 20 Oct. 15/3 Fleet Admiral William Halsey, the only five-star officer who will be called. 1951 I. SHAW *Troubled Air* xxii. 380 It doesn't make any difference..whether a five-star general or an ambassador to England sat next to you. 1953 P. FRANKAU *Winged Horse* IV. iii. 261 We always celebrate the moment..with a five-star quarrel. 1959 *Times* 10 June 12/6 An island where a five-star hotel caters for people who like deep-sea fishing. 1970 *New Scientist* 2 Apr. 5/2 Oysters, salmon and other five-star seafood. 1914 Five-to-two [see *big-mouthed* s.v. *BIG *a.* B.2]. 1932 P. P. *Rhyming Slang* 21 *Five to two*, Jew. 1941 G. KERSH *They die with their Boots Clean* I. 40 A Five is a Five-to-Two, or Jew. 1948 E. WAUGH *Loved One* 7 They respect us. Your five-to-two is a judge of quality. 1929 *Times* 9 July 15/3 Soviet Industries. Failure of Five-Year Plan. *Ibid.*, A special conference to consider how far Soviet organizations have succeeded during the past eight months in carrying out the five-year economic plan. 1932 *Ann. Reg. 1931* II. 220 The Cabinet [of Romania]..drew up a 'five-year plan' for agriculture. 1938 *Time* 10 Jan. 20/1 The second Five-Year Plan ended last week, and the Soviet Union entered 1938 by starting on a third. 1969 N. W. PIRIE *Food Resources* i. 36 The Indian Government is sometimes condemned because the fourth Five Year Plan..envisages only a 1 per cent increase.

five-finger. Add: **1. d.** Any of several New Zealand trees with five-lobed leaves.

1926 *Trans. N.Z. Inst.* LVI. 701 *Nothopanax arboreum*, five-finger. 1960 B. CRUMP *Good Keen Man* 32 There wasn't a live five-finger or mangeo left in the valley.

five-fingered, *a.* (Later examples.)

1896 *Westm. Gaz.* 8 Feb. 2/1 *Uraster rubens*, the Five-fingered one. 1910 W. B. YEATS *Green Helmet* 21 An old five-fingered trick to tumble you out of the place.

fivepence. a. *as fine as fivepence* (examples). See also FIPPENCE.

1564 BULLEIN *Dial. agst. Fever* (E.E.T.S.) 62 Out of the countree..as fine as fippence! 1662 I.T. *Grim Collier Croydon* II. in *Gratiæ Theatrales* sig. G12ᵛ, As a man would say, finer than Five pence, or..more proud than a Peacock. 1672 WYCHERLEY *Love in Wood* v. 80 His Mistress is as fine as fippence, in her embroidered Satens. 1793 W. B. STEVENS *Jrnl.* 21 Feb. (1965) 69 *Unde derivatur* as fine as Fivepence *vulgo dict* Fippence? 1805 E. CAVANAGH *Let.* 4 Oct. in M. & C. Wilmot *Russ. Jrnls.* (1934) II. 187 All sorts of Music playing *as fine as five Pence!*

fivepenny, *a.* (Examples.)

1799 in *Deb. 7th Congr. U.S.* 2 Sess. 1410, I was likewise informed that this was done, and a five-penny-bit each paid freely for a copy. 18.. [see TENPENCE]. 1968 *Guardian* 24 Apr. 3/3 The woman at the till took the new fivepenny piece. 1971 A. ROSS *Huddersfield Job* 51 A fistful of fivepenny pieces.

fiver. 1. (N. Amer. examples.)

1843 *Spirit of Times* 7 Jan. 536/2 If any man heard more than a double X bet, he was in a livelier crowd than we met with. But in '*fivers*' something was done. 1931 M. DE LA ROCHE *Finch's Fortune* xxiv. 315 Renny won a fiver off me.

fivesome. Add: **B.** *sb. Golf.* A round in which five players take part.

1928 *Collier's* 10 Nov. 6/2 The idea was not to play a fivesome on a crowded course.

fix, *sb.* Add: **1.** (Earlier and later examples.) Also, condition, state; (working) order. *U.S.*

1816 HORRY & WEEMS *Life Gen. F. Marion* (ed. 4) xiv. 124 They are in a mighty good fix. 1818 E. P. FORDHAM *Pers. Narr. Trav.* (1906) 167 The fire went out and it got quite dark. I was in a bad *fix*, as they say in the backwoods. 1834 D. CROCKETT *Narr. Life* vi. 50, I believe..we should all have been genteelly licked that time, for we were in a devil of a fix. 1834 W. A. CARRUTHERS *Kentuckian in N.Y.* I. 29, I couldn't get my hands in no sort of a comfortable fix. 1836 *Pleasant Peregrinations* 50 (Th.), Tables and settees are put into a sleeping fix in the twinkling of a bedpost. 1839 *Spirit of Times* 27 Apr. 90/3 The filly is a keener, but looked out of fix. 1839 C. BRONTË *Let.* 14 Aug. in W. Gérin *C. Brontë* (1967) x. 153 It so happens that I can get no conveyance..so I am in a fix. 1852 C. H. WILEY *Life in South* 126 I'll be delighted to be in your company in any fix (*i.e.* costume). 1868 H. WOODRUFF *Trotting Horse* xi. 113 In getting a whole stable of horses into fix to trot races, there will seldom be two whose treatment during their preparation ought to be the same. 1895 *Outing* (U.S.) XXVI. 356/1 George's rod is splintered, and the Kid's reel out of fix. 1931 *Kansas City Times* 7 Dec. 16 What a fix this old world might have been in if our boys had not made it safe for democracy.

3. A reliable indication of the position of a ship, aircraft, etc., obtained by determining the bearings, visually or by radio, of objects whose position is known (as fixed points on land, or celestial objects); the position so obtained; also, the action of obtaining a 'fix'.

1902 *Encycl. Brit.* XXXIII. 97/2 The 'station-pointer' is the instrument used in plotting fixes. 1919 H. SHAW *Textbk. Aeronaut.* xviii. 208 An 'absolute' fix is one in which all the observations are taken together, while if a considerable interval of time is allowed to elapse between the taking of the observations, a 'running' fix is obtained, which is not so reliable. 1925 F. C. BOWEN *King's Navy* 182 On a well-lighted coast night navigation can be easier than by day since there will always be lights from which to take compass bearings and so to get a 'fix'. 1935 *Travel* Sept. 8/1 Throughout the flight position reports will be made every thirty minutes, based on these 'fixes', radio bearings and dead reckoning. 1942 J. A. HAMMERTON *ABC of RAF* (ed. 2) 110/1 When the landmark is recognized, then the airman is said to have obtained a 'visual fix'; the word 'fix' merely means the position of the aircraft. *Ibid.* 110/2 If the weather is bad and clouds hide the ground the navigator cannot obtain a visual fix, but there are other kinds of fix. He may obtain a 'radio fix' by getting in touch with two wireless stations. *Ibid.*, With clear sky above, the navigator may use the stars and get an 'astro' fix. 1950 A. P. HERBERT *Independent Member* xxxi. 196 If two or three reported the same splash from different angles..there might even be an accurate 'fix'. 1958 *Observer* 12 Oct. 1/1 The Jodrell Bank radio telescope has been following the American moon rocket out into space. After tracking ended..65 'fixes' of the rocket's position had been made. 1963 *Times* 14 Jan. 13/7 There was only one fix on the trawler's chart for that day and no time was given for the fix. 1971 *Daily Tel.* 17 June 17/4 Using radio-direction finding equipment he obtained a 'fix' from the signals and traced the transmitter to Westmoreland Road, Bromley.

4. *slang* (orig. *U.S.*). A dose of a narcotic drug. Also **fix-up.** Cf. *FIX *v.* 16 c.

[1867 W. H. DIXON *New Amer.* I. 191 Claret-cobbler..eye-opener, fix-ups, or any other Yankee deception in the shape of liquor.] 1934 *Flynn's* 21 Apr. 113/2 A package of narcotics is called a deck or a bindle; a shot is a geezer,..or fix-up. 1936 *Amer. Speech* XI. 121/1 *Fix-up*, a ration of dope, especially one which has just been taken. 1938 *Ibid.* XIII. 184/1 *Fix*, var. of *fix-up*. A ration of narcotics, especially one to be injected. 1949 N. ALGREN *Man with Golden Arm* (1959) I. 57 'Don't vomit, student,' he taunted Frankie to remind him of the first fix he'd had. **1958** *Oxf. Mail* 29 July 6/7 A weird scene where the dope peddlers gather to beat up Johnny, who gets more into debt with each 'fix'. **1966** *Evening News* 17 Nov. 10/5 Hardly a week goes by without a report of drug-taking and 'fix' parties among teenagers.

5. A bribe; bribery; an illicit arrangement (e.g. amongst politicians, or between policemen and criminals). *U.S. slang.*

1929 *Flynn's* 20 Apr. 793/2, I can put the fix on for you so you can run your stuff over the line and straight into Chicago. **1930** E. D. SULLIVAN *Chicago Surrenders* (1931) ii. 48 It's impossible to scare tough hombres who can get $55 a barrel for beer that costs them $7. There's plenty leeway for the 'fix'. **1950** H. E. GOLDIN *Dict. Amer. Underworld Lingo* 70/2 *Fix.* 1. An agreement, secured through bribery, chicanery, intimidation, whereby a criminal indictment is quashed, or the severity of a sentence or of a charge lessened... 2. Any arrangement by which laws, rules, or regulations are circumvented..'his mob got a license (police permission) to hustle (steal) on the cannon (picking pockets) here. The fix is in solid.' **1953** W. P. MCGIVERN *Big Heat* x. 124 He had come up through the ranks of a society that was founded on the fix.

fix, *v.* Add: **2. e.** *Genetics.* To establish (a character, or the gene responsible for it) as a permanent property of subsequent generations.

1900 *Jrnl. R. Hort. Soc.* XXIV. 142 Seeds were again selected from the plants producing smooth seed and planted the third year, and so on through five generations, when the character was fully fixed, and all the plants came true, producing only the smooth black seed. **1902** BATESON & SAUNDERS *Rep. Evol. Committee R. Soc.* I. 131 It has long been known to breeders that certain forms cannot be fixed by selection indefinitely continued. **1905** *Jrnl. R. Hort. Soc.* XXIX. 418 Whenever it is desired in a cross-bred strain to fix a *dominant* character selection must always be made of *single* families containing no recessive members. **1957** *Ann. Math. Statistics* XXVIII. 891 The gene A will ultimately be fixed in the population or completely lost from it. **1969** *Times* 17 Feb. 6/7 The breed was created after 17 years of 'fixing' a type from a beef Shorthorn–Highland cross. **1970** R. GORER *Devel. Garden Flowers* i. 34 The recessive genes are always the easiest to fix, once they have been brought out.

3. a. Also *refl.*

1838 DICKENS *O. Twist* xliv, A mind unable..to detach itself from old..associations, though enabled to fix itself steadily on one object. **1904** L. T. MEADE *Love Triumphant* IV. vii, Her eyes fixed themselves on Leonora's face.

4. c. *trans.* Of a plant or micro-organism: to assimilate (the nitrogen or carbon dioxide of the atmosphere) by causing it to become combined in a non-gaseous metabolizable form. Hence, to cause (an element, esp. nitrogen) to form a compound, whether gaseous or not, as the first step in some biological or industrial process.

1850 *Jrnl. Hort. Soc. Lond.* V. 59 For a given quantity of water evaporated the amount..of the *nitrogenous* proximates fixed is..about *twice* as great in the Leguminosæ as in the Graminaceæ. **1862** *Phil. Trans. R. Soc.* CLI. 448 The Nitrogen might enter directly into the organism of the plant, provided its green parts were adapted to fix it. **1887** *Chem. News* 4 Mar. 104/2 According to the author's [*sc.* Berthelot's] experiments arable soil continually fixes free atmospheric nitrogen. **1906** *Jrnl. Soc. Chem. Industry* 30 June 568/2 The electrochemical processes which have been proposed for fixing atmospheric nitrogen are..very numerous. **1928** F. A. ERNST *Fixation of Atmospheric Nitrogen* i. 9 Calcium carbide at red heat will absorb free nitrogen as a sponge will water, thus fixing the nitrogen in the form of calcium cyanamide. **1930** BUCHANAN & FULMER *Physiol. & Biochem. Bacteria* III. xvii. 174 The nitrogen fixed by microörganisms is used at once in the synthesis of the cell proteins. **1931** H. G. WELLS et al. *Science of Life* VI. iv. 558 The green plant is good at fixing carbon, the bacterium at fixing nitrogen. **1940** *Jrnl. Biol. Chem.* CXXXV. 790 *Escherichia coli* and *Propionibacterium pentosaceum* fix CO_2 with formation of succinic acid. **1946** *Nature* 28 Sept. 447/2 Application of phosphate fertilizers together with any substances which..stimulate the soil microflora will tend to increase the total amount of phosphorus fixed. **1947** J. R. PORTER *Bacterial Chem. & Physiol.* x. 831 The most important bacteria capable of fixing atmospheric nitrogen belong to the genera *Rhizobium* (symbiotic root-nodule bacteria) and *Azotobacter* (nonsymbiotic bacteria). **1971** *Sci. Amer.* Sept. 62/3 Photosynthesis fixes carbon in the leaf and stores solar energy in the form of carbohydrate.

d. To preserve and harden biological material, esp. before microscopic examination.

1878 *Jrnl. R. Microsc. Soc.* I. 189 The property which osmic acid has of fixing the histological elements instantaneously in their actual form. **1969** BROWN & BERTKE *Textbk. Cytol.* iii. 16/1 Formaldehyde..is probably the most widely used chemical for fixing tissues.

e. *Immunol.* To bring about the fixation of (complement).

In early use occas. const. *to* and used as if belonging to sense 1.

1902 *Jrnl. Hygiene* II. 257 The plague bacillus in the presence of the anti-plague immune body of the horse fixed complement of the normal guinea-pig and caused its disappearance from the mixture. **1903** *Lancet* 15 Aug. 447/2 In Tube B the complement becomes fixed to the treated corpuscles before sufficient immune-body can pass to the untreated corpuscles; as a matter of fact, immune-body does pass over, but by the time this has occurred all the complement has been fixed. **1964** M. HYNES *Med. Bacteriol.* (ed. 8) vii. 85 As little as 0·00001 ml. of human serum..will fix complement in the presence of 0·1 ml. of human antiserum. **1970** PASSMORE & ROBSON *Compan. Med. Stud.* II. xxii. 12/1 Both IgG and IgM molecules have the capacity to fix complement.

14. b. Add to def.: *spec.* to prepare (food or drink); *to fix the table* (see quot. 1842[2]); *to fix one's face*, etc.: to put on or rearrange one's make-up, etc.; *to fix out* (examples). Also with *off* and *over*, and const. *for* (doing something). (Further examples.)

1725 S. WILLARD in H. S. Nourse *Early Rec. Lancaster, Mass.* (1884) 237, I fixed the men out with stores. **1783** JAS. SMITH *Tour* 1 Dec. in *Ohio State Archaeol. & Hist. Q.* (1907) XVI. 360 After having fixed up our luggage and taken breakfast we started from Capt. Owsley's. **1804** W. CLARK in Lewis & Clark *Orig. Jrnls.* (1905) VII. 301 Captain Lewis..has been detained at St. Louis to fix off the Osage chiefs. **1839** F. TROLLOPE *Dom. Manners* (1949) 414 You must fix me a drink. **1842** DICKENS *Amer. Notes* II. ii. 46 You call upon a gentleman in a country town, and his help informs you that he is 'fixing himself' just now, but will be down directly: by which you are to understand that he is dressing. *Ibid.*, You inquire..whether breakfast will be ready soon, and he tells you..they were 'fixing the tables': in other words, laying the cloth. **1853** J. G. BALDWIN *Flush Times Alabama* 99 Tom had unfortunately fixed him for visiting his mother on crutches. **1856** *Knickerbocker* June XLVII. 617 Only point me out your traps, and I'll send them up to the hotel, and fix you off all as square as a box. **1873** C. M. YONGE *Pillars of House* IV. xlv. 278 He examined the last roll of proof..and said..'I can fix off a sheet for you.' **1883** 'MARK TWAIN' *Life on Mississippi* 286 The feud..could have been fixed up, easy enough. **1889** —— *Connecticut Yankee* 21 We fixed up an agreement. **1893** W. TIREBUCK *Sweetheart Gwen* II. iii. 223 It's the weekly meeting and I fix the table. **1908** R. W. CHAMBERS *Firing Line* xxi. 367 Use a sponge and cold water and fix your hair and put on your shoes. **1917** C. MATHEWSON *Second Base Sloan* 279 Tell him to show that to the man at the ticket office and he will fix him out. **1928** A. WAUGH *Last Chukka* 87 Why don't we go into the club now, and fix it up? **1933** A. CHRISTIE *Ld. Edgware Dies* ii. 23 Come in here and talk to me while I fix my face. **1936** WODEHOUSE *Laughing Gas* ii. 23 They come to take me out to some party, and they find me in my kitchen in a gingham wrapper, fixing a Welsh rarebit. **1941** J. P. MARQUAND *H. M. Pulham, Esq.* xxiv. 385 When we did have a guest, Gladys had to move..into the old nursery, which we never had got around to fixing over. **1947** AUDEN *Age of Anxiety* (1948) v. 103 Cutting sandwiches and fixing drinks. **1953** N. GORDIMER *Lying Days* II. vii. 69 In my room..I fixed my hair this way and that. **1962** K. ORVIS *Damned & Destroyed* xiii. 88, I should go and fix my face. **1969** A. GLYN *Dragon Variation* iii. 74 She..went to the bathroom, bathed her face and fixed it.

c. For 'U.S.' read 'orig. and chiefly U.S.' *to fix* (a person): to deal with, settle or 'do for' (a person); to kill (a person). (Further examples.)

1836 *Col. Crockett's Exploits* 125 The people in Texas would have nothing to do with that affair, nohow they could fix it. **1842** DICKENS *Amer. Notes* II. ii. 46 You beg a porter to collect your luggage, and he entreats you not to be uneasy, for he'll 'fix it presently'. **1860** HOTTEN *Slang Dict.* (ed. 2) 140 To fix one's flint for him, *i.e.*, to settle his hash. **1874** 'MARK TWAIN' *Let.* (1917) I. 216 I'll close the door against them all—which will 'fix' all of the lot. **1875** *Chicago Tribune* 25 Sept. 7/1 McDonald..said, 'I'll fix you, Fiddler Neary.' He drew a weapon. **1876** C. C. ROBINSON *Gloss. Mid-Yorksh.* 42/1 To 'fix' the flint of any person, is to serve him out. **1942** WODEHOUSE *Money in Bank* (1946) xii. 106 'Don't you worry. I'll fix him.'..'How do you mean, fix?' 'Just fix.' 'You wouldn't croak him?' Mrs Molloy laughed merrily at the whimsical thought. **1957** *Times Lit. Suppl.* 8 Nov. 670/3 He tries to 'fix' her by another [lawsuit], but she wins that too. Unhappily, Seymour 'fixes' it so that her besotted husband gets custody of the younger child.

d. For 'U.S.' read 'orig. U.S.' Add to def.: usu. by illegal means, esp. bribery. (Further examples.)

1790 in *Jrnl. Wm. Maclay* (1927) 248 It is expected of us that we should fix the Governor of Pennsylvania. **1872** G. P. BURNHAM *Mem. U.S. Secret Service* 72 When Biebusch saw *this* man in Court, whom he fancied he had 'fixed' for certain, the criminal wilted. **1881** *Standard* 7 Sept. 5/2 It is true that they talk of 'fixing' a horse, but they also use 'nobbling' in the same sense. **1908** *Atlantic Monthly* Aug. 230/2 A jockey or a pugilist is 'fixed'. **1941** L. A. G. STRONG *Bay* ix. 229 She had ignored the difficulty of fixing Muriel. **1959** *Daily Tel.* 10 Dec. 20/3 He was told that a [driving] test could be 'fixed' for £10. **1960** *Guardian* 18 Oct. 8/2 The mere suspicion that matches were being 'fixed' would be disastrous for the game.

e. To mend, repair. orig. *U.S.*

a **1762** S. NILES *Wars* in *Mass. Hist. Soc. Coll.* (1861) II. 401 A number of hands came to fix our whale-boats. **1870** 'MARK TWAIN' *Sk. New & Old* (1875) 20 [I] finally took the watch to another watchmaker...He fixed it, and gave it a fresh start. **1947** 'N. SHUTE' *Chequer Board* iii. 59 They fixed the vicar's Austin Seven. **1949** F. SARGESON *I saw in my Dream* 37 Arnold was fixing his motor bike.

f. To castrate, sterilize (an animal).

1930 D. H. LAWRENCE *Nettles* 21 Is he a gentleman or a lady?—Neither, my dear! I had him fixed. It saves him from so many undesirable associations. **1962** *Coast to Coast 1961–62* 13 Many years ago he had chosen Silvanus from a litter of kittens, and Brett had 'fixed' it, as he had 'fixed' so many male animals on the farm. **1970** *Washington Post* 30 Sept. B7 'If you fix a stallion,' said Dawnita, 'something happens. He just doesn't have as much action.'

16. *intr.* **a.** To intend; to arrange, get ready, make preparations, *for* or *to do* something. Also with *out* and *up*. *U.S.*

1716 B. CHURCH *Hist. Philip's War* (1865) I. 140 He fixes for another Expedition. **1779** D. LIVERMORE in *Coll. New Hampsh. Hist. Soc.* (1850) VI. 335 Troops are busy in clearing and fixing for laying the foundations of the huts. **1829/30–1943** in WENTWORTH *Amer. Dial. Dict.* **1854–5** in N. E. Eliason *Tarheel Talk* (1956) iv. 136 Aunt Lizy is just fixing to go to church. **1871** MRS. STOWE *Old Town Fireside T.* 55 He was a fixin' out for the voyage. **1875** 'MARK TWAIN' *Speeches* (1923) 55 You fix up for the drought. **1907** *Springfield* (Mass.) *Weekly Republ.* 22 Aug. 6 What a pretty night! The moon is fixing to shine! **1914** G. ATHERTON *Perch of Devil* I. 32, I meet..schoolgirls..so painted up they look as if they was fixin'..to be bad. **1935** J. B. PRIESTLEY *Wonder Hero* iv. § 3. 121, I may be able to fix up for you both to go out to supper afterwards. **1970** H. E. ROBERTS *Third Ear* 6/2 *Fixin' to*, about to do something; to be ready or to intend to.

b. (Usually with *up*.) To put oneself in proper trim; to dress up; to spruce up.

1783 JAS. SMITH *Tour* 7 Dec. in *Ohio State Archaeol. & Hist. Q.* (1907) XVI. 361 We fixed up and started from our encampment as soon as it was well light. **1834** D. CROCKETT *Narr. Life* vii. 51, I fixed up, and joined old Major Russell again. **1845** S. JUDD *Margaret* III. 393 Pa and Ma..were for fixing up a little. *a* **1852** F. M. WHITCHER *Widow Bedott Papers* (1883) xxvi. 116 She looked as if she'd fixed in a wonderful hurry. **1873** J. H. BEADLE *Undevel. West* x. 177 When we can build larger houses it will be time to fix up. **1929** W. FAULKNER *Sartoris* III. iii. 202 You better go to the bathroom and fix up. **1964** MRS. L. B. JOHNSON *White House Diary* 26 Feb. (1970) 75 They were readying a press conference. ..I fixed up a bit and went down with Lynda.

c. *intr.* and *trans.* To inject (oneself) with narcotics. (Cf. *FIX *sb.* 4.) *slang* (orig. *U.S.*).

1938 *Amer. Speech* XIII. 184/1 *To fix*, to take narcotics. **1962** K. ORVIS *Damned & Destroyed* ix. 59 No wonder you're setting there so smug and relaxed. You fixed this morning. You're floating! **1963** A. TROCCHI *Cain's Book* 78 The one vital coil in him is the bitter knowledge that he can choose to fix again. **1967** M. M. GLATT et al. *Drug Scene* vii. 87 At first I 'fixed' only once a week, then more often, and after about six months I was addicted. **1969** *Guardian* 8 July 18/1 The Association for the Prevention of Addiction's centre..shelters addicts and allows them to 'fix' themselves from 10.30 a.m. to 10.30 p.m.

fixate, *v.* Add: **3.** *trans.* To direct the eyes upon, concentrate the gaze directly on. Hence **fixa·ted** *ppl. a.*

1889 *Amer. Jrnl. Psychol.* II. 639 Carefully fixate the point marked, keeping the eye entirely from wandering. **1924** R. M. OGDEN tr. *Koffka's Growth of Mind* iii. § 5. 71 In fixation, the eye is turned until the fixated object falls upon the place of clearest vision. **1931** *Brit. Jrnl. Psychol.* July 9 Let a subject fixate a point between two discs of different shades of grey. **1951** G. HUMPHREY *Thinking* ix. 268 Instructions were: to fixate no special point, but if fixation is necessary to direct attention towards the middle of the dividing line, [etc.]. **1971** *Sci. Amer.* June 36/3 The subject must move his eyes and look around the picture, fixating each part he wants to see clearly.

b. *Psychol.* Orig., in Freudian theory, to cause (a component of the libido) to be arrested at an immature stage leading a person to abnormal attachments to people or things, etc.; hence, to cause (a person) to react automatically to stimuli in terms which relate to a previous strong emotional experience; to establish (a response) in this way. Usu. *pass.* or as **fixated** *ppl. a.*, established or responding in this way; also *loosely*, obsessed with. Cf. *FIXATION 3 b.

1926 W. MCDOUGALL *Outl. Abnormal Psychol.* 133 According to this theory [*sc.* the theory of the Œdipus complex] the *libido*..of every infant normally becomes fixated upon the parent of the opposite sex. **1945** 'G. ORWELL' *Crit. Ess.* (1951) 185 It is clear that for many years he remained 'fixated' on his old school. **1951** N. E. MILLER in S. S. Stevens *Handbk. Exper. Psychol.* 443 When animals that have learned a specific habit.. are given a few electric shocks at the choice point, it 'fixates' their behaviour. **1952** W. J. H. SPROTT *Social Psychol.* IX. 180 Furthermore..frustration may give rise to a response..which gets 'fixated' and is repeated in the frustrating situation over and over again, irrespective of its uselessness. **1955** *Bull. Atomic Sci.* Jan. 14/1 As a nation, we became fixated to the transitory moment of our victory. **1957** C. M. ANDERSON *Beyond Freud* i. 13 The person who is 'fixated at the anal level' has traits supposedly characteristic of the child from one to three years. **1958** *Times Lit. Suppl.* 21 Mar. 159/3 The build-up is good: a rich, crippled husband, fixated on his only son; a nymphomaniac wife, [etc.]. **1962** C. E. BUXTON in Hilgard *Introd. Psychol.* (ed. 3) XVII. 480/1 An individual may in some sense have remained immature by being fixated or caught at one stage of development.

fixation. Add: **1.** (Further examples.)

1953 *Brit. Commonw. Forest Terminol.* I. 56 *Fixation*, the process of stabilizing soil, particularly shifting sand, against erosion. **1968** *Gloss. Terms Timber Preservation* (*B.S.I.*) 21 *Fixation*, the act or state of fixing a preservative in the timber so that it will not leach. **1971** *Novum Testamentum* Jan. 56 The fixation in writing of a cycle of oral tradition. **1971** *Nature* 2 Apr. 289/1 Species formation through genetic divergence and its fixation by reproductive isolating mechanisms is a slow process.

2. d. The process of fixing nitrogen or

another substance as part of a biological or industrial process; see *FIX *v.* 4 c. Cf. **nitrogen fixation.*

1850 *Jrnl. Hort. Soc. Lond.* V. 57 The suspicion that the water evaporated had a more definite quantitative relationship to the fixation of the *non*-nitrogenous than to that of the nitrogenous constituents of the plants. **1862** *Phil. Trans. R. Soc.* CLI. 468 The question whether or not the assimilation of free Nitrogen by plants may account for all, or a part, of the otherwise unexplained fixation, is seen to be left in a dilemma almost inexplicable. **1888** *Ann. Bot.* I. 351 Experiments to determine the fixation of atmospheric nitrogen by vegetable substances. **1914** J. KNOX *Fixation of Atmospheric Nitrogen* II. 73 A great advantage of the NH_3 synthesis for the fixation of nitrogen is that the amount of energy required for the process is small. **1927** S. A. WAKSMAN *Princ. Soil Microbiol.* xxii. 558 Fixation of nitrogen in the soil is carried on largely by bacteria. **1930** BUCHANAN & FULMER *Physiol. & Biochem. Bacteria* III. xvii. 178 In aerobic nitrogen and oxygen fixation there is a competition between nitrogen and oxygen to act as hydrogen acceptors. **1941** *Jrnl. Biol. Chem.* CXXXIX. 375 The fermentation of glucose by propionic acid bacteria was accompanied by fixation of $C^{13}O_2$ in the succinate. **1966** *New Statesman* 2 Dec. 824/3 A celebrated chemical process, on which all food supplies ultimately depend, is the 'fixation' of nitrogen from the air to fertilise the soil or the sea.

e. The process of fixing biological specimens (cf. *FIX *v.* 4 d).

1891 *Jrnl. R. Microsc. Soc.* 287 Sublimate fixation produced excellent results. **1929** C. E. MCCLUNG *Handbk. Microsc. Technique* i. 8 Living protoplasm is a gelatinous or semi-fluid material, and after fixation it becomes a solid. **1969** BROWN & BERTKE *Textbk. Cytology* iii. 15/2 The most important single step in the preparation of tissues is fixation.

f. *Immunol.* = **complement-fixation.*

1905 *Centralblatt f. Bakteriol.* (*Originale*) XXXIX. 603 (*heading*) The fixation of alexines by specific serum precipitates. **1970** PASSMORE & ROBSON *Compan. Med. Studies* II. xxii. 12 The activation or fixation of complement..depends upon changes in configuration of immunoglobulin molecules when they are brought into close apposition during reaction with antigen.

3. a. The concentration of the gaze upon some object for a given time with the intention of holding the retinal image upon the area of direct vision. Cf. *FIXATE *v.* 3.

1889 H. H. ELLIS tr. *Moll's Hypnotism* (1890) i. 2 Indian yogis and fakirs..throw themselves into the hypnotic state by means of fixation of the gaze. **1896** G. F. STOUT *Anal. Psychol.* I. 214 For the most part, however, the fixation of images is not accompanied by overt movement or by any very appreciable muscular strain. **1924** [see *FIXATE *v.* 3]. **1941** *Nature* 13 Sept. 321/1 Reading ordinary type an average adult reader traverses a line of print 3½ in. long in about six stages or fixations. **1951** [see *FIXATE *v.* 3]. **1971** *Sci. Amer.* June 35/1 During normal viewing of stationary objects the eyes alternate between fixations, when they are aimed at a fixed point in the visual field, and rapid movements.

b. *Psychol.* In Freudian theory, the arresting of the development of a libidinal component at a pregenital stage, so that psychosexual emotions are 'fixed' at that point. Also *loosely*, an obsession, an *idée fixe*. Cf. *FIXATE *v.* 3 b.

1910 A. A. BRILL tr. *Freud's Three Contribs. to Sexual Theory* i. 11 A phase of very intense but short-lived fixation on the woman (usually on the mother). *Ibid.* 27 We find feelings of inversion in the unconscious psychic life, fixation of libido on persons of the same sex. **1925** A. & J. STRACHEY tr. *Freud's Coll. Papers* III. iv. 447 In my *Drei Abhandlungen zur Sexualtheorie* I have expressed the opinion that each stage in the development of psychosexuality affords a possibility for the occurrence of a 'fixation' and thus for the laying down of a disposition to illness in later life. *Ibid.* 453 Fixation can be described in this way. One instinct or instinctual component fails to accompany the rest along the anticipated normal path of development. **1946** O. FENICHEL *Psychoanal. Theory of Neurosis* I. v. 65 They represent, rather, pathological cases with special oral-sadistic fixations. **1947** A. HUXLEY *Let.* 9 Mar. (1969) 567 Marlow is one of those classical cases, so dear to psychoanalysts, with a fixation on his mother, about which he feels subconsciously guilty, so that he can't associate sex with respectability. **1957** F. ALEXANDER *Psychoanal. & Psychotherapy* v. 103 According to this view there is a reciprocal relationship between fixation and regression. **1963** *Daily Tel.* 15 May 25/8 A young man with a fixation about boats. **1970** *Ibid.* 21 Feb. 8/3 Victorian taste had a fixation with the Gothic.

c. *Med.* The immobilization and holding in place of a limb or joint or the parts of a fractured bone; the holding in place of a displaced or floating organ by means of sutures.

1897 W. J. WALSHAM *Surg.* (ed. 6) 829/1 (Index), Fixation of kidney. **1906** WALSHAM & SPENCER *Surg.* (ed. 9) III. 254 Fixation of a fracture by open operation. **1908** ROSE & CARLESS *Man. Surg.* (ed. 7) xix. 476 The fixation of the fracture in a good position is provided for by the application of suitable splints. **1926** *Jrnl. Bone & Joint Surg.* VIII. 496 (*caption*) Vanadium (author's design) steel plate with transfixion of bone, giving maximum fixation and security. **1948** F. W. HOLDSWORTH in *Brit. Surg. Pract.* IV. 192 Internal fixation is, therefore, advisable by plating the fracture of the radius with a 4-screwed vitallium plate. **1951** DORLAND *Med. Dict.* (ed. 22) 566 *Internal fixation*, the fastening together and fixation of the ends of a fractured bone by means of wires, plates, screws or nails applied directly to the fractured bone. **1957** R. NISSEN in Mulholland *Current Surg. Managem.* 57 (*title*) Repair of esophageal hiatal hernia by fixation to the abdominal wall. **1966** J. S. BATCHELOR in R. Furlong *Fractures & Dislocations* 215, 85 fractures treated by early fixation within six days of injury gave a non-union of 23 per cent. **1968** S. TAYLOR et al. *Short Textbk. Surg.* (ed. 2) xxv. 379 In the past, fixation of the kidney (nephropexy) was performed for such symptoms as aching in the loins. *Ibid.* xxx. 449 Pathological Fracture... Treatment includes immobilization by internal fixation of the fracture if that is feasible.

4. *attrib.* and *Comb.* **fixation-point,** the point at which an eye is directed, so that its image falls on the most sensitive part of the retina; also, the most sensitive part of the retina.

1890 BILLINGS *Med. Dict.*, *Fixation-point*.., the point of sight—*i.e.* the point of which the retinal image is on the centre of clearest vision; the crossing-point in outer space of the lines of regard. **1909** E. B. TITCHENER *Text-bk. Psychol.* I. 81 The object first becomes visible as a patch of black;..finally, as it approaches the fixation-point, it appears in its true colour. **1932** W. S. DUKE-ELDER *Text-bk. Ophthalmol.* I. xiii. 580 A position in which the eyes are both directed upon the same fixation point situated straight ahead and infinitely far away. **1950** *Sci. News* XV. 24 Seven different fixation points were discovered, for the following spectral regions. **1971** *Nature* 11 June 395/1 The subject fixated a central point in the pre-exposure field and after a 'ready' signal the stimulus was presented briefly to the left or right of the fixation point.

fixative, *sb.* Add: **1.** (Additional examples.)

1930 R. CAMPBELL *Adamastor* 64 Speed, motion, flight!..Fixative of blurred effects. **1959** *Times* 3 Aug. 9/6 Moustache fixatives.

2. Any substance or preparation used to fix biological specimens. Cf. *FIX *v.* 4 d.

1891 *Jrnl. R. Microsc. Soc.* 287 As a fixative, the author has the highest opinion of the value of osmic acid. **1929** C. E. MCCLUNG *Handbk. Microsc. Technique* i. 8 Some fixatives preserve nuclear conditions especially well, others those of the cytoplasm. **1969** BROWN & BERTKE *Textbk. Cytol.* iii. 16/1 Acetic acid is a rapid penetrant but a poor fixative.

3. A liquid that serves to reduce the rate of evaporation of more volatile components in a mixture of odoriferous liquids, esp. a perfume.

1923 E. J. PARRY tr. *Durvelle's Prep. Perfumes & Cosmetics* iv. 123 Fixatives..are bodies which serve to render more permanent odours of a fugitive nature, such as those of most flowers... Amongst the classic fixatives are ambergris, civet and musk. **1949** R. W. MONCRIEFF *Chem. Perfumery Materials* xiii. 203 The fixative need not itself be strongly odorous, for benzyl benzoate is an excellent fixative and yet has only a faint scent. **1971** *Nature* 16 Apr. 433/2 Deer musk is a direct preputial gland secretion and a fixative for secondary odours.

fixed, *ppl. a.* Add: **2.** *fixed idea* (earlier example).

1829 H. C. ROBINSON *Diary* 13 Aug. (1967) 102 [Goethe] repeated..one of his fixed ideas that it is by the most laborious collection of facts that even a poetical view of nature is to be corrected.

6. *fixed point* (*attrib.* examples); cf. *point-constable*, *-policeman* (POINT *sb.*[1] D. 14) and POINT-DUTY; (*b*) *Physics*, a temperature of some well-defined and reproducible physical process, such as a change of phase of a pure substance, used in defining a scale of temperature; (*c*) *Math.*, a point that is unchanged by a given transformation or by each of a given set of transformations; (*d*) *Broadcasting* (see quot. 1941); (*e*) *Computers*, designating a method of representing numbers by a single sequence of digits with a fixed or constant position for the radix point; usu. *attrib.*

1896 *Daily News* 15 Feb. 3/1 'Fixed point' constables. **1900** *Daily Chron.* 15 Aug. 5/1 City Police on fixed-point duty. **1904** *Ibid.* 16 Sept. 4/5 Fixed-point constables, whose imperious arms will later indicate the right of way.

1778 *Phil. Trans. R. Soc.* LXVII. 844 In the following pages we have thrown together the practical rules, which we would recommend to be observed in adjusting the fixed points of thermometers. **1849** R. V. DIXON *Treat. Heat* I. 10 Sir Isaac Newton first pointed out the advantage to be obtained by this method of graduating temperature scales, and at the same time proposed as the fixed points in the graduation the temperatures of melting ice and boiling water. **1966** *Units & Standards of Measurement: Temperature* (H.M.S.O.) (ed. 2) 3 The International Practical Scale of Temperature is based on six reproducible temperatures (defining fixed points) to which numerical values are assigned and on formulae establishing the relation between the temperature and the indications of instruments calibrated by means of the values assigned to the six defining fixed points.

1893 A. R. FORSYTH *Theory Functions Complex Variable* xix. 514 If we assume that *w* and *z* are two points in the same plane, then there will in general be two different points which are unaltered by the transformation; they are called the *fixed* (or *double*) points of the transformation. **1963** G. F. SIMMONS *Introd. Topol.* 337 A topological space *X* is called a fixed point space if every continuous mapping *f* of *X* into itself has a fixed point. **1968** E. T. COPSON *Metric Spaces* viii. 111 The problem of solving an algebraic equation $f(z)=0$ can be expressed as a fixed point problem. The relation $w=z+f(z)$ maps the complex plane into itself; and the fixed points are the zeros of $f(z)$.

1941 *B.B.C. Gloss. Broadc. Terms* 13 *Fixed point*, time set apart over a continuous period for the broadcast, at regular intervals, of programmes of a similar type over a particular network, or for successive contributions to a regular series. **1969** *Radio Times* 18/25 Dec. 94/2 There are not many fixed-points in transmission times, but obviously they have to be strictly adhered to.

1948 *Math. Tables & Other Aids to Computation* III. 318 The position of the radix point in the fixed-point representation is a matter of some importance since it determines what combinations of operand values are permissible for multiplication and division. **1955** *Jrnl. Assoc. Comput. Mach.* Jan. 57 Both floating binary point and fixed point operation are included. **1964** F. L. WESTWATER *Electronic Computers* ix. 143 Soon, 'libraries' of standard subroutines were available for each particular computer and..sets of subroutines to be performed in 'floating-point' rather than the fixed-point mode arithmetic. **1970** *Math. Rev.* XXXIX. 215/1 This paper examines the distribution of the binary digits of a number representation that is supplied to a conversion routine in decimal form (all in fixed point).

c. *fixed asset*: an asset which cannot be promptly converted into cash (cf. LIQUID *a.* 6); freq. *pl.*; *fixed odds*: used *attrib.* of a bet on association football results that is paid off at predetermined odds, as opp. to a bet on a football pool; *fixed trust*, an investment trust or unit trust in which the investment of funds is restricted to a fixed list of securities.

1898 L. R. DICKSEE *Auditing* (ed. 3) vii. 275 (*heading*) Fixed assets. *Ibid.* 276 It would have seemed more natural to have placed Plant before Stock-in-trade, as being, properly speaking, a 'fixed asset'. **1907** *Act 7 Edw. VII* c. 50 § 21 Every company..shall include..a summary of..its assets, giving such particulars as will disclose..how the values of the fixed assets have been arrived at. **1930** *Economist* 8 Nov. 865/1 The 'fixed trust'..holds a fixed number of common stocks and participation in the trust takes the form of certificates authenticated by a trust company and specifying a fractional interest in a block or unit of the fixed common stocks held by the trust company. These fixed investment trusts first came into public notice about 1924. **1935** *Ibid.* 6 Apr. 795/2 The stream of new investment which has come under fixed trust control has been directed predominantly into the shares of well-established..companies. **1950** *Sporting Investor* Feb. 20 (Advt.), The superb methods for the smaller pools and fixed-odds coupons provide the steady solid income for the small or large investor. **1951** R. W. JONES *Thomson's Dict. Banking* (ed. 10) 280/2 *Fixed assets*, assets (such as land, buildings, plant, machinery), which are not turned into cash, but are used indirectly for the purpose of providing the income of a business. **1959** *Times* 14 Jan. 13/6 Early Unit Trusts had a fixed unit of investment, as their name, Fixed Trusts, implied. It was of the essence of those trusts that unit certificates were issued only against successive deposits of a fixed portfolio of securities, the unit of investment. **1964** *Daily Tel.* 15 Apr. 1/6 (*heading*) Fixed odds tax shocks bookmakers. *Ibid.*, The turn-over on fixed odds football betting was well over £60 million last year.

d. *fixed-head*: (*a*) of a car body, having a fixed roof (opp. **drop-head* (b)); (*b*) of a car engine, having a fixed cylinder-head; *fixed light* (*Naut.* and *Aeronaut.*): see quot. 1960.

1889 L. DELBOS *Naut. Terms* 31 Fixed light. **1948** R. DE KERCHOVE *Int. Maritime Dict.* (1958) 264 *Fixed light*, a coastal light in which the illuminating apparatus gives a continuous light of uniform intensity. **1960** *Guide Civil Land Aerodrome Lighting* (*B.S.I.*) 9 *Fixed light*, a light having constant luminous intensity when observed from a fixed direction. **1962** *Times* 14 Mar. 15/4 A full width rear bench seat in the fixed-head coupé. **1963** BIRD & HUTTON-STOTT *Veteran Motor Car* 96 Delahaye..never made fixed-head engines and never, apparently, had trouble with their detachable heads. **1971** *Daily Tel.* 20 Oct. 6 The new 350 SLC fixed head coupé—a stretched version of the sports car with..room for five.

7. (Further examples.)

1882 R. BITHELL *Counting-House Dict.* s.v. *Trust* 309 It is recommended that one half-yearly coupon should be *fixed* [bearing a fixed rate of interest] and the second variable. **1935** *Planning* III. LII. 7 The Board is financed by the issue of fixed-interest stock which carries no voting rights. **1971** *Daily Tel.* 19 Oct. 19 The demand for fixed interest stocks is still running strongly.

9. Corrupted, bribed, 'squared'; tampered with; of a sporting contest, having the result dishonestly prearranged. (Cf. FIX *v.* 14 d in Dict. and Suppl.) *slang* (orig. *U.S.*).

a **1889** *San Francisco News Letter* (Farmer), His friends on the grand jury..acted precisely as fixed jurors had been known to act. **1901** *Denver Republican* 26 Aug. 3/4 'Fixed' roulette wheels, 'squeeze' faro boxes, loaded dice and marked cards. **1931** L. STEFFENS *Autobiogr.* I. v. 37 Being in with the stables, I soon began to hear about 'fixed races'. **1960** *Guardian* 18 Oct. 8/2 (*heading*) 'Fixed' football.

10. In predicative use: situated materially or financially. *colloq.* (orig. *U.S.*).

1844 [see FIX *v.* 8 c]. **1873** J. H. BEADLE *Undevel. West* 791 From these figures I think they should be better fixed than they are. **1875** *Chicago Tribune* 25 Aug. 8/1 He was very well 'fixed' both in money and in whisky. **1904** H. R. MARTIN *Tillie* 47 I'm well fixed. I got money plenty. **1906** *Springfield* (Mass.) *Weekly Republ.* 25 Jan. 1 The new Connecticut senators are much better fixed financially than their predecessors. **1963** M. LAURENCE in R. Weaver *Canad. Short Stories* 2nd Ser. (1968) 134 Sabina had once had a wealthy lover—well, not wealthy, perhaps, but certainly nicely fixed. **1970** *Guardian* 31 Dec. 11/4 Ali will come out of this fight very well fixed.

11. Special comb.: **fixed bayonet,** a bayonet fixed to the mouth of a musket or rifle; also *attrib.*; **fixed focus** *Photogr.*, a focus on a

camera that is fixed by the design and cannot be adjusted by the user; also *attrib.*; **fixed income,** an unvarying income derived from fixed-interest investments, a pension, etc., as distinct from a salary or wage; **fixed-pitch airscrew, propeller** *Aeronaut.*, a propeller with blades whose pitch cannot be varied; **fixed price,** (*a*) the price charged for a set meal at a restaurant (cf. Fr. *prix fixe*); also *attrib.*; (*b*) used *attrib.* of a 'bazaar' type shop at which all articles are sold at or below a stated price; (*c*) a net price not subject to discount; **fixed seat,** a rowing-seat without a slide; **fixed-wing aircraft, plane,** an aeroplane of the conventional type as opposed to a rotating-wing aircraft such as a helicopter.

1802 C. JAMES *Mil. Dict.* s.v. Manual § vi, The first motion of the charge is the position which the soldier will ..take, in order to advance on an enemy, whom it is intended to attack with fixed bayonets. **1815** J. SIMPSON *Visit Flanders* 173 Eight battalions of the guard debouched with fixed bayonets. **1858** *Leisure Hour* 660/1 We essay to enter the waiting-room but are stopped by a ..soldier, armed with musket and fixed bayonet. **1908** *Daily Chron.* 3 Oct. 4/6 The Royal Marines..enjoy the same fixed bayonet rights in the City. **1909** *Ibid.* 1 Oct. 4/6 Other fixed-bayonet corps in the City are the Honourable Artillery Company,..the 'Buffs', or East Kent Regiment, and the 'Royal Fusiliers' (City of London Regiment). **1892** *Photogr. Ann.* II. 320 Those who object to fixed focus cameras. **1904** *Westm. Gaz.* 27 Aug. 14/2 It will be found that with the majority of lenses the best fixed focus is at about a distance of 20 ft. **1951** G. H. SEWELL *Amateur Film-Making* (ed. 2) iii. 27 The simplest and most inexpensive cameras are equipped with what is known as 'fixed-focus' lenses. **1970** *Which?* July 217/2 Most of the lenses were fixed focus, that is to say you could not vary the distance setting. **1858** TROLLOPE *Three Clerks* III. xii. 213 You will not object to state whether or no you possess any fixed income? **1929** A. HUXLEY *Do what you Will* 110 So much middle-class pride on such small fixed incomes. **1933** B. ELLINGER *This Money Business* xi. 107 All people who are in receipt of fixed incomes (whether it be from investments in Government securities or in debentures, or landlords in receipt of long-term fixed rents, or old age pensioners) benefit from the fall in prices. **1967** *Listener* 16 Mar. 359/2, I am a fixed income pensioner. **1931** *R.A.F. Quart.* July 397 Various types of fixed pitch airscrew were tried. **1934** *Flight* 8 Feb. 124/2 As a fixed-pitch airscrew does not possess satisfactory characteristics for both take off and speed in high-speed aircraft, a variable-pitch airscrew is usually fitted to the latest Northrop models. **1968** MILLER & SAWERS *Techn. Devel. Mod. Aviation* iii. 68 The performance suffered from the use of fixed-pitch propellers, so the Hamilton–Standard variable pitch propeller..was quickly adopted. **1907** *Baedeker's Paris & Environs* (ed. 16) 17 Among the *Brasseries* and *Tavernes*..some are elegant establishments à la carte, while others (charges indicated) have fixed prices. **1909** *Bradshaw's Railway Guide* Apr. 1097 All 'fixed price' meals served in Restaurant. **1933** D. C. E. PELL *Life's Enchanted Cup* xvi. 211 A special dining-room for those who wished for lunch at a fixed price of 3.50 francs. **1935** *Economist* 6 Apr. 798/2 Though the principle of the 'chain chemists' has allowed Boots to build up an excellent business.. it does not seem to offer the opportunity for rapidly expanding profits which, say, Woolworth and Marks and Spencer have discovered in the fixed-price store. **1960** *Farmer & Stockbreeder* 22 Mar. 104 With the lowest price in its class, low running costs, and low fixed-price maintenance and parts costs the Fordson Dexta is a great cost saver. **1965** J. L. HANSON *Dict. Econ.* 180/1 For most consumers' goods the custom of the fixed or specified price has replaced the older custom of haggling over price. **1970** N. FREELING *Kitchen Bk.* iii. 27 There was nothing so vulgar as a fixed-price menu in this hotel. Everything was à la carte. **1889** W. B. WOODGATE *Rowing* 103 The mechanical power gained by a sliding seat is so great that even if he who uses it sets at defiance all recognised principles of fixed-seat rowing, he can still command more pace than if he adhered to fixed-seat work. *Ibid.*, Oarsmen of the fixed-seat school. **1907** *Daily Chron.* 27 July 9/3 It would be a good thing if all junior races could be rowed on fixed seats. Many novices who promised well have lost all style and cultivated wrong methods by having rowed races on sliding seats before knowing how to work on a fixed seat. **1950** *Chambers's Encycl.* XII. 15/2 The boats first used were in-rigged with fixed seats and clinker-built. **1949** *Aero Digest* Sept. 120 Picking up fuel supplies dropped by parachutes from fixed-wing aircraft. **1955** *Oxf. Jun. Encycl.* VIII. 4/1 For fixed-wing aircraft..the varying requirements of speed, altitude, and range affect the choice of engine. **1958** *Times* 1 Mar. 7/3 It [*sc.* a helicopter] may be slower than the fixed wing aircraft. **1966** *Times* 19 July 11/1 So far..B.E.A. has bought fixed-wing types [of American aircraft].

fixer. Add: **1.** (Further examples.) Now, esp. one who arranges or adjusts matters (often illicitly).

1900 'FLYNT' & WALTON *Powers that Prey* III. i. 64 Do you know what fixers are travelin'? **1909** W. IRWIN *Confess. Con Man* ii. 52 At the head of the outfit stood the 'fixer', whose job it was to bribe or stall city officials so that the gamblers could proceed with reasonable security. **1914** JACKSON & HELLYER *Vocab. Criminal Slang* 33 *Fixer*,..one who acts as go-between for thieves and bribe takers. Example: 'If you get a rumble, send for Jones, the mouthpiece; he's a sure-shot fixer and can square anything short of murder.' **1926** J. BLACK *You can't Win* xii. 155 Lawyers haggling at the desk about releases for prisoners, 'fixers', hawk-eyed and rapacious, lurked about. **1930** *Observer* 14 Sept. 15 The Americans have a word for the professional collaborator: they call him a 'fixer'; and his business is to remove the technical defects from a play. **1931** *Economist* 15 Aug. 302/2 A corps of lawyers and lobbyists who enjoyed a high reputation at Ottawa and Quebec as political 'fixers'. **1952** *Country Life* 12 Dec. 1959/2 The most successful dog..has the commanding eye, enabling him to hypnotise the sheep, which earns him the name of fixer. **1959** *Listener* 25 Mar. 561/1 The second [group in Parliament] are the onlookers, the fixers, and the rank and file. **1971** *Guardian* 21 Jan. 3/3 Victor Louis who was the principal contact man and fixer.

2. (Further examples.) Also in *Photogr.*

1922 C. BEATON *Diary* 28 Dec. (1961) 29, I..went to Kodak's and bought a portrait attachment, some gaslight paper, developers and fixers. **1935** KINGSBURY & JOHANNSEN *Histol. Technique* 1 A fixer may be defined as a fluid (or gas) in which the living, or at least very fresh, tissue is placed in order to preserve the structure of its elements as nearly as possible as in life. **1941** J. D. CORRINGTON *Working with Microscope* 161 Zenker's fluid, one of the most efficient and widely used of all fixers in animal histology. **1968** *Gloss. Terms Offset Lithogr. Printing* (*B.S.I.*) 15 *Fixer*, a solution which changes undeveloped light-sensitive material into a water-soluble form for subsequent removal from the photographic layer by washing.

fixing, *vbl. sb.* Add: **1.** (Further examples.)

1910 *Jrnl. R. Hort. Soc.* XXXVI. 28 This enables the breeder to dispense entirely with the old and laborious method of so-called 'fixing' by continuous selection. **1929** C. E. MCCLUNG *Handbk. Microscop. Techn.* i. 7 Fixing is the process of preserving, by means of coagulation, the normal structural characters of organs, tissues and cells. **1939** *Times* 10 July 13/1 Gold to the amount of £183,000 was dealt in at the fixing. **1965** J. L. HANSON *Dict. Econ.* 202/1 *Gold fixing*, determining the price of gold on the London Gold Market. *Ibid.* 264/1 The traditional 'gold-fixing' is carried out [at the premises of N. M. Rothschild in London] by representatives of four firms of bullion dealers. **1967** *Times* 18 Mar. 16/5 At the fixing in London the gold price was lowered 1d. to 251s. 5½d.

2. a. (Earlier and *transf.* examples.)

1820–1 R. FLOWER *Let.* in R. G. Thwaites *Early Western Trav.* (1904) X. 126 'There wife,' said he, 'did you ever see such fixings?' He felt the paper, looked in a mirror.., and gazed with amazement. **1840** *Kentucky Rifle* (Danville, Ky.) 31 Oct. 110/2 Let every man that can play a fiddle or blow a horn, bring his fixins along and make music. **1847** J. S. ROBB *Streaks of Squatter Life* 31 Throw yourself wide on the literary fixins and poetry, for the galls. **1864** BAGEHOT *Coll. Works* (1965) II. 333 There is a setting of surroundings—as the Americans would say, of *fixings*—without which the reality is not itself. **1957** *N.Z. Timber Jrnl.* Mar. 52/1 Fixings, certain items like plugs, pallets, nogs and backings used to fix joinery. **1963** WODEHOUSE *Stiff Upper Lip, Jeeves* v. 40 She was undeniably an eyeful, being slim, svelte and bountifully equipped with golden hair and all the fixings.

3. *fixing agent, fluid.*

1855 T. F. HARDWICH *Man. Photogr. Chem.* v. 52 The Iodide and Bromide of Potassium have both been employed as fixing agents. **1878** W. ABNEY *Treat. Photogr.* iv. 28 In determining the fixing agent to employ in silver printing. **1919** *Brit. Jrnl. Photogr. Alm.* 248 The fixing agent is hypo. **1936** C. J. WALLIS *Pract. Biol.* 6 Immerse the object in several times its own volume of the fixing agent. **1944** B. J. ORBAN *Oral Histol. & Embryol.* xvi. 326 The length of time necessary for a fixing agent to act upon a tissue varies. **1968** C. E. ENGEL *Photogr. for Scientist* v. 254 The fixing agent must not affect the silver of the developed image. **1927** H. E. JORDAN *Textbk. Histol.* (ed. 4) xx. 727 An excellent fixing fluid for general embryologic work.

fixity. Add: **2. d.** Fixedness (of look).

1895 *Cornhill Mag.* Nov. 519 Dark, restless eyes, that could..fasten upon one with a disagreeable fixity.

fixture. 2. b. (Earlier examples.)

1767 in *Archives State New Jersey* XX. 547 Tables, chests of drawers, fixtures, &c. **1849** N. KINGSLEY *Diary* (1914) 66 Saw a man that had died, and the fixtures and mode of preparation for burial.

5. Also *transf.* and *fig.*; *fixture-card* (later examples); so *fixture-list*.

1890 E. DOWSON *Let.* 11 Feb. (1967) 136, I hope to have it complete for your criticism when you come here on Thursday. For you will come—must come.. Let me have a card to-morrow to acquaint me if it is a fixture. **1905** *Strand Mag.* Sept. 298/1 We always take it last on our fixture-list. **1914** C. MACKENZIE *Sinister St.* II. III. vii. 638 A gilt mirror..stuck round with the fixture cards of the university and the college. **1923** [see *BET *sb.* 1]. **1946** M. C. SELF *Horseman's Encycl.* 453 The 'fixture card' is sent to all members of a hunt and tells them what coverts are to be drawn on what days. **1955** *Times* 5 Aug. 4/5 Although their fixture list includes as many as 13 schools, they have not been beaten by another school since July, 1953. **1964** C. WILLOCK *Enormous Zoo* ix. 161 The annual fixture in which one tribe comes over to steal its neighbours' cows, the return match being played for the attackers' women.

fix-up (fi·ks,ʌp). *U.S.* [f. verbal phrase *to fix up*: see FIX *v.* 8 and 14 b.] Something 'fixed up'; a contrivance or 'get-up'. See also *FIX *sb.* 4.

1832 *Polit. Examiner* (Shelbyville, Ky.) 8 Dec. 4/1 She says Mr. Bunker sit down, well I thought I would whilst she was getting her fixups off. **1855** *Knickerbocker* XLVI. 84 The old gentleman looked at me very disdainfully, and then in his most staid accents drawled out: 'Well, if this isn't one of the curiousest fix-ups ever I did see!' **1861** *N.Y. Tribune* 27 Nov. (Chipman's Bartlett), The 'Albany Argus', still hoping for some sort of a compromise or fix-up with the rebels, says [etc.]. **1867** W. H. DIXON *New Amer.* I. 191 Claret-cobbler,.. eye-opener, fix-ups, or any other Yankee deception in the shape of liquor. **1873** J. MILLER *Life Modocs* (1876) x. 149 The lady who has the least amount of natural hair has invariably the largest amount of artificial fix-ups on her head.

fizgig. Add: **1.** Also *attrib.* or as *adj.* = flighty.

1928 GALSWORTHY *Swan Song* III. xii. 308 Like all these 'fizz-gig' young moderns, she was just fluttering without basic purpose or direction.

4. (Later example.)

1947 *Yr. Bk. Arts N.Z.* III. 166 Steal their canoes, Smash their fishing fiz-gigs.

6. An informer. *Austral. slang.*

1902 *Daily Chron.* 21 Aug. 5/2 Each detective [in Melbourne] had a criminal in his employ, who was known as a 'fizz-gig'. **1953** BAKER *Australia Speaks* v. 124 A stool pigeon or informer, otherwise known as a fizgig, fizzer, shelf and topoff.

fizz, fiz, *sb.* Add: **1.** (Earlier example.)

1812 J. & H. SMITH *Rejected Addr.* 69 The water when he was baptized gave a fizz.

3. Also any alcoholic or non-alcoholic effervescing drink.

1889 *Cent. Dict.*, *Fizz, fiz*,..a light frothy liquid; specifically, in the United States, soda-water or other effervescent water. **1891** Gin fiz [see GIN *sb.*[2] 2]. **1906** MRS. BEETON *Househ. Managem.* xlviii. 1512 *Silver fizz*,.. 1 wineglass of gin, the juice of ½ a lemon, the white of 1 egg, 1 teaspoonful of icing sugar, a pinch of carbonate of soda, pounded ice. **1942** V. M. COTTRELL *Lost Cave of Pukerangi* ix. 81 A bottle of 'fizz' each. **1955** G. FREEMAN *Liberty Man* I. i. 17 I'll 'ave a coke...No I won't, I'll 'ave a raspberry fizz. **1958** A. L. SIMON *Dict. Wines* 76/2 *Fizzes*, American 'long' drinks made up of various kinds of spirits and some sugar or syrup..served in tall glasses which are then filled with a syphon.

b. **fizzbang** = whizz-bang (see after WHIZZ *int.* and *adv.* b).

1916 'BOYD CABLE' *Action Front* 67 He has a confused memory..of drumming rifle fire..and, then, immediately after, the rush and crash of a couple of German 'Fizz-Bang' shells. *Ibid.* 186 This cellar roof is too thin to stop an ordinary Fizzbang.

fizz, *v.* Add: **d.** Also with *out.* Cf. FIZZLE *v.* 3.

1847 *Punch* 4 Sept. 87/2 Jenny Lind..has created a *fureur*, but it has fizzed out. **1941** BAKER *Dict. Austral. Slang* 29 *Fizz out on*, to let down, fail in a promise.

fizzer. Add: **1. b.** *Cricket.* A very fast ball; also, one that deviates with unexpected speed after pitching. *colloq.*

1904 *Daily Chron.* 9 May 7/3 It will be good fun watching Tom keeping his fingers out of the way of a springing 'fizzer' of Lockwood. **1929** A. CONAN DOYLE *Maracot Deep* 251 Challen sent down another over of fizzers. **1956** R. ALSTON *Test Commentary* 44 Lock's first ball.. was an absolute 'fizzer' which broke sharply.

3. Usu. in phr. *on a* (or *the*) *fizzer.* A charge-sheet. *Services' slang.*

a **1935** T. E. LAWRENCE *Mint* (1950) I. xi. 39, I got back after..twelve, and they shoved me on the fizzer! **1940** *Daily Mail* 7 Sept. 3/8 Here are some current military phrases, interpreted...'On a fizzer'—On a charge. **1966** *New Society* 9 June 25/1 Feeling I was on a fizzer (army talk for a disciplinary charge).

fizzle, *sb.* Add: **2.** Also with *out.*

1861 O. W. NORTON *Army Lett.* (1903) 23 The Erie Regiment is one grand fizzle out. **1958** *Spectator* 8 Aug. 185/2 A Suez-type fizzle-out.

fizzle, *v.* **3.** *to fizzle out* (earlier U.S. examples). Also with *away.*

a **1848** *Cincinnati Gaz.* (Bartlett), The factious and revolutionary action of the fifteen has..disgraced the actors, and fizzled out! **1854** *Olympia* (Wash.) *Pioneer* 15 Apr. (Th.), The Stellacoom gold excitement has entirely fizzled out. **1866** *Richmond Enquirer* (De Vere), The enterprise fizzled out in the most contemptible manner. **1910** R. BROOKE *Coll. Poems* (1928) Mem. p. li, I've several times started to write to you a notable.. letter, but my life has been too jerky to admit of much connected thought lately, so the letter always fizzled away.

fizzy, *a.* Delete *rare*[−1] and add later examples.

1955 *Times* 20 Aug. 7/6 With their N.A.A.F.I.'s and their bus stops, their parlour games and their fizzy lemonades, the soldiers strike a staid and respectable note. **1963** WODEHOUSE *Stiff Upper Lip, Jeeves* xviii. 143 A very different Bassett from the fizzy rejoicer who had exited so short a while before.

Also *sb.*, a fizzy drink, e.g. champagne.

1896 KIPLING *Seven Seas* 209 The Captain stood a limberful of fizzy—Somethin' Brutt. **1970** 'M. UNDERWOOD' *Shem's Demise* v. 35 Can I have something to drink, Mum?..Is there any fizzy?

fl. = L. *floruit* he flourished (FLOURISH *v.* 4).

1879 LEWIS & SHORT *Latin Dict.* p. xi/1, F. Vegetius Renatus, writer on the art of war, fl. A.D. 386. **1905** F. H. COLLINS *Author & Printer* 124/1 *fl.*, *floruit* (flourished). **1959** *Webster's Biogr. Dict.* 808/1 Kay.., John. fl. 1733–1764. English inventor of flying..shuttle.

flab, *sb.* Add: **2.** *slang.* Fat, flabbiness; *spec.* at Christ's Hospital, butter. (See also quot. 1959.)

1923 *Glasgow Herald* 15 Nov. 8 Other terms in every day use [at Christ's Hospital] are 'flab', butter [etc.]. **1958** *Globe Mag.* 27 Sept. 17/5 He ballooned to 175 pounds of flab and was a poor advertisement for his product.

1959 I. & P. OPIE *Lore & Lang. Schoolch.* ix. 168 The unfortunate fat boy..is known as..flab. **1966** K. GILES *Big Greed* 136 She looks pretty good..no flab round the thighs yet.

flabby, *a.* Add: **4.** *Comb.* **a.** In parasynthetic adjs., as *flabby-breasted*, *-jowled*, *-mouthed.*

1965 H. GOLD *Man who was not with It* xxix. 266 A gutty, chicken-necked, flabby-breasted manager. **1932** W. FAULKNER *Light in August* xiii. 290 Flabbyjowled and darkcaverneyed. **1905** *Daily Chron.* 1 June 5/7 Those blear-eyed, flabby-mouthed swillers.

b. With pr. pple. forming adj., as *flabby-looking.*

1895 J. ASHBY-STERRY *Tale of Thames* 43/2 A flabby-looking, unhealthy person. **1947** W. DE LA MARE *Coll. Stories for Children* 201 Rather flabby-looking mackerel.

flack (flæk). *slang* (chiefly *U.S.*). [Origin unkn.] A press agent; a publicity man.

1946 'S. STERLING' *Where there's Smoke* xiii. 99 That publicity flack is here. **1961** A. BERKMAN *Singers' Gloss. Show Business* 23 *Flack*, a member of the Publicity Department (usually of a motion picture studio); press agent. **1966** L. CHARBONNEAU *Way Out* xiv. 102 And all of a sudden my flack friend believes in his own fairy tales. **1968** C. DRUMMOND *Death & Leaping Ladies* v. 120 They were booked to do ten matches in Mexico City; Bull, their flack, had lined up the opposition.

flacket, *sb.*[1] (Later example.)

1936 W. JAMES *Gangways & Corridors* v. 51 A verse my grandfather used to repeat to me..When I was a boy No bigger than a flacket Half a yard of flannel Would make me a jacket.

flacket, *v.* (Later U.S. and dial. examples.)

1863 A. D. WHITNEY *Faith Gartney* v. 45 You go flacketting out, bareheaded, into the streets, after a topping jade like that. **1885** J. SPILLING *Daisy Dimple* ix. 72, I see her go past flackerting with him last night.

‖ **fladbrod, -bröd** (fla·tbrö, flā·brö). [Da. (Norw.) *fladbröd*, mod.Norw. *flatbröd*, lit. flat bread.] A type of thin unleavened bread eaten in Norway.

1799 MALTHUS *Jrnl.* 12 June (1966) 60 The Norwegians who are used to the flad brod made of oatmeal always find the rye bread disagree with them. **1881** P. B. DU CHAILLU *Land Midnight Sun* I. 394 *Fladbröd* is made from an unfermented dough of barley and oat-meal, often mixed with pea-flour. **1903** B. HARRADEN *Kath. Frensham* II. v. 144 To-morrow Mette makes Fladbröd. **1963** 'G. CARR' *Lewker in Norway* iv. 82 Two packets of the thin but nourishing Norwegian biscuit-bread called *fladbrod*.

flag, *sb.*[1] Add: **4.** *flag-grass* (U.S.), *-pond*, *-root*; **flag-basket** (earlier example); **flag-lily,** the common yellow flag, *Iris pseudacorus*, and other irises.

1747 H. PUREFOY *Let.* 9 Dec. (1931) II. 280 Pray bring a flag Basket or Wallet to take them [*sc.* books] away with you. **1848** W. H. EMORY *Notes Milit. Reconn.* 92 It [*sc.* the island] was overgrown with willow, cane, Gila grass, flag grass, &c. **1884** 'C. E. CRADDOCK' *In Tenn. Mts.* I. 18 Among their roots flag-lilies..and devil-in-the-bush mingled in a floral mosaic. **1922** A. S. MACMILLAN *Pop. Names Flowers Somerset* 107 Flag Lily. Yellow Iris, *Iris Pseudacorus*. **1652** *Rec. Providence, R.I.* (1893) II. 64 John Field shall have the Flagge-pond. **1680** *Portsmouth* (R.I.) *Rec.* 204 The flag pond where the fence now stands. **1837** T. C. HALIBURTON *Clockmaker* 1st Ser. ix. 66 He chases [the horses]..over ditches, creeks, wire holes, and flag ponds. **1707** J. MORTIMER *Husb.* (1708) 287 At D is a spade..the Edges of which are as sharp as a knife, which makes it easie to cut Flag-roots. **1881** S. P. MCLEAN *Cape Cod Folks* v. 107 Grandma fed him with bits of unsweetened flag-root.

flag, *sb.*[2] Add: **5.** *flag-paved* adj.

1895 *Daily News* 21 Nov. 6/2 The street is flag-paved. **1926** W. J. LOCKE *Old Bridge* II. vii, A narrow flag-paved street.

flag, *sb.*[4] Add: **1. c.** *to keep the flag flying*: to refuse to haul down one's flag and surrender; to carry on the fight; chiefly *fig.*; *to show the flag*: (of one of H.M. ships) to make an official visit to a foreign port or elsewhere, showing the White Ensign; also *transf.* and *fig.*; hence *flag-showing* vbl. sb. and ppl. a.

1914 J. W. STALKER (*song title*) Keep the old flag flying. **1918** A. HURD *Brit. Fleet Gt. War* iv. 48 It was only..by releasing 11,000 or 12,000 trained officers and men from non-fighting ships—vessels that 'showed the flag', to quote the phrase of the moment—that it became possible ..to obtain crews for what was to become the Grand Fleet. **1919** BEERBOHM *Seven Men* 20 Neither he nor his work received the slightest encouragement; but he persisted in behaving as a personage: always he kept his dingy little flag flying. **1931** *Times Lit. Suppl.* 1 Oct. 740/3 Professor Warfield kept the flag flying in the theological seminary of Princeton. **1937** PARTRIDGE *Dict. Slang* 281/1 *Show the flag*, to put in an appearance, just to show that one is there. **1957** D. MACINTYRE *Jutland* iii. 36 There had been flag-showing cruises. **1959** *Listener* 14 May 826/2 'Showing the flag' means a British ship going to a foreign port. **1963** *Times* 7 Feb. 18/6 This was a genuine effort on the bank's part to show the flag at a time when they thought it should be shown. A series of six British products would be advertised. **1965** A. NICOL *Truly Married Woman* 29 They had formed a literary club to keep the flag of culture flying.

d. A metal plate bearing the words 'For Hire' affixed to the meter of a taxi.

The flag is raised when the vehicle is disengaged, and when engaged is lowered to start the meter and register the fare due for the distance travelled.

1909 *Daily Chron.* 26 June 6/2 Taxi-cab 'flag up' case... A taxicab driver..was found guilty of defrauding his employers..by driving a cab of the company with the flag up, thus putting into his own pocket money that should have gone to the company. **1910** *Punch* 14 Dec. 421/1 Or in the pouring rain..a taxi will go by with the flag up. **1924** C. MACKENZIE *Old Men of Sea* i. 3, I would walk by the kerb and peer with futile optimism at the drooping flags of the many taxis. **1969** 'J. MORRIS' *Fever Grass* i. 5 The driver..did not lean across to lift the little red metal flag of the meter mounted on the dashboard by the left front window.

e. *Sporting.* A flag used chiefly to indicate the start or finish of a race; also, a device on a chess-clock which falls when the time-limit is reached. So *to drop the flag*, to give the signal for the start or finish; also, *the flag falls.*

1856 'STONEHENGE' *Man. Brit. Sport* viii. 203 The Duties of the Flag-Steward are to..see that the flagman hoists the right flag...The Field-Stewards..should have one or two flagmen with blue flags. **1890** 'R. BOLDREWOOD' *Col. Reformer* xx, He will be there, or thereabouts, when the flag falls, I'll lay. **1890** in A. E. T. Watson *Turf* (1898) 232 Every horse shall be considered as having started which is under the Starter's orders when the advance flag has been raised. **1895** MANSON *Sporting Dict.* s.v. *Distance Judges*, When the first horse reaches the winning post, one of the judges there drops a flag. **1898** in A. E. T. Watson *Turf* iv. 113 It had been supposed by many that no horse of this age could win with so heavy a burden; but there was never any doubt as to the result after the flag had fallen. **1925** E. F. NORTON *Fight for Everest, 1924* 200 It would be a great help to the leader of a future expedition in making up his mind when to 'drop the flag'. **1930** *Brit. Chess Mag.* I. 48 In a time-scramble..the Hungarian's flag actually fell. **1951** 'ASSIAC' *Adv. in Chess* III. ii. 93 Reshevsky..would remain quite unperturbable by the nerve-racking need to make half a dozen moves or more with the 'flag' on his clock about to drop in a matter of seconds. **1969** A. GLYN *Dragon Variation* ii. 49 Carl's clock flag had now fallen, but it didn't matter any more. The game was over.

f. *flag of convenience*: a foreign flag under which a ship is registered in order to avoid certain duties, charges, etc.

1956 *Times* 20 Aug. 7/2 'We Greeks didn't invent the so-called flag of convenience,' he said, somewhat indignantly. **1957** *Britannica Bk. of Yr.* 416/2 The phenomenal growth of merchant fleets registered under the flags of such non-maritime countries as Liberia and Panama began [in 1956] to cause alarm... The ownership of such vessels, under what came to be known as 'flags of convenience', was vested mainly in U.S. and Greek companies and citizens of other nationalities. Owners of ships registered in such countries were able to avoid the high rates of taxation applicable in most of the traditional maritime countries, or, in the case of U.S. companies, the handicaps of U.S. shipping legislation. **1971** B. CALLISON *Plague of Sailors* 94 'Owners?' I queried. 'Greek firm—it's a flag of convenience set-up—called Manentis Shipping.'

g. A statement of the name, ownership, etc., of a publication, or of its name alone, as printed on the editorial page or the front page.

1956 E. C. ARNOLD *Functional Newspaper Design* vi. 108 Some papers have dropped the article *The* from their flag. **1967** *Boston Globe* 18 May 27/5 The New York Times recently made a sensational change in its 'flag' across the top of Page One.

h. *Computers.* A symbol or symbols used to indicate some property of the data in a record.

1959 *Jrnl. Assoc. Comput. Mach.* VI. 147 It is necessary to define the information with flag words inserted directly in the physical records. *Ibid.* 148 There are a number of end flags which indicate the end of a string of information. **1967** COX & GROSE *Organiz. Bibl. Rec. by Computer* 151 The fields and their 'flags' are specified..to ensure consistency in input.

2. d. *pl.* A flag-lieutenant. *Naut. slang.*

1929 F. C. BOWEN *Sea Slang* 49 *Flags*, a Flag Lieutenant. **1936** *S.P.E. Tract* XLVII. 257 His Admiral and perhaps others will continue to call him 'the Flag Lieutenant' (familiarly 'Flags').

7. *flag-planter*, *-planting*; *flag-bedecked*, *-decked*, *-hung* adjs.; *flag-wise* adv.; **flag-day,** (*a*) *U.S.*, the anniversary of the adoption by Congress of the Stars and Stripes as the American national flag on June 14th, 1777; (*b*) a day on which money is raised for a cause by the sale of small paper flags or other tokens which are worn as evidence that the wearer has contributed; **flag discrimination** *Comm.*, the application of differential duties or charges according to a ship's nationality; **flag-fall,** the falling or dropping of a flag to indicate the start of a race (see *1 e); **flag-flying,** (*a*) the flying of flags; (*b*) *colloq.*, overbidding at Bridge; so *flag-flier*; (*c*) *slang* (see quot. 1889); **flag line** *U.S.*, a transport line flying the flag of a specified country; **flagpole sitter,** one who sits on top of a flagpole, for exhibition, etc.; **flag-signal** *v.*, to signal by means of flags; so **flag-signaller;** **flag-wagger** *slang*, (*a*) a flag-signaller; (*b*) = *flag-waver*; **flag-wagging,** (*a*) recent examples of sense 'flag-signalling'; also *attrib.*; (*b*) = *flag-waving*; also *attrib.*

1904 *Daily Chron.* 12 Aug. 5/7 The flag-bedecked town. **1906** *Ibid.* 9 May 7/4 The flag-bedecked fleet. **1937** KOESTLER *Spanish Testament* i. 29 Bending over a flag-bedecked map with Colonel Questa and a German airman. **1894** *Chicago Tribune* 17 June 1/7 American Flag Day has come to stay. **1905** *N.Y. Even. Post* 10 June 9 Flags shall fly from the City Hall on June 14 in observance of Flag Day. **1914** *Scotsman* 5 Oct. 10/3 The Flag Day effort organised to help the Belgian Relief Fund. **1919** G. B. SHAW *Heartbreak House* p. xvii, The passionate penny collecting of the Flag Days was brought under some sort of regulation. **1899** *Daily News* 18 July 6/3 In flag-decked cages. **1926** M. LEINSTER *Dew on Leaf* 41 Flag-decked floating temple. **1928** *Britain's Industr. Future* (*Liberal Ind. Inquiry*) I. v. § 2. 50 They pointed out the disadvantages of indirect methods of Protection, such as subsidies, flag discrimination, [etc.]. **1959** *Listener* 28 May 920/1 Other governments assist their National Flag lines by ship-building subventions, operating subsidies, and flag discrimination. **1899** *Westm. Gaz.* 16 May 5/3 At Newmarket,..in two consecutive races, the favourite was practically out of the race at flagfall. **1922** E. WALLACE *Flying Fifty-Five* xxvi. 153 But Besse o' the Barn was not favourite at flag-fall. **1927** *Observer* 29 May 25 The more sober and sane flag-fliers who calculate risks. **1889** BARRÈRE & LELAND *Dict. Slang*, *Flag flying* (tailors) is used in reference to a bill posted up when hands are required. **1904** *Westm. Gaz.* 10 Aug. 2/3 There was some diversity in the flag-flying on the various official and semi-official buildings. **1917** E. BERGHOLT *Royal Auction Bridge* 101 'Flag-flying'. In the early days of Auction, it was considered a very heroic thing, when you saw that the opponents would make game on their call, to rush in with an overbid that you were sure would fail, in order to keep the game alive. **1918** *Ibid.* (ed. 2) 152 As this is an instructive example of 'flag-flying' it is as well to compare the two results in figures. **1928** *Daily Express* 21 May 3/7 Jack, with a pitying smile for Sam's heroic flag-flying, doubled—and Sam made a grand slam. **1947** S. HARRIS *Fund. Princ. Contract Bridge* I. iii. 31 The business double ..can obviously be used with great effect against reckless flag-flying. **1897** E. L. VOYNICH *Gadfly* III. viii, The sunlit blaze of carpeted street and flag-hung walls. **1905** *Daily Chron.* 25 Dec. 3/5 The Rev. Michael Adler preached a sermon from the flag-hung pulpit. **1944** B. HERSHEY *Skyways of Tomorrow* v. 83 To this observer it becomes clear that unlimited competition for routes by various United States flag lines would tend very strongly toward weakening the ability of any United States aviation company to compete. **1901** *Westm. Gaz.* 28 Oct. 2/2 The Frenchman..is an indefatigable flag-planter. **1899** *Daily News* 12 July 4/4 Foreign politics in these regions, not content with pinpricks and flag-plantings. **1901** *Westm. Gaz.* 10 Oct. 3/1 This plan of flag-planting is distinctly ingenious. We shall have to speak of planting out flags (instead of pegging out claims) for posterity. **1931** *Kansas City Star* 18 Sept., A man who would do that would do most anything, even to being a flagpole sitter. **1939** A. KEITH *Land below Wind* x. 171 The flagpole sitter rests on the top of his flagpole. **1930** *Daily Express* 6 Sept. 5/5 To make the robot swing his arms and go through the flag-signaller's alphabet. **1888** KIPLING *Soldiers Three* (1889) 15 As if he were flag-signallin' to t' world at large. **1910** H. G. WELLS *Mr. Polly* vi. 130 His mind passed to Mrs. Larkins and the bonnet that was to gain such a hold upon him; it seemed to be flag-signalling as she advanced. **1919** *Athenæum* 11 July 582/2 Flag-wagger. **1966** *Guardian* 30 Sept. 6/8 An imperialist, a flagwagger. **1971** *Daily Tel.* 18 June 14/6 No mere flag wagger, Mr Marten adopted the cool advocacy of a barrister presenting a well-studied brief. **1898** *Westm. Gaz.* 21 Sept. 2/2 Flag-taking (like 'flag-wagging') is more exhilarating than remunerative. **1908** *Ibid.* 20 Nov. 2/2 Flag-wagging rhetoric. **1915** G. ADAM *Behind Scenes at Front* 130 The old name of Army Signals still exists, but flag-wagging is to Signals what Euclid is to mathematics. **1916** 'BOYD CABLE' *Action Front* 152 Wally and me was both in the flag-wagging class. **1928** *Daily Express* 6 June 13 To live among them without flag-wagging or publicity. **1956** BANNERMAN *Birds Brit. Isles* V. 121 The approaching birds scattered..and no amount of frantic flag-wagging by the flankers would turn them. **1958** *Times* 1 Feb. 3/3 That is not a threat, is not flag-wagging, and it is not bluff. **1849** D. ROCK *Ch. of Fathers* II. I. vii. 490 Figures.. fastened flag-wise upon staves. **1906** *Westm. Gaz.* 8 June 2/1 One leaf left, flagwise, on its battered stem.

flag, *v.*[4] Add: **1. b.** To mark with a small flag or tag so that relevant items may be readily found. Also *transf.*

1934 H. NICOLSON *Curzon: Last Phase* 74 Instead of being placed in a jacket of its own, each paper as it arrived was affixed to the top of its own file, and these files were encased in a large folder. Any previous paper mentioned in the minutes had to be 'flagged'. It was this flagging process which caused such pain and irritation. The flags consisted of tabs of thick scarlet paper marked 'A' 'B' 'C' 'D' and so on through the letters of the alphabet. These tabs had to be affixed to any back page in the file to which reference was made in the current minute. **1966** *New Statesman* 28 Jan. 113/3 This phrase.. purports to embody *Brown and Jones*, 2 B & C 1827, which I have flagged for your Lordship. **1967** *Amer. Speech* XLII. 70 The arrangement follows the method and apparatus of Thompson's revised *Motif-Index*, with the 'new' variants of motifs flagged by an asterisk.

c. In *Computers*: see *FLAG *sb.*[4] 1 h.

1959 *Jrnl. Assoc. Comput. Mach.* VI. 128 Symbolic coding—in which only those instructions and data which are referred to by other instructions need be flagged. **1966** C. J. SIPPL *Computer Dict. & Handbk.* (1967) 129/2 *Flag operand*, the third operand of a symbolic instruction, designating which digits of the object-level instruction will be flagged (IBM 1620).

2. a. *spec.* To stop (a train) by signalling

with a flag. Hence, to stop (a vehicle, person, etc.) by waving or signalling. Also *absol.* So *to flag down, in.*
1856 *N.Y. Herald* 12 Jan. 1/3, I flagged the Albany express train..with my white flag. **1871** *Scribner's Monthly* II. 433 Old Tom, who flagged at the Cherry street crossing. **1886** [in Dict.]. **1899** A. H. Quinn *Pennsylvania Stories* 168 At Broad Street the outfit was flagged by a Sergeant. **1915** Wodehouse *Something Fresh* iii. 63 George, that nice, fat carver is wheeling his truck this way. Flag him, and make him give me some more of that mutton. **1932** W. Faulkner *Light in August* xii. 270 And I flagged that car with my right hand. **1932** *Kansas City Times* 18 Feb. 22 Fellows who flag a newspaper man down in order to..pay a subscription. **1940** R. Stout *Over my Dead Body* xi. 149 A taxi appeared and I flagged it. **1943** N. Coward *Middle East Diary* 23 Sept. (1944) 100 The car broke down..however we flagged a passing lorry..and whirled off. **1945** E. Bowen *Demon Lover* 93 Eric, do you think you could flag the *maître d'hôtel*? **1954** L. Klemantaski tr. *Fraichard's Le Mans Story* viii. 80 Faroux flags in Chinetti's 2 litre Ferrari. **1957** S. Moss *In Track of Speed* vi. 86 His pit attendants..flagged him in after the race had been in progress for some time. **1966** *Listener* 6 Jan. 23/1, I was driving along Holland Park Avenue..when I was flagged down by three women. **1970** 'H. Carmichael' *Remote Control* ii. 22 Mrs. Melville managed at last to flag a passing taxi.

flagellate, *a.* Add: **1. a.** (Earlier example.)
1867 *Mem. Boston Soc. Nat. Hist.* I. 306 The whole question..hinges upon the determination as to the animal or vegetable nature of the Monad-like, or so-called flagellate infusoria.
B. *sb.* A microscopic protozoan organism of the class Mastigophora (or Flagellata), characterized by the possession at some stage of its life of one or more flagella that are used for locomotion.
1879 *Q. Jrnl. Microsc. Sci.* XIX. 100 Sometimes the movements of the flagellum are energetic, and the organism then begins to move like a Flagellate, by help of its flagellum. **1897** H. M. Bernard in A. H. Miles *Concise Knowl. Nat. Hist.* 718 It is impossible to draw any hard and fast line between the lowest plant and animal cells, and..such simple Flagellates may be regarded as belonging to a border land. **1924** H. J. Van Cleave *Invertebr. Zool.* ii. 28 Relatively constant body form with usually one or two chromatophores characterize these small flagellates. **1949** C. A. Hoare *Handbk. Med. Protozool.* v. 111 In addition to the flagellates of the alimentary and genital tracts,..human beings may be infected with flagellates living in the blood and cells of the reticulo-endothelial system. **1963** H. Sandon *Ess. Protozool.* ii. 25 If we could trace the ancestry of all multicellular plants and animals back beyond the time of the earliest creatures capable of leaving fossil remains, most people agree that we should come to organisms which, if they lived today, would be included among the flagellates.

flagellated, *a.* (Earlier example.)
1874 *Monthly Microsc. Jrnl.* XI. 70 Free-swimming and flagellated monads.

flagellation. Add: **2.** *Biol.* **a.** The arrangement of flagella on an organism.
1893 J. Tuckey tr. *Hatschek's Amphioxus* 164 The flagellation of the body.
b. = *exflagellation.
1898 *Jrnl. Exper. Med.* III. 94 The process of flagellation presented by the elongate organism is remarkable. **1926** C. M. Wenyon *Protozool.* II. ii. 881 In the typical coccidia the male gametocyte produces male gametes after a relatively slow process of nuclear multiplication, while in the hæmosporidia the male gametes are formed by a violent process known as flagellation or exflagellation, which occurs in the stomach of the invertebrate.

flagellative, *a.* (Further example.)
1903 *Daily Chron.* 19 Feb. 3/5 All the officers actively concerned in inflicting the late illegal flagellative punishments..should be promptly dealt with.

flagellin (flădʒeˑlin). *Biochem.* [f. Flagell(um + -in¹.] A fibrous protein isolated from bacterial flagella.
1955 W. T. Astbury et al. in *Symp. Soc. Exper. Biol.* IX. 292 We propose, therefore, to call this new member of the k-m-e-f group 'flagellin'. **1969** S. T. Lyles *Biol. Microorganisms* viii. 191 Flagella are uniform throughout their length and are composed of fibrous protein called flagellin.

flagellist. (Further example.)
1926 H. Nicolson *Let.* 21 July (1966) 31 They were followed by half-naked flagellists.

flagellomania (flæːdʒĕlomēⁱˑniă). [f. Flagell(um + -o + -mania.] Enthusiasm for flogging. Hence **fla:gellomaˑniac** *sb.* and *a.*, (one who is) enthusiastically in favour of flogging.
1895 G. B. Shaw in *Daily Chron.* 24 Feb. 8/5 Flagellomania has been victorious by seven votes to five on the Industrial Schools Committee. **1899** —— in *Humanity* May 136/2 The male flagellomaniac—who is sometimes, unfortunately, a judge—craves intensely for the flogging of women. **1908** *Humanitarian* Sept. 66/2 We are constantly assured by the flagellomaniac section of the Press that crime is 'stamped out' by the 'cat'. **1917** G. B. Shaw in *New Republic* 6 Jan., Any newspaper can get up a flagellomaniac garotting scare. **1952** R. Hughes *Swinburne's Lesbia Brandon* 553 It is not mere flagellation in the void,..as is usually the case with flagellomaniacs.

flageolet². (Earlier and later examples.)
1877 E. S. Dallas *Kettner's Bk. Table* 243 Haricot beans.—These are the seeds: first, the flageolets, which are green and are to be had fresh from July to October, or dried always. **1951** E. David *French Country Cooking* 235 Pour the whole contents of a tin of *flageolets* into a pan.

flaggy, *a.*¹ **2.** (Later example.)
1831 W. Howitt *Bk. Seasons* iv. 94 The large flaggy nests of the water-hen.

flagstone. Add: **2. flagstone artist** = *pavement-artist* (see Pavement *sb.* 4).
1861 Mayhew *Lond. Labour* Extra vol. (1862) 436/2 They arrested these flag-stone artists with others. **1891** Kipling *Light that Failed* iv. 54 They believed I was a self-taught flagstone artist.

flail, *sb.* Add: **5. flail(-type) harvester,** a type of harvesting-machine for forage-grass (see quot.); **flail-joint** *Med.*, a joint showing grossly excessive mobility; **flail tank,** a type of tank used for clearing a mine-field.
1959 *Farmer & Stockbreeder* 12 May 70 (*heading*) With a flail harvester. *Ibid.* 70/1 The true flail-type harvester has a horizontal rotor to which is attached a number of free-swinging flails or cutters. [These flails] cut the grass by high speed impact. **1876** *Trans. Clin. Soc.* IX. 173 A flail joint, *i.e.* union by a fibrous bond, more or less long, between the bones of thigh and leg. **1967** A. R. Shands et al. *Handbk. Orthopaedic Surg.* x. 190 Useful procedures in the surgical treatment of poliomyelitis and other types of flaccid paralysis..restore stability to flail joints. **1944** *Hutchinson's Pict. Hist. War* 12 Apr.–12 Sept. 423 (*caption*) The enemy has sown mines and the flail tank in the distance is clearing them away.
B. *adj.* [f. the sb. used *attrib.* as in **flail-joint* (see above).] Of a part of the body, esp. a joint: exhibiting grossly excessive mobility as a result of the loss or absence of normal muscular control.
1876 [see **flail-joint* above]. **1919** R. C. Elmslie *After Treatment Wounds & Injuries* vi. 61 A flail but mobile hip joint can be supported by..a Thomas Caliper splint. *Ibid.* xiv. 206 A flail condition of the hip joint results from removal of the head of the femur. **1959** A. G. Apley *System of Orthopaedics & Fractures* viii. 78 Where the muscles controlling a joint are all equally weakened..the joint becomes flail. **1961** G. Perkins *Orthopaedics* xx. 300 A Charcot's elbow is also flail, but it is unlike the flail elbow caused by surgical excision. **1968** A. B. Ferguson *Orthopaedic Surg.* (ed. 3) vii. 605 The foot, if completely flail, must be supported on both sides, as well as anteriorly and posteriorly, by a double upright brace. **1968** S. Taylor et al. *Short Textbk. Surg.* (ed. 2) xvi. 202 When a number of ribs are doubly fractured and a segment of the chest wall cannot be used in respiration it is usually referred to as a flail chest.

flail, *v.* Add: **4.** *intr.* To move in the manner of a flail. Also *fig.*
1874 J. S. Blackie in A. M. Stoddart *J. S. Blackie* (1895) II. xvii. 99 Carlyle..is flailing about him in the same one-sided magnificently unreasonable way that you know. **1951** M. McLuhan *Mech. Bride* 122/2 She comes flailing along, head back, toes pointed.

flair¹. Add: **2.** Also in weakened senses: (*a*) special aptitude or ability; (*b*) liking, taste, enthusiasm. (Both senses freq. const. *for.*)
1925 S. Lewis *Martin Arrowsmith* ix. 106 You have a real flair for investigative science. **1926** Fowler *Mod. Eng. Usage* 182/1 Mrs. —— has homely accomplishments; a f[lair] for cooking goes with her f[lair] for writing. **1932** *McCall's Mag.* July 96 Her flair for adventure and conquest was rising at the thought of great, strange cities. **1942** *Burlington Mag.* June 134/1 One explanation offered by the psychologist is that this special insight or 'flair' is only found in the repressed or frustrated artist. **1955** *Times* 17 June 3/5 The 'flair' for producing unexpected lobs. **1959** T. S. Eliot *Elder Statesman* iii. 98 Michael's head is well screwed on. He's got brains, he's got flair.

flair (flēᵊɹ), *v.* [f. Flair¹.] *trans.* To scent. Also *absol.* and *fig.*
1921 *Glasgow Herald* 22 Jan. 4 With a strange and peculiar zest we flair survivals of phrase and intonation lurking in the speech..of very aged persons. **1923** J. Masefield *Poems* 560 The sharp nose flaired them heedfully. *Ibid.* 563 His ears were cocked and his keen nose flaired. *Ibid.* 565 The fox drew in and flaired with his muzzle. *Ibid.* 596 He flaired the air, then he padded out.

flaith (flā). *Irish Hist.* [Irish.] The chief of an ancient family or *fine.*
1861 E. O'Curry *Lect. MS. Materials Anc. Ir. Hist.* ix. 202 The hereditary proprietors, the *Flaiths* (pronounced 'flahs'), or landlords. **1876** *Encycl. Brit.* V. 799/2 An *aire* whose family held the same land for three generations was called a *flaith*. *Ibid.* 800/1 The rank of a *flaith* depended upon the number of his *ceiles*. **1898** J. Heron *Celtic Ch.* 15 The 'nobles' or 'flaiths' possessed not only cattle but deis.

flak (flæk). [G., f. the initials of the elements of *fliegerabwehrkanone* 'pilot-defence-gun'.] An anti-aircraft gun; also (the usual sense in Eng.), anti-aircraft fire. Also *attrib.* So **flak-happy** *a.*, mentally affected by flak (cf. **bomb-happy* adj., *-happy); **flak jacket,** a protective jacket of heavy fabric containing metal strips or plates; **flak ship,** a German anti-aircraft vessel.
1938 *Jane's Fighting Ships* 218/2 [On a German vessel] *A.A. guns.* 4·1 inch, 3·5 inch on H.A. mounts ('Flak'). **1940** *Times Weekly* 23 Apr. 2 Heavy and accurate anti-aircraft fire from the flak ships. **1940** *Times* 16 Sept. 2/3 The word 'Flak' is probably used in every Bomber Command pilot's report after a raid on Germany. **1940** *Flight* 19 Sept. 224/1 One of the British bombers which raided Berlin..had been damaged when diving through heavy *Flak* fire. **1941** D. Garnett *War in Air* x. 99 Every motorised column of troops..had its own 'Flak' detachment of anti-aircraft guns. **1941** *Times* 30 Sept. 4/7 Blenheim and Beaufort aircraft of Coastal Command flew through intense flak. **1943** in *Amer. Speech* (1944) XIX. 60 *Flak Happy.* **1943** *Hutchinson's Pict. Hist. War* 12 May–3 Aug. 50 One of Hitler's hush-hush flak-towers designed for high elevation defence against raiding aircraft. **1944** *Amer. Speech* XIX. 10 *Flak-happy*..was used to designate the mental confusion of an airman exposed to severe *flak*. **1956** W. A. Heflin *U.S.A.F. Dict.* 207/1 Flak curtain...Flak jacket. *Ibid.* 207/2 Flak suit...Flak vest. **1957** J. Braine *Room at Top* xxviii. 230 The bomb-aimer got a faceful of flak. **1961** J. Heller *Catch-22* (1962) xiv. 141 He left his flak suit and parachute there. **1968** *Observer* 4 Feb. 1/1 No man ventures out without a gun and a flak jacket. **1970** *Guardian* 10 July 11/1 Experienced reporters here [*sc.* in Belfast]..have been sending to London for flak jackets. **1971** *Daily Tel.* 5 Aug. 8/7 The burst of orgasms coming as incessantly as a barrage of flak leave one a little sex-shocked.

flake, *sb.*¹ Add: **6. flake-yard** (example).
1856 J. Reynolds *Peter Gott* (Bartlett), The owners of vessels have a flake-yard in the vicinity of the landing-places, to which the fish are carried on being landed.

flake, *sb.*² Add: **4. c.** *pl.* Short for *cornflakes* (*corn *sb.*¹ 11).
1951 *Good Housek. Home Encycl.* 371/2 Crisp flakes with cold milk and sugar.
11. c. *attrib.* and *Comb.* uses in *Archæol.* (see sense 4 in Dict.); **flake culture,** a prehistoric culture using flake implements.
1865 Flake-knife [see Dict. 11 b]. **1924** M. C. Burkitt *Our Forerunners* 81 If small flakes have been taken off round a flint core which is then split in half, the result is a flat under-surface..on the other side of which are flake scars. **1926, 1935** [see *core *sb.*¹ 5]. **1927** Peake & Fleure *Hunters & Artists* iv. 42 Flake implements, or those formed by working up the edges of the flakes struck from a core, only came into gradual use in Acheulian times, and even then were not common. **1928** D. A. E. Garrod in *Proc. Prehist. Soc. E. Anglia* V. iii. 266 From what I have called the 'blade-culture'-group we turn to the great cycle of 'flake-industries'. **1935, 1943** [see *blade *sb.* 6 b]. **1937** [see *biface]. **1937** Garrod & Bate *Stone Age Mt. Carmel* I. i. iii. 32 Flake-scrapers..are flakes with scraper retouch round some part of the edge. The majority are rough and shapeless. **1947** J. & C. Hawkes *Prehist. Brit.* (ed. 2) i. 10 White flake cultures are predominantly Eastern, extending right across Asia, the core cultures have an African bias. **1957** Childe *Dawn Europ. Civilization* (ed. 6) i. 11 Flake-axes..mounted as adze-blades in perforated antler sleeves. **1957** L. MacNeice *Visitations* 29 Flake-tool; core-tool. **1959** *Antiquity* XXXIII. 17 A flake-blade industry of Neolithic type.

flake (flēⁱk), *sb.*⁷ [Perh. f. Flake *sb.*²] A name under which dogfish is marketed for food.
1906 *Daily Chron.* 1 Mar. 5/5 A meeting of the Sea Fisheries Committee..had approved of the change of the name from dogfish to flake, and after the dinner the company unanimously affirmed that flake was a most excellent..form of food. **1932** *Times Lit. Suppl.* 7 Apr. 242/4 There is also long-line fishing for dog-fish (renamed 'flake' for marketing). **1959** *Chambers's Encycl.* XII. 458/2 Spotted Dog-fish and Spiny Dog-fish..are sold as 'flake' or 'rock salmon'.

flake, *v.*¹ **3. b.** Further examples in *Archæol.* (see quot. 1879 in Dict.).
1954 S. Piggott *Neolithic Cultures* x. 288 The axes were flaked or chipped on the spot in large quantities. **1955** *Sci. Amer.* May 110/2 Large pebbles were flaked to give a cutting edge. Flaking is a kind of chipping or peeling, analogous to the whittling of wood.

flake, *v.*² Restrict †*Obs.* to senses in Dict. and add: **2.** *to flake (out)*: to faint, fall asleep (from exhaustion, drunkenness, etc.). So *flaked (out)* ppl. adj., exhausted; unconscious, asleep. *colloq.*
1942 L. Kennedy *Sub-Lieutenant* vi. 39 During the week's [P.T.] course, two of them broke their ankles; the others usually flaked out from exhaustion before the end of the afternoon. **1943** 'H. Green' *Caught* 21 There is a man flaked out at your feet. **1953** F. Robb *Sea Hunters* xiii. 200 'Olley, where's old Drum?' 'Flaked out.' **1958** H. Johnston *Phantom Limb* viii. 71 'Can we go to bed soon?' she asked. 'I'm absolutely flaked out.' **1960** B. Crump *Good Keen Man* 139, I flaked out more thoroughly than a man who is blind drunk. **1961** S. Price *Just for Record* vii. 60 When it was over I was flaked. **1971** *Guardian Weekly* 17 July 14/3 The nuns are still there, looking a bit flaked out now but bearing up.

flaking (flēⁱˑkiŋ), *vbl. sb.* [f. Flake *v.*¹ + -ing¹.] The action of the vb.; the condition of being flaked or flecked; *spec.* an adventitious appearance of light flecks on animals. Also *attrib.*
1829 G. Griffin *Collegians* xxxix. 184 I'm only sorry I didn't give him a thing more..an' that was a good flaking. **1841** [see Flake *v.*¹ 6]. **1879** [see Flake *v.*¹ 3]. **1902** W. Bateson et al. *Rep. Evol. Comm. Roy. Soc.* i. 47 Flower-colour of First Cross-bred Generation...In connection with the question of colour one point of interest may be briefly mentioned here, viz., the occurrence of 'flaking'...Such flaking was recorded in thirty-one out of thirty-nine combinations in the first year. **1904** Goodchild & Tweney *Technol. & Sci. Dict.* 224/1 *Flaking*, a defect in distemper in which the paint flakes off, owing to imperfect adhesion with the surface to which it is applied. **1920** *Chambers's Jrnl.* 591/2 The automatic carrier once more picks them up, to transfer them to the flaking-machine. **1921** *Ibid.* 173/2 The entire flint armoury of Philip Bentley, along with his fire-drills and flaking-tools. **1954** S. Piggott *Neolithic Cultures* x. 293 A small flaking-site at Stake Pass near the Langdale Pikes in Westmorland.

flaky, *a.* Add: **2. b.** *spec.* Of pastry: consisting when baked of thin delicate flakes or layers. So **flakiness.**
1837 [see Dict., sense 2]. **1853** Dickens in *Househ. Words* (Christmas No.) 1/1 Look at the pie-crust alone. There's no flakiness in it. It's solid—like damp lead. **1857** —— *Dorrit* II. xxxiv. 619 A pie as far from flaky as the present. **1865** *Harper's Mag.* Apr. 610/1 Crisp 'short-cakes' or 'flaky' pie-crust. **1904** *Daily Chron.* 14 Sept. 8/4 A pie that has not made up its mind whether it is to be short or flaky is not worth eating. **1943** A. Simon *Conc. Encycl. Gastron.* IV. 92/1 Indian flaky pastry. **1969** *Harper's Bazaar* Oct. 12/2 Quails wrapped in ham and cooked in flaky pastry.

flam, *sb.*² Delete def. and substitute: (See quot. 1819.) Add later example.
1931 G. Jacob *Orchestral Technique* vii. 71 Very characteristic of the side-drum are the strokes known as 'the flam' and the 'drag'. Flam.

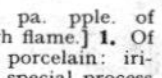

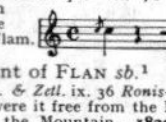

flam (flæm), *sb.*⁴ Variant of Flan *sb.*¹
1711 R. Sibbald *Descr. Ork. & Zetl.* ix. 36 Ronis-voe..could Harbour many Ships, were it free from the Flams of Wind, which come from the Mountain. **1820** [see Flan *sb.*¹]. **1903** *Northern Ensign* 28 July 2/1 'Er's a flam o' win' doon 'e shimley.

‖ **flambé** (flaⁿbe), *a.* [Fr., pa. pple. of *flamber* to singe, pass through flame.] **1.** Of a certain type of Chinese porcelain: iridescent from the effects of a special process of firing, or from the irregular application of glaze; of a (usu. reddish or bluish) glaze so applied. Also as *sb.*, a piece of porcelain decorated in this way.
1886 S. W. Bushell *Chinese Porcelain* 4 Prince Kung one day admired a glazed Buddha from the ruins of the Summer Palace, taking it for old flambée porcelain. **1888** *Harper's Mag.* Oct. 658 The comparison of these *flambé* vases with onyx or precious stones is to the advantage of the brilliant porcelain. **1904** *Daily Chron.* 31 May 3/1 A dozen specimens of 'flambé', which exhibit the splashed reds, browns, and purples, due to a special process of firing. **1904** E. Dillon *Porcelain* 42 In the case of the *flambé* or 'transmutation' glazes, the strange caprices of colour have their origin, in part at least, in the contrast of the red sub-oxide and the green silicate of copper. **1959** G. Savage *Antique Coll. Handbk.* 80 The reign of Ch'ien Lung saw the manufacture of *flambé* glazes in great variety.
2. *Cookery.* Applied to a dish covered with spirit and served alight.
[**1906** Mrs. Beeton *Househ. Managem.* lxii. 1659 *Flamber* (Fr.). To singe poultry or game. To cover a pudding or omelet with spirit and set it alight.] **1914** N. Newnham-Davis *Gourmet's Guide to London* lvi. 363, I enjoyed the *sole Monico*,..a woodcock *flambé* and a salad. **1935** M. Morphy *Recipes of all Nations* 87 Crêpes Suzettes..are usually served *flambées*—hot Curaçoa being poured over them and set alight just before serving. **1958** *Times* 16 June 11/5 Americans..have made more use of spectacular flambé dishes in their homes.

flambeau. 2. (Later examples.)
1936 R. Campbell *Mithraic Emblems* 25 Silent and vertical and dim The lunar flambeau of a prayer. **1939** A. E. Housman *Coll. Poems* 107 The chestnut casts his flambeaux.

flamboyance. (Further examples.)
1909 *Daily Chron.* 2 July 4/7 Its architectural flamboyance. **1940** W. S. Maugham *Books & You* i. 26 The flamboyance of Elizabethan English.

flamdoodle (flæmdūˑd'l). *U.S.* [An arbitrary formation.] = Flapdoodle *sb.* 2.
1888 *N.Y. Sun* (Farmer), We..planted [=buried] Uncle George in ship-shape and proper manner. We wasn't goin' to have any highfalutin' flamdoodle business over him. **1921** H. Crane *Let.* 28 Jan. (1965) 52 M. Ray will allow the Dada theories and other flamdoodle of this section run him off the track.

flame, *sb.* Add: **3. b.** (Example in *sing.*)
1959 I. & P. Opie *Lore & Lang. Schoolch.* ix. 170 Red heads attract a barrage of nicknames:..fire head, flame, flarey, [etc.].
5. c. The colour of flame, flame-red.
1711 tr. *H. van Oosten's Dutch Gardener* (ed. 2) iii. xiv. 151 The Fire..in this Plant [*sc.* tulip], leaves only its own Colour, which is Flame or Gold. **1921** *Queen* 13 Aug. 198 The buds are of extraordinarily deep colour with a suggestion of flame. **1923** *Daily Mail* 16 July 16 In Peach, Brown, Mastic, Royal, Flame, Gold. **1971** *Rose Ann.* 192 Mojave. 1954. Deep orange and reddish flame. *Ibid.* 197 Shepherd's Delight. 1958. Flame, orange and yellow.
9. a. *flame-light* (later examples).
1921 W. de la Mare *Crossings* 47 A fire-place, its brass and steel merrily twinkling in the flame-light. **1923** D. H. Lawrence *Birds, Beasts & Flowers* 170 Three great elephants..in the torch-light, Slowly sailing in gorgeous apparel through the flame-light.
b. *flame-throwing.*
1952 C. Day Lewis tr. *Virgil's Aeneid* vi. 125 The flame-throwing Chimaera.
c. *flame-burnt, -flushed.*
1841 *Civil Engin. & Arch. Jrnl.* IV. 340/2 When the bricks were intended to be..flame-burnt,..no addition was made to the clay. **1917** D. H. Lawrence *Look!* 101 We near the flame-burnt porches. **1923** —— *Birds, Beasts & Flowers* 66 Flame-flushed, enraged, splendid salvia. **1937** W. de la Mare *This Year, Next Year*, Its drowsy eye Fixed on the flame-flushed company.
10. flame float, a buoyant pyrotechnic device used for illumination or as a signal, etc.; **flame gun,** a flame-throwing gun, used to destroy weeds, etc.; **flame machine, projector** = **flame-thrower* (a); **flame manometer** (see Manometer); **flame-plating** *Engineering* (see quots.); so (as a back-formation) **flame-plate** *v.*; **flame-proof** *v. trans.*, to render flame-proof (Webster 1934); so **flame-proofing** *vbl. sb.* and *ppl. a.*; *flame-proofed* ppl. adj.; **flame-red** *a.* and *sb.*, (of) a vivid orange-red shade; **flame-resistant, -retardant** *adjs.*, not readily inflammable; so *flame-resistance*; **flame spectrum,** a spectrum of the light produced when a substance is vaporized in an otherwise non-luminous flame; **flame-thrower** [cf. G. *flammenwerfer*], (*a*) a weapon consisting essentially of a reservoir from which a long spray of flame can be ejected against the enemy; (*b*) = **flame gun*; **flame-trap** (see quot. 1949¹); **flame-ware,** a type of cooking equipment, often of glass, that can withstand the heat of an open flame.
1942 T. Rattigan *Flare Path* ii. ii. 135 It got in the way of a flame float I was throwing out. **1943** 'T. Dudley-Gordon' *Coastal Command* xvii. 167 They hurriedly signalled her [*sc.* the ship] to the spot, dropping flame floats all round the U-boat. **1946** *Jrnl. R. Aeronaut. Soc.* L. 49/2 At night, flame floats were dropped to observe wind. **1931** *Times Lit. Suppl.* 1 Oct. 743/3 Flame guns which had at first been..used against them [*sc.* locusts]. **1948** *Times* 20 Mar. 6/7 Even a flame-gun has been devised for their [weeds'] destruction. **1952** E. R. Janes *Flower Garden* 51 Seedlings..looking as if they had been scorched by a flame gun. **1917** *Times* 19 Feb. 7/2 The Germans have used flame machines for the first time in the Balkans. **1954** *Steel* 5 Apr. 130/3 Productive life of steel core rods has been increased..by having them flame-plated. **1959** *Engineering* 9 Jan. 59/1 A leading American cable company has saved over $1,500 per year..by having its aluminium alloy pulleys flame-plated to increase their wear resistance. **1954** *Steel* 5 Apr. 130/3 Flame-plating is a method of applying tungsten carbide to metal parts...Low temperature deposition is a major advantage of this new process. **1956** *Jrnl. Iron & Steel Inst.* CLXXXIV. 100/1 Flame-plating is a process whereby layers of tungsten carbide/cobalt..may be applied, by powder deposition, to a wide variety of base metals. **1959** *Engineering* 9 Jan. 59/1 The flame-plating process is a surfacing operation in which particles of tungsten carbide are blasted on to the surface of the workpiece. **1915** *War Illustr.* 4 Sept. 70 German 'Flammenwerfer' (flame-projector) in action. **1955** *Sci. News Let.* 27 Aug. 143/3 Chemicals..used..for flame-proofing cotton. **1962** *Listener* 25 Jan. 198/2 Most electric blankets are now made of flame-proofed fabric. **1929** *Chem. Abstr.* 4353 (*heading*) Waterproofing and flameproofing composition for use on textile fabrics. **1938** *Encycl. Brit. Bk. of Yr.* 332/1 A new flame-proofing material for textile fabrics and paper was announced. **1963** A. J. Hall *Textile Sci.* V. 257 The flameproofing of textile materials has in recent years aroused considerable attention. **1382** Wyclif *Lev.* xiv. 4 Flawm reed silk. **1906** L. Claremont *Gem-Cutter's Craft* 77 The beautiful 'flame red' variety which displays the extraordinary effect of a burning coal. **1956** G. Durrell *My Family* ii. 28 Roses dropped petals that seemed as big and smooth as saucers, flame-red,..glossy and unwrinkled. **1959** *B.S.I. News* May 6/1 The durability of flame-resistance of such treated fabrics can be seriously affected by washing. **1947** *Amer. Dyestuff Reporter* XXXVI. 135/2 A flame-resistant or flame-retardant fabric is one which 'exhibits appreciable resistance to afterflaming'. **1959** *Listener* 16 Apr. 695/2 A textile that is claimed to be flame resistant. **1961** *B.S.I. News* Nov. 9/2 The law makes it an offence to sell fabrics and garments as flame-proof or flame-resistant unless they comply with these standards. **1947** Flame-retardant [see **flame-resistant*]. **1966** *Listener* 15 Dec. 893/2 Thanks to the flame-retardant finish, which public health authorities insist on, the paper dress will char but not flare up when a lighted match is held to it. **1862** *Chem. News* 26 Apr. 234/1 The broad band in the flame-spectrum of calcium named Ca β, is replaced in the spectrum of the intense calcium-spark by five green lines. **1907** *Proc. R. Soc.* A. LXXIX. 243 The spectra of the glow from metallic vapours have been photographed repeatedly in the oxy-hydrogen flame spectra of all the alkali metal salts. **1962** R. E. Dodd *Chem. Spectroscopy* i. 12 Flame spectra are..either line spectra of neutral atoms or bands of very closely spaced lines arising from molecules. **1917** P. Gibbs *Battles of Somme* 178 There were eight of these flame-throwers brought against the Sussex lads. **1944** *Living off Land* vii. 154 To start a backfire rapidly under difficult conditions, flame-throwers are used. **1945** *Reader's Digest* Oct. 84/1 Flame throwers kill the weeds. **1956** A. H. Compton *Atomic Quest* 303 Those burned by flame throwers. **1932** *Jrnl. R. Aeronaut. Soc.* XXXVI. 854 It seemed..that flame traps were one of the things they could leave off for the moment in their struggle against the weight of engines. **1949** *Gloss. Aeronaut. Terms* (*B.S.I.*) ii. 11 *Flame trap* [*Induction flame damper*], a device fitted in the induction system to prevent the passage of flame in the event of a 'backfire' or 'blow back'. **1949** *Hansard Commons* 15 Mar. 1932 Can he tell the House whether the problem of flame traps for jet aircraft at night has been overcome? **1938** *Trade Marks Jrnl.* 23 Nov. 1420 Flameware. Made in JP Pyrex Brand JP England. Registration of this Trade Mark shall give no right to the exclusive use, either separately or in combination, of the letters 'J.P.' and the word 'Flameware'. **1941** C. J. Phillips *Glass* xii. 281 Pyrex-brand Flameware is made from a special glass of relatively low expansion which has been treated to give it added mechanical strength and resistance to thermal shock. **1967** *House & Garden* Jan. 8/3 A complete range of oven- and flameware in multicolour..design. *Ibid.* Mar. 79/1 There is a steady demand for good oven-ware and flameware.

flame, *v.* Add: **1. d.** *to flame out*: spec. of a jet engine, to cease operation through extinction of the flame in the combustion chamber. So *flame-out* sb.
1950 *Nat. Geogr. Mag.* Sept. 307/2 If that boy..had blasted his throttle forward fast, he would have got what we call a 'flameout'. **1951** *Britannica Bk. of Yr.* 686/1 *Flame-out*, of a jet plane, to exhaust its supply of fuel. **1957** *Wall St. Jrnl.* 15 July 7/4 Curtiss-Wright Corp. announced it has successfully conducted tests with fuels which ignite on contact with air and which can re-light jet engines which have 'flamed-out' at high altitude. **1957** *Aeroplane* 1 Feb. 149/1 When the aircraft was about 90 minutes out of Salisbury, all four engines began 'bumping' and four flame-outs occurred in the space of five minutes. **1960** *Guardian* 30 June 10/4 The nature of the fuel gave the engine a strong tendency to 'flame out'. **1965** *New Scientist* 27 May 577/2 Some of the combustion troubles associated with flame velocities which cause flame-outs in jet engines, especially at great heights.
5. Later example of *fig.* use.
1942 T. S. Eliot *Little Gidding* i. 7 The brief sun flames the ice, on pond and ditches.

flamenco (flămeˑŋko). [Sp., = Flamingo.] A Spanish gipsy style of singing or dancing; a song or dance in this style.
1896 *Daily News* 8 June 4/2 A few steps from the flamenco, or wild Spanish dance. **1923** *Blackw. Mag.* July 28/1 Gramophone records of flamenco-singing. *Ibid.* Oct. 503/2 Spanish flamenco music. **1925** *Ibid.* Jan. 80/2 The nasal eastern flamenco which gives to Spanish singing its most marked character. **1950** *Ballet Ann.* IV. 32 There is no clear distinction in people's minds between *flamenco* dancing and the classical or 'school' dancing of Spain. **1952** B. Ulanov *Hist. Jazz in Amer.* ii. 11 The intensity..resembles..flamenco dances in Spain. **1959** F. Newton *Jazz Scene* i. 14 There is a great difference between the first flamenco songs of the 1860s and the flamenco of today.

flaming, *vbl. sb.* (Additional *concr.* example.)
1703 tr. *H. van Oosten's Dutch Gardener* iii. ii. 153 Concerning the Striping or Flaming, it must be low, beginning from the bottom of the Flower.

flaming, *ppl. a.* Add: **1. c.** *flaming onions*, an anti-aircraft projectile consisting of about ten balls of fire shot upwards in succession, so called from its resemblance to a string of onions (see also quot. 1943). *Services' slang.*
1918 V. Drake *Above Battle* vi. 118 These 'flaming onions' were huge rockets of balls of phosphorus which burst at the top of their climb and spread out into a great umbrella shaped curtain of flame, and sank slowly down, the idea being that any machine in its path would be instantly set fire to by the burning phosphorus. **1925** Fraser & Gibbons *Soldier & Sailor Words*, *Flaming onions*, a colloquial name for a German anti-aircraft projectile. **1928** C. F. S. Gamble *N. Sea Air Station* xxii. 384 One of the flying-boats..came into a heavy barrage of 'flaming onions'. **1943** Hunt & Pringle *Service Slang* 32 *Flaming onions*, tracer fire from the ground.
2. Used euphemistically for a profane epithet. Also as *adv.*
1895 'G. Mortimer' *Like Stars that Fall* xv, Yes, by God, I'll get flaming drunk. **1922** D. H. Lawrence *England, my England* 230 I've never been patient to no flaming doctor, and hope I never shall be. **1936** *Punch* 10 June 646/1 Some flaming person has gone and stolen my flaming bicycle-pump, flame it all! **1944** *Coast to Coast 1943* 188 It was lovely work Patsy, but, stone the flamin' crows! you can't do that at your age. **1946** *Coast to Coast 1945* 163 Brand new line, and you're too flaming lazy to get the lot! **1960** *Analog Science Fact/Fiction* Oct. 174/1 Miss Fulton says..'Not bad, but Chandler has done better...' Too flaming right he has. **1966** 'J. Hackston' *Father clears Out* 61 To hell with the flaming place. **1969** *Private Eye* 9 May 14 He's saved my life if he only flamin' knew it.
4. b. *flaming youth* [Shakes. *Ham.* iii. iv. 84]: a collective term for young persons, alluding to their characteristically vigorous and unrestrained behaviour or ways. Chiefly *U.S.*
1923 'W. Fabian' (*title*) Flaming youth. **1937** *Time* 6 Sept. 40/2 The traditional belief that flaming youth is guilty of most drunken driving accidents was last week scouted by the results of a six-month survey. **1960** J. Mitford *Hons & Rebels* viii. 58 The Bright Young People had faded from the London scene..as their

transatlantic counterparts, the Flaming Youth, had no doubt disappeared from the American scene with the passing of the 'twenties. **1962** *Amer. Speech* XXXVII. 27 The hectic era of Prohibition, bootleggers, flappers, and flaming youth. *Ibid.* 31 The age of flaming youth.

‖ **Flamingant** (flamiŋgaⁿ), *sb.* and *a.* [Fr., = Flemish-speaking, f. **flameng*, fem. *flamenge*, ad. Du. *Vlaming* Fleming¹.] **a.** *sb.* An advocate of the recognition of Flemish as an official language of Belgium, or of the exclusive use of Flemish in certain parts of Belgium. **b.** *attrib.* or as *adj.* Of or pertaining to such advocates or their policies. Hence **Flaˑmingantism,** the policy of furthering the use of Flemish.
1920 *Contemp. Rev.* Jan. 86 While the chauvinists were badly beaten, the Parliamentary power of both Socialism and Flamingantism was increased tremendously. **1921** *Ibid.* Nov. 700 The aim of the Flamingant programme has been to obtain equal rights, both in theory and practice for the Flemish language. **1922** *Ibid.* Feb. 244 They [*sc.* the Walloons] consider the Flamingants to be unpatriotic and 'pro-German'. *Ibid.* Dec. 786 The result is that the Flamingant officials are rapidly promoted. **1964** P. Geyl *Hist. Low Countries* ix. 199 The Flamingants (as the conscious and active Flemings were called) indulged freely in romanticism. *Ibid.* 200 In Antwerp..the town hall was brought under Flamingant sway. *Ibid.* 201 Militant Flamingantism had been able to come to power in some municipalities.

flamingo. Add: **1. b.** The deep pink colour of the flamingo.
1897 *Westm. Gaz.* 5 June 7/1 The poor little flamingo-caped lassies. **1923** *Daily Mail* 29 Jan. 1/4 In shades of Powder Blue,..Cyclamen,..Flamingo, Pink. *Ibid.* 12 June 15 Favourite Colour of the Season is Flamingo. **1927** V. Woolf *To Lighthouse* i. 41 Lovely evenings, with all their flamingo clouds and blue and silver.

flammability. Delete †*Obs.* and add recent examples. Revived in modern use to avoid the possible ambiguity of *inflammability*, in which the prefix *in-* might be taken for a negative (In- *pref.*³). So also **flammable** *a.*, now freq. in commercial use.
1942 *Jrnl. R. Aeronaut. Soc.* XLVI. 471 That intense combustion in Diesel engines probably occurs only in regions in which local concentration of fuel at least equals a certain minimum value analogous to the lower limit of flammability. **1959** *Gloss. Packaging Terms* (*B.S.I.*) 10 In order to avoid any possible ambiguity, it is the Institution's policy to encourage the use of the terms 'flammable' and 'non-flammable' rather than 'inflammable' and 'non-inflammable'. **1959** *Scotsman* 1 May 7/6 Scientists at the Government's research establishment at Boreham Wood are carrying out flammability tests on some of the materials used in Glasgow tramcars. **1961** *Times* 1 Mar. 21/7 The flammability of clothing fabrics has caused much concern. **1963** *B.S.I. News* July 10/1 *Inflammability* is unquestionably capable of being misunderstood...The fire research interests on the..committee were particularly anxious that we should standardize on *flammability* in order to prevent such misunderstanding. **1970** *Which?* May 136/2 The plastic linings provided for the kitchen areas..will do little to stop your tent catching fire, because they also are highly flammable.

‖ **flammenwerfer** (flamənvɛˑrfər). [G., f. *flamme* Flame *sb.* + *werfer* thrower, mortar, f. *werfen* to throw.] = **flame-thrower* (a).
1915 *War Illustr.* 4 Sept. 70 German 'Flammenwerfer' (flame-projector) in action. **1917** P. Gibbs *Battles of Somme* 178 It was against the Sussex men that the Germans used their 'flammenwerfer' or flame-jets. **1957** G. Barker *Coll. Poems 1930–55* 130 The flammenwerfer and the fish And also I acknowledge him creator.

flammulated (flæˑmiulēⁱtĕd), *a.* [f. mod.L. *flammulatus*, f. L. *flammula*, dim. of *flamma* Flame.] Of a reddish colour, ruddy. So **flammulation** (flæmiulēⁱˑʃən), a small flame-like marking.
1860 *Ibis* II. 41 The nearest ally of the present species [*sc. Malacoptila veræ pacis*] is Lafresnaye's *Malacoptila panamensis*, from which it may be distinguished by the absence of any flammulations below. **1872** E. Coues *N. Amer. Birds* 203 Flammulated owl. Above grayish-brown, obscurely streaked with black.

flan (flæn), *sb.*⁵ [Fr.: see Flawn.] An open tart containing fruit or other filling.
1846 A. Soyer *Gastronomic Regenerator* 502 A Flan of Puff Paste...Have a plain round or oval flan mould. **1858** Thackeray *Virginians* ix, The sweets and *flans*, Madam Esmond prepared herself. **1906** Mrs. Beeton *Househ. Managem.* 897 A 6-inch diameter flan or paste-ring. *Ibid.* 899 There are two ways of making a flan without the aid of a ring. **1951** *Good Housek. Home Encycl.* 344/1 Line the flan ring..with the pastry.

flanch, *v.* Add: With *up*: to slope inwards towards the top; applied especially to the outsides of chimney-shafts.
1833 Loudon *Encycl. Archit.* § 234 Each flue to have a Roman cement chimney shaft..flanched up (sloped in a way to throw off wet).

flanching, *vbl. sb.* [See Flanch *v.*] **a.** The action or state of spreading outwards (see Dict.). **b.** The sloping fillet of cement or mortar in which the base of a chimney-pot is bedded. Also called *flanched work.*
1833 Loudon *Encycl. Archit.* § 234 Sections of the flanchings. **1904** Goodchild & Tweney *Technol. & Sci. Dict.* 226/2 *Flaunching* or *flaunc[h]ed work*, the cement fillet round the bed of a chimney pot.

Flanders. Add: **2.** *Flanders field, make, mud.* **b. Flanders baby, doll,** a small doll manufactured in the Low Countries to display fashionable dress, or for use in a puppet show; **Flanders brick** (earlier example); **Flanders poppy,** a poppy of Flanders, the emblem of the Allied soldiers who fell in the war of 1914–18; also, one of the artificial poppies worn in Britain on Remembrance Sunday, in November.
1823 J. Galt *Entail* I. xix. 156 Yon Flanders baby is no for a poor man's wife. **1899** A. M. Earle *Child Life in Colonial Days* xviii. 365 'Flanders babies' had a cherished old age. **1969** E. H. Pinto *Treen* 206 Leslie Daiken considered that Dutch dolls, known in 17th- and 18th-century England as 'Flanders babies', and in America as 'peg dolls', really originated in the Thuringian Forest in Germany. *a* **1700** Evelyn *Diary* an. 1670 22 July (1955) III. 555 Here my Lord & his Partner had built two or 3 roomes with flanders white brick, very hard. **1801** *Monthly Mirror* Aug. 139 Her stage appearance..might be mistaken for a Flander's doll, moved by wires. **1915** J. McCrae in *Punch* 8 Dec. 468/3 In Flanders fields the poppies blow Between the crosses, row on row. **1799** Malthus *Jrnl.* 29 May (1966) 36 The country girls have a little of the Flanders make in their persons. **1918** Kipling in *Metropolitan* Dec. 46/2 One man lurched in—helmet, Flanders mud, accoutrements and all. **1933** A. G. MacDonell *England, their England* i. 13 An eleven-inch or eight-inch howitzer, both fortunately rare in Flanders mud. **1921** *Times* 21 Oct. 13/6 Australia, Canada, France, and the United States, as well as Newfoundland, have adopted the Flanders poppy as the national remembrance flower. *Ibid.*, All will be able to buy a Flanders poppy. *Ibid.* 29 Oct. 11/5 The King..has expressed his desire to include Flanders poppies in his wreath to be placed on the Cenotaph on that day. **1971** *Guardian* 21 May 2/8 Mr Heath..laid a wreath on the tomb of the unknown soldier beneath the Arc de Triomphe..a simple circle of Flanders poppies.

Flandrian (flæˑndriăn), *a.* and *sb.* [f. Flanders + -ian.] **A.** *adj.* **a.** Of or pertaining to Flanders or its inhabitants. *rare.* **b.** *Geol.* Of, pertaining to, or designating the period since the retreat of the ice sheet and the rise in sea-level at the end of the last glaciation in north-western Europe. **B.** *sb.* **a.** An inhabitant of Flanders. *rare.* **b.** *Geol.* The Flandrian period.
1637 J. Shirley *Lady of Pleasure* ii. D2. 62 *Celestina.* But some one noble bloud or lusty kindred, Claps in, with his gilt coach, and Flandrian trotters, And hurries her away to be a Countesse. **1670** in J. W. Draper *Cent. Broadside Elegies* (1928) No. 52, Proud France! no more thy Flandrian Conquests borst [*sic*]. **1886** J. H. Blunt *Dict. Sects* 163/2 *Flandrians*, Flemings. **1934** R. A. Daly *Changing World of Ice Age* v. 169 Table XXII..includes the Flandrian stage, named by him [*sc.* Dubois] to represent the time and processes involved by the melting of the last great Pleistocene ice-caps. **1937** Wooldridge & Morgan *Physical Basis Geogr.* xxiii. 421 The evidence of emergence during the last glaciation, followed by the great submergence, which has been called by Dubois the 'Flandrian transgression'. **1952** *Proc. Prehist. Soc.* XVIII. 109 It would be particularly useful to associate the three stages of the Flandrian transgression with mesolithic or neolithic industries. **1958** F. E. Zeuner *Dating Past* (ed. 4) 100 Interglacial beach deposits..to the east of Calais, round the corner where the Flandrian plains begin. **1968** R. G. West *Pleistocene Geol. & Biol.* xii. 281 Many raised bogs show the development of *Sphagnum* peats during the middle Flandrian, above earlier lake sediments. **1971** *Nature* 1 Jan. 43/2 The survival of the Teesdale rarities through the Flandrian period.

flâneur. Add: (Earlier, later, and *transf.* examples.) Hence **flane, flâne, flané, flâné** *v. intr.*, to saunter, to laze.
1854 *Harper's Mag.* Aug. 411/2 Did you ever fail to waste at least two hours of every sunshiny day, in the long-ago time when you played the *flaneur*, in the metropolitan city, with looking at shop-windows? **1876** L. Troubridge *Life amongst Troubridges* (1966) xi. 143 Shopped the whole morning—*flanéd* down Regent Street. **1894** G. du Maurier *Trilby* III. viii. 155 They are going to laze and flane about the boulevards. **1896** G. B. Shaw *Our Theatres in Nineties* (1932) II. 217 The boundary which separates the clever *flâneur* from the dramatist. **1897** G. du Maurier *Martian* iv. 175 To his great surprise he saw Bonzig leisurely flâning about. **1938** H. G. Wells *Apropos of Dolores* i. 13 In Paris, in London I have been a happy flâneur; I have flâné-d in New York and Washington and most of the great cities of Europe. **1954** I. Murdoch *Under Net* xv. 203 The fishermen were fishing, and the *flâneurs* were flaning. **1969** *Computers & Humanities* IV. 29 The electronic age may yet see every man a *flaneur.*

flange, *sb.* Add: **3. c.** *transf.* Of natural objects: a rim or fan which stands out from the main part of the object. Also *fig.*
1880 'Mark Twain' *Tramp Abroad* 393 Stepping on an outlying flange of her foot. **1897** M. Kingsley *Trav. W. Afr.* 607 The brown water..striking a ridge of higher rock..flew up in a lovely flange some twelve feet or so high. **1930** Auden *Poems* 25 Nowise withdrawn by doubting flinch Nor joined to any by belief's firm flange.
4. *flange-hilted* adj.
1930 V. G. Childe *Bronze Age* iii. 86 Early versions of the flange-hilted leaf-shaped sword..are very common in Northern and Central Europe.

flangeless (flæˑndʒlĕs), *a.* [f. Flange *sb.* + -less.] Having no flange.
1903 W. M. Camp *Notes Track Constr.* v. 247 With all but four of the drivers flangeless. **1927** T. Woodhouse *Artificial Silk* 116 This machine is intended to beam warps on flangeless beams.

flank, *sb.*¹ Add: **1. b.** (Earlier example.)
1747 Mrs. Glasse *Art of Cookery* xxi. 160 A Bullock...The Hind Quarter..the Thin and Thick-flank.
9. *flank-guard*; **flank angle** *Mech.* (see quot. 1954); **flank forward** *Rugby Football* (orig. *S. Afr.*), a wing forward; so *flank*, used *absol.*
1951 *Engineering* 21 Sept. 369/3 A screw thread has..seven elements of shape and size: major diameter, minor diameter, effective diameter, pitch, flank angles, [etc.]. **1954** *Defs. for Use in Mech. Engin.* (*B.S.I.*) 17 *Flank angles*, the angles between the individual flanks and the perpendicular to the axis of the thread measured in an axial plane section. **1937** *Rand Daily Mail* 10 Apr. 18/2 Of the flank or back rankers Strachan cannot be left out. *Ibid.* 25 June 22/1 Van der Berg will probably be one of the flank forwards. **1956** V. Jenkins *Lions Rampant* i. 19 Scotland's Greenwood, at flank-forward, saw to it that his country's honour was not besmirched. **1960** E. S. & W. J. Higham *High Speed Rugby* p. xx, We have adopted the name 'flank' instead of the more common 'wing forward' as being less likely to cause confusion between wing forward and wing three-quarter. *Ibid.* xii. 154 The Flanks (wing forwards): These are the open play specialists. **1901** 'Linesman' *Words by Eyewitness* (1902) 305 Here they are spotted by the self-constituted British flank-guard. **1918** E. S. Farrow *Dict. Mil. Terms* 234 *Flank guard*, a detachment detailed to cover the flank of a column marching past, or across the front of an enemy. **1930** *Nation* 6 Dec. 327/2 To Hutchinson..the *Shelleyans* are indebted for the fullest one-volume edition, and the *Elians* for the best cheap substitute for, or flankguard to, Mr. Lucas's *Works of Charles and Mary Lamb.*

flanker, *sb.*¹ Add: **3. c.** A trick, a swindle. *slang* (orig. *Mil.*).
1923 Manchon *Le Slang* 124 Flanker..3. une échappatoire; he's done us a flanker, il nous a trompés, nous a attrapés, ou (rare) nous a échappé. **1937** Partridge *Dict. Slang* 282/1 Do a flanker..; work a flanker. **1962** *Observer* 27 May 3/8 An Army spokesman said: '..he has certainly pulled a flanker on the Army'. **1962** B. Knox *Little Drops of Blood* ii. 47 This bloke wasn't content wi' just fiddling the h.p. He'd been workin' another flanker.
d. = *flank forward* (*flank *sb.*¹ 9). Also in American and Canadian football (see quot. 1961²).
1953 H. Muller *Tot Siens to Test Rugby* 180 My first-choice flanker would be Des O'Brien. **1961** *Auckland Weekly News* 5 July 53/1 As a wingforward and later a flanker he will always be remembered. **1961** J. S. Salak *Dict. Amer. Sports* 165 *Flanker* (football), an offensive end or back who lines up five or more yards wide of his own end. **1962** *Times* 12 June 3/5 Pask for J. Douglas as the blindside flanker in the pack. **1968** [see *end *sb.* 3 g]. **1970** *Toronto Daily Star* 24 Sept. 18/7 Waivers were asked on flanker Tom Bland and nobody picked him up.

flanking, *vbl. sb.* (see under Flank *v.*) Add: *flanking party* (see quots.).
1802 C. James *Mil. Dict.*, *Flanking-party*, a select body of men on foot or on horseback, whose object is to harrass and perplex the enemy, to get upon his wings, or by any manœuvre to hang upon the flank of an opposing force. **1888** Kipling *Departm. Ditties* (1890) 79 Sent out a flanking-party. **1918** E. S. Farrow *Dict. Mil. Terms* 234 *Flanking party*, any body of men detached from the main army to act upon the flanks of an enemy.

flanky (flæˑŋki), *a.* *Tanning.* [f. Flank *sb.*¹ + -y¹.] Of a skin: loose and coarse.
1903 L. A. Flemming *Pract. Tanning* 116 Some classes of skins are naturally more flanky than others.

flannel, *sb.* Add: **1. e.** A piece of flannel (or other fabric) for washing the face or hands, etc., or washing the floor, etc.
1819 Keats *Let.* 20 Sept. (1958) II. 201 The door steps always fresh from the flannel. **1906** Mrs Beeton *Househ. Managem.* 1911 If there is much cheesy-looking substance on the body [of a newly born child] it may be removed with a little sweet oil, and then well soaped with a soft flannel. **1914** D. H. Lawrence *Widowing of Mrs. Holroyd* iii. 87 She takes a flannel and soap and towel. **1926** I. M. Peacocke *His Kid Brother* 10 He gasped and spluttered, as the inexorable soapy 'flannel' passed over his rosy face. **1945** Mencken *Amer. Lang. Suppl.* (1948) II. 486 Several English correspondents say that the English for *wash-rag* is really *flannel*, but others deny it. **1962** M. Duffy *That's how it Was* xxiii. 194 'I've soaped the flannel.' I took it from her and..washed the pitiful back.
f. *slang.* Nonsense, 'hot air'; flattery, unnecessary ostentation; (see also quot. 1943). (Cf. *flannel-mouth* (*c*) s.v. sense *6 and *flannel *v.* d.)
1927 *Daily Express* 11 Oct. 3/4 One day his sister died sudden. Up he comes to ask for fourteen days' leave..'to mourn over the body..according to the

Jewish faith'...The padre wired to a rabbi..and..it was all flannel..just flannel from beginning to end. **1943** Hunt & Pringle *Service Slang* 32 *Flannel*, honeyed words or small gifts made to N.C.O.s with intent to ask favours. **1945** *Penguin New Writing* XXIII. 49 The ship's company know what is coming. Jimmy the One is going to give us a pep talk. Tons of flannel. **1958** *Sunday Times* 9 Feb. 4/3 The only book I've read about flying that isn't flannel. **1970** *Daily Tel.* 17 Apr. 4 (Advt.), This coupon will bring you our 'all facts—no flannel' brochure telling you all about us.
2. b. *spec.* Flannel trousers (see also quots. in Dict.).
1911 L. J. Vance *Cynthia* 145 The Brazilian had changed to a costume of white flannels, white shirt of silk, and white leather shoes. **1935** *Discovery* May 148/1 Two pairs of flannels were insufficient to prevent their [*sc.* mosquitoes'] drawing blood. **1958** I. Murdoch *The Bell* i. 19 He wore dark grey flannels and a white open-necked shirt.
c. *sing.* A person who has got his flannels as a member of the Harrow cricket or football team.
1901 *Harrovian* 30 Mar. 35/1 In addition to the good batsmen we have among our old flannels, it seems likely that the Eleven will be stronger in batting than in bowling. **1923** *Daily Mail* 13 July 11 Unfortunately for Harrow, they have not the same eleven available. In fact, only two old 'flannels' are left.
5. a. (Further examples.)
1758 *Newport* (R.I.) *Mercury* 26 Dec. 3/1 An old brown Jacket, a Flannel Jacket and Flannel Shirts. **1905** *Grand. Mag.* I. 116 The Duke's flannel suit. **1908** [see *Ascot]. **1915** T. S. Eliot *Prufrock* (1917) 15, I shall wear white flannel trousers, and walk upon the beach. **1946** J. Irving *Royal Navalese* 76 'Have you got flannel ears?' is a rather brutal way of saying 'Can't you pay attention?'
6. *flannel-clad, -slacked, -suited* adjs.; *flannel dance*; **flannel-cake** (later U.S. examples); **flannel flower,** (*b*) an Australian plant, *Actinotus helianthi*, the involucre of which resembles a snipped piece of white flannel; **flannelgraph,** a sheet of flannel to which (paper or cloth) cut-outs will adhere, used as a teaching aid; **flannel-mouth,** (*a*) a large species of cat-fish common in North American rivers, esp. the Mississippi and the Great Lakes; (*b*) any of various grunts of the genus *Hæmulon* or *Bathystoma* found off the south Atlantic coast of North America and having the inside of the mouth red, esp. the tomtate (*B. aurolineatum*); (*c*) *U.S. fig.*, used contemptuously of a person; in later quots. *spec.* a braggart or empty talker, a flatterer (cf. *1 f and *flannel *v.* d); so *flannel-mouthed* adj.
1847 'H. Franco' *Trippings Tom Pepper* I. 112 A very delicate species of food, which I tasted then for the first time, called flannel cakes. **1909** 'O. Henry' *Options* (1916) 212 We..then parted, after Château Margaux, Irish stew, flannel-cakes, [etc.]. **1885** Warren & Cleverly *Wand. Beetle* 59 The Beetle and her flannel-clad crew. **1898** *19th Cent.* Dec. 993 Its [*sc.* Oxford's] sun-burnt, flannel-clad youth. **1948** J. Betjeman *Coll. Poems* (1958) 153 Alfred Brown Remembers courting days in Gospel Oak And takes her to the Flannel Dance. **1895** J. H. Maiden *Flowering Plants N.S.W.* i. 9 We only know one truly local name for this plant, and that is the 'Flannel Flower'. **1911** A. E. Mack *Bush Days* 29 The summer flowers—the Christmas bells, Christmas bush, and flannel flowers. **1944** *Guatemala News* Oct.–Nov. 28 An interest aroused among the Sunday School teachers in the use of flannelgraph materials. **1955** E. Blishen *Roaring Boys* iii. 119 The flannelgraph..was a large sheet of flannel that, after some difficulty with drawing-pins, he managed to attach to the blackboard. **1968** *Listener* 27 June 844/2 Glimpses of the evangelism of the street corner, of 'flannelgraph' classes for children. **1882** Jordan & Gilbert in *Bull. U.S. Nat. Mus.* XVI. 108 *I. lacustris* (Walbaum)..Cat-fish of the Lakes; Great Fork-tailed Cat; Mississippi Cat; Florida Cat; Flannel-mouth Cat...Abundant in all large bodies of water. One of the largest of the cat-fishes, reaching a weight of 100 pounds. *Ibid.* 882 On page 108, instead of *Ictalurus lacustris*, read:..*I. nigricans* (Le S.). **1884** G. B. Goode *Nat. Hist. Aquatic Anim.* I § 133. 398 Several species of small fish, belonging to the genus *Diabasis*..are distinguished by the brilliant red color of the inside of the mouth and throat, from which they have sometimes been called Red Mouths, or Flannel Mouths. *Ibid.*, The Red-mouth Grunt, *Diabasis aurolineatus*, is probably the 'Flannel-mouthed Porgy', familiar to Florida fishermen. **1912** T. Dreiser *Financier* 8 You do, and I'll kick your head off, you flannel mouth. **1929** A. Ellis *Life* 202 George starts to complain that it was run by a bunch of 'flannel mouths'. **1933** *Amer. Speech* VIII. iv. 51/1 [In Nebraska] a *flannel mouth* was one who talked much with little sense or who was a braggart. **1966** *Publ. Amer. Dial. Soc.* 1964 XLII. 39 *Chaw-mouth*..refers to the Irishman's talkativeness and parallels the more common *flannel-mouth*. **1893** *Funk's Stand. Dict.* I. 690/3 *Flannel-mouthed.* 1. Having a large mouth. 2. (Slang.) Talking thickly or with a brogue; also, loud-mouthed. **1934** Webster *Flannel-mouthed*, orig., talking thickly with or as with a brogue; now, smooth-spoken; oily-tongued;—often contemptuous. **1948** J. Betjeman *Coll. Poems* (1958) 180 Her sturdy legs were flannel-slack'd. **1917** S. McKenna *Sonia* vii. § 3. 297 A group of barristers, flannel-suited for the Long Vacation.

flannel, *v.* Add: **c.** *intr.* To put on flannels.
1919 W. T. Grenfell *Labrador Doctor* (1920) ii. 19 Each boy..had to 'flannel' and run round the Aylesbury Arms.
d. *slang.* To flatter, curry favour; so **flannelling** *vbl. sb.* (Cf. *flannel *sb.* 1 f.)
1941 *New Statesman* 30 Aug. 218/3 (*list of war slang*) *Flannel*—To flatter. **1945** Partridge *Dict. R.A.F. Slang* 26 *To flannel*; *flannelling*: (to use) flattering or wheedling words. **1957** J. Braine *Room at Top* v. 46, I managed to flannel him into the belief that I approved of his particular brand of efficiency.

flannelled, *ppl. a.* Add: (Earlier example.) Also, wearing flannel trousers.
1778 H. More *Let.* (1834) I. i. iv. 136 The doctor..consented on condition that I should be well furred and flannelled. **1898** *19th Cent.* XLIV. 994 The single..trotting figures of flannelled men. **1902** Kipling *Five Nations* (1903) 135 Then ye contented your souls With the flannelled fools at the wicket or the muddied oafs at the goals. **1967** T. C. Worsley (*title*) Flannelled fool.

flap, *sb.* Add: **2. b.** A consonant sound produced by a flapping motion of the tongue.
[**1867** A. M. Bell *Visible Speech* 57 The subsequent removal of the centre obstruction is attended with a slightly percussive *flap*, which is the essential characteristic of the class of 'Divided' Consonants.] **1887** *Encycl. Brit.* XXII. 383/1 Glides to and from Flaps.—Flaps are consonants where there is a slack organ which flaps with the breath as it passes. **1888** H. Sweet *Hist. Eng. Sounds* 12 The most distinct glide-consonants are the flaps, of which the Norwegian 'thick' *l* is an example. **1950** D. Jones *Phoneme* p. xiii, Single flap tongue-tip r. **1968** P. M. Postal *Aspects Phonol. Theory* ii. 24 Both are pronounced with the same voiced flap [D] as medial consonant.
c. *colloq.* *fig.* A state of worry, agitation, fuss, or excitement. Esp. in phrases, *to be in, get in(to), a flap.* Also *spec.*, an alert (military sense). (Cf. *flap *v.* 10 b; *unflappable *a.*)
1916 *In Northern Mists* xliv. 188 Sometimes our departures are more than usually sudden and unexpected. The whole proceeding then becomes what is known amongst us as 'a Flap'. **1925** Fraser & Gibbons *Soldier & Sailor Words* 95 *A flap*: the familiar Navy term for the sudden 'liveliness' on board ship on the arrival of an emergency order involving general activity at extreme high pressure...Also, an airman's term for an air-raid. **1936** *Punch* 2 Sept. 265/1 At the moment there is a flap because the price of bread is going up. **1939** *Ibid.* 30 Aug. 231 Now don't go and get into a flap or anything, Mother, but Joan's broken her arm. **1940** 'Gun Buster' *Return via Dunkirk* i. x. 78 Then happened one of our monthly 'flaps' (invasion scares). **1942** W. Simpson *One of our Pilots is Safe* ii. 39 From then on until the 10th May, when the blitz started in Holland and Belgium, we were subjected to a series of 'flaps'. *Ibid.* 40 Get your flight up right away, there's a flap on. **1945** E. Waugh *Brideshead Revisited* 14 'D' you think it's the real thing?' 'No.' 'Just a flap?' 'Yes.' **1949** H. Pakington *Young Wm. Washbourne* xx. 186 There was a sense of something afoot. 'What's the flap?' asked William of the man whose watch he was relieving. 'Only our battle-cruisers engaging the enemy,' replied the other. **1956** W. Slim *Defeat into Victory* xi. 236 Everything was working smoothly, there was no flap. **1960** *Cambr. Rev.* 7 May 510/2 It is quite untrue to say that emotionally vulnerable patients who 'get into a flap' over exams will subsequently panic when a real situation threatens them in later life.
5. e. *Aeronaut.* (Also *wing-flap*) = *aileron. (The word has been applied to various kinds of hinged or sliding sections at the leading or trailing edge of the wing or tail of an aircraft, the primary function of which is to modify the lift.)
1906 *Sci. Amer.* 18 Aug. 116/3 One part of the wing..is formed of a series of longitudinal flaps, fixed at the edges to a wire gauze network, so that the flap is made to close when the wing is brought down, but keeps open when the wing is raised. **1909** *Flight* 27 Nov. 755/2 Small triangular planes are fitted above the extremities of the wings...The rudder is put over for steering to the right and simultaneously the flap above the extremity of the right-hand wing-tip is raised. **1911** *Reports & Mem.* (*Adv. Comm. Aeronaut.*) Nov. 100 It would be useful to know..whether the wing flaps should be additional to the width of the wing,..or whether the extra efficiency of warping indicates the abandonment of flaps altogether. **1921** *Aeronaut. Jrnl.* June 274 Flap Experiments with Slotted Aerofoil. An increase in the lift coefficient can be obtained by the use of a plane with flaps and altering the angle of incidence of these flaps. **1930** *Engineering* 14 Feb. 189/2 The vertical top and bottom flaps, which form the rudders, and the horizontal port and starboard flaps, which form the elevators, have each a span of about 44 ft. **1941** A. O. Pollard *Bombers over Reich* xi. 154 With flaps out of action. **1971** *Air Enthusiast* June 22/1 Possibilities of modifying the wing leading edge profile or using a leading-edge flap were eventually discarded.
6. d. A loose covering for the lower part of the abdomen. *U.S.*
[**1701** C. Wolley *Jrnl. N.Y.* (1860) 29 A piece of Cloth about a yard and a half long, put between their groins, tied with a Snake's Skin about their middle, and hanging down with a flap before.] **1813** *Niles' Weekly Reg.* V. 270/1 The fort was attacked by 725 Indians..; they were entirely naked, except a flap. **1846** J. J. Hooper *Adv. Simon Suggs* 113 Generally there are two or three hundred [Indians]..engaged in the sport at once; all naked except the 'flap'. **1919** C. G. Raht *Romance Davis Mts.* 54 The clothing..consisted usually of leggings..the breech clout, or 'flap'.
10. *flap-board, ear(s), -shutter, -top; flap-down* adj.
1833 Loudon *Encycl. Archit.* § 2031 A flap-board. **1899** W. C. Morrow *Bohem. Paris* 269 The bleeding trunk of the victim lying upon the flap-board. **1937** *Archit. Rev.* LXXXII. 57/1 The drinks-cupboard is in mahogany, and has a rubber-faced flap-down serving-board. **1922** Joyce *Ulysses* 170 Look at his mouth. Could whistle in his own ear. Flap ears to match. **1959** I. & P. Opie *Lore & Lang. Schoolch.* x. 183 Nosey people are known as: Flap-ears, Keyhole Kates, Nosey-Parkers [etc.]. **1867** Sutton & Dawson *Dict. Photogr.* 156 Some use flap-shutters in front of the lens. **1927** R. A. Freeman *Certain Dr. Thorndyke* xv. 225 A large, old-fashioned flap-top desk.
11. flap table, one with a hinged flap, a leaf table.
1833 Loudon *Encycl. Archit.* § 613 An ironing-board, or flap table. **1924** Kipling *Debits & Credits* (1926) 183 His pitch-pine-lined caravan, with..its flap-table.

flap, *v.* Add: **7.** *fig.* With *ears*: to listen attentively. *intr.* (less commonly *trans.*). *colloq.*
1925 Wodehouse *Carry on, Jeeves* i. 26 It was the work of a moment with me to..dive into a bush..and stand there with my ears flapping. **1937** M. Allingham *Dancers in Mourning* ix. 124 The next thing Konrad heard—and he must have sat there with his ears flappin'—was Sutane sayin', 'I don't want you.' **1955** M. Gilbert *Sky High* vi. 80 We never..ask any questions but..we manage to keep our ears flapping. **1959** G. Mitchell *Man who grew Tomatoes* x. 137 I'm sure he followed us. Didn't you notice him flapping his ears when you were talking to the receptionist? **1967** 'M. Erskine' *Case with 3 Husbands* ii. 33 You get on and explain them...Harris, I can see, has his ears positively flapping.
10. b. *intr. colloq.* To speak (anxiously) *about*; to be upset; to become agitated; to fuss; to panic. (Cf. *flap *sb.* 2 c.) Also (*trans.*) *to flap one's mouth.*
[**1870** G. Meredith *Let.* 27 Jan. (1970) I. 415 Out flaps the big girl with a whinny, Fire! Fire!] **1910** H. G. Wells *Mr. Polly* vii. 166 You go flapping your silly mouth about me, and I'll give you a poke in the eye. **1912** Kipling *Diversity of Creatures* (1917) 40 'We're a nice lot to flap about governing the Planet,' De Forest laughed. **1927** J. Elder *Thomasina Toddy* ix, 'Anne's a jolly nice person,' she said to Stella. 'Not bad. Nothing to flap about,' said Stella. *Ibid.* xxii, 'It's beastly, being so little,' she said crossly. Anne..said 'But it *doesn't* matter, Tom. You mustn't let it. It's silly to flap about things which can't be remedied.' **1943** Hunt & Pringle *Service Slang* 32 A person who can't 'cope' or who is very nervous is told to stop 'flapping'. **1952** J. Cannan *Body in Beck* viii. 171 My mamma flaps about my climbing, he said, and I said..it's no more dangerous than crossing the Corn. **1959** J. Verney *Friday's Tunnel* iii. 39 Mummy..burst into tears. I put my arm round her waist. 'Please don't flap.'

flapjack. Delete 'Now *dial.* or *U.S.*' and add: **1.** (N.Z. examples.)
1871 M. A. Barker *Christmas Cake in Four Quarters* 294 We told Munro we'd have his flap-jacks for second course. **1945** A. P. Harper *Camping & Bushcraft in N.Z.* i. 21 To make a flap-jack, mix some flour and baking powder into a thick paste.
c. A biscuit usu. containing rolled oats, syrup, etc. (Not known to some correspondents in S. England.)
1935 M. Struan *Pop. Home Cookery* 234 *Date Flapjacks*...Cream the butter and sugar and when they are soft work in the rolled oats. Press half the mixture in a well-buttered tin, and spread with the dates. **1942** *Radiation Cookery Bk.* (ed. 24) 149 Flap Jacks. Ingredients. 6 oz. butter or margarine. 8 oz. rolled oats. 6 oz. Demerara sugar. Pinch of salt. **1950** *Good Housek. Tea-Time Fare* 34 Syrup Flapjack...Melt the margarine, sugar and syrup together, add the rolled oats..and bake it in a moderately hot oven. When the flapjack is golden-brown, remove it from oven. **1962** *Guardian* 17 Nov. 12/6 A proper recipe for genuine oats-and-treacle flapjacks.
3. A vanity case for face-powder.
1934 *Punch* 16 May 553/3 A flapjack with a powder-puff Might well be indiscreet [as a present]. **1941** S. Gibbons *Rich House* vii. 75 Slowly opening her hand-bag and taking out her flapjack.

flapless (flæˑplĕs), *a.* [f. Flap *sb.* + -less.] Without a flap.
1916 Joyce *Portrait of Artist* iv. 195 His Norfolk coat with the flapless sidepockets. **1928** *Daily Express* 2 May 8 Your tailor may insist..that pocket flaps break the line of the waist and will cut your lounge jacket with jetted, or flapless, pockets.

flapper, *sb.*¹ Add: **3.** (Earlier example.)
1747 Mrs. Glasse *Art of Cookery* xxi. 162 Wild Ducks, called Flappers or Moulters.
4. e. *Racing slang.* = *flapping *vbl. sb.* 4.
1928 *Weekly Dispatch* 24 June 2 No flapper meetings for me.

flapper (flæˑpəɹ), *sb.*² *slang.* [Commonly supposed to be a fig. use of sense 3 ('young wild duck or partridge') of Flapper *sb.* in Dict. (cf. the G. equivalent *backfisch* perch, fish for frying); but the earlier use ('immoral young girl') app. connects the word with mod. north. dial. (Northumb. and Durham) *flap* ('an unsteady young woman', Halliwell); see Flap *sb.* 9.] †**1.** (See quots.) *Obs.*
1889 Barrère & Leland *Dict. Slang*, *Flippers, flappers*, very young girls trained to vice. **1893** Farmer *Slang*, *Flapper*..(3) A very young prostitute. **1909** J. R. Ware *Passing Eng.*, *Flapper*, a very immoral young girl in her early 'teens'.

2. A girl in her late teens, orig. one with her hair down in a pigtail; a young woman, esp. with an implication of flightiness or lack of decorum. *slang* or *colloq.*

1888 B. Lowsley *Berkshire Gloss., Vlapper*,..applied in joke to a girl of the bread-and-butter age. [**1893** Farmer *Slang, Flapper*. 2...A little girl.] **1903** D. F. T. Coke *Sandford of Merton* iii, There's a stunning flapper. **1905** D. Sladen *Playing Game* i. ix, A red-faced flapper, with a lot of freckles and a pigtail. **1906** *Varsity* 18 Oct. 23/1 Here we were in tight uniforms stepping out to raucous bugles beneath the eyes of many 'flappers'. **1909** *Tatler* 30 June 149 The first appearance of a 'flapper' at a ladies' golf championship was in 1895,..in these two long-haired, long-legged colleens were the two most famous lady golfers the world has yet produced. **1915** *Home Chat* 6 Nov. 237/1 She was the jolliest flapper I had seen, with her long plait of hair down her back. **1927** *Punch* 30 Nov. 591 'Flapper' is the popular press catch-word for an adult woman worker, aged twenty-one to thirty, when it is a question of giving her the vote under the same conditions as men of the same age. **1928** *Ibid.* 30 May 605 Attention was called in the Upper House to the conspicuous absence of the Peer who had violently attacked what he was pleased to call the 'Flappers' Vote' in his Press. **1929** H. A. Vachell *Virgin* i. 22 She had behaved like a flapper. **1936** F. Clune *Roaming round Darling* ix. 76 A modern flapper, wearing slacks and a beret, riding a man's bicycle, passed.

b. *attrib.*, as *flapper cousin*, etc.; **flapper-bracket, -seat,** a seat at the back of a bicycle to accommodate a young woman; **flapper vote,** a contemptuous expression for the parliamentary vote which was granted to women of 21 years and over by Act of Parliament in 1928; so *flapper voter*.

1909 *Tatler* 30 June 149/2 The flapper brigade is a force [at golf] which grows every year. **1916** B. Ruck *Girls at his Billet* ii, As long as [I] can persuade her to let me take her out on the flapper-bracket of my motor-bike. **1917** *Church Q. Rev.* July 317 Educated India..is still possibly at the flapper age, a little awkward..but full of the joy of life. **1921** S. Thompson *Rough Crossing* ii. § 3 The ..attention bestowed..by her 'flapper' cousins on these ordinary, pleasant-faced young men. **1923** U. L. Silberrad *Lett. J. Armiter* xiii, She was thrown off the flapper-seat of a motor-cycle. **1928** *Hansard Commons* 29 Mar. 1414 As to all this talk about the flapper vote, I want to know whether the flapper vote is to keep off the register these 2,000,000 women who are merely excluded at the present moment by a technicality. **1928** *John Blunt* 11 Aug. 2/1 Ten thousand new flapper voters. **1933** O. Sitwell *Miracle on Sinai* 353 An old..opponent of—as it had been called—the Flapper Vote. **1957** *Observer* 8 Sept. 7/2 Mrs. Rosenthal, who launched the 'Maiden-form' brassiere thirty-three years ago at the height of the flapper era.., sent me over from New York the original sketch of this most modest, flattened bust-line.

Hence **fla·pperdom,** (*a*) = **flapperhood*; (*b*) flappers collectively; **fla·pperhood, fla·pperism,** the condition of being a flapper; **fla·pperish** *a.*, pertaining to or characteristic of a flapper or flappers.

1907 'I. Hay' *Pip* i. vi. 157 She was in the last stages of what slangy young men call 'flapperdom', and her hair was gathered on the nape of her neck with a big black bow. *Ibid.* ii. vii. 216 The flapper going so far as to ask her two admirers for a quotation of odds—in the current coin of flapperdom, chocolates. **1922** M. Sadleir *Excurs. Vict. Bibliogr.* 5 Brought up on Jane Austen, Scott, and Dickens, I read, during my years of flapperdom, Marryat, Trollope, and Wilkie Collins. **1905** D. Sladen *Playing Game* ii. ix, That was during her childish beauty, before she passed into red-faced flapperhood. **1921** *Times Lit. Suppl.* 16 June 377/2 The first full-grown, full-blown stories which their mothers considered suitable for their years of flapperhood. **1920** W. J. Locke *House of Baltazar* xvii, Her inconsequence and flapperish immaturity. **1927** *Manch. Guardian Weekly* 2 Dec. Suppl. p. xiii/1 This array of flapperish literature, which makes our desk look younger than it has done for years. **1961** I. Murdoch *Severed Head* xxvii. 235 'Oh, I *do* love you!' She embraced me in a flapperish manner, lifting one high-heeled foot impetuously behind her. **1909** *Tatler* 30 June 149/2 Whilst the elder [sister]..wears her hair on high, the younger..has still a year or more of flapperism. **1927** *Sunday Express* 14 Aug. 4 She represents the essence of youth and flapperism.

flapping, *vbl. sb.* Add: **3. b.** *Aeronaut.* The angular up-and-down oscillation of the blade of a helicopter about its hinge. Also *attrib.*

1937 *Jrnl. R. Aeronaut. Soc.* XLI. 820 Although the airscrew may be at an angle, I will call the first flapping 'vertical flapping' and the second 'horizontal flapping'. **1940** H. E. Baughman *Aviation Dict.* 80/1 *Flapping angle*, the difference between the coning angle and the instantaneous angle of the span axis of a blade of a rotary wing system. **1949** *Aircraft Engin.* Feb. 33/1 Don Juan de la Cierva, father of rotating wing aircraft, introduced a very ingenious but simple invention, the flapping blade. **1950** *Gloss. Aeronaut. Terms (B.S.I.)* I. 44 *Flapping angle*, the angle between the tip-path plane and the plane normal to the hub axis. *Ibid.* 45 *Flapping hinge*, a pivot which allows the zenithal angle of the blade to be varied with respect to the rotor head. **1955** Liptrot & Woods *Rotorcraft* v. 43 In forward flight the advancing blade moves upwards about the hinge..the retreating side moves downward...It is this motion about the mean coning position, due to this dissymmetry of lift, which is known as 'flapping'.

4. *Racing slang.* A form of racing which is not subject to Jockey Club or National Hunt Committee regulations, or, in greyhound racing, to those of the National Greyhound Racing Club.

1911 *Queen* 8 Apr. 581/1 In racing parlance there are three sorts of racing, 'the flat', 'over the sticks', and 'flapping'. The first is the spring, summer, and autumn sport, the second is the winter sport of steeplechasing, and the third either form of racing which takes place neither under Jockey Club nor National Hunt regulations. **1916** *Daily Express* 9 Sept. 3/5 There was trouble at the 'flapping' meeting at Blaydon..on Saturday. **1928** *Daily Tel.* 14 Feb. 11/5 'Flapping Meetings'..will not be exempted by the bill from the provisions of the Betting Act, 1853. **1947** F. Tomlinson in *Police Jrnl.* July–Sept. (*title*) The 'Flapping Track' [of a greyhound-racing stadium]. **1955** *Daily Tel.* 27 Apr. 7/3 The Inspr. said: 'By injecting them [*sc.* greyhounds] with this stuff?' and Selby said 'Yes.' Asked if he realised it was illegal, Selby replied: 'Never. I only do it at flapping tracks.' **1969** C. Drummond *Odds on Death* vi. 136 The old 'flapping' meetings.

flapping, *ppl. a.* Add: Applied *spec.* to the upward and downward movement of the wings of birds and, formerly, of flying machines. So *flapping flight*, etc.

a **1857** G. Cayley *Let.* in C. H. Gibbs-Smith *Cayley's Aeronaut.* (1962) xiii. 135 You must mind, when you couple the flapping wings to the handle that works them, that the connecting rod takes [?] into the butt end of the wing rod at such a distance from the hinge of the wing as to allow the hand of the person working [it] to go through two feet, whilst the tip end of the wing just completes its full range. **1899** *Aeronaut. Jrnl.* July 59/1 (*title*) On Flapping Flight of Aeroplanes. *Ibid.*, The older mathematical investigation by Navier of the problem of flapping flight, seems to be quite discredited. **1906** *Sci. Amer.* 18 Aug. 117 (*caption*) The Florencie Orthopter, or Flapping-Wing Machine. **1909** *Flight* 20 Feb. 99/2 The question is..what fundamental qualities of a flapping-flight machine assimilates its wings to those of the bird. **1921** *Ibid.* 15 Sept. 621/2 (*title*) Problem of Flapping Flight. **1954** Bannerman *Birds Brit. Isles* III. 133 The feathers are raised, the tail widely spread and raised and lowered at various angles..and the flapping flight indulged in.

flappy, *a.* Add: **3.** That flaps.

1905 Chesterton *Club of Queer Trades* iii. 91 He rose.. flapping like a seal... He flapped a plaid shawl over his..arm..he flapped his eyelids...He was a..clergyman, of a flappy and floppy type. **1908** *Daily Chron.* 9 Oct. 7/1 Sailors, with bare feet and flappy blue trousers. **1924** *Scribner's Mag.* Aug. 200/2 A flappy little bag of gray and silver beads. **1965** M. Bradbury *Stepping Westward* iv. 205 Old ladies in flappy dresses.

flare, *sb.*[1] Add: **1.** Also *fig.* in other senses; *spec.*, a sudden or loud noise, a fanfare.

1834 Dickens *Let.* 3 Sept. (1965) I. 40 An unpremeditated flare at the English Opera House last night with the ladies. **1916** E. Pound *Lustra* 97 Since he died My wit and worth are cobwebs brushed aside In the full flare of grief. **1928** *Daily Tel.* 17 Jan. 10/7 After the service was over, amid a flare of trumpets, the procession returned through the Cathedral. **1946** Mezzrow & Wolfe *Really Blues* xv. 264 Flares were always in the right places, to help build up the pulse. **1962** A. Nisbett *Technique Sound Studio* 242 s.v. *Bass*, The most efficient way of producing bass is an acoustic exponential horn, preferably with a low rate of flare.

b. *Astr.* A sudden increase in brightness of part of the sun as seen at certain visible and ultra-violet wavelengths. Also *solar flare*.

1937 *News Service Bull., School Ed.* (Carnegie Inst. Washington) 24 Oct. 153/3 Bright flares of hydrogen light were visible on the Sun. *Ibid.*, It is believed that all of these fade-outs occur simultaneously with solar eruptions—absence of solar observations some times accounting for the failure of any astronomer to report observation of a flare. **1956** H. Spencer Jones in A. Pryce-Jones *New Outl. Mod. Knowl.* 121 A violent disturbance will not infrequently occur, in which a localized region of the Sun becomes intensely bright. Coincident with this eruption or *flare*, as it is termed, there is a sudden and complete fading on short-wave radio transmissions in channels passing over the sunlit face of the Earth. **1963** H. J. & E. v. P. Smith *Solar Flares* p. xii, Flares occur in the solar atmosphere and are closely associated with sunspots and other aspects of solar activity.

c. *Astr.* Any sudden and short-lived increase in the overall brightness of a star other than the sun. So *flare star*, a star in which flares occur from time to time.

1949 *Publ. Astron. Soc. Pacific* LXI. 179 The plate with the Selected Area shows the star in the process of one of its flares. **1951** *Ibid.* LXIII. 142 Krüger 60 B..is the only flare star for which the mass is known. **1956** C. Payne-Gaposchkin *Introd. Astron.* xi. 264 A solar flare brightens the whole light of the sun only slightly, but on a cool star a flare may more than double the star's total brightness. **1964** *New Scientist* 19 Mar. 760/1 A flare star is a particular type of variable star which produces an almost undetectable burst of radio emission at the same time as it flares up to become visually brighter.

2. Also in military, aeronautical, and general use; also *attrib.* (see quots. 1918 and 1956 and *flare path).

1912 *Aeroplane* 5 Dec. 565/1 Were the flares lighted in order that our own airship should know where to alight? **1915** *Times* 14 Apr. 7/6 The enemy fires some flares across to us, and this is a good sign, for it shows they are afraid of an attack from us. **1916** 'Boyd Cable' *Action Front* 48 Magnesium flares. **1918** E. S. Farrow *Dict. Mil. Terms* 235 *Flare*, an unsteady, dazzling light used as an illumination and in signaling: in aëronautics, a guide for landing. *Flare lights*, lights used in combination with obstacles, either protected or screened, to prevent the enemy removing them. They are screened in rear so that the defenders may remain in shadow. *Flare pistol*, a large pistol, which looks like a sawed-off shotgun, from which flares are fired. **1925** N. E. Odell in E. F. Norton *Fight for Everest, 1924* 134 We watched till return, or even an indication by flare of distress. **1941** *Times Weekly* 15 Oct. 6 Flares dropped by British aircraft hung almost stationary in the air, causing a bright glow over the French coast. **1942** *Aeroplane Spotter* 23 Apr. 101 Flare chutes. **1943** T. Horsley *Find, Fix & Strike* 83 In conjunction with dive bombing, and in connexion with it, their [Albacores'] suitability as flare droppers was equally well marked. **1944** *Gen. Ord. Dict. Add.*, *Flare*,..container of combustible material, dropped from aircraft to illuminate target area etc. **1956** W. A. Heflin *U.S.A.F. Dict.* 208/1 *Flare chute*, a parachute attached to a flare for letting the flare down slowly. **1965** *Observer* 11 July 1/2 Americans 'flare-ship' aircraft..continually illuminated our positions at night.

3. Delete 'See quot. 1868.' and substitute: (See quot. 1968.). (Further examples.)

1867 Sutton & Dawson *Dict. Photogr.* (ed. 2) 119 *Flare*, stray light falling upon the sensitive plate during its exposure in the camera. When certain forms of double or triple compound lenses are used, and the camera is turned towards a strong light, a circular spot of flare is sometimes seen in the centre of the ground glass. **1968** *Gloss. Terms Offset Lithogr. Printing (B.S.I.)* 14 *Flare*, non-image light that reaches the light-sensitive material in the camera, usually caused by surface reflections in the optical system, extraneous light and/or reflection from sources other than the original. **1971** *Amateur Photographer* 13 Jan. 54/2 Of his dozen or so entries about to every light source, the inbuilt photographic lenses..constitute considerable flare—caused by internal lens reflections.

4. b. *transf.* A gradual widening or spreading outwards; also, that part which spreads. Also *attrib.*

a **1910** 'O. Henry' *Trimmed Lamp* (1916) 6 Her skirt is shoddy, but has the correct flare. **1966** Stanford & Forsyth *Hist. Music* 166 The shawm-player prefers to place his lips..on the widened rim or 'flare' at the top of the pipe. **1929** *Star* 21 Aug. 2/4 Cut..with the new smart waist line and full flare skirt.

c. = *flare-out 2.

1967 J. Fay *Helicopter* (ed. 2) viii. 120 Whereas at high speed a flare could..produce a momentary climb the only practical result of a flare at low speed would be to change the fuselage attitude. **1967** D. P. Davies *Handling Big Jets* xii. 213 Keep the stabiliser in trim so that full elevator effectiveness is available for the flare. **1969** L. Kemp *Brit. G.I. in Vietnam* iii. 53 Gradually the nose came up for the flare.

flare, *v.* Add: **4. c.** *intr.* To open or spread outwards, as the sides of a bowl, a skirt, the mouth of a horn. Hence **flared** *ppl. a.*

1835 C. F. Hoffman *Winter in Far West* II. 24 Finally, it [*sc.* the vault] flares upward, so that the edges of the arch themselves the projecting face of the cliff. **1857**, etc. [see *flaring *ppl. a.* 3]. **1899** *Daily News* 27 Feb. 6/6 The fronts of the short coat flare open to show a gathered lace vest. **1928** *Daily Mail* 23 July 17 Delightful Lace Tunic with the new Flared Skirt. **1930** *Times* 17 Mar. 15/6 A short slightly flared above the hem. **1968** J. Ironside *Fashion Alphabet* 44 A culotte skirt..is any skirt which..hangs usually one flared from the hip.

d. *intr.* To make the glide path of an aircraft about to land gradually less steep until it is parallel to the ground; to raise the nose of an aircraft when doing this; also *trans.*, to cause (an aircraft) to descend in this way. Also with *out*. Hence **flared** *ppl. a.*

1935 *Jrnl. R. Aeronaut. Soc.* XXXIX. 869 Unexpectedly high lift and drag force coefficients were developed in the later stages of the flared landings. **1944** W. Langewiesche *Stick & Rudder* xii. 215 Because of this steeper glide, the heavily wing-loaded ship has more upward turning to do in the flare-out. This makes it necessary to flare out even earlier, even higher. **1951** *Jrnl. R. Aeronaut. Soc.* LV. 326/1 The pilot shuts off power, glides down on to the deck, flares out just before touch down. **1967** J. Fay *Helicopter* (ed. 2) viii. 120 Flaring.. **1967** D. P. Davies *Handling Big Jets* vii. 213 From the threshold on, simply flare (if necessary), reduce to idle thrust, push off drift (if necessary), then land. **1969** L. Kemp *Brit. G.I. in Vietnam* iii. 72 Hayes followed the routine procedure of 'flaring' the helicopter before descent—letting the tail drop and the nose come up to reduce airspeed—and then dropping her vertically with the collective.

6. b. Also with *out*, quasi-*trans.* with quoted words. (Cf. Flare-out.)

1907 *Smart Set* Feb. 95 'Your name is Oughelttree,' he suddenly flared out.

flareless (flɛəˈlɛs), *a.* [f. Flare *sb.*[1] + -less.] **a.** Not producing a flare (Flare *sb.*[1] 1). **b.** Of a garment: having no flare (*flare *sb.*[1] 4 b).

1932 *N.Y. Times* 21 Feb. (*headline*) Japan using new smokeless and flareless powder. **1954** W. Faulkner *Fable* (1955) 32 The simple flareless breeches.

flare-out. Add: **2.** *Aeronaut.* A lessening of the steepness of the glide path of an aircraft about to land (see *flare *v.* 4 d).

1944 [see *flare *v.* 4 d]. **1958** *Times* 17 Oct. 3/4 An automatic landing is completed with the aid of a Standard Telephones radio altimeter which gives an accurate measurement of height above ground and is used for the 'flare-out' (levelling off to land). **1963** *Spectator* 22 Mar. 351 With BLEU the flare-out is beautifully smooth and accurate. The judgement of the touchdown point is exact.

flare path. *Aeronaut.* [*flare *sb.*[1] 2.] A line of lights on an airfield or elsewhere to guide aircraft in taking off or coming in to land; an illuminated runway.

1919 R. H. Reece *Night Bombing with Bedouins* 56 The answer received, the machine taxies..turns, hurtles down the flare-path and leaves the ground. **1936** *Jrnl. R. Aeronaut. Soc.* XL. 91 On a fine clear night the flare path is visible for about 50 miles. **1942** T. Rattigan (*title*) Flare path. **1943** R. Hillary *Last Enemy* ii. 46 Across about a hundred yards from us lay the flare path, a straight line of dimly glowing light.

flare-up. 2. Delete '(not in dignified use)' and add further examples.

1880 Rossetti *Let.* 1 Sept. (1965) I. 92 *The Guardian*..contains a flare-up review of *The Game*. **1960** *Times* 2 Mar. 11/2 A border flare-up not directly concerned with the water issue. **1970** F. McKenna *Gloss. Railwaymen's Talk* 35 A flare-up usually took place on a Saturday night when the men enjoyed themselves after a week of tension and hard work.

flaring, *ppl. a.* Add: **3.** Also of a pan or dish (cf. *flare *v.* 4 c).

1857 *Trans. Ill. Agric. Soc.* II. 158, I..allow a little of the milk to pass along into a large flaring pan. *a* **1877** Knight *Dict. Mech.* I. 876/1 *Flaring*...Increasing in diameter upwards as of an upwardly expanding pan. *Ibid.* III. 2633/2 *Trumpet*..The flaring mouth of a railway-car draw-head. **1940** *Montgomery Ward Catal.* 434/2 Flaring Pails—Tin..6 quarts [etc.]..14 quarts.

flaser (flɑːzə). *Geol.* [a. G. *flaser*, dial. form of *fladder* streak, vein.] Used *attrib.* to denote the presence of lenses of little-altered rock in a streaky parent rock that has been metamorphosed by shearing under pressure; it is found mainly in gabbro, gneiss, and granite; so *flaser-gabbro, granite, rock, schist, structure*.

1886 C. Lapworth *Page's Geol.* (ed. 12) 108 A curious veiny or banded structure (flaser structure). **1888** J. J. H. Teall *Brit. Petrogr.* vii. 177 In speaking of the foliated gabbros it is convenient to recognise two principal types which may be designated by the terms *flaser-gabbro* and *gabbro-schist*. *Ibid.* 245 In some of these a flaser structure has been developed by intercrystal movement. **1891** G. A. J. Cole *Aids Pract. Geol.* 210 This structure..appears to arise by interaction of the minerals when subjected to earth-pressures ('Flaser-gabbros', etc.). **1930** Peach & Horne *Geol. Scot.* 18 Flaser structure with abundant phacoids of felspar and hornblende. **1951** Turner & Verhoogen *Ign. & Metamorph. Petrol.* xxi. 143 There is continuous transition from mylonites to a group of rocks variously referred to as augen schists, mylonite gneisses, flaser rocks, and blastomylonites. **1954** H. Williams et al. *Petrogr.* XI. 204 The terms flaser gabbro and flaser granite have long been applied by European petrologists to cataclasites in which granulated streaks surround lenses..and between streamlined eyes of undestroyed gabbro or granite.

flash, *sb.*[2] Add: **1. d.** A brief telegraphic news dispatch, usually as a preliminary to a fuller report; a brief item of broadcast news. So *news flash*; orig. *U.S.* (in telegraphic sense).

1857 *Richmond* (Va.) *D. Whig* 31 Aug. 3/1 The first flash came across the ocean by the Submarine Telegraph at noon to-day. **1904** *Post Express* (Rochester, N.Y.) 12 Sept. 1 News Flashes from All Over. **1933** *Evening Standard* 19 Apr. 6/2 The '[Central] Newsagency' had the news..the tape machines..in this form. Flash 11.28 p.m. Moscow Trial. **1934** H. N. Rose *Thesaurus of Slang* vii. 481/2 Brief News Bulletin..a flash. **1938** *Manch. Guardian Weekly* 21 Oct. Suppl. There was little hope that..a..news flash would break in..but her voice at all once recorded. 'Flash!' a masculine announcer put in. **1940** F. Fleming *Flying Visit* 118 The Censorship, after passing a news agency 'flash' stating that the *New York Morning Post* had published an amazing dispatch in which its London correspondent alleged (etc.). **1966** *New Statesman* 1 Nov. 604/1 It often got out 'flashes' quicker which enabled us to..save valuable time on big stories.

e. *Cinematography.* Exposure of a scene; a scene momentarily shown on the screen.

1913 E. W. Sargent *Technique of Photoplay* (ed. 2) ii. 14 We know what is in the letter, so just a flash about three feet long is used. **1922** A. C. Lescarboura *Cinema Handbk.* i. 53 *Flash*, a short scene, usually three to five feet. **1944** *Ann. Reg.* 1943 344 Propaganda shorts..came in a steady flow..their length varying from a flash to five minutes.

f. = **flash-lamp* (b).

1913 D. E. Adams in F. H. Harris *Dartmouth out o' Doors* 40 A pocket flash was the only light on hand. **1943** R. Chandler *Lady in Lake* (1944) xxxiv. 178 'Got a flash?' 'No.' I said: 'There's one in the car pocket on the left side.' Shorty fumbled around and metal clicked and the white beam of a flashlight came on.

g. A flash-light photograph; also, = **flash-gun, flash-lamp* (a).

1945 Wakefield & Smith *Synchronized Flashlight Photogr.* v. 72 The flash is fixed to the camera. **1959** J. Cary *Captive & Free* lxiii. 287 A camera man held up his reflector and took a flash. **1963** L. Deighton *Horse under Water* xvi. 66 He brought it [*sc.* a camera] complete with flash and a green filter. **1971** *Amateur Photographer* 13 Jan. 42 A man came into the shop and said he would like to buy a small electronic flash he had seen in the window.

h. The brief pleasurable sensation received immediately after an injection of certain narcotic drugs. *slang.*

1967 M. M. Glatt et al. *Drug Scene* iii. 39 He no longer got a 'kick' or 'flash' from taking drugs. **1970** *Observer* 3 May 3/3 The pleasure comes apparently from the half-dreamlike state between consciousness and sleep which the addict calls his 'flash' or 'buzz'. **1971** *Oz* XXXV. 40/1 More & more people started shooting it to get the flash all the real hip suckers were talking about.

7. b. A patch of cloth sewn on a military uniform, usu. on the upper arm or shoulder, with a device to indicate the unit or country, etc., to which the wearer belongs.

1918 (*title*) Flashes of 51st Division Prior to Aug. 18 (*chart in Imperial War Museum Libr.*). **1927** W. Deeping *Kitty* xi. 142 A captain wearing the ribbon of the Military Cross, and black and white chess-board flashes. **1943** *Stars & Stripes* 15 June 3/5 Here's how the British and American armies describe different items: *American*..Insignia, shoulder, sleeve. *British*..Divisional sign or flash. **1944** *Times* 5 July 5/5 One cannot fail to notice the interest the Germans display in shoulder-flashes bearing the name Australia worn by the few A.I.F. officers who are here. *Ibid.*, They nearly all ask where the rest of the Australians are when they see my flashes. **1953** C. Day Lewis tr. *Virgil's Aeneid* ii. 47 Change shields with these dead Greeks, put on their badges and flashes!

14. b. flash boiler, generator = Flasher 6; **flash bomb** (see quot. 1940); **flash-bulb** *Photogr.*, a glass bulb producing the light used for taking flash-light photographs; **flash burn,** a burn caused by sudden intense heat, esp. that generated by a nuclear explosion; **flash-butt welding** *Metallurgy* (see quot. 1958); also **flash weld(ing)** (cf. **butt-weld(ed)*); **flash card** (see quot. 1945); **flash colour,** a patch of bright colour on an animal's body which is visible only when the animal is in motion; so *flash-coloured* adj.; **flash-dry** *v. trans.*, to dry in a very short time; so *flash-drying* vbl. sb.; **flash flood** [cf. sense 11 a, b in Dict.] *Phys. Geogr.*, a sudden, destructive flood; so *flash flooding* vbl. sb.; **flash-gun** *Photogr.*, a device that can be attached to a camera to hold and operate a flash-bulb; **flash-lamp,** (*b*) a portable electric lamp which produces a light by the pressure of a button, etc.; **flash-light,** (*c*) chiefly *U.S.* = **flash-lamp* (b); also as *v. trans.*, to photograph by flash-light (also *fig.*); hence *flash-lighting* vbl. sb.; **flash meter,** a device similar to the shutter of a camera which permits momentary exposure of slides for teaching purposes; **flash pasteurization,** a method of pasteurization in which the substance is suddenly raised to a higher temperature than in normal pasteurization but for a shorter period; also *flash process*; **flash-photolysis** *Chem.*, the use of a very short, intense flash of light to bring about chemical decomposition or dissociation; so **flash-photolyse** *v.*, to decompose by this means; *flash-photolysed* ppl. adj.; **flash-point,** (*b*) *fig.*, a point of climax, indignation, etc. (cf. *boiling-point*); **flash powder,** powder used in flash-light photography; also extended uses; **flash roasting** *Metallurgy* (see quot. 1958); so *flash-roast* v. trans.; **flash spectroscopy** *Chem.*, spectroscopic examination of rapid chemical reactions initiated by a very short, intense flash of light; **flash spectrum,** (*a*) a spectrum of the chromosphere which appears at the beginning and end of totality of a solar eclipse; (*b*) a spectrum of the reactants and reaction products obtained in flash spectroscopy; **flash-spotting** *Mil.*, the locating and reporting of hostile battery positions by observation of their gun-flashes; hence *flash-spotter*; **flash steam generator** = **flash generator*; **flash tube** *Photogr.*, a tube, filled usually with xenon under reduced pressure, by means of which a flash is produced when an electrical current is suddenly passed through the gas.

1902 *Lockwood's Dict. Mech. Engin.* (ed. 3), *Flash Boiler*, a rapidly steaming boiler in which the steam is generated in coils of small tubes. **1906** *Daily Chron.* 3 Mar. 3/6 Water is converted into steam in a tubular boiler, called a flash boiler. **1940** *Flight* 26 Dec. b/2 Flash bombs are dropped to illumine the target. **1935** *News Chron. Amateur Photogr.* ii. iii. 200 The flashlight need not be situated directly behind the camera..particularly if a flash-bulb is available. **1937** *Pop. Sci.* Aug. 86/1 (*heading*) Using two flash bulbs for portrait photos. *Ibid.*, Any flash-bulb holder, operated by batteries, may be easily adapted. **1939** Henney & Dudley *Handbk. Photogr.* iv. 93 Synchronized flash guns are devices which enable the photographer to fire off a flash bulb at the same instant the shutter of his camera is opened. **1954** X. Fielding *Hide & Seek* ix. 113 Half a dozen sheep stood there motionless for a second, as though posing for a flash-bulb photograph. **1946** *Nature* 3 Aug. 152/1 An attack by atomic bombs would, no doubt, cause some casualties by 'flashburn', although even ordinary clothing appears to offer substantial protection against it. **1951** *Ann. Reg.* 1950 420 The three major effects [of atomic warfare] were blast, flash burn, and radiological. **1933** *Welding Ind.* Aug. 224/2 On flash butt welding machines, the transformer..inside the housing is exposed to the damage which may be caused by the unavoidable considerable spattering. **1958** A. D. Merriman *Dict. Metallurgy* 95/1 *Flash-butt welding*, a resistance welding process in which an arc is struck and maintained between the joint members until welding heat is attained. The current is then shut off and the weld made by forcing the parts together. **1923** *Cumulative Book Index* 159/1 Flash cards for rapid word drills. **1945** C. V. Good *Dict. Educ.* 173/2 *Flash card*, a card of heavy cardboard having on it written or printed letters, words, phrases, numerals, or combinations of numerals for computation; used as an aid to learning, the teacher holding each card up for the class to see for a brief interval. **1963** Weyrich & Schuller *Audio-Visual Materials* iii. 55 Flash cards..ultimately create relationships and understandings of the symbols and the objects for which these symbols stand. **1964** *Listener* 12 Nov. 775/2 The raising of children with 'flash-cards': what do you think of this?—two seconds to answer. **1928** T. H. Savory *Brit. Spiders* vii. 160 Coloration may assist concealment..by what have been called 'flash-colours'. **1935** —— *Spiders* 28 When the spider has been disturbed and runs away, its colouration..quickly moving bright points are conspicuous. Then suddenly it stops and draws in its legs: the bright colours are hidden, and the spider becomes almost invisible—a method of protection known as 'flash colouring'. **1950** *Time* 26 June 10 Big heating ovens..flash-dry the ink almost instantly. **1960** *Times* 20 Sept. (Pure Food Suppl.) p. xx, Because the food is flash-dried in a high vacuum the moisture takes nothing away with it. **1946** *Nature* 10 Aug. 194/1 Considerable progress has been made in..spray-drying, flash-drying and drum-drying have been developed with considerable success. **1940** *Words* Apr. 55/1 Harry Burns..died in a 'flash' flood which raced down Wheeling creek. **1939** *Nature* 17 June 1028/1 A two feet deep band of boulders and sand, deposited in a period of flash flooding. **1913** W. E. Dommett *Motor Car Mech.* 141 Flash generators are really a particular case of water-tube boilers, in which the tubes serve not only for the production of steam but simultaneously as super-heaters. **1930** *Pop. Sci.* July 46/3 (*caption*) New flash gun attaches to camera and is used with an ordinary flashlight cartridge. **1939** Flash gun [see **flash-bulb*]. **1959** *Which?* May 25/2 The cameras tested could be used with flashguns—an advantage for anyone wanting to take indoor pictures. **1908** *Model Engin. & Electrician* 11 June 570/2 Flashlamp batteries. **1914** G. W. Young *From the Trenches* xi. 274 Electric flash-lamps. **1928** *Daily Express* 11 Oct. 8 Tom shone his flashlamp on the knife. **1901** *Field & Stream* Jan. 774/2 The Comet Baby Flash light. **1912** A. Bennett *Grand Babylon Hotel* xiii. 144 Racksole had photographed the corpse by flashlight. **1909** *Westm. Gaz.* 19 Mar. 3/1 To attempt to flashlight him at night was rather too serious a project. Kipling *Diversity of Creatures* (1917) 207 We were studying the interior of a soul, flash-lighted by the dread of 'losing its position'. **1939** F. Hume [illegible] 213 A gold-handled umbrella with a bachelor-girl flash-light in its tip. **1926** E. Caldwell *Trouble in July* xii. 179 Several men rushed inside, flashlighting the room. **1958** 'A. Gilbert' *Death against Clock* 65 Carry a flash-light by any chance? **1969** H. Barnes *Oceanogr. & Marine Biol.* 185 Electronic flash-lighting with a G.E.C. flash-meter is again employed, the camera mechanism being synchronized to the flash. **1957** D. T. Herman et al. in Saporta & Bastian *Psycholinguistics* (1961) 558/2 A Keystone model 1045 projector equipped with an Ilex flashmeter was used. **1952** J. W. Brown et al. *A-V Instructional Materials Man.* iv. 157 Keystone overhead projector with flashmeter attachment. **1921** M. Mortenson *Managem. Dairy Plants* iii. 150 It requires approximately 17 per cent more heat for flash pasteurization than for vat pasteurization. **1927** H. E. Ross *Care & Handling of Milk* viii. 101 Flash pasteurization is often used in pasteurizing milk and cream for manufacturing purposes where high temperatures may be used without injuring the product. **1940** *Economist* 23 Dec. 798/1 Ordinary pasteurisation imparts an objectionable cooked flavour...But there are certain methods such as flash pasteurisation, film evaporation and freezing which are not open to these objections. **1950** G. Porter in *Proc. R. Soc.* A. CC. 284 A new technique of flash photolysis and spectroscopy has been developed. **1962** R. E. Dodd *Chem. Spectroscopy* x. 310 An important kinetic technique which relies upon spectrophotometry is flash photolysis. **1963** G. Herzberg *Spectra & Struct. Simple Free Radicals* 8 (*caption*) Apparatus for the study of absorption spectra of free radicals in the vacuum ultraviolet by flash photolysis. *Ibid.* 13 Almost the first absorption spectrum of flash-photolyzed diazomethane showed a new transient feature, which turned out to be the spectrum of CH_2. **1971** *Nature* 1 Jan. 41/1 The technique was checked by flash photolysing a 3:1 mixture of hydrogen and nitrogen dioxide. **1955** *Times* 26 Aug. 3/3 What actually happens is that the idealist is horrified by what he learns of police methods, his social conscience is brought to the flash-point of concentration by a love affair. **1958** *Economist* 13 Sept. 817/1 China's tactics seemed to be to build up heat to near flash-point and then suddenly lower the temperature. **1963** *Ann. Reg.* 1962 132 The flashpoint of East–West relations in 1962 was reached on 22 October. **1889** *Brit. Jrnl. Photogr.* 805 The characteristic of flash powder is that it contains within itself the elements by which the flash is produced. **1952** Granville *Dict. Theatr. Terms* 77 *Flash powder*, a chemical substance through which is passed an electric current producing a flash and a cloud of smoke. Used for explosive effects in war plays. **1966** F. H. Brightman *Oxf. Bk. Flowerless Plants* 56/2 The spores form a very fine, bright-yellow powder called 'lycopodium powder', which was formerly used as a constituent of 'flash powder'. **1910** Ayers & Johnson in *Bull. U.S. Dept. Agric. Bur. Animal Ind.* cxxvi. 14 Experiments were made, using both the 'flash' process, which consists of heating flowing milk for from thirty to forty seconds and then cooling, and the 'holder' process, where the milk is heated in a tank and held for thirty minutes before cooling. **1951** *Engineering* 20 July 75/2 Hydrogen sulphide in town's gas can be obtained by flash-roasting iron pyrites. **1926** D. M. Liddell *Handbk. Non-Ferrous Metallurgy* II. 1428/2 (*Index*) Flash roasting. **1958** A. D. Merriman *Dict. Metallurgy* 95/1 *Flash roasting*, a process for removing sulphur from ores by blowing the pulverised concentrates through a combustion chamber. **1952** *Proc. R. Soc.* A. CCX. 440 Flash spectroscopy makes possible the recording of absorption spectra in a time less than 50 microseconds. **1953** *Ibid.* CCXVI. 165 The explosive oxidation of acetylene, initiated homogeneously by the flash photolysis of a small quantity of nitrogen dioxide, has been investigated by flash spectroscopy. **1899** C. A. Young *Gen. Astron.* 267 The 'flash-spectrum' was successfully photographed at a number of stations. **1900** *Proc. R. Soc.* LXVII. 373 The centre of the flash spectrum arcs was..midway between the edges of the spectrum in the photographs obtained at mid-eclipse. **1926** H. Macpherson *Mod. Astron.* 32 Confirmatory observations of this 'flash spectrum', as the phenomenon of reversal from dark to bright was called, were secured at subsequent eclipses. **1950** *Proc. R. Soc.* A. CC. 299 Effect of capacity on flash spectra of krypton and hydrogen. **1930** Blunden *De Bello Germanico* 78, I believe our observers or flash-spotters used it. **1922** *Encycl. Brit.* XXX. 253/1 Any gun which fired at night within direct view of the enemy was liable to be marked down by the 'flash-spotting' section... It was therefore necessary to introduce flashless powder. **1907** *Westm. Gaz.* 19 Nov. 4/2 The flash steam generator, which is of the Serpollet type. **1945** Wakefield & Smith *Synchronized Flashlight Photogr.* ix. 109 The light-producing unit is a coiled glass tube filled with pure xenon... The flash-tube. **1946** *Electronic Engin.* 113 (*title*) Electronic flash tubes in high-speed photography of explosions. **1957** T. L. J. Bentley *Man. Miniat. Camera* (ed. 5) v. 74 The practical uses of electronic flash sets are governed by the characteristics of the modern xenon-filled flash tube. **1959** Flash-tube [see **flash-lighting*]. **1959** *Jrnl. Iron & Steel Inst.* CXCII. 401/2 Flash welds gave the highest and most consistent results. **1933** *Welding Ind.* Feb. 4/1 Butt welding may be divided into (*a*) direct or upset method..(*b*) flash welding, in which the material is brought together and an arc is drawn between the parts to be welded. This arc plays along the material in a continuous flash and the pieces are moved together during this flashing period and when in the plastic state the current is interrupted and pressure applied. **1943** *Jrnl. R. Aeronaut. Soc.* XLVII. 289 Flash welding is a type of butt weld in which the two parts to be joined are connected to the secondary terminals of a low-voltage high-current transformer, are then brought into close proximity, and the voltage applied.

15. Excess metal or plastic that is forced between facing surfaces as the two halves of a two-part die or mould close up, forming a thin projection on the resulting object.

1910 *Encycl. Brit.* X. 665/1 A large amount of metal is squeezed out beyond the concavity of the forging dies... There are two methods adopted for removing this 'fin', or 'flash' as it is termed. **1936** H. W. Rowell *Technol. Plastics* xxi. 158 If the width [of the land] is made greater to reduce pounds pressure per sq. in., the thickness of flash may exceed the desirable ·004 in. **1952** J. Wulff et al. *Metall. for Engin.* xxvii. 521 Excess metal, called flash, is squeezed from the die faces at the parting line and is subsequently trimmed or ground from the part. **1966** Walker & Martin *Injection Moulding of Plastics* iv. 132 Moulds for nylon usually have to be well bedded together to prevent flash.

16. A thin layer (of glass, chromium, etc.). (Cf. Flash *v.*[1] 14 a, *c.)

1909 Webster, *Flash*.., a layer of glass flashed on. **1950** J. Osborne *Dental Mech.* (ed. 3) xxii. 369 One important modification must be made, however, to prevent any flash being present at the all-important fitting edge. **1959** *Times* 11 Nov. 6/3 What matters most with chromium plating is the layer of nickel under the top 'flash' of chromium. **1961** *B.S.I. News* Apr. 9/2 The durability of chromium-plating depends largely on the thickness and quality of the layer of nickel which is applied under the final 'flash' of chromium.

flash, *v.*[1] Add: **8. e.** *to flash a glance, a look, one's eyes.*

1886 'M. Gray' *Silence of Dean Maitland* i. ix, Cyril flashed upon him one of his droll glances, and laughed. **1888** Mrs. H. Ward *R. Elsmere* xiii, She flashed a quick, defiant look at him. **1903** R. Langbridge *Flame & Flood* xvii, The young man..flashed his insolent eyes.. at her.

f. *trans.* and *intr.* To switch (lights on a motor vehicle) on and off in order to communicate a warning, a direction, etc.; to signal in this way (*trans.* and *intr.*).

1957 J. Kerouac *On Road* (1958) iii. ix. 233 The guy shot by us..and tooted his horn and flashed the tail lights for challenge. **1962** L. Deighton *Ipcress File* xxviii. 180, I flashed the headlights and got an answering signal from the brake actuated red rear lights of a vehicle parked there. **1963** *Guardian* 23 Jan. 10/3 After having my headlights adjusted, I started driving with them dipped. Having been 'flashed' by large numbers of drivers I can only assume that they are..annoyed. **1965** 'W. Haggard' *Hard Sell* ix. 100 The Merc suddenly flashed him and he drew in. **1971** *Daily Tel.* (Colour Suppl.) 22 Oct. 23/4 Hold straight until you are quite sure you are in charge, flashing your brake lights if necessary..so as to warn following traffic. *Ibid.* 26, I suspect he is the same man who flashes me frenziedly when I dare to use my headlights before ten o'clock on a summer's night.

9. b. Also with *round*.

1853 Kingsley *Hypatia* v, The steel-clad apparition suddenly flashed round, and vanished. **1893** M. E. Mann *In Summer Shade* xvi, 'He must be paid.' 'How?' demanded Mary, flashing round upon him. **1903** R. Langbridge *Flame & Flood* iii, Susette flashed round upon him with a brilliant smile.

c. *to flash back*: to jump back, as when a flame in a Bunsen burner retreats down the tube and burns at the air-inlet; to 'light back'.

1902 *Encycl. Brit.* XXVIII. 596/2 A still further addition of air causes the mixture to become so highly explosive that it flashes back into the tube of the burner.

10. a. quasi-*trans.* with quoted words.

1875 *Harper's Mag.* Aug. 415/1 One day she flashed out upon Tom Saymour 'Vote!—why should I?' **1886** 'M. Gray' *Silence of Dean Maitland* i. v, 'The whole village looking on and not lifting a finger—the cowards!' Lilian flashed out. **1907** *Munsey's Mag.* Nov. 169 'I intend to see Varani—alone,' she flashed.

12. b. *Cinemat.* *trans.* and *intr.* To show abruptly on the screen; *intr.*, to change abruptly *to* (another scene).

1913 E. W. Sargent *Technique of Photoplay* (ed. 2) ii. 15 A one word leader flashed on the screen that said 'Later'. **1935** H. G. Wells *Things to Come* x. 93 Flash the date A.D. 2054. **1936** —— *Man who could work Miracles* (film ed.) vii. 43 Flash to a brief bright scene in a San Francisco hospital.

14. c. *Photogr.* To cover *over* with a very thin layer.

1903 *Nature* 29 Jan. 301/2 The so-formed negative is sprung from the wax, cleaned and polished, and flashed over with a very thin layer of nickel in a nickel bath.

flash, *v.*[2] Delete *dial.* and add later examples.

1957 *N.Z. Timber Jrnl.* Mar. 52/1 To flash means to make a watertight joint with sheet lead and other metal known as flashings, at the intersection, e.g. between chimney and roof. **1964** J. S. Scott *Dict. Building* 130 *Flash*, to make a weathertight joint, called a flashing.

fla·shback, *sb.* [f. the verbal phr. **to flash back*.] **1.** (See quots. and cf. *flash *v.*[1] 9 c.)

1903 *Motoring Ann.* 306 The highly inflammable vapour of petrol, and a 'flash-back', resulted in the total destruction of the car. **1963** *Times* 27 Feb. 4/5 Three electricians were taken to hospital with burns after a flashback and fire in the power house at the Premier Flour Mills, Silvertown, E.

2. *Cinemat.* A scene which is a return to a previous action in the film, a *cut-back; hence, a revival of the memory of past events, as in a pictorial or written presentation. (See *flash *sb.*[2] 1 e, and **v.*[1] 12 b.)

1916 *Variety* 13 Oct. 28/4 In other words the whole thing is a flash-back of the episodes leading up to her marriage. **1928** J. Gallishaw *Only Two Ways to write Story* i. vii. 177 With *Sunk* the method of presentation was chronological...In the case of *Paradise Island* the method is reversed. The order instead of being chronological is anti-chronological: it is the flash-back method. **1934** H. G. Wells *Exper. Autobiogr.* II. vii. 487 When goddesses and Sea Ladies vanish and a flash back to the ancestral chimpanzee abolishes the magic caverns of Venus, human beings arrive. **1947** *Times* 1 Nov. 6/4 The film relates, in a prolonged flash-back, how the innocent Indian became corrupted by bewildering contact with those supposed to be his superiors in civilization. **1957** *Times Lit. Suppl.* 26 July 453/2 In his new novel..[he] uses with enviable ease a complicated system of flash-backs.

flasher. Add: **2. b.** An automatic device for alternately lighting and extinguishing incandescent lamps, as in advertising and warning signs; such a sign or signal itself.

1909 *Install. News* III. 127 The effectiveness of the fixed pattern is greatly enhanced by the addition of a 'flasher'. **1928** *Publisher's Weekly* 30 June, Inset, Electric flashers 7×11" (loaned for special displays). **1932** *Flight* 1 July 613/2 A flasher is incorporated in the L.T. circuit and the beacon is usually operated on a flashing code sequence. **1944** *Times* 23 Aug. 5/4 A new technical term.. comes from the Underground, where there are to be more thermal flashers..[i.e.] illuminated signs urging passengers to pass along the platform. **1958** *Observer* 17 Aug. 15/6 Triggers under the steering-wheel work the self-cancelling indicators and the headlamp flasher. **1966** T. Wisdom *High-Performance Driving* ix. 97 Never take for granted.. that the car..with turn-indicator flashing left is actually proposing to turn left. The driver may well have left his 'flasher' on many corners ago. **1971** *Gloss. Electrotechnical, Power Terms (B.S.I.)* i. iii. 16 *Flasher relay*, relay in which the contact units make and break with a self-determined periodicity. **1971** K. Royce *Concrete Boot* iii. 36 The police closed in. They did it well. Plain cars and vans; no sirens or flashers.

8. *Cricket.* (See quot. 1936.)

1936 *Daily Herald* 24 Dec. 15/6 The hitter is a left-hand batsman, but is inclined to be a 'flasher' (one apt to chase balls outside the off-stump). **1956** R. Alston *Test Commentary* 14 If the regular opening batsmen fail, I'd try Benaud, he's no more of a 'flasher' than Trumper.

flash-house. (Later example.)

1922 Joyce *Ulysses* 434 You won't get a virgin in the flash houses.

flashing, *ppl. a.* Add: **1. c.** Hence, applied to a direction indicator on a motor vehicle which signals with flashes of light.

1932 *Daily Express* 20 Sept. 3 Everyone's talking about the new flashing Direction Indicator which only Morris cars carry. This device gives safer signalling. **1959** '*Motor*' *Man.* (ed. 36) vi. 183 Another change since the war has been the progressive displacement of the semaphore-type of direction indicator by flashing light signals, and this latter system is now nearly universal on modern cars...Flashing turn indicators are operated either by a switch of the self-cancelling type mounted on the steering column, or by an independent switch mounted within easy reach of the driver.

fla·shless, *a.* Emitting or producing no flash, *esp.* of gunpowder which ignites without flashing; without a flash (various senses).

1908 *Westm. Gaz.* 2 Mar. 6/2 The latest American artillery wonder is..a 'noiseless, smokeless, flashless, colourless, fumeless gun', which, it is claimed, can throw two million bullets an hour. **1922** [see *flash-spotting* s.v. *flash *sb.*[2] 14 b]. **1929** R. Graves *Good-bye to All That* xi. 123 Before the war there were the two line battalions and the depot; the affiliated and flash-less territorials. **1945** *Times* 10 Oct. 3/4 General Marshall, however, gives the Germans two points: their triple-purpose 88 mm. gun and their powder, flashless as well as smokeless.

fla·sh-over. Also **flashover.** [f. verbal phr. *to flash over* (see Flash *v.*[1] 5).] An accidental arc discharge from a conductor when its voltage is greater than that which the insulation can withstand.

1892 S. P. Thompson *Dyn.-Electr. Mach.* (ed. 4) 88 Commutators of the ordinary sort with thin mica insulation between the bars..are easily short-circuited by the flash-over. **1931** *Nature* 21 Feb. 290 It was..found out that flashovers occurred in the strings of suspended insulators in industrial areas. **1950** *Engineering* 20 Jan. 79/2 Lightning currents which flowed to earth after the tower or earth wire had been struck, did not build up..high voltages..and so cause flashover. **1962** H. C. Weston *Sight, Light & Work* (ed. 2) ix. 253 Similar radiation is emitted when a 'flash over' occurs between electricity conductors other than welding electrodes.

flask, *sb.*[2] Add: **4. b.** *Dentistry.* A sectional metal container for holding a denture during vulcanization.

1859 *Brit. Jrnl. Dental Sci.* Apr. 463/1 In 1846, the necessity became imperative for what are now called in America 'Putnam's vulcanizing flasks'. **1860** J. Richardson *Pract. Treat. Mech. Dentistry* xv. 362 In forming the matrix, a vulcanizing flask is used. *Ibid.* 364 On separating the flask, the teeth, with the wax and temporary plate, will be found attached to the section of the matrix last formed. **1911** G. H. Wilson *Man. Dental Prosthetics* vi. 258 The edges of the denture as it comes from the flask are rough and irregular from the extension of excess vulcanite. **1963** J. Osborne *Dental Mechanics* (ed. 5) xix. 378 Whilst ordinary denture flasks may be used for acrylic teeth, small flasks are easier to handle and less wasteful in plaster.

flask, *v.*[2] Add: **c.** *spec.* in *Dentistry*, to place (a denture) in a flask and surround with plaster ready for vulcanizing. So **flasked** *ppl. a.*, **fla·sking** *vbl. sb.* and *ppl. a.*

1873 O. Coles *Man. Dental Mech.* x. 194 The teeth and bands having been properly adjusted to the model, and to each other, the case is ready for flasking. *Ibid.* 199 When sinking a repair in the flask cover everything but the portion that will require packing, and instead of flasking in the lower portion use the upper section. **1916** J. H. Prothero *Prosthetic Dentistry* (ed. 2) xiii. 194 The flasked case..will then present three separate openings, leading from the base of the crucible to the matrix. **1927** D. M. Shaw *Dental Prosthetic Mech.* vi. 75 The first portion of the flasking plaster is brought up on the gum nearly to the necks of the teeth. **1963** J. Osborne *Dental Mechanics* (ed. 5) xix. 379 Porcelain patterns must be flasked so that they are half-way into the flasking plaster and may thus be removed easily after flasking is completed.

flasqued (flɑskt), *a.* *Her.* [f. Flasque + -ed[2].] Having flasques.

1881 in R. S. Ferguson *Church Plate Carlisle Dioc.* (1882) 52 Small Roman n in a shaped escutcheon, the sides flanched, or flasqued.

flat, *sb.*[2] Add: **2.** In recent use, not necessarily a suite or a complete residence: also used even of one room with shared access to others.

3. *flatdom, -land, -lord* [after Landlord *sb.* 1], *-mate*; *flat-breaking, -dwelling, -hunting, -letting, -warming*.

1936 J. Curtis *Gilt Kid* xxx. 289 These crimes of housebreaking and flatbreaking are far too common. **1926** *Glasgow Herald* 2 Mar. 9 The centre of aristocratic flatdom has also the reputation of being the abiding place of all that is best in dogdom. **1911** *Tariff Reform League: Rep. Labour & Soc. Cond. Germany* III. 173 In the attics of these 'flat' dwellings there is a store for each tenant. **1937** *Sunday Dispatch* 24 Jan. 2/1 So we resigned ourselves to flat-dwelling for ever. **1971** *News of World* 26 Sept. 8/7 Cats with access to gardens like to doze themselves regularly. It is hard on the flat dwelling cat when he cannot do so. **1920** J. Joyce *Let.* 11 Oct. (1966) III. 24 Am very busy flat hunting. **1924** A. Christie *Poirot Investigates* iii. 73 A horrid thing to do..but you know what flat-hunting is. **1889** E. Dowson *Let.* 16 Nov. (1967) 117, I shall be spending the next 40 hours or so in Flatland. **1901** *Daily News* 20 Apr. 4/5 An interesting study in flat-land was provided yesterday at the Royal Courts of Justice, when the owner of a block of flats sued a tenant for a quarter's rent. **1971** *Rand Daily Mail* 3 Apr. 1/8 Within three years Johannesburg's congested flatland will have gained nearly 6500 new flats. **1906** *Westm. Gaz.* 8 Feb. 5/1 Another firm of estate agents.. stated that the outlook in the flat-letting business was anything but cheerful. **1909** *Ibid.* 2 June 2/1 Its pious pretence is to warn simple flatlords against non-payers of rent. **1960** *Woman* 5 Mar. 71/1 All the colourful tales your flat-mates tell you. **1965** K. Giles *Some Beasts no More* i. 18 One day he doesn't come home and his little flatmate eventually gets round to telling the coppers. **1942** *Penguin New Writing* XV. 70 Felix was giving a flat-warming.

flat, *a.*, *adv.*, and *sb.*[3] Add: **A.** *adj.* **2. h.** Of relatively small curvature or inclination. *Golf.* Of a club: having the head at a very obtuse angle to the shaft; of a swing of the club: not upright, oblique.

1857 H. B. Farnie *Golfer's Man.* (1947) v. 27 Regarding the lie of a club for effecting distance, whether it should be flat or upright, little can be said..the rule being, the longer the club, the flatter the lie. **1887** *Jamieson's Scot. Dict.* Suppl., *Flat*, adj. A term in golfing, applied to a club of which the head is at a very obtuse angle to the shaft. **1888** *Lockwood's Dict. Mech. Engin.*, *Flat sweep*, a flat sweep or curve signifies one that is relatively of less curvature than others with which it may be compared. **1909** P. A. Vaile *Mod. Golf* i. 17 [One who plays an upright swing] will keep longer in the line of the ball's flight to the hole, and in the plane of its flight, than one who stands away from the ball and uses a flat swing. *Ibid.* iii. 32 A club with a lie which is too flat.

3. g. *flat tyre* (U.S. *tire*): (*a*) a deflated or punctured tyre; also *ellipt.* as *flat*; (*b*) *U.S.* a dull and spiritless person. Also *flat wheel*.

1923 Wodehouse *Inimitable Jeeves* xvi. 218 I'm bound to say it isn't very often I find my own existence getting a flat tyre. **1925** H. L. Foster *Trop. Tramp Tourists* xvi. 300 You think you're the berries, don't you? Well, you might have been once, but you're a flat-tire these days! You can't make the grade! **1927** *New Republic* 26 Jan. 277/2 He's a flat tire. **1929** 'C. Walt' *Love in Chicago* xv. 211 Stopping at the crossroads to see if I had a flat. **1934** J. M. Cain *Postman always rings Twice* i. 12, I was in the filling station, fixing flats. **1942** 'N. Shute' *Pied Piper* 81 The driver wrestled to jack up the bus and get the flat wheel off.

h. Of the frequency response of an amplifier or other electronic device: uniform (over a certain range of frequencies); of a device: having such a response; amplifying, attenuating, or reproducing equally signals of all frequencies.

1926 *Encycl. Brit.* XXVI. 281/2 A properly designed system is 'flat', i.e., indiscriminatory, over a sufficiently wide auditory band. **1949** Frayne & Wolfe *Elem. Sound Recording* xxix. 604 With this machine a frequency response flat within 2 db is obtained from 30 to 15,000 cycles. **1958** J. Tall *Techn. Magn. Recording* vi. 86 A flat amplifier is one that amplifies all frequencies equally. **1970** J. Earl *Tuners & Amplifiers* ii. 57 Trimmers across the bass and treble tone controls..make it possible to balance them for a 'flat' response when the controls are at the centre setting.

4. b. *Engraving.* Wanting in sharpness; applied to a pull or impression of a plate which has received only the flat impression of the press without the overlay used to develop light and shade.

1888 C. T. Jacobi *Printers' Voc.* 44 *Flat*, an expression used to indicate excessive flatness in an illustration owing to want of light and shade in overlaying. **1888** Flat pull [see Dict., 15]. **1897** Singer & Strang *Etching* 175 A 'flat' proof of a block as it comes from the photo-engraver generally shows itself to be in need of some touching up.

c. Of paint, lacquer, or varnish: lustreless, dull. (Cf. Flat *a.*, etc., A. 4, C. 12, Flat *v.*[2] 8 a, Flatted *ppl. a.* 5.)

1896 in Dict. s.v. Flat *v.*[2] 8 a, To cover (a surface) with flat, i.e. lustreless, paint. **1935** H. R. Simonds *Finishing Metal Products* xxv. 263 There may be gloss or flat paint, or an intermediate semigloss frequently known as egg-shell paint. **1940** R. C. Martin *Lacquer & Synthetic Enamel Finishes* xii. 370 Clear or flat lacquers may then be used to a finish. **1951** H. W. Chatfield *Gloss. Terms Paint Trades* 115 Flat varnishes, lacquers, enamels, etc. **1953, 1958** [see *egg-shell c]. **1971** *Sci. Amer.* Sept. 224/3 Coat the inside of the box with flat black paint.

d. *Photogr.* Wanting in contrast.

1901 G. E. Brown *Finishing Negative* vii. 66 Increasing contrasts..gives a 'snap' to otherwise flat negatives which is often very welcome. **1923** *Kodak Mag.* Mar. 36 From the negative least developed we shall get a print flat and grey with little difference between high light and shadow. **1953** T. L. J. Bentley *Man. Miniat. Camera* (ed. 4) viii. 113 A negative may be so flat and deficient in printing density that straightforward printing will not yield an enlargement of acceptable quality.

6. d. Impecunious, penniless. *U.S. slang.* (Cf. *flat broke* s.v. Flat *adv.* 2.)

1833 *Sk. & Eccentr. D. Crockett* (1834) 60 Retiring to bed, comfortably situated, he awoke next morning flat without a dollar. **1930** *Times Lit. Suppl.* 4 Sept. 698/2 Satisfying his desires freely when he can, starving when he is 'flat'.

e. *to leave* (a person) *flat*, to 'drop' suddenly and completely; to go away from.

1902 G. V. Hobart *It's up to You* ii. 37 Then they both chuckled and left me flat. **1919** in *Saucy Stories* Aug. 83/2 She got up enough spunk to leave him flat on Broadway. Lost him in the crowd...Refused to see him when he showed up. **1930** Wodehouse *Very Good, Jeeves!* iii. 85 He buzzed off, leaving me flat. **1942** T. Rattigan *Flare Path* i. i. 121 You meant my Johnny's going to leave me flat the minute the war's over.

11. Also in *Combs.*, as *flat-sounding, -vowelled*.

1920 'K. Mansfield' *Lett.* (1928) II. 3 His *flat-sounding* voice. **1936** 'M. Franklin' *All that Swagger* i. 10 The haw-haw, flat-vowelled Public School English.

12. b. *Phonetics.* Of a vowel: (see quot. 1934[3]). Also, of a sound: characterized by the downward shift of higher frequencies.

1901 H. Sweet in *Maître Phonétique* 145, if wij dis-'tiŋgwiʃ bi'twijn 'mikst' pə'ziʃən ən 'flæt' ʃeip əv ðə tʌŋ, wij ʃəd nætʃərəli koːl ðə θrij njuw siəriz bæk-flæt bæk-mikst frʌnt-mikst. **1934** H. C. Wyld in *S.P.E. Tract* xxxix. 607 Another long vowel [ā] (low-flat-tense). *Ibid.* 608 This vowel..is the mid-flat-tense. *Ibid.* 609 The tongue may be so used that neither back nor front predominates, but the whole tongue, which lies evenly in the mouth, is raised or lowered. Vowels so formed are called 'mixed' by Sweet, but I owe to him also the term 'flat' which I prefer as more descriptive. The vowel [ā] in *bird* is low-flat. **1952** R. Jakobson et al. *Prelim. Speech Analysis* 31 Flat vs. Plain... Flattening manifests itself by a downward shift of a set of formants.

c. *Gram.* Not distinguished by a characteristic ending, as an adverb which has the same form as an adjective or substantive, or a substantive used as an adjective.

Flat adverbs of modern English often go back to an Old English form ending in *-e*.

1871 J. Earle *Philol. Eng. Tongue* 361 The Flat Adverb is simply a substantive or an adjective placed in an adverbial position. **1901** Greenough & Kittredge *Words* (1902) 199 Not all of our 'flat adverbs' actually go back to such *-e* forms. **1965** *English Studies* XLVI. 356 The 'flat-adverbs' (like e.g. *fast*).

13. (Examples.)

1841 *N.Y. Standard* Jan. (Th.), *Flat*, without interest, in brokers' slang. **1870** *Congress. Globe* 25 Jan. 733/2 [Certificates] have been sold 'flat'..that is to say, without taking the interest into account. **1870** J. K. Medbery *Men & Myst. Wall St.* 61 Stock can almost always be obtained by borrowers, either flat, i.e. with no interest on either side, or with interest at market rates for the money advanced. **1885** *Harper's Mag.* Nov. 843/2 To lend 'flat' means without interest.

b. *Comm.* Unvarying, fixed, uniform; of a standard amount; not varying with changed conditions; without excess or diminution for particular cases. Also quasi-*adv.*

1898 *Engineering Mag.* XVI. 38 Three costs are kept,—the flat cost (including labor and material only), the factory cost (factory expense added to flat cost), and total cost (including all expenses of every kind). **1902** *Encycl. Brit.* XXXII. 163/2 The statistics as to the street railway earnings in America are based upon the universal practice there of charging a 'flat' 5 cent fare for the whole trip. **1903** *Westm. Gaz.* 30 July 10/2 The Tube Railway, on which there was what was known as a 'flat fare'—that was, a fare of twopence for the whole way. **1907** *N.Y. Even. Post* (semi-weekly ed.) 11 Feb. 4 A company.. will take in exactly as much money if the whole lot pay fare at two cents flat. **1908** *Daily Chron.* 21 Feb. 2/6 They had found..that the 'flat' rate system—the fixed annual rate—was unsound. **1920** *Westm. Gaz.* 1 Apr. 4/2 There..ought not to be any flat rate for all classes of horse-drawn vehicles. **1928** *Britain's Industr. Future* (*Liberal Ind. Inq.*) iii. xvi. 193 The majority of workers are paid at flat time-rates. **1950** T. H. Marshall *Citizenship & Social Class* 55 Flat-rate benefits do not reduce the gaps between different incomes. **1958** *Ann. Reg.* 1957 92 Labour offered a flat-rate reduction of £100 to every taxpayer. **1963** *Times* 24 May p. vii/3 The 'six-penny tube', or to give it its correct name—the flat fare—if adopted by London Transport, would lead straight to bankruptcy.

14. a. *flat-capped, -chested* (later examples), *coated* (of animals), *-cut, -edged, -heeled* (later example), *-leaved, -pearled*.

1947 J. Mulgan *Report on Experience* 18 Tenement houses crowded with pale, flat-capped working men. **1961** H. E. Bates *Now sleeps Crimson Petal* 25 A muscular flat-capped skittles player who drove a brewers' dray. **1929** Flat-chested [see *chemicalize *v.*]. **1939** M. Dickens *One Pair of Hands* vi. 97 It sounds so governessy and flat-chested. **1872** 'Stonehenge' *Dogs Brit. Isl.* (ed. 2) 89 The flat-coated or short-coated small St. John's or Labrador breed [of retriever]. **1902** C. J. Cornish *Naturalist on Thames* 109 Our sheep..their wild ancestors, the active and flat-coated animals which still feed on the stony mountain-tops. **1948** C. L. B. Hubbard *Dogs in Brit.* 227 Another of the lesser-known varieties is the Flat-coated Retriever. **1922** Joyce *Ulysses* 178 A flatcut suit of herringbone tweed. **1923** D. H. Lawrence *Kangaroo* xiv. 311 They were walking home in a whirl of the coldest, most flat-edged wind they had ever known. **1890** Kipling *Barrack-r. Ballads* (1892) 183 The hand of every honest man flat-heeled across your mouth. **1926** D. H. Lawrence *Sun* i. 5 The flat-leaved cactus called prickly pear. **1924** E. Sitwell *Sleeping Beauty* vi. 28 Upon the flat-pearled and fantastic shore.

b. *flat-lying* (later examples).

1949 E. Pound *Pisan Cantos* lxxiv. 24 A nice little town in the Tyrol in a wide flat-lying valley. **1965** G. J. Williams *Econ. Geol. N.Z.* vi. 55/2 (*caption*) Flat-lying lodes of the Golden Point group.

c. With *adjs.*, as *flat-icy, -sleek*.

1923 D. H. Lawrence *Kangaroo* xiv. 308 So, in the flat-icy wind..they crouched. **1922** E. Sitwell *Façade* ii And finer Their black hair seemed (flat-sleek to see) Than the leaves of the springing Bohea.

15. flat back, (*c*) a book whose back is flat when the book is closed; (*d*) various other technical senses (see quots.); **flat bed** [Bed *sb.* 11], a bed or frame with a horizontal surface; *spec.* (freq. *attrib.* or as *adj.*): (*a*) a printing machine with a flat printing surface; (*b*) a trailer with no top or sides; (*c*) on a sewing machine, the flat surface on which the fabric rests; **flat bug,** any of the family Aradidæ of bugs, which are very flat and live chiefly under the bark of trees; **flat chasing** (see quot. 1960); also **flat-chased** *ppl. a.* [Chased *ppl. a.*[2]]; **flat-coil,** a pond-snail of the genus *Planorbis*, having its shell coiled in a plane; so *flat-coiled* ppl. a.; **flat-compounded** *a.*, *Electr. Engin.* [*compound *v.* 2 f] (see quot. 1940); **flat-crown,** (*b*) a popular name for two trees, *Albizzia gummifera* and *A. adianthifolia*, both found in southern Africa; **flat-earth** *a.*, of or pertaining to the theory that the earth is flat; so *flat-earther, flat-earth-man*; **flat-fell seam** (see quot. 1964); **flat field** *Photogr.* (see quots. 1904 and 1918); **flat film** *Photogr.*, film on a card or sheet (opp. *roll film*); also *ellipt.* as *flat*; **flat finish** (see quot. 1940); **flat-four** *a.*, of a four-cylinder engine in which two cylinders are placed on each side of the crankshaft and all the cylinders are parallel to the ground; **flat joint,** (*a*) *Building* (see quot. 1904); (*b*) *U.S. slang* (see quot. 1914); **flat-knitting,** a knitting process in which the needles on which the yarn is spun are set in a straight line; so **flat-knit** *a.*, of a fabric made by flat-knitting; **flat-minded** *a.* (see quot. 1928); **flat pea**: see Pea[1] 3; **flat pointing** *Building* (see quot. 1940); hence *flat-joint pointing*; **flat-ring** *a.*, denoting an armature taking the form of a flat ring; **flat-rolled** *ppl. a.*, formed by rolling between smooth cylindrical rollers; **flat-sawn** *a.* (see quot. 1957); **flat silver** *N. Amer.*, knives, forks, spoons, and other eating or serving utensils made of or plated with silver (Webster 1961); **flat-skein work** *Basket-making* (see quot. 1943 and Skein *sb.*[2] 1); **flat slab** *Building*, a concrete slab reinforced in two or more directions to enable it to be supported by columns, etc., without the use of beams or girders; also *attrib.*, as *flat-slab construction*; **flat sour,** fermentation of tinned products by the action of micro-organisms which produce acid but not gas, and so do not cause distension of the tin; also (with hyphen) *attrib.*; **flat space,** Euclidean space; **flat spin** *Aeronaut.*, a spin in which an aircraft descends in tight circles while not departing greatly from a horizontal attitude; *fig.*, a frenzy of agitation, a worried confusion of mind; **flat spot** (see quot. 1940); **flat-tail mullet,** an Australian fish (*Liza argentea*); also *flat-tailed mullet*; **flat-top,** (*b*) *U.S. slang*, an aircraft-carrier; also *baby flat-top*, a smaller (cargo, etc.) vessel converted into a carrier; (*c*) used *attrib.* of a style of hairdressing; **flat tuning** *Radio* [*tuning *vbl. sb.*] (see quot. 1940); **flat turn** *Aeronaut.* (see quot. 1935); **flat-ware** (examples); **flat water** *local*, patches of oily water in the sea, indicating the presence of pilchards; **flat work** *Laundry* (see quot. 1928).

1904 Goodchild & Tweney *Technol. & Sci. Dict.* 226/1 *Flat backs*, whole bound or half bound books whose backs have the leather firmly glued or pasted to them. **1940** *Chambers's Techn. Dict.* 339/1 *Flat-back* (*Moulding*), a pattern having a flat upper surface at the joint of the mould, so lying wholly within the drag or bottom half. **1957** Mankowitz & Haggar *Encycl. Eng. Pott. & Porc.* 53/1 Intended for display on the mantelpiece; often of a 'flat-back' type, modelled and decorated on one side only. **1963** C. R. Cowell et al. *Inlays, Crowns & Bridges* xii. 140 'Steele's' flatbacks—these are useful for patients with close bites. **1875** J. Southward *Dict. Typogr.* (ed. 2) 48 Hoe's Machines...The forme of type is locked-up in the bed by means of screws, by which the type is held as securely as in the ordinary manner upon a flat bed. **1886–7** *Proc. Inst. Civil Engineers* LXXXIX. 247 In the second class is the single-cylinder machine—a cylinder impressing a forme of type on a flat bed, and printing one side only. **1892** J. Southward *Princ. & Progr. Printg. Machin.* 108 The machine printed 3,145 sheets in the hour—a rate that is altogether unprecedented for a flat bed cylinder machine. **1906** F. H. Hitchcock *Building a Book* 121 In the latter part of 1812, the first flat-bed cylinder press was erected by them [*sc.* Koenig and Bauer] in Bensley's office. **1927** E. St. John *Pract. Hints Presswk.* p. xvi, For the general run of commercial work the two-revolution flat-bed cylinder press is preferred. **1932** C. C. Knights *Printing* 69 As the name implies the flat-bed machine has a flat surface or bed upon which the material to be printed is laid. **1959** *Times* 14 Jan. 12/4 The future will see photocomposition allied to letterpress printing, both flatbed and rotary, on an increasing scale. 1960 M. Spark *Ballad of Peckham Rye* iv. 67 They are advertising for ten twin-needle flat-bed machinists. 1961 *Amer. Speech* XXXVI. 272 *Flat bed*, a truck, semitrailer, or trailer with no sides or top. 1970 *Which?* Aug. 237/2 All the [sewing] machines we tested this time had a flat bed. 1971 P. Driscoll *White Lie Assignment* x. 82 Two long flatbed trailers were parked with canvas-covered cylinders on them. 1895 J. H. & A. B. Comstock *Man. Study of Insects* xiv. 139 The flat-bugs .. are the flattest of all bugs, the body appearing as if it had been stepped upon. 1921 *Trans. Amer. Entom. Soc.* XLVII. 1 Every entomologist is in a general way familiar with the 'flatbugs' of the Hemipterous genus *Aradus*, which are often met with beneath the dead bark of trees, but no systematic treatment of the numerous..species has ever been attempted. 1923 W. E. Britton *Guide Insects Connecticut* iv. 11 It is believed the Aradidæ or 'flat bugs', are predatory on insects and other small animals. 1959 Southwood & Leston *Land & Water Bugs* ii. 13 *A. cinnamomeus* differs from other flatbugs whose habits are known in not being a fungus feeder, for it lives on pine sap. 1956 G. Taylor *Silver* iv. 73 Flat-chased or embossed in such low relief as to be almost indistinguishable from flat-chasing. 1960 H. Hayward *Antique Coll.* 119/1 *Flat chasing*, surface decoration in low relief on precious metal, produced by hammering with small blunt tools. The characteristic feature of the process is that no metal is removed. 1901 E. Step *Shell Life* xiii. 234 The *Skenea planorbis*, whose reddish or tawny shell at first sight looks like one of the small fresh-water Flat-coiled Shells (*Planorbis*), whence its specific name. *Ibid.* xvii. 319 The next section of these Pond-snails comprises the Flat-coils (*Planorbis*), made familiar by the typical species—the Ram's-horn or Trumpet-snail. 1915 R. Lankester *Divers. Nat.* xxxiv. 346 The flat-coiled pond-snail, Planorbis. 1926 A. E. Ellis *Brit. Snails* ii. 118 The snails of this family [*sc.* Planorbidæ] are popularly called Ram's-horns, Flat-coils, or Trumpet Snails. 1909 Webster I. 828/1 Flat-compounded. 1940 *Chambers's Techn. Dict.* 339/1 *Flat-compounded*, said of a compound-wound generator the series winding of which has been so designed that the voltage remains constant at all loads between no-load and full-load. 1868 J. Chapman *Trav. Int. S. Afr.* II. 451 The umbrella-like Flat-crown, common in the Berea-bush near D'Urban. 1887 C. A. Moloney *Sk. Forestry W. Afr.* 346 Flatcrown of Natal. 1897 'Mark Twain' *More Tramps Abroad* lxviii, The 'flat-crown' (should be flat-roof)—half a dozen naked branches, full of elbows, slant upward like artificial supports, and fling a roof of delicate foliage out in a horizontal platform as flat as a floor. 1950 *Cape Argus* 18 Mar. (Mag. Section) 7/7 The Cape ebony, white stinkwood, flatcrown, essenhout and umzimbiti, trees that yield beautiful timber for furniture-making, grow in profusion in every kloof. 1905 *Westm. Gaz.* 25 Feb. 3/2 This Flat-Earth Society. 1909 *Ibid.* 21 Sept. 4/1 A lifelong upholder of the flat-earth theory. 1922 A. S. Eddington *Theory of Relativity* 26 Those who adhered to the flat-earth theory must hold that the flat map gives the true size of Greenland. 1934 *Punch* 21 Nov. 562/1 Without being a bigoted flat-earther, he [*sc.* Mercator] perceived the nuisance..of fiddling about with globes..in order to discover the South Seas. 1963 *Times* 11 May 6/2 His treatment of appoggiaturas was that of a Flat Earther with spasmodic doubts. 1908 G. B. Shaw *Fabian Essays* p. xii, Fewer votes than one would have thought possible for any human candidate, were he even a flat-earth-man. 1951 Auden *Nones* (1952) 47 Lovers of small numbers go benignly potty,.. are Millerites, Baconians, Flat-Earth-Men. 1939 M. B. Picken *Lang. Fashion* 58/2 *Flat fell seam*, flat, sturdy seam. 1964 *McCall's Sewing* ii. 28/2 *Flat-fell*, seam used on shirts, slacks and other tailored garments in which one seam is trimmed and the other stitched over it. Gives a flat, finished seam on both sides of the garment. 1841 R. Hunt *Pop. Treat. Art of Photogr.* 80 A photographic camera should possess, according to Sir John Herschel, 'the three qualities of a flat field, a sharp focus, ..and a perfect achromaticity'. 1878 W. Abney *Treat. Photogr.* 205 Since the manufacture of non-distorting doublets giving a fairly flat field has been perfected. 1893 *Jrnl. Soc. Arts* XLI. 384/2 The efforts of opticians.. are being continually put forth in the direction of the attainment of..as near approach as possible to a 'flat field'. 1904 Goodchild & Tweney *Technol. & Sci. Dict.* 226/1 A lens is said to give a flat field when the image of a distant object is equally in focus, whatever part of the screen it occupies. 1918 *Photo-Miniature* Mar. 20 *Flat field*, applied to a lens which, when photographing a flat subject, e.g., a painting, gives equal definition in all parts of the plate. 1966 LaCour & Lathrop *Photo Technol.* x. 115/1 Enlarging lenses are designed to produce an extremely flat field. 1950 *Rev. Documentation* XVII. 134 (*title*) The Microcopy on flat film as an aid in documentation. 1958 *Engineering* 31 Jan. 155/1 The two basic types—roll-film and 'flats', the latter including micro-cards and micro-sheet (or micro-fiches). 1913 *Sat. Evening Post* 12 Apr. 43 (Advt.), The beautiful, modern flat finish for interior walls and ceilings. 1940 *Chambers's Techn. Dict.* 339/1 *Flat finish*, a non-glossy finish, showing no brilliancy of surface. 1959 '*Motor*' *Man.* (ed. 36) 38 They [*sc.* each pair of cylinders] are at 180 degrees and the cylinders are parallel with the ground. This is known as the flat-four type. 1961 *New Scientist* 19 Jan. 162/2 The main features of the Ferguson 'flat four' design are aimed at obtaining the utmost rigidity of the crankcase. 1825 J. Nicholson *Oper. Mech.* 555 In one kind of pointing, the courses are simply marked with the end of a trowel, called *flat-joint pointing*. 1904 Goodchild & Tweney *Technol. & Sci. Dict.* 126/1 *Flat joint*, a mortar joint flush with the face of the wall. 1914 Jackson & Hellyer *Vocab. Criminal Slang* 34 *Flat joint*. Current amongst open-air sure-thing men who operate at circus gatherings, fairs, carnivals, any gaming establishment...The 'Shells'; 'three card monte'; the 'eight die case'..are all grafting flat joints. The term is derived from the essentiality in all of these crooked devices of a counter or other flat area across or upon which the swindle may be conducted. 1963 Mencken *Amer. Lang.* xi. 731 Carnival workers, and especially *strong-joint* or *flat-joint* operators, have a more or less secret argot. 1963 A. J. Hall *Textile Sci.* iii. 149 Large amounts of flat-knit fabric are produced with straight bar knitting machines. 1969 *Sears Catal.* Spring/Summer 19 Seamless stretch tights knit of nylon...Flat knit heels and toes. 1939 M. B. Picken *Lang. Fashion* 58/2 *Flat knitting*, type of knitting done in flat form. 1885 G. Meredith *Diana of Crossways* I. i. 15 One is not astonished at her appearing an 'actress' to the flat-minded. 1928 *Funk's Stand. Dict.* I. 937/3 *Flat-minded*, lacking mental power, imagination, or feeling; devoid of prominent characteristics. 1891–3 *Dict. Techn. & Trade Terms of Arch. Design* 110/1 What is called 'flat pointing' is done by marking the joints of the brickwork with a flat trowel. 1900 *Eng. Dial. Dict.* II. 386/1 *Kentish*. In flat-pointing the mortar is smeared up against, and on to, the edges of each brick. 1940 *Chambers's Techn. Dict.* 339/2 *Flat pointing*, the method of pointing, used for uncovered internal wall surfaces, in which the stopping is formed into a smooth flat joint in the plane of the wall. 1884 S. P. Thompson *Dyn.-Electr. Mach.* iii. 29 The flat-ring armature may be said to present a distinct type from those in which the ring tends to the cylindrical form. 1893 Hawkins & Wallis *Dynamo* 122 A second magnet..can be presented to the other face of the flat-ring core. 1902 *Encycl. Brit.* XXVII. 577/2 The discoidal or flat-ring method. 1935 H. L. Campbell *Working of Steel* iii. 29 Steel is supplied in the form of strips, sheets, plates, and bars. The following definitions apply to these classes of flat-rolled steel. 1962 *Times* 8 Feb. 3/1 Stainless flat-rolled products. 1882 W. D. Hay *Brighter Britain* I. v. 120 Rough split sections of the great logs..fixed in the ground..so as to bring their flat-sawn tops upon a uniform level. 1957 *N.Z. Timber Jrnl.* Mar. 52/1 *Flat sawn*, timber cut tangentially to the annual rings and giving flat grain. 1928 E. Post *Etiquette* 626 The most complete list of flat silver possible. 1968 *Canad. Antiques Coll.* June 10/3 Of what we call 'flat silver' there were of course spoons. 1912 T. Okey *Introd. Basket-Making* xii. 143 Except for Flat Skein work, a purely local industry, skeins are chiefly used by the ordinary basket-maker for handling, and for siding up light work. 1943 A. G. Knock *Willow Basketry* 25 What can be called Flat-Skein Work is being used, each stroke lying as nearly as possible flatly upon the one underneath. 1906 C. S. Hill in Buel & Hill *Reinforced Concrete Constr.* (ed. 2) ii. vii. 175 (*heading*) Flat slab construction. *Ibid.*, The arrangement of the reinforcement in flat-slab floors differs with the form of reinforcement used and with the form of slab arrangement. *Ibid.* 176 The accompanying drawings..show flat slab constructions with Columbian bars. 1960 K. Billig *Struct. Concrete* ii. xiv. 537 Because of the absence of exposed corners, flat-slab construction is less vulnerable in case of fire than beam-and-girder construction. 1926 *Delineator* July 52 In the canning of greens, asparagus, peas, beans and corn, flat sour seems to be responsible for more failures than any other one factor. 1943 J. G. Baumgartner *Canned Foods* iv. 63 The facultative anaerobic group of 'flat-sour' organisms are so called because they attack carbohydrates with resultant acid..formation. *Ibid.* 64 Products containing sugar or starch are particularly liable to undergo severe spoilage by the 'flat-sour' organisms. [1873 W. K. Clifford tr. Riemann in *Nature* 1 May 16/1 These manifoldnesses in which the square of the line-element may be expressed as the sum of the squares of complete differentials I will call *flat*.] 1883 *Encycl. Brit.* XV. 664/1 The space with which we are familiar..has been called flat space or homaloidal space to distinguish it from other spaces in which the curvature is not zero. 1949 Synge & Schild *Tensor Calculus* viii. 295 A space in which the curvature tensor vanishes identically is called *flat*. 1953 B. Spain *Tensor Calculus* v. 56 In a flat space the property of parallelism is independent of the choice of a curve. 1917 'Contact' *Airman's Outings* iv. 104 Suddenly the machine quivered, swung to the left, and nearly put itself in a flat spin. 1919 W. H. Downing *Digger Dial.* 23 *Flat spin*, to be in difficulties. Only applied to an airman. 'On a flat spin', in a bad position. 1928 *Daily Mail* 7 May 6/4 When a person becomes excited or confused, aviators say 'He went into a flat spin'. 1930 *Punch* 30 Apr. 500 Getting into a flat spin over the perishing spelling. 1957 M. Spark *Comforters* vi. 126 It is possible for a man matured in religion by half a century of punctilious observance.. to go into a flat spin when faced with some trouble which does not come within a familiar category. 1967 D. Piggott *Gliding* (ed. 2) xv. 90 Recovery from a flat spin is slow and unpredictable. 1935 C. G. Burge *Compl. Bk. Aviation* 307/2 '*Flat spot*', a term applied to a particular form of hesitation when the throttle is opened from the slow running position. 1940 *Chambers's Techn. Dict.* 339/2 *Flat spot*, in a carburettor, a point during increase of air flow (resulting from increased throttle opening or speed) at which the air-fuel ratio becomes so weak as to prevent good acceleration. 1962 *Which?* Apr. (Suppl.) 74/1 Drivers complained repeatedly of a carburation 'flat spot'. 1896 F. G. Aflalo *Nat. Hist. Austral.* 232 The Flat-Tailed Mullet is also met with in estuaries. 1908 D. G. Stead *Edible Fishes N.S.W.* 43 In form the Flat-tail Mullet is more compressed or slab-sided than the Sea Mullet. 1951 T. C. Roughley *Fish Austral.* 35 The flat-tail mullet is found in all states except Tasmania. 1943 F. Pratt *Navy has Wings* 190 'Scratch one flat top,' Commander Dixon's voice had shouted..through the ship's radio. 1943 *Time* 22 Nov. 26/3 That beats a previous high scorer: the escort carrier 'B'.., another 'baby flat-top'. 1955 C. S. Forester *Good Shepherd* 170 Escort vessels and destroyers and baby flat-tops were coming off the ways as fast as America and England and Canada could build them. 1956 L. S. Trusty *Art & Sci. Barbering* 93 The principal feature of the Flat Top style is the flat top...The top should be visualized as flat and smooth as the bristles of a brush and in length from 1 and ½ to ¾ of an inch. 1957 *N.Y. Times* 2 June vi. 26/1 A stiff version [of the crew cut] is the Flat Top Crew. 1933 'R. Stranger' *Dict. Wireless Terms* 72 *Flat tuning*. A receiver is said to possess flat tuning when a station can be heard over a wide range of movement of the condenser dial. 1940 *Chambers's Techn. Dict.* 339/2 *Flat tuning*, inability of a tuning system to discriminate sharply between signals having different frequencies. 1934 V. M. Yeates *Winged Victory* i. v. 86 An Avro would do anything you wanted..even do a flat turn just for fun if you kicked the rudder with decision. 1935 P. W. F. Mills *Elem. Practical Flying* vi. 88 Turns made without any bank, or flat turns, as they are called. 1851 *Illustr. Catal. Gt. Exhib.* iii. iii. 719/1 Plates, dishes, saucers, &c., termed 'flat ware', are made from moulds which form the inside of the article, the exterior being given by 'profiles' of the required outline, made of fired clay, glazed. 1895 *Montgomery Ward Catal.* 188 Solid Sterling Flat Ware... Tea Spoons..Dessert Forks..Sugar Shells..Butter Knives. 1901 *N.Y. Even. Post* 7 May 4/5 A complete line of Rogers Flatware. 1914 G. Atherton *Perch of Devil* ii. 241 A magnificent silver service, from many dozens of 'flat ware', to silver platters. 1952 M. McCarthy *Groves of Academe* (1953) ix. 198 She seemed to fix her eyes on the flatware and napery with the same hypnotised effort that dragged her fork to her lips and back again. 1927 *Glasgow Herald* 10 Sept. 4 We cruise back and forward watching for signs. The chief of these is 'flat' water, as the men call the smooth, oily patches that so puzzle landward folk. [1906 *Westm. Gaz.* 14 Nov. 9/1 Speaking generally, there has certainly been no rise during recent years, especially in what we call 'flat' articles—tablecloths, and such things.] 1921 *Electrician* 11 Mar. 304/2 In the United States.. some women use their power wringer as a cold mangle for the smaller pieces of 'flat-work'. 1928 *Funk's Stand. Dict.* I. 937/3 *Flat work* (Laundry), articles that are not to be starched, as, sheets or pillow-cases, in distinction from starched articles, as, waists, collars, etc.

B. *adv.* **2.** Delete 'Now *rare*' and add further examples.

1838 Dickens *Nickleby* lxiv. 621, I be not the mun to crow..so I tell 'ee flat. 1842 *Spirit of Times* 21 May 138/1 Every friend of Old Whitenose would have been flat broke! 1933 W. S. Maugham *Sheppey* (1952) ii. 244 You haven't turned it [*sc.* the governor's invitation] down flat?

3. b. Exactly, precisely, not exceeding the stated value: used of amounts, distances, and the like. orig. *U.S.*

1909 Webster 827/3 *Flat*,..Without excess; exactly; due;—used chiefly of numbers or quantities; as, to run a hundred yards in ten seconds *flat*. 1945 *Sat. Rev.* 4 Aug. 22 This one, for instance,..all you fiendishly clever people will solve in no time flat. 1962 *Guardian* 5 July 1/5 In no time flat the very voice of Mrs Chichester..was on the horn. 1969 J. Clarke *Foxon's Hole* viii. 49 It took her about two minutes flat to step into jeans and sweater.

C. *absol.* and *sb.*[3] **2. e.** A flat space or flattened surface; *spec.* a flat place on the tyre or wheel of a vehicle, or the flat space on a commutator caused by sparking or irregularity of rotation.

1873 J. H. Beadle *Undevel. West* xxvii. 576 Up the cliffs, where caves open inward, flats have been worked upon the rock. 1893 Hawkins & Wallis *Dynamo* 391 Occasionally, one or two segments in a commutator wear down below the general cylindrical surface of the rest, and form what is known as a 'flat'. 1906 *Westm. Gaz.* 3 July 10/2 Flats on the wheels may to a great extent be avoided by skilful driving, but there is a potential 'flat' in every skid. 1906 Goodchild & Tweney *Technol. & Sci. Dict.* 808/1 A violent application of the brakes may cause several pounds' worth of damage by grinding a 'flat' on the tyres. 1930 *Engineering* 30 May 708/2 It was suggested that flats in railway wheels could, with advantage, be restricted to a depth of 3 mm. (0·118 in.). 1966 H. Sheppard *Dict. Railway Slang* (ed. 2) 5 *Flat*, worn part on wheel tyre due to skidding, or, in London Transport, due to excessive braking.

f. In full *optical flat*. A block or lamina (usually of glass) with one or more surfaces made accurately plane and smooth, any unevenness or departure from a perfect plane being small compared with the wavelength of light.

1897 *Astrophysical Jrnl.* V. 134 The second..requires two large optical flats, each about one and one-half times the aperture of the telescope itself. 1932 Hardy & Perrin *Princ. Optics* xvi. 345 Round flats are edged exactly like lenses but rectangular ones are edged somewhat differently. 1957 R. S. Longhurst *Geom. & Physical Optics* viii. 135 If an optical flat is placed in contact with a shallow convex spherical surface, a thin air film of varying thickness results. 1971 *Nature* 30 Apr. 575/1 Silver chloride sheet..was made as flat as possible by compression between glass optical flats.

5. (Further examples.)

1765–75 P. Pond in C. M. Gates *Five Fur Traders* (1933) 53 The wind took the Canew up in the Air—Leat hir fall on the frozen flat. 1811 J. Farey *Gen. View Agric. Derbysh.* I. i. 133 Alluvial flat of loam or sandy loam has accumulated upon the Gravel. 1857 R. B. Paul *Lett. fr. Canterbury, N.Z.* iv. 68 The flat on which Mr. Gebbie's house stands. 1888 'R. Boldrewood' *Robbery under Arms* (1889) i. 6 Here it widened out into a large, well-grassed flat. 1944 *Living off Land* iii. 54 These mulga flats contain rocky boulders. 1968 K. Weatherly *Roo Shooter* 130 The doe usually fed on the plateau, but.., she sometimes came down on to the flats.

8. g. (Further examples.)

1804 J. Roberts *Penn. Farmer* 55 It is made like a gate, with five bars or flats. 1891 *Century Mag.* Feb. 526/2 The Mexican system of crushing grain by hand on the *metate*, as the flat under the millstone of the Mexicans and native Californians is called. 1962 A. Nisbett *Technique Sound Studio* 253 *Flat*. (*a*) On a stylus, this is a surface of wear which appears on the two sides of the tip after some period of use...(*b*) On the rubber tyre of an idler wheel, a 'flat' is an indentation which may form if the idler is left 'parked' in contact with the drive spindle, or other surface. 1967 E. Chambers *Photolitho-Offset* 272 *Flat*, a number of negatives stripped-up or assembled in position for printing-down on to a single sheet of metal.

9. f. (Examples of *flat*, *flattie*, a flat-heeled shoe.)

1937 *Amer. Speech* XII. 236/2 She was just a brown-haired kid who wore flatties and went barelegged. 1938 *Times* 11 Mar. 19/4 With your suit, coloured shoes and bags are favourites. Particularly chic are the crocodile 'flats' in cornflower-blue, [etc.]. 1950 A. Lomax *Mr. Jelly Roll* (1952) i. 19 They wore what they called the St. Louis Flats and the Chicago Flats, made with cork soles and without heels and with gambler designs on the toes. 1955 'C. H. Rolph' *Women of Streets* vi. 77 I'm wearing flatties now to rest my feet. 1970 *Guardian* 2 May 1/2 Mary Wilson..saying she liked to 'put on my flatties and head scarf and walk for miles'.

11. (Earlier and later examples.) Also phr., *to join the flats*: to make into a consistent whole, to give unity.

[1746 Garrick *Let.* 11 Dec. in *Corresp.* (1831) I. 46 He had built up the stage, but as nobody came there, he shut in a flat scene to hide it.] 1795 F. Reynolds *Rage* ii. ii. 27 An elegant Apartment leading to Lady Sarah's Dressing-Room—the Door in the Flat. 1901 *Daily Chron.* 21 Aug. 3/4 The 'flats' of her career, so to speak, are not quite joined. 1908 *Ibid.* 29 Apr. 3/3 The 'flats' of the new edition are not very well 'joined'. 1921 G. B. Shaw *Pen Portraits* (1932) 175 Really, Henry Arthur [Jones], you might at least join your flats. 1923 —— *Shaw on Theatre* (1958) 161 A pit without stalls, which jeered mercilessly when the flats would not join. 1932 E. V. Lucas *Reading, Writing & Remembering* iii. 66 He [*sc.* Mr. Asquith] gave the reporters less work in making him grammatical and fluent than any other speaker. There was no need to join his flats. 1957 *Oxf. Compan. Theatre* (ed. 2) 264/1 The frame of an English flat consists of four 3×1 in. timbers, of which the two vertical side-pieces are the Stiles, and the others, the top and bottom Rails. *Ibid.* 265/1 In the mid-nineteenth century..it was..used in the phrase 'a pair of flats', and was confined to the two separate halves of a back scene... Farther back, the word is used only adjectivally, and the full term is Flat Scene.

flat, *v.*[2] **7.** Delete † and add later U.S. examples.

1868 *Harper's Mag.* Aug. 429/2 A bull..commenced to bellow, which awoke our friend, who..exclaimed—dreaming, of course, that some member of his class was exercising his vocal organ—'I say, you have flatted your A, and it won't do!' 1895 *North Amer. Rev.* July 11 When a person has a poor ear for music, he will flat and sharp right along. 1944 W. Apel *Harvard Dict. Mus.* 332/2 Variants in which some of the original tones are flatted.

11. To saw lengthwise through the thickness of a plank, deal, or batten, so reducing the width.

1883 M. P. Bale *Saw-Mills* 333 *Flatting*, sawing through the flat or thinnest way of boards. 1945 J. W. Bush in N. W. Kay *Pract. Carpenter & Joiner* iv. 68 This machine is most useful for..flatting, that is, resawing deals into scantlings.

flat-cap. 2. (Later example.)

1922 Joyce *Ulysses* 391 He..hankered about the coffee-houses and low taverns with crimps, ostlers, bookies, Paul's men, runners, flatcaps, waistcoateers, [etc.].

flat-foot. Add: **1. b.** A condition in draught-horses in which the hoof is large and very sloping, with a flat sole and large prominent frog.

1894 D. Roberge *Foot of Horse* 58 Thoroughbred horses are great sufferers from overgrowth of hoof, although the *form* of his foot is preferable to that of the flat foot or the convex form. 1903 *Dis. Horse* (U.S. Dept. Agric.) 372 Flatfoot is that condition in which the sole has little or no convexity. 1906 H. C. Reeks *Dis. Horse's Foot* 144 Flat-foot is undoubtedly a congenital defect, and is seen commonly in horses of a heavy, lymphatic type.

2. b. A foot-soldier, an infantryman. (Cf. *flat feet*, quot. 1873 s.v. Flat *a.* 15.)

1889 Barrère & Leland *Dict. Slang* I. 348/2 *Flat-feet*, a foot-soldier; applied generally to the Foot-guards. 1903 *Westm. Gaz.* 13 Feb. 2/2 To increase the paper strength of the Army by..'flat-foots' who could not march. 1955 *Archivum Linguisticum* VII. 68 The flatfoot or infantrymen [*sic*].

c. A sailor. *slang.*

1897 *Daily News* 3 Sept. 5/6 When some marine reads it, he'll say, 'It's easy to see that a flat-foot has put that in the paper.' 1898 *Tit-Bits* 30 Apr. 81/3 The blue-jacket himself is a 'flatfoot'. 1904 Kipling *Traffics & Discov.* 73 E's a flat-foot, a indigo-blue matlow. 1909 J. R. Ware *Passing Eng.* 134/1 *Flat-foot*, a young sailor less than twenty-one.

d. A policeman, a plain-clothes man. (Cf. *flatty[2] 3.) *slang* (orig. *U.S.*).

1913 A. Stringer *Shadow* ii. 48 By the time he had fought his way up to the office of Second Deputy he no longer resented being known as..a 'flat foot'. 1932 J. T. Farrell *Studs Lonigan* (1936) iv. 73 He got sore as a boil and stepped up to the lousy flatfoot. 1933 E. C. Vivian *Ladies in Case* iii. 49 Spike laid out a flat-foot—pleeceman, I mean. 1938 J. Curtis *They drive by Night* xxiii. 262, I don't like no flatfoots myself, but I'll help them get anyone of that sort. 1943 P. Cheyney *You can always Duck* iii. 52 This guy's a flatfoot—one of Mr. Hoover's little boys—a Special Agent of the Federal Bureau of Investigation. 1948 C. Day Lewis *Otterbury Incident* vii. 83 Suppose the flatfeet got to hear of it? 1959 I. & P. Opie *Lore & Lang. Schoolch.* xvii. 369 His name is P.C. Wallace but we call him old Walrus. When he has gone round the corner we call names such as slob, natter knob, or flatfeet.

e. A person (with flat feet). *colloq.*

1922 Joyce *Ulysses* 375 Some flatfoot tramp on it in the morning.

So **fla·t-foot** *v. intr.*, to walk in a flat-footed manner.

1932 J. T. Farrell *Young Lonigan* iv. 194 He flat-footed it back to the shade. 1966 *Listener* 27 Jan. 147/1 'Fit for an Alba..' he said, flat-footing round one of Europe's most extravagant palaces.

flat-footed, *a.* Add: **2.** No longer restricted to the U.S. (Earlier U.S. examples and later examples of British use.) Also, dead, insipid, maladroit.

1828 A. Royall *Black Bk.* II. 114 He was one of your right down flat-footed ox-drivers. 1834 *Knickerbocker* III. 35, I haint got no shoes, tis true, but I stand flat-footed and damn the man who can move me one inch. 1899 *Westm. Gaz.* 6 Nov. 2/3 A flat-footed, commonplace scribbler of heroic verse. 1902 *Daily Chron.* 14 Jan. 3/2 His 'enigmatic smile' and his flat-footed compliments. 1923 *Times Lit. Suppl.* 4 Jan. 4/2 Histories of literature ..in which the erudite..mind traverses with flat-footed thoroughness the country it has mapped out for itself. 1931 W. G. McAdoo *Crowded Years* 367 Germany came out flat-footed with the belligerent warning..that she would engage in unrestricted submarine warfare. 1957 R. Campbell *Coll. Poems* II. 111 Weird blue-stockings with damp, flatfooted minds. 1962 *Daily Mail* 19 Jan. 3/8 It is time Associated-Rediffusion presented something less flat-footed and insipid.

b. *U.S.* Unready, not 'on one's toes'.

1912 in *Amer. Speech* (1951) XXVI. 31/1 (Baseball terminology) *Flat-footed*, unprepared, caught napping. 1928 *Funk's Stand. Dict.*, *Flat-footed* (slang). *Racing*. Standing still; unprepared: said of a horse when the jockey is not on the qui vive and expecting a start. 1940 *Topeka Jrnl.* 14 Nov. 1/4 (AP), The Italians were caught flat-footed..and from that moment the story was one of deadly Greek bayonet charges. 1955 H. Roth *Sleeper* ix. 68 A group of people who lie so shamelessly that they are constantly being caught flat-footed. 1963 J. Joesten *They call it Intelligence* i. v. 51 The C.I.A...was caught flatfooted by the military coup in Baghdad.

flat-footedly, *adv.* Also *fig.*

1886 J. A. Logan *Gt. Conspiracy* 660 The old Rebel leaders..came out flat-footedly again with the 'demand that all Custom-house taxation shall be only for revenue'. 1949 'N. R. Nash' *Young & Fair* ii. i. 50, I couldn't come out flatfootedly against the Vidge? 1963 *Times Lit. Suppl.* 18 Jan. 40/4 A flat-footedly dramatic definition of time and place.

flat-head. Add: **1. b.** A fool, simpleton. *dial.* and *slang*.

1862 C. C. Robinson *Dialect of Leeds* 50 Ah tell'd him 'at he hedn't t'sense 'at he wur born wi'—a big flathead. 1884 'Mark Twain' *Huck. Finn* xxiii. 229 Greenhorns, flatheads! 1902 *Westm. Gaz.* 31 May 2/1 'G'n, ye little flathead,' he said admonishingly, 'folks don't get headaches here.' 1922 Wodehouse *Let.* 29 Dec. in *Performing Flea* (1953) 19 Your stuff is really too good for the ordinary magazine, so that when you don't have a plot which is all right for the flatheads anyway, editors are apt to turn you down. 1941 Baker *Austral. Slang* 29 *'Flathead*, a simpleton or fool. 1966 *New Statesman* 6 May 654/2 Gobbledygook is the defence of the American intellectual aware of the hostile mockery of the surrounding flatheads.

flat-headed, *a.* Add: **c. flat-headed borer** *U.S.*, a larva of any beetle of the family Buprestidæ, which bores in the bark and sap-wood of trees.

1882 M. Treat *Injurious Insects* 144 The Flat-Headed Apple-Tree Borer..may at once be recognized by it [*sic*] anterior ends being enormously enlarged and flattened. *Ibid.* 145 This Flat-headed borer is far more common with our Eastern brethren. 1926 E. O. Essig *Insects Western N. Amer.* xxiii. 395 The larvæ [of the family Buprestidæ] are known as flat-headed borers because of the greatly enlarged and flattened thoracic segments. 1968 *Gloss. Terms Timber Preservation* (*B.S.I.*) 13 *Buprestidae*, a family of beetles, often brightly coloured or of metallic brilliance, which bore in or under the bark or in the sap-wood of dying trees or timber; called 'flat-headed borers' in the U.S.A.

flat-iron, *sb.* Add: **1. b.** A kind of boat (see quot. 1961).

1886 *Outing* (U.S.) VIII. 58/1 There are..the 'pumpkin-seed', and the 'flat-iron' models. 1902 *Westm. Gaz.* 18 Aug. 5/2 Steaming down Southampton Water at half-past seven in the 'Albion', a craft which yachtsmen are fond of designating a 'flat-iron'. 1959 *Times* 16 Mar. (Suppl., Port of London) p. viii/3 At Battersea, a strange craft may be seen...This is a 'flat-iron', a coastal collier bringing fuel to the power stations and gasworks. 1961 F. H. Burgess *Dict. Sailing* 90 *Flat-iron*. Any vessel of shallow draft, abnormally wide in the beam, and with low upper-works.

2. Applied to a building, from its triangular shape.

1866 Larwood & Hotten *Hist. Signboards* xiv. 468 The house being wedgeshaped, has an entry at each side. Such a house in London is often called by the vulgar a 'Flat-iron'. 1893 Farmer & Henley *Slang* III. 16/2 *Flat-iron*, a corner public house. 1903 *Westm. Gaz.* 8 June 2/1 The Flatiron Building, it seems, did not exhaust all the possibilities of ingenuity in skyscraper construction. 1929 D. H. Lawrence *Pansies* 76 Then you will hit the Flat-iron Building and flatten it out. 1931 *Times Lit. Suppl.* 10 Sept. 677/3 A 'flat-iron' building is monotonous in a sense in which the Escorial is not.

flatland. Designate sense in Dict. as sense 2 and add: **1.** Also **flat land.** A region of flat land. orig. *U.S.* (See also *Flat *sb.*[2] 3.)

1735 *Springfield Rec.* II. 505 In the Said Ledge Hill then over Rocky flatland we found a Small white marked & Renewed it. 1836 J. Hall *Statistics of West* 31 That [water] which overflows the flat lands, will be stagnant, or flow gently backward in eddies. 1892 A. E. Lee *Hist. Columbus, Ohio* I. 697 The rain poured down in sheets, inundating the flat-land. 1907 W. H. Koebel *Return of Joe* 167 A couple of thousand feet below us was spread the stretch of flatland. 1934 J. L. Myres in E. Eyre *Europ. Civilization* I. 89 Westward towards the Atlantic from the lofty core of Central Asia, the north-west quadrant of the Old World land-mass extends in three great belts, a Northern and a Southern Flatland, separated by a.. Mountain-zone...The two flatlands consist of relatively steady blocks of old crust. 1959 *Times* 18 June (Suppl., Queen in Canada) p. vii/7 The flatlands of Prince Edward county. 1961 *Guardian* 25 March 6/4 Its rich wheat fields..do show what can be done in the Sardinian flatlands. 1970 *Guardian* 18 July 1/8 Newark sprawls across the industrial flatlands that lie across the Hudson River from New York.

flatlet (flæ·tlėt). [f. Flat *sb.*[2] + -let.] A small flat, usu. of one or two rooms.

1925 *Glasgow Herald* 21 Mar. 7 There is a large scheme evolving by which big well-built houses..are adapted to provide one-two-roomed flatlets at a very moderate rental. 1931 W. Holtby *Poor Caroline* vi, The experimental pioneer life in a two-roomed flatlet at Haverstock Hill. 1935 *Times* 9 Oct. 1/7 Wanted. Flatlet houses in really good letting districts. 1950 L. A. G. Strong *Which I Never* 110 Full of tall houses which..were now let out in flats, flatlets and bed-sitting-rooms.

flatness. Add: **3.** (Example in *Photogr.*) Cf. *Flat *a.* 4 d.

1889 E. J. Wall *Dict. Photogr.* 71 *Flatness*, a want of vigour and contrast in the negative and resulting prints, due to under- or over-exposure, or to the use of too strong or too weak a developer.

b. *flatness of field* (see quots. and *flat field* s.v. *Flat *a.* 15).

1867 [see Dict., 2]. 1878 W. Abney *Treat. Photogr.* 206 On the distance of the diaphragm from the lens is dependent the amount of distortion, as is also the size of the picture which the lens is capable of defining; whilst at the same time the flatness of the field is also in a great measure due to a large distance being maintained between them. 1884 *Encycl. Brit.* XVII. 805/2 To get tolerable definition and flatness of field a stop must be added. 1919 *Brit. Jrnl. Photogr. Alm.* 238 Flatness of field means the property of giving equal definition in all parts of the plate when photographing a flat subject with the lens pointed squarely to it.

flat out, *advb. phr.* [Cf. Flat *adv.* 2.]

1. Using or involving all one's (or its) strength or resources; at top speed; = *all out *advb. phr.* 4. Also as *adj.*

Several of the quots. illustrate common Australian colloq. phrases.

1932 *Daily Dispatch* (Manch.) 25 July 6/4 Driving flat out. 1942 *We speak from Air* 51, I followed my bombs down until I was just above the ground again, and then I beat it, flat out, across the roof-tops of Nantes. *Ibid.*, We raced over the chimney-pots, our engines flat out. 1943 C. H. Ward-Jackson *Piece of Cake* 29 *Flat out*. At full speed. Borrowed from motor-racing parlance early in the Great War, full speed being attained on the straight flat run of the track. 1944 J. Devanny *By Tropic Sea & Jungle* 227 The young [rat-kangaroo]..ran straight, as flat out as a lizard drinking. 1945 *Coast to Coast 1944* 157 I'm flat out!..Flat out, like a lizard on a log. 1946 E. S. Tompkins *Speed Camera* 37 There is much more going on..than there is in straight flat-out running. 1957 *Economist* 16 Nov. 581/2 If the unions decide on war, they will have a strong incentive to choose the inexpensive guerrilla type rather than a flat-out effort. 1959 Baker *Drum* 110 Ideas of lying flatly (on one's face, not one's back) and of travelling or working at great speed are recorded in the phrases 'flat out like a lizard drinking' and 'flat out like a lizard on a log'. 1962 *Listener* 29 Mar. 542/1 A flat-out strategic war. 1969 *Private Eye* 23 May 14 I've been flat out, like a lizard drinking, lugging your kiddies round the back paddock. 1969 *New Yorker* 14 June 44/2 Running flat out, Graebner hits a superb hard backhand.

2. Phr. *to be flat out for* (something): to direct one's attention, energies, etc., towards securing, effecting, doing, etc. (something). (Emphatic form of *to be out for*: *out *adv.*).

1930 *Morning Post* 7 Aug. 10/1 Everyone is 'flat out' for flying—my trouble is to persuade anyone to do anything else. 1943 C. H. Ward-Jackson *Piece of Cake* 29 *Flat out for*, in favour of.

flat-out, *sb. U.S.* [f. verbal phr. *to flat out* (see Flat *v.*[2] 4 b).] A failure, a fiasco.

a 1870 in D.A.E., *To flat out*..is heard also as a noun. 'It was a complete flat-out.' 'He made a flat-out.'—N[ew] E[ngland]. 1886 *Century Mag.* Mar. 727/1, I didn't know it was goin' to be s'ch a perfect flat-out.

flatstone (flæ·tstōun). [f. Flat *a.* + Stone *sb.*] **1.** and **2.** (See Dict. s.v. Flat *a.* 15.)

3. A horizontal grave-stone.

1855 *Parl. Papers 1854–5* XLI. i. 52 Flatstone, not exceeding 6 ft. 6 in. by 3 ft. *Ibid.* 53 Memorial, consisting of a flat stone over such grave, in remembrance of one person..£1 10s.

flatted, *ppl. a.* Add: **6.** *Mus.* Of a note: lowered by one semitone. *U.S.*

1938 *Oxf. Compan. Mus.* 322 *Flat*...There is a slight difference of usage in the language of Britain and the United States—in the former 'to flatten' and 'flattened'; in the latter 'to flat' and 'flatted'. 1949 L. Feather *Inside Be-Bop* ii. 50 The 'blue' (flatted) seventh was used incidentally. 1956 —— *Encycl. Jazz* 27 In California in 1940 Oscar Moore, guitarist with the King Cole trio, ended the group's first Decca record, *Sweet Lorraine*, on a ninth chord with a flatted fifth. 1958 A. Jacobs *New Dict. Mus.* 127 *Flattened seventh* (U.S., *flatted seventh*), the lowering of the seventh degree of the scale by a semitone.

flatted (flæ·tėd), *a.*[2] [f. Flat *sb.*[2] + -ed[2].] Divided into, constructed as, or consisting partly of, flats.

1913 *Chambers's Etym. Dict.* Suppl., *Concierge*,.. a door-keeper, esp. in a flatted house. 1921 *Glasgow Herald* 28 Jan. 10 English visitors..are sometimes astonished at our flatted houses. 1941 *When we build Again* (Bournville Village Trust) 107 Many of these numerous small factories are grouped in districts. From the planner's point of view this..makes possible a consideration of the job in terms of units. The solution suggested here for this state of affairs is the 'flatted' factory. 1957 *Times* 22 Aug. 8/7 Some form of 'flatted factories' to cater for the smaller businesses which need more room. 1962 D. Tench *Law for Consumers* vii. 106 The purchase and sale of flatted property has for a long time been a feature of Scottish life. 1970 D. Craig *Young Men may Die* xiv. 93 Opposite is another old flatted mansion.

flatten, *v.* Add: **2. b.** *intr.* Aeronautics. *to flatten out*: to bring an aeroplane into a position parallel with the ground. Also, of the aeroplane: to assume such a position.

1913 *Aeroplane* 17 Apr. 453/1 Apparently he tried to flatten out too quickly. 1914 H. Rosher *In R.N.A.S.* (1916) 37, I just managed to flatten out and straighten up a little as I hit the ground sideways. 1917 C. C. Turner *Aircraft of To-Day* vii. 116 When the sea is calm the pilot often finds it anything but easy to see when to flatten out to 'land'. 1950 [see *approach *v.* 13].

4. *fig.* Also with *down*.

1924 *History* Oct. 209 The general average flattened down to a moderate thirteen shillings. 1971 *Guardian* 11 June 19/3 Furniture and silver prices have flattened out lately.

flattened, *ppl. a.* (Further examples.)

1929 St. John Ervine *First Mrs. Fraser* i. 9 Ninian.. has a conventional English public school accent; colourless, unvaried, with flattened vowels and elided consonants. 1948 *Penguin Music Mag.* Feb. 45 Its typical scale includes the flattened third and flattened seventh (over-flattened by any characteristic Blues singer). 1968 *Gloss. Terms Mechanized & Hand Sheet Metal Work* (*B.S.I.*) 28 *Flattened can*, a complete unflanged can body. 1969 [see *blue *a.* 3 d].

flatter (flæ·təɹ), *sb.*[4] [f. Flat *sb.*[3] 4 b + -er[1].] One who couples and uncouples trucks at a flat in a coal-mine.

1894 *Times* 11 Oct. 4/6 After two years or so..the trapper is advanced to the rank of a flatter. His occupation then consists in coupling or uncoupling the trucks or wagons. 1921 *Dict. Occup. Terms* (1927) § 043 *Flat keeper*, flat coupler-on; flatman, flatter; couples full and uncouples empty tubs at 'flat' or working face terminus of haulage system.

flatter, *v.*[1] Add: **7.** *absol.* (Later examples.)

1913 *Field* 15 Nov. 1046/2 Two furlongs from home Maiden Erlegh looked most dangerous, but he flattered only to deceive. 1928 *Daily Express* 2 Aug. 12 Smirke.. came through well..to settle Goodwin and Dakota, who had flattered in the run home.

9. b. To show to the best advantage, make effective, emphasize the good points of.

1904 *Westm. Gaz.* 11 Feb. 4/2 A dark blue velvet will flatter diamonds remarkably. *Ibid.* 27 Aug. 2/3 The wickets at Nottingham..flatter batsmen enormously. 1909 *Ibid.* 27 Feb. 15/1 A good white shoe worn with a white dress flatters the foot immensely.

flattering, *ppl. a.* Add: **3. b.** *flattering unction*: see Unction 5 b.

flattie: see *flatty[2], and *flat *sb.*[3] 9 f.

flatting, *vbl. sb.* **3.** Delete †*Obs.* and add later U.S. and *transf.* examples.

1895 *N. Amer. Rev.* July 11 When a person has a poor ear for words, the result is a literary flatting and sharping; you perceive what he is intending to say, but you also perceive that he doesn't say it. 1956 Jakobson & Halle in Saporta & Bastian *Psycholinguistics* (1961) 348/1 Secondary tonality features (flatting and/or sharping).

flatty[2] (flæ·ti). *colloq.* Also **flattie.** [f. Flat *a.* + -y[6].] **1.** A flatfish.

1892 O. Heslop *Northumberland Words* I. 291 *Flatty*, a flatfish. 1950 W. Holtby *South Riding* vi. iv. 384 They used to go spiking for flaties in the mud. 1971 *Angling Times* 10 June 25/4 A 2 lb 1¼ oz flounder was the best 'flattie' in Teignmouth SAS's first contest.

2. A flat-bottomed boat.

1892 *Otago Daily Times* 26 Mar., We..accepted the use of a flat-bottomed boat, of the species known in the district as 'flatties'. 1908 E. J. Banfield *Confessions of Beachcomber* i. ii. 60 A huge black beacon waddled along.. followed by a roughly-built 'flatty'. 1937 Partridge *Dict. Slang*, *Flattie*, *flatty*, a small flat-bottomed sailing-boat: coll., esp. among boys: from ca. 1860. 1939 *Yachting* June 109 It's the ideal building material for rowboats, folding boats, kyaks and 'flatties'. 1946 K. Tennant *Lost Haven* (1947) ii. 31 Jamaica was breathing heavily with the effort of rowing the flattie against the tide. 1970 *Motor Boat & Yachting* 16 Oct. 38/3 To be successful a flattie should have a length/beam ratio of 4 to 5, the latter figure giving the faster boat. A typical American flattie of the 1800s was 20ft. in length with 5ft. beam at deck level and 4ft. 3in. at the bottom.

3. A policeman. Cf. *flat-foot 2 d.

1899 'J. Flynt' *Tramping with Tramps* 393 *Flatty*, a policeman; synonymous with 'bull'. 1900 'Flynt' & Walton *Powers that Prey* iv. 51 The 'flatties' in uniforms surrounded the place. 1911 *Sessions Papers* 13 Feb. 359 One of the men says, 'Look out, here comes a "flatty"' (meaning a policeman). 1922 Joyce *Ulysses* 487 The squeak is out. A split is gone for the flatties. 1928 E. Wallace *Gunner* xvii. 133 That is what you tell the flattie if there is any fuss. 1934 D. L. Sayers *Nine Tailors* 216 It's damned slack of you flatties not to have dug me out earlier. 1949 Wodehouse *Mating Season* xix. 166 'You know Dobbs?' 'The flatty?' 'Our village constable, yes.'

flatwoods (flæ·twudz). *U.S.* [f. Flat *a.* + Wood *sb.*[1]] Low-lying timber land; applied especially to types of well-wooded land in Ohio, Alabama, and Florida.

1841 in *Amer. Speech* (1940) XV. 179/1 Running up under the foot of the ridge to a large span oak tree and a forked sugar standing in a flat woods. 1888 'C. E. Craddock' *Keedon Bluffs* 169 The cattle..had been..driven home to the farms in the 'flat-woods'. 1908 R. W. Chambers *Firing Line* ix, Where have you been for a week? In the flat-woods. *Ibid.* x, Those lemon-tinted butterflies which haunt the Florida flat-woods. 1946 G. Wilson *Fidelity Folks* 95 We got to see funny people from up the creek and out in the Flatwoods.

Flaubertian (flobɛ̄·ɹtiăn), *a.* [f. the name of Gustave *Flaubert* (1821–1880), French novelist.] Of, pertaining to, or characteristic of Flaubert.

1926 *Glasgow Herald* 30 Sept. 5 The Flaubertian earnest and emotional intensity with which Conrad wrote his books. 1932 *Week-end Rev.* 13 Aug. 187/1 Flaubertian or Gauterian manifestos on art. 1959 *Listener* 1 Oct. 541/1 The novelist with the almost Flaubertian regard for form. 1963 *Punch* 27 Mar. 462/3 A Flaubertian sense of perfection in literature and in life.

flaunching: see *flanching *vbl. sb.*

‖ **flautato** (flautā·to). *Mus.* [It.; lit. 'fluted'.] = Flautando.

1842 J. F. Warner *Univ. Dict. Mus. Terms* p. xl/2 *Flageoletto*, when employed in relation to stringed instruments, denotes a particular way of making the strings sound, namely in such a manner as to elicit from them their natural accessory tones... The words *Flautino*, *Flautato*, *Flautando*..are used in the same sense. 1876 [see Flautando]. 1935 *Chambers's Encycl.* IV. 728/1 In violin music flautato or flautando is a direction to play harmonics, which have a flute-like quality of tone.

flavanthrone (flēivæ·n-, flēi·vænþrōun). *Chem.* Also **-ene.** [*flavanthrene* ad. G. *flavanthren*, f. Flav(o- + Anthr(a- + -ene; *flavanthrone* f. *flavanthrene* after *anthrone (see -one).] An anthraquinone derivative, $C_{28}H_{12}N_2O_2$, isolated as yellowish brown needles and used as a yellow vat dye.

1902 in *Proc. R. Inst.* (1906) XVII. 107 Flavanthrene, a yellow dye-stuff, which has not yet left the laboratory of its inventor, has a blue leuco-compound. 1902 *Jrnl. Soc. Dyers & Colourists* XVIII. 188/1 The new colouring matter which was referred to by Prof. O. N. Witt in his recent lecture at the Royal Institution as *Flavanthrene*, does not appear to have come into the market yet. 1920 F. A. Mason tr. *Georgievics's Text-bk. Dye Chem.* (ed. 2) 452 Flavanthrone..is sold as Indanthrene Yellow G (B.A.S.F.), Chloranthrene Yellow (B.D.L.), and Caledon Yellow G (Sol.). 1952 K. Venkataraman *Chem. Synthetic Dyes* II. xxxii. 992 Flavanthrone.., which dyes orange-yellow shades from the blue vat, continues to be an important dye (partly because it is a nontenderer and has good light fastness). 1961 Cockett & Hilton *Dyeing Cellulosic Fibres* i. 13 In 1901, the first anthraquinone vat dye, Indanthrone (Indanthrene Blue B), was discovered by Bohn...This was followed in the same year, and from the same chemist, by Flavanthrone (Indanthrene Yellow G) and many other variations.

Flavian (flēi·viăn), *a.* and *sb.* [ad. L. *Flavi-ānus*.] **A.** *adj.* Of or pertaining to the Roman gens Flavia. **B.** *sb.* A member of this gens.

1598 R. Grenewey tr. *Tacitus' Ann.* iii. 123 The captaines also of the Flauian side helped. 1776 Gibbon *Decl. & F.* I. iii. 76 Such a prince [*sc.* Vespasian] consulted his true interest by the association of a son, whose more splendid..character might turn the public attention, from the obscure origin, to the future glories of the Flavian house. 1796 *Encycl. Brit.* XVI. 425/1 [Domitian] distributed great rewards; sitting as president himself, adorned with a purple robe and crown, with the priests of Jupiter and the college of Flavian priests about him. 1867 C. M. Yonge *Pupils of St. John* viii. 139 Beyond this colossus lay the great Flavian Amphitheatre, called from it the Colosseum. 1872 Church & Brodribb *Pliny's Lett.* i. 4 With its close, which witnessed the establishment of the Flavian dynasty, began a more hopeful era. 1882 [see *Claudian *a.* and *sb.*]. 1900 [see *Antonine B. 2]. 1903 *Jewish Encycl.* V. 406/2 An early branch of the imperial Flavian house was at one time inclined toward Judaism and Christianity. 1956 M. Hadas *Hist. Rome* (1958) v. 93 (*heading*) Flavians and Antonines. 1956 M. J. Costelloe tr. *Herling & Kirschbaum's Roman Catacombs* (1960) ii. 36 The domestic tragedy within the imperial household of the Flavians..reached its climax in the assassination of the emperor. 1959 *Chambers's Encycl.* XI. 787 Vespasian established the Flavian dynasty.

flavin. Add: **2.** *Biochem.* Also **flavine.** [ad. G. *flavine* (R. Kuhn et al. 1933, in *Ber. d. Deut. Chem. Ges.* LXVI. B. 320).] Any of the derivatives of isoalloxazine, many of which are important biologically, esp. the cofactors flavin mononucleotide and flavin adenine dinucleotide. (See *riboflavin.)

1933 *Chem. Abstr.* XXVII. 2167 A water-sol. nitrogenous pigment having a yellow-green fluorescence can be isolated from natural sources of *vitamin* B_2... The group name flavin is proposed. 1967 J. P. Lambooy in R. C. Elderfield *Heterocyclic Compounds* IX. ii. 174 Since the majority of the flavins possess polyhydric side chains, there are some unique reactions which such functional groups undergo.

flavine (flēi·vīn). [f. L. *flāv-us* yellow + -ine[5].] **1.** Var. Flavin 1 and *2.

2. *Pharm.* Any of an ill-defined series of yellow acridine derivatives used as antiseptics, esp. the chloride or other salt of 3,6-diamino-acridine. Cf. *acriflavine.

1917 C. H. Browning et al. in *Brit. Med. Jrnl.* 20 Jan. 75/2 On account of its trypanocidal action the compound was called trypaflavine, but, as its range of use promises to be much wider, we shall refer to it simply as flavine. 1951 A. Grollman *Pharmacol. & Therapeutics* xxv. 520 These flavines or acridine dyes possess considerable antiseptic and bactericidal properties.

flavone (flēi·-, flæ·vōun). *Chem.* [ad. G. *flavon* (von Kostanecki & Tambor 1895, in *Ber. d. Deut. Chem. Ges.* XXVIII. 2302), f. Flav(o- + -one.] **a.** A colourless crystalline tricyclic compound, $C_{15}H_{10}O_2$; 2-phenyl-benzo-1,4-pyrone. **b.** Any of the derivatives of this compound, many of which are plant pigments; a flavonoid.

1897 *Jrnl. Chem. Soc.* LXXII. i. 425 Flavone, $C_{15}H_{10}O_2$, is formed by the condensation of ketocumaran with benzaldehyde, and separates in almost colourless crystals melting at 108°. 1938 G. H. Richter *Textbk. Org. Chem.* 664 The powder on the flower stalks, leaves, and seed capsules of many varieties of the primula is chiefly flavone. 1965 Bell & Coombe tr. *Strasburger's Textbk. Bot.* (new ed.) 21 In many genera in which there are blue- or red-flowered species containing anthocyanins, other species have yellow flowers containing flavones.

flavonoid (flēi·-, flæ·vŏnoid). *Chem.* [f. *flavon(e + -oid.] = *flavone b.

1949 *Jrnl. Pharmacol. & Exper. Therap.* XCV. 399 At a recent symposium on vitamin P it was suggested..that the generic term 'Flavonoids' be used to refer to the flavonols, flavanones and related compounds. 1962 *Lancet* 27 Jan. 194/1 A pharmacist who is a hæmophiliac had noted that by taking hesperidin chalcone (a flavonoid) he could ward off hæmorrhagic episodes. 1968 *New Scientist* 25 Jan. 188/1 Pigments found in higher plants include..the yellow, blue, or red flavonoids. 1970 *Watsonia* VIII. 168 Flavonoids may have important physiological functions and are not simply waste products of metabolism.

flavonol (flēi·-, flæ·vŏnǫl). *Chem.* [a. G. *flavonol* (von Kostanecki & Tambor 1895, in *Ber. d. Deut. Chem. Ges.* XXVIII. 2303), f. *flavon(e + -ol.] **a.** 3-Hydroxyflavone, $C_{15}H_{10}O_3$, a colourless crystalline flavone derivative. **b.** Any of the derivatives of this compound, many of which are yellow plant pigments.

1898 *Jrnl. Chem. Soc.* LXXIV. i. 327 Morin differs from quercetin and other flavonol derivatives in colour. 1922 *Encycl. Brit.* XXX. 478/1 A number of flavonol compounds has been found to exist in plants in the sugar free condition. 1949 [see *flavonoid]. 1951 A. Grollman *Pharmacol. & Therapeutics* xxvii. 611 The therapeutic value of rutin and other flavonols (hesperidin, eriodictin, citrin) is still disputed. 1959 E. H. Rodd *Chem. Carbon Compounds* IVb. viii. 906 Many of the ubiquitous yellow colouring materials found in nature are derivatives of flavonol, 3-hydroxyflavone.

flavoprotein, flavo-protein (flēivoprōu·tīn). *Biochem.* [f. Flavo- + Protein.] Any of a group of conjugated proteins having flavin mononucleotide or flavin adenine dinucleotide as cofactor and involved in oxidation-reduction reactions in the cell.

1934 *Chem. Abstr.* XXVIII. 3765 Flavins and flavoproteins as vitamin B_2. 1954 *New Biol.* XVII. 63 For these reactions a series of catalysts is necessary (cytochromes, flavoproteins, cozymase, etc.) which appear to be common to all types of cell. 1966 D. Keilin *Hist. Cell Respiration* xii. 287 With substrates other than succinate it appears that flavoprotein acts as the terminal oxidase.

flavour, *v.* Add: **1. b.** To have the flavour *of*, to savour.

1887 M. Corelli *Thelma* i. xii, A strange sickening sense of unrest that flavoured of despair. 1897 *Westm. Gaz.* 4 Mar. 3/3 Though they flavour more of antiquity and the early Victorian era than of novelty.

flavourful (flēi·vəɹfŭl), *a.* [f. Flavour *sb.* + -ful.] Full of flavour.

1927 *Daily Express* 5 Oct. 5/2 One of the most nourishing and flavourful vegetables. 1968 *Globe & Mail* (Toronto) 17 Feb. 28 The oysters are beautifully succulent and flavorful. 1971 *Nat. Geogr.* May 715/2 Attracted partly by flavorful Norman ways, hordes of British vacationists crowd the [Channel] isles in July and August.

flax, *sb.* Add: **2. b.** (Earlier example.)

1832 A. Earle *Narr. Res. N.Z.* (1966) 64 Their food is always eaten out of little baskets, rudely woven of green flax.

7. a. *flax-stick*, *-swamp* (N.Z.).

1871 C. L. Money *Knocking about in N.Z.* iv. 52 *Mogue*[s is the] Maori name for a raupo or flax-stick raft. 1933 L. G. D. Acland in *Press* (Christchurch) 14 Oct. 15/7 The dried reed stalks..are called f[lax]-sticks. 1871 M. A. Barker *Christmas Cake in Four Quarters* iv. i. 251 A view of downs rolling into a narrow gully, the flax swamp of which formed a natural boundary to one bit of the kitchen garden. 1926 H. Guthrie-Smith *Tutira* (ed. 2) 283 The destruction by fire of flax-swamps.

d. *flax-dressing* (examples), *raising* vbl. sb.

1766 *Complete Farmer* s.v. *Flax* m 3/1 They laid out money for breeding apprentices to flax-raising and flax-dressing. 1874 A. Bathgate *Colonial Experiences* v. 52 Flax dressing is now a permanent industry.

e. Referring to the colour of the flower of the flax, as *flax-blue*, *flax-flower blue*.

1899 *Daily News* 20 May 8/6 Flax-flower blue, a lovely shade that seems to reflect the summer sky itself. 1900 *Ibid.* 28 July 6/7 Flax-blue Irish linen.

8. flax-honey *N.Z.*, honey from bees feeding on flowers of *Phormium tenax*.

1950 G. Dempsey *Wind from Sea* ii. 20 There's plenty of flax-honey..the season has been too dry for clover honey.

flax-seed, flaxseed. Add: **3.** (sense 1) *flax-seed tea*.

1904 E. Glasgow *Deliverance* 101 Polly makes her hot flaxseed tea every two hours. 1922 Joyce *Ulysses* 89 Only measles. Flaxseed tea. Scarlatina influenza epidemics.

flaxy, flaxie (flæ·ksi), *sb. N.Z.* [f. Flax *sb.* + -y[6].] A worker in a flax-mill.

1927 J. Devanny *Old Savage* 51 They would..mingle with the 'flaxies' across the inlet. 1928 —— *Dawn Beloved* iii. 29 No one came by but a young flaxie.

flea, *sb.* Add: **6. flea-bag**, also, a soldier's sleeping-bag; also applied to a shabby building, place, etc.; **flea-bug** *U.S.* = *flea-beetle*; **flea circus** orig. *U.S.*, a show of performing fleas; **flea-hopper** *U.S.*, any of various jumping species of bugs of the family Miridæ, esp. (*a*) the garden flea-hopper (genus *Halticus*), a small black bug that sucks sap from garden plants; (*b*) the cotton flea-hopper, *Psallus seriatus*, a small green sucking insect that attacks cotton; **flea market** *colloq.* [cf. Fr. *marché aux puces*, in Paris], term applied jocularly to a street market; **flea pit** *colloq.*, an allegedly verminous place of public assembly, e.g. a cinema.

1915 H. Rosher *In R.N.A.S.* (1916) 41, I am going to invest in a Jaeger flea bag. 1930 R. Pertwee *Pursuit* i. xi. 58 He snaked his feet into his flea bag. 1932 D. Runyon *Guys & Dolls* v. 94 His room on the top of an old fleabag in Eighth Avenue. 1958 E. Dundy *Dud Avocado* iii. iv. 249 God, how I hated Paris! Paris was one big flea-bag. 1961 *John o' London's* 14 Sept. 296/4 Drab three-storey buildings that now serve as fleabag hotels. 1877 *Rep. Vermont Board Agric.* IV. 158 It is very lively in its movements, and is sometimes called flea-bug. 1928 *Amer. Mag.* May 67/1 Professor Heckler's Trained Fleas... Tiny golden carts rattled merrily across the strip of white blotting paper. The first act in the flea circus was under way. 1932 *Screenland* Apr. 89/1 Shooting galleries, the flea circus, and ending the spree by having their pictures made. 1936 Auden & Isherwood *Ascent of F6* (1937) ii. iii, When I was at school, I tried to keep a flea circus. 1950 *Oxf. Jun. Encycl.* IX. 369/1 'Flea Circuses', which are still to be seen at fairs—the main attraction usually being a chariot race between fleas attached by wire to tiny aluminium chariots. 1953 Flea circus [see *Cheap Jack* s.v. *cheap *a.* D]. 1902 L. O. Howard *Insect Bk.* 301 *Halticus uhleri* Giard, known as the 'garden flea-hopper', is common in gardens and is injurious to flowers and vegetables. 1926 *Jrnl. Econ. Ent.* Feb. 106 The natural hosts of the cotton flea-hopper are various species of Croton. 1920 *Jrnl. Agric. Res.* 15 Mar. 485 The cotton flea hopper, *Psallus seriatus* Reuter,..has only recently become important as a cotton pest. 1959 Southwood & Leston *Land & Water Bugs Brit. Isles* ix. 244 The North American garden fleahopper *Halticus bracteatus* (Say), is often a serious pest of clovers or lucerne. 1922 G. S. Dougherty *In Europe* 130 It is called the 'Flea' Market because there are so many second hand articles sold of all kinds that they are believed to gather fleas. 1960 N. Mitford *Don't tell Alfred* xx. 213 He must learn to clean and crate and pack the object as well as to discover it and purchase it and resell it. From flea-market to Jayne Wrightsman's boudoir. 1970 *New Yorker* 15 Aug. 62/2 The preservation of the open-air flea market. 1937 *Daily Herald* 3 Feb. 12/4 Even the patrons of these palaces [*sc.* cinemas] referred to them as 'flea-pits'. *Ibid.*, A peaked service cap with the name of the flea pit written on the band in gold braid. 1971 *Ink* 12 June 14/4 He went to a fleapit cinema.

flea, *v.* (Later examples.)

1920 T. S. Eliot *Ara vos Prec* 15 Who clipped the lion's wings And flea'd his rump and pared his claws? 1930 E. Waugh *Labels* 100 A pet monkey..fleaed its rump on the terrace. 1932 Auden *Orators* i. 21 The dog fleaing itself in the hot dust. 1937 *Sunday Times* 18 Apr. 8/4 She had the kind of vitality which is ready to start an open-air meeting at Hampstead with an audience of a policeman and a dog fleaing itself.

flea-bit, *a.* (Later U.S. examples.)

1830 O. W. Holmes *Poems* (1852) 181 The small-voiced pug-dog welcomes in the sun, And flea-bit mongrels ..give answer all. 1917 E. Pound *Lustra* 195 Sheep! jabbing the wool upon their flea-bit backs.

flea-bitten, *a.* Add: **3.** Also *fig.*

1917 D. H. Lawrence *Look!* 89 As to whether you've found a fortune In me, or a flea-bitten fate. 1971 *Guardian* 3 July 11/1 Gwalior town itself is quite as seedy and flea-bitten as any other central Indian town.

flead: see Fleed.

fleasome (flī·səm), *a. joc.* (in association with Gleesome *a.*). [f. Flea *sb.* + -some[1].] Full of fleas.

1853 J. Payn *Poems* 79 (*On our Dog Jock*) A gleesome fleasome affectionate beast. 1886 Kipling *Departmental Ditties* 38 Gentle *Bandar*, an Inscrutable Decree, Makes thee a gleesome fleasome Thou, and me a wretched Me. 1924 *Blackw. Mag.* Feb. 220/1 The still fleasome but now gleesome puppy.

flèche. Add: **3.** One of the twenty-four points on a backgammon board; = Point *sb.*[1] B. 3 g.

1863 G. F. Pardon *Hoyle's Games Modernized* 331 Backgammon is played..on a board divided into sections, and figured with twenty-four points or flèches, of different colours, placed alternately. 1867 *Bohn's Hand-bk. Games* 381 Backgammon is played..upon a quadrangular.. board, on which are figured 24 points or *flèches*, of two colours, placed alternately. *Ibid.*, The points or flèches are numbered from 1 to 12 of each colour [*sc.* in a diagram]. 1897 R. F. Foster *Compl. Hoyle* 617 *Flèches*, the points upon a backgammon board.

4. *Fencing*. (See quot. 1961.)

1928 *Daily Tel.* 17 July 17/7 With the épée Walter met his most dangerous opponent..in the first round, and beat him with a rapid flèche attack on the body. 1930 *Morning Post* 14 July 15 In the concluding series Armstrong's pace embarrassed Mowlam into an ill-timed flèche. 1961 F. C. Avis *Sportsman's Gloss.* 192/2 *Flèche*,..used in fencing to mean a running attack.

fléchette (fleʃe·t). [Fr., dim. of *flèche* arrow.] A missile resembling a dart, dropped from aircraft. (*Disused.*)

1915 F. A. Talbot *Aëroplanes & Dirigibles* 141 Another missile which has been introduced by the French airmen.. is the steel arrow, or 'fléchette' as it is called. 1915 Grahame-White & Harper *Aircraft in Gt. War* 247 The first use of 'flechettes' in the war was at the beginning of September.

fleck (flek), *v.*[2] U.S. var. Flick *v.*[2]

1887 A. W. Tourgée *Button's Inn* 124 'You handle that as if there were eggs in it,' laughed the other, flecking the whip. 1889 'C. E. Craddock' *Despot of Broomsedge Cove* xxiii. 424 He experimented with some delicate flecking touches of the bow. 1918 C. Sandburg *Cornhuskers* 50 He lived flecking lint off coat lapels.

fled, *ppl. a.* Add: **2.** Of pottery: liable to crack at a late stage of manufacture by a too rapid change of temperature during or after a firing. *fled ware*, ware cracked after the bisque firing.

1904 in Goodchild & Tweney *Technol. & Sci. Dict.*

fledge (fledʒ), *sb.*[2] [f. Fledge *v.* (sense 4).] A feather- or down-like covering.

1915 D. H. Lawrence *Rainbow* iv. 102 He now had a thick fledge on his upper lip, a black, finely-shaped line. 1923 —— *Birds, Beasts & Flowers* 44 The deep, soft fledge of Sicilian winter-green.

fleece, *sb.* Add: **3. e.** (Further examples.)

1957 M. B. Picken *Fashion Dict.* 133/1 *Fleece*, heavy erect-pile coat fabric of all wool or wool and hair of llama, alpaca, camel, vicuna or cashmere goat. 1964 *McCall's Sewing* iv. 57/1 *Fleece*, a heavy-weight woollen with very long nap used for coats. It is inclined to be bulky, and may be difficult to sew.

6. fleece-picker *N.Z.*, one who picks up fleeces in a shearing-shed; **fleece-wool** (later examples).

1861 H. W. Harper *Lett. fr. N.Z. 1857–1911* (1914) 20 July 54 Shearing, it happens, is in full swing, so there are a number of extra men, besides the shepherds of the station, shearers, fleece-pickers, wool sorters, and 'rouse-abouts'. 1892 W. E. Swanton *Notes on N.Z.* ii. 96 There are the boys to pick up the fleeces, one fleece picker to every four or five shearers. 1945 Baker *Austral. Lang.* iii. 63 Woolshed workers include the fleece-picker or fleecy. 1954 E. C. Studholme *Te Waimate* (ed. 2) xv. 130 Fleece-pickers received 15s. a week, wool-rollers 25s. to 30s. [in 1882]. 1891 R. Wallace *Rural Econ. Austral. & N.Z.* xxix. 385 The years 1880, 1881, and 1882 were good years, when unsorted wool ranged between about 11*d.* and 1*s.* old. per lb., and fleece wool rose to close upon 1*s.* 6*d.* per lb. 1921 *Daily Colonist* (Victoria, B.C.) 8 Oct. 15/2 The cost of putting an extremely good value in fleece wools has made it difficult for the puller to operate. 1951 L. Acland *Early Canterbury Runs* 378 *Fleece*..The main part of a sheep's wool, which is picked up in one, skirted, and rolled. This is sold as fleece wool, as opposed to locks, bellies, and pieces.

fleece-o, fleecy. = *fleece-picker*. *N.Z.*

1894 E. Wilson *In Land of Tui* xv. 244 Meanwhile the shorn fleeces are carried to a large table in the wool-shed by a man appropriately named 'Fleecy', and are spread out by four wool-pickers, who tear away the bad parts and fold the fleeces square, passing them to the classer. 1909 C. Owen *Philip Loveluck* 173 A young Maori boy acted as 'Fleece oh!'..he flicked the fleece up in his arms when it was cut from the sheep, grasped it by its four ends, and flung it out for Loveluck to fold. 1933 'E. Milton' *Waimana* iii. iv. 195 Two male Maori experts..were engaged with the wife of Kuru..to act as 'Fleeço'. 1956 G. Bowen *Wool Away!* (ed. 2) vii. 88 The 'fleece-o', whose work is to pick the fleece off the shearing board when shorn, to throw it on the wool-table, and also to keep the board swept clean.

fleed. (Earlier examples.)

1845 E. Acton *Mod. Cookery* xi. 237 The leaf-fat, or fleed. *Ibid.* xvi. 347 Flead crust. *Ibid.* xxiii. 523 Fleed or flead cakes..very much served as a tea-cake at the tables of the superior order of Kentish farmers.

fleet, *sb.*[1] Add: **1.** *fleet in being*: a phrase first used by the Earl of Torrington after the engagement off Beachy Head in 1690 to describe a fleet which, though inferior to that of the enemy, is able to hamper his movements.

[1690 Earl of Torrington *Sp. to Ho. Comm.* (1710) 29 Most Men were in fear that the French wou'd invade; but I was always of another Opinion,.. for I always said, that whilst we had a Fleet in being, they wou'd not dare to make an Attempt.] 1898 Kipling (*title*) A Fleet in Being. 1899 *McClure's Mag.* Jan. 237/2 It is indeed as a threat to communications that the fleet in being is chiefly formidable. 1902 *Encycl. Brit.* XXXII. 501/1 Of late years controversy has raged round this phrase, 'a fleet in being' and the strategic principle which it expresses. 1964 D. Macintyre *Battle for Mediterranean* i. 31 A lack of desire on the part of the Italians to risk their fleet..is..probable, a policy which accorded with the theory of maintenance of a 'fleet in being'.

d. *Fleet Air Arm*, the branch of the air force formed to operate with the fleet. (First formed 1923; came under the Admiralty's control in 1938.) Abbrev. *F.A.A.*

1923 *Rel. Navy & Air Force* 2 in *Parl. Papers* (Cmd. 1938) XV. 827, 1. *Navy and Air Force*..Naval officers belonging to the Fleet Air Arm are therefore to be attached to the Air Service on the nomination of the Admiralty. 1933 *Jrnl. R. Aeronaut. Soc.* XXXVII. 337 The references made..to the activities of the Fleet Air arm were fully justified...It was surprising how tiny a speck on the ocean an aircraft carrier could appear when one had to return to it after a sea reconnaissance. 1939 *Navy List* Sept. p. xiii, F.A.A.—Officers serving with or undergoing training in the Fleet Air Arm. 1940 E. C. Shepherd *Brit. Air Power* 27 But for the Fleet Air Arm, which is exclusively controlled by the Admiralty, the whole of British Air Power is thus under the direction of a single staff—the Air Staff—and is commanded by R.A.F. 1953 *Times* 21 May 8/2 The Admiralty has decided to reintroduce the term 'Fleet Air Arm' after a lapse of seven years, during which the air forces of the Royal Navy have been known officially as 'Naval Aviation'.

2. Revived in the sense of: a number of vehicles or aircraft forming a definite group or 'unit'.

1881 W. D. Hay *300 Years Hence* x. 248 The fleet of the Avengers sweeps onward through the air. 1889 *Kansas Times & Star* 7 May, A fleet of hacks was cruising around to take them to their..homes. 1905 *Daily Chron.* 11 Sept. 5/6 At five o'clock, when the motor fleet drew up. 1908 H. G. Wells *War in Air* iii. § 5, A fleet of airships. 1911 *Chambers's Jrnl.* 767/2 The vast sum of money..expended on..a single Dreadnought might better be devoted to creating a whole fleet of dirigibles and aeroplanes. 1915 W. J. Locke *Jaffery* iii, Barbara has gone away with the Daimler,..and as I don't keep a fleet of cars, I had to choose between this and the donkey-cart. 1924 F. J. Haskin *Amer. Govt.* (rev. ed.) 433 Large fleets of trucks and automobiles. 1967 *Boston Sunday Globe* 23 Apr. 8/1 A separate electrically powered fleet [of trains] will be cutting the time between Manhattan and Washington, D.C. 1971 *Guardian* 22 Feb. 12/4 By accident I joined what drivers described as a company with one of the worst fleets of transport in the kingdom.

4. *fleet-action, engineer, -man.*

1901 J. Blake *How Sailors Fight* iv. 91 As the first phase of a fleet action the captains of the various warships meet together on the admiral's flagship. 1910 *Daily Chron.* 17 Mar. 3/5 The fleet action of the future will..develop into an aggregation of duels between opposing battle units. 1901 J. Blake *How Sailors Fight* ii. 44 A fleet engineer is always on duty in South Wales, and it is his business to recommend to the Government the coal they shall include in their contracts. 1904 *Daily Chron.* 1 Feb. 3/3 It will encourage straight shooting among the fleetmen. 1905 *Ibid.* 19 July 5/6 The magic personality of British fleetmen. 1909 *London Mag.* Aug. 605/2 To cheer the fleetmen as they march through the streets.

fleet, *sb.*[2] Add: **2. b.** *Fleet Street*, a street in London devoted largely to the production and publication of daily newspapers and periodical journals; hence *allusively*, the national newspapers generally, the journalistic press, journalism. So *Fleet Streeter*.

1882 C. Pebody *Eng. Journalism* ii. 19 Fleet Street to-day, with its energy, enterprise, and intelligence, is a characteristic representation of the whole spirit of the English Press. 1893 Farmer & Henley *Slang* III. 19/2 *Fleet-Street*, the estate of journalism. *Ibid.*, *Fleet-Streeter*, a journalist of the baser sort. 1904 J. R. Robinson *50 Yrs. Fleet St.* 236 The passion for letter-writing to newspapers is recognised in Fleet Street as a distinct form of mental aberration. 1905 H. Leach *Fleet St.* 87 A faculty for quick and perfect condensation is one of the most valuable possessions of the Fleet Street man. *Ibid.* 143 Fleet Street has its particular specialists for several of the courts of law. *Ibid.* 189 Their [*sc.* editors'] differences show how impossible it is for Fleet Street to tell what the morrow will bring forth for it. 1920 [see *DOWNING STREET]. 1927 *Scots Observer* 12 Mar. 11/3 Shabbiness and flyblownness represent a Fleet Street tradition that has been broken. 1962 *John o' London's* 22 Feb. 177/3 The event is a natural for the copy-starved Fleet Streeters. 1969 *Times* 10 Dec. 11/4 The underminers in Fleet Street: *voilà l'ennemi.*

fleet, *a.*¹ Add: **4.** *fleet-feathered* adj.; **fleet-foot** *a.*: also *absol.*; **fleet-hound,** in later use (see quot.).

?1862–63 G. M. HOPKINS *Poems* (1948) 177 Divinity of air, fleet-feather'd gales. 1940 C. DAY LEWIS tr. *Virgil's Georg.* IV. 88 Arethusa the fleetfoot, her arrows at last laid by. 1888 H. DALZIEL *Brit. Dogs* (ed. 2) I. ii. 47 The Deerhound..is also named the Rough Greyhound, and the Northern, or Fleet-hound.

fleetful (flī·tful). [f. FLEET *sb.*¹ + -FUL 2.] As many as would make a fleet; *transf.* a large number.

1899 *Westm. Gaz.* 21 Feb. 2/2 The public would have paid for a fleetful of ships. 1908 P. DINNEEN tr. *Keating's Hist. Irel.* II. 165 The king gave him a fleetful, that is, two thousand two hundred, and they put out to sea. 1923 S. GWYNN *Hist. Irel.* v. 45 The 'fleetfuls' of foreign students who came to attend surprisingly advanced teaching.

Flem (flem). An abbreviation of FLEMING¹.

1909 *Daily Chron.* 20 Aug. 3/3 Rubens..This exuberant, robust, vivacious Flem. 1925 *Contemp. Rev.* Mar. 312 The Flems outnumber the Walloons in Belgium.

flemingin (flemi·ndʒin). *Chem.* [f. mod.L. *Fleming-ia*, a genus of plants, f. the name of John *Fleming* (1747–1829), British physician and botanist + -IN¹.] A crystalline compound, $C_{12}H_{12}O_3$, obtained from *Flemingia* (= *Moghania*) *congesta* and used esp. as a yellow dye for silk.

1898 A. G. PERKIN in *Jrnl. Chem. Soc.* LXXIII. 661 *Flemingin*, which is the name proposed for the substance thus obtained, is a dull, orange-red, crystalline powder, having a faint lustre. 1943 A. H. COOK tr. *Mayer's Chem. Natural Coloring Matters* iv. 258 Flemingin is obtained from waras.

flesh, *sb.* Add: **1. e.** Also with *crawl* (cf. CRAWL *v.*¹ 6).

1876 'MARK TWAIN' *Tom Sawyer* 61 It makes my flesh crawl to hear you.

12. a. *flesh-burst, -case, -creep, -disguise, -flower, -hotpot* (nonce formation on FLESH-POT), *-rose, -scape, -scent; flesh-stuff, -tint* (later examples); *flesh-tone.*

1876 G. M. HOPKINS *Wr. Deutschland* (1918) 60 How a lush-kept plush-capped sloe Will, mouthed to flesh-burst, Gush! 1922 JOYCE *Ulysses* 189 For years in this fleshcase a shesoul dwelt. 1904 *Daily Chron.* 17 Dec. 3/3 The fight between Tyler and the huge ape is worth the money, and Mr. Fritz Bergen's illustrations of it will throw in several extra flesh-creeps. 1910 J. MASEFIELD *Ballads & Poems* 64 Arrayed in some new flesh-disguise. 1862 G. M. HOPKINS *Vision of Mermaids* (1929) Diadem'd Like an Assyrian prince, with buds unsheath'd From flesh-flowers of the rock. 1922 JOYCE *Ulysses* 560 Her sleeve, falling from gracing arms, reveals a white fleshflower of vaccination. *Ibid.* 502 Fleshhotpots of Egypt to hanker after. 1955 C. TOMLINSON *Necklace* 21 It moves with equal certainty Through a register of palm-greens and flesh-rose. 1949 S. SPENDER *Edge of Being* 21 A fleshscape woven of fiery fleece. 1923 D. H. LAWRENCE *Birds, Beasts & Flowers* 41 The flesh-scent of this wicked tree. 1935 G. BARKER *Janus* 42 His face..no more than a mounding of flesh-stuff. 1967 E. SHORT *Embroidery & Fabric Collage* iv. 97 In figure work they depicted form with line rather than shading, and avoided naturalistic flesh tints. 1931 H. READ *Meaning of Art* II. 128 Flowers and the flower-like flesh-tones of a woman's or a child's body.

b. *flesh-creeper, -creeping* ppl. adj.; *flesh-devouring, -eating* (later examples).

1887 H. BAUMANN *Londinismen* 56/1 *Flesh-creeper*, scherzhaft: Schauerroman Gruselgeschichte. 1932 T. S. ELIOT *Sel. Essays* 321 He is a little of the religious spell-binder..the flesh-creeper, the sorcerer of emotional orgy. 1959 *Times* 14 Feb. 9/7 The moving *Funérailles* which he played simply as a flesh-creeper. 1896 *Westm. Gaz.* 17 Mar. 3/1 The flesh-creeping announcement that our old friend Osman Digna, who has died so often, is advancing with a considerable force in the direction of Kassala. 1936 *Times Lit. Suppl.* 7 Mar. 195/3 Mr. Burke gets the right flesh-creeping quality into some of the scenes. 1936 R. CAMPBELL *Mithraic Emblems* 28 The flesh-devouring bird of time. 1953 J. S. HUXLEY *Evol. in Action* iii. 72 One original tiny flesh-eating creature.

c. *flesh-bound, -petalled, -untrammelled, -winged.*

1877 G. M. HOPKINS *Poems* (1918) 31 Man's spirit will be flesh-bound when found at best, But uncumbered. 1909 W. JAMES *Pluralistic Universe* vi. 252 Turn your face toward sensation, that flesh-bound thing which rationalism has always loaded with abuse. 1949 S. SPENDER *Edge of Being* 36 This flesh-petalled tree. 1917 J. MASEFIELD *Lollingdon Downs* 57 Bodiless joy of flesh-untrammelled mind. 1947 S. SPENDER *Poems of Dedication* 53 Legendary heroes..On flesh-winged ships fluttered from their island.

d. *flesh-fragrant, -rosy.*

1923 D. H. LAWRENCE *Birds, Beasts & Flowers* 15 Gods..half-sinisterly flesh-fragrant As if with sweat. 1957 L. DURRELL *Bitter Lemons* 102 There was a fine fire of flesh-rosy carob wood.

13. flesh-split, that part of a split hide or skin which is nearest the flesh.

1897 C. T. DAVIS *Manuf. Leather* (ed. 2) 429 Flesh splits are splits which are finished on the flesh side. The usual weight of the flesh splits is from 1½ to 3½ pounds. 1904 GOODCHILD & TWENEY *Technol. & Sci. Dict.* 37/2 s.v. *Bag hides*, The lower or Flesh Split is used for insoles and stiffeners in boots.

flesh, *v.* **4. b.** Delete †*Obs.* and add later examples.

1909 H. G. WELLS *Tono-Bungay* II. i. 103 We've fleshed ourselves a bit, eh? *a* 1909 *Century Mag.* (Webster), A rangy bay that ought to be fleshed for six months.

flesh-coloured, *a.* (Earlier and later examples.)

1703 tr. *H. van Oosten's Dutch Gardener* II. lxix. 111 There are three sorts of common Pionies...Second, the Flesh Colour'd. 1966 J. S. COX *Illustr. Dict. Hairdressing* 58/2 Flesh-coloured hair net.

fleshly, *a.* **1.** (Later examples.)

1881 W. S. GILBERT *Patience, Dramatis Personae*.. Reginald Bunthorne (a Fleshly Poet). 1882 [see *ÆSTHETIC *a.* 4]. 1961 *New English Bible John* i. 13 Not born of any human stock, or by the fleshly desire of a human father.

fleshmonger. Restrict †*Obs.* to senses in Dict. and add: **3.** A slave-dealer.

1845 F. DOUGLASS *Life* (1846) 4 Cruel as the deed may strike any one to be, for a man to sell his own children to human flesh-mongers, it is often the dictate of humanity for him to do so.

Fletcherian (fletʃīə·riăn), *a.* [f. the name of John *Fletcher* (1579–1625), English dramatist + -IAN.] Of, pertaining to, or characteristic of Fletcher or his works.

1850 *Gentl. Mag.* Aug. 119/2 Many passages were strongly marked with the favourite Fletcherian cadence. 1907 *Westm. Gaz.* 3 Apr. 4/1 It is curious..that FitzGerald should have made his translations of Calderon so strongly Fletcherian in style. 1921 *Spectator* 19 Feb. 236/2 We all know how this new flower flourished and ramped in the Fletcherian garden and was taken up by the poet's dramatic imitators. 1947 *Scrutiny* XIV. 318 Anxious to fit the play [sc. *Henry VIII*] in as the final goal of the Shakespearian progress..he is driven to explain away the limp 'Fletcherian' verse.

Fletcherism (fle·tʃəriz'm). [f. the name of Horace *Fletcher* (1849–1919), American author + -ISM.] The practice of thorough mastication advocated by Fletcher. So **Fle·tcherite,** a follower of Fletcher; **fle·tcherize** *v. trans.*, to masticate thoroughly. Also *fig.*

1903 *Lit. Digest* 28 Nov. 739/1 It is now proposed to speak of the 'Fletcherizing' of food that is thoroughly chewed. 1904 *Daily Chron.* 31 Oct. 4/7 The Fletcherites preach the gospel of chewing. 1906 *Suburban Country Life* (Boston) Aug. 101/1 His investigation of what has been known as Fletcherism. 1907 *Practitioner* June 852 The Fletcherites, who, so far from not giving two bites to a cherry, insist on thirty-two to a mashed potato. 1909 H. JAMES in W. D. Orcutt *Quest Perf. Bk.* (1926) 88 It is impossible save in a long talk to make you understand how the blessed Fletcherism..lulled me, charmed me, beguiled me. 1910 'O. HENRY' *Strictly Business* (1917) xvii. 182 Annette Fletcherized large numbers of romantic novels. 1922 WODEHOUSE *Adv. Sally* ii. 41 The raffish mongrel was apparently endeavouring to fletcherize a complete stranger of the Sealyham family. 1925 'J. DOYLE' *Marmosite's Miscellany* 8 The rank spiny grass and the remotest flower..I am garlanding together in a glorious wreath To fletcherise at leisure.

Fletton (fle·tən). Also **fletton.** [The name of a town near Peterborough.] Used *attrib.* or *ellipt.* to designate a type of brick made by a semi-dry process, orig. from the Oxford clay found near Fletton.

1908 C. F. MITCHELL *Brickwork & Masonry* (ed. 2) xiv. 390 The average Fletton brick, when immersed in water for 24 hours, will absorb 20 per cent of its weight of water. 1927 *Daily Tel.* 24 May 3/3 The imported bricks cannot compete..in our area with Fletton bricks. 1934 *Archit. Rev.* LXXVI. 122/2 The rear elevation has rustic flettons distempered. 1967 *Times* 9 Feb. 16/1 London Brick, the largest fletton producer in Britain. 1971 *Daily Tel.* (Colour Suppl.) 28 May 21/4 Nearly half the bricks made in Britain are 'flettons' from this bed of clay, and nearly half the flettons come from the Bedfordshire brickfield.

fleur. Add: **2. b.** In names of artificial silk materials, as *fleur de chine, fleur de soie* (also *fleursoie*).

1927 *Daily Express* 5 Apr. 3 It is practically impossible to tell the newest material 'fleur de Chine' from crêpe de Chine. 1928 *Times* 9 May 10/4 A draped gown of parchment fleur de soie, trimmed with old lace. 1928 *Daily Express* 23 July 5/2 Fleursoie, a soft firm material rather like triple georgette.

3. fleur de coin (də kwæ̃) *Numismatics*, mint or perfect condition.

1892, 1966 [see F.D.C. (s.v. *F III. 3)]. 1931 *N. & Q.* 12 Dec. 430/2 The expression 'mint state' has passed on from our fraternity...The strange old French expression *fleur de coin* has generally been substituted for it, as expressing that exquisite gloss which characterizes virgin new-born specimens.

Fleuss (flois). The name of H. A. *Fleuss*, British inventor, used *attrib.* or in the possessive of apparatus designed by him.

1882 *Rep. Insp. Mines 1881* 463 Fleuss' breathing apparatus and lamp. 1883 *Encycl. Brit.* XVI. 467/1 Persons equipped with the Fleuss breathing apparatus can now enter mines after explosions. 1902 *Ibid.* XXXII. 190/1 The small Fleuss machine will produce about 1½ lb of ice in one operation of 20 minutes. 1904 GOODCHILD & TWENEY *Technol. & Sci. Dict.* s.v. *Air Pumps*, A great improvement in mechanical pumps has been made of recent years in what is called the Fleuss Pump. 1906 *Ibid.* s.v. *Tyres*, The Fleuss Tyre possesses no inner tube, but the edges of the cover overlap and fit together closely enough to retain the air. 1909 *Westm. Gaz.* 22 Feb. 8/1 'The Fleuss,' he added, 'is the prototype of all apparatus using compressed oxygen and caustic soda, and it has already proved its value in mine rescue-work.'

flex, *v.* (Further examples.)

1912 L. J. VANCE *Destroying Angel* xiii. 160 Tensing and flexing his tired muscles while his eyes shifted quickly from one quarter to another. 1926 R. MACAULAY *Crewe Train* II. ix. 167 Flexed her toes against the rock's level face, and drew herself up. 1947 R. ALLEN *Home made Banners* iii. 16 It dedicated its soul to the war and flexed its muscles for the war, and then there was a pause.

flex (fleks), *sb.*¹ *Math.* [f. L. *flexus*, f. *flectere* to bend.] A point of inflexion (see INFLEXION, INFLECTION 3).

1866 A. CAYLEY in *Q. Jrnl. Pure & Appl. Math.* VII. 213 The stationary ineunt, or cusp, considered as..a particular case of the double ineunt, is a spinode; to render this notation symmetrical, we require certain new terms, say link, as the correlative to node, and flex as the correlative to cusp...The ordinary singularities of a plane curve would thus be the node, the cusp, the link, and the flex. 1950 H. G. FORDER *Geometry* ix. 152 The line which joins two flexes of a cubic goes through a third flex.

flex (fleks), *sb.*² [Abbrev. of *FLEXIBLE *sb.*] Flexible low-current electric cable consisting of two or more stranded conductors separately insulated, used esp. for connections to electric lamps and portable domestic appliances.

1905 C. C. METCALFE *Pract. Electr. Wiring* vii. 104 Flexible cord, generally known in the wiring trade as 'Flex', consists of two flexible conductors laid together, each insulated from the other. 1907 *Install. News* July 10/2 A flat piece of vulcanite..provided with three holes, the bottom two being used to take the flex. 1922 *Daily Mail* 18 Dec. 1 Electric iron, complete with flex cord and adaptor. 1966 *Electricity at Work* Sept., In a 2 core flex there is no earth core. 1967 P. HONEY *Househ. Electr.* 65 Never run flexes under a carpet.

flexed, *ppl. a.* Add: **b.** *Archæol.* Of a corpse: buried with the legs drawn up under the chin. Hence also of a burial in which the corpse is in this position.

1915 A. KEITH *Antiquity of Man* vi. 117 The body had been laid on its back,..with knees, thighs, and elbows flexed. 1934 V. G. CHILDE *New Light Most Anc. East* ix. 274 The bodies had been interred extended or flexed. 1952 [see *EXTENDED *ppl. a.* 1 b]. 1959 J. D. CLARK *Prehist. S. Afr.* ix. 249 The burial at Driekops Eiland was flexed, lying on its side and facing south.

flexibility. 2. (Further examples.)

1961 *Oxf. Univ. Gaz.* 19 Oct. 176/1 The principle of flexibility in fixing stipends within scales. 1962 *Listener* 29 Mar. 540/2 It follows then, doesn't it, that a civil defence programme is not meant primarily to add to the flexibility of American strategy?

flexible, *a.* Add: **1. b.** In modern mechanical and electrical usage (see quots.).

1859 W. J. M. RANKINE *Man. Steam Engine* iv. 126 (*caption*) Flexible tube and diaphragm valves. *a* 1877 KNIGHT *Dict. Mech.* I. 882/1 Thirion's flexible coupling is used for conveying power from one shaft to another when they are not in line. 1888 *Lockwood's Dict. Mech. Engin.* 144 *Flexible crank shaft*, a crank shaft in which the strains due to the rigidity of an unyielding mass compelled to revolve under conditions of strain due to the want of alignment of its bearings, are reduced and minimised by the introduction of flexible joints in its length. 1909 *Install. News* II. 186/1 Flexible cord surface wiring. 1909 *Daily Chron.* 24 Aug. 7/5 Another advantage of the 'flexible' engine..is the power it gives the driver to avoid accidents and collisions. 1918 E. S. FARROW *Dict. Mil. Terms, Flexible*, a term applied to a dirigible balloon in which a flexible gas container is held in shape only by the pressure of gas within and to which the load is hung. This characterizes the whole non-rigid system of airships. 1965 *Economist* 27 Mar. 1412/3 Flexible or 'black top' road design is also being studied. 1967 *Gloss. Sanitation Terms (B.S.I.)* 28 *Flexible joint*, a joint which permits some movement of the jointed parts out of their original alignment or which permits the jointing of parts which are not truly aligned. 1970 A. BYERS *Home Lighting* i. 22 Floor standards, table lamps,..and permanent lighting may all be supplied from plugs and socket outlets using a flexible cord.

4. d. *flexible response*: see quot. 1966. Cf. *controlled response* (*CONTROLLED *ppl. a.* 2).

1963 *Time* (Internat. ed.) 26 July 15/3 Pentagon planning now puts relatively more emphasis than it did a few years ago on 'graduated' or 'flexible' responses. 1966 SCHWARZ & HADIK *Strategic Terminol.* 71 *Flexible response*, strategy based on the capability of reacting appropriately across the entire spectrum of possible challenge from atomic war to infiltration and subversion. 1967 *Punch* 26 Apr. 597/2 Both parties will have embarked on a policy of manufacturing anti-ballistic missiles

on a scale which will cause defence costs to escalate out of all proportion, and possibly even to destroy the philosophy of flexible response, owing to the necessity to saturate enemy defences at the first stroke. 1968 *Economist* 10 Feb. 14/2 Last year Nato at long last adopted a strategy of flexible response in place of massive retaliation.

B. *ellipt.* as *sb.* = *FLEX *sb.*²

1896 H. J. DOWSING *Griffin's Electr. Engin. Price-Bk.* (ed. 2) 254 In factories, mills, etc., where mechanical strength is of importance, a pendant fitting is made with iron armoured wire in place of the concentric flexible. 1907 *Install. News* May 11/1 Conductors (excepting flexibles) may be enclosed in steel conduits. 1925 *G.E.C. Publications: Wires & Cables* 77 Motor car ignition flexibles. 1942 MEARES & NEALE *Electr. Engin. Pract.* (ed. 5) II. xxii. 277 The flexible often gets caught as the sweeper is rolled along. 1971 *Engineering* Apr. 104/1 (Advt.), Braided flexibles.

flexing (fle·ksiŋ), *vbl. sb.* [f. FLEX *v.* + -ING¹.] The action of the verb FLEX. Also *transf.*

1902 *Westm. Gaz.* 9 May 10/1 Those earth movements which have resulted in the flexing and fracturing of strata. 1920 *Blackw. Mag.* July 71/1 Never a tremor of wing except just the flexing and twisting of the tips for balance. 1968 *Listener* 6 June 752/3 His critical flexings are limited to some he-man anecdotes and a lengthy interview.

flexography (fleksǫ·grăfi). [f. *flexo-* (f. L. *flexus*, f. *flectere* to bend) + -GRAPHY.] A rotary letterpress technique using rubber or plastic plates and aniline inks for printing on fabrics, plastics, metal foils, and other materials, as well as on paper; also called *aniline printing.* So **flexogra·phic** *a.*

1952 *Print* (U.S.) Nov. 51/1 An informal poll..indicated many drawbacks to the term *Aniline* printing...Three suggestions were presented: Flexographic Process, Permatone Process, or Rotopake Process. *Flexographic* was chosen as the new official description. 1954 *Penrose Ann.* XLVIII. 151 The process known as Flexography dates back to the time when the first rubber stamp was moulded. 1954 *Brit. Printer* LXVI. 36/1 The aniline process, or 'flexographic' printing, as it is now commonly termed in America, has its origins in the paper converting field rather than in the general printing industry. 1960 *News Chron.* 7 Sept. 8/3 Flexographic printing. Competent charge hand required by leading firm of film converters. 1966 *Print* (Wynkyn de Worde Soc.) iv. 55 To these three conventional processes may be added aniline or flexographic printing, which is a form of rotary letterpress, using a soft rubber plate and very thin inks, predominantly for printing packaging materials. 1970 *Brit. Printer* July 69/2 Screen process is now competing with offset-litho and flexography in some large packaging applications. *Ibid.* 85/1 Three models..are available in the new range of flexographic presses.

Flexowriter (fle·ksoraitəɹ). Also **flexowriter.** The proprietary name of a kind of electric typewriter incorporating a tape-punch and a tape-reader.

1955 *Trade Marks Jrnl.* 21 Sept. 927/1 *Flexowriter.* Appliances consisting of a typewriter and a duplicator in one unit for office use. 1956 I. A. HERRMAN *Man. Office Reprod.* xiii. 142 A Flexowriter punches codes in the edge of cards or in tape, and reads edge-punched cards or punched tape for automatic operation. 1964 C. DENT *Quantity Surveying by Computer* iii. 31 If a computer has facilities for accepting eight-channel tape, then the additional expense of Flexowriter input and output with its greater flexibility may be justified. 1968 *Bodl. Libr. Rec.* VIII. 63 The entries were typed on the flexowriter at RSL on envelopes of a standard size.

fli:bbertigi·bbet, *v.* [f. FLIBBERTIGIBBET *sb.*] *intr.* To play the flibbertigibbet; to gad about frivolously.

1921 GALSWORTHY *To Let* I. i, His daughter *would* flibberty-gibbet all over the place like most young women since the War.

‖ **flic** (flik, flik). [Fr.] A French policeman.

1899 E. DOWSON *Let. c.* 13 June (1967) 413 The 'flics' are a joy for ever. Each one has a different face & each one is inevitably and transparently a policeman in plain clothes. 1918 C. B. JORDAN tr. *Ibañez's Four Horsemen Apocalypse* II. i. 168 He hated the *flics*, the Paris police. 1941 KOESTLER *Scum of Earth* II. i. 60 On the platform..sat a row of *flics*, rifles between knees. 1964 WODEHOUSE *Frozen Assets* iii. 51 Just as I was about to strike him on the mazzard, this *flic* intervened, and his was the mazzard I struck.

flick, *sb.*¹ Add: **1. d.** *Cricket.* A quick turn of the batsman's or bowler's wrist in playing or delivering the ball; also, a turn of the ball. *Hockey.* A stroke played with such an action; so *flick shot, stroke.*

1897 K. S. RANJITSINHJI *Jubilee Bk. Cricket* 73 There is a certain amount of 'flick' from the fingers, but this is quite different from the twist of the slow bowler. *Ibid.* 84 Spin imparted by wrist 'flick'. 1903 JEPHSON in H. G. Hutchinson *Cricket* 91 The varying 'flicks' or 'whips' of the wrist. *Ibid.* 112 Two or three long loose strides, two at a trot, and an arm swinging round like a flail, a good length, great pace, and on any wicket at times a considerable flick back from the off. 1957 *Encycl. Brit.* XI. s.v. *Hockey* (*caption*, facing p. 616) Start of flick stroke. 1959 *Times* 6 Mar. 16/2 Hodgetts..placed his flick poorly. 1962 *Times* 10 Dec. 4/6 A flick shot.

e. A film; also in *pl.*, the cinema. *colloq.* (Cf. *FLICKER *sb.*³ 3.)

1926 E. WALLACE *Square Emerald* xv, We'll occupy the afternoon with a 'flick'. I love the movies—especially the romantic ones. 1927 W. E. COLLINSON *Contemp. Eng.* 113 We all know the word movies, but still use pictures or cinema [si'nima] in preference to the American term... Mr. Titley adds the slang flicks or flickers, unknown to me. 1931 *Repertory* (Oxford) 18 Apr. 9/1 Hand in hand with the Muse of the Theatre departs the appropriate feminine deity who presides over the 'flicks'. 1936 *Times* 29 Feb. 13/4 The present flick-minded generation may see rather than read its Victorian classics. 1949 F. SWINNERTON *Doctor's Wife comes to Stay* 112 He would take her to the theatre, the ballet, the flicks. 1959 J. BRAINE *Vodi* xviii. 215 Where shall we go this afternoon anyway? Tanbury and tea at the Raynton, then a flick?

4. *Comb.* **flick-knife,** a weapon with the blade held in the handle by a catch which can be released with a flick of the finger; **flick roll** *Aeronaut.* (see quot. 1950); also as *vb.*

1957 *Times* 15 Nov. 13/1 Mr. Justice Streatfeild said at York Assizes yesterday: 'What an invention of the Devil is the "flick" knife, which unhappily so often features in crimes of violence in this country, often committed by young people.' 1958 *New Statesman* 6 Sept. 261/2 Young thugs who have poured into the area carrying flick-knives, bicycle chains and petrol-bombs. 1967 *Spectator* 4 Aug. 138/3 There's only one flick knife on show; but the view of New York's educational system has seldom looked bleaker. 1928 O. STEWART *Aerobatics* 6 In the flick roll the machine again turns over sideways but, instead of doing so gently and lazily, it does so quickly with a violent wriggle. *Ibid.* 9 Some aeroplanes require a good deal of firm handling before they can be induced to flick roll. 1950 *Gloss. Aeronaut. Terms (B.S.I.)* I. 9 *Flick roll*, a rapidly executed roll in which the autorotative tendency of the wings is aided to some extent by the rolling moment due to the use of rudder at high angles of incidence. 1956 D. BARNHAM *One Man's Window* 65, I have never seen flick rolls used in combat before.

flick, *sb.*² **2.** (Later examples.)

1920 'SAPPER' *Bull-Dog Drummond* iii. 87 All that I have, dear old flick, is yours for the asking. 1930 G. MANNING-SANDERS *Burnt Man* xiv. 168 'Why, hullo Ambrose, old flick!' cried Joe.

flick, *v.*² Add: **2. c.** *Cricket.* (*a*) Of the batsman: to play (the ball) with a slight turn of the wrist.

1897 *Encycl. Sport* I. 223/2 The two wrists moving the bat with a sort of 'flicking' movement. 1903 P. F. WARNER in H. G. Hutchinson *Cricket* 62 Short and straight balls, if they do not get up to any height, may be flicked round on the on side by a quick turn of the wrist. 1927 M. A. NOBLE *Those 'Ashes'* 176 Gregory..was all at sea to Larwood, whom he flicked three times dangerously through the slips.

(*b*) Of the bowler: to deliver (a ball) with a flicking movement of the wrist. Also *intr.*, said of the ball so delivered.

1903 JEPHSON in H. G. Hutchinson *Cricket* 110 Bowling with a long bouncing run, he can make the ball flick higher and faster than any other bowler. *Ibid.* 116 He 'flicks' the ball as we have all seen many a wrathful billiard-player do when returning the white from a most unexpected pocket. 1920 E. R. WILSON in P. F. Warner *Cricket* 102 He [*sc.* Simpson-Hayward, the underhand bowler] 'flicked' the cricket ball as if it were a billiard ball.

flicker, *sb.*³ Add: **2. b.** A rapid, rhythmic variation in the degree or quality of illumination which is perceptible to the eye; also, the visual sensation caused by this.

1892 E. S. FERRY in *Amer. Jrnl. Sci.* CXLIV. 195 A vibration of the lamp or of the diffraction-grating produces a flicker in the field of view that cannot be distinguished from the appearance produced by a too slow rotation of the sectored disc. 1912 *Phil. Mag.* 6th Ser. XXIV. 370 The fovea is more sensitive to red flicker, the periphery to blue. 1929 C. MURCHISON *Found. Exper. Psychol.* iv. 200 Chromatic flicker can be evoked by alternating, upon the same retinal area, two stimuli which differ in wave-length composition but which are photometrically equal. 1944 W. D. WRIGHT *Measurem. Colour* iv. 107 The sensation of flicker due to alternations in colour vanishes at a lower frequency of alternation than the flicker due to difference in brightness. 1962 H. C. WESTON *Sight, Light & Work* (ed. 2) v. 161 There is ordinarily no visible flicker from fluorescent lamps, although sometimes it is apparent at their extremities as seen by peripheral vision.

c. *spec.* in *Cinemat.* and *Television*, a succession of sudden, abrupt changes in a picture, such as occurs when the number of frames per second is too small to produce persistence of vision.

1899 H. V. HOPWOOD *Living Pictures* vi. 208 A continual rattle impinging on the ear tends to intensify irritation caused to the eye by flicker on the screen. 1923 F. A. TALBOT *Moving Pictures* 7 The regularity of this recurring damping wave, described as 'flicker', set up severe eye-strain. 1933 *Discovery* May 157/1 The ultra-short waves enable images with much greater detail and almost complete absence of flicker to be transmitted. 1953 H. A. CHINN *Telev. Broadc.* i. 6 The minimum acceptable frame frequency is that required for the elimination of flicker. 1968 *Times* 18 Apr. 2/8 (*headline*) TV flicker led to death.

d. A rapid variation in the quality of a sound analogous to visual flicker (see *2 b above).

1922 *Physical Rev.* XX. 332 It has been found possible ..to make use of the perception of flicker in an alternation phonometer. Let two frequencies alternate in the ear at a suitable rate, [etc.]. 1934 C. MURCHISON *Handbk. Gen. Exper. Psychol.* xvi. 897 The former used a flicker method, the tones to be compared being presented to the ears alternately at the rate of about 25 alternations a second. 1953 C. E. OSGOOD *Method & Theory Exper. Psychol.* iv. 146 Simultaneous auditory flicker..was found to increase the pronouncedness of visual flicker if it was already present.

3. *pl.* = *FLICK *sb.*¹ 1 e. *colloq.*

1927 [see *FLICK *sb.*¹ 1 e]. 1930 W. DE LA MARE *Desert Islands* 2 Some of the best of the 'flickers' or 'movies'. 1938 G. GREENE *Brighton Rock* I. iii. 43 What about you and me going across to the flickers? 1969 GISH & PINCHOT *L. Gish* iii. 31 Mother, guess who we saw acting in 'flickers'!

4. *attrib.*, as *flicker frequency, method*; **flicker effect** *Electronics*, a relatively low-frequency random fluctuation in the current emitted by a cathode in a thermionic valve, caused by changes in the state of the emitting surface; also, a similar effect in transistors and other solid-state devices; **flicker fusion,** the apparent steadiness of a regularly varying source of light when the frequency of the variation is sufficiently great; so **flicker-fusion frequency,** the lowest rate of variation at which flicker is not perceptible; also called *critical flicker* (or *fusion*) *frequency*; **flicker noise** *Electronics*, noise due to the flicker effect; **flicker photometer,** an instrument for measuring the relative intensities of light from two different sources, esp. sources of different colours, by measuring the flicker or observing the absence of flicker when the field of view is illuminated alternately by the two sources of light; so **flicker photometry.**

1926 W. SCHOTTKY in *Physical Rev.* XXVIII. 77 If we had to do with emission of light instead of electrons, we would speak of a chaotic variation of light intensity taking place over the surface of the cathode, a phenomenon which we should describe by the word 'flicker'. If..Johnson's explanation of the phenomenon is the correct one, then we may use the analogy and call the new effect the 'flicker effect'. 1907 *Nat. Electr. Light Assoc. 30th Convention* I. 337 The most sensitive flicker frequency for small ranges of flicker was in the neighborhood of 2.5 cycles per second. 1922 *Jrnl. Optical Soc. Amer.* VI. 7 The substantial independence of critical flicker frequency upon chroma is..the basis of the critical frequency method of heterochromatic photometry. 1936 *Jrnl. Gen. Physiol.* XIX. 514 Flicker fusion depends upon the suppression of intensity discrimination. 1944 *Federation Proc.* III. 6 (*heading*) Changes in flicker fusion frequency (F.F.F.) under experimental stress. 1962 H. C. WESTON *Sight, Light & Work* (ed. 2) v. 161 The fluctuation in light output of electric lamps caused by a 50-cycle alternating current..is faster than the usual 'flicker fusion frequency' for the bright-adapted eye. 1897 *Jrnl. Physiol.* XXII. 143 The most marked discrepancy between the band and flicker methods was in the case of red. 1934 Flicker method [see *2 d above]. 1947 *Proc. Physical Soc.* LIX. 366 At low frequencies (<10² to 10⁴ c.p.s.) flicker noise may be tens or even hundreds of times greater than Johnson or shot noise. 1960 D. A. BELL *Electr. Noise* xii. 274 When working with semiconductors there is always a risk of flicker noise being introduced at the electrode contacts. 1896 F. P. WHITMAN in *Physical Rev.* III. 241 (*heading*) On the photometry of differently colored lights and the 'flicker' photometer. 1955 J. W. WENTWORTH *Color Telev. Engin.* iii. 59 The best instrument for measuring the luminances of 'colored' surfaces is the flicker photometer. 1897 *Jrnl. Physiol.* XXI. 47 According to the rules of flicker-photometry..flicker should disappear from the whole surface of the disc at the same minimal speed.

flicker, *v.* Add: **7. b.** To cause to move in a fitful and unsteady manner; to indicate by a flicker.

1903 R. LANGBRIDGE *Flame & Flood* viii, He watched her eye-lashes flicker dismissal. *Ibid.* xxii, There was yet a partial inhumanity which licked its lips..which..burned as fiercely on the side of justice as injustice,..flickering an equal encouragement to 'I *can't* go!' and—'I must!' 1907 *Munsey's Mag.* Dec. 308/1 [They] entered the castle..; the torches flickering weird shadows as they walked between them. 1923 D. H. LAWRENCE *Birds, Beasts & Flowers* 113 He lifted his head from his drinking..And flickered his two-forked tongue from his lips.

flickering, *vbl. sb.* (Further example.)

1968 *Times* 18 Apr. 2/8 A boy's susceptibility to the flickering of a television set led to his death.

flickerless (fli·kəɹlės), *a.* [f. FLICKER *sb.*³ + -LESS.] Without flickers, producing no flicker.

1922 M. J. WRIGLEY *Film* 117 Mr. Martin Harper..has the credit to his name of producing the 'Extralite' flickerless shutter.

flickery (fli·kəri), *a.* [f. FLICKER *v.* or *sb.*³ + -Y.] = FLICKERING *ppl. a.*

1893 in *Funk's Stand. Dict.* 1922 *Daily Mail* 21 Nov. 10 A flickery resemblance of his old self. 1950 F. STARK *Traveller's Prelude* 197 One wonders if he will live through the night, his pulse is so flickery. 1953 R. BRADBURY *Fahrenheit 451* (1954) 133 Running like an ancient flickery Keystone Comedy.

flicky (fli·ki), *a.*² [f. FLICK *sb.*¹ + -Y¹.] Of or pertaining to a flick, jerky.

1897 K. S. RANJITSINHJI *Jubilee Bk. Cricket* 127 There is something distinctly jerky and flicky in this projection of the elbow and straightening of the arm.

flight, *sb.*¹ Add: **1. f.** The action or technique of travelling through the air or space in an aircraft or spacecraft or in a balloon; the movement through air or space of such a machine.

1784 *Universal Mag.* Sept. 358 We again took our flight [in a balloon], and ascended to near 1200 feet. 1879 *Encycl. Brit.* IX. 308/1 It is not necessary to enter upon a history of artificial flight. 1909 *Daily Chron.* 19 Mar. 4/5 Lilienthal..shares with Pilcher..and Le Bris, the honour of being the pioneer of modern experiments in gliding flight. 1916 H. BARBER *Aeroplane Speaks* 1 The Elementary Principles of Flight. 1935 C. G. BURGE *Compl. Bk. Aviation* 312/2 It [*sc.* the tail plane] will be set at such an angle of attack that it carries no load in normal flight. 1951 A. C. CLARKE *Exploration of Space* ii. 9 The first difficulty one encounters in trying to envisage interplanetary flight is that of scale. 1955 *Oxf. Jun. Encycl.* IV. 35/1 The modern balloon dates from 1783, when two kinds of lighter-than-air methods of flight were introduced almost at once. 1971 P. J. MCMAHON *Aircraft Propulsion* xiii. 350 Reference to any textbook on principles of flight will show that the induced drag is related to the lift of a wing.

g. An instance of flight (sense *f above); a (usu.) numbered regularly timed journey by air-line from place to place; a journey through the air or through space; a voyage by an aircraft, balloon, or spacecraft through the air or through space.

1785 [See FLIGHT *sb.*¹ 2]. 1835 *Nautical Mag.* Oct. 613 It is related that the aërial ship took her first flight in a waggon. 1842 *Household Words* IV. 98/2 Exertions he had undergone in preparing for the flight. 1868 *Aeronaut. Soc. Catal.* 8 Flying machine—which being attached to the body, enables a person to take short flights. 1904 *Pall Mall Mag.* Jan. 19/2 Nor is it an easy task to calculate an airship's speed. On these flights up and down the Mediterranean coast, [etc.]. 1909 *Aero* 25 May 14/1 Mr. J. T. C. Moore-Brabazon has made several short flights with the *Bird of Passage* at Shellness. 1912 *Aeroplane* 26 Dec. 637/2 Their Yuletide present takes the form of a ticket for a passenger flight at Hendon. 1922 *Encycl. Brit.* XXX. 57/1 The longest flight by an N.S. airship was 101 hours. 1928 *Times* 22 Mar., [The King of Afghanistan's] first flight in any civil aircraft, and his only flight since he visited Europe. 1940 CRUMP & MAUL *Our Airliners* ii. 28 Transcontinental and Western Air announces that 'Flight 10..will arrive in three minutes at Gate Number 11'. 1951 N. BALCHIN *Way through Wood* viii. 108 We..happened just to catch a flight, and were at Heathrow by seven. 1962 F. I. ORDWAY et al. *Basic Astronautics* xiii. 547 During the 32-hr flight Dr. Simons wore a full-pressure suit and remained seated in the tiny gondola. *Ibid.* 550 Manned balloon flights. 1968 *Times* 10 Dec. 6/7 The Surveyor flights which soft-landed on the moon. 1969 *Listener* 20 Feb. 232/2 The pictures you brought back from the Moon were not as good as those taken on an unmanned flight—the American Lunar Orbiter flight.

h. A Royal Air Force unit consisting of about five or six machines; the members of such a unit.

1914 H. ROSHER *In R.N.A.S.* (1916) 25 We shall not get our squadron together until the end of January... We may, however, go over in pieces, a flight at a time. 1915 *War Illustr.* 27 Nov. 345 This branch of the Army is organized in what are called wings, divided into squadrons, and subdivided into flights. 1917 *Blackw. Mag.* July 121/1 Twelve officers flew to France with the flight to which I belonged. 1932 W. S. CHURCHILL *Thoughts & Advent.* 181 The art of flying was in its childhood [1911–15]... Even the nomenclature had to be invented, and I may claim myself to have added the words 'seaplane' and 'flight' (of aeroplanes) to the dictionary. *a* 1935 T. E. LAWRENCE *Mint* (1950) II. i. 101 The flight staggers off parade to drop bonelessly into bed. 1959 *Chambers's Encycl.* I. 189/1 The main formations of the R.A.F. are the flight, the squadron, the wing and the group. There are three flights to a squadron.

6. d. *Cricket.* The trajectory and pace of the ball in its flight before pitching; also, the art of controlling these. (Cf. *FLIGHT *v.* 7.)

[1897 K. S. RANJITSINHJI *Jubilee Bk. Cricket* iii. 64 The bowler must learn to manipulate and deliver the ball in such a way that, after pitching, it deviates from its original line of flight.] 1903 JEPHSON in H. G. Hutchinson *Cricket* 88 He..possesses a wonderful command of length, ..a considerable variation of flight. 1911 P. F. WARNER *Bk. Cricket* 74 Every slow left-hander should cultivate 'flight'. 1924 N. CARDUS *Days in Sun* 48 Even the changeful flight of a Lohmann could not hold back our Haywards and our Frys from mastery. 1963 A. ROSS *Australia 63* v. 113 His..control of flight..had rescued a match that seemed to be drifting far out of England's reach.

10. c. The tail of a dart.

1938 A. & G. D'EGVILLE *Darts* 61 The flights, or tails, may be of paper, linen, tin, cardboard or feathers. 1950 *Oxf. Jun. Encycl.* IX. 176 There are many variations in the darts themselves, from heavy brass darts with paper or plastic flights, to light, sometimes unweighted, wooden darts with feather flights. 1968 N. E. WILLIAMSON *Darts* i. 15 Several friends of mine, all good dart players, will swear by the canes and paper flights, but used with brass barrels. *Ibid.* iii. 26 It is also advisable to make sure that you have a suitable box or container in which to place your darts when not in use in order to preserve the flights.

13. d. (Earlier U.S. example.)

1813 *Niles' Weekly Register* V. Add. A. 6/2 It was extended with flights to draw the meal towards the Hopper.

14. *flight bag, shed, time* (later examples), *way.*

1943 REDDING & LEYSHON *Skyways to Berlin* 95 A flight bag and a musette lay beside the foot locker. 1965 *Family Circle* Oct. 31/2 Get a handy, colorful flight bag. 1934 'E. CAMBRIDGE' *Sycamore Tree* IV. vi. 260 Howell was gazetted..to an aerodrome..where the flight sheds looked no longer than hen houses on the wide green plain. 1959 F. D. ADAMS *Aeronaut. Dict.* 76 *Flight time*, time spent in flight or in flying operations, measured, when exactness is required, between specified instances, as between the commencement of the take-off run and the end of the landing run. 1971 *Guardian* 3 June 5/8 The flight time of 65 minutes allowed some..of the economy-class passengers to climb..to the first class lounge. 1933 *Jrnl. R. Aeronaut. Soc.* XXXVII. 3 It will become necessary clearly to mark flightways to be used even on an 'all way' field. 1963 *Times* 28 Feb. 13/5 The Battle of Britain fighter station at Hornchurch, Essex, was sold yesterday for £517,000 by public auction...The last lot, comprising 38 acres of open grassland and flightways, was sold to a gravel company for £61,000.

b. In titles of officers of various ranks in the Royal Air Force. Also *ellipt.* = *flight sergeant.*

1914 *Times* 19 Nov. 3/4 Royal Naval Air Service. Flight Lieuts...reappointed as Acting Flight Commanders. 1916 H. BARBER *Aeroplane Speaks* 27 The Flight-Sergeant is awaiting the Pilot's orders. 1917 *Blackw. Mag.* Mar. 380/2 'Good-night, you chaps,' said one of the flight-commanders. *Ibid.* May 800/1 The good work of my pilot had brought him a flight commandership. 1922 *Man. Seamanship (H.M.S.O.)* I. 11 Distinction lace worn by officers of the Royal Air Force...R.A.F. Flight Lieutenant. R.N. Lieutenant. 1928 C. F. S. GAMBLE *N. Sea Air Station* iv. 75 Flight Commander (relative rank of Lieutenant, R.N.). Flight Lieutenant (relative rank of Lieutenant, R.N.). Flight Sub-Lieutenant (relative rank of Sub-Lieutenant, R.N.). 1943 C. H. WARD-JACKSON *Piece of Cake* 30 *Flight*, flight sergeant. 1959 *Chambers's Encycl.* XI. 519 Army and Royal Marines, Captain. Royal Air Force, Flight Lieutenant (Flight Officer, W.R.A.F.). *Ibid.*, Army and Royal Marines, Colour Sergeant. Royal Air Force, Flight-Sergeant. 1960 T. RATTIGAN *Ross* I. ii. We'd nearly finished fatigue, Flight.

15. flight call, (*a*) the call made by a bird during flight; (*b*) an announcement at an airport to passengers for a particular flight, informing them that they may board the aircraft; **flight control** (see quot. 1959); **flight crew** (see quot. 1965); **flight deck,** (*a*) of an aircraft-carrier: the deck on which aircraft take off and land; (*b*) of an aeroplane: the part accommodating the pilot, navigator, etc.; **flight engineer** (see quot. 1965); **flight envelope** *Aeronaut.*, the set of limiting combinations of speed and altitude, or speed and range, etc., possible for a particular kind of aircraft or aero-engine; **flight-line,** (*a*) the direction of flight by birds, esp. during migration; (*b*) *Aeronaut.* (see quot. 1956); **flight net,** a net used for the capture of birds; so *flight-netter, -netting*; **flight note** = **flight call* (a); **flight-number,** the identifying number of a flight (cf. *1 g above); **flight path** (see quot. 1919); the planned course of an aircraft or space vehicle from point to point; also *transf.*; **flight plan** *Aeronaut.*, the prearranged scheme for a particular flight; so **flight-plan** *v. intr.*, **flight-planning** *vbl. sb.*; **flight recorder,** a device in an aircraft which records the relevant technical details of each flight, in order to assist investigation in the event of an accident; hence *flight recording* vbl. sb.; **flight refuelling,** refuelling of an aircraft whilst in flight; so **flight-refuel** *v. trans.*; **flight simulator,** an apparatus designed to simulate the actual conditions of flight, used esp. by airline pilots in training; **flight-test** *v. trans.*, to test an aircraft, missile, etc., during flight; so **flight-testing** *vbl. sb.*

1937 *Brit. Birds* XXXI. 98 When disturbed, they flew up without travelling far, producing a whinnying and metallic flight-call. 1959 BANNERMAN *Birds Brit. Isles* VIII. 63 Birds flying about the island.., calling with a soft squeaky note..not unlike the flight call of the stormy petrel. 1969 'J. MUNRO' *Innocent Bystanders* vi. 86 They were in the departure lounge, waiting for their flight call. 1937 *Discovery* Oct. 277/2 Since the reaction thrust, acting from the extreme rear of the plane, is in no way different from the pull of the aero engine in its nose, and remains in fixed relation to the aeroplane axis, stability and flight-control are not interfered with. 1959 F. D. ADAMS *Aeronaut. Dict.* 75/1 *Flight control*, 1. *In plural.* Controls for guiding or trimming an aircraft, missile, [etc.]. 2. A general if somewhat loose term applied to any activity or organization that directs and controls the movement of aircraft. 1969 *Sunday Times* 13 July 13/2 *Flight Control System*, a system that serves to maintain stability and control during the flight. 1951 *Philippine Air Lines Timetable* 15 May 2 P.A.L.'s experienced flight crews, with skilled American pilots and trained cabin attendants, are your assurance of a perfect trip. 1958 'N. SHUTE' *Rainbow & Rose* i. 14 The flight crew were getting ready to take off the freighter. 1965 *Gloss. Aeronaut. Terms (B.S.I.)* § 16 *Flight crew* (operating crew), those members of the aircrew whose primary concern is the operation and navigation of the aircraft and its safety in flight. 1924 *Sci. Amer.* Oct. 248/2 On the flight deck there are various devices for checking the speed of landing airplanes. 1936 *Meccano Mag.* June 325 The members of the flight deck party are rushing out to seize the machine as she comes to a standstill. 1949 *Aeronautics* Nov. 44/1 As one might expect, the Brabazon's flight deck is spacious and well laid out. 1958 *Oxf. Mail* 19 July 1/2 Rebel snipers fired on one of the planes..and hit it twice. One shot went through the flight deck. 1964 D. MACINTYRE *Battle for Mediterranean* iii. 47 But for her armoured flight deck which absorbed the shock of some of the hits, the *Illustrious* must have been sunk. 1971 *Guardian* 29 June 6/4 Airlines have been warned to review flight deck procedure after an air miss..between a BOAC VC10 and an El Al Boeing. 1938 *Flight* 21 July 67/2 Opposite the captain's desk are the flight engineer's quarters. 1939 *Meccano Mag.* Oct. 570/3 The radio operator occupies a position behind them, and a little farther aft is the flight engineer, both on the port side of the cabin. 1958 'N. SHUTE' *Rainbow & Rose* vi. 226, I was talking to Dick Scott, my new flight engineer. 1965 *Gloss. Aeronaut. Terms (B.S.I.)* § 16 *Flight engineer*, a member of the flight crew responsible for engineering duties. 1944 *Jrnl. R. Aeronaut. Soc.* XLVIII. 488 The best known of the envelope cases is the 'flight envelope', which is in general use in this country and in the United States...The 'flight envelope' covers all possible conditions of symmetrical manœuvring flight instead of the few isolated points specified in the previous system. 1966 *McGraw-Hill Encycl. Sci. & Technol.* I. 177 (*caption*) Airplane flight envelope, speed and range, medium altitude. 1966 D. STINTON *Anat. of Aeroplane* ii. 10 The operational environment of an aircraft lies within a boundary, drawn on a basis of speed and height, called the flight-envelope... The outline marks the limit of performance in one particular configuration. 1933 *Brit. Birds* XXVI. 366 The migratory flight-line of the others appears to lie well off the shores of west Wales. 1956 W. A. HEFLIN *U.S.A.F. Dict.* 210/2 *Flight line*, on an airfield, a general area including the hangars and the ramps and other grounds between and surrounding the hangars where aircraft are parked, serviced, etc. 1958 BANNERMAN *Birds Brit. Isles* VII. 147 The majority of the large flocks..travel by two main flight-lines. 1971 *Country Life* 25 Feb. 407/2 In America, the native geese regularly move on well-plotted flight lines. 1889 *Leisure Hour* 675/2 The birds caught in the flight nets are sold to a dealer. 1897 *Pearson's Mag.* Feb. 213 (*heading*) Flight-netting for wildfowl. *Ibid.* 216/2 The birds when captured are packed into large bags, and are carried in this way to the flight-netter's home. 1960 E. ENNION *House on Shore* vi. 78 The flight-net has been reintroduced recently ..to 'control' oystercatchers in Morecambe Bay. 1937 *Discovery* Feb. 47/2 He [*sc.* an owl] has..a loud flight-note ..like the quacking of a duck. 1961 BANNERMAN *Birds Brit. Isles* IX. 77 J. T. Nichols, who considers this the 'flight note' of the transient bird. 1949 *Internat. Air Transport Assoc. Bull.* Dec. 136 *Flight number*, which is equivalent to the term line number, means the numerical designation of a flight. 1953 C. DAY LEWIS *Italian Visit* i. 21 We have Ten minutes until our flight-number is called. 1955 E. BOWEN *World of Love* xi. 204, *I* don't even know where he's coming from... Harris will have the flight-number. 1911 *Chambers's Jrnl.* I. 55/2 Instruments which will exhibit the angle of the flight-path. 1919 W. B. FARADAY *Gloss. Aeronaut. Terms* 52 *Flight path*, the path of the centre of gravity of an aircraft with reference to the air. 1967 *Sunday Mail Mag.* (Brisbane) 1 Oct. 5/1 The peregrine flew towards it...Every bird in her flightpath froze to the ground. 1969 *Times* 19 Feb. 13/6 Another experiment..will measure the mass and distance of Mars simply by tracking the spacecraft and measuring the effect the planet exerts on its flight path. *Ibid.* 10 Mar. 10/1 Fortunately none of the flight paths used by the birds crosses the airport. 1936 M. B. GARBER *Mod. Mil. Dict.* 135 *Flight plan*, a plan for an aerial flight, setting forth the probable time of departure, direction of flight.., with estimated time of arrival. 1940 *Meccano Mag.* July 322/3 These flight plans cannot be standardised like a railway timetable, because of the great effect changes in weather have on the aircraft. 1945 *Aeroplane* 30 Nov. 638/2 Aircraft ex-Baltimore could always 'flight-plan' for Stephenville. 1971 *Guardian* 3 July 1/7 The pilot filed a flight plan for Trinidad with the airport tower. 1959 *Listener* 26 Feb. 372/1 Good flight planning—to use a term already long in use in terrestrial air navigation—is going to be of the highest value [in space navigation]. 1971 *Flying* Apr. S5/1 Not enough pilots seem ready to lay out hard cash for weather and flight-planning information. 1948 *Shell Aviation News* cxv. 5/1 The Civil Aeronautics Board is now requiring all U.S. certified airlines to install flight recorders on their aircraft by 30th June, 1948. 1964 Flight recorder [see *black box* (*BLACK *a.* 19)]. 1971 *Times* 4 Oct. 1/3 Vital clues to the cause of the crash.. were gained yesterday when the flight recorder was decoded. 1962 *Flight Internat.* LXXXI. 945/1 The relatively new art of flight recording. 1963 *Economist* 23 Nov. 733/1 TSR 2 can be flight-refuelled. 1939 *Air Ann. Brit. Emp.* 50 Flight refuelling..is likely to become a prominent means of allowing flight at very high wing loadings. 1947 *Shell Aviation News* CXIII. 14/1 It is fairly generally known that a series of flight-refuelling trials has recently taken place, in conjunction with British South American Airways. 1950 *Jrnl. R. Aeronaut. Soc.* LIV. 600/1 One recent development which should help both to raise the standard of training and reduce the time and cost of non-revenue flying involved in crew training is the 'flight simulator'. This ground equipment is designed to reproduce exactly the control cabin of the selected aircraft type, with all instruments and controls. 1961 *Engineering* 12 May 677/1 Control of an aircraft in bad flying conditions can now be tested on the ground by a machine..known as a 'rocking cockpit' or..flight simulator. 1931 C. SPRIGG *Airship* xiv. 210 Before the rigid airship can be used on a commercial service it must be thoroughly flight-tested under every variety of conditions. 1958 *Observer* 9 Nov. 1/8 This is a new solid fuel rocket that had been flight-tested only once before. 1969 *Times* 3 June (Suppl.) p. iii/5 He [*sc.* Neil Armstrong] became a Nasa research pilot in 1955 and flight-tested the X-15 rocket plane. 1943 *Sci. News Let.* 30 Jan. 73/3 (*heading*) Flight testing advances win award for MacClain. 1961 *Shell Aviation News* Dec. 2/2 With the coming of auto-observers and telemetry, 'test flying' has been gradually and unobtrusively ousted by 'flight testing'.

flight, *sb.*² Add: **1. f.** *Economics.* The selling of a particular currency by foreign holders, e.g. in anticipation of a fall in its value; the

withdrawal of investments from a particular country.

1923 *Ann. Reg. 1922* II. 88 The instability of money caused a continuous flight from the mark in Germany, the normal consequence of inflation. 1930 *Economist* 27 Sept. 554/1 For a few days at the beginning of the week there was an incipient flight from the mark, the French withdrawing funds. 1938 *Times* 18 Feb. 21/2 The ease with which..'flight' money is prepared to exchange its refuge—either from one currency to another, or from currencies into gold or from gold into currencies. 1948 G. CROWTHER *Outl. Money* (ed. 2) vii. 205 There is.. no less money [in London] as the result of a 'flight of foreign capital'. All that has happened is that..a smaller proportion..of the total of British money belongs to foreigners. 1967 *Times* 4 Aug. 19/8 It is a fallacy to conclude that such sales [by London banks] must represent a flight from the pound by UK residents.

flight, *v.* Add: **3. b.** (Earlier example.)

1873 *Young Englishwoman* Nov. 531/1 The habit of the wild fowl is this...At evening they 'flight' to the uplands.

7. *Cricket.* To vary the trajectory and pace of (the ball) in its flight before pitching. Also *intr.*

1912 P. F. WARNER *Eng. v. Austral.* iv. 33 Woolley kept an excellent length, and 'flighted' the ball. 1925 *Country Life* 18 July 95/1 This power of varying the trajectory of the ball and its pace is known as 'flighting' the ball. 1955 *Times* 16 Aug. 3/3 He flighted and turned from leg just enough on what had previously seemed to be an easy paced surface. 1961 *Times* 16 May 4 Cowper had been bowling slow flighted off-breaks.

flighted, *ppl. a.* **1.** Delete 'Only in *drowsy-flighted*.' and add example of *feeble-flighted.*

1906 *Westm. Gaz.* 5 Jan. 4/2 Short-winged and feeble-flighted little birds.

3. Of steps: arranged in flights.

1929 A. CLARKE *Later Poems* (1961) 19 There Are flighted steps to climb.

flightless, *a.* (Later examples, not referring to birds.)

1953 E. SITWELL *Gardeners & Astronomers* 1 The entity of primal, flightless, winged Stupidity. 1967 M. E. HALE *Biol. Lichens* vii. 101 Large flightless weevils in the mossy forests of New Guinea. 1971 *Nature* 23 Apr. 482/1 BOAC this week found no difficulty in cramming the passengers from a flightless 747 into a conventional jet aircraft, with plenty of room to spare.

flim² (flim). *slang.* [abbrev. of FLIMSY *sb.* 1.] A £5 note.

1870 *Chambers's Jrnl.* 9 July 448/1 'What would it be worth?' 'A flim, Sam.' 1954 'N. BLAKE' *Whisper in Gloom* I. ii. 28 They..offer Bert..a flim for his boat.

flim-flam, *v.* Delete † from sense a and add later examples of both senses.

1903 O. KILDARE *My Mamie Rose* 72 'Flim-flamming'... Doubling a bill in a number of them and counting each end of it as one separate bill. 1904 E. ROBINS *Magnetic North* II. 144 But I didn't know *you'd* get flim-flammed out o' your boots. 1911 H. QUICK *Yellowstone Nights* xi. 290 Financial ingenuity has found a way to flim flam the devil himself. 1929 D. HAMMETT *Dain Curse* (1930) xii. 127 His success in flimflamming his followers had gone to his head. 1934 J. M. CAIN *Postman always rings Twice* x. 117 We've been flim-flammed, Cora. 1963 *Punch* 1 May 620/1 Marketing practices that smacked of flimflamming the public. 1971 *Publishers' Weekly* 22 Mar. 33 One expects somehow to be flim-flammed on a vacuum cleaner.

flim-flammer. (Earlier and later U.S. examples.)

1881 *Nat. Police Gaz.* 12 Nov. 10/2 Among the numberless small swindles of the metropolis there is none more curious than that of the flim-flammer. 1968 *Sat. Rev.* 27 Apr. 62/1 Peddlers, guides, flim-flammers and other friends of the tourist are forbidden there.

So **flim-flammery.**

1898 W. J. LOCKE *Idols* xiv. 201 Sick of your flim-flammeries of philosophy. 1960 *Observer* 17 Jan. 12/6, I got a taste of his flimflammery the first time I met him.

flimmer, *v.* (Later U.S. examples of extended use.)

1918 C. SANDBURG *Cornhuskers* 17 Canada thistle blue and flimmering larkspur blue. *Ibid.* 114 The shafts across her bed are flimmering.

flimsy, *sb.* Add: **2.** (Earlier and later examples.) Also, a sheet of thin paper, esp. that used on a typewriter for taking carbon copies; a document on thin paper.

1857 J. E. RITCHIE *Night Side London* 202 The reporters —or, rather, the penny-a-liners—who write on 'flimsey', and leave 'copy' on spec. at all the daily paper offices. 1888 C. T. JACOBI *Printers' Vocab.* 44 *Flimsy*, thin paper, such as..telegraph forms. 1909 *Daily Chron.* 10 Mar. 7/6 He had sent to the defendants 'flimsies' of the original reports. 1916 J. BUCHAN *Greenmantle* i. 1, I had just finished breakfast..when I got Bullivant's telegram...I flung him the flimsy with the blue strip pasted down on it. 1929 F. C. BOWEN *Sea Slang* 50 *Flimsy*, an officer's report at the end of a commission or when leaving a man-of-war. 1930 D. L. SAYERS *Strong Poison* xiv. 175 She dragged the cover off the typewriter..shook the top sheet, carbons, and flimsies together as a terrier shakes a rat. 1952 M. TRIPP *Faith is Windsock* vi. 95 Craig scrambled among his flimsies and maps. *Ibid.* xi. 163 Wafer-like sheets of paper with transparent celluloid covering, known as flimsies, were distributed by the bombing leader. 1958 M. DICKENS *Man Overboard* xiii. 193 The character references supplied by the Flimsies of his Confidential Reports. 1960 'N. SHUTE' *Trustee from Toolroom* iii. 47 'I told you that I had an answer to that cable...' He passed the flimsy to Keith.

flinch (flinʃ), *sb.* [f. FLINCH *v.*¹] Flinching.

1832 J. P. KENNEDY *Swallow Barn* I. xx. 205 Thar's no flinch in me, you may depend upon it. 1922 A. S. M. HUTCHINSON *This Freedom* II. ii. 91 But she came back bravely from her flinch. 1930 AUDEN *Poems* 25 Nowise withdrawn by doubting flinch Nor joined to any by belief's firm flange.

flinders, *sb. pl.* (Later examples.)

1949 E. POUND *Pisan Cantos* lxxx. 86 And the Osservanza is broken And the best de la Robbia busted to flinders. 1971 W. BURROUGHS *Speed* 42 About noon the transmission went all to flinders.

Flinders² (fli·ndəɹz). The name of Captain Matthew *Flinders* (1774–1814), English navigator and explorer, used *attrib.* in **Flinders bar(s),** a soft iron bar or bundle of rods, placed vertically near a ship's compass to correct deviation due to magnetic induction and to lessen the heeling-error.

1881 W. THOMSON in *Jrnl. R. United Serv. Inst.* XXIV. 408 The Flinders bar essentially corrects..the constituent of the heeling error, which has its maximum values on the east and west courses. 1895 F. J. EVANS *Elem. Man. Deviations Compass in Iron Ships* (ed. 9) xi. 119 Sir W. Thomson in his patent compass has adopted a malleable iron bar..which he has called a 'Flinders' bar' after its accomplished inventor. 1902 *Encycl. Brit.* XXVII. 181/2 A method of correcting deviation by means of a bar of vertical iron so placed as to correct the deviation nearly in all latitudes. This bar, now known as a 'Flinders bar', is still in general use. 1924 R. CLEMENTS *Gipsy of Horn* (1925) 217 The binnacle was useless..for the Flinders-bars had gone with the sky-light. 1959 *Chambers's Encycl.* VIII. 822/1 A 'flinders' bar, a soft iron bar mounted vertically on the binnacle, will..correct that part of the deviation that is due to the induced vertical fields.

flindosa (flindō̆ᵘ·zā). *Austral.* Also **flindosy.** [Corruption of *Flindersia*, a genus of trees, f. the name of Captain Matthew *Flinders*: see prec.] An Australian hardwood rain-forest tree, *Flindersia australis*, also known as Australian teak.

1862 *Internat. Exhib. Catal. Prod. N. S. Wales* 28/2 Flindersia Australis. *Cedrelaceæ.* A large-sized tree... Timber valuable for staves... Ash, Beech, and Flindosa. *Ibid.* 30/1 Flindersia Australis var. *Cedrelaceæ*...Timber used for house building, and occasionally for staves... Flindosa. 1884 A. NILSON *Timber Trees N.S.W.* 47 *C[ryptocarya] obovata.*—Sycamore; Flindosa. *Ibid.* 80 *F[lindersia] australis.*—Ash; Cugerie; Bulboro; Flindosa. 1907 J. H. MAIDEN *Forest Flora N.S.W.* II. 151, I have also seen it [*sc. Flindersia australis*] labelled 'Flindosa', a name more strictly applied to *F. Schottiana.* 1908 *Ibid.* III. 112 *Cryptocarya sp.* Called Flindosa at Brisbane Water. 1913 *Ibid.* V. 101, I have received it [*sc. Litsæa reticulata*] under the name of 'Flindosa'. 1929 W. D. FRANCIS *Austral. Rain-Forest Trees* 157 Flindersia australis R. Br. Crow's Ash, Teak, Flindosy, Flindersia. 1965 *Austral. Encycl.* IV. 111 *F[lindersia] australis* is a large tree..known generally throughout Queensland as crow's ash, but in some places as 'flindosy' or 'flindowzee'.

fling, *sb.* **4. a.** *Highland fling*: also *fig.*

1866 'MARK TWAIN' *Lett. fr. Hawaii* (1967) 32 Mrs. Jones..felt something grab her cheek; she dropped the sponge and out popped a scorpion...Well, she just got up and danced the Highland Fling for two hours and a half.

flinkite (fli·ŋkəit). *Min.* [ad. G. *flinkit* (A. Hamberg 1889, in *Geol. Fören. Förhandl.* XI. 213), f. the name of Gustav *Flink* (1849–1931), Swedish mineralogist + -ITE¹.] A basic arsenate of manganese known only as greenish-brown crystals from Sweden.

1891 *Jrnl. Chem. Soc.* LX. I. 20 Flinkite, a hydrated manganese arsenate, occurs in the Harstigen mine, Pajsberg, Wermland. 1967 *Amer. Mineralogist* LII. 1603 Flinkite, $Mn_2^{2+}Mn^{3+}(OH)_4(AsO_4)$. *Ibid.* 1604 Both retzian and flinkite have formulae suggesting isotypy.

flint, *sb.* Add: **2. b.** Also, a piece of metal (usu. an alloy of misch metal with a base metal such as iron) used to produce a spark for the ignition of the fuel in a cigarette-lighter.

1929 *Sears, Roebuck Catal.* Spring–Summer 471 Extra flints. For all lighters using a round flint. 1944 *N. & Q.* 30 Dec. 294/1 The *flint* used in petrol lighters is a cerium-iron alloy, and has very questionably taken over the name given..to lumps of silicon oxide found in chalk. 1971 *Grattan's Catal., Autumn/Winter 1971–2* 615 Ronson 'Comet VC40' Gas Lighter... Flick-top design for easy access to flints.

8. c. ellipt. for *flint corn* (see *10).

1802 J. DRAYTON *View South-Carolina* 137 The flint is more hard and nourishing. 1857 *Trans. Ill. Agric. Soc.* III. 63, I plant the white flint. It is rather on the gourd seed order. 1883 *Encycl. Brit.* XV. 309/2 The 'Flint' varieties are most common east of Lake Erie and north of Maryland. 1909 [see *DENT *ppl. a.* 3]. 1947 *Ann. Missouri Bot. Garden* Feb. 17 There is as yet little exact evidence as to how completely the gene combinations introduced from the northern flints have been broken up in modern dent corns.

9. b. *flint-chipper.* **c.** *flint-headed* adj. (earlier U.S. example). **d.** *flint-eyed* adj., *-grey* (later examples).

1872 *Amer. Naturalist* VI. 208 Unfinished specimens suggest that variety was continually aimed at by the 'flint chipper'. 1898 *Westm. Gaz.* 30 Mar. 1/3 A tall, solemn, shy, flint-eyed young man. 1933 W. DE LA MARE *Fleeting* 51 A still quiet challenge Fills her dark, her flint-grey eyes. 1951 S. SPENDER *World within World* 126 The sun-bathers [were] lying in profusion on the flint-grey grass. 1843 'R. CARLTON' *New Purchase* I. 2 Shadows of branching antlers and flint-headed arrows caused many a darkness in his path.

10. flint corn, the name of certain varieties of maize, esp. *Zea mays* var. *indurata*, having very hard kernels; **flint-gun** (earlier U.S. example); **flint paper** (see quot. 1962); also *ellipt.*; **flintwood** (example).

1705 R. BEVERLEY *Virginia* (1722) 126 The one [grain] looks as smooth, and as full as the early ripe Corn, and this they call *Flint-Corn.* 1838 H. COLMAN *Rep. Agric. Mass.* 20 The best kinds of flint corn weigh 60 lbs. to the bushel. 1872 E. EGGLESTON *End of World* viii. 60 The relative merits of 'gourd-seed' and 'flint' corn. 1950 *New Biol.* VIII. 37 Flint corns (type of maize) have hard flinty kernels with a little soft starch inside. 1837 W. IRVING *Capt. Bonneville* (1895) II. 133 Their flint guns were at fault, and missed fire. 1916 H. A. MADDOX *Paper* x. 137 Flint papers and box enamels are burnished in the friction glazer. 1920 R. W. SINDALL *Paper Technol.* (ed. 3) xviii. 295 *Flint.*—Papers, usually coated, to which an extra polish has been imparted by friction with a long flat stone moving quickly to and fro across the surface of the paper. 1962 F. T. DAY *Introd. to Paper* 115/1 *Flint papers*, a base paper coated one side with a colour and afterwards hard burnished or flint glazed to produce a high gloss, water-proof surface. 1967 E. G. LOEBER *Suppl. Labarre's Dict. Paper* 25 *Flint*, also called *flint (glazed) paper. Ibid., Flint paper*,..an abrasive paper..coated with ground flint-stone. 1884 A. NILSON *Timber Trees N.S.W.* 135 *Flintwood*, Eucalyptus pilularis.

flip, *sb.*¹ Add: **3. flip-iron** *U.S.* = *flip-dog.*

1869 MRS. STOWE *Oldtown Folks* xxxvii. 480 Draw the flip-iron from the fire and stir the foaming bowl. 1947 F. D. DOWNEY *Our Lusty Forefathers* 7 Shaking flip-irons in each other's face, as the saying went, 'at loggerheads'.

flip, *sb.*² Add: **2. b.** *Gunnery.* The springing of the barrel of a gun at the moment of discharge.

1896 W. W. GREENER *Gun* (ed. 6) xxiii. 546 Recoil.. 'Jump' and 'flip' are secondary movements—vertical and lateral respectively. *Ibid.* 564 *Flip* is a term used to denote the lateral deflection of a rifle barrel due to the same or similar causes as the vertical deflection termed 'jump'. 1903 *Kynoch Jrnl.* June–July 101/2 The flip, or springing of the barrel may entirely counteract the increase in jump. 1904 *Ibid.* Jan.–Mar. 63 The flip was evidently not entirely vertical since the direction varied as well as the elevation. 1905 *Ibid.* Apr.–June 80 There is the probability that the 'jump' or 'flip' of the rifles will not be quite the same. 1925 G. BURRARD *Notes on Sporting Rifles* (ed. 2) 75 They do not take into account the effect of jump or flip. This varies in every individual rifle and barrel. 1960 *Times* 5 Mar. 9/5 A recoil carriage which allows full recoil of the gun, even at the pressures used, and permits 'flip'.

4. A short flight in an aircraft; also, a trip in another form of conveyance; a quick tour on foot. *colloq.*

1914 H. ROSHER *In R.N.A.S.* (1916) 28 It was much too foggy for my trip to Hartlepool yesterday afternoon, but I went for a short flip around. 1920 *Blackw. Mag.* Feb. 167/2 The *Moewe* arrived, untouched, on 4th March, after a two months' 'flip'. 1929 *Daily Express* 7 Nov. 2/1, I went to Croydon and went up again for a short 'flip'. 1943 HUNT & PRINGLE *Service Slang* 32 *Flip*, a trip in an aeroplane or in a car. 1958 *Times* 16 June 12/4 He..promised her a flip in his five-metre yacht. 1959 *Vogue* Oct. 120 A quick flip around Fenwicks always delights us.

5. Applied to a person (see quots.). *slang.*

1942 BERREY & VAN DEN BARK *Amer. Thes. Slang* § 398.3 *Impudent person*,..flip. *Ibid.* § 410.3 *Flip*, a forward or flippant person. 1955 N. SHAPIRO *Hear me talkin' to Ya* 347 He's not a flip as far as business is concerned. 1958 *New Statesman* 6 Sept. 294/3 If you are not cool, you can be put down; you have lost your will and confidence. You are a flip and you are beat. 1961 'I. Ross' *Old Students never Die* (1963) iv. 60 'She's a flip... Nuts,' he translated, 'Loony. Off her rocker.'

6. Abbrev. of **flip side.*

1960 *Melody Maker* 31 Dec. 6/1 The same can be said of the flip, also featured in the film.

7. flip jump, a toe jump in figure skating; **flip side,** the reverse, or less important, side of a gramophone record; also *transf.*

1940 S. HENIE *Wings on my Feet* 163 A flip jump..is done by placing the right toe-point to the right back outside edge at the finish of a left three. In this case, the left foot swings around, but on a close arc to the right, and you land as in a Salchow. 1960 M. V. OWEN *Fun Figure Skating* vii. 138 If you do add the half-turn, you will be doing a full 'flip jump', which lands on the ROB edge in the exact way that the waltz jump landed. 1964 'J. NOEL' *Figure Skating for Beginners* ix. 92 The flip jump itself used to be known as the toe salchow or spot salchow. 1968 *Daily Tel.* 6 Dec. 15/6 She fell after attempting a double flip jump, but otherwise was always elegant and technically correct. 1949 *Down Beat* 11 Mar. 14 The flip side (*South*) will be a shade slower but with the same general routine. 1962 *Spectator* 14 Dec.

945 A slice off the flipside. 1968 Times Lit. Suppl. 11 Apr. 377/1 The actual readings are on the flip side. 1971 Guardian 25 Mar. 14/4 A dogged courage..is the flip side of Mr Heath's best known vice. 1971 Sunday Times (Johannesburg) 28 Mar. 7/2 The manoeuvring, the muscling, the promoting and the hustling which is the flip-side of showbiz.

flip, a.[1] Add: (Further examples.) Now esp., glib, flippant.

1924 Drama Mag. (Chicago) Feb. 177/2 Doris is flip, exaggerated, and stagey—off the stage. 1935 Time 24 June 38/2 He is entirely too flip and smart-alecky. 1944 Auden For Time Being (1945) 110 Every martyrdom an occasion for flip cracks and sententious oratory. 1947 'N. Shute' Chequer Board x. 290 He coloured hotly, and wished desperately for eloquence that he might make some flip and smart rejoinder, but no inspiration came. 1958 R. Williams in N. Mackenzie Conviction 78 How can anyone..use these new flip words for any attachment to learning or the arts? 1963 Listener 21 Mar. 529/2, I find it difficult to reconcile the authorship of the flip, alert opening scenes with the awful feyness of what followed. 1969 N. Cohn AWopBopaLooBop (1970) xii. 104 The musicians fitted themselves sensibly to the situation—they kept things light and flip and sexy. 1970 Times 9 May 9/5 The word 'schizophrenia' is flung about today with flip facility.

flip, v. Add: **1.** (Later absol. examples.)

1945 J. Steinbeck Cannery Row xxvii. 118 They had to flip for who would go to the party first. 1971 'A. Blaisdell' Practice to Deceive x. 146 You want to flip for who does the report?

8. To fly in an aircraft. colloq. or slang.

1915 H. Rosher In R.N.A.S. (1916) 41, I crashed into the atmosphere first thing this morning and flipped around for 55 minutes.

9. In full to flip one's lid, wig. To be or become wildly excited or enthusiastic; to go wild, lose one's head. slang (orig. U.S.).

1950 Neurotica Autumn 44 If I'm not right back don't flip. 1951 Jrnl. (Ithaca) 30 Jan. 6/6 Present war emergencies..have been too much for local government officials. I fear they have flipped their lids. 1951 N.Y. Times Bk. Rev. 2 Dec. 50/2 The funniest book of the lot is enough to make a reader 'flip' or 'flip his lid'. 1952 C. Brossard Who walk in Darkness viii. 53 He flipped his wig when it was finished and they took him to a sanatorium. 1961 R. Bloch Blood runs Cold (1963) 156 Mitch and his crowd didn't flip for jazz, but he'd come on strong with the bingos. 1960 B. Crump Good Keen Man 183 As he spoke one of the dogs sank his teeth into a tender part and the bull flipped his lid completely. 1960 Time & Tide 24 Dec. 1599/1 Does he make me flip my wig! 1967 Boston Globe 18 May 18/1 (Advt.), Our food and service are great. Our decor's delightful. Your club treasurer will flip over our low rates. 1969 'R. Macdonald' Goodbye Look iii. 23 She's a phoney blonde...I can't understand why he would flip over her.

flip-flap, sb. Add: **3. d.** In a place of amusement or the amusement section of an exhibition, etc., a machine with long moving arms by which passengers are raised on platforms (see quot. 1908).

1908 Daily Chron. 3 Apr. 7/2 The..huge steel arms [are] 150 ft. in length, much like the main shaft of a crane, greatly magnified...These arms will be slowly raised until their extremities cross in the air, 150 ft. above the level, and then each will complete the semi-circle. Suspended from..the hand of each steel arm will be a car containing passengers... The flip-flap is built on the cantilever principle with heavy counter weights. 1922 C. Sidgwick Victorian xxii, He is going to have millions of Japanese lanterns and a Flip-Flap and an open air café like at Earl's Court. 1937 Evening News 2 Feb. 8/6 Those who were young enough..for such delights thirty years ago must have been surprised to learn that the Flip-Flap of Franco-British Exhibition fame is still in existence.

flip-flop, sb. Restrict 'nonce-uses' to senses in Dict. and add: **c.** A somersault. Cf. Flip-flap sb. 3 a. U.S.

1902 G. H. Lorimer Lett. Merchant xvii. 245 And when a fellow's turning flip-flops up among the clouds, he's naturally going to have the farmers gaping at him. 1929 Liberty 30 Nov. 43/1 Turning hand-springs and flip-flops all over the sawdust covered floor. 1946 Gunnison (Colo.) News-Champion 2 May 1/2 The convertible Ford coupe.. missed a narrow bridge..turned a flip-flop in the air and came to a stop right side up in waist-deep waters. 1969 New Yorker 12 Apr. 100/2 Every time Lind or any other astronaut opens his mouth, the entire space industry turns flip-flops.

d. As advb.

1904 H. G. Wells Food of Gods ii. i. 145 She..passed, flip-flop, within three yards of them.

e. Electronics. Either of two types of electronic switching circuit: (a) one that passes from a stable to an unstable state and back again in response to a triggering pulse; (b) one that has two stable states and makes a single transition from one to the other in response to a triggering pulse.

1935 Wireless Engineer XII. 606/2 The earliest accounts of the device, which is sometimes called the 'flip-flop', are by Eccles and Jordan. 1946 Math. Tables & other Aids to Computation II. 100 The basic electronic memory device of the ENIAC is the flip-flop or trigger. 1964 F. L. Westwater Electronic Computers ii. 27 If a pulse is supplied to the lowest flip-flop in the counter the number 1 will be registered. 1966 New Scientist 3 Nov. 242/3 The cell contains its own version of an electronic 'flip-flop' circuit, a two-tube circuit which is stable when either tube is conducting, but not when both are. 1971 New Scientist 11 Mar. 551/3 Flip-flops are widely used in computers.

f. A plastic or rubber sandal consisting of a flat sole and straps. Also attrib.

1970 Observer 15 Mar. 6/3 Milligan has a beard and wears flip-flops with jeans. 1971 M. Polland Package to Spain xii. 152 'She had flip-flops with white daisies on the front of them.' 'Flip-flops?'..'Sandals. Beach sandals all made of plastic.' 1971 W. J. Burley Guilt Edged i. 18 She wore..blue jeans and flip-flop sandals.

Hence **flip-flop** v. intr., to go, proceed, act, etc., in a 'flip-flop' manner; also trans., to turn (something) over. So **flip-flopping** vbl. sb.

1897 Outing (U.S.) XXX. 176/1, I could hear a vigorous flip-flopping going on beyond the weeds, and I knew the captive was a trout. 1924 W. Deeping Three Rooms xxxv. 313 He [sc. a person distempering a wall] grew quite jaunty and confident, flip-flopping with considerable dexterity. 1940 F. D. Davison in B. James Austral. Short Stories (1963) 59 Wallabies flip-flopped out of your way as you rode. 1968 J. D. Watson Double Helix xxvi. 197 Both pairs could be flipflopped over and still have their glycosidic bonds facing in the same direction.

flipper, sb.[2] Add: **1. b.** A rubber attachment to the foot used for underwater swimming, esp. by frogmen.

1945 Newsweek 17 Sept 113 (Advt.), He'd organized a spear-fishing party, and this is the proper regalia—glass-front mask, flipper shoes, and a..spear. 1953 New Yorker 29 Aug. 17/2 Such paraphernalia..as flippers,..Aqua-Lungs, and snorkels. 1959 Spectator 28 Aug. 251/1 He pulled on rubber flippers to join his wife in the water. 1970 New Yorker 19 Sept. 104/2 Clumping around in frogman flippers.

3. (Example.)

1928 A. E. Krows Equipm. Stage Prod. iii. 40 The wing-piece for an exterior setting is usually..hinged in two parts... If it is very tall, a small extra section.. called a 'flipper', is hinged to one of the larger parts.

4*. Cricket colloq. (See quot. 1967.)

1959 J. Fingleton Four Chukkas to Australia 36 He [sc. Benaud] varied his attack with a leg-break..a well-disguised bosie and an excellent flipper. 1967 Partridge Dict. Slang Suppl., Flipper, a top-spinner delivered by the bowler with an extra flip of the fingers.

flipper, v. [f. Flipper sb.[2]] intr. To move by means of flippers.

1955 A. Ross Australia 55 ix. 122, I saw green turtles surfacing sadly in pairs, soon flippering away as if in disappointment at the weather. 1969 Look & Learn 15 Mar. 35/1 Knowing that his harpoon gun was useless against two such killers, Frankie flippered for the raft.

flipping (fli·piŋ), adv. and ppl. a. slang. [Flip v.] Used as a substitute for a strong expletive. Usually derog. (Cf. *blinking ppl. a. 4.)

1911 D. H. Lawrence White Peacock ii. ix. 347 'Ain't it flippin' 'ot?' drawled Creswell. 1948 C. Day Lewis Otterbury Incident i. 2 Flipping heroes, ain't we all? 1954 A. Heckstall-Smith Eighteen Months xii. 145, I suggested that..he might take up the game professionally. He shook his head. 'Too much flippin' trainin',' he sneered. 1959 Spectator 20 Nov. 713/3 While terms of approval.. change rapidly with the fashion, terms of disapproval (blinking..flippin' awful..) show very little alteration. 1971 Guardian 24 Aug. 5/1 They wax indignant about pornography but when it comes to doing anything about it they are bone flipping lazy.

flit, sb. Add: **a.** spec. do a flit, to decamp.

1952 M. Tripp Faith is Windsock vii. 108 They'll say I got windy and did a flit. 1964 R. Braddon Year Angry Rabbit (1967) ii. 16 A fourth [daughter]..had cunningly got herself pregnant by the one pathologist in the team who was too honourable to do a flit. 1970 Sunday Truth (Brisbane) 27 Dec. 16/3 They live on the generosity of the small country storekeeper, then do a flit.

c. (See quots. 1942.) Also attrib. or as adj. U.S. slang.

1942 Berrey & Van den Bark Amer. Thes. Slang § 405.2 Effeminate man,..Fauntleroy, flit, fuddyduddy. Ibid. § 508.4 Male homosexual,.. flit, four-letter man, fruit, fruiter. 1951 J. D. Salinger Catcher in Rye xix. 170 Sometimes it was hard to believe, the people he said were flits and lesbians. 1964 E. Lacy Pity Honest ii. 30 'Could he have been on the flit side?' 'Doubt that,..not that I'm any authority on queers.'

Flit (flit), sb.[2] [f. the vb.] The proprietary name of an insecticide. Flit gun, a syringe intended for use in spraying insecticides.

1923 Official Gaz. U.S. Pat. Office 27 Nov. 723/1 Standard Oil Company (New Jersey).. Flit. Insecticide. Claims use since May 17, 1923. 1926 Trade Marks Jrnl. 17 Mar. 628 Flit, chemical substances used for Agricultural, Horticultural, Veterinary and Sanitary purposes. Standard Oil Company..New Jersey. Ibid. 10 Nov. 2513 Flit, Sprayers, Squirt-guns and Atomizers..Standard Oil Company..New Jersey. 1927 Blackw. Mag. Sept. 310/2 She had had the foresight to buy a metal spray and a tin of insecticide called Flit. With Flit we had come scatheless through Persia. Ibid., Even our Flit gun ceased to interest him. 1932 E. Hemingway Death in Afternoon xvi. 186 You ought to spray him with flit. 1948 C. Day Lewis Otterbury Incident iv. 43 They had a pail of muddy water..and one of those Flit guns loaded with it. 1958 Times 7 Aug. 3/4 They only had a Flit-gun, the sort of thing we used in India to keep the mosquitoes off.

flitch, sb.[1] Add: **3. c.** In full flitch-plate. A strengthening plate added to a beam, girder, or any woodwork.

1888 Lockwood's Dict. Mech. Engin., Flitch, a plate of metal or of wood bolted to an otherwise weak and unstayed beam or structure in order to strengthen and support it. Flitch Plate, a broad thin plate or rolled bar used in building up flitch beams or plated work. 1912 Motor Man. (ed. 14) iii. 89 To build the frame up of stout section ash and bolt on steel 'flitch' plates to the sides. 1950 Engineering 6 Jan. 3/3 The engine is reinforced by a channel-section flitch over most of its length.

fliting, flyting, vbl. sb. **1. b.** (Later examples.)

1934 A. Huxley Beyond Mexique Bay 21 The proceedings ended with a 'flyting'. Three of the singers..proceeded to improvise stanzas of derision at one another's expense. 1948 English Studies XXIX. 166 The fliting between Unferth and Beowulf. 1959 A. G. Brodeur Art of Beowulf 144 Flytings are either exchanges of rude wit, rough games, or invective preceding a fight. 1962 G. K. Hunter John Lyly vi. 335 In the 'flytings' between Katherine and Petruchio the exchanges wear the guise of wit. 1968 Listener 25 Apr. 525/3 Beckett had anticipated the sequence in the flyting in Waiting for Godot.

flivver (fli·vəɹ), sb. slang (orig. U.S.). [Of obscure origin.]

1. A cheap motor car or aeroplane. Also, 'a destroyer of 750 tons or less' (Funk's Stand. Dict. 1928).

1910 W. A. Fraser Red Meekins (1921) 22 You stick to me an' you'll be travellin' 'round the country in a flivver. 1919 F. Hurst Humoresque 296 Was his little Sid fool enough to beat it all the way over here in a flivver for eight bucks the round trip? 1920 Glasgow Herald 21 July 9 May be I will disguise the Shamrock as a 'flivver' (as the Ford car is known here). 1924 'Digit' Confess. 20th Cent. Hobo 11 Flivver, Ford automobile. 1924 W. M. Raine Troubled Waters vi. 58 Rowan McCoy drove his new car—it was a flivver, though they did not call it that in those days. 1926 Ladies' Home Jrnl. Apr. 39 Won't it be amusing when we can..step into our little up-shooting flivvers at the back door? 1927 Punch 2 Feb. 135/1 He has successfully fought the villain Trust; his workmen all own 'flivvers'; there has never been a strike. 1936 Beaver Mar. 9/1 So-called 'flea' or 'flivver' types [of aeroplanes]. 1956 S. Bellow Seize Day (1957) i. 22 He had driven a painted flivver.

2. A person or thing that has a damaging or deleterious influence; a failure.

1915 H. L. Wilson Ruggles of Red Gap (1917) xiii. 230 That Jackson lad has offered me about ten thousand of them vegetable cigarettes, but I'll have to throw him down. He's the human flivver. Put him in a car of dressed beef and he'd freeze it between here and Spokane. 1915 Dialect Notes IV. 233 Flivver, a hoax; also, a failure.

Hence **fli·vver** v. intr., (a) to fail, to come short of success; (b) to travel in a flivver.

1912 L. J. Vance Destroying Angel vi. 74 If the production flivvers, I'll need that thirty cents. 1927 Bulletin 11 Apr. 14/1, I was finding the desert a bit flat when you flivvered in.

flix-weed, now usu. **flixweed,** see Flux sb. 13.

float, sb. Add: **1. c.** Finance. An operation of floating a currency. Cf. *float v. 1 d.

1971 Daily Tel. 10 May 14 Between the two German mark 'floats'—in Oct. 1969 and today—the international monetary scene has been comparatively calm. 1971 Guardian 18 Aug. 18/3 A float against the dollar by a unified European block might provide the necessary breathing space. 1971 Economist 4 Sept. 13/1 The Bank of Japan is operating energetically to prevent the float from being anything except a bogus one.

8. e. A structure fitted to the alighting gear of an aircraft to enable it to float on water.

1897 Strand Mag. June 717/1 The conical vessel in front is an empty float, whose use is to keep the whole from sinking if it should fall in the water. 1909 Flight 30 Jan. 63/1 Delagrange..has ordered a set of special floats for his aeroplane. 1913 Aeroplane 17 Apr. 453 Labouret.. 'stalled' the machine, fell over sideways, and smashed.. the right float. Ibid. 455 The concertina floats on this machine are evidently a great success. 1928 C. F. S. Gamble N. Sea Air Station 13 Experiments with various types of floats and flotation bags for aeroplanes. 1941 W. Nelson Airplane Lofting i. 11 Float bottoms are designed to give small water resistance, and the float as a whole is shaped to give as little air drag as is consistent with its other functions. 1958 R. D. Blacker Basic Aeronaut. Sci. viii. 130/2 The particular model..is amphibious, inasmuch as it has wheels which are retracted into the floats by the pilot for water landings.

9. a. Also, in a petrol engine, a device which floats on the petrol in the float-chamber of the carburettor and regulates the supply so that the level remains constant.

1901 Motor-Car World Mar. 42/1 The Float. 1902 A. C. Harmsworth et al. Motors vii. 112, F is a float with wire attached to indicate the height of the petrol. 1912 Motor Man. (ed. 14) 10 When the float sinks it opens a small valve and allows the petrol to flow in till a certain level is reached. The float then rises and closes the valve. 1967 K. Ullyett 1100 Companion viii. 122 Fuel in the float chamber is maintained at constant level by float and needle in the conventional way.

10. (Earlier examples.)

1829 J. R. Planché Paris & London i. v. 24 A diagonal view of the stage of the Odeon is seen through the wings.. At the end of the float a section of the audience and part of the theatre is visible. 1840 A. Bunn Stage I. iii. 54 That mysterious line of light across the stage, (yclept in theatrical phraseology the float).

14. c. ellipt. = milk-float (*milk sb. 10).

1971 Daily Tel. 19 Apr. 2/6 Postmen and milkmen should share deliveries from two-man floats.

22. (Later example.)

1948 E. N. Dick Dixie Frontier 74 Quantities of affidavits were printed and distributed to paid agents, who with the co-operation of ignorant or corrupt justices of the peace secured floats wholesale for speculators.

23*. A sum of money in a shop, etc., used to provide change, small payments, etc., at the start of business; a shop till or its contents (slang); a small loan. Also attrib.

1902 W. H. Chantrey Theatre Accounts ii. 24 There seems no necessity in a Theatre to have a Petty Cash Account, as each week the treasurer will reimburse all the small amounts which have been laid out by the various members of the staff, and if necessary a 'float' could be provided to meet this expenditure meanwhile. 1931 Police Jrnl. Oct. 503 The other day a thief..stole (knocked off) the contents of the till (the float). 1955 Times 8 July 6/7 Mrs. Foote produced £57 from the shop till, which she said was her 'float'. 1966 L. Southworth Felon in Disguise xi. 159 Larceny of cash from the pub float of the 'Crosby Arms'. 1967 K. Giles Death in Diamonds ix. 170 The branch offices carry a float account because the salesmen get a portion of their commission the following morning after a sale. 1967 V. Canning Python Project ii. 18 'That's a lot of money to keep around the flat.' 'My husband always said one should have a substantial cash float, just in case.' 1969 Times 1 Apr. 6/1 (Advt.), When you need a quick £10 float, at any time, you go to a branch with a cash dispenser in the wall.

b. The amount of money represented by cheques, etc., in transit. Chiefly U.S.

1915 H. P. Willis Federal Reserve xi. 228 There is, in short, a so-called 'float' which represents the volume of checks afloat in the mails at any time and not liquidated. 1924 W. O. Scroggs Cent. Banking Progress 273 The amount of these checks continually in transit, the 'float', was estimated at about $300,000,000. 1930 Economist 18 Oct. 707/2 A..'float', that is, clearing house checks and exchanges in transit, the volume of which depends largely on stock market activity. 1931 T. E. Gregory in W. Rose Outl. Mod. Knowl. xv. 652 As the volume of speculation grows, the 'float', i.e. the sums due at any moment, also increases.

24. float-chamber, a small chamber in the carburettor of a petrol engine from which petrol, maintained at a constant level by the action of a float, is supplied to the jets; **float-feed,** a device for controlling the feed of a liquid by means of a float; also attrib.; **float glass,** glass manufactured by the *float process; **float-gold** (earlier U.S. example); **float needle,** a thin rod attached to a float (sense *9 a) which by passing into or out of the inlet to the float-chamber allows less or more petrol to enter it; **float-plane** = *float-seaplane; **float process,** a process for making flat glass in which the glass is drawn in a continuous sheet from the melting tank and made to float on the surface of molten metal in a controlled atmosphere while it hardens; **float road** U.S. (see quot. 1905); **float-seaplane,** a seaplane equipped with floats.

1901 Motor-Car World Mar. 42/1 Should petrol run over at the jet when the pressure is on, give the spindle in the centre of the float-chamber a few turns with a screwdriver to grind in the needle valve. When the engine is not running keep the spring on the float needle. 1967 K. Ullyett 1100 Companion viii. 125 It is a compact, dustproof carburetter, with concentric float chamber. 1902 A. C. Harmsworth et al. Motors vii. 115 The method of supplying petrol to the carburetter is on the same principle.., float feed. 1907 Westm. Gaz. 28 Nov. 4/1 The carburetter is of the float-feed type. Ibid. 5 Dec. 4/2 Carburetter..of the usual float-feed spray type. 1959 Times 21 Jan. 10/3 The advantages of float glass..include its freedom from distortion. 1962 Guardian 25 June 4/6 Float glass is an entirely new glass which combines the best qualities of both plate and sheet glass. 1873 J. Miller Life amongst Modocs xvi. 204 They had found only a few bars with float gold. 1901 Float needle [see *float-chamber above]. 1965 Punch 8 Dec. 824/1 Bits of radio aerial, carburettor float needles, and an AA route from Leicester to Devizes. 1922 Flight XIV. 126/2 Seaplanes include Float Planes and Flying Boats, denoting, respectively, seaplanes fitted with floats or hull. 1939 War Weekly 24 Nov. 140/4 It is almost impossible to make a floatplane anything like as fast as a landplane. 1958 Listener 20 Nov. 818/2 Nothing can be more hazardous than landing a float-plane among sea ice. 1965 New Scientist 27 May 577/1 The Organization of the USSR State Committee on Aviation Technology has patented a device..for use on cargo floatplanes. 1959 Economist 24 Jan. 346/1 In the float process, a continuous ribbon of molten glass is fed out of the furnace across the surface of a tank of molten metal in a controlled atmosphere, gradually being allowed to cool. 1964 Guardian 6 Aug. 11/8 The float process's main advantage is that it cuts costs considerably, uses less labour, and takes up less than half the space of a production line using the grinding polishing method. 1970 Physics Bull. Apr. 153/1 The development of the float process for the manufacture of a continuous ribbon of flat glass has provided a rich variety of novel technological and scientific problems. 1901 F. A. Montgomery Reminisc. Mississippian 115 Next morning I determined to follow an old float road in which we found ourselves. 1905 Terms Forestry & Logging 37 Float road, a channel cleared in a swamp and used to float cypress logs from the woods to the boom at the river or mill. 1919 W. B. Faraday Gloss. Aeronaut. Terms 72 Float seaplane, an aeroplane provided with floats for alighting on water. 1928 C. F. S. Gamble N. Sea Air Station xiii. 212 The flying-boat is not so fast, nor has it the same ceiling, as float-seaplanes.

float, v. Add: **1. d.** transf. Finance. Of a currency: to fluctuate as regards its international exchange rate. Also trans., to arrange for (a currency) so to fluctuate.

1965 Guardian 30 Sept. 1/8 The decision to let the mark float was forced on the German authorities by a sudden inflow of funds. 1965 J. L. Hanson Dict. Economics & Commerce 182/1 To permit the pound to 'float'..would.. be a return to free exchange rates. 1970 Daily Tel. 2 June 20/7 Foreign exchange markets went into a flurry of activity as a result of the decision to float the Canadian dollar. 1971 Ibid. 10 May 14 If the mark floats high, sterling and the franc may face large inflows of foreign currency. 1971 Economist 4 Sept. 3/2 Japan nominally floats the yen, but really keeps it fixed.

2. d. Electr. Of a part of an electrical circuit: to be unconnected to a source of fixed potential.

1931 Proc. R. Soc. A. CXXXI. 688 The grid of the first valve floats at a potential just over 2 volts negative with respect to the negative end of the filament. 1947 F. G. Spreadbury Electronics ii. 94 Instead of allowing the probe to float, a potential difference is maintained between it and either the cathode or anode. 1945 R. C. Walker Electronic Equipment iii. 54 A floating grid acquires a negative potential by collecting electrons from the cathode emission, and its final potential is in equilibrium with the cathode stream. 1967 Electronics 6 Mar. 118/2 A floating input that can be operated up to 500 v above ground.

5. d. To move or proceed, esp. in a leisurely or casual way; to wander from place to place. slang (orig. U.S.).

1901 'H. McHugh' John Henry 10 I'm sitting on the sofa ..when my lady friend floats into the arena. 1909 R. A. Wason Happy Hawkins 96 So I just floated, punchin' cows most o' the time but not runnin' very long over the same range. 1931 'D. Stiff' Milk & Honey Route i. 15 The hobo really floats, which explains the name 'floater', by which he is often labelled. 1935 G. Ingram Cockney Cavalcade 232 Come on. Let's float. 1936 J. Curtis Gilt Kid v. 53, I floated because I got fed up and wanted her to turn in thanking me. 1959 I. & P. Opie Lore & Lang. Schoolch. x. 192 Juvenile language is well stocked..with expressions inviting a person's departure,.. flit, float away, [etc.]. 1970 Globe Mag. (Toronto) 26 Sept. 18/2 Some immigrants are bound to float on, but is Canada really doing very well for itself when a chartered accountant like Chris Mawhood gets turned off?

9. (Earlier and later U.S. examples.)

1871 J. Burroughs Wake-Robin (1884) 106 Our guide proposed to conduct us to a lake..where we could float for deer. 1885 Outing (U.S.) Oct. VII. 80/2 'Kill any deer over there?' 'No,' said Carl, 'we floated two nights, but it was terrible foggy.'

floatation, flotation. Add: **1.** attrib. Applied to any device that gives buoyancy.

1928 [see *float sb. 8 e]. 1940 Chambers's Techn. Dict. 343/1 Flotation gear, a system of air or gas bags, sometimes with hydrovanes, to enable a land plane, in an emergency, to land and remain afloat, on water. 1958 Times 1 Sept. 4/6 A..general-purpose missile..is being shown complete with parachutes and flotation bag; the latter inflates automatically when the missile falls into the sea. 1963 Times 17 May 14/1 Divers jumped from the helicopters and attached the flotation collar to the capsule. 1971 Times 17 July 5/4 Under your seat there is a floatation device.

b. The separation of particulate material, esp. pulverized ore, by utilizing the varying capacity for floating on a given liquid that results from differences in the surface properties of the particles (rather than from differences in density); freq. attrib. So **froth flotation,** flotation in which the separation is enhanced by using a liquid to which a frothing agent has been added and bubbling air through it, so forming a froth in which certain of the particles collect.

1908 Westm. Gaz. 1 Aug. 15/1, 11,775 tons of dump tailings were treated by the flotation process. 1925 Jrnl. Iron & Steel Inst. CXI. 471 The application of flotation-methods to existing coal-washery-plants, and the cost and working expenses of such appliances, are discussed. Ibid. 472 The article describes the application of froth-flotation processes to the recovery of coal from the sludge. 1927 Daily Express 15 July 10/1 The original ore is readily amenable to treatment by flotation. 1951 Kirk & Othmer Encycl. Chem. Technol. VI. 595 Separations made by flotation include not only the concentration of base-metal ores but also widely divergent applications such as the separation of ink from repulped paper stock, of peas from pea pods, [etc.]. 1968 Coulson & Richardson Chem. Engin. (ed. 2) II. xvii. 704 Froth flotation is widely used in the metallurgical industries where, generally, the ore is difficult to wet and the residual earth is readily wetted. 1970 Materials & Technol. III. ii. 117 Many flotation plants treat over 20,000 tonnes of ore per day.

floater. Add: **1. a.** spec. (a) A dead body found floating in water. U.S. slang.

1890 J. A. Riis How Other Half Lives xix. 230 'Floaters' come ashore every now and then with pockets turned inside out. 1963 J. Mitford Amer. Way Death iii. 53 Floaters..are another matter; a person who has been in the Bay for a week or more..will decompose more rapidly.

(b) A golf-ball capable of floating on water.

1897 Westm. Gaz. 30 Mar. 9/2 There are bournes from which no ball, not e'en a floater, returns. 1927 Daily Express 29 Sept. 9 Many leading American professionals ask that the 'floater' be adopted as the official standard ball.

(c) A mine adrift.

1916 'Taffrail' Carry On 54 Almost every day 'floaters', which have broken adrift from their moorings, are solemnly sunk by rifle fire.

(d) A piece of float-ore.

1921 Chambers's Jrnl. 508/1 The molybdenite-seeker next proceeds to work in earnest by breaking all the loose 'floaters' or detached boulders, and collecting all the flakes that are set free in bags. 1928 Sunday Dispatch 25 Nov. 3/5 The chance discovery..of a large 'floater', or piece of gold-bearing quartz. 1950 K. S. Prichard Winged Seeds 21 But I reckon the lode these floaters came down from 's not far off.

(e) (See quot. 1933.) Prison slang.

1933 Punch 25 Oct. 456/1 A 'floater' is an old magazine, book 'or even a newspaper' which is smuggled irregularly from cell to cell. 1958 F. Norman Bang to Rights 97 It's [sc. a book] a floater so you can sling it if you think you are going to get a turn over.

3. b. Insurance. A policy in general terms, esp. covering portable goods. (Cf. *floating ppl. a. 5 b.)

1900 Policy-holder 6 June 441/2 The Norwich Union is largely interested..by specific amounts and by floaters... Messrs. John McNairn and Co. had a £20,000 floater from the North British.

4. a. (Earlier U.S. and later U.K. examples.) In U.K., without any suggestion of corrupt practice.

1847 Knickerbocker XXIX. 329 Early the next morning the 'floaters' were marched in single file with votes in hand, to the ballot box. 1896 Westm. Gaz. 22 Feb. 5/2 The 'floaters' should be mostly credited to Mr. Morley. Many people at Montrose believe he is certain to have a majority of 1,500. 1959 Spectator 25 Sept. 394/1, I find it hard to believe that many floaters were impressed by the Conservatives' first TV broadcast. 1963 Punch 6 Feb. 194/1 A charming Tory supporter..may well sway incalculable numbers of 'floaters' when polling day comes.

b. (Earlier and later examples.) Also, one who frequently changes his job; a temporary employee.

1859 T. S. Woodward Reminisc. (1939) 49 He was a floater..but he located him a tract in the fork of Coosa and Tallapoosa. 1873 J. H. Beadle Undevel. West xxiii. 455 There are clerks, agents..and perhaps fifty 'floaters', making up the American population. 1878 —— Western Wilds iii. 45 A man..failed, lost hope, and sank into a 'floater'. 1909 Webster, Floater, one who takes temporary employment; specif., a substitute teacher. U.S. 1923 J. D. Hackett in Managem. Engin. May, Floater, a person who habitually leaves one occupation and goes to another for the sake of variety. 1927 W. T. Root in C. Johnson Negro in Amer. Civ. (1931) 321 The larger number of unmarried 'floaters' drifting into the city. 1931 [see *float v. 5 d]. 1931 G. Irwin Amer. Tramp & Underworld Slang 76 Floater, a migratory worker, one who moves from place to place, but who has some excuse for this in that he works occasionally. 1934 Sun (Baltimore) 11 Jan. 1/5 He denied that the order marked the establishment of a policy designed to prevent the employment of transient or 'floater' labor. 1967 L. Deighton Expensive Place xv. 104 'The murdered girl was working for us'.. 'A floater?'..'No. Permanent.' 1969 Daily Tel. 24 Oct. 16/5 There are only a score of vacancies to be filled, and these are of no interest to a number of young 'floaters' in and out of jobs as delivery boys, petrol pump attendants, car washers. 1971 H. C. Rae Marksman ii. vi. 149 [He's a] Detective Inspector; a floater, I think they call it. He circulates from department to department.

d. An official order to leave a town or district; a sentence suspended on condition that the offender leaves the area. U.S. slang.

1914 Jackson & Hellyer Vocab. Criminal Slang 34 Floater,..A suspended sentence; a mandatory order to quit a community or locality. Example: 'The rap wasn't strong enough, so they took a floater.' 1926 J. Black You can't Win vi. 69, I was just after gettin' a six months' floater out of Denver. 1952 J. Steinbeck East of Eden 334 There's a permanent order in the Sheriff's office.. that if I..admit I'm your wife I'll get a floater out of the county and out of the state.

5. A mistake, 'bloomer'. slang.

1913 A. Lunn Harrovians iv. 78 There is no phrase for a faux pas at Harrow... It is only when he reaches the university that he realizes that such banter is often a 'floater', and for this handy expression he has no parallel in school slang. 1925 A. Huxley Those Barren Leaves i. i. 8 What she called, in her jovial undergraduatish moments, a 'floater'. 1929 Wodehouse in H. Cotton Legion Bk. 110 It's just when our intentions are best that we always make the most poisonous floaters. 1938 E. Waugh Scoop i. ii. 33 Have a cigarette or—had he made a floater?—or do you prefer your churchwarden? 1967 A. Wilson No Laughing Matter iii. 312 I've as good as said that we don't want your money... Just the sort of floater I would make, babbling on.

floating, ppl. a. Add: **1.** (Further examples.)

1819 P. Wakefield Excurs. N. Amer. (ed. 3) 25 There is one boat which, without exaggeration, may be denominated a floating palace. 1832 J. Lindley Introd. Bot. i. 84 Stomata are found..in floating leaves upon the latter [sc. the upper side] only. 1838 Southern Lit. Messenger IV. 26/1 The noise of steam, and the dissonant voices of the crowd subside, and give place to the regular thump of the floating hotel [sc. a steamer]. 1843 Knickerbocker XXII. 85 The Knickerbocker-Steamer, that floating palace of the Hudson. 1872 B. Jerrold London xxi. 187 Soothing a

convict's last moments in the floating infirmary off Woolwich Dockyard. 1912 Chambers's Jrnl. 184/2 A floating factory means an old steamer of three to four thousand tons, a size which has annually been increased. ..By using the larger size of steamers the owners are able to fit them with plant not only for trying out the blubber ..but also with plant for turning the carcasses into oil and guano. 1933 Discovery July 213/1 In 1930 floating factory ships, with their consorts the whale-catchers, were to be found encircling three-quarters of the Antarctic continent. 1966 Guardian 14 Mar. 18/5 Two floating hotels..to carry tourists upstream from Cairo to Aswan. 1966 Economist 22 Oct. 366/1 The Poles started out as specialists in modern trawlers, floating fish factories and base ships—with reinforced hulls for operation in Arctic waters—and they now rank second in this specialised field to Japan. 1967 C. D. Sculthorpe Biol. Aquat. Vasc. Plants iv. 72 Floating-leaved hydrophytes are generally unable to withstand severe winds and turbulent waters.

3. b. (See quot.)

1888 Lockwood's Dict. Mech. Engin. s.v., When the lever of a weigh-bridge or of a testing machine is in equal balance it is said to be floating.

4. Esp. in floating population.

1835 Niles' Register 27 June 289/1 The remark of a distinguished politician..in the convention of 1822, as to the power of what he called the 'Floating Population' over the result of elections. 1876 [in Dict.]. 1933 'G. Orwell' Down & Out i. 7 The lodgers were a floating population..who used to turn up..stay a week and then disappear again. 1962 Lancet 26 May 1119/1 More people come to London. There is a greater floating population. 1965 B. Sweet-Escott Baker St. Irreg. i. 35 Innumerable other departments were represented—..Electra House,.. the B.B.C., and a floating population from the Ministry of Information.

5. b. Of a marine insurance policy: (see quot. 1959).

1839 Southern Lit. Messenger V. 7/1 The packets, by their 'floating policies of insurance' offer another inducement to shippers. 1902 Encycl. Brit. XXIX. 527/2 As it frequently happens that merchants desire to have all their shipments covered, by whatever vessel they may come, they require insurance in general terms; such a policy is termed a floating policy. 1959 Jowitt Dict. Eng. Law I. 814/2 Floating policy, a marine insurance policy issued to cover all shipments of goods by the assured by any vessel between specified places.

c. Of a currency, or its exchange rate: fluctuating, not fixed. Cf. *float v. 1 d.

1964 H. G. Grubel World Monetary Reform 442/2 Floating exchange rates. 1970 P. Einzig (title) The case against floating exchanges. 1971 M. A. G. van Meerhaeghe Economics xii. 419 After 1931 several monetary units became floating currencies. 1971 Guardian 17 Aug. 3/6 The cumulative effect of the floating mark (now tantamount to an upward revaluation of more than 8 per cent) and the 10 per cent import surcharge means that German exports there will be 18 per cent dearer.

6. floating axle, a live axle in which the revolving part turns the wheels while the weight of the vehicle is carried on the ends of a fixed axle housing; **floating battery** (later example); **floating crane,** a crane mounted on a pontoon; **floating derrick** (see Derrick 2 d); **floating drydock** = floating dock; **floating floor** (see quot. 1963); **floating mill** U.S., a mill so constructed as to float in a river and be worked by the current; also attrib.; **floating point** Computers, designating a method of representing numbers by two sequences of digits, one sequence being the significant digits of the number and the other indicating the position of the radix point; usu. attrib.; **floating voter,** a voter who has not attached himself to any political party; in U.S. spec. = Floater 4 a; also **floating vote,** the vote of such a person; also collect.

1907 Westm. Gaz. 26 Nov. 4/3 The famous 'floating' back axle. 1866 E. A. Pollard Southern Hist. War I. 60 The floating battery dismounted two of the parapet guns [of Fort Sumter]. 1903 Work 27 June 331/2 Another crane in use in Germany is called the floating crane, and is built on a large, massive flat boat. 1969 Jane's Freight Containers 1968–69 216 (caption) Port of Copenhagen. Floating crane 'Herkules'. 180 tons capacity also used for containers. 1838 E. Flagg Far West I. 135 Upon the river-bank..stands the 'Floating Dry Dock'. 1921 Daily Colonist (Victoria, B.C.) 8 Apr. 19/4 Extensive improvements have been made in the floating drydock, increasing its efficiency. 1970 Encycl. Brit. VII. 712/2 The floating dry dock is a barge-like structure that can be partially submerged so that a vessel can enter it. The dock is then raised in the water and pumped dry to permit work on the vessel. 1934 Builder CXLVII. 552/1 The patent 'Cullum' soundproofing system..has suitable insulating value. The standard thickness of this floating floor is 3½ in. 1963 Gloss. Build. Terms (B.S.I.) 18 Floating floor, the upper portion of a floor including the flooring, when this is supported on a resilient layer or mountings to provide insulation against sound or vibration. 1970 New Yorker 8 Aug. 46/3 What is particularly effective against noise is a so-called floating floor. This consists of a main concrete floor four centimeters thick on top of which lies a one-centimeter-thick layer of glass wool, with another four-centimeter layer of concrete. 1796 A. Ellicott in C. V. Mathews A. Ellicott (1908) 138 These floating mills are erected upon two, or more, large canoes or boats, and anchored out in a strong current. 1817 S. R. Brown Western Gaz. 52 It has a post office, and a floating mill anchored abreast of the town. 1888 Lockwood's Dict. Mech. Engin. 145 Floating Mill Wheel, a water wheel, having its bearings in a boat moored in the stream of a rapidly flowing river, which turns the wheel and performs work for which it is suitable. 1948 E. N. Dick Dixie Frontier 250 Some of the first water-power mills were 'floating mills'. These were built on two large dugout canoes. An undershot wheel was placed between them. 1948 Math. Tables & Other Aids to Computation III. 318 Floating-point operation greatly reduces the need for scale factors, but complicates the operations of addition and subtraction. 1954 Jrnl. Assoc. Computing Machinery Oct. 194 The Type 704 is the first large-scale commercially available computer to employ fully automatic floating point arithmetic commands. 1964 F. L. Westwater Electronic Computers ix. 143 Sets of subroutines to be performed in 'floating-point' rather than the fixed-point mode arithmetic. 1847 Knickerbocker XXIX. 328 Such constitute what in common parlance is called the 'floating vote'. 1856 Cincinnati Enquirer 7 May 4/1 Fremont, having been a successful explorer and trapper, can discover and trap a great many floating votes. 1935 H. Nicolson Let. 13 Nov. (1966) I. 223 There is always the slender chance that the floating vote may swing away from Labour and that I shall creep in. 1971 Daily Tel. 19 July 2/2 He has been fighting a vigorous campaign to mop up any floating votes among the 77 delegates at the union's fortnight-long conference. 1905 D. G. Phillips Plum Tree 14 Those stiff-armed men were the 'floating voters' of that ward of Pulaski. 1955 Observer 1 May 11 The floating voter is not a unique creature. 1958 Economist 13 Sept. 828/2 That now well-known and inoffensive favourite of the touchy floating voter, Mr. Aneurin Bevan.

b. In the names of various aquatic plants (see quots.). U.S.

1814 J. Bigelow Florula Bostoniensis, Alopecurus geniculatus, floating foxtail grass. 1847 Floating glyceria [see manna grass, Manna[1] 9]. 1934 Nat. Geogr. Mag. LXV. 598 The dainty white display of floating-heart nods over its own reflection in the water. 1940 N. C. Fassett Man. Aquat. Plants 42 Floating Moss. Ibid. 55 Floating Brownleaf. Floating-leaf Pondweed.

floating-out. [f. vbl. phrase to float out (see Float v. 11).] The action of floating a ship out of dock. Also attrib.

1896 Daily News 18 Sept. 6/7 The floating-out of dock of the battleship Illustrious at Chatham Dockyard. 1899 Ibid. 29 June 5/5 Her floating-out weight was 1,500 tons.

Flobert (flōu·bəɹt, ‖ flobę̄r). The name of a French armourer, N. Flobert (1819–94), used attrib. and ellipt. to describe a cartridge or breech-loading rifle designed by him.

1890 Webster, Flobert, a small cartridge designed for target shooting. Ibid., Flobert rifle, a rifle adapted to the use of floberts. 1895 Montgomery Ward Catal. 461/2 Flobert Rifle, oiled walnut pistol grip stock. Ibid. 463/3 Flobert Rifle Parts...Flobert Hammers, Remington action...Flobert Breech Blocks. 1904 Kynoch Jrnl. Jan.–Mar. 33 The Flobert cartridge is notable as being an evolution of the percussion cap. 1918 E. S. Farrow Dict. Mil. Terms 237 Flobert, a small cartridge used in target shooting, sometimes called ball cap. Ibid., Flobert-Gras Rifle, a breech-loading rifle, having several valuable improvements on the Chassepot, used in France 1874–1885. 1950 Publ. Amer. Dial. Soc. xiv. 29 Flobert, flaubert [flobət] n., any small rifle, twenty-two caliber. Florence, S.C. Obsolescent? Once a popular make of rifle.

floc (flǫk). [f. Floc(culus.] A flocculent mass of fine particles and colloidal material.

1921 H. P. Eddy in Jrnl. Western Soc. Engin. XXVI. 259 When sewage containing a suitable supply of dissolved oxygen is agitated..the fine suspended matter and the colloidal substances here gather into small aggregations called flocculi or floc. [Note] The word 'floc' as used herein, signifies a single sponge-like mass or an aggregation of such masses. 1947 H. E. Babbitt Sewerage & Sewage Treatm. (ed. 6) xix. 396 As the floc settles through the liquid it sweeps other settleable particles down with it. 1959 J. Clegg Freshwater Life (ed. 2) xix. 324 The flocs in the activated sludge system perform the same function of purification as the zooglea layer on the percolating filter.

flocculation. (Later example.)

1961 Lancet 16 Sept. 631/1 Flocculation tests [were] negative.

flocculus. Add: **3.** Astr. **a.** Applied to the wisps of luminosity in a nebula.

1826 Herschel in Mem. Astron. Soc. Lond. II. 491 The flocculi [in the nebula of Orion], instead of being generally round, are drawn out into little wisps.

b. One of the small cloudy wisps or masses on the sun's surface, revealed when the sun is photographed with the spectroheliograph.

1903 G. E. Hale in Publ. Yerkes Observ. III. i. 5, I now propose the name flocculi for the regions on the Sun's disk which are shown only on photographs made with the spectroheliograph. Ibid. 14 It is necessary to speak of calcium flocculi, hydrogen flocculi etc. as the..forms of the various vapors in the same part of the disk are not identical. 1915 —— Ten Yrs. Wk. Mount. Observ. 20 The spectroheliograph..discloses extensive clouds of calcium, hydrogen, iron, and other vapors, which..are recorded (as flocculi). 1930 R. H. Baker Astron. vii. 291 In the calcium spectroheliograms the flocculi are generally bright, and they are especially conspicuous in the sun-spot zones. 1953 L. H. Aller Astrophysics ix. 341 Elsewhere on the disk are often seen large sinuous dark filaments or dark flocculi, prominences seen in projection. 1967 [see *fine structure 2].

flock, sb.[1] Add: **5. flock-book,** a list of pedigrees of sheep; **flock pigeon,** an Australian species of pigeon, Histriophaps histrionica, usu. seen in large flocks; the harlequin bronze-wing.

1912 W. Deeping Sincerity xxvi. 195 Crabbe kept a complete register of his tenants..their ages, resources, infirmities, and characters, like the records of sheep in a 'flock-book'. 1950 N.Z. Jrnl. Agric. Feb. 144/1 Stud sheep which are entered in flock books. 1887 G. J. Broinowski Birds Austral. III. Pl. 111, Leucomelana Norfolciensis..White-headed Fruit-Pigeon... This bird, more commonly known among settlers as the 'Flock' Pigeon, is found generally along the whole of the north-eastern sea-coast. 1959 Observer 17 May 8/4 The flock-pigeon, a plump bronze-wing formerly thought to be on the brink of extinction. 1966 N. W. Cayley What Bird is That? (ed. 4) 23 Topknot pigeon...Also called Flock Pigeon. Usually seen in flocks, frequenting brushes.

flock, v.[1] **4. a.** Delete † and add later Jamaican example.

1943 L. Bennett Jamaican Humour in Dialect 21 But dose bwoys ovah dere Wi' flock we like dese gals flock De soljas ovah here.

flog, v. Add: **1. c.** fig. Also (freq. intr.) in slang use: (i) to proceed by violent or painful effort; (ii) to obtain, usu. by violent effort.

1925 Fraser & Gibbons Soldier & Sailor Words 96 To flog,.. to walk; go on foot.—e.g., 'There was no train so we flogged it.' 1936 Geogr. Jrnl. LXXXVII. 166 We had to flog our way through snow up to our waists. 1943 N. Coward Middle East Diary 26 Aug. (1964) 64 Visualised himself flogging through the provinces in Shakespearian Rep. indefinitely. 1959 B. Goolden For Richer, for Poorer viii. 121 A sports model out of which they could flog eighty with ease. 1964 Times 11 Feb. 11/6 [Lorry drivers] are being encouraged to 'flog on' even in bad weather.

d. Phr. to flog a dead horse: see Horse sb. 18.

2. b. (Later example, const. out.)

1924 Kipling Debits & Credits (1926) 132, I went to bed; for I was fair flogged out.

c. slang (orig. Mil.). To sell or offer for sale, orig. illicitly. So **flogging** vbl. sb.

1919 War Terms in Athenæum 1 Aug. 695/2 'Flogging', the illegal disposal of Army goods. 1925 Fraser & Gibbons Soldier & Sailor Words 96 To flog, to sell something not the vendor's own to dispose of. 1935 'D. Hume' Gaol Gates are Open 10 If a crook disposes of stolen property through a fence he is 'fencing', but if it is sold through any other channel it is 'flogging'. 1951 G. Hanley Consul at Sunset 125 He was dead...His kit was collected and flogged to those who would buy it in the mess. 1966 J. Porter Sour Cream x. 134 Filching state property and flogging it to the eager populace is a common enough crime in the Soviet Union as it was with us during the war. 1967 M. Drabble Jerusalem the Golden v. 112 Let's go..and look at the ghastly thing that Martin flogged us.

flood, sb. Add: **6*.** Usu. in pl., colloq. abbrev. of *flood-lamp, -light.

1930 Punch 19 Feb. 198/2 Spool is the dramatic society's honorary electrician. His conversation was full of floods, floats, spots, battens, and dimmers. 1933 P. Godfrey Back-Stage i. 14 The up-stage O.P. flood isn't properly masked. 1967 Punch 16 Aug. 242/3 For Figaro and Verdi's Macbeth at the first [Edinburgh] festival twenty-one years ago, John Christie had to bring in a lighting bridge and sixty floods and perches from Glyndebourne.

7. flood-bank, control.

1928 Manch. Guardian Weekly 31 Aug. 178/3 The building of floodbanks, or levees, along the river banks. 1945 Finito! Po Valley Campaign 23 A battalion fought its way to the near floodbank. 1943 J. S. Huxley TVA iii. 22 Power as a by-product of flood control. 1957 G. E. Hutchinson Treat. Limnol. I. i. 116 The whole system has recently been modified by flood-control works.

8. flood-lamp = *flood-light; **flood-light,** a light providing a beam of intense illumination; the illumination so provided; also attrib. and fig.; hence as v. trans., to illuminate by means of flood-lights; **flood-lighting** vbl. sb., the action of illuminating with flood-lights; also, flood-lights collectively; so **flood-lit** ppl. a.; **flood-plain** (earlier and later examples); **flood-way,** (a) a piece of flooded road or path; (b) a channel constructed for the purpose of taking the flood-waters of a river.

1916 Amer. Year Bk. 563 The line of demarcation between search-lamps and so-called 'flood lamps' is a narrow one. 1933 P. Godfrey Back-Stage iv. 43 Lighting towers..each capable of carrying a dozen 1000-watt flood lamps to supplement the battens overhead in lighting the scene. 1957 Economist 21 Sept. 912/2 The officious display of security, the floodlamps and police dogs that surrounded the campaign train might have been better avoided. 1923 M. Luckiesh Light & Color 256 Tall or isolated buildings are particularly attractive when flood-lighted. 1924 J. F. Hobart Tulley's Handbk. (ed. 7) III. 814 Another means of illumination is by floodlights which are mounted at a distance from the space to be lighted. 1925 A. E. Newton Greatest Bk. in World 82 From either side of the proscenium arch two flood-lights played upon the actors. 1925 Cady & Dates Illum. Engin. 410 The light may be projected from flood-light projectors. 1928 Daily Tel. 29 May 10/5 From thirty flood-light projectors the Royal Pavilion was bathed in amber, red, and green. 1928 C. F. S. Gamble N. Sea Air Station ix. 131 By flood-lighting the sky with flares placed on the tops of balloons. 1930 Aberdeen Press & Jrnl. 29 Mar. 7/1 It would be useless to turn a 'flood-light' on to any article which cannot stand this searching test. 1930 Flood light [see *approach sb. 13]. 1917 Electr. News 26, 48 A Resumé of

Flood-lighting. 1923 J. W. T. Walsh Elem. Princ. Lighting 188 The illumination of open spaces..may often be carried out satisfactorily by a floodlighting equipment. 1927 M. Borden Flamingo ii. ii, It was, of course, most beautiful at night against the dark, when Peter's flood-lighting streamed over it to make it translucent as ice. 1955 Times 2 June 3/4 In view of the need to conserve fuel in the emergency, the Ministry of Works have discontinued the floodlighting of public buildings in London. 1928 Daily Express 6 Aug. 15/1 Its walls, brilliantly flood-lit,.. are made almost entirely of sheets of glass. 1934 H. Nicolson Curzon: Last Phase 40 The flood-lit self-righteousness, the timid imprecisions, the appalling amateurishness of democratic diplomacy. 1958 W. T. O'Dea Social Hist. Lighting vii. 175 The floodlit office building or factory is an advertisement. 1873 J. H. Beadle Undevel. West xxx. 656 West of the 'Buckskin' was a singular flood plain some six miles wide. 1957 G. E. Hutchinson Treat. Limnol. I. i. 104 The local rise in the water table intensified solution and so produced in the flood plain very striking karstic forms. 1971 Nat. Geogr. Sept. 407 These are but puddles compared to the deluges that once ripped across the lower Missouri's flood-plain. 1889 F. E. Gretton Memory's Harkback 108 Near Tewkesbury and Upton there was a nasty bit of flood-way. 1915 E. Pound Cathay 25 The heart turns to travel so that he then thinks On flood-ways to be far departing. 1928 Manch. Guardian Weekly 31 Aug. 178/3 The construction of a 'floodway' leaving the main stream [of the Mississippi] at Bird Point, near Cairo.

flood, v. Add: **1.** (Later fig. examples in Econ.)

1936 Discovery Oct. 320/2 Foreign pins still continued to flood the English market. 1944 Ann. Reg. 1943 265 The presence of large armies..flooded the country with money. 1951 E. Paul Springtime in Paris iv. 75 Where shall I be when rationing ends, and beef begins flooding the market?

3. b. To drive out by floods.

1910 Westm. Gaz. 21 June 8/3 The flooded-out attendants.

4. d. To become flooded.

1908 Daily Report 24 Aug. 9/1 A carburettor which persistently floods is not uncommon. 1912 Motor Man. 13 A persistent tendency to flood..due to a punctured float.

flooded, ppl. a. Add: **flooded gum** Austral., any of several eucalypts growing in damp soil.

1847 F. W. L. Leichhardt Jrnl. Overland Exped. Austral. x. 324 The latter part of the stage was again over a large box-flat, intersected by shallow grassy depressions, timbered with flooded-gum. 1884 A. Nilson Timber Trees N.S.W. 136 Flooded gum, Eucalyptus rostrata, saligna, viminalis, resinifera. 1933 D. G. Stead Tree Bk. 24 A splendid tree is the Sydney Blue Gum, also called Flooded Gum. This tree grows to more than 150 feet high in the eastern forests of New South Wales and southern Queensland. 1965 Austral. Encycl. III 405/1 Large Sydney blue gums or flooded gums (E. saligna) at Tamborine Mountain.

floody, a. Delete †Obs. and add later examples.

1818 Keats Let. 13 Mar. (1931) I. 119 Devonshire..is a splashy, rainy, misty,..floody, muddy, slipshod County. 1964 Penguin Bk. Austral. Ballads 201 The creek was full and bloody floody.

floor, sb.[1] Add: **1. a.** Phr. to mop or wipe the floor with: see Mop v.[2] 1 b, Wipe v. 9 e.

c. fig. A minimum, esp. of prices or wages. Cf. *ceiling vbl. sb. 6 d.

1938 Reader's Digest Sept. 1 Even fair price ceilings and quality floors won't answer real needs unless an adequate supply of goods is made available. 1941 Time 21 July 70/3 Excuse for the silver-buying program and its artificial price floor..was to keep Western miners at work. 1949 Economist 24 Sept. 670/1 Price-floors were set for bituminous coal. 1959 Ibid. 11 Apr. 106/2 A floor of only £12 a week on the wages of British artists.

d. spec. The floor of a studio where films or television programmes are shot; hence used allusively: (a) a film or television studio; (b) in phr. on the floor, of a film: in production.

1937 M. Robinson Continuity Girl v. 89 Rene was to be continuity. Mr. Kimmins asked me to come on the floor as her assistant. 1948 Ann. Reg. 1947 443 The inauguration of the modern Metro-Goldwyn-Mayer floors at Elstree. 1948 Observer 22 Feb. 5/1 At Elstree, only one film, the Guinea Pig, is on the floor. 1950 'E. Crispin' Fresh Hearses i. 15 'How far has it [sc. a film] got?' 'It's not on the floor yet... I mean that they haven't actually started making it yet.' 1957 M. Kennedy Heroes of Clone i. i. 13 I'll make a shooting script... It'll be something you can go on the floor with. 1971 R. Busby Deadlock xiii. 200 [He's] down there on the floor. We're doing one for the can and then he goes out live.

4. c. from the floor: of a question, speech, etc.: delivered by an individual (member, spectator, etc.), as opp. to the governing body, the 'platform'.

1966 Rep. Comm. Inquiry Univ. Oxf. I. 234 Resolutions moved not by Council, but from the floor.

6. Also, a dance-floor. Cf. sense 4 in Dict.

1839 W. Carleton Fardorougha (1848) iv. 46 Answer Mrs. Fogarty, statin' fedher you'll take a month's larnin' on the flure. 1884 L. Troubridge Life amongst Troubridges (1966) 171 We all went to the New Club Ball.. such a floor, such music, and such a partner. 1938 [see *cabaret[1] 2 b]. 1967 R. Rendell New Lease Death viii. 82 'Hard to make conversation when you're dancing.'.. 'Like "Don't you think this is a good floor"?'

7. b. the floor (Cricket colloq.): the ground. So to put a catch on the floor (Cricket colloq.): to fail to hold it.

1903 Strand Mag. XXV. 624/2 A large majority of them [sc. catches] were 'put on the floor'. 1960 Times 14 June 16/1 With the field drawn tight around the bat and catches being snapped up off the floor.

10. Delete ? Obs. and add later example.

1969 Jane's Freight Containers 1968–69 18/3 Freight Container Components... Floor. Component supporting the payload.

14. floor-slab, -space, -stone; floor-mounted adj.

1962 Times 25 May 18/5 The cranked, floor-mounted gear lever. 1936 Discovery Feb. 56/1 The two principal floor-slabs..were of a specially hard kind of granite. 1963 Gloss. Build. Terms (B.S.I.) 18 Floor slab, a slab forming the continuous loadbearing structure of a floor and spanning between supports or laid on the ground. 1876 J. S. Ingram Centenn. Exposition v. 150 It occupied about one-seventh of the entire floor-space in that structure. 1930 Times Educ. Suppl. 23 Aug. 363/2 The small floor-space of the war museum. 1927 Blackw. Mag. Apr. 527 Little of the floorstone remains. 1956 E. Pound tr. Sophocles' Women of Trachis 30 Seemed to corrode of itself. Ate itself up, there on the floor-stones.

15. floor-board, hence as vb.: to press (the accelerator pedal of a motor vehicle) down until it reaches the floor; to accelerate, drive very fast; **floor lamp** U.S., a tall lamp designed to stand on the floor; **floor-leader** U.S., a leader in debate, esp. in legislative assemblies; **floor-length** a., reaching to the floor; **floor man,** one who helps to attract customers to a mock auction; **floor manager,** (a) U.S., a 'master of ceremonies' at a dance; (b) orig. U.S., a shop-walker; (c) U.S., one who organizes support for a candidate in the hall of a political convention; (d) in television production: see quot. 1961; **floor pattern** (see quot. 1964); **floor polish,** a manufactured substance for rendering floors glossy; hence floor-polisher; **floor show,** an entertainment presented on the floor of a restaurant, night-club, etc.; **floor-waiter,** a waiter who serves on one floor of a hotel; **floor-walker** (earlier and later examples).

1942 Berrey & Van den Bark Amer. Thes. Slang § 728/4 Drive fast..floorboard, give 'er the gas. 1948 Mencken Amer. Lang. (Suppl. II) xi. 719 Of the more original words and phrases of the truckmen I offer a few specimens:— floor-boarding... Running at high speed. 1971 Scope (S. Afr.) 19 Mar. 59/2 (Advt.), You slice her into second gear—all clear ahead—and floorboard the pedal. 1971 Islander (Victoria, B.C.) 16 May 10/3 As he floor-boarded the throttle he noticed his quarry had turned a corner. 1907 Yesterday's Shopping (1969) 255 Polished brass telescopic floor lamps. 1940 C. McCullers Heart is a Lonely Hunter (1943) i. i. 5 He walked into a store one day and hauled out a floor lamp without paying for it. 1965 J. M. Cain Magician's Wife (1966) ii. 18 He.. went into the living room, and without turning the floor lamps on, sat down in a chair by a window and stared out at the gathering dusk. 1899 Congress. Rec. 11 Feb. 1764/2 Congress has witnessed very few more successful floor leaders. 1954 Encounter Mar. 59/1 Congressman John McCormack, Democratic floor-leader in the House. 1939 Ottawa Jrnl. 26 June 8/4 The bride wore a floor-length gown of white chiffon over taffeta. 1967 E. Short Embroidery & Fabric Collage iii. 74 A round table in a bedroom or drawing-room may sometimes have a floor-length cover permanently in position. 1928 Daily Express 3 Mar. 7 Meader and Solomons were what is known as 'floor men', or..'pitch gatherers'. 1887 Harper's Mag. May 967/1 Jerry, as one of the floor-managers, was gorgeous. 1892 Ibid. Feb. 439/1 Like the floor-walkers in the stores, they're all floor or aisle managers now. 1913 J. London Valley of Moon iii. xiii, An' here's you making' rough-house at a dance, an' I'm the floor manager, an' I gotta put you out. 1924 W. S. Hayward Retail Handbk. 74 The floor manager is the first person to arrive in the department and the last to leave. 1930 J. B. Priestley Angel Pavement vi. 292 Tells me she's had some bother with the buyer or floor manager. 1953 Manch. Guardian Weekly 3 Sept. 3 The convention floor manager's hard-won knowledge. 1960 O. Skilbeck ABC of Film & TV 55 Floor manager. 1961 G. Millerson Technique Telev. Production 14 The floor-manager is the director's contact man on the studio floor, and checks staging, action and performers on his behalf. 1966 S. Jackman Davidson Affair i. 17, I..saw the floor-manager's hand drop to cue me in..and turned to face the camera. 1943 M. Mayo Amer. Square Dance 11 It is well also to picture clearly in the mind what the dance will look like and what the floor pattern will be. 1958 J. Winearls Mod. Dance 130 (caption) Floor Pattern. 1964 W. G. Raffé Dict. Dance 374/2 Floor pattern: the track, or footsteps, traced by the dancer on the stage. 1907 Yesterday's Shopping (1969) 18/2 Floor polish. 1926 Daily Colonist (Victoria, B.C.) 2 July 4/5 (Advt.), Floor polishes, you may say, are pretty much alike. But try 'Poliflor' just once and you will immediately realize what a vast difference there can be. 1895 Army & Navy Co-op. Soc. Price List 184/2 Floor Polisher, extra large, with swivel joint, including handle. 1897 Westm. Gaz. 7 Dec. 10/2 One man only will be allowed on the premises—the floor-polisher. 1939–40 Army & Navy Stores Catal. 114/1 Floor polisher... Double sided. One side for applying wax. One side for polishing. 1927 Stage Year Bk. 69 The opportunities and scope for employment [in vaudeville] of an artist are practically unlimited, apart from the large field in musical productions, cabaret floor shows, etc. 1931 Durante & Kefoed Night Clubs 4 A floor show with six principals and no chorus. 1959 Times 2 Mar. 12/7 Snippets from a night-club floor-show of resplendent tattiness. 1930 A. Bennett Imperial Palace xvii. 109 Once he had memorably discovered fourth-floor silver in the fifth-floor service-room..which disconcerted all the floor-waiters. 1967 G. Greene May we borrow your Husband? 19 Peter's mislaid his tie. He thinks the floor-waiter has purloined it. 1876 Scribner's Monthly Feb. 599/2, I next went into a store a few doors further up Broadway. When I entered I approached the floor-walker. 1942 E. Paul Narrow St. iv. 32 The tenant of this room and bath was a floor-walker from the Samaritaine.

floor, v. **3. a.** (Earlier example.)

1830 Coleridge Table T. 8 July (1884), The other day I was what you would call floored by a Jew.

floor-cloth. 2. (Later examples.)

1951 Catal. of Exhibits, South Bank Exhib., Festival of Britain 136/1 Dishcloths & floorcloths. 1960 M. Spark Bachelors x. 154 Carrie, you have wiped the oven with the floor cloth.

floorer. Add: **1. b.** In the game of skittles, a throw with the first of three balls which floors all the pins.

1881 Cassell's Bk. Sports 128 The only play that will secure a floorer is to throw the ball with a good round-handed swing, imparting to the ball what in billiards would be called a 'side', and so that the edge of the ball alights on the right-handed shoulder of the front pin. 1928 Daily Express 31 Mar. 3/4 A scroll on which are inscribed the names of members who have scored floorers. 1930 A. P. Herbert Water Gipsies xxi, A stranger to the game could not have told how it happened, but in less than a second the frame was clear—not one pin standing— a 'floorer'. 1960 Observer 17 Jan. 3/5 A chap..said he had done nineteen floorers in a row.

flooring, vbl. sb. **5.** flooring-board (earlier example), -machine.

1733 W. Ellis Chiltern & Vale F. vii. 95 So it will in Flooring-boards, notwithstanding they are nailed down ever so fast. 1847 H. Howe Hist. Coll. Ohio 372 There are now in operation within the corporate limits..2 flooring machines. 1957 N.Z. Timber Jrnl. Mar. 52/1 Flooring machine, a machine for preparing floorboards.

‖ **floose** (flūs). Also **floos, flus(e), fulus.** [See Fluce sb.] A small coin of north Africa, Arabia, India, and neighbouring countries.

1625 in W. Foster Eng. Factories India (1909) III. 71 There money goeth by the caffala, of which 9½ maketh a doller and 60 fluse to a caffala. 1858 Simmonds Dict. Trade, Floose, the tenth part of a danim, a petty money of Bussorah and other parts of Arabia. 1877 H. M. Elliot's Hist. India VII. 345 They [sc. the Portuguese] also use bits of copper which they call buzurg, and four of these buzurgs pass for a fulús. 1906 Daily Chron. 2 May 7/6 To-day a well-dressed man, who sees a European passing in the street, extends his hand, and mutters something about 'flus'. 1907 Daily Mail 18 Jan. 6/4 Often have I bestowed upon him a floos for bread.

floosie, floozie (flū·zi). colloq. [Orig. unkn., but cf. dial. floosy adj., fluffy, soft (f. floose, see Floss[2]) and *flossy a.] A girl or woman, esp. one of disreputable character.

1911 C. B. Chrysler White Slavery iii. 30 Tell that floosie to cut out that yelping. 1927 [see *baby sb. 1 c]. 1944 B. Marshall All Glorious Within xix. 144 In the bathroom scene..she has her bath and sings Ave Maria just to show she ain't no floosie. 1952 H. W. Tilman Nepal Himalaya ii. xv. 183 The men dallied with a Siren, in less classical language, a 'floozie' who sat on a roof combing her wonderfully straight black hair. 1958 B. Nichols Sweet & Twenties vii. 91 The usual anaemic floosies in pastel shades. 1959 B. Goolden For Richer, for Poorer i. 11 Why the floozies?

flop, sb. Add: **1. c.** Something loose and pendulous; = Flap sb. 4.

1900 in Eng. Dial. Dict. II. 419/1. 1902 Westm. Gaz. 11 Sept. 3/2 She achieves another immense flop with the back of the brim well pinned in position over the knob of hair which..flops on her neck. 1909 J. R. Ware Passing Eng. 134/2 Flop. When the lower classes of women adopted the 'cretin' or 'poodle' style of wearing the hair low down over the forehead, they gave it this name. 1933 V. Woolf Writer's Diary 9 May (1953) 200 A little servant girl with honest eyes, hair brushed in a flop.

4. b. A turn-round; a sudden change of policy or party. U.S.

1880 N.Y. World 22 Nov. 5/1 Mr. Skinner's apparent flop on the railroad question is injuring his chances in the Speakership struggle. 1904 Springfield (Mass.) Weekly Republ. 7 Oct. 2 That a flop by the most militant of the unionists is under contemplation has been denied. 1911 H. S. Harrison Queed xviii. 230 So ran the editorial, which was offensively headed 'West's Fatal Flop'. 1929 Collier's 5 Jan. 41/1 It was basically a 'flop'.

c. A failure, collapse, or decline. Also, a person or enterprise (esp. a play, etc.) that is a failure. slang.

1893 Farmer Slang, Flop. 2...A sudden fall or 'flop' down. 3...A collapse or breakdown. 1899 Westm. Gaz. 28 Jan. 6/2 There has been a flop in Trunks, but Canadas have been good. 1927 Sunday Express 15 May 5/7 Nearly all the American turns prove a flop. Yet they think they can command the earth. 1930 Publishers' Weekly 18 Oct. 1851 These authors every once in a while write a flop. 1931 Discovery Nov. 372/2 Fokker's first invention was a 'flop'. 1934 Times Lit. Suppl. 7 June 406/3 She, too, is a common type—the Hollywood flop. 1936 Amer. Speech XI. 221 If it [sc. a play] just somehow doesn't click or register, it's doomed to be another flop. **1945** L. A. G. Strong *Othello's Occupation* 121 He's pretty wobbly, professionally speaking. He's had two flops in the suburbs. **1957** *Economist* 5 Oct. 24/2 As a gesture of defiance Argentina's one-day general strike last week was a flop. **1969** *Times* 7 Nov. 3/1 Neil Simon..has had eight Broadway hits..and the question everyone is asking..is whether he's got a flop in him.

d. A 'flabby' or 'soft' person. *slang.*

1909 H. G. Wells *Tono-Bungay* II. iv. 171 All the little, soft feminine hands, the nervous ugly males, the hands of the flops, and the hands of the snatchers! **1923** *Glasgow Herald* 12 Dec. 10 If that little flop..believes he can play fast and loose with the moral consciousness of this nation. **1936** 'F. O'Connor' *Bones of Contention* 70 She was a great flop of a woman.

e. *U.S. slang.* A bed; a place to rest or sleep; = **flop-house.*

1910 D. Ranney *Autobiogr.* iv. 70 You can get a bed in a lodging-house for ten cents, or if you have only seven cents you can get a 'flop'. **1913** E. A. Brown *Broke* iii. 28 Say, Jack, can you tell a fellow where he can find a free flop? **1916** *Amer. Mag.* May 14/1 She said to tell you this ain't no hobos' flop, neither. **1925** *Lit. Digest* 11 July 50/1 You better go around to one of the missions. There's a couple of 'em will give you a flop for nothing. **1930** J. Dos Passos *42nd Parallel* 75 They couldn't find any-place that looked as if it would give them a flop for thirty-five cents. **1955** *Publ. Amer. Dialect Soc.* xxiv. 120 So we go up to my flop.

5. *flop-eared* (earlier U.S. example); **flop-house** *slang* (orig. *U.S.*), a doss-house.

1846 J. J. Hooper *Adv. Simon Suggs* ii. 28 You..gnatty, flop-eared varmint! **1923** N. Anderson *Hobo* iii. 30 'Flop-houses' are nearly all alike. Guests sleep on the floor or in bare, wooden bunks. The only privilege they buy is the privilege to lie down somewhere in a warm room. **1927** *Scots Observer* 26 Mar. 10/3 The lowest of the derelicts spent the night..in a 'flophouse' (which is worse than the lowest 'model'). **1930** *Harper's Mag.* July 133 The Welfare Council of New York had to charter an old barge..as an overflow flop house. **1941** Wyndham Lewis *Let.* 3 Sept. (1963) 297 If I don't do something to break out of the net, I shall end my days in a Toronto flophouse. **1964** S. Bellow *Herzog* (1965) 249 Get out! I leave you nothing! ..Croak in a flophouse.

flop, *adv.* and *int.* Add: Also *fig.*

1930 *Daily Express* 6 Sept. 4/2 Every one adopts a 'wait and see' policy, and business goes 'flop'.

flop, *v.* Add: **2. b.** Also without *over*; and *trans.*, to cause to change sides; to bring over. *U.S.*

1884 *Puck* 6 Aug. 359/1 It is not the Independents who have 'flopped' this time. It is the Republican Party that has 'flopped' from honesty to dishonesty. **1894** *Daily Ardmoreite* (Ardmore, Okla.) 18 Jan. 1/4 The purported change was..a fake to enable that canine barnacle, Soule, to flop his politics. **1904** *Omaha Bee* 3 Sept. 6 A number of New York newspapers have flopped to the support of Parker. **1904** *Springfield* (Mass.) *Weekly Republ.* 16 Dec. 8 Mr. Roche flopped the Boston Pilot to the support of the republican candidate. **1926** C. R. Cooper *Oklahoma* 123 Hurriedly lawmakers who had been opposed to it 'flopped' to the other side.

c. *spec.* To sleep. *slang* (orig. *U.S.*).

1907 J. London *Road* (1914) 107 'Kip', 'doss', 'flop', 'pound your ear', all mean the same thing; namely, to sleep. **1926** J. Black *You can't Win* vi. 66 It was time to 'flop'. They took off their shoes and coats. **1936** W. A. Gape *Half a Million Tramps* x. 301 Where the hell are you going to 'flop' tonight? **1959** N. Mailer *Advts. for Myself* (1961) 84 They're filthy..diseased..what's the town mean, why aren't they put in the coop where they belong, why should they be flopping so near our house in a meadow?

d. *fig.* To collapse, fail (cf. *flop *sb.* 4 c). *slang.*

1919 Wodehouse *Damsel in Distress* viii, The summer-time number [in a theatre] flopped on the second night. **1928** *Observer* 15 July 15/1 If..the play 'flops' after a run of..three or four nights. **1936** P. Fleming *News from Tartary* 28 She published a book on that journey, which flopped. **1967** M. Reynolds *After Some Tomorrow* 61 Lenin supposedly tried to apply the teachings of Marx to Russia—and flopped.

flopper (flǫ·pəɹ). *U.S.* [f. Flop *v.* 2 and *2 b.] **1.** *Criminals' slang.* A perpetrator of any of several kinds of frauds.

1876 B. Harte *G. Conroy* II. vi. ii. 312 It is worthy of a short-card sharp and a keno flopper. **1914** Jackson & Hellyer *Vocab. Criminal Slang* 35 *Flopper*, in general use by money changers, switchers (substituters); flim-flammers. **1960** F. Gibney *Operators* v. 137 'Floppers' fall down on 'slippery' floors in..supermarkets, tumble deftly in front of slow-moving automobiles...Then they submit their claims to the stores or the insurance companies.

2. One who deserts to the opposite side in politics.

1880 *Cleveland Leader* 8 June 1/7 On the twenty-fifth ballot the Florida flopper went to Sherman. **1904** *Courier Jrnl.* (Louisville, Ky.) 17 Aug. 4 There are always floppers. The mere circumstance that somebody deserts his party and goes over to the other proves nothing.

flopperoo (flǫ:pĕrū·). *N. Amer. colloq.* Also **floperoo.** [f. *flop *v.* 2 d; cf. *-eroo.] A flop, failure.

1936 in *Amer. Speech* (1937) XII. 18/2 John Oliver in the Richmond *News-Leader* for Dec. 26, 1936, terms Georgia Tech the '*greatest flopperoo*' in football of the sports season. **1945** Lou Shelly *Jive Talk Dict.* 11 Flopperoo,..a failure. **1951** M. McLuhan *Mech. Bride* (1967) 58/2 The intellectually creative men with whom the future of mankind always rests will be regarded only as floperoos. **1970** R. Jeffries *Dead Man's Bluff* xviii. 163 His case was a real floperoo.

floppy, *a.* Add: Also *fig.*

1905 D. Sladen *Playing Game* ix, She's such a young heifer—she's at the floppy sentimental age.

flopsy bunny (flǫ·psi bʌ·ni). [f. Flop *v.* + -sy.] One of a group of rabbits in the children's stories of Beatrix Potter (1866–1943); hence, a sentimental designation of a rabbit. Also *transf.*

[**1902** B. Potter *Tale of Peter Rabbit* 9 Once upon a time there were four little Rabbits, and their names were—Flopsy, Mopsy, Cotton-tail, and Peter.] **1909** —— (*title*) The tale of the Flopsy Bunnies. **1954** 'N. Blake' *Whisper in Gloom* ix. 128 A nauseating tale about some flopsy bunnies. **1960** *News Chron.* 30 June 6/4 Maybe a juxtaposition of Flopsy Bunny sentiment and harsh reality is just the kind of shock the human race needs. **1966** M. Waddell *Otley* vii. 62 Her face looked pretty when she was asleep, like a sexy flopsy bunny. **1971** 'A. Gilbert' *Tenant for Tomb* vi. 96 All this Robin Redbreast malarky..it's all on a par with the Flopsy Bunnies in coloured hats and white pinnies.

flora. Add: **3. b.** The plants or plant life of any particular type of environment.

1874 D. Brandis (*title*) The forest flora of North-West and Central India. **1880** A. R. Wallace *Island Life* II. xxiv. 507 The discussion of a series of typical Insular Faunas and Floras with a view to explain the interesting phenomena they present. **1908** *Jrnl. Biol. Chem.* V. 285 The gas ratio is not an especially important characteristic in mixed fecal flora. *Ibid.* 296 The influence of these organisms upon the intestinal flora of mice. **1909** Groom & Balfour tr. *Warming's Oecol. Plants* ix. lxvii. 257 In depressions lying within the subglacial tract where snow remains for a long time, one finds characteristic, greasy mud, which sustains a vegetation of its own—Öttli's snow-patch flora. **1939** A. Huxley *After Many a Summer* I. v. 65 He began to talk..about fatty alcohols and the intestinal flora of carp. **1971** *Nature* 8 Jan. 120/1 The resident flora of the external auditory canal.

floral, *a.* Add: **3.** *floral clock* (see quot. 1962); *floral tribute*: a gift of flowers at a funeral; a wreath.

1925 *Ward Lock's Pict. Guide Edin.* (ed. 8) 62 Below it, in summer, is a beautiful floral clock of rare design. **1962** E. Bruton *Dict. Clocks* 74 *Floral clock*, large public clock set out in bedding plants on the ground. Some strike; a few are cuckoo clocks. **1963** *Listener* 14 Feb. 302/2 The road to the floral clock lay wide open. **1887** *Nuneaton Chron. In Memoriam C. N. Newdegate* 13 Apr., Floral Tributes. Prior to the funeral, a large number of.. wreaths were received. **1963** C. E. Vulliamy (*title*) Floral tribute. *Ibid.* i. 17 Those pompous laurels and lilies, the floral tribute of a person or persons unknown. **1971** *Times* 31 July 2/2 Mrs. Meena Jayant Madhavani with her children..gratefully thank..those who arranged floral tributes following the passing away of our Mr. Jayant in New Delhi.

floral (flōə·răl), *sb.* [f. the adj.] † **1.** A dancer at the Floralia, a Roman festival in honour of the goddess Flora. *Obs.*

a **1658** Lovelace *Lucasta* (1659) 13 So Cato sometimes the nak'd Florals saw.

2. A fabric with a floral design.

1897 *Sears, Roebuck Catal.* 222/1 Persians, plaids, checks, brocades, dots, floral. **1953** A. Upfield *Murder must Wait* x. 91 The other women..wear dizzy florals. **1960** *Guardian* 8 Feb. 6/7 The initial designs..are all conventional florals.

Floreal, *sb.* (Earlier example.)

1802 C. Wilmot *Let.* 25 Apr. (1920) 56 Sunday 25th April, 1802, 5me Florial.

floreat (flǭ·ri,æt). [L. *floreat* may he (it) flourish, 3rd sing. pres. subj. of *florēre* to flower, flourish.] Used in conjunction with a name to indicate the hope that the named person, institution, etc., may prosper; *spec.* in *Floreat Etona*, the motto of Eton College.

1888 C. A. Wilkinson *Reminisc. Eton* xxxiv. 340 Join with one heart and voice in the old shout, 'Floreat Etona'. **1898** Lady Butler in W. Meynell *Life & Work Lady Butler* (*caption*, facing p. 1) Floreat Etona! **1899** L. Cust *Hist. Eton Coll.* xii. 294 When Robert Elwes rode to certain death against the Boer bullets at Laing's neck.. his last words were 'Floreat Etona'. **1965** J. P. Carstairs *Concrete Kimono* ii. 17 All decent fellow *Floreat Etona*, and all that jazz! **1967** 'J. Ashford' *Forget what you Saw* xiii. 109 My years at school weren't wasted. *Floreat Etona!* **1969** *Listener* 26 June 899/3 *Floreat* Barry Gold-water.

floreated, *ppl. a.*: see Floriated *ppl. a.*

florencite (flǫ·rĕnsəit). *Min.* [f. the name of W. *Florence*, who made a preliminary examination of the mineral: see -ite¹.] A basic phosphate of cerium (and sometimes other rare earths) and aluminium.

1899 *Nature* 30 Nov. 119/1 Dr. E. Hussak and Mr. G. T. Prior gave an account of a new Brazilian mineral, Florencite, a hydrated phosphate of aluminium and cerium earths. **1962** *Mineral. Mag.* XXXIII. 285 Florencite occurs at Kangankunde as pink rhombohedra about 0·01 cm long.

florentium (flŏre·nʃiə̆m). *Chem.* [a. It. *florentium* (also *florenzio*) (Rolla & Fernandes 1924: see *Gazz. chim. ital.* (1926) LVI. 862–4 and *Atti Accad. Lincei: Rend.* (*Sci. fisiche,* etc.) (1926) IV. 498–500), f. *Florentia*, ancient name of Florence, Italy + -ium.] A disused name for the element *promethium.

1927 *Chem. Abstr.* XXI. 1209 The exact text of the original sealed note..shows that it contains..the proposal to name the new element *florentium* (*florenzio*), with the symbol Fr. **1939** *Nature* 9 Dec. 959/1 The existing data about No. 61, which is called illinium or florentium, are derived from X-ray spectral lines in the *L*- and *K*-series. **1962** Cotton & Wilkinson *Adv. Inorg. Chem.* xxxi. 871 As early as 1926, several groups of workers reported optical and X-ray evidence for the existence of element 61 in various lanthanide concentrates, and the names Illinium, Il, and Florentium, Fl, were proposed by workers at the Universities of Illinois and Florence.

floribunda (flōəribʌ·ndă, flǫ-). [mod.L., f. *floribundus* flowering freely, f. L. *flōr(i)-*, *flōs* flower + *-bundus* (as in *moribundus*), influenced in meaning by *abundus* copious.] A plant bearing flowers in dense clusters, *esp.* a type of rose formerly described as a hybrid polyantha. Freq. *attrib.*

1898 *Daily News* 6 Dec. 5/3 Various primulas are making a pleasant show, especially the vivid little yellow floribunda. **1945** *Rose Ann.* 15 Fresh names have been found for the new group; in this country Hybrid Polyanthas, and in America Floribunda. **1956** B. Park *Collins Guide to Roses* ix. 123 In recent years other species have been bred into this class of rose so that the term hybrid polyantha has been dropped, and floribunda has taken its place. **1959** *Times* 28 Mar., It is usually sufficient to cut these floribunda roses. **1969** E. B. Le Grice *Rose Growing* vii. 91 Climbing sports of a perpetual flowering floribunda are not themselves necessarily perpetual. **1970** *Times* 28 Sept. 8/6 A hybrid tea or floribunda rose should have at least two shoots.

florid, *a.* Add: **3. a.** *florid counterpoint* (earlier example).

1740 Grassineau *Mus. Dict.* 77 Florid Descant and Counterpoint.

Florida. Add: **Florida moss** = Long-beard 2, *Spanish beard*; **Florida orange,** any of several varieties of orange grown in Florida; also *ellipt.*; **Florida water** (earlier and later examples).

1840 *Picayune* (New Orleans) 28 July 4/1 Lavender and Florida waters; perfumed toilet and pearl powders. **1861** Mrs. Beeton *Bk. Househ. Managem.* xxiii. 528 When the bird is roasted,.. surround it with Florida oranges. **1888** G. Trumbull *Names of Birds* 75 A large bunch of 'Florida moss'. **1892** S. Hale *Lett.* (1919) 270 Oranges are delicious...I think they must be Floridas. **1895** *Montgomery Ward Catal.* 259/2 Florida Water is either used as a perfume..or as a lotion. **1923** A. Ward *Encycl. Food* 360 The Florida orange is typically sweeter and less 'sprightly' than the California. *Ibid.*, The best Louisiana oranges are upheld by many as of equal rank with the Floridas. **1924** *Contemp. Rev.* Jan. 95 Giant trees hung with florida moss. **1967** J. McPhee *Oranges* i. 9 Californians say that if you want to eat a Florida orange you have to get into a bathtub first. **1970** *Harper's Bazaar* Oct. 120/2 A bottle of Florida water, 24s 6d.

Floridan (flǫ·ridăn), *a.* and *sb.* [f. Florida + -an.] = *Floridian *a.* and *sb.*

1763 W. Roberts *First Discov. Florida* 3 The size, firmness,.. and longevity of the Floridan Indians. **1763** T. Robinson *Let.* 6 June *Ibid.* 101 The Floridans are a manly well-shaped race. **1856** W. Phillips *Conquest of Kansas* 320 Shortly after the sack of Lawrence, Colonel Titus, the Floridan, offered three hundred dollars for his head. **1900** *Proc. U.S. Nat. Mus.* XXIII. 287 Passing through the Suwanee Strait (now the neck of the Floridan peninsula).

floridean (flǫri·dĭăn), *a.* Also **Floridean.** [f. mod.L. *Florideæ* (f. L. *flōridus* Florid) + -an.] Of or pertaining to the Florideæ (cf. Florideous *a.*) (now often considered a sub-class of the Rhodophyceæ or red algæ); *floridean starch*, a reserve polysaccharide found in many red algæ.

1902 *Encycl. Brit.* XXV. 272/1 The colourless granules of *Florideæ*..have been called floridean-starch. **1935** J. E. Tilden *Algae* ii. 26 The empty shells of the Bryozoa contained large numbers of *Ph[ormidium] persinicum* of a Floridean (Rhodophycean) red color. **1956** *Nature* 25 Feb. 366/1 The purification of Floridean starch. **1970** G. O. Aspinall *Polysaccharides* iv. 66 Floridean starch, which is found in granular form in various red seaweeds, notably *Dilsea edulis*, contains no amylose, and the branched polysaccharide has an average unit chain length of 18–19.

Floridian (flǫri·diăn), *sb.* and *a.* [f. Florida + -ian.] **A.** *sb.* **a.** A Floridan Indian. *Obs.* exc. *Hist.* **b.** A native or inhabitant of Florida. **B.** *adj.* Of, pertaining to, or associated with Florida.

1589 Hakluyt *Princ. Navig.* 541 The Floridians haue pieces of Unicornes hornes, which they were about their necks. **1777** in C. F. Jenkins *B. Gwinnett* (1926) 218 The inroads & depradations of the Floridians. **1819** *Western Rev.* I. 307 In the basin of the St. Lawrence or even in the Floridian waters, a total difference of [fish] inhabitants

may be detected. **1827** J. L. Williams *View W. Florida* 76 The barking..of a congregation of half starved whelps, is music to the ear of a native Floridian. **1859** D. G. Brinton (*title*) Notes on the Floridian peninsula, its literary history, Indian tribes and antiquities. **1945** W. L. McAtee *John & Joe* 14 There are..no fewer than 10 species and sub-species of birds in the Floridian region. **1957** G. E. Hutchinson *Treat. Limnol.* I. i. 102 The Floridian Peninsula..is now covered with a considerable thickness of Tertiary sediments.

florigen (flǭ·ridʒĕn, flōə-). *Bot.* [f. L. *flōr(i)-*, *flōs* flower + -gen.] An unidentified hormone supposed to induce flowering in plants.

1936 M. Kh. Chailakhyan in *Compt. Rend. Acad. Sci. URSS* XIII. 83 We may term this blossom forming or blossom hormone 'florigen', meaning 'blossom-former'. **1966** *Amer. Jrnl. Sci.* XXIX. 427/1 Proposed bioassays for florigens cannot be tested because of the lack of isolated florigens.

Florisbad (flǭ·risbæd). The name of a village near Bloemfontein, South Africa, applied *attrib.* to (the remains of) a primitive hominid discovered there.

1935 J. F. Drever in *Proc. Sect. Sci. Kon. Akad. Wetensch. Amsterdam* XXXVIII. 124 Until..the classification of the Hominidae is revised and modernised, the status of the Florisbad Man will be most suitably expressed by giving it the value of a sub-genus and calling it: Homo (Africanthropus) helmei. **1940** *Jrnl. R. Anthropol. Inst.* 18 The Florisbad skull. **1948** A. L. Kroeber *Anthropol.* (ed. 2) iii. 105 The fossil forms nearest to Rhodesian man..is the Florisbad skull..is probably the nearest to being Palaeoanthropic in type among the more recent South African skeletons. **1959** J. D. Clark *Prehist. S. Afr.* ii. 27 In 1932 Professor Dreyer found the Florisbad skull during excavations in the mineral spring there. *Ibid.* iv. 87 At the time that Florisbad Man was living at the site the vegetation must have been very like that of the Karoo and Middleveld today.

floristic (flōri·stik), *a.* and *sb.* [f. Flora: see -istic.] **A.** *adj.* Of or pertaining to the study of plants with reference to their distribution. **B.** *sb. pl.* That branch of phytogeography which deals with the distribution and abundance of plants. So **flori·stically** *adv.*

1898 Pound & Clements *Phytogeogr. Nebraska* p. iii, The authors..directed the work of the Survey with a view to..a report..in which the floral covering of the state should be treated from the phytogeographical standpoint, instead of mere monographs dealing with it from a floristic standpoint. *Ibid.* v. 215 Phytography..dealing not only with floristic but distribution, but..with morphology, histology, and ecology as well. **1909** Groom & Balfour tr. *Warming's Oecol. Plants* i Floristic plant-geography is concerned with..The conception of a 'Flora', that is, a list of species growing within a larger or smaller area...The division of the earth's surface into floristic regions..according to their affinities...The sub-division of the larger natural floristic tracts—floristic kingdoms—into smaller natural tracts. *Ibid.* 145 Grassy surfaces lining a valley differ floristically according to the aspect. **1918** L. Huxley *Life J. D. Hooker* II. 214 This great floristic work was fairly rounded off by his completion of the 'Ceylon Flora'. **1958** V. G. Childe *Most Anc. East* iii. 50 To find a floristic and faunistic environment comparable to that encountered by the ancient Egyptians one must travel far upstream into the monsoon zone. **1964** V. J. Chapman *Coastal Veget.* i. 9 On the Continent, another system of vegetation analysis is employed. This is based essentially on floristics. **1969** *Nature* 29 Nov. 846/2 The floristic and faunistic richness of old grasslands. **1970** *Watsonia* VIII. 99 Hundreds of plants of this fairly distinct species being scattered over a large area of otherwise floristically rather poor rock.

floristry. (Later examples.)

1963 *John o' London's* 20 Mar. 255/4 Violet Stevenson, well known for her broadcasts on floristry and flower arrangement. **1969** *Oxford Times* 17 Oct. 30/5 (Advt.), Courses in Floristry.

florule (flǭ·riūl). Anglicized form of Florula.

1849 *Amer. Jrnl. Sci.* ... In view of the fact that in both countries [sc. Portugal and America] a number of distinct horizons showing the progressive change in the flora throughout that period have yielded fossil plants in such a way that, if the Portuguese beds were as fully developed as are the American ones, each of these florules might also be compared, the subject becomes rather fascinating.

flossy, *a.* Add: Also *fig.* Saucy, impertinent, 'fresh'; fancy, showy. *N. Amer. colloq.*

1889 *Road* (Denver, Colo.) 28 Dec. 4/3 Phil, we have got it in for you if you don't quit being so flossy. **1895** W. C. Gore in *Inlander* Dec. 113 Flossy, beautiful, stylish. **1900** Ade *More Fables* (1902) 136 He'd show you if you could get Flossy with a Lady, even though she Works. **1903** A. H. Lewis *Boss* 122 He's as flossy a proposition as ever came down the pike. **1922** H. Titus *Timber* i. 14 Do you list that with your references? Your luck with these flossy young petticoats? **1958** *Fisherman* (Vancouver) 15 Aug. 2/3 The flossy propaganda issued by the CMA.

Hence **flossied up** *ppl. a.*, dressed up.

1943 *Penguin New Writing* XVIII. 63 There was a tremendous crowd going, all flossied up for a day out. **1946** F. Sargeson *That Summer* 57, I..was all flossied up. **1957** I. Murdoch *Sandcastle* i. 10, I suppose I'll have to dress. She's sure to be all flossied up.

flot³ (flǫt, flo). [Fr., lit. 'wave'.] A trimming of lace or loops of ribbon, arranged in overlapping rows.

1872 *Young Englishwoman* Nov. 595/2 A *flot* of mauve ribbon falling over the chignon. **1882** Caulfeild & Saward *Dict. Needlework*, *Flôts*, a French term, used to signify successive loops of ribbon or lace arranged to lie over-lapping one another in rows, so as to resemble the flow of small waves...What is called a Flôt-bow is made after the same style. **1902** *Daily Chron.* 12 Apr. 8/3 The long flots of frills.

flouncy (flau·nsi), *a.* Also **-ey.** [f. Flounce *sb.*² + -y¹.] Having flounces, flounced.

1821 M. Wilmot *Let.* 17 Jan. (1935) 93 This gauzy, flowncy, furbelow, flybysky place. **1900** *Westm. Gaz.* 20 Sept. 3/2 A deep-kilted flouncy chiffon. **1906** *Daily Chron.* 31 Aug. 5/7 Floppy hats and skirts, all 'fluffy' and 'flouncy', have come into vogue again. **1909** H. G. Wells *Tono-Bungay* I. i. 19 Shiny and flouncey clothing. **1927** *Sunday at Home* June 525/2 Flouncy petticoats..are giving way in Spain to modern styles.

flour, *sb.* Add: **3.** *flour bag, -dredger* (examples), *loaf, man, miller, -shoot.*

1806 S. Mickle *Diary* 28 June in F. H. Stewart *Notes Old Glouc. County, N.J.* (1917) I. 181 Marked a number of my flour bags with oil and lamp black. **1872** B. Jerrold *London* vii. 71 The whole scene, from thimble-rigger to the peer armed with flour-bags. **1950** *N.Z. Jrnl. Agric.* Jan. 27/3 For the domestic beekeeper an old flour bag makes an ideal strainer. **1867** 'T. Lackland' *Homespun* I. 125 Spoons, and knives, and rolling-pins, and flour-dredgers. **1939–40** *Army & Navy Stores Catal.* 165/3 Flour dredgers. **1828** A. Sherburne *Mem.* ii. 52 He with-drew and sent us in some flour loaves. **1743** W. Ellis *Mod. Husb.* IV. I. 7 The Flour-men do not care to buy this Yellow Wheat in Summer. **1848** *Knickerbocker* XXXI. 221 The butcher's bill, the coal man's bill, the flour man's bill, the house rent, were all quickly settled. **1921** *Daily Colonist* (Victoria, B.C.) 26 Mar. 1/4 Flour millers have asked for a compensatory duty on flour. **1880** Hardy *Trumpet-Major* xxxii, The miller entered the mill as if he were simply staying up to grind.. But he continually left the flour-shoot to go outside and walk round.

4. flour-worm, the larva of any one of the flour-beetles or flour-moths.

1880 Hardy *Trumpet-Major* xvi, Such abundance of water that the old-established death-watches, wood-lice, and flour-worms were all drowned.

flour, *v.* Add: **3.** Also *trans.*

1882 *Rep. Prec. Metals* (U.S.) 648 The action of pounding is likely..to flour the gold as well as the quicksilver.

flouring, *vbl. sb.* **1.** (Earlier and later U.S. examples.) Also *flouring mill-stone.*

1797 *Southampton* (N.Y.) *Rec.* III. 353 John Jermain [shall] have privilege..of erecting a grist mill..for flouring or for packing. **1797** *N.Y. State Soc. Arts* I. 375 Possessing some flouring mills, I was naturally led to converse at times with my millers..on the economy of water in grinding. **1842** *Amer. Pioneer* I. 204 In the city and its vicinity are twenty-five pairs of flouring mill-stones. **1932** *Sun* (Baltimore) 19 Sept. 4/3 Fire destroyed the Alpine buckwheat flouring mill, which was to have resumed operations next week.

flourishing, *ppl. a.* Add: **3.** Also in weakened or trivial use.

1855 Geo. Eliot *Let.* 7 May (1954) II. 201, I am extremely well and jolly..I hope you are all equally flourishing. **1942** N. Balchin *Darkness falls from Air* i. 9 He said, 'How's Marcia?'..I just said, 'Oh, flourishing.'

floury, *a.* Add: **d. floury miller,** *Abricta curvicosta*, an Australian cicada whose body is covered with white down.

1904 *Proc. Linn. Soc. N.S.W.* XXIX. 600 This is a common species in the neighbourhood of Sydney in mid-summer, and is known as the 'Floury Miller' on account of the quantity of silvery pubescence covering the body, which makes it look as though it had been dusted with flour. **1925** *Illustr. Austral. Encycl.* I. 269/1 Well-known species [of cicadas] are the Red-eye..and the Floury Miller.., both names well describing the insect's appearance. **1970** T. E. Woodward et al. in *Insects of Australia* (C.S.I.R.O.) xxvi. 413/1 Many striking cicadas occur in Australia...Among these are the 'double drummer', *Thopha saccata* (F.), the 'floury miller', *Abricta curvicosta* Germ., and the 'green Monday', *Cyclochila australasiae* (Don.).

flow, *sb.*¹ Add: **1. e.** A gradual deformation of a solid (as rock or a metal) under stress in which it suffers a permanent change in shape without fracture or loss of cohesion between its parts.

1889 *Bull. U.S. Geol. Surv.* No. 55. 68 The elaborate and exhaustive series of experiments made by Henri Tresca on 'the Flow of Solids'. **1897** *Geol. Mag.* Nov. 513 Some Experiments on the Flow of Rocks. *Ibid.* 514 The conditions of pressure to which the marble is subjected are those in the 'zone of flow' of the earth's crust. **1932** F. F. Grout *Petrography* vii. 402 The visible deformation of rocks near the surface of the earth is mostly by fracturing and only in very weak rocks, such as clays, by flow. **1959** A. G. Guy *Elem. Physical Metall.* (ed. 2) ix. 322 We might expect plastic flow to begin when the maximum shear stress reaches a certain value. **1965** A. Holmes *Princ. Physical Geol.* (ed. 2) viii. 170 Slate is thus an example of a rock on which a new 'grain' has been impressed—partly by the mechanical effects of flow, partly by the growth of new minerals which have similarly accommodated themselves to the direction of flow. **1971** M. J. Manjoine in H. Liebowitz *Fracture* III. iv. 275 In polycrystalline materials, the initiation and propagation of fracture are usually preceded or accompanied by plastic flow, even though this flow may be small.

3. c. = **honey-flow.*

1951 E. Crane *Dict. Beekeeping Terms* 22 Main flow. Miellée principale. Haupttracht. **1952** H. Mace *Bee-keeper's Man.* xvi. 87 In summers of continued drought, clover is soon over on light soil and the flow may not continue more than two or three weeks. **1953** R. Graves *Poems* 21 In the red West, Where bees come thronging to the apple flow.

9. *flow pattern, -rate*; **flow-blue,** a blue colour applied to pottery or porcelain which diffuses readily through the glaze; **flow chart,** a diagram showing the movement of goods, materials, or personnel in any complex system of activities (as an industrial plant) and the sequence of operations they perform or processes they undergo; also, a diagram in which conventional symbols show the sequence of actual or possible operations and decisions in a data-processing system or computer program, esp. one that is more detailed than a block diagram; hence *flow(-)charting* vbl. sb.; **flow diagram** = **flow chart*; **flow-line,** (*b*) *pl.* [cf. F. *lignes d'écoulement*], the lines that appear on the surface of wrought metal when it is polished and etched, indicating the directions of flow and elongation of the metal during working; (*c*) (also *flowline*), any of the inter-related routes followed by goods, materials, etc., in passing through the various stages of manufacture or treatment; a route depicted on a flow chart; **flow(-line) production,** the continual passage of goods from one machine or piece of equipment to another in the successive stages of production; **flow-meter,** an instrument for measuring rate of flow (of gas, liquid fuel, etc.); **flow-pipe,** the pipe by which hot water leaves the boiler in a system of heating (see also quot. 1967); **flow-sheet, flowsheet** = **flow chart*; hence *flow-sheeting* vbl. sb.; **flow-structure** *Geol.*, the structure in igneous rock produced by the flow of the molten mass before solidification.

1961 Webster, Flow-blue. **1962** K. Shaw *Ceramic Colours* iv. 42 Flow Blues depend for their formation on the volatilisation of chlorides which combine with the cobalt compound of the underglaze colour..during the glost firing. **1967** J. P. Cushion *Eng. China Coll.* ii. 71 The early prints tend to be of a rather dark and blurred blue, rather aptly named by the American collectors as 'flow-blue'. **1970** *Globe & Mail* (Toronto) 25 Sept. 37/4 (Advt.), Primitive scales, andirons, some flow blue china, glass. **1920** C. E. Knoeppel *Graphic Production Control* xii. 136 What should be considered in making up these flow charts are [etc.]. **1949** G. R. Terry *Office Managem.* xxviii. 637 The two types of procedure flow charts are paper distribution and paper correlation. **1966** *Digital Computer Needs* 156/2 A program ..usually includes the preparation of a flow chart showing, diagrammatically, the desired sequence of discriminations and actions. **1968** *Brit. Med. Bull.* XXIV. 242/2 (*caption*) Flow chart showing scope of automation required in the treatment of a patient in a radiotherapy department. **1964** T. W. McRae *Impact of Computers on Accounting* i. 27 The technique of flow charting is especially useful for pointing out improbable or unusual exceptions which have been omitted from the programme. **1943** *Industr. & Engin. Chem.* July 769/2 A simplified flow diagram illustrating the first commercial design of a Fluid Catalyst cracking plant. **1947** Goldstine & Von Neumann *Planning & Coding of Problems for Electronic Computing Instrument* vii. 3 We therefore propose to begin the planning of a coded sequence by laying out a schematic of the course of C [control] through that sequence...This schematic is the flow diagram of C. **1949** D. R. Hartree *Calculating Instruments & Machines* viii. 112 A method of indicating the structure of the sequence of operating instructions by means of a 'flow diagram' representing the control sequence. **1960** R. M. Currie *Work Study* 60 The flow diagram is a drawing, substantially to scale, of the working area, showing the location of the various activities identified by their numbered symbols. **1963** *Times Rev. Industry* May 83/1 A flow diagram..sets out every step of the calculation telling the computer exactly what to do wherever an alternative course presents itself. **1912** P. A. Amos *Processes Flour Manuf.* xxi. 171 Care in 'block-spacing'..before filling in the flow lines of the stock will keep the figures clear. **1913** G. H. Gulliver *Metallic Alloys* (ed. 2) vii. 231 When the section is vigorously etched..the surface shows alternate dark and light striations called flow-lines. **1950** *Engineering* 3 Nov. 334/1 Flow-line production is a particular aspect of mass production. **1956** W. D. Hargreaves in D. L. Linton *Sheffield* 294 An entirely new plant..provides for the continuous flow-line production of railway axles. **1959** B. Chalmers *Physical Metallurgy* vii. 348 The strings of inclusions delineate the 'flow lines' of a forging, and indicate the directions and regions of weakness. **1960** *Times Rev. Industry* Apr. 48/1 A simultaneous attack on building layout, handling, flow lines, placing of equipment, and actual methods of construction is rarely carried out. **1920** *Flight* XII. 353/1 A petrol flowmeter should be fitted so that the engineer can see at a glance his fuel consumption. **1925** Odell in E. F. Norton *Fight for Everest, 1924* 362 The.. ..flow-meter..was connected up with the rubber tubing conveying the gas from the cylinders to the mouthpiece. **1952** *Electronic Engin.* XXIV. 162 The flowmeter to be described was designed for continuously recording the rate of flow of blood to the lungs of an animal. **1950** *Sci.*

News XV. 141 Plate 13 shows a typical photograph of the flow pattern in one of the planes of the model side-blown converter, showing particularly the flow of 'gases' above the 'steel' surface. **1955** *Times* 13 June 9/6 'Flow patterns' can be quickly calculated of air movements over the whole of north-western Europe, for interpreting millions of varied and detailed reports of meteorological observations from all over the Northern Hemisphere. **1904** Goodchild & Tweney *Technol. & Sci. Dict.* 229/2 *Flow* or *flow pipe*, the pipe by which water leaves a boiler. **1967** *Gloss. Sanitation Terms* (*B.S.I.*) 34 *Flow pipe*, a pipe in a primary hot water circuit in which water moves away from the boiler, or a pipe in a secondary hot water circuit in which water moves away from the hot-water storage vessel. **1937** *Times* 13 Apr. p. xii/2 The layout of the Wolseley factory has been scientifically planned in accordance with modern flow production methods. **1955** *Times* 22 June 7/5 The common aim is to achieve in the building of ships a rhythm corresponding to the 'flow production' of, say, a motor-car factory. **1960** *Times* 2 Dec. 17/2 Any significant leak leads to a reduction in flow-rate at the place of the leak. **1962** *Lancet* 27 Jan. 182/2 The peak expiratory flow-rate was measured with a Wright's peak flow-meter. **1912** P. A. Amos *Processes Flour Manuf.* xxi. 171 'Flow-sheets', or diagrams, illustrating the course through which any material travels whilst undergoing treatment in manufacture. **1932** Auden *Orators* i. 22 Designs for the flow sheet of a mill. **1963** *Times Rev. Industry* May 83/2 He can..put the flowsheet information into machine code which entails a detailed time-consuming reproduction of the problem in computer language. **1964** M. Gowing *Britain & Atomic Energy* xii. 336 This involved much basic design work and flowsheeting. **1890** Flow-structure [see *fluxion-structure*]. **1893** Flow-structure [see Fluidal *a.*]. **1903** *Athenæum* 11 July 65/2 A flow-structure has been developed in the matrix. **1968** R. A. Lyttleton *Myst. Solar Syst.* vi. 187 The solidified flow-structure within the tektite.

flow, *v.* Add: **2. d.** Substitute for def.: Of a solid: to suffer a permanent (i.e. non-elastic) change in shape under stress without fracturing or rupturing. (Examples.)

1887 *Encycl. Brit.* XXII. 595/2 When the stress is sufficiently increased..the substance then assumes what may be called a completely plastic state; it flows under the applied stress like a viscous liquid. **1888** Flowing metal [see Flowing *ppl. a.* 1]. **1894–5** Van Hise in *16th Ann. Rep. U.S. Geol. Surv.* i. 594 Whether rocks flow or fracture is in many cases largely dependent on the rapidity of deformation. **1897** *Geol. Mag.* Nov. 514 The experiments therefore show that limestone..does possess a certain degree of plasticity, and can be made to 'flow'. **1901** *Phil. Trans. R. Soc.* A. CXCV. 398 Many limestones under pressure in the earth's crust flow precisely as metals do. **1914** Ries & Watson *Engin. Geol.* iii. 192 When subjected to stresses of sufficient intensity, rocks are deformed either by fracturing or by flowing. **1932** F. F. Grout *Petrography* vii. 402 At moderate and great depths in the crust, competent rocks yield elastically up to the elastic limit and then fracture; and weaker ones recrystallize and flow. **1971** M. J. Manjoine in H. Liebowitz *Fracture* III. iv. 278 In this region, the material can flow more rapidly at a lower stress.

3. a. Also, of animals.

1890 'R. Boldrewood' *Col. Reformer* xix, He..confined himself to riding..round the cattle on the camp, preventing them from flowing out in unnecessary directions.

8. d. (Later U.S. example.)

1906 *Amer. Naturalist* June 446, I observed a tree which flowed little sap and continued flowing after the other trees had ceased.

flowage. Add: (Earlier and later U.S. examples.)

1830 *Massachusetts Spy* 3 Feb. (Th.), The flowage, which would be occasioned by a dam to turn the water into the Feeder. **1936** *Sun* (Baltimore) 5 Oct. 4/6 Army engineers, faced with the task of completing the Maryland portion of the new waterway.., said today 633 acres [of land] would be purchased outright. Only flowage rights will be bought on 308 acres. **1971** *Capital Times* (Madison, Wis.) 28 Aug. 11/7 Campsites and picnic areas near the flowage..brought in more than $1 million of tourism to the Hayward area annually.

b. *Geol.* The flow (*flow *sb.*¹ 1 e) of rock.

1894–5 Van Hise in *16th Ann. Rep. U.S. Geol. Surv.* i. 594 Even in homogeneous rocks, the zone of fracture and the zone of flowage are not sharply separated from each other. **1921** *Science* 4 Mar. 196/2 While rock flowage and rock fracture constitute two distinct types of deformation, there is almost complete gradation between the two. **1965** G. J. Williams *Econ. Geol. N.Z.* xviii. 302/2 Upper seams tend to be lenticular and rarely of workable thickness, though local thickenings were found in areas of complex structure owing to flowage of the coal along structural lines.

flower, *sb.* Add: **3.** *no flowers* (*by request*): an intimation that flowers are not desired at a funeral; hence, no outward sign of mourning or regret.

1900 C. Hanbury *Let.* 5 Aug. in *Autobiogr.* (1901) xv. 224 When the Home-going is, I want to say with more than common emphasis—'by desire'—no flowers; also no bell tolling. **1923** H. C. Bailey *Mr. Fortune's Practice* iii. 84 No flowers, by request. Don't be an owl. **1928** E. Wallace *Gunner* xxii, I shall..explain to the police why I shot you, and there will be no flowers from Scotland Yard. **1970** *Times* 17 Nov. 20/2 Cremation at 11.30 a.m. at Exeter Crematorium, Thursday 19th November. No flowers by request.

12. a. *flower-jar, -seed, -shop.* **b.** *flower-gathering* vbl. sb. **d.** *flower-bright, -sweet* adjs.

1906 *Daily Chron.* 11 Apr. 8/1 The..flower-bright stretches of the great park. **1952** C. Day Lewis tr. *Virgil's Aeneid* xii. 278 Led by the queen's daughter, who tore at her flower-bright tresses And rose-petal cheeks. **1933** R. Tuve *Seasons & Months* iv. 163 A flower-gathering scene is similar in temper. **1835** *Southern Lit. Messenger* I. 239 Getting some water from a flower jar, I sprinkled it over her face. **1863** B. Taylor *H. Thurston* ii. 35 Mrs. Babb..had gathered..the chrysanthemums..and stuck them into an old glass flower-jar. **1807** Jane Austen *Let.* 20 Feb. (1932) I. 182 You are recommended to bring away some flower-seeds. **1866** A. D. Whitney *L. Goldthwaite* vii, There were flower-seeds—bags labelled 'Petunia'. **1867** *Amer. Naturalist* May I. 154 Branches of this early gem may now be purchased..at the flower-shops of Boston. **1865** Swinburne *Chastelard* i. i, I love her for sweet eyes or brows or hair, For the smooth temples, where God touching her Made blue with sweeter veins the flower-sweet white. **1908** Hardy *Dynasts* iii. v. ii. 431 To put it so is flower-sweet of you. **1925** J. Gregory *Bab of Backwoods* 34 Flower-sweet attributes of girlhood.

13. flower arrangement, the decorative arrangement of flowers; hence (as a back-formation) **flower-arrange** *v. intr.*; also **flower-arranger, flower-arranging** *vbl. sb.*; **flower-beetle** *U.S.*, a beetle which feeds upon flowers, esp. one belonging to the sub-family Cetoniinæ; **flower-box,** a box in which flowers are grown, e.g. one placed outside a window, a window-box; **flower-bud,** an unopened flower, as distinguished from a leaf-bud; **flower decoration** = **flower arrangement*; **flower-fly,** a dipterous insect which frequents flowers, esp. one belonging to the family Syrphidæ; **flower-printed** *a.*, bearing a floral design; **flower room,** a work-room in which flowers are arranged.

1955 in *Amer. Speech* (1956) XXXI. 84 J— is flower-arranging. **1900** G. Jekyll *Home & Garden* xiii. 141 The elaborate system of flower arrangement practised by the Japanese. **1933** A. L. Sadler *Art Flower Arrangem. Japan* iv. 54 The names of the three elements of a flower arrangement are taken from the three powers of nature, Heaven, Earth and Man. **1962** I. Murdoch *Unofficial Rose* xxxv. 336 She noted..which buds would be ready.. for her flower-arrangement at the Women's Institute. **1966** *Times* 17 Sept. 12/1 (*caption*) Flower arrangers make great use of shrubs, both for their flowers and their foliage. **1967** *Listener* 16 Mar. 357/3 It is no wonder that flower arranging is a difficult business. **1842** T. W. Harris *Insects Injur. Veget.* 22 The tree-beetles..during the same period of their lives, frequent flowers, and are called flower-beetles. **1895** J. H. & A. Comstock *Man. Stud. Insects* xxi. 564 The flower-beetles are so called because many of them are often seen feeding upon pollen and flying from flower to flower. **1955** Borror & Delong *Introd. Study Insects* xxii. 390 The flower beetles are principally pollen feeders. **1876** J. S. Ingram *Centenn. Exposition* xii. 398 The general description of tiles made by this firm are here enumerated. Encaustic and tesselated tile pavement;..flower-boxes, etc. **1871** *Leisure Hour* 478/1 One kind [of caterpillar]..manages to enclose a young flower-bud betwixt the leaves. **1907** G. Jekyll (*title*) Flower decoration in the house. **1937** C. Spry *Flowers in House & Garden* 143 If one wants to achieve a facility with flower decorations there should be endless experiment of arranging them. **1967** *Everyman Encycl.* V. 338/1 Flower decoration. The arrangement of F[lower]s for interior adornment has achieved..popularity. **1842** T. W. Harris *Insects Injur. Veget.* 414 Flower-flies..are easily distinguished from the preceding flies..by the smaller size of their winglets. **1951** C. N. Colyer *Flies Brit. Isles* xii. 154 (*heading*) *Syrphidae*, hover-flies, drone flies, flower flies. **1952** R. S. Thomas *Acre of Land* 31 He cannot read the flower-printed book Of nature. **1970** *Daily Tel.* 17 Apr. 17 Maxi dress with sash belt, in turquoise flower-printed cotton. **1873** *Young Englishwoman* Nov. 562 A flower room. **1967** M. Summerton *Memory of Darkness* vii. 95, I retired to the flower room in which Fenton had dumped enough ill-assorted blooms to stock a florist's. **1968** A. Christie *By Pricking of my Thumbs* vii. 85 'A flower room, I shouldn't wonder,' he said. 'Where people used to do the flowers. See?'

b. flower children, people, term applied to the adherents of a sub-group of hippies, so called from their habit of wearing or carrying flowers as symbols of peace and love. Also **flower power,** an expression used to describe the beliefs and conventions of this sub-group.

1967 *Times* 13 July 2/6 The term flower children has been given to devotees of mind-expanding drugs in California. **1967** *Guardian* 25 July 6/6 Beat-reared, Greenwich-nurtured teenagers are running away..to be flower people. **1967** *Listener* 3 Aug. 131/3 A new world consisting very largely of the hippies, the flower-power generation. **1967** *Time* 25 Aug. 38 In came the beautiful people on four motorcycles, right into the ballroom, oozing with flower-power. **1969** N. Cohn *AWopBopaLooBop* (1970) xxiv. 225 As fads go, Flower Power was less than impressive...London was content mostly to ape California. Everyone wore kaftans and beads and bells. Everyone spoke in hushed tones of San Francisco and Monterey, of acid and Love and the Maharishi. **1969** *Listener* 5 June 793/2 The political innocence of the Hornsey flower children. **1970** *Globe Mag.* (Toronto) 26 Sept. 14/1 The flower children..arrive by the hundreds to find thousands like themselves sleeping in vacant lots. **1971** K. Amis *Girl, 20* iv. 152 Just another fun thing and now thing, like these clothes they all wear and theatre in the nude and flower power and environmental art.

flower, *v.* **1. b.** Add: Also with *into.*

1865 M. C. Harris *Christine* xxxii, It seemed as if earth had suddenly flowered into a paradise. *Ibid.* xxxvi, The evil in him had flowered out into luxurious growth.

flowery, *a.* Add: **1. c.** *Flowery Empire, Kingdom, Land,* or *Nation* [transl. Chinese *hwa kwo*]: China. *Flowery Land* also = Florida.

1847 R. Fortune *Three Years' Wand.* i. 2 Was this.. the 'flowery land',..of which I had heard so much in England? **1859** Mayne Reid *Oçeola* i, The Flowery Land. **1862** *Once a Week* 26 July 138/1 Celestials of the 'flowery' empire. **1867** F. S. Cozzens *Sayings* i. 4 Silk, sir, came from the Flowery Nation. **1870** *Macm. Mag.* XXIII. 172/1 Nor has Christianity..ever been able to obtain a solid footing in the Flowery Land [*sc.* China]. **1901** *Daily Colonist* (Victoria, B.C.) 30 Oct. 8/4 Two of the big Chinese companies have had men at work at Ross Bay cemetery..exhuming the bodies of their dead countrymen for shipment home to China where all loyal citizens of the Flowery Kingdom wish to have their last resting place.

d. *flowery dell*, rhyming slang for a prison cell. Also *ellipt.*

[**1859** Hotten *Dict. Mod. Slang* 40 *Flowery*, lodging, or house entertainment.] **1925** E. Jervis *25 Yrs. in Six Prisons* i. 16, I talked to him of his 'flowery' ('flowery dell' rhymes with 'cell', hence 'flowery'). **1938** F. D. Sharpe *Sharpe of Flying Squad* xxiv. 248, I bangs and rattles on the door of the Flowery until the screw comes. **1970** T. Clayton *Men in Prison* vi. 128 Found aht on the Moor,..that if you have a new play to read weekends in the flowery..you can kid yourself you're having a Saturday night aht.

flown, *ppl. a.*¹ Add: **b.** *flown cover,* in *Philately*, an envelope or cover dispatched by airmail.

1930 *Times Educ. Suppl.* 16 Aug. p. iv/1 A new form of collecting is taking the imagination of those who are interested in flying. It is the collecting of what are called 'flown covers'. **1936** *Times* 3 Jan. 14/1 A collection of postage stamps and flown covers..are also on show. **1962** M. Savill tr. *Schenk's Romance Postage Stamps* iii. 132 The expeditionary forces in Korea..sent their field post letters with their national airmail stamps, and so the Korean battles are..documented..by genuine flown covers.

flowstone (flōu·stōun). *Geol.* orig. *U.S.* [f. Flow *v.* + Stone *sb.*] Rock, or a rock formation, deposited by water flowing in a thin sheet.

1925 *U.S. Geol. Surv. Bull.* 760 110 Deposits of calcium carbonate have also accumulated against the walls in many places where water trickles from the rock..To distinguish this material from that deposited by dripping water, it has been called 'flowstone'. *Ibid.* (Plate XXV, *caption*) Because of the resemblance of this material to ice, it is popularly known as flowstone. **1953** G. T. Warwick in C. Cullingford *Brit. Caving* iv. 63 Very similar to these [dripstones] are those formed from thin films of water flowing over the surfaces of a cave, which are called flowstone deposits by the Americans. **1958** *Times Lit. Suppl.* 25 July 426/3 The reader who hardly knows what a pothole or a flowstone is. **1969** J. D. Hanwell in C. Cullingford *Man. Caving Techniques* xiii. 241 Massive barriers of stalagmite and flowstone can often block an obvious way on. **1971** *Islander* (Victoria, B.C.) 24 Jan. 5/2 Stalactites formed by water flowing along a wall or crack are flowstones.

flu, 'flu: now the usual spelling of Flue *sb.*⁶

1893 [see Flue *sb.*⁶]. **1911** *Chambers's Jrnl.* Apr. 239/2 We naturally ask ourselves what season—the 'flu' season, or was it the festive season? **1935** G. Barker *Janus* 151 She's in bed; she's got the 'flu. **1951** Auden *Nones* (1952) 28 Little birds with scarlet legs, Sitting on their speckled eggs, Eye each flu-infected city. **1957** *Times Lit. Suppl.* 15 Nov. p. ii/3 When the place is snowbound and the staff laid low with flu, the girls take over.

fluavil (flū·ăvil). *Chem.* [ad. F. *fluavile* (Payen 1852, in *Compt. Rend.* XXXV. 118), f. L. *flu-ere* to flow + *fl-āv-us* yellow: see -il, -ile.] A yellow amorphous resin found in gutta-percha.

1864 Watts *Dict. Chem.* II. 669 Fluavil, a constituent of gutta percha. **1910** *Encycl. Brit.* XII. 744/1 Albane $C_{10}H_{16}O$ and fluavil $C_{20}H_{32}O$..can be separated from the pure gutta by the use of solvents. **1933** *Chem. Abstr.* XXXIII. 8447 Compns. which are suitable for sealing cans, etc., are produced substantially free from the balata hydrocarbon but comprising vulcanized rubber latex and fluavil substantially unchanged by interaction with S.

flub (flʌb), *v. U.S. colloq.* [Etym. unknown.] *trans.* To botch, bungle. Also *intr.* Hence as *sb.*, something badly or clumsily performed; a slip-up.

1924 P. Marks *Plastic Age* xii. 122, I have the feeling.. that I have flubbed this talk. **1926** Wodehouse *Heart of Goof* iv. 142 The spectacle of a flubber flubbing ahead of us on the fairway inclines to make us flub as well. **1942** Berrey & Van den Bark *Amer. Thes. Slang* § 578.18 *Play inexpertly* [music],..flub. *Ibid.* § 711.8 Make a bad shot [golf]..flub. **1952** J. Steinbeck *East of Eden* 495 In my younger days I played tennis...A servant..could pick up his master's flubs at doubles. **1952** S. Kauffmann *Philanderer* (1953) iv. 70 They'll bring someone else in for the other job if you flub it. **1959** N. Mailer *Advts. for Myself* (1961) 76 Al watched him flub balls all over the table. **1969** *TV Times* (Brisbane) 13 Aug. 8/3 Sullivan twitches. He mumbles. He flubs.

flubdub (flʌ·b,dʌb). Chiefly *U.S.* [Fanciful.] Bombastic or inept language.

1888 *Detroit Free Press* Aug. (Farmer), By swiping out the flub-dub and guff, I guess we'll have room to put in the points. **1902** Kipling *Traffics & Discov.* (1904) 16 Any God's quantity of fuss and flubdub to bury a man. **1904** *Rochester Post-Express* 13 July 4 There is an immense amount of flubdub and nonsense and gush in this sort of talk. **1908** *Herald* (Buenos Aires) in *Daily Chron.* 30 June 4/7 All the oratory and political flub-dub that was ever mouthed or printed. **1928** *Daily Express* 6 Aug. 8/7 Psychic research..whether mere 'flub-dub'—like some fortune-telling [etc.]. **1952** *Daily Mirror* (N.Y.) 8 July 11/4 Maybe Mike Todd or Berle should take over the management of the conventions...They would remove much of the amateur flub-dub.

fluctuational (flʌktiu,ēi·ʃənăl), *a.* [f. Fluctuation + -al.] Of or pertaining to fluctuation.

1913 W. Bateson *Mendel's Princ. Heredity* 113 Negative results which suggest that these features are largely fluctuational.

fludrocortisone (flū:drokǭ·ɹtizōun). *Pharm.* [f. **flu(orohy)drocortisone.*] A synthetic derivative ($C_{21}H_{29}FO_5$) of hydrocortisone which causes retention of sodium and loss of potassium and which is given orally in the treatment of adrenocortical deficiency and to supplement cortisone in the treatment of Addison's disease; also *ellipt.* for the acetate, the form in which the drug is usu. given.

1958 *Clin. Sci.* XVII. 205 These compounds differed from hydrocortisone only by the presence of an additional halogen atom, and fludrocortisone (9α-fluorohydrocortisone acetate) has proved to be the most active of the series. **1961** *Lancet* 25 Mar. 632/1 In the adrenalectomised subjects, aldosterone by mouth had 1/25th to 1/30th the potency of fludrocortisone in the maintenance of sodium balance. **1965** Goodman & Gilman *Pharmacol. Basis Therapeutics* (ed. 3) lxxii. 1633 Fludrocortisone acetate.. is the most potent synthetic mineralocorticoid. **1969** R. George in J. A. Bevan *Essent. Pharmacol.* xxxiv. 361 Fludrocortisone is used extensively as an anti-inflammatory agent but only in topical preparations.

flue, *sb.*³ Add: **6.** *flue-tube*; *flue-like* adj.; **flue-boiler** (earlier example); **flue-cure** *v. trans.*, to cure (tobacco) by using artificial heat introduced by flues; so *flue-cured* ppl. adj., *-curing* vbl. sb.; **flue-dust,** dust which collects in the flue of a furnace, *spec.* of a metallurgical furnace, and which contains valuable particles of metal, etc.; **flue-gas,** any mixture of gases from the flues of chemical or smelting factories; **flue-surface** (see quot.).

1859 *Trans. Ill. Agric. Soc.* III. 522 My boiling was not conducted under as favorable auspices..as in the experiment previously alluded to with the flue boiler. **1909** *Cent. Dict.* Suppl., Flue-cure. **1905** G. M. Odlum *Culture of Tobacco* 99 In case of the flue-cured tobaccos, these barns would be too large to properly maintain the heat necessary. **1931** *Times* (Trade & Engin. Suppl.) 5 Sept. 535/4 The progress has been confined..to the production of bright flue-cured tobacco, principally in Norfolk, Oxford, and Elgin counties in Ontario. **1966** *Times* 23 May 16/5 The flue-cured crop represents only a small fraction of the Malawi tobacco crop. **1886** C. G. W. Lock *Tobacco* iv. 220 We..have no space to describe the different methods of 'curing' tobacco, as, for instance,.. 'flue-curing', 'open-fire-curing', &c. **1923** *Glasgow Herald* 23 June 10 Each of these growers has a flue-curing barn. **1857** R. S. Burn *Steam-Engine* (ed. 2) 77 Each chamber.. is also connected with the bottom of the boiler by a series of vertical flue passages,..which ..allow the flue-dust to precipitate into the spaces beneath. **1902** *Encycl. Brit.* XXV. 42/2 Both kinds of gas, when issuing from the burner, hold in mechanical suspension a considerable quantity of 'flue-dust'. **1917** *Nature* C. 92/1 The flue-dust of blast-furnaces. **1951** *Engineering* 20 July 75/2 Behaviour of aggregates of small particles such as..boiler flue dusts. **1898** *Jrnl. Chem. Soc.* LXXIV. ii. 188 Flue Gases in Relation to Furnace Efficiency. **1900** *Engineering Mag.* XIX. 760/1 The now easy process of flue-gas analysis. **1958** *Chambers's Techn. Dict.* Suppl. 980/1 *Flue gas temperature*, temperature of flue gases at the point in the flue where it leaves the furnace. **1908** *Westm. Gaz.* 9 Mar. 8/2 The bursting in of the door..sent the flames up through the flue-like staircase. *a* **1877** Knight *Dict. Mech.* I. 891/1 *Flue-surface* (Steam-engine), the area of surface of the boiler which is exposed to the action of the flame and heated gases after they have left the fire-chamber or furnace. The heating surface of a boiler is made up of the fire-surface and flue-surface. **1878** *Design & Work* IV. 335/1 Smoke and soot were discharged in such abundance as speedily to choke the flue-tubes.

flued, *ppl. a.*¹ Delete 'Only in comb.' (Earlier and later examples.)

1824 *Greenhouse Comp.* I. 114 All of them may be better grown and flowered in a flued pit. **1842** *Florist's Jrnl.* III. 87 In a green-house well raised, seed is produced better perhaps than even on the south side of a flued wall. **1963** *B.S.I. News* May 15/2 Flued convector heaters. **1969** *Sci. Jrnl.* Feb. 25/2 The burner works only in flued appliances where chimney draught can aspire the necessary air.

flueless (flū·lės), *a.* [f. Flue *sb.*³ + -less.] Without a flue.

1903 *Daily Chron.* 27 Jan. 3/7 Flueless fires. **1927** *Glasgow Herald* 25 Mar. 10 Flueless grates.

fluence² (flū·ĕns), aphæretic form of Influence *sb.*, occurring esp. in phr. *to put the fluence on* (a person), to apply mysterious, magical, or hypnotic power to (a person).

1909 J. R. Ware *Passing Eng.* 203/2 *Put on the flooence*, attract, subdue, overcome by mental force. **1923** Wodehouse *Inimitable Jeeves* iii. 31 She was always able to turn me inside out with a single glance, and I haven't come out from under the 'fluence yet. **1937** D. Jones *In Parenthesis* iv. 68 Put the fluence on..drownd the bastards on Christmass Day in the Morning. **1957** A. E. Coppard *It's Me, O Lord!* ii. 21 It was avouched..that if you rubbed the juice of a lemon on the palm of your hand you were armoured against suffering..and as long as the 'fluence' lasted other canes broke too. **1958** M. Procter *Man in Ambush* vii. 82 If ever I saw a girl trying to put the 'fluence on a fellow it was Tess. **1965** E. Bruton *Wicked Saint* viii. 105 Put the 'fluence on him and we'll be away.

fluff, *sb.*¹ Add: **1. d.** Usually *bit of fluff* (cf. *bit *sb.*² 4 h): a young woman. *colloq.*

1903 'Marjoribanks' *Fluff-hunters* 11 The guard was about to whistle, when a bit of fluff was blown up the platform, and, before Webster had had time to send up a petition for a safe journey, it (the fluff) had come to rest on the corner seat opposite him. **1919** W. Deeping *Second Youth* xxiv. 207 Got a little party on, you know, two bits of fashionable fluff. **1937** W. S. Maugham *Theatre* xiv. 120 It was strangely flattering for a woman to be treated as a little bit of fluff that you just tumbled on to a bed. **1962** Wodehouse *Service with Smile* ii. 39 His ball of fluff will be there.

3. b. Add to def.: Also, a mistake made when broadcasting, playing music, etc.

1937 *Variety Radio Directory 1937/38* 342 *Fluff*, missing a cue, or muffing a gag. **1962** A. Nisbett *Technique Sound Studio* 253 *Fluff*, a small accidental error in operational work or in speech. **1967** *Word Study* Oct. 2/2 The announcer's fluff: 'Kids, be sure Mother is stopped [stocked] up with Kellogg's Pep.' **1970** *Daily Tel.* 28 Oct. 12/7 In spite of a surprising number of fluffs, there was some excellent playing.

c. A bungling of a stroke or movement in games; a mistake, a blunder.

1928 *Daily Express* 10 May 10/2 Walter Hagen..hit a [golf] ball no more than a couple of yards...There is nothing malicious in the way the mind fixes on that two-yard 'fluff'. **1960** *Times* 21 Mar. 3/3 In addition he achieved four astonishing place kicks, which made his costly fluff against France unbelievable. **1970** *Guardian* 5 Aug. 18/8 Naturally, the Manson defence lawyers leaped on the fluff as an outrage.

fluff, *v.*¹ Add: **3.** (Earlier example.)

1875 L. Troubridge *Life amongst Troubridges* (1966) x. 103 Her black hair combed right down over her forehead and fluffed out behind.

5. d. Add to def.: Also, to make mistakes in broadcasting, playing music, etc.; to mispronounce, play wrongly (a phrase, note, etc.).

1936 H. Herd *Press Days* i. 20, I had a good natural memory and could be relied upon [in reciting] not to fluff my lines. **1946** *Penguin Mus. Mag.* Dec. 71 But, if you are in his orchestra, don't fluff that phrase next morning. **1955** *Times* 11 May 15/1 He loomed nearer in a closeup to tell viewers (fluffing his lines a little) about the decision of the three Great Powers to invite Russia to join in four-Power talks. **1960** C. Day Lewis *Buried Day* ix. 186, I had kept fluffing when I practised them [*sc.* songs]. **1967** *Word Study* Oct. 3/1 Steve Allen fluffed one time in advertising Dunhills: 'They're *much*, *much* wilder [milder].'

e. To bungle (esp. a stroke or movement in games); to make a mistake; to fail (an examination). (Cf. *duff *v.*²)

1902 *Westm. Gaz.* 30 Apr. 7/3 Blain played first, and topped his shot into the bunker. Hilton then played, half-fluffed his. **1909** *Ibid.* 10 June 12/3 He skied his second into the rough at the third, and then fluffed his approach. **1915** H. L. Wilson *Ruggles of Red Gap* (1917) xv. 264 If it proves that I have fluffed my catch, rely upon me to use extreme measures. **1944** 'N. Shute' *Pastoral* i. 18, I fluffed the twelfth and lost a ball. **1955** *Amer. Speech* XXX. 303 *Fluff*, flunk (an exam), or make a mistake of any kind. **1959** G. Jenkins *Twist of Sand* viii. 148 There's a new attack plan. I don't want it fluffed, like the other. **1971** *Daily Tel.* 6 Aug. 11/5 Many school-children..awaiting their summer exam results have now got this particular worry. Has mother fluffed hers—or has she got through? **1971** *Times* 17 Sept. 6/4 Palmer fluffed it because there was a hedge where his backswing should have gone.

f. *intr.* To bluff, to lie; also *trans.*, to falsify (accounts, etc.).

1902 E. Nesbit *Five Children & It* x. 268 We must pretend like mad. Like that game of cards where you pretend you've got aces when you haven't. Fluffing, they call it. **1941** M. Allingham *Traitor's Purse* xiv. 159 You're up to something you don't want me to know about... Don't fluff. **1955** M. Gilbert *Sky High* xv. 209 Watch him..he's fluffing... He's acting. Dangerous man. Don't relax. **1957** *Times Lit. Suppl.* 11 Oct. 603/2 If he had chosen to fluff the accounts, or to render none, he would not have exposed himself.

g. Of railway porters: (see quots.).

1923 in J. Manchon *Le Slang* 127. **1926** *Westm. Gaz.* 7 Jan., 'Fluffing'—that is, unauthorized tip-hunting by railway men—must stop, by order of the National Union of Railwaymen. **1966** H. Sheppard *Dict. Railway Slang* (ed. 2) 5 *Fluffing*, selection of rich passengers by off duty employees.

fluffed (flʌft), *ppl. a.* [f. Fluff *sb.*¹ or *v.*¹ + -ed.] **a.** Having or covered with fluff or soft hairs; puffed *out* like a ball of fluff. **b.** Of a bungled shot at golf (cf. *fluff *v.*¹ 5 e). **c.** Of a badly-played piece of music (cf. *fluff *v.*¹ 5 d).

1906 *Westm. Gaz.* 6 Jan. 14/1 A pair of little sedge-warblers busy feeding a fluffed-out young cuckoo. **1909** M. B. Saunders *Litany Lane* i. xi, A dark-fluffed, dimple-fisted citizen [*sc.* a baby]. **1923** *Glasgow Herald* 25 June 9 There were fluffed strokes near the green. **1965** *New Statesman* 9 Apr. 586/2 The Mahler performance..had all the marks of under-rehearsal: changed character and contrast..and every now and again a fluffed entry.

fluffer (flʌ·fəɹ). [f. Fluff *sb.*¹ + -er¹.] A worker on a railway system (*spec.* the London Underground) employed to clear the track of refuse (see also quot. 1964).

1956 in *Shorter Oxf. Dict.* Add. **1962** *Guardian* 19 Jan. 9/6 They call them fluffers..because their job is to remove the fluff and dust that gets..on to the rails. **1964** *Daily Mirror* 30 Jan. 18/5 Fluffers...That's the name given to the gangs of women employed to clean the walls of London Underground tunnels during the early hours of the morning.

fluffily (flʌ·fili), *adv.* [f. Fluffy *a.* + -ly².] In a fluffy manner or condition, like fluff.

1906 W. Churchill *Coniston* i. xiii. 166 Miss Cassandra was arrayed fluffily in cool, pink lawn. **1923** *Weekly Dispatch* 14 Jan. 15 Beautiful pink plumes..swing fluffily from the waist on either side. **1927** *Scots Observer* 30 Apr. 11/3 One was a fluffily pretty stranger. **1946** G. Millar *Horned Pigeon* xvi. 251 She was a blonde, fluffily pretty in grotesquely high wooden-soled shoes.

fluffment (flʌ·fmĕnt). *dial.* and *U.S.* [f. Fluff *sb.*¹ or *v.*¹ + -ment.] Something having a light or loose texture (*lit.* and *fig.*).

1890 *Century Mag.* Apr. 817 That is but a fluffment of talk, my Lady...I never heard tell in my time of nurture being stronger than nature. **1899** Dickinson & Prevost *Gloss. Cumbld.*, *Fluffment*, light and loose talk, or material. [Ex.] Her dress o' fluffment an' leace. **1928** *Funk's Stand. Dict.*, *Fluffment*, a bustling manner or fussy style of dress.

fluffy, *a.* Add: **1. c.** *fig.*, often with reference to personal character or intellect.

1898 *Westm. Gaz.* 16 Apr. 1/3 Celia is strong-minded. You would not think so to look at her: she is what I call a fluffy girl. **1904** E. F. Benson *Challoners* xiv, Begin instantly without playing any fluffy arpeggios. **1927** A. P. Herbert *Plain Jane* 10, I like them fluffy,..With downy soft eyebrows and artful blue eyes,..With fluffy complexions, like plums on a wall, And fluffy opinions, and no brains at all. **1964** *Punch* 23 Dec. 967/3 His fluffy wife prepared to help him by seducing the boss.

5. *Comb.*: *fluffy-brained, -minded* adjs.

1905 *Author* 1 Feb. 150 Has he thrown you over for *her*, the fluffy-brained thing? **1902** *Westm. Gaz.* 23 Oct. 3/1 A fluffy-minded woman. **1935** Wodehouse *Blandings Castle* i. 12 The ninth Earl of Emsworth was a fluffy-minded and amiable old gentleman with a fondness for new toys.

flügelhorn (flū·g'lhǫɹn). Also **flu-.** [G., f. *flügel* wing + *horn* horn.] A brass wind instrument. Also *flugel* used *ellipt.*

1854 *Times* 20 Oct. 1/2 Henry Distin's Flugel Horn Union, at 9.15. **1878** Stainer & Barrett *Dict. Mus. Terms* 171/2 *Flügel-horn*, a bugle. A valve-horn. **1926** Whiteman & McBride *Jazz* iv. 105 These extra ones included accordion, bass tuba, flugelhorns, euphonium, [etc.]. **1954** *Grove's Dict. Mus.* (ed. 5) III. 166/1 In English brass bands..a 'flugel' (as it is usually called) is an essential constituent. **1958** R. Vaughan Williams in M. Kennedy *Works R.V.W.* (1964) 633 The usual symphony orchestra is used [in Vaughan Williams' ninth symphony], with the addition of three saxophones and flügel horn. This beautiful and neglected instrument is not usually allowed in the select circles of the orchestra and has been banished to the brass band, where it is allowed to indulge in the bad habit of vibrato to its heart's content. **1965** *Melody Maker* 10 July 12/5 Ian Carr swopped between muted trumpet and occasional flugelhorn.

fluid, *a.* and *sb.* Add: **A.** *adj.* **1. a.** *fluid extract* (U.S.), a concentrated solution (usu. in alcohol) of the active principle of a vegetable drug prepared to a standard strength (see esp. quot. 1965); freq. as *fluidextract.*

1851 Wood & Bache *Dispens. U.S.A.* (ed. 9) 991 Mix thoroughly with the resulting Fluid Extract the Tincture of Ginger. **1885** *Buck's Handbk. Med. Sci.* I. 783/1 The Aromatic Fluid Extract (*Extractum Aromaticum Fluidum*, U.S. Ph.). **1935** C. Solomon *Prescription Writing & Formulary* ii. xi. 102 Fluidextracts (called liquidextracts in the B.P.) are liquid alcoholic preparations of drugs so prepared that 1 cc. contains 1 gram of the drug (or, 1 minim contains 1 grain). **1951** A. Grollman *Pharmacol. & Therapeutics* i. 38 The tinctures and fluid extracts are the most commonly used liquid preparations. **1965** *Pharmacopeia U.S.A.* (ed. 17) 787/2 Fluidextracts are liquid preparations of vegetable drugs, containing alcohol as a solvent or as a preservative, or both, and so made that each ml. contains the therapeutic constituents of 1 Gm. of the standard drug that it represents.

b. (Later examples.)

1949 Koestler *Promise & Fulfilment* xiv. 159 This phase of fluid guerilla fighting ended on April 2. **1956** A. Huxley *Let.* 10 Jan. (1969) 784 Fluid staging, as I remember, used to be all the rage in Germany and Belgium thirty-five years ago. **1959** *Times* 11 June 3/3 The position, as the scores suggest, is decidedly fluid.

B. *sb.* **3.** (Also in sense of *fluidic *a.* 3) **fluid amplifier,** a fluidic device in which small changes in a low-energy flow of fluid produce corresponding changes in a much larger flow; **fluid circuit**, a system of tubes, nozzles, and cavities designed to perform a fluidic function in a way analogous to an electrical circuit; **fluid clutch** = *fluid drive; **fluid compression**, compression of steel while in a fluid state; so *fluid-compressed* adj.; **fluid coupling**, a device that makes use of oil or some other liquid to transmit torque from one shaft to another; **fluid drive**, a transmission in a motor vehicle, etc., in which a fluid coupling is used to transmit the power from the engine to the gears; **fluid flywheel** = *fluid coupling; **fluid logic**, the performance of logical operations by fluidic devices; fluidics; **fluid pressure**, pressure of a fluid, or resembling that of a fluid, being equal in all directions about a point and acting perpendicularly to any surface.

1960 *Product Engin.* 14 Mar. 17/1 Oil or air circuits may soon compete for control applications previously thought suitable only for electronic and electrical controls. Reason: a simple new fluid amplifier just unveiled at the Army's Diamond Ordnance Fuse Laboratories here. 1963 *S.A.E. Jrnl.* Aug. 38 Fluid amplifiers perform electronic-like functions. 1964 *Control Engin.* Sept. 92/2 (*heading*) Fabricating pure fluid circuits. 1965 *New Scientist* 9 Dec. 719/1 Last year..an F-101B fighter was flown with stability against yaw under fluid-circuit control. 1966 *Ibid.* 24 Mar. 766/3 Fluid circuits, the gas or liquid analogues of electronic circuits, are much slower than their electronic counterparts. 1951 *Engineering* 26 Oct. 533/3 The transmission assembly..incorporates a fluid clutch. 1963 R. F. Webb *Motorists' Dict.* 58 A fluid clutch is..operated by engine speeds through two propellers (turbos) facing each other in a fluid filled container. 1888 *Lockwood's Dict. Mech. Engin.* 146 Fluid Compressed Steel. 1884 W. H. Greenwood *Steel & Iron* xxii. 510 The moulds employed in Sir Joseph Whitworth's process of fluid compression are of special construction. 1940 *Automobile Engin.* 50/3 An oil pump maintains pressure in the fluid coupling. 1966 *McGraw-Hill Encycl. Sci. & Technol.* V. 335/1 Because the fluid flow cannot change abruptly, a fluid coupling absorbs rather than transmits shock loads. 1941 *Automobile Engin.* XXXI. 49/1 Fluid drive is available as special equipment on Dodges with conventional transmission and clutch. 1968 *Jane's Surface Skimmer Systems 1967–68* 31/1 The two propulsion engines drive the propellers through a fluid drive system, a reversing gearbox and 1:1 ratio vee box. 1930 *Autocar* 5 Sept. 1 (Advt.), The Daimler transmission system, comprising the Daimler fluid flywheel and a self-changing silent four-speed gearbox. 1967 E. Rudinger *Consumer's Car Gloss.* (ed. 2) 44 The fluid flywheel was used originally in place of a clutch in cars with a pre-selector gearbox..and it is now used in some types of automatic transmission. 1960 J. R. Greenwood *B.S. Thesis, Mass. Inst. Tech.* (*title*) The design and development of a fluid logic element. 1963 *Trans. Soc. Instrument Technol.* XV. 123/2 Fluid devices..without movable parts originated much more recently and the interest in fluid logic is due mainly to these types of devices. 1965 *Times* 23 July 17/3 In a fluid logic system a switch is activated when a small jet of air impinges on another's path. 1845 *Rep. Brit. Assoc. 1844* I. 348 To show how..each of the individual particles of water..shall unite in the production of an aggregate motion consistent with the continuity of mass and with the laws of fluid pressure,—this is a problem which belongs to the mathematician. 1858 W. J. M. Rankine *Man. Appl. Mech.* 100 The term fluid pressure is used to denote a thrust which is normal and equally intense in all directions round a point. 1910 *Encycl. Brit.* XIV. 117/1 This mechanical axiom of the normality of fluid pressure is the foundation of the mathematical theory of hydrostatics. 1940 J. D. Jevons *Metallurgy of Deep Drawing & Pressing* ix. 311 With fluid-pressure actuation the applied pressure cannot rise above a value determined by a release valve.

fluidible (flū·idib'l), *a.* [f. Fluid + -ible.] Capable of change of shape like a fluid under pressure.

1908 J. W. Gregory *Geog.* 5 The shape of the earth depends on the fluidible nature of its interior. 1926 *Encycl. Brit.* II. 172 Under the influence of the intense heat and pressure, the material must at that depth be capable of flow into any cavities like a fluid, so that it has been described as elastico-rigid, or 'fluidible'.

fluidic, *a.* Add: **3.** [Back-formation from next.] Designed or operating in accordance with the principles and techniques of fluidics; of or pertaining to fluidics.

1960 [see *fluidics *sb. pl.*]. 1965 *Missiles & Rockets* 8 Feb. 18/3 In the first fluidic circuits..a fluid stream of relatively high energy was directed across a cavity. 1965 Humphrey & Tarumoto *Fluidics* i. iii. 40/2 Fluidic systems often..have few if any moving parts. 1968 *McGraw-Hill Yearbk. Sci. & Technol.* 12/2 Common to all of the many different types of fluidic devices is the principle that a stream of fluid flowing through a small channel..can be deflected from its initial path when hit by a smaller control jet. 1968 *Sci. Jrnl.* Dec. 53/1 The fluidic system uses a flow of fluid instead of a flow of electrons (electric current). 1969 *Nature* 20 Dec. 1193/1 When the control flows are provided by a feedback from the efflux of the power jet, a fluidic oscillator is set up.

fluidics (flu,i·diks), *sb. pl.* [f. Fluid *sb.* after *electronics, mathematics*, etc.] A field of technology concerned with using small interacting flows and jets of fluid in systems of tubes, nozzles, and cavities that have few or no moving parts, to carry out operations characteristic of electronics, such as amplification and switching.

Quot. 1960 represents a different sense.

[1960 *Compressed Air & Hydraulics* Apr. 149/1 [Abstract of a paper by H. Masheder read at the 2nd European Fluid Power Conf., 28 Apr. 1960.] This paper examines future trends in automatic fluid power control in the light of the following factors: the reasons for the unique position occupied by fluidics; the stability of this position and whether it is likely to be threatened by competitive non-fluidic systems; [etc.]. *Ibid.* 171/2 One prominent engineer [sc. Masheder]..suggests 'fluidics' with which to designate all aspects of the behaviour of liquids and gases.] 1965 *Missiles & Rockets* 8 Feb. 18/1 In five short years, with an investment of some $30 million, the United States has brought into being a new technology—fluidics. *Ibid.* 18/2 The term 'fluidics' as used throughout this report refers to that field of technology that deals with the use of fluids, either gaseous or liquid, in motion to perform functions such as signal or power amplification, temperature or rate sensing, logic or computation, and control. Inherent in the term is the concept of achieving amplification or gain—and often, the absence of moving parts. 1965 Humphrey & Tarumoto *Fluidics* III. xiv. 139 Fluidics will have significant applications in control valves; temperature, pressure and flow sensing and control systems;..and in small, low speed process computers. 1966 *Times* 6 Dec. 20/6 The technique of fluidics..is based on a method of controlling the flow of either gases or liquids just as electron flow is controlled and amplified in electronic circuits. 1967 *New Scientist* 19 Jan. 143/1 The analogy between fluidics and electronics has been taken one step further with the introduction of a planar turbulence amplifier. 1970 *Engineering* 13 Mar. 268 Moving-part fluidics using membrane elements is little practised in the West.

fluidization (flū:idaizēi·ʃən). [f. Fluidize *v.* + -ation.] The process of fluidizing; the state of being fluidized. (In quot. 1932 = Fluidity 1 b.)

1932 *Private Let. J. Lonke* (G. & C. Merriam Co. files) 3 Feb., The correction of the present maldistribution and the restoration of the fluidization of the world's gold supply. 1947 *Chem. Engin. Progr.* XLIII. 429/2 If the gas velocity is relatively low the appearance of the column of solids will be like that of a boiling liquid, and the condition may be referred to as 'dense phase' fluidization. 1954 D. L. Reynolds in *Amer. Jrnl. Sci.* CCLII. 577 As a volcanic process fluidization is exemplified by the uprushing gas phase at Vesuvius. 1955 *Science News* XXXVI. 74 Even and regular transport and feeding of the finest coal powder can be successfully brought about by fluidization. 1960 *Times Rev. Industry* Jan. 39/1 Compressed air is sometimes employed to discharge difficult material from storage bins by 'Fluidization'. 1962 E. M. Shoemaker et al. in A. E. J. Engel et al. *Petrologic Stud.* 344 Intricate mixing, rounding, and polishing of debris derived from depth..suggest fluidization.

fluidize, *v.* Add: **2.** To cause (a mass of finely divided solid) to assume fluidity and other characteristics of a liquid by passing a current of gas, vapour, or liquid upwards through it.

1945 *Industr. & Engin. Chem.* Apr. 332/2 The density of the powder is related to the air rate necessary to fluidize it. 1950 *Petroleum Refiner* Sept. 191/1 Air serves to transfer catalyst to the regenerator and to fluidize the catalyst bed there. 1965 *Sci. Jrnl.* June 65/1 Powder is transferred by pipe..and is fluidized whenever it is necessary to lift it vertically. 1967 P. J. Wyllie *Ultramafic & Related Rocks* viii. 241/1 In the Mesozoic, the massive kimberlite was invaded by an influx of gases, fluidized, and carried upwards at low temperatures. 1968 Coulson & Richardson *Chem. Engin.* (ed. 2) II. vi. 196 Systems fluidised with a liquid do not tend to give rise to bubbles.

So **flu·idized** *ppl. a.*, esp. in *fluidized bed*; **flu·idizing** *ppl. a.* and *vbl. sb.* Also **flu·idizer**, an apparatus in which fluidization is carried out.

1943 *Industr. & Engin. Chem.* July 768/2 A fluidized mass can be circulated by application of the gas lift principle. 1947 *Chem. Engin. Progr.* XLIII. 435/1 Drop in pressure due to gas travel though a bed of solids is practically independent of its velocity when smooth, dense phase fluidizing action takes place. 1949 *Industr. & Engin. Chem.* June 1135/2 Heat transfer between the fluidized solid and the fluidizing gas. *Ibid.* 1227 (*heading*) Reaction of methane with copper oxide in a fluidized bed. 1954 D. L. Reynolds in *Amer. Jrnl. Sci.* CCLII. 580 The conduit of the volcano was eroded and widened by fluidized solid particles. 1955 *Science News* XXXVI. 69 (*caption*) Prototype of fluidizer. 1959 *Times Rev. Industry* June 55/1 The growing importance of fluidizing techniques for the movement of materials in bulk. 1963 *Jrnl. Pharm. Sci.* LII. 284 Fluidized bed drying of tablet granulations is at least 15 times faster than tray drying procedures. 1963 *Economist* 31 Aug. 760/2 Everything from fluidised-bed coating with plastics to electrophoretic deposition. 1964 *New Scientist* 24 Sept. 759/3 There are some interesting arguments in favour of the fluidiser as an auxiliary means of cooking. 1971 *Guardian* 4 June 9/4 Suspended in a fluidised bed like pingpong balls in a funfair.

fluidly, *adv.* (Later example.)

1971 *Oxford Times* 5 Nov. 30 Movement is..confidently expressed by the additional use of fluidly handled wash.

fluidra(ch)m, fluigram, contracted forms of *fluid dra(ch)m, fluid gram(me)*.

1880 O. Oldberg in *Med. Rec.* (N.Y.) 14 Aug. 178/2 We would have to learn *four* new units:..in fact, only two..are important. One is the Gram,..and the other the (cubic centimeter, which, as suggested by Mr. Alfred B. Taylor, should be called a) Fluigram, equal to about 15 minims, or ¼ fluid-drachm. 1889 *Buck's Handbk. Med. Sci.* VII. 731/2 In medicine and pharmacy the gram and the cubic centimeter are the most important metric units, and to express their close relationship the cubic centimeter has been very appropriately termed a 'Fluigram'. 1961 *Stedman's Med. Dict.* (ed. 21) 1827/1 The minim, fluidrachm, and fluidounce of the U.S. apothecaries' measure are slightly larger than the corresponding denominations in the Imperial (British) measure.

fluke, *sb.*[3] Add: Also *attrib.* Phr. *a fluke of wind* (examples).

1906 'Q' *Mayor of Troy* x. 139 The mass huddled together, rubbing flanks, swaying this way and that in the pressure of panic as corn is swayed by flukes of summer wind. 1909 *Daily Chron.* 12 Oct. 8/1 It was no fluke victory. 1955 *Times* 3 Aug. 4/5 Because of the calms and flukes of wind yesterday numbers of vessels retired from their respective events.

fluke, *v.*[2] **2.** Add examples of extended uses. Also *intr.*

1897 N. Gould *Seeing him Through* xiii, Even if he managed to fluke home in this trial. 1904 *Daily Chron.* 5 May 7/2 Even if Mr. P. F. Warner's XI. were fluked out of the game. 1929 *Star* 21 Aug. 16/2 Marvellous handicap horses that may fluke the City and Suburban.

fluked (flūkt), *a.*[2] [f. Fluke *sb.*[1] + -ed[2].] Infested with flukes, fluky.

1855 [see Fluky *a.*[1]]. 1925 *Glasgow Herald* 20 June 4 Badly fluked sheep should be destroyed.

fluky, *a.*[2] **1.** (Earlier example.)

1867 *Australasian* 2 Mar. 268/1 Cosstick..playing a very flukey innings.

flume, *sb.* Add: **3. a.** (Further examples.)

1940 *Chambers's Techn. Dict.* 343/2 *Flume*, a metal chute used for the distribution of concrete from a placing plant. 1957 *N.Z. Timber Jrnl.* Mar. 52/2 *Flume*, an inclined trough with running water for transporting logs. 1959 *B.S.I. News* Nov. 25 Notches, weirs and flumes for flow measurement.

c. (Earlier example.)

1865 *Eastern Slope* (Washoe, Nev.) 23 Dec. 3/1 The great Stockholder..has in the classic language of the mines, 'gone up the flume'.

4. *flume-water.*

1897 B. Harraden *Hilda Strafford* 59, I wish I hadn't filled up my reservoir so full with flume-water.

fluming, *vbl. sb.* (see after Flume *v.*). Add examples of action in sense 1 of the vb.

1851 *San Francisco Picayune* 23 Sept. 2/5 There is another fluming company..that will commence operations this week. 1869 *Trans. N.Z. Inst.* II. 372 [The oldest drifts] can only be worked by bringing water to bear on them by a system of 'fluming'.

flummadiddle. Add: Also **flummer-**, **flummy-**, **fuma-**. **2.** (Earlier and later examples.)

1850 in Wentworth *Amer. Dial. Dict.* s.v., Fumadiddle, flummydiddle. 1854 M. J. Holmes *Tempest & Sunshine* iv. 51 What does she want of any more flummerdiddle notions? 1882 *Century Mag.* Oct. 837 Well, see all that flummer-diddle he got off about it. 1941 *Sat. Even. Post* 25 Oct. 35/3 An' does you try any fumadiddles, you is right away gwine happen to a catastrophe.

flummery. Add: **2. b.** *dial.* and *U.S. collect.* Trifles, useless trappings or ornaments.

1879 *Dewsbre back at Mooin Olmenac* 5 (E.D.D.), For fear t'sun sud fade yor carpets and yor bits o' flummery. 1883 'Mark Twain' *Life on Mississippi* 406 The bridal chamber whose pretentious flummery was overawing.

flummocky (flʌ·məki), *a.* *dial.* and *U.S. dial.* Also **-ucky**, **flommocky**, **-ucky**. [f. *flummock* sb. or vb.: see Flummox *v.*] Confused, muddled; untidy, slovenly. (Cf. Slummocky.)

1834 W. A. Carruthers *Kentuckian in N.Y.* II. 215 Hang me if I don't think he's a little flumucky altogether about the head. 1884– in dial. glossaries (Cheshire, Shropshire, Warwick).

flummox, *v.* **3.** (Earlier example.)

1839 D. P. Thompson *Green Mountain Boys* xxiv. 256 Well, if he should flummux at such a chance, I know of a chap..who'll agree to take his place.

flunk, *sb.* **1.** (Later examples.)

1904 *N.Y. Even. Post* 6 Jan. 5 A sprinter and football player has received a flunk in one study and a condition in another. 1948 *Time* 16 Feb. 94/3 This time there were twice as many flunks.

flunk, *v.* Add: **1.** (Later examples.)

1850 H. C. Watson *Camp-fires Revol.* 414 They were, of course, exposed to the fire of the red-coats..but they didn't flunk a bit. 1894 P. L. Ford *Hon. Peter Stirling* (1898) 355 What will people say of me on November fourth, if my regiment flunks on September thirtieth? 1901 *Munsey's Mag.* XXV. 408/2 It looks pretty middling tough, and it won't do to try it and flunk. 1910 J. Hart *Vigilante Girl* xxi. 294, I don't mean that he's flunking, for he's no coward. 1971 *Sunday Times* (Johannesburg) 28 Mar. 7/1 Sinatra himself said: 'I've flunked out with women more often than not. Like most men, I don't understand them.'

b. (Later examples.) Also *trans.*, to fail (an examination, etc.); *to flunk out*: to be dismissed from a school or university for failing examinations.

1899 A. H. Quinn *Pennsylvania Stories* 166 He never attracted attention by his scholarship, but yet he drifted along somehow without flunking. 1920 F. Scott Fitzgerald *This Side of Paradise* (1921) 35 He'll fail his exams, tutor all summer..and flunk out in the middle of the freshman year. 1923 R. D. Paine *Comr. Rolling Ocean* vi. 99 He tutored for Princeton and flunked in freshman year. 1924 P. Marks *Plastic Age* xviii. 202, I don't..chase around with filthy bags or flunk my courses. 1936 L. C. Douglas *White Banners* xvi. 342 He was working hard to take a calculus examination that he had flunked..two years ago. 1951 *Reader's Digest* May 12/1 He flunked out of various high schools, not because he was too stupid. 1968 *Listener* 27 June 841/1 The scene is..Columbia University, where a number of young Second World War vets and/or non-combatants are making gestures at working for degrees or just hanging around after flunking out. 1970 *Times* 12 Aug. 5/7, I was utterly, deeply, completely depressed and flunked my A levels.

2. (Examples.)

1843 *Yale Lit. Mag.* IX. 61 That day poor Fullman was *flunked*, and was never again reinstated in the good graces of our officer. 1893 W. K. Post *Harvard Stories* 231 That was all very well for him, who..never got 'flunked'. 1899 A. H. Quinn *Pennsylvania Stories* 40 He..finally flunked him in his finals. 1910 *N.Y. Even. Post* 29 Nov. 8 Examining boards may 'flunk' an officer in his first examination. 1966 *Word Study* Feb. 2/2 For if English teachers had always based their grades in English on the moral probity of their students' private lives, they would have had to flunk such naughty boys as Christopher Marlowe, James Boswell, Dylan Thomas, and Baltimore's own Edgar Allan Poe.

flunker (flʌ·ŋkəɹ). *U.S.* [f. Flunk *v.* + -er[1].] **a.** One who fails in an examination. **b.** One who causes candidates to fail.

1895 W. C. Gore in *Inlander* Nov. 62 *Flunker*, one who habitually fails in recitations. 1910 O. Johnson *Varmint* i. 14 'What had he done to you?' said Jimmy, winking at Mr. Hopkins,..master of the Latin line and distinguished flunker of boys. 1948 *Chicago Daily News* 2 Nov. 10/5 College flunkers.

flunkey, *sb.*[1] Add: **1. c.** (See quots.) *N. Amer.*

1906 *N.Y. Even. Post* 12 Sept. 7 'Flunkeys' in the Northwest do not wear uniforms; their work is to act as assistant cooks in mining and lumber camps. 1931 'D. Stiff' *Milk & Honey Route* 205 *Flunky*, camp waiter. Always male. A woman is a *hasher*. 1956 J. S. Gowland *Sikanska Trail* 177 'You're a flunkey,' he said. 'Report to the cook.' 1970 *Islander* (Victoria, B.C.) 17 May 6/1 The first pay slips the waitresses received listed them as 'flunkies'. 'We didn't like this,' Mrs. Kusha says. 'The men employed in the cookhouse were flunkies, but we had been hired as waitresses.'

Hence **flunkeyia·na**, the sayings or characteristics of flunkeys.

1854 *Punch* XXVI. 44 (*heading*), *Flunkeyiana*—a fact. *Flunkey* (*out of place*). 'There's just one question I should like to ask your Ladyship—Ham I engaged for Work, or ham I engaged for Ornament?' 1928 J. Gore in E. Eden *Semi-Attached Couple* p. xi, Miss Eden..could..draw..the characters of servants with brilliance of touch and a knowledge of flunkeiana which Thackeray might have envied.

fluonomist (flūǫ·nŏmist). [f. Flu(e *sb.*[3] + -*onomist* as in *economist*.] (See quots.)

1947 *Christian Century* LXIV. 943 The 'fluonomist' is simply our old and much respected neighbor the chimney sweep. 1948 *Word Study* Oct. 11/1 [citing London *Times*] A sweep entitled himself 'fluonomist' and put up his prices. 1965 *N.Z. Woman's Weekly* 4 Oct. 103/1 Our chimney sweep calls himself a fluonomist!

fluoranthene (fluŏræ·nþīn). *Chem.* [ad. G. *fluoranthen* (Fittig & Gebhard 1877, in *Ber. d. Deut. Chem. Ges.* X. 2143), f. Fluor(ene + Phen)anth(r)ene.] An aromatic hydrocarbon, $C_{16}H_{10}$, crystallizing as colourless needles or plates and found in coal tar; also, any of the derivatives of this compound.

1878 *Jrnl. Chem. Soc.* XXXIV. 431 (*heading*) Fluoranthrene [*sic*], a new coal-tar hydrocarbon. 1947 *Jrnl. Amer. Chem. Soc.* LXIX. 505/1 Pure synthetic fluoranthene is now readily available. *Ibid.* 507/1 The synthesis..of other compounds containing the fluoranthene nucleus is being investigated. 1952 *Chem. Rev.* I. 485 The solutions of many fluoranthenes fluoresce blue or blue-green in ultraviolet light.

fluorescence. Add: Also *attrib.*, as **fluorescence spectrum**, the spectrum of the light emitted by a substance when excited to fluorescence.

1885 J. & C. Lassell tr. *Schellen's Spectrum Anal.* (ed. 2) lxi. 242 The visible part of the spectrum is then removed from sight, and the fluorescence spectrum alone occupies the field. 1962 R. E. Dodd *Chem. Spectroscopy* ii. 80 The fluorescence spectrum appears at a characteristic wavelength and does not shift with the wavelength of the exciting light.

fluorescent, *a.* Add: **b.** *fluorescent lamp*, a lamp in which light is produced largely by fluorescence; *esp.* a tubular electric-discharge lamp containing mercury vapour, ultra-violet radiation from which causes the fluorescence of a coating of phosphor on the inside of the tube; also *fluorescent light, lighting, tube*; *fluorescent screen*, an opaque or transparent screen coated on one side with a fluorescent material and used for displaying images produced when it is struck by X-rays, electrons, or other ionizing radiation.

1896 T. A. Edison *U.S. Pat.* 865,367 (1907), Fluorescent Electric Lamp... The object I have in mind is to produce light by fluorescence... Fluorescent lamps made in accordance with this invention may be operated singly or may be worked together in series. 1938 *Trans. Illum. Engin. Soc.* (London) III. 126/2 The tubular fluorescent lamp. *Ibid.*, The low-pressure mercury vapour fluorescent tube..is singularly comparable with the sodium lamp. 1966 D. F. Galouye *Lost Perception* xiv. 147 He regained consciousness in the glare of fluorescent lights strung along an acoustical tile ceiling. 1971 *Nature* 1 Jan. 46/1 The light intensity was 40 Wm^{-2} from warm white fluorescent lights. 1942 C. L. Amick *Fluorescent Lighting Man.* i. 2 From a lamp and fixture market that did not exist 4 years ago, fluorescent lighting has grown to an estimated $250,000,000 business for 1941. 1953 'M. Innes' *Christmas at Candleshoe* i. 11 With a flicker and a ping a bar of fluorescent lighting has snapped on. 1961 L. Mumford *City in History* x. 298 The American supermarket, with its ghastly fluorescent lighting. 1863 G. G. Stokes in *Phil. Trans. R. Soc.* CLII. 615 On forming an image of the spark between a sharp and a flat electrode..and receiving it on a fluorescent screen, the flat electrode gave the brighter of the two images. 1901 F. H. Williams *Roentgen Rays* iii. 99 The X-ray examination by the photograph is better, as a rule, than that by the fluorescent screen for all parts of the body except the trunk. 1956 M. Slurzberg et al. *Ess. Television* x. 361 The act of converging the beam of electrons to produce a pinpoint of light on the fluorescent screen is called focusing.

fluorescer (flūŏre·səɹ). [f. Fluoresce *v.* + -er[1].] A fluorescent substance.

1904 *Nature* 31 Mar. 523/1 The most powerful fluorescer towards the α radiations is Sidot's hexagonal blende, a crystallised form of zinc sulphide. 1959 *Guardian* 28 Sept. 3/5 A pair of pastel blue cotton sheets..had changed colour... The fluorescer added to synthetic detergents was the cause. 1967 M. Donbrow *Instrum. Meth. Analyt. Chem.* II. viii. 216 'Xanthene' dyes are usually good fluorescers.

fluorescing (flūŏre·siŋ), *ppl. a.* [f. Fluoresce *v.* + -ing[2].] That fluoresces, fluorescent; that causes fluorescence.

1896 *Westm. Gaz.* 27 May 6/3 Mr. Edison by coating the interior surface of a Crookes' tube with crystals of a new fluorescing substance has succeeded in causing the X rays to change to light. 1924 *Discovery* June 97/2 A fluorescing screen. 1957 *Technology* July 168/4 Fluorescing agents, some of which, added to detergents, tend to intensify the colours..of cellulose fabrics such as cotton. 1966 *McGraw-Hill Encycl. Sci. & Technol.* V. 350/2 At very low fluorescent intensities, the intensity is almost directly proportional to the concentration of the fluorescing material.

fluorian (flū·ŏriăn), *a.* *Min.* [f. Fluor(ine + *-ian.] Of a mineral: having a (small) proportion of a constituent element replaced by fluorine.

1930 W. T. Schaller in *Amer. Mineralogist* XV. 571 The adjectival endings thus formed for the names of all the chemical elements are given below... Fluorine—fluorian. 1959 W. H. Dennen *Princ. Mineral.* vii. 363 Chemical varieties [of phlogopite]: Fluorian, ferroan, manganoan.

fluoridate (flū·ŏridēit), *v.* [Back-formation from next.] *trans.* **a.** To add traces of a fluoride or other source of fluoride ions to (water, tooth-paste, food, etc.). Also *absol.* (Cf. *fluoridation 2 a.)

1949 *Bull. Amer. Assoc. Public Health Dentists* Aug. 12/1 Many cities throughout the nation..have taken or are taking steps to fluoridate their water supplies. 1951 *Jrnl. Amer. Water Works Assoc.* XLIII. 644/1 Some questions have been raised concerning the advisability of fluoridating a communal water supply because of the objection to so-called medication of water supplies. 1952 *Public Health Rep.* (U.S.) Jan. 41/1 It will take about 14 years after a community starts fluoridating before a two-thirds reduction can be made in the caries attack rate of all children. 1958 G. B. Denton *Vocab. Dentistry & Oral Sci.* iv. 182 An opinion favoring the use of *fluoridate* was adopted by the *Journal of the American Dental Association* for its style in 1950, and the publicity given to it was probably largely responsible for establishing the use of that term to supersede *fluorinate*. 1964 Goose & Hartles *Princ. Preventive Dentistry* vi. 93 Tablets containing about 1 mg of fluoride have been used..as a means of fluoridating the daily water supply at home. 1969 *Brit. Dental Jrnl.* CXXVII. 605/1 There is no evidence of a general unwillingness on the part of water undertakings to fluoridate when asked to do so by the local health authorities.

b. To treat (teeth) with a preparation containing fluoride. (Cf. *fluoridation 2 b.)

1963 *Brit. Dental Jrnl.* CXIV. 219 If teeth are fluoridated topically with strong solutions, mostly protective fluorides are formed initially. 1966 *Helv. Odontol. Acta* X. 29 Investigations to study some aspects of the processes occurring at the enamel surface fluoridated topically under various conditions.

Hence **flu·oridated** *ppl. a.*

1950 *Amer. Mag.* May 46 (*heading*) They drink away their toothaches; fluoridated water. 1955 *Science News* XXXVIII. 41 The use of fluoridated toothpaste, tooth powders, mouthwashes, pastilles, and even chewing gum, has been canvassed. 1964 Goose & Hartles *Princ. Preventive Dentistry* vi. 94 The effect of the consumption of fluoridated salt on the permanent teeth of children. 1965 *Helv. Odontol. Acta* IX. 148 The effect of the oral environment on the fluoride content of topically fluoridated enamel.

fluoridation (flū:ŏridēi·ʃən). [f. Fluoride: see -ation.] **1.** *Min.* The process by which a mineral absorbs fluorine.

1904 *Monogr. U.S. Geol. Surv.* XLVII. 206 Fluoridation is the addition of fluorine, forming fluorides. 1953 *Chem. & Engin. News* 4 May 1905/1 The earlier, geologic sense of 'fluoridation', namely, introduction of fluorine into a mineral, would seem to be more properly fluorination.

2. a. The addition of traces of a fluoride or other source of fluoride ions to drinking-water for the prevention or control of dental caries.

1949 *Jrnl. Amer. Water Works Assoc.* XLI. ii. 576/1 The present demands for the fluoridation of water supplies are coming from the public and the press. 1955 *Sci. Amer.* Apr. 2/2, I believe that a big reason why fluoridation failed to win acceptance was the hasty and somewhat authoritarian way in which it had been introduced. 1958 *Spectator* 6 June 737/2 In Britain many leading dentists and physicians are opposing fluoridation. 1960 *Times* 10 Feb. 7/3 A Bill to introduce the compulsory fluoridation of water. 1966 McKeown & Lowe *Introd. Social Med.* xiii. 139 In the British studies there was a sixty-six per cent reduction in caries among three year old children after the introduction of fluoridation. 1971 *Daily Tel.* 3 June 2/1 Fluoridation of water supplies..has been shown to prevent dental decay.

b. = *fluoridization b.

1963 *Brit. Dental Jrnl.* CXIV. 216 Topical fluoridation procedures should be discussed only as an alternative to water fluoridation if..a community cannot benefit from water fluoridation. 1965 *Helv. Odontol. Acta* IX. 150 The fluoride content of the outermost enamel layer immediately after topical fluoridation ranged between 1385 and 2800 ppm.

fluorida·tionist. [-ist 3.] One who advocates the fluoridation of public water supplies.

1951 *Jrnl. Amer. Water Works Assoc.* XLIII. 1 'Medication', or even worse 'mass medication', has been plaguing fluoridationists almost ever since it was introduced innocently enough way back when. 1967 *Brit. Dental Jrnl.* CXXIII. 414/1 Dr. George Walbott, the well-known allergist and anti-fluoridationist is organising an international society for fluoride research.

fluoridization (flū:ŏridəizēi·ʃən). [f. Fluorid(e + -ization.] **a.** = *fluoridation 2 a.

1939 *Jrnl. Amer. Water Wks. Assoc.* XXXI. 1929 Every use of water must be examined as well as current water treatment practices before fluoridization can be begun. 1947 *Amer. Jrnl. Public Health* XXXVII. 1156 Fluoridization of community water supplies. 1949 *Bull. Amer. Assoc. Public Health Dentists* Aug. 12/2 Investigators are agreed that the term 'fluorination' of water supplies is incorrect and that the proper term is 'fluoridization'. 1966 *Listener* 6 Oct. 499/3 The evidence upon which the Ministry of Health recommends the fluoridization of drinking water.

b. (See quot. 1963.)

1951 *Sci. News Let.* 19 May 310/3 Fluoridation for the water supply, fluoridization for the treatment procedure when the dentist puts a fluoride on the children's teeth. 1951 *Jrnl. Dental Res.* XXX. 181 In cases of rampant caries, fluoridization of the teeth would simply provide a delaying action. 1963 C. O. Boucher *Current Clinical Dental Terminol.* 151/2 *Fluoridization*, the topical application of a solution of a fluoride to teeth.

Hence **flu·oridize** *v. trans.*; **flu·oridized** *ppl. a.*

1951 *Sci. News Let.* 19 May 310/3 Your town may fluoridate your drinking water, but your dentist will fluoridize your children's teeth. 1951 *Jrnl. Dental Res.* XXX. 178 The normal enamel contained 126·5 ppm. F, while the fluoridized enamel contained 1,434·6 ppm F. *Ibid.* 251 They found increasingly more low lactobacillus counts among children from the city with artificially fluoridized water. 1952 *Public Health Rep.* (U.S.) Jan. 39/1 It is not necessary to continue the use of fluoridized water after the enamel has been calcified.

fluorimeter, var. *fluorometer 2.

fluorinate (flū·ŏrinēit), *v. trans.* [f. Fluorin(e + -ate[3].] **1.** *Chem.* To introduce an atom of fluorine into (a molecule or compound).

1931 *Chem. Abstr.* XXV. 3643 The..unsatd. residues condense to compds. of higher mol. wt. and are also in part fluorinated by the $[PhIF_2]$ still present. 1943 A. L. Henne in H. Gilman *Org. Chem.* (ed. 2) I. xi. 946 Hydrocarbons can be fluorinated in the gaseous form over copper gauze by means of fluorine diluted by nitrogen. 1961 L. F. & M. Fieser *Adv. Org. Chem.* x. 356 A method in which hydrocarbons are fluorinated to saturated fluorocarbons by cobalt trifluoride was used on a small industrial scale during the war to produce fluorocarbon lubricants.

2. = *fluoridate *v.* a.

1945 *Fluorine in Dental Health* (N.Y. Inst. Clin. Oral Path.) 40 A method to control..dental caries—by fluorinating a communal water supply that is deficient in content of fluorine. 1958 [see *fluoridate *v.* a]. 1968 Passmore & Robson *Compan. Med. Studies* I. v. 9 Local authorities in Europe and America have fluorinated their water with a consequent improvement in the teeth of their children.

Hence **flu·orinated**, **flu·orinating** *ppl. adjs.*

a. (sense 1.) 1932 *Industr. & Engin. Chem.* XXIV. 331/2 The series of fluorinated products, C_2F_6, C_2F_5Cl, and $C_2F_4Cl_2$. 1935 P. H. Groggins *Unit Processes in Org. Synthesis* iv. 180 Hydrofluoric acid, antimony trifluoride, and boron trifluoride are the fluorinating agents most frequently employed. 1966 Cotton & Wilkinson *Adv. Inorg. Chem.* (ed. 2) xiv. 388 Sulfur tetrafluoride is a particularly useful and selective fluorinating agent for organic oxygen compounds. 1969 H. B. Hass in S. A. Miller *Ethylene* x. 861 The production of fluorinated refrigerants and propellants for aerosol bombs consumes about 80 per cent of the carbon tetrachloride produced in the United States.

b. (sense 2.) 1945 *Fluorine in Dental Health* (N.Y. Inst. Clin. Oral Path.) 44 Will artificially fluorinated water produce in humans the results..caused by waters in which fluorine is found naturally? 1951 *Jrnl. Dental Res.* XXX. 380 Onset of use of fluorinated waters has to occur within the first six years of life in order to produce an appreciable protection against caries.

fluorination (flū:ŏrinēi·ʃən). [f. Fluorin(e + -ation.] **1.** *Chem.* The action or process of fluorinating a compound. (See *fluorinate *v.* 1.)

1931 *Chem. Abstr.* XXV. 3642 (*heading*) Experiments on the fluorination of organic compounds. 1946 *Nature* 9 Nov. 673/1 Disproportionation occurs during the fluorination (whence the possibility of preparing $(C_2H_5O)_2SiF_2$ from $(C_2H_5O)_3SiCl$). 1966 *McGraw-Hill Encycl. Sci. & Technol.* VI. 328/1 Fluorination reactions occur with enormous vigor.

2. = *fluoridation 2 a.

1943 *Jrnl. Amer. Water Wks. Assoc.* XXXV. 1179/1 Theoretically, the idea of fluorination of the domestic water supply for the reduction of dental caries prevalence appears sound. 1949 [see *fluoridization a]. 1950 *Lancet* 9 Dec. 779/2 Dr. A. P. Black..suggested that fluorination of drinking-water was 'the biggest step in public health in the last hundred years'. 1958 G. B. Denton *Vocab. Dentistry & Oral Sci.* iv. 182 *Fluorinate* and *fluorination* were commonly employed for the addition of fluoride to drinking water from 1943–1949.

fluorine. Add: **2.** *attrib.*, as **fluorine dating, method, test**, etc., a method of determining the relative age of organic remains by measuring the amount of fluorine that has been absorbed from surrounding ground-water.

1949 Oakley & Montagu in *Bull. Brit. Mus. (N.H.), Geol.* I. ii. 29 The authors..independently reached similar conclusions with regard to the probable dating of the [Galley Hill] skeleton and..prepared a joint report on their findings. One author (K.P.O.) has prepared the introductory sections, the account of the geology, and of the fluorine dating. 1959 J. D. Clark *Prehist. S. Afr.* iv. 83 Fluorine and uranium tests have shown that the skull fragments are of the same age as the faunal remains found with them. 1968 R. G. West *Pleistocene Geol. & Biol.* ix. 160 The fluorine method is useful for the relative dating of animal skeletal remains found in sand and gravel.

fluoro- (fluŏro). **1.** Used as comb. form of Fluorine (rarely of Fluoride), chiefly in the names of chemical compounds, as **fluoroace·tamide**, $CH_2F{\cdot}CONH_2$, a stable, toxic, fluorinated derivative of acetamide with strong insecticidal properties; **flu:orohydroco·rtisone** = *fludrocortisone; **flu·oroform** [ad. F. *fluoroforme* (Meslans 1890, in *Compt. Rend.* CX. 717): see Chloroform], a gas (CHF_3) that is almost completely inert both chemically and physiologically and is the fluorine analogue of chloroform; **flu·orotype** *Photogr.* [-type], an old positive process in which paper sensitized with sodium fluoride was used.

1909 *Jrnl. Chem. Soc.* XCVI. ii. 297 The author has determined the heats of formation of the following compounds...Fluoroacetamide, 249·55 cal. 1958 *Nature* 28 June 1810/2 Fluoroacetamide is as effective as the more dangerous sodium fluoroacetate as a systemic insecticide. 1964 *New Statesman* 20 Mar. 438/3 Last month the government banned the sale of fluoroacetamide as an insect control. 1954 Fried & Sabo in *Jrnl. Amer. Chem. Soc.* LXXVI. 1455/2 9α-Fluorohydrocortisone acetate..was obtained in about 50% yield. *Ibid.* 1456/1 Deacetylation of I [*sc.* 9α-fluorohydrocortisone acetate] with sodium methylate afforded 9α-fluorohydrocortisone. 1970 W. Modell *Drugs of Choice 1970–71* xxxii. 511 All patients with Addison's disease now can be satisfactorily managed with oral therapy using 10 mg. of hydrocortisone 2 to 3 times a day, plus 0·1 mg. of 9-alpha-fluorohydrocortisone (fludrocortisone) a day or every other day. 1890 *Jrnl. Chem. Soc.* LVIII. ii. 724 This gas is fluoroform, CHF_3. 1950 N. V. Sidgwick *Chem. Elements* II. vii. 1130 A guinea-pig put into a 50 per cent. air-fluoroform mixture..did not know that the gas was there. 1844 R. Hunt *Res. Light* 106 It has been found that the fluate of soda has the property of quickening the sensibility of brominated papers to a very remarkable extent; and from this quality a new process, which I would distinguish by the name of the Fluorotype, results. 1955 H. & A. Gernsheim *Hist. Photogr.* iii. xi. 124 Fluorotype derives its name from the fluoride of sodium used in preparing the paper...The exposure was only half a minute and the picture was developed with protosulphate of iron.

2. Used as combining form of Fluorescence, as in *fluorometer, etc.

fluorocarbon (flū:ŏrokā·ɹbǫn). *Chem.* [f. *fluoro- + Carbon, after *hydrocarbon*.] Any of a large class of synthetic, chemically stable compounds of carbon and fluorine analogous to the hydrocarbons (see also quot. 1962).

1937 *Jrnl. Amer. Chem. Soc.* LIX. 1407/1 From a reaction mixture of carbon and fluorine..fluorocarbons have been isolated. 1950 J. H. Simons *Fluorine Chem.* I. xii. 402 Substances made from or protected by fluorocarbons are free from decay and insect damage. 1951 *Sci. News Letter* 15 Sept. 165 The relatively new chemicals known as fluorocarbons when used as a cooling spray greatly increase the efficiency of electrical transformers. 1959 *Times* 27 Apr. (Rubber Industry Suppl.) p. vi/7 The newer fluoro-carbon polymers have valuable properties in heat and fluid resistance. 1962 A. J. Rudge *Manuf. Fluorine* vii. 57 In the technical literature there has developed a tendency to use the term 'fluorocarbon' to include compounds containing elements additional to carbon and fluorine, e.g. CF_2Cl_2. This practice is confusing and is to be deprecated. 1964 R. E. Banks *Fluorocarbons* v. 136 Like their aliphatic relatives, aromatic fluorocarbons are colourless. 1966 *New Scientist* 24 Nov. 456/2 Fluorocarbons such as polytetrafluoroethylene (PTFE) and related polymers.

fluorometer (fluŏrǫ·mītəɹ). [f. *fluoro- + -meter.] **1.** A device used to aid in the fluoroscopic location of one object within another.

1897 *N.Y. Tribune* 9 Feb. 3/4 In a series of scientific experiments with the new fluorometer, invented by John Dennis of this city [*sc.* Rochester], Professor A. L. Arey demonstrated definitely the angles and direction of the Roentgen rays with regard to their source. 1898 *Sci. Amer.* 12 Feb. 101/1 It is the province of the 'fluorometer' to enable observers to form an exact and certain diagnosis of the presence of bullets, needles, calculi or any other substance which is comparatively more dense in its fluoroscopic shadow than the subject in which it is contained. 1899 D. Walsh *Röntgen Rays in Med. Work* (ed. 2) 97 In America the 'Dennis Fluorometer' is in vogue.

2. Also **fluorimeter**. Any apparatus or instrument for measuring the intensity of fluorescence or the duration of its afterglow. Hence **fluoro-**, **fluorime·tric** *adjs.*, of, pertaining to, or employing the fluorometer or fluorometry; **fluoro-**, **fluorime·trically** *advs.*; **fluoro·-**, **fluori·metry**, the use of the fluorometer.

1913 *Chem. Abstr.* VII. 2905 The instrument described, termed a fluorometer. 1918 *Ibid.* XII. 450 (*heading*) New fluorometric apparatus for the determination of X-rays. 1920 *Ibid.* XIV. 2584 (*heading*) A fluorimeter. 1920 *Jrnl. Amer. Chem. Soc.* XLII. 1351 The natural term to designate the new method could be either 'fluorometry' or 'fluoremetry' [*sic*]. 1942 *Electronic Engin.* XV. 127 A supersonic cell fluorometer..for the measurement of the rise and decay of luminescence in phosphors. 1953 Bowen & Wokes *Fluorescence of Solutions* viii. 58 The term 'fluorimeter'..is now generally accepted in Great Britain. Alternative terms such as 'fluorometer' and 'fluorophotometer' are in use in America. 1961 *Lancet* 7 Oct. 793/1 The urine was..estimated fluorimetrically. 1962 T. O. Sippel in A. Pirie *Lens Metabolism in Rel. Cataract* 368 The oxidized and reduced forms of diphospho- and triphosphopyridine nucleotide were measured fluorometrically. *Ibid.*, Readings were made on a photomultiplier fluorometer.

fluorophor, fluorophore (flū·ŏrofǭəɹ). Also **fluorphor.** [a. G. *fluorophor* (R. E. Meyer 1897, in *Zeitschr. f. physiol. Chem.* XXIV. 508), f. *fluoro- + -phore.] **a.** An atomic group the presence of which in a molecule causes it to be fluorescent. **b.** A fluorescent substance.

1898 *Jrnl. Chem. Soc.* LXXIV. ii. 105 In order..that the fluorescence may be developed, it is further necessary that the fluorphore be situated between two heavy atomic groups, usually benzene nuclei. 1939 Radley & Grant *Fluorescence Anal. U.V. Light* (ed. 3) ii. xiv. 290 Some fluorescent compounds of simple composition but containing no 'fluorophors' are known. 1962 S. Udenfriend *Fluorescence Assay* ii. 33 Nonfluorescent steroids are converted to fluorophores by dehydration in concentrated sulfuric acids. 1966 *McGraw-Hill Encycl. Sci. & Technol.* VII. 611/1 Other terms sometimes used synonymously with phosphor are luminophor..or fluorphor.

fluorophotometer (flū:ŏrofotǫ·mītəɹ). Also **flu:ophoto·meter**, **flu:orphoto·meter.** [f. *fluoro- + Photometer.] A fluorometer (sense 2) incorporating a photometer. Hence **flu:orophotome·tric** *a.*, of, pertaining to, or employing a fluorophotometer or fluorophotometry; **flu:orophoto·metry**, the use of the fluorophotometer.

1928 *Jrnl. Sci. Instrum.* V. 273 (*heading*) A simple ultraviolet fluorophotometer. 1946 *Nature* 28 Sept. 451/2 The result of the 'fluorimetry' or 'fluorphotometry' of samples will depend on their previous history as regards exposure to light. 1950 *Arch. Ind. Hyg. & Occup. Med.* II. 311 Fluorophotometric analysis for uranium. 1953 Fluorophotometer [see *fluorometer 2]. 1969 G. W. Ewing *Instrum. Meth. Chem. Analysis* (ed. 3) iv. 107 Instruments for the measurement of fluorescence are known as fluorimeters (sometimes fluorometers or fluophotometers).

fluoroscope (flū·ŏroskōup). [f. *fluoro- + -scope.] An apparatus which incorporates a fluorescent screen and is used in conjunction with an X-ray machine to produce a visible image of a body placed between the screen and the source of the rays. Hence **fluorosco·pic** *a.*, formed or done by means of a fluoroscope or fluoroscopy; pertaining to the fluoroscope or fluoroscopy; **fluorosco·pically** *adv.*; **fluoro·scopy**, the use of the fluoroscope; an examination by means of a fluoroscope.

1896 *Lancaster* (Pa.) *New Era* 2 Apr. 2 He [*sc.* Edison] calls his instrument the Fluoroscope. 1896 *Boston Med. & Surg. Jrnl.* 1 Oct. 336/1 A fluoroscopic examination of the heart. *Ibid.* 335/1 The constant motion of the heart and diaphragm interfere with the use of radiography but renders fluoroscopy all the more valuable. 1897 *Chem. News* 24 Sept. 158/1 (*heading*) Photography of the fluoroscopic image. 1908 *Practitioner* Sept. 437 Fluoroscopic examination of the thorax was also negative. 1940 G. L. Clark *Applied X-Rays* (ed. 3) ix. 166 A typical unit for continuous fluoroscopic inspection of..food products on the conveyor belt. 1959 *Medicamundi* V. 4 (*heading*) Fluoroscopically controlled cholangiography with the image intensifier. 1961 A. Taylor *X-Ray Metallogr.* iii. 32 The smaller castings made from light alloys are usually examined for major defects by fluoroscopic inspection. 1970 *Nature* 18 July 296/1 In many hospitals the records will be sufficiently accurate to determine the number of fluoroscopies per patient treated, as well as radiation dose per fluoroscopy.

fluorosis (fluŏrōu·sis). *Path.* [ad. F. *fluorose* (H. Christiani 1927, in *Compt. Rend. Sixième Congrès Chim. Ind.* 164/1), f. Fluor- + -osis.] Poisoning by fluorine or a fluorine compound; any condition caused by such poisoning.

1927 *Chem. Abstr.* XXI. 3404 An investigation which resulted in characterizing a new disease, fluorosis. 1936 *Nature* 16 May 828/2 Fluorine from chemical works or resulting from volcanic activity can get into soil and pasture and cause fluorosis in cattle. 1958 *Spectator* 6 June 737/2 Crippling fluorosis in *natural* fluoride areas at or near the so-called safe concentration has been admitted by some of the top American proponents themselves. 1971 *Daily Tel.* (Colour Suppl.) 28 May 21/4 Cattle nearby have in the past suffered from fluorosis, a condition not unlike rheumatoid arthritis in which the joints seize up.

fluorphor, var. *fluorophor.

flurry, *sb.* Add: **2. c.** A sudden burst of activity (in the stock-market).

1876 *Fur, Fin & Feather* Sept. 129 The prospect of a flurry in stocks..is sure to strip the island of visitors. 1889 *Kansas Times & Star* 2 Nov., The big flurry in the stock market yesterday should cause no particular alarm. 1907 E. S. Field *Six-Cylinder Courtship* 80 A column..sandwiched in between The Latest Armenian Atrocities and the Unprecedented Flurry in Chewing Gum. 1971 *Daily Tel.* 25 Aug. 14 A flurry of speculative activity saw BSA 3 up at 26p.

flurry, *v.* Add: **2.** Also *transf.* and *fig.*

1917 W. Owen *Let.* 13 Dec. (1967) 515, I was flurrying round like any Mrs. Smith de Smith when 'Company' is expected. 1920 D. Lindsay *Voy. Arcturus* xii. 145 The freezing wind, flurrying across the desert, drove the fine particles of sand painfully against their faces.

flus(e, var. *floose.

flush, *sb.*[2] Add: **8. b.** Of a lavatory, its plumbing, etc., as *flush toilet*. Also occas. *ellipt.* Cf. Flush *v.*[2] 3, *flush-box, -tank*.

1908 *Sears, Roebuck Catal.* 604/1 The closet..is furnished with a positive flush valve. *Ibid.* 605/3 China push button flush in front of tank. 1950 *Ann. Rep. Smithsonian Inst. 1949* 220 Flush toilets, bathing and laundry, street cleaning, and fire protection require an average of about 40 to 75 gallons per day per capita. 1961 L. Mumford *City in Hist.* viii. 216 Flush toilets both wasted potential fertilizer and polluted the streams. 1965 G. McInnes *Road to Gundagai* iii. 38 We were lucky to have arrived after the installation of 'the flush'. 1967 *Guardian* 28 Apr. 4/6 A notable change from chemical to flush lavatories, both inside and outside caravans. 1967 *Gloss. Sanitation Terms* (*B.S.I.*) 60 *Flush valve* (flushing valve), a valve, for controlling the flushing of a W.C. pan.

flush, *a.*[1] Add: **5. b.** Also *absol.*

1799 A. Young *Agric. Lincoln.* 30 Two inch six-panel bead and flush. 1842 Gwilt *Archit.* § 2131 The panels are surrounded with a bead formed on the edge of the framing, and the work is called bead and flush. 1885 *Spons' Mech. Own Bk.* 379 When it [*sc.* a piece of copebead stuff] lies close in the cheek, and also close at the mitres, it receives a little glue, and is nailed on with ⅞-in. fine brads, 3 or 4 to each. These are punched below the flush. 1937 [see *bead *sb.* 6 a].

c. (Additional examples.)

1933 *Jane's Fighting Ships* 527 Note raised C.T. [conning tower] and flush deck forward. 1940 *Illustr. London News* CXCVII. 359 (*caption*) The structure of an American 'flush-deck' destroyer, typical of the 50 transferred to Great Britain.

e. *Typogr.* Not indented or protruding.

1900 H. Hart *Cent. Typogr.* 139/2 The ascending letters are cut flush at the top of the body. 1961 T. Landau *Encycl. Librarianship* (ed. 2) 144/1 *Flush*, 'even with', *i.e.* the left margin.

7. flush-binder, one who cuts the pages of a book flush with the cover; so *flush hand*; **flush-decker**, a flush-decked ship; **flush ring** (see quot. 1967[1]).

1881 *Instr. Census Clerks* (1885) Index, Flush binder. 1902 *Daily Chron.* 28 Apr. 11/4 Flush Binders wanted, used to piece work. 1921 *Dict. Occup. Terms* (1927) § 532 *Binder, flush*; glues up books and affixes boards;..cuts book flush with edge of cover at top, bottom and fore-edge, by hand or power-operated guillotine. 1937 *Jane's Fighting Ships* 505 *United States*..boats by various yards (Flush Deckers). 1957 *Ibid.* 285 In 1953–54 were modernised by Canada for submarine warfare and converted to flush-deckers. 1904 *Daily Chron.* 12 Jan. 10/6 Flush Hands wanted. 1961 Webster, Flush ring. 1967 *Gloss. Terms Builders' Hardware* (*B.S.I.*) iv. 13 *Flush ring*, a flush handle consisting of a metal ring lying in a circular recess in a mounting plate. *Ibid.* iii. 8 *Flush ring catch* (showcase catch), a flush fitting catch having a spring bolt withdrawn by lifting a ring shaped handle from a recess in the body of the catch.

flush, *v.*[1] Add: **2. b.** *transf.* and *fig.* To reveal; to bring into the open; to drive *out*.

1950 [see *earth *sb.*[1] 4]. 1958 *Spectator* 1 Aug. 176/1 After being flushed from his rural retreat in England by an unfortunate affair with a literary horsewoman, he [etc.]. 1971 *Scope* (S. Afr.) 19 Mar. 22/1 When we flushed them out of the old city of Jerusalem..we really knew fear.

flush, *v.*[2] Add: **11.** *trans.* To fatten *up* (sheep); to stimulate (ewes) with generous diet at the breeding season.

1764 *Museum Rust.* III. xxxiii. 148, I had a fine piece of turneps, with which I intended to flush up five score sheep. 1886 C. Scott *Sheep-farming* 74 Nor is this the only evil of 'flushing' the ewes when they are put to the rams. From actual test we are convinced that ewes which have been flushed one year are never so prolific the next. 1923 *Discovery* Sept. 243/2 Many flockmasters..practised the methods of 'flushing' or artificially stimulating their ewes by means of an extra supply of special food at the approach of the 'tupping'..season.

flush, *v.*[3] Add: **1.** Also with *up*.

1883 *Specif. Alnwick & Cornhill Rlwy.* 3 The whole of the work is to be flushed up with mortar or cement.

flute, *sb.*[1] **3. a.** Delete †*Obs.* and add later examples.

1924 *Glasgow Herald* 6 Oct. 8 'Sect', or German champagne..is to be drunk only from French-fashioned flutes or tall glasses. 1959 *House & Garden* Dec./Jan. 36/3 Deep wine goblets or tall champagne flutes. 1964 *Observer* 14 June 33/5 A champagne 'flute', tall enough for a quarter or a fifth of a bottle.

5. (Later examples in *Geol.*)

1935 *Jrnl. Geol.* XLIII. 729 Swiftly moving sand or silt-laden water currents sometimes abrade grooves, elongated in the direction of flow, in stream boulders and in the bedrock of the stream channel...In this article the grooves are termed 'flutes', and the process of their formation is called 'stream fluting'. 1963 D. W. & E. E. Humphries tr. *Termier's Erosion & Sedimentation* 408 The *lapies* of karst scenery where small ridges or flutes of limestone are dissected and isolated by a surficial flow of water.

flute, *v.* Add: **3. b.** *intr.* To hang or jut *out* in flutings.

1896 *Sun* 11 Dec. 1/7 Arranged with the usual side-pieces, which flute out gracefully to the lower edge. 1908 M. & J. Findlater *Crossriggs* xxxi, Her skirts fluted out about her like the ruffled petals of a flower.

‖ **flûte-à-bec** (flütabεk). [Fr., lit. 'flute with beak'.] An old form of flute, also called English flute, having a mouthpiece with a fipple; now, a recorder.

1797 *Encycl. Brit.* VII. 310/1 This [*sc.* the flute] is a very ancient instrument. It was at first called the flute *à bec*. 1825 [see Flauto]. 1876 Stainer & Barrett *Dict. Mus. Terms* 171 The *flûte à bec* was used so commonly in England that it was called on the continent *flûte d'Angleterre*. 1879 *Encycl. Brit.* IX. 351/1 The flageolet is a smaller variety of the old *flute-à-bec*. 1930 *Daily Express* 8 Sept. 6/3 The half-forgotten flute-a-becs, bassethorns..and oboes d'amore. 1959 *Collins' Music Encycl.* 250/1 *Flûte à bec, flûte douce*, recorder. 1959 A. Rowland-Jones *Recorder Technique* iii. 30 The wind-way of a recorder is formed by the insertion of an incised plug..into the blown end of the tube, which is generally made beak-shaped (whence 'Flûte à bec') to fit between the lips.

‖ **flûte d'amour** (flütdamür). [Fr., lit. 'flute of love'.] An old form of flute with a pitch a third below that of the flute in D.

1876 Stainer & Barrett *Dict. Mus. Terms* 173/2 *Flûte d'amour*, a low-toned flute, an A flute, sounding a minor third below the notes actually written. It is now obsolete. 1961 J. A. Macgillivray in A. Baines *Mus. Instruments* x. 249 The..*flûte d'amour* in A seems to have been used mainly for variety in flute recitals.

flutey, var. Fluty *a.*

1929 E. Bowen *Joining Charles* 7 The blackbirds with a wild flutey note. 1938 E. Waugh *Scoop* iii. ii. 274 Mr. Salter's voice sounded curiously flutey and querulous.

flutter, *sb.* Add: **1. c.** *Med.* Abnormal contractions of a muscular organ that are very rapid but regular.

1910 *Heart* II. 177 (*heading*) Auricular flutter & fibrillation. *Ibid.* 182 We cannot say how long the rapid auricular flutter lasted at that time. 1920 T. Lewis *Mechanism & Graph. Repres. of Heart Beat* xxii. 263 We may divide the simple paroxysm from the attack of flutter arbitrarily by defining the latter as a new rhythm whose rate surpasses 200 and may reach 350 per minute. 1936 *Jrnl. Amer. Med. Assoc.* 21 Mar. 992/1 (*heading*) Diaphragmatic flutter with symptoms of angina pectoris. 1957 *Encycl. Brit.* XI. 311/2 Auricular flutter and auricular fibrillation are close cousins. 1966 Dunlop & Alstead *Textbk. Med. Treatm.* (ed. 10) 625 If flutter recurs, a second course of digitalis may succeed in establishing normal rhythm. 1969 Crofton & Douglas *Respiratory Dis.* xxxvi. 637/1 The rapid contraction of the diaphragm (up to 100/min or more) known as diaphragmatic tic may be a feature of encephalitis but more commonly is a hysterical phenomenon...When the rate is faster the term diaphragmatic flutter is used.

d. Abnormal oscillation of a wing or other part of an aircraft.

1911 H. T. Wright in Grahame-White & Harper *Aeroplane* 258 Another experiment was tried recently to illustrate 'Propeller flutter'. 1916 H. Barber *Aeroplane Speaks* 123 Propeller 'flutter' or vibration, may be due to faulty pitch angle, balance, camber, or surface area. 1927 *Aeronaut. Research Committee Rep. & Mem. 1041* (1925) 1 Since its commencement the investigation of flutter and vibration in aeroplanes has proved to be a much more complex problem than was at first realised. 1930 *Flight* 3 Jan. 28/1 It is certain that wing flutter would have occurred. 1933 *Jrnl. R. Aeronaut. Soc.* XXXVII. 497 It speaks volumes for the reliability and safety of the wooden airscrew that it is not called upon to undergo any endurance test, and only very occasionally a spinning tower test when flutter is suspected. 1937 *Ibid.* XLI. 233 Later appeared a statement that the design criterion for cantilever wings, which determined flutter speed and critical reversal speed, was the torque stiffness. 1960 *Aeroplane* XCVIII. 192/1 Blades can self-excite in much the same way as aircraft wings flutter. 'Choking flutter' can occur at the high axial velocities obtaining at the back end of the compressor.

e. Rapid tonguing in playing wind instruments. Usu. *attrib.* and *Comb.*, as *flutter-tongue, -tongued* ppl. adj., *-tonguing* vbl. sb.

1926 Whiteman & McBride *Jazz* iv. 110 A flutter-tongued, drunken whoop of an introduction that had the audience rocking. *Ibid.* ix. 201 The flutter tongue in the brasses is rather like a covey of quail flying out from ambush. *Ibid.* 205 Schoenberg is also the father of the flutter on the trombone—that is, very rapid tonguing on the same note. 1944 W. Apel *Harvard Dict. Mus.* (1960) 754/1 A special type of tonguing, called *Flatterzunge* or flutter-tonguing, has been introduced by R. Strauss. It calls for a rolling movement of the tongue, as if pronouncing d-r-r-r. 1962 N. Del Mar *R. Strauss* v. 159 The whistling of the wind is given by harp glissandi and flutes fluttertonguing with the addition of a Wind Machine. 1967 *Spectator* 22 Dec. 787/3 His slow glissandos, his 'fluttertongue' effects (familiar enough as a flautist's device).

f. A rapid fluctuation in the pitch or loudness of a sound (in sound reproduction usu. the former).

1931 A. Nadell *Projecting Sound Pict.* ii. 43 Any condition that allows the film to be pulled past the sound aperture with the least trace of intermittent motion will cause flutter. 1933 C. W. Glover *Pract. Acoustics for Constructor* x. 126 The reflection of sound between parallel walls gives rise to a characteristic phenomenon known as flutter. 1946 *Electronic Engin.* XVIII. 331 Records of a combined playing of a violin and piano—the most exacting test for 'wow' and 'flutter'. 1948 P. M. Morse *Vibration & Sound* (ed. 2) vi. 262 Proper auditorium design is aimed, in part, at the elimination of flutter echoes. 1955 *Gloss. Acoust. Terms* (*B.S.I.*) 36 *Flutter echo*, a rapid multiple echo of even rate. 1962 A. Nisbett *Technique Sound Studio* iv. 92 Once wow or flutter have actually been recorded on to a tape or disc, I know of no means of curing it. 1967 *Reader's Digest Pocket Compan.* 74 An annoying flutter, caused by bad acoustics in the..concert hall, was covered up by adding reverberation.

4. (Further examples.) Now usu. used of speculation or betting on a small scale.

1900 J. Robinson *Life Time S. Afr.* 366 The opportunities of the share market were open to the humblest operator. Men and youths, women and maidens, all could have a 'flutter'. 1930 *Cambridge Daily News* 24 Sept. 7/6 The shares ought to be worth a mild flutter at round 8s. 6d. 1967 *Listener* 13 Apr. 479/1 Money-raising efforts..like selling charity Christmas cards, a rake-off from football pools, and other flutters of various kinds.

5. flutter-mill *U.S.*, a mill worked by a flutter-wheel; **flutter-wheel** (earlier U.S. example).

1866 C. H. Smith *Bill Arp* 85 The Choctaw children built their flutter mills. 1938 M. K. Rawlings *Yearling* i. 5 His knife was snug in his pocket;..he had planned as long ago as Christmas, to make himself a flutter-mill. 1817 *Index of Patents* (1874) 559 Flutter-wheels, letting water on.

flutter, *v.* Add: **6. b.** *to flutter the dove-cot(e)s*: to alarm, or cause excited discussion among, quiet people. Cf. quot. 1664 in Dict.

1853 Lytton *My Novel* III. ix. xiii. 69 Nor did the great Roman general more nervously 'flutter the dove-cots in Corioli', than did the advance of the supposed X.Y. agitate the bosoms of Lord Spendquick and his sympathising friends. 1864 [see Dict.]. 1893 [see Dove-cot, -cote]. 1940 C. P. Snow *Strangers & Brothers* ii. 24 'How are they taking it?' said George. 'It's fluttered the dove-cotes.'

flutteration. Add later U.S. example.

1896 J. C. Harris *Sister Jane* 220, I heard a big flutteration at the top, and out came a pair of wood-peckers.

fluvio-. Add: **flu:vio-gla·cial** *a.*, pertaining to or produced by the action of streams which have their source in glacial ice, or the combined action of rivers and glaciers; **flu:vio-lacu·strine** *a.*, pertaining to or produced by the action of both rivers and lakes; **fluviology** (flūviǫ·lŏdʒi), potamology; the facts and conditions relating to a river or river-system.

1894 *Geol. Mag.* Jan. 30 The younger fluvio-glacial gravel deposits of the Limmat valley. 1970 R. J. Small *Study of Landforms* xi. 389 A number of different fluvio-glacial features are found at or very close to the margins of an ice-sheet. The best known of these are karnes, karne-terraces and eskers. 1864 *Q. Jrnl. Geol. Soc. London* XX. 383 The fluvio-lacustrine series, consisting of deposits from lakes fed by mountain-rivers. 1963 D. W. & E. E. Humphries tr. *Termier's Erosion & Sedimentation* ix. 180 It [*sc.* tangue] is a fluvio-lacustrine silt, which has been reworked by the sea. 1904 *Amer. Geol.* Jan. 43 The modern science of fluviology. 1924 *Glasgow Herald* 13 Dec. 8 An unexpected addition to the fluviology of Glasgow.

flux, *sb.* Add: **9. a.** Also used with reference to other forms of matter and energy that can be regarded as flowing, such as radiant energy, particles, etc. (Later examples.)

1911 *Encycl. Brit.* XXI. 525/2 Across this surface there will pass a definite amount of radiant energy, in other words a definite total luminous flux. 1947 *Science* 9 May 491/1 A graphite uranium pile, a 'hot' laboratory.., still another pile with 100 times the neutron flux of the first, [etc.]. 1957 *New Scientist* 23 May 32/2 The thermal neutron flux in such a reactor will be on the average about 2×10^{13} neutrons per square centimetre per second. 1963 R. W. Ditchburn *Light* (ed. 2) x. 397 A flux of one watt of radiation of this wavelength constitutes a luminous flux of 692 lumens. 1964 C. J. Maiden in LeGalley & McKee *Space Explor.* ix. 242 Recently spacecraft have become available to make direct measurements of the flux of micrometeorites in space. 1967 Condon & Odishaw *Handbk. Physics* (ed. 2) ix. 280 A definite flux of $(1{\cdot}5 \pm 0{\cdot}8) \times 10^{-4}$ per cm^2 per sec for γ rays of a few hundred MeV has been reported..from the direction of the constellation Cygnus. 1971 *Nature* 7 May 11/1 The flux of hydrogen atoms in the solar wind.

b. *Electricity* and *Magnetism*. (The number of) lines of magnetic induction (*magnetic flux*) or electric displacement (*electric flux*); the quantity of flux through any surface is equal to the integral of the normal component of the induction or displacement over the surface.

1873 J. C. Maxwell *Treat. Electr. & Magn.* I. 11 Electric and magnetic induction, and electric currents, belong to the second class, being defined with reference to areas. When we wish to indicate this fact, we shall refer to them as Fluxes. 1898 J. A. Fleming *Magnets & Electr. Curr.* iii. 65 If..a disc of iron is placed in a uniform field of magnetic force, the flux concentrates itself in the iron. 1911 W. F. Magie *Princ. Physics* xix. 452 The theorems concerning electric flux are similar to those concerning magnetic flux. 1933 E. Mallett *Vectors for Engineers* iv. 52 The e.m.f. produced by the alternating flux in the primary coil. 1957 B. I. & B. Bleaney *Electr. & Magn.* vi. 147 If the two coils are closely wound..then the ratio of the two fluxes N_1 and N_2 will just be equal to the ratio of the number of turns n_1, n_2 on the two coils. 1969 L. Young *Syst. Units Electr. & Magn.* iv. 42 The flux 'lines' spread out over space to produce a certain electric flux density or electric displacement.

13. *flux-linkage, -turn*; **flux density**, the quantity of flux passing through unit area in a plane normal to the direction of the flux; *esp.* magnetic induction; **flux(-)gate, fluxgate** (**magnetometer**), a kind of magnetometer used esp. in aerial surveys which consists essentially of one or more soft magnetic cores each surrounded by primary and secondary windings, the signal produced in the latter representing in phase and magnitude the direction and magnitude of the external magnetic field; **flux(-)gate compass**, an aeronautical compass incorporating a gyroscopically controlled flux-gate; **flux line**, (*a*) one of the lines conceived of as representing by their direction and density the direction and strength of either magnetic induction or electric displacement; (*b*) (see quot. 1962); **fluxmeter, flux meter** [ad. F. *fluxmètre* (M. E. Grassot 1904, in *Jrnl. de Physique* 4me Sér. III. 696)], an instrument for measuring changes in magnetic flux.

1898 J. A. Fleming *Magnets & Electr. Curr.* iii. 63 When an iron circuit is strongly magnetised it may have across its section a magnetic flux density as great as..20,000 lines or units of induction per square centimetre. 1934 *Discovery* Oct. 301/2 The 2-in. speech coil attached to the 11-in. cone works in a flux density of 11,500. 1967 Condon & Odishaw *Handbk. Physics* (ed. 2) vi. 48 The luminous intensity of a source of light in any given direction is the solid angular luminous flux density in the direction in question. 1946 M. Davidson *Gyroscope* III. iv. 218 The gyro used to stabilise the Flux-gate compass unit..consists of a capacitor type split-phase four-pole induction motor. 1947 *Canadian Jrnl. Res.* A. XXV. 125 (*heading*) Flux-gate magnetometers. 1962

F. I. Ordway et al. *Basic Astronautics* iv. 156 A later satellite, Explorer 6, carried search coil and flux gate magnetometers. **1967** *New Scientist* 27 Apr. 217/3 The fluxgate is an electronic device that normally measures variations in field strength. **1898** J. A. Fleming *Magnets & Electr. Curr.* iii. 62 Where the flux density is large the flux lines are closely packed. **1925** Hodkin & Cousen *Textbk. Glass Technol.* xxiii. 306 Fireclay blocks, intended for use as 'flux-line' blocks, should be made of the same clay mixture as pots. **1962** *Gloss. Terms Glass Industry (B.S.I.)* 19 *Fluxline*, 1. The level of the molten glass surface in a tank. 2. The boundary line between unmelted batch and clear glass in a tank. **1933** E. Mallett *Vectors for Electrical Engineers* iv. 52 The flux linkages..through the primary coil are made up of two parts,..due to the current in the primary, and..due to the current in the secondary. **1904** *Sci. Abstr.* B. VII. 741 (*heading*) Grassot fluxmeter. **1933** *Proc. R. Soc.* CXXXIX. 619 The field was explored by means of a large search coil..connected to a fluxmeter. **1960** Jerrard & McNeill *Theoretic. & Exper. Physics* xv. 461 The Grassot fluxmeter..is a robust form of moving-coil ballistic galvanometer used for measuring magnetic flux. **1932** E. B. Moullin *Princ. Electromagn.* iii. 122 The scale of the flux meter can be engraved to read directly in flux turns.

fly, *sb.*[1] Add: **1. e.** *a fly in the ointment* [after Eccl. x. 1]: some small or trifling circumstance which spoils the enjoyment of a thing, or detracts from its agreeableness; (*to*) *drink with the flies* (Austral. and N.Z. colloq.): see quot. 1943.

1833 Lamb *Elia* 2nd Ser. (1840) 3 A Poor Relation—is the most irrelevant thing in nature,—..a lion in your path,—a frog in your chamber,—a fly in your ointment. **1914** *Scotsman* 10 Sept. 9/1 The present situation is not without its 'fly in the ointment' for those motorists who have patriotically lent the assistance of their cars to the military authorities. **1928** *Daily Express* 11 May 10/7 The insurance of school fees has now become so general that it is as well to point out to parents that there is a rather large fly in the ointment. **1936** A. Huxley *Olive Tree* 5 There is only one fly in the ointment offered by commercial propagandists; they want your money. **1940** *Bulletin* (Sydney) 3 Jan. 35/3 He never needs to buy whisky, either, though a natural antipathy to drinking with the flies saves the officer's supply. **1943** Baker *Dict. Austral. Slang* (ed. 2) 28 *Drink with the flies*: a drink consumed without the company of others. Also, to drink alone.

f. *there are no flies on*: (*a*) there is no lack of activity or astuteness in (a person); there is no fault to be found with, there are no blemishes in; (*b*) there is nothing dishonest or 'shady' about (a transaction). So *to have no flies on*, etc. *slang* (orig. *Austral.* or *U.S.*).

The earliest examples indicate that the phrase was prob. orig. applied to cattle that are so active that flies do not settle on them.

1848 H. W. Haygarth *Recoll. Bush Life Austral.* ix. 101 'It's lucky we got them,' said Amos; 'there were "no flies" about that black bull.' *Note.* This expression is very common in Australia...Anything particularly good is said by the class of men we are here describing to have 'no flies' about it. **1868** J. Diprose *St. Clement Danes* 99 To this celebrated pugilist [*sc.* Deaf Burke] is attributed the old story of the 'flies in the gin-and-water', and hence the term 'no flies' became prevalent. [The story follows.] **1888** *Missouri Republ.* 24 Feb. (Farmer *Amer.*), Persons who are capable of descending to New York and Boston English are fully justified in saying that there are no flies on St. Louis or the St. Louis delegation either. **1888** *Detroit Free Press* 25 Aug. (Farmer & Henley), There ain't no flies on him, signifies, that he is not quiet long enough for moss to grow on his heels, that he is wide awake. **1893** Farmer & Henley *Slang* s.v. *Flies*, *There are no flies on me, on him*, etc.,..'I am dealing honestly with you'; 'he is genuine, and is not humbugging'. In America, the expression is used of (1) a man of quick parts..(2) a person of superior breeding or descent. **1897** R. M. Stuart *Simpkinsville* i. 18 They wasn't no flies on his shape, nor his rig, nor his manners neither. **1898** *Strand Mag.* May 516, I kin put you in the way of making your pile, I kin. This is a *bona-fide* offer. No flies on *my* business. **1900** G. Bonner *Hard-Pan* iii. 83 There are no flies on your little sister. **1928** Galsworthy *Swan Song* II. vi. 156 There are no flies on your uncle. **1948** C. Day Lewis *Otterbury Incident* iv. 43 There are no flies on Rickie. I'm pretty sure he'd spotted how his own shoes got dirty. **1961** *Observer* 23 Apr. 18/2 There are no flies on Benaud. If England start bowling their overs slowly, no one will have to draw his attention to it.

g. *like flies*: in large numbers or huge quantities.

[**1595** Shakes. *3 Hen. VI* II. vi. 8 The common people swarme like summer flies.] **1934** A. Hyder *Black-Girl, White-Lady* xxi. 309 'Yo' kills niggers?' 'Like flies,' Charley assured her. **1958** R. Liddell *Morea* III. ii. 245 Now, at the end of 1955, people were marrying like flies; it is unlucky to marry in a leap year.

3. b. *ellipt.* The tsetse fly. *S. Afr.*

1856 C. J. Andersson *Lake Ngami* 488 The natives, who are well acquainted with localities frequented by the fly. **1864** T. Baines *Explor. S.W. Afr.* [424] They have lost all their oxen by the fly. **1877** —— *Gold Regions S.E. Afr.* 57 Keeping a sharp look out upon our own and each others horses to see that no 'fly' settled on them.

10. a. *fly-bottle* [tr. G. *fliegenglas*], *-country* (earlier example), *-screen* (examples).

1953 G. E. M. Anscombe tr. *Wittgenstein's Philos. Investig.* § 309 What is your aim in philosophy?—To shew the fly the way out of the fly-bottle. **1967** *Philos. Q.* XVII. 363, I would like an illustration of a flybottle, which people describe variously. **1969** *Proc. Arist. Soc.* Suppl. Vol. XLIII. 108 If the exit from the fly-bottle is an entry into too restrictive a cupboard, it seems necessary to find a way out of the cupboard as well. **1864** T. Baines *Explor. S.W. Afr.* iv. 68 Treachery of guides purposely sent by the chiefs to lead him into the fly country. **1903** *Work* 1 Aug. 412/3 Inside this lining will be laid a thin coating of tinfoil, on which will be placed what will look like a strong netting or fly-screen, except that the wires of which it is composed will be carefully joined and highly polished. **1952** 'N. Shute' *Far Country* 18 She went and rang the dinner bell outside the flyscreen door. **1971** *Sunday Times* (Johannesburg) (Business Sect.) 28 Mar. 5/2 Adjustable aluminium louvres..may also be fitted with flyscreens.

b. *fly-caster* (examples); *fly-proof* adj.

1897 *Outing* (U.S.) XXX. 221/2 It is not every fly-caster who can properly manipulate a live or dead minnow, a crawfish, or a spoon. **1926** *Daily Colonist* (Victoria, B.C.) 9 July 16/1 Earl C. McFarland..nationally known fly-caster and sportsman, died..today. **1817** *Niles' Reg.* XII. 284/1, I found several of the other kinds of wheat..growing on some farms in the same field with the fly-proof or Lawler wheat. **1944** *Living off Land* ii. 34 Keep the smokehouse fly-proof by covering the top.

11. fly-blister (example); **fly-brush** (earlier U.S. example); **fly-dope** *N. Amer.*, a liniment used as a protection against flies; **fly-eater,** an eater of flies, *spec.* any of several small Australian fly-catchers of the genus *Gerygone*; **fly-fan,** a fly-flick, or a motor-driven fan for driving away flies; **fly-flick,** an instrument for killing or driving away flies; **fly-fungus,** (*a*) the fly-agaric, *Amanita muscaria*; (*b*) the house-fly fungus, *Entomophthora muscæ*; **fly-mould** = *fly-fungus* (*b*); **fly-paper** (later examples); (*b*) *Flypaper Act* (slang), the Prevention of Crimes Act, 1909; so *to be on the flypaper*, to be subject to this Act, to be a criminal known to the police; **fly-speck** (U.S. and *fig.* examples); (*b*) a plant disease, esp. of apples and pears, caused by the fungus *Leptothyrium pomi*; also *attrib.*; **fly-specked** *a.* (earlier U.S. example); **fly-strike,** a skin disease of sheep, caused by the maggots of blow-flies, esp. those of the genera *Lucilia* and *Calliphora*; hence **fly-struck** *a.*; **fly-swat, -swatter** = *swatter; **fly-swish** = *fly-whisk*; **fly-weight** *Boxing*, a weight of 8 stone or less; a pugilist of such a weight; also *transf.*; **fly-whisk** (later examples); **fly-wire,** screening to exclude flies, esp. *fly-wire door*.

1869 'Mark Twain' *Innoc. Abr.* xvii. 169 A Venus with a fly-blister on her breast. **1841** G. Catlin *Indians* I. 113 The Indian's fly brush was made of the buffalo tail. **1897** *Outing* (U.S.) XXX. 377/1 The fly-dope, or preventive against the attacks of insects. **1903** S. E. White *Forest* ix. 106 Each and every man..heralds the particular merits of his own fly-dope. **1895** W. O. Legge in *Rep. Australasian Assoc. Adv. Sci.* VI. 447 [The habits and habitats of the genus as] applied to *Gerygone* suggested the term Fly-eater, as distinguished from Fly-catcher. **1906** *Westm. Gaz.* 30 May 12/1 Starlings [are] great fly-eaters. **1930** W. de la Mare *Desert Islands* 212 Owl, kingfisher, robin, fly-eater. **1958** P. C. Morrison *Austral. Bird Bk.* 143 *Brown warbler*, flyeater, G[erygone] richmondii. **1891** A. H. Keane tr. *Junker's Trav. Afr.* II. viii. 276 His hand also grasped, not a warlike assegai, but a peaceful fly-fan! *a* **1911** D. G. Phillips *Susan Lenox* (1917) I. xvii. 300 They went into a large restaurant with fly fans speeding. **1914** D. Fraser *Winning Prim. People* iv. 45 He carried nothing in his hand but a fly-flick. **1822** *Mem. Wernerian Soc. Edinb.* IV. 343 The plant commonly known by the name of the fly-fungus (from its property of destroying flies when steeped in milk), has made some noise of late on the Continent. **1910** *Encycl. Brit.* I. 780/1 *Amanita muscaria*, the fly fungus, formerly known as *Agaricus muscarius*. **1912** L. O. Howard *House Fly* ii. 64 With the common house fly fungus (*Empusa muscae*) a slight change in the amount of atmospheric moisture is sufficient to bring about germination. **1952** C. J. Alexopoulos *Introd. Mycol.* vii. 177 The most familiar of the Entomophthorales is *Entomophthora muscae*, commonly called the fly fungus. **1908** *Westm. Gaz.* 8 Aug. 16/3 Flies..may be seen..lying about in odd corners covered with a mouldy grey fur, which is called fly-mould. **1910** F. Martyn *Burglar in Baulk* 8 *On the flypaper*, subject to the Crimes Prevention Act. **1933** C. E. Leach *On Top of Underworld* x. 139 *Flypaper Act*..Prevention of Crimes Act. **1936** J. Curtis *Gilt Kid* ii. 23 He gets nicked for suspect and being on the flypaper, he gets a stretch in the Ville. **1941** *Coast to Coast 1941* 76 A spiral fly paper hung from the ceiling. **1971** *Islander* (Victoria, B.C.) 22 Aug. 3/1 From the..ceiling hung a few well encrusted fly papers. **1865** Mrs. Stowe *House & Home Papers'* 290, I would shut my eyes on fly-specks, and open them on the beauties of Nature. **1907** *Westm. Gaz.* 15 July 2/1 Even your Caesar Borgia is but a fly-speck in the infinite. **1909** R. A. Wason *Happy Hawkins* 113 Lookin' like a fly-speck on a new tablecloth. **1939** H. Wormald *Dis. Fruits & Hops* v. 103 Frequently associated with Sooty Blotch, but sometimes occurring alone, are groups of black, circular dots which from their size and appearance are known as Fly Specks. **1943** Baker *Dict. Austral. Slang* (ed. 3) 75 *The speck*, Tasmania. Also, 'the Fly Speck'. **1952** E. Ramsden tr. *Gram & Weber's Plant Dis.* 126/2 The fly speck fungus produces groups of round black dots on apples and sometimes on pears and plums. **1961** *Trans. Brit. Mycol. Soc.* XLIV. 304 The Micropeltaceae or 'fly-speck' fungi are treated by Prof. Batista as one of the seven families of the Microthyriales. **1872** 'Mark Twain' *Roughing It* 43 Only one cruet left..a stopperless, fly-specked, broken-necked thing. [**1890** J. H. Steel *Treat. Dis. Sheep* x. 325 Even a dog is able to tell when sheep have been struck by the fly. **1937** A. Fraser *Sheep Farming* xv. 143 When the weather favours strike..the maggot fly may cause hundreds of pounds worth of damage.] **1950** H. G. Belschner *Sheep Managem.* vi. 178 The treatment of fly-strike in sheep should be undertaken as soon as possible. *Ibid.*, The only equipment necessary to dress a fly-struck sheep is a pair of shears, a suitable dressing, and a small horsehair brush. **1965** C. R. W. Spedding *Sheep Prod.* viii. 210 When the eggs hatch, the maggots burrow under the skin and this is described as 'fly strike'. **1922** C. R. Edmonds *Dis. Anim. S. Afr.* II. 310 This trouble [*sc.* sheep blowfly] is usually designated as 'fly-blown sheep' or 'fly-struck sheep'. **1934** *Bulletin* (Sydney) 18 Apr. 28/1 A fresh paddock will work wonders even if the flock in it includes fly-struck sheep. **1937** A. N. Lyons *Tom, Dick & Harriet* xxvi. 225 He carried a fly-swat..and flourished this..in a menacing manner. **1917** *Blackw. Mag.* May 753/1 Fly-papers and fly-swatters proved illusive. **1937** *Sunday Express* 30 May 17/3 Then he took to 'fly-swatters' and things for protecting food from the fly menace. **1922** *Blackw. Mag.* Apr. 462/1 An old fly-swish made of palm-leaves. **1927** *World Dominion* Oct. 318 His inevitable fly-swish under his arm. **1947** J. Stevenson-Hamilton *Wild Life S. Afr.* iv. 42 The trunk is indeed everything to the elephant...Furnished with a good large bunch of long grass it makes a better fly-swish than do most animals' own tails. **1911** *Boxing* 9 Sept. 454/3 Kleber reigns over the fly-wt championship. **1913** J. G. B. Lynch *Comp. Amat. Boxer* 234 Championships. Standard Weights. Fly Weight, 8 stone and under. **1914** —— *Promin. Pugilists* 158 He [*sc.* Jimmy Wilde] is far below the fly-weight limit. **1944** Auden *For Time Being* (1945) 12 A fly-weight hermit in a dream Of gardens. **1955** *Times* 12 May 4/3 They fell behind again when Gnr. Lloyd was beaten in one round by the opposing flyweight. **1930** E. Waugh *Labels* iii. 65 He lunged out with a spanner and rapped an old man on the knuckles who was trying to sell us a fly-whisk. **1953** Dylan Thomas *Under Milk Wood* (1954) 52 Mrs Ogmore-Pritchard belches in a teeny hanky and chases the sunlight with a flywhisk. **1963** W. Soyinka *Lion & Jewel* 25 Fly-whisk and whisky. **1952** 'N. Shute' *Far Country* v. 140 The house itself had deep verandas on two sides and fly-wire doors. **1969** *Southerly* XXIX. 305, I..pushed open the fly-wire door then walked into the kitchen.

b. fly-agaric (earlier example).

1788 J. Bolton *Hist. Fung.* I. 27 Fly Agaric... The root is large, and bulb-shaped.

fly, *sb.*[2] Add: **1. d.** Phr. *to give* (it) *a fly*, to make an attempt, to have a go. Also, *to have a fly at* (1941 Baker, *Dict. Austral. Slang* 29). *Austral.*

1919 Downing *Digger Dial.* 24 *To give it a fly*, to make an attempt. **1926** K. S. Prichard *Working Bullocks* xiv. 129 'They say you're going to race him at Blue Flowers, Red?' 'Might give him a fly,' Red admitted. **1931** V. Palmer *Separate Lives* 114 Righto! I'm willing to give it a fly if that's settled. **1934** T. Wood *Cobbers* ii. 19 They come in for miles..to give it [*sc.* betting at trots] a fly.

e. *Football.* = **fly-half.*

1921 E. H. D. Sewell *Rugby Football* 195 L. A. Phillips was a combination half, who was really equally at home as scrum or fly.

2. b. (Earlier examples.) Also in *Football* and *Hockey*.

1856 *Spirit of Times* 8 Nov. 165/1 Several fine catches..made on the fly. **1868** [see *catch *sb.*[1] 1 c]. **1887** Dr. Irvine in G. A. Hutchison *Football* iii. 91 If he throws forward, his opponent may make his mark, if he manages to catch the ball on the fly. **1960** T. McLean *Kings of Rugby* xiii. 173 One such run..led to Dawson putting a hefty boot to the ball on the fly. **1961** J. S. Salak *Dict. Amer. Sports* 226 *Hitting on the fly* (field hockey), hitting an oncoming ball without first stopping or fielding it.

4. a. *spec.* (freq. *pl.*). The piece of cloth that hides the fastening at the front of a pair of trousers; also, the fastening itself.

1941 I. Baird *He rides the Sky* 234 A pair of tennis shorts with zipper fly. **1942** E. Paul *Narrow St.* i. 6 The professor..turned toward the *pissoir*, unbuttoning his fly en route. **1952** 'Vigilans' *Chamber of Horrors* 27 The words *button one's fly* are offensive only to the prurient. **1953** M. Dickens *No More Meadows* i. 49 Champ, your flies are undone again. That boy! He'll be arrested yet. **1959** R. Fuller *Ruined Boys* II. viii. 128 'Your flies are undone,' said Matley primly.

7. a. (sense 4 a) *fly-button*; *fly-front* (later examples).

1895 *Montgomery Ward Catal.* 86/1 Metal Snap Fly Buttons. **1933** J. E. Liberty *Pract. Tailoring* vi. 87 The fly buttons..should be carefully marked first by laying the fly edge along the catch seam and marking over from the holes on to the button catch. **1961** A. Wilson *Old Men at Zoo* ii. 81 You've been state nursed from the cradle, without ever learning to do your own flybuttons up. **1897** *Sears, Roebuck Catal.* 175/2 Men's fancy fly front corduroy vests will be all the style for the coming season. **1961** *Sunday Express* 12 Mar. 16/7 Simple trench coat..fly-front.

c. instrumental, as (sense 4 a) *fly-fronted* adj.

1901 in C. W. Cunnington *Eng. Women's Clothing* (1952) ii. 43 A fly-fronted coat. **1929** *Evening News* 18 Nov. 6/5 Look at the above Overcoat, the Fly-fronted Chester, which is a specially dignified style, and is much worn by business and professional men.

8. fly ash, ash resulting from the burning of pulverized coal; **fly boy** *U.S. slang*, a member of the Air Force, esp. a pilot; **fly camp,** a temporary camp; **fly-cruise,** a holiday involving an air journey to the place where a sea-cruise begins; also *attrib.*; **fly-frame,** (*a*) *Cotton-spinning* = **flyer frame*; (*b*) a machine used

in the polishing and grinding of glass; **fly-half** *Football*, the half-back who stands off from the scrum-half; = *stand-off half*; **fly-kick,** a kick (esp. in Rugby football) made while moving at rapid speed; so **fly-kick** *v. trans.* and *intr.*; **fly-nut,** a screw nut having wings or projections which enable it to be tightened by hand; **fly-post** *v. trans.* and *intr.*, to display (posters) rapidly in unauthorized places; *trans.*, to cover with bills so posted; so *fly-posting* vbl. sb.; **fly stitch** (see quots.).

1931 *Electr. World* 23 May 961/1 The solution of the fly-ash problem lies in reducing the amount of ash that is discharged to the atmosphere to the point where it is no longer objectionable. **1937** *Jrnl. Amer. Concrete Inst.* VIII. 578 The residue from the burning of powdered coal is commonly designated as 'fly ash'. **1956** *Science News* XL. 83 Fly-ash produced in boilers fired by pulverized fuel can be used in making building materials. **1967** *New Scientist* 15 June 650/1 Many schemes have been proposed for using up the millions of tons of 'fly-ash', a dust-like residue produced by coal-burning plant. **1946** *Amer. Speech* XXI. 248 Airforce flying personnel are sometimes labelled bird-men or flyboys. **1948** *Life* 1 Nov. 87/2 The generals are no full-throttle 'fly-boys'. **1954** Berrey & Van den Bark *Amer. Thes. Slang* § 756/2 Cloud hopper, eagle, eagle chaser, fly-boy, manbird, sky chauffeur. **1939** *Amer. Speech* XIV. 236, I heard the branch camps attached to one of our mountain divisions of the CCC referred to as fly camps. **1959** *Tararua* (N.Z.) XIII. 49 Fly-camp for a small, subsidiary, temporary camp..suggests a camp with merely a fly as shelter. **1964** *Imperial Oil Review* (Toronto) Dec. 5/2 Outside on the river bank..Hughes moved one helicopter..off to the nearest fly camp, 35 miles southwest. **1968** *Times* 14 Sept. 24/3 Sorting through the information I had gathered on 'fly-cruise' holidays, I came upon what was for me a new aspect of the regulations. **1970** *Daily Tel.* 19 Sept. 11 Flying to join a ship for a couple of weeks' cruising holiday in the Mediterranean had gradually won popularity...This year several firms are offering Caribbean fly-cruises. **1971** *Sunday Times* 3 Jan. 72/1 The increasingly popular fly-cruise holidays. **1835** Fly frame [see Roving *vbl. sb.*[2] 3 a]. **1890** W. J. Gordon *Foundry* 131 The huge sheets of glass are..ground with sand under a heavy iron 'fly-frame', said to have been invented by James Watt. **1940** *Chambers's Techn. Dict.* 344/2 *Fly frames*, a series of machines used to attenuate roving in preparation for the spinning frame. **1918** J. E. Raphael *Mod. Rugby Footb.* 127 It is the business of the three-quarters not to force the fly half forward, but to stand back also. **1921** E. H. D. Sewell *Rugby Football* 60 He [*sc.* the scrum half] knows by the signal which way his fly-half is going to run. **1955** *Times* 1 Aug. 3/3 Scrum half Williams and Morgan at fly-half. **1906** Gallaher & Stead *Compl. Rugby Footballer* 208 Ninety-nine successful fly-kicks do not atone for one failure. **1921** E. H. D. Sewell *Rugby Football* 92 Whether by punt, drop, or 'field' or fly-kick does not matter. **1930** *Daily Express* 6 Nov. 17/7 Hunt suddenly turned defence into attack by fly-kicking. **1959** *Observer* 28 June 29/3 A pass which Dawson fly-kicked some 30 or 40 yards down the field. **1825** J. Nicholson *Oper. Mech.* 454 The small finding screw with a fly-nut. **1896** *Palmer Tyres* 22 Fly nut for opening valve. **1953** Flood & West *Dict. Sci. & Techn. Words* (ed. 2) 138/2 *Fly-nut*, nut with wings so that it can be screwed on by hand. **1903** *Daily Mail* 25 Feb. 3/7 A young advance agent for a theatrical company who went out 'fly-posting' in the dead of night and pasted bills on private property. **1906** *Daily Chron.* 21 July 6/6 A staff of men..to fly-post the constituency during the night with Tariff Reform literature. **1961** *Times* 21 June 13/6 The special triumph of what was known as 'fly posting' was to obliterate some rival announcement. **1970** *Daily Tel.* (Colour Suppl.) 11 Dec. 28/4 Intricate Art Nouveau offerings which he once used to help to fly-post stealthily all over London. **1934** M. Thomas *Dict. Embroidery Stitches* 101 Fly stitch is..a kind of open detached chain stitch. **1960** G. Lewis *Handbk. Crafts* 32 Fly stitch or Y stitch...The needle comes up at the top of the left arm of the V part of the stitch. It is inserted at the top of the right arm of the V and comes out at the base of the V over the loop of the working thread. The tail of the Y can be as long or short as desired.

fly, *a.* Add: **3. fly cop** *slang* (see quots. 1859 and 1962); *U.S.*, a detective, a plain-clothes policeman; **fly-flat** (earlier and later examples in extended use); **fly-pitch** *slang*, a street pitch; hence **fly-pitcher,** one who operates a fly-pitch, a street-trader.

1859 Matsell *Vocabulum* 34 *Fly-cop*, sharp officer; ..an officer that understands his business. **1872** *Chicago Tribune* 27 Dec. 8/3 Blacklegs, gamblers, 'fly cops', and sports. **1889** Barrère & Leland *Dict. Slang* I. 376/1 *Fly-cop* (thieves'), a sharp policeman. **1900** Fly cop [see *chair *sb.*[1] 1 d]. *a* **1906** 'O. Henry' *Trimmed Lamp* (1916) 227 It was purely the still panic produced by the sound of the ax of the fly cop, Conscience, hammering at the gambling-house doors of the Heart. **1929** Wodehouse *Gentleman of Leisure* xviii. 149 'A detective?' 'Dat's right. A fly cop.' **1962** *John o' London's* 25 Jan. 82/2 A motorised police officer is..a *fly cop.* **1864** *Revelations Lady Detective* 126, I have an appointment with a fly flat (*i.e.* a clever fool). **1907** *Daily Chron.* 1 Apr. 4/4 The criminal classes..always speak of fools who think they are wise as 'fly-flats'. **1938** J. Cary *Castle Corner* x. 559 'I don't see why we should consider the speculators.' 'A lot of fly-flats who thought they could beat us at the game.' **1934** *Evening News* 9 July 11/2 There are the 'fly pitches', spots..where the cheap-jacks take their stand... The most famous is at the foot of the Irving statue at the side of the National Portrait Gallery. Whenever you see a really big crowd collected at this spot you can be sure that one or other of the familiar fly-pitchers of London is doing his stuff. It may be the coloured Prince of Tipsters, proving vehemently that he..holds the secret of making money out of racing. **1939** J. B. Priestley *Let People Sing* x. 256 'Bin workin' a fly pitch in the Sat'day gaff,' said Micky Barnet. **1965** *Sun* 26 Oct. 7/6 Charlie Byass..is one of the two dozen or so licensed chestnut men in London. There are others, called 'fly pitchers', who sell without a licence.

fly, *v.*[1] Add: **1. g.** To travel by aircraft. Also *trans.*, to cover, traverse, or perform by aircraft (also said of the machine). Also, *to fly in, out*, to arrive, depart, by air.

[**1826** W. Hone *Every-Day Bk.* I. 1462 He had purposed, by a pair of patent wings,..to fly from one of the Dover cliffs down into the town of Calais.] **1884** *Illustr. London News* 26 Jan. 91/3 Many successful and unsuccessful attempts have been made to fly the Channel. **1909** *Flight* 20 Feb. 100/2 To bring together other would-be aviators who, like himself, are primarily anxious to learn to fly. **1909** *Punch* 4 Aug. 73/1 On the spot where the first man to fly the Channel alighted from his aeroplane. **1909** *Captain* Nov. p. xii (Advt.), Model of the Machine which flew the Channel. **1911** Grahame-White & Harper *Aeroplane* 265 As their skill and confidence increased, they began to fly in breezes. **1916** H. Barber *Aeroplane Speaks* 44 If he were flying over the shortest route to his destination. *Ibid.* 45 The Pilot and Observer fly on and on. **1917** *Brit. Dominions Year Bk.* 242 We who still regard flying the Straits of Dover as something of a feat. **1931** *Times* 2 Mar., The Sahara route which is now being flown by the French. **1958** 'N. Shute' *Rainbow & Rose* i. 4 He was flying a courier service. **1966** *Listener* 21 July 103/3 An expert who had flown in from Venezuela to do the job. **1967** *Daily Tel.* 14 Mar. 20/7 An Indian, aged 100,..applied..for a ticket home to Delhi on the 'Fly Now Pay Later' plan. **1968** *Globe & Mail* (Toronto) 17 Feb. B5/5 The [a]erial surveys flown by Texas Gulf in 1960. **1970** *Observer* 20 Sept. 26/1 The pilot [is] determined to fly no more World War Two bombing missions.

h. Of an aircraft or spacecraft: to travel through the air or through space.

1848 *Chambers's Edin. Jrnl.* 6 May 302/2 To construct a machine to fly by mechanical powers alone. **1902** *Aeronaut. World* (U.S.) I. 65/2 S. Byerley says he has been working on a novel flying machine, which, he is confident will fly when finished. **1959** *Chambers's Encycl.* I. 112/1 The faster a given aeroplane flies, the less is the power required to produce its lift. **1971** *Daily Tel.* (Colour Suppl.) 12 Nov. 8/1 Everything depends on three robot spacecraft..which have been flying towards Mars for the past 22 weeks.

i. Of pigeons: to fly from (a certain place).

1898 *Daily News* 21 Nov. 7/1 There were also young birds that had, to use the technical term, 'flown Berwick'. **1907** E. Chamberlain *Homing Pigeon* 178 The best birds in the Barrowford lofts have all flown Nantes as yearlings, and subsequently flown Nantes and Marennes again and again.

5. a. *to fly a kite*: also, to try something out; to act in an exploratory manner. Also *go fly a kite* = 'go away', 'be off'. *colloq.* (chiefly *U.S.*).

1928 *Sat. Even. Post* 7 Jan. 21/2 Now would you like for mamma to go on reading to you about this Polark soldier and his girl friend, or would you like to go fly a kite. **1937** Partridge *Dict. Slang* 291/2 *Fly a, the, kite*,..to test public opinion by tentative measures. **1942** 'H. Pentecost' *I'll sing at your Funeral* (1945) vii. 64 'He suggested,' said Bradley, with a sigh, 'that I go fly a kite!' **1951** C. P. Snow *Masters* xv. 137 At the beginning Brown had, as he used to say, 'flown a kite' for compromise, now Chrystal joined him. **1961** *Lebende Sprachen* VI. 99/2 Go jump in the lake,..go fly a kite; DDT (drop dead twice). **1971** *Nature* 26 Feb. 584/1 The solution, the committee suggests—and it is plainly flying a kite and not laying down policy—may be a more selective way of choosing the departments to which studentships will be allocated.

b. Also, to conduct or pilot (an aircraft).

1902 *Young Engineer* I. 265/1 He, in his indignation, undertook to fly his 'Illustrissimo', as he termed his balloon, with the whole of the Holy Office. **1911** Grahame-White & Harper *Aeroplane* 264 A machine that..will require no more skill and nerve to fly it than does the driving of a car. **1917** *Brit. Dominions Year Bk.* 243 It calls for considerable courage..to fly heavy seaplanes a hundred miles or so inland. **1920** *Act* 10 & 11 *Geo. V* c. 80 § 10 (1), Where an aircraft is flown in such a manner as to be the cause of unnecessary danger. **1928** *Manch. Guardian Weekly* 21 Sept. 224/3 Señor Juan de la Cierva, the inventor,..flew one of these machines to Paris from Croydon on Tuesday. **1968** Miller & Sawers *Techn. Devel. Mod. Aviation* iv. 99 The prototype DC-1 was flown.

c. To convey by aircraft.

1928 *Daily Express* 18 May 2/6 More than £26,000,000 worth of gold coin and bullion has been flown in and out of Croydon during the past three years. **1930** C. R. Samson *Fights & Flights* 209 You will fly me to Calais, and I will repair 1241. **1958** 'N. Shute' *Rainbow & Rose* i. 17 You come to fly him out? **1971** *Daily Tel.* 9 Nov. 15/8 Cremation is taking place in Australia today. The ashes will be flown back to England.

7. Of a fox: to break covert.

1897 *Encycl. Sport* I. 546/1 They force the fox through the cover in such a determined fashion as will make him 'fly or die'. **1902** *Daily Chron.* 14 Nov. 8/5 The pack had to wait some time before they could get their fox to fly.

e. (Earlier U.S. examples.)

1831 S. Smith *Life & Writings J. Downing* 151, I flew round and washed my face and hands. **1833** J. Hall *Legends of West* 88 She flew around among the folks mighty peart. **1839** C. M. Kirkland *New Home* ii. 16 Come, gals! fly round, and let's git some supper.

f. *to fly the coop*: to escape or elope; to leave suddenly; *to fly the track*: to turn from the usual or expected course. *U.S. colloq.*

1847 *Congress. Globe* 4 Feb. 322/2, I had been accused of flying the track on the creed of the Democratic party. **1901** 'B. Burgundy' *Toothsome Tales* (1902) 68 Chester..used to..toast her until she was on the edge of hysterics, then he'd fly the coop in quest of a new cargo of thirst-erasers. **1906** 'O. Henry' *Four Million* (1916) 123 She'll never regret flyin' the coop with Chunk McGowan. **1909** —— *Options* (1916) 157 One of them..'roller-coasters' flew the track and killed a man. **1966** H. Kemelman *Saturday the Rabbi went Hungry* (1967) xxi. 206 This man..ran off, flew the coop, beat it.

9. h. *Baseball.* To hit a fly-ball (Fly *sb.*[2] 8). Chiefly with *out.* (Pa. t. usu. **flied.**)

1893 *Chicago Tribune* 3 July 7/3 Kittridge flied out to Brodie. **1904** *St. Louis Globe-Democrat* 4 July 12/3 Wallace flew to Lush for the third out. **1912** C. Mathewson *Pitching in a Pinch* ix. 202 Sheckard flied out to Seymour, Kling being held on second base. **1948** *Durant* (Okla.) *Daily Democrat* 2 July 4/4 Baker then flied to center and neither runner was able to advance.

flyable, *a.* Add: **2.** Of weather: suitable for flying. Of goods: transportable by air. Of an aircraft: that is capable of flying or of being flown.

1936 J. Grierson *High Failure* xii. 271 McNab arranged..to take me out to Robert Bruce [name of aeroplane], provided the weather should be flyable. **1945** *Times* 3 Aug. 5/6 Forecasters who are amazingly expert at assessing how long the weather will remain flyable at night at any particular airfield in the country. **1960** R. W. Marks *Dymaxion World of B. Fuller* 151/4 The more ephemeralization advances the more flyable becomes any one cargo. **1963** *Economist* 30 Nov. 928/3 A real flyable aeroplane. **1970** *Daily Tel.* (Colour Suppl.) 9 Oct. 61/1 A corrugated silver Ford Tri-Motor in flyable condition.

fly-about, *a.* Add: **b.** Of horses: skittish.

1889 F. E. Gretton *Memory's Harkback* 110 Two little blood-chestnuts..very fly-about, and disposed to bolt.

fly-away, *a.* and *sb.* Add: **A.** adj. *fly-away grass* (example).

1856 *Rep. Mass. Board Agric.* I. 29 Hair Grass, or Fly Away Grass,..is another species belonging to the genus.

B. *sb.* (Earlier example.)

1804 M. Wilmot *Let.* 24 Jan. (1934) I. 78 She was not amongst the fly aways but return'd for the Ball.

fly·-back. [f. verbal phr. *to fly back* (see Fly *v.*[1] 9 c).] **1.** In a stop-watch or chronograph, the return of the hands to zero. Usu. *attrib.*

1888 *Phil. Mag.* 5th Ser. XXV. 263 The fly-back chronograph enables time to be measured to ⅕th second, without taking the eye for one moment off the object to be watched. **1903** F. J. Garrard *Watch Repairing* xix. 200 (*caption*) Fly-back work of chronograph. **1923** W. I. Milham *Time & Timekeepers* xv. 250 A double split-second fly back stop watch..gets out of order quite easily. **1929** G. H. Baillie *Watches* x. 233 A watch with fly-back minute hand, which..flies back to 0 as soon as it reaches 60. **1953** W. J. Gazeley *Watch & Clock Making & Repairing* xiv. 315 Another arrangement met with in some stop watches..is where a separate push-piece is used for the fly-back action.

2. a. The return of the scanning spot in a cathode-ray tube to the starting point after the end of a line or frame. Freq. *attrib.* **b.** The portion of each cycle of a saw-tooth wave-form in which the signal diminishes.

1934 *Jrnl. Inst. Electr. Engin.* LXXV. 66/1 The arrangement..constitutes a time base which provides a scanning sweep..and a rapid fly-back. **1948** T. Soller et al. *Cathode Ray Tube Displays* iv. 133 The flat portion [of the waveform] is the 'quiescent' voltage and the portion that follows the rise is the 'return' or 'flyback'. **1950** *Electronic Engin.* XXII. 55 The flyback time in these tests was about one-seventh of the sweep time. **1957** Amos & Birkinshaw *Television Engin.* I. i. 15 During the time occupied by each horizontal traverse and fly-back the beam moves downwards.

fly-blow, *v.* Add: So **fly-blowing** *vbl. sb.*

1891 R. Wallace *Rural Econ. Austral. & N.Z.* ii. 42 Sheep do not require to be dipped in this district, as there are no ticks, no foot-rot, and little fly-blowing.

fly-blown, *ppl. a.* Add: **2. b.** (Earlier Austral. example.) Also *N.Z.*

1853 C. R. Read *What I heard, saw & did at Austral. Gold Fields* 51 Being 'fly-blown' is a Colonial term for being 'done *up*'. **1864** C.R. Thatcher *Songs of War* 20 When they know you're *fly-blown*, You're then left alone By the damsels inside the Bazaar. **1921** H. Foston *At Front* vii. 56 [The men] usually returned to the works fly-blown.

fly-by (flai·bai). [f. verbal phr. *to fly by* (see Fly *v.*[1] 1).] **a.** = *fly-past. orig. *U.S.*

1953 *Tandemeer* 11 Dec. 2 On each day of the show there will be spectacular aerial flybys of jet planes. **1970** *Courier-Mail* (Brisbane) 29 Apr. 2/8 As for the fly-by, it was now surmised that having once passed overhead, the same planes had flown out of sight, circled and returned again and again.

b. In space flights: a close approach to a celestial body for the purpose of observation; a spacecraft that makes such an approach.

1960 *Missiles & Rockets* 31 Oct. 15/2 First *Mariner* mission will be a fly-by of Venus in 1962. **1961** *Ibid.* 2 Jan. 14/1 One of the major questions in the fly-by program is money. **1961** *Flight* 26 Oct. 653/2 Additional manœuvres carried out near to the target could reduce the dispersions still further so that it might be possible to accomplish close fly-bys, orbiters, or soft landings. **1962** F. I. Ordway et al. *Basic Astronautics* v. 235 Fly-bys would be equipped with magnetometers to detect the general interplanetary field and Jupiter's field. **1967** *Listener* 15 June 792/2 The surface [of Mars] was photographed [in 1965] at close range by the United States fly-by probe Mariner 4. **1969** *New Scientist* 27 Feb. 439/1 The first of a pair of *Mariner* spacecraft should now be on its way to carry out an observation fly-by of Mars.

fly-by-night. Add: **1.** (Additional *attrib.* examples; also later examples in sense of quot. 1796.)

1903 G. S. Wasson *Cap'n Simeon's Store* 72 Blowed ef 't wa'n't downright horrid the works them two ole fly-by-nights was into them days! **1914** J. London *Let.* 25 June (1966) 425, I have no patience with fly-by-night philosophers such as Bergson. **1925** A. Huxley *Those Barren Leaves* II. v. 151 When the next full moon invites New bugaboos and fly-by-nights. **1938** 'N. Shute' *Ruined City* (ed. 2) i, 5 We've come to you rather than to one of them fly-by-night financial houses because we're prudent business folk. **1958** E. Dundy *Dud Avocado* I. vi. 96 This is no fly-by-night proposition. **1971** *Guardian* 12 July 7/3 It is not all heart in the mini-cab world. Far too many are fly-by-night hustlers.

fly-catcher. Add: **4.** In the war of 1914–18, a fast type of aeroplane. *colloq.*

1924 *Glasgow Herald* 28 July 9 The royal yacht had three 'fly-catchers'. **1925** Fraser & Gibbons *Soldier & Sailor Words*, *Fly-catchers*, a fast type of aeroplane, officially rated as 'Fleet Fighters'. **1928** *Times* 23 Mar. 19/2 They were engaged in turn by the 'Flycatchers' from the Furious.

flyer, flier. Add: **1. c.** An aviator.

[**1679** tr. *R. Hooke's Philos. Coll.* I. 17 A Tail will help both to support, and also to steer or guide the Flyer.] **1934** Webster, *Flier, flyer*, an airman. **1936** F. Clune *Roaming round Darling* ix. 79 In the dining-room were two lady fliers, dressed in blue. **1953** C. Day Lewis *Italian Visit* ii. 31 Recall how flyers from a raid returning, Lightened of one death, were elected for another.

d. A flying machine.

1880 *15th Rep. Aeronaut. Soc.* 21 June 22 If you will leave the balloon to me I will not trespass on your flyers. **1903** M. Wright in M. W. McFarland *Wright Pap.* (1953) I. 393, I spend the day largely in getting typewriter copies of the description of the Wright flyer. **1908** H. G. Wells *War in Air* viii. 254 Next only to the Butteridge machine these were certainly the most efficient heavier-than-air fliers that had ever appeared. **1964** *Listener* 14 May 780/1 Ordinary motorists will not mind, because about this time they will all possess small private flyers.

2. (Additional examples.)

1842 *Spirit of Times* 4 June 163 We hope the Rocket of our old friend Dr. W. may be such another 'flyer'; he certainly 'goes off' well. **1877** *Coursing Calendar 1876* 79 Mr. Brocklebank's Bombard although not a flyer, he will turn out a useful dog. **1888** *St. Louis Globe Democrat* 2 Mar. (Farmer), In spite of the strike, passenger trains excepting what are known as the flyers, are running with reasonable regularity. **1919** F. Hurst *Humoresque* 131 You ought to see the flier a friend of mine has got. A Mercury Six [motor-car] with a limousine top like a grand-opera box. **1922** E. Wallace *Flying Fifty-Five* xi. 67 Fifty-Five is a flyer...He did the five furlongs in fifty-eight and a fifth seconds. **1948** C. L. B. Hubbard *Dogs in Brit.* 35 Heredity and environment count a great deal in whether a puppy will be a 'flyer' or a 'bad-doer'. **1971** *Daily Tel.* 21 July 11/2 The normal carburettor-fed 2002 is one of the fastest cars in its class: with the new Kugelfischer fuel injection system it is a real flyer.

b. A fast kangaroo. *Austral. colloq.*

1848 H. W. Haygarth *Recoll. Bush Life Austral.* xi. 118 At first starting, a young male or female kangaroo, called in the colony 'a flyer', can leave both horse and hound far behind. **1861** T. McCombie *Austral. Sks.* 172 The settlers designate the old kangaroos as 'old men' and 'old women', the full-grown animals are named 'Flyers'. **1968** K. Weatherly *Roo Shooter* 16 Two young fliers had got through the fence surrounding the 'turkey nest' tank.

c. *fig.* A person exceptional in some (specified) way.

1930 T. Thurston *Man in Black Hat* xvi. 278 We had an extraordinarily pretty girl here in the bar. Real flier, she was. **1936** Wodehouse *Laughing Gas* iii. 30, I am no flier in the way of looks. **1939** J. Cary *Mister Johnson* 140 He's not a flier at the accounts.

d. *Cricket.* A ball pitched short that flies up sharply; a 'bouncer'.

1913 *Cricket* 305 Every bowler pitches short sometimes, and when he does so, the resultant rib-roaster..flier..call it what you will, is no more than an ordinary risk. **1954** J. H. Fingleton *Ashes crown Year* 58 May ducked a flier from Archer and was hit on the back. **1962** *Times* 1 May 4/2 Nobody was the victim of a flyer.

3. e. Also *attrib.*, as *flyer frame.*

1825 J. Nicholson *Oper. Mech.* 383 A roving-frame of a different construction has received more general adoption..; it is called the bobbin and flier roving-frame. **1957** *Encycl. Brit.* VI. 562/1 Following on the last head of drawframes the process becomes one principally of attenuation. To this end the material is passed through one, two, three or four passages of machines collectively known as flyer frames.

4. d. (Later examples.) Also, *spec.* one issued by the police. *U.S.*

1952 Berrey & Van den Bark *Amer. Thes. Slang* (ed. 2) § 477.4 *Flyer*,..a warrant for arrest. **1955** D. W. Maurer in *Publ. Amer. Dial. Soc.* xxiv. 149 Is there a *flyer* out for this man? This is a police circular carrying photographs, fingerprints. **1962** K. Orvis *Damned & Destroyed* xiv. 94 Boards covered with tacked-up police flyers. **1965** E. McBain *Doll* (1966) iii. 38 Leslie offered him the agency flyer on Tinha, the composite that went to all photographers, advertising agency art directors, and prospective clients. **1966** D. Shannon *With a Vengeance* (1968) i. 13 They had also an unidentified corpse, and all the paperwork on that: the flyers sent out with the description. **1966** *Word Study* Feb. 6/2 This sentence from a recent advertising flier of *Fortune*.

e. A flying buttress.

1912 F. Bond *Cathedrals* 73 A second set of fliers was built above the aisle roof.

5. b. (Earlier and later examples.)

1846 *Spirit of Times* 11 July 229/3 Lend me a quarter..just for a flyer. **1848** W. Armstrong *Stocks* 37 A. B. Neilson occasionally sells fifty or a hundred shares of stock by way of a flyer. **1870** J. K. Medbery *Men & Myst. Wall St.* 136 A 'flyer', is a small side operation, not employing one's whole capital. It is nearly equivalent to what is ordinarily known as a venture. **1966** *Newsweek* 2 May 64/1 More and more 'Aunt Janes' are deciding to take a flyer, growing impatient with their slow-moving quality stocks as they see high-flyers double and triple in value.

fly-in (flai·in). [f. verbal phr. *to fly in* (see Fly *v.*[1] 1).] **a.** The action or an act of delivering troops, goods, etc., by air to a specified place. **b.** A service, entertainment, etc., provided for persons who have arrived by air. Also *attrib.*

1943 W. S. Churchill 6 Dec. in *Second World War* (1952) V. 368 Medium bombers should be..posted in Cyrenaica for 'softening' action against enemy airfields and shipping and to cover the fly-in of the British fighter squadrons. **1952** *N.Y. Times* 9 Mar. 2/7 Not eligible for this prize are small planes landing at a combination drive-in and fly-in. **1956** W. Slim *Defeat into Victory* 220 The fly-in of the 26th Division to Indaw. *Ibid.*, Wingate was dissatisfied with the rate of fly-in. **1960** in *Lebende Sprachen* (1963) VIII. 33/2 There are also fly-in banks at airports (a teller meets your helicopter or plane right on the runway) and sail-in banks for people in boats. **1966** *Times* 28 Feb. (Canada Suppl.) p. xii/5 He saw a column of smoke rising from a clearing, and recognized it as a beef barbecue 'fly-in' of Western Club pilots.

flying, *vbl. sb.* Add: **1. b.** The action of guiding or piloting an aircraft or spacecraft, or of travelling in one.

1834 *Amer. Railroad Jrnl.* III. 791/1 Such is my plan for flying. **1864** *R. Chambers' Bk. Days* II. 722/1 This extraordinary aërial locomotive is perhaps one of the most curious of these apparatuses for the purpose of flying. **1909** *Flight* 27 Feb. 113/1 In the present stage of flight the art of learning how is almost as great as that of the art of flying itself. *Ibid.* 23 Oct. 668/2 As a result of the good flying on Saturday afternoon, a large attendance was attracted to the racecourse on Monday afternoon. **1916** H. Barber *Aeroplane Speaks* 48 When..involuntary descents [were] often a part of 'cross-country flying. **1930** P. White *How to fly Airplane* xxiii. 306 Cross-country flying includes all flights over land..more than a mile or two away from the home field. **1947** D. J. Ingells *They tamed Sky* i. 2 This is push-button flying, automatic flight. **1960** C. H. Gibbs-Smith *Aeroplane* 1 Flying has never appeared to its devotees as a mere method of transportation. **1962** J. Glenn in *Into Orbit* 5 There is nothing spooky or supernatural about flying in space. **1967** *Drive* Spring 106 He loathes flying, endures sea travel, is happy in a car only when at the wheel.

3. (Some of these collocations could with equal or greater justification be listed under the *ppl. a.*: for convenience they are all grouped here, since all of them fall within or appertain to the subject of aviation. Cf. sense *1 b above.) *flying boot, circus* (see *circus 2 d), *coat, ground, meeting, school, squadron, suit, time, velocity, week*; **flying bedstead,** an experimental aircraft, shaped like a bedstead, for testing methods of vertical take-off; **flying boat,** (*a*) an experimental flying machine built by J. P. Blanchard; (*b*) a boat-shaped car on a merry-go-round; (*c*) a form of seaplane having a boat-like fuselage; **flying bomb,** a pilotless jet-propelled aeroplane with an explosive warhead, first used by the Germans against England in June 1944; **flying boxcar,** a bomber or a large freight-carrying aircraft; **flying corps,** a unit of aircraft for military or naval purposes; the Royal Flying Corps was the precursor of the Royal Air Force; **flying deck** = **flight deck* (a); **flying doctor,** a doctor who habitually uses an aeroplane for visiting patients in areas remote from his headquarters; also *attrib.*, as *flying doctor service*, etc.; **flying field** = *airfield; **flying hour** (see quot. 1956); **flying machine,** a machine or contrivance capable of being flown in the air; usu. restricted to a heavier-than-air machine dependent on its motors for maintenance and progression in the air (see also Flying *ppl. a.* 4 b in Dict. and Suppl.); **flying officer,** a rank in the Royal Air Force; **flying saucer,** the fanciful name given to various unidentified disc- or saucer-shaped objects reported as appearing in the sky; **flying speed,** the speed of an aircraft at take-off or in normal level flight; **flying test bed, bench,** an aircraft fitted with apparatus for measuring its performance in the air; **flying wing** (see quots.).

1955 *Sci. News Let.* 4 June 357/2 Rolls-Royce, in Britain, has designed a model dubbed the 'flying bedstead'. It is thrust into the air vertically by two jet nozzles pointed downward in the front and rear. **1957** *Times* 29 Nov. 10/6 The 'flying bedstead' is..a tubular structure on which two 5,000 lb. thrust Rolls-Royce Nene turbo-jet engines are mounted. **1784** *Universal Mag.* LXXV. 221/1 Mr. Blanchard..during the late war, formed a flying boat, which he intended for carrying the dispatches from Brest to Paris, but as this did not answer his expectations, he was obliged to..relinquish the idea of elevating himself above the clouds. **1854** A. C. Mowatt *Autobiogr. Actress* 108 In a single night, the booths, the fairy houses, the circus, flying boats, cars, horses disappear. **1903** O. Chanute *Let.* 13 Mar. in M. W. McFarland *Wright Pap.* (1953) I. 301 Today I spent with Wm. Kress who experimented with a flying boat last year. **1913** *Aeroplane* 24 Apr. 470/1 The death of Louis Gaudart showed what a deathtrap the ordinary type of flying-boat is. **1917** C. C. Turner *Aircraft of To-day* xix. 277 The twin-engined Curtiss flying-boat 'America'. **1928** *Manch. Guardian Weekly* 31 Aug. 178/3 A flying-boat service between Sydney and Wellington. **1934** 'R. Crompton' *William—the Gangster* iii. 49 Riding upon merry-go-rounds, flying in flying-boats, sliding down helter-skelters. **1958** *Times* 18 July 6/5 Britain's only remaining air line using flying-boats, Aquila Airways, are to cease operations on October 1. **1971** *Daily Tel.* (Colour Suppl.) 16 July 5/3 One of the Sunderland flying boats between Cairo and India: a yacht of the sky, landing on the sea or inland lakes as it hopped from point to point. **1944** *Times* 3 July 4/5 Strong measures to counter the flying bombs have been continued throughout the week-end. *Ibid.* 5 July 4/5 The Prime Minister intended to make a statement on Thursday on the flying-bomb attack. **1946** Flying bomb [see *buzz *sb.*[1] 5]. **1935** *Times* 9 Oct. 9/4 They are also showing flying boots for women lined in sheepskin and covered in willow calf and crêpe rubber. **1944** G. Netherwood *Desert Squadron* vi. 49 An airman wearing a badge in the form of a silver flying boot ..the insignia of the famous 'Late Arrivals'' Club, a club formed from airmen who..after having been posted as missing, returned safely to their units. **1932** *Beaver* Mar. 393 The new Junkers freight plane 'JU-52', 'flying boxcar'. **1941** *Amer. Speech* XVI. 165/2 *Flying boxcar*, a bomber. **1958** *Oxf. Mail* 19 July 1/2 A fleet of 'flying boxcars' bringing jeeps, bazookas and..rifles. **1931** *Chambers's Jrnl.* Sept. 545/1 Outside the squadron office was gathered a small group of pilots, in their sheepskin thigh-boots and heavy leather flying-coats. **1913** Flying Corps [see **flying officer*]. **1914** *Scotsman* 15 Sept. 5/7 The success attained by the Royal Flying Corps. **1926** T. E. Lawrence *Seven Pillars* (1935) II. xxi. 134 It was decided to lay out on the beach by the aerodrome a small position, capable of being held..by the..Flying Corps. **1931** *Jrnl. R. Aeronaut. Soc.* XXV. 373 The 'Vindex'..was the first ship to be fitted with a forward hangar and flying deck. **1954** P. K. Kemp *Fleet Air Arm* 86 The first carrier..was fitted with a flying deck forward. **1926** *Lancet* 17 Apr. 824/1 (*heading*) Flying Doctors. **1932** *Ibid.* 24 Dec. 1396/2 In May 1928 Dr K. St. Vincent Welch, the first 'flying doctor', set up practice at Cloncurry in Queensland. **1944** F. Clune *Red Heart* 7 There are six Flying Doctor Bases in Arid Australia. **1966** *Daily Tel.* 27 Sept. 13/3 Dr. Anne Spoerry, of the East African Flying Doctor Service. **1971** *Daily Tel.* 19 Jan. 10/5 The Falkland Islands Government maintains a 'flying doctor' service. **1927** J. M. Saunders *Wings* (1928) ii. 91 The giant searchlights from the tower played over the dark corners of the flying field. **1958** P. Kemp *No Colours or Crest* ix. 183 A disused flying field pitted with craters. **1909** *Westm. Gaz.* 23 Oct. 2/1 Those who, in spite of the half-gale which was blowing, went to the flying-ground. **1918** F. H. Colvin *Aircraft Mech. Handbk.* xxii. 305 To keep all the available machines of one squadron in commission for 800 flying hours. **1956** W. A. Heflin *U.S.A.F. Dict.* 214/1 *Flying hour*, an hour spent in flight (which includes takeoff and landing) by a person, aircraft, or piece of equipment. **1736** *Gentl. Mag.* Oct. 617/1 One James Todd..at the Theatre in Covent-Garden, fell from the upper Stage, in a flying Machine, the Wires breaking. **1769** *Oxf. Mag.* Sept. 108 (*caption*) A New Flying Machine upon Dr. Musgrave's Plan, that moves with yᵉ same Rapidity as Mr. Moore's Machine without Horses. **1797** *Encycl. Brit.* I. 198/2 In the year 1709..as we were informed by a letter published in France in 1784, a Portuguese projector, Friar Gusman, applied to the king for encouragement to his invention of a flying machine. **1808** G. Cayley *Aeronaut. & Misc. Note-Bk.* (1933) 3 Apr. 63 It obliges the centre of gravity of flying machines to be much forwarder of the centre of bulk than could be supposed a priori. **1834** *Amer. Railroad Jrnl.* III. 790/1 (*caption*) Travelling Balloon, or Flying Machine. **1865** *Reader* 1 Apr. 375/3 Until the time shall arrive when an aerostat or flying machine can be introduced with something like success. **1881** Flying machine [in Dict.]. **1895** *Nature* 1 Aug. 321/2 A large party of scientific men..to witness a trial of the celebrated flying machine. **1908** H. G. Wells *War in Air* VIII. § 2, It was not in their airships, but..in their flying-machines proper, that the strength of the Asiatics lay. **1909** *Westm. Gaz.* 11 Jan. 9/1 The regulation of flying-machine races. **1959** *Times* 2 Apr. 4/6 The challenge of the missile does not mean that the flying machine is outdated. On the contrary, there will always be great need for the work of the manned aircraft, which it alone can perform. **1970** *Daily Tel.* 30 Oct. 11/2, I was at school when I saw my first real flying machine. **1909** *Flight* 20 Nov. 749/1 A flying meeting to be held on the

Napoule aerodrome. **1913** *Aeroplane* 17 Apr. 446 Royal Flying Corps, Military Wing.—To be flying officers and to be seconded. **1922** *Man. Seamanship* (H.M.S.O.) I. i. 11 Distinctive Lace worn by members of the Royal Air Force...Flying Officer, Observation Officer. **1947** *Times* 8 July 4/4 During the past fortnight reports that dish-like objects, nicknamed 'flying saucers', have been seen travelling through the air at great speed..have come from the United States and Canada. **1948** *Jrnl. Brit. Interplanetary Soc.* VII. 199, I haven't examined the details carefully, but the 'flying saucers' bear all the hall-marks of mass-suggestion. **1953** Leslie & Adamski *Flying Saucers have Landed* I. i. 13 Ever since the cliché 'flying saucer' was coined, the greatest and most exciting mystery of our age has been automatically reduced to the level of a music hall joke. **1965** *New Society* 9 Sept. 14/2 When Kenneth Arnold saw something from his airplane near Mount Rainier in June 1947, he gave them the happy name of flying saucers. **1894** *Proc. Internat. Conf. Aerial Navig. 1893* 307 Just as the rider and swimmer go to a school for their instruction, so we must go to a flying school for ours. **1920** *Act* 10 & 11 *Geo. V* c. 80 § 7 (I), Any aerodrome, flying school, or landing ground. **1917** 'Contact' *Airman's Outings* 39 The throttle was again opened full out as the bus raced into the wind until flying speed had been attained, when it skimmed gently from the ground. **1785** H. Walpole *Let.* 24 June (1905) XIII. 278 France has conceived hopes of annihilating our Pyrenees by these flying squadrons [*sc.* balloons]. **1917** 'Contact' *Airman's Outings* p. xii, A record for the casualties of any one flying squadron during any three months since the war began. **1935** D. Pilley *Climbing Days* iii. 52 My brother John, looking magnificent in a Teddy Bear Jaeger flying-suit. **1936** *Discovery* Sept. 285/1 Upper atmosphere flights employing..special flying suits have actually been made. **1947** *Shell Aviation News* CXII. 11/1 This year the development of turbine installations in flying test beds has been taken a step further. **1949** *Jrnl. R. Aeronaut. Soc.* LIII. 135/1 Neither the turbo-propeller nor the turbo-reaction engine had actually flown so far. A flying test bench was being prepared, which would take either engine. This test bench was a four-engined machine which had been fitted with a nacelle above the fuselage. **1966** *Times* 10 Sept. 1/5 A Vulcan V-bomber, modified to act as a 'flying test bed' for the Anglo-French Concord's..engine, made a maiden flight yesterday. **1946** *Mod. Lang. Notes* Nov. 442 Travel time, flying time, running time. **1956** W. A. Heflin *U.S.A.F. Dict.* 214/2 *Flying time*, the elapsed time spent in flight or in associated actions by aircraft..or by persons. **1918** E. S. Farrow *Dict. Mil. Terms*, *Flying velocity*, the speed requisite to raise an aëroplane from the ground. **1909** *Flight* 20 Nov. 749/2 The flying week to be held there by the Aero Club of France. **1937** *Jrnl. R. Aeronaut. Soc.* XLI. 162 The final logical result..is..the flying wing with the engines, passengers or military load housed inside...There are still certain difficulties which will prevent us from achieving..a mere flying wing. **1948** *Science News* VII (*caption* to pl. 44) The Flying Wing...An aircraft without fuselage or tail-plane.

flying, *ppl. a.* (See note at *flying *vbl. sb.* 3.) Add: **1. b. flying coachman,** the regent honey-eater, *Zanthomiza phrygia*; **flying mouse,** the smallest gliding marsupial, *Acrobates pygmæus*, of the family Phalangeridæ.

1918 J. A. Leach *Austral. Bird Bk.* (ed. 4) 170 Regent Honey-eater (Warty-faced), [or] Turkey-Bird (Mock Regent), [or] Flying Coachman..one of the most beautiful of birds. **1934** *Bulletin* (Sydney) 16 May 20/4 The..flying-mouse is common in the wide belt of scrub which stretches across S.A. and Western Vic.

3. Also *flying-boom, -iron*; *flying deck*, 'a deck supported at the side by railings, stanchions, or the like' (Webster 1934); *flying-kites*, see Kite *sb.* 5; *flying ring*, each of a number of rings suspended in pairs or rows by swinging ropes used in acrobatics and gymnastics; *flying trapeze*, = Trapeze 1.

1864 P. Paterson *Glimpses Real Life* xiii. 130 If the bounding youth be swinging to and fro on the flying trapeze. **1883** *Man. Seamanship for Boys* 24 The given diameter of a flying-boom is at the boom-iron. *Ibid.* 23 The extreme end [of the jib-boom] is eight-square and an iron hoop driven on to receive the flying iron. **1896** *Daily News* 17 Dec. 6/1 The boats are relegated to a flying deck overhead. **1901** *Ibid.* 20 Feb. 6/7 Flying-ring performers, a troupe consisting of two men and a girl. **1919** C. Mackenzie *Sylvia & Michael* iii. 77 A fellow with a moustache like a flying trapeze. **1934** W. Saroyan (*title*) The Daring Young Man on the Flying Trapeze & Other Stories. **1953** *Britannica Bk. Year* 639/2 *Flying boom*, a system used to refuel aircraft in flight. **1965** E. Bruton *Wicked Saint* iii. 35 Heidi's mother continued in a ground job with the circus..but for ever hankered after the flying trapeze.

4. *flying jump*, = flying leap; *flying start*: in a race, (*a*) a start in which the competitors pass the starting-point at full speed; (*b*) a start by one competitor prior to the starting signal being given; (*c*) *transf.* and *fig.*

1851 *Fraser's Mag.* XLIII. 658/2 A flying start being the rule. **1893** Kipling *Many Invent.* 35 You know how you take a flying jump on to a fellow's head at school, when he snores in the dormitory. **1899** *Harmsworth Mag.* July 531/2 The cycle is next best with a flying start record of 1 m. 35⅗ sec. **1921** E. H. D. Sewell *Rugby Football* 61 The easiest way to make an opening is to steal a march on your opposing outside half by getting a flying start. **1924** G. Lambton *Men & Horses* 233 In the Knowsley Nursery, at Liverpool,..thanks to a flying start, she was only just caught close home. **1928** *Daily Mail* 7 Aug. 15/5 [He] set up a new record for the mile (flying start) with a speed of 40 miles per hour. **1933** L. A. G. Strong *Sea Wall* 245 How are the likes of you and me to get a good flying start with nothing to kick off from? **1958** *Spectator* 31 Jan. 137/2 It got him off to a flying start; in it were the seeds of his future Christian life. **1962** *Listener* 7 June 983/1 He climbed slowly in Elizabeth's reign, despite his flying start as the son of a Lord Keeper.

b. *flying machine* (earlier example); *Flying Scotchman* (*Scotsman*): see Scotchman b.

1764 *Ipswich Jrnl.* 18 Feb. 4/3 A new Flying Machine, to accommodate Passengers by..the nearest Way for London, will set out..on Mondays, Wednesdays, and Fridays.

c. *flying visit* (earlier examples).

1797 R. M. Roche *Childr. of Abbey* II. iv. 52 Amanda supposed, as usual, she only came to pay a flying visit. **1825** H. Wilson *Mem.* II. 175 He paid me a flying visit.

e. (*a*) *flying squadron*: transf. and fig. applied to any body of persons organized for rapid movement or action.

1670 [see Dict., 4 d *transf.*]. **1875** *Temple Bar* June 197 A constant system of espionage carried on by the beautiful demoiselles of her Court, whom Brantôme has immortalised under the term of the Queen-Mother's 'flying squadron'. **1901** *Girls' Realm* 930/2 Flying Squadron Race.

(*b*) *flying squad*: a detachment of a police force organized for rapid movement. Also = **flying squadron* transf. & fig.

1927 *Daily Tel.* 10 May 6 Fatally injured by being knocked down by a motor-van of the Flying Squad. **1929** J. B. Priestley *Good Companions* II. iv. 347 What with photographs and finger-prints and telegraphs and wireless and flying squads!..Not a dog's chance! **1938** *Lancet* 109/2 A 'Flying Squad' in London...Islington is shortly to have the first maternity emergency unit to be established in London. **1939** T. S. Eliot *Old Possum's Pract. Cats* 33 He's the bafflement of Scotland Yard, the Flying Squad's despair. **1961** *Daily Tel.* 1 Mar. 24/1 To speed justice, more frequent assizes, continuous quarter sessions and a 'flying squad' of judges to hear cases during the Long Vacation are recommended to-day. **1966** *Guardian* 13 Dec. 5/4 In 1920..he [*sc.* W. Hambrook] was put in charge of a team of three inspectors, four sergeants, and four detective constables, together with two drivers...Within a few days the team had been named the Flying Squad.

f. *flying flock, flying* (*ewe*) *stock* (see quot. 1886); opp. *running stock* (Running *ppl. a.* 17 d). Also *flying stock* (in Austral. and N.Z.) (see quots. 1891). So *flying herd* (of cows).

1837 W. Youatt *Sheep* xv. 519 In a breeding stock this [examination of ewes and lambs] is absolutely necessary, but in a flying stock, or that in which the ewes and the lambs are usually sold before the termination of the year, this may be dispensed with. **1844** H. Stephens *Bk. Farm* III. 1105 The ewes are sold in autumn to be fed on turnips, and the lambs are disposed of, after being weaned, to rearers of sheep who breed none. Such a one is called a flying-flock...Flying-stock require no drafting. **1886** C. Scott *Sheep-farming* 29 A flying ewe stock is one where the ewes are bought in annually, and sold out again after bringing a lamb. **1891** R. Wallace *Rural Econ. Austral. & N.Z.* xviii. 255 Walhallow Estate [Australia]..is largely used for fattening purposes, and a 'flying' stock is maintained in addition to the regular and permanent station flock. *Ibid.* 259 [On Edendale Estate, Southland, N.Z.] The sheep kept are largely a 'flying' or bought-in stock for fattening and marketing. **1950** J. G. Davis *Dict. Dairying* 437 Other herds are replenished by purchasing newly calved cows. These are called 'flying herds'.

g. Football. (*a*) *flying man*, in the Eton field game (see quot. 1898); (*b*) *flying half, man*, earlier terms for *fly-half* (see *fly *sb.*[2] 8).

1864 B. Hemyng *Eton School Days* xxiii. 255 He was a little too heavy for 'flying man'. **1898** *Encycl. Sport* II. 143/1 In eleven a-side games the bully is formed of four players...Immediately behind is 'flying man', usually the most skilful forward on the side. **1906** Gallaher & Stead *Compl. Rugby Footballer* iv. 64 Two half backs, one being delegated to attention to the scrum, and the other being by way of a flying half. *a* **1914** J. E. Raphael *Mod. Rugby Football* (1918) 127 With the flying man standing as far back as the scrum half can throw the ball full pitch, it is extremely difficult to bottle him.

h. *flying shear*, a device for shearing a long, continuous length of metal into short pieces without arresting its forward motion; so *flying-sheared* adj.

1902 *Trans. Amer. Soc. Mech. Engin.* XXIII. 468 The success or failure of the whole equipment was entirely dependent on something so little demonstrated..as the Flying Shear. **1958** *Jrnl. Iron & Steel Inst.* CLXXXVIII. 30/2 The flying shear limits the finishing speed and reduces output. *Ibid.*, Flying-sheared sheets must be cooled quickly.

i. *flying spot*: a small spot of light that is made to move rapidly over an object, the reflected or transmitted light from successive parts of it being converted by a photo-electric tube or cell into an electrical signal that can be made to reproduce an image of the object. Usu. *attrib.*, as *flying-spot scanner, scanning.*

1933 *Television* Oct. 342/1 The transmitter utilises the principle of indirect or 'flying spot' scanning by means of an aperture disc. **1960** *How TV Works* vi. 34/2 For many purposes..reproduction of higher quality is achieved by means of a 'flying spot' machine in which each frame of an ordinary positive film is rapidly scanned by a flying spot of light. **1965** *Math. in Biol. & Med.* (*Med. Res. Council*) iv. 192 The focus and intensity controls and high-voltage supplies are associated with the 1·5-mil-spot cathode ray tube that produces the short-persistence flying spot, with a sweep diameter of 10 cm. **1968** *Brit. Med. Bull.* XXIV. 191/2 They are scanned line by line (television-wise) by a 'flying-spot scanner'.

7. Building. *flying façade*, a façade that rises above the level of the roof of a building; *flying shore*, a horizontal shore used to provide temporary support between two walls.

1894 J. P. Allen *Pract. Building Construct.* xxii. 355 End houses which cannot be strutted from another building, are dealt with by means of flying shores. **1936** P. E. Thomas *Mod. Building Pract.* II. 167 A 'flying shore' is horizontal, and incapable of taking any vertical load, its function being to keep the wall plumb. **1961** S. Sitwell *Golden Wall & Mirador* viii. 161 *Espesaña* denotes a bell-wall or flying façade, rising above and often doubling or trebling the height of the building. **1962** A. E. Bridgwood *Newnes Carpentry & Joining* III. 5 The elevation of one of a series of flying shores offering support to the walls of four floors is shown.

flying fox. Substitute for def.: A member of the genus *Pteropus* of fruit-eating bats, found in India, Madagascar, south-east Asia, and Australia. (Later examples.)

1877 W. S. Dallas in *Cassell's Nat. Hist.* I. 268 The Flying Fox drinks by lapping. **1910** *Encycl. Brit.* X. 586/1 The flying-foxes are the largest of the bats. **1965** D. Morris *Mammals* 98 The largest bats of this group are the Flying Foxes of the Pacific region. **1969** E. C. Rolls *They all ran Wild* xvii. 390 The party shot fifty-five flying-foxes.

2. A carrier operated by cables across a gorge, etc. *Austral.* and *N.Z.*

1936 F. Clune *Roaming round Darling* vii. 64 The mountain-sides near by are vast stores of black marble which is quarried out and brought down by flying foxes (buckets run on overhead wires). **1948** Partridge *Dict. Forces' Slang*, *Flying Fox*, a device used by the Australians in the Pacific to speed up the advance by slinging heavy equipment across rivers and other obstacles by ropes. **1965** G. J. Williams *Econ. Geol. N.Z.* x. 157/1 A certain amount of serpentinous rather than nephritic material had been brought down by flying-fox and cut into slabs. **1969** *Landfall* XXIII. 59 Adam had rigged up two flying-foxes—two long wires across the valleys on which he sent the hay flying across to the ferny slopes on the other side where they fed the cattle.

fly-leaf. (Earlier example.)

1832 J. Constable *Let.* 28 Mar. (1966) IV. 371, I have bothered Sparrow out of a fly leaf, & I like it very well, preserving the original title.

flyless (flai·lės), *a.* [f. Fly *sb.*[1] + -less.] Without flies. Hence **fly·lessness.**

1904 *Daily Chron.* 28 July 4/7 You leave Brindisi..in a perfectly fly-less ship. **1906** *Ibid.* 2 Oct. 4/7 The window-panes of a hitherto flyless house. **1918** W. Hutchinson *Doctor in War* (1919) xviii. 254 This miracle of flylessness was achieved..even in most of the Ospedali da Campo and Ospedaletti da Campo in the..camps just behind the Front.

flyness (flai·nės). [f. Fly *a.* + -ness.] The quality or fact of being 'fly'; wide-awakeness.

1888 *Inter-Ocean* (Chicago) 8 Mar. (Farmer *Amer.*), Each conviction is a lesson to the youthful politician that flyness and smartness cannot be pleaded in mitigation of contempt of honor. **1925** *United Free Ch. Mission Rec.* June 247 Free from any of that holy flyness, with which, justly or not, ecclesiastics are sometimes credited.

fly-off (flai·ǫf). [f. verbal phr. *to fly off* (see Fly *v.*[1] 9 c).] The action of flying off. Also *attrib.* or as *adj.*, designating a brake in a motor vehicle that requires a manual operation to put it on or off but not to keep it on.

1901 H. G. Wells *First Men in Moon* xx. 180, I had reckoned that..the tangential 'fly off' of the moon's spin would be at least twenty-eight times less than the earth's. **1959** *Motor* 19 Aug. 7/1 Good features of the control include..a central fly-off handbrake. **1970** *Motoring Which?* Apr. 47/2 Drivers liked the DB5's fly-off design. You pressed the button only if you wanted to set the brake on, not to release it.

fly-over (flai·ōuvəɹ). [f. verbal phr. *to fly over* (see Fly *v.*[1] 9 c).] **1.** A railway or road bridge over another (e.g. a local over a main) line or road. Also *fig.* Similarly **fly-under,** a line or road under another. Both words also *attrib.*

1901 *Daily News* 5 Jan. 6/5 The junction for the Aldershot branch..is being..rearranged on the 'fly-over' system, that is, the down line..remains as it was, but a new one..is being brought over the top of the main line by means of a skew bridge...This 'fly-over'..will abolish a fruitful source of delay. **1930** *Engineering* 7 Nov. 596/1 At Cogload Junction a fly-over bridge..will be built. **1937** *Daily Express* 7 Apr. 8/4 Whitehall's idea now is that great motoring highways be driven across the parks from one end to the other. They would meet at fly-overs, where one road would be built over the other. **1937** *Archit. Rev.* LXXXI. 158 (*caption*) Fly-over [road] crossing. **1938** *Times* 26 Jan. 14/4 The council also approved schemes for the construction of a fly-over junction on the Sutton by-pass road at its junction with the Merton–Belmont road. *Ibid.* 8 Nov. 11/6 The scheme has involved the construction of a 'fly-under' bridge under the present electric lines near North Acton. **1961** *Daily Tel.* 30 Aug. 15/3 'Fly-under' on old railway. *Ibid.*, A stretch of abandoned railway line is to be converted into a 'fly-under' junction. **1962** W. Nowottny *Lang. Poets Use* viii. 183 This kind of symbolic language can use vocabulary that is apparently very simple..but conceals a tracery of fly-overs from literal to metaphorical terrain. 1963 [see *ELEVATED ppl. a. 1]. 1970 Courier-Mail (Brisbane) 15 May 2/3 The fly-over freeway leading from Tokyo International Airport to the centre of the world's biggest city.

2. = *FLY-PAST; the passage of an aircraft over (an area).
1931 Times 5 Sept. 10/7 The fly-over, though robbed of the excitement of a race, should nevertheless afford a fine spectacle. 1953 Time 30 Mar. 25/1 On his way to England, four other Britons had been killed during a 60-plane 'flyover' staged at Gibraltar. 1971 Daily Tel. 24 Apr. 2/6 Concorde's approach and fly-over noise was less than that of the..Boeing 707.

fly-past (flaɪ·pɑːst). [f. verbal phr. *to fly past* (see *FLY v.[1] 1 g), after *march past*.] The action of flying past, or forming part of a procession of aircraft; also, a fly-by (sense b).
1914 Flight 24 July 791 The machines which took part in the 'fly past' were three of the big 160 h.p. Gnome-Short machines. 1928 Daily Express 12 Oct. 1 At the end of the display when a fly-past of five different types of bombers was taking place. 1956 Times 8 May 9/5 From the balcony of Buckingham Palace, the Queen will take the salute at a fly-past of RAF aircraft. 1967 Times 19 Oct. 11/2 The American fly-past [of Venus].. cannot compete in results with the Russian soft landing. 1971 Daily Tel. 21 Aug. 4/5 A 21-gun salute and a fly past by South African air force jets completed his send-off.

fly-sheet. Add: (Earlier example.)
1833 W. C. MACREADY Diary 23 July (1912) I. 51 Received letter from Serle, written on the fly-sheet of a circular from the Society of Dramatic Authors.

2. A canvas cover pitched outside and over a tent to provide protection against the weather.
1926–7 Army & Navy Stores Catal. 1000/2 Ridge tents ..fitted with outer fly sheet forming double roof. 1946 J. E. Q. BARFORD Climbing in Brit. ii. 33 A fly-sheet is desirable in heavy rain, but none have so far been designed which will stand up to a strong wind.

flyting, var. FLITING vbl. sb.

fly-trap. Add: **1. b.** *fig.*
1909 'O. HENRY' Options (1916) 62 Old Jerome was lingering long after breakfast..before setting forth to his down town fly-trap. 1925 W. DEEPING Sorrell & Son iv, The 'Cubby Hole' of the Angel Inn was a pivot, a fly-trap, a cave into which all sorts of male things crowded, and drank.

fly-under: see *FLY-OVER 1.

flyway. [f. FLY v.[1] + WAY sb.[1] 7.] The route taken by birds during migration; also in extended use (see quot. 1948[2]).
1891 Outing (U.S.) Dec. 222 But a few years ago San Diego Bay was gay with life as we rowed across it to the mouth of the Spanish Bight, which was then a favorite 'fly way' of this bird. 1948 Brit. Birds XLI. 279 The Yukon delta described as perhaps the richest breeding ground and migration flyway of arctic birds ever discovered. 1948 Conservation in Action V. 8/1 The term 'flyway' denotes a vast geographic region that is occupied by definite bird populations and contains both breeding and wintering grounds, connected by..migration routes. 1962 Times 26 Feb. (Canada Suppl.) p. xv/6 Wildfowl are abundant in the prairie provinces, on the central flyway. 1971 Country Life 25 Feb. 435/2 It [sc. Ireland] is indeed the terminus of two wildfowl flyways.

fly-wheel. Add: Also *attrib.* and *Comb.*
1896 Daily News 18 Nov. 7/7 The great problem was to get the required fly-wheel effect without too great a weight. 1909 Westm. Gaz. 30 Nov. 5/2 The flywheel effect is increased owing to the kinetic energy of the mass of the revolving liquid medium. 1963 BIRD & HUTTON-STOTT Veteran Motor Car 123 The flywheel-fed splash lubrication. 1971 Flying Apr. 72/2 The weight of the fuel at the wingtips created a 'flywheel effect'.

foam, sb. Add: **1. d.** A mass or layer of foam used in fire-fighting, produced by adding a foaming agent to a flow of water or by other means; also, a foaming agent so used. Also *attrib.*, as *foam carpet, extinguisher, generator*.
1906 A. G. LAURENT Brit. Pat. No. 18,357 15 Aug., The invention has for its object a hand fire extinguishing apparatus..that..sprays foam instead of liquid. 1933 Meccano Mag. Feb. 99/3 A fire engine in use at Dagenham carries practically all the equipment likely to be required by a fire brigade. This includes a chemical 'foam' generator. 1934 WEBSTER, Foam extinguisher, a fire extinguisher discharging a frothy substance which covers and extinguishes by excluding air from the burning surface. 1957 Encycl. Brit. IX. 271/1 Foam is also used in the fire service in the form of a dry powder which employs a foam generator to mix the powder with water being discharged by the fire pump. 1960 Daily Tel. 12 Aug. 18/4 'Foam carpet' for runway...A device for carpeting a runway with foam has been developed by the Engineering Flight at the R.A.F. Station, Waddington, Lincs. 1966 McGraw-Hill Encycl. Sci. & Technol. V 279/1 On oil fires water is effective only as a foam, or in some cases as a fog. 1970 Oxford Mail 3 Aug. 1/7 The pilot was reported to have turned down the RAF offer, which would have given him a foam cushion to land on. Ibid., The cost of laying an anti-fire foam carpet on the runway.

e. Rubber or plastic in the form of a cellular mass similar to foam in structure. Usu. *attrib.*, as *foam mattress, plastic, rubber, sheet*. Also **foamed** *ppl. a.*, made or existing in this form.
1937 Jrnl. R. Aeronaut. Soc. XLI. 1090 Some foam plastics are mentioned for the isolation against noise and the protection against cold. 1943 Air News 15 July 8/4 (caption) New uses for foamed rubber. 1945 Life 1 Oct. 129/1 Foam rubber..will soon be used to soften millions of U.S. beds and chairs. 1948 Daily Mail 30 Aug. 3/6 The chief new outlet in the rubber industry is in 'foam' rubber. 1951 Good Housek. Home Encycl. 108/2 Latex foam padding. 1955 Americana Ann. 125/2 Addition of water to the resin during processing causes elimination of carbon dioxide with the production of a foamed plastic. 1959 B.S.I. News Apr. 4/2 To work economically, this method of heating requires really efficient thermal insulation, and this is to be provided by extensive use of plastics foam sheets. 1959 Economist 6 June 940/2 Protected by space suits and foam mattresses. 1959 Observer 27 Sept. 4/3 'Expanded neoprene' (a kind of foamed rubber). 1961 B.S.I. News Aug. 19/2 The B.S.I. recently drew the attention of the Board of Trade to a flock-filled pillow which was described as being made of 'foam'. 1962 Which? July 198/1 The fifth type (of life-jacket)..used non-absorbent foam plastic to provide buoyancy. 1966 McGraw-Hill Encycl. Sci. & Technol. X. 494a/2 The rigid foams are employed as the core in structural laminates, such as in airplane wings. 1971 Brit. Med. Bull. XXVII. 75/2 Alternative materials for insulation, such as..foam plastics and cork.

5. a. (*a*) *foam bath, -fleck, -fleece, -flower* (see also *5 b); (*c*) *foam-backed, -filled, -upholstered* adjs.; also *foam-cold, -pale* adjs.
1963 Punch 10 Apr. p. xiv, Men's wear..includes foam-backed car coats. 1938 L. MACNEICE I Crossed Minch II. xvi. 232 She says that you're using astringents and taking foam baths. 1969 A. LASKI Dominant Fifth v. 181 She wanted..to go home, take off her shoes, soak in a foam bath. 1929 BLUNDEN Near & Far 39 This foam-cold vale. 1959 Sunday Times 21 June 19/4 A chair that folds into a neat parcel...Foam-filled cushion and back. 1895 M. PEMBERTON Impregnable City iv. 23 The foam-flecks breaking upon the face. 1876 G. M. HOPKINS Wr. Deutschland (1918) 16 The cobbled foam-fleece. 1928 BLUNDEN Retreat 56 Where befouled the foam-fleece stays. 1866 SWINBURNE Poems & Ballads 61 And the sea moving saw before it move One moon-flower making all the foam-flowers fair. 1878 Ibid. 2nd Ser. 29 Look forth from the flowers to the sea; For the foam-flowers endure when the rose-blossoms wither. 1951 W. DE LA MARE Winged Chariot 31 The foam-flowers of sea-wilderness. 1895 W. B. YEATS Poems 41 From the foam-pale distance. 1961 Housewife Apr. 67/1 Chairs have foam-upholstered seats.

b. foam-back, foamback (see quot. 1963); **foam flower,** a member of the genus *Tiarella*, esp. *T. cordifolia*, a small perennial herb native to North America; = *false mitre-wort* (MITRE-WORT); **foam glass** (see quot. 1962).
1961 Drapery & Fashion Weekly 2 Mar. 1 Foambacks will change the trade's whole way of life. 1963 A. J. HALL Textile Sci. iii. 157 Foam-back fabrics. A new development in the manufacture of fabric is to apply to the back of a woven or knitted fabric a very thin layer of synthetic foam sheeting...The presence of the foam layer confers great warmth and induces the garment to retain its shape better during wear. 1895 W. ROBINSON Eng. Flower Garden (ed. 4) II. 788/1 Some real treasures, for no apparent reason, are overlooked. Such has been the fate of the lovely little Foam Flower, a hardy plant of rapid increase. 1908 Suburban Country Life July 20/2 The foam flower, the clintonia, [etc.]..make a carpet in a certain bit of woods. 1962 Amat. Gardening 27 Jan. 4 Tiarella cordifolia and the taller T. Wherryi are good weed suppressors..their feathery appearance has earned them the name Foam Flower. 1948 Archit. Rev. CIV. 268 Insulation for walls, floors and ceilings is 6 in. cork, and for the roof it is foam glass and composition. 1962 Gloss. Terms Glass Ind. (B.S.I.) 33 Foam glass, a rigid multi-cellular insulating block fabricated from glass of high durability and comprising a homogeneous mass of closed cells.

foamed (fōᵘmd), *ppl. a.* [See at end of FOAM v.] **1.** Covered with or as with foam.
1820 Far-foamed sands [in Dict.].

2. Having or made to have a cellular structure like that of foam.
1943, etc. [see *FOAM sb. 1 e]. 1959 Engineering 6 Feb. 183/3 (heading) Foamed glass insulation.

b. *foamed slag*, the product of blast-furnace slag subjected to a special treatment.
1937 Discovery Feb. 58/2 An example in this country of the use of foamed slag concrete blocks is to be found in the Fire Testing Station at Elstree. 1956 Gloss. Terms Concrete (B.S.I.) 5 Foamed slag, a lightweight cellular material [aggregate]..manufactured by treating molten blastfurnace slag with water or other suitable medium to produce a cellular dry product.

fob, sb.[3] Restrict *dial.* to senses in Dict. and add: **c.** *Soap-making.* The scum or froth which rises to the top of the semi-liquid soap during a certain process of manufacture.
1857 W. MILLER Elem. Chem. III. vi. § 1. 373 The fob is skimmed off, and the semi-solid pasty mass of soap is transferred to the frames. 1860 Ure's Dict. Arts (ed. 5) III. 714. 1884 A. WATT Soap-Making v. 47 A black foam or 'fob' appears on the surface, which only ceases when the materials are completely saturated with alkali. 1904 GOODCHILD & TWENEY Technol. & Sci. Dict. s.v. Fitting, A solid crust or Fob of frothy soap.

fob, v.[1] Add: **1.** Also, *to fob* (a person) *of* or *out of* (something). *colloq.*
1843 Punch V. 152/2 Kings and ministers have fobbed us of our renown. 1854 Ibid. XXVII. 232/1 Noodles who complain..of being fobbed..out of various sums from £25 to £90.

fob, v.[3] (s.v. FOB sb.[3]). Add: So **fo·bbing** *vbl. sb.*
1898 G. H. HURST Soaps 229 The material may rise up in the form of froth very considerably, this proceeding being known to soap-makers as fobbing...It is this liability to fob that renders it undesirable that a soap pan should be fully charged to start with. 1938 Thorpe's Dict. Appl. Chem. (ed. 4) II. 95/2 Some brewers add only a small proportion of the hops as the copper is filling up—they help to prevent 'fobbing'.

focal, a. Add: **3.** *focal aperture, capacity*: see quots.; *focal plane*: also often attrib., as *focal plane shutter*, a roller-blind shutter with a wide slit that moves across the front of the plate or film.
1905 Westm. Gaz. 21 Oct. 18/2 In photographing rapidly moving objects the focal-plane shutter does give unmistakable distortion. 1906 Ibid. 8 Sept. 14/2 In a piece of apparatus so complex as the focal-plane reflecting camera. 1918 Photo-Miniature XV. Mar. (Gloss.), Focal aperture (relative aperture), a term expressing the speed of a lens...Focal capacity, an item in the specification of a camera denoting the greatest focal length of lens which can be used. Ibid., Focal-plane shutter, an exposure shutter of the roller-blind or curtain type, placed as nearly in contact with plate or film as possible. 1921 H. G. PONTING Gt. White South 58 To make focal-plane photographic exposures. 1950 [see between-lens shutter s.v. *BETWEEN adv. 4].

focalize, v. Add: **3.** *Med.* To confine to a certain focus (FOCUS sb. 4). Also *intr.* for *pass.* Also **focalization** (later examples relating to sense *3 of the vb.); **fo·calized** *ppl. a.*
1906 Practitioner Nov. 589 The severer disturbances of the liver, those in which the intoxication..focalises in the liver. 1906 [see *DETOXICATE v.]. 1914 V. HORSLEY in S. Paget Sir V. Horsley (1919) II. 196 Every case of focalised epilepsy. 1937 G. W. ALLPORT Personality xi. 295 We are left with a concept of trait as a generalized and focalized neuropsychic system (peculiar to the individual). 1951 H. A. MURRAY in Parsons & Shils Toward Gen. Theory Action 452 The focalizations and reëvaluations that occur in the mind of a hungry man. 1959 S. ARIETI Amer. Handbk. Psychiatry I. I. iv. 103 A need is a general disposition which commonly becomes associated (through 'focalization', or 'canalization', as Murphy would say) with a number of specific entities.

fo'c'sle: see FORECASTLE.

focus, sb. Add: **2. e.** *depth of focus* (examples): now usu. expressed as a distance, and (*a*) used as a synonym of *depth of field* (see *FIELD sb. 16 c); (*b*) distinguished from this and used with the sense: the distance between the two extreme axial points behind a lens at which an image is judged to be in focus.
1860 C. WALDACK Treat. Photogr. ii. 22 The most essential conditions required in a portrait tube, are, rapidity of action, sharpness, and depth of focus...It is for want of depth of focus, that in so many portraits the hands..are out of shape. 1911, 1939 [see depth of field s.v. *FIELD sb. 16 c]. 1940 MORTIMER & SOWERBY Wall's Dict. Photogr. (ed. 15) 183 Depth of focus. Synonyms: Depth of Field, Depth of Definition. 1948 A. L. M. SOWERBY Dict. Photogr. (ed. 17) 179 Depth of focus, strictly, means range of movement of lens that is permissible before image of object focused upon becomes noticeably unsharp. More usually applied to range of distances over which an object can move without becoming noticeably unsharp... In this sense, synonyms are 'Depth of Field' and 'Depth of Definition'. 1958 M. L. HALL Newnes Compl. Amat. Photogr. 48 Another type of depth of focus scale is engraved upon the lens barrel. 1961 G. MILLERSON Technique Telev. Prod. iii. 35 The near and far distances..give us the depth of field (often erroneously called depth of focus—a different matter entirely). 1966 McGraw-Hill Encycl. Sci. & Technol. II. 431/1 There may be some latitude in focusing in which there is no apparent change in sharpness. This distance is known as the depth of focus.

focused, *ppl. a.* Add: **1. b.** *transf.* and *fig.*
1892 J. D. HOOD Waterspouts Yorksh. Wolds 25 The force of the focussed body of water made a triple rent down the slope of the hill. 1909 Milton Memorial Lect. 99 The beautiful, but not strictly focussed scenery of 'Lycidas'. 1934 H. G. WELLS Exper. Autobiogr. I. iii. § 2. 95 Yet impossible as it is to get any focussed clearness and exactitude here, it is equally impossible to ignore this phase of completed puerility.

focuser. Add: **2.** An electrostatic or magnetic device for focusing particles.
1954 Physical Rev. 2nd Ser. XCV. 284/1 Some selection of rotational states could be effected by such a focuser. Similarly, a focuser using magnetic fields would allow spectroscopy of atoms. 1958 Sci. News L. 60 The stream of molecules..passes through the focuser, which consists of four cylindrical electrodes...Those in the higher of the two energy states are deflected toward the axis of the focuser. 1963 G. TROUP Masers & Lasers (ed. 2) viii. 120 An electrostatic focuser and state separator..was used to provide a beam of excited molecules.

fodder, sb. Add: **4.** *fodder-crop, -cutter, -plant* (earlier U.S. example), *-rack, -stack*; **fodder-corn,** also *U.S.*, maize used as fodder.
1856 in D.A. s.v., Hauled up two loads wood & 1 of fodder corn. 1947 Reader's Digest Jan. 59/2 Barns and stable-loft bulging with hay, grain and fodder corn. 1850 Rep. U.S. Comm. Patents 1849: Agric. 158 It is..a fodder-crop and an improver of the land. 1867 Rep. Mass. Board Agric. I. 297 Hay and fodder cutters have become quite indispensable. 1848 Rep. U.S. Comm. Patents 1847 229 The perfection which has been obtained in the root culture and of the fodder-plants. 1902 Daily Chron. 9 Sept. 3/1 When he [sc. a horse] happens to be confronted with an overhead fodder-rack..he raises his head to snatch a mouthful and then lowers it to the pendent position. 1961 Countryman LVIII. 464, I doubt if a single Corsican flock knows the luxury of fodder-racks for the night. a 1738 W. BYRD Hist. Dividing Line (1929) 305 When it rain'd, or was colder than Ordinary, the whole Family took refuge in a Fodder Stack (not far from their roofless house). 1835 A. B. LONGSTREET Georgia Scenes (1871) 24 He commanded all the corn-cribs and fodder-stacks in Georgia. 1890 Century Mag. Dec. 284 The fodder stacks.. might conceal dozens of guerrillas.

foddered, *ppl. a.* (Later U.S. example.)
1864 Rep. Agric. Soc. Maine 52 It will require the attraction of provender..to bring them [sc. sheep] back to their foddered rack.

foehn, var. FÖHN (in Dict. and Suppl.).

fœtalization, fetalization (fiːtălaizēⁱ·ʃən). [ad. G. *fetalisation* (L. Bolk *Das Problem der Menschwerdung* (1926) 25).] (See quot. 1957.) So **fœ·talized, fetalized** *a.*
1930 G. R. DE BEER Embryol. & Evol. viii. 59 The principle here described..is that to which Bolk has applied the term foetalization. 1937 Nature 2 Jan. 37/1 The Bushman..shows features, in which it [sc. the skull] resembles the juvenile skull. It is here suggested that this 'fœtalization'..is not due to an adult retention of juvenile features, but is a precocious assumption of the adult. 1940 AUDEN in I Believe 19 The distinguishing mark of man as an animal is his plastic, unspecialized 'foetalized' nature. 1957 L. EISELEY Immense Journey 130 'Foetalization' or 'pedomorphism', as it is termed, means simply the retention, into adult life, of bodily characters which at some earlier stage of evolutionary history were actually only infantile. Ibid. 133 It [sc. man] is not..the most foetalized race nor the largest brained.

fog, sb.[2] Add: **6. b.** *fog-lamp, -light, -whistle* (example). **d. fog-buoy** (see quots.); **fog-horn** *v. intr.*, to speak with a loud, penetrating voice.
1899 F. T. BULLEN Way Navy 31 A fog-buoy—a barrel towed at the end of a rope at the distance each ship should keep astern of the other. 1939 War Illustr. 29 Dec. 532/2 In foggy weather..every ship tows from the stern on the end of a long line a brightly-painted wooden cask, which is known as a fog-buoy. 1923 A. HUXLEY Antic Hay i. 4 The Reverend Pelvey, M.A., foghorning away from behind the imperial bird. 1937 Times 13 Apr. p. xxii/4 In contributing to safety on the road by day or night, the makers of direction indicators and fog and anti-dazzle lamps have played a notable part. 1962 Fog-light [see *ACCESSORY B. sb. 1]. 1859 First Impressions New World i. 3 We had a good deal of fog when off Newfoundland, which obliged us to use the fog-whistle frequently.

foggage. Add: **1. b.** = FOG sb.[1] 1.
1775 J. WATSON Hist. Halifax 538 Fog, the Grass which grows in meadows after mowing...It is otherwise called Feg, or Fogage. 1834 W. YOUATT Cattle 118 Highland stots..are turned out on the foggage. 1844 H. STEPHENS Bk. Farm III. 841 A foggage field. 1886 C. SCOTT Sheep-farming 169 Lambs..will do very well.. if they have the run of a good aftermath or foggage. 1909 Daily Chron. 28 Jan. 4/7 Devonshire did not recognise what Yorkshire meant by 'fog'—the second crop of grass, the aftermath, known also as 'foggage' and 'eddish'.

fogging, *vbl. sb.*[2] Add: **1.** Also used *attrib.* (= fog-signalling), as *fogging duty, post.*
1898 Daily News 5 Jan. 2 A platelayer..proceeding to his post on 'fogging' duty, was knocked down. 1902 Westm. Gaz. 5 Nov. 12/1 At the different fogging posts on their system. 1907 Ibid. 29 Oct. 9/1 The light of the home signal was out when he arrived at his fogging post.

foggy, a. Add: **5. b.** Used negatively in superl., with ellipsis of *idea, notion.*
1917 Punch 22 Aug. 145 Uncle. '..Wonder who she is.' Niece. 'Haven't the foggiest. Must be pre-war.' 1933 D. L. SAYERS Murder must Advertise viii. 147 'Who was the benefactor?' 'I don't know. Do you, Mr. Bredon?' 'Haven't the foggiest.' 1951 J. B. PRIESTLEY Festival at Farbridge II. iii. 375 'Is that a good idea?' asked Laura. 'My dear, I haven't the foggiest.' 1966 I. JEFFERIES House Surgeon iii. 27 'D'you know where he's to be found at this time?' 'Haven't the foggiest.' 1967 P. MCGERR Murder is Absurd i. 14 'Then you've no idea what his play's about?' 'Not the foggiest,' she said cheerfully.

fogie, var. FOGY.

fogou (fōᵘ·gu). [ad. Cornish *fogo, fougo* a cave, underground chamber. Cf. VOGAL, VUG.] A Cornish souterrain or earth-house.
1889 Chambers's Encycl. III. 490 Cornwall abounds in ..menhirs, cromlechs, artificial caves or fogous, cairns and barrows. 1891 J. H. PEARCE Esther Pentreath IV. ix. 299 He do know the owld adits and fogous..p'raps he's shiltrin' in wan o' they somewhere. 1932 Times Lit. Suppl. 11 Aug. 565/1 The fogous, or covered tunnels, are unknown in the rest of South Britain or on the Continent. 1962 Listener 5 Apr. 591/1 If..you visit the enclosure which contains the fogou, the underground structure, that is, which is called Carn Euny. 1963 Field Archæol. (Ordnance Survey) (ed. 4) 65 Earth-houses...The other part of Britain where similar underground structures are found is Cornwall. Here they are known as fogous...Several Cornish forts..have a fogou which acts as a kind of narrow postern giving access to the ditch from inside the defences. 1970 N. CHADWICK Celts v. 126 The souterrain, sometimes referred to in Cornwall as a fogou.

fogy. 4. (Earlier U.S. examples.)
1852 Democr. Rev. XXX. 207 The fogy atmosphere of Washington makes cowards of you all. 1865 Trans. Ill. Agric. Soc. 1862–4 V. 691 We flaunt our labor-saving farm machinery in the faces of slow and fogy Germany.
¶ See also *old-fogyish, -fogyism* s.v. OLD a. D. 2.

föhn. Add: **1.** (Earlier example.)
1861 A. CATLOW Sketching Rambles I. ii. 57 The most violent tempests are raised by the Föhn, or south wind.

2. Also **foehn.** A warm dry katabatic wind developing on the lee side of a mountain range in response to air moving across the range. Also *föhn wind.*
1889 W. FERREL Pop. Treat. Winds vi. 333 Temporary foehns are produced whenever there is a cyclone so situated as to draw a current of air over a high mountain range. 1910 Encycl. Brit. VI. 523/2 Greenland föhn winds..blow down warm and dry, raising the temperature even 30° or 40° above the winter mean, and melting the snow. 1951 F. DEFANT in T. F. Malone Compend. Meteorol. 667/1 In North America the foehn, known as the chinook, has a climatic influence on a very wide belt east of the Rocky Mountains. 1957 J. K. CHARLESWORTH Quaternary Era II. xxxi. 678 Foehn winds, varying in strength, duration and range as in modern Greenland.. probably blew throughout the year.

foie gras, foie-gras (fwa,gra). *colloq.* Short for *pâté de foie gras*: see PÂTÉ 1.
1818 T. MOORE Fudge Fam. Paris 23 The liver of the unfortunate goose is enlarged, in order to produce that richest of all dainties, the foie gras. 1865 'OUIDA' Strathmore I. i. 5 The host and the guests..lounged;..peeling an apricot, or cutting into a foie gras. 1913 H. CARADOC City of Plain iii, Sussex here helped himself to two foie-gras sandwiches. 1946 G. MILLAR Horned Pigeon xx. 307 They were foie gras sandwiches. 1958 L. VAN DER POST Lost World of Kalahari i. 15 The nocturnal hippopotamus whose sweet lard meant more to him than foie-gras to any gourmet. 1970 New Yorker 3 Oct. 43/2 Mothers perennially sending caviar and foie gras to their sons on duty.

foil, sb.[1] Add: **4. d.** Used as a wrapping, container, etc., for food.
1946 Steelways Sept. 30 Aluminum foil is the star in frozen foods packaging. 1959 Woman 30 May 3/4 The little foil trays in which tarts are sold are useful afterwards for placing beneath medicine bottles. 1961 Harrods Food News 3 Cover the pan with two or three layers of foil. 1969 Daily Tel. 27 Mar. 17/4 Cut and trim the toast bread and set it in table position (tightly foil-wrapped to keep it perfectly moist) with the butter.

foi·l-borne, foilborne, a. [*BORNE ppl. a. 2 b.] Of a craft: raised out of the water by lift provided by hydrofoils. Of the motion, etc., of a craft: taking place while it is so supported.
1961 Hovering Craft & Hydrofoil Oct. 6/2 The foils will be lowered, the main engine started, and the craft can proceed foilborne at 60 knots. Ibid., Driftwood and other floating debris has little effect on foilborne boats. 1964 Weekly News (Auckland) 8 July 31 (caption) The Manu-wai..settles heavily as her foil-borne hull re-enters the water. 1967 Jane's Surface Skimmer Systems 1967–68 95/1 An automatic control system is installed that utilises craft motion and height inputs to maintain foilborne flight. Ibid. 91/2 Foilborne propulsion. 1968 New Scientist 17 Oct. 138/2 One civil application would be to inject additive round the hull of a hydrofoil while it was accelerating to become foil-borne.

foilist (foi·list). [f. FOIL sb.[5] + -IST.] One who fences with a foil.
1907 Times 2 Apr., I implore you..to withdraw the word 'foilist' from circulation before the authority of The Times has given it..currency. 1932 Times Lit. Suppl. 24 Mar. 220/2 The beginner..should be able to develop into a more than competent foilist. 1958 Oxf. Mag. 13 Mar. 377/1 The foilists..defeated the strong Cambridge foil team.

foist, sb.[3] Add: **d.** *U.S.* = *FICE, FIST sb.[2] 3.
1851 T. A. BURKE Polly Peablossom's Wedding 18 [The sounding of the horn] was sufficient invitation to every hound, foist, and 'cur of low degree' to join in the chorus.

Fokker (fǫ·kəɹ). [Name of A. H. G. *Fokker* (1890–1939), a Dutch engineer, the inventor.] Orig. a German tractor monoplane characterized by its speed and climbing power. Later used *attrib.* to designate aircraft manufactured by the Fokker company in the Netherlands and elsewhere.
1913 Aeroplane 23 Jan. 78/1 The..successful flying of the Rumpler 'Taube', the 'Albatros', and the 'Fokker' in Germany. 1917 'CONTACT' Airman's Outings 41 An aerodrome just east of the wood was the home of the Fokker star, Boelcke. 1930 C. R. SAMSON Fights & Flights 324 A German Fokker, which..was more than a match for a slow old Short with only a Lewis gun. 1941 W. S. CHURCHILL Secret Session Speeches (1946) 27 Air attack upon our Western Approaches by means of long-range Fokker Wulf machines. 1966 Times 9 Aug. 13/6 This company of 5,000 employees has an international reputation, based primarily on the popularity of the F27, the Fokker Friendship.

folacin (fōᵘ·lăsin, fǫ·lăsin). *Biochem.* Also **-ine.** [f. *FOL(IC *ac*(*id* + -IN[1].] = *folic acid (pteroylmonoglutamic acid).
1949 Jrnl. Laboratory & Clinical Med. XXXIV. 501 Evidence indicating rapid metabolic liberation of folacin (folic acid, pteroylglutamic acid) from teropterin (pteroyltriglutamic acid). 1955 Chem. & Engin. News 6 June 2433/1 The pteroylglutamic acids may be designated generically as folic acids. The pure substance hitherto known as folic acid, folacine, or vitamin B_c shall be named pteroylmonoglutamic acid.

folboat, var. *fold boat* (*FOLD sb.[3] 7).

fold, sb.[3] Add: **1. b.** (Further examples.)
1888 ROLLESTON & JACKSON Forms Anim. Life 294 Two 'mantle' folds, one dorsal, the other ventral, line the corresponding valves...A 'brachial' groove bounded on its inner side by a prominent fold or lip. Ibid. 692 They [sc. the valves of Coelomate Metazoa] are lined by two mantle folds or extensions of the body walls. 1895 Cambr. Nat. Hist. III. 264 The folds or plaits on the columella, which are often characteristic of the genus or even family (e.g. Fasciolariidae, Mitridae, Turbinellidae) are not merely external, but continue down the whole spire. 1942 L. H. HYMAN Compar. Vertebr. Anat. (ed. 2) xi. 289 The tonsillar fossa is bounded in front and behind by low folds. 1949 A. S. ROMER Vertebr. Body x. 294 In jawed fishes, amphibians and reptiles, the lips are in general small and unimportant skin folds. 1964 W. W. BALLARD Compar. Anat. & Embryol. xi. 193 The amnion is usually formed through the agency of two crescentic amniotic folds.

c. *spec.* an undulation or gentle curve of the ground; a slight hill or hollow; the general grouping of heights and hollows.
1869 G. M. HOPKINS Jrnl. 27 June (1959) 191 Br. Sidgreaves has heard the high ridges of a field called folds and the hollow between the drip. 1901 Daily News 19 Jan. 5/3 Four companies of the Rifle Brigade were concealed behind a fold of ground. 1904 A. E. FLETCHER Gainsborough 88 The fold of the fields and the grouping of trees. 1930 Morning Post 17 June 20/5 Delightful house in a fold of the Downs.

h. *Geol.* A bend in rock strata, esp. one having a wave-like form.
1863 J. D. DANA Man. Geol. 104 One fold or flexure in the rocks may succeed to another, or they may form interrupted series. 1885 C. LYELL Elem. Geol. (ed. 4) 61 Stratified rocks have their strata usually bent into parallel folds forming anticlinal and synclinal curves. 1937 [see *FAN sb.[1] 10 f]. 1965 A. HOLMES Princ. Physical Geol. (ed. 2) ix. 212 Folds range in intensity from broad and gentle undulations to tightly compressed plications in which the dips of the beds are almost parallel, except near the hinge-lines.

7. *attrib.* **fold boat** = *FALTBOAT; also in contracted form *folboat*; **fold(-)mountain,** a mountain formed directly by folding, or one in which the strata are extensively folded; so *fold ridge.*
1938 Times Lit. Suppl. 24 Sept. 616/2 With his wife he made a journey in a fold-boat..down the Danube. 1946 R. CAPELL Simiomata I. 19 The Marines entered the harbour in their little collapsible Folboats, attached explosive charges to two destroyers and two cargo boats, and retired. 1955 'N. SHUTE' Requiem for Wren 128 An M.T.B...with a folboat on board, a sort of kayak built of waterproof canvas on a wooden frame. 1969 Publ. Amer. Dial. Soc. LI. 6 Fold boat, kayak made of skeleton frame covered with a skin of rubberized cloth. It folds into a small package. 1908 J. W. GREGORY Geogr. 34 Fold-mountains arise from the crumpling of bands of the earth's crust by lateral pressure. Such fold-mountains resemble the wrinkles formed when a table-cloth is pushed across a table. 1965 A. HOLMES Princ. Physical Geol. (ed. 2) xxx. 1111 The axes of the major structures of a range of fold mountains..are generally roughly parallel to the trend of the range. 1936 Discovery Jan. 20/1 The late Miocene-Pliocene fold ridge which extended to New Caledonia, and in the other direction formed the Himalayas and the European Alps.

fold, v.[1] Add: **1. e.** Now esp. with *up*.
1914 Daily Mail 21 Feb. 10/1 One of the little seats that folded up when they were not required. 1939–40 Army & Navy Stores Catal. 759/1 This roll-up chair.. folds up small enough to go in a large pocket.

5. Delete † and add later examples. Esp. with *up*.
1911 A. BENNETT Honeymoon I. 15 Cedric: The mater folded up like that? Flora: ..Naturally she folded up. She only needs proper treatment. 1928 J. P. MCEVOY Showgirl 199 The Main Stem hears under cover that 'Get Your Girl' may fold up soon for lack of suitable house. 1936 Variety 24 June 66/1 First tour of the Draegermen.. proved a box office flop, and folded here. 1937 N. COWARD Present Indicative v. vi. 191 In spite of excellent press notices..the play folded up at the end of eight weeks. 1939 E. B. WHITE Quo Vadimus? I. 47 Nutshell folded up, because, an expert said, the name was too long; but half a dozen others sprang up to take its place. 1942 Penguin New Writing XV. 12 Yes, but when he gets outside he'll fold up. 1958 I. MURDOCH Bell xxvi. 315 He was sorry.. to hear that Imber was folding up. 1965 G. MELLY Owning-Up vi. 71 The band folded in December 1961. 1971 Guardian 21 Jan. 1/1 (headline) Skyways folds up. 1971 Sunday Times (Colour Suppl.) 24 Oct. 31/1 This generous subsidy could not go on for ever and when it was withdrawn the magazine folded.

6. In mod. use freq. with *together*.
1771 H. MACKENZIE Man Feel. 223 He folded his hands together. 1853 KINGSLEY Hypatia x, Her hands folded together before her. 1894 B. HARRADEN Varying Moods 60 She folded her hands together on the buckle of her waist-belt.

8. Also const. certain advs. and prepositions.
1891 [see *ENEMY sb. and a. B.2]. 1906 K. TRASK Night & Morn. 34 Take me, Leonidas, to thy strong arms—.. fold me from the whole wide world. 1915 H. MACKAY London, One Nov. 46 The mystery of things the twilights fold away.

e. Const. *in. Cookery.* To add (an ingredient) gently by lifting a mixture with a spoon, etc., so as to enclose it without stirring or beating.
1915 D. C. PEEL Learning to Cook xiv. 172 Add the whipped whites last, and..fold them in lightly. 1933 W. G. R. FRANCILLON Good Cookery xv. 316 Fold in the slightly warmed flour. 1958 Listener 20 Nov. 855/2 Add the yolks and cheese to the mixture, then stiffly beat the whites of the eggs and fold in gently.

10. foldaway *a.*, adapted to be folded away; **fold-out** *sb.*, an oversize page in a book, magazine, etc., which has to be unfolded by the reader; also as *adj.*
1960 Farmer & Stockbreeder Suppl. 26 Jan. 4/1 A wash boiler with all refinements except a pump but with a fold-away hand wringer. 1967 Gloss. Caravan Terms (B.S.I.) 3 Foldaway bed, a bed designed to be lowered from and retract into a cupboard or wall or to fold on itself to make an upright furniture unit. 1969 Jane's Freight Containers 1968–69 577/3 Foldaway guides to aid truck location. 1971 Guardian 24 Aug. 7/3 An 84-year-old woman..was trapped in her foldaway bed for 13 hours. 1961 WEBSTER, Foldout. 1963 Amer. Speech XXXVIII. 125 It might be well to include a fold-out map at the end of each volume. 1967 Time 17 Mar. 7 The prurient appeal of an overripe foldout is no worse than the peekaboo enticement of gossip about 'People'. 1968 Listener 30 May 705/1 A coffee-table book, with Topolski drawings, wide margins and a fold-out frontispiece.

folder, sb. Add: **h.** A folding case or cover for loose papers.
1911 H. S. HARRISON Queed 329 West went to a filing cabinet..pulled out a large folder marked, Reformatory. 1922 Brit. Stationer 11 Dec. 66 Folders for name and number system. 1930 A. PALMER Company Secretarial Pract. 182 To file the copies in a folder. 1950 E. BOWEN Coll. Impressions p. vi, These are impressions..which, by force of continuing to be written and of having been kept in a folder, have collected themselves.

folding, *vbl. sb.*[1] Add: **3. folding strength,** the strength of paper when subjected to continuous alternate creasing.
1936 Discovery May 156/2 The percentage decrease in the folding-strength of a paper after it has been heated for 72 hours at 105°C. is taken as a measure of its permanence on ageing. 1963 R. R. A. HIGHAM Handbk. Papermaking iii. 79 The folding strength of a paper is dependent on the fibre length and flexibility of the sheet.

folding, *ppl. a.* Add: **1. b.** *folding money*: paper money. *colloq.* (chiefly *U.S.*).
1930– in WENTWORTH Amer. Dial. Dict. (1944) 225/1 Foldin' money, paper money, bills;—humorous. W.Va. & general. 1940 J. O'HARA Pal Joey 162 Any time the socialites go out they leave there [sic] folding money at home. 1968 L. DEIGHTON Only when I Larf iv. 47 I'd walked straight out of the bank carrying a bag full of folding money. 1970 Daily Tel. (Colour Suppl.) 15 May 10/2 They used to ask me to prepare the cars and then work, say, in the pits at Le Mans. This gave me a bit of folding money.

folic (fōᵘ·lik, fǫ·lik), *a.* *Biochem.* [f. L. *fol-ium* leaf + -IC.] *folic acid*: pteroyl-(mono)glutamic acid, a vitamin of the B group found in a wide variety of organisms, the deficiency of which in man is associated with anæmia; also, any of the other pteroylglutamic acids.
1941 K. MITCHELL et al. in Jrnl. Amer. Chem. Soc. LXIII. 2284/1 Using Streptococcus lactis R as a test organism, we have obtained..an acid nutrilite...It is especially abundant in green leaves...Because of this fact, and since we have obtained what appears to be a nearly pure chemical entity, we suggest the name folic acid. 1949 H. W. FLOREY et al. Antibiotics II. xlvii. 1448 The synthesis of pteroylglutamic acid (folic acid). 1960 A. HUXLEY Let. 23 June (1969) 892 She is doing very well on the B12+Folic Acid which I prescribed..as a cure for her dizzy spells. 1961 W. J. E. JESSOP Fearon's Introd. Biochem. (ed. 4) xiii. 201 All the folic acids are effective in the treatment of..non pernicious anæmias. 1961 Lancet 26 Aug. 483/1 The complex relation between vitamin B12 and folic acid is still far from being completely understood.

‖ **folie** (foli), sb. [Fr.; see FOLLY.] Madness, insanity, mania. Chiefly in various pathological terms (see quots.).
1803 Edin. Rev. II. 163 The maniac appears perfectly composed, makes the best and most correct answers to the questions put to him; but, if set at liberty, immediately becomes furious. This kind of derangement..is what is vulgarly called folie raisonnante. 1880 Encycl. Brit. XIII. 109/2 To this condition foreign authorities have applied the term folie circulaire...After intervals of comparative sanity, the patient manifests symptoms which run their course through the prodromal, the acute, and the demented stages, on again to recovery. 1890 BILLINGS Med. Dict., Folie du doute, doubting insanity. Ibid., Folie des grandeurs, delirium of grandeur. 1913 DORLAND Med. Dict. (ed. 7), Folie à deux, communicated insanity. 1914 W. J. LOCKE Jaffery viii, This is absurd. It's megalomania—la folie des grandeurs. 1915 A. HUXLEY Let. Oct. (1969) 80 A new growing country, swelled with its own pride, filled by its growing pains with an immense folie de grandeur. 1923 —— On Margin 181 The baroque folie de grandeur. 1960 Guardian 8 July 13/6 Friedrich Grunwald..denied having 'folie de grandeur'—being a big head who has gone round the bend. 1961 Listener 24 Aug. 290/2 You..are afflicted with folie de doute (was it really 10,000,000 cells per minute, or 10,000, and if so, why?). 1962 Sci. Amer. Aug. 66/3 In twins, as in other pairs of individuals who are emotionally close to each other, the condition known as folie à deux has been commonly observed. The two partners tend to develop shared delusions and symptoms. 1962 John o' London's 27 Sept. 307/4 A brilliantly observed clinical study of the folie à deux indulged in by a pair of chronic fantasists.

folinic (fōᵘli·nik, fǫl-), *a.* *Biochem.* [f. *FOL(IC *a.* + -IN + -IC.] *folinic acid*: a derivative of folic acid, believed to be the form in which the latter acts as a coenzyme.
1949 T. J. BOND et al. in Jrnl. Amer. Chem. Soc. LXXI. 3852/2 On the basis of structure and functional relationship to folic acid, this active principle has been termed folinic acid. 1962 New Scientist 17 May 364/3 To counteract the damaging effects of the methotrexate that spilled over into the general circulation, a biological antagonist, folinic acid (leucovorin), was injected intramuscularly. 1966 E. S. WEST et al. Textbk. Biochem. (ed. 4) xviii. 814 Folinic acid, citrovorum factor (CF), and leucovorin are names for a naturally occurring derivative of folic acid... It is N[5]-formyl-5,6,7,8-tetrahydropteroylglutamic acid.

foliot. Add: **3.** [Fr., 14th cent.: see Dict.] A type of clock escapement consisting of a bar with adjustable weights on the ends. (*Disused.*)
1899 F. J. BRITTEN Old Clocks & Watches 23 Froissart has left a descriptive eulogium of a clock, written in 1370...In this the controlling medium is referred to as a 'foliot', which was doubtless the straight armed balance with weights. 1948 A. L. KROEBER Anthropol. (ed. 2) xi. 453 The first clock escapements, large and heavy, were of the type called 'foliot balance and verge escape'. 1958 New Scientist 11 Dec. 1481/1 (caption) The Dover Castle clock, showing the verge-and-foliot escapement. 1962 E. BRUTON Dict. Clocks 152 Many years ago it had been converted to pendulum. In 1956 it was restored to foliot..and set going again inside the cathedral.

folk. Add: **3. c.** *folk of peace* [mistranslation of Gael. *daoine sidhe*, lit. people of the fairy hill (cf. Ir. *bean sidhe* BANSHEE), by confusion with *sithe*, gen. of *sith* peace]: fairy folk, fairies. *Sc.*
[1691 R. KIRK Secret Commonwealth (1815) 30 The Seers avouch, that severalls who go to the Siths, (or People at Rest, and, in respect of us, in Peace,) before the natural Period of their Lyfe expyre, do frequently appear to them. 1841 H. MILLER Old Red Sandstone xi. 215 'Not of the race of Adam,' said the creature, turning for a moment in its saddle: 'the People of Peace shall never more be seen in Scotland.' 1900 J. G. CAMPBELL Superstitions Highlands & Islands Scotland i. 1 The Fairy or Elfin people, or, as they are called both in Irish and Scottish Gaelic, the sith people, that is, 'the people of peace', the 'still folk', or 'silently-moving' people.] 1875 Encycl. Brit. II. 203/2 The Highlanders call them the folk of peace. 1893 R. L. STEVENSON Catriona I. i, I am nameless, like the Folk of Peace.

5. b. Add to def.: traditional, of the common (local) people, esp. opp. sophisticated, cosmopolitan, as *folk-art, -artist, -belief* (examples), *-comedy, -culture, -drama, -epic, -hero, -legend, -life, -literature* (example), *-medicine, -mind, -museum, -name* (examples), *-poem, -poetry, -rhyme, -tale* (later example), *-tradition.*
1921 Art & Archæol. XI. 185 (title) Folk art. 1938 Encycl. Brit. Bk. of Yr. 476/2 A quiet growth of more spontaneous painting, some of it idyllic and reminiscent of folk art. 1967 E. SHORT Embroidery & Fabric Collage ii. 41 A more recent tradition in dress which could be classed as folk art is that of the pearly kings and queens. 1934 Hound & Horn July–Sept. 589 The spirit of a folk school of music so excites the folk artist that [etc.]. 1959 Times 11 Dec. 16/2 He [sc. Barrie] remains a folk-artist. 1892 G. L. GOMME Ethnol. in Folklore v. 123 These ghastly ceremonies throw much light on the old folk-belief as to the dead. 1922 W. B. YEATS Trembling of Veil 243 Folk-belief, tales of the fairies. 1933 E. K. CHAMBERS Eng. Folk-Play 216 A folk-belief may..explain the singular passage..in which the Fool..beholds his own face. 1952 D. HOFFMAN Paul Bunyan iv. 73 The authors..can dramatize folk comedy by contrasting it with the standards of cultivated society. 1936 Discovery Mar. 95/2 The volume is an important contribution to the study of European folk-culture. 1963 AUDEN Dyer's Hand 300 The beast in it [sc. fable] may be a folk-culture hero whose qualities of courage or cunning are to be imitated. 1964 GOULD & KOLB Dict. Social Sci. 272/1 A folk culture..is a culture in which behaviour is highly conventionalized, personal, based on kinship, and controlled informally, traditionally, and through the 'sacred'. 1917 W. B. YEATS in Lett. J. Joyce (1966) II. 405 We do not play the folk drama very well. 1961 Times 23 May 15/1 Ballad and folk-drama. 1904 C. G. CHILD Beowulf p. v, Great indeed would have been our loss, if..the only remaining folk-epic of the Germanic peoples, had perished in doing menial service to grocer or soap-seller. 1950 John o' London's 24 Nov. 629/1 A great folk-epic about the Cid Campeador. 1899 Folk-lore X. IV. 407 The folk-heroes Hari Chand and Raja Amba. 1927 J. L. BROOKS (title of unpubl. thesis) Paul Bunyan: American Folk Hero. 1932 Q. D. LEAVIS Fiction & Reading Public II. i. 92 The traditional folk-heroes like Crispine and Crispianus, Simon Eyre, and the Six Worthy Yeomen of the West. 1960 20th Cent. Dec. 557 Blues-singers..sanctified by white jazz intellectuals into folk-heroes. 1971 P. DRISCOLL White Lie Assignment xiii. 103 A guerilla in the new folk-hero style. 1909 A. HERBERT Isle of Man vii. 101 Around ..the golden plover..is hung one of the prettiest of the folk-legends abounding in the isle. 1923 W. B. YEATS Plays & Controversies 210 We thought we could bring the old folk-life to Dublin. 1955 A. LOMAX Mister Jelly Roll (ed. 3) p. xv, It is within the folklife of these Creoles that the emotional character of hot jazz is to be found. 1966 G. E. EVANS Pattern under Plough 21 Folk-life implies a holistic approach whose main definition is not in method but in the field of study. 1893 W. B. YEATS Celtic Twilight 201 Irish folk-literature. 1898 E. CLODD Tom Tit Tot vi. 61 Folk-medicine, the wide world through, is full of prescriptions based on sympathetic or anti-pathetic magic. 1931 Times Lit. Suppl. 6 Aug. 605/1 In this book an attempt is made to resuscitate and standardize much forgotten folk medicine. 1968 Times 3 Dec. 10/7 Yet another piece of folk medicine seems to have been vindicated in the laboratory. 1899 Folk-Lore X. IV. 385 The surest way therefore of projecting oneself into the folk-mind..is..to take up the various points as they have seemed to grow one out of the other in folk-logic and processes of thought. 1924 P. C. BUCK Scope of Music ix. 116 The composer, if it is the work of one man, or the folk-mind, if it is the result of the friction of use, knew better. 1957 Times Lit. Suppl. 1 Nov. 656/3 It is hard..to think of Loki as other than fundamentally malignant, and very hard to think of the folk-mind, even over centuries, turning him into a sympathetic comic figure. 1936 M. ALLIS Eng. Prelude xxiv. 258 In Strangers' Hall and Suckling House Norwich offers the most complete, if not the only, folk museums in England. 1966 P. V. PRICE France 126 Collections of objects connected with wine and food..are frequently found in folk museums. 1924 MAWER & STENTON Introd. Surv. Eng. Place-Names iii. 50 These names are important as a link between the folk-names and place-names restricted to some particular spot. 1960 P. H. REANEY Orig. Eng. Place-Names vi. 99 The earliest place-names created by the Anglo-Saxons were not originally place-names in the strict sense of the word; they were folk-names. 1940 Horizon Mar. 170 A real folk-poem, it was in its way a work of art. 1963 Times 18 July 5/3 A suite of old folk-poems about Charity as exemplified in the tale of Dives and Lazarus. 1892 S. A. BROOKE Hist. Early Eng. Lit. I. 90 As to the spears singing..that is a common phrase in ancient folk-poetry. 1903 L. F. ANDERSON Anglo-Saxon Scop 10 Mone..states his belief that we have in the..passage a reference to Germanic folk-poetry. 1923 A. HUXLEY On Margin 54 The Folk Poetry of 1920 may best be classified according to subject-matter. 1965 G. MCINNES Road to Gundagai xvi. 280 The earthy folk-poetry of C. J. Dennis. 1889 Chambers's Jrnl. VI. 670/1 English folk-rhymes are very numerous and curious. 1959 I. & P. OPIE Lore & Lang. Schoolch. xv. 333 The future is worked out with the aid of an old folk rhyme. 1935 R. GIRVAN Beowulf & 7th Cent. 82 In Beowulf the folk-tale is the story: all the rest is incidental. 1950 M. J. C. HODGART Ballads v. 110 There were poets of taste who were close enough to folk-tradition to be able to adapt it without making it look literary. 1971 Guardian 3 July 8/5 We [sc. Londoners]'ve got so much more folk tradition than most places.

6. folk-blues [*BLUES], the original 'blues' of the Negroes of the southern U.S., as opposed to composed imitations; so *folk-jazz*; **folk-dance,** a dance of popular origin; the music for such a dance; also **folk-dance** *v. intr.*, **folk-dancer; folk-dancing** *vbl. sb.*; **folk high school** [tr. Da. *folkehøjskole*], a school of adult education, originating in Denmark, now also in other Scandinavian countries; **folk-law** (usu. in *pl.*), a customary law of the people, applied esp. to the *Leges Barbarorum*, the laws of the Germanic peoples; **folk-memory,** recollection of the past persisting among a people or group; **folk-music,** music of popular origin; hence **folk-musician**; so *folk-tune*; **fo·lknik** [after *beatnik], a devotee of folk-music; also *attrib.*; **folk-play,** a traditional type of play; so *folk-player*; **folk-psychology** [tr. G. *völkerpsychologie*] = ETHNOPSYCHOLOGY; so *folk-psychologist*; **folk-singer,** a singer of folk-songs; so **folk-singing; folkway** (usu. in *pl.*), the traditional behaviour of a people or group; **folk weave, folkweave** (see quot. 1960); also *attrib.*
1926 A. NILES in W. C. Handy Blues 3 Some folk-blues were love songs. 1959 'F. NEWTON' Jazz Scene vi. 99 The anonymity and impersonal grandeur of the folk-blues. 1909 E. BURCHENAL (title) Folk-dances and singing games. 1912 C. J. SHARP Folk Dancing in Schools 4 The three main types of folk-dance found in England are:— (1) The Morris Dance. (2) The Sword Dance. (3) The Country Dance. 1927 Observer 2 Oct. 19/4 He defied anyone who folk-danced to be unhappy. 1954 M. EWER Heart Untouched v. 77 Do they folk-dance? Do they make ghastly things in raffia? 1967 CHUJOY & MANCHESTER Dance Encycl. 370/2 In N.Y. alone there are enough folk dance sessions to offer anyone an opportunity to join in folk-dancing almost every day of the week. 1936 Discovery Dec. 396/2 The Portuguese folk-singers, folk-players, and folk-dancers go through their traditional performances with a complete lack of self consciousness. 1907 Folk-dancing [see *folk-singer]. 1908 Amer. Physical Educ. Rev. Oct. 375 The place of folk dancing. 1927 Observer 2 Oct. 19/4 If there was more singing of the old songs and more folk-dancing. 1914 U.S. Bureau of Education Bull. no. XXII. 1914 (title) The Danish Folk High Schools. 1949 E. L. ALLEN Bishop Grundtvig vii. 82 A large proportion of the heads of folk high-schools to-day are men with a theological training. 1966 D. JENKINS Educated Society iii. 125 The Folk High School movement obviously has achieved a great success in the predominantly agrarian communities of Scandinavia. 1950 A. LOMAX Mister Jelly Roll (1952) 185 Linking the folk-jazz half-world to the super-respectable and stuffy world of the music business. 1884 Folk-law [see Dict. 5 b]. 1898 E. JENKS Law & Politics Middle Ages ii. 32 While France and Germany have their feudal laws.. England is still in the twilight of the folk-laws. 1962 H. R. LOYN Anglo-Saxon England iv. 175 It is not a mere quibble..to question a sharp cleavage between ideas of royal law and folk-law. 1908 W. JOHNSON (title) Folk-memory; or, The continuity of British archæology. 1939 R. G. COLLINGWOOD Autobiogr. xi. 143 'Folk-memory'..the transmission by example and precept of certain ways of thinking and acting from generation to generation. 1960 K. M. KENYON Archæol. in Holy Land viii. 208 The length of folk memory, though it may be reasonably accurate as to the occurrence of important events, is short as regards chronological exactitude. 1889 G. B. SHAW London Music 1888–89 (1937) 79 All good 'folk music' is as international as the story of Jack the Giant Killer. 1907 C. J. SHARP Eng. Folk-Song p. ix, It is impossible to say how far the folk-music that has survived in a county like Somerset is..truly representative of English folk-song as a whole. 1971 Guardian 3 July 8/1 Cecil Sharp deliberately looked for English folk music in other countries, especially America. 1907 C. J. SHARP Eng. Folk-Song 34 The folk-musician..is under no such temptation [to make music for the sake of making it]. 1941 L. MACNEICE Poetry of Yeats viii. 165 A poet of the folk-musician type. 1958 in Amer. Speech (1966) XLI. 139 In Greenwich Village..lies the Folklore Center..near the door is the lettering 'Israel G. Young'... Invariably, folkniks (to quote Izzy) are present, for this den is the meeting place..for New York's up and coming folksingers and hangers on. 1961 McLean's 25 Mar. 18/1 He is exhibiting with increasing frequency in the circle of Toronto restaurants and coffee houses known locally as folknik spots. 1963 Observer 8 Sept. 12/6 Folkniks.. are the adherents, mostly teenaged, of the new-style folk-song movement—an austere group who prefer their songs unaccompanied...In the spreading folknik clubs, amplifiers are forbidden. 1905 Westm. Gaz. 7 Apr. 12/1 This pageant, which takes the form of a folk-play specially written and invented by Mr. Louis N. Parker, deals with the chief events in the history of the town. 1933 E. K. CHAMBERS (title) The English folk play. 1969 in Halpert & Story Christmas Mumming in Newfoundland 59 An attempt to write the detailed history of the folk play must await the results of current investigations. 1936 Folk-player [see *folk-dancer]. 1889 Cent. Dict., Folk-psychology. 1918 R. S. WOODWORTH Dynamic Psychol. i. 12 In the sixties, there was even published for several years in Germany a journal of 'folk psychology'. Ibid. 13 The methods and presuppositions of the older folk psychologists have not stood the test of time. 1957 J. S. HUXLEY Relig. without Revelation v. 108 Explanations of the origin of religion in terms of animism or magic or folk psychology. 1898 Folk-lore IX. I. 49 The sagas..died out because there were no folk-singers qualified to present them. 1907 C. J. SHARP Eng. Folk-Song p. viii, Only those, perhaps, who have been brought into close contact with the old folk-singers of to-day, can fully realize how intimately folk-singing and folk-dancing have ..been bound up with the social life of the English village. 1927 Observer 12 June 13/2 Altogether more than fifty folk-singers and craft workers hailing from all parts of French Canada were gathered together. 1961 Guardian 9 May 7/2 The white folk-singers of the southern back-woods. 1931 Times Lit. Suppl. 30 Apr. 344/4 The free rhythm and florid ornament..characteristic of Greek folk-singing. 1934 W. B. YEATS King of Gt. Clock Tower 18 Her method was 'folk-singing' or allied to it. 1971 Guardian 3 July 8/2 Purists..regard the folk-singing.. as being 'impure'. 1907 C. J. SHARP Eng. Folk-Song p. x, The folk-tune presents many problems of absorbing interest to the musical theorist. 1914 —— Folk Singing in Schools 5 The folk-tunes which have recently been collected from the English peasantry. 1957 MANVELL & HUNTLEY Technique Film Music ii. 51 A glimpse of a folk-tune is sometimes heard on French horns. 1907 W. G. SUMNER Folkways p. iii, I formed the word 'folk-ways' on the analogy of words already in use in sociology. ..Folkways are habits of the individual and customs of the society which arise from efforts to satisfy needs... Then they become regulative for succeeding generations and take on the character of a social force. 1925 W. P. MONTAGUE Ways of Knowing v. 142 There are cases when a 'folkway' or social habit becomes actively evil. 1959 G. D. MITCHELL Sociol. v. 78 Customs may be differentiated into folkways and mores. 1964 GOULD & KOLB Dict. Social Sci. 273/2 Folkways are the learned shared behaviour common to a people. 1971 E. FENWICK Impeccable People iii. 17 He had learned to swallow the folkways of Parsons Point. 1938 Decorative Art 50 The gaily striped folk-weaves on sofas. 1939 Soft Furnishing in Workroom 10 Brown/Beige Folkweave Fabric..for Loose Cover. 1949 E. COXHEAD Wind in West v. 123 Her three-piece leatherette suite, her beige and orange folk-weave curtains. 1960 Textile Terms & Defs. (ed. 4) 70 Folk weave, n. or adj., a term applied to any construction which, when used in loosely woven fabrics made from coarse yarns, gives a rough and irregular surface effect.

b. Short for *folk-music*. So *folk-rock*, folk-music with a strong beat; *folk club*, etc.
1963 Observer 8 Sept. 12/7 MacColl learnt folk at his mother's knee, felt people should know their own music. 1965 New Society 20 May 26/3 There are..as many as 300 folk clubs..all over Britain..most of them grouped around a resident singer or group. 1966 Time 1 July 50 The folk-rock movement. 1966 Guardian 22 Dec. 4/7 In the pop/folk field the best new release is by The Incredible String Band. 1969 Down Beat 17 Apr. 24/2 The lyrics are characterized by the self-consciously whimsical or ironic or nouveau-romantic trends of folk-rock at its worst. 1971 Melody Maker 18 Sept. 36/1 The Crown is one of the foremost folk clubs in Edinburgh. Ibid. 36/3 It's still one of the best places in the country for folk in terms of performance and audience reaction.

folkish (fōᵘ·kiʃ), *a.* [f. FOLK + -ISH.] Of the common people; traditional, unsophisticated.
1938 D. DAVIDSON Attack on Leviathan 226 Semi-literary but folkish creations like Leatherstocking, Rip Van Winkle, Huck Finn. 1951 M. COOPER Russian Opera iv. 40 Most striking is the warm, folkish lyricism [of the opera]. 1962 Listener 25 Oct. 695/1 The generally folkish vein of much of the vocal melody. 1968 Vogue 15 Apr. 23/1 Covent Garden's production..is folkish, parochial in style.

folk-lore. Now usu. unhyphenated. Add: **b.** Recently in extended use: popular fantasy or belief.
1954 Economist 11 Sept. 8/1 Some scientists..add to the store of folk lore by asking whether the occasional flying saucer is not a Russian technician who has missed his bearings on the return to earth. 1959 J. BRAINE Vodi xvii. 211 She was always the Little Woman, almost the Pretty Little Nursie of masculine folklore. 1959 Listener 5 Nov. 763/1 Finland..has a special niche in American folklore, under the caption: 'Brave little Finland'. 1971 Nature 29 Jan. 293/2 The Santa Barbara oil spill..may not be entirely worthy of its place in contemporary ecological folklore.

fo·lkloric, *a.* [f. FOLK-LOR(E + -IC.] Of or pertaining to folklore.
1883 C. S. BURNE Shrops. Folk-Lore xxxviii. 618 Thus two independent lines of research, philological and folk-loric, combine to support the historic evidence. 1952 G. SARTON Hist. Science I. iv. 106 The intimate knowledge of Aegean culture that we owe primarily to a great variety of monuments is confirmed..by folkloric remains in the Aegean area. 1969 Daily Tel. (Colour Suppl.) 10 Jan. 32/3 A folkloric dance group from a mining town.

fo·lklorish, *a.* [f. FOLK-LOR(E + -ISH.] Resembling folklore.
1926 Sat. Rev. 10 July 914 This..piece of humor..is, on the surface, anything but folk-lorish. 1944 BLUNDEN Cricket Country 164 Bowlers who insist on being symbolical or folk-lorish will not be given a second over. 1964 F. WHITE West of Rhone xiii. 125 The Langue d'Oc.. dwindled to a patois... Latterly there has been a folk-lorish revival of the patois, with..poetry competitions at the jeux floraux.

folk-song (fōᵘ·k,sǫŋ). Earlier **folk's song.** [f. FOLK + SONG sb., after G. *volkslied.*] A song originating from the common people; also, a modern imitation of such a song; also *collect. sing.*
1847 Howitt's Jrnl. 17 Apr. 212/1 The Three Little Roses. A German Folk's song. Translated by Mary Howitt. 1870 Folk-Song [see FOLK 5 b]. 1871 Cornh. Mag. Nov. 573 In a large collection of folk-songs from all the chief Dravidian languages, not one may be called immoral, not one cheerful. 1875 Encycl. Brit. III. 283/1 It was not till the publication of Allan Ramsay's Ever-green and Tea Table Miscellany, and of Bishop Percy's Reliques, that a serious effort was made to recover Scotch and English folk-songs from the recitation of the old people who still knew them by heart. 1883 Grove's Dict. Mus. III. 584/2 In some countries of Europe the development of the Song can be followed from the primitive form of folk-song to the highest type of artistic composition. 1907 C. J. SHARP (title) English Folk Song. 1964 GOULD & KOLB Dict. Social Sci. 272/2 Folklore or folksong..belong to such small, scattered, isolated groups. 1965 M. MORSE Unattached v. 180 Folk-song concerts. 1966 Listener 18 Aug. 236/2 That passage from Ecclesiastes which Pete Seeger has turned into a beautiful folk song.
Hence **fo·lk-songish, -songy** *adjs.*, resembling or relating to folk-song.
1925 P. A. SCHOLES 2nd Bk. Gramophone Rec. 154 The dignity of Jupiter in a fine folk-songish strain. 1934 C. LAMBERT Music Ho! iii. 151 Vaughan Williams..whose style..without necessarily being folk-songy in the picturesque way, is intimately connected with the inflections and mood of English folk music. 1959 Times 11 May 3/7 I'm sure that the folk-songy sounds, the flattened leading notes and so on, are as artificial as a Tudor Tea Shoppe.

folksy (fōᵘ·ksi), *a.* orig. *U.S.* [f. *folks*, pl. of FOLK + -Y[1].] **1.** Sociable; also, unpretentiously companionable; informal, casual.
1852 L. CLAPPE Let. from Calif. 15 Mar. (1922) 199 'They ain't kinder like eour folks', or, as that universal Aunt Somebody used so expressively to observe, 'Some-how, they ain't folksy!' 1868 A. D. WHITNEY P. Strong's Outings 178 So pleasant and folksy. 1890 Harper's Mag. Dec. 138/2 I'd kind o' thought 'twould be folksy an'.. cheerful. 1892 R. T. COOKE Huckleberries 331 I'm real folksy; grasshoppers ain't no neighbours to me. 1942 D. POWELL Time to be Born (1943) v. 117 The public really doesn't like its idols to be folksy. 1953 John o' London's 23 Jan. 76/4 A homesy, folksy, knowing account of a small town boy called Jason.

2. Having the characteristics of folk-art, culture, etc.; *esp.* ostensibly or artificially folkish.
1947 A. EINSTEIN Mus. Romantic Era vi. 55 Mozart is not 'folksy': he is always himself. 1949 New Statesman 15 Oct. 422/1 A chorus of youths and maidens, wearing a specially designed fancy dress of folksy cut. 1958 Manch. Guardian 20 Sept. 4/6 There lingers around the island of Lundy..an agreeable aura of folksy romance, eminently conducive to fables and premonitions. 1964 M. ANDERSON Saint at Stake xx. 174 Even the least 'folksy' modern doctors admit that in such cases 'magic' is sometimes effective.
Hence **fo·lksiness,** the state or quality of being 'folksy' (in either sense). orig. *U.S.*
1931 P. W. SLOSSON Gt. Crusade viii. 227 The democratic geniality and 'folksiness' which Michigan..ranked as the chief of human virtues. 1956 Ess. Divers Hands XXVIII. 80 There is a smack of folk-lore about them, but no earnest, bespectacled folksiness. 1961 Times 9 Feb. 3/1 The folksiness of choruses and peasant dances recalls Der Freischütz.

folky (fōᵘ·ki), *a.* [f. FOLK + -Y[1].] Characteristic of the common people; = *FOLKSY a. 2. Hence **fo·lkiness,** the state or quality of being folky.
1914 A. H. SIDGWICK Prom. Ticket 41 Schubert could have written..folk-songs, much better and much folkier than anything collected from embarrassed rustics. 1936 Scrutiny V. 151 One admits the futility and sentimentality, at this date, of the folky composer. 1938 L. MACNEICE I Crossed Minch i. 8 That more blatant surface-charm or glamour—that sugared folkiness—which we are led to expect in out-of-the-way places. 1940 Archit. Rev. LXXXVII. 87/2 What must be avoided at all costs is folky architecture. 1964 Listener 20 Feb. 330/3 Conventional Slav folkiness. 1970 E. LEE Music of People x. 197 The Seekers were beginning to gain a success with a 'Folky Pop' style.

follicle. Add: **4.** Comb. *follicle-stimulating* adj. (see quot. 1956).
1948 Biol. Abstr. XXII. 2389/1 In the ♀ only the follicle-stimulating hormone can exert its effect. 1956 New Gould Med. Dict. 559/1 Follicle-stimulating hormone, an adenohypophyseal hormone which stimulates follicular growth in the ovary, and spermatogenesis in the testis... Abbreviated FSH.

follicular, *a.* Add: **2. c.** *follicular hormone*, a natural œstrogenic hormone, now usually called œstrone.
1923 Jrnl. Amer. Med. Assoc. LXXXI. 821/1 The conclusion seems justified that this follicular hormone is the cause of estrual or mating instincts. 1947 A. T. CAMERON Recent Adv. Endocrinol. vi. 181 Oestrone (follicular hormone)..has marked oestrogenic properties.

follis (fǫ·lis). Pl. **folles** (-ēz). [L.] A bronze or copper coin introduced by Diocletian in 296 A.D. and again used in Byzantine currency c 800 A.D.
1885 H. A. GRUEBER in S. Lane-Poole Coins & Medals iii. 60 In copper two new pieces were issued, called the follis and the denarius. 1908 W. WROTH Imperial Byzantine Coins I. p. lxxviii, The chief denominations are marked M (the follis). K, I, E, being pieces of 40, 20, 10, and 5 nummia. 1962 R. A. G. CARSON Coins 176 For both Augusti and Caesars the mints in Britain and Gaul and the Eastern mints continued to issue folles with the Genio reverse. Ibid. 211 Theophilus also replaced the old 40 nummia coin by a new bronze coinage of follis and half-pieces with types. 1971 Ashmolean Mus. Rep. of Visitors 1970 46 The Byzantine coins from the same source include a 'consular' follis of Heraclius and his father.

follow, sb. Add: **1. b.** A supplementary portion (in a restaurant); also *pl.*, = *AFTERS.
1910 A. A. MILNE Day's Play 213 At most restaurants you can get a second help of anything for half-price, and that is technically called a 'follow'. 1946 G. MILLAR Horned Pigeon ii. 31 Robeson..made us some kind of stew; and the 'follows'..were tinned 'yellow cling' peaches.

3. (Earlier example.)
1876 Baily's Mag. Aug. 100 Against which [score] Yorkshire could only put 208..which, unfortunately, did not prevent a 'follow on'.

b. *attrib.*; also in general use.
1897 Badminton Mag. Apr. 441 The original 'follow-on' limit was 100. 1899 W. G. GRACE Cricket. Remin. 229 The compulsory follow-on innings. 1960 Farmer & Stockbreeder 15 Mar. 94/1 (Advt.), This new booklet contains advice about 'follow-on' feeding. 1964 Financ. Times 12 Mar. 19/1 A large follow-on order for..engines. 1964 Economist 28 Mar. 1280/2 The follow-on rate [for electricity]. 1971 Nature 19 Mar. 143/2 It recommended that follow-on Vikings should be held in abeyance until the results of the first missions are analysed.

4. follow-through. *Golf, Cricket*, etc. The action or an act of following through (see *FOLLOW v. 20*). Also *transf.*
1897 Encycl. Sport I. 465/1 Both force and direction are imparted by what is technically known as the 'follow through'. 1904 Daily Chron. 28 Apr. 3/2 It is..worthy of note that after the ball has departed, when the follow-through is nearly completed, Vardon's gaze is still fixed on the spot whence it has flown. 1920 D. J. KNIGHT in P. F. Warner Cricket 27 Another great factor of the batsman's art is what is known as the follow through. 1924 F. G. LOWE Lawn Tennis 46 The wrist brings the striking face square with the ball, and after impact gradually turns the striking face over until at the finish of the 'follow through' it almost faces the ground. 1931 Times Lit. Suppl. 17 Sept. 699/3 The upward swing of a bait-rod (that gentle easy follow-through). 1959 Daily Mail 20 Feb. 10/4 Bowlers cutting up the pitch with their follow-through. 1964 WODEHOUSE Frozen

Assets i. 14 The Sergeant stamped some more papers. He had a wristy follow-through which at any other moment Jerry would have admired.

b. *fig.* or *gen.*

1926 *Socialist Rev.* Jan. 308 He should be familiar with what is meant by 'follow-through Departments', for the condition of a patient five years after he leaves is a good test of the Hospital's efficiency. **1955** *Times* 25 July 5/6 The follow-through from this beginning by our respective Governments will be decisive in the measure of this conference. **1959** *Times* 11 June 16/3 Kaffirs..eased later because of lack of 'follow through' buying. **1970** *Daily Tel.* 15 June 2/5 The test of the agreement will be on the success of the productivity follow-through.

5. follow-up. The action of following up; the pursuit or prosecution of something begun or attempted; see also quot. 1942. Also *attrib.*

1923 J. D. Hackett in *Managem. Engin.* May, *Follow-up*, methods used by the personnel department to maintain friendly relations with employees. **1929** *Melody Maker* Jan. p. iii (Advt.), Sensational follow-up song in America, to *That's My Weakness Now.* **1929** *Sat. Even. Post* 14 Dec. 13/2 It's the follow-up of that injunction gag. **1931** *Brit. Jrnl. Psychol.* July 87 A 'follow up' investigation of one thousand particularly gifted children. **1938** W. S. Churchill *Into Battle* (1941) 68 A follow-up department in the Ministry to see exactly what was the state of production in the different firms. **1942** *Ann. Internal Med.* XVI. 655 Many cases have repeated hospital admissions and form a long line of 'follow-ups' in the out-patient department. **1950** W. Hammond *Cricketers' School* xvi. 149 The umpire himself may be temporarily unsighted by the bowler's follow-up. **1956** C. Willock *Death at Flight* xvii. 227 Mr. Goss had briefed this team with all the relevant facts necessary for the follow-up story. **1957** *Technology* Apr. 70/3 The manufacturer's attack on a market needs firepower..; but, as in war, there must also be consolidation, the 'follow-up'. **1959** B. Wootton *Social Sci. & Social Path.* ii. 60 Except for the occasional follow-ups of rehoused populations. **1968** *Brit. Med. Bull.* XXIV. 208/1 A nation-wide follow-up system for samples of patients classified in field surveys. **1968** *New Scientist* 21 Nov. 417/1 Follow-ups of persons who have attempted suicide are not carried through sufficiently.

follow, *v.* Add: **1.** Also with advs., e.g. *about, out.*

1750 Goadby *Apol. Life Bampfylde Moore Carew* (ed. 2) v. 48 Parson Bryant followed him out. **1860** W. Collins *Woman in White* xi, I opened the door for her in silence, and followed her out. **1865** M. C. Harris *Christine* xxx, Richard followed his brother slowly out into the path. **1877** A. Sewell *Black Beauty* xlviii, I used to come to him in the field and follow him about. **1910** E. M. Albanesi *For Love of Anne Lambart* 59 He follows me about like a dog.

2. d. To provide (a thing) *with* a sequel (cf. *21 c) or a successor.

1671 Head & Kirkman *Eng. Rogue* iv. viii. 125 At the first blow, I thought he had cut me in two, following that with three or four more. **1901** *Daily Chron.* 14 Dec. 8/2 He had arranged to follow 'Iris' with..'My Lady Virtue'. **1907** *Smart Set* Jan. 72 Her efforts to follow 'Anchored' with other stories.

†**e.** Of a side at *Cricket*: *to follow their innings*, to follow on (see Dict. sense 19 d). *Obs.*

1815 *Suffolk Chron.* 2 Sept. 4/4 The latter, immediately following their first innings, obtained 13. **1854** J. Pycroft *Cricket Field* (ed. 2) xi. 251 The M.C.C...in playing Surrey followed their innings, being headed by 106. **1894** *Laws of Cricket* § 53 The side which goes in second shall follow their innings if they have scored 120 runs less than the opposite side in a three days' match, or 80 runs in a two days' match.

3. c. To attend (the body of a deceased person) to the grave; (*colloq.*) to attend a person's funeral. Also *absol.*

1814 *New Monthly Mag.* Feb. 103/1 Behind the waggon followed the chief mourner, who was his own riding horse, attached by the bridle. **1819** C. Wolfe in *Rem.* (1827) 155 Last night I helped to lay poor M— in his coffin, and followed him this morning to his grave. **1820** *Kaleidoscope* New Ser. I. 142/3 His brothers..agreed to follow the body to the grave. **1831** J. Banim *Smuggler* I. xi. 201 They renewed their cries for 'Hood! to follow in the funeral!' **1897** Hall Caine *Christian* i. viii, Gimme a black cloth on the corfin, my dear, and mind yer tell 'im to foller. **1902** *Westm. Gaz.* 18 Dec. 12/2 In Norfolk it is customary to speak of attending a funeral as 'following' the remains. **1940** Brahms & Simon *Don't, Mr. Disraeli* ii. 21 Defiantly Henrietta sent a wreath to the funeral. It did not follow her friend to the grave.

15. b. Const. *on.*

1888 Mrs. H. Ward *R. Elsmere* i. iii, That state which so often follows on the long confinement of illness. **1903** R. Langbridge *Flame & Flood* ii, A rich-souled organ poured out its absolution; following on the voice of the violin.

19. d. (Earlier example.)

1865 F. Lillywhite *Guide to Cricketers* 86 Surrey 'followed on', but left only 23 runs for Oxford to get to win.

20*. follow through. *intr. Golf*, etc. To continue the stroke, after the ball has been struck, to the full extent of the swing. (Cf. *Follow *sb.* 4.)

[**1895** H. G. Hutchinson *Golf* (ed. 5) iv. 87 The difficulty..of getting the club to follow easily through after the ball.] **1897** *Encycl. Sport* I. 464/1 The player should have the habit, so important in this 'following through', of regarding the ball merely as a point *through* which the club head is to pass in the course of its circuit. **1909** P. A. Vaile *Mod. Golf* xiv. 197 The timing of the stroke, so that the head of the driver..runs into the line of flight of the ball, hits it truly, and then follows through in a true vertical plane. **1967** *Know the Game: Tennis* 31 The racket is following through.

21. follow up. c. = sense *2 d.

1795 T. Paine *Age of Reason* ii. 66 Matthew..follows up this part of the story of the guard..with a second part. **1905** W. Bodie *Bodie Bk.* 173, I must explain these two statements, and then follow them up by a third.

22. *Comb.* **follow-me-lads** *sb. pl.* [cf. F. *suivez-moi-jeune-homme*], curls or ribbons hanging loosely over the shoulder; also *sing.*, *U.S. colloq.*, a furbelow in a woman's costume; **follow-spot,** a spotlight that follows a performer on the stage; also *fig.*

1862 *Times* 21 Jan. 10/1 Vagrant ringlets straying over the shoulder, better known by the name of 'follow me, lads'. **1872** *Spectator* (Farmer), 'Follow-me-lads' are not in themselves very pretty, though, like any other fashion, they become the Princess. **1874** Hotten *Slang Dict.*, *Follow-me-lads*, curls hanging over a lady's shoulder. **1901** Greenough & Kittredge *Words* (1902) 190 Kiss-me-quick, hug-me-tight, follow-me-lads,—names for articles of female attire. **1929** K. H. Brown *The Father* i, Miss Evelina Amberley, in all her frills and laces and follow-me-lads, as scalloped and frilled and fluted as her name. *Ibid.* ix, The knife sawed through Aunt Euphemia's follow-me-lad, slashed out a thick chunk of somber plush. **1957** J. Osborne *Entertainer* i. 12 The lighting is..bright and hard, or a simple follow-spot. **1960** Lawrence & Lee *Gang's All Here* 7 The follow spot is preparing to seek the man who will go to 1600 Pennsylvania Avenue.

fo·llowing, *pres. pple.* of Follow *v.*, loosely used as quasi-prep.: as a sequel to, in succession to (an event), after.

1947 *Evening News* 11 Dec., The prologue was written by the company following an incident witnessed by them during anti-Jewish demonstrations following the hanging of two British soldiers in Palestine. **1948** E. Gowers *Plain Words* 56 Perhaps the fight against *following* as a preposition ought to be regarded as lost. **1968** *Observer* 24 Mar. 6/4 Used car prices are going up, following the Budget.

following, *ppl. a.* Add: **6.** In various technical usages.

1839 Ure *Dict. Arts* 979 When a coal has a following or roof stone, which regularly separates with the coal. **1883** Gresley *Gloss. Coal-m.*, *Following dirt*, loose shale, etc., in a thin bed forming the roof of a coal seam. *Ibid.*, *Following-up bank*, a breadth of about 6 yards of coal taken off on either side of a leading bank. **1888** *Lockwood's Dict. Mech. Engin.*, *Following edge*, that edge of the blade of a screw propeller which leaves the water behind it, as distinguished from the leading edge. *Following joints.* The rings of which cylindrical boilers are built, being lap jointed, fit one within the other. Instead, therefore, of being parallel cylinders, they are necessarily frustra of cones, and as the joints all lap in one direction, they are termed following joints. *Following steady*, a steady which is attached to the back of the side rest of a lathe, and which embraces the work behind or after the tool and follows it along with the rest. **1967** *Gloss. Mining Terms* (*B.S.I.*) xi. 8 *Following roof*, a layer of roof which falls as coal is excavated, or soon after.

folly, *sb.*¹ Add: **5. b.** *pl.* A revue notable for the glamour of its female performers; used esp. as a title, as *Ziegfeld Follies*; also, the female members of such a revue.

[**1880** J. Stirling tr. *Zola's Nana* v. 104 Maria Blond—a girl of fifteen—frightfully thin and frightfully vicious, who had just made her début at the Folies.] **1908** *Theatre Mag.* (U.S.) Aug. 201/1 In novelty of ideas, variety, talent of performers and general smartness of production, 'Follies of 1908' is fully up to the standard of the best this enterprising..young manager [*sc.* Ziegfeld] has yet attempted. **1917** R. W. Lardner *Gullible's Travels* (1926) iv. 202 They wouldn't nobody of ever mistook the women for *Follies* chorus girls. **1919** F. Hurst *Humoresque* 300 The Moncrieff Follies—twenty-four of them, not counting two specialty acts and a pair of whistling Pierrots—burst forth into frolic. **1929** H. Miles tr. *Morand's Black Magic* i. 71 Head-dresses worthy of the Ziegfeld Follies. **1957** 'Gypsy Rose Lee' *Gypsy* xxxiv. 309 Julie Bryan, my understudy in the *Follies* (less experienced than the maid), went to work as featured strip teaser for the Minskys. **1965** P. Ziegfeld *Ziegfeld's Girl* ii. 43 The first *Ziegfeld Follies* opened at the Apollo Theater in Atlantic City and came to New York in July of 1907.

Folsom (fōᵘ·lsəm). The name of a village in north-eastern New Mexico, U.S.A., applied *attrib.* to the remains of a prehistoric culture first found there, esp. to a type of projectile point (see quots.).

1928 H. J. Cook in *Sci. Amer.* July 39/3 The fact that these arrowheads were found under the conditions they were, makes it possible to designate it as evidence of a definite cultural stage, for which the name 'Folsom Culture' has been suggested. **1931** *Proc. Colorado Mus. Nat. Hist.* X. ii. 12 Fragments with long and broad lengthwise grooves..with often long and sharp base points, are really the only ones that could be called 'Folsom points', strictu sensu, that is to say, points of same shape and technique as those actually found at Folsom and..made by the same ancient people. **1948** A. L. Kroeber *Anthropol.* xvi. 678 The typical fine 'Folsom points' are projectile heads made by pressure chipping. *Ibid.* 679 The true or specific Folsom culture..seems more likely to have been glacial rather than post-glacial; certainly terminal Pleistocene..rather than initial or middle. **1954** *Ann. Reg. 1953* 387 In the New World, after 27 years of searching, the first bone (a rib) of Folsom Man was unearthed in New Mexico.

Fomorian (fomōᵊ·riăn). *Celtic Mythol.* Pl. also **Fomori.** [f. Ir. *fomor* pirate, monster, f. *fo* under + *mor* monster (cogn. w. Mare *sb.*²) = Gael. *famhair* + -ian.] In Irish legend, one of a race of pirates or giants, perhaps originally representing the gods of death and darkness. In Scottish legend, a giant. Also as *adj.*

1876 *Encycl. Brit.* V. 300/2 That the Fomorian and *Tuatha Dé Danaan* contests are mythological there can be no doubt. **1891** W. B. Yeats *John Sherman & Dhoya* 173 One evening Fomorian galleys had entered the Bay of the Red Cataract. **1898** J. Heron *Celtic Church* 5 Partholonians, Nemedians, and Fomorians, or sea-rovers, are..the first..to take possession [of Ireland]. **1901** J. Rhŷs *Celtic Folklore* II. vii. 435 The Irish Morc..and another called Conaing are represented in the legendary history of early Erin as the naval leaders of the Fomori. **1921** L. Spence *Introd. Mythol.* xi. 294 Among these warring elements the Fomorians are a race of Titans. **1926** *Contemp. Rev.* Mar. 352 Scottish folk lore knows the giants as Fomorians. Tale after tale tells of their great boulder-throwing contests on the Scottish hills. **1959** Aldington & Ames tr. *Larousse Encycl. Mythol.* 236/1 After a battle with the Fomorians in which Conann and many Fomorian followers were killed, the remnants of the people of Nemed fled from the country. **1964** E. O. G. Turville-Petre *Myth & Religion of North* vii. 160 This second battle has been seen as one between the Tuatha Dé Danann, gods of light, life, day, and the Fomorians, gods of death and darkness.

‖**fons et origo** (fǫnz et ǒrəi·go). [L.] The source and origin (*of*).

[**1809**, etc. examples of *fons et origo mali*: see Stanford's Dict. Anglicised Words.] **1873** Browning *Red Cott. Nt.-Cap* iii. 202 Never mind the cause, *Fons et origo* of the malady: Apply the drug with courage! **1890** W. James *Princ. Psychol.* II. xx. 189 Many writers..hold that the consciousness of active muscular motion, aware of its own amount, is the *fons et origo* of all spatial measurement. **1897** *Westm. Gaz.* 12 Nov. 2/2 If we may reserve our opinion on the one point which was the *fons et origo* of Liberal Forwardism—the Eastern Question. **1927** A. H. McNeile *Introd. N.T.* 285 Papias, the *fons et origo* of many problems. **1930** *Times Lit. Suppl.* 20 Nov. 955/3 The *fons et origo* of the downfall of the West. **1966** *Listener* 12 May 701/1 That conjunction of..emotion and expression, that is the *fons et origo* of his music.

food, *sb.* Add: **2. b.** *transf.*, as in *skin food.*

1898 H. A. Browning *Beauty Culture* vii. 134 Let me, however, warn you to study your skin (and not to choose its 'food' hastily or casually). *Ibid.* x. 221 The face..is smeared with skin-food. **1907** *Yesterday's Shopping* (1969) 537/1 Skin food. Massage & complexion cream. *Ibid.* 538/3 Squaw hair food. **1908** *Queen* 30 May 932/1 The introducer of the Russian skin food Valaze is Dr. Lyruski. **1912** *Ibid.* 30 Nov. 1006/1 It is a skin food in the truest sense and really works wonders with the skin. **1942** N. Marsh *Death & Dancing Footman* ii. 43 A fat lot of good 'Hersey's Skin Food' is to your middle-aged charms.

7. a. *food-crank, -faddist, -fish* (earlier example), *habit, parcel, product, queue, -shortage, -supply, tax, -ticket.*

1906 J. Condon *Let.* 17 Nov. (1966) 220 Don't think I am some sort of a food-crank. **1951** R. Campbell *Light on Dark Horse* 150 Any Fabian food-crank. **1910** *Daily Chron.* 14 Apr. 4/2 The 'food faddists' or 'food reformers'. **1927** W. E. Collinson *Contemp. Eng.* 38 Some of the food-faddists went beyond the vegetarian stage and became fruitarians. **1953** A. Christie *Pocket Full of Rye* viii. 47 One of those food faddists who'll eat any mortal thing so long as it isn't cooked. **1970** *Guardian* 15 Aug. 6/6 Top pundits tend to make fun of food faddists. **1865** J. G. Bertram (*title*) The harvest of the sea. A contribution to the natural and economic history of the British food fishes. **1927** *Maclean's Mag.* 1 June 32/1 Food habits are important. **1919** E. H. Jones *Road to En-dor* v. 52 He was comfortable enough himself (thanks to the contents of our food parcels). **1946** J. B. Priestley *Bright Day* x. 310 He'd been sending some food parcels to his sister in England. **1967** *Guardian* 5 Sept. 7/5 The poorest depending on cash grants and food parcels from Turkey. **1897** *Daily News* 6 Mar. 6/4 The Select Committee on food products adulteration. **1906** *Daily Chron.* 5 June 5/4 To enable Government inspectors to supervise, from hoof to can, the preparation of meat food-products. **1940** *Manch. Guardian Weekly* 5 Jan. 3 There is nothing more irksome and more damaging to a smooth-running war system than local shortages and food queues. **1931** J. S. Huxley *What dare I Think?* iv. 135 Food-shortage..will..bring about an equilibrium. **1965** M. Hilton tr. Meuvret in *Pop. in Hist.* xxi. 511 Movements undertaken to escape regions experiencing food-shortage. **1881** W. D. Hay *300 Years Hence* i. 4 Inquiries about..food-supply and so on. **1957** P. Worsley *Trumpet shall Sound* vi. 120 The Japanese organized work-teams..to cultivate huge gardens for their own and the natives' food-supply. **1906** Food-tax [see **food-taxer* below]. **1913** *Punch* 22 Jan. 67 Food taxes. **1907** *Westm. Gaz.* 12 Oct. 13/1 The food-ticket is invaluable, when money might prove a danger. **1909** *Ibid.* 30 Jan. 2/2, I felt also bound to refuse the gift of a food-ticket until their cases had been investigated. **1937** Koestler *Spanish Testament* ii. 55 The revolutionary committees issued food tickets which the shopkeepers were obliged to honour.

b. *food-chopper, -mixer, -taxer; food-collecting, -gathering* (see also *8), *-getting, -taxing.*

1908 *Sears, Roebuck Catal.* 468/1 Enterprise food chopper..will chop raw meat, cooked meat, vegetables. **1911** *Daily Colonist* (Victoria, B.C.) 6 Apr. 3/1 (Advt.), Meant to save and sure to please bargains we are offering Food Choppers, $1.75 kind..$1.25. **1911** J. A. Thomson *Biol. Seasons* iii. 294 The marked shortening of the daylight hours available for food-collecting. **1932** E. Step *Bees, Wasps, Ants* 4 When she [*sc.* the bee] is not making food-collecting excursions, she sits upon the cell. **1927** Haldane & Huxley *Animal Biol.* xii. 288 The last task undertaken [by bees] before going out food-gathering is that of sentry-duty. **1941** J. S. Huxley *Uniqueness of Man* xiv. 278 Practical activities of communal existence such as food-getting and war. **1959** *Observer* 19 Apr. 14/3 There are over a dozen brands of food mixers on the market. **1961** *Times* 26 Apr. 25/4 Domestic appliances such as..food-mixers. **1903** *Westm. Gaz.* 19 Aug. 5/1 Mr. Arnold-Forster a food-taxer. **1906** *Daily Chron.* 13 Feb. 4/2 There was something for the Food-taxers also, for..the matter of food-taxes is 'not a question of principle'. **1905** *Ibid.* 25 Mar. 7/4 This food-taxing policy.

8. food-call, the cry of a bird for food; also *transf.*; **food-card,** a card used in the rationing of food to indicate the amount of food allowed to a person for a specified period of time; **food-chain** *Ecology*, a series of organisms each dependent upon another for food, esp. by direct predation; **food-controller,** an official having control of food supplies; **food-cycle** *Ecology*, an interdependent group of food-chains in a community; **food-gatherer,** *spec.* in *Anthropol.*, one who obtains food from natural sources rather than through agriculture, etc.; so **food-gathering** *sb.* and *attrib.*; **food-lift,** a lift for the conveyance of food; **food poisoning,** any illness caused by the presence in food of harmful bacteria or toxic substances (as bacterial toxins or poisons from inedible plants); **food stamp** (see quot. 1967); **food-value,** value as food; *spec.* in dietetics, the relative nourishing power assigned to foods; also *fig.*; **food-vessel** *Archæol.*, a type of prehistoric pottery found in northern England (see quot. 1963); applied *attrib.* to the culture characterized by such pottery; also (*rare*) *food-vase*; **food-web** = **food-cycle.*

1926 T. E. Lawrence *Seven Pillars* (1935) iii. xxxv. 207 Down the visible wind in the misted valley came the food-call of Turkish bugles. **1949** *Brit. Birds* XLII. 236 She may utter food-calls as during incubation. **1956** Bannerman *Birds Brit. Isles* V. 196 The 'food calls' of the young..are uttered in excited chorus. **1918** *Times* 6 Feb. 8/2 Food cards taken out for children educated at boarding schools. **1923** E. A. Ross *Russian Soviet Republ.* 113 Bread- and food-cards of four different colors were issued to four class divisions of the population. *Ibid.* 114 In the spring of 1920 there were only eight thousand adults in Petrograd who had not taken out food-cards, i.e. had not gone to work. **1927** C. Elton *Anim. Ecol.* v. 56 There are, in fact, chains of animals linked together by food, and all dependent in the long run upon plants. We refer to these as 'food-chains', and to all the food-chains in a community as the 'food-cycle'. **1959** J. Clegg *Freshwater Life* (ed. 2) iv. 65 The algae provide the first link in the food-chain, utilising simple chemical substances in the water to build up their own structures; thereby making available abundant food for the lower animals. **1968** *Times* 17 Dec. 10/5 Birds of prey are particularly vulnerable..because of their position at the end of a food chain. For example, worms eat contaminated vegetation, sparrows eat worms, and peregrines eat sparrows. **1916** *Act 6 & 7 Geo. V* c. 68 § 3 For the purpose of economising and maintaining the food supply of the country during the present war, it shall be lawful for His Majesty to appoint a Minister of Food under the title of Food Controller. **1927** Food-cycle [see **food-chain*]. **1963** J. E. G. Raymont *Plankton & Productivity in Oceans* xviii. 542 Food cycles in the oceans commence with the synthesis of organic material by the phytoplankton. **1928** C. Dawson *Age of Gods* iii. 49 Man was entirely at the mercy of nature—a mere scavenger who eked out a miserable existence as a food-gatherer and an eater of shell-fish. **1949** W. F. Albright *Archaeol. of Palestine* iii. 61 The Natufians were still essentially food-gatherers, in spite of their discovery of the cultivation of grain. **1960** K. M. Kenyon *Archaeol. in Holy Land* i. 19 The men of the period were dependent for their existence on the food they could gather by hunting, fishing and other natural sources; they were food-gatherers. **1926** V. G. Childe *Aryans* v. 103 Men who had made the great advance from a food-gathering to a food-producing economy. **1936** *Proc. Prehist. Soc.* II. 253 People dependent upon hunting, fishing and food-gathering tend everywhere to produce the same kind of art. **1958** *Listener* 30 Oct. 689/2 The food-gathering state when man was, economically at least, a savage like the early Eskimo. **1898** A. Bennett *Man fr. North* xvi. 139 The screen which hid the food-lift. **1922** Joyce *Ulysses* 167 A diner..stared towards the foodlift. **1887** *Practitioner* Apr. 302 (*heading*) Report on food poisoning at a wedding breakfast. **1917** E. O. Jordan *Food Poisoning* i. 2 Most attacks of food poisoning are usually of a slight and apparently temporary nature. **1951** Whitby & Hynes *Med. Bacteriol.* (ed. 5) ii. 15 Potassium permanganate is much used in the tropics to kill food-poisoning bacteria on vegetables. **1951** E. W. H. Cruikshank *Food & Nutrition* (ed. 2) xiv. 285 Food poisoning may also occur..by the ingestion of poisonous plants. **1970** *Daily Tel.* 6 May 19/8 A doctor is only compelled to notify the health authorities of certain infectious diseases and cases of food poisoning. **1962** *Economist* 28 Apr. 363/2 The 146,000 certified [American] needy now receiving food-stamps. **1967** *Ibid.* 15 July 214/3 The other 'anti-hunger' device, the Food Stamp Programme, allows families to purchase stamps at substantially less than their face value and use them to buy food at their local shops. **1970** *Guardian* 31 Mar. 11/1 The flower children fled to communes in New Mexico where pot is plentiful and Federal food stamps prevent starvation. **1899** E. G. White *Counsel to Editors* (1939) xxi. 104 Everything that the imaginative mind can think of is woven into the book, and presented to the world as mental food. But very often it has no food value. **1907** *Chambers's Jrnl.* 29 June 495 The York Health and Housing Reform Association has published a table of food-values. **1909** *Ibid.* Jan. 6/2 The average Chinese and Japanese diet is rather richer in food-values than the average American. **1915** *Lit. Digest* (N.Y.) 4 Sept. 479/2 (Advt.), A nourishing and appetizing first course like..Tomato Soup..contributes rich food-value. **1960** C. Storr *Marianne & Mark* iii. 35 Aunt Pamela, though..no expert on food values, was immensely keen on providing a good mixed diet. **1871** *Archaeologia* XLIII. 385 One of the four food-vases..is ornamented with fine punctures at the bottom. **1963** H. N. Savory in Foster & Alcock *Culture & Environment* iii. 34 The rim of a Corded Beaker, part of a jar with a rim deeply bevelled internally, which Piggott hailed as a possible ancestor of the southern 'Food-vase'. **1871** *Archaeologia* XLIII. 378 Food vessels are rare in the barrows of Wiltshire and the South of England. **1930** F. Elgee *Early Man in N.E. Yorks.* viii. 68, I shall first describe the food-vessel skeleton burials. *Ibid.* 70 Burials in which no pottery was present, but which are nevertheless characteristic of the food-vessel period. **1963** E. S. Wood *Collins Field Guide Archæol.* i. iv. 62 In the North the 'Food Vessel' culture absorbed very obviously many of the traditions of the secondary neolithic and Beaker peoples, and produced a somewhat coarse pottery, with neck and shoulder, and much ornament in bands and panels, from which it takes its name. **1961** *Estuarine Bull.* VI. 12 (*title*) Sand shrimp: cross-link in an estuarine food web. **1971** *Nature* 1 Jan. 14/1 Important links in the food web of the sea.

foo-foo (fū·fū). Also **foofoo, fou-fou.** [Of West Afr. origin: recorded in Twi, Ewe, Wolof, etc. Cf. Cuban Sp. *fufú.*] A kind of dough made out of plantains: a traditional food of Negroes on both sides of the Atlantic.

These are variant spellings of Fufu (q.v.).

1826 A. Barclay *Pract. View Slavery W.I.* 437 A negro..would greatly prefer his own good substantial dish of *foofoo*, composed of eddoes, ochras, and mashed plantains. **1851** *Illustr. Catal. Gt. Exhib.* iv. i. 977/2 This mass [of plantain boiled whole], beaten in a mortar, constitutes the *foo-foo* of the negroes [of British Guiana]. **1858** Simmonds *Dict. Trade*, *Foo-foo*, a negro name for dough made from plantains; the fruit being boiled and then pounded in a mortar. **1899** J. Rodway *In Guiana Wilds* 54 There were fou-fou soup, pepper-pot, barbecued meat, and piles of oranges. **1924** *Glasgow Herald* 8 Nov. 5 The women pounded their 'fou-fou' in the courtyard of the [Ashanti] village. **1964** E. Huxley *Back Street New Worlds* xiv. 141 The Shepherd's Bush market has a shop devoted wholly to West African foods..like..garden eggs and fou-fou, edwene and dried snails.

fool, *sb.*¹ and *a.* Add: **B.** *adj.* (Later examples.) Freq. in U.S.

1805 L. Dow *Jrnl.* (1806) II. i. 76, I showed the contrast of a gentleman and a fool deist. **1854** M. J. Holmes *Tempest & Sunshine* ii. 25 Tempest..can hardly wait till I'm dead before she spends my money on fool fixins. **1862** S. Hale *Lett.* (1919) 13 Everybody talking such fool nonsense as sometimes almost to prevent digestion. **1896** M. Corelli *Mighty Atom* xvi. 335 My fool tears a-flowin' on her coffin. **1902** W. N. Harben *Abner Daniel* 2 Oh, Alan, don't you see he's goin' to ruin us with his fool notions? **1912** R. A. Wason *Friar Tuck* xxiii. 165 It was the foolest lookin' group I was ever part of. **1924** W. M. Raine *Troubled Waters* xxiii. 245 You've heard that fool story about Norma and Mac. **1932** E. Wilson *Devil take the Hindmost* ix. 104 The local banks have failed through the speculations of some fool gambler. **1951** 'J. Wyndham' *Day of Triffids* iv. 85 You never can tell what fool carelessness may go on.

fool, *v.* Add: **2. a.** *to fool along* (U.S.): to proceed slowly or aimlessly; also *fig.*; *to fool (a)round* (earlier U.S. examples).

1837 A. Greene *Glance at New York* (Farmer), He mustn't come foolin' round my gal, or I'll give him fits. **1866** C. H. Smith *Bill Arp* 44 You get a government contract for a few thousand pounds and you fool along with it, selling what you make to these drug men at a bigger price. **1870** *Galaxy* May 726/2 What did Abel come fooling around there for? **1884** 'Mark Twain' *Huck. Finn* xxxiii. 338 You turn back and fool along slow, so as to get to the house about the time you ought to. **1934** H. Vines *This Green Thicket World* 20 They fooled along and did not much try to reach the ferryman's house.

fooler (fū·lər). [f. Fool *v.* 3 + -er¹.] A person or thing that 'fools' one.

1909 R. A. Wason *Happy Hawkins* 316 They finally located a mine that looked good-natured an' generous; but it was a fooler.

foolishment (fū·lĭʃmĕnt). [f. Foolish *a.* + -ment] = Foolishness.

1918 *N.Y. Evening World* 4 Jan. 18/1 (*title of verse*) Foolishment. **1928** M. Chapman *Happy Mountain* (1929) xxviii. 306 Don't you pay heed to Dena's foolishment. **1966** *Sunday Mail* (Brisbane) 18 Dec. 29/3 There's nothing nicer than a good foolishment.

fool-proof (fū·lprūf), *a.* orig. *U.S.* [f. Fool *sb.*¹ + Proof *a.* 1 b.] Proof against even the incompetence of a fool; simple and straightforward so as to respond even to the most inexperienced or careless handling; safeguarded against every sort of accident.

1902 A. C. Harmsworth et al. *Motors* 309 The car..is comparatively 'fool-proof'. **1904** *Westm. Gaz.* 24 Oct. 2/2 The car is so 'simple' that my daughters drive it—'fool-proof' the Americans call it. **1926** W. R. Inge *Lay Thoughts* 220 Everywhere we find the same demand to make life easy, safe, and fool-proof. **1928** Galsworthy *Swan Song* iii. ii. 231 A base of operations with a fool-proof title was essential. **1960** *Farmer & Stockbreeder* 19 Jan. (Suppl.) 37/2 It should be as foolproof and as easy to manage as possible, even to the most amateur poultry keeper. **1968** *Times* 24 Oct. 7/7 The cost of making nuclear reactors absolutely foolproof would outweigh their economic advantages.

foolscapped (fū·lzkæpt), *pa. pple.* and *ppl. a.* [f. Fool's-cap + -ed².] Furnished with a fool's-cap.

1909 A. Noyes in *Westm. Gaz.* 26 Aug. 1/3 Poor fool's-capped scholars. **1918** *Hist. Amer. Lit.* I. 265 He was never laurelled like Byron, never foolscapped like Keats by critics or public.

foot, *sb.* Add: **1. d.** In the colloq. exclamation *my foot!* (also *your foot!*), expressing a contemptuous contradiction.

1923 R. Crothers *Mary the Third* ii. ii. 69 *Mother*: She was honest enough to tell me that...*Father*: Honest your foot! She's fooled you—deceived you. **1925** N. Coward *Hay Fever* iii, *Judith*: It's so silly to get cross at criticism—it indicates a small mind. *David*: Small mind my foot! **1928** D. L. Sayers *Lord Peter views Body* xi. 262 'I thought he was doing a motor-tour.' 'Motor-tour your foot!' said the Inspector, with more energy than politeness. **1945** L. A. G. Strong *Othello's Occupation* 72 Cooperation my foot. You're trying to trap me into admitting a motive for doing the old girl in. **1961** H. E. Bates *Day of Tortoise* 55 'But it's a serious matter for you.' 'Serious my foot. Why should I worry?'

6. (Further examples.)

1830 S. Fox *Menologium* p. vi, In these compositions..trochaic feet predominate. **1888** A. S. Cook *Judith* p. i, A normal hemistich contains two metrical feet. **1942** J. C. Pope *Rhythm of Beowulf* 12 Sievers was only borrowing mistakes from contemporary metrical theory when he marked the 'feet' of his five types.

7. d. As a measure of coal gas: the amount of gas contained in one cubic foot of space.

1838 *Penny Cycl.* XI. 88/2 A sufficient quantity of gas was turned on to give a light equal to that of a mould candle; the consumption in this case was a foot and a half per hour. **1879** *Encycl. Brit.* X. 99/2 A burner passing 7 feet of gas per hour.

26. c. *feet of clay*: see *Clay *sb.* 4 c.

27. *to carry off his feet* (earlier example); *to have one's feet on the ground*: to base oneself on realities; to be practical.

1854 Mrs. Gaskell *Let.* 27 Oct. (1966) 316 I'll enclose you two pieces of Mr S. Gaskell to show how *he's* carried off his feet. **1950** 'P. Woodruff' *Island of Chamba* i. 21 El Hadramauti..is a bit of a theorist and H.M. for all his oddity has his feet very firmly on the ground.

28. *to put one's foot down*: (*b*) when driving a motor vehicle: to accelerate; *not* (or *never*) *to put a foot wrong*: to make no mistakes; hence *not to put a foot right*: to make many mistakes.

1948 'N. Shute' *No Highway* iv. 108 His superiors could rest content that Mr. Symes would never put a foot wrong. **1959** I. Jefferies *13 Days* viii. 98, I didn't feel I ought to put a foot wrong with Watson, the Ordnance Captain. **1959** *Listener* 22 Jan. 152/1 He never put a foot wrong. He was extraordinarily good in the House. **1961** H. Nicolson *Let.* 1 June (1968) III. 395 The Americans..blame..the 'diplomatists', meaning thereby their intelligence services. 'Why can't we put a foot right?' they wail. **1962** J. Braine *Life at Top* ii. 38, I put my foot down and the Zephyr gathered speed up the slope.

29. *to catch* (a person) *on the wrong foot*: to catch unawares; *to get* (or *set*, etc.) *off on the right foot*: to start successfully; similarly *to get* (or *start*, etc.) *off on the wrong foot*: to start unsuccessfully; to fail to establish good relations.

1909 R. Beach *Silver Horde* xiii. 173, I want to see you get off on the right foot; I'd feel bad if you fell down. **1925** *Country Gentleman* (U.S.) Sept. 11/2, I know I got off on the wrong foot. It was manifest in the faces and general demeanor of the grave and reverend Senators. **1937** N. Coward *Present Indicative* vi. xi. 262 To me a round of applause..even though it be conventional rather than spontaneous, almost always sets my performance off on the right foot. **1947** C. Witting *Let X be Murderer* iv. 43 They caught me on the wrong foot, Glad. What did the old fool want to go and ring up for? **1949** R. Chandler *Little Sister* xxxii. 232, I got off on the wrong foot. After that I just had to take my lumps. **1955** *Times* 12 May 11/2 Starting the Colonial Development Corporation off on the wrong foot. **1956** A. Wilson *Anglo-Saxon Attitudes* i. iii. 71 'Oh, he's brilliant all right,' said Robin...'But he obviously gets off on the wrong foot with his colleagues.' **1958** *Listener* 7 Aug. 195/2 There is that vast number of marriages..where the marriage takes place to give the baby a name. The whole thing starts on the wrong foot. **1958** *Times* 18 Dec. 11/4 The least athletically inclined are frequently finding themselves..bowled out, tackled or caught on the wrong foot. **1960** H. Innes *Doomed Oasis* ii. iii. 120 The relationship hadn't been at all easy at first. 'They started off on the wrong foot, you see.' **1961** *Times* 9 May 13/3 The Commons-peerage question must be disposed of before these wider discussions can be got off on the right foot.

30. d. *to have* or *set one's foot on the neck of*: (fig.) to hold completely in subjection: see Neck *sb.*¹ 3 a.

34. a. *foot-wear* (later examples).

1922 *Daily Mail* 1 Nov. 8 Women and girls, with their short skirts, neat footwear, and other prevailing fashions. **1954** F. C. Avis *Boxing Ref. Dict.* 43 *Footwear*, regulation boots.

e. esp. in names of speed and control appliances on vehicles, as *foot-accelerator, -brake, -braking, -clutch, -starter*; also *foot-acted, operated* adjs.

1888 *Lockwood's Dict. Mech. Engin.*, Foot Brake. **1905** *Westm. Gaz.* 17 Oct. 4/2 Many motorists who find the working of a foot-clutch trying. **1908** *Ibid.* 16 Jan. 4/2 A foot accelerator is also fitted. *Ibid.* 30 Jan. 4/1 A large foot-operated contracting brake. **1908** *Daily Chron.* 21 Nov. 9/4 The three-speed gear in association with a foot-acted brake. **1909** *Ibid.* 27 Feb. 7/7 Foot braking is less fatiguing for prolonged spells of application. **1925** *Morris Owner's Man.* 11 Both foot and hand brakes operate on drums fitted to the back wheels. **1959** *Times* 2 Oct. 11/3 The foot-operated dip switch. **1967** *Jane's Surface Skimmer Systems 1967–68* 9/1 The engine is controlled by a foot-operated accelerator. **1971** *Engineering* Apr. 69/2 A hydraulic master cylinder unit, coupled to a pivoted foot-operated lever.

35. foot-candle, a unit of illumination equivalent to the illumination of a surface all of which is at a distance of one foot from a point source of light having a luminous intensity of one candela (or formerly one international candle), corresponding to a luminous flux density of one lumen per square foot; **foot-coal,** read: an underlying stratum of coal; add later examples; **foot-dragging,** a deliberate delay or slowness (cf. *Drag *v.* 1 b); **foot-drop** *Path.*, a permanently extended position of the foot, due to paralysis of the flexor muscles; **foot-lambert,** a unit of luminance equal to the average luminance of a surface emitting or reflecting one lumen per square foot; **foot-locker** *U.S.*, a small trunk or chest; **foot-log** *U.S.*, a log used as a foot-bridge; **foot-pad,** also a device on a space vehicle; **foot-pound-second,** used to designate a system of units based upon the foot, the pound, and the second as units of length, force, and time respectively; **foot-ring,** the circular rim on the base of a plate, vase, etc.; **foot-rot,** (*b*) a fungal disease of plants, affecting the base of the stem; **foot-rule,** also *fig.*; **foot run** [Run *sb.*¹ 13 a], a length of one foot measured lengthwise along a material or structure, esp. one conceived of as having a potentially variable length; **foot-scraper** = Scraper 5; **foot-wall** (earlier Austral. example); **foot-work,** (*d*) in other games, dancing, etc.: agility, sureness, and accurate placing of the feet; also *fig.*

1906 *Illum. Engin.* I. 66/1 Owing largely to the loose and ambiguous manner in which the various terms expressing light-measurements are used..the term 'foot-candle'..expresses but a vague idea to the average architect and engineer. **1949** H. C. Weston *Sight, Light, & Efficiency* v. 150 The term foot-candle..is still in common use..but lumen per sq. ft. is now the preferred expression. **1963** J. M. Fraser *Psychol.* (ed. 2) xvii. 224 For certain jobs..up to four hundred foot-candles (the equivalent of bright sunlight) have been tried. **1712** Foot-coal [in Dict.]. **1840** *Knickerbocker* XV. 105 About a foot from the bottom of every vein, there is a layer of earth...This divides it into 'foot-coal' and 'upper coal'. **1879** G. F. Jackson *Shropsh. Word Bk.* 91 Foot coal. **1966** *New Statesman* 16 Dec. 896/1 One likely result of all this foot-dragging is that the Nato Council will fail this week to seize what might have been an excellent opportunity to simplify the arrangements for European defence. **1969** *Guardian* 31 Jan. 10/2 There is no university now which does not have some appeals machinery..though there has been some foot-dragging on other issues. **1908** C. W. Daniels in Hutchison & Collier *Index of Treatment* (ed. 4) 119 When the 'foot drop' is extreme, a cradle should be used to prevent..increasing the deformity. **1920** *Glasgow Herald* 8 July 4 Conditions affecting the feet..e.g. foot-drop, corns and contracted toes, clawfoot. **1950** E. D. W. Hauser *Dis. Foot* (ed. 2) v. 84 In most instances the paralysis that causes a valgoplanus is associated with paralysis of the dorsiflexors, which means that foot drop is present. **1925** *Trans. Illum. Engin. Sci.* XX. 631 The Foot-lambert is the average brightness of any surface emitting or reflecting one lumen per square foot, or the uniform brightness of a perfectly diffusing surface emitting or reflecting one lumen per square foot. **1942** *Jrnl. Aeronaut. Sci.* IX. 263/1 A brightness level of 0·01 foot-lambert is..comparable in order of magnitude with the brightnesses of roads and highways under moonlight. **1958** *Van Nostrand's Sci. Encycl.* (ed. 3) 686/2 A foot candle is a unit of incident light and a foot lambert is a unit of emitted or reflected light. **1943** *Harper's* June 16 They sit obscurely on foot lockers during the daytime, when they must keep off their bunks. **1943** *Infantry Jrnl.* Aug. 51 Others went to town and came in late stumbling against footlockers and cursing. **1969** *Eugene* (*Oreg.*) *Register-Guard* 3 Dec. 1A/3 Foot lockers, cabinets and standard doorways are painted in combinations of the bright yellow and orange. **1969** *Sears, Roebuck Catal.* 298 *Metal-covered Footlocker.* Features a removable full-length molded plastic tray. Sheet steel covers sturdy plywood frame, fiberboard..top and bottom. **1845** W. T. Porter *Big Bear Arkansas* 130, I husseled off..Jem to the foot log. **1945** B. A. Botkin *Lay my Burden Down* 252 Go to the mill and cross on a foot log. **1966** *New Scientist* 30 June 835/1 The mechanical properties seen in *Surveyor's*..photographs..show that the spacecraft's footpads penetrated the surface for one inch. **1969** *Times* 22 July (Moon Rep.) p. i/1 I'm at the foot of the ladder. The L.M. footpads are only depressed in the surface about one or two inches. **1892** Foot-pound-second [see F.P.S. s.v. *F III. 3]. **1968** *Van Nostrand's Sci. Encycl.* (ed. 4) 1912/2 The length-force-time system to be discussed here is the foot-pound-second system. **1937** *Proc. Prehist. Soc.* III. 52 A large vessel with slightly raised base and foot-ring. **1952** G. Savage *18th Cent. Eng. Porc.* xxiii. 309 Champion's plates are more numerous than those of Cookworthy...Mostly they have a double foot-ring to prevent the centre of the plate from sagging downward during firing. **1967** *Antiquaries Jrnl.* XLVII. 229 Large platter with thick wall and scarcely defined rim...The foot-ring is low but carefully moulded. **1899** G. Massee *Text-bk. Plant Dis.* 333 This disease is known as 'mal-di-gomma' in Italy, and 'foot-rot' in Florida. **1926** H. H. Hume *Cultiv. Citrus Fruits* xxix. 462 Its history in Europe extends back to about 1845, and foot-rot worked destruction in the groves of the Azores some years previous to that time. **1952** E. Ramsden tr. *Gram & Weber's Plant Dis.* 68/2 Foot rot is a similar condition in older plants. *Ibid.* 69/1 Sclerotia are rarely to be found on damping off or foot-rot lesions. **1903** *Westm. Gaz.* 16 June 2/2 The advantage of having a foot-rule, so to speak, by which to test agreements for purchase. **1904** *Daily Chron.* 12 May 3/2 Mr. Richard Bagot's work may not always satisfy the critical foot-rule. **1908** *Westm. Gaz.* 9 June 1/2 It was Mr. Chamberlain who had pointed to exports as a foot-rule with which he wanted us to measure up our trade as a whole. **1837** Foot run [see Run *sb.*¹ 13 a]. **1869** W. Richardson *Timber Trades Price Bk.* 1 (*heading*) The price per foot run and its equivalent per 120·12 ft., irrespective of thickness or width. **1968** *Bodl. Libr. Rec.* VIII. 61 The installation..provides 12,500 foot-run of shelving for books up to 12 inches in height. **1872** *Harper's Mag.* XLIV. 547/2 Foot-scrapers and mats were doubled at all the approaches. **1938** J. Steinbeck *Long Valley* 115 On the front and back porches foot-scrapers and cocoa-fibre mats kept dirt out of the house. **1860** *Mining Surveyors' Rep.* (*Victoria Dept. Mines*) xii. 213 Slabs..being also placed longitudinally on the foot-wall to save the wear of the oxhide buckets. **1908** *Daily Chron.* 29 Jan. 9/2 (Wrestling) Vallotton..showed fine form, his footwork being wonderfully smart. *Ibid.* 25 June 3/3 (Tennis) It is foot-work that wins. **1921** A. W. Myers *20 Yrs. Lawn Tennis* 167 So well controlled was her footwork. **1929** Wodehouse *Mr. Mulliner Speaking* ii. 62 In face of the danger, his footwork, always impressive, took on a new agility. **1963** *Times* 8 May 5/5 Each offered brilliant foot-work [in dancing].

b. With adv.: **foot-up** *Rugby football*, in scrummaging, the illegal lifting of either foot by any member of the front row of forwards on either side before the ball is put in the scrummage. Also *attrib.*

1921 E. H. D. Sewell *Rugby Football* 361 Inadvertent offside, foot-up,..are..absolutely unavoidable at times. **1927** Wakefield & Marshall *Rugger* 183 The forwards..merely leant up against one another while the front row tried trick hooking and foot-up tactics. *Ibid.* 185 He must be careful..not to be penalised..for foot-up. **1963** *Times* 7 Mar. 3/5 For a foot-up offence, MacCormac..got three points for the Pay Corps.

foot, *v.* Add: **2.** Also of a ship: to move or sail with speed. Also with *it.* (In windward sailing, denoting speed as distinguished from pointing.)

1899 *Daily News* 4 Oct. 3/1 Shamrock, under clever handling, and footing splendidly, again took the lead. **1901** *Daily Chron.* 27 Sept. 5/7 His boat seemed to be footing it better. **1905** *Ibid.* 19 May 5/5 The latter boat was closely pursued by Hamburg, which was footing splendidly.

9. c. (Earlier U.S. example.)

1819 E. Evans *Pedestrious Tour* in R. G. Thwaites *Early Western Travels* (1907) VIII. 183 My dogs..helped themselves to the first repast presented, leaving their master to foot the bills.

footage (fu·tėdʒ). [f. Foot *sb.* + -age.]

1. *Mining.* A piece-work system of paying miners by the running foot of work; the amount paid; also, the amount mined.

1892 *Rep. Mineral Industries* (11th Census U.S.) 220 Two systems are employed in extracting the ore, (1) the footage system, and (2) the tribute system. The footage system is usually employed in new labores. **1909** *Westm. Gaz.* 9 Dec. 10/4 The output was 15,800 tons, or 2·15 tons per head, and the development footage 3,700. **1923** *Glasgow Herald* 31 May 16 The fathoms broken in the stopes were 234,960 against 192,784, the development footage 104,447 against 87,159, the tons milled 3,786,666 as compared with 3,447,736. **1927** *Sunday Times* 23 Jan. 5 The Development Footage sampled totalled 5,530 feet.

2. The length in linear feet of cinematographic or television film used in photographing a scene, etc. Also *attrib.*

1916 'B. M. Bower' *Phantom Herd* ii. 22 He visualized a stampede and the probable amount of footage it would require. **1918** H. Croy *How Motion Pictures are Made* v. 128 Directors..craftily working to keep the production expense as low as possible, do not altogether forget the footage possibilities of an exterior love-scene. *Ibid.* vi. 150 The amounts are added up and a footage rate determined. **1950** O. Skilbeck *ABC of Film & TV* 57 *Footage counter.* All ciné cameras are fitted with a meter showing the amount of film run off. **1967** *Spectator* 28 July 101/1 NBC decided to puff the footage up into a fifty minute documentary. **1970** *Which?* July 217/1 They all had a footage counter so that you could work out how much film you had left.

football. Add: **4. football coupon,** a coupon used in an entry for a football pool; **football pool,** an organized system of betting on the results of football matches; also, loosely, = **football coupon.*

1918 *Methodist Times* 5 Dec. 9/1 The amendment of the Gambling Laws, particularly in relation to football coupons, [etc.]..is long overdue. **1940** S. O'Casey *Star turns Red* i. 5 The concentration necessary to choose the right teams for entry in his football coupons. **1929** *Times Index* Jan.–Mar. 59/2 Football 'pool'. **1936** *Economist* 7 Mar. 517/2 We may..put..the total 'rake off' of football pool promoters..at not less than 30 per cent. of the amounts staked. **1957** M. Spark *Comforters* i. 6 Louisa Jepp sat at the table writing out her football pools. **1959** *Chambers's Encycl.* VI. 159/2 In 1954 football pools were estimated to be worth about £10 million to the post office. **1971** R. Roberts *Classic Slum* x. 183 Football pools 'nationalised' a local form of gambling. Before their introduction bookmakers supplied men in factory and workshop with fixed odds football coupons weekly.

footer, *sb.*¹ Add: **3. b.** (Examples.)

1863 *Boy's Own Vol.* July 36 A peculiar fashion of their own [at Harrow] which prompts them to call football 'footer'. **1896** *Westm. Gaz.* 12 Dec. 1/3 Who'd have thought of finding the old smug at a footer match? **1925** W. Deeping *Sorrell & Son* xvi, To perform on the footer field with a lot of young louts. **1945** E. Waugh *Brideshead Revisited* i. i. 30, I had to change for *F-f-footer.*

6. (See quot.)

1927 T. Woodhouse *Artificial Silk* 95 In the manufacture of stockings..two machines are used...One of these machines, termed the 'legger', knits the upper and longer part of the stocking, whereas the other machine, termed the 'footer', knits the remainder of the stocking.

foot-fault (fu·tfǫlt). *Lawn Tennis.* [f. Foot *sb.* + Fault *sb.*] A fault or infringement of the rules made by overstepping the base-line while serving. Hence **foo·t-fault** *v. intr.*, to make a foot-fault; *trans.*, to record a foot-fault against (a player). **foo·t-faulter, foo·t-faulting** *vbl. sb.*

1886 J. Dwight *Lawn Tennis* i. i. 5 Frequent foot-faulting in a match spoils your service altogether. *Ibid.* ii. v. 69 It is nearly impossible to call foot faults and to watch the service line too. **1889** W. M. Brownlee *Lawn-Tennis* iii. v. 144 Be careful to put the toe of the left foot on the line, and you are not likely to be foot-faulted. **1902** *Westm. Gaz.* 6 Oct. 4/3 The ex-champion was annoyed at being several times pulled up for 'foot-faulting'. *Ibid.* 23 Oct. 6/2 The New Lawn Tennis Laws. The Foot-Fault Difficulty Abolished...The feet of more than one 'crack' had to be carefully watched by relentless linesmen to detect the slightest inclination to foot-fault. **1909** *Ibid.* 9 Feb. 12/2 The ruling..that players who..lift the left foot before throwing up the ball, or while doing it, and then do the same with the right foot before the ball leaves the racket, are foot-faulters. **1921** A. W. Myers *20 Yrs. Lawn Tennis* 114 Larned..was the foot-fault judge in the Davis Cup doubles...When he foot-faulted,..he signalled to the umpire with his hand and the man in the chair called the foot-fault. **1922** W. T. Tilden *It's All in the Game* 35 In his anxiety he foot-faulted and his first serve was wasted. **1927** *Daily Express* 6 June 1 Tilden was foot-faulted by a special umpire in the fifth game of the third set.

foot-folk (fu·tfōᵘk). Delete †*Obs.* and add: In mod. use a new formation, prob. partly after G. *fussvolk*, as in quot. 1859 in Dict.

1876 Morris *Æneids* vii. 793 A cloud of foot-folk follow him. **1903** *Daily Chron.* 11 Dec. 3/4 The British footfolk at close grips are the very devil. **1905** *Ibid.* 24 June 4/6 The Scottish army consisted almost exclusively of foot-folk.

foo·t-gear. [Foot *sb.* 1.] Boots, shoes, or similar covering for the feet.

1837 [see Foot *sb.* 34 a]. **1874** 'H. Churton' *Toinette* (1881) ix. 101 The heavy brogans and a part of the gray stockings which formed her foot-gear. **1891** *Scribner's Mag.* X. 279 The proper foot-gear is the canvas shoe with rubber sole. **1904** S. E. White *Silent Places* viii. 81 A man's standing..can be accurately gauged by the magnificence of his foot-gear.

Foot-guards, footguards. For 'Now..Guards' read: The Footguards now comprise the Grenadier Guards, the Coldstream Guards, the Scots Guards, the Irish Guards, and the Welsh Guards.

foot-hill (fu·thil). orig. *U.S.* [Foot *sb.* 18.] A hill forming a lower eminence at the foot of a mountain or mountain-range.

1850 *Western Star* (Milwaukie, Ore.) 21 Nov. 3/2 The most productive portions..have been in the foot-hills. **1870** W. Gladden in *Scribner's Monthly* I. 153 The road leading to it passes over and among a series of conical foot-hills that rest against the base of the mountains. **1873** J. H. Beadle *Undevel. West* xv. 253 In California, everything under two thousand feet high is called a hill; if it leads up to a mountain, a foothill. **1879** [see Foot *sb.* 35]. **1882** *Rep. Prec. Metals U.S.* 262 The mineral belt is in the western foot-hills of the Verde range of mountains. **1916** J. B. Thoburn *Hist. Oklahoma* II. 4 The Black Mesa is in reality one of the foot-hills of the Rocky mountains. **1971** *Nature* 4 June 330/2 This bandicoot is widespread in the forested foothills and lowlands of western New Guinea. **1971** *Daily Tel.* 19 July 7/1 Its primary object is the exploration of Hadley Rille, one of the Moon's mysterious 'ditches' in the foothills of the Apennine range.

footie, var. *Footy *sb.*

footing, *vbl. sb.* **17.** *footing-place* (later example; also *fig.*).

a **1828** D. Wordsworth *Tour Continent* in *Jrnls.* (1941) II. 288 A gulph..towards the edge of which our footing-place..sloped directly. **1891** O. Wilde *Intentions* 26 Facts are not merely finding a footing-place in history, but they are usurping the domain of Fancy.

footler (fū·tlər). [f. Footle *v.* + -er¹.] One who footles.

1904 Chesterton *Nap. Notting Hill* iv. iii. 242 Drop your public-house flag, you footler! **1923** *Daily Mail* 22 June 7 Urbane persiflage directed against 'footlers', bores, and busybodies. **1923** U. L. Silberrad *Lett. J. Armiter* ix. 191 The rest are just footlers who'd piffle and gas. **1940** 'G. Orwell' *Inside Whale* 133 A novelist who simply disregards the major public events of the moment is generally either a footler or a plain idiot.

footlights, *sb. pl.* Add: *across the footlights*: see *Across B. *prep.* 2 b.

footling (fū·tliŋ), *ppl. a.* [f. Footle *v.* + -ing².] That footles or trifles; 'drivelling', 'blithering', trivial; hence of things.

1896 [O.E.D. s.v. Footle *v.*]. **1905** G. B. Shaw in *Daily Chron.* 15 Mar. 6/3 They are paraphrases of great works, made by footling people. **1907** *Daily Chron.* 2 Apr. 4/4 That dreary room..with its footling little lace window curtains. **1923** A. Bennett *Things that have interested Me* II. ii. 20 A perfectly footling popular play which fails to be popular. **1930** T. Thurston *Man in Black Hat* x. 181, I had heard it before from fatuous idealists and footling divines. **1930** W. S. Maugham *Breadwinner* i. 46 You do talk the most footling rot. **1960** *Guardian* 14 Dec. 8/7 Wondering why he should do 'footling' experiments with ball bearings.

foo·t-loose, *a.* orig. *U.S.* [Cf. Foot *sb.* 35.] Free to act as one pleases; not hampered by any ties. (See also quot. *a* 1699 s.v. Foot *sb.* 35.)

1873 J. H. Beadle *Undevel. West* xxxi. 669 All my friends who were 'footloose' had the 'Arizona fever'. **1904** F. Lynde *Grafters* xxvii. 347 If the mine should happen to explode..it'll be a comfort to have a foot-loose friend or two on the outside to pick up the pieces of us. **1910** W. M. Raine *B. O'Connor* 52 We went in together for all the kinds of spreeing that young fellows who are footloose are likely to do. **1924** —— *Troubled Waters* x. 104 Oh, she's nice enough, when she isn't a little divvle. The trouble is she isn't foot-loose. **1943** Forshaw & Abercrombie *County of London Plan* § 46 The London Docks..are not directly affected by the Plan: this is an industry that is not 'foot-loose'. **1959** T. S. Eliot *Elder Statesman* i. 25 You see, I'm a widower, like you, Dick. So I'm pretty footloose. **1961** *Guardian* 6 Feb. 8/4 A migrant group is likely to include..some who are merely footloose, who will settle nowhere. **1971** E. Candy *Words for Murder Perhaps* xi. 125 I'm travelling around, you see. Footloose and fancy free, you might call me. I've no special ties anywhere.

footnote. Add: *transf.* and *fig.*

1892 R. L. Stevenson (*title*) A footnote to history. Eight years of trouble in Samoa. **1935** E. V. Lucas in *Punch* 4 Sept. 269/3 What a glorious possession this Abbey is and what a footnote to history, science and art! **1942** *Sphere* 15 Aug. 220/2 A significant footnote to China's great struggle. **1965** *Listener* 3 June 834/1 Sir Colin Coote's breezy, affectionate life of his friend Walter Elliot provides a useful footnote.

foot-path, footpath. Add: **1. b.** A pavement.

1813 J. Lambert *Trav. N. Amer.* (ed. 2) II. 56 The street [*sc.* Broadway, New York] is well paved, and the foot-paths are chiefly bricked. **1908** *Westm. Gaz.* 21 Sept. 8/3 His car struck the footpath-face..and the car staggered badly on the road. **1943** K. Tennant *Ride on Stranger* vi. 50 The city swarm poured over the footpaths and across the taxi's front wheels.

footsie, var. *Footy *sb.* 1.

foot-slog (fu·tslǫg), *v.* [f. Foot *sb.* + Slog *v.* 2.] *intr.* To go on foot; to tramp, march. Also with *it.* Hence **foo·t-slog** *sb.*, a tramp, march; **foo·t-slo:gger,** a foot-soldier, infantry-man, pedestrian; **foo·t-slo:gging** *vbl. sb.* and *ppl. a.*

1894 C. H. Donovan *With Wilson in Matabeleland* ix. 198 Some of the Johannesburg 'Foot-Sloggers'..were doing damage in the village. **1899** Kipling *Five Nations* (1903) 185 We're foot—slog—slog—slog—sloggin' over Africa! **1906** *Macm. Mag.* Apr. 472 Must get in and foot-slog it, that's all. **1916** 'Peter' *Trench Yarns* iv. 34 I'm not a Shop gunner, y'know. I'm only a miserable foot-slogger! **1921** E. R. G. R. Evans *South with Scott* vi. 89 There were people who preferred foot-slogging to ski at any time. **1925** B. Travers *Mischief* viii, Another long foot-slog..in squelching shoes. **1927** *Chambers's Jrnl.* June 392/1, I..had to foot-slog until I could find a British Consul to borrow from. **1928** *Observer* 26 Feb. 6/3 A world where people travel by preference footslogging. **1928** *Sunday Express* 28 Oct. 14/3 A long, long way to Berlin, especially for the poor bally foot-sloggers who did the fighting and the dying. **1929** *Daily Tel.* 1 Jan. 9/6 Do coroners think that footsloggers willingly ask for trouble? **1939** *War Illustr.* 7 Oct. 98/3 Their speed..was not that of foot-slogging infantry but of motorized columns.

footstep. Add: **6.** *attrib.*: **footstep bearing** = sense 5 d.

1887 D. A. Low *Introd. Machine Draw.* 110 Plan and sectional elevation of a footstep bearing for an upright shaft. **1888** *Lockwood's Dict. Mech. Engin.* 147 Footstep bearings are..made adjustable by placing the bearing in the centre of a ring casting. **1959** *Chambers's Encycl.* II. 175/2 A 'footstep bearing' is a thrust bearing at the bottom of a vertical shaft to support the revolving parts.

footstool. 1. c. *U.S.* (Earlier and later examples.)

1821 T. Dwight *Trav. New Eng.* (1823) III. 231 We felt a total superiority to all the humble beings who were creeping on the footstool beneath us. **1859** *Congress. Globe* 18 Feb. 1122/1, I would give that power [of declaring war] to no President—none that has ever stood upon this footstool. **1906** *N.Y. Globe* 20 Feb. 8 This New York of ours, regarded by many the wickedest city on the footstool.

footsy, var. *Footy *sb.* 1.

footy, -ie (fu·ti), *sb. colloq.* **1.** [Jocular dim. of Foot *sb.*: see -y⁶.] Amorous play with the feet; also *transf.* and *fig.* Also **footsy, footsie,** and redupl.

1935 S. Lewis *It can't happen Here* xix. 215 Lindy and you, playing footie-footie these last couple years. **1944** G. Fowler *Good Night, Sweet Prince* (1949) ii. iv. 131, I played footsie with her during Don José's first seduction by Carmen. **1945** S. Lewis *Cass Timberlane* (1946) xl. 291 You got to do something about Bradd and your wife. Town's beginning to talk. They're playing a little too much footie-footie. **1947** K. Jaediker *Tall, Dark, & Dead* ii. 36, I have neither the inclination nor the time to play footie with the police. **1948** A. Keith *Three came Home* iii. 61 The eight guards..lolling in the women's barrack playing footsy-footsy. **1955** D. W. Maurer in *Publ. Amer. Dialect Soc.* xxiv. 136 Sometimes the party out of power retains one or two fixers who play 'footsie' with the *fixer* for the party which is in power. **1959** J. Thurber *Years with Ross* iii. 49 In [a drawing]..showing a man and his wife and another woman at a table..the designing minx was playing footy-footy with the husband. **1959** K. Waterhouse *Billy Liar* iii. 48 This was the sequence and rhythm of daylight love-play as she knew it, a kind of oral footy-footy that was the nearest she could get to intimate conversation. **1963** *Economist* 11 May 537/3 Pakistan is..despite recent games of footsie with Peking, a staunchly anti-communist-ally. **1971** *Ink* 12 June 14/2 The real trouble started when he started playing footsie with the real capitalists.

2. Dim. of Football. esp. *Austral.* and *N.Z.*

1940 *Bulletin* (Sydney) 31 Jan. 40/2 When school came out I played footy, and picked a row with a kid and had a fight. **1947** 'A. P. Gaskell' *Big Game* 14, I cleaned my footy boots. **1949** [see *Clean *v.* 6 b]. **1949** *Hilltop* (N.Z.) I. 8, I got footie practice tonight. **1966** F. Shaw et al. *Lern Yerself Scouse* 44 A game a footee, a game of football. **1967** *Southerly* XXVII. 297 Evans..strides off with her to ask race-goers, cinema queues and footy fans to sign peace petitions.

foozled (fū·z'ld), *ppl. a.* [f. Foozle *v.* + -ed¹.] Bungled; esp. of a stroke in *Golf.*

1899 *Westm. Gaz.* 25 Aug. 3/1 Foozled drives or missed approaches. **1909** *Ibid.* 22 Jan. 5/2 A round freely interspersed with foozled tee shots.

So **foo·zling** *vbl. sb.*, bungling.

1927 J. Adams *Errors in School* 187 In his playing the pupil finds no lack of errors, mistakes, foozlings, call them what you will. **1965** G. McInnes *Road to Gundagai* xiv. 259 The rest of the eighteen holes were a miserable exhibition of foozling, duffing, [etc.].

for, *prep.* Add: **8.** Also, in preparation for or anticipation of (the stated time of a dinner, etc.).

1900 *Athenæum* 5 May 545/1 The Annual Dinner of the Society will take place on Wednesday, May 16..at 7, for 7.30 P.M. **1929** E. Wallace *Iron Grip* viii, You can please yourself about accepting..No. 109 Grosvenor Place, and the hour is 7.30 for 8. **1970** J. Burke *Four Stars for Danger* iii. 49 'Seven-thirty,' he was saying, 'for eight. Right?'

d. *For Sale*: used *attrib.* or as adj. phr. Cf. also Sale *sb.*² 2 e.

1938 R. Field *All this & Heaven Too* (1939) xxxviii. 488 The best houses in the neighbourhood suddenly stood empty with 'To Let' or 'For Sale' signs on their walls. **1965** K. Giles *Some Beasts no More* ii. 42 There was a For Sale notice which alluded to a fine residence and ninety acres of land.

9. c. (Earlier example of *for the life of him.*)

1809 [see Life *sb.* 3 c].

13. e. *to be for it* (orig. *Mil. slang*): to have one's name on the crime sheet, i.e. to be marked down for punishment or trouble; hence, to be in for trouble.

1909 *Captain* Apr. 14/2 'He'll give the whole show away?' 'Then, I suppose, we'll all be for it.' **1915** 'I. Hay' *First Hundred Thousand* xvi. 214 No more trial-trips; no more chaperoning! This time, we decided, we were 'for it'. **1915** *War Illustr.* 21 Aug. 22/1 Then it is that he realises so acutely that if anything happens to his pilot he is 'for it', as the current flying phrase has it. **1917** P. Macgill *Brown Brethren* iv. 68 If Captain Thorley finds 'im missin' he'll be for it. **1919** *Athenæum* 25 July 664/1 'You're for it', *i.e.*, for the 'orderly room'. *Ibid.* 1 Aug. 695/1 'On the peg', to be charged with a 'crime'. 'You'll be for it', the sergeant's threat of 'the peg'. **1926** *Punch* 28 Apr. 466/1 Major Atlee..declared that if the discipline of a unit broke down it was the officer in charge who should be 'for it'. **1928** [see *Bonzo]. **1940** *War Illustr.* 5 Jan. 567/1 If a bomb comes, one is 'for it'.

19. g. With an adjective, in pleonastic use, as *for free*, for no charge, without payment; *for real*, real. Also in such phrases used *attrib.* Chiefly *U.S.*

1887 in *Amer. Speech* (1950) XXV. 39/2 When a for-true doctor come to see him. **1900**, etc. [see *Fair *sb.*³ 1 c]. **1942** in Wentworth *Amer. Dial. Dict.* (1944) 228/1 Railroads don't haul trash for free. **1943** *Amer. Speech* XVIII. 47 It might be reasonable to assume that 'for free' results from the confusion of 'free' and 'for nothing'. **1954** W. M. Miller *Conditionally Human* (1963) 72 Don't worry, Richard. This time it's for real. **1957** *New Yorker* 21 Sept. 33/2 He said psychiatrists had been enthusiastic, patents have been applied for, and it's for real. **1957** G. Smith *Friends* 147 Back home we pay if we're ill...You don't expect to be ill for free. **1958** K. Amis *I like it Here* xi. 133 Bowen tried to buy some drinks, conscious of having been fed and made drunk for free. **1960** J. Kirkwood *There must be a Pony* (1961) xii. 95 A good guy; a movie cop..; a for-real cop.

forage, *sb.* Add: **4. forage-cap** (earlier U.S. example); **forage harvester,** an implement for harvesting forage grass (see quot. 1944); **forage-master** (later U.S. example).

1827 J. F. Cooper *Prairie* I. x, He wore a forage-cap of fine blue cloth. **1944** C. Culpin *Farm Machinery* (ed. 2) x. 226 The Allis Chalmers forage harvester..is a machine which cuts the standing crop, conveys it..to a chopping cylinder, chops it into short lengths, and blows it into a trailer alongside. **1960** *Farmer & Stockbreeder* 12 Jan. 79/1 The forage-harvester..is the best thing that has hit grass farming; just as the binder revolutionized the corn-field. **1861** *U.S. Army Regul.* 149 No wagon-master or forage-master shall be interested..in any wagon or other means of transport employed by the United States.

foramen. Add: Also with defining name, as **foramen of Magendie** [described by François *Magendie* (1783–1855), French physiologist], the median aperture in the roof of the fourth ventricle of the brain; **foramen of Monro** [described by Alexander *Monro* (1733–1817), Scottish anatomist], a foramen in the brain connecting the third ventricle with each of the two lateral ventricles; the interventricular foramen; **foramen of Winslow** [described by Jakob Benignus *Winslow* (1669–1760), Danish anatomist], a narrow passage connecting the lesser sac and the greater sac of the peritoneal cavity of the abdomen; the epiploic foramen.

1826 J. & C. Bell *Anat. & Physiol. Human Body* (ed. 6) III. 274 My reader must now find his way into the marsupium, or purse of the omentum, viz. the porta omenti, the celebrated foramen of Winslow. **1837** J. Quain *Elem. Anat.* (ed. 4) 901 (Index), Foramen of Monro. **1882** *Quain's Elem. Anat.* (ed. 9) II. 290 The roof or posterior wall [of the fourth ventricle]..is not quite complete, for there is a hole in it termed the foramen of Majendie [*sic*]. **1907** *Practitioner* Aug. 267 A hernia into the foramen of Winslow was discovered. **1910** *Ibid.* July 50 The orifice in the ependyma roofing the fourth ventricle, which is called the foramen of Magendie. **1951** J. T. & O. T. Lewis tr. *Houssay's Human Physiol.* xxv. 226/2 Obstruction of Monro's foramen provokes dilatation in the ventricle occluded.

forastero (fǫræstę̄·ro). [a. Sp. *forastero*, strange.] Any of a group of varieties of the cacao tree, *Theobroma cacao.*

1858 L. A. A. de Verteuil *Trinidad* vii. 262 The cacao forastero is hardier, more robust, and yields more abundantly. **1908, 1929** [see *Criollo]. **1953** E. M. Chatt *Cocoa* ii. 16 Criollo means 'native' and Forastero 'foreign'. The contradistinction arose in Venezuela after the introduction of cacao trees from the West Indies, when the terms were used to denote the local and imported kinds of cacao.

forb (fōᵊɪb). [f. Gr. φορβή fodder, forage (φέρβειν to feed).] A herbaceous plant of a kind other than grass: applied chiefly to any broad-leaved herbs growing naturally on grassland.

1924 Clements & Weaver *Exper. Vegetation* i. 21 The term 'forb' is here used for herbs other than grasses in order to afford a clear cut distinction and at the same time avoid an awkward phrase. **1952** A. W. Sampson *Range Managem.* x. 193 Forbs such as the clovers, filaree, dandelion, and cow parsnip contribute in no small way to the hay and pasture crops. **1960** N. Polunin *Introd. Plant Geogr.* xiii. 399 On well-drained banks there may be Grasses and some hardy but attractive forbs. **1970** *Nature* 25 July 340/2 High numbers of mixed-feeding or forb-feeding species may not produce economically significant damage.

forbesite (fǭ·ɪbzəit). *Min.* [ad. G. *forbesit* (A. Kenngott *Uebersicht d. Resultate min. Forschungen 1862–1865* (1868) 47), f. the name of David *Forbes* (1828–76), English geologist and explorer, who first analysed it: see -ite¹.] A hydrated arsenate of nickel and cobalt found in the Atacama desert, Chile, as greyish-white crusts.

1868 J. D. Dana *Syst. Min.* (ed. 5) 560 D. Forbes describes..a mineral occurring in the desert of Atacama in veins in a decomposed dioryte...Kenngott names it Forbesite. **1935** J. W. Mellor *Inorg. & Theoret. Chem.* XIV. 424 The cobalt minerals include..Forbesite, $H_2(Ni,Co)_2(AsO_4)_2.8H_2O$.

forbidden, *ppl. a.* Add: **b.** *forbidden line,* a spectral line produced by a forbidden transition; *forbidden transition,* 'a transition between two states of a quantum-mechanical system (as a molecule, atom, or nucleus) that does not conform to some selection rules, esp. those for electric dipole radiation from an unperturbed system.

1923 H. L. Brose tr. *Sommerfeld's Atomic Struct. & Spectral Lines* vi. 366 The forbidden lines..belong to transitions in which n_i jumps by two or..by three units. *Ibid.* 367 In the combinations (sp_i) in the H.S. or the II N.S. no forbidden transitions occur. **1939** J. W. T. Spinks tr. *Herzberg's Molecular Spectra & Molecular Struct.* I. v. 305 Another forbidden transition, also involving magnetic dipole radiation, has been observed for oxygen. **1957** *Encycl. Brit.* II. 591/1 The two strong green lines in the spectra of the gaseous nebulae prove to be forbidden lines of doubly ionized oxygen.

force, *sb.*^{1} Add: **2.** With reference to the force of wind described by numbers in the Beaufort scale.

1867 [see Dict.]. **1933, 1961** [see *Beaufort scale]. **1963** *Listener* 21 Mar. 528/3 The wind is not only there, but assuming the proportions of a force-ten hurricane.

4. a. *pl.* In *attrib.* use or in the possessive, esp. during the war of 1939–1945.

1942 *New Statesman* 3 Jan., I see that a new version of this feature is now to begin in the new year on the Forces Programme. **1943** E. Olivier *Night Thoughts of Country Landlady* iii. 25 The very inferior music often produced in the B.B.C. Forces' Programme. **1945** *News Review* 10 May, If you're a Forces bride you will be given a travelling warrant for the whole journey from your British home to your new home in America. **1945** *Manch. Guardian* 18 July, A statement in a 'forces' newspaper. **1952** Granville *Dict. Theatr. Terms* 80 *The Forces' sweetheart.* Vera Lynn, the British vaudeville and radio singer, was a great favourite with the troops in the second world war, hence the sobriquet. **1959** *Times Lit. Suppl.* 2 Oct. 556/4 The man and the girl,.. whom we observe listening intently to Forces Favourites.

d. *U.S.* (See quots.)

1807 C. W. Janson *Stranger in Amer.* 309 Force, is here employed when speaking of the number of slaves employed in field labour on each plantation. **1834** W. G. Simms *Guy Rivers* II. 97 The force of the traveller—for such is the term by which the number of his slaves are understood—was small. **1837** H. Martineau *Society in Amer.* I. II. 344 All the 'force' that could be collected on a hasty summons,—that is, almost every able-bodied man in the city and neighbourhood, was sent out with axes to build us a bridge. **1871** Schele de Vere *Americanisms* (1872) 475 *Force* is a common name for a gang of laborers, whether they are Irishmen at work on a railway, or negroes employed on a plantation. **1899** *Monthly South Dakotan* (Mitchell) I. 138 A high wind..showered down hundreds of bushels of apples, [and] one is confronted by the alternative of sending for the 'force' to pick them up on Sunday or letting the sun scald and ruin them.

7. e. *Austral.* and *N.Z.* (See quots.)

1933 *Press* (Christchurch) 21 Oct. 15/7 *Force,* the power of dogs to move sheep... Huntaways are sometimes spoken of as *forcing dogs*; but the term *f[orce]* is also applied to the ability of a heading dog to pull sheep. **1960** Baker *Drum* 110 *Force,* the ability of a sheepdog to control a mob of sheep, esp. without legging, i.e., leg-biting. A good dog is said to have a lot of force.

15. b. *Tennis.* (See quot. 1890^{2}.)

1662 [see Dict., 5 d]. **1890** J. M. Heathcote *Tennis* 50 The Force is the usual resource of a player who must try to win at very 'close chase', or who returns a ball which comes 'fair-off' from the end-wall. *Ibid.* 124 *Force,* a stroke played, either direct or boasted, for the dedans with some strength. **1927** *Daily Tel.* 26 Apr. 17/1 Some admirable tennis was seen, with good returns, short chases, and accurate forces. **1955** *Times* 2 May 4/1 Dear went all out for winning openings, making a severe attack on the *dedans*—he scored with 11 forces during the two sets.

23. force cup, a rubber cup attached to a handle which by creating a vacuum in a blocked drain is used to clear it; **force(-)field,** a field of force (see *field *sb.* 17); esp. in *Science Fiction,* one that acts as an invisible barrier; **force-land** *v.* (see *forced *ppl. a.* 2 d); **force-out,** in *Baseball,* the obligatory retirement of a base runner at the base he is forced to run to by a following base runner.

1907 *Yesterday's Shopping* (1969) 118/3 Force Cup. For cleansing stopped pipes, drains, &c. **1951** *Good Housek. Home Encycl.* 257/1 Try using a rubber force cup with a vigorous up-and-down movement. **1960** D. V. Davis *Domestic Encycl.* i. 54 An emergency force cup can be made by cutting a piece out of an old rubber ball, placing the pole over the sink outlet and squeezing the ball several times. **1920** *Rep. Brit. Assoc. 1920* 236 Each atom must form the centre of an electromagnetic field of force. These force fields were first dealt with by Humphreys. **1926** *Bull. Nat. Res. Counc.* LIV. 294 A clear understanding of the form of the orbit and quantum conditions for central force fields is often essential in the theoretical interpretation of spectra. **1944** F. Brown in B. Aldiss *Introd. SF* (1964) 74 'There is a barrier.' A force-field, of course. **1962** F. I. Ordway et al. *Basic Astronautics* iv. 120 There are four groups of phenomena: (1) interstellar and interplanetary particulate matter, (2) energetic particles, (3) electromagnetic radiations, and (4) force fields. **1964** *Observer* 13 Dec. 34/7 An electronic bird-repeller..that will send out a science-fiction type 'force field' to keep birds away. **1896** H. Chadwick *Spalding's Base Ball Guide* 76 The result being a force-out play to second, if not a double play. **1926** *N.Y. Times* 11 Oct. 24/1 His grounder to Bell was turned into a forceout of Ruth at second while Combs dashed on to third. **1968** *Washington Post* 4 July C 2/4 Wills..took third on Gene Alley's single and scored on Roberto Clemente's force out.

force, *v.*^{1} Add: **3. d.** *intr. Austral.* and *N.Z.* Of a sheep-dog: to move sheep. Cf. *force *sb.*^{1} 7 e.

1920 Paton & Reid in J. B. Cramsie *Managem. Sheep Austral.* v. 29 Close working [by a dog] in the open paddock is to be avoided because this means forcing, and forcing means over-heated sheep. **1934** J. Lilico *Sheep Dog Mem.* 27 [The dogs] would head, lead, hunt away, force and back though..they were best at rouseabout work.

5. Examples of use in *Cricket.*

1897 *Encycl. Sport* I. 226/2 Under such conditions the batsmen should run a few risks to score runs quickly (in cricketers' parlance, 'to force the game') while they have the chance. **1904** P. F. Warner *How We recovered Ashes* xiv. 276 Hopkins was evidently bent on forcing the hitting. **1908** W. E. W. Collins *County Cricketer's Diary* x. 168 An attempt..to force the game on a ground that does not lend itself to forcing tactics. **1963** A. Ross *Australia 63* iii. 79 Dexter forced him through mid-wicket. **1970** *Times* 19 Aug. 6/3 They all pitched a little short, for fear of being driven, and because of this they were forced and hooked.

7. b. *to force down*: to compel (an aircraft) to land.

1917 *War Illustr.* 15 Dec. 359 The German machine..was 'forced down' on the French front in an intact state, and its airmen were made prisoners. **1958** *Times* 30 June 10/5 U.S. aircraft forced down by Soviet fighters.

8. b. *Tennis.* To use the force stroke (see prec. 15 b).

1890 J. M. Heathcote *Tennis* 52 It is impossible to force as severely, difficult to force as accurately, with a back-hand as with a fore-hand stroke.

9. *to force* (a person) *on, upon* (something). Delete † and add later examples.

1856 De Quincey *Confess.* 238 Nervous irritation forced me..upon frightful excesses; but terror from anomalous symptoms sooner or later forced me back. *Ibid.* 269 The..riotous prodigality of life naturally forces the mind more powerfully upon the antagonist thought of death. **1903** R. Langbridge *Flame & Flood* xxiv, Her lack of money had forced her back upon the most respectable costume which she had.

forced, *ppl. a.* Add: **2.** *forced labour* (later examples); also as *attrib. phr.*

1931 *Economist* 17 Jan. 105/2 Some of the worst features of the report..are those relating to what is euphemistically described as the official recruitment of labour, for export, mainly to Fernando Po, and to forced labour conditions. **1941** Koestler *Scum of Earth* 276 They..were kept in the Moroccan forced-labour battalions to work in mines and quarries. **1942** *Ann. Reg. 1941* 175 Trials of people for kindness towards prisoners of war and foreigners doing forced labour. **1960** *Guardian* 15 Nov. 8/5 Lord Altrincham's call for conscription seems to be for both military service and civil jobs such as road making. The latter is ordinarily termed forced labour.

b. (Earlier example.)

1847 H. Staunton *Chess-Player's Handbk.* I. ii. 22 When a player has only one legal move at command, it is said to be a *forced move.*

c. *forced feeding* = *force-feeding* (see *force-feed *v.*).

1901 G. C. Watson *Farm Poultry* ix. 161 It is quite possible to injure the digestion of laying hens by a system of forced feeding. **1923** H. A. Roberts *Commercial Poultry Raising* xv. 207 Forced feeding must not be carried to the point where it affects and impairs stamina and health.

d. *forced landing*: the unpremeditated landing of an aircraft in an emergency; hence *force-land* v. intr., to make a forced landing.

1917 'Contact' *Airman's Outings* 8 That aerial highwayman the forced landing. **1921** *Flight* XIII. 440/2 They left Croydon..in a Goliath, which had, unfortunately, to force-land at Amiens. **1928** [see *crash *v.* 6 b]. **1936** *Jrnl. R. Aeronaut. Soc.* XL. 28 An engine failure is defined as engine trouble (apart from fuel system) causing a forced landing. **1958** *Oxf. Mail* 21 Jan. 1/1 A United States Navy plane..force-landed about 450 miles from its destination. **1961** L. van der Post *Heart of Hunter* viii. 120 Some R.A.F. men..had made a forced landing in the desert.

e. *forced-choice*: used *attrib.* of a question, technique, etc., in which the participant must choose between a number of pre-arranged alternatives.

1944 Shipley & Graham *Summary Res. on Personal Inventory* 1 Its items..are cast in forced-choice form to promote valid answering. **1957** E. Bott *Family & Social Network* vi. 170 Answers to forced-choice questions need not represent the subject's attitudes. **1964** D. B. Fry in D. Abercrombie et al. *Daniel Jones* 62 The listener is placed in a 'forced choice' situation. **1967** J. F. Corso in H. Helson et al. *Contemp. Approaches Psychol.* vii. 304 To predict performance in a forced-choice test, an extension of the decision model is required. **1970** *Jrnl. Gen. Psychol.* Jan. 96 These statements were mimeographed and presented in the form of a forced-choice true and false test. **1970** *Language* XLVI. 304 Despite their close agreement in the 'forced-choice' selections, the preceptive reaction of UK and US subjects shows them sharply polarized in the judgment test.

7. Of a draught: produced artificially; so *forced-air, ventilation*; *forced (feed) lubrication,* force-feed lubrication (see *force-feed *sb.*).

1865 Webster s.v. *Draught, Forced draught,* the draft produced by a blower, as by compressing the air beneath a fire. **1885** *Marine Engineer* VII. 39/1 (*heading*) On the Application of moderate forced draught to the furnaces of small steam vessels. **1887** *Encycl. Brit.* XXII. 496/2 Where forced draught has been substituted for chimney draught. **1898** *Westm. Gaz.* 18 May 2/3 In our service we are satisfied with a forced draught trial of four hours as a rule. **1901** *Feilden's Mag.* IV. 441/2 The forced-air draught fans... Forced-air draught is preferable to steam blast. **1907** J. G. Horner *Encycl. Pract. Engin.* V. 40/2 It is hardly possible to overrate the importance of forced lubrication in modern practice. **1907** *Westm. Gaz.* 19 Nov. 4/2 Forced feed lubrication. **1909** *Ibid.* 12 Aug. 4/2 Designing an engine on the forced-induction principle. **1930** *Engineering* 28 Feb. 280/3 Forced lubrication is employed for the main engine and for the turbo-blower. **1960** *Farmer & Stockbreeder* 19 Jan. 108/1 The forced-draught cabinet incubators. **1971** *Sci. Amer.* May 73/3 In a forced-draft [cooling] tower the fan is at the bottom and pushes the air up through the tower.

‖ **force de frappe** (fors də frap). [Fr., lit. 'striking force'.] A striking force; *spec.* the French independent nuclear striking force. Also *attrib.*

1962 *Observer* 6 May 10/2 Britain has had her 'independent deterrent' for several years, and General de Gaulle is now trying to give France her own 'force de frappe'. **1962** *Spectator* 31 Aug. 294 Useless objects of conspicuous consumption like steel mills in underdeveloped countries or *forces de frappe* elsewhere. **1965** A. Werth *De Gaulle* x. 344 De Gaulle continued to look upon his *force de frappe* as absolutely essential to France's independence and her status as a great power. **1968** S. Serfaty *France, de Gaulle & Europe* vi. 119 While the left is staunchly against the force de frappe, judged 'inefficient, ruinous and dangerous', it remains nonetheless true that the French atomic forces originated during leftist governments. **1968** *Listener* 4 July 2/1 Their leaders were evidently trained and equipped as a *force de frappe*, and acted in accordance with the precepts of Debray's guerrilla manual.

force-fee·d, *v.* [f. Force *sb.*^{1} 23 + Feed *v.*] To feed by force; also *fig.* So **force-fee·ding** *vbl. sb.*

1909 Webster, *Force feeding,* feeding (poultry, etc.) by pushing food forcibly down the throat, as by a cramming machine. **1912** D. F. Laurie *Poultry Foods & Feeding* v. 83 Machine or force feeding is the most..satisfactory method of feeding large numbers of fattening chickens. **1938** S. Beckett *Murphy* 182 Otherwise he had to be force-fed. **1962** *Times* 15 Feb. 15/5 We had..put her [*sc.* a dolphin] in a small side tank in order to force-feed her. **1964** *Economist* 8 Feb. 504/2 The feeders who buy the animals and force-feed them nearly two tons of grain. **1964** D. Varaday *Gara-Yaka* vi. 52 He brought a jug of lukewarm beef tea, and the task of force-feeding began. **1971** *Guardian* 26 Aug. 10/5 Being force-fed information is a habit one can break.

force-fee·d, *sb.* [f. Force *sb.*^{1} 23 + Feed *sb.*, or f. prec.] A supply, esp. of lubricant, that is maintained under applied force or pressure. Usu. *attrib.*

1918 V. W. Pagé *Aviation Engines* vii. 19 When a force-feed lubricating system is used, the oil..is thrown off at a tangent to the crank-pin circle in all directions. **1921** W. H. Berry *Mod. Motor Car Pract.* i. 44 This system of partial force-feed has been developed into a true forced-feed system. **1938** C. Culpin *Farm Machinery* viii. 116 It has the advantage of all force-feed drills, and gives a more continuous flow of seed than the external-feed type. **1966** *McGraw-Hill Encycl. Sci. & Technol.* VII. 607/1 In the force-feed system the pump pressure insures a steady replacement of the oil which has been heated by friction work with fresh cool oil.

‖ **force majeure** (fors maʒœr). [Fr., lit. 'superior strength'.] Irresistible force or overwhelming power.

[**1858** Simmonds *Dict. Trade, Force-majeure,* a French commercial term for unavoidable accidents in the transport of goods, from superior force, the act of God, &c.] **1883** *Academy* 8 Sept. 158/1 Tyranny, upheld by law, will generally be 'tempered' by outrage, so long as a *force majeure* prevents its being met in any other way. **1886** *Macm. Mag.* Sept. 342/1 They [*sc.* politicians] will not combine except under *force majeure.* **1902** *Encycl. Brit.* XXV. 112/1 The expression 'act of God'..is not synonymous with *force majeure*; but it includes every loss by *force majeure* in which human agency, by act or negligence, has had no part. **1907** W. De Morgan *Alice-for-Short* i, Hindered from..determined effort by a *force-majeure* trying to the temper but heroically endured. **1916** 'Peter' *Trench Yarns* v. 51 *Force majeure* being on George's side, the transaction was accomplished to the accompaniment of awful threats as to George's lurid future. **1930** W. de la Mare *Poems for Children* p. xxix, Whose right is often founded solely on *force majeure.* **1941** A. L. Rowse *Tudor Cornwall* xi. 265 It was thought..that the mayor had yielded the town by treachery, but later they learned that it was rather by *force majeure.* **1959** G. Mitchell *Man who grew Tomatoes* iii. 39 One cannot be a respected and highly respectable Civil Servant for twenty years without learning to bow to the *force majeure* of public opinion.

forcer^{2}. Add: **1. b.** One who produces forced crops.

1789 J. Abercrombie (*title*) The complete kitchen gardener and hot-bed forcer. **1902** *Westm. Gaz.* 4 Apr. 8/2 There are forcers who have a dozen large hothouses devoted solely to Easter lilies. **1905** *Ibid.* 21 Feb. 12/1 A 'forcer' in Kent gets a living from strawberry-forcing on less than half an acre of land.

forcing, *vbl. sb.* Add: **3. a. forcing pen** (*Austral.* and *N.Z.*), a pen into which sheep are forced

or driven in order to guide them to a certain point, as in a drafting yard; a crush-pen.

1923 W. Perry et al. *Sheepfarming in N.Z.* iii. 16 [The sheep yards] should consist of receiving yards at each end, forcing pen, and drafting race. **1950** *N.Z. Jrnl. Agric.* July 5/2 If the forcing pens and..check pens run up a slight incline, it is quite satisfactory to have the rest of the yards on level ground. **1965** J. S. Gunn *Terminol. Shearing Indust.* I. 28 *Forcing pen,* a narrow area through which sheep are forced on their way to the sheep shower.

forcing, *ppl. a.* Add: (Examples of use in *Cricket* (cf. Force *v.*^{1} 5) and in *Bridge.*)

1888 Steel & Lyttelton *Cricket* ii. 57 The ball most to be dreaded for the forcing stroke is the hanging ball, which stops and does not come on evenly and fast to the bat. **1891** W. G. Grace *Cricket* viii. 236 There are many other points to be considered, such as knowing when to play a slow, patient game, or a forcing game. **1930** E. Culbertson *Contract Bridge Blue Bk.* vii. 95 The Forcing principle is a logical and necessary extension of the Approach principle with the added proviso that..both partners are required to keep the bidding *unconditionally* open until the game bid is selected. The reason for the *Forcing* bids is: a majority of strong hands contain a choice between..suit bids and no-trump; a game..is assured provided *neither partner passes* before the best fitting bid is selected. **1933** C. Vandyck *Contract Contracted* iii. 28 The Forcing Take-Out is a jump bid in a new suit. *Ibid.* iv. 32 The Forcing Call for the Opening Hand is variously known as the Forcing Two, Demand Two and Big Two. **1935** *Times* 4 Nov. 4/7 With no flourish or eccentricity, he was a forcing batsman who scored at an immense rate. **1939** N. de V. Hart *Bridge Players' Bedside Bk.* 137 Clearly he has made a forcing pass, leaving it to North to double or to go on in Hearts or Spades. **1947** S. Harris *Fund. Princ. Contract Bridge* iii. 27 The only way in which East can make a forcing bid over North is to double informatively, or overbid North in his own suit. **1959** Reese & Dormer *Bridge Player's Dict.* 95 Traditional Acol players do not regard a change of suit by responder as forcing unless it is combined with a jump. **1970** *Globe & Mail* (Toronto) 26 Sept. 51/4 Partner must have at least two aces for his jump raise (forcing).

b. *spec.* **forcing dog** *Austral.* and *N.Z.*, a dog with the power to move sheep in the direction it wishes.

1933 [see *force *sb.*^{1} 7 e].

forcing-pump. = Force-pump 2.

1838 *Civil Engin. & Arch. Jrnl.* I. 322/1 A substitute for the Forcing-pump in supplying Steam Boilers.

Ford (fōəɹd), *sb.*^{2} The name of Henry *Ford* (1863–1947), American manufacturer of motor vehicles, used to designate the products of the company he founded. Also *fig.* (see quots. 1964, 1968).

1914 *Scotsman* 14 Oct. 12/1 Ford Chassis £105. **1918** *Automobile Engineer* Mar. 85/2 The minimum-priced car, represented by the Ford. **1920** *Blackw. Mag.* Jan. 105/2 It is the sort of road that even a Ford would hesitate to tackle. **1925** A. P. Herbert *Laughing Ann* 48 Like the starting of a Ford. **1927** A. Huxley *Let.* 8 Oct. (1969) 292, I don't know whether one's interests are better looked after by an individual or a Ford factory. **1937** [see *anti-freeze *a.* and *sb.*]. **1958** B. Nichols *Sweet & Twenties* vi. 83 She saw a Ford van in the window. **1964** W. L. Goodman *Hist. Woodworking Tools* 100 The Stanley-type plane may be described..as the Ford of woodworking tools. **1968** J. Ironside *Fashion Alphabet* 87 *Ford,* a best-seller, a winner. A line which continues from season to season.

ford, *v.* Add: **1. b.** (Later U.S. example.)

1903 A. Adams *Log of Cowboy* vii. 88 You can hardly imagine what a difference there is in fording this herd, between a cool cloudy day and a clear hot one.

fore-. Add: **2. a.** *fore-suffer* (later example).

1923 T. S. Eliot *Waste Land* iii. 17 And I Tiresias have foresuffered all.

c. *fore-poling.*

1881 R. W. Raymond *Gloss. Mining Terms, Fore-poling,* a method of securing drifts in progress through quicksand by driving ahead poles, lath, boards, slabs, etc., to prevent the inflow of the quicksand on the sides and top, the face being protected by breastboards. **1967** *Gloss. Mining Terms* (*B.S.I.*) xi. 8 *Forepoling,* the use of girders or bars projecting forward in cantilever as temporary supports beneath newly exposed roof.

3. a. *fore-rib* (earlier example).

1861 Mrs. Beeton *Bk. Househ. Managem.* xii. 276 *Fore quarter.* Five ribs, called the fore-rib.—This is considered the primest roasting piece [of beef].

5. fore-bitter, a sea song (see quots.); **fore-caddie,** a caddie who goes in advance of the players to watch where the balls fall; **fore-car,** an obsolete form of motor-cycle having the passenger's seat in front; **foredeep** *Geol.*, a trench or elongated depression along the edge of an orogenic belt, often beneath the sea; **fore-dune,** the part of a dune system nearest to the sea; **fore-hock,** the hock of the fore-leg of pork or bacon; **fore-kidney** = Pronephros; **fore-milk,** (*a*) the colostrum (Dorland, 1901); (*b*) the first milk drawn from a cow at each milking; **fore-pitch,** ? projection; also *fig.* (G. M. Hopkins); **foreplay,** stimulation or love-play preceding sexual intercourse; **fore-pleasure** [tr. G. *vorlust* (S. Freud *Drei Abhandlungen zur Sexualtheorie* (1905) iii. 56)], pleasure induced by sexual stimulation; **fore-stone,** also, the front cross-piece of a blast-furnace.

1906 C. Bridge in C. Stone *Sea Songs* p. iv, In the Royal Navy the term sea song was unknown. What landsmen would have so designated, blue-jackets called 'Fore-bitters'. **1962** A. G. Course *Dict. Naut. Terms* 79 *Fore bitters,* sea songs sung round the fore bitts.. Sea shanties were never sung on these occasions. **1792** *Scots Mag.* LIV. 223/2 Dickson was then performing the duty of what is now commonly called a fore-cadie. **1926** *Glasgow Herald* 27 Dec. 6 An intelligent forecaddy..can materially enhance his prospects. **1904** *Motor Cycle* 11 Apr. 348/1 One of the most highly developed forecars we have yet seen. **1908** *Westm. Gaz.* 21 May 4/3 My machine was a three-wheeler, known in those days as a forecar. **1963** Bird & Hutton-Stott *Veteran Motor Car* 90 Tricar, fore-car and light car proper. **1909** H. B. C. Sollas tr. *Suess's Face of Earth* IV. v. ix. 295 These depressions mark the subsidence of the foreland beneath the recent folds. They have received the name of 'fore-troughs'. But.. their two sides are of different structure, and it therefore seems advisable to call them 'fore-deeps'. **1968** R. W. Fairbridge *Encycl. Geomorphol.* 35 'Pacific' tectonics are those associated with contemporary orthogeosynclines, island arcs, oceanic trenches and active foredeeps. **1921** L. Cockayne *Veget. N.Z.* II. iv. 73 A typical, highly developed dune-area in New Zealand consists of a fairly even-topped low wall of sand the foredune which rises from the upper strand beyond reach of an ordinary tide. **1964** V. J. Chapman *Coastal Veget.* vi. 140 In the absence of sand couch grass, the fore-dunes are generally colonized by the Lyme grass. **1923** R. E. Davies *Pigs & Bacon Curing* 29 The side may be made into smaller cuts by either of the following divisions:—First: fore hock, thick streaky, thin streaky, flank, gammon, corner, long loin, back and ribs, and collar. **1892** J. A. Thomson *Outl. Zool.* 404 The pronephros or fore-kidney persists. **1904** *Daily Chron.* 15 July 3/7 Thousands of bacteria have been found in the 'fore' milk, whereas the 'middle' milk was.. germ-free. **1908** *Stratford-upon-Avon Herald* 24 July 7/1 The rejection of foremilk and the discontinuance of wet milking. **1949** *New Biol.* VII. 21 During any one milking time the fat-content of the milk increases very rapidly from the thin 'fore-milk' to the rich 'strippings'. **1870** G. M. Hopkins *Jrnl.* 19 Dec. (1959) 201 Each [icicle] like a forepitch of the shape of the piece of potsherd it grew on, like a tooth to its root for instance. **1871** *Ibid.* Mar. 204 Possibly each tuft [of cloud] in forepitch or in origin is quained and a crystal. **1881**—— *Sermons* (1959) II. viii. 196 There is..in the works of creation..the order of intention, ..not only intention in understanding and intention in will but also intention or forepitch of execution, of power or activity. **1929** J. B. Eggen in Calverton & Schmalhausen *Sex in Civilization* 594 The difference between perversion and fore-play. **1950** M. Mead *Male and Female* II. iii. 73 Love-affairs are matters of the eyes, foreplay is almost completely concentrated in a glance-exchanging courtship. **1953** A. C. Kinsey et al. *Sex. Behav. Hum. Female* II. ix. 364 Many persons..feel that the intensity of the ultimate orgasm is heightened by extended foreplay. **1961** R. Liswood *Doc. Speaks* (1963) iv. 64 When there is adequate foreplay..intercourse is more satisfying. **1910** A. A. Brill tr. *Freud's Three Contributions to Theory of Sex* iii. 62 In contradistinction to the end-pleasure, or pleasure of gratification of sexual activity, we can properly designate the first as fore-pleasure. The fore-pleasure is then the same as that furnished by the infantile sexual impulse, though on a reduced scale. **1925** J. Riviere et al. tr. *Freud's Coll. Papers* IV. 183 The increment of pleasure which is offered us in order to release yet greater pleasure arising from deeper sources in the mind is called an 'incitement premium' or technically, 'fore-pleasure'. **1951** C. Berg *Unconscious Significance of Hair* vii. 58 Lingering at the stage of 'fore pleasure' relations to the sexual object. **1839** Ure *Dict. Arts* 756 The front of the furnace is open for about 12 inches from the lower part of the front cross-piece called fore-stone. **1884** Fore-stone [see *work-stone* s.v. Work *sb.* 34].

fore-and-aft, *a.* Add: *spec.* Applied to a field service cap.

1940 *War Illustr.* 5 Jan. 571/2 Boys in the same battle-kit that the British Army wears, but with a maple-leaf badge in their fore-and-aft caps. **1950** A. Baron *There's no Home* viii. 87 He wore a British soldier's fore-and-aft cap with the flaps buttoned under his chin.

forecast, *sb.* Add: **1. b.** Recent examples of use spec. with regard to the weather.

1930 *Observer* 14 Sept. 17 The previous forecast of no serious gale. **1955** *Ann. Reg. 1954* 2 It proved to be one of those occasions when the forecast says 'widespread rain' and all that follows is a display of cloud. **1957** *Times* 11 May 7/1 Forecasts for period 6 a.m. to midnight.

forecastable (fōəɹkɑ·stăb'l), *a.* [f. Forecast *v.* + -able.] That may be forecast.

1897 'Mark Twain' *Following Equator* 546 In India, the annual man-killings by snakes are..as forecastable as the tiger-average and suicide-average. **1965** *Philos.* XL. 340 This must not be..forecastable within the calendar.

foreclosure. Add: attrib., *foreclosure action.*

1905 *Westm. Gaz.* 15 May 2/1 [The] suggestion of a foreclosure without a foreclosure action. **1929** *Encycl. Brit.* IX. 494/2 A foreclosure action is brought by the mortgagee against the mortgagor in the chancery division of the High Court in England. **1959** Jowitt *Dict. Eng. Law* I. 818/1 Foreclosure actions are now comparatively rare in practice.

foreconscious (fōəˑrkǫ·nʃəs), *a.* (*sb.*). *Psychol.* [tr. G. *vorbewusst* (see *preconscious *a.*), f. Fore- + Conscious *a.*] Pertaining to that part of the mind, below the threshold of immediate conscious attention, whose memories and perceptions can re-enter the conscious field; = *preconscious *a.* Also as *sb*

1915 C. R. Payne tr. *Pfister's Psychoanalytic Method* 47 Freud distinguishes foreconscious ideas which lack only the conscious investment of energy, from the real unconscious, but attributes to this distinction more practical than theoretical value. **1915** Stedman *Med. Dict.* (ed. 3) 348/1 *Fore-conscious,* noting an unconscious mental process which becomes conscious only on the fulfillment of certain conditions. **1920** T. P. Nunn *Education* (1923) 49 The complex may be on the 'fore-conscious' level—that is the ideas belonging to its activity, though forgotten, may be capable of being recalled. **1921** *Sat. Westm. Gaz.* 1 Oct. 16/2 The fore-conscious or marginal region of the mind. **1924** E. E. Cummings *Let.* 27 Feb. (1969) 105 Many voices answer..but not mine because *gas-bill* flashed thru our hero's foreconscious. **1942** J. G. Miller *Unconsciousness* i. 20 Preconscious and foreconscious. These are psychoanalytic terms used synonymously. **1957** R. L. Munroe *Schools of Psychoanalytic Thought* II. iii. 82 The perceptual conscious is what we are actually aware of at any given moment. This rather narrow field extends into the preconscious or foreconscious (the terms are interchangeable), which includes the host of immediate perceptions and memories available to us if our attention requires them.

forecourt. Add: *spec.* the petrol-dispensing part of a filling-station.

1958 *Motor* 24 Sept. 239 Most motorists patronize a particular garage when they are on their home ground, having found that the forecourt service is efficient and.. reflects good workmanship in the premises behind. **1961** *Oxford Mail* 4 Oct. 2/5 Girl or boy to train as forecourt attendant. **1969** *Guardian* 11 Aug. 5/6 Stamps are 'paid' for on the forecourt by artificially high prices.

fore-edge. Add: *fore-edge painting* (see quot. 1960).

1912 A. J. Philip *Business of Bookbinding* 199 *Foredge painting,* pictures found, usually in old books, on the fore-edge of books. **1938** *Times Lit. Suppl.* 17 Dec. 808/2 A seventeenth-century English Bible..with a signed fore-edge painting. **1960** G. A. Glaister *Gloss. Bk.* 143/2 *Fore-edge painting,* a water colour painted on the fanned out fore-edge of a book. **1969** *Canad. Antiques Coll.* Apr. 23/1 Fore-edge paintings in rare books provide not only delight to collectors but mysteries in more ways than one.

fore-end. Add: **1. c.** = **fore-hock.*

1906 Mrs. Beeton *Househ. Managem.* 628 Fore end.. weight about 17 lb. **1951** *Good Housek. Home Encycl.* 343/1 For boiling, choose a piece of fore-end or gammon. **1960** *Farmer & Stockbreeder* 16 Feb. (Suppl.) 11/2 Good length, well-shaped hams..fine fore-end and head.

d. *pl.* The forward space in a submarine, used for the storage of torpedoes and as living-quarters for members of the crew; also, the submarine ratings quartered there. *colloq.*

1940 G. Hackforth-Jones *Submarine Flotilla* i. 24 A torpedo..had slipped through its carrying-band and plunged into the 'fore-ends'. **1942** —— *One-One-One* i. 7 Petty Officer John Rodgers, the Torpedo Gunner's Mate, was browsing over his torpedo tubes in the 'Fore-ends' of the *Plunger*. *Ibid.*, Those [torpedoes] ready for firing lying concealed in the tubes..were the pride of the 'Fore-ends' crew. **1958** J. D. Drummond *H.M. U-Boat* vii. 105 Naval Intelligence..indicated that the 'Salmon and Gluckstein'—as the fore-ends lightly nicknamed the enemy heavy units—were ready to make a dash for it. *Ibid.* viii. 118 The long-suffering Scots in the fore-ends.

3. *Comb.* **fore-end loader** *Agric.*, a loader attached to the front of a tractor or other vehicle; a *fore-loader.

1958 *Times* 15 Dec. 2/5 The indiscriminate use of fore-end loaders can add materially to the dirt tare. **1964** J. C. Turner *Farm Machinery* ix. 82 When using a fore-end loader the following..should be remembered...The load should not be lifted higher than necessary.

foreground, *sb.* Add: **3.** *foreground music* (opp. *background music*: see *background *sb.* 1 e).

1961 *Time* 21 July 36/1 When his wife asks where the cash came from, he mumbles something about the stock market and adds, as cellos groan ominously in what ought to be called the film's foreground music, 'I made a killing.' **1967** L. Deighton *London Dossier* 30 Deafening foreground music is provided by a giant jukebox.

fo·reground, *v.* [f. the sb.] *trans.* To place in the foreground. Hence **fo·regrounding** *vbl. sb.*; *spec.* in *Linguistics* [rendering Czech *aktualisace* modernization (Havránek and Weingart *Spisovná čeština a jazyková kultura* (1932))], the use of unorthodox or unexpected devices in language.

1892 'Mark Twain' *Amer. Claimant* xvi. 153 We could do a prodigious trade [in portrait-painting] with the women if we could foreground the things they like, but they don't give a damn for artillery. **1959** R. Quirk in Quirk & Smith *Teaching of English* i. 45 The Prague School notion of what has been translated as 'foregrounding'..is defined by Mukařovský as 'the aesthetically intentional distortion of the linguistic components' in relation to the normal standard language on the one hand and to 'the traditional aesthetic canon' on the other. **1962** S. R. Levin *Ling. Struct. Poetry* ii. 17 Foregrounded linguistic elements.. call attention to themselves. **1964** P. L. Garvin *Prague School Reader* p. viii, *Automatization* refers to the stimulus normally expected in a social situation; *foregrounding*—in Czech *aktualisace*—on the other hand refers to a stimulus not culturally expected in a social situation and hence capable of provoking special attention.

forehand, *a.* and *sb.* Add: **A.** *adj.* **5.** *Lawn Tennis.* **a.** Said of a stroke made by holding the racket with the palm of the hand turned forwards. **b.** Applied to the area of a court in which a player uses the forehand stroke.

1889 H. W. W. Wilberforce *Lawn Tennis* 21 When about to make a forehand stroke, the feet should be apart. *Ibid.* 36 Supposing it [*sc.* the return] comes to the player's forehand, he may..return it in the direction it came but at an acuter angle. *Ibid.* 37 From the forehand court the return will be either—(*a*) Down the line...(*b*) Across the court. **1890** C. G. Heathcote *Tennis* 212 The moment it is seen that the ball will come to the fore-hand, the left foot should be brought across and in front of the right. **1921** A. W. Myers *20 Yrs. Lawn Tennis* 9 He attacked the Australian's backhand corner (the forehand corner of a right-handed player) very adroitly. **1929** *Morn. Post* 13 July 16/2 The Forehand Grip. **1958** *Everyman's Encycl.* VII. 663/1 In the forehand drive the hitting arm is extended fully.

B. *sb.* **3.** *Lawn Tennis.* A forehand stroke. *on the forehand*: forehanded.

1908 H. S. Mahony in A. W. Myers *Compl. Lawn Tennis Player* 246 Of their ground-play, especially on the forehand, there is nothing to be said but praise. **1924** G. W. Hillyard *40 Yrs. Lawn Tennis* 69 He..never did acquire the right method of hitting the ball on the forehand. **1934** *Times* 26 July 5/4 His forehand became utterly wild... It was on the forehand that he made the losing stroke.

C. *adv. Lawn Tennis.* (See quot.)

1925 S. Lenglen *Lawn Tennis* 24 The ball is cut forehand when it is struck below the centre of its circumference and across from right to left.

forehanded, *a.* Add: **3.** *Lawn Tennis.* With a forehand stroke. Also *adv.*

1889, 1890 [see *back-handed *a.* 2]. **1908** A. W. Myers *Compl. Lawn Tennis Player* 129 Lawford attempted to take them forehanded. **1922** [see *back-handed *a.* 2].

foreha·ndedness. *U.S.* [f. Forehanded *a.* 2.] The quality of being 'forehanded'.

1840 R. H. Dana *Bef. Mast* (1854) 262 Regular habits, forehandedness (if I may use the word) in worldly affairs. **1897** A. T. Mahan *Life Nelson* II. 277 The incident shows at once the forehandedness of Nelson. **1905** *Springfield* (Mass.) *Weekly Republ.* 8 Dec. 1 The attempt..to persuade people to do their Christmas shopping early appears to be making headway. Such forehandedness is in the interest of the buyers. **1937** *Times Lit. Suppl.* 13 Feb. 99/3 His [*sc.* Ezra Pound's] forehandedness makes this new volume particularly exciting.

forehander (fōəˑɹhæ·ndəɹ). *Lawn Tennis.* [f. Forehand + -er^{1}.] A forehand stroke.

1922 A. E. Crawley *Lawn Tennis Do's & Don'ts* 25 If you hold [the racket] across your chest, you may be too late for a forehander.

foreign, *a.* and *sb.* Add: **A.** *adj.* **5.** *spec.* in phr. *foreign body.*

1767 [in Dict.]. **1898** *Daily News* 30 Nov. 8/1 The presence of the 'foreign body'—a disguising euphemism for half an ounce..of lead entering a man at great velocity. **1961** *Brit. Med. Dict.* 577/1 *Foreign body,* a substance present in any part of the body in which normally it is not found, and usually of external origin. **1964** S. Duke-Elder *Parsons' Dis. Eye* (ed. 14) xxvi. 375 Foreign bodies,..particles of coal dust, emery, steel, etc.—may pitch upon the conjunctiva or..cornea.

8. *Foreign legion*: see Legion 1 b.

c. *foreign devil* [tr. Chin. *yang kuei-tzu* (also, formerly, *fan kuei*(*-tzu*), and other regional forms)]: a term of contempt for a foreigner (esp. a European) in China; also *transf.*

[**1842** *China as it Was* viii. 51 Be it remarked, that the term Fan-Qui, signifying literally 'barbarian wanderer', or 'outlandish demon' is applied invariably by the subjects of the 'Celestial Empire' to all foreigners.] **1860** *Englishman in China* 237 All rush to their doors, and press to see the '*Fan Qui*', or foreign devils; though I must here add that but once was that name applied to us, and really they were all of them most civil. **1889** G. B. Shaw *Fabian Ess. Socialism* 174 A Frenchman or Scotchman was a natural enemy: a Muscovite was a foreign devil. **1926** A. Huxley *Jesting Pilate* III. 248 The trees might just be saying, 'Foreign Devil, Foreign Devil', and repeating it monotonously, mile after mile. **1937** E. Snow *Red Star over China* I. iv. 43 It was the perfect setting for the blotting-out of a too inquisitive foreign devil. **1965** P. Ordway *Night of Reckoning* (1967) ii. 39 Most of us foreign devils [in Spain] found seasonal rentals a very necessary source of additional income. **1969** V. G. Kiernan *Lords of Human Kind* v. 167 Against these new barbarians China was building a new Great Wall, of hatred. Wherever they went they were saluted with cries of 'foreign devil'.

9. *spec.* opp. *inland,* as *foreign bill, exchange* (see Exchange *sb.* 4 and quots.).

1691 [see Exchange *sb.* 4]. **1849** [see Inland *a.* 2]. **1940** *Manch. Guardian Weekly* 5 Jan. 3 Behind all these questions of supply lie the inexorable limitations of foreign exchange and shipping space. **1948** G. Crowther *Outl. Money* (ed. 2) vii. 202 (*heading*) The Foreign Exchanges. *Ibid.* 210 The foreign exchange market was originally the 'foreign-bills-of-exchange market'. It was the place where bills payable in New York, Shanghai, Buenos Aires and half a hundred other places could be bought. **1951** J. R. Winton *Dict. Econ. Terms* (ed. 3) 38 *Foreign exchange,* strictly, this term denotes the transaction of exchanging one currency for another but, in practice, it is often used to refer to actual currency, e.g. foreign currency notes or the means of obtaining such through travellers' cheques or letters of credit. **1971** *Daily Tel.* 10 May 14/2 The currency of any other country is 'foreign exchange'.

10. (Earlier examples.) Also in *Journalism.*

1611 Coryat *Crudities* Epist. Dedic., The Observations of my trauels..I hope..will be very delectable to every Reader that loueth to heare of forraine affaires. **1623** J. Chamberlain *Let.* 21 Nov. (1939) II. 527 The selected commissioners for forain affaires sit much. **1848** Thackeray *Pendennis* I. xxx, Look! here comes the Foreign Express galloping in. *Ibid.*, Mr. Doolan..is foreign sub-editor, and sees the mail on the newspaper sheet before he goes to his own. **1870** *Nation* (N.Y.) 27 Jan. 54/2 There is..a column of telegrams from Europe in every issue of the daily papers, which editors are accustomed to refer to as furnishing the latest 'foreign intelligence'. **1877** *Harper's Mag.* Dec. 53/2 Foreign correspondence in the hands of a foreign editor. **1959** C. Ogburn *Marauders* (1960) ii. 47 A reporter and a writer on the foreign desk of *The Chicago Daily News.* **1967** [see *city page* s.v. *city 9]. **1971** *Guardian* 22 July 11/8 Foreign correspondents during the previous days used, perforce, comments from the angry hippies.

14. Further examples of combs. of the type of *foreign-owned* (= owned by a foreigner, of foreign ownership). Also *foreign aid* [*aid *sb.* 3 c], *-born* (= born abroad), *-language* attrib.

1958 Foreign-aid [see *aid *sb.* 3 c]. **1968** B. Turner *Circle of Squares* viii. 51 The foreign aid we have lashed out so lavishly. **1969** *New Yorker* 31 May 68/3 Because foreign-aid resources have been so limited, Indonesia's economic-development efforts..have been confined largely to food production. **1856** J. P. Hambleton *H. A. Wise* 57 Both the Alien and Sedition laws were intended for the oppression of foreign born citizens. **1940** J. H. Jagger *English in Future* 124 Native-born with one parent foreign-born. **1964** P. F. Anson *Bishops at Large* ix. 440 A foreign-born Mexican citizen. **1898** *Westm. Gaz.* 16 June 1/3 A foreign-controlled department of public works. **1898** *Daily News* 5 Nov. 6/3 The commander of the foreign drilled troops. **1905** *Daily Chron.* 22 Nov. 6/3 Our corn is part home-grown and part foreign-grown. **1933** Bloomfield *Lang.* iii. 54 The factor of foreign-language speaking does not lend itself to measurement. **1938** *Times Lit. Suppl.* 20 Aug. 514/3 The kind of Americans of foreign-language origin who are members of the population of the United States. **1959** *Encounter* Dec. 56/2 The BBC set new standards for foreign-language broadcasts. **1971** *Incorporated Linguist* X. 39/1 No one would claim that British foreign-language policy towards Europe is or should be principally determined by the extent to which English is known and used in Europe. **1906** *Daily Chron.* 13 Aug. 4/7 The foreign-trained fitters and machinists.

B. *sb.* **4. b.** ellipt. for *foreign language, foreign parts, foreign service,* etc.

1906 R. Brooke *Let.* 3 Feb. (1968) 39 The Calendar has appeared. There are about eight foreigns [*sc.* matches against teams from other schools]. **1907** A. Quiller-Couch *Major Vigoureux* ix. 90, I was thinking that—she being from foreign and the Islands the first place she've touched at—I might pick up a bravish order in the way of fresh milk and eggs. **1955** G. Freeman *Liberty Man* I. iii. 49 'On leave, Jack?' 'Yeah... Just back from foreign.' **1962** *Times* 13 Feb. 12/6, I had tried to explain I'd just arrived from 'foreign' [*sc.* foreign service]. **1971** P. O'Donnell *Impossible Virgin* v. 103 He was a foreigner, and he babbled in foreign.

foreigner. Add: **1. c.** *pl.* Foreign stocks and shares.

1898 *Westm. Gaz.* 14 Feb. 9/1 Foreigners are quiet. **1904** *Ibid.* 10 May 11/1 The Account Changes in Foreigners. **1964** *Financ. Times* 3 Mar. 19/2 Among Foreigners, Baltic bonds attracted buying interest.

2. (Later examples.)

1966 O. Norton *School of Liars* viii. 129, I know very little about him. He's a foreigner in the village. **1968** M. Allingham *Cargo of Eagles* xiv. 157 He was a poor silly foreigner from Kent. **1968** J. Drummond *Gantry Episode* ii. 23, I want the lot of you foreigners out of Gantry... I don't like interference.

fo·relady. *U.S.* = Forewoman b.

1889 Farmer *Americanisms, Forelady,* a forewoman. **1904** *N.Y. Even. Post* 6 Oct. 3 (Advt.), Help Wanted. Designer and Forelady for long-established waist factory near Chicago. *a* **1910** 'O. Henry' *Trimmed Lamp* (1916) 15 Two of the most 'refined' women in the store—a forelady and a cashier—had a few 'swell gentlemen friends'. **1906** U. Sinclair *Jungle* x. 125 Miss Henderson, the forelady in her department. **1909** *Public Ledger* (Philadelphia) 24 June 14/1 Umbrella forelady, to take charge of finishers in large umbrella factory. **1923** M. Watts *L. Nichols* 191 She had advanced..to..the position of forewoman—again, pardon!—of forelady at Stein and Merkel's.

foreland. Add: **2. a.** Esp. in *Physical Geogr.*, land deposited by the action of the sea in front of a coast, usu. with no intervening water; also, such land forming a cape (cf. sense 1). (Later examples.)

1896 *Bull. Geol. Soc. Amer.* VII. 400 At a later stage transportation of material alongshore begins and the waste from the edge and bottom of the land, together with the river sediment, is built out at certain points in front of the older mainland in deposits of various shapes, which are appropriately grouped together under the general term forelands. **1959** C. A. M. King *Beaches & Coasts* viii. 261 The Paekakariki coast north of Wellington, where a wide sandy foreland fills in a broad bay in the wide northern part of Cook Strait.

4*. *Geol.* A firm unyielding block of the earth's crust which is opposed to or partially surrounds an orogenic belt and towards which the folding is inclined.

1907 W. B. Scott *Introd. Geol.* (ed. 2) xxiii. 506 In folded mountain ranges three zones may be distinguished: (1) A rigid, unyielding mass which is not folded, (2) the zone of folding, (3) the zone of diminishing action, where the folding gradually dies away or ends in a fault... The side of the range toward which the overturned folds incline is called the foreland, and may be either the unfolded mass or the zone of diminishing action. **1909** [see *back-land 3]. **1937** Wooldridge & Morgan *Physical Basis Geogr.* vi. 76 The African 'hinterland' is believed to have moved northward towards the European 'foreland'. **1968** R. W. Fairbridge *Encycl. Geomorphol.* 734/1 The Mediterranean-Alpine-Himalayan belt is double sided, the folds tending to be overthrust against forelands..both to the north and to the south, while in between lie block-faulted collapsed regions. **1970** *Nature* 28 Nov. 838/2 All of these 'orogenies' have resulted in thrusts and overturned folds directed towards the foreland.

fo·re-loader. [Loader^{1} 1 c (*a*).] A loader mounted on the front of a tractor or other vehicle. Cf. **fore-end loader.*

1954 *Farmers Weekly* 4 June 65/2, I bought a fore-loader for use with the hydraulic lift of a modern tractor. **1958** *Times* 27 Oct. 15/4 The obvious loading tool is the tractor-mounted foreloader, to be found on most farms today. **1960** *Farmer & Stockbreeder* 8 Mar. 112/2 This makes it possible for a tractor and foreloader to go in for cleaning out the manure. **1963** *Times* 14 Jan. 13/2 Silage is machine-chopped and blown into the feed truck, meal and potatoes being added by foreloader.

foremanship. (Earlier and later examples.)

1853 'Mark Twain' *Let.* (1917) I. 24 If he cannot get a foremanship. **1941** H. J. Massingham *Remembrance* viii. 74 The craftsman..drilled the thin slabs with the slat-pick..in tune with the foremanship of the seasons.

forerunner. Add: **1. d.** *Skiing.* [tr. G. *vorläufer.*] A skier who runs over the course as a preliminary to a skiing race.

1949 P. Cummings *Dict. Sports, Forerunner* (Skiing), one who breaks a trail first before a competition starts. **1964** *Times* 31 Jan. 4/1 Three forerunners then came down to open the course.

foreseeable, *a.* Freq. in phr. *foreseeable future.*

1932 *Nation* June 613 The population problem..does not seem likely to be solved in the foreseeable future. **1958** *Spectator* 22 Aug. 241/3 Nobody is going, within the foreseeable future, to lay a finger on the trade unions. **1959** *Cambr. Rev.* 2 May 469/1, I cannot picture fewer motorists in the city during the foreseeable future. **1965** *New Statesman* 30 Apr. 672/3 The sad fact is that, in the readily foreseeable future, governments have to persuade us to ration ourselves to one child each on the average.

foreset, *ppl. a.* Add: **2.** (fōəˑɹset). *Geol.* [directly f. Fore- *prefix* + Set *ppl. a.*] In a delta or similar deposit, designating a series of inclined layers of sediment formed by successive deposition on the advancing forward slope and usu. situated between bottom-set beds and top-set beds. Also *ellipt.*

1905, 1942 [see **bottom-set bed* s.v. *bottom *sb.* 19]. **1939** W. H. Twenhofel *Princ. Sedimentation* xiv. 507 Foreset lengths are largely directly controlled by quantity of sands and inversely by velocity of currents. Foresets may range in length to 30 meters or more but about 3 meters is an average high maximum. **1957** G. E. Hutchinson *Treat. Limnol.* I. ii. 183 As more and more foreset beds are laid down, the mouth of the river is carried out into the lake. **1971** *Nature* 13 Aug. 448/1 The foreset slopes of major deltas.

fo·re-shock. *Seismology.* [Fore- 5.] A lesser shock preceding the principal shock of an earthquake.

1902 *Q. Jrnl. Geol. Soc.* LVIII. 393 (*heading*) Foreshocks of 1901. **1959** *Times* 3 Jan. 4/1 It is not uncommon to have a preliminary or 'foreshock' followed by the main shock. **1970** *Nature* 18 July 317/2 Small local foreshocks and aftershocks which often precede and follow larger earthquakes.

foresight. 5. (Earlier examples.)

1806 M. Lewis in *Orig. Jrnls. Lewis & Clark Exped.* (1905) IV. 169 Labuish,..having by some accedent lost the fore sight of his gun shot a great number of times but killed only the number mentioned. **1847** [see Sight *sb.*^{1} 14 b].

foresighted, *ppl. a.* Add: Hence **fo·resightedness.**

1909 'Mark Twain' *Speeches* (1910) 179 You complimented Mr. Rogers on his foresightedness. **1924** *Glasgow Herald* 10 Apr. 9 Knowing the foresightedness of the Prime Minister the consensus of opinion is that he has taken an opportune moment to test the feeling of the country. **1926** *Proc. U.S. Naval Inst.* Jan. 116 No limit should be placed on the foresightedness of a commander.

forespent, *ppl. a.* Delete †*Obs.* and add later example.

1884 G. B. Shaw *Let.* 15 Apr. (1965) 83 The additional ten hours depend for their table producing power on the forespent other ten hours.

forest, *sb.* Add: **4. a.** *forest fire* (also *fig.*), *-floor* (earlier example), *-land* (later examples), *pathology, reserve.*

1878 F. B. Hough *Rep. Forestry* I. 158 The frequent occurrence of forest-fires along railroad-lines. **1958** *Spectator* 8 Aug. 183/2 He was forced to intervene in the island to protect Turkish nationals, to prevent indirect

aggression, and to put out a neighbouring forest fire. **1849** Thoreau *Week Concord Riv.* 233 Ere the black bear haunted Thy red forest-floor. **1805** King in *Hist. Rec. Austral.* 1st Ser. V. 586 Forest land: [land which] abounds with Grass and is the only Ground which is fit to Graze; according to the local distinction, the Grass is the discriminating Character and not the Trees, for by making use of the Former it is clearly understood as different from a Brush or Scrub. **1936** *Discovery* Apr. 107 A typical Finnish scene of water and forest-land. **1968** G. Jones *Hist. Vikings* IV. ii. 383 He headed into the forestlands of Dalarna. **1944** *Forestry Terminol.* (Soc. Amer. Foresters) 3/2 Forest pathology borders on a number of related fields, such as forestry, plant pathology, mycology. **1882** *North Amer. Rev.* Oct. 400 Preserving certain portions..as Government forest reserves. **1945** *Craig* (Colo.) *Empire-Courier* 25 July 2/4 There's forest reserve country up there that's just waiting for you.

d. *forest-born* (later examples).

1837 *Southern Lit. Messenger* III. 238 The walls..once resounded with the accents of the forest-born Demosthenes. **1841** H. S. Foote *Texas & Texans* I. 120 It was in fact perfectly natural..that 'forest-born' orators [should have come forward] to rouse..the spirit of resistance.

5. forest-bed (earlier examples); **forest mahogany,** a name used for several species of the genus *Eucalyptus,* especially *E. resinifera*; **forest marble** (earlier example); **forest red gum,** *Eucalyptus tereticornis*; **forest shrew,** a name used for several African shrews of the genus *Myosorex.*

1840 Forest bed [see *Cromer]. **1861** *Geologist* IV. 70 The dark sandy clay, known as the Forest bed, from the abundance..of stems and trunks of trees found in and on it. **1884** A. Nilson *Timber Trees N.S.W.* 10 The most valuable and best-known species of *Eucalyptus* are those called..'Red or Forest Mahogany'. **1889** F. von Mueller *Eucalyptographia* I, s.v. *Eucalyptus resinifera,* It bears the colonial name of Red or Forest-Mahogany, which appellations are very inaptly given, inasmuch as the wood bears no real similarity to that of the true West Indian Mahogany. **1858** Forest Marble [see *Bathonian *a.* 2]. **1904** J. H. Maiden *Forest Flora N.S.W.* II. 1 'Forest Red Gum'... This species is..usually found in open forest country, hence I recommend the adoption of the prefix 'Forest' to Red Gum, the name by which it is very commonly known, with the view to save confusion. **1931** E. Maxwell *Afforestation in Southern Lands* lxiv. 273 This other Red Gum, the Forest Red Gum, will grow under conditions that the River Red Gum will not. **1957** *Forest Trees Austral.* (Commonw. Forestry & Timber Bur.) 86 Forest red gum..extends beyond the shores of Australia to the drier parts of Papua. **1958** G. Durrell *Encounters with Animals* i. 25 If anything lives to eat, this forest shrew does.

fore-stage. Restrict †*Obs.* to sense in Dict. and add: **2.** *Theatre.* That part of the stage which lies nearest to the audience, freq. extending in front of the curtain.

1923 G. B. Shaw *Shaw on Theatre* (1958) 163 Mr. Granville-Barker..had reconstructed the London and American stages on which he worked by building a forestage out into the auditorium. **1934** T. S. Eliot *Rock* i. 31 On the fore-stage, an agitator is addressing a tattered crowd. **1958** *Times* 22 Oct. 6/3 A stepped forestage that gives access to a semi-circular apron, placed, in the manner of a Greek orchestra, immediately before it.

forestall, *sb.* Add: **4.** In the writings of Gerard Manley Hopkins (see quots.).

c **1883** G. M. Hopkins *Sermons* (1959) II. iii. 150 The.. act (of consenting to grace) is really necessary... But if after this we are left to ourselves for a leisurely and deliberate avowal or disavowal of this 'forestall' (as I shall call it), [etc.]. *Ibid.* 155 Remark that prayer understood in this sense, this sigh or aspiration or stirring of the spirit towards God, is a *forestall* of the thing to be done, as on the other side grace prevenient is God's forestall of the same.

fore-stick. (Earlier U.S. examples.)

1793 *Massachusetts Spy* 7 Mar. (Th.), He found his companion lying in a large body of live coals, her head on the backlog and knees on the forestick. **1821** [see *chunk *sb.*^{1} 1 b].

foreteach, *v.* Delete †*Obs.* and add later examples. (Still *rare.*)

1876 Morris *Æneids* x. 843 The father's soul foretaught of ill, afar their wail he knew. **1909** tr. *Sermons of St. Bernard on Advent & Christmas* 152 They are foretaught by the Holy Spirit.

forete·llable, *a.* [f. Foretell *v.* + -able.] Capable of being foretold.

1912 F. von Hügel *Eternal Life* 212 The very range and slowness of such an immense, assumedly necessary, foretellable evolution. **1927** H. G. Wells in *Sunday Express* 20 Feb. 12/3 A foretellable disaster.

foretop. 3. (Later U.S. examples, of sheep.)

1866 *Ret. Agric. Soc. Maine* 149 They all had the large foretop on the forehead. **1874** *Rep. Vermont Board Agric.* II. 416 When young they should have good foretop, and the skin a light pink.

forge, *sb.* Add: **6. b. forge-master,** one having charge or ownership of a forge; also *fig.*; **forge-scale** = *hammer-scale* (Hammer *sb.*^{1} 7).

a **1628** Forge-master [see Forge *sb.* 6 a]. **1814** M. Birkbeck *Journey through France* 71 The sale of the ore to the forge-masters. **1959** *Times Rev. Industry* July 75/2 Wire-drawers, forgemasters and other consumers of semi-finished steel products. **1969** *Daily Tel.* 24 Jan. 4/6 Crankshaft makers and forgemasters. **1883** *Specif. Alnwick & Cornhill Rlwy.* 9 To substitute a portion of forge scales or cinders. **1940** *Chambers's Techn. Dict.* 349/1 *Forge scale,* the iron oxide coating which forms on iron and steel during forging.

forget, *sb.*^{2}, var. Forgett.

1886 [see *fourchette 1 d]. **1891** [see Forgett]. **1921** *Dict. Occup. Terms* (1927) § 429 Side pieces for fingers of glove (called fourchettes or forgets).

forget, *v.* Add: **1.** Used *colloq.* in admonitory phrases. Phr. *forget it*: take no more notice of it, don't mention it.

1888 *Detroit Free Press* 6 Oct. (Farmer *Amer.*), You don't want to fool with those Quakers any, and don't you forget it. **1890** R. K. Fox *Slang Dict. N.Y.* 48 'And don't you forget it' [and other slang expressions] are all, or nearly all, Californianisms. **1903** R. L. McCardell *Conversat. Chorus Girl* 91, I gave him the laugh, and said, 'Forget it!' **1909** E. C. Hall *Aunt Jane of Kentucky* ii. 32 He was the Rev. Lemuel Page, and don't you forgit it. **1915** *Let.* in C. Mackenzie *My Life & Times* (1966) V. 13 The next bloody army I join is the Salvation bloody army and don't you forget it. **1930** R. Macaulay *Staying with Relations* xx. 302 Well, then, forget it.. I sure don't want to wait any more. **1932** [see *aw *int.*]. **1951** 'A. Garve' *Murder in Moscow* xxii. 181 'By the way, Jeff—thanks!' 'Forget it.' **1960** 'W. Haggard' *Closed Circuit* ix. 111 'I hope I'm not interrupting you..' 'Forget it.' **1970** W. Smith *Gold Mine* xxxi. 78 She was genuinely puzzled by the question. ..'What's he got to do with it?' 'All right, forget it.'

3. Also, *to forget about*: not to recall the facts concerning; not to remember to take action in the matter of. *colloq.*

1897 N. Gould *Seeing him Through* xix, He had forgotten about that, it was such a long time ago. *a* **1915** R. Brooke *Mem.* in *Coll. Poems* (1918) p. xxv, I often wonder whether I haven't written several of them myself under a pseudonym, and forgotten about it.

forget-me-not. Add: **4.** *forget-me-not brooch, ring*; also *ellipt.*

1836 Dickens *Sk. Boz* II. 154 A small gold chain and a 'Forget me not' ring, the girl's property. **1844** C. M. Yonge *Abbeychurch* vi. 118 Jane gave me the pretty forget-me-not brooch I wore yesterday. **1854** —— *Heartsease* I. I. v. 85 There—that forget-me-not—the first ring I ever had. **1863** A. D. Whitney *Faith Gartney* xx. 131 There was the forget-me-not ring lying in her box of ornaments. *Ibid.* xxiv. 170 You must wear this, now, and keep the forget-me-not for a guard.

forging, *vbl. sb.*^{2} [f. Forge *v.*^{2} + -ing^{1}.] = Clicking *vbl. sb.* b.

1831 [see Clicking *vbl. sb.* b]. **1892** in *Funk's Stand. Dict.*, In *forging,* a horse merely hits one of his forward shoes with his hind shoes, making a disagreeable noise. **1963** Bloodgood & Santini *Horseman's Dict.* 9 *Forging* or *clicking*: when the foreshoe strikes the hind shoe on the same side.

forgivable, *a.* Add: Hence **forgi·vableness,** the quality of being forgivable; **forgi·vably** *adv.*, in a manner that is excusable or deserves forgiveness.

1898 *Expositor* Aug. 105 When general unbelief prevails in the forgivableness of transgression, it is a truth worth proclaiming. **1926** *Spectator* 29 May 917/2 The quality which distinguishes his great prototype Pepys, we mean the quality of forgivableness. **1926** H. W. Phillips *Mod. Foreign Exch.* 31 Then came the reaction from the strain and privation of the war, expressing itself quite forgivably in an orgy of spending. **1927** *Sunday Times* 13 Mar. 6/4 [The part] was quite ludicrously, if forgivably, presented by a substitute. **1968** *Times Lit. Suppl.* 2 May 460/3 'Voices'..forgivably crowd the earlier note-books.

forgottenness (fǫɹgǫ·t'n,nės). [f. Forgotten *ppl. a.* + -ness.] The state of being forgotten; oblivion.

1924 *Brit. Weekly* 4 Sept 483/4 Archbishop Leighton's writings are apt to be relegated to forgottenness to-day. **1924** L. M. Watt *Prayers Public Worship* 159 Awake them from the lethargy of forgottenness. **1947** *Sat. Rev. Lit.* 22 Nov. 17 No one has fewer illusions about that forgottenness.

forint (fǫ·rint). [Hungarian, f. It. *fiorino*: see Florin.] The principal monetary unit of Hungary; a coin of this denomination.

1946 *Times* 31 Aug. 7/4 The National Bank of Hungary's buying rate for sterling remittances from the United Kingdom is at present 46.96 forint = £1. **1947** [see *filler^{2}]. **1962** R. A. G. Carson *Coins* 383 Hungary, declared a republic in 1946, began a new coinage with a new unit, the forint consisting of 100 filler.

fork, *sb.* Add: **10. d.** The part of a bicycle frame in which the (front or back) wheel revolves; also *attrib.*, as *fork-blade, -crown, -end, -head.* Also *pl.*

1871 *English Mechanic* XIII. 144 Can any reader give me a description of socket and fork of 'Ariel' bicycle? **1880** *Scott. Football Ann.* (Advt.), The Howe Bicycle, all the latest improvements, hollow oval fork, &c. **1892** *Work* 27 Aug. 379/2 Front mud-guard should be fixed to fork crown with a bolt and nut. **1898** *Cycling* 48 The fork ends. **1902** *Captain* VII. 83/1 The fork head of the Raleigh. **1908** *Daily Chron.* 7 Nov. 7/1 In a fog it is better to carry one's lamp on the fork-blade than high up. **1957** *Encycl. Brit.* III. 544/2 The fork crown (at the top of the forks) is fixed to the steering column. **1952** *Cycling* ('Know the Game') (1964) 17 The wheel revolves on a spindle contained in the hub and this is usually secured in the forks of the cycle by means of two ordinary nuts. **1968** [see *frame *sb.* 11 f].

12. g. *Chess.* A simultaneous attack on two pieces, esp. by a knight.

1656 Beale *Royall Game of Chesse Play* vii, Forke is, that when you see two of the enemies Noble-men standing in the same ranke, and but one house betwixt them, advance a pawne, guarded with an other, unto the middle house before them both, and you may commonly take one of them. **1688** R. Holme *Armoury* (1905) 66/2 A Fork or dilemma, is a way of takeing a chesse man, by runing vp a pawn to the rank next two great men of the aduerse part standing in one rank with a house betweene them, where if one be saued, the other wilbe taken. **1761** E. Hoyle *Ess. Chess* A4, Take care of a Fork or a Check by Discovery, or a Stale mat. **1764** R. Lambe *Hist. Chess* 98 A Fork is, when an adverse Pawn must take one of your Pieces, by standing upon a Square of the next Line, between two of them. **1890** R. F. Green *Chess* v. 17 A frequent and fatal fork. The White Knight..attacks both the Black King and Queen. **1969** A. Glyn *Dragon Variation* ix. 293 Carl was threatening a Knight fork against Boghossian's Queen and other Rook.

16. fork-lift (truck, etc.**),** a vehicle fitted with a pronged device in front for lifting and carrying heavy goods; also *fork-lift* ellipt.

1946 *Engin. Index 1945* 557/2 Use of carrier fork lift trucks capable of handling filled cable reels. **1950** *Jrnl. R. Aeronaut. Soc.* LIV. 521/1 Fork lifts, and in exceptional cases loading ramps, were used for the bulky awkward items. **1955** *Radio Times* 30 Sept., Modern package-handling machines—fork-lifts they are called. **1958** *Times* 1 July p. iv/7 The tractor is fitted with a fork lift loader to lift 35 cwt. capacity pallets. **1968** *Times* 11 Mar. p. ii (*caption*), A heavy load..handled by a forklift truck.

b. fork supper (also *-buffet, -dinner, -lunch*(*eon,* etc.), a meal served at a buffet, etc., consisting of food suitable for eating with fork alone, making the provision of set places at table unnecessary.

1940 N. Mitford *Pigeon Pie* i. 12 He came to all the week-end parties, tea parties, fork luncheons. **1949** *Antiquity* XXIII. 121 After a sumptuous fork-luncheon they all set off for Amiens. **1952** A. Colby *Beauty Bk.* 230 The buffet supper..is a 'fork supper'. **1957** H. Croome *Forgotten Place* vii. 89 We'll have a party. Just a fork supper, and a bit of dancing. **1958** *TV Times* 18 July 18/3 The party is always on a Sunday or Monday—fork buffets, of course. **1965** *Courier-Mail* (Brisbane) 9 Oct. 17/11 A fork dinner for 40 young friends. **1971** *Times* 22 Jan. 14/3 Aspirants..would be well-advised to practise the art of conversing while wielding wine glass and loaded plate, because the plan is to hold a two-hour fork luncheon before the selection meeting.

fork, *v.* Add: **1. c.** *Chess. trans.* To attack (two pieces) simultaneously with the same piece.

1745 E. Hoyle *Piquet & Chess* 61 Take care that no guarded Pawn of your Adversary's fork two of your Pieces. **1891** E. Freeborough *Chess Endings* 116 The Queen may be forked by a diverging check, with the Kt, preceded by B-Kt 7 ch. **1891** R. B. Swinton *Chess for Beginners* 73 Sometimes a Pawn can play a shrewd trick by 'forking' two pieces. **1899** E. E. Cunnington *Mod. Chess Primer* 20 There is nothing, in chess, more dangerous than to allow one of your opponent's men to..'fork' two of your men by one move. **1922** A. Emery *Elem. Chess* 68 When a knight checks, and at the same time attacks a man which can be captured with advantage on the next move the latter is said to be 'forked'.

5. (Earlier examples with *over.*) Also *to fork up.*

(*a*) **1839** *Observer & Reporter* May 18 The gambler should fork over his illgotten gains. **1849** D. Nason *Jrnl.* 113 As he was the biggest man I had to fork over $1·25. **1866** J. C. Gregg *Life in Army* xv. 132 Every person..forks over his picayune.

(*b*) **1836** *Franklin Repos.* (Chambersburg, Pa.) 29 Mar. 1/2 'Fork up, and that instantly, or take the contents of this,' he added fiercely, as he thrust the cold barrel of the pistol against the supplicant's cheek. **1839** *Observer & Reporter* Nov. 23 Well then, fork up, and be quick. **1843** *Punch* V. 86/2 'Oblige me with that hod', and 'Have the kindness to hand me that gimlet', are phrases which might be well substituted for a request to 'chuck' or 'fork up' 'this here' and 'that 'ere'. **1857** *Quinland* II. III. iii. 167 The plethoric dog of a New Jersey Jew has got the 'tin', and will fork up as soon as the matter is fixed. **1871** *Scribner's Monthly* I. 601 So the governor forked this up, though it's my selection entirely. **1918** E. Pound in *Lett. J. Joyce* (1966) II. 424, I will fork up the remaining £20 of the fifty promised. **1932** H. Crane *Let.* Jan. (1965) 395 The family will just have to fork up a loan or something for me.

10. To bestride or mount (a horse). *U.S.*

1903 A. Adams *Log of Cowboy* xix. 295 So fork that swimming horse of yours and wet your big toe again in the North Platte. **1920** J. M. Hunter *Trail Drivers of Texas* 245 Throw your rope and whatever it falls on, fork him.

forlornity (fǫɹlǫ·ɹnĭti). orig. *U.S.* [f. Forlorn *a.* + -ity.] **a.** Forlornness. **b.** A forlorn person.

a **1870** In *D.A.E.* s.v., *Forlornity,* forlorn condition. **1878** 'Mark Twain' *Let.* (1917) I. 341 Livy and Clara sat down forlorn, and cried... Last night the forlornities had all disappeared. **1904** B. von Hutten *Pam* III. ii, 'Oh!' she added, breaking into rueful laughter at the sight of his fat forlornity. **1917** H. T. Comstock *Man thou Gavest* 320 Thomas explained and apologized for the admittance of the two 'forlornities', as he called them. **1922** *Contemp. Rev.* Oct. 489 In their rusty forlornity.

form, *sb.* Add: **5. c.** (Further examples.) Also in extended uses in *Linguistics*. Cf. *linguistic form.

1921 E. Sapir *Lang.* iv. 63 The evolution of forms like *teeth* and *geese*. **1926** Bloomfield in *Language* II. 155 The vocal features common to same or partly same utterances are forms. *Ibid.*, Thus a form is a recurrent vocal feature which has meaning. **1933** —— *Lang.* 168 A form like *John* or *run*,..without, for instance, any specification as to final-pitch, is, properly speaking, not a real linguistic form, but only a lexical form; a linguistic form, as actually uttered, always contains a grammatical form. **1953** J. B. Carroll *Study of Lang.* ii. 49 The comparative linguist can attempt to trace back the forms of a given language to the forms of another, older language. *Ibid.*, Historical linguists have prepared lists of 'reconstructed' forms. **1962** E. F. Haden et al. *Resonance-Theory Linguistics* ii. 15 Two language entities, between which there is a state of Resonance, may be found to be 'sames' as to their Form. **1966** J. M. Sinclair in C. E. Bazell *In Memory of J. R. Firth* 430 A form, in this article, is a stretch of language which has not yet been assigned a lexical status.

d. *Math.* A homogeneous polynomial in two or more variables; a quantic.

1859 G. Salmon *Less. Mod. Higher Alg.* xii. 88 A quadratic form can be reduced in an infinity of ways to a sum of squares, yet the number of positive and negative squares in this sum is fixed. **1903** Grace & Young *Alg. Invariants* i. 4 The transformation of the binary form $a_0x_1^2 + 2a_1x_1x_2 + a_2x_2^2$. **1928** H. W. Turnbull *Theory of Determinants* iii. 31 It is a linear homogeneous form in *n* arguments. *Ibid.* viii. 133 A homogeneous polynomial is a form or quantic. **1953** F. Blum tr. *van der Waerden's Mod. Algebra* (ed. 2) I. iii. 48 A polynomial is said to be homogeneous or to be a form if all of its terms are of the same degree.

e. *Librarianship.* Used *attrib.* in *form-catalogue*, *-class*, etc., to denote a catalogue or catalogue entry in which books of a certain kind (poetry, almanacs, fiction, etc.) are listed together.

1876 C. A. Cutter *Rules Dict. Catal.*, Contents, Form-catalogue. *Ibid.* 14 *Form-entry*, registry under the name of the kind of literature to which the book belongs. *Ibid.* 49 In the catalogues of libraries consisting chiefly of English books, if it is thought most convenient to make form-entries under the headings Poetry, Drama, Fiction, it may be done. *Ibid.*, There is no reason but want of room why only collections should be entered under form-headings. *Ibid.*, In the case of English fiction a form-list is of such constant use that nearly all libraries have separate fiction catalogues. **1913** J. H. Quinn *Library Catal.* 30 Form-Catalogue is one in which the entries are arranged according to the forms of literature and the languages in which the books are written, either alphabetically or according to the relations of the forms to one another. **1966** T. Landau *Encycl. Librarianship* (ed. 3) 105/2 Form Classes or divisions are used to contain those works which are required more for the way in which they are written or presented than their subject content.

11. *a matter of form*: a point of formal procedure; orig. a legal phrase; hence *colloq.* = a merely formal affair; a point of ordinary routine.

1711 [see Form *sb.* 15]. **1787** [in Dict.]. **1824** H. J. Stephen *Treat. Princ. Pleading* ii. § 1. 254 As the party has no option in accepting the issue, when well tendered, and as the similiter may in that case be added for him, the acceptance of the issue when well tendered, may be considered as a mere matter of *form*.

b. In somewhat more *colloq.* use: the state of affairs, what is happening or going on, the position; the correct procedure.

1934 E. Waugh *Handful of Dust* 21 'I'm going to Hetton tomorrow.'..'What's the form?' 'Very quiet and enjoyable.' **1936** P. Fleming *News from Tartary* II. ii. 76, I sent a wire to..Reuter's Correspondent at the capital..who..would be sure to know the form. **1940** N. Mitford *Pigeon Pie* xii. 192 She read somewhere that this was the form in Imperial Russia. **1951** 'N. Shute' *Round Bend* 87 He came along to the hotel each morning and evening to find out the form and when I wanted him. **1958** J. Cannan *And be a Villain* i. 36 Eve, you know the form—telephone for a doctor.

16. c. *slang.* (Without preceding article.) A 'police record'; a criminal conviction.

1958 F. Norman *Bang to Rights* III. 151 You can get at least a five..for getting captured with a shooter especially if you'v got a bit of form behind you. **1960** 'M. Underwood' *Death by Misadventure* vi. 91 He has form for false pretences, mostly small stuff. **1964** J. Prescot *Case for Court* vi. 107 There's loads of form written down on my conviction card...Juvenile court, police court, quarter sessions and assizes—I've seen 'em all.

18. b. A temporary structure for containing fresh concrete and giving it the required shape while it sets.

1908 T. Potter *Concrete* (ed. 3) xvi. 281 There are no set methods for making forms; a common-sense joiner will soon develop ideas of his own..especially if he sees castings turned out a few times. **1964** *Economist* 6 June 1151/1 A 'slip form paver'..pulls along behind it its own shuttering or 'forms' which keep the concrete from spreading sideways as it sets.

22. (sense 3) *form-fitting* adj.; (sense 6 b) *form-room*; (sense 12 b) *form-filling* vbl. sb.; **form-board,** (*a*) a device used in intelligence tests; (*b*) = *form *sb.* 18 b; **form book** [Form *sb.* 16], a record of the performances of a race-horse; also *transf.*; **form-class,** (*a*) *Forestry* (see quot. 1905); (*b*) *Linguistics*, a class of linguistic forms having some feature in common, such as being usable in the same position within a given construction, or being spoken with exclamatory final pitch; **form-criticism** *Theol.* [tr. G. *formgeschichte*], a method of literary criticism mainly applied to the Bible, and carried out by first classifying passages as belonging to certain forms (e.g. sayings, myths), and then tracing the early history of these forms with the aim of discovering the original form and relating this to its historical setting; so **form-critic,** one who engages in form-criticism; *form-critical* adj.; **form drag** *Aeronaut.*, the drag on a moving body that depends directly on its shape and is due to the unequal pressure over its surface that results from the disturbance of the fluid; **form factor** *Forestry* (see quots.); **form-genus** *Biol.*, a collective group of **form-species,** showing morphological similarities but not necessarily a genetic relationship; **form-historical** *adj.* = *form-critical* adj.; also *form-historian*, *-history*; **form letter,** a standardized letter, esp. one that can be sent to correspondents who inquire about routine matters or topics of frequent occurrence; **form-line** *Cartography*, (usu. *pl.*) lines drawn on a map to indicate the estimated configuration or elevation between the contour lines; **form quality** [tr. G. *gestaltqualität* (C. von Ehrenfels 1890, in *Vierteljahrsschrift f. wissensch. Philos.* XIV. 256] (see quot. 1901); **form-species**: see *form-genus*; **form-symbol,** a symbol designating the form of a crystal (see Form *sb.* 1 c); **form-word,** also *gen.*, a word expressing a formal or grammatical feature; a function word; = *functor 2; **formwork,** timber, steel, etc. made up into a form or set of forms for concrete (see *form *sb.* 18 b).

1917 L. M. Terman et al. *Stanford Rev. Binet-Simon Scale* vii. 142 In the form-board test, the younger subjects were a little superior to the older. **1947** P. L. Harriman *Dict. Psychol.* 141 *Form-board*, a measure of intelligence which consists of blocks to be fitted into a recess. **1952** *Brit. Jrnl. Psychol.* Feb. 79 Two versions of the test, of which the first is a 'formboard' requiring the actual placing of a piece to complete the pattern. **1958** *Archit. Rev.* CXIII. 222/1 Formboards are laid between their flanges, mesh reinforcement is laid over tees and formboard, and gypsum concrete is pumped on to make a roof of about 3 in. thick. **1923** Wodehouse *Inimitable Jeeves* xiv. 180 The race went by the form-book all right. **1934** —— *Right Ho, Jeeves* i. 14 Gussie Fink-Nottle, against all the ruling of the form book, might have fallen in love. **1955** T. Rattigan *Separate Tables* i, According to this form book, Marston Lad is worth a bob or two each way. **1962** *John o' London's* 1 Feb. 115/1 Her acting range, when you go through the form-book, though deep, is also somewhat narrow. **1905** *Terms Forestry & Logging* 12 *Form class*, all trees in a stand so similar in form that the same form factor is applicable in determining their actual volume. **1921** E. Sapir *Lang.* Index 252 Form-classes. **1932** *Forestry* VI. 143 Each tree was placed in its form-class according to the value of the girth quotient. **1933** Bloomfield *Lang.* ix. 146 All English substantives belong to a form-class. *Ibid.* xii. 202 The form-class of limiting adjectives is much smaller than that of descriptive adjectives. **1958** C. F. Hockett *Course in Mod. Ling.* xviii. 162 A class of forms which have similar privileges of occurrence in building larger forms is a form-class. **1962** Householder & Saporta *Probl. Lexicogr.* 280 The principle of form-class equivalence, that glosses should be perfect and complete translations in respect to grammatical class. **1970** A. Cameron et al. *Computers & O.E. Concordances* 48 Through the form classes, the user has an index to various class correlations—agreement, concord, [etc.]. **1933** V. Taylor *Format. Gospel Trad.* i. 19 The Form-Critics..appear to proceed as if they had lighted upon a method which supersedes all others. **1951** H. Gardner *Business of Criticism* II. ii. 105 The fundamental question which the form-critic asks is 'Why was the story told?' or 'What is the point of the story?' **1933** V. Taylor *Format. Gospel Trad.* i. 19 The historical sketch will have served the purpose..of introducing the leading members of the Form-critical School. **1957** D. E. Nineham *Stud. Gospels* p. x, He seemed largely to accept Bultmann's very negative assessment of the historical value of the Gospels as seen through form-critical eyes. **1928** B. S. Easton *Gospel before Gospels* ii. 32 With Dibelius form-history is raised to the rank of a distinct discipline..in his hands 'form-history' becomes 'form-criticism'. **1935** C. H. Dodd *Parables of Kingdom* iv. 111 The most recent school of Gospel criticism, that of *Formgeschichte*, or 'Form-criticism', has taught us that in order to understand rightly any passage in the gospels we must enquire into the 'setting in life'..in which the tradition underlying that passage took form. **1951** N. Annan *L. Stephen* vi. 183 Some critics declare that form-criticism, which discloses the tradition of the early Church, rather than source-criticism, is the most fruitful method of interpreting the evidence. **1963** C. H. Dodd *Hist. Trad. in Fourth Gosp.* 6 The application of form-criticism opened up new lines of approach. **1931** T. G. Whitlock *Appl. Aerodynam.* iii. 38 Form drag..is not a frictional or rubbing force acting tangentially to the surface, like skin friction. **1937** *Jrnl. R. Aeronaut. Soc.* XLI. 1120 Air resistance consists of the skin friction and of the drag due to the inertia of the medium ('form drag' according to modern terminology). **1951** D. O. Dommasch et al. *Airplane Aerodynam.* vi. 146 The various types of drag..are as follows: (1) form or pressure drag; (2) skin-friction drag; (3) induced drag; and (4) wave drag. **1970** A. C. Kermode *Flight without Formulae* 63 That part of the drag which is due to the shape or 'form' of a body, and which can be reduced by streamlining, is called form drag. **1895** W. Schlich *Man. Forestry* III. i. 36 Under 'form factor' is understood the proportion which exists between the volume of a tree and that of a regularly shaped body which has the same base and height as the tree. **1953** *Brit. Commonw. For. Term.* I. 63 *Form factor*, the ratio of the volume of a tree or its part to the product of its basal area and height. **1857** Dickens *Dorrit* II. viii. 387 The work of form-filling,..memorandum-making. **1960** *Guardian* 3 Feb. 6/4 A minimum of fuss but plenty of form-filling. **1897** *Sears, Roebuck Catal.* 237/2 Balbriggan Drawers... Elastic and form fitting. **1970** J. Sangster *Touchfeather* iv. 87 A girl like me, with a trim little form-fitting uniform to wear, has absolutely no place where she can conceal a gun. **1873** *Q. Jrnl. Microsc. Sci.* XIII. 411 He is careful to retain a complete and concise enumeration of the various 'artificial species', 'form-genera', or 'phases', in which the natural species may manifest themselves. **1886** *Encycl. Brit.* XXI. 399/2 Billroth came forward in 1874 with the startling view that the various 'form-species' and 'form-genera' are only different states of one and the same organism. **1935** E. A. Bessey *Text-bk. Mycol.* xiv. 375 Since the genera based on asexual structures do not necessarily indicate true relationships of the included species the term 'form genus' was suggested for such groups by Schroeter. **1960** H. H. Swinnerton *Fossils* x. 58 These genera are therefore not based upon genetic affinity but upon outstanding similarities of form. They are in fact morphological or form genera marking grades of evolution. **1928** B. S. Easton *Gospel before Gospels* ii. 31 The Synoptic material, obviously, offers a tempting field to the form-historian. *Ibid.* iii. 59 In 1924 such an examination appeared, *The Form-Historical Method*, by Licentiate Emil Fascher. **1934** F. C. Grant in *Bultmann's Form Criticism* I. 7 To grasp the main outlines of the form-historical method. **1954** *Scott. Jrnl. Theol.* VII. 339 There is considerable evidence on form-historical principles themselves. **1909** *Sat. Even. Post* 13 Feb. 8/1 The credit man..had a perfect passion for form letters. **1921** *Daily Colonist* (Victoria, B.C.) 11 Mar. 16/7 The original bill and the polite note are as nothing compared with the 'form' letter sent out by firms which make a business of debt collecting. **1922** S. Lewis *Babbitt* iii. § 3 The fortnightly form-letter, to be mimeographed and sent out to a thousand 'prospects'. **1932** Crooks & Dawson *Etheridge's Dict. Typewriting* (ed. 3) 147 *Form letters*..are printed forms or skeleton letters which are allotted numbers for identification by the typists when they are instructed..to send stock letters in reply to inquiries which do not warrant separate letters. **1968** *Globe & Mail* (Toronto) *Mag.* 17 Feb. 5 A form letter to boys who send in asking for hockey photos. **1935** *Geogr. Jrnl.* LXXXVI. 253 The topography sketched in by the use of formlines at intervals of approximately 200 feet. **1951** F. Stark *Beyond Euphrates* 231, I am still in hectic work over my map, which has to have all the heights and form-lines fixed by Tuesday. **1961** L. D. Stamp *Gloss. Geogr. Terms* 125/1 Many cartographers are careful to draw a distinction between 'contours' which are based on instrumental survey and 'form-lines' which are sketched in from general observations. **1901** Baldwin *Dict. Philos.* I. 391/2 *Form quality*, that which characterizes a mental whole as being of a particular form, as being formed, or as having relations of parts. **1938** R. S. Woodworth *Exper. Psychol.* xxv. 623 The form quality of the tune does not reside in the constituent notes but in their pattern and mutual relations. **1948** *Mind* LVII. 23 We can..apprehend directly such form-qualities as sketchiness and schematicity. **1955** F. H. Allport *Theories of Perception* xviii. 498 Form-qualities are one type of such universals, for..they transcend the particular receptor pattern made by the stimulus-object of the moment. **1875** *Cliftonian* Apr. 40 Let the admirer of antiquity confine his excitement to the form-room. **1897** H. W. Bleackley *Short Innings* xv, The bell..signified that all the boys should be in their form rooms. **1907** C. L. Thomson *Teaching English* 11 The form-room library, under the control of the form mistress. **1959** *Dana's Man. Mineral.* (ed. 17) ii. 27 Miller indices also may be used as form symbols and are then enclosed in braces. **1889** *Cent. Dict.*, *Formword*,..a word showing relation only or chiefly; an independent word performing an office such as in other languages..is performed by the formative parts of words. **1892** H. Sweet *New Eng. Gram.* I. 22 We call such words as *the* and *is* form-words, because they are words in form only. When a form-word is entirely devoid of meaning, we may call it an empty word, as opposed to full words such as *earth* and *round*. **1937–8** *Proc. Arist. Soc.* XXXVIII. 196 Such form-words as *all, some, this, not,* and *implies*. **1964** *English Studies* XLV. 241 Such formwords as *thou, yon(der),* and *there*. **1967** [see *functor 2]. **1918** Hool & Johnson *Concrete Engineers' Handbk.* ii. 94 Formwork, of course, should in every case leave the finished concrete true to line and surface. **1943** J. S. Huxley *TVA* 83 (*caption*) The arrangement of rough timber formwork into which the concrete is poured. **1962** *Engineering* 31 Aug. 267/1 A method of using precision-made steel formwork for the placing *in situ* of structural concrete in room width tunnel sections.

form, *v.* Add: **1. g.** *Electr.* (i) [after F. *former* (G. Planté 1872, in *Compt. Rend.* LXXIV. 593).] To convert electrolytically the surface of (a positive or †negative plate of a lead-acid accumulator) into its active form by passing a direct current through it in both directions alternately.

1881 S. P. Thompson *Storage Electr.* 11/2 M. Planté..minutely describes the process of 'forming' the cell. *Ibid.*, The charges may last several hours, and by the end of several months, the cell will be well 'formed'. **1932** Althouse & Turnquist *Mod. Storage Battery Pract.* iii. 73 After the active materials have been pasted into the plates, the plates are formed by a series of charges and discharges. **1964** G. Smith *Storage Batteries* ii. 23 Planté negatives have been obsolete for many years, and it is

modern practice to form Planté positives against plain lead sheets, or 'dummies'.

(ii) To subject (a semiconductor device or some kinds of rectifier) to a relatively large current or voltage in order to produce or modify permanently certain electrical characteristics.

1926 [implied by *forming vbl. sb. b]. **1949** Bardeen & Brattain in *Physical Rev.* 2nd Ser. LXXV. 1210/1 When the two points were connected as emitter and collector, and the collector was electrically formed, transistor action was obtained. **1953** R. A. Harvey *Battery Chargers* iv. 92 Once the [selenium] rectifier is formed it remains reasonably stable. **1962** L. P. Hunter *Handbk. Semiconductor Electronics* (ed. 2) viii. 15 A procedure similar to that of forming a point contact can be used to create small-area PN junctions under a metal contact.

8. c. Esp. in the orders *form fours!* and *form two deep!*

1889 *Infantry Drill* 152 The companies that will form the side faces will form fours in the required direction. *Ibid.* 154 On the words Form Two-Deep, the original rear-rank men will take one pace to the rear. **1915** 'I. Hay' *First Hundred Thousand* i, 'Squoad—'Shun! Move to the right in fours. Forrm—fourrrs!'..'On the command "form fours", odd numbers will stand fast.'..'Forrm—two deep!'

form-. Add: **formal** (fǭ·ɹmæl) *sb.*², any of the acetals derived from formaldehyde, having the general formula $CH_2(OR)_2$; *spec.* Methylal, $CH_2(OCH_3)_2$; **formamide** (earlier examples); **forma·midine,** the postulated compound, $NH_2·CH:NH$, which is known only as its salts; **forma·nilide,** a white crystalline compound, $C_6H_5·NH·COH$, used as an antipyretic, analgesic, and local anæsthetic.

1900 Dorland *Med. Dict.*, *Formal*, methylol or methylene dimethylate..: a clear, fragrant liquid. It is anesthetic and hypnotic. **1952** E. H. Rodd *Chem. Carbon Compounds* I. xvi. 1031 A mixture of the isomeric formals is obtained from glycerol, trioxymethylene and conc. HCl at 100°. **1852** Formamide [see *formanilide]. **1863** A. W. Hofmann in *Jrnl. Chem. Soc.* XVI. 72 Half-way between formate of ammonium and hydrocyanic acid, theory justifies us in searching for formamide. *Ibid.* 73 This substance, which is equally soluble in water, alcohol and ether is formamide. **1882** *Jrnl. Chem. Soc.* XLI. 266 The solid therefore obtained by the action of the substance, 2HCN + 3HCl, upon alcohol is the hydrochloride of the base..formamidine. **1952** *Jrnl. Appl. Chem.* II. 203 It is now possible to report a simple method for the synthesis of formamidine hydrochloride. **1852** W. Gregory *Handbk. Org. Chem.* 337 Formanilide, the analogous compound to which, formamide, is not known. **1966** Millar & Springall *Sidgwick's Org. Chem. Nitrogen* (ed. 3) vii. 224 The formylation of aromatic amines proceeds very readily; formanilide, H·CONHPh, is rapidly formed on boiling dilute formic acid with aniline.

formal, *a.* and *sb.* Add: **1. d.** *formal concept* [tr. G. *formaler begriff*]: a concept of logic, free from the descriptive content that would restrict it to any particular subject-matter (see quots.); *formal implication* (see quots.).

1903 B. Russell *Princ. Math.* § 15 The relation of formal implication..holds between propositional functions when the one implies the other for all values of the variable. **1922** tr. *Wittgenstein's Tractatus* 4. 126 In the sense in which we speak of formal properties we can now speak also of formal concepts. (I introduce this expression in order to make clear the confusion of formal concepts with proper concepts which runs through the whole of the old logic.) **1932** Lewis & Langford *Symbolic Logic* 101 The relation of formal implication is transitive, like material implication. **1939** G. Ryle in *Mind* XLVIII. 149 To use the language of Kant and Wittgenstein, we were trying to treat formal concepts as if they were 'proper' or material concepts. *Ibid.* 151 A formal concept is one which may have a place in a proposition about any subject-matter you please, and some..will be present in any proposition...'Not', 'exists', 'some', 'other',..and many others are not peculiar to any special topics. **1955** A. N. Prior *Formal Logic* 197 The Russellian 'formal' implication is simply universal material implication. **1965** E. J. Lemmon *Beginning Logic* iv. 154 The term 'formal implication' was coined by Russell to describe the universal quantification over a material implication: i.e. a proposition of the form '(x) (Fx→Gx)'.

B. *sb. esp.* (An) evening dress; an engagement at which such dress is worn. orig. and chiefly *U.S.*

1941 *Amer. Speech* XVI. 96 They are all right for casuals or spectator wear, but I'd get that formal in either Waltz Blue or Coronado Coral. **1946** *New Yorker* 28 Sept. 27/2 The neighboring children were strange, with..their queer talk of dates, and formals, and going steady. **1948** *Ibid.* 27 Mar. 29/1 The initiates wore evening dresses (formals, formals, Emily reminded herself to say). **1961** M. Beadle *These Ruins are Inhabited* (1963) v. 71 If a dinner invitation does not specify 'informal', guests assume that black ties will be worn. Young ladies then wear short formals, and old ladies wear floor-length evening dresses. **1968** 'A. York' *Predator* vi. 86 She looked sufficiently virginal to be attending her first formal.

formaldehyde. Substitute for def. (in Dict. s.v. Form-): A choking gas, CH_2O, usually prepared as an aqueous solution, and used as a preservative, an antiseptic, and in large quantities as a raw material in the chemical and plastics industries.

1872 *Jrnl. Chem. Soc.* XXV. 234 (*heading*) Reduction of formic acid to formaldehyde and methyl alcohol. **1894** [in Dict.]. **1947** R. L. Wakeman *Chem. Commercial Plastics* iii. 28 Formaldehyde, made by oxidation of methanol, is an essential component of all of the large volume thermosetting resins now produced. **1957** *New Scientist* 9 May 11/2 Salk and his colleagues..had shown that poliomyelitis viruses could be grown in tissue culture in bulk and treated with formaldehyde to make a vaccine. **1958** *Times Rev. Industry* Feb. 77/1 Monsanto Chemicals (Australia) has announced that its latest formaldehyde unit will double the capacity of the plant laid down some years ago.

formalin (fǭ·ɹmălin). Formerly also **-ine.** [f. Formal(dehyde + -in¹.] An aqueous formaldehyde solution, stabilized with methanol, containing 37 per cent by weight of formaldehyde. Hence **fo:rmaliniza·tion,** treatment with formalin; **fo·rmalinize** *v. trans.*, to treat with formalin; **fo·rmalinized** *ppl. a.*

1893 *Brit. Pat.* 7038, The new antiseptic material, herein called formalin,..contains 40% of pure formaldehyde combined with 60% of water. **1894** [see *formol]. **1901** *Westm. Gaz.* 25 July 5/3 The formalin treatment of consumption. **1907** *Practitioner* Apr. 570 The use of formalinised gut is advised. **1958** *Listener* 23 Jan. 153/2 The vaccine used in Britain and America is 'dead'—that is to say the virus is killed with formalin before it is made up for use. **1960** *Proc. Soc. Exper. Biol. & Med.* CIII. 158/2 After formalinization, these cells were not subject to hemolysis. **1964** M. Hynes *Med. Bacteriol.* (ed. 8) xi. 145 Toxoid, which can be prepared by formalinizing the purified toxin, may be used for active immunization of the human subject.

formalism. Add: **2. b.** *Theology.* (See quot. 1957.)

1931 K. E. Kirk *Vision of God* i. 8 Formalism, as we have defined it hitherto,—the demand for a definite rule of life—has rigorism as one of its branches. **1937** O. Wyon tr. *Brunner's Divine Imperative* vii. 65 The fact that the knowledge of God has become obscure also leads to the fatal dilemma—either the formalism of the legalistic ethic..or eudaemonistic materialism. **1957** *Oxf. Dict. Chr. Ch.* 513 *Formalism*...Of theories of ethics which look for the ground of moral action in the form of the moral law alone..without reference to any specific purposes or values which it is desired through the action to achieve or realize.

3. *Math.* **a.** The conception of pure mathematics as the manipulation according to certain formal rules of symbols that are intrinsically meaningless.

1913 tr. L. E. J. Brouwer in *Bull. Amer. Math. Soc.* XX. 82 On what grounds the conviction of the unassailable exactness of mathematical laws is based has for centuries been an object of philosophical research, and two points of view may here be distinguished, intuitionism (largely French) and formalism (largely German). **1933** M. Black *Nature of Math.* 148 'Formalism' has always been the working attitude of a group of practising mathematicians rather than a fully explicit philosophy. **1940** E. T. Bell *Devel. Math.* xxiii. 527 Formalism denies logicalism and seeks to controvert the conclusions of intuitionism. **1970** H. Weyl *Appreciation* in C. Reid *Hilbert* 270 Hilbert's formalism restores the principle of the excluded middle which was the main target of Brouwer's criticism.

b. (See quot. 1940.)

1940 E. T. Bell *Devel. Math.* xiii. 262 'Formalism' in analysis means manipulation of formulas involving infinite processes without sufficient attention to convergence and mathematical existence. **1947** Courant & Robbins *What is Math.?* (ed. 4) p. vii, There exists the danger of frustration and disillusionment unless students and teachers try to look beyond mathematical formalism and manipulation and to grasp the real essence of mathematics. **1964** H. Eves *Introd. Hist. Math.* (rev. ed.) xii. 360 Euler's work represents the outstanding example of eighteenth-century formalism.

4. *Theatre.* A movement originating *c* 1890 in Russia which, at first a reaction against excessive naturalism in stage production, after the Revolution led to excessive emphasis on symbolism and stylization.

1927 J. Dolman *Art Play Prod.* xviii. 396 Most of the attempts at formalism today are really compromises—accidental compromises with stylization..or intentional compromises with realism or symbolism. **1931** Selden & Sellman *Stage Scenery* i. 28 Among the more popular styles..are: Realism,..Stylization,..Formalism, the mode of using forms which mark a place for action but do not attempt to represent or suggest any given locality. **1966** A. Nicoll *Devel. Theatre* (ed. 5) i. 3 We are here [*sc.* in the Nō plays] confronted by a theatre in which formalism is dominant. **1967** A. S. Gillette *Introd. Scenic Design* xii. 165 Theatres with conventional proscenium arches have been adapted to formalism by the substitution of drapery backgrounds for scenery.

5. a. The theory held by a Russian literary group in existence between 1916 and 1930 that technique and form are both the means to and the goal of artistic creation. **b.** Subsequently, a term often used pejoratively in Communist criticism to denote an artist's concentration on form at the expense of social reality and content.

1943 V. Ughet tr. *Miliukov's Outl. Russ. Culture* II. 118 It [*sc.* Socialist Realism] was directed against 'formalism' with its tendency to degenerate into mere aesthetic preoccupation with literary forms. **1955** V. Erlich *Russ. Formalism* I. i. 4 Formalism was..the first critical movement in Russia which attacked in systematic fashion the problems of rhythm and meter, of style and composition. **1957** C. Hunt *Guide to Communist Jargon* xxi. 73 The *Political Dictionary* defines formalism as 'putting to the forefront the outer side of a question, the detachment of form from content'. *Ibid.* 75 The charge of formalism will commonly mean that a novelist has devoted too much attention to plot, characterization and description, and that his work lacks the requisite inspirational quality. **1962** *Listener* 19 July 115/1 The best circles in the West are now almost exclusively preoccupied with technical means for their own sake, with what their colleagues on the wrong side of the Iron Curtain would describe as 'formalism'.

6. [cf. G. *formalismus* in same sense (Hilbert et al. 1928, in *Math. Ann.* XCVIII. 1–30).] A particular mathematical theory or mode of description of a physical situation or effect.

1934 *Ann. Math.* XXXV. 29 (*heading*) On an algebraic generalization of the quantum mechanical formalism. **1955** J. A. Wheeler in W. Pauli *Niels Bohr* 177 These circumstances make it appropriate to discuss the spontaneous fission rates of even very heavy nuclei in terms of the liquid drop formalism. **1968** C. G. Kuper *Introd. Theory Superconductivity* xii. 193 In the field-theoretical approach to the theory of solids, quasiparticles enter the formalism in precisely the same way as 'physical' particles enter the formalism of elementary particle physics. **1970** *Nature* 12 Sept. 1087/2 A fresh attempt to impose a simple formalism on the kinetics and equilibrium of the oxygenation reaction.

formalist. Add: **5.** A follower or advocate of *formalism (3, 4, 5).

1913 tr. L. E. J. Brouwer in *Bull. Amer. Math. Soc.* XX. 83 For the formalist..mathematical exactness consists merely in the method of developing the series of relations, and is independent of the significance one might want to give to the relations or the entities which they relate. **1926** *Encycl. Brit.* III. 635/2 The 'Formalists' held that the naturalistic theatre was not art, but a soulless attempt to photograph life. **1934** P. A. Markov *Soviet Theatre* vii. 119 The 'formalists'..stressed the need for freeing the actor from the bonds of realism or æstheticism. **1941** Courant & Robbins *What is Math.?* iv. 216 The clash between the intuitionists and the formalists has been much publicized by passionate partisans of these schools. **1954** A. J. Ayer *Philos. Essays* ix. 221 The formalist aims to get rid of numbers by construing statements about numbers as statements about signs. **1955** V. Erlich *Russ. Formalism* I. vi. 95 By focussing sharply on the specific aims and methods of literary scholarship, the Formalists had induced their Marxian opponents to get off the high horse of dialectical generalizations. **1967** A. S. Gillette *Introd. Scenic Design* xii. 165 The goal of the formalists originally was to return to a completely functional theatre. **1971** *Sci. Amer.* Aug. 92/2 A Formalist is likely to say that the real-number line is whatever we define it to be.

B. *adj.* Of or pertaining to formalism, in various senses; formalistic.

1900 B. Russell *Philos. Leibniz* xiv. 170 But as a method of pursuing philosophy, it had the formalist defect which results from a belief in analytic propositions, and which led Spinoza to employ a geometrical method. **1926** *Proc. Lond. Math. Soc.* 2nd Ser. XXV. 339, I hold that mathematics is part of logic, and so belong to..the logical school as opposed to the formalist and intuitionist schools. **1931** K. E. Kirk *Vision of God* III. iii. 132 The characteristic dangers of Judaism..were exactly the dangers that formalist codification tends to foster. **1955** V. Erlich *Russ. Formalism* I. iii. 47 In Petersburg..the first steps of the Formalist movement were marked by a close alliance with the poetic avant-garde. **1967** G. H. Hamilton *Painting & Sculpt. in Europe* 5 The scope of strictly formalist criticism has undoubtedly been enlarged by surrealism and by the doctrines of social realism. **1970** *Times Lit. Suppl.* 23 July 787/2 The Formalist philosophy was an exemplary one in several respects.

formalistic, *a.* (Later examples.)

1941 Courant & Robbins *What is Math.?* ii. 88 In some way or other,..even under the most uncompromising formalistic, logical, or postulational aspect, constructive intuition always remains the vital element in mathematics. **1957** C. Hunt *Guide to Communist Jargon* xxi. 77 The resolution [of 10 Feb. 1948] went on to censure for their 'formalistic perversions and anti-democratic tendencies' all the more prominent Soviet composers. **1962** *Listener* 19 July 115/2 The Sixth [Symphony of Shostakovich] was denounced as empty and 'formalistic'.

formalize, *v.* Add: So **fo·rmalizable** *a.*, capable of being formalized.

1944 H. Reichenbach in P. A. Schilpp *Philos. B. Russell* 27 This formulation given at a later stage was anticipated by Russell's original distinction of formalizable and non-formalizable parts of logic. **1965** N. Chomsky *Aspects of Theory of Syntax* i. 19 No adequate formalizable techniques are known for obtaining reliable information concerning the facts of linguistic structure.

formant (fǭ·ɹmănt). [a. G. *formant* (Hermann and Matthias 1894, in *Arch. f. d. ges. Physiol.* LVIII. 262), f. L. *formant-*, *-ans* pres. pple. of *formāre* to Form: see -ant¹.] **1.** *Phonetics.* The characteristic pitch of a vowel-sound; *spec.* one of several characteristic bands of resonance, a combination of which determines the distinctive sound-quality of a vowel (or *transf.*, of a musical instrument).

1901 *Nature* 26 Dec. 187/2 A vowel, according to him [*sc.* Hermann], is a special acoustic phenomenon, depending on the intermittent production of a special partial, or 'formant' or 'characteristique'. The pitch of the 'formant' may vary a little without altering the character of the

vowel. **1930** *Times Lit. Suppl.* 3 July 545/1 Later investigators have developed a theory of the vowel as a *formant*, which, though occupying a definite region of pitch, nevertheless manages somehow to affect the whole sound-complex. **1952** *Electronic Engin.* XXIV. 368 The peaks of the transmission curve are the resonant frequencies, and generate formants, which have been defined as the frequencies of energy concentration of the speech signal. **1956** Jakobson & Halle in Saporta & Bastian *Psycholinguistics* (1961) 349/1 The clear-cut formant structure of a vowel. **1957** A. C. Baines *Woodwind Instrum.* 34 Harmonics that fall within certain zones of pitch ('formants') tend to dominate. **1964** J. B. Carroll *Lang. & Thought* ii. 12 Vowel quality is directly correlated with the relative positions, on the frequency scale, of two or three formants, or bands of strong resonance overtones, present in the frequency spectrum of a sound. **1968** J. Lyons *Introd. Theoret. Ling.* iii. 110 The acoustic analysis of vowels in terms of their constituent formants..correlates quite well with their description in terms of the articulatory dimensions.

2. *Philol.* An affix to a root or stem, forming a longer root or stem. (Sometimes used more loosely.)

1935 G. K. Zipf *Psycho-Biol. Lang.* (1965) iv. 144 The stem (root plus formant) to which endings are added. **1956** J. Whatmough *Language* viii. 136 Some morphemes occur only as parts of words, e.g. English *-ing* in *loving*, or *-nt-* in Latin *amantis*.., and these are known as 'formants'.

‖ **forma pauperis** (fǭ·ɹmă pǭ·pĕris). [L.] = *in forma pauperis* (s.v. ‖In 4). Also † *under forma pauperis*.

1627 Middleton et al. *Widow* II. i, I scorn to get thee under *forma pauperis*. **1633** T. Adams *Exp. 2 Peter* ii. 6. 621 Doe they not sue for their inheritance in heaven, *Formâ pauperis*; refusing to give the least scrap of their superfluitie for eternall life? **1907** *Daily Chron.* 4 Mar. 3/5 In the Divorce Court..fees to the lawyers are supposed to be paid even in formâ pauperis causes. **1959** Jowitt *Dict. Eng. Law* II. 947/2 Appeals may still be taken *in forma pauperis* to the House of Lords. The Appeal (Forma Pauperis) Act, 1893, enables the House to refuse leave to appeal.

format, *sb.* Delete ‖. Pronunc. now usu. (fǭ·ɹmæt). Add: **b.** *transf.*

1936 O. S. Puckle tr. *von Ardenne's Telev. Reception* ii. 42 The photograph shows the evenness of the illumination and the almost perfect rectangularity of the picture format. **1962** *Listener* 1 Feb. 224/1 The low relief and the often square format of the canvas also establish the pictures as firmly constructed panels. **1971** *Daily Tel.* 13 Oct. 6/7 The Brick Development Association proposed a single metric brick format which was adopted by the industry. It was slightly smaller than the imperial format: 225 mm × 112·5 mm × 75. **1971** D. Potter *Brit. Eliz. Stamps* iii. 32 The first vertical format double-size stamp was issued.

2. A style or manner of arrangement or presentation; a mode of procedure.

1955 *Times* 3 June 9/2 It is a tribute to the attractive format..that one does not realize that the report runs to 83 pages. **1957** *N.Y. Times* 29 Dec. iv. 3/2 It is so fundamental that the President must never appear before the American people in a minor role that the choice of this format is inexplicable. **1958** *Spectator* 1 Aug. 167/1 The principal performer..had to write the scripts herself, and when..she attempted to heed criticism and alter the format, she was told that the show, however bad, must go on. **1958** *Listener* 23 Oct. 648/2 Goldberg has developed from the sectional, chopped-up forms of de Kooning to a large centralized format. **1963** *Amer. Speech* XXXVIII. 165 The slang items in this study are subdivided into two major groupings...The format followed supplies first the denotative word or phrase which was used to elicit the slang responses. Next, [etc.]. **1967** *Times* 6 Jan. 15/1 The usefulness of the I[ndustrial] R[eorganization] C[orporation] will be in its ability to help the two sides agree upon a new supply format. **1970** *Daily Tel.* 19 Nov. 14/4 This programme format, consisting of a report followed by a public meeting arranged for television, never seems to me satisfactory. **1971** *Homes & Gardens* Oct. 175/2 The noise generated by tape play-back systems..is the most serious drawback of the small cassette format.

b. *Computers.* A particular arrangement of data or characters in a record, instruction, word, etc., in a form that can be processed or stored by a computer.

1955 *Computers & Automation* Jan. 16/2 Logical operations of verifying format, insuring proper placement of significant characters,..are all performed in less than ½ second. **1959** *Jrnl. Assoc. Comput. Mach.* VI. 325 One such typical program..might involve the conversion and rearrangement of data from the format in which it is stored on external units to a format compatible with internal use, e.g. the conversion of characters from decimal to binary form. **1963** B. F. Green *Digital Computers in Research* v. 84 An associated format statement..specifies the precise manner in which the quantities appear in the input or output medium. **1968** *Amer. Documentation* Jan. 74/2 A format routine converts scanner output to element form for entry in the central data file. **1970** *Nature* 4 July 7/2 A scheme for coding rock and mineral names into a format suitable for computers.

format (fǭ·ɹmæt), *v.* Pa. t. and pa. pple. **formatted**; pres. pple. **formatting.** [f. the sb.] *trans.* To arrange or put into a format: used chiefly in connection with *Computers* (see *format *sb.* 2 b). Also **fo·rmatted** *ppl. a.*, **fo·rmatting** *vbl. sb.*

1964 *CIS Gloss. Automated Typesetting* (Composition Information Services, Calif.) 14 Book formatting. **1965** *Adv. Computers* VI. 7 These systems have come to be known as formatted file systems, since the input data are arranged into various formats for ease of storage and retrieval. **1968** *Guardian* 23 Sept. 9/2 We formatted this Top 100 Sounds, covering the top sounds both east and west of the Atlantic. **1969** *Computers & Humanities* IV. 98 Little or no composition control data or formatting punctuation is encoded with the input data. *Ibid.* 99 We format pages internally and then pass them to an output module tailored to a specific output device. **1970** A. Cameron et al. *Computers & O.E. Concordances* 48 A concordance, however, if formatted along appropriate lines..can shorten this mind-wearying labour tremendously. **1970** *McGraw-Hill Yearbk. Sci. & Technol.* 160/1 The principal formatting function in the graphic arts is justification, which is frequently coupled with hyphenation.

formate (fǭ·ɹmē¹t), *v.*² *Aeronaut.* [Back-formation from *formation 4 b.] *intr.* Of an aircraft or its pilot: to take up formation *with*, to fly in formation. Also const. *on*. Also (*rare*) of a boat: to proceed in formation.

1929 *Times* 6 Mar. 5/6 He saw Mr. Sealy-Allin 15 minutes before the collision and attempted to formate with him. **1933** *Aeroplane* 19 Apr. 690/1 The Avros formated by themselves. **1934** T. E. Lawrence *Lett.* (1938) 805 We are all up to the teeth in 5 more target boats. Three or four are [to] formate from here to Bridlington. **1934** V. M. Yeates *Winged Victory* xiii. 106 He kept getting in everybody's way. If there was one thing a pilot ought to do properly, it was to formate. **1935** *Flight* 23 May 549/1 It is possible to recollect a number of occasions when, as passenger in a big machine, one has been 'formated' on by odd and sometimes gay pilots. **1942** *We speak from Air* xvi. 51 The other Beauforts formated on me until we separated near the target.

formation. Add: **1. b.** *Electr.* The action or process of forming an accumulator plate, a semiconductor device, etc. (see *form *v.* 1 g).

1881 S. P. Thompson *Storage Electr.* 12/1 The amount of charge will..depend on the size of the plates and on the degree to which the process of 'formation' has been carried out. **1926** L. B. W. Jolley *A.C. Rectification* (ed. 2) xv. 372 During the process of formation the oxide film is being deposited on the anode. **1964** G. Smith *Storage Batteries* ii. 22 A forming agent is added to the sulphuric acid, which ..accelerates the production of the lead dioxide coating when current is passed through the plate during formation.

3. b. The disposition of fibres in a sheet of paper.

1937 E. J. Labarre *Dict. Paper* 110/1 *Formation*, the disposition or texture of the fibres in a sheet of paper. **1952** J. P. Casey *Pulp & Paper* I. viii. 417 Formation..is a highly important property and influences many of the properties of the final paper, such as strength and appearance. **1953** J. N. Stephenson *Pulp & Paper Manuf.* III. 119 Formation is that quality which describes the structure of a sheet of paper, the manner in which the fibers are interlaced.

4. b. *Aeronaut.* The orderly disposition of a number of aircraft in flight.

1914 *Engineering* 4 Dec. 680/3 Formation Flying...In the handling of a single squadron, the formation adopted may evidently be fairly elastic. **1917** 'Contact' *Airman's Outings* p. xiv, Dotted everwhere, singly or in formations of two, three, four, or six, were..the single-seater fighting scouts. **1931** H. A. Jones *War in Air* III. iv. 296 They were made to..practise..flying in formation. **1941** E. C. Shepherd *Military Aeroplane* 4 Anti-aircraft fire can..break up the formations. **1963** *Amer. Speech* XXXVIII. 117 The bomber then comes slowly up behind the tanker and flies a close formation with it. **1971** D. N. James *Gloster Aircraft* 211 Three Gladiators of No. 87 Squadron flew tied-together formation aerobatic sequences with cord connecting their interplane struts.

5. b. *Ecol.* [a. G. *formation* (A. Grisebach 1838, in *Linnaea* XII. 160).] A community formed by groups of plants which have adapted themselves to similar climatic conditions.

1898 *Bot. Gaz.* XXV. 394 The grass formations are more or less intermediate between the sand hill and the foothill formations. **1909** Groom & Balfour tr. *Warming's Oecology of Plants* xxxv. 140 A formation may then be defined as a community of species,..which have become associated together by definite external..characters of the habitat. **1929** Weaver & Clements *Plant Ecol.* i. 6 No formation is uniform throughout its entire extent. **1960** N. Polunin *Introd. Plant Geogr.* xi. 333 Each formation usually covers a wide area involving various conditions.

6. formation-rule *Logic*, one of a set of rules together specifying which combinations of symbols are to count as well-formed formulas in a given system. (Opp. *transformation rule.)

1937 A. Smeaton tr. *Carnap's Logical Syntax Lang.* § 1. 2 The difference between syntactical rules in the narrower sense and the logical rules of deduction is only the difference between formation rules and transformation rules. **1948** H. Reichenbach *Elem. Symb. Logic* i. 16 Formation rules..tell us under what conditions a set of signs is meaningful. **1965** P. Caws *Philos. of Sci.* xvi. 119 Formation-rules..specify how primitive or other terms may be put together to form acceptable expressions. **1966** *Amer. Philos. Q.* III. 7/1 Speech activities..proceed in accordance with formation rules.

Also **forma·tionally** *adv.*, in respect of formation, as regards formation.

1898 *Bot. Gaz.* XXV. 382 The forests of Mexico..are both floristically and formationally distinct from the northern forests. **1953** C. E. Bazell *Linguistic Form* 51 Phonemically the form [*wished*] is on a par with the formationally underivable *fist*, etc.

formative, *sb.* Add: (Further examples.) Also *gen.*, a formative agent.

1907 Galsworthy *Country House* I. iv, That essential formative of character, east wind. **1953** C. E. Bazell *Linguistic Form* i. 8 A unit of formation, commonly called (*allo-*) *morph* in America and *morpheme* in Europe (outside Copenhagen), and for which the present writer has proposed the term *formative*. **1965** N. Chomsky *Aspects of Theory of Syntax* i. 3 The rules that specify the well-formed strings of minimal syntactically functioning units (*formatives*). **1965** B. M. H. Strang *Metaphors & Models* 18 In the first stage a terminal *string* is generated by applying an ordered set of rules, the formula *F*, to some of a finite repertoire or set, Σ, of formatives.

‖ **formatore** (fǭɹmatō·re). [It., agent-n. f. *formare* to form.] A modeller in wax or plaster; a technical assistant who repairs or restores pottery, metalwork, etc.

1862 *Catal. Internat. Exhib., Brit.* II. No. 5451 Formatore to the Science and Art Department, the Trustees of the British Museum, and the Royal Academy of Arts. **1899** *Proc. Zool. Soc.* 776 Assistant formatore in the Museum.

former, *sb.*¹ Add: **3. d.** *Electr.* A frame or core about which a coil is wound.

1891 J. W. Urquhart *Dynamo Constr.* v. 76 The first matter to attend to in winding the bobbins is to ensure that there shall be no possibility of metallic contact between the wire and..the metal of the reel or former. **1910** *Encycl. Brit.* VIII. 772/1 The coils are either wound by hand directly on to the armature core, or are shaped on formers prior to being inserted in the armature slots. **1943** *Electronic Engin.* XVI. 80/2 A high quality ceramic coil former..for aerial coupling coils.

e. *Aeronaut.* A transverse member that strengthens and gives shape to a wing or fuselage.

1919 H. Shaw *Text-bk. Aeronaut.* ix. 110 In order to maintain the front portion of the upper surface [of the wing] it is customary to place false ribs or formers from the leading edge to the main spar. **1930** *Flight* 28 Nov. 1384/2 The fuselage formers or frames..are attached to the longerons by angle-section brackets. **1951** 'N. Shute' *Round Bend* 261 The great empty cabin, floored with duralumin, with bare stringers and formers supporting the outer skin of the walls. **1966** D. Stinton *Anat. of Aeroplane* 203 The skin is usually formed of metal sheets riveted, or spot-welded, to metal frames, formers and bulkheads.

4. *attrib.* and *Comb.*, as **former-wound** *a.*, of an armature coil, wound on a former before being mounted.

1902 *Encycl. Brit.* XXVII. 583/2 The former-wound coils have the advantage that they are perfectly symmetrical and interchangeable. **1906** A. Russell *Treat. Altern. Curr.* II. 328 The winding of the stator of a polyphase motor is simple. It may be made up of rectangular former-wound coils, that is, coils which are wound into shape on a rectangular wooden block before being fixed on the stator. **1940** [see *bar-wound adj.].

former, *a.* Add: **1. e.** Used to designate a former holder of an office; = British *ex-*. *U.S.*

1905 *N.Y. Herald* 5 Feb. 47 Former President Cleveland is among the arrivals of the week at the Lakewood Hotel. **1919** Mencken *Amer. Lang.* iv. 118 Such clumsy quasi-titles as ex-United States Senator,..and former Chief of the Fire Department. **1948** *Minneapolis Morn. Tribune* 28 Sept. 1/6 Sen. George Wilson, threatened by Democratic former Sen. Guy M. Gillette, was frankly warned by Dewey's advisers to intensify his campaign. **1970** *Encycl. Brit.* XXII. 650/1 President Hayes..appointed former Senator David M. Key..to his cabinet.

-former. [f. Form *sb.* 6 b + -er¹.] In schools, a member of a specified class or form, as *sixth-former*.

1869 Third-former [see Third *a.* 5]. **1927** J. Elder *Thomasina Toddy* xi, Sixth and Fifth formers who were in the Elevens. *Ibid.* xii, They gained on the Sixth-formers' boat. **1928** *Daily Express* 31 Aug. 15/2 One of them inquired of a small fourth-former whether he had seen any suspicious characters about. **1960** D. Potter *Glittering Coffin* vi. 82 The choice of grammar school sixth-formers would be Oxford or Cambridge first. **1969** *Listener* 20 Feb. 252/1 A more likable, fourth-former-to-fifth-former note.

‖ **Formgeschichte** (fǭ·ɹm,gəʃi:χ⁾tə). [G., f. *form* form + *geschichte* history.] = *form-criticism.

1923 *Harvard Theol. Rev.* XVI. 92 Two monographs on special phases of gospel 'Formgeschichte'. **1932** *Expos. Times* XLIII. 396/1 The 'investigations'..of the German school of *Formgeschichte*. **1935** [see *form-criticism]. **1954** *Scott. Jrnl. Theol.* VII. 334 The work of R. Bultmann and M. Dibelius in the field of Formgeschichte in the Gospels. **1970** *Computers & Humanities* IV. 199 It offers a scientific method of crossing..the quagmire of *Formgeschichte* which has seemed to be the limit of Synoptic studies.

Formica² (fǭɹmai·kă). Also **formica.** The proprietary name of a hard, durable plastic laminate used esp. as a decorative surfacing material. Also *attrib.* and *Comb.*

1922 *Official Gaz.* (U.S. Pat. Off.) 10 Jan. 409/2 The Formica Insulation Company, Cincinnati, Ohio...*Formica*. Electric Insulating Compound. **1946** *Ibid.* 12 Mar. 179/2 The Formica Insulation Company, Cincinnati, Ohio...*Formica*. For laminated sheets of wood, fabric, or paper impregnated with synthetic resin and consolidated under heat and pressure, for use on table tops, furniture and wall panelling. **1957** *Oxf. Mail* Suppl. 16 Oct. 6/5 Red

upholstered chairs with chromium legs and tables in formica. **1958** *House & Garden* Feb. 91 (Advt.), Furniture surfaced for ever with Formica decorative laminates. **1958** *Woman's Own* 5 Mar. 11/4 Formica-topped units link the automatic electric cooker (with eye-level grill) and sink unit into an ideal one-level working surface. **1959** H. Hobson *Mission House Murder* i. 9 One tinselly formica-and-chrome establishment. **1960** *Guardian* 25 Apr. 2/4 A Formica-covered table top.

forming, *vbl. sb.* Add: **b.** *Electr.* = *formation 1 b.

1902 *Encycl. Brit.* XXV. 26/2 Faure..greatly shortened the time required for 'forming' by giving the plates a preliminary coating of red lead. **1926** *Physical Rev.* 2nd Ser. XXVII. 813 No forming is necessary so the rectifier is immediately operative when the alternating voltage is applied. **1956** L. P. Hunter *Handbk. Semiconductor Electronics* iv. 17 The transistor before forming shows a collector resistance of the order of 10^5 ohms...After forming, the collector resistance is usually about 2×10^4 ohms. **1962** Simpson & Richards *Junction Transistors* viii. 181 The other is reverse-biased and forms the collector. It is given a special heat treatment called 'forming' which consists of passing a large current through it in the reverse direction for a short time.

forming, *ppl. a.* Add: **c.** That forms or brings about formation (see *form *v.* 1 g).

1914 H. Pender *Amer. Handbk. Electr. Engineers* 105 After the forming charge the plates are dried and are ready for the market. **1962** L. P. Hunter *Handbk. Semiconductor Electronics* (ed. 2) iv. 42 The forming pulse..heats the germanium in the neighborhood of the collector point to a relatively high temperature. **1964** [see *formation 1 b].

formol (fǭ·ɹmǫl). [Former trade name, app. arbitrarily f. Formaldehyde.] = *formalin. Also *attrib.*, as **formol titration** *Biochem.*, a method of estimating amino acids involving the use of formalin. Hence **fo·rmolize** *v. trans.*, = *formalinize *v.*; **fo·rmolized** *ppl. a.*

1894 *Brit. Med. Jrnl.* 22 Dec. 1429/2 Formol was introduced by Professor Leber; it was supplied under two names—formol or formaline. **1910** *Chem. Abstr.* IV. 606 The conclusion of Soerensen, that amino acid N can be determined by means of the formol titration..is confirmed. **1932** *Proc. Soc. Exper. Biol. & Med.* XXIX. 456 We..succeeded in immunizing guinea pigs with formolized Rickettsia suspensions. **1949** Miles & Pirie *Nature Bacterial Surface* i. 7 The partly lysed bacteria were formolized and washed repeatedly on the centrifuge. **1961** Greenstein & Winitz *Chem. Amino Acids* I. iv. 516 (*heading*) Formol titration constants of amino acids and peptides at 30°. **1964** *Bull. World Health Org.* XXX. 46/1 The cells were formolized by a modification of Csizmas' (1960) method.

Formosa (fǭɹmōu·să). [See next.] Used *attrib.* in names of products of Formosa, esp. *Formosa tea* (also *ellipt.*).

1889 *Jrnl. Soc. Chem. Ind.* 30 Nov. 931/1 Formosa camphor in the Hong Kong market rose to 30 dols. per picul. **1892** [see Pekoe *v.*]. **1896** J. D. Clark *Formosa* 58 The fear lest Formosa Tea, being converted into Congou by the Indian methods of preparation..should thereby lose the distinguished position it occupies as Oolong. **1951** W. G. Copsey *Law's Grocer's Man.* 157/1 *Formosa tea*. This is a distinctive leaf tea known as Formosa Oolong. It is semi-fermented in character and variegated in colour.

Formosan (fǭɹmōu·săn), *a.* and *sb.* [f. *Formosa*, orig. fem. (sc. *insula* island) of L. *formōsus* beautiful; a name (orig. Portuguese) of the island Taiwan, off the south-east coast of China: see -an.] **A.** *adj.* Of or pertaining to the island of Formosa, its inhabitants, or their language. **B.** *sb.* **a.** A native or inhabitant of Formosa. **b.** The language of the indigenous Formosans.

1650 H. Jessei tr. *Sibellius's Conv. 5,900 E. Indians* 4 M. Junius being carryed by the good hand of God to the Formosan Island, in the East-Indies, preached in Dutch. *Ibid.* 8 M. Junius..translated certaine Psalmes into the Formosan Islanders Language. **1704** G. Psalmanazar *Descr. Formosa* 122 The Formosan Alphabet. *Ibid.* 335 The Lord's Prayer, Apostle's Creed, and the Ten Commandments, translated by the Author into Formosan. **1707** —— (*title*) A dialogue between a Japonese and a Formosan, about some points of the religion of the time. *Ibid.* (Pref.), You'll see a Formosan, like a willing Prisoner, endeavouring to bring his friend unto the same Condition as he is himself. **1731** R. Millar *Hist. Propag. Christ.* II. viii. 318 Mr. Robert Junius..translated certain psalms into the Formosan language. **1879** *Encycl. Brit.* IX. 416/2 The Formosan fauna has been but partially ascertained. **1915** W. Campbell *Sks. fr. Formosa* xvi. 122 Some of the Formosan tribes boil down every head brought in to a thick jelly, from which thin oblong cakes are made, for being nibbled to inspire fresh courage. **1933** Bloomfield *Lang.* iv. 71 The languages of the great islands of the East, such as Formosan, Javanese (20 millions), Sundanese (6½ millions) [etc.]. **1955** H. Kurnitz *Let.* 29 Dec. in G. Marx *Groucho Lett.* (1967) 250 The Krasnas had a Xmas party with Chinese food, which..was cooked by some Formosan tribesmen. **1962** *Times* 20 Aug. 9/4 The Kuomintang has no appeal to the younger generation in Formosa whether they are children of those who came from the mainland or native Formosans.

formose (fǭ·ɹmōuz, -s), *sb. Chem.* [a. G. *formose* (O. Loew 1886, in *Jrnl. f. prakt. Chemie* XXXIII. 328), f. Form(aldehyde + -ose².] A mixture of hexose sugars, originally thought to be a single compound, produced by the condensation of formaldehyde in the presence of weak alkalis.

1886 *Jrnl. Chem. Soc.* L. 610 Formose, $C_6H_{12}O_6$, is obtained by shaking a 3·5 to 4 per cent. solution of formaldehyde with an excess of milk of lime. **1957** J. C. Sowden in W. Pigman *Carbohydrates* ii. 103 Formaldehyde condenses in the presence of weak bases to form a complex mixture of sugars called formose or methose. **1969** *New Scientist* 13 Nov. 342/3 NASA scientists are studying the formose sugar synthesis.

formula. Add: **1. c.** A form of words serving to reconcile different aims, opinions, or points of view.

1905 Asquith in *Westm. Gaz.* 9 Oct. 2/3 The recovery of freedom of negotiation, the reloading of the big revolver, the summoning of an open conference, and all the other temporising formulæ of the Balfourian school. **1928** Galsworthy *Swan Song* I. i. 5 We shall sit and glower at each-other, and use the word 'formula' at stated intervals. **1940** Wodehouse *Quick Service* i. 18 It was plain that this girl and he were poles apart and could never hope to find a formula. **1971** A. Bullock *20th Cent.* 51/1 At the Yalta Conference (February 1945) verbal formulae were found to disguise growing differences between the Russians and the Anglo-Americans.

d. *Literary Criticism.* In various technical and semi-technical uses (see quots.).

1888 A. S. Cook *Judith* p. lii, Rime and various forms of assonance are occasionally employed by Old English poets, sometimes for the purpose of uniting more closely the two halves of the same line,..rarely in formulas or compounds within the same hemistich. **1903** L. F. Anderson *Anglo-Saxon Scop* 39 Grein gives four examples of the occurrence of the formula ['*singan and secgan*'] in Anglo-Saxon. **1921** B. Tarkington *Let.* 26 Mar. (1959) 49 The formulas that have prevailed are now all familiar to the audience. **1932** M. Parry in *Harvard Stud. Class. Philol.* XLIII. 8 Each idea to be expressed in the poetry has its formula for each metrical need. **1934** H. J. Rose *Handbk. Gk. Lit.* ii. 31 There are embedded in it [*sc.* the style of Homer] many formulae, epithets and turns of expression which strongly indicate..the existence of an age of balladry before the date of the epics. **1935** A. C. Bartlett *Larger Rhet. Patterns Anglo-Saxon Poetry* 91 Although not every Anglo-Saxon poem has a formal introductory passage, there is a type of opening sentence which occurs so often as to allow its being called a formula. **1953** *Speculum* XXVIII. 446 The unlettered singer ordinarily composing rapidly and extempore before a live audience, must and does call upon ready-made language, upon a vast reservoir of formulas filling just measures of verse. **1958** T. B. L. Webster *From Mycenae to Homer* iii. 89 Homer preserves much of this manner [*sc.* of second-millennium poetry] in his noun-epithet formulae, typical scenes, formulae for opening and closing of speeches, refrain lines, etc. **1965** M. Hodgart *Faber Bk. Ballads* 14 A formula is a theme expressed in identical or almost identical words and metre. **1967** C. L. Wrenn *Study of O.E. Lit.* iii. 47 *Andreas*, a heroic hagiographical poem of the early ninth century, begins with exactly the same exordial formula..as does *Beowulf*. **1971** *English Studies* LII. 350, I..fail to see why syntactic, idiomatic and even single semantic units must be called 'formulas' because they occur in OE poetry.

4. *Motor Racing.* The class or specification of a racing car, usu. expressed in terms of engine capacity. Also *attrib.*

1927 *Autocar* 20 May 850/2 Half an hour later we knew for certain that we had won the race on formula. **1939** *Motor Sport* June 181/1 Last year..Indianapolis was run under the Grand Prix formula laid down by the A.I.A.C.R., specifying engine limits of 3-litres supercharged and 4½-litres unsupercharged. **1958** *Times* 31 Oct. 15/4 There is little doubt that the new Grand Prix formula recommended by the C.S.I. will be accepted by the F.I.A. at their next meeting. **1965** *Listener* 3 June 841/1 This was the weekend of the Monaco Grand Prix, when the faded Edwardian *pensions* seemed stunned and shaken by the noise of Formula One racing engines.

formulable (fǭ·ɹmiŭlab'l), *a.* [f. Formul(ate + -able.] That may or can be formulated; capable of formulation.

1891 W. James *Let.* 21 Sept. in R. B. Perry *Tht. & Char. of W. J.* (1935) II. 174 The way in which you squeeze the last drops of formulable truth out of the facts is admirable. **1902** M. H. Liddell *Introd. Sci. Study Eng. Poetry* p. vii, Certain fundamental laws as definitive and formulable in the one case as in the others. **1934** T. S. Eliot *Eliz. Ess.* 151 If we may use the word 'purpose' for something more profound than any formulable purpose can be. **1939** *Mind* XLVIII. 202 The class *k* of non-analytic declarative sentences formulable in a given language.

formulaic, *a.* (Later examples.)

1933 M. Parry in *Trans. Amer. Philol. Assoc.* LXIV. 179 (*title*) Whole formulaic verses in Greek and South-slavic heroic song. **1952** C. M. Bowra *Heroic Poetry* vi. 222 Not all heroic poetry shows these two kinds of formulaic elements on the same scale. **1953** *Speculum* XXVIII. 446 Parry, aided by Lord, demonstrated that the characteristic feature of all orally composed poetry is its totally formulaic character. **1963** R. Quirk in Brown & Foote *Early Eng. & Norse Studies* 171 While formulaic utterances and habitual collocations are the necessary starting point in the study of the early alliterative poetry, they are *only* a starting point. **1967** C. L. Wrenn *Study of O.E. Lit.* iii. 51 *Word-hord onlēac* is a formulaic phrase from oral tradition. **1971** *New Statesman* 16 Apr. 529/2 There is something in the Homeric poems that carries them far beyond the standards of the oral, formulaic poetry of the Southern Slavs.

Hence **fo:rmula·ically** *adv.*

1957 *Archivum Linguisticum* IX. ii. 96 *Cena* (with which *yantar* was paired off formulaically). **1963** J. Lyons *Structural Semantics* vi. 111 This convention..makes for easier reading of syntactic equivalences, formulaically expressed. **1966** S. B. Greenfield in E. G. Stanley *Continuations & Beginnings* 153 The first-person speaker recounts graphically (however formulaically) his tribulations as an exile upon the sea.

formularism (fǭ·ɹmiŭlări:z'm). [f. Formular(y + -ism.] Rigid adherence to or dependence on formularies. So **fo·rmularist.**

1896 Hardy *Jude* II. i. 95 The well-known three, the enthusiast, the poet, and the formularist. **1927** *Observer* 30 Oct. 7 The cut-and-dried formularism of the Socialist theory.

formulatable (fǭ·ɹmiŭlēi·tab'l), *a.* [f. Formulat(e *v.* + -able.] = *formulable *a.*

1895 H. James *Notebk.* 22 Dec. (1947) 239 Everything is formulated and formulatable to the child. **1923** *Spectator* 29 Sept. 426/1 Reduction to a small number of formulatable principles. **1946** C. I. Lewis *Analysis of Knowledge* II. viii. 204 Only by confining statements to an intent thus formulatable in expressive terms can anything be proved conclusively by single experiences.

formulate, *v.* Add examples of more extended use.

1893 A. Conan Doyle *Mem. S. Holmes* 12, I listened..to the dialogue of the two detectives. Gregory was formulating his theory. **1906** E. Phillpotts *Portreeve* I. x, She..stood for the moment powerless to formulate a clear pathway through this thicket of ideas. **1907** *Smart Set* Mar. 134 He resolved to send her some volumes of Ruskin and to formulate a graduated course of reading. **1910** E. M. Albanesi *For Love of Anne Lambart* 91 Anne began to put into movement the scheme that had formulated itself in her mind. **1924** A. D. Sedgwick *Little French Girl* I. vii. 69 It was strange to her, as she spoke, to feel how deeply she knew all this about Giles, though she had never before formulated it to herself.

Hence **fo·rmulator.**

1919 K. Pearson in C. Goring *Eng. Convict* 12 The use by the formulator of a thoroughly unscientific method does not justify us in rejecting his hypothesis. **1966** *New Statesman* 13 May 682/1 The three officials..should not be regarded..simply as civil servants. Stan Morris..is more a formulator of policy.

formulation. Add: **b.** A material or mixture prepared according to a particular formula.

1958 *New Scientist* 19 June 208/1 Different formulations of rubber vary..in their performance as cushioning materials. **1959** *Ibid.* 2 Apr. 744/2 The preparation of the liquid mix from which plastic tools are moulded is said to be simplified by a new series of pre-measured formulations. **1964** *Times Rev. Industry* Feb. 7/1 Between 40 and 50 new drugs and formulations are marketed every year and each one will involve the consideration of masses of data.

formylate (fǭ·ɹmilē¹t), *v. Chem.* [f. Formyl + -ate³.] To introduce one or more formyl groups into (a compound or molecule). So **fo·rmylated, -ing** *ppl. adjs.*; also **fo·rmylatable** *a.*

1931 *Jrnl. Amer. Chem. Soc.* LIII. 3491 The scheme of formylating by distilling a mixture of toluene, formic acid and methionine. **1964** *Jrnl. Molec. Biol.* VIII. 836 Artificially formylated [^{35}S]methionyl-RNA was prepared. **1965** *Ibid.* XIV. 64 The methionyl-sRNA synthetase and the formylating enzyme were separated by precipitation. *Ibid.* 65 Methionine was formylated after esterification to sRNA. **1970** *Nature* 29 Aug. 918 Reticulocyte initiation factors bind a formylatable methionine tRNA but not a non-formylatable tRNA.

formylation (fǭɹmilē¹·ʃən). *Chem.* [f. as prec. + -ation.] The introduction of one or more formyl groups into a compound or molecule by means of a chemical reaction.

1930 *Jrnl. Biol. Chem.* LXXXVI. 695 (*heading*) The formylation of amino acids. **1962** *Lancet* 29 Dec. 1381/1 The introduction of the no. 2 carbon to form the purine ring proceeds via formylation of a.i.c.a.r. to the 4-formamido compound. **1964** *Jrnl. Molec. Biol.* VIII. 835 The formylation of the free α-amino group of methionine ..takes place.

forra(r)der: see *forwarder *a.* and *adv.*

Forstner bit (fǭ·ɹstnəɹ). [The name of Benjamin *Forstner*, who patented the invention in 1874 and 1883 (U.S. patents 155,148 and 280,026).] A type of wood-drilling bit.

1902 G. L. Sutcliffe *Mod. Carpenter* II. 131 (*caption*) Forstner Auger bit. **1931** Reid & Higgins *Fund. Woodworking Trades* viii. 90 Forstner bits..are guided by their circular rim instead of a spur in the centre... They are especially adapted for delicate work. **1966** A. W. Lewis *Gloss. Woodworking Terms* 8 Forstner bit, bit with a very short centre point and sharp cylindrical rim.

forsythia (fǭɹsai·þiă). [mod.L. (M. Vahl *Enumeratio Plantarum* (1805) I. 39), f. the name of William *Forsyth* (1737–1804), English botanist: see -ia¹.] A plant of the genus of spring-flowering shrubs so named,

native to Albania and eastern Asia and belonging to the family Oleaceæ; cultivated for their bright yellow, bell-shaped flowers.
1846 *Jrnl. R. Hort. Soc.* I. 227 The species obtained by Mr. Fortune is very distinct from the original Forsythia. **1851** *Curtis's Bot. Mag.* LXXVII. 4587 The original *Forsythia*, established on a Chinese plant cultivated in Japan..appears to have been introduced into Holland in 1833. **1893** [see **golden bell*]. **1905** *Daily Chron.* 2 Aug. 3/1 Daffodils in the grass under drooping Forsythia. **1927** E. H. M. Cox *Evol. Garden* xiii. 250 The Forsythias with bright yellow flowers. **1965** S. C. Emberton *Shrub Gardening* ix. 52 The gardens along the line bubble over with sugar-pink almond blossom and irrepressible yellow forsythias.

fort, *sb.*[1] Add: **1. b.** Phr. *to hold the fort*: to act as a temporary substitute; also, to remain at one's post, to maintain one's position, to 'cope'.
c **1870** P. P. Bliss *Sacred Songs & Solos sung by Ira D. Sankey* (1874) 3 'Hold the fort, for I am coming,' Jesus signals still; Wave the answer back to heaven, 'By Thy grace we will.' **1928** D. H. Lawrence *Phoenix II* (1968) 601 Thirty-six years ago men..still believed in the fight for life and the fun of it. 'Hold the fort, for I am coming.' **1936** E. S. Gardner *Case Sleepwalker's Niece* x. 73 In the meantime I want to put in a telephone call. You hold the fort. **1941** A. L. Rowse *Tudor Cornwall* 341 Elizabeth and her archbishops..had held the fort until their church had come..to have an ethos of its own. **1960** J. Stroud *Shorn Lamb* viii. 107 I'm going out for an hour or so, can you hold the fort? **1962** J. Wain *Strike Father Dead* v. 243, I shan't need more than ten minutes, if you can hold the fort that long. **1966** *Times* 22 Sept. 11/5 Mr. George Brown's determination to hold the Foreign Office fort and to sally forth as little as possible.

fort, *v.* (Later U.S. examples.)
1834 D. Crockett *Narr. Life* vii. 52 The fort was built right in the middle of a large old field, and in it the people had been forted so long and so quietly, that they didn't apprehend any danger at all. **1838** *Southern Lit. Messenger* IV. 295 The settlers were sometimes forted. **1851** W. De Hass *Hist. W. Virginia* 315 Of those forted at Grave creek, was a William McIntosh, wife and child. **1853** B. Young in *Jrnl. Discourses* I. 165 Suppose we should say to all the wards in this city, the time has come to fort up. **1878** J. H. Beadle *Western Wilds* xii. 186 They lit on the Yorker and his friends and druv 'em back into Miller's Store, when they forted and held their own. **1905** G. E. Cole *Early Oregon* iv. 53 We started on and arriving at Grave creek, found the people 'forted up'. **1948** E. N. Dick *Dixie Frontier* 267 Castleman warned the pioneers to 'fort'.

forte, *sb.* Now often pronounced (fǭ·ɹti) or (fǭ·ɹte).

forte-piano. Add: **A.** *adj.* Also as *sb.*, such a succession of notes or chords.
1917 G. B. Shaw *How to become Mus. Critic* (1960) 294 Sir Thomas Beecham..seems to have no conception of the dynamic range of Mozart's effects, of the fierceness of his *fortepianos*.
B. *sb.* For def. read: An early name of the Pianoforte. Delete †. Add further examples.
1948 *Penguin Music Mag.* Oct. 25 How many domestic pianists would relish having to revert to the thin-toned *forte-piano* of Beethoven's day? **1961** *Proc. R. Mus. Assoc.* LXXXVII. 43 It is well known that J. C. Bach preferred the forte piano to the harpsichord. It is, indeed, the ideal instrument for accompanying.

Fortescue, fortescue (fǭ·ɹtėskiū). *Austral.* [perh. alteration of the pop. name *forty skewer* after the proper name *Fortescue*.] A scorpænid fish, *Centropogon australis*.
1874 E. S. Hill in J. E. Tenison-Woods *Fish N.S.W.* (1882) 49 The scorpion or Fortescue..bear that name no doubt in memory of the pain they have hitherto inflicted; and for its number and array of prickles it enjoys in this country the *alias* 'Forty skewer' or 'Fortescure'. **1882** J. E. Tenison-Woods *Ibid.*, The 'Fortescue'..is a common fish in Port Jackson, with a very long dagger-shaped preorbital spine. **1896** F. G. Aflalo *Nat. Hist. Austral.* 228 Fortescue (*Pentaroge*)..is a terrible pest, lurking among the *débris* in the nets and all but invisible. **1951** T. C. Roughley *Fish Austral.* 129 Most fishermen on the eastern Australian coast are familiar with..the fortesque (*Centropogon australis*). **1968** G. P. Whitley et al. *Animals of World: Austral.* 21/2 Several related fishes such as the..Fortescue (*Centropogon australis*) are less severe in their toxic effects.

forthcoming, *ppl. a.* **2.** Delete *rare* and add later examples. Add to def.: Also, informative, responsive.
1942 *Punch* 8 July 12/1 The War Minister..got into a brisk row with most of the House when he was not very forthcoming on the question of dive-bombers. **1971** J. Aiken *Nightly Deadshade* v. 54 'Some of your colleagues,' I say, 'are not so forthcoming as you seem to wish to be.'

forthcomingness. (Later example.)
1969 *Daily Tel.* (Colour Suppl.) 7 Feb. 30/1 The people have that unsophisticated forthcomingness tempered by a subtle irony.

forthtell, *v.* Add: So **forthte·ller.**
1920 *Glasgow Herald* 17 July 4 Mr Wells takes himself much more seriously as historian than as humorous novelist, or 'forthteller' of scientific development. **1964** G. A. Williamson *World of Josephus* iii. 63 A prophet was not primarily a foreteller, but a forthteller.

fortification. Add: **2. b.** The addition of nutrients, usually vitamins, to food.
1939 *Milbank Mem. Fund Q.* XVII. 241 (*heading*) The public health aspects of the fortification of foods with vitamins and minerals. **1941** *Lancet* 20 Dec. 773/1 Fortification must be considered solely in relation to the proven need..for a given nutrient. **1969** *New Scientist* 9 Nov. 328/3 Fortification of cereals and cereal flours with proteins offers the best immediate prospect for alleviating the world's shortage of protein foods.

fortified, *ppl. a.* Add: **2. a.** Of food: supplied with added nutrients.
1910 C. Watson *Food & Feeding* iv. 36 Milk fortified with starch. *Ibid.* viii. 139 A 'fortified cocoa' can be prepared as follows:—cocoa..milk sugar..water..cream. **1940** *Nature* 27 July 117/1 The white loaf would be fortified with a supplement of vitamin B1 (aneurin) and a calcium salt. **1970** *Guardian* 7 Aug. 3/8 Modern bakeries..have pushed out fortified bread to tickle the palate.
b. Of wines: strengthened with alcohol. Cf. Fortify *v.* 4.
c **1906** *Harmsworth Encycl.* VIII. 6320/2 Madeira is a sherry type of fortified wine. **1954** R. W. Schery *Plants for Man* xx. 525/2 Fortified wines are 'cocktail' or 'dessert' wines such as port, sherry, and vermouth, in which the alcoholic content is supplemented by the addition of brandy..to a strength unobtainable by simple fermentation. **1968** *Daily Tel.* (Colour Suppl.) 13 Dec. 41/1 So often in statistics no distinction is made between table wine, fortified wine, and aperitif wine.

fortify, *v.* Add: **4. b.** To add nutrients, usually vitamins, to (food).
1940 *Nature* 27 July 118/1 There would be two outstanding examples of fortified foodstuffs, bread and margarine. **1942** *Ann. Reg.* 1940 345 Recognition..of the need to fortify various products such as bread, margarine etc., with vitamins. **1942** *Endeavour* I. 30/1 1 gram of calcium is sufficient to fortify 4 oz. of margarine. **1969** *New Scientist* 9 Nov. 328/3 By 1969..all emergency shipments of grain will be fortified to the optimal extent either with amino acids or with protein concentrates.

Fortin (fǭ·ɹtin). The name of J. Fortin (1750–1831), French physicist, used *attrib.* or in the possessive to designate a type of mercury barometer invented by him.
1875 *Encycl. Brit.* III. 382/2 (*caption*) Fortin's Barometer. **1930** *Meteorol. Gloss.* (Met. Office) (ed. 2) 80 *Fortin barometer*, a portable form of mercurial barometer in which the zero of the scale is fixed by a pointer, inside the cistern. By making the cistern partly of leather and providing an adjusting screw, the level of mercury in the cistern can be brought up to the scale zero before each reading. **1970** *Encycl. Brit.* III. 181/1 Since the Fortin barometer is not a primary instrument, its instrumental errors must be determined by comparison with a standard barometer.

fortis. Add: **B.** *adj.* Of one of two or more homorganic consonants: strongly articulated. Opp. *lenis*. So also as *sb.* (pl. **-es**), such a consonant.
1909 Webster, Fortis *a.* **1932** G. K. Zipf *Sel. Stud. Rel. Freq. Lang.* i. 4 The fortis tenuis aspirata is decidedly more conspicuous than the fortis..tenuis. **1933** Bloomfield *Lang.* vi. 99 In English the unvoiced stops are aspirated fortes. iv. 32 A voiceless/voiced pair such as [s,z] are distinguished not only by the presence or absence of voice but also by the degree of breath and muscular effort involved in the articulation...The fortis consonants which are usually voiced tend to be articulated with relatively weak energy, whereas those which are always voiceless are relatively strong. Thus, it may be important to define [s]..as strong or *fortis* [s] and [z] as weak or lenis. **1964** Jakobson & Halle in D. Abercrombie et al. *Daniel Jones* 101 The distinctive mark in a fortis-lenis pair. **1965** *Language* XLI. 480 The stop is of greater duration and fortis articulation. **1969** *Jrnl. Lancs. Dial. Soc.* 14 The weakening of the fortis consonants in Jestetten was a phonetic change which threatened the stability of the whole consonant system.

fortissimo, *adv.* Add: **B.** *sb.* A very loud passage or point; *fig.* a high pitch of excitement.
1856 Geo. Eliot *Jrnl.* in *Lett.* (1954) II. 243 It was a crescendo of delight when we found a 'Strawberry', and a *fortissimo* when I..saw..an Anthea Cereus. **1885** G. B. Shaw *How to become Mus. Critic* (1960) 69 He [sc. Massenet] pours forth all his energy in a screeching, grinding, rasping *fortissimo* of extraordinary exuberance and vigor. **1928** *Observer* 5 Feb. 12/5 The hard and sudden fortissimos. **1928** *Daily Express* 17 Mar. 9/7 As this jazz harmony reached its fortissimo. **1931** E. Sackville-West *Simpson* III. xiii. Flinging a fortissimo of emotion—whether of anger or of delight—with sudden and deafening violence on to everything which twitched the edge of his soul. *a* **1963** C. S. Lewis *Poems* (1964) 54 Those who have seen no lions must revere A bull for Pan's *fortissimo*.

Fortran (fǭ·ɹtræn). *Computers.* Also **FORTRAN, fortran.** [f. *for(mula tran(slation).*] A high-level programming language used chiefly for scientific and mathematical calculations. Freq. *attrib.*
1956 *Computers & Automation* Nov. 9/2 More recently, John Backus' group at IBM has prepared FORTRAN (FORmula TRANslation) for the IBM-704 computer. FORTRAN will translate into computer language a program written very close [*sic*] the language of the mathematician or scientist. **1957** J. W. Backus et al. in *Proc. Western Joint Computer Conf.* Feb. 188 (*heading*) The FORTRAN automatic coding system. *Ibid.*, The programmer attended a one-day course on FORTRAN and..then programmed the job in four hours using 47 FORTRAN statements. **1966** *Listener* 15 Dec. 893/2 The technologist can always write his fortran and the literate composer his iambics. **1967** *Electronics* 6 Mar. 140/1 The Norden program features..Fortran language, which permits it to be used on any computer having a Fortran compiler. **1969** P. B. Jordain *Condensed Computer Encycl.* 223 The principal dialects today are FORTRAN II (now obsolescent) and FORTRAN IV.

fortress, *sb.* Add: **2. b.** *fortress-castle, -church, -city, -palace, -prison, -town.* Also in phrases of the type *Fortress America, Europe* [after G. *Festung Europa*].
1951 *N.Y. Herald Tribune* 25 Feb. 24 The central thesis..is the idea that fortress America could survive alone. **1960** *Times* 13 Jan. 12/7 American interest in civil defence shelters might be interpreted as implying American abandonment of the concept of 'Fortress America'. **1966** *Guardian* 22 Apr. 18/3 A retreat to a 'fortress America' concept. **1906** *Westm. Gaz.* 16 Aug. 2/3 No Imperial fortress-castle is strong enough to Germanise the vast plains where peer and peasant alike are still passionately Polish. **1963** *Times* 9 Feb. 11/2 Fortress-church. **1909** *Westm. Gaz.* 24 Feb. 1/3 The Byzantine fortress-city of Jericho. **1942** *Nation* 19 Dec. 683 It was the Italian press which first launched the idea of Fortress Europe. **1944** *Sat. Even. Post* 27 May 12/1 We shall still face the real problem of Fortress Europe, the penetration of the German mind. **1971** *Guardian* 22 July 4/1 The Prime Minister..said..'I have never seen a European policy as a policy of withdrawal into a fortress Europe.' **1905** *Westm. Gaz.* 12 Sept. 3/1 The gloomy fortress-palace. **1935** J. Thomas *No Banners* viii. 70 A fire-gutted convent which had been transformed under the Franco régime into a fortress-prison. **1937** *Discovery* Aug. 250/2 The fortress-town..enclosed an area of about 270 by 200 metres.

fortune, *sb.* Add: **4. d.** *a small fortune*: also used hyperbolically to designate any large sum of money.
1874 Hardy *Far fr. Madding Crowd* I. vi. 65 He played on with spirit, and in half an hour had earned in pence what was a small fortune to a destitute man. **1912** T. Dreiser *Financier* iv. 37 If he had been able to buy them all and dispose of them all as readily as he had his soap, he would have made a small fortune. **1951** E. Paul *Springtime in Paris* xi. 33 Gobelin tapestries on the walls and old French rugs on the floors, each one worth a small fortune. **1964** D. Mayo *Island of Sin* xi. 56 It's one of the finest known islands in the group, and Doreen pays a small fortune to keep it that way.
9. *fortune-favoured* adj.; *fortune-seeker*; **fortune cookie** *N. Amer.*, a dessert, freq. served in Chinese restaurants, made from a thin dough folded and cooked around a slip of paper bearing a prediction or maxim.
1962 K. Orvis *Damned & Destroyed* xx. 149, I didn't need to break a fortune-cookie to know I was a loser. **1970** *Toronto Daily Star* 24 Sept. 6/3 Chinese fortune cookies. **1898** *Daily News* 6 Oct. 6/7 The more fortune-favoured individuals. **1919** W. J. Locke *Jaffery* i. He had put poor old Jaffery and fortune-favoured me in the shade. **1906** *Daily Chron.* 5 May 4/7 A vast throng of fortune-seekers.

forty, *a.* and *sb.* Add: **A.** *adj.* **c.** *the forty hours* (also qualifying *devotion*, etc.; It. *le quarant'ore*): in the R.C. Church, the continuous exposition of the Host for forty hours, used as an occasion of special devotion or intercession.
1759 A. Butler *Lives Saints* IV. 560 The saint..ordered the forty hours prayer for his recovery. **1849** R. H. Dyer *Moves Cath.* I X. iii. 70 The devotion of the forty hours prayer instituted by a poor Capuchin friar, Joseph of Milan. **1869** A. T. Drane *Life Mother Margaret Hallahan* vii. 180 During the Octave of Corpus Christi this year the Devotion of the Forty Hours was for the first time celebrated in St. Catherine's Convent. **1922** *Cath. Encycl. Suppl.* 19/1 The Forty Hours' Adoration. **1967** *New Cath. Encycl.* VI. 1006/1 Forty hours devotion. A continuous period of prayer before the Blessed Sacrament, begun and terminated with a solemn High Mass, processions where possible, Litany of the Saints, and special prayers. *Ibid.*, Forty Hours has come to be a devotion simply honoring the Blessed Sacrament rather than a means of making reparation or of petitioning for peace.
B. *sb.* **5.** One fourth of a quarter section of land, comprising forty acres. *U.S.*
1845 C. M. Kirkland *Western Clearings* 2 Eighties and forties..are plain enough when one is habituated to them. **1873** E. Eggleston *Myst. Metrop.* i. 19 It was just so many quarter sections, 'eighties', and 'forties' to be bought low. **1902** S. E. White *Blazed Trail* i. 13 The men who were to fell the trees, Radway distributed along the boundary of a 'forty'. **1913** G. Stratton-Porter *Laddie* (1917) xiv. 276, I had thought we would commence on the east forty when planning the work [of ploughing]. **1947** *Pacific Discovery* Jan.–Feb. 5/1 This was wilderness, as distinct from the back forty.
6. A period of forty minutes' play.
1913 *Field* 25 Oct. 904/2 Light forwards are bound to be worn down in two 'forties' by heavier.
7. A 'crook', thief, sharper (in quot. 1879, a convict). *Austral. slang.*
1879 Mrs. C. Cook *Comic Hist. N.S.W.*, Fifteen of the 'Forties' became free of the colony this June. **1882** *Sydney Slang Dict.* 8/2 *The Forties*, the worst types of 'the talent' who get up rows in a mob,..and sometimes assault and rob, either in barrooms or the streets. **1885** *Australasian Printers' Keepsake* 116 Ah, them were jolly days indeed, Long ere the Vandemonian swarm had come,

..Or ere the 'Forty' had capsized our trade. **1927** M. M. Bennett *Christison* xxii. 194 Their numbers swelled with rowdies and 'forties'—gambling sharpers who travelled from shed to shed making five pounds by cheating for every five shillings they earned.
C. 1. forty-acre *U.S.* and *N.Z.*, a section of land comprising forty acres (cf. *B. 5); **forty-footer,** a forty-foot yacht; **forty-rod whisky** (earlier, *ellipt.*, and N.Z. examples); **forty skewer**: see *Fortescue.
1742 *New Hampshire Probate Records* III. 94, I give to my Grand Son..one Forty Acre Lot. **1860** in A. F. Ridgway *Voices from Auckland* 48 The Forty-acre men will ruin the country. **1869** J. May *Guide to Farming in N.Z.* 42 We were lately on the sections of two 'forty-acre men'. **1877** C. B. George *40 Yrs. on Rail* xi. 227, 'I live just over there', pointing to his house across a forty-acre lot. **1943** C. Crow *Gt. Amer. Customer* 185 There was no reason why a farmer could not plant a whole forty-acre field in wheat. **1902** *Essex Inst. Hist. Coll.* XXXVIII. 256 The schooner-yacht 'Excelsior' was one of the earliest of the 'forty-footers'. **1863** W. H. Russell *My Diary North & South* II. 11 Their cries for water were incessant to allay the internal fires caused by '40 rod' and '60 rod'. **1869** 'Mark Twain' *Sks. New & Old* (1875) 70 Trading for forty-rod whiskey..has played the everlasting mischief. **1873** J. H. Beadle *Undevel. West* xiii, The standard drink is whisky—'stone fence', 'forty-rod', and 'tarantula-juice'. **1916** 'Anzac' *On Anzac Trail* v. 77 Shebangs [in Cairo] where they sell you whisky that takes the lining of your throat down with it..a soothing liquid that licks 'forty-rod', 'chained lightning', or 'Cape smoke' to the back of creation. **1919** Forty-rod [see *boot-legging]. **1948** *Daily Oklahoman* (Okla. City) 7 June 8/1 The mere possession of a few gills of forty rod is not counted as an ample offset to planned assassination.
2. forty-eight, (*b*) *pl.* a sheet of a book folded into forty-eight leaves; (*c*) the forty-eight preludes and fugues of J. S. Bach; **forty-eightmo,** the size of a book in forty-eights.
1808 C. Stower *Printers' Gram.* 192 A Half Sheet of Forty-eights, with Two Signatures. **1839** T. C. Hansard *Print. & Type-Founding* 168 Forty-eights to be paid 2s. per sheet extra. **1873** H. C. Banister *Music* (1889) III. xxvi. 206 Bach's Fugue in C♯ Minor, No. 4 of the 48. **1888** C. T. Jacobi *Printers' Vocab.*, *Forty-eightmo*, a sheet of paper folded into forty-eight leaves. **1902** *Westm. Gaz.* 29 Dec. 1/3 Those famous fugues and preludes known to all musicians as the 'Forty-Eight'. **1965** G. Hughes *Handbk. Gt. Composers* vii. 45 At times the..counterpoint seems to spring from the interweaving of rhythmic figures rather than of melodic phrases—as in the prelude in F sharp major from the second book of the 'forty-eight'.
b. forty-niner (earlier example).
1853 *Mt. Echo* (Downieville, Calif.) 12 Feb. 1/1 Speeches were made by some of the worthy old forty-nin-ers.

forty-five. Add: **c.** A revolver of ·45 calibre. *U.S.*
1881 G. W. Romspert *Western Echo* 285 We now used the six-shooters,—known as forty-fives,—and which were also deadly weapons. **1926** J. Black *You can't Win* iii. 21 He owned the latest pattern of repeating rifle and a couple of 'forty-fives'.
d. A 45 r.p.m. 7-inch microgroove gramophone record; also written **45**.
[**1949** *Gramophone* Nov. 108/2 Single, double and treble speed motors fitted with heavy non-magnetic turntables, suitable for high fidelity reproduction of 78, 45 and 33⅓ r.p.m. records.] **1950** *Ibid.* Dec. 162/2, I have had an opportunity of hearing the Rachmaninov No. 2 and the Tchaikovsky No. 1 Pianoforte Concerto on these 45 discs. **1968** R. Clapperton *No News on Monday* vii. 82, I did a forty-five for Sunny Coast Records once that had as much impact as a handful of sand.

forward, *a.*, *adv.*, and *sb.* Add: **A.** *adj.* **3. d.** *forward batsman, block, cut, drive, glance, lunge, play* (earlier example), *player, point, stroke*. Cf. *backward *a.* 1.
1876 *Haygarth's Cricket Scores & Biogr.* 1855–75 v. 38 A steady, forward batsman. **1849** *Boy's Own Bk.* 85 Smother it by the forward block. **1851** J. Pycroft *Cricket Field* vii. 145 The Forward Cut..sends the ball between point and middle wicket. **1836** E. Jesse *Angler's Rambles* 301 He would draw diagrams of the angles of incidence and reflexion: shew..the exact length which he could cover in a forward drive. **1903** P. F. Warner in H. G. Hutchinson *Cricket* iii. 64 The back glance and the forward glance have practically taken the place of the leg hit. **1861** 'Old Cricketer' *Cricket-Bat* 66 The forward cut, also called the forward lounge, is an off hit. **1828** G. T. Knight in W. Denison *Cricket: Sk. Players* (1846) 41 The method then [*c* 1804] introduced was running in at the ball, hard hitting, and a bold forward play. **1851** J. Pycroft *Cricket Field* vii. 126 The best forward players may err. **1870** Forward player [see *backward *a.* 1]. **1851** J. Pycroft *Cricket Field* xi. 222 With a fine forward player, a near Middle-wicket or forward point often snaps up a catch. **1897** Ranjitsinhji in *Daily News* 10 Aug. 6/1 In order to make a forward-stroke effectively aggressive, the batsman must throw himself into the correct attitude. **1906** A. E. Knight *Compl. Cricketer* ii. 66 A forward stroke..generally known in our time as the 'Push stroke'.
4. Also of delivery, exchange, and material.
1930 M. Clark *Home Trade* 107 'Forward delivery' [means] that the goods will be delivered..at a future date. **1930** [see *forward *adv.* 1 c]. **1957** *Economist* 21 Dec. 1076/2 If they could offset that reduction cheaply by larger purchases of forward rubber they might do so, but rubber three months forward commands a premium of about ½d. per lb. **1962** *Ibid.* 28 Apr. 367/2 A forward exchange operation is a contract to exchange one currency for another at a future date at a rate of exchange fixed now.
B. *adv.* **1. c.** (Additional example.)
1930 *Economist* 23 Aug. 350/1 It by no means follows that the policy of buying forward will always operate to the buyer's advantage, for unless a foreign buyer simultaneously covers himself by a corresponding forward exchange operation, he may find that an adverse movement of the exchange has destroyed his profit on the shipment.
3. b. (Further examples of *Cricket* usage.)
1851 J. Pycroft *Cricket Field* x. 191 A good ball..is pushed forward to middle wicket. *Ibid.*, Shortleg is standing sharp in forward.
5. *to put forward* (earlier *refl.* example).
1836 Newman *Parochial Sermons* (ed. 2) II. xxiv. 325 Now, it is plain that there are two sorts of men in the world;—those who put themselves forward, and speak much; and those who retire.
11. a. *forward-looking* (later examples). **c.** *forward-looker.*
1919 Mencken *Amer. Lang.* ix. 302 The forward-looker, whose belief in the continuity..of the evolutionary process takes on the virulence of a religious faith. **1923** [see *Babbitt]. **1933** P. Godfrey *Back-Stage* xiii. 164 It is this vanguard which makes the studio or art theatre possible. These serious people are prompted by feelings of responsibility towards the rest of humanity. They are the forward-lookers. **1843** H. D. Thoreau *Let.* 8 June in R. B. Perry *Tht. & Char. W. James* (1935) I. 48 He [sc. H. James, sen.] is a refreshing, forward-looking and forward-moving man. **1922** S. Lewis *Babbitt* xvi. 209 Weeklies and monthlies..as practical and forward-looking, as the real-estate columns. **1928** F. Hurst *President is Born* xxii. 235 A man born into his time as snugly as a bug into his rug. Forward-looking. Backward-heeding. Un-neurotic. Normal. **1962** *BSI News* June 25/1 A forward-looking attitude prevails.
C. *sb.* **4.** Also in various other games.
1895 J. M. Brown *Polo* 78 The two 'forwards', No. 1 and No. 2, should work together, and, if possible, be exactly the same class of player. **1897** *Encycl. Sport* I. 517/2 All the forwards should make a point of taking passes with their sticks. **1929** *Encycl. Brit.* III. 182/2 The duty of the forwards [in basket-ball] is primarily to make goals. **1956** *Men's Hockey* (Know the Game Series) (rev. ed.) 17/2 At long corners the defence come out in two waves, the forwards in the first wave each making for his attacking number. **1968** Eagleson & McKie *Terminol. Austral. Nat. Football* II. 7 *Forward*, a player occupying a forward position.

forwardal (fǭɹwǭ·ɹdăl). [f. Forward *v.* + -al.] = Forwarding *vbl. sb.*
1911 *Supt. Govt. Printing India* Form No. 283 Books, The book..is available, price Rs. — inclusive of forwardal charges.

forwarder (fǭ·ɹwəɹdəɹ), *a.* and *adv.* [compar. of Forward *a.* and *adv.*] Further forward. Often in jocular colloq. use in the (orig. dial. or vulgar) form **forra(r)der** (fǫ·rədəɹ), in the phr. (*to get*) *no* or *any forra(r)der*.
1796 F. Burney *Camilla* x. vii, She ventured to bend forwarder, to take a view of the side wings. **1813** Jane Austen *Pride & Prej.* III. xiii, Kitty is forwarder than either of us. **1814** —— *Mansf. Park* I. xviii, Without his being much the forwarder. **1860** H. Spencer *First Princ.* (1863) ii. 33 And so on in an infinite series, leaving us at last no forwarder than at first. **1888** 'R. Boldrewood' *Robbery under Arms* II. vii. 143 Every week..we found ourselves a lot forrader. **1898** *Daily Tel.* 15 Dec. (Ware), Whether the Liberal Forwards will get any 'forrader' over the light claret [etc.]. **1918** W. Owen *Let.* 6 Mar. (1967) 537 Am no forrader with my Chest of Drawers. The man won't sell as it is, & says he has no time yet to work on it. **1966** *Times* 10 Oct. 1/2 The case of Gerald Brooke, the British lecturer sentenced in Russia, was discussed but, in the words of a British source, 'We got no forrarder.'

forwarding, *vbl. sb.* Add: **3.** *forwarding agency, establishment, house, mechanism, yard.*
1837 W. Jenkins *Ohio Gaz.* 126 It has..ten heavy forwarding houses, connected with lake and canal transportation. **1847** H. Howe *Hist. Coll. Ohio* 43 The harbor of Ashtabula..has several forwarding establishments. *Ibid.* 331 Toledo..has..9 forwarding and commission houses. **1897** *Outing* (U.S.) XXX. 197/2 The tourist should look after his own mount..and he should not send it by an independent forwarding agency. **1950** *N.Z. Jrnl. Agric.* July 7/3 The receiving yards [for sheep] usually communicate with the forwarding yards or pens. **1968** *Gloss. Terms Offset Lithogr. Printing* (*B.S.I.*) 34 *Forwarding mechanism*, a device on a machine or feeder by which the top sheet is advanced in its normal direction of travel.

forward scattering. a. *Physics.* Scattering in which the direction of the scattered radiation or particles makes an angle of less than ninety degrees with the original direction. **b.** *Telecommunications.* Scattering or reflection of high-frequency radio waves by irregularities in the troposphere or ionosphere so that some of their energy returns to earth beyond the horizon of the point of transmission. Hence (as a back-formation) **forward-scatter** *v. trans.*; also **forward-scattered** *ppl. a.*
1950 *Proc. IRE* XXXVIII. 406/1 This formula for forward scattering applies no matter what may be the relation of the scale of turbulence to the wavelength. **1952** *Physical Rev.* LXXXVI. 1 (*heading*) Forward scattering of light by a Coulomb field. **1953** *Jrnl. Geophysical Res.* LVIII. 90 The amplitude of the forward-scattered signal exceeds that of an equivalent back-scattered signal. **1954** *Proc. IRE* XLII. 530/1 Echoes forward-scattered from individual meteor ionization trails. **1967** Rodberg & Thaler *Introd. Quantum Theory of Scattering* xii. 370 For a large target of uniform particle density..only forward scattering will occur. **1968** W. G. Burrows *V.H.F. Radio Wave Propag.* 6 The propagation of a VHF wave beyond the radio horizon by forward scattering.
So **forward(-)scatter** *sb.*, (*a*) forward scattering (sense *b); (*b*) radio waves that have returned to earth after being scattered in this way. Also *attrib.*
1954 *Canad. Jrnl. Physics* XXXII. 450 Reflections from meteoric trails can support, intermittently, high frequency radio transmissions well beyond the line-of-sight.. The Stamford workers have developed a theory of this forward-scatter transmission. **1956** *Ann. Reg.* 1955 409 Waves reflected from the E-layer..and sent on by 'forward scatter'. **1958** *Times* 17 Jan. 6/6 Experiments are to be carried out in Britain to ascertain whether television pictures can be transmitted by the 'forward scatter' technique over distances of 200 miles or more. **1960** *Gloss. Terms Telecomm.* (*B.S.I.*) 131 A signal received by forward scattering is often from tropospheric..

fossa², foussa (fǫ·să, fū·să). Also **fosa.** [Malagasy: see Fossane.] **a.** A mammal (*Cryptoprocta ferox*), related to both cats and civets, which is the largest carnivore found in Madagascar. **b.** *Fossa fossa* (popular names Fossane, *foussa*, and *fanaloka*), a monotypic genus of civets also found in Madagascar.
1838 W. Ellis *Hist. Madagascar* I. ii. 39 Animals of the badger kind, found..in the Sakalava country; it is called fosa. **1885** *Encycl. Brit.* XV. 435/2 *C. ferox*, the 'foussa' of the Malagasy. **1887** Chisholm tr. *Vogt & Specht's Nat. Hist. Anim.* I. 184 The Fossa of the Malagasy..Cryptoprocta ferox. **1916** *Ann. & Mag. Nat. Hist.* 8th Ser. XVII. 413 The Fossa (*Cryptoprocta*) has probably received more anatomical attention than any single genus of Carnivora. **1954** F. Harper *Extinct & Vanishing Mammals of Old World* 249 Fossane..is presumably a diminutive of fossa and is used in English as a translation of it. **1964** E. P. Walker et al. *Mammals of World* II. 1262/1 Fossas walk in a flat-footed manner on the soles of their feet. *Ibid.* 1238/2 Falanoucs, Fossanes, Malagasy Civets... The fact that the generic name of this animal [*Fossa*] is the same as the vernacular name of another Madagascar mammal, the fossa (*Cryptoprocta*) should not lead to confusion, as they are distinctly different animals.

fosse. Add: **2. b.** In full *fosse-road*. = *Fosse-way.
1130–35—*a* **1450** [in M.E.D.]. **1882** *Encycl. Brit.* XIV. 424/2 The Fosse road from Lincolnshire..passes by Leicester to Watling Street. **1906** The Via Devana..joins the Fosse at Leicester on its way from Colchester. **1937** Collingwood & Myres *Roman Brit.* (ed. 2) vi. 91 The Fosse begins at Seaton..on the Devonshire coast. **1965** C. Cochrane *Lost Roads of Wessex* i. 14 No Roman way has been listed as located between Salisbury and the southern part of the Fosse. *Ibid.* vi. 119 For a short distance the Roman quarrystone paving of the Foss was intact.

Fosse-way, Fosse Way (fǫ·swēi). *Hist.* Also **Foss Way, Foss-way.** [f. Fosse 2 + Way *sb.*¹] One of the four great Roman roads in Britain, so called from the ditch or fosse on each side; it probably ran from Axminster to Lincoln, via Bath and Leicester.
1610 Holland *Camden's Brit.* i. 562 Thus much of the Feldon, or champion part, which that ancient Fosse-way (a thing that would not be overpassed) cutteth overthwart. *c* **1703** C. Fiennes *Journeys* (1947) iv. vi. 33, I came..to High Crosse which is esteemed the middle of England, where the two great Roads meete that divides the kingdom in the Saxons tyme in 4 parts, the Whatling Streete.., and the Fosse Way. **1797** *Encycl. Brit.* V. 26/1 The fosse-way, which comes from Scotland, passes through this county and town to Totness in Devonshire. **1838** *Penny Cycl.* X. 381/1 From Ilchester the Fosse Way continued its course in a pretty direct line to Ischalis (Ilchester), of which town it forms the principal street. **1876** *Encycl. Brit.* V. 792/2 The junction of the Fosse Way with the Ermin and Icknield streets. **1937** Collingwood & Myres *Roman Brit.* (ed. 2) iv. 94 The Fosse Way shows that..the south-east of Britain as far as the Fosse Way had passed by 442 under Saxon dominion, there is other fifth-century evidence which is hard to reconcile with such an idea. **1965** Timperley & Brill *Anc. Trackways of Wessex* xxii. 131 As the trackway came near to the Avon it may have been..joined by the Fosse Way for the rest of its journey to join up with the Cotswold Ridgeway.

fossick, *v.* Add: **1.** *to fossick about* (earlier N.Z. example).
1874 A. Bathgate *Colonial Experiences* xi. 138 [These unsocial miners], so long as they can make a living by 'fossicking' about, are happy.
2. (Additional examples.)
1938 F. S. Anthony in D. M. Davin *N.Z. Short Stories* (1953) 213 Why not bring them cows up there and let them fossick for tucker in the bush. **1956** E. Hiscock *Around World in Wanderer III* xiii. 180 A place where people go to fossick, i.e. to gather shells and coral at low water or on the reef. **1968** *Times* 11 Sept. 11/5 Unpaved backstreets..where old crones fossicked.
3. (Earlier example.)
1857 C. R. Thatcher *Goldrush Songster* (1958) 77 Next morning I well fossicked it.

fossicking, *vbl. sb.* and *ppl. a.* (Earlier and later examples.)

1853 E. Clacy *Lady's Visit Gold Diggings Austral.* vii. 123 William was the one whose fossicking-knife first brought its hidden beauties to light. **1916** J. B. Cooper *Coo-oo-ee* xiv. 213 All the same, I don't think Hawley got much colour for all his 'fossickin'. **1936** F. Clune *Roaming round Darling* xxv. 273 Gorrie..had given up cooking and taken to fossicking, as a mate of his found a gold nugget on Christmas Eve worth £130. **1970** K. Giles *Death in Church* iv. 77 No suspicion of illegal fossicking or buying.

fossil, *a.* and *sb.* Add: **A.** *adj.* **1. c.** *fossil fuel*: combustible material obtained from below ground and formed during the geological past; now esp. contrasted with sources of nuclear energy. Also *attrib.*
1835 [J. Holland] (*title*) The history and description of fossil fuel, the collieries and coal trade of Great Britain. **1854** [see Fossil *a.* and *sb.* A. 1]. **1921** *Sci. Amer.* 14 May 292/3 The early exhaustion of our fossil fuels will require the use of such other sources of power as water, wind, and sun. **1949** *Science* 4 Feb. 107/2 In view of the eventual exhaustion of fossil fuels, it is of interest to know to what extent water power can be depended upon to replace them. **1952** Ayres & Scarlott *Energy Sources* i. 4 The combustion of fossil fuels or the accelerated nuclear fission of elements. **1955** *Times* 9 Aug. 8/2 This would exhaust the known reserves of coal, oil and natural gas (the so-called fossil fuels) within less than a century. **1969** R. B. Fuller *Operating Man. Spaceship Earth* viii. 122 Physical energy..is ever increasingly deposited as a fossil-fuel savings account aboard our Spaceship Earth through photosynthesis and..topsoil fossilization.
Hence *fossil-fuelled* adj.
1964 J. B. McClure in *Trans. World Power Conf.* (Lausanne) Ia. 326 The early application of higher efficiency in fossil fueled processes is limited by declining fuel costs..and high capital costs. **1967** *Technology Week* 23 Jan. 34/2 The nuclear power plants currently being constructed will be competitive with fossil-fueled plants. **1970** *Times* 10 Apr. 31/5 The bugbear of short range with existing battery electric systems could be overcome in the short term..by offering a small fossil-fuelled engine alternator.

fossildom (fǫ·sĭldəm). [f. Fossil *sb.* + -dom.] The condition or character of being a fossil or a lifeless piece of antiquity.
1898 *Daily News* 3 Dec. 5/7 Spurious vulgar fossildom secretly urges members to oppose this non-party measure. **1905** J. Rickaby *Development* 48 Protestantism, even Anglican Protestantism, is essentially fossildom.

fossor. Add: **2.** [ad. mod.L. *Fossores* (P. A. Latreille 1817, in *Nouveau Dictionnaire d'Histoire Naturelle* X. 287), name of a group of Hymenoptera.] One of the burrowing Hymenoptera, the digger-wasps, formerly described as Fossores.
1899 *Cambr. Nat. Hist.* VI. 95 The apterous condition of the females of the Multillides and Thynnides is very anomalous in the Fossors. **1919** *Jrnl. E. Afr. Nat. Hist. Soc.* XV. 440 The Fossors are not true wasps. *Ibid.* 441 Each Fossor lives entirely an independent life.

fossorial, *a.* and *sb.* Add: **A.** *adj.* **1.** (Earlier and later examples of *fossorial Hymenoptera*.) Cf. prec.
1837 W. E. Shuckard (*title*) Essay on the indigenous fossorial Hymenoptera. **1880** *Notes from Leyden Museum* II. 225 (*title*) On two new exotic species of fossorial Hymenoptera. **1935** *Bull. Soc. Sci. Nat. Maroc* XV. 1 (*title*) On some fossorial Hymenoptera from the Soudan.

foster-mother. Add: **2.** An apparatus for rearing chickens hatched in an incubator.
1907 *Yesterday's Shopping* (1969) p. xxxv/4 Foster mothers. **1911** J. W. Hurst *Successful Incubation* 78 A cheap form of ordinary lamp should never be used, as is sometimes attempted..by makers of home-made brooders and foster-mothers. **1914** *Poultry* 2 Jan. 1042 When my Chicks are in your Foster-Mothers I can rest comfortably at night, and so can the Chicks. **1960** *Farmer & Stockbreeder* 19 Jan. Suppl. 37/3 There is still much to be said for the old foster-mother as this not only provides a brooder but a house as well.

Foucault (fuko, fū·kōᵘ). The name of J. B. L. *Foucault* (1819–1868), French physicist, used *attrib.* and in the possessive to designate experiments or discoveries made, or apparatus designed, by him; as *Foucault current*, an eddy current; *Foucault* (*knife-edge*) *method, test*, a method of testing lenses and mirrors for surface irregularities; *Foucault('s) pendulum*, a pendulum designed to demonstrate the rotation of the earth by the rotation of its plane of oscillation; *Foucault prism*, a polarizing prism resembling a Nicol prism but having the two halves separated by a film of air.
1852 *Phil. Mag.* 4th Ser. III. 331 (*heading*) A mathematical theory of M. Foucault's pendulum experiment. **1882** L. Wright *Light* xi. 251 A large Foucault prism is only about one-fourth the cost of a Nicol of the same aperture. **1883** Atkinson tr. *Ganot's Physics* (ed. 11) 869 The currents..produced in solid conductors and which are converted into heat, are often spoken of as Foucault currents. **1923** Glazebrook *Dict. Appl. Physics* IV. 329/1 For the investigation of spherical aberration errors Foucault's original knife-edge method is still frequently employed. **1930** L. C. Martin *Introd. Appl. Optics* I. 204 In case a prism is wanted for purposes of ultra-violet polarimetry, the Foucault type..is employed. *Ibid.* 257 (*caption*) The Foucault knife-edge test. **1931** *Discovery* May 164/1 A knowledge of the rotation of our world might be gained by such means as the Foucault pendulum or the gyro-compass. **1952** F. Twyman *Prism & Lens Making* (ed. 2) xi. 373 Correcting mirrors by local polishing with the help of the Foucault test. **1953** P. Kemp *A.C. Electr. Engin.* (ed. 8) x. 96 These e.m.f.s..set up local circulating currents known as Foucault or Eddy Currents and give rise to a loss of power. **1966** *McGraw-Hill Encycl. Sci. & Technol.* IX. 618 A Foucault pendulum is a spherical pendulum suspended so that its plane of oscillation is free to rotate.

‖ **fouetté** (fuete), *sb. Ballet.* [Fr., pa. pple. of *fouetter* to whip.] A step in which the dancer stands on one point and executes a rapid sideways 'whipping' movement with the free leg, freq. turning on the point at the same time.
1830 R. Barton tr. *Blasis's Code of Terpsichore* II. 102 Afterwards come the..attitudes..the *grands fouettés*, facing and revolving. **1922** Beaumont & Idzikowski *Man. Class. Theatr. Dancing* II. iv. 80 Execute a *Fouetté à terre en dedans*. **1932** A. Haskell *Ballet in Eng.* iii. 20 It is straightforward, strong, helpful music, a powerful aid in the famous (notorious?) 32 fouettés. **1947** *Ballet Ann.* I. 96 In the early classical ballets where we have the repetitive execution of steps (the *fouetté* is a good example), the pace will vary with the dancer. **1959** *Times* 2 Mar. 12/5 Technically, apart from a miscalculation in the *fouettés* of the third act, her dancing was smooth.

fougade. Delete † and add later example.
1904 *Daily Chron.* 4 June 5/5 The explosion of a number of mines and fougades.

foul, *a.*, *adv.*, and *sb.* Add: **A.** *adj.* **1. b.** *foul brood* (earlier example). Hence *foul-broody* adj., infected with foul brood.
1863 *Jrnl. Hort.* 28 July 78/1 It..occurred to me that all the mischief might possibly arise from that fatal scourge of continental and American apiaries denominated 'foul brood'. **1875** J. Hunter *Man. Bee-keeping* 193 Twenty stocks..were foul-broody, and I lost them all.
3. *foul case* (see quots.); *foul papers*, a draft or working manuscript, as opposed to a fair copy.
c **1625** in *Library* (1925) 4th Ser. VI. 152 The Booke where by it was first Acted from is lost: and this hath beene transcribed from the fowle papers of the Authors wch were found. **1659** [in Dict.]. **1729** J. Gay *Polly* Pref. p. iii, I take this occasion in the most solemn manner to affirm, that the very copy I delivered to Mr. Rich was written in my own hand some months before at the Bath from my own first foul blotted papers. **1872** W. Blades *Shaks. & Typogr.* 74 The compositor works with two of these cases slanting up in front of him, and, when from a shake, a slip, or any other accident, the letters become misplaced, the result is technically known as a 'foul case'. **1955** *Essays & Studies* VIII. 3 Investigators have worked out theories about how we can recognize 'foul papers' ('author's manuscript in its last stage before the making of a fair copy', to quote a recent definition). **1959** F. Bowers *Textual & Lit. Crit.* iii. 75 Formerly, both quartos were thought to be printed from Shakespeare's foul papers. **1963** Kenneison & Spilman *Dict. Printing* 73 *Foul case*, a case of type in which the separate pieces of type have been distributed into the wrong compartments.
13. b. Revolting, disgusting. *slang.*
1911 D. Coke *Wilson's* vi. 59, I was stopping somebody making a foul row. **1930** A. P. Herbert *Water Gipsies* xxi, 'Hullo, Gordon! My dear, it's a *foul* night. Arctic!' She shivered charmingly. **1930** *Punch* 20 Aug. 200/1 'How foul!' I said..'How dare the creature long to meet me!' **1963** 'J. le Carré' *Spy who came in fr. Cold* xxv. 212 Mundt won. London won—that's the point. It was a foul, foul operation. But it's paid off, and that's the only rule.
14. (Earlier *Austral.* example of *foul riding*.)
1882 *Sydney Slang Dict.* 4/1 *Foul-riding*, boring a competitor against the rails.
B. *sb.* **3. b.** *foul-up*: a state of muddle or confusion. orig. *U.S.*
1953 'S. Ransome' *Drag Dark* (1954) vii. 69, I couldn't judge how far I could rely on you to find the way out of this foul-up. **1958** 'P. Bryant' *Two Hours to Doom* 30 Second weapon to be used if there's any foul up with the first. **1967** *Observer* 14 May 2/8 He traces the foul-up back to 1953.
C. *adv.* **6. foul-hooked** *a.*, of a fish, hooked anywhere on the body except in the mouth.
1898 *Westm. Gaz.* 6 Oct. 4/2 Without doubt this fish was foul-hooked, though it was surely a heavy one. **1959** *Cape Times* 9 Mar. 3/6 Galjoen, haarders, and white steenbras, most of which were foul hooked, were landed. **1971** *Country Life* 11 Mar. 533/3 If there is a fish in the run, casts subsequent to the first will tend to fall across it and disturb it, if indeed it is not pricked or foul-hooked.

foul, *v.*¹ Add: **1. b.** Also with *up*.
1922 D. H. Lawrence *England, my England* 238 He could smell the cold, rotten clay that fouled up into the water. **1960** E. L. Delmar-Morgan *Cruising Yacht Equip. & Navig.* xii. 146 Tendency of a two-stroke to 'oil up' or to 'foul up'.
3. d. Also with *up*: to spoil, (cause) to bungle or muddle (something or someone). Chiefly *U.S.*
1947 *Harper's Mag.* May 425/1 'Wing Ding' has a tricky opening where the sax and trumpet play against each other, accenting different beats...Sparrow came in a beat late, fouling it up. **1950** 'S. Ransome' *Deadly Miss Ashley* xii. 143 It would foul us up for fair this time. **1951** J. D. Salinger *Catcher in Rye* ix. 80 Boy, I *really* fouled that up. **1954** J. & W. Hawkins *Death Watch* (1959) ii. 55, I got all fouled up in the chain of command. **1958** S. Ellin *Eighth Circle* (1959) II. ix. 100 You've got fine bone structure, but look at the way you're fouling it up. **1971** N. Freeling *Over High Side* III. 201 Everybody still has their mouth far too close to the utterly fouled-up microphone.
5. b. Also *fig.* with *up*. Chiefly *U.S.*
1957 P. Frank *Seven Days to Never* (1959) viii. iv. 184 Somewhere in his career he had fouled up. **1958** 'P. Bryant' *Two Hours to Doom* 16 A man marked for high rank..if he did not foul up on the way.

foul-mouth, *sb.* Delete †*Obs.* and add later example.
1966 Auden *About House* 39 A foul-mouth gets the cold Shoulder.

found, *ppl. a.* Add: **3.** *absol.* in *pl.* Advertisements of found articles, usually in phr. *losts and founds*.
1913 G. B. Dibblee *Newspaper* 126 The most important groups of classified advertising are as follows: financial, theatrical, public notices, losts and founds, educational.

foundation. Add: **6. c.** *pl.* [tr. G. *grundlagen*, etc.] The underlying principles or logical basis (of a subject), esp. as a separate matter for study.
1817 Coleridge *Biog. Lit.* x. 194, I retired to a cottage in Somersetshire..and devoted my thoughts and studies to the foundations of religions and morals. **1897** B. Russell (*title*) An essay on the foundations of geometry. **1931** F. P. Ramsey (*title*) The foundations of mathematics. **1941** Courant & Robbins *What is Math.?* ii. 87 Paradoxes like this have led Russell and others to a systematic study of the foundations of mathematics and logic. **1944** H. Reichenbach (*title*) Philosophic foundations of quantum mechanics. **1965** A. S. & E. H. Luchins *Logical Found. Math.* xviii. 377 Despite the popularity of the term *foundations*, there are wide differences of opinion as to the importance of foundations for psychology.
7. (Earlier examples.) **b.** *attrib.* **c.** = **foundation garment*.
1845 [see *body *sb.* 7 b]. **1865** M. Eyre *Lady's Walks* xvii. 195 The materials to be used are..neatly tacked on the silk—any foundation could..be used; merino, cloth, or whatever material was thought best suited for curtains. **1882** [in Dict.]. **1907** *Westm. Gaz.* 9 Mar. 13/2 Such a coat is..mounted on a foundation silk. **1939** P. J. Marquand *Wickford Point* 370 She said that she had heard that I worked in a department store and she wanted to know about foundations. **1960** [see *backless *a.* b].
d. A base for cosmetics. Used esp. *attrib.*, as *foundation cream, lotion, make-up*.
[**1911** G. Stratton-Porter *Harvester* xiii. 279 She found a box of cream and rubbed it on for a foundation. Then she opened some pink powder, and carefully dusted her cheeks.] **1929** D. L. Moore *Pandora's Let. Box* x. 179 A thick layer of powder..may be spread on a foundation cream having almost the consistency of paste. **1935** 'E. Ann' *Beauty Adorned* x. 86 A greaseless foundation-cream as a base. **1939–40** *Army & Navy Stores Catal.* 434/3 Foundation Cream—3/6. Foundation Lotion—4/6. **1942** D. Powell *Time to be Born* (1943) x. 227 Miss Finkelstein..checked on her Chinese yellow foundation make-up. **1952** C. W. Cunnington *Eng. Women's Clothing* viii. 265 Starting with foundation lotion over cherry-coloured cream rouge..this is followed by cameo powder and cherry lipstick. **1959** *News Chron.* 12 Aug. 5/5, I don't use foundation now. It dries my skin. **1965** [see *cleansing *ppl. a.*]. **1966** *Guardian* 25 Apr. 6/2 Lipsticks (and foundations) in rich as well as pale colours for the darker woman.
8. b. Also applied to persons, etc. belonging to or associated with the founding of a society or institution. Cf. *founder *sb.*² 5.
1900 *Westm. Gaz.* 28 June 1/1 The Royal herd and flock at Sandringham were established in 1863, when the foundation animals were selected from some of the best stocks in England. **1928** *Daily Mail* 31 July 8/5 The subscription will be 5s. a year for foundation members. **1965** Doughty & Wahl in *Rossetti Lett.* I. 41 James Collinson, ..a foundation member of the P.R.B. **1970** *Morning Star* 26 Aug. 5 A foundation member of the Communist Party, Tommy Strudwick,..has died in New Zealand.
c. foundation deposit, an object placed in the foundation of a building; **foundation garment,** a woman's corset, or corset-substitute; **foundation-stone** (earlier example of lit. use); **foundation-trench** (see quot. 1954).
1955 L. Woolley *Alalakh* ii. 84 It is tempting to regard these concealed 'cupboards' as intended to receive foundation-deposits. **1962** D. Harden *Phoenicians* 225 As a foundation-deposit under one wall there was a late geometric amphora with twisted handles. **1927** *Daily Express* 12 Sept. 5/5 These are best described as camibockers, plus a skirt, and are quite the most sensible foundation garment. **1952** M. McCarthy *Groves of Academe* (1953) iv. 54 Courses tailored to the individual need, like their own foundation-garments. **1957** *Times* 25 Nov. 11/1 It is wise first of all to give thought to the question of the right foundation garments. Unconditioned muscles tire less with the support of a well-fitted brassière and pantie-belt. **1856** J. A. Symonds *Let.* 28 June (1967) I. 74 The laying of the Foundation Stone of the New Aisle. **1942** *Oxoniensia* VII. 28 The purpose of this enclosure is obscure. The rectangular sides suggest a foundation-

trench for a masonry building, but no trace of masonry or even rubble was found in the section. **1954** M. Wheeler *Archaeol. fr. Earth* (1955) vi. 72 A trench, known as the 'foundation-trench', is cut along the line of the proposed wall in order that the foundations of the latter may rest upon the solid sub-soil rather than upon the relatively unstable soil that is commonly found on the surface.

founder, *sb.*² Add: **5. founder member,** a person belonging to or associated with the founding of a society or institution. Cf. *foundation 8 b.
1909 *Westm. Gaz.* 23 Apr. 9/3 The committee of the Aero Club of the United Kingdom have decided that after the first thousand members have been elected, such thousand members being founder members, the subscription will be raised and an entrance fee charged. **1960** *Farmer & Stockbreeder* 8 Mar. 63/2 A founder-member of the British Charollais Committee..said many farmers would like to see Charollais imported.

fou·nding, *ppl. a.* [f. Found *v.*²] Associated with or marking the establishment of (something specified); that originated or created. Spec. *founding father* (freq. with capital initials), an American statesman of the Revolutionary period, esp. a member of the American Constitutional Convention of 1787; also *transf.*; *founding member* = **founder member*.
1903 *Westm. Gaz.* 17 Nov. 10/1 Founding members are now being elected for the Ladies' Military and Naval Club. **1914** K. B. Umbreit (*title*) Founding fathers: men who shaped our tradition. **1953** *Manch. Guardian Weekly* 8 Oct. 5/1 The Hyde Park–Kenwood Community Conference was formed. Many of the 'founding-fathers' of the conference were Quakers. **1957** C. Hunt *Guide to Communist Jargon* viii. 21 The invitation to the founding Congress of the Comintern. **1958** C. Baker *Friend in Power* i. 14 The Founding Fathers, eternized in dark oils, looked benignly down from the white and gold walls. **1959** *Ann. Reg.* 1958 180 Article 4 provided that the control commission should consist of seven members, one representative each from the three founding nuclear Powers [etc.]. **1959** *Listener* 5 Mar. 419/2 It is possible, too, that some of the founding families [in Tarentum] may have enjoyed political privileges. **1961** *Ibid.* 7 Sept. 336/1 Sociologists tend to accept the distinction stressed by their Founding Father, Max Weber, between fact and value. **1966** *Times* 28 Feb. (Canada Suppl.) p. v/1 One third of Canadians are not of the founding races. **1967** *Observer* (Colour Suppl.) 15 Oct. 44/2 It was with abuses of this sort ..in mind that Lord Keynes and the founding fathers of the new regime at Covent Garden set to work in 1945. **1969** P. A. Robinson *Sexual Radicals* (1970) II. 84 Roheim..rejected the naïve attempts of the founding fathers to explain culture in terms of a simple-minded rationalistic psychology.

fountain, *sb.* Add: **2. b.** A metal vessel containing aerated water for drinking; a box containing ice and a coil through which aerated water is drawn (= **soda-fountain*). *U.S.*
1843 'R. Carlton' *New Purchase* I. 52, I shall make no attempt to record their..puns—good things of the sort, like soda-water, had better be taken at the fountain. **1873** J. H. Beadle *Undevel. West* xxxi. 675 A drink equal to the best soda from fountains. **1936** *Amer. Speech* XI. 38 The craze for this sort of fountain entertainment seems to be on the wane. **1955** H. Roth *Sleeper* xvi. 131 He had neglected to tip the fountain man.
6. b. fountain-pen (earlier examples).
1710 M. Henry *Expos. Bible, Zechariah* iv. 2 Without any further Care they [sc. lamps] received Oil as fast as they wasted it, (as in those which we call Fountain Inkhorns, or Fountain Pens). **1789** F. Burney *Diary* 18 Aug. (1842) 51, I took my fountain pen, and wrote my rough journal. **1833** *Mechanics' Mag.* XIX. 319/1 The introduction of this paper will go far to supersede the use of fountain-pens of all kinds.

fou·ntain, *v.* [f. the sb.] **a.** *intr.* To rise like the waters of a fountain. **b.** *trans.* To cause to well up in the manner of a fountain.
1903 Hardy *Dynasts* I. I. iii. 27 Soon stars will shut.. And sunbeams fountain forth. **1948** G. H. Johnston *Death takes Small Bites* vii. 174 And then a high log broke and dropped into the ashes of the fire and the sparks fountained. **1967** D. Knight *Turning On* 94 Fire fountained from the church spire, red sparks floating on the wind. **1969** *Daily Tel.* (Colour Suppl.) 16 May 45/3 Seven tons of water fountained up at 125 m.p.h. to a height of over 400 feet. *Ibid.* 28 Nov. 8/1 A water main.. suddenly fountained 20,000 gallons of water down the Haymarket.

fountaining (fau·ntėniŋ), *ppl. a.* [f. Fountain + -ing².] = Founting *ppl. a.*
1883 G. Meredith *Poet. Wks.* (1912) 218 A chariot, cleaving the storm, Clove the fountaining lake with a plough. **1923** *Glasgow Herald* 5 Sept. 8 When the sun plays on the foam the impression of innumerable fountaining pearls is conveyed. *Ibid.* 11 Dec. 6 The pale fountaining gold of the clumps of tall windlestraws. **1955** S. Spender *Coll. Poems 1928–1953* 131 This early summer prepares its feasts In the garden..Fountaining bird-song, criss-crossed with bees.

four, *a.* and *sb.* Add: **A.** *adj.* **2. d.** *on all fours* (*with*). (Examples of use of sense 2 of All fours *phr.*)
1877, 1883 [see All fours *phr.* 2]. **1909** H. G. Wells *Tono-Bungay* II. iii. § 2. 191 'It's just on all fours with the rest of things,' he remarked; 'only more so. You needn't think you're anything out of the way.' **1931** *Economist* 8 Aug. 277/1 The railways maintain that conditions in Great Britain and America are not on all fours. **1940** W. S. Churchill *Into Battle* (1941) 186 It was only natural that the Admiralty, since this war began, should have brought this precedent—although it is not exactly on all fours, and there are some differences—..to the notice of His Majesty's Government. **1961** *Ann. Reg.* 1960 49 It was impossible to make an agreement exactly on all fours with the Anglo-American bomber base agreement.
B. *sb.* **1. b.** The figure (4) denoting this number.
2. f. The four players of a game, esp. of Bridge. Also *fig.*
1895 Kipling *Day's Work* (1898) 348 Can't I get him to make up a four at tennis with the Hammon girls? **1905** H. A. Vachell *Hill* vii. 144 Bridge...We could get up a four in this house. **1924** C. Mackenzie *Old Men of Sea* vii. 113 Cosway was called away to make up a four at bridge. **1971** *Times* 9 Aug. 4 A four in the East-West bridge game.
3. d. *pl.* (in form *fourses*). A light meal taken in the fields at four o'clock in the afternoon. *local.* Cf. Bever *sb.* 3.
1823 E. Moor *Suffolk Words* 132 Foorzes. **1849** [see *elevens]. **1887** E. S. Suffling *Land of Broads* (ed. 2) 266 At 4 p.m., when they have their 'fourses'. **1895, 1923** [see *elevens]. **1953** A. Jobson *Household & Country Crafts* ii. 28 Another speciality..was the harvest-cake or biscuit, also known as a bever cake... This was a most welcome addition to the 'fourses' provided by the farmer's wife for the harvest field. **1966** G. E. Evans *Pattern under Plough* xv. 154 A substitute harvest horn..did service to summon up the women and children to bring their *elevenses* and *fourses* into the harvest field.
C. *Comb.* **1. a.** *four-cylinder, -day, -door, -mile, -story* (earlier U.S. example).
1902 *Westm. Gaz.* 8 Sept. 7/3 Quadruple expansion four-cylinder engines of 4,000 horse-power. **1936** *Discovery* Feb. 37 Improved four-cylinder compound locomotive with poppet-valves and double blast-pipe. **1902** *Daily Chron.* 1 Sept. 5/2 The half-yearly four-day closing of the British Museum reading room. **1909** *Westm. Gaz.* 3 Sept. 12/1 Thus making the first four-day westward voyage. **1935** *Discovery* Aug. 221/1 The five-day week has already arrived, the four-day week will come and there will be increasing leisure to be employed. **1960** *News Chron.* 9 Apr. 1/3 Thousands of men stayed away from work.. trying..to enforce a four-day week. **1957** P. Frank *Seven Days to Never* i. 13 A four-door sedan. **1897** S. L. Hinde *Fall of Congo Arabs* 115 The whole crowd jumped into the river, here about a hundred yards wide, with a four-mile current. **1903** Kipling *Five Nations* 90 From the Four-mile Radius roughly to the plains of Hindustan. **1835** C. F. Hoffman *Winter in West* I. 95 [In Detroit] there are..many four-story stores erecting.
2. four-ale, also *attrib.*; (*b*) a four-ale bar; **four-ball** *a.*, defining a foursome at golf in which four balls are used, the best ball on each side counting at each hole; also *absol.*; **four-baller,** a golfer playing in a four-ball match; **four-by-two,** (*a*) *Mil. colloq.*, the cloth attached to a pull-through; (*b*) also, rhyming slang for a Jew; **four-chromatic** *a.* = **four-colour* adj.; **four-colour** *a.*, having, or pertaining to the use of, four colours; *four-colour problem*, the as yet unsolved problem of proving as a mathematical theorem that on any plane map only four colours are needed to give different colours to any regions that have a common boundary; hence **four-coloration, four-colour** *v. trans.*, **four-colouring** *vbl. sb.*; **four-cycle** *a.* = *four-stroke *a.*; **four-dimensional** *a.* (earlier example); **four-figure** *a.*, (*a*) consisting of four figures, i.e. a thousand or over (but less than ten thousand), a thousand pounds or over, etc.; (*b*) evaluated to or containing four significant figures or four decimal places; **four-four time** *Mus.*, time or rhythm consisting of four crotchets in a bar (also *ellipt.*); **four freedoms**: see *freedom 4; **four-high** *a.*, having four rolls one over another, as a rolling-mill (cf. *two-high* s.v. Two B. IV. 2); **four hundred** *U.S.*, the highest society of a locality; **four-letter** *a.*, consisting of four letters; applied esp. to any of several monosyllabic English words, referring to the sexual or excretory functions or organs of the human body, that are conventionally excluded from polite use; *four-letter man*, an obnoxious person; **four-minute mile,** a mile run by an athlete in four minutes or less (first achieved by R. Bannister in 1954); hence **four-minute miler,** an athlete who performs this feat; freq. used hyperbolically; **four paws** *U.S. logging* (see quot. 1905); **four-place** *a.* = **four-figure* adj. (*b*); **four-point assay** *Chem.* (see quot.); **four-point bearing** *Naut.* (see quot. 1948); **four-seater,** a car with seating accommodation for four people, also *attrib.*; **four-spot,** a playing card with four spots; **four-star** *a.*, designating a (high) rank in a grading system for hotels, etc., usu. with four or five grades; also *transf.*, of a high degree of excellence; also *four-starred* adj.; **four-vector** *Math.* [ad. G. *vierervektor* (A. Sommerfeld 1910, in *Ann. d. Physik* 4th Ser. XXXII. 750)], a vector defined by four scalar components, esp. a space-time vector in the theory of relativity; **four-walled** *a.*, having or enclosed within four walls; **four walls,** the walls of a room or a house; hence, *allusively*, in reference to confinement within the limits of a (small) building.
1902 *Daily Chron.* 23 Sept. 6/7 He had been to a 'four-ale club'. **1930** *Ibid.* 6 Nov. 5/7 The man had been in his [public-]house—in the four-ale bar. **1953** Dylan Thomas *Under Milk Wood* (1954) 2 Night in the four-ale, quiet as a domino. **1966** 'H. Carmichael' *Suicide Clause* vi. 73 A kindred spirit beside me in the four-ale bar. **1904** *Westm. Gaz.* 25 Mar. 4/1 The players in a four-ball match. **1909** *Ibid.* 8 Mar. 12/2 Four-ball foursomes. **1969** *Times* 25 Sept. 15/1 In the wake of that extraordinary feature of American golf, the four-ball, gang mowers at last got to work on the fairways. **1927** *Observer* 31 July 15/5 A selfish 'four-baller', intent solely on his own pleasure, has blocked the way. **1925** Fraser & Gibbons *Soldier & Sailor Words* 97 *Four by two*, the term for the piece of flannel issued as a 'pull through' for cleaning a rifle. From its size, four inches by two. **1930** P. Macdonald *Link* 88 Cross maintains that the oil-bottle and four-by-two must have been to clean the rifle. **1936** J. Curtis *Gilt Kid* vii. 69 That's all right so long as it's not ham... I'm a four-be two, you see. **1968** L. Berg *Risinghill* 106 It was like the rhyming slang 'four-by-twos' for 'Jews', used..perhaps with an undertone of lurking contempt. **1970** E. McGirr *Death pays Wages* iii. 60 'This Marx, was he a four by two?' demanded Quimple. 'Pardon?' 'A Jew, sir, a Jew.' **1952** G. A. Dirac in *Jrnl. London Math. Soc.* XXVII. 85 (*title*) A property of 4-chromatic graphs and some remarks on critical graphs. **1962** A. Doig tr. *Berge's Theory of Graphs* xxi. 214 The following theorem which has never yet been proved is known as the four-colour problem: every planar graph is 4-chromatic. **1967** O. Ore *Four-Color Probl.* xii. 210 This..can be extended..to give a 4-coloration of G. [**1879** Cayley in *Proc. R. Geogr. Soc.* I. 260 The theorem then is, that if an area be partitioned in any manner into areas, these can be, with four colours only, coloured in such wise that in every case two attached areas have distinct colours.] **1879** A. B. Kempe in *Amer. Jrnl. Math.* II. 200, I will conclude with a theorem...It is one of which I long endeavoured to obtain an independent proof, as a means of solving the four-colour problem. **1897** P. J. Heawood in *Q. Jrnl. Math.* XXIX. 270 (*title*) On the four-colour map theorem. **1931** *Everyman's Encycl.* IV. 209/1 The early years of the twentieth century have seen many improvements in working the three-colour block process, while in America especially the four-colour process has been developed. **1941** Courant & Robbins *What is Math.?* v. 247 The four color theorem has indeed been proved for all maps containing less than thirty-eight regions. *Ibid.*, In the four color problem the maps may be drawn either in the plane or on the surface of a sphere. **1963** E. L. Johnson in *Calif. Univ. Operations Res. Center Rep.* No. 28 (*title*) A proof of four-coloring the edges of a regular three-degree graph. *Ibid.* 8 The edges of the reduced graph G¹ can be 4-colored. **1967** E. Chambers *Photolitho-Offset* xv. 236 The demand for colour, especially four-colour half-tone work, either for catalogue or carton use, has brought about the development of four and six-colour machines. **1971** *Daily Tel.* (Colour Suppl.) 8 Jan. 21/4 The best four-colour lithographic printers in the country. **1963** G. A. Dirac in *Proc. London Math. Soc.* XIII. 195 Theorem 2 establishes the truth of the four-colour conjecture for a new class of planar graphs.. and..furnishes a procedure for verifying that a given suitable planar graph is 4-colourable without having to find a 4-colouring for the whole of it. **1967** O. Ore *Four-Color Probl.* viii. 107 To define a four coloring of the faces in G. **1909** *Westm. Gaz.* 25 Mar. 4/1 It is open to question if a two-cycle engine will ever be produced which will show the same economy of fuel as the four-cycle. **1924** *Times* 29 Nov. (Trade & Engin. Suppl.) 250/4 Four-cycle double-acting engines. **1950** *Engineering* 19 May 577/3 Fuel within this range..can be burnt successfully in four-cycle [diesel] engines. **1880** Four-dimensional [see Dimensional *a.* 2]. **1842** *Penny Cycl.* XXIII. 499/2 Four-figure logarithms on a card. **1877** *Porcupine* 17 Feb. 741/1 The same modest four-figure price. **1968** Fox & Mayers *Computing Methods for Scientists & Engineers* iii. 51 The fifth digit in our four-figure computation. **1970** *Daily Tel.* 8 June 4/3 The newcomers should help to send Dame Patricia back to Westminster with a good four-figure majority. **1826** J. F. Danneley *Mus. Gram.* (Index) 100 Four-four time. **1853** C. Engel *Pianist's Handbk.* 49 The *Allemande*, in the style of an old German dance, in 4/4 time, in moderate movement. **1880** [see Compound *a.* 2 f]. **1959** *Listener* 5 Feb. 264/2 Alternating bars of five-four and four-four. **1966** *Crescendo* Dec. 26/1 The time signature 4/4 seems to be at its most productive rhythmically when 12/8 is imposed on it. *Ibid.*, To superimpose 18/8 on to 4/4. **1878** *Iron Age* 5 Dec. 3/5 The use of Bleckly's four-high wire mill..has been attended with very satisfactory results. **1928** *Jrnl. Iron & Steel Inst.* CXVII. 840 The chief object of the four-high mill is to provide an exceedingly rigid working roll of relatively small working diameter. **1958** *Times* 7 Jan. 14/2 Heavy, medium and light plates all rolled in two 4-high mills. *Ibid.*, The new slabbing mill and the 4-high plate mills. **1888** *N.Y. Tribune* 8 Apr. 5/1 Not all of 'the 400' have yet returned from the South. **1948** *Coronet* Aug. 36/1 To social strivers she is Queen of the 400. **1923** J. Manchon *Le Slang* 265 *Shit*,..un type embêtant... L'euph. est *four-letter-man*. **1924** J. Sutherland *Circle of Stars* xxiii. 236 Carter isn't that kind of a four letter man if he does soak. **1927** C. S. Lewis *Let.* 12 Dec. (1966) 122 Louis the Pious was 'a man of blameless and virtuous habits'—tho' every other sentence in the chapter makes it clear that he was a four letter man. **1934** *Amer. Speech* IX. 264/1 The obscene 'four-letter words' of the English language are not cant or slang or dialect, but belong to the oldest and best established element in the English vocabu-

lary. *Ibid.* 267/1 For most people, the bare word forms of these four-letter words have become sexual fetishes. **1935** E. Hemingway *Green Hills Africa* (1936) ii. iii. 97 Ashamed at having been a four-letter man about books. **1947** A. Huxley *Let.* 9 Mar. (1969) 568 She would bring him to amorous life again by re-assuming her cockney accent.. going very nearly to the point of murmuring four-letter words into his ear. **1960** *Times* 7 Nov. 17/4 Having regard to the state of current writing, it seems that the prosecution against *Lady Chatterley* can only have been launched on the ground that the book contained so-called four-letter words. **1962** I. Murdoch *Unofficial Rose* xvii. 164 Felix regarded Randall as a four-letter man of the first order. **1969** N. Cohn *AWopBopaLooBop* (1970) xx. 191 He was heckled. Immediately, he exploded in a rash of four-letter words and the curtain came down. **1955** T. Sterling *Evil of Day* iv. 40 These people would believe I was going to die if they saw me running the four minute mile. **1958** *Sunday Times* 30 Nov. 38/6 His phenomenal series of four-minute miles. **1963** *Listener* 31 Jan. 209/2 A 4-minute miler could take 3.6 seconds off his time under the influence of amphetamine. **1969** J. Wainwright *Take-Over Men* i. 16 He broke the four-minute mile to position her chair. **1905** *Terms Forestry & Logging* 35 *Double couplers*, two coupling grabs joined by a short cable, used for fastening logs together. Syn.: four paws. **1957** *Brit. Commonw. Forest Terminol.* ii. 76 *Four paws*,..used for fastening a skidding chain or cable to large logs. **1888** *Encycl. Brit.* XXIII. 13/1 A four-place table. **1927** Carey & Grace (*title*) Four-place mathematical tables with forced decimals. **1940** A. L. Bacharach et al. in *Biochem. Jrnl.* XXXIV. 1464 The comparison of activity with that of a stable standard preparation..can be done by means of the 'four-point assay'; two doses of the standard and two doses of test material are used. **1926** G. Bradford *Gloss. Sea Terms* 19/1 Bow and beam bearing..is also called a four point bearing. **1948** R. de Kerchove *Internat. Maritime Dict.* 278/2 *Four-point bearing*, the simplest and most commonly used method of determining the distance off a terrestrial object when coasting. The object is observed when it is four points (45°) on the bow and again when it is abeam. **1909** *Westm. Gaz.* 16 Nov. 4/2 The famous 15-h.p. four-seater touring-car. **1922** *Times* 20 June 8/5 The increased convenience and comfort of a four-seater. **1878** B. F. Taylor *Between Gates* 193 We are a four-spot of dirty spades. **1920** C. E. Mulford *J. Nelson* x. 100 I'm layin' down as fine a pair of four-spots as I've ever held. **1921** *Automobile Assoc. Ann. Rep.* 4 Upwards of 2,000 firms carry the official appointment of the Association. During the past year the work of inspecting and appointing, on a 'four-star' classification basis, has continued. **1952** E. Grierson *Reputation for Song* xvii. 138 The 'Golden Fleece'..was a four-star hotel. **1958** K. Amis *I like it Here* xiii. 155 A four-star monastery with chanting and flagellation. **1959** *Sunday Express* 21 June 6/6 Oscar Heinrich's life is four-star reading for anyone who [etc.]. **1968** *Listener* 1 Aug. 159/1 No one has ever shown that a car which will run happily, without pinking, on, say, four-star petrol will run any better—faster, or more economically—on five-star. **1970** *Daily Tel.* 30 June 3/1 Ladbroke's, the bookmakers, announced yesterday that they are going to build a four-star hotel in the centre of Leeds. **1950** A. Christie *Murder is Announced* viii. 85 George, it's my own particular, one and only, four starred Pussy. The super Pussy of all old Pussies. **1914** L. Silberstein *Theory of Relativity* v. 148 The length, thus defined, of a four-vector may be either real, or purely imaginary, or nil, according as we have..a space-like, a time-like, or a singular vector. **1952** E. G. Ramberg tr. *Sommerfeld's Electrodynam.* iii. 213 The four-vector attains a meaning in the four-dimensional world which is independent of the choice of the coordinate system. **1968** *Physics Bull.* Nov. 370/2 The four-vector potential of the electromagnetic field. **1905** *Daily Chron.* 18 Aug. 3/7 A four-walled prison life. **1908** C. W. Wallace *Children Chapel Blackfriars* 7 The four-walled coach-court of the public inn. [**1834** tr. *Jacquemont's Journ. India* I. 156, I was much more comfortable under it..than within four naked walls.] **1849** N. P. Willis *Rural Lett.* 17 Between five in the morning and 'flower-shut' I feel as if four walls and a ceiling would stop my breath. **1922** 'E. Percy' (*title*) If four walls told. **1930** Harrison & Cartwright (*title*) Within four walls.

Fourcault (fūə·ɹko, -kōᵘ). The name of Émile *Fourcault*, Belgian inventor (1862–1919), used *attrib.* to designate a continuous process for making sheet glass in which the glass is drawn vertically upwards through a slot in a debiteuse. Also *Fourcault machine*, etc.

1908 W. Rosenhain *Glass Manuf.* 359 (Index), Fourcault process. **1948** L. M. Angus-Butterworth *Manuf. Glass* xvii. 158 The great disadvantage of the Fourcault process was that it was necessary to continue drawing the sheet up vertically after it left the furnace. **1949** P. Davis *Devel. Amer. Glass Industry* viii. 192, 44 per cent of total window-glass production in 1929 was made by the Fourcault machine. **1965** *Glass Technol.* VI. 27A The one-compartment Fourcault tanks..must necessarily have a certain length to permit the glass to cool down sufficiently on its way towards the drawing compartment.

fourchette. Add: **1. d.** *Glove-making.* The forked piece between two adjacent fingers of a glove; a forgett. Also *Comb.*, as *fourchette-cutter, -puncher.*

1862 *Catal. Internat. Exhib., Brit.* II. No. 4942 The Children's [gloves] have an improved fourchette. **1881** *Instr. Census Clerks* (1885) 76 Thumb and Fourchette Cutter and Puncher. **1886** *Chambers's Jrnl.* 226/2 He cut pieces for the thumbs and fourchettes or sides of the fingers—usually pronounced 'forgets'. **1957** M. B. Picken *Fashion Dict.* 136/1 *Fourchette*, small forked piece of leather or other material set in between fingers of well-fitted gloves.

2. *Card games.* (See quots.)

1885 R. A. Proctor *Whist* 241 King and Knave make the *fourchette* to the Queen led; Queen and ten make the *fourchette* to the Knave; and so on. **1898** 'Cavendish' *Princ. Whist* 102 When you have the *fourchette* it is almost always right to cover. **1906** 'Cut-Cavendish' *Compl. Bridge Player* 131 A fourchette is the combination of cards in one hand immediately above and below the card led, such as knave, nine, sitting over the ten. It applies to the cards of the second player over the leader, and to the cards of the third player over the second. **1907** *Westm. Gaz.* 18 May 14/1 In the unusual case of holding a hand that is full of fourchettes, the middle heart is the correct opening.

fourcrœa: see *furcræa.

Fourdrinier (fuəɹdri·nie, fūɹdrinīə·ɹ). The name of Henry (1766–1854) and Sealy (d. 1847) *Fourdrinier*, British printers, used attrib. (†or in the possessive) to denote esp. the paper-making machine invented by them, and also the wire cloth used for draining the pulp in the machine.

[**1825** J. Nicholson *Oper. Mech.* 369 The machines which are now almost universally employed..the invention of Messrs. Fourdrinier.] **1837** *Rep. Sel. Comm. Fourdrinier's Pat.* 9 in *Parl. Papers* XX. 35 Do you understand the working of the machine which is called Fourdrinier's Patent Machine for making paper? **1839** Ure *Dict. Arts* II. 930 One of the Fourdrinier machines made at London by Mr. Donkin. **1845** *Encycl. Metrop.* VIII. 766 Fourdrinier's machine.—This was the first successful attempt at supplying machinery to perform the duties of the deckle or mould. **1874** *Art of Paper-Making* 164 They certainly well deserve to be immortalised in the name of the present Fourdrinier. *Ibid.* 182 It is this shaking movement, though it is very trifling (about ½ inch), which makes the Fourdrinier paper superior to that made on a cylinder-machine. *a* **1877** Knight *Dict. Mech.* I. 912/1 *Fourdrinier-machine*, a paper-making machine, the first to make a continuous web. **1964** *Gloss. Paper Terms (B.S.I.)* 16 *Fourdrinier machine*, a machine for the production of a continuous sheet of paper or board which is formed by drainage on an endless wire cloth unit known as the 'Fourdrinier wire part'. **1966** H. Williamson *Methods Bk. Design* (ed. 2) xviii. 295 Almost any Fourdrinier is huge; the biggest is colossal. *Ibid.* 298 It is the drying cylinders which extend the Fourdrinier machine to its enormous size.

fourer (fōə·rəɹ). *Cricket.* [f. Four + -er¹.] A hit from which four runs are scored. (No longer current.)

1875 *Field* 8 May 458/1 Boys made a 'fourer'. **1901** R. H. Lyttelton *Out-door Games* 14 The public only applaud him [*sc.* the batsman] when he hits a fourer. **1927** *Observer* 29 May 28/5 At the other end I'll make old Brown bowl full half-volleys wide of the off-stump and have an outer ring of three to save the fourers.

four flush, *sb.* *U.S.* [Flush *sb.*³ 1.] In poker, a flush containing only four (instead of five) cards, and hence almost worthless; a bob-tail flush. Hence *attrib.* or as *adj.*, lacking in genuineness. So **four-flush** *v. intr.*, to act in a 'bluffing' or fraudulent manner; chiefly as **four-flushing** *vbl. sb.* and *ppl. a.*

1887 J. W. Keller *Draw Poker* 35 If in opening a pot a player finds in his hand a pair and a four flush, or four straight, he may break his pair. **1896** Ade *Artie* iv. 37 'I thought he was going to fight.' 'Not that boy. He was four-flushin'.' **1899** —— *Fables in Slang* (1911) 12 She always saw the same old line of Four-Flush Drummers from Chicago and St. Louis. **1901** 'J. Flynt' *World of Graft* 46 That's what I call four-flushin'. **1901** [see *call-down* s.v. *call *sb.* 15]. **1904** *N.Y. Even. Post* 20 Feb. 10 Mayor Harrison's assertion that the Sunday closing law is a 'four-flush' law..one that was meant to deceive and not meant for actual enforcement. **1942** W. Stegner *Mormon Country* 286 A four-flushing holdup man named Gunplay Maxwell.

Hence **fou·r-flusher,** a pretender, braggart, humbug.

1904 'No. 1500' *Life in Sing Sing* xiii. 255 *Four-flusher*, one who poses for effect. **1910** *Outlook* 25 June 376 Many doubted his intention, and one newspaper called him a 'four-flusher'. **1916** 'B. M. Bower' *Phantom Herd* ii. 30 Some of the four-flushers..kept eyeing that bluff. **1920** A. B. Baxter *Parts Men Play* xxi. § 1, Vot you mean, you great big fourflusher? **1923** R. D. Paine *Comrades of Rolling Ocean* vi. 106, I could have sized up this Maddigan four-flusher if he had merely walked across the campus in my freshman year. **1924** W. M. Raine *Troubled Waters* ii. 22 You're a fine bunch of four-flushers. **1944** L. A. G. Strong *Director* xiii. 112 You shouldn't let these four-flushers come it over you.

fou·r-foo:ter. [f. Four *a.* + Foot *sb.* + -er¹.] A creature having four feet; a quadruped.

1906 R. Whiteing *Ring in New* xxxviii. 281 They [*sc.* tigers] are four-footers, we are two—perhaps that's all the difference. **1920** Galsworthy *In Chancery* ii. x, 'Strange life a dog's,' said Jolyon suddenly; 'the only four-footer with rudiments of altruism, and a sense of God!'

Fourier (fūə·rie). The name of J. B. J. *Fourier* (1768–1830), French mathematician, used *attrib.* or in the possessive to designate certain principles enunciated by him and many mathematical expressions and techniques arising out of his work, as *Fourier analysis*, the analysis of a periodic function into a number of simple harmonic functions or, more generally, into a series of functions from any orthonormal set; *Fourier's law*, that any non-sinusoidal periodic vibration can be regarded as the sum of a number of sinusoidal vibrations each having a frequency that is an integral multiple of some fundamental frequency; *Fourier('s) series*, a series of the form $\frac{1}{2}a_0 + (a_1 \cos x + b_1 \sin x) + (a_2 \cos 2x + b_2 \sin 2x) + \ldots$, where the constants a_0, a_1, b_1, etc. are defined in terms of a function $f(x)$ to which the series may converge; *Fourier's theorem*, (*a*) that if a function $f(x)$ satisfies certain conditions within the interval $-\pi \leqslant x \leqslant \pi$, it can be represented within that interval by a Fourier series; (*b*) (see quot. 1880); *Fourier transform*, a function $f(x)$ related to a given function $g(t)$ by the equation $(2\pi)^{\frac{1}{2}} f(x) = \int_{-\infty}^{\infty} g(t)e^{\pm itx}\,dt$, used to represent a non-periodic function by a spectrum of sinusoidal functions. Also *Fourier coefficient, expansion, integral, transformation*, etc.

1834 *Rep. Brit. Assoc. 1833* 343 If the interval of the roots be determined, by the application of Fourier's theorem of the succession of signs of the original function X and its derivatives. **1842** A. De Morgan *Differential & Integral Calculus* xx. 641 In applying Fourier's theorem ..to discontinuous functions, we find that at the point where the discontinuity takes place, and a function which generally can have but one value might be expected to have two, it takes neither, and gives only the mean between them. **1877** Rayleigh *Theory of Sound* I. ii. 24 The pre-eminent importance of Fourier's series in Acoustics. **1880** G. S. Carr *Synopsis Pure & Appl. Math.* I. i. 134 Fourier's Theorem.—Fourier's functions are..$f(x)$, $f'(x)$, $f''(x)$...$f^n(x)$...As x increases, Fourier's functions lose *one* change of sign for each root of the equation $f(x) = 0$, through which x passes, and r changes of sign for r repeated roots. **1884** A. Daniell *Text Bk. Princ. Physics* v. 127 Longitudinal vibrations of a string or rod..whose ends are held fixed obey the same principles as transverse vibrations. Fourier's law holds good. **1902** E. T. Whittaker *Course Mod. Anal.* vii. 152 The question arises..whether the Fourier expansion is *unique*. **1911** *Proc. R. Soc.* A. LXXXV. 14 We can also sum the series of the products of the Fourier coefficients of two such functions. **1912** *Phil. Mag.* 6th Ser. XXIV. 866 Fourier's integral. **1923** *Proc. Cambr. Philos. Soc.* XXI. 463 The notion of Fourier transforms arises from Fourier's integral formula,..which gives ..reciprocal relations..connecting the two functions $f(x)$ and $F(x)$. **1929** V. Bush *Operational Circuit Analysis* x. 186 Direct operational methods may be regarded as shorthand processes of evaluating and tabulating the results of Fourier analysis. **1936** *Discovery* Apr. 114/2 All sound-waves (except those from a flute, closed organ-pipe, etc.) are composed of many combined vibrations whose composition follows Fourier's law. **1957** R. S. Longhurst *Geom. & Physical Optics* xi. 226 The Fraunhofer pattern is the Fourier transform of the amplitude across the diffracting aperture and vice versa. **1963** R. W. Ditchburn *Light* (ed. 2) iv. 89 The 'top-hat curve' shown in fig. 4.6 can be represented by an appropriate Fourier series..for all values of x_0 because the curve to be represented is periodic. **1964** *Oceanogr. & Marine Biol.* II. 14 The correlation coefficient $f_u(\tau)$ is related by a Fourier transformation to the spectrum function $F_u(n)$. **1965** Pearson & Maler *Introd. Circuit Anal.* ix. 439 The Fourier transform..is useful in analyzing pulses from a frequency standpoint. **1967** Condon & Odishaw *Handbk. Physics* (ed. 2) ii. iii. 26/2 The part of $F(t)$ effective in exciting the oscillator is the component in its Fourier integral representation associated with the natural frequency of the oscillator.

Hence vbl. uses, as *Fourier-analyse, -transform* vbs.; *Fourier-transformed* ppl. adj.

1962 W. B. Thompson *Introd. Plasma Physics* vi. 108 To get a numerical value, the curve may be Fourier analysed, and the effect of each harmonic component discovered. **1967** *Oceanogr. & Marine Biol.* V. 24 For several North Sea ports..continuous sea-level records were Fourier-analysed to determine if any non-tidal periods were present. **1968** C. G. Kuper *Introd. Theory Superconductivity* iv. 57 These Fourier-transformed quantities. **1970** *Nature* 5 Sept. 1032/1 The Fourier coefficients..could then be inversely Fourier transformed to reconstruct the specimen in three dimensions.

four-in-hand. Add: **1. b.** A kind of neck-tie of the variety designed to be tied in a loose knot with hanging ends. Chiefly *U.S.*

1892 Four-in-hand tie [see Dict. 3]. **1902** S. Hale *Let.* (1919) 374 The young men in Tuxedos, four-in-hands, panamas. **1909** 'O. Henry' *Roads of Destiny* x. 166 While trying to figure out Jimmy's method of tying his four-in-hand he cordially gave information. **1909** *Public Ledger* (Philadelphia) 24 June 5/1 Knitted and French Four-in-hands, made of fiber and short lengths of silk. **1919** T. K. Holmes *Man fr. Tall Timber* xix. 224 She had fashioned.. a crimson four-in-hand to be tied at the low, rolling collar of her blouse. **1942** E. Paul *Narrow St.* ii. 12 His twinkling blue eyes..noted my American felt hat and four-in-hand tie. **1967** *Boston* (Mass.) *Herald* 1 Apr. 21/2 The traditional four-in-hand or ascot.

‖ **fou rire** (fu rir). [Fr., lit. 'mad laughter'.] (A fit of) wild or uncontrollable laughter.

[**1897** P. G. Hamerton *Autobiogr.* vii. 51 The eldest boy in our class was liable to fits of perfectly uncontrollable laughter (what the French call *le fou rire*).] **1914** R. Brooke *Let.* 3 Sept. (1968) 613 Murmurs of subdued applause from me. Eddie [Marsh] had a *fou rire.* **1919** G. Saintsbury *Hist. Fr. Novel* II. iv. 137 But I own that an access of *fou rire* once came upon me when I was told in a printed page that *La Chartreuse de Parme* was a 'very lively and very amusing book'. **1924** M. MacCarthy *Nineteenth-Cent. Childhood* vii. 75 Adela and I exchange glances... We recover quickly from our hidden *fou-rire.* **1963** P. H. Johnson *Night & Silence* xxvi. 194 Halfway up the stairs she..fell into a *fou rire.*

fourm(e, var. Form *sb.* 21.

c **1290**, etc. in M.E.D. *c* **1386, 1576** [see Form *sb.* 21]. **1589** [see Form *sb.* 21 b]. **1916** Blunden *Harbingers* 63 Strange streams Flow flagging in the undescribed deep fourms Of creatures born the first of all. **1952** R. Campbell tr. *Poems of Baudelaire* 77 Whereon as in a fourm you would fill out And mould your hair.

Fournier (fūə·ɹnie). The name of P. S. *Fournier* (1712–68), French engraver and type-founder, used to designate a style of type designed by him and revived in modern use. *Fournier('s) point*, a unit of measurement of type.

1902 T. L. De Vinne *Pract. Typogr.* 146 Attempts have been made to return to the Fournier point by making it in accord with the new measures. *Ibid.* 155 In the Fournier system 1000 points made 35 centimetres. **1927** in O. Simon *Introd. Typogr.* (ed. 2, 1963) v. 92 Fournier, a specimen of a classic old face designed by Pierre Simon Fournier (*Fournier-le-jeune*) Paris ca. 1750 & now first revived by the Lanston Monotype Corporation Ld...1927. **1930** H. Carter *Fournier on Typefounding* p. xxxv, The traditional 'Fournier point', long used in Belgium, was ·013728 English inch. **1935, 1945** [see *Bembo]. **1962** D. B. Updike *Printing Types* (ed. 3) I. ii. 32 Fournier's point is scarcely used nowadays, except in Belgium. **1963** O. Simon *Introd. Typogr.* (ed. 2) iii. 13 Fournier, Plantin and Imprint are..more successful in their smaller sizes.

fourpenny, *a.* Add: **1. c.** *fourpenny one*: a blow, hit; also, a scolding. *colloq.*

1936 *Evening News* 29 Feb. 11/6 The chairman rang the lift bell... Great was our mirth ..when the little messenger ..said: 'Blimey, guv'nor,..George won't 'alf give you a fourpenny one when he comes down.' **1953** N. Balchin *Sundry Creditors* 89 On the last bend someone went hard into the retaining fence and took what looked like a fourpenny one. **1964** N. Freeling *Double-Barrel* ii. xi. 73, I think he got mad because he gave her a real fourpenny one. I bet she has a black eye.

fourreau (fu·ro). [F. *fourreau*, lit. sheath, scabbard.] A tight-fitting dress; an underslip as part of a dress. (Cf. Slip *sb.*³ 4 c.)

1872 *Young Englishwoman* Dec. 645/1 The fourreau dress, reminding one of modes of the First Empire. **1908** [see *sheath gown* s.v. Sheath¹]. **1913** in C. W. Cunnington *Eng. Women's Clothing* (1952) iii. 118 Evening gown.. mounted on flesh pink marquisette with a fourreau of sea-green charmeuse. **1921** *Glasgow Herald* 22 Jan. 4 A fourreau of black charmeuse, with which..is worn a simple over-dress of crepe de chine. **1927** *Observer* 10 Apr. 22 The whole dress is also slit from throat to hem, permitting a glimpse of a contrasting fourreau. **1964** O. Coburn tr. *Braun-Ronsdorf's Wheel of Fashion* 112/1 In such a sheath dress (called *fourreau*), which fitted the hips tightly..a lady could scarcely move.

foursome, *a.* and *sb.* Add: **B.** *sb.* **2.** A company, party, or dance of four persons.

1926 [see *eightsome *a.*]. **1926** *Amer. Speech* I. 630/2 The expressions *onesome, twosome, threesome, foursome, fivesome*, are coming to be used in other fields than golf, as a 'threesome at tennis', a 'foursome for bridge', a 'fivesome for dinner'. **1971** *Petticoat* 17 July 15/1 Your friend wants you to make up a foursome with her boyfriend's unknown friend. **1971** G. Mitchell *Lament for Leto* i. 26 Do you play bridge?.. I want to make up a foursome.

fou·r-stroke, *a.* [Four *a.* + Stroke *sb.*¹] In internal combustion engines, designating a cycle of operations which consists of four strokes (intake, compression, combustion, and exhaust); also *four-stroke engine, working.*

1900 W. W. Beaumont *Motor Vehicles* xi. 190 The two cylinders give an impulse for each revolution with a four-stroke cycle. **1902** P. N. Hasluck *Automobile* vi. 113 Petrol motors employed on automobiles belong..to the compression and explosion type, and the Beau de Rochas four-stroke cycle of operations, known in Great Britain as the Otto cycle, is employed almost exclusively. **1906** *Motor Boat* 12 July 14/1 A four-stroke engine. **1951** *Engineering* 6 July 2/1 Valve..drives are more heavily stressed in two-stroke working than in four-stroke working.

fourth, *a.* and *sb.* Add: **A.** *adj.* **2. b.** *Fourth of July* (earlier examples). Hence *Fourth-of-Julyism.* (Also later examples of *Fourth.*)

(*a*) [**1777** *Jrnls. Cont. Congr.* VIII. 625 There is due..a bill for material, workmanship, &c. furnished for the fire works on the 4 July, the sum of 102 69/90 dollars.] **1779** *New Haven Colony Hist. Soc. Coll.* VI. 317 Toasts were drank, to wit:..the Fourth of July, '76: The memorable era of American Independence. **1807** M. L. Weems *Lett.* (1929) II. 369 From these reflections..we may collect some good fourth-of-July ideas. **1834** *Southern Lit. Messenger* I. 156, I do not know that the celebration of a Fourth of July in a country village has ever been thought worthy of appearing in print. **1853** *Ibid.* XIX. 473/1 Fourth-of-Julyisms fled to the stump or the national anniversary barbecues. **1874** B. F. Taylor *World on Wheels* ii. i. 183 A Fourth-of-Julyism has somehow become an object of contempt.

(*b*) **1830** S. Breck *Recoll.* (1877) 91 On the Fourth, being a national holiday, there was a great parade on the Common. **1854** W. G. Simms *Southward Ho!* xiii. 253 Ordinarily admirable, our dinner on the glorious Fourth was worthy of the occasion. **1872** E. Eggleston *End of World* xi. 76 No more electin' presidents.., no more Fourths,..no more nothin. **1909** 'O. Henry' *Roads of Destiny* xiii. 206 It reminds me of a Fourth I helped to celebrate down in Salvador. **1967** D. Francis *Blood Sport* viii. 94 I'm glad you could take me at such short notice, with the Fourth coming up this weekend.

B. *sb.* **6.** A person who comes in to complete a party of four in a game or at a social event.

1803 *Lett. Miss Riversdale* I. 155 He..insisted upon Marlow's leaving with him to come and make up a rubber for his sister, knowing he could depend upon me for a fourth. **1841** Dickens *Let.* 10 Feb. (1969) II. 209 Browne ..dines with me... The gay and festive Thompson also joins us. Will you make a fourth? **1865** —— *Mut. Friend* I. ii. iv. 196 [He] implores to be asked to..make a fourth at the play. **1902** E. Glyn *Refl. Ambrosine* 159 The men played Bridge. Augustus made one of the fourths. **1959** T. S. Eliot *Elder Statesman* ii. 43 I'll feel more confidence after a fortnight..of people not staring..or wanting a fourth at bridge.

7. The fourth forward gear of a motor vehicle.

1909 J. S. V. Bickford *Faults & how to find Them* § 1173 Changing down. By this we mean changing from a high speed to a lower (fourth to third, third to second etc.). **1968** *Autocar* 12 Sept. 30/1 By 4,500 rpm (73 mph in fourth, almost 80 mph in fifth) this resonance has practically died away.

8. A place in the fourth class in an examination list.

[**1911** *Encycl. Brit.* X. 42/1 In the 15th century the candidates for the mastership of arts were divided into three classes..while a fourth, which was not published, contained the names of those who failed.] **1914** C. Mackenzie *Sinister St.* II. iii. xi. 714, I shall never get my Blue. I shall get a fourth in Greats. *Ibid.* xii. 746 If he stays up ten years he'll never get a Fourth. **1945** E. Waugh *Brideshead Revisited* i. i. 24 You want either a first or a fourth. There is no value in anything between. **1946** C. L. Wrenn *Word & Symbol* (1967) 163 The consciousness in the minds of the Oxford authorities that Sweet had obtained a 'fourth' in Greats.

C. *Comb.* **fourth arm** (see *arm *sb.*² 9); **fourth dimension,** a supposed or assumed dimension, additional to length, breadth, and thickness (see Dimension 3 *note*); hence **fourth-dimensional** *a.*, of or relating to the fourth dimension, *fig.* superhuman, extraordinary; hence *fourth-dimensionalism*; **fourth estate** (see Estate *sb.* 7 b); **fourth leader,** from 1922 to 1966, the fourth leading article in *The Times*, usually of a light or humorous nature; **fourth (cranial) nerve** *Anat.*, either of the trochlear nerves (see Trochlear *a.* 1), together also called the *fourth pair of nerves*; **fourth party** *Eng. Hist.*, a group of politicians (1880–85) led by Lord Randolph Churchill, forming a party independent of the three existing political parties; **fourth position** *Dancing* (see quot. 1957); **fourth ventricle** *Anat.*, a rhomboidal cavity situated between the medulla oblongata and the pons Varolii in front and the cerebellum behind; **fourth wall** *Theatr.*, the proscenium opening through which the audience sees the performance.

1875 G. H. Lewes *Problems* II. 279 We have no experiences out of which a fourth dimension could be constructed. **1895** W. D. Howells *My Lit. Passions* 202 That fourth dimension of the poem which is not yet made palpable or visible. **1904** B. Russell in *Mind* XIII. 574 The merit of speculations on the fourth dimension..is chiefly that they stimulate the imagination, and free the intellect from the shackles of the actual. **1934** *Discovery* Aug. 239/1 The Fourth and higher Dimensions. **1964** C. Chaplin *Autobiogr.* xvii. 303 Wells said that as a struggling young writer he had written one of the first scientific articles touching on the fourth dimension. **1902** E. Carpenter *Civilisation* ii. 70 An immaterial mediation or a fourth-dimensional mediation..would simply remove the problem out of the regions of scientific analysis. **1908** *Westm. Gaz.* 28 July 2/1 A forcible proof of superhuman or fourth-dimensional power. **1921** *Quest* Oct. 55 These few crude remarks on fourth-dimensionalism and the time-enigma are ventured solely as the tentative opinions of a layman. [**1946** *Times* 28 June 5/6 Those of your readers who are under a perpetual obligation to the charm of the fourth leading article may like to be reminded of Dr. Johnson's praise of the Scottish breakfast.] **1949** 'M. Innes' *Journeying Boy* vi. 64 He picked up *The Times*... He was mildly diverted by the fourth leader. **1964** E. S. Turner *How to measure a Woman* 165 *The Times*..has allowed its Fourth Leader writer, more than once, to drool over the temptation the alarm provides. **1966** *Times* 6 Oct. 9/6 'If it was not so tragic it would make a fine fourth leader for *The Times*,' said a seasoned observer of the Commonwealth Parliamentary Association conferences that meet round the world annually. **1681** S. Pordage tr. *Remaining Med. Works of T. Willis* 140 The fourth conjugation of nerves (which we call rightly the fourth by order and succession, although it is accounted the eighth and last by Fallopius) hath a diverse origine from all the rest. **1685** S. Collins *Syst. Anat.* II. Tab. xlviii, The Pathetic, or Fourth pair of Nerves. **1858** Gray *Anat.* 478 The Fourth, or trochlear nerve, is the smallest of the cranial nerves. **1963** Brock & Krieger *Basis Clin. Neurol.* (ed. 4) xii. 189 The superior oblique muscle [of the eye] is innervated by the fourth nerve. **1880** H. Wolff *Let.* 29 Sept. in W. S. Churchill *Ld. R. Churchill* (1906) I. 166 My dear Randolph,—After you left yesterday I received two very handsome tributes to the Fourth Party. **1897** J. McCarthy *Hist. Own Times from 1880 to Jubilee* 27 Now came Lord Randolph with his new group, having its distinct individual purpose, and it claimed to be recognised as the Fourth party. **1966** R. S. Churchill *Winston Churchill* I. iii. 44 This led to the formation of the Fourth Party—Lord Randolph Churchill, Sir Henry Drummond Wolff, Mr John Gorst, with Mr Arthur Balfour, Lord Salisbury's nephew, in loose attendance. **1884** D. Anderson *Compl. Ball-Room Guide* 10 *Fourth position*, put out right foot in a straight line with left toe, right heel about four inches from left toe. **1957** G. B. L. Wilson *Dict. Ballet* 124 In the fourth position.. one foot is placed before the other in an extension of the first position with the weight evenly distributed, both feet being fully turned out and separated by a distance of about one foot. **1578** J. Banister *Hist. Man* i. 100ᵛ The fourth ventricle is not much capable, and is comprehended of the thinne membran. **1685** S. Collins *Syst. Anat.* II. Tab. 1, The Scobs which makes the fourth ventricle. **1848** *Quain's Elem. Anat.* (ed. 5) II. 723 The fourth ventricle, or ventricle of the cerebellum. **1962** *Gray's Anat.* (ed. 33) 956 The cavity of the rhombencephalon is expanded to form the fourth ventricle. **1807** L. Hunt *Crit. Ess.* 60 The stage appears to be his room, of which the audience compose the fourth wall. **1910** E. F. Spence *Our Stage* v. 111 That fourth wall, the existence of which Mr. Jerome K. Jerome rather quaintly..suggested by the fender and fireirons laid in front of the footlights. **1959** *Listener* 4 June 997/2 His music-hall experience made him aware of the artificiality of the 'fourth wall' which in the legitimate theatre separates the stage from the auditorium: the music-hall comedian speaks directly to the audience. **1967** *Times* 19 Apr. 6/4 In the last act, Andrey..breaks the fourth wall and delivers his assault..straight out to the audience.

fou·r-wheel, *a.* **1.** Of a horse-drawn or motor vehicle: = Four-wheeled *a.*

1740 W. Ellis *Mod. Husb.* i. 131 Four-wheel Carriages.. go easier, when the Weight..is laid on the hinder Wheels. **1777** P. Thicknesse *Year's Journey* I. iii. 16, I have sold my four-wheel post-chaise. **1820** J. S. Fry *Ess. Constr. Wheel-Carriages* vi. 75 A four-wheel Coach, as at present constructed. **1902** P. N. Hasluck *Automobile* xviii. 441 The Decauville voiturette is the first of the four-wheel voiturettes to be described.

2. Of the braking or driving system of a motor vehicle: operating on all the four wheels.

1924 *Nation* 29 Oct. 461 Vast multitudes are entirely content, entirely occupied with their four-wheel brakes and their radios. **1926** *Encycl. Brit.* I. 229 Another vehicle, of a different type, is the '4-wheel drive' lorry, in which the wheels are driven independently. **1956** 'N. Shute' *Beyond Black Stump* ii. 60 A big four-wheel-drive truck. **1969** *Times* 12 May 16/2 A four-wheel drive forward control tractor.

foussa, var. *fossa².

fovea. Add: *fovea centralis*, the fovea of the retina.

1858 Gray *Anat.* 560 In the centre of the posterior part of the globe, the retina presents a yellow spot..having a central depression, *fovea centralis.* **1924** R. M. Ogden tr. *Koffka's Growth of Mind* iii. 71 The eye is turned until the fixated object falls upon the place of clearest vision lying at the centre of the retina (the *fovea centralis*). **1970** T. S. & C. R. Leeson *Histol.* (ed. 2) xx. 488 The fovea centralis is a shallow, rounded pit lying 4 mm. to the temporal side of the optic disk.

foveal, *a.* (Examples.)

1904 *Westm. Gaz.* 15 June 1/3 In the *fovea centralis* there are cones only, so that in direct or foveal vision the rods are out of play. **1931** *Brit. Jrnl. Psychol.* Jan. 286 Some persons possess a quite appreciable proportion of foveal rods. **1962** H. C. Weston *Sight, Light & Work* (ed. 2) i. 4 Each of these foveal cones has its own separate nerve path to the brain. **1966** M. Kabrisky *Proposed Model Visual Information Processing Human Brain* ii. 19 The 100,000 foveal receptors are allocated about 100,000 channels in the optic nerve, leaving the rest of the 125 million receptors only about 900,000 channels.

fowl, *sb.* Add: **1. d.** A troublesome sailor, one unamenable to discipline. *slang.*

1937 in Partridge *Slang Dict.* **1938** 'Giraldus' *Merry Matloe Again* 145, I was a 'fowl' of the first water. I was always getting 'run-in', always in trouble and had no zeal for the Navy whatsoever.

5. c. fowl paralysis, Marek's disease, a type of cancer affecting poultry; **fowl pest,** (*a*) = **fowl plague*; (*b*) = **Newcastle disease*; **fowl plague,** an acute, highly contagious virus disease of the domestic fowl and other birds that is usually fatal; **fowl pox,** a virus disease of the domestic fowl and other birds, in which lesions appear on feather-free parts of the body or on the mucous membranes of the mouth, nose, or throat.

1932 *Ann. Reg. Scientific Poultry Breeders' Assoc.* 75 There seems little doubt that there is a special condition affecting poultry which, for want of a better name, is termed fowl paralysis. **1960** *Farmer & Stockbreeder* Suppl. 22 Mar. 1/2 An outstanding feature is the resistance to leucosis and fowl paralysis. **1909** Webster, Fowl pest. **1912** J. R. Mohler et al. tr. *Hutyra & Marek's Spec. Path. Dis. Dom. Animals* I. i. xiii. 290 Fowl pest is an acute, contagious, infectious disease of fowls. **1953** 'M. Innes' *Christmas at Candleshoe* x. 110 The other gentlemen had fallen into a grave discourse of fowl-pest, hard-pad and foot-and-mouth disease. **1970** *Times* 21 Oct. 10/1 The current upsurge of fowl pest is at its most destructive in the factory farming units. **1937** *Poultry World Ann.* 82 Fowl plague or pest is a very highly infectious disease caused by a virus...The so-called Newcastle Disease closely resembles Fowl plague in many respects, but is caused by a different virus. **1970** *Q. Poultry Bull.* (Brit. Oil & Cake Mills) Dec. 13 The Fowl Pest Order of 1936 with amendments lays down the legislation governing both Newcastle Disease and Fowl Plague. **1908** M. H. Hayes tr. *Friedberger & Fröhner's Vet. Path.* (ed. 6) II. 425 The causative agents in fowl-pox belong to the group of so-called filtrable agents of infection. **1932** Gaiger & Davies *Vet. Path. & Bact.* xxvii. 388 Fowl-pox occurs naturally in fowls, turkeys and other birds. **1953** L. Robinson *Mod. Poultry Husbandry* xx. 551 Fowl pox is extremely contagious.

fowling, *vbl. sb.* Add: **2.** *fowling-gun.*

1647 *Archives of Maryland* IV. 323 Tho. Greene.. demandeth of Richard Span a fowling Gunne. **1847** *Knickerbocker* XXX. 239 Reader, in your younger days you have, no doubt, left your fowling gun slyly hid in a fence corner.

fox, *sb.* Add: **2. c.** An attractive woman. *U.S. slang.*

1963 R. I. McDavid in Mencken *Amer. Lang.* xi. 745 A cat in hot pursuit of a chick or fox is said to have his nose wide open. **1964** L. Hairston in J. H. Clarke *Harlem* 290 Daddy, she was a real fox! **1967** Wentworth & Flexner *Dict. Amer. Slang* (Suppl.) 684 *Fox*, an energetic, attractive, desirable girl, esp. one having a happy and alert personality. Orig. Negro use. Pop. into white use during 1963 by heavyweight boxer Cassius Clay. **1970** R. H. Greenan *Nightmare in Colour* (1971) lviii. 181 A fox is a girl. A fox is a chick, you see?

15. d. similative, as *fox-red* adj.

1910 Galsworthy *Inn Tranq.* (1912) 53 The wet fox-red drifts under the beech-trees. **1926** D. H. Lawrence *David* x. 70 Thou hast spied out every hair in his beard. Is he not fox-red? **1956** G. Durrell *My Family* vii. 87 A tall, square Venetian mansion, with..green shutters and a fox-red roof.

16. b. fox-snake (for *Coluber vulpinus* read *Elaphe vulpina*): add examples.

1857 *Trans. Ill. Agric. Soc.* II. 658 They are also found in the stomachs of the milk-snake,..and of the large fox-snake (Scotophis vulpinus). **1958** R. Conant *Field Guide Reptiles U.S.* 157 *Fox snake*... Ground color varies from yellowish to light brown, and the dark spots and blotches from chocolate to black.

fox, *v.* **2. c.** Delete †*Obs.* and add later examples.

1835 A. B. Longstreet *Georgia Scenes* 63 Has he foxed you? **1906** E. Phillpotts *Portreeve* i. xv, Don't fox yourself to think that. That's mad. *Ibid.* ii. xviii, Love makes a man cunning...I foxed you fifty times. **1942** *Times* 1 June 4/3 'We had the guns absolutely foxed,' a pilot said. **1963** *Times* 25 Feb. 13/5, I like to imagine some householder in the next century trying to piece together our lives and interests. Will she be foxed by literature addressed to Mr. H.-T. (my husband)?

foxaline (fǫ·ksălīn). [Fanciful formation on Fox *sb.*] Imitation fox-fur.

1907 *Ladies' Field* 12 Jan. (Advt.), White Foxaline long Stole. **1923** *Daily Mail* 16 Apr. 4 White Foxaline Tie. **1927** *Daily Express* 13 Sept. 5/2 Grey Chinese cat (sometimes called foxaline). **1930** R. Macaulay *Staying with Relations* ii. 18 One of those mythical animals imagined by the fur trade, like the foxaline, the pony-seal, and the rabbit-skunk. **1963** *Retail Trading-Standards Assoc. Bull.* Aug.–Sept. 3/1 *Animal*, hare flank... *Accepted trade name*, foxaline.

fox-fire. (Further examples.)

1824 J. Doddridge *Notes Virginia & Pennsylv.* 290 If they had seen any thing like fire, between that and the fort, it must have been fox fire. **1832** J. P. Kennedy *Swallow Barn* I. xxviii. 311 The fox-fire—as the country people call it—glowed hideously from the cold and matted bosom of the marsh. **1872** E. Eggleston *End of World* xxxii. 210 The 'fox-fire', rotting logs glowing with a faint luminosity, startled her several times. **1920** J. M. Hunter *Trail Drivers of Texas* 149 There is where you could see phosphorescence (fox fire) on our horses' ears.

fo·x-hole. [See Hole *sb.* 1 b; the compound *fox(e)hole* (OE. *fox-hol*) appears in early local designations, e.g. in Domesday Book, and persists in place-names.] A hole in the ground used by a soldier for protection; a slit trench. Also *transf.* and *fig.*; also used *attrib.*, as **foxhole circuit** (see quot. 1946²).

1919 *Red Cross Mag.* Apr. 29/1 The bitter weeks of the Argonne when the same Yank lay hungry, cold, wet, and exhausted in some insufficient fox-hole. **1928** Blunden *Undertones of War* xii. 143 Those dead men in field grey overcoats at the entrances, and others flung down by their last 'foxholes' near by. **1942** *Hutchinson's Pict. Hist. of War* 18 Mar.–9 June 130 Slit trenches used for protection by troops were called 'fox-holes' by American soldiers fighting against the Japanese on Bataan peninsula. **1944** *Newsweek* 31 July 60 The performers who were selected to make up the first entertainment..wave, as part of USO-Camp Show's 'Foxhole Circuit'. **1946** W. V. Thomas in Hawkins & Boyd *War Report* xviii. 351 Foxholes with the Germans manning bazookas. **1946** *Amer. Speech* XXI. 223/1 *Foxhole circuit*, a theater circuit at the fighting fronts played by American troupes. **1955** *Sci. News Let.* 19 Feb. 116/1 Simple foxhole shelters provide 'surprising protection' from this invisible radioactivity. **1957** *Economist* 28 Sept. 1003/2 The arrangement..might get trade flowing without abandoning the principle of reunion through elections, and yet it might conceivably inveigle the communists out of their foxholes. **1964** D. F. Dowd in I. L. Horowitz *New Sociol.* 59 The academic profession in America is the social critic's refuge; even, in extreme cases, his foxhole. **1968** *Times* 4 Apr. 8/6 The pound is in the same foxhole as the dollar.

foxie, foxy (fǫ·ksi). *Austral.* and *N.Z. colloq.* [Abbrev. of *fox-terrier* + -y⁶.] A fox-terrier.

1906 E. Dyson *Fact'ry 'Ands* xviii. 246 Like er bally foxie after er rat. **1933** *Bulletin* (Sydney) 13 Sept. 21/1 The first dog I loved lived long—a fat old foxie older than I. **1948** D. W. Ballantyne *Cunninghams* (1963) ii. xv. 192 The dog—a dirty-coated foxie—whimpered across the kitchen. **1966** *Weekly News* (N.Z.) 27 Apr. 40 (*caption*) The smooth-haired 'foxie' probably developed later than the wire-haired type.

fox-mark (fǫ·ksmāɹk). [f. Fox *v.* 4 + Mark *sb.*¹] A brown spot or stain on a print, book, etc., caused by damp affecting impurities in the paper. Hence *fox-marking*; also *fox(e)y* adj., *foxiness.*

1880 W. Blades *Enemies of Books* ii. 23 It grieves the heart to see 'foxey' stains common in his most superb works. **1931** *Book-Collectors' Q.* June–Aug. 13 A recent attempt was made to get at the cause of fox-marks. *Ibid.*, One portion of a leaf separated into two parts was sent to one chemist, and the other portion to another chemist, to report on the fox-marking. **1935** C. A. Mitchell *Docum. & Sci. Exam.* 24 The so-called 'foxy' appearance of paper caused by the oxidising action of mould fungi or chemical processes of oxidation. **1937** E. J. Labarre *Dict. Paper* 149/2 *Foxed, foxing* or *foxy*, stains, specks or spots in paper, e.g. prints, books, mostly due to mould or paper-mildew. **1956** *Bookman's Conc. Dict.* 117/2 *Foxiness*, decay or spottiness of book pages. **1958** *Oxf. Mail* 19 Mar. Suppl. 21/4 First a word on 'fox-marks'. These brown stains..are caused by damp. **1960** G. A. Glaister *Gloss. Bk.* 28/2 Fox marks and certain apparently indelible stains on printed paper can be removed by bleaching.

fo·x-trot, *sb.* **1.** (See Fox *sb.* 16 and add earlier and later examples.)

1872 F. M. A. Roe *Army Lett.* (1909) 70 He has a fox trot, which is wonderfully easy. **1894** R. Kipling *Day's Work* (1898) 58 Would you consider a fox-trot, an' single-foot, an' rack, an' pace, *an'* amble, distinctions not worth distinguishin'? **1946** M. C. Self *Horseman's Encycl.* 134 *Fox trot*, a slow, shuffling trot, the fox-trot is one of the gaits permitted in a five-gaited saddler as a 'slow-gait'.

2. A modern dance, of American origin, consisting chiefly of alternating measures of long and short steps; also, a piece of music suitable as an accompaniment for the fox-trot.

1915 *Truth* 17 Mar. 1/5 A new dance, the 'Fox-trot', a relation of..'Ragtime'. **1915** *Victor Record Catal.* May, Dance records... Fox trots. **1917** S. B. Leacock *Frenzied Fiction* (1919) v. 70 The others were dancing the fox-trot to the victrola on the piazza. **1919** G. D'Egville *How & what to Dance* (1922) 55 The Fox-Trot is a dance of many steps, and to the casual observer everybody seems to have different ones. **1919** E. Scott *All about Latest Dances* 68 The true basis of the American Fox-Trot is an alternation of four slow and four or eight quick movements, depending on the step chosen. **1923** —— *A.B.C. of Dancing* 84 The foxtrot is not a dance in the sense that the waltz and polka are dances because it has no distinctive rhythm and no characteristic step or figure. **1928** *Melody Maker* Feb. 127/1 You have just heard a fox-trot, 'I call her honey because she sticks to me'. **1946** R. Capell *Simiomata* ii. 48 Kirou remembers Macaskie singing foxtrots.

Hence **fo·x-trot** *v. intr.*, to dance a fox-trot; **fo·x-trotter,** one who fox-trots; **fo·x-trotting** *vbl. sb.*

1916 *Daily Colonist* (Victoria, B.C.) 11 July 3/7 Such [folk] dances should not be classed with fox-trotting, bunny-hugging, Wilson gliding, [etc.]. **1919** *Punch* 26 Feb. 166/2 I've fox-trotted in Stranraer, Irish-jigged in Mullingar. **1919** E. Scott *All about Latest Dances* 72 Some fox-trotters spin round at intervals on the soles of their feet. **1923** R. D. Paine *Comrades of Rolling Ocean* v. 75 A dozen couples..fox-trotted in a clear space of the forward deck. **1928** F. N. Hart *Bellamy Trial* i. 18 He seems to have spent most of his time perfecting his golf game and his fox-trotting abilities. **1967** V. Nabokov *Speak, Memory* (ed. 2) xiii. 254 On the Cunard liner *Pannonia* which left Greece on May 18, 1919..I learned to foxtrot.

foxy, *a.* Add: **1.** (Later examples of sense 'crafty, cunning'.)

1895 W. C. Gore in *Inlander* Dec. 110 *Foxy*, cunning, wary, clever, tricky, politic. 'That's a foxy trick.' 'The French ambassador is a foxy old bird.' **1946** G. Millar *Horned Pigeon* xx. 335 And he is much too foxy to let a poor girl even smell at his antecedents. He is so clever that he speaks even his native language without accent. **1957** Mezzrow & Wolfe *Really Blues* (Gloss.) 373 *Foxy*, sly, clever.

b. Of weather: misleadingly bright.

1876 in *Eng. Dial. Dict.* **1909** *Westm. Gaz.* 10 July 3/1 Fine 'foxy' days, when hot sunshine burns the breeze up, don't last. **1933** 'L. Luard' *All Hands* 240 'Tis foxy old weather.

3. (Additional example.) Cf. also *fox-mark.

1817 W. Tucker *Family Dyer & Scourer* iii. 28 Should the silks appear rusty, or what is known to the dyers by the name of copper burst, or foxy, it is customary to [etc.].

6. *foxy-featured, -looking* adjs.

1899 S. R. Crockett *Black Douglas* x. 67 A red, foxy-featured man, with mean and shifty eyes. **1924** J. Masefield *Sard Harker* 56 A middle-aged, foxy-looking man.

foxy, var. *foxie.

foyer. Add: **1. b.** Hearth, home.

1908 *Daily Chron.* 28 Jan. 4/6 The love of parents for their children..makes the every-day life of the foyer lovely. **1959** R. Postgate *Every Man is God* xxiii. 210 Happy families are centripetal; they have a heart, a centre to which all the rest are attracted..there is a warm *foyer* of love to hold that group together.

2. Also, the entrance hall of a hotel, restaurant, theatre, etc.

1901 *Dialect Notes* II. 140. **1910** *Bradshaw's Railway Guide* Apr. 1116 The Restaurant with Foyer is one of the prettiest Dining Rooms in London. **1915** 'Bartimeus' *Tall Ship* iv. 77 There were at least half a dozen mothers in the *foyer* of the big..hotel. **1938** E. Bowen *Death of Heart* iii. iv. 380 She had proposed to meet him in the entrance foyer of Quayne and Merrett's.

3. = Focus *sb.* 5 b.

1799 Malthus *Jrnl.* 29 May (1966) 36 The Emperor of Russia had..been..misinformed in supposing that Hamburgh was the foyer of Democrats. **1939** A. J. Toynbee *Study Hist.* VI. 36 A political *foyer* for acts of religious creation.

frabjous (fræ·bdʒəs), *a.* A nonsense-word invented by 'Lewis Carroll' (C. L. Dodgson), app. intended to suggest 'fair' and 'joyous'; used vaguely by others in various contextual senses. Hence **fra·bjously** *adv.*

1872 'L. Carroll' *Through Looking-Glass* i. 24 O frabjous day! Callooh! Callay! **1899** Kipling *Stalky & Co.* 144 Oh, you frabjous asses! *Ibid.* 244 You're a dirty little schoolboy. Besides bein' frabjously immoral. **1935** 'D. Smith' in *Famous Plays 1935–36* 171 I'm frabjously late. **1937** D. L. Sayers *Busman's Honeymoon* iv. 96 Oh, frabjous day!.. All my life I have waited to hear those exquisite words, *Peter darling, The sweep's come.* **1970** *New Yorker* 10 Jan. 6/1 A frabjous sort of place in a somewhat vorpal neighborhood.

fraction, *sb.* Add: **6.** A portion of a section (sense 2 e (b)) of land. *U.S.*

1789 in C. Cist *Cincinnati* (1841) 209 The whole of the township and fraction to be surveyed. **1815** D. Drake *Cincinnati* vi. 202 The principal wall or embankment, encloses an entire block of lots and some fractions. **1837** J. M. Peck *Gaz. Illinois* i. 77 Fractions are parts of quarter sections intersected by streams or confined claims. **1847** in H. Howe *Hist. Coll. Ohio* 206 Matthias Denman.. had purchased the fraction of land on the bank of the Ohio, and the entire section adjoining it on the north. **1901** S. E. White *Westerners* xviii. 156 Thar is a lode.. over on the J. G. fraction that's shore th' purtiest bit of quartz lead you ever see.

7. Any one of the portions, differing in physical or chemical properties, into which a mixture may be separated, esp. by physical methods.

1857 *Phil. Mag.* 4th Ser. XIII. 134, I selected the fraction boiling in the fifteenth rectification at 240°F. to make a preliminary experiment upon. **1873** *Ibid.* XLV. 129 In such a case each fraction of the distillate will have the same composition as every other fraction and as the original mixture. **1904** [see *fractionally *adv.*]. **1913** A. R. Warnes *Coal Tar Distill.* 52 When it is not desired to recover the anthracene, the heavy creosote and anthracene oil fractions are sometimes not separated, but worked as one fraction. **1955** *Sci. Amer.* Aug. 92/3 Electrophoresis is the special name given to the technique of separating molecular mixtures into fractions. **1958** *New Scientist* 9 Jan. 12/1 The purification of these liquids gives rise to aromatic fractions suitable for use as gasoline components. **1960** L. Picken *Organization of Cells* v. 188 Five different particle fractions were..separated from the homogenate at five different speeds and times of centrifugation. **1971** *Nature* 13 Aug. 455/2 Textural variation depends on changes in the proportions of sand and clay with the silt fraction remaining fairly constant.

8. In Communist use: see quots. 1922 and 1927. Also, a deviant or schismatic group. Hence **fra·ctionism,** the views or policies of such a group; **fra·ctionist,** a member or adherent of such a group. Also (all forms) *transf.* Cf. *fractional *a.* c.

1922 *Communist Party Gt. Brit.: Rep. Organisation* ii. 25 Besides his area group, every member will have some special task. He may be on the Speakers' Group, or working in the Local Labour Party Fraction. *Ibid.* 26 A Fraction is a Party organisation inside a representative or delegate body. (It is also used for a grouping of all the Communists and their followers inside a trade union or similar organisation.) **1927** *Communist Party Training (Communist Party Gt. Brit.)* 117 A fraction is composed of all Party members inside any unit or representative committee of a non-party organisation, united for the purpose of conducting Party work in same. **1949** *Times* 10 Jan. 6/1 Other new words recently collected from this paper include *fractionist* (akin to deviationist). **1952** *New Yorker* 25 Oct. 106/2 Among Marty's other criminal fractionisms were that he thought American imperialism in France was of only secondary importance. **1952** M. McCarthy *Groves of Academe* (1953) xiii. 262 In the fraction, I was given the assignment of recruiting Mulcahy to the Party. **1954** *Encounter* Feb. 48/2 Last came the rank and file, whose work consisted of making speeches, distributing leaflets, attending party and fraction meetings. **1954** Koestler *Invis. Writing* xxiii. 255 It is essential to..discredit a person or fraction 'in the eyes of the masses' before he or they are liquidated. **1958** *New Statesman* 1 Feb. 136/2 The fact is that Conservatives, Liberals, Socialists and Roman Catholics all hold 'fraction meetings' and decide what their policy is to be in the trade unions. **1959** B. & R. North tr. *Duverger's Pol. Parties* (ed. 2) i. iii. 174 The development of fractions is not a sign of the liberty of members..: rather does it point to differences of opinion between members of the ruling class.

fractional, *a.* Add: *fractional currency* (U.S. examples); *fractional note* (*N. Amer.*), a note in fractional currency; *fractional section* = *fraction *sb.* 6.

1815 D. Drake *Cincinnati* 129 Cincinnati is built upon one entire and two fractional sections. **1863** *Stat. at Large U.S.A.* (1864) XII. 711 In lieu of postage and revenue stamps for fractional currency, and of fractional notes, commonly called postage currency,..the Secretary of the Treasury may issue fractional notes of like amounts in such form as he may deem expedient. **1878** *Congress. Rec.* 29 Jan. 638/2 [Silver] is not like our fractional notes, a promise to pay. It is payment. *Ibid.* 4 Feb. 727/2 The fractional currency had become so ragged and so dirty that people were willing to surrender it for almost anything. **1903** R. Stiles *4 Yrs. under Marse Robert* 63 In the South during the war banks, municipalities, companies, and..individuals issued fractional notes or shin plasters which passed as currency supplementary to the Treasury notes issued by the Confederate Government. **1946** J. T. Adams *Album Amer. Hist.* III. 135 Later (July, 1862) Congress legalized small change notes by providing for the issuance of fractional currency in denominations of 3, 5, 10, 15, 25 and 50 cents.

b. *Chem.* Designating a process in which different fractions of a mixture are separated in consequence of their differing physical properties. So *fractional crystallization, distillation* (see Distillation 3), *precipitation*, etc.

1857 W. A. Miller *Elem. Chem.* III. i. 9 This process of *fractional* distillation is well adapted to the separation of liquids. **1886** *Chem. News* 10 Sept. 131/1 Fractional crystallisation has yielded new results with didymium. *Ibid.*, Working with the samarskite earths, fractional precipitation with oxalic acid separates first erbia, holmia, and thulia, then terbia, and lastly yttria. **1897** G. F. Becker in *Amer. Jrnl. Sci.* CLIV. 257 (*heading*) Fractional crystallization of rocks. *Ibid.* 258 The process.. usually known as fractional crystallization..has been employed in the purification of compounds ever since chemistry was pursued. **1949** Alexander & Johnson *Colloid Sci.* II. xxvii. 797 Fractional solution may also be employed, a series of solvents of increasing solvent power being used to extract the unfractionated polymer. **1950** N. V. Sidgwick *Chem. Elements* I. 456 Fractional precipitation of the hydroxides is useful in isolating the very weakly basic $Sc(OH)_3$ or the strongly basic $La(OH)_3$. **1960** Turner & Verhoogen *Ign. & Metamorph. Petrol.* (ed. 2) iv. 85 The nature of the liquid fraction of a crystallizing magma at a given moment depends on.. the degree to which fractional crystallization has already been effective. **1966** *McGraw-Hill Encycl. Sci. & Technol.* XI. 118 Fractional distillation to remove iron, lead, and cadmium from crude zinc. **1968** Coulson & Richardson *Chem. Engin.* (ed. 2) II. xiv. 577 It is often possible to separate a mixture of two soluble salts by fractional crystallization.

c. In Communist use: of, or pertaining to, a fraction (sense *8). Hence **fra·ctionalism,** the doctrine or policy of a fraction; an instance of deviating from the official line of the Communist Party; **fra·ctionalist,** an adherent or supporter of fractionalism; also *attrib.* or as *adj.* Also (all forms) *transf.* Cf. *fractionism, -ist (s.v. *fraction *sb.* 8).

1929 *Communist Party Gt. Brit.: New Line* 77 Greater attention must be given to..the strengthening of all fractional activities within the trade unions. **1941** Koestler *Scum* 157 The *abstentionisme* of the Chamber was intended to be a subtle fractional device; in fact it was a declaration of irresponsibility. **1945** —— *Yogi & Commissar* 203 Behind the façade of Popular-Front Committees..the air was thick with Comintern-intrigue and fractional conspiracy. **1950** E. H. Carr *Bolshevik Rev.* I. viii. 202 The criticism of individuals or even of groups would be tolerated within the party, but the opposition must not organize: that would be to commit the sin of 'fractionalism'. **1953** *Newsweek* 2 Mar. 38/3 Two months ago he was finally ousted from the French Communist Party as a 'fractionalist'. **1959** *Oxf. Mail* 28 Jan. 1/7 Mr. Podgorny..said the 'fractional activity' of the group had been exposed. **1959** *Times* 29 Jan. 8/3 The anti-party group members were accused of 'fractionalism'. **1960** *Spectator* 19 Feb. 241 These last he dismissed as 'fractionalist' because, he claimed, they subordinated the national interest to that of fractions of the nation in coalition.

fractionally *adv.*, (also), in *Chem.*, so as to separate into fractions.

1871 *Jrnl. Chem. Soc.* XXIV. 405 Bromotetracodeine is conveniently obtained as hydrobromide by..dissolving the residue in the smallest possible quantity of weak hydrobromic acid, and fractionally precipitating by cautious addition of stronger acid. **1904** F. Soddy *Radio-Activity* 17 If the active barium chloride was fractionally crystallised, the activity tended to concentrate in the least soluble fraction. **1954** Reilly & Rae *Physico-Chem. Methods* (ed. 5) II. 253 Intimate mixtures are fractionally disintegrated.

fractionalize (fræ·kʃənălaiz), *v.* [f. Fractional *a.* + -ize.] *trans.* To break up or separate into distinct parts or fractions. Hence **fra·ctionalizing** *vbl. sb.* Also **fra:ctionaliza·tion,** an act of, or the process of, fractionalizing.

1931 *Economist* 5 Sept. 420/2 Their report underlines even more categorically..the disadvantages which result from the fractionalising of the economic life of Europe. **1945** *Annals Amer. Acad. Pol. & Soc. Sci.* Jan. 55/1 Hampered by political fractionalization..the countries of eastern Europe have nevertheless made rapid progress. **1947** *Amer. Friend* 24 July 286/1 It is so easy to sectionalize and fractionalize humanity. **1955** *N.Y. Times* 17 June 14/7 John B. Hollister..said he thought it would be a mistake to begin an 'early fractionalization' of foreign aid activities. **1957** L. F. R. Williams *State of Israel* 158 The Israel system has the disadvantage of so fractionalizing the representation in the legislature that no party has yet succeeded in gaining an absolute majority. *Ibid.* 159 The fractionalization of representation in the Knesset between some fourteen groups. **1967** *Times Rev. Industry* July 89/1 It [*sc.* information] is fractionalised; in other words, it is scattered about between different departments and different people.

fractionate, *v.* Add: **b.** *gen.* = prec.

1961 in Webster. **1963** C. D. Simak *They walked like Men* xxvi. 155 Might there not be no more than a single one of them, with that one fractionated into units the size of bowling balls? **1965** *Listener* 12 Aug. 233/2 We spend much of our time fractionating living matter. We break it up with sonic oscillations..or by chilling it in solid carbon dioxide and hitting it with a hammer. **1970** *Sci. Jrnl.* Apr. 5/1 Pollution matters beat even space activities for the extent to which they are fractionated between different departments—no less than 12 are involved.

2. *intr.* To break up into fractions.

1957 G. E. Hutchinson *Treat. Limnol.* I. viii. 543 If the drop fractionates, small daughter drops may be carried to the top of the cloud in an updraft.

Hence **fra·ctionating** *ppl. a.* and *vbl. sb.*; *fractionating column*, a long vertical chamber for fractionating liquids in which vapour from a still passes upwards to a condenser and mixes intimately with falling condensate on the way, so becoming enriched in the lower-boiling fractions.

1908 *Jrnl. Chem. Soc.* XCIV. ii. 17 (*heading*) A Fractionating Column... A condenser is provided inside the top of the..apparatus, and the distilled liquid is collected at the lower part, being returned to the distillation flask or run off as desired. **1922** C. S. Robinson *Elem. Fract. Distill.* iii. 17 The second general method by which fractionating can be obtained, is fractional condensation. **1952** Smith & Stinson *Fuels & Combustion* i. 22 The vapors are passed into the fractionating column where the oil is separated into several fractions. **1970** R. M. E. Diamant *Appl. Chem. Engineers* (ed. 2) xiv. 248 The degree of separation of the liquid into upper and lower fractions depends upon the number of plates or the length of the fractionating column.

fractionation. (See under Fractionate *v.*) Add examples of uses other than in Chemistry.

1926 *Encycl. Brit.* II. 164/1 The production of an innumerable series of colour-forms, as in the sweet pea, is almost certainly due to the fractionation of the colour-complex. **1934** H. C. Warren *Dict. Psychol.* 109/2 *Fractionation*, a piecemeal method of describing a conscious experience, in which the situation is repeated exactly.. a number of times with varied instructions. **1951** G. Humphrey *Thinking* 68 The subject might be required to notice..what happened in consciousness before the stimulus word appeared. This was the method of fractionation. **1963** *Language* XXXIX. 11 'Position' in a sentence is defined by successive fractionations, i.e. when a sentence has just two positions, each of which may be occupied by a word or a phrase, and when each constituent phrase in turn may be divided into two positions. **1966** *Rep. Comm. Inquiry Univ. Oxf.* I. 135 The growth and present size of most departments now means that these duties are far beyond the grasp of a single head, and fractionation has become essential to well-being and progress.

fracto- (fræ·kto), used as comb. form of L. *fractus* broken, in Meteorology (see quots.).

1896 *Internat. Cloud-Atlas (Internat. Meteorol. Committee)* 18 (*heading*) International designations of cloud-forms... Fr. Cu...Fracto-cumulus. Fr. N...Fracto-nimbus. Fr. S...Fracto-stratus. **1917** A. G. McAdie *Princ. Aërography* 115 The true cumulus..is often torn by strong winds,..('fracto-cumulus'). *Ibid.*, If the mass of nimbus is torn up into small patches..they may be called 'fracto-nimbus', the 'scud' of the sailors. **1926** —— *Man & Weather* 68 The word *fracto*..is now in general use to designate a cloud form in which the mass is broken into small divisions. **1956** J. C. Swayne *Conc. Gloss. Geogr. Terms* 62 *Fracto-nimbus cloud* (*scud*), ragged fragments of cloud drifting rapidly in a strong wind often under rain clouds.

fracturable (fræ·ktiŭrăb'l, -tʃərăb'l), *a.* [f. Fracture *v.* + -able.] That can be fractured.

1897 *Kodak News* May 6 Nor should there be anything easily fracturable. **1911** H. S. Harrison *Queed* xxxi, He might fracture all of the decalogue that was refinedly fracturable.

fracture, *sb.* Add: **6.** *Geol.* (of fracture in the earth's crust): *fracture-line, -system, -zone.*

1925 J. Joly *Surface-Hist. Earth* viii. 140 The long rift valley of South Australia, mainly meridional in direction, as well as meridionally-directed fracture-lines along the eastern side of that continent. **1922** *Encycl. Brit.* XXXI. 214/2 Rectilinear and regularly intersecting fracture-systems. **1965** A. Holmes *Princ. Physical Geol.* (ed. 2) xxvi. 939 By far the most astonishing structures of the East Pacific Rise are the E–W fracture zones which have sliced the Rise into long crustal slabs.

fracture, *v.* Add: **1. b.** To impress, excite, amuse greatly. *U.S. slang.*

1946 H. Brown *Sound of Hunting* ii. 116 This guy Muller fractures me. **1951** M. Shulman *Many Loves of Dobie Gillis* (1953) 41 We're a riot, hey. We play all kinds of funny stuff. We fracture the people. **1955** L. Feather *Encycl. Jazz* (1956) x. 346 *Fracture*, impress, excite, amuse. **1966** *Crescendo* Aug. 20/2, I know he fractured you the same as he did me. **1970** H. E. Roberts *Third Ear* 7/1 *Fracture v.*, to make one laugh.

3. *Phonology.* (*trans.*) To subject to fracture.

1889 Ellis *E.E. Pronunc.* v. 496 The peculiar manner of fracturing the vowels in the A-, A', Æ, E-, EA', O' words by prefixing an accented (ee) or (ii) and reducing the vowel itself to indistinct (ə).

frænulum. Add: Also **frenulum** (frе·niŭlə̆m). Hence **fre·nate, fre·nular** *adjs.*

2. *Ent.* A bristle or group of bristles attached to the base of the hind wing in many Lepidoptera, and interlocking with a process on the front wing, thus uniting the two wings of a side in flight.

1892 G. F. Hampson *Fauna Brit. India, Moths* I. 6 The fore and hind wings of the majority of families of moths are united by a 'frenulum' arising from the base of the costa of the hind wing and fitting into a 'retinaculum' on the fore wing. **1898** A. S. Packard *Text-bk. Entomol.* 122 In many Lepidoptera they [*sc.* the wings] are loosely connected by the loop and frenulum. **1925** A. D. Imms *Gen. Textbk. Entomol.* iii. 418 Among Heteroneura two main types of wing-coupling apparatus are also evident, viz: the frenate and the amplexiform. In the frenate type a sexual difference is very noticeable. **1969** R. F. Chapman *Insects* viii. 175 Many other Lepidoptera have the frenulum well developed. *Ibid.*, In Micropterygidæ the jugum is folded under the fore wings and holds the frenular bristles.

frag (fræg), *v.* *U.S. Mil. slang.* [Abbrev. of **fragmentation* (*grenade*).] To throw a fragmentation grenade at one's superior officer, esp. one who is considered over-zealous in his desire for combat. So **fra·gging** *vbl. sb.*

1970 *Daily Colonist* (Victoria, B.C.) 9 Dec. 5/4 Another spectacular technique is 'fragging'. 'To frag' is a term meaning to use a fragmentation grenade 'to cool the ardor of any officer or NCO too eager to make contact with the enemy'. **1971** *Courier-Mail* (Brisbane) 19 May 1/3 There were 209 fragging incidents last year, according to the Army and 34 deaths were listed as probably due to these. **1972** *Sat. Rev.* 8 Jan. 12/1 Fragging is a macabre ritual of Vietnam in which American enlisted men attempt to murder their superiors. The word comes from the nickname for hand grenades, a weapon popular with enlisted men because the evidence is destroyed with the consummation of the crime.

fragile, *a.* The usual pronunciation in Britain is now (fræ·dʒail).

fragment (fræ·gmĕnt, frægme·nt), *v.* [f. the sb.] *trans.* and *intr.* To break or separate into fragments.

1818 Keats *Endymion* iii. 845 For what poor mortals fragment up, as mere As marble was there lavish. **1901** H. McIntosh *Is Christ Infallible?* i. iii. 76 They vainly dream that they can fragment and vivisect the Spirit's embodiment and environment of Christ's teaching. **1919** *Proc. Class. Assoc.* 27 Specialism..has fragmented the specialities themselves in a way that makes the outlook hazardous. **1961** *Time* 7 July 30 The Revolutionary Council..was fragmenting in despair.

fragmentation. Add: (Further examples.) Also **fragmentation bomb, grenade,** one designed to disintegrate into small fragments on explosion.

1918 *Aviation* 15 Nov. 502/2 Deliveries of the 'demolition' and 'fragmentation' bombs..were made in April. **1920** H. G. Wells *Outl. Hist.* I. vi. xxx. 380/1 The state of Europe beneath its political fragmentation was a social disorder. **1943** *Roof over Britain* ii. 10 The fragmentation of metals under certain conditions. **1944** *Times* 10 Feb. 4/6 An airfield was hit with fragmentation bombs. **1957** F. Krollmann *Langenscheidts Fachwörterbuch Wehrwesen* 647/1 Fragmentation shell..fragmentation grenade. **1958** *Spectator* 15 Aug. 213/1 The fragmentation of French political parties. **1959** *Listener* 23 July 152/2 The music is a study in fragmentation. **1970** *Peace News* 3 Apr. 5 (*caption*) This Vietnamese woman was the victim of a fragmentation bomb (made in the USA). **1970** Fragmentation grenade [see *frag *v.*]. **1971** *Daily Tel.* 22 Apr. 4/5 Incidents of American soldiers in Vietnam tossing fragmentation grenades at officers more than doubled, to 209, last year.

fragmented, *pa. pple.* and *ppl. a.* (Earlier and later examples.)

a **1817** Jane Austen *Northanger Abbey* (1818) II. ix. 172 In the shape of some fragmented journal, continued to the last gasp. **1955** *Times* 24 June 10/7 The necessity of avoiding such isolation..was felt more profoundly than elsewhere in a Europe fragmented by history. **1964** G. L. Cohen *What's Wrong with Hospitals?* v. 99 With responsibility for the expectant mother fragmented or frantically shuffled between three authorities (hospital, local health, general practice).

fragmentize, *v.* Add: So **fra:gmentiza·tion.**

1951 *Yale Rev.* Spring 514 The majority of the population, up in arms against a fragmentization of their native land. **1968** *New Scientist* 21 Mar. 617/1 The French Minister for Science, M. Maurice Schumann objected, we have been told, to the use of the phrase 'fragmentization of Europe' in the draft report of last week's OECD meeting.

'fraid (frēɪd), *a. colloq.* [aphet. f. Afraid *ppl. a.*] (I'm) afraid. So (chiefly *Children's*) phr. *fraid cat* (also *fraidy cat*), a coward. Cf. Frayed *ppl. a.*[1]

1888 Kipling *Story of Gadsbys* (1889) 85 'Fraid you won't be entered in the Stud Book correctly unless you go Home? **1895** —— *Day's Work* (1898) 195 Wish I could, Lizzie. 'Fraid I can't. **1908** G. Jekyll *Children & Gardens* (caption, facing p. 64), 'Fraid I've dropped a stitch! *c* **1910–23** in Wentworth *Amer. Dial. Dict.* (1944) 'Fraidy cat. **1912** *Dialect Notes* III. 576 *Fraid-cat*, a coward. **1942** C. Morley *Thorofare* xlix. 286 You're an old fraid-cat. **1945** N. L. McClung *Stream runs Fast* iv. 26 He might not have been so composed, if he had known he had two fraidy-cats to take care of him. **1968** C. Aird *Henrietta Who?* i. 4 'Accident?' shouted the man at the wheel... 'Fraid so,' shouted back Ford. **1968** D. Gray *Died in Red* xiii. 65 'So you won't go on working here?' 'No. 'Fraid not.'

frail, *sb.*[1] Add: **3.** *frail basket.*

1815 A. Constable *Let.* 27 Jan. in *J. Constable's Correspondence* (1962) I. 112 The frail basket in which the turkey is, is your own. **1900** A. Morrison *Cunning Murrell* i. 9 Over his shoulder he carried a large gingham umbrella..and from its handle hung a frail basket.

frail, *sb.*[3] *slang* (chiefly *U.S.*). [Subst. use of Frail *a.*] A woman.

1908 H. Green *Maison de Shine* 50 Aw, the frails is all the same... A guy comes along and shoots that old con about how he's the grandest thing on earth, an' the wisest of 'em fall. **1926** *Amer. Speech* I. 462/1 The Apollo Theater in London prints the following glossary of slang in its program as a guide to 'Is Zat So?' *Dame, Frail*,.. Girl. **1931** E. Linklater *Juan in Amer.* II. xvi. 177 Bullets whistling through the air to..threaten widowhood for the ravished frail. **1945** P. Cheyney *I'll say she Does* v. 141 She's a swell dish—a lovely piece of frail. **1970** K. Platt *Pushbutton Butterfly* (1971) iv. 36 A smaller soggy shape was huddled behind him...An Angel and his frail challenging another night.

frail, *v. U.S. dial.* [prob. f. Eng. dial. *frail* flail.] *trans.* To beat, thrash. Hence **frai·ling** *vbl. sb.*

1851 M. L. Byrn *Arkansaw Doctor* 82 (Th.), The old man plainly told her..he would frail her worse than a dog would a pole-cat. *Ibid.* 123 (Th.), He..did not like the thought of getting a frailing for it. **1890** J. C. Harris in *Century Mag.* Dec. 287/1 He uped an' frailed me out, an' got the gal to boot. **1896** *Peterson Mag.* Jan. 89/2 A frailing with a dead branch left him [*sc.* a mule] unmoved. **1901** W. Churchill *Crisis* I. xii. 97 Reckon I'd frail 'em good ef he catched hold of me with his black hands. **1932** W. Faulkner *Light in August* (1933) xi. 229 I'll frail the tar out of you.

Fraktur (fræ·ktūɹ). *Typogr.* [G.] A German style of black-letter.

[**1886** T. L. De Vinne *Historic Printing Types* ii. 18 Considering its angularity, the name, Fractur, which Germans give to their modern German character is well chosen.] **1904** —— *Mod. Meth. Bk. Composition* 253 Though black and forbidding, fraktur is a thin character. **1926** S. Morison *Type Designs* 11 [Gutenberg's] informal type is..interesting..mainly as the ancestor of the *fraktur* and *schwabacher* later employed for vernacular and less important uses. **1965** *Amer. Speech* XL. 106 In handwriting and printing, the German *Fraktur* script was used exclusively [amongst German-Americans]. **1969** H. Carter *View Early Typogr.* iii. 57 The Schwabacher was the normal vehicle for German from 1490 until, about 1540, it was outdone in favour by Fraktur. *Ibid.* iv. 69 The Fraktur..is a revolution within the gothic, expressing, as it does, a courtly taste of the highest secular majesty. **1970** *Times Lit. Suppl.* 7 Aug. 884/5 Among the vernaculars, the German *Fraktur* still survives.

framboise. Restrict †*Obs.* to sense in Dict. and add: **B.** *adj.* Of raspberry colour. Also *absol.*

1904 *Daily Chron.* 24 Sept. 8/3 The soft..framboise tones. **1959** *Times* 28 Sept. 13/2 It [*sc.* a suit] can also be obtained in size 18..in framboise.

frame, *sb.* **2.** Delete †*Obs.* and add later examples in *U.S. slang*; esp. = *frame-up.

1914 Jackson & Hellyer *Vocab. Criminal Slang* 35 *Frame*,..a prearranged plan of action; a secret implying sinister intention; a 'frame-up'. **1924** G. Bronson-Howard *Devil's Chaplain* ii. 36 This louse woulda beat us to the frame if I hadn't plugged him. **1948** 'J. Evans' *Halo for Satan* (1949) xiii. 175 He..wasn't a killer but just the victim of a frame.

4. d. In full *frame of reference*: (i) A system of co-ordinate axes in relation to which position may be defined and motion conceived of as taking place.

1897 A. E. H. Love *Theoret. Mech.* i. 5 A set of lines of reference such as *OA, OB, OC* with respect to which the position of a point *P* can be determined will be called a frame of reference. *Ibid.* xiii. 360 'Acceleration', and by consequence 'force', have no meaning except as dependent on a frame; 'acceleration' means 'acceleration relative to a frame', and similarly with force. **1928** A. S. Eddington *Nature Physical World* iii. 61 The particular frame in which we are relatively at rest has a symmetry with respect to us which other frames do not possess. **1942** Synge & Griffith *Princ. Mech.* i. 12 Latitude and longitude define position on the earth's surface; we are here using the earth as a frame of reference. **1965** J. D. North *Measure of Universe* xvi. 362 The triply-infinite set of inertial frames connected by the Lorentz transformations is often referred to as a new 'Absolute'. **1967** Rodberg & Thaler *Introd. Quantum Theory of Scattering* x. 260 The *S* matrix we have been using.. describes collisions in an arbitrary reference frame.

(ii) *fig.* and *transf.* (Always as *frame of reference.*) A set of standards, beliefs, or assumptions governing perceptual or logical evaluation or social behaviour.

1924 *Sci. Amer.* Dec. 401/1 This is the mind's 'frame of reference' for the facts of the outer world. **1933** G. B. Shaw *Polit. Madhouse Amer.* 16 If I may borrow an expression from my friend Professor Archibald Henderson, who is a mathematician, he had no frame of reference. He had no scientific postulates of any kind. He was in the air. **1936** M. Sherif *Psychol. Soc. Norms* ii. 9 We shall give concrete examples to illustrate the existence of norms or frames of reference which are different from those that are taken by western civilization to be as 'natural' as air or water. **1949** Koestler *Insight & Outlook* xxvii. 366 In each pair two universes clash, two self-contained frames of reference, two hierarchies of values intersect. **1957** V. G. Childe *Dawn Europ. Civilization* (ed. 6) x. 176 But his [*sc.* Montelius's] disciples and imitators have clumsily extended his system beyond the regions for which it was devised and have used it as a frame of reference into which cultural phenomena in Central Europe, South Russia, and even Turkestan must be fitted! **1958** *Listener* 6 Nov. 724/1 There is as yet no assured frame of reference for comparisons between the scrolls and the New Testament. **1967** A. Arent *Gravedigger's Funeral* xii. 192 That was the thing about Jenny. Her frame of reference. She was solid.

9. b. *Austral.* and *U.S.* An emaciated animal; *spec.* in *pl.* in *Austral.*, draught cattle. (Cf. sense 11 in Dict.)

1880 *Bradstreet's* 29 Sept. 3/4 The north British farmers are finding it profitable to import what the American dealers graphically call 'frames' to feed for market. **1903** 'T. Collins' *Such is Life* 249 By-the-way, there's four of your frames left—out near those coolibahs. **1934** *Bulletin* (Sydney) 10 Oct. 21/4 No poorer or weaker old frames ever travelled the Birdsville stock route.

10. (Later U.S. example.)

1904 W. N. Harben *Georgians* 299 Half a dozen expert workmen were putting up the frame of a two-storied building on massive pine sills.

b. (U.S. examples.) Also = Frame-house 2.

1732 *Rec. Early Hist. Boston* (1885) XII. 40 Henry Gibbs has very lately Incroched on the Towns Land on Dock Square by Erecting Frames thereon. **1841** C. Cist *Cincinnati* 41 Of these last [buildings] 200 were brick and 121 frames. **1892** 'Mark Twain' *Amer. Claimant* 28 The 'mansion'..was a rickety old two-story frame.

c. *ellipt.* as adj. (= Framed *ppl. a.*) or sb. (= Framework 1). *U.S.*

1790 *Pennsylvania Packet* 3 Apr. 3/3 A good frame loaf bread bake house with one oven. **1816** U. Brown *Jrnl.* in *Maryland Hist. Mag.* (1915) X. 268 The 3rd House is Log and frame with Double Porches. **1836** J. Hall *Statistics of West* iv. 66 The dwelling houses are usually large edifices of brick or frame. **1872** 'Mark Twain' *Roughing It* (1882) xiii. 74 Block after block of trim dwellings, built of 'frame' and sunburnt brick. **1897** J. L. Allen *Choir Invisible* ii. 12 Rude dwellings of logs now giving way to others of frame and of brick. **1924** C. E. Mulford *Rustlers' Valley* iv. 38 The street was a busy one in front of a line of lighted buildings, frame, one-story shacks all.

11. e. That part of a pair of spectacles which encloses the lenses and holds them in their proper position. Also *attrib.*

1847 *Penny Cycl.* XXII. 328/2 Where side-pieces are added to the frame..the instrument is denominated simply a pair of spectacles. **1895** *Montgomery Ward Catal.* 203/2 Steel frame spectacles. **1939–40** *Army & Navy Stores Catal.* 921/2 Lenses fitted to any of the above frames or customers' own frames. **1969** *Which?* Feb. 45 Two-thirds of those buying spectacles pay several pounds extra on private frames.

f. The fixed part of a bicycle. Hence *frame-bag* (a bag for carrying articles, fixed within the frame).

1871 *English Mechanic* 10 Feb. 491/3 The great ease in riding..the makers attribute to the form of the frame. **1897** *Westm. Gaz.* 14 Aug. 7/2 In the frame bag was a package containing blocks of writing paper. **1898** *Cycling* 8 Lacking the elaborate plant of the great firms—especially that for frame building—they are apt to fit their frames together untruly. **1968** *Which?* Aug. 233 We measured the strength of front forks (the parts of the frame which are likely to bend or crack first), pedal cranks and carriers. **1971** *Oxford Mail* 6 Oct. 16/1 (Advt.), Boys bicycles, 16 in. frame, relatively little used.

g. In Pool, the triangular form used in setting up the balls; also, the balls as set up, or the round of play required to pocket them all; similarly in Skittles and Tenpin Bowling; also, one of the several innings forming a game.

1890 C. C. Moore *Games of Pool* 79 These balls are placed in the form of a triangle upon the table, as in Pyramids, a wooden frame or 'triangle' being employed to save trouble and insure correctness. **1897** R. F. Foster *Compl. Hoyle* 608 (Tenpins) Each player rolls ten frames or innings. **1910** *Hints on Skittles, Offic. Rules* 23 The Frame shall be 4-ft. 6-in. square, the Plate circular, 3-in. in diameter, and 22½-in. from centre to centre. *Ibid.* 24 The ball is not dead until it is motionless on, or off the frame. **1929** *Encycl. Brit.* III. 978/1 The teams roll one frame (2 balls) on one alley, and for the next frame alternate and use the other alley. **1930** A. P. Herbert *Water-Gipsies* xxi, There were thirty competitors for the Cup, and each of them played seven frames. **1958** *Economist* 20 Dec. 1085/1 If he succeeds in spilling all ten pins with one ball, this is called a 'strike', and the bowler is credited with ten points in that particular section of the game (each such section is called a 'frame'). **1968** Mrs. L. B. Johnson *White House Diary* 8 Oct. (1970) 718, I..rolled two balls that went in the gutter for my second frame. **1971** J. Wainwright *Last Buccaneer* i. 10 He could play snooker. It was a three-frame match and he had already won the first frame.

h. The horizontal skeleton of a motor vehicle that supports the body, the engine, and the various mechanisms; also, the structural framework of an aircraft (now usu. *air frame* (*air *sb.*[1] B. III. 6)).

1900 W. W. Beaumont *Motor Vehicles* I. 232 The frame is built up of tubes, and is carried on double-leaf springs on the front axle, which carries stud axles at its ends with forwardly projecting steering arms. **1902** A. C. Harmsworth et al. *Motors* ii. x, A frame which has some diagonal stays or parts which act as diagonals is very desirable, though few car frames are so made. **1909** *Flight* 20 Feb. 103 *Frame*, in French, the term 'chassis' is sometimes used, but more often the word 'fuselage' on account of the bodies of most monoplanes being spindle-shaped. *Ibid.*, *Half-Elliptic Frames*, a frame of the fusiform type which has been curtailed at the middle. **1921** *Chambers's Jrnl.* 341/1 The rattle and jolt of the little frame-car and the noise of the motor made conversation..inaudible. **1922** *Encycl. Brit.* XXX. 24/1 The body [of an aeroplane] is most often a frame of wood compression members and wire bracing. *Ibid.*, To these spars are attached transverse ribs which give the shape of the wing and a light wood edge completes the frame. **1929** *Encycl. Brit.* XV. 893/2 The type of frame which has become practically standard consists of two longitudinal beams.. with cross members spanning between them at intervals. ..The frame with the mechanical parts of the car assembled upon it is called the chassis. **1959** '*Motor*' *Man.* (ed. 36) i. 13 The cruciform bracing could be applied to a frame only to the rear of the engine and gearbox. **1968** G. N. Georgano *Compl. Encycl. Motorcars* 574 Based on light and simple tubular frames, the cars were easily adaptable for various racing classes.

i. *Aeronaut.* A transverse structural member in a fuselage, wing, etc., which follows the outline of the part and supports the longitudinal members and the skin.

1925 E. H. Lewitt *Rigid Airship* i. 13 All the loads carried by the ship are supported by this corridor, which transmits their weight to the transverse frames. **1930**, **1966** [see *former *sb.*[1] 3 e]. **1933** W. Munro *Marine Aircraft Design* v. 74 The scantlings of a wing float frame are given in Fig. 55. The frames are spaced about 8 in. apart. **1966** *McGraw-Hill Encycl. Sci. & Technol.* V. 587/2 Most..stiffened-shell fuselages use both frames and stringers to support the skin.

12. b. *Cinematography.* One of the series of separate pictures on a film.

1916 E. W. Sargent *Technique Photoplay* (ed. 3) 362 *Frame.* A single photograph in the roll of film. A picture one inch wide by three-fourths of an inch high. There are sixteen of these frames to the foot. **1918** H. Croy *How Motion Pictures are Made* 32 The light revealed the postures of the dancer as she appeared on the 'frames' in front of the viewing aperture. **1936** *Discovery* June 192/2 Red figures underneath the black shutter speeds show the standard ciné speeds from 8 to 96 frames per second. **1966** *Listener* 24 Feb. 286/2 The very last (and most horrifying) frame. **1971** *Sci. Amer.* Oct. 35/2 Frame-by-frame analysis of motion pictures of the head movements and eye movements.

c. *Television.* A single complete image or picture built up from a series of lines; formerly also = *field *sb.* 16 d. Also *attrib.*, as *frame frequency.*

[**1928** *Television* I. ii. 10/3 All that is required is simply an adjustment to bring the picture properly into its 'frame'.] **1935** E. H. Robinson *Televiewing* iii. 49 This number of complete 'frames' or pictures is barely sufficient to prevent flicker in the receiver. **1936** O. S. Puckle tr. *von Ardenne's Telev. Reception* i. 5 The Marconi–E.M.I. television system transmits 25 complete pictures per second each of 405 total lines. These lines are interlaced so that the frame and flicker frequency is 50 per second. **1943** [see *field *sb.* 16 d]. **1943** *Gloss. Terms Telecomm.* (*B.S.I.*) 77 In interlaced scanning the frame-frequency is an integral multiple of the picture-frequency. **1955** J. W. Wentworth *Color Telev. Engin.* ix. 249 It takes two fields to produce a complete television image or frame. In a field-sequential color system, it takes 6 fields or 3 frames to produce a complete color picture, since each line must eventually be scanned in all three colors. **1971** *Sci. Amer.* Sept. 183/2 Television stations broadcast 30 frames per second, each frame containing 525 scan lines.

13. d. An open box of slats in which bees build and which can be removed from the hive. Also *attrib.*

1673 *Phil. Trans. R. Soc.* VIII. 6097 The Frame for the Bees to fasten their Work upon. **1875** J. Hunter *Man. Bee-keeping* 137 Select a frame of comb. **1881** *Gardening Illustr.* III. 123 Remove the top, and four or five of the frames, so as to let the bees drop in the hive. *Ibid.*, The bees will run up into the frame hive. **1908** Kipling *Actions & Reactions* (1909) 98 Melissa found a far-off frame so messed and mishandled by abandoned cell-building experiments that, for very shame, the bees never went there. **1955** A. C. Waine *Background to Bee-keeping* v. 46 Part of this problem was solved by the blind Swiss naturalist Huber (1789) who induced his bees to build their combs within wooden frames each of which could be conveniently handled. **1971** D. Galton *Bee-keeping in Russia* ii. 10 The movable frames..now in standard use in Russia as elsewhere.

15. (sense 10) *frame barn, construction, dwelling, shop*; (sense 11) *frame tent*; (sense 13 c) *frame yard.*

1753 in *New Jersey Archives* (1897) 1st Ser. XIX. 284

To be Sold...Plantation..there is on it..a large frame barn, and stables. **1825** *Colonial Advocate* (Toronto) 5 Jan. 1/4 For Sale...A Frame Barn with Stabling, Forty by Forty-two. **1831** J. M. Peck *Guide Emigrants* 182 He may put up a frame barn. **1847** Carlyle *Lett.* 2 Oct. (1904) II. 50 You will be very wise to get that frame barn you speak of. **1935** *Archit. Rev.* LXXVII. 64/3 All Wright houses look like frame-construction houses. **1956** J. M. Richards in A. Pryce-Jones *New Outl. Mod. Knowledge* 381 Rhythmic façade patterns that frame-construction naturally promotes. **1760** in *New Jersey Archives* (1898) 1st Ser. XX. 425 To be sold, a certain Tract of Land,..with three good frame Dwelling-houses. **1832** J. P. Kennedy *Swallow Barn* I. xix. 192 Half a dozen frame dwellings..were scattered over the landscape. **1913** J. London *Valley of Moon* iii. xiv, He had just begun work on a small frame dwelling. **1894** 'Mark Twain' *Pudd'nhead Wilson* i, Two or three brick stores..towered above interjected bunches of little frame shops. **1874** A. Bathgate *Colonial Experiences* x. 134 We reached the township of Maori Point..consisting of..a few 'frame tents', as fixed tents usually lined with green baize or drugget are called. **1961** *Spectator* 23 June 933 Frame tents are what liken the camping fields of Europe to Agincourt. **1963** *Guardian* 21 Jan. 5/2 The frame tent.. is built on a folding tubular frame, with four or five uprights. **1970** *Which?* May 132/1 A frame tent usually consists of two tents. *Ibid.* 136/1 The number of people a frame tent can sleep is determined by the width of the inner tent. **1899** *Daily News* 28 Oct. 7/6 The early chrysanthemums..are well worth a visit. Ask for the 'frame-yard'. **1933** *Jrnl. R. Hort. Soc.* LVIII. i. 13 The plants..should be..left on the beds for two or three days before removing to the frame yard.

16. frame aerial *Radio*, an aerial composed of a rectangle or loop of wire, adapted for directional reception; **frame story**, a story which serves as a framework within which a number of other stories are told.

1921 *Wireless World* 10 Dec. 562/2, I can receive the Dutch Concert...on a..frame aerial. **1924** *Wireless Weekly* 8 Oct. 745/1 A good range of receivers, including reflex sets using small frame aerials. **1960** *Gloss. Terms Telecomm.* (*B.S.I.*) 122 *Loop aerial* (frame aerial), a closed circuit designed to be used as an aerial. **1924** F. Edgerton *Panchatantra Reconstructed* II. i. 4 Each of the five books contains not only a primary story, which we call the 'frame-story', but also at least one, and usually several, 'emboxt' stories. **1960** J. V. Cunningham *Trad. & Poetic Struct.* 60 For the second feature—the general idea of a frame story—it is agreed that no particular model need be sought.

frame, *v.* Add: **8. e.** *to frame up*: to prearrange (an event) surreptitiously and with sinister intent; to plan in secret; to fake the result of (a contest, etc.). *U.S. slang.*

1906 A. H. Lewis *Confess. Detective* ii. 26 One of the gang..cried out to his fellows: 'Hold off! He's pulled his cannister; an' if you crowd him he's framed it up to do Red.' **1910** E. A. Walcott *Open Door* vii. 86 'An' then he frames up dis job on me,' said Jimmy bitterly. **1913** C. E. Mulford *Coming of Cassidy* vii. 119 As long as this deal was framed up, we'll say it was this mornin'. **1919** *Detective Story Mag.* 25 Nov. XXVIII. 5 If you give me the signal I'll frame up something. **1923** R. D. Paine *Comrades of Rolling Ocean* vii. 121 All I need is a little work with your catcher, to frame up signals and so on.

10. To concoct a false charge or accusation against; to devise a scheme or plot with regard to; to make the victim of a 'frame-up'. *slang* (orig. *U.S.*).

1922 E. Titus *Timber* xxvi. 234 So they were after Bryant were they? They were framing him? **1926** C. E. Mulford *Cassidy's Protégé* iv. 40 He had seen honest men framed, and guilty men let off for political reasons. **1926** J. Black *You can't Win* xxii. 347 The police.. knew I was trying to frame myself out; they began framing me in. **1927** *Observer* 6 Nov. 21 Her heart has been given to Tom Mix, who, in consequence, is 'framed'. **1931** *Daily Tel.* 19 Jan. 11/6 They're for ever after me all the time, trying to frame me. **1956** R. Braddon *Nancy Wake* vii. 70 If they were prepared to lie about Marseille then obviously they intended to frame her.

framed, *ppl. a.* Add: *spec.* in U.S. of houses. (Cf. Frame *sb.* 10, 15.)

1639 [in Dict.]. **1751** in *New Jersey Archives* (1897) 1st Ser. XIX. 68 To be sold, a plantation containing..a good fram'd Barn. **1784** *Maryland Jrnl.* 11 May, Advt. (Th.), A large framed House almost as good as new. **1823** E. James *Long's Exped. Rocky Mts.* I. 82 This town.. contained..several framed dwellings of two stories. **1837** J. M. Peck *Gaz. Illinois* ii. 133 The rough..cabin is giving place to comfortable framed or brick tenements.

fra·me-up. *colloq.* (orig. *U.S.*). [See *frame *v.* 8 e, 10.] Anything that has been prearranged or concocted, esp. with a sinister intent; a conspiracy or plot, e.g. for the purpose of incriminating a person on false evidence.

1900 'Flynt' & Walton *Powers that Prey* 141 He could arrange a 'frame-up', and relieve 'Soapy' of the stolen pocketbook, after 'Soapy' had lifted it from his victim's pocket. **1913** C. E. Mulford *Coming of Cassidy* vii. 116 The crookedness would not come more than once in a deal if the frame-up was 'single-odd'. **1919** *New Appeal* (Girard, Kans.) 11 Jan. 1/5 They were the victims of a frameup inspired by class hatred. **1924** *Westm. Gaz.* 12 Feb., The Government, he said, were guilty of manufacturing a plot which never existed. He claimed that the plot was from beginning to end a 'frame-up'. **1956** A. L. Rowse *Early Churchills* xi. 214 Their signatures were very cleverly forged. Coming at such a moment it looks like a frame-up. **1971** *It* 2–16 June 5/2 While serving a six month sentence..Ian learned a lot about frame ups, about prison conditions.

framework. Add: **1. b.** *framework of reference*, †*axes* = *frame of reference (now chiefly in sense (ii)).

1910 E. T. Whittaker *Hist. Theories Aether & Electr.* xii. 446 The physicist..finds it convenient to construct a framework of axes in space and time for the purposes of fitting his experiences into an orderly arrangement. **1914** L. Silberstein *Theory of Relativity* i. 4 We are compelled to give up our earth as a system of reference and replace it..by a framework of axes pointing from an initial point fixed in the sun. *Ibid.* 3 The earth loses its privilege as a framework of reference. **1931** *Amer. Jrnl. Orthopsychiatry* Oct. 493 Clearly the framework of reference within which a department of life is thought partly determines the basic attitude. **1959** *Listener* 14 May 857/1 The idea of a Luddite phase is clearly important in his framework of reference.

framing, *vbl. sb.* **5.** *framing-timber* (later examples).

1940 *Chambers's Techn. Dict.* 353/1 *Framing timber*, the Australian equivalent of carcassing timber. **1957** *N.Z. Timber Jrnl.* Mar. 52/2 *Framing timber*, wood used in the carcassing of buildings. **1968** *Gloss. Formwork Terms* (*B.S.I.*) 9 *Carcassing timber* (framing timber), timber used for any structural purpose in the support of the forms, but not normally in contact with the concrete.

franchisal (fræ·ntʃizăl, -tʃəiz-), *a.* [f. Franchise *sb.* + -al.] Of or belonging to the franchise.

1897 F. W. Maitland *Domesday Bk.* 80 With feudal justice therefore we contrast 'franchisal' justice. **1954** *Bull. Inst. Hist. Res.* XXVII. 31 There was no court, royal, communal or franchisal, which did not come within the purview of the central courts of common law.

franchise, *sb.* Add: **2. c.** *Marine Insurance.* A percentage below which the underwriter incurs no responsibility.

The term was borrowed from French practice, but the English application differs from the French; cf. **1857** M. Hopkins *Handbk. Average* 140 In some foreign policies the warranty has a rather different signification. It not only requires that the damage should amount to a certain proportion of the value of the goods insured, but it gives to the assured, in case of Average, only that portion of the loss which exceeds the agreed limit, which is there called the *franchise*, or *affranchisement*.

1895 W. Gow *Marine Insur.* 192 It has been arranged that all claims falling short of a certain amount or percentage should not attach to the policy covering the goods. This amount or percentage is termed the *franchise*. *Ibid.* 195 Nowadays in England when the franchise is once reached, the whole amount of average including the franchise is paid by the underwriter. **1929** V. Dover *Marine Insur.* (ed. 3) 240 Once the franchise is reached the policy pays in full. *Ibid.* 330 Unless a franchise of 5 *per cent.* is attained. **1962** R. H. Brown *Dict. Marine Insur. Terms* 103 *Franchise*, an amount or percentage specified in the policy which must be reached before a claim is payable... In some foreign insurances the term may be used to denote a deductible franchise.

d. The authorization granted to an individual or group by a company to sell its products or services in a particular area. Hence *franchisee, franchiser, franchisor.*

1959 *Listener* 2 Apr. 600/3 Paul had obtained the barber shop franchise in a local hotel. **1966** *Economist* 9 July 148/1 Franchisers, to be successful, must to some degree tell their franchisees what to sell and at what price. **1967** *Times Rev. Industry* Aug. 77/2 In the United States..some 1,500 franchisors..handle about 35 per cent of total retail sales. *Ibid.*, The car distributor with the exclusive local franchise. *Ibid.* 77/3 The management, marketing, product and service know-how of the franchise organization. **1970** *Guardian* 31 July 9/7 The endless seekers of franchises on the [pop festival] site. Drinks, hot dogs, records..are a ready source of cover.

franchised, *ppl. a.* Add: **4.** Of a company: possessing special powers or rights conferred on the ground of public utility. Also, possessing a franchise (sense *2 d). *U.S.*

1908 *Munsey's Mag.* Nov. 165/2 His attitude toward the franchised and privileged monopolies of Washington. **1968** *Globe & Mail* (Toronto) 5 Feb. 24/8 (Advt.), For Sale. Franchized dry cleaning plant, operating in busy well located Oshawa Shopping Centre. **1970** *Motoring Which?* Oct. 150/1 Cars were generally serviced better.. at franchised garages.

franchising, *vbl. sb.* Delete †*Obs.* and add later examples. (Cf. *franchise *sb.* 2 d.)

1966 *Economist* 9 July 147/3 Congress is listening to arguments that any curbs on franchising would hurt small businessmen. **1967** *Times Rev. Industry* Aug. 77/3 Franchising, as the Wimpy bar idea in Britain illustrates, has become a method of very quickly establishing a distributive network for a product or service.

Francis I: see *François Premier.

francium (fræ·nsiəm). *Chem.* [mod.L. (M. Perey 1946, in *Jrnl. de Chim. phys.* XLIII. 157), f. *Franc*(*e*: see -ium.] A radioactive metallic element that is the heaviest member of the alkali-metal series and is chemically similar to cæsium; all its isotopes have short half-lives and only one (francium 223) occurs naturally, being produced by the radioactive decay of actinium 227. Symbol Fr; atomic number 87.

1946 *Jrnl. & Proc. R. Inst. Chem.* Dec. 277 For element 87,..the name Francium (Fr) has been suggested. **1951** J. R. Partington *Gen. & Inorg. Chem.* (ed. 2) xii. 323A It was at first called *actinium-K* but has been renamed *francium* (Fr.). **1963** A. G. Maddock in *Mellor's Compreh. Treat. Inorg. & Theor. Chem.* II. Suppl. III. ii. vi. 2516 Francium-211 may also have a half-life longer than five minutes and there is some evidence that ^{210}Fr and ^{211}Fr may have half-lives near one minute. *Ibid.* 2518 Asimov has calculated that the earth's crust contains 24·5 g. of francium. **1968** C. A. Hampel *Encycl. Chem. Elem.* 224/1 Francium exists in aqueous solution as a large singly-charged ion with small tendency to form complex ions.

franckeite (fræ·ŋkəˌait). *Min.* [ad. G. *franckeit* (A. W. Stelzner 1893, in *Neues Jahrb. f. Min.* II. 124), f. the name of Carl and Ernest *Francke*, mining engineers: see -ite[1].] A sulphostannate of lead, $Pb_5Sn_3Sb_2S_{14}$, forming lustrous greyish-black orthorhombic crystals.

1893 *Jrnl. Chem. Soc.* LXIV. ii. 576 This new mineral, which is of special interest as indicating the character of the occurrence of tin in the Bolivian mines..is named franckeite by the author in honour of C. and E. Francke, two mining engineers specially interested in Bolivian mining. **1968** [see *cylindrite].

‖ **franco** (fræ·ŋko), *a.* [It. (*porto*) *franco* free (carriage).] Free of any postal or delivery charge.

1873 *Young Englishwoman* May 256/2 The Malle des Indes sends out patterns by post *franco*. **1962** H. O. Beecheno *Introd. Bus. Stud.* xii. 113 A *free, franco* or *rendu* quotation includes all costs up to and including delivery to the purchaser's place of business. **1969** J. L. Hanson *Dict. Econ.* (ed. 3) 211/1 *Franco*, a term used in foreign trade transactions, it is a price quotation which includes not only the cost of the commodity but also insurance, freight, and all delivery charges to the importer's warehouse.

Franco-. Add: *Franco-Canadian* adj. and sb., *-Irish*, *-Prussian* (examples).

1837 *Niagara Reporter* 24 Sept. 2/3 A meeting..of the Franco-Canadian clique. **1841** *Montreal Transcript* 30 Oct. 2/4 The vanity of the Franco-Canadians. **1845** *Knickerbocker* XXV. 59 The ever-noisy sailors, with their strange Franco-Canadian *patois*, 'made the air vocal with sweet sounds'. **1881** W. J. Rattray *Scot in Brit. N. Amer.* II. 404 Under the Constitutional Act of 1791, the Franco-Canadians were, if not quite satisfied, at least tranquil and submissive. **1969** *Publ. Amer. Dial. Soc.* LII. 12 Almost all Franco-Canadians learn English. **1931** H. S. Weaver *Let.* 19 July in Joyce *Lett.* (1966) III. 223 A small Franco-Irish contingent of influential people. **1945** R. Hargreaves *Enemy at Gate* 65 The combined Franco-Irish force. **1871** 'E. Perkins' (*title*) The Franco-Prussian war in a nutshell. **1966** V. L. Lidtke *Outlawed Party* iv. 114 Remembering his opposition to the Franco-Prussian War, no one could justifiably accuse him of surrendering to militarism.

‖ **François Premier** (frańswa prəmye). Also Francis I. The name of Francis I, King of France (1515–1547), used adjectivally to designate the styles in architecture, furniture, etc., characteristic of his reign.

1860 *Dict. Archit.* (Archit. Publ. Soc.) II. 88/2 *Francois Premier* (*Style*). This term properly applies to that portion of *Renaissance* architecture seen at Chenonceaux and Chambord. **1878** Mrs. B. Palliser tr. *Jacquemart's Hist. Furnit.* iv. v. 401 (*caption*) Nef of rock crystal, cut and engraved. Italian workmanship of the Francis I. period. **1910** *Encycl. Brit.* II. 413/2 Between the last phase of Flamboyant Gothic and the introduction of the pure Italian Revival there existed a transitional period, known generally as the 'Francis I. style'. **1923** F. M. Atkinson tr. *Félice's Fr. Furnit. Middle Ages* ii. i. 49 This 'François Ier' art is at bottom truly French, vivid, varied, full of gaiety and fancy. **1961** L. A. Boger *Compl. Guide Furnit. Styles* vi. 79/2 The Francois I style was essentially a transitional style in furniture design in which Italian Renaissance ornament and detail were found on Gothic forms. **1967** J. Porter *Chinks in Curtain* iv. 43 The exterior of the château was mock François Premier but the room..was pure Second Empire.

Francomania (fræŋkomēɪ·niă). [See -mania.] A craze or excessive liking for France and for things French. Hence **Francoma·niac.**

1899 *Daily News* 8 May 8/2 From 1750 to the Revolution, Anglo-mania in France was fostered by Francomania in England. **1908** *Daily Chron.* 4 Mar. 6/6 That Francomaniac, Frederick the Great. **1930** *Social Sci. Abstr.* 1000 The Francomania of the Restoration.

Franconian (fræːŋkōu·niăn), *a.* and *sb.* [f. *Franconia* (see below) + -an.] **A.** *adj.* Of or pertaining to (the inhabitants of) Franconia, a region of Germany bordering the river Main and in medieval times a duchy. **B.** *sb.* **a.** An inhabitant of Franconia. **b.** (See quot. 1954.)

1805 *Times* 7 Nov. 2/1 An army, consisting of Silesian and Franconian regiments..is on its march to Franconia. **1812** C. Butler *Hist. Rev. Emp. Germany* iii. 59 The principal events in the history of the latter princes of the Franconian line, and of all the princes of the Suabian line, were produced or influenced by the contests between the popes and the emperors. **1847** R. B. Paul *Hist. Germany* xii. 81 Both were Franconian princes, and grandsons of Conrad. **1849** *Art-Jrnl.* XI. 108/2 The

procession is arranged according to the four tribes of Bavarians, Suavians, Franconians and Saxons. **1888** [see Germanic *a.* 2]. **1928** J. W. Thompson *Feudal Germany* xvi. 607 Carinthia and Styria (which was separated from Carinthia in 1035) were mainly colonized by Franconians and Rhinelanders. **1946** 'O. Marx' tr. *Valentin's German People* i. 6 The East Frisians and the Franconians on the Lower Rhine resemble the Dutch. **1954** Pei & Gaynor *Dict. Linguistics* 76 *Franconian*, a group of medieval West Germanic dialects, combining the characteristics of Low and High German, and subdivided into Lower Franconian..and Middle and Upper Franconian. **1959** *Chambers's Encycl.* II. 165/2 In the 6th century the Bavarians expanded their territory..to the north beyond the Danube as far as the Franconian Jura.

Francophilia (fræŋkofi·liă). [f. Francophil(e *a.* and *sb.* + -ia[1].] Friendliness to France.

1945 *N.Y. Herald-Tribune* (Books) 16 Dec. 3 'The French flu' (Francophilia), a disease which ravages the mind. **1962** *Economist* 9 June 963/2 M. Houphouët-Boigny..neatly combines Francophilia with power.

Francophobia (fræŋkofōu·biă). [See -phobia.] Dread or dislike of France or the French, tending to become an obsession.

1887 [see -phobia]. **1900** *Fortn. Rev.* May 722 To deny Frenchmen the quality of courage in the face of danger is really to push Francophobia beyond the bounds of common sense. **1928** *Manch. Guardian Weekly* 10 Aug. 104/1 In his indignation M. Jacques Bainville, of the 'Liberté', brings a direct charge of Francophobia. **1961** *Spectator* 20 Oct. 532 Federalist Francophobia led to the passage of the Alien and Sedition Laws.

francophone (fræ·ŋkofōun), *sb.* and *a.* Also with capital initial. [f. Franco- + Gr. φωνή voice.] **A.** *sb.* A French-speaking person. **B.** *adj.* French-speaking. Hence **francopho·nia**; **francopho·nic** *a.*

1900 tr. *Deniker's Races of Man* xiii. 508 In Canada two-thirds of the white population are Anglophones, and the rest Francophones. **1962** *Economist* 23 June 1192/1 The steady economic waltz of the general's francophonic clients. *Ibid.* 29 Dec. 1280/2 The theoretically quadrilingual, but in practice largely Francophone, community. **1968** *New Statesman* 12 Jan. 33/3 It is not exactly France's fault..that the francophone African states are mostly vast in area but poor in population and in natural resources. **1969** *Daily Colonist* (Victoria, B.C.) 5 July 5/1 The study employs the terms anglophones, francophones and 'others' to denote respectively those who speak English, French or another language in their homes. **1969** *Maclean's* Aug. 1/1 The man behind this sudden surge of Francophonia..is the paper's new editor-in-chief, Frank Walker. **1971** *Guardian* 5 June 10/6 The 80 million Francophones on whom President Pompidou rests his case for linguistic parity.

franc-tireur. (Earlier and later examples.)

1808 W. W. Wynn *Let.* 31 Dec. in R. Leighton *Corresp. Lady Williams Wynn* (1920) xi. 130 He [*sc.* Lord Ebrington] is gone a *franc-tireur* with General Cameron. **1870** *Daily Globe* 23 Nov. 1/6 Uhlans and Francs-Tireurs. **1910** *Encycl. Brit.* XI. 16 The Germans treated captured francs-tireurs as irresponsible non-combatants found with arms in their hands and usually exacted the death penalty. **1955** J. Thomas *No Banners* xx. 193 He..went to meet a *Franc-Tireur* who was on the run after killing several Germans. **1968** *Listener* 15 Feb. 197/3 The *Washington Post*..has been bombarded with letters expressing horror at that photograph of the Saigon police chief..executing a Vietcong *franc-tireur*.

frangipane. 2. Substitute for def.: A fragrant ornamental shrub or tree of the genus *Plumeria*, esp. *P. rubra.* (Further examples.)

[**1842** *Curtis's Bot. Mag.* LXVIII. 3952 It is from this circumstance [the white juice], probably, that the French call the species of this Genus 'Franchipanier', *Franchipane* being coagulated milk.] **1871** J. Smith *Domestic Bot.* 292 *P. rubra.* in the West Indies being called Red Jasmine, as also 'Frangipane'—a name also given to the sweet-smelling flowers of *P. acuminata.*

frangipa(n)ni. (Further examples.) *Frangipani* is now the usual spelling of Frangipane.

1882 J. Smith *Dict. Pop. Names Plants* 180 In the West Indies *Plumeria rubra* and *alba*, shrubs of the Dogbane Family (Apocynaceæ) are called Frangipanni on account of their deliciously-scented flowers. **1898** *Jrnl. R. Hort. Soc.* XXII. 159 Frangipani (*Plumiera* [sic] *rubra*, *P. alba*, *P. fragrantissima* &c.).—Commonly planted on graves in Borneo and Malayan islands. **1904** J. Cameron *Firminger's Man. Garden. India* (ed. 5) iv. 497 P[lumeria] rubra.—The Frangipani Plant. Introduced from Jamaica. **1934** H. B. Sharpe in A. J. Jex-Blake *Gardening E. Afr.* xvii. 252, I planted out everywhere my pink frangipani. **1965** A. Thomas *Gardening in Hot Countries* vii. 100 Other kinds of *Plumieria* [sic]..have fainter perfumes and may be more pleasing at close quarters than ordinary frangipani.

‖ **franglais** (frańgle). [Blend of Fr. *fran*(*çais* + *an*)*glais.*] A corrupt version of the French language produced by the indiscriminate introduction of words and phrases of English and American origin. Also *transf.* and as *adj.*

[**1959** M. Rat in R. Etiemble *Parlez-vous Franglais* (1964) ii. 34 Faudra-t-il appeler bientôt *franglais* ce français émaillé de vocables britanniques que la mode actuelle nous impose?] **1964** *Economist* 25 Apr. 354/1 'Franglais'—what the professor [*sc.* Etiemble] sees as French bastardised and ruined by Anglo-Saxon..borrowings. **1964** *Cambr. Rev.* 24 Oct. 45/2 The 'New Britain'.. will be a world of tab collars, Mary Quant, and academic *franglais*, no doubt. **1965** *New Statesman* 7 May 735/1 At a time when young English writers and journalists are struggling to strike Gallicisms out of their equipment, the French complain that *le franglais* is taking over theirs. **1967** *Guardian* 8 Feb. 6/5 Upholders of Gallic linguistic purity lament the encroachment of what they call 'franglais'. **1969** N. Freeling *Tsing-Boum* xi. 80 'Alors bye-bye,' she was saying... That awful way French women had on the phone, using idiotic franglais phrases like 'because le job'.

frank (fræŋk), *sb.*[7] *U.S.* Short for *frankfurter.

1936 E. E. Cummings *Let.* 9 May (1969) 145 Sand there shall be with our franks. **1957** J. Kerouac *On Road* (1958) i. 9 We had a farewell meal of franks and beans. **1959** *Consumer Reports* Sept. 455/3 Slitting the franks may adversely affect the device's ability to cook them properly. **1968** *Washington Post* 5 July A17 (Advt.), Safeway Skinless All Meat Franks. 2 lb. pkg. 99c.

frank, *v.*[2] Add: **1.** Revived in later use: to mark (a letter, etc.) with a sign (in lieu of an affixed postage stamp) by means of a franking machine. Chiefly as **franked** *ppl. a.*

1927 *Times* 25 Feb. 10/6 The posting of franked letters at a date subsequent to that marked on the envelope is a violation of the terms of the licence. **1928** M. Rittenberg *Mail-Order made Easy* xiv. 179 Franked or metered mail is accepted under certain conditions. **1971** D. Potter *Brit. Eliz. Stamps* xiii. 141 Letters and packets are franked with a handstamp or machine impression.

franked *ppl. a.*: spec. *franked income* (see quot. 1965).

1955 *Times* 2 July 9/7 Profits tax absorbed £6,000 less, mainly because the proportion of franked income was greater this year. **1965** J. L. Hanson *Dict. Econ.* 188/1 *Franked income*, a term used of a company which derives income from the profits of another company which have been subjected already to profits tax. Such income is said to be 'franked' and so is not liable to profits tax a second time.

Frankenstein (fræ·ŋkənstəin). The name of the title-character of Mrs. Shelley's romance *Frankenstein* (1818), who constructed a human monster and endowed it with life. Commonly misused allusively as a typical name for a monster who is a terror to his originator and ends by destroying him. Also *attrib.* Hence **Frankenstei·nian** *a.*

1838 Gladstone in *Murray's Handbk. Sicily* (1864) p. xlvi, They [*sc.* mules] really seem like Frankensteins of the animal creation. **1889** S. Webb in G. B. Shaw *Fabian Ess.* 38 The landlord and the capitalist are both finding that the steam-engine is a Frankenstein which they had better not have raised. **1907** *Sat. Rev.* 6 Apr. 414/1 Is Great Britain creating for herself something of a Frankenstein monster on the Nile? **1931** R. Campbell *Georgiad* i. 14 No sooner was our Frankenstein set free Than for a name he racks his nimble wits. **1958** I. Asimov *Naked Sun* xiv. 172 Do you know robots started with a Frankenstein complex against them? They were suspect. Men distrusted and feared robots. **1965** J. Wainwright *Death in Sleeping City* ii. 107 Like some Frankensteinian monster, the Police Machine moved forward. **1971** B. Callison *Plague of Sailors* i. 51 The Frankensteinian exhibit in the alcohol bath. **1971** *Daily Tel.* 3 May 1/4 There are now growing indications that the Nationalists in South Africa have created a political Frankenstein which is pointing the way to a non-White political revival.

Frankenthal (fræ·ŋkentāl). [The name of a town in Germany, in the Bavarian Palatinate.] The designation of a porcelain made at Frankenthal from the middle to the end of the eighteenth century.

1863 W. Chaffers *Marks Pott. & Porc.* 183 The stock and utensils of the Frankenthal manufactory were purchased in 1800 by M. Von Recum. **1887** *Sale Catal. Porc. & Decorative Objects* (*Christie, Manson, & Woods*) 4 Feb. 4 A large Frankenthal group of Chinese figures at an arbour. **1925** W. W. Worster tr. *Hannover's Pott. & Porc.* III. ix. 215 The porcelain body of Frankenthal, which during the best period was brought from Passau, is as a rule of so soft a whiteness as to resemble at times the French soft porcelain. **1947** W. B. Honey *German Porc.* 25 The Frankenthal table-wares and vases of the best period were..often painted with elaborate figure-subjects of a kind seldom attempted elsewhere in Germany. **1960** H. Hayward *Antique Coll.* 123/1 Frankenthal porcelain figures often have rich pierced, scrolled Rococo bases and sometimes include trellis work.

franker. Add: Also, an instrument for franking postal matter.

1925 *Glasgow Herald* 5 Jan. 8 There are addressers, duplicators,..postal frankers and cheque writers.

frankfurter (fræ·ŋkfəːtəɹ). Also **frankfurt** (fræ·ŋkfəɹt). [G. *Frankfurter wurst* Frankfurt sausage. Cf. Frankfort.] A highly seasoned smoked beef and pork sausage, originally made at Frankfurt am Main.

[**1877** E. S. Dallas *Kettner's Bk. of Table* 317 In the absence of Chorizos use Frankfort sausages.] **1894** *San Francisco Midwinter Appeal* 17 Feb. 2/1 Four bits for a Frankfurter seems rather steep. **1902** G. H. Lorimer *Lett. Merchant* x. 127 Only last week the head of our sausage department started to put out a tin-tag brand of frankfurts. **1908** —— *J. Spurlock* vii. 147, I told the Major that he had undoubtedly got hold of the hottest dog in the frankfurter can. **1919** F. Hurst *Humoresque* 262 Screaming peanut- and frankfurter-stands. **1949** F. Maclean *Eastern Approaches* i. vii. 80 We would congregate round a roaring open fire in the *dacha* and Avis would dispense frankfurters and peanut butter. **1964** H. Kemelman *Friday the Rabbi slept Late* (1965) vii. 66 They offered..baked beans and frankfurts. **1971** *Sunday Times* (Colour Suppl.) 27 June 50/2 Frankfurters and Viennas are the world's best-known boiling sausages.

fra·nking machine. [f. Franking *vbl. sb.* + Machine *sb.*] An officially authorized machine, introduced by the British Post Office in 1922, used by large concerns for 'stamping' letters, etc., with a sign (in lieu of an affixed postage stamp); it simultaneously records the cost of postage (this being periodically checked and collected by the Post Office). Also **franking stamp.**

1927 *Times* 22 Feb. 10/4 Remembering to have read of a franking machine called a postage meter being recently issued to the public, I wrote to the P.M.G. **1929** *Encycl. Brit.* XVIII. 311/1 Another method of avoiding the use of postage stamps is the employment of postal franking machines, which are now licensed by the post office. These machines impress the correspondence with a red franking stamp which is accepted as the equivalent of a postage stamp. **1961** *Lebende Sprachen* VI. 70/1 *Postage meter machine* (B[ritish] E[nglish]: franking machine).

Franklin[2] (fræ·ŋklin). [f. the name of Benjamin *Franklin* (1706–1790).] †**1.** *U.S.* A lightning-conductor. Also *Franklin's rod. Obs.*

1818 J. Palmer *Jrnl. Trav. U.S.* 104 Franklins, or conductors, are a certain safe-guard, and generally used. *a* **1862** S. Breck *Recoll.* (1877) ii. 71 Our professor of natural philosophy was desirous to erect on the old lofty tower..a lightning-rod, very properly called at that time a 'Franklin'. **1910** *Encycl. Brit.* XI. 30/1 He [*sc.* Benjamin Franklin] planned the lightning-rod (long known as 'Franklin's rod'), which he described and recommended to the public in 1753.

2. N. Amer. *Franklin stove*, a kind of iron fireplace invented by Franklin; also, a free-standing stove for heating a room. Also *ellipt.*

1787 *Massachusetts Centinel* 3 Nov. 3/3 An assortment of Franklin stoves. **1797** *Upper Canada Gaz.* (York, Ont.) 4 Nov. 3/3 For Sale—A Franklin Stove Complete. **1818** J. Palmer *Jrnl. Trav. U.S.* 15 In the best room some have an iron fireplace..called a Franklin; these look very neat, and will much sooner heat a room than the open fire-place. **1897** B. F. Taylor *Summer-Savory* x. 95 Within the wide door was the bar-room, with a great hospitable Franklin. **1949** *Los Angeles Times* (Home Mag.) 3 Apr. 20/2 In the living room, an old Franklin stove substitutes for a fireplace. **1954** C. Bruce *Channel Shore* 28 Anse..threw a cigarette butt at the Franklin stove. **1970** *Globe & Mail* (Toronto) 26 Sept. 13/1 A small Franklin stove..and a number of Canadian antiques add charm.

frankly, *adv.* **3.** (Later examples with ellipsis of *to speak.*)

1888 Kipling *Under Deodars* (1890) 52 *She.*—.. Can you wonder that I'm disinclined for amusement? *He.*—Frankly, I do. **1960** L. R. Banks *L-Shaped Room* xiv. 196 The one [*sc.* board] under his front said, '*Are you ready to meet your maker??*' 'Frankly, yes,' Toby answered it. **1961** A. Christie *Pale Horse* i. 20 'But you didn't come here to talk to me about my books.' 'Frankly no.'

frantic, *a.* Add: **2. b.** *colloq.* In exaggerated use = 'terrific', 'awful'.

1908 *Punch* 26 Feb. 152/2 I'm working simply most awfully hard for Ireland just now. I've bought a ghastly heap of poplin..and a frantic lot of Limerick lace.

frap, *v.* Add: **1. a.** (Later example.)

1896 Kipling *Seven Seas* 23 Once we frapped a ship, and she laboured woundily.

2. (Later example.)

1962 G. Compton *Too Many Murderers* xx. 162 They saw where Jason left his dinghy, and they fraped theirs out on the same line.

frape[3], **frap.** Add: **2.** (See quot. 1963.)

1901 *Pall Mall Mag.* Oct. 183 The boy crept down through the moonlit garden to the dinghy which Billy had left on its frape under the cliff. **1905** 'Q' *Shining Ferry* iii. xviii. 186 Through this ring...Mr. Hosken had run a frape, on which he kept his blue boat. **1963** *Amer. Speech* XXXVIII. 299 *Frape*, a rope with blocks for mooring a boat.

frappé, *a.* Add: Also as *sb.*, an iced drink, a soft water-ice served in a glass, etc. Also as *v. trans.*, to put (a drink) on ice, to ice (a drink). Also *transf.*

1890 B. Hall *Turnover Club* 230 The guests had rubbed the luckless Agent against the ice-box until he was thoroughly frapped. **1907** J. Masefield *Tarpaulin Muster* xxii. 211 'Put a quart of champagne on ice and bring it up here,' said the boss. 'See you don't *frappé* it too cold.' **1922** *Glasgow Herald* 13 May 6 Besides untold recipes for cobblers, coolers, highballs, frappés, daisies, sangarees. **1936** *Punch* 15 July 70/1 There are Frappés, there are Parfaits, there are Sundaes. **1955** J. G. Davis *Dict. Dairying* (ed. 2) 517 *Frappés, soufflés*, etc., are essentially water ices.

Frascati (fræskā·ti). The name of a district in Latium, SE. of Rome, used for the name of a wine (usu. white) produced there.

[**1908** N. Young *Murray's Handbk. Rome* (ed. 17) 29 The white wine of Frascati is light, sparkling, and rather sweet.] **1935** A. L. Simon *Wines & Liqueurs from A to Z* 24 *Frascati*, one of the best white wines of the Latium (Italy). **1968** 'A. York' *Predator* v. 61 Two tumblers of red Frascati. **1970** *House & Garden* May 140/3 The province of Lazio grows the famous white Frascati, a wine that can be anything from quite dry to very sweet. **1970** 'O. John' *Diamond Dress* xii. 139 I'd been looking forward to some delicious spaghetti alla carbonara and a bottle of Frascati.

Frasnian (fræ·zniăn), *a. Geol.* Formerly also -ien. [ad. F. *Frasnien* (J. Gosselet 1879, in *Ann. Soc. géol. du Nord* VI. 396), f. *Frasne*, var. *Frasnes*, name of a village and commune in southern Belgium.] Name of the lower of the two stages constituting the Upper Devonian in Europe; of or pertaining to this stage or the geological period during which it was deposited. Also *absol.*

1882 A. Geikie *Text-bk. Geol.* 701 In the Belgian and Eifelian tracts they have been subdivided as under:.. Fammenien..Frasnien. **1894** *Geol. Mag.* Oct. 474 The Frasnien series in the North of France and Belgium. **1963** D. W. & E. E. Humphries tr. *Termier's Erosion & Sedimentation* xiii. 288 The Frasnian reefs on the edge of the Dinant basin. *Ibid.* 289 The bioherms of the Frasnian form a subsidising series up to 1,300 feet thick.

frat (fræt), *sb.*[1] *U.S. College slang.* **a.** Abbreviated form of *fraternity 7. Also *attrib.* **b.** A member of a fraternity.

1895 W. C. Gore in *Inlander* Nov. 64 *Frat*, 1. A fraternity. 2. A member of a fraternity. **1899** A. H. Quinn *Pennsylvania Stories* 186 But any Frat would have wanted Harington. **1902** *Chicago Record-Herald* 7 Sept. i. 1/3 Alfalfa Eta Fraternity Frat House. **1906** 'N. Newkirk' *Recoll. Gold Cure Candidate* 117 The fraternities..have been pointing out that I would never amount to much as a student until I became a Frat man. **1909** *Daily Chron.* 9 Feb. 4/4 The Frat-man is one of a group of about fifteen students... Frat-houses may be expensive or cheap. **1926** *Ladies' Home Jrnl.* Nov. 12 It's a frat house. **1947** *Chicago Tribune* 2 July 19 Too bad he couldn't go to college. He doesn't know what proms and frats an' trig an' hot dates are all about. **1951** M. Shulman *Many Loves of Dobie Gillis* (1953) 210 Some frat men say he stole their loving cup. **1967** *Punch* 13 Sept. 377/3 The only Frank Lloyd Wright building on my campus was a frat. house.

frat, *v.* and *sb.*[2]: see *fratting *vbl. sb.*

fraternal, *a.* Add: **b.** *fraternal order*, a brotherhood or friendly society. *U.S.*

1889 *Kansas Times & Star* 7 June, The Iron Hall, a western fraternal order, is in session here this week. **1905** D. G. Phillips *Plum Tree* 267 Local machine leaders of Scarborough's party, with corruptible labor and fraternal order leaders.

c. *fraternal polyandry*, a form of polyandry in which brothers hold a wife in common; *fraternal twin*, a dizygotic twin.

1896 W. Crooke *Tribes & Castes N.-W. Prov. & Oudh* II. 445 On this fraternal polyandry, see Westermarck. **1904** H. H. Wilder in *Amer. Jrnl. Anat.* III. 389 Corresponding to this hypothesis..we may designate these two types [of twin] respectively as *Fraternal* and *Duplicate*, thus doing away with the misleading and inapplicable terms 'identical' and 'homologous' as applied to the one type, and furnishing a distinguishing term for the other, which seems thus far to have remained without a name. **1917** H. H. Newman *Biol. of Twins* i. 8 Biologists have for some time recognized at least two distinct types of human twins: *fraternal* and *duplicate*. *Fraternal* twins may or may not be same-sexed, are usually no more alike than are brothers and sisters, and are believed to be dizygotic, derived from two fertilized eggs. **1921** E. Westermarck *Hist. Human Marriage* (ed. 5) III. xxix. 116 Fraternal polyandry is more or less common in vast districts of the Himalayan region..chiefly among people of Tibetan affinities. *Ibid.* 122 The mountaineers of the Himalayas are not the only people in Northern India that practise fraternal polyandry. **1937** R. H. Lowie *Hist. Ethnol. Theory* (1938) v. 47 Patrilineal descent results from patrilocal residence with fraternal polyandry. **1938** *Brit. Jrnl. Psychol.* Jan. 356 Comparing the separated identical twins with the fraternal twins it was found that in some of the physical characteristics,..separated identicals were more alike than fraternals. **1941** J. S. Huxley *Uniqueness of Man* ii. 48 Fraternal twins of like sex, though as we would expect they show considerably less resemblance than identical twins, are more alike than pairs of brothers or sisters born at different times. **1953** Beals & Hoijer *Introd. Anthropol.* (1956) xiv. 429 The ideal pattern of marriage in Toda culture is fraternal polyandry, which dictates that when a woman marries a man she becomes, in theory at least, the wife of all his brothers, both the living and those as yet unborn.

B. *sb.* A fraternal twin.

1911 *Jrnl. Morphol.* XXII. 859 Were it possible in a number of cases to determine by examination of the placental relationships of new-born twins whether they were duplicates or fraternals..we would have facts from which we could with confidence draw conclusions. **1935** R. S. Woodworth *Psychol.* (ed. 10) vii. 144 The question is whether identical twins are more alike mentally than fraternals or than siblings. **1939** *Nature* 18 Nov. 870/1 The quintuplets were unlike at birth but have become.. more similar than fraternals.

fraternity. Add: **7.** A social association of the students or alumni of a college or university, usually having a name consisting of three Greek letters, as 'Phi Beta Kappa'. Also *attrib. U.S.*

1777 in F. W. Shepardson *Phi Beta Kappa* (1915) 9 [At the January meeting of 1777..a mode of initiation was reported.] 'I, A.B. do swear..to prove true, just, and deeply attached to this our growing fraternity.' **1844** in A. P. Jacobs *Psi Upsilon Epitome* (1884) 180 Catalogue of the Psi Upsilon Fraternity. **1879** W. R. Baird (*title*) Baird's Manual of American college fraternities. **1899** A. H. Quinn *Pennsylvania Stories* 59 He could not help seeing that the Fraternity men were in general the best dressed. **1901** H. D. Sheldon *Student Life* 224 In general, the fraternities dominate the smaller colleges,..although an occasional revolt on the part of the non-fraternity men sometimes occurs. **1902** J. Corbin *American at Oxford* 54 In Balliol there are three debating clubs, and they are of course in some sense rivals. Like the fraternities in an American college, they look over the freshmen each year pretty closely. *Ibid.* 269 The fraternity houses so widely diffused in America offer almost a counterpart of the halls of the golden age of the mediaeval university. **1905** A. H. Rice *Sandy* 125 Annette counted her fraternity pins. **1927** *Sat. Even. Post* 24 Dec. 19/3 There was heart-breaking rivalry among us to secure fraternity pins from the boys who possessed them. **1967** *Punch* 13 Sept. 377/3 On my own campus, as at many middle-western universities, the fraternity system had great power and prestige.

8. Used by Galton for: the brothers (and sisters) of a family collectively.

1889 F. Galton *Nat. Inheritance* 234 A Fraternity consists of the brothers of a family, and of the sisters after the qualities of the latter have been transmuted to their Male Equivalents. **1900** K. Pearson *Gram. Sci.* (ed. 2) xi. 459 The intensity of parental correlation is about ·3 to ·5, of grandparental about ·15 to ·3, and of fraternal about ·4 to ·6, the latter correlation being somewhat reduced when the 'fraternity' consists of members of opposite sexes.

fraternization. Add: **b.** *spec.* (cf. quot. 1851 in Dict.). The cultivation, subject to military discipline, of friendly relations by occupying troops with local inhabitants; relations with local inhabitants that contravene military discipline: used esp. of relations with German women after the war of 1939–45. Also *transf.*

1944 *New Statesman* 7 Oct. 233/1 The administration of occupied Germany..must necessarily be 'tough'...To judge by his Proclamation No. 1 to the German people, and by his orders forbidding fraternisation, General Eisenhower is going to present his subordinate commanders and staffs with some knotty problems. **1945** *Ibid.* 21 Apr. 252/2 At present it is not clear whether rape [by occupying troops] is a crime to be punished by death, or whether it should be classified as fraternisation, for which the penalty is a fine, or non-fraternisation, which is a laudable act. **1945** *Spectator* 10 Aug. 130/1 'Fraternisation'...I write to challenge the implication in 'Serving Officer's' letter that to frequent brothels is to 'be human'. **1946** *Amer. Speech* XXI. 251, I once saw a soldier who had philanthropically sold a prisoner of war a pack of cigarettes for $12 charged with 'fraternization'. **1949** Koestler *Promise & Fulfilment* vii. 88 The famous order of the day of the G.O.C. Palestine..which..imposed a ban on fraternization with the Jews. **1969** *Guardian* 29 Aug. 1/6 In Olomouc there are still many girls..whose heads have been shaved for too intimate a fraternisation with their Russian brothers.

fraternize, *v.* Add: **1. b.** *spec.* To cultivate friendly relations *with* (troops of an opposing army); to practise *fraternization. Hence **fra·ternizing** *vbl. sb.*

1897 G. B. Shaw *Crude Criminology* in *Doctors' Delusions* (1932) 272 The whole army might..realize that they had no quarrel with the enemy and fraternize with them. **1915** *Sphere* 9 Jan. 31 It has puzzled our French allies.. that while we are in a war to the death with the German Empire our soldiers and theirs should have been able to fraternise on Christmas Day... Soldiers of the rival armies exchanged sweets, cigars, and cigarettes, and sang carols and songs in unison. **1928** H. Williamson *Pathway* x. 228 The German and British staffs both issued orders, about the same time, that any man found fraternizing with the enemy, would be court-martialled. **1937** Koestler *Spanish Testament* iii. 72 Soldiers from the rebel ranks fraternising with and discussing the situation with the Government troops. *Ibid.* ix. 179 And then ensued a perfect orgy of fraternising and eating. **1944** *New Statesman* 17 June 400/1 The War Office regulation..has laid down that British soldiers must be polite to coloured [U.S.] troops, but not fraternise with them, or offer them drinks. **1945** in *Amer. Speech* (1947) XXII. 147/2 (*heading*) Fraternizing in Germany. **1946** *Britannica Bk. of Yr.* 832/2 *Fraternize*, to deal socially with members of an enemy nation, especially with women; and, *specifically*, sexually. **1957** *New Yorker* 26 Oct. 144/2 The fantastic amount of fraternizing with the enemy in the Korean prison camps.

fratting, *vbl. sb. slang.* [Short for *fraternizing *vbl. sb.*] Friendly relations between British and American soldiers and German women in the occupied parts of Western Germany after the war of 1939–45. Also **frat** *v. intr.*, to *fraternize; **frat** *sb.*[2], a woman met by 'fratting'; also as short for 'fratting'; **fratter**, one who 'frats'. Also *transf.*

1945 *New Statesman* 14 July 20/3 'Fraternisation' has become a word denoting sexual intercourse ('frats' and 'fratting' are new Army words). **1945** *Manch. Guardian* 18 July 8/4 (*caption*) Frat or Non-frat. **1946** J. Irving *Royal Navalese* 81 *A piece of Frat*, Wren-language for any attractive young woman—ex-enemy—in occupied territory. **1946** S. Spender *European Witness* i. 13 At the messes most of the conversation was about 'fratting'. **1949** G. Cotterell *Randle in Springtime* i. 8 Then, take my frat I go with, what harm did she ever do? *Ibid.* ii. ii. 45 You see all the men here go fratting and it simply isn't fair on us girls...I can't see what they see in these German women. *Ibid.* v. 285 So he's married...I bet she doesn't know what a shameless old fratter you were. **1955** *Times* 11 May 15/3 To an outside observer, brotherly love is not the most striking feature of the Labour Party at this moment—apart, perhaps, from a little 'fratting' with a certain Liberal lady, with an obvious eye to the main chance. **1957** M. K. Joseph *I'll soldier no More* (1958) 13 'He was fratting, wasn't he?' 'Sure—dark piece, lives up the Ludwigstrasse.' *Ibid.*, Maybe he'd've rather stayed with his frat. **1965** J. Fleming *Nothing is Number* ii. iv. 78 He did frat, I mean make friends a bit with the people you're thinking of.

fraud, *sb.* Add: **6.** *fraud squad*; **fraud order** *U.S.*, an official order prohibiting the delivery of letters to a firm or individual suspected of making illegal use of the postal service.

1905 Calkins & Holden *Art Mod. Advertising* 258 It is often impossible to prosecute the advertisers, and the most the post-office department can do is to issue what is known as a fraud order. Such an order peremptorily and without redress stops the mail of the advertiser. **1931** C. Kelly *U.S. Postal Policy* 150 Under a 'fraud order' all mail directed to such persons or company is stamped 'Fraudulent' on the outside and returned to the sender. **1967** *Economist* 7 Jan. 49/2 There is also the Fraud Squad, of gallant lay policemen undertaking (astonishingly well under the circumstances) inquiries that need the most sophisticated legal and financial expertise. **1971** *Times* 27 Sept. 3/3 Post Office investigators and members of Scotland Yard's Fraud Squad are searching for a gang of expert forgers.

fraudsman. Delete † *nonce-wd.* and add later examples.

1958 *Times* 9 Oct. 3/5 In all, the officer questioned 32 witnesses who testified to seeing certain of the 'fraudsmen'. **1966** M. Procter *His Weight in Gold* xi. 93 Like nearly all fraudsmen he had a lot of personal charm. **1971** *Daily Express* 22 Mar. 13/3 You cannot..bar your mind against a fraudsman.

‖ **Frauendienst** (frau·əndinst). [Title of a work by U. von Lichtenstein (d. *c* 1275); f. G. *frauen* pl. of *frau* woman + *dienst* service.] Exaggerated chivalry towards women. Also *transf.*

[**1879** *Encycl. Brit.* X. 525/2 As the expression of all that was fantastic and ridiculous in the age of chivalry, must be mentioned the *Frauendienst* of Ulrich von Lichtenstein, a work which was written about the middle of the 13th century... The author relates the amazing tasks imposed on him by his mistress.] **1932** *Essays & Studies* XVII. 66 Chaucer is emphasizing that parody, or imitation, or rivalry..of the Christian religion which was inherent in traditional *Frauendienst*. **1941** Auden *New Year Let.* iii. 69 And his goal The Frauendienst of his weak soul. **1963** —— *Dyer's Hand* 212 We find neither outright sexual passion nor sentimental *Frauendienst*.

fraught, *ppl. a.* Add: **4.** Distressed; distressing.

1966 *Listener* 3 Mar. 330/3 All that had gone before led me to expect an end more fraught. **1967** P. Purser *Twentymen* i. 8 Sleep didn't often withhold her favours from me but if she did it was always when the next day was going to be particularly fraught. **1970** D. Francis *Rat Race* ix. 121 Don't look so fraught...They said it was clear there now.

Fraunhofer (frau·nhōufəɹ). The name of Joseph von *Fraunhofer* (1787–1826), Bavarian optician and physicist, used *attrib.* or in the possessive to designate certain phenomena investigated by him, as *Fraunhofer('s) diffraction*, diffraction in which the diffraction pattern is a linear function of the variation in phase across the diffracting aperture or object; *Fraunhofer('s) lines*, the dark lines in the spectra of the sun and other stars; *Fraunhofer spectrum*, a solar or stellar spectrum containing Fraunhofer lines.

1837 Fraunhofer's lines [see Line *sb.*[2] 8 c]. **1840** J. F. W. Herschel in *Phil. Trans. R. Soc.* CXXX. 17 The orange-red rays, or the less refrangible half of the portion from C to D in Fraunhofer's spectrum. **1843** *Phil. Mag.* 3rd Ser. XXII. 361 The spectrum which results is given in great purity, and Fraunhofer's lines are quite apparent. **1888** *Encycl. Brit.* XXIV. 430/1 (*heading*) Fraunhofer's diffraction phenomena. **1905** R. W. Wood *Physical Optics* vii. 161 (*heading*) Mathematical treatment of Fraunhofer diffraction phenomena. **1937** *Discovery* Apr. 109/2 The Fraunhofer spectrum line, C, of hydrogen in the red at wavelength 6563 Ångström units (also known as $H\alpha$). **1963** R. W. Ditchburn *Light* (ed. 2) viii. 291 A Fraunhofer diffraction pattern is formed in the focal plane of the lens. **1966** *McGraw-Hill Encycl. Sci. & Technol.* XIII. 267 Of the 92 natural elements, 64 are represented in the Fraunhofer spectrum. **1968** R. A. Lyttleton *Myst. Solar Syst.* iv. 141 When a comet is quiescent,..the detectable light consists of a continuum crossed by the familiar solar Fraunhöfer [*sic*] lines.

frawn. Add: Also **fraughan, fraun.**

1869 P. Kennedy *Evenings on Duffrey* xix. 208 They might..gather..fraughans in the woods. **1892** J. Barlow

Bog-Land Studies 69 Wee frauns. 1906 Temple Bar June 553 They trod the heather and fraughan beneath them. 1914 'G. A. BIRMINGHAM' *Lost Tribes* xi, Under their shelter fraughan bushes nestled, their black fruit already ripening.

frayed, *ppl. a.*[2] (Later *fig.* examples.)

1934 *Discovery* Dec. 345/1 Super-sensitivity to sounds normally arises with frayed nerves due to worry or illness. 1966 J. PORTER *Sour Cream* i. 7 The grandiose schemes of my youth had got more than a bit frayed round the edges.

Frazerian (frĕⁱzīᵊ·riăn), *a.* Of or pertaining to Sir James George *Frazer*, Scottish anthropologist (1854–1941), or his work. Also as *sb.*, a follower or adherent of Frazer.

1932 *Times Lit. Suppl.* 1 Dec. 915/1 The King of the Shilluk has long been entitled to rank as a classical case of the Frazerian embodiment of the tribal luck whose duty it is to vacate his office by death before the vigour that he must pass on..can suffer diminution. 1937 A. HUXLEY *Ends & Means* v. 38 Two peoples may have what is, according to Frazerian ideas, the same custom. 1962 *Listener* 16 Aug. 258/2 We learnt that an untrained laboratory worm which eats a trained one takes over its responses—though..this principle doesn't apply to human beings, so that any eager Frazerian contemplating a supper of Bronowski à la mode had better think again. 1968 *Int. Encycl. Soc. Sci.* V. 552/1 Many anthropologists thus became 'enslaved', as Seligman was and as Malinowski once claimed to have been, by Frazerian anthropology. 1970 E. LEACH *Lévi-Strauss* i. 19 The ethnographic observations on which Lévi-Strauss, like his Frazerian predecessors, has chosen to rely.

frazzle, *v.* Now *slang* or *colloq.* (Earlier U.S. example.) Also *intr.* and in ppl. adj. (*frazzled-out*).

1872 *Congress. Globe* 30 May, App. 577/3 The ends of the switches were all frazzled. 1896 J. C. HARRIS *Sister Jane* 344 He's the genuine article, guaranteed not to rip in the seams or frazzle at the sleeves. 1912 J. H. MOORE *Ethics & Educ.* 34 Many a frazzled-out member of society owes his failure in life to no greater misdemeanour than the mere failure to make connection with his calling. 1912 J. LONDON *Son of Sun* viii. 285 Loose ends of rope stood out stiffly horizontal, and, when a whipping gave, the loose end frazzled and blew away. 1912 *Chambers's Jrnl.* Mar. 194/1 For bed a mud kang with a frazzled mat on it. 1927 J. DEVANNY *Old Savage* 43 His fight had left him 'frazzled', as he expressed it. 1960 *Guardian* 6 Jan. 1/7 The insistence of frazzled parents that merry-making and goodwill to men have got to stop somewhere.

frazzle, *sb.* Add examples of phr. (orig. U.S.) *to a frazzle* in fig. expressions denoting complete exhaustion or extinction.

1865 GORDON in W. C. Church *Ulysses Grant* (1897) 318 Tell General Lee, I have fought my corps to a frazzle. 1880 J. C. HARRIS *Uncle Remus* (1881) xi, Brer Fox dun know Brer Rabbit uv ole, en he know dat sorter game done wo' ter a frazzle. 1905 *Washington Star* 24 Nov. 22 The Becham machine whipped Blackburn to a frazzle, giving him the first real defeat he had ever experienced. 1908 *Westm. Gaz.* 3 Nov. 7/3 Mr. Roosevelt, when asked for his opinion on the result, said, 'We have got them beaten to a "frazzle".' 1916 *Daily Colonist* (Victoria, B.C.) 29 July 4/4 As a briber and corrupter, Mr. Brewster..has everyone else 'beat to a frazzle'. 1927 *Daily Express* 20 June 3/4 They run motor races on their sea wall at Lowestoft. They..have Canute beaten to a frazzle. 1935 T. E. LAWRENCE *Let.* 15 Feb. (1938) 856 Some of those lovely worm-driven water hose clips... They beat [*name omitted*] clips to a frazzle. 1937 *Daily Herald* 3 Feb. 15/6 Listens with such inscrutability that he's got the Chinese licked to a frazzle. 1966 C. BERMANT *Diary of Old Man* 111 There he goes again, coughing himself to a frazzle.

freak, *sb.*[1] Add: **4. c.** One who 'freaks out' (*FREAK *v.* 3); a drug addict (see also quots.).

1967 *Atavar* (Boston) 1–14 Sept. 17/1 The life expectancy of the average speed-freak..is less than five years. *Ibid.* Dec. 4/1 Some of us are beginning to wonder who are the 'freaks' in this world and who are the 'straight' people. 1969 R. R. LINGEMAN *Drugs from A to Z* 79 Freak...One who prefers a certain kind of drug, as in *acid freak* or *meth freak*...By extension, one who is obsessed with a certain way of thinking as in 'political freak'. 1970 C. MAJOR *Dict. Afro-Amer. Slang* 55 Freak, one who practices socially unaccepted forms of sexual love; strong believer in something. 1971 *Oz* May 7/2 Its hills and valleys are full of hippies, and freaks, camping along the river beds. 1971 *It* 9–23 Sept. 5/5 Power freaks like Ted Heath and union leader Vic Feather. 1971 *Ink* 19 Oct. 7/1 An ideological community of 25 freaks plus guru in Copenhagen. *Ibid.* 7/3 Far from there being any noticeable improvement in the quality of relationships as practised among freaks, I would say there has been a distinct deterioration compared even with the most miserable standards of the straight world.

5. Also quasi-*adj.* to denote something abnormal or capriciously irregular; **freak show**, at a fair, etc., a sideshow featuring freaks (sense 4 b).

1887 Freak show [in Dict.]. 1898 *Daily News* 17 Mar. 6/5 'The yellow kid', a personification of 'freak' or sensational journalism. 1907 *Westm. Gaz.* 26 Sept. 7/2 The boats which have been built for this race of recent years are freak boats pure and simple. 1907 *Daily Chron.* 5 Oct. 4/4 Conditions in America seem particularly favourable to the propagation of freak religions. 1908 *Westm. Gaz.* 7 Mar. 8/1 The production of freak fruits, such as white blackberries..and seedless oranges. 1928 [see *COME *v.* 24 d]. 1939 G. GREENE *Lawless Roads* i. 30 A freak show in a little booth. 1951 R. CAMPBELL *Light on Dark Horse* x. 141 These Berghens take a delight in freak-flying. 1963 *Times* 8 May 16/3 The freaker the hands before the goulash the more normal the distribution after the redeal. 1968 A. TAYLOR *Honour in Shallow Cup* i. 8 They'd been having freak storms there.

freak, *v.* Add: **3.** *to freak out* (occas. without *out*): to undergo an intense emotional experience, to become stimulated, to rave, esp. under the influence of hallucinatory drugs. Also *trans.*, to cause (a person) to be aroused or stimulated in such a way. (Also in more trivial uses.) So **freaked-out** *a.*, affected thus; **freaking-out** *vbl. sb.*

1965 *Village Voice* (N.Y.) 2/1 (Advt.), Grand Opening!!! Freak with the Fugs!!! The East Side's Most Infinite Hallucination in Person. 1966 *Life* 25 Mar. 33/4 When my husband and I want to take a trip together..I just put a little acid in the kids' orange juice..and let them spend the day freaking out in the woods. 1967 *Oxf. Mail* 3 Mar. 4 'Freak out, baby,' goes the latest rebel war whoop. ..Frank Zappa..answers: 'On a personal level, freaking out is a process whereby an individual casts off outmoded and restricted standards of thinking, dress and social etiquette.' 1967 *Atavar* (Boston) 7–13 July 13/2 (*heading*) Freaked-out in the Federal Building. 1968 *It* 1–14 Nov. 8/4 Suppose..that total freedom could be granted now, today. Have you thought of what it would mean? Freedom to freak-out, yes; freedom to do your thing, sure. 1969 *Gandalf's Garden* iv. 9/1 He was the first guy I had ever met who used his music to influence people, to turn them on, or freak them out. 1969 *Daily Tel.* 21 Nov. 4/7 Some 'freaked-out' teenagers cannot move, let alone sit up straight when they attend classes. 1970 *It* 27 Feb.–13 Mar. 11/1 Aage leaps about and shouts at them and freaks them out. 1970 *Times* 5 May 9/3 The full African look..complete with enormous freaked out 'Hair' wig. 1970 *Nature* 23 May 704/1 One question asked the respondents how often they had seen other people 'freak out', that is, have intense, transient emotional upsets.

frea·k-out. [f. the vbl. phr. *to freak out*: see *FREAK *v.* 3.] An intense emotional experience, a 'rave-up', esp. one resulting from the use of hallucinatory drugs. (Also in more trivial uses.)

Quot. 1749, an isolated use, is better analysed as a use of the sb. *freak* (sense 3) plus the verbal phr. *to have out* 'to bring to a conclusion' (cf. OUT *adv.* 7 b).

[1749 J. CLELAND *Mem. Woman Pleasure* II. 198 She had had her freak out, and had pretty plentifully drowned her curiosity in a glut of pleasure.] 1966 *Daily Tel.* 10 Aug. 13/3 The tape-recorder picked up the horrifying moans and shrieks of one man who had made 33 pleasurable 'trips' with LSD and was encountering his first 'freakout' or bad LSD experience. 1967 *Spectator* 11 Aug. 158/1 This morning he had got in quick while the others were still half asleep from paper rounds or recovering from Sunday night freak-outs. 1968 L. DEIGHTON *Only when I Larf* iv. 48 That helicopter trip is a futuristic freak-out. 1968 J. IRONSIDE *Fashion Alphabet* 194 Mop (also called Freak-out)...Introduced in 1967 by the way-out young... A wild looking bush with curls, worn by both girls and boys. 1970 *Toronto Daily Star* 24 Sept. 1/9 They give the impression 'freak outs' happen every week.

freaky, *a.* **2.** In senses corresponding to *FREAK *v.* 3. Also as quasi-*adv.*

1966 YOUNG & HIXSON *LSD on Campus* p. vi, I think it would do everybody good to take LSD. But soon it's gonna get pretty freaky. 1967 A. K. BAER et al. *Study 18 LSD Users on Sunset Strip* 208 Everybody in the car was positive he was on an acid trip. He was freaky. 1969 *Observer* 7 Dec. 25/1 'Live freaky, die freaky,' was the judgment of a neighbour in Benedict Canyon on last August's Sharon Tate massacre. 1969 *Gandalf's Garden* vi. 10/3 Freaky-straights, either ordinary-looking people with fanatical ideas on one particular theme, like the 'Flat Earth Society', or people whose appearance is very weird but whose minds are channelled into one usual line of thought... The lack of light in the eyes betrays the freaky-straight.

freckly, *a.* (Later example.)

1935 HUXLEY & HADDON *We Europeans* v. 156 Skin Colour. Light, Medium, Swarthy, Freckly.

free, *a.* Add: **1. d.** Colloq. phr. *to be free, white and (over) twenty-one*: to be a free agent.

1929 J. BUCHAN *Courts of Morning* II. xiv. 346 We're all of us free, white, twenty-one, and hairy-chested, and we know how to be kind to a pretty girl. 1952 D. AMES *Murder, Maestro, Please* xix. 138 She's free, white and—no, I guess she's not twenty-one. 1958 J. CANNAN *And be a Villain* iii. 59 You're free, white and twenty-one. He couldn't *make* you go there. 1962 M. CARLETON *Dread Sunset* (1963) v. 108 What could I do when she insisted?.. She was free, white and, heaven knows, well over twenty-one!

2. (Further examples.)

1769 [see PRESS *sb.*[1] 13 f]. 1829 W. E. CHANNING *Wks.* (1886) 633/1 Through a free press, all public measures should be brought before the tribunal of the people. 1888 E. BELLAMY *Looking Backward* xv. 229 A free newspaper press..was a redeeming incident of the old system. 1946 *Observer* 10 Feb. 3/3 Britain would have a better chance of recapturing her share of the world markets with a 'free' economy than with nationalised industries. 1948 J. M. MURRY (*title*) The free society. 1951 I. SHAW *Troubled Air* xxi. 347 The benefits of a free society extended from one end of the economic spectrum to the others. 1968 *Guardian* 23 Aug. 9/1 Surprise grew into something like amazement at BBC monitoring stations yesterday as 'free' Czechoslovak broadcasts continued to be transmitted on eight normal wavebands. 1971 P. WORSTHORNE *Socialist Myth* vii. 143 The inevitably fissiparous tendencies of a free society.

b. Colloq. phr. *it's a free country*: a catch-phrase asserting a person's rights as an individual, implying that the action proposed is not illegal.

1885 'F. ANSTEY' *Tinted Venus* iv. 51 It's a free country, this is! 1911 A. BENNETT *Card* i. 17 Well, *you* go and do it. It's a free country. 1939 N. STREATFEILD *Luke* 244 Why stop with her if she was always nagging about her soul? That's what gets me. After all it's a free country. 1962 'H. CALVIN' *System* i. 13 'A girl could do worse,' she said sullenly. 'It's a free country, isn't it?' 1971 R. J. WHITE *Second-hand Tomb* i. 19 'Nuns and photographers.' 'Only one of each,' said Margaret. 'And it's a free country.'

c. In recent specific collocations. (i) Used in titles to denote those who continued resistance to Germany in the 1939–45 war after the capitulation of their respective countries.

1940 *Times* (Weekly ed.) 27 Nov. 11 He dwelt on the importance of the war in the Mediterranean theatre and of the part that Free France could play there in snatching triumph out of defeat. 1941 *Ibid.* 5 Feb. 15 Gen. de Gaulle, leader of the Free French forces. 1941 C. GRAVES *Life Line* 4 To-day we have the scattered remnants of the Free French, Free Dutch, Free Polish, and Free Norwegian Fleets to render certain assistance. 1944 H. G. WELLS *'42 to '44* 150 Vice-Admiral Muselier was appointed Commander of the Free French naval and air forces. 1965 B. SWEET-ESCOTT *Baker St. Irreg.* ii. 52 What the refugees themselves seemed to have in mind was that we should recognize them as a 'Free Rumanian' or a 'Free Bulgarian' movement. 1965 A. J. P. TAYLOR *Eng. Hist. 1914–1945* xiv. 527 German aeroplanes were using air-bases in Syria: Wavell must intervene there also, in co-operation with the Free French.

(ii) *free world*: a name used of themselves collectively by non-communist countries; so *free Europe*.

1950 *Time* 17 July 32/3 A branch of the National Committee for a Free Europe founded last year by a group of private U.S. citizens. 1955 *Ann. Reg. 1954* 241 There was the Council of Free Czechoslovakia..the 'rats and jackals' of Radio Free Europe. 1955 *Bull. Atomic Sci.* Feb. 42/3 The Soviet World and the Free World are running neck and neck in the training of scientists. 1957 *BBC Handbk.* 169 Throughout free Europe from Norway to the Turco-Soviet frontier. 1958 *Listener* 7 Aug. 185/2 The scale of human values..which distinguishes the free world from the Communist world. 1963 *Guardian* 19 Apr. 11/2 The free-world countries involved were Britain and Greece.

(iii) In various other collocations, e.g. of or pertaining to subversive movements inside a country.

1968 *N.Y. Rev. Books* 11 July 34/3 We have been unimpressed by the supposedly free universities established on the peripheries of major universities... What is wanted are really new ways of learning, not additional courses taught by Marxists and acidheads. 1969 *Guardian* 16 Sept. 11/3 The absurd romanticism of Free Belfast.

8. f. *spec.* in *Cricket*. Applied to (one who adopts) an unrestrained style of batting.

1851 J. PYCROFT *Cricket Field* iv. 59 In olden time the freest hitter was the best batsman. *Ibid.* x. 203 Many a man..whose talent lies in defence, tries free hitting, and between the two proves good for nothing. 1860 *Baily's Mag.* Aug. 387 His 45 and 30 were made in a fine free style. 1885 *Punch* 19 Sept. 143/2 Behind the stumps unbeatable, free bat, and slashing field.

10. c. *Philol.* Designating a linguistic form that can be used in isolation. Opp. *BOUND *ppl. a.*[2] 4 b.

1926, 1957 [see *BOUND *ppl. a.*[2] 4 b].

d. *Phonetics.* Of a vowel: not followed by a consonant in the same syllable. Opp. *CHECKED *ppl. a.*[1] 1 b.

1946 PRIEBSCH & COLLINSON *German Lang.* (ed. 2) I. iii. 76 Vowels in absolute final position ('free' vowels). 1962 [see *CHECKED *ppl. a.*[1] 1 b].

14. b. *Physics* and *Chem.* [after G. *frei* in the same sense (P. Drude in *Zeitschr. f. phys. Chem.* (1894) XV. 79, *Ann. d. Physik* (1900) I. 572).] Of an electron: not bound (to an atom, molecule, etc.), and therefore able to move unrestrictedly under the influence of electric and magnetic fields. (Occas. used of other particles, but the sense becomes indistinguishable from sense 16.)

1906 E. E. FOURNIER D'ALBE *Electron Theory* x. 193 They are propagated through space with the velocity of light, and if any matter containing free electrons or positive atoms intervenes, their rate of propagation is lessened. 1907 N. R. CAMPBELL *Mod. Electr. Theory* iii. 70 These 'bound' electrons, as they may be called in distinction to the 'free' electrons which are subject to no restraining force, take no part in electrostatic actions. 1956 N. F. RAMSEY *Molec. Beams* viii. 203 (*heading*) Nuclear and molecular interactions in free molecules. 1958 *Listener* 25 Dec. 1071/1 The number of free electrons in the ionosphere. 1959 *Chambers's Encycl.* V. 127/2 The electron theory of metals..ascribes their conductivity to the presence of free electrons. 1965 PHILLIPS & WILLIAMS *Inorg. Chem.* I. vi. 196 Such a situation corresponds to the so-called 'free-electron theory'.

16. (Later examples.) *free radical*: see *FREE *a.* D. 2.

1929 *Chem. Abstr.* XXIII. 5159 It is concluded..that the reactive substance in all these expts. is free methyl. 1937 *Discovery* July 199/2 With hydrogen atoms attached

to all the free bonds. 1952 *Science News* XXVI. 57 A covalent bond may..break by homolytic fission, each of the electrons separating with one of the atoms, giving two free atoms, e.g.: H—Cl→H— + Cl—, or—if the atoms..have other atoms bound to them—free radicals. 1953 R. W. GURNEY *Ionic Processes in Solution* iv. 64 There will be a dissociative equilibrium in the solution between the free ions and the neutral ion pairs. 1965 PHILLIPS & WILLIAMS *Inorg. Chem.* I. x. 347 Reactions which involved free atoms. 1970 D. F. SHAW *Introd. Electronics* (ed. 2) ix. 181 We thus have a diatomic system with two valence electrons whose energy levels are slightly different from those in the free atomic state.

31. *free-market, -port* attrib.

1905 *Westm. Gaz.* 8 Sept. 4/1 The great advantage a Free-Port system has over 'Protection'. 1907 *Ibid.* 5 Apr. 3/2 What Preferentialists ask from the masses in England is a price above the free-market price. 1947 V. A. DEMANT *What is happening to Us?* i. 8 The approach to a completely free-market economy began to wreck what any genuine conservative would want to conserve. 1967 *Boston Sunday Herald* 26 Mar. vi. 7/5 (Advt.), Buy silver, china, watches, etc., in our free-port shops right in the hotel.

C. *adv.* **b.** Colloq. phr. *free, gratis, (and) for nothing*. Phr. *for free*: see *FOR *prep.* 19 g.

1841 DICKENS *Let.* 30 June (1969) II. 317, I have declined to be brought in, free gratis for nothing and qualified to boot, for a Scotch county that's going a-begging. 1850 KINGSLEY *A. Locke* I. ii. 32 Spittoons, as you see, perwided free gracious [*sic*] for nothing. 1893 E. F. BENSON *Dodo* II. xi. 222 No charge for mixed metaphors. Supplied free, gratis, *and* for nothing. 1941 H. G. WELLS *You can't be too Careful* III. viii. 146, I was to be sent to France, free gratis and for nothing for six months. 1965 S. JEPSON *Third Possibility* xii. 89 The man you're going to meet can give you more of it free, gratis and for nothing than you can use in a lifetime.

D. 1. a. *free-floating, -flowering, -flowing, -flying* (so *free-fly* v., nonce-use), *-growing, -ranging, -running*.

1921 *Jrnl. Ecol.* IX. 241 A suggestion as to factors influencing the distribution of free-floating vegetation. 1926 W. McDOUGALL *Outl. Abnormal Psychol.* xviii. 316 If this accident had not taken place, the free-floating fear would have broken out in a phobia for some other object. 1936 *Discovery* Jan. 3/1 Ultra-microscopic particles of living protoplasm were the germs from which visible free-floating organisms were derived. 1965 *Times Lit. Suppl.* 25 Nov. 1058/3 The more free-floating culture of bohemia. 1824 *Greenhouse Compan.* 68 Free-flowering plants of different degrees of beauty. 1952 A. G. L. HELLYER *Sanders' Encycl. Gardening* (ed. 22) 223 A large, rounded, free-flowering bush is formed. 1934 WEBSTER, Free-flowing. 1960 *Farmer & Stockbreeder* 16 Feb. 109 In the Nicholas Liver Fluke Drench, hexachloroethane particles are suspended in a free-flowing liquid and this ensures accurate dosing. 1922 JOYCE *Ulysses* 256 They threw young heads back..to let freefly their laughter. 1915 E. R. LANKESTER *Divers. Nat.* 222 Fully-formed free-flying state. 1961 BANNERMAN *Birds Brit. Isles* X. 290 Only three chicks were reared to the free-flying stage. 1824 LOUDON *Encycl. Gard.* (ed. 2) 396 The species of stocks for fruit-trees are divided into what are called free-growing and dwarfing stocks. 1902 *Daily Chron.* 1 Apr. 2/1 The free-growing heralds of spring [*sc.* daffodils]. 1952 A. G. L. HELLYER *Sanders' Encycl. Gardening* (ed. 22), L[ycopodium] *Billardieri*, free-growing creeper, New Zealand. 1942 C. R. CARPENTER in *Jrnl. Compar. Psychol.* XXXIII. 113 (*title*) Sexual behavior of free ranging rhesus monkeys. 1948 V. MASSEY *On being Canadian* iii. 43 Free-ranging experimentation in ideas. 1958 *Observer* 19 Oct. 17/4 They [*sc.* geese] demand an open-air, free-ranging life. 1963 *Times* 7 Feb. 8/4 The Government..are anxious that the Commons should work for a free-ranging debate. 1940 *Chambers's Techn. Dict.* 353/2 *Free-running speed*, the speed which a vehicle or train will attain when propelled by a constant tractive effort. 1960 *Farmer & Stockbreeder* 15 Mar. 48/1 (Advt.), Free-running baler and binder twine. 1962 SIMPSON & RICHARDS *Junction Transistors* xvi. 404 The astable circuit is an oscillator whose output is a flat-topped wave... It is often called a free-running multivibrator or..simply a multivibrator.

b. *free-agency* (earlier example).

1754 FIELDING *Voy. Lisbon* (1755) 129, I would rob him of nothing but that free-agency which is the cause of all the corruption..of human nature.

2. **free activity**, used *attrib.* of a type of school or method of teaching in which children learn through their own efforts and experiments and not through instruction by a teacher; also, usu. in *pl.* (not *attrib.*), such activities; **free association** *Psychol.*, in an experiment based on the association of ideas, an association freely made by the person undergoing the test without suggestion or control on the part of the experimenter; hence **free-associational, free-associative** *adjs.*; **free-associate** *v. intr.*; **free cinema** (see quots. 1956); so *free film*; **free city**, (*a*) in the Middle Ages, a sovereign city-state in Germany; (*b*) a semi-autonomous city under the authority of an international body; **free-cutting** *a. Metallurgy*, applied to a metal with good machining properties, esp. when these are due to small quantities of some additional substance; **free diver**, a skin-diver; so **free-diving** *vbl. sb.*, skin-diving; **free drop**, a parachute descent made with a free parachute; **free enterprise**, the freedom of private business from state control; hence **free enterpriser**, an advocate or follower of such a system; **free expression**, the uninhibited expression of one's thoughts, feelings, creative capacities, etc.; **free fall**, the movement, of a body under no forces other than gravity, there being neither thrust nor (appreciable) drag acting on it; as (*a*) the flight of a ballistic missile after the driving power is cut off; (*b*) in a parachute descent, the part of the fall before the parachute opens (drag on the parachutist being neglected); (*c*) the flight of a spacecraft in space when there is no thrust from the engines, and any occupants of it experience weightlessness; phr. *in free fall*, moving or flying in these conditions; hence **free-fall** *v.*; **free film**, see *free cinema*; **free flight**, *spec.* (*a*) the flight of an unmoored balloon, or of a glider released from its towing-rope; (*b*) flight of an aircraft, missile, etc., in free fall; (*c*) used *attrib.* to designate a wind tunnel in which the model is not mounted but supported by aerodynamic forces like an aircraft in flight; **free food**, food imported free of tax or duty; also *attrib.*; **free-fooder**, a politician who opposes taxes on food; **free-form** *attrib.*, spec. of an irregular shape or structure; also *ellipt.*; **free gift**, *spec.* an object given away without charge to promote sales (cf. sense 21 b in Dict.); **free gold**, gold occurring naturally in a pure state or uncombined with other substances; **free (public) house** (see sense 31, quot. 1858, in Dict.); **free jump** *Parachuting*, = *free drop*; **free kick** (see KICK *sb.*[1] 1); **free library**: see LIBRARY[1] 1 b; **free list**, a list of persons from whom, or things on which, payment is not required (see also quot. 1870); so **free list** *v. trans.*; **free-loader** *slang* (orig. *U.S.*), one who eats or drinks without expense to himself, a sponger; so **free-loading** *vbl. sb.* and *ppl. a.*, and (as a back-formation) **free-load** *v. intr.*; **free love** (earlier and later examples); **free lover** (earlier U.S. example); **free lunch**, a lunch given gratis, esp. by bar-keepers to attract customers; so **free-luncher**; **free paper** *U.S.*, (*pl.*) documents proclaiming the status of a manumitted slave; **free parachute**, one released by the parachutist and not by a static line attached to the aircraft; **free pass**, authority to travel on a railway, etc., or to enter a place of entertainment without payment; **free path** *Physics*, (*a*) the distance which a molecule or other particle traverses without encountering another particle and without colliding with the walls of the containing vessel; (*b*) the distance a sound wave travels between successive reflections from the walls of an enclosure; usu. as *mean free path* (in both senses); **free period** (see quot. 1961); **free place**, a place in a secondary school awarded free to a scholar from an elementary school; also *attrib.*; hence **free-placer**, one who holds a free place; **free radical** *Chem.*, an uncharged atom or group of atoms having one or more unpaired electrons, esp. when these normally form part of a bond; **free range**, (*a*) *U.S.*, free pasturage; (*b*) used esp. *attrib.* of chickens given freedom to range for food (opp. *BATTERY 12* c); so *free-range egg*, etc.; **free return** *Astronaut.*, the positioning of a spacecraft on to the correct return flight path by planetary gravitation; **free silver** *U.S.*, the free coinage of silver bullion at government mints; also, belief in or advocacy of such a policy; **free skating**, a competitive programme of variable skating figures performed to music; **free speech** (cf. sense 25 in Dict.); **free-standing** *a.*, standing alone; not supported by a structural framework; **Free Stater**, a native or inhabitant of a free state (see Dict.), as the Orange Free State or Irish Free State; one supporting such a state; **free stock**, plants grown from seed to be used as rootstocks in grafting; **free-style** *a.*, = FREE *a.* 9; applied *spec.* to a swimming race in which the style of stroke used is left to the competitor's choice; also *absol.*; **free union** [F. *union libre*], cohabitation of a couple without marriage; **free vector** *Math.*, a vector of which only the magnitude and direction are specified, not the position or line of action; **free verse** = *VERS LIBRE; so **free-verser**; **free vote**, a Parliamentary vote not made subject to party discipline.

1941 N. CATTY *Learning & Teaching in Junior School* ii. 34 Other periods are devoted primarily to children's individual work and free activities. 1942 *Brit. Jrnl. Psychol.* Oct. 140 Four 'experimental' ('free activity') schools were paired with 'control' schools..in which the teaching was of the formal type. 1965 W. LAMB *Posture & Gesture* viii. 106 'Free activity' methods in the teaching of physical education give scope for much emotional and chaotic work. 1941 *Mind* L. 80 If you start free-associating from any item in consciousness you will be bound to reach a sexual item sooner or later. 1954 *Amer. Jrnl. Psychol.* LXVII. 244, I will read *four* words slowly and..you are to free-associate to the last word only. 1971 *Daily Tel.* (Colour Suppl.) 2 Apr. 7/1 He is not very interested in probing the subconscious of his patients, or allowing them to free-associate about childhood miseries. 1899 W. JAMES *Talks to Teachers* 219 Stated technically, the law is this: that strong feeling about one's self tends to arrest the free association of one's objective ideas and motor processes. 1905 E. B. TITCHENER *Exper. Psychol.* II. i. 192 We show him a word; he is to react when the word has suggested something, no matter what. The word *sea* may arouse the idea of land or water or ships or some particular sea or some particular incident at sea,—anything it likes. Associations of this sort are termed, technically, *free* associations. 1964 GOULD & KOLB *Dict. Social Sci.* 557/1 Freud..induced his patients to disclose their thoughts without censorship, allowing ideas to enter their minds in undirected free association. 1941 *Brit. Jrnl. Psychol.* July 41 The responses given..were very full and largely free-associational in content. 1960 *Guardian* 9 Sept. 7/3 A solid chunk of free-associational near-verse. 1958 *Times Lit. Suppl.* 24 Jan. 44/1 A time when literary fantasy of a free-associative, surrealistic kind was by no means the commonplace that it is today. 1970 *New Yorker* 29 Aug. 24/1 He..diverted the rest of the hour into a lengthy free-associative screen about certain rhapsodic sexual events. 1956 *Living Cinema* I. 9 The phrase 'Free Cinema' coined by the National Film Theatre is a happy thought, for it can be used..to cover words like 'avant-garde' and 'experimental'. *Ibid.*, Free Cinema..can..include any film in which the maker has succeeded in breaking the chains..of financial or ideological control. 1963 *Listener* 14 Feb. 300/1 *Saturday Night and Sunday Morning*, *A Kind of Loving*, and other products of Free Cinema are, at best, admirable documentaries. 1617 Free city [see IMPERIAL 10]. 1621 P. HEYLYN *Microcosmus* 145 The citties of this country are of 3 sorts...The third sorte are the Free or Imperiall citties...These free citties are in number 60. 1665 Free city [see HANSEATIC *a.*]. 1797 *Encycl. Brit.* VIII. 284/1 [Hamburg] was declared a free imperial city by a decree of the aulic. 1919 *Times* 28 June (Suppl. Treaty Versailles) p.vii/6 (Article 102) The Principal Allied and Associated Powers undertake to establish the town of Danzig..as a Free City. It will be placed under the protection of the League of Nations. 1958 *Listener* 4 Dec. 936/1 A proposal that..the western sector [of Berlin] should be made a 'free city' with its own government. 1959 *Chambers's Encycl.* VI. 261 Less secure was the position of the 'free cities' whose autonomy was not protected by imperial privileges. Among them..were flourishing communities, such as Brunswick, Lüneburg, Magdeburg and Emden. 1965 A. J. P. TAYLOR *Eng. Hist. 1914–1945* iv. 134 Thanks to him [*sc.* Lloyd George], Danzig became a Free City, instead of being annexed to Poland. 1927 *Mech. Engin.* XLIX. 163/2 The writer..asks what is free-cutting steel, and how does it differ from steel difficult to machine? The latter is tough or ductile, and..it is often most difficult to cut good threads in such material, but..free-cutting steel is 'free-breaking' steel. 1949 R. T. ROLFE *Dict. Metallogr.* (ed. 2) 108 *Free-cutting quality*, the quality possessed by a specific grade of a particular material whereby the machinability..is increased, usually by the addition of a further constituent. 1958 *Jrnl. Iron & Steel Inst.* CXC. 89/3 In the field of free-cutting steels, special mention is made of leaded steels as a means of improving the machinability of low-carbon, high-nickel carburizing steels. 1963 L. DEIGHTON *Horse under Water* xxvi. 109 A professional salvage free-diver. 1963 *Harper's Bazaar* Jan. 30/2 Breathing in from containers strapped to their backs, experienced 'free divers' can go as deep as 130 ft. for 15 minutes. 1955 *Sci. News Let.* 2 July 13/1 Free-diving has given marine biologists a revolutionary method of specimen collecting. 1957 G. CLARK *Archaeol. & Society* (ed. 3) ii. 48 The development of the aqualung and of the technique of free-diving by M. Cousteau and his associates has recently made it possible to envisage submarine archaeology as a field of purposeful research as well as of lucky finds. 1940 *War Illustr.* 26 Jan. 20/2 Not until 1919 did the first successful 'free' drop take place. 1944 *Jane's All World's Aircraft 1943–44* 25/1 After seven free drops [the parachutist] qualifies for his parachute badge. 1890 A. MARSHALL *Princ. Econ.* I. I. ii. 30 The growth of free enterprise in England. 1938 *Newsweek* 3 Oct. 37/3 Management leaders representing the world's democratic countries agreed that free enterprise, not government control, is the key to better times. 1957 L. F. R. WILLIAMS *State of Israel* 103 The so-called middle-class villages—based on free-enterprise farming and hired labour. 1943 *Lincoln* (Nebr.) *Jrnl.* 16 Aug., Free enterprisers in this country include the big industrialists. 1967 *Listener* 8 June 752/1 Conservatives and free-enterprisers of every colour. 1943 H. READ *Educ. through Art* v. 109 Play is the most obvious form of free expression in children and there has been a persistent attempt on the part of anthropologists and psychologists to identify all forms of free expression with play. 1958 *Spectator* 4 July 14/1 First there were filmed excerpts from the studio, with its presiding talent, Lee Strasberg, analysing free expression exercises. 1967 M. DRABBLE *Jerusalem Golden* vi. 149 The efforts of the American teacher at their nursery group to make the children paint with free expression. 1967 D. P. CAREW *Many Years, Many Girls* vii. 140 A 'free expression' afternoon was instituted during which we could please ourselves what we did. 1919 R. D. GODDARD *Method of Reaching Extreme Altitudes* 59 The time of descent [of a rocket] will also be short; but free fall can be satisfactorily prevented by a suitable parachute. 1920 *Flight* XII. 210/1 With parachutes having problematical opening..the uncertainty of the free-fall is such that no one ever dreams of making a practice drop with it at

less than 2,000 ft. 1930 P. WHITE *How to fly Airplane* xxii. 304 The usual and safest time to open the parachute is immediately after clearing the ship, unless an emergency arises in which it is necessary to make a long free fall before pulling the rip cord. 1953 *Authentic Science Fiction* 15 Feb. 49 She wanted me to explain to you that..she was weightless at that point, 'as I'm in free fall'. 1955 *Times* 25 Aug. 6/3 She claimed a world record for a 'free fall' jump, having descended approximately 26,238 ft. before she pulled the ripcord. 1956 W. A. HEFLIN *U.S.A.F. Dict.* 221 Free-fall, to make a free fall. 1958 C. C. ADAMS et al. *Space Flight* 142 In the free fall (zero gravity of orbit), telescopes (particularly radio telescopes) of fantastic size could be assembled. 1959 W. GOLDING (*title*) Free fall. 1959 *Time* 17 Aug. 15/3, I free-fell an eternity. 1961 *Listener* 9 Nov. 765/1 The occupants of the projectile would have been in free fall, and therefore weightless, from the moment of firing. 1962 F. I. ORDWAY et al. *Basic Astronautics* viii. 346 An interplanetary vehicle after its freefall back to Earth will have an approach velocity of 25,000 mph or more. 1971 *Daily Tel.* 16 Aug. 9/7 When he pulled the cord he had been free-falling for 5,000 ft. 1958 R. HOGGART in N. Mackenzie *Conviction* 137 The 'free' films made by such men as Lindsay Anderson and Karel Reisz. 1922 *Flight* XIV. 147/2 It is probable that the resistance of the supports in free flight can be cut down to one-half the minimum drag of the wing. 1939 *Jrnl. R. Aeronaut. Soc.* XLIII. 792 With background experience in development of a free-spinning tunnel..the Committee felt that the attempt to develop a free-flight tunnel was amply justified. 1954 *N.Y. Times Mag.* 29 Aug. 49/1 The Honest John free-flight artillery rocket is this [pre-set] type of missile. 1959 *Listener* 12 Feb. 282/1 The condition of weightlessness during free-flight. 1970 R. TURNILL *Lang. Space* 48 An Apollo spacecraft is also in 'free flight' all the way to the moon and back once it has been injected into the required trajectory. 1903 *Westm. Gaz.* 17 June 7/2 The free food Unionists. 1903 *National Rev.* Aug. 893 Though calling themselves Free Fooders, they are not in favour of Free Food. 1967 R. S. CHURCHILL *Winston Churchill* II. ii. 62 Churchill was a prime mover in what was ultimately to be named the Free Food League. 1903 DUKE OF DEVONSHIRE *Let.* 10 Oct. in J. Amery *Life J. Chamberlain* (1969) VI. cix. 482, I am not sure whether you and the other Free Fooders do not wish to take up a more hostile attitude..than I. 1905 *Spectator* 7 Jan. 13/2 Such a supposition..is surely strange as coming from convinced Free-Fooders. 1969 J. AMERY *Life J. Chamberlain* VI. cxi. 559 This amendment was not one which the 'Free Fooders' would be likely to oppose. 1952 in N. Mailer *Advts. for Myself* (1961) 157 Sam resents..the slender coffee table, a free-form poised like a spider on wire legs. 1957 'P. QUENTIN' *Suspicious Circ.* iv. 37, I walked..towards the free-form swimming-pool. 1960 M. MILLAR *Stranger in my Grave* xv. 161 A half-finished free-form table which Jim was making. 1968 *Economist* 17 Feb. 76/1 All these conglomerates have 'free-form' management structures, with a small central staff supervising the profitability of a host of disparate subsidiaries. 1969 *Sunday Times* (Colour Suppl.) 12 Jan. 15/3 In most of the communities there is free-form music, an uncoordinated bedlam of bells, guitars, drums and oriental chant. 1909 *Daily Chron.* 28 July 7/6 (Advt.), Make use of the Free Coupon printed here and you will receive..a free-gift parcel containing a Bottle of Guy's Tonic. 1933 *Punch* 19 July 61/1 She..collected over a thousand Free Gift coupons which she has exchanged for a portable radio set. 1965 *Guardian* 19 May 7/1 Gimmicks and the offering of free gifts to promote sales were condemned. 1897 'MARK TWAIN' *Following Equator* 687 No way of getting anything out of the rock but the coarser-grained 'free' gold. 1901 *Daily Colonist* (Victoria, B.C.) 11 Oct. 8/4 The rock is a free-milling quartzite, plentifully impregnated with free gold in specks, plainly visible to the naked eye. 1960 B. RAMSEY *Barkerville* 21 A small vein which showed substantial values and several sizeable fragments of free gold. 1894 G. MOORE *Esther Waters* xxx. 236 The 'King's Head'..had..one thing in its favour—it was a free house. 1927 W. E. COLLINSON *Contemp. Eng.* 80 Tied houses and free houses. 1930 *Daily Express* 16 Aug. 10/2 To bring about in large regional areas the compulsory merging of brewery companies, including 'free' houses. 1959 'O. MILLS' *Stairway to Murder* ii. 18 The notification 'Free House' beside the inn-sign. 1930 D. VERRILL *Aircraft Bk. for Boys* xi. 189 A diagram of a 'free jump' from a plane, in which the jumper counts five (or more) before pulling the rip-cord and releasing the parachute. 1806 *Let.* 26 Dec. in L. Sumbel *Mem.* (1811) III. 212 He has received Mr. Thomas Sheridan's directions to put her name and friend on the free-list. 1833 *Deb. Congress U.S.* 20 Feb. 1749 The gentleman moved to strike the article of cotton out from the free list of imports. 1845 *Ann. Rep. Treas. U.S.* 6 An adequate revenue will still be produced, and permit the addition to the free list of salt and guano. 1855 M. THOMSON *Doesticks* xxvi. 229 The Bowery Theatre..where they announce a grand 'benefit' five nights in the week, for the purpose of cutting off the free list. 1856 [see FREE *a.* 32]. 1870 J. K. MEDBERY *Men & Myst. Wall St.* 20 The securities..are divided into two classes, known respectively as the Regular and the Free List. 1909 *Daily Chron.* 13 Apr. 5/2 Cocoa was free listed on the ground that..it was a common necessity of life. 1969 *Listener* 13 Nov. 680/3 The *Spectator*'s reviewer..has been dropped from the free list at that theatre. 1967 B. NORMAN *Matter of Mandrake* viii. 56 Free-load, Neddy Masters had said. Get a little drunk. 1968 *Word Study* Dec. 5/2 A leech is one who attaches himself to another person in order to borrow or to freeload or merely to relieve his loneliness. 1947 *Time* 24 Mar. 63 Even men who don't frequent saloons would come in to see a hockey game. And free loaders were no problem; most people bought at least a few beers while they watched. 1948 R. CHANDLER *Let.* 20 Sept. (1966) 165 We didn't have this cat seventeen years in order for some freeloader to say God forgive him he'd even take a piece about her for his goddam parish magazine. 1951 *N.Y. Times* 15 Apr. VII. 23/3 Congressmen are great freeloaders. 1960 *News Chron.* 12 July 4/4 The weirdest assortment of beatniks, drunks, romantics, poets, free-loaders, millionaires. 1967 *Melody Maker* 23 Dec. 7 A Freeloader is one who has discovered that you can drink yourself silly for absolutely no expense if you attend all the receptions. 1956 J. CANNON *Who struck John?* 123 [The art of] free loading in the fight racket has diminished. 1964 S. BELLOW *Herzog* (1965) 124 My lousy, free-loading bohemian family, all chisellers. 1971 *Daily Tel.* 1 July 7/3 Since the food cost an average of £1·75 a head, she thought this was freeloading at the taxpayers' expense. [1814 *Theatr. Inquis.* IV. 384 Mr. and Mrs. Kemble, as Freelove and Lady Eliza,..were beyond all praise.] 1822 A. CUNNINGHAM *Tales Eng. & Sc. Peasantry* II. 73 [Her] thoughts had been weaned..from free love to religion. 1971 *Guardian* 23 July 7/2 A young person today was bombarded on all sides by invitations to free love. 1858 *Baltimore Sun* 28 June (Bartlett), Abolitionists, spiritualists, and free lovers. 1854 *Wide West* (San Francisco) 26 Nov. 2/3 The excitement during the week on the subject of the 'free lunches' has been of the most intense character. 1889 *Kansas Times & Star* 25 June, Two..Saloon keepers here were notified by the police department to discontinue the free lunches at their places. 1905 *Daily Chron.* 9 Sept. 6/7 The free-lunch system in Manchester is dying out. 1908 G. H. LORIMER *J. Spurlock* v. 82, I had mapped out the most complete little free-lunch route in New York City. 1909 'O. HENRY' *Roads of Destiny* iv. 58 [His] memories of the old hotel are limited to his having been kicked out from its free-lunch counter in 1873. 1927 H. CRANE *Let.* 16 Feb. (1965) 286 If you want the good old beer, the old free-lunch counter and everything thrown in—for 15¢ a glass. 1876 *Ventura Free Press* (San Buenaventura, Calif.) 8 Jan. 1/6 A healthy free luncher..commenced operations with a sandwich. 1898 *Daily News* 15 July 6/6, I felt flattered myself, a mere free-luncher, when I saw the great Joseph, with the income of a prince, thus deigning to superintend my viands. 1838 in J. S. Buckingham *America* (1841) I. 282 Henry has relations..some of them free, and likely he has free papers. 1881 *Century Mag.* Nov. 126/1 It was the custom in the state of Maryland to require the free colored people to have what were called free papers. 1933 *Flight* 30 Mar. 302 (*heading*) Free parachute jumping. 1935 C. G. BURGE *Compl. Bk. Aviation* 489/2 *Free parachute*, a parachute whose pack is secured only to the body of the user, the release of which is dependent on some act on his part other than falling from the aircraft. 1942 A. M. LOW *Parachutes* iii. 39 In the 'free' parachute the airman pulls the rip-cord himself. 1857 *Trans. Ill. Agric. Soc.* II. 2 The railroad company sent free passes over their whole line. 1894 Free pass [see FREE *a.* 32]. 1879 *English Mechanic* 5 Sept. 639/1 This dark space is found to increase and diminish as the vacuum is varied, and is found to be the mean free path of the molecules of the residual gas. 1940 J. H. JEANS *Introd. Kinetic Theory Gases* v. 131 Viscosity and conduction of heat can be explained in terms of the collisions of gas molecules, and of the free paths which the molecules describe between collisions. 1947 *Proc. Physical Soc.* LIX. 536 For rooms of the usual shapes the mean free path of sound is independent of the shape. 1966 *McGraw-Hill Encycl. Sci. & Technol.* VIII. 187 The types of mean free paths which are used most frequently are for elastic collisions of molecules in a gas, of electrons in a crystal, of phonons in a crystal, and of neutrons in a moderator. 1945 C. V. GOOD *Dict. Educ.* 292/1 *Period, free*,..designating the time in a regular school day during which a teacher or a pupil has no definitely assigned duties. 1961 *Where?* Winter 1960–61, 14 *Free period*, (1) time in a school curriculum for private or unsupervised study. (2) time when a teacher is not allocated a class. 1965 'O. MILLS' *Dusty Death* xx. 196 Old Cowley knows I skive off during that free period. 1968 'P. HOBSON' *Titty's Dead* ii. 22 'So sorry if I'm late. But it's my free period.'.. She taught classics and she was the Senior Mistress. 1907 *Hansard, Commons* 15 May 1054 These free places..would be for Public Elementary School children who would not be asked to compete with children outside but who would only be asked to pass a qualifying examination. 1909 *Daily Chron.* 28 July 5/5 The distinction between fee-paying and free-place scholars. 1920 *Rep. Departm. Committee on Scholarships* 2 in *Parl. Papers* (Cmd. 968) XV. 385 To qualify as a 'free place' pupil the child must have attended a Public Elementary School for a certain period previously. *Ibid.* 35 The existing provision of free places in secondary schools appears to us inadequate. 1926 *Rep. Consult. Cttee. Educ. of Adolescent* 133 The Free Place Examination is conducted in writing. 1921 G. SAMPSON *English for the English* 44 Scholarship children or 'free-placers'. 1961 *Economist* 16 Dec. 1119/1 These 'free-placers' are sometimes brighter than the fee-payers. [1886 E. F. SMITH tr. *Richter's Chem. Carbon Compounds* 200 The free acid radicals, like all monovalent groups, cannot exist free.] 1900 *Jrnl. Amer. Chem. Soc.* XXII. 768 We have to deal here with a free radical, triphenylmethyl. 1948 W. A. WATERS *Chem. Free Radicals* (ed. 2) i. 4 The great chemical activity of the free radicals is to be associated with available combining energy of the odd electron. 1954 *Sci. Amer.* Sept. 86/3 Fragments of molecules known as free radicals. 1971 *Nature* 12 Feb. 500/1 The presence of free radicals in cigarette smoke was demonstrated by Lyons et al. 1912 MULFORD & CLAY *Buck Peters* 186 Outlying free range had been thoroughly combed. 1947 *Steamboat* (Colo.) *Pilot* 13 Feb. 8/4 Then sheep commenced to come for a share of the free range. 1950 *N.Z. Jrnl. Agric.* Dec. 550 Pullets are usually given good free range, but frequently cockerels are not so well treated. 1960 *Times* 2 May 8/6 At least one farmer with a roadside trade knows that increasing numbers of motorist shoppers are looking for eggs produced by free-range hens. *Ibid.*, His blackboard announces two prices which show the premium he expects to be paid for the free-range egg. 1968 *Times* 29 Nov. 13/1 There seems to be no scientific basis for the commonly held belief that free-range eggs have more flavour and are more nutritious than battery eggs. 1964 R. H. BATTIN *Astronaut. Guidance* 392 Free-return trajectories. 1969 *Daily Tel.* (Colour Suppl.) 10 Jan. 21 (*caption*) Moonbug retro-fires to achieve 'free-return' elliptical orbit. 1970 *Times* 15 Apr. 10/2 As the vehicle travelled round to the back of the moon, the effect of the lunar gravitation field was such as to swing the ship round on to a homeward trajectory. This is what the flight dynamics specialists call a free return trajectory. [1890 *Nation* (N.Y.) 1 May 346/1 The latest bill..provides for..the purchase by the Treasury of 4,500,000 ounces of 'free silver' per month.] 1895 *Chicago Tribune* 6 Apr. 1 Free Silver or Ruin, Hinrichsen will force his Fiat Money campaign. 1900 'MARK TWAIN' *Speeches* (1910) 192, I am in favor of..the gold standard and free silver. 1942 R. G. LILLARD *Desert Challenge* 50 Free silver was the symbol..of the agricultural West and South against the financial East. 1902 *Encycl. Brit.* XXXII. 642/2 Free skating,..affording scope for the performance of dance steps and brilliant individual figures. 1970 *Radio Times* 29 Jan. 54 Double and treble Axel-Paulsen Jumps are among the most difficult and spectacular manœuvres in free-skating. 1848 *N.Y. Weekly Tribune* 15 July 3/4 To address the citizens..upon the subject of Free Soil and Free Speech. 1859 GEO. ELIOT *Let.* 18 Sept. (1954) III. 154, I begin to think silence the only good thing..if the inevitable result of free speech is, that we must fall..into complaint and accusation. 1943 J. S. HUXLEY *Evol. Ethics* vii. 59 The suppression of free speech and inquiry. 1876 J. FERGUSSON *Hist. Indian Archit.* I. v. 121 A free-standing building. 1936 *Archit. Rev.* LXXIX. 14 (*caption*) It shows the free-standing structural column which occurs in the hall of each flat. 1951 *Good Housek. Home Encycl.* 232/1 Both free-standing and built-in models..are available. 1963 W. F. GRIMES in Foster & Alcock *Culture & Environment* v. 95 A ring of free-standing stones set within or mounted upon a circular bank. 1899 *Daily Chron.* 27 Sept. 5/3 The sickly sentiment of the Free Staters. 1940 L. A. G. STRONG *Sun on Water* 206 When the troubles came, it was Johnny..who turned Republican, and Denis a fanatical Free Stater. 1852 G. W. JOHNSON *Cottage Gardeners' Dict.* 851/2 Free Stocks are such as are raised from the seed, layers, &c., of any of the cultivated varieties of fruit-trees, and others. 1955 R. ATKINSON *Growing Apples* ii. 20 A few seedlings—called 'free' or 'crab' stocks and grown from cider apple or wild crab pips—are still sometimes used. 1934 WEBSTER, Free style. 1950 *Oxf. Jun. Encycl.* IX. 454/2 In free-style and breast-stroke races the competitors start with a dive. 1957 N. FRYE *Anat. Criticism* 91 The free-style allegories. 1958 P. GAMMOND *Decca Bk. Jazz* x. 125 He helped to keep free-style jazz alive with his Clambake Seven. 1960 M. WOODHOUSE *Tree Frog* ii. 10 She..swam free style for the West of England. 1931 *Times Lit. Suppl.* 1 Oct. 742/4 Marriage, companionate marriage, free unions, are considered intimately. 1919 L. SILBERSTEIN *Elem. Vector Algebra* i. 2 It is then obviously convenient *not* to include position among the determining characteristics of a vector. Such vectors, in distinction from localized ones, are called free vectors. 1964 E. Œ. WOLSTENHOLME *Elem. Vectors* i. 2 When the term *vector* is used, it is assumed that it refers to a free vector. 1908 A. NOYES *W. Morris* 119 The so-called 'free-verse' experiments, with abrupt and meaningless jerks or bumps. 1914 C. MACKENZIE *Sinister St.* II. III. ix. 682, I hate Free Thought, Free Love and Free Verse. 1926 W. R. INGE *Lay Thoughts* 31 A cubist or a free-verse writer. 1926 *Glasgow Herald* 1 Feb. 8 Those among the free-versers who are not purely imbecile are disgruntled, sarcastic, and gloomy. 1931 *Economist* 25 Apr. 885/1 The House was accorded the opportunity of a 'free vote', which it gave..in favour of the Bill. 1955 *Times* 30 June 11/3 In a free vote..the House of Commons signified a desire that legislation should be introduced to improve the financial position of junior Ministers. 1971 *Times* 20 Oct. 1/1 Mr Heath's dramatic decision on Monday to concede a free vote to Conservative backbenchers placed Opposition leaders in an extremely difficult position.

free-and-easiness. (Later examples.)

1912 BEERBOHM *Christmas Garland* 36 The general free-and-easiness. 1959 H. HAMILTON *Answer in Negative* i. 9, I came to Fleet Street..and I still like the free-and-easiness of it.

freebie (frī·bi), *a.* and *sb.* *U.S. colloq.* Also **freebee, freeby.** [Arbitrarily f. FREE *a.*] **A.** *adj.* Free, without charge. **B.** *sb.* Something that is provided free.

1942 BERREY & VAN DEN BARK *Amer. Thes. Slang* § 551/17 Free..free gratis for nothing, freebee, freeby, [etc.]. 1946 MEZZROW & WOLFE *Really Blues* xiv. 256 It's the brakeman who throws freebie passengers off. *Ibid.* 373 *Freebie*, handout, something gratis. 1954 L. ARMSTRONG *Satchmo* (1955) vi. 81 That meal was a freebie and didn't cost me anything. 1962 E. LACY *Freeloaders* iv. 83 She'll write 'free' on the slip... They come in for the freebie and end up buying more copies.

free-board. Add: **2.** *transf.* and *fig.*

1896 *McClure's Mag.* Dec. 142/2, I saw Mrs. McPhee swell and swell under her garance-coloured gown. There is no small free-board to Janet McPhee, nor is garance any subdued tint. 1925 J. JOLY *Surface-Hist. Earth* iii. 54 The change of buoyancy of the magma which must arise if the substratum changes from the solid to the liquid state... How will this change affect the 'freeboard' of the continents? When we melt basalt in the laboratory we easily float granite in the fluid rock. It floats with a good free-board, and appears to preserve its buoyancy indefinitely.

Free Church. Add: **3.** (With small initials.) A church in which the seats are free. *U.S.*

1835 J. MARTIN *Gaz. Virginia* 126 It contains several dwelling houses—one free church—one common school. 1835 A. A. PHELPS in *Life W. L. Garrison* (1885) II. 63 He has since been requested to preach the sermon to the Free Church in Hartford. 1860 *Massachusetts Acts & Resolves* c. 181 An act to incorporate the trustees of the free church of Saint Mary, for sailors.

freedom. Add: **4.** *four freedoms*: those propounded by Franklin D. Roosevelt (see quot. 1941).

1941 F. D. ROOSEVELT in *N.Y. Times* 7 Jan. 4/7 In the future days..we look forward to a world founded upon four essential human freedoms. The first is freedom of speech and expression... The second is freedom of every person to worship God in his own way... The third is freedom from want... The fourth is freedom from fear. 1944 H. G. WELLS *'42 to '44* 69 Our government's good faith in promising mankind a regenerated world, the 'four freedoms' and so forth after the war. 1950 A. HUXLEY *Themes & Variations* 246 To talk about the Rights of Man and the Four Freedoms in connection..with India is merely a cruel joke. 1964 R. CHURCH *Voyage Home* iii. 35 The assumption of the Government..was that..grateful civilians would see to it that the ex-soldiers be eased back with sympathy into an economic way of life supported on the newly-painted pillars of the four freedoms.

10. *degree of freedom.* **a.** (Earlier and later examples.)

1867 THOMSON & TAIT *Treat. Nat. Philos.* I. ii. 130 A free point has *three* degrees of freedom, inasmuch as the most general displacement which it can take is resolvable into three, parallel respectively to any three directions, and independent of each other... If the point be constrained to remain always on a given surface, *one* degree of constraint is introduced, or there are left but *two* degrees of freedom. 1958 *Van Nostrand's Sci. Encycl.* (ed. 3) 610/2 When heat energy is imparted to a pure diatomic gas, ⅕ of it goes into each degree of molecular freedom. Three of these degrees of freedom are concerned with motions of translation, so that ⅗ of the energy takes this form.

b. *Physical Chem.* An independent capability of a system to vary without altering the number of phases and components present.

1904 A. FINDLAY *Phase Rule* ii. 15 We shall therefore define the number of degrees of freedom of a system as the number of the variable factors, temperature, pressure, and concentration of the components, which must be arbitrarily fixed in order that the condition of the system may be perfectly defined. 1948 GLASSTONE *Physical Chem.* (ed. 2) vi. 475 A system consisting of one phase only, e.g., solid, liquid or gaseous, of water has two degrees of freedom,..for..it is necessary to specify both temperature *and* pressure to define completely the state of the system. 1966 TER HAAR & WERGELAND *Elem. Thermodynam.* vi. 82 Gibbs' phase rule... The number of phases plus the number of degrees of freedom exceeds the number of components by 2.

c. *Statistics.* A property of a statistical distribution or of a statistic, in number equal either to the number of values that can be arbitrarily and independently assigned to the distribution, or to the number of independent and unrestricted quantities contributing to the statistic.

1922 R. A. FISHER in *Jrnl. R. Statistical Soc.* LXXXV. 88 We shall show that Elderton's Tables of Goodness of Fit..may still be applied, but that the value of n' with which the table should be entered is not now equal to the number of cells, but to one more than the number of degrees of freedom in the distribution. 1950 YULE & KENDALL *Introd. Theory Statistics* (ed. 14) xxi. 485 In the quantity $\Sigma(x-m)^2$ there are n independent contributions of the type $(x-m)^2$, and hence we may say that n is the number of degrees of freedom of that estimate; but in the quantity $\Sigma(x-\bar{x})^2$ we have used the data to estimate $\bar{x}$, and hence the number of degrees of freedom is lowered by unity, i.e. equals $n-1$. 1961 H. JEFFREYS *Theory of Probability* (ed. 3) iii. 145 The estimate of any unknown from the observations directly relating to it may be of very doubtful accuracy on account of the small number of degrees of freedom. 1969 KENDALL & STUART *Adv. Theory of Statistics* (ed. 3) I. xvi. 375 It is thus natural to speak of the number of degrees of freedom, ν, of a function such as χ^2, meaning thereby that it is distributed as the sum of squares of ν independent standardized normal variates.

15. *freedom-loving* (earlier and later examples); **freedom box** (see quot. 1960); **freedom fighter,** one who takes part in a resistance movement against the established political system of a country; also *transf.*; **freedom ride** *U.S.*, an organized ride (in buses, etc.) by people demonstrating against racial segregation; so **freedom rider,** one who takes part in such a demonstration of racial protest; similarly *freedom march, walk*.

1911 C. J. JACKSON *Illustr. Hist. Eng. Plate* II. xxix. 913 Lord Carbery owns an interesting silver-gilt freedom-box..which was presented by the borough of Bandon-Bridge to George, the third Baron Carbery in 1760. 1960 H. HAYWARD *Antique Coll.* 123/2 *Freedom box*, small circular or oblong box, silver-gilt, when not gold, presented with a script conferring the freedom of a town. Particularly popular in Ireland from the late 18th to the early 19th cent. 1942 J. LEHMANN *40 Poems* 32 Their freedom-fighters staining red the snow. 1958 *Listener* 13 Nov. 786/1 The Hungarian 'freedom fighters'. 1962 *Listener* 20 Dec. 1042/2 The Information Minister [in Jakarta] had urged the press to support the freedom fighters in northern Borneo. 1964 *Ann. Reg. 1963* 113 Mr Obote had played a prominent part in the Addis Ababa conference..offering training grounds for 'freedom fighters' against South Africa. 1969 *Daily Nation* (Nairobi) 31 Oct. 16/1 Freedom fighters ring Portuguese territories and Rhodesia with the support of African governments, if not always full acknowledgment. 1969 M. PUGH *Last Place Left* vii. 45 She was a freedom fighter in the sex war. 1841 H. S. FOOTE *Texas & Texans* I. 102 The descendants of..freedom-loving ancestors in Texas nobly resolved to imperil their lives. 1942 W. S. CHURCHILL *End of Beginning* (1943) 122 We shall..build a sure peace for all freedom-loving peoples. 1965 *Listener* 20 May 755/2 The Alabama freedom march. 1961 *N.Y. Times* 20 Aug. 2E/6 Freedom Rides..are expected to continue until bus-terminal segregation ends. 1961 *Guardian* 25 May 1/3 Anti-segregation 'freedom riders' arrived here by bus from Montgomery, Alabama. 1961 *Economist* 27 May 890/1 The flow of 'Freedom Riders'—mostly young Negroes, with a few white sympathizers, who are determined to challenge illegal discrimination in the defiant heart of the old South. 1967 *Listener* 17 Aug. 211/2 The old days when Freedom Riders were widely applauded..have gone. 1967 *Word Study* Dec. 7/2 There were..'freedom walks' across Southern states in 1962 and 1963.

free-for-all, *a.* and *sb.* orig. *U.S.* [FREE *a.* 10 b.] **a.** *adj.* Open to all. **b.** *sb.* A fight, etc., in which anyone may take part.

1881 [see FREE *a.* 10 b]. 1902 S. E. WHITE *Blazed Trail* xxvii. 175 In a free-for-all knock-down-and-drag-out, kicking, gouging, and biting are all legitimate. 1918 C. E. MULFORD *Man fr. Bar-20* vi. 56 From the noise it appeared to be turning into a free-for-all. 1923 L. J. VANCE *Baroque* xiv. 85 The girl quietly lifted the lower sash, letting in..cacophony of a domestic free-for-all in one of the neighboring flats. 1931 *Amer. Speech* VII. 49 [Lumberjacks'] 'free-for-all' when the 'shanty men' join the 'bull pen boys' around the box stove. 1941 L. A. G. STRONG *Bay* ii. 56 To judge by the sounds, it was a regular free-for-all. 1955 A. L. ROWSE *Expansion Eliz. Eng.* viii. 296 The development of privateering from a promiscuous free-for-all scrimmage..into a regular business. 1971 *Guardian* 12 July 7/6 The sort of free-for-all tariff system we saw at London Airport in the sixties.

freeha·ndedness. [f. FREE-HANDED *a.*] Open-handedness, liberality.

1888 J. R. LOWELL *Lit. & Pol. Addresses* (1914) 214 The power of the political boss is built up..by his free-handedness in distributing the property of other people. 1910 M. S. WATTS *N. Burke* 386 [He] certainly could not be accused of encouraging the Mexicans in their extortionate practices by any particular free-handedness on his own part. 1928 T. J. GRAYSON *Investment Trusts* 68 The facility and freehandedness with which they distribute profits. 1940 K. ROBERTS *O. Wiswell* 192 Here's another thing you'll have to consider, Mr. Wiswell: that's freehandedness with rebels.

free lance. Add: **2.** In recent usage, a person working for himself and not for an employer; freq. *attrib.*; also of occupations or work performed by free lances.

1882 J. HATTON *Journalistic London* v. 106 The name of Grenville-Murray..might be associated with clever work on many other English as well as French journals. The free lance *par excellence* of journalism was laid to rest..during..1881. 1901 *Westm. Gaz.* 7 Mar. 9/1 Someone who calls himself a free-lance journalist. 1912 W. OWEN *Let.* 3 June (1967) 139 There entertained my guest, the Preacher—a funny old man who is a free-lance (as he vaunts) and answered the Vicar's advertisement. 1927 CARR-SAUNDERS & JONES *Soc. Struct. Eng. & Wales* 62 Free-lance professional men, doctors and barristers for instance. *Ibid.* 75 When members of a free-lance profession take salaried positions. 1950 *Science News* XV. 7 If they had to rely on the free-lance articles which come in they could close down tomorrow. 1962 MCLUHAN *Gutenberg Galaxy* 74 Leopold Bloom..is a free-lance ad salesman.

Hence **free-lance** *v. intr.*, to act as a free lance; **free-lancer**, a free lance; **free-lancing** *vbl. sb.* and *ppl. a.*

1903 A. BENNETT *Truth abt. Author* v. 60 What in Fleet Street is called 'free-lancing'. 1904 *Westm. Gaz.* 6 May 2/3 Lord Londonderry..has done a bit of free-lancing himself in his non-Ministerial days. 1907 *Ibid.* 27 Mar. 4/2 Some free-lancing Parliamentary iconoclast. 1909 *Daily Chron.* 7 Apr. 4/7 If the clergy were allowed to free-lance in each other's parishes. 1915 W. J. LOCKE *Jaffery* i, He had a terrible time for a dozen years or so, taking pupils, acting, free-lancing in journalism. 1937 *Times* 30 Dec. 6/3 My conviction that one could do more for the general cause of good architecture from within that body [*sc.* the R.I.B.A.] than as a rebel, free-lancing outside it. 1955 L. FEATHER *Encycl. Jazz* i. 28 Mary Lou Williams, free-lancing along the Street. 1966 *New Statesman* 11 Mar. 357/4 (Advt.), *Journalism*: occasional freelancer..seeks full-time position.

free-living, *a.* Add: **1.** (Further example.)

1935 G. GREENE *Basement Room* 40 Any child in those free-living parts might be expected to play truant.

2. (Examples.)

1912 J. S. HUXLEY *Individ. in Anim. Kingdom* ii. 38 Free-living animals, the Protozoa. 1915 E. R. LANKESTER *Divers. Nat.* 113 It is this change from the life of a free-living shrimp to that of a living lump, adherent by its head to rocks or floating logs. 1967 *Oceanogr. & Marine Biol.* V. 352 (*caption*) Free-living polyzoa.

free-trader. Add: **1. c.** A trader not in the service of a company. *U.S.*

1837 W. IRVING *Adv. Capt. Bonneville* (1895) I. 66 Waylaying and dogging the caravans of the free traders, and murdering the solitary trapper. 1902 S. E. WHITE *Conjuror's House* iv. 39 Brooding on his imprisonment the Free Trader forgot his surroundings. 1945 R. W. SERVICE *Ploughman* 429 McTosh is a free trader..and hates the Company.

3. (Earlier U.S. example.)

1832 *Deb. Congress U.S.* 6 June 3305 He admits a fact which the free traders have always found it difficult to explain, that goods do not usually rise under an increase of duties.

free·way. orig. *U.S.* [WAY *sb.*[1] 1.] **a.** A thoroughfare with restricted access (see quots. 1930, 1936). **b.** An express highway.

1930 *Amer. City* Feb. 95 A freeway is a strip of public land, dedicated to movement, over which the abutting owners have no right of light, air or access. 1935 *Harvard City Planning Studies* VII. 40 Such master plan shall show desirable streets, highways, freeways..and other planning features. 1936 E. M. BASSETT *Zoning* (new ed.) 83 The increase of freeways, which are thoroughfares to which the abutting owners have no right of direct access, will help to solve the problem of too extensive business districts along main arteries. 1939 *Sun* (Baltimore) 16 Jan. 5/7 Construction of freeways, either elevated or on the ground, would be necessary to facilitate movement of traffic in and out of Washington. 1943 *Life* 22 Nov. 118/2 Almost everybody in Los Angeles has some pet plan for rebuilding large parts of the city after the war and opening up new 'freeways' (express auto highways). 1946 *Archit. Rev.* XCIX. 127/3 The internal by-pass or 'freeway' is a feature that will demand high technical skill. 1951 *Engineering* 17 Aug. 196/1 Distinction between a freeway and an ordinary highway is in the..access to the freeway...'Owners of abutting lands have no..or only restricted right..of access.' 1959 *Guardian* 18 Sept. 7/2 The freeways...Mile after mile of smooth, three or four-lane road, unbroken by intersection or traffic lights..minimum permissible speed about 40 m.p.h.

free wheel, free-wheel. [FREE *a.* 14.] The rear wheel of a bicycle arranged so that it can rotate freely while the pedals remain stationary; in a machine other than a bicycle, a wheel, propeller, etc. that can run free of a clutch or other connection with the machine itself; also *attrib.* and *absol.* (= a bicycle with a free wheel).

1899 *Westm. Gaz.* 17 Nov. 7/1 The free-wheel device..is..*the* novelty of the show. 1900 *Isle of Man Exam.* 3 Nov. 7 The recent free wheel contests. 1901 *Westm. Gaz.* 18 Sept. 8/1 Injuries sustained while riding a free-wheel machine down Knockholt Hill. 1902 A. C. HARMSWORTH et al. *Motors* x. 217 Free-wheel steering gear. 1902 E. NESBIT *Five Children & It* ix. 240 There..stood a bicycle—a beautiful new free-wheel. 1912 W. OWEN *Let.* 6 Feb. (1967) 114 My bicycle is quite disabled now—the free-wheel having worn out. 1930 *Engineering* 7 Feb. 163/1 Free wheels or over-running clutches are referred to. 1958 *Ibid.* 7 Mar. 298/1 There was a free-wheel arrangement between the air turbine and the alternator.

Hence **free-wheel** *v. intr.*, to ride with the wheel that is normally driven rotating freely; to operate as a free wheel; **free-wheeled** *a.*, having a free wheel; **free-wheeler,** a free-wheel bicycle; **free-wheeling** *vbl. sb.* and *ppl. a.* Also *transf.* (of lawn-mowers, etc.).

1900 *Captain* III. 80/2 All these free wheeled machines made a clicking noise. 1903 *Motor Cycle* 31 Mar. 10/2 There is nothing incongruous in the sight of a club of bicyclists—some of them propelling their bicycles by muscular power at all times, except when free-wheeling downhill. 1905 *Mosquito* Aug. 4 They thoroughly enjoyed the ride coming back and free-wheeled the whole way. 1908 E. M. SNEYD-KYNNERSLEY *H.M.I.* viii. 78 It was the first year of 'free-wheelers'. 1932 *Oxford Times* 23 Sept. 22/3 Free-wheeling and syncro-mesh gears have for some time been almost universal on cars built in the United States. 1935 *Jrnl. R. Aeronaut. Soc.* XXXIX. 1025 The drag of the propeller when locked in position, or free-wheeling. *Ibid.* 1026 Above 13° blade-angle setting the free-wheeling propeller has less drag. 1936 A. BRUNEL *Film Production* 16 Their principal camera developed a strange habit of 'free-wheeling' intermittently. 1970 *Which?* Mar. 83/1 The cutting cylinders of most of the mowers free-wheeled.

fig.

1911 'I. HAY' *Safety Match* iii. 48 For a moment he was silent—free-wheeling, so to speak, over the pulverised remains of Mr Winch. 1931 M. DE LA BÉDOYÈRE *Drift of Democracy* iii. 49 Romanticism glides along free-wheeling wheresoever the mood suggests. 1944 AUDEN *Sea & Mirror* (1945) ii. 45 Had you..really left me alone to go my whole free-wheeling way to disorder. 1956 M. STEARNS *Story of Jazz* (1957) vii. 67 In a free-wheeling music such as jazz, a musician is judged by his capacity for sustained and swinging improvisation. 1958 *Times* 24 Nov. p. viii/5 At the other end of the scale from free-wheeling Quebec are its neighbours on the east. 1966 *Listener* 20 Jan. 111/3 He saw drama as something half-way between a poem and a prose narrative, a balance between formalism and free-wheeling. 1967 A. BATTERSBY *Network Analysis* (ed. 2) xiv. 237 He must then ask what is the maximum period for which the project can be allowed to free-wheel without serious departures from the plan occurring.

free will. Add: **3. b.** *Free Will Baptist*: a member of a North American sect, of Welsh origin, of Arminian Baptists.

1732 SWIFT *Advantages Repeal. Sacr. Test* 6 Of the three Judges on each Bench, the first may be a Presbyterian, the second a Free-Will Baptist, and the third a churchman. 1816 W. BENTLEY *Diary* 19 May (1914) IV. 389 In Salem, the Free Will Baptists had an Immersion. 1823 *Baptist Mag.* IV. 32 Some who usually assembled with those denominated Free-will Baptists. 1847 H. HOWE *Hist. Coll. Ohio* 348 The village contains..1 Free Will Baptist, 1 Methodist and 1 Universalist church. 1889 'MARK TWAIN' *Connecticut Yankee* xxxi. 398 My first wife was a Free Will Baptist. 1966 R. G. TORBET *Hist. Baptists* (rev. ed.) x. 285 The Freewill Baptist periodical, *The Morning Star*, became so vocal against slavery that the printing establishment..was twice refused the privilege of incorporation.

freeze, *sb.*[1] Add: Also in specific *fig.* uses, as: (i) *to do a freeze*: to be overlooked or ignored. *Austral.* and *N.Z. slang*.

c 1926 'MIXER' *Transport Workers' Song Bk.* 23 'I know,' said one, 'I did a freeze Till I tumbled to the lurk.' **1941** Baker *Dict. Austral. Slang* 30 *Do a freeze*, to be overlooked, ignored.

(ii) The fixing or establishing of assets, dividends, military strength, etc., at a certain level or figure. Cf. **wage-freeze* and *freeze *v.* 5 e, f.

1942 *Business Week* 11 Apr. 88/1 In wartime there is much to be said for a general price, wage, and profit freeze. **1948** *Electronics* Nov. 132/1 Television Application Freeze Announced. Recent action by the FCC temporarily halted any further authorization of new television stations...The freeze would remain in effect long enough for the commission to decide whether certain changes should be made. **1951** *Economist* 24 Nov. 1245 In the first twelve months since the end of the so-called freeze, wage rates have increased by 10 per cent. **1959** *Daily Tel.* 23 Mar. 14/4 Mr. Macmillan's plan for a controlled and inspected 'freeze' of forces in a prescribed area. **1962** *Daily Tel.* 26 Oct. 1/3 He has accepted the proposal..for a 'freeze' of two to three weeks. This will involve him in the suspension of all arms shipments to Cuba. **1965** *New Statesman* 9 Apr. 560/3 A socialist government should actively support a new nuclear freeze in Europe.

2. *Cinemat.* and *Television.* A shot in which the movement is arrested by printing the same frame many times. Also **freeze-frame, -shot, frozen-frame.** Cf. *freeze *v.* 4 f.

1960 O. Skilbeck *ABC of Film & TV* 59 *Freeze frame*, T.V. term meaning a briefly Frozen Shot after the Jingle to allow ample time for Change over at the end of a T.V. 'Commercial'. **1965** L. Halliwell *Filmgoer's Compan.* 157 *Freeze frame*, a printing device whereby the action appears to 'freeze' into a still, this being accomplished by printing one frame many times. **1966** *New Statesman* 3 June 819/2 Daisy..breaks down but recovers in a frozen-frame finale. (Incidentally, *The Moving Target* ends on a freeze too: perhaps some more conventional way of signing off a story might be returned to.) **1969** *New Yorker* 17 May 127/1 The sound track uses the creak of the prison doors..and, finally, to accompany a freeze-shot of the start of a massacre, the Haydn tune that the Hapsburgs adopted as their anthem. *Ibid.* 20 Dec. 36/3 The freeze-frame of the dream resumes. **1970** A. Fowles *Dupe Negative* xiv. 190, I crashed berserkly through the door and a freeze frame was stamped obscenely on my mind.

freeze, *v.* Add: **3. b.** (U.K. examples.)

1897 *Westm. Gaz.* 23 June 7/1 Londoners, when they get hold of a good thing, like to 'freeze on to' it. **1935** Wodehouse *Blandings Castle* i. 25 You won't mind if I freeze on to the two-seater for the nonce? **1960** M. Stewart *My Brother Michael* xv. 188 We'll freeze on to those facts, and let the rest develop as it will.

4. d. *to freeze out*: (of a plant) to die through freezing. *U.S.*

1872 *Trans. Dep. Agric. Illinois* 73 They [*sc.* strawberry plants] dry out and freeze out worse in a loose and well aerated sand. **1872** *Rep. Vermont Board Agric.* 128 Alsike luxuriates in damp soils, and will not freeze out as red clover.

e. To make oneself suddenly rigid or motionless.

1848 [see Dict., sense 2 b *fig.*]. **1865** *Detroit Tribune* 6 Oct. 3/1 The raiders remained in the back room some minutes without making any demonstration, and Smith in the meantime 'froze' to the door latch. **1908** S. E. White *Riverman* iii. 27 But Orde..had frozen in an attitude of attentive listening. **1916** H. Titus *I Conquered* ix. 109 Of a sudden the horse froze, stopped his breathing. **1931** D. L. Sayers *Five Red Herrings* xxviii. 334 The Chief Constable hurriedly snatched up the rug and froze. **1933** *Brit. Birds* XXVII. 130 It 'froze' here for about five minutes and then started moving its head around. **1959** *Listener* 5 Mar. 414/2 Whenever a sentry appeared, they froze. **1969** I. & P. Opie *Children's Games* vi. 193 As the person in front turns round, the players 'freeze', for if he sees anyone moving..he sends that player back.

f. *Cinemat.* (See quot. 1960.) Cf. *freeze *sb.*[1] 2.

1960 O. Skilbeck *ABC of Film & TV* 59 *Freeze*, to arrest movement by successively Printing one Frame of Negative. Done, for instance, to extend a Shot for Optical purposes, as in a Title background; or for comedy effect. **1965** *Movie* Spring 29/1 Oval masking, 'freezing', multiple-image, slow-motion. **1965** [see *freeze *sb.*[1] 2].

5. e. To make (assets, credits, etc.) unrealizable. Also (*nonce-use*) *intr.*, to become unrealizable.

1922 *Ann. Reg. 1921* ii. 71 Credits granted by banks and financial houses to merchants have been frozen in enormous amounts. **1931** *Economist* 1 Aug. 219/1 In so far as the President's plan hints at definite freezing of existing bank credit for an agreed period, it is not acceptable to bankers. **1936** Auden *Look, Stranger!* 43 Europe grew anxious about her health, Combines tottered, credits froze. **1941** *Times* (Weekly ed.) 30 July 5/2 The Chinese Government have officially requested the British Government to 'freeze' Chinese assets. **1942** *Ann. Reg. 1941* 63 Great Britain..promptly 'froze' Japanese assets. **1966** *Listener* 27 Oct. 606/2 They froze his money in the bank.

f. To fix (wages, prices, resources, etc.) at a stated level.

1933 H. L. Ickes *Diary* 12 Oct. (1955) I. 106 This contemplates the freezing of prices at their present level. **1940** *Economist* 16 Mar. 455/2 In addition to the reduction in wages, prices and employment were 'frozen'. **1944** *Ann. Reg. 1943* 287 There should be less political difficulty in 'freezing' wages. **1956** *Ann. Reg. 1955* 135 Military budgets should be 'frozen' at the 1 January 1955 level.

g. To make immobile or inflexible; to arrest at a certain stage of development, etc.

1936 J. Gunther *Inside Europe* ix. 142 It would, by 'freezing' the present borders, prevent *Anschluss*, union of Germany and Austria. **1941** *Time* (Air Exp. Ed.) 26 May 22/3 General Electric..had..frozen its models of receivers. **1949** 'G. Orwell' *Nineteen Eighty-Four* ii. 204 The purpose of all of them was to arrest progress and freeze history at a chosen moment. **1958** *New Statesman* 12 Apr. 454/1 Co-op representation is to be 'frozen' at something like the present level of 20 MPs and 10 prospective candidates. **1958** *Spectator* 8 Aug. 185/1 This attempt to freeze frontiers and governments would be absurd coming from anyone. **1969** *Daily Tel.* 18 Jan. 1/6 A Federal Court judge in New York yesterday froze action on the merger of the Atlantic Richfield Company and Sinclair Oil Corporation.

7. b. (Earlier U.S. examples.)

1861 G. K. Wilder *Diary* (MS.) 20 July, We finally froze him out. **1867** 'Mark Twain' *Amer. Drolleries* (1875) 62 They would let that man go on and pay assessments,..and then they would close in on him and freeze him out. **1869** C. L. Brace *New West* v. 69 They can..lay assessments to bring a stock down to the lowest point, thus 'freezing out' the unhappy stockholders.

freeze-drying, *vbl. sb.* (Stress variable.) [Freeze *v.*] A method of drying foodstuffs, blood plasma, pharmaceuticals, etc. while retaining their physical structure, the material being frozen and then warmed in a high vacuum so that the ice sublimes. So **freeze-dried** *a.*; (as a back-formation) **freeze-dry** *v. trans.*

1944 *Nature* 22 Apr. 485/2 Many biological materials can be most conveniently preserved..if they are dried from the frozen state. The success of the 'freeze-drying' procedure appears to be chiefly related to the fact that the resulting 'solid state' prevents the concentration and aggregation of the molecules of protein. **1946** *Ibid.* 7 Sept. 349/1 Freeze-dried milk benefited from the addition of cystine. **1949** E. W. Flosdorf *Freeze-Drying* i. 6 In 1933 in the author's laboratory, the first products for actual clinical use were freeze-dried. **1956** *Nature* 14 Jan. 85/2 Freeze-dried bovine uterine cervical secretions. **1957** *Times* 11 Nov. 13/1 The new 'freeze-dried' BCG tuberculosis vaccine...The virus in this case is live, as opposed to the poliomyelitis virus, which is 'inactivated'. **1959** *Times* 24 Sept. 7/3 Officers sample new freeze-dried foods. **1962** *Engineering* 26 Jan. 133 Freeze-drying, AFD, continues..to hold the imagination with its possibilities for preserving the purity and flavour of perishable foodstuffs. **1962** V. N. Orekhovich et al. in A. Pirie *Lens Metabolism Rel. Cataract* 325 The solution was dialysed and then freeze-dried. **1963** *Daily Tel.* 29 Aug. 11/2 The valve came from the body of a man who died..about a month ago...The valve was then 'freeze-dried' and stored. **1967** *New Scientist* 9 Feb. 352/3 More recently, freeze-dried coffee extracts have appeared on the market.

freeze-out. *U.S.* [Freeze *v.* 7.] **1.** In full *freeze-out poker.* A variety of the game of poker in which the players, as fast as they lose their staked capital, drop out, all the money going to the player last remaining.

1856 *Butte Record* (Oroville, Calif.) 25 Oct. 1/6 He was..playing 'freeze-out' for the whiskey. **1877** *Harper's Monthly* Oct. 799 (Bartlett), They doant do nuthin' but drink whiskey and play frease aout poker. **1889** Farmer *Americanisms* s.v., In *freeze-out* poker..no player, when his money is exhausted, can borrow, or continue in the game on credit under any circumstances. **1907** B. Tarkington *His own People* iv. 61 I'll put it up against that tin automobile of yours, divide chips even and play you freeze-out for it. **1945** R. W. Service *Ploughman* 214 I'll play you for one of them there geese. Stud poker or freeze out.

2. An act of 'freezing' or forcing out.

1884 [see Freeze *sb.*[1] 1]. **1904** F. Lynde *Grafters* xiv. 190 By that time enough of the stock will have changed hands..and the freeze-out will be a fact accomplished.

freezer. Add: **1.** (Earlier and later examples.) Also, a refrigerated room; a compartment in a refrigerator or freezing machine. So in *Comb.*

1847 in *Missouri Hist. Rev.* (1942) XXXVI. 121 An article called a 'Freezer', which consists of a cylindrical jar, made of block tin, and fitted with a close cover [for ice-cream manufacture]. **1924** *U.S. Dept. Agric. Dept. Bull.* No. 1246. 9 Most packing and cold-storage plants are also equipped with what are known as 'freezers', which are refrigerating rooms in which the temperature can be lowered to 5° or 10°F.—sometimes lower. **1949** *Gloss. Terms Refrigeration* (*B.S.I.*) 9 *Freezer*, a low-temperature cold store, normally maintained below 20°F. **1959** *B.S.I. News* Apr. 17/1 The British Standards covering electric refrigerators and food freezers..have now been revised. **1959** *Housewife* June 129 A really big Kelvinator..gives you a roomier freezer compartment. **1961** *Times* 9 Aug. 5/2 The freezer trawler can stay on the fishing grounds until her holds are full. **1970** *Kenya Farmer* Feb. 15/1 A new freezer and chiller block.

3. (Earlier and later N.Z. examples.)

1889 Williams & Reeves *Colonial Couplets* 21 Be they [*sc.* sheep] freezers or crawlers or wethers or ewes. **1907** W. H. Koebel *Return of Joe* 177 Many were the pressings and pinches bestowed on each indignant animal before he was decided worthy of the honour of becoming a 'freezer'.

4. In *Chasing*, a punch for producing a frosted groundwork. (Cf. Frieze *v.*[2] 3.) Also **freezing tool.**

1887 L. L. Haslope *Repoussé Work* 23 A small punch, called a freezing tool, which produces a small star. **1898** T. B. Wigley *Art Goldsm. & Jeweller* 79 Punches of various shapes, called chasing tools...Freezer. Mat. Dead Mat. Hair Mat.

freeze-up. [f. the vbl. phr. *to freeze up*: see Freeze *v.* 5 b.] The condition of being stopped by frost or ice; a period in which land or water is frozen, *esp.* so as to prevent travel; an area so affected: a frozen condition (as of a water tank, engine-cylinder jacket, etc.). Also *transf.* and *fig.*

1876 *Oregon Weekly Tribune* (The Dalles) 29 Jan. 3/2 We hope to see the day when..all the inhabitants east of the Cascades will not be detrimentally affected by any freeze-up which may occur. **1879** [see Freeze *sb.*[1]]. **1882** *Golden* (N.M.) *Retort* 28 July 3/2 He says a freeze-up occurred from insufficient fluxing, but thinks the smelter will start up again soon. **1904** J. Lynch *3 Yrs. Klondike* 129 A couple of steam-engines had been..brought to Dawson last October just before the freeze-up. **1915** J. London *Lost Face* 116 The freeze-up came on when we were at the mouth of Henderson Creek. **1912** *Motor Man.* (ed. 14) 177 If the motor-house is continually heated..there will be no risk of a freeze-up. **1942** Wyndham Lewis *Let.* 30 Apr. (1963) 320 Meanwhile we are heading for an economic freeze-up. **1948** A. L. Rand *Mammals Eastern Rockies* 2/3 Spreading out over the..forested plains after the freeze-up. **1963** *Daily Tel.* 28 Jan. 18/2 (*headline*) 'Freeze-up' in talks with Treasury. **1964** *Ann. Reg. 1963* p. xvii, One of the longest and most severe winters on record.., with a freeze-up that lasted in Britain well into March.

freezing, *vbl. sb.* Add: **2.** *freezing chamber* (earlier example); **freezing process** (see quot. 1967); **freezing works** *Austral.* and *N.Z.*, an establishment in which animals are killed and the carcasses prepared and frozen for export.

1889 E. Wakefield *N.Z. after 50 Yrs.* vi. 130 The sheep..are skilfully slaughtered..and trucked down to the hulk, the whole interior of which is a freezing chamber. **1889** *Science* XIV. 142/1 The Pœtsch freezing process in mining operations. A brief description of the freezing process devised by Herman Pœtsch for sinking shafts in quicksands and other difficult ground was given. **1905** T. C. Futers *Mech. Engin. Collieries* I. 73 (*caption*) Freezing process of shaft sinking. **1967** *Gloss. Mining Terms* (*B.S.I.*) ix. 9 *Freezing process*, a method of consolidating water-bearing strata, to prepare it for shaft sinking, in which a freezing agent (usually brine) is circulated through suitably disposed boreholes drilled into the strata around the site of the shaft. **1889** E. Wakefield *N.Z. after 50 Yrs.* vi. 130 The solution of all their difficulties..was found to lie in having freezing works on shore, near the place of shipment. **1891** R. Wallace *Rural Econ. Austral. & N.Z.* xxxv. 464 For the shipment of the Queensland supply, freezing works are in process of construction in Brisbane, Rockhampton, and Townsville. **1906** E. W. Elkington *Adrift in N.Z.* v. 72 Cattle and sheep in their thousands..eventually find their way, *via* the freezing works, to our own dinner-tables. **1963** N. Hilliard *Piece of Land* 28 Within smelling distance of the freezing works and the stock-yards.

freezing, *ppl. a.* Add: **1.** Of drizzle, fog, rain, etc.

1797 *Encycl. Brit.* VII. 460 *Freezing Rain, or Raining Ice*, a very uncommon kind of shower. **1940** S. Petterssen *Weather Analysis & Forecasting* i. 40 In the United States weather reports, rain and/or drizzle that produce glazed frost are reported as 'freezing rain'. **1958** *Polar Record* May 92 *Freezing drizzle*, drizzle, the drops of which freeze on impact with the ground or with objects on the earth's surface or with aircraft in flight. **1963** *Meteorol. Gloss.* (*Met. Office*) (ed. 4) *Freezing drizzle, fog, rain*, supercooled water drops of drizzle (or fog or rain) which freeze on impact with the ground (in aviation with an aircraft) to form glazed frost or, in the case of the smaller droplets which comprise fog, rime. **1968** G. R. Rumney *Climatol.* v. 76/1 An ice storm, or freezing rain, occurs when liquid drops of rain, having a temperature slightly below freezing, fall upon solid surfaces of the earth..whose temperature is well below freezing.

freezy, *a.* Add: Freezing cold.

1902 O. Wister *Virginian* xxiii, Thought it looked pretty freezy out where yu' war riding. **1968** W. Garner *Deep, Deep Freeze* ix. 111 She said, 'Wow! What's a pretty girl like me doing in a freezy old place like this?'

freight, *sb.* Add: **4.** For *U.S.* read 'orig. *U.S.*' (Earlier and later examples.)

1861 *Remin. Life Railroad Engineer* 123 He was running the Night Express,..while I was running the through freight. **1922** M. B. Houston *Witch Man* i. 14 There were four trains a day in Innessburg, not counting the freights. **1955** E. Hillary *High Adventure* ix. 158 Feeling a little like an express train overhauling a slow freight.

5. (Chiefly N. Amer., exc. in the language of Containers.) *freight agent, car* (earlier and later examples), *container, depot, elevator, forwarder, house, locomotive, rate, terminal, train* (earlier and later examples), *wagon, yard*. Also **frei·ghtliner,** a train carrying goods in containers.

1843 *Rep. Western R.R.* 17 Freight-agent. **1875** *N. Amer. Rev.* CXX. 403 He has been promoted to the office of freight-agent. **1944** *Sears, Roebuck Catal.* 900 If there is a freight agent at your station, you pay freight charges when shipment is received. **1833** *Amer. Railroad Jrnl.* II. 325/2, 10 freight cars. **1947** *Chicago Daily News* 17 Jan. 4/2 A freight car and caboose were destroyed by fire. **1967** Freight car [see *car *sb.*[1] 2 a]. **1969** *Jane's Freight Containers 1968–69* 12/1 A freight container is an article of transport equipment. **1841** *Spirit of Times* 25 Sept. 354/1 Freight depot. **1904** W. N. Harben

Georgians i The long, brick freight depot. **1906** *Springfield Weekly Republ.* 4 Oct. 7 What the English call a 'railway goods station', and what we call a freight depot. **1903** Ade *In Babel* 18 He had patented certain devices which were used by all makers of passenger and freight elevators. **1911** *Daily Colonist* (Victoria, B.C.) 21 Apr. 16/3 Expedition combined with safety, can only be secured by means of a freight elevator. **1968** *Globe & Mail* (Toronto) 3 Feb. B5/2 'Do you know that only three apartment buildings in Toronto have freight elevators?' he asked. 'We sometimes have to put furniture on top of elevators to get it into buildings.' **1922** G. T. Stufflebeam *Traffic Dict.* 64 *Freight forwarder*, one who is located at a port, attends to documentation, booking space, securing marine insurance and other detail in connection with export shipments. **1968** *Economist* 14 Sept. p. xxv/1 There is more pressure on the British freight forwarders than there is on their continental counterparts. **1969** *Jane's Freight Containers 1968–69* 28/1 The members of the Canadian International Freight Forwarders Association Inc. **1970** *Times* 2 June (Container Suppl.) p. iv/3 The increasing range of forwarding work, the changing conditions under which it is carried out, and the new physical equipment—containers, in particular—have called for the new and wider description as freight forwarders. **1848** *Hunt's Merch. Mag.* XVIII. 383 The Worcester Railroad..has a freight-house in Boston. **1918** *Essex Inst. Hist. Coll.* LIV. 218 They should be run as one road, thus doing away with the expensive separate staffs, repair shops, freight houses, etc. **1948** *Daily Ardmoreite* (Ardmore, Okla.) 28 July 13/2 The car was shunted to a siding near the freight house yesterday afternoon. **1965** *Times* 12 July 14 (*caption*) A typical Freightliner..will be loaded..by the latest equipment...The container is lifted off the road transport..and immediately transferred on to the rail wagon. **1966** *Daily Tel.* 11 Aug. 22/7 British Rail is to build 60 refrigerated glass fibre containers to carry meat and perishable goods in its freightliner trains. **1969** *Lebende Sprachen* XIV. 65/2 A grain terminal and a British Railways freightliner terminal are to be built at Tilbury. **1934** Webster, Freight locomotive. **1935** *Discovery* Nov. 326/1 The freight locomotives were designed to haul considerably heavier trains than the passenger locomotives. **1848** *Hunt's Merch. Mag.* XVIII. 101 Freight rates—Iron, 25 cents per 100 pounds. **1881** *Chicago Times* 12 Mar., A freight-rate war between the eastern lines of railroads is regarded..as unavoidable. **1907** *Westm. Gaz.* 20 Nov. 10/2 Startling freight-rate experiments of the German railways. **1943** J. S. Huxley *TVA* xii. 114 The inter-territorial freight-rate problem of the U.S.A. **1967** *Economist* 18 Feb. 596/1 The dispute between the railway unions and the management of British Railways over freight terminals is like an old and hardy sore. **1967** *Punch* 12 July 66/1 A new international freight terminal. **1845** *Ann. Rep. Boston & Worc. R.R.*, Freight-train. **1887** M. Roberts *Western Avernus* 35 We found there was a freight train leaving this town at one o'clock. **1968** B. Foster *Changing Eng. Lang.* i. 59 It has recently become fashionable to speak of a goods train as a *freight train* in the American fashion. **1832** *Amer. Railroad Jrnl.* I. 85/3 The snow has not prevented the running of carriages or freight waggons. **1889** K. Munroe *Golden Days* ix. 91 Holstead and Thurston soon had their modest outfit..on board an Oregon freight wagon. **1910** J. Hart *Vigilante Girl* viii. 122 Behind these were..waiting stage coaches, and behind them freight wagons. **1881** *Chicago Times* 14 May, Twelve days ago the railroad switchmen in all but one of the freight-yards in this city struck for a slight increase of their wages. **1903** *N.Y. Times* 24 Sept. 5 Two youths..were seen prowling about the freight yards. **1961** L. Mumford *City in History* xv. 461 A waste of freight yards and marshalling yards.

freight, *v.* Add: **2.** Also *absol.*

1855 *N.Y. Weekly Tribune* 6 Jan. 1/2 The Ohio Railroads are freighting heavily. **1931** *Randolph Enterprise* (Elkins, W.Va.) 8 Jan. 4/5 He lived at Brownsville, Nebraska, and freighted from there to Fort Riley.

freightage. 3. (Earlier U.S. example.)

1872 'Mark Twain' *Roughing It* xvii. 137 High freights and bewildering distances of freightage.

freighter. 2. Also, one who owns or conducts a freight wagon or train of wagons. *local U.S.*

1852 *Knickerbocker* Mar. 224 The freighters were as impatient of delay as those Æneas saw crowding the shores of Styx. **1874** B. F. Taylor *World on Wheels* i. v. 43 You meet now and then a 'freighter', as the ox-expressmen of plain and prairie are called. **1907** S. E. White *Arizona Nights* i. xvi. 229 It happened to be..a freighter without the fear of God in his soul.

3. (Earlier U.S. example.)

1839 *Southern Lit. Messenger* V. 5/2 The Great Britain is now sailing as a mere freighter, and larger vessels are sailing as packets.

b. A freight-wagon. *U.S.*

1885 *Century Mag.* Nov. 65/1 Heavily loaded freighters were lurching in. **1921** C. E. Mulford *Bar-20 Three* xiii. 147 Pete Jarvis was proud of his new sixteen-foot freighter. **1929** *Randolph Enterprise* (Elkins, W.Va.) 14 Nov. 1/3 The Conestoga Wagon, known as the freighter, hauled travelers too.

c. A freight-carrying aircraft.

1920 *Flight* XII. 560/1 Owing to increased quantities of heavy and bulky freight being now carried to and from Paris by the Handley Page aeroplanes, a number of specially-constructed freighters will soon be added to the machines employed. **1935** *Ibid.* XXVIII. 657 The D.L.H. freighter, a Ju. 52, is just about to leave on its nightly journey to Berlin. **1959** *Listener* 15 Jan. 95/1 The Royal Air Force..is also receiving a 'substantial' number of short-range freighters.

frenate, *a.*: see *frænulum.

French, *a.* and *sb.* Add: **A.** *adj.* **1. d.** French-Canadian.

1673 *Hudson's Bay Co. Minutes* (1942) 59 That Mr. Kirke bee desired to treate for provideing Such french goodes as may be necessary. **1842** *Montreal Transcript* 15 Nov. 2/2 The murderer of lieut. Weir..could be returned as member for any French County in Lower Canada. **1963** W. L. Morton *Kingdom of Canada* 334 The attitude..spread to the French element in Red River.

2. b. With the implication of 'spiciness'. *French kissing* (see quot. 1965); so *French kiss sb.* and *v.*

[**1749** Fielding *Tom Jones* V. xiii. ix. 75, I am so far from desiring to exhibit such Pictures to the Public, that I would wish to draw a Curtain over those..in certain French novels.] **1842** R. Browning *Bells & Pomegranates* III. 6/2 My scrofulous French novel On grey paper with blunt type! **1850** Thackeray *Pendennis* II. xxx. 305 Young De Boots..recognised you as the man who..did business one-third in money,..and one-third in French prints. **1922** Galsworthy *Windows* iii. 28 Mother, I'd no idea you were so—French. **1923** J. Manchon *Le Slang* 130 *French kiss*, baiser très appuyé. **1926** Galsworthy *Silver Spoon* ii. viii. 181, I don't advise you to read it; it's very French. **1930** J. Dos Passos *42nd Parallel* v. 390 She taught him how to frenchkiss and would stroke his hair. **1931** 'D. Stiff' *Milk & Honey Route* xiv. 152 At one time the hobo enjoyed almost exclusively the 'French post cards' (called 'American cards' in France). **1959** G. Endore *Detour through Devon* 166 And in our slang, French means 'perverted'. **1960** V. Nabokov *Invitation to Beheading* xix. 192 A cursory inspection of the prison collection of French postcards. **1960** J. Updike *Rabbit, Run* (1961) 147 Sometimes just French kissing.., sloppy tongues and nobody can breathe. **1965** W. Young *Eros Denied* xxvii. 271 In England..we call..kissing with the tongue in the other's mouth French kissing. **1967** A. Wilson *No Laughing Matter* v. 450, I might give you a nice French kiss.

3. French bed (see quot. 1965); hence *French bedstead*; **French blue,** (*b*) *colloq.* (see quot. 1964); **French brace,** (*a*) a type of breast-drill; (*b*) in a theatre, a hinged brace (see quot. 1967); **French-chalk** *v. trans.*, to clean or mark with French chalk; **French clock,** a clock made in France; usu. applied to an elaborately decorated clock of the eighteenth century; **French cricket,** an informal type of cricket without stumps in which a player is out if the 'bowler' succeeds in getting the ball past the bat so that it hits the legs of the batsman; **French curve,** a template used for drawing curved lines; **French defence,** a defence in Chess in which Black replies to an opening move of P-K4 by White with P-K3; also *ellipt.*; **French door** *N. Amer.* = *French window*; **French dressing,** a salad-dressing consisting of vinegar and oil, usually with added flavouring; **French 'flu,** excessive fondness for all things French (an expression first used by A. Koestler); **French fried potatoes,** potato chips (see *chip *sb.*[1] 2 b); also *French fried(s)*, used *absol.*, and *French fries*; **French knot** (see quot. 1964); **French letter** *colloq.*, a contraceptive and disease-preventing sheath, = *condom; **French maid,** a lady's maid of French origin, freq. employed in the Victorian and Edwardian eras as a status symbol; **French sewing** (see quots.); **French toast,** any of various kinds of toasted bread (see quots.).

1825 H. Wilson *Mem.* IV. 38, I pointed..towards the small French bed. **1844** B. Disraeli *Coningsby* II. iv. vii. 77 Under ordinary circumstances a French bed and a brasier of charcoal alone remained for Villebecque, who was equal to the occasion. **1852** Ruskin *Let.* 25 Mar. in M. Lutyens *Effie in Venice* (1965) ii. 288 Tell Mr Snell not to put a four poster—but a large French bed in the bedroom at Herne hill. **1853** C. Brontë *Villette* I. i. 4 My own French bed in its shady recess. **1965** J. Aronson *Encycl. Furnit.* (1966) 234/1 *French bed*, roll-end bedstead without posts. **1967** D. Mackenzie *Death is Friend* 23 The French bed had a velvet padded headboard. **1837** Dickens *Pickwick* xii. 115 The large man..condensed himself into the limits of a dwarfish French bedstead. **1842** —— *Amer. Notes* I. iii. 142 Our bedroom was spacious and airy..having no curtains to the French bedstead or to the window. **1868** C. L. Eastlake *Hints Household Taste* viii. 185 To the four-poster succeeded..the French bedstead, of which the head and foot-piece were in shape and size alike, and over which two curtains fell, sometimes from a pole fixed at the side, and sometimes from a small circular canopy attached to the ceiling. **1878** H. James *Europeans* I. iv. 131 She can have the large north-east room. And the French bedstead. **1901** Mrs. A. Praga *How to Furnish* i. 29 We must reserve £2 5s., for..a black and brass French bedstead,..a double-woven steel wire mattress, a wool mattress, bolster, and two feather pillows. **1964** *Lancet* 29 Aug. 452/1 'French blues' (the name for a non-proprietary mixture of amphetamine and a barbiturate). **1967** *Daily Tel.* 21 Feb. 15/7 A schoolgirl of 14 said at Birmingham yesterday that she took 10 'French blue' pep pills she had bought outside the Midnight City dance club in the town. **1968** 'J. Welcome' *Hell is where you find It* xiv. 166 Some say French blues kill the itch to drink. With me they didn't. **1846** C. Holtzapffel *Turning* II. xxv. 561 The French brace..is also constructed in iron, with a pair of equal bevil pinions. **1947** *Gloss. Techn. Theatr. Terms* (*Strand Electr. Co.*) 5 *French brace*, a tall wooden right-angled triangular frame, hinged to the back of a flat or other piece. **1948** H. Melvill *Designing & Painting Scenery* ii. 18 The brace (called in this instance a 'French brace') can swing out into position, or go flat against the groundrow when not in use. **1967** *Oxf. Compan. Theatre* 912/2 A variant [of the stage brace], composed of a right-angled triangle of wood hinged to the flat, opened out and weighted, is known as a French Brace. **1969** E. H. Pinto *Treen* 380 The more complex French brace or breast drill, came into the English wood-worker's toolchest just before 1900, but it was illustrated in Bergeron more than 100 years earlier. **1933** P. Godfrey *Back-Stage* xiii. 167 The floor of the stage, which was French-chalked for dancing to the gramophone. **1963** J. Osborne *Dental Mechanics* (ed. 5) iv. 61 The waxed-up model is French-chalked. **1841** *Fraser's Mag.* XXIV. 717/2 On the mantelpiece, huge French clocks. **1864** J. S. Le Fanu *Uncle Silas* I. xxiv. 290 There was a little French clock over the mantlepiece in the school-room. **1964** I. Oellrichs *Death in Chilly Corner* iv. 30 The little old French clock buzzed three times to indicate three o'clock. **1964** J. Symons *End of Solomon Grundy* ii. i. 69 She..went to look at the French clock in the sitting-room. **1967** W. Edey *French Clocks* 9 It is in this capacity, as a decorative art rather than as a work of science, that French clocks excel. **1926** E. Bowen *Ann Lee's* 221 They went back to the garden and played French-cricket there with a tennis-ball. [**1935** *Encycl. Sports* 296/2 *French cricket*, game played with a cricket bat and ball in a confined space. It is played by any number, without sides, and there is no scoring. Each batsman plays until he is out, and is then succeeded by the next.] **1959** I. & P. Opie *Lore & Lang. Schoolch.* viii. 137 An informal game of cricket, such as French cricket or Stump cricket. **1885** *Army & Navy Co-op. Soc. Price List* 720 French Curves, Pear wood, assorted patterns. **1969** R. Mayer *Dict. Art Terms & Techniques* 154 French curves..are also called irregular curves and drawing curves. **1971** D. Clark *Sick to Death* i. 18 An artist's drawing-board...A long T-square..a straight edge with a metal ribbon on one side; a french-curve template. **1889** Freeborough & Ranken *Chess Openings* v. 239 The French Defence dates from Lucena (1497). **1895** J. Mason *Art of Chess* 292 The French Defence is often adopted in anticipation of the eternal Lopez, and may be considered one of the best methods of eluding that wearisome game. **1964** *Listener* 1 Oct. 531/2 Peter Lee..specializes in the French Defence. In this game..superior opening knowledge gave him a quick win. **1965** H. Evans et al. *Korn's Mod. Chess Openings* (ed. 10) 145 The French derives its name from a correspondence game between London and Paris in 1834, even though Lucena had examined it in 1497. **1971** *Oxford Mail* 27 Oct. 16/8 Fischer..opened..with his favourite move and Petrosian went into a French defence. **1923** *Southern Calif. Hardwood Manuf. Co. Catal.* Feb. 97 French Doors while intended primarily for outside use—porches, sun rooms etc., are now used to good advantage for inter-room use where lighting or heating is a reason for separating two rooms. **1926** *Daily Colonist* (Victoria, B.C.) 18 July 11/5 (Advt.), Thrown wide between two small rooms, French doors allow you the often convenient space of a large room. **1968** *Globe & Mail* (Toronto) 13 Feb. 30/1 (Advt.), Living room with..French doors to screened veranda. **1900** *Ladies' Home Jrnl.* June 40/3 French Dressing is a mixture of oil and vinegar in the proportions of six tablespoonfuls of oil to two of vinegar. **1923** H. Crane *Let.* 5 Dec. (1965) 159, I made some fine lettuce salad with onions and peppers and French dressing. **1945** *La Junta* (Colo.) *Tribune-Democrat* 15 Feb. 2/6 French Fried Onions Chef's Salad with French Dressing. **1962** *Woman's Own* 17 Mar. 14/3 Tossed green salad with French dressing. **1943** A. Koestler in *Yogi & Commissar* (1945) 21 The managerial class on Parnassus..have lately been affected by a new outbreak of that recurrent epidemic, the French 'Flu. **1959** *Guardian* 11 Dec. 6/6 Arthur Koestler once diagnosed French 'flu as an unreasoned admiration for everything Gallic. **1966** J. Fowles *Magus* xv. 83 'Voila.' 'Very nice.' I determined to stamp out the French 'flu before it spread. **1894** 'O. Henry' *Rolling Stones* (1916) 150 Our countries are great friends. We have given you Lafayette and French fried potatoes. **1913** F. H. Burnett *T. Tembarom* xxii. 287 Beefsteak and French fried potatoes were the favourites. **1918** in F. A. Pottle *Stretchers* (1930) 289 After looking around a while we found the Cafe Riche and had a fine steak with French fries. **1944** *This Week Mag.* 18 Mar. 12/2 Sirloin steak I get them, with French frieds and mushrooms. **1951** C. Armstrong *Black-Eyed Stranger* (1952) ii. 10 Bring me two lamb chops, French fries, cup of coffee. **1956** 'A. Gilbert' *Riddle of Lady* ix. 130 An excellent steak..and what the Nell Gwynn called French fried potatoes and Crook called chips. **1958** G. Greene *Our Man in Havana* v. iii. 213 Roast turkey, cranberry sauce, sausages and carrots and French fried. *c* **1885, 1899** French knot [see Knot *sb.*[1] 1 b]. **1960** G. Lewis *Handbk. Crafts* 29 Bumpy stitches like French knots which require great care in pressing to prevent them being flattened are better not used. **1964** *McCall's Sewing* ii. 29/1 *French knot*, decorative hand-stitch in which the thread is twisted around the needle and brought down through the fabric at almost the same spot to form a small dot. *c* **1856** *Paul Pry* in C. Pearl *Girl with Swansdown Seat* (1955) vi. 256 French letters..prevent the spread of venereal contagion in casual intercourse between the sexes, and in the marriage state, the increase of the family. *c* **1857** *Amours, Intrigues, & Adv. Musical Student* 144 (Advt.), How many industrious men have been crushed by poverty..through ignorance of the existence of French Letters. **1886** R. F. Burton *Arab. Nts.* X. 239 Of the penis succedaneus..every kind abounds, varying from a stuffed 'French letter' to a cone of ribbed horn. *c* **1888–94** *My Secret Life* III. 324 What Jenny's sister paid for French letters I dont know, I used to pay nine pence each. **1960** B. Askwith *Tangled Web* 162, I daresay he would have liked to give me a baby—but he always used a French letter. **1960** M. Spark *Bachelors* x. 181 They fumble about with their french letters or they tear open their horrible little packets of contraceptives like kids with sweets. **1968** J. R. Ackerley *My Father & Myself* vi. 49 My elder brother Peter was the accident. 'Your father happened to have run out of french letters that day.' **1848** Thackeray *Vanity Fair* lv. 491 She went up stairs and dressed herself this time without the aid of her French maid. **1853** Dickens *Bleak Ho.* xviii. 178, I was conscious of being distressed even by the observation of the French maid. **1899** Wilde *Importance of being Earnest* iii. 132 A thoroughly experienced French maid produces a really marvellous result in a very brief space of time. **1905** F. H. Burnett

Little Princess ii. 17 One or two of them had even caught a glimpse of her French maid, Mariette, who had arrived the evening before. **1912** A. N. Lyons *Clara* viii. 73 I'm only a little dis'eartened because the motor's broken down and my French maid forgot to air the curling-tongs. **1960** B. Keaton *Wonderf. World of Slapstick* (1967) 55 This little lady carried thirty-six pieces of luggage..and an entourage consisting of a pianist, a chauffeur, a footman, and two French maids. **1923** H. A. Maddox *Dict. Stationery* 31 *French sewing*, a method of sewing small books and pamphlet work without tapes, the cover usually being drawn on. **1963** Kenneison & Spilman *Dict. Printing* 75 *French sewing*, in binding, the method of machine-sewing without tapes, or in hand-sewing, the method of sewing a book without the use of a sewing frame, the sections of the book being sewn at the edge of the bench. **1660** R. May *Accomplisht Cook* vi. 162 *French Toasts.* Cut French Bread, and toast it in pretty thick toasts on a clean gridiron, and serve them steeped in claret, sack, or any wine, with sugar and juyce of orange. **1882** F. E. Owens *Cook Bk.* 128 French toast. Make a batter of two eggs, one-half cup of milk, pinch of salt, and teaspoon of cornstarch. Dip thin slices of bread in and fry brown in a well-buttered frying pan. **1892** T. F. Garrett *Encycl. Pract. Cookery* I. 192/1 French Toast. Beat up one egg in a basin with a little salt and 1 teaspoonful of milk, and in this dip some thin slices of Bread..; then plunge the Bread into a frying-pan of boiling fat, and fry to a light brown. Take them out, drain them...Spreading over them with stewed rhubarb, or other fruit in season, is a great improvement. **1924** *Western Daily Press* (Bristol) 20 Mar. 9/5 A piece of bread and butter toasted on the dry side is said to be French toast.

b. French cuff, a double cuff formed by turning back a long cuff band and fastening it with buttons or links; **French heel** (later examples); **French pleat,** (*a*) a type of pleat at the top of a curtain (see quot. 1964[1]); (*b*) = **French roll*; **French roll,** a woman's hairstyle, in which the hair is tucked into a vertical roll down the back of the head; **French seam** (see quot. 1968); **French tack** (see quot. 1964); **French twist,** = **French roll*.

1916 *Daily Colonist* (Victoria, B.C.) 22 July 12/5 (Advt.), Each Shirt is finished with French cuffs and separate soft collar. **1964** *McCall's Sewing* xi. 171/1 Apply the French cuff to the lower edge of sleeve. **1967** 'L. Egan' *Nameless Ones* xv. 187 A fancy sports shirt with French cuffs and gold cuff links. **1922** Joyce *Ulysses* 353 With her high crooked French heels on her. **1928** M. de la Roche *Whiteoaks* (1929) xii. 157 Four girls approached abreast, wearing French heels and flesh-coloured stockings. **1948** Derieux & Stevenson *Compl. Bk. Interior Decorating* 147 To simplify making of French pleats two products are available. **1964** *McCall's Sewing* xvi. 284/1 *French pleats.* Divide fullness of pleat fold into three equal, smaller folds. At the bottom of the heading, sew by hand through the three pleats on the right side of the fabric. Pull thread up tight and fasten securely. Top edge of heading is not pressed into pleats. **1964** G. Lyall *Most Dangerous Game* ix. 64 She had..silky fair hair pulled back and coiled in a French pleat. **1966** P. Carlon *Running Woman* viii. 69 Long hair could be worn in a French pleat, a chignon, at times. **1968** R. W. Semmens in F. H. Radford *Art & Craft Hairdressing* (ed. 5) I. vii. 116/2 *French Rolls and Pleats.* This is a popular hair style suitable for longer hair. **1941** F. E. Wall *Princ. & Pract. Beauty Culture* 423 French Roll. This coiffure is particularly becoming to the small woman who wishes to appear taller, but it is equally suitable to a taller woman with a thin face. **1962** in R. Corson *Fashions in Hair* (1965) xii. 630 The bouffant, a hairdo filched from eighteenth-century France, whose tortured variations—mushroom flip, French twist, chemise, French roll..—began sprouting a few years back. **1965** J. Caird *Murder Reflected* ii. 19 She had hair of true wheat-gold, worn up in a French roll. **1968** J. Ironside *Fashion Alphabet* 192 With the advent of the 'bouffant' hair-do, the French roll was an easy adaptation, the teased hair being brushed up from the neck and made into a thick roll down the back of the head, thicker at the top. *c* **1890** tr. *T. de Dillmont's Encycl. Needlework* 8 For joining such stuffs as fray, use the so-called French seam. **1895** *Montgomery Ward Catal.* 37/1 Ladies' Shirt Waists..all felled French seams. **1956** *And so to Sew* 118/2 Do not use french seams where perfect fitting is necessary or where a flat finish is required. **1967** *Guardian* 23 May 6/5 Bogged down among the plackets and French seams of school dressmaking. **1968** J. Ironside *Fashion Alphabet* 99 *French seam*, a seam made by sewing the two fabrics together on the right side; then turning and sewing on the wrong side, enclosing the raw edges. **1934** *Vogue's Guide Smart Dressmaking* 39 The lining is finished separately from the coat at lower edge, with French tacks at the seams. **1960** *Vogue Pattern Bk.* Early Autumn 57 Catch the seams together near the hemline with loose French tacks. **1964** *McCall's Sewing* ii. 29/1 *French tack*, a thread bar fastening used to hold two pieces of a garment together loosely. Used to attach lining to coat at hemline. **1877** E. Creer *Lessons in Hairdressing* iii. 48 That [coiffure] which is called a 'French twist' has two different forms and uses. According to some authorities it is made with the back hair, and fixed by means of a comb at the crown of the head...Another 'French twist'..is also known by the names of '*Torsade Dondel*', '*Torsade Repoussée*', '*Torsade Gouffrée*'. It is made with a long, thin strand of hair, and is hollow when completed. **1892** I. Mallon in R. Corson *Fashions in Hair* (1965) xii. 494 Women who wear their hair very plainly part it in the back and turn it over like a French twist, drawing it up to the top of the head. **1962** J. Fleming *When I grow Rich* viii. 102 She..did her hair up into a French twist at the back. **1966** J. S. Cox *Illustr. Dict. Hairdressing* 61/1 *French twist*,..the same as French Roll.

4. French bulldog, a smaller variety of the English bulldog.

1894 R. B. Lee *Mod. Dogs* ix. 215, I am surprised that there are show committees who will provide classes for these French bulldogs to be called 'toys', going up to 23 lb. in weight. **1922** R. Leighton *Complete Bk. of Dog* xx. 308 It was from the English variety of pygmy Bull-dogs that the now fashionable French Bulldog was evolved. **1952** E. F. Davies *Illyrian Venture* iv. 73 Our shopping expedition produced a French bulldog.

5. French pink *U.S.*, the cornflower, *Centaurea cyanus*, or a species of *Armeria*; **French walnut** = Common walnut (Walnut[1] 2 b).

1863 S. K. Holmes *Jrnl.* 12 July (1955) 226 The prairie is a mass of moving purple plumes, 'French Pinks', the natives [*sc.* Texans] call them. **1896** S. O. Jewett *Country of Pointed Firs* xv. 132 There was little trace of her flower-garden except a single faded sprig of much-enduring French pinks. **1914** L. H. Bailey *Stand. Cycl. Hort.* II. 711/2 Centaurea..cyanus..Bluebottle. Bluet. Bachelor's Button..Cornflower. Ragged Sailor. French Pink. **1841** *Southern Lit. Messenger* VII. 221/1 [I] rode into the yard; found large old French walnut there. **1966** A. W. Lewis *Gloss. Woodworking Terms* 107 French Walnut.

7. *French-heeled* adj. Also, esp. in the sense 'French and ——', *French-American, -British, -Creole, -provincial, -Spanish, -Swiss.* See also *French Canadian *sb.* and *a.*

1915 W. Owen *Let.* 1 Aug. (1967) 350 The fellow is a Radio-Telegraphist on a French-American Liner. **1934** Dylan Thomas *Let.* 16 Mar. (1966) 96 The French-American Transitionists. **1612** W. Strachey *Trav. Virginia* (1953) i. x. 121, I haue drunck often of the rath wyne, which..people haue made full as good as your French-Brittish wyne. **1868** G. Rose *Great Country* 195 The French Creole ladies. **1956** M. Stearns *Story of Jazz* (1957) v. 54 The slaves are apparently singing a French-Creole tune in the French-Creole patois. **1915** J. London *Let.* 18 Sept. (1966) 460, I should like to see you in your French heeled slippers. **1951** R. Chandler *Let.* 5 Feb. (1962) i. 28, I live in a French Provincial chateau on Mulholland Drive. **1956** 'E. S. Aarons' *Assignment Treason* (1967) x. 78 There was French provincial furniture, softly glowing with centuries of hand polishing. **1970** *Globe & Mail* (Toronto) 25 Sept. 34/1 (Advt.), Lovely 4-bedroom French provincial home with bath, den, family room, [etc.]. **1934** Wyndham Lewis *Let.* 3 July (1963) 219 The french-spanish border. **1921** H. Crane *Let.* 16 May (1965) 56 There is a French-Swiss artist here.

B. *absol.* and *sb.* **1. b.** Used euphemistically for 'bad language', esp. in the phr. *excuse* (or *pardon*) *my French!*

1895 [see Durned]. **1909** J. R. Ware *Passing Eng.* 171/1 *Loosing French*, violent language in English. **1922** Joyce *Ulysses* 446 Bad French I got for my pains. **1936** M. Harrison *All Trees were Green* ii. 104 A bloody sight better (pardon the French!) than most. **1940** *S.P.E. Tract* iv. 181 *Excuse my French!* (forgive me my strong language). **1955** M. McCarthy *Charmed Life* (1956) ii. 52 'Damn fool,' he said, vehemently, 'pardon my French.' **1961** J. O'Donovan *Middle Tree* ii. 12 A kick in the arse, smartly administered... Excuse my French! **1966** A. La Bern *Goodbye Piccadilly* xxv. 220 Well I'll be buggered. Excuse my French.

3. Dry vermouth. Usu. in phr. *gin and French.*

1938 M. Allingham *Fashion in Shrouds* xxii. 420 There's some gin and French and a little Advocaat..in the cupboard. **1949** A. Wilson *Wrong Set* 208 'How about that gin..?' he asked. 'It's in the shaker,..with some French.' **1967** M. Gilbert *Dust & Heat* 14 He was drinking double gins with single Frenches in them.

4. Fellatio. Also as *v. trans.* and *intr.*, to practise fellatio or cunnilingus (with). *slang.*

1958 Murtagh & Harris *Cast First Stone* (Gloss.), *French, Frenchy*, fellatio. **1958** in Partridge *Dict. Underworld* (1968) 821/1 French, *v.* **1965** W. Young *Eros Denied* xxvii. 271 In England..we call..cunt-licking Frenching. *c* **1965** in Partridge *Dict. Slang Suppl.* (1970) 1146 She thought he was asleep, and Frenched him. **1967** Wentworth & Flexner *Dict. Amer. Slang* 684 *French*, to commit fellatio or cunnilingus. **1968** B. Turner *Sex Trap* vii. 52 You're not queer, are you?..I just thought you might want a spot of French if you were. *Ibid.* viii. 64 You can be whipped or caned..or you can have French for another pound. **1969** T. Parker *Twisting Lane* 203 There's two things I won't let her do though, that's French and sadism.

french, *v.* Add: **6.** Of tobacco, cotton, or other plants: to become diseased and distorted. (Cf. Frenchman 4.)

1852 G. N. Jones *Florida Plant. Rec.* (1927) 67 My Corn Crop Looks better than it did Last year; this time it have Frenched a Little I see, in Places. **1888** *Congress. Rec.* 17 May 4069 Then [the cotton-plant] begins to blight, then comes frenching and the shedding of squares and formes.

French Canadian, *sb.* and *a.* **A.** *sb.* **1.** A French settler in Canada or a French-speaking descendant of such a settler; a Canadian of French ancestry.

1758 *Descr. Cape Breton* 4 These *French Canadians*..he represents as an indolent People. **1775** J. Adair *Hist. Amer. Indians* 153 The French Canadians are highly censurable..for debauching our peacable northern Indians, with their infernal catechism. **1838** *Westm. Rev.* XXXI. 405 The Hindoos are a more ignorant and more passive people than the French Canadians. **1876** [see Canadian *a.* and *sb.*]. **1935** Huxley & Haddon *We Europeans* viii. 261 The stock from which the French Canadians sprang. **1967** *Times* 28 Feb. (Canada Suppl.) 24 If a national break-up were to occur, French-Canadians would face a substantial loss. **1968** *Globe & Mail* (Toronto) 17 Feb. 6/2 Sections of the Quebec press have questioned if this is the proper time in our history for a French Canadian to seek the Prime Ministership.

2. The language of this people.

1929 S. Brown tr. *Constantin-Weyer's Man scans his Past* 203 He hailed me in French Canadian. **1960** *Amer. Speech* XXXV. 229 In French Canadian the high vowels [i y u] are generally unvoiced in unstressed (nonfinal) position.

B. *adj.* (Now usu. with hyphen.) Of or pertaining to the above or to French Canada.

1775 J. Adair *Hist. Amer. Indians* 164 The black tribe, the French Canadian priests. **1822** *Kingston* (Ont.) *Chron.* 23 Nov. 3/4 The pages of some French Canadian papers. **1853** S. Strickland *27 Yrs. Canada West* I. 39 One pair of our horses were French Canadian. **1886, 1897** [see *English-Canadian* s.v. *English *a.* 2 e]. **1930** *Times Lit. Suppl.* 6 Feb. 106/4 The French-Canadian politicians. **1966** *Times* 28 Feb. (Canada Suppl.) p. vi/4 A sharp revival of French-Canadian nationalism. **1969** Halpert & Story *Christmas Mumming in Newfoundland* 112 We may contrast this with the French-Canadian pattern of mumming at Lent.

So **French Canadianism,** (advocacy of) the cultural distinctiveness of French Canada or political separatism in Quebec.

1916 A. Bridle *Sons of Canada* 33 One might have naively surmised that..he had renounced some of his French Canadianism. **1962** I. Fleming *Spy who loved Me* ii. 28 My family had belonged to the very innermost citadel of French-Canadianism.

Frencher. (Earlier U.S. example.)

1826 J. F. Cooper *Last of Mohicans* xx, Why did not the grand Frencher [*sc.* Montcalm]..bury the tomahawks of the Hurons?

Frenchie, var. Frenchy *sb.*

1905 Baroness Orczy *Scarlet Pimpernel* xii. 117 We seek him here, we seek him there, Those Frenchies seek him everywhere. **1916** 'Boyd Cable' *Action Front* 89 Now I wonder who it is. A Frenchie by his tunic sleeve. **1955** *Times* 23 May 3/1 Bobbing up now here, now there, he seeks these Frenchies everywhere in Lisbon. **1965** H. Gold *Man who was not with It* i. 3 There he is on the midway, Grack the Frenchie, talking for his countstore.

Frenchman. Add: **5.** A knife used in pointing brickwork.

1885 *Spons' Mechanics' Own Bk.* 591 This Frenchman is simply an old dinner-knife ground to a point, the tip of which is turned down square to form a hook. **1934** T. S. Eliot *Rock* i. 22 Fancy anyone not knowin' that a Frenchman's a pointin' tool!

6. The French or red-legged partridge, *Alectoris rufa.*

1893 *Baily's Mag.* Oct. 258/1 The red leg, the very much-abused 'Frenchman' of the English fields. **1920** F. M. Ogilvie *Field Observ. Brit. Birds* 58 A Frenchman..has a loud, combative call. **1965** P. Wayre *Wind in Reeds* iii. 41 If they are Frenchmen, they may refuse to fly and simply run in front of the dog.

Frenchy, *a.* (*adv.*) and *sb.* Add: **A.** *adj.* (Earlier and later examples.)

a **1828** D. Wordsworth *Tour Continent* in *Jrnls.* (1941) II. 318 A long Village—frenchy in the better style. **1950** J. Cannan *Murder Included* iii. 57 People will suspect Bunny just because she's Frenchy. **1967** C. O. Skinner *Madame Sarah* viii. 147 Word had gotten about that the plays in her repertoire were 'Frenchy', meaning naughty.

B. *sb.* Also, a French Canadian.

1891 J. Oxley *Chore-Boy of Camp Kippewa* 74, I wouldn't give him for half a dozen of those *parlez vous* Frenchies. **1966** *Maclean's* 2 May 50/4, I was constantly laughed at, pointed at and corrected, as a stupid Frenchy.

frenular *a.*, **frenulum:** see *frænulum.

Freon (frī·ǫn). Also **freon.** Proprietary name for any of a group of partially or completely halogenated simple hydrocarbons containing fluorine and usu. chlorine or bromine, used esp. as refrigerants and aerosol propellants.

1932 *Industr. & Engin. Chem.* XXIV. 620 Freon..was developed in a deliberate attempt to produce, by chemical arrangement, a refrigerant..which still has properties especially favorable to economic operation. **1941** *Archit. Rev.* XC. 52 Cooling is provided by refrigerating compressors, using Freon gas, a safe and harmless refrigerant. **1945** *Electronic Engin.* XVII. 762 Insulation is provided by freon gas. **1950** J. D. Park in J. H. Simons *Fluorine Chem.* I. xv. 527 The structures of most of the 'Freons' having from one to five carbon atoms. **1959** *Trade Marks Jrnl.* 25 Feb. 220/2 *Freon*, fluorinated hydrocarbons being chemical substances for use as propellents for aerosols. E. J. Du Pont de Nemours and Company. **1966** *McGraw-Hill Encycl. Sci. & Technol.* V. 501/1 The most popular freons are Freon-11 and Freon-12. **1967** *Jane's Surface Skimmer Systems 1967–68* 106/1 The pilothouse..and berthing spaces are air-conditioned during the cooling season by a 13 ton capacity freon type compressor system. **1968** *New Scientist* 25 July 197/1 Freons, the halogenated varieties of methane and ethane..are pretty inert chemically.

frequency. Add: **4. b.** (Further examples.) In *Telecommunications*, the number of cycles per second of the carrier wave of a transmission; hence *loosely*, a frequency band, a channel. Also, a signal, or a component of a signal, having a frequency of a certain value or in a certain range.

1835 *Rep. Brit. Assoc. 1834* 315 The colour of the light is supposed to be determined by the frequency of the

ethereal vibrations, or by the length of the wave. **1943** C. L. Boltz *Basic Radio* viii. 133 A succession of such circuits will attenuate the higher frequencies so much that their amplitude is negligible. **1948** P. M. Morse *Vibration & Sound* (ed. 2) vi. 226 Corresponding to the physical quantities intensity and frequency are the physiological..quantities loudness and pitch. **1958** *Radio Times* 14 Feb. 3/2 To make sure that the BBC keeps strictly to the frequencies (or wavelengths), both in sound and television broadcasting, allotted to it by international agreement. **1958** 'N. Shute' *Rainbow & Rose* i. 7 We've been monitoring their frequency. **1962** *Which?* Feb. 41/2 Sound travels in the form of waves—rather like radio waves, but of a far lower frequency. *Ibid.* 42/1 The other VHF radios would reproduce frequencies higher than 4,200 cycles per second, but only at a lower volume. **1971** *Daily Tel.* 7 Jan. 14/6 Enough radio frequencies could be made available on the medium wave to satisfy commercial radio and the existing BBC local stations.

c. *Statistics.* The number of times an event or character occurs in a given sample; also (the *relative* or *proportionate frequency*), this number expressed as a proportion of the total possible number of occurrences.

1854 G. Boole *Invest. Laws of Thought* xvii. 268 For if we substitute..we shall form the expression of that event whose probability constitutes the numerator of *c*, and the ratio of the frequency of this event to that of the former one..will give the value of *c*. **1857** *Phil. Mag.* 4th Ser. XIV. 61, I have commenced collecting statistics showing the relative frequency with which the different simple letters, and various compounds of simple letters, occur in the English language. **1875** F. Galton in *Ibid.* XLIX. 37 We may work backwards, and, from the relative frequency of occurrence of various magnitudes, derive a knowledge of the true relative values of those magnitudes, expressed in units of probable error. **1937** L. H. C. Tippett *Methods of Statistics* (ed. 2) i. 28 Frequencies and proportionate frequencies underlie nearly all methods of statistical representation. **1943** M. G. Kendall *Adv. Theory Statistics* I. vii. 165 The second approach seeks to define probability in terms of the relative frequency of events. **1950** W. Feller *Introd. Probability Theory* I. i. 18 The frequency of multiple aces in the hands at bridge. **1955** M. Loève *Probability Theory* i. 5 In a game of dice..'double-six' occurs about once in 36 times, that is, its observed frequencies cluster about 1/36. The number 1/36 is a permanent numerical property of 'double-six' under the conditions of the game, and the observed frequencies are to be thought of as measurements of the property. **1965** A. H. Roberts *Statistical Ling. Anal. Amer. Eng.* v. 60 These 8,103 words had a total frequency of 10,503,788 or ⅞ of the total frequencies of occurrence.

d. *Ecol.* A measurement of the way individuals of a species are distributed in a community.

1913 C. Raunkiaer in *Jrnl. Ecol.* I. 24 Each species acquires the same degree of frequency as in a shoot-counting method if the frequency be expressed in whole numbers from 1 to 5. **1921** *Jrnl. Ecol.* IX. 97 If we know the absolute degree of frequency of a species (that is the number of individuals belonging to one species growing on a large area Y) we can calculate the probable occurrence on every area (y) which is smaller than Y. **1932** *Ecol. Monogr.* ii. 478 Frequency is concerned with the uniformity with which the plants of a species are distributed throughout a plant community. *Ibid.* 479 The percentage is called the 'frequency index' of the species. **1964** K. A. Kershaw *Quantitative & Dynamic Ecology* i. 16 The frequency of a species is a measure of the chance of finding it with any one throw of a quadrat in a given area. Thus, if a species has a frequency of 10 per cent then it should occur once in every ten quadrats examined.

5. *attrib.* and *Comb.*, as (sense 4 b) *frequency characteristic, correction, -dependent* adj., *discriminator, divider, doubler, indicator, meter, multiplier, -multiplying* ppl. adj., *raiser, range, -selective* adj., *teller, transformer*; (sense *4 c) *frequency-count.*

1926 Frequency characteristic [see **frequency response*]. **1957** *BBC Handbk.* 49 Sound circuits have to be equalized in frequency characteristic. **1962** A. Nisbett *Technique Sound Studio* i. 17 (*caption*) Between studio, continuity and transmitter there may be..switching centres, boosting amplifiers and frequency correction networks on land-lines, etc. *Ibid.* 254 *Frequency correction*, the change in the frequency characteristics of a signal which is required to restore it to its original form. **1956** J. Whatmough *Language* v. 73 The 20,000 most frequently occurring words (as ascertained by previously made frequency counts). **1962** W. B. Thompson *Introd. Plasma Physics* ii. 14 Thus the plasma appears as a dielectric with a frequency-dependent dielectric coefficient. **1935** Frequency discriminator [see *discriminator 2 b]. **1943** *Gloss. Terms Telecomm.* (*B.S.I.*) 73 *Frequency discriminator*, a device for producing an output voltage approximately proportional to the deviation of frequency of a signal from a predetermined value. **1944** *Electronic Engin.* XVII. 196 The frequency discriminator..has only recently been finding its place in commercial radio receivers in this country. **1938** S. R. Roget *Dict. Electr. Terms* (ed. 3) 141/1 Frequency divider. **1952** Terman & Pettit *Electronic Measurem.* (ed. 2) v. 196 By means of a chain of subharmonic generators (frequency dividers), one obtains frequencies of 10,000 cycles, 1000 cycles, and 100 cycles which have the same degree of precision as the standard frequency. **1924** S. R. Roget *Dict. Electr. Terms* 94/2 Frequency doubler. **1950** P. Parker *Electronics* xi. 374 A class C amplifier is sometimes used as a frequency doubler, by tuning the tank circuit to the second harmonic of the signal frequency. **1906** J. A. Fleming *Princ. Electr. Wave Telegr.* i. 42 One well-known form of frequency indicator is that due to Mr. Campbell. *Ibid.*, In addition to the voltmeters and ammeters..a frequency meter ought to be provided. **1957** E. L. Ginzton *Microwave Measurem.* viii. 389 The frequency of the unknown signal can be measured with moderate accuracy by an instrument called the heterodyne frequency meter. **1933** K. Henney *Radio Engin. Handbk.* xviii. 462 Present practice favors the use of static frequency multipliers where it is desired to use an alternator of comparatively low frequency. **1966** *McGraw-Hill Encycl. Sci. & Technol.* V 515/1 A lower frequency is generated in a crystal-controlled oscillator..and a frequency multiplier is employed to obtain the high frequency desired. **1962** Simpson & Richards *Junction Transistors* xviii. 457 Using a frequency-multiplying circuit to increase both the carrier and deviation frequencies. **1893** *Jrnl. Soc. Arts* 5 May 624/1 We want..a frequency-raiser.., which shall transform the frequency, even as the induction-coil transforms the voltage. **1926** G. G. Blake *Hist. Radio Telegr.* 234 The main object of the foregoing frequency raisers has been to obtain an increase of frequency from comparatively low frequency alternators. **1936** *Summ. Doct. Diss. N.W. Univ.* IV. 180 Within any response-band the glottal note was lowered as the supply-tube was lengthened, the successive response-bands, or optima, covering similar frequency-ranges. **1962** *Which?* Feb. 41/2 We found personal radios had a limited frequency range, from about 220 to 3,900 cycles per second. **1946** *Nature* 23 Nov. 759/1 Frequency-selective attenuation by interstellar matter. **1962** A. Nisbett *Technique Sound Studio* 245 Marked acoustic colouration in a studio may be due..to frequency-selective excessive absorption of sound. **1906** J. A. Fleming *Princ. Electr. Wave Telegr.* i. 43 Many other forms of frequency teller have been developed for practical use in connection with transformer working. **1941** S. R. Roget *Dict. Electr. Terms* (ed. 4) 144/2 *Frequency teller*, a name used, particularly in radio-communication for a frequency meter. **1893** *Jrnl. Soc. Arts* 5 May 624/1 The frequency-transformers hitherto used in the laboratory for purposes of research..are not practical for electric-lighting purposes. **1940** *Chambers's Techn. Dict.* 355/2 *Frequency transformer*, a static piece of apparatus (e.g. a transformer or mercury-arc convertor) which receives power at one frequency and delivers it at another frequency.

6. Special comb. **a.** (sense 4 b) **frequency band** = *band *sb.*[2] 14; **frequency changer,** a machine, circuit, or component that changes an alternating current of one frequency into one of another frequency; *spec.* in a superheterodyne receiver, the circuit or component that combines the incoming signal with that from the local oscillator to produce the intermediate frequency; **frequency converter,** a frequency changer; **frequency distortion,** distortion of a signal in which components of different frequencies are amplified or attenuated to different extents; **frequency diversity,** the use of channels of different frequencies in a diversity system of communication; **frequency response,** the relationship between the output-input ratio of a device or system designed to transform or transmit oscillatory signals and the frequency of the signal; **frequency shift,** used *attrib.* (also *absol.*) to designate a method of radio telegraphy in which the carrier frequency takes either of two approximately equal values according as the signal is a 'space' or a 'mark'; **frequency spectrum,** the entire range of frequencies of electromagnetic waves, sound waves, etc.; also, the distribution of the energy of a wave-form among its different Fourier components. Also *frequency modulation.

1922 *Bell Syst. Techn. Jrnl.* July 117 The method consists in analyzing the speech waves as impressed on a condenser transmitter, using a tuned circuit to transmit narrow frequency bands of energy. **1929** J. H. Morecroft *Elem. Radio Communication* iii. 82 The normal frequency band for broadcasting purposes is..from 550 to 1500 kilocycles. **1956** Tibbs & Johnstone *Frequency Modulation Engin.* (ed. 2) v. 106 Of the three major frequency bands, the v.h.f. band has been developed rapidly. **1902** Sheldon & Mason *A.C. Machines* vii. 157 The synchronous motor necessary to drive the frequency changer. **1965** *BBC Handbk.* 116 For best results on short waves, a receiver should incorporate a tuned radio-frequency amplifier preceding the frequency-changer stage. **1909** *Cent. Dict. Suppl.* I. 292/3 Frequency-converter. **1912** Bedell & Pierce *Direct & Altern. Current Man.* (ed. 2) ix. 291 The usual form of frequency changer or frequency converter consists merely of an induction motor and a separate driving motor. **1957** A. C. Clarke *Deep Range* (1968) xii. 106 This was a simple enough task for the sub's frequency converters; if he wished, Franklin could tune in to any sounds from almost a million cycles a second down to vibrations as sluggish as the slow opening of an ancient, rusty door. **1959** K. Henney *Radio Engin. Handbk.* (ed. 5) ix. 9 Semiconductor diodes are used over a very wide frequency range (a-f to microwave) as frequency converters. **1932** F. E. Terman *Radio Engin.* v. 121 Frequency distortion tends to be greater as the amplification per stage is increased. **1962** A. Nisbett *Technique Sound Studio* xii. 214 Quite severe frequency distortion can and does go practically unnoticed by 99% of the general public. **1931** *Proc. Inst. Radio Engin.* XIX. 531 Frequency diversity is effective due to the difference of fading times for frequencies differing by as little as several hundred cycles. **1964** *Electronics Weekly* 8 July 7/4 Frequency diversity is a standard feature of most microwave links... If fading occurs in the 'working' channel the telephone traffic..is automatically switched to the other channel. **1926** *Jrnl. Inst. Electr. Engin.* LXIV. 1023/2 Perfect reproduction of the originating wave-form necessitates a system with a uniform frequency response or frequency characteristic at all volume-levels. **1932** C. L. Boltz *Everyman's Wireless* xi. 225 The frequency-response curve..shows a very severe resonance at just over 500 cycles. **1971** *Which?* Jan. 19/1 Almost all microphones had at least as good a frequency response as the tape recorders they were to be used with. **1944** *Electronics* Nov. 126/2 The bandwidth required by frequency-shift transmission is no greater than that required by the carrier make and break system. *Ibid.* 127/3 Frequency shift can result in a much lower bandwidth than carrier make-break. **1959** *Chambers's Encycl.* XI. 471/1 A method of keying which is extensively employed with teleprinters is that known as frequency-shift (or carrier shift), in which the sender radiates continuously. **1955** F. E. Terman et al. *Electronic & Radio Engin.* (ed. 4) xvii. 591 When the instantaneous frequency of a frequency-modulated wave is varied in a more complex manner..the frequency spectrum becomes very complicated. **1962** *Rep. Comm. Broadc. 1960* 5 in *Parl. Papers 1961–2* (Cmnd. 1753) IX. 259 The division by international agreement of the frequency spectrum into bands allocated to particular services forms part of the International Radio Regulations. **1962** A. Nisbett *Technique Sound Studio* 269 *Scale*, division of the audio frequency spectrum by musical intervals (i.e. frequency ratios). **1968** C. W. McMullen *Commun. Theory Princ.* i. 10 The shape and dimensions of the pulse signal determine the frequency spectrum.

b. (sense *4 c) **frequency curve,** a frequency diagram in the form of a continuous curve, with the variable taking any of a continuous range of values for each member of a large population; **frequency diagram,** a diagram in which frequency of occurrence is plotted against the value of the variable; **frequency distribution,** a classification of the members of a population according to the value assumed for each member by some variable; a diagram or table showing the frequency with which a variable takes each of its possible values; **frequency polygon,** a frequency diagram containing a (small) finite number of points that are joined to form a line composed of a number of straight segments.

1893 K. Pearson in *Nature* 26 Oct. 615/2 The asymmetrical character of certain frequency curves in physical and biological measurements. **1911** G. U. Yule *Introd. Theory Statistics* vi. 76 Such an ideal limit to the frequency-polygon or histogram is termed a frequency-curve. **1932** J. S. Huxley *Probl. Rel. Growth* vii. i. 210 The frequency-curve for female body-length is unimodal. **1925** R. A. Fisher *Statistical Methods* ii. 37 Fig. 4 is a frequency diagram illustrating the distribution in stature of 1375 women. **1931** L. H. C. Tippett *Methods of Statistics* ii. 23 Frequency diagrams are useful as giving a visual impression of the characteristics of a sample. **1895** K. Pearson in *Phil. Trans. R. Soc.* A. CLXXXVI. 412 A method is given of expressing any frequency distribution by a series of differences of inverse factorials with arbitrary constants. **1924** Whittaker & Robinson *Calc. of Observations* viii. 165 The type of frequency distribution which is most familiar to the worker in experimental science is the distribution of the measures obtained by repeated measurements of the same observed quantity. **1948** *New Biol.* IV. 29 We cannot interpret the change in mean until we have examined the type of scatter of the individual heights around the mean, that is what the statistician describes as the frequency distribution of heights. **1965** A. H. Roberts *Statistical Ling. Anal. Amer. Eng.* v. 44 Table 3 and Appendix VIII show the joint frequency distribution of word length by syllable and phoneme number. **1897** K. Pearson *Chances of Death* I. 273 If the tops of these lines be joined we obtain a frequency polygon. **1966** E. B. Mode *Elem. Probability & Statistics* ix. 123 The frequency polygon..gives a picture of the way in which frequency of occurrence varies over the complete gamut of values.

frequency modulation. *Electr.* **a.** Modulation of a wave by variation of its frequency. **b.** The system of transmission of radio waves, etc., using such modulation. Abbrev. F.M. Cf. *amplitude modulation* (*amplitude 6 d).

1922 J. R. Carson in *Proc. Inst. Radio Engin.* X. 59 It has been proposed..to employ an apparently radically different system of modulation which may be termed frequency modulation as distinguished from amplitude modulation, in the belief that the former system makes possible the transmission of signals by a narrower range of transmitted frequencies. **1943** *Electronic Engin.* XV. 440/1 The ideal detector for frequency modulation is one which not only faithfully reproduces the true form of the f.m. components, but also is unresponsive to amplitude fluctuations of the carrier signal. **1955** *Times* 29 July 5/4 The present system of amplitude modulation in the v.h.f. maritime services should be changed to one of frequency modulation. **1958** *N.Z. Listener* 10 Oct. 21/1 The possibility of introducing FM (frequency modulation), a method of broadcasting that gives high fidelity reception. **1962** *Which?* Feb. 36/2 On the VHF band the FM (frequency modulation) system is used.

Hence **frequency-modulated** *ppl. a.*; (as a back-formation) **frequency-modulate** *v.*

1936 R. Glasgow *Princ. Radio Engin.* xi. 314 In order for a phase- or a frequency-modulated signal to become audible, it must be converted into an amplitude-modulated signal. **1950** *Gloss. Terms Radar* (*B.S.I.*) 6 *Frequency-modulated radar*, a form of radar in which the radiated wave is frequency-modulated and the returning echo beats with the wave being radiated, thus enabling the range to be measured. **1962** F. I. Ordway et al. *Basic Astronautics* vii. 322 The radio frequency carrier wave is modulated by a group of subcarriers... The subcarriers are frequency-modulated by the characteristics of the intelligence.

frequentist (frī·kwĕntist). [f. *frequent-*, stem of adjs., etc., related to FREQUENCY + -IST.] One who believes that the probability of an event should be defined as the limit of its relative frequency in a large number of trials. Also *attrib.* or as *adj.*
1949 M. G. KENDALL in *Biometrika* XXXVI. 104 It might be thought that the differences between the frequentists and the non-frequentists (if I may call them such) are largely due to the differences of the domains which they purport to cover. 1965 I. HACKING *Logic of Statistical Inference* xii. 227 Neither frequentists nor subjectivists have been right about probability. 1969 D. V. LINDLEY *Introd. Probability & Statistics* II. p. xi, This is not the place to criticize in detail the defects of the purely frequentist approach.

fresco, *sb.* Add: **2. a.** Also *transf.*
1883 'OUIDA' (*title*) Frescoes, etc., dramatic sketches. 1899 *Daily News* 28 Mar. 6/1 Florence has often been sketched before, putting Browning aside with his astounding fresco-music. 1933 A. MACLEISH (*title*) Frescoes for Mr. Rockefeller's city. 1966 *Listener* 23 June 924/2 This impressive musical fresco for string orchestra.
c. fresco buono = *BUON FRESCO; **fresco secco** = SECCO *b. sb.*
1843 W. B. S. TAYLOR *Man. Fresco & Encaustic Painting* vi. 122 Fresco secco..cannot be placed in the same elevated rank as fresco buono. 1886 H. C. STANDAGE *Artists' Man. Pigments* xii. 104 Describe the difference between 'fresco secco' and 'fresco buono'. 1957 *Fresco secco* [see *BUON FRESCO].

fresh, *a.*[1], *adv.*, and *sb.*[1] Add: **A.** *adj.* **6.** *fresh air* used in attrib. phrases, as *fresh-air fiend* or *maniac*, etc.
1884 *N.Y. Tribune* 2 July 7/1 The work of the Fresh Air Fund..sending children for a week or two from rural homes in unhealthy quarters of the city to healthful villages and farms. 1908 *Daily Chron.* 6 July 4/4 The fresh-air cure has been..very much boomed of late years. ..One result of this has been the evolution of what I may term the fresh-air maniac. 1927 W. E. COLLINSON *Contemp. Eng.* 99 Before the war we had our fresh air fiends..and the hatless brigade. 1950 N. CARDUS *Second Innings* 109 He went rambles all over the likes—the first of the 'fresh-air' fiends.
10. c. Of a cow: yielding a renewed or greatly increased supply of milk; coming into milk. *U.S.*
1884 *Vermont Agric. Rep.* VIII. 29 The cows will go dry for a time during the hot weather in summer and be fresh in fall. 1896 *Ibid.* XV. 67 This [inoculating of cream] may be done by using a 'starter' made from cream of the skim-milk of a fresh cow. 1971 *Independent* (Deerfield, Wis.) 23 Sept. 22/4 (Advt.), Fresh, springing, bred back cows and heifers.
14*. [Perhaps influenced by G. *frech* saucy, impudent.] Forward, impertinent, free in behaviour. orig. *U.S.*
1848 BARTLETT *Dict. Amer.* App., Fresh, forward; as 'don't make yourself too fresh here'; that is to say, 'not quite so much at home'. 1887 F. FRANCIS *Saddle & Moccasin* 136 What's the matter, then? Has Piggy been too 'fresh'? 1902 H. L. WILSON *Spenders* xxiii. 270 And when she goes out and says that isn't right they tell her she's too fresh. 1904 'A. DALE' *Wanted: a Cook* 190, I smiled, and was about to speak, when she rose, and in a loud voice, cried, 'Say, you're too fresh! Where d'ye think ye are?' 1908 G. H. LORIMER *J. Spurlock* ii. 26 That [remark] was pretty fresh, and my only excuse for doing it was that I couldn't think of anything fresher. 1923 WODEHOUSE *Adv. Sally* xiii. 156 I'm going to show that guy up this afternoon... He's been getting too fresh. 1928 S. VINES *Humours Unreconciled* iii. 41 A woman who does that sort of thing has no business to turn one down as soon as one gets a little bit fresh. 1932 H. NICOLSON *Public Faces* viii. 233 'Those Britishers,' mumbled the President eventually, having taken a large gulp of iced water, 'are getting fresh.' 1953 *Manch. Guardian Weekly* 20 Aug. 7/1 Anybody try any fancy stuff, or they got fresh, and they let you have it.
B. *adv.* **2. fresh-find** *v. trans.*, to find (a deer) after the scent has been lost; hence **fresh-found** *ppl. a.*
1780 in C. P. Collyns *Chase Wild Red Deer* (1862) 193 He was then found lying in a rush-bed. 1799 *Ibid.* 209 Here they fresh found him. 1895 in J. Fortescue *Rec. Stag-hunting Exmoor* (1887) 189 Still persevered in hopes of fresh finding him in Haddon. *Ibid.* 190 We had fresh found our deer. 1899 *Westm. Gaz.* 18 Aug. 3/1 A clever huntsman..usually succeeds in fresh-finding his deer. 1922 JOYCE *Ulysses* 571 After him, freshfound, the hue and cry zigzag gallops.

fresh, *v.* **1.** Delete †*Obs.* and add further examples in sense a. Also with *up*.
1835 J. P. KENNEDY *Horse-Shoe Robinson* I. 66 Put a sprinkling of salt in a bucket o' water..it sort of freshes the creatur up like. 1897 KIPLING *Capt. Cour.* 260 The fresh air will fresh Mrs. Cheyne up. 1910 *Westm. Gaz.* 16 Apr. 16/2 The rains have freshed the trout streams.

freshen, *v.* Add: **I. d.** Of a cow: to become fresh (see *FRESH *a.* 10 c). *U.S.*
1915 J. LONDON *Let.* 26 Jan. (1966) 446 Get Timms'..number of freshening cows. 1931 *Randolph* (W.Va.) *Enterprise* 9 Apr. 2/2, I have for sale 2 year old Jersey heifers to freshen in April and two Jersey cows,..one of them fresh now.
e. To wash one's hands and face, tidy one's hair and clothes, etc. Const. *up*. Chiefly *U.S.*
1961 in WEBSTER. 1962 'A. GILBERT' *No Dust in Attic* vi. 73 She thought she'd run along and freshen up. 1971 'D. HALLIDAY' *Dolly & Doctor Bird* vii. 92 They all foregather..for drinks before dinner. Would you like to freshen up and we'll take you?
3. *to freshen the nip* (examples; also *fig.*: cf. NIP *sb.*[2] b).
1827 J. F. COOPER *Red Rover* (1881) iii. 51 Profiting by the occasion 'to freshen his nip', as he quaintly called swallowing a pint of rum and water, he continued his narrative. 1842 *Southern Lit. Messenger* VIII. 764/1 After going into the saloon [grog-shop] to 'freshen the nip'—as they professionally called taking a glass of brandy and water—they led me into the upper tier of boxes. 1859 (see NIP *sb.*[2] 1 c). 1961 F. H. BURGESS *Dict. Sailing* 96 *Freshen the nip*, veer or haul slightly so that a rope may bear on a fresh place.

freshman. Add: **2. a.** Applied also to female students (in their first year). *rare.*
1897–8 *Vassar Coll. Catal.* 50 Freshman Class. Adair, Barbara. Affeld, Antoinette [etc.]. 1971 *Scotsman* 20 May 21/8 The tall, 19-year-old Glasgow University 'freshman' faces her first major test of the season.
b. freshman class *U.S.* (earlier examples).
1805 D. McCLURE *Diary* (1899) 8, I..was examined & admitted into the Freshman Class at Yale College. 1832 *Coll. New H. Hist. Soc.* III. 9 He was, in 1751, admitted a member of the freshman class in Harvard University at the age of twelve years. 1842 *Knickerbocker* XIX. 433 From time immemorial a playful animosity has existed between the freshman and sophomore classes.

freshness. Add: *spec.* Forwardness, impertinence. Cf. *FRESH *a.* 14*.
1892 'MARK TWAIN' *Amer. Claimant* 160 The mob began to lose its temper, and to get insulted, I had brought upon itself by its own too rash freshness. 1901 *Munsey's Mag.* XXIV. 791/1 He had once heartily 'jarraped' a new hand who had exhibited an unpleasing 'freshness' when speaking to her. 1928 J. C. LINCOLN *Silas Bradford's Boy* 13 The captain's dignity was slightly ruffled by what he considered freshness on the part of his nephew.

freshwater, *a.* Add: **1. b.** *U.S.* (See quot. 1925.)
1840 O. W. HOLMES *E. Venner* vii, A Sophomore from one of the fresh-water colleges. 1881 *Harper's Mag.* Jan. 224/1 'There is enough to send him through college.' 'To a fresh-water college?' 'Why not, for a fresh-water boy? He will always live in the West.' 1903 C. T. BRADY *Bishop* xi. 230 He had just entered the preparatory class of a little Eastern Fresh-water college. 1925 G. P. KRAPP *English Lang. in Amer.* I. 135 One speaks also..of regions further inland with the qualifying adjective freshwater, as in *freshwater towns* or *freshwater colleges*, the adjective carrying with it some implication of rusticity and provincialism. 1963 *Punch* 17 Apr. 548/1 A very great improvement in the standard and aims of even quite small 'fresh-water colleges'.

freshwoman. Delete † and 'imaginary' and add later examples. Still *rare.*
1874 *Scribner's Monthly* II. 347 To bring them where they can enter as Freshmen, or Freshwomen. 1885 *Academy* 21 Nov. 347 A fresh-woman—if that is the girl-equivalent of fresh-man—is to play the second lady.

Fresnel (frenĕ·l). [The name of A. J. *Fresnel* (1788–1827), French physicist and engineer.]
1. Used *attrib.* and in the possessive to designate apparatus, phenomena, and concepts relating to his work in optics, as *Fresnel biprism* = *BIPRISM; *Fresnel diffraction*, diffraction in which the diffraction pattern is a non-linear function of the variation in phase across the diffracting aperture or object; *Fresnel's formulæ*, two formulæ giving the proportion of linearly polarized light reflected from a plane surface in terms of the angles of incidence and refraction (see quot. 1957); *Fresnel lens*, a lens consisting of a number of concentric annular sections, each of different curvature and so designed that a parallel beam relatively free from spherical aberration can be produced; *Fresnel's integrals*, the integrals $\int_0^v \cos \frac{1}{2}\pi t^2 dt$ and $\int_0^v \sin \frac{1}{2}\pi t^2 dt$, used in the theory of Fresnel diffraction; *Fresnel('s) mirror* or *mirrors*, two plane mirrors set together at an angle of just less than 180 degrees; *Fresnel's rhomb*, a glass parallelepiped of such a shape that light can be passed through it to undergo two total internal reflections and emerge parallel to its initial direction.
1830 D. BREWSTER in *Phil. Trans. R. Soc.* CXX. 73, I am persuaded that the formulæ of Fresnel are accurate expressions of the phenomena under every variation of incidence and refractive power. *Ibid.* 77 M. Fresnel's general formula has been adapted to this species of rays.] 1835 *Rep. Brit. Assoc.* 1834 333 M. Poisson applied Fresnel's integral to the case of diffraction by an opaque circular disc. *Ibid.* 370 The parallelopiped thus constructed, and which is known under the name of Fresnel's rhomb, is of essential service in experiments on circular and elliptic polarization. 1848 A. STEVENSON *Acc. Skerryvore Lighthouse* II. 257 The divergence..may be described as the angle which the flame subtends at the principal focus of the lens, the maximum of which, produced at the vertex of Fresnel's great lens,..is about 5° 9'. 1849 G. G. STOKES in *Camb. & Dublin Math. Jrnl.* IV. 9 There are three particular angles of incidence..for which special results are deducible from Fresnel's formulæ. 1854 Fresnel's rhomb [see RHOMB 2 b]. 1871 tr. *Lommel's Nature of Light* (1875) xv. 207 (*caption*) Fresnel's mirror. *a* 1884 KNIGHT *Dict. Mech. Suppl.* 356/1 Fresnel lens, a lens consisting of a central portion of spherical section and surrounding rings, so adapted as to direct the rays practically parallel. 1890, 1904 Fresnel's biprism [see *BIPRISM]. 1905 R. W. WOOD *Physical Optics* vii. 195 (*heading*) Fresnel diffraction phenomena. 1957 JENKINS & WHITE *Fund. Optics* viii. 173 Since Fresnel diffraction is the easiest to observe, it was historically the first type to be investigated. 1937 G. S. MONK *Light* i. 19 A much more direct method..is the Fresnel biprism. 1957 *Encycl. Brit.* XIV. 611/1 If *i* is the angle of incidence and *r* that of refraction, the fraction reflected is $\sin^2(i-r)/\sin^2(i+r)$ or $\tan^2(i-r)/\tan^2(i+r)$, according to the direction of polarization. These expressions are usually called Fresnel's sine and tangent formulae. 1957 *Oxf. Compan. Theatre* (ed. 2) 476/1 The Fresnel lens spotlight or step-lens spotlight seems to have had more development and use in the U.S.A. than in England. 1959 BORN & WOLF *Princ. Optics* i. 50 One may also invert the procedure and prepare, by means of Fresnel's rhomb, linearly polarized light from elliptically polarized light. 1962 *McGraw-Hill Encycl. Sci. & Technol.* VII. 185 Another way of splitting the light from the source is the Fresnel double mirror.
2. (With lower-case initial letter.) A name occas. used by spectroscopists for a unit of frequency equal to 10^{12} Hz (10^{12} cycles per second).
1939 W. R. BRODE *Chem. Spectroscopy* viii. 191 The choice of the fresnel as a unit for recording visible and ultraviolet data is very satisfactory in that the units are not unwieldy. 1951 *Nature* 3 Mar. 367/2 Frequencies expressed in sec^{-1} involve large powers of ten ($\times 10^{14}$), while the fresnel ($=10^{12} sec^{-1}$) has never become popular. 1960 DECKER & CORRINGTON in W. G. BERL *Physical Methods in Chem. Anal.* (ed. 2) I. 194 The limits of the visible spectrum in frequency are from 750 to 400 fresnel units.

fret, *v.*[1] Add: **9.** Often const. *about, after, at, over, upon.*
a 1790 B. FRANKLIN *Autobiogr.* (1909) 79 Fretting about the money Collins had got from me. 1837 E. G. G. HOWARD *Old Commodore* III. 69 Timothy began to fret upon it. 1854 W. COLLINS *Hide & Seek* I. vi. 247 Don't forget the letter, sir, for I shan't fret so much about it when once I've got that! 1865 R. C. HARRIS *Christine* xi, She went through life..fretting at her lot. 1899 SHEEL & BEARDSLEY *King Washington* 224 In vain the captain fretted over the delay.
10. a. (Later example.)
1897 W. J. SYKES *Princ. & Pract. Brewing* 481 Often the secondary fermentation becomes unduly excited; the beer is then said to 'fret' or 'fret up'.

frettage[2] (fre·tédʒ). [f. FRET *v.*[1] + -AGE.] = *FRETTING *vbl. sb.*[1] 1 c; *frettage corrosion*, fretting corrosion.
1938 *Times* 29 June 11/3 The Engineering Department itself is attacking the puzzling problem of 'frettage corrosion', which, when vibration is present, causes a fine reddish-brown dust to appear between surfaces usually regarded as fitting so tightly that no relative movement is possible. 1960 U. R. EVANS *Corrosion & Oxidation of Metals* xviii. 738 In aircraft, frettage has been found between plates of bolted or riveted assemblages. *Ibid.*, The presence of oxygen..increases damage due to frettage, which in presence of oxygen is called fretting oxidation or fretting corrosion.

fretting, *vbl. sb.*[1] Add: **1. c.** Damage suffered by two metal surfaces when clamped or otherwise held together, owing to slight relative motion to and fro; *fretting corrosion*, fretting accompanied by a chemical change of the surfaces.
1939 G. A. TOMLINSON et al. in *Proc. Inst. Mech. Engineers* CXLI. 235/2 It appears desirable to denote the present action by the term 'fretting corrosion'. *Ibid.* 223/1 Although..chemical action accompanies fretting corrosion, the process nevertheless is almost certainly not one of corrosion as ordinarily understood. 1948 H. H. UHLIG *Corrosion Handbk.* III. 590 Increasing trouble from fretting corrosion can be expected as machine parts are operated at higher stresses. 1960 *New Scientist* 22 Sept. 778/3 This variety is termed fretting corrosion, and it is differentiated from simple fretting by the fact that the abraded particles undergo chemical change, generally as a result of oxidation.

fretty, *ppl. a.* Add: **3.** (Later example.) Also of beer.
1940 H. L. HIND *Brewing* II. xxxv. 870 Some of the smallest yeasts..are [illegible] and the beer [illegible] it will usually be found entirely difficult to remove them from a fretting beer.

fretty, *a.*[2] Add: **c.** Of beer: characterized by fretting or secondary fermentation (see FRET *v.*[1] 10, FRETTING *vbl. sb.*[1] 2).
1897 W. J. SYKES *Princ. & Pract. Brewing* 382 Beers produced from such contaminated waters show a great tendency, especially marked in hot weather, to become 'fretty', go turbid, turn sour.

Freudian (froi·diăn), *a.* (*sb.*). [See -IAN.] Of or pertaining to Sigmund *Freud* (1856–1939),

Austrian specialist in neurology and founder of psychoanalysis, or his teaching; spec. *Freudian censor* = *CENSOR *sb.* 4; *Freudian slip*, an unintentional mistake that seems to reveal a subconscious intention. Also *sb.*, a follower or adherent of Freud. Hence **Freu·dianism, Freu·dism**, the teaching or system of Freud; a characteristic specimen of this; **Freu·dianly** *adv.*, in a Freudian manner.
1910 *Amer. Jrnl. Psychol.* Apr. 289, I recently dreamt that I was traveling to Italy on my way to the next Freudian Congress (which is to be held in March). 1914 A. A. BRILL tr. *Freud's Psychopathol. Everyday Life* iii. 44 The Freudians will continue their researches for the causes of mental diseases. 1915 E. B. HOLT *Freudian Wish* v. 76 The idea has gone abroad that the term 'Freudian' is somehow synonymous with 'sexual'. 1920 B. LOW *Psycho-Anal.* 10 The Freudian theory and technique, and these alone, constitute Psycho-Analysis. 1921 *19th Cent.* Mar. 477 The attitude of the new school towards Freudism. 1922 R. S. WOODWORTH *Psychol.* xix. 505 The Freudian would shake his head at our interpretation of the lightning dream. 1923 *Times Lit. Suppl.* 12 Apr. 245/2 One or two lapses into Freudism. 1924 E. & C. PAUL tr. *Wittels' Sigmund Freud* 225 Thinkers who have no intimate connexion with Freudianism. 1927 A. HUXLEY *Proper Studies* 95 The Freudian censor is a real person with lodgings inside the skull. 1943 *Mind* LII. 78 He distinguishes very properly between the psycho-analytical technique..and what he calls 'Freudism'—the hypotheses and 'philosophy' which has been built up on the revelations for which the technique is responsible. 1955 M. McCARTHY *Charmed Life* (1956) ii. 37 He did not see quite how it worked Freudianly. 1958 *Times Lit. Suppl.* 19 Sept. 533/4 M. Sartre's own existentialism..making it, potentially at least, much more nourishing than the other two forms of *Weltanschauung* which have divided the mind and heart of Europe, Marxism and Freudianism. 1959 *Listener* 13 Aug. 257/2 A Freudian slip of the tongue at a dinner party. 1963 'N. BLAKE' *Deadly Joker* iii. 51 It was an odd little slip of the tongue..They call them Freudian slips nowadays. 1964 A. BURGESS *How Marry Miles?* xv. 95 A girl friend of mine..is married to a strict Freudian-type analyst. 1966 *Guardian* 25 May 8/5 Fine..But when we explain these flaws nowadays, Freudianly or otherwise. 1971 *New Yorker* 27 Feb. 57/1 It is such a primitive kind of Freudianism that..it hardly seems Freudian at all now.

‖ **friagem** (fri·āʒem). [Pg., cold spell.] A spell of unusually cold weather in Brazil, caused by the presence of Antarctic air.
1924 W. G. KENDREW *Clim. Cont.* xl. 327 On the middle Amazon there is usually a cool spell of 5 or 6 days in May or June, brought by a south wind called 'friagem' (ed. 3 (1937), a south wind; it is called 'friagem'). 1931 A. A. MILLER *Climatology* vi. 130 Cold waves known as 'friagems'. 1944 HAURWITZ & AUSTIN *Climatology* x. 225 Throughout the world, invasions of markedly cold air into equatorial regions are usually rare, and consequently the *friagem* is a very interesting meteorological phenomenon. 1968 G. R. RUMNEY *Climatology* xv. 288 Occasional outbursts of unusually cold, relatively dry air..originating over the expanded frozen surfaces around Antarctica, make their way much farther north than in normal and sometimes actually cross the equator. ..it is called a friagem.

friar, *sb.* Add: **9.b. friar's balsam** (earlier and later examples); also inhaled and used internally as an expectorant.
1753 W. LEWIS *New Dispensatory* 427/2 *Balsamum commendatoris*... This balsam has been inserted in some foreign pharmacopoeias..under the titles of..Balsam of Berne, Wade's balsam, Friar's balsam, Jesuit's drops, &c. 1772 [see MOROCK]. 1831 R. COX *Adv. Columbia River* vi. 78 The wound was dressed with friar's balsam and lint. 1959 W. GOLDING *Free Fall* i. 30 Then they realized of course that they had given him poison instead of friar's balsam... They had pulled and pulled but the spoon wouldn't come out of his mouth. 1963 *Brit. Pharm. Codex* 1261 Tincture of benzoin, compound... Friars' balsam. 1967 *Listener* 28 Sept. 419/1 For congested noses, Friar's Balsam... You inhale this—remember that nostalgic paraphernalia of cloths and steam?

frib: see *FRIBBY *a.*

fribble, *sb.* and *a.* Add: **A.** *sb.* **1.** Also, a woman of this type.
1874 M. CLARKE *His Natural Life* (1875) I. i. xi. 162 Flirt, fribble, and adventurer as she was. 1912 D. H. LAWRENCE *Let.* 14 Nov. (1932) 76 William gives his sex to a fribble.

fribby (fri·bi), *a.* (*sb.*). *Austral.* and *N.Z.* [Origin unknown.] Applied to small short locks of wool. Also as *sb.* (usu. *pl.*), such locks. Also **frib** (usu. *pl.*), short wool pieces and second cuts.
1900 A. HAWKESWORTH *Austral. Sheep & Wool* 180 A fleece is said to be fribby when a great number of second cuts or fribs fall out when it is shaken or in the process of rolling. 1915 J. R. MACDONALD *N.Z. Sheepfarming* xxvi. 69 When the fleece is placed on the table.. the stained and fribby pieces should be taken off the edges. 1929 H. B. SMITH *Sheep & Wool Industry Austral. & N.Z.* (ed. 3) 209 *Fribby*, short locky pieces of wool such as second cuts and small black yolky locks from crutch and under fore-legs of sheep. 1951 L. G. D. ACLAND *Early Canterbury Runs* 379 *Fribby*. Perhaps more a wool trade term than a station one. The yolky locks round the points taken off by the roller from a decently skirted fleece.

friction, *sb.* Add: **1. b.** *Hairdressing.* A massage of the scalp.
1931 G. A. FOAN *Art & Craft Hairdressing* ii. 114/2 Frictions are very popular in the gentleman's saloon, and may be considered as invigorating and beneficial in that they tone up the debilitated scalp. 1948 *Hairdressing & Beauty Culture* i. 10 A friction is a service which is greatly beneficial to the scalp and hair, particularly after an oil or a wet shampoo. 1961 S. J. COX *Illustr. Dict. Hairdressing* 61/2 *Friction*, a massage movement in which the fingers press and rub the scalp surface, imparting their effect in depth.
5. *friction-brush*; **friction-ball** (earlier U.S. example); **friction-drive**, a transmission of power by means of friction-gearing; **friction drum** (see quot. 1960); **friction-glazing**, the process of producing a high polish on paper by passing it through calender rollers that are revolving at unequal speeds; also *attrib.*; hence **friction-glazed** *a.*, **-glazer**; **friction head** [HEAD *sb.* 17 b], the head that goes to overcome the frictional resistance of a liquid flowing in a pipe; **friction welding**, a welding technique in which the necessary heat is produced by first rotating one component mechanically while pressing it against the other, which is held stationary; hence, any bonding of surfaces as a result of frictional heating.
1813 *Niles' Weekly Reg.* IV. 111/2 The wheels of both boats and carriages were provided with double ratchets reversed, or friction cups and balls. 1895 *Montgomery Ward Catal.* 103/2 Superior friction brushes made from genuine bristles. 1939–40 *Army & Navy Stores Catal.* 104/3 Body friction brush... Healthy exercise for the skin. 1907 *Motor Boat* 19 Sept. 19/1 The change-speed gear is a friction drive and a good brake for lowering. 1927 T. WOODHOUSE *Artificial Silk* 54 By these means, and a suitable combined belt, wheel, and friction drive, the trough can be tilted, when desired. 1957 *Times* 18 Nov. 11/1 An electric friction-drive car for which no key is required for ignition. 1960 C. WINICK *Dict. Anthropol.* 177/2 *Friction drum*, a drum with a string or stick attached to the center of the membrane. The fingers are resined or moistened and drawn along the string or stick and the resulting vibrations are transmitted to the membrane. Friction drums are often used ceremonially. 1907 CROSS & BEVAN *Text-bk. Paper-Making* (ed. 3) x. 271 The 'friction-glazed' or 'burnished' surface..is used chiefly for certain kinds of strong wrapping papers, and for certain coated papers. 1962 F. T. DAY *Introd. to Paper* vi. 65 The body paper may vary considerably from plain writings and printings, to coated friction-glazed papers, such as enamels, chromos and surfaces. 1899 H. R. A. HIGHAM *Handbk. Paper-making* viii. 213 Friction glazers are used to produce a very high finish on single-sided coated papers which contain wax in the coating. 1898 *Paper Makers' Handbk.* 20 *Friction-glazing calender*, a machine consisting of several rolls of unequal diameter working on one another and kept in position by very strong upright frames. 1888 CROSS & BEVAN *Text-bk. Paper-Making* xi. 169 Another method, known as 'friction-glazing', employed for giving a very high finish, is to pass it between a very large paper roll and a smaller iron one, the latter revolving at a much greater speed than the former. 1962 F. T. DAY *Introd. to Paper* v. 54 The finishing is then carried out by..friction-glazing in the case of flint and enamel papers. 1880 M. MERRIMAN *Treat. Hydraulics* vii. 207 The friction head is summed in the large main. 1963 A. C. TWORT *Textbk. Water Supply* x. 274 (*caption*) Friction head loss through various fittings, etc. 1959 *Times Rev. Industry* Mar. 48/1 Friction welding in its simplest form was one of the first methods used by the Germans for the welding of thermoplastic pipes. 1964 *Ann. Rep.* 1963 388 Another new way of treating metals, friction-welding, was evolved in the Soviet Union some years ago. 1971 C. CHANDLER *Ceramics in Mod. World* vi. 173 (*caption*) Silicon powder (A) is isostatically pressed, causing weak friction-welding at junctions (B).

frictionally, *adv.* Add: By means or by way of friction.
1927 T. WOODHOUSE *Artificial Silk* 90 These vertical discs impart motion frictionally to horizontal discs.

fridge, frig (fridʒ). Colloq. abbrev. of REFRIGERATOR 2. **b. fridge-heater**, a machine that uses the heat extracted by a refrigerator to provide hot water.
The proprietary name *Frigidaire* may also have contributed to the currency of the shortened form *frig*. Cf. also *frigerator* in *D.A.* (quots. 1886, 1909).
1926 E. F. SPANNER *Broken Trident* xvi. 181 Best part of our stuff has gone in the frig. 1935 [illegible] working, the mercury will climb like a rocket. 1935 C. BROOKS *Frame-Up* xix. 243 Do you mean that you keep a dead body in a fridge waiting for the right moment to..? 1939 M. DICKENS *One Pair of Hands* xii. 198 Your frig is out of order and the trifle hasn't got cold. [illegible] vi. 90 In the fridge was salmon, raspberries and considerable quantities of butter, milk and cheese. 1955 G. GREENE *Quiet American* 90 We haven't a frig—we send out for ice. 1957 *Archit. Rev.* CXXI. 459/2 Also on the ground floor is a demonstration larder for the new Ferranti 'fridge-heater' with an adjacent tea-kitchen that serves the entire premises; the fridge-heater cools the larder and at the same time produces hot water for the tea-kitchen. 1960 [see *BLINK *sb.*[2] 1 d]. 1971 *Islander* (Victoria, B.C.) 21 Mar. 2/2 We usually landed more than could be eaten fresh and having no fridge, dressed the surplus fillets with smoked salt and hung them up to dry.

fried, *ppl. a.* Add: **b.** *fried-fish shop*, a shop that sells fried fish, usually with fried (chipped) potatoes. Also, formerly, *fried-fish warehouse.*
1838 DICKENS *O. Twist* II. xxvi. 92 Field Lane..has its barber, its coffee-shop..and its fried-fish warehouse. 1898 J. D. BRAYSHAW *Slum Silhouettes* 25 I've jist seed Liz Dukeson, the donah at the fried-fish shop. 1939 'G. ORWELL' *Coming up for Air* IV. vi. 275 Flats, pubs, fried-fish shops, picture-houses, on and on for twenty miles.
2. Drunk. *slang.*
1926 in WENTWORTH & FLEXNER *Dict. Amer. Slang* (1960) 201/1 Princeton has completed the idiom of the cuisine by adding *fried* to *boiled* and *stewed*, meaning intoxicated. 1954 N. COWARD *Future Indefinite* iv. 193 After a gay reunion party..I retired to be slightly fried, blissfully happy. 1966 'D. SHANNON' *With a Vengeance* (1968) vi. 82, I don't know nothin' about how he got took off... I also heard he was fried that night.

Friedel-Crafts (frī·dĕl kra·fts). *Chem.* The names of Charles *Friedel* (1832–99), French chemist, and J. M. *Crafts* (1832–1917), American chemist, used *attrib.* with reference to a method, discovered by them, of alkylating or acylating aromatic hydrocarbons or ketones by the use of anhydrous aluminium chloride (or a similar substance) as a catalyst.
1892 *Jrnl. Chem. Soc.* LXII. I. 337 Friedel-Crafts' synthesis. 1900 *Ibid.* LXXVIII. 1006 The Friedel-Crafts reaction is among all synthetical methods the one most commonly used. 1927 *Industr. & Engin. Chem.* XIX. 1063 The summer of 1927 marks the fiftieth anniversary of the discovery of the Friedel-Crafts reaction. 1937 *Jrnl. Chem. Soc.* 504 Three main possibilities in the Friedel-Crafts mechanism. 1964 *New Scientist* 16 Mar. 686/1 Mixtures of the compounds containing 17–32 per cent of chlorine can be polymerized with heat and a Friedel-Crafts catalyst. 1964 N. G. CLARK *Mod. Org. Chem.* xix. 256 Friedel-Crafts alkylation [illegible] in a number of important aspects.

Friedreich (frī·draik, -ɔiχ). The name of Nikolaus *Friedreich* (1825–82), German neurologist, used *attrib.* and in the possessive to designate: **a.** Hereditary locomotor ataxia, an inherited disease of the central nervous system marked by an unsteady gait and an inability to co-ordinate voluntary movements; also its accompanying deformity, *Friedreich('s) foot.*
1883 *Brit. Med. Jrnl.* 31 Mar. 627/2 The term 'hereditary ataxy' is, therefore, a misnomer. 'Family ataxy', although more correct, does not stand for all..under these circumstances, we may provisionally accept Browne's [? an error for Brousse's] proposal to call the malady in question 'Friedreich's disease'. 1890 *Brain* XIII. 503 Tabes is an affection of adult age; Friedreich's ataxy a disease of childhood. 1934 S. A. K. WILSON *Neurology* II. xlix. 950 'Friedreich [sic] foot' may have been present from the outset. 1964 S. DUKE-ELDER *Parsons' Dis. Eye* (ed. 14) xxxiv. 545 In Hereditary Ataxia (Friedreich's Disease) optic atrophy and paralyses of the ocular muscles are very rare. 1970 *Sunday Times* 13 Dec. 24/6 (Advt.), Friedreich's ataxia is a crippling disease of the Nervous System which starts in childhood or adolescence.
b. *Paramyoclonus multiplex*, a rare disease characterized by involuntary twitching of the muscles, esp. of the limbs.
1889 *Brain* XI. 417 Friedreich's paramyoclonus, a fibrillar twitching insufficient to move the limb. 1903 *Jrnl. Nerv. & Mental Dis.* XXX. 365 In his opinion, the distinguishing feature of myoclonus of the Friedreich type was the occurrence of contractions of individual muscles not under the control of the will. 1967 *Brain* XC. 673 Similar families have been reported previously, often under the title of Friedreich's paramyoclonus multiplex.

friend, *sb.* Add: **4.** Delete †*Obs.* and add later examples.
1961 PARTRIDGE *Dict. Slang Suppl.* 1097/2 *Friend*, the man who keeps a harlot as his mistress. 1967 *Guardian* 11 Dec. 1/8 The boy's mother..was joined..by a man described as her 'friend'. The mother is apparently divorced.
5. Usu. in *pl.* A supporter of an institution or the like, contributing help, money, etc.
1926 (*title*) Friends of the Bodleian. First Annual Report 1925–26. *Ibid.* title-p. verso, In presenting this first Annual Report of the Friends of the Bodleian, the Secretary would like to point out that a considerable part of the year has..been spent in launching the Society... However, the Friends are now a considerable body. 1927 *Times* 20 July 17/6 The Dean and Chapter of Canterbury are forming a society of men and women to be known as 'The Friends of Canterbury Cathedral'. *Ibid.*, The Prince of Wales, on the special ground of the link between Canterbury Cathedral and Edward the Black Prince, is the first Friend entered on the roll. 1963 *Guardian* 21 Apr. 7/3 The twenty more or less aristocratic Friends of the Tate Gallery. 1963 *Times* 3 June 10/4 He also initiated 'The Friends of Bristol Cathedral', of which Queen Mary became the first Friend-in-Chief. 1971 *Guardian* 7 July 24 The Friends of the Lake District, a conservationist body which has already fought Manchester Corporation over its water extraction plans.

friendly, *a.* (*sb.*) and *adv.* Add: **A.** *adj.* **3. c.** (Further examples.)

1903 WODEHOUSE *Tales of St. Austin's* 18 Merevale's were playing a 'friendly' with the School-House, and.. Harrison had been pressed into service as umpire. *Ibid.* 22 Lucky the game was only a friendly. 1963 *Times* 11 May 3/6 He has appeared in a number of first-team friendlies. 1971 *Daily Tel.* 2 Aug. 3/7 Fighting broke out on the terraces during the Port Vale 'friendly' match against Manchester City at Vale Park.
B. *sb.* (Earlier N.Z. examples.)
1861 *Let.* 18 May in *Richmond–Atkinson Papers* (1960) I. xii. 707 No natives about except the 'friendlies' at Poutoko and Hauranga. 1869 B. Y. ASHWELL *Let.* 8 May IV. 554, 200 Friendlies went with Heta.

Friesian, var. of FRISIAN *a.* and *sb.*, as the name of a breed of cattle.
1923 R. WALLACE *Farm Live Stock* (ed. 5) 222 The general type of the Friesian is that of a large dairy animal. *Ibid.* Index, Friesian (or Fresian) British cattle. 1955 *Times* 6 July 7/1 The British Friesians are good. 1969 *Listener* 30 Jan. 159/1 Friesian cows have been found more profitable than others.

frie·zing, *vbl. sb.*[2] [f. FRIEZE *sb.*[2] or *v.*[2]] Frieze-work. Also *attrib.*
1769, *c* 1850 [see FRIEZE *v.*[2] 2]. 1899 *Westm. Gaz.* 17 May 5/2 The large amount of window space will contrast pleasingly with the plain friezing above. 1909 WEBSTER, *Friezing machine*, a machine for making edge moldings on woodwork.

friezy, *a.* Add: **2.** Resembling frieze.
1891 J. H. PEARCE *Esther Pentreath* I. iii, Covering him over with Tom's freezy coat. 1892 R. O. HESLOP *Northumbld. Words* s.v., A friezy coat is made of a kind of rough home-spun yarn. 1901 *Westm. Gaz.* 25 July 3/1 The Highwayman coat,..formed of a friezy tweed or homespun.

frig (frig), *v.* Restrict †*Obs.* to senses in Dict. and add: **2*.** Freq. used with euphemistic force. **a.** *trans.* and *intr.* = *FUCK *v.* 1. **b.** To masturbate.
1598 FLORIO *Worlde of Wordes* 139 *Fricciare*, to frig, to wriggle, to tickle. 1680 ROCHESTER *Poems on Several Occasions* (1950) 14 Poor pensive Lover, in this place, Would Frigg upon his Mothers Face. 1680 OLDHAM in *Ibid.* 131 There Punk, perhaps, may thy brave works rehearse, Frigging the senseless thing, with Hand, and Verse. *c* 1684 *Sodom* IV. 120, I caught one frigging with a curs bob tayle. *a* 1749 A. ROBERTSON *Poems* (1751) 83 So to a House of Office..a School-Boy does repair, To..fr— his P— there. 1858 H. SILVER *Diary* 21 Oct. in G. Ray *Thackeray* (1955) I. 452 Thackeray says one of the first orders he rec^d [at Charterhouse] was 'Come & frig me'. 1865 E. SELLON *New Epicurean* (1875) 11, I frigged and kissed their fragrant cunnies. *Ibid.* 22 The next minute I had flung her back on the hay, and was frigging away at her maidenhead. *c* 1888–94 *My Secret Life* II. 268, I have frigged myself in the streets before entering my house, sooner than fuck her. *Ibid.*, I got up,..frigged my prick, probed her. *a* 1935 T. E. LAWRENCE *Mint* (1950) II. x. 128 'If you ask me, Sir,' replied Taffy unabashed 'like a pack of skeletons frigging on a tin roof.' 1946 MEZZROW & WOLFE *Really Blues* 380 *High-pressure romancing* (find 'em, fool 'em, frig 'em and forget 'em).
c. *fig.* Also used as a coarse expletive. Cf. *FUCK *v.* 2.
1905 S. JOYCE in *Lett. J. Joyce* (1966) II. 104 Cosgrave says it's unfair for you to frig the one idea about love, which he had before he met you. 1936 S. KINGSLEY *Dead End* I. 33 *Spit*: Frig you! 1958 'E. MCBAIN' *Killer's Choice* (1960) 10 'He's telling us politely to go to hell.' 'Well, frig him,' Monoghan said. 1970 L. MEYNELL *Curious Crime of Miss Julia Blossom* xi. 153 'And what about the rent?' 'Frig the rent.'
d. [cf. sense 1 and *FUCK *v.* 3.] Const. *about, around*: to muck about, fool around (*with*). Const. *off*: to go away, make off.
1933 J. MASEFIELD *Conway* 211 *Frig about*, to fool around. 1940 HEMINGWAY *For whom Bell Tolls* xx. 272 We do not let the gypsy nor others frig with it [*sc.* a gun]. 1943 *Penguin New Writing* XVII. 87 You get an order, and you're frigging around like a pack of schoolgirls for five blinking minutes. 1955 N. BALCHIN *Fall of Sparrow* 166 So we've got to *move* people. See? Not keep them about here frigging around being tested for things. 1960 D. FEARON *Murder-on-Thames* xx. 188 It's not so easy to frig about with a stiff corpse and not leave bruise marks. 1965 H. C. RAE *Skinner* IV. xiii. 261 'Frig off,' he said, swinging towards the door. 1967 'J. PALMER' *Above & Below* iv. 44 He asked to be sent to a fighting unit where he could do something more useful..than frig about with papers.
Hence **frigging** *vbl. sb.* (also *ppl. a.* and quasi-*adv.*), **frigger**[1]: further examples (in above senses).
In quot. 1736 Friga is a proper name applied to a hermaphrodite lover, but the pun is transparent.
1707 *Indictment J. Marshall for publ. School of Love* (P.R.O. 6 Anne KB 28/24/9) My lovely Phil. is..so well versed in the various manners of fucking and frigging as the captain of the virtuosas. 1736 'F. SCHEFFER' *Toast* IV. 177 And surpriz'd (well-a-day!) the great Friga and Frow In a Posture—the Muse must not venture to shew! *Ibid.* 180 Ill presage! And had Friga but made a right Use Of a Hint so instructive, and op'd her own Sluice; Swelling high the Imp's Pool in a Torrent had run. *c* 1888–94 *My Secret Life* I. 93 Having come to the conclusion..that frigging makes people mad. 1922 JOYCE *Ulysses* 725 Doing that frigging drawing out the thing by the hour. 1930 J. DOS PASSOS *42nd Parallel* Newsreel v. 63 Station agent's so friggin' tough in this dump. *Ibid.* 77 When does the friggin' boat go? 1944 J. B. PRIESTLEY *Daylight on Saturday* xii. 78 This shunting frigging new arrangement..has got every flaming thing foxed up. 1951 E. MITTELHOLZER *Shadows move among Them* I. xv. 145 'You have the right spirit—and that's no frigging pun.' 'Hendrik! Please! No obscenity!' 1953 W. R. BURNETT *Vanity Row* xiv. 100 I'm god-damned frigging tired of getting shoved around by you. 1959 K. WATERHOUSE *Billy Liar* 40 Take your frigging mucky hands off my pullover. 1962 *Observer* 4 Mar. 1/6 We cannot go on with this piecemeal frigging around...Now we want to see finality brought about.

frig (frig), *sb.* [f. the vb.] In senses corresponding to the verb (sense 2*).
c 1888–94 *My Secret Life* V. 171, I pulled out my prick and with two or three frigs spent in a spasm of pain and pleasure. 1941 BAKER *Dict. Austral. Slang* 30 *Frigg-up*, a confusion, muddle. 1955 M. McCARTHY *Charmed Life* (1956) iii. 70, I don't give a frig about Sinnott's heredity.

frig: see *FRIDGE.

frigate. Add: **2. b.** Also, since 1943, a naval escort vessel, a large corvette.
1955 *Times* 10 May 7/1 The Leopard, a multi-purpose frigate with a heavy anti-aircraft armament, has an extreme length of 340 ft. and a beam of 40 ft. 1959 *Chambers's Encycl.* VI. 88/1 *Frigate*... The name was reintroduced during the second world war for a new type of convoy-escort vessel.
4. frigate mackerel, *Auxis rochei* or *A. thazard*, fishes of the family Scombridæ, found in warm seas.
1884 G. B. GOODE *Nat. Hist. Aquat. Anim.* I. III. 306 It is not unusual in the Bermudas, where it is called the 'Frigate Mackerel', a name not inappropriate for adoption in this country [*sc.* the U.S.]. 1902 JORDAN & EVERMANN *Amer. Food & Game Fishes* I. 277 The only species of this genus is the frigate mackerel. 1933 *Bulletin* (Sydney) 19 Apr. 20/4 The rare capture on the coast of Otago ..of a fish known in Australia as the frigate mackerel. 1965 A. J. MCCLANE *Standard Fishing Encycl.* 374/1 Frigate mackerels are generally small, seldom reaching a length of 2 feet.

frigger[2] (fri·gəɹ). [Etym. unknown.] A small glass ornament or testing sample.
1923 H. J. POWELL *Glass-Making in England* vii. 100 'Friggers' may be seen as chimney-piece ornaments in glass-blowers' homes in every glass-making district. 1938 A. FLEMING *Scott. & Jacobite Glass* ix. 107 Friggers have ever been a feature among East Coast glass-blowers who can produce whole 'Zoos' of little glass animals. 1964 *Punch* 29 July 175/1 The best 'glassmakers' frigger' of all—a flip-flop.

frighten, *v.* Add: **b.** with *off* (adv. and prep.).
1861 GEO. ELIOT *Silas M.* xiv. 261 If you can't bring your mind to frighten her off touching things, you must.. keep 'em out of her way. 1904 L. T. MEADE *Love Triumphant* IV. x, I frightened those fellows off, didn't I, Beaufort?

frightener. Add: *spec.* a member of a criminal gang who intimidates the victims of its activities, esp. in phr. *to put the frighteners in, on*: to intimidate. *slang.*
1958 F. NORMAN *Bang to Rights* i. 23 Spud Murphy gave him a very strong pull, and put the frighteners on him. 1962 R. COOK *Crust on its Uppers* ii. 18 They were just slag punters and couldn't pay even when we put the frighteners in. 1962 *Daily Tel.* 19 Dec. 15/6 Soho 'frighteners'—gangsters who try to extort money from club owners—were told..at the Old Bailey..that they faced severe punishment. 1966 A. PRIOR *Operators* vi. 64 His job had been to put the frighteners on various shopkeepers. *Ibid.*, The other man and himself had got a quid a piece out of it, the traditional ten per cent always paid to the frighteners. 1971 R. BUSBY *Deadlock* ix. 146 Some firm's trying to put the frighteners on, so I give him a bit of protection.

frighteningly (frəi·t'niŋli), *adv.* [f. FRIGHTENING *ppl. a.* + -LY[2].] In a manner, or to an extent, that frightens.
c 1854 [see Dict. s.v. FRIGHTEN *v.*]. 1906 W. J. LOCKE *Beloved Vagabond* xi, The backward vista down the years is too frighteningly long. 1921 GALSWORTHY *To Let* I. ii, She was frighteningly self-willed. 1922 A. S. M. HUTCHINSON *This Freedom* IV. i. 251 Her mother had scalded her hand and had cried out to her, frighteningly. 1952 E. HOBSBAWM in *Granta* 15 Nov., Most of us—at any rate the frighteningly large number which gets into print—specialise in the tedium of telling you how mistaken we were about politics.

frightfully, *adv.* Add: **2. b.** Now freq. without depreciatory reference, but merely = 'awfully', greatly, very. *colloq.* or *slang.*
1875 *Punch* 24 July 31/1 Everything is..'How deliciously frightful!' 'How frightfully charming!' 1928 GALSWORTHY *Swan Song* I. v. 39 With more jaw, and deeper set eyes, but frightfully like Jon! *Ibid.* III. vii. 308, I don't know that I approve of it frightfully. 1958 P. KEMP *No Colours or Crest* ii. 14, I say, you know, it's frightfully nice of you chaps to go on this show.

frightfulness. Add: Used during the war of 1914–18 to render G. *schrecklichkeit*, implying a deliberate policy of terrorizing the enemy (esp. non-combatants) as a military resource. Also *transf.*
1914 R. BROOKE *Let.* 11 Nov. (1968) 632 Belgium is.. the country where three civilians have been killed to every one soldier. That damnable policy of 'frightfulness' succeeded for a time. 1915 D. O. BARNETT *Lett.* 115 We are having a quiet time to-day, without any frightfulness for a change. 1915 P. GIBBS *Soul of War* 155 It was only when special orders for 'frightfulness' had been issued,.. that the rank and file of the enemy's army committed its brutalities. 1924 J. S. C. BRIDGE *Hist. France* II. 119 Attributing to the French the deliberate pursuit of a policy of 'frightfulness'. 1945 R. HARGREAVES *Enemy at Gate* 70 The impression persists..that his preference was for the methods of 'frightfulness'. But if terrorisation was the besiegers' object, their sanguine hopes were doomed to condign disappointment. 1968 G. JONES *Hist. Vikings* III. iii. 212 Ragnar attacked the smaller Frankish force..and took 111 prisoners. These as a deliberate exercise in 'frightfulness' he hanged..in full view of the second Frankish division.

frigid, *a.* **2.** Delete †*Obs.* and add later examples. Now usu. applied to women who are sexually unresponsive.
1893 J. A. SYMONDS *Life Michelangelo* II. 384 The whole weight of argument..leaves the impression on our mind that he was a man of physically frigid temperament, extremely sensitive to beauty of the male type. 1927 W. M. GALLICHAN *Sexual Apathy* i. 11 The frigid wife sheds an unconscious influence upon her children and those around her. 1953 J. S. VAN TESLAAR tr. *Stekel's Frigidity in Woman* v. 96 (*heading*) Psychology of the frigid woman. *Ibid.* 97 On their part they remain frigid during the act. 1962 C. ALLEN *Textbk. Psychosexual Disorders* v. xiv. 279 Women who are frigid frequently show great fear of snakes and often dream of them.

frigidity. Add: **2.** (Later examples.) Now usu. lack of sexual response in a woman.
1897 H. ELLIS *Stud. Psychol. Sex* I. v. 113 The lad in this case..has recently consulted me for *impotentia coeundi*, manifesting a frigidity for women. 1903 *Ibid.* III. 162 In dealing with the characteristics of the sexual impulse in women..we have also to consider the prevalence of frigidity, or sexual anæsthesia. 1963 A. HERON *Towards Quaker View of Sex* 65 But 'frigidity'..implies more than failure to have orgasm: it is the inability to enjoy love-making and penetration. 1970 *N. & Q.* Dec. 450/2 This relation between sexual frigidity and wrath does not seem to occur in the standard popular treatments of the Sins.

Frigidaire (fridʒidē^ə·ɹ). [Quasi-Fr., ad. L. *frigidarium.*] The proprietary name of a brand of refrigerator. (In quot. 1929 *fig.*)
1926 *Publishers' Weekly* 18 Sept. 964/2 Vacuum cleaners, frigidaires, radios. 1929 *Melody Maker* Jan. 25 We'd make a lovely pair If you would only care But you're a Frigidaire. 1930 *Morning Post* 17 June 20/3 Frigidaire fitments. 1931 *Trade Marks Jrnl.* 16 Dec. 1644/1 *Frigidaire.* Refrigerating machines and parts thereof. Frigidaire Corporation..City of Dayton, State of Ohio, United States of America. 1940 AUDEN *Another Time* 96 And had everything necessary to the Modern Man, A gramophone, a radio, a car and a frigidaire. 1946 A. CHRISTIE *Come, tell me how you Live* 9 But I was thinking of a plan To kill a millionaire And hide the body in a van Or some large Frigidaire.

‖ **frigorifico** (frigŏri·fiko). [Amer. Sp. *frigorífico*, lit. refrigerator.] In S. America, a meat-packing plant, esp. one concerned with the freezing of carcasses for export.
1917 L. E. ELLIOTT *Brazil Today & Tomorrow* v. 210 Opened in 1913, the *frigorifico* first supplied chilled meat to the city of S. Paulo. 1928 *Daily Tel.* 20 Mar. 10/7 He [*sc.* Lord Bledisloe] found..that except in the frigorificos and in the corrales..adjoining them there was little or no systematic attempt to control the dissemination of the [foot-and-mouth] disease. 1931 *Discovery* July 218/1 We passed several cattle barges on our journey up the river, most of them taking cattle to the frigorificos near Rosario. 1960 H. S. FERNS *Brit. & Argentine* xiii. 417 In 1884 the Sansinena brothers converted their *saladero* to a *frigorífico*.

frikkadel (fri·kădĕl). *S. Afr.* Also **fricadel, frickadel, frik(k)adel, frikkadelle, frikkedel.** [Afrikaans. Cf. FRICANDEL, -ELLE.] A ball of minced meat, fried or baked; a rissole.
1870 [see *BOBOTIE]. 1913 PETTMAN *Africanderisms* 178 *Frikadels* or *Frikadeletjes*, balls of minced meat and vegetables either fried or baked. 1926 O. SCHREINER *From Man to Man* 330 He bent low over his plate, shovelling in the frikkedel and gravy with his knife. 1946 *Cape Times* 5 Oct. (Week-end Mag.) 2/5 Fish frikkadel. 1966 E. PALMER *Plains of Camdeboo* vii. 122 Eating frikkadelle laced with rosemary and drinking coffee.

frill, *sb.*[1] Add: **1. c.** Esp. in pigeons (see quots. 1879, 1957); hence, a pigeon having a frill (cf. *frill-back* in 5).
1860 B. P. BRENT *Pigeon Bk.* 53 The eye..is of a pearl or gravel colour..very different from the Turbit's,..and the frill is rarely so long. 1879 L. WRIGHT *Pract. Pigeon Keeper* xiv. 165 The fourth and last property is frill. *Ibid.*, If the frill, or group of re-curved feathers on the breast is to extend [etc.]. 1890 *Daily News* 8 Jan. 2/4 Oriental frills, imported 20 years ago from Turkey and Smyrna. 1957 C. OSMAN *Racing Pigeons* i. 24 The same family [*sc.* the Git group of pigeons] occasionally has frills, feathers running across the breast of the bird and curling over backwards.
d. orig. *U.S.* (Earlier and later examples.)
a 1845 SYDNEY SMITH in S. Holland *Mem.* (ed. 3, 1855) I. ix. 407 Mr. —— has great good sense, but I never met a manner more entirely without frill. 1865 'MARK TWAIN' in Harte & 'Twain' *Sks. Sixties* (1926) 189 You put on

as many frills and make as much fuss..as if it were..a first-class power among nations. 1870 *Sacramento paper* (De Vere), I can't bear his talk, it's all frills. 1871 SCHELE DE VERE *Americanisms* (1872) 603 *Frills*, denotes, in California and the West generally, any assumption of style. 1885 'MARK TWAIN' *Huck. Finn* v, You've put on considerable many frills since I been away. *Ibid.*, He cussed me for putting on frills, and trying to be better than him. 1891 E. KINGLAKE *Austral. at Home* 157 Do not put on what the Americans call 'frill'. 1892 KIPLING *Barrack-R. Ballads* 12 It's the commissariat camel puttin' on 'is bloomin' frills! 1900 G. BONNER *Hard-Pan* vi. 194 She suffered from none of that rancor which the boarder who is suspected of 'putting on frills' is liable to arouse. 1928 GALSWORTHY *Swan Song* iii. 23 The first book was born too still for anything. A sort of 'African Farm', without the spiritual frills—if you remember it.
e. *fig.* A thing or feature which is merely ornamental; an embellishment. Usually in a derogatory sense.
1893 FARMER *Slang, Frills*, swagger; conceit; also accomplishments (as music, languages, etc.); and culture; *cf.*, Man with no frills. 1904 *N.Y. Tribune* 13 Oct. 6 The Board of Education should be encouraged in its tendency to lop off a few 'frills' from the curriculum and add the time saved to the substantial parts of the course. 1909 *Westm. Gaz.* 4 Aug. 4/1 A full-fledged Parliament.., a Speaker, a Mace, and all the frills and furbelows still considered essential to Parliamentary Government. 1916 A. QUILLER-COUCH *Art of Writing* vii. 133 The editor of a mining paper in Denver, U.S.A., boldly the other day laid down this law, that niceties of language were mere 'frills'. 1919 L. F. CODY *Buffalo Bill* 21 There were no frills about Will Cody's story as he told it to me. 1949 F. SWINNERTON *Doctor's Wife comes to Stay* 121 A simple, good-natured soul, who wants to be painted just as he is..'No frills.' 1962 PARKER & ALLERTON *Courage of Convictions* iv. 168 He was a pig-headed authoritarian without frills.

frilling, *vbl. sb.* Add: **1. b.** *fig.*
1899 C. J. C. HYNE *Further Adv. Capt. Kettle* vi, Great masses of foliage growing to the crown of the splintered heights, with a surf frilling. 1903 *Westm. Gaz.* 2 Mar. 8/2 A mid-day meal of stewed steak and frillings.

frilly, *a.* (under FRILL *sb.*[1]). Add: Full of frills; resembling frills. Also *fig.* **b.** *sb. pl.* Frilled undergarments. *colloq.* Hence **fri·lliness**, frilly character; also *fig.*
1890 E. CUSTER *Following Guidon* xiv. 202 Our opera-glasses looked just a little 'frilly' in such a place, but they were really useful. 1899 *Westm. Gaz.* 21 Sept. 3/2 Fluffy, frilly wraps. 1900 *Ibid.* 19 July 3/1 Capelines.. more or less frilly and floppy. 1902 *Ibid.* 20 Nov. 3/1 The effect was delightfully rippling and frilly. 1904 B. VON HUTTEN *Pam* I. iii, Pamela in a frilly nightgown. 1911 H. S. HARRISON *Queed* vi. 75, I am a clerk,..and office girl. My official title of course, is a little more frilly. 1922 *Glasgow Herald* 22 Mar. 10 [He] insisted upon the need for a thorough elementary education.. even if less attention were paid to the 'frilly' subjects than at present.
1900 E. GLYN *Visits Eliz.* 16 Lady Doraine and that horrid Smith creature made a place for me in the empty hammock between them, and, as I knew my 'frillies' were all right, I hammocked too. 1917 H. JENKINS *Bindle* vi. 87 In the bed was a figure..rising up from a foam of frillies. 1927 *Daily Express* 27 June 13/4 There are bargains to be found from fur coats to 'frillies'. 1970 S. J. PERELMAN *Baby, it's Cold Inside* 97, I am..shopping for some feminine 'frillies'.
1909 *Westm. Gaz.* 3 July 15/1 Once more everything is tending towards paniers, frilliness, flounces, and full skirts. 1927 *Daily Express* 27 Apr. 3/2 There was..more coy frilliness..about the evening dresses than last year. *Ibid.* 23 Sept. 3/3 The food was nourishing and English— ..none of your Frenchified frillinesses.

fringe, *sb.* Add: **2. b.** Also in wider use: an outer edge or margin of any kind, material or immaterial; an outer limit of a country, area, or population. Also *attrib.* = existing on the edge or margin of an area or region. Cf. also *CELTIC *a.* 2.
1898 Q. VICTORIA in *Daily News* 20 Dec. 5/4 All these people ask is to be allowed to do their daily task in peace..and to have a little fringe of play. 1902 D. G. HOGARTH *Nearer East* 162 The density of this Fringe population depends on fertility. *Ibid.* 181 It is necessary to detach the Arabs, who are found in the Arabian and Mesopotamian Fringe, from the central Semitic group. *Ibid.* 185 The outer desert ring has up to now proved effective to separate this loose unity from the Fringe populations. *Ibid.* 221 This system of railways..will serve to neutralise the immemorial tendency of this Fringe to display upon a superficial area..a *congeries* of human groups bitterly antagonistic in tradition and hope. 1903 [see MARGINAL *a.* 3]. 1926 J. GEDDIE (*title*) The Fringes of Edinburgh. 1935 *Amer. Speech* X. 10/2 Information..regarding the language spoken on the fringes of society. 1943 *Times* 21 Dec. 2/3 Altogether 35 sneak-raiders were destroyed in 'fringe target' attacks alone during the first half of the year. 1947 in *Amer. Speech* (1950) XXV. 65/1 'Fringe' Parking. 1951 *Electronic Engin.* XXIII. 262 He is resident in a fringe area. 1958 *Spectator* 14 Feb. 195/3 It may have won him a few fringe votes in Rochdale. 1958 *Times* 11 Nov. 4/3 Certain 'fringe' events arranged in Bath during the festival, notably the Festival Ball, [etc.]. 1960 *Guardian* 23 June 8/1 This is not part of the festival, but one of the 'fringe' entertainments. 1960 G. MURRAY in *Spectator* 28 Oct. 639, I have..adopted the term 'fringe' to indicate those medical practices which are not available to the patient in the NHS as a matter of course. *Ibid.*, I have not attempted to assess the merits of the treatments that fringe medicine can provide. 1962 *Guardian* 4 July 7/1 The posturings of a bunch of fringe-lunatics. 1966 *Times* 9 Nov. 13/5 Fringe Londoners like to keep the odd pig or two in their outbuildings. 1971 *Ink* 12 June 12/4 Events include discussions, rock groups, fringe theatre.
c. (Earlier example.) Also *Grecian fringe.*
1876 *Queen* 29 July (Advt.), Curled or waved fringes for the front hair. 1878 *Cassell's Fam. Mag.* IV. 175/2 None of that affected 'Grecian fringe' with which modern 'girls of the period' strive to hide what little forehead they possess.
g. Substitute for def.: A band or strip of contrasting brightness or darkness produced by the diffraction or interference of light and usually seen as one of a series. (Further examples).
1827–8 J. F. W. HERSCHEL in *Encycl. Metrop.* IV. 468 On the whole, the lamina would appear marked all over with dark and bright alternating fringes, just as we see it actually does. 1866 E. ATKINSON tr. *Ganot's Elem. Treat. Physics* (ed. 2) VII. viii. 491 From the fact that the dark fringes disappear when one of the beams is intercepted, it is concluded that they arise from the interference of the two pencils which cross obliquely. *Ibid.* 493 A series of alternate dark and light bands or fringes are seen parallel to the line of shadow. 1923 GLAZEBROOK *Dict. Appl. Physics* IV. 144/2 The simplest application of interference fringes is the testing of optical surfaces by means of test plates. 1963 R. W. DITCHBURN *Light* (ed. 2) v. 127 Coloured fringes are often seen when a thin film of transparent material is viewed by reflected light. The film may be a layer of oil on water or on the surface of a road. 1966 W. J. SMITH *Mod. Optical Engin.* xiv. 416 The accuracy of the fit between work and gage is described in terms of the number of fringes seen when the gage is placed in contact with the work.
h. A strip of false colour between regions of contrasting brightness in an optical image, or surrounding the representation of an object on a colour film or in a projected colour picture (cf. *FRINGING *vbl. sb.* b).
1891 H. T. WOOD *Light* vii. 105 Newton..arrived at the conclusion..that an achromatic object-glass, or a glass not showing these coloured fringes, was impossible. 1936 A. B. KLEIN *Colour Cinematogr.* 322 A fringe may be caused by parallax, error in printing registration, or by movement in the object which has taken place between the exposure of color-separation negatives. 1943 A. COX *Optics* 220 When there is any appreciable amount of lateral chromatic aberration present one side of the image has a red fringe, and the other has a deep blue fringe. 1963 R. W. DITCHBURN *Light* (ed. 2) vii. 259 When white light is used, coloured fringes appear at the boundaries between light and dark regions in an image.
3. fringe benefit orig. *U.S.*, a perquisite or benefit of some kind provided by an employer to supplement a money wage or salary; also *transf.*; **fringe-net**, a net intended to confine a fringe (2 c) of hair; hence **fringe-netted** *ppl. a.*; **fringe-variation** *Chess* (see quot. 1907).
1952 *Newsweek* 18 Feb. 74/2 For its 650,000 members it had asked..several cents' worth of fringe benefits. 1955 *Times* 6 June 10/3 Many of the gains that labour has been getting through new contracts are not so much increases in money wages as in so-called 'fringe' benefits—such as larger pensions and welfare payments. 1962 *Times* 12 Nov. 11/3 The term 'fringe benefit' was apparently first recorded..in an announcement by the United States War Labour Board during the Second World War. 1969 *Times* 30 Apr. 24/4 (Advt.), Successful candidates can expect to receive salaries based on their experience and ability and fringe benefits which include 18 working days holiday a year, generous contributory pension scheme, free life insurance to the value of two years salary and car purchase at reduced rates. 1899 *Westm. Gaz.* 10 June 2/1 She was now engrossed with a refractory fringe-net. 1909 M. B. SAUNDERS *Litany Lane* I. vi, Her round face, tightened up in a stiff frame of fringe-nets. 1966 J. S. COX *Illustr. Dict. Hairdressing* 61/2 *Fringe net*, a net of human hair or, exceptionally, of silk or nylon used to confine a fringe of hair. 1909 M. B. SAUNDERS *Litany Lane* I. x, Provincial mayoress..with fringe-netted hair. 1898 *Brit. Chess Mag.* 472 It would not be very difficult to find quite a goodly collection of two-movers in which every specimen contained the weakness of a 'fringe' variation. 1907 S. BLACKBURNE *Terms & Themes Chess Probl.* I. 54 A variation which does not spring naturally from the position, but has been tacked on by the employment of men which have otherwise no effect upon the solution, is known as a 'Fringe Variation'.

fringe, *v.* Add: **4.** *intr.* To spread like a fringe *away, off, out, over*, etc.
1858 GEO. ELIOT *Sc. Cler. Life* II. III. iv. 113 The Bridge Way was an irregular, straggling street, where the town fringed off raggedly into the Whitlow Road. 1875 'MARK TWAIN' in *Atlantic Monthly* Mar. 286/2 Do you see where the line fringes out at the upper end and begins to fade away? 1877 *Ibid.* Nov. 591 Its [*sc.* the town's] borders fringed off and thinned away among the cedar forest. 1899 C. P. ALLEN *What is Liberalism?* viii. 48 Its limits fringe away from it like a huge penumbra. 1901 *Westm. Gaz.* 15 Aug. 2/1 Exmoor..is for the most part in the west of Somerset, but it fringes over into the north-eastern part of Devon.

fringer (fri·ndʒəɹ). [f. FRINGE *sb.* + -ER[1].] One who is on the fringe (sense *2 b).
1938 *Times Lit. Suppl.* 26 Feb. 139/3 Picked up a bit of a living by introducing snob-conscious bourgeois to social fringers. 1952 J. MASEFIELD *So Long to Learn* 100 There were so many ways, cliques and coteries; there were..Celtic fringers. 1962 *Spectator* 26 Jan. 97, I knew he would denounce even the most casual approach to a 'fringer'.

fringing, *vbl. sb.* Add: **b.** The appearance of fringes of false colour in a projected colour picture as a result of imperfect registration of component images.
1912 F. A. TALBOT *Moving Pictures* xxvi. 298 Another disconcerting feature..is the apparent duplication of the outlines of figures near the camera...'Fringing', as this defect is called, is difficult to eliminate. 1921 *Conquest* Oct. 511/2 The moment motion of any rapidity occurs the objectionable defect of 'fringing' makes itself apparent. 1936 A. B. KLEIN *Colour Cinematogr.* iv. 244 The writer has photographed the ripples on a stream without obvious signs of fringing. 1969 *Focal Encycl. Film & Telev.* 301/1 Fringing is particularly objectionable in colour synthesis, since the fringes will then be of different colours from adjacent parts of the images.
c. The spreading of magnetic or electric lines of force at the edge of a region in which they otherwise form a simple pattern amenable to calculation.
1893 D. C. JACKSON *Text-bk. Electro-Magn.* I. v. 145 To this should be added about 10 per cent to allow for the 'fringing' of the lines of force at the corners of the pole pieces. 1943 *Electronic Engin.* XVI. 254 As we approach the edge of the plates the field becomes non-uniform and we have wide curvature of the lines of force. This is technically known as 'fringing'. 1957 AMOS & BIRKINSHAW *Television Engin.* (ed. 2) I. x. 262 The expression is approximate because it was assumed that the field between the plates was parallel whereas in fact it tends to spread at the edges... This effect is known as fringing. 1963 B. FOZARD *Instrumentation Nucl. Reactors* iii. 26 Fringing effects in the electric field at the ends of the electrodes tend to cause the magnitude of the sensitive volume to vary with applied voltage.

fringing, *ppl. a.* Add: *fringing forest*: see quot. 1926.
1903 W. R. FISHER tr. *Schimper's Plant-Geogr.* II. i. 177 We find the banks of rivers and lakes stocked with woods as far as infiltration extends. These are often mere brush-woods, but not unfrequently developed as luxuriant forests (fringing forests) not inferior to those of the best forest climate. 1926 TANSLEY & CHIPP *Study of Vegetation* x. 208 Fringing forest..occurs as a fringe of forest or woodland along the banks of perennial rivers. 1960 N. POLUNIN *Introd. Plant Geogr.* xiv. 439 There often occur, along rivers, 'fringing forests' which are evergreen and otherwise reminiscent of the rain forest.

frippet (fri·pĕt). *slang.* [Origin unknown.] A frivolous or showy young woman.
1908 D. H. LAWRENCE *Let.* 15 May (1962) I. 11 The girl will have a soul—she will not be a frippet. 1927 *Blackw. Mag.* May 673/2 'Frippets—three in number,' was the..diagnosis. 1933 W. CHETHAM STRODE *Sometimes Even Now* III. ii. 89 I'll invite him—and he can bring all his 'lovelys' and 'frippets' with him. 1945 E. TAYLOR *At Mrs. Lippincote's* xxiii. 196 'Mistress!' he thought... It was like the swine of a man to use such a word for what he and Edwards would have called a bit of a frippet. 1960 W. SHEED *Middle Class Education* (1961) 50 Did you have a rewarding life experience last night with that rather enticing bit of frippet? 1971 D. CLARK *Sick to Death* vi. 123 Quite a nice bit of frippet. But too young for me. I like a mature woman.

Frisbee, frisbee (fri·zbi). Also **frisby.** [See quot. 1970.] A concave plastic disc which spins when thrown into the air and is used in a catching game.
The word *Frisbee* is a proprietary name in the U.S. (see quot. 1959).
1957 *Newsweek* 8 July 85 The object of the game is simply for one player to toss the Frisbee, or disk, into the air and try to keep it from his opponent's grasp. 1959 *Official Gaz.* (U.S. Pat. Office) 26 May TM 168/2 Frisbee. 679,186. Wham-O Mfg. Co. For: Toy Flying Saucers for Toss Games in Class 22. 1967 *Wanganui* (N.Z.) *Photo News* 11 Feb. 35 Frisbees are plastic dishes which perform all manner of gyrations when tossed in the air. 1969 *Guardian* 5 July 9/2 The frisbee..is a plastic disc about the size of a frying pan that in the hands of the experts out boomerangs the boomerang. 1970 *Sports Illustrated* 3 Aug. 48/1 Thirteen years ago the Wham-O Manufacturing Company of San Gabriel, Calif...brought out the first Frisbee. Wham-O purchased the rights from a Los Angeles building inspector named Fred Morrison, who in turn had been inspired by the airworthy pie tins of the Frisbie bakery in Bridgeport, Conn. (which went out of business in March of 1958). He changed the spelling to avoid legal problems. 1971 *Islander* (Victoria, B.C.) 27 June 6/3 Gunny-sacked racers hobble about and frisbee circles form.

Frisesomorum (frəi:sesŏmō^ə·rŏm). *Logic.* [Formed by the letters signifying the propositions of the syllogistic mood.] The mnemonic term for that indirect mood of the first figure of syllogisms in which the major premiss is particular and affirmative, the minor universal and negative, and the conclusion particular and negative. Also called **Frisesmo.**
1599 BLUNDEVIL *Logike* 121 Celantes: Dabitis: Fapesmo: Frisesomorum. 1685 tr. *Arnauld & Nicole's Logic* III. viii. 35 These five modes are generally denoted by these words, Baralipton, Celantes, Dabitis, Fapesmo, Frisesomorum [printed Trisesomorum]. 1849 H. L. MANSEL *Aldrich* 65 The five indirect moods of the first figure were called Baralip, Celantes, Dabitis, Fapesmo, Frisesmo. 1884 J. N. KEYNES *Formal Logic* 199 Similarly, *Fapesmo* and *Frisesomorum* (the *Fesapo* and *Fresison* of Figure 4) have no corresponding direct moods. 1937 I. CHOQUETTE tr. *Maritain's Introd. Logic* iii. 189 When

these conventional words have more than three syllables, *e.g.*, *Frisesomorum*, the extra syllables are but stopgaps, and have no symbolic meaning. **1960** J. Roland et al. tr. *Dopp's Formal Logic* viii. 172 The moods similar to *Dabitis* and *Frisesomorum* (4th figure)..cannot, therefore, be valid.

frisk, *v.* **4. a.** For def. read: To search (a person or place); esp. to run the hand rapidly over (a person or his clothing), in a search for a concealed weapon, stolen goods, etc. *slang.*
1789 G. Parker *Life's Painter* xv. 179 They *frisk* him, that is, search him. **1812** [in Dict.]. **1920** I. Ostrander *How Many Cards?* 65 When you frisk this man Hill at Headquarters, if you find a pair of gloves on him keep them aside till I get there. **1929** C. E. Merriam *Chicago* 289 To take the police away from the task of frisking hip pockets and inspecting refrigerators. **1930** E. V. Knox in *Punch* 26 Feb. 236 His arm was twisted during the course of the interview, and his person was frisked on the chance of finding a gun. **1937** 'N. Blake' *There's Trouble Brewing* iii. 52 'Frisk him, Sergeant.'..The cleaner had to submit to a thorough examination of his pockets and person. **1940** R. Stout *Over my Dead Body* xiii. 173 The place has been frisked by someone in a hurry. **1955** *Times* 25 June 6/4 At a local exhibition several similar discs were handed to members of the public to hide in their clothing before being 'frisked' with a geiger counter. **1959** 'A. Gilbert' *Death takes Wife* xv. 198 Put down money for a joint you didn't frisk in advance. **1970** *New Statesman* 23 Oct. 520/2 Showing his teeth in a vicious snarl as they frisk him and open his jacket to feel under his arm.

frisking, *vbl. sb.* (Later examples in sense *4 a of vb.)
1913 H. A. Franck *Zone Policeman 88* v. 140 We..gave them and their baggage such a 'frisking' as befalls a kaffir leaving a South African diamond mine. **1938** M. Allingham *Fashion in Shrouds* xix. 342 She only admitted one person at a time into the room, and then only after a keen visual 'frisking'. **1958** S. Ellin *Eighth Circle* II. xviii. 141 It was a professional frisking, down to the way his wrist watch..and pen were removed and examined. **1968** *New Scientist* 9 May 288/3 These [uses] included alarm devices in the entrances to banks and aircraft and as a police tool for the 'frisking' of the public. **1971** *Times* 7 May 2/3 (*headline*) Judge rejects protest over frisking.

‖ **frisson** (frisoṅ). [Fr., shiver, thrill.] An emotional thrill.
1777 H. Walpole *Let.* 8 Oct. (1904) X. 130, I tore open the sacred box with..little reverence... No holy *frisson*, no involuntary tear warned me. **1903** A. Bennett *Truth abt. Author* xii. 156 The cult of the literary *frisson*. **1920** *Public Opinion* 24 Sept. 290/1 There had been a frisson of horror because the enemy was over the Marne. **1956** B. Russell *Portraits from Memory* 209 They rather enjoyed the *frisson* that its horrors gave them. **1967** *Times* 24 May 8/4 She provides that indescribable *frisson* which is the prerogative of only a few singers in each generation. **1971** *Observer* 31 Oct. 11/4 The dramatist seemed to have projected him on stage first to grip the audience with a frisson of alarm and impending melodrama.

frisure. Delete †*Obs.* and add later example.
1927 *Daily Express* 4 Oct. 9/6 Other resolutions declared that bobbed and shingled heads should have a higher frisure than formerly.

frit, *a.* Dial. and colloq. pa. pple. of Fright *v.* 2 a.
1821 [see Fright *v.* 2 a]. **1908** D. H. Lawrence *Let.* 31 Dec. (1962) I. 44 The woman is..a wee bit frit. **1960** *Oxford Mail* 4 Aug. 1/5 When I heard he was dead I got frit. **1970** *New Society* 19 Nov. 897/3, I was frit, cold and bored.

‖ **fritto misto** (fri·to mi·sto). [It., mixed fry (Fry *sb.*[2] 2).] A mixed grill.
1903 N. Newnham-Davis *Gourmet's Guide to Europe* ix. 158 The *Minestrone* of Milan and Genoa differ and so does the *Fritto Misto* of Rome and Turin. **1933** H. G. Wells *Bulpington of Blup* vii. 277 They were eating a fritto misto..in the *Isola Bella*. **1958** A. Wilson *Middle Age of Mrs Eliot* II. 220 Well cooked escalopes, or fritto misto. **1961** M. Howard *Surgeon's Dilemma* ii. 48 'What are you thinking about?' '*Spaghetti Bolognese* and *Fritto misti della Mare*.'

Fritz[1] (frits). German nickname for *Friedrich* (= Frederick). Hence, esp. in the 1914–18 war, used for: a German, *esp.* a German soldier (as typical of the German army); also, a German shell, aircraft, submarine, etc. Also *attrib.*
[**1883** G. Meredith *Let.* 1 Feb. (1970) II. 684 Many of your friends have bitterly deplored the German Invasion... But..let each have his time—Fritz after Alphonse.] **1915** D. O. Barnett *Let.* 11 June 174 By that time, of course, Fritz had made himself scarce. **1915** *Sphere* 28 Aug. 229/1 'Fritz' is the name our men have given to the latest air creation of Germany. **1916** *Daily Mail* 1 Nov. 4/4 An effective bombardment of the enemy's lines or a successful trench raid [is] described by Tommy as 'strafing the Fritzes'. **1919** G. K. Rose *2/4th Oxf. & Bucks Lt. Infty.* 54 At night it seemed probable that a patrol of a few brave men could crawl right up to the German wire and listen, or by setting foot in them enquire whether 'Fritz' was at home in his trenches or no. **1919** J. B. Morton *Barber of Putney* xiv. 228 Anyone'd think you was the first bloke that ever done a Fritz in. **1932** J. Dos Passos *Nineteen Nineteen* 377 The Fritz officer..shouted in English. **1955** J. Thomas *No Banners* xxvi. 261, I gathered he was more of a *collaborateur* than anything else. He praised you Fritzes up to the skies.

fritz[2] (frits). *slang* (orig. and chiefly *U.S.*). [Origin unknown.] Phr. *on the fritz*: out of order, defective, unsatisfactory; *to put on the fritz* (also *to put the fritz on*): to spoil, destroy, put a stop to.
1903 R. L. McCardell *Conversat. Chorus Girl* 15 They gave an open air [performance] that put our opera house show on the Fritz. **1906** H. Green *At Actors' Boarding House* 359 What with me ketchin' 'em cookin' spaghetti on the gas an' tearin' up the bedspreads to use fur makeup towels, they're puttin' the place on the fritz! **1924** Wodehouse *Bill the Conqueror* v. 122 Everything's on the fritz nowadays. **1962** *Guardian* 11 Apr. 9/2 It appeared, for an awful moment, that a cue had failed, that the teleprompter was on the fritz. **1968** R. H. R. Smithies *Shoplifter* (1969) vii. 151 It's Mother's plan to put the fritz on shoplifting.

Friulian (fri,ū·liăn), *a.* and *sb.* [f. *Friuli*, a district in N.E. Italy: see -an.] **a.** *adj.* Of or pertaining to Friuli or its inhabitants. **b.** *sb.* A native or inhabitant of Friuli; the Rhæto-Romance dialect spoken by the people of Friuli.
1880 *Encycl. Brit.* XIII. 492/2 *Eastern Section or Friulian Region.*—Here there still exists a flourishing 'Ladinity', but at the same time it tends towards Italian. **1910** C. Hulton tr. *Villari's Mediæval Italy* i. 4 In Italy the Friulian Marches, eventually extending to the Adige, were constituted. **1959** *Chambers's Encycl.* VI. 89/2 Friuli is rich in both maize and wine, and its inhabitants ..are mostly Italians, but some speak Slovenian and some Friulian, a Romansch dialect. **1963** *Times* 12 Feb. 8/6 The Friulians can hardly have been happy that Trieste is to be the regional capital. **1966** *Listener* 2 June 814/3 The recent *Elegie* for voice, violin, clarinet, and guitar (1962) on the Friulian poems of Franco Gironcoli. **1968** *Encycl. Brit.* XVIII. 1101/1 The RR dialects are spoken in three major areas: (1) In Friuli, the region between the Alps and the head of the Adriatic sea... This idiom is called Friulian.

frizziness (fri·zinės). [f. Frizzy *a.* + -ness.] Frizzy style or character.
1906 W. J. Locke *Beloved Vagabond* iii. 36, I criticised her straight Teutonic fringe and fanfaronaded on the captivating frizziness of Janna's hair. **1912** L. A. Harker *Mr. Wycherly's Wards* xvii, Your type is severe and classical; 'frizziness' would be quite dreadful and incongruous. **1955** *N.Y. Times* 14 Aug. III. 11 Some domestic supplies [of hair]..proved unsatisfactory because of frizziness.

fro, *v.* Add: Hence **fro·ing** *vbl. sb.* (See To and fro *phr.* E.)

Fröbel (frö·bėl). Also **Froebel**. The name of F. W. A. *Fröbel* (1782–1852), German teacher, used *attrib.* or in the possessive to designate the system of child education introduced by him, or a school following this system. (Cf. Kindergarten.) Hence **Froebe·lian** *a.*, of or pertaining to his system; as *sb.*, an adherent of his system; **Froe·belism**, the system of education introduced by Fröbel.
1873 *Jrnl. Women's Educ. Union* I. 144/2 Those who are interested in the Froebelian system of education will be glad to know that the Ecole Normale..will be re-opened in September. **1875** *Ibid.* III. 31 Froebel Society for the Promotion of the Kindergarten System.—An Association has been formed with the object of promoting ..Kindergarten work. **1876** E. Shirreff *Kinder-Garten* v. 40 It is..an object of the Fröbel method to hinder the ripening of the reasoning and critical faculty without corresponding practical activity. **1879** *Encycl. Brit.* IX. 794/1 The uncle and nephew [*sc.* F. W. A. and Karl Froebel] differed so widely that the new 'Froebelians' were the enemies of the old. *Ibid.* 795/1 The late Joseph Payne advocated Froebelism in a pamphlet, *Froebel and the Kindergarten System.* **1881** H. Barnard (*title*) Papers on Froebel's Kindergarten. **1904** G. S. Hall *Adolescence* I. iii. 171 Froebelian influence in manual training reaches through the eight school years. **1958** *Observer* 23 Feb. 8/4 If anything distinguishes Montessori from the Froebelians ..it is the fact that she stands definitely for work and not for play. **1959** *Chambers's Encycl.* VI. 90/2 The central organizing body for Froebel education in England is the National Froebel Foundation. **1969** L. C. Schiller in M. Ash *Who are the Progressives Now?* 154 In the middle of the [nineteenth] century a new influence arrived in England, the Froebel Kindergarten. *Ibid.* 155 In Froebel Colleges..the principles and practice of Friedrich Froebel were expounded.

frock, *sb.* Add: **5. c.** *transf.* A *député* or politician.
[**1917** H. Wilson *Diary* 24 Apr. (1927) I. xviii. 342 It makes it all the more necessary to put off the next meeting of the Frock Coats till after that date.] **1919** *Ibid.* 28 Mar. II. xxvii. 177, I think the Frocks have gone mad. They sit and talk all day. **1921** W. Orpen *Onlooker in France* xiv. 100 A little mass of black frock-coated figures—'frocks' as we called them. **1926** *Blackw. Mag.* Oct. 455/2 These deputies were not as the 'frocks' of 1914–18, remarkable only for their ignorance of matters military. **1928** *Observer* 26 Aug. 5 Contemptible as a soldier, Lord George was impossible as a 'frock'.
6. frock dress, (*b*) a form of court dress (see quot. 1875).
1875 *Dress worn by Gentlemen at Court* 3 Frock Dress is worn at Dinners and Evening Parties, where uniform is not worn, and consists of Dress Coat lappelled, Waistcoat, Breeches or Pantaloons, with white Cravat. **1903** *Westm. Gaz.* 17 Nov. 1/1 'Windsor Uniform—Frock dress' is the order for evening attire at Court this week.

Froebel: see *Fröbel.

frog[1]. Add: **3. b.** = Froggy *sb.* 2. Also, the French language. Also *attrib.* or as *adj.*
1778 F. Burney *Evelina* I. xiv. 79 Hark you, Mrs Frog..you may lie in the mud till some of your Monseers come to help you out of it. **1845** F. A. Kemble *Let.* 15 Dec. in *Rec. Later Life* (1882) III. 110 Surely I shall always be able, go where I will, among frogs or maccaronis, to procure *sucre noir*, or *inchiostro nero*. **1914** R. Brooke *Let.* July (1968) 601 Could we go on Friday to the Frog-Art show at Grosvenor House? From the First Frog to Cézanne. **1932** J. Dos Passos *Nineteen Nineteen* 55 Even the dogs looked like frog dogs. **1938** S. V. Benét *Thirteen O'Clock* 234 But there'd be the nuisance of learning frog-talk and the passage there and back. **1955** W. Faulkner *Fable* 333 Ask him...You can speak Frog. **1962** I. Murdoch *Unofficial Rose* vii. 84 Not that I want you to marry a frog, but she sounded quite a nice girl. **1970** *Private Eye* 27 Mar. 16, I dunno about the no hard feeling's bit—from what I hear about them frog sheilahs!
4. b. Colloq. phr. *frog in the throat*: (temporary) hoarseness; an irritation in the throat.
1909 in *Cent. Dict.* Suppl. **1933** F. Richards *Old Soldiers never Die* xvii. 223 One was speaking very thickly and the other lost his temper and told him to pull the bloody frog out of his throat. **1962** A. Nisbett *Technique Sound Studio* vii. 125 If the speaker has a frog in his throat..it will not do any harm to leave in the cough which clears it.
7. *frog-spear.*
1891 *Fur, Fin & Feather* Mar. 196 If the tourist likes frogs' legs..a frog spear is handy but not necessary.
8. frog eye, a fungal disease of plants indicated by spots on the leaves, *esp.* a tobacco disease caused by *Cercospora nicotianæ* or an American disease of apple and other trees caused by *Physalospora obtusa*; also *attrib.*; also **frog's eye**; **frog-face,** a face like that of a frog; *Path.*, a type of facial deformity usu. caused by a tumour in the region of the nose.
1914 *Jrnl. Agric. Res.* II. 57 Enlargements, which give to the disease the common name of 'frog-eye', are usually in alternating rings or zones of brown and gray. *Ibid.* 66 (*caption*) Typical spots of the frog-eye disease. **1926** F. D. Heald *Man. Plant Dis.* xxii. 585 The leaf attacks [of *Physalospora cydoniæ*] are referred to as leaf spot, leaf blight, brown spot and frog eye. **1950** C. Westcott *Plant Dis. Handbk.* iv. 308 Frog-eye leaf spot, general on apple, [etc.]. **1971** K. M. Graham *Plant Dis. Fiji* 210 Frog eye is common on tobacco wherever it is grown... The frog eye fungus persists in crop refuse. **1872** Geo. Eliot *Middlem.* iv. xxxix. 189 A little round head with bulging eyes—a sort of frog-face. **1884** M. Mackenzie *Man. Dis. Throat & Nose* II. 385 The most marked symptoms [of enchondromata of the nose] are obstruction of the nasal passages, and deformity in advanced cases amounting to 'frog-face'. **1948** *Ann. Surg.* CXXVII. 522 As the growth expands and advances, the floor of the orbit is elevated (unilateral or bilateral), producing 'frog-face' deformity. **1911** *Encycl. Brit.* XXVI. 1038/1 'Frog's eye', or 'leaf spot', denotes the occurrence of small white specks on the leaf. This disease is probably bacterial in origin.
b. frog-lily *U.S.*, the American yellow water-lily, *Nuphar advena*; also called Spatter-dock and cow-lily.
1869 J. G. Fuller *Flower-Gatherers* 204 It flourishes best in dull, stagnant pools, and is often called the *Frog-lily.* **1931** W. N. Clute *Common Names of Plants* 111 The frog lily (*Nymphaea advena*) is better named, for frogs delight to rest on its round floating leaves.

frog[2]. Add: **b. frog-band,** a band running from above the wall below the coronary band to join the frog; **frog-pad** = Cushion *sb.* 4 b.
1831 W. Youatt *Horse* p. vi, The Anatomy of the Foot... The coronary ligament: the coronary ring: the frog-band. **1908** *Animal Managem.* 217 A firm, soft cushion, sometimes called the 'frog pad' or 'cushion of the heels'.

frog[4]. (Earlier U.S. example.)
1847 *Rep. U.S. Comm. Patents 1846* 95 Frogs are used having guards or grinders on their outsides, and double inclined planes up and down, by which the wheels are guided to the right track.

froggish, *a.* (Earlier U.S. example.)
a **1861** T. Winthrop *John Brent* (1883) xxii. 194 Thanks La Grenouille, rover of the wilderness, for thy froggish instinct.

froggy, *sb.* Add: **2.** Also **froggee.** (Later examples.)
1883 *Referee* 15 July 7/3 While Ned from Boulogne says '*Oui, mon brave*, The Froggies must answer for Tamatave.' **1887** W. S. Gilbert *Ruddigore* I. 11 Froggee answers with a shout As he sees us go about. **1955** G. Greene *Quiet American* iv. ii. 241 Don't go, Fowler... I can't talk to those Froggies. **1965** *Guardian* 3 Sept. 9/6 A group of stage-type Limeys spend a weekend in France where they mix with a series of stage-type Froggies.

froggy, *a.* Add: **3.** *slang.* French.
1937 in Partridge *Dict. Slang.* **1953** G. Kersh *Brighton Monster* 187 We specialised in French brandy: stopped the Froggy boats in mid-Channel, grabbed the cargo. **1959** *Listener* 24 Dec. 1108/2 Where is this froggy speech of mine? **1962** I. Murdoch *Unofficial Rose* viii. 84 What about that froggy girl, the one you met in Singapore?

froglet (frǫ·glėt). [See -let.] A small or young frog.
1907 W. De Morgan *Alice-for-Short* xxi, Just one chirp of a nightingale or trill of a flute from a froglet. **1926** J. S. Huxley *Ess. Pop. Sci.* xvii. 196 The brusque transformation of tadpole into froglet. **1956** *Nature* 18 Feb. 342/2 By August..the froglets from this spawn were approximately 1½ cm. long.

fro·gman. [Frog[1] 1.] A man wearing a close-fitting suit of rubber or the like, with goggles and flippers, and equipped with a self-contained supply of oxygen to enable him to swim and operate under water; so **fro·gwoman.** **b.** (See quot. 1962.)
1945 *Sphere* 22 Sept. 359/3 (*caption*) The complete apparatus worn by the 'frog-men'. **1947** *Nature* 18 Jan. 88/2 An attempt to obtain samples of blood and tissues from the whale carcase..will be made by a scientific man working in a self-contained oxygen swim-suit ('frog man suit'). **1949** *N.Y. Times* 25 Dec. E7/5 'Frogmen', as they are called, wear self-contained breathing apparatus and web-like rubber shoes, used in the war, and swim like fish without stirring up mud, so that fish can be stalked with cameras. **1955** *Times* 6 June 4/2 Royal Navy frogmen flown by helicopter from Gosport searched the wreckage and recovered the four bodies. **1962** D. E. Barnes *Newnes Conc. Encycl. Nucl. Energy* 266/1 Frogmen is the name given to those who work in completely enclosing rubber suits, known as frogsuits, which are used to give protection against airborne radioactive material. *Ibid.* 266/2 (*caption*) Frogman suit. **1963** *Daily Tel.* 22 July 11/8 Teams of frogmen and frogwomen from all parts of Britain searched the sea and two lakes yesterday for a trail of clues leading to three sunken 'treasure' chests.

frog-march, frog's-march, *v.* Now usually, to hustle (a person) forward after seizing him from behind and pinning his arms together. Add further examples.
1931 J. Ferguson *Death comes to Perigord* viii. 97 Cæsar slewed him round, and forcing both arms behind his back, got ready to frog-march him to the door. **1935** D. L. Sayers *Gaudy Night* xxi. 446 Someone seized her round the neck from behind and frog's-marched her off to the cellar. **1952** B. Hamilton *So Sad, so Fresh* xii. 84 We took complete charge..frogs-marching Marya to the piano. **1965** *Times Lit. Suppl.* 19 Aug. 715/4 When I was at school in Yorkshire to be frog-marched meant that one's arms were secured behind one's back by a heftier fellow than oneself and one was propelled forward by the tormentor's knee being applied regularly and forcibly to one's behind. **1969** R. F. Delderfield *Come Home Charlie* viii. 89 He..took me by the collar and the seat of my pants and frogmarched me the length of the café.

frog-spawn. Add: **2. b.** *colloq.* Tapioca or sago pudding.
1959 I. & P. Opie *Lore & Lang. Schoolch.* ix. 163 Tapioca or sago is.., very commonly, 'frog spawn'. **1963** *New Society* 22 Aug. 5/1 The well-known fly cemetery and frogspawn are still o.k.-descriptions of currant puddings and tapioca.

Fröhlich (frö·lix). Also **Froehlich.** The name of Alfred *Fröhlich*, Austrian neurologist (1871–1953), used *attrib.* and in the possessive to designate a syndrome characterized esp. by obesity and sexual infantilism and caused by a tumour in the pituitary gland.
1909 *Jrnl. Amer. Med. Assoc.* 24 July 255/1 Delayed adolescence of the Fröhlich type. **1912** H. Cushing *Pituitary Body* 337 (*index*) Fröhlich's syndrome. **1964** A. Grollman *Clin. Endocrinol.* vi. 89/2 The chief symptoms observed in Fröhlich's syndrome are obesity, faulty skeletal development and genital hypoplasia.

frolicky, *a.* Delete †*Obs.* and add later U.S. example.
1883 E. W. Nye *Baled Hay* 88 They was just frolicky and gay because they felt good.

front, *sb.* (and *a.*). Add: **3. b.** Also *fig.* Outward appearance or aspect; façade; *spec.* a bluff. Cf. sense *7 g.
1900 'Flynt' & Walton *Powers that Prey* 181 It riles a bloke's sense o' justice to be accused false an' helps him to put up a front. **1934** J. O'Hara *Appointment in Samarra* (1935) 27 There was dancing and a hat-check girl and waiters in uniform and all that front. **1949** E. Coxhead *Wind in West* vii. 193 While he still put a good front on the affair, she said nothing. **1952** A. Baron *With Hope, Farewell* 124 It was only a front. He was scared stiff. **1953** R. Lehmann *Echoing Grove* 136 Will you kindly assist me to preserve a front till Monday?.. My parents have been through enough—we've got to put a face on it.
5. c. (Later examples.)
1944 *Hutchinson's Pict. Hist. War* 27 Oct. '43–11 Apr. '44 237 A 20-mile advance on the south of the bulge which opened up a 'breakthrough' front 60 miles in width. **1967** J. Marshall-Cornwall *Napoleon* iv. 50 Only 25,000 were available as a mobile field force, and these were extended..on a front of about 30 miles.
f. *transf.* With preceding epithet, an organized sector of activity, as *domestic front, home front*, etc.
1919 *Punch's Hist. Gt. War* 19 The trials..on the home front. **1929** *Nation* 4 Dec. 696/1 Gleb's victory on the economic front is somewhat spoiled by his partial defeat on the 'domestic front'. **1934** A. Huxley *Beyond Mexique Bay* 6 The amusement front had its duly appointed commissar. **1938** *Ann. Reg. 1937* 67 Sir A. Sinclair complained that the Government was concentrating its energies too much on preparations for attack, to the neglect of what was commonly known as 'the home front'. **1941** *Punch* 3 Sept. 211/3 My sister..writes..of the many..economical dishes she is now able to prepare as a result of the B.B.C. talks on the Kitchen Front. **1959** *Encounter* July 79/2 Myths of Soviet home-front propaganda. **1969** *Times* 6 Jan. 7/8 But the industry is fighting back on the marketing as well as the political front.
g. *transf.* An organized body of political forces.
1926 D. L. Sayers *Clouds of Witness* vii. 157 A woman ..looking like a personification of the United Front of the 'Internationale'. **1934** *Ann. Reg. 1933* II. 171 A uniform National-Socialist organisation embracing all German workers was called into existence under the designation 'The German Workers' Front'. **1936** E. A. Peers *Spanish Tragedy 1930–36* iv. 188 And combine under him [*sc.* Sr. Azaña] they did—Republican Left, Republican Union, Socialists, Syndicalists, Anarchists, Marxists and Communists—forming, for the purpose of the election campaign, a united phalanx, a 'Popular Front', as they called it: *Frente Popular.* **1940** *Amer. Speech* XV. 453/2 He [*sc.* Dimitrov]..urged members of the Communist Party to organize 'Popular Fronts' in the democracies. **1968** *Listener* 15 Aug. 195/2 The Popular Front for the Liberation of Palestine.
7. e. *spec.* with *the*: the promenade of a seaside resort, often with adjoining gardens. Cf. Sea-front 2.
1904 *Ward, Lock's Guide Isle of Man* 44 The Queen's Promenade..is the part of the Front most favoured by visitors with a taste for quiet. **1920** *Glasgow Herald* 17 July 7 For them the 'front', palpitating with cheerful humanity, is Elysium. **1938** G. Greene *Brighton Rock* I. i. 20 A blow along the front'll do you good. *Ibid.* iii. 42 I'd like to have asked them why he left me like that, to go scampering down the front in that sun.
f. (Earlier example.)
1806 G. F. Cooke *Diary* 6 Dec. in W. Dunlap *Mem. G. F. C.* (1813) I. 328 Went to the theatre..passed Mr Rae into the front of the house.
g. A person, organization, etc., that serves as a cover for subversive or illegal activities. So *front man, organization.* orig. *U.S.*
1905 *McClure's Mag.* XXIV. 346 For Brayton was the front, not the head of the System. **1926** J. Black *You can't Win* iv. 27 The store was but a 'front' or blind for a poker game and other games in the back room. **1934** H. N. Rose *Thesaurus of Slang* 25/1 Representative Who Poses as the 'Big Shot' of a Gang (n. phr.): *the front man.* **1938** H. Asbury *Sucker's Progr.* 345 Their agent and front man was the Chief of Police. **1940** in *Amer. Speech* (1941) XVI. 146/2 Foreign 'isms'..masquerading behind 'front' organizations. **1940** *Time* 29 Jan. 23/3 Department of Justice investigators believe that Earl Browder is a mere front-man. **1949** M. Miller *Sure Thing* (1950) 67 It's a front; the Commies control it. *Ibid.* 70 'I attended a camp of the Youth League of America.' 'You knew..that was merely a front for the Communist Party.' **1951** J. Cornish *Provincials* 213, I clare say it was all a 'front' for spy activities. **1959** 'M. Erskine' *House of Enchantress* ix. 130 He was.. respectable-looking and meek,..just the type to make an excellent front for Madame Rosario. **1960** *Spectator* 16 Sept. 408 He becomes the nark and front-man for an unscrupulous white landlord in a slum-house area. **1965** *Ibid.* 19 Feb. 220/1 A Communist 'front' organisation formed and financed by the Communist régime in North Vietnam.
h. *Meteorol.* A bounding surface or a transition zone between two air masses at different temperatures; also, the line on the ground that marks the lower edge of this surface; so *cold, warm front*: the forward boundary of a mass of advancing cold, or warm, air.
1921 Bjerknes & Solberg in *Geofysiske Publikationer* II. iii. 12 In the first case, the boundary line at the ground will be the front of advancing cold air, or, to introduce a shorter expression, a '*cold front*'. In the latter case, the boundary line will be the front of advancing warm air, or simply a '*warm front*'. **1923** N. Shaw *Air & its Ways* vii. 74 The polar front is regarded as being a bank of air with stream lines..over which the equatorial air is advancing gradually upward by motion directly transverse to the line of motion in the front. **1938** *Nature* 29 Oct. 804/1 Any pressure system, such as a cyclone, an anticyclone, a trough or a front. **1956** *Weather* May 147 The front is not a surface but a zone of rapid temperature transition between air masses. **1957** *Times* 11 May 7/1 A depression near Ireland will be slow moving and associated weak fronts will move slowly N.E. to E. districts. **1970** R. W. Longley *Elem. Meteorol.* x. 223 A frontal surface is the bounding surface between two air masses...The front is the intersection of such a frontal surface with the ground.
9. e. The front part of a woman's garment.
1801 Jane Austen *Let.* 5 May (1932) I. 125 It is to be a round gown, with a jacket and a frock front..to open at the side. *Ibid.*, The front is sloped round to the bosom and drawn in. **1889** *Daily News* 23 July 7/2 The travelling mantle..buttons the whole way down the front, and is provided with over-fronts which fall straight from the shoulders. *Ibid.*, These fronts are lined with yellow and pink..surah. **1932** E. Bowen *To North* iv. 38 One wore frills down her front, she was going to have a baby.
11. = *front-pipe* (see *14).
1879 *Organ Voicing* 12 Zinc is frequently used for basses and 'fronts'.
b. *front of*: in front of. *U.S.*
1843 'R. Carlton' *New Purchase* xv. 110 Front of the fire-place was the parlour. **1871** Mrs. Stowe *Sam Lawson* 45 Wall, she was a standin' front o' this. **1896** S. Jewett *Country of Pointed Firs* 107 He used to..throw a little bundle 'way up the green slope front o' the house.
c. As a command: to the front, forward.
1907 *N.Y. Even. Post* (semi-weekly ed.) 13 May 6 The register clerk [at a Shanghai hotel] assigns you to a room, and instead of 'Front!' he shouts 'Boy!'
13. In various uses or less technical uses.
1897 *Encycl. Sport* I. 267/2 It ruins the tyres, and, if applied to the front wheel, puts a great strain on the front forks. *Ibid.* 493/1 It often happens in heavy shooting that the recoil of the second barrel causes the front trigger to cut the first joint of the fore-finger. **1902** *Captain* VII. 474 Crabb front rim brake. **1907** *Daily Chron.* 11 Nov. 7/4 The front axle being..used for both steering and driving alike. **1908** *Westm. Gaz.* 18 Jan. 2/4 Front-pressure over the area of a railway carriage must be from 25lb. to 35lb. per sq. foot before the stability of the train is imperilled. **1908** *Daily Chron.* 3 July 6/4 This front-cover picture. **1932** *Morris Man.* 68 When the front axle is off the ground, the pedal should be depressed. **1967** *Gloss. Mining Terms (B.S.I.)* xx. 8 Front abutment pressure.
b. *Phonetics.* Applied to sounds in the formation of which the fore-part of the tongue touches or is raised towards the hard palate.
1867 A. M. Bell *Vis. Speech* 52 Front. The Tongue contracted from the front part between it and the roof of the mouth. Front-Mixed. The Front and the Point of the Tongue both raised. *Ibid.* 58 The 'Front-divided' Consonant has its side apertures within the palatal arch. **1888** H. Sweet *Hist. Eng. Sounds* 2 Front vowels are rounded by the lips only. **1918, 1962** [see *clear *a.* 13 b]. **1933, 1965** [see *central *a.* 1 d].
14. front-action *a.* (see quots.); **front-bencher,** an occupant of a front bench, a leading member of the Government or Opposition; **front brake** = *front-wheel brake*; **front cloth** *Theatr.*, a painted cloth before which a scene is played while the stage is set for another scene behind it; **front-end loader,** a tractor or haulage vehicle which has a shovelling or loading implement attached to the front; **front flight** = first flight (see Flight *sb.*[1] 8 d); also *attrib.*; **front foot,** a linear foot of frontage (cf. *foot front* in 11); **front line** = Front *sb.* 5; also *spec.* the musicians in a jazz band other than the rhythm section; **front-loader,** a machine, esp. a washing machine, designed to be loaded from the front, as distinct from one loaded from the top, etc.; **front man,** *spec.* (orig. *U.S.*), (i) see sense *7 g; (ii) the leader of a band; **front matter** *Printing* (orig. *U.S.*), all matter (title-page, preface, table of contents, etc.) in a book that precedes the text; the prelims; **front name** (*U.S.*) (earlier example); **front office,** orig. *U.S.*, a main or head office; *spec.* police headquarters; **front page,** the front outside page of a newspaper; often *attrib.* to indicate an important or striking piece of news; so **front-page** *v.* trans. (orig. and chiefly *U.S.*), to feature on the front page; **front-pager,** one who is worthy of being featured on the front page; a celebrity; **front-piece** *Theatr.*, a small play acted in front of the curtain; **front-pipe,** each of the row of pipes which form the front of an organ, often gilded or otherwise decorated; **front rank,** the first or foremost rank; also *attrib.*; **front-ranker,** a person (ship, etc.) of the highest class or of leading position; **front-runner** (orig. *U.S.*), '(i) a contestant who runs best when in the lead; also, one who can set his own fast pace; (ii) the leading contestant in a competition' (Webster 1961); so **front-running**; **front trench** *Mil.*, the trench nearest the enemy; **front wheel,** the foremost or either of the foremost wheels upon which a vehicle runs; also *attrib.*, as *front-wheel brake, drive.*
1881 W. W. Greener *Gun* 209 Back-action locks..tend to weaken the stock at the grip more than front-action locks. **1907** *Ibid.* (ed. 8) 144 If the mainspring [of the lock]..is placed before the tumbler, it is 'front-action' or 'bar'. **1907** *Westm. Gaz.* 31 Aug. 1/3 Each Unionist Front-Bencher will have to do a double or treble turn. **1919** G. B. Shaw *Heartbreak House* p. ix, Where were our front benchers to nest if not here? **1968** *Listener* 16 May 642/1 Would you say that it can be in opposition as a backbencher, or indeed even as a frontbencher, is virtually a role of impotence? **1959** *Motor Man.* [illegible] the ends of the front number plate to the front brake assemblies. **1959** *Motor Man.* (ed. 36) v. 137 (*caption*) Details of a Lockheed hydraulic two-leading-shoe front brake. **1884** J. Hatton *Irving's Impressions of America* II. xi. 268 Every scene is a set, except two, and they are front cloths. **1896** G. B. Shaw *Our Theatres in Nineties* (1932) II. 32 Long enough to allow the carpenters time to set the most elaborate water-scene behind the front cloth. **1958** B. Nichols *Sweet & Twenties* 141 Those most precious of all items to the revue writer, the 'front-cloth numbers', which can be played without props or scenery.

1959 C. Ogburn *Marauders* (1960) iii. 92 Bulldozers, power shovels, earth-movers and front-end loaders. **1960** *Farmer & Stockbreeder* 8 Mar. 76/3 They had mounted a cutter-bar driven by hydraulic motor, on to a tractor front-end loader. **1969** *Jane's Freight Containers 1968–69* 83/2 Stevedoring contractors' equipment at Port Brownsville includes approximately..14 front end bulk loaders, [etc.]. **1899** *Westm. Gaz.* 1 Dec. 4/2 The field gradually tailed off and only the front-flight men were able to keep on terms. **1902** *Daily Chron.* 5 Dec. 3/4 A department of the chase upon which front-flight men of the shires may be inclined to look down. **1812** *Deb. Congress U.S.* 4 May (1853) 2288 [The city of Washington] shall have power to cause [street improvements] to be done at any expense not exceeding two dollars and fifty cents per front foot. **1865** *Harper's Mag.* Aug. 319/1 Men bought town lots for $400 a front foot. **1925** B. Snyder *Real Estate Handbk.* 341 The land value map is designed to show the value of the land per front foot. *Ibid.*, These front-foot values are called unit values. **1915** 'I. Hay' *First Hundred Thousand* xviii. 248 That sudden disturbance in the front-line trench. *Ibid.* 251 Our front-line parapet. **1917** F. M. Ford *Let.* 1 Jan. (1965) 81, I hope to get to Mesopotamia as I am not fit for the front line. **1920** G. K. Rose *2/4th Oxf. & Bucks Lt. Infty.* 36 My company supplied parties to carry wire and stakes up to the front line. **1931** *Times Lit. Suppl.* 12 Feb. 116/4 From a vast mass of published diaries..by officers and men whose duties brought them constantly into the front line, Professor Cru, himself a front-line soldier, has studied the fundamental nature of war. **1955** *Jazzbook 1955* 18 A coloured front line..and a mixed white and coloured rhythm section. **1965** G. Melly *Owning-Up* vi. 72 The front line were blowing through their mouth pieces to warm them up sufficiently to play in tune. **1960** *Farmer & Stockbreeder* 12 Jan. 84 (Advt.), Independent operation of front-loaders, dozers, etc. **1970** *Guardian* 17 Nov. 9/2 The best laundromat machines.. are the front-loaders. **1937** *Amer. Speech* XII. 46/1 *Front man*, the leader of the band. **1946** R. Blesh *Shining Trumpets* (1949) xii. 279 To hire New Orleans players and then leave them free to play jazz, never occurred to the natty swing 'front men'. **1959** 'F. Newton' *Jazz Scene* xi. 186 A good and permanent band is normally run by a martinet, or a 'natural' front man with an eye to the public. **1909** *Cent. Dict.* Suppl., Front matter. **1920** *Publishers' Weekly* 4 Oct. 1660/2 We have cast-off your manuscript and estimate it will make 8 pages of front-matter and 248 pages of text. **1966** H. Williamson *Methods Bk. Design* (ed. 2) xii. 176 American designers sometimes use the terms 'front matter' and 'end matter'. **1877** Bartlett *Dict. Amer.* (ed. 4), *Front name*, Christian name. 'The familiar manner in which the telegraph handles my front name', i.e. in calling him Ben. **1900** 'J. Flynt' *Notes Itinerant Policeman* 73 The capture dwindles down to a request on the part of the chief or his officer that the man shall go to the 'front office'. **1935** Wodehouse *Blandings Castle* xii. 302 The Front Office has just sent out a communication to all writers. **1966** *Punch* 1 June 818/1 This is the sort of thing that can happen when the 'front office' is dubious about a film's popular appeal. **1902** *Out West* Jan. 39 Bearing on its front page a picture of the murderer, and the 'story'. **1917** *Writer's Bulletin* Mar. 56/2 He pounds the typewriter keys And in the distance clearly sees A front page story and a raise. **1917** Wodehouse *Uneasy Money* xii. 134 'Why, we may all be murdered in our beds!' he cried. 'Front page stuff!' said Roscoe Sherriff, with gleaming eyes. **1929** *Times* 4 Feb. 13/5 The recent serious illness of the King has been 'front page news' from the beginning. **1929** M. Lief *Hangover* 54 Most of these society dames front-paged their fed-uppance with tea-fights and garden soirées. **1957** *N.Z. News* 17 Dec. 2/1 *The New York Times* front-paged Labour's victory. **1958** *Times Lit. Suppl.* 4 July 377/3 It is most gratifying to see a front-page article..devoted to an intelligent and knowledgeable discussion of contemporary Russian poets. **1970** *Daily Tel.* 1 Jan. 5/8 Mr Barber is too good a journalist not to give the whole story breathless front-page excitement. **1899** *Daily News* 28 Feb. 4/7 A dazzling array of eminent 'front-pagers'. **1934** A. Huxley *Beyond Mexique Bay* 3 The gay and charming front-pagers who go on winter cruises are, in the main, elderly people. *a* **1889** *Evening News* (Barrère & Leland), At the Gaiety..a farce, 'Lot 49', by Mr. Fisher, as a front piece to 'Frankenstein'. **1907** *Daily Chron.* 9 Nov. 4/4 Thirty-four full-length plays and seven front-pieces. **1855** E. J. Hopkins *Organ* 73 Tin does not soon become tarnished; hence its peculiar appropriateness for ungilded 'front pipes'. **1905** T. Elliston *Organs & Tuning* 393 The front pipes to be of stout V.M. zinc silvered with aluminium leaf. **1954** *Grove's Dict. Mus.* (ed. 5) VI. 290/2 The front pipes were made of tin, those inside of lead. **1872** G. Meredith *Let.* 15 Nov. (1970) I. 472 The suspicion that Burton did not do what he said is unworthy, considering the things he has undoubtedly accomplished, and which place him in the front rank of adventurous travellers. **1897** *Sears, Roebuck Catal.* 579 This rifle..stands in the 'front rank' with the very best target rifles of this and other countries. **1899** *Westm. Gaz.* 10 Mar. 4/3 Mr. Fox, another front-rank costumier. **1935** *Discovery* Nov. 321 British manufacturers are in the front rank in this branch of scientific industry. **1905** *Westm. Gaz.* 25 Mar. 15/1 The eagerness which was displayed by some 'front-rankers' to get a 'line' of the Ceylon Pearl Syndicate's underwriting. **1914** in E. C. Barnes *Alfred Yarrow* (1923) xxvii. 260 The *Firedrake* and *Lurcher* were looked upon as two front-rankers. **1928** *Manch. Guardian Weekly* 7 Dec. Suppl. p. vi/1 It is a long time since a year's verse list was led by so many front rankers. **1963** *Times* 9 Feb. 10/5 London's light operatic conductors, at a time when Ivan Caryll, Herman Finck, Jimmy Glover, and Alfred Dove were front-rankers. **1940** *Time* 4 Nov. 71 He has won most of his subsequent newspaper clippings by front-running for the U.S.'s No. 1 anti-Wall Street financier. **1958** *Times* 24 Oct. 17/6 Eldon tried to front-run without the necessary strength and experience. **1914** *Automobile* 27 Aug. 390/2 Always a front-runner, he soon opened a gap on the field which looked to be a safe one. **1952** *Birmingham* (Ala.) *News* 5 May 12/1 Not a front runner, he is a dark horse who might come in first should Taft and Eisenhower cancel each other out. **1960** *Times* 29 June 17/7 There can be no doubt that the emergence of a few brave front-runners has lifted the middle distance running out of the doldrums. **1970** W. Smith *Gold Mine* vii. 18 He had joined C.R.C. a mere twelve years previously..and now he was the front runner. **1950** *N.Y. Times* 11 June 53/2 Wade beat Pearman by nine yards in the 880 with a front-running race. **1951** *Life* 1 Oct. 32/1 Taft is the strongest single Republican and the front-running candidate for the '52 nomination. **1916** 'Boyd Cable' *Action Front* 143 The stretcher-bearers carried their burden into the front trench. **1878** *Design & Work* IV. 218/1, 52 in. Wolverhampton..with lamp and pouch, front wheel brake. **1897** Front wheel [see sense *13 above]. **1900** *Captain* III. 463/2 Gamage's Holborn front wheel rim brake. **1902** A. C. Harmsworth et al. *Motors* x. 218 The automobilist should frequently jack up the front of his car so that the front wheels are free of the ground. **1904** Goodchild & Tweney *Technol. & Sci. Dict.* 147/2 The front wheel brake usually carries two blocks on a horseshoe-shaped clip. **1908** *Westm. Gaz.* 3 Nov. 4/1 The Allen-Liversidge system of front-wheel braking. **1928** *Daily Mail* 13 Aug. 12/6 Will the new principle of front wheel drive prove a success? **1968** N. Fleming *Counter Paradise* iii. 42 'Great heap,' Jake grinned...'Yes..and it's got front-wheel drive.' **1971** *Guardian* 18 Feb. 2/6 The Renault 17..is a front-wheel drive model.

front, *v.*[1] Add: **7. b.** To serve as a 'front' (see *front *sb.* 7 g). *slang* (orig. *U.S.*).
1932 J. Sayre *Rackety Rax* vii. 55 You'll have to front for us, knowin' the collegiate racket and all. **1939** R. Chandler *Big Sleep* xxvi. 232 Why should I front for that twist? **1939** *Nation* 5 Aug. 134/2 *America*, accusing us of 'fronting' for the Semites and Communists. **1951** *Manch. Guardian Weekly* 1 Mar. 3 Mr. Churchill agreed to 'front' for the quick pride of the Royal Navy. **1959** M. Ainsworth *Murder is Catching* vii. 90 Was he merely doing his job..? Or was he fronting for Pender? **1971** N. Freeling *Over High Side* I. 41 To..help him out occasionally I have fronted for him—a telephone call. And I'm bound to say he helped me.
c. To lead (a band). Also *intr.* orig. *U.S.*
1936 *Amer. Mercury* May p. x/2 *Baton weaver*, the joe personality who fronts the band. **1937** *Amer. Speech* XII. 46/2 Ted is fronting for Smith's old band. **1946** P. Fischer in *Jazzways* 48/2 Hampton was with the Les Hite Orchestra, occasionally 'fronted' by Louis Armstrong. **1949** L. Feather *Inside Be-Bop* iii. 26 Coleman Hawkins, who was fronting Clarke's band, copyrighted the tune. **1958** P. Gammond *Decca Bk. Jazz* x. 128 The remnants of Isham Jones's Orchestra were taken over..by Woody Herman, a clarinet player with a taste for jazz and a talent for fronting a band.
11. *Phonetics.* To pronounce with the tongue in a front position, i.e. touching or raised towards the hard palate; to palatalize. Also *intr.* Hence **fronted** *ppl. a.*, **fronting** *vbl. sb.*
1888 H. Sweet *Hist. Eng. Sounds* 36 The fronting is carried out most fully with the point nasals and stops. *Ibid.*, These fronted consonants again in their turn influence a preceding sound. *Ibid.* 37 [This sound] fronts the preceding ʃ. **1907** H. C. Wyld *Hist. Study Mother Tongue* viii. 160 A natural inference is that..*e* being a front vowel, fronted the preceding consonant. **1929** *Encycl. Brit.* I. 1/2 These [changes] are due to fronting.. or to rounding. **1939** *Trans. Philol. Soc. 1939* 89 In OE, Germ *a* appears as *æ*. Some dialects have *e*, but..the fronting first produced *æ*. **1964** *Language* XL. 31 Perhaps we can see a reason why /a/ should front.

frontage (frɒ·ntėdʒ), *v.* [f. the sb.] *trans.* To face; to have the front towards.
1914 N. Munro *New Road* ii. 17 Narrow, broken lanes with all the gable-ends of the abutting buildings frontaging the thoroughfare. **1958** *Times* 15 Mar. 7/5 His ratable value increased because he now frontages a made-up road.

frontal, *a.* Add: **2. c.** [after Da. *frontal* (J. Lange 1892 as in next).] Of or pertaining to the façade of a building; also in *Gr. Art*, pertaining to front or full-face view of a sculptured object (see next); similarly of a naked human body.
1893 *Funk's Stand. Dict.* s.v., The frontal effect of a fine building. **1905** P. Gardner *Gram. Gr. Art* v. 57 One finds figures stooping, or kneeling, or in a variety of other attitudes; but the frontal law still holds. **1971** W. J. Burley *Guilt Edged* iii. 46 Full frontal nudity on the stage. **1971** *Times* 12 Aug. 8/2 His [*sc.* Linnæus's] sketch..of *Andromeda polifolia* shows a naked virgin in full frontal exposure. **1971** *Times* 23 Aug. 10/4 The many publications which print explicit colour photographs of full frontal female nudes..and in some cases male nudes.

frontality (frɒntæ·lĭti). [tr. Da. *frontalitet* (J. Lange, *Billedkunstens Fremstilling græske Kunst* (1892) 54), f. Frontal *a.* + -ity: so G. *frontalität*, F. *frontalité*.] A principle in sculpture, according to which the figure is carved or moulded as seen from the full front.
In the strictest style, the figure is divided by an imaginary plane into two symmetrical halves, with no bending to either side.
1905 P. Gardner *Gram. Gk. Art* v. 56 Lange has expounded in detail his theory of frontality in early art. **1910** E. A. Gardner *Six Greek Sculptors* 34 The bold modelling of the figure, and the skill with which the twist of the body is rendered, make it worthy of notice in its departure from the rigid 'frontality' of early art. **1920** *Q. Rev.* July 39 A new decorative style of portraiture, of which the leading characteristic was a return to the frontal view of primitive Greek art. That is to say, 'frontality' became, as in the archaic periods, the basic principle of composition. **1938** *Burlington Mag.* June 307/2 Rousseau found in frontality a compositional device used by the primitives. **1958** *Times Lit. Suppl.* 21 Mar. 148/4 The well-known panels in San Vitale at Ravenna depicting Justinian and Theodora with their retinues (which illustrate the replacement of the Hellenistic three-quarter pose of the head by full-faced frontality).

frontally (frɒ·ntăli), *adv.* [f. Frontal *a.* + -ly[2].] By a frontal attack. Also *fig.*
1900 *Westm. Gaz.* 2 Feb. 7/1 Thus Buller, in order to relieve Ladysmith, must frontally force a tremendous position held by 10,000 or 12,000 of the best riflemen in the world. **1961** M. Howard *Franco-Prussian War* iv. 152 While his 5th Division attacked the enemy frontally from the south, 6th Division was to sweep round to the west. **1971** *Listener* 5 Aug. 181/1 The political impact of student discontent is touched on, rather than tackled frontally.

frontier, *sb.* and *a.* Add: **A.** *sb.* **4.** Later examples in sense 'the border or extremity conterminous with that of another'.
1956 E.-J. Finbert *Israel* 2 [Israel's] land frontiers are spread over 951 k. (591 miles). **1964** *Ann. Reg. 1963* 306 A U.N. observer would be sent to the Yemen-Saudi frontier.
b. (Earlier examples.) Also in specific use (see quot. 1894 and D.A.).
1676 W. Berkeley *Let.* 1 Apr. in W. E. Washburn *Governor & Rebel* (1957) 184 We have now such a strength on the frontiers of al our Plantations that we cannot feare them [*sc.* the Indians] if they were ten times more in number then they are. **1756** [see Frontier B. *adj.* 1]. **1842** *Amer. Pioneer* I. 226 Of all the men I ever knew he was the best qualified to live on a frontier where there were savages..to contend with. **1894** F. J. Turner *Frontier in Amer. Hist.* 3 What is the frontier?..In the census reports it is treated as the margin of that settlement which has a density of two or more to the square mile.
B. *adj.* **1.** (Further examples.) Also, characteristic of people living at a frontier; pioneering; primitive.
1611 Coryat *Crudities* 638 This Bommel is the farthest frontier towne westward of Gelderland. **1840** J. S. Mill in *Westm. Rev.* XXXIV. 489 The comparatively petty interest of some frontier dispute. **1854** Thoreau *Walden* 14 It would be some advantage to live a primitive and frontier life, though in the midst of an outward civilization, if only to learn what are the gross necessaries of life. **1902** S. Weyman *In Kings' Byways* i. 137 In the summer of 1706..one of the Austrian captains sat down before the frontier town of Huymonde, in Spanish Flanders, and prepared to take it. **1958** *Spectator* 14 Feb. 206/3 The connection in him between his 'frontier' mentality and his secret craving for respectability. **1960** *Guardian* 8 Dec. 8/2, I don't know about university education. In 'frontier' societies it is perhaps not a good thing. **1963** A. Heron *Towards Quaker View of Sex* 64 In frontier conditions or settled agricultural societies children represent an economic asset. **1968** Mrs. L. B. Johnson *White House Diary* 2 Oct. (1970) 714 There..came into being today the North Cascades National Park in the State of Washington —an area that's called the American Alps and is still very much frontier.

fro·ntierless, *a.* [See -less.] Having no frontier or dividing line.
1925 *Inner Life* ix. 165 The English layman as he comes to receive the Holy Communion in his parish Church, finds himself caught up into a Fellowship which is frontierless. **1935** G. Greene *England made Me* iv. 131 One believed in a new frontierless world.

frontierman, frontiersman. Add: (Earlier example.) *Legion of Frontiersmen*, an organization intended to enrol for irregular service men with colonial or frontier experience.
1782 St. Jean de Crèvecœur *Lett. Amer. Farmer* xii. 270 (*heading*) Distresses of a frontier man. **1905** *Times* 11 Apr. 8/1 A corporation to be called the 'Legion of Frontiersmen'. **1906** *Daily Chron.* 21 May 7/1 Mr. Roger Pocock, founder and hon. secretary of the Legion of Frontiersmen. **1930** *Times* 8 Nov. 8/7 A movement has been set on foot for the formation of a Maritime Command of the Legion of Frontiersmen.
2. *transf.* and *fig.*
1912 A. Conan Doyle *Lost World* ii. 24 I'm a frontiersman from the extreme edge of the Knowable. **1963** *Guardian* 13 Mar. 13/1 Thanks to these four men, and many another British literary frontiersman who has followed in their pioneering footsteps.

frontogenesis (frɒ:nto,dʒe·nėsis). *Meteorol.* [ad. G. *frontogenese* (T. Bergeron *Über die dreidimensional verknüpfende Wetteranalyse* (1928) I. 15), f. Fronto- + Genesis.] The formation or development of fronts.
1934 D. Brunt *Physical & Dynamical Meteorol.* xviii. 333 Bergeron has suggested the name *frontolysis* to denote the smoothing out of a front from discontinuity to continuity, and the name *frontogenesis* to denote the reverse process. **1939** *Geogr. Jrnl.* XCIV. 145 The primary factor in the formation of fronts (called frontogenesis) consists of the air movements which bring cold and warm masses into juxtaposition and form a diffuse front. **1967** *Jrnl. Atmospheric Sci.* Nov. 627/1 The time scale for the frontogenesis process must be small since fronts often form in less than 24 hr.
Hence **fro:ntogene·tic** *a.*
1966 *Jrnl. Atmospheric Sci.* Sept. 455/2 A paper.. presented with the purpose of determining the frontogenetic potentialities of deformation fields.

front yard. *U.S.* [Front *sb.* (*a.*) 13.] A piece of ground or garden in front of a house.

1767 J. Browne *Diary* 122 My pump in the front yard froze. **1835** J. H. Ingraham *South-West* II. 114 Southerners, with the exception of the cultivation of a few plants in a front yard, pay little regard to horticulture. **1897** 'Mark Twain' *Following Equator* xxix. 282 There cannot be another town in the world that has no..weed-grown front-yards of the poor. **1908** *Westm. Gaz.* 29 May 1/3 'The front-yard', as my old farming friend in the train informed me, was mostly grass. **1916** H. L. Wilson *Somewhere in Red Gap* ii. 69 The Piedmont Queen dahlia bulbs I'd ordered for the front yard.

frost, *sb.* Add: **7. a.** *frost-free, -proof* adjs.
1926 *Nat. Geogr. Mag.* May 513/1 (*heading*) Over mountain roads to a frostfree thermal belt. **1946** *Nature* 20 July 78/1 Most suitable frost-free sites are gentle slopes. **1896** T. W. Sanders *Encycl. Gardening* (ed. 2) 1 Lift tubers..& store in frost-proof place. **1960** *Farmer & Stockbreeder* 23 Feb. 72/2 The store was made frost-proof by lining the inner wall with glass-fibre.
c. frost-crack, a vertical split in a tree-trunk caused by the stress created as the wood freezes; **frost-flower,** an ice-crystal resembling a flower; **frost-grape** (earlier U.S. example); **frost-heave, -heaving,** uplift of a soil surface caused by expansion of water beneath the surface on freezing; so **frost-heaved** *a.*; **frost-hollow** (see quots.); **frost pocket,** a small low-lying area affected by frost (cf. Pocket *sb.* 8 c); **frost-ring,** a ring-shaped zone of tissue damaged by frost in the trunk of a tree; **frost-thrust** *Geol.*, (*a*) *sb.* = *frost-thrusting; (*b*) *adj.*, that results from or has been subjected to frost-thrusting; **frost-thrusting** *Geol.*, movement of soil during freezing, often with consequent lateral or vertical movement of partly buried rocks.
1894 T. Laslett *Timber & Timber Trees* (ed. 2) iv. 61 An expansion may result at a certain stage of the freezing of the inner woody cylinder, and the consequence is a frost-crack. **1960** Kramer & Kozlowski *Physiol. Trees* xvi. 490 Frost cracks are more common in forest trees than in orchard trees. **1847** C. Brontë *J. Eyre* I. iv. 47, I fell to breathing on the frost-flowers with which the window was fretted. **1923** E. Sitwell *Bucolic Comedies* 78 The frost-flowers upon the window-panes. **1789** *Trans. Amer. Philos. Soc.* I. 261 The frost or winter grape is known to every body. **1946** *Amer. Jrnl. Sci.* CCXLIV. 623 Thus the general use by geologists of..'frost-heave' and 'frost-heaved' is not completely supported by dictionary definitions. **1957** J. K. Charlesworth *Quaternary Era* I. xxvii. 567 Geologically, it [*sc.* permafrost] is most important in giving rise to frost thrust..and frost heave. **1967** F. A. Cook in Hamelin & Cook *Illustr. Gloss. Periglacial Phenomena* ix. 193 Frost-heaved congelifracts, involutions and injection plugs. **1929** *Jrnl. Geol.* XXXVII. 430 The maximum amount of frost heaving that occurs during cold winters..is not known. **1930** *Ibid.* XXXVIII. 303 (*title*) The mechanics of frost heaving. **1895** W. R. Fisher *Schlich's Man. Forestry* IV. iv. 431 Damp, low-lying places with stagnating air..termed frost-hollows. **1953** H. L. Edlin *Forester's Handbk.* v. 75 Valley bottoms and slight depressions of the ground in which cold air collects on still, cloudless nights..are known as frost-hollows. **1931** *Forestry* V. 118 The leaders of Norway Spruce in small low-lying areas, 'frost pockets', are killed back. **1932** Fuller & Conard tr. *Braun-Blanquet's Plant Sociol.* v. 86 The frost pockets (*Frostlöcher*) of the southwestern Swiss Jura are well known and shunned. **1968** R. Amberley *Incitement to Murder* ii. 29 Is the more exposed land, farther down..also yours?..There is a frost pocket there, no doubt. **1929** T. Thomson tr. *Büsgen's Struct. & Life Forest Trees* v. 146 The frost rings did not extend beyond the fourth year. **1960** Kramer & Kozlowski *Physiol. Trees* ii. 16 Frost rings are most often found near the center of a tree because young shoots are more susceptible to frost injury than are old stems. **1946** K. Bryan in *Amer. Jrnl. Sci.* CCXLIV. 625 There is no common expression for the lateral thrust resulting from expansion although horizontal as contrasted with vertical frost-thrust would sufficiently carry the meaning. **1951** [see *frost-thrusting]. **1957** [see *frost-heave]. **1951** *Jrnl. Geol.* LIX. 65 The writer has found no terms descriptive of these particular phenomena. Consequently the following terminology is used: *Frost-thrusting* is used for the process whereby large blocks of rock have been raised, commonly vertically, or have changed position with respect to the rest of the outcrop, owing to the frost action. *Frost-thrust blocks* refers to the blocks of rock which have been moved by frost-thrusting. **1954** *Amer. Jrnl. Sci.* CCLII. 55 From their structures we conclude that frost thrusting —lateral soil movement during freezing—plays a major role in their development.

froth, *sb.* Add: **4. c. froth-blower** *joc.*, a beer-drinker: adopted as the title of a certain charitable organization.
1905 *Daily Chron.* 25 Mar. 6/4 The quaintly-named 'Froth Blowers'..are a little group of twenty-four working men. **1927** *Times* 25 June 9/7 A company limited by guarantee under the title of 'Ye Ancient Order of Froth-Blowers, Limited', has been registered to take over all or part of the property and liabilities of the incorporated Ancient Order of Froth-Blowers. **1928** *Times* 8 Feb. 5/3 In 1924 Mr. Herbert Temple..desired to benefit certain charities. He invented the name 'Ancient Order of Frothblowers'. **1928** 'Sapper' *Female of Species* xvii, Now then, boys, once again—and all together—Froth Blowers for ever. **1931** E. Waugh *Remote People* i. 12 He had Froth Blowers' cuff-links, and a Rotarian wheel in his buttonhole.

frothy, *a.* Add: **2. c.** Used to describe very light, tenuous dress-material. So **fro·thiness.**
1901 *Westm. Gaz.* 10 Jan. 3/2 In evening frocks and tea-gowns there is still a very marked liking for frilled frothy edges. **1930** *Times* 13 Mar. 11/6 The outstanding feature is the frothiness of the dresses made from perishable materials with their innumerable frills, flounces, and furbelows.

‖ **frottage** (frǫtā·ʒ). [a. F. *frottage* rubbing, friction, f. *frotter* to rub (see Frot *v.*).]
1. *Psychiatry.* (See quot. 1933.)
1933 H. Ellis *Psychol. Sex* ii. 37 The special perversion of frottage..consists in a desire to bring the clothed body, and usually though not exclusively the genital region, into close contact with the clothed body of a woman. **1939** G. R. Scott *Encycl. Sex* s.v., Like fetichism, too, frottage is evidently a morbid development of the normal sexual excitatory effects of touching or contact with the opposite sex. **1955** H. T. Moore in D. H. Lawrence *Sex, Lit. & Censorship* 27 The entire day's catalogue of voyeurism, frottage, and various other kinds of aberrant gratification.
2. *Art.* The technique or process of taking a rubbing from an uneven surface, e.g. grained wood or sacking, as a point of departure for a work of art.
1935 [see *collage]. **1959** H. Read *Conc. Hist. Mod. Painting* iv. 138 Max Ernst..also invented the 'frottage'— that is to say, designs composed of 'rubbings' of various rough surfaces. **1961** *Spectator* 29 Sept. 428 Evolved his own special techniques of frottage (a process not unlike brass-rubbing) and collage. **1967** G. H. Hamilton *Painting & Sculpture in Europe* vii. 265 The discovery of.. *frottage* (rubbing), occurred on the rainy evening of 10 August 1925 in a room by the sea.

‖ **frottola** (frǫ·tǫlă). *Mus.* Pl. **frottole.** [It.] A form of Italian popular song common esp. in the 15th and 16th centuries.
1854 R. Brown tr. *Giustiniano's Despatches* I. 83 The..word..'Frottola'..is used by the novellist Sachetti, ..and applied to certain compositions of his in verse... The word is now merely employed to signify a comic song. **1875** A. W. Ward *Hist. Eng. Dram. Lit.* I. ii. 128 The *frottola* (literally a comic ditty) marks a step in advance. Here types take the place of abstractions. **1900** *Grove's Dict. Mus.* I. 566/1 The words of the Frottole were often comic (in fact the word is a synonym for a joke) but still oftener extremely sentimental. **1960** *Times* 2 Sept. 13/6 Second came the more complete Hispanization of the *villancio* (the equivalent of the Italian frottola). **1970** *Computers & Humanities* V. 109 The thematic indices of printed sources of the frottola and the Palestrina repertory.

‖ **frotton** (frǫtoṅ). [Fr., f. *frotter* to rub. The termination *-on* denotes a tool or implement.] In early block-printing, a dabber or burnisher used to press the damp sheet upon the inked block.
1876 T. L. De Vinne *Invention of Printing* 83 How were these images printed? Almost every author who has written on printing has said that they were printed by friction, with a tool known as the frotton, which has been described as a small cushion of cloth stuffed with wool. **1893** E. G. Duff *Early Printed Bks.* 17 It consists of three sheets of paper, each of which contains an impression from a block... They are printed by means of the frotton in light-coloured ink. **1911** *Encycl. Brit.* XXVII. 511/2 Then a sheet of damp paper was laid upon it [*sc.* the block], and the back of the paper was carefully rubbed with some kind of dabber or burnisher, usually called a *frotton*, till an impression from the ridges of the carved block had been transferred to the paper. **1960** G. A. Glaister *Gloss. Bk.* 149 *Frotton printing*, a method of taking an impression from type by rubbing the verso of a leaf laid upon it. This method was used before the invention of the press.

frou-frou, *v.* [f. the sb.] *intr.* To move about with a rustle of draperies. Only in *frou-frouing.*
1905 *Truth* 18 May 1289/2 Frou-frouing femininities. **1909** M. B. Saunders *Litany Lane* II. xvii, She found herself floating and frou-frouing up Majorson's monastic stairs. **1958** A. Wilson *Middle Age of Mrs. Eliot* II. 166 Up those stairs should have come frou-frouing skirts.

frowst (fraust), *sb. colloq.* Also **froust.** [Back-formation from Frowsty *a.*]
1. At Harrow School: see quots.
1880 *Hugh Russell at Harrow* 12 Can't you let a fellow take his frowst the very second morning? *Ibid.*, Gloss., *Frowst*, extra time in bed on Sundays, saints' days and whole holidays. **1905** H. A. Vachell *Hill* v. 113 Lying in bed in the morning when there is no First School is a 'frowst'. By a subtle law of association, an armchair is also a 'frowst'. **1923** Galsworthy *Captures* 60 Ruding.. looked right down on me in my old 'froust', as we called arm-chairs.
2. The close and fusty air of a room, etc., which is over-warm or over-crowded and without adequate ventilation.
1907 'J. Halsham' *Lonewood Corner* 130 For the manufacture of froust trust the elementary schools! **1921** *Blackw. Mag.* Apr. 453/1 Poor Colin..had to listen in candle-lit frowst [in a dug-out] to the banalities talked by the Brigadier. **1929** D. Coke *Monkey Tree* vii, He liked the cheap room and its genial frowst. **1946** H. Croome *Faithless Mirror* 70 Inside, a magnificently fruity frowst greeted them, strongly flavoured with wood smoke.

frowst, *v.* Substitute for def.: To take pleasure in staying in a warm, close, 'fuggy' atmosphere. Hence **fro·wster.**
1917 'I. Hay' *Carrying On* iii. 68 In winter it is much warmer below the earth than upon its surface, and Thomas Atkins is the most confirmed 'frowster' in the world. **1929** 'A. Berkeley' *Wychford Poisoning Case* xxiii. 280 'Hallo, you frowsters!' cried Sheila, bursting without warning into the room.

frowstiness (frau·stinės). [f. Frowsty *a.* + -ness.] Frowsty quality; fustiness, stuffiness.
1923 *Daily Mail* 21 June 10 To clear his lungs of their over-night 'frowstiness'. **1926** W. J. Locke *Old Bridge* iii. ix, The old frowstiness of dirt had given place to the fragrance of cleanliness. **1960** *Guardian* 24 Feb. 12/6 The froustiness of last night's supper at breakfast-time.

frowsty, *a.* Add: (Later non-dialectal examples.) Also in *Comb.*
1922 D. H. Lawrence *Aaron's Rod* xiv. 195 A frowsty-looking man. **1924** F. Hurst *Lummox* § 3. 10 She was kind enough in a frowsty way. **1927** *Manch. Guardian Weekly* 28 Jan. 75/2 The frousty, self-conscious hedonism of the swagger music-hall. **1958** P. Kemp *No Colours or Crest* iii. 25 Neither daylight nor fresh air could penetrate to lighten the frowsty atmosphere of our dormitories.

frowzled (frau·z'ld), *ppl. a.* Rumpled, tousled, dishevelled, frowzy. Also **frowzly** (frau·zli) *a.* (in quots. *frowsly*), in similar sense.
1808 M. Wilmot *Jrnl.* 20 Sept. (1934) III. 376 [His] enthusiasm caught fire & made him..rub his frouzled Crop till every hair stood erect. **1872** S. Hale *Lett.* (1919) 111 Both these Fräuleins had short frowsly hair. **1901** *Harper's Mag.* CII. 665/1 Look at the poor thing's hair! Only see how frowsly it is.

frozen, *ppl. a.* Add: **1.** *spec.* Of food: preserved by refrigeration to below freezing point.
1872 [in Dict.]. **1891**, etc. [see *chilled *ppl. a.* 2 b]. **1933** *Discovery* Apr. 127 The authors described some new experiments designed to retain the full flavour and colour of frozen fruit. **1950** *Archit. Rev.* CVIII. 142/3 A 'Kelvinator' dual freeze unit—i.e., for ordinary refrigeration and frozen food storage. **1957** *Daily Mail* 5 Sept. 9/4 Frozen-food producers came to the rescue with frozen fried fish and now the latest, frozen chips. **1967** *Listener* 20 Apr. 533/3 The same frozen scampi and vegetables.
b. *the frozen limit* (colloq.): the hard and fast limit; the *ne plus ultra* of what is objectionable or unendurable. See *limit *sb.*
1917 W. H. L. Watson *Adv. Despatch Rider* x. 216, I don't mind their machine-guns, but their Minnenwerfer are the frozen limit! **1920** 'Sapper' *Bull-Dog Drummond* i. § 1, Only his eyes redeemed his face from being what is known in the vernacular as the Frozen Limit.
c. *Billiards.* Used to designate a ball at rest in close contact with another ball or a cushion.
1904 S. A. Mussabini *Mannock's Billiards* II. 275 When the object is frozen to the side cushion. *Ibid.* 287 Here we have the red ball 'frozen' or 'tight up' against an end cushion. *Ibid.* 293 A ball 'frozen' on a cushion. **1961** J. S. Salak *Dict. Amer. Sports* 188 When object balls are frozen they remain in play as they are.
d. Of credits, assets, etc.: impossible to liquidate or realize at maturity or some other given time. Also *fig.* (Opp. Liquid *a.* 6; cf. *freeze *v.* 5 e.)
1922 *Daily Mail* 16 Dec. 9 Germany has an immense quantity of 'frozen credits' locked up in this country. **1923** *Ibid.* 9 Jan. 7 A large part of British assets..were temporarily 'frozen'. **1929** *Encycl. Brit.* IX. 876/2 *Frozen credit*, credit (generally bank-loans) which has been extended but which the creditors find it impossible or highly inexpedient to collect at maturity or at any given time... In such a case the bank will often renew or extend the loan, and this credit is said to be *frozen*, a term signifying the opposite of 'liquid' as applied to credit, capital or assets. **1930** *Times* 24 Mar. 23/5 The opportunity to accomplish this liquidation of over $10,000,000 of frozen assets. **1937** Wodehouse *Ld. Emsworth & Others* iv. 158 Angus McTavish was the sort of man who, just by going about looking like a frozen asset, takes all the edge and zip out of a girl's game. **1959** *Daily Tel.* 2 May 6/2 Property sequestrated or, to use the current jargon, 'frozen'.
3. *frozen-faced* adj.
1921 S. Adams *Success* vii. 212 Old frozen-faced Willis Enderby. **1925** T. Dreiser *Amer. Trag.* (1926) II. III. ix. 150 His frozen-faced terror. **1964** *Daily Tel.* 6 Jan. 1/2 Even the most frozen-faced Cardinals in Vatican City were taken by surprise in viewing..the overpowering.. enthusiasm the Pope encountered.

fructan (frɒ·ktæn). *Chem.* [f. Fruct(ose + *-an*, as in *glucosan.*] Any of a group of polysaccharides that are composed of fructose residues and occur widely in plants, esp. as reserve foods; a fructosan.
1953 Whistler & Smart *Polysaccharide Chem.* i. 17 The polysaccharides termed fructosans in the literature are hence incorrectly named. The preferred name fructans is employed in this book. **1957** Whistler & Corbett in W. Pigman *Carbohydrates* xii. 683 Dahlia tubers and the tubers of the Jerusalem artichoke are rich in fructans. **1964** D. D. Davies et al. *Plant Biochem.* iii. 145 Most fructans also contain very small amounts of glucose.

fructivorous, *a.* (Later examples.)
1947 J. Stevenson-Hamilton *Wild Life S. Afr.* xxxi. 261 Being already both fructivorous and insectivorous, the species [*sc.* the baboon] may be at the early beginning of becoming carnivorous also. **1970** *Watsonia* VIII. 67 Red berries for fructivorous birds.

fructolysis (frʊktǫ·lɪsɪs). *Biochem.* [f. FRUCTO(SE + LYSIS.] (See quot. 1943.)
1932 DICKENS & GREVILLE in *Biochem. Jrnl.* XXVI. 1252 For the sake of brevity, we shall use the following terms. *Glycolysis*. General expression denoting the breakdown of carbohydrate to lactic acid.. *Glucolysis*. The glycolysis in the presence of added glucose. *Fructolysis*. The glycolysis in the presence of added fructose. 1943 *Thorpe's Dict. Appl. Chem.* (ed. 4) VI. 74/1 'Glucolysis' has been the name given to the phenomenon of glucose breakdown to lactic acid, and 'fructolysis' has been applied when the sugar in question has been fructose.

fructosan (frʊ·ktosæn). *Chem.* [f. FRUCTOS(E + -an, as in *glucosan*.] = *FRUCTAN.
1928 *Chem. Abstr.* XXII. 4480, 9 g. of a yellow trifructosan..is obtained. *Ibid.* 4481 Heating inulin for 6 hrs. at 140° produces 51% of a difructosan. *Ibid.*, A monofructosan. *Ibid.* 5849 (Index), Fructosans, and acetates. 1931 E. C. MILLER *Plant Physiol.* viii. 412 The hexosans may be further classified according to the type of hexose that they yield on hydrolysis into the dextrosans (glucosans), the levulosans (fructosans), the mannans, and the galactans. 1940 *New Phytologist* XXXIX. 185 While the term fructosan should properly apply to the whole group of fructose polymers including inulin, it is frequently used to designate only this second class of polymers found in monocotyledons. 1953 [see *FRUCTAN]. 1966 P. MCDONALD et al. *Anim. Nutrition* ii. 18 Many grasses contain appreciable quantities of fructosans.

frug (frʊg, frūg). [Origin unknown.] A modern dance. Hence as *v. intr.*
1964 in *Amer. Speech* (1965) XL. 143 Discotheque dresses make dancing the frug..and the Watusi a delight because they move with the beat. 1965 *Daily Express* 27 Sept. 3/5 Other guests..were wearing themselves out with the latest dance craze, the Frug. 1965 J. PHILIPS *Twisted People* i. 10 The frug, the Watusi, the surf, the monkey—strange dances in which the partners do not touch, do not talk. 1965 *Punch* 22 Dec. 930/2 They were frugging around to a new Who waxing. 1967 *Spectator* 29 Sept. 350/1 Mr. George Brown went to New York to dance the frug and tell the United Nations about Vietnam. 1968 J. C. HOLMES *Nothing More to Declare* 11 Energetic kids, with their guitars and placards, who frug and demonstrate so indefatigably everywhere these days. 1968 J. UPDIKE *Couples* (1969) iv. 365 Bernadette..heard a frug record put on the phonograph and held wide her arms. 1971 *Courier Mail* (Brisbane) 1 Feb. 16/8 They frugged and danced.

fruit, *sb.* Add: **2. e.** *old (tin of) fruit*: a term of familiar address. (Cf. *old bean*.) *slang.*
1923 *Daily Mail* 8 Feb. 6 Was she simply bursting to address him Gaily as her 'dear old tin of fruit'? 1928 *Ibid.* 25 July 10/6 Then their politeness. No slapping a friend on the back with a 'What cheer, old fruit?' 1951 T. RATTIGAN *Who is Sylvia?* i. 212 You don't mind me asking, did you, old fruit? 1968 K. BIRD *Smash Glass Image* xiii. 162 *Adiós*, old fruit. *Hasta luego.* Go and jump in the nearest lake.

7*. a. A dupe, an 'easy mark'. **b.** A male homosexual. *slang* (orig. *U.S.*).
1895 W. C. GORE in *Inlander* Dec. 111 *Fruit*, one who can be easily deceived. 1913 *Punch* 22 Jan. 72/2 It was a flaw in the new play that its mugs were such 'easy fruit'. 1931 G. IRWIN *Amer. Tramp & Underworld Slang* 81 *Fruit*, an 'easy mark'. A girl or woman willing to oblige. Probably..from the fact that they are 'easy picking'. 1935 N. ERSINE *Underworld & Prison Slang* 38 *Fruit*, a sexual pervert. 1957 K. MARTIN *Aubade* v. 79 The way I'm acting anyone would think I was a fruit. Gary probably is. He looks like one. 1970 *Guardian* 13 Feb. 9/5 He is a fruit, which means..that he is a queer. 1971 *Rolling Stone* 24 June 3/2 John Mendelsohn did an excellent job acting like a fruit.

8. a. *fruit-farm, -juice, -lot, -pulp, -year.* **b.** *fruit-farmer, -farming, -grower* (earlier examples), *-growing* sb. and ppl. adj. (earlier examples), *-raising*.
1872 *Trans. Dep. Agric. Illinois* IX. 65 The first consideration in the establishment of a fruit farm is accessibility to market. 1921 E. M. CLOWES *On Wallaby* iv. 92 Some neighbouring station, dairy, or fruit-farm. 1872 *Trans. Dep. Agric. Illinois* IX. 66 The fruit farmer can raise cheap pork in his apple and peach orchards. 1890 W. BOOTH *In Darkest Eng.* II. iii. 133 Fruit farming affords a great opening for female labour. 1825 G. BLISS (*title*) The fruit grower's instructor. 1854 *Trans. Illinois Agric. Soc.* I. 70 The fruit and fruit growers of seven states were represented. 1850 *Rep. U.S. Comm. Patents Agric.* 1849 106 This perhaps is the best fruit-growing district of the State. 1854 *Trans. Illinois Agric. Soc.* I. 135 Fruit growing..has not increased any faster than the growing of other agricultural produce. 1872 *Trans. Dep. Agric. Illinois* IX. 66 A soil adapted to the growth of forest trees is..the best for fruit growing. 1905 J. F. FRASER *Canada* iii. 32 This jut of land is as rich for fruit growing as Kent itself. 1880 *Jrnl. Chem. Soc.* XXXVIII. 354 Behaviour of fruit-juices of different ages with reagents. 1951 AUDEN *Nones* (1952) 64 The unamerican survivor Hears angels drinking fruit-juice with their wives. 1912 *Chambers's Jrnl.* Mar. 173/2 It is very beautiful up behind the fruit-lot among the rocks and the pine-trees. 1887 *Colon. & Indian Exhib., Rep. Col. Sect.* 131 Importing a large quantity of fresh fruit, and what is called 'fruit-pulp' from Tasmania. 1887 C. A. MOLONEY *Sk. Forestry W. Afr.* 339 The fruit-pulp is eaten and also prepared into a pleasant beverage. 1906 *Westm. Gaz.* 28 Aug. 6/1 To avail themselves of any cheap and defective fruit-pulp for the making of jam. 1854 *Trans. Illinois Agric. Soc.* I. 500 The fact that fruit raising here is attended by drawbacks. 1891 *Harper's Mag.* Jan. 168/2 California has much to learn about fruit-raising. 1742 W. ELLIS *Timber-Tree* (ed. 3) ii. xl. 192 When they sell well, as they do in plentiful Fruit-years. 1811 R. SUTCLIFF *Trav. N. Amer.* ii. 27 This was likely to be a very abundant fruit year.

9. fruit bark beetle = *fruit tree (bark) beetle; **fruit bat**, (read:) a member of the suborder Megachiroptera, which includes the flying-foxes (*Pteropus*) and other fruit-eating bats; add earlier and later examples; **fruit-body**, the part of a fungus that bears the spores and spore-producing organs; **fruit-cake**, (*a*) (examples); **fruit cocktail**, a preparation of fruit used as an appetizer or refreshment; **fruit cup**, a preparation of fruit used as an appetizer or dessert; **fruit drop** [DROP *sb.* 10 e], a fruit-flavoured sweet; **fruit-gatherer**, an implement for gathering fruit from tall trees; **fruit gum** [GUM *sb.* 1 h], a fruit-flavoured gum; **fruit-jelly**, a fruit-flavoured table-jelly; **fruit machine**, a coin- or token-operated gaming machine which pays out according to the combination of symbols (often representations of fruit) appearing on the edges of wheels spun by the operation of a lever; also *transf.* in various *slang* uses; **fruit salad**, (*a*) fruits, usually uncooked, cut up and mixed together, often served with sugar, cream, etc.; (*b*) an array of service ribbons and decorations (*Services' slang*); **fruit salts**, effervescent health salts (also formerly in *sing.*); **fruit tree (bark) beetle**, *Scolytus rugulosus*, which burrows beneath the bark of fruit-trees; **fruitwood**, the wood of fruit-trees, esp. pear, used to make furniture.
1892 *Insect Life* IV. 293 The topics treated are the Fruit Bark-beetle (*Scolytus rugulosus*); [etc.]. 1951 *Dict. Gardening* (R. Hort. Soc.) II. 843/1 Fruit-bark Beetle, *Scolytus rugulosus*, attacks a number of fruit trees, especially Apple and Plum. 1877 W. S. DALLAS in *Cassell's Nat. Hist.* I. 169 The young African Fruit Bats born in the Zoological Gardens were covered with short, smooth hair. 1905 *Westm. Gaz.* 8 Nov. 10/2 The Indian fruit bat..is not a novelty in the menagerie. 1936 *Discovery* Oct. 307/1 Before the advent of the white man and his cultivated fruits, these great fruit bats lived on scrub figs, berries, ti-tree and gum blossoms for the honey they contained. 1966 V. SERVENTY *Continent in Danger* iv. 86 Some of the 'batteries' or camps of the fruit bats..number hundreds of thousands of individuals. 1912 C. THOM in C. E. Marshall *Microbiol.* i. 13 The systems of classification used are largely based upon the types of sexual fruit bodies produced. 1927 GWYNNE-VAUGHAN & BARNES *Struct. & Develop. Fungi* 1 In relation to the fruit bodies of higher forms, they [sc. the hyphae] become woven into a dense mass. 1968 *Gloss. Terms Timber Preservation (B.S.I.)* 10 *Fruit(ing) body*, in wood-destroying fungi, a structure that bears the spore-producing organs and spores, commonly a conspicuous bracket, toadstool or compacted sheet, with pores, gills, spines, etc., bearing the spores. 1969 *New Scientist* 27 Nov. 451/1 Only when fruitbodies are formed does the maximum conversion of compost nutrients into edible food take place. 1854 E. RUSKIN *Let.* 27 Feb. in M. Lutyens *Millais & Ruskins* (1967) 143 At night Old Anne gets out wine and fruit cake and they sit and gossip. 1967 J. POTTER *Foul Play* xiii. 153 Through a mouthful of fruit cake he congratulated her crumbily. 1922 *N.Y. Hotel Rev.* 18 Mar. 62 Fruit cocktail. 1928 *Sat. Even. Post* 12 May 107/1 Mr. Montgomery had taken a morsel of fruit cocktail. 1938 *Fortnum & Mason Catal.* 32/1 Fruits in syrup.. Fruit Cocktail..per tin 1/8. 1968 C. DRUMMOND *Death & Leaping Ladies* i. 22, I ended up with eighteen fruit cocktails and..was left with eleven sure-fire hits. 1931 *N. & Q.* 4 Apr. 241/2 Where fruit-cup, ices and wonderful cakes were served. 1960 N. MAILER *Advts. for Myself* (1961) 358 You Americans always eat the last course first..Fruit cup with sherbet, if you please. 1907 *Yesterday's Shopping* (1969) 50 Acidulated fruit drops. 1935 G. GREENE *Basement Room* 40 A sticky fruit drop in his hand. 1959 I. & P. OPIE *Lore & Lang. Schoolch.* ix. 166 'Lollies' is becoming a general term..for..humbugs, large aniseed balls, and fruit drops. 1971 *Guardian* 18 Aug. 4/3 Fruit-drops, lollipops, a stick of chewing-gum. 1847 *Rep. U.S. Comm. Patents* 1846 19 A fruit-gatherer, of very ingenious and simple construction, has been patented. 1910 *Daily Chron.* 10 Feb. 1/5 A fruit-gatherer attached to a pole and armed with cutting teeth. 1938 G. GREENE *Brighton Rock* vii. vii. 328 The packets of fruit gums came dropping out. 1960 E. W. HILDICK *Boy at Window* xiv. 108, I asked him for a box and he simply tipped the rolls of fruit gum out of this. 1971 *Times Educ. Suppl.* 25 June 20/3 A skyver..offered the teacher a fruit gum. 1846 A. SOYER *Gastronomic Regenerator* p. xxi, All fruit jellies [should be] as near as possible to the colour of the fruit. 1875 *Encycl. Brit.* I. 172/1 Jams, marmalades, and fruit jellies. 1933 *Times* 7 Apr. 4/2 Committed to trial..on a charge of receiving 20 automatic 'fruit' machines..knowing them to be stolen. 1943 HUNT & PRINGLE *Service Slang* 33 *Fruit machine*, an anti-aircraft predictor. 1957 *Economist* 30 Nov. 781/2 Permission to install the minor gambling devices known as 'fruit machines'..is by county option. 1959 G. JENKINS *Twist of Sand* vii. 110 The 'fruit machine' fed by information from two officers, gave the course and speed of the warships. 1965 *Listener* 2 Sept. 342/3 There are three juke boxes; pin tables; fruit machines. 1861 MRS. BEETON *Bk. Househ. Managem.* 798 Fruit salads are made by stripping the fruit from the stalks, piling it on a dish, and sprinkling over it finely-pounded sugar. 1943 HUNT & PRINGLE *Service Slang* 33 *Fruit salad*, ..large collection of medal ribbons which runs to three or more rows. 1955 'N. SHUTE' *Requiem for Wren* iii. 74 A red-faced old gentleman with..a fruit salad of medal ribbons on his chest. 1961 *Times* 19 Apr. 13/3 The ribbons worn on the chest and colloquially called 'fruit salad'. 1964 B. FALK *Peacock Cookery Bk.* 136 A fruit salad can be made from any mixture of fresh fruit in season together with some tinned fruit, or from fresh fruit alone. 1889 *Illustr. London News* 31 July 33 (Advt.), For health and longevity, use Eno's 'Fruit Salt'. 1892 E. LYTTON *Let.* 18 Nov. in E. Lutyens *Blessed Girl* (1953) ix. 179 When he comes down cured, Sir Augustus puts it down to Eno's Fruit Salts. 1939 D. L. SAYERS *In Teeth of Evidence* 64 Mr. Loveday had gone in to borrow a dose of fruit salts from Mr. Turnbull. 1898 E. A. ORMEROD *Handbk. Insects Injur. Fruits* 197 (*heading*) Fruit-tree Bark Beetle. 1921 T. W. SANDERS *Fruit Foes* i. 30 Fruit Tree Beetle (Scolytus rugulosus).—Both the beetle and the larvæ are injurious to fruit trees. 1927 MACQUOID & EDWARDS *Dict. Eng. Furnit.* III. 29 In common with other fruit woods, it [sc. pear wood] has been used from a very early period. 1939 E. WENHAM *Old Furnit. for Mod. Rooms* i. 7 Bobbin-turned chairs of oak, or of some fruitwood. 1947 J. C. S. BROUGH *Timbers for Woodwork* xvi. 134 One of several 'fruitwoods' it [sc. apple wood] was in former days employed largely for domestic commodities. 1964 J. GLOAG *Englishman's Chair* ix. 194 Elm seats, ash or yew bows, with arms and turned work in fruit wood or yew. 1971 *Times* 19 June 12/5 The marquetry of flowers, birds and grotesque masks is in ivory and fruitwood.

fruitarian. Add: Also as *adj.*
1902 H. BEGBIE *Adv. Sir John Sparrow* x. 150 The form of eating which I practise..is not vegetarian, but fruitarian. 1903 *Sci. Amer.* 10 Oct. 255/2 Fruits contain little protein, and nuts are relied on in the fruitarian plan of eating to balance the ration.
Hence **fruita·rianism**, the principles or practice of fruitarians.
1902 H. BEGBIE *Adv. Sir John Sparrow* xiv. 221 We flee from before the face of vegetarianism, fruitarianism, theosophy, and religious manias. 1908 *Practitioner* Mar. 401 Vegetarianism, fruitarianism, and even zomotherapy may suit a few. 1930 E. T. THURSTON *Man in Black Hat* viii. 149 The medical profession at the moment approves of fruitarianism.

fruiting, *ppl. a.* Add: **fruiting body** = *fruit-body; also occas. applied to spore-producing bacteria.
1918 HAWLEY & HAWES *Man. Forestry Northeast. U.S.* vii. 125 These fruiting bodies are the only part of the fungus visible to the naked eye. 1944 *Proc. Soc. Exp. Biol. & Med.* LVI. 18 The production of fruiting bodies in aqueous media is a new observation. 1962 W. P. K. FINDLAY *Preservation of Timber* ii. 16 A fruiting body of one of the larger bracket fungi may produce over eight hundred million spores per hour.

fruition. ¶ Now a standard usage, esp. *transf.* and *fig.* (Later examples.)
1936 G. B. STERN *Monogram* III. 202 The words you have written for publication..should be a fastidiously selected portion of your mind and experience which has slowly grown to fruition and importance. 1958 *Times Lit. Suppl.* 8 Aug. 447/3 Sir Edmund Chambers's monumental labours on Shakespeare and the Shakespearian stage were based largely on the resources of the London Library, without which they could never have been brought to fruition, even if they had been attempted. 1959 *Times* 7 July 3/6 This process..has now reached full fruition with the standardization of body shells for a whole range of models (of motor cars). 1968 *Times* 28 Nov. 14/1 A project for reviving the undiscovered burial chambers thought to exist within the pyramid of Khephren at Giza, Cairo, is shortly to come to fruition.

fruity, *a.* Add: **2.** (Earlier example.)
1844 DICKENS *Mart. Chuz.* xii. 151 Whether he would wish to try a fruity port with greater body.
3. *colloq.* Full of rich or strong quality; highly interesting, attractive, or suggestive. Cf. JUICY *a.* 2, SPICY *a.* 7.
1900 T. HOPKINS *Silent Gate* ii. 45 When pulled up short, his language was of the Dials, fruity. 1915 T. BURKE *Nights in Town* 337 A popular murder, fruity, cleverly done, and with a sex interest. 1921 WODEHOUSE *Indiscretions of Archie* 299 It's here now. The dickens of a fruity picture. 1925 *Weekly Westminster* 19 Sept. 522/3 Mr. John Garside's Young Launcelot is more intelligent, even if it is like the work of a fruity comedian without his fruitiness. 1928 S. VINES *Humours Unreconciled* xv. 201 An unusually 'fruity' political scandal connected with bribery. 1938 G. HEYER *Blunt Instrument* xi. 204 It might strike him as a pretty fruity idea to do in his victims as clumsily as he could. 1958 WODEHOUSE *Cocktail Time* xviii. 151 Some minutes later, a fruity voice caressed his ear. Albert Peasemarch's mentor, Coggs, had advised making the telephone-answering voice as fruity as possible in the tradition of the great butlers of the past. 1959 *Oxf. Mag.* 4 June 452/1 Angela Fedlar would have been wonderful if she could have been twice as fruity and three times as loud; she sounded exciting but far away. 1969 *Country Life* 25 Dec. 1697 A design as robust as it is ostentatious; Renaissance at its fruitiest, Elizabethan at its most exuberant, are gaily mixed together.

frumpiness (frʊ·mpinés). [f. FRUMPY *a.* + -NESS.] The quality or condition of a frump.
1912 C. N. & A. M. WILLIAMSON *Heather Moon* i. v. 59 Aline tried to think that she was the weirdest frump in the world... The thing was to hurry her away in all her frumpiness. 1924 *Sunday at Home* Feb. 258/1, I am going to take you up and save you from frumpiness and spinsterhood.

frumpishly (frʊ·mpɪʃli), *adv.* [-LY².] In a frumpish manner; like a frump or dowdy.
1927 *Daily Tel.* 23 Aug. 8/5 The middle-aged matron refuses definitely to take a back seat. She does not feel a frump. Why, then, should she dress frumpishly?

frustrated, *ppl. a.* Add: **1.** *spec.* Of persons. (Later examples.)

1943 J. S. HUXLEY *Evol. Ethics* iii. 22 The individual if not grossly frustrated or oppressed, can adjust himself..to the ethical standards of his society. 1957 *Times Lit. Suppl.* 1 Nov. 651/1 A whole class of discontented and frustrated young men who, after returning from the war, found inadequate scope for their energies.
2. Of goods, etc.: delayed or prevented from being dispatched.
1941 *Times* (Weekly ed.) 30 July 17/3 We have recently had the unsatisfactory business of 'frustrated exports'.. Exports to several South American countries were 'frustrated' by official decree. 1952 *Economist* 31 May 612/1 The 'frustrated exports' that are already individually licensed for sale at home.

fruticetum (frūtɪsī·təm). [a. L. *fruticetum* a place full of shrubs or bushes, f. *frutex* shrub, bush.] A collection of shrubs; cf. ARBORETUM.
1838 J. C. LOUDON *Arboretum et Fruticetum Britannicum* I. p. xiv/2 (*summary*) Planters of British arboretums and fruticetums since the commencement of the 19th century. 1936 *Discovery* Mar. 86/1 There is no separate arboretum, fruticetum, and herbaceous garden.

fry, *v.*¹ Add: **1.** With *up*: to 'hot up' (cold food) in a frying-pan.
1899 G. B. BURGIN *Bread of Tears* i. v, 'What are you cooking there, Fenella? It smells good.' 'Frying up the cold fowl,' said Fenella.
2. c. To execute in the electric chair; also *intr.*, to be executed thus. *U.S. slang.*
1929 *Flynn's* 17 Aug. 702/2 I'll fry for it, I suppose—that's the law, Doc. 1934 A. MERRITT *Burn Witch Burn!* v. 68 They'll laugh themselves sick an' fry us at Sing Sing. 1956 'J. WYNDHAM' *Seeds of Time* 107 You'll hang or you'll fry, every one of you.

fryer. Add: **3.** A chicken suitable for frying. *U.S.*
1923 *Dialect Notes* V. 207 *Frier*, a chicken of frying size. 1940 C. MCCULLERS *Heart is Lonely Hunter* (1943) I. iii. 41 They haves [*sic*] a mule and..from twenty to twenty-five laying hens and fryers. 1957 J. STEINBECK *Short Reign of Pippin IV* 67 H. W...was selling eggs and fryers to the Army and Navy.

frying, *vbl. sb.* Add: **2. frying-basket**, a vessel of metal basket-work used for frying fish, etc.
1877 E. S. DALLAS *Kettner's Bk. Table* 490 Put them [sc. whitebait] into a wire frying-basket. 1951 *Good Housek. Home Encycl.* 393/2 Put the balls into a frying basket and fry in deep fat.

fry·-pan. [f. FRY *v.*¹] = FRYING-PAN.
1832 *Chambers's Edin. Jrnl.* I. 130/1 To the ropes..are attached fry-pans, children's toys, and other light articles. a 1862 THOREAU *Maine Woods* (1864) 323 Three tin plates, a fry-pan. 1918 H. BINDLOSS *Agatha's Fortune* xxi. 192 Two fry pans stood upon the logs. 1959 *Sunday Times* 30 Aug. 14/4 Sunbeam are bringing out a new deep lid for their plug-in frypan. 1963 *Amer. Speech* XXXVIII. 270 The term *fry pan* rarely occurs before the 1950s. When it does, it is often as the *double fry* or *omelette pan*.. But the advent of the electric fryers marks a revival of *fry pans*. 1970 *Toronto Daily Star* 24 Sept. 8/3 (Advt.), Toastess electric frypan.

fry-up (frəɪ·ʊp). [f. *to fry up* (see *FRY *v.*¹ 1).] An easily prepared and quickly cooked dish of fried food, esp. of cold food heated up in a frying-pan; also, the preparation of such a dish.
1967 'M. ERSKINE' *Case with Three Husbands* vii. 96 The two men decided finally on a mixed grill..and a fry-up of cold potatoes. 1968 'E. TREVOR' *Place for Wicked* xix. 258 Are you hungry?..We'll have a fry-up, shall we? 1969 J. WAINWRIGHT *Take-Over Men* xii. 203 Then lunch. More often than not a 'fry-up'—I became a dab hand with a frying-pan.

fuchsia. Add: **d.** A red colour like that of the fuchsia flower, fuchsia-red.
1923 *Daily Mail* 5 Mar. 13 Colours: Peach, Apple, Apricot, Mauve, Fuchsia, Periwinkle. 1930 *Times* 29 Nov. 16 A colour scheme of fuchsia and gold predominates. 1963 *New Yorker* 29 June 64/1 (Advt.), Mandarin Coat—or Dress..Red, Black, Fuchsia.

fuchsinophil, -phile (fūk-, fiūksi·nǫfil, -fəil), *a. Biol.* [f. FUCHSIN(E + -O + -PHIL, -PHILE.] Readily stained with fuchsine; produced by staining with fuchsine. Also **fu:chsinophi·lic** *a.*, in the same sense. Hence **fu:chsinophi·lia**, affinity for fuchsine.
1900 DORLAND *Med. Dict.* 268/1 Fuchsinophil. 1904 *Jrnl. R. Microsc. Soc.* 639 (*heading*) The fuchsinophile granules of spinal ganglia cells. 1910 H. W. ARMIT tr. *Ehrlich & Lazarus' Anæmia* I. i. iii. 102 The fact that these cells cannot possibly be lymphocytes is proved by the complete absence of fuchsinophile granules. 1931 *Amer. Jrnl. Anat.* XLVIII. 165 Many of the fuchsinophilic or eosinophilic necrotic cells have disappeared. 1936 H. D. ROLLESTON *Endocrine Organs* vi. 310 In female foetuses this fuchsinophil reaction did not last so long. *Ibid.*, The suppression of the fuchsinophil stain in female foetuses may be due to inhibition exerted by the pituitary. 1940 *Amer. Jrnl. Anat.* LXVII. 190 After Orth's fixation the fuchsinophilia varies considerably from gland to gland. 1967 *Pathol. & Microbiol.* XXX. 402 A marked fuchsinophilia lasting up to 30 minutes was observed in the heart muscle fibers. 1970 H. SELYE *Exper. Cardiovasc. Dis.* II. iv. 665 Fuchsinophilic degeneration is a pre-necrotic change affecting certain myocardial fibers in damaged areas. *Ibid.*, The fuchsinophilic fibers or fiber segments. *Ibid.*, Fuchsinophilic staining.

fucidin (fiū·sidin). *Pharm.* Also **Fucidine** (proprietary name). [Formed by alteration of *FUSIDI(C *a.* + -INE⁵.] The sodium salt of fusidic acid, used as an oral antibiotic esp. against staphylococci. Also used *loosely* for *fusidic acid.*
1961 *Trade Marks Jrnl.* 30 Aug. 1189/2 Fucidine. 1962 *Lancet* 8 Sept. 478/1 A small controlled trial of the new antibiotic 'Fucidin'..in patients whose burns were colonised by *Staph. aureus.* 1966 DUNLOP & ALSTEAD *Textbk. Med. Treatm.* (ed. 10) 73 The use of fucidin should only be seriously considered for a patient with severe staphylococcal disease who is able to swallow and who is known to be hypersensitive to penicillin. 1968 GARROD & O'GRADY *Antibiotic & Chemotherapy* (ed. 2) xi. 200 Fucidin is active against species of Gram-positive bacteria and the Gram-negative cocci. 1970 *To-day's Drugs* (Brit. Med. Assoc.) 21 It is very rare to find a fucidin-resistant staphylococcus.

fuck (fʊk), *v.* Also 6 **fuk**, 7– **f—k**, etc. [Early mod.E. *fuck*, *fuk*, answering to a ME. type *fuken (wk. vb.) not found; ulterior etym. unknown. Synonymous G. *ficken* cannot be shown to be related.]
For centuries, and still by the great majority, regarded as a taboo-word; until recent times not often recorded in print but frequent in coarse speech.
1. *intr.* To copulate. *trans.* (Rarely used with female subject.) To copulate with; to have sexual connection with.
a 1503 DUNBAR *Poems* lxxv. 13 Be his feiris he wald haue fukkit. 1535 LYNDESAY *Satyre* 1363 Bischops..may fuck thair fill and be vnmaryit. 1535–36 — *Answer to Kingis Flyting* 49 Ay fukkand lyke ane furious Fornicatour. 1598 FLORIO *Worlde of Wordes* 137/1 *Fottere*, to iape, to sard, to fucke, to swive, to occupy. c 1650 in Hales & Furnivall *Percy's Folio MS.* (1867) 99 Which made him to haue a mighty mind To clipp, kisse, & to ffuck. 1680 ROCHESTER *Poems on Several Occasions* (1950) 14 Much Wine had past with grave discourse, Of who Fucks who, and who does worse. c 1684 *Sodom* II. 30 Hee Fucks to please his will, but I for need. 1707 'MADAM B[RAN]LE' *Fifteen Plagues of Maidenhead* 4 But I poor Virgin never shall be F—. a 1750 A. ROBERTSON *Poems* (1750) 256 But she gave proof that she could f—k. c 1800 BURNS *Merry Muses* (1911) 71 You can f—k where'er you please. c 1863 PHILO CUNNUS *Festival of Passions* II. 54 That female nation who being encamped near their enemies were fucked, during a truce, by the hostile party. 1869 ROSSETTI *Let.* 15 Sept. (1965) II. 743 If Byron f—d his sister he f—d her and there an end. c 1888–94 *My Secret Life* IV. 64 Then a dread came over me. I had fucked a common street nymph. 1922 JOYCE *Ulysses* 765 His wife is fucked yes and damn well fucked too. 1928 D. H. LAWRENCE *Lady Chatterley* iv. 44 Fellows with swaying waists fucking little jazz girls. 1967 D. WRIGHT tr. *R. Queneau's Between Blue & Blue* v. 48 'Well, Lamélie,' says Cidrolin, 'while you're waiting to get married, do you want to be entertained or educated?' 'No, Dad, what I want to do is to fuck.' 1971 *Ink* 19 Oct. 15/3, I don't want to fuck anyone, and I don't want to be fucked either.
2. Used profanely in imprecations and exclamations as the coarsest equivalent of DAMN *v.* 5.
1922 JOYCE *Ulysses* 587 God fuck old Bennett! 1929 F. MANNING *Middle Parts of Fortune* II. xv. 379 'Fuck the bloody thing!' he said fiercely under his breath. 1955 S. BECKETT *Molloy* 69 Fuck the son of a bitch. 1959 F. KING *So Hurt & Humiliated* 151 'Suppose any of the neighbours were to look out and see them.' 'Oh, f— the neighbours!' 'Really, Henry!' 1969 'J. MORRIS' *Fever Grass* ii. 24 Why don't you..tell whoever it is to go fuck themselves?
3. Const. with various adverbs: *fuck about*, to fool about, mess about; *fuck off*, to go away, make off; *fuck up*, to ruin, spoil, mess up. Also **fu·ck-up** *sb.*, a mess, muddle.
1929 F. MANNING *Middle Parts of Fortune* I. ii. 31 They kept 'em fuckin' about the camp, while they sent us over the bloody top. *Ibid.* iii. 37 As soon as a bit o' shrapnel comes their way, [they] fuck off 'ome jildy, tout sweet. 1944 *Amer. Speech* XIX. 108 Go on, now, fuck off! 1958 S. BECKETT *Malone Dies* 116 She wants to know if you're the one in charge. Fuck off, said Lemuel. 1958 J. O'HARA *From Terrace* (1959) 257 Such a Goddam fuck-up. 1967 R. SHAW *Man in Glass Booth* (1969) xvii. 114 'Can't be too specific,' answered the old man. 'Don't want to fuck up my case.' 1968 M. RICHLER *Cocksure* iv. 29 I'm sorry about this fuck-up, Mr Griffin. 1968 J. UPDIKE *Couples* iv. 294 This fucks up our party, doesn't it? 1969 *It* 11–24 Apr. 12/3 The..neatly planned plot to fuck up their transport scene.
Hence **fu·cking** *vbl. sb.* Also as *ppl. a.* and *adv.*, used esp. as a mere intensive.
a 1568 A. SCOTT *Poems* iv. 55 Thir foure, the suth to sane, Enforsis thame to fucking. 1680 ROCHESTER *Poems on Several Occasions* (1950) 30 Through all the Town, the common Fucking Post, On whom each Whore, relieves her tingling Cunt. 1707 [see *FRIGGING *vbl. sb.*]. c 1888–94 *My Secret Life* III. 228 This house had but eight rooms, and two mere closets to let out for fucking. *Ibid.* VIII. 307 She was..a magnificent bit of fucking flesh, but nothing more. 1893 FARMER & HENLEY *Slang* III. 80/2 *Fucking*.. Adj., A qualification of extreme contumely. Adv. Intensitive and expletive; a more violent form of *bloody.* 1922 JOYCE *Ulysses* 580 I'll wring the neck of any bugger says a word against my fucking king. 1929 F. MANNING *Middle Parts of Fortune* I. ii. 23 Blown to fuckin' bits as soon as we got out of the trench, poor bugger. 1930 J. DOS PASSOS *42nd Parallel* 94 It was a fucking shame it was Freddy hit you. 1938 DYLAN THOMAS *Let.* 31 Aug. (1966) 208 None using obscene words, none..to do with fucking. 1939 — *Let.* 29 Sept. (1966) 240 I'll give Dent the whole fucking works. 1960 D. LESSING *In Pursuit of English* 12 'What the f—ing hell's that for?' my father said. 1969 AUDEN *City without Walls* 49 I'm so bored with the whole fucking crowd of you I could scream! 1971 *It* 2–16 June 18/3 The Youngbloods..are so fucking good they can do spontaneous albums.

fuck (fʊk), *sb.* [f. prec.] **1.** An act of copulation.
1680 ANON. in *Rochester's Poems on Several Occasions* (1950) 37 Thus was I Rook'd of Twelve substantial Fucks. c 1684 *Sodom* (Epilogue spoken by Fuckadilla) 1. 19 A little fuck can't stay our appetite. c 1800 BURNS *Merry Muses* (1911) 68 When maukin bucks, at early f—ks, In dewy glens are seen, sir. c 1888–94 *My Secret Life* III. 139, I was dying with want of a fuck. 1928 D. H. LAWRENCE *Lady Chatterley* xviii. 342 A lily-livered hound with never a fuck in him. 1965 E. J. HOWARD *After Julius* iii. 38 Eat well, don't smoke, and a fuck was equal to a five-mile walk.
b. *concr.* A person (usu. a woman) considered in sexual terms.
1874 *Lett. fr. Friend in Paris* II. 168, I had always held that dear mamma was the best fuck in the family, and in every way a most desirable and splendid creature. 1969 S. GREENLEE *Spook who sat by Door* ix. 77 An aborted marriage to a favourite fuck. 1969 'J. MORRIS' *Fever Grass* ii. 26 She was a good fuck... She was great in bed.
2. Phr. *not to give* (or *care*) *a fuck*: not to care in the slightest.
1929 F. MANNING *Middle Parts of Fortune* I. v. 87 'They don't care a fuck 'ow us'ns live,' said little Martlow bitterly. 1934 H. MILLER *Tropic of Cancer* (1948) 37 Nobody gives a fuck about her except to use her. 1962 I. MURDOCH *Unofficial Rose* vi. 63 Not that I care a fuck. 1971 *Ink* 12 June 3/2 We don't give a fuck if we have to stand around all day doing nothing.
3. Various other casual, intensive, etc., uses.
1959 W. BURROUGHS *Naked Lunch* 63 How in the fuck should I know? *Ibid.* 96 You may be a tedious old fuck yourself some day. 1966 D. HOLBROOK *Flesh Wounds* 129 Driver speed up. Come on, for fuck's sake. 1968 *Anarchy* Oct. 314 A lad of 13 who has had his hand up for some time trying to attract the chairman's attention says 'Oh, for fuck's sake.' 1969 *Oz* May 13/1 (Advt.), Pete Quesnal, late of St. Nicholas, where the fuck are you. 1970 G. LORD *Marshmallow Pie* xv. 136 What the fuck do you think you're doing?

fu·cker. [f. *FUCK *v.* + -ER¹.] One who fucks. Also in extended use as a general term of abuse.
1598 FLORIO *Worlde of Wordes* 137/1 *Fottitore*, a iaper, a sarder, a swiver, a fucker, an occupier. 1893 FARMER & HENLEY *Slang* III. 80/2 *Fucker*. 1. A lover. 2. A term of endearment, admiration, derision, etc. 1922 JOYCE *Ulysses* 584 I'll wring the bastard fucker's bleeding blasted fucking windpipe! 1928 D. H. LAWRENCE *Lady Chatterley* xviii. 334 I'm not just my lady's fucker, after all. 1929 F. MANNING *Middle Parts of Fortune* II. xii. 268 'Laugh, you silly fuckers!' he cried in vehement rage. 1961 A. WILSON *Old Men at Zoo* v. 276 'We'll get you, you fucker!' Barley was shouting.

fuck-up: see *FUCK *v.* 3.

fuck-wind, *sb.*: see WINDFUCKER.

fucoxanthin (fiū·kozæ·nþin). *Chem.* Also 9 -ine. [f. FUC(US + -O + XANTHINE.] A brown carotenoid pigment, $C_{40}H_{54}O_6$, occurring in and generally characteristic of the brown algæ.
1873 H. C. SORBY in *Proc. R. Soc.* XXI. 461 *Fucoxanthine*. This is the name I propose for the principal colouring-matter of *Fucei* and other olive *Algæ*. 1917 HAAS & HILL *Introd. Chem. Plant Products* (ed. 2) v. 229 Carotin, Xanthophyll, and Fucoxanthin..are known collectively as the Carotinoids. 1949 *New Biology* VII. 90 The masking pigment [of chlorophyll] in the brown algae is called fucoxanthin, and in the cases of the reds phycoerythrin. 1965 T. W. GOODWIN *Chem. & Biochem. Plant Pigments* iv. 133 Three main classes of the division Phaeophyta all characteristically synthesize fucoxanthin.

fu·ddydud. Used occas. as a shortened form of next.
1914 *Dialect Notes* IV. 72 *Fuddydud*.., fussy person. 1952 M. MCCARTHY *Groves of Academe* (1953) ix. 167 'You old fuddydud,' she finally teased him. 1967 *Listener* 2 Nov. 584/2 The Robust Individualist is too often a whimsical fuddydud born out of his time.

fuddy-duddy (fʊ·di,dʊ·di). *slang.* [Origin unknown.] An old-fashioned person; an ineffectual old fogy. Also *attrib.* or as *adj.*
[1899 W. DICKINSON *Cumberland Gloss.* 106/2 *Duddy fuddiel*, a ragged fellow.] 1904 in H. WENTWORTH *Amer. Dial. Dict.* (1944) 236/2. 1907 *Dialect Notes* III. 244 Fuddy-duddy, n. and adj. 1938 D. MOFFAT *Mott Family in France* xvii. 191 Mr. Mott wandered along the Promenade..looking at the damned old fuddy-duddies who cluttered it up. 1940 *Harper's* Dec. 1a A great many people..believe its [American army] officer class is caste-ridden and fuddy-duddy. 1951 'N. SHUTE' *Round Bend* 310 People may call the Sheikh of Khulal an old fuddy-duddy, but he's an important man in these parts. 1957 *Times* 23 Aug. 9/4 What was once called prudish

and is now sneered at as being 'fuddy-duddy'. 1958 *Observer* 23 Mar. 14/4 Deserted country-house with fuddy-duddy owner. 1964 L. NKOSI *Rhythm of Violence* ii. ii. 30 Some priests can be..more militant than some fuddy-duddies who pass as politicians! 1970 D. BALSDON *Oxford Now & Then* 134 So much for 'a straight run to the grave'. How fuddy-duddy they all were.

fudge, *int.* and *sb.* Add: **B.** *sb.* **4.** A patch of print, esp. a piece of late news, inserted in a newspaper page; also, a machine or cylinder for inserting such patches. Also *attrib.*, as *fudge-cylinder, -plate, -shaft, -space, -unit*; **fudge-box** (see quot. 1929).
1899 *Daily News* 23 Sept. 5/1 The blank space left for 'fudge' while the evening paper is being printed off is caused by the presence of the empty fudge-box. 1902 *Daily Mail* 27 June 7/5 After the paper has passed through the big cylinders that print the entire sheets, it goes past the small cylinder which prints the contents of the 'fudge box' in the vacant space. 1910 E. WALLACE *Nine Bears* vi. 45 'Issued at 4.10,' he said, glancing at the 'fudge' space, where the result of a race had been printed. 1929 *Encycl. Brit.* XVIII. 508/1 This late news is printed into the blank spaces by a contrivance called a 'fudge box' which is circular in form and into which are secured linotype slugs. The 'fudge' is fastened on to an auxiliary cylinder equipped with inking mechanism. 1964 *Gloss. Letterpress Rotary Print. Terms (B.S.I.)* 13 Fudge cylinder, plate, shaft, unit. 1967 *Evening News* 13 July 4/4 (*caption*) You must excuse me while I break off to supply the results..in the fudge spaces for our 3.30 fudge.
5. A soft-grained sweetmeat prepared by boiling together milk, sugar, butter, etc. orig. *U.S.*
1896 W. C. GORE in *Inlander* Jan. 147 *Fudges*, a kind of chocolate bonbons. 1902 *Queen* 3 May 763/1 The greatest 'stunt' among college students is to make *Fudge*. *Ibid.*, Nut Fudges... Fruit Fudge. 1905 *Buffalo Express* 16 Jan. 2 The overturning of an alcohol lamp over which some girls were cooking fudge. 1960 *Good Housek. Cookery Bk.* (ed. 5) 423/1 The fudge is poured into an oiled tin or caramel bars and cut up as soon as it is firm. 1971 *Islander* (Victoria, B.C.) 18 July 2/1 A batch of white fudge.

Fuegian (fu,ī·djiăn, fwē·djiăn), *a.* and *sb.* [f. the name *Tierra del Fuego*, lit. 'land of fire' + -IAN.] **A.** *adj.* Of or pertaining to Tierra del Fuego, or its inhabitants. **B.** *sb.* One of a primitive people inhabiting this S. American archipelago.
1825 J. WEDDELL *Voy. towards S. Pole* vi. 148 It was not long before the Fuegians arrived without hearing; and soon made themselves known by a singing noise. *Ibid.* 161 According to the Fuegian mode of salutation, they soon commenced shouting. 1872 C. M. YONGE *Pioneers & Founders* x. 273 The Fuegians are as degraded a people as any on the face of the earth. *Ibid.* 274 The poor little ship..drifting near the Fuegian coast. 1890 W. JAMES *Princ. Psychol.* II. xix. 110 The Fuegians, in Darwin's voyage, wondered at the small boats, but took the big ship as a 'matter of course'. 1900 J. G. FRAZER *Golden Bough* (ed. 2) I. i. 120 Fuegian wizards throw shells against the wind to make it drop. 1918 G. B. SHAW in *W.E.A. Educ. Yearbk.* 1918 42 A certificate classing you with a Fuegian or with Plato or Shakespear. 1931 *Times Lit. Suppl.* 5 Mar. 167/1 Tasmanian or Fuegian equivalents of eternity, omniscience and so on. 1969 *Nature* 6 Dec. 974/1 Once round the Horn, the three pathetic Fuegians who had spent a year or so in Britain..landed.

Fuehrer: see *FÜHRER.

fuel, *sb.* Add: **2*.** Specific senses, related more or less closely to senses 1 and 2.
Senses b and c below are properly regarded as uses of senses 1 and 2 in specific contexts. Senses a and d represent two extensions in sense, more or less corresponding to similar extensions in the meaning of *burn* vb. and *combustion*.
a. Food, regarded as that which supplies the body with energy; those constituents of food which are utilized by the body to produce energy. (Usu. as a conscious metaphor.)
1876 A. H. CHURCH *Food* i. i. 1 In the case of the human body we likewise have, first, a material structure; secondly, fuel, in the form of our daily rations of food. 1902 W. G. THOMPSON *Pract. Dietetics* (ed. 2) i. 9 [If water is withheld], preventing the transportation of the fuel and of tissues to various parts of the body, death follows in about two to seven days or more. 1936 J. J. BOGERT *Nutrition & Physical Fitness* viii. 121 Calorie values are especially useful in thinking of or talking of food as body fuel. 1951 MCLESTER & DARBY *Nutrition & Diet* (ed. 6) iii. 39 In the calculation of fuel requirements, the physician will not go far astray if he adds to the physiological basal metabolism some 10 per cent to cover the cost of the specific dynamic action of the ingested food. 1964 A. Z. BAKER *Dietetics & Nutrition* ii. 21 When other energy-releasing foods are lacking, as in starvation, body proteins are used as fuel. 1968 PASSMORE & ROBSON *Compan. Med. Studies* I. iv. 2 The fuels of the body are carbohydrate, fat, protein and ethyl alcohol, taken in the diet. 1970 FISHER & BENDER *Value of Food* ii. 19 When fuel, for example glucose, is burned oxygen is used and carbon dioxide is formed.
b. (A kind of) liquid or other material which by its combustion with air in an internal combustion engine provides power.
1886 D. CLERK *Gas Engine* i. 23 The fuel..is, in the gas engine, introduced directly to the motive cylinder and burned there. It is indeed part of the working fluid. 1902 A. C. HARMSWORTH et al. *Motors* xii. 243 Instead of the combustible or fuel being burned..in the cylinder of the engine, as in the internal combustion engine of the petrol car, it is burned under a boiler. c 1915 *Autocar Handbk.* (ed. 6) iii. 61 The function of the carburetter is to convert the liquid fuel (petrol in most cases) into a gas. 1928 V. W. PAGÉ *Aviation Engines* v. 110 The power obtained from the gas-engine depends upon the combustion of fuel in the cylinders. 1955 *Times* 26 Aug. 4/5 In a conventional jet engine all the air is compressed and then heated by the injection of burning fuel. 1963 C. CAMPBELL *Sports Car Engine* iv. 61 The commercial motor fuels sold to-day are all blended to conform to a very narrow specification.
c. (A kind of) material which reacts with an oxidizer to produce thrust (in a rocket engine) or electricity (in a fuel cell). Also *loosely*, a propellant.
1922 *Trans. Faraday Soc.* XVII. 467 If..natural fuel consumed at metallic zinc—coal or oil—could be burnt in primary cells to give electrical power directly. 1929 R. H. GODDARD *Papers* (1970) II. 663 Construction and tests of rocket having same fuel capacity as large rocket previously used in tests of lifting power. *Ibid.* 675 The ascent [of the rocket] was about 90 feet, the horizontal distance being much greater, but there was fuel enough to last about half a minute. 1947 W. LEY *Rockets & Space Travel* (1948) x. 236 By far the more distant future one may speculate on monatomic hydrogen as a rocket fuel. 1953 E. BURGESS *Rocket Propulsion* i. 14 Essentially then, the rocket consists of two tanks containing a fuel and an oxidizer and these two propellants are forced into a combustion chamber. 1962 A. SHEPARD in *Into Orbit* 29 At 5:30 a.m., with the liquid oxygen fuel already loaded aboard the booster, the technicians..declared a hold. 1966 *McGraw-Hill Encycl. Sci. & Technol.* V. 551/1 Modern fuel cells use gaseous fuels, either H_2 or CO or mixtures of these gases. The oxidant may be gaseous O_2 or air. 1970 G. M. BROOKLEY et al. in G. A. Partel *Space Engin.* 167 Current operational propellants are based on hydrogen and hydrogen-rich compounds of carbon and nitrogen as fuels and oxygen or oxygen-based oxidizers.
d. *Nuclear Sci.* (A kind of) material used as a source of energy in a nuclear reactor; material that can support a self-sustaining chain reaction.
1946 *Edison Electric Inst. Bull.* Jan. 23/3 Large power and heating plants using atomic fuel. 1948 *Nature* 28 Aug. 318/1 Uranium 235, the only material occurring..'primary', nuclear fuel must be made to breed 'secondary' fuel from more abundant materials: either plutonium from uranium 238 or U233 from thorium. 1950 GLASSTONE *Sourcebk. Atomic Energy* xiv. 402/2 In the conception of a breeder reactor, the fertile material..may form a blanket or jacket surrounding the reactor core. 1962 F. I. ORDWAY et al. *Basic Astronautics* x. 421 Less than 1 percent of the uranium fuel will be fissioned in present rocket engine reactors. 1963 B. FOZARD *Instrumentation Nucl. Reactors* xiii. 158 For a given core arrangement the reactivity depends upon the balance between the amount of neutron-producing material (fuel) and the amount of neutron-absorbing material (absorber).
3. a. *fuel consumption, tank.*
1890 W. ROBINSON *Gas & Petroleum Engines* i. 6 The total fuel consumption..was only 1·21 lb. per indicated horse-power per hour. 1907 WOODWARD & PRESTON tr. *Sorel's Carburetting & Combustion in Alcohol Engines* i. 28 It is probable that for any given engine the fuel consumption is also affected by the quantity and temperature of cooling water used. 1971 *Sat. Stemmer's Today* Sept. 4/2 Devices for cutting fuel consumption are frequently advertised as bringing about substantial savings in motoring costs. 1900 A. H. GOLDINGHAM *Oil Engines* i. 3 The oil is stored under pressure in the fuel-tank. 1933 *Economist* 9 Dec. 1144/2 The question of fuel pressure systems..has enabled the fuel tank to be located in a non-vulnerable position. 1959 *Observer* 1 Mar. 11/5 With two fuel tanks in the rear wings leave space for a big, deep trunk.
b. fuel cell, a primary cell which consumes fuel continuously and converts its chemical energy directly into electrical energy; **fuel element**, an assemblage of nuclear fuel with other materials to form a unit for use in a reactor; also *fuel rod*; **fuel food**, food that is rich in fats or carbohydrates and therefore provides the body with energy, in contrast to food that is of value chiefly because of the vitamins or trace elements it contains; **fuel injection**, the direct introduction of fuel under pressure into the combustion chamber or its intakes in an internal combustion engine; so **fuel injector**, the nozzle through which the fuel is forced, with its associated valve mechanism; **fuel oil**, oil used as fuel in an engine or furnace; **fuel-value**, the value of a combustible article as fuel; (*b*) the value of food as a source of energy; the amount of energy obtainable by the body from a given quantity of food.
1922 *Trans. Faraday Soc.* XVII. 482 Fuel cells may be classified as—(1) Direct fuel-cells burning solid fuel.. (2) Semi-direct fuel-cells burning gaseous fuel... (3) Indirect cells of (a) Oxidation-reduction Type.. (b) Metal Anode Type. 1956 *Engineer* 20 July 93 (*heading*) High-pressure hydrogen-oxygen fuel cell. 1960 *New Scientist* 13 Nov. (Energy Suppl.) 16/2 Both the Gemini and Apollo programmes relied on fuel cells of different kinds for their power. 1951 *Nucleonics* Nov. 20/1 The fission products gradually build up inside the fuel elements. 1970 *Nature* 26 Dec. 1245/1 Among the recent innovations is the introduction of fuel elements in the form of large blocks of graphite with machined holes to carry tubular pins. 1905 *Daily Chron.* 14 July 4/4 At the present period of year there is comparatively little demand for fuel-food. 1944 J. S. HUXLEY *On Living in Rev.* xiii. 134 The civilian population..will have been going short of vitamins and fuel-foods. 1900 A. H. GOLDINGHAM *Oil Engines* 53 This method of fuel injection forms the subject-matter of U.S. patent 654,583, granted to the writer May 29, 1900. 1941 *Nature* 26 July 105/1 The Minister of Aircraft Production..mentioned that where the British aero engine industry uses the carburettor, the Germans have adopted the fuel injection system. 1966 *McGraw-Hill Encycl. Sci. & Technol.* V. 555/2 In engines with continuous combustion, such as gas turbines and liquid-fuel rockets, fuel injection is necessary because the pumping action of piston-type engines is unavailable to draw fuel into the combustion chamber. 1914 HISCOX & PAGE *Gas, Gasoline & Oil-Engines* (ed. 21) xvii. 456 The fuel injectors are of the orthodox pattern with oil-operated valves. 1962 F. I. ORDWAY et al. *Basic Astronautics* x. 401 The combustion chamber contains the fuel injectors, spark plugs, flame holder, and so forth. 1883 *Chambers's Jrnl.* 3 Mar. 140/2 A contract..for the supply of all fuel oil required for the steamers. 1968 *Observer* 26 Sept. 11/4 He had returned to Williamstown in *Anzac* on the last of her fuel oil. 1947 C. GOODMAN *Sci. & Engin. Nuclear Power* i. 319 Solid fuel rods, clad with a non-corrosive protective coating. 1886 *Lett. fr. Donegal* 36 Bog once 'cut out' does not grow again, and the fuel-value is consequently lost to the land. 1902 W. G. THOMPSON *Pract. Dietetics* (ed. 2) i. 13 A day laborer requires 0·28 pound of protein per diem plus enough fat and carbohydrate to yield a total fuel value of 3,500 calories. 1938 A. D. CALLOW *Food & Health* 18 Water cannot be burnt in the body, and therefore has no fuel-value. 1949 L. J. BOGERT *Nutrition & Physical Fitness* viii. 122 In general, the foods with high fuel value will be found to be those which are either rich in fat or low in water content. 1964 M. H. MITCHELL *Comp. Nutrition Man & Dom. Anim.* II. xviii. 681 In terms commonly used in human nutrition, the calorie or fuel value of the available energy in the diet would be 2745 cal.

fuel, *v.* **1.** (Later examples.)
1950 F. GAYNOR *Encycl. Atomic Energy* 62 The Los Alamos fast-neutron reactor..is 'fuelled' by plutonium. 1950 *Sci. Amer.* Mar. 11/2 In the President's first announcement of the Hiroshima atomic bomb, it was stated that the bomb drew its energy from the same source that fuels the sun and the stars. 1962 A. SHEPARD in *Into Orbit* 99 The crews started filling the Redstone's tanks with the liquid oxygen shortly after midnight. 1970 *Guardian* 15 Apr. 5/6 Students who raided administrative files discovered that outside complaints had been recorded... This fuelled suspicion at other universities.

fuelling (fiū·ĕliŋ), *vbl. sb.* [f. FUEL *v.* + -ING¹.] The action of laying in, or furnishing with, fuel; supply or storage of fuel. Also *attrib.* and *transf.*
1921 *Cent.* June 1063 At each port, facilities for fuelling and for receiving the various supplies which a fleet requires must..be forthcoming. *Ibid.* 1068 There are numerous minor ports available as temporary fuelling bases. 1927 *Glasgow Herald* 14 July 11 The broad Imperial development of the science of fuelling. 1949 W. G. MOORE *Dict. Geogr.* (1950) 67 *Fuelling station*, a port or seaport within a port, which supplies fuel to the ships calling there. 1961 *Times* 9 Sept. 10/4 Her commander..had to put his ship in for fuelling three times in the course of a fortnight's cruise. 1967 *Guardian* 24 Oct. 4/5 The outlook for commerce is delicately poised between changing fashions..and the fuelling factors—salary increases and expected price rises.

fufu. (Earlier and later examples.) (See also *FOO-FOO.)
1740 C. LESLIE *New Hist. Jamaica* ii. 33 And boil it with beaten Maiz or Indian Corn, (which they call Fu Fu). 1930 *Discovery* Mar. 79/1 The various vegetables used for making the soups or sauces which flavour their *fu-fu*, a kind of dough. 1968 *Guardian* 24 Oct. 4/5 Quantities of yellow gari and the more expensive fufu.

fug (fʌg), *sb. colloq.* (orig. *dial.* and *School slang*). [? Related to FOGO. Cf. *FUGGY *a.*] A thick, close, stuffy atmosphere, esp. that of a room overcrowded and with little or no ventilation. Also **fug-footer, -soccer, -socker** *School* and *University slang*, indoor football.
1888 E. F. BENSON *Sk. fr. Marlborough* i. 16 Seating himself in the most comfortable chair, as a consolation for the prevailing fug. 1905 C. RANGER-GULL *Harvest of Love* i. 10 He met a group of school-boys..They had been playing 'fug-soccer' in the racquet courts. 1914 C. MACKENZIE *Sinister St.* II. iii. viii. 665 Nigel had booked himself to play fug-socker with three hearty Trinity men. 1927 *Tall Ship* iv. 78 We get up quite a good fug in our casemate at night. *Ibid.* xi. 175 'Fug!' he exclaimed. 'What a fug!' And elevated his nose with a sniff. 1923 U. L. SILBERRAD *Lett. J. Armiter* x. 214 Can you smell the old cold pipe..a cosy fug. 1930 W. DEEPING *Kitty* xvii. [sc. a sickroom] smelt like a greenhouse, full of soft fug. 1940 M. MARPLES *Pub. Sch. Slang* 85 *Fug-footer* (Harrow, 1884+), indoor football. 1948 G. H. JOHNSTON *Death takes Small Bites* i. 29 The fug of the room. 1968 J. E. MATTHEWS *Brew's Youth & Youth Groups* (ed. 2) i. 48 There are only three necessities—light,..warmth, and fug.
Hence **fug** *v. intr.*, to stay in a stuffy atmosphere. Also with *it*. Hence **fugged** *ppl. a.*, stuffy, thick.
1889 BARRÈRE & LELAND *Dict. Slang* I. 387/2 *To fug* (Shrewsbury), to stay in a close, stuffy room. 1901 W. T. GRENFELL *Labrador Doctor* (1920) ii. 36 Every one was 'fugging' in the house or had gone to bed. 1921 A. S. M. HUTCHINSON *If Winter Comes* ii. v. 122, I like it a jolly

sight better than fugging up in those carriages with all that gassing crowd. 1971 H. C. RAE *Marksman* i. v. 45 The room was stuffy, fugged with tobacco smoke.

fugal, *a.*: see also *FUGUE *sb.* 2.

fugato, *adv.* Add examples of *sb.* and *attrib.*
1876 STAINER & BARRETT *Dict. Mus. Terms*, *Fugato*, in the fugue style; a composition containing fugal imitation, but which is not a strict fugue form. 1891 E. PROUT *Fugue* § 358 Passages of imitation, provided that all the voices take part in them, will very often be also fugato passages. *Ibid.* § 359 The freedom allowed in fugato. 1903 *Times* 5 May 4/3 Fugato entry for the lower strings. 1906 *Grove's Dict. Mus.* II. 118/1 Beethoven was particularly fond of the fugato. 1968 *Listener* 26 Sept. 418/3 The least well-managed moment was the fugato in the middle.

fuggle (fʊ·g'l). Also **Fuggle's**. A variety of hops.
1888 *Westm. Gaz.* 5 Sept. 2/3 All good sound hops are eagerly sought after. 1902 *Times* ... 1905 *Westm. Gaz.* 25 Sept. 9/2 The growers of 'Fuggles' and other comparatively hardy kinds. 1930 *Encycl. Brit.* XIII. 679/2 The variety known as Fuggle's, a heavy-cropping though slightly coarse hop, has been much planted in the Weald of Kent.

fuggy (fʊ·gi), *a. colloq.* (orig. *dial.* and *School slang*). [f. *FUG *sb.* + -Y¹; or *FUG may be a back-formation from this. Cf. FOG *sb.*² and FOGGY *a.*] Of the air in a room: close, stuffy, fusty, and smelly, from want of ventilation. Of persons: addicted to living in such an atmosphere.
1888 E. F. BENSON *Sk. fr. Marlborough* i. 15 How beastly fuggy this place is. *Ibid.* vi. 58 He was rude enough to say that I was a fuggy beast. 1889 BARRÈRE & LELAND *Dict. Slang, Fuggy* (Shrewsbury), stuffy. 1900 *Globe* 12 Jan. 3/1 They missed the warm 'fuggy' heat of their old cabins. 1921 *Chambers's Jrnl.* Feb. 189/1 In the fuggy comfort of the engineers' mess. 1923 U. L. SILBERRAD *Lett. J. Armiter* iv. § 3. 102 He was a fuggy old man.

fughetta (fiugĕ·tă). *Mus.* [It., dim. of *fuga* FUGUE. Cf. G. *fughette*.] A short, condensed fugue. Also *attrib.*
1876 STAINER & BARRETT *Dict. Mus. Terms*, *Fughetta*. 1891 E. PROUT *Fugue* § 351 In the form most frequently met with, a fughetta is an abridged fugue. *Ibid.* § 356 Another variety of the fughetta form. 1898 *Westm. Gaz.* 29 Dec. 10/2 The fughetta..on the first subject of the first movement. 1947 A. EINSTEIN *Mus. Romantic Era* viii. 157 In..Schumann's Seven Pieces in Fughetta Form, Op. 126 (1853). 1959 *Times* 21 Sept. 5/5 The fughetta was particularly well played.

fugue, *sb.* Add: *Comb.* (Additional examples.)
2. *Psychiatry.* A flight from one's own identity, often involving travel to some unconsciously desired locality. It is a dissociative reaction to shock or emotional stress in a neurotic, during which all awareness of personal identity is lost though the person's outward behaviour may appear rational. On recovery, memory of events during the state is totally repressed but may become conscious under hypnosis or psycho-analysis. A fugue may also be part of an epileptic or hysterical seizure. Also *attrib.*, as *fugue state*. So **fu·gal** *a.*
1901 C. R. CORSON tr. *Janet's Mental State Hystericals* 422 Those long flights [*fugues*],..those strange excursions, accomplished automatically, of which the patient has not the least recollection. 1923 OGDEN & RICHARDS *Meaning of Meaning* vi. 220 'The Unconscious' is what causes dreams, fugues, psychoses. 1946 J. LAIRD *Our Minds & Their Bodies* iv. 86 There is a palpable difference between behaviour in somnambulism, or in a fugue, or in a panic, or in 'shock', and ordinary human conduct. 1940 C. S. MYERS *Shell Shock* 89 When the 'shock' is severe, it may be followed by..fugal automatism, or stupor, on recovery from which the patient can recall none of the last events that happened. 1965 LANDIS & BOLLES *Textbk. Abnormal Psychol.* vii. 94 A middle-aged embezzler recalled events that had occurred during a fugue state more than twenty years earlier, although he had been unable to remember them during the entire twenty years. 1961 *Lancet* 29 July 240/1 She had also been a voluntary patient in a mental hospital..once after an epileptic fugue. 1965 ROSEN & GREGORY *Abnormal Psychol.* ii. xii. 241 A fugue is a combination of amnesia and physical flight. The individual flees from his customary surroundings; what he is really trying to escape is his own fear. 1969 ULLMANN & KRASNER *Psychol. Approach to Abnormal Behavior* ii. xv. 289/2 Two days before the fugue, an anonymous letter addressed him as a homosexual.

Führer (fū·rəɹ). Also **Fuehrer.** [G., leader.] Part of the title (*Führer und Reichskanzler*) assumed by Adolf Hitler (see *HITLER) in 1934 as head of the German Reich, after the model of *DUCE. Also *transf.* and *attrib.*
1934 *N. & Q.* CLXVII. 254/1 Since the Führer refuses permission for this, it is thought probable the German stamps will show the swastika. 1936 A. HUXLEY *Eyeless in Gaza* viii. 91 Stupidity has come back, as a king—no; as an emperor, as a divine Führer of all the Aryans. 1937 — *Ends & Means* v. 34 They yet persist in taking the same roads as are taken by the Duces and Fuehrers. 1943 'G. ORWELL' in *Partisan Rev.* July–Aug. 347 Some..declare that he has a 'fuehrer complex' and..would split it rather than share authority. 1944 H. G. WELLS *'42 to '44* 183 Galton had the mental disposition of a Fascist and was all for fuehrers and duces. 1948 E. E. CUMMINGS *Let.* 24 Sept. (1969) 186 Presently a smallish mob of muckers..materialized, & their führer-commissar suggested a friendly snowfight. 1959 *Sunday Times* 14 June 22/5 The cold and pointless ferocity of the gang fights came out vividly in interviews..with two or three young fuehrers. 1967 *Spectator* 29 Dec. 819/3 Mr. Gordon Walker..issued his führer order that the Bloomsbury site be scrapped.
Hence **Führer-prinzip** (or **principle**): see quot. 1950.
1937 AUDEN *Let.* in Auden & MacNeice *Lett. fr. Iceland* v. 57 The Teutonic Führer-Prinzip would have appealed to you. 1940 *Ann. Reg. 1939* 148 The tendency to adopt the 'führer' principle would threaten religious life and freedom. 1950 THEIMER & CAMPBELL *Encycl. World Politics* 176/2 The *Fuehrer Prinzip* was the principle that the fuehrer had the right to command and the people the duty to obey him. 1959 *Chambers's Encycl.* XI. 51/2 Fascism was inherently militaristic and anti-democratic, with the 'Führer-principle' as its dominant political idea.

Fulah (fū·lă), *sb.* and *a.* Also **Foulah, Ful, Fula. A.** *sb.* **a.** One of a Sudanese people of partly non-Negro extraction. **b.** The language of this people. **B.** *adj.* Of or pertaining to this people or their language. Also called **Fulbe, Fullatah**, etc. Cf. next.
1832 *Jrnl. R. Geogr. Soc.* II. 283 The principal internal trade is maintained with the Foulahs. 1854 E. NORRIS *MacBrair's Gram. Fulah Lang.* 1 The pronunciation of the Fulah language is rather guttural. *Ibid.* 4 Terminations expressing the size of objects are very common in Fulah. 1875 *Encycl. Brit.* I. 263/1 The *Foulahs* or *Fellatas* occupy the central parts of the Soudan. 1884 *Ibid.* XVII. 318/1 Nor is it possible to regard the..Fulah, and many other Soudanese tongues as fragments or offshoots of Bantu. 1887 *Ibid.* XXII. 278/2 Recent research has shown..that the Fulahs and Nubians differ fundamentally in speech and physique, the former being of Caucasic and the latter of Negro type. 1901 *Ibid.* XXIX. 234/1 Most of the kings and of the ruling class are not Hausas, but belong to a tribe variously called Fulah, Fulani, Fulbe, or Fullatah. Whilst the Hausas are a quiet commercial people, the Fulahs are a race of soldiers and rulers. 1921 E. SAPIR *Lang.* 75 Ful, an African language of the Soudan. 1938 J. CARY *Castle Corner* 351 The Emir was a high bred Fulbe. 1957 M. BANTON *W. Afr. City* ii. 34 After the French colonies had joined the Allies, Fula traders appeared in greater numbers.

Fulani (fū·lăni, fūlā·ni), *sb.* and *a.* [Native name.] = *FULAH *sb.* and *a.*; *spec.* the Fulah of northern Nigeria and adjacent territories.
1855 S. CROWTHER *Jrnl. Exped. Niger* 147 We had not seen a real Filani, till we came to the neighbourhood of Hamaruwa. 1860 R. CAMPBELL *Pilgr. Motherland* ix. 105 Every person in Ilorin is said to speak both the Aku and Fulanee languages. 1901 [see *FULAH *sb.* and *a.*]. 1931 *Times Lit. Suppl.* 26 Feb. 145/1 The Fulani form perhaps the one group that is really of ambiguous status. 1932 J. CARY *Aissa Saved* i. 6 A half-bred Fulani girl with big soft eyes and a fine golden skin. 1957 *Encycl. Brit.* I. 312/1 The most important non-Negro group to be found in the region is the Fulani (Fula, Peul), a people of white Mediterranean race closely related to the Berbers. 1960 *Guardian* 15 July 14/3 Although the ruling classes [in northern Nigeria] are still Fulani, they early on became assimilated with the Hausas. 1960 G. DURRELL *Zoo in my Luggage* vi. 150 A herd of the long horned Fulani cattle..surrounded us...They were huge, beautiful beasts of a dark chocolate brown, with enormous melting eyes and a massive spread of white horns, sometimes as much as five feet from tip to tip.

Fulbright (fu·lbrəit). The name of Senator William *Fulbright*, of Arkansas, U.S.A., designating the *Fulbright* Act (Public Law 584 of the 79th Congress), of 1 August 1946, which authorized agreements with foreign countries by which local currencies acquired by the American government from the sale of surplus war property might be used for financing higher learning; hence applied *attrib.* to a grant, etc., awarded by virtue of such agreements, or to a person holding such a grant or position, as *Fulbright professor, scholar*. Also *ellipt.* Hence **Fu·lbrighter** *colloq.*, the holder of a Fulbright grant.
1951 *Times Educ. Suppl.* 23 Mar. 229/2, 23 visiting American lecturers and 20 research scholars are..here with Fulbright awards. 1952 M. MCCARTHY *Groves of Academe* (1953) x. 197 Percy copped a Fulbright to lecture on Amiel in Lebanon. 1957 *Times Lit. Suppl.* 8 Nov. 680/1 Professor Jayne's year of research in England, supported by Fulbright and Guggenheim funds. 1958 *Times* 22 Sept. 11/6 A party of some 200 Americans disembark... If you..asked who they were, you might well be given the laconic reply: 'Fulbrighters.' 1958 S. ELLIN *Eighth Circle* (1959) II. ii. 28 He went to Oxford on a Fulbright. 1959 *Encounter* Oct. 38/2 Some stranded Fulbrights, a Mexican pundit, a Lapp. 1959 *Times Lit. Suppl.* 6 Nov. 647/1 A Fulbright scholar.

fulcrum. Add: **2. c.** *Zool.* The stem or median part of the incus of the mastax of certain rotifers.
1886 C. T. HUDSON *Rotifera* I. 118 The trophi consist mainly of two ribbed rami, attached to a long narrow plate (the fulcrum). 1896 *Cambr. Nat. Hist.* II. 210 In the ventral wall of the gizzard of most Ploima is a median piece, the *fulcrum*, from which run forwards and upwards two pieces, the *rami*, which are hinged on the fulcrum. The Y-shaped structure formed of these three pieces is called the *incus* (anvil). 1951 L. H. HYMAN *Invertebrates* III. xiii. 88 The fulcrum, the median, usually posterior piece, is commonly a thin plate lying in the sagittal plane of the body. 1967 P. A. MEGLITSCH *Invert. Zool.* viii. 261/1 The fulcrum and rami together form the incus.
d. In trilobites (see quot. 1909).
1900 C. R. EASTMAN tr. *Zittel's Text-bk. Palæontol.* I. 612 The distal portion [of the pleura], beginning at the fulcrum, may continue of equal thickness. 1909 *Cambr. Nat. Hist.* IV. 234 At some distance from the axis the pleurae are bent downwards and backwards. The point where this bend occurs is called the 'fulcrum'; it divides the pleura into an internal and an external part. 1953 SHROCK & TWENHOFEL *Princ. Invert. Paleontol.* (ed. 2) xiii. 583 (*caption*) Fulcrum.
e. *Ent.* In Diptera, a chitinous portion of the pharynx.
1910 C. G. HEWITT *House-Flies* 12 The Fulcrum.—This chitinous portion of the pharynx..lies on the lower part of the head and in the rostrum. 1925 A. D. IMMS *Gen. Textbk. Ent.* 597 Situated within this region is a complex framework of chitin known as the *fulcrum*, which forms a kind of case enclosing the pharynx, and is present in almost all Diptera. 1951 L. S. WEST *Housefly* iii. 51 Its [sc. the rostrum's] internal support consists chiefly of a large stirrup-shaped sclerite, the fulcrum.

fulfil, *v.* Add: **5.** *refl.* Of a person: to work out one's destiny; to develop one's gifts and character to the full. (A development of Tennyson's use: see Dict.)
1920 R. MACAULAY *Potterism* III. i. § 8. 117 In what place, under what conditions, would Oliver Hobart now fulfil himself, now carry on the work so faithfully begun on earth? *Ibid.* VI. v. § 8. 259 Jane would, no doubt, fulfil herself in the course of time, make an adequate figure in the world she loved. 1971 *Times* 16 Oct. 8/7 The belief that a degree is evidence of 'ability' and that without one you..have not 'fulfilled' yourself.

fulgorid (fʊ·lgǒrid), *sb.* and *a. Ent.* [ad. mod. L. *Fulgoridæ*, f. the generic name *Fulgora*, a. L. *Fulgora* goddess of lightning, f. *fulgur* lightning: see -ID³.] **A.** *sb.* An insect of the homopterous family Fulgoridæ which includes the lantern-flies. **B.** *adj.* Belonging to this family.
1893 in *Funk's Stand. Dict.* 1899 *Cambr. Nat. Hist.* VI. 576 The wax of Fulgorids is used by the Chinese for candles. 1913 *Zoologist* 4th Ser. XVII. 281 Some notes on a Luminous South African Fulgorid Insect (Rhinortha Guttata). *Ibid.* 289 With approach of cold weather all signs of the Fulgorid parasites disappeared. 1923 H. M. LEFROY *Man. Ent.* 259 Some Fulgorids are quite small, others are very large: the Lantern-flies are included in this family; they were so called because the front of the head, extending into a bladder-like structure, was believed to be luminous. 1926 E. O. ESSIG *Insects Western N. Amer.* 218 The date fulgorid..is 4–6 mm. long, reddish-brown in the nymphal stage and very dark brown in the adult stage. 1954 BORROR & DELONG *Introd. Study Insects* xx. 260 None of the fulgorids is known to be luminescent.

fulguration. Add: **3.** *Med.* [a. F. *fulguration* (S. Pozzi 1907, in *Bull. de l'Acad. de Méd.* 30 July 193).] The destruction of tissues, esp. tumours, by means of electric sparks.
1907 *Daily Chron.* 11 Oct. 5/6 Dr. Hart [sc. de Keating-Hart] treats cancer by ordinary surgical means, utilising fulguration as a valuable healing agent. 1962 *Lancet* 31 Mar. 693/1 Fulguration [of the bladder neck] aims at destroying infected glands and pseudo-polyps but may cure by destroying part of the internal sphincter.

full, *a.*, *sb.*³, and *adv.* Add: **A.** *adj.* **4. d.** *as full as a tick* (see *TICK *sb.*¹). Also, in this and similar comparisons, extremely drunk. *Austral.* and *N.Z. slang.*
1892 *Dialect Notes* I. 210 *Full as a tick*, drunk. 1915 *Bulletin* (Sydney) 11 Mar. 44/2 We both got full as ticks. 1930 *Ibid.* 2 Apr. 51/4 'I met Mace down at Cafferty's,' he told his father... 'Full as an egg, and inclined to be nasty!' 1947 P. NEWTON *Wayleggo* (1949) ix. 103 By the time the dance was under way I was as 'full as a bull' and ripe for anything. 1949 D. M. DAVIN *Roads from Home* I. v. 74 Wasn't he in here this afternoon and as full as a tick? 1960 N. HILLIARD in C. K. Stead *N.Z. Short Stories* (1966) 237, I noticed he was full as a bull in no time.
11. *full and by*: also *fig.*
1930 *New Statesman* 5 July 412/1 And, take it full and bye, it is one of the very few things to be honoured in our nature.
12. a. *full-page* (later examples), *-scale, -size, -term.*
1911 A. BENNETT *Card* xi. 245 The only hotel in the Five Towns seriously pretending to be 'first-class' in the full-page advertisement sense. 1954 M. RICHERT *Painting in Brit.: Middle Ages* iv. 81 The few uncoloured full-page miniatures which precede a copy of the gospels. 1933 *Discovery* Aug. 257/1 Full-scale experiments in flight are of necessity difficult and take a long time. 1957 *Times Lit. Suppl.* 20 Dec. 778/1 Since Alexander Henderson..

FULL

there has been no full-scale monograph on the wines of antiquity. **1832** in A. Adburgham *Shops & Shopping* (1964) iv. 40 Full-size Paper Patterns. **1957** *Times Survey Brit. Aviation* Sept. 7 A full-size mock-up of the Vanguard's cockpit. **1907** W. J. Maloney tr. Budin (*title*) The nursling: the feeding and hygiene of premature and full-term infants. **1949** M. Mead *Male & Female* viii. 179 They smother in economic details the memory of their miscarriages, as if these had been full-term children.

c. *full-bellied* (later examples), *-leaved* (later examples), *-lipped*, *-winged* (later example).

1909 *Daily Chron.* 2 June 5/6 The full-bellied barges of 600 tons freightage. **1937** N. Coward *Present Indicative* VIII. 333 A full-bellied roar of laughter. **1864** G. M. Hopkins *Poems* (1948) 131 Now while the full-leaved hursts unalter'd stand. **1917** J. Masefield *Lollingdon Downs* 74 The full-leaved summer bore no fruit. **1859** Bagehot in *National Review* IX. 383 Lancelot, the great knight of many exploits and full-lipped enjoyment. **1932** E. Hemingway *Death in Afternoon* ix. 244 He had a full-lipped face. **1957** Bannerman *Birds Brit. Isles* VI. 251 There has been a mixed flock of *full-winged* snow and blue geese at Woburn for fifty years at least.

d. **full board**, (*a*) [*board *sb.* 2 d] *Austral.* and *N.Z.*, a full complement of shearers on the board; (*b*) the provision of a bed and all meals: an arrangement offered by hotels, boarding houses, etc.; **full-bound** *a. Bookbinding*, bound entirely in leather; **full-cell process**, Bethell's process (cf. *Bethell); **full character** (see quot.); **full-choke**, a gun or gun-barrel with the maximum amount of choke-boring; **full-cream** *attrib.*, consisting of or made from unskimmed milk; **full employment** (see quot. 1948); **full-forward**, in Australian National Football, one of three players (esp. the centrally positioned one) who constitute the front forward line and stand closest to the opposing goal; **full hand** *Poker* = *full house 2; **full-mouth**, (*b*) a full-mouthed animal; **full pitch, toss** *Cricket*, a ball pitched right up to the batsman; also as *advb. phr.*, without the ball having first touched the ground; so **full-pitched** *a.*; **full professor** orig. *U.S.*, a professor with the highest ranking position on the staff of a university or college; **full score** *Mus.*, a score in which the parts for all voices and instruments are given on separate staves; **full snipe**, a popular name for the common snipe, *Gallinago gallinago*; **full word** *Linguistics*, a word that has an independent meaning; also *attrib.*; spec. in *Chinese Grammar* (see quot. 1954).

1894 E. Wilson *In Land of Tui* xv. 242 At one side [of the shearing shed], down the whole length of the building, the shearers work, thirteen of whom are called a 'full board'. **1910** *Bradshaw's Railway Guide* Apr. 1149/2 Very comfortable rooms from 3/-, and full board from 7/6. **1941** Baker *N.Z. Slang* v. 39 The use of *board* for the floor of a shearing shed is also slang. Whence comes *a full board*. **1971** *Oxford Mail* 11 Oct. 11/4 (Advt.), Accommodation urgently reqd for daughter aged 15. North Oxford preferred, bed and breakfast or full board. **1880** J. W. Zaehnsdorf *Art of Bookbinding* 171 When the sides and back of a volume are covered with leather it is said to be full-bound. **1946** H. Whetton *Pract. Printing & Binding* xxxi. 381 The book may be full-bound..that is, the leather may fully cover the boards. **1915** H. F. Weiss *Preserv. Struct. Timber* v. 57 The Bethell or full-cell process is considered the standard process of treating timber with creosote. **1940** H. Trotter *Man. Indian Forest Utiliz.* ix. 202 The full-cell process has been known for over 80 years. **1962** Full cell process [see *Bethell]. **1863** J. Edkins *Gram. Chinese Lang.* (ed. 2) II. iii. 105 They call significant words..full characters, while the auxiliary words or those which are non-significant, they term..empty characters. **1881** W. W. Greener *Gun* 387 A full choke is constricted to the extent of 30 to 40,000ths of an inch. **1892** —— *Breech-loader* 134 The 16-bore full-choke, with barrels 30 inches in length. **1895** *Outing* (U.S.) XXVII. 65/1, I have ruined too many fine birds with the full-choke to want to use it any more. **1964** H. L. Peterson *Encycl. Firearms* 52/1 Full chokes are used for long-range shooting, modified chokes or cylinder bores for upland birds and skeet shooting. **1881** Full-cream [see Full *a.* 12 a]. **1929** *Punch* 8 May p. viii, Fresh, full-cream milk. **1970** *Guardian* 14 Apr. 10/3 We needed full cream milk powder for the babies. **1835** J. Loudon *Philanthropy* 107 If the self same cause which gives cheap food, gives full employment to labour [etc.]. **1846** Ld. Beauvale in C'tess of Airlie *Lady Palmerston* (1922) II. xvi. 102 By God's blessing there is full employment with high wages. **1940** *Economist* 20 Jan. 88/1 Optimistic assumptions of the effect that the attainment of 'full employment' will have on the revenue. **1948** G. Crowther *Outl. Money* (ed. 2) v. 138 'Full employment' does not necessarily mean that every man and woman has a job. It means only that there are no more supplies of idle labour or idle capital of the sorts that are actually being demanded. **1965** *Austral. Encycl.* IV. 135 The full-forward of each team, in particular, is chosen for his ability to take the ball high in the air. **1969** *Sun-Herald* (Sydney) 13 July 48/1 Hawthorn full-forward Peter Hudson kicked five goals although heavily guarded by Essendon's defenders. **1850** *Bohn's Handbk. Games* 382 Full hand. **1950** *Hoyle's Games* (ed. 20) I. 118 With Full Hands the higher threes win, e.g. 3 threes and 2 fours are better than 3 twos and pair aces. **1950** *N.Z. Jrnl. Agric.* Oct. 351/1 No sheep are carried beyond full mouth; at this age the cast ewes are disposed of privately. **1959** S. J. Baker *Drum* II. 111 *Full-mouth*, an eight-tooth sheep. **1843** 'Wykehamist' *Pract. Hints Cricket* 15 The player being *enticed* to play to it as to a full-pitch. **1895** H. G. Hutchinson *P. Steele* i. 29 One or two [balls] went over the wicket altogether, and were taken, full pitch, by the wicket-keeper. *Ibid.*, He..sent..ball after ball full-pitched. **1929** Full-pitch [see *carry *v.* 9 b]. **1934** J. W. Burgess *Reminisc. Amer. Scholar* iii. 42 The faculty of the college was at that time a strong body of teachers, most of the n being full professors of long experience and high standing. **1947** *Partisan Rev.* XIV. 474 The academic hierarchy, from instructor up to full professor, enforces caution on the imaginative or adventurous thinker. **1966** P. Green tr. *Escarpit's Novel Computer* iii. 116 Statisticians..do a numerical break-down of the full professors..and various other species of academic fauna. **1971** *Nature* 4 June 275/1 Even Berkeley is only able to boast that two per cent of its full professors were women in 1970. **1876** Stainer & Barrett *Dict. Mus. Terms*, Full score. **1946** *Penguin Music Mag.* Dec. 15 Purcell never saw an Italian opera... He ma·· just possibly have read a full score—we know that P pys possessed one. **1885** F. S. Mitchell *Birds Lancs.* 185 Common Snipe... Local Names.—Full Snipe, Lady Snipe. **1887** A. C. Smith *Birds Wilts.* 431 The provincial names of these three species accurately describe their relative size; the Jack or Half Snipe weighing about 2 ozs., the Common, Whole, or Full Snipe 4 ozs., and the Great or Double Snipe 8 ozs. **1913** H. K. Swann *Dict. Eng. & Folk-Names Brit. Birds* 90 The Common Snipe is also sometimes termed Full Snipe to distinguish it from the Jack (or Half) Snipe. **1956** *Jrnl. Lancs. Dial. Soc.* Jan. 15 *Snipe*,..full snipe, scalp snipe: Heywood. **1826** *Manch. Guardian* 5 Aug. 4/3 Barker, the Nottingham bowler,..so far lost his temper as to give a full toss, as we thought, at the face of the player, instead of his wicket. **1906** A. E. Knight *Compl. Cricketer* iii. 113 Over-pitch the ball and it presents the easiest of full tosses. **1892** H. Sweet *New Eng. Gram.* I. 22. § 58 When a form-word is entirely devoid of meaning, we may call it an *empty word*, as opposed to *full words* such as *earth* and *round*. **1929** Grattan & Gurrey *Our Living Lang.* 98 Words which have an independent meaning of their own. These are known as Full-words. **1933** Bloomfield *Lang.* xii. 199 The parts of speech [in Chinese] are *full words* and *particles*. **1934** Priebsch & Collinson *German Lang.* iii. 249 Other full-word suffixes are *-bar*, *-haft*. **1954** Pei & Gaynor *Dict. Linguistics* 78 *Full word*, in Chinese grammar, any word which expresses a concept or idea or designates a person, object, quality, etc.—i.e., verbs, substantives and adjectives. **1966** J. Buse in C. E. Bazell *In Memory of J. R. Firth* 53 Full words consist of a single free morpheme..with or without one or more of the following affixes.

C. *adv.* **2. c. full out.** Delete '*Obs.* exc. *dial.*' and add: now esp., at full power, at top speed; also *attrib.* or as *adj.*

1917 'Contact' *Airman's Outings* 46 With nose down and engine full out, we raced towards the lines and safety. **1933** *Jrnl. R. Aeronaut. Soc.* XXXVII. 877 On its official trials the machine did a full-out speed, while flying level at 19,000 ft., of 121 m.p.h., its cruising speed. **1938** *Times* 2 Feb. 14/5 It sounded as if the engine was 'full out' when the machine struck the ground. **1938** *Sunday Express* 13 Nov., Alvis was working *full out* to supply the demand. **1942** *Ann. Reg. 1941* 120 The Government [of Canada] had pledged themselves to a full-out war effort. **1971** *Financial Mail* (Johannesburg) 26 Feb. 652/3 A dedicated Mr Botha himself runs the Department full out.

5. b. *full-fledged* (earlier and later examples); **full-fashioned** = *fully-fashioned* adj.

1883 Full-fashioned [in Dict.]. **1927** T. Woodhouse *Artificial Silk* 92 Originally, the spring beard needles were used in full-fashioned (flat) machines. *Ibid.* 95 Needles have to be taken out of action when it is desired to decrease the width of any part of the full-fashioned fabric... This possibility of effecting various changes in the width according to the amount of material required for various parts of the human body, gave rise to the term 'full-fashioned'. **1937** *Times* 29 Nov. 38/2 'What does "full-fashioned" mean?'..'It means that a stocking is made to fit the contours of the leg.' **1883** 'Mark Twain' *Life on Mississippi* 246, I was a pilot now, full-fledged. **1961** *Brno Studies* III. 92 A 'full-fledged' verb.

full, *sb.*[4] *local* (Kent). [Prob. a use of Full *sb.*[3]] A ridge of shingle or sand pushed or cast up by the tide. So **full** *v.*[4] *trans.*, to form such a ridge on (the beach).

1846 F. Drew *Geol. Folkestone & Rye* 16 The beach is gradually 'fulled', that is pebbles are heaped up in front of the breaker, generally forming a small ridge. **1847** *Proc. Inst. Civil Engin.* VI. 476 No certain record has ever been kept of the increase of the coast line; but from the best existing data, it appears to be about two yards annually, and allowing the accumulation to have been rather more rapid at first, say three yards per annum, a period of about nineteen hundred years will have elapsed, since the sea first left the original 'Full' at Lydd. **1864** *Ibid.* XXIII. 195 The spit..is formed of parallel 'fulls' of beach. **1902** Ld. Avebury *Scenery of Eng.* 163 It [*sc.* North Weir Point] consists of a series of curved concentric ridges or 'fulls'..forming a projecting cape or 'Ness'. **1964** V. J. Chapman *Coastal Veget.* viii. 205 At Dungeness the ridges, known as 'fulls', generally bear Curled dock.

full, *v.*[2] Add: **1. b.** (Further examples.) Also said of the tide.

1794 E. Drinker *Jrnl.* 9 Sept. (1889) 237 The moon fulled this morning about 8 or 10 o'clock. **1878** B. F. Taylor *Between Gates* 40 It is as if a poor little aster should full like the moon and be a dahlia. **1912** L. J. Vance *Destroying Angel* xix, He should be able to catch the tide just as it was nearing high water. Allowing it to swing him north-west until it fulled, he ought to be a third of the way across by the time it slackened. **1938** W. de la Mare *Memory* 32 Fulling moon aloft doth ride.

full-blood, *a.* Add: **b.** (Earlier U.S. examples, of animals.)

1812 *Niles' Weekly Reg.* II. 408/1 His full blood merino ram lamb. **1850** *Rep. U.S. Comm. Patents Agric. 1849* 88 Sheep..ranging in quality from half to full-blood merino.

B. *sb.* A pure-bred person or animal.

1883 R. B. Sage *Scenes Rocky Mts.* xx. 166 They [*sc.* half-breed children] were more beautiful..than the same number of full-bloods,—either of whites or Indians. **1864** *Ret. Agric. Soc. Maine* 9 Those [sheep] exhibited were mostly Spanish Merinos..most full bloods. **1873** J. H. Beadle *Undevel. West* xii. 198 One may travel for days in the Territory, and never see a full-blood. **1888** *Vermont Agric. Rep.* X. 38, I breed Jerseys; have no full-bloods but high grades. **1964** *Observer* 12 Jan. 6/8 Gathering opinion throughout Australia that the nation's 20,000 full-bloods cannot be indefinitely treated as a lesser breed.

full-blooded, *a.* (Earlier Amer. examples.)

1774 P. V. Fithian *Jrnl.* 8 Jan. (1900) I. 89 Balantine, either to shew himself a true full-blooded Buck, or out of mere wantonness..turned the Bones..into many improper and indecent postures. **1786** *Maryland Jrnl.* 31 Mar. (Th.), A number of full-blooded Colts and Fillies. **1801** *Steele Papers* I. 218 The present first, altho' a full blooded Yankee, as we call him in these parts, knows the importance of this place. **1810** *Massachusetts Spy* 25 Apr. 3/1 Forty-five full-blooded Merino Sheep.

Hence **full-bloo·dedly** *adv.*, forcefully, whole-heartedly.

1931 *Times Lit. Suppl.* 19 Feb. 137/4 The story is full-bloodedly crowded with sudden deaths. **1958** *Listener* 18 Dec. 1030/1 The difficulty of putting Lorca across full-bloodedly.

full dress. b. (Earlier example, and further examples of *fig.* use.)

1761 *Ipswich Jrnl.* 24 Jan. 4/1 (Advt.), I have laid in an entire fresh Stock in the Peruke Way, and..will make..full-dress Bobs, from one Pound ten to one Pound five shillings the piece. **1926** Fowler *Mod. Eng. Usage* 155/1 Those who like a full-dress word better than a plain one continue to use..it. **1942** *Amer. Speech* XVII. 271/1 It amounted to a full-dress war argument once it was finished. **1962** J. Joesten *They call it Intelligence* ix. 71 He (Rudolf Abel) had become a full-dress member of the Communist Party.

fullerphone (fu·ləfəʊn). [f. name of Major A. C. Fuller + *-phone* in *telephone*.] A telegraphic instrument used in war-signalling.

1920 *Glasgow Herald* 2 Mar. 7 The claim of Major A. C. Fuller in respect of the 'Fullerphone'. **1922** *Encycl. Brit.* XXXI. 491/2 The fullerphone is a telegraph instrument, the essential point of which is the changing at the receiving end of a steady current into an interrupted current of audible frequency, while at the same time the current in the line remains steady. **1928** Blunden *Undertones of War* xxii. 253 The mechanism of the 'fullerphone' or 'power buzzer'.

full-face. [See Full *a.* 8 a, quots. 1702, 1876, 1895.] **1.** *attrib. phr.* = *full-faced *a.* 2.

1909 *Daily Chron.* 4 Mar. 1/3 The Halcyon was struck a heavy full face blow on the port quarter. **1937** R. H. Wilenski *Mod. Movement in Art* 133 The full-face eye in profile heads in Egyptian art.

2. *Printing.* A full-faced letter or fount of such letters (cf. next).

1892 [see Face *sb.* 21]. **1923** J. J. Little *Bk. Types, etc.* (N.Y.) 409 Bold Face.—A fullface letter similar to the roman, containing both hair lines and heavy strokes. **1963** Kenneth & Spilman *Dict. Printing* 77 *Full face*, a fount of capitals designed to occupy the complete body size, as there are no descenders to provide for.

full-faced, *a.* Add: **3.** *Printing.* Designating letters, chiefly capitals, which have a face occupying the complete body size.

1824 J. Johnson *Typogr.* II. i. 10 Being cast in all the various sizes, both Full-faced and Open. **1841** W. Savage *Dict. Art Printing* 247 A full faced letter is considerably larger in proportion than a letter of the regular face upon the same body. **1888** C. T. Jacobi *Printers' Vocab.* 49 *Full-faced letter*, a fount of capitals which has no beard on the top of the shank, occupying the whole depth of the body.

full house. 1. An assembly or audience which fills the building in which a performance is given or a meeting is held; also in extended use. Also, a session of a legislative or deliberative body, in which all or most of the members are present in their usual capacity (*Funk's Stand. Dict.* 1893).

[Cf. 1662–3 s.v. House *sb.*[1] 4 g, and 1720 s.v. Full *a.* 1.] [illegible] It brought full Houses for thirty-eight Nights. **1828** *Ohio* 20 Mar. 164/1 A party of itinerant Maromeros (or rope dancers) held their exhibition in the large walled yard..to about eight hundred people; which was considered as a very 'full house'. [illegible] when Spurs visit Old Trafford.

2. *Poker.* A hand containing three of a kind and a pair (next in value below four of a kind). Also *fig.*

1887 *Puck* (U.S.) 7 Sept. 21/2 Noah drew to pairs and got a full house [in the ark]. **1908** C. E. Mulford *Orphan* xxi. 267 You two make a pair of aces what can beat any full-house ever got together. **1922** —— *Tex* iii. 33 Tex wondered what the crowd would say if he should lean over and pull a royal flush out of Williams' ear, or a full-house from the nephew's nose. **1929** E. Linklater *Poet's Pub* xxv. 274 He had filled the kitty roof-high, bluffed the four-ace-players, scared the full-house-holders. **1963** G. F. Hervey *Handbk. Card Games* 238 If the 3 of Diamonds is discarded, the odds against improving to a full house is about 11 to 1.

full length. 2. *attrib.* and *ellipt.* (Earlier examples.)

1760 Sterne *Tr. Shandy* I. xxiii. 171 One of these you will see drawing a full-length character *against the light*. **1782** T. Pennant *Journ. Chester to London* 383 In the hall is a full-length of the unfortunate Mary Queen of Scots.

fullness, fulness. 8. (Earlier example.)

1801 Jane Austen *Let.* 5 May (1932) 127 No fulness appears either in the body or the flap; the back is quite plain.

full-ri·gger. [See Rigger[1] 4.] A full-rigged vessel. Also *fig.*

1899 C. J. Cutcliffe Hyne *Further Adv. Captain Kettle* vii, Image nodded towards the deserted vessel. 'Fine full-rigger, hasn't she been?' **1907** *Daily Chron.* 27 Dec. 3/2 A well-found full-rigger of 600 tons. **1928** *Ibid.* 17 Dec. 3/3 Instead of being the fine tall rigger that it might have been, it is, comparatively, only a topsail schooner.

full time. The total number of hours normally allotted to daily or weekly work, etc. Chiefly *attrib.* (hyphened) and advb., esp. in sense 'that occupies all one's time, that engages one to the exclusion of other activities'.

1898 *Daily News* 13 Dec. 5/7 The half-time system..does irretrievable hurt to the full-time scholars as well. **1911** *Rep. Labour & Social Conditions in Germany* III. 89 Full-time employment. *Ibid.* 97 Full time is worked in the Tobacco factory. [illegible] All the factories we visited were running full time. **1921** *Act* 11 & 12 *Geo. V* c. 51 § 77 (2) Any young person..shown..to be under suitable and efficient full-time instruction. **1926** in Fowler *Mod. Eng. Usage* 737/2 That motherhood is a full-time job with work-while mothers will readily admit. **1927** Bowley & Stamp *Nat. Income 1924* 31 The general average income of wage-earners in full-time work. **1930** A. Flexner *Universities* 87 So-called 'full-time' units, consisting of groups on salary devoting themselves solely to teaching, research, and the care of hospital patients. **1955** *Times* 23 July 3/6 But though the universities will probably be able to satisfy most of the demand for full-time degree courses they will certainly not be able to supply industry with all the technologists it requires. **1959** *Manch. Guardian* 6 Aug. 3/5 Mr. Wesker will become a full-time professional writer. **1968** *Times* 17 May 66 The Institute ..now employs 575 full-time civilian analysts. **1971** *Engineering* Apr. 55/2 It is the full-time business of these agencies.

full-timer. Add: **2.** One who works full-time.

1868 *Fortn. Rev.* Oct. 430 At thirteen the boy 'passes the doctor' (*i.e.* obtains a medical certificate of age), and becomes a 'full-timer'..amenable to the same rules, and subject to the same hours of labour, as the adult operative. **1877** H. E. Manning *Misc.* II. ii. 96 Is it possible for a child to be educated who becomes a full-timer at ten or even twelve years of age? **1974** *Times Lit. Suppl.* 21 June [Ital. Suppl.] p. v/4 If a full-timer, he began his duties before the lady rose, when he might have to air her linen.

fully, *adv.* Add: **a. fully-fashioned** *a.*, of a garment (esp. a stocking): shaped to fit closely to the body; also *transf.*; **fully-fledged** *a.* = full-fledged (see Full *adv.* 5 b).

1923 *Army & Navy Stores Catal.* 649 Cashmere hose, fully fashioned. **1945** E. Dabeley in L. Mall *B. D., Broadcaster* (1966) 41 A woman friend..wearing what looked to me like fully-fashioned silk stockings. **1946** *Picture Post* 11 May 17/2 Pure silk stockings cost about 15s. a pair, and fully-fashioned 'mixtures' from 12s. to over 60s. **1959** *Times* 18 May 13/4 More fully-fashioned stockings are being sold in Britain to-day than ever before. **1963** A. J. Hall *Textile Sci.* iii. 149 One must be made of a fully-fashioned knitting machine. [illegible] The contract plates has since been compulsorily converted into fully-fledged corporate bodies. **1939** *Discovery* Aug. 462/1 It was thought that a gap intervened between the Old and the New Stone Ages, during which man retreated from Northern Europe to return later fully-fledged as neolithic man. **1957** C. Hunt *Guide to Communist Jargon* xxvi. 92 Fully-fledged nations..possessing a common language, territory, economic life and national character. **1971** *Engineering* Apr. 27/2 Management and marketing are now fully-fledged subjects in their own right.

fully (fu·li), *v. slang.* [f. the adv., in phr. 'fully committed for trial'.] *trans.* To commit (a person) for trial. Hence **fu·llied** *ppl. a.*

1849 *Sessions Papers* 1 Feb. 324 The prisoner said..be expected either to be *turned up* or *fully'd*..—those are cant expressions, meaning either to be discharged, or committed for trial. **1859** Matsell *Vocabulum*, *Fullied*, committed for trial. **1926** E. Wallace *More Educated Evans* iv. 97 We found a lot of stolen property in his rooms, and he was fullied. **1936** J. Curtis *Gilt Kid* xxiii. 281 They'll fully me to the Old Bailey, I reckon.

fulminate, *v.* Add: **10.** *Path.* Of a disease: to develop suddenly and severely. (Cf. *fulminating *ppl. a.* 3.)

1910 *Practitioner* June 744 A gland presumably tuberculous... Sooner or later such a gland almost always fulminates, that is to say, rapidly bursts its capsule and allows the broken-down contents to invade the surrounding lymphoid tissues.

fulminating, *ppl. a.* Add: **1.** (Earlier U.S. example of *fulminating powder*.)

1804 T. G. Fessenden *Terrible Tractoration* 142 Sound Discord's jarring tocsin louder, Than Howard's fulminating powder.

3. *Path.* Of a disease: coming on suddenly with intense severity; foudroyant; = Fulminant *a.* 2.

1900 in Dorland *Med. Dict.* **1908** *Brit. Med. Jrnl.* 22 Aug. 477/1 Two cases of fulminating pyorrhœa alveolaris specifica. **1910** *Practitioner* Feb. 264 Fulminating cases of infection with virulent organisms. **1964** M. Hynes *Med. Bacteriol.* (ed. 8) x. 135 A fulminating gastro-enteritis which is commonly fatal. **1970** R. M. Goodman *Genetic Disorders Man* xvii. 571/2 The disease [*sc.* galactosemia] may be fulminating and result in early death.

fulvo-, used as combining form of Fulvous *a.*, meaning 'having a reddish-yellow hue'.

1879 W. A. Leighton *Lichen-flora Gt. Brit.* (ed. 3) 512 *Fulvo-testaceus*, tawny-yellow-brown. **1887** W. Phillips *Brit. Discomycetes* 356 Exterior fulvo-rufous. **1943** H. A. Dade *Colour Terminol. Biol.* 5 Armeniaca: apricot colour; lightly greyed orange-scarlet; = Fries's terms fulvo-cinnamomeus and helvolo-alutaceus.

fumagillin (fiū·mădʒi·lin). *Biochem.* [f. mod.L. *Aspergillus fumigatus* (see def.) by rearrangement of some of its elements: see -in[1].] An unstable colourless crystalline compound, $C_{26}H_{34}O_7$, which is produced by the growth of the fungus *Aspergillus fumigatus* and which has antibiotic activity against some viruses and protozoa.

1951 Eble & Hanson in *Antibiotics & Chemotherapy* I. 54 The present paper concerns the isolation and properties of a new antibiotic compound which we call fumagillin. [illegible] Encouraging results were achieved using fumagillin in chronic intestinal amoebiasis in man.

fumagine (fiū·mădʒin, -in). *Bot.* [Fr., f. *Fumago*.] A black superficial mould on plants, caused by fungi once grouped under the name *Fumago*, and associated with the honey-dew produced by certain insect pests.

1879 G. B. Buckton *Monogr. Brit. Aphides* II. 20 Passerini remarks that *Rhopalosiphum dianthi* is one of the most destructive Aphides in foreign greenhouses. They there give rise to a kind of mould on the plants they infest, to which the French give the name *Fumagine*. **1913** D. Grant tr. *Bourcart's Insecticides* 393 Fumagine is the term applied to the black coating which appears on certain plants infested by plant lice or cochineals (scale insects). This coating is formed by the black mycelium of a fungus which lives solely on the saccharine liquid, the honey-dew, which the insects project on the leaves.

fumaroid (fiū·măroid), *a. Chem.* [f. Fumar(ic *a.* + -oid.] Resembling fumaric acid in having a *trans* configuration in geometrical isomerism.

1894 [illegible] Many cases of stereoisomerism are believed to be explicable by formulæ resembling those given above, so that the expressions maleinoid and fumaroid structure are used. **1938** A. J. Mee tr. *Karrer's Org. Chem.* xvii. 256 It thus became common to call *cis*-ethylenic compounds, maleinoid, and *trans*-ethylenic derivatives, fumaroid forms. **1964** *Internat. Encycl. Chem. Sci.* 480/1 *Fumaroid form*, the axial symmetric or *trans* form of ethylene geometrical isomers, named from fumaric acid.

fumarolic (fiūmărǫ·lik), *a.* [f. Fumarole + -ic.] Of or belonging to a fumarole; formed by a fumarole.

1903 *Science* 3 Apr. 543 The placing of various ore deposits of many well-known districts in such classes as fumarolic, solfataric, pneumatolytic, etc...seemed to the present writer to be premature. **1944** C. A. Cotton *Volcanoes* xii. 203 In this case the fumarolic activity is on a grand scale. [illegible] Having showed the gypsum to be prominent in altered deposits at the sites of extinct or dying fumarolic activity.

fumaroyl (fiū·măroil). *Chem.* [f. as next + -oyl.] A bivalent radical, –CO·CH:CH·CO–, derived from fumaric acid.

1952 *Chem. Abstr.* XLVI. 10945/1 (Index), Fumaroyl chloride. **1965** *Nomencl. Org. Chem.* (I.U.P.A.C.) 136 Fumaroyl (preferred to *trans*-butenedioyl). **1970** R. W. McGilvery *Biochem.* xvii. 387 Another oxidase..cleaves the ring to form the *cis*-unsaturated compound, C-maleoylacetoacetate. An isomerase converts this to the *trans*-compound, C-fumarylacetoacetate.

fumaryl (fiū·măril, -ǝil). *Chem.* [f. Fumar(ic *a.* + -yl.] = prec.

1864 Watts *Dict. Chem.* II. 747 Chloride of Fumaryl. **1965** *Jrnl. Pharmacol.* LXI. 1168 There are marked differences in biological potency between maleyl and fumaryl dicholines compared with succinyl dicholine.

fumble, *v.* Add: **6.** Forming combs., as *fumble-fisted*, *-footed* adjs.

1847 Halliwell, *Fumble-fisted*, very awkward in handling things. *Suffolk.* **1877** A. Sewell *Black Beauty* xxxi. 149, I don't know what is the matter with this horse, he goes very fumble-footed. **1926** A. B. Smith *Studies & Caprices* 150 Music which even the most fumble-fisted can play with pleasure.

fume, *sb.* Add: **8. fume-chamber, -cupboard, -hood, -pipe**, ventilation contrivances for getting rid of noxious gases generated in laboratory work.

1905 *Strand Mag.* Apr. 422/1 It is..fitted up with.. 'muffles'..fume chambers, [etc.]. **1913** *Oxf. Univ. Gaz.* 4 June 943/2 (Apparatus of the pharmacological laboratory) Fume chamber. **1921** A. E. Munby *Laboratories* 35 Fume cupboards or draught closets..consist of a wood-framed glazed case in which the experiment is performed, with some special means of ventilation. **1938** *Jrnl. R. Aeronaut. Soc.* XLII. 744 The bath is placed in a fume cupboard with forced ventilation. **1921** A. E. Munby *Laboratories* 34 Fume hoods on the benches are sometimes provided in the form of a metal or wood tube or boxing with a small hood 9 or so in. square attached to a ventilating trunk in the bench. **1965** *Economist* 25 Dec. 1437/3 (*caption*) Fume hood. **1921** A. E. Munby *Laboratories* 48 Fume pipe for benches or fixture table.

fume, *v.* Add: **6. b.** *quasi-trans.* with a sentence or words as obj.: to utter irritatedly. Also with *away*: to pass or spend (time) fuming.

1897 W. W. Jacobs *Skipper's Wooing* xii, Glover fumed the afternoon away. **1907** *Munsey's Mag.* Dec. 392/1 'What are you doing—guying us?' fumed Bailey. **1908** *Smart Set* Sept. 76/2 'Good Call, Titcomb,' fumed Parmalee..'it's a chamber of horrors.'

fumed, *ppl. a.* (at end of Fume *v.*) Add: *fumed oak*, oak which has been darkened by exposure to ammonia vapour. (Cf. Fume *v.* 8 and Fumigated *ppl. a.*)

1902 *Idler* Nov. 355 Bookcase in Fumed Oak. **1910** *Encycl. Brit.* XI. 903/1 'Fumigated' or 'fumed' oak. **1915** T. Burke *Nights in Town* 110 There is one of those ubiquitous fumed-oak bookcases. **1926** *Punch* 23 Dec. 726/2 But Mrs. Twankey and her son Jack now live..in a semi-detached villa, where their fumed-oak and silver-plated butter-dishes are as good as anything to be seen in their neighbours. **1936** N. Coward *To-night at 8.30* ii. 33 (*title*) Fumed oak. A comedy in two scenes. *Ibid.* 35 There is a fumed oak dining-room suite consisting of a table, and six chairs. *Ibid.* [illegible] There was a bit more to it [*sc.* life] than refinement and fumed oak and lace curtains. **1958** B. Wright *Remodelling Old Furniture* vi. 140 Fumed oak should be coated with clear polish.

fumerole, var. Fumarole.

fumet[1]. Delete † *Obs.* and add later examples. Also a concentrated fish stock used for flavouring.

1877 E. S. Dallas *Kettner's Bk. of Table* 290 Pheasant... requires to be kept till the fumet is fully developed. **1906** Mrs. Beeton *Househ. Managem.* 1116 *Fumet*, the flavour or essence of game, fish, used to impart a rich flavour to certain dishes. **1936** Lucas & Hume [illegible] **1965** *House & Garden* Dec. 84/3 Fumet ..is very highly concentrated stock of, usually, game or fish, used to flavour other dishes or sauces.

fumigacin (fiūmigǟ·sin, fiumi·gāsin). *Biochem.* [f. mod.L. *fumig(a-tus* (see *fumigatin) + *-cin* as in *actinomycin*: see -in[1].] **a.** An antibiotic produced by the fungus *Aspergillus fumigatus*, now considered to be a mixture of the antibiotics helvolic acid and gliotoxin. **b.** = *helvolic acid.

1942 [see *clavacin]. **1944** *Jrnl. Biol. Chem.* CLII. 479 The crystalline material formerly described as fumigacin is a mixture of fumigacin and gliotoxin. Pure fumigacin..appears to be identical with the helvolic acid recently isolated. **1949** H. W. Florey et al. *Antibiotics* I. vii. 339 Fumigacin...appeared to differ..from helvolic acid. Further investigations..showed that fumigacin was, in fact, a mixture of helvolic acid and gliotoxin. **1963** J. Ramsbottom *Mushrooms & Toadstools* xxiii. 289 Fumigacin (helvolic acid)..has all the necessary qualities except that bacteria readily acquire resistance to it. **1959** [see *clavacin].

fumigant, *a.* and *sb.* **B.** *sb.* Delete *rare* and add later examples.

1940 *Chambers's Techn. Dict.* 359/1 Examples of fumigants are hydrogen cyanide and ethylene oxide. **1952** *Oxf. Jun. Encycl.* VI. 337/1 The farmer and gardener have a number of methods of control they can use against pests, apart from..spraying with..Fumigants. **1971** *Listener* 15 Apr. 493/2 Dr Bannerman had put microbes in the privy to test the fumigant.

fumigatin (fiumigǟ·tin). *Biochem.* [f. mod.L. *fumigāt-us*, specific epithet of the fungus *Aspergillus fumigatus*, f. L. *fūmigāt-* ppl. stem of *fūmigāre* (see Fumigate *v.*) + -in[1].] A maroon crystalline compound, $C_8H_8O_4$, with antibiotic activity, obtained synthetically and from *Aspergillus fumigatus*.

1938 Anslow & Raistrick in *Biochem. Jrnl.* XXXII. 695 A hitherto undescribed mould metabolic product fumigatin, $C_8H_8O_4$..[has] been isolated from cultures of *Aspergillus fumigatus* Fresenius. **1946** *Nature* 17 Aug. 241/1 Fumigatin, spinulosin, helvolic acid, and gliotoxin are the four metabolic products of *A. fumigatus* which show considerable antibiotic activity. **1967** E. Paryski tr. *Korzybski's Antibiotics* II. III. 1273 Fumigatin, $C_8H_8O_4$, is a quinone containing one methoxyl and one methyl group: 3-hydroxy-4-methoxytoluquinone.

fumigatorium (fiū:migātō^ə·riǝm). *U.S.* [f. Fumigat(e *v.* + -orium.] An air-tight container or building in which fumigation, esp. of plants, takes place.

1902 W. G. Johnson *Fumigation Methods* xi. 97 *A fumigatorium* is a house or room constructed or adapted for the fumigation of nursery stock or other materials. **1903** *Florida Experiment Station Bull.* LXIII. 651 To kill the white fly larvae in an air-tight fumigatorium, I commenced by using one gramme of KCN to 15 cubic feet of space. **1909** *U.S. Dept. Agric. Bur. Ent. Bull.* LXXXIV. 24 A larger fumigatorium was constructed. **1919** *U.S. Dept. Agric. Farmers' Bull.* MXXIX. 36 A fumigatorium is a container in which anything can be fumigated.

fun, *sb.* Add: **2.** Also, a source or cause of amusement or pleasure.

1934 *Punch* 9 May 526/1 A Rector in an unapostolic fury is rather fun. **1954** *Economist* 20 Mar., His book has all the charm of science fiction; it is enormous fun. **1958** *Listener* 25 Dec. 1085/1 The clothes were Jacobean, and fun to wear.

b. *like fun* (earlier and later examples); *for the fun of the thing*: for amusement; *to have fun* (*with*): to enjoy (a process); *spec.* to have sexual intercourse.

1833 C. A. Davis *Lett. J. Downing* (1834) 24 And began to laugh like fun. **1848** Mrs. Gaskell *M. Barton* I. v. 73 Carsons' mill is blazing away like fun. **1877** *Independent* 19 July 15/2 Little Tad commissioned lieutenant by Stanton, 'just for the fun of the thing'. **1893** Farmer & Henley *Slang* III. 86/2 *To have* (or *do*) *a bit of fun*, to procure or enjoy the sexual favour. **1903** Beerbohm *Around Theatres* (1924) I. 425 Amateur mimes..go in for private theatricals..just for the fun of the thing. **1958** *Times Lit. Suppl.* 7 Feb. 73/4 The clerks..get their own back by unmasking frauds and..having fun with the low standard of French commercial honesty. **1961** M. Dickens *Heart of London* II. 198 Ambrosia had pushed Edgar and the girl in there with the admonition to have some fun, dears.

c. Exciting goings-on. Also *fun and games*, *freq.* used ironically; *spec.* amatory play. *colloq.*

1879 [see *Chinkey, etc.]. **1897** *Daily News* 13 Sept. 7/1 The engineer officers who are engaged in carrying out some of the Sirdar's plans get much more than their fair share of 'the fun'. **1898** *Westm. Gaz.* 28 Oct. 3/1 It is possible that there may be rare fun by-and-by on the Nile. **1920** 'Sapper' *Bull-Dog Drummond* vi. 155 We've had lots of fun and games since I last saw you. **1940** N. Mitford *Pigeon Pie* iii. 66 Farther on, however, you come to jolly fun and games—great notices. **1948** Partridge *Dict. Forces' Slang* 78 *Fun and games*, any sort of brush with the enemy at sea. **1948** 'N. Shute' *No Highway* iii. 70 'Fun and games,' he said. 'The boffin's going mad.' **1952** E. Grierson *Reputation for Song* xxix. 260 Beneath the orderly conduct of her bar there was always present the possibility of 'fun and games'. **1954** C. Armstrong *Better to Eat You* ii. 22 If it happened because somebody is having fun-and-games with Miss Sarah Shepherd, somebody is going to be sorry. **1966** J. Porter *Sour Cream* v. 59, I headed the car in the direction of the coast road. We had the usual fun and games with the local drivers. **1970** *Globe & Mail* (Toronto) 26 Sept. B3/3 Mr. Brown also expects the fun and games of tax haven subsidiaries to disappear with the new legislation.

3. a. *attrib.* and *Comb.*, passing into *adj.* with the sense 'amusing, entertaining, enjoyable'.

a **1846** B. R. Haydon *Autobiogr.* (1927) III. xvii. 358 There was a room at Holly House called the 'fun-room', without chair or table. It was for dancing and romping. **1853** N. P. Willis (*title*) Fun jottings; or, Laughs I have taken pen to. **1908** *Daily Chron.* 26 Dec. 4/4 The side-show is blossoming out again at all points of the compass in 'fun towns' and the like. **1959** J. Osborne *World of Paul Slickey* I. vii. 61 You'll always be a Fun Person. **1962** *Sat. Even. Post* 13 Oct. 69 Some fur coats are, however, just for fun. The 'fun fur' coat has given a big boost to the industry. **1962** *Sunday Express* 16 Dec. 18/5 Nowadays you can't rustle up enough fun people for a small party any more. **1965** *New Statesman* 7 May 712/3 Millions have sampled the delights of St Tropez and St Moritz and a lot more of the so-called 'fun places' for themselves. **1965** *Punch* 18 Aug. 254/3 The cheap 'fun furs' acclaimed as so young and amusing. **1968** A. Diment *Bang Bang Birds* x. 186, I was remembering Marianne and the fun times we have had. **1968** J. Ironside *Fashion Alphabet* 151 The young have taken to 'fun' furs which may be rabbit dyed to any colour under the sun. **1969** *Listener* 13 Feb. 221/3 Much better to give St Katharine Docks to Joan Littlewood for a fun palace. **1971** *New Yorker* 8 May 107 We have the Osborns, the Beals, the Hartungs, the Falmers, and us. Now let's think of someone fun.

b. Special comb., as **fun fair**, a fair (or that part of a fair) which is devoted to amusements and side-shows; **funfest** [*fest] chiefly *U.S.*, a gathering for the purposes of amusement; **fun-maker**, a comedian, humorist, jester.

1925 A. Huxley *Those Barren Leaves* II. iii. 106, I cannot claim to bring every attraction of the Fun Fair into your place of labour—only the switchback, the water-shoot and the mountain railway. **1951** A. Baron *Rosie Hogarth* 16 Fun fairs discharge their screech and blare upon the passer-by. **1955** *Times* 18 July 5/3 The fun fair at the Festival Gardens, Battersea, opened bashfully.. for its first Sunday session yesterday. There was no music. Nor was there any beer. **1918** in *Dialect Notes* V. 11 (*headline*) Lincoln High invites to fun-fest. **1922** S. Lewis *Babbitt* xiii. 172 You will get 111% on your kale in this fun-fest. **1962** J. D. MacDonald *Key to Suite* x. 140 The convention..wasn't..a fun-fest, a week of broads and bottles. **1963** *Punch* 4 Dec. 803/1 A riotous funfest. **1904** *Daily Chron.* 12 Nov. 5/6 Mr. James Welch to be Prominent Fun-maker at Drury Lane. **1906** R. L. Ramsay *Skelton's Magnyf.* p. xcvii, The two fun-makers of the morality are the brothers Fancy and Folly. **1936** *Variety* 15 July 14/2 At last those three ace fun makers.. get a chance to go to town in a big time laugh hit of their own.

function, *sb.* Add: **3. c.** *spec.* in *Philology.* So **function word** (see quot. 1940).

1894 O. Jespersen *Progress in Lang.* v. 135 Syntax is nothing but the theory of the functions, *i.e.*, meanings, of the grammatical forms. **1919** E. Classen *Outl. Hist. Eng. Lang.* ii. 35 One of the consequences of the fixing of the word order was that each word in the sentence had its particular syntactic function according to its exact position, and that the survivals of free word order which were to be found in Old English disappeared in Middle English. **1926** L. Bloomfield in *Language* II. 159 The positions in which a form occurs are its *functions*. **1933** —— *Lang.* xvi. 265 The lexical form in any actual utterance, as a concrete linguistic form, is always accompanied by some grammatical form: it appears in some function, and these privileges of occurrence make up, collectively, the grammatical *function* of the lexical form. *Ibid.*, Lexical forms which have any function in common, belong to a common form-class. **1940** C. C. Fries *Amer. Eng. Gram.* 109 By a function word I mean a word that has little or no meaning apart from the grammatical idea it expresses. **1963** F. T. Visser *Hist. Syntax* I. iii. 189 The verb is not a content-word, but a function-word, and is traditionally called 'copula'. **1967** [see *functor 2].

d. *Computers.* Any of the basic operations in a computer, esp. one that corresponds to a single instruction.

1947 D. R. Hartree *Calculating Machines* 11 Consider, therefore, the functions required in the operation of a digital calculating machine. *Ibid.* 12 The components carrying out these functions may not all be physically distinct; for example a single unit may act both as an adding unit and a memory unit. **1952** *Math. Tables & other Aids to Computation* VI. 170 A 'left-shift' function is available. This shifts the entire product one place to the left and may be called by a digit impulse supplied to its function gate. **1962** *Gloss. Terms Autom. Data Processing (B.S.I.)* 37 *Function part* (operation part, function digits, function number), that part of an instruction which specifies the operation to be performed. **1964** F. L. Westwater *Electronic Computers* i. 10 A calculation can be carried out if the necessary functions are carried out in the correct order.

6. function space, a topological space the elements of which are functions.

1931 H. P. Robertson tr. *Weyl's Theory of Groups & Quantum Mech.* i. 32 The totality of such functions $\psi(s)$ therefore constitute a linear 'function space' of continuously infinite dimensions. **1961** R. A. Silverman tr. *Smirnov's Linear Algebra & Group Theory* v. 218 So far, we have considered the space *H*, where a vector is defined as an infinite set of numbers... We now consider a function space *F*, where continuously varying functions of one or several variables play the role of vectors. **1968** E. T. Copson *Metric Spaces* iv. 57 The simplest function space consists of all functions $x(t)$ continuous on a closed interval $[a, b]$.

function, *v.* **1.** Add later examples of the meanings derived immediately from F. *fonctionner*.

1897 G. Allen *Evol. Idea of God* iii. 46 We..know.. consciousness ceases altogether at death, when the brain no longer functions. **1907** *Daily Chron.* 21 Aug. 4/7 This joy does not 'function', as the French so charmingly say, over the present sample of autumnal weather in a month that should strictly be August. **1918** *Times* 18 Apr. 8/3 The agencies of obstruction and party intrigue which will immediately begin to function in every section of the political arena. **1924** Galsworthy *White Monkey* I. viii, There are rules of the game which must be observed, if society is to function at all. **1965** M. Spark *Mandelbaum Gate* iii. 50 Freddy functioned on with his letter, as he had done for thirty years of his natural history, a letter a week.

functional, *a.* Add: **1. b.** Relating to the system which specializes and divides the functions of managers, workers, or employees in a business, factory, etc. *U.S.*

1903 F. W. Taylor *Shop Management* § 234 'Functional management' consists in so dividing the work of management that each man from the assistant superintendent down shall have as few functions as possible to perform. *Ibid.* § 240 The four functional bosses who are a part of the planning department. *Ibid.* § 245 Functional Foremanship. **1911** —— *Princ. Sci. Management* 65 It is necessary, therefore, to provide teachers (called functional foremen) to see that the workmen both understand and carry out these written instructions. Under functional management, the old-fashioned single foreman is superseded by eight different men, each one of whom has his own special duties. **1930** M. Clark *Home Trade* 198 Foremen are of the type termed 'functional'.

c. As related to the arts, esp. to architecture: designating work executed with a view to its utilitarian purpose; also, of artists, builders, etc.: concerned with the use intended for their product, not with traditional or other theories of design.

1928 G. H. Edgell *Amer. Archit. To-day* iv. 301 These ..stand for the two types of daring engineering with architectural beauty and of pre-eminent monumentality with perfect functional fitness. **1931** C. H. Reilly in W. Rose *Outl. Mod. Knowl.* 997 Their buildings are in the first and last place functional and..with them form follows function both within and without. **1937** H. Read *Art & Society* vii. 261 (*heading*) Functional Art. **1938** *Encycl. Brit. Bk. of Yr.* 54/2 Even before the World War.. structural requirements, *i.e.* 'functional' plan, more often conflicted than harmonized with the 'architectural' design of a building. **1956** A. Wilson *Anglo-Saxon Att.* I. iii. 81 Light industry was embodied in a splendid compromise between functional and more conventional English taste. **1957** J. Braine *Room at Top* vii. 72 It [*sc.* the house] was 1930 functional in white concrete, with a flat roof. **1966** *Illustr. London News* 30 July 29/1 The outside of this building is functional and severe.

2. a. Also of a mental disorder: having no discernible organic cause.

1920 *Proc. R. Soc. Medicine* XIII. III. *Psychiatry* 32 Functional disorders of particular systems and communities of neurones. **1926** W. McDougall *Outl. Abnormal Psychol.* i. 1 There are two great classes of disorders of our mental life, those that are directly due to organic lesions of the nervous system and those which seem to imply no such lesion, no gross injury to the structure of the brain, and which are therefore called 'functional disorders'. **1960** Hinsie & Campbell *Psychiatric Dict.* 310/2 A functional mental disorder is one which, insofar as knowledge permits, stems from the psyche. **1962** *Listener* 21 June 1082/3 The patient complains of, and suffers, the appropriate pains; but there are no, or inadequate, physiological grounds for them; in current medical jargon they are 'functional'. **1971** *Brit. Med. Bull.* XXVII. 77/1 The term 'functional psychosis' has no precise meaning.

c. *Social Sciences.* Of or pertaining to a particular mental or social function. Applied to an approach to the study of the interrelations of particular phenomena within a given framework or structure, freq. in contrast to a structural approach.

1884 W. James in *Mind* IX. 19 The contrast is really between two *aspects*, in which all mental facts without exception may be taken; their structural aspect, as being subjective, and their functional aspect as being cognitions. **1898** *Philos. Rev.* VII. 451 There is, however, a functional psychology, over and above this psychology of structure. **1921** *Amer. Jrnl. Psychol.* XXXII. 519 Functional psychology, in this sense, is especially American, and the psychology of act especially German. *Ibid.* 533 If the view of my book is accepted, both 'functional' and 'structural', as qualifications of 'psychology', are now obsolete terms. **1926** B. Malinowski in *Encycl. Brit.* Suppl. I. 132/2 *The Functional Analysis of Culture.*—This type of theory aims at the explanation of anthropological facts at all levels of development by their function, by the part which they play within the integral system of culture, by the manner in which they are related to each other within the system, and by the manner in which this system is related to the physical surroundings. **1937** G. W. Allport *Personality* (1938) v. 133 The boundaries between the functional systems are weak in infancy, causing the child to react as a whole. *Ibid.* vii. 191 For convenience of discussion this new principle [of growth] may be christened the functional autonomy of motives. **1944** B. Malinowski *Sci. Theory of Culture* x. 115 The cogency of the functional approach consists in the fact that it does not pretend to forecast exactly how a problem posed for a culture will be solved. **1948** T. G. Andrews *Methods of Psychol.* i. 10 Experiments of the functional type are designed to answer the question *how*. **1951** R. C. Sheldon in Parsons & Shils *Toward Gen. Theory of Action* I. 35 Such difficulties are at the bottom of the dissatisfaction that many social scientists feel with the functional approach in social science, which assumes that every action has the function of promoting the maintenance of a system. **1951** E. E. Evans-Pritchard *Social Anthropol.* iii. 56 Functional anthropology, with its emphasis on the concept of social system and hence on the need for systematic studies of the social life of primitive peoples as they are today,..not only separated..social anthropology from ethnology; it also brought together the theoretical study of institutions and the observational study of primitive social life. **1968** *Encycl. Soc. Sci.* VI. 21/2 Structural-functional analysis is not new in either the social or the natural sciences... The only new aspect of it is its formidable new name, structural-functional analysis. *Ibid.* 23/2 What general conditions must be met (i.e. functional requisites) if the unit is to persist in its setting without change.

d. Pertaining to or serving a function (opp. *functionless*): see *transf.* examples in Dict. after sense 2 b. (Also further examples of *transf.* use.) Also, practical, utilitarian.

1890 W. James *Princ. Psychol.* I. x. 401 If there be such a meaning..it alone can make clear to us why such finite human streams of thought are called into existence in such functional dependence upon brains. **1943** *Mind* LII. 360 The whole point is to discover unexpected functional dependences, and this can only be done by operating in unusual conditions and with unusual material. **1950** *Times Lit. Suppl.* 13 Jan. 26/2 In a good poem imagery and rhythm are functional, not merely decorative. **1960** C. Winick *Dict. Anthropol.* 535/1 The instability in human speech is caused by the varying needs of society. These needs are manifest in functional changes of the elements, eventually resulting in modifications of the linguistic structure. **1961** E. Nagel *Structure of Sci.* iv. 73 The more sophisticated modern notions of cause as invariable functional dependence. **1962** J. Glenn in *Into Orbit* 40 It was an extremely difficult vehicle to build... It was not perfect; but it was functional.

4. *Logic.* **functional calculus** = *predicate calculus.

1933 M. Black *Nature of Math.* 63 These new symbols are both primitive in the functional calculus. **1951** *Mind* LX. 265 The basic notions of the lower functional calculus.

B. *sb. Math.* [ad. F. *fonctionnelle* in same sense (M. Fréchet 1906, in *Rendiconti del Circ. matem. di Palermo* XXII. 38; Fréchet ascribes the word to Hadamard (1903), who, however, used it only as an adj., in *opération fonctionnelle*).] A function the value of which depends

on the whole form of another function; **functional analysis**, the analysis of functionals.

1917 *Proc. Nat. Acad. Sci.* III. 640 (*heading*) On bilinear and *n*-linear functionals. **1918** G. C. Evans *Functionals & their Applications* i. 1 The maximum of a function is a functional of that function. **1945** E. T. Bell *Devel. Math.* (ed. 2) xxii. 535 Volterra used the terminology 'functions depending on other functions', and the special case of 'functions of lines', for what were subsequently called functionals..by Hadamard. *Ibid.*, $F[x(t)]$ is called a functional of $x(t)$ when its value depends on all the values assumed by $x(t)$. **1948** E. Hille (*title*) Functional analysis and semi-groups. **1966** R. Stone *Math. in Soc. Sci.* i. 11 Often when taking a decision.. what we have to maximize (or minimize) is a functional rather than a function. **1968** E. T. Copson *Metric Spaces* Pref., The theory of the topology of metric spaces..is not only the basis of functional analysis but also unifies many branches of classical analysis.

fu·nctionalism. [f. Functional *a.* + -ism.] **1.** *Social Sciences.* The method of studying, or the theory of, the functional interactions and adaptations of particular phenomena within a given framework or structure.

1914 J. B. Watson *Behavior* i. 9 We advance the view that behaviorism is the only consistent and logical functionalism. **1932** *Encycl. Social Sci.* VI. 524/1 The reason is probably the fact that the whole trend of the social process since Darwin has been in the direction of functionalism and might be described as its verification and validation. **1933** P. Radin *Method & Theory Ethnol.* vi. 176 Why a new term like functionalism should be applied to it is not at all evident. **1937** R. H. Lowie *Hist. Ethnol. Theory* xiii. 235 Malinowski's functionalism is avowedly antidistributional, antihistorical, and treats each culture as a closed system except insofar as its elements correspond to vital biological urges. **1944** B. Malinowski *Sci. Theory of Culture* x. 117 The type of criticism levelled against functionalism, to the effect that it never can prove why a specific form..of table implement or theological concept, is prevalent in a culture, derives from the prescientific craving for first causes. **1951** R. B. MacLeod in Rohrer & Sherif *Social Psychol. at Crossroads* 224 Functionalism..has settled down to the 'middle of the road' American psychology. **1956** J. H. M. Beattie in A. Pryce-Jones *New Outl. Mod. Knowl.* 259 But basic to all forms of functionalism is the view that the facts of social life may not usefully be thought of as a collection of separate elements. **1963** C. Jacobson tr. *Lévi-Strauss's Struct. Anthropol.* v. xv. 290 Therefore, historico-geographical concerns should not be excluded from the field of structural studies, as was generally implied by the widely accepted opposition between 'diffusionism' and 'functionalism'.

2. Regard for the function and purpose of a building as regulating its shape and style. Also in extended use.

1930 *Observer* 29 June 20 Sugar-cube architecture, or to be impressively abstract 'Functionalism'. *Ibid.* 5 Oct. 19 This is what is called the architecture of functionalism. The architectural form arises purely out of the purpose of the building. **1935** *Fortnightly* Apr. 410 We are now told that 'Functionalism' is the one and only test of art. **1941** *Burlington Mag.* Feb. 66/2 His definition of modern 'Functionalism' will not satisfy all readers, though it will make many of them chuckle: 'The theory that any building which could be made to stand up at all was All Right.' **1955** S. Spender *Making of Poem* i. 18 Functionalism is the philistinism of people who talk about a work of art as a 'well-done job' like any other piece of plumbing. **1957** E. H. Gombrich *Story of Art* xxvii. 421 The theories for which the Bauhaus stood are sometimes condensed in the slogan of 'functionalism'—the belief that if something is only designed to fit its purpose we can let beauty look after itself.

fu·nctionalist. [f. Functional *a.* + -ist.] **1.** *Social Sciences.* An adherent or advocate of the functional approach (see *functional *a.* 2 c). Also *attrib.* or as *adj.*

1914 J. B. Watson *Behavior* i. 8 The difference between functional psychology and structural psychology, as the functionalists have so far stated the case, is unintelligible. *Ibid.*, The terms sensation, perception, affection, emotion, volition are used as much by the functionalist as by the structuralist. **1931** R. S. Woodworth *Contemp. Schools Psychol.* iii. 48 This attempt of the functionalists to revise the definition of psychology by lining it up with biology. **1933** P. Radin *Method & Theory Ethnol.* ii. 24 The so-called functionalist school of Radcliffe-Brown and Malinowski. **1937** R. H. Lowie *Hist. Ethnol. Theory* xiii. 236 There is only one natural unit for the ethnologist—the culture of all humanity at all periods and in all places; only when the functionalist has..defined his particular culture within that frame of reference, does he know what he is talking about. **1951** R. B. MacLeod in Rohrer & Sherif *Social Psychol. at Crossroads* 224 For the functionalist perceiving is also a function, one of the organism's ways of adjusting itself. **1952** W. J. H. Sprott *Social Psychol.* 221 The 'Functionalists'—such as Malinowski and Radcliffe-Brown—come to our aid. At least every society must provide means of satisfying the basic needs of its members, and we can therefore explain many of their institutions in terms of the function they perform to this end. **1961** *Oxf. Mag.* 1 June 387/1 Professor Evans-Pritchard's criticisms of 'functionalist' (i.e. unhistorical) anthropology have been made on earlier occasions. **1970** E. Leach *Lévi-Strauss* i. 7 Most..social anthropologists..claim to be 'functionalists'; broadly speaking they are anthropologists in the style and tradition of Malinowski. In contrast, Claude Lévi-Strauss is a social anthropologist in the tradition of Frazer.

2. An adherent of functionalism in design. Also *attrib.* or as *adj.*, exhibiting functionalism.

1932 H. Robertson *Mod. Archit. Design* iv. 140 The functionalist, in theory at any rate, seeks in design only the most effective expression of the *use* of any object, whether it be a house to live in, or a chair to sit on. **1934** *Archit. Rev.* LXXV. 9 Having been first designed as a quasi 'free classic' building, it was, as it were, driven to re-designing itself in a more functionalist fashion by sheer pressure of circumstances. **1939** O. Lancaster *Homes Sweet Homes* 78 Recent compliance on the part of the insular British with the extreme dictates of the continental functionalists.

functionalize, *v.* Add: *U.S.* In the theory of business management: to distribute or assign (work) with due regard to the special function of the individual worker. Hence **fu:nctionaliza·tion**, the system governing such assignment of work.

1923 R. H. Lansburgh *Industr. Management* 55 Functionalization has brought with it basic changes in the structure of industrial organizations. *Ibid.* 60 These functionalized foremen. *Ibid.* 63 Functionalized departments working through one foreman. **1925** W. H. Leffingwell *Office Management* 108 As business grows ever larger and becomes increasingly functionalized and specialized. *Ibid.* 118 The functionalization of all industrial departments. **1958** *Times Lit. Suppl.* 4 Apr. 186/3 Collectivization and functionalization, in short, are facts: or rather, series of events, because they have happened historically and uncritically.

functionally, *adv.* (Later examples.)

1927 E. A. Sonnenschein *Soul of Grammar* 3 Two cases might differ functionally without differing morphologically. **1934** H. Read *Art & Industry* 11 C. 91/1 For sitting upright, actually the stool or the bench is all that is functionally necessary. **1951** E. E. Evans-Pritchard *Social Anthropol.* ii. 22 [Montesquieu] had the idea of everything in a society and its ambient being functionally related to everything else. **1971** *Nature* 30 Apr. 578/3 The indications are clear that the ancestral middle Miocene hominoids were structurally and functionally advanced in this locomotor style.

functor (fɒ·ŋktəɹ). [f. Funct(ion *sb.* + -or, after type *factor*.] **1.** *Logic.* A function or operator.

1937 A. Smeaton tr. *Carnap's Logical Syntax Lang.* § 3. p. 14 In order to express properties or relations of position by means of numbers, we shall use functors. *Ibid.*, Besides such descriptive functors, we make use also of logical functors. For example: 'sum (3, 4)' has the meaning: '3 plus 4'. **1947** H. Reichenbach *Elem. Symb. Logic* § 54. p. 312 Among the functors, the unique mathematical functions are of particular importance. **1951** J. Łukasiewicz *Aristotle's Syllogistic* iv. 78 The principle of my notation is to write the functors before the arguments. In this way I can avoid brackets. **1956** E. H. Hutten *Lang. Mod. Physics* vi. 232 Newton's first law..finds a 'natural' translation as, e.g. (x)(Rx v Ux⊃∼Fx), if the functors R, U, and F mean 'being at rest', 'being in uniform motion', and 'being acted upon by a force', respectively. **1963** *Med. Ævum* XXXII. 144 One should have '*can* exist of itself'; this modal functor.. is vital to the course of the..exposition.

2. *Linguistics.* = *form-word; also used more widely, to include similar affixes, etc.

1958 C. F. Hockett *Course in Mod. Ling.* xxxi. 264 There are at least three types of grammatical forms which are to be classed as functors. **1964** R. A. Hall *Introd. Ling.* iii. 15 These elements which indicate grammatical functions are termed 'function-words' or 'functors'. **1965** *Listener* 17 June 903/2 In themselves prepositions (which may be described as structure-words or functors) have no lexical meanings. **1967** R. A. Waldron *Sense & Sense Devel.* ii. 45 Even individual words like *and*, *but*, *or*,.. (which are variously known as function words, form words, operators, functors, or kenemes)..are defined grammatically rather than, or as well as, lexically.

Hence **functo·rial** *a.*

1951 J. Łukasiewicz *Aristotle's Syllogistic* v. 132 All these may be called functorial propositions, since in all of them there occurs a propositional functor, like 'if—then', 'or', 'and'. **1955** A. N. Prior *Formal Logic* I. iv. 94 If quantifiers and functorial variables are introduced, we may obtain just as rich a system.

fund, *sb.* Add: **8.** *fund-raising* adj. and sb.; so *fund-raiser*.

1957 V. Packard *Hidden Persuaders* i. 8 A great many advertising men, publicists, fund raisers, personnel experts, and political leaders..still do a straightforward job. **1961** *Guardian* 26 Jan. 22/6 Money should..be available for..a..memorial..if the Press gave the fund-raisers its backing. **1940** *Struct. Amer. Econ.* (Nat. Resources Planning Board) II. 39/1 The entire present contributory basis of the social security system could be reexamined with a view..to merging fund-raising for social security with fund raising for all other governmental purposes. **1954** Koestler *Invis. Writing* xx. 224 Willy asked me to..write a fund-raising pamphlet about it. **1959** *Times* 30 Dec. 9/1 A worldly young Rabbi who is more interested in fund-raising than in religious rites. **1964** A. Wykes *Gambling* x. 242 The fund-raising potentials of the state lottery. **1970** *New Yorker* 14 Nov. 44/2 Marriage and fatherhood eventually nudged him into public relations and fund-raising.

fund, *v.* Add: **5.** *trans.* To supply with funds, pay (a person); to finance (a position or enterprise).

1900 *Westm. Gaz.* 6 July 6/3 The War Office had..given the London Scottish commander to understand that they ..would 'fund'..the 320 men who were in readiness to join the emergency camp. **1966** *New Statesman* 7 Jan. 27/4 (Advt.), These posts are funded from a Hayter Committee Grant. **1970** *Sci. Jrnl.* Jan. 28/4 We work in a system in which research projects are funded by grants. **1970** *Sunday Times* 8 Mar. 60/8 An average local radio station.. costs £80,000 a year to run. Eventually all will be funded wholly by the BBC, at an annual cost..of some £3,200,000.

fundament. Add: **1. c.** *Geogr.* Any landscape before colonization by man in general or by any particular group of men.

1928 J. B. Leighly in *Univ. Calif. Publ. Geogr.* III. 3 The forces which condition and shape a cultural landscape are many and of varied origin, each fluctuating in intensity through time... The natural fundament which they modify, on which they erect their proper structures, is similarly varied from place to place, itself changing through natural processes. **1934** *Ann. Assoc. Amer. Geogr.* XXIV. 80 By an extension of its dictionary meaning, fundament is used to indicate the foundation on which the works of man have been built. Fundament may be defined as the face of the earth as it existed before the entrance of man into the scene. **1954** *Ibid.* XLIV. 248 What the fundament or 'natural' Calumet originally looked like.

fundamental, *a.* Add: **4.** *fundamental complex* (see quot. 1961).

1893 A. Geikie *Text-bk. Geol.* (ed. 3) II. VI. I. 715 The pre-Cambrian rocks..may be divided into two great series. At the base lies a vast mass of gneisses, schists, and eruptive rocks, which, known as the 'Fundamental Complex', is regarded as the oldest of the whole. **1910** *Encycl. Brit.* II. 361/1 The so-called 'fundamental complex', an assemblage of acid, basic and intermediate irruptive rocks, associated together in a complex of extraordinary intricacy. **1961** L. D. Stamp *Gloss. Geogr. Terms* 202/2 *Fundamental complex*, in geology the rocks of the 'original' crust of the earth formerly applied to the great areas of pre-Cambrian crystalline rocks. It is still used although it is now recognized that probably no part represents the 'original' crust of the earth.

fundamentalism (fɒndăme·ntăliz'm). [f. Fundamental *a.* + -ism.] A religious movement, which orig. became active among various Protestant bodies in the United States after the war of 1914–1918, based on strict adherence to certain tenets (e.g. the literal inerrancy of Scripture) held to be fundamental to the Christian faith; the beliefs of this movement; opp. *liberalism* and *modernism*.

1923 *Daily Mail* 24 May 8 Mr. William Jennings Bryan ..has been exerting the full force of his great eloquence in a campaign on behalf of what is termed 'Fundamentalism'. **1925** K. Lake *Relig. Yesterday & To-morrow* 63 There has been in America some surprise at the sudden rise of Fundamentalism in the last five years. **1927** *Observer* 5 June 5/3 Fundamentalism and the Klux Klan are signs of alarm on behalf of the older ideals. **1955** *Times* 25 Aug. 14/1 'Fundamentalism'..appears to have been used first in connexion with the (American) Northern Baptist Convention of 1920 to describe the more conservative delegates who desired 'to restate, reaffirm, and re-emphasize the fundamentals of our New Testament faith'. *Ibid.*, Now 'fundamentalism'..appears to describe the bigoted rejection of all Biblical criticism, a mechanical view of inspiration and an excessively literalist interpretation of scripture.

So **fundame·ntalist**, an adherent of fundamentalism. Also *attrib.* or as *adj.*, and *transf.*

1922 *Contemp. Rev.* July 20 The fundamentalist creed. *Ibid.* 21 The Fundamentalists have been fortunate in their non-ministerial leader [*sc.* W. J. Bryan]. **1925** K. Lake *Relig. Yesterday & To-morrow* 60 The most energetic.. group, but the least well educated, is the Fundamentalist. *Ibid.* 62 The Fundamentalists have zeal, but it is certainly not according to knowledge. **1926** H. F. Osborn *Evol. & Relig. in Educ.* 12 The fundamentalist movement.. sought to re-establish the Biblical literalism of the time of Cromwell, Milton, and the Puritans. **1955** *Times* 25 Aug. 14/1 The dangers of the new fundamentalist movement. One of the encouraging developments for Christian teachers to-day is the new relationship which is growing up between scientific and religious thought. **1969** *New Yorker* 14 June 45/1 I've never been a flashy stylist, like Arthur. I'm a fundamentalist. Arthur is a bachelor. I am married and conservative.

fundamentum. Short for *fundamentum relationis.

[**1599** Blundevil *Logike* 33 Here note that of the schoolemen the thing from which the application is made, is called in latine *fundamentum*, in English the foundation.] **1883** F. H. Bradley *Princ. Logic* II. II. i. 295 What is the fact or phenomenon constituting the *fundamentum* of this relation? **1907** W. James *Meaning of Truth* (1909) vii. 163 This relation, like all relations, has its *fundamentum*. *Ibid.*, In the case of the relation between 'heir' and 'legacy' the *fundamentum* is a world in which there was a testator, and in which there is now a will and an executor.

‖ **fundamentum divisionis** (fɒndăme·ntǝm divizĭǒu·nis). *Logic.* Pl. **-menta** (-me·ntă). [L., = foundation or basis of division.] The principle or basis of logical division of a genus into its constituent species.

1849 W. Thomson *Outl. Laws of Thought* (ed. 2) 134 The division must be made according to one principle (*fundamentum divisionis*). **1870** W. S. Jevons *Elem. Lessons Logic* xii. 105 The size of the books is in this case the ground, basis, or principle of division, commonly called the *Fundamentum Divisionis*. **1906** H. W. B. Joseph *Introd. Logic* v. 104 The *fundamentum divisionis*, the principle or basis of a division, is that aspect of the genus, in respect of which the species are differentiated. **1956** J. O. Urmson *Philos. Analysis* v. 68 The distinction 'positive-negative' has a different *fundamentum divisionis* from the distinction 'particular-general'.

fundamentum relationis (fʊndǽme·ntəm rɪlēiʃiōu·nis). *Logic.* [L., = foundation of the relation.] Those elements of the objective world that constitute the terms of a relation. (See also *FUNDAMENTUM.)

1843 MILL *Logic* I. ii. § 7. I. 55 The series of events may be said to *constitute* the relation; the schoolmen called it the foundation of the relation, *fundamentum relationis*. **1906** J. N. KEYNES *Formal Logic* (ed. 4) I. iv. 64 The fact or facts constituting the ground of both correlative names is called the *fundamentum relationis*. For example,..in the case of husband and wife, the facts which constitute the marriage tie. **1907** W. JAMES *Let.* 4 Aug. in R. B. PERRY *Tht. & Char. W. J.* (1935) II. 540 The *fundamentum relationis* of the fact that the idea may mean and point to that reality, and know it truly, is to be found in the enveloping experiences and nowhere else.

fundatrix (fʊndēi·triks). Restrict †*Obs.* to sense in Dict. and add: **2.** *Ent.* The founding female of a colony of aphids, which produces young parthenogenetically.

[**1875** G. B. BUCKTON *Monogr. Brit. Aphides* I. 76 This hibernation of the foundress of a colony..is somewhat exceptional.] **1907** W. R. FISHER *Schlich's Man. Forestry* (ed. 2) IV. 362 The wingless parthenogenetic ♀, stem-mother or *fundatrix*, hibernates alone on spruce buds. **1923** H. M. LEFROY *Entomol.* 272 These first females are known as fundatrices, or stem-mothers. **1936** *Forestry* X. 134 A primary [host], on which the fundatrix vera or foundress,..and the gallicolae or gall-dwellers are reared. **1952** ESSIG & ABERNATHY *Aphid Genus Periphyllus* 41 The mature fundatrix is wingless, small, dark brown. **1971** *Nature* 2 July 14/1 Each fertilized female..lays one egg which later hatches to give the fundatrix—the first wingless female of the cycle.

fundi (fʊ·ndi). [Native African name.] A West African grass, *Digitaria exilis*, cultivated for its seed, which resembles millet; hungry rice.

1858, 1887 [see HUNGRY *a.* 4]. **1915** *Nature* XCVI. 350/2 Fundi, *Digitaria exilis*, has been known for some time as a cultivated cereal... It is cooked by being thrown into boiling water or used like porridge. **1957** M. BANTON *W. Afr. City* iii. 45 *April*... Farms are prepared for fundi, ginger and groundnuts.

fundiform (fʊ·ndifǭɹm), *a.* *Anat.* [ad. mod.L. *fundiform-is* (A. Retzius 1841, in *Arch. f. Anat., Physiol. und wiss. Med.* 499, translating G. *schleuderförmig* shaped like a sling), f. L. *funda* sling: see -FORM.] *fundiform ligament*: a ligament having the shape of a sling; *spec.* (*a*) a ligament on the front of the ankle which encloses the tendons of two extensor muscles of the leg; (*b*) a ligament which forms a loop enclosing the root of the penis and partly supports the weight of the latter.

1854 E. WILSON *Anatomist's Vade Mecum* (ed. 6) iv. 305 The tendons of the extensor longus digitorum and peroneus tertius have also a separate loop connected with the cruciform ligament, the sling ligament (fundiforme) of Retzius. **1900** DORLAND *Med. Dict.* 356/2 *Fundiform ligament*, a portion of the anterior annular ligament of the ankle. **1962** *Gray's Anat.* (ed. 33) 1530 The fundiform ligament..springs from the lower side of the linea alba and splits into two lamellae which pass one on each side of the penis and unite below with the septum of the scrotum.

funeral, *sb.* Add: **1. c.** *none of your* (*our*, etc.) *funeral*: no affair of yours (ours, etc.), nothing to do with you (us, etc.); *your* (etc.) *funeral*: your (etc.) affair or concern (often with an implication of unpleasant consequences). orig. *U.S. slang.*

1854 *Oregon Weekly Times* 25 Nov. (Th.), A boy said to an outsider who was making a great ado during some impressive mortuary ceremonies, 'What are you crying about? it's none of your funeral.' **1867** *Trans. Ill. Agric. Soc.* VI. 167 As this was 'none of our funeral', of course the party of the other part was provided with all the regalia of the occasion. **1895** *Century Mag.* Sept. 674/1 We don't know for certain it was them, and it's none of our funeral, anyhow. **1870** *Congress. Globe* 12 Apr. 2780/3 Mr. Painter then addressing you said, 'This is not my funeral.' **1877** *Hartford Times* 17 Oct. (Farmer), Oh, that isn't my funeral, I want you to understand. **1908** S. E. WHITE *Riverman* vii. 60 However, it's your funeral. Come on, if you want to. **1917** S. GRAHAM *Priest of Ideal* i, But although we go into black it is not our funeral. **1921** A. S. M. HUTCHINSON *If Winter Comes* IV. i. 254 However his wife was his funeral, not mine, and I said nothing. **1930** SAYERS & 'EUSTACE' *Documents in Case* 72 But how he can ever imagine that it will sell...But that's his funeral. **1948** *Sat. Even. Post* 26 June 82/4, I assured LaGuardia that it was 'his funeral, and not mine'. **1952** G. W. BRACE *Spire* (1953) xii. 114, I refuse to involve myself. It is your problem and your funeral.

3. (Delete †*Obs.*) Now *U.S.* Also, a funeral service.

1853 in N. E. Eliason *Tarheel Talk* (1956) 273 Was her funerald preached at the burying, if not, I would like to know when it will be. **1871** SCHELE DE VERE *Americanisms* (1872) 238 This led to the custom, still prevailing in densely-settled districts, and especially among the freedmen of the South, to have *funerals*, i.e. special funeral sermons preached some time after the death of the person. **1905** *N.Y. Even. Post.* 9 Mar. 8 An official funeral will be held in the Senate chamber to-morrow at two o'clock, and the body will be taken to Tennessee to-morrow evening. **1922** M. B. HOUSTON *Witch Man* xi. 131 The cemetery to which Wilda had fared as much as twice for many of the graves within it, first when the sleeper was laid there and later—sometimes there was the lapse of a year—when his funeral was preached.

6. *funeral director, home, parlour.* (orig. *U.S.*)

1886 J. A. PORTER *New Stand. Guide Washington* 178 General Furnishing Undertaker and Funeral Director. **1963** *New Statesman* 12 Apr. 516/3 'The reason why we have changed the name,' said Peter Beaty, secretary of the National Association of Funeral Directors, 'is for the services rendered.' **1964** *English Studies* XLV. (Suppl.) 23 Britain's 4,400 undertakers, or funeral directors, as they prefer to call themselves. **1936** MENCKEN *Amer. Lang.* (ed. 4) vi. 287 When it [sc. the mortician's business] is achieved the patient is put into a *casket* and stored in the *reposing room* or *slumber-room* of a *funeral-home*. **1968** *Globe & Mail* (Toronto) 17 Feb. 53/5 (Advt.), Hallowell Funeral Home. **1927** *Sat. Even. Post* 12 Mar. 6/3 Rows of wooden folding chairs, often rented from funeral parlors. **1938** AUDEN & ISHERWOOD *On Frontier* I. i. 24 Our doctors will see to your health, and our funeral parlours will bury you. **1963** J. MITFORD *Amer. Way Death* i. 29 The nomenclature has gradually changed. From 'undertaker' *tout court* to 'funeral parlor' to 'funeral home' to 'chapel' has been the linguistic progression.

fungation (fʊŋgēi·ʃən). *Path.* [f. as FUNGATE *v.*: see -ATION.] The formation of a fungus-like growth.

1908 *Practitioner* Sept. 395 These palliative operations are performed to prolong life, to prevent fungation, to give relief to pain. **1910** *Ibid.* June 764 Owing to the fungation, this was the most unfavourable of the cases. **1949** *Year Bk. Gen. Surg.* 187 In special circumstances, such as fungation or a large tumor, simple mastectomy may be indicated.

fungicidal (fʊndʒisəi·dăl), *a.* [f. FUNGICID(E + -AL.] Of the nature of, or acting as, a fungicide; characteristic of a fungicide.

1905 G. F. STRAWSON *Insects & Fungi Injurious to Plants* II. 60 The quantity of copper sulphate decomposed is small and not wasted, as the resulting precipitate has fungicidal qualities. **1909** B. M. DUGGAR *Fungous Dis. Plants* vi. 87 In the case of some of the powdery mildews the use of any fungicidal sprays or dusts may be beneficial. **1930** *Nature* 13 Dec. 921 The fungicidal power of wood preservatives. **1942** [see *FUNGISTATIC *a.*]. **1969** B. E. J. WHEELER *Introd. Plant Dis.* xx. 333 In recent years a number of heterocyclic nitrogen compounds have been developed with marked fungicidal and bactericidal activity.

fungistasis (fʊndʒistēi·sis). [f. *fungi-*, combining form of FUNGUS *sb.* + Gr. στάσις stopping.] Inhibition of the growth of fungi.

1930 *Arch. Dermatol. & Syphilol.* XXII. 225 (*heading*) Fungistasis in ringworm of the toes and of the feet. **1970** *Nature* 14 Mar. 1071/1 Fungistasis is usually considered to result from biological activity in soil.

fungistatic (fʊndʒistæ·tik), *a.* [f. as prec. + STATIC *a.* and *sb.*] Inhibiting the growth of fungi. Hence **fungista·tically** *adv.*

1922 *Arch. Dermatol. & Syphilol.* VI. 754 We have endeavored by a fungistatic (restraining) test and a fungicidal (killing) test to determine the affinities of various commonly employed medicaments and of certain laboratory stains for three different, pathogenic molds. **1941** *Experiment Station Record* Feb. 161 A branched chain acid, in general, proved less effective fungistatically than the corresponding straight chain acid. **1942** *Contrib. Boyce Thompson Inst.* XII. 451 Protective fungicides can function in their initial action fungistatically though in time the effect may be lethal or fungicidal. **1946** [see *ACETYLENIC *a.*]. **1971** *Times* 25 Oct. 12/8 It [*sc.* a new paint] comprises a fungistatic high polymer with mycologically active reagents.

fungitoxicity (fʊndʒitǫksi·sĭti). [f. as prec. + TOXICITY.] Toxicity to fungi. So **fungito·xic** *a.*, toxic to fungi; **fungito·xicant**, any substance with fungitoxic action.

1951 *Contrib. Boyce Thompson Inst.* XVI. 332 The fungitoxicity of phenothiazone. *Ibid.*, Many of them are quite fungitoxic. **1953** *Nature* 1 Aug. 199/1 The finding by Neilson-Jones of some fungitoxicity in the more normal Oxshott Heath soil is also significant. **1958** V. W. COCHRANE *Physiol. of Fungi* xiv. 461 Sulfanilamide is, however, not strongly fungitoxic. *Ibid.* 463 None of these four types of reaction is certainly known to be responsible for the action of fungitoxicants. **1970** *Nature* 19 Sept. 1267/2 There is..no correlation between the therapeutic effect of the compounds *in vivo* and their fungitoxicity *in vitro*.

fungus, *sb.* Add: **1. c.** A beard. Also *face-fungus. slang.*

1925 WODEHOUSE *Sam the Sudden* xiii. 89 Where did you get the fungus? **1936** [see *BEAVER¹]. **1937** 'R. CROMPTON' *William—the Showman* x. 240 'Is it to be me or that ass with the fungus on his cheeks?' demanded Richard belligerently. **1959** H. HOBSON *Mission House Murder* ii. 14 In addition to the chin-fungus he'd put on a little weight.

5. *fungus-eater, -flora, -hunt, -hunter, -ring*; *fungus-eating, -plagued* adjs.; **fungus-garden** (see quot.); **fungus-gnat, -midge** *Ent.*, a fly of the dipterous family Mycetophilidæ.

1899 G. H. CARPENTER *Insects* 304 Fungus-Eaters... A considerable number of small Beetles together with the grubs of many Flies and Midges find their sustenance in fungi. *Ibid.* Index 397/1 Fungus-eating insects. **1892** G. MASSEE (*title*) British fungus-flora. **1924** J. A. THOMSON *Science Old & New* xvi. 88 The fungus-gardens of the termites are seen at their best in Ceylon, and the characteristic feature is the construction of a maze of chewed wood with labyrinthine passages, on the walls of which the fungi grow. **1884** J. S. KINGSLEY *Stand. Nat. Hist.* (1888) II. 407 The family *Mycetophilidæ*, commonly called Fungus Gnats. **1955** *Sci. Amer.* May 104/3 In caves of the Great Smokies in Tennessee and North Carolina, there is a large fly, the fungus gnat (*Platyura fultoni*), which is luminous in its larval stage. **1899** BURNAND in *Pall Mall Mag.* XVIII. 536 We espied the Professor..evidently enjoying a fungus-hunt. **1886** P. ROBINSON *Valley of Teetotum Trees* 134 A common object of the country in 'the fall of the year'..is the fungus. And scarcely less familiar, in wooded districts especially, the fungus-hunter. **1899** G. H. CARPENTER *Insects* 256 The *Mycetophilidæ* or Fungus-midges. **1918** W. BEEBE *Jungle Peace* (1919) viii. 178 Light-starved and fungus-plagued, the shrub and saplings are stunted and weak. **1907** WOODRUFFE-PEACOCK *Pasture & Meadow Anal.* 5 The flora of fungus-rings..should always be most carefully noted the season through. **1963** *Field Archæol.* (Ordnance Survey) (ed. 4) 7 The fungus rings which suggest barrow circles.

funicle. Add: **c.** *Ent.* = FUNICULUS 4.

1881 L. O. HOWARD in *Rep. U.S. Dept. Agric. 1880* III. 361 The club is rather large..as long as or longer than the last four funicle joints together. **1888** J. H. COMSTOCK *Introd. Ent.* ii. 11 The funicle is that part of the clavola between the club and the ring-joints; or, when the latter are not specialized, between the club and the pedicel. **1923** A. D. MACGILLIVRAY *Guide Ext. Insect Anat.* iii. 79 In those species..where the distal end of the flagellum is clavate, the enlarged segments are known as the club and the segments between the club and the ring-segments as the funicle. **1925** A. D. IMMS *Gen. Textbk. Ent.* 14 In some insects, particularly among Hymenoptera, the flagellum is divisible into the ring-joints, the funicle, and the club...The funicle comprises those joints which intervene between the ring-joints and the club,..or between the latter and the pedicel in cases when the ring-joints are not differentiated. **1937** [see *FUNICULUS 4 b].

‖ **funiculaire** (fünikülęr). [Fr.: see FUNICULAR *a.*] A funicular railway.

1907 *Westm. Gaz.* 9 Oct. 2/1 We never go up anything except by funiculaire. **1927** *Observer* 18 Dec. 9/4 Every attention is paid to the comfort of travellers, even in the funiculaires.

funicular, *a.* Add: **B.** *sb.* A funicular railway (Webster 1909).

1911 A. BENNETT *Card* xi. 280 He had been to Montreux and missed the funicular back. *Ibid.* 284 We all saw the Countess off in the funicular at three o'clock. **1926** H. BENNETT tr. *Leroux's Adv. Coquette* vi, A number of travellers set down by the funicular gazed from a respectful distance at the actress's enthusiasm. **1957** P. KEMP *Mine were of Trouble* x. 187 The disused funicular that climbed La Rhûne, a prominent mountain nearby. **1971** *Daily Tel.* 30 Jan. 7/2 The resort has..four funiculars and one chairlift.

funi·culared, *ppl. a.* [f. *FUNICULAR *sb.*] Provided with a funicular railway.

1933 *Times Lit. Suppl.* 30 Mar. 211/2 The Arctic figures as going the way of Africa and funiculared mountains.

funicularize (fiuni·kiŭlărəiz), *v.* [f. FUNICULAR *a.* + -IZE.] *trans.* To provide with a funicular railway.

1927 R. CAPELL in H. J. Foss *Heritage of Music* 216 The Alps..are now mostly funicularised. **1936** *Times Lit. Suppl.* 9 May 397/3 The Alps before they were funicularized for Winter Sports.

funiculus. Add: **4. b.** *Ent.* (See quots.)

1826 KIRBY & SPENCE *Introd. Entomol.* III. xxxiii. 389 The Funiculus, a small cartilaginous cord, passing through a minute orifice of the *Postfrænum*, just above the point where the footstalk is fixed, to an opposite hole above it. **1895** *Cambr. Nat. Hist.* V. i. 492 The petiole, besides articulating..with the propodeum by means of certain prominences and notches, is also connected therewith by means of a slender ligament placed on its dorsal aspect and called the funiculus. **1923** A. D. MACGILLIVRAY *Guide Ext. Insect Anat.* iii. 79 When the ring-segments are wanting and the antenna is geniculate, the segments between the club and the scape are known as the funiculus. **1937** J. R. DE LA TORRE-BUENO *Gloss. Ent.* 110 Funicle, funicule, funiculus, in the insect antenna, that part of the clavola between the club and the ring-joints (Comstock).

funk, *sb.*² Read: *Obs.* exc. *U.S. dial.*

1917 *Dialect Notes* IV. 412 Open up the door and let the funk out. **1937–40** in WENTWORTH *Amer. Dial. Dict.* (1944) 237/2.

2. Music that is 'funky' (see quots. and *FUNKY *a.*³ 2). *slang* (orig. *U.S.*).

1959 *Jazz* Fall 292 You can even try to put too much 'funk' in a thing. **1960** *Melody Maker* 31 Dec., The American jazz public is 'funk'-crazy. **1961** *Sunday Times* 5 Feb. 36/5 The contemporary jazz cult of 'blues roots'—otherwise described as 'soul' or 'funk'. **1962** *Radio Times* 10 May 42/2 *Funk*, basically, 'smelly'—signifies return of modern jazzmen to earthy roughage of blues and New Orleans, but rephrased with modern techniques; similar to soul, only more extrovert. **1970** C. MAJOR *Dict. Afro-Amer. Slang* 56 *Funk*, 'soul' quality in black music, melancholy mood of the blues.

funk, *v.*² Add: **1.** Also const. *at.*

1841 *Punch* 13 Nov. 213/2 Funking at the rejection of a clever man,..determining to take prussic acid in the event of being refused.., the student finds his first ordeal approach. **1857** T. HOOD *Pen & Pencil Pict.* 144, I have seen him out with the governor's hounds: he funked at the first hedge. **1859** *Punch* 23 July 34/2 Louis Napoleon, who had gone to war on a pledge that the Austrians should be driven out of Italy, had 'funked' at the Quadrilateral. **1863** READE *Hard Cash* xxv, I began to funk again at his knowing that;..I was flustered, ye see.

5. funksticks, also *transf.*, a coward; also **funkstick.**

1897 R. BADEN-POWELL *Matabele Campaign* xvii. 438 A nervous man is forty thousand times worse than a frightened woman, especially when..he has any number of drink-fuddled 'funk-sticks' ready to echo his alarm. **1916** *National Rev.* No. 382. 527 The rising predicted by Bernhardi and feared by all funksticks. **1930** A. E. W. MASON *Dean's Elbow* vi. 68 She thought of William Mardyke and his timidities. 'He'll never do that. What did you call him?' 'A funkstick.'

funk-hole. *slang.* [f. FUNK *sb.*¹] A place of safety into which one can retreat; esp. in war, as a dug-out. Also *transf.* and *fig.*, e.g. an employment which is used as a pretext for evading military service.

1900 *Daily News* 20 Nov. 3/2 The Funk Holes which the besieged residents had mined in the river bank. **1914** *Daily Mail* 4 Dec. 8/3, I am sitting in my 'funk hole' lined with straw. **1918** W. J. LOCKE *Rough Road* xx. 245 'J. M. T. and I have looked Death in the face many a time—and really he's a poor raw-head and bloody-bones sort of Bogey; don't you think so, old chap?' 'It all depends on whether you've got a funk-hole handy.' **1920** *Blackw. Mag.* May 608/1 Grain-pits that afford excellent ready-made funk-holes. **1922** C. E. MONTAGUE *Disenchantment* xiv. 188 Some of our higher commanders would use their A.D.C. rooms as funk-holes to shelter..their distant cousin the marquis. **1928** *Sunday Dispatch* 20 Sept. 2/2 Jim might have stayed! A few hours away from the office wouldn't matter. Lovely funk-holes, offices! **1932** 'A. BRIDGE' *Peking Picnic* v. 48 This place was one of her favourite funk-holes. **1946** R. CAMPBELL *Talking Bronco* 15 To jobs in ministries they'll fly, And funk-holes in the B.B.C. **1959** J. D. CLARK *Prehist. S. Afr.* ix. 219 Deep, dark caves were never occupied except very occasionally as refuges or 'funk holes'.

funkia (fʊ·ŋkiă). Also **funckia.** [mod.L., ad. *funckia* (K. Sprengel 1817, in *Anleitung zur Kenntnis d. Gewächse* (ed. 2) I. 246), f. the name of H. C. *Funck* (1771–1839), Prussian botanist + -IA¹.] A member of the genus of liliaceous plants from Japan once so named, but now called *Hosta*, having racemes of drooping white or lilac bell-shaped flowers; a plantain-lily.

1839 *Curtis's Bot. Mag.* LXV. 3657 This interesting species of Funckia was received from Belgium. **1883** W. ROBINSON *Eng. Flower Garden* 132/2 The bold and striking foliage of some of the strongest plain-leaved section of Funkias renders them very effective as edging plants for large beds. **1927** E. H. M. COX *Evol. Garden* v. 78 Funkias, or Plantain Lilies, are among the best foliage plants of the herbaceous border. **1970** *Countryman* LXXV. 291 We tend to think of hostas as recent introductions, but this variety [*sc. Hosta undulata variegata*] was brought from Japan as long ago as 1834...But then, of course, they were called funkias.

funky, *a.*³ Delete †*Obs.* and add later examples in sense 'mouldy, old, musty'. Also, smelling strong or bad. *U.S. dial.*

1906 *Dialect Notes* III. 118 This butter's funky. **1909** *Ibid.* 396 This room smells funky. **1929** T. WOLFE *Look Homeward* iv. 41 [A Negress's] strong smell, black and funky. **a 1938** —— *Web & Rock* (1947) 25 Breathe in the funky nigger stench. **1962** J. BALDWIN *Another Country* (1963) I. i. 12 They knew..why his hair was nappy, his armpits funky.

2. Of jazz or similar music: down-to-earth and uncomplicated; emotional; having the qualities of the blues. Also in extended use: 'in', 'swinging', fashionable. *slang* (orig. *U.S.*).

1954 *Time* 8 Nov. 42 *Funky*, authentic, swinging. **1957** [see *BOP *sb.*²]. **1959** 'F. NEWTON' *Jazz Scene* vi. 119 Critics are on the search for something a little more like the old, original, passion-laden blues: the trade-name which has been suggested for it is 'funky' (literally: 'smelly', i.e. symbolising the return from the upper atmosphere to physical, down-to-earth, reality). **1960** *Melody Maker* 31 Dec., Horace [Silver] recalls that the use of the word funk in the modern sense goes back to his composition 'Opus de Funk'. 'When you put a lot of little blues inflections in the solos, people would say you were really funky, by which they just mean bluesy, and that is how I came up with the title. So the critics started to label with me as funky.' **1962** *John o' London's* 10 May 457/1 He has a strong blues feeling and he doesn't overdo the fashionable 'funky' business. **1969** *Sat. Rev.* 27 Sept. 25/1 You can't hype kids into buying things they don't want, even with people in funky clothes soft-selling them. **1970** *Sunday Times* 8 Feb. 15/1 John and Yoko: a life-style that's getting funkier all the time. **1971** *Frendz* 21 May 17/1 *Brown Sugar* and *Bitch* are Jagger at his foxy, dirty, funky best.

funnel, *sb.*¹ Add: **7. b. funnel-beaker** *Archæol.*, a type of beaker with a narrow neck; also *attrib.* of a culture characterized by the use of beakers of this type; **funnel web,** a funnel-shaped spider's web; **funnel-web spider** (*a*) *U.S.*, a spider of the family Agelenidæ; (*b*) *Austral.*, a poisonous spider of the family Dipluridæ; also *ellipt.*; **funnel-web tarantula** = *funnel-web spider (b); **funnel-web weaver** = *funnel-web spider (a).

[**1929** V. G. CHILDE *Danube in Pre-History* vii. 117 A relatively closed group is constituted by collared flasks, funnel-necked beakers, and amphorae.] **1954** R. W. EHRICH *Rel. Chronol. Old World Archeol.* viii. 118 There is a penetration southward from the northern European plain of various groups of people who had..funnel beakers,..and the like. **1955** *Proc. Prehist. Soc.* XXI. 120 It does not seem that they have expanded west of the Vistula in the territory occupied by the Funnel-Beaker population. **1957** V. G. CHILDE *Dawn Europ. Civilization* (ed. 6) ix. 159 A herding group who sometimes decorated their funnel-beakers..with cord imprints had cleared tracts of Denmark and Southern Sweden for pasture in Late Atlantic times. *Ibid.* x. 176 The subdivision of Early Neolithic was originally based on the typology of the funnel-beaker—the most distinctive vase in the dominant culture which is usually called after it..the Funnel-Beaker culture. **1895** J. H. & A. B. COMSTOCK *Man. Stud. Insects* ii. 30 The greater number of the webs seen at such times [*sc.* in the early morning] are of the form we term funnel-webs. **1912** J. H. COMSTOCK *Spider Bk.* iv. 193 The principal part of a funnel-web is sheet-like in structure. **1949** W. J. GERTSCH *Amer. Spiders* x. 216 The funnel web of the agelenids is little changed from the silken cell of their forebears. **1912** J. H. COMSTOCK *Spider Bk.* vii. 582 The members of this family [*sc.* Agelenidæ] spin sheet-like webs, which are usually furnished with a tubular retreat; this suggests the common name funnel-web spiders for the family. **1933** *Bulletin* (Sydney) 15 Mar. 21 Three people have been killed recently from the bite of a spider, known as the atrax or funnel-web spider. **1954** BORROR & DELONG *Introd. Study Insects* xxx. 797 The web of the funnel-web spiders is somewhat sheetlike, but is shaped like a funnel. *Ibid.* 805 The Agelenidae, or funnel-web spiders, are a large group. **1956** S. HOPE *Diggers' Paradise* xxii. 203 The funnel-web lives almost exclusively in the Greater Sydney zone. **1970** *Times* 30 Dec. 4/5 She was bitten yesterday by a funnel web spider about an inch long. **1912** J. H. COMSTOCK *Spider Bk.* vi. 247 The funnel-web tarantulas resemble the two preceding families. **1954** BORROR & DELONG *Introd. Study Insects* xxx. 798 The Dipluridae are the funnel-web tarantulas. **1895** J. H. & A. B. COMSTOCK *Man. Stud. Insects* ii. 31 The funnel-web weavers..are long-legged, brown spiders.

funnel, *v.* Restrict † to senses in Dict. and add: **2.** *trans.* and *intr.* To guide or move through a funnel. Also *transf.* and *fig.*

1901 KIPLING *Kim* xiii. 330 Jammed into a corner between cliffs that funnelled and focused every wandering blast. **1923** —— *Land & Sea* 182 That wind's funnelling badly in the valley. **1936** L. C. DOUGLAS *White Banners* x. 214 Crowds like swollen rivers funnelled through station exits. **1958** *Sunday Times* 12 Oct. 1/3 All the data is funneled into the Air Force Data Reduction Centre at Englewood, near Los Angeles. **1961** *Times* 20 Feb. 14/3 [soccer] They made less impression..on Vowels and his colleagues, funnelling back. **1964** W. MCCORD in I. L. Horowitz *New Sociol.* 431 The popular policy of funneling resources into primary education must be viewed with some caution. **1971** *Sci. Amer.* June 65/1 Three magnetic spectrometers, which funnel the electrons into a system of detectors.

funniosity (fʊniǫ·sĭti). *jocular.* [f. FUNNY *a.* + -OSITY.] Comicality, jocularity; also, something comical, a comicality.

1890 in W. T. Vincent *Recoll. F. Leslie* (1894) II. xxiv. 117 Then the conversation reverted for a brief space to that very choice mixture of funniosities. **1900** *Westm. Gaz.* 5 July 3/2 There is no end to the freaks and funniosities of the cricket field. **1920** S. GRAHAM *Childr. of Slaves* xiv. 271 The mind stocked with music-hall funniosity and pseudo-cynicism. **1921** —— *Europe* x. 138 Shylock's cleverness and intellectual assurance was obscured by funniosities.

funny, *a.* Add: **1. b.** *funny business*, action (on the part of a clown or actor) intended to excite laughter; hence, jesting, nonsense; also *slang*, fooling or monkeying about; deceitful or underhand practices; similarly *funny stuff*, *funny dope*; *funny column, paper*, a (section of a) newspaper containing humorous matter or illustrations. orig. *U.S.*

1860 W. HANCOCK *Emigrant's Five Years Free States Amer.* iv. 93 The 'funny' column of any American journal. **1874** in L. Hearn *Amer. Miscellany* (1924) I. 20 The said *Giglampz* is not a funny paper. **1888** E. W. NYE *Baled Hay* 38 There was no funny business in his nature. **1890** *Century Mag.* Dec. 303 She even ventured on the funny column, for it was not Sunday. **1891** E. S. ELLIS *Check No. 2134* xiv. 93, I hope we'll get through without any more funny business. **1915** FROEST & DILNOT *Crime Club* x, The blue barrel of a revolver showed in the electric light. 'No funny business!' he warned them. 'You guys can't play it on me.' *Ibid.* xii, Especially if you try to put any of the funny dope over on me. **1930** *London Mercury* Feb. 324 He'll be out and about in a fortnight. Till then we will visit him together—and no funny stuff! **1936** 'I. HAY' *Housemaster* xv. 191 She seemed to be labouring under the idea that he, Victor, was trying to start some funny business with her. **1946** WODEHOUSE *Money in Bank* xii. 105 He is far too scared of our hostess to try any funny stuff on her. **1960** O. MANNING *Great Fortune* III. xviii. 210 Our permits..are issued on the understanding that we do not get mixed up in any funny business. **1968** *Globe & Mail Mag.* (Toronto) 13 Jan. 15/3 Despite an early apprenticeship to the funny papers McArthur was a subtle humorist. **1971** D. POTTER *Brit. Eliz. Stamps* iii. 44, 1968 was even quieter. Only four sets, no strange se-tenant strips or funny business.

2. b. *funny-peculiar*, a colloquialism introduced to distinguish sense 2 in Dict. from sense 1 (*funny-ha-ha*), the two antithetic expressions freq. appearing together.

1938 'I. HAY' *Housemaster* III. 78 *Chris.* That's funny. *Button.* What do you mean, funny? Funny-peculiar, or funny ha-ha? **1942** *Horizon* July 69 He says, 'Funny-haha or funny-peculiar?' and I says, 'Funny-peculiar.' **1946** S. GIBBONS *Westwood* vii. 95 You said her *conscience* forced her into being a you-know-what. It sounds awfully funny-peculiar. **1955** M. MCCARTHY *Charmed Life* (1956) iii. 58 His art-school training rendered him funny ha-ha to the cognoscenti. **1959** J. VERNEY *Friday's Tunnel* xxv. 227 John Gubbins leant forward, smiling in a funny-peculiar not funny-ha-ha way. **1965** *Times Lit. Suppl.* 8 Apr. 281/3 Whether it is also amusing or merely funny-peculiar will depend upon the reader. **1970** 'D. SHANNON' *Unexpected Death* (1971) ii. 24 He thought the blonde was rather funny..in the sense of funny-peculiar.

4. funny-bone, also *fig.*; **funny-face,** a *joc.* and *colloq.* form of address; **funny farm** *slang*, a mental hospital; **funny party** *Naut.*, a ship's concert party.

1902 *Daily Chron.* 12 June 3/3 Two principal figures and a few carefully careless scratches—that is all Mr. Raven-Hill uses in the pointing of his joke, but he hits the universal funny-bone with his pencil. **1965** G. MCINNES *Road to Gundagai* x. 161 Kennedy..hit our funny-bones because he was a man who used long words with an extremely sober face. **1927** I. GERSHWIN (*song title*) Funny face. **1930** SAYERS & 'EUSTACE' *Documents in Case* 38 Ever and ever yours, funny-face, old dear. **1943** A. CHRISTIE *Moving Finger* v. 52 'It's all right, funny face,' I said. **1963** J. N. HARRIS *Weird World Wes Beattie* v. 61 Before Baldwin Ogilvy agrees to locking his client away in the funny farm, he might like to investigate the whole bang shoot. **1969** E. AMBLER *Intercom Conspiracy* (1970) ix. 173 *Intercom* was described as 'the Batman of the funny-farm set' and its editor as 'the Lone Ranger of the lunatic fringe'. **1911** 'GUNS' & 'THEELUKER' *Middle Watch Musings* 137 Nearly every ship has a 'funny' party. **1917** 'TAFFRAIL' *Sub* vii. 178 Once a year came the squadron regatta and sports, while at intervals our 'Funny Party', or pierrot troupe, gave an entertainment.

B. *sb. pl.* Comic illustrations, etc.; *spec.* comic strips, or the section of a newspaper devoted to these (D.A.). Hence, funny persons, books, etc.; jokes. *rare* in *sing.* orig. *U.S.*

1852 *Lantern* (N.Y.) II. 114/1 Keeping our dear public advised of all operations in the 'Funnies'. **1920** C. SANDBURG *Smoke & Steel* 33 About the funnies in the papers. **1922** —— *Slabs of Sunburnt West* 35 Turning among headlines, date lines, funnies, ads. **1925** A. HUXLEY *Those Barren Leaves* I. i. 8 The funnies and the fuzzy-wuzzies—in a word, the artists. **1936** WODEHOUSE *Laughing Gas* xxvi. 272 We've only read the movie section and the funnies. **1941** BELLOC *Silence of Sea* v. 33 What some modern funnies call 'wish-fulfilment'. **1952** in WENTWORTH & FLEXNER *Dict. Amer. Slang* (1960) 205/2 Cut the funnies an' git a can-opener. **1959** *Punch* 30 Dec. 682/2 Makers of pantomimes engaged the ripest music-hall funnies they could find. **1961** *John o' London's* 14 Sept. 307/1 The best bet would have been to ship this Mexican funny [*sc.* a film] straight out on circuit. **1961** 'B. WELLS' *Day Earth caught Fire* viii. 118 No funnies, Dave. Something's going on. **1970** *New Yorker* 17 Oct. 148/2, I hear he's a lawyer now, restricted, I suppose, to sneaking in a funny now and then in his summation to the jury.

funster. (Earlier example.)

1788 *Gentl. Mag.* LVIII. 4/1 Mr. T—ſle, a Fellow of the same college, and a *fellow-funster* also,..arrived when the Dr. was out upon his visits.

fur, *sb.*¹ Add: **2. b.** *to make the fur fly*: for U.S. slang read orig. U.S. slang. Also *the fur flies*, etc. (Add earlier and later examples.)

[**1663** BUTLER *Hud.* I. iii. 184 I'le make the fur Flie 'bout the eares of the old Cur.] **1814** *Niles' Reg.* VI. 67/2 Smugglers look out, or you will soon see 'the fur fly'. **1834** D. CROCKETT *Narr. Life* ii. 11, I knew very well that I was in a devil of a hobble, for my father had been taking a few horns, and was in a good condition to make the fur fly. **1955** M. GILBERT *Sky High* v. 70 There was a slip-up... There was some fur flew about *that*, I can tell you. **1963** V. GIELGUD *Goggle-Box Affair* viii. 73 A good deal of fur's always flying about Gargantua.

9. *fur company, -farm, -farmer, -farming, -hunting, -sewer, trade* (earlier and later examples), *-trader* (earlier example), *-trading*; *fur-bordered, -bound, -muffled, -topped, -trimmed, -wrapped* adjs.

1903 *Westm. Gaz.* 8 Oct. 4/2 The fur-bordered, serrated lace collar. **1898** *Ibid.* 18 Nov. 3/2 A fur-bound coat. **1818** *Niles' Reg.* XV. 19/2 A small canal has been opened..by the North West fur company. **1880** *Scribner's Monthly* May 125/1 Trappers and hunters for the fur companies would have thought themselves in paradise could they have seen our stores in '74. **1914** *Outing* (U.S.) Dec. 345/1 A manager of a fur farm must have a liking for animals. *Ibid.* 343/1 The trade must look to the fur-farmer for a considerable portion of its supply. **1961** *New Scientist* 23 Feb. 460/3 Many types of animals being bred at fur-farms are particularly sensitive to the..noise created by subsonic aircraft, and the sudden nature of sonic bangs might cause still greater losses than the fur-farmers have so far experienced. **1911** *Daily Colonist* (Victoria, B.C.) 12 Apr. 5/4 This is nothing less than fur farming on an extensive scale, the breeding of marten being specialized. **1923** J. C. SACHS *Furs & Fur Trade* 92 Fur farming..is one of the oldest of occupations. **1819** L. A. ANSPACH *Hist. Newfoundland* xiv. 378 This fur-hunting employs a great number of persons. **1919** W. T. GRENFELL *Labrador Doctor* (1920) xiii. 238 The only trouble with..fur-hunting is that its very nature limits its supply. **1905** *Westm. Gaz.* 16 Jan. 2/1 Sleighs with fur-muffled occupants. **1896** MRS. H. WARD *Sir G. Tressady* xi. 227 Tailoresses and shirtmakers and fur-sewers. **1928** *Daily Express* 11 Oct. 5/5 Russian boot..fur-topped, fleece-lined. **1732** L. ARMSTRONG *Let.* 11 Sept. in *Calendar of State Papers, America & W. Indies* (1939) 255 They..engross the whole management of the fur trade. **1807** P. GASS *Jrnl.* 65 This and Hudson's Bay Company..carry on almost the whole of the fur trade in that extensive country. **1911** *North Amer. Rev.* Mar. 396 Astor's experience in the fur trade..enabled him to plan on a vast scale. **1815** *Kingston* (Ontario) *Gaz.* 8 July 1/3 Conveying goods for the fur traders. **1841** G. CATLIN *Indians* II. 29 St. Louis..is the great depot of all the Fur Trading Companies to the Upper Missouri and Rocky Mountains. **1911** FLETCHER & KIPLING *School Hist. Eng.* ix. 169 We had a whale-fishing and fur-trading station in Hudson Bay. **1898** *Westm. Gaz.* 4 Nov. 3/1 The Common Councillors..in their fur-trimmed robes. **1895** KIPLING *2nd Jungle Book* 151 He and his master..hunted together,..fur-wrapped boy and savage,..yellow brute.

10. fur-bearer, an animal that yields a fur of (commercial) value; so *fur-bearing* adj.; **fur-cloth, -fabric** (see quot. 1928); **fur felt,** a felt fabric, deeply napped to give a fur-like effect; **fur-fever, -moth** (see quots.).

1906 E. INGERSOLL *Life Anim.: Mammals* 162 The Fur Bearers..the martens, weasels, badgers, ratels, skunks, otters and their kin of the family Mustelidæ. **1941** J. S. HUXLEY *Uniqueness of Man* viii. 186 Direct destruction may be for commercial gain, as with whales, egrets, or fur-bearers. **1876** J. BURROUGHS *Winter Sunshine* (1883) iv. 91 The fox..furnishes, perhaps, the only instance..of a fur-bearing animal..that actually increases in the face of the means that are used for its extermination. **1936** *Discovery* Jan. 30/2 To slaughter fur-bearing animals for Broadway shops. **1928** *Daily Express* 23 June 5/1 Fur-cloth is specially made to imitate fur, and can be used for all furry animals. **1944** M. LASKI *Love on Supertax* v. 57 Four little boys in everyday clothes topped by fur-cloth Cossack hats. **1938** *Decorative Art* 57/1 Armchairs with imitation fur fabric upholstery. **1952** *Vogue* Dec. 126 (Advt.), For coats: velours, fancy and check tweeds, fur fabrics. **1968** *Guardian* 22 Oct. 7/3 Gabardine coats lined with a light-weight fur fabric. **1897** *Sears, Roebuck Catal.* 232/1 The newest shape, black stiff, fur felt hat. **1971** *Guardian* 4 Jan. 7/3 When Mr Patey's firm makes an Ascot hat it is a proper 'pullover'—grey fur felt pulled over a gossamer frame. **1905** *Pearson's Mag.* July 102/2 Bronchial catarrh, and 'fur fever'..are both caused by inhaling this 'fluff' and dust. **1842** T. W. HARRIS *Insects Injur. Veget.* 360 The fur-moth (*Tinea pellionella*). **1938** L. HUNTER *Domestic Pests* xvii. 154 The case-bearing clothes-moth (*Tinea pellionella*), sometimes called the fur moth or the single spotted clothes-moth.

furaldehyde (fiuræ·ldĭhəid). *Chem.* [Abbrev. of *FURFURALDEHYDE.] Either of the two aldehyde isomers of the compound C_4H_3O·CHO derived from furan, *esp.* 2-furaldehyde or *FURFURAL.

1916 *Chem. Abstr.* X. 3508/1 (*index*) 2-Furaldehyde. **1953** DUNLOP & PETERS *Furans* i. 22 The aldehyde (VII) related to furoic acid would be named 2-furaldehyde, or simply furaldehyde. Its common name is furfural.., a non-systematic name, yet one so widely used that it would be unwise to make a change.

furan (fiūᵊ·ræn). *Chem.* Also furane (-ēin). [Abbrev. of *FURFURAN.] **a.** A heterocyclic compound having a five-membered ring, $(CH)_4O$, and isolated as a colourless liquid with an ethereal odour. **b.** Any compound derived from furan that contains a furan ring. Also *attrib.*, as furan nucleus or ring, a ring of four carbon atoms and one oxygen atom; so *furan-ringed* adj.; **furan resin,** any of various resins that are polymers of a furan, esp. of furfuryl alcohol or furfural.

1894 G. MCGOWAN tr. *Bernthsen's Text-bk. Org. Chem.* (ed. 2) xv. 329 The furfurane derivatives are frequently termed 'furane' derivatives for shortness' sake. **1899** *Jrnl. Chem. Soc.* LXXV. II. 1201/2 (*index*) Furfuran (furan). **1907** *Ibid.* XCII. I. 146 (*heading*) Reduction of the furan nucleus. **1930** *Jrnl. Amer. Chem. Soc.* LII. 1697 Glycosides which possess the furan ring are very readily hydrolyzed. *Ibid.*, The furan-ringed methylglycosides. **1935** C. ELLIS *Chem. Synthetic Resins* II. 1516/2 (*index*) Furan resins. **1951** R. S. MORRELL in Morrell & Langton *Synthetic Resins & Allied Plastics* (ed. 3) xiii. 409 Furanes may be used in fungicides and fly-repellent preparations. **1953** DUNLOP & PETERS *Furans* i. 3 Furans substitute with greater ease than their benzene analogs. **1969** J. A. BRYDSON *Plastics Materials* (ed. 2) xxiv. 511 The furan resins have never achieved importance for mouldings or laminates but..have found value in anti-corrosion applications.

furanose (fiūᵊ·rănōuz, -s). *Chem.* [f. *FURAN + -OSE².] A structure containing a furan ring thought to be sometimes assumed by monosaccharide sugars that contain four or more carbon atoms; also, a sugar having this structure. Freq. *attrib.*, as *furanose form, ring*, etc.

1927 GOODYEAR & HAWORTH in *Jrnl. Chem. Soc.* II. 3140 (*caption*) Furanose (a tetrose). *Ibid.* 3141 The corresponding labile or γ-isomerides may be described as arabofuranose, xylo-furanose etc. **1948** W. W. PIGMAN *Chem. Carbohydrates* ii. 52 The basic types of the furanoses are the pentoses. **1957** E. V. MILLER *Chem. Plants* i. 6 Glucose and other hexoses may exist in the form of a pyranose ring or a furanose ring, the former being the more stable form. **1963** E. PERCIVAL in Florkin & Stotz *Comprehensive Biochem.* V. i. 24 Furanose forms of unsubstituted sugars have never been isolated although in solution they are considered to exist in equilibrium with the open-chain and pyranose ring forms.

furanoside (fiuræ·nŏsəid). *Chem.* [f. *FURANOS(E + *-IDE.] Any glycoside in the furanose form.

1932 *Jrnl. Chem. Soc.* 2255 The crystalline furanosides are reasonably stable to neutral aqueous permanganate. **1934** E. F. & K. F. ARMSTRONG *Carbohydrates* (ed. 5) xvi. 193 Haworth compares the relative velocity constant values for the hydrolysis of sucrose and furanosides with certain pyranosides, proving that sucrose is a typical furanose in this characteristic. **1966** C. A. MARSH in G. J. Dutton *Glucuronic Acid* i. 43 The electrophoretic mobilities on paper of the pyranoside and furanoside forms of both 2-naphthyl β-D-glucoside and 2-naphthyl β-D-glucosiduronic acid.

furca. Add: **2.** *Ent.* An apodeme or process in the thorax of many insects.

1895 *Cambr. Nat. Hist.* V. I. 103 Entothorax (apophysis or furca). **1910** C. G. HEWITT *House Fly* 9 The theca rests on a triradiate chitinous sclerite—the furca, which consists of a median, slightly convex rod. **1925** A. D. IMMS *Gen. Textbk. Ent.* 47 The endosternites (apophyses of some writers) are commonly represented by the *furcæ*; each furca is a median apodeme, unpaired at its base, with two free distal arms. *Ibid.* 598 The *theca*..articulates distally with a short rod or *furca*. **1969** R. F. CHAPMAN *Insects* viii. 131 In higher insects the two apophyses arise together in the midline and only separate internally, forming a Y-shaped furca.

b. = *FURCULA 3.

1939 H. WOMERSLEY *Prim. Insects S. Austral.* 81 In most species [of Collembola], the fourth ventral segment carries a pair of large, partially fused, appendages which form the characteristic spring or furca. **1969** R. F. CHAPMAN *Insects* ix. 154 The jump is produced by the sudden release of the furca.

3. *Zool.* A pair of divergent processes on the last abdominal segment of certain crustaceans.

1903 J. S. KINGSLEY tr. *Hertwig's Man. Zool.* 420 The terminal abdominal segment [of the *Copepoda*] is two-forked, forming the 'furca'. **1932** BORRADAILE & POTTS *Invertebrata* xii. 303 On the telson [of Crustacea] usually..is a pair of caudal rami forming the caudal furca. **1956** *Nature* 11 Feb. 289/2 In the arrangement of the setæ on uropods and furca the new species is similar to *Th*[*ermobathynella*] *leleupi*.

furcal (fʊ·ɹkăl), *a.* *Zool.* [f. L. *furca* fork + -AL.] Forked, furcate; esp. of or pertaining to a furca.

1851 *Trans. Linn. Soc.* XX. 434 Some traces of the entrances into these furcal bones exist in the sternal plates of *Perla*. **1895** *Cambr. Nat. Hist.* V. I. 399 [In Perlidæ] the metasternum..has on each side a peculiar slit; similar orifices exist on the other sterna...Newport..says that they are blind invaginations of the integument; he calls them the sternal or furcal orifices. **1969** R. F. CHAPMAN *Insects* ix. 154 It is probable that the mechanism of furcal extension is not the same in all Collembola.

furciferine (fʊɹsi·fĕrəin), *a.* *Zool.* [f. L. *furcifer* (see FURCIFEROUS) + -INE¹.]. Of or pertaining to western South American deer of the genus *Hippocamelus*, which have forked antlers and were formerly grouped in the genus *Furcifer*.

1891 FLOWER & LYDEKKER *Introd. Stud. Mammals* 329 The Furciferine group includes *Cervus chilensis*. **1966** *Punch* 25 May 775/3 Chile should also be all right with its *huemal*, an elusive furciferine deer which only exists in the high mountains of the country.

furcræa (fʊɹkrī·ă). Also four-, -crœa, -croya. [mod.L. (A. Ventenat 1793, in *Bulletin des Sciences de la Société Philomatique* I. 65), f. the name of the French chemist A. F. de *Fourcroy* (1755–1809).] A tropical American plant of the genus so named, closely related to *Agave* and belonging to the family Amaryllidaceæ.

1821 *Curtis's Bot. Mag.* XLVIII. 2250 (*heading*) Gigantic Furcrœa. **1900** L. H. BAILEY *Cycl. Amer. Hort.* s.v., As a rule, Furcræas bear fruit not more than once, and then die without producing suckers. **1901** *Westm. Gaz.* 3 July 7/1 Poles of the furcroea aloe for the construction of recreation-rooms and school houses. **1963** ROBERTSON & GOODING *Bot. Caribbean* v. 45 (*caption*) Bulbil of *Furcraea*. *Ibid.* xxiii. 213 Other members [of the Agavaceæ] familiar in the West Indies are..*Furcraea* spp. (Maypole) with creamy white flowers.

furcula. Add: (Further examples.)

1903 J. S. KINGSLEY tr. *Hertwig's Man. Zool.* 605 Clavicles and furcula are united directly or by ligaments to the broad sternum. **1954** G. C. KENT *Compar. Anat. Vertebrates* viii. 253 The two clavicles unite in the midline to form a furcula (wishbone).

2. *Embryol.* A process from which the epiglottis is developed.

1887 A. C. HADDON *Introd. Study Embryol.* vi. 183 Behind this [projection] are a pair of folds (furcula), which eventually will form the epiglottis. **1926** JORDAN & KINDRED *Textbk. Embryol.* xv. 196 The crest of the furcula becomes elevated and develops into the epiglottis.

3. *Ent.* A forked appendage at the end of the abdomen in springtails.

[**1871** *Amer. Naturalist* V. 10 The spring consists of a pair of three-jointed appendages..forming a fork.] **1906** J. W. FOLSOM *Entomology* ii. 68 *Appendages*...In Collembola..those of the fourth segment form the characteristic leaping organ, or furcula. **1909** *Jrnl. Econ. Bot.* IV. 9 Furcula somewhat broader than in *S*[*minthurus*] *aureus*. **1932** BORRADAILE & POTTS *Invertebrata* xiv. 408 By contraction of the extensor muscles of the furcula the latter is pulled down out of contact with the hamula and the animal is propelled forwards into the air. **1964** J. T. SALMON *Index to Collembola* I. 111 Re-examination showed this species to have a well developed furcula.

furfural (fʊ·ɹfiŭræl). *Chem.* [f. FURFUR(OL + *-AL².] An aldehyde of furan, C_4H_3O·CHO, a colourless liquid which darkens in air and which is used chiefly as a solvent and in the manufacture of furan derivatives and synthetic resins; 2-furaldehyde; = FURFUROL; **furfural resin** = *furan resin.

1879 H. WATTS *Dict. Chem.* (ed. 2) VIII. I. 832 Furfural or furfurol, $C_5H_4O_2$. Pyromucic aldehyde. **1901** *Daily Chron.* 11 May 2/2 In the case of whisky the chief cause of mischief is furfural, from which old matured whisky is free. **1939** H. R. SIMONDS *Industr. Plastics* (1940) iii. 74 The furfural resins are dark in color and therefore are used only with great difficulty for producing light colored plastics. **1960** *Times Rev. Industry* May 82/3 By the hydrolysis of the waste products [from the manufacture of cellulose] furfural—an important raw material for the oil processing industries and an intermediate in the manufacture of nylon—may be obtained. **1969** J. A. BRYDSON *Plastics Materials* (ed. 2) xxiv. 511 Furfural occurs in the free state in many plants but is obtained commercially by degradation of hemi-cellulose constituents present in these plants.

furfuraldehyde (fʊɹfiŭræ·ldĭhəid). *Chem.* [f. FURFUR(OL + ALDEHYDE.] = prec.

1879 *Jrnl. Chem. Soc.* XXXVI. 137 A preparation of glacial acetic acid..which had been observed..to give a deep red coloration with aniline..he finds to contain furfuraldehyde. **1969** L. K. ARNOLD *Introd. Plastics* xvi. 172 Furfural (furfuraldehyde) is used in phenol-furfural resins and as an intermediate in nylon 6,6.

furfuran (fʊ·ɹfiŭræn). *Chem.* Also -ane. [a. G. *furfuran* (A. Baeyer 1877, in *Ber. d. Deut. Chem. Ges.* X. 1361), f. FURFUR(OL + -AN, -ANE.] = *FURAN.

1877 *Jrnl. Chem. Soc.* II. 745 The author [*sc.* A. Baeyer] proposes for the group C_4H_4O (Limpricht's *Tetraphenol*) the name 'furfuran'. **1895** *Bloxam's Chem.* (ed. 8) 744 Furfurane, C_4H_4O, is found in the first runnings of the distillation of wood-tar. **1899** [see *FURAN]. **1923** C. ELLIS *Synthetic Resins & Plastics* xi. 202 The cyclic compound, furfurane, may be looked upon as a member of the ethylene oxide series. **1958** PACKER & VAUGHAN *Org. Chem.* XXVII. 937 Furan (furane, furfuran) is a colourless liquid.

furfuryl (fʊ·ɹfiŭril, -əil). *Chem.* [f. FURFUR(OL + -YL.] **a.** The univalent radical $C_4H_3O·CH_2$— containing a furan ring. †**b.** = *FURYL. *Obs.* **furfuryl alcohol,** a colourless liquid, $C_4H_3O·CH_2OH$, darkening in air and used industrially as a solvent and in the manufacture of synthetic resins.

1873 *Jrnl. Chem. Soc.* XXVI. 626 Furfuryl alcohol, $C_5H_6O_2$, is produced, together with pyromucic acid, by the action of alcoholic potash and furfurol. **1884** ROSCOE & SCHORLEMMER *Treat. Chem.* III. II. 614 Constitution of the Furfuryl Compounds.—These contain the group C_4H_3O, which in accordance with Baeyer's suggestion is termed 'furfur'. **1958** *Spectator* 20 June 804/3 The stuff which, apparently, gives coffee its flavour—a chemical called furfuryl mercaptan.

furiant (fūᵊ·riănt, fiūᵊ·riănt). *Mus.* [Czech.] A type of Bohemian dance, or its music, in quick triple time with frequently-shifting accents.

1881 GROVE *Dict. Mus.* III. 614/2 Bohemia is preeminently rich in dances (such as the *beseda*, *dudák*, *furiant*, [etc.]). **1902** *Encycl. Brit.* XXVII. 554/1 The fiddlers who scraped out their 'furiants' and other wild dances. **1943** *Times* 11 Nov. 6/5 The ballet,..with Miss Sasha Machov to teach them the authentic style in polka and furiant. **1967** *Listener* 13 Apr. 505/2 An example of just that sort of thing is Dvořák's *Czech Suite* of 1879. Its five movements include a polka, a bucolic minuet, and a *furiant* finale.

furious, *a.* **1. d.** Add to def.: Also *advb.*, rapidly, uproariously, noisily. (Add further examples. For FUN 3 in quot. 1790 read: FUN *sb.* 2.)

1851 H. MELVILLE *Moby Dick* II. xlii. 284 He swam so fast and furious. **1859** A. ROBERTSON *Let.* 10 Feb. in R. Fulford *Dearest Child* (1964) 162 The dancing was resumed, and..they were going on at it 'fast and furious'. **1899** *Tit-Bits* 29 Apr. 103/1 Pictures came fast and furious. **1931** *Week-end Rev.* 21 Nov. 654/2 The fun is fast and furious.

furnace, *sb.* Add: **5. furnace spectrum,** the spectrum of the light emitted by a substance when heated in an electric furnace. So *furnace line*.

[**1905** *Astrophys. Jrnl.* XXI. 256 The oven spectrum shows new groups of bands in the spectra of Ca, Sr, Ba, and Cu.] **1911** *Contribs. Mt. Wilson Solar Observ.* III. 6 The quality of the furnace lines for measurement is in general good. *Ibid.* 16 The..strong lines..are given by the core of the arc, and appear in the furnace spectrum very faintly at the highest temperatures. **1922** A. S. EDDINGTON in *Encycl. Brit.* XXX. 298/2 The 'enhanced lines' of strontium 4077 and 4215 are relatively strong in stars of high luminosity; whereas the 'furnace lines' of strontium 4607 and calcium 4455 behave in the reverse manner. **1943** *Astrophys. Jrnl.* XCVIII. 33 For the furnace spectrum, dysprosium was vaporized in the carbon-tube vacuum furnace at a temperature near 2600°C.

furnish, *sb.* Add: **d.** The materials from which paper is manufactured.

1920 CROSS & BEVAN *Paper-Making* 374 The characters of these sorted rags are taken into account in the composition of the paper-maker's furnish. **1929** *Penrose's Ann.* XXXI. 99 A good proportion of cotton and/or linen in the furnish of a paper, as well as high chemical purity, are essential for durability. **1969** M. KILBY *Write on Both Sides* i. 31 'Depends what furnish you're working to.' 'Furnish?' 'That's what we call the formula of a paper.'

furnish, *v.* **10. a.** (Later examples.)

1850 F. E. SMEDLEY *F. Fairlegh* iv, Clayton had..slain a sufficient number of victims to furnish forth pies for the supply of the whole mess. **1860** GEO. ELIOT *Bro. Jacob* ii, What housewife..would not think shame to furnish forth her table with articles that were not home-cooked? **1903** R. LANGBRIDGE *Flame & Flood* xvi, So she would..order..the best that the 'cuiseen'..could furnish forth.

furnisher. Add: **2.** In textile printing, a revolving brush or roller that supplies the colour.

1897 C. F. S. ROTHWELL *Printing Textile Fabrics* I. 29 The furnisher generally revolves the same way as the printing roller, and consists then of the ordinary cloth-covered wooden roller; but when pigment colours, or any colour that has a tendency to stick in the engraving, are being printed it is usual to use a brush furnisher and to work it in the opposite way. **1898** F. H. THORP *Outl. Industr. Chem.* I. 495 The color is fed to the print roll from the color box by a revolving cylindrical brush called the 'furnisher', which dips into the color paste. **1961** BLACKSHAW & BRIGHTMAN *Dict. Dyeing* 29 The Brush Furnisher is used for print pastes containing a high solids content of heavy density ingredients such as zinc oxide.

furnishing, *vbl. sb.* **1. a.** (Earlier and later *attrib.* examples.)

1833 *Knickerbocker* I. 157 Gardiner's magnificent furnishing establishment already totters on its base. **1848** E. BRYANT *California* i. 14 These I obtained at reasonable rates, of Messrs. Wilson & Clarke, who keep a general furnishing store for these expeditions. **1860** GEO. ELIOT *Let.* 20 Dec. in J. W. Cross *Life* (1885) II. xi. 282 Our curtains are not up and our oil-cloth is not down. Such is life, seen from the furnishing point of view! **1967** E. SHORT *Embroidery & Fabric Collage* i. 18 Colours which shocked a few years ago are now accepted, and commonly found on furnishing and dress fabric.

furniture. Add: **6. a.** *spec.* The mountings of a rifle.

1852 *Househ. Words* 13 Mar. 582/1 Such close fittings as those of the furniture of guns to the stocks. **1892** W. W. GREENER *Breech-Loader* 14 The furniture of the stock consists of the heel-plate, trigger-guard, &c.

7. b. Applied in the book trade to well-bound volumes and 'standard' sets which serve to fill and adorn the shelves of a private library. (Cf. *furniture-picture* in 10 in Dict.)

1928 *Periodical* No. 143. 25 Its [*sc.* the O.E.D.'s] claims as 'furniture' are not despicable. **1928** *Times Lit. Suppl.* 5 Apr. 260/3 The big 'furniture' books in folio and quarto which long formed the most conspicuous feature of private libraries. **1928** *Publishers' Circular* 6 Oct. 461 Much of it just furniture, and yet there were many very nice items.

10. *furniture-polish* (examples), *-van* (earlier example); **furniture beetle,** a small wood-boring beetle of the family Anobiidæ, esp. *Anobium punctatum*; **furniture cream,** a creamy substance used for polishing furniture.

1915 *Proc. R. Phys. Soc. Edinb.* XIX. 220 The Furniture beetles are comprised in the two families Anobiidæ and Lyctidæ. **1938** L. HUNTER *Domestic Pests* xvi. 121 The common furniture beetles emerge from the pupal stage just beneath the surface of the wood. **1959** E. F. LINSSEN *Beetles Brit. Is.* II. 62 The most annoying of all household Furniture Beetles is *Anobium punctatum*. **1971** *Times* 28 Jan. 12/5 He can point to the exact spot where the furniture beetle may be found at work on a damaged specimen beech. **1873** E. SPON *Workshop Receipts* 86/2 *Furniture Cream.*—1. Yellow wax, 4 oz.; yellow soap, 2 oz.; water, 50 oz.; boil and add boiled oil and oil of turpentine, each 5 oz. **1904** J. VAIZEY *More about Pixie* viii. 90 Sukey mixed the lettuce with furniture cream instead of salad-dressing. **1962** G. BUTLER *Coffin in Oxford* x. 138 A smell of furniture cream, lightly scented with lavender. **1833** J. NEAL *Down-Easters* I. 15 Plastering the foot all over with his furniture polish and wrapping it up. **1891** KIPLING *Light that Failed* iv. 59 If they want furniture-polish, let them have furniture-polish, so long as they pay for it. **1951** *Good Housek. Home Encycl.* 222/2 *Liquid Furniture Polish.* ½ pint turpentine. ½ pint linseed oil. ½ pint methylated spirit. ½ pint vinegar. **1858** C. M. YONGE *Christmas Mummers* i. 7 A large scarlet vehicle, painted in yellow letters, with the words, 'Postlethwayte's Furniture Van'.

furnitureless (fʊ·ɹnitiŭɹlĕs), *a.* [f. FURNITURE + -LESS.] Having no furniture.

1887 W. JAMES *Let.* 5 Feb. (1920) I. 262 My heart was heavy in my breast that so rich a nature..should have nothing but that furnitureless cabin within and snow and sky without, to live upon. **1890** W. BOOTH *In Darkest Eng.* II. v. 167 We found them both out of work, home furnitureless, in debt. **1927** *Glasgow Herald* 8 Nov. 11 Marlborough House is still furnitureless, and is a place of ladders and dust-sheets.

furor. Add: **4.** Sometimes with L. adj. added to define the nature of the 'frenzy', as *furor academicus, biographicus, papisticus, poeticus, teutonicus*. Also *furor scribendi*.

1850 KINGSLEY *A. Locke* I. xx. 278 Mackaye grumbled at my writing so much, and so fast, and sneered about the *furor scribendi*. **a 1873** MILL *Ess. Relig.* (1874) 33 The fire of London, which is believed to have had so salutary an effect on the healthiness of the city, would have produced that effect just as much if it had been really the work of the 'furor papisticus' so long commemorated on the Monument. **1922** A. HUXLEY *Let.* 8 June (1969) 207 His parents-in-law elect are extremely averse to being anything more than allied—on the score, I gather, of poor Robert's..furious *furor poeticus*. **1928** *Daily Tel.* 11 Sept. 12/1 Once upon a time Macaulay complained of the furor biographicus or lues Boswelliana which makes biographies an orgy of praise. **1960** *Times* 2 Dec. 20/6 From the early *furor teutonicus* to the present mood of lyric vitality, the achievement..is impressive. **1963** P. H. JOHNSON *Night & Silence* v. 27 The man was obviously wild with excitement, with *furor academicus*. **1964** F. BOWERS *Bibliogr. & Textual Crit.* v. iv. 151 A balance that is sometimes neglected in the *furor poeticus* of textual speculation. **1964** *Economist* 31 Oct. 507/1 The obsessive *furor scribendi* which..drove him to proliferate lectures, addresses and prefaces.

furore. Add: (Earlier examples.)

1790 E. WYNNE *Diary* 15 Feb. (1935) I. ii. 34 Went to the opera... They made a great furore for Mrs Banti. **1831** J. C. YOUNG *Diary* 16 June in *Mem. C. M. Young* (1871) I. vi. 208, I heard Paganini. The *furore* there has been about this man has bordered on fatuity.

2. Uproar, disturbance, fury.

1946 H. MILLER *Let.* 7 Oct. in Durrell & Miller *Private Corr.* 231 Girodias, Gallimard and Denoël will all be brought to trial in a few months for publishing French versions of the *Tropics* and *Black Spring*. A real shindig!...A tremendous furore. They now talk about 'Le Cas Miller', as they talked once of the Dreyfus affair. **1947** I. BROWN *Say Word* 54 Consider Furore. Nowadays, especially in the Press, it often has a totally incorrect meaning. We read that so-and-so's speech caused a furore, i.e. an uproar of resentment. **1948** H. ACTON *Mem. Aesthete* v. 114 My 'Conversazione of Musical Instruments', which was to create a furore when I recited it at Oxford. **1970** E. O'BRIEN *Pagan Place* II. 124 Your father laughed recalling fist-fights about such issues as the best goalie in the county...One thing he always made a point of was to stand a round of drinks after the furore had died down.

furphy (fʊ·ɹfi). *Austral. slang.* [Associated with *Furphy carts*, water and sanitary carts used in the 1914–18 war, manufactured at a foundry established by the Furphy family at Shepparton, Vict.] A false report or rumour; an absurd story. Also *attrib.*

1916 *Anzac Bk.* 56/1 These furphies are the very devil. *Ibid.*, Furphy was the name of the contractor which was written large upon the rubbish carts that he supplied to the Melbourne camps. The name was transferred to a certain class of news item, very common since the war, which flourished greatly upon all the beaches. **1916** C. J. DENNIS *Moods of Ginger Mick* 122 Soljerin's me game. That's no furphy. **1919** W. H. DOWNING *Digger Dial.* 25 *Furphy-king*, a retailer of rumours. **1933** *Bulletin* (Sydney) 12 Apr. 21 The persistent rumour that they were introduced to check ragwort is a furphy. **1942** C. BARRETT *On Wallaby* iv. 70 Tim broke the..news... It was no furphy this time. We really were going..to somewhere in Sinai. **1945** L. GLASSOP *We were Rats* III. xlv. 247 We did not believe it when they told us...'Mick,' [Eddie] said earnestly, 'it's a furphy. It's the same old furphy..an'..it's not funny any more.' **1960** 'J. WYNDHAM' *Trouble with Lichen* xii. 153 The furphy round the House is that *they* believe in the thing so solidly that they've convinced their husbands.

furriered (fʊ·riəɹd), *pa. pple.* [-ED¹.] Made or treated by a furrier.

1923 *Daily Mail* 2 July 4 This regal looking garment..Lined and Furriered beautifully. **1927** *Observer* 20 Nov. 20 French fur productions. Furriered from hundreds of perfect pelts. **1929** *Daily News* 5 Oct. 1, 8 Fitch Skins..furriered in Paris.

furriery. Substitute for entry: †**1.** *pl.* Furs collectively. *Obs.*

1784 KING in King & Cook *Voy.* III. VI. vi. 340 No labour can ever be turned to so good account as what is employed upon their furrieries. The animals, therefore, which supply these, come next to be considered. **1799** W. TOOKE *View Russian Empire* III. 51 For smaller furrieries and edgings, the skins of the Marten, the Squirrel, the Ermine, the Rabbit and the Marmotte are the choicest. [**1828–32** WEBSTER, *Furriery*, furs in general. *Tooke.*]

2. The art of dressing and making up furs; furrier's work; the trade or business of a furrier.

1920 C. J. ROSENBERG *Furs & Furriery* p. v, There was no work obtainable dealing with the practical side of Furriery. **1929** *Daily News* 5 Oct. 1 A stole which cannot fail to please—the quality and furriery are *sans reproche*. *Ibid.*, The finest points of furriery are embodied in this charming Coat. **1959** *Chambers's Encycl.* VI. 127/2 The Rex coat is ideal for furriery, being produced in a variety of colours.

furrin (fʊ·rin), *a.* Humorous or dialectal perversion of *foreign*.

1895 MRS. H. WARD *Story of Bessie Costrell* iii. 52 'Ee's a bishop, they ses—someun from furrin parts. **1915** F. H. BURNETT *Lost Prince* vi. 60 It'd be likely to be in some o' these furrin places. England'd be too far. **1950** L. DURRELL *Spirit of Place* (1969) 104 Dear Anne, I have been owing you a letter From furrin parts. **1969** E. MCGIRR *Entry of Death* vii. 145 May I present Sergeant Griffid. I'll spell it on account of it being hard on furrin ears.

furriner (fʊ·rinəɹ). Humorous or dialectal perversion of *foreigner*.

1849 C. BRONTË *Shirley* I. vii. 191 As a furriner,..you have not that understanding of huz and wer ways. **1852** E. TWISLETON *Let.* 29 Oct. (1928) iv. 61 This is what it is to have married a 'furriner' and left one's sisters. **1924** J. BUCHAN *Three Hostages* xvii. 242 The worst of them furriners..is that you can't never be sure what they thinks of you. **1968** D. HOPKINSON *Incense-Tree* ii. 17 They left the beaches to 'furriners' like ourselves.

furring, *vbl. sb.* Add: **4.** The business of collecting furs or hunting furred animals; fur-trading. Also *attrib.*

1778 G. CARTWRIGHT *Jrnl.* (1792) II. 373, I fitted out Joseph Tero for a furring voyage to White-bear River. **1849** *Jrnl. Bp. Newfoundland's Voy.* 67 Sixty miles within that bay the Hudson's Bay Company have a furring establishment. *Ibid.*, The settled inhabitants..are supported by the salmon fishery, by furring, and by killing seals. **1856** W. E. CORMACK *Narr. Journey Newfoundland* (1874) VIII. 73 The extent of their salmon and cod fisheries, and of their furring, was noticed when speaking of the occupation collectively of the inhabitants of St. George's Bay. **1920** GRENFELL & SPALDING *Le Petit Nord* (1921) 62 In the winter the northern people move up the bays and go 'furring'.

furrowed, *ppl. a.* Add: in *Masonry* (see quots.).

1904 C. F. MITCHELL *Brickwork & Masonry* 128 *Furrowed Work*... This labour, used to accentuate quoins, consists in sinking a draught about the four sides of the face of a stone, leaving the central portion projecting about ⅛ inch, in which a number of vertical grooves about ⅛ inch wide are sunk. **1929** E. G. WARLAND *Mod. Pract. Masonry* 24 *Furrowed Surfaces.*—Small flutings, from ¼ to ⅜ in. wide, are worked vertically or horizontally across the surface.

further, *a.* Add: **2.** *further education*, formal education organized for adults, or for young people who have left school. Also *attrib.*

[**1898** G. BALFOUR *Educ. Syst. Gt. Brit. & Ireland* p. xxii, No adequate system of intermediate schools exists to receive, as a matter of routine, those elementary pupils who are fitted for further education.] **1913** J. H. WHITEHOUSE *Nat. Syst. Educ.* ii. 6 The writer suggests..the conferring on local authorities of the power to enforce further education compulsorily for children between the ages of 14 and 18. **1937** G. A. N. LOWNDES *Silent Social Revolution* II. 95 The growth of our modern system—or rather systems—of technical and further education drawing their strength both from the secondary and (since 1926) the Hadow senior schools. **1944** *Act* 7 & 8 *Geo. VI* c. 31 § 7 The statutory system of public education shall be organised in three progressive stages to be known as primary education, secondary education, and further education. **1962** J. BRAINE *Life at Top* i. 11 Now he was dead and Sindram Grange had been taken over by the County as a Further Education Centre. **1970** *Daily Tel.* 28 Aug. 2/2 The establishment and development of polytechnics as bridges between further and higher education.

furunculosis (fiurʊŋkiŭlōu·sis). *Path.* [mod. L., f. L. *furunculus* FURUNCLE: see -OSIS.]

1. A diseased condition marked by the appearance of a crop of boils.

1886 *Med. & Surg. Reporter* LV. 609 (*heading*) Furunculosis. **1908** *Practitioner* June 858 Eczema, with complicating furunculosis. **1946** MACKEE & CIPOLLARO *Skin Dis. Children* (ed. 2) ii. 25 Localized recurrent furunculosis is best treated with roentgen rays. **1968** NAYLOR & ROOK in A. Rook et al. *Textbk. Dermatol.* I. xx. 607/1 Epidemics of furunculosis attributable to specific strains of staphylococci have occurred.

2. [ad. G. *furunkulose* (Emmerich & Weibel 1894, in *Archiv für Hygiene* XXI. 5).] A disease of salmonid fishes caused by the bacterium *Aeromonas salmonicida*.

1912 MASTERMAN & ARKWRIGHT *Rep. Epidemic Salmonidæ 1911* 13 The epizootic disease which occurred amongst the salmon and trout in the south-west of England during the summer of 1911 presented the same symptoms as the disease which occurs in Germany, especially Bavaria, and in France, and is called Furunculosis of the Salmonidae. **1930** *Discovery* Aug. 268/2 'Furunculosis'..affects salmon and trout rivers where it may cause great havoc among the fish. The disease is caused by a bacterium, which cannot survive in absolutely pure water but arises in water polluted by organic matter. **1959** J. W. JONES *Salmon* 165 One of the most important factors in causing epidemics of furunculosis is water temperature.

furyl (fiūᵊ·ril, -əil). *Chem.* [Abbrev. of *FURFURYL.] The univalent radical C_4H_3O— based on furan. Cf. *FURFURYL.

1916 *Chem. Abstr.* X. 3400 Organic radicals...Furyl (2-, 3-),

O·CH:CH·CH:C–.

1963 F. M. DEAN *Naturally Occurring Oxygen Ring Compounds* i. 3 Furyl carbinols behave as allylic alcohols in suffering hydrogenolysis to alkyl furans. **1966** *Nomenclature Org. Chem.* (I.U.P.A.C.) (ed. 2) 67 Univalent radicals derived from heterocyclic compounds by removal of hydrogen from a ring are in principle named by adding 'yl' to the names of the parent compounds. . . The following exceptions are retained: furyl, [etc.].

fusain. Add: **2.** (fiū·zein). A friable, porous type of coal that occurs as dull layers or masses in bituminous seams; 'mineral charcoal'.
1883 *Jrnl. Chem. Soc.* XLIV. 941 Fusain is a variety of coal, resembling wood-charcoal in appearance. Some stalks, the interior of which is composed of fusain, are covered with a bark which has been converted into coal. 1913 *Proc. Amer. Philos. Soc.* LII. 89 The unaided eye can discern many features of coal in the bed; it can group types into glance, matt, cannel, fusain. 1919 [see *CLARAIN]. 1935 *Nature* 23 Nov. 818/2 Quite moderate variation in the amount of fusain in the coal has important effects on its utility. 1961 W. Francis *Coal* (ed. 2) v. 296 Fusain differs very considerably in constitution and properties from the remaining ingredients of coal. *Ibid.*, The proportion of fusain in most coals is small, generally less than 3%.

fuse (fiūz), *sb.*² *Electr.* [f. Fuse *v.*²] In full *safety fuse*. A strip or wire of easily fusible metal (or a device containing this) inserted in an electric circuit, which melts (or 'blows') and thus interrupts the circuit when the current increases beyond a certain safe strength. Also *fig.* (see *BLOW *v.*¹ 19 c).
1884 R. Hammond *Electr. Light* v. 56 Conductor Joined with Fuse. *Ibid.* 58 Showing Safety Fuses in Circuit. 1890 J. W. Urquhart *Electr. Light Fitting* 160 The main object, then, of a safety fuse or cut-out . . is to prevent accidental overheating. *Ibid.* 165 [see *b]. 1923 Wodehouse *Inimitable Jeeves* ii. 28 That scheme of yours . . has blown out a fuse. 1969 *Which?* Sept. 282/2 When you buy a plug, it will normally have a 13-amp fuse inside.
b. *fuse-block, -board, -box, -carrier, -holder, -mounting, -plate, -plug*, various contrivances for holding a fuse or a number of fuses; **fuse-wire**, wire used to make fuses.
1885 J. Dredge *Electr. Illum.* II. 325 A fuse box containing six fuses. 1890 J. W. Urquhart *Electr. Light Fitting* 162 The fuse plate may easily be removed and replaced by others. *Ibid.* 165 In such cases it is considered safer to assemble all the fuses upon a fuse board. 1891 H. E. Swift *U.S. Patent 455,366*, Fuse-wire cut outs or 'fuse-blocks', as they are commonly called, have also been used in connection with incandescent lamps, said fuse-blocks being located on the ceiling. 1892 F. C. Allsop *Pract. Electr. Light Fitting* 57 The reason why lead or lead-tin alloy is preferred for the fuse-wire of a cut-out. 1893 W. P. Maycock *Electr. Lighting* III. xv. 220. 384 A S.P. fuse block. *Ibid.* 386 The fuse wire is fitted in what is called a fuse plug. 1894 D. Salomons *Electr. Light Install.* (ed. 7) II. 219 Another improvement . . is that of placing under the fuse binding-screw a washer carrying a steady pin, which passes loose into the fuse block. *Ibid.* III. 161 It is almost impossible to remove the fuse-box cover or replace the fuse without removing this box completely. 1899 W. P. Maycock *Electr. Wiring* 185 The removable fuse carrier is of porcelain. *Ibid.* 245 The fuse-holders are of porcelain. 1899 H. M. Leaf *Internal Wiring* 85 A guide for determining . . the sizes of fuse wires that may be employed for cut-outs. 1914 S. C. Batstone *Electr.-Light Fitting* 104 Porcelain Tubular Fuse Carrier. 1936 M. Kennedy *Together & Apart* i. 47 She snipped off lengths of fuse wire. 1964 R. F. Ficchi *Electrical Interference* vi. 78 The fuse-mounting and relay assembly is usually mounted close to the power source.

fuse, *v.*² Add: **2. c.** Of an electric light, appliance, etc.: to fail or be extinguished owing to the melting of a fuse. Also *trans.*, to cause (a circuit, etc.) so to fail.
1930 *Daily Express* 6 Nov. 2/2 Two minutes before he arrived . . the lights in the building fused. 1940 G. D. H. & M. Cole *Murder at Munition Works* xx. 192, I was in his shop once when the light fused. 1951 *Good Housek. Home Encycl.* 80/2 A heater should never be run from a lighting plug, as this is apt to fuse the circuit.

fuse (fiūz), *v.*³ [f. *fuse sb.*²] *trans.* To insert a fuse in (a circuit), to furnish with a fuse.
1914 D. C. Shafer *Harper's Everyday Electricity* ix. 90 The method of fusing a house circuit is best shown in Fig. 12. 1954 F. Wiseman *Penguin Handyman* (ed. 3) i. 11 A circuit supplying this [current] should be fused with a 5-amp fuse wire.
Hence **fused** *ppl. a.*
1934 C. Haslett *Electr. Handbk. for Women* ix. 153 Fused plugs are available and are often found very useful. 1962 *B.S.I. News* Mar. 24/2 Socket-outlets requiring fused plugs must not accept unfused plugs. 1967 *Do It Yourself* Nov. 1283/2 Feed the circuit from a fused spur box situated in a conspicuous position on the landing.

fused, *ppl. a.* Add: **c.** *fused participle*: a participle regarded as being joined grammatically with a preceding noun or pronoun, rather than as a gerund that requires the possessive, or as an ordinary participle qualifying the noun (see quot. 1926).
1906 H. W. & F. G. Fowler *King's Eng.* ii. 119 The mistake is caused by certain types of sentence in which a real, not a fused participle is so used that the noun and its (unfused) participle give a sense hardly distinguishable from a possessive noun and a gerund. *Ibid.* 120 Long fused-participle phrases are a variety of abstract expression, and as such to be deprecated. 1926 Fowler *Mod. Eng. Usage* 205/2 *Fused participle* is a name given to the construction exemplified in its simplest form by 'I like *you pleading* poverty'. . . The name was invented . . for the purpose of labelling & so making recognizable & avoidable a usage considered by the inventor to be rapidly corrupting modern English style. 1942 Partridge *Usage & Abusage* 124/2 Idiomatically, the *excuse me doing* form is generally understood to be positive; but it is much rarer than the *excuse my doing* form, especially since ca. 1920, when the fused participle doctrine began to ravage the land. 1954 E. Gowers *Compl. Plain Words* ix. 157 What they are not agreed about is whether it is also correct to . . write 'the Bill getting a second reading surprised everyone'. If that is a legitimate grammatical construction, the subject of the sentence, which cannot be *Bill* by itself, or *getting* by itself, must be a fusion of the two. Hence the name 'fused participle'.

fuselage (fiū·zeläʒ, -edʒ). [Fr., f. *fuseler* to shape like a spindle, f. *fuseau* spindle: see Fusil¹ and -age. So called from its spindle-like shape.] The elongated body of an aeroplane, to which the wings and tail unit are attached and which (in modern aircraft) contains the crew and the passengers or cargo. (In early use given slightly varying meanings: see quots.)
1909 *Flight* 19 June 366/2 The aeroplane itself was considerably damaged, the *fuselage* which carries the elevating-plane in front, being completely broken. 1910 R. Ferris *How it Flies* xx. 461 *Fuselage*, the framework of the body of an aeroplane. 1913 A. E. Berriman *Aviation* 249 Another French word at present in common use is *fuselage*, meaning the girder-like backbone employed in modern aeroplane design. This member also forms the body of the machine. 1913 *Times* 14 Apr. 4/1 The number of machines which have broken their fuselage—to say nothing of their wings. 1913 A. E. Verrill *Harper's Aircraft Bk.* xi. 120 The parts of an aeroplane are mainly the frame, or 'chassis'; the body, or 'fuselage'; the wings, or 'planes'; [etc.]. 1918 Cowley & Levy *Aeronautics* viii. 141 The fuselage connecting the body and the tail system exists purely to provide support for the latter. 1932 E. Bowen *To North* xvi. 175 Catching the full summer sun on its wings and fuselage . . the plane evenly passed to the coast. 1960 C. H. Gibbs-Smith *Aeroplane* I. xvi. 129 The construction of a large conventional all-metal aeroplane of today . . is stressed skin, with [in the fuselage] the metal skin laid over a multiple system of ring-line formers.

fusidic (fiūsi·dik), *a.* *Biochem.* [f. mod.L. *Fusid-ium* (see def.), f. L. *fūs-us* spindle + Conidium: see -ic.] *fusidic acid*: a steroid, $C_{31}H_{48}O_6$, with antibiotic properties, orig. isolated from a strain of the fungus *Fusidium coccineum* and esp. effective against *Staphylococcus aureus*.
1962 W. O. Godtfredsen et al. in *Nature* 10 Mar. 987/1 From the fermentation broth of a strain of *Fusidium* a hitherto unrecorded antibiotic, for which the name 'fusidic acid' is proposed, has been isolated. 1964 *Lancet* 15 Dec. 1278/1 Success with fusidic acid in a highly resistant case of staphylococcal septicaemia. 1967 Gottlieb & Shaw *Antibiotics* II. 144 Fusidic acid, $C_{31}H_{48}O_6$, a metabolite of *Fusidium coccineum* and *Cephalosporium lamellaecola*. *Ibid.* 145 Fusidic acid, sometimes called 'Antibiotic ZN-6', is identical with 'Ramycin' from *Mucor ramannianus*.

fusimotor (fiū·zimōᵘtəɹ), *a.* *Physiol.* [f. L. *fūs-us* spindle + Motor *a.* 2]. Of, pertaining to, or designating the motor neurones with fibres of small diameter which innervate the intrafusal fibres (spindles) of muscle.
1958 Hunt & Paintal in *Jrnl. Physiol.* CXLIII. 195 The motor fibres to the spindle have been designated γ efferents by Leksell (1945) and small motor fibres by Kuffler et al. (1951). Both designations have possible ambiguity. . . We therefore propose the term fusimotor with reference to neurones that are motor to the muscle spindle. 1962 P. B. C. Matthews in *Q. Jrnl. Exper. Physiol.* XLVII. 327 There exist fusimotor fibres of two functionally distinct kinds, which for the purposes of the present description, and for the want of better terms, will be called static and dynamic fusimotor fibres. 1968 *Jrnl. Physiol.* CXCV. 420 It has been proposed that voluntary movement may be initiated and sustained by the activation of fusimotor neurones . . operating the muscle spindles and monosynaptic reflex as a follow-up length servo. 1969 *Proc. R. Soc.* B. CLXXIII. 162 It is difficult to see how such a segmental motor apparatus could allow the cortex to initiate precise movements of the hand, employing only the fusimotor route.

fusion. Add: **3. c.** (i) *Psychol.* and *Physiol.* [tr. G. *verschmelzung* (J. F. Herbart *Psychol. als Wiss.* (1824) I. 200).] A blending together of separate simultaneous sensations into a new complex experience or qualitative perception; the process whereby a succession of similar stimuli produces a continuous response or the sensation of a continuous stimulus.
1890 W. James *Text-bk. Psychol.* 57 The so-called Fusion of Sensations in Hearing. 1903 G. F. Stout *Groundwork Psychol.* 45 They [*sc.* sensations] may combine . . like the bitterness, sweetness, and aroma of a cup of coffee. . . The first of these modes of union is called *fusion* or blending. . . Fusion is characterised by the absence of any definite order among the constituents of the sensation complex. 1911 E. B. Titchener *Text-bk. Psychol.* II. 351 The classical instance of the qualitative perception is the tonal fusion. 1946 D. P. C. Lloyd in J. F. Fulton *Howell's Textbk. Physiol.* (ed. 15) ii. 37 The frequency of stimulation necessary to promote full mechanical fusion of the contraction response is . . different for different muscles. 1952 H. H. Emsley *Visual Optics* (ed. 5) I. ii. 29 The final meticulous adjustment of the eyes is carried out under the compelling desire for fusion of the two unocular images into a single perceptual image in the cortex.
(ii) *Psychiatry.* [tr. G. *mischung* (Freud *Das Ich und das Es* (1923) iv. 50).] In Freudian theory, the union and balance of life and death instincts which exist in normal persons.
1927, 1946 [see *defusion]. 1953 *Hibbert Jrnl.* July 332 Freud noted how masochism and sadism both arise from fusion of the life and death instincts. *Ibid.*, This peculiar fusion, with the death instinct preponderating, lies at the base of the desires . . for sexual self-mutilation.
d. *Nuclear Sci.* The formation of a heavier, more complex nucleus by the coming together of two or more lighter ones, usu. accompanied by the release of relatively large amounts of energy; also, the utilization of this process as a source of energy.
1947 *Sci. News Lett.* 7 June 358 (*heading*) Atom fusion gives energy. 1952 *Economist* 22 Nov. 541/1 This may have been a hybrid bomb, part atom, part hydrogen, but enough to prove that the scientists have solved the problem of releasing energy by nuclear fusion. 1957 *Rev. Mod. Physics* XXIX. 565/1 It has been suggested . . that the fusion of helium plays an important role in energy generation and element synthesis in the red-giant stage of the star's evolution. 1958 *Engineering* 21 Feb. 227/3 Fission and fusion, at least in the initial stages, may well be complementary, both being necessary for a properly integrated nuclear power programme. 1964 R. H. Baker *Astron.* (ed. 8) xvi. 479 The fusion of hydrogen into helium in the cores of these more massive stars may be principally by means of the carbon cycle. 1968 M. S. Livingston *Particle Physics* i. 5 Another special type of interaction is the fusion of light-weight nuclei such as ^2H . . into the more stable He^4 nuclei plus extra neutrons, with the release of several million electron volts per particle. 1969 J. Dougall tr. *Born's Atomic Physics* (ed. 8) x. 356 The only example of a man-made thermonuclear fusion process is in the form of the hydrogen bomb, where the high temperatures are produced by means of an initiating atomic (i.e. fission) explosion.
4. *attrib.* and *Comb.*, as *fusion energy, reaction, reactor, weapon*; **fusion bomb**, a bomb in which the energy released is derived from an uncontrolled process of nuclear fusion; *spec.* a hydrogen bomb; (critical) **fusion frequency** [tr. G. *verschmelzungsfrequenz* (J. v. Kries 1903, in *Zeitschr. f. Psychol. u. Physiol. d. Sinnesorgane* XXXII. 115)] *Physiol.* = *flicker-fusion frequency*; **fusion nucleus** *Biol.*, a cell nucleus that results from the fusion of two or more nuclei; **fusion welding**, a welding technique in which the metal is melted and joined without the application of pressure; so **fusion weld** *sb.*, **-welded** *ppl. a.*
1950 *Sci. Amer.* Mar. 13/2 The designer of a fusion bomb clearly would start with a fission bomb of uranium or plutonium, the explosion of which would produce the high temperature required for the thermonuclear fusion reaction. 1955 *Sci. News* XXXVIII. 8 He [*sc.* Dr. Bhabha] predicted that a way would be found 'for liberating fusion energy in a controlled manner within the next two decades'. 1962 J. P. C. Southall tr. *Helmholtz's Treat. Physiol. Optics* II. 374 Porter compared the fusion frequencies for a series of different intensities of illumination. 1970 T. Cornsweet *Visual Perception* xiv. 393 If the frequency of flashing of a light is gradually increased from a low rate, there will be some frequency for which the light will just look steady . . and this is called the critical fusion frequency, or CFF. 1904 *Ann. Bot.* XVIII. 345 The fusion-nucleus then increased in size. 1928 E. B. Wilson *Cell* (ed. 3) v. 400 The pronuclei conjugate immediately after the entrance of the sperm and apparently fuse completely to form a fusion-nucleus. 1941 Johannsen & Butt *Embryol. Insects* ii. 9 The zygote, or fusion nucleus, in which the diploid number of chromosomes has been restored. 1950 F. Gaynor *Encycl. Atomic Energy* 129 Fusion reactions go on constantly in the interiors of the stars . . and form the basic energy sources of these stars. 1960 T. Frisch in *Atlantic* Oct. 76 The fusion reactor, in so far as we can see at present—will have to be operated at a temperature of many million degrees, and the production, control, and containment of such temperatures . . is a stupendous problem. 1955 *Time* 25 July 61/3 If a fusion reactor works with deuterium, the fuel will be cheap and abundant. 1955 *Bull. Atomic Sci.* Mar. 94/1 The presence of fission or fusion weapons in another major conflict seems inevitable. 1963 H. Kahn *On Escalation* vi. 96 'Clean' fusion weapons, leaving little or no residual fallout. 1930 *Engineering* 14 Mar. 364/1 With fusion welds, there is no difficulty in obtaining metal stronger than the plate. *Ibid.*, It has been customary to speak of the strength of a fusion-welded butt joint as a percentage of the plate strength. 1959 *B.S.I. News* Nov. 7/2 Fusion-welded pressure vessels. 1928 *Nature* 11 Apr. 505/2 The oxy-acetylene flame is most generally used for fusion welding, owing to its high temperature. 1958 *Engineering* 18 Apr. 501/1 The equipment includes an a.c./d.c. set for fusion welding under argon . . and a resistance welding machine.

fusional (fiū·ʒənăl), *a.* [f. Fusion + -al.] Of or pertaining to fusion; *spec.* designating a class of languages (see quots.).
1921 E. Sapir *Lang.* vi. 144 'Fusional' and 'symbolic' contrast with 'agglutinative', which is not on a par with 'inflective' at all. What are we to do with the fusional and symbolic languages that do not express relational concepts in the word but leave them to the sentence? *Ibid.* 148 By a 'fusional-agglutinative' language we would understand one that fuses its derivational elements but allows a greater independence to those that indicate relations. 1932 W. L. Graff *Lang.* ix. 333 When the union of radical and nonradical parts [of a word] is so great that a considerable alteration of either or both occurs, we obtain another structural type altogether, namely the *fusional* one. 1957 *Encycl. Brit.* XIII. 700/2 Fusional languages (the term inflected is also used) are the languages of the classical European tradition, such as Latin or Greek. Here grammatical endings are numerous, but it is not easy to isolate out the separate functions fused in each. 1964 R. H. Robins *Gen. Ling.* 332 Words in which several grammatical categories are marked by word forms in which it is difficult . . to assign each category to a specific and serially identifiable morphemic section are instances of fusional word structure (the term *inflecting*, or *inflectional*, is the older and traditional name for the typological class here called fusional . .). 1964 S. Duke-Elder *Parsons' Dis. Eye* (ed. 14) xxviii. 451 The involuntary reflexes which depend on vision (fixation, fusional movement, convergence, etc.).

fuss, *sb.* Add: **1. a.** *to make a fuss of* or *over* (†*with*): to pamper; to treat with an excessive display of affection or attention.
1814 Jane Austen *Mansf. Park* III. xiv. 265, I thought little of his illness at first. I looked upon him as the sort of person to be made a fuss with in any trifling disorder. 1928 *Daily Mail* 13 Aug. 10/2 He asked her if she were not going to let him make a fuss of her before he went. 1931 E. O'Neill *Mourning becomes Electra* I. iii. 72 His mother was sheet with him, while Marie, she made a fuss over him and petted him. *a* 1953 — *Long Day's Journey* (1956) I. 37 You like to get us worried so we'll make a fuss over you. *Ibid.* II. ii. 78 You want to be petted and spoiled and made a fuss over.
4. *Comb.*, as **fuss-box, -budget, -pot**, a person who fusses.
1901 R. Meinertzhagen *Army Diary* 30 Apr. (1960) 21 Major Mainwaring . . is a grumpy old fuss-box. 1939 D. L. Sayers *In Teeth of Evidence* 256 What an old fuss-box you are, darling. 1904 *Dialect Notes* II. 397 *Fuss-budget*, a nervous, fidgety person. 1936 M. Mitchell *Gone with Wind* xxvii. 476 Her husband was old fuss-budget Frank. 1966 *Punch* 30 Nov. 822/2 The two Mrs. Oswalds are visible characters, the younger played very artfully by Sarah Miles and the elder, on its terms of chatterbox fussbudget, vividly impersonated by Bessie Love. 1923 G. O'Donovan *Vocations* i. 16 You *are* a fuss-pot. First you won't and then you will. 1951 J. B. Priestley *Festival at Farbridge* II. ii. 249 In the woman's opinion, the gentleman was a fusspot but wonderfully free with his money. 1966 M. R. D. Foot *SOE in France* iii. 50 Some were foolhardy, some were fusspots.

fusticate, *v.* (see at end of Fusty *a.*). Add: Also *intr.*, to stay in a close stuffy atmosphere.
1901 L. Silberrad *Lett. J. Armiter* iv. 102 'Are you going to stir?' . . ('I feel sure he meant "fusticating", though he was too polite to say it) 'in here all afternoon?' he asked.

fusula (fiū·ziŭlă). *Zool.* Also fusule. Pl. -æ. [mod. L., incorrectly f. L. *fūsus* spindle.] (See quots.)
1909 A. E. Shipley in *Cambr. Nat. Hist.* IV. 325 These spinnerets . . are movable turrets on which are mounted the 'fusulae' or projections whence the tubes from the spinning glands open. *Ibid.* 326 In some spiders the fusulae are all much alike, but usually a few very much larger than the rest are noticeable under the microscope, and these are often alluded to as 'spigots'. 1928 T. H. Savory *Biology of Spiders* ii. 41 The smaller ones [*sc.* tubes], called spools or fusulae, consist either of a cylindrical basal portion traversed by a long thin tube or of a slightly conical base with a curved thin tube. 1951 Locket & Millidge *Brit. Spiders* I. 28 The *fusulae* from which the silk is produced are situated on the tips or on the apical segments of the spinners.

||**fusuma** (fū·sŭma). [Jap.] A sliding screen, covered with paper on both sides, used to separate one room from another in a Japanese house.
1880 I. L. Bird *Unbeaten Tracks Japan* I. ix. 88 The large tea-houses contain the possibilities for a number of rooms, which can be extemporised at once by sliding paper panels, called *fusuma*, along grooves in the floor and in the ceiling or cross-beams. 1886 E. S. Morse *Jap. Homes* 126 The *fusuma* forming the movable partitions between the rooms are covered on both sides with thick paper. 1890 B. H. Chamberlain *Things Japanese* 24 The rooms are divided from each other by opaque paper screens, called *fusuma* or *karakami*, which run in grooves at the top and bottom. 1905 *Westm. Gaz.* 23 Sept. 10/2 Sliding the fusuma, the screen of opaque paper separating the living- from the sleeping-room, Sada entered the chamber.

fut (fʌt). [Echoic.] = Phut.
1898 J. M. Falkner *Moonfleet* ix. 125 There came a flash of fire . . and a fut, fut, fut, of bullets in the turf.
Hence **fut** *v. intr.*, of a bullet: to land with a dull abrupt sound.
1901 *Westm. Gaz.* 3 Jan. 2/1 The bullets . . came more thickly now, squealing over our heads and futting on the ground between the horses' hoofs.

||**futah** (fū·tă). Also footah, fotah, futu. [ad. Arab. *fūṭa* cloth used as a waist-wrapper.] A kind of material orig. imported to Arab countries from India; hence, an article or garment made from this material, *spec.* a kind of loin-cloth or short skirt worn by Arabs.
[1616 B. Farie *Let.* 26 May in W. Foster *Lett. Received by E. India Co.* (1900) IV. 306 As for the sales of the Surratt clothing you sent along in the Sea Adventure, I refer you to the advice of Mr. Sairs, entreating you that if you have . . lunges and footaes of all sorts that will not vend in Japon, to ship them away for this place.] 1738 T. Shaw *Travels or Observations Barbary & Levant* 293 But when the Women are at Home and in Private, then they lay aside their Hykes and sometimes their Tunicks, and, instead of Drawers, find only a Towel about their Loyns. . . This is called both in Barbary and the Levant, a Footah. 1836 E. W. Lane *Mod. Egyptians* I. 175 A napkin ('footah') is given to each person. 1853 J. Richardson *Mission to Cent. Afr.* I. v. 67 A few [of the Touaregs] sport a red *fotah*, or turban. 1923 *Other Lands* Oct. 17/1 The fūtah is tied round the waist, and reaches nearly to the knees. 1936 F. Stark *S. Gates Arabia* iii. 32 They wore their futahs (loin-cloths) as long as a kilt or even a skirt. 1962 *Times* 12 Jan. 12/6 The tribesmen in their kilts, or *futas* as they call them. 1966 'S. Harvester' *Treacherous Road* x. 90 A knee-length *futah*, a sort of kilt, of printed or white cotton.

futilitarianism (fiū:tilitē°·riăniz'm). [f. Futilitarian + -ism.] Futilitarian policy.
1921 *Glasgow Herald* 3 Feb. 8 To associate the activities of the Church with the Labour Party would be one more example of the Higher Futilitarianism with which ecclesiastics in all ages have been tempted to experiment. 1924 C. E. Montague *Right Place* x. 153 At least for some eager and absorbed hours your true rambler has washed all that futilitarianism out of his soul.

||**futon** (fū·tǫn). [Jap.] A Japanese bed-quilt.
1876 *Trans. Asiatic Soc. Japan* IV. 172 Those . . who . . are tired of tinned meats and live *futons*. 1886 E. S. Morse *Jap. Homes* 212 The *futons*, or comforters, are . . hung over the balcony rail to air. 1891 Chamberlain & Mason *Handbk. Japan* 8 Beds are still rare; but good quilts (*futon*) are laid down on the mats. 1959 *Encounter* Jan. 20/2 Their *futon*—the wadded quilt stuffed with cotton-wool which serves the Japanese for a bed.

futtah (fʌ·tă). *N.Z.* Also futter, whata, etc. Early spellings representing the pronunciation of Maori *whata*, a food-store raised on posts.
1834 E. Markham *N.Z. or Recollections of It* 37, I have seen a Wutter or Platform 80 feet above ground in an immense Tree all the Branches cut off And a Platform well secured, and the Potatoes on it and thatched over. 1878 E. S. Elwell *Boy Colonists* 100 Walker . . afterwards gave him another pair of trousers from the futtah. 1888 A. H. Duncan *Wakatipians* iii. 22 The buildings at the home-station . . were represented by a 'futter' and a long narrow hut. 1891 *Rep. Austral. Assoc. Advancem. Sci.* III. 378 The men gathered the food and stored it in *whatas* or store-rooms, which were attached to every chief's compound, and built on tall posts to protect the contents from damp and rats. 1892 *N.Z. Alpine Jrnl.* I. ii. 95 Our second [camp] was in George's Creek, where we built a 'futtah' for the provisions. 1954 E. C. Studholme *Te Waimate* (ed. 2) III. v. 260 Most stations . . had a futtah—a corruption of the Maori word, whata, meaning 'a raised store-house in which food is kept'.

futural (fiū·tiŭrăl). *a.* [ad. med.L. *futūrālis* (see Futurality).] Of or pertaining to the future; spec. *Gram.*, having a future sense.
1906 J. H. Moulton *Gram. N.T. Gk.* I. 120 Futural presents . . have no lack of durativity about them. 1953 *Jrnl. Philos.* 27 Aug. 553, I suppose positive results for all possible tests, present as well as futural.

future, *a.* and *sb.* Add: **A.** *adj.* **1. b.** *future life* (examples).
1776 Gibbon *Decl. & F.* I. xv. 450 The doctrine of a future life, improved by every additional circumstance which could give weight and efficacy to that important truth. 1814 J. Kenrick (*title*) The necessity of revelation to teach the doctrine of a future life. 1879 *Psychol. Rev.* I. 335 Those who believe in God and a Future Life. 1945 B. Russell *Hist. Western Philos.* II. ii. 330 As regards the doctrine of a future life, in the West it was first taught by the Orphics and thence adopted by Greek philosophers.
c. (Philos.) *future contingent*: a proposition which is neither true nor false, because it is about the future; sometimes, the future event itself, that may or may not occur. Also *attrib.*
1623, 1788 [see Contingent *sb.* 2]. 1885 J. Veitch *Institutes* § 463 This happens especially in future contingents. 1957 A. N. Prior *Time & Modality* 120 We assign to future contingents a third truth-value. 1969 N. Rescher *Many-Valued Logic* ii. 148 The so-called future-contingent propositions . . involve the outcome of free human choices.
B. *sb.* **3. b.** *spec.* A prospective condition of success, prosperity, etc.; (*there is*) *no future in* (something): there is no prospect of success or advancement in (it).
1891 Wilde *Pict. Dorian Gray* xv. 268, I like men who have a future, and women who have a past. 1921 [see *futuristic *a.*]. 1926 Galsworthy *Silver Spoon* II. x. 199 No future in it! 1932 *Week-end Rev.* 7 May 586/2 Miss Stone certainly has a future as a novelist. 1948 'N. Shute' *No Highway* ii. 40 We can't hide things up. There's no future in that. 1960 L. Cooper *Accomplices* I. vi. 62 You can't rob a bank or bash a chap's head in and then say to the police, Oh it doesn't matter I'm leaving here next week. (You can, I suppose, if you feel that way, but there's no future in it.) 1968 *Listener* 28 Nov. 714/2 I've known Lowry since 1919, when we were both working in the life class in Manchester School of Art, and I predicted a future for Lowry. 1971 'H. Carmichael' *Most Deadly Hate* v. 62 There's no future in being an accessory after the fact in a murder case.
7. *attrib.*, as future(s) research, study = *futurology.
1969 *New Scientist* 18 Dec. 588/2 A conflict about the conduct of futures research in West Germany has blown up among the futurologists themselves. 1971 *Observer* 10 Jan. 21/4 'The Institute for the Future' . . is trying to get futures study out of the gee-whiz stage and give it a much more intellectually solid basis. 1971 *Sunday Times* (Johannesburg) (Business Section) 28 Mar. 10/2 (Advt.), Duties will include so called 'future' research into the implications of rapid technological change for *Man*.

futurism (fiū·tiŭriz'm, -tʃər-). [f. Future *a.* + -ism, after It. *futurismo*, F. *futurisme*.] An art-movement, originating in Italy, characterized by violent departure from traditional forms, the avowed aim being to express movement and growth in objects, not their appearance at some particular moment. Also applied to similar tendencies in literature and music.
1909 *Daily Chron.* 5 May 6/6 'Futurism' is the declaration of the new school of literature grounded by the International Review 'Poesia'. 1915 W. H. Wright *Mod. Painting* 276 Marinetti, a poet, is the spiritual (and monetary) father of Futurism. 1921 Galsworthy *To Let* II. ii, Paul Post—that painter a little in advance of Futurism. 1937 N. Slonimsky *Mus. since 1900* (1938) p. xviii, Futurism was proclaimed in a manifesto published in the Paris *Figaro* on 20 February 1909. It aimed at complete annihilation of all accepted forms in favor of a future music created according to some imagined law of machine-like perfection. 1963 *Cambr. Rev.* 4 May 405/1 Voznesensky perpetuates . . one specific trend in Russian poetry—the school of futurism, represented by Khlebnikov and Mayakovsky, Tsvetayeva, and Pasternak. 1967 E. Short *Embroidery & Fabric Collage* iv. 89 In a comparatively short time Fauvism, Cubism, Expressionism, Dadaism, Futurism, and Surrealism threw new light on the treatment of subject matter.

futurist, *sb.* (*a.*). Add: **2.** One who has regard to or studies the future; a believer in human progress.
1846 [see small-type note in Dict.]. 1936 G. K. Chesterton *Autobiogr.* 25 The one thing forbidden to such futurists was Looking Backwards. 1956 A. Toynbee *Historian's Approach to Religion* vi. 79 The futurists are revolutionaries who consciously and deliberately set out to break with a disintegrating social past in order to create a new society. 1964 E. M. Forster in *Granta* 15 Feb. 9/1 *Traction Engine*: I shall pass often enough in the future. *Punt*: A Futurist! Better and better. 1971 *Observer* 10 Jan. 21/7 A man whose name had often been mentioned respectfully by other futurists.
3. [After It. *futurista*, F. *futuriste*.] An adherent of futurism. Also *attrib.* or as *adj.*
1911 W. J. Locke *Clementina Wing* xxii. 278 After that they had gone to see the New Futurists. 1914 *Star* 16 Dec. 6/2 Small bullet-proof shields, . . painted in cubist patterns in futurist colours. 1915 W. H. Wright *Mod. Painting* 272 The famous Futurist statement that 'a running horse has not four legs, but twenty'. 1916 'Boyd Cable' *Action Front* 128 Erratic daubs of bright colours laid on after the most approved Futurist style. 1924 C. Hamilton *Prisoners of Hope* 129 The walls . . were covered with the raw and confused handiwork of the people who called themselves futurists because they had never been taught how to paint. 1958 *Times Lit. Suppl.* 31 Jan. 58/1 Trotsky writes superbly on the literary 'fellow-travellers', Alexander Blok, and the Futurists in particular.

futuristic (fiūtiŭri·stik, -tʃər-), *a.* [f. prec. + -ic.] **1.** Having the characteristics of futurism; avant-garde; ultra-modern.
1915 W. H. Wright *Mod. Painting* 257 His [*sc.* Gleizes'] well-known L'Homme au Balcon appears to us today as most Futuristic in conception. 1921 *Queen* 10 Sept. 326 Quaint futuristic or jazz embroidery. 1921 Galsworthy *To Let* I. i, What was the use of going in to look at this crazy, futuristic stuff with the view of seeing whether it had any future? 1928 *Melody Maker* Feb. 182/2 Some pretty futuristic harmony. 1968 J. Ironside *Fashion Alphabet* 16 Garments, both futuristic and yet based on armour and chain-mail.
2. Of or pertaining to the future; predicting for the future.
1958 *Listener* 18 Dec. 1021/2 It may well be that only when optical telescopes can be carried in earth satellites or erected on the moon will it be possible to look back into the past to this extent. Before the advent of such futuristic enterprises it seems likely that the great radio telescopes will give us the answer we require. 1969 *N.Y. Rev. Books* 16 Jan. 32/4 Goldbloom has merely invented a novel kind of anachronism that might be called 'futuristic' or 'projective'.

futurology (fiūtiŭrǫ·lŏdʒi, -tʃər-). [f. Future *sb.* + -ology.] The forecasting of the future on a systematic basis, esp. by the study of present-day trends in human affairs. Hence **futuro·logist**.
1946 A. Huxley *Let.* 29 Mar. (1969) 542 Thank you for . . the interesting enclosure on 'Teaching the Future'. I think that 'futurology' might be a very good thing. 1967 *Listener* 23 Mar. 397/1 Futurology, the systematic study of trends which enable us to forecast the shape of things to come. *Ibid.* 397/2 Here we are with the automation process which, according to well-known futurologists, will lead to enforced leisure for a large part of the population. 1969 *New Scientist* 11 Dec. 570/2 Futurology . . cannot be turned into a respectable 'hard' science merely by getting the economists and the technologists to put some numbers to it. 1970 *Financial Times* 13 Apr. 7/1 American futurologist Herman Kahn is wrong in seeing the 21st century as belonging to Japan.

futz (fʌts), *v.* *U.S. slang.* [Origin uncertain; perh. alteration of Yiddish *arumfartzen* (*Amer. Speech* (1943) XVIII. 43).] *intr.* To loaf, waste time, mess *around*.

1932 J. T. Farrell *Studs Lonigan* (1936) iii. 56 Studs kept futzing around until Helen Shire came out with her soccer ball. 1941 B. Schulberg *What makes Sammy Run?* vii. 114 Sammy said, '. . Maybe you ought to write a play too.' . . 'I've been working on one for a couple of years,' I said. 'Don't futz around with it too long,' Sammy said. 1968 N. Benchley *Welcome to Xanadu* vi. 130 It's bad for your blood pressure to futz around like this.

fuzz (fʌz), *sb.*³ *slang* (orig. *U.S.*). [Origin uncertain.] A policeman or detective (see also quot. 1931); freq. *collect.*, the police.
1929 E. Booth *Stealing through Life* xii. 301 Don't run, and rank yourself—the fuzz don't know what's doin' yet. 1931 G. Irwin *Amer. Tramp & Underworld Slang* 81 *Fuzz*, a detective; a prison guard or turnkey. Here it is likely that 'fuzz' was originally 'fuss', one hard to please or over-particular. 1938 D. Runyon *Take it Easy* ii. 33 A race-track fuzz catches up with him. 1959 *Observer* 1 Nov. 7/5 'Fuzz broke up two of my poetry readings last night.' . . Fuzz means police. 1961 Wodehouse *Ice in Bedroom* x. 75 The society of coppers, peelers, flatfeet, rozzers and what are known in the newest argot of her native land as 'the fuzz' always affected her with an unpleasant breathlessness. 1964 E. Ambler *Kind of Anger* iii. 76 If the fuzz get excited we may have to produce her. 1969 *Times* 25 July 5/2 (*heading*) Fuzz sock it to the mean cats. 1971 Wodehouse *Much Obliged, Jeeves* xv. 165 If the fuzz search my room, I'm sunk.

fuzzy, *a.* **5. fuzzy-wuzzy,** add to def.: hence as a slang term for a coloured native of other countries, such as Fiji and New Guinea.
1940 N. Mitford *Pigeon Pie* iii. 41 There are Chinks and Japs and Fuzzy Wuzzies and Ice Creamers and Dagos, and so on. *Ibid.* xiii. 208 You brutish Hun. . . Savage—Kaffir—fuzzy-wuzzy. 1965 *Listener* 30 Dec. 1067/1 England came up against a tough opponent after a series of easy victories against Tibetans, Zulus, and other relatively speaking unarmed fuzzy-wuzzies.

fuzzy (fʌ·zi), *sb. slang.* = *fuzzy-wuzzy* (Fuzzy *a.* 5 in Dict. and Suppl.).
1890 Kipling *Barrack-r. Ballads* (1892) 10 The Fuzzy was the finest o' the lot. 1923 — *Land & Sea Tales* 11 A wounded Sudanese—what our soldiers used to call a 'fuzzy'. 1926 R. Macaulay *Crewe Train* II. iii, Those infernal fuzzies of yours have upset the lamas.

fyke, *sb.* Add: (Earlier and later examples.)
1832 in T. F. De Voe *Market Assistant* (1867) 197 We discovered a sea-dog stealing bass from a fuik of a bass-net. 1942 C. Weygandt *Plenty of Pennsylvania* 45 They set fykes for the up-stream run of suckers in March.
b. *fyke-net* (earlier example).
1842 *Nat. Hist. N.Y., Zool.* I. 54 Mr. Everson . . has taken them . . in the fyke-nets.

G

G. III. b. Substitute for def.: In *Physics* *g* (or *G*) is the symbol for the acceleration due to gravity = about 9·8 m. (32 ft.) per second per second at sea level; also used to denote gravity generally, or the associated forces. Also *attrib.*, as *g-force, -stress*; *G-suit.
1785 T. Parkinson *Syst. Mech.* xiv. 244 If *G* be the force of gravity, the axis and all diameters of the parabola, are perpendicular to the horizon. 1806 G. Gregory *Dict. Arts & Sci.* I. 870/2 M. Krafft gives a formula for the proportion of gravity in different latitudes on the earth's surface, which is this: $y = (1 + 0.0052848 \sin^2 \lambda)g$; where *g* denotes the gravity at the equator, and *y* the gravity under any other latitude λ. 1825 *Phil. Mag.* LXVI. 110, *l* denotes the height of the homogeneous atmosphere in feet, and *g* the gravitating force. 1869 *Ibid.* 4th Ser. XXXVIII. 83, M is the mass of a unit volume of a fluid, and *g* the accelerating force of gravity. 1928 N. Macmillan *Art of Flying* x. 138 Sustained high manœuvre loadings . . are referred to for brevity as 1G, 2G, or 6G. The question of G affects the pilot as well as the aeroplane. 1931 *Jrnl. R. Aeronaut. Soc.* XXXV. 1042 A figure of 3 or 4g is common in fast turns and in loops. 1938 *Lancet* 31 Dec. 1508/1 This was seen especially during looping, the maximal *g* stress occurring at the point of pulling up from the dive. 1945 [see *anti-g *a.*]. 1959 *Daily Tel.* 23 Feb. 11/7 Protective clothing, or 'space suits' will be needed to help guard a man against the heavy G forces imposed on him because of the great thrust upwards that a space vehicle will develop. 1959 Kaye & Laby *Tables of Physical & Chem. Constants* (ed. 12) 9 A standard value for the gravitational acceleration *g* at the Earth's surface is necessary for fixing the values of certain derived units and standards in the C.G.S. and M.K.S. systems.
f. Miscellaneous abbreviations. g., gramme; g, the 'general factor' of intelligence (see quots.); g, Ger. *gerade*, 'even', used in *Physics* in the designation of functions (especially wave-functions) which do not change sign on inversion through the origin, and of atomic states, etc., represented by such functions; G = *giga-; G. = *grand *sb.* 8 (*U.S. slang*); G-*agent*, a type of nerve-gas (orig. *U.S.*); G.A.W., guaranteed annual wage (*U.S.*); G.B., g.b., grand bounce, *i.e.* dismissal, ejection (*U.S. colloq.*, *Obs.*); G.B., Great Britain (esp. as a sign carried on a motor vehicle to indicate that it is registered in Britain); G.B.E., Grand Cross of the British Empire; G.B.H., grievous bodily harm; G.B.P., g.b.p., great British public; G.C., George Cross; G.C.E., General Certificate of Education; G.C.I., ground-control(led) interception; G.C.I.E., Grand Commander of the Indian Empire; G.C.M.G., Grand Commander of St. Michael and St. George; G.D.P., gross domestic product; G.D.R., German Democratic Republic; GeV [f. *g(iga- + *electron volt], a thousand million (10^9) electron volts; G.F.S., Girls' Friendly Society; G.H., growth hormone; G.H.Q., general headquarters; G.L.C., Greater London Council; G.M., g.m. [shortening of *G.M.T.], humorous substitute for *a.m.* or *p.m.*; G.M.T., Greenwich mean time; G.N.P., gross national product; G.O.C., General Officer Commanding; G.O.M., Grand Old Man, a name given to W. E. Gladstone; also *transf.*; G.O.P. (see *Grand Old Party 2); G.P., general practitioner; G.P., grand(e) passion; G.P.I., g.p.i., general paralysis of the insane; G-*plan* [f. the initial letter of the manufacturer's name + Plan *sb.*], the proprietary trade-mark of a type of furniture manufactured by Messrs. E. Gomme, Ltd.; G.P.O., General Post Office; G.P.U. = *Ogpu; G.S., general service; G.S.R. = *galvanic skin response* or *reflex*; G.T., *gran turismo*; G.T.T. (orig. G.T.), gone to Texas; absconded (*U.S.*); G.V. (*slang*), governor; G.W.(R.), Great Western (Railway). See also *G.A.T.T., *G.I., *G-man, *G string.
1885 A. Macfarlane *Physical Arithmetic* iv. 142 The authorized abbreviation for gramme is *g.*, but it is customary with English writers to use *gm.*, in order to distinguish between gramme and grain. 1902 *Encycl. Brit.* XXXIII. 808/2, 1 gramme (g.) = 15.4323564 grains or 0.7716 scruple. 1915 E. Webb in *Brit. Jrnl. Psychol.* Monogr. Suppl. III. iv. 38 The marks for these tests together with the above coefficients of their 'saturation' by the general factor (*g*) ought to furnish a measure of '*g*' itself. 1927 C. Spearman *Abilities of Man* vi. 75 The 'general factor' . . denoted by the letter *g* . . is so named because, although varying freely from individual to individual, it remains the same for any one individual in respect of all the correlated abilities. 1930 R. S. Mulliken in *Physical Rev.* 2nd Ser. XXXVI. 617 The subscripts *g* (German *gerade*) for even terms and *u* (German *ungerade*) for odd terms are recommended. 1962 R. E. Dodd *Chem. Spectroscopy* iv. 220 In homonuclear diatomic molecules there is also a prohibition against g–g or u–u transitions. 1928 J. O'Connor *Broadway Racketeers* xvii. 182 They had me in the bag for nearly ten G's before I pulled the string and let the joint go blooey. 1937 D. Runyon *More than Somewhat* ii. 35, I am going to stake them to a few G's. 1971 A. Curry *Shack-up* vi. 86 He'd probably drop me a few G's for the names of the guys in London. 1956 *Rep. Properties War Gases* (*Edgewood Arsenal, Md.*) I (*title*) G-agents. 1968 *New Scientist* 29 Feb. 465/1 The G-agents (nerve gases) . . are quick killers. 1955 *Times* 7 June 7/2 The 'guaranteed annual wage'—G.A.W.—has for long been a major objective in the programmes of the American trade union movement. 1880 *News & Press* (Cimarron, N.M.) 23 Dec. 1/7 Well, I've got the g.b. . . I've been fired! 1917 G. B. McCutcheon *Green Fancy* 157 Guess I'm going to get the G.B. 'fore long. 1920 *Michelin Guide to Great Britain* (ed. 5) 807 To be carried in addition to the ordinary GB plaque. 1926 *Automobile Assoc. Handbk.* iv. 157 An oval nationality plate bearing the letters G.B. . . must be carried at the rear of every car or motor cycle of British registration. 1937 M. Sharp *Nutmeg Tree* xii. 144 There were four cars standing outside, but only one with a G.B. plate. 1968–9 *A.A. Continental Handbk.* 29 A plate (GB for Great Britain) . . must be fixed . . next to the rear registration plate. . . GB plates cost 2*s.* 6*d.* 1918 *Whitaker's Alm.* 143 G.B.E., Knights Grand Cross or Dames Grand Cross [of the British Empire]. 1958 F. Norman *Bang to Rights* III. 166 Bottles was doing a lagging G.B.H.! (Grievious Bodily Harm). 1966 L. Southworth *Felon in Disguise* ii. 40 He did have a couple of arrests. One for malicious damage and the other for G.B.H. 1925 G. Frankau *Life—and Erica* xv. 189 It's no good making the G.B.P. laugh *at* you. 1928 *Punch* 8 Feb. 150/3 The g.b.p. does like a wedding. 1941 H. G. Wells *You can't be too Careful* 5 'You got the G.C.,' said young Jewler. 'We're all proud of you.' 1951 *Times Educ. Suppl.* 27 July 597/4 The travel agency that can offer visitors to the Channel Islands duty-free scent and tobacco and also a G.C.E. 1966 *Rep. Comm. Inquiry Univ. Oxf.* I. 64 The results of GCE examinations. 1945 *Electronic Engin.* XVII. 683 With the Battle of Britain by night must be associated . . the use of G.C.I. or Ground Control Interception. 1947 Crowther & Whiddington *Science at War* Pl. vi (*caption*), The guiding of fighter aircraft to the enemy in the air by G.C.I. (Ground Controlled Interception). 1889 *Whitaker's Alm.* 108 The Most Eminent Order of the Indian Empire . . Knights Grand Commdrs. G.C.I.E. 1871 *Ibid.* 70 Lord Lisgar, G.C.M.G. 1962 *Economist* 20 Oct. 284/1 Investment might have had to . . constitute 24 per cent of GDP by 1966. 1958 *Listener* 18 Dec. 1026/2 Herr Grotewohl ended by saying that the G.D.R.'s main task was to demonstrate the 'superiority' of the socialist system. 1971 *Guardian* 8 Sept. 10/6 At least 100,000 dachas are now estimated to exist in the GDR. 1952 *Proc. Physical Soc.* B. LXV. 307 It was recommended that the use of the word 'billion' should be avoided in scientific literature. The use was recommended, for example, of either 10^9ev or gev (as agreed by IUPAP in 1948). 1954 *Nature* 12 June 1127/2 The bevatron has produced protons of energy 5 GeV. 1969 *Times* 5 Feb. 13/8 The new generation of even larger proton accelerators, like the proposed 300 GeV machine at Cern, will produce particles of sufficient energy to make it possible to observe some of the really short-lived resonances. 1888 C. M. Yonge *Our New Mistress* xvii. 161 She wanted me to help her in . . her G.F.S. Bible class. 1939 A. Thirkell *Brandons* x. 264 She had been brought up . . on mothers' meetings, G.F.S. meetings . . and the hundred activities of the Vicarage. 1952 *Metabolism* July 356 The growth hormone used was Lot No. GH-3. 1961 *Lancet* 29 July 235/2 The diabetogenic effect of G.H. in laboratory animals is . . well established. 1856 F. Nightingale in C. Woodham-Smith *F. Nightingale* (1950) xii. 275 G.H.Q. feeding their horses on the biscuits the men could not eat. 1914 *Isle of Man Weekly Times* 21 Nov. 7/4, I have been moved higher up, my address now being 'Adv. G.H.Q.' 1935 Wodehouse *Blandings Castle* ii. 37 The housekeeper's room at Blandings Castle, G.H.Q. of the domestic staff. 1962 *Times* 23 Nov. 5/2 The inner London education authority will consist of members of the G.L.C. 1901 'Linesman' *Words by Eyewitness* (1902) 246 Merely referring to it as such and such an hour G.M. 1929 S. Lewis *Dodsworth* xxxii, I bet I never went to bed before three g.m. once, the whole way over! 1928 W. M. Smart *Sun, Stars & Universe* iv. 49 The true Greenwich Mean Time (G.M.T.). 1961 *Listener* 26 Oct. 657/1 The limited size of a British market with a G.N.P. of some £25 billion. 1969 *Guardian* 25 July 10/4 The Apollo project . . uses only 1 per cent of the American GNP. 1883 *Army Regulations* II. 45 Office of G.O.C. Camp, Colchester. 1968 *Listener* 8 Feb. 167/1 There was a massive press conference held by the British GOC, General Keightley. 1884 *Punch* 23 Feb. 95/2 G.O.M. been looking worn and irritable since Session opened. 1903 A. H. Beavan *Tube, Train*, etc. i. 2 This is by no means the oldest steam-engine at work in the kingdom, the doyen being one built as far back as 1767. . . It is said that this G.O.M. is more economical than many of the modern engines. 1944 *Gramophone* July 27/1 (*heading*) Coward of Sheffield England's G.O.M. of music. 1968 *Listener* 5 Sept. 316/3 The stonemason's son . . who was eventually elevated to the position of GOM of letters. 1887 *Graphic* 30 Apr. 447/2 The ordinary general practitioner—the 'G.P.', as he is familiarly called. 1942 R. A. Knox *In Soft Garments* viii. 60 Some ass of a French G.P. didn't know his own business. 1960 *Times* 17 Mar. 13/6 Rural G.P.s evidently like the prospect more than others. 1935 D. L. Sayers *Gaudy Night* ii. 30 She wormed round rather. Had a sort of G.P. for Miss Shaw. 1922 Joyce *Ulysses* 8 That fellow . . says you have g.p.i. . . General paralysis of the insane. 1930 *Daily Express* 8 Sept. 1/1 Pathological experiments . . have enabled 'G.P.I.' to be diagnosed in its incipient stages. 1969 King & Nicol *Venereal Dis.* (ed. 2) v. 62 The clinical manifestations of GPI do not usually appear until 10 years or more after infection. 1957 *Times* 4 Nov. 13/2 Some of the mass-production G-plan designs are good, but have a sameness. 1967 E. Wymark *As Good as Gold* ii. 33 The flat was strictly modern inside. The hall tended to G-plan. 1869 *All Year Round* 23 Jan. 180/1 The hard-worked servant of the G.P.O. who handed in the 'comic' missive. 1955 *Times* 7 May 9/6 Could the G.P.O. be persuaded to undertake an operational research into the use of Post Offices? 1925 G. Popoff *Tcheka* 238 The third period of the Tcheka dates from February 6, 1922. On this date the Tcheka was 'dissolved'; the 'G.P.U.', however, was immediately set up in its place. 1941 Koestler *Scum of Earth* 122 The dark silhouette of the . . G.P.U. agent had replaced the . . symbols of the struggle for a happier world. 1918 F. M. Ford *Let.* 13 Jan. (1965) 86 The doctors won't pass me G.S., confound them. 1954 J. Masters *Bhowani Junction* vii. 62 She got on the push bike. It was a heavy ugly thing . . what the Army calls a G.S. bicycle. 1927 *Jrnl. Exper. Psychol.* X. 206 The mucous surfaces manifest no electrical charges corresponding to the G.S.R. 1964 *Philos. Rev.* LXXIII. 204 He exhibits no change in GSR, perspiration, and so forth. 1966 *Publ. Amer. Dial. Soc.* 1964 XLII. 5 *Gran turismo* . . Also GT. 1968 W. Garner *Deep, Deep Freeze* xx. 192 He saw the red Ferrari 330 GT coupé in his rear mirror. 1839 *Daily Eastern Argus* (Portland, Maine) 17 May 2/2 G.T.—This is said to be a common mode of making Sheriff's returns in the South West. It means, 'Gone to Texas'. 1839 *Georgia Messenger* (Macon, Ga.) 1 Aug. 2/6 G.T.T.—General Nathaniel Smith . . has fled to Texas, with from $70,000 to $100,000 of Uncle Sam's money in his pocket. 1884 (*title*) G.T.T. Gone to Texas. Letters from our boys, ed. by Thomas Hughes. 1949 *Sat. Even. Post* 4 June 30/2 That famous old initialed forwarding address: 'G.T.T.' 1905 H. G. Wells *Kipps* v. 106 'I suppose the G.V.—' began Kipps. 'He knows,' said the housekeeper. 1879 C. L. Dodgson *Let.* 24 Oct. in S. D. Collingwood *Life & Lett. L. Carroll* (1898) xi. 413 The young lady, who was travelling on the G.W. Railway. 1889 E. Dowson *Let.* 18 Oct. (1967) 109, I bought it at Truro coming up the G.W.R. 1934 *Discovery* Nov. 314/2 Many expresses on the L.M.S. and G.W.R. now load up to 500 tons or over. 1970 R. B. Wilson (*title*) Go Great Western. A history of GWR publicity.

Ga (gā). Pl. Ga, Gas. [Native name.] The name of a Negro people of Ghana and their language. Also *attrib.* or as *adj.*
1858 J. Zimmerman (*title*) A grammatical sketch of the Akra—or Gã—language. 1879 *Encycl. Brit.* X. 756/1 The south-eastern corner of the Gold Coast is occupied by another language known as the Ga or Acra, which comprises the Ga proper and the Adangme and Crobo dialects. 1884 *Ibid.* XVII. 318/1 *Ewe Group*: Acra (Ga), Fantee, [etc.]. 1955 P. Strevens *Papers in Lang.* (1965) ix. 114 The languages spoken in this area are almost entirely confined to members of the Kwa Larger Unit, comprising such languages as Twi, Ga, Fante, [etc.]. 1962 [see *Ashanti]. 1963 *Listener* 7 Feb. 231/2 African languages are being studied as the modern languages of Europe are studied in European universities. A Ga student learns Akan. 1969 *Times* 22 Oct. (Ghana Suppl.) p. i/3 Not one Ewe among them, and only one Ga, the Minister for Housing. *Ibid.* p. ii/2 The Ga carved out a kingdom for themselves.

gabardine. A form of Gaberdine used *spec.* for a twill-woven cloth, usu. of fine worsted.
The two spellings are used interchangeably.
1904 *Ladies' Field* 14 May 426/1 Gabardine, a material of flax and cotton, with a wool lining, both gabardine and lining being waterproofed. 1908 *Times* (weekly ed.) 14 Aug. p. iii/2 Gabardine is a material which has many qualities not to be overlooked when it is a question of sporting dresses. 1923 G. C. Denny *Fabrics* 46 Gabardine or gaberdine (wool). A firm material similar to whip cord. 1938 'E. Queen' *Four of Hearts* (1939) xi. 153 She looked lovely . . in a tailored gabardine suit.

gabbart. Add: **2.** (Freq. **gabbard.**) A support used in erecting a scaffold.
1897 *Ardrossan & Saltcoats Herald* 31 Dec. 3 The position of one of the gabbard legs. *Ibid.*, The second gabbard from the north wall, not far from the centre of the scaffold. 1962 *Engineering* 13 Apr. 489/1 Peine tower cranes . . are being used, as well as two 10-ton derricks on gabbards.

gabbiness (gæ·binės). [f. Gabby *a.* + -ness.] Garrulousness; talkativeness.
1952 S. Kauffmann *Philanderer* (1953) ii. 27 Perry Downing reduced that restraint to a minimum because of his gabbiness and vulgarity. 1961 *Guardian* 10 Oct. 7/3 The gabbiness of the Chicago musicians, who wrote prolifically about themselves.

gabbro. Add: Hence **gabbroi·tic** (-o͡i-) *a.* = Gabbroic *a.*

1893 *Jrnl. Geol.* I. 595 Diabases from dykes..in which the gabbroitic as well as the diabasic structures are well exhibited. **1895** *Ibid.* III. 1 The rarity of olivine in the granulitic rocks is in marked contrast with its abundance in the gabbroitic variety.

gabbroid (gæ·broid), *a. Petrol.* [f. GABBR(O + -OID.] Resembling gabbro; *spec.* belonging to the petrographic clan containing the gabbro and dolerite families, the rocks of which contain a high proportion of ferromagnesian minerals. Also as *sb.* Hence **gabbroi·dal** *a.*

1900 *Amer. Geologist* XXVI. 151 (*heading*) Mineralogical and petrographic study of the gabbroid rocks of Minnesota, and more particularly of the plagioclasytes. **1904** CHAMBERLIN & SALISBURY *Geol.* (1905) I. 432 The term *gabbroids* may be used to include the dark crystalline rocks in which the ferromagnesian minerals predominate, as the diorites, gabbros, dolerites, peridotites, etc. *Ibid.*, The granitoids are usually acidic and the gabbroids usually basic. **1931** *Mineral. Abstr.* IV. 401 Chemical analyses of the differentiation products of a gabbroidal magma, viz. diorite-pegmatite, diorite-aplite [etc.]. **1940** *Geogr. Jrnl.* XCV. 394 The association of gabbroid and granophyric rocks. **1965** G. J. WILLIAMS *Econ. Geol. N.Z.* x. 160/2 These lenses..were invaded by gabbroidal and peridotitic dykes conformably with the bedding planes.

gabby, *a.* Now in more general use. Add further examples.

1899 ADE *Fables in Slang* 124 He was a Gabby Young Man, and he could Articulate at all Times, whether he had anything to Say or not. *a* **1911** D. G. PHILLIPS *Susan Lenox* (1917) II. viii. 214, I was awful gabby downstairs. Yes—I told him. **1949** 'J. TEY' *Brat Farrar* xix. 176 Never very 'gabby' at any time. A quiet boy? Yes, a nice quiet boy. **1958** M. ALLINGHAM *Hide my Eyes* 125 As for the public, they're interfering, gabby, can't keep their traps shut. **1965** *Spectator* 22 Jan. 108/2 Hubert Humphrey shows himself a gabby and literate and battling soul. **1970** *New Yorker* 28 Nov. 116/2 The women..were mostly political and gabby.

gabfest (gæ·bfest). *slang* (chiefly *U.S.*). [f. GAB *sb.*² + *FEST.] A gathering for talk; a spell of talking; a prolonged conference or conversation.

1897 *Boston Transcript* 7 Jan. 15/7 A Chicago paper speaks of the speechmaking on Andrew Jackson's day as 'the Democratic gabfest'. **1910** W. M. RAINE *B. O'Connor* 101 I'll leave you and your jelly fish Scotty to your gabfest. **1942** J. D. CARR *Seat of Scornful* xi. 147 When can I come and see you for a bit of gabfest about this business? **1960** *Spectator* 30 Sept. 470 A shambles as big as the Labour gabfest. **1971** D. CORY *Sunburst* v. 78 A watching brief on the Common Market gabfest.

gable, *sb.*¹ Add: **4. gable-topped** *a.*, topped by a gable, having a gable-shaped top.

1842 *Civil Engin. & Arch. Jrnl.* V. 81/1 Some..architects..give us a gable-topped and an elliptic bed-topped window alternately. **1903** LD. R. GOWER *Rec. & Remin.* 214 A small gable-topped building. **1923** R. G. COLLINGWOOD *Roman Brit.* 86 A gable-topped canopy.

Gaboon, gaboon (găbū·n). Also **Gabun.** [The name of a province and river in West Africa.] A hardwood from the tree *Aucoumea klaineana*. Also *attrib.* **Gaboon adder** or **viper** [tr. F. *échidnée du Gabon* (Duméril & Bibron, in *Erpétologie générale* (1854) VII. 1428)], a snake, *Bitis gabonica*.

1910 *Encycl. Brit.* VIII. 843/2 The heart-wood is very black and hard and is known as black ebony,..and Gabun, Lagos, Calabar or Niger ebony. **1913** *Chambers's Jrnl.* Aug. 621/1 Gaboon mahogany is another timber which the wood-worker regards with disdain, owing to its softness. **1920** A. L. HOWARD *Man. Timbers of World* 140 Gaboon. *Boswellia Klaineana*...This..useful wood..fills a place of importance which justifies a title which would give it individuality, though it should not be called mahogany. *Ibid.*, Gaboon has been tried for the purpose of making aeroplane propellers. **1928** *Daily Tel.* 29 May 6 Specimens of the green mamba and gaboon viper from West Africa. **1954** J. A. PRINGLE *Common Snakes* 26 Gaboon Adder...This snake is widely distributed in the Western forest of tropical Africa..but its occurrence in the Union is restricted to the northern tips of Zululand and South West Africa. **1960** *Guardian* 29 Sept. 9/1 Another very pleasant thing about the classrooms is the row of gaboon-faced sliding doors. **1965** R. & D. MORRIS *Men & Snakes* i. 16 The Gaboon viper has a pair of enlarged scales between its nostrils which stick out like two little horns.

Gadarene (gæ·dărīn), *a.* Also † **Gadarean.** [ad. late L. *Gadarenus*, f. Gr. *Γαδαρηνός* inhabitant of Gadara, a town of ancient Palestine near the Sea of Galilee.] Of or pertaining to Gadara; used allusively, esp. of a headlong rush or flight, with reference to the story in Matt. viii. 28 of the swine that rushed down a steep cliff into the sea and drowned.

1820 Gadarean [see PIG *sb.*¹ 2 a]. **1899** KIPLING *Stalky & Co.* 210 'This man,' said M'Turk, with conviction, 'is *the* Gadarene Swine.' **1919** *Brit. Manufacturer* Nov. 30/2 In such case we are a debtor nation, and our path is likely to lead us downhill like the Gadarene swine. *a* **1930** D. H. LAWRENCE *Phoenix* (1936) v. 600 A vast mob which may at any moment start to bolt down to the precipice Gadarene-wise. **1952** [see *BOUZOUKI].

gadbush, var. *GODBUSH.

gaddi: see *GADI.

gadget (gæ·dʒėt). Also formerly **gadjet.** [Origin obscure. First known in use among seafaring men, and said by several correspondents to have been current *c* 1870, and by a few as far back as the fifties of the nineteenth century, but not found in print before 1886.

One of the most plausible etymological suggestions is F. *gâchette*, which is or has been applied to various pieces of mechanism, e.g. in a lock and in a gun; it is a dim. of *gâche* staple (of a lock), wall-staple or hook. The possibility of connexion with F. *engager* to engage (one thing with another) has also been suggested; cf. dial. F. *gagée* tool, instrument. Derivation from GAUGE is improbable.]

Used as an indefinite or general name for: a comparatively small fitting, contrivance, or piece of mechanism.

1886 R. BROWN *Spunyarn & Spindrift* xxxi. 378 Then the names of all the other things on board a ship! I don't know half of them yet; even the sailors forget at times, and if the exact name of anything they want happens to slip from their memory, they call it a chicken-fixing, or a gadjet, or a gill-guy, or a timmey-noggy, or a wim-wom—just *pro tem.*, you know. **1902** KIPLING *Traffics & Discov.* (1904) 179 Steam gadgets always take him that way. *Ibid.* 190 You've certainly got the hang of her steamin' gadgets in quick time. **1907** *Motor Boat* 4 July 439/1 There will be no harm in pointing out a few of the requirements of a Whitehead torpedo, which will enable the reader to appreciate the number of 'gadgets' which are involved. **1911** C. E. W. BEAN *'Dreadnought' of Darling* ii. 30 He had known every stick in their sides and every gadget on their decks. **1914** KIPLING *New Army* (1915) 39 They have installed decent cooking ranges and gas, and the men have already made themselves all sorts of handy little labour-saving gadgets. **1918** *Glasgow Even. News* 14 Sept. 6 There are a variety of gadjets connected with a motor-car; an aeroplane is replete with them—thermometer, barometer, altimeter, and the pilot only knows what besides. **1923** 'B. M. BOWER' *Parowan Bonanza* i. 12 Solid ledge of gold...Knock it off in chunks with a single-jack and gadget. **1927** *Glasgow Herald* 18 Aug. 9 Innumerable motor car and wireless gadgets. **1951** *Good Housek. Home Encycl.* 46/1 Too many kitchen gadgets are discarded because so much time is required for their cleaning.

b. *spec.* in local use. A winch or similar mechanical gear for discharging vessels; also, a craft equipped with such gear.

1899 *Bristol Times & Mirror* 10 June 3/8 The gadget which was used in the discharge of vessels was being towed down the Harbour...The man who was steering the gadget rather lost his head...There was ample room for the gadget to have passed through if it had been steered properly. **1931** *Ibid.* 16 May, Pulley gear was rigged to the stays or spars of the sailing ship over the hold, and the heaving rope passed to the winch drum of the gadget lying alongside.

c. *transf.* and *gen.* An accessory or adjunct; a knick-knack or gewgaw.

1915 'BARTIMEUS' *Naval Occas.* 142 Look here, old lady, here's a gadget I got for you—he fumbled with the tissue paper enclosing a little leather case. **1922** *Blackw. Mag.* Apr. 421/1, I had to work with concordances, glossaries, all sorts of gadgets. **1925** *Ibid.* Sept. 423 Another waiter offered a selection of 'gadgets',—the appetising morsels of anchovy, stuffed-olive, or pâté.

d. *Glass-making.* A spring-clip used for gripping the foot of a wine or other footed glass when it is being shaped.

1918 P. MARSON *Glass & Glass Man.* 83 The servitor has now done his part of the work, and the glass is handed to the workman. It is then cracked off, and the foot caught by a spring clip arrangement attached to a pontil, called a 'gadget'. **1923** H. J. POWELL *Glassmaking in England* 43 *Gadget*, a spring-clip, attached to a puntee, to hold foot of wine glass, while the bowl is being finished, in order to prevent puntee-mark.

2. *attrib.* and *Comb.*

1924 *Illustr. London News* 24 May 931/2 A paradise for the 'gadget-loving' boy. **1941** *Amer. Speech* XVI. 316/2 *Gadget bag*, a case for camera accessories. **1945** Gadget-ridden [see *chromium-plated* s.v. *CHROMIUM 2]. **1946** *Consumers' Guide* (U.S.) May 16/1 Job-hungry GI's and gadget happy housewives make shining targets for swindlers. **1949** 'J. TEY' *Brat Farrar* i. 8 Ulysses..was 'without doubt..a gadget-contriver'. **1964** E. A. NIDA *Toward Sci. Transl.* xii. 252 Of course, MT does capture the imagination, especially of the gadget-minded public.

Hence **gadgetee·r**, a person given to the use or invention of gadgets; **ga·dgetry**, gadgets collectively; the use of gadgets; **ga·dgety** *a.* having the qualities or characteristics of a gadget; fitted with a gadget or gadgets.

1920 *Christian World* 19 Aug. 4/1 The door of my bedroom in one great American hotel was a blazing triumph of sheer gadgetry. **1937** *Night & Day* 26 Aug. 24/1 The Tanova self-supporting sock looked pathetically gadgety on the plaster legs in the window. **1938** *Reader's Digest* Mar. 93/1 Such Yankee ingenuity usually brings the gadgeteer to grief. **1957** *London Mag.* Aug. 52 It is a scarcely individualized appetite that desires gadgetry and goods, and calls them Utopia. **1958** *Times* 22 July 9/7 The temple of gadgetry known as the underground control room. **1962** *Punch* 19 Sept. 397/1 A good gadgety camera. **1964** M. GOWING *Britain & Atomic Energy* ii. 82 They did not even commit the error ascribed to 'gadgeteers', of believing they had a monopoly of their invention. **1967** *Daily Tel.* 29 Mar. 12/7 Despite the latest gadgetry—echo-sounders, freezer trawlers, automatic gear—the commercial sea fisherman remains basically a hunter.

‖ **gadi** (gā·di), **gaddi** (gɒ·di). Also **gadhi, gâdi, guddy, -ee.** [Marathi *gādī*, Bengali *gadi*, Hind. *gaddi*, lit. cushion.] The cushioned throne of an Indian ruler; *transf.* = THRONE *sb.* 5 a.

1855 H. H. WILSON *Gloss. Judicial & Revenue Terms.* **1879** E. ARNOLD *Light of Asia* II. 333 The marriage feast was kept, as Sâkyas use, The golden gadi set, the carpet spread. **1886** YULE & BURNELL *Hobson-Jobson* s.v. *Guddy*, *Gaddee*, 'To be placed on the *guddee*' is to succeed to the Kingdom. **1906** *Westm. Gaz.* 6 Sept. 4/3 The Jâm died in 1895, whereupon this child was installed on the gâdi of Jamnagar by Colonel Hancock, then 'Agent to the Governor' in Kathiawar. **1923** *Blackw. Mag.* Jan. 10/1 He had had a very narrow escape of being deposed from the *gadi*. **1927** *Times* (weekly ed.) 29 Sept. 346/2 The Nawab of Bhopal succeeded to *gadi* on the death of the Begum his mother. **1960** J. MASTERS *Venus of Konpara* 18 He succeeded to the gaddi, the cushion which was the Indian symbol of rule. **1961** *Times* 23 Jan. (India Suppl.) p. viii/1 The welfare state has succeeded to the *gaddi* of Raja Sansar Chand.

Gaditan, *a.* and *sb.* Add: Also **Gaditane.** Delete † *Obs.* and add later example.

1967 'LA MERI' *Sp. Dancing* (ed. 2) x. 123 The gaditanes used them when they conquered fashionable Rome.

Gaditanian, *a.* and *sb.* (Earlier and later examples.)

1845 R. FORD *Hand-bk. Trav. Spain* II. ii. 18 Gaditanian dances..are more marked by energy than by grace. **1967** 'LA MERI' *Sp. Dancing* (ed. 2) v. 66 Some say the dances of Cadiz in Gaditanian days come under this style.

gadolinium (gædoli·niɒm). *Chem.* [mod.L. (J. C. G. de Marignac 1886, in *Compt. Rend.* CII. 902), f. GADOLIN(ITE + -IUM.] A metallic element of the lanthanide series which is present in monazite, gadolinite, and other rare earth minerals and forms a series of mostly colourless salts in which it is trivalent; it resembles steel in appearance and is strongly magnetic below room temperature. Symbol Gd, atomic number 64.

1886 *Jrnl. Chem. Soc.* L. 667 Marignac gives the name *Gadolinium* (Gd) to the substance which has hitherto been provisionally distinguished as Yα. **1920** *Ibid.* CXVIII. II. 317 The fractional crystallisation of the acetates effects a rapid separation of gadolinium from samarium and didymium. **1924** J. W. MELLOR *Comprehensive Treat. Inorg. & Theor. Chem.* V. xxxviii. 686 The terbium family of rare earths is arbitrarily taken to include the three elements, europium, Eu; gadolinium, Gd; and terbium, Tb. **1938** R. W. LAWSON tr. *Hevesy & Paneth's Man. Radioactivity* (ed. 2) x. 119 Gadolinium..absorbs slow neutrons very strongly. **1959** *Electronic Engin.* XXXI. 589/2 New kinds of ferromagnetic ferrites are continually being discovered: gadolinium ferrite types have been investigated. **1961** WOYSKI & SILVERNAIL in Spedding & Daane *Rare Earths* xxi. 517 Gadolinium..has been used to obtain extremely low temperatures, the classic example being Giauque's experiments.., in which a temperature of 0·29 K was obtained by the adiabatic demagnetization of gadolinium sulfate octahydrate.

Gaekwar (gai·kwāɹ). Also **Guickwar, Guicowar.** [ad. Marathi *gāekwāḍ*, lit. cowherd.] The title of the native ruler of Baroda in India.

1813 J. FORBES *Oriental Mem.* II. xviii. 84 These princes were all styled Guickwar, in addition to their family name...The word literally means a cow-keeper, which, although a low employment in general, has, in this noble family among the Hindoos, who venerate that animal, become a title of great importance. **1854** E. THORNTON *Gazetteer India* I. 266 Baroda,..the capital of the territory of the prince called the Guicowar. **1881** W. W. HUNTER *Imp. Gazetteer India* I. 451 The Gáekwár of Baroda is entitled to a salute of 21 guns. **1921** *Glasgow Herald* 26 Nov. 7 The Maharaja—better known here [*sc.* Baroda] and in England as the Gaekwar. **1953** B. SUBBARAO *Baroda through Ages* III. 129 The capture of Baroda in 1734..inaugurates a new era in the history of Baroda under the Gaekwars.

Gaeldom. Add to def.: Also, Gaelic culture or civilization; the Gaelic people. (Further examples.)

1890 D. HYDE *Beside Fire* p. xxx, This story..was carried by some Irish bard..to the Gaeldom of Scotland. **1898** *Folk-Lore* IX. I. 46 Until 300 years ago there was practical identity between the mythical and heroic literature of the two main divisions of Gaeldom—Ireland and Scotland. **1925** D. CORKERY *Hidden Ireland* x. 300 It was a Gael who wrote this poem, a Gael whose mind was nurtured in Gaeldom. **1930** W. B. YEATS *Diary* 20 Oct. (1944) 56 The devout Catholicism and enthusiastic Gaeldom. **1959** *Listener* 20 Aug. 292/2 A reader..might well wonder whether Gaeldom, having 'declined', is now dead. **1963** *Economist* 19 Oct. 252/1 Scottish Gaeldom's great annual festival.

Gaelic, *a.* and *sb.* Add: **A.** *adj.* **2.** *Gaelic coffee*, coffee with cream and Irish whisky.

1952 *Coffee Trade News* May 8 As one of the Scots Quartette who were introduced to Gaelic coffee in Dublin last week, I share their curiosity as to whether it would prove as popular in Scotland as in the land of its invention. **1967** J. B. PRIESTLEY *It's Old Country* iii. 29 She invited him to take Gaelic coffee in her sitting room.

Gaeltacht (gēⁱ·ltəχt). [Ir.] A region in Ireland in which Irish is the vernacular; also, these regions collectively. Also *attrib.*

1929 A. DE BLÁCAM *Gaelic Lit. Surveyed* i. 7 He who dwells in the Gaeltacht learns to relish the neat turn of phrase, the proverbial allusion, and the apt retort. **1956** B. INGLIS *Story of Ireland* IV. 232 The number of Irish speakers in the country had continued to shrink, and the 'Gaeltacht' areas—where Irish was the vernacular—were losing their Irish speakers fast. **1969** *Guardian* 16 July 7/1 She would like..to move back to the West, to the Gaeltacht where Irish is spoken as the native language.

Gaertner¹ (gēᵊ·ɹtnəɹ). Also **Gärtner.** The name of August *Gaertner*, German bacteriologist (1848–1934), used attrib. or in the possessive to designate *Salmonella enteritidis*.

1899 MUIR & RITCHIE *Man. Bacteriol.* (ed. 2) 331 During the last few years, in some epidemics of meat-poisoning, similar bacilli differing slightly from Gaertner's bacillus have been isolated. **1928** *Jordan & Falk's Bacteriol. & Immunol.* 444 The term 'Gärtner bacillus' came to be used by some writers..as a general term synonymous with 'paratyphoid bacillus'. **1928** *Daily Express* 11 Aug. 7/1 Death was due to heart failure following infection by the bacillus gaertner from ham purchased in the police canteen. **1930** *Brit. Med. Jrnl.* 22 Mar. 546/1 The patient's serum after operation contained no agglutinins for the organism, isolated, a feature not unusual in Gaertner infections.

Gaertner². *Med.* Also **Gärtner.** [The name of G. *Gaertner*, Austrian physician (1855–1937).] *Gaertner's phenomenon*, the degree of fullness of the veins of the arm as it is raised to varying heights as indicating the degree of pressure in the right auricle. *Gaertner('s) tonometer*, an instrument for measuring blood-pressure by means of a compressing ring applied to the finger.

1901 *Boston Med. & Surg. Jrnl.* 8 Aug. 150/1 The blood pressure taken in the finger by the Gärtner tonometer might approximate the mean pressure..but such an approximation would be accidental and uncertain. **1903** G. W. CRILE *Blood-Pressure* 310 Gaertner's tonometer..consists of a pneumatic ring about 1 cm. in height and 2½ cm. in diameter. *Ibid.* 354 With the Gaertner tonometer he noted that..the pressure fell once from 120 to 90. **1915** STEDMAN *Med. Dict.* (ed. 3) *G[aertner]'s vein phenomenon*, varying fulness of the veins of the arm as the limb is raised to different heights, affording an index to the amount of pressure in the right auricle. **1928** G. W. NORRIS et al. *Blood-Pressure* (ed. 4) iv. 380 Gärtner's phenomenon. **1934** J. F. H. DALLY *High Blood Press.* (ed. 3) xv. 245 Gärtner's tonometer (1899) afforded another index of systolic pressure by the point at which colour returned to the skin of the blanched finger after removal of compression.

Gaertner, erron. var. *GARTNER.

Gætulian, var. *GETULIAN *a.* and *sb.*

gaff, *sb.*¹ Add: **5.** (sense 2) *gaff-rigged* adj.

1933 'L. LUARD' *All Hands* 256 A gaff-rigged ketch. **1961** *Times* 5 Aug. 7/7 They made the crossing in a gaff-rigged yawl.

gaff, *sb.*² Add: **3.** *to stand the gaff*, to receive severe treatment, criticism, etc. Similarly *to give, take*, etc., *the gaff. U.S. slang.*

1896 ADE *Artie* xii. 111 If he gets the gaff he'll be flat on his back. **1899** A. THOMAS *Arizona* 124 Will they stand the gaff? Will they set sixty hours in the saddle..? **1903** *McClure's Mag.* Oct. 563 'Good,' they cheer, when you find fault; 'give us the gaff. We deserve it and it does us good.' **1910** W. M. RAINE *B. O'Connor* 244 Neil has got to stand the gaff for what he's done. **1924** —— *Troubled Waters* xviii. 194 Just because he shuts his mouth and stands the gaff.

gaff, *sb.*⁴ Add: **3.** A house, shop, or other building.

1932 G. S. MONCRIEFF *Café Bar* viii. 73 He went back to his gaff and broke into the gas meter. **1936** J. CURTIS *Gilt Kid* ii. 20, I had it off last week.., not a big job, just a little snout gaff. **1938** G. GREENE *Brighton Rock* III. iii. 125 The barred and battlemented Salvation Army gaff. *Ibid.* v. i. 188 It's the best road-house this side of London...You can't pick 'em [*sc.* girls] up in this gaff. **1961** J. MACLAREN-ROSS *Doomsday Book* I. viii. 86, I was keeping an eye on the gaff—seen you going in. **1971** *Fremdsprachen* XV. 63 Two new words..for a flat or dwelling are 'pad' and 'gaff'.

b. *spec.* A brothel; a prostitute's room.

1947 *New Statesman* 10 May 330/3 In most cases the girls do not live where they work (they call this their 'gaff'). **1959** *Streetwalker* i. 14 Take a cab for the five minutes drive back to your gaff, or flat.

gaff, *v.*² Add: **2.** *trans.* To deceive or trick; to make (a game or device) crooked or dishonest. *slang* (orig. *U.S.*).

1935 N. ERSINE *Underworld & Prison Slang* 39 *Gaff*, to cheat. **1934** J. PROSKAUER *Suckers All* 15 He was..putting in slot machines for a racketeer, and 'gaffing' them..to increase the house percentage. *Ibid.* v. 77 There *was* a way of gaffing every game. **1963** *New Scientist* 6 June 553 Roulette wheels can of course be crookedly fixed ('gaffed' is the trade term). **1965** H. GOLD *Man who was not with It* xv. 135, I want to play you straight fifty-fifty, not gaff you for fifty-fifty.

gaffe (gæf). Also **gaff.** [Fr. (see GAFF *sb.*²).] A blunder, an instance of clumsy stupidity, a 'faux pas'.

1909 *Pall Mall Gaz.* 12 Apr. 1/1 These two gentlemen, whose weather predictions are still listened to with some deference, have made a bad 'gaffe', to use a popular slang expression. **1922** *Blackw. Mag.* Mar. 303/1, I had obviously said the wrong thing, committed a gaffe. **1924** GALSWORTHY *White Monkey* I. ii, 'He's a bit romantic, of course.' 'Oh! Have I made a gaff?' 'Not a bit; jolly good shot.' **1928** C. MACKENZIE *Extremes Meet* 207 Knowing nothing, you might easily make a bad gaffe. **1959** *Economist* 6 June 922/2 It would be too silly if this newspaper *gaffe* induced..any future prime minister to relapse yet further into the grave political vice of showing immutable loyalty to his friends.

gag, *sb.*¹ Add: **1. b.** (Earlier U.S. examples of modern sense.) Also *attrib.*

1798 *Aurora* (Phila.) 1 Aug. (Th.), Query have the Cherokees any gag-bill? **1840** J. Q. ADAMS *Mem.* X. 273 Then came Atherton, of New Hampshire, the man of the mongrel gag. **1861** *Congress. Globe* App. 49/1 The Pacific railroad bill, just passed through this House under the 'gag', and in violation of the constitution.

3. c. A joke; a humorous remark, situation, action, etc. Cf. GAG *sb.*², esp. quots. 1819 and 1880.

1863 H. MORFORD *Sprees & Splashes* i. 12 All play-goers will remember..a piece of pure nonsense by Brougham, in which an old hunks is intruded upon by his daughter's lover in disguise, the whole farce crowded full of gags. **1923** WODEHOUSE *Inimitable Jeeves* xv. 195 That..curate wanted to give the public some rotten little fairy play..without one good laugh or the semblance of a gag in it. **1929** M. LIEF *Hangover* 235 Maybe I can tell you about the gag one of our better-known critics pulled in his review of Dillingham's new musical comedy..he wrote: 'Eulalia Duncan sang so off-key last night that she had great difficulty moving her vowels.' **1965** *New Statesman* 16 Apr. 607/1 Braid Scots is spoken only by comedians, as a gag. *Ibid.* 621/1 Chekhov, still a one-word gag about gloom to P. G. Wodehouse and Noel Coward.

4. gag-bill, a theatrical bill summarizing the sensational episodes of a play; **gag-book,** a book containing gags; a joke-book; **gag-man** (orig. *U.S.*), a deviser or writer of gags; also, a comedian; so *gag-writer, -writing*.

1885 J. K. JEROME *On Stage* 147 The old man has got the knack of making out good gag bills. **1909** B. WILLIAMS *Actor's Story* xix. 259 Not having a 'gag' book I had to think how to adapt a scene suitable for my particular line of business. **1961** *Guardian* 23 June 7/5 Peter de Vries's latest novel..reads like a gag-book. **1928** *Sunday Express* 15 Jan. 4 'Gag men' have long flourished in America. One of them thought of making the villain sit on a red-hot stove. **1928** *Collier's* 29 Dec. 28/3 'What you need is a smart gag man,' I say. **1951** DYLAN THOMAS *Let.* 12 Apr. (1966) 357, I am about to take on a new job; co-writing, with the best gagman in England..a new comic series for the radio. **1957** *Observer* 29 Sept. 12/3 A quiet, mike-clutching gagman, in a discreetly elegant grey suit. **1959** *Listener* 23 July 124/1 A team of gag-writers. **1928** H. CRANE *Let.* 27 Mar. (1965) 321 Meanwhile there are mechanical jobs such as title-writing, gag-writing, 'continuity' writing, etc. **1959** *Spectator* 11 Sept. 342/2 A young undertaker's clerk given to gag-writing.

gag (gæg), *sb.*³ *U.S.* [Local name.] A large serranoid fish, *Mycteroperca microlepis*, found off the coasts of the southern United States.

1884 G. B. GOODE *Fisheries U.S.* I. 413 There appear to be..at Key West, as well as in Bermuda, various local forms closely related to this [*sc.* the rock-fish], one of which is known by the name 'Gag'. **1896** JORDAN & EVERMANN *Fishes Amer.* 1177 *Mycteroperca Microlepis* (Gag). **1950** R. A. DAHNE *Salt-Water Fishing* viii. 155 The gags seem to be increasing in popularity with many salt-water anglers.

gag, *v.*¹ **1. b.** Delete *Obs.* exc. *dial.* and add later examples.

1908 *Practitioner* Sept. 367 The operation causes a little gagging and retching at first, but the patient soon becomes accustomed to the feeling of the cords in the throat. **1939** H. HODGE *Cab, Sir?* 49 The stench of stale beer and cheap scent makes me gag. **1945** J. STEINBECK *Cannery Row* xvii. 70 The idea gagged him, but he couldn't let it alone. **1963** M. DUGGAN in C. K. Stead *N.Z. Stories* (1966) 97 Suppose you gag a little at the sugar coating, its the same old fundamental toffee, underneath.

5. c. *intr.* To tell a joke or jokes. Also *trans.*, with the joke as object.

1942 in BERREY & VAN DEN BARK *Amer. Thes. Slang* § 281/11. **1959** I. & P. OPIE *Lore & Lang. Schoolch.* iii. 53 'Do you know where Smudger takes his girl?' gags the would-be comic.

gaga (gā·gā, gæ·gā), *a.* and *sb. slang.* Also **ga-ga.** [a. F. *gaga* a senile person; senile.] **A.** *adj.* Doting, exhibiting senile decay; mad, 'dotty'; fatuous. Also as *adv.*

[**1905** *Daily Chron.* 18 Mar. 8/6 'Ah, you English,' quoth Mr. De Vries not so long ago, after a round of the London theatres, 'you like to laugh—ga-ga!'..Is not that the pathetic cry of our present drama, 'Ga-ga!'] **1920** F. M. FORD *Let.* 26 July (1965) 116 The V.G.F. [*sc.* Victorian Great Figure] must be gaga! **1921** M. BARING *Passing By* 210 Sir Arthur is quite gaga and took me for George the whole evening. **1926** E. FERBER *Show Boat* xix. 385 Nola darling, you've just gone gaga, that's all. What do you mean by staying down there in that wretched malarial heat! **1927** *Sunday Express* 13 Mar. 4 The conventional pictures of a young man and a young woman looking 'ga-ga' at each other. **1929** W. J. LOCKE *Ancestor Jorico* xviii, 'But why did he leave the half-million to his son, in his will?' '*Gaga*, my dear Binkie. Just *gaga*. Senile, if you'd like it better.' **1956** C. P. SNOW *Homecomings* xvii. 116 Hiding behind his smoke-screen of platitudes like an amiable old man already a bit ga-ga. **1961** A. WILSON *Old Men at Zoo* iii. 158 If Godmanchester was so gaga that he blabbed like this, then our prospects were alarming.

B. *sb.* A doting or senile person; a madman.

1938 S. BECKETT *Murphy* xi. 240 Mr. Endon..the most biddable little gaga in the entire institution. **1941** KOESTLER *Scum of Earth* 180 Couldn't understand what he said...Disastrous old gaga.

‖ **gagaku** (ga·gaku). [Jap., f. *ga* authentic, graceful, noble + *gaku* music.] A type of Japanese instrumental music performed chiefly on ceremonial occasions at the Imperial court and closely resembling the secular music of ancient China during the Tang period.

[**1893** F. T. PIGGOTT *Mus. Japan* I. 14 The new Musical Bureau was christened '*Gagaku Rio*': *Ga* meaning 'tasteful', and therefore, in old days as in new, an equivalent to something but lately introduced from foreign parts.] **1936** K. SUNAGA *Jap. Mus.* ii. 24 *Gagaku* is the special form of music performed in the ceremonies of the Imperial Court, as well as music and dancing performed on such other occasions as banquets and entertainments. **1968** *Daily Tel.* 15 Nov. 19/8 During the 30-minute acceptance ceremony in Tokyo this morning, court musicians, wearing resplendent period costumes, played age-old instruments to produce ancient music known in Japan as 'gagaku'.

‖ **gage d'amour** (gaʒ damur). [Fr.] A pledge of love; a love-token.

1768 STERNE *Sent. Journ.* II. 144 His faith-less mistress had given his *gage d'amour* to one of the Count's footmen. **1838** DICKENS *Let.* 27 July (1965) I. 422 It is many months since I sent you a slight gage d'amour. **1873** TROLLOPE *Eust. Diamonds* xxxi. 60 That gage d'amour which you carry on your finger. **1930** 'R. CROMPTON' *William's Happy Days* vii. 157 He'd picked up a glove that she'd left in church last Sunday, and was treasuring it as a kind of *gage d'amour*.

gaggle, *sb.* Add: **1.** Also *transf.*, a group of people or things, esp. a disorderly assemblage; *spec.* (*slang*) a group of aircraft.

1946 G. GIBSON *Enemy Coast Ahead* 206 We started off first in squadrons, then in wings and finally in a sort of formation known as a group gaggle, meaning a flock of geese. **1946** E. C. CHEESMAN *Brief Glory* vi. 73 Ferry pilots had to fly in 'gaggles' to make it easier for the Observer Corps. **1956** J. E. JOHNSON *Wing Leader* i. 13 We curved across Berlin, sparred cautiously with large gaggles of Russian fighters. **1966** *Listener* 8 Sept. 354/1 There is hardly a modern skyscraper in midtown that does not have its gaggle of sightseers. **1971** *Islander* (Victoria, B.C.) 21 Mar. 14/3 A gaggle of sparsely inhabited islands.

gagster (gæ·gstəɹ). *colloq.* [f. GAG *sb.*¹ + -STER.] One who makes 'gags' or jokes; a gag-writer or comedian.

1935 E. WEEKLEY *Something about Words* 65 The very modern gagster, at the theatre. **1959** *Listener* 5 Nov. 797/3 Music-hall gagsters. **1961** *Observer* 16 July 5/3 He writes his own material (another break with the Bob Hope generation and their teams of gagsters).

gah (gā), *int.* [Echoic.] An exclamation indicating impatience or exasperation.

1917 R. NICHOLS *Ardours & Endurances* I. 38 The German line Vanishes in confusion, smoke. Cries, and cry Of our men, '*Gah, yer swine!*' **1936** 'N. BLAKE' *Thou Shell of Death* iii. 50 Gah! Walks like a jaguar with the gripes. **1947** —— *Minute for Murder* x. 236 'Affair'—gah, what a word! **1953** A. MILLER *Crucible* (1956) I. 20 Gah! I'd almost forgot how strong you are, John Proctor! **1966** M. PROCTER *His Weight in Gold* viii. 66 'Gah!' he told himself. 'Ham Wayman, you're going daft.'

gaiassa (gaiˌæ·sā). Also 9 **caiash.** [Arab. *ḳayyāsa*.] A high-stemmed vessel with lateen sails used on the Nile for carrying freight.

1818 H. LIGHT *Trav. Egypt* II. i. 123, I began to descend the Nile in a small caiash or barge. **1906** *Westm. Gaz.* 21 Mar. 3/2 One of the most heterodox of boats to European eyes is the 'Gaiassa' of the Nile. **1938** *Times Lit. Suppl.* 31 Dec. 829/1 The craft in question being, presumably, *gaiassas*.

‖ **gaieté de cœur** (gete də kœr). [Fr.] Light-heartedness; playfulness.

1728 VANBRUGH & CIBBER *Prov. Husb.* I. i. 4 Take a frolicksome Supper at an India House—perhaps, in her *Gayeté de Cœur* toast a pretty Fellow. **1747** H. WALPOLE *Let.* 24 Nov. (1903) II. 295, I did not mention returning to Florence out of *gaieté de cœur*. **1886** C. M. YONGE *Chantry House* I. xiii. 119 Emily..in youthful *gaieté de cœur* had got a little tired of her. **1936** *Times Lit. Suppl.* 4 July 558/2 The latest ebullition of his [*sc.* the Regent's] *gaieté de coeur* is to make the French Ambassador's wife drunk at table.

gaiety. Add: **1. a.** Esp. in phr. *the gaiety of nations*, the cheerfulness or pleasure of a large number of people; general gaiety or amusement (freq. used hyperbolically or ironically).

1779 JOHNSON *Life of Edmund Smith* in *Prefaces* IV. 61 That stroke of death [Garrick's], which has eclipsed the gaiety of nations, and impoverished the publick stock of harmless pleasure. **1967** *Guardian* 15 Dec. 7/4 Nowadays..musical comedies all seem to hearken back to Old Vienna or Gay Paree, and they don't exactly contribute to the gaiety of nations. **1968** *Listener* 29 Aug. 264/2 Analysts have added substantially to the gaiety of nations, by the cheerful and ingenious inventiveness with which they have produced fantastic and highly risible analyses of all and sundry.

4. (With capital initial.) The name of a former London theatre famous, esp. in the 1890s, for its musical shows, used *attrib.* of features characteristic of these shows. **Gaiety girl,** a performer in such a show; also *transf.*

1890 G. B. SHAW *London Mus. 1888–89* (1937) 314 These can aspire to kicking through a Gaiety *pas de quatre*. **1898** *People* 27 Feb., One of the brightest and most graceful of 'Gaiety' girls. **1910** G. K. CHESTERTON *G. B. Shaw* 143 Shaw..introduced the Gaiety Girl, but did not represent her life as all gaiety. **1946** BRAHMS & SIMON *Trottie True* ii. 20 There was nothing she could do to stop the man nipping off to a stage door and taking a Gaiety Girl out to supper. **1963** *Listener* 10 Jan. 58/1 The black stockings and swirling skirts of the Gaiety Girls. **1965** C. FREMLIN *Jealous One* xx. 156 This gaiety-girl business is all fairly new. It's not natural to her. She's one of Nature's frumps, really. **1970** *Guardian* 19 Dec. 6/2 There were Gaiety Girls on garlanded swings.

gain, *sb.*² Add: **2. d.** *Electronics.* An increase in power, voltage, or current, expressed as the ratio of the increased quantity to the original quantity or (more commonly) as the logarithm of this; **gain control** (see quot. 1930).

1922 GLAZEBROOK *Dict. Appl. Physics* II. 862/2 The 'loss' in such a case will actually be a 'gain'. That is, in such a case greater current will flow into the receiving circuit when the apparatus is inserted than when it is omitted. **1929** K. HENNEY *Princ. Radio* xiii. 309 There is a voltage step-up in the transformer, therefore some voltage gain may be secured by its use. **1930** *Sel. Gloss. Motion Pict. Techn.* (*Acad. Motion Pict., Hollywood*), *Gain control*, device for varying the gain of an amplifier. **1936** R. S. GLASGOW *Princ. Radio Engin.* vii. 171 Since the decibel is a logarithmic unit, the total gain of an amplifier can be conveniently found by adding together the gains of the individual stages. **1950** *Sci. News* XV. 19 The platinum electrode..is connected to a valve amplifier of high gain. **1959** K. HENNEY *Radio Engin. Handbk.* (ed. 5) v. 29 The effectiveness of the coupled circuit of Fig. 44 is determined by its gain and its selectivity. **1961** N. H. CROWHURST *High Fidelity Sound Engin.* iii. 79 The effect of a gain control is to alter volume or loudness. **1962** A. NISBETT *Technique Sound Studio* ii. 38 The speech may be at such a low level that considerable gain has to be used, and noise from the amplifier becomes apparent. **1970** D. F. SHAW *Introd. Electronics* (ed. 2) xii. 267 The current gain in a junction transistor is normally less than unity.

gain, *v.*² Add: **3. d.** Of a clock, watch, etc.: to become fast (FAST *a.* 8 c); to indicate a time ahead of the correct time. Also *trans.*, to run fast by the amount of (a specified period).

1861 *Horological Jrnl.* June 123/2 When the watch is in a horizontal position, the weight of the balance is supported on a single point, hence its greater freedom of motion and tendency to gain. *Ibid.* Nov. 36 Columns 3 to 7 show the mean daily rate for each of the five weeks, *gaining* when no sign is used, and *losing* when the – sign is used. **1863** *Ibid.* July 121/1 In the same temperature..it was gaining two seconds and one tenth. **1870** 'MARK TWAIN' in *Galaxy* X. 882/2 My beautiful new watch had run eighteen months without losing or gaining. **1917** H. E. DUDENEY *Amusements in Math.* 10/1 Does that watch gain or lose, and how much per hour? *Ibid.* 153/2 It gains $\frac{5}{7}$ of a minute in 65 minutes. **1946** D. DE CARLE *Pract. Watch Repairing* 154 It is possible to make the watch gain by making the balance heavy at its lowest point. *Ibid.* 266 The watch may suddenly gain a few seconds due to the coils of the balance spring sticking together. **1970** *Encycl. Brit.* XXIII. 267/2 The watch would tend to gain.

gaine (gēⁱn). [Fr., = sheath.] A metal tube attached to a fuse (see quots.).

1918 *Nature* 14 Nov. 218/1 The gaine is a metal tube screwed to the fuse, which enters a cavity in the filling and makes good contact with it. **1932** LLOYD GEORGE *War Mem.* II. xix. 592 A gaine..is a tube filled with explosive, attached under the nose-cap of a high-explosive shell...Its purpose is to ensure that the detonation of the fuse in the nose-cap shall effectively detonate the contents of the shell.

Gainsborough (gēⁱ·nzbŏrə). [The name of Thomas *Gainsborough* (1727–88), English painter.] **1.** A portrait painted by Gainsborough.

1798 J. CONSTABLE *Let.* 2 Dec. (1964) II. 16 Consequently I have not taken advantage of your friendship or Mr. Strutt's good nature respecting the Gainsborough. **1894** W. ARMSTRONG *Gainsborough* 83 As refutations of Sir Joshua we could name many Gainsboroughs which beat the *Blue Boy*. **1903** A. B. CHAMBERLAIN *Gainsborough* 132 She [*sc.* Signora Grassi] called her house a 'painted Paradise' because of the Gainsboroughs hanging there. **1958** E. WATERHOUSE *Gainsborough* 13 An almost unknown painter some of whose works..are curiously similar to early Gainsboroughs.

2. A large broad-brimmed hat of the type worn by women in Gainsborough's portraits. In full, *Gainsborough hat*.

1878 *Cassell's Fam. Mag.* Aug. 569/2 The..wearers of the Gainsborough, Rembrandt, and beef-eater hats. **1884** [see *CART-WHEEL *sb.* 5]. **1904** *Westm. Gaz.* 12 Aug. 5/1 Extemporised Gainsboroughs. **1928** *Amer. Speech* IV. 92 One remembers the Gainsborough hat.

gait, *sb.*¹ Add: **a.** (Further *fig.* examples.) Esp. in phr. *to go one's* (*own*) *gait*, to go one's own way; to pursue one's own course. (For literal senses of this phr. see GANG *v.*¹ 1 b, GATE *sb.*²)

1865 'MARK TWAIN' *Celebr. Jumping Frog* (1867) 37 Preachin' was his nateral gait. **1922** J. B. PRIESTLEY *Papers from Lilliput* 31 Caring little whether he is still a shepherd or metamorphosed into a fisherman or cobbler, so long as he is still with us, going his own fantastic gait. **1940** H. READ *Annals of Innocence* II. ii. 82 These are qualities to be enjoyed by non-poetic people: the poet must go his own gait. **1958** *Times* 4 Oct. 9/5 Miss Watts, whose voice is of the right weight and gait for Bach.

‖ **gaita** (gai·tā). [Sp.] A Spanish musical instrument resembling the bagpipes. So **gaite·ro,** a piper.

1846 R. FORD *Gatherings from Spain* vii. 77 The tune of the *gaita*, or Moorish bagpipe. **1922** *Glasgow Herald* 14 Aug. 4 Within the deep, wide recess of the chimney the son of the house is singing to the accompaniment of the 'gaita'. **1927** *Chambers's Jrnl.* 29 Oct. 758/2 Their little *gaitas* (bagpipes) tasselled with the national colours of red and yellow. *Ibid.*, A famous band of *gaiteros*, or pipers. **1967** 'LA MERI' *Sp. Dancing* (ed. 2) ii. 32 The *gaita*..is a bagpipe, green in color, which is the same as the better-known bagpipes of Scotland and Ireland.

gaiter, *sb.*¹ Add: **3. gaiter-boot, -shoe** (orig. *U.S.*) = sense 2 in Dict.

1840 *Picayune* (New Orleans) 30 July 2/4 Many of the ladies of Philadelphia..now wear..gaiter boots, with little straps of black leather. **1849** N. P. WILLIS *Rural Lett.* 230 Dandies strolling and stealing an occasional look at their loose *demi-saison* pantaloons and gaiter-shoes. **1875** MRS. STOWE *We & Neighbors* x. 106 Looking as if they never had heard of a French hat or a pair of gaiter-boots. **1932** 'A. BRIDGE' *Peking Picnic* x. 111 Blue trousers tucked into high white gaiter boots.

‖ **gaize** (gēⁱz). *Geol.* [Fr.] A fine-grained micaceous and calcareous sandstone found in Cretaceous beds in northern France and southern England.

1885 A. GEIKIE *Text-bk. Geol.* (ed. 2) VI. 832 A porous calcareous and argillaceous sandstone known as *Gaize*, containing a large percentage of silica soluble in alkali. **1949** KIRK & OTHMER *Encycl. Chem. Technol.* III. 430 Along the valleys of the Meuse and Ardennes in France is an extensive formation of..gaize. This material..is used as a pozzolanic admixture with portland cement. The mixture is known as *gaize cement*. **1955** G. G. WOODFORD tr. *Gignoux's Stratigr. Geol.* viii. 419 In the southern part of the Ardennes..we already know that its base is composed of the top of the *gaize* with *Mortoniceras rostratum*.

gal¹ = GIRL. (Earlier examples.)

1795 B. DEARBORN *Columbian Gram.* 135 Improprieties, commonly called Vulgarisms, [include]..Gal for girl. **1824** J. WIGHT *Mornings at Bow St.* 132 *Gal*—cockney for *girl*. **1837** DICKENS *Pickw.* iv. 39 My daughters, gentlemen—my gals these are. *Ibid.* xxxii. 343 You *are* a nice gal and nothin' but it.

gal² (gæl). [f. the name *Galileo*: see GALILEAN *a.*²] A unit used in expressing the value of the acceleration due to gravity and equal to one centimetre per second per second.

1914 O. KLOTZ in *Nature* 13 Aug. 611/2 So long ago as 1909 Wiechert used the term 'gal' in the report for the Göttingen earthquake station for that unit, being the first syllable of Galileo...Others, as well as myself, have used 'gal' or rather 'milligal' in analyses of earthquakes...Dyne is the unit of force, gal the unit of acceleration. **1935** *Nature* 23 Mar. 471/1 A particular case as example in which an error of 0.01 gal in '*g*' would lead to an error of 40 ft. in geoidal elevation. **1937** *Geogr. Jrnl.* LXXXIX. 92 Unfortunately the old Helmert (1901) is used instead of the international formula adopted at Stockholm in 1930, in which the equatorial value of '*g*' is greater by 0.019 gals (= cm. per sec. per sec.). **1955** *Sci. Amer.* Sept. 165/3 Painstaking pendulum measurements have established the value of gravity as 981.274 gal. **1971** *Geophysical Jrnl.* XXII. 449 With the general adoption of the International System of Units..the c.g.s. unit of acceleration, the gal (or Gal), is doomed to obsolescence.

gala¹. Add: Now more usually pronounced (gā·lă). **4.** *gala dinner*, *dress* (earlier example), *night* (earlier example), *performance*, *show*.

1766 S. SHARP *Lett. from Italy* xxi. 86 Some of these persons of quality..order the undertaker of the opera to let out their boxes when they do not go themselves, and often stay at home purposely on *gala* nights. **1805** C. WILMOT *Let.* 29 Aug. in *Russ. Jrnls.* (1934) II. 177 There is hardly any of the Costume of Women, & what there is not in their *Gala dress*. **1925** R. GRAVES *Welchman's Hose* 35 A Frenchman was struck dead by a meteorite, That was the sort of gala-show it was! **1932** E. SITWELL *Bath* 238 Lord Chesterfield..speaks of one of these gala-performances. **1934** A. HUXLEY *Beyond Mexique Bay* 6 He..decreed that our dinner, every third or fourth day should be called a Gala Dinner. **1966** *Illustr. London News* 30 July 22 Princess Muna went with the King, the Duke of Gloucester..and the Duchess to the Royal Opera house, where they saw a gala performance of Giselle.

Also as *adj.*, festive, gay. Chiefly *N. Amer.*

1964 *McCall's Sewing* iii. 36/2 When you want to feel really dressed up and ready for a big night, try the real evening fabrics. You'll feel more gala. **1968** *Globe & Mail* (Toronto) 17 Feb. 25/7 The whole thing will be very gala and a lot of fun.

galabiya (galā·biya). Also **gal(l)abea, -beeah, -bia, -bieh, -biyah, -biyeh, -bya.** [= Arab. *jallābiya*, pop. equivalent of *jilbāb*.] A smock-like garment worn in Arabic-speaking Mediterranean countries.

1725 J. WINDUS *Journey to Mequinez* 29 A Gelebia..is made of a course and thick wrought woollen Stuff, without Sleeves, but Holes to put their Arms thro', it reaches to their Knees, and hangs loose about their Bodies like a Sack. **1820** J. G. JACKSON *Account of Timbuctoo* 200 Their costume is a *jelabea*, and a belt, without shoes or head dress. **1839** J. DAVIDSON *Notes Trav. Afr.* 12 The *jelábiyah* are merely coarse sacks, with holes in them for the arms to pass through. **1892** D. A. CAMERON *Arabic-Engl. Vocab.* 45 'Galábiya'. **1900** A. CONAN DOYLE *Green Flag* 328 Tell them to undo the man's galabeeah. **1916** V. HORSLEY in S. Paget *Sir V. Horsley* (1919) 321 The small boys..are given the head-dress and a few feet of stuff for a galabea. **1921** *Chambers's Jrnl.* 26 Mar. 265/1 A figure in a blue *galabieh* and a white-turbaned tarboosh. **1965** *Economist* 2 Jan. 25/1 The [Egyptian] working man in the towns..rustles a few more notes in his trouser (no longer, let us note in passing, his galabiya) pocket.

galactan (gălæ·ktæn). *Chem.* [a. G. *galactan* (C. Scheibler 1882, in *Neue Zeitschr. f. Rübenzucker-Industrie* VIII. 277/1), adapted from *galactin* (f. Gr. *γαλακτ-*, *γάλα* milk) to conform with G. *dextran* *DEXTRAN.] Any of a group of polysaccharides that yield galactose on hydrolysis and occur in many plants, esp. algæ, lichens, mosses, and conifers, and a few animals.

1886 *Jrnl. Chem. Soc.* L. 609 This carbohydrate [from the seed of *Lupinus luteus*] very closely resembles galactin (or adopting Scheibler's nomenclature galactan), obtained by Muntz from the seed of the lucerne. **1912** C. A. BROWNE *Handbk. Sugar Analysis* xix. 599 In the galactans d-galactose shows the same tendency to form combinations with other sugars..; there are arabogalactans, xylogalactans, mannogalactans, glucogalactans and other combinations. **1963** FLORKIN & STOTZ *Comprehensive Biochem.* V. vii. 222 Galactans of animal origin are rare. The snail *Helix pomatia* and a few other cold-blooded animals produce a highly branched levorotatory galactan containing D-galactose and a lesser amount of L-galactose.

galactase (gălæ·ktēⁱz, -ēⁱs). *Biochem.* [f. GALACT(O- + *-ASE.] A proteolytic enzyme present in the milk of many animals.

1898 *Proc. Amer. Assoc.* XLVII. 232 The authors [*sc.* S. M. Babcock and H. L. Russell]..discovered an inherent proteolytic ferment in milk, to which they have given the name galactase. **1901** C. A. MITCHELL tr. *Oppenheimer's Ferments* x. 118 Galactase..is stated to work best in weakly alkaline media, and to resemble trypsin, though not to be identical with it. **1937** A. L. & K. G. B. WINTON *Struct. Foods* III. I. 185 The part played by enzymes in the ripening of cheese is complex. Some of these are derived from the milk or starter, others are formed by micro-organisms. Galactase..belongs to the former class.

galactic, *a.* Add: **2.** Also, of or pertaining to another galaxy or to galaxies in general; *galactic belt*, the Milky Way; *galactic concentration*, the increase in the number of stars from the galactic poles to the galactic equator; *galactic co-ordinates*, spherical co-ordinates used in astronomy, comprising galactic latitude and longitude; *galactic equator*, a great circle passing as nearly as possible through the middle of the Milky Way; *galactic latitude*, latitude measured from the galactic equator; *galactic longitude*, longitude measured from some reference point on the galactic equator, usu. defined either by the direction of the centre of the galaxy or by the intersection of the galactic and the celestial equators; *galactic plane*, the plane defined by the galactic equator; *galactic pole* (examples); *galactic system*, the galaxy containing the sun and most of the visible stars; *galactic zone* = *galactic belt*.

1849 J. F. W. HERSCHEL *Outl. Astron.* xv. 534 The following table, expressing the densities of the stars at the respective distances, 1, 2, 3, &c., from the galactic plane. *Ibid.* 535 The law of the visible distribution of stars over the southern galactic hemisphere, or that half of the celestial surface which has the south galactic pole for its center. **1878** S. NEWCOMB *Pop. Astron.* IV. 417 Telescopic stars..are least numerous in the regions most distant from the galactic belt. **1888** C. A. YOUNG *Text-bk. Gen. Astron.* xxi. 510 The 'galactic plane' 'is to sidereal what the ecliptic is to planetary astronomy, a plane of ultimate reference, the ground plan of the sidereal system'. **1890** A. M. CLERKE *Syst. Stars* xxiii. 355 The bright spaces of the galactic zone are commonly surrounded and set off by dark winding channels. *Ibid.* 359 A regular progression of density from the galactic poles to the galactic equator. *Ibid.*, Two or three hundred 'beginning, or gathering clusters', might be pointed out in the galactic system. **1906** *Rep. Brit. Assoc. 1905* 260 Mean values were computed for different galactic latitudes by combining the results of regions at equal distances from the Milky Way. **1914** A. S. EDDINGTON *Stellar Movements* viii. 165 The stars with proper motions greater than 10″ per century show no galactic concentration; they are all comparatively near to us. *Ibid.* ix. 196 A systematic dependence on galactic longitude in some of the zones. **1926** *Encycl. Brit.* I. 252/1 In plan the system is elongated with its axis in galactic longitude 325°. **1930** R. H. BAKER *Astron.* xi. 442 The galactic equator is the great circle halfway between the galactic poles. **1931** W. M. SMART *Text-bk. Spherical Astron.* xi. 276 The galactic co-ordinates of S. **1936** E. HUBBLE *Realm Nebulæ* vi. 125 The galactic system is a member of a typical, small group of nebulæ which is isolated in the general field. **1965** J. B. SYKES tr. *Ogorodnikov's Dynamics Stellar Syst.* i. 25 The choice of any particular coordinate system is a matter of convenience. In stellar astronomy the most frequently used is that of galactic coordinates, defined by the galactic equator and the galactic poles. **1971** *Nature* 18 June 437/1 Results from sounding rocket experiments..have revealed the existence of cosmic X-ray sources at high galactic latitude.

galacto- Add: **gala·ctophore,** a galactophorous duct; **gala:ctopoie·sis** [Gr. *ποίησις* production], the production of milk (see also quot. 1961).

1904 G. S. HALL *Adolescence* I. vi. 420 Both the galactophores or glands and the supportive areolar tissue develop rapidly. **1961** *Lancet* 29 July 241/2 Saline solutions of the human pituitary fractions were injected, via the teat galactophore openings. **1842** DUNGLISON *Dict. Med. Sci.* (ed. 3) 319/1 *Galactopoiesis*, galactosis. **1948** *New Biol.* IV. 127 A distinction has..been drawn between lactogenesis—the power of initiating milk secretion—and galactopoiesis—the power of maintaining the flow of milk once it has been started. **1961** KON & COWIE *Milk* I. iv. 164 The term 'galactopoiesis', i.e., the enhancement of an established milk secretion, is used by some research workers in the more general sense of the maintenance of milk secretion.

galactonic (gælæ̆ktǫ·nik), *a. Chem.* [tr. G. *galactonsäure* galactonic acid (H. Kiliani 1885, in *Ber. d. Deut. Chem. Ges.* XVIII. 1551), f. GALACT(OSE + *-ONIC.] *galactonic acid*: a monobasic sugar acid, $CH_2OH(CHOH)_4COOH$.

1885 *Jrnl. Chem. Soc.* XLVIII. 967 *Galactonic acid*, $C_6H_{12}O_7$, is best prepared by the action of bromine on an aqueous solution of lactose. **1956** *Biochem. Jrnl.* LXIV. 21/2 The enzymic reduction of derivatives of D-galacturonic acid to derivatives of L-galactonic acid has been demonstrated in peas. **1957** *Jrnl. Biol. Chem.* CCXXVII. 748 An initial oxidation of galactose to galactonic acid is followed by a conversion of galactonic acid..to yield pyruvic acid.

galactosæmia (gălæ:ktosī·miă). *Med.* Also **galactosemia.** [f. GALACTOS(E + Gr. *αἷμα* blood: see -IA¹.] **a.** The presence of galactose in the blood. **b.** A metabolic disorder (of which galactose in the blood is a symptom) resulting from a hereditary inability to metabolize galactose.

1934 *Biol. Abstr.* VIII. 1431/1 [There is a] need in clinical studies of galactosemias and galactosurias, of regulating the proportions of glucose and galactose. **1945** BRUCK & RAPOPORT in *Amer. Jrnl. Dis. Child.* LXX. 267 (*heading*) Galactosemia in an Infant with Cataracts. *Ibid.* 272/1 The ability of the liver to convert and to store galactose seems to be impaired but not completely lost in galactosemia. **1958** *Observer* 5 Jan. 9/1 She has a condition called 'congenital galactosaemia', which is an inherited inability to break down another sugar found in milk, known as galactose. **1970** R. M. GOODMAN *Genetic Disorders Man* xvii. 872/2 The presence of galactose in the blood or urine is not a reliable indication of galactosemia.

galactosamine (gæ:læ̆ktōᵘ·sămin). *Chem.* [ad. G. *galaktosamin* (Schulz & Ditthorn 1900, in *Zeitschr. Phys. Chem.* XXIX. 373), f. GALACTOSE + AMINE.] Any amino derivative of galactose; *spec.* 2-amino-2-deoxy-D-galactose (chondrosamine), $C_6H_{13}NO_5$.

1900 *Jrnl. Chem. Soc.* LXXVIII. I. 478 (*heading*) Galactosamine, a new amino-sugar. **1956** *Nature* 28 Jan. 186/2 A substance reacting as a hexosamine but which is not glucosamine or galactosamine has been found. **1969** *New Scientist* 23 Jan. 187/2 Mucopolysaccharide contains two amino sugars in large amounts, glucosamine and galactosamine.

galactoside (gălæ·ktŏsəid). *Chem.* [a. F. *galactoside* (M. Berthelot *Chim. org. fondée sur la Synthèse* (1860) II. III. iii. 286), f. GALACTOSE + -IDE.] Any glycoside in which the sugar is galactose. Hence **gala:ctosi·dase,** any enzyme that hydrolyses a galactoside to galactose and an aglycone; **galactosi·dic** *a.*, of, pertaining to, or characterized by a galactoside.

1862 H. WATTS tr. *Gmelin's Hand-bk. Chem.* XV. 317 He [*sc.* Berthelot] then subdivides the saccharides into—glucosides..; levulosides..; galactosides. **1902** *Jrnl. Chem. Soc.* LXXXII. I. 263 *Tetra-acetyl-β-phenolgalactoside*, $C_{20}H_{24}O_{10}$, crystallises from dilute alcohol or from benzene in stout prisms...This galactoside is not hydrolysed by yeast extract. **1917** *Chem. Abstr.* XI. 2819 (*heading*) Detection of β-galactosidase in the vegetable kingdom. **1950** *Jrnl. Bacteriol.* LX. 390 Several of the problems of the enzyme chemistry of galactosidase..require the use of preparations more highly purified than can be readily accomplished with bacterial enzymes. **1952** *Biol. Abstr.* 2533/2 The formation of beta-galactosidase (I) by *E. coli* is induced exclusively by substances possessing an intact galactosidic radical. **1956** *Nature* 10 Mar. 478/2 Only enzymes capable of splitting α-glucoside and α-galactoside linkages were required to be present.

galacturonic (gălæ:ktiŭrǫ·nik), *a. Chem.* [ad. G. *galacturonsäure* galacturonic acid (F. Ehrlich 1917, in *Chemiker-Zeitung* 7 Mar. 197/2), f. GALACT(OSE + *URONIC *a.*] *galacturonic acid*: the uronic acid derived from galactose.

1917 *Chem. Abstr.* XI. 2583 E[hrlich] isolated from the decompn. products of pectin *d*-galacturonic acid, a new isomer of glucuronic acid of the animal kingdom, and the *d*-form of E. Fischer's racemic aldehyde mucic acid. **1930** *Jrnl. Biol. Chem.* LXXXVI. 491 Flax pectin..is made up of 61 per cent of galacturonic acid, and the pectic acid from orange peelings..is composed of 73.7 per cent of galacturonic acid. **1957** *Sci. News* XLV. 82 Hexuronic acids such as glucuronic acid and galacturonic acid with the molecular formula $C_6H_{10}O_7$.

galah. Add: **2.** *Austral. slang.* A fool, simpleton.

1944 L. GLASSOP *We were Rats* xxv. 142 When will these galahs wake up? **1949** *Geogr. Mag.* Feb. 374 A foolish person is often called a *galah*, after the name of the garrulous bird. **1960** *Encounter* XIV. 28 Don't be a galah, Art and I are old drinking cobbers. **1965** E. BROWN *Big Man* vii. 56, I have to report for special duty because of this Pommie galah. **1971** *Guardian* 9 Sept. 9/5 He has visited the mineral belt and met the..larrikins, galahs and dills who gravitate to such a scene.

Galahad (gæ·lăhæd). [The name of the noblest knight of the Round Table in Arthurian legend.] A man having the qualities of the legendary Sir Galahad; one characterized by nobility, integrity, courteousness, etc. Also *attrib.*

1870 C. M. YONGE *Caged Lion* xv. 263 A Galahad would he strive to be. **1911** G. B. SHAW *Getting Married* Pref. 145 Our bachelors and widowers would no longer be Galahads. **1921** D. H. LAWRENCE *Sea & Sardinia* 76 We have set forth, politically, on such a high and Galahad quest of holy liberty. **1967** 'E. TREVOR' *Freebooters* xxi. 223 He'd thought he was just a tin Galahad trying to share Sgt Proctor's martyrdom by yapping about it.

Galalith (gæ·lăliþ). Also written with lower-case initial letter. [f. Gr. *γάλα* milk + *λίθος* stone.] A proprietary name for a product resembling celluloid made from the casein of milk with the addition of other substances, used as a substitute for horn, celluloid, ivory, etc.

1901 *Trade Marks Jrnl.* 20 Mar. 296 Galalith. Goods, included in Class 50.., made of Celluloid-like Material consisting of an Animal Substance (viz., Dried Curds) Hardened by means of Formaldehyde. Vereinigte Gummiwaaren-Fabriken Harburg-Wien..in the United Kingdom, c/o Haseltine, Lake, & Co.,..London. **1922** *Daily Mail* 13 Nov. 1 A number of curved galalith handles, in all white or antique colour. **1934** [see *CONSTRUCTIVIST *a.* and *sb.*]. **1963** G. S. BRADY *Materials Handbk.* (ed. 9) 148 The first commercial casein plastic was called *Galalith*, meaning *milkstone*.

galangin (gălæ·ŋgin, gălæ·ndʒin). *Chem.* [a. G. *galangin* (E. Jahns 1881, in *Ber. d. Deut. Chem. Ges.* XIV. 2385), f. GALANG(A + -IN¹.] A derivative of flavone, 3,5,7-trihydroxyflavone, present in a species of galingale (*Alpinia officinarum*).

1882 *Jrnl. Chem. Soc.* XLII. 209 Galangin is deposited from a solution in absolute alcohol in six-sided plates. **1918** PERKIN & EVEREST *Nat. Organic Colouring Matters* vii. 177 Galangin, $C_{15}H_{10}O_5$, the second constituent of galanga root, crystallizes in yellowish-white needles, melting-point 214–215°, soluble in alkaline solutions with a yellow colour. **1961** *Encycl. Brit.* II. 27/1 Among the better known of the anthoxanthins..may be mentioned..galangin.

‖ **galant** (galañ, gălā·nt), *a.* and *sb. Mus.* [a. Fr. and G. *galant* (see GALLANT *a.* and *sb.*).] **A.** *adj.* Designating or pertaining to a light and elegant style of eighteenth-century music, esp. that of harpsichord composers, contrasted with baroque music. **B.** *sb.* This style of music.

1949 *Scrutiny* XVI. 76 A coherently consistent style like classical baroque or Mozartian *galant*. **1953** *Times Lit. Suppl.* 20 Feb., The reinstatement of medieval, Baroque, or 'galant' conditions. **1959** D. COOKE *Lang. Mus.* v. 232 To their ears, nurtured on the normal 'tasteful' classical style of the *galant* and *rococo* periods, Mozart's exploitation of the expressive power of music..appeared to be wild and perverse experiments.

‖ **galanterie** (galañtəri). [Fr., see GALLANTRY.] † **1.** = GALLANTRY 2 a. *Obs.*

1616 T. ROE *Jrnl.* 1 Nov. (1899) II. 320, 10,000 horse, many in Cloth of Gould.., all in Galanterie.

2. *Mus.* (See quot. 1938). (Also with Ger. pl. **galanterien.**)

1911 *Encycl. Brit.* XXVI. 52/1 Then come the *galanteries*, from one to three in number. **1938** *Oxf. Compan. Music* 351/2 In the classical suite of the earlier eighteenth century the Galanterien are those movements which are not essential to the scheme and may perhaps be looked upon as interpolations of light relief.

3. Courtesy, politeness, esp. to women.

1920 D. H. LAWRENCE *Lost Girl* vii. 133 He had an amiable and truly chivalrous *galanterie*. **1959** *Times* 3 June 7/3 Eighteenth-century *galanterie*.

Galatian (gălēⁱ·ʃiăn), *sb.* and *a.* [f. *Galatia* + -AN.] **A.** *sb.* **a.** A native or inhabitant of Galatia, an ancient country of central Asia Minor. **b.** *pl.* St. Paul's Epistle to the Galatians. **B.** *adj.* Of or pertaining to Galatia or its people.

1587 J. PRIME *Exposition upon St. Paul to Galathians* 15 The Galathians were deeply plunged in the sea, the fault was most in the mariners. **1611** BIBLE *Galat.* iii. 1, O foolish Galatians [1388 Wyclif: Galathies; 1526 Tindale: Galathyans], who hath bewitched you? **1842** BROWNING *Bells & Pomegranates* III. 6/2 There's a great text in Galatians, Once you trip on it, entails Twenty-nine distinct damnations. **1853** J. BROWN *Exposition Epistle of Paul to Galatians* v. ii. 208 Some of the Galatian converts had yielded to the Judaising teachers. **1879** *Encycl. Brit.* X. 19/2 Unless the churches of Derbe, Lystra, &c., be regarded as Galatian, we are left in ignorance of the names..of the churches addressed. *Ibid.* 20/1 The Galatians..continued to make use of their ancient dialect. **1912** C. W. EMMET *St. Paul's Epistle to Galatians* p. xxi, Romans was in fact originally written as a circular letter addressed to mixed Churches..at the same time as Galatians. **1950** R. HEARD *Introd. N.T.* xviii. 182 Who were the Galatians? This is one of the classical controversies of New Testament criticism. Two answers are possible. The first..is that they were the inhabitants of North Galatia... The second answer, more generally accepted by recent critics, is that the 'Galatians' were not Galatians by race..but Christians of those towns in the southern part of the Roman province of Galatia which Paul and Barnabas had visited on their first missionary journey. **1957** *Oxf. Dict. Chr. Ch.* 535/1 St. Paul wrote this letter to his Galatian converts.

galax (gæ·læks). *Bot.* [mod.L. (Linnæus *Species Plantarum* (1753) I. 200); cf. Gr. *γαλαξίας* GALAXY.] An evergreen plant of the genus so called of the family Diapensiaceæ, native to the south-eastern United States, and having white flowers and shiny leaves.

1804 *Bot. Mag.* XX. 754 (*heading*) Galax Aphylla. Carolina Galax...Michaux found it on the high mountains of Carolina, flowering in May. **1894** W. FALCONER in W. Robinson *Wild Garden* (ed. 4) viii. 84 The sombre-hued Pyrola and Galax. **1898** B. TORREY in *Atlantic Monthly* Apr. 458/2, I could only conjecture the plant to be galax. **1922** M. B. HOUSTON *Witch Man* xi, The darkly polished galax leaves. **1959** C. DORMON *Flowers Native to Deep South* 88 Everyone has seen the shining, stiff leaves of Galax in florists' wreaths and bouquets.

galaxy, *sb.* Add: **1. b.** Any of the numerous large groups of stars and other matter that exist in space as independent systems; *spec.* (often with capital initial) the group that contains the solar system and whose plane of maximum star density gives rise to the observed Milky Way.

1848 J. P. NICHOL *Stellar Universe* iii. 59 Superb groups or galaxies separated from each other by gulfs so awful, that they surpass the distances which divide star from star...Amid this system of clusters, floats the galaxy whose glories more nearly surround us. **1888** C. A. YOUNG *Gen. Astron.* xxi. 503 The belief that these star-clusters are *stellar universes*,—'galaxies', like the group of stars to which the writers supposed the sun to belong. **1930** R. H. BAKER *Astron.* xii. 480 The existence of galaxies beyond the Milky Way was finally demonstrated by Hubble in 1925. **1966** F. HOYLE *Galaxies, Nuclei & Quasars* i. 1 The best known galaxy is naturally the one we live in...In total mass and in size it seems to be much like the nearest of the large external galaxies, the nebula in Andromeda. **1968** D. S. EVANS *Observation Mod. Astron.* vii. 228 The spiral character of the Galaxy cannot be so marked as that in some external galaxies.

galbe (gælb). Also **galb.** [Fr. (cf. GARB *sb.*² 1 b).] A contour, outline or profile (see also quots.).

1899 *Cent. Dict.*, *Galbe*,..in *art*, the general outline or form of any rounded object, as a head or vase; especially, in architecture, the curved form of a column, a Doric capital, or other similar feature. **1909** J. R. WARE *Passing Eng.* 138/2 *Galbe* (Thieves'), profile of a violent character, and even applied to any eccentricity of shape above the knees. **1926** R. FRY *Transformations* 71 Look..at the neck of the jar..the clear salience of the lip which is exactly enough..to give a perfect close to the curves of the galb that leads up to it. **1936** H. READ *Meaning of Art* (ed. 2) II. 101 On pottery..it is found in the *galbe* or outline which the pot makes.

GALBRAITHIAN

Galbraithian (gælbrēi·þiăn), *a.* and *sb.* [f. the name of J. K. *Galbraith* (born 1908), U.S. economist and diplomat + -IAN.] **A.** *adj.* Of, pertaining to, or characteristic of J. K. Galbraith and his writings. **B.** *sb.* One who favours or is influenced by the writings of J. K. Galbraith.

1960 *Guardian* 15 June 2/4 The familiar Galbraithian indictment of 'private opulence and public squalor'. **1961** *Times Lit. Suppl.* 27 Jan. 51/2 The contemporary (Galbraithian or 'affluent') conception of the welfare state. **1966** *Economist* 23 Apr. 374/2 He is a Galbraithian in his belief that wealth..is outrunning welfare, social justice and the means of a good life.

gale, *sb.*[3] Add: **1. a.** In restricted use, applied to a wind having a velocity within certain limits (see quots.).

1801 [see STORM *sb.* 1 b]. **1899** *Westm. Gaz.* 24 Jan. 4/3 A gale is not a gale until it has reached Force 7 on the Beaufort scale, though many people lightly class all heavy winds as gales. **1923** N. SHAW *Forecasting Weather* (ed. 2) 456 As a result of the investigation of 1905 we now classify winds with velocity above 75 miles per hour as hurricane winds, those with velocity between 64 and 75 miles per hour as storm winds, and those between 39 and 63 as gales. **1963** *Meteorol. Gloss.* (*Met. Office*) 109 *Gale*, a wind of a speed between 34 and 40 knots (force 8 on the Beaufort scale of wind force, where it was originally described as 'fresh gale'), at a free exposure 10 metres (33 feet) above ground. *Ibid.*, Statistics of gales refer to the attainment of mean speeds of 34 knots or over.

3. *gale force, -warning.*

1902 *Daily Chron.* 4 Mar. 5/2 Southerly winds were blowing in all parts of the British Isles yesterday, reaching gale force on the west coast of Ireland. **1967** C. B. CHRISTESEN in *Coast to Coast 1965–6* 32 The impact of gale-force winds was such that at first we thought the ship had struck a rock. **1918** *Meteorol. Gloss.* (*Met. Office*) 128 Gale-warning. Notice of threatening atmospherical disturbances on or near the coasts of the British Islands are issued by telegraph from the Meteorological Office to a number of ports and fishery-stations. **1940** C. DAY LEWIS tr. *Georgics of Virgil* I. 26 Another gale-warning often is given by shooting stars. **1969** C. R. BURGESS *Suppl. Meteorol. for Seamen* 2 These forecasts include Gale warnings, General weather synopsis and expected changes within the next 24 hours.

Galego, var. *GALLEGO.

galenobismutite (gălī:nobi·zmiutəit). *Min.* Also **-muthite.** [ad. Sw. *galenobismutit* (H. Sjögren 1878, in *Geol. För. Stockh. Förh.* IV. 110), f. GALEN(A + -O + BISMUTH + -ITE[1].] A sulphide of lead and bismuth usu. occurring as light grey columnar or fibrous masses.

1880 *Jrnl. Chem. Soc.* XXXVIII. 14 Bismuth occurs in Wermland..as the new mineral galenobismuthite, $PbS.Bi_2S_3$. **1962** *Acta Crystallogr.* XV. 691 (*heading*) A redetermination of the crystal structure of galenobismutite, $PbBi_2S_4$.

‖ **galère** (galę̄r). [Fr., lit. galley; also *fig.* (cf. GALLEY *sb.* 1 b).] A coterie, circle; a (usu. undesirable) set of people; an unpleasant place or situation.

1756 CHESTERFIELD *Let.* 26 Nov. (1777) II. 435, I most frequently..congratulate..myself for having got out of that *galère*. **1842** MILL *Let.* 11 Mar. in *Wks.* (1963) XIII. 507 Everybody now condemns the folly of involving ourselves in that *galère*. **1895** S. WEYMAN *Mem. Minister of France* iii. 121 'Why, your excellency,' he cried, in a tone of boundless surprise, 'what are you doing in this *galère*?' **1961** D. G. JAMES *M. Arnold* i. 15 If it comes to 'ruling ideas' what was Newman doing in a *galère* which contained Goethe and Sainte-Beuve? **1963** *Times* 26 Sept. 17/5 The head of the Security Service gave this ruling on February 1:—'Until further notice no approach should be made to anyone in the Ward *galère*.'

‖ **Galgenhumor, galgenhumor** (ga·lgĕnhūmǫr). [G., f. *galgen* gallows + *humor* humour.] = *gallows-humour* (*GALLOWS *sb.* 8).

1948 MENCKEN *Amer. Lang.* Suppl. II. 727 Not a few of these terms show *Galgenhumor*, e.g., *meat-wagon* for an ambulance. **1959** *Observer* 26 Apr. 25/4 Berliners, with their *Galgenhumor* and *Schnoddrigkeit*. **1963** AUDEN *Dyer's Hand* 523 The Gravedigger's song in Hamlet is..an expression of the *galgenhumor* which suits his particular mystery.

Galician (găli·þiăn, găli·ʃiăn), *a.*[1] and *sb.*[1] [f. *Galicia* + -AN.] **A.** *adj.* Of or pertaining to Galicia, a province in north-west Spain, or its inhabitants. **B.** *sb.* An inhabitant of Galicia; also, the language of Galicia.

1749 U. RHYS *Account Spain & Portugal* 23 The *Galicians* make good Soldiers; and are pleased with the Profession. **1809** tr. *A. de Laborde's View Spain* II. 428 This road is frequented by..a great many Galician workmen. *Ibid.* 456 The Galician who serves either his master, or the public, or in the army, is contented to appear a slave. **1823** T. ROSS tr. *Bouterwek's Hist. Span. Lit.* I. 13 The vulgar idiom spoken by the Galician water-carriers in Madrid. **1828** *Encycl. Metrop.* (1845) XIX. 438/1 The Galicians, or Galegos as the Spaniards call them, are a grave and sober people. **1927** *Chambers's Jrnl.* 29 Oct. 759/1 Her head is the Galician woman's carry-all. **1960** W. D. ELCOCK *Romance Lang.* v. 428 Galician, in the meantime, degenerated to a patois status; it is still widely spoken and practised as a literary cult by local enthusiasts. *Ibid.*, Probably the earliest specimen of Galician-Portuguese to have survived is an act of partition dated 1230. **1966** *Tablet* 29 Jan. 144/2 Other Spanish languages, such as Basque and Galician. **1969** *Daily Tel.* (Colour Suppl.) 10 Jan. 32/1 The Asturians and Galicians dance in strong colours, kicking with controlled but fierce abandon.

Galician (găli·ʃiăn), *a.*[2] and *sb.*[2] [f. *Galicia* + -AN.] **A.** *adj.* Of or pertaining to Galicia, a region of Poland and West Russia, or its inhabitants. **B.** *sb.* An inhabitant of Galicia. Also used loosely in Canada to denote a Central European immigrant.

1835 *Penny Cycl.* III. 131/2 The Galician soil is no where so productive as in the districts of Zloczoff and Stanislawoff. **1903** *Eye Opener* (High River, Alta.) 25 July 3/3 The lost tribes of Israel, in the dishevelled shapes of Galicians and Doukhobors. **1909** *Chambers's Jrnl.* Mar. 154/1 It is customary in the West to classify the various contingents which come from east-central Europe as Galicians. **1918** C. G. MONTEFIORE *Lib. Judaism & Hellenism* v. 246 It has been too rashly assumed that the Russian, Polish and Galician 'masses' must be for ever wedded to Orthodox Judaism. **1927** *Times* (weekly ed.) 16 June 664/4 He took charge of a Red-Cross unit on the Galician front. **1960** S. BECKER tr. *Schwarz-Bart's Last of Just* (1961) III. 93 The young Galician followed him with a constrained smile.

Galignani (gælinʸā·ni). *Obs.* exc. *Hist.* [f. the name of its founder, Giovanni Antonio *Galignani* (1752–1821).] Colloquial appellation of the English-language newspaper *Galignani's Messenger*, founded in 1814 and published daily in Paris until 1884 (and continued until 1904 as the *Daily Messenger*).

1822 M. EDGEWORTH *Let.* 5 Jan. (1971) 309 In Florence..there is..a sort of coffee-house, where many English newspapers are openly read daily—..Galignani and others. **1837** H. C. ROBINSON *Diary* 2 Aug. (1967) 180, I found opportunity today..to read some *Galignanis*. They contained an account of the borough elections. **1848** THACKERAY *Van. Fair* lxvi. 695 Jos was in his great chair dozing over Galignani. **1868** GEO. ELIOT *Let.* 24 July (1956) IV. 460 We have been chiefly in regions innocent even of Galignani. **1880** *Ibid.* 12 May VII. 277 We read in Galignani yesterday that Sir W. Harcourt is turned out of Oxford!

Galilean, *a.*[1] and *sb.*[1] Add: Also **Galilæan.** **A.** *adj.* (Later examples.) Also, Christian.

1671 MILTON *P.R.* III. 233 Thy life hath yet been private, most part spent At home, scarce view'd the *Gallilean* Towns. **1821** SHELLEY *Hellas* (1822) l. 550 Every Islamite who made his dogs Fat with the flesh of Galilean slaves. **1927** W. B. YEATS *Resurrection* in *Adelphi* IV. 729 He walked that room and issued thence In Galilæan turbulence. **1958** A. TOYNBEE *East to West* 210 The southern face of the Galilaean highlands is blurred.

B. *sb.* (Later examples.)

1811 SHELLEY *Let.* 24 Apr. (1964) I. 66 The Galilean is not a favorite of mine. **1866** SWINBURNE *Poems & Ballads* 78 Wilt thou yet take all, Galilean? *Ibid.* 79 Thou hast conquered, O pale Galilean; the world has grown grey from thy breath. **1957** *Encycl. Brit.* IX. 976/1 Deborah, Jonah, Elisha and perhaps Hosea were Galileans.

Galilean, *a.*[2] (and *sb.*[2]). Also **Galileian.** Add to def.: Discovered by Galileo, as *Galilean satellite*, any of the largest four moons of the planet Jupiter; also, pertaining to or arising out of the work of Galileo.

1757 W. EMERSON *Doctrine of Fluxions* (ed. 2) p. viii, Let a heavy Body descend through a perpendicular Height of $16\frac{1}{12}$ Feet in one Second of Time, according to the Gallilean Hypothesis of Gravity. **1903** J. J. FAHIE *Galileo* v. 94 From about 1637, Francesco Fontana of Naples also began to turn out good glasses [*sc.* lenses] of the Galilean pattern. **1911** *Encycl. Brit.* XV. 564/1 In apparent brightness each of the four Galilean satellites may be roughly classed as of the sixth magnitude. **1944** *Chambers's Techn. Dict.* Suppl. 961/1 *Galilean binoculars*, binoculars in which the objectives are of the usual doublet telescope objective type and the eyepieces are negative lenses. **1954** A. R. HALL *Scientific Revolution* vi. 168 By its attention to actual phenomena Galilean science was made real and experiential. **1970** *Nature* 25 Apr. 316/1 It should be possible by 1974 to obtain radar echoes from the Galilean satellites of Jupiter.

b. *Physics.* Pertaining to the properties of space and time assumed in classical physics; *Galilean transformation*, a transformation of co-ordinates in which the classical laws of motion remain unchanged.

1910 *Sci. Abstr.* XIII. 261 Mathematically these principles are represented by an invariance or covariance with respect to certain transformations. These transformations are (1) for Newtonian mechanics the 'Galilean transformation'..; (2) for Lorentz's electrodynamics the 'Lorentz transformation'. **1918** A. S. EDDINGTON *Rep. Relativity Theory Gravitation* ii. 18 The laws of mechanics and electrodynamics are usually enunciated with respect to 'unaccelerated rectangular axes', or, as they are often called, 'Galilean axes'. *Ibid.* v. 48 The path of a particle in Galilean co-ordinates (*i.e.*, under no forces) is a straight line. **1920** R. W. LAWSON tr. *Einstein's Relativity* iv. 11 A system of co-ordinates of which the state of motion is such that the law of inertia holds relative to it is called a 'Galileian system of co-ordinates'. **1922** E. P. ADAMS tr. *Einstein's Meaning Relativity* iii. 65 There are finite regions, where, with respect to a suitably chosen space of reference, material particles move freely without acceleration, and in which the laws of the special theory of relativity..hold with remarkable accuracy. Such regions we shall call 'Galilean regions'. **1960** R. M. PALTER *Whitehead's Philos. Sci.* viii. 166 Newton's laws of motion are covariant with respect to Galilean transformations (i.e., transformations which carry one inertial system into another).

B. *sb.* One who holds or supports Galileo's views.

1925 A. N. WHITEHEAD *Sci. & Mod. World* (1926) viii. 186 The difference is very analogous to that between the Galileans and the Aristotelians: Aristotle said 'rest' where Galileo added 'or uniform motion in a straight line'.

galipine (gæ·lipīn). *Chem.* Also **-in,** † **-eine.** [ad. It. *galipeina* (Körner & Böhringer 1883, in *Gazzetta Chim. Ital.* XIII. 365), f. mod.L. *Galipea*, a generic name of the tree producing Angustura bark.] An alkaloid, $C_{20}H_{21}NO_3$, obtained from Angustura bark. Similarly **galipoidine** (gælipoi·dīn) [ad. G. *galipoidin* (Tröger & Runne 1911, in *Arch. Pharm.* CCXLIX. 184)], an alkaloid, $C_{19}H_{15}NO_4$; **galipoline** (gæli·polīn) [ad. G. *galipolin* (Späth & Papaioanou 1929, in *Monatsh. Chem.* LII. 129)], an alkaloid, $C_{19}H_{19}NO_3$.

1884 *Jrnl. Chem. Soc.* XLVI. 341 In the mother-liquors from which the cusparine was originally precipitated..another alkaloid is found, to which the authors have given the name of *galipeine*, $C_{20}H_{21}NO_3$. **1911** *Ibid.* C. i. 482 Angostura bark contains..a new alkaloid, now named galipoidine. **1912** *Ibid.* CII. i. 896 Galipine is colourless when pure, and yields colourless salts; the yellow colour usually ascribed to the salts is due to the presence of impurities. **1913** T. A. HENRY *Plant Alkaloids* 404 Galipine, $C_{20}H_{21}O_3N$, crystallises from alcohol or ether in prisms. *Ibid.*, Galipoidine, $C_{19}H_{15}O_4N$, m.pt. 235°, is sparingly soluble in most organic solvents. **1929** *Chem. Abstr.* Oct. 4703 Extn. of angostura bark with EtOH for 8 days and working up the alkali-sol. portion gives galipolin..which gives galipin on methylation. **1939** T. A. HENRY *Plant Alkaloids* (ed. 3) 399 Galipoline..contains two methoxyl groups, and on methylation by diazomethane yields galipine. **1968** I. L. FINAR *Org. Chem.* (ed. 4) II. xxv. 843 Galipine..has been isolated from angostura bark, e.g., cusparine, galipine, galipoline.

galjoen (galyu·n). *S. Afr.* [Afrikaans, = Du. *galjoen* galleon.] The sea-fish *Coracinus capensis* (family Coracinidæ); also any of several related fish of the family Coracinidæ.

[**1853** L. PAPPE *Edible Fishes Cape of Good Hope* 23 *Dipterodon capensis*...(*Galjoenvisch*, Galleon-fish).] This fish, more plentiful..of the Colony, is highly esteemed. **1902** *Trans. S. Afr. Philos. Soc.* XI. 221 The Galjoen also can readily be supposed to have derived its name from its resemblance in shape to the high built three-decker..called..by the Dutch 'Galjoen' or 'Galleon'. **1949** L. G. GREEN *In Land of Afternoon* (1950) xviii. 226 In fine weather many a galjoen is landed on those rocks. **1971** *Cape Times* 20 Aug. 22/3 One angler..took nine galjoen.

gall, *sb.*[1] Add: **1. b.** *to dip one's pen in gall*, to write with virulence and rancour. (Cf. quot. 1614 in sense 3 a.)

Probably derived from instances like those in quots. 1601, 1605, where there appears to be a pun on GALL *sb.*[3] (the oak-gall, which is used in the manufacture of ink).

[**1601** SHAKS. *Twel. N.* III. ii. 52 Let there be gall enough in thy inke, though thou write with a goose-pen, no matter: about it. **1605** —— ... *Ier.* What, is your pen foule? *Hor.* No, Father, cleaner then Lorenzos soule; Thats dipt in inck made of the enuious gall: Else had my pen no cause to write at all.] **1750** J. DOVE *Creed founded on Truth* 15, I shall omit the Consideration of the particular Reasons of these Differences, because I would not dip my pen in Gall. **1828** *Imperial Mag.* Apr. 360/1 In the same spirit of bitter enmity..the Doctor has dipped his pen in gall, to blast the memory of this great man. **1934** *Rev. Reviews* V. 376/1 In the *Contemporary Review* for April an anonymous writer dips his pen in gall in order to depict the German Emperor. **1946** W. S. MAUGHAM *Then & Now* 228 His pen had been dipped in gall and as he wrote he chuckled with malice.

4. For *U.S. slang* read: orig. *U.S. slang.* (Add earlier and later examples.)

1882 *Denver Republican* 29 Jan. 4/1 There is only one word which thoroughly expresses the quality of Dr. Anderson's communication. That word is the strong expression, 'gall'. **1936** 'I. HAY' *Housemaster* xvi. 220 And what do you think they had the gall to do then? **1948** WODEHOUSE *Spring Fever* xv. 153 He was a young man abundantly equipped with what he called *sang froid* and people who did not like him usually alluded to as gall.

8. gall-sickness, (*b*) the name [tr. Du. *galziekte*] given in South Africa to diseases of the liver in cattle, sheep, and goats.

1875 J. NOBLE *Descrip. Handbk. Cape Colony* 259 The 'gal zeickte' or gall sickness is also a common disease. **1896** R. WALLACE *Farm. Industr. Cape Col.* 288 Deaths in Cape Colony from gall-sickness. **1953** *Official Yr. Bk. Union S. Afr. 1950* XXVI. I. xix. 914 *Anaplasma marginale*, the cause of gallsickness in cattle.

gall, *sb.*[3] Add: **2. gall-berry, gallberry** *U.S.*, a holly (*Ilex glabra* or *I. coriacea*); also

attrib.; **gall-bush** *U.S.*, the gall-berry bush; **gall-wasp**, a gall-producing, hymenopterous insect of the family Cynipidæ.

1709 J. LAWSON *New Voyage Carolina* 90 Gall-Berry-Tree, bearing a black Berry, with which the Women dye their Cloathes and Yarn black. **1834** J. J. AUDUBON *Ornith. Biogr.* II. 191 The holly,..the gall-berry, and the poke, are those which they first attack. **1901** C. MOHR *Plant Life Alabama* 816 With gallberry bushes for the undergrowth. **1938** M. K. RAWLINGS *Yearling* i. 7 Open galberry flats spread without obstructions. *Ibid.* iv. 40 The gallberry bushes. **1962** KUHN & GODFREY *Trees of N. Florida* 205 The large or sweet gallberry (*Ilex coriacea*) is more often seen as a shrub than a tree. It is not uncommonly associated with the shrubby, bitter gallberry, *Ilex glabra* (L.) Gray. **1728** in *N. Carolina Col. Rec.* (1886) II. 802 They measured..16 chains and 30 links to a Gall Bush. **1835** J. MARTIN *Gaz. Virginia* 41 An ever-green shrub, called the gall-bush,..bears a berry which dies a black color like the gall of an oak—and hence its name. **1879** *Encycl. Brit.* X. 44/1 Among the Hymenoptera are the gall-wasps. **1925** *Glasgow Herald* 28 June 4 The rose gall-wasp (Rhodites). **1965** L. H. NEWMAN *Man & Insects* i. 86 Many of the gall wasps have alternating generations.

gall-, see *GALLO-[2].

Galla (gæ·lă). Pl. **Galla, Gallas.** A member of a group of Hamitic peoples inhabiting equatorial Africa, allied to the Ethiopians in language and origin; also, their language. Also *attrib.* or as *adj.*

1875 *Encycl. Brit.* I. 263/2 The next great branch of the Ethiopic race comprehends the Galla, who occupy an immense tract in Eastern Africa... Our knowledge of them is chiefly confined to those Gallas who border on Abyssinia. **1878** K. JOHNSTON *Africa* xiv. 256 The Somali and Galla people are so closely related as they are hostilely disposed towards each other. **1892** A. S. WHITE *Devel. Africa* (ed. 2) 101 As agriculturists and herdsmen, and in the industrial arts, the Galla bordering on Abyssinia and the Somali of the coast are the most advanced. **1895** H. H. JOHNSTON *Hist.* ... The typical Gallas of Kaffa and surrounding regions are perhaps the finest people in all Africa. **1941** ... The Galla love of roaming. **1920** *Blackw. Mag.* May 678/2 The poor old Galla vendor clucking with rage. **1941** W. L. GRANT *Lang.* ... One of the most important [Cushitic] dialects are..Somali, Galla, etc. **1970** *Encycl. Brit.* ... The Galla..used to make a pilgrimage to the Abba Muda, 'father of anointing'..the personification of Galla law and tradition.

gallacetophenone (gælæ·sĭto,fīnō·n, gæ,lăsĭto,fī·nōn). *Chem.* [ad. G. *gallacetophenon* (Nencki & Sieber 1881, in *Jrnl. prakt. Chem.* XXIII. 537), f. *GALL- + *ACETOPHENONE.] A yellow crystalline compound, $C_6H_2(OH)_3\cdot COCH_3$, formerly used in ointments for skin-diseases and as a mordant dye-stuff.

1881 *Jrnl. Chem. Soc.* XL. 813 Gallacetophenone is obtained on heating pyrogallol with glacial acetic acid and zinc chloride at 145°. **1893** E. KNECHT et al. *Man. Dyeing* II. vii. 620 Alizarin yellow C is gallacetophenone and is produced by the action of glacial acetic acid and chloride of zinc on pyrogallol. **1904** H. W. STELWAGON *Treat. Dis. Skin* (ed. 3) III. 238 Gallacetophenone is likewise employed in this disease [*sc.* psoriasis], in the form of an ointment. **1952** K. VENKATARAMAN *Chem. Synthetic Dyes* II. xxvi. 798 Alizarin Yellow C..one of the earliest synthetic mordant dyes, was gallacetophenone.

gallamine (gæ·lămīn, -in). *Chem.* Also **-ine.** [f. *GALL- + AMINE.] **1.** *gallamine blue*, a basic mordant dye of the oxazine series.

1889 *Jrnl. Soc. Dyers & Colourists* 25 Nov. 171/1 Gallamine blue produces on wool, mordanted with chrome and tartar, a blue which somewhat resembles in shade that obtained with gallocyanine, with which gallamine blue is indeed closely related. **1905** *Jrnl. Chem. Soc.* LXXXVIII. i. 831 The yield of gallamine-blue, $C_{15}H_{13}O_4N_3Cl$, obtained by treating gallamide in aqueous solution with nitrosodimethylaniline, is largely increased by using an excess of the latter. **1935** *Jrnl. Comparative Neurol.* LXI. 120 The gallamin blue method is thus of a great importance as a neurological stain. **1952** K. VENKATARAMAN *Chem. Synthetic Dyes* II. xxv. 782 The Gallocyanine derivative prepared from gallamide is marketed as the soluble bisulfite compound, Gallamine Blue.

2. In full *gallamine triethiodide*. A whitish powder, $C_6H_3[O\cdot(CH_2)_2\cdot N(C_2H_5)_3]_3I_3$, having a slightly bitter taste and used as a neuromuscular blocking drug.

1951 *Proc. R. Soc. Med.* XLIV. 375 Flaxedil, or Gallamine triethiodide, as it is now named by the British Pharmacopoeial Commission, is a British production of a synthetic curarizing agent,..which was first synthesized in France. **1961** *Lancet* 26 Aug. 486/2 The man was given a small dose of atropine by mouth and 1.5 ml. gallamine triethiodide ('Flaxedil') intravenously. **1963** A. H. DOUTHWAITE *Hale-White's Materia Medica* (ed. 32) 210 Gallamine is used chiefly to obtain adequate muscular relaxation for surgical operations carried out under general anæsthesia.

gallanilide (gælæ·nilǝid). *Chem.* [f. *GALL- + ANILIDE.] An anilide of gallic acid, $C_6H_2(OH)_3\cdot CONH\cdot C_6H_5$, used in the manufacture of some dyes.

1883 *Jrnl. Chem. Soc.* XLIV. 335 Gallanilide is deposited as a crystalline mass when digallic acid is dissolved in aniline. **1916** E. KNECHT et al. *Man. Dyeing* (ed. 3) II. vii. 620 Gallanilide blue is produced by the action of gallanilide on nitrosodimethylaniline and subsequent treatment with aniline. **1924** *Chem. Abstr.* XVIII. 2009 The ferric salt..of gallanilide..is a dark violet powder. **1952** K. VENKATARAMAN *Chem. Synthetic Dyes* II. xxv. 783 The Gallocyanine derivative prepared from gallamide is marketed as the soluble bisulfite compound, Gallamine Blue... The gallanilide analog is Gallanil Violet.

gallant, *a.* Add: **1. c.** *Mus.* = *GALANT *a.*

1925 *Musical Q.* XI. 396 (*title*) The 'gallant' style of music. *Ibid.*, Gallant, the recent classic homophonous style of an amatory and dramatic music whose evolution during the seventeenth and eighteenth centuries runs nearly parallel with that of rococo architecture.

gall-darned, var. *GOLDARNED *ppl. a.*

Gallegan (gæl'ī·găn), *a.* and *sb.* [f. next + -AN.] = *GALICIAN *a.*[1] and *sb.*[1]

1845 R. FORD *Hand-bk. Trav. Spain* I. I. xv. 59 A Gallegan is the best groom. **1922** *Glasgow Herald* 14 Aug. 4 The 'arrival'..of the Gallegans. **1927** *Chambers's Jrnl.* 759/1 The pipers..break off their metal to chant an impromptu couplet in the Gallegan dialect.

Gallego, gallego (gælē·go). Also **Galego, Gallejo.** [Sp.] = *GALICIAN *sb.*[1] (In quots. 1600, a Galician ship.)

1600 HAKLUYT *Voy.* III. 631 We had before lost sight of a small Galego on the coast of Spaine. *Ibid.* 633 An old Galego which I caused to be fashioned like a galley. **1656** BLOUNT *Glossogr.*, *Gallego* (Sp.), a man of Galitia. **1811** J. CARR *Descr. Trav. Spain* 50 The Gallejos (pronounced Galleyos) or Galicians are a remarkably fine athletic race of men. **1828** [see *GALICIAN *a.*[1] and *sb.*[1]]. **1846** THACKERAY *Cornhill to Cairo* i. 6 A little..boat, rowed by three ragged gallegos. **1925** *Chambers's Jrnl.* 10 Oct. 709/1 The Gallegos are for the most part miserably poor. **1927** P. KEMP *Mine were of Trouble* v. 85 General Franco is a gallego (Galician), with all the obstinacy and subtlety of that race.

gallein. (Earlier and later examples.)

1871 *Jrnl. Chem. Soc.* IX. 700 Gallein is decomposed by heat like hæmatein, carbonisation taking place. **1953** H. W. GRIMMER tr. H. Gilman *Org. Chem.* III. iv. 317 This dye [*sc.* Coerulein] is obtained from Gallein by treatment with concentrated sulfuric acid at 200°.

gallery, *sb.* Add: † **2. c.** *Aeronaut.* An enclosed platform attached to a balloon to carry passengers. *Obs.*

1784 *London Chron.* LVI. 1/3 Yesterday..an air balloon..was let off from Versailles..There was a large gallery fixed to it, in which were M. Charles, M. James, M. Montgolfier, and an Officer of the army. **1905** G. BACON *Balloons* ii. 28 Charles had arranged a gallery to carry the ... a car shaped like a boat was suspended from the net.

3. e. The part of a Friends' meeting-house occupied by the ministers or elders; *gallery Friend*, a Quaker minister or elder.

1802 M. PATTERSON *Recorder* i. 121 The galleries of London. **1913** *Jrnl. Friends' Hist. Soc.* Jan. 2 Jane Wigham..the second wife of John Wigham, Tertius, was also a gallery Friend. **1921** R. M. JONES *Later Per. Quakerism* I. 58 The phrases so characteristic of Molinos, Guyon, Fénelon..were heard everywhere in Quaker 'galleries'.

4. c. At écarté, the spectators who are betting on either player and are allowed to offer suggestions.

1890 'BERKELEY' *Écarté & Euchre* 28 French Écarté. If a player chooses to join in a game of Écarté, it is generally arranged in the following manner. Two of the number sit down to play a game in the usual way,.. and the remainder, called 'The Gallery', are allowed to take part in the game to the extent of betting on the player of their choice, and advising him, if necessary. **1897** R. F. FOSTER *Complete Hoyle* 255 Any person in the gallery is allowed to draw attention to errors in the score, and may advise the player he is backing, or even play out the game for him.

d. A group of spectators at a golf match or other game or sport. Also *transf.*

1890 H. G. HUTCHINSON *Hints on Golf* (ed. 6) 71 If you rise to such heights of golfing powers as to attract a gallery. **1894** *Strand Mag.* VIII. 661/2 One can do very well without a gallery when one is trying a new experiment on 'ski'. **1899** *Captain* II. 65/2 The gallery of white-robed spectators. **1906** *Westm. Gaz.* 21 Sept. 4/2 It is virtually impossible for a player attended by a big 'gallery' to lose his ball. **1925** J. BUCHAN *John Macnab* iii. 59 Sir Archie was aware that his style of jumping was not graceful and he was discomposed by this sudden gallery. **1970** *New Yorker* 20 Oct. 183/1 The gallery had virtually won him the fourth set with a huge surge of support.

5. b. A shooting-gallery (see SHOOTING *vbl. sb.* 8 b).

1848 MRS. GASKELL *M. Barton* II. ii. 33 Some workman with whom her son had made some arrangement about shooting at the gallery. **1897** *Sears, Roebuck Catal.* 779/3 Gallery Targets.

6. (See quot.)

1950 *Manch. Guardian Weekly* 12 Oct. 15/1 In general in the United States a 'gallery' is a place that shows paintings and sells them, a 'museum' is one that simply shows them.

12. a. (sense 3 c) *gallery boy, girl, god, goddess, queue, ticket.* **b. gallery camera** (see quot. 1968); **gallery forest** = **fringing forest*; **gallery grave,** a tunnel-shaped megalithic tomb; **gallery-play,** playing to the gallery (see sense 4 a in Dict.); showy play designed to gain applause; also *fig.*

1887 *Courier-Journal* (Louisville, Ky.) 3 Feb. 1/4 The gallery boys stood up and again and again clapped their hands. **1911** BEERBOHM *Zuleika D.* ii. 15 All the gallery-boys..were scornful of the sweethearts wedged between them. **1964** E. CHAMBERS *Camera & Process Work* iv. 26 (*caption*) Elevation and plan of a gallery camera. **1968** *Gloss. Terms Offset Lithogr. Printing* (*B.S.I.*) 13 *Gallery camera*, a camera in which the photographic material is transported from the darkroom to the camera in a dark slide. **1920** A. H. UNWIN *W. Afr. Forests* vii. 130 *Hexalobus morepetalus*..Grows to large tree in Gallery Forests—small in Savannah. **1937** *Discovery* Apr. 100/2 The savannah country, into which the gallery forests, with their accompanying forest-frequenting birds, shoot out long tongues—remnants of a once continuous forested area. **1927** E. WALLACE *Feathered Serpent* xiii. 169 You'll probably be pulled up by gallery girls who want autographs. **1851** Gallery god [see GOD *sb.* 4]. **1812** Gallery Goddess [see GOD *sb.* 4]. **1937** *Proc. Prehist. Soc.* III. 86 The brilliance of the Breton passage grave culture has long obscured the existence of gallery graves in Brittany. **1963** G. DANIEL in Foster & Alcock *Culture & Environment* ii. 14 Suggesting that it was in the area from south Finistère to the Vendée were found the tombs most likely to be the prototypes of the Gallery Graves and Transepted Gallery Graves and Rectangular Chambers of south Wales. **1871** F. GALE *Echoes Cricket Fields* v. 23 Frequent useless appeals to the umpire, a practice which Nyren condemned as being what we now call 'gallery play'. **1899** *Westm. Gaz.* 31 Jan. 7/2 A weak opposition, evidently disposed to indulge in 'gallery' play. *a* **1914** J. E. RAPHAEL *Mod. Rugby Football* (1918) vi. 107 'Gallery play' should only be resorted to as a last desperate resource. **1916** 'BOYD CABLE' *Action Front* 121 You've..done good work for your first show; don't spoil it with rank gallery play. **1923** KIPLING *Land & Sea T.* 21 In the Great War there was very little suspicion, or chance, of gallery-play for the V.C. **1934** A. P. HERBERT *Holy Deadlock* 143 Nobody in the gallery queue would have recognised their Mary Moon. **1789** *Loiterer* 18 Apr. 5 A woman who pawned her prayer book for a gallery ticket. **1833** R. DYER *9 Yrs. of Actor's Life* 137 His actors being..the companions of any who are likely to take a gallery ticket.

galley, *sb.* Add: **1. b.** (Earlier and later examples.) Cf. *GALÈRE.

1859 DICKENS *T. Two Cities* v. 21 What the devil do *you* do in that galley there! **1909** G. B. SHAW in *Nation* 16 Jan. 601/1 Asking incredulously what the devil he is doing in *that* galley. **1948** R. SCOTT-JAMES in F. M. Ford *Man could stand Up* Pref. 5 Was he not that dextrous, fanciful, almost dilettante writer..? What was he doing in this galley?

5. b. A galley-proof; = SLIP *sb.*[2] 10 d.

1890 in WEBSTER. **1934** T. R. COWARD in G. Gross *Publishers on Publishing* (1961) 149 When the corrections are made, the galleys go back to the printer and are made into page proofs. **1951** S. JENNETT *Making of Books* I. vi. 88 The page proofs come to the reader, and must be checked against the corrected galleys, to see that all the corrections have been carried out. **1971** *Times Lit. Suppl.* 20 Aug. 999/1, I *have* had galleys from Penguin Books, but more usually the finished product, fresh misprints and all.

8. galley-yarn = *galley-packet.*

1874 HOTTEN *Slang Dict.* 172 *Gally-yarn*, a sailor's term for a hoaxing story. **1884** HENLEY & STEVENSON *Admiral Guinea* III. iv, You the one overtaken and denounced; and you spin me a galley yarn like that! **1905** *Daily Chron.* 11 July 3/4 Mr. Baring-Gould has gathered up all the old galley-yarns.

galley-west (gæ·liwest), *adv. colloq.* (orig. and chiefly *U.S.*). Also **galleywest, gally-west.** [An alteration of Eng. dial. *colly-west*, *-weston*: see E.D.D. and *COLLYWESTON 2.] Phr. *to knock galley-west*, to knock sideways or askew; to bring to confusion; to knock out or dispose of completely.

1875 'MARK TWAIN' *Lett.* (1917) I. 250 Your verdict has knocked what little [critical penetration] I did have gally-west! **1884** —— *Huck. Finn* xxxvii. 382 Then she grabbed up the basket and slammed it across the house and knocked the cat galleywest. **1891** M. E. RYAN *Pagan of Alleghanies* xiv. 184 Here you come with your theories of hue and knock my serenity galley-west. **1892** KIPLING *Lett. of Travel* (1920) 66 It knocks pleasant domestic arrangements galleywest. **1936** J. DOS PASSOS *Big Money* 77 A wave had knocked him galleywest. **1938** S. CHASE *Tyranny of Words* viii. 87 Einstein shattered a whole cosmology of concepts. Let us not be knocked galley-west again, says Bridgman.

gallica (gæ·likă). [f. L. *Gallicus* GALLIC *a.*[1]] A species of rose, *Rosa gallica*; a plant or flower of a variety of this species.

1848 W. PAUL *Rose Garden* I. iv. 51 Summer Roses: Provence, 8. Moss, 10. Damask, 6. Alba, 6. Gallica, 75. **1869** S. R. HOLE *Bk. about Roses* xi. 160 These variegated Roses..are affiliated in the catalogues to the family of Gallicas. *Ibid.* 161 Let us suppose that they could distinguish at any distance the Gallica from the Provence Rose. **1912** E. WILLMOTT *Genus Rosa* II. 329 The *gallica* Rose having been common in gardens throughout Europe from very early times. **1962** I. MURDOCH *Unofficial Rose* xvi. 154 Despatched to the rose slopes to cut a bunch of gallicas.

‖ **Gallice, gallice** (gæ·lisi), *adv.* [L., 'in Gaulish'.] In French.

1889 in *Cent. Dict.* **1897** G. DU MAURIER *Martian* I. 44 He had received from Barty a sore eye too—*gallicè*, 'un pochon'.

gallied, *pa. pple.* and *ppl. a.* (see GALLY *v.*). (Earlier and later examples.)

1787 GROSE *Prov. Gloss.*, *Gallied*, frightened. Exm. **1851** H. MELVILLE *Moby Dick* III. i. 9 That strange perplexity of inert irresolution, which, when the fishermen perceive it in the whale, they say he is *gallied*. **1888** 'R. BOLDREWOOD' *Robbery under Arms* II. iii. 48 They seemed awfully gallied about being stuck up and robbed of it [*sc.* gold]. **1903** KIPLING *Five Nations* 14 Our gallied whales are blind! **1946** K. TENNANT *Lost Haven* (1947) xix. 318 Look here, Alec. I was wondering if you could come and have a look at a little job I've got that I'm a bit gallied about.

Gallio (gæ·lio). The name of a Roman proconsul of Achaia, whose refusal to take action is recorded in Acts xviii. 17 ('And Gallio cared for none of those things'), applied *gen.* to one who is indifferent. Hence **Ga·llio-like** *a.* and *adv.*; **Gallionic** (gæliǫ·nik) *a.*, characteristic of a Gallio, indifferent.

1850 [see CARER]. **1855** MACAULAY *Hist. Eng.* III. xiii. 293 Unhappily, Scotland was ruled, not by pious Josiahs, but by careless Gallios. **1865** L. STEPHEN *Sk. from Cambridge* xii. 137 Our prevailing tone is what I should venture to describe as one of quiet good sense, and what fanatics would consider to be only fit for careless Gallios. *Ibid.* 140 A Gallio is generally a pleasant companion. **1884** S. J. REID *Sydney Smith* x. 240 Gallio-like, he cared for none of these things. **1909** *Westm. Gaz.* 8 Mar. 4/3 Gallio-like bargain. **1920** *Q. Rev.* Oct. 403 Japan seemed Gallionic all the time. **1926** *Glasgow Herald* 20 Mar. 8 The Gallionic attitude of the soldiers along the various lines of the route.

Gallipoli (gæli·pǫli). The name of a town in Apulia, Italy, used to designate the olive-oil exported from there. *Gallipoli soap*, soap made from this oil.

1822 C. LILLIE *Brit. Perfumer* xl. 223 Gallipoly Soap. This, like the French soap, above described, comes in large cakes, which generally smell so strong of bad oil, as hardly to be borne. **1839** URE *Dict. Arts* 541 *Gallipoli oil*, is a coarse olive oil, containing more or less mucilage. **1868** F. H. JOYNSON *Metals* 120 Common rosin, melted with a little gallipoli oil and spirits of turpentine. **1884** W. S. B. MCLAREN *Spinning* (ed. 2) 45 If..oil is used..it should be the best quality of olive. Gallipoli is the name for the very best.

gallisin (gæ·lisin). *Chem.* Also **-ine.** [a. G. *gallisin* (Schmitt & Cobenzl 1884, in *Ber. d. Deut. Chem. Ges.* XVII. 1003), f. *gallisieren*: see GALLIZE *v.* and -IN[1].] An amorphous unfermentable substance obtained from commercial glucose, now regarded as a mixture.

1884 *Jrnl. Chem. Soc.* XLVI. 981 The authors [*sc.* C. Schmitt and A. Cobenzl]..have succeeded in separating from fermented glucose (prepared from potato-starch) an intensely hygroscopic substance which they have named gallisin. **1890** *Chem. News* 14 Mar. 132/2 In the analysis of eleven samples of commercial starch-sugar the quantities of gallisine found were from 22.49 to 6.82 per cent. **1938** *Thorpe's Dict. Appl. Chem.* (ed. 4) II. 299/1 The product variously known as isomaltose, gallisin, revertose is now regarded as a mixture of maltose with other substances. **1965** K. MAYER tr. *Staněk's Oligosaccharides* vi. 108 The substances so obtained and designated by various names (such as gallisin, revertose, δ-dextrose, and dextrinose) were undoubtedly not chemical individuals.

gallo-[2] (gæ·lo), also before a vowel **gall-**, combining form of GALLIC *a.*[2] in the names of compounds derived from gallic acid or related substances, as *gallanilide, gallein, gallocyanine, gallo-nitrate, gallo-tannate*, etc.

gallocyanine (gæ:losəi·ănīn, -in). *Chem.* Also **-in.** [a. F. *gallocyanine* (H. Kœchlin 1882, in *Bull. Soc. Chim.* XXXVIII. 97, 162), f. *GALLO-[2] + CYANINE.] A green crystalline substance, $C_{15}H_{12}N_2O_5$, obtained from gallic acid and used as a bluish-violet mordant dye for wool and cotton and in calico-printing; also, any of a group of structurally similar dyes of the oxazine series that are derived from gallic acid.

1883 H. KŒCHLIN in *Chem. News* 13 Apr. 170/1 By allowing an alcoholic solution of nitroso-dimethylaniline hydrochlorate to react in heat upon the tannins, as upon gallic acid, there are obtained violet colouring-matters, verging more or less upon a blue according to the nature of the tannin employed. These colouring-matters..the author names gallo-cyanines. **1893** E. KNECHT et al. *Man. Dyeing* II. vii. 624 Gallocyanin yields a bright blue-violet shade, which resists light, soap, and acids very well, but is not fast to soda. **1956** HICKINBOTTOM & GARWOOD in E. H. Rodd *Chem. Carbon Compounds* IIIB. xii. 781 The so-called gallocyanine dyes, mordant dyes of the oxazine series, largely used in calico printing. **1958** J. R. BAKER *Princ. Biol. Microtechnique* xii. 242 A basic dye may be used..to obtain an approximation to the amount of DNA in a nucleus or of RNA in a nucleolus or in the cytoplasm. A mordant-dye, chrome alum/gallocyanine..is particularly recommended for this purpose. **1960** HOWARD & RAMAGE in E. H. Rodd *Chem. Carbon Compounds* IVc. xvii. 1546 The gallocyanines..are important for application in calico printing, with chromium salts and tannin mordant.

gallonage (gæ·lənēdʒ). [f. GALLON + -AGE; after TONNAGE *sb.*] An amount in gallons; a quantity of liquid produced or sold.

1909 in *Cent. Dict.* Suppl. **1930** *Economist* 11 Oct. 662/1 Stocks of both crude and gasoline in the hands of the producers, the refiners and the distributors are so huge that the struggle for 'gallonage' is growing hotter daily. **1936** A. G. STREET *Gentleman of Party* iv. 80 The figures on the churn tickets..showed the daily gallonage. **1967** *Times Rev. Industry* Mar. 59/3 Shell obviously does not reveal the extra gallonage achieved. **1970** *Country Life* 1 Oct. 855/1 In terms of gallonage on which duty has been paid..the total of 13 million gallons up to June is only about 7 per cent down.

gallop, *sb.* Add: **1. c.** A track designed or suited for the galloping or exercising of horses.

1848 TROLLOPE *Kellys & O'Kelleys* II. ii. 45 They've proper gallops there, which we haven't. **1923** in F. Siltzer *Newmarket* App. 269 Horses not completing the full length of a gallop..must at once walk off the gallop to the nearest Walking Ground. **1927** *Times* 6 July 10/5 Scotland Lodge Estate, 1,009 acres,..including the residence, stud farm, and gallops. **1935** *Proc. Prehist. Soc.* I. 16 Along the main ridge of the spur..is a broad gallop. **1971** *Country Life* 18 Feb. 381/4 Had anyone been out on the Newmarket gallops earlier..they would have seen him at work tirelessly.

galloper. Add: **1. b.** A wooden horse on a merry-go-round; a roundabout with such horses on it.

1945 *Archit. Rev.* XCVII. 50/2 These gallopers continued without competitors for more than thirty years. *Ibid.*, The real thrill of the galloper had to await the successful results of the patient toil of the steam engineers before it could be realized. **1968** D. BRAITHWAITE *Fairground Archit.* iii. 55 Initially rating second in popularity to the 'Gallopers', the 'Scenic Railway' attained a higher point in terms of fairground architecture. *Ibid.* vii. 122 Both 'Galloper' and switchback had a fixed centre.

Ga:llo-Ro·man, *a.* and *sb.* [GALLO-.] **A.** *adj.* Belonging to Gaul when it formed part of the Roman Empire. **B.** *sb.* An inhabitant of Gaul under Roman rule; also, the language of these people.

1841 T. ARNOLD *Lect. Mod. Hist.* 33 Throughout the south of France the population is predominantly..of Gallo-Roman origin. **1861** W. H. ANDERTON tr. *Dupanloup's Sermon Cath. Ireland* 8 Saint Patrick, that young Gallo-Roman whom we had sent to her [*sc.* Ireland]. **1879** *Encycl. Brit.* IX. 528/1 These were also the days of what is called 'the Gallo-Roman empire'. *Ibid.* 528/2 The Visigoths and Gallo-Romans defeated the terrible hordes of Attila. **1933** [see *CATALAN]. **1934** M. K. POPE *From Lat. to Mod. Fr.* I. II. 11 Thus the Romance languages in their early stage, Gallo-Roman, Hispano-Roman, Italo-Roman, etc., slowly took shape. The Gallo-Roman styled himself *romanus*. **1962** H. R. LOYN *Anglo-Saxon England* i. 16 Names of townships that trace unbroken descent from the name of their eponymous Gallo-Roman estate-owners.

Galloway. Add: **2.** Also *attrib.*

1849 W. & J. DEANS in J. Deans *Pioneers of Canterbury* (1937) app. 284 We would strongly advise that a few bulls of the Galloway breed should be sent out. **1862** [see *ABERDEEN 1]. **1960** *Times* 25 Jan. 19/1 When choosing our Galloway calves in 1948 we had this point in view.

gallows, *sb.* Add: **6.** Freq. in the form *gallus* in the U.S.

1837, 1896 [in Dict.]. **1932** W. FAULKNER *Light in August* xviii. 404 Even if I dont wear no tin star on my galluses. **1932** E. CALDWELL *Tobacco Road* xiv. 166 He stepped into his overalls, put one arm through a gallus. **1942** O. NASH *Good Intentions* 145 To supply each of my pairs of pants with its own set of galluses. **1957** R. A. HEINLEIN *Door into Summer* (1967) iii. 51 Oh, I'm a gallus-snapper when I get started; you should see me wear women's hats at parties.

8. gallows-humour, grim, ironical humour; 'sick' humour; cf. *GALGENHUMOR.

1901 W. D. HOWELLS in *North Amer. Rev.* Nov. 710 The honors are not quite so easy in the matter of gallows-humor. **1935** *Archit. Rev.* LXXVII. 31/1 These may not be altogether unintentional touches of what the Germans call 'gallows-humour'. **1958** *Times* 17 July 4/3 The explosive vitality, full-blooded sentiment, and gallows humour that pervade the play.

Gallup (gæ·lǝp). The name of the American statistician George Horace *Gallup* (born 1901), used *attrib.* (chiefly in *Gallup poll*) to denote an assessment of public opinion made by ascertaining the opinions of a representative cross-section of the people; a public-opinion poll. Hence occas. **Gallup-poll** *v. intr.*, to take a Gallup poll; also *v. trans.*, to receive (a certain number of votes) in a Gallup poll.

1940 *Illustr. London News* CXCVII. 565 According to the latest U.S. Gallup poll as we go to press, President Roosevelt is assured of a total of 410 electoral votes. **1941** *Electronic Engin.* XIV. 383 It would be interesting to have a Gallup Survey..of reasons why manufacturers have not adopted electronic aids. **1948** A. TOYNBEE *Civilization on Trial* ii. 16 If it were made the subject of a world-wide Gallup Poll, there would be no unanimity in the answer. **1951** *Listener* 23 Aug. 297/1 A kind of gallup-poll mentality. **1951** KOESTLER *Age of Longing* I. x. 181 You don't have to go around gallup-polling to see that the people of France have only one wish. **1963** C. IRVING et al. *Scandal '63* xx. 222 Neville Chamberlain..Gallup-polled 32 per cent in 1940. **1966** *Observer* 30 Oct. 27/1 The age of Gallup Polls and TV documentaries.

gallus: delete 'obs.' (Cf. *GALLOWS *sb.* 6.)

Galois (galwa). The name of É. *Galois* (1811–32), French mathematician, used *attrib.* to designate various concepts in algebra arising out of his work, as *Galois field*, a field with a finite number of elements; *Galois function, group, resolvent, theory.* So **Galoi·sian** *a.*

1893 E. H. MOORE in *Bull. N.Y. Math. Soc.* III. 75 The Galois-field of order $s = q^n$, $GF[q^n]$, consists of the system of q^n incongruous classes (modulis q, $F_n(\xi)$) of rational integral functions of ξ with integral coefficients. *Ibid.*, Galois theory. **1898** *Bull. Amer. Math. Soc.* IV. 335 This is precisely the Galoisian scheme of solution. **1899** *Ibid.* V. 237 Galois group. **1900** J. PIERPONT in *Ann. Math.* 2nd Ser. I. 124 Any one of the equations $G_\kappa(t) = 0$ is called a Galoisian Resolvent of (A) for R. *Ibid.* 125 We shall call for convenience of notation take $G_1 = 0$. We shall call this *the* Galoisian Resolvent. **1901** BURNSIDE & PANTON *Theory of Equations* (ed. 4) II. xx. 246 This function, ψ_1, is called the Galois Function. *Ibid.* 257 The equation of the Nth degree whose roots are these N values, viz., $\Psi(z) \equiv (z-\psi_1)(z-\psi_2)..(z-\psi_N) = 0$, is called the Galois resolvent. **1937** R. D. CARMICHAEL *Introd. Groups of Finite Order* ix. 255 A particular Galois field, so far as its abstract properties are concerned, is completely determined by its order. **1963** G. D. MOSTOW et al. *Fund. Struct. Algebra* x. 297 Galois theory shows that the roots of the polynomial can be obtained from the coefficients a_i by ordinary field operations $(+, -, \times, \div)$ and root extractions if and only if the Galois group is solvable. **1965** PATTERSON & RUTHERFORD *Elem. Abstract Algebra* iii. 97 It can be shown that for each prime number p and each positive integer n, there is essentially only one field consisting of p^n elements; this..is called a *Galois field*.

galon, var. GALLOON.

1877 in A. Adburgham *Shops & Shopping* (1964) xvi. 178 A light blue cashmere, trimmed with diagonal bands of oriental galon and cream lace. **1890** [see GALLOON]. **1923** *Daily Mail* 1 Mar. 15 A practical day gown..girdled with wool galon. **1939–40** *Army & Navy Stores Catal.* 1077/3 Galon for trimming.

galoot. 2. For *U.S.* read orig. *U.S.* and add further examples.

1898 *Bulletin* (Sydney) 6 Aug. 32/4 'Who's that galoot 'as lost his 'orse, Sis?' Bendigo was asking. **1944** A. CLARKE *Coll. Plays* (1963) 223 Big galoots in hobnailed boots! **1946** I. L. IDRIESS *In Crocodile Land* xxx. 209 We'll lock this silly galoot up. **1966** *New Statesman* 8 Apr. 497/3 The galoots who blew up Nelson's Pillar.

galpon (gælpǭᵘn). [Local name.] In South America, a building given to the use of labourers on a farm.

1894 *Proc. Zool. Soc.* 305 Behind some wood piled up at the side of the big galpon. **1918** W. H. HUDSON *Far Away & Long Ago* xiv. 193 My father went to the *galpon*, the big barnlike building used for storing wood, hides, and horsehair. **1961** G. DURRELL *Whispering Land* (1965) iii. 67 A small white *estancia*..and to left of it a large Dutch barn or *galpón*.

Galsworthian (gǭlzwɜ̄·ɪðiăn), *a.* [f. the name of John *Galsworthy* (1867–1933), English novelist and playwright.] Of, pertaining to, or characteristic of John Galsworthy or his work.

1912 *Eng. Rev.* June 494 A grim, almost Galsworthian, unpleasantness. **1928** *Daily Express* 24 May 10/6 A tight-lipped Galsworthian pessimist like the Duke of Wellington or Lord Kitchener. **1936** *Times Lit. Suppl.* 3 Oct. 789/1 For her fellow-mortals..she has an almost Galsworthian compassion.

Galton (gǭ·ltǝn). The name of Sir Francis *Galton* (1822–1911), English scientist and anthropologist, used in: *Galton's law*, the formula proposed by him to account for ancestral heredity, by which the parents each contribute a quarter of the characters to their offspring, the grandparents each contribute an eighth of the characters, and so on; also, the tendency of offspring of outstanding parentage to regress to, or below, the average for the species; so *Galton's curve*, etc.; *Galton('s) whistle* (see quots.). Hence **Galto·nian** *a.*

1889 A. R. WALLACE *Darwinism* xiv. 414 'Panmixia', or free intercrossing, will co-operate with Galton's law of 'regression towards mediocrity', and the result will be that..the organ in question will rapidly decrease till it reaches a mean value considerably below the mean of the progeny that has usually been produced each year. **1890** W. JAMES *Princ. Psychol.* I. x. 298 One may, I

think, without fear of being upset by any future Galtonian circulars, believe that all men must single out from the rest of what they call themselves some central principle. **1903** *Daily Chron.* 31 July 5/2 'Galton's law', ..now one of the cardinal principles of biology. **1904** J. A. & M. R. THOMSON tr. *Weismann's Evol. Theory* II. 206 Galton's curve of frequency of variations. **1904** GOODCHILD & TWENEY *Technol. & Sci. Dict.* 245/1 *Galton's whistle*, a sounding pipe, resembling in principle a very small closed organ pipe, producing a note of very high frequency. Used in experiments on the limit of audition. **1907** V. L. KELLOGG *Darwinism To-day* 71 Variation..is subject to Galton's law of regression. **1908** J. A. THOMSON *Heredity* ix. 335 Mendelian formulæ apply to the progeny of known crosses or hybrids, while Galtonian formulæ apply to intra-racial heredity. **1931** *Discovery* Apr. 106/2 It might be a Galton whistle, which is too high-pitched to be audible to man. **1953** A. HUXLEY *Let.* 21 June (1969) 676 Do Galtonian visualizers react in a different way from non-visualizers.

galumph, *v.* Add to def.: Now usu., to gallop heavily; to bound or move clumsily or noisily. Hence **galu·mphing** *vbl. sb.* (Add further examples.)

1881 *Punch* 27 Aug. 94/2 The [H.M.S.] *Hercules* got up steam and went on her way westward galumphing. **1901** *Westm. Gaz.* 15 Aug. 2/2 A postman in uniform galumphed about on a farm-horse. **1903** *Daily Chron.* 31 Oct. 8/1 There would be such a galumphing up their stairs that peace and security would forsake them. **1930** C. MACKENZIE *April Fools* xii. 271 Viola..had slept through the stifled cries of her parents beneath the bedclothes when Beyle [*sc.* a bull-dog] was galumphing round their room. **1965** S. RAVEN *Friends in Low Places* vi. 129 In the hall was a galumphing lass with a lot of jerseys and a po face.

galumptious: see *GOLUPTIOUS *a.*

galvanic, *a.* Add: **a.** *galvanic battery* (earlier example); *galvanic skin response* (or *reflex*), the rapid variation in the electrical conductivity of the skin as a measure of the effect of an emotional stimulus on autonomic activity.

1801 *Phil. Mag.* X. 93 He [*sc.* Cruickshank] employs a galvanic battery of his own invention. **1927** C. W. DARROW in *Jrnl. Exper. Psychol.* X. 197 Employing the term *galvanic skin reflex* (galvanische Hautreflex) because it is less misleading than the term psychogalvanic reflex. **1962** *Listener* 19 Apr. 691/2 With special instruments you can measure electrical changes in the skin when a person is emotionally upset or aroused, changes which we call the galvanic skin response, and which are made use of in the so-called 'lie detector'.

galvanotaxis (gæ:lvănǝtæ·ksis). *Biol.* [mod. L., f. GALVANO- + Gr. τάξις arrangement.] The disposition exhibited by certain organisms of movement in relation to the direction of an electric current or field. Hence **ga:lvanota·ctic** *a.*

1899 F. S. LEE tr. *Verworn's Gen. Physiol.* v. 455 Galvanotaxis. It is characteristic of the galvanic current that it always calls out phenomena of polar excitation. **1905** Galvanotactic [see *ANODIC *a.* 3]. **1961** *New Scientist* 3 Aug. 278 The voltage gradient across the width of the fish is sufficient to induce galvanotaxis and turn it in the direction of the positive electrode. **1965** *Ibid.* 29 Apr. 303/1 The orientation of the whole organism during exposure to direct current—the phenomenon known as 'galvanotactic' or 'electrotactic' response.

galvanotropism (gæ:lvănotrǫ·piz'm). *Biol.* Add to def.: movement produced in plants or animals by the stimulus of an electric current. Hence **ga:lvanotro·pic** *a.*

1895 *Science Progress* IV. 100 The essential fact upon which the galvanotropic effect turns is that current down the animal has a quieting action, current up the animal an exciting action. *Ibid.* 101 Fertilised frog's eggs..exhibit remarkable galvanotropic effects. *Ibid.*, There is in many cases a remarkable orientation in relation to the current—'galvanotropism'. **1918** J. LOEB *Forced Movements* iv. 41 Galvanotropic reactions are found throughout the whole animal kingdom. *Ibid.* 42 Phenomena of galvanotropism can be observed also in infusorians. **1936** *Discovery* June 198/1 The planarian *Leptoplana variabilis* is negatively phototropic and kathodically galvanotropic. **1959** T. H. SAVORY *Instinctive Living* iii. 28 Some of them are synonyms—e.g. electrotropism and galvanotropism.

Galvayne (gæ·lvē̆ⁱn). The name of Sidney *Galvayne*, a writer on horses, used *attrib.* to designate a method or devices for breaking in a horse; so as *v. trans.*, to break in (a horse) by Galvayne's method; so **Galva·yning** *vbl. sb.* Also *Galvayne's mark* (see quot. 1908).

1905 S. GALVAYNE *20th Cent. Bk. Horse* 23 'Galvayning' is really a scientific utilization of the animal's strength against itself... The 'Galvayne' position is attained by tying the animal's head round with the halter-shank, or the 'Galvayne' strap, to its tail. *Ibid.* 29 Once you have got the horse properly 'Galvayned', watch it carefully for a little while before proceeding further. *Ibid.* 32 If the animal be 'Galvayned' on the 'near' side, the trainer must work on the 'off'. **1908** *Animal Managem.* 43 At ten years 'Galvayne's mark', a depression on the outer side of the upper corner incisor, appears. **1911** *Encycl. Brit.* XIII. 725/1 Galvayning is accomplished by bending the horse's neck round at an angle of thirty-five to forty degrees and tieing the halter to the tail. **1942** 'M. INNES' *Daffodil Affair* I. v. 31 Well I always suspected rareying with Daffodil—though, mind you, it may have been galvayning all the same.

galziekte (gălzi·ktǝ ‖ χa·lzīktǝ). Also **gall-ziekte.** [Du.] = *gall-sickness* (*GALL *sb.*[1] 8). Also *attrib.*

1871 J. MACKENZIE *10 Yrs. N. Orange River* xiii. 261 Examination of the horse after death throws little light upon the subject. The Dutch call it 'gall-ziekte'; the English, inflammation. **1875** [see *gall-sickness* s.v. *GALL *sb.*[1] 8]. **1912** *Agric. Jrnl. Union S. Afr.* Aug. 177 A strong smelling weed, often sent to us under the name of *galziekte bosch*, and considered by many farmers to be a cure for galziekte.

gamahuche (gæ·măhūʃ), *v. slang.* Also **gamaruche.** [ad. Fr. *gamahucher.*] *trans.* To practise fellatio or cunnilingus (with); also *intr.* Also as *sb.* Hence **gamahucher.**

1865 E. SELLON *New Epicurean* (1875) 13 'Quick, quick, Blanche!' cried Cerise, 'come and gamahuche the gentleman.' **1867** —— *Ups & Downs of Life* 91 in H. S. Ashbee *Index Librorum Prohibitorum* (1877) 389 Augusta would strip naked, place herself in any attitude, let me gamahuche her, would gamahuche in her turn. **1868** *Index Expurgatorius of Martial* 15 Lesbia was a gamahucher. *Ibid.* 33 They agreed wonderfully well in both being gamahuchers. *Ibid.* 47 So I think no objection he'll raise, To a gamahuche even from you. **1879–80** *Pearl* (1970) 271 You may frig and gamahuche and try every plan, But fair fucking's the pride of an Englishman. **1888** P. PERRET *Tableaux Vivants* ix. 73 My dear, do you know, this is my only ambition! To gamahuche a lady of fashion! *c* **1888–94** *My Secret Life* X. 5 She gave me a gamahuche for a few minutes. **1893** FARMER & HENLEY *Slang* III. 107 Gamaruche. **1968** PARTRIDGE *Dict. Underworld* (ed. 3) 854 *French, go down, nosh*, are prostitutes'..verbs..for 'to gamaruche' a man.

Gambia (gæ·mbiă). The name of a state of western Africa, used *attrib.* to designate a form of trypanosomiasis prevalent there.

1905 *Proc. R. Soc.* LXXIV. 389 These experiments go to show that the two diseases—Gambia Fever and Sleeping Sickness—..are distinct. **1908** *Practitioner* Feb. 224 The trypanosomata of Gambia fever (trypanosomiasis).

Gambian (gæ·mbiăn), *a.* and *sb.* [f. *Gambia* (see prec.) + -AN.] **A.** *adj.* Of or pertaining to Gambia. **B.** *sb.* A native or inhabitant of Gambia.

1906 F. B. ARCHER *Gambia Colony* i. 23 It certainly merits more than a passing mention in any record..of Gambian history. **1913** DORLAND *Med. Dict.* (ed. 7), Gambian horse sickness. **1943** K. W. BLACKBURNE *Devel. & Welfare Gambia* i. 6 It is a question whether, in the case of Gambians, this [consent to international transfer of territory] would be forthcoming. **1957** C. MACINNES *City of Spades* I. v. 30 They talked in their barbarian Gambian language. *Ibid.* II. xi. 172, I..waited ..in case those two Gambians might start some fight again. **1966** *Transition* (Uganda) XXVI. 20/1 In the work of his contemporary, the Gambian poet Lenrie Peters, the emphasis is slightly different.

gamble, *v.* **2. a.** Delete *rare* and add further examples.

1922 JOYCE *Ulysses* 763 When do you ever see women rolling around drunk like they do or gambling every penny they have and losing it on horses. **1930** *Publishers' Weekly* 8 Feb. 706 He would not have gambled his money upon them by adding them to his list.

b. Also (*a*) with *off*, and (*b*) without *away*.

1848 THACKERAY *Van. Fair* lxiv, When she got her money she gambled; when she had gambled it she was put to shifts to live. **1874** 'H. CHURTON' *Toinette* xii, Loyd probably traded her off, perhaps gambled her off, in some drunken spree.

gambling, *vbl. sb.* **b.** *gambling-club, -joint, -machine, -school, -table.*

1966 *Listener* 10 Mar. 361/1 A picture of gambling clubs in Manchester. **1901** S. E. WHITE *Westerners* xiii. 94 Bunco men can clean him out in a gambling joint. **1925** B. TRAVERS *Mischief* v, Who does Captain Dumfoil expect to find running a gambling joint? The Archbishop of Canterbury? **1935** AUDEN & ISHERWOOD *Dog beneath Skin* II. iii. 101 Gambling-machines and switchbacks. **1935** A. J. CRONIN *Stars look Down* I. ii. 17 Some colliers..that made up the gambling school in ordinary times—squatted upon their hunkers against the wall. **1852** M. EASTMAN *Aunt Phillis's Cabin* 210 He fancied he would find happiness..at the gambling table. **1857** C. KINGSLEY *Two Y. Ago* I. i. 26 He's..croupier at a gambling-table. **1891** H. CAMPBELL *Darkness & Daylight* (1895) XXXIII. 639 Whenever they have money, no matter how obtained, they generally drop the most of it at the gambling-tables.

gambo. For *Monmouthshire dial.* read *Eng. dialects of Wales and W. Midlands dial.* Add to def.: a simple kind of cart or trolley. Also *attrib.* and *Comb.* (Add further examples.)

1878 *N. & Q.* 5th Ser. X. 105/1 Some Radnorshire Words...*Gambo*, a cart of the very simplest construction. **1887** F. T. HAVERGAL *Herefordshire Words* 16/1 *Gambo*, *Gamber*, or *Gambrel*, a cart with sides only; no front or back. **1945** DYLAN THOMAS *Let.* 30 July (1966) 278 There's his brown hen cluck in the gambo-swished mud. **1950** —— in *Botteghe Oscure* VI. 336 Butter fat goosegirls, bounced in a gambo bed. **1951** —— *Let.* 28 May 360 The word *gambo* means a farm-cart. **1963** W. H. BOORE *Valley & Shadow* v. 21 Ten coaches, apart from the hearse, and a gambo from the farm to carry the overflowering of tribute. **1967** *Listener* 16 Feb. 229/2, I unearthed a two-wheeled hay-wagon, locally known as a gambo.

gambrel. Add: **4.** *gambrel-joint*; *gambrel-roof, -roofed a.* (earlier U.S. examples).

1876 J. BURROUGHS *Winter-Sunshine* iv. 104 A fox struggling with a trap which held him by the hind leg above the gambrel-joint! [**1737** in *Old-Time New Eng.* (1926) July 21 One Tenement two Stories upright, with a Gambering Roof.] **1765** *Massachusetts Gaz.* 19 Dec. (Th.), A large building with two upright Stories and a Gambrel Roof. **1779** in *Mass. Hist. Soc.* 2nd Ser. II. 466 The [Indian] Queens Pallace was a gambril roofed house. **1824** *Microscope* 21 Feb. (Th.), In a gambrel-roof'd house,..She dwelt with a heart void of care.

game, *sb.* Add: **3. b.** Delete † *Obs.* and add later examples; *spec.* signifying sexual intercourse.

1938 G. GREENE *Brighton Rock* VI. ii. 259 What mattered was the game. The two main characters made their stately progress towards the bed sheets. **1964** *Mademoiselle* Sept. 164 There's a new dance, the Ska—like The Game set to music.

4. c. (See also PLAY *v.* 16 b, and cf. 1854 *fig.* quot. under sense 4 a in Dict.)

d. *pl.* In Scotland, a number of contests in athletics, piping, and dancing held esp. in various Highland centres; a meeting for the purpose of holding such contests; freq. in *Highland games.*

1822 *Inverness Jrnl.* 4 Oct., One of the spectators at the Highland games, yesterday, was plundered of a gold watch. **1831** *Aberdeen Jrnl.* 31 Aug., The Games were contested with a spirit which would have pleased our friend the Ettrick Shepherd himself. **1905** *Westm. Gaz.* 15 Oct. 7/2 The competitors have..piped their way up to the games ground. **1948** A. RAEBURN *This is Scotland* 54 There are balls and Highland Games, where tossing the caber, throwing the hammer, Highland dancing and other professional spectacles are arranged.

e. *pl.* Athletics or sports as organized in a school, college, etc. Freq. *attrib.* (see sense *16 c).

1895 KIPLING *Day's Work* (1898) 343 He blossomed in full glory as head of the school, ex-officio captain of the games. **1899** —— *Stalky & Co.* 66 King and Macrea, fellow house-masters, had borne it in upon him that by games, and games alone, was salvation wrought. **1934** LD. BERNERS *First Childhood* xi. 105 At my preparatory school, it [*sc.* Latin grammar] ceased to be a game (we did also games themselves). **1960** J. BETJEMAN *Summoned by Bells* vii. 67 Greatest dread of all, the dread of games!

f. *The Game*: a form of charades.

1940 S. LEWIS *Bethel Merriday* xiv. 119 That monstrous form of charades called 'The Game'. **1960** S. FOOTE *Emergency Exit* vii. 51 They played 'The Game' after dinner. **1962** E. ROOSEVELT *Autobiogr.* xxx. 230 We played The Game—a form of charades. Queen Elizabeth..chose the words that the rest of us were called upon to act out.

5. a. Colloq. phr. *two can play at that game*: others can act in a similar way (usu. said as a threat of retaliation for unfair dealing).

1826 SCOTT *Woodst.* II. iii. 79 I'll show you that two can play at the game of wrestling. **1842** *Knickerbocker* Apr. 355 I'll show him that *two* can play at that game. **1911** BEERBOHM *Zuleika D.* xv. 230 The Socratic manner is not a game at which two can play. **1955** G. GREENE *Quiet American* I. iv. 60 They must have been caught in the cross-fire, trying to get back, and I suppose every one of us along the bank was thinking, 'Two can play at that game.'

e. *the game*: thieving, housebreaking; freq. in phr. *on the game.* Thieves' slang.

1739 W. UDALL in *Ordinary of Newgate's Account* II. 14/1, I, and others went out again upon the Old Game, till I was taken up for Clacking the Doctor. **1811** *Lex. Balatron.* sig. G2, *Game*, any mode of robbing. **1839** H. BRANDON *Poverty, Mendicity & Crime* Gloss., *On the game*, thieving. **1905** *Daily Chron.* 14 Apr. 6/6 Pacey pressed me to go out 'on the game'.

f. *the game*: prostitution; usu. in phr. *on the game.* (Cf. quot. 1606 for sense 3 b in Dict.) *slang.*

1898 *Daily News* 21 July 8/6 The prosecutrix pestered her to 'go on the game', i.e. the streets. **1911** F. HARRIS *Women of Shakespeare* 194 The phrase of the prostitute to-day on the streets of London is: 'I'm on the game'. **1958** [see *BRASS *sb.* 2 g]. **1963** T. & P. MORRIS *Pentonville* viii. 191 This is almost certainly an underestimate of the number who are either married to prostitutes or have women 'on the game' as an additional source of income. **1969** T. PARKER *Twisting Lane* 201 Betty's on the game, isn't she? Has she got you at it too?

6. b. *game, set, and match*, a complete and decisive victory. (From the use in Lawn Tennis.)

1968 *Listener* 15 Aug. 210/2 This seemed to be game, set and match to the garage. **1969** *Sunday Express* 6 Apr. 4 It was game, set, and match to the passenger.

d. (Later examples in Chess.)

1917 J. DU MONT tr. *Lasker's Chess Strategy* iv. 24 The mobility of the White Queen..begins to have a threatening effect on Black's game. **1952** E. LASKER *Chess Secrets* 271 Had I kept my Bishop back at QB1, and castled instead, this advance would have given me a satisfactory game.

f. A person's performance in a particular

game; the normal standard of one's play; *to be on* (or *off*) *one's game*, to be playing well (or badly); to be in (or out of) form. Cf. quot. 1885 under sense 6 c in Dict.
1851 J. PYCROFT *Cricket Field* ii. 30 What can surpass a hardly-contested match when you have been manfully playing an up-hill game. 1873 J. BLACKWOOD *Let.* 7 June in *Geo. Eliot's Lett.* (1955) V. 421, I..backed Tommy when most folk thought he was off his game! 1887 F. GALE *Game of Cricket* 59 We had played..together, and knew each other's game. 1891 H. G. HUTCHINSON *Hints on Golf* 16 What am I doing wrong, Tom? I'm quite off my game. *Ibid.* 52 If you are one of the many golfers who overrate their game. 1904 S. A. MUSSABINI *Mannock's Billiards* i. 23 It is wonderful how strength of nerve improves with the strength of one's game. 1920 *Westm. Gaz.* 16 Oct. 2/2 Their disregard of the recognised rules was accentuated by the fact that neither man was on his game.

8. h. *Chess.* (i) A method of play, esp. a series of initial moves; cf. *close game* (*CLOSE *a.* and *adv.* A. 2 c), **open game*.
1750, etc. [see *CLOSE *a.* and *adv.* A. 2 c]. 1845 *Souvenir Bristol Chess Club* 59 This move produces the opening known by the name of 'The two Knights' game'. 1894 J. MASON *Princ. Chess* iv. 188 The Scotch Game may or may not be a Gambit. 1952 E. LASKER *Chess Secrets* 414 To steer into his favorite game, the Vienna variation of the King's Gambit. 1968 BOTT & MORRISON *More Chess for Children* 158 (*heading*) Four knights' game.
(ii) A sequence of moves forming a recognized stage in the play, esp. in *end-game* (END *sb.* 25), *middle game*.
1884, etc. [see END *sb.* 25 in Dict. and Suppl.]. 1949 H. GOLOMBEK *World Chess Championship 1948* 219 He was quite outplayed in the early middle game. 1952 E. LASKER *Chess Secrets* 273 The clearest understanding of opening and endgame strategy was not sufficient.. unless it was matched by masterly middle game tactics.

16. a. *game-drive, -land, -park, -path, -pie* (earlier example), *reserve, -shot.*
1895 W. C. SCULLY *Kafir Stories* 98 These game-drives were fraught with considerable danger. 1931 J. MOCKFORD *Khama* vii. 60 The rich tribute in hides, ivory, and ostrich-feathers which the Masarwa bushmen and other vassals occupying the game-land..brought annually to Shoshong. 1958 *Listener* 16 Jan. 102/1 But unspoilt gamelands, such as those in the Congo, afford unrivalled opportunities for a study of natural balances. 1897 J. L. ALLEN *Choir Invisible* xiii, The great neutral game-park of the Northern and the Southern Indians. 1963 A. SMITH *Throw out Two Hands* x. 111 Even the Serengeti, the main game park, had closed its gates. 1856 C. J. ANDERSSON *Lake Ngami* xxxi. 404, I struck into a large 'game path'. 1935 L. G. GREEN *Great Afr. Mysteries* xiv. 181, I saw the first crocodile, waiting at a gamepath to grip its victim by the nose. 1827 H. G. LEWIS *Let.* 13 Oct. in J. Constable *Corr.* (1966) IV. 70 This accompanys a French game pye, made by my Cook. 1921 *Daily Colonist* (Victoria, B.C.) 27 Mar. 27/2 These [Queen Charlotte] islands would make the greatest game reserve in the world. 1962 J. G. BENNETT *Witness* xix. 231 My visit to the game reserves in Kenya and other parts of Africa. 1901 KIPLING *Kim* xiii. 336 Hurree was no game-shot,—the snick of a trigger made him change colour. 1927 A. CONAN DOYLE *Case Bk. S. Holmes* 82 The famous game-shot, sportsman, and man-about-town was a big, swarthy fellow.
b. (senses 3 and 4) *game-playing* sb. and adj., *game-shy* adj.; (sense 11) *game-dealer, -finding* sb. and adj., *game-proof* adj.
1876 J. S. INGRAM *Centenn. Exposition* 691 One of the oldest game-dealers in the West. 1922 R. LEIGHTON *Compl. Bk. Dog* xi. 159 A better dog for game-finding than could have been found in any other part of the world. 1960 *Times* 24 Sept. 9/4 A game-finding dog. 1908 *Westm. Gaz.* 18 Apr. 3/1 Centralisation will mean larger public-houses and greater attractions owing to the crowds, game-playing,..and late hours. 1909 *Ibid.* 8 Mar. 4/3 They were at 'famous' game-playing schools. 1947 AUDEN *Age of Anxiety* (1948) v. 111 Barns and shrubberies For game-playing gangs. 1970 *Guardian* 29 Apr. 9/6 The long-term objective of Women's Liberation ..is an end of 'game-playing' between men and women. 1908 *Daily Chron.* 21 Aug. 7/3 A nine-foot game-proof fence of woven buffalo wire. 1963 *Times* 10 May 10/7 A game-proof ditch between the forest boundaries and the settlement schemes. 1932 AUDEN *Orators* III. 92 To work-shy and game-shy kind.
c. *attrib.* and *Comb.* uses of the plural, esp. in sense *4 e, as *games-mad* adj., *games-mania, -master, -mistress, -room, -worship.*
1960 C. DAY LEWIS *Buried Day* vi. 108 A games-mad school. *Ibid.* 128 The Sherborne games-mania. 1902 *Daily Chron.* 16 July 7/1 Special games-masters and mistresses. 1965 G. MCINNES *Road to Gundagai* x. 159 Our combined sergeant, wet-nurse, games-master, schoolteacher and doctor. 1907 *Westm. Gaz.* 18 July 10/2 The home supplies also a temperance bar..and games rooms. 1959 *House & Garden* June 44 The house is on three floors. At garden level is a large gamesroom. 1960 C. DAY LEWIS *Buried Day* iv. 68 Though we were encouraged to enjoy our games, there was no games-worship.

17. game ball *pred. a.* (*Anglo-Irish slang*), excellent, fine; **game-board** = BOARD *sb.* 2 c; **game-book** (see quot. 1807); **game-chick,** a game-chicken; **game chips,** very thin fried chipped potatoes served with game (sense 11 b); **game-forcing** *a.*, in *Bridge*, denoting a bid intended to instruct one's partner to continue the bidding until a contract is reached that will win a game; **game-goer,** in *Bridge*, a bid that undertakes to win a game; so **game-going** *a.*; **game-piece,** (*a*) = PIECE *sb.* 12; (*b*) a painting of game (sense 11); **game-rhyme,** a rhyme used in a game; **game(s) theory,** the mathematical theory of situations of conflict (such as occur in games of skill, in economics, and in war), esp. as used to determine the most advantageous strategy for any of the participants; hence **game-theoretic, -theoretical** *adjs.*; **game-theorist**; **game-warden** (see WARDEN *sb.*¹ 7 b).
1916 JOYCE *Portrait of Artist* iii. 119 That's game ball. We can scut the whole hour. 1922 —— *Ulysses* 224 That'll do, game ball, Blazes Boylan said. 1934 *Discovery* Oct. 287/2 Though found in Central Ireland, this game-board betrays Celto-Norse influences in the border pattern. 1807 SOUTHEY *Lett. from England* I. xxv. 293 A game book, as it is called, is one of the regular publications, wherein the sportsman may keep an account of all the game he kills, the time when, the place where. 1964 R. PERRY *World of Tiger* i. 10 On a first examination of the game-books it seems a miracle that there could have been any tigers left alive in India by the end of that [sc. the nineteenth] century. 1937 *Brit. Birds* XXXI. 166 A special game-chick and poultry investigation. 1951 *Good Housek. Home Encycl.* 478/2 Accompaniments vary slightly for different birds, but thin gravy, game chips and fried crumbs are usual. 1961 G. SMITH *Business of Loving* xi. 222 Some pheasant with game chips. 1945 PHILLIPS & REESE *How to play Bridge* v. 28 Game-forcing bids of unlimited strength. A jump bid in a new suit, such as Two Spades over One Heart, or Three Clubs over One Diamond. 1929 M. C. WORK *Compl. Contract Bridge* i. 6 The dealer's first concern when he picks up his hand is to get his side into a game-going contract if game is in their cards. *Ibid.* 7 Dealer having made a bid that is not a game-goer. 1929 *Daily Express* 3 Jan. 15/5 How these game-pieces..were played in olden days. 1956 HEDSTRÖM & TAYLOR tr. *Bergström's Dutch Still-Life Painting* vii. 247 The Dutch game-piece, with which we shall..include pictures of dead domestic birds, is a late branch of still-life painting. 1927 W. E. COLLINSON *Contemp. Eng.* 13 Old game-rhymes from the district. 1959 I. & P. OPIE *Lore & Lang. Schoolch.* vii. 112 Shirley Temple and Deanna Durbin are two of the stars most often mentioned in game rhymes. 1962 *Antioch Rev.* XXII. 8 What are we to think of a civilization which has not been able to talk about killing almost everybody except in prudential and game-theoretic terms? 1962 *Listener* 19 Apr. 693/2 Weapons systems and game-theoretical studies. 1964 K. WINETROUT in I. L. Horowitz *New Sociol.* 156 As a mathematician, he may become a game-theorist with Herman Kahn and others. [1928 J. VON NEUMANN in *Math. Ann.* C. 295 (*heading*) Zur Theorie der Gesellschaftsspiele.] 1954 J. D. WILLIAMS *Compleat Strategyst* i. 13 One-person games are uninteresting, from the Game Theory point of view... Most of the work done to date in Game Theory deals with the two-person game. 1957 LUCE & RAIFFA *Games & Decisions* i. 10 Game Theory is a model for situations of conflict among several people, in which two principal modes of resolution are collusion and conciliation. 1959 I. JEFFERIES *13 Days* iii. 38, I don't like cards, but I did some games-theory once and I could handle the well-known ones. 1961 *Listener* 12 Oct. 558/1 The 'game theory', as it is now being called, is being used to construct quite sophisticated theoretical frameworks to a variety of natural phenomena...Its mathematics can.. be used to deal with matters not always thought of as games, such as military situations, labour-management relationships and foreign trade. 1969 B. SUTTON-SMITH in R. D. Abrahams *Jump-Rope Rhymes* p. ix, Much of the work of the world is carried on through mathematical games theory. 1971 G. K. ROBERTS *Dict. Political Anal.* 86 Game theory is a useful analytical technique, and has been employed in the study of..campaign tactics in an election, coalition formation,..etc.

game, *a.*¹ Add: **a.** Esp. in Australia in colloq. phr. (*as*) *game as Ned Kelly*: very spirited or brave.
1941 BAKER *Dict. Austral. Slang* 30 *Game as Ned Kelly*, highly courageous; willing to tackle big odds. 1956 S. HOPE *Diggers' Paradise* 89 A common expression on the lips of Aussies of both sexes to describe someone engaged in a risky enterprise is 'he's as game as Ned Kelly'. 1969 *Sun-Herald* (Sydney) 13 July 13/3, I asked him [sc. Mick Jagger] if he had heard the expression 'as game as Ned Kelly'. He said he had.

gamelan (gæ·məlæn). Also **gamelang.** [Javanese, f. *gamel* to handle.] A type of orchestra common in the East Indies, esp. in Java, consisting mainly of percussion instruments with some woodwind and string ones. Also *transf.* and *attrib.*
1817 T. S. RAFFLES *Hist. Java* I. viii. 469 The musical instruments of the Javans are peculiar. Several of them are necessary to compose a *gámelan*, set, or band. 1869 A. R. WALLACE *Malay Archip.* I. vii. 161 A native band or Gamelang, was playing almost all the evening. 1949 C. MCPHEE in *Musical Q.* XXXV. 250 Just at what point the gamelan emerged from a small ensemble of gongs into the elaborately organized orchestra it is today is not known. 1965 *New Statesman* 10 Sept. 370/2 The wild music that Boulez scores for his gamelan orchestra of xylophones, harps, celesta, vibraphone, piano and multitudinous bells. 1967 J. CLEARY *Long Pursuit* viii. 180 Behind him half a dozen musicians had begun to play, two of them on percussion instruments like a xylophone, the others with gongs, cymbals and a drum. 'The gamelan, that is the orchestra, plays what the Darlan tells it.'
b. A percussion instrument resembling a xylophone, used in the East Indies.
1934 in WEBSTER. 1967 *Listener* 10 Aug. 170/1 A woman singer and a gamelan-player moved through the Pullman coaches.

gamesman (gēⁱ·mzmæ̆n). [f. GAME *sb.* + MAN *sb.*¹; cf. next.] **1.** = MAN *sb.*¹ 15.
1931 *Antiquity* V. 464 Gamesmen of the same form and material have been unearthed by Watelin at Kish.
2. One who engages in games or sports; *spec.* one who is skilled in gamesmanship.
1947 S. POTTER *Gamesmanship* ii. 21 Psychological tendency, if not temporal necessity, will cause him to drive faster, and—behold! now the gamesman can widen his field and bring in carmanship. 1959 *Punch* 19 Aug. 37/1 If Henry V sending those tennis balls to the King of France wasn't a gamesman I don't know one. 1961 *Times* 8 Mar. 17/7 C. W. Brasher, a member of Achilles, which has had more than its fair share of gamesmen, has admitted that he could bring a mile time of 4 mins. 15 sec. down by 10 sec. with a series of excuses..about track, spikes, and wind. 1963 *Punch* 21 Aug. 264/3 One..wonders how the modern gamesman ever has time to play any games. 1969 *Observer* 21 Dec. 6/1 Academics ..can't even think like..Nasser because he's a leader and the gamesmen are just pretending to be.

gamesmanship (gēⁱ·mzmænʃip). [f. GAME *sb.* + *-MANSHIP.] Skill in winning games, esp. by means that barely qualify as legitimate. Also *transf.*
1947 S. POTTER (*title*) The theory & practice of gamesmanship or the art of winning games without actually cheating. 1952 E. GRIERSON *Reputation for Song* ii. 11 This was so like Laura, with whom a hand she could not play was a hand wasted, but Mr. Clarke, a practitioner of 'gamesmanship' himself, would not be rushed. 1960 *Guardian* 23 Mar. 9/3 Mr. Smirnov..assumed a more confidential manner towards the end of his essay in gamesmanship. 1967 *Times* 8 Apr. 13/6 Sportsmanship.. is a switched-off word that has lost ground to gamesmanship.

gamester. 1. a. Delete † *Obs.* and add later examples.
1832 P. EGAN *Bk. Sports* 340/2 The great interest attached to cricket-matches..will warrant something more than the mere comparison of numbers and names of the gamesters. 1942 BERREY & VAN DEN BARK *Amer. Thes. Slang* § 636/1 Athlete; player,..gamester. 1965 *Eng. Studies* XLVI. 464 Gamesters are cheered by *pepsters* and *rallisters*.
6. For GAME *sb.* 14 read GAME *sb.* 12. (Add further example.)
1957 N. F. TICEHURST *Mute Swan in Eng.* 121 *Gamester, gamster*, an owner of a game of swan.

gameta·ngium. Pl. gametangia. Now the commoner form of GAMETANGE.
1886 [see GAMETANGE]. 1952 *New Biol.* XIII. 103 A transverse wall is formed a short distance behind each tip, so dividing each swollen branch into two portions, the distal portion being known as a gametangium. 1964 *Oceanogr. & Marine Biol.* II. 198 Gametangia have hitherto been unknown in the genus *Udotea*.

gamete. Add: Also *Comb.*
1914 GEDDES & THOMSON *Sex* ii. 45 These foci of gamete-making began to be enclosed..and nourished by adjacent tissue. 1927 HALDANE & HUXLEY *Animal Biol.* ii. 63 Before gamete-formation, the two members of such a pair segregate. 1963 R. P. DALES *Annelids* viii. 164 The posterior gamete-producing segments.

gametic (gæme·tik, -ī·tik), *a. Biol.* [f. GAMETE + -IC.] Of or pertaining to gametes.
1905 R. C. PUNNETT *Mendelism* 25 The theory of gametic purity can be further tested by deducing from it the results which should follow from crossing the heterozygote with either of the homozygotes. *Ibid.* 64 Gametic coupling. 1955 *Sci. News Let.* 2 Apr. 210/1 The third system is 'gametic selection'. This means that the germ cell carrying the disease is more likely to result in offspring than other cells. 1962 *Lancet* 29 Dec. 1383/2 It is difficult to see how selection against a given gametic genotype could occur at all on the female side.

gameto-. Add: **game·tocyst,** a cyst containing two associated gametocytes; **game·tocyte,** a cell that gives rise to gametes; **gameto·ge·nesis,** the formation of gametes; **gametoge·nic, -o·genous** *adjs.*, giving rise to gametes or to reproductive cells; **gameto·geny, -o·gony** = *gametogenesis; **game·tophore** *Bot.*, a modified branch or filament bearing sexual organs; so **gametopho·ric** *a.*; **gametophy·tic** *a.*, pertaining to or occurring on a gametophyte.
1926 C. M. WENYON *Protozool.* I. i. 134 A cyst may be formed around two gametocytes, as in the case of gregarines. It is then distinguished as a gametocyst. 1955 HUTNER & LWOFF *Biochem. & Physiol. Protozoa* II. 190 The gametes escape from the gametocyst of *Trichonympha* after it is ingested by the roach. 1899 *Nature* 3 Aug. 322/2 The parasites [of malaria] reach their highest development within the vertebrate host, and become either (*a*) sporocytes or (*b*) gametocytes. 1925 E. B. WILSON *Cell* (ed. 3) vii. 594 The nucleus of each gametocyte now divides successively to form a number.. of gamete-nuclei which pass to the periphery and here are budded off..to form minute true gametes. 1966 *New Scientist* 1 Sept. 482/1 The original, parent parasite in the red cell may produce..a merozoite..[or] a gametocyte. 1900 B. D. JACKSON *Gloss. Bot. Terms* 108/1 *Gametogenesis*, the production of gametes. 1946 *Nature* 14 Dec. 879/1 In the adult *Drosophila*..mustard gas produces a selective action on gametogenesis. 1966 F. A. E. CREW *Found. Genetics* viii. 154 The essential features of gametogenesis are the pairing or conjugation of the two members of each of homologous chromosomes, their separation without any longitudinal splitting and their passage into the nuclei of the two daughter cells. 1900 B. D. JACKSON *Gloss. Bot. Terms* 108/1 *Gametogenic*, giving rise to gametes. 1903 J. B. FARMER et al. in *Proc. R. Soc.* LXXII. 500 The gametogenic tissues which are destined to become the reproductive elements. *Ibid.*, The term *gametogenic*, as here proposed, is also intended to include the primary sporogenous tissue of plants. 1971 *Nature* 2 July 65/1 The gametogenic events of the living testis. 1903 E. A. MINCHIN in E. R. Lankester *Treat. Zool.* I. ii. 210 The gametocytes are termed gametogenous monoints, the formation of gametes being regarded as a special case of monogony. 1900 B. D. JACKSON *Gloss. Bot. Terms* 108/1 *Gametogeny*, the production of gametes. 1971 *Nature* 9 Apr. 408 By the fourth week gametogeny has been completed and mating has begun. 1910 H. B. FANTHAM in *Proc. Zool. Soc.* II. 678 The processes leading to the formation of the gametes may be termed gametogony. 1969 *New Scientist* 8 May 306/1 A tiny protozoan that divides its time between a sexual cycle (gametogony)..and two asexual cycles. 1895 D. H. CAMPBELL *Struct. & Devel. Mosses & Ferns* i. 2 The gametophyte itself may show two well-marked phases, the protonema and the gametophore. *Ibid.* vi. 155 The sexual organs are borne either separately or together at the summit of the gametophoric branches. 1957 H. C. BOLD *Morphol. Plants* xvi. 290 The leafy *Sphagnum* plant sometimes is called the leafy gametophore.., inasmuch as it bears the sex organs when mature. 1905 D. H. CAMPBELL *Struct. & Devel. Mosses & Ferns* (ed. 2) 640/2 Gametophytic buds. 1919 F. O. BOWER *Bot. Living Plant* III. xxi. 341 This gametophytic budding is less common here [sc. in ferns] than in the Bryophytes. 1959 FOSTER & GIFFORD *Compar. Morphol. Vascular Plants* ii. 20 The alternation of sporophytic and gametophytic generations in the life cycle.

gametoid (gæ·mĭtoid), *sb.* and *a. Biol.* [f. GAMET(E + -OID.] **A.** *sb.* (See quot. 1953.) **B.** *adj.* Having the form or function of a gamete; gamete-like.
1891 [see ZYGOTOID]. 1903 *Proc. R. Soc.* LXXII. 503 The evidence justifies us in..correlating the appearance of these 'gametoid' neoplasms with the result of a stimulus which has changed the normal somatic course of cell development into that characteristic of reproductive (not embryonic) tissue. 1908 *Practitioner* Feb. 226 Neither gametoid cells, nor flagellated forms of this parasite, have been observed. 1953 I. F. & W. D. HENDERSON *Dict. Sci. Terms* (ed. 5) 165/2 *Gametoid*, a structure behaving like a gamete, as apocytes uniting to form a zygotoid.

Gamgee, gamgee (gæ·mdʒi). The name of Joseph Sampson *Gamgee* (1828–1886), English surgeon, used attrib. (usu. in *Gamgee tissue*) or absol. to designate a wound-dressing invented by him, consisting of a thickness of cotton-wool between two thin layers of gauze.
1895 *Army & Navy Co-op. Soc. Price List* 699/2 Gamgee Tissue..per lb 2/6. 1906 *Practitioner* Nov. 674 The application of a baked or scorched pad of Gamgee tissue. 1924 *Glasgow Herald* 8 Apr. 10 Strips of emergency field-dressing gamgee. 1937 'A. BRIDGE' *Enchanter's Nightshade* xxvi. 347 The complicated apparatus of poultices and gamgee jackets.

gamine (gami·n). [Fr.] A female gamin; an attractively pert, mischievous or elfish girl or young woman, usually small and slim. Also *attrib.* or quasi-*adj.*
1899 C. SCOTT *Drama of Yesterday* II. 243 A veritable *gamine*, there is no other word which so thoroughly expresses her. 1921 *Glasgow Herald* 23 Apr. 4 The 'Coquette' is an unscrupulous milliner's assistant, a regular 'gamine', who plays off a lover against a husband. 1925 D. H. LAWRENCE *St. Mawr* 8 He was fascinated by Lou's quaint aplomb..her *gamine* knowingness. 1957 'D. RUTHERFORD' *Long Echo* vii. 131 The actress..who had lately adopted the gamine style of hair-cut. 1962 I. MURDOCH *Unofficial Rose* xxvi. 250 The puckish *gamine* who, alas, sometimes enjoyed teasing and tormenting him.

‖**gaminerie** (gaminri, gămī·nəri). [Fr.] The behaviour or characteristics of a gamin or gamine.
1917 J. AGATE *Buzz, Buzz!* 221 With the interment of his little friend [sc. a monkey] Anatole buried for ever his own *gaminerie*. 1930 *Observer* 1 June 15 Amused by the rich *gaminerie* of Miss H. 1960 A. POWELL *Casanova's Chinese Restaurant* 51 Determination not to be jockeyed out of either his *gaminerie* or accustomed manner of ordering his own life.
So **gamine·sque** *a.*, resembling or characteristic of a gamin or gamine.
1928 *Manch. Guardian Weekly* 17 Aug. 135/2 She lets fly a full handful of her best gaminesque tricks. 1940 H. G. WELLS *Babes in Darkling Wood* I. i. 26 Our young people cast off the cares of the world abruptly and became gaminesque.

gaming, *vbl. sb.* Add: **3.** *gaming board.*
1932 *Discovery* Nov. 341/2 A magnificent gaming board is unique among Irish examples of Norse culture. 1938 *Burlington Mag.* July 40/1 Inlaid gaming boards.

gamma. Add: **1. b.** An examiner's third-class mark. Also *transf.* and *attrib.*
1932 A. HUXLEY *Brave New World* xi. 188 The fifty-six..machines were being manipulated by fifty-six aquiline and ginger Gammas. *Ibid.* xii. 204 A creature with a Gamma-Minus physique. 1956 L. E. JONES *Edwardian Youth* iv. 98 A Second Class in Greats, thanks to 'Gammas' for Greek and Latin composition. 1958 [see *BETA 3]. 1959 E. WAUGH *R. Knox* iv. 94 To qualify for a first it was necessary to get alphas in seven papers... A gamma cancelled an alpha. Betas were neutral. 1961 M. BEADLE *These Ruins are Inhabited* (1963) vi. 85 He's my gamma card...That's what the fellows call it. Because the lowest grade is gamma.
c. Used as a symbol for various quantities, etc., in science, etc.:
(i) *Metallurgy.* (*a*) Applied to one of a series of allotropic forms of a metal, as *gamma iron*, the allotrope of iron stable between 910°C and 1,403°C, characterized by a face-centred cubic crystal structure. (*b*) Applied to a solid solution in a range of alloys, as *gamma brass*, the third of a series of alloys of copper and zinc.
1896 F. OSMOND in *Jrnl. Iron & Steel Inst.* XLIX. 180 It was necessary to distinguish at least three molecular states of iron, which were respectively stable within certain limits of temperature: α below 700°, β between a range of 750° to 860°, γ above 860°. *Ibid.* 187 The distinction established between γ and β iron..is not an arbitrary, but an experimental one. 1902, etc. [see *ALPHA 3 d]. 1904 E. S. SHEPHERD in *Jrnl. Physical Chem.* VIII. 435 Copper and zinc unite to form six series of solid solutions, which have been distinguished [in this paper] as α, β, γ, δ, ε, and η. 1925 *Phil. Mag.* L. 311 Some authors have supposed γ-brass to be $CuZn_2$, while others believe that it is a solid solution of zinc in Cu_2Zn_3. 1956 G. S. BRADY *Materials Handbk.* (ed. 8) 116 Gamma brass, with the zinc above 45%,..is not easily worked either hot or cold. 1963 W. H. DENNIS *Metallurgy Ferrous Metals* i. 4 From 1,403°C to its melting point it again changes from the f.c.c. gamma iron to b.c.c. delta iron.
(ii) *Physics. gamma rays* or *radiation*, electromagnetic radiation of very short wavelength emitted by radioactive substances, orig. regarded as the third and most penetrating kind of radiation emitted by radium but now known to be identical with very short X-rays. So *gamma ray* or *particle*, a single photon of gamma radiation.
1903 RUTHERFORD in *Phil. Mag.* Feb. 177 The γ rays, which are non-deviable by a magnetic field, and which are of a very penetrating character. 1904 R. J. STRUTT *Becquerel Rays* iii. 83, I have succeeded in observing the γ-radiation from 10 milligrammes of radium bromide. 1926 R. W. LAWSON tr. *Hevesy & Paneth's Man. Radioactivity* iv. 49 A γ-'particle' produces 1·5 pairs of ions per unit of length. 1930 J. BUCKINGHAM *Matter & Radiation* 31 The longest X-rays have the properties of the shortest ultra-violet rays and the longest Gamma-rays merge into the shortest X-rays. 1942 S. TOLANSKY *Introd. Atomic Physics* xvi. 272 The nucleus, being usually left in an unstable condition after the departure of the β-particle, reverts back to a more stable state, the surplus energy being radiated as a γ-ray. 1957 [see *BETA 2 f]. 1958 *Listener* 30 Oct. 704/1 Food-preservation by gamma radiation. 1961 *Aeroplane* C. 391/1 This will be the first attempt at 'gamma ray astronomy' from a satellite. Scientists are now limited in their study of extra-terrestrial gamma radiation because their measurements are affected by existing radiation in the Earth's atmosphere. 1969 *Times* 29 Sept. 12/6 What is almost certainly the discovery of the first gamma ray star has now been reported.
Hence ellipt. for *gamma rays* in *gamma counter, emitter, irradiation*, etc.; so *gamma emitting* adj.
1929 *Brit. Med. Jrnl.* 14 Sept. 508/1 (*heading*) The effect of gamma irradiation on cell division. 1949 W. S. EASTWOOD *Surv. Pile Made Isotopes* 1 Sources of this kind have previously been confined to the natural gamma emitting elements. 1949 *Non-Destructive Testing* VIII. ii. 23/1 It is standard practice in gamma radiography to place the film between sheets of lead when making an exposure. 1951 *Nucleonics* IX. i. 50 British industry has never made wide use of the natural gamma emitters radium and radon for radiography. 1953 *Ann. Reg. 1952* 403 Gamma radiography (the method of photographing through solid objects such as metals). 1955 *Gloss. Terms Radiology (B.S.I.)* 6 *Gamma emitter*, an atom whose radioactive decay process is associated with the emission of gamma rays. 1961 *Lancet* 16 Sept. 633/2 Rb is a hard gamma emitter. 1971 *Nature* 4 June 324/1 Radioactivity was measured in a well-type gamma counter.
(iii) *Photogr.* The gradient of the straight-line portion of the characteristic curve of a photographic emulsion, taken as measuring the contrast of the developed image compared with that of the scene photographed; hence *loosely*, contrast. Similarly in *Television*, a measure of the contrast of the transmitted picture compared with that of the scene televised.
1903 *Photogr. Jrnl.* XLIII. 48 The relation between the development factor (afterwards called γ) and the time of development (*t*). 1935 *Discovery* Mar. 85/1 When the plate is removed from the [compound] developer it generally has a lower contrast (gamma) than if it had been normally developed. 1937 A. T. WITTS *Telev. Cycl.* 65 *Intensity contrast*, the contrast in detail in a picture, sometimes referred to as the gamma of the picture. 1961 G. MILLERSON *Technique Telev. Production* iii. 48 From the angle of the slope, the curve's gamma can be deduced (tan α)...A low gamma device accepts a wide contrast range, but compresses it to fit reproduction limits. 1962 *Unesco Bull. Libr.* XVI. 16 A high value of gamma is desirable in photographic reproduction of lines, as in the pages of a printed book.
(iv) A unit of magnetic field strength, 10^{-5} oersted.
1903 *Nature* 5 Nov. 6/1 There was a movement of the declination needle to the west through about 34′, and a diminution of 240 γ in the horizontal force. 1967 *New Scientist* 7 Dec. 617/2 The lunar magnetic field is no bigger than two gammas.
(v) A unit of mass, one millionth part of a gramme.
1931 *Industrial & Engin. Chem.* (*Analytical Ed.*) 15 July 314/2 The value obtained was 4·0 γ copper per gram (1 γ=0·001 mg.) for the combined filtrates. 1940 *Ibid.* 15 June 359/2 The Committee on Nomenclature, Spelling, and Pronunciation of the American Chemical Society has approved the following recommendation of the Division of Microchemistry: For 0·001 milligram the term 'microgram', designated by the symbol γ (the word 'gamma' should not be used as a substitute for 'microgram'). 1956 A. HUXLEY *Let.* 23 Sept. (1969) 807 Fifty gamma of LSD were sufficient to produce in me virtually the full effect of the standard dose. 1963 JERRARD & MCNEILL *Dict. Sci. Units* 55 In 1937 it was claimed that the gamma was being superseded by the microgramme in the United Kingdom whereas the gamma reigned supreme in the U.S.A. and continental Europe.
(vi) *gamma globulin* (see quot. 1957).
1937 A. TISELIUS in *Biochem. Jrnl.* XXXI. 1466 The fastest of these components could be identified with serum albumin. The other three..are more or less completely precipitated by half saturation with ammonium sulphate. They will therefore be named α, β, and γ serum-globulin. *Ibid.* 1473 The β- and γ-globulins gave rather diffuse boundaries. 1957 *Dorland's Med. Dict.* (ed. 23) 563/2 *Gamma g[lobulin]s*, globulins of plasma which in neutral or alkaline solutions have the slowest electrophoretic mobility...Most antibodies are gamma globulins. 1960 M. E. FLOREY *Clin. Appl. Antibiotics* II. vii. 202 Gamma globulin has been considered to play a part in enhancing the effect of antibiotic therapy. 1970 *Nature* 7 Nov. 509/2 'Immunoglobulin' thus superseded 'γ-globulin'.

gamma grass, var. GAMA GRASS.

Gammexane (gæme·ksēⁱn, gæmĕksēⁱn). Also **gammexane.** [f. *gamma*-hexachlorocyclohexane, the systematic chemical name.] The proprietary name of an isomer of benzene hexachloride used as an insecticide.
The proprietary term was registered as *Gammexan* in 1951 (*Trade Marks Jrnl.* 7 Oct. p. 920).
1945 *Trade Marks Jrnl.* 7 Nov. 541/1 *Gammexane*, pharmaceutical substances for human use and for veterinary use; sanitary substances, insecticides, preparations for destroying vermin. Imperial Chemical Industries Limited. 1945 [see *BENZENE 3]. 1961 C. H. DOUGLAS-TODD *Pop. Whippet* 148 Gammexane powder.. will kill all living insects on a dog.

gammy, *a.* Add: **3.** Also, disabled through injury or pain.
1893 (in Dict.). 1913 *Punch* 23 Apr. 331 Gammy thumb! The rest o' the squad ain't got gammy thumbs, has they? 1947 D. M. DAVIN *Gorse blooms Pale* 80 That gammy foot of mine.

gamont (gæ·mǫnt). *Biol.* [a. G. *gamont* (M. Hartmann 1904, in *Biol. Centralbl.* XXIV. 25), f. GAM(ETE + Gr. ὄντ-, ὤν, pres. pple. of εἶναι to be, exist.] In Protozoa, a cell that produces gametes.
1912 E. A. MINCHIN *Introd. Stud. Protozoa* ix. 181 It would perhaps be better to replace the terms 'schizont' and 'sporont' by 'agamont' and 'gamont' respectively. 1957 *New Biol.* XXIV. 27 Typically there are two distinct generations [in foraminifera], namely a sexual one (the gamont) which alternates with an asexual one (or schizont). 1964 V. A. DOGIEL *Gen. Protozool.* (ed. 2) vii. 149 In *Adelina* a male and female gamont are invested in a thin membrane, under which the female gamont is transformed into a female gamete, while the male gives rise to several gametes.

gampless (gæ·mplĕs), *a.* [f. GAMP *sb.* 2.] = UMBRELLALESS *a.*
1883 M. E. MANN *In Summer Shade* I. ix. 208 Any old gampless woman of the congregation. 1899 *Westm. Gaz.* 2 Oct. 6/2 To provide 'gampless' seatholders with umbrellas.

-gamy, *suffix*, f. Gr. γάμος marriage + -Y³, appended to Gr. stems to form sbs. with the senses: (*a*) 'marriage (of the type specified)', as in ENDOGAMY, EXOGAMY, *HYPERGAMY; (*b*) '(such a) means of fertilization or reproduction', as in ALLOGAMY, CLEISTOGAMY, *POROGAMY. Cf. BIGAMY, MONOGAMY, POLYGAMY.

Gamza (gæ·mză, gɒ·mză). Also **Gumza.** [Bulgarian.] A dark red grape of Bulgaria, or the red wine made from it. Also *attrib.*
1959 *World Crops* XI. 111/1 The most universally grown is the *Gumza red grape*. The typical Gumza red wines are made mainly in northern Bulgaria. Gumza grapes are grown in some neighbouring countries. 1959 *Wine & Spirit Trade Rev.* 13 Feb. 32/2 *Mavroude* and *Gamza* [are] dark reds. 1961 *Spectator* 7 Apr. 495 The red Gamza, which is fresh and flavoury, rather like a Beaujolais.

ganaderia (gænădīᵊ·riă). [Sp., f. *ganado* livestock, cattle.] A cattle-ranch or stock farm.
1860 MAYNE REID *Odd People* 62 A 'ganaderia' of cattle, or a plantation of cocoa trees [in S. America]. 1960 *Guardian* 16 Sept. 16/6 At Méjanes, he has turned his ganaderia into a luxury holiday centre.

gander, *sb.* Add: **2. c.** A look or glance (see quot. 1914). *slang* (orig. *U.S.*). Cf. next.
1914 JACKSON & HELLYER *Vocab. Criminal Slang* 36 *Gander*, an inquisitorial glance; a searching look; an impertinent gazing or staring. Also the simple act of looking or seeing...'Take a gander at this dump as we pass...' 1937 D. RUNYON *More than Somewhat* i. 16 Now I am taking many a gander around the bedroom to see if I can case the box of letters. 1943 [see *GANDER *v.* 1 b]. 1959 A. LEJEUNE *Crowded & Dangerous* viii. 91, I was sitting..taking a gander at what's running at 'urst Park. 1971 *Sci. Amer.* Oct. 74 (Advt.), Take a gander at the see-through door below. See that corrugated piece of steel?

gander, *v.* **1. b.** Add to def.: (See also quots.) *dial.* or *colloq.* (now chiefly *U.S.*).
1887 T. DARLINGTON *Folk-Speech S. Cheshire* 206 *Gonder*, to stretch the neck like a *gander*, to stand at gaze. 'What a't gonderin' theer fur?' 1903 *Cincinnati Enquirer* 9 May 13/1 *Gander*, to stretch or rubber your neck. 1935 *Amer. Speech* X. 17/2 To wander about looking for someone or something. Modern *to gander*. 1939 *Ibid.* XIV. 239/2 *To gander*, to examine. 1943 HUNT & PRINGLE *Service Slang* 33 *Gander*, a look through the mail, a glance over another's shoulder at a letter or paper. To perpetrate this long-necked nuisance.

Gandharva (gændā·ɹvă). *Hindu Mythol.* Also **Gandharba, -arwa.** [Skr., cf. Gr. κένταυρος: see CENTAUR.] (See quot. 1876.)
1846 H. H. WILSON *Sk. Relig. Sects Hindus* ii. 17 A few of the *Tántrikas* worship the *Siddhas*, or Genii,.. the same class furnishes occasional votaries of the.. *Gandharbas*, and even of the..goblins and ghosts. 1856 M. MÜLLER *Chips* (1867) II. xvi. 102 Then the Gandharvas sent a flash of lightning. 1876 *Encycl. Brit.* IV. 208/2 The retinue of Indra consists chiefly of the *Gandharvas*..a class of genii, considered in the epics as celestial musicians. 1934 *Burlington Mag.* Sept. 169/1 Their decoration..may show Gandharwas flying or playing musical instruments. 1956 K. CLARK *Nude* vii. 276 In India..they appear in the 6th century..as flying gandharvas. 1968 *Indian Mus. Jrnl.* V. 8 The human Gandharva-s (genii).

Gandhi (gæ·ndi). The name of M. K. *Gandhi* (1869–1948), Indian political leader and social reformer, applied *attrib.* to a close-fitting white cap with a wide band encircling the head.
1921 E. M. FORSTER *Let.* 19 Sept. in *Hill of Devi* (1953) 125 Crowds of the Nagpur people..wearing the white Gandhi cap. 1963 M. MALIM *Pagoda Tree* xv. 94 A portentous Hindu in a snowy *dhoti*, a fawn cotton jacket and a 'Gandhi cap'. 1967 H. R. F. KEATING *Inspector Ghote caught in Meshes* x. 142 Here and there a turban or a Gandhi cap or the vivid splash of a sari.

Gandhian (gæ·ndiăn), *a.* [f. prec. + -IAN.] Of, pertaining to, or characteristic of Gandhi. So **Gandhi-e·sque** *a.*, resembling or characteristic of Gandhi and his principles; **Ga·ndhism, Ga·ndhi-ism,** Gandhian principles or actions; **Ga·ndhist,** a follower of Gandhi; also *attrib.*
1921 *Daily Tel.* 7 Mar. 7/1 Some..have resolutely opposed the Gandhian policy. 1921 *Public Opinion* 2 Dec. 554/3 The inherent selfishness of Ghandi-ism. 1921 *Glasgow Herald* 2 Dec. 8 It is..undesirable..to mention 'Ghandhism' just now in this picturesque Rajput city. *Ibid.* 12 Dec. 9 Protests from some few Gandhists. 1928 *Observer* 15 Jan. 4 The general Gandhist idea. 1954 M. LOWRY *Let.* 10 May (1967) 367 Endeavouring to remedy this in the approved Gandhi-esque manner. 1954 *Encounter* Dec. 11/2 Gandhism..visualises ultimately a stateless society. 1958 J. V. BONDURANT *Conquest of Violence* i. 6 Prominent among Gandhian principles are non-violence, adherence to truth, and dignity of labor. 1963 *Guardian* 23 Feb. 4/7 Simple, authoritative and very Gandhi-esque. 1966 *Listener* 6 Jan. 14/2, I detected a desire to return to Gandhi-ism: 'Let us do without aid, without American wheat; let us starve if necessary, but regain our economic independence and self-respect.' 1971 *Hindustan Times Weekly* (New Delhi) 4 Apr. 3/6 The oldest survivors of the freedom struggle and Gandhian methods.

gandoura (gændūᵊ·ră). Also **gandourah, gandura(h).** [ad. Algerian Arab. *gandūra*, classical Arab. *ḳandūra*.] A long, loose gown worn mainly in the Near East and North Africa.
1851 *Illustr. Catal. Gt. Exhib.* IV. I. 1262/2 Gandoura, made of wool and silk. 1902 *Daily Chron.* 12 June 7/5 Wearing only his rich white gandurah. 1927 *Glasgow Herald* 17 Sept. 4 The Cadi in spotless burnous and fine white linen gandoura. 1951 R. SENHOUSE tr. *Colette's Chéri* (1963) 131 The coloured lining of the white gandoura she put on was suffused with a vague pink.

gandy dancer (gæ·ndi dɑ·nsəɹ). *slang* (orig. *U.S.*). [Orig. uncertain.] A railroad maintenance-worker or section-hand. Hence **gandy dancing.**
1923 N. ANDERSON *Hobo* vi. 93 A 'gandy dancer' is a man who works on the railroad track tamping ties. 1929 *Amer. Speech* V. 172 Gandy dancing is not considered a very honorable profession. 1933 *Ibid.* VIII. 26/2 *Gandy dancer*, section hand. (From the rhythmic up-and-down motion of workers pumping a handcar.) 1957 J. KEROUAC *On Road* (1958) III. vi. 215 Working in a railroad gandy-dancing cookshack. 1959 J. THURBER *Years with Ross* iv. 63 They discussed the parlance of railroading—deadhead, highball, whistle stop, gandy dancer. 1970 F. MCKENNA *Gloss. Railwaymen's Talk* 35 Footplatemen have a great regard for gandy dancers, the men who keep the rail safe for the train to run over.

gang, *sb.*¹ Add: **10. a.** (Examples without depreciatory connotations.) *colloq.*
1945 A. KOBER *Parm Me* 28 The bunch is waiting at my house. I thought I'd call fa you and take you over, so that we'd meet the gang. 1955 G. FREEMAN *Liberty Man* I. i. 15 All the gang would be there, and she'd be ever so proud of him.
12. (senses 9 and 10) *gang-boss, -fight, -life, -man* (earlier and later examples), *-war, -warfare*; **gang-bang, -shag** *slang* (orig. *U.S.*), an occasion on which several men have sexual intercourse one after another with one woman; hence as *v. trans.* and *intr.* See also *GANGLAND.
1953 *San Francisco News* 14 Oct. 23/3 'Wonder gear', it developed, is an attractive girl, in Navy lingo, and 'gang-bang' is a juicy party involving wine, women and song. 1957 *Nation* 23 Feb. 161/1 Sometimes he lapses into pages of terrifying gibberish that sound like a tape recording of a gang bang with everybody full of pod, juice and bennies all at once. 1967 G. LEGMAN *Fake Revolt* 28 Wife-swapping, husband-ditching, gang-banging, and the rest. *Ibid.* 30 The gang-bang ethic. 1968 B. TURNER *Sex Trap* xxi. 193 What's the next arrangement to be? A gang-bang for the whole Vice Squad? 1969 *Guardian* 16 Aug. 7/3 A pretty 18-year-old girl..used to 'stuff' herself with heroin and let herself be 'gang-banged' all the time. 1887 C. B. GEORGE *40 Yrs. on Rail* ix. 193, I was in charge of a construction train, being engineer, conductor and gang-boss combined. 1923 'R. DALY' *Enchanted Isl.* xii. 109 Tulagi, the chief gang-boss, will keep the other boys up to their work [on the copra plantation]. 1931 *Times Lit. Suppl.* 18 June 485/1 The modern 'gang-boss' and his henchmen in the boot-legging and hijacking world in America. 1932 H. SIMPSON *Boomerang* x. 271 A man left lying insensible in the wake of a gang-fight. 1958 *New Statesman* 20 Dec. 880/3 The films have conditioned us to demanding a high degree of realism in this kind of gang-fight. 1949 M. MEAD *Male & Female* xv. 309 This asocial gang-life of boys provides a basis for the adult criminal world in America. 1867 J. E. WHITE in E. R. Pike *Human Doc. Vict. Golden Age* (1967) 218 The gangman acts as foreman or overlooker. 1928 *Daily Express* 10 Oct. 6 We can hide you away where all the gangmen in London won't find you. 1955 E. WAUGH *Officers & Gentlemen* 153 The gangmen of the departments closed in for the kill. 1964 A. SWINSON *Six Minutes to Sunset* ii. 42 At midnight the railway between Chheharta and Khasa was torn up, the job being done expertly with gangmen's tools. 1927 F. M. THRASHER *Gang* xiii. 237 The gang shag includes boys from sixteen to twenty-two years of age. It is a party carried on with one woman by from fifteen to thirty boys from one gang or club. A mattress in the alley usually suffices for this purpose. 1932 *Amer. Speech* VII. 253 A cycle of gang war pictures. 1959 WYNDHAM LEWIS *Let.* 12 Dec. (1963) 514 The people were no more troubled than Chicagoans are by gang-wars. 1934 WEBSTER, Gang warfare. 1962 *John o'London's* 8 Mar. 235/1 Gang-warfare and teenage problems.

gang, *v.*² Add: **1. b.** To arrange (implements or instruments) in gangs (GANG *sb.*¹ 8 b). Hence **ganged** *ppl. a.*
1900 *Yearbk. U.S. Dept. Agric.* 540 The plows are usually ganged, two to one frame. 1938 *Nature* 3 Sept. 446/1 The use of a motor driving the main ganged variable condenser. 1942 *Electronic Engin.* XV. 9 The tuning..is carried out by..variometers ganged together and controlled by a single knob. 1964 *Economist* 6 June 1149/2 Adaptation..made it feasible for disc-plows to be 'ganged' so that several could be pulled at once. 1971 *Hi-Fi Sound* Feb. 41/2 (Advt.), Rotary knob ganged controls for tone, balance and volume.
2. Also, to go in company *with*.
1928 W. A. WHITE *Masks in Pageant* 348 He was frail [in his boyhood] and never ganged with his fellows.
b. *to gang up* (*with, on*): to form a gang or combine (with, against). *colloq.* (orig. *U.S.*).
1925 T. A. BOYD *Points of Honor* 50 Then I set out to kick hell out of 'im. I'd a done it too if they hadn't a ganged up on me. 1940 *Nature* 6 July 36/2 The old folks..and the youngsters..are 'ganging up' on the half of the population that does the work. 1942 BERREY & VAN DEN BARK *Amer. Thes. Slang* § 343/6 Affiliate or associate with,..*gang up with*. 1942 D. POWELL *Time to be Born* (1943) xiv. 332 Things must have gone wrong with her!..Maybe he's got people to gang up against her. 1944 *Hutchinson's Pict. Hist. War* 27 Oct. 1943—11 Apr. 1944 365 A change in U-boat tactics was essential if they were to gang-up on and strike at our convoys. 1946 'P. QUENTIN' *Puzzle for Fiends* viii. 64 Thought he was Jesus Christ..with a lot of Satans ganging up on him. 1951 *Ann. Reg. 1950* 47 A fear that Germany and France might 'gang-up' in Europe without Britain. 1959 'J. BYROM' *Take Only as Directed* i. 16 He'll get David to gang up with him and stop me sailing.

ganga, var. GANJA (in Dict. and Suppl.).

gangan. (Further examples in form *gang-gang*.)
1891 E. KINGLAKE *Australian at Home* 124 You hear the sweet note of the wonga pigeon or the hoarse scream of the ganggang. 1965 *Austral. Encycl.* II. 440/1 The smallest cockatoo is the gang-gang..of south-eastern Australia and Tasmania.

gangdom (gæ·ŋdəm). [f. GANG *sb.*¹ + -DOM.] = *GANGLAND.
1926 *Springfield* (Mass.) *Union* 20 Apr. 7 A definite inquiry into gangdom's manipulations. 1930 *Observer* 19 Oct. 17 A mere commonplace of New York gangdom. 1935 J. T. FARRELL *Judgment Day* (1936) iii. 515 Lonigan and..Gallagher together leaped up the career of gangdom's adventurous ladder to fame. 1935 *Amer. Speech* X. 10/1 An age when highly organized gangdom had no voice in the government.

gang-gang, var. GANGAN (in Dict. and Suppl.).

gangland (gæ·ŋlænd). orig. *U.S.* [f. GANG *sb.*¹ + LAND *sb.*] The domain of gangsters; the underworld; gangs or gangsters collectively. Also *attrib.*
1912 A. H. LEWIS *Apaches N.Y.* 15 The first lesson of Gangland is never to inform nor give evidence. 1928 *Daily Tel.* 28 Aug. 7/5 A bomb, known in gangland parlance as a 'pineapple'. 1932 E. WALLACE *When Gangs came to London* viii. 64 Gangland does not talk scandal even of its worst enemies. 1966 *Observer* 25 Sept. 1/5 (*headline*) Police stop gangland carve-up. 1971 *Daily Tel.* 15 Sept. 1/1 Police believe a gangland killing may be connected with the discovery early yesterday of a bloodstained car at Chigwell, Essex.

gangliectomy (gæŋglie·ktŏmi). *Surg.* [f. GANGLI(ON + *-ECTOMY.] = *GANGLIONECTOMY.
1903 *Alienist & Neurol.* Feb. 20 That great operation on the nervous system, trigeminal gangliectomy. 1970 *Excerpta Med. Ser. IX* XXIV. 642/1 Lumbar gangliectomy for obliterative diseases in a stage of destruction was performed on 92 patients.

gangling, *ppl. a.* (Further examples.)
1935 A. SQUIRE *Sing Sing Doctor* v. 63 Tall and gangling, Fay had a long-jawed horse face. 1943 C. S. FORESTER *Ship* 176 Tall and gangling and clean-shaven. 1953 'N. SHUTE' *In Wet* x. 350 A black haired, lean, gangling lad of twenty. 1955 *Times* 18 Dec. 3/2 The massive gangling figure of Mr — hurtled out of his room.

ganglion. Add: **6. ganglion-blocking** *a.*, preventing the transmission of nerve impulses across the synapse in a ganglion. Cf. *GANGLIONIC *a.*
1951 *Brit. Med. Jrnl.* 14 Apr. 775/2 When using ganglion-blocking agents..it is necessary to remember that the response to drugs and procedures affecting the blood pressure will necessarily be exaggerated, as the mechanisms maintaining a normal blood pressure are weakened. 1962 H. BURN *Drugs, Med. & Man* vi. 65 In the presence of one of these substances, which are called ganglion-blocking drugs, the acetylcholine when liberated is unable to stimulate the next fibre.

ganglionectomy (gæ:ŋgliŏne·ktŏmi). *Surg.* [f. GANGLION + *-ECTOMY.] Excision of a ganglion.
1925 *Jrnl. Amer. Med. Assoc.* 20 June 1908/2 (*title*) Treatment of Raynaud's disease by lumbar ramisection and ganglionectomy. 1937 *Brit. Jrnl. Surg.* XXXIV. 788 Sympathetic ganglionectomy is the usual surgical treatment of Raynaud's syndrome and other vasospastic states of the extremities. 1970 GURDJIAN & THOMAS *Operative Neurosurg.* (ed. 3) 463 Leriche used cervical and stellate ganglionectomy with some favorable results.

ganglionic, *a.* Add: **b.** *ganglionic blocking* = *ganglion-blocking adj.
1946 ACHESON & MOE in *Jrnl. Pharmacol. & Exper. Therap.* LXXXVII. 236 We therefore propose that the predominant action of tetraethylammonium ion, which accounts for its vasodepressor and cardiodecelerator effects, be called simply a ganglionic blocking effect. 1961 A. GOTH *Med. Pharmacol.* vii. 99 The majority of the ganglionic blocking drugs are quaternary ammonium compounds.

ganglioside (gæ·ŋgliosaid). *Biochem.* [a. G. *gangliosid* (E. Klenk 1942, in *Zeitschr. f. physiol. Chem.* CCLXXIII. 76), f. GANGLI(ON + *-OSIDE.] Any of a group of glycolipids present chiefly in the grey matter of man and some animals and distinguished from cerebrosides by the presence of neuraminic acid among the products of their hydrolysis.
1943 *Chem. Abstr.* XXXVII. 5743 (*heading*) The gangliosides and cerebrosides of cattle spleen. 1955 R. J. ROSSITER in K. A. C. Elliott et al. *Neurochem.* ii. 20 In the brains of patients suffering from Tay-Sach's disease the cerebrosides are largely replaced by gangliosides. 1969 *New Scientist* 2 Jan. 32/2 Tetanus toxin is adsorbed by nerves, in particular by the chemical components known as gangliosides.

gangly (gæ·ŋgli), *a.* orig. *U.S.* [Alteration of GANGLING *ppl. a.*: cf. -Y¹.] = GANGLING *ppl. a.*
1872 'MARK TWAIN' *Roughing It* vii. 34, I should have shot that long gangly lubber they called Hank, if I could have done it without crippling six or seven other people. 1950 G. GREENE *Third Man* i. 7 His long gangly legs. 1965 *Punch* 11 Aug. 215/2 How..does his gangly stature and personality still impose itself?

gang-plank. For *U.S.* read orig. *U.S.* Also **gangplank.** (Add earlier and later examples.)
1846 *Knickerbocker* XXVII. 469 The last bell rang; the gangplank was drawn in-board; the hawsers were cast off. 1961 F. M. BURGESS *Dict. Sailing* 100 *Gang-plank*, a long narrow plank, fitted with treads, to enable people to walk over the bow of a beached boat.

gangsman. Add: **3.** = next 1.
1923 *Glasgow Herald* 30 Aug. 7 The complete story of the gangsmen and their crimes.

gangster (gæ·ŋstəɹ). orig. *U.S.* [f. GANG *sb.*¹ + -STER.] **1.** A member of a gang of criminals. Also *attrib.* and *Comb.*
1896 *Columbus* (Ohio) *Even. Dispatch* 10 Apr. 4/2 The gangster may play all sorts of pranks with the ballot box, but in its own good time the latter will get even by kicking the gangster into the gutter. 1911 *N.Y. Even. Post* 17 June 6 A musician, returning to his home, was knocked down and killed by a group of gangsters. 1928 *Daily Express* 31 July 8/2 Finding the three Brighton gangsters guilty of murder. 1935 WODEHOUSE *Blandings Castle* x. 253 Kind of tough and ugly he looks, like something out of a gangster film. 1938 F. D. SHARPE *Sharpe of Flying Squad* xv. 164 London at that time was in some danger of becoming a gangster-ridden city like Chicago. 1948 F. R. LEAVIS *Great Tradition* i. 4 One applauds the determination to explode the gangster-hero.
2. A member of a gang of workmen. *rare.*
1927 *Daily Express* 20 June 1/5 Gangsters followed with new ballast and new track.
Hence **ga·ngsterdom** = *GANGLAND; **ga·ngsterish** *a.*, resembling or characteristic of a gangster or gangsters; **ga·ngsterism,** the actions or methods of gangsters.
1923 *Nation* 26 Dec. 743/1 The autobiography of one who has come up in the world from sneak-thiefery and gangsterdom. 1927 *New Masses* Mar. 19/4 Gangsterism reigned supreme, both in and out of the convention hall. 1932 *New Statesman* 7 May 581/1 The alternative is a steady growth of gangsterdom. 1934 *Times* 31 July 10/2 Political 'gangsterism'..is only an extreme manifestation of the temper which is being allowed to dominate the whole political outlook. 1945 G. MILLAR *Maquis* viii. 165 The Resistance in the Haute-Saône followed more gangsterish lines than that of..the Doubs. 1947 *Penguin New Writing* XXXI. 68 He walked with a manly, indifferent saunter, eyes half-closed in gangsterish confidence. 1958 *Times Lit. Suppl.* 31 Jan. 67/2 More tales of American night-life and gangsterdom between the wars. 1962 *Daily Tel.* 19 Mar. 13/4 The economic gangsterism of capitalism.

gang-up (gæ·ŋʌp). *colloq.* [f. *GANG *v.*² 2 b.] An act of 'ganging up'; a meeting of a gang; *spec.* = *gang-bang. (See also quot. 1936.)
1936 *Metronome* Feb. 21/3 *Gang up*, medley. 1949 *Time* 27 June 29 Congress was..harassed by a Communist gang-up with Bose. 1951 E. WATERS *Eye on Sparrow* vi, This is the gang-up. Men like that put you to sleep with their drops. Then one man after another goes in and takes you. 1957 *Economist* 7 Dec. 876/2 Trying to bring about the closest possible pre-conference accord between the British and German governments without giving other Nato partners the impression of an Anglo-German 'gang-up'. 1967 G. STEINER *Lang. & Silence* 137 The narrative of an S.A. gang-up in a beer hall is unforgettable.

gangway. Add: **3. f.** Used interjectionally, as a demand to clear the way.
1925 FRASER & GIBBONS *Soldier & Sailor Words* 101 *Gangway, make way for a naval officer!* An expression, heard sometimes among New Army men in the War,.. meaning 'Get out of the way', 'Stand back', 'Clear a passage'. 'Gangway!' is ordinarily a common warning call on board ship, meaning as above. 1942 O. NASH *Good Intentions* 56 Gangway, you motoring proletariat. 1946 E. O'NEILL *Iceman Cometh* (1947) IV. 219 Gangway for two good whores!
5. (sense 2 a) *gangway seat.*
1924 GALSWORTHY *White Monkey* I. iii. 19 They occupied those gangway seats in the front row of the dress circle of which Fleur had a sort of natural monopoly. 1939 G. GREENE *19 Stories* (1947) 155 He sat down in a gangway seat. 1971 E. PAUL *Reluctant Cloak & Dagger Man* x. 123, I walked up the plane and saw that Tanya had a window seat. The gangway seat next to her was empty.

ganister. Add: Also *attrib.* and *Comb.*
a 1835 PHILLIPS & DAUBENY in *Encycl. Metrop.* (1845) VI. 587/2 The ganister or calliard rocks of the coal strata. 1855 J. PHILLIPS *Man. Geol.* 191 Ganister measures. 1890 W. J. GORDON *Foundry* 107 The ganister lining of the converter is replaced by bricks of magnesian limestone. 1902 *Daily Chron.* 12 May 3/1 Ganister crushing. 1956 W. EDWARDS in D. L. Linton *Sheffield* 13 Only the Soft Bed and Ganister Coal are locally fit to work.

ganja. (Later examples of form *ganga*.)
1957 C. MACINNES *City of Spades* I. xii. 98 Many good men..have lived inside penitentiaries on account of that goddamned ganga. 1959 A. SALKEY *Quality of Violence* viii. 117 He..poured out some *ganga powder* and started to fix the powder..along a section of a dry banana leaf. *Ibid.* 118 If you want to smoke the *ganga*, at least you could offer me some, too.

gannet. Add: **2.** A greedy person, esp. a greedy seaman. *slang.*
1929 F. C. BOWEN *Sea Slang* 56 *Gannet*, a greedy seaman. 1950 P. TEMPEST *Lag's Lexicon* 91 The bet may be on how many plates of porridge one 'gannet' can put away at a sitting. 1969 *Gandalf's Garden* IV. 34/3 (Advt.), Welcome to our humble guest house all holiday-makers, meditators, yogis or vegetarian gannets.

gannetry (gæ·nĕtri). [f. GANNET + -RY.] A breeding-place for gannets.
1913 J. H. GURNEY *Gannet* xi. 324 Let us..fix a date and suppose a census taken at every Gannetry on the 1st May, 1910. 1960 WILLIAMSON & BOYD *St. Kilda Summer* 113 St. Kilda has much the biggest gannetry in the northern hemisphere, and..more than one-fifth of the North Atlantic gannets nest there. 1971 *Stornoway Gaz.* 10 July 4/8 The world's greatest gannetry, in which is estimated 44,000 pairs of gannets breed.

ganodont (gæ·nodǫnt). *Palæont.* [f. mod.L. *Ganodonta* (J. L. Wortman 1896, in *Bull. Amer. Mus. Nat. Hist.* VIII. 259), f. Gr. γάνος brightness + ὀδόντ-, ὀδούς tooth.] = *TÆNIODONT.
1897 *Natural Sci.* Sept. 153 The Eocene Ganodonts of the northern hemisphere.

ganophyllite (gænofi·ləit). *Min.* [ad. G. *ganophyllit* (A. Hamberg 1890, in *Geol. Fören. Stockholm Förh.* XII. 586), f. Gr. γάνος brightness + φύλλον leaf: see -ITE¹.] A brown hydrous silicate of manganese and aluminium.
1892 *Jrnl. Chem. Soc.* LXII. II. 1412 *Ganophyllite*.. occurs in fairly large, monosymmetric crystals. 1948 *Mineral. Mag.* June 347 Ganophyllite..readily lost water when heated.

ganosis (gănōᵘ·sis). [a. Gr. γάνωσις, f. γάνος brightness, f. γάνουν to polish.] (See quots.)
[1896 E. SELLERS in K. Jex-Blake tr. *Pliny's Hist. Art* 157/2 The process [sc. *circumlitio*, a kind of painting] must be kept distinct from the γάνωσις or toning down of the whole statue.] 1911 *Encycl. Brit.* XX. 485/2 When the plaster wall surface has been carefully smoothed it must be anointed with a mixture of wax, resin and oil, which is to be driven in by heat, and then polished till the surface shines like a mirror. This is a classical process referred to by Vitruvius under the name 'ganosis', as applied to the nude parts of marble statues, possibly to tone down the cold whiteness of the material. 1962 R. C. HAGGAR *Dict. Art Terms* 147/2 *Ganosis*, the removal of the dead-white surface of Greek sculpture by applying beeswax and polishing with a cloth, which gives a warm, buttery complexion to the marble.

gansey (gæ·nzi). Also **gansy, ganzee, ganzey, ganzie, ganzy.** Dial. var. GUERNSEY 2 a; a jersey.
1886 J. J. H. BURGESS *Shetland Sketches* 68 He rubbit aff da shute wi' da sleeve o' his gansey. 1922 JOYCE *Ulysses* 469 In workman's corduroy overalls, black gansy with red floating tie and apache cap. 1935 S. BECKETT *Echo's Bones*, Scarlet and blue ganzies. 1961 *Guardian* 13 Dec. 7/3 He had on an old cap, no muffler, and a faded red ganzey under a fraying jacket. 1969 J. WOOD *Three Blind Mice* vi. 82 Who ever went tae sea in those latitudes without his sea-boots and a decent ganzie on his back?

ganted, U.S. var. GAUNTED *a.*
1887 *Scribner's Mag.* Oct. 475/2 Looks right puny an' ga'nted. 1909 R. A. WASON *Happy Hawkins* 333 His face was pale..an' he was ganted down in weight a little.

gantline. Add: (Earlier and later examples.) Now the commoner form.
1840 [see GIRT-LINE]. 1938 J. MASEFIELD *Dead Ned* 45 They'd have laid me out and stripped me to a bare gantline, but for him.

gantry. Add: **2.** (Further *attrib.* examples.)
1888 *Lockwood's Dict. Mech. Engin.* 157 *Gantry crane*, an overhead travelling crane carried on a gantry. 1943 J. S. HUXLEY *TVA* 23 Gantry cranes swung the concrete from the mixing barges to the forms. 1958 *Engineering* 7 Mar. 298/3 The gantry tower, which is used in preparing the vehicle for flight, is being moved away from the launching area.
b. A structure crossing several railway-tracks to accommodate signals.
1889 [in Dict.]. 1902 *Encycl. Brit.* XXXII. 144/1 In cases where room must be economized, signals are usually placed on narrow overhead bridges or 'gantries' spanning a number of tracks. 1939 R. S. LYONS *Wonders Mod. Industry* iii. 35 Ten rail-joints, then Exe Station; ten more rail-joints, then signal gantry, bridge. 1958 DAY & COOPER *Railway Signalling Systems* viii. 124 The train stops must always be beside the line to which they refer, but the signals, with which they must be connected may be on a gantry or bracket some distance from that line.

Gantt (gænt). The name of Henry Laurence *Gantt* (1861–1919), American management consultant, used *attrib.* to designate a chart in which a series of horizontal lines shows the amount of work done or production completed in certain periods of time in relation to the amount planned for those periods.
1918 L. P. ALFORD in *Industrial Management* Feb. 100/2 The supply division of the Ordnance Bureau of the Army..has its information in regard to requirements and progress in graphical form on the Gantt charts. 1923 W. CLARK *Gantt Chart* i. 5 In the Gantt chart a division of space represents both an amount of time and an amount of work to be done in that time. 1958 J. F. MAGEE *Production Planning & Inventory Control* ix. 223 Gantt charts make use of a horizontal scale marked in time units. A series of horizontal lines..is used to display data.

gaon (gā·,ōᵘn). Pl. **gaonim, gaons, geonim.** [Heb. *gā'ôn* (pl. *geônîm*) excellence (Ps. xlvii. 4), pride (Prov. xvi. 18).] Used as an honorific for the heads of Jewish academies in Babylonia, Palestine, Syria, and Egypt from the 6th to the 11th centuries; later in Spain, Italy and in the eighteenth century in Vilna applied to any person outstandingly gifted in language and art; (after 19th- and 20th-century Yiddish usage) a genius, prodigy. Hence **gao·nic** *a.*
1780 *Encycl. Brit.* V. 3178/2 *Gaons*, a certain order of Jewish doctors, who appeared in the East, after the closing of the talmud. 1892 ZANGWILL *Childr. Ghetto* (1893) I. vii. 71 The light of his generation, the great Gaon..Rav Shemuel. *Ibid.* 72 It was like the Wilna Gaon come again. 1934 WEBSTER, Gaonic. 1935 S. LEWIS *It can't happen Here* xviii. 191 You Geonim never lap up anything but snow cooled nectar. 1941 G. G. SCHOLEM *Major Trends in Jewish Mysticism* ix. 341 The 'Gaon' Elijah of Vilna, the outstanding leader of Lithuanian Jewry. *Ibid.* 352 Assuming that these passages are subsequent Gaonic interpolations. 1964 H. KEMELMAN *Friday the Rabbi slept Late* (1965) i. 23 'What do you think of the Rabbi now?'..'A gaon,..a regular gaon.' 1968 *New Scientist* 23 May 385/3 The responsa written by Jewish legal experts in Medieval Spain since the period of the *gaonim* (spiritual heads of communities).

gap, *sb.*¹ Add: **6. a.** Also, a disparity, inequality or imbalance; a break in deductive continuity; a (usu. undesirable) difference in development, condition, understanding, etc.; in modern use freq. qualified by a preceding sb., as in *credibility gap, dollar gap, generation gap, missile gap* (see under the first elements). Cf. senses 6 b and 7.
1930 *Economist* 13 Sept. 477/1 The widening gap between income and expenditure disclosed by the latest revenue return. 1940 *Ibid.* 30 Nov. 662/1 The 'gap' in the Budget is of the order of £2,000 millions a year. 1952 *Sat. Rev.* 20 Sept. 9/3 *Gap* ought to be removed from public speech, as in the dollar *gap*, the export *gap*, inflationary and deflationary *gap*. 1956 A. J. AYER *Probl. Knowl.* ii. 80 They accept the existence of the gap between evidence and conclusion, but they hold that it can be bridged by a legitimate process of inductive reasoning. 1959 *Times Lit. Suppl.* 22 May 305/1 The Gap that has come to enjoy the dignity of capital letters, namely, the gap that has opened up in the past hundred years between men of science and students of the humanities. 1964 *Philos. Rev.* LXXIII. 165 (*title*) The Gap between 'Is' and 'Should'. 1967 *Nature* 4 Mar. 851/1 The phenomenon which has become known as the technology gap is more a matter of management than of technology. 1968 *Listener* 29 Aug. 273/3 It's this imbalance which gives the Soviet Union such an advantage in its dealings with the West, and it's far more important than any 'gap' in technology or arms. 1969 *Mind* LXXVIII. 27 He deals with the logical gap by jumping it. 1971 *Nature* 25 June 481/1 The 'technology gap'—the supposed disparity between the United States and the rest of the industrialized world in the capacity to exploit advanced technology.
b. Also, *to bridge* or *close a gap*.
1879 [see BRIDGE *v.*¹ 1 d]. 1927 *Cherwell* 25 June 238/1 A few days ago a London journalist was commenting.. on the cleavage between..the Oxford Set and the young officers of the Guards...However, Mr. Drury Lowe is the one person..who has successfully bridged this gap, if gap there be. 1944 *Ann. Reg. 1943* 50 The production of coal..had fully kept pace with consumption and in fact somewhat exceeded it, so closing the previous 'gap' of eleven million tons annually. 1950 S. E. TOULMIN *Exam. Place of Reason* xv. 223 All his experience and wisdom are needed to bridge the gap between facts and values. 1957 *BBC Handbk.* 84 'Forces' Requests'..is an attempt to close the gap between National Servicemen all over the world and their families at home in Britain. 1961 *Ann. Reg. 1960* 500 Efforts to 'bridge the gap' between the Common Market and the European Free Trade Association continued.
d. *Aeronaut.* The vertical distance between the upper and lower wings of a biplane.
1909 *Hazell's Annual 1910* 487/2 Another tailless biplane is the Cody, the main planes of which have..a gap of 9 ft. in the centre 1915 W. E. DOMMETT *Aeroplanes & Airships* 33 The gap between the planes, which are staggered, is 4 ft. 9 in. 1916 H. BARBER *Aeroplane Speaks* 63 There will be..loss of efficiency unless the gap between the top and bottom surfaces is

equal to not less than 1½ times the chord. **1936** B. Jones *Elem. Pract. Aerodynam.* vi. 99 Struts act as columns, and, if the gap is unduly big, the struts are excessively long and the structure is weak. **1949** J. W. Vale *Aviation Mechanic's Aircraft Man.* i. 25 The greater the gap, the smaller the interference of air flow between the upper and lower wing.

9. *gap-filling* sb. and adj.; **gap-bed**, the bed of a gap-bed lathe; a gap-bed lathe; **gap-bridge**, a bridge or piece of casting which closes the gap in a gap-bed lathe when not in use; **gap-rope** (see quot.).

1888 *Lockwood's Dict. Mech. Engin.* 157 *Gap bed*, a lathe bed having a portion recessed out in front of and below the headstock, to receive work larger in diameter than the height of the lathe centres, under ordinary conditions, will allow of. **1905** *Daily Chron.* 12 Sept. 1/3 (Advt.), Lathe, 7 in. centres, 12 ft. gap bed. **1955** C. T. Bower *Bk. of Lathe* ii. 44 The gap bed is very popular with British lathe manufacturers, but it has never been thought desirable by American makers. *Ibid.* 45 General purpose lathes..can have gap beds so that they can swing the unexpected big jobs. **1888** *Lockwood's Dict. Mech. Engin.*, Gap bridge. **1890** W. James *Princ. Psychol.* I. ix. 265 All the rest..is *internodal* consciousness, giving us the sense of continuity, but having no significance apart from its mere gap-filling function. **1958** *Times Lit. Suppl.* 14 Mar. 142/1 Thinking appears in an obvious sense as an exercise in gap-filling. **1964** C. Dent *Quantity Surveying by Computer* vi. 81 The whole of the vocabulary for that trade in which the heading occurs is now read on to the drum, in readiness for the gap-filling subroutine which is to follow. **1883** *Man. Seamanship for Boys* 60 Foot-Line or Gap Rope is a rope's end rove through a block at the bowsprit cap, and bent to an eyelet-hole in the foot of the jib, or to the clew of the jib.

gap, *v.* Add: **c.** To make a gap or breach in or between; to open (a gap or passage). Cf. last quot. for Gapped *ppl. a.* 2.

1893 *Funk's Stand. Dict.*, *Gap*,..to make an opening or cause a hiatus or breach in. **1916** 'Boyd Cable' *Action Front* 95 Eager hands tore down the sandbags to gap a passage for them. **1925** *Brit. Weekly* 5 Mar. 545/3 Demolishing two-thirds of the houses, and 'gapping' and 'loop-holing' the remainder. **1927** *Daily Express* 29 July 1/2 The sowing, 'gapping', 'singling', hoeing, and other processes. **1959** *Motor* 2 Sept. 76/1 Distributor points properly gapped.

d. *intr.* To be broken through at intervals; to have gaps; to gape open.

1948 D. Ballantyne *Cunninghams* 22 The cottage veranda gapped in places. **1951** in M. McLuhan *Mech. Bride* 74 Because it fits so well, the sides don't gap.

gape, *sb.* Add: **6.** *Comb.*: **gape-worm**, the worm that causes the gapes (see 3 a).

1873 L. Wright *Bk. Poultry* 196 The egg of the gape-worm. **1895** *Montgomery Ward Catal.* 217/3 Gape Worm Extractor..quickly removes, without injury to the chick, the worms..from the windpipe. **1935** *Discovery* Sept. 266 *Syngamus trachea*, the Gape Worm of poultry and other birds.

gaper. Add: **5.** An easy catch in Cricket. *colloq.*

1903 Wodehouse *Prefect's Uncle* viii. 117 Seeing you miss a gaper like that right into your hands made me think I was playing stump-cricket with the juniors. **1963** *Times* 7 Feb. 3/3 Certain of the younger members of the side were dropping some regular 'gapers'.

gapped, *ppl. a.* Add: **3.** *Mus.* Designating a scale or mode with less than seven notes, esp. the pentatonic scale.

1910 D. Macdonald *Irish Music* 3 The Ancient Irish used the gapped scale in many of their airs. **1933** *Times Lit. Suppl.* 2 Mar. 139/1 The gaps in the gapped modes are what gives such modal tunes their decisive character. **1962** *Listener* 5 July 37/2 The suggestion of gapped scales.

garage (gæ·rāʒ, gæ·rėdʒ), *sb.* [F. *garage*, f. *garer* to shelter.] **1. a.** A building, either private or public, intended for the storage and shelter of motor vehicles while not in use. **b.** A commercial establishment that sells petrol, oil, and similar products and freq. also undertakes the repair and servicing of motor vehicles. Also *attrib.* and *Comb.*

1902 *Daily Mail* 11 Jan. 6/7 The new 'garage' founded by Mr. Harrington Moore, hon. secretary of the Automobile Club...The 'garage', which is situated at the City end of Queen Victoria-street,..has accommodation for eighty cars. **1902** *Times* 1 Mar. 4/3 Motor Garage. **1902** A. C. Harmsworth et al. *Motors* 65 The simplification of the motor engine and the establishment of *garages*. *Ibid.* 306 In stations or 'garages' where a number of cars are kept. **1909** *Westm. Gaz.* 13 Apr. 4/2 A Midland firm of garage proprietors. *Ibid.* 18 Nov. 5/3 Customers who need garage room for business and private vehicles. **1934** P. A. Reynolds *Garage Organiz. & Managem.* i. 4 The third type of garage is..the Full Service business, which caters, in addition to the services already mentioned, for repairs of every description. **1940** C. S. Lewis *Let.* 1 Jan. (1966) 174 Garage-hands and motor touts. **1958** [see *forecourt]. **1965** *New Society* 11 Nov. 8/3 Whether it is garage or street parking or car ports..most people want their cars conveniently near their homes... The garage court is sometimes advanced as a compromise. **1971** 'H. Calvin' *Poison Chasers* i. 9 The garage man.. had come into the public bar. **1971** *Oxford Times* 3 Dec. 24 (Advt.), Excellently maintained detached freehold House..Separate access to brick garage.

2. A hangar for the storage of aircraft. (Disused.)

1909 A. Berget *Conquest of Air* II. vii. 280 The construction of *garages*, landing stations and shelters.

garage (gæ·rāʒ, gæ·rėdʒ), *v.* [f. prec.] *trans.* To place (a motor vehicle) in a garage for storage or repairs; to accommodate (a motor vehicle) at a garage. Also *absol.*

1906 *Daily Chron.* 26 May 3/7 They will garage your car.., wash it, clean it, adjust it, repair it, keep it always at 'concert pitch'. **1907** (*title*) Where to garage. **1928** Galsworthy *Swan Song* vii. 55 Having garaged his car, he remembered that he had not lunched. **1972** *Daily Tel.* 3 Feb. 15/3 We hope it will not be too long before Mr Hayward can..garage his car.

Hence **ga·raging** *vbl. sb.*, garage accommodation.

1926 *Glasgow Herald* 23 Mar. 9 One or two of the great London stores which had provided garaging for the use of their customers. **1936** *Evening Standard* 30 May 14/1 To save himself 7s. 6d. a week garaging. **1968–9** *A.A. Continental Handbk.* 375/2 There are facilities for garaging, minor repairs, and refuelling.

garagist (gæ·rāʒist). Also ‖ **garagiste.** [Fr. *garagiste.*] The owner or proprietor of a commercial garage; an employee of a garage.

1928 *Observer* 11 Mar. 12/2 The wayside garagist who sold the tins of petrol. **1958** *Spectator* 22 Aug. 247/3 The can was filled up by an Old Etonian *garagiste*. **1961** P. Ustinov *Loser* iv. 71 The local mechanic and only garagist for kilometers round. **1967** *Punch* 20 Sept. 426/1, I had not expected to meet with so much opposition and dismay when I casually mentioned my intention to the garagists. **1967** *Observer* (Colour Suppl.) 22 Oct. 33/3, I arrived at a French garage..one night and announced that the engine was falling out. The *garagiste* greeted the news with the tired smile of those accustomed to dealing with the ignorant.

garamity, var. *gor-amity.

Garamond (gæ·rămǫnd). *Typogr.* [f. the name of Claude *Garamond* (died 1561), a French typefounder.] Any of a class of typefaces cut by Garamond or based on his design or that of Jean Jannon. Also *attrib.*

[**1780** *Encycl. Brit.* V. 3178/2 The small Roman was, by way of excellence, known among the printers of these nations by the name of *Garamond's small Roman*.] **1868** H. N. Humphreys *Hist. Art Printing* ix. 144 The small Roman letters of this skilful engraver became technically known, not only in France, but in other countries, as Garamond type. **1926** 'P. Beaujon' in *Fleuron* V. 160 The seventeenth century..produced..[a] roman and italic *derived* from Garamont's designs..which..survives as the 'original Garamond'...There can be little doubt that this is the design of Jean Jannon, dating from 1615. **1928** *Scholartis Press Catal.* June, Printed by the Westminster Press in 14 point Garamond. **1929** *Encycl. Brit.* XVIII. 509/2 The 'Garamond' type steadily acquired influence. **1931** *Times Lit. Suppl.* 25 June, Suppl. p. i/3, The brilliant Garamond setting for Louys' 'Roi Pausole'. *Ibid.*, p. i/4 Jules Meynial shows..a catalogue well printed in handset Garamond. **1962** D. B. Updike *Printing Types* (ed. 3) II. xxiii. 234 While the Cloister or the Garamond—both brought out by the American Type Founders Company—may not be absolutely necessary to an office, a type of this historic class should be selected. **1966** H. Williamson *Methods Bk. Design* (ed. 2) viii. 85 The Garamond series was the first in the ambitious programme of matrix production undertaken by Monotype in 1922. *Ibid.*, Jannon, at the outset of his career, used Garamond types, and although he based his design on them it was new in several ways.

‖ **garance** (garaǹs). [Fr., see Garancin.] = Garancin.

1896 [see *free-board 2]. **1931** *Times Lit. Suppl.* 22 Jan. 56/3 Names of the dyes and juices..garance and arzica.

garbage, *sb.* Add: **5.** *garbage can, collection, collector, disposal* (all chiefly N. Amer.); **garbage disposal unit** (or **disposer**), a device, usually fitted to the waste-pipe of a kitchen-sink, which grinds up and disposes of small amounts of kitchen waste, esp. the remains of food.

1906 *N.Y. Even. Post* 23 Aug. 2 The landlords are not providing the tenants with garbage and ash cans. **1914** Kipling *Lett. of Travel* (1920) 231 Every conceivable subject from the disposal of a garbage-can to that of a corpse. **1967** E. M. Noblet in *Coast to Coast 1965–6* 164 A pitiless tipping into the street of the..garbage-can contents of human lives. **1911** *Daily Colonist* (Victoria, B.C.) 20 Apr. 6/3 The new system of garbage collection will be inaugurated as soon as a sufficient number of garbage cans are secured. **1968** *Globe & Mail* (Toronto) 3 Feb. 10/6 Other services such as mail, transportation and garbage collection have come to a halt. **1924** *Collier's* 19 Jan. 19/3 They determine whether the garbage collector or the street cleaner is responsible for the removal of dead cats from blind alleys. **1953** *Manch. Guardian Weekly* 8 Oct. 5 City garbage collectors had been especially indolent. **1900** *Engineering Mag.* XIX. 732 Garbage disposal and water supply. **1949** *Official Gaz.* (U.S. Pat. Off.) 6 Sept. 73/1 Electrically-driven, Garbage Disposal Unit. **1954** *Good Housek. Home Encycl.* 113/1 An electric garbage disposal unit can be fitted into the waste-pipe of some sinks. **1958** *Times* 13 Jan. 11/2 There is a garbage disposal unit for attachment to the sink which will eliminate all food waste such as small bones, vegetable scraps and peelings, fibrous material. **1940** S. Lewis *Bethel Merriday* xxiv. 245 We're paying for.. the new electric garbage disposer.

garbanzo (gāɹbæ·nzo). Also 8 **garvanzo.** [Sp.: see Calavance.] The chick-pea, *Cicer arietinum*.

1759 tr. *Venegas's Nat. & Civ. Hist. Calif.* I. 45 The same success attended the experiments made with.. garvanzo, or a kind of pease. **1841** G. Borrow *Zincali* I. vi. 104 A puchéra, the ingredients of which were beef, bacon, garbanzos, and berdolaga. **1918** *Chambers's Jrnl.* Jan. 33 The fat Castilian garbanzos. **1925** *Glasgow Herald* 31 July 6 These garbanzos or chichis give the peculiar flavour to soups. **1933** H. Allen *Anthony Adverse* VIII. lxi. 998 The ever-present garbanzo or chick-pea. **1950** C. S. Forester *Mr. Midshipman Hornblower* 275 To taste salt beef and biscuit again, and never again to taste beans and garbanzos.

Garbo[1] (gā·ɹbo). The name of the film actress Greta *Garbo* (real name Greta Gustafsson; born 1905), used in various locutions (chiefly *colloq.*) as: *to do a Garbo*, to behave like Greta Garbo; to be aloof or retiring; to seek seclusion; **Garboesque, Garboic, Garboish, Garbo-like** *adjs.*, resembling or characteristic of Greta Garbo; **Garbo hat**, a felt hat with a wide floppy brim.

1934 [see *do v. 11 d]. **1934** *Tit-Bits* 31 Mar. 12/2 'To do a Garbo'..means to be proud, aloof and unbending. **1937** *Times* 18 Oct. 15/4 We all fully understand the necessity for these Garboic tactics. **1942** Jespersen *Mod. Eng. Gram.* VI. xix. 326 Thus we have *Casanovesque*.., *Garboesque* (NP [= in a newspaper] 1936). **1955** A. West *Heritage* vi. 255 She was wearing a Garboish hat that framed her entirely unmade-up face. **1958** *Vogue* Apr. 134 (*caption*) A grey flannel Garbo pull-on hat. **1958** J. Courage in C. K. Stead *N.Z. Short Stories* (1966) 32 The night-sister and her Garbo-like face. **1962** N. Freeling *Love in Amsterdam* II. 74 Her features.. had, indeed, the Garboesque cast that is exceedingly attractive. **1963** —— *Because of Cats* vii. 105 Housewives..in playsuits and sunglasses and Garbo hats, languidly pushing energetic toddlers—also in Garbo hats—in go-carts. **1967** N. Marsh *Death at Dolphin* ii. 38 He loathes publicity and does a Garbo. **1970** R. Johnston *Black Camels* ii. 27 You really are in your Garbo mood, Ben.

garbo[2] (gā·ɹbo). *Austral. slang.* [f. Garbage *sb.* + Austral. dim. suff. *-o[2].] A dustman, a collector of rubbish.

1953 Baker *Australia Speaks* iv. 105 *Garbo*, a garbage man. **1967** *Sunday Truth* (Brisbane) 19 Mar. 13/1 The wage of the 'garbo' is in excess of $40 a week. **1969** *Sunday Mirror* (Brisbane) 33 The garbos of Shrewsbury.. are tired of wearing tattered jackets and trousers while emptying loads of old rubbish. **1971** *Guardian* 27 May 13/6 Australian garbos probably could not compete with English dustmen in the length and scale of their strikes.

garbologist (gāɹbǫ·lŏdʒist). Also **garbiologist.** [f. Garbage *sb.* + -ologist.] A dustman.

1965 *N.Z. Woman's Weekly* 4 Oct. 103/1 RCA tells us of their local 'garbiologists'. **1966** *New Scientist* 13 Jan. 97/3 One dustman in court last week called himself a garbologist. **1968** *Radio Times* 26 Sept. 23/4 Waste is now big technicological business. No wonder British dustmen are campaigning to have themselves officially renamed 'garbologists'.

garcinia (gāɹsi·niă). *Bot.* [mod.L. (Linnæus *Genera Plantarum* (1737) 343), f. the name of Laurent *Garcin* (1683–1751), French botanist and traveller, + -ia.] A member of the genus so named of tropical, evergreen trees, belonging to the family Guttiferæ and found in Asia, Africa, and Polynesia; some species produce gamboge, and another is the mangosteen.

1760 J. Lee *Introd. Bot.* xi. 25 In respect to its *Apex* or *Top*, it [*sc.* the calyx] is either *Acute, sharp*, as in *Primula*..*Obtuse, blunt*, as in *Nymphæa* and *Garcinia*; or with one of its Indents *lopped off*, as in *Verbena*. **1886** G. Nicholson *Illustr. Dict. Gardening* II. 43/2 Garcinias thrive in a peat and loam compost. **1890** W. J. Gordon *Foundry* 126 The garcinia that yields the gamboge. **1951** *Dict. Gardening* (R. Hort. Soc.) II. 858/1 Garcinias need stove conditions and a compost of loam and peat.

garçon. Add: (Earlier and later examples.) Also used *attrib.* or *Comb.* to designate a type of short hair-style for women, the Eton crop.

1788 A. Young *Jrnl.* 15 Sept. in *Travels* (1792) I. 87, I thought only of my poor mare being squeezed a night amongst the Bretagne garrans; sixpence, however, to the garçon, had effects marvelously to her ease. **1829** H. Foote *Compan. to Theatres* 142 Each spectator is expected to take refreshment, for which purpose marble tables, etc are in all parts of the theatre, and *garçons* are at hand to anticipate the wishes of the company. **1850** E. Ruskin *Let.* 9 Mar. in M. Lutyens *Effie in Venice* (1965) I. 153 Such a parting..with all the garçons, chambermaids, gondoliers. **1942** E. Paul *Narrow St.* iii. 20 Georges, the *garçon*, was serving drinks to a couple of officials and an oyster vendor. **1956** A. Wilson *Anglo-Saxon Att.* I. iii. 87 A young woman with black garçon-cut hair.

‖ **garçonnière** (garsonʸę̄r'). [Fr.] A bachelor's rooms or flat.

1927 *New Statesman* 15 Oct. 17 A young rake who.. turns his capercailzie hut into a *garçonnière*. **1938** H. G. Wells *Apropos of Dolores* iii. 124 In Paris, in her world, there is no such thing as an innocent garconnière. **1945** E. Waugh *Brideshead Revisited* viii. 178 Two emancipated American girls who shared a garconnière.

‖ **garde champêtre** (gard ʃampę̄tr'). [Fr., lit. 'rural guard'.] A rural policeman; a gamekeeper.

1814 M. Birkbeck *Journey through France* 97 In the country, the Garde champetres, a revolutionary institution, are the great means..of crushing them [*sc.* crimes] in the egg. **1831** *Edin. Rev.* LIV. 309 The village poacher will find him [*sc.* the farmer] a more active *garde champêtre* than any keeper. **1957** G. Bellairs *Death in High Provence* iii. 29 The *garde champêtre*, the mayor's runabout, mainly concerned with duties as game and forest warden.

‖ **Garde Mobile** (gard mobil). [Fr.] A French military force, now chiefly engaged in police activity; also (usu. with lower-case initials), a member of this force.

1848 *Illustr. London News* 1 July 428/3 The great barricade was stormed by National Guards, Cuirassiers, Republican Guards, and Gardes Mobiles. **1848** *Observer* 2 July 3/1 The combat was altogether fought by the army, the Garde Mobile, and by the shopkeepers and middle classes of Paris. **1857** C. M. Yonge *Dynevor Terr.* I. xx. 334 The brave boy *gardes mobiles* came crowding round him, all black with powder. **1935** G. Greene *Basement Room* 87 Praying secretly to the Virgin to send an angel, to send the police, to send the Gardes Mobiles.

garden, *sb.* Add: **1. d.** *to cultivate one's garden* [after Voltaire *Candide* (1759) xxx, 'Il faut cultiver notre jardin'], to attend to one's own affairs.

[**1759** tr. *Voltaire's Candid* xxx. 132 All that is very well, answered Candid, but let us take care of our garden.] **1931** O. W. Holmes *Let.* 7 May (1964) 324 It would be a bore to be ill abroad and I suppose there would be a good chance of being so. I will cultivate my garden. **1933** L. Strachey *Characters & Comm.* I. ii. 13 The precept 'Il faut cultiver notre jardin' has come down to the degenerate descendants of Candide in the form of 'Have an eye to the main chance'—a very different exhortation. The twentieth century has learned to cultivate its garden so well that it makes a profit of ten per cent. **1955** R. M. MacIver *Pursuit of Happiness* iv. 35 We must cultivate our own garden and find the joy of doing it in our own hearts.

e. Colloq. phr. *everything in the garden is lovely*, the situation is perfectly satisfactory; all is well.

[**1909** J. R. Ware *Passing Eng.* 125/1 *Everything is nice in your garden*, a gentle protest against self-laudation...This is said to be derived from one of the young princesses (probably a daughter of the Princess Beatrice) who made this reply when something in her garden at Osborne was praised by Her Majesty.] **1910** 'G. B. Lancaster' *Jim of Ranges* v. 110 You're all right, 'n I'm aw right..'n ev'y thing'n the garden's lovely. **1924** C. Mackenzie *Old Men of Sea* v. 71 The financial outlook's lovely. It's lovely, Mr. Marsham. Yes, everything in the garden's lovely, as Marie Lloyd used to sing. **1939** O. Lancaster *Homes Sweet Homes* 56 Everything in the garden, to use a contemporary phrase, was lovely. And then..the prevailing optimism received a shock. **1941** A. Christie *Evil under Sun* ii. 30 Then everything in the garden—or shall we say at the seaside?—is lovely, Mademoiselle. **1969** *Guardian* 13 Feb. 4/3 But not everything in the garden was lovely and the public, in particular, was not sure what higher education was for.

f. *to lead* (someone) *up the garden* (*-path*), to lead on, entice; mislead, deceive. *colloq.*

1925 E. Mannin *Sounding Brass* I. viii, They're cheats, that's wot women are! Lead you up the garden and then go snivellin' around 'cos wot's natcheral 'as 'appened to 'em. **1927** G. D. H. & M. Cole *Murder at Crome House* ix. 94 To lead Flint up the garden-path and relieve him of his cash. **1957** I. Murdoch *Sandcastle* ii. 29 I'm going to lead her up the garden. I've got her thoroughly foxed so far. **1963** *Times* 29 Jan. 12/6, I made these confessions to lead the police up the garden path.

4. a. *garden-book, -boy, -hat, -rake, -seed* (earlier example), *-shears* (earlier example), *soil, umbrella.*

1903 *Bookman* XXV. 61/1 The volume now before us comes opportunely as a justification of what has grown to be called a 'garden book'. **1906** A. Bennett *Let.* 6 Mar. (1966) I. 70, I expect another is his garden-book. **1937** A. Lamplugh (*title*) The garden book. **1936** *Discovery* Aug. 243/1 The common Kenya garden-boy—whom I have known to plant a cherished seedling upside down, and then thoughtfully water it. **1951** D. Lessing *This was Old Chief's Country* i. 12 Working in our house as servants were always three natives: cook, houseboy, garden boy. **1857** Dickens *Dorrit* I. xxviii. 246 In her simple garden-hat. **1967** E. Taylor *Second Thursday* iii. 39 Anna's garden hat..lay at the bottom of the steps. **1743** W. Ellis *Mod. Husb.* (1744) Aug. xvi. 83 A man with a Garden-rake raked in Cole-seed all over it. **1858** J. A. Warder *Hedges & Evergreens* 60 A light garden-rake. **1605** J. Rosier *Voy. G. Waymouth* (1887) 105 We..digged a plot of ground, wherein, amongst some garden seeds, we sowed peaze and barley. **1629** J. Parkinson *Paradisi* iii. 5 Thrift..may be kept, being cut with a paire of Garden sheeres. **1848** *Cultivator* V. 213, I..potted it in common garden soil. **1875** Mrs. Stowe *We & Neighbors* ii. 20 Catnip..which grew in a bit of garden soil back of the house. **1939–40** *Army & Navy Stores Catal.* 63/2 Loams, garden soils. **1907** *Ibid.* 238/3 Garden Umbrella..with brass jointed stem, 6 ft. long..with spike to screw into the ground. **1970** *Times* 28 May 8/1 In Sweden..one household in 15 owns..a garden umbrella. **1971** *E. Afr. Standard* (Nairobi) 10 Apr. 2/1 (Advt.), Garden umbrellas—picnic sets—safari mattresses.

5. a. *garden vegetable.*

1818 T. G. Fessenden *Ladies Monitor* 171 Provincial words [include]:..*bete* for beet, a garden vegetable. **1841** *N.Y. Hist. Soc. Coll.* 2nd Ser. I. 152 The Netherlanders.. have introduced every kind of garden vegetables.

6. garden carpet, (*a*) a species of Geometer moth (cf. *carpet-moth*); (*b*) (see quots. 1960 and 1963); **garden centre**, a place where gardening requirements are sold; **garden city**, (*a*) a name given to Chicago and certain other cities (freq. with capital initials); (*b*) a town of limited size, surrounded by a wide rural or agricultural belt, and planned so as to combine the advantages of town and country; **Garden Colony**, a name given to Natal, on account of the wealth and beauty of its flora; **garden county** *U.S.*, the county (of a state) most resembling a garden; the most fertile and beautiful part; **garden-craft**, the management of gardens; gardening; **garden egg**, a Jamaican name for the egg-plant or its fruit; **garden gnome**, a figure of a gnome used as a garden ornament; **garden magic**, magic practised by certain primitive tribes as an aid in the cultivation of their gardens; **garden-patch** *U.S.* = *garden-ground* (b); **garden-path**, a path in a garden (see also sense *1 f above); **garden-pea** (see sense 5 a in Dict.); *spec.* (see quot. 1960); **garden-room**, (*a*) space in a garden; (*b*) a room with a door or window opening out on to a garden; **garden sauce** (**sars, sarse, sass**) *U.S.*, garden vegetables (see Sauce *sb.* 4 a); **garden spot** *U.S.*, (*a*) a garden or a place suitable for a garden; (*b*) a region notable for its fertility, climate, etc.; **Garden State** *U.S.*, a name given to various American states, esp. New Jersey; **garden suburb**, a suburb organized on the same lines as, or having some of the characteristics of, a garden city; **garden town** = **garden city* (b); **garden-truck** *U.S.*, garden vegetables; **garden village**, a village organized on the same lines as, or having some of the characteristics of, a garden city.

1908 R. South *Moths Brit. Is.* II. 195 (*caption*) Garden Carpet at rest. **1928** *Sunday Express* 8 July 10/4 A fine 'firdausi', or garden carpet. **1960** H. Hayward *Antique Coll.* 127/1 *Garden carpets*, Persian carpets, showing formal designs with figures and animals. **1963** *Times* 8 June 1/7 There is reason to believe that the first important carpet was known as the Garden Carpet. When the sons of the Prophet took Ctesiphon in A.D. 637 they discovered a vast woven fabric of over one thousand square yards in area,..in which the motive was a garden in the full beauty of spring...The garden carpet is a symbol—the emblems of life and death passing to eternity in an endless chain. **1965** C. Kelway *Gardening on Sand* ix. 117 Look out for them [*sc.* dwarf trees] at the now-popular garden centres where many may be seen growing in containers ready for taking home. **1971** *Guardian* 25 Sept. 7/1 Plant container-grown Roses, Flowering Shrubs, Conifers..Lymm Garden Centre—on A56. **1848** *Gem of Prairie* (Chicago) 25 Nov. 3/3 My rambles round the Garden city. **1872** Schele de Vere *Americanisms* 665 Savannah, in Georgia,..claiming the name of Garden City. **1898** E. Howard *To-morrow* i. 13 Garden City, the Town-country magnet, which it is intended to build. **1902** —— (*title*) Garden cities of tomorrow. **1903** *Times* 29 Aug. 4/5 The Garden City Pioneer Company (Limited) has acquired about 4,000 acres of land near Hitchin on which to build the first garden city. **1914** *Scotsman* 6 Oct. 7/2 There are at present in existence a number of schemes such as the First Garden City at Letchworth, and the various garden suburbs which have been successfully started in various parts of the country. **1927** *Daily Tel.* 1 May 9/4 The Federal Capital City of Australia..its splendid design on 'garden city' lines. **1966** W. L. Creese (*title*) The search for environment. The garden city: before and after. **1896** 'S. Cumberland' *What I think of S. Afr.* xv. 175 Natal is called the 'Garden Colony'. **1902** *Westm. Gaz.* 28 July 10/1 It is the Garden Colony of South Africa. **1872** *Newton Kansan* 21 Nov. 2/3 This bids fair to be the favored spot in the garden county. **1891** J. D. Sedding (*title*) Garden craft, old and new. **1904** E. Wharton *Italian Villas* 5 To enjoy and appreciate the Italian garden-craft one must always bear in mind that it is independent of floriculture. **1927** E. S. Rohde (*title*) Garden-craft in the Bible, and other essays. **1811** W. J. Titford *Sks. Hortus Bot. Amer.* p. xiii, Garden Egg. (*Solanum Melongena.*) Cut in slices, parboiled and fried, resembles fried eggs. **1964** E. Huxley *Back Street New Worlds* xiv. 141 The Shepherd's Bush market has a shop devoted wholly to West African foods..like..garden eggs and fou-fou. **1969** *Daily Tel.* 11 Jan. 14/1 Melongene is our own familiar aubergine known in Jamaica as 'garden egg'. **1946** *Woman & Beauty* Dec. 31 Garden gnomes—*some* people like them. **1966** M. Kelly *Dead Corse* vii. 100 The owners of plastic flowers and garden gnomes, and plaster teal taking wing across the wall of the lounge-diner. **1929** B. Malinowski *Sexual Life Savages* ii. 36 In garden magic, the officiator plays an economically and socially important role and is the organizer and director of the work. **1957** P. Worsley *Trumpet shall Sound* iii. 71 The cult has come a long way from simple garden-magic. **1832** J. P. Kennedy *Swallow Barn* II. 224 Little garden-patches..where cymblings, cucumbers..flourished. **1924** R. Cummins *Sky-High Corral* 17 The fat cook puffed in from the garden patch. **1870** 'F. Fern' *Ginger-Snaps* 162 The irresolute hand.. could only lead them up and down that narrow garden path. **1932** B. Nichols (*title*) Down the garden path. **1629** J. Parkinson *Paradisi* II. lii. 522 Garden Pease are for the most part the greatest and sweetest kinds. **1681, 1832** Garden pea [see sense 5 a in Dict.]. **1951** E. David *French Country Cooking* 231 Put the contents of a large tin of garden peas through a sieve. **1960** *Which?* June 120/1 Canned peas come in two forms, 'garden' and 'processed'. 'Garden' peas are canned when they are fresh...'Processed' peas are dried peas which are soaked to bring them back to their ordinary size before they go through much the same canning procedure as garden peas. **1969** V. C. Clinton-Baddeley *Only Matter of Time* 37 Alice, bearing potatoes..and 'garden' peas which had not long been emancipated from the deep freeze. **1750** Garden-room [see sense 4 a in Dict.]. **1892** Stevenson & Osbourne *Wrecker* xx. 317 The garden-room, whither Lady Ann had now retired. **1961** *Times* 6 May 11/3 This exquisite garden-room is the most perfect of its kind in Italy. **1971** *Country Life* 24 June 1593/1 At the end of this wall Captain Radcliffe built a terrace and portico or garden room of stone in the Grecian style. **1791** *Amer. Museum* X. 179 For want of garden sauce, they..eat more flesh than is consistent with their health. **1833** J. Neal *Down-Easters* I. 91, I wanted cabbage or potaters or most any sort o' garden sarse. **1869** S. Bowles *Our New West* xi. 231 A load of grain, pork, or 'garden sass'. **1873** J. H. Beadle *Undevel. West* xxix. 504 The country West does not contain arable land enough to supply garden-sauce to an average population. **1921** C. Greer-Petrie *Angeline at Seelbach* 11 The next thing he fotch us was some gyarden sass and some biled beef. **1687** T. Faunce in *Rec. Town Plymouth* (1889) I. 190 We are bounded by goodman Watsons garden spot. **1813** *Niles' Reg.* IV. 317/2 The Rapids of the Miami may justly be termed the 'garden spot' of the territory. **1884** 'C. E. Craddock' *In Tenn. Mts.* I. 53 She was welcomed to a chair and a view of the weed-grown 'garden-spot'. **1898** T. N. Page *Red Rock* 298 It's the garden spot of the world—the money's jest layin' round to waste on the ground. **1865** *Daily Morning Chron.* (Washington, D.C.) 29 Sept. 2/3 Verily, Illinois is justly called 'The Garden State'. **1871** Schele de Vere *Americanisms* 659 Kansas is often called the Garden State, from the beautiful appearance of rolling prairies and vast cultivated fields. **1948** *Sat. Even. Post* 20 Nov. 57/2 When the first nip of frost chills the New Jersey air, cooks in the Garden State revive this recipe. **1967** Mrs. L. B. Johnson *White House Diary* 25 June (1970) 537 Betty Hughes was the tour guide...Her state [*sc.* New Jersey] is called the 'Garden State', and they have many truck farmers. **1905** S. A. Barnett *Let.* 28 Jan. in H. Barnett *Canon Barnett* (1918) II. 191 My wife has had a very busy week, interviewing people *re* the Garden Suburb. **1914** Garden suburb [see *garden city*, above]. **1953** C. Day Lewis *Italian Visit* ii. 42 Here is one corner of a foreign field That is for ever garden suburb. **1915** P. Geddes *Cities in Evolution* ii. 54 The conditions for labour and its real wages, in the innumerable garden-towns and villages which are springing up. **1807** P. Gass *Jrnl.* 51 The Rees..had left..some garden truck, such as squashes. **1923** M. Watts *L. Nichols* 292 A wagon-load of garden-truck. **1915** Garden village [see *garden town* above]. **1933** *Archit. Rev.* LXXIV. 120 He may preserve as much as he can and create a garden village which will be an addition to the neighbourhood.

garden, *v.* Add: **1. b.** *Cricket.* Of a batsman: to remedy any unevenness in the pitch by clearing away loose fragments, patting the ground flat, etc. *colloq.*

1897 *Encycl. Sport* I. 226/2 Fragments of grass and turf should be removed...The process of clearing the ground of *débris* is known to cricketers as 'gardening'. **1956** N. Cardus *Close of Play* 15 They both attended to the turf every ball, 'gardening' assiduously, and how thoroughly Sutcliffe would pat the earth.

gardening, *vbl. sb.* **3.** (Further examples.)

1716 H. Stevenson *Young Gard'ner's Director* p. iii, In all the Gard'ning Books. **1843** *Ainsworth's Mag.* IV. 473 She seemed ashamed neither of her gardening gloves..nor her quizzical companion. **1932** D. C. Minter *Mod. Needlecraft* 220/2 Remnants of cretonne..made up into a sewing or gardening apron. **1934** A. Thirkell *Wild Strawberries* xiii. 277 The schoolroom, where Lady Emily kept some of her valuable old gardening books. **1964** E. Bowen *Little Girls* III. v. 206 Gardening gloves crammed bulkily into a pocket.

Gardner (gā·ɹdnəɹ). The name of Captain M. W. *Gardner*, used attrib. and absol. to designate a type of machine gun invented by him, which has two to five barrels side by side.

1884 *Illustr. London News* LXXXIV. 222/1 The force was drawn up..with three Gardners on the left and three Gatlings on the right. **1899** *Daily News* 30 Jan. 5/1 The workshops were turning out two Gardiners [*sic*] and two Maxims every month. **1903** A. Conan Doyle *Green Flag* 11 Caught in the blast of lead from a Gardner.

‖ **gare** (gar), *sb.*[5] [Fr.] **a.** A dock-basin on a river or canal. **b.** A railway station. **c.** A pier, wharf, or the like.

1869 S. Northcote *Diary* 17 Nov. in A. Lang *Life* (1890) I. xi. 360 Two or three *gares* have been hollowed out, where vessels may lie. **1870** E. G. E. Ward *Jrnl.* 15 July in *Outside Paris* (1871) 2, I ran to the Gare of the Railway, and bought a paper just come from Paris. **1885** H. James *Little Tour in France* (1900) xxxviii. 250 The whole despotic *gare*: the deadly *salle d'attente*, the insufferable delays over one's luggage, the porterless platform. **1902** *Encycl. Brit.* XXVII. 708/2 Commander Edwards, R.N., proceeded down the Canal, taking possession of the *gares* and dredgers. **1906** *Daily Chron.* 10 Sept. 5/6 When seen at the gare he was wearing a bowler hat and had a grey beard. **1912** *Q. Rev.* Oct. 315 Gares, or mooring stations, are now provided.

‖ **gargouillade** (garguiyad). *Ballet.* [Fr., f. *gargouiller* to gurgle, bubble.] (See quots. 1957 and 1961.)

1930 Craske & Beaumont *Theory & Pract. Allegro Class. Ballet* 33 (*heading*) Gargouillade dehors. *Ibid.* 34 (*heading*) Gargouillade volé. *Ibid.* 76 Execute a *Gargouillade en dehors*. **1957** G. B. L. Wilson *Dict. Ballet* 132 *Gargouillade*, a brilliant series of steps in which the left leg describes two small circular movements in the air.. before the point of the left foot is drawn up to the knee of the supporting leg. **1959** *Times* 9 Jan. 6/2 He threw off the turns and leaps and *gargouillades*..with marvellous *éclat*. **1961** Webster, *Gargouillade*, a pas de chat with a double rond de jambe.

gargoylism (gā·ɹgoiliz'm). [f. Gargoyle + -ism.] **1.** Grotesqueness. *rare.*[-1]

1902 *Athenæum* 14 June 747/2 He is an artist in sensible nonsense, a master of gargoylism, a priest of the grotesque.

2. *Med.* A syndrome characterized by mental deficiency and skeletal deformities, including an abnormally large head, short limbs, and a protruding abdomen; Hurler's syndrome.

1936 Ellis et al. in *Q. Jrnl. Med.* New Ser. V. 119 The association of a peculiar type of osseous dystrophy, with congenital clouding of the cornea,..[etc.] has been noted sufficiently often to justify its being regarded as a definite syndrome...We have chosen the name 'Gargoylism' to describe it. **1965** R. H. Durham *Encycl. Med. Syndromes* 262 The term 'gargoylism' has been applied to this syndrome because the gross disfiguration resembles the gargoyles of Gothic architecture. **1970** C. G. Tedeschi *Neuropath.* iii. 67/2 In contrast, the skull deformity of gargoylism belongs to the dolichocephalic variety.

Garibaldi. Add: Now usu. pronounced (gæribǭ·ldi).

3. In full *Garibaldi biscuit*. A sandwich biscuit containing a paste of currants.

1896 J. T. Law *Grocer's Man.* 87 *Garibaldi*..with currants inside. **1909** H. G. Wells *Tono-Bungay* III. iii. 276 Instead of offering me a Garibaldi biscuit, she asked me with that faint lisp of hers, to 'have some squashed flies, George'. **1951** C. V. Wedgwood *Last of Radicals* i. 30 A cup of tea, with 'Garibaldi' biscuits ('squashed flies' we called them).

Garibaldian (gæribǭ·ldiăn, -bæ·l-), *a.* and *sb.* [f. Garibaldi + -an.] **A.** *adj.* Of, pertaining to, or supporting Garibaldi. **B.** *sb.* An adherent of Garibaldi. Also **Garibaldi·no** *a.* and *sb.*, **Gariba·ldist** *sb.*

1860 *Universe* 8 Dec. 1/2 The Garibaldian General Türr. **1861** E. Dicey *Cavour* xxiv, The Garibaldian Dictatorship. **1863** *All Year Round* 2 May 222/1 The Garibaldists are coming! **1864** *Dublin Rev.* July 144 Mr. Shaen, a zealous Garibaldian..openly maintained that there had been 'a plot'. **1892** O'Clery *Making of Italy* 121 The numerous expeditions which followed for the purpose of reinforcing..the Garibaldian army. **1904** Conrad *Nostromo* iv. 20 There were three doors in the front of the house, and each afternoon the Garibaldino could be seen at one or another of them. **1965** 'C. Hibbert' *Garibaldi & his Enemies* I. iv. 67 The fear of the Garibaldians..led the inhabitants to desert the town. *Ibid.* viii. 112 Anita had abandoned her Garibaldino uniform some days before.

‖ **garigue, garrigue** (garig). [Fr.] In the south of France, uncultivated land of a calcareous soil overgrown with low scrub; also, the vegetation found on such land.

1896 *Smithsonian Rep.* 414 Uncultivated lands, there called the garrigues. **1903** W. R. Fisher tr. *Schimper's Plant-Geogr.* III. v. 516 Such waste tracts, in South France termed '*garigues*', usually occur only on calcareous soil. **1914** A. G. Tansley in H. S. Thompson *Flowering Plants of Riviera* 6 In rocky places where the pines are not well developed or are absent altogether the limestone scrub is very open, and there is much bare rock between the shrubs. This type of vegetation is called *garigue*. **1963** G. Bellairs *Death in Wasteland* iv. 46 Part of it is *garrigue*, rocky uncultivated wasteland covered in prickly oaks. **1965** Polunin & Huxley *Flowers of Mediterranean* 11 Extensive areas of the hottest and driest terrain are covered with garigue.

garland, *sb.* Add: **9. Garland Day, Friday, Sunday** (see quots.).

1836 A. E. Bray *Descr. Tamar & Tavy* II. xxx. 289 Amongst the little boys, this day [*i.e.* 29th May, date of the Restoration 1660] goes by the name of *garland day*. **1933** *Irish Press* 31 July 1/7 For fourteen centuries pilgrims have come to Croagh Patrick on Garland Sunday. **1955** D. D. C. P. Mould *Irish Pilgrimage* ix. 134 Croagh Patrick attracts enormous crowds..the pilgrimage goes..on the last Sunday of July, the so-called Garland or Garlic Sunday. **1959** I. & P. Opie *Lore & Lang. Schoolch.* xii. 262 At Abbotsbury in Dorset 13 May has long been 'Garland Day'. The children customarily carry round a large flower garland on a pole, and show it at front doors. **1960** *Cath. Herald* 22 July 8/1 Hawkers will not be allowed..near Croagh Patrick from Garland Friday [*i.e.* the Friday before Garland Sunday] until Reek Sunday.

garlander (gā·ɹlăndəɹ). *rare.* [f. Garland *sb.* + -er[1].] One who carries a garland on Garland Day.

1939 F. Thompson *Lark Rise* xiii. 234 One of the most dependable of the older girls, who was made responsible for the behaviour of the garlanders. *Ibid.* 237 Sometimes the garlanders would forsake the road for stiles and footpaths.

garlic, *sb.* Add: **3. a.** *garlic bread, press, salt, sausage.*

1951 E. J. Marvel *Cook it Ahead* 3 (*heading*) Garlic Bread. **1960** 'I. T. Ross' *Murder out of School* xiv. 181 She brought a basketful of garlic bread from the oven. **1958** *Catal. County Stores Taunton* June 12 Garlic Press. **1968** *House & Garden* May 15/2 Garlic press..useful all-metal kitchen utensil for pressing garlic..11/6. *c* **1938** *Fortnum & Mason Catal.* 52/2 Salt..Garlic—per bot. 1/6. **1958** *Catal. County Stores Taunton* June 12 Garlic or Onion Salt. **1905** *Daily Chron.* 5 Jan. 4/5 The remnants of [horse] flesh—which even the garlic-sausage manufacturers cannot use. **1930** E. Waugh *Labels* 44 In the haversack on his back he carries a map and garlic sausage. **1968** P. Dickinson *Weathermonger* vi. 69 They..ate garlic sausage, processed cheese, bread and tomatoes.

garn (gāɹn), *int.* Colloq. (chiefly Cockney) pronunciation of *go on* (see Go *v.* 84 j), often used to express disbelief or ridicule of a statement.

1886 in H. Baumann *Londinismen* 62/1. **1888** J. Runciman *Chequers* 80 Garn, you farthin' face! **1914** G. B. Shaw *Pygmalion* I, in *Nash's Mag.* Nov. 152/2 The Note Taker (*whipping out his book*). Heavens! what a sound!..Ah—ah—ah—ow—ow—ow—oo! The Flower Girl (*tickled by the performance, and laughing in spite of herself*). Garn! **1922** 'R. Crompton' *Just-William* viii. 166 Garn! S'ours! We found it. **1925** *Glasgow Herald* 9 Jan. 8 He complained that if he used such words as 'garn' or 'struth' he was accused of vulgarity, whereas were he capable of imitating the peculiar sounds such as were heard from those north of the Tweed he would be able to move in any society. **1968** A. Holden *Death after School* iii. 22 'Garn,' called out someone, 'tell us somefing we don't know!'

Garnet[5] (gā·ɹnėt). *slang.* [f. the name of Sir *Garnet* Wolseley (1833–1913), leader of several successful military expeditions.] Phr. (*all*) *Sir Garnet*, highly satisfactory; all right.

1894 *Sporting Times* 1 Sept. 1/4 The start was all 'Sir Garnet'. **1913** C. Mackenzie *Sinister St.* I. I. iv. 61 That's All Sir Garnet, and don't you make no mistake. **1941** H. G. Wells *You can't be too Careful* I. i. 17 'All's well with the Missus, Tewler?' he asked. 'All Sir Garnet, Sir,' said Mr Richard Tewler. **1958** 'A. Gilbert' *Death against Clock* 111 She'd been knocked out..and her heart not being quite Sir Garnet did the rest.

Garnett (gā·ɹnėt), *sb.* [f. as the vb.] A 'Garnett's machine' (see Garnett *v.*); **Garnett tooth**, a form of saw-tooth used in these machines.

1884 W. S. B. McLaren *Spinning* ix. 189 The machines by which the first three of these materials are made are various. The one for opening the looser waste is called a Garnett. **1892** J. Nasmith *Students' Cotton Spinning* 132 It is customary to cover the licker-in with a set of teeth...These are called 'Garnett' teeth.

garniture. Add: **1. c.** *garniture de cheminée* (see quot. 1960).

1900 F. Litchfield *Pott. & Porc.* iv. 39 The famous Sèvres *garnitures de cheminées*. **1960** R. G. Haggar *Conc. Encycl. Cont. Pott. & Porc.* 203/2 *Garniture de cheminée*, a set of five vases, three baluster-shaped with covers, and two beakers with trumpet-shaped apertures called 'flutes', decorated *en suite*.

Garo (gā·ro). Also **Garrow.** A member of one of a group of Mongoloid tribes of the Garo Hills, Assam; also, their language. Also *attrib.*

1872 E. T. Dalton *Descr. Ethnol. Bengal* 88 The Garo apology for a robe does not reach half down the thigh. **1880** *Encycl. Brit.* XII. 777/2 The following are the twenty principal dialects of the Tibeto-Burman group: —(1) Cáchári or Bodo, (2) Garo, [etc.]. **1883** *Ibid.* XV. 295/1 Of the aboriginal tribes [of Maimansinh] the most numerous are the Hajongs (24,936) and Gáros (10,997). **1898** C. M. Yonge *J. Keble's Parishes* v. 54 Among the Garrows, a people of Hindustan, the youngest *daughter* inherits the property. **1912** A. Gordon *Life A. H. Charteris* 339 The plains at the foot of the mountains—called the Dooars—are peopled by Mechis, Rajbunsis, and Garos. **1959** *Chambers's Encycl.* VII. 482 The Assam-Burmese group is represented in eastern India, chiefly in Assam. There is very great linguistic diversity in this area. Some of the more important languages are..Gārō, [etc.].

garope, garoup(h)a, varr. Grouper.

garrapata, var. *carrapato.

garri (gæ·ri). *W. Afr.* Also **gari, garry.** = Cassava 2.

1926 P. A. Talbot *Peoples S. Nigeria* III. xxxvii. 908 Cassava grated up into 'Garri', or beaten into 'fufu', is becoming more and more general. **1946** Forde & Scott *Native Economies of Nigeria* ii. 106 The high incidence of pellagra in association with a gari (cassava flour) diet was due not only to the lack of the anti-pellagric vitamin factor..but to a lack of organic sulphur. **1962** *New Scientist* 27 Sept. 660 Gari, a meal made by fermenting cassava root. **1964** J. P. Clark *Three Plays* 115 His.. sentence to a six-months' diet On beans and garri. **1968** *Guardian* 4 July 1/1 His official supply of nine bags of garry (an inert vegetable staple), two fish, and two cups of salt.

garrigue: see *garigue.

garrison, *sb.* Add: **6.** *garrison duty*; **garrison cap** *U.S.*, a peakless cap worn esp. as part of a military uniform; **garrison state** (see quot. 1954).

1947 S. Bellow *Victim* i. 6 A Negro janitor in a garrison cap was washing the stairs. **1971** *U.S. Army Reg. 670–5* (*Uniform & Insignia, Male Personnel*) Jan. 14 (*caption*) Distinctive unit insignia worn on garrison cap. **1813** Southey *Nelson* i. 37 The few who were able to perform garrison duty. **1937** H. D. Lasswell in *China Q.* II. 643 In the garrison state the specialist on violence is at the helm, and organized economic and social life is systematically subordinated to the fighting forces. **1941** —— in *Amer. Jrnl. Sociology* XLVI. 455 We are moving toward a world of 'garrison states'—a world in which the specialists in violence are the most powerful group in society. **1954** Webster Add., *Garrison state*, a state dominated by military rather than by civilian personnel and policies, esp. one whose military preparations threaten to convert it into a totalitarian state. **1959** *Listener* 26 Mar. 553/2 Cultural activities, on the one hand, tend to become a commercial part of an overdeveloped capitalist economy or, on the other, an official part of the Science Machine of the Garrison State.

‖ **garrocha** (găro·tʃă). [Sp.] (See quots.)

1846 R. Ford *Gatherings from Spain* xxii. 301 The bull..well remembers this *garrocha*, or goad, by which herdsmen enforce discipline. **1882** H. de Windt *On Equator* ix. 136 Each [picador] was armed with a *garrocha*, or spear, the blade of which..is only about an inch long, as the picadores are not allowed to kill the bull, but merely to irritate and goad him. **1910** *Encycl. Brit.* IV. 790 As the bull attacks the horse, the *picador*, who is armed with a short-pointed, stout pike (*garrocha*), thrusts this into the bull's back with all his force. **1957** R. Campbell *Portugal* 110 The mounted *campinos* and grazers are skilled in the use of the *garrocha* or cattle-spear. **1967** McCormick & Mascareñas *Compl. Aficionado* vii. 215 Mounted men pursue the calves in order to upset them with the blunt pole, or garrocha, which they carry.

garrupa, var. Grouper.

Garry (gæ·ri). The name of Nicholas *Garry* (1781–1856), an officer of the Hudson's Bay Company, used attrib. to designate a species of oak, *Quercus garryana*, native to south-western British Columbia and the Pacific coast to the south.

1908 G. B. Sudworth *Forest Trees Pacific Slope* 283 Next to valley oak, Garry oak, known most commonly as 'white oak', is the largest oak in the Pacific coast region. **1926** *Daily Colonist* (Victoria, B.C.) 25 July 17/1 The Garry oaks rivetted his attention as the most beautiful trees he had ever seen. **1952** D. F. Putnam *Canad. Regions* xxi. 465/1 Around Victoria..such trees as madrona and Garry oak indicate a similarity to some Mediterranean environments.

garrya (gæ·riă). *Bot.* [mod.L. (D. Douglas 1834, in *Bot. Reg.* XX. 1686), f. the name of Nicholas *Garry* (see prec.).] An evergreen shrub of the genus so called, esp. *Garrya elliptica*, which is native to California and Oregon, and cultivated for the ornamental catkins it bears during the winter.

1834 *Bot. Reg.* XX. 1686 Garrya elliptica. Elliptic-leaved Garrya. **1854** *Explor. Route to Pacific* (*U.S. War Dept.*) *Botany* IV. 136 Colonel Frémont found on the Upper Sacramento, 'above the Great Cañon', in 1846, a Garrya nearly allied to this species. **1893** A. D. Webster *Hardy Ornamental Flowering Trees & Shrubs* 50 There are male and female plants of the Garrya, and..the former is the more ornamental. **1937** C. H. Middleton *Winter-flowering Plants* 61 Young Garryas do not transplant readily. **1963** *Oxf. Bk. Garden Flowers* p. viii, Garrya does best if grown on a wall or sunny bank.

Garshuni (gāɹʃū·ni). Also **Carshuni, Karshuni.** [ad. Arab. *karšūnī*.] Arabic written in Syriac characters.

1856 T. H. Horne *Introd. Holy Script.* (ed. 10) IV. 26 The propaganda at Rome issued an edition in 1703 in Syriac and Carshuni (i.e. Arabic in Syriac letters) for the use of the Maronites. **1902** *Daily Chron.* 13 Feb. 5/1 His accomplishments in this direction included the Karshuni alphabet, in which very few Orientalist professors are skilled. **1910** *Cath. Encycl.* IX. 685 The Maronite is a Syrian Rite, Syriac being the liturgical language, though the Gospel is read in Arabic for the benefit of the people. Many of the priests, who are not sufficiently learned to perform the Liturgy in Syriac, use Arabic instead, but Arabic written in Syriac characters (*Karshuni*). **1912** F. J. Bliss *Relig. Mod. Syria & Palestine* iii. 124 The books containing these are written or printed in Syriac character, this combination of Arabic words and Syrian form being known as Karshûni. **1922** J. Hastings *Encycl. Relig.* XII. 168/1 In Syria proper..many of the prayers are said in an Arabic translation, so as to be intelligible to the people; they are then written in Syriac characters, and this combination of Syriac and Arabic is called Carshuni (Syr. *garshūnī*). **1948** D. Diringer *Alphabet* iv. 269 There are rare instances of Arabic being written in non-Arabic scripts, for instance, in *garshuni*, or *karshuni*, which is the Syriac script adapted to Arabic. **1957** P. K. Hitti *Lebanon in History* xxxi. 457 Introduced probably by some Maronite student of Rome, this establishment put out in 1610 the Arabic Psalms in Syriac characters (Garshūni).

garter, *sb.* Add: **7.** See also Prick *v.* 29; *to fly the garter*: see Fly *v.*[1] 4 b; *to have one's guts for garters*: see *gut *sb.* 1 b.

8. garter belt, a suspender-belt; **garter-snake**, (*b*) *S. Afr.* the name of various banded snakes, as *Elaps lacteus*; **garter-stitch**, the simplest stitch in knitting, orig. used in making garters; also called *plain knitting*.

1959 W. Brown *Cry Kill* viii. 76 Naked except for her dark nylon stockings and a garter belt. **1963** *Time* 4 Jan. 40/3 No girdle or garter belt was needed [for leotards]. **1971** 'D. Shannon' *Whim to Kill* ii. 30 Her white nylon panties and garter-belt had been ripped off. **1789** W. Paterson *Narr. Journeys Country of Hottentots* 163 The Kouse Band, or Garter Snake, is another of the poisonous reptiles of that country. **1910** *East London Dispatch* 10 June 6/2 The *Garter Snakes* and the Vipers are the only snakes of South Africa which permit themselves generally to be closely approached without evincing much concern. **1931** R. L. Ditmars *Snakes of World* xiii. 165 Garter snakes or Coral snakes. **1947** J. Stevenson-Hamilton *Wild Life S. Afr.* xxxvi. 328 The various species of garter snakes (*Elaphechis* and *Homorelaps*) likewise are akin to the true cobra. **1909** *Daily Chron.* 8 Dec. 9/4 Sixty stitches are loosely cast on, and sixty rows of garter-stitch knitted. **1970** M. Hamilton-Hunt *Knitting Dict.* 8 Garter stitch is used for this corner.

garth[1]. **2.** Delete †*Obs.* and add later examples.

1894 J. C. Atkinson *Memorials of Old Whitby* 43 A *garth* is a fence of any kind,—earth, stones, turves, posts and rails. **1908** W. G. Collingwood *Scand. Brit.* 193 He made his *bær*, a group of buildings, in the *tún*, or homefield, which he manured and mowed for hay, and surrounded with a garth.

Gartner (gā·ɹtnəɹ). Also (erron.) **Gärtner, Gaertner.** The name of Herman Treschow *Gartner* (1785–1827), Danish anatomist, used to designate certain anatomical structures distinguished or described by him; as *Gartner's canal* or *duct*, a vestigial part of the Wolffian duct in some female mammals, running from the parovarium through the broad ligament to the vagina. Hence **Gartne·rian** *a.*; esp. *Gartnerian cyst* (see quot. 1900).

1874 Dunglison *Dict. Med. Sci.* 447/2 *Gärtner, Canals or Ducts of.* The excretory ducts of the Wolffian bodies disappear in the human female, but in some of the lower animals, as the pig, they persist, and receive this name. **1876** *Quain's Elem. Anat.* (ed. 8) II. 821 In the sow and several ruminants..the subdivided upper tubular part or epoophoron has disappeared, and the main part (middle part of the Wolffian duct) remains in the duct of Gaertner, a strong, slightly undulated tube. **1900** Dorland *Med. Dict.* 188/1 *Gartnerian cyst*, a cystic tumor developed from Gärtner's duct. **1901** C. H. Roberts *Outl. Gynæcol. Path.* ii. 36 The Gartnerian (parovarian) duct can be actually recognised in some animals (cows and pigs) and traced to the vagina giving rise to cysts in those animals. **1908** J. Campbell *Obstetr. & Gynaecol.* 233 In removing Gärtnerian cysts, damage to the ureters must be guarded against. **1957** M. F. Campbell *Princ. Urology* vi. 125 Persistence of Gartner's duct explains ectopic ureteral opening into the vagina, cervix, or uterus. **1968** Passmore & Robson *Compan. Med. Stud.* I. xxxvii. 58/2 The rest of the mesonephric duct degenerates, although part of it may persist in vestigial form as Gartner's duct.

Gärtner: see *Gaertner[1], *Gaertner[2].

‖ **garua** (gārū·ă). [Peruvian Sp.] = *camanchaca.

1840 *Penny Cycl.* XVIII. 3/1 From June to November, the sky is covered with a kind of fog, which is called the *garua*. **1847** T. Ross tr. *Von Tschudi's Trav. Peru* ix. 248 In the oases the garuas are much heavier than in the adjacent wastes. **1914** C. R. Enock *Ecuador* 195 The Peruvian coast..receives..a light mist-drizzle, known as *garua*. **1927** W. G. Kendrew *Clim. Cont.* (ed. 2) 313 At Lima..heavy showers are rare, the usual rain being a fine drizzle or a wet mist called Garua.

garuda (ga·rŭda). Also **Garuda, Garuṛa.** [Skr. *garuḍá.*] **a.** In Indian mythology, the name of a fabulous bird, half-eagle, half-man, ridden by the god Vishnu. **b.** The eagle in the official seal of Indonesia.

1882 M. Lockwood tr. *Lenormant's Beginnings Hist.* (1883) ii. 89 There is a singular analogy between these symbolic beings and the *Garuda*, or rather *Garudas*, of the Aryans of India, genii, half men and half eagles. **1924** E. Best *Maori* II. xxi. 574 The myth and representations of Vishnu and the two *garuda* birds that represent the powers of Good and Evil. **1952** W. D. P. Hill *Holy Lake of Acts of Rāma* 510 Garuṛa..has the head, wings, talons and beak of an eagle or vulture and the body and limbs of a man. **1964** *Telegraph* (Brisbane) 19 Nov. 8 (*heading*) Cage me a garuda! *Ibid.*, The mythical garuda bird, Indonesia's national emblem.

gas, *sb.*[1] **3. c.** For 'hydrogen' read 'hydrogen, helium', and for 'balloon' read 'balloon or airship'.

f. Any of various gases used in warfare to cause poisoning, asphyxiation, irritation, etc., to the enemy; freq. preceded by defining word, as *asphyxiant, asphyxiating, lachrymatory, mustard, nerve, poison, tear gas.*

1900 F. W. Holls *Peace Conf. Hague 1899* vii. 337 Concerning the prohibition of the use of projectiles containing asphyxiating gas. **1915** D. O. Barnett *Lett.* 28 Apr. 124 You may like to know we've got a way of competing with asphyxiating gas. **1915** *War Illustr.* 8 May 282/1 Germans held back by their own gas. **1921** [see *asphyxiant *a.* and *sb.*]. **1959** *Chambers's Encycl.* III. 334/1 Gas was first used on a large scale in the first world war.

4. (Earlier example.) Also, a jet or jets of gas used for heating or cooking.

1851 Mrs. Gaskell *Let.* *c* 28 Mar. (1966) 147 They acted in the outer lobby, under the gas. **1896** E. Turner *Little Larrikin* xiv. 160 Several contrivances she saw about. One over the gas, for instance—a piece of hoop-iron, bent hook-wise.

5. b. Colloq. phr. *all* (or *everything*) *is gas and gaiters*, everything is very satisfactory; (*all*) *gas and gaiters*, pompousness, verbosity.

1839 Dickens *Nickleby* xlix. 489 She is come at last —at last—and all is gas and gaiters! **1923** A. C. Benson *Trefoil* iii. 26 My father was profoundly irritated by him, and..said something..about 'gas and gaiters' which seemed to us a harsh description of so pretty a man. **1925** A. Christie *Secret of Chimneys* xiii. 124 I've only got to get hold of dear old Stylptitch's Reminiscences.. and all will be gas and gaiters. **1932** G. B. Shaw *Adv. Black Girl* 67 Its [*sc.* the Bible's] one great love poem is the only one that can satisfy a man who is really in love. Shelley's Epipsychidion is, in comparison, literary gas and gaiters. **1961** Wodehouse *Ice in Bedroom* ii. 21 She cries 'Oh, Freddie darling!' and flings herself into his arms, and all is gas and gaiters again.

c. Fun; a joke. *Anglo-Irish slang.*

1914 Joyce *Dubliners* 24 He told me he had brought it [*sc.* a catapult] to have some gas with the birds. Mahony used slang freely. **1962** E. O'Brien *Lonely Girl* ii. 20 'Let's do it for gas,' Baba said. **1965** M. Kenyon *May you die in Ireland* xxiii. 197 Gas is Irish for joke—gag, fun, having a wonderful time.

d. (Esp. prec. by indef. article.) Something or someone that is very pleasing, exciting, impressive, admirable, etc. Also *attrib.* or quasi-*adj.* Cf. *gasser 2. *slang* (orig. *U.S.*).

1957 J. Baldwin in *Partisan Rev.* III. 357 Brand-new pianos certainly were a gas. **1962** *Austral. Women's Wkly.* Suppl. 24 Oct. 3/2 *Gas*, anything which is very good. **1963** *Guardian* 5 July 11/3 You can listen to his old records now and they are still a gas. **1963** L. Deighton *Horse under Water* xxxii. 124 'Yeah,' he said, 'got me an E-type Jag: Cambridge blue—wire wheels—it's a gas.' **1964** 'J. Welcome' *Hard to Handle* vii. 79 I'm bound to say I thought it rather gas at the time. **1967** *Melody Maker* 27 May 10/6 Altogether an indefinite sound except for a gas build-up in the middle. **1971** *Frendz* 21 May 16/1 The Stones..were a screaming, speeding, sexy gas.

6. a. (sense *3 f) *gas alert, attack, bomb, -bombardment, war, warfare*; also *gas fire* (later example); **b.** *gas-driven, -filled, -heated* adjs.

1918 E. S. Farrow *Dict. Mil. Terms* 256 When conditions of atmosphere and wind are favorable for a gas attack preparations are made to meet it. This is known as *gas alert*. **1916** *King's Royal Rifle Corps Chron. 1915* 20 The Germans opened a burning gas attack upon the front of the trenches we had captured from them ten days before. **1971** *Daily Tel.* 13 Sept. 1/5 Hundreds of New York State police prepared last night for a gas attack on 1,000 rebellious prisoners holding 38 hostages at Attica prison. **1915** *Sphere* 1 May 107/2 The Germans throw their gas bombs with a species of sling. **1920** G. K. Rose *2/4th Oxf. & Bucks. Lt. Infty.* 203 Throughout the night of August 7/8..a heavy gas-bombardment was kept up. **1899** *Westm. Gaz.* 23 June 8/3 Gas-driven cars, similar to those running from Blackpool to Lytham. *a* **1918** W. Owen *Poems* (1963) 95 Gas-driven busses. **1907** *Jrnl. Soc. Arts* LV. 596/2 Gas-filled airships have lately been so improved. **1919** *Conquest* I. 25/3 A modern half-watt gas-filled lamp. **1932** *Discovery* Oct. 319/2 The..introduction of thermionic valves containing mercury, called 'gas-filled relays' and 'thyratrons'. **1971** F. Raphael *Who were you with Last Night?* 113 She's slowly turning blue in a gas-filled room. **1910** *Encycl. Brit.* XIII. 160/1 Gas fires, as a substitute for the open coal fire, have many points in their favour. *a* **1877** Knight *Dict. Mech.* I. 946/1 Gas-heated furnace. **1963** *B.S.I. News* Apr. 31 (*heading*) Gas-heated catering equipment. **1931** 'Miles' (*title*) The gas war of 1940. **1916** W. Owen *Let.* Aug. (1967) 402, I am now as well up in *Gas* Warfare as can be.

7. gas alarm, (*c*) *U.S.* an alarm device operated by gas; (*d*) a warning of the presence of poisonous gas; **gas amplification**, a process in which, in a strong electric field, an ion produced in a gas by ionizing radiation gives rise to further ions; also, the factor by which the total ionization is increased by this process; **gas analysis**, the measurement or estimation of the quantity of different gases in a mixture; = Gasometry; so *gas analyser*; **gas arc**, an electric arc in any of the rare gases under high pressure; also, a lamp using a gas arc; **gas-balloon** = Balloon *sb.*[1] 6; **gas barrel** (see quot. 1904); **gas-bill** (*b*) (examples); **gas-cape**, a cape for protection against poison gas; **gas-cell**, a cell containing gas in an airship; **gas centrifuge**, a centrifuge for partially separating a mixture of gases, esp. gases that are isotopes of a single element or compounds of them; hence **gas-centrifuging** *vbl. sb.*; **gas chamber**, (*b*) a chamber in which gas masks are tested or demonstrated; (*c*) esp., one of the chambers used by the Germans in the 1939–45 war for exterminating groups of human beings by gas-poisoning; any place of execution by gas-poisoning; **gas chromatogram, chromatograph** (see *chromatogram, *chromatograph); **gas chromatography**, chromatography analogous to liquid chromatography but using a carrier gas as the moving phase, which carries the gaseous or vaporized sample through a column containing the solid or liquid stationary phase so that the sample emerges separated into its constituents; hence **gas-chromatographic** *a.*, relating to or involving gas chromatography; **gas-cooker**, a cooker in which the heat is supplied by the burning of gas; a gas-stove; so *gas-cooking*; **gas-cooled** *a.*, of a nuclear reactor, having a gaseous coolant; **gas-cylinder**, a cylinder in which pressurized gas is stored; **gas-detector**, substitute for def.: a device for detecting the presence of poisonous gas; **gas-discharge**, applied *attrib.* to a gas-filled lamp, tube, etc., designed so that electric discharges may be produced in it; **gas dynamics**, the study of the dynamic properties of gases; **gas-escape**, a leakage of gas; **gas-fired** *a.*, heated by the combustion of gas; **gas-firing**, (*b*) a gas-fired system of heating; **gas-focused** *a.*, applied to a cathode-ray tube that makes use of the ionization produced in residual gas to focus the beam of electrons; **gas gangrene**, a rapidly spreading form of gangrene marked by the evolution of gas and usually resulting from the infection of deep wounds by *Clostridium* species; **gas-gong**, a gong to give warning of a gas attack; **gas-gun**, (*b*) a gun using gas as a propellant or as fuel; **gas helmet** = **gas mask*; **gas-house** (orig. *U.S.*), a gas-works, or a building forming part of a gas-works; also *attrib.* (U.S.), designating a run-down area or its inhabitants (see also quot. 1926); **gas kinetics**, the study of the kinetic properties of gases; so **gas-kinetic** *a.*; **gas laws** *Physics*, a group of laws (as Boyle's law, Graham's law) that describe the physical properties of gases in general; **gas-lighter**, (*a*) a device for igniting gas; (*b*) a cigarette-lighter in which the fuel is a gas; **gas-liquid** *a.*, designating or pertaining to a chromatographic process in which the moving phase is gaseous and the stationary phase is liquid; **gas-mantle** = Mantle *sb.* 5 g; **gas mask**, a mask used as a protection against poisonous gas; hence *gas-masked* adj.; **gas-meter** (additional examples); **gas-office**, the office of a gas-company; **gas officer**, an army officer responsible for the precautionary measures taken against gas attacks; **gas oil** (see quot. 1949); **gas oven** (further examples, with allusion to its use as a means of committing suicide); (*b*) = **gas chamber* (c); **gas-pipe chair**, a cantilever chair with legs and arms resembling gas-pipes; **gas pliers** (see quot. 1940); **gas poker**, a hollow poker perforated with holes through which gas can be made to flow, and which on being lit provides heat for kindling a fire; **gas-ring** (*b*) (examples); **gas shell** (see quot. 1918); **gas show** [Show *sb.*[1] 5 c], a detected escape of natural gas at the surface, taken as indicating that oil may be present underground; **gas-spurt**, little heaps or lumps occurring on the surface of some strata, thought to be due to the intermittent escape of gas from decomposing organic matter; **gas starter** (see quot. 1935); **gas storage**, the preservation of fruit and vegetables in carbon dioxide or other gases; so *gas stored* adj.; **gas-stove** (additional examples); **gas thread**, a standard form of screw thread of relatively fine pitch, used on metal tubes; **gas trap** = Trap *sb.*[1] 8 a; **gas-turbine**, a turbine in which the motive power is derived from a flow of gas; also (now more commonly), an internal combustion engine in which air is compressed, heated by combustion with fuel, and the expansion of the resulting hot gases used to power a turbine; also *attrib.*; **gas-vacuole** [a. G. *gasvacuole* (H. Klebahn 1895, in *Flora* LXXX. 252)], a type of vacuole found in certain bacteria and blue-green algæ, containing gas; **gas van,** a mobile gas chamber (senses b and c above); **gas welding,** fusion welding in which the metal is heated by the combustion of a gas; hence *gas-welded* adj.

1866 *Rep. Comm. Patents 1864* (U.S.) I. 364 The construction of a gas alarm, for the protection of property. **1915** D. O. Barnett *Lett.* 183 There was a false gas alarm last night. **1938** Gas-alarm [see **gas-detector* below]. **1933** A. F. Collins *Exper. Telev.* iv. 82 This great increase in sensitivity and current-carrying power is due to what is called gas amplification. **1950** D. H. Wilkinson *Ionization Chambers* vi. 143 When fewer electrons are released in the initial ionization..the 'signal' must obviously be increased, and gas amplification may be resorted to. **1956** *Nature* 25 Feb. 391/1 The latter process, particularly photo-electric effects and positive ion bombardment of the cathode, have set an upper limit..to the gas-amplification which has been practicable in single-stage devices. *a* **1884** Knight *Dict. Mech.* Suppl. 378/1 *Gas analyzer*, an instrument for determining the presence and quantity of the gases obtained by the destructive distillation of coal. **1956** *Nature* 25 Aug. 407/1 A gas analyser depending on the variation of the velocity of sound with the nature of the gas. **1859** *Proc. R. Soc.* IX. 218 In Bunsen's admirable method of gas analysis, considerable time and trouble are expended in observing the exact temperature and pressure to which the gas is subjected at the time of measurement. **1887** W. Dittmar *Quant. Chem. Analysis* 155 Gas Analysis, in the customary acceptance of the term, comprises only the *gas volumetric* methods for the analysis of gaseous bodies. **1933** H. A. Daynes (*title*) Gas analysis by measurement of thermal conductivity. **1938** *Thorpe's Dict. Appl. Chem.* (ed. 4) II. 687/1 The sensitiveness of spectroscopy as a method of gas analysis varies greatly according to the gas under test. **1949** *Trans. Illum. Engin. Soc.* XIV. 19 The radiation from the Gas Arc is characterised by an intense continuum extending from the ultra violet through the visible region into the infra red. **1951** W. R. Stevens *Princ. Lighting* v. 117 Xenon may be used in either air-cooled or water-cooled lamps (for example, the 'gas arc'). **1960** H. Cotton *Princ. Illumination* xv. 390 The gas arc is the name given to the discharge at heavy current-density through one of the rare gases at high pressure. *Ibid.*, Under gas-arc conditions there is an intense continuum covering the whole visible range. **1881** W. D. Hay *300 Years Hence* ix. 187 Supporting his car with a proportionately large gas-balloon. **1914** H. M. Buist *Aircraft in German War* i. 19 An airship is any form of gas balloon fitted with motors, propellers and steering gear. **1904** Goodchild & Tweney *Technol. & Sci. Dict.* 247/1 *Gas barrel*, the wrought iron tube used for conducting gas from the large mains..into buildings, and up to the point from which it is distributed by 'compo' pipes. **1937** *Archit. Rev.* LXXXI. 12 The heating in the cloakrooms is by means of heating coils extending underneath boot lockers (which are constructed of welded gas-barrel) and behind seats under clothes pegs. **1882** 'Mark Twain' *Stolen White Elephant* 22 Three month's unpaid gas-bills taken. **1969** *Listener* 9 Jan. 59/1 In these first weeks of the New Year the TV consumer is receiving his quota of new programmes, which arrive as inevitably as the quarterly gas bill. **1941** *Illustr. London News* CXCVIII. 487 (*caption*) The men are all wearing gas-capes. **1948** A. Baron *From City from Plough* 165 They sat in their shallow pits at the roadsides, huddled in their waterproof gas-capes. **1928** C. F. S. Gamble *N. Sea Air Station* xix. 334 Above us the gas-cells are hanging limply down. **1950** *Chem. Abstr.* XLIV. 6694 (*heading*) Convection processes in the gas centrifuge. **1955** *Sci. Abstr.* A. LVIII. 391/1 The fractions of xenon drawn off at the two ends of the rotor of a gas centrifuge are enriched respectively in the heavy and light isotopes. **1960** *Times* 19 Oct. 15/3 The production of enriched uranium by gas-centrifuging. **1967** *New Scientist* 6 Apr. 17/1 Capital investment in a gas centrifuge plant would probably be favourable with that of the gaseous diffusion plant. **1938** *Times* 12 Jan. 11/6 The only way to convince is to multiply the 40 Home Office gas-vans and allow anybody to put on a mask and enter the gas chamber. **1945** *Daily Mirror* 27 Sept. 8 The Germans knew what to do with women with white hair, or too exhausted to work. For them was the gas-chamber. **1953** *Encounter* Nov. 26/1 In the gas chambers of Auschwitz, Belsen, and other extermination camps, around six million human beings were put to death during the closing phase of the Second World War. **1969** *Listener* 3 July 11/3, I wanted to find out..what would have happened if Huey Newton had been sent to the gas chamber. **1952** J. Griffiths et al. in *Analyst* LXXVII. 897 (*title*) Adsorption and partition methods. Gas chromatography. *Ibid.*, Recent work on various gas chromatographic techniques in the authors' laboratory is described. **1961** *Listener* 7 Dec. 977/1 What is now called 'gas chromatography' allows mixtures of vapours to be analysed with the same degree of subtlety that is possible when liquids are being analysed by 'paper chromatography'. **1965** Phillips & Williams *Inorg. Chem.* I. xi. 416 More recently, gas chromatographic methods have been employed in which the hydrides are transferred in streams of oxygen-free hydrogen or nitrogen. **1967** *New Scientist* 25 May 463/1 Present gas chromatography units (in which the components of a mixture are separated by diffusing at different rates through a porous solid). **1884** Gas cooker [see Cooker 1]. **1935** *Discovery* Dec. 361/1 The allied gas, butane, compressed similarly in cylinders, is now in frequent use for gas-cookers etc. **1886** *Chambers's Jrnl.* 5th Ser. III. 63/2 Six hundred pounds is saved yearly since the introduction of gas-cooking. **1949** C. Goodman *Sci. & Engin. Nuclear Power* II. x. 141 It is obvious that split-flow would be advantageous for gas-cooled reactors..It would not be advantageous for liquid-cooled reactors. **1955** *Sci. Amer.* Oct. 58 Gas cooled reactor being built by British at Calder Hall is described as 'the slow-speed reciprocating engine of the reactor world'. **1965** *Listener* 2 Sept. 342/3 This will use the now famous advanced gas-cooled atomic reactor which can produce electricity more cheaply than any other in the world. **1904** Goodchild & Tweney *Technol. & Sci. Dict.* 247/1 *Gas cylinders.* Compressed gases are sent out in steel cylinders. **1923** Kipling *Irish Guards in Great War* I. 120 Recessions which they had been cutting out in the trenches for the reception of gas-cylinders. **1886** *Cassell's Fam. Mag.* Feb. 189/1 (*heading*) A miner's gas detector. **1938** *Times* 11 Feb. 11/4 The sentry sat watching his gas-detector, and when it changed colour he sounded the gas-alarm. **1934** Webster, Gas-discharge lamp. **1940** *Chambers's Techn. Dict.* 367/1 *Gas-discharge tube.* Generally, any tube in which an electric discharge takes place through a gas. Specially, a tube comprising a heated cathode, a grid, and an anode enclosed in an atmosphere of gas at low pressure, generally mercury vapour, argon, or neon. **1960** *Times* 11 Mar. 16/1 Considered as a light source, they [*sc.* glow-worms] are about three times as efficient as a gas-discharge lamp. **1969** *Times* 29 May 12/8 Based on an orange-red wavelength of light emitted by an isotope of the inert gas krypton, when stimulated in a standardized gas discharge tube, the system gives precision to about one part in 100 million. **1949** O. G. Sutton *Science of Flight* 138 Flow problems in which density variations are produced or accompanied by large variations in velocity are said to belong to the science of *gas dynamics*. **1890** *Monthly Packet* Christmas 184 She certainly was not troubled with either gas-escapes or School Boards. **1940** W. Empson *Gathering Storm* 46 He makes no leak we ought to mend Or gas-escape that should not blow. **1889** Groves & Thorp *Chem. Technol.* I. 546 The gas-fired boiler showed an evaporation of from 8·6 to 9·2 lbs. of water per lb. of coal. **1967** *Times* 17 Jan. 10/6 Heating units for gas-fired..central-heating systems. **1961** *Listener* 12 Oct. 583/1 Dwellers in urban districts may find gas-firing has much to commend it. **1934** *Discovery* Nov. 324/2 Two classes of the [vacuum] tubes are made, hard and soft or gas-focused tubes. **1959** Rider & Uslan *Encycl. Cathode-Ray Oscill.* (ed. 2) i. 11/1 One of the first (1922) cathode-ray tubes developed..was the Western Electric 224, a gas-focused type of tube. **1914** *Jrnl. R. Army Med. Corps* XXIII. 514 There are some ten different organisms that have been isolated from cases of gas gangrene in man. **1918** *Nomencl. Dis.* (ed. 5) 238 Bacillus œdematis maligni. ..The cause of malignant œdema and some cases of gas gangrene. **1939** Auden & Isherwood *Journey to War* 93 The sweet stench of gas-gangrene from a rotting leg. **1944** *Ann. Reg. 1943* 359 The routine use of the sulphonamides have made wound infections and gas gangrene relatively infrequent. **1924** R. H. Mottram *Spanish Farm* III. 230 The gas-gong hanging from a nail. **1896** W. W. Greener *Gun & its Development* (ed. 6) 756/1 Gas guns. **1955** L. Wesley *Air-Guns & Air-Pistols* ii. 18 The Giffard gas-gun created much interest when it first appeared on the English market. **1962** *Daily Tel.* 15 Nov. 1/7 A liquid or powder had been sprayed over them from a 'gas gun'. **1915** *Punch* 15 Dec. 488/3 Will officers please state how many Gas Helmets they possess? **1916** A. Bennett *Jrnl.* 1 June (1932) II. 164 The 'gas officer' came down yesterday to inspect gas-helmet efficiency of troops. **1880** G. N. Lamphere *U.S. Govt.* 227/1 The buildings connected with the hospital..are..a gas-house [etc.]. **1923** T. S. Eliot *Waste Land* iii. 15 Fishing in the dull canal..round behind the gashouse. **1926** Wood & Goddard *Dict. Amer. Slang* 20 *Gas-house district*, district in city unsuitable for living; abode of gangsters. **1952** B. Cerf *Good for Laugh* 163 On the mound pitching for the old St. Louis Cardinal gas-house gang. **1929** J. A. Ratcliffe *Physical Princ. Wireless* ii. 21 In a metal we suppose that there exist free electrons, which move about hither and thither in random directions with 'gas-kinetic' velocities. **1937** *Discovery* Dec. 394/2 The main topics include..gas kinetics, magnetic and electric moments. **1899** J. Walker *Introd. Physical Chem.* iv. 29 These simple gas laws.. apply in strictness only to ideal or 'perfect' gases. **1966** Barnard & Mansell *Fund. Physical Chem.* iii. 108 These Gas Laws are really generalizations about the behaviour of gases to which most conform only approximately. **1971** J. Howlett in B. de Ferranti *Living with Computer* ii. 12 The gas laws tell how the volume of a gas varies with pressure and temperature. **1856** *Rep. Comm. Patents 1855* (U.S.) I. 514 Improved Gas-Lighter. **1874** Gas-lighter [see Gas *sb.* 6 c]. **1962** R. Jeffries *Exhibit No. 13* ii. 15 He flicked open a gas lighter. **1966** J. B. Priestley *Salt is Leaving* iii. 45 He..lit his pipe again, using what seemed to be an outsize gas lighter. **1951** *Biochem. Jrnl.* XLVIII. p. vii/1 A gas-liquid partition chromatogram of similar efficiency to a liquid-liquid partition chromatogram should lead to a better separation of the two compounds. **1952** *Ibid.* L. 679/2 The theory of gas-liquid chromatography differs from that of liquid-liquid chromatography..only by virtue of the fact that the mobile phase is compressible and thus produces a gradient of gas velocity down the column. **1963** *Times* 22 May (Margarine Suppl.) p. vi/3 The analysis of these fatty acids has been revolutionized by the development of gas-liquid chromatography. **1900** *Jrnl. Soc. Arts* XLVIII. 460/2 The Incandescent Gas Mantle and its Use. **1909** H. G. Wells *Tono-Bungay* III. i. 282 The cerium and thorium alone were worth the money he extracted for the gas-mantles then in vogue. **1915** *War Illustr.* 4 Sept. 69 French soldiers wearing anti-poison gas masks and respirators. **1917** *Ibid.* 17 Feb. 15/3 An Austro-Hungarian infantryman with gas-mask. **1971** *Daily Tel.* 13 Sept. 1/5 Following an appeal to the rioters to free their captives, guards in watchtowers began putting on gas masks. **1940** *War Illustr.* 5 Jan. 560 The gas-masked gun crew are ready for all eventualities. **1865** J. A. Symonds *Let.* 5 Oct. (1967) I. 575 On reading our gas meter I found that a huge quantity must have been wasted. **1906** M. H. Baillie-Scott *Houses & Gardens* xl. 116 Rows of little houses..each with its gas-meter under the stairs. **1825** *Edin. Jrnl. Sci.* July 88 The boy was then sent to the gas-office to give information there of the smell. **1882** 'Mark Twain' *Stolen White Elephant* 22 Gas office broken open here during night. **1916** Gas officer [see **gas helmet*]. **1916** W. Owen *Let. c* 23 July (1967) 401, I *may*, thus, become Gas Officer. **1901** *Jrnl. Soc. Chem. Ind.* Dec. 1257/1 Crude oil is now used instead of gas oil in the making of gas. **1949** *Thorpe's Dict. Appl. Chem.* (ed. 4) IX. 390/2 Gas oil is the term broadly applied to petroleum distillates which distil between kerosene and lubricating oils..The term gas oil has been broadly applied to these distillates because of their use in carburetting water gas to raise its heating and illuminating value. However, they are widely used as domestic fuel oils and are in demand as cracking-plant charging oils. **1960** *Which?* Jan. 5/1 Larger burners, especially those for central heating boilers, use 'gas oil', sometimes marketed under proprietary names such as 'Domestic Fuel Oil'. **1922** Joyce *Ulysses* 482 Many..women also commit suicide by..asphyxiating themselves by placing their heads in gas ovens. **1939** G. Greene *Confid. Agent* I. ii. 77 You've got to choose some line of action and live by it. Otherwise nothing matters at all. You probably end with a gas oven. **1945** 'G. Orwell' *England your England* (1953) 54 Was it true about the German gas ovens in Poland? **1968** J. R. Ackerley *My Father & Myself* xv. 163 It was the kind of shock that people must receive when some old friend..goes home and puts his head in the gasoven. **1937** 'G. Orwell' *Road to Wigan Pier* xii. 223 No one..who has not looked at a gaspipe chair and reflected that the machine is the enemy of life. *a* **1942** L. G. Blochman *See you at Morgue* (1946) iii. 30 One of the chrome-plated gaspipe chairs..in Pen's living-room. **1894** Gas pliers [see *monkey-wrench* s.v. Monkey *sb.* 17 a]. **1940** *Chambers's Techn. Dict.* 367/2 *Gas pliers*, stout pliers with narrow jaws, the gripping faces of which are concave and serrated, to provide a secure grip on pipes, nipples, etc. **1940** H. Brighouse in *Best One-Act Plays 1940* 49 It's a gas-poker...You light it and put it under the coals. **1962** *Which?* Feb. 60/1 The fire lighters using gas were of two kinds: the *gas poker*, which is put into the fuel, and the *underbar fire lighters*, which are used under the grate. **1901** *Westm. Gaz.* 18 Dec. 1/3 You stirred them for five minutes on the gas-ring. **1954** T. S. Eliot *Confid. Clerk* I. 20 Cooking a sausage on a gas ring. **1915** D. O. Barnett *Lett.* 204 Coming through a district where they'd been using gas-shells. **1918** E. S. Farrow *Dict. Mil. Terms* 257 *Gas shells*, ordinary shells from which the greater part of explosive has been taken and replaced by a poisonous liquid, which, when the shell bursts, turns into a poisonous vapor or gas. **1940** *Chambers's Techn. Dict.* 367/2 Gas show. **1964** *Times Rev. Industry* Apr. 82/1 Shell once had a concession in the [Nile] delta and obtained some 'gas shows'. **1879** *Encycl. Brit.* X. 294/1 Gas-spurts. **1932** *Jrnl. R. Aeronaut. Soc.* XXXVI. 853 The R.A.E. Mark I. gas starter was still doing very good service, particularly in civil aviation. **1935** C. G. Burge *Compl. Bk. Aviation* 585/2 *Gas starter*, a device whereby an engine is rotated for starting, by supplying gaseous combustible mixture under pressure to the cylinders during the power stroke. **1935** *Economist* 19 Nov. 949/1 Remarkable progress has been made in the 'gas' storage of home-grown orchard fruit—i.e. storage in an atmosphere whose carbon dioxide content has been appreciably increased. **1935** *Ibid.* 25 May 1230/1 English gas-stored Bramleys and imported boxes of apples have sold well. **1853** F. Nightingale *Let.* Dec. in C. Woodham-Smith *F. Nightingale* (1950) vi. 122 The whole flue of a new gas stove came down the second time of using it. **1861** Mrs. Beeton *Bk. Househ. Managem.* 266 (*caption*) Gas-stove. **1919** V. Woolf *Night & Day* x. 128 They sat.. talking together over the gas-stove in Ralph's bedroom. **1967** K. Giles *Death & Mr Prettyman* viii. 157 Persuade Mrs Jewel to leave some stewy soup on the gas stove. **1888** *Lockwood's Dict. Mech. Engin.* 158 *Gas thread*, a screw thread of fine pitch employed for wrought iron and brass tubes. **1902** Gas thread [see Thread *sb.* 5]. **1882** 'M. Harland' *Eve's Daughters* 81 Nature scorns the use of patent gas-traps. **1904** *Proc. Inst. Mech. Engin.* Oct. 1061 (*heading*) A scientific investigation into the possibilities of gas-turbines. *Ibid.* 1125 He did not see any immediate future for a gas-turbine, except in possibly utilizing the exhaust gases from reciprocating engines. **1930** *Engineering* 10 Jan. 38/2 The early development of what may possibly be the next important form of heat engine, the gas turbine. **1935** *Jrnl. R. Aeronaut. Soc.* XXXIX. 53 This invention relates to an aeroplane propelled by a constant pressure gas turbine, the exhaust gases of which are exhausted towards the rear at so high a velocity that their recoiling effect sustains or wholly replaces the propeller. **1948** *Jane's Fighting Ships 1947–48* 5 The first time gas turbines have ever been employed for the purpose of marine propulsion. **1957** *Times* 21 Dec. 5/3 Travel in the B.O.A.C. Britannia is.. luxurious, and the low level of noise and vibration, which is a feature of gas turbine engined aircraft, makes a long journey less tiring. **1902** *Encycl. Brit.* XXV. 265/1 A gas-vacuole..may be made to vary its volume with varying pressure. **1968** *New Scientist* 21 Nov. 437/1 Most planktonic blue-green algae possess gas-vacuoles, and the buoyancy that they confer tends to lead to the accumulation of the algae at the surface of natural waters in dense growths. **1938** Gas-van [see **gas chamber* above]. **1946** *Ann. Reg. 1945* 185 Mass-murder by shooting, the gas-van or the crematorium. **1951** *Archit. Rev.* CX. 394/2 The ring main is of cast iron; risers and supply pipes are of copper, gas-welded. *a* **1877** Knight *Dict. Mech.* I. 958 *Gas welding-furnace*, a heating jet, or cluster of jets, to heat pieces of metal locally, in order to bring them to a welding temperature. **1927** *Jrnl. Iron & Steel Inst.* CXV. 911 Gas welding has many applications, particularly to chrome-molybdenum steel.

gas (gæs), *sb.*^{2} *colloq.* (orig. *U.S.*). Abbrev. of *Gasolene (=petrol); *to step, tramp* or *tread on the gas, to give it* (or *her*) *the gas*: to accelerate a motor vehicle by pressing down the accelerator pedal; also *fig.*, to put on speed; to hurry.

1905 R. Beach *Pardners* (1912) v. 125, I turned the gas into the tug, blowin' for the Wells Street Bridge. **1911** *Daily Colonist* (Victoria, B.C.) 22 Apr. 8/6 (Advt.), Having installed a new portable gasoline wagon we can supply you with 'gas' in record time from our Central Garage. **1916** H. L. Wilson *Somewhere in Red Gap* vii. 289 Once she'd tramped on the gas of a ninety-horse-power racer and socked him against a stone wall. **1918** E. M. Roberts *Flying Fighter* xvi. 236 We were to land the machines [*sc.* aeroplanes] to be refilled with gas and oil. **1920** C. W. Hoffman in C. L. Gregory *Condensed Course Motion Pict. Photogr.* xxii. 322 Any suggestion to him that—'He get a move on', or 'Step on the gas'. **1924** W. M. Raine *Troubled Waters* viii. 81 Next moment she was pushing home the brake and shutting off the gas. **1926** A. Huxley *Jesting Pilate* iv. 297 Jazz it up. Keep moving. Step on the gas. **1929** *Times* 6 Nov. 15/4 The time may come when, to speed up the cause, he may tread on the gas with the best of them. **1932** *Atlantic Monthly* Mar. 314/1 The old Ford rocked and rattled as I gave it the gas. **1942** L. Rich *We took to Woods* (1948) 66, I remember shoving for dear life while Ralph gave her the gas. **1955** L. P. Hartley *Perfect Woman* xxviii. 249 Something that can keep people going and make them feel their value, which..is what we want to feel—without stepping on the gas and running people down. **1963** H. Garner in R. Weaver *Canad. Short Stories* (1968) 2nd Ser. 36 He claimed he didn't have enough gas in his car to drive him to the hospital. **1965** Priestley & Wisdom *Good Driving* iii. 27 More gas means more speed. Less gas means less speed.

b. *attrib.* and *Comb.*, as *gas-pump, rationing, tank*; **gas boat,** a boat driven by a petrol-engine; **gas pedal,** an accelerator pedal; **gas station,** a filling-station.

1910 *Our Navy* (U.S.) III. xi. 21/1 The gas boat bumps and mutters. **1959** *Native Voice* (Vancouver) Jan. 2/5 Chief Assu had four fine sons, each of whom he set up with a house and gas boat. **1961** A. Miller *Misfits* xi. 116 Guido turns sharply with them—the truck leaning dangerously—and works brake and gas pedal simultaneously. **1925** F. Scott Fitzgerald *Great Gatsby* i. 21 Garages, where new red gas-pumps sat out in pools of light. **1968** *Globe & Mail* (Toronto) 17 Feb. 47/5 (Advt.), B-A Gas pumps,..good highway, busy fishing. **1951** in M. McLuhan *Mech. Bride* 116/1 You willingly put up with gas rationing. **1932** E. Wilson *Devil take Hindmost* viii. 65 The city..cannot afford to have the gas stations go broke. **1957** J. Kerouac *On Road* (1958) 149 Dean was wearing his gas-station coveralls. **1926** *Daily Colonist* (Victoria, B.C.) 7 July 5/3 An extra ninety-gallon gas tank has been installed in the Yorktown, giving her a non-stop cruising radius of 800 miles.

gas, *v.* Add: **1. c.** To inflate with gas.

1922 *Encycl. Brit.* XXX. 48/2 Airship harbours have facilities for gassing airships with hydrogen. **1934** J. A. Sinclair *Airships* x. 204 The whole work of gassing the ship fell to the crew.

4. (Further examples.) Also, to be subjected to a gas attack.

1914 Glaister & Logan *Gas Poisoning in Mining* viii. 145 Recognising the characteristic action of carbon monoxide as causing powerlessness of the limbs long before, perhaps, total loss of consciousness supervenes, it can readily be understood that men are sometimes placed in most perilous positions by being gassed. **1915** *Times* 7 May 9/4 The men in hospital..who were 'gassed' ..on Hill 60. **1915** D. O. Barnett *Let.* 9 June 168 Young had to go off this morning to a village a few miles back to be gassed. He (and a lot of staff men) were put in a trench and given a dose, with respirators on of course. *Ibid.* 14 June 178 They say that round Wipers the German infantry is rotten..[and] won't advance if there is any fire to hold them up, that is, unless we're gassed out. **1919** G. K. Rose *2/4th Oxf. & Bucks Lt. Infty.* 203 The Colonel,..Regimental Sergeant-Major and many signallers and runners, all found that they were gassed. **1922** *Daily Mail* 11 Nov. 7, 18 Girls 'Gassed'. A number of employees were overcome by fumes..through a stopper flying from a cylinder of ammonia gas. *Ibid.* 14 Nov. 7 Army Officer Gassed in his Bath...Accidental Death, the result of poisoning by a gas escape from a geyser. *Ibid.* 20 Nov. 7 Residents of Horley, Surrey, complain of being 'gassed' by the fumes released in the breaking up of mustard gas shells at a dump near Gatwick Racecourse, about a mile and a half away.

5. b. (Earlier U.S. example.)

1852 F. A. Buck *Let.* 1 Jan. (1930) 95, I have been.. doing the agreeable to the customers at the Hermitage Hotel..and gassing with travellers generally.

6. Of a storage battery or dry cell: to give off gas.

1902 E. J. Wade *Second. Batteries* 332 When once the surfaces of a pure lead anode in an electrolyte of dilute sulphuric acid are peroxidised and gassing. *Ibid.* 335 As a rule the positives commence to gas almost immediately the current passes. **1907** R. W. Vicarey *Storage Batteries* 7 The charge must be continued until every cell in the whole battery has become milky or 'gasses freely'. **1950** G. W. Vinal *Primary Batteries* iv. 123 That dry cells do gas is well established, but..as the public normally uses them such gassing is negligible and the cause of no concern. **1964** G. Smith *Storage Batteries* iii. 45 Most of the charge is now being used in dissociating the water of the sulphuric acid solution into hydrogen and oxygen, and the cell begins to gas freely.

7. *trans.* To impress or please enormously; to excite or thrill. Cf. *Gas *sb.*^{1} 5 d. *slang* (orig. *U.S.*).

1949 L. Feather *Inside Be-Bop* vi. 44 Woody, after a year's retirement, had decided to come back with a band that would 'gas' everyone. **1958** *Sunday Express* 19 Oct. 7/6 Duke Ellington said the visit 'gassed' him—a swing expression which means that he was thrilled. **1967** *Crescendo* Mar. 6/2 A..cadenza at the end of 'Watermelon man' which really gassed me.

gas (gæs), *v.*^{2} *colloq.* (orig. *U.S.*). [f. *Gas *sb.*^{2}] **1.** *trans.* To supply (a motor car) with petrol. Freq. const. *up*.

1934 J. M. Cain *Postman always rings Twice* i. 18, I went to gas up a car. **1959** 'J. R. Macdonald' *Galton Case* (1960) x. 78 The attendant..was busy gassing a pickup truck. **1962** E. Ambler *Light of Day* ii. 25 You'd gone to gas up the car.

2. *intr.* To fill *up* the petrol tank of a motor car.

1962 E. S. Gardner *Case of Blonde Bonanza* (1967) xiv. 161 One of the city police picked up Dillard at a service station..where he was gassing up. **1967** W. Murray *Sweet Ride* x. 170 They'll know at the station, on account of they gas up there sometimes.

gas-bag. Add: **1. b.** A balloon or airship; a bag inflated with gas, forming part of an airship or balloon.

1852 *Illustr. London News* 18 Sept. 224/2 Above us reeled the monster gas-bag like a monster peg-top. **1877** *Design & Work* III. 602 Science! indeed, to talk of propelling a flaccid gas-bag! **1897** *Aeronaut. Jrnl.* Jan. 6/2 The machine flew against the wind. There was nothing of the balloon nature about it. There were no gas bags to uphold it. **1912** *Rev. Reviews* XLVI. 61 Inside the frames go the gas-bags, sixteen or so in number. **1932** *Ann. Reg. 1931* 32 The liability of the gas-bags to chafing was the only weak spot in the airship. **1963** A. Smith *Throw out Two Hands* (1966) iv. 43 Somehow I was unprepared for looking into the gas-bag above us,.. and for seeing nothing.

Gascon, *sb.* **1.** *a.* (Later examples.)

1849 Thackeray *Pendennis* I. xxvii. 262 He is a Gascon, and comes from the borders of Spain. **1959** *Chambers's Encycl.* VI. 176/2 The Gascons were virtually independent under their national dukes until 1032.

‖ **gaseosa** (gæsiōu·sä). [Sp., = soda-water.] An effervescing drink. Also *attrib.*

1927 D. H. Lawrence *Mornings in Mexico* 70 He had had a bottle of *gazoosa* [*sic*], fizz, with us. **1932** E. Hemingway *Death in Afternoon* iii. 35 The vendors of cold beer and gaseosas. **1939** G. Greene *Lawless Roads* viii. 199 There were stacks of empty 'gaseosa' bottles.

gaseous, *a.* Add: **1. b.** *gaseous diffusion*, the diffusion of gas (see Diffusion 5), esp. through a porous wall or membrane as a means of partially separating the gases in a mixture. Also *attrib.*

1846 *Phil. Trans. R. Soc.* CXXXVI. 623 The full elucidation of the transpiration of mixed gases, must await..the further extension of our knowledge of the laws of gaseous diffusion. **1922** *Jrnl. Amer. Chem. Soc.* XLIV. 41 A possible cause of loss of efficiency in gaseous diffusion is local convective action, due to unevenness in the texture of the membrane. **1958** W. K. Mansfield *Elem. Nucl. Physics* i. 11 Separation using this fact occurs in gaseous diffusion plants.

c. *gaseous gangrene* = *gas gangrene* (see *Gas *sb.*^{1} 7).

1889 *Pacific Med. Jrnl.* XXXII. 469 That septic disease now generally called malignant œdema..is also called..fulgurating gaseous gangrene. **1965** *Dorland's Med. Dict.* (ed. 24) 316/2 C[*lostridium*] *novyi*, a species which is an important cause of gaseous gangrene, being reported in 32 to 48 per cent of cases.

gash (gæʃ), *sb.*^{3} *slang.* Also **gashing, gashion.** [Origin unknown; cf. Eng. dial. *gaishen* a skeleton, a silly-looking person, an obstacle (see E.D.D.).] Something superfluous or extra; waste, rubbish, garbage. Also *attrib.* and *Comb.*

1925 Fraser & Gibbons *Soldier & Sailor Words* 103 *Gashions*, extra of anything. **1943** Baker *Dict. Austral. Slang* (ed. 3) 34 *Gash*, a second helping of food; any surplus or residue. **1945** 'Tackline' *Holiday Sailor* 62 Gash can mean so many things. It is rubbish, waste... It can be supernumerary—a gash-hand who hasn't any particular job to do. **1953** Scott & Fisher *Thousand Geese* v. 54 James planned to tidy up the camp—dig a gash-pit. **1960** *Times* 2 Apr. 8/7 A disgusted stoker is emptying a bottle of best Demerara down the gash-shute. **1965** P. Wayre *Wind in Reeds* xiv. 194 She [*sc.* a bear] has an unerring nose for what in the Navy was called 'the gash bucket', and..invariably went straight to the sink and turned out the contents of the garbage pail underneath.

gash (gæʃ), *a.*^{4} *slang.* [Origin unknown; cf. *Gash *sb.*^{3}] Superfluous, extra, spare; free.

1945 [see *Gash *sb.*^{3}]. **1946** R. Harling *Steep Atlantick Stream* vii. 190 You can grab an oilskin from the wardroom and there's a gash lifebelt there. **1967** G. Hammond *Mud in his Eye* vii. 82 I've always said the big one [*sc.* shed] was gash, and the harbour might as well have the rent. **1968** *Guardian* 15 June 8/2 Would he like to go to London, all expenses paid..? Of course he would; a good gash trip. **1971** M. Russell *Deadline* ix. 103 Cop-shop's stuffed with gash CID apprentices.

gashing, gashion: see *Gash *sb.*^{3}

gasifier (gæ·sifəi,əɹ). [f. Gasify *v.* + -er^1.] Any apparatus for manufacturing gas (see also quot. 1959).

1883 J. W. Ingham *Brit. Pat. 2339* p. 2 An apparatus which I designate a Gasifier consisting of retorts chambers or pipes maintained at an elevated temperature and containing what is known as Soda Lime. **1959** *Times Rev. Industry* Jan. 60/1 A gasifier consists..of a single cylinder containing a pair of opposed pistons which reciprocate freely without connecting rods or crankshaft. On the working stroke they compress air...The air compressed, together with the exhaust gases, provides a flow of medium temperature gas which is used to drive a gas turbine. **1961** *New Scientist* 22 June 728/3 Steam and oxygen at high pressure are introduced at the lower end of the gasifier.

gasket. Add: **2. b.** A flat sheet or ring of some relatively soft material made to be placed between adjoining metal surfaces so as to seal the joint against the pressure of gas or liquid; *spec.* one inserted between the cylinder-head and the cylinder-block in an internal-combustion engine.

1903 R. J. Mecredy *Dict. Motoring* 123 *Gasket*, a packing washer of asbestos, rubber, or other material used to make liquid, steam, or air-tight joints. **1915** *Autocar Handbk.* (ed. 6) ii. 59 This head is held down by a number of bolts, and a copper and asbestos gasket is interposed, so that a gas and watertight joint is obtained. **1925** *Morris Owner's Man.* 34 The cylinder head lifted, showing the valves, ports, water-ways and the gasket. **1946** Kuns & Plumridge *Automobile Engines* 46 The cylinder-head gasket, of course, is a specially prepared gasket in all cases, of copper, bronze, and asbestos. **1963** C. Campbell *Sports Car Engine* xiii. 291 Solid copper gaskets should always be used on an engine used for racing. **1969** *Which?* Aug. 229/2 Replace rubber plugs and gaskets [in pressure cookers] if they show signs of age.

gas-light. Also **gaslight.** Add: **c. gaslight paper,** a photographic printing-paper sufficiently insensitive to be usable in weak artificial light, without the necessity for a dark-room.

1906 R. C. Bayley *Compl. Photogr.* 243 Gaslight papers, as their name implies, are sufficiently insensitive to be worked in gaslight. **1963** John & Field *Textbk. Photogr. Chem.* ii. 24 Contact papers have long been known under the alternative name of 'gaslight' papers, since they can be handled cautiously in a weak domestic gas or electric light.

gas-man. Also **gasman.** Add: **1. c.** A workman who installs or repairs equipment for supplying household gas.

1873 A. T. Ritchie *Let.* (1924) viii. 151 Two gas-men in one room, a carpenter in another. **1971** *Guardian* 21 June 9/5 The North Thames Gas Board..dump and collect meters in vans and other gasmen have to use bicycles.

† **2. b.** A man who lights gas-lamps in the street. *Obs.*

1849 Thackeray *Pendennis* I. xxxix. 379 The gasman came and lighted the lamps before Sir Francis's door. **1854** M. Cummins *Lamplighter* 520 The gas-man came quickly up the street, lit, as by an electric torch, the bright burners. **1863** S. C. Massett *Drifting About* 50 The narrow and gloomy passage slightly illuminated by one or two jets of gas, which 'Tom the Gasman'..had a few moments before lighted.

gasolene, gasoline. Add: (Earlier and later examples.) Now usu. = Petrol *sb.* 3 (chiefly *U.S.*).

1865 *Appleton's Ann. Cycl. 1864* 669/2 Many refiners [of petroleum] separate first of all the lightest naphtha..; to this the name of *gasolene* has been given. **1881** *Harper's Mag.* May 815/1 The street hawkers, with gasoline torches, are crying their wares. **1895** *New Rev.* Oct. 390 Thirteen [vehicles] were driven by petroleum (or gasolene) motors. *Ibid.* 392 Of the petroleum vehicle..it may be said that it owes much of its extreme lightness to its modesty in the matter of fuel. A few pints of gasolene or rectified petroleum will suffice it for five or six hours. **1896** *Cosmopolitan* XX. 420/2 The practicability of the gasolene road vehicle. **1908** *Westm. Gaz.* 4 Aug. 10/1 The winner became unconscious..through the gasolene fumes. **1914** Duchess of Sutherland *Six Weeks at War* iii. 39 The Germans seemed to be doing the whole of this campaign on gasolene—that is to say, they use motor transport throughout. **1926** *Daily Colonist* (Victoria, B.C.) 13 July 1/3 Albert..attended the gasoline station of Corfield Motors. **1951** *Manch. Guardian Weekly* 15 Feb. 5 Gasoline, kerosene, and other petroleum products are smuggled out. **1959** K. Davis *Human Society* 627 A change from horse-powered to gasoline-powered vehicles.

gasper. Add: **2.** A cigarette, esp. a cheap or inferior one. *slang.*

1914 *Isis* 14 Feb. 5/2 Not only does he prefer the Irish-grown 'gasper' to all others, but he provides them for his Saxon guest. **1916** *Punch* 11 Oct. 261 'D'yer fink Jellicoe gits 'is share?' 'Course not, stoopid. Admirals don't smoke gaspers!' **1921** S. P. B. Mais *Why we should Read* 32 Why should one prefer a Corona cigar to a 'gasper'? **1965** *Listener* 21 Oct. 624/3 'Gasper' commercials are with us still a every peak viewing hour.

gaspereau (gæ·spĕro, gæspĕrōu·). *Canad.* Also **gaspereaux, gasperot.** [Canadian F. *gaspareau, gasparot.*] = Ale-wife^2.

1703 tr. *Lahontan's New Voy. N. Amer.* I. 243 Gasperots, a small Fish like a Herring. **1890** in Webster. **1912** *Sea Fisheries E. Canada* (*Commission of Conservation*) 105 The gaspereaux..is mostly used as a bait fish. **1938** *Nature* 5 Mar. 421/1 The gaspereau (Pomolobus pseudoharengus) spawns in the spring.

gaspergou (gæspəɹgū·). *local U.S.* Also **gaspar-goo, gaspergoo.** [f. Louisiana Fr. *casseburgau*, f. *casser* to break + *burgau* a species of shellfish.] The freshwater drum-fish, *Aplodinotus grunniens.*

1809 F. Cuming in R. G. Thwaites *Early Western Trav.* (1904) IV. li. 330 A fine dish of gaspar-goo, the best fish I had yet tasted of the produce of the Mississippi. **1831** W. B. Dewees *Lett. fr. Texas* (1852) xv. 137 Up the country our rivers abound with various kinds of fish, such as cat, buffalo, perch and gaspergoos. **1947** B. W. Dalrymple *Panfish* 344 For years they've been catching Sheepshead and Gaspergou.

gas-proof (gæ·s,prūf), *a.* [f. Gas *sb.*^{1} + Proof *a.* 1 b.] Impervious to poisonous gas.

1909 *Westm. Gaz.* 17 Feb. 7/2 Experts in the use of gas-proof apparatus. **1939** W. H. Davies *Loneliest Mountain* 17 Is it the Baby, three weeks old, That wears a gas-proof mask?

Hence as *v. trans.*, to render gas-proof; so **gas-proofed** *ppl. a.*, **gas-proofing** *vbl. sb.*

1936 *Lancet* 19 Dec. 1466/2 The [first-aid] posts would be structurally protected as far as possible and should certainly be gas-proofed. **1938** *Times* 11 Jan. 13/6 Will the materials be ready for gas-proofing rooms? *Ibid.* 12 Jan. 12/1 Southampton has a model gas-proofed house on a new estate. *Ibid.* 10 Mar. 15/4 A sudden interest in the gas-proofing of rooms.

gassed (gæst), *ppl. a.* [f. Gas *v.* 4.] **1.** Affected by poisonous gas.

1915 *Morn. Post* 29 May 7/7 The death of the gassed victims around Ypres. **1918** E. M. Roberts *Flying Fighter* 28 Several thousands of other wounded and gassed men followed us.

2. Drunk; intoxicated. *slang.*

1925 Fraser & Gibbons *Soldier & Sailor Words* 103 *Gassed*, drunk. **1956** Wallis & Blair *Thunder Above* (1959) xiii. 136 He's getting gassed already. We'd better steer him out of here. **1960** *Daily Mail* 22 Aug. 4/6 When I'm with people I laugh so much..they figure I'm 'gassed'. But I'm not. I don't drink.

gasser. Add: **2.** = *Gas *sb.*^{1} 5 d. *slang* (orig. *U.S.*).

1944 C. Calloway *Hepsters Dict.* 7. **1945** L. Shelly *Jive Talk Dict.* 11 *Gasser*, show-stopper. **1955** L. Feather *Encycl. Jazz* 11, I wrote a piece of music for this scene... It was a gasser; real great, I confess it. **1962** *Radio Times* 17 May 43/3 Should he [*sc.* a jazz musician] 'blow' with feeling, or great excitement..he is..considered by one and all to be a 'gasser'. **1970** *Sunday Truth* (Brisbane) 26 Apr. 14/2 (*heading*) Ron's Friday night show was a gasser.

‖ **gasthaus** (ga·sthaus). Also **Gasthaus.** [G., = Guest house.] A small German inn or hotel.

1834 F. B. Head *Bubbles from Brunnens* 254 At one of these solemn villages the postilion stopped at a 'gast-haus' to bait his horses. **1841** *Fraser's Mag.* June 710/1 If he has dined at an inn or restaurant, gasthaus, posada, albergo, or what not, [he] invariably inserts into his log-book the bill of fare. **1874** R. H. Busk *Tirol* 86 Its principal inn..claims to be not merely a *Gasthaus*, but a *Gasthof*. **1936** R. Lehmann *Weather in Streets* II. 252 Stopping in the evening at a small gasthaus. **1969** *Daily Tel.* 13 June 16/7 Most Europeans..have always drunk in places which offer fairly broadly-based catering services, like the café or *gasthaus*.

‖ **gasthof** (ga·sthōf). Also **Gasthof.** [G., f. *gast* guest + *hof* hotel, large house.] A German hotel (usu. a larger establishment than a *Gasthaus).

1832 J. H. Stocqueler *15 Months' Pilgr. Khuzistan* II. iv. 106 A spacious gasthof, or public house. **1865** 'Ouida' *Strathmore* v, A Bohemian Gasthof is about the only place..where you see the doctrine of equality in absolute..practice. **1874** [see prec.]. **1960** B. K. Wilson *Lovely Summer* ix. 107 Time to stop at a wayside *Gasthof* for delicious hot chocolate topped with cream. **1969** A. Glyn *Dragon Variation* viii. 228 The boy and the rest of the British team were in a gasthof near the university, students' rooms.

gastric, *a.* Add: **c.** *gastric mill*, a framework consisting of movable calcareous or chitinous plates in the stomach of certain crustacea.

1897 Parker & Haswell *Text-bk. Zool.* I. 547 A gastric mill is present in Malacostraca. **1927** Haldane & Huxley *Anim. Biol.* i. 7 Gastric mill, with grinding teeth and straining apparatus. **1967** *Oceanogr. & Marine Biol.* xv. 377 The masticatory apparatus or gastric mill.

gastrin (gæ·strin). *Physiol.* [f. Gastric *a.* + -in^1.] A hormone that stimulates gastric secretion.

1905 J. S. Edkins in *Proc. R. Soc.* B. LXXVI. 376 The active principle, which may be called 'gastrin', is not destroyed by boiling. **1955** *Sci. News Let.* 7 May 293/1 Gastric juice production is also caused by a hormone called gastrin, formed in the lower third of the stomach, the antrum, when this part of the stomach comes in contact with food. **1968** *New Scientist* 25 July 199/2 The hormone gastrin..is produced by the wall of the stomach itself.

gastro-. Add: **gastrodi·aphane** [Gr. διαφανής translucent], an instrument for illuminating the inside of the stomach in order that its outline may be seen through the abdominal wall; so **gastrodia·phany; ga:stro-duodeno·stomy,** the surgical creation of a passage between the stomach and the duodenum; **ga:stro-entero·logy,** the branch of medicine dealing with the stomach and intestines and their pathology; hence **ga:stro-enterolo·gical** *a.*; **ga:stro-entero·logist; ga:stro-gastro·stomy** (see quot. 1900); **ga:stro-ileo·stomy,** the surgical creation of a passage between the stomach and the ileum; **ga:stro-jeju·nal** *a.*, pertaining to or involving both the stomach and the jejunum; **ga:stro-jejuno·stomy,** the surgical creation of a passage between the stomach and the jejunum; **gastro-œsopha·geal** *a.*, pertaining to or involving both the stomach and the œsophagus; **ga·stropexy** [see *-pexy], an operation for restoring a prolapsed stomach to its proper position by suturing it to the abdominal wall; **gastroplica·tion** [L. *plicāre* to fold], an operation for reducing the capacity of the stomach by sewing together folds made in the stomach wall; **gastropto·sis** [Ptosis], a downward displacement of the stomach; **gastrosta·xis** [Gr. στάξις a dripping], oozing of blood through the mucous membrane of the stomach.

1892 M. Manges tr. *Ewald's Dis. Stomach* ii. 63 The use of the gastrodiaphane has been suggested by Einhorn. This instrument consists of a small electric light, which is introduced into the stomach; the contours of this organ are outlined by the light shining through the gastric wall and the abdominal parietes. **1894** Gould *Dict. Med.*, *Gastrodiaphany*, a method of exploration of the stomach by means of the gastrodiaphane. **1896** A. E. Maylard *Treat. Surg. Alim. Canal* 156 Gastrodiaphany..has received up to the present but a very limited trial. **1904** *Nature* 4 Aug. 316/2 This includes the ordinary clinical methods and the examination by the Röntgen rays and by so-called gastro-diaphany. **1890** Billings *Med. Dict.*, Gastroduodenostomy. **1903** *Lancet* 29 Aug. 591/2 We rapidly considered the relative advantage of pyloroplasty..and gastro-duodenostomy. **1927** F. F. Imianitoff tr. *Rachet's Pract. Gastroscopy* p. vi, Gastro-enterological endoscopy. **1935** *Jrnl. Amer. Med. Assoc.* 2 Feb. 402/2 This dictum of the gastro-enterologists. **1904** *Appleton's Med. Dict.*, *Gastroenterology*, the pathology, etc., of the stomach and intestines. **1908** *Practitioner* Feb. 283 The student of gastro-enterology. **1952** F. A. Jones *Mod. Trends Gastro-enterol.* p. xiii, In many countries, as in Great Britain, gastro-enterology is not an established speciality..but remains within the sphere of general medicine and surgery. **1900** Dorland *Med. Dict.*, *Gastrogastrostomy*, the formation of an anastomosis between the pyloric and cardiac ends of the stomach, performed for hour-glass contraction of the stomach. **1902** *Encycl. Medica* XI. 454 *Gastro-anastomosis or Gastro-gastrostomy* (Wölfler)..consists in making a free communication between the two compartments of an hour-glass stomach. **1921** J. Sherren *Lect. Surg. Stomach & Duodenum* iii. 35 If there is a double hour-glass stomach,..double gastrojejunostomy should be performed if this operation is indicated. Plastic operations on the stomach, such as gastro-gastrostomy, should not be done. **1896** A. E. Maylard *Treat. Surg. Alim. Canal* 249 When the junction is with the ileum, it is strictly a gastro-ileostomy. **1957** H. A. Davis *Princ. Surg. Physiol.* xvii. 453 The unintentional performance of a gastro-ileostomy often results in the development of an ileal ulcer or an ileitis. **1908** *Practitioner* Sept. 450 The original posterior gastro-jejunal opening had closed. **1891** *Lancet* 2 May 988/2 The growth having much increased and the stomach being greatly dilated, gastro-jejunostomy was performed. **1921** Gastrojejunostomy [see **gastro-gastrostomy* above]. **1889** *Cent. Dict.*, Gastro-esophageal. **1962** *Lancet* 1 Dec. 1178/2 He has made a particular study of gastroœsophageal disorders in children. **1896** A. E. Maylard *Treat. Surg. Alim. Canal* 225 An operation termed gastropexy has been performed by Duret for certain cases of displacement and dilatation. **1908** *Practitioner* Sept. 440 For ptosis and atony of the stomach, gastro-enterostomy is preferable to gastropexy or gastroplication alone. **1896** A. E. Maylard *Treat. Surg. Alim. Canal* 248 Brandt also records a successful case [*sc.* of gastrorrhaphy] under the title of gastroplication. **1903** Riegel & Stockton *Dis. Stomach* I. 284 The operation of Bircher consists in reducing the capacity of the stomach by constructing folds in the stomach-wall and sewing these reduplications together (gastroplication). **1893** Dunglison *Dict. Med. Sci.* (ed. 21) *Gastroptosis*, falling or dragging down of the stomach. **1900** *Encycl. Medica* V. 207 In dilatation and gastroptosis the location of the pain approaches nearer to the umbilicus. **1909** Osler & McCrae *Syst. Med.* V. 608 Gastroptosis, although one of the less frequently recognized of the visceral ptoses, is one of the most important as regards symptomatology. **1927** S. Wyard *Handbk. Dis. Stomach* vii. 170 Gastroptosis..occurs as an isolated phenomenon or in association with a downward displacement of certain of the other viscera as well, e.g. intestines, kidneys, etc. **1906** W. H. White in *Lancet* 3 Nov. 1190/1 Sir Cooper Perry kindly suggested to me that the disease might be called gastrostaxis. **1929** Hurst & Stewart *Gastric & Duodenal Ulcer* III. iii. 82 Many cases of so-called 'gastrostaxis', i.e. gastric hæmorrhage without easily recognizable lesion, are due to bleeding from innumerable microscopic lesions. **1960** Jones & Gummer *Clinical Gastroenterol.* xiii. 395 Patients with small acute ulcers, diffuse erosive gastritis, and also gastrostaxis with diffuse mucosal bleeding.

gastroscope (gæ·stroskōup). [f. Gastro- + -scope.] An instrument which makes possible a visual inspection of the interior of the stomach when passed into it via the mouth and œsophagus.

1888 *Lancet* 7 July 35/1 A special means of examination introduced by Mikulicz—viz., the direct inspection of the gastric mucous membrane by means of a specially constructed gastroscope, to be termed *gastroscopy*. **1956** J. J. Spira *Gastro-duodenal Ulcer* v. xxii. 323 The duodenal lesion is obviously out of the area which can be seen through the flexible gastroscope. **1965** S. Wolf *Stomach* x. 209 The gastric mucous membrane as seen through the gastroscope is red and thickened.

Hence as *v. trans.*, to examine (a patient) with a gastroscope.

1937 R. Schindler *Gastroscopy* 5 Many patients were gastroscoped repeatedly. **1970** H. M. Spiro *Clin. Gastroenterol.* xv. 255 Some physicians recommend that all patients with upper digestive symptoms..be gastroscoped.

gastroscopy. Add: **2.** The practice and techniques of using the gastroscope; an examination of the stomach by means of the gastroscope.

1888 [see prec.]. **1927** F. F. Imianitoff tr. *Rachet's Pract. Gastroscopy* p. v, We have had the opportunity of watching several gastroscopies. **1956** J. J. Spira *Gastro-duodenal Ulcer* v. xxii. 323 Gastroscopy is restricted by its limited field of vision but may be useful as contributory evidence of a gastric lesion. *Ibid.*, 3 deaths in an unknown number of gastroscopies. **1967** S. Taylor et al. *Short Textbk. Surg.* xx. 267 Gastroscopy may help decide if a gastric ulcer is simple or malignant.

Hence **gastrosco·pic** *a.*; **gastrosco·pically** *adv.*, by means of a gastroscope; **gastro·scopist,** one skilled in the use of the gastroscope.

1888 *Lancet* 7 July 35/2 One case of pyloric cancer (not examined gastroscopically). **1892** M. Manges tr. *Ewald's Dis. Stomach* ii. 63 Gastroscopic examination. **1937** R. Schindler *Gastroscopy* 1 In most of the fundamental matters the gastroscopists are in complete accord. **1956** J. J. Spira *Gastro-duodenal Ulcer* v. xxii. 323 Gastroscopic examination is particularly useful in diagnosing mucosal atrophy..and early tumours. **1958** H. Ogilvie in Blackburn & Lawrie *Textbk. Surg.* xx. 533 The opinions offered even by expert gastroscopists on the nature of the lesion are more often wrong than right.

gastrozooid (gæstro,zōu·oid). *Zool.* Also **gastero-.** [f. Gastro- + Zooid *sb.*] A nutritive zooid in colony-forming Hydrozoa and Thaliacea.

1879 H. N. Moseley in *Phil. Trans. R. Soc.* CLXIX. 429 In all Stylasteridæ two forms of zooids are present. One form has a mouth and gastric cavity, the other is devoid of these and has a purely tentacular function. For the former the term 'gastrozooid' is here adopted, and for the latter that of 'dactylozooid'. **1932** Borradaile & Potts *Invertebrata* v. 147 The manubrium has come to lie outside the medusa bell as the gastrozooid. *Ibid.* xix. 608 These [buds] of the lateral row become gasterozooids which gather food for the community. **1961** *Encycl. Brit.* XII. 3/2 In certain colonies [*sc.* of polyps and medusæ], such as those of *Millepora* and *Hydractinia*.., the hydranths are not all alike. Some of them (*gastrozooids*) possess mouths as well as tentacles. **1967** E. J. W. Barrington *Invert. Struct. & Function* xxi. 449 The differentiation of the colonial polyps into gastrozooids, concerned with feeding, and gonozooids (blastozooids) concerned with reproduction.

gat^2 (gæt). *slang* (orig. *U.S.*). Also **gatt.** [Short for Gatling.] A revolver or other gun.

1904 'No. 1500' *Life in Sing Sing* xiii. 248/2 *Gatt*, a revolver. **1920** [see **cap-pistol*]. **1928** E. Wallace *Gunner* xviii. 146 Why was a pistol called a gat, he wondered. Probably it was an abbreviation of gatling, and was obviously an Americanism. **1931** Wodehouse *Big Money* xiii. 283 He produced the gat..and poised it in an unsteady but resolute grasp. 'Hands up!' he said. **1966** *Punch* 1 June 792/3 The dive he took us to..was as bad as anything Akim Tamaroff ever ran through with his gat blazing.

gatch (gætʃ). [ad. Pers. *gach.*] A type of plaster used by Persian craftsmen.

1886 *Century Mag.* XXXII. 718 By the aid of *gatch* or plaster of Paris, the artisan of Teherân often transforms these mud structures into dreams of loveliness. **1934** F. Stark *Valleys of Assassins* 109 A slender piece of column moulded in a pattern of overlaid leaves, and probably once intended to be covered with metal, for it was made of very fragile 'gatch'.

gate, *sb.*^{1} Add: **5. b.** *to get the gate*, to be dismissed, rejected or jilted; so *to give* (someone) *the gate. slang* (orig. and chiefly *U.S.*).

1918 *Diary* 26 Mar. in *Wine, Women & War* (1926) 52 Col. Y— slated to get the gate. **1918** H. C. Witwer *From Baseball to Boches* iv. 143 'I wanna speak to you in private, Jeanne,' I says to her. 'Give this hick the gate!' **1922** S. Lewis *Babbitt* viii. 119 If any of us were to indulge in it here, he'd get the gate so fast it would make his head swim. **1930** Wodehouse *Very Good, Jeeves!* x. 263, I don't believe there's a female in the world who could see Uncle George fairly often in those waistcoats without feeling that it was due to her better self to give him the gate. **1951** *Sat. Even. Post* 8 Dec. 44/1 There's no reason why he should be fired..or given the gate. **1971** E. Fenwick *Impeccable People* viii. 47 She billed you for an extra month because Monnie gave her the gate.

c. The mouth. *slang.*

1937 in Partridge *Dict. Slang* 318/1. **1955** P. Wildeblood *Against Law* 98 Eats sweets...Unwraps them as slowly as he can, and then pops them into his gate, all gloating like. **1963** 'A. Garve' *Sea Monks* iii. 85 Macey pointed the gun at him...'You keep your gate shut, big boy.' **1966** B. Naughton *Alfie* viii. 54 Shut your big ugly gate at once.

d. [See quot. 1937, but perh. a shortening of *alligator.*] A person, esp. a jazz-musician; = *Cat *sb.*^{1} 2 c; freq. used as a form of address. *U.S. slang.*

1937 L. Armstrong *Swing that Music* vii. 77 When I was a kid..they started calling me 'Gate-mouth'...I started calling the other boys 'Gate' too...Then I got used to saying it and when I got into Kid Ory's band when the boys were all swinging good and hot, I would sing out, 'Swing it, Gate'..and now 'Gate' is a word swing players use when they call out to one another in their own language. **1939** *Collier's* 8 Apr. 33/3 You've handicapped your tunes with stuff no gate wants to play. **1952** B. Ulanov *Hist. Jazz in Amer.* (1958) 351 *Gate*, once (and occasionally used after the swing era) synonymous with jazz musician.

6. c. *spec.* A starting-gate.

1928 *Daily Mail* 25 July 14/1 Drift has hitherto been a disappointment. She is bad at the gate. **1963** *Sunday Mirror* (Sydney) 20 Jan. 43/1 Gili..was actually the first to hit the ground when the gates opened.

d. *Cricket.* The space between the bat and the batsman's body; formerly also, the wicket.

1851 W. Bolland *Cricket Notes* v. 108 His 'Gates' fell as flat as my hand. *Ibid.* 109 Upon the earliest appearance of the game in Ireland, the people applied the word Gate instead of Wicket. **1958** *Times* 17 Dec. 3/1 He [*sc.* the batsman] is on his way back to the pavilion bowled through the 'gate'. **1966** E. R. Dexter *Ted Dexter Declares* ii. 22 If you're out, bowled through the gate.

e. A numbered entrance to the apron of an airport.

1956 J. Stroud *Famous Airports of World* v. 70 Each building will have one or two of these 'fingers' [*sc.* finger-like walkways] and each of these will have eight or more gates leading to the aircraft. **1969** J. Elliot *Duel* I. iv. 79 We're called at gate two; we might as well go over. **1971** F. Raphael *Who were you with Last Night?* 168 We were standing around like passengers who've had their flight called but haven't been told which gate number.

8. b. (Earlier example.) *U.S.*

1815 D. Drake *Cincinnati* iii. 145 The engine drives four saws in separate *gates*.

d. An arrangement of slots, usu. in the shape of the letter **H**, through which the lever of a gear-box is moved to engage different gears. Freq. *attrib.* and *Comb.*

1906 *Westm. Gaz.* 23 Jan. 4/2 The advantage of the gate change is..its simplicity and certainty. **1906** *Daily Chron.* 14 Nov. 9/3 The 'gate' system of gear changing. **1908** *Westm. Gaz.* 28 May 4/2 The two motions that an ordinary gate-quadrant necessitates. *Ibid.* 14 Nov. 15/2 Four speeds operated through a gate. **1928** *Daily Express* 11 Oct. 1/2 The gear lever can be shifted through the gate at will according to the road and traffic. **1956** C. Monday *Centre Lathe Work* i. 29 A yoke holds *B* and *C* in contact, and is extended to form an operating handle which fits into a gate on the outside of the gearbox. **1957** S. Moss *In Track of Speed* xi. 137, I fumbled the five speed gear box until I got used to it. The reason for this was that the gate change was back to front to what was usual. **1967** A. Diment *Dolly Dolly Spy* i. 8 Only put the throttles through the gate in a dire emergency.

e. The mechanism in a cinematographic camera or projector that holds each frame momentarily behind the lens.

1909 *Statutory Rules & Orders* 12 The film gate shall be of massive construction and provided with ample heat radiating surface. **1915** B. E. Jones *Cinematogr. Bk.* i. 6 The ratio of speed at which the film passes the gate will be found to be 50 ft. in one minute. *Ibid.* iii. 14 The function of the gate is to keep the film flat and steady during exposure. **1951** G. H. Sewell *Amateur Film-Making* (ed. 2) iii. 22 The camera is an instrument whose function is to hold a supply of film, feed this to a position behind the lens known as the 'gate' of the instrument, move it intermittently past this gate, [etc.].

f. *Electronics.* An electrical signal that is used to trigger or control the passage of other signals in a circuit; a gate pulse.

1946 *Jrnl. Appl. Physics* XVII. 954/1 The ideal system would be to have an infinitesimal gate come right at the maximum of the rectified pulse. **1947** H. B. Abajian in J. S. Hall *Radar Aids to Navig.* III. vii. 205 The timing circuits of the system..provide a rectangular pulse, called a 'gate', coincident in time with the desired signal to switch on a normally cutoff section of the receiver. **1959** Rider & Uslan *Encycl. Cathode-Ray Oscill.* (ed. 2) vii. 50/2 The control of the negative gate is such that for recurrent sweeps it is periodically applied to the pentode, where it causes a longer period of tube nonconduction as compared to tube conduction.

g. *Electronics.* A circuit with one output and a number of inputs, the output signal of which is determined by the combination of signals applied to the inputs.

1946 *Electronics* May 144/3 It is with the gate circuit, or switching unit, that the present discussion deals... The gate shown was simply substituted for the existing gate of the chronograph. **1948** *Ibid.* Sept. 114/3 A gate is essentially a device having two inputs and one output. ..Logically the gate detects the AND concept. **1964** C. Dent *Quantity Surveying by Computer* iii. 23 Multiplication and division are carried out in a similar manner, circuits employing such devices as 'gates'..and 'flip-flops', being used to control and hold the binary digits.

h. *Electronics.* The material in a unipolar transistor forming the electrodes to which the input signal is applied, corresponding to the grid of a thermionic valve.

1952 W. Shockley in *Proc. Inst. Radio Engineers* Nov. 1368/1 The principles of operation of the unipolar transistor are substantially different from those of the bipolar types. For this reason, it seems appropriate to consider choosing a new set of names for the three terminals...The choice selected is 'source'.., 'drain'.., and 'gate' for the control electrodes that modulate the channel. One reason for selecting 'gate'..is that the subscript '*g*' is reminiscent of 'grid' and the analogy is close between the two. **1959** J. N. Shive *Semiconductor Devices* xiii. 237 The gates act like the control grid wires in that the voltage applied to the gates determines the width of the channel through which source-to-drain current may flow. **1970** D. F. Shaw *Introd. Electronics* (ed. 2) xii. 304 In normal operation the gate [of a field effect transistor] must always be reverse biassed with respect to source and drain.

13. gate circuit = *Gate *sb.*^{1} 8 g; **gate city** *U.S.*, a city placed at the entrance to a district; *spec.* (with capital initials) applied to Atlanta, Keokuk, Kansas City, etc.; **gate-fine,** at a university, a fine imposed for failing to be within the College gates by a certain hour; **gatefold,** 'a folded insert (as a map) in a book or other publication larger in some dimension than the page' (1961 Webster); also, a similar piece of folded paper or other material; **gate-leg, -legged** *a.*, designating a table with a leg or legs set in a gate-like frame which may be swung back to allow the leaves to be shut down; **gate-lodge,** a lodge (Lodge *sb.* 3) at the gate of the grounds belonging to a mansion, etc.; **gate-money,** also *transf.*; (*b*) money charged on a gate-bill; **gate pulse** *Electronics*, a pulse that allows a gate circuit to pass a signal; = *Gate *sb.*^{1} 8 f; **gate receipts** = *gate-money*; **gate-table,** a gate-leg table; **gate-valve,** a valve in which a sliding part controls the extent of the aperture; **gate-wheel,** a wheel operating the gates protecting a level crossing.

1945 *Electronics* Dec. 135/1 The timing circuit is controlled by the counter and its associated gate circuit. **1859** *Ladies' Repository* XIX. 51/2 Keokuk, Iowa, is the 'Gate City'—a translation, I believe, of its Indian name. **1865** *Atlanta Daily Intelligencer* 1 Oct. 2/1 Atlanta, the 'Gate City' of the South. **1887** *Courier-Journal* (Louisville, Ky.) 8 May 9/16 He enters the new South through the gate city. Louisville is to the South what Chicago is to the Northwest. **1909** 'O. Henry' *Roads of Destiny* 348 In the Gate City of the South the Confederate Veterans were reuniting. **1898** *Chambers's Dict.* 377/2 *Gate-fine*, the fine imposed for disobedience to such orders. **1963** *Nation* 4 May 18/1 They were even less real than the gatefold girls in 'Playboy'. **1963** 'E. McBain' *Ten Plus One* (1964) ii. 19 There were seventy dollars in cash in the wallet...Carella found the photographs in the gatefold. **1965** *Spectator* 5 Mar. 308/3 The Consumer Council's eight-page gatefold leaflet on 'How to say "No" to a doorstep salesman'. **1968** *Punch* 6 Mar. 325/2, I don't think we are in danger of seeing a Modigliani in the gatefold pin-up section of *Playboy*. **1905** A. Hayden *Chats Old Furnit.* 25 *Gate-leg table.* This term is self-explanatory. The legs of this class of table open like a gate. They belong to Jacobean days and are sometimes spoken of as Cromwellian tables. **1924** 'J. Sutherland' *Circle of Stars* viii. 64 An oak gate-leg table holding books and flowers. **1962** *Williamsburg Reproductions Catal.* 6/2 Among the most admired furnishings in the Governor's Palace is the oval gate-leg table in the supper room. **1903** *Connoisseur* V. xix (Advt.), Oak gate-legged table. **1919** W. Deeping *Second Youth* iii. 28 A gate-legged table in the hall. **1971** D. Francis *Bonecrack* i. 13 The furniture..consisted of an uninspiring three piece suite.., a gate-legged table, a standard lamp. **1922** Joyce *Ulysses* 697 A baronial hall with gatelodge and carriage drive. **1938** L. MacNeice *Earth Compels* 8 A Yorkshire terrier ran in and out by the gate-lodge. **1889** G. B. Shaw *London Music 1888–89* (1937) 118 A host of strangers who never did and never will contribute a farthing to the gate money upon which music depends for her living. **1920** *Hibbert Jrnl.* July 715 Now that paintings may only be seen behind barriers by paying gate-money, there is 'no demand for that sort of thing'. **1961** E. Williams *George* xix. 302 After..9.20 nobody's allowed out, and if we are out already, on our battels—our bills—we're charged gate-money, so much every half-hour, mounting till 12.20. **1945** *Electronics* Dec. 136/2 Upon firing the first round, the final trigger circuit shifts its state of equilibrium, initiating a negative gate pulse. **1948** Woodward & Williams in J. A. Pierce et al. *Loran* II. vii. 227 The gate-pulse mixer is a double-triode coincidence mixer. **1914** G. V. Hobart *Boobs* v. 80 He..proceeded to cover the gate receipts with eager mitts. **1959** *Daily Tel.* 27 June 1/7 He now expected the crowd to reach 35,000 and gate receipts £178,500. **1904** *Daily Chron.* 4 Feb. 8/1 Choose a gate table in preference to any other, since after meals are over it is but a small matter to fold this and place it at one side. **1906** *Mag. Fine Arts* Apr. 443 Spiral-legged Gate-table. *a* **1884** Knight *Dict. Mech.* Suppl. 394/2 Jenkins' gate valve has a disk of compressible packing held to its seat by a wedge-shaped follower. **1958** *Engineering* 11 Apr. 469/1 Hydraulically-actuated wedge gate valves..have successfully met the rigorous specification. **1903** *Jrnl. Inst. Electr. Engin.* XXXII. 620 In all cases where the gates and signals are worked from one point the gate-wheel is interlocked with the signal levers.

gate, *v.*^{1} Add: **2.** *Electronics.* To subject to the action of a gate (see *Gate *sb.*^{1} 8 f, g). **a.** To select those parts of (a signal) that occur within given time intervals or have amplitudes within given limits. **b.** To switch by means of a gate circuit. Usu. as **ga·ted** *ppl. a.*, **ga·ting** *vbl. sb.* and *ppl. a.*

1946 *Jrnl. Appl. Physics* XVII. 953/2 Gating under such circumstances would result in a distortion of the pulse by shearing it fore and aft. *Ibid.* 968/1 The gated background audio-noise voltage. **1947** H. B. Abajian in J. S. Hall *Radar Aids to Navig.* III. vii. 205 (*caption*) Block diagram of the gating system. **1950** Lawson & Uhlenbeck *Threshold Signals* ix. 253 In the aural method of detection one usually gates or strobes the incoming signals. **1957** R. F. Shea *Transistor Circuit Engin.* x. 317 A transistor switching circuit which may be used to gate an input signal from a common input terminal to one of two output terminals. **1959** J. N. Shive *Semiconductor Devices* xiv. 255 (*caption*) This gating circuit not only performs the gating function but amplifies the gated signal as well. **1966** *McGraw-Hill Encycl. Sci. & Technol.* XI. 207/1 The counting circuit can operate a gating circuit which passes the clock pulses directly to the cathode-ray tube. **1968** P. A. P. Moran *Introd. Probability Theory* i. 44 An electric source producing random impulses..was gated (switched on and off) once every second.

gate (gēit), *v.*^{4} *Founding.* [f. Gate *sb.*^{4}] *trans.* To provide (a mould, etc.) with a gate or gates (see Gate *sb.*^{4} 1 b). Also *intr.*

1901 *Shop & Foundry Practice* (*Colliery Engineer Co.*) IV. xl. 5 It is usually better to gate a mold for a heavy casting as near the bottom as possible. **1950** J. S. Campbell *Casting & Forming Processes* ix. 77 For many metals such as aluminum and magnesium it is good practice to gate into the castings at several places as a help in avoiding hot spots and in filling the mold cavity.

Hence **ga·ted** *ppl. a.*^{2}, provided with a gate or gates; **ga·ting** *vbl. sb.*^{2}, the action of providing gates; the arrangement of gates, runners, etc., through which molten metal enters a mould.

1906 F. W. Barrows *Pract. Pattern-Making* I. i. 16 The moulder..suggested the gating of the patterns, having noticed the castings as he dumped the mould, with their runners tying them all together, making one pattern of the flask full. *Ibid.* IV. i. 228 The moulding-machine manufacturers advertise that you can use your present patterns, whether of wood or metal, single or gated, on 'our' machines. **1907** J. G. Horner *Encycl. Pract. Engin.* V. 88/2 *Gating*, signifies the channels, and formation of the same by which liquid metal enters a mould after it leaves the pouring basin. **1931** *Foundry Trade Jrnl.* 28 May 369/2 This gating is not so simple as the top one, and it requires a heavier weight of sprues. **1948** *Metals Handbk.* (*Amer. Soc. Metals*) 6/2 *Gated pattern*, mold patterns designed to permit two or more castings, joined by connecting channels, to be made in one mold.

gâteau. Add: (Earlier and later examples.) Also **gateau,** and with pronunc. gæ·to. Now usu. a large rich cake often filled with cream, or cream and fruit, and highly decorated; also, meat or fish baked and served in the form of a cake (see also quot. 1861).

1845 E. Acton *Mod. Cook.* xviii. 473 (*heading*) A French Rice Pudding, or Gateau de Riz. *Ibid.* 474 It will..turn out..having the appearance of a cake...An admirable variety of this Gâteau is made with cocoa-nut flavoured milk. **1849** Thackeray *Pendennis* I. xxiv. 232 A basket containing little gateaux of apricots. **1861** Mrs. Beeton *Bk. Househ. Managem.* 45 *Gâteau*, a cake, correctly speaking; but used sometimes to denote a pudding and a kind of tart. *Ibid.* 680 French rice pudding, or gateau de riz. **1906** *Ibid.* lxvii. 1754 Gateau of minced meat. **1932** *Edinburgh Bk. Plain Cookery Recipes* 51 Gâteau of Fish. **1955** M. Hastings *Cork & Serpent* v. 64 Bread and butter, tea-cakes, and gateaux. **1960** C. Dale *Spring of Love* i. 16 Have something else—some gateau, an ice-cream?

gate-crasher (gēi·tkræ:ʃəɹ). [f. Gate *sb.*^{1} + *Crasher 2.] One who enters a sports ground or a private party, reception, entertainment, etc., without an invitation or ticket; also *transf.* and *fig.* Hence **ga·te-crash** *v. intr.* and *trans.*, to enter (a party, etc.) as a gatecrasher; **ga·te-crashing** *vbl. sb.* and *ppl. a.*

1927 *Daily News* 28 June 5/3 'One-eyed Connolly', the champion American 'gate crasher' (one who gains admittance to big sporting events without payment). *Ibid.*, The Committee of the White Rose Ball..dealt severely with a few cases of 'gate crashing'. **1928** *Sunday Dispatch* 15 July 1 The doom of the society gate-crasher is irretrievably fixed. *Ibid.* 5 Aug. 3/2 He was arrested for 'gate-crashing' over the frontier from Canada to America without a passport. **1929** *Daily Mail* 18 July 7/4 (*headline*) Gate-crashing woman. *Ibid.*, Helly Cozzonis..was the gate-crasher at the Mansion House reception. **1930** *Morn. Post* 19 Aug. 12/1 A 'Political Gatecrasher'. **1931** E. Mannin *Ragged Banners* viii. 80 Geoffrey Hayes is giving a party to-night—shall we gate-crash? **1933** D. L. Sayers *Murder must Advertise* xi. 192 I'm sick of Slinker's parties..let's go and gate-crash something really virtuous. *Ibid.* 198 'I had the pleasure of entertaining your cousin there one night.' 'Did he gate-crash on you? That is exactly what he would do.' **1957** D. Robins *Noble One* vii. 73 'Afraid I'm a gate-crasher,' he said. 'Your father heard me knock and let me in.' **1963** *Times* 23 May 4/7 The British women can be said to have gate-crashed the semi-final round.

gate-post. Add: **b.** Colloq. phr. *between* (or †*betwixt*) *you* (*and*) *me and the gate-post*, in strict confidence. Cf. *Bed-post, Post *sb.*^{1} 1 c.

1871 E. Eggleston *Hoosier Schoolmaster* (1872) xiv. 82 A-twix you and me and the gate-post, don't you never believe nothing that Mirandy Means says. **1875** Browning *Inn Album* ii. 39 A secret's safe 'Twixt you, me and the gate-post! 1884 'C. E. Craddock' *Where Battle was Fought* 132 Between you and me and the gate-post, old Walter Percy is a fool about everything in this world except money. 1959 P. H. Johnson *Humbler Creation* xxxix. 265 Strictly between you and me and the gate-post, Colonel, I don't care for them.

Gath: see Tell *v.* 3 b.

Gatha (gā·ta). Also with lower-case initial. [Zend; = Skr. *gāthā* song, verse, stanza.] Any of a number of psalms or versified sermons that form part of the Avesta. Also *attrib.*, esp. with reference to their language. Hence **Gathic** (gā·tik) *a.* [cf. F. *gathique*], of or pertaining to the gathas or the language in which they are written; *sb.*, the language itself, being the more archaic form of the Avestic language. Also **Gathaic** (gātā·ik) *a.*

1862 M. Haug *Ess. Parsees* iii. 136 The first Gâtha contains verses, each of which contains 48 syllables. 1867 —— *Old Zand-Pahlavi Gloss.* p. xliii, The difference of the Gâtha dialect (*gâsânîk*) from the common Avesta language is shown in the pronouns of the first and second persons. 1891 K. E. Kanga *Pract Gram. Avesta Lang.* p. iv, The examples, both Avestaic and Gâthâic..are such as are actually met with in the sacred writings. 1902 L. H. Mills (*title*) A dictionary of the Gâthic language of the Zend Avesta. 1910 E. C. Sykes *Persia* 21 The entire copies of the Avesta and Gathas are believed to have been destroyed when Alexander burnt one of the palaces of Persepolis. 1961 *Times* 13 Apr. 14/7 As soon as they see the moon they worship her by chanting a gatha (psalm) in her honour. 1968 *Language* XLIV. i. 2 It is only the Later Avestan form; Gathic instead has..the aorist imperative.

gather, *sb.*[1] Add: **1. c.** The action of 'gathering' a ball (see Gather *v.* 4 d).

1921 E. H. D. Sewell *Rugby Football* iii. facing p. 64 (*caption*) A 'gather' at full-speed..of a well-heeled ball.

3. b. *Glass-making.* A mass of molten glass on the end of a punty or blowing-iron.

1934 in Webster. 1944 *Electronic Engin.* XVII. 101 A portion of hot glass, called a gather, is made to drop into a mould in which the loose pins have previously been assembled. 1967 C. Gaskin *Edge of Glass* viii. 187 They blew a gather of glass into a mould.

gather, *v.* **4. d.** Add to def.: Esp. to pick up (*a*) a ball in fielding at cricket or in rugby football, or (*b*) a shot bird.

1838 *Sheffield Independent* 29 Sept., Gathered the ball. 1891 *Hurlingham Club Rules* 14 All birds must be gathered by the dog or trapper, and no member shall have the right to gather his own bird. 1892 W. W. Greener *Breech-Loader* 225 If a wounded bird struggles in front of you from a companion's gun, drop it if you can. Explain to the first shooter that you did so to save time in gathering it. 1925 D. J. Knight *More Compleat Cricketer* 12 The ball must be gathered in the palm of the hand. *Ibid.*, Gather it off the ground with a kind of sweeping movement.

gathering, *vbl. sb.* Add: **4. a.** *spec.* An assembly organized annually in various parts of the Scottish Highlands for contests in athletics, dancing, piping, etc.; = *game *sb.* 4 d.

1828 *Edin. Evening Courant* 16 Aug., The Athole Gathering or Highland Meeting was held..on Wednesday the 6th instant. 1850 Queen Victoria *Jrnl.* 12 Sept. (1868) 123 We..went..to the Gathering at the *Castle of Braemar*. 1947 Duke of Hamilton in H. W. Meikle *Scotland* xxxi. 244 Of the greater 'gatherings' the northern meeting at Inverness, the Braemar Gathering on Deeside, and the Oban Games are the outstanding examples. 1969 M. Pugh *Last Place Left* xvi. 113 Always wanted to see a Gathering through local eyes.

6*. *Glass-making.* = *gather *sb.*[1] 3 b.

1908 W. Rosenhain *Glass Manuf.* vi. 88 The introduction of the gathering into the molten glass is each time liable to produce air bells which would spoil the whole mass of glass. 1925 Hodkin & Cousen *Textbk. Glass Technol.* xxxi. 403 It is not necessary to blow down the pipe, when this method is used, before the gathering is brought from the furnace.

gating: see Gate *v.*[1], *gate *v.*[4]

'gator (gē·tər). Also **gater, gator.** Colloq. abbrev. of Alligator[2]. orig. *U.S.*

1844 *Knickerbocker* XXIII. 407 The 'gator is n't what you may call a han'some critter. 1886 *Outing* (U.S.) Apr. 60/2 The natives often feed their dogs and hogs boiled 'gator meat. 1890 C. W. Butler in G. O. Shields *Big Game N. Amer.* 563 This is not based upon the dog's love of 'Gator-hunting, but upon the 'Gator's love of dog-hunting. 1933 *Bulletin* (Sydney) 20 Sept. 33/2 Halifax (where the local pubkeeper used to keep a couple of 'gators chained up in the back yard like dogs). 1948 G. B. Shaw *Buoyant Billions* IV. 52 You cannot charm the rattlers and gators as I can. 1967 C. O. Skinner *Madame Sarah* viii. 168 Mr. Bell..managed to get hold of a baby 'gator, which was dumped into one of the lakes.

G.A.T.T., GATT, Gatt (gæt). [f. the initials of *General Agreement on Tariffs and Trade.*] A convention or organization established in 1947 to agree common rules for tariffs and to reduce trade restrictions.

1948 *Hansard Commons* 29 Oct. 3 Multilateral tariff negotiations should be held..to afford an opportunity to certain additional countries to adhere to the G.A.T.T. 1955 *Ann. Reg. 1954* 148 G.A.T.T.'s objective was to raise the standard of living of all peoples by means of the freest possible exchange of goods. 1957 *Economist* 19 Oct. 198/1 They have got to convince the members..that the common market does not conflict with GATT. *Ibid.* 2 Nov. 420/1 The question whether the common market treaty was compatible with Gatt.

gattine (gæti·n). Also **gattina.** [Fr.] A disease of the silkworm; = Pébrine.

1825 F. Rozier in tr. *Dandolo's Art of Rearing Silk-worms* xii. 281 M. Dandolo..makes some observations on two diseases..called in Lombardy *calcinaccio*..and *gattine*. *Ibid.* 283 The change of the nature of the silk-worm, when subject to the *gattina*,..is a real animal disease. *Ibid.*, By gattina is generally understood a worm that cannot accomplish the prescribed functions of nature. 1884 *Encycl. Brit.* XVII. 613/2 Owing to the ravages of *gattine*, the silk industry has greatly declined since 1864. 1888 E. A. Butler *Silkworms* v. 72 The most terrible of the diseases of the silkworm is *pébrine*, or pepper disease, which was formerly called *gattine*. 1954 E. Attwood *Louis Pasteur* x. 56 The silkworms were dying of a strange disease: they were covered with tiny black or brown spots, and they grew stiff and made little movements with their feet like a cat trying to scratch. One of the names given to this disease was gattine, from a word meaning a small cat.

‖**gau** (gau). [G.] A territorial and administrative division of ancient Germany, including several villages or communities; in the Middle Ages, a larger division, over which, under Frankish rule, was placed a graf.

The word is a frequent final element of place-names, as Breisgau, Oberammergau.

1845 B. Thorpe tr. *Lappenberg's Hist. Eng.* I. 88 In the time of Charlemagne, they [*sc.* the Saxons] were in possession of..the 'gaus', or districts, of the later Upper and Lower Saxony. 1874 Stubbs *Const. Hist.* §26 The four [*sc.* marks] were in A.D. 804 made into a Gau, in which the archbishop of Bremen had the royal rights of Heerbann and Blutbann. 1916 E. G. A. Holmes *Nemesis of Docility* i. 19 The counts..were supposed to..keep order in their own *gaus* or 'shires'. 1959 *Chambers's Encycl.* VI. 257 The people looked for protection to the local ruler rather than the reigning prince, to the count of the district (Gau) and to the tribal duke.

gaucheness. Delete *rare* and add further examples.

1936 Auden *Look, Stranger!* 55 O can you see precisely in our gaucheness The neighbour's strongest wish, to serve and love? 1937 *Discovery* July 226/1 Sixty-one excellent plates atone for any 'gaucheness' of the style. 1967 M. Wilding in *Coast to Coast 1965–6* 253 She liked him for his gaucheness.

gaudy, *a.*[1] For †*Obs.* read *arch.* and add later example.

1902 W. de la Mare *Songs of Childhood* 29 There's a ring gaudy-green in the dell.

gaudy, *a.*[2] **4.** *gaudy-coloured* (later examples).

1824 'H. L. Howard' *Joseph & Brethren* II. 123 The gaudy-colour'd story of his mind. 1921 W. de la Mare *Crossings* 67 Muffled up in a gaudy-coloured shawl.

gaudy-night (s.v. Gaudy-day). (Later examples.)

1935 D. L. Sayers (*title*) Gaudy night. 1963 Lockhart & Woodhouse *Rhodes* xxiii. 404 It happened to be the College's annual Gaudy night.

‖**gauffrage** (gōfrā·ʒ). *Printing.* [ad. F. *gaufrage* embossing, f. *gaufrer* (see Goffer *v.*).] = *blind printing (blind *a.* 16).

1904 [see *blind printing*]. 1927 H. Hubbard *Colour Block Print Making* 172 The Japanese call this 'gauffrage', but it is known amongst western artists as 'blind printing'. 1960 H. Hayward *Antique Coll.* 127/2 *Gauffrage*, blind-printing, producing an embossed effect without colour, and used in Japan from the late 18th cent. as a technique in making prints.

gaufre, var. Gofer.

gauge, *sb.* Add: **1. c.** *Physics.* [tr. G. *maßstab* (H. Weyl 1918, in *Sitzungsber. d. Preuss. Akad. d. Wissensch.* 30 May 475).] A concept introduced by Weyl as a measure of the vector field that in his cosmology related length and position, represented mathematically by a potential function; hence, any function introduced as an additional term into the equations of the potentials of a field such that the derived equations of observable physical quantities are unaltered by the introduction. Freq. *attrib.*

1920 A. S. Eddington *Space, Time & Gravit.* xi. 169 The only possible course is to lay down (1) a mesh-system filling all the space and time considered, (2) a definite unit of interval, or gauge, *at every point of space and time*. *Ibid.* 176 The radius of curvature of the world provides a natural and absolute gauge at every point; and it will presumably introduce the greatest possible symmetry into our laws if the observer chooses this, or some definite fraction of it, as his gauge. 1923 —— *Math. Theory of Relativity* vii. 202 Change of gauge is a generalisation of change of unit in physical equations, the unit being no longer a constant but an arbitrary function of position. 1940 W. Pauli in *Physical Rev.* 15 Oct. 718/1 By 'gauge-transformation of the first kind' we understand a transformation $U \rightarrow Ue^{i\alpha}$ $U^* \rightarrow U^*e^{-i\alpha}$ with an arbitrary space and time function α. By 'gauge-transformation of the second kind' we understand a transformation of the type $\phi_k \rightarrow \phi_k - i(\partial\alpha/\partial x_k)/\epsilon$ as for those of the electromagnetic potentials. 1959 J. Aharoni *Special Theory of Relativity* i. 29 We draw an arbitrary Cartesian system *x*, *t* and prescribe along the axes two gauges of geometrical length g_1 and g_2, one to represent a unit of distance, the other a unit of time. 1961 *Encycl. Dict. Physics* III. 409/2 The gauge is chosen for convenience in a particular problem to bring about symmetry (gauge invariance) or a convenient form of equations involving the potentials. 1965 R. Adler et al. *Introd. Gen. Relativity* iv. 106 The four-vector function ϕ_μ has no *direct* physical meaning; only its four-dimensional curl has physical meaning. It is thus clear that we may make a so-called gauge transformation on ϕ_μ; that is, we can add an arbitrary four-dimensional divergence..to ϕ_μ without altering $F_{\mu\nu}$ and therefore without altering the physical situation. *Ibid.* xiii. 405 Quantities and relations that do not change under gauge transformation are called gauge invariants.

15. gauge-field *Astr.*, a restricted area of the sky photographed for the purpose of gauging the number and density of the stars in that region; **gauge function** *Math.* (see quot. 1965); **gauge-work** = *gauged work* (see Gauged *ppl. a.* 2).

1891 *Smithsonian Inst. Rep.* 107 A photographic gauge-field on a small scale. 1937 J. von Neumann *Coll. Wks.* (1962) IV. 208 A familiar symmetric gauge function is

$$\phi_p(u_1,..u_n) = \left(\sum_{i=1}^{n} |u_i|^p\right)^{1/p}$$

for $1 \leqslant p < +\infty$. 1965 *Pacific Jrnl. Math.* XV. 242 A function ϕ on a complex vector space is called a gauge function if $\phi(u) > 0$ when $u \neq 0$, $\phi(\alpha u) = |\alpha|\phi(u)$ for complex α, $\phi(u+v) \leqslant \phi(u) + \phi(v)$. 1906 *Daily Chron.* 23 Aug. 2/5 Gage-work window headings. 1920 *Conquest* Apr. 269/2 These slips of steel..are much used in gauge work for building up bars of any desired thickness.

gauleiter (gau·ləitəɹ). Also with capital initial. [G., f. *gau* *gau + *leiter* leader.] A political official controlling a district under Nazi rule; also *transf.* and *fig.*, a local or petty tyrant.

1936 A. Huxley *Eyeless in Gaza* xxxi. 413 Tyranny under commissars, tyranny under *Gauleiters*—it doesn't seem to make much difference. 1943 *Penguin New Writing* XVI. 9 Mr. Scollop, with his draper's shop and his councillorship, is by way of being the gauleiter of this town. 1949 L. P. Hartley *Boat* 299 Two bullies, gauleiters almost, from whom everyone fled in terror. 1958 *Times Lit. Suppl.* 4 Apr. 186/4 The Nazi gauleiters in occupied countries. 1970 *New Statesman* 23 Oct. 526/1 His friend Karl Hanke, the Gauleiter of Lower Silesia, visited him in a state of shock after he had been to Auschwitz concentration camp.

Gaullism (gǭ·liz'm). [ad. F. *Gaullisme*, f. the name of General Charles de *Gaulle* (1890–1970), French military and political leader: see -ism.] The principles and policies associated with General de Gaulle; adherence to or support of these principles. Also *transf.* (Cf. next.)

1950 *Chambers's Encycl.* IV. 61/1 The dual threats of communism and Gaullism. 1951 *20th Cent.* Nov. 400 (*title*) Aspects of Gaullism. 1955 *Times* 25 June 7/5 It is regrettable to consider either the history of the Armistice of 1940 or that of Gaullism and 'Free France' only with humour. 1959 *New Statesman* 7 Feb. 174/1 'Gaullism'..is a growing factor in Italian politics. 1968 *Listener* 21 Mar. 377/1 He was terrified by the effect Gaullism was having, and might have, on his own countrymen.

Gaullist (gǭ·list), *a.* and *sb.* Also **Gaulliste.** [ad. F. *Gaulliste*, f. as prec.: see -ist.] **A.** *adj.* Supporting or pertaining to the principles and policies of General de Gaulle (see prec.). **B.** *sb.* One who supports General de Gaulle and his principles. Also *transf.* Cf. *de Gaullist *a.* and *sb.*

1941 H. Nicolson *Diary* 20 Jan. (1967) II. 139 De Gaulle says that he had received a long letter of the most Gaulliste nature. 1947 Koestler in *Partisan Rev.* XIV. 344 In the name of what principle do you prefer the Communist jail to Gaullist restricted freedom? 1952 *Ann. Reg. 1951* 206 Proportional representation would increase Communist and Gaullist representation. *Ibid.* 209 Gaullists and Communists were opposed to it and Socialists hesitant. 1958 [see *band-wagoner]. 1960 *New Statesman* 2 Apr. 471/1 In French eyes, Khrushchev has allowed himself to appear first and foremost as a Gaullist of the left. 1971 *Guardian* 7 Sept. 2/5 The election of June 1968, when it was hard for any candidate running under Gaullist colours to miss.

Gauloise (gǭ·lwāz). Also **Gaulois,** and with lower-case initial. [A proprietary name, f. Fr. *gaulois, -oise* Gallic.] A cigarette of a popular French brand.

1920 *Trade Marks Jrnl.* 17 Mar. 528, 20 Cigarettes Gauloises. 1926–7 *Army & Navy Stores Catal.* 76/2 Imported French cigarettes. Gauloises. 1933 *Trade Marks Jrnl.* 23 Aug. 1033 *Gauloises.* Cigarettes. Service d'Exploitation Industrielle des Tabacs..Ministère des Finances, Palais du Louvre. 1933 'G. Orwell' *Down & Out* v. 35, I even squandered two francs fifty on a packet of Gaulois Bleu, in anticipation of my wages. 1958 R. Godden *Greengage Summer* iv. 49 It's a Gauloise, a French cigarette. 1959 *Listener* 20 Aug. 285/3 She smoked Gauloises, but she preferred black Russian cigarettes with gold tips. 1960 *20th Cent.* Sept. 259 He sat in his truck smoking gauloises.

gauntleted, *a.* Add: Of a glove: having a gauntlet (see Gauntlet *sb.*[1] 2 b).

1902 C. Major *D. Vernon* 38 He was drawing on his gauntleted gloves. 1924 'J. Sutherland' *Circle of Stars* iv. 37 Her loose gauntleted gloves of wash-leather.

gaup, gawp, *v.* Add: (Later examples of *gawp*, which is now the commoner spelling.) Also = Gawk *v.*

1942 D. Powell *Time to be Born* (1943) vii. 175 You sit there gawping at him like some little goon. 1952 J. Cannan *Body in Beck* vii. 134 I'm here to get on with the job, not to gawp at the clients. 1959 J. Burke *Echo of Barbara* v. 42 Flauntin' yourself before a lot of gawpin' idiots.

gauss. Substitute for def.: The electromagnetic unit of magnetic induction (flux density) in the C.G.S. system, defined as the induction that exerts a force of one dyne on each centimetre of a straight wire carrying one e.m.u. (10 amp.) of current, when the induction is perpendicular to the wire. (In the International System of Units the gauss is replaced by the tesla, or weber per square metre, equal to 10,000 gauss.)

The name has been suggested for various magnetic units. The original sense was the C.G.S. unit of magnetic field strength (later called an *oersted*), and, because in the C.G.S. system the electromagnetic units of field strength and of induction have the same dimensions and *in vacuo* are numerically the same, the word has freq. been used for both indiscriminately.

1891 L. Clark *Dict. Metric & Other Useful Measures* 38 *Gauss*, a unit [*printed* unity] of intensity of magnetic field, proposed by Prof. S. P. Thompson. $=10^8$ C.G.S. units. Other values have also been suggested. 1893 *Electrician* 29 Sept. 579/1 Some attempt was made to secure the adoption of the name of 'gauss' for the C.G.S. unit of magnetic field. 1895 *Rep. Brit. Assoc.* 196 The Committee [*sc.* the British Association Committee on Electrical Standards]..recommend for tentative adoption the following terminology...That the c.g.s. unit of magnetic potential or of magneto-motive force be called a *gauss*...Hence the number of gausses round any closed curve linked on an electric circuit is equal to 1·2566 the number of ampère-turns in this circuit. 1913 C. W. C. Barlow *Electr. & Magn.* xii. 241 The Gauss is the unit of Magnetic Induction and also of Field, the Gilbert of M.M.F...These names are not in common use. 1930 *Nature* 16 Aug. 252/2 The committee on nomenclature [of the International Electrotechnical Commission]..adopted the following names for the magnetic C.G.S. units. The unit of magnetic flux was called the 'maxwell' and the unit of flux density the 'gauss'. 1947 *Science News* IV. 116 The gauss is the unit of the magnetic field. It exerts unit force (1 dyne) on a unit magnetic pole placed in it. 1955 *Sci. Amer.* May 56/3 The earth's field near the poles is ·7 gauss. 1962 A. R. W. Hayes *Revision Physics* v. 170 In free space an applied field *H* oersted produces a flux density *H* gauss; in a magnetic medium the flux density is modified to a value *B* gauss which is the sum of the original flux and that due to magnetization of the medium. 1965 A. H. Morrish *Physical Princ. Magn.* i. 2 The cgs unit of magnetic field is the oersted, although the term gauss is still frequently used.

Gaussian, *a.* Add: (Earlier and later examples.)

1874 *Rep. Brit. Assoc. 1873* 75 Gaussian logarithms have for their object to facilitate the finding of the logarithms of the sum and difference of two numbers whose logarithms are known, the numbers being themselves unknown. 1897 A. G. Webster *Theory Electr. & Magn.* II. ix. 367 We may, when dealing with electrical quantities, assume that the dimensions of the electrical inductivity are zero. This gives the electrostatic system of units. We may.., when dealing with magnetic quantities, assume that the dimensions of the magnetic inductivity are zero. This gives the magnetic system. Both these systems are due to Gauss, and when we use both systems for their respective kinds of quantities, we shall say that the quantities are measured in *Gaussian* units. 1957 B. & B. I. Bleaney *Electr. & Magn.* xxiii. 646 In electromagnetic theory Maxwell's equations involve both electrical and magnetic units, and in the c.g.s. system they are generally written in mixed or Gaussian units.

b. *Statistics.* Designating a curve, frequency distribution, or statistical process, etc., that is described mathematically by a function of the form $\exp(-x^2/2\sigma^2)/\sqrt{(2\pi\sigma^2)}$, where *x* is the variable.

1905 K. Pearson in *Biometrika* IV. 173 Many of the other remedies which have been proposed to supplement what I venture to call the universally recognised inadequacy of the Gaussian law..cannot..effectively describe the chief deviations from the Gaussian distribution. *Ibid.* 189 In writing for Germans I naturally spoke of the Gaussian curve. 1920 *Biometrika* XIII. 10 To get the Gaussian or normal curve we must..replace differences by differentials and we have $(dy/dx)/y = -x/\sigma_0^2$. *Ibid.* 15 The so-called Gaussian or normal curve was first introduced into statistics as a *rough and ready* solution for the sum of a certain number of terms in a definite hypergeometrical series. 1968 P. A. P. Moran *Introd. Probability Theory* ix. 424 The homogeneous additive process in which the distribution of $x(t_1) - x(t_2)$ is always a normal distribution is known by various names as the Gaussian, normal, Wiener, or Brownian process.

gauze, *sb.* Add: **3. gauze ring** = *crape ring* (*crape *sb.* 3 b).

1867 G. F. Chambers *Descriptive Astron.* vii. viii. 709 As a rule the gauze ring must not be expected to be seen with any aperture below 4 inches. 1964 D. H. Menzel *Field Guide Stars* ix. 297 Toward the inner edge the grains again thin out, producing a partially transparent ring, the gauze, or crepe, ring.

gauze, *v.* Add to def.: To cover with or as with gauze; to veil. Also *intr.*, to become gauzy or misty.

1902 B. Baynton *Bush Studies* 45 The wide plain gauzed into a sea on which the hut floated lonely. 1938 E. Bowen *Death of Heart* ii. vii. 292 Thickets of hazel gauzed over the distances inside.

gauzily (gǭ·zili), *adv.* [f. Gauzy *a.* + -ly[2].] In a light, delicate manner; so as to resemble gauze.

1903 *Daily Chron.* 16 May 3/6 The black chiffon..very gauzily and daintily treated. 1934 I. W. Hutchison *North to Rime-Ringed Sun* vii. 64 In the placid gold water..the clouds were gauzily mirrored. 1970 *Daily Tel.* 21 Sept. 10, I don't see myself..drifting gauzily about like Botticelli's Primavera. I just want a dress.

gavage (gævā·ʒ). [Fr.] A method of forcible feeding by the use of a force-pump and a tube passed into the stomach.

1889 *Buck's Handbk. Med. Sci.* VIII. 102/2 Dr. Turnier, of Paris, has advocated the method of gavage in infants prematurely born. 1905 Osler *Princ. Med.* (ed. 6) 511 When there is persistent anorexia, gavage may be necessary. 1963 L. Durrell in *Holiday* Jan. 70/1 In goose country, where the practice of force-feeding geese (*gavage*) is in operation, there are always a goodly number of casualties.

gavel (gæ·vĕl), *v.*[4] orig. and chiefly *U.S.* [f. Gavel *sb.*[4]] To hammer with, or as with, a gavel (Gavel *sb.*[4] b).

1925 T. Dreiser *Amer. Tragedy* (1926) II. ii. xxv. 313 And at once Oberwaltzer gaveling for order and ordering the arrest of the offender. 1934 *Words* Nov. 5/2 *To gavel* and *to yes* are new, and..obviously American. 1959 *Time* 23 Feb. 20/1 'The faces will be different,' said Arkansas' John McClellan as he gaveled his Senate labor-racketeering committee into session. 1966 D. F. Galouye *Lost Perception* v. 53 Radcliff strode..onstage, the determined *thuds* of his heels gavelling order among the assembly.

‖**gavroche** (gavrɔʃ). [Fr., f. the name of a gamin in Victor Hugo's *Les Misérables.*] A street urchin or gamin, esp. in Paris.

1876 *Times* 2 Nov. 4/4 A Norwegian *gavroche* is balancing himself by a miracle of adroitness on the dorsal extremities of the slender shafts. 1882 *Pall Mall Gaz.* 8 Dec. 4/1 'Mo-sieu'..is..pronounced as 'm'sieu' in 1882, or, if one wishes to talk as a real *gavroche*, 'm'seu'. 1921 *Blackw. Mag.* Feb. 251/1 Beneath the demure exterior the Parisian *gavroche* lurked. 1964 *Guardian* 16 June 9/2 Paris..adopted her as a gavroche of the theatrical boulevards.

gawblimey, -blimy, varr. *gorblimey.

Gawd (gǭd). Also **gaw, gawd.** Vulgar or slang pronunciation of God *sb.* So **Gawd-forbid** = *God-forbid* (*God *sb.* 8 d); **Gawd-help-us, Gawdelpus,** a helpless or exasperating person.

1877 G. Meredith *Let.* 24 Nov. (1970) I. 553 Your Black Christian..appears to have been blest by *Gawd* recently. 1892 I. Zangwill *Childr. Ghetto* I. x. 223 'Is your Daniel in?' 'Yes.' 'Tank Gawd! I mean, can I see him?' 1896 A. Morrison *Child of Jago* xxviii. 272 Run, for Gawd's sake, or the woman'll croak! 1908 H. G. Wells *War in Air* iv. 115 'Gaw!' said Bert. 'We're going up!' 1911 L. Stone *Jonah* I. ix. 104 Gawd, 'e's stiffened 'im! 1912 A. N. Lyons *Clara* xv. 162 Look at all the Gawdelpuses ya see on the trams—cow-spirited, cocoa-faced makeshifts. 1928 'Sapper' *Female of Species* v. 76 'A lounge lizard. A ball-room snake. What matter that his Black Bottom is the best in London.' 'My Gawd! sir,' gasped the other. 'His 'ow much?' 1931 Wodehouse *If I were You* ix. 98 He had always been an ugly sort of Gawd-help-us. 1936 J. Curtis *Gilt Kid* vii. 73 Why don't you take off your gawd-forbid? We're passing the Cenotaph. 1937 Partridge *Dict. Slang* 319/1 *Gawd forbid*, a variant of *God forbid* [= kid]. 1955 M. Allingham *Beckoning Lady* xiv. 204 You take 'Er Ladyship and the Gawd-ferbid to the party. 1961 Wodehouse *Ice in Bedroom* iii. 25 A potbellied baggy-trousered Gawd-help-us. 1967 L. White *Crimshaw Memo.* (1968) v. 93 Gawd knows I got enough problems.

gawk, *v.* For *dial. U.S.* read '*colloq.*, orig. *U.S.* or *dial.*' (Add further examples.)

1905 *Westm. Gaz.* 30 Mar. 8/1 Many of the audience turned round to 'gawk'—as he phrased it—at a poor fainting girl. 1959 'M. Ainsworth' *Murder is Catching* vii. 82 Everyone will come and gawk. 1965 C. D. Eby *Siege of Alcázar* (1966) ii. 58 Gawking in wonder at the falling bombs.

gawkily (gǭ·kili), *adv.* [f. Gawky *a.* + -ly[2].] In a gawky or ungainly manner; like a gawk.

1812 L. M. Hawkins *C'tess & Gertrude* IV. 39 'I suppose you read a great deal then,' said Lady Elma gawkily. 1928 *Daily Express* 13 Feb. 3 He is tall and almost gawkily built. 1928 *Oxf. Poetry* 38 These boughs..stretch gawkily above the cross-roads. 1944 W. S. Maugham *Razor's Edge* iv. 153 She held herself gawkily as though she didn't know how to cope with the length of her limbs.

gawp (gǭp), *sb. dial.* and *colloq.* [f. Gaup, gawp *v.*] A fool or simpleton. (See also Gaupy, gaup.)

1926 L. Coutts *Lyrics* 41 I've affin seen conceit in men But ee're the warst! Ye gawp! 1937 L. A. G. Strong *Swift Shadow* 155 Standing by like gawps to see bloody murder done. 1960 R. Williams *Border Country* 95 Country's not going to be run by a lot of young gawps.

gay, *a.* Add: **1. f.** Forward, impertinent, too free in conduct, over-familiar; usu. in phr. *to get gay. U.S. slang.*

1896 W. C. Gore in *Inlander* Jan. 147 *Get gay*, to joke boisterously; to show off; to act 'smart'. 1899 Ade *Fables in Slang* 109 The Copper, perceiving that he had come very near getting Gay with our First Families, apologized for Cutting In. 1901 Merwin & Webster *Calumet 'K'* xii. 226 He got gay one day. I warned him once, and then I threw him off the distributing floor. 1911 J. F. Wilson *Land Claimers* vi. 80 And I wouldn't get gay round her. 1915 Wodehouse *Something Fresh* iv, The flush on the little man's face deepened. 'Are you trying to get gay with me?' he demanded dangerously.

g. Of a dog's tail: carried high or erect.

1927 W. H. Dowling in C. C. Sanderson *Pedigree Dogs* 334 *Tail.*—Should be carried proudly, curved or plumed in a tight curl over and close to the back (never gay as in a 'Peke'). 1952 C. L. B. Hubbard *Pembr. Corgi Handbk.* 111 *Gay tail*, one which from root to tip is carried over the horizontal.

2. a. Esp. in *gay dog*, a man given to revelling or self-indulgence; *gay Lothario*: see Lothario.

1900 G. Swift *Somerley* 54 Oh! that first kiss! how proud of it we are, what gay dogs we feel! 1910 S. Kaye-Smith *Spell Land* xix. 221 He felt rather a gay dog.

c. Of a person: homosexual. Of a place: frequented by homosexuals. *slang.*

1935 N. Ersine *Underworld & Prison Slang* 39 *Geycat*, ..a homosexual boy. 1951 E. Lambert *Sleeping-House Party* vii. 74 In a way it was an odd threesome. It occurred to me that Esther rather hung round our two gay boys. 1955 P. Wildeblood *Against Law* I. 23 Most of the officers at the station had been 'gay'..an American euphemism for homosexual. *Ibid.* III. 105 The place [*sc.* a prison] is *packed* with gay people who are in for something else. 1960 [see *bent *ppl. a.* 5 c]. 1963 A. Heron *Towards Quaker View of Sex* iii. 24 These may form the 'queer' society; these will frequent 'gay' bars. *absol.* 1966 A. Firth *Tall, Balding, Thirty-Five* xv. 194 Would he ever dare, even if he wanted to, join the shrill freemasonry of the London gay? 1968 *Globe & Mail Magazine* (Toronto) 13 Jan. 6/1 A coffee shop frequented by the gay.

9. *gay-hearted* adj.; **gay cat** *U.S. slang*, a young or inexperienced tramp; a hobo who accepts occasional work; (see also sense *2 c); **gay deceiver,** (*a*) a deceitful rake (Rake *sb.*[5]); (*b*) *pl. slang* = *falsies *sb. pl.*; **Gay Gordons,** (*a*) (see quot. 1925); (*b*) a Scottish dance popular in old-time and modern dancing.

1897 'J. Flynt' in *Forum* Feb. 741 Nothing arouses his [*sc.* the hobo's] scorn more than the dilettante, or 'gay-cat', as he calls him. 1901 J. London *Let.* 6 Dec. (1966) 126 Wyckoff is a gay cat. That was his rating when he wandered over the States. 1926 J. Black *You can't Win* vi. 74 He must have been an awful gay cat to get into the end of a carload of planed lumber. It's suicide. 1950 R. Chandler *Let.* 18 May (1966) 78 A gay-cat is a young punk who runs with an older tramp and there is always a connotation of homosexuality. Again, he could be a 'look-out' (outside man) or a 'finder' (finger or finger man), but that is a derived or occasional meaning and not exact. 1803 G. Colman *Love laughs at Locksmiths* (1823) II. 25 Says he, 'I am a handsome man, but I'm a gay deceiver.' 1809 Gay deceiver [see Gay *a.* 2 a]. 1898 J. D. Brayshaw *Slum Silhouettes* 44 Ah! he's a gay decaver, is Billy, like all the min. Sure I wouldn't trust my ould gran'mother wid him. 1942 D. Powell *Time to be Born* (1943) x. 232 Her pink sweater..clung properly to the seductive curves of her Gay Deceivers. 1962 *Guardian* 12 Mar. 4/7 False bosoms..were known as 'gay deceivers'. 1925 Fraser & Gibbons *Soldier & Sailor Words* 185 *Gay Gordons, The*, The Gordon Highlanders. In particular the 2nd Battalion, the 92nd Highlanders. 1947 J. R. Gillespie *Old Tyme Dancing* 32 (*heading*) The Gay Gordons. 1955 J. I. M. Stewart *Guardians* III. iv. 238 The music of the Samba and the Gay Gordons. 1966 *Crescendo* Dec. 27/1 A musical chore to be dealt with grudgingly like the Veleta or the Gay Gordons. 1853 Whittier *Panorama* (1856) 33 Gay-hearted Health. 1947 W. de la Mare *Coll. Stories for Children* 44 Fairies, sly, small, gay-hearted.

gay, *v.* Delete †*Obs.* and add later examples of sense b (now usu. with *up*). Hence **gayed** *ppl. a.*

1960 *News Chron.* 2 May 6/6 The charming effect of houses gayed up with these flower-filled boxes. 1968 'E. Trevor' *Place for Wicked* iii. 39 She called in with a geranium in a pot for them because they'd never think about gaying the place up for themselves.

gāyatrī (gā·yatri). [Skr., f. *gā* to sing.] **a.** An ancient twenty-four-syllable metre. **b.** A hymn, etc., composed in this metre; *esp.* the

verse of the Rig-veda repeated daily as a prayer by Brahmins.

1843 *Penny Cycl.* XXVI. 177/1 The principal metres used in the hymns of the Vedas are the Jagatî, Gâyatrî, and Trishtup. 1845 *Encycl. Metrop.* XVI. 351/1 His father.., guided by the priest,..pronounces three times the Gâyatrî; 'let us meditate on the adorable light of the Divine Ruler, may it guide our intellects!' 1865 Max Müller *Chips* (1867) I. 19 The famous hymn, the Gâyatrî, which now for more than three thousand years has been the daily prayer of every Brahman. 1886 *Encycl. Brit.* XXI. 275/2 The *gâyatrî*..and other chiefly octosyllabic metres. 1916 A. A. Macdonell *Vedic Gram.* 446 Gâyatrî triplets. 1965 *Language* XLI. 11 *Yaja* occurs at the end of a gāyatrī.

Gay-Pay-Oo (gē¹,pē¹,ū·). Also **Gay-Pay-U.** Phonetic representation of G.P.U. as pronounced in Russian (see *G III. f).

1927 *Observer* 11 Dec. 21/2 The Gay-Pay-OO, or Political Police. 1928 *Punch* 18 Jan. 57/2 The Gay-Pay-Oo, it seems, is the name of the reorganised Russian Cheka. 1934 G. B. Shaw *Pref.* 359 The Russians were forced to set up an Inquisition or Star Chamber, called at first the Cheka and now the Gay Pay Oo (Ogpu). 1936 D. Powell *Turn, Magic Wheel* III. 225 He yanked words out of the corner of his mouth..as..cautiously as if Gay-Pay-U men were all about.

gazabo (găzē¹·bo). *slang* (orig. and chiefly *U.S.*). Also **gazebo.** [Perh. f. Sp. *gazapo* a sly fellow; cf. Gazebo.] A fellow, 'guy'; often with a pejorative connotation.

1896 Ade *Artie* v. 44 Who does I meet comin' out o' the house but a cheap gazabo that was with her the first time I see her. 1910 W. M. Raine *B. O'Connor* 39 A big gazabo in a red wig held up Frost, the engineer. 1934 B. Penton *Landtakers* (1935) I. i. 14 Aw, leave the old gazebo. He's drunk. 1953 H. Miller *Plexus* (1963) vii. 262 But there was one thing he seldom did, queer gazabo that he was—he seldom asked questions.

gazania (găzē¹·niă). *Bot.* [mod.L. *Gazania* (G. Gaertner *De Fructibus Plantarum* (1791) II. 451), f. the name of Theodorus *Gaza* (1398–1478), a Greek scholar.] A plant of the genus of South African herbs so named, belonging to the family Compositæ and having showy yellow or orange flowers.

1813 W. T. Aiton *Hortus Kewensis* (ed. 2) V. 140 Great-flower'd Gazania. 1844 Mrs. Loudon *Ladies' Flower-Garden Ornam. Perennials* II. 29 The Peacock Gazania..only expands its flowers in broad daylight. *Ibid.*, The Golden-Flowered Gazania..is a very pretty little plant, with golden yellow flowers. 1921 G. Jekyll *Colour Schemes* 63 A few belated Orange Lilies have their colour nearly repeated by the Gazanias next to the path. 1966 E. Palmer *Plains of Camdeboo* xvii. 280 There are Nemesias and Gazanias.

gazob (gă·zǫb). *Austral. slang.* [Origin unknown; perh. f. *gazabo.] A fool; a blunderer.

1906 E. Dyson *Fact'ry 'Ands* xii. 162 But I thought barrer-pushin' was er game fer gazobs? 1916 C. J. Dennis *Songs Sentimental Bloke* 42 Ar! but 'e makes me sick! A fair gazob! 1966 Baker *Austral. Lang.* (ed. 2) vi. 135 Fools of one kind and another..*gazob, gimp, gup*, [etc.].

gazoo, gazooka, varr. Kazoo.

1927 *Bulletin* 24 Aug., A new musical atrocity is reported to be coming to Scotland. It is the Gazoo. This instrument of ear-torture is simply an adaptation of the primitive 'comb and tissue paper'. 1938 'R. Hyde' *Nor Years Condemn* x. 195 The Maoris could make the dish-like gazookas of tin and wire sound like guitars. 1959 *Manch. Guardian* 26 Aug. 5/2 There is..a new outbreak of kazooing in South Wales...On the history of gazooka (as it is called in South Wales) Mr. Gwyn Thomas..is an expert.

gazook (găzū·k). *slang.* [Origin unknown, but cf. *gazabo and *gazob.] A fool; a stupid or unpleasant person.

1928 M. H. Weseen *Crowell's Dict. Eng. Gram.* 266 *Gazook*, slang name applied contemptuously to any person who is disliked. 1936 B. Penton *Inheritors* I. ii. 15 Look at that poor gazook, Sambo. He'd call your old man God Almighty even if he starved him to death.

gazoomph (găzu·mf), *v. slang.* Also **gasumph, gazump(h), gezumph.** [Origin uncertain.] *trans.* To swindle; *spec.* (normally with the spelling *gazump*) to act improperly in the sale of houses, etc. (see quots. 1971). Also as *sb.*, a swindle. Hence **gazoo·mpher,** a swindler.

1928 *Daily Express* 19 Dec. 2/7 'Gazoomphing the sarker' is a method of parting a rich man from his money. An article is auctioned over and over again, and the money bid each time is added to it. *Ibid.*, I 'gazoomphed' a friend of mine with complete success last night. 1932 G. S. Moncrieff *Café Bar* xvii. 185 'Ere 'e is..parasitin' on people all day..and then 'e objects to a little *gasumph*! *Ibid.* 187 That's all we are, gasumphers an' researchers of tripe. 1934 P. Allingham *Cheapjack* xv. 189 Grafters speak a language comprised of every possible type of slang... Quite a number of words are Yiddish. These include 'gezumph', which means to cheat or to overcharge. *Ibid.* xxiii. 301 The whole gezumphing world seemed to have gathered together. *Ibid.* 319 *Gezumpher*, a swindler. 1961 *Daily Mail* 25 Feb. 9/2 M.P.s had admitted that they had been 'gazoomphed' by fast-talking racketeers. 1969 Y. Carter *Mr. Campion's Farthing* xiii. 121 I've never known an offer from you that wasn't a *gezumph*. 1971 *Guardian* 8 Nov. 13/2 'Gazumping'—a system of profiteering by double selling and pushing prices up—is creeping into the property market... The word is car trade slang for selling to one buyer and then, as values rise, to a second buyer. 1971 *Times* 6 Dec. 4/1 A Bill to be presented to the Commons next week..would outlaw the raising of the price of a house before contracts are exchanged. The practice is known as 'gazumping'.

‖**gazpacho** (gaþpā·tʃo, gasp-). Also **gaspacho.** [Sp.] A cold Spanish soup (see quot. 1845).

1845 R. Ford *Hand-bk. Trav. Spain* I. i. 69 *Gazpacho*..is a cold vegetable soup, and is composed of onions, garlic, cucumbers, *pepinos*, pimientas, all chopped up very small and mixed with crumbs of bread, and then put into a bowl of oil, vinegar, and fresh water. 1903 Conrad & Hueffer *Romance* I. iv, He spent his days cooking extraordinary messes,..making substitutes for eternal gaspachos. 1963 R. Carrier *Great Dishes of World* 56/2 *Gazpacho* is traditionally served accompanied by individual small bowls of raw vegetables and garlic *croûtons*. 1969 *New Yorker* 20 Sept. 164/2 Gazpacho..should rest and 'marry' for a time before being eaten.

gazump: see *gazoomph *v.*

geanticline (dʒi,æ·ntiklain). *Geol.* [f. Gr. γῆ earth + Anticline.] = Geanticlinal *sb.*

1895 J. D. Dana *Man. Geol.* (ed. 4) 389 A mountain range includes all the mountain ridges made over the area and border of a single geanticline. 1925 J. Joly *Surface-Hist. Earth* vii, The first Rocky Mountains geanticline came into existence in closing Carboniferous (early Pennsylvanian) time. 1957 G. E. Hutchinson *Treat. Limnol.* I. i. 14 Lake Poso..is probably a remnant of an old surface that escaped the general Pleistocene uplift which affected two geanticlines east and west of the lake.

gear, *sb.* Add: **1. a.** Now common in colloq. use.

1963 *Sunday Express* 22 Dec. 14/2 A designer of singularly *avant garde* 'gear' (men's wear to the uninitiated). 1966 *Listener* 3 Mar. 327/2 There was no dressing up—in case, presumably, the customers fled at the sight of the queer gear.

5. e. *that's* (or *it's*) *the gear*: an expression of approval. Hence as *adj.*, good, excellent, 'great'. Also as *int. slang.*

1925 Fraser & Gibbons *Soldier & Sailor Words* 103 *Gear*, apparatus generally... Also used as a colloquial term for anything giving satisfaction—*e.g.*, 'That's it, that's the gear!' 1951 *Sunday Pictorial* 29 Oct., If a guy is 'gear', as they call a smart boy, he will dress in a single-breasted..drape jacket. 1963 *Guardian* 8 Oct. 9/1 The Liverpool Sound..put expressions like 'it's the gear' into the mouths of debs...How long has it been since a native expression ousted a transatlantic jargon import like gear did to crazy and judy to chick? 1963 *Today* 30 Nov. 15 (*caption*) They're gear! The Beatles leave for London after their triumphant tour of Sweden. 1964 J. Burke *Hard Day's Night* iii. 71 'Gear!' John jumped up from his seat. 'Come on, girls, let's have a bit of a dance.' *Ibid.* iv. 89 Once we even all sat down and wrote those letters saying how gear she was and all that rubbish. 1967 W. Murray *Sweet Ride* vi. 88 Choo Choo, you're boss! Fab! Gear! Bitchin'!

6. d. *Aeronaut.* Short for *landing gear. *U.S.*

1931 J. E. Younger *Airplane Constr. & Repair* iii. 48 Such a device is called a retractable landing gear. The method is quite practicable provided the pilot does not forget to unfold the gear before landing. 1936 *Jrnl. R. Aeronaut. Soc.* XL. 275 Call for a maximum permissible time for raising or lowering the undercarriage, although I understand that the U.S. Dept. of Commerce require that a gear should be lowered in 60 seconds or less. 1963 *Amer. Speech* XXXVIII. 118 This is accomplished by adding power, raising the gear and flaps, and climbing back to a safe altitude.

7. b. Any of the several sets of gear-wheels in a motor vehicle, bicycle, etc., which can be used to alter the relation between the speed of the engine or driving mechanism and the speed or torque of the driven wheels; also, the relation between the number of revolutions made by a motor vehicle or cycle-wheel and the number of revolutions made by the gear-wheels; *high* (or *low*) *gear*: a gear affording a high (or low) ratio between the speed of the driven wheels and the speed of the driving mechanism, and consequently a high (or low) speed to the vehicle itself; *top gear*: see *s.v.* Top *sb.*[1] 32.

1888 [see sense 7 in Dict.]. 1896 F. T. Bidlake *Cycling* 66 High gear for downhill, low gear for up. 1897 *Encycl. Sport* I. 268/1 The higher the gear, the further the machine travels at each stroke of the pedals. 1902 H. Sturmey in A. C. Harmsworth et al. *Motors* x. 203 When putting a lower gear into operation, as is necessary when climbing a hill. *Ibid.* 204 If the calculated speed of the second gear is, say, eighteen miles per hour. 1907 'I. Hay' *Pip* II. ix. 283 Sometimes slowing through a village or changing gear up a hill. 1909, 1910 [see *top-gear* (b) s.v. Top *sb.*[1] 32]. 1923 W. E. Dommett *Motor Car Mech.* (ed. 2) 185 On the top gear with the dog clutches engaged the speeds are equal. 1930 *Strand Mag.* May 348/1 Trout retired, top gear, into the distance. 1958 *Times* 19 Aug. 11/6 The three-speed gearbox with a comparatively low second gear which encourages the driver to stay in top.

IV. *gear-oil, -shaft, -sleeve*; *gear-driven* adj.; **gear-box, -case** (further examples, esp. in connection with motor vehicles); hence *gear-box-less* adj.; **gear-change,** (*a*) the action of changing gear; (*b*) = *gear-lever*; so *gear-changing*; **gear-lever,** the lever by means of which one changes gear; **gear-pump,** a pump in which liquid is drawn in by one gear-wheel and expelled by another gear-wheel in mesh with the first; **gear ratio,** the ratio between the rates at which the last and the first wheels of a train of gears rotate; **gear-shift** (orig. *U.S.*) = *gear-lever*.

1902 W. W. Beaumont in A. C. Harmsworth et al. *Motors* x. 208 Many cars have..a secondary frame to which the motor and gear-box, &c., are attached. 1909 *Chambers's Jrnl.* Jan. 62/2 There is a centrifugal governor mounted on or driven from one of the rotating shafts behind the gear-box, and its speed of rotation is always in fixed relation to that of the road-wheels. 1957 *Times* 2 July (Agric. Suppl.) p. viii/4 Other new features [of the Ferguson 35 tractor] include the six-speed gear-box. 1909 *Westm. Gaz.* 14 Sept. 4/3 A gear-box-less 'Sheffield-Simplex' car. 1963 Bird & Hutton-Stott *Veteran Motor Car* 45, A 45 h.p. 'gearbox-less' car came on the market. 1896 G. B. Shaw *Let.* 6 Mar. (1965) 608 Their net price, new, with gear case, saddle &c, complete is £25. 1902 H. Sturmey in A. C. Harmsworth et al. *Motors* x. 206 The gear-case should be kept sufficiently full of lubricant to enable the lower edges of the gear wheels to be constantly passing through it. 1912 *Motor Man.* (ed. 14) v. 169 (*heading*) A method for obtaining a 'silent' gear change. 1955 *Times* 5 July 5/7 The excellent gearchange..is effected by a short central lever. 1967 *Autocar* 5 Oct. 73/3 With the gearchange switched from the left of the cockpit to the right. 1909 *Westm. Gaz.* 23 Feb. 4/1 The beginner..can practise the art of gear-changing on the hill. 1906 *Ibid.* 26 Mar. 10/2 A gear-driven car. 1906 *Daily Chron.* 14 Nov. 9/3 The water circulates by a gear-driven pump. 1904 *Motor Man.* (ed. 6) ix. 97 Put the gear lever on to the next speed. a 1951 D. Stewart in *Austral. Short Stories* (1951) 409 Placing his hand on her knee as, doubtless in error for the gear-lever, he so often does. 1910 *Blackw. Mag.* Nov. 606/1 A black and glutinous compound, which I afterwards learned to recognise as gear-oil. 1922 *Automobile Engineer Ref. Bk.* 202/1 Gear pump. 1933 Rutherford in *Proc. R. Soc.* A. CXLI. 261 The plug..is cooled by a rapid stream of transformer oil supplied by a gear-pump through insulating tubes. 1909 Gear ratio [see *top-drive* s.v. Top *sb.*[1] 32]. 1915 Young & Aston *Complete Motorist* (ed. 8) iii. 81 An arrangement of toothed pinions, which allow two, three or four different gear ratios to be employed as the case may be. 1959 '*Motor*' *Man.* (ed. 36) iv. 74 The gear ratios usually quoted..take also into account the reduction in the final drive. *Ibid.* 91 Reference was made to an overdrive unit...Its purpose is to provide an extra gear ratio, higher than direct drive, so that quiet and economical cruising is possible. 1904 A. B. F. Young *Complete Motorist* iii. 59 Different sets of gearing, by means of which the gear shaft transmits its motion to the road wheels in different ratios. 1927 *Glasgow Herald* 30 Nov. 11 The new model..will be of the standard selective gearshaft type. 1926 *Amer. Speech* I. 686/2 American: gear shift... English: change speed lever. 1959 *Observer* 12 Apr. 3/3 The steering-column gear-shift..is partly cable-operated. 1908 *Westm. Gaz.* 20 Nov. 4/2 It is arranged with divided gear-sleeve to enable short shafts to be used.

gear, *v.* Add: **4. a.** Also *fig.*, to adjust, correct or co-ordinate; *spec.* to adjust or adapt (something) *to* a particular system, situation, etc.; to bring into dependence, conformity or harmony; freq. in *pass.*

1900 J. London *Let.* 21 Jan. (1966) 85 The tongue is too often geared at too high a pitch to adequately carry on its labors. 1937 *Time* 22 Feb. 26 That part of *John Meade's Woman* which is geared to these phenomena is an effectively written, well-photographed slice of U.S. industrial history. 1945 *Reader's Digest* July 13/1 Every country's production had been geared to an intricate system of German demands and supplies. 1958 *Spectator* 22 Aug. 249/1 This might have been effective in an entire [ballet] production geared to the Bolshoi interpretation. 1962 *Sunday Times* (Colour Suppl.) 10 June 10 Authors', composers' and other royalties are often closely geared to recording rights. 1969 *Times* 31 Jan. 11/8 The rate of the addition reaction is somehow geared to the rate of growth of the cell as a whole.

gear, *a.* See *gear *sb.* 5 e.

geared, *ppl. a.* Add: **3. b.** *Finance.* Of a company or its capital (see *gearing *vbl. sb.* 3 b).

1930 *Economist* 25 Oct. 768/2 Ordinary stocks..by reason of the 'high geared' capital structure of the companies..must always bear the jolts and jars of trade fluctuations. 1940 *Ibid.* 6 July 14/1 The relatively high income thus secured was indispensable for paying generous debenture interest and preference dividends on a highly geared paid-up capital. 1965 [see *gearing *vbl. sb.* 3 b].

gearing, *vbl. sb.* Add: **3. b.** *Finance.* (See quots.) Also *attrib.*

1932 H. Parkinson *Sci. Investment* ix. 88 Among the stocks shown in the table, the high gearing of the railways is conspicuous. 1936 *Economist* 14 Mar. 594/1 The 'gearing ratio' is the proportion borne by the total net assets of the company to the net assets available for the common stock. 1959 *Jrnl. Inst. Bankers* Apr. 149 The simplest form of gearing is the issue of two classes of shares of which one class gives a preferential right to a

fixed percentage dividend before the holders of the other class receive anything. 1965 Seldon & Pennance *Everyman's Dict. Econ.* 193 *Gearing*, the proportion of a company's annual income allocated to 'prior charges' (interest on debentures and preference dividends), the remainder going to the ordinary shareholders. Where it is high the company is said to be highly geared, and conversely.

4. *gearing-beam, -wheel.*

1835 J. H. Ingraham *South-West* II. 288 Two gearing beams are laid across, for supporting the machinery. 1835 Ure *Philos. Manuf.* 109 Two eighty-horse steam-engines..are mounted with their great geering-wheels on the ground floor. 1876 Preece & Sivewright *Telegraphy* 81 Made to revolve by means of the 'gearing' or driving wheels.

gearless, *a.* (Further examples.)

1906 *Westm. Gaz.* 15 May 4/2 The long-promised six-cylinder 'gearless' Napier car. 1951 *Catal. Exhibits, S. Bank Exhib., Festival of Brit.* 183/1 Gearless passenger lift to carry 20 persons. 1966 *Listener* 6 Oct. 515/3 Dr Donald Firth with his gearless car and revolutionary machine tools deserves far more urgent attention.

geaster (dʒi,æ·stəɹ). [mod.L., f. Gr. γῆ earth + ἀστήρ star.] An earth-star (see Earth- II).

1887 W. Phillips *Man. Brit. Discomycetes* 213 Splitting into several lobes, like a *Geaster*.

gebang (dʒi̇bæ·ŋ). Also **gebanga.** [Mal. *gĕbang*.] In full *gebang palm.* A Malaysian fan palm, *Corypha elata*, the leaves of which are used for basket-work and thatching, while the pith of the trunk yields a sago.

1817 T. S. Raffles *Hist. Java* I. i. 37 Another species of palm, the *gebáng*, also yields valuable ropes. 1846 Lindley *Veget. Kingd.* 138 Employed..in weaving into baskets and bags the young leaves of the Gebang Palm. 1952 'W. March' *October Isl.* x. 125 Lontar, gebanga, rattan and nipa. 1966 E. J. H. Corner *Nat. Hist. Palms* xii. 300 The most widespread gebang palm of Malaysia, *C. elata*, is distributed from Bengal..to north Australia.

gebel, var. *jebel.

‖**Gebrauchsmusik** (gəbrau·χsmūzi:k). Also **gebrauchmusik.** [G., f. *gebrauch* use + *musik* music.] Music intended primarily for practical use and performance, esp. music suitable for amateur groups and domestic playing.

1930 *Mod. Music* Feb.–Mar. 17 Present day 'Gebrauchsmusik' seems to be principally music for the dance, the sound films, [etc.]. 1932 *Music & Lett.* XIII. 43 Hindemith..is young and impetuous enough to have one or two bees in his musical bonnet. These are known variously as *Gebrauchmusik, Gemeinschaftsmusik*,..and so on. 1934 C. Lambert *Music Ho!* IV. 249 Hindemith's music has been somewhat unduly saddled with the description of Gebrauchsmusik—bread-and-butter music, workaday, or utility music are perhaps the best equivalents. 1947 A. Einstein *Mus. Romantic Era* ii. 11 Until late in the 18th century there was scarcely any music conceivable that was not utilitarian, *Gebrauchsmusik*, or that would not have served an immediate purpose. 1952 P. Hindemith *Composer's World* p. viii, A quarter of a century ago, in a discussion with German choral conductors, I pointed out the danger of an esoteric isolationism in music by using the term *Gebrauchsmusik*..that ugly term showed a power of penetration and a vigor that would be desirable for worthier formulations. 1962 *Times* 14 Dec. 15/2 The now unfashionable category of *Gebrauchsmusik*.

gedackt, gedact (gəda·kt). Also **gedeckt.** [G. *gedackt*, old pa. pple. of *decken* to cover.] An organ flue-stop having its pipes closed at the top. So *gedackt-work* [G. *gedacktwerk*], such stops collectively; *lieblich gedackt* (see quot. 1938).

1855 E. J. Hopkins *Organ* xxxviii. 304 To a 'Gedact of 8-feet tone' he introduced a 'Rohrflöte of 4-feet tone'. *Ibid.* 364 Lieblich Gedact. 1880 Grove *Dict. Mus.* II. 601/2 In Schulze's organ the Gedact (No. 4) was formed of stopped wood pipes that produced a fuller tone than the usual Stopped Diapason. 1904 Goodchild & Tweney *Technol. & Sci. Dict.* 252/2 *Gedact work*, those organ stops belonging to the flue work, which consist of closed pipes. 1938 *Oxf. Compan. Mus.* 668/2 *Lieblich Gedact* (literally 'lovely and covered' or 'lovely and lidded'). A Gedact of 8- or 4-foot length and pitch.

‖**Gedankenexperiment** (gəda·ŋkən,eksperime:nt). *Philos.* [G., f. *gedanke* thought + *experiment* experiment.] An experiment carried out only in imagination or thought; an appeal to imagined experience; a *thought-experiment.

1958 N. R. Hanson *Patterns of Discovery* vi. 137 The first presentation involves a *gedankenexperiment*; it supposes a super-microscope more powerful than any electronic microscope. 1966 *Amer. Philos. Q.* III. 306/2 The child in his [*sc.* Ayer's] *Gedankenexperiment* is taught the use of psychological concepts by a recorded voice.

Gedinnian (dʒədi·niăn), *a. Geol.* Also †**Gedinnien.** [ad. F. *gedinnien* (A. H. Dumont 1848, in *Mém. de l'Acad. R. des Sciences de Belgique* XXII. 4), f. *Gedinne*, name of a village in Belgium: see -ian.] Pertaining to or designating the lowest division of the Devonian system in Europe, or the epoch or age during which it was deposited. Also *absol.*

1853 *Q. Jrnl. Geol. Soc.* IX. 25 The 'Gedinnian system' must be entirely separated from the Rhenane series with which..it has no organic remains in common. 1882 A. Geikie *Text-bk. Geol.* vi. ii. iii. 702 Gedinnien, comprising an upper group of shales and sandstones and a lower group of fossiliferous shales, quartzo-phyllades, quartzites, and conglomerates. 1928 E. Neaverson *Stratigr. Palaeont.* xi. 255 The Gedinnian fauna of Belgium is not well known, since it is often represented only by fragments of ostracoderm fish. 1955 G. G. Woodford tr. *Gignoux's Stratigraphic Geol.* iv. 119 The Gedinnian and Coblenzian are completely missing and the Devonian begins with a basal conglomerate..directly overlain by sandstones.

gee (dʒī), *sb.*[3] *slang.* [Origin unknown; cf. next, and *gee *v.*[2] 1 c.] The accomplice of a cheapjack or showman. Also *gee-man*.

1898 *Bulletin* (Sydney) 17 Dec. (Red page), A gee is their outside confederate, who 'gees up' the mugs for them. 1928 *Daily Express* 19 Dec. 2/7 'Ricks' or 'gees': people who mingle with the crowd to arouse their enthusiasm. 1941 K. Tennant *Battlers* xiii. 141 'I'm geeing for him, and I'll fix it.' The busker's spirits fell again. In the show world a 'gee-man' or 'micky finn' was socially on the level of a duck's feet. He is the man who goes out in the crowd and touts for custom. 1959 *News Chron.* 16 Nov. 5/3 Strategically placed in the crowd, the 'gee men' started the bidding going.

gee (dʒī), *sb.*[4] *U.S. slang.* [f. the pronunciation of the initial letter.] = Guy *sb.*[2] 3 d.

1921 P. & T. Casey *Gay-Cat* 302/2 Gee, guy, gun, mug, plug, stiff, etc.—a fellow. 1930 [see *go *v.* 58 d]. 1931 *Flynn's* 24 Oct. 656/2 They tell me you've got a mortgage on the dump...I could find a gee with free money to stake you. 1968 S. Challis *Death on Quiet Beach* xii. 177 'Just a minute, this ain't O'Brien.' 'No. This is some other gee.'

gee (dʒī), *sb.*[5] The name of the letter G; *spec.* in *U.S. slang*, a thousand dollars (cf. *G III. f).

1926 *Evening Standard* 12 July 3/2 If one branch of English society drops its initial aitches, and another branch ignores its terminal gees. 1936 L. Duncan *Over Wall* i. 21 A thousand-dollar bill was a Gee. 1946 M. Taylor in 'D. Stanley' *Treas. Sports Humor* 149 There's a hundred gees at stake.

gee (gī, dʒī), *sb.*[6] *slang* (orig. *U.S.*). [Origin uncertain; perh. f. Ghee.] Opium or some similar drug. Also *attrib.* Hence **geed-up** *a.*, drugged.

1936 *Amer. Speech* XI. 121/1 *Gee-yen*, opium which precipitates in very small quantities in the stem of the pipe. 1938 *Ibid.* XIII. 184/2 *Gee*,..smoking opium, especially refined or reworked opium. *Ibid.* 185/1 Geed up. *Ibid.*, *Gee-stick*, an opium pipe. 1939 *Flynn's* 18 Mar. 56/2 She fell from the doorway, either not knowing there were two steps down into the bathroom or else forgetting it on account of being so geed up. 1953 Anslinger & Tompkins *Traffic in Narcotics* 308 *Gee*, drugs, especially opium.

gee (dʒī), *sb.*[7] *Aeronaut.* Also **G.** [f. initial letter of *grid.*] A navigational radar system developed in the war of 1939–45 to guide bombers to their targets. Also *attrib.*

1945 *Electronic Engin.* XVII. 685/1 Gee. The navigational system which enabled our bombers to know exactly where they were at any time *en route* to or from Germany. *Ibid.* 713 'G' or 'Gee' as it is more often spelt as an abbreviation for 'Grid', relating to the grid map references used by R.A.F. navigators. 1947 Crowther & Whiddington *Science at War* Pl. x (*caption*), The two boxes shown here contain the essential Gee receiver used on aircraft. 1959 *Times Lit. Suppl.* 3 Apr. 194/4 It could give the bombers a set of lattice-lines from which the navigator might infallibly determine his position on the way to the target up to a distance of about 400 miles. This was Gee.

gee (dʒī), *v.*[2] [f. Gee *int.*] **1.** *trans.* **a.** *Sc.* = Jee *v.* 2.

1813 G. MacIndoe *Wandering Muse* 114 For me, I never geed my noddle, Nor car'd I Snip, or Tib a boddle. 1931 H. S. Roberton *Curdies* xxvi. 106 The same men widna gee their ginger to put a bane-teeth comb through a wean's heid to stop it frae scartin'.

b. To direct (a horse, etc.) by the call of 'gee'. Also *absol.*

1845 E. J. Wakefield *Adv. N.Z.* II. 133 He *geed* the bullocks and ploughed on. 1867 'T. Lackland' *Homespun* II. 194 The regiment is somehow got back, by hawing and geeing, into line. *Ibid.* III. 286 The man..is turning up the sod with the gleaming share..while he 'gees' and 'haws' the yoke of cattle.

c. To incite, encourage; *spec.* to entice or encourage (the public) to patronize side-shows, etc., at a fair (cf. *gee *sb.*[3]). Also, to make (a person or animal) move more quickly. Freq. const. *up*. Cf. Gee-(h)up *v. slang.*

1898 [see *gee *sb.*[3]]. 1932 'Ex-Convict No. ——' *Dartmoor from Within* iv. 86 Yells of encouragement were shouted from the cell windows by men in my own hall—'geeing' him on. 1936 J. Curtis *Gilt Kid* xxiii. 231 He could flash a few oncers before her eyes if he wanted to gee her up. 1941 [see *gee *sb.*[3]]. 1956 R. Fuller *Image of Society* v. 141 The directors of the company must be gee'd up. 1958 F. Norman *Bang to Rights* 64 With us still shouting and geeing the twirls up. *Ibid.* 130 My life I'm not geeing you up. 1967 L. Deighton *Expensive Place* ix. 67, I tried to decide whether she was geeing me up.

2. *intr. dial.* (chiefly *Sc.*) = Jee *v.* 1.

1835 D. Webster *Original Sc. Rhymes* 116 Gloomy clouds may dim the air, But winna mak my fancy gee. 1891 R. P. Chope *Dial Hartland* 47 To *gee* or *gee round* is to turn towards the right.

gee (dʒī), *int.*[2] orig. *U.S.* [Prob. a shortening of *Jesus!* (or *Jerusalem!*); cf. *geewhillikins *int.*, *gee whiz(z *int.*] An exclamation of surprise or enthusiasm; also used simply for emphasis. Cf. Jee *int.*, *Jeez(e *int.*

1895 S. Crane *Red Badge of Courage* ii. 28 Gee rod! how we will thump 'em! 1901 S. E. White *Westerners* ii. 11 'Gee Christmas!' ejaculated Billy, and laughed loudly. 1902 *Captain* VII. 40 Gee! I thought the hair would come out of my head. 1906 *Amer. Illustr. Mag.* Apr. 701 'I guess he's tall and thin and homely and dark, gee,' he yowled explosively. 1909 E. Underhill *Runaway Place* 134 'Gee, I believe you're right!' he exclaimed. 1927 H. A. Vachell *Dew of Sea* 269 But gee! if she is downing you, she offs it. 1953 'N. Shute' *In Wet* 91 'Gee,' said Wing Commander Dewar, 'this thing'll drive me nuts.'

Geechee (gī·tʃi). *U.S. dial.* [f. the name of the *Ogeechee* River, Georgia.] (See quot. 1934.) Also, a derogatory term for a Negro of the southern United States. Cf. *Gullah.

1926 *Nat. Geogr.* Sept. 278 Among the negroes living on the Ogeechee River a patois, developed in ante bellum days, has persisted...The origin of 'Geechee', as the patois is called, is explained by the fact that slaves employed on the old rice plantations were more or less isolated and rarely conversed with their white owners, with the result that their knowledge of English words was slight and the pronunciation of them was bizarre. The 'Geechee' negro speaks in a sort of staccato and always seems excited when talking. 1934 Webster, *Geechee.* A dialect, originally of Negro slaves on the Ogeechee river, Georgia, formed of English and native African words. 2. One who speaks Geechee. 1940 E. Caldwell *Trouble in July* xiii. 208 He sounds like one of those Geechee niggers...That breed'll do anything to keep them from working. 1945 *New Yorker* 8 Sept. 20/2 Creecy was a Geechee..blacker than the soot in the fire-place.

geek, var. Geck *sb.*[1]; *spec.* in *U.S. slang* (see quots. 1954 and 1961).

1876 [see Geck *sb.*[1]]. 1916 *Wells Fargo Messenger* Oct. 29/2 A new Wells agent struck our town the other week, and say—you never saw a more enthusiastic geek! 1954 Webster Add., *Geek*, a carnival 'wild man' whose act usually includes biting the head off a live chicken or snake. 1961 *Times Lit. Suppl.* 27 Jan. 62/2 He picks up a waitress, a simple girl, and enslaves a 'geek', a dumb sideshow stooge whose daily routine consists of being exhibited in a pit which he has to dig for himself.

geelbek (gī·lbek, ‖χiə·lbek). *S. Afr.* Also **geelbec, geelbeck.** [Afrikaans, f. *geel* yellow + *bek* mouth, beak.] **1.** A marine fish, *Atractoscion æquidens*, of the family Sciænidæ, which has bright yellow edges to jaws and gill-cover and is found off the southern coasts of Africa and also of Australia, where it is called teraglin.

1853 L. Pappe *Edible Fishes Cape of Good Hope* 16 *Otolithus Æquidens*..(*Geelbeck.*)..Common along the whole coast. 1900 *Trans. S. Afr. Phil. Soc.* 216 The name Cape Salmon..is now appropriated almost exclusively by the Geelbek. 1931 *Times Lit. Suppl.* 16 Apr. 301/2 Like so many of the Cape fish,..the geelbek..gets his name from the original Dutch settlers. 1953 J. L. B. Smith *Sea Fishes S. Afr.* (rev. ed.) 227 Sometimes the tails of cured Kobs are trimmed by dishonest vendors so as to resemble those of the more costly Geelbek.

2. A yellow-billed wild duck, *Anas undulata*, found in Eastern and Southern Africa.

1867 E. L. Layard *Birds S. Afr.* 352 A[*nas*] *Xanthorhyncha*..*Geelbec* of Colonists and Sportsmen. 1901 W. T. Black *Fish River Bush* i. 21 An occasional shot may be got at a Geelbec duck (*Anas*) as it flies low along the bush lining the river banks. 1957 McLachlan & Liversidge *Roberts's Birds S. Afr.* 48 Yellowbill. Geelbek.

‖**geelhout** (χīə·lhōut). *S. Afr.* Also formerly **geele-houtt.** [Afrikaans, f. *geel* yellow + *hout* wood.] = Yellow-wood.

1790 E. Helme tr. *Le Vaillant's Trav. Afr.* II. xiv. 288, I also remarked the Geele-Houtt, (yellow-wood) so called from its colour: which..is very serviceable for planks, beams, and rafters. 1801 J. Barrow *Trav. S. Afr.* I. iii. 133 The most common was the *geel hout* or yellow wood, (*taxus elongatus*) erroneously called by Thunberg the *ilex crocea*. 1834 [see Yellow-wood]. 1953 *Cape Argus* 25 Feb. 9/4 Eighty trees, including jakkals-bessie, geelhout, wild fig,..have been planted along Table Bay boulevard. 1961 Palmer & Pitman *Trees S. Afr.* 122 The common yellowwood, Outeniqua yellow-wood,..geelhout.

gee-string, var. *G-string 2.

geewhillikins (dʒī,wi·lĭkinz), *int.* orig. *U.S.* Also **ge-, je-, -whil(l)iken(s), -whil(l)ikin(s), -whit(t)aker(s).** [Perh. a fanciful substitute for *Jerusalem!*, but cf. next.] = *GEE *int.*[2]

1851 'A MISSOURIAN' in T. A. Burke *Polly Peablossom's Wedding*, etc. 52 Jewhilliken, how he could whip er nigger! 1856 *Town Talk* (San Francisco) 20 July 1/1 Gee-whitaker! what a kurchy she made, and bowed so low that I nearly fell outer my dickey onto the floor. 1857 *Knickerbocker* Nov. 435 (Th.), And great Gewhilikins! wasn't the snow peppering down! 1865 R. H. NEWELL *Orpheus C. Kerr Papers* ii. 31 Geewhillikins! how the critter did squawk and squeal. 1872 SCHELE DE VERE *Americanisms* 611 *Jerusalem!*, a favorite New England exclamation...In the West it is, as usual, improved to suit the louder taste of the people, and becomes Jewhillikin. 1884 'MARK TWAIN' *Huck. Finn* xxxviii. 386 'Geewhillikins,' I says, 'but what does the rest of it mean?' 1906 E. NESBIT *Railway Children* xii. 251 When I tried to get up again I couldn't stand, so I sat down. Gee whillikins! it does hurt, though. 1932 E. WEEKLEY *Words & Names* viii. 112 Of similar origin [*sc.* euphemistic substitution] are *gee-whittakers, gee-whiz*, [etc.]. 1941 C. S. FORESTER *Capt. from Connecticut* xx. 261 'Geewhillikins, sir,' said Hubbard; the dark mobile face lengthened in surprise.

gee whiz(z (dʒī,whi·z), *int.* orig. *U.S.* Also **gee whitz, gee wiz,** and with hyphen. [Prob. a corruption of prec. or a euphemism for *Jesus!*: cf. *GEE *int.*[2] and WHIZZ, WHIZ *int.*] = *GEE *int.*[2]

1885 W. T. GRAY *Bad Boy at Home* 8 (at back) Gee wiz! but they do put on stile in New York. 1888 *Judge* (U.S.) 1 Sept. 334 Geewhitz! 1904 *N.Y. Times* 21 Apr. 9 He appeared in the saloon and shouted: 'Gee whiz! The gas pipes is all leakin'.' 1908 *Westm. Gaz.* 9 May 6/1 The 'Raid on Prosperity' is what he calls his statement for the defence, and, gee-whiz!..he goes for the Raiders in a way that is positively refreshing. 1940 'R. CROMPTON' *William & Evacuees* vi. 164 'Gee whiz!' breathed William in ecstasy. 1957 J. D. SALINGER *Franny & Zooey* (1962) 86 Well, gee whizz. I'm only trying to make polite bathroom talk.

Geez (gi·ez, gīz). Also **Geëz, Gheez, Giez, Giz.** [Ethiopic.] The ancient language of Ethiopia, still used as a liturgical language; = ETHIOPIC *a.* 2. Also *attrib.*

1790 J. BRUCE *Trav.* I. 424 Speaking of the Ethiopic at present, I mean only the Geez language, the language of the Shepherds, and of the books. *Ibid.* 425 The Geez is exceedingly harsh and unharmonious. 1839 *Penny Cycl.* X. 52/2 The antient Ethiopian, or Geez. 1867 [see ETHIOPIC 2]. 1883 R. N. CUST *Sk. Mod. Lang. Afr.* I. 74 The Written Characters of the old Ethiopic, or Gíz. 1908 *Daily Chron.* 6 Mar. 3/5 Their Bible, which is in Gheez,..contains all the books of our own Old Testament, except the Book of Esther. 1920 *Blackw. Mag.* May 676/1 A book in the Geez language. 1957 *Listener* 14 Nov. 774/1 Their recital of Geez texts and the copying of old manuscripts on the lives of Ethiopian saints. *Ibid.*, The priests are satisfied if the ordinary man says his prayers in Geez which he does not understand.

geez, geeze, varr. *JEEZ(E *int.*

geezer. Also **geyser.** Add: (Further examples.) Now usu. of men, not necessarily elderly; a chap, fellow.

1904 'No. 1500' *Life in Sing Sing* 248/2 *Geezer*, a fellow. 1910 *Punch* 5 Oct. 243 'Old yer blooming rah, an' give the old geyser [*sc.* a lady singer] a charnst, cawn't yer? 1914 *Dialect Notes* IV. 201 The old geezer wouldn't let us play ball in his pasture. 1938 G. GREENE *Brighton Rock* II. i. 74 You're a grand little geezer. *Ibid.* II. ii. 90 A geezer can't have an alibi for every minute of the day. *Ibid.* IV. ii. 267 I'll tell you a thing or two, you bloody little geezer. 1958 [see *BARON 2 c]. 1958 J. SYMONS *Gigantic Shadow* vii. 37 There's a geezer I know named Twisty Dodds, kind of a small-time crook you might call him. 1960 [see *BIRD *sb.* 1 d]. 1965 [see *CHINA[1] 7].

gefuffle (gĕfʊ·f'l), var. CURFUFFLE, *KERFUFFLE.

1943 B. J. HURREN *Eastern Med.* ii. 32 The gefuffle created in the search for this stick was unbelievable. 1959 J. VERNEY *Friday's Tunnel* xxvi. 234 He made such a gefuffle of safety precautions and locked doors I couldn't imagine what was up.

‖**gefüllte fish** (gĕfi·ltə fiʃ). Also **gefil(l)te, gefuellte, gefulte, gefülte fish** or **fisch.** [Yiddish, a. G. *gefüllt* filled, stuffed, f. *füllen* FILL *v.*] A Jewish dish of stewed or baked stuffed fish or fish-cakes, boiled in a fish- or vegetable-broth.

1892 I. ZANGWILL *Childr. Ghetto* I. i. iv. 114 There is even *gefüllte Fisch*, which is stuffed fish without bones. 1916 J. LAIT *Beef, Iron, & Wine* (1917) 15 That crowd of tired *hoi-polloi* fighting for a spot to stand to get their corned beef and cabbage, or *gefuellte* fish. 1931 D. RUNYON *Guys & Dolls* (1932) iv. 71, I am sitting in Mindy's restaurant putting on the gefillte fish. 1941 L. G. BLOCHMAN *See You at Morgue* (1946) ix. 61 Don't your Old Lady make gefüllte fish any more for Shabbath dinner? 1959 H. PINTER *Birthday Party* II. 26 The nicest piece of gefilte fish you could wish to find on a plate. 1962 *Coast to Coast 1961–62* 113 It was Friday, so I ran into *gefüllte* fish.

Gegenschein, gegenschein (gē·gənʃain). *Astr.* [G., f. *gegen* opposite + *schein* shine, glow; first used technically by T. J. C. A. Brorsen 1854, in *Unterhaltungen im Gebiete der Astron.* 17 May 156, after an earlier use in the same connection by F. H. A. von Humboldt *c* 1803, published in his *Kosmos* (1845) I. 144.] A patch of very faint nebulous light occurring in the ecliptic opposite the sun.

1880 *Amer. Jrnl. Sci.* CXX. 442 The gegenschein is an extremely faint spot of light some 7° in diameter, lying in the zodiacal band. 1888 [see *COUNTER-GLOW]. 1968 *Observer* 22 Dec. 3/1 The astronauts..have a heavy programme of photography, including an attempt to photograph the 'Gegenschein', the image of the sun faintly reflected back from dust and tenuous gas in space. 1971 *Nature* 30 July 323/1 Observations of the gegenschein are made best away from the Milky Way in February, March and April.

geggie (ge·gi). *Sc.* Also **gaggie.** [f. GAG *sb.*[1]] A travelling theatrical show, usu. held in a tent. Cf. GAFF *sb.*[4]

1898 D. WILLOX *Poems & Sk.* 16 Glasgow Fair was then held in all its glory..and..young Willox sometimes neglected his duties to witness the glories of the 'geggie'. 1922 *Glasgow Herald* 17 July, The 'geggie', which to the children of the present day is little more than a name, if it is even that. 1931 A. J. CRONIN *Hatter's Castle* I. i. 23 This is the first night of Levenford Fair. I saw the start o' the stinking geggies on my way home. 1952 T. JOHNSTON *Memories* i. 14 The gaggie shed or theatre was built of wooden partitions.

‖**Geheimrat** (gəhai·mrāt). Also **Geheimer Rat.** [G.] In Germany, a privy councillor (see also quot. 1911).

1837 H. C. ROBINSON *Diary* 27 July (1967) 180, I called on Geheimrat Schlosser and had an interesting chat with him. 1911 *Encycl. Brit.* XXII. 373/2 The title..of *Geheimer Rat* is simply very generally, *e.g.* in Prussia, given to high officials, usually with the addition of the branch of the service to which they belong, *e.g.* *Geheimer Finanzrat.* 1936 A. HUXLEY *Eyeless in Gaza* liii. 601 The sofas where the Geheimrats slept off the effects of noonday eating. 1943 H. L. MENCKEN *Heathen Days* vii. 91 Thus a plain Mister became Doctor, a Doctor became Professor, and a Professor became *Geheimrat.* 1971 AUDEN *Academic Graffiti* 19 The *Geheimrat* in Goethe Made him all the curter With *Leute* who were leery Of his Colour Theory.

Geiger (gai·gəɹ). Also with small initial letter. [The name of Hans *Geiger* (1882–1945), German physicist.] **Geiger counter** [*COUNTER *sb.*[2] 3 b], an instrument for detecting and counting ionizing radiation, used esp. for measuring radioactivity; it consists essentially of a wire anode surrounded by a cylindrical cathode in a chamber containing gas at low pressure. Also *Geiger-Müller counter.*

1924 *Sci. Abstr.* A. XXVII. 295 (*title*) The Geiger counter as a sensitive detector of X-radiation. 1926 *Physical Rev.* XXVII. 111 A study of Geiger counting chambers has been made. 1930 *Ibid.* XXXV. 651 (*title*) Multiple coincidences of Geiger-Müller tube-counters. 1945 *Electronic Engin.* XVII. 612/2 An improved circuit..for reading the rate of pulses from a Geiger-Müller counter. 1951 *New Biol.* X. 34 A single particle or γ-ray passing through a Geiger-Muller counter triggers off an electrical discharge which can be amplified to work an electrical and mechanical relay counter. 1956 'N. SHUTE' *Beyond Black Stump* 1 A number of substances that are trapped in the earth's centre will influence a Geiger counter sufficiently to set it clicking. 1957 M. LOWRY *Let.* 4 Apr. (1967) 406 When the last old sourdough has traded in his divining rod and gold sifting pan for a geiger counter.

b. *fig.*

1957 *London Mag.* June 65 We do not feel inclined to treat political beliefs as geiger counters by which we can measure the significance of events as they occur. 1959 *Times Lit. Suppl.* 24 Apr. 246/5 Dr. Simon runs his Geiger counter over the writings of Bernard Shaw to detect the educationist beneath the playwright and pamphleteer. 1961 *John o'London's* 12 Oct. 423/3 A sadistic gangster story..at times, unless my moral geiger counter deceived me, rather slanted, sympathy-wise, towards the gang.

geikielite (gī·kilait). *Min.* [f. the name of Sir Archibald *Geikie* (1835–1924), Scottish geologist + -LITE.] A titanate of magnesium, with some magnesium replaced by iron, known only as brownish or bluish black rolled pebbles from Ceylon.

1892 *Nature* 27 Oct. 620/2 The pebbles were handed over to Mr. Allan Dick for chemical investigation. Quantitative analysis proved the mineral to be essentially magnesium titanate...To this interesting new species Mr. Dick..gave the name Geikielite, in honour of Sir Archibald Geikie,..in whose laboratory the analysis had been made. 1906 *Jrnl. Chem. Soc.* XC. II. 459 The ferric oxide shown in these analyses is in some cases primary (isomorphously replacing the geikielite molecule). 1944 C. PALACHE et al. *Dana's Syst. Min.* (ed. 7) I. 536 Ilmenite is $FeTiO_3$ with as much as 54 atomic per cent of Mg..and grading into geikielite, $(Mg,Fe)TiO_3$.

geilfine (gēⁱ·lfinĭ). *Irish Hist.* Also **geilfhine.** [Ir.] One of the four divisions of the *fine*, comprising the chief and his four nearest male relatives. Cf. FINE *sb.*[2], *IARFINE, *INDFINE.

1869 W. N. HANCOCK et al. tr. *Anc. Laws Irel.* II. 161 The 'Geilfine'-tribe relationship in the direct line, such as the father, and the son, and the grandson, and the great grandson, and the great great grandson to the fifth generation. 1875 H. S. MAINE *Lect. Early Hist. Instit.* 211 A Geilfine class may consist of a father and four sons who are not in the same degree. 1882 C. G. WALPOLE *Short Hist. Kingdom Irel.* i. 9 The father and the four sons formed a family group of five households, which went by the name of the *Geilfine.* 1903 P. W. JOYCE *Soc. Hist. Anc. Irel.* I. vii. 188 The *Gelfine* system, under which four groups of persons, all nearly related to each other, held four adjacent tracts of land as a sort of common property. 1967 F. J. BYRNE in Moody & Martin *Course Irish Hist.* iii. 49 The brehon lawyers drew up a very elaborate scheme of the different degrees of relationship. The *geilfhine*, sometimes called *deirbfhine* was the normal family group—basically the relationship between a man and his brothers; but it was extended over five generations to include his own children, his father's brothers, his grandfather's brothers, and even his great-grandfather and *his* brothers!

‖**geis** (geʃ, gēⁱʃ, giʃ). Also **gaysh, geas.** Pl. **geasa, geise.** [Ir.] In Irish folklore: a solemn injunction, prohibition, or taboo; a moral obligation.

1880 S. FERGUSON *Poems* 63 This journey at this season was ill-timed, As made in violation of the *gaysh.* 1899 D. HYDE *Lit. Hist. Irel.* 344 He thought he saw Gradh son of Lir upon the plain, and it was a *geis* (tabu) to him to see that. *Ibid.* 373 Every man who entered the Fenian ranks had four *geasa* (gassa, *i.e.*, tabus) laid upon him. 1928 *Observer* 22 Jan. 5/4 Apparently a man could be either:—(1) Born under a 'geis' prohibiting certain actions on his part, or (2) Laid under 'geis' either at birth or any time during his life, either by divine or human agency. 1965 *New Statesman* 23 July 129/2 In a sense which most Irish people will know, this put Fallon under a *geas*, a moral compulsion, to say his bit.

geisha. Substitute for def.: A Japanese girl whose profession is to entertain men by dancing and singing; loosely, a Japanese prostitute. (Add further examples.)

1910 *Encycl. Brit.* XI. 553/2 *Geisha*..., strictly the name of the professional dancing and singing girls of Japan. The word is, however, often loosely used for the girls and women inhabiting Shin Yoshiwara, the prostitutes' quarter of Tokyo. 1922 JOYCE *Ulysses* 95 And they call me the jewel of Asia, Of Asia, The geisha. 1939 B. H. CHAMBERLAIN *Things Japanese* (ed. 6) 460 Up-to-date garb and manners make the geisha appear prim and old-fashioned.

attrib. 1887 [see KIMONO]. 1895 'C. HOLLAND' *My Japanese Wife* i. 2 A real mousmé, not a *geisha* girl with a past, an ambiguous present, and a who-knows-what future. 1910 *Chambers's Jrnl.* Sept. 617/1, I have heard of the geisha girls of Japan. 1947 J. BERTRAM *Shadow of War* VII. iii. 221 Omori..had enjoyed a certain reputation as a home of rather superior geisha-houses. If not quite a red-light district..., it was down in the guide-books as a 'pleasure-quarter' of Tokyo.

Geissler (gai·sləɹ). The name of Heinrich *Geissler* (1814–79), German mechanic and glass-blower, used attrib. (†or in the possessive) to designate certain apparatus invented by him, as **Geissler tube,** a sealed glass tube containing gas at low pressure and a pair of electrodes so designed that a luminous discharge can be produced between them.

1863 E. ATKINSON tr. *Ganot's Physics* § 740. 706 Plücker has found that the light in Geissler's tubes does not depend on the substance of the electrodes. 1870 *Chem. News* 28 Jan. 45/2 Experiments made with perfectly pure gases enclosed in so-called Geissler tubes. 1883 *Encycl. Brit.* XVI. 31/1 As it takes a height of about 30 inches of mercury to balance the pressure of the atmosphere, a Geisler pump necessarily is a somewhat long-legged and unwieldy instrument. *Ibid.*, Even a Geisler's stop-cock requires to be lubricated to be absolutely gas-tight. 1930 W. M. DEANS tr. *Pohl's Electr. & Magn.* 219 Geissler tubes are being used to an increasing degree in practice for lighting. 1966 *McGraw-Hill Encycl. Sci. & Technol.* XII. 589/1 Modified forms of Geissler tubes containing helium, neon, argon, or mercury have become familiar to everyone as luminous signs or fluorescent tube lamps.

‖**Geist, geist** (gəist). [G., = spirit: see GHOST *sb.*] Spirit; spirituality; intellectuality; intelligence.

1871 M. ARNOLD *Friendship's Garland* i. 8 In Berlin we oppose 'Geist'—*intelligence*, as you or the French might say—to 'Ungeist'. *Ibid.* 9 France has 'Geist' in her democracy, and Prussia in her education. 1883 *19th Cent.* Aug. 274 So that the heaven to which we must look forward will be a heaven of pure *Geist.* 1958 *Times Lit. Suppl.* 10 Jan. 19/1 Papers acquire a certain temperament with time. They have their own *Geist*, their own quirks of personality. 1964 *Economist* 12 Dec. 1256/1 The main *geist* which rules the Treasury is that of amenability.

‖**Geisteswissenschaft** (gəi·stĕsvi:sĕnʃaft). Pl. **Geisteswissenschaften.** [G., f. *geist* spirit (see GHOST *sb.*) + *wissenschaft* learning (*wissen* to know).] Usu. in *pl.*, the arts or humanities, opp. 'the sciences'. So **Gei·steswi:ssenschaftler,** one who studies the arts or humanities.

1909 W. M. URBAN *Valuation* i. 10 Psychology..is the fundamental *Geisteswissenschaft.* 1925 *Phil. Rev.* XXXIV. 334 Dilthey held that for the mental sciences, which he called, following Mill, the *Geisteswissenschaften*, the natural scientific method is insufficient. 1931 M. R. COHEN *Reason & Nature* 371 We shall here concern ourselves only with the values which are the study of the *Geisteswissenschaften.* 1934 M. BLACK tr. *Carnap's Unity of Sci.* 31 Science..includes, in addition to the 'natural' sciences, Psychology and the *Geisteswissenschaften* (social sciences) generally. 1938 *Mind* XLVII. 114 Artists and 'Geisteswissenschaftler'..wishing to have a concentrated glimpse of the whole breadth and wealth of human thought.

gel (dʒel), *sb.* [Orig. a suffix f. the first syllable of *gelatin* (as in *ALCOGEL, *HYDROGEL) and later used as an independent word.] A semi-solid colloidal system consisting of a solid dispersed in a liquid. Also *attrib.*

The invention of this word and of *sol* is ascribed by later writers to Thomas Graham (died 1869).

1899 W. B. HARDY in *Jrnl. Physiol.* XXIV. 164 Graham's nomenclature is as follows: The fluid state, colloidal solution, is the 'sol', the solid state the 'gel'. The fluid constituent is indicated by a prefix. Thus an aqueous solution of gelatine is a 'hydrosol', and on setting it becomes a 'hydrogel'. 1904 W. C. D. WHETHAM *Recent Devel. Phys. Sci.* iv. 137 The mass will solidify under certain conditions to form a solid which may be called a gel. 1918 *Nature* 28 Mar. 66/1 A very minute amount of electrolyte added to a stable colloidal solution may bring about complete precipitation or flocculation of the sol, the sol separating out in a gelatinous form known as a gel. 1922 *Encycl. Brit.* XXX. 591/1 On digestion with water the cellular tissue is transformed into a series of hydrated gel-products. 1926 *Ibid.* Suppl. I. 785 The Brownian movement affords a criterion of whether protoplasm is in the condition of a sol or a gel. 1930 *Engineering* 31 Jan. 142/3 Cement could be regarded as what in colloid chemistry is known as a rigid gel, which changes its physical properties..according to the variations of its moisture content. 1930 *Nature* 15 Mar. 432 Structure in very permeable collodion gel films and its significance in filtration problems. 1931 *Oxf. Univ. Gaz.* 17 June 693/2 The equilibria between (*a*) silica gel and water vapour, (*b*) silica gel and alcohol vapour, (*c*) iron oxide gel and alcohol. 1951 WILLIAMS & ALBERTY in Taylor & Glasstone *Treat. Physical Chem.* (ed. 3) II. v. 675 There is at present almost universal acceptance of the view that gels possess a ramifying, more or less coherent, framework that retains the liquid component and confers elasticity and rigidity upon the system as a whole. 1966 *McGraw-Hill Encycl. Sci. & Technol.* VI. 93/1 Commonly, gels have a low solid content, for example, 2–5% for ferric oxide, and as little as 0·1% for coagulated blood.

gel (dʒel), *v.* [f. the sb.] *intr.* To become a gel. Hence **gelled** *ppl. a.*, **ge·lling** *vbl. sb.* (Cf. JELL *v.*)

1917 *Sci. Amer. Suppl.* 22 Sept. 191/2 Ligno-cellulose fibre..does not gel so readily by cold mechanical treatment as does cellulose. 1924 H. B. WEISER in R. H. Bogue *Theory & Applic. Colloidal Behavior* I. xv. 390 A sol containing 1 per cent of pure gelatin does not gel until around 10°. 1933 FISCHER & HOOKER *Lyophilic Colloids* I. ix. 159 These explanations..do nothing to make clear why the silicic acid so frequently gels before being precipitated, or the relationship of the gelling process to the precipitation. 1953 KIRK & OTHMER *Encycl. Chem. Technol.* XI. 162 It causes the ink to body up and finally to gel completely. 1970 *Nature* 25 July 371/1 The mixture gelled at T_e °C. After 5 h..the cell was warmed a few degrees to make the gelled oil transparent.

fig. 1958 *Observer* 30 Mar. 14/3 The combination of drawing-room and documentary failed to gel. 1966 *Listener* 14 Apr. 549/1, I recognize the moment As when an awareness gels.

gelada (dʒe·lădă). [Native name.] In full *gelada baboon.* An Ethiopian baboon, *Theropithecus gelada*, characterized by a heavy mane in the adult male, and by a tufted tail.

1843 J. E. GRAY *List of Mammalia in Brit. Mus.* 9 The Gelada. 1878 *Proc. Zool. Soc.* 1 A Family of Gelada Baboons. 1893 R. LYDEKKER *Royal Nat. Hist.* I. I. iv. 125 The aspect of the gelada forcibly suggests a large black poodle dog, with an unusually abundant mane. 1906 *Westm. Gaz.* 27 Aug. 10/2 A fine specimen of the Dusky Gelada. 1967 *Listener* 6 Apr. 459/3 Other ground-dwelling monkeys such as the patas monkey, the gelada baboon and the hamadryas baboon. *Ibid.*, The gelada is especially interesting.

Gelalæan (dʒelălī·ăn), *a.* Also **Gelalean, Jalalæan, Jel(l)alæan.** [f. Arab. *Jalāl ad-dīn.*] Of or pertaining to Gelal-ed-Din, 'Glory of the Faith', a title of Malek Shah, Sultan of Khorasan, and reformer of the Persian calendar in 1079.

1780 *Encycl. Brit.* VI. 4009 *Gelalean* or *Jellalæan* Kalendar, is a correction of the Persian kalendar, made by order of sultan Gelaleddan, in the 467th year of the Hegira; of Christ 1089. 1838 H. NICOLAS *Chronol. Hist.* (ed. 2) 25 The Gelalæan era..March 14. A.D. 1079. 1866 BRANDE & COX *Dict. Sci.* etc., *Gelalæan Era*, the era of Yezdegird. 1910 *Encycl. Brit.* VI. 317/1 Era of Yazdegerd, or Persian or Jelalaean Era.

Gelasian (gĕlēⁱ·ziăn), *a.* [f. *Gelasius* (see below): see -IAN.] Of or pertaining to Pope Gelasius I (492–6 A.D.); esp. applied to liturgical books or prayers attributed to him.

a 1773 A. BUTLER *Feasts Cath. Ch.* (1839) II. i. 59 In the Gelasian liturgy..only five Sundays occur in Advent. 1832 *Archaeologia* XXIV. 10 The whole ancient Roman liturgy is comprised in the three books called the Leonine, Gelasian, and Gregorian Sacramentaries. 1915 F. E. BRIGHTMAN *Engl. Rite* p. x, [The Leonine Sacramentary] contains much official matter, some of which recurs in the later Gelasian and Gregorian books.

gelate (dʒelēⁱ·t), *v.* [Back-formation f. *GELATION[2].] *intr.* = *GEL *v.* So **gela·ted** *ppl. a.*

1915 [see *GELATION[2]]. 1948 *New Biol.* V. 33 A reversible liquefaction of any gelated regions of the protoplasm. 1951 *Ibid.* X. 17 When this plasmasol reaches the tip of the pseudopodium we can imagine it gelating to form new pseudopodial wall.

gelatin, gelatine. Add: **3.** *gelatine film, size.*

1878 *Design & Work* 26 Jan. 121/2 The gelatin film is then moistened with cold water. 1889 *Judge* (U.S.) 22 June 180/2 The gelatine film..is one two-thousandths of an inch in thickness. 1934 H. HILER *Notes Technique Painting* iii. 179 To the second series, the albumins, belong..the gelatine-size glues.

gelatinase (dʒelæ·tinēⁱz, -ēⁱs). *Biochem.* [f. GELATIN + *-ASE.] An enzyme which liquefies gelatin, occurring among bacteria, yeasts, etc.

1911 *Jrnl. Chem. Soc.* C. II. 810 The pancreas gelatinase differs from gelatinase *B. prodigiosus*, by the action of which glycine is formed. 1912 *Ibid.* CII. II. 283 The Gelatinase of Bacillus prodigiosus...The most active gelatinase preparations are obtained from bouillon cultures of the organism. 1949 H. W. FLOREY et al. *Antibiotics* I. xi. 514 In addition to the antibiotic, gelatinase was secreted. Metabolism solutions containing gelatinase lysed suspensions of heat-killing gram negative bacteria.

gelation[2] (dʒelēⁱ·ʃən). [f. *GEL *sb.* + -ATION.] The process of becoming a gel.

1915 W. W. TAYLOR *Chem. Colloids* 11 There seems no escape from the adoption of the following terms:—gel, to gelate, gelation, sol, to solate, solation. 1950 *Thorpe's Dict. Appl. Chem.* (ed. 4) X. 56/2 Gelation is not necessarily due to the formation of very large molecules. *Ibid.*, Gelation occurs only when there is the possibility of unlimited growth in three dimensions. 1954 *Electronic Engin.* XXVI. 141 If gelation takes place before maximum expansion has been reached, the product will be much denser than expected.

‖**Gelehrte(r** (gĕlēᵊ·ɹtəɹ). Also **gelehrte(r.** Pl. **-ehrten.** [G., f. *gelehrt* learned, *lehren* to instruct.] A learned person; a scholar or savant.

1836 MILL *Let.* 9 Jan. in *Wks.* (1963) XII. 292 Heaven knows whether our stupid public will read or buy a book by a German *gelehrte* who is neither prince nor minister & whom they never heard of. 1854 GEO. ELIOT *Jrnl.* Aug.–Oct. (1884) I. vi. 271 Sauppe is also a *Gelehrter*, Director of the Gymnasium, and Editor of a series of Classics. 1875 *Nation* 9 Sept. 169/2 These *Gelehrten* are not true Baconians. 1939 WYNDHAM LEWIS *Let.* Jan. (1963) 261 The theoretic basis was never very clearly understood, it seems, even by the Gelehrter. 1961 L. NAMIER *Eng. in Age of Amer. Revol.* (ed. 2) 15 No German *Gelehrter* was ever made a baron or a count.

gelly, var. *JELLY *sb.*[2]

gelt, *sb.*[2] For 'now only *dial.*' read 'now only *slang*', and add later examples.

1892 I. ZANGWILL *Childr. Ghetto* II. I. xxii. 165 Fourteen *Shtibbur's* a lot of *Gelt.* 1960 *Observer* 24 Jan. 7/1 Had I ever heard the underworld saying: No grass ever grasses for gelt alone? 1968 C. DRUMMOND *Death & Leaping Ladies* iv. 80 'The gelt?' said Reed...'Four thousand dollars,' said Miss Pocket.

‖**Gemeinschaft** (gĕmai·nʃaft). Also with lower-case initial. [G., f. *gemein* common, general + *-schaft* -SHIP.] A social relationship between individuals based on affection, kinship, or membership of a community, as within a family or group of friends; contrasted with *GESELLSCHAFT. So *gemeinschaft-like* adj. [G. *gemeinschaftlich*].

[1887 F. TÖNNIES (*title*) Gemeinschaft und Gesellschaft.] 1937 T. PARSONS *Struct. Social Action* xvii. 687 Both *Gemeinschaft* and *Gesellschaft* are what are sometimes referred to as positive types of social relationship. *Ibid.* 688 *Gemeinschaft*..is a broader relationship of solidarity over a rather undefined general area of life and interests. 1940 C. P. LOOMIS tr. *Tönnies's Concepts Sociol.* 18 The essence of both *Gemeinschaft* and *Gesellschaft* is found interwoven in all kinds of associations. 1961 L. MUMFORD *City in History* x. 310 Both primary and secondary groups, both *Gemeinschaft* and *Gesellschaft*, took on the same urban pattern. 1964 GOULD & KOLB *Dict. Social Sci.* 281/2 *Gemeinschaft* is an ideal-type concept, and as such is most correctly applied in describing or analyzing social systems in its adjectival form, *Gemeinschaft*-like. *Gemeinschaft*-like social systems are those in which *Wesenwille* (natural or essential will) has primacy.

gemellion (dʒeme·liŏn). Also **gemellione.** [ad. med.L. *gemellio*, f. L. *gemellus* a twin.] One of a pair of basins used for washing the hands before meals, the water being poured over the hands from one basin and caught by the other; hence, any decorative basin; *spec.* in liturgical use (see quot. 1960).

1889 in *Cent. Dict.* 1920 *Proc. Soc. Antiq.* 18 Mar. 131 The enamelled basins known as gemellions. 1960 H. HAYWARD *Antique Coll.* 127/2 *Gemellion*, this term describes one of a pair of basins used for the liturgical washing of the hands at the Mass.

geminate, *a.* and *sb.* **A.** *adj.* Delete 'Now only *Nat. Hist.*' and add further examples.

1964 R. H. ROBINS *Gen. Ling.* iii. 103 Consonants can be long, or geminate, when the closure or obstruction is held momentarily before release. 1967 V. NABOKOV *Speak, Memory* (ed. 2) vii. 141 Spacious windows alternated with narrower ones, single or geminate, and some of these were of frosted glass.

B. *sb.* (Further examples.)

1958 *Archivum Linguisticum* X. 61 The question of 'subsidiary' syllables, perceptibility as against sonority, whispered speech, diphthongs and geminates. 1965 *Language* XLI. 449 Intervocalic *p t k* indicate a so-called short geminate; *pp tt kk* stand for overlong geminates.

gemmologist (dʒemǫ·lŏdʒist). [f. GEMMOLOGY + -IST.] One who is skilled in gemmology.

1931 (*title of periodical*) The Gemmologist. 1937 R. WEBSTER (*title*) The gemmologist's pocket compendium. 1959 *Listener* 23 Apr. 706/2 The diamond is known to the gemmologist as the hardest..of all minerals. 1960 *Guardian* 11 July 4/2 A gemmologist's two-year course..includes the study of crystallography.

gemmology. (Later example.)

1970 *Daily Tel.* 21 Sept. 19 My son is interested in gemmology and would like to be a diamond cutter.

gemmule. 2. Add to def.: In Darwin's theory of pangenesis, one of the hypothetical units conceived of as capable of reproducing the part from which it is thrown off.

1871 [in Dict.]. 1875 [see PANGENETIC *a.*]. 1913 *Q. Rev.* Oct. 376 No one now accepts Darwin's theory of gemmules or pangenesis. 1952 C. P. BLACKER *Eugenics* ii. 45 According to this provisional hypothesis,..the cells of the body throw off minute particles called *gemmules* which circulate freely throughout the blood-stream and multiply there.

‖**gemütlich** (gəmǖ·tliχʸ), *a.* [G.] Pleasant, cheerful; cosy, snug, homely; genial, good-natured.

1852 QUEEN VICTORIA *Jrnl.* 11 Oct. (1868) 141 The view was so beautiful over the dear hills; the day so fine; the whole so *gemüthlich.* 1858 —— *Let.* 7 Apr. in R. Fulford *Dearest Child* (1964) 86, I like him very much, he is pleasant, gemütlich, accomplished. 1867 W. JAMES *Let.* 24 July in R. B. Perry *Tht. & Char. W.J.* (1935) I. 239 Young Thies is a great wit, to judge by a saying of his the other night that 'the Americans were almost as *gemütlich* as the Germans, at a fire.' 1934 A. HUXLEY *Beyond Mexique Bay* 180 Germany is to be made as remote from Europe as New Guinea...The result promises to be extremely *gemütlich.* 1958 A. WILSON *Middle Age of Mrs. Eliot* III. 427 A very roly-poly, chuckling gemütlich middle-aged German. 1960 *Times* 11 June 11/6 Those yellow Biedermeier villas which give the town its gemütlich atmosphere. 1965 T. CAPOTE *In Cold Blood* iv. 206 Their five-room apartment, with its *gemütlich* mélange of plump hassocks and squashy chairs.

‖**Gemütlichkeit, gemütlichkeit** (gəmǖ·tliχʸkait). [G., cf. prec.] The quality of being *gemütlich*; geniality; cosiness; cheerfulness.

1892 W. JAMES *Let.* 18 Dec. (1920) I. 332 When *Gemüthlichkeit* is banished from the world, it will still survive in this dear and shabby old country [*sc.* Italy]. 1912 W. E. HOCKING *Meaning of God* xxviii. 406 We must have..music with beer, Gemütlichkeit, and a fine outlook. 1929 A. HUXLEY *Holy Face* 2 The *gemüthlichkeit*, the prettiness, the cosy sublimities of the Lake District. 1959 *Oxf. Mag.* 11 June 464/1 The 'cosiness' and 'Gemütlichkeit' of the Bodleian. 1969 H. MACINNES *Salzburg Connection* i. 9 The other lake offered..folk music and dancing and general Gemütlichkeit. 1971 *Times* 6 Nov. 14/5 The double stress on God's 'terrible' demands on men and on his loving-kindness revealed in Christ gives German piety a quality which not even Catholic *Gemütlichkeit* quite overcomes.

gen (dʒen), *sb. slang* (orig. *Services'*). [Perh. abbrev. of *general* in the official phrase 'for the general information of all ranks', or possibly from part of the words *genuine* or *intelligence.*] Information; facts. Also *attrib.*

1940 MICHIE & GRAEBNER *Their Finest Hour* iv. 60 Operations room, where I got my 'Gen' (R.A.F. slang for information, or instructions). 1943 C. H. WARD-JACKSON *Piece of Cake* 31 *Gen book*, note-book kept for jotting down useful bits of information. *Ibid.*, *Genwallah*, orderly room sergeant, or anyone else well-couched in..King's Regulations and Air Council Instructions [etc.]. 1944 *Penguin New Writing* XXII. 41 My mate Tich..gives me the 'gen' of the last nine days. 1950 M. KENNEDY *Feast* 46 He pretends to have the gen on the visitors, but no visitor would know so much as that. 1958 G. MITCHELL *Spotted Hemlock* ix. 84 I'll get you a brochure...Last season's will give you all the gen you need. 1970 *Daily Tel.* 10 Dec. 8/6 A vast amount of 'gen' is included, and this will be invaluable for settling arguments.

Hence as *v. intr.* (see quot. 1943); also *trans.*, to inform (const. *up*); so **ge:nned-u·p** *a.*, informed; well supplied with information or facts.

1943 C. H. WARD-JACKSON *Piece of Cake* 31 *Gen-up, to*, to learn quickly, to swot—prior to a trade test board or some equally binding ordeal. 1945 PARTRIDGE *Dict.* *R.A.F. Slang* 29 *Genned up*, well supplied with information. 1958 E. HYAMS *Taking it Easy* I. i. 5 He wanted information; I had it. I was in a position to, as we said then, gen him up. I genned him up.

-gen. Add: **2.** The suffix is also occasionally used in terms denoting plant tissues that give rise to particular kinds of cells, as *calyptrogen, dermatogen, phellogen.*

3. *Geol.* In the form **-gene,** used in terms indicating the type, method, or place of formation, as *tectogene.*

gena (dʒī·nă). *Anat.* Pl. **genæ.** [L., see GENAL *a.*, GENIAL *a.*[2]] The cheek or lateral part of the head, *esp.* of insects.

1826 KIRBY & SPENCE *Introd. Ent.* III. xxxiv. 488 The cheeks of insects (*Genæ*) usually surround the anterior part of the eyes, and lie between them and the mandibles or their representatives. 1887 MARSHALL & HURST *Pract. Zool.* ix. 153 The genæ, or 'cheeks', are a pair of vertical plates covering the sides of the head, behind and below the eyes. 1911 *Encycl. Brit.* XXVII. 283/1 A lateral unsegmented plate, the *gena*, which carries the eyes [in Trilobites]. 1957 RICHARDS & DAVIES *Imms's Gen. Textbk. Ent.* (ed. 9) I. 20 The *gena*..forms the whole of the lateral area below and posterior to the eyes on each side. *Ibid.* III. 587 The *genae* (parafacials or cheeks) comprise the region lying between the face and the anterior margin of the eye on either side. 1964 C. J. GOODNIGHT et al. *Gen. Zool.* x. 198/1 The dorsal region [of the head of the grasshopper] is known as the vertex; the lateral, the genae, or cheeks.

genal, *a.* (Earlier example.)

1877 *Encycl. Brit.* VI. 660/1 (*caption*) Another of the same stage, in which the genal or cheek spines are developed.

gendarme. Add: **2. c.** A policeman. *slang.*

1906 H. GREEN *At Actors' Boarding House* 204 She'll be cabling the old guy to set the gendarmes on us as we leave the boat. 1931 H. CRANE *Let.* 30 Mar. 367, I am to sail to *Mexico* (damn the gendarmes!) next Saturday.

gendarmery, gendarmerie. Add: **2. b.** The head-quarters of the gendarmes; the police-station.

1945 G. MILLAR *Maquis* vi. 131 We left our bicycles in the cabbage patch behind the Gendarmerie. 1955 J. THOMAS *No Banners* ix. 78 Well, try the *Gendarmerie.* They've got cells there. *Ibid.* xx. 191 Patrice was a British officer who had the bad luck to land by parachute on the roof of a *gendarmerie.* 1967 G. M. WILSON *Cake for Caroline* v. 61 Ring the gendarmerie...Ask somebody to meet me up there.

gene (dʒīn). *Biol.* [a. G. *gen* (W. Johannsen *Elem. d. exacten Erblichkeitslehre* (1909) 124), irreg. f. Gr. γεν- (see -GEN).] Each of the units of heredity which (except for polygenes) may be regarded as the controlling agents in the expression of single phenotypic characters and are usu. segments of a chromosome at fixed positions relative to each other; they were orig. defined as ultimate units of mutation and recombination, but are now best regarded as sequences of nucleotides within nucleic acid molecules each of which determines the primary structure of some protein or polypeptide molecule.

The term *gene* is used indiscriminately for both a genetic locus and an allele.

1911 W. JOHANNSEN in *Amer. Naturalist* XLV. 132, I have proposed the terms 'gene' and 'genotype'..to be used in the science of genetics. The 'gene' is nothing but a very applicable little word, easily combined with others, and hence it may be useful as an expression for the 'unit-factors', 'elements' or 'allelomorphs' in the gametes, demonstrated by modern Mendelian researches. 1917 T. H. MORGAN in *Ibid.* LI. 520 The linkage relations of genes..have a very important bearing on the problem of the localization of genes in the germ plasm. 1925 *Nation* 3 Oct. 19/1 If..in a given kind of plant there is a single 'factor' for flower colour and the flowers are blue or yellow, the gamete will bear either the blue factor (gene) or the yellow one, not both. 1930 G. R. DE BEER *Embryol. & Evol.* iii. 21 By introducing into an egg containing a weak female-producing gene a sperm containing a strong male-producing gene (as may be done by crossing moths of different races), it is possible to convert would-be females into males more or less completely. 1946 *Nature* 6 July 30/2 The evidence indicates that resistance to blight is controlled by major genes, though minor genes determine the degree of susceptibility in susceptible varieties. 1955 *Sci. Amer.* July 74/2 Heredity is determined by the chromosomes, the threadlike bodies in the nucleus of the cell, and by their subunits the genes. 1964 G. H. HAGGIS et al. *Introd. Molecular Biol.* x. 260 The functional genes turn out to occupy a much greater chromosomal length than the genes defined by mutation or recombination. 1971 *Sci. Amer.* Aug. 52/1 When we speak of genes, we usually have in mind the hereditary material—the DNA—in the chromosomes of the cell nucleus. Yet genes are also found outside the nucleus in the cytoplasm, notably in association with chloroplasts and mitochondria.

2. *attrib.* and *Comb.*, as **gene frequency** (see quot. 1968); **gene mutation,** a mutation in which the change is confined to a single gene; **gene pool,** the stock of different genes in a breeding population.

1930 R. A. FISHER *Genet. Theory Nat. Selection* i. 10 Even relatively intense selection will change the ratio *p*:*q* of the gene frequencies relatively slowly. 1968 R. RIEGER *Gloss. Genetics & Cytogenetics* 170 *Gene frequency*, the proportion of one particular type of allele to the total of all alleles at this genetic locus in a breeding population. 1928 *Daily Tel.* 11 Sept. 8/2 Whether X-rays could produce 'gene mutations'. 1930 R. A. FISHER *Genet. Theory Nat. Selection* iii. 49 Gene-mutation..consists in a change in a single hereditary particle, or gene, into a gene of a new type. 1950 T. DOBZHANSKY in *Amer. Naturalist* LXXXIV. 404 Every sexual species accordingly possesses a gene pool, in which each gene may be represented by a certain number of alleles, and each chromosome by one or more structural variants. 1959 *New Biol.* XXVIII. 90 Sexual reproduction provides such a mechanism in allowing gene-recombination; but..if unlimited in its scope it would inhibit diversification by randomizing the 'gene pool' in each generation. 1970 *Times* 29 Dec. 10/6 There is a necessity to conserve a wide range of different genes in the 'gene pool' from which new strains can be selected.

‖**gêne** (ʒĕn). [Fr., f. OF. *gehine* torture (*gehir* to confess under torture): cf. OHG. *jehan* to speak, say.] Constraint, embarrassment, discomfort.

1817 H. C. CAMPBELL *Jrnl.* 17 Oct. in *Journey to Florence* (1951) 108, I dislike seeing such things in a crowd particularly in an English one. There is a certain *gêne* which they always occasion. 1853 LYTTON *My Novel* II. VII. xxi. 283 This will prevent a great deal of *gêne* and constraint. 1914 WODEHOUSE *Man Upstairs* 257 The advent of an unbidden guest rarely fails to produce a certain *gêne.* 1953 R. A. KNOX *Off Record* p. ix, He is condemned to listen to one side of a telephone conversation, with something of the *gêne* which that exercise entails.

‖**gêné** (ʒene), *a.* Also fem. **gênée.** [Fr., pa. pple. of *gêner* to embarrass; cf. prec.] Constrained, embarrassed, discomforted. Also as *v. trans.*, to make (someone) feel *gêné*; so **gêné(e)d** *ppl. a.*

1806 *Wynne Diaries* (1940) III. 301 Mrs. Bishop seemed gênée, we played at Loo, and were very merry all the Evening. 1817 H. C. CAMPBELL *Jrnl.* 4 Oct. in *Journey to Florence* (1951) 102 A few people came in the evening. I felt geneed not having been in company for some time. 1823 BYRON *Don Juan* XIII. ciii. 106 But none were 'gêné'. 1851 E. RUSKIN *Let.* 25 Nov. in M. Lutyens *Effie in Venice* (1965) 219 He is very good and well bred and..they don't see how gened he is. 1929 E. BOWEN *Last Sept.* I. viii. 91 She looked gênée, dispirited.

genecology (dʒen-, dʒīnikǫ·lŏdʒi). *Biol.* [f. Gr. γέν-ος race + *ECOLOGY.] The study of genetic differences between related species, and populations of a species, in relation to the environment. Hence **genecolo·gic, geneco-lo·gical** *adjs.*

1923 G. TURESSON in *Hereditas* IV. 172 It seems appropriate for several reasons to denote this study of species-ecology by the term *genecology*..as distinct from the ecology of the individual organism, for which the old term autecology seems to be the adequate expression. 1928 B. D. JACKSON *Gloss. Bot. Terms* (ed. 4) 439/2 Genecological. 1947 R. F. DAUBENMIRE *Plants & Environment* x. 366 (*heading*) The genecologic classification. 1963 DAVIS & HEYWOOD *Princ. Angiosperm Taxon.* xii. 398 Genecological differentiation has been found to be especially common in trees. *Ibid.* 401 The basis of genecology..is the observation that this intra-specific variability can be demonstrated to be of genetic origin and is not randomly dispersed throughout the species range, but is distributed in such a way that neighbouring plants tend to resemble one another.

generable, *a.* **1.** (Later example.)

1962 S. R. LEVIN *Ling. Struct. Poetry* ii. 11 Many poetic sequences are generable by the kind of grammar constructed for ordinary language, but some are not.

general, *a.* and *sb.* Add: **A. 1. a.** *general health*, the state of health of the body as a whole, or of a community.

1819 in A. Clarke *Ess. Dis. Skin* (1821) v. 97 It does not seem to hurt my general health; my appetite is good. 1870 *Food Jrnl.* 1 Mar. 67 So numerous are the forms which charitable relief assumes, that it would be next to impossible to analyse their separate influence on the general health. 1879 *Man. Artill. Exerc.* III. 69 Attentions to the comforts of the men will..add to their general health. 1899 *Daily News* 11 May 4/1 Although my general health, to use a well-known phrase, is wonderfully good, I seem indeed..to fear being kept here too long. 1903 *Westm. Gaz.* 26 Mar. 9/2 His leg wound did not heal well, and this, with his poor general health, greatly depressed him. 1951 E. M. GRAVELIUS *Brit. Red Cross Nursing Man.* (ed. 9) xx. 240 The general health plays an important part in the recovery from skin complaints.

2. a. *general American*, a form of U.S. speech without marked dialectal or regional characteristics; *General Certificate of Education*, an examination set by each of a group of examining boards for pupils in secondary schools in England and Wales; also, the certificate awarded to those who pass this examination; cf. *G.C.E.* (s.v. *G III); *general circulation*, (*a*) (see quot. 1928); (*b*) 'circulation, as of a newspaper, among readers not confined to a narrow class in business or interests' (Webster 1909); *general headquarters*, the headquarters of the commander-in-chief; *general meeting*, a meeting which all members of a society or other organization may attend; *general quarters*, in the navy, the stationing of all hands, and the making of preparations, as if for an encounter with the enemy; also *attrib.* (as *general-quarter*); *general reserve* (see quot.); *general semantics*, 'a doctrine and educational discipline due to Alfred Korzybski (1879–1950) intended to improve the habits of response of human beings to their environment and one another esp. by training in the better and more critical uses of words and other symbols' (Webster 1961); *general strike*, (*a*) a strike of all the workers of one industry; (*b*) a concerted strike by workmen of all or most of the important trades and occupations of a country with a view to securing some common object by the stoppage of business; *spec.* that in Britain in 1926.

1934 WEBSTER, Introd. p. xxvi/1 In America three main types of cultivated speech are distinguishable:..(3) the variously named Western, Midwestern, or General American. *Ibid.*, The fact that more speakers in the English-speaking world habitually use the General American than any other single type cannot vitiate the standing of the Southern British..for the educated Englishman. 1947 C. K. THOMAS *Introd. Phonetics Amer. Eng.* xxii. 171 Eastern New England, the New York City area, the South, and the General American area clearly speak according to..different standards. *Ibid.* 172 The New York City area..has a population verging on ten million; General American,..a population of over eighty million. 1964 *Listener* 16 Apr. 634/3 What is called general American—a more-or-less middle-western accent. 1947 *Examinations Secondary Schools* (Rep. Second. School Exam. Council) 6 It is because the 'Ordinary' and 'Advanced' papers have fundamentally the same qualifying purpose..and because we wish them to be available to candidates who are not 'at school' that we have proposed the single certificate and offered the title 'General Certificate of Education'. *Ibid.* 8 A 'General Certificate of Education' should be awarded showing the subjects (and the level—'Ordinary' or 'Advanced'—in each subject) in which the candidate has satisfied the examiners. 1955 *Times* 28 June 5/3 It is hoped to establish courses leading to the General Certificate of Education, for pupils who need the G.C.E. to qualify for professional training. 1899 *Q. Jrnl. R. Meteorol. Soc.* XXV. 166 Schemes and discussions concerning the general circulation of the atmosphere. 1926 N. SHAW *Man. Meteorol.* I. 291 Dove rendered a signal service to the observational representation of the general circulation by producing monthly maps of isotherms of the globe. 1928 D. BRUNT *Meteorol.* iv. 26 When we draw charts on which are represented, at a large number of stations, the most frequently occurring wind directions..., we find the movement of the winds of the globe form a system which is much simpler in appearance than might have been expected by one accustomed only to the variability of weather in the British Isles. This system is called the 'general circulation' of the atmosphere. 1859 *War in Italy* 54 The infantry of the guard followed general head-quarters to Castenedolo. *Ibid.*, General head-quarters..moved to Montechiaro. 1914 *Times* 3 Oct. 8/2 The Press Bureau..issued the following descriptive account, which has been communicated by an eye-witness present with General Headquarters. 1782 R. GOADBY *Life B. M. Carew* 11 Their general meetings at stated times, which they are all obliged to be present at. 1812 *Dramatic Censor for 1811* 419 The Committee, therefore, might be left..to call a General Meeting when they might deem such a proceeding necessary. 1900 *Westm. Gaz.* 5 Mar. 9/2 On every 'general-quarter' day, in my last ship. 1919 W. LANG *Sea Lawyer's Log* vi. 61 When the bugle sounds 'General Quarters', the prelude to action. 1918 E. S. FARROW *Dict. Mil. Terms* 259 *General reserve*, a reserve retained in the hands of the general officer commanding of the whole force until required. 1933 A. KORZYBSKI (*title*) Science and sanity. An introduction to non-Aristotelian systems and general semantics. 1951 *Essays & Studies* IV. 119 The term 'General Semantics' was used by the late Alfred Korzybski for a kind of linguistic therapy quite unrelated to technical linguistics. 1810, 1830 General strike [see STRIKE *sb.* 9]. 1902 *Encycl. Brit.* XXXIII. 26/1 In 1891 a general strike took place in the German printing trade. 1924 J. F. BRYANT *Gandhi & Indianisation* 72 His methods were tinged with the ideas of Passive Resistance and the General Strike. 1926 *Hansard, Commons* 3 May 71, I do not think all the leaders when they assented to ordering a general strike fully realised that they were threatening the basis of ordered government. 1952 *Chambers's Encycl. World Survey* 215 A general strike of textile workers took place in Bombay and towards the end of the period notice of a general strike on the Indian railways was given. 1954 B. & R. NORTH tr. *Duverger's Pol. Parties* I. i. 15 General Strikes of 1891 and 1893 in Belgium; of 1902 and 1908 in Sweden. 1971 A. BULLOCK *20th Cent.* iii. 72/1 The failure of the General Strike in 1926 marked the defeat of the militants in the British Labour movement.

b. (*b*) *general post* (earlier and later examples). Also *fig.*, a general and rapid exchange or interchange of positions, etc.

1707 [see POST *sb.*[2] 5 a]. 1838 DICKENS *Nickleby* i. 2 There came one morning, by the general post, a black-bordered letter. 1889 K. GREENAWAY *Bk. Games* 63 General Post. One person is selected as 'postman' and blindfolded, the others all take the names of different places, except one, who is chosen the leader, and has a written list of the places chosen by the players. 1898 A. B. GOMME *Games for Parlour & Playgr.* 51 An occasional call of 'General Post' by the leader, when all players must change their seats, gives a good chance to the blind man. 1941 A. L. ROWSE *Tudor Cornwall* xii. 307 They were..instituted to other livings; in effect, it meant a sort of general post of the affected clergy. 1954 M. BERESFORD *Lost Villages* vi. 213 In this general post the land-use of much of the Midlands, the grassy shires, could be considered afresh. 1969 I. & P. OPIE *Children's Games* vi. 209 The party game known variously as 'General Post', 'Move All', [etc.].

6. a. *general dealer* (earlier examples); *general practitioner* (earlier example); see also PRACTITIONER 1 b.

1830 J. F. COOPER *Water Witch* II. i, A man whose misdeeds in commerce are as universally noted as the stoppage of a general dealer. 1844 G. RAYMOND *Mem. R. W. Elliston* I. xiv. 348 The general practitioner had an exceedingly pretty wife. 1859 DICKENS in *All Year Round* 13 Dec. 3/1 The general-dealer opposite..is opening his shop.

7. a. *general knowledge*, knowledge of miscellaneous facts, information, etc. (cf. quot. 1860 under sense 8 a in Dict.); *general public*, the ordinary people; = PUBLIC *sb.* 1 b; *general (theory of) relativity*: see *RELATIVITY.

1820 HAZLITT in *London Mag.* II. 250/2 Books of liberal taste and general knowledge. 1834 G. CRABBE JR. in *Poetical Wks. G. Crabbe* I. iv. 97 The fund of general knowledge which my father gradually showed..much surprised his patron. 1854 J. E. MILLAIS *Let.* 10 May in M. Lutyens *Millais & Ruskins* (1967) 210 This is what a number of comfortable, portly, philosophers will say merely in direct opposition of the general public. 1877 General public [in Dict.]. 1906 *Daily Chron.* 25 Jan. 4/7 A Scriptural general knowledge paper. 1934 *Discovery* Nov. 317/2 When the general public finds that the railways are providing local services much faster and no dearer than motor bus companies,..it will begin to return..to the railway. 1938 F. B. YOUNG *Dr. Bradley Remembers* (1940) iii. 124 That same Act of Parliament..had decreed that a medical student, before registration, must first pass an examination in General Knowledge. 1952 G. RAVERAT *Period Piece* iv. 65 When I went away to school, [I] was asked in a General Knowledge Paper, which were my three favourite composers.

b. *general shop, store* (examples).

1835 J. MARTIN *Descr. Virginia* 134 A neat village..containing 16 dwelling houses, 3 general stores, 2 groceries. 1836 General shop [see STORE *sb.* 12 a]. 1851 *Lyttelton Times* (N.Z.) 17 May 7/3 He has opened a General Store. 1861 DICKENS *Gt. Expect.* I. vii. 92 Mr. Wopsle's great-aunt..kept—in the same room—a little general shop. 1865 —— *Mut. Fr.* IV. vi. 206 At the general shop, at the butcher's and at the public house. 1948 *Bangor* (Maine) *Daily News* 28 July 1 Shopping at the general store here to replenish his food supply.

c. Freq. in phrases used *attrib.*, as *general-fish, -produce, -purpose, -purposes, -utility.*

1860 *Leisure Hour* 10 May 294 These are the 'general utility' men, as they are sometimes facetiously called by those whose genius is not quite so versatile. 1888 J. C. HARRIS *Free Joe*, etc. 127 One of the many 'general-utility' men that improved methods enable the high schools and colleges to turn out. 1894 *Country Gentlemen's Catal.* 230/2 Patent general purpose drill. 1909 *Westm. Gaz.* 11 Feb. 3/3 The largest fruit and general produce merchants in New York. *Ibid.* 3 Apr. 16/4 As the president of this excellent club, he spoke of the good trout and general-fish waters it leased on the Surrey Wey. *Ibid.* 28 Aug. 16/4 The Thames is in good condition, and general-fish anglers are promised a continuance of sport. 1911 F. O. BOWER *Plant-Life* 58 It served as a general-purposes shoot. 1923 KIPLING *Land & Sea T.* 143 A general-utility shed. 1933 *Meccano Mag.* Mar. 193/1 It [*sc.* the 'Tiger' engine] has been designed for use in general purpose and torpedo-carrying aircraft. 1937 B. H. L. HART *Europe in Arms* x. 131 The plans of the General Staff were dominated by the idea of contributing a general-purpose force of three army corps to join the French field armies. 1966 *Listener* 3 Nov. 668/1 German's generation was content with one general-purposes 'olde' manner. 1968 FOX & MAYERS *Computing Methods* i. 7 The Runge-Kutta method..is a useful general-purpose routine for non-linear first-order equations.

B. 7. c. *General February, Fevrier, January, Janvier* or *Winter*: personifications of wintry months, alluding to their effect upon military campaigns, etc.

1855 *Punch* 10 Mar. 95 Russia has two generals in whom she can confide—Generals Janvier and Fevrier. 1908 KIPLING *Lett. to Family* (1920) 159 Here, General January will stiffen him up. 1919 *Mr. Punch's Hist. Great War* 20 'General Janvier' is doing his worst, but our men are sticking it out though slush and slime. 1926 B. PARES *Hist. Russia* xviii. 339 Nicholas had placed his hopes in 'General January' and 'General February', that is to say in the severity of a Russian winter. 1966 A. FIRTH *Tall, Balding, Thirty-Five* v. 63 When the crunch came in 1941 he [*sc.* Stalin] owed a good deal to General Winter. 1967 D. G. CHANDLER *Campaigns Napoleon* XIV. lxxi. 815 Time was playing into the hands of the Russians by bringing 'General Winter' even closer.

d. The head of the Salvation Army; *spec.* General William Booth, its founder.

1883 *All about Salv. Army* 26 If the General were to be removed by death to-morrow, his successor, without a minute's delay, would step into his position. 1884 W. BOOTH *General's Lett.* (1890) 18 Go to the crowd of sinners, or spot them individually..because out of them you may make Lieutenants, and Captains, and Majors, and Generals. 1886 —— *Orders & Regul. Salv. Army* 165 The General must and will appoint his own successor—each successive General doing the same. 1959 *Chambers's Encycl.* XII. 177/1 The general..no longer has the right of nominating his successor.

general hospital. [app. after F. *hôpital général*, the name of the hospital founded at Bicêtre, France, in 1656.] **a.** A hospital which does not confine itself to patients suffering from one particular class of disease or to those of a particular sex or age-group.

1737 (*title*) The plan and elevation of a new general hospital intended to be erected at Bath for the reception of 150 poor strangers. 1757 A. R. *Curiosities of Paris* iv. 62 General Hospital, is a small Distance from the City...They receive all Kinds of Patients upon Application, such as Women with Child, Foundlings, Orphans, and Lunatics. 1794 *Medical Extracts* II. 62 The air in the lowest ward in the General Hospital. 1803 *Guide Watering & Sea-Bathing Places* 44 The General Hospital..is open to the sick poor of the United Kingdom,..the inhabitants of Bath alone excepted. 1869 [see GENERAL *a.* 2 a]. 1880 *Encycl. Brit.* XII. 302/1 Hospitals are usually divided into General and Special. In General Hospitals cases of all kinds are admitted in some, whilst in others certain classes are excluded. 1966 R. P. SLOAN *Today's Hospital* i. 7 The trend is away from the 'special disease' hospital and toward the 'general' hospital assuming responsibility directly or indirectly for all physical or mental disabilities.

b. A military hospital receiving the sick and wounded from field hospitals.

1854 F. NIGHTINGALE *Let.* 15 Dec. in C. Woodham-Smith *F. Nightingale* (1950) ix. 183 The proportion of Roman Catholics..is already making an outcry...Dr. Menzies has declared that he will have two only in the General Hospital. 1863 *Good Words* Nov. 817/1 The army hospitals near Washington are of course full, and a general hospital has been lately established at Gettysburg. 1899 *Daily News* 27 Nov. 5/3 Large hospitals..known as 'general hospitals' (or 'base hospitals'), each accommodating 500 men and 20 officers...A large staff is necessary for the working of each 'general hospital'. 1914 *Times* 30 Oct. 10/1, 3rd Southern General Hospital, Oxford. 1964 W. R. M. DREW in F. N. L. Poynter *Evol. Hospitals in Brit.* 165 Altogether, there were nineteen bearer companies, twenty-eight field hospitals..and twenty-two general hospitals.

generalism. Add: **c.** The fact or quality of generalizing; the actions, principles or qualities of a generalist (opp. SPECIALISM 1).

1908 *Times Lit. Suppl.* 13 Aug. 260/3 The essays..were, in their day, too good specimens of the best generalism to be of permanent value. 1963 *Times* 1 May 10/3 The Common Market was encouraging a European particularism..at just the time in history when speedy communications..required a free world generalism. 1968 *New Scientist* 3 Oct. 31/2 There is a need for a new brand of 'science-based generalism'.

generalist. Add later examples of sense b.

1961 *Economist* 2 Dec. 938/1 The complacent belief that a well trained 'generalist' could turn his hand to anything. 1961 L. MUMFORD *City in Hist.* v. 123 Gifted generalists like Aristotle. 1964 R. WILKINSON *Gentlemanly Power* vi. 71 The backbencher was a generalist also; Parliament has made less use of specialized committees than have either the French or American legislature. *Ibid.* 72 In British bureaucracy, the generalist backgrounds of the top men have shown an interesting relation to professional procedure. 1968 *Listener* 26 Sept. 395/2 I've been asking some large questions—when you are a generalist you learn to look towards big patterns. 1971 *Physics Bull.* Apr. 231/2 Scheme for an initial two year generalist course for the majority of students backed up by specialist courses for the further studies of the few.

generalizability (dʒe:nĕrălaiˑzăbi·lĭti). [f. GENERALIZABLE *a.* + -ILITY.] The fact or quality of being generalizable.

1951 T. PARSONS et al. in Parsons & Shils *Toward Gen. Theory Action* II. iv. 202 The very fact that they [*sc.* rewards] become the objects of competing claims..is in part evidence of their generalizability to cover the claims of different individuals. 1967 *Language* XLIII. 749 He has provided us with no criteria for determining generalizability from sets of pattern correlations of the type he considers.

generate, *v.* Add: **2. d.** *Math.* and *Linguistics.* To produce (a set or sequence of items) by certain specified operations or by the repeated application of rules to some basic items.

1896 *Bull. Amer. Math. Soc.* III. 115 If *G* is generated by substitutions of order *p*α it is self-conjugate. 1903 B. RUSSELL *Princ. Math.* xxiv. 200 The simplest method of generating a series is as follows. 1947 BIRKHOFF & MACLANE *Surv. Mod. Algebra* vi. 138 The group G is cyclic if it contains some one element x whose powers exhaust G; this element is said to generate the group. *Ibid.* xiv. 373, $x^2+4x+2=0$ has a root $-2+\sqrt{2}$ which generates the same field $R(\sqrt{2})$, for any number in the field can be expressed in terms of this new generator. 1956 N. CHOMSKY in *IRE Trans. Information Theory* IT-2 Sept. 113/1 We investigate several conceptions of linguistic structure to determine whether or not they can provide simple and 'revealing' grammars that generate all of the sentences of English and only these. 1957 —— *Syntactic Structures* ii. 13 The grammar of L will thus be a device that generates all of the grammatical sequences of L and none of the ungrammatical ones. 1958 J. T. CULBERTSON *Math. & Logic Digital Devices* iv. 45 This rule for generating the integers applies to all systems regardless of radix. 1959 *Word* XV. 237 Even in this grammar, however, it would probably be foolish to generate such small and unproductive sets as the Turkish personal pronouns. 1965 PATTERSON & RUTHERFORD *Elem. Abstract Algebra* ii. 44 Every element *x* of a finite group *G* generates a cyclic subgroup whose order is a factor of the order of *G.* 1968 *Language* XLIV. 57 If the optional Noun deletion rule does not apply, *ma maison* is generated instead of *la mienne.* 1968 J. LYONS *Introd. Theoret. Ling.* iv. 156 When we say that a grammar generates the sentences of a language we imply that it constitutes a system of rules..which..yield..a decision-procedure for any combination.

generating, *ppl. a.* Add: *generating station*, a power station for the generation of electricity.

1898 *Daily News* 13 June 7/3 The Company are practically reconstructing their electrical generating station. 1959 *Chambers's Encycl.* V. 93/1 The capacities of some of the largest generating stations.

generation. Add: **2. b.** *spec.* The production of electricity. Also *attrib.*

1886 W. WORMELL tr. *Urbanitzky's Electr. in Service of Man* 95 The generation of electricity is to be explained by the mere contact of bodies with each other. 1900 *Westm. Gaz.* 7 Mar. 9/1 The generation and distribution expenses for the year. 1901 *Ibid.* 14 Nov. 8/3 The generation plant for the first section of the new electric tramways. 1943 J. S. HUXLEY *TVA* i. 9 Flood-control could be readily tied up..with the profitable generation of electric power.

c. *Linguistics.* The process of deriving the grammatical sequences of a language from a basis that constitutes the grammar.

1959 *Word* XV. 234 Following through with a very condensed generation sequence (in which most irrelevant rules and choices are omitted), we might have something like this. *Ibid.* 237 A short sample idiom-generation might look like this. 1963 J. LYONS *Structural Semantics* ii. 31 In the learning and use of language there are two complementary factors to be reckoned with. These I shall call generation and recall...By 'generation' [I mean] the construction of a form by the individual speaker from elements which are themselves taken from 'storage'. 1965 N. CHOMSKY *Aspects of Theory of Syntax* i. 60 Both weak and strong generation are determined by the procedure.

4. b. first- (or second-, etc.) **generation** *a.*, designating a member of the first (or second, etc.) generation of a family, *spec.* of descendants of immigrant parents, esp. in the United States; also, designating a naturalized immigrant (or a child, etc., of a naturalized immigrant). Also *transf.* and *fig.*

1896 S. A. BARNETT *Let.* Sept. in H. Barnett *Canon Barnett* (1918) II. 119 There are the usual Americans. One 'first-generation man', as he calls himself..has made a great fortune. 1946 J. O'HARA in *55 Short Stories from New Yorker* (1952) 199 Francis had his place at the bar, at the far corner, and it was his so long as he was present. First-generation Jimmy and second-generation Jimmy had seen to that. 1951 M. MCLUHAN *Mech. Bride* 67/1 First-generation immigrants who quickly made good. *Ibid.* 67/2 The father is just such a second-generation type. 1953 E. COXHEAD *Midlanders* vi. 153 Herself a second-generation college girl, she now under-valued the freedoms the pioneers had won. 1956 *Nature* 10 Mar. 489/2 Second-generation inbreds...Fourth-generation inbreds...F_1 hybrids between third-generation inbreds. 1960 *Guardian* 29 July 4/4 Deriabin, born in 1921, was a second generation Communist. 1960 *Ibid.* 5 Nov. 3/6 A second-generation Kenyan whose father was one of the pioneer settlers in the White Highlands. 1962 *Chem. Engin. Progress* Oct. 44 (*caption*) First generation (left) and improved (right) void-free laminates of phenolic resin and graphite fabric. 1963 *Computer Jrnl.* VI. 144/1 It has recently become conventional to distinguish between 'first-generation computers' and 'second-generation computers'. 1968 L. BLACK *Outbreak* ix. 86 The number of notifications [of smallpox] will rise sharply, as second and third generation cases emerge from the incubation period.

7. *attrib.* and *Comb.*, as *generation-conscious* adj.; **generation gap** (cf. *GAP *sb.*[1] 6 a).

1930 R. MACAULAY *Staying with Relations* iv. 57 'I remember an air raid,' said Julian. 'They woke me up and carried me down to the basement. I am the air raid generation.' 'They're terribly generation-conscious,' Adrian explained. 1934 R. CAMPBELL *Broken Record* 50 Quarrelling with my father made me generation-conscious. 1967 *Boston Globe* 18 May 10/1 He acknowledged that the 'generation gap' is difficult both for the younger and the older generations. 1968 *Guardian* 25 Oct. 14/6 There is a generation gap in attitude but not in the way this family treat each other. 1969 W. GARNER *Us or Them War* xxi. 167 He said, 'Patti, whatever becomes of the generation gap?' She said, 'I jumped across...' 1971 K. DICK *Ivy & Stevie* 54 Generally, poets in her [*sc.* Stevie Smith's] age group..do not escape criticism from what is journalistically termed 'generation-gap' censure.

generational, *a.* (Further examples.)

1947 WELLEK & WARREN *Theory Lit.* xviii. 259 The largest reputations survive generational tastes: Chaucer, Spenser, Shakespeare..and Tennyson have a permanent..position. 1956 WYNDHAM LEWIS *Red Priest* xxvi. 217 You are younger than I am, and probably are in a different generational stream. 1964 *Punch* 9 Sept. 382/2 The generational gap is even more extended at student level. 1971 *Guardian* 9 Feb. 8/7 The generational outbursts against humanity.

generative, *a.* Add: **2. b.** *Linguistics.* Able to generate (in sense *2 d of the vb.); concerned with generation (*2 c); *generative grammar*: see quots. 1964 and 1965.

GENERATIVIST

1959 *Word* XV. 233 A generative grammar, as Chomsky has shown, may be conveniently arranged in the form of a series of equation-like rules. **1960** *Language* XXXVI. 360 (*title*) The place of intonation in a generative grammar of English. **1964** E. Bach *Introd. Transformational Gram.* ii. 13 A (*generative*) grammar of a language is a theory or set of statements which tells us in a formal and explicit way which strings of the basic elements of the language are permitted. **1964** E. A. Nida *Toward Sci. Transl.* iv. 60 A generative grammar is based upon certain fundamental kernel sentences, out of which the language builds up its elaborate structure by various techniques of permutation, replacement, addition and deletion. **1965** N. Chomsky *Aspects of Theory of Syntax* i. 8 By a generative grammar I mean simply a system of rules that in some explicit and well-defined way assigns structural descriptions to sentences. **1968** J. Lyons *Introd. Theoret. Ling.* iv. 139 Any linguistic description which has this capacity of describing actual utterances as members of a larger class of potential utterances, is said to be generative.

ge·nerativist. [f. Generativ(e *a.* + -ist.] One who employs the methods of generative grammar.
1965 *Amer. Speech* XL. 289 This seems to allow generativists a token amount of equal time, but it is clear that they are 'the other party'. **1967** *Word* XXIII. 47 Since generativists began to turn to the concept of deep grammar as input to the transformational rules that produce 'surface' sentences.

generator. Add: **4. a.** *Geom.* = Generant A. b.
1863 *Phil. Trans. R. Soc.* CLIII. 455 The nodal generating lines or Nodal Generator. **1893** N. F. Dupuis *Elem. Synthetic Solid Geom.* i. 7 The variable line N is called the generator, and the fixed guiding lines are directors. **1959** *Chambers's Encycl.* III. 837/1 Analytically, it is convenient to regard the generators of a cone (i.e. the lines joining the vertex to points on the base circle) as extending to infinity in both directions.
b. *Algebra.* Any of a subset of the elements of a set in terms of which all the other elements of the set can be represented, using specified operations.
1894 *Bull. Amer. Math. Soc.* I. 63 A substitution σ of Γ' is determined by the elements *a'*, *b'*, *c'* which it makes correspond to the generators *a*, *b*, *c*. **1940** C. C. MacDuffee *Introd. Abstract Algebra* ii. 53 A cyclic group has as its elements the powers of a single generator. **1947** Birkhoff & MacLane *Surv. Mod. Algebra* xiv. 373 Any number in the field can be expressed in terms of this new generator.
c. *Computers.* A routine that enables a computer to construct from a set of parameters other routines or sub-routines with specific applications. Also *attrib.*
1953 *Computers & Automation* May 4 Editing is but one phase of the commercial and logistic problems which lend themselves to generator techniques. **1956** Berkeley & Wainwright *Computers* viii. 344/2 *Generator*, a computer program which generates coding. **1958** Gotlieb & Hume *High-Speed Data Processing* xiv. 293 Generators have also been written for editing, re-run procedures, tape checking, and moving records. **1962** Huskey & Korn *Computer Handbk.* xvii. 19 If memory space is not a problem the input information can be reduced to reasonable size by devising a generator code which is usually cyclic in character and can produce the linear code.

generic, *a.* (*sb.*) Add: **a.** *generic image*, a mental image representing a class or genus of objects, whether formed (as is usually supposed) by blending images of several particular members of that class or by preventing an image from becoming fully determinate.
1817 Coleridge *Biog. Lit.* v. 98 Des Cartes..showed, in what sense not only general terms, but generic images (under the name of abstract ideas) actually existed. **1878** W. James *Noteb.* in R. B. Perry *Tht. & Char. of W.J.* (1935) II. 80 Generic images..will be remembered as further instances of facts persistently denied by empiricists. **1882** F. Galton in *Proc. R. Inst.* IX. 166 The generic images that arise before the mind's eye..are analogues of these composite pictures. **1918** J. Ward *Psychol. Princ.* xii. 299 The generic image (*Gemeinbild* of German psychologists) constitutes the connecting link between ideation and conception. **1953** H. H. Price *Thinking & Exper.* ix. 292 Both Locke and Kant were talking about generic images, though they did not know it.
c. *sb.* A generic word; *spec.* (see quot. 1961).
1961 Webster, *Generic*, an element of a compound proper name that is general and often lowercased (as *river* in 'Mississippi River' and *store* in 'XYZ Store'). **1962** Burrill & Bonsack in Householder & Saporta *Probl. in Lexicogr.* 184 The words *hope* and *folly*, encountered in some names, have a topographic meaning and are used as generics in England. **1964** *Language* XL. 49 Generics may be considered the name given to a particular interpretation of the definite article.

genericness (dʒĕne·riknės). [f. Generic *a.* + -ness.] Generic quality or characteristics.
1939 P. Christophersen *Articles* 33 *The* represents an aggregating genericness,..*a* is a singularizing form. **1939** *Mind* XLVIII. 150 Scales of kinds or sorts, which scales exhibit differences in degree of generic-ness or specific-ness.

generous, *a.* Add: **7.** *Comb.*, as *generous-hearted*, *-lipped*, *-natured*, *-souled* adjs.
1813 Jane Austen *Pride & Prej.* III. i. 10 He was always the sweetest-tempered, most generous-hearted, boy in the world. **1856** Whittier *Panorama* 6 Wise-thoughted age, and generous-hearted youth. **1924** M. A. Lowndes *Terriford Myst.* iii. 35 Her generous-lipped mouth was too large for beauty. *Ibid.* vi. 68 She was the most devoted and generous-natured of wives to me. **1907** *Daily Chron.* 9 Nov. 8/5 Like all generous-souled men, her grandfather ran to extremes.

genetic, *a.* (and *sb.*) Add: **1. e.** *genetic psychology* (see quot. 1909[1]).
1909 *Cent. Dict.* Suppl. (s.v. *psychology*), *Genetic psychology*, that division of psychology which deals with the development of mind in the individual and with its evolution in the race. **1909** W. M. Urban *Valuation* iii. 72 How such presuppositions arise is..a problem of genetic psychology. **1947** O. Barfield in *Essays presented to C. Williams* 106 What I am talking about is not poetic diction, but etymology or philosophy or even genetic psychology.
f. *genetic fallacy*: the fallacy of judging the value of something, or the truth of a belief, by its origin.
1934 Cohen & Nagel *Introd. Logic* xix. 388 (*heading*) The genetic fallacy. **1941** *Mind* L. 386 The 'scientific method of interpreting Spinoza's philosophy' must avoid both the 'normative fallacy' and the 'genetic fallacy'. **1959** I. G. MacCaffrey *Paradise Lost as 'Myth'* 210 Milton never committed the genetic fallacy which claims that good and evil are rendered indistinguishable when they are seen to have a common source. **1965** *Philos.* XL. 351 To commit a Genetic Fallacy, the fallacy of supposing that an opinion is discredited when its causal origins are revealed.
g. Of or pertaining to genetics or genes; *genetic code*: the system by which nucleic acid molecules store genetic information, now known to operate by means of triplets of nucleotides read in sequence.
1908 W. Bateson *Methods & Scope Genetics* 11 The conception..of the individual as composed of what we call presences and absences of all the possible ingredients ..is the basis of all progress in genetic analysis. *Ibid.* 46 At last by genetic methods we are beginning to obtain such facts. **1936** *Discovery* May 161/1 Recently attention has been paid to..the interaction of genes with, what may be termed, the genetic environment. **1939** C. D. Darlington (*title*) Evolution of genetic systems. **1941** J. S. Huxley *Uniqueness of Man* ii. 42 One and the same genetic outfit will give different effects in different environments. *Ibid.* 43 Our ignorance of the precise genetic constitution of human populations. **1947** H. J. Muller in *Proc. R. Soc.* B. CXXXIV. 30 Practically every mutation, even a 'small' and non-lethal one, with the rarest of exceptions, requires finally a genetic death, that is, a failure to live or to breed, somewhere along the line of its descent, if the population would remain genetically at par. For each mutation, then, a genetic death. **1959** *New Biol.* XXVIII. 23 Diverse non-adaptive evolutionary changes as may be brought about in small isolated populations by the phenomenon of 'genetic drift'. **1961** S. A. Sand in *Amer. Naturalist* XCV. 242 Our concept of a genetic system storing information should not be restricted to the denotation of 'gene' or to limited aspects of DNA structure...We would then recognize more than one type of genetic code and several classes of genetic change. **1962** *Nature* 31 Mar. 1268/1 The correspondence between groups of nucleotides and amino-acids has come to be known as the 'genetic code'. **1963** *Times* 12 Jan. 8/1 New experiments have suggested that viruses may, in effect, be genes or 'genetic messengers', the report says. **1969** *Times* 6 Jan. 10/7 If man is indeed carrying a large number of such inactive or 'nonsense' genes, the implication is that he is using only a small fraction of the genetic information that his D.N.A. is capable of storing. **1969** *New Scientist* 28 Aug. 415/2 The day may be approaching when genetic engineering may make it possible to make a plant to order. **1970** *Guardian* 8 May 3/7 Most couples who seek genetic counselling come after they have had one or more defective children. **1971** *Ibid.* 8 Oct. 13/1 Human genetic engineering aimed at the elimination of genetic diseases.
B. *sb. pl.* **a.** (See sense 3 in Dict.)
b. That branch of biology which is concerned with the study of natural development when not complicated by human interference.
1897 L. F. Ward *Outl. Sociol.* 180 But there is a shorter adjective form *telic*, which is preferable to teleological and possesses the advantage of being converted into the name of a science, *telics*, as proposed by Dr. Small. These two words may be conveniently set over against *genetic* and *genetics*.
c. The scientific study of heredity and variation.
1905 W. Bateson *Let.* 18 Apr. in B. Bateson *W. Bateson* (1928) 93 The best title would, I think, be 'The Quick Professorship of the study of Heredity'. No single word in common use quite gives this meaning..., and if it were desirable to coin one, 'Genetics' might do. **1906** —— in *Nature* 14 June 146/1 May it be suggested that the branch of science should now receive a distinctive name? The physiology of heredity and variation is a definite branch of science...To avoid further periphrasis, then, let us say genetics. **1907** *Daily Chron.* 23 Feb. 9/3 The.. International Conference on Genetics. **1908** W. Bateson *Methods & Scope Genetics* Pref., The physiology of Heredity and Variation, a study now spoken of as Genetics. **1930** R. A. Fisher *Genet. Theory Nat. Selection* p. viii, That an independent study of natural selection is now possible is principally due to the great advance which our generation has seen in the science of genetics. **1949** Darlington & Mather *Elem. Genetics* 15 These inborn causes..have to be defined as materials or processes whose behaviour and effects we can predict and control. This is the aim and scope of genetics. **1965** Peacocke & Drysdale *Molec. Basis of Heredity* i. 3 The development of modern genetics dates only from the rediscovery of Mendel's paper in 1900.

genetical, *a.* (Further examples, in sense of *genetic *a.* 1 g.)
1910 *Amer. Naturalist* XLIV. 108 (*title*) Genetical studies on Œnothera. Notes on the behavior of certain hybrids of Œnothera in the first generation. **1922** *Ibid.* LVI. 59 Genetical and cytological proof was obtained that these intersexes in *Drosophila* possess two X-chromosomes and three sets of autosomes. **1925** *Jrnl. Bot.* LXIII. 360 A greater knowledge of modern genetical work would save taxonomists from falling into some common errors. **1930** R. A. Fisher (*title*) The genetical theory of natural selection. **1959** *Listener* 26 Nov. 920/2 Nucleic acids are the substances that embody genetical information. **1971** *Nature* 30 Apr. 552/3 The pathogen is now open to much more detailed genetical and physiological examination.

genetically, *adv.* Add: **b.** By the agency of genes; according to genetics.
1902 Bateson & Saunders *Rep. Evol. Comm. R. Soc.* I. 134 An organism can be strictly defined as genetically sure if all its gametes when united with similar gametes reproduce the parent identically. **1932** S. Zuckerman *Soc. Life Monkeys* ii. 27 If an impulse to live a herd life is a genetically determined response. **1968** *Times* 17 Oct. 18/5 An attempt to determine whether the capacity of the rats to convert a high proportion of cyclamate into chemicals is inherited genetically.

geneticist (dʒĕne·tisist). [f. *genetic *sb.* c + -ist.] An expert in, or student of, genetics.
1913 in Dorland *Med. Dict.* (ed. 7). **1924** E. W. MacBride *Study Hered.* vii. 186 This theory..uncritically accepted by many 'geneticists'. **1930** G. R. de Beer *Embryol. & Evol.* 20 The geneticist who studies this distribution in hereditary transmission is really following the genes through two cell-generations only. **1932** *Discovery* Mar. 73/2 Plant geneticists learnt how to build up a plant almost to a specification. **1950** A. Huxley *Themes & Variations* i. 74 The issue between Soviet geneticists and the geneticists of the West is similar in essence to that which divided the Pelagians from the Augustinians. **1971** A. E. H. Emery *Elem. Med. Genetics* (ed. 2) ix. 162 We will..limit our discussion to those problems which face the medical geneticist.

genetics: see *genetic *a.* (and *sb.*) B.

Geneva[2]. Add: **Geneva Convention** (see quot. 1910[2]); **Geneva Protocol** (see quot. 1970); **Geneva watch**, a Swiss-made watch; also *ellipt.*
1844 C. M. Yonge *Abbeychurch* ii. 23 Above it hung a small Geneva watch. [**1865** *Parl. Papers* LVII. 473 (*title*) Accession of the British Government to the Convention signed at Geneva, August 22, 1864, for the Amelioration of the Condition of the Wounded in Armies in the Field.] **1880** L. Hertslet *Treaties & Conventions Gt. Brit. & Foreign Powers* XIV. 1164 The following is a list of the Parties who signed, or have acceded to, the Geneva Convention up to this date. **1900** *Daily News* 21 Mar. 5/6 'Explosive bullets'..were..expressly condemned by the Geneva Convention. **1910** A. Bennett *Clayhanger* III. xiv. 422 Edwin..drew out the silver Geneva. **1910** *Encycl. Brit.* XI. 592/1 *Geneva Convention*, an international agreement for the purpose of improving the condition of wounded soldiers of armies in the field, originally adopted at an international conference held at Geneva, Switzerland, in 1864, and afterwards replaced by the convention of July 6, 1906, also adopted at Geneva. This later agreement is the one now known as the Geneva Convention. **1933** [see *bacteriological *a.* b]. **1959** *Listener* 26 Feb. 392/1 The tradition of the Geneva convention. **1970** *Sat. Rev.* 6 June 28/3 The Geneva Protocol of 1925..banned the use in war of all 'asphyxiating, poisonous, or other gases' and of 'bacteriological methods of warfare'.
b. *Mech.* **Geneva drive, mechanism**, etc., a mechanism for converting a continuous rotatory motion into an intermittent rotatory motion, consisting of a driving wheel to which is attached a pin that engages with one of several radial slots in the driven wheel for part of each revolution; also *ellipt.*; **Geneva stop**, a stop (Stop *sb.*[2] 9 d) consisting of a pair of gears in which the driven wheel lacks cogs or slots on part of its circumference.
1873 Geneva stop [see Stop *sb.*[2] 9 d]. **1913** F. A. Talbot *Pract. Cinematogr.* iii. 26 The early types of camera were fitted with what is known as the Geneva stop movement. **1918** F. D. Jones *Mechanisms* vi. 154 The general type of intermittent gearing..is commonly known as the 'Geneva wheel', because of the similarity to the well-known Geneva stop used to prevent the over-winding of springs. **1949** *Tool Engineers Handbk.* (Amer. Soc. Tool Engineers) lxxviii. 1143 A special geneva drive is utilized to impart intermittent motion to the circular electrodes. **1953** L. J. Wheeler *Princ. Cinematogr.* vi. 179 The Geneva Movement, or Maltese Cross,..is now used almost exclusively throughout the industry. **1967** S. B. Tuttle *Mechanisms for Engin. Design* iii. 35 The Geneva mechanism. *Ibid.* 37 The internal Geneva produces intermittent circular motion in which the dwell period is less than 40% of the elapsed time. **1970** R. M. Phelan *Fund. Mech. Design* (ed. 3) xii. 286 The Geneva wheel..is satisfactory for relatively high-speed operation. *Ibid.* 287 The Geneva stop is a useful variation of the

Geneva wheel in which one of the slots in the driven member is left closed, thus permitting only a certain number of revolutions..before the pin hits the closed slot and stops the motion.

Genevese, *a.* and *sb.* Add: **A.** *adj.* (Earlier examples); *spec.* designating a type of sauce for fish.
1826 E. Craven *Mem. Margravine of Anspach* II. ii. 44 His house was built by a Genevese architect named Billion. **1845** E. Acton *Mod. Cookery* Index, p. xv/1 Genevese Sauce, or *Sauce Genevoise*. **1857** C. Brontë *Professor* xix. 67, I have one object before me now—to get that Genevese girl for my wife.

Genevois. Pl. **Genevois.** Delete †*Obs.* and add later examples. Also as *adj.* = Genevese *a.*
1765 J. Conyers *Let.* in E. Hamilton *Mordaunts* (1965) viii. 181 The Genevois being lovers of order and decency. **1845** [see prec.]. **1959** *Times* 17 Nov. 16/5 The year in which the earlier Genevois, Jean Etienne Liotard, so detested by Reynolds, died. **1968** D. Torr *Treason Line* 100 The knot of early comers, all of them prominent Genevois bores.

genic (dʒi·nik, dʒe·nik), *a.* [f. *gene + -ic.] Of or pertaining to genes.
1922 C. B. Bridges in *Amer. Naturalist* LVI. 57 Comparison..between the effects of haploidy for an autosome and the effects normally present in diœcious sex shows that they have similar genic bases—namely, each is due to differences in the ratio between two aggregates of genes. **1925** *Ibid.* LIX. 129 Each character of an individual is the index of the point of balance in effectiveness of a large but unknown number of genes... This conception of 'genic balance' was applied to the sex characters of the intersexes. **1937** *Nature* 11 Sept. 450/2 Differences in chromosome constitution..may..mean a difference in genic balance. **1956** *New Biol.* XX. 32 All developmental processes..are to some extent under genic control.

-genic (dʒe·nik), *suffix* [f. -gen + -ic], forming adjectives with the meaning: 'of, pertaining to, or relating to generation or production'; *spec.* (*a*) generating, producing, as *carcinogenic, *epeirogenic, Pathogenic, Pyrogenic; (*b*) generated, produced, originating in, as *blastogenic, *cryptogenic.
b. After *photogenic *a.* used to form adjectives with the meaning 'well suited for reproduction or dissemination in a particular medium', as in *radiogenic, *telegenic.

genicular, *a.* Delete *rare*$^{-1}$ and add: Also, of or pertaining to the knee or a genu. (Later examples.)
1902 H. Morris *Human Anat.* (ed. 3) v. 756 These fibres..at the genu constitute the genicular bundle. **1913** *Gray's Anat.* (ed. 18) 853 The point where it [*sc.* the facial nerve] changes its direction is named the geniculum; it presents a reddish gangliform swelling, the genicular ganglion, or nucleus of the sensory root of the nerve. **1951** P. Thorek *Anat. in Surgery* xlv. 837 The middle genicular arteries enter the knee joint through the posterior ligament. **1967** G. M. Wyburn et al. *Conc. Anat.* vi. 161/1 The descending genicular artery arises distally and descends to the knee joint.

geniculum (dʒĕni·kiŭlŏm). *Anat.* [L., dim. of *genū* Genu.] A small genu; an angular knee-like or knot-like structure.
1889 *Buck's Handbk. Med. Sci.* VIII. 131/1 The thalami and geniculums project caudad beyond the intersegmental line. **1913** [see prec.]. **1967** Anson & Donaldson *Surg. Anat. Temporal Bone & Ear* II. 115 The fascial canal..passes horizontally lateralward, then bends at a right angle, forming the geniculum.

genin (dʒe·nin). *Chem.* [The ending of Saponin, Saligenin, used as a generic word.] **a.** Any of various steroids that occur as aglycones in certain glycosides present in some plants and toad venoms. **b.** Occas. used as the name of specific compounds.
[**1874** O. Schmiedeberg in *Arch. f. exper. Path. u. Pharm.* III. 24 So bildet sich ein schön krystallisirender Körper, den man in Analogie mit dem Sapogenin Digitogenin nennen kann.] **1915** *Chem. Abstr.* IX. 1336 Genin, $C_{21}H_{30}O_5$, wartlets from 15 parts of b.96% alc. **1925** *Jrnl. Chem. Soc.* CXXVIII. i. 1295 The genin is converted by cold, concentrated hydrochloric acid into 'dianhydrogitoxigenin', identical with digitaligenin from 'Digitalinum verum'. **1927** *Jrnl. Biol. Chem.* LXXIV. 789 The toxicities of the unhydrogenated 'genins' were of themselves not of the highest order. **1938** *Thorpe's Dict. Appl. Chem.* (ed. 4) II. 381/1 All these drugs are nonnitrogenous glycosides of plant origin yielding on hydrolysis various sugars and a number of structurally similar aglycones or 'genins'. **1959** L. F. & M. Fieser *Steroids* i. 4 The glycoside of which digitogenin is the aglycone, or genin, is a saponin. **1964** C. W. Shoppee *Chem. Steroids* (ed. 2) v. 367 Four aglycones: genin B, $C_{21}H_{30}O_4$, genin D, $C_{21}H_{30}O_5$, genin E, $C_{21}H_{34-32}O_5$ and genin F, $C_{21}H_{34}O_5$, isolated from *D. grandiflora*.., are probably digitenols.

Genist (dʒi·nist). [ad. late L. *Genistæ* sb. pl., f. L. *gen-* to be born.] One of a sect of ancient Jews who took no foreign wives during the Babylonian captivity, and who therefore claim to be pure-blooded descendants of Abraham. So **Ge·nite.**
1613 Purchas *Pilgrimage* II. viii. 128 He there nameth ..diuers other sects if they may beare that name: as the Genites or Genists, which stood vpon their stocke and kindred. **1882** F. W. Farrar *Early Chr.* II. 342 Even down to the fifth century there continued to be..'Genists', or Jews by race.

genistein (dʒĕni·ste,in, -stīn). *Chem.* [f. Genist(a + *-ein.] An isoflavone derivative, $C_{15}H_{10}O_5$, that is present in dyer's broom, *Genista tinctoria*, and some other plants and is a weak yellow colouring matter. Also called *prunetol*.
1900 Perkin & Newbury in *Jrnl. Chem. Soc.* LXXV. 833 This new colouring matter, for which the name *genistein* is proposed, crystallises in long, glistening, colourless needles. **1943** A. H. Cook tr. *Mayer's Chem. Natural Coloring Matters* iv. 195 Genistein..is contained in dyer's broom (*Genista tinctoria*), together with luteolin, and in soybeans (*Soja hispida*) as the 7-glucoside. **1953** *Jrnl. Chem. Soc.* 871 Genistein (5:7:4'-trihydroxyisoflavone) is a pro-œstrogen, responsible for most of the œstrogenic activity of subterranean clover.

genitivally (dʒenitəi·văli), *adv.* [f. Genitival *a.* + -ly[2].] In the genitive case; as a genitive.
1893 in *Funk's Stand. Dict.* **1966** G. N. Leech *Eng. in Advertising* xiv. 133 In 'Kellogg's Rice Krispies great new card series', a whole product name 'Kellogg's Rice Krispies' is used genitivally, itself containing a further genitive ('Kellogg's').

genitive, *a.* and *sb.* Add: **B.** *sb.* **genitive absolute**, a construction in Greek similar to the Latin ablative absolute.
1860 W. W. Goodwin *Syntax Greek Verb* vi. 297 We sometimes find the Participle in the Genitive Absolute with ὡς. **1897** A. N. Jannaris *Hist. Greek Gram.* III. 499 The Greek genitive absolute substantially corresponds to the Latin ablative absolute.

ge·nitively, *adv. rare.* [f. Genitive *a.* + -ly[2].] As a genitive.
1846 Monier Williams *Sanscr. Gram.* ix. 161 Dependent Compounds, or Compounds Dependent in Case.. Genitively Dependent. Or those in which the relation of the first word to the last is equivalent to that of a genitive.

genitor[2]. Add: **b.** *spec.* in *Anthropol.* (See quots.)
1949 E. E. Evans-Pritchard in M. Fortes *Social Structure* 86 When, as often happens among the Nuer, the physiological father, the *genitor*, is a different man from the sociological father, the *pater*, his sons will not marry into his minimal lineage. **1963** *Brit. Jrnl. Sociol.* XIV. 24 We use the words *genitor* and *pater* to distinguish between the begetter of a child and its legal father.

geniture. Transfer †*Obs.* to senses 2 to 6 and add later examples to sense 1.
1916 A. Quiller-Couch *Art of Writing* viii. 145 A man's lineage and geniture being reckoned, as a rule, among the things he cannot reasonably be asked to amend. **1931** A. J. Cronin *Hatter's Castle* 439 She had failed him in everything.., in the very geniture of her children.

genizah (geni·zā). Also **geniza.** Pl. **genizoth.** [Heb., lit., a hiding, hiding-place, f. *gānaz* to set aside, hide.] A store-room or repository for damaged, discarded, or heretical books and papers and sacred relics, attached to many synagogues; also, the contents of a genizah.
1897 *Times* 7 Aug. 11/5 The honour of discovering the Genizah belongs to the 'nameless' dealers in antiquities of Cairo. **1903** *Jew. Encycl.* V. 612/1 The term 'genizah' is now applied almost exclusively to the hoard at the old synagogue of Fostat near Cairo. **1926** *Glasgow Herald* 24 Feb. 8 All the old prayer books or torn Bibles are reverently carried to the synagogues, where they are placed in what is called a Genizah...In these ancient Genizahs have been found some priceless old MSS. **1968** *Economist* 10 Aug. 36/1 Devout Jews in the Middle Ages believed that pieces of paper which bore the name of God should not be burned but be put aside in a special room or *geniza*.

gennemic (dʒĕne·mik), *a.* [f. Gr. γέννημα product + -ic.] In phonetics, of or pertaining to speech sounds after they have been uttered. So **genne·mically** *adv.*
[**1921** G. Panconcelli-Calzia *Experimentelle Phonetik* 7 Die experimentelle Phonetik behandelt ihren Stoff zuerst in bezug auf die erzeugende Tätigkeit, d.h. genetisch (ἡ γένεσις), und dann in bezug auf das durch diese Tätigkeit entstandene Erzeugnis, d.h. gennemisch (τὸ γέννημα).] **1949** R-M. S. Heffner *Gen. Phonetics* i. 3 The phonetician, in his effort to delimit and to describe the several distinct constituent speech sounds, can examine the movements of the speech organs which produce the sounds (genetic investigation), or he may examine the sounds as acoustic phenomena after they have been produced (gennemic investigation). **1953** L. F. Brosnahan *Some O.E. Sound Changes* 22 A vowel sound, considered gennemically, consists of a series of 'distinctive frequency regions'. **1967** *Word* XXIII. 254 These terms have been divided into three groups..(*a*) the first column refers to genetic ('articulatory'), (*b*) the second to gennemic (acoustic) and (*c*) the third to energemic (auditory) phonetics.

Genoa. Add: **Genoa jib**, a large jib used on racing yachts. Also *ellipt.*
1932 *Yachting* Aug. 66/2 Genoa jibs, as cut at present, are a doubtful blessing in ocean racing. *Ibid.* 74/2 Genoas and jib topsails were of very little use...One boat, *Vamare*, split her Genoa at the start. **1933** *Rudder* Apr. 51/1 You will find your Snipe class boat greatly improved in both speed and stability if you equip her with the new Genoa jib. **1961** P. Moyes *Sunken Sailor* iv. 50, I want a proper complement of sails. Storm jib, beating jib, a Genoa for reaching and a spinnaker.

genocidal (dʒenŏsəi·dăl, dʒe·n-), *a.* [f. next + -al.] Of, pertaining to, or involving genocide.
1948 *Sat. Rev. Lit.* 3 Jan. 8 Belligerents would probably be producing atomic bombs—unless they found other genocidal weapons. **1951** *Scottish Jrnl. Theol.* IV. 419 A fundamentally dishonest casuistry to justify genocidal warfare. **1958** *Observer* 20 July 21/3 Scientists far more deeply involved in these genocidal problems. **1968** *Sunday Tel.* 20 Oct. 8/4 In Zanzibar this has involved a total and genocidal revolution.

genocide (dʒe·nŏsəid). [f. Gr. γένο-ς (see Genus) + -cide 2.] The deliberate and systematic extermination of an ethnic or national group.
1944 R. Lemkin *Axis Rule in Occupied Europe* ix. 79 By 'genocide' we mean the destruction of a nation or of an ethnic group. **1945** *Sunday Times* 21 Oct. 7 The United Nations' indictment of the 24 Nazi leaders has brought a new word into the language—genocide. It occurs in Count 3, where it is stated that all the defendants 'conducted deliberate and systematic genocide—namely, the extermination of racial and national groups...' **1951** *Amer. Jrnl. Psychiatry* Feb. 595/1 Genocide as defined by the United Nations is the direct physical destruction of another racial or national group. **1962** *Listener* 20 Sept. 452/2 One of the things the seventeenth century never achieved was genocide. **1969** *Peace News* 13 June 4/1 The government are..conducting cultural genocide by destroying this Scottish (Gaidhlig) community.

Genoese, *a.* and *sb.* **A.** *adj.* (Earlier and later examples.)
1741 M. W. Montagu *Let.* 15 Feb. (1966) II. 226 A Genoese Abbé, who..is resolved never to return to Genoa. **1849** *Art Jrnl.* XI. 45/2 One of those light-sailing Genoese boats. **1967** *Times* 22 Mar. 13/7 The last remaining vines of the Coronata valley which used to provide the exact white wine to accompany Genoese fish.
B. *sb.* Also *sing.*, a native or inhabitant of Genoa; also, the Genoese dialect.
1818 W. Marsden tr. *Trav. of Marco Polo* II. lxxvii. 560 There is a diversity of dialect, similar to what is found between the Genoese, the Milanese, the Florentine, and the dialects of other Italian states. **1878** [see *Emilian *a.* and *sb.*]. **1905** [see Tuscanize *v.*]. **1950** E. Whelpton *By Italian Shores* i. 20 Columbus, a Genoese of Jewish origin, discovered America, though he went there on behalf of the king of Spain.

genome (dʒi·nōᵘm). *Biol.* Formerly also **genom** (-nŏm). [a. G. *genom* (H. Winkler *Verbreitung u. Ursache d. Parthenogenesis* (1920) iv. 165), irreg. f. *gen* *gene + *chromosom* *chromosome.] A haploid set of chromosomes; the sum-total of the genes in such a set.
1930 *Cytologia* I. 14 Chromosomes from different sets (or genoms) of *Triticum vulgare* show affinity toward each other. **1930** [see *allopolyploidy]. **1932** *Proc. 6th Int. Congr. Genetics* I. 275 The inviability of deficient genomes in the haploid generation serves to some extent as an alternative distinction between mutation and deficiency. *Ibid.* II. 5 There are two species having genoms resembling *C. neglecta*. **1952** C. P. Blacker *Eugenics* x. 243 The appearance of such terms as gene-complex and genome (denoting a set of chromosomes as a working unity) testify to the movement towards holism in genetics. **1965** A. M. Srb et al. *Gen. Genetics* (ed. 2) vii. 190 Among organisms with chromosomes, each species has a characteristic set of genes, or genome. In diploids a genome is found in each normal gamete. It consists of a full set of the different kinds of chromosomes. **1970** *Sci. Amer.* Oct. 19/1 The human genome..consists of perhaps as many as 10 million genes.

genomere (dʒe·nomiᵊɹ). *Biol. Obs.* exc. *Hist.* [f. *gen(e + -o + *-mere.] A hypothetical component of a gene.
1928 W. H. Eyster in *Zeitschr. f. indukt. Abstamm. u. Vererb.* (Suppl.) I. 666 The gene is an organization of more elementary genetic units, the genomeres...The term genomere was suggested by Dr. P. W. Whiting. **1939** C. H. Waddington *Introd. Mod. Genetics* xvi. 373 Eyster suggested that this might be due to the sorting out of sub-units (genomeres) of which the gene was supposed to consist. **1966** E. A. Carlson *Gene* xxvi. 247 Demerec's enthusiasm for the study of the gene was not discouraged to the point of withdrawal when he was unable to find support for his genomere hypothesis in *Delphinium*.

genonema, genoneme (dʒenŏnī·mă, dʒe·nŏnīm). *Biol.* [f. *gene + -o + Gr. νῆμα thread.] = *chromonema.

1934 N. Koltzoff in *Science* LXXX. 313/1 The size of the chromosomes in the salivary glands is determined through the multiplication of *genonemes*. By this term I designate the axial thread of the chromosome, in which the geneticists locate the linear combination of genes; the cytologists call it generally the 'axoneme' or 'chromonema'. **1937** *Ann. Reg.* 1936 53 Chromosome study of plants tended to confirm the spiral chromonemata theory but giant salivary chromosomes seem to contain ring-shaped chromomeres connected by oblique genonemes. **1965** L. C. Dunn *Short Hist. Genetics* xiii. 137 The logical next step, asking what kind of molecule composed the gene string ('genonema', a new term for chromosome).

genotype (dʒe·nŏtəip), *sb.*[1] *Biol.* [f. Gr. γένος Genus + -type.] The type-species of a genus.
1897 C. Schuchert in *Science* 23 Apr. 639/2 Genotype applies to any typical material of the type species of a genus. **1923** *Times Lit. Suppl.* 15 Mar. 171/2 It is.. necessary to fix on a single species as the standard of each genus, new or old; it is called the genotype. **1926** F. A. Bather in S. S. Buckman *Type Ammonites* VI. 6 In establishing a new genus an author should fix on one species as the genotype (or genoholotype). **1953** E. Mayr et al. *Methods Syst. Zool.* xiv. 265 The International Commission (Paris, 1948) recommends that the term *genotype* not be used because of possible confusion with the same word as used in genetics.

genotype (dʒe·nŏtəip), *sb.*[2] *Biol.* [ad. G. *genotypus* (W. Johannsen *Elem. d. exakten Erblichkeitslehre* (1909) ix. 130), f. *gene + -o + -type.] The genetic constitution of an individual, esp. as distinguished from its phenotype; the sum-total of the genes in an individual or group. Hence **genoty·pic, -ty·pical** *adjs.*; **genoty·pically** *adv.*
1910 *Science* XXXII. 588/2 The general program will consist of a symposium on the subject of 'Genotypes or pure lines of Johannsen'. **1911** [see *gene]. **1922** *Hereditas* III. 211 The genotypical response of the plant species to the habitat. **1925** D. F. Jones *Genetics Plant & Anim. Improvement* xv. 503 No one can injure a breed more than by permitting genotypically inferior animals to enter. **1930** R. A. Fisher *Genet. Theory Nat. Selection* 9 The genotypes are probably unequally fitted, at least to a slight extent, to their task of survival and reproduction. *Ibid.* 81 A change in the genotypic constitution of the species. **1964** G. H. Haggis et al. *Introd. Molecular Biol.* x. 245 It is necessary to distinguish between the assemblage of outwardly recognizable traits by which an individual is defined, known as its phenotype, and the assemblage of inherited factors which determine these traits, known as its genotype. **1971** *Nature* 2 Apr. 279/2 Skin grafts exchanged between genotypically identical siblings survive significantly longer than grafts made between non-identical siblings.

genotype (dʒe·nŏtəip), *v.* [f. prec. sb.] *trans.* To investigate the genotype of.
1961 *Lancet* 9 Sept. 602/1 When the mother has rhesus antibodies..the husband should be genotyped. **1971** *Nature* 2 Apr. 279/2 Some monkey families could be genotyped and the antigens were found to segregate in a pattern consistent with a single complex system.

-genous (dʒĕnəs), *suffix*, [f. -gen + -ous] = *-genic, as (*a*) Alkaligenous, Calcigenous, Kaligenous, Pathogenous, Pyrogenous 2; (*b*) Pyrogenous 1.

genre. Add: **1. b.** *spec.* A particular style or category of works of art; esp. a type of literary work characterized by a particular form, style, or purpose.
1770 C. Jenner *Let.* 5 May in *Private Corr. D. Garrick* (1831) I. 384 With regard to the *genre*, I am of opinion that an English audience will not relish it so well as a more characteristic kind of comedy. **1790** A. Young *Jrnl.* 15 Jan. in *Travels* (1892) I. 301 It is a *genre* little interesting when the works of the great Italian artists are at hand. **1843, 1880** [in Dict., sense 1]. **1856** Geo. Eliot *Ess.* (1884) 84 In every *genre* of writing it [*sc.* wit] preserves a man from sinking into the *genre ennuyeux*. **1882** G. Saintsbury *Short Hist. Fr. Lit.* 50 A better notion of the *genre* may perhaps be obtained from a short view of the subjects of some of the principal of those Fabliaux whose subjects are capable of description. **1967** *Radio Times* 13 Apr. 10/5 *Laike Moussike*, the new genre which in the last eight years has given a new impetus..to Greek popular music.
2. a. (Earlier example.)
1861 C. M. Yonge *Young Step-Mother* xvii. 232 'I used to be very fond of drawing.' '*Genre* is my style.'
b. (Further examples.) Also *transf.*, of music and literature.
1849 *Art Jrnl.* XI. 59/3 This picture is certainly one of the masterpieces of the English school of *genre* painting. *Ibid.* 108/1 Dietz, a genre-painter of merit. **1920** G. B. Shaw *How to become Mus. Critic* (1960) 310 It would be so much easier if Cockaigne were genre music, with the Westminster chimes, snatches of Yip-i-addy, and a march of the costermongers to Covent Garden. **1931** *Times Lit. Suppl.* 21 May 406/3 The story settles down for a time into the methods of a *genre* novel of Shropshire life. **1934** C. Lambert *Music Ho!* iii. 200 They are.. satirical genre pieces—over in a flash, but unerringly pinning down some particular aspect of popular music, whether foxtrot, tango or tarantella. **1937** *Burlington Mag.* Sept. 139/1 Frans Hals's genre-pictures. **1959** *Times* 13 Jan. 3/3 All three are genre-portrait groups.

Genro (ge·nro). *Obs.* exc. *Hist.* [Jap., = principal elders, f. *gen* root + *ro* old.] The 'elder statesmen' of Japan, a former body of retired statesmen who were at times informally consulted by the emperor. Also, a member of this body.
1876 E. M. Satow tr. *Shōzan Yashi's Kinsé Shiriaku* i. 10 He was generally nick-named 'the swaggering Chief Minister' (Bakko Genrô). **1880** E. J. Reed *Japan* I. 364 The second of the governing bodies of the state is the Genro-In (house of seniors), or senate. **1921**, etc. [see *elder statesman* (*elder *a.* 1 c)].

‖ **gens de (la) robe** (ʒaṅdəlarob), *sb. pl.* [Fr., lit. 'men of the (long) robe'.] Lawyers.
1679 H. Savile *Let.* 5 June in *Savile Correspondence* (Camden Soc., 1858) 93 All Protestants are turn'd out of all places except just the *gens de robe*. **1850** C. M. Yonge *Kenneth* xxii. 272 His own future..arrayed itself ..in sombre colours, as regarding English lawyers, with all a young French soldier's contempt for the *gens de la robe*. **1886** *Athenæum* 10 July 47/2 Henri de Mesmes.. was one of not a few *gens de robe* of whom L'Hôpital is the best known. **1967** L. Norton tr. *Mem. St. Simon* ix. 130 The *gens de la robe* (men of the long robe) were despised because they sold their services to the state for money.

gent, *sb.* Add to note: Used in tradesmen's notices.
1901 *Daily Chron.* 10 Sept. 9/1 Hair Dresser.—First-class gent's junior wanted. **1921** *Dict. Occup. Terms* (1927) § 920 Hairdresser's assistant; gents' or ladies' hand.
b. *pl.* = *gentleman 4 e. *colloq.*
[**1933** R. Strachey *Many Happy Returns* I. 41 He let me build the Palace Pier too, and the gents' place on the front.] **1938** G. Greene *Brighton Rock* I. ii. 28 He's just gone into the gents to have a wash. **1958** B. Hamilton *Too Much of Water* viii. 75 The Gents on the lower deck.. consisted of three water-closets, a urinal, a..shower, and two small rooms containing large baths. **1965** *Crescendo* Oct. 7/1 It was confirmed by conversation overheard in the interval along the aisles or in the Gents. **1968** *Listener* 1 Aug. 141/1 The only place you could talk to Harrer was in the gents, and even there not for long.

genteelism (dʒenti·liz'm). [f. Genteel *a.* + -ism.] Genteel behaviour, attitudes or characteristics; *spec.* (see quot. 1926); also, a genteel word or expression; a genteel euphemism.
1908 *Westm. Gaz.* 24 Jan. 3/1 They are the marks of 'genteelism', as distinguished from gentility. **1926** Fowler *Mod. Eng. Usage* 212/2 By *genteelism* is here to be understood the substituting, for the ordinary natural word that first suggests itself to the mind, of a synonym that is thought to be less soiled by the lips of the common herd, less familiar, less plebeian, less vulgar, less improper. *Ibid.*, The small selection of genteelisms offered below. **1934** *S.P.E. Tract* XLIII. 156 A collection of chocolates and sweets to which even the worst enemy of genteelisms could scarcely deny the epithet *recherchée*. **1949** *Archit. Rev.* CV. 248 To dismiss entirely the whole field of publicity when landscaping new towns would seem to be an act of genteelism reminiscent of the days when the designer ignored everything that didn't fall into line with his own private taste. **1957** E. Gowers *H. W. Fowler* 9 When King Richard..asks for a mirror and Bolingbroke tells someone to go and fetch a looking-glass, the usurper was inflicting..the final humiliation of being corrected for using a genteelism. **1960** *Spectator* 19 Aug. 275 There is no end to the flood of genteelism that is eroding the language.

gentian. Add: **1. c.** A bitter liquor made from the root of gentian. Also *gentiane*, *gentian brandy*, *gentian spirit*.
1857 L. Pappe *Floræ Capensis* p. vi, Different kinds of Sebæa..could..serve as a good substitute for Gentian. **1892** *Cornhill Mag.* XVIII. 615 The gentian spirit may be said to be the very elixir of life to the mountain folk. *Ibid.* 616 The smell of gentian brandy is not pleasant, especially if new, but with age the spirit greatly improves, mellows, and loses its disagreeable aroma. **1902** J. T. Law *Grocer's Man.* 408/2 *Gentian Spirit*, a bitter liquor made from gentian root. **1966** P. V. Price *France* 65 A pre-prandial digestive of a herby type, such as gentian or *anisette*. **1967** I. Marder *Paris Bit* ix. 151 'I haven't been able to think of a blue drink. Can you?' 'How about *gentiane*?'
2. b. quasi-*adj.* Having the blue colour of the gentian; gentian-blue.
1908 *Westm. Gaz.* 7 Dec. 2/3 A gentian sky untinged By any haze its noontide blue outstretches. **1967** S. Pakenham *Sixty Miles from England* xv. 204 Lady Blanche, in her white mantilla and pussy-cat bow of gentian silk.

gentianose (dʒe·nʃiănōᵘs, -ōᵘz). *Chem.* [a. G. *gentianose* (A. Meyer 1882, in *Zeitschr. f. physiol. Chem.* VI. 135), f. Gentian + -ose[2].] A trisaccharide found in the roots of various species of gentian.
1883 *Jrnl. Chem. Soc.* XLIV. 811 The other properties of gentianose would appear to place it near cane-sugar. **1948** Kirk & Othmer *Encycl. Chem. Technol.* II. 879 Gentianose may be hydrolyzed by invertase or by dilute acids to D-fructose and gentiobiose.

gentle, *a.* and *sb.* Add: **A. 1. e.** Enchanted or haunted by fairies. Also *gentle people*, the fairies. (Cf. Gentry 3 b.) *dial.*
1823 S. McSkimin *Carrickfergus* (ed. 2) IV. 258 The large hawthorns growing singly..are deemed sacred to fairies, and are hence called *gentle thorns*. **1877** *Cornhill Mag.* Feb. 174 Woe betide the foolhardy person who ventures to raise an axe against one of these 'gentle bushes', as they are called. *Ibid.* 178 All the land was of excellent quality except half an acre of rocky ground, which was 'allowed' to be a very 'gentle place'. **1893** W. B. Yeats *Celtic Twilight* 94 In some more than commonly 'gentle' place. **1907** J. Masefield *Tarpaulin Muster* xiii. 136 There's many places here is gentle. **1913** E. M. Wright *Rustic Speech* xii. 207 Collective names [for Fairies] are: the Fair Folk..the Gentle People, or Gentry (Irel.).
8. Freq. in phr. *a gentle hint.*
1896 J. C. Harris *Sister Jane* 32 Taking this as a gentle hint, I went out. **1922** Joyce *Ulysses* 354 She just gave a gentle hint about its being late.

gentle, *v.* Add: **2. a.** (Additional examples.)
1908 *Daily Chron.* 3 Feb. 3/3 Tragic circumstances that are put with all the vigour of a strong hand, yet gentled by a beautiful thread of love. **1936** 'R. Hyde' *Passport to Hell* 23 Horse-dung trodden into the mire and yet gentled with the smell of warm straw. **1958** *Engineering* 28 Feb. 258/1 Whether a few rounds in the ring did in fact gentle his condition I was never able to discover.
3. *intr.* To become gentle. *rare.*
1912 J. Masefield in *Eng. Rev.* Oct. 388 When the light gentles and the wind is soft. **1953** V. Buchanan-Gould *Vast Heritage* i. 2 For a short time..the African world gentled to soft femineity.
Hence **ge·ntling** *ppl. a.*
1906 *Westm. Gaz.* 10 Sept. 2/3 Above the vale, with its gentling tale Of a life that is lived for right. **1952** E. Hyams *Soil & Civilization* 167 The gentling influence of the Atlantic allows deciduous trees to grow farther north. **1966** *New Statesman* 4 Feb. 172/2 Within his gentling limits, Mr Sucksdorff seems authentic.

gentleman. Add: **4. a.** *spec.* in *Cricket.* A non-professional player (opp. Player 2 c). Also *transf.*
1806 in F. Lillywhite *Cricket Scores & Biographies 1746–1826* (1862) I. 328 This being the first match between the Gentlemen and the Players. **1884** [in Dict.]. **1891** [see Player 2 c]. **1966** *Listener* 25 Aug. 265/2 The social split..which I have described as the amateur and gentleman *versus* the professional and player. **1971** A. Price *Alamut Ambush* xii. 140 That calculated..amateurishness of his—the flouting of the rules to prove that he was a gentleman rather than a player.
e. *pl.* Designation of a public convenience for male persons. Freq. *gentlemen's*, and with capital initial. Cf. *gent *sb.* b.
1929 'H. Green' *Living* xviii. 213 'You go and leave them in the Gentlemen.' 'Leave 'em in the lavatory?' **1933** S. Frazer *Acorned Hog* 78 Over on that platform's the general waiting-room,..and over there's the Gentlemen's, and, any'ow, everythink's written up. **1934** E. Waugh *Handful of Dust* iii. 117 'I tell you what I must do, is to telephone. Where is it?' 'D'you mean really the telephone or the gentlemen's?' **1941** J. Cary *Herself Surprised* xliv. 108 There are quays there and lamps and some squares of grass; a ladies and gentlemen, and a cinema.
5. a. Also a euphemism for a smuggler.
1906 Kipling *Puck of Pook's Hill* 251 Watch the wall, my darling, while the Gentlemen go by! **1929** G. Greene *Man Within* I. iv. 92 'You know who I am?'..'One of the Gentlemen.' **1966** E. & M. A. Radford *Death of 'Gentleman'* xxii. 188 A smuggler..engaged in the Gentlemen's trade.
c. *gentleman of the road*: (*a*) a highwayman; also *ellipt.* as *gentleman*; (*b*) a commercial traveller; (*c*) a gypsy; †*the gentleman that* (or *who*) *pays the rent* (Obs.): a pig.
1728 Gay *Begg. Op.* III. xvi. 57 The fine Gentlemen imitate the Gentlemen of the Road. **1778** J. Wedgwood *Let.* 19 Mar. (1965) 218, I got some knowledge of the gentlemen..and sent some of my people in pursuit of them who soon brought me in two of the robbers. **1809** [see Road *sb.* 5 b]. **1855** Mrs. Gaskell *Squire's Story* in *Novels & Tales* (1873) IV. 228 Kate Hearn's husband collected his rents on the highway, like many another 'gentleman' of the day. **1869** Trollope *Vicar of Bullhampton* v. xxix. 184 They had no coffee room at the Bull, and strangers who came that way were of necessity shown into that in which the gentlemen of the road were wont to relax themselves. **1907** G. B. Shaw *J. Bull's Other Island* iv. 80 The gentleman that pays the rint... They call a pig that in England. **1969** K. Giles *Death cracks Bottle* viii. 96 'A survival of the gentlemen of the road—all little cads..today.'..'I should say..that Mr Rumly is the best in the business of selling wines.' **1969** *Radio Times* 3 Apr. 9/1 Gentlemen of the Road—a study of the Gypsy way of life.
7. a. *gentleman-cadet*; *gentleman-cricketer*, *-player*.
1775 [see Cadet 3]. **1840** in A. Ponsonby *H. Ponsonby* (1942) 19 Every one of the Masters speaks of Gentleman Cadet Ponsonby in very favourable terms. **1953** E. Smith *Guide Eng. Traditions* 7 Up to 1939 those who gained high enough positions in the annual entrance examinations received (at their own expense) special military training as Gentleman Cadets at Sandhurst and Woolwich. **1885** F. Gale *Life R. Grimston* vi. 64 The M.C.C. was a small club..containing only from three to four hundred members, who were..all good gentlemen cricketers. **1823** M. R. Mitford in *Lady's Mag.* July 386/1 A set match at Lord's ground for money..between a certain number of gentlemen players, as they are called—people who make a trade of that noble sport. **1832** P. Egan *Bk. Sports* 346/1 Who can beat twenty-two, with some wickets to spare, Of the gentlemen-players. **1906** A. E. Knight *Compl. Cricketer* ix. 299 The real gentleman player has no love for these miserable.. labels and distinctions.

b. gentleman friend, a beau; a boy-friend.
1829 M. B. Smith *Let.* 27 Nov. in *Forty Yrs. Washington Soc.* (1906) 307 We have at least 6 or 7 young gentlemen friends, who are frequently with us. **1894** Somerville & 'Ross' *Real Charlotte* III. xxxvi. 50 She respected him, an emotion not hitherto awakened by a varied experience of 'gentlemen friends'. **1935** Marsh & Jellett *Nursing-Home Murder* xii. 187, I hope you were not very shocked at what my gentleman-friend said..? **1970** R. Rendell *Guilty Thing Surprised* v. 58 My gentleman friend come in about seven.
c. gentlemen's (or **-man's**) **agreement** orig. *U.S.*, an agreement which is not enforceable at law, and which is only binding as a matter of honour; **Gentleman's Relish**, the proprietary name of a savoury paste (Paste *sb.* 1 d).
1929 Wodehouse *Mr. Mulliner Speaking* vi. 201 What we had better do is to have a gentleman's agreement. **1930** *Times Lit. Suppl.* 27 Feb. 153/3 By a gentleman's agreement—if one may use such a term when speaking of ruffians who now have not even courage to commend them—rival gangs enjoyed the monopoly of trade in different districts. **1931** J. K. Winkler *Morgan the Magnificent* vii. 107 In 1886..was the first of a series of memorable dinner-table conferences. At them were formulated so-called 'gentlemen's agreements'. **1971** Lord Avon in *Listener* 7 Oct. 472/1 I'd made an agreement with Mussolini which he'd dubbed 'the Gentlemen's Agreement', by which we were both going to observe certain conditions about Spain. **1907** *Trade Marks Jrnl.* 17 Apr. 676 *Patum Peperium The Gentleman's Relish*... For Toast, Biscuits, &c. London. Potted Meat. Charles Augustus Osborn..Manufacturer. **1918** H. G. Wells *Joan & Peter* ix. 291 Perhaps a sandwich, Gentleman's Relish or shrimp paste. **1950** *Trade Marks Jrnl.* 6 Dec. 1033/1 *The Gentleman's Relish.* Food pastes consisting principally of anchovy extracts. C. Osborn & Company Limited..Manufacturers. **1966** 'K. Nicholson' *Hook, Line & Sinker* v. 61 There was woodcock pâté, Gentleman's Relish, home-made black cherry jam.

gentlemanism (s.v. Gentleman). (Example.)
1898 G. B. Shaw in *Sat. Rev.* 29 Jan. 139/1 The dream-fed gentlemanism of the age which Shakespear inaugurated in English literature.

gentlewomanliness. (Earlier example.)
1808 M. Wilmot *Jrnl.* 16 Aug. (1934) III. 368 He had felt..the Gentlewomanliness of her manners.

genu. Add: **b.** In certain mod.L. terms, with the sense 'knee', as **genu recurvatum**, 'a backward curvation of the knee-joint' (Dorland 1900); **genu valgum**, knock-knee; **genu varum**, bow-legs.
In classical Latin *valgus*, *vārus* meant respectively 'bow-legged', 'knock-kneed'.
1887 *Encycl. Brit.* XXII. 691/2 During the last few years..other deformities, such as knock-knee or *genu valgum* and bow-leg or *genu varum*, have been remedied by operation. **1910** *Practitioner* Mar. 348 Genu-valgum (with genu-recurvatum). **1936** Bailey & Love *Short Pract. Surg.* (ed. 3) xli. 920 Genu Varum is less common than knock-knee, and is usually due to rickets. **1954** J. Cyriax *Textbk. Orthopaedic Med.* (ed. 2) I. xxi. 559 Nowadays genu valgum due to vitamin-D deficiency is almost as rare as venal rickets. **1968** S. Taylor et al. *Short Textbk. Surg.* (ed. 2) xlii. 563 Patients are rarely disabled by genu recurvatum, and operative treatment is therefore seldom needed.

genuine, *a.*[1] Add: **4. a.** (*the*) *genuine article*: see *article *sb.* 14 d.

genus. Add: **5.** *genus irritabile* (*vatum*) [after Horace, *Ep.* II. ii. 102], the irritable or over-sensitive race or class (of poets).
1721 Swift *Let. to Young Poet* 23 These of your Profession have been call'd *Genus irritabile vatum*. **1818** Keats *Let.* (1958) I. 386 That matter..which is accounted so acceptable in the 'genus irritabile'. **1838** Dickens *Let.* (1965) I. 398 Your great patience is so exemplary and peculiar that I fear you will never make an author, as you can never be one of the *genus irritabile*. **1940** 'N. Blake' *Malice in Wonderland* vii. 86 The Mad Hatter realized he was of the *genus irritabile* of singers, so he doped the piano for him, knowing it would make him fly off the handle.

geo, gio. Add: Also, in wider use: a long, narrow, steep-sided cleft or inlet formed by erosion in coastal cliffs, and typically represented by the geos of Orkney.
1934 E. Linklater *Magnus Merriman* xxxiv. 346 The waves..drove another [*sc.* trawler]..hard ashore in a Westray geo. **1960** Williamson & Boyd *St. Kilda Summer* x. 102 The route rises to the right following a series of ledges into the geo. **1961** L. D. Stamp *Gloss. Geogr. Terms* 208/1 *Geo*..has been adopted by geomorphologists to describe coastal clefts, often marking joints, faults or dikes from which material has been removed by wave action. **1970** R. J. Small *Study of Landforms* xii. 449 In plan the cliffs are usually complex and regular, with inlets and geos developed along joints and faults.

geo-. Add: **geobota·nic** *a.*, geobotanical; **geobo·tany** = Phytogeography; so **geobo·tanist**; **ge:ochrono·metry**, (*a*) an extension of geometry conceived as taking time into account as the fourth dimension; the 'geometry' of space-time; (*b*) absolute geochronology, in which events are assigned (approximate) dates in relation to the present instead of to other events; **geocoro·na**, an envelope of gas surrounding the earth, resembling the sun's corona and consisting chiefly of ionized hydrogen; **geocra·tic** *a.* [Gr. -κρατία rule + -ic], (*a*) applied to earth-movements which reduce the area of the earth's surface covered by water: opp. *hydrocratic *a.*; (*b*) of or pertaining to the predominant influence of the natural environment on man; **ge:odyna·mics**, the study of geodynamic forces; **ge:omorpho·geny**, the science dealing with the genesis of the physical features of the earth's surface; so **ge:omorphoge·nic** *a.*, **ge:omorpho·genist**; **ge:oplana·rian** [L. *plānus* flat + *-arian], one who believes the earth to be flat, a 'flat-earther'; **geopote·ntial**, the work that must be done against gravity to raise unit mass to a given point from sea level; **ge·osphere**, any of the more or less spherical concentric regions that together constitute the earth and its atmosphere; **geosta·tionary** *a.*, of, pertaining to, or designating an artificial satellite that revolves round the earth in one day and hence remains above a fixed point on the earth's surface; **geostra·tegy**, strategy as applied to the problems of geo-politics, 'global strategy'; hence **ge:ostrate·gic(al)** *adjs.*; **geota·xis** *Biol.*, a taxis (see Taxis 6) in which the external stimulus is the force of gravity; so **geota·ctic** *a.*; **geote·chnic** *a.*, of or pertaining to geotechnics; **geote·chnics**, the art of modifying and adapting the physical nature of the earth to the needs of man; **geotechno·logy**, 'the application of scientific methods and engineering techniques to the exploitation and utilization of natural resources (as mineral resources)' (Webster 1961); **geothermal** (later examples).
1904 *Pop. Sci. Monthly* May 71 The immense region.. on geo-botanic maps..has not the uniformity which one would be inclined to attribute to it. **1901** *U.S. Dept. Agric. Bur. Plant Industry Bull.* III. 18 The most thorough investigations have been given to the Chernozem soils by Russian geo-botanists. **1960** *Times* 24 Sept. 19/2 A geobotanist has been included..on all major geological expeditions. **1904** *Pop. Sci. Monthly* May 68 (*title*) The geology and geo-botany of Asia. **1956** *Nature* 17 Mar. 520/2 Geobotany, geochemistry and geophysics of all the central and eastern African lakes. **1923** C. D. Broad *Sci. Thought* xii. 457 A sense-history and the physical world are both four-dimensional spatio-temporal wholes, and we must therefore talk of their geo-chronometry rather than their geometry. **1949** *Mind* LVIII. 219 The alternative 'geochronometries' developed in mathematical physics convey little or no information about the ultimate nature of time. **1960** *Bull. Geol. Soc. Amer.* LXXI. 223/1 Despite the major advances in the technique of geochronometry the establishment of the absolute age of the geological, i.e. paleontological, time scale has proceeded very slowly. **1970** *Nature* 2 May 473/1 In practice.. potassium-argon geochronometry does not always reveal the initial age of crystallization. **1960** *Aeroplane* 8 July 53/1 Analysis of data obtained from the Soviet space-probe, Lunik II,..has shown that the Earth is enveloped in a 'geo-corona' of ionized gas. **1962** *New Scientist* 12 July 94/3 This region composed essentially of protons..is called variously the exosphere, hydrogen geocorona or.. the magnetosphere. **1969** *Nature* 20 Dec. 1187/1 The solar Lyman-α radiation..is scattered by the geocorona into the dark hemisphere of the Earth. **1898** *Geogr. Jrnl.* Feb. 133 Hydrocratic and geocratic movements alternated during Jurassic times. **1951** G. Taylor *Geogr. 20th Cent.* i. 5 Humboldt..thus developed what I have been accustomed to call a '*Geocratic*' type of geography, which suggests that the earth (i.e. Nature) itself plays a great part in determining the type of life which develops in a particular area. **1885** *Nature* 22 Oct. 609/2 Full scope was given to seismology, vulcanology, and geodynamics. **1958** A. E. Scheidegger (*title*) Principles of geodynamics. **1896** *Nature* 18 June 147/1 After the geomorphogenic introduction, two lessons are given to geological principles. **1904** *Amer. Geologist* Mar. 159 Very few of the geomorphogenists have carried their new science forward into a geographical relation. **1894** A. C. Lawson in *Univ. Calif. Bull. Dept. Geol.* I. viii. 241 (*title*) The geomorphogeny of the coast of Northern California. **1909** W. B. Scott *Introd. Geol.* (ed. 2) 435 It would be an advantage in clearness and precision of nomenclature, if Geomorphogeny could be substituted [for Physiography or Physiographical Geography]. **1930** *Proc. Arist. Soc.* XXX. 114, I am thinking, say, of the earth as flat, as when I want to refute a geoplanarian. **1914** V. Bjerknes in *Q. Jrnl. Meteorol. Soc.* XL. 161 It should be borne in mind that in dynamical meteorology gravity-potential (or *geo-potential* as it is now proposed to call it) has to be used as a *co-ordinate*. **1939** *Meteorol. Gloss.* (*Met. Office*) (ed. 3) 97 The zero of potential is taken as at sea level...Points with the same geopotential may be said to be at the same level. By using geopotential rather than height for specifying the position of parts of the atmosphere, the consideration of the air movements is simplified. **1970** *Nature* 9 May 494/2 The theory of the determination of the geopotential from satellite tracking data is not fully understood. **1898** W. J. McGee in *Nat. Geogr. Mag.* IX. 436 The atmosphere..is one of the geospheres, the outermost of the four. *Ibid.* 437 The earth has an interior portion much denser than the known exterior; and this.. may conveniently be called a centrosphere—the innermost of the four geospheres. **1913** J. Murray *Ocean* x. 227 Our earth..is composed of concentric spheres or shells of matter in the gaseous, liquid, and solid or 'trans-solid' states. These have been called Geospheres, viz., the atmosphere, the hydrosphere, the lithosphere, the biosphere, the tektosphere, and the great centrosphere. **1961** *Aeroplane* CI. 16/2 Raising a communication satellite from a low-circular orbit into a geostationary orbit at 22,300 miles. **1967** *New Scientist* 25 May 456/3 How difficult, if not impossible, it is to ensure that a so-called geostationary satellite is truly stationary. **1968** *New Statesman* 13 Dec. 828/1 The existence of geo-stationary satellites and the enormous investment in world-wide communications will increase the flow of information and disseminate it on a scale that almost defies the imagination. **1944** G. B. Cressey *Asia's Lands & Peoples* ii. 32 The function of geostrategy is to understand a nation's problems and potential and to suggest a program of internal development and international cooperation that will be of mutual value. **1957** *Encycl. Brit.* X. 182H/1 In theory a branch of geopolitics,..geo-strategy treated warfare as total, embracing the entire populations and resources of the contesting states...It helped to make Germany the first country to realize that airpower could take a position alongside seapower and landpower. **1958** *New Statesman* 26 Apr. 517/1 Pearl Harbour, indeed, provides the point of departure for American geostrategy. **1899** *Natural Sci.* Apr. 329 The negatively geotactic organism should become positively geotactic in solutions of greater specific gravity than its own. *Ibid.*, The tendency that some Infusorians have to collect near the surface of the water in which they live has been regarded as a reaction to the force of gravity,—a negative geotaxis. **1908** C. Davenport *Exper. Morphol.* v. 117 On warm days the typical geotactic phenomena are often absent. **1962** *New Scientist* 6 Dec. 545/2 Common observation indicates that moths fly towards a light ('positive phototaxis') and flies climb up a window pane ('negative geotaxis')... The response of the fruitfly *Drosophila melanogaster* to gravity ('geotaxis') was tested in an ingenious vertical maze. **1964** *Oceanogr. & Marine Biol.* II. 476 The marked geotactic orientation of lamellibranch molluscs is well known. *Ibid.*, The animals orient with a positive geotaxis in the sand or mud. **1914** Geddes & Thomson *Sex* x. 241 Our aims are not only synthetic, as men-philosophers say, but applied—that is geotechnic, as with practical women, who, as the anthropologists confess, had the first word in cultivation. **1924** *Glasgow Herald* 15 Nov. 4/2 Man.. indulges in big geotechnic operations such as cutting a Panama Canal. **1927** A. Defries *Interpreter* ix. 217 Neotechnics has its physical science, geotechnics its vital sciences, its synthetic aims. **1968** *New Scientist* 19 Sept. 607/1 The geotechnics division of the station is well-known for its studies of civil engineering enterprises. **1942** *Sci. News Let.* 12 Dec. 370 A new word, '*geotechnology*', has been coined to include all the mineral arts and sciences from metallurgy to ceramics. **1961** *Times* 7 Mar. 2/5 Engineering Pedologist/Geologist..with..postgraduate research experience in pedology, sedimentary geology or geotechnology. **1940** *Chambers's Techn. Dict.*, 374/1 *Geothermal gradient* (Mining), the rate at which the temperature of the earth's crust increases with depth. **1955** *Times* 5 July p. ii/3 An interesting possibility is the exploitation of geothermal steam—steam generated naturally below the surface of the earth. *Ibid.* 14 July 14/6 The large heavy water plant which will form part of the scheme for using the steam from the geothermal springs in New Zealand. **1971** *Nature* 29 Jan. 300 An unsuccessful attempt to develop geothermal power in a known geothermal area.

geochemistry (dʒi,oke·mistri). [f. Geo- + Chemistry; cf. G. *geochemie* (C. F. Schönbein 1838, in *Ann. d. Physik u. Chem.* XLV. 281).] The chemistry of the earth; the study of the chemical composition of the earth.
1903 *Carnegie Inst. Yearbk.* 1902 27 The problems of geophysics and geochemistry involve the applications of pure physics and pure chemistry from the minutest parts of the earth to the mass of the earth as a whole. **1952** *New Biol.* XIII. 122 The science of geochemistry, with its emphasis on the dynamic exchange of molecules and ions between lithosphere, hydrosphere, biosphere, and atmosphere.
So **geoche·mical** *a.*, of or pertaining to geochemistry; **geoche·mically** *adv.*; **geoche·mist**, an expert in, or student of, geochemistry.
1888 *Jrnl. Soc. Chem. Industry* VII. 338 (*title*) Geochemical studies. **1918** Webster Add., Geochemist. **1946** *Nature* 6 July 31/1 Some conclusions which may be of interest to oil- and water-geologists, geophysicists and geochemists. **1957** *Times* 20 Dec. 13/1 Developments in geochemical prospecting—the use of chemical analyses of soil to detect underlying deposits of minerals. **1957** G. E. Hutchinson *Treat. Limnol.* I. viii. 559 This Rb: K ratio is by no means geochemically impossible. **1965** G. J. Williams *Econ. Geol. N.Z.* x. 156/2 Miss X. F. Williams has reported copper and nickel geochemically in this region. **1970** *Nature* 17 Oct. 273/2 Geochemical studies indicate that the weathering of 100 tons of average continental surface rock supplies about 0·9 ton of sodium to the hydrological cycle.

geochronology (dʒi:okrŏnǫ·lŏdʒi). [f. Geo- + Chronology.] The chronology of the earth; the measurement of geological time and the ordering of past geological events. (The term *geochrone* introduced by Prof. H. S. Williams to designate a unit of geological time does not seem to have been widely adopted.)
1893 H. S. Williams in *Jrnl. Geol.* I. 294 In all these studies in which the geological time-scale is applied to the evolution of the earth.., the time concerned is not human chronology but is what may be called geochronology.

For this purpose we need a standard time-unit or geochrone. **1934** *Discovery* Mar. 66/2 The high upper limits are supported by the geochronology of the Swedish geologist, de Geer. **1957** G. Clark *Archæol. & Society* (ed. 3) v. 133 Geochronology, the chronology based on the natural changes recorded in the geological sequence, depends on many branches of natural science. **1965** F. J. Monkhouse *Dict. Geogr.* 1/2 *Absolute age*, in geochronology, the dating of rocks in actual terms of years.

Hence **ge:ochronolo·gical** *a.*, of or pertaining to geochronology; **ge:ochronolo·gically** *adv.*; **ge:ochrono·logist**, an expert in, or student of, geochronology.

1934 *Antiquity* VIII. 245 A geochronological investigation of the ice-lake sediments. **1936** *Proc. Prehist. Soc.* II. 169 The absolute geochronological scale which has been established by Scandinavian workers. **1958** F. E. Zeuner *Dating Past* (ed. 4) 4 There are several geochronological methods, each capable of covering not more than a limited range of time. *Ibid.* iv. 109 Fromm's (1938) geochronologically dated pollen-diagrams from Angermanland provide the remainder of dates in the Scandinavian sequence, and Welten's work in Switzerland may become important as a second pollen-time-scale. **1960** *New Scientist* 14 July 137/3 The latest method in the repertoire of the geochronologist is the rubidium-strontium method. **1970** *Nature* 24 Oct. 320/1 The matching of discrete geochronological zones across the boundaries of continents thought to be adjacent before the onset of continental drift. *Ibid.* 320/2 Other pre-drift configurations have not been so well documented geochronologically.

geodesic, *a.* Add: *geodesic curve*, a geodesic line on a curved surface; also *ellipt.* as *geodesic*; *geodesic dome*, a dome built according to the principles of geodetic construction (see quot. 1959) enunciated by the American designer and architect, R. Buckminster Fuller (b. 1895).

1881 Geodesic curve [in Dict.]. **1950** [see next]. **1959** *Times* 20 Mar. 3/5 The geodesic dome combines the structural advantages of the sphere (which encloses the most space within the least surface, and is strongest against internal pressure) with those of the tetrahedron (which encloses least space with most surface and has the greatest stiffness against external pressure). **1960** R. W. Marks *Dymaxion World of B. Fuller* 58/1 In modern geometry, as we have seen, any arc of a great circle is called a 'geodesic'. **1968** *Listener* 26 Sept. 394/1 The key invention that emerged from these speculations was the geodesic dome..made up of a triangulated web of short struts. He developed the geodesic principle with countless experimental structures.

geodetic, *a.* and *sb.* Add: **A.** *adj.* **b.** *geodetic construction* (see quot. 1950); *geodetic dome* = **geodesic dome*.

1936 *Jrnl. Aeronaut. Sci.* Dec. 78/1 The multitudinous fittings employed in the geodetic construction..add much to the weight and cost of the structure. **1950** *Gloss. Aeronaut. Terms (B.S.I.)* 3105 *Geodetic construction*, a method of making curved space frames in which the particular structural members follow geodesics in the surface, the curves being designed in such a manner that the forces set up in the members are either tension or compression. **1958** *New Scientist* 15 May 27/2 The airship embodied Wallis's first essay in geodetic construction, and it was a complete success. **1959** *Times Lit. Suppl.* 6 Nov. p. xxx/5 With his geodetic domes and his synergetic geometry..he [*sc.* Buckminster Fuller] is designing ..methods of enclosing space that others may one day make into an architecture. **1970** K. Platt *Pushbutton Butterfly* (1971) viii. 84 The house..was a series of geodetic domes clustered igloo-like beneath the main house.

Geoffroy (ʒo·frwa). [tr. *Felis geoffroyi* (D'Orbigny & Gervais 1844, in *Extr. Proc.-Verb. Soc. Philomathique Paris* 40), f. the name of Étienne *Geoffroy* Saint-Hilaire (1772–1844) or his son Isidore (1805–1861), both French zoologists.] *Geoffroy's cat*: a South American species of cat, *Felis geoffroyi*.

1870 *Proc. Zool. Soc.* LIII. 796 A fine specimen of Geoffroy's Cat (*Felis geoffroii*),..obtained from Paraguay. **1906** *Westm. Gaz.* 30 Apr. 4/3 A nice example of Geoffroy's Cat may now be seen in the Small Mammals' House at the Zoo. **1961** G. Durrell *Whispering Land* (1965) v. 124 It was a baby Geoffroy's cat, a small species of wild cat which is getting increasingly rare in South America. Its basic colouring was a pale fawny yellow, and it was dappled all over with neat, dark brown spots.

geography. Add: **1. b.** *linguistic geography* (see *linguistic *a.*); *dialect geography* (see *dialect 2 b).

d. *the geography* (*of the house*), the arrangement and position of the rooms, staircases, and other internal features of a house; hence as a jocular euphemism for lavatory, water-closet. *colloq.*

1864 C. M. Yonge *Trial* II. xiii. 239 The little gentleman showed himself minutely acquainted with the whole geography of the house, knew all the rooms and pictures. **1920** 'Sapper' *Bull-Dog Drummond* ii. 65 He wanted to get the geography of the house firmly imprinted on his mind. **1927** R. Graves *Lars Porsena* (ed. 2) 68 For a man to show a woman the way to the lavatory..an evasive phrase had to be used:..'Have you been shown the geography of the house?' **1930** A. Lyall *It isn't Done* 59 It is all very baffling for the uninitiated foreigner..who when his host offers to 'show him the geography of the house' finds that his tour begins and ends with the smallest..room. **1958** 'A. Bridge' *Portuguese Escape* ix. 154 Will you excuse me if *I* show you the geography?.. There—that's the gentlemen's bathroom. **1963** L. Meynell *Virgin Luck* v. 117 'That's the bathroom,' she said, explaining the upstairs geography of the place, 'with the loo just beyond it.' **1967** *Listener* 21 Dec. 802/2 The Business Man Jocular: 'I say, where's the geography, old son?'

geologic, *a.* Add: **1. a.** *geologic time* = *geological time* (see next). Chiefly *U.S.*

1955 *Sci. Amer.* Mar. 82/1 Throughout geologic time, as far back as we can read the rocks, the continents have repeatedly been invaded by the oceans. **1968** Dayhoff & Eck *Atlas Protein Sequence & Structure 1967–68* 74/1 *Geologic time*, a time scale related to, and comparable with, the age of the earth. It is sometimes used vaguely to refer to times of many millions of years, sometimes specifically referring to the dated occurrences of various geologic events or eras.

geological, *a.* Add: *geological time*: time as measured in terms of geology; also, the time which has elapsed since the formation of the earth, or the stretch of time between the formation of the earth and the beginning of the historical period.

1875 J. Croll *Climate & Time* xix. 326 It is..impossible to form an adequate conception of the length of geological time. **1904** Goodchild & Tweney *Technol. & Sci. Dict.* 253/2 *Geological time*, a chronological measure (of a somewhat indefinite character) which bears the same kind of relation to the measures of time used in history that the distances of the stellar bodies do to the ordinary standard of terrestrial measurement. **1911** *Jrnl. Chem. Soc.* C. II. 570 The Measurement of Geological Time. **1923** L. D. Stamp *Introd. Stratigr.* i. 5 Geological time is divided into five great Eras, and into a number of Periods. **1971** S. Smith in K. Dick *Ivy & Stevie* 50 It is heaven to think of geological time.

geomagnetism (dʒiˌomæ·gnĕtiz'm). [f. Geo- + Magnetism.] The study of the magnetic properties of the earth and related phenomena; terrestrial magnetism.

1938 S. Chapman in *Terrestr. Magn.* XLIII. 321 Workers on the science of the Earth's magnetism.. should regularly adopt the title *geomagnetism*, and the corresponding adjective *geomagnetic*, in place of the more usual *terrestrial magnetism* and *terrestrial magnetic*. **1940** Chapman & Bartels (*title*) Geomagnetism. **1956** *Encycl. Brit. Ann.* 354/2 Geomagnetism has broad and basic implications in the study of the ionosphere, radio-wave propagation, [etc.]. **1968** W. M. Kaula *Introd. Planetary Physics* iii. 137 The fields..are..represented as vector spherical harmonics, with spheroidal (in geomagnetism called 'poloidal') and toroidal parts.

So **geomagne·tic** *a.*, of or pertaining to geomagnetism; **geomagne·tically** *adv.*; **geomagneti·cian, geoma·gnetist**, an expert in, or student of, geomagnetism.

1903 *Sci. Amer. Suppl.* 19 Sept. 23177/1 The geomagnetist Paulins..essayed to supply the..organs of the plants with this life and vigor-giving excitement. **1904** *Nature* 21 Apr. 581/1 Everyone who has ever been engaged in geomagnetic investigations. **1938** *Geogr. Jrnl.* XCII. 386 The position of the Earth's magnetic axis when the distorting effects caused by local conditions of the Earth's crust are eliminated..is known as the Geomagnetic Pole. **1955** *Sci. Amer.* Sept. 152/1 As a famous geomagnetician of the early 19th century, Christopher Hansteen, truly said: 'The earth speaks of its internal movements through the silent voice of the magnetic needle.' **1956** *Nature* 7 Jan. 29 (*heading*) Observations of whistling atmospherics at geomagnetically conjugate points. **1970** *Ibid.* 23 May 740/1 The contribution of ionospheric currents should predominate at both stations on days which are not geomagnetically disturbed.

geometric, *a.* Add: **c.** Designating or pertaining to a style of English architecture preceding or corresponding to the decorated style (see Decorated *ppl. a.* b).

1889 *Cent. Dict.* s.v. *decorated*, The Decorated style has been divided into two periods: namely, the Early or Geometric Decorated period, in which the ornament consists especially of simple curves and lines and combinations of them; and the Decorated style proper. **1899** R. Glazier *Man. Hist. Ornament* 39 Decorated or geometric period. *Ibid.*, The aisle windows with mullions and bold geometric tracery. **1957** *Times Lit. Suppl.* 25 Oct. 636/3 So far as architecture is concerned the book covers three periods, Early English, Geometric, and Early Decorated.

d. Designating a period, the 'Geometric Age', of ancient Greek culture lasting from *c* 900 to *c* 700 B.C., or objects belonging to that period, esp. the pottery, characterized by the use of geometrical forms in decoration.

1902 *Encycl. Brit.* XXV. 572/1 In the remains of the Geometric Age we may trace the influence of the Dorians. *Ibid.* 572/2 (*caption*) Geometric vase from Rhodes. **1939** J. D. S. Pendlebury *Archaeol. Crete* vi. 319 The Geometric pottery of Crete never attained the high standard of the Attic school. *Ibid.* 323 The most probable dating for the end of the true Geometric Period is about 700 in the East, and about 750 in Crete. **1950** H. L. Lorimer *Homer & Monuments* p. viii, The salient features of the Late Geometric age with which the Early Iron Age terminates had become known..through the exploration of the Dipylon cemetery. *Ibid.* ii. 70 Vases in the shape of a pomegranate are fairly frequent in Greek Geometric art.

geometrical, *a.* Add: **1. a.** *spec.* = *geometric *a.* c.

1817 T. Rickman *Architecture* 74 The figures..are all worked with the same moulding, and do not always regularly join each other, but touch only at points. This may be called geometrical tracery. **1848** [see Decorated *ppl. a.* b]. **1849** E. Sharpe *Treat. Decorated Window Tracery in Eng.* I. ii. 8, I propose..to name these three styles of Window tracery, Geometrical, Curvilinear, and Rectilinear. *Ibid.* II. i. 89 Towards the close of the Geometrical Period there occurred some singular attempts at originality in the designs of Window Tracery. **1875** *Encycl. Brit.* II. 425/2 Edward I, 1272 to 1307. Transition from Early Pointed to Complete, or Geometrical Pointed. **1942** N. Pevsner *Outl. Europ. Archit.* iii. 51 The kind of tracery which is called flowing as against the geometrical tracery of 1230 to about 1275.

c. *geometrical optics*: the branch of optics which deals with the geometrical analysis of the paths of light in refraction and reflection.

1838 W. N. Griffin *Treat. Optics* i. 1 In Geometrical Optics the circumstances of the transmission and modification of light are computed on certain laws established by experiment; in Physical Optics these laws are accounted for on hypotheses of the structure of bodies. **1936** *Discovery* Nov. 364/1 The Physical Society's report on the teaching of Geometrical Optics.

geometrid, *a.* and *sb.* (Earlier and later examples.)

1876 A. S. Packard (*title*) A monograph of the Geometrid moths or Phalænidæ of the United States. *Ibid.* 37 The times of appearance of our geometrids. **1963** V. Nabokov *Gift* i. 30 The four lovely gauze wings Of the softest Geometrid moth in the world. *Ibid.* ii. 109 A tropical geometrid coloured in perfect imitation of a species of butterfly infinitely removed from it in nature's system.

geometrize, *v.* Add: (Earlier and later examples.)

1603 Holland *Plutarch's Mor.* 768 The said matter.. refusing to be thus geometrized, that is to say, reduced to some finit and determinate limits. **1661** Blount *Glossogr.* (ed. 2) sig. S6v/2 To *geometrize*, to play the Geometrician, to hold a due proportion, to observe order. **1931** E. H. W. Meyerstein *Let.* 28 Oct. (1959) 137 Even to try to geometrize *faith* was the inspiration of a Prophet.

Hence **geo:metriza·tion.**

1927 *Observer* 13 Nov. 15/5 Geometrisation is carried much further by Mr. Claude Flight in his Futurist essay entitled 'Holland'..a jumble of curves, straight lines, and acute or obtuse angles. **1931** *Jrnl. Philos.* XXVIII. 19 The real basis of Meyersonian causalism is in geometrization...He seems satisfied..if he can show relativity to be a 'reduction' of physics to geometry. **1933** *Antiquity* VII. 365 The progressive geometrization to which their ceramic decoration is subject.

geometry. Add: **1. b.** Applied to the relative arrangement of objects or constituent parts, as specified by geometrical quantities.

1933 H. B. Howard *Stresses in Aeroplane Structures* viii. 175 The loads in such frames do not depend on the elastic properties of the members, provided only that the extensions of those members do not materially alter the geometry of the frame. **1934** *Physical Rev.* XLV. 598/1 Earlier results..showed the same trend.., but with a slightly lower value of collision area, as would be expected from the geometry. **1955** *Rev. Sci. Instr.* XXVI. 126/1 This was a 'good geometry' experiment, so that an aperture of, say, 90 percent of the total opening gave a photocell reading of 80 percent. **1962** *Jrnl. Geophys. Res.* LXVII. 5077/2 Next, we consider the geometry of the experiment...The spin axis of the spacecraft..was at an angle of approximately 52·50° to the *z* axis...At the same time, the angle between..the projection of the spin vector in the *xy* plane and the positive *x* direction was 43°. **1970** *Sci. Jrnl.* Apr. 56 (*caption*) Leyland 2S/350R powerplant is here illustrated in the form of exterior and cutaway views, slightly simplified but having the correct geometry. **1971** *E. Afr. Standard* (Nairobi) 10 Apr. 8/6 Suspension is fully independent all round, incorporating automatic ride-height control and anti-lift geometry.

geomorphic, *a.* Restrict *nonce-wd.* to sense in Dict. and add: **2.** Of or pertaining to the natural features of the earth's surface; geomorphological.

1893 [see next]. **1894** *Bull. Dept. Geol. Univ. Calif.* I. 242 The present geomorphy is the result of..the advance in the new geomorphic cycle to a stage of late adolescence or early maturity. **1934** *Jrnl. R. Anthropol. Inst.* LXIV. 344 Geomorphic studies in Uganda reveal an arid climate in late Pliocene times. **1954** W. D. Thornbury *Princ. Geomorphol.* i. 11 His [*sc.* W. M. Davis's] concept of the geomorphic cycle..is the idea that in the evolution of landscapes there is a systematic sequence of land forms. **1969** *Nature* 15 Mar. 1005/2 These [projects] include..the preparation of geomorphic, geological and vegetation maps.

Hence **geomo·rphically** *adv.*

1969 *Nature* 15 Mar. 1005/1 Geomorphically, Aldabra consists of a peripheral narrow ridge 8 m above sea level, with a lower terrace cut into its seaward face.

geomorphology. Substitute for def.: The branch of geology dealing with the origin, evolution, and configuration of the natural features of the earth's surface or a particular region of it.

1893 W. J. McGee in *Congrès Géol. Internat. 1891* 199 The systematic examination of land forms and their interpretation as records of geologic history introduces a new branch of geologic science, called 'physical geography' or 'physiography' by different writers, which has been designated 'geomorphic geology' by Powell and the 'new geology' or 'geomorphology' by the writer; but the term 'geomorphy', first employed in a somewhat different connection by Sir William Dawson, though never extensively used with this meaning, is preferable. **1898** J. Geikie *Earth Sculpture* p. viii, Prof. A. De Lapparent's *Leçons de Géographie Physique*—a most instructive and comprehensive outline of geomorphology. **1931** L. D. Stamp in W. Rose *Outl. Mod. Knowl.* 818 The underlying structure, in so far as it determines the character of the land surface, is obviously important, and at the present day forms a subject of its own known as geomorphology. **1932** J. A. Steers (*title*) The unstable earth. Some recent views in geomorphology. **1946** *Nature* 31 Aug. 300/1 He has made considerable contributions to the geomorphology of Wales. **1946** F. E. Zeuner *Dating Past* IV. xi. 339 The reaching by an area of the senile stage of geomorphology requires many millions of years. **1969** *Nature* 15 Mar. 1005/1 The results of the first eight months of the expedition..chiefly concerned..the distribution of terrestrial groups, with basic studies of geomorphology, climate and vegetation.

Hence **ge:omorpholo·gical** *a.*, of or pertaining to geomorphology; **ge:omorpholo·gically** *adv.*; **geomorpho·logist**, an expert in, or student of, geomorphology.

1896 *Nature* 26 Nov. 76 (*title*) Geomorphological speculation. **1928** *Funk's Stand. Dict.*, Geomorphologist. **1935** *Geogr. Jrnl.* LXXXVI. 98 The group of glaciers which are geomorphologically characterized by extending in a continuous sheet in which the ice moves outwards in all directions. **1936** *Ibid.* LXXXVIII. 301 Geographers, geologists and geomorphologists..want accurate maps. **1946** F. E. Zeuner *Dating Past* III. v. 113 Ridges of pressure-moraines are formed which sometimes are most prominent geomorphological features. **1960** B. W. Sparks *Geomorphol.* i. 1 The geomorphologist does not study all aspects of the evolution of landforms. *Ibid.* 3 Geomorphological laboratories help greatly in understanding some processes. **1970** *Nature* 4 July 97/1 French geomorphologists are very interested in the effect of climate on landforms.

geomorphy. Add: (Later examples.) So **geomo·rphist**, a geomorphologist.

1893 [see *geomorphology]. **1894** [see *geomorphic *a.* 2]. **1904** *Amer. Geologist* Mar. 175 The geomorphist who is satisfied with the study of land forms as a finality.

geonoma (dʒiˌǫ·nŏmă). *Bot.* [mod.L. (C. L. Willdenow 1804, in *Mémoires de l'Académie Royale des Sciences de Berlin* 37), f. Gr. γεωνόμος a colonist, in allusion to its rapid propagation.] A tree of the genus of small palms so named, natives of South and Central America.

1849 R. Spruce *Notes of Botanist on Amazon & Andes* (1908) I. i. 11 The roofs were..made by tying several of the broad flat fronds of a small palm called Ubím (Geonoma) on to a stick so as to closely overlap each other. **1871** *Jrnl. Linn. Soc. Bot.* XI. 104 The fine Geonomas discovered by Wendland in Central America are not always represented in the Kew Herbarium by specimens perfect enough to enable one to classify them with certainty. **1910** *Daily Chron.* 12 Mar. 8/4 Other graceful plants for our rooms are the Latania, the Grevillea, the Kentia, the Geonoma, the Rhapis, and the Dracæna. **1966** E. J. H. Corner *Nat. Hist. Palms* xii. 307 If one wanted to study thoroughly the minutiae of palm-evolution.. *Geonoma* should be the choice.

geophone (dʒi·ofōun). [f. Geo- + *-phone.] A device or instrument used to detect vibrations such as sound-waves or shock-waves in the ground.

1919 *Engin. & Min. Jrnl.* 17 May 872/1 The geophone, a 'listening' instrument invented by the French to detect enemy sapping and underground mining operations..is now being used by the Bureau of Mines..in establishing the location of miners who have been entombed after a disaster. **1922** *Encycl. Brit.* XXXII. 526/2 The *geophone* is an instrument for direction-finding of sounds proceeding through the earth, and its particular use during the war was for localizing the sound of picks, etc. used in tunnelling and land mining. **1933** *Discovery* Dec. 375/2 The geophones..such as were used by the sappers in the war to localize enemy saps. **1953** *Sci. News* XXIX. 15 Ashore, use is made of moving coil instruments, called geophones, which respond to the velocity with which the ground surface moves after an explosion. **1965** *New Scientist* 4 Feb. 271/1 The largest seismic array in the world..will contain 525 geophones (a kind of seismometer).

geophysical (dʒiˌofi·zikăl), *a.* [See Geo- and cf. *geophysics.] Of or pertaining to geophysics.

1888, 1894 [see Geo-]. **1946** *Nature* 28 Dec. 931/2 (*title*) Geophysical prospecting and English oilfields. **1955** *Sci. Amer.* Sept. 49/1 During the Geophysical Year scientists of many disciplines will observe all the large-scale aspects of the earth: its interior, the oceans and glaciers, the lower and upper atmosphere, gravity and magnetism, and the extraterrestrial forces which profoundly influence these interacting features. **1970** *Earth-Sci. Rev.* VI. 275 Geophysical zones (e.g., electric-log zone, velocity zone, radioactivity zone).

geophysicist (dʒiˌofi·zisist). [f. next + -ist.] An expert in, or student of, geophysics.

1903 *Amer. Jrnl. Sci.* Dec. 402, I shall, leaving the consideration of the pre-fossil period to the geo-physicist, mention but briefly several associated theories of an astronomical or physical nature. **1934** *Times* 18 Sept. 9/1 The Swedish geophysicist who introduced one method of locating metals by electrical means. **1957** *Technology* Dec. 359/3 Work carried out by..a geophysicist..made use of an optical technique for measuring the temperature in..the upper atmosphere. **1970** *Sci. Jrnl.* May 32/3 The geochemical theories..do not seem to meet the geophysicist's requirement that extra mass must find its way into the basins to give a gravity excess.

geophysics (dʒiˌofi·ziks). [f. Geo- + Physics.] The science or study of the physics of the earth, esp. of its crust; the application of the principles, methods, and techniques of physics to the study of the earth.

1889 in *Cent. Dict.* **1906** *Nature* 12 July 258/1 The problems of geo-physics..are..intimately connected with each other. **1929** Eve & Keys (*title*) Applied geophysics in the search for minerals. **1964** *Oceanogr. & Marine Biol.* II. 37 This is usually the case for other types of density current in geophysics.

geophyte (dʒi·ofəit). *Bot.* [ad. G. *geophyt* (F. W. C. Areschoug 1896, in *Acta Univ. Lundensis* XXXI. 1), f. Geo- + -phyte.] A plant that produces its perennating buds below the surface of the ground. Hence **geophy·tic** *a.*

1900 B. D. Jackson *Gloss. Bot. Terms* 110/2 *Geophytes*, applied by Areschoug to those plants which produce underground buds, without partial development there. **1913** *Jrnl. Ecol.* I. 17 Cryptophytes include plants with their dormant parts subterranean in the case of geophytes with bulbs, rhizomes, tubers on stem and root, and root-buds. **1952** P. W. Richards *Tropical Rain Forest* i. 10 While in temperate forests the herbaceous ground flora includes a large proportion of..geophytes, in the rain forest..geophytes are only rarely present. **1956** *Nature* 4 Feb. 214/2 The Pteridophyta..include only such plants as show a geophytic habit or are aquatic annuals.

geo-politics, geopolitics (dʒiˌopǫ·litiks). [ad. Sw. *geopolitisk* geopolitical (R. Kjellén 1900, in *Ymer 1899* XIX. 283), f. Geo(graphy + Politics.] **a.** The influence of geography on the political character of states, their history, institutions, and esp. relations with other states; also, the study of this influence. **b.** A pseudo-science developed in National-Socialist Germany, which regarded the state as an organism with 'powers of action independent of and superior to the human groups or individuals who constitute it' (*Encycl. Brit.*, 1957).

1904 E. Reich *Found. Mod. Europe* i. 8 History, in Europe, and still more outside Europe, is written largely, if not wholly in characters of that geography, or, as we prefer to call it, geo-politics, that has..determined the trend and tenor of decisive events. **1931** *Geogr. Jrnl.* LXXVIII. 186 The rôle he assigns to 'geopolitics' is..the estimation of the strength and direction of geographical, economic, and political influences in the past, and their probable development in the future. **1942** H. Beukema in A. Dorpalen *World of Gen. Haushofer* p. xii, Geopolitics is the science of the earth relationships of political processes. It is based on the broad foundations of geography, especially on political geography... Geopolitics sets out to furnish the tools for political action and the directives for political life as a whole. Thus, geopolitics becomes an art, namely, the art of guiding practical politics. **1961** *Listener* 16 Mar. 475/2 In terms of geo-politics, China and the United States are less compatible powers than the Soviet Union and the United States.

So **geo-poli·tical** *a.*, of or pertaining to geo-politics; **geo-poli·tically** *adv.*; **geo-politi·cian**, a supporter or exponent of geo-politics.

1902 E. Reich in *Monthly Rev.* Nov. 124 Geographical, or rather geo-political, influences of an abiding character. *Ibid.* 131 History is made by minorities of men, if by majorities of geo-political factors. **1938** *German Life & Lett.* II. 261 Geo-politically the South Tyrol under Austrian rule, represented an enormous fortified bastion. **1942** A. Dorpalen *World of Gen. Haushofer* i. 16 Germany had to be schooled geopolitically. *Ibid.* ii. 75 Geopolitical action is to rely on a careful, detailed analysis of any given political situation. This implies a minute study of geographic, social, economic, and many other factors. **1942** D. Whittlesey *German Strategy of World Conquest* 96 This attribute the geopoliticians claim for Nazi Germany with eager pride. **1945** *Times* 21 Aug. 3/2 He emphasized Norway's geopolitical significance in the Scandinavian sphere. *Ibid.* 15 Dec. 4/7 The Nuremberg Tribunal to-day entered upon the maze of geo-political theories by which the Nazi State..attempted to change the face of the conquered territories of Europe. **1946** 'G. Orwell' *James Burnham* 15 If totalitarianism triumphs and the dreams of the geo-politicians come true, Britain will disappear as a world power. **1957** J. Masters *Far, Far the Mountain Peak* 227 That mission was being studied in secret on a geopolitical level.

Geordie. Add: Also **Geordy, Jordie. 2. d.** A native or inhabitant of Tyneside. Also in slightly *transf.* senses. Also *attrib.* or as *adj. colloq.* **e.** A Scotsman. *Austral.* and *N.Z. colloq.*

1866 C. Nordhoff *Young Man-of-War's Man* iv. 69 The sailors belonging to the ports on the north-eastern coast of England are called Jordies. **1872** T. & G. Allan *Tyneside Songs* (1891) 416 Where's a' his funny sayin's, that set a' the Geordies in a roar? **1890** 'R. Boldrewood' *Miner's Right* I. ix. 227 Whose yer friend; a Geordie, most like? **1892** R. O. Heslop *Northumberland Words* I. 196 The men who went from the lower Tyneside to work at the pits in South Tynedale were always called 'Geordies' by the people there. **1943** *Amer. Speech* XVIII. 89 [In New Zealand] a Scotsman is a *Geordie*, and an Irishman, as in vulgar American, a *Mick*. These synonyms are also current in Australia. **1955** 'C. H. Rolph' *Women of Streets* iv. 56 A large rough Geordie woman in her mid-thirties. **1959** 'M. Ainsworth' *Murder is Catching* 11 He had a faint Geordie twang. *Ibid.* 12 'Have a heart,' said the Geordie. **1971** *Listener* 12 Aug. 201/2 There's a people's culture in Geordieland—it was the last place in England to have its own circuit of music halls.

George. Add: **6.** Also simply *George!*

1888 A. C. Gunter *Mr. Potter* i, George! isn't it horribly lonely?

b. Colloq. phr. *let George do it*: let someone else do the work or take the responsibility. orig. *U.S.*

1910 *Bookman* May 293/2 What's going to happen when Lovey asks papa to hold Snookums and that hitherto devoted parent replies, 'Let George do it.' **1942** Wodehouse *Money in Bank* (1946) xvi. 140 He was not familiar with the fine old slogan, Let George Do It. **1948** *Chicago Tribune* (Grafic Mag.) 10 Oct. 8/1 Producers have a way of saying 'Let George do it' whenever a particularly difficult villain role turns up. **1971** P. G. Wodehouse in *N.Y. Times Encycl. Almanac 1971* 448 It is the old, old story. Overconfidence. We tell ourselves, 'Oh, I can't be bothered getting a divorce. They'll be plenty without me. Let George do it.'

c. Used as a familiar form of address to a stranger; *spec.* in Services' slang: (*a*) an airman; (*b*) an automatic pilot in an aircraft.

1925 Fraser & Gibbons *Soldier & Sailor Words* 104 *George*, a colloquial term for an airman, corresponding to 'Jack' for bluejacket, and 'Tommy' for a soldier. Also used sometimes in the Air Force in addressing any stranger. **1931** *Flight* 30 Oct. 1082 'George' is the automatic pilot about which there has been so much talk lately. **1943** C. H. Ward-Jackson *Piece of Cake* 31 In the early days of the Great War soldiers and sailors came to refer to any airman as 'George'...Airmen thus came to address any strange fellow by the same name. **1961** *Shell Aviation News* No. 282, 5/2 Even if powered controls do all the physical work, if 'George' does all the 'steering',..the test pilot is the man who bears the responsibility, and is still irreplaceable.

7. George Cross, George Medal, decorations for gallantry in civilian life instituted by King George VI; **George guinea** = sense 4 b.

1905 *Westm. Gaz.* 25 May 7/3 The finding of George guineas. **1940** *N. & Q.* 5 Oct. 235/1 In the broadcast.. which he delivered on Sept. 23, His Majesty the King.. announced that..he has decided to create at once a new mark of honour for men and women in all walks of civilian life. 'I propose,' he said, 'to give my name to this new distinction, which will consist of the George Cross, which will rank next to the Victoria Cross, and the George Medal for wider distribution.' **1941** H. Brighouse in J. W. Marriott *Best One-Act Plays of 1940* 293 *Philip.* You can take it as official that you'll be recommended for the George Cross. *Alfred.* Crimes! That's the civilian V.C. **1959** *Daily Tel.* 24 Apr. 13/4 (*heading*) George Medal man gaoled. **1966** *Whitaker's Almanack* 772/1 Malta was..besieged in the last war and again withstood the attacks...In recognition of the part played by the Maltese people, King George VI awarded the George Cross to the island.

georgette (dʒǫɹdʒe·t, ʒǫɹʒe·t). [Fr., f. the name of Mme *Georgette* de la Plante, French modiste.] A thin, semi-transparent, plain-woven crêpe made from fine, hard-twisted silk or other yarns. Also *georgette crêpe.*

1915 *Sphere* 22 May p. vi/1 A deep yoke and front composed of plain forget-me-not blue *chic* Georgette. **1920** *Glasgow Herald* 29 Nov. 15 Silk georgette. **1922** *Daily Mail* 10 Nov. 8 The bride's gown of gold embroidered georgette. **1927** A. Miller *Colfax Bk.-Plate* xiii. 161 My negligée..was made of georgette and satin. **1960** *Which?* Feb. 42/1 *Georgette*, fine, light, open, plain woven fabric with a crêpe yarn. Often silk, but other fibres as well.

Georgian, *a.*[1] (and *sb.*[2]). Add: **A.** *adj.* **1.** *spec.* Of or resembling the style of architecture, esp. domestic architecture, characteristic of the reigns of the first four Georges (1714–1830). *Georgian group*, the name of a society formed to advocate the preservation of examples of this style.

1875 Geo. Eliot *Let.* 14 Aug. (1955) VI. 165 Our house here is rather a fine old red brick Georgian place. **1896** B. & B. F. Fletcher *Hist. Archit.* 280 At this time, there grew up a vernacular style, most of the less important houses being erected in the useful and modest Queen Anne and Georgian type of square house. **1940** N. Mitford *Pigeon Pie* vi. 99 They were going to pull it down and build a block of flats. (The Georgian Group, wrapped in dreams of Federal Union, stirred in its sleep on hearing this.) **1945** J. Summerson *Georgian London* v. 53 With the general adoption of parapet-roofs and sash-windows with recessed frames, the characteristic Georgian town house had arrived. **1961** W. Gérin *B. Brontë* xvi. 236 A double-fronted Georgian house in the centre of the village. **1967** *Observer* 9 July 32/3 She wants to buy a house in Regent's Park, St. John's Wood or Chelsea—modern, but in Georgian style.

b. Belonging to or characteristic of the reign of George V (1910–36) or VI (1936–52).

In literary criticism applied *spec.* to writers or writing of the first years of George V's reign, and esp. to contributors to five anthologies of 'Georgian Poetry' published between 1912 and 1922.

1910 P. Gibbs in *Lady's Realm* July 272 Under the new regime of Georgian England. **1912** E. H. Marsh (*title*) Georgian Poetry 1911–1912. **1921** R. Lynd in *Anthol. Mod. Verse* Introd. p. xxx, The Georgian poets, like the Lake poets, are re-establishing the claim of familiar experiences to poetical treatment in familiar language. **1924** V. Woolf *Mr. Bennett & Mrs. Brown* 19 The Georgian writer had to begin by throwing away the method that was in use at the moment. **1927** [see *Edwardian *a.* 3]. **1963** J. I. M. Stewart *Eight Mod. Writers* i. 14 The Georgian poets..included a writer of unique vision in Walter de la Mare, a genius in D. H. Lawrence, and a dark horse in Robert Graves.

3. *Georgian green* (see quot. 1949).

1934 *Historical Colours* (Thos. Parsons & Sons) 39 It is likely that William and Mary as well as Georgian greens were derived..from the vert pomme or apple green of our French neighbours..as intercourse in those times (*circa* 1757) was greater than generally credited. **1942** J. Cary *To be a Pilgrim* lxxvi. 171 She showed me the paint. 'Is this the colour, uncle. Georgian green they called it.' **1949** *British Colour Council Dict. Colours Int. Decoration* III. 12 *Georgian green*, one of the most typical of the slightly yellowish greens found in wall decoration, hangings and woven upholstery fabrics of many kinds during the period covered by the first four Georges, Kings of England, 1714–1830.

B. *sb.* One belonging to the time of the Georges, kings of England; *spec.* a writer of the early part of the reign of George V (see A. 1 b above).

1901 *Sketch* 28 Aug. 249/2 The arid stucco of the unimaginative Georgians and Early Victorians. **1913** R. Brooke *Let.* 24 July (1968) 493 Send photographs of a Georgian or two..to Canadian & American papers, when the book [*sc.* *Georgian Poetry*] goes for review. **1916** *Q. Rev.* Oct. 386 From such a catastrophe, the humour, commonsense, and the artistic judgment of the best of the new 'Georgians' will assuredly save their generation. **1924** V. Woolf *Mr. Bennett & Mrs. Brown* 4 Mr. Forster, Mr. Lawrence, Mr. Strachey, Mr. Joyce, and Mr. Eliot I will call the Georgians. **1943** *English Studies* XXV. 1 After the War the Georgians still commanded a large public.

Georgianism (dʒǫ·ɹdʒiăniz'm). [f. Georgian *a.*[1] + -ism.] The qualities or characteristics of Georgian architecture, poetry, etc.

1914 R. Brooke *Let.* 6 July (1968) 597 Marsh has got Georgianism on the brain, & will shortly issue a series of Georgian poker-work. **1929** *Daily Tel.* 15 Jan. 9/2 (*caption*) The best traditions of Georgianism inspire this most modern of..dining-rooms. **1933** *Scrutiny* Dec. 300 It isn't the wordy debility of Georgianism that needs reacting against to-day. **1954** *Essays & Studies* VII. 83 'Georgianism' has for some time now been a term of implicit rebuke. **1970** *English Studies* LI. 271 That most fitting showpiece of Central Georgianism, his 'Weekend' sequence.

geostrophic (dʒiˌostrǫ·fik), *a.* [f. Geo- + Gr. στροφή a turning, f. στρέφειν to turn + -ic.] Of, pertaining to, or caused by the Coriolis force (see *Coriolis); usu. applied to a wind (or a current of water) in which there is a balance between the Coriolis force and the horizontal pressure gradient (cf. *gradient wind* s.v. *gradient *a.* and *sb.* B. 6).

1916 N. Shaw in *Nature* 4 May 210/1 The motion of air ..is persistent because the pressure-gradient is balanced by the centrifugal action of the earth's rotation, which we may call the geostrophic component, and of the curvature of the path over the earth's surface, which we call the cyclostrophic component. *Ibid.* 211/2 We call the wind computed according to the first term the geostrophic wind, and regard it as generally representing the actual wind of temperate and polar regions. **1928** D. Brunt *Meteorol.* vii. 66 The value of the gradient wind which is derived when the curvature of the path is neglected is called the *geostrophic wind*. **1933** *Jrnl. R. Aeronaut. Soc.* XXXVII. 314 The ground wind is rarely the same in direction as the geostrophic wind which blows at 1,500 feet above the surface, and in which the airman is more likely to be flying. **1957** *Times Survey Brit. Aviation* Sept. 7/3 Up to about 20,000 ft. the classical 'geostrophic balance' between the pressure and motion fields holds fairly well. **1966** [see *gradient wind* s.v. *gradient *a.* and *sb.* B. 6]. **1970** *Sci. Jrnl.* Mar. 56/1 In most ocean currents the horizontal pressure gradient is found to be in balance with coriolis force. Such currents are called geostrophic or 'Earth-balanced' currents.

geosyncline (dʒiˌosi·ŋkləin). *Geol.* [Back-formation from Geosynclinal *sb.*] = Geosynclinal *sb.*

1895 J. D. Dana *Man. Geol.* (ed. 4) 389 A mountain system includes all ranges in a region made in different, more or less independent, geosynclines at the same epoch. **1925** J. Joly *Surface-Hist. Earth* v. 85 The geosyncline wherein the Cordilleras of N. America were cradled extended from the Gulf of Mexico to the Arctic Ocean. **1954** M. P. Billings *Struct. Geol.* (ed. 2) iii. 55 A geosyncline is a large basin in which many thousands of feet of sediments accumulate. **1965** A. Holmes *Princ. Physical Geol.* (ed. 2) xxx. 1122 The modified form *geosyncline* came into general use.

gephyrocercal (dʒe:firo,sɜ̄·ɹkăl), *a.* *Ichthyol.* [f. Gr. γέφυρα bridge + κέρκος tail + -al.] Having the caudal fin formed by the meeting of the dorsal and anal fins. So **gephyrocercy** (dʒe:firo,sɜ̄·ɹsi), gephyrocercal condition.

1886 J. A. Ryder in *U.S. Comm. Fish Rep. 1884* 983 *Gephyrocercy* and *gephyrocercal* are terms which will apply to the type of caudal structure appearing normally in only a few forms, such as *Mola* and *Fierasfer*. *Ibid.* 991 Gephyrocercy.—This type of tail appears to be normally met with in only two forms, of Teleosts, namely, *Mola* and *Fierasfer*. *Ibid.*, The interval so bridged by a secondary process of development leads to the formation of what we may call a gephyrocercal tail. **1939** *Nature* 18 Mar. 455/2 The gephyrocercal tail with protruding axial supplement, the normal first dorsal,..and the form of the dermal armour of the head, are all typically cœlacanthid. **1955** E. H. Colbert *Evol. Vertebrates* v. 68 Both these modern genera [of lungfishes] are characterized by gephyrocercal tails.

Geraldton wax (dʒe·rāldtǫn wæks). *Austral.* [f. *Geraldton*, a town in W. Australia + Wax *sb.*[1]] An evergreen shrub of Western Australia, *Chamælaucium uncinatum*, of the family Myrtaceæ. Also *attrib.*

1920 *Jrnl. & Proc. R. Soc. W. Austral.* VI. 42 Geraldton Wax-plant, Ornamental shrub. **1921** E. H. Pelloe *Wildflowers W. Austral.* iv. 42 From Geraldton..southward almost to Fremantle blooms..the well-known 'Geraldton Wax-flower'. **1934** *Bulletin* (Sydney) 14 Nov. 26/3 The Geraldton wax especially can be trimmed to any shape to get a better effect, since the blooms are carried along the stems. **1955** A. Ross *Australia 55* 44 The foggy mauve of Geraldton wax.

geranial (dʒĕrē̆·niăl). *Chem.* [a. G. *geranial* (F. W. Semmler 1891, in *Ber. d. Deut. Chem. Ges.* XXIV. 201), f. *geraniumaldehyde.*] A fragrant colourless oil present in many essential oils (as lemon-grass oil) and used extensively in perfumery as citral, of which it is the *trans* form; it is the aldehyde, $C_{10}H_{16}O$, corresponding to geraniol.

1899 *Jrnl. Chem. Soc.* LXXVI. I. 67 Crystals are obtained which yield geranial..when they are freed from oil. **1953** E. H. Rodd *Chem. Carbon Compounds* IIB. xii. 498 The aldehyde citral..occurs in two geometrical forms, citral-a (geranial) and citral-b (neral).

geraniol (dʒĕrē̆·niǫl). *Chem.* [a. G. *geraniol* (O. Jacobsen 1871, in *Ann. d. Chem. und Pharm.* CLVII. 234), f. Gerani(um + -ol.] A colourless oil with a fragrance of roses that is present in many essential oils (as geranium oil and rose oil) and is used extensively in perfumery; it is an open-chain primary alcohol of the terpene series, $C_{10}H_{18}O$.

1871 *Jrnl. Chem. Soc.* XXIV. 261 By repeated fractional distillation..a liquid was obtained..called by the author geraniol. **1906**, etc. [see *citronellal*, *-ol* s.v. *citronella]. **1923** *Nature* 29 Dec. 952/1 Of active principles, australol, geraniol, citral, and piperitol gave coefficients of 22·5, 21, 19·5, and 13 respectively. **1949** E. Guenther *Essential Oils* II. 173 Geraniol serves extensively in the perfume, cosmetic, soap, and flavor industries. **1971** *Sci. Amer.* May 6/2 Chemical stereoisomers, whose molecules are mirror-images of one another, often taste and smell different from one another. For example, geraniol smells like roses while its stereoisomer smells like fresh oil.

gerbera (dʒɜ̄·ɹbĕră, gɜ̄·ɹbĕră). *Bot.* [mod.L. (J. F. Gronovius 1737, in Linnæus *Corollarium Generum Plantarum* 16), f. the name of T. *Gerber* (died 1743), German naturalist.] A plant of the genus so named, of the family Compositæ; *spec.* the Transvaal daisy. Also *gerbera daisy.*

1889 *Garden* 12 Oct. 341/1 The Gerbera may be planted outside for the summer. As a pot plant it is perfectly charming. **1949** H. E. Bates *Jacaranda Tree* i. 6 Fringes of heliotrope and gerbera daisy, mauve and pink and orange. **1956** J. Hearne *Stranger at Gate* xv. 119 He got up and shifted the hose..to a bed of gerberas. **1960** [see *Barberton].

gerenuk (ge·renŭk). Also **geranook, gerenook.** [Somali name.] A long-necked gazelle-like antelope of East Africa, *Litocranius walleri.*

1895 H. G. C. Swayne *17 Trips through Somaliland* 312 The Gerenuk is the commonest and most widely distributed of the Somali Antelopes. **1895** [see *dik-dik]. **1915** Roosevelt & Heller *Life-hist. Afr. Game Animals* II. xviii. 610 The gerenuk was first described by Sir Victor Brooke [in 1878]. **1965** *New Scientist* 26 Aug. 505/1 Arid-country species like gerenuks never need to drink water. **1970** *Ibid.* 24 Sept. 639/2 A famous zoo breeding group of gerenuk or giraffe-necked antelope are now failing to be as productive.

geriatric (dʒeriˌæ·trik), *a.* [f. as next.] Of or pertaining to geriatrics; designed for use by old people.

1926 *Med. Rev. of Revs.* XXXII. 193/1 We will publish a special geriatric number which will cover the field of diseases of old age. **1947** *Times* 20 June 5/4 Small, long-stay annexes, under the direct control of the 'geriatric' departments, for patients in need of constant nursing care. **1956** W. Hobson *Mod. Trends in Geriatrics* xiv. 345 The organization of a geriatric unit requires..a flexible attitude of mind towards the rearrangement of the hospital. **1963** *Times* 16 Feb. 5/6 A new and deceptive form of restraint which has appeared in certain mental hospitals to fill the vacuum left by the strait-jacket—the geriatric chair.

geriatrics (dʒeriˌæ·triks). [f. Gr. γῆρας old age + ἰατρικός (see Iatric *a.*), after Pædiatrics.] The branch of medicine, or of social science, dealing with the health of old people.

1909 I. L. Nascher in *N.Y. Med. Jrnl.* 21 Aug. 358/2 Geriatrics, from geras, old age, and iatrikos, relating to the physician, is a term I would suggest as an addition to our vocabulary, to cover the same field in old age that is covered by the term paediatrics in childhood. **1919** M. W. Thewlis (*title*) Geriatrics. A treatise on senile conditions, diseases of advanced life, and care of the aged. **1939** *Nature* 25 Nov. 902/2 Courses in geriatrics should be established in the medical schools to give future physicians a better understanding of the effects of mental attitudes on the bodily ailments of the aged. **1967** *New Scientist* 19 Jan. 160/2 There is in Britain only one university chair in geriatrics, and none in gerontology.

Hence **geriatri·cian**, an expert in geriatrics.

1934 in Webster. **1968** *Guardian* 9 May 7/2 Most progressive geriatricians now want to keep old people out of hospital beds at all costs.

germ, *sb.* Add: **5.** *germ-caused, -free, -proof* adjs.

1931 J. S. Huxley *What dare I Think?* i. 10 The numerous brood of germ-caused diseases. **1933** D. L. Sayers *Murder must Advertise* iv. 69 We spend our whole time asking intimate questions of perfect strangers... 'Are you *Sure* that your Toilet-Paper is Germ-free?' **1951** M. McLuhan *Mech. Bride* 61/1 When scrubbed, deloused, germ-free,..then she is lovely to love. **1969** *Times* 28 Jan. 6/7 The apparatus for the germ-free birth..is designed to make possible a cure for a deficiency disease of the body's immune defence system. **1902** *Daily Chron.* 12 July 5/2 A germ-proof house is the latest addition to the hygienic terrors of life. **1940** *Illustr. London News* 23 Mar. 376/1 (*caption*) Nurses in germproof clothing sealing the bottles.

6. germ bomb, a bomb containing germs, used in germ warfare; **germ-centre** [tr. G. *keimcentrum* (W. Flemming 1884, in *Arch. f. mikrosk. Anat.* Sept. 55)], a group of pale-staining cells forming the central area of a lymphatic nodule; **germ-line** = **germ-track*; **germ-plasm,** add to def.: *spec.* [tr. G. *keimplasma* (A. Weismann *Über die Vererbung* (1883) 15)] that part of the germ-cell which, according to Weismann's theory of heredity, bears the factors determining the transmission of characters from parent to offspring and is itself transmitted unchanged from generation to generation; = Id, Idioplasm; also *germ-plasma*; **germ-track** (see quot. 1949); **germ warfare,** the deliberate dissemination of disease-germs among an enemy as a weapon of war; = **bacteriological warfare.*

1934 *Archit. Rev.* LXXV. 98/2 But I visualize poison, explosive, incendiary and germ bombs in an air-fleet attack. **1954** *Ann. Reg. 1953* 133 American airmen who, as prisoners, had confessed to using 'germ bombs' had retracted their confessions. **1898** E. A. Schäfer *Essentials Histology* (ed. 5) xxv. 152 Active multiplication of the lymph-cells by karyokinesis is..the cause of the formation of nodules in the tissue...On this account the nodules are sometimes termed *germ-centres*. **1906** *Practitioner* Nov. 708 The tonsils are composed of lymphoid tissue, which, besides being diffused throughout the whole organ, is collected at various intervals into small nodules, in which the lymphocytes are closely packed... These nodules are sometimes spoken of as *germ-centres*. **1956** Yoffey & Courtice *Lymphatics, Lymph & Lymphoid Tissue* (ed. 2) v. 269 There is increased activity of the lymphoid tissues in response to bacterial injections, and..the main site of the reaction is the germ centre. **1925** E. B. Wilson *Cell* (ed. 3) iv. 312 In some animals the germ-cells are set aside from the somatic cells at a very early stage in the ontogeny, so that we can actually trace their line of ontogenic descent or *germ-line* backwards to early stages of development. **1968** R. Rieger *Gloss. Genetics & Cytogenetics* 204 *Germ line*, a lineage of 'generative' cells..ancestral to the gametes..which, during the development of an organism (particularly in animals), are set aside as potential gamete-forming tissues. **1889** S. Schönland tr. *Weismann's Continuity of Germ-Plasm* in *Ess. upon Heredity* 168, I propose to call it the theory of 'The Continuity of the Germ-plasm', for it is founded upon the idea that heredity is brought about by the transference from one generation to another, of a substance with a definite chemical, and above all, molecular constitution. I have called this substance 'germ-plasm'. **1890** Billings *Med. Dict.* I. 577/2 *Germ plasma*, the substance from which the sexual products are developed. According to the theory of Weismann, a portion of the ovum is preserved, unaltered in constitution, to give rise to the sexual bodies, while the rest of the ovum is altered to produce the tissues of the body. **1893** Germ-plasm [see Id]. **1905** *Westm. Gaz.* 30 Mar. 2/1 The germ-plasm is the hereditary substance of the germ-cell, or the primary constituent substance of the whole organism. **1957** *Encycl. Brit.* XIII. 608/1 The continuity of the germ-plasm..suffices to account for the inheritance of the great majority of characters common to all individuals of a species without the need of postulating contributions from the body-cells. **1893** Parker & Rönnfeldt tr. *Weismann's Germ-Plasm* vi. 184 The transmission of the germ-plasm from the ovum to the place of origin of the reproductive cells..takes place in a regular manner, through perfectly definite series of cells which I call *germ-tracks*. **1906** *Rep. Brit. Assoc. 1905* 434 In *Ascaris*..there are two chromosomes in the fertilised egg, and in all the nuclei in the 'germ-track'. **1949** Darlington & Mather *Elem. Genetics* 394 *Germ track*, the lineage of cells in the development of an organism, particularly in animals, which are potential ancestors of germ cells, as opposed to somatic cells. **1938** *Harper's Monthly* Mar. 367 Lurid descriptions of death rays, rocket planes, germ warfare. **1940** Blunden *Poems 1930–1940* 228 Revival of germ warfare. **1953** *Encounter* Oct. 15/2 The stories of germ warfare in Korea. **1963** *Ann. Reg. 1962* 395 Porton, a British Government Research establishment set up 20 years before to study germ warfare.

German, *a.*[2] and *sb.*[2] Add: **A.** *adj.* **1. a.** *German Ocean* (earlier examples); also *German Sea.*

1635 J. von Langeren *Direction for Eng. Traveller*, 'Northfolke' Germain-sea-N[orth]. *a* **1661** Fuller *Worthies* (1662) I. 317 Essex hath..the German Ocean on the East. *Ibid.* II. 307 This had..the German-Sea at the same distance. **1814** [see Ocean *sb.* 2].

4. a. German band, an instrumental band of street musicians, properly of German extraction; **German Catholic**: see Catholic *sb.* 3 b in Dict. and Suppl.; **German collie** (see quot. 1933); **German sausage** (earlier examples); **German sheep-dog, shepherd (dog),** a strong breed of dog, originally from northern Europe, frequently used as a guard-dog and for police work; = *Alsatian *a.* and *sb.* B. 2; **German wool** (examples).

[**1830** Lady Williams Wynn *Let.* 27 July (1920) 377 The German Band is all disbanded..& throughout His Household he has dismissed every Foreigner.] **1851** J. D. Lewis *Across Atlantic* 100 Immediately after dinner, an excellent German band, of a dozen performers, plays for a couple of hours. **1898** H. G. Hutchinson *Golfing Pilgr.* 25 The German bands of our Margates and our Ramsgates. **1906** B. Kennedy *Wander Pict.* 190 At first I thought it was a Salvation Army band, but it turned out to be an awful German band that was making the town hideous. **1933** L. G. D. Acland in *Press* (Christchurch, N.Z.) 21 Oct. 15/7 *German collie*, a variety of sheepdog which is blue (black and white spots). He usually has one or both wall eyes. **1934** J. Lilico *Sheep Dog Mem.* 3 She was what is known in New Zealand as a German collie but on the Borders they were called 'Bilton Blues'. [**1837** Dickens *Pickw.* xxxi. 334 The ham (which was also from the German sausage-shop round the corner).] **1839** C. Sinclair *Holiday House* xv. 334 Perigord pies, German sausages, cold fowls, pastry, and fruit. **1850** German sausage [see Sausage 1]. **1922, 1926** German sheepdog, Shepherd Dog [see *Alsatian *a.* and *sb.* B. 2]. **1959** H. Neilsen *Fifth Caller* i. 17 A man.. decided to take his German shepherd for a walk. **1847** A. Brontë *Agnes Grey* vii. 103 The elder girl..was trifling over a piece of canvas and a basket of German wools. **1858** Geo. Eliot *Scenes Clerical Life* II. 88 Beaded urn-rugs and chair-covers in German wool.

b. German ivy (see Ivy *sb.* 2), **larch.**

1838 H. Colman *1st Rep. Agric. Mass.* 116 The Larch referred to in the text and there called the German Larch, is the common or White Larch.

B. *sb.* **3. a.** (Earlier U.S. examples.)

1863 A. D. Whitney *Faith Gartney* x, Who danced with who at the 'German' last night. *Ibid.* xiii, It was very agreeable..to dance the German with the nicest partner in the Monday class.

c. (Earlier example.)

1865 Dickens *Mut. Fr.* I. ii. viii. 240 Circumstances.. interpose obstacles between yourself and small germans.

C. *German-occupied* adj.; also **German-Italian** *a.*, **German-Swiss** *a.* and *sb.*; **German-American,** an American of German ancestry; also *attrib.* or as *adj.*

1824 I. Candler *Summary View Amer.* 333 Natives of Germany all told the German Americans of their bad language. **1880** *Harper's Mag.* Sept. 567 A German-American School..is kept over a disreputable little grog-shop. **1919** *Mr. Punch's Hist. Great War* 134 Perhaps he over-estimated the strength of the German-American and Pacifist elements. **1941** Wyndham Lewis *Let.* Jan. (1963) 284 You, as a German-American, can feel little sympathy for poor old England. **1936** H. Nicolson *Diary* 4 July (1966) 268 The French foresee and dread a German-Italian alliance. **1940** W. S. Churchill *Into Battle* (1941) 169 In German-occupied Poland the most hideous form of terrorism prevails. **1964** M. Gowing *Britain & Atomic Energy* vi. 180 Outside German-occupied Europe, the main source of supply open to Germany was Portugal. **1919** C. Mackenzie *Sylvia & Michael* i. 31 She asked if the other two women spoke English, and Miss Savage told her that one was a German-Swiss. **1934** E. Waugh *Ninety-two Days* v. 119 The monk was a German-Swiss. **1955** J. Thomas *No Banners* xx. 190 Two elderly German-Swiss gentlemen.

Germanic, *a.*[1] and *sb.* **A.** *adj.* **2.** (Further examples.)

Now tending to supersede *Teutonic* in scholarly work.

a **1773** A. Butler *Moveable Feasts Cath. Ch.* (1839) III. 80 Roman conquerors often took names..from countries which they had subdued,..as the African,.. the Germanic,..&c. **1774** R. Henry *Hist. Gt. Brit.* II. iii. 215 But though these Germanic nations differed very much from one another..yet they appear to have sprung from the same origin. **1842** B. Thorpe *Codex Exoniensis* 513 To suppose it a translation from another Germanic dialect, would be giving..credit for a kind of knowledge hardly..in existence among our simple forefathers. **1894** *Trans. Philol. Soc. 1891–4* 383 Germanic verse occasionally allows the freedom that syllables which in the spoken language are short, are, for the purposes of versification, treated as long. **1909** F. B. Gummere *Oldest Eng. Epic* p. viii, These two poems [*sc.* *Deor* and *Widsith*]..contain many references to persons and stories of Germanic heroic legends. **1928** W. W. Lawrence *Beowulf & Epic Trad.* 4 Anglo-Saxon verse..was.. rooted in the traditions of professional singers, the main features of whose craft were shared by the poets of the other peoples of Germanic stock. **1948** *Mod. Philol.* XLVI. 73, I shall not try to relate this meter to that of other Anglo-Saxon poetry or of the poetry in the other early Germanic dialects. **1963** R. W. V. Elliott in S. B. Greenfield *Stud. in Honor of A. G. Brodeur* 64 Originality of invention was not the Germanic scop's aim.

germanic (dʒəɹmæ·nik), *a.*[2] *Chem.* [f. Germanium + -ic.] Of, pertaining to, or containing germanium in the quadrivalent state. So **germanous** (-mē̆·nəs) *a.*, of, pertaining to, or containing germanium in the bivalent state; †also **germa·nious** (*obs.*).

1888 Bloxam *Chem.* (ed. 6) 306 Germanium..forms two series of compounds, the germanious and germanic. **1895** *Ibid.* (ed. 8) 418 White germanic sulphide, GeS_2... Germanous oxide, GeO. **1962** P. J. & B. Durrant *Introd. Adv. Inorg. Chem.* xviii. 621 Germanous sulphide, GeS, is made by the reduction of germanic sulphide, GeS_2, with germanium. **1966** *Kirk-Othmer Encycl. Chem. Technol.* (ed. 2) X. 522 Ionic germanic salts of the common inorganic oxy-acids..are not known.

Germanicism (dʒəɹmæ·nisiz'm). [f. Germanic *a.*[1] + -ism.] A Germanic characteristic, idiom, etc.; Germanic quality.

1905 *Daily Chron.* 14 Sept. 3/2 Mr. Zilliacus's translator, who gives us..sch's for j's, w's for v's, and other Germanicisms. **1947** Partridge *Usage & Abusage* 94/2 Attempting to rejuvenate it [*sc.* the literary language] with Gallicisms, Germanicisms, Grecisms and Latinisms. **1964** *Listener* 2 Apr. 559/2 The style in which he painted and the school to which he so loosely belonged were tainted with the labels of vulgarity and Germanicism— the two qualities the English can least abide.

germanium. Substitute for def.: A hard, brittle, semi-metallic element, greyish-white in colour, which has chemical properties intermediate between those of silicon and tin and is used esp. as a semiconductor in transistors and other solid-state devices. Atomic number 32; symbol Ge. (Add further examples.)

1948 *Bell Lab. Record* XXVI. 323/1 In semi-conductors, such as silicon and germanium,..there may be as few as one current-carrying electron for every million atoms. **1954** H. R. Harner in C. A. Hampel *Rare Metals Handbk.* ix. 163 Commercial recovery of germanium has..been chiefly from zinc ores, germanite, and flue dust (from coals). *Ibid.*, Production of germanium in England was started in 1950. **1955** *Sci. News Let.* 19 Feb. 119/3 The germanium transistor, versatile little crystal substitute for the radio tube. **1959** [see *argyrodite]. **1970** D. F. Shaw *Introd. Electronics* (ed. 2) xi. 236 To manufacture a *P–N* junction diode using germanium, the crystal is grown from *N*-type germanium (i.e. germanium with arsenic or antimony as impurity).

Germano-. Add: **Germa·nophil(e)** *a.* and *sb.* (see -phil, -phile); so **Germano·phily**; **Germa·nophobe** *a.* and *sb.* (see -phobe); **Germanophobia** (earlier example).

1898 *Daily News* 14 Dec. 5/7 M. Barthélemy-St.-Hilaire, 'the only Germanophile Frenchman then living'. **1906** *Outlook* 15 Sept. 338/2 Those whom the *Deutsche Revue* describes as 'very Germanophil Ministers'. **1911** *Q. Rev.* July 254 The editor of one of the greatest Russian newspapers, who is at one time a strong Germanophil, and at another an equally strong Germanophobe. **1946** R. Capell *Simiomata* III. 204 The second prime minister of the occupation..was..a convinced Greek Germanophil. **1961** *Guardian* 29 Sept. 6/4 The irrationalism and anti-rationalism of the Germanophiles. **1910** *Westm. Gaz.* 16 Feb. 8/2 That champion of germanophily. **1903** *N.Y. Tribune* 15 Apr. 8/3 The extreme Germanophobes of England. **1904** *Daily Chron.* 22 Oct. 3/2 In this story he is neither Francophil nor Germanophobe. **1905** *Ibid.* 10 Jan. 5/5 The Germanophobe attitude of the English Press. **1964** *Economist* 12 Dec. 1225/2 A noisy handful of Germanophobes. **1887** Germanophobia [see -phobia].

germarium (dʒəɹmē̆ə·riɒm). *Zool.* Pl. **-aria.** [mod.L., f. L. *germen* germ.] In certain invertebrates, the ovary proper, as distinguished from the vitellarium; also, the corresponding part of a testis.

1877 T. H. Huxley *Man. Anat. Invert.* 178 A single or double *germarium*, having nearly the same structure as the ovary of *Macrostomum*. **1888** Rolleston & Jackson *Forms Anim. Life* 646 The germarium is always single, usually globular or elongate, and even folded on itself. **1925** A. D. Imms *Gen. Textbk. Ent.* 145 The germarium is the region containing the primordial germ cells or spermatogonia which undergo multiplication. **1970** D. Gilmour et al. in *Insects Austral.* (C.S.I.R.O.) ii. 69 The germarium is situated at the apex of each ovariole.

Germinal, *sb.* (Earlier example.)

1802 C. Wilmot *Diary* 25 Apr. (1920) 60 Three days in this month of Germinal, it is the custom for all the world to drive four miles out of Town in State, and return back again.

germinal, *a.* Add: **d.** *fig.* Capable of development; productive of new ideas, influences, etc.; = Seminal *a.* 4 a.

1934 H. G. Wells *Exper. Autobiogr.* I. iii. 136 The European country houses and chateaux that were so alive and germinal, mentally, in the seventeenth and eighteenth centuries, stand now mere empty shells. **1935** J. R. Moore *Defoe & Mod. Econ. Theory* ii. 5 He was one of the rare germinal minds in literature. **1936** C. S. Lewis *Allegory of Love* IV. i. 157 The *Romance of the Rose* is one of the most 'successful' books, in the vulgar sense, that have ever been written...As a germinal book during these centuries [*sc.* the later Middle Ages] it ranks second to none except the Bible and the *Consolation of Philosophy.*

germy (dʒɜ̄·ɹmi), *a.* [f. Germ *sb.* + -y[1].] Full of germs; unclean, polluted.

1912 P. A. Amos *Processes Flour Manuf.* xxi. 202 Flaking..is especially noticeable upon soft, damp, or germy stock. **1933** Dylan Thomas *Let.* (1966) 70 Remember this: each rose is wormy. And every lovely woman's germy. **1935** M. de la Roche *Young Renny* xiv. 123 Mary has washed this child's offering. It will put these new germy ideas in his head. **1970** *Guardian* 6 Apr. 9/2 A natural slut can languish unwashed, hairy, germy and odorous.

gerontocratic (dʒerǫ:ntokræ·tik), *a.* [f. Gerontocracy + -ic.] Of, pertaining to, or characteristic of a gerontocracy.

1950 *African Studies* Dec. 159 Some anthropologists are of the opinion that gerontocratic age-groups are an invention of the Negro. **1957** P. Worsley *Trumpet shall Sound* ii. 40 Many youths looked forward to migrant labour as an exciting liberation from parental and gerontocratic control. **1957** V. W. Turner *Schism & Continuity in Afr. Soc.* iii. 81 Grandparents..belong to the generation which exerts gerontocratic political and jural authority in the matrilineal village.

gerontology (dʒe:rǫntǫ·lŏdʒi). [f. Gr. γεροντ-, γέρων old man + -o + -logy.] The scientific study of old age and of the process of ageing.

1903 [see Thanatology]. **1954** *Medical Press* 25 Aug. 180/1 The science of gerontology..includes the medical and biologic, the psychologic and sociologic, the economic and philosophic aspects of ageing. **1967** [see *geriatrics].

Hence **gero:ntolo·gical** *a.*, of or pertaining to gerontology; **ge:ronto·logist**, an expert in, or student of, gerontology.

1941 *Time* 22 Sept. 48 First problem facing the gerontologists is to get funds for research. **1944** *Science* 8 Dec. 508 Thus McCay, referring to gerontological work, expressed the need [etc.]. **1954** *Medical Press* 25 Aug. 180/1 No such science as a 'science of ageing' has actually been constituted—even, apparently, where most attention has been given to gerontological studies. **1955** *New Biol.* XVIII. 29 The gerontologist, with the prolongation of human life in mind, is interested in something..which is in no sense comparable with the evolution of sight. **1966** *Listener* 21 Apr. 573/1 Muscle aging is of particular interest, however, for gerontologists—people, that is, who are working on the nature and control of age changes. **1967** *New Scientist* 19 Jan. 160/1 The relatively small numbers engaged in gerontological research in Britain.

gerontomorphic (dʒĕrǫ:ntomǫ·ɹfik), *a.* [f. Gr. γεροντ-, γέρων old man + -o + μορφή form + -ic.] Of, pertaining to, or designating anatomical specialization most fully represented in the mature male of a species.

1939 C. S. Coon *Races of Europe* iv. 85 It represents a gerontomorphic or sexually differentiated Mediterranean or Galley Hill form. **1959** J. D. Clark *Prehist. S. Afr.* iv. 86 These two fossils..are, however, not the only remains of this gerontomorphic, proto-Australoid type from Africa, for fragments of three fossil crania were found in 1935. *Ibid.* 88 'Proto-Australoid' type with fully adult anatomical features (gerontomorphic).

gerontophil (dʒĕrǫ·ntofil), *a.* [f. Gr. γεροντ-, γέρων old man + -o + -phil.] Loving or favouring old people, esp. old men; desiring sexual relations with old people. So as *sb.* So **gerontophi·lia**; **gerontophi·lic** *a.*; **geronto·philism**; **geronto·phily.**

1918 E. Jones *Papers on Psycho-Analysis* (ed. 2) xxxviii. 655 Some of the resultants of the 'grand-father-complex' may now be mentioned. The most striking is the tendency to gerontophilia—i.e., a special fondness for old people. **1937** M. Hirschfeld *Sex Anomalies* II. v. 90 The third category of psycho-sexual infantilism, in which the individual seeks the object of his desire.. among older people (gerontophilia). **1939** Joyce *Finnegans Wake* 115 Speaking anent Tiberias and other incestuish salacities among gerontophils. **1945** E. Waugh *Brideshead Revisited* I. vii. 162 Among Julia's friends there was a kind of gerontophilic snobbery; young men were held to be gauche and pimply. **1959** *Listener* 23 July 146/3 She seems to share the gerontophilism which Mlle Sagan has made so popular. **1963** *Times Lit. Suppl.* 18 Jan. 35/1 The English..are notoriously gerontophil, and the surest road to success for a writer or a painter.. is to survive to a venerable age. **1964** *Listener* 15 Oct. 603/1 Sado-masochism, gerontophily, a whiff of incest. **1965** *Ibid.* 25 Nov. 869/1 Robin, one of those gerontophil types described by Proust as being so fortunately provided by Nature for the exclusive gratification of old men. **1965** F. Sargeson *Memoirs of Peon* v. 113 It's not that I suspect you of gerontophilia.

gerrymander, *sb.* Add: (Earlier examples.) Also, one elected by gerrymandering. Also *attrib.*

1812 *Columbian Centinel* 23 May 2/3 The sensibility of the good people of Massachusetts is..awakened to this

'Gerrymander'. 1812 *Massachusetts Spy* 4 Nov. (Th.), Gerrymander Senate. 1813 *Ibid.* 12 May (Th.), An official statement of the returns of voters for senators give[s] twenty nine friends of peace, and eleven gerrymanders.

gerrymander, *v.* (Earlier examples.)
1812 *Salem Gaz.* 22 Dec. 2/4 So much..for *War* and *Gerrymandering*. 1812 *N.Y. Post* 28 Dec. 3/1 They attempted also to *Gerrymander* the State for the choice of Representatives to Congress.

gertcha (gɜ̄·ɪtʃă), *int.* Also **gercha, gertcher.** Vulgar corruption of *get away* (or *along*) *with you*, etc., used esp. as a derisive expression of disbelief.
1937 PARTRIDGE *Dict. Slang* 323/1 *Gertcher*, get out of it, you! 1939 'J. BELL' *Death at Half-Term* vii. 132 'Go down to the Old Vic sometime and see the real thing for yourself.' 'Gercha!' said Inspector Mitchell. 1949 J. B. PRIESTLEY *Delight* xxxii. 89 'One of the most energetic and prolific of our authors...' Gertcha! 1963 'G. CARR' *Lewker in Norway* ii. 30 'Gertcha!' The orator..elbowed him away.

Gervais (ʒɛrve). [f. the name of Charles *Gervais*, French cheese-maker (1830–92).] In full *Gervais cheese*. The proprietary name of a soft, creamy cheese.
1896 LONG & BENSON *Cheese* v. 60 The Gervais cheese is a delicate little luxury produced..by M. Gervais and M. Pommel...Gervais is a mixture of cream and milk. 1902 [see *BONDON]. 1950 J. G. DAVIS *Dict. Dairying* 120 *Gervais cheese*. This is a popular French soft cheese and is usually made from two parts of whole milk and one of thin cream. 1951 E. DAVID *French Country Cooking* 197 Pound 6 ozs. of *Petit Suisse* (Pommel) or *Demi-Sel* (Gervais) cheeses with ½ teacup of cream or milk.

Gerzean (gəɪzi·ăn), *a. Archæol.* [f. El *Gerzeh*, name of a district in Egypt + -AN.] Of, pertaining to, or designating the middle period of the ancient pre-Dynastic culture in Egypt.
1925 *Catal. Antiquities at Badari* (*Brit. Sch. Archæol. in Egypt*) 3 Approximate Dates. 13,000 B.C...Badarian Age, 10,000 B.C...Amratian Age, 9,000 B.C...Gerzean Age. 1928 BRUNTON & CATON-THOMPSON *Badarian Civilisation* I. i. 1 By 'Predynastic' is meant the age.. which has been subdivided into the three divisions of Early (or Amratian), Middle (or Gerzean), and Late (or Semainian). 1949 W. F. ALBRIGHT *Archæol. of Palestine* iv. 70 There was a large village and cemetery of Middle Gerzean date, roughly from about the third quarter of the fourth millennium.

‖ **Gesamtkunstwerk** (gəza·mt,kunst,veɪk). [G., f. *gesamt* total + *kunstwerk* work of art.] In the æsthetic theory of Richard Wagner (1813–83), an ideal work of art in which drama, music, and other performing arts are integrated and each is subservient to the whole.
1939 B. FLES tr. *Křenek's Music Here & Now* 223 Wagner went so far as to lower his orchestra into a cavity below the audience's line of vision to emphasize the illusionary character of his *Gesamtkunstwerk*, or 'symbiosis of the arts'. 1947 A. EINSTEIN *Mus. Romantic Era* xix. 356 In his *Gesamtkunstwerk* all the individual arts were supposed to give up something of their own nature in order to create a higher unity. 1948 L. SPITZER *Linguistics & Lit. Hist.* iv. 160 The nephew..lends his voice to imitating an orchestra..and impersonating a Wagnerian-like *Gesamtkunstwerk*. 1966 *Listener* 6 Oct. 517/3 Whether Gascoyne saw his poem in the light of a *Gesamtkunstwerk* I'm not qualified to say.

‖ **Gesellschaft** (gĕze·lʃɑft). Also with lower-case initial. [G., f. *gesell(e)* companion + *-schaft* -SHIP.] A social relationship between individuals based on duty to society or to an organization; contrasted with *GEMEINSCHAFT. So *gesellschaft-like* adj. [G. *gesellschaftlich*].
1887, etc. [see *GEMEINSCHAFT]. 1964 GOULD & KOLB *Dict. Soc. Sci.* 286/1 *Gesellschaft*-like social systems are those in which rational will (*Kürwille*) has primacy.

gesnera. Add: Also **gesneria** (the spelling preferred by Linnæus and now the accepted form).
[1737 LINNÆUS *Genera Plantarum* 179 *Gesneria*. Gesnera *Plum*.] 1845 *Bot. Mag.* LXXI. Tab. 4152 *Schiede's Gesneria*..is another lovely addition to the many beautiful *Gesnerias* now cultivated in our stoves. 1901 G. NICHOLSON *Dict. Gard.* Suppl. II. 389/1 *Gesnera*. According to the 'Index Kewensis' the correct spelling is that of Linnæus—*Gesneria*. 1956 *Dict. Gardening* (R. Hort. Soc.) (ed. 2) II. 886/1 The Gesnerias being tropical, need stove conditions. 1961 *Amat. Gardening* 14 Oct. Suppl. 23/3 The gesneras are related to the gloxinias and require very much the same conditions.

gestagen (dʒe·stădʒĕn). Also **gestogen.** [f. GESTA(TION + -GEN.] Any substance, such as the sex hormone progesterone, having progestational effects. Hence **gestage·nic** *a.*
1948 K. MIESCHER in *Rec. Progr. Hormone Res.* III. 47 The class of sexogens comprises the estrogens, the androgens, and the gestagens, as we propose to call the compounds with progestational action. 1949 L. F. & M. FIESER *Nat. Prod. related to Phenanthrene* (ed. 3) iv. 300 There are two types of female sex hormones, exemplified by the estrogens: estradiol, estrone, and estriol; and by progesterone, the sole natural progestational hormone or gestogen. 1958 *Jrnl. Clin. Endocrinol. & Metabolism* XVIII. 338 It has been generally supposed that progesterone is the only naturally occurring substance with a primarily gestagenic effect. 1962 *Lancet* 2 June 1176/1 The addition of œstrogens to the commercial preparations of orally active gestagens has enhanced their efficacy in contraception.

Gestalt, gestalt (gəʃta·lt). [G., = form, shape.] A 'shape', 'configuration', or 'structure' which as an object of perception forms a specific whole or unity incapable of expression simply in terms of its parts (*e.g.* a melody in distinction from the notes that make it up); cf. *CONFIGURATION 6. Freq. *attrib.*, as **Gestalt psychology,** a school of psychology which holds that perceptions, reactions, etc., are Gestalts; also *ellipt.*
[1890 C. VON EHRENFELS in *Vierteljahrsschrift für wissenschaftliche Philosophie* XIV. 249 (*title*) Ueber 'Gestaltqualitäten'.] 1922 K. KOFFKA in *Psychol. Bull.* XIX. 531 The *Gestalt*-psychologists proper. *Ibid.* 574 The *Gestalt* theory is fundamentally incompatible with the associationist's principles. 1924 tr. K. Koffka in *Psyche* V. 80 Gestalt-Psychology has so far got a number of important answers to its questions. *Ibid.* 81 Gestalt-Theory. *Ibid.* 84 Prof. Wertheimer, in his lectures, has treated personality as a Gestalt. 1926 *Encycl. Brit.* Suppl. I. 45/1 The work of the Gestalt school with its stress upon the unity of psychic processes. *Note*, The Gestalt theorists. 1930 W. KÖHLER (*title*) Gestalt psychology. 1931 M. BELGION *Human Parrot* i. 15 The Behaviourists and the apostles of *Gestalt*. 1936 A. J. AYER *Lang., Truth & Logic* ii. 57 The Gestalt psychologists who of all men talk most constantly about genuine wholes. 1941 AUDEN *New Year Let.* I. 19 A true gestalt where indiscrete Perceptions and extensions meet. 1959 *Times* 13 Feb. 13/5 Webern..eliminates not only rhythm but the *Gestalt* of a melodic line and all traces of coherence by tonality. 1962 [see *AFTER-IMAGE].

Hence **Gesta·ltism,** Gestalt psychology; = *CONFIGURATIONISM; **Gesta·ltist,** one who accepts or practises the principles of Gestalt psychology; a Gestalt psychologist; also as *adj.* Also in loan-words from German: **Gestaltpsychologie,** Gestalt psychology; **Gestaltqualität,** the quality of a Gestalt; **Gestalttheorie,** Gestaltism.
1909 E. B. TITCHENER *Lect. Exper. Psychol. of Thought-Processes* I. i. 32 We may speak of general attributes of sensation, as Ebbinghaus does; or we may speak of *Gestaltqualität*, form of combination, funded character. 1925 I. A. RICHARDS *Princ. Lit. Crit.* iv. 25 There are very many problems of psychology, from those with which some of the exponents of *Gestalt theorie* are grappling to those by which psycho-analysts are bewildered. *Ibid.* xxii. 183 The exponents of *Gestalt-psychologie*. 1931 *Psyche* July 6 The rise of Gestaltists, hormic psychologists and many other varieties. 1933 *Times Lit. Suppl.* 16 Nov. 786/2 His own fivefold scheme of opposed viewpoints (Structure and Function, Association and *Gestaltqualitat*, Introspection and Behaviourism, Mechanism and Teleology, Conscious and Unconscious). 1938 *Mind* XLVII. 377 If Associationism may be regarded as Psychological Enemy No. 1, cannot Gestaltism put in a claim to be at any rate No. 2? 1938 *Amer. Speech* XIII. 295 Mr. Firth's gestaltist chapter on meaning ('Context of Situation') makes excellent sense. 1938 *Brit. Jrnl. Psychol.* July 77 Prof. Spearman's general theory of visual perception must of necessity lead him into conflict with the protagonists of the *Gestalttheorie*. 1951 H. McCLOY *Alias Basil Willing* vii. 65 The Gestaltists believed in studying a complete picture of every psychiatric situation.

Gestapo (gĕstā·po, ge-). [G., acronym f. the initial letters of *Ge*heime *Sta*ats*po*lizei, Secret State-Police, set up by Hermann Göring in Prussia, 1933, and extended to the whole of Germany in January 1934.] The secret police of the Nazi regime in Germany. Also *transf.*
1934 *New Republic* 18 July 249/2 The names and significance of the..semi-military organizations in Germany..are:..The *Geheime Staats Polizei*, usually called the *Gestapo*, the secret state police of Prussia. 1937 *N.Y. Times Mag.* 21 Nov. 1/3 The spokesmen of the regime,..are taking every opportunity..to ridicule the idea that every German servant girl abroad is a disguised Gestapo agent or a spy. 1940 *Mind* XLIX. 222 Morality..becoming nothing more than subservience to the decrees of a dictator with his Ogpu or Gestapo at his back. 1944 G. B. SHAW *Everybody's Political What's What?* xxiv. 224 The municipal statesman sends his sanitary Gestapo into an unhealthy private house and prosecutes the tenant. 1966 L. P. DAVIES *Psychogeist* xii. 109 My private Gestapo has already brought the news.

gesticulatingly (dʒesti·kiŭlēɪ:tiŋli), *adv.* [f. GESTICULATING *ppl. a.* + -LY².] With gesticulations. Also **gesti·cula:tively** *adv.*
1893 F. ADAMS *New Egypt* 22 Strolling about, noisily and gesticulatingly. 1898 E. P. EVANS *Evol. Ethics* vii. 223 He can..express the number of objects lying before him gesticulatively with his fingers.

gestogen, var. *GESTAGEN.

gestural, *a.* Add: (Further examples.) *spec.* Designating or pertaining to the theory that human speech originated in oral imitation of bodily gestures. Hence **ge·sturally** *adv.*
1930 R. PAGET *Babel* iii. 84 The American twang..is mainly due to a tightening of the pharynx, and has nothing to commend it on gestural or phonetic grounds. *Ibid.*, Each word should, so far as possible, be gesturally appropriate to its meaning. 1941 D. EFRON *Gesture & Environment* II. iii. 95 This gestural vocabulary comprehends no less than 125 manual 'words', implying definite meaningful associations. 1949 *Trans. Philol. Soc. 1948* 49 The Gestural Theory is largely linked with the name of Sir Richard Paget. 1957 D. L. BOLINGER in *Publ. Amer. Dial. Soc.* XXVIII. 11 *Wonder* is a borderline case when it precedes. Gesturally it is often treated as a Q[uestion], but intonationally it usually is not. 1966 *Listener* 2 June 814/3 The *Sette peccati*..represents each of the seven deadly sins gesturally on the stage.

gesture, *sb.* Add: **4. b.** *transf.* and *fig.*; *spec.* [after F. *geste*; cf. *BEAU GESTE] a move or course of action undertaken as an expression of feeling or as a formality; esp. a demonstration of friendly feeling, usu. with the purpose of eliciting a favourable response from another.
1916 *Daily News* 2 Feb. 4/4 The cost of museums and galleries ought to be considered as part of the cost of the war...To shut them is a mean and shabby gesture before the whole world. 1921 *Times* 18 Oct. 10/4 The gift of your Medal of Honour to a British comrade in arms, whose tomb in Westminster Abbey stands for all our best endeavour and hardest sacrifice in the war, is a gesture of friendly sympathy and good will which we will not forget. 1921 *Daily News* 9 Nov., The hope that Sir James Craig might make a generous gesture. *Ibid.* 24 Nov., You cannot quite get that gesture from Mr. Balfour. 1922 *Daily Mail* 9 Nov. 9 So far as the movement against Prohibition is concerned, the victory of Mr. Edwards, Governor of New Jersey, is only a gesture. As Governor he promised to make the State as wet as the Atlantic. *Ibid.* 16 Dec. 9 The United States Cabinet to-day sat..to consider a world gesture which is intended ..to assist Europe and to allay discontent at home. 1922 *Westm. Gaz.* 20 Dec., The semi-official gestures of Greece towards a reconciliation with this country. 1933 BLOOMFIELD *Lang.* ix. 147 Vocal gestures, serving an inferior type of communication, occur not only outside of speech, as in an inarticulate outcry, but also in combination with speech-forms. 1959 *Listener* 8 Oct. 563/2, I do not advocate, instead, an imitation of the gestures of the new 'Holy Trinity' of European music: Stockhausen, Boulez, and Nono. 1963 *Ibid.* 7 Mar. 418/2 The Lijnbaan ..would be a very long, completely straight two-storey street for pedestrians were it not for a single formal gesture which acts like a magic wand, providing canopies across the Lijnbaan as well as along it. 1964 *Ann. Reg. 1963* 253 France did not sign the test ban treaty, described..as 'a purely platonic gesture'.

5. gesture theory, a theory of the origin of language (see *GESTURAL *a.*); hence *gesture-theorist.*
1930 R. PAGET *Babel* ii. 54 The gesture theory of human speech is not new. *Ibid.* 62 To the gesture-theorist it is a natural consequence of the fact that every tongue- and lip-gesture can be construed in a variety of ways.

‖ **gesundheit** (gĕzu·nt,həit), *int.* [G., lit. 'health'.] An exclamation used to wish good health to a person, esp. to someone who sneezes.
1914 *Everybody's* Feb. 484 'Saved your life,' he murmured mechanically, as one suffixes 'Gesundheit' to a sneeze. 1942 O. NASH *Good Intentions* 124 Mr. Weaver said 'A cashew', and the man said '*Gesundheit*'. 1959 H. PINTER *Birthday Party* II. 37 *Goldberg* (lifting his glass). Gezunteheit [1960 ed.: Gesundheit]. 1961 L. PAYNE *Nose on my Face* xvi. 252 Saunders sneezed suddenly. '*Gesundheit*,' said Jim.

get, *sb.*¹ Add: **1. c.** The action of returning the ball, esp. a difficult shot, in lawn tennis. *colloq.*
1927 *Daily Tel.* 22 Mar. 15/6 One does not remember seeing Hake play better, and some of his gets were most spectacular. 1969 *Sunday Times* 6 July 20/1 He was broken only once, in the third game as a result of an amazing 'get' by the champion.

2. b. (Further examples.) Also *spec.* a bastard; hence as a general term of abuse: a fool, idiot. (Cf. *GIT.) Now *dial.* and *slang.*
1706 in W. Cramond *Court Bks. Regality of Grant* (1897) 20 Gregor Burgess protested against the said Allane that called him a witch gyt or bratt. 1822 J. GALT *Provost* ix. 65 A donsie mother that could gie no name to her gets. 1880 W. H. PATTERSON *Gloss. Antrim & Down* 43 *Get*, an opprobrious term used in scolding matches. 1908 J. MASEFIELD *Capt. Margaret* xi. 325 He's a mother's joy, the Portuguese drummer's get. 1922 JOYCE *Ulysses* 319 The bloody thicklugged sons of whores' gets! 1934 'L. G. GIBBON' *Grey Granite* 126 The woman said of all the whoreson's gets she'd ever met he was the worst. 1940 *Daily Mail* 7 Sept. 3/8 Here are some current military phrases interpreted:..*get*, chump, fool. 1965 *Listener* 24 June 949/1, I would..define him as a daft old get. 1967 'H. CALVIN' *Nice Friendly Town* viii. 101 Put something on him, the stupid get!

4. A getaway; a hasty retreat; esp. in phr. *to do* (or *make*) *a get*. Cf. GET *v.* 31 d. *Austral.* and *N.Z. slang.*

1898 *Bulletin* (Sydney) 28 May 31/2 Their inquisitiveness..compelled Jim to kill his stud-sluts and growing stock, and do a timely 'get'. 1906 E. DYSON *Fact'ry 'Ands* ix. 117 They thort his jills had done er get. 1909 T. H. THOMPSON *Ballads about Business* 92, I..prepared to do a get. 1914 A. A. GRACE *Tale of Timber Town* v. 32, I must make a git. So-long.

get, *sb.*² Delete † and add later example.
1957 T. HUGHES *Hawk in Rain* 33 Grubbing his get among your lilies.

get (get), *sb.*³ Also **gett.** [Aramaic.] Among the Jews, a written 'bill of divorcement' prepared according to a prescribed form; also, the divorce itself.
1892 ZANGWILL *Childr. Ghetto* I. I. iv. 122 'He must give her *Gett*!' 'Of course!..I divorce her at once!' 1960 L. P. GARTNER *Jewish Immigrant in England* vi. 168 Social pressure and legal adjustments in the *ketubah* (marriage document) could force the most recalcitrant of husbands to grant his estranged wife a *get* (divorce). 1963 *Listener* 17 Jan. 123/1 The husband delivered a Jewish letter of divorce, called a *gett*, to his wife.

get, *v.* Add: *Etym. note.* The pa. pple. is also used *colloq.* with omission of (*I*) *have.* Cf. *GOTCHA, GOTCHER, *GOTTA.
1849 *Knickerbocker* XXXIV. 12 They got no principles. They got no platform to stand onto. 1857 *Quinland* I. 1 Got an hour to spare—thought I'd just run in and see what you were all about. 1884 [see *GET-OUT 1]. 1887 M. E. WILKINS *Humble Romance* 370 What you got there, grandma? 1911 R. D. SAUNDERS *Col. Todhunter* i. 11 Oh, of course, you got to laugh at me. 1911 J. F. WILSON *Land Claimers* ix. 118 But I got several plans, and I need ye. 1941 [see sense *14 d]. 1967 [see *GAWD].

1. e. *I wish you may get it, don't you wish you may get it?*: ironical colloq. expressions implying the speaker's doubt of, or lack of desire for, another's success.
1836 DICKENS *Sk. Boz* I. 42 An 'I wish you may get it' sort of expression in his eye. 1837 —— *Pickw.* xxvi. 274 'But the plaintiff must get it,' resumed Mrs. Cluppins. ..'Vell,' said Sam...'All I can say is, that I wish you may get it.' 1842 BARHAM *Ingol. Leg.* 2nd Ser. 245 Ah, ha! my good friend!—don't you wish you may get it? 1848 THACKERAY *Van. Fair* xiii, 'There's one of the greatest men in the kingdom wants some.' 'Does he?' growled the senior. 'Wish he may get it.' 1851 MAYHEW *Lond. Labour* I. 1. 56 I've heard people say when I've cried 'all a-growing' on a fine-ish day, 'Aye, now summer's a-coming.' I wish you may get it, says I to myself; for I've studied the seasons. 1857 HUGHES *Tom Brown* I. ix, Don't you wish you may get it?

5. b. *to get anything* (or *something*) *on* (a person): to gain or possess incriminating information about (someone); to have an advantage over. Cf. *ON *prep.* 21 b, d. *colloq.* (orig. *U.S.*).
1919 *Detective Story Mag.* 25 Nov. 129 He gave me the slip...Maybe it's just as well since I haven't got anything on him yet. 1923 L. J. VANCE *Baroque* vii. 42 You haven't got any thing on me. 1946 T. JONES *Skinny Angel* 85 Those fellows are trying to get something on someone. 1960 'W. HAGGARD' *Closed Circuit* iii. 31 Get something on the men who counted. Then you could do almost as you pleased. It was astonishing how most of the men who counted had something to hide.

e. *Racing.* To hold out for, to stay (a specified distance).
1898 A. E. T. WATSON *Turf* vii. 148 There are not a few horses that cannot fairly 'get' even five furlongs. 1898 *Daily News* 17 Oct. 3/3 He will..be opposed by plenty of candidates who can get the Cambridgeshire course. 1907 *Daily Chron.* 14 Nov. 3/3 Only a wonder of a horse can 'get' those four miles and a half of ditches and fences.

6. b. In various games: to make (a certain score); to score (so many points, runs, goals, etc.); in *Cards*, to take (so many tricks). Also, in *Cricket*, to take (a wicket), to take the wicket of.
1548, 1553 [see GOAL *sb.* 3 a]. 1710 [see ODD *a.* 1]. 1731 in H. T. Waghorn *Cricket Scores* (1899) 4 The Duke's hands came in first, and got 79 before they were out. 1778 *Miss Wicket & Miss Trigger* (*caption of print*), Miss Trigger you see is an excellent Shot, And forty five Notches Miss Wicket's just got. 1857 HUGHES *Tom Brown* II. viii. 387 We haven't got the best wicket yet. *Ibid.* 397 Only seventeen runs to get with four wickets—the game is all but ours! 1901 *Encycl. Sport* I. 231/2 Many a bad ball gets a wicket. 1912 A. A. LILLEY *Twenty-four Years Cricket* (1914) x. 164 The substantial support Trumper received..left us 194 to get to win. 1930 C. V. GRIMMETT (*title*) Getting wickets. 1971 *Sunday Express* 31 Oct. 31/3 He could not get the goal he sought so eagerly.

7. d. To understand (a person or statement). Also *absol. colloq.* (orig. *U.S.*).
[1892 'MARK TWAIN' *Amer. Claimant* xiii. 101, I don't know that I quite get the bearings of your position.] 1907 M. C. HARRIS *Tents of Wickedness* I. iii. 33 'I don't get her,' she murmured, as if Leonora was a telephone number. 1913 J. LONDON *Valley of Moon* I. vii, When I go after anything I get it, an' if anything gets in between it gets hurt. D'ye get that? 1918 WODEHOUSE *Piccadilly Jim* xi. 114, I get you not, friend. Supply a few footnotes. 1937 'J. BELL' *Murder in Hospital* vii. 136 'I'd go about it rather quietly if I were you...' 'I get you,' said Thornton. 1948 —— *Wonderful Mrs. Marriott* xxi. 273 Oh, I get. The Condover Court lady. 1956 I. BROMIGE *Enchanted Garden* II. ii. 93 Fiona broke into peals of laughter and became quite helpless for a few moments. 'Don't get it,' said Julian. 1966 'M. INNES' *Change of Heir* ii. 14 Okay, okay. I get. Norval. My name is Norval.

e. To notice, look at (a person, esp. one who is conceited or laughable); usu. as *imp.* with a pronoun as object. *colloq.*
1958 *News Chron.* 22 May 4/4 If he is conceited the girls mutter get *yew*! 1967 H. DALMAS *Fowler Formula* (1968) i. 16 It was almost like hearing himself say, 'Get me! I had a special invitation to the Universal party this afternoon.'

12. d. For U.S. vulgar read orig. U.S., and add earlier and later examples.
1772 in *D.A.* 1802 *Methodist New Connexion Mag.* Nov. 432 A number, too, are wrought upon in the usual way, and hopefully get religion without any of these extraordinary appearances. 1952 *Manch. Guardian Weekly* 9 Oct. 7 It is sad news for his publishers that he has got religion.

14. c. *to have got 'em* (*bad*): to have the D.T.'s, to have 'the horrors'; also in milder sense, to have a fit of nerves. *slang.*
1893 FARMER & HENLEY *Slang* III. 188/1 A very sick person, especially a patient in the horrors, is said to have *got 'em bad*. 1936 P. M. CLARK *Autobiogr. Old Drifter* xiii. 184 Another fellow who 'got 'em' was 'Taffy'. He got 'em so badly one night that he ran from the Old Drift, clad only in his nightshirt.

d. *to have got it bad*(*ly*): to have fallen in love; to be infatuated. *slang.*
1911 G. B. SHAW *Getting Married* in *Doctor's Dilemma*, etc. 263 You seem to have got it pretty bad. 1921 W. J. LOCKE *Mountebank* xiii. 163 'She's got it rather badly,' Charles murmured to me. 1941 WEBSTER & ELLINGTON (*song-title*) I got it bad and that ain't good. 1969 D. CLARK *Nobody's Perfect* v. 148 Take it from me he's got it badly. He couldn't even hear me mention your name without wanting to talk about you.

16. c. Also *to get it in the neck*: see *NECK *sb.*¹ 1 e.

e. *to get his* (or *theirs*): to be killed. *slang.*
a 1910 'O. HENRY' *Rolling Stones* (1913) iii. 65 Clifford Wainwright being shot by a squad of soldiers...Oh, yes, it was rum that did it. He backslided and got his. 1913 KIPLING *Diversity of Creatures* (1917) 288 He'd got *his*. I knew it by the way the head rolled in my hands. 1928 E. WALLACE *Flying Squad* xiii. 110 He'll get his one of these days. 1938 F. D. SHARPE *Sharpe of Flying Squad* viii. 107 The other women leave her alone because they know that if they don't—they'll get theirs from Johnny. 1959 N. MAILER *Advts. for Myself* (1961) 66 He was going to get his, come two three four hours. That was all right, of course, you didn't live forever.

21. a. (Further examples.) Also, to puzzle, perplex, nonplus. So *to get* (someone) *where one wants him* (or *her*): to have at one's mercy; to render subservient, dependent, etc.
1868 *Harper's Mag.* Mar. 538/2 Scratching his head a minute, Benjamin F. replied: 'Well, I confess your Honor's got me there!' 1906 W. CHURCHILL *Coniston* I. xiv. 171 'What's the name of your gal?' 'Well,' said Mr. Hopkins, 'I guess you've got me. We did christen her Lily, but she didn't turn out exactly Lily...I guess her name's Cassandra.' 1936 W. DE LA MARE *Wind blows Over* 32 'That's Mistaken Point,' he said. 'Why was it mistaken?' He shook his head...'That's got me, miss,' he replied. 1939 A. THIRKELL *Before Lunch* iii. 76 You only want to get her where you want her. Most people are like that. 1968 A. CLARKE *Darkened Room* iii. 41 Silly young fool, I thought angrily; he's got you where he wants you.

c. To succeed in taking or catching (a person or animal); *spec.* (orig. *U.S.*) to succeed in killing or injuring; to shoot or kill.
1853 'P. PAXTON' *Yankee in Texas* 118 [A Texan] does not kill his game, he *saves* or *gets it*, or *makes it come*. 1887 F. FRANCIS *Saddle & Mocassin* viii. 138 They'll get you one of these days, Colonel, when you are driving around in your wagon. 1899 B. TARKINGTON *Gentleman fr. Indiana* ix. 160 Wiley,..you don't think they've got him? 1900 E. GLYN *Visits Eliz.* 50 She did not hit any rabbits, but she got a gardener in the leg. 1908 *Daily Chron.* 16 Sept. 7/5 This climate is sure to get a white man sooner or later. 1917 'S. ROHMER' *Hand of Fu-Manchu* (1920) viii. 65, I turned, dizzily, to see Fletcher sinking to his knees, one hand clutching his breast. 'She got me...with the knife,' he whispered. 1932 [see *AGENT *sb.* 4 d]. 1951 R. CAMPBELL *Light on Dark Horse* xxiii. 334 He never told me whether he was implicated in the attempt to 'get' Guillermo, or not.

d. To exercise, worry, annoy. *colloq.* (orig. *U.S.*).
1867 B. HARTE *Condensed Novels* 280 To have let bigger things go by, and to be taken in by this cheap trick..is what gets me. 1904 W. H. SMITH *Promoters* xii. 190, I wish to the Lord he hadn't been so quick about it. That's what gets me. 1926 W. DEEPING *Sorrell & Son* xxix, 'Do you think he minds?' 'I know he doesn't. But it gets me.' 1960 B. COBB *Don't lie to Police* ix. 149 It got me—her talking that way.

e. To enthral, attract, appeal to; to touch emotionally; to obsess. *colloq.*
1913 R. BROOKE *Let.* 20 Nov. (1968) 532 It's the *Rhythm* that gets you. 1916 *To-Day* 11 Nov. 50/3 The 'curtain' on that dog walking across the stage and sticking his cold nose into the hand of his lonely master always used to get me. 1928 F. N. HART *Bellamy Trial* i. 6 'It [*sc.* murder]'s always interested me more than anything else.'..'Well, don't let it get you. I'd just keep it as a hobby.' 1938 G. GREENE *Brighton Rock* II. i. 68 'It gets you,' the Boy said, 'it gets you,' surrendering himself to the huge brazen suggestion. 1958 B. HAMILTON *Too Much of Water* xi. 243 'How was it that the '*Dichterliebe*' made you cry?' 'I can't answer that. Somehow it got me.'

22. b. To obtain an audible signal from (a radio or television transmitter or station); to 'pick up'.
1899 *Windsor Mag.* X. 145/2 Another station can always get us in a few minutes. 1921 *Wireless World* IX. 571/1 To ensure that the signals be picked up if it is at all possible to get them. *Ibid.* 586/2 If it be possible to get the Dutch concerts..with this three-valve set. 1924 *Wireless Ann.* 25 Hello, did you get me? 8 XY standing by. 1947 M. LOWRY *Under Volcano* vi. 162 Hugh..turned the radio dial back and forth, trying to get San Antonio. 1965 H. C. WOODRUFF *Short Wave Listener's Guide* (chapter for U.K. readers), Although one may expect to be able to get the majority of the stations in this list, the frequencies..may well be different from those quoted..for American reception.

c. To get in touch by telephone with (a person or place).
1907 [see sense *7 d, above]. 1908 G. B. SHAW *Let.* 1 July (1956) 126, I have just telephoned Stella...I tried to get Mrs. Pat herself...I also tried to get Forbes..but he was out. 1958 WODEHOUSE *Cocktail Time* xxi. 174 He took up the receiver...'I hear you've been trying to get me. What's your trouble?'

d. To answer (a telephone, door-bell, etc.); usu. *to get it.*
1941 THURBER & NUGENT *Male Animal* I. 29 (Doorbell rings.).. *Tommy* (going to door). I'll get it.

24. *to have got to* (further examples); also (orig. *U.S.*) in the sense of 'must', 'to be certainly'.
1865 'L. CARROLL' *Alice in Wonderland* iv. 54 The first thing I've got to do is to grow to my right size again. *Ibid.* 57 I'd nearly forgotten that I've got to grow up again. 1875 'MARK TWAIN' in *Atlantic Monthly* Mar. 283/2 This has got to be learned. 1919 E. JORDAN *Girl in Mirror* (1925) iv. 79 You'll have to see me every day. I've got to look after you. 1968 *Amer. N. & Q.* Mar. 104/1 If there's cockfighting today it's got to be in the bluegrass around Lexington, Kentucky. 1968 L. O'DONNELL *Face of Crime* (1969) ix. 125 Now I was really appalled. 'You've got to be kidding.' 1969 R. AIRTH *Snatch!* vi. 61 He said to Morland, 'This guy has got to be a comedian.'

26. (Later *poet.* example not referring to animals.)
1923 R. GRAVES *Feather Bed* 23 We are his sons Got on she-furies of our Northern gales.

27. d. *to get* (someone) *at it*: to have (a person) 'on', to make fun of. *slang.*
1958 F. NORMAN *Bang to Rights* III. 136 You see I did this on perpose just to get her at it. *Ibid.* 151 He had half sused that the boggie was getting him at it.

31. c. *to get nowhere, not to get anywhere*: not to reach any goal or object; to make no progress; to achieve nothing; *to get somewhere*: to be successful; to make some progress.
1923 H. C. WITWER *Fighting Blood* i. 18 I'm going to *get* somewhere! Right now I ain't got no more idea than a baby of what I'm going to be. 1925 *New Yorker* 14 Nov. 13/3 If he was sharper, this Sandburg, he'd get nowhere. 1932 W. CATHER *Obscure Destinies* i. 84 Mrs. Rosen felt that she was not getting anywhere. 1940 E. H. W. MEYERSTEIN *Let.* 4 June (1959) 242 Even when Jews 'get somewhere'—if they marry Englishwomen they are condemned by their wives. 1960 C. DAY LEWIS *Buried Day* viii. 174 Unless you get inside their ring you will get nowhere. 1961 *Guardian* 4 Nov. 7/6 Talking alone will get nowhere.

d. For *U.S. colloq.* or *slang* read *colloq.* (orig. *U.S.*). Also = sense 72 g in Dict. (Add earlier and later examples.)
1864 *Harper's Mag.* Oct. 565/2 Belaboring the mules till he was tired, and telling them to 'git' till he was hoarse. 1874 'MARK TWAIN' *Sk.* 12/1 Then he says, 'one-two-three-git!' 1888 'R. BOLDREWOOD' *Robbery under Arms* II. xi. 190 'I reckon you're bound to git.' 'Yes, Bill, sharp's the word.' 1892 H. NISBET *Bushranger's Sweetheart* xxiii. 176 None of your damned impertinence. Get. 1907 *Daily Chron.* 21 Nov. 5/1 In other words, as the Americans said, 'That fleet can get.' 1938 G. GREENE *Brighton Rock* II. i. 81 Turn out the light and get. 1959 'D. BUCKINGHAM' *Wind Tunnel* viii. 72, I want you out of the way—so git. And what's more—git quickly. 1967 K. GILES *Death & Mr. Prettyman* ii. 62 Anybody in a room either gets or pays for another twenty-four hours.

g. To reach the point or stage where; freq. in *U.S.* const. *so. colloq.*
1906 E. DYSON *Fact'ry 'Ands* viii. 98, I got I could pick 'em out in me sleep. 1944 E. S. GARDNER *Case of Crooked Candle* (1947) xviii. 193 You get so you know your way around. 1967 *Boston Sunday Globe* 23 Apr. B. 41/5 It's getting so now that real estate news is getting 'daily space'..to match the Vietnam war stories.

32. c. *to get going*: to begin; to start talking, acting, etc., vigorously; to get into full swing; to 'get a move on'. Also *trans.*, to start; to render (someone) excited, talkative, etc. See also *to get cracking* s.v. *CRACK *v.* 22 b.
1897 O. W. HOLMES *Pollock–Holmes Lett.* (1942) I. 77 He is really fine when he gets going on the Church of England. 1898 E. N. WESTCOTT *David Harum* 391 David is not only living, but appears almost no older than when we first knew him, and still just as likely to 'git goin'' on occasion. 1920 S. LEWIS *Main St.* xxviii. 326 She kidded him along, and got him going. 1932 'A. BRIDGE' *Peking Picnic* iv. 38 She's rather a character, you know, when you get her going. 1956 A. H. COMPTON *Atomic Quest* i. 8 If this task is as important as you men say.. we must get going. *Ibid.* iii. 189 To get the Hanford plant going.

33. *to get left*: see LEAVE *v.*¹ 7 d; *to get lost* (*slang*, orig. *U.S.*), to go away; to take oneself off (freq. *imp.*); *to get next to*: see *NEXT *a.* 13 c.

1947 F. WAKEMAN *Hucksters* v. 66 If Kimberly were to walk in tomorrow..I'd tell him to get lost. 1959 *Times* 25 Sept. 9/2 Tell him to get lost. 1961 'B. WELLS' *Day Earth caught Fire* ix. 145 'Cut that out, man,' the beatnik said...'Get lost, man,' replied Pete. 1962 'H. CALVIN' *System* xiii. 179 The last time Carabine came in I told him to get lost. 1967 WODEHOUSE *Company for Henry* v. 80 Can I have a word with you?..In private. Get lost, young Jane.

b. *how —— can you get?*: a colloq. phr. implying that the person referred to has an extreme amount or an excess of the quality described by the adjective.
1951 H. WOUK *Caine Mutiny* VII. 488 How unconscious can you get? Don't you know to-day's Navy Day? 1966 'A. GARVE' *Murderer's Fen* II. iii. 84 Damn it, the writer himself admits he isn't sure.—How vague can you get? 1967 J. FLEMING *No Bones about It* 80 Ben! How old-fashioned can you get? 1968 'P. HOBSON' *Titty's Dead* xii. 125 There's been an affair...And I never twigged it. How dim can you get?

VI. With preps.

35*. **get across —.** See *ACROSS B. 1 c and 2 b.

35.** **get around —.** = *get round* — (42 a, b). *U.S.*
1849 G. A. F. RUXTON *Life in Far West* 89 One from the Land of Cakes..sought to 'get around' (in trade) a right 'smart' Yankee, but couldn't 'shine'. 1875 MRS. STOWE *We & Neighbors* iii. 38 Eva is my girl; I sha'n't let anyone get around her. 1875 'MARK TWAIN' in *Atlantic Monthly* Mar. 283/2 This has got to be learned; there isn't any getting around it. 1894 —— *Those Twins* iv. 362 There is no getting around proof like that.

36. **get at —. c.** Also, to solicit or pester; to try to influence.
1952 W. SPROTT *Social Psychol.* (1964) vii. 123 We are all 'propaganda conscious' in the sense that we put up a resistance if we feel we are being 'got at'. 1958 *Times Lit. Suppl.* 31 Jan. 57/1 We resent, as the Victorians did not, being 'got at' by the social or religious moralist.

d. (Earlier and later examples.)
1823 J. CONSTABLE *Let.* 2 Aug. (1964) II. 283, I fear my great coat is got at by moths, as I find my father's is that I am come down here with. 1891 *Ally Sloper's Half Holiday* 3 Jan. 7/1 'Your family don't seem to get on, missie?'..'*On!* Who're ye gettin' at?' 1895 *Punch* 14 Dec. 277/1 Smart women..delight In 'getting at' you in a shameful way. 1957 J. OSBORNE *Entertainer* iii. 30 Don't look hurt. I'm not getting at you. I love you very much.

e. To begin; to start work on; to turn one's attention to. *U.S. colloq.*
1884 'MARK TWAIN' *Lett. to Publishers* 14 Apr. (1967) 173 Get at your canvassing early, and drive it with all your might. 1923 H. CRANE *Let.* 6 Feb. (1965) 118, I have been so rushed around..that I have not yet got at the review for your study.

f. To mean or intend; to hint, imply; usu. in phr. *what are you getting at? colloq.*
1899 D. BELASCO *Naughty Anthony* II, in *Heart of Maryland* (1941) 294 What are you getting at? What do you refer to when you call me the husband? 1921 *Collier's* 26 Mar. 22/1 'Say, what are you gettin' at?' says the kid, interested at last. 1931 N. COWARD *Post Mortem* vi. 75, I wish I knew what you were getting at.

36*. **get by —.** To succeed in getting past (someone); to evade. *colloq.*
1904 S. E. WHITE *Blazed Trail Stories* II. v. 199 How he had gotten by the office boy Brown could not conceive. 1919 H. CRANE *Let.* 13 Dec. (1965) 27, I am thoroughly confident about the thing itself since it has got by the particular, hierarchic Josephson.

37*. **get inside —.** To penetrate; to investigate closely; to achieve a deep or intimate understanding of.
1875 *Athenæum* 14 Aug. 222/2 More important..is the power of getting inside a character and revealing it to the public. 1959 *Listener* 16 July 112/3 No need to stress the good qualities of William Parsons as a vocalist or his ability to get inside a song and really interpret it. *Ibid.* 23 July 146/1 No one, not even Thurber, ever got inside Harold Ross. *Ibid.* 150/1 Mr. Denis Constanduros's *As Far as the Flagstaff*..got inside the doldrum life of a retired engineer.

38. **get into —. d.** Also, to become interested, involved, or absorbed in; to specialize in (sometimes with mixture of sense 38 e).
1938 E. HEMINGWAY *Fifth Column* (1939) 220 'What are you reading?' '*Richard Feverel.*' 'I couldn't get into it.' 1966 'C. KEITH' *Elusive Epicure* (1968) v. 69 He did advise me one time to get into Early American antique furniture. 1969 *It* 11–24 Apr. 11/2 What sort of things are you getting into musically now?

g. To take possession of; to 'come over'.
1876 'MARK TWAIN' *Tom Sawyer* iii. 37 All through supper his spirits were so high that his aunt wondered 'what had got into the child'. 1937 I. BAIRD *John* xiv. 163 You, too? Why, what's got into you tonight? 1946 D. STIVENS in *Austral. Short Stories* (1951) 386, I dunno what's got into you to-night.

h. To have sexual intercourse with (a woman).
c 1888–94 *My Secret Life* in S. Marcus *The Other Victorians* (1966) iv. 166, I felt as if I was wicked in getting into her, almost as if I was going to poke my mother. 1922 F. HARRIS *My Life & Loves* I. iii. 61 Again I dreamed of Lucille and again I was trying, trying in vain to get into her when again the spasm of pleasure overtook me. 1957 J. KEROUAC *On Road* (1958) 44 I've just got to get into her sister Mary tonight.

39. **get off —. c.** (Earlier example.)
1835 J. CONSTABLE *Let.* 12 Sept. (1967) V. 27, I wish I could get off going there to lecture.

40. **get on —. g.** To affect in such a way as to harass or obsess; to become a source of worry to; esp. in phr. *to get on one's nerves* (see *NERVE *sb.* 8 e).
1920 R. MACAULAY *Potterism* III. ii. 127 'Never mind Arthur,' she said. 'I wouldn't let him get on my mind if I were you, mother.'

40*. **get outside —.** To eat (occas., to drink). Also, *to get outside of. slang.*
1886 GREEN & HALL *Jack in Box* 15 Here, get outside some grub. 1886 P. G. EBBUTT *Emigr. Life Kansas* 182 Directly he got outside of a few glasses of whisky, his manner was very different. 1909 S. WATSON *Wops the Waif* xi. 26 So git outside your scran as quick as yer knows how. 1927 WODEHOUSE *Meet Mr. Mulliner* viii. 256 He lay there in a sort of delirium, picturing himself getting outside a medium-cooked steak smothered in onions. 1967 D. CAMPBELL in *Coast to Coast 1965–6* 21 It takes me half an hour to get outside the mixed grill and the ice-cream and coffee.

41. **get over —. b.** (Earlier and later examples.)
1712 C. MORDAUNT *Let.* in E. Hamilton *Mordaunts* (1965) iv. 83, I hope she may be in no great danger but got over the Measles as easily as I did. 1966 *Listener* 13 Jan. 77/1 'I shall never get over her death..,' he said, over eighty years later.

f. (Earlier example.)
1840 H. COCKTON *Life Valentine Vox* viii. 49 And as the old boy's not always exactly wide awake, he's to be got over just in the same way.

g. *to get over the footlights* = to get across (see *ACROSS B. 2 b). *U.S.*
1915 *Munsey's Mag.* Aug. 515/1 Shaw was generally considered altogether too wild to stand a chance of getting over the footlights.

44. **get to —. b.** To bribe. *U.S. slang.*
1927 *Dialect Notes* V. 447 *Get to one*, to bribe. 1930 E. D. SULLIVAN *Chicago Surrenders* i. 10 Gangsters can't operate on a satisfactory scale anywhere until they have 'got to someone'.

c. To worry, depress, or obsess; = sense *40 g above. *U.S. slang.*
1961 in WEBSTER. 1968 *New Yorker* 28 Dec. 42/2 It depresses me, but I don't let it get to me. *Ibid.*, You can't excuse yourself that way, any more than you can let drunks and such get to you.

d. To get across to (an audience, etc.) (see *ACROSS B. 2 b). *U.S. colloq.*
1968 *Globe & Mail* (Toronto) 13 Jan. 28/6 They didn't even realize that they were hearing a great man in Tea-garden..even though we always got to them by the end of the evening.

VII. With adverbs.

49. **get about. b.** (Earlier example.)
1816 JANE AUSTEN *Emma* III. xii. 219 Mr. Weston.. did not conceive..that it would be of any consequence; for 'such things..always got about'.

50*. **get across.** See *ACROSS A. 4 and B. 2 b.
1913, etc. [see *ACROSS B. 2 b]. 1923 U. L. SILBERRAD *Lett. J. Armiter* vi. 148 Sorry—my fault—one fails to get across. 1928 *Observer* 1 Apr. 6 His verse..in spite of all the efforts of his friends and admirers..has not really 'got across' eleven years after his death. *Ibid.* 17 June 8 Something, somehow, fails to get across. For it cannot be denied that the actual story..is a little slow. 1930 *Times* 24 Mar. 15/5 He can 'get his words across' as Gilbert intended.

53. **get along. b.** (Earlier U.S. example.)
1830 [see *ALONG *adv.* 2 a].

53*. **get around. a.** *to get around to* = to get round to (sense *67 c).
1887 M. E. WILKINS *Humble Romance* 35 There has been a good many things I haven't got around to. 1936, 1952 [see *AROUND *adv.* 4].

b. To go round; to circulate; *spec.* to go out a great deal; to visit many places. Chiefly *U.S.*
1928 *Amer. Speech* III. 219 *Get around*, to..have many desirable dates. 'Mary Jane sure did get around last semester.' 1951 M. MCLUHAN *Mech. Bride* 60/2 The news got around fast. 1959 *Times Lit. Suppl.* 20 Mar. 159/3 Still, Mr. Donnelly has got around...He makes his way to places like Tashkent, Samarkand and Alma Ata.

54. **get away. a.** Also (usu. with a negative), to disregard or escape *from* (a fact, implication, etc.).
1912 T. DREISER *Financier* lvii. 601 A jail is a jail; and there is no getting away from that. 1930 A. CHRISTIE *Murder at Vicarage* xi. 84 It's his pistol—you can't get away from that.

b. Also as a colloq. expression of astonishment or incredulity = 'go on', you don't say (so).
1848 THACKERAY *Van. Fair* xxii. 190 'Don't trifle with her affections, you Don Juan!' 'Get away,' said Jos Sedley, quite pleased. 1960 H. PINTER *Dumb Waiter* 124 *Ben.* The lorry started and ran over him. *Gus.* Go on! *Ben.* That's what it says here. *Gus.* Get away. 1969 D. CORY *Night Hawk* 16 'Do you speak Spanish?' 'Of course I do. I *am* Spanish.' 'Get away.' 'I am. I can prove it.'

c. (Earlier and later examples.) Also (*colloq.*, orig. *U.S.*), to carry off successfully; to succeed in winning or stealing; to do (something) with impunity; freq. in phr. *to get away with it*: to succeed in what one tries; to act without being detected or punished; so *to get away with murder*: to get away with anything; to do whatever one wishes.
1878 J. H. BEADLE *Western Wilds* ii. 41 More'n once the robbers would tackle some gritty man that was handy with his 'barkers', an' he'd get away with two or three of 'em. 1886 *Boston Jrnl.* 18 Dec. 2/4 They got away with the pennant three successive seasons. 1892 *Congress. Rec.* 13 Dec. 122/2 [These gentlemen] will have to be content with the pitiful $240,000 that they have already 'got away with'. 1908 *Dialect Notes* III. 314 You can't get away with me. 1912 *Maclean's Mag.* Oct. 56/2 In the Elizabethan days you could assault the watch..and have a jolly set-to with the blades in any convenient angle of a wall and 'get away with it'. 1921 *Collier's* 26 Mar. 25/1 The Kid loved her enough for her to get away with murder—which he undoubtedly did. 1923 A. HUXLEY *Antic Hay* x. 145 He had no sense of time or of order. But he got away with it, as he liked to say. 1926 *Amer. Speech* I. 292/2 This stable has been getting away with murder. 1939 *Chatelaine* Sept. 19/2 Usually she's young enough and amusing enough to 'get away with murder'. 1945 E. WAUGH *Brideshead Revisited* 12 He would sometimes say of the ways of the Army in pay and supply and the use of 'man-hours': 'They couldn't get away with that in business.' 1958 *Times* 6 Oct. 3/3 A film set in Ireland and relying upon whimsical comedy can get away, if not with murder, at least with weaknesses. 1967 J. CAIRD *Murder Scholastic* viii. 98 George gets away with murder...I mean, he does things in his own way. 1968 *Listener* 8 Aug. 164/3 Nobody can quite believe that Mr Dubcek is being allowed to get away with it.

d. *spec.* in *Cricket.* To hit (the ball) past the fieldsmen, so as to make a run or runs; also with the bowler as object.
1868 *Baily's Mag.* July 128 The two first-named.. exhibited splendid defence; but they could not get the ball away. 1903 C. F. WOOD in H. G. Hutchinson *Cricket* xii. 379 Scoring is out of the question. You may stop the ball as long as your patience lasts, but you can't get it away. 1955 *Times* 12 May 4/4 Phillips was one of five freshmen to bowl. He is tall and an awkward man to get away by reason of his length.

f. *intr.* To begin; to start quickly; esp. of a plant, to start growing vigorously or well.
1930 *Forestry* IV. 113 If the oak got away quickly it would soon be out of reach of the weeds. 1933 *Jrnl. R. Hort. Soc.* LVIII. 99 You wish to ensure the roots getting away quickly into the new soil. 1950 *N.Z. Jrnl. Agric.* Oct. 295/1 There would be danger of fires getting away on sunny faces when the vegetation was very dry. 1957 *Jrnl. R. Hort. Soc.* LXXXII. 370 If the ground is broken into rubble the roots can get away and the plant will flourish. 1960 F. C. STERN *Chalk Garden* xv. 163 It is advisable to put them in as small plants as they get away much quicker than large plants. 1967 R. MACKAY *House & Day* 86 'The trees are quite big.'..'Yes. They've got away well now.'

55. **get back. b.** *to get one's own back*: to revenge oneself; to get even with someone.
1910 J. DRISCOLL *Ringcraft* 17 He wanted to get his own back, and..he fancied he saw his chance. 1914 G. B. SHAW *Pygmalion* IV, *Higgins*...You have wounded me to the heart. *Liza*...I'm glad. I've got a little of my own back. 1920 'IXION' *Motor Cycle Remin.* 124 Whenever he met me I was able to get my own back. 1922 *Westm. Gaz.* 28 Nov., Busby said that he did it to 'get his own back'.

c. *to get back at* (or *on*): to retort or retaliate upon. *colloq.* (orig. *U.S.*).
1888 *Chicago Inter-Ocean* (Farmer), The open letter writers are getting back at Sam for his fondness for tobacco. 1907 *Daily Chron.* 17 Oct. 3/3 You cannot afford to be rude to a journalist. Some day he will get back on you. 1923 WODEHOUSE *Inimitable Jeeves* ix. 90 A lesser man might easily have snatched at the chance of getting back at me a bit by loosing Cyril into my bedchamber. 1972 *Guardian* 11 Feb. 1/6 For most of question time..Mr Thorpe tried to get back at the..Prime Minister.

56*. **get by.** *intr.* To be successful in escaping or evading something; to succeed, get along, 'manage'; to pass muster, be acceptable; to get away *with*. *colloq.* (orig. *U.S.*).
1908 J. M. SULLIVAN *Crim. Slang* 12 *Getting by*, living without doing any hard work. 1918 in *Wine, Women & War* (1926) 24 Absorbing what's useful in foreign methods ..just getting by, myself...Too damned technical. 1922 H. CRANE *Let.* 2 Apr. (1965) 83 It has enough in it to 'get by' on the first reading with a rather pleasing effect. 1926 *S.P.E. Tract* XXIV. 122 That chap could get by with murder. 1939 WODEHOUSE *Uncle Fred in Springtime* i. 12 Polly thinks I can get by all right. 1952 A. WILSON *Hemlock & After* I. i. 20 *Our* old bus will get by with a new engine. 1968 *Listener* 14 Nov. 663/3 The pseudo-metaphysical jargon that gets by as art criticism today.

57. **get down. c.** *to get down on* (someone): to develop a dislike for or grudge against; to be hostile or oppressive to. *U.S. colloq.*
1875 'M. QUAD' *Quad's Odds* 381 The adult male population of the village got down on John Anderson Tompkins. 1898 E. N. WESTCOTT *David Harum* 105 Dave got down on him fer some little thing or other.

d. *intr.* To settle down *to* (something); to turn one's attention *to*; freq. in phr. *to get down to it*: to get started; to begin work seriously or energetically. See also *to get down to brass tacks* (*BRASS *sb.* 5 b), *to get down to cases* (*CASE *sb.*¹ 1 d).
1892 'MARK TWAIN' *Amer. Claimant* xiv. 108 You've got to get right down to it and amuse your mind. 1924 *Sunday Times* 30 Mar. 6/2 The Bishop of Beauvais, the Earl of Warwick, and Chaplain de Stogumber assemble round a table and 'get down to it'. 1930 J. B. PRIESTLEY *Angel Pavement* viii. 413 Then come back here, bring your notebook, and we'll get down to it. 1957 K. M. KENYON *Digging up Jericho* 39 The first stages of a dig.. start long before one actually gets down to excavating.

e. *trans.* To depress or weary (someone). Cf. DOWN *adv.* 18.
1930 C. V. GRIMMETT *Getting Wickets* v. 115 There is quite enough to worry about on the field without allowing the troubles of the game to 'get you down' when you leave it. 1932 N. COWARD *Cavalcade* III. ii. 137 Blues, Twentieth Century Blues, are getting me down. 1953 'N. SHUTE' *In Wet* v. 171 It's just being cooped up in the office gets you down a bit.

60. **get in. b.** (Earlier example.)
1700 J. VERNEY *Let.* 10 Nov. in M. M. Verney *Verney Lett.* (1930) I. x. 159 The Coll. may carry it for the County, & Sir T. Lee get in at Aylesbury.

d. *spec.* To bring in or buy; to get a stock of.
1869 MRS. H. WOOD *Roland Yorke* II. xx. 125 She [*sc.* the landlady] gets things in for us, and wants to be paid for them. 1893 *Chambers's Jrnl.* 1 July 414/2, I fetched water, got in sticks, cleaned boots. 1907 R. BROOKE *Let.* July (1968) 92 The Mrs Chaffey, the lady who will land us, wanted to know what food to get in. 1932 A. J. WORRALL *Eng. Idioms* ix. 72 Most shopkeepers are getting in their Christmas goods. 1962 *Oxford Mail* 22 June 4/7 So long as I can earn enough to pay the rent and get in the odd bottle, I'm happy.

j. (Earlier example.)
1838 MRS. GASKELL *Let.* 17 July (1966) 19, I had engaged a girl..to help in all the extra work, & meant to get all sorts of things in.

l. *to get in a word* (earlier example).
1835 H. C. ROBINSON *Diary* 12 Nov. (1967) 143 Rogers ..said in his sneering way: 'Can Mrs. Masquerier get in a word?'

m. *to get in bad* (or *wrong*): to incur dislike; to get into trouble; also *trans.*, to bring (someone) into disfavour; to get (someone) into trouble; cf. *BAD B. 1 c; so *to get in good*. Freq. const. *with*. *colloq.* (orig. *U.S.*).
1910 O. JOHNSON *Varmint* xi. 152 Dink, you're getting in wrong again. 1913 *Dialect Notes* IV. I. 26 *Get in bad*, to make a mistake or a false move. 1920 F. SCOTT FITZGERALD *This Side of Paradise* (1921) I. iv. 139 People are beginning to think he's odd... He certainly is getting in wrong. 1921 S. FORD *Inez & Trilby May* iii. 56 So much prattle about a rich uncle who couldn't be produced was bound to get us in wrong sooner or later. 1928 *Observer* 19 Feb. 16/2 Young Woodley..prefers poetry to cricket. That, of course, 'gets him in bad' with his house-master. 1928 *Sunday Dispatch* 22 July 22/3 It will be he who will get in bad with the fans. 1931 *Kansas City Star* 7 Nov., The husband, jumping at a chance to 'get in good' came home from work the next day with a bundle of books. 1966 *Listener* 27 Oct. 622/2 The speech he made in Chicago which got him in bad with the organizers of Negro protest marches.

62. **get off. a.** Also, to succeed in falling asleep; to fall asleep. Cf. sense *62 m.
1922 V. WOOLF *Jacob's Room* i. 17, I thought he'd never get off—such a hurricane. 1934 L. A. G. STRONG *Corporal Tune* III. iv. 267 If you find you can't get off tonight..don't lie awake. Ring your bell, and ask sister to give you my 'A' drink. 1969 A. LASKI *Dominant Fifth* v. 182 The doctor gave me some very good sleeping-pills and said I must take one every evening, and so I did, though not until I'm sure Tess has got off.

h. (Earlier and later examples.) Also *intr.*, to get married or engaged to be married.
1710 [see OFF *adv.* 3 b]. 1801 M. EDGEWORTH *Belinda* I. ii. 41 There's no less than six of her nieces, whom she has *got off* within these four winters—Not one of 'em now, that has not made a catch-match. 1868 F. C. BURNAND *Hit & 'Miss'* i. 14 You should have got her off as she's a daughter—Why, noblemen in numbers must have sought her. 1915 B. RUCK *Courtship of Rosamond Fayre* iii, 'Miss Urquhart's got off herself.' 'She has and she hasn't. Her chap's always away.' 1923 E. BOWEN *Encounters* 116 It had been difficult to get poor Cicely off.

i. (Earlier U.S. example.)
1849 *Yale Lit. Mag.* XIV. 187 There is the writing of one who tried to 'get off', as the boys say, something comic on every occasion.

j. *where one gets off*: the point beyond which one is not competent, entitled, or required to go; esp. in phr. *to tell* (someone) *where he gets* (or *to get*) *off*: to rebuke for presumption or interference; to 'tell off'. *colloq.* (orig. *U.S.*).
1900 ADE *More Fables* 163 He said he was a Gentleman, and that no Cheap Skate in a Plug Hat could tell him where to Get Off. 1922 S. LEWIS *Babbitt* vii. 93 Once in a while I got to assert my authority, and...I told him just exactly where he got off. 1932 A. J. WORRALL *Eng. Idioms* 73, I told him where he got off. 1953 J. TRENCH *Docken Dead* vi. 90 I'm sure you knew how to deal with the police. Told them where they got off, I expect. 1963 D. LESSING *Man & Two Women* 128 If just for once she told us where to get off.

k. *intr.* To become acquainted or friendly *with* (one of the opposite sex), esp. with amorous intentions. *colloq.*
1915 [see *CLICK *v.*¹ 1 d]. 1925 FRASER & GIBBONS *Soldier & Sailor Words* 104 *Get off with, to*, to make the acquaintance of or 'pick up' with anyone, usually some girl, without the formality of an introduction. 1925 F. LONSDALE *Spring Cleaning* I. 13 What fun it would be if one of us could get off with him. 1936 AUDEN *Look, Stranger!* 35 The lady who admires us, you Have thought you're getting off with too.

l. Of a jazz musician: to improvise skilfully. *U.S. slang.*
1933 *Fortune* Aug. 47/1 Returning to Trombonist Brown, he can *get off, swing it, sock it, smear it*, or *go to town* (all of which mean syncopate to beat the band). 1955 R. BLESH *Shining Trumpets* (ed. 3) xii. 289 The present-day solo is esteemed modern and full of ideas in direct proportion to the more unrecognizable it makes the melody. Such 'getting off' conceals lack of true invention.

m. *trans.* To succeed in getting (a child) to go to sleep. *colloq.* Cf. sense *62 a.
1951 N. MITFORD *Blessing* I. iii. 25 Well, I only hope he won't overexcite the poor little fellow. You know what it's like getting him off, evenings. 1968 A. LASKI *Keeper* ii. 23 Gavin's been playing up; teeth; I think she may have just about got him off.

n. Used as an exclamation expressing impatience or incredulity; = sense *54 b. *colloq.*
1958 J. WAIN *Contenders* 29 'Get off,' I said. I should explain that 'Get off' is an expression much used in North Staffordshire as an ironic rejoinder to obvious remarks.

63. **get on. b.** Often *to get a move on*: see *MOVE *sb.* 6.

d. Also *intr.*
1836 *Spirit of Times* (N.Y.) 5 Mar. 20/1 Other parties were anxious to 'get on' at this price, but could not succeed.

f. Freq. in phr. *to get on with it*: to continue with one's affairs; to pursue one's course.
1932 R. FRASER *Marriage in Heaven* II. vi. 161 I've always just let people get on with it, especially men, if they didn't like what I said or did. 1955 J. BINGHAM *Paton Street Case* v. 91, I started out with some idea of serving the community and bunk like that, and now the community can get on with it as far as I'm concerned. 1962 *Listener* 8 Feb. 242/1 The only thing for France was to get out and leave Guinea to get on with it on its own.

g. (Earlier and later examples of: *to get on in the world.*)
1809 [see ON *adv.* 9 a]. 1883 [see WORLD *sb.* 17 b]. 1911 G. B. SHAW *Getting Married* Pref., in *Doctor's Dilemma*, etc. 124 It used to be said that members of large families get on in the world.

l. With *to*: to grasp the meaning, truth or significance of; to understand; to detect or find. *colloq.* (orig. *U.S.*).
1880 *Chicago Inter-Ocean* 2 June 6/3 The visitors taking kindly to Ward's curves, Dunlap and McCormick especially getting on to him in fine style. 1889 J. W. RILEY *Pipes o' Pan* 28 Get onto that position for a poet! 1893 'JOHNSTON SMITH' *Maggie* xv. 130 Do yehs want people ter get onto me? 1923 WODEHOUSE *Inimitable Jeeves* ix. 97, I knew there wasn't a chance of my being able to work this stage wheeze in London without somebody getting on to it and tipping off the guv'nor. 1930 J. B. PRIESTLEY *Angel Pavement* vi. 277 That was a bit of smart thinking on your part... There aren't many men about here who could have got on to it like that. 1940 E. PERCY in *Best One-Act Plays 1940* 70 It's very fortunate I got on to it in time. I'm sure I've saved Ann a great deal of unhappiness.

m. With *to*: to get into touch or communication with (someone).
1895 W. B. YEATS *Let.* 3 Mar. (1954) II. 252, I am beginning to think of getting on at Roscommon to Douglas Hyde. I shall go from that to Dublin. 1955 *Times* 30 June 9/5 Then later I read the body was to be exhumed. This thing got on my nerves, so I got on to the police.

64. **get out. a.** *to get out from under*: see *UNDER *adv.* 4 b.

h. (Later examples.) Also *intr.*
1846 GEO. ELIOT *Let.* Mar. in J. W. Cross *Life* (1885) I. ii. 141, I wish we could get the book out in May. 1870 D. G. ROSSETTI *Let.* 3 Feb. (1965) II. 787, I suppose I cannot get out till April.

j. *Cricket.* To put out, dismiss (a batsman or side). Also *intr.*, to be put out. So *to get oneself out*: to be dismissed, to be got out, freq. with the implication that one is oneself largely to blame.
1833 J. NYREN *Young Cricketer's Tutor* 89 They were devilish troublesome customers to get out. 1836 [see OUT *adv.* 4 c]. 1897 K. S. RANJITSINHJI *Jubilee Bk. Cricket* iv. 178 People get themselves out off slow bowling more often than the bowler gets them out. 1908 E. P. OPPENHEIM *Missioner* I. vi. 62 Stephen is in now... If he gets out, the match is over. 1912 A. A. LILLEY *24 Yrs. Cricket* x. 137 He [*sc.* Victor Trumper]..never gave one the remotest suggestion that he would ever get out. 1926 J. B. HOBBS *My Cricket Mem.* xvi. 214 We did well to get them out for this total.

k. To succeed in solving or finishing (a puzzle, game, etc.). *colloq.*
1924 B. DALTON *Games of Patience* 34 *Lady Betty*... The game..is not easy to get out. 1928 R. KNOX *Footsteps at Lock* xvi. 158 He had 'got it out'. 'The cipher?' 'No, the patience.' 1931 N. COWARD *Post Mortem* ii. 16 Lady Cavan is seated at a bridge table playing Canfield Patience... *Lady C.* I got it out yesterday. 1951 C. P. SNOW *Masters* III. xli. 328 I've got it out!..I've got the answer to the slow neutron business.

66. **get over. c.** (Earlier example.) Also, *to get it over with.*
1813 JANE AUSTEN *Let.* 15 Sept. (1952) 323 At nine we are to set off for Grafton House, and get that over before breakfast. 1935 *Punch* 4 Sept. 262/2 Already from Australia I hear of 'meet up', 'rest up', and 'get it over with'. 1947 R. ALLEN *Home Made Banners* iii. 18 Figure I might as well sign up tomorrow and get it over with.

e. *trans.* and *intr.* = *get across* (see *ACROSS B. 2 b).
1916 *Picture-Play* III. 122 If he works from characters and uses..bits of effective business to 'get his plays over', he [*sc.* the scenario writer] should keep such material in handy files. 1920 WODEHOUSE *Jill the Reckless* xviii. 261 Dramatic critics..were telling each other that 'The Rose of America' was just another of those things but it had apparently got over. 1921 H. A. VACHELL *Blinkers* viii. § 3 Mrs. Merrytree, delighted to perceive that she had, in stage parlance, 'got over', held the situation firmly. 1928 *Sunday Express* 29 Apr. 5/6 Her friendliness 'gets over'.

67. **get round. c.** With *to*: to succeed in finding the time, energy, or inclination for (doing something); to come to the point of dealing with.
[1902 W. D. HOWELLS *Lit. & Life* 155 The high banks of the seasonable Christmas-week snow, which the street-cleaners had heaped up there till they could get round to it with their carts.] 1946 K. TENNANT *Lost Haven* (1947) xiv. 221 Everything in Lost Haven was put off until someone should have enough time to 'get round to it'. 1961 J. SEYMOUR *Fat of Land* viii. 106 Our neighbour Richard cans hares, but we never got round to that. 1967 K. GILES *Death in Diamonds* viii. 145 He must take Elizabeth there for a weekend, he resolved, with a slight undertone of sadness at the thought he would probably never get round to it.

68. **get through. e.** (Earlier example.)
1839 A. CONSTABLE *Let.* 4 Feb. in *J. Constable's Corr.* (1962) 306 Your Uncle Golding's affairs I have not got through with yet.

f. *trans.* To secure the implementation of (a bill or other political measure).
1873 'MARK TWAIN' & WARNER *Gilded Age* xx. 190 The Senator..favored the appropriation and he gave the Colonel..to understand that he would endeavor to get it through.

g. *intr.* To establish communication by radio or telephone. Also *trans.*, to send or receive (a message) by radio or telephone.
1895 A. R. BENNETT *Teleph. Syst. Europe* 11 The delay and uncertainty in getting through would probably deter him from using the telephone at all. 1902 BEERBOHM in *Sat. Rev.* 27 Dec. 805/1 We feel..that he has rung up a messenger-boy after failing to 'get through' on the telephone. 1916 'BOYD CABLE' *Action Front* 188 The signallers leaped to their instruments, buzzed off the call, and getting through, rattled their messages off. *Ibid.* 189 They haven't had time since they got my message through. 1954 G. DURRELL *3 Singles to Adv.* iv. 85, I tried to contact McTurk to let him know that we were coming, but I could not get through.

h. With *to*: to reach the attention or understanding of (someone); to communicate with. *colloq.*
1961 in WEBSTER. 1962 J. BRAINE *Life at Top* iii. 54 He had defeated me; I couldn't think up any way to get through to him. 1969 A. HUNTER *Gently Coloured* ii. 14 You don't have to answer them, but you can do. Am I getting through to you, Osgood?

70. **get together. b.** (Later examples.) Also, to confer; to meet in friendly conference; to agree.
1816 JANE AUSTEN *Emma* II. iii. 47 It is such a happiness when good people get together. 1889 *Judge* (U.S.) 10 Aug. 282/1 Five Men..are to be hanged on the same day. In other words, they will follow Mr. Dana's advice and get together. 1889 *Puck* (U.S.) 14 Aug. 418/2 'I saw you conferring with Congressman Shouter this morning.' ..'Why, yes; he said that we ought to get together.' 1904 *N.Y. Times* 23 Dec. 1 The jury was unable to get together, and the Presiding Justice had ordered them locked up for the night. 1923 *Illustr. London News* 1 Sept. 418/3 So widely divergent are the standpoints that I wonder the police authorities do not get together..and formulate a standard practice.

c. *trans.* In *Rowing*, to cause (a crew) to work together. Also *intr.*
1876 E. D. BRICKWOOD *Boat Racing* I. viii. 97 If the progress made by the crew is satisfactory, and they have got well together, a regular racing outrigger may be substituted for the tub. 1888 W. B. WOODGATE *Boating* xii. 170 The other days are long-course days of long grinds, to get men together, and to reduce weight. 1898 *Encycl. Sport* II. 280/2 As the day of the race..draws near, the attention of the coach must be given entirely towards getting the crew absolutely together.

d. To organize, harmonize, put in order. *slang* (orig. *U.S.*).
1962 *Down Beat* 12 Apr. 22, I guess I was on my way in '57, when I started to get myself together musically. 1969 *It* 4–17 July 10/3 The promoters just hadn't got things together at all and would have made an immense loss.

72. **get up. a.** Colloq. phrases: *to get up and get* (U.S.), *to get up and go*: to start moving quickly or acting energetically; to make haste; *to get up early* (*in the morning*): to be alert, wide awake, or quick.
1864 B. COTTON *Songster* 10 Monsieur will be invited to just 'get up and get'. 1870 'MARK TWAIN' *Lett. to Publishers* (1967) 39 Have written first four chapters of the book, and I tell you the 'Innocents Abroad' will have to get up early to beat it. 1884 [see EARLY *adv.* 1 a]. 1903 J. FOX *Little Shepherd* xxii, A voice bellowed from the rear..'Git up and git, boys!' That was the order for the charge. 1940 F. L. ALLEN *Since Yesterday* iv. 79 They were exhibiting the same emotional willingness to get up and go, they knew not where, that was being exhibited in Germany by multitudes of men and women. 1960 J. MORTIMER *Lunch Hour* 150 Well, you didn't pin it on Sammy Noles. Oh, no. Sammy gets up too early in the morning for that little carry on.

h. (Earlier example.)
1828 *Sporting Mag.* Feb. 244/2 Straight-armed bowlers are invariably slow bowlers. Their balls, indeed, get up fast, but they never come fast to the long stop.

VIII. *Comb.* **73. b.** **get-off** (earlier example); (*b*) = *TAKE-OFF *sb.* 3 b; (*c*) the action of 'getting off' in jazz (see *62 l); an improvisation or 'break'; also *attrib.*; **get-on,** one who 'gets on'; a successful person; also *attrib.*; **get-there** (orig. *U.S.*), energy, ambition; also *attrib.*

1832 *Chambers's Jrnl.* I. 121/2 As a get-off, she commences a eulogy on her butter. **1898** E. N. Westcott *David Harum* xix. 169 He hain't got much 'git there' in his make-up. **1901** *Daily Chron.* 22 June 10/5 Their style of rowing..is certainly the 'get there' style. **1908** *Westm. Gaz.* 10 Mar. 14/2 A little weary of this 'get on' gospel being continually dinned in their ears. **1908** *Daily Chron.* 13 Mar. 4/6 Prophetically numbered by Landor among the 'ons' who are get-ons. **1916** H. Barber *Aeroplane Speaks* 50 The Pilot turns the Aeroplane in order to face the wind and thus secure a quick 'get-off'. **1932** *Melody Maker* July 593 There is an abundance of trumpet-playing of the first order from the local 'get-off' man. **1935** *Vanity Fair* Nov. 71 Breaks are sometimes known as *get-offs* or *take-offs*. **1956** M. Stearns *Story of Jazz* (1957) xvi. 180 The formula consisted of importing one or two 'hot' soloists, or 'get-off' men, letting them take a chorus once in a while.

d. The *intr.* verb with an adj., forming compound adjectives, as **get-tough; get-well,** esp. designating cards or other forms of message sent to a sick person to express good wishes for his recovery.

1959 *Daily Mail* 1 Apr. 7/2 The United States wants to force its 'get tough' line on the other Western Powers. **1960** *News Chron.* 14 Oct. 1/5 The Government's get-tough policy follows a growing demand..to 'teach the natives a lesson'. **1971** *Guardian* 19 Nov. 14/3 Criminal Justice Bills tend to be miscellaneous collections of proposals...The current one..is a sweet-and-sour concoction, with several 'get tough' provisions. **1956** B. Cleary *Fifteen* (1962) x. 145 She looked over the get-well cards in Woodment's stationery store. **1966** *Guardian* 29 Dec. 7/4 Mrs. Kennedy..sent him a 'get well' telegram.

geta (gē·tă), *sb. pl.* [Jap., f. *ge* lower, under + *ta* footwear.] Wooden shoes worn by the Japanese, with thongs between the big toe and the small toes.

1884 tr. *Rein's Japan* 416 He leaves his geta or zōri at the door, so as not to soil the beautiful mats. **1897** *Outing* (U.S.) XXX. 214/2 Their wooden geta clattering and slapping their soles as they go. **1904** *Daily Chron.* 13 June 8/5 Some of the 'geta' worn by little girls are painted in many colours. **1905** *Westm. Gaz.* 16 Mar. 2/1 The Yorozu are artistic to the stilts of their wooden geta. **1947** J. Bertram *Shadow of War* 334 If the youngest tot stumbled in his *geta* he was bullied unmercifully.

get-away, getaway (ge·tăwē). [Get *v.* 73.] The action of getting away; *spec.* (*a*) the breaking cover of a fox; (*b*) an escape; a method or chance of escape; esp. of thieves with their booty (often *to make one's get-away*); (*c*) the start of a race. Also *attrib.*, esp. designating a vehicle in which thieves make their get-away.

1852, 1890 [see Get *v.* 73]. **1893** L. W. Moore *His Own Story* xlviii. 622 Only three [prisoners] had made their escape. Mrs. Keene had prevented a clean 'get-away' for many more. **1907** C. E. Mulford *Bar-20* xv. 166 We'll have to make our get-away plumb sudden or we'll never go. **1922** *Punch* 1 Nov. 423/1 Quickly removing the ball from my toe I crawled under the net and made a get-away in the referee's car. **1923** *Motor Cycling* 26 Sept. 657/2 No one failed to start, although in general the getaways were not so fast or neat as in the case of the trade riders. **1924** [see *bean *v.*]. **1930** E. H. Lavine *Third Degree* (1931) iv. 38 He will make his getaway before the fun starts. *Ibid.* xi. 129 The stolen get-away Cadillac car. **1966** M. R. D. Foot *SOE in France* x. 325 Robert Benoist, arrested in the street three days after his brother, made the first sensational get-away of a splendid run of escapes. **1968** 'L. Marshall' *Blood on Blotter* xxvi. 174 You provided the job, you cased it, took care of the law, provided the get-away car.

Gethsemane (geþse·mănĭ). [ad. L. (Vulgate) *Gethsemani*, Gr. Γεθσημανῆ, ad. Aramaic *gath shemānī(m)* oil-press.] The name of a 'garden' or enclosure on the Mount of Olives, scene of the agony of Christ (*Matt.* xxvi. 36–46). Hence **a.** A representation in painting or sculpture of Christ's agony in the garden. **b.** A scene of spiritual or mental anguish; an instance of such anguish.

1901 *Westm. Gaz.* 26 July 2/1 A great, sad, German-looking Gethsemane, its big white figures shining mysteriously behind a grating. **1913** D. H. Lawrence *Love Poems* 16 And I at the foot of her cross did suffer My own gethsemane. **1923** —— *Birds, Beasts & Flowers* 53 See it [*sc.* an almond-tree] come forth in blossom.. Sweating his drops of blood through the long-nighted Gethsemane Into blossom.

get-out. *colloq.* [f. Get *v.* 64 and 65.]

1. Phr. *as* or *like* (*all*) *get-out*, used to indicate a high degree of something.

1838 J. C. Neal *Charcoal Sks.* 12 We look as elegant and as beautiful as get out. **1869** W. C. Hazlitt *Eng. Proverbs* 60 As common as get out. *Cornw.* **1881** J. Sargisson *Joe Scoap's Jurneh* 119 He glooart at meh as impident as *git oot*. **1884** 'Mark Twain' *Huck. Finn* xxxviii, We got to dig in like all git-out. **1916** H. L. Wilson *Somewhere in Red Gap* v. 197 When I got into the parlour she had them on, pleased as all get-out. **1959** M. Steen *Tower* I. ii. 35 By then I was as drunk as all-get-out. **1967** V. Canning *Python Project* vii. 146 They looked as comfortable as all get-out.

2. An escape from a difficult position; an evasion, withdrawal.

1899 E. W. Hornung *Amateur Cracksman* 115, I ask you, was there ever a better get-out? **1909** *Westm. Gaz.* 4 June 7/3 I'm entirely in your hands. I have got no get-out. **1932** *Economist* 30 Jan. 221/2 An orderly 'get-out' from their German commitments. **1968** *Listener* 4 Jan. 27/2 The lads didn't know what they wanted to say and fell on the magic and mystery idea as an unanswerable get-out.

3. *Theatr.* The total weekly cost of a theatrical production; orig., the ability of a touring company to leave a town after paying all expenses; the amount of box-office takings required to assure this. Also *attrib.*

1952 Granville *Dict. Theatr. Terms* 88 *Get out*, the ability to leave a town with all expenses paid... *Get-out figure*, the weekly running cost of a touring show. **1959** P. Bull *I know Face* iii. 54 In actual fact we should have been sitting pretty, as the get-out was only £663 a week. *Ibid.* 55, I was able to get the 'get-out' figure down to £500 a week. **1961** *Times* 3 Jan. 3/7 The 'get-out' for Brecht's *Galileo* (cast of 40) was £1,920 a week, for eight performances a week.

get-rich-quick, *a.* orig. *U.S.* [Get *v.* 33 and Quick *adv.*] Characterized by attempts at, or hopes of, acquiring wealth rapidly. Hence **get-rich-quicker,** a person who desires to make quick profits.

[**1902** O. Wister *Virginian* xxxi, He's disturbed over getting rich quick and being a big man in the Territory.] **1904** 'O. Henry' in *McClure's Mag.* Aug. 360/1 A get-rich-quick—excuse me—gang giving back the boodle! Oh, no. **1910** *Daily Chron.* 22 Apr. 1/6 Tales of the get-rich-quick order are quite common. **1914** *Chambers's Jrnl.* Feb. 144/1 California passed from the 'get rich quickers' into the hands of scientific exploiters. **1940** J. Buchan *Memory Hold-the-Door* v. 114 The get-rich-quick offscourings of European capitals. **1957** *Sat. Rev.* 8 June 25 As a businessman, yes; as a private enterpriser, a get-rich-quicker, of course. **1960** *Times* 5 Dec. 4/2 The Government..had made things far too easy for the get-rich-quick boys. **1971** D. Potter *Brit. Eliz. Stamps* i. 15 Stamps were solemnly discussed in the financial columns, ..and get-rich-quick operators joined in the *mêlée*.

gett, var. *get *sb.*[3]

getter, *sb.* Add: **1. c.** A substance used to remove residual gas from an evacuated enclosure by chemical or other action. Also *attrib.* and *Comb.*

1922 *Proc. Inst. Radio Engin.* X. 469 By 'getter' we mean the substances such as phosphorus, arsenic, sulphur and so on. These substances, when applied to metals and volatilized by heating them, trap the gases. **1943** J. Yarwood *High Vacuum Technique* iv. 38 These 'getters' during volatilization combine with the residual gases in the vessel and they are fixed as chemical compounds deposited on the walls of the tube. **1944** *Light Metals* Jan. 34/1 Getters are used in all radio valves.. and in cathode-ray-oscillograph bulbs. **1959** *New Scientist* 26 Mar. 708/3 In other types of getter-ion pumps, the getter film is produced by sputtering—that is, by knocking atoms out of the bulk getter metal by bombarding it with the ions of the residual gas.

getter (ge·təɹ), *v.* [f. prec.] *trans.* **a.** To remove (gas) by means of a getter. **b.** To evacuate (an enclosure) by means of a getter. So **ge·ttered** *ppl. a.*; **ge·ttering** *vbl. sb.*, freq. *attrib.*

1922 *Brit. Pat.* 187,055 3/1 The present invention.. depends upon the process of 'gettering'. **1925** F. H. Newman *Prod. & Measurem. Low Pressures* vii. 133 In the 'gettered' lamp a solution of red phosphorus in alcohol was painted on the leads below the filament. **1933** *Phil. Mag.* XVI. 673 Gettered hydrogen may be re-evolved from magnesium owing to a 'vapour pressure' exerted by the cleaned-up gas. *Ibid.* 674 Hydrogen may be gettered by and reliberated from magnesium. **1943** J. Yarwood *High Vacuum Technique* v. 51 Gettering is carried out at the lowest possible pressure, and the magnesium or barium pellet reduces the pressure in the valve to less than 10^{-5} mm. Hg. **1967** Ward & Bunn *Introd. High Vacuum Technol.* iii. 65 The gettering process..can obviously be used to remove gas from a system and hence produce a pumping action.

ge·t-toge:ther. *colloq.* (orig. *U.S.*). [Get *v.* 70 b.] A meeting, gathering; an informal conference; *esp.*, an informal social gathering. Also *attrib.* Hence **ge:t-toge·therness.**

1911 *Springfield Weekly Republican* 9 Feb. 12 It was the biggest get-together the organization has ever held. **1932** *News Chron.* 23 Sept. 5/2 All passengers greatly enjoyed the Get Together Dance last night. **1935** Wodehouse *Luck of Bodkins* xv. 173 Brother Bodkin said as much last time we were having one of our get-togethers. **1936** 'J. Tey' *Shilling for Candles* xv. 169 The article.. finished with a real Hopkins touch of get-togetherness. **1958** *Punch* 1 Jan. 60/3 Accepting one of his hostess's 'specials' at one of her now-famous 'get-together' parties. **1959** J. Arden *Party* I. i, in J. C. Trewin *Plays of Year* XVIII. 238 You know Richard, how he really loves a get-together. A party, eh, on Wednesday? **1962** *Times Lit. Suppl.* 30 Nov. 938/3 An era of political and ecclesiastical get-togetherness. **1966** J. Caird *Perturbing Spirit* xv. 161 She normally dislikes crowds and jollity and get-togetherness.

Getulian (gĭtiū·liăn), *a.* and *sb.* Also **Gætulian, Getulan.** [f. L. *Gætūli, Gētūli*, ad. Gr. Γαιτοῦλοι, perh. from Berber root.] **A.** *adj.* Of or pertaining to the ancient nomadic Berber people inhabiting the desert region to the south and east of Numidia, or to the region under their control.

1607 Topsell *Four-f. Beasts* 659 This Getulian or Barbarian Squirrell, is of mixt colour, as it were betwixt black and red. **1697** Dryden *Virgil's Æneis* iv. 56 Getulian Cities here are spread around. **1850** W. Smith *Dict. Gr. & Roman Biogr.* II. 636/1 Some of the Gaetulian tribes were at the same time subjected to his sway. **1935** A. Toynbee *Study Hist.* I. 241 An infiltration of 'Getulan' Nomadic hunters from North-West Africa. **1950** *Chambers's Encycl.* X. 126/1 A new kingdom of Mauritania now limited Numidia in the west, as did a confederacy of free Gaetulian tribes in the south.

b. *Archæol.* (See quots.)

1927 Peake & Fleure *Hunters & Artists* iv. 40 It [the Capsian industry]..seems to occur in two cultural regions, an eastern or *Getulian*, and a western or *Ibero-Maurusian*, each of which displays characteristic differences. **1932** *Antiquity* June 193 In North Africa the Upper Palaeolithic industries are known as *Capsian* (Early and Late, or *Getulian*).

B. *sb.* One of the Getulian people.

1607 Topsell *Four-f. Beasts* 201 The *Getulians* fought with them with Speares and Dartes. **1854** W. Smith *Dict. Gr. & Roman Biogr.* I. 925/2 The ancients clearly recognised the distinction between the Gaetulians and the Negro peoples who dwelt S. of them. **1910** *Encycl. Brit.* I. 360/1 The Berber tribes..bore in ancient times the generic names of Numidians, Gaetulians and Moors or Maurusiani. **1928** V. G. Childe *Most Anc. East* iii. 48 And others [*sc.* migrants], conveniently termed Getulans, remained in North Africa.

get-up. Add: **4.** Inclination to get up and be active; energy, enterprise, determination. Also *get-up-and-get, get-up-and-go*, etc. (cf. *get *v.* 72 a). *colloq.* (orig. *U.S.*).

1841 'Dow, Jr.' *Short Patent Sermons* 86 It flats right down, and stays there, like a junk of dough—no get up to it. **1863** *Ladies' Repository* Aug. 477/1 In vain I tried to convince him that there was some 'get-up' in the animal. **1884** G. W. Peck *Peck's Boss Book* 183 The adjutant..felt that his position demanded a horse that had some git-up-and-git. **1889** *Kansas Times & Star* 2 Dec., It's not wealth, nor birth, nor state, But get-up-and-get, that makes cities great. **1907** N. Munro *Daft Days* xvi. 142, I wish to goodness folk here had a little git-up-and-go to them! **1910** W. M. Raine *B. O'Connor* 223 When it comes to the get-up-and-hustle, she's there. **1915** Wodehouse *Something Fresh* x, He'll make a name for himself one of these days. He's got get-up in him. **1947** J. Steinbeck *Wayward Bus* 123 It's a pleasure to talk with a man with some get-up and go. **1962** *Punch* 11 Apr. 573/2 Tortoises are not easy to race because they are devoid of get-up-and-go. **1964** Mrs. L. B. Johnson *White House Diary* 12 Jan. (1970) 42 Lyndon went to church...I am sorry I did not have the get-up-and-get to go with him.

geyser. Also pronounced (gī·zəɹ). Add: **2.** (Earlier and later examples.) Also for the heating of water for use in wash-basins, sinks, etc. (See also *geezer.)

1878 *Gas Engineer* Feb. 184 (Advt.), The instantaneous water heater; or Maughan's Patent Geyser..so constructed that any quantity of hot water can be drawn from it with the utmost facility. **1920** L. C. Hale *American's London* viii. 109 The aristocratic landlady was telling me of the advantage of her own particular geezer... I moved closer to descry the lettering on the cylinder, and lo! it was a geyser. I suppose the word is universally mispronounced over here because they have not been brought up in a geyser country. **1929** *S.P.E. Tract* xxxii. 400 The mechanical device for heating bath-water made *geyser* a household word, and though the introducers gave it the vowel of *grey*, the pronunciation as in *key* gained ground. **1965** G. McInnes *Road to Gundagai* v. 83 A crude precursor of the 'Ascot' gas heater—known to various Lady Helps as the geezer, the gayzer or the guyzer.

gezumph, var. *gazoomph *v.*

ghaffir (gā·fīᵊr). Also **ghafir,** etc. [ad. Arab. *ġafīr*.] A native Egyptian policeman; a guardian, watchman.

a **1817** J. L. Burckhardt *Notes on Bedouins* (1831) I. 366 The Arab guardian was called *ghafeyr*. **1856** R. F. Burton *Narr. Pilgr. El-Medinah* I. 231 Without guns or gunners, it [*sc.* the fort] is occupied by about a dozen Fellahs, who act as hereditary 'ghafirs', (guardians). **1883** D. M. Wallace *Egypt* 216 At the bottom of the communal organisation stand the Gaffirs or village watchmen. **1923** *Glasgow Herald* 19 Feb. 9 The whole distance was guarded..by 'ghaffirs', or watchmen. **1928** *Blackw. Mag.* Feb. 248/1 A canal watchman, known as a Ghaffir. **1938** W. B. Ziff *Rape of Palestine* 421 Even the Jewish *Ghaffirs* and supernumerary constables were usually armed with truncheons only. **1971** G. Household *Doom's Caravan* iv. 169 The naval stores were..guarded by Arab ghaffirs.

Ghanaian (gānē·ăn), *sb.* and *a.* Also (as short-lived variants) **Ghanan, Ghananian, Ghanian.** [f. *Ghana* + -ian.] **A.** *sb.* A native or inhabitant of Ghana, a West African state formerly known as the Gold Coast. **B.** *adj.* Of, pertaining to, or characteristic of Ghana or its people.

1949 *Accra Even. News* in B. Timothy *K. Nkrumah* (1955) vii. 81 Kwame Nkrumah..Born a true Ghanaian for Ghana. **1957** *Evening Standard* 9 Apr. 4/6 After dinner the Ghanians, wearing national costume, will give a dancing display. **1957** *Economist* 31 Aug. 671/1 The two Muslims were Ghanaians. **1957** *Times* 17 Dec. 9/6 The C.P.P.'s greatest handicap has been failure to capture the sympathy of the Ghanaian middle classes. **1964** P. Strevens *Papers in Lang.* (1965) vi. 81 There are well-known accents of English..as a second language (such as Ghanaian's English accent or Urdu-speaker's English accent). **1966** *Daily Tel.* 29 Dec. 9/1 A ship, ordered by the Nkrumah régime and apparently not wanted by the present Ghanaian Government. **1966** *Transition* (Uganda) XXVI. 17/2 But by the tragedy of his domestic excesses *after* independence, Nkrumah fell short of becoming a great Ghanaian.

ghanta (ga·ntă). Also **gunté.** [Hind. *ghaṇṭā*.] A bell or gong, used as an instrument in Indian music.

1865 *Proc. R. Irish Acad.* IX. I. 108 (*Gunté*).—Bell. As a musical instrument, the bell is used somewhat in the same manner as the cymbals. **1954** *Grove's Dict. Mus.* (ed. 5) IV. 459/2 The conch (*śaṅkha*), of few but mellow notes, is used in the temple, as also the bell (*ghaṇṭa*).

ghap (gāp). Also **ghaap, ghab, guaap, ngaap.** [Nama name.] A South African carrion flower belonging to the genus *Trichocaulon*, esp. *T. piliferum*.

1819 *Rees Cycl.* XXXIII. (s.v. *Stapelia pilifera*), The Hottentots are said to eat it, knowing it by the name of *Guaap*. **1857** L. Pappe *Floræ Capensis* (ed. 2) 29 *Stapelia pilifera*..is eaten by the natives, who call it *Guaap*, for the purpose of quenching their thirst. **1878** *Trans. S. Afr. Philos. Soc.* I. 24 Here we find stunted Mesembryanthemums, numerous kinds of Euphorbias, the peculiar *Ghap* (Stapelia) in at least a half dozen varieties. **1913** Pettman *Africanderisms* 186 *Ghab*..*Stapelia pilifera* (as well as others of these Carrion-flowers as they are called), is known by this name in the Karoo. [**1932** Watt & Breyer-Brandwijk *Medicinal & Poisonous Plants S. Afr.* 151 Natives formerly ate the stem of *Stapelia pilifera* L., Nama *guaap*, as a thirst quencher. The Namas still use it for this purpose.] **1962** *Ibid.* (ed. 2) 138 The Trichocaulons, known generically to the Hottentots as *ngaap* (the spiny-stemmed species) are mostly edible in the raw state.

gharana (garā·nă). Also **gharwana.** [Hind. *gharānā*.] In Indian music, a school of players who practise a particular style of interpretation.

1957 O. Gosvami *Story Indian Mus.* xxiv. 260 These [groups] came to be known as *Gharwanas* or schools of interpretations [*sic*]. **1967** B. V. Keskar *Indian Mus.* i. 20 The tendency to run down the music of other *gharanas* increased. There was great rivalry between these various *gharanas*. **1968** *Indian Mus. Jrnl.* V. 33 Though based on the traditional Gwalior *Gharānā* he evolved his own individual style. **1969** *Cultural News fr. India* Nov. 30 The active society Raag Rang brought us Mushtaq Hussain Khan's son Ishtiaq and grandson Sharafat in a splendid programme of the Rampur gharana.

gharial, gharrial, varr. Gavial.

1864 [see Gavial]. **1923** *Chambers's Jrnl.* Dec. 811/1, I had constant opportunity of seeing and observing the habits of the gharial. **1930** *Times Educ. Suppl.* 29 Nov. p. iv/3 The Zoo has received five young Eastern gharials from Borneo.

Ghassulian (gæsū·liăn), *a.* *Archæol.* [ad. Fr. *Ghassoulien* (R. Neuville 1930, in *Jrnl. Palestine Oriental Soc.* X. 202), f. the site Teleilat el-*Ghassul* near Jericho in Jordan: see -ian.] Of, pertaining to, or designating a chalcolithic culture of which remains have been found at Teleilat el-Ghassul. Also as *sb.*, an inhabitant of this area in the chalcolithic age.

1931 *Bull. Amer. Sch. Oriental Res.* Apr. 15 Then comes the characteristic Chalcolithic of Palestine, the Ghassulian. **1949** W. F. Albright *Archaeol. of Palestine* ii. 45 The new Ghassulian culture was at first dated much too low by the excavators. *Ibid.* iii. 64 The Ghassulian graves excavated by Stekelis were dug in the ground and lined with stones. **1952** V. G. Childe *New Light on Most Anc. East* (ed. 4) xi. 229 The Ghassulians in Transjordan were buried individually in stone cists, sometimes covered by round barrows supported by a circular kerb of boulders. **1966** R. de Vaux in *Cambr. Anc. Hist.* (ed. 3, 1970) I. ix(b). 523 The name 'Ghassulian' has sometimes been applied to everything which is regarded as 'Chalcolithic' in Palestine. This usage is inaccurate, but there are certain sites, especially in the southern half of the country, which were related to Ghassūl and were almost contemporary with it.

ghastly, *a.* Add: **1. b.** Also applied to persons.

1931 F. Grierson *Mystery in Red* iv. 70 Don't be a ghastly idiot. **1960** N. Mitford *Don't tell Alfred* vii. 76 Then he'd never have heard of this ghastly Yanky. *Ibid.* 81, I bought her [*sc.* a tortoise] from those ghastly children. **1969** C. Fremlin *Possession* i. 9, I know his mother, and she's *frightful*! She really is, Clare, I promise you: she's *ghastly*!

ghaut, ghat. Add: **4.** In full *burning-ghat*. A level spot at the top of a river ghat on which Hindus burn their dead; a funeral pyre.

1877 *Encycl. Brit.* VI. 567/1 To erect cinerators on the burning ghat or ground. **1885** Kipling *Life's Handicap* (1891) 328 A woman's corpse going down to the burning-ghat. **1908** *Daily Chron.* 14 Nov. 5/4 At the ghat, or funeral pyre, a mixture of milk and cocoanut water.. was poured into the mouth of the corpse. **1937** L. Bromfield *Rains Came* I. i. 12 He came frequently along these paths to the ghats after nightfall. There was a kind of macabre beauty about the place, and in the spectacle of the cremation itself there was a kind of faith and certainty. **1964** A. Swinson *Six Minutes to Sunset* ii. 33 The corpses were taken to the burning *ghats* outside the city.

ghawazee (găwā·zi), *sb. pl.* Also **ghawâzi, ghowazee.** [ad. Arab. *ġawāzī*, pl. of next.] Egyptian dancing-girls.

1799 W. G. Browne *Trav. Afr.* vi. 86 The last are the female dancers or *ghawasié*. **1836** E. W. Lane *Mod. Egypt* I. 240 The *Ghawa'zee* (or public dancing-girls). *Ibid.* II. 95 The Ghawa'zee perform, unveiled, in the public streets, even to amuse the rabble. **1845** R. Ford *Hand-bk. Trav. Spain* I. 188 The *ghowazee* of the Egyptians. **1924** 'S. Rohmer' *Grey Face* iv, Her hair, which was dressed in the fashion of the dancing girls who pose for ever upon the Egyptian monuments, and which is preserved in life to this day by the ghawâzi of Keneh. **1964** W. G. Raffe *Dict. of Dance* 198/1 *Ghawazee* or *Ghazeeyeh* (Egypt, Modern Arabian), the professional public dancing girls.

ghazeeyeh (gazī·ye). Also **ghazie.** [ad. Arab. *ġāziya*.] An Egyptian dancing-girl.

1819 T. Hope *Anastasius* I. xv. 325 A knot of ghazie distorted their limbs into as uncouth postures as if they had been frogs themselves. **1877** *Encycl. Brit.* VII. 727/1 The Ghawâzee (sing. Ghâzeeyeh) form a separate class. **1964** [see prec.].

gheblee, var. *gibli.

ghee. Add to def.: Also made from cow's milk. Add examples of the form *ghi*.

1873 H. Blochmann tr. *Abul Fazl 'Allami's Ain i Akbari* I. 130 Most of them get..4s. of g'hí, and half a *man* of rice. **1939** M. Pallis *Peaks & Lamas* ii. 21 In most districts the village shop can be relied on for flour, rice, lentils, *ghi* (clarified butter), sugar, onions and pepper. **1957** *Encycl. Brit.* X. 323/2 *Ghee*...The best is prepared from butter of the milk of cows, the less esteemed from that of buffaloes. **1959** *Times* 31 Dec. 10/6 A ramshackle bus which smelt of packed humanity, spiced with garlic, *ghi*, [etc.]. **1967** Singha & Massey *Indian Dances* iv. 59 A small diva (oil-lamp) which burnt ghee (clarified butter) was placed on a salver. **1971** *Illustr. Weekly India* 4 Apr. 31/3, I tried to retaliate by objecting to his eating spiced food cooked in ghee.

Gheez, var. *Geez.

Gheg (geg). Also **Geg, Gegde, Gheghe, Ghegide, Ngeghe.** A people of Northern Albania; a member of this people; also, the language spoken by this people. Also *attrib.* and as *adj.*

1812 Byron *Ch. Har.* 129 Though they have some cavalry amongst the Gegdes, I never saw a good Arnaout horseman. **1835** W. M. Leake *Trav. N. Greece* I. ii. 61 The only important divisions of the Albanians are four: the Ngéghe, Tóshke, Liápe and Tjame. **1851** E. Lear *Jrnls. Landscape Painter in Albania* 110 These Gheghes are all armed. *Ibid.* 125 The Gheghe Albanians..are still allowed the privilege of carrying arms. **1891** *Trans. Philol. Soc. 1888–90* 355 The Gheg or Central Albanian of Albania. **1954** Pei & Gaynor *Dict. Linguistics* 83 *Gheg*, one of the two predominant dialects of Albanian, spoken in northern Albania and parts of Dalmatia. **1958** P. Kemp *No Colours or Crest* v. 77 The rivalry of Gheg and Tosk and the age-long hatred of both for their Slav and Greek neighbours. *Ibid.* 78 A stocky, granite-faced, illiterate Gheg from Krujë in the Mati country. *Ibid.*, Gheg tribesmen from Kossovo and his own territory of Mati. **1971** P. Driscoll *White Lie Assignment* ix. 73 My method has kept me safe for twenty years against stupid Ghegs like him.

Ghent (gent). *Hort.* The name of a city in Belgium (Flemish *Gent*, French *Gand*), used *attrib.* to designate any of a number of hybrid azaleas first developed by P. Mortier of Ghent between 1804 and 1834.

1841 J. Loudon *Ladies' Companion* 23/1 The kinds called the Ghent Azaleas, are hybrids and varieties raised in Belgium. **1882** Lindley & Paxton *Paxton's Flower Garden* I. 65 Blooming somewhat later than the Ghent varieties. **1889** W. Robinson *Eng. Flower Garden* (ed. 2) 769/2 The hardy Azaleas, called Ghent azaleas, have sprung chiefly from the wild azaleas that grow in North America. **1963** *Oxf. Bk. Garden Flowers* 56 *Azalea pontica*..is one of the species used in developing the first 'Ghent' azaleas.

gherao (gerau·). [f. Hind. *gherna* to surround, besiege.] A form of harassment in labour disputes in India and Pakistan, whereby workers detain their employers or managers on the premises, refusing to let them depart until their claims are granted. Hence as *v. trans.*, to detain (a person) in this manner.

1967 *Statesman* (Calcutta) 7 Apr. 1 (*heading*) Gheraos must be declared illegal, say employers. *Ibid.*, There had been 'gheraos' on minor issues like the quality and quantity of food to be supplied by a canteen. **1967** *Economist* 9 Dec. 1049/2 Hospitals were gheraoed; so were the Calcutta city government and the West Bengal assembly. **1968** *Ann. Reg. 1967* II. iv. 79 Industrial labour diminished production by frequent organization of *gheraos* or forceful confinement of the managers. **1969** *Times* 22 Mar. 1/5 Pakistan workers..have 'gheraoed' (put pressure upon) two Scottish executives. **1969** *National Herald* (New Delhi) 29 July 3/2 They threatened Academic Council members with gherao. **1969** *Amrita Bazar Patrika* 5 Aug. 5/5 More than 50 students of Bengal Veterinary College gheraoed the Director of Veterinary Services..and other officers of the department in the New Secretariat building for three hours on Monday.

ghetto, *sb.* Add: [Perh. f. It. *getto* foundry, as the first ghetto founded in Venice in 1516 was on the site of a foundry.] **2.** *transf.* and *fig.* A quarter in a city, esp. a thickly populated slum area, inhabited by a minority group or groups, usu. as a result of economic or social pressures; an area, etc., occupied by an isolated group; an isolated or segregated group, community, or area.

1892 I. Zangwill *Children of Ghetto* I. I. 2 The particular Ghetto that is the dark background upon which our pictures will be cast is of voluntary formation. **1897** [in Dict.]. **1908** J. London *Martin Eden* (1910) xxxvi. 310 They dismounted and plunged off to the right into the heart of the working-class ghetto. **1909** *Westm. Gaz.* 20 Aug. 8/3 The people..have grown superior to the banal excitement and cheap attractions of the densely crowded areas. The day on which the tramways went over Westminster Bridge recorded the unlocking of the London ghettos. **1937** *Times* 6 Oct. 13/7 Part of the benches [in the Warsaw Polytechnic] have been marked for students belonging to a union almost exclusively controlled by 'Aryans', and others for the Jewish students' union, while a few seats for non-union students are left unmarked...The establishment of the 'bench ghetto' is an important precedent, unknown even in Germany. **1957** *Times Lit. Suppl.* 29 Nov. 713/3 On records the Coloured jazz musicians still played largely for their race; in life they played for the immigrants into the Negro city ghettoes. **1961** L. Mumford *City in History* xvi. 493 The suburb..was a segregated community..a sort of green ghetto dedicated to the elite. **1961** *Listener* 7 Dec. 1000/3 [The television programme] 'Bookstand', (November 30), capriciously shifted to the Tuesday ghetto, had one memorable item. **1966** *Ibid.* 29 Sept. 454/1 The ghetto is one of two in which most of Chicago's Negroes, who make up a quarter of the city's population, are forced to live. **1968** *N.Y. Rev. Books* 11 July 34/1 The 'breakdown of the Catholic ghetto' is a good thing, but the breakdown of intellectual ghettos at M.I.T. and Harvard might be, educationally, an even better thing.

3. *attrib.* and *Comb.*

1892 I. Zangwill *Childr. Ghetto* (1893) 2 The Ghetto.. becomes only a swarming-place for the poor and the ignorant...Such people are their own Ghetto gates. **1903** *Daily Chron.* 5 Aug. 5/2 What he calls the Ghetto face and the Ghetto eye, observable enough in immigrants, cannot be detected after a generation or two on American soil. **1908** J. London *Martin Eden* (1910) xxxviii. 328 Tell them why you don't want Socialism. Tell them what you think about them and their ghetto ethics. **1941** Koestler *Scum of Earth* 48 The country which was the first to introduce yellow ghetto benches in its schools. **1949** —— *Promise & Fulfilment* II. iv. 251 The victory of the new type of Israelis grown on Palestine soil over the obstinate fanaticism of ghetto-bred politicians. *Ibid.* III. I. 294 The same ghetto-heritage of suspicion. **1968** *Guardian* 24 Oct. 10/2 Those ancestors of today's ghetto-dwellers came to areas where, unlike the South, there was no discrimination. **1969** *Ibid.* 17 Sept. 10/2 Catholics are not going to abandon the ghetto mentality which the events of the past month have created unless Stormont shows some belated signs of having the stomach for dealing with Protestant extremists. **1971** *Radio Times* 16 Sept. 37/5 Social workers are becoming increasingly worried by the 'ghetto mentality' in the deprived areas of our cities.

ghetto (ge·to), *v.* [f. the sb.] *trans.* To put or keep (people) in a ghetto. So **ghe·ttoed** *ppl. a.*; **ghe·ttoize** *v. trans.*; also **ghettoiza·tion.**

1936 *Times* 15 Feb. 11/3 Jews, who are ghettoed under the racial legislation. **1939** *Jewish Standard* (Toronto) Apr. 13 An element of ghettoization, self demoralization. **1939** *Canadian Jewish Chron.* 4 Aug. 3 Arcand's attempt ..to ghettoize a minority. **1964** M. A. Galamison in J. H. Clarke *Harlem* 225 It is a land of ghettoized human beings: men denied creative work. **1965** *Listener* 15 Apr. 543/1 We don't ghetto them or keep them out of work. *Ibid.* 545/2 The immigrant can be ghettoed psychologically. **1970** *Guardian* 27 Oct. 11/8 Skilled white workers are moving South from the ghettoed and polluted cities of the North.

ghilgai, var. *gilgai.

ghillie, var. Gillie[1].

Ghilzai (gi·lzəi). Also **Ghilji.** The name of one of the most famous of the tribes of Afghanistan. Also *attrib.*

1826 Leyden & Erskine tr. *Mem. Zehir-Ed-Din* 220 Plundering and beating up the quarters of the Ghiljis. **1842** C. Masson *Narr. var. Journeys* II. 207 The Ghiljis, although considered, and calling themselves, Afghâns.. are undoubtedly a mixed race. **1875** *Encycl. Brit.* I. 235/1 On the British invasion the Ghilzais showed a rooted hostility to the foreigner. **1920** *Blackw. Mag.* Aug. 172/2 Information at hand pointed to the Ghilzais from Afghanistan. **1971** *Whitaker's Almanack* 815/2 The most numerous race [of Afghanistan] is the Pathan.., the main divisions being the Durranis, from whom the Royal family springs, and the Ghilzais.

Ghiordes (gi,ǭ·ɹdĕz). The name of a town (Gördes) in western Turkey, used *attrib.* or as *adj.* to designate a fine type of Turkish rug, or a kind of knot used in weaving some oriental carpets (see quots.).

1900 J. K. Mumford *Oriental Rugs* x. 147 In some Ghiordes rugs the main border is made up of a pattern which..suggests a comb. *Ibid.* 159 The Ghiordes knot is used [in new Meles rugs]; the warp is of two- or three-strand wool. **1957** *Encycl. Brit.* IV. 918/1 There are two different kinds of knot employed, the Ghiordes or Turkish, and the Senne or Persian; the tufts of yarn in the former coming in pairs between the two warp threads. **1967** R. Jeffries *Deadly Marriage* ii. 28 Cabbot.., correctly identified it as a Ghiordes prayer mat.

gholam (golā·m). Also **gholaum, ghulam, goulam.** [Arab.] A courier, messenger.

1820 M. Tancoigne *Narr. Journey Persia* 251 The military household of the king of Persia..is composed of a certain number of cavalry, who are called goulams, or slaves. *Ibid.* 252 The goulams always escort the king when he appears in public, or when he leaves the city to hunt. **1840** J. B. Fraser *Trav. Koordistan* I. i. 21 'What am *I* to do then, your Excellency?' roared out some Gholaum or Beg. **1882** E. O'Donovan *Merv Oasis* I. 319 We had with us a *gholam*, or courier, belonging to the British Legation at Teheran.

ghoont (gūnt). Also **ghounte, goont, gunt.** [Hind.] A Himalayan pony.

1625 Purchas *Pilgrims* I. iv. iv. 438 Heere is the great breed of a small kind of Horse, called *Gunts*, a true trauelling scalecliffe beast. **1834** tr. *Jacquemont's Journ. India* II. 12 In Cashmere, I shall buy, without regard to price, the best *ghounte* in Tibet. **1858** Simmonds *Dict. Trade*, *Ghoont*, a small surefooted Indian pony, used in the mountain ranges as a pack or saddle-horse. **1871** E. Balfour *Cycl. India* (ed. 2) II. 312/2 During winter the ghoont live on the roots of the stunted bushes, and are very expert at scraping the snow from off them with their fore feet.

Ghoork(h)a, varr. *Gurkha.

ghost, *sb.* Add: **3. b.** Philos. *the ghost in the machine*: Gilbert Ryle's name for the mind viewed as separate from the body (see quots.).

1949 G. Ryle *Concept of Mind* i. 15 Such in outline is the official theory. I shall often speak of it, with deliberate abusiveness, as 'the dogma of the Ghost in the machine'. *Ibid.* 22 The dogma of the Ghost in the machine..maintains that there exist both bodies and minds; that there are mechanical causes of corporeal movements and mental causes of corporeal movements. **1960** J. O. Urmson *Conc. Encycl. Western Philos.* 350/1 We are inclined to construe the concept of mind as of an extra object situated in the body and controlling it by a set of unwitnessable activities; this is what he [*sc.* Ryle] calls the dogma of the ghost (the mind) in the machine (the body). Ryle regards this picture as totally misleading. **1961** *Mind* LXX. 103 Certainly he [*sc.* Teilhard de Chardin] imports a ghost, the entelechy or *élan vital* of an earlier terminology, into the Mendelian machine. **1967** Koestler (*title*) The ghost in the machine.

8. b. (Earlier examples.) Also *fig.*

1716, etc. [see Lay *v.*[1] 3 b]. **1826** [see Raise *v.*[1] 20 a]. **1833** R. Dyer *9 Yrs. of Actor's Life* 53 If I played with applause, it was a matter of indifference whether 'the ghost' walked on Saturday or not.

10. c. An impression of a signature made by folding the paper over while the ink is still wet.

1929 *Sotheby's Catal.* 4–7 Nov. 82 On the back of one 'ghost' there is a note in the Author's hand.

11. d. *Spectroscopy.* A spurious spectral line produced by periodic errors in a diffraction grating.

1879 C. S. Peirce in *Amer. Jrnl. Math.* II. 330 (*heading*) On the ghosts in Rutherford's diffraction-grating. *Ibid.* 338 The positions of some of these 'ghosts', or repetitions of the principal spectrum, have been carefully measured. **1882** [see sense 11 b in Dict.]. **1963** R. W. Ditchburn *Light* (ed. 2) vi. 188 Every fifth line [of a diffraction grating] slightly displaced has the effect of superposing a 'grating' of spacing five times the true spacing and gives a 'ghost' whose apparent wavelength is λ/5.

e. *Biol.* A cell wall or cell membrane, esp. of a red blood corpuscle, that has lost its protoplasmic contents; also, in extended use, a bacteriophage with an empty 'head'.

1890 *Jrnl. Chem. Soc.* LVII. 499 The parenchymatous tissue of the endosperm portions..is completely disintegrated, the cell-walls either entirely disappearing or remaining in a much swollen and altered form as mere 'ghosts'. **1902** *Jrnl. Exper. Med.* VI. 267 Whether this increase of permeability persists when the corpuscles have been reduced to ghosts by the escape of hæmoglobin I am unable to say. **1951** *Jrnl. Bacteriol.* LXI. 753 The ghosts have a host range specificity similar to that of the virus from which they were derived. [*Note*] The word 'ghost' has been used because of its obvious relationship to the red cell ghost produced in a somewhat similar manner. It consists of the deflated head and the tail. **1959** Garen & Kozloff in Burnet & Stanley *Viruses* II. 207 Ghosts of a T1..and a staphylococcal phage have been seen in electron micrographs. **1970** *Nature* 31 Oct. 415/1 At least 50 per cent of the protein in red cell ghosts is present in this form.

f. *Metallurgy.* A faint band on the surface of steel due to the segregation of certain of its constituents.

1905 *Trans. Inst. Naval Archit.* XLVII. II. 369 All [these micrographs], however, showed lines which are technically known as 'ghosts', namely lines of metal high in sulphur and phosphorus. *Ibid.* 380 The well-marked 'ghosts' in the micro-structure. **1945** Greaves & Wrighton *Pract. Microsc. Metallogr.* x. 173 A ghost is due to local segregation of impurities during the solidification of the ingot.

g. *Television.* A displaced repeated image on a television screen caused by a duplicate signal travelling by a longer path.

1927 *Bell Syst. Techn. Jrnl.* VI. 646 When marked fading occurred, the normally clear [television] reproduction was accompanied by 'ghosts' or additional images which faded in and out. **1929** Sheldon & Grisewood *Television* xiii. 143 The 'ghosts' represented energy which had traveled from the transmitter upward to the Heaviside..whence it was reflected to the receiver. **1967** *Guardian* 24 June 1/3 After a rain storm of static and a flutter of television 'ghosts' we were able to recognise LBJ.

h. *Cinemat.* (See quots.)

1940 *Chambers's Techn. Dict.* 375/2 *Ghost*, vertical streaks on high-lights in a projected picture, arising from incorrect phasing of the rotary shutter with respect to the moving film. **1959** W. S. Sharps *Dict. Cinematogr.* 100/1 In film projection, a ghost is an extension above or below the main images on the screen and is caused by faulty synchronization between the shutter and intermittent mechanisms of the projector.

i. A spurious signal on a radar screen that does not correspond to a target at the indicated location.

1945 *Amer. Speech* XX. 309/2 *Ghost*, echoes which do not follow normal characteristics and for which definite targets cannot be found. **1961** D. J. Povejsil et al. *Airborne Radar* vi. 331 In a two-PRF system, two targets would yield four possible range values: two correct ranges and two 'ghosts'. **1969** Barton & Ward *Handbk. Radar Measurement* i. 12 Common examples are ambiguities in range and Doppler measurements..and the ghosts produced by scanning with two fan beams which are oriented at right angles to each other.

14. d. *ghost-white* adj.; *ghost-wise* adv.

1893 Kipling *Seven Seas* (1896) 61 A sloop-of-war, ghost-white and very near. **1954** L. MacNeice *Autumn Sequel* III. 27 Two great swans, ghost-white. **1905** *Daily Chron.* 16 May 4/5 He weeps, ghost-wise, at the auctioneer's bids in his bedroom. **1924** *Public Opinion* 11 July 48/3 Your nearness gathers ghostwise down the room.

e. ghost-family, the family of a ghost-marriage; **ghost-form** = *ghost-word*; **ghost-gum** *Austral.*, a species of *Eucalyptus* (cf. White-gum[2]); **ghost image,** applied to various kinds of spurious or false images (see quots. and sense 11 in Dict. and Suppl.); **ghost line** = sense *11 f; **ghost-marriage** (see quots.); **ghost-raiser,** one who 'raises a ghost' (see 8 b); **ghost-soul,** in folk-lore, a double or apparition of a person; **ghost town** orig. *U.S.*, a town partially or completely devoid of its inhabitants; **ghost train,** (*b*) a train of cars at a fun fair that travels through dark tunnels in which there are ghost-like effects; (*c*) a train run during the night to keep the track clear in periods of heavy snowfall or severe frost; **ghost writer** orig. *U.S.*, a hack writer who does work for which another person takes the credit (cf. sense 13 in Dict.); hence **ghost-write** *v. trans.* and *intr.* = *ghost *v.* 4; *ghost-writing* vbl. sb.; *ghost-written* ppl. adj.

1951 E. E. Evans-Pritchard *Kinship & Marriage among Nuer* iii. 110 The family that develops out of a ghost-marriage may be called a ghost-family in acknowledgement of the ghostly status of the pater of the children. It consists of a ghost, his wife, his children born in the union of marriage, and the kinsman who begat these children and acts as father to them. **1933** Bloomfield *Lang.* xxvii. 487 Chaucer's phrase *in derring do that longeth to a knight* 'in daring to do what is proper for a knight', was misunderstood by Spenser, who took *derring-do* to be a compound meaning 'brave actions' and succeeded in introducing this ghost-form into our elevated language. **1954** Y. Hillman in *Coast to Coast* 45 It was a good pelt and as white as a ghost gum. **1954** G. Dutton *Ibid.* 152 The chalky ghost-gums. **1959** *Listener* 7 May 785/1 On the slopes, and in the valleys too, grow ghost-gums—a species of eucalyptus with drooping, dark-green leaves and smooth sculpted trunks of an exquisite pallor. **1902** *Daily Chron.* 19 June 4/4 'The "Grubs" Collimating Telescope Gun Sight', an optical contrivance arranged so that a vertical or ghost image is made to appear as if projected upon the object aimed at. **1904** J. McIntosh *Photogr. Ref. Bk.* (ed. 2) ii. 43 The ghost image is a reflection of the bright object in front of the camera reflected from the back lens upon the front lens, and finally transmitted to the plate. **1940** *Chambers's Techn. Dict.* 375/2 *Ghost image*, the image arising from a mirror when the rays have experienced reflection within the glass between the surface and the silvering. **1953** L. J. Wheeler *Princ. Cinematography* vi. 185 If the shutter is out of phase with the pull-down a 'ghost image' will be seen on the screen. **1926** *Jrnl. Iron & Steel Inst.* CXIII. 109 When the ascending segregates are impeded by the growing crystals, they tend to elongate and form typical stringlike formations of segregate known as 'ghost lines'. **1951** E. E. Evans-Pritchard *Kinship & Marriage among Nuer* iii. 109 A very common feature of Nuer social life is a union I have called ghost-marriage. If a man dies without legal male heirs, a kinsman of his..ought to take a wife to his name. **1964** Gould & Kolb *Dict. Soc. Sci.* 388/1 *Ghost marriage*, the custom of marrying a girl to the name of a dead brother or other relative. **1886** J. K. Jerome *Idle Thoughts* 166 Memory is a rare ghost raiser. **1908** *Daily Chron.* 30 June 3/1 An efficient new ghost-raiser presents himself. **1871** E. B. Tylor *Prim. Cult.* I. 451 The notion of a ghost-soul animating man while in the body. **1924** W. B. Selbie *Psychol. Relig.* 29 As formulated by Professor Tylor this theory of a quasi-material ghost-soul is to be regarded as the typical and almost universal source of religious ideas and practices. [**1875** *Cincinnati Enquirer* 2 July 5/1 The deserted mining towns, like the ghosts of their departed prosperity.] **1931** G. F. Willison *Here they dug Gold* 71 Today all lie ghost towns smelling of the long slow processes of ruin and decay. **1944** F. Clune *Red Heart* 6 Then Broken Hill will be a Ghost Town—like White Cliffs which once boomed on opals. **1952** H. Innes *Campbell's Kingdom* I. iii. 51 Come Lucky was a rotten clutter of empty shacks. It was my first sight of a ghost town. **1967** *Times* 18 July 1/3 Qantara, after the Middle East war, has been turned into a ghost town, as not more than 1,000 of its 8,000 people remain. [**1925** A. Ridley (*title of play*) The Ghost Train.] **1936** Auden & Isherwood *Ascent of F6* II. iii. 100 The Ghost Train and the switchback did not always disappoint. **1958** A. Sillitoe *Sat. Night* xi. 157 Winnie clamoured for the Ghost Train. **1968** *Daily Express* 10 Jan. 1 (*headline*) Ghost trains fight big freeze. **1927** *Editor & Publisher* 18 June 9/1 When he isn't ghost-writing [he] is a member of the European staff of the New York Times. **1932** *New Republic* 10 Feb. 347 The autobiographical boloney ghost-written by Samuel Crowther for Ford. **1938** J. Hilton *To you, Mr. Chips* i. 37 The kind..whose autobiography, written or ghost-written, exudes the main idea that 'school made him what he was'. **1951** Wodehouse *Old Reliable* iii. 40 I'm ghostwriting the story of her life. **1956** Auden *Making, Knowing & Judging* 9 And end up by ghost-writing poems for him which he was too busy to start or finish. **1927** *Lit. Digest* 30 July 27 A ghost-writer may do all the work. **1928** J. P. McEvoy *Show Girl* 77, I told you I was one of them there ghost writers doing my bit for belle lettres. **1964** G. A. Williamson *World of Josephus* xvii. 261 Ghost writers are always with us. **1927** *Editor & Publisher* 18 June 9/1 Ghost-writing originated very naturally in the field of sports. **1964** M. McLuhan *Understanding Media* II. xxi. 212 Modern ghost-writing, teletype, and wire services create an insubstantial world of 'pseudo-events'.

ghost, *v.* Add: **3. b.** Of a sailing vessel: to make relatively good progress when there is very little wind.

1891 [in Dict., sense 3]. **1921** *Glasgow Herald* 6 July 8 The Lady Anne's remarkable facility for what is known as ghosting was undoubtedly the means of falsifying all forecasts of the issues. **1926** D. Byrne *Brother Saul* xvii. 450 Every inch of sail was set as they ghosted along. **1954** W. G. Arnott *Orwell Estuary* i. 1 We ghosted into the harbour in the gloom of an autumn evening. **1962** A. G. Course *Dict. Naut. Terms* 87 The British clippers, such as the *Cutty Sark*, were the best ever at ghosting. **1971** E. Fenwick *Impeccable People* iv. 24 They took an evening sail down the coast..and came ghosting back near shore in the quiet dark.

4. *trans.* and *intr.* To write (something) as a ghost-writer; to act as a 'ghost' (see Ghost *sb.* 13).

1922 *Glasgow Herald* 11 Oct. 3 'A certain general' for whom he did some 'ghosting'. **1925** T. E. Lawrence *Let.* 21 Apr. (1938) 474 A very distinguished person's wife once asked me if I would care to edit or 'ghost', her husband's diary. **1932** *Times Lit. Suppl.* 1 Sept. 609/1 The articles which he 'ghosted' for Jerry Turnbull, a millionaire with literary ambitions. **1954** Koestler *Invis. Writing* 338, I enjoyed fussing over her, ghosting her dispatches. **1959** M. Cumberland *Murmurs in Rue Morgue* viii. 57 A man of talent who has failed—to the point where he is ready to ghost for a man like Galotin.

ghoster (gōᵘ·stəɹ). *Naut.* [f. *ghost *v.* 3 b + -er[1].] (See quots.)

1956 E. C. Hiscock *Around World in Wanderer III* ii. 20 Such is the property of really light large sails, the ghoster gave us steerage way. *Ibid.* 267 *Ghoster*, a sail similar to a genoa but made of very light material for use in light airs. **1962** *Guardian* 28 June 5, I had my 'ghoster' (large Genoa rig) set this afternoon.

ghostiness (gōᵘ·stinės). [f. Ghosty *a.* + -ness.] The quality or state of being ghosty.

1895 G. Macdonald *Lilith* xi. 69 What a night I found it! How shall I make my reader share with me its wild ghostiness? **1904** *Westm. Gaz.* 1 Jan. 2/2 The ghostiness of these faded, frail old men gliding and whispering about in the old house.

ghostish, *a.* Delete †*Obs.*[−0] and add examples.

1911 D. H. Lawrence *White Peacock* II. iii. 245 It's rather ghostish to have the road suddenly smudged out. *Ibid.* vi. 306 You do see yourself a bit ghostish.

ghou·lishness. [-ness.] Ghoulish nature or quality.

1910 H. Walker *Lit. Victorian Era* II. vii. 505 Her poetry altogether escapes the taint of ghoulishness. **1928** *Weekly Dispatch* 24 June 12/4 What is there in poor human nature..that creates this ghoulishness? **1954** *New Yorker* 15 May 92 One of the best-known facts about him..has gained currency simply by virtue of its ghoulishness.

ghou·lism. [-ism.] Ghoulish characteristics or quality.

1959 I. & P. Opie *Lore & Lang. Schoolch.* ii. 32 *Ghoulism*. When children are about ten years old they enter a period in which the outward material facts about death seem extraordinarily funny. **1959** *20th Cent.* Dec. 426 In ghoulism the necrophiliac traffic is one way, as it were.

ghur(r)ial, ghuryal, varr. Gavial.

1835 [see Gavial]. **1878** P. Robinson *In my Indian Garden* 83 The ghurial is of a finer breed. **1905** *Kynoch Jrnl.* July–Sept. 152 On the sandbars Ghurrials in thousands basked in the sun.

G.I. (dʒi·,əi). (Stress variable.) **1.** Abbrev. of *galvanized iron*; used chiefly in *G.I. can*: (*a*) *slang*, a German artillery shell; (*b*) a galvanized-iron can; a rubbish-bin.

1928 *Papers Mich. Acad. Sci. & Arts* X. 294 *G.I. cans*, German heavy shells, so called because of their similarity to the galvanized cans found around hospitals. **1929** F. A. Pottle *Stretchers* (1930) 38 An army 'G.I. can' is identical with a civilian ash can. **1930** *Amer. Speech* V. 381 *G.I. can*, a German shell. *Ibid.* 384 *G.I. can or boiler*, galvanized iron utensil made for army cooking. **1951** *Word Study* May 8/1 Navy men of World War II will recall that on every ship and station big garbage cans were invariably referred to as G.I. cans.

2. Abbrev. of *government* (or *general*) *issue*. **a.** Used *attrib.* of equipment designed or provided for members of the armed forces of the U.S. and hence applied to things belonging to or associated with American servicemen; *G.I. bride*, a foreign woman married by an American serviceman while he is on duty abroad; *G.I. Joe*, an American soldier.

1936 *Amer. Speech* XI. 51 *G.I.* soap is government issue soap. **1940** *Ibid.* XV. 211/1 The boys are equipped with *Government Issue* materials—abbreviated to *G.I.* These include *G.I. shoes* and *clothes*, *G.I. trucks* and, if necessary, *G.I. pills*. *Ibid.*, An official camp dance is a *G.I. stomp*. **1942** E. Colby *Army Talk* 92 The girls whom kindly chaperones bring to an army dance..they call..'G.I. girls', and the dance a 'G.I. Hop'. The soldier calls peacetime maneuvers a 'G.I. War' and a close military haircut a 'G.I. haircut'. **1942** D. Breger in *Yank* 17 June 24 (*strip-cartoon title*) G.I. Joe. **1945** *News Chron.* 17 May 2 Four G.I. Joes calmly playing dice. **1945** *Evening Standard* 20 Dec. 3/2 (*headline*) Ships soon to take G.I. brides. **1947** Auden *Age of Anxiety* (1948) i. 26 Bringing to all John Doakes and G.I. Joes tidings of great joy. **1956** A. H. Compton *Atomic Quest* v. 337 The eagerness with which young men who had served in World War II accepted the educational opportunities offered by the GI Bill of Rights was not surprising. **1961** *Guardian* 18 May 8/3 A group of GI brides was asking Tracy for his autograph.

b. As *sb.* An enlisted member of a U.S. armed force.

1943 *Time* 15 Nov. 11, I am surprised that you ask for 'a name for soldiers who also serve—behind the front'... That term, of course, is 'G.I.' **1944** *Spectator* 5 May 400/2 Relations between British other ranks and their equivalents, the American G.I., notoriously vary. **1945** *Manch. Guardian* 8 Aug. 3 The appearance of American soldiers in uniform, 'G.I.'s' or officers, as 'guest' booksellers in British bookshops. **1957** J. Braine *Room at Top* vi. 55 The G.I. who drove the jeep he crashed into.

giant, *sb.* and *a.* Add: **A.** *sb.* **2. c.** *Astr.* One of the class of larger diffuse stars, as distinguished from the dwarfs (see *dwarf *sb.* 2 b).

1912 *Proc. Amer. Philos. Soc.* LI. 573 The naked-eye stars of this class are all 'giants'. **1925** F. J. M. Stratton *Astronomical Physics* ix. 126 For a given spectral type, the giants are redder or have lower effective temperatures than the dwarfs. **1956** H. S. Jones in A. Pryce-Jones *New Outl. Mod. Knowl.* 114 Some stars are so large that their diameters are several hundreds of times greater than the Sun's... Such stars are called giants.

6. giant-disc *v. trans.* (N.Z.), to cultivate by means of a machine with very large disc-cutters; **giant fibre** *Zool.*, an enlarged and modified nerve-fibre esp. in certain invertebrates; **giant racer,** a large switchback at a fun fair; also *fig.*; **giant('s) stride** (earlier and later examples) (cf. quot. 1862 for B. 1 a).

1951 *Landfall* V. 175 Billy's going to cut out that manuka so I can giant-disc it and put in a crop of turnips. **1963** *Weekly News* (Auckland) 3 July 37/2 We climbed a hill giant-disced to grey dust with chopped vegetation showing through. **1888** Giant-fibre [see *neurochord* s.v. Neuro-]. **1897** Parker & Haswell *Text-bk. Zool.* I. x. 438 Running longitudinally through the ventral cord in many forms are certain *giant fibres* of very large size. **1963** R. P. Dales *Annelids* vi. 111 Giant fibres have been known for some time, but their nervous nature was not at first appreciated. *Ibid.* 116 The giant-fibre system has been evolved for the rapid conduction of impulses. **1934** G. Greene *It's Battlefield* iv. 203 Marriage was the switch-back, the giant racer,..the guarantee that one would never be alone. **1949** 'J. Tey' *Brat Farrar* ix. 79 Lost in contemplation of the Giant Racer. **1863** Crown Princess of Prussia *Let.* 11 May in R. Fulford *Dearest Mama* (1968) 210 Having a swing, a see-saw and a giant stride put up for little and big children. **1906** M. H. Baillie-Scott *Houses & Gardens* 121 The woodwork of the swing and giant's stride is painted in gay colours. **1963** C. Mackenzie *My Life & Times* II. 29 A giant-stride is a column of wood from a revolving horizontal wheel at the top of which depend about a dozen ropes with small wooden bars at the end of them.

B. *adj.* **1. b.** (Further examples.)

1911 C. E. W. Bean *'Dreadnought' of Darling* xiii. 130 The giant emu, giant kangaroo, alligator, tortoise, and giant wombat. **1937** *Discovery* Jan. 27/1 The first Giant Panda to be captured alive was taken by Mrs Ruth Harkness, on a recent expedition into the frontier country between South-Western China and Tibet. *Ibid.* Oct. 308/2 The giant squid is certainly the largest invertebrate animal. **1947** I. L. Idriess *Isles of Despair* xxxv. 234 We must show you the giant clam. **1947** J. Stevenson-Hamilton *Wild Life S. Afr.* xxxiv. 300 The giant kingfisher (*Megaceryle masima*), attains a total length of close on eighteen inches. **1969** *Nature* 21 June 1126/1 In the wild, giant pandas are usually alone; except for mothers with cubs they have rarely been seen in company.

c. Applied to a star (see the sb., sense *2 c).

1912 *Proc. Amer. Philos. Soc.* LI. 571 The existence of these two series was first pointed out by Hertzsprung, who has called them by the very convenient names of 'giant' and 'dwarf' stars—the former being of course the brighter. **1913** *Observatory* Aug. 328 Giant stars must either have low density or great surface-brightness, and the reverse is true of the dwarf stars. **1959** *Listener* 3 Dec. 971/1 We have red giant stars which are accompanied by bluish-green companions.

giantesque (dʒəi,ănte·sk), *a.* [f. Giant *sb.* + -esque, after F. *gigantesque*.] Having characteristics of a giant, gigantic.

1909 in Webster. **1929** *Sunday Express* 20 Jan. 9 You gloat over their giantesque sins and sorrows.

giantism. Add: **c.** *transf.* and *fig.*

1936 M. Allis *Eng. Prelude* xxiv. 176 London remains the great metropolis, but those of the Midlands give an impression of giantism London lacks. **1952** *Archit. Rev.* CXI. 127/3 Giantism in classical architecture is a well-known phenomenon. **1961** L. Mumford *City in History* xv. 456 The more units in a given area, the more efficient was the source of power: hence the tendency towards giantism. **1967** *Economist* 7 Jan. 4 Before succumbing to a philosophy of industrial giantism you must answer this question.

gib (dʒib, gib), *sb.*[5] [Of obscure origin; cf. Jib *sb.*[2], Gibbet *sb.*[1]] **1.** A piece of wood or metal employed to keep something else, *e.g.* some part of a machine, in place. **a.** A bolt, pin, or wedge for insertion in a hole, to fasten the adjoining parts more tightly together.

Gib and cotter, gib and key: a two-part contrivance, consisting of a fixed and a movable wedge, used to tighten the strap at the end of a connecting rod.

1795 W. Felton *Treat. Carriages* II. 221 *Gib*, a small half round wedge, which keeps the pole from rising. **1838** *Civil Engin. & Arch. Jrnl.* I. 150/1 The corners were secured by a strong iron strap with a gib and key. **1839** R. S. Robinson *Naut. Steam Eng.* 78 The rectangular hole for the gib and cutter. *a* **1865** Smyth *Sailor's Word-bk.* (1867) 339 *Gib*, a forelock. **1880** *Encycl. Brit.* XIII. 732/1 The pin is sometimes rectangular in section, and tapered or parallel lengthwise. 'Gibs' and 'cottars' are examples of the latter. **1887** D. A. Low *Introd. Machine Draw.* 49 The gib is provided with horns at its ends to keep it in its place. **1888** *Lockwood's Dict. Mech. Engin.* 91 The gib and cottar arrangement is that in which the cottar is formed in two parts, whose outer faces are parallel with each other, the requisite taper being given to their inner or sliding faces.

b. A bearing surface to reduce friction, or a wedge to take up wear at a sliding contact.

1873 J. Richards *Treat. Wood-working Machines* 13 Gibs of lignum vitæ will be found to wear well. *a* **1877** Knight *Dict. Mech.* 650/1 The box has two taper-cheeks *a a* and two taper-gibs *b b* adjustable by screws, so as to set up the boxing to the wrist and the cheeks to the guides in the cross-head.

c. (See quot.)

1902 *Encycl. Brit.* XXX. 402/2 The extractor [in a Maxim gun] is recessed to take a movable plate termed a 'gib', behind which is a spring.

2. *Mining.* A piece of wood used to support the roof of a coalmine.

1847 in Halliwell. **1879** G. F. Jackson *Shropsh. Word-bk.* 172 *Gib*, a wooden prop used to support the coal when being 'holed'. **1883** Gresley *Gloss. Coal-m.*, *Gib*, a short prop of timber by which the coal is supported whilst being holed or undermined.

3. *attrib.* and *Comb.*: **gib-head,** a head (of a rod, pin, etc.) with projecting shoulder(s); **gib-headed key** (see quot.).

1854 Ronalds & Richardson *Chem. Technol.* (ed. 2) I. 152 The opposite ends of the connecting-rods are provided..with gib-heads. **1887** D. A. Low *Introd. Machine Draw.* 23 When the point of a key is inaccessible the other end is provided with gib-head. **1888** *Lockwood's Dict. Mech. Engin.*, *Gib-headed key*, a key having a set-off standing at right angles with the thicker end, for convenience of drawing it back in situations where the use of a drift is not practicable.

gib (dʒib, gib), *v.*[3] [f. prec.] *trans.* To provide or secure with a gib or gibs. Hence **gibbed** *ppl. a.*

a **1884** Knight *Dict. Mech.* Suppl. 398/1 *Gibbed lathe*, a lathe, the carriage of which has a bar which grips beneath the overhang of the bed, beneath which it is secured by a wedge known as a *gib*, to prevent the riding up of the carriage. **1893** *Funk's Stand. Dict.*, *Gibbed way*, a guiding surface provided with gibs for adjustment.

gibber, *sb.*[3] Add: **b.** *gibber-plain*.

1933 *Bulletin* (Sydney) 27 Sept. 20/4 A well-found homestead..set in a howling wilderness of bare gibber plain. **1953** *Sci. News* XXVII. 9 In cases where the initial rock surface is harder..stone-desert surfaces are commonly formed. These are known as hamada in Africa, gobi in Mongolia, and gibber-plains in Australia.

gibberellic (dʒibəre·lik), *a.* [f. next + -ic.] *gibberellic acid*: one of the gibberellins, a tetracyclic lactonic acid ($C_{18}H_{21}O_4COOH$) produced by the fungus *Gibberella fujikuroi*.

1954 Curtis & Cross in *Chem. & Ind.* 28 Aug. 1066/1 In attempting to isolate gibberellin A from culture filtrates of this fungus we have instead isolated another compound with similar biological properties. We propose to call this compound gibberellic acid. **1956** *Wall St. Jrnl.* 13 Dec. 1/1 In tests, gibberellic acid made such plants as sunflowers, roses, petunias and asters grow up to three times the height of untreated plants in three to four weeks. **1962** C. A. West in Florkin & Stotz *Comprehensive Biochem.* XI. ix. 152 An exchange of samples and a comparison of properties soon showed the identity of gibberellic acid, gibberellin A_3 and gibberellin X. **1968** *New Scientist* 21 Mar. 651/1 Gibberellic acid, the principal gibberellin, is known to be responsible for inducing the synthesis of the enzyme..in barley seed..that breaks down the starch food reserve within the seed during germination.

gibberellin (dʒibəre·lin). [f. mod.L. *Gibberell-a* (generic name of the fungus *G. fujikuroi*, from cultures of which gibberellin was first isolated), dim. of the generic name *Gibbera*, f. L. *gibber* hump: see -in[1].] Any of numerous compounds which are chemically related to gibberellic acid and which are present in many higher plants as growth regulators; their characteristic effects include elongation of the stem and other parts of the plant and the promotion of germination and flowering.

1939 Yabuta & Hayashi in *Jrnl. Agric. Chem. Soc. Japan* (*Eng. Abstr.*) Mar. 41 (*heading*) Isolation of 'gibberellin', the active principle which makes the rice seedlings grow slenderly. **1958** *Science News* XLVII. 9 Plants treated with a few micrograms of gibberellin became much longer. **1961** *Q. Rev.* XV. 59 The gibberellins are used to stimulate swelling of fruits, e.g., grapes and tomato. **1965** Bell & Coombe tr. *Strasburger's Textbk. Bot.* (new ed.) III. 498 *Gibberella fujikuroi*, which causes a disease of rice in Japan, yields the growth substance gibberellin, now much used in plant physiological researches. **1968** *New Scientist* 21 Mar. 651/1 Some 23 gibberellins, each with a slightly different physiological action and chemical structure, have so far been recognized.

gibberish, *sb.* and *a.* Add: Now freq. with pronunc. (dʒi·bəriʃ).

Gibbonian (gibōᵘ·niăn), *a.* [f. the name *Gibbon* (see def.) + -ian.] Relating to or resembling the style or opinions of the historian Edward Gibbon (1737–94).

1794 *Crit. Rev.* XI. 60 That sneering infidelity, which after times, perhaps, will distinguish by the epithet of Gibbonian. **1892** A. Birrell *Res Judicatae* 46 Is it not sometimes a relief to exchange the quips and cranks of some of our modern writers..for the stately roll of the Gibbonian sentence? **1939** A. Toynbee *Study Hist.* IV. 149 Great Britain..could congratulate herself, in Gibbonian language, upon having survived..a fluctuation in the Balance of Power. **1970** *Daily Tel.* 6 Aug. 6/2 [J. G.] Frazer's Gibbonian style [in *The Golden Bough*].. does not make for exactly easy reading.

gibli (gi·bli). Also **gheblee, gibleh, qibli,** etc. [ad. Arab. *ḳiblī* south wind.] A local name in Libya for the sirocco.

1821 G. F. Lyon *Narr. Trav. N. Afr.* 94 Nothing can be more overpowering than the south wind, El Gibli. **1848** J. Richardson *Travels* I. 17 A *ghiblee* day. The wind from The Desert is coming with a vengeance. **1853** —— *Narr. Mission Cent. Afr.* 61 Under the enervating influence of the gheblee, or hot wind. **1857** H. Barth *Trav. N. & Cent. Afr.* 652 A heavy gibleh. **1902** *Encycl. Brit.* XXXIII. 481/2 It is very wind-swept and parched in summer by the terrible south wind (*qibli* or 'sirocco'). **1927** W. Kendrew *Clim. Cont.* (ed. 2) xxxi. 240 The Sirocco..has naturally received local names, in south-east Spain levèche, in Algeria sirocco, in Tunis chili, in Tripoli gibli, [etc.]. **1966** *Punch* 29 June 965/3 Her steadfastly sunny outlook even in the middle of a *ghibli* (khamsin or sirocco, by any other name as blinding and painful). **1968** *Encycl. Brit.* VII. 792/1 The khamsin of Egypt (or gibleh of Libya).

Gibraltar. Add: **4. Gibraltar ape,** the Barbary ape, *Macaca sylvana*, esp. a member of the colony of these animals living on the rock of Gibraltar.

1932 S. Zuckerman *Soc. Life Monkeys* iii. 49 The young of the Gibraltar ape, *Macaca sylvana*, are born in spring and early summer. **1968** *Sunday Tel.* 20 Oct. 3/5 He has had to be content with Gibraltar apes.

Gibson[1] (gi·bsən). The name of C. Dana *Gibson* (1867–1944), American artist and illustrator, used *attrib.* to designate a type of girl fashionable in the late 1890s and early 1900s, or dress, etc., characteristic of this type of girl.

1901 *Cosmopolitan* June 120/1 It is a saying among artists that nine out of ten models who come..seeking employment say they are the original 'Gibson girl' or the 'Diana of the Garden'. **1902** *Encycl. Brit.* XXVI. 590/1 American men..impart as much importance and humour to his [*sc.* Gibson's] pages as his 'Gibson girls' give radiance. **1904** *Daily Chron.* 11 Apr. 8/4 The lean-limbed, lantern-jawed American..accompanied by a Gibson girl, who calls him 'poppa'. **1905** *Ibid.* 1 Mar. 6/6 A Gibson girl..is a lady with a perfect figure and a perfectly-fitting dress...Miss Hardren, a handsome brunette in black, with a black picture hat, criticised the

Gibson dress unsparingly. *Ibid.* 13 Mar. 8/1 Gibson pleats starting from the shoulder line and brought in symmetrical lines to the waist. **1942** D. Powell *Time to be Born* (1943) x. 221 Miss Finkelstein..had swept her sleek black locks into a Gibson girl pompadour.

Gibson² (gi·bsən). In full *Gibson cocktail.* A cocktail consisting of gin and vermouth garnished with pearl onions.
1930 H. Craddock *Savoy Cocktail Bk.* i. 75 Gibson cocktail. ½ French Vermouth. ½ Gin. **1951** H. McCloy *Alias Basil Willing* x. 121 Two double Gibsons, very dry..icy, oyster-white liquid that was almost pure gin with only a dash of vermouth. **1968** 'J. Welcome' *Hell is where you find It* iv. 51 He busied himself with a martini jug. I supposed that he was making a Gibson.

giddap (gi·dæp, gidæ·p), *v.* (Chiefly in *imp.*) *colloq.* (orig. *U.S.*). Also **giddy-ap, -up.** [Colloq. pronunc. of *get up*.] **A.** *int.* = Get *v.* 72 g. **B.** *v. intr.* To move quickly; also *trans.*, to urge (a horse) forward.
1925 T. Dreiser *Amer. Trag.* (1926) I. i. xviii. 133 'Giddap, horsey,' she played. 'Giddap.' **1938** D. Runyon *Furthermore* xiv. 293 Princess O'Hara..tells Gallant Godfrey to giddap, and Gallant Godfrey is giddapping very nicely. **1938** O. Nash *I'm Stranger here Myself* 136 Then giddy-ap, Napoleon! Giddy-ap, Gideon! **1948** D. W. Ballantyne *Cunninghams* (1963) i. xviii. 82 Another joker giddaped a horse. **1954** *Manch. Guardian Weekly* 24 June 7 He harnessed the brawling Pennsylvanians and with a twinkling 'giddy up' drove them into the Eisenhower camp.

giddea. Add: **2.** A long spear made from this wood, used by Australian Aborigines.
1878 *Catal. Objects Ethno-typical Art in Nat. Gallery, Melbourne* 46 *Gid-jee*, hardwood spear, with fragments of quartz set in gum on two sides and grass-tree stem. Total length, 7 feet 8 inches. **1966** W. S. Ramson *Austral. Eng.* vi. 117 *Gidgee* (also *gidya* or *giddea*)..is used..of a long spear made from the wood of this tree.

giddy, *a.* Add: **3.** *to play the giddy goat*: see Goat 3 b in Dict. and Suppl.; *to play the giddy ox*: to behave foolishly or frivolously; to play the fool.
1892 *Ally Sloper's Half-Holiday* 19 Mar. 91/2 Fanny Robinson was flighty; she played the giddy ox—I mean, heifer. **1915** W. S. Maugham *Of Human Bondage* xiii. 47 Don't play the giddy ox. **1922** Joyce *Ulysses* 9, I don't want to be debagged! Don't you play the giddy ox with me!
b. Used (often ironically) as an intensive; also used in the expression of surprise *my giddy aunt* (cf. *Aunt 5).
1896 Kipling *Seven Seas* 171 'E's a kind of a giddy harumfrodite—soldier an' sailor too! **1899** —— *Stalky & Co.* 28 King'll have to prove his charges up to the giddy hilt. *Ibid.* 112, I spoke the giddy truth... I said I didn't know. *Ibid.* 142 Hullo. What's the giddy jest? **1915** J. Buchan *39 Steps* iv. 97 A giddy lot Scudder's friends cared for peace and reform. **1919** 'W. N. P. Barbellion' *Jrnl. Disappointed Man* 268 Oh my giddy aunt! **1924** A. J. Small *Frozen Gold* iii. 92 The giddy part of it is that our strike isn't up north at all.

giddy-ap, -up: see *giddap *v.*

Gideon (gi·diən). *U.S.* [f. the name of an Israelite leader (see Judges vi. 11 ff.).] A member of a Christian organization of American commercial travellers, founded in 1899. *Gideon Bible*, a Bible purchased by this organization and placed in a hotel room, Pullman car, etc.
1906 *Springfield Weekly Republican* 13 Sept. 11 The religiously-minded commercial travelers, known as Gideons, have four prosperous camps in cities of New York state. **1920** *Atlantic Monthly* July 88 He replied in words that may still be read, thanks to the Gideons, in any hotel room: **1922** *Bookman* Feb. 552/2 He has stolen, to date, fifty-eight Gideon Bibles, twenty-seven from the Hotel Astor alone, and presented them to friends. **1956** 'N. Shute' *Beyond Black Stump* 17 He had nothing to read till he discovered the Gideon Bible in the drawer of the bedside table. **1961** K. Vonnegut *Sirens of Titan* (1967) iii. 53 He took the Gideon Bible that was in his room, and he started with the first sentence in Genesis.

gidgee, var. Giddea, gidya.

gift, *sb.* Add: **3. e.** In kindergartens, one of a series of educative toys designed to develop the child's powers of observation, etc.
1855 E. von Wickerode tr. *von Marenholtz's Woman's Educ. Mission* 19 The little gymnastic games for the hands and fingers connected with the 'First Gift', answer the triple end of amusing, occupying, and educating the infant. **1892** C. M. Yonge *That Stick* II. xxxix. 195 The elder children..were busied..in building up coloured cubes, 'gifts' in Kindergarten parlance. **1892** —— *Old Woman's Outlook* 80 Children are supposed to learn multiplication rationally by proof on the abacus frame, or by the 'gifts' of the Kindergarten. **1905** J. H. Boardman *Educ. Ideas of Froebel & Pestalozzi* iv. 48 According to the general definition of the term, there are altogether twenty Gifts, although most Kindergartners now limit the name to the first six or seven.
9. b. *gift-giving*. Also **gift-book** (earlier examples); **gift coupon**, a coupon issued with certain commodities, a specified number of which entitles the holder to a free gift; **gift-exchange** (see quot. 1963); **gift shop** orig. *U.S.*, a shop dealing in articles suitable for gifts; also with pseudo-archaic spelling, *gifte-shoppe*; **gift-token** or **voucher**, a voucher (intended to serve as a gift) for a sum of money to pay for the purchase of an article; **gift-wrap** *v. trans.* (orig. *U.S.*), to wrap (an article intended as a gift) attractively; also *absol.* and *fig.*; hence *gift-wrapped* ppl. adj.; *gift-wrapping* vbl. sb. (also *attrib.*).
1834 *Knickerbocker* III. 113 It is, I believe, your standing gift-book. **1842** (*title*) The royal gift book for the young. **1848** D. Vedder (*title*) The pictorial giftbook of lays and lithography. **1931** *Morning Post* 18 Feb. 6/5 The 'Gift' Coupon system. **1933** D. L. Sayers *Murder must Advertise* iv. 54 It only needed the alteration of a sentence and the introduction of a panel about gift-coupons. **1936** R. Linton *Study of Man* 144 All trade was phrased in terms of gift exchange. **1951** R. Firth *Elem. Social Organiz.* i. 21 Such diverse social relations as buying and selling, gift-exchange, [etc.]. **1963** *Brit. Jrnl. Sociol.* XIV. 26 *Gift exchange*..describes a type of transaction which formally consists in the making of a gift and its repayment by another. **1937** M. Mead *Cooperation & Competition among Primitive Peoples* i. 22 All this importation is phrased as gift giving between devoted friends. **1949** —— *Male & Female* 408 The fish were then exchanged in a gift-giving context. **1959** I. & P. Opie *Lore & Lang. Schoolch.* xii. 236 Sometimes..the older children take advantage of the gift-giving and play tricks. **1918** C. Wells *Vicky Van* i. 10 Little faddly prize bags of gift-shop novelties are her stakes. **1929** Gift shop [see *art *v.*² 5]. **1932** E. Bowen *To North* xxi. 229 There are too many shops... Especially gift shops. **1948** J. Cannan *Little I Understood* xi. 136 Expensive and financially unstable gifte shoppes. **1963** *She* Dec. 51 Best leg forward in 1964 with a gift token for nylon stockings. *Ibid.*, Make life easier for a hard-working house-wife whose feet are killing her with a Scholl Gift Voucher entitling her to one or a set of foot massaging sessions. **1969** *Times* 17 Dec. 18/5 (Advt.), Who's lucky? Everybody who gets a gift voucher for two tickets. **1936** *Amer. Speech* XI. 101/1 During the holiday season many department stores advertised, 'We Gift-Wrap Here.' **1948** in *Amer. Speech* (1956) XXXI. 210 Ready to gift-wrap your package. **1958** *Economist* 20 Dec. 1083/2 Whatever he buys will be professionally gift-wrapped for him. **1969** *New Yorker* 11 Oct. 146/3 (Advt.), We gift wrap and ship everywhere. **1964** *Punch* 16 Dec. 936/3 People..pile them [*sc.* cars] up with gift-wrapped presents. **1949** *Word Study* May 8/1 (*caption to cartoon*) Gift Wrapping. **1963** M. McCarthy *Group* xi. 240 They also sold.. Valentines and gift-wrapping paper.

gift-over (gi·ft,ōuvəɹ). [f. Gift *sb.* + Over *adv.*] The act of making over as a gift.
1927 *Daily Mail* 8 July 7/1 That the gift-over, by which the property might pass away from the children on account of the son's marriage, was void. **1929** *Glasgow Herald* 8 July 8 The gift-over whereby the estate would go elsewhere than to the children.

gig (gig), *sb.*⁶ *colloq.* [Origin unknown.] An engagement for a musician or musicians playing jazz, dance-music, etc.; *spec.* a 'one-night stand'; also, the place of such a performance. Also *transf.* and *attrib.* Hence **gi·gster**, one who does 'gigs'.
1926 *Melody Maker* Sept. 7 One popular 'gig' band makes use of a nicely printed booklet. **1927** *Ibid.* May 457/3 This seven-piece combination does many 'gigs' in S.E. London, but is hoping to secure a resident engagement at Leamington in the near future. **1934** S. R. Nelson *All about Jazz* vi. 113 Jack runs numerous bands which play 'gig' work—i.e. private engagements or public work. In his office, he has a file in which some hundreds of 'gig' musicians are listed. **1939** *Melody Maker* 9 Sept., When King George died there was terrible confusion, especially among gigsters, as to whether they should fulfil their gigs or not. **1964** L. Hairston in J. H. Clarke *Harlem* 287 Pa—knockin' hisself out on a mail-handler gig at the Post Office where the pay is so lousy he's gotta work a part-time gig. **1965** G. Melly *Owning-Up* vii. 80 Another Proustian gig was the Civic Hall, Nantwich. We played there fairly regularly right through the 'fifties. **1969** *Observer* 12 Jan. 31/5 Leading groups will be given two hours in which to play what they want, without the limitations imposed by commercial gigs.

gig, *v.*² **2.** (Earlier examples.)
1815 *Niles' Weekly Reg.* 16 Sept. 36/1 The carriages run upon cast racks, are propelled by the improved short hand and gigged backwards by bevel wheels, in the manner of the best mills. **1874** W. M. Baines *Narr. E. Crewe* viii. 180 This carriage [to the frame-saw] could be 'gigged' backwards or forwards or fed forwards.

gig (gig), *v.*⁷ *colloq.* [f. *gig *sb.*⁶] *intr.* To do a 'gig' or 'gigs' (see *gig *sb.*⁶); freq. to *gig around* (see quot. 1939). Hence **gi·gging** *vbl. sb.*
1939 C. E. Smith in Ramsey & Smith *Jazzmen* (1940) xiii. 267 To gig around meant to play for small parties, week-end engagements, and the like. **1949** L. Feather *Inside Be-Bop* III. 77 [He] gigs around New York with Chubby Jackson, Lennie Tristano, Benny Goodman. *Ibid.* 92 Settling in California, [he] gigged with Boyd Raeburn, [etc.]. **1952** B. Ulanov *Hist. Jazz in Amer.* (1958) xviii. 227 He had the usual Gotham gigging beginning. **1959** 'F. Newton' *Jazz Scene* xii. 222 He had to earn his living as a blueprint inspector, occasionally gigging in his spare time. **1967** *Crescendo* Feb. 12/2 Buy Professor Jacko's 'Gig-Book For All Occasions'. Gives melody line, chords and starting note for complete evening's gigging.

giga- (dʒ-, gəi·gă; dʒ-, gi·gă), *pref.* An arbitrary derivative of Gr. γίγας giant, prefixed to the names of units in the metric system to form the names of units 1000 million (10^9) times greater. Abbrev. G.
1951 *Symbols, Signs & Abbrev.* (R. Soc.) 15/1 Prefixes to abbreviations for the names of units indicating multiples..giga (× 10^9). **1960** *Gloss. Terms Telecomm.* (*B.S.I.*) 179 Frequencies of 1,000 Mc/s and above are sometimes expressed in gigacycles per second (Gc/s). 1 Gc/s = 1000 Mc/s. **1964** *New Scientist* 30 Jan. 286/1 An extraordinarily high output (from megawatts to gigawatts). **1966** *Electronics* 3 Oct. 171 It will relay aircraft transmissions..over existing 4-gigahertz and 6-Ghz satellite frequencies. **1966** Auden *About House* 17 Translated in a nano-second To a c.c. of poisonous nothing In a giga-death. **1968** *Nature* 19 Oct. 311/2 CIPM at its meeting in 1958 recommended the prefix giga (pron. jī'ga; symbol G) for the multiple 10^9.

gigantism. Add: (Further examples.) *spec.* **a.** In man, excessive size due to an increase in the supply of growth hormone caused by overactivity of the anterior lobe of the pituitary gland. **b.** In plants, excessive size due to polyploidy. Also *fig.* Cf. Giantism.
1899 *Q. Rev.* CXC. 279 Disease may..lead to.. acromegaly, or gigantism. **1927** *Bot. Gaz.* LXXXIV. III. 314 An aberrant form of *Potentilla anserina* was characterized by gigantism, sterile pollen, and lack of fruit formation. **1932** C. D. Darlington *Chromosomes & Plant-Breeding* xi. 57 Gigantism is a normal and perhaps universal property of those tetraploids which can be compared directly with the diploids from which they have arisen. **1932** J. S. Huxley *Prob. Rel. Growth* v. iv. 130 Some cases of dwarfing gigantism. *Ibid.* 131 Simple pituitary gigantism is associated with relatively long limbs. **1944** *Science* 16 June 481/1 In the case of *Gigantopithecus* the gigantism reaches a new climax. **1953** *Encounter* Nov. 9/1 The prosperous gigantism of the Japanese papers is based..on the high..literacy rate of the population. **1965** C. D. Darlington *Cytology* i. vi. 222 Simple doubling [of chromosomes]..is usually accompanied by more or less gigantism. **1970** R. Gorer *Devel. Garden Flowers* ii. 42 Gigantism is one of the commonest phenomena in cultivated plants.

gigantoblast (dʒəigæ·ntoblast). *Med.* [f. Gr. γιγαντο-, γίγας giant + βλαστός embryo, germ.] A particularly large erythroblast.
1898 A. C. Coles *Blood* 176 Large nucleated erythrocytes—megaloblasts or gigantoblasts..are occasionally seen. **1935** Whitby & Britton *Disorders of Blood* ix. 180 Gigantoblasts, or very large cells of megaloblastic type, may be met with.

gigantopithecus (dʒəigæ:ntopi·þĭkŏs, -þi·kŏs). [mod.L. (G. H. R. von Koenigswald 1935, in *Proc. Sect. Sci. K. Nederl. Akad. Wetensch.* XXXVIII. 874), f. Gr. γιγαντο-, γίγας giant + πίθηκος ape.] A large fossil primate, sometimes considered hominid, belonging to the genus so called, which is known from bones found in China.
1943 *Palaeont. Sinica* (Ser. D) x. 271 *Gigantopithecus* is to be considered not only as true anthropoid but as a type very close to the hominids. **1944** *Science* 16 June 481/2 We seemingly do not know more of the provenance of the *Gigantopithecus* teeth than the fact that they were gathered from drawers of a chemist's shop. **1950** D. P. Quiring *Funct. Anat. Vertebr.* xiii. 509 The third lower molars of Gigantopithecus have a mass about six times larger than those of modern man. **1965** M. H. Day *Guide to Fossil Man* III. 263 The fossil bones discovered with the new mandibles confirm that *Gigantopithecus* formed a part of this fauna. *Ibid.* 262 (*heading*) The 'Gigantopithecus' Teeth.

gigas (dʒəi·gās), *a.* *Bot.* [a. Gr. γίγας: see Giant.] Of or designating a polyploid form of a plant which is larger and more vigorous than the normal form.
1915 *Amer. Naturalist* XLIX. 703 Both gigas forms are more persistently biennial in habit than their parents. **1939** Sinnott & Dunn *Princ. Genetics* (ed. 3) xiv. 321 A case in mosses (*Bryum*) where an autotetraploid (2n gametophyte, 4n sporophyte) showed gigas characters. **1946** *Nature* 22 June 843/2 Isolations were made from these large-spored colonies and 'gigas' lines established. **1967** Briggs & Knowles *Introd. Plant Breeding* xxi. 268 Plant organs may show some increase in size, termed *gigas* characteristics. **1970** R. Gorer *Devel. Garden Flowers* i. 39 The leaf, or any other part of a gigas plant, contains not only larger cells than the wild type, but also more of them.

‖ **gigerium** (dʒidʒĭə·riŏm). [sing. of L. *gigeria* cooked entrails of poultry.] = Gizzard 1 a.
1875 *Encycl. Brit.* III. 726/1 A proventriculus, connected by a narrow neck with the gizzard (*gigerium*). **1884** E. Coues *N. Amer. Birds* (ed. 2) II. 213 The gizzard, *gigerium*, or muscular division of the stomach. **1956** A. N. Worden *Funct. Anat. Birds* vi. 31 The gizzard (known also as the *gigerium* or *ventriculus bulbosus*) or second part of the stomach, usually follows immediately after the proventriculus.

giggering. *Book-binding.* See *jigger *v.*⁵

giggle, *sb.* Add: **2. b.** An amusing person or thing; a joke; fun; *no giggle*: no joke (see Joke *sb.* 3). *colloq.*

1936 J. Curtis *Gilt Kid* x. 108 It's no giggle being in the nick. **1958** F. Norman *Bang to Rights* 117 There was one geezer who was a right giggle. **1959** C. MacInnes *Absolute Beginners* 152 Wiz told the tale as I've just done, for giggles, but even he didn't seem to think it all that laughable, I could see. **1963** 'A. Garve' *Sea Monks* vi. 169 Prob'ly have the Home Secretary on the line next... What a giggle, eh? **1963** *Sunday Express* 29 Sept. 17/1 As the gang burst in..one warned: 'This is no giggle, I will shoot you.' **1966** D. Francis *Flying Finish* ii. 21 It's all very well you taking on Peters's job for a bit of a giggle but you surely can't mean to go on with it permanently? **1968** J. Rathbone *Hand Out* iv. 24 He enjoyed the course, which was mostly rather a giggle.
3. [Cf. Gaggle *sb.* 1.] A group of girls or young children. *colloq.*
1940 B. Ruck *Pennies from Heaven* xxix. 236 He had picked her out of the whole giggle of Society débutantes. **1957** J. Braine *Room at Top* iii. 28 A giggle of schoolgirls round a pile of brightly-coloured rayon underwear. **1967** *Evening News* 12 Sept. 8/3 At Mitcham..a blue M.G.B...impressed a giggle of schoolgirls.
4. *attrib.* and *Comb.*, as **giggle-house** *Austral.* and *N.Z. slang*, a mental hospital; **giggle-pants, -suit** *Austral. Services' slang*, working trousers, clothes; overalls; **giggle-water** *slang*, intoxicating liquor.
1919 W. H. Downing *Digger Dial.* 26 *Giggle-house*, lunatic asylum. **1943** N. Marsh *Colour Scheme* iv. 72 When I've taken over this joint the resemblance to a giggle-house will fade out. **1944** K. Levis in *Meanjin Papers* III. i. 32 The Lieutenant..was a young fellow with deep blue eyes, and fresh-creased giggle pants. **1942** A. G. Mitchell in *Southerly* Apr., *Gigglesuit*, a fatigue dress. **1945** J. B. Blair in *Coast to Coast* 133 Always smart and spruce in his Field Service uniform, Wang never looked his best in giggle-suits. **1929** *Amer. Speech* IV. 386 Some of the common names for whiskey..*giggle water*, *nose paint*, [etc.]. **1946** G. Hackforth-Jones *Sixteen Bells* i. i. 32 Drop o' gin'll go down nicely on top of that giggle-water [*sc.* champagne cocktails]. **1962** *John o' London's* 14 June 571/1 Giggle-water is any unseamanlike drink.

gigglish, *a.* Delete †*Obs.*⁻¹ and add later example.
1900 E. Glyn *Visits Eliz.* 22 [She] got rather gigglish.

gigmanity. (Further examples.)
1931 *Times Lit. Suppl.* 12 Feb. 112/3 The sculptor's abandonment of exotic delights and infidelity to return to swaggering Sicilian gigmanity. **1966** *New Statesman* 16 Dec. 908/3 The style..was suitable for the full-souled, for the gigmanity and for those in office.

gigolo (ʒi·gŏlo, dʒi·gŏlo). [Fr., formed as masc. correlative of *gigole* tall thin woman, woman of the streets or public dance-halls.] A professional male dancing-partner or escort; a 'kept' man (see Kept *ppl. a.* 1 a), esp. a young man supported financially by an older woman in return for his attentions.
1922 *Woman's Home Companion* Nov. 7/1 A gigolo, generally speaking, is a man who lives off women's money. In the mad year 1922..a gigolo, definitely speaking, designated one of those incredible and pathetic male creatures,..who, for ten francs..would dance with any woman wishing to dance..in the cafés, hotels, and restaurants of France. **1927** *Daily News* 21 May 5/4 The audience was delighted with the grannies' dance with gigolos—as lounge lizards are called. **1927** *Daily Express* 24 Oct. 10/4 The Riviera wakes up... Well-known mannequins, dance partners, gigolos, and barmen..have once more returned to their place in the sun. **1928** *Daily Tel.* 27 Mar. 6 Similar indulgence, perhaps, would not be extended to her adoption of a 'gigolo'. **1933** N. Coward *Design for Living* i. 21 I'd forgotten about your French accent and the the way you move your hands, and the way you dance. A sleek little gigolo! **1961** P. Ustinov *Loser* x. 226 The role of Captain Val di Sarat..has been consigned to a superannuated gigolo.

Gila, gila (hī·lă). [The name of a river in New Mexico and Arizona.] In full *Gila* (or *gila*) *monster*. A large venomous lizard, *Heloderma suspectum*, found in the south-western United States.
1877 H. C. Hodge *Arizona* 226 There is one variety [of saurian lizard], however, peculiar to Arizona..and locally known as the Gila monster. **1890** *Chambers's Jrnl.* 8 Mar. 158/2 The Gila Monster..which lives in the valleys and sandy plains of Arizona and Sonora, is called by the native Mexicans *Escupion*, which means 'Spitter'. **1902** C. E. Mulford *Bar-20* v. 60, I only had th' main show—Gilas, rattlers an' toads. **1930** *Times Educ. Suppl.* 9 Aug. p. iv/2 The Zoo has bought six examples of the Arizona poisonous lizard known as the Gila monster. **1969** A. Bellairs *Life of Reptiles* I. vi. 228 Others such as the Gila monster and many geckos and snakes are principally nocturnal.

gilbert² (gi·lbəɹt). *Physics.* [The name of William *Gilbert* (1544–1603), English physician and natural philosopher.] The electromagnetic unit of magnetomotive force in the C.G.S. system of units, equal to $10/4\pi$ ampere-turns.
1893 *Trans. Amer. Inst. Electr. Engin.* X. 364 The four units—gilbert, weber, oersted and gauss—have been suggested by the Committee of the Institute. **1930** *Nature* 16 Aug. 252/2 The committee on nomenclature [of the International Electrotechnical Commission].. adopted the following names for the magnetic C.G.S. units... Magnetomotive force... The 'gilbert'. **1967** *Handbk. Chem. & Physics* (ed. 48) F-79 Magnetizing force is measured by the space rate of variation of magnetic potential and as such its unit may be the gilbert per centimeter.

Gilbertese (gilbəɹtī·z), *a.* and *sb.* [f. *Gilbert* (see below) + -ese.] **A.** *adj.* Of or pertaining to the Gilbert Islands in the mid-Pacific. **B.** *sb.* **a.** Collectively, the people of the Gilbert Islands; also, one of these people. **b.** The language of the Gilbert Islands.
1908 H. Bingham (*title*) A Gilbertese-English dictionary. *Ibid.* Pref., Such Gilbertese as desire to acquire a knowledge of English. *Ibid.*, The spiritual and social uplift of the Gilbertese people. **1945** H. Luke *From South Seas Diary* ix. 101 A South Gilbertese, neither young nor beautiful, but none the less plagued by a jealous husband. *Ibid.* 112 Mrs Maude is doing valuable work in developing Gilbertese arts and crafts. **1964** M. Dickson *World Elsewhere* v. 195 I've been staying among the Gilbertese who've been settled here.

Gilbertian (gilbɔ̄·ɹtiăn, gilbɔ̄·ɹʃən), *a.* [f. the name of W. S. *Gilbert* (1836–1911), librettist of the Gilbert and Sullivan operas + -ian.] Of, pertaining to, or characteristic of W. S. Gilbert or his work; *spec.* resembling or reminiscent of the ludicrous or paradoxical situations characteristic of the Gilbert and Sullivan operas. Hence **Gilbe·rtianism.**
1879 *Scribner's Monthly* Oct. 909/1 'Thespis' ran one hundred nights. Of course the plot unfolds a Gilbertian conceit. **1887** *Graphic* 29 Jan. 107 All do their duty, and, to borrow a Gilbertian phrase, 'do it very well'. **1891** *Strand Mag.* Oct. 331/2 There is a perfect home farm on the Gilbertian land. **1894** P. Fitzgerald *Savoy Opera* 13 The 'Gilbertian' topic of the English traveller 'turning up his nose' at everything he sees abroad. *Ibid.* 14 A sort of 'Gilbertian humour'. **1906** *Daily Chron.* 16 Nov. 6/7 The Gilbertian question whether a Lord Chancellor could bring himself to justice for contempt of his own court..is equalled in the quality of Gilbertianism by the puzzle of Baton Rouge, Indiana. **1918** T. H. Ward *Eng. Poets* V. 540 His peculiar quality of topsy-turvydom, which has perhaps added the word 'Gilbertian' to the language. **1929** *Times* 18 July 15/2 A Gilbertian situation arose in which a Government pledged to attack and an Opposition pledged to defend private enterprise simply exchanged roles. **1961** *Sunday Times* 30 Apr. 12/6 Might one deduce a coming Gilbertian revival? There seems no reason why that difficult Gilbertian morality should not appeal to the connoisseur like contemporary wax-fruit and beadwork.

Gilderoy (gi·ldəroi). The name of a Scottish robber, used in the colloq. phr. *higher than Gilderoy's kite*: extremely high; out of sight. Chiefly *U.S.*
1869 'Mark Twain' *Innoc. Abr.* (1870) xxv. 192 The first time she took her new toy into action she got it knocked higher than Gilderoy's kite—to use the language of the Pilgrims. **1888** E. C. Brewer in *N. & Q.* 7th Ser. V. 357/1 To be 'hung higher than Gilderoy's kite' means to be punished more severely than the very worst of criminals... The ballad says: Of Gilderoy sae fraid they ware They bound him mickle strong, Tull Edenburrow they led him thair, And on a gallows hong; They hong him high abone the rest, He was so trim a boy. They 'hong him high abone the rest', because his crimes were deemed to be more heinous. So high he hung, he looked like 'a kite in the air'. **1903** G. Brown *How to beat Game* 53 This theory, however, like many others, is often knocked higher than Gilderoy's kite when put to the test of practice. **1934** H. Corey *Crime at Cobb's House* iii. 35 The colt threw him higher'n Gilderoy's kite.

gilgai (gi·lgəi). *Austral.* Also **ghilgai, gilgie, gilguy.** [Native name.] A saucer-like depression forming a natural reservoir for rain-water. Also *attrib.*
1898 Mrs. K. L. Parker *More Austral. Legendary Tales* 7 A..shower fell,..filling the gilguy holes on the plain. **1898** Morris *Austral Eng.* 160/1 Ghilgais vary from 20 to 100 yards in diameter, and are from five to ten feet deep. **1930** V. Palmer *Men are Human* v. 41 They watched the gilgais turn to mud on the blacksoil plains. **1933** *Bulletin* (Sydney) 28 June 20/3 They found the old girl bogged to the belly in a gilgai. **1959** A. Upfield *Bony & Black Virgin* xv. 135 Bony was led to a gilgie hole amid the scrub trees.

gilguy. Add: **2.** A 'thingummy', a 'what-d'ye-call-it'. *Naval slang.*
1886 [see *gadget]. **1918** *Chambers's Jrnl.* May 291/2 The placing on board of the various engines, machines, and fittings,—boilers, turbines,..and a hundred and one other gilguys. **1926** [see *gimmick].

gill, *sb.*¹ Add: **2. d.** *Aeronaut.* (See quot. 1949.)
1949 *Gloss. Aeronaut. Terms* (*B.S.I.*) II. 13 *Gills*, a set of movable flaps at the rear of a cowling to control the flow of air. **1971** D. N. James *Gloster Aircraft* 229 The engine was an 840 hp Bristol Mercury IX..with a leading-edge exhaust collector ring and controllable cooling gills.
5. b. gill-book, the lamellate respiratory organs of the king crabs which form the order Xiphosura; **gill-pore**, a small pore between the gill-pouch and the exterior in acorn worms of the class Enteropneusta; **gill-pouch**, (*a*) a pouch into which the gills of acorn worms open; (*b*) a pouch enclosing the gills in fishes of the subclass Cyclostomata; (*c*) *Embryol.*, a pouch present during part of the development of all vertebrate embryos.
1881 *Q. Jrnl. Microsc. Sci.* XXI. 541 The lamellæ of the gill-books of Limulus are..delicate flattened bags with a setose free border. **1902** *Encycl. Brit.* XXV. 521/2 The leaves (some 150 in number) of the gill-book..correspond to the tooth-like processes of the pectens of Scorpio. *Ibid.*, The gill-books of Limulus. **1932** Borradaile & Potts *Invertebrata* xv. 445 The gill books are stated to be the most primitive respiratory organs. **1902** *Encycl. Brit.* XXVI. 85/1 Each gill-slit may be said to open into its own..gill-pouch; this in its turn opens to the exterior by a minute gill-pore. **1967** P. A. Meglitsch *Invert. Zool.* xii. 427/1 The gill pores are usually guarded by sphincter muscles. **1888** Gill-pouch [see Pouch *sb.* 3 a]. **1888** Rolleston & Jackson *Forms Anim. Life* 590 New gill-pouches appear to be constantly added throughout life. **1895** B. Dean *Fishes Living & Fossil* ii. 18 There has been a general tendency to press closely together the gill pouches. **1914** W. E. Kellicott *Outl. Chordate Devel.* iv. 269 There is a pair of lateral extensions of the fore-gut [of the chick]..marking the positions of the future first gill pouches. **1931** J. R. Norman *Hist. Fishes* iii. 37 The gill-pouches are large in the lamprey. **1932** Borradaile & Potts *Invertebrata* xix. 663 A pair of collar pores open backward from the collar cavities into the first gill pouch. **1959** L. H. Hyman *Invertebrates* V. xvii. 111 In *Stereobalanus canadensis*..all the gill pouches on each side are fused. **1960** B. I. Balinsky *Introd. Embryol.* i. 10 The formation of gill pouches in the ontogenetic development of all vertebrates. **1962** K. F. Lagler et al. *Ichthyology* iii. 82 Some hag-fishes have numerous gill pouches on a side.

gill, jill, *sb.*⁴ **1.** Delete †*Obs.* and add later examples.
1938 *Times Lit. Suppl.* 8 Oct. 638/4 She was a rolling-stone and an ignorant Jill-of-all-trades. **1942** Berrey & Van den Bark *Amer. Thes. Slang* § 382/2 A female, esp. a girl or young woman,..*gill*,..*Jill*.
b. A female ferret, polecat, or weasel. *colloq.* or *dial.*
1851 *N. & Q.* III. 461/2 The name by which the male ferret is known in the midland counties is the *hob*; the female is called the *jill*. **1875** G. C. Davies *Rambles School Field-Club* xxviii. 210 A..'hob', or male ferret, and..a 'jill', or female. **1902** W. E. de Winton in C. J. Cornish *Naturalist on Thames* 72 The female, or 'Jill', changes her entire coat directly she has young.
2. a. See also Jack *sb.*¹ 1 b.

gillenia (dʒili·niă, gili·niă). *Bot.* [mod.L. (C. Moench *Methodas Plantas Suppl.* (1802) 286), f. the name of Dr. Arnold *Gille*, or *Gillenius*, a seventeenth-century German botanist.] A member of a small genus of herbaceous plants belonging to the family Rosaceæ; also the root of these plants, used as an emetic.
1820 J. Bigelow *Amer. Med. Bot.* III. 17 Some authors have attributed a tonic power to the Gillenia, when administered in small doses. **1848** Dunglison *Med. Lex.* (ed. 7) 390/2 The root of this shrub,—*Gillenia*,..is a safe and efficacious emetic. **1901** W. Robinson *Eng. Flower Garden* (ed. 8) 575 The Gillenias have a distinct and delicate beauty all their own. **1951** *Dict. Gardening* (R. Hort. Soc.) II. 891/1 Gillenias are hardy, easily grown plants.

gillie, *sb.*¹ Add: **3.** A type of shoe (see quots.).
1934 Webster, *Gillies*, low-cut sport shoes or ties with laces wound about the ankles. **1939** A. Keith *Land below Wind* xiii. 220 We had tried lace boots, gillies, short riding boots. **1950** H. McCloy *Through Glass Darkly* (1951) vi. 57 And brown shoes. The sort with no tongues and criss-cross laces that they call 'gillies'.

gillie (gi·li), *v.* [f. Gillie¹.] *intr.* To act as gillie. Hence **gi·llying** *vbl. sb.*
1905 *Daily Chron.* 2 Mar. 3/2 Contriving during the long vacation of six months to make enough money—by teaching, gillying or field-labour—to keep himself at the University for another year. **1926** *Chambers's Jrnl.* Mar. 153/1 Archie usually gillied for me. **1927** *Scots Observer* 9 Apr. 10/1 Nor is it intended to have classes in game-keeping and ghillie-ing.

gillion (gi·liən, dʒi·liən). [Blend of *giga- and Million.] A name sometimes used for 1000 million by writers wishing to avoid the ambiguity of 'billion'.
1961 N. W. Pirie in *New Scientist* 11 May 331/3 A name is needed for 10^9... Because G is the symbol, gillion would be the logical word. **1966** V. A. Firsoff *Facing Universe* III. ii. 95 The number of galaxies within the observable universe..is estimated at about 3 gillion (3×10^9). **1971** *Nature* 12 Mar. 133/2 The difficulty with N.W. Pirie's proposal of 'gillion' and 'tillion'..is—where do we go after 'giant' and 'monster'.

gilsonite (gi·lsənəit). Also **Gilsonite.** [f. the name of S. H. *Gilson* (see quot. 1888): see -ite¹.] The proprietary name of an exceptionally pure variety of asphalt, found in Utah and Colorado as a brittle, brilliant black solid with a high softening temperature; it resembles glance pitch and grahamite.
1888 J. M. Locke in *Trans. Amer. Inst. Mining Eng.* XVI. 162 Gilsonite or Uintahite, a new variety of Asphaltum from the Uintah mountains, Utah. The discovery of this asphaltum was made by S. H. Gilson, of

Salt Lake, and since then the material has borne the local name of Gilsonite. **1921** *Brit. Mus. Return* 180 Basalt coated with asphalt, and gilsonite, from Vera Cruz, Mexico. **1929** H. Abraham *Asphalts* (ed. 3) x. 166 Gilsonite is one of the most valuable asphalts for manufacturing paints and varnishes. **1938** J. S. Miller in A. E. Dunstan et al. *Sci. Petroleum* IV. 2724/1 Because of its brown undertone, Gilsonite is used widely in printing-inks. **1960** K. Stern in *Industr. Min. & Rocks* (ed. 3) xxxiii. 635/1 Gilsonite's high dielectric strength, heat-resisting qualities and its resistance to acids and alkalies makes it of great value as a waterproofing and weather-resisting agent.

gilt, *sb.*¹ Add: **3. b.** A gilt-edged security.
1936 W. B. M. Ferguson *Somewhere off Borneo* vi. 37 If he could get free board and travel.., it would mean just so much more towards another 'gilt' for his private collection. **1953** *Economist* 14 Nov. 508/1 (*heading*) Spotlight on gilts. **1967** *Times Rev. Industry* June 93/1 They went into equities largely because the scope for gilts had been mostly exhausted by three cuts in Bank rate.

gilt, *sb.*² Delete *Obs.* exc. *dial.* and add later examples.
1959 *Chambers's Encycl.* X. 724/2 Well-grown gilts can be put to the boar at eight months. **1967** *Oxf. Times* 27 Oct. 11/6 A Large White pig from Burford Grammar School's farm was very highly commended in the carcase class (gilts).

gilt, *ppl. a.* Add: **3. a.** *gilt-latten*; **gilt-bronze** (see quot. 1889); also *fig.*; cf. Ormolu; **gilt-edge** *a.* = *gilt-edged*; also *absol.*, a gilt-edged security; **gilt tooling** (see Tooling *vbl. sb.* 2 b); **gilt toy** (see quot. 1862); **giltwood** *a.*, made of wood and decorated with gilding.
1889 *Cent. Dict.*, *Gilt-bronze*, a gilded metal much used for decorative objects, either real bronze, or often brass, latten, or some similar yellow metal. **1906** F. Treves *Highways & Byw. Dorset* iv. 43 A track made of gilt-bronze moss. **1936** *Burlington Mag.* Nov. p. xix/2 Various candelabra in gilt-bronze. **1818** Keats *Let.* 21 Sept. (1931) I. 236 A dissertation on letter writing... Gilt edge [paper] Dandies in general, male female and literary. **1895** *Brewer's Dict. Phr. & Fable* 519/2 *Gilt-edge investments*, a phrase introduced in the last quarter of the 19th century..for investments in which no risks are incurred. **1901** *Westm. Gaz.* 28 Aug. 7/1 The Demand for Gilt-Edges. **1905** G. Wilson *Ann. Glover Incorp.* 24 What is termed a gilt-edge security. **1900** S. R. Gardiner *Student's Hist. Eng.* 314 Gilt-latten effigies of Richard Beauchamp. **1907** *Yesterday's Shopping* (1969) 442/1 Russia leather, with gilt tooling. **1930** G. H. Bushnell *University Librarianship* viii. 50 Marking books permanently by gilt tooling. **1839** Ure *Dict. Arts* 571 Supposing the articles desired to be gilded be brass or copper buttons, or small articles for gilt toys, or ornaments of dress. **1851** *Illustr. Catal. Gt. Exhib.* III. III. 629/1 Much of the Birmingham jewellery and gilt toys are produced by means of dies or steel blocks, with impressions of articles to be sunk therein. **1862** *Chambers's Encycl.* IV. 754/2 *Gilt toys*, this term is known in trade as a designation for small articles which are gilded, but is chiefly applied to the cheap jewellery which is almost exclusively manufactured at Birmingham. **1939** *Burlington Mag.* LXXV. 32/1 A set of giltwood chairs by Thomas Chippendale. **1958** *Times* 2 Dec. 20/7 A pair of Italian giltwood side tables with rustic tree supports, 18th century. **1970** *Daily Tel.* 18 Apr. 10/5 A suite of 14 pieces of George I giltwood furniture was bought..for £19,000.
b. *gilt-framed*, *-tooled*; **gilt-edged**, delete 'in commercial slang' and add earlier *fig.* examples; also used to denote something of very high quality, richness, elegance, etc.; also *absol.*, gilt-edged securities.
1867 *Galaxy* July 278 He decidedly should not honor any more drafts unless secured by previous deposits of specie, gilt-edged paper endorsed with two good names, or U.S. or solvent State stocks. **1872** *Vermont Board Agric. Rep.* I. 163 The high prices reported to be paid in the Boston market for what is called 'gilt-edged butter'. **1897** Kipling *Capt. Cour.* x. 230 Pastors of great, gilt-edged congregations, at the seaside for a rest. **1913** E. Wharton *Custom of Country* ix. 113, I don't want to break into your gilt-edged crowd. **1930** *Economist* 10 May 1038/2 Gilt-edged were fairly steady after a shaky start, but little business was done. **1952** Koestler *Arrow in Blue* ii. 16 Dr. Bedoe's eyes flashed and he broke into his gilt-edged smile. **1958** *Times* 11 Jan. 12/1 While gilt-edged took the news of the new £500m. Exchequer stock issue in their stride, the equity markets turned dull on the Stock Exchange yesterday. **1965** *New Statesman* 9 Apr. 588/2 There is to be no [tax] exemption..for gilt-edged. **1854** Thoreau *Walden* 48 A..gilt-framed looking-glass. **1936** *Discovery* Dec. 386/2 The Coffee House, where once the little gilt-framed poster had found so fit a home. **1926–7** *Army & Navy Stores Catal.* 431/2 Leather photograph frames...Crushed calf, gilt tooled. **1936** *Burlington Mag.* Apr. 198/2 Small gilt-tooled ornament.

Gilyak (gi·lyæk). A people of the northern part of Sakhalin Island or the adjacent part of Siberia; a member of this people; the language of this people. Also *attrib.* or as *adj.*
1858 *Jrnl. R. Geogr. Soc.* XXVIII. 389 One hundred and seventy miles below Pul, one finds the dwelling places of the Gilyaks. *Ibid.*, The Gilyak villages passed were Mangal, Dengdala,..and Tyr. **1861** E. G. Ravenstein *Russians on Amur* xx. 391 A Gilyak would not..permit any fire to be taken from his hut. **1890** J. G. Frazer *Golden Bough* II. iii. 105 The Gilyaks, a Tunguzian people of Eastern Siberia, hold a bear festival. *Ibid.* 107 The Gilyak custom of leading him from house to house. **1933** Bloomfield *Lang.* iv. 70 *Gilyak* is spoken in the northern part of Sakhalin Island and round the mouth of the Amur River. **1956** J. Whatmough *Language* 32 Grouped together with Kamchadal and Gilyak, as 'Hyperborean'. **1958** E. A. Armstrong *Folklore of Birds* i. 6 The Ainu and the Gilyak treat the bears they slay with great reverence. **1968** Chomsky & Halle *Sound Pattern Eng.* 305 In Ubykh, as in many other languages, such as Gilyak..the difference between velar and uvular points of articulation is paralleled by the difference between nonstrident and strident.

gimbal. Add: Also pronounced (gi·mbăl).

gimbri (gi·mbri). Also **gunibri, gunibry.** [ad. Arab. *gunbrī*.] A small Moorish guitar played by plucking the strings with a piece of dry palmetto leaf; also, a player of this instrument.
1876 Stainer & Barrett *Dict. Mus. Terms* 214/1 The resonance-body of the Gunibry is made of a bottle-pumpkin cut longwise, and covered with sheep skin. **1894** G. Montbard *Among Moors* 7 He sings in a sad broken voice,..accompanying the song on a sort of square guitar, a *gimbry*. **1903** *Westm. Gaz.* 18 Feb. 2/3 The slender fanatical singer, Whose fingers were skilled on the gimbri. **1907** F. Campbell *Shepherd of Stars* xv. 177 The Gimbri wagged his [head] from side to side. **1954** *Grove's Dict. Mus.* (ed. 5) V. 506/2 Those [instruments] of popular acceptance are the pandore (*gunibri*), [etc.].

gimlet, *sb.*¹ Add: **1. b.** *fig.*
1925 Hardy *Human Shows* 34 In the waves they bore their gimlets of light. **1946** *John Bull* 18 May 14/4 Crowthers was..a little gimlet of a man with glasses.
c. A cocktail, usu. consisting of gin and lime-juice.
1928 D. B. Wesson *I'll never be Cured* III. 73 The 'Gimlet' we were introduced to..at the Golf Club: and it proved to be the well and flavorably known ricky, but described as 'gin, a spot of lime, and soda'. **1937** N. Coward *Present Indicative* IX. 378 Standing about in the ward-room accepting with gracious melancholy 'gimlet' after 'gimlet'. **1953** R. Chandler *Long Good-Bye* (1959) iii. 18 A real gimlet is half gin and half Rose's Lime Juice and nothing else.
3. gimlet-eyed *a.* (later examples).
1948 Wyndham Lewis *Let.* 30 Jan. (1963) 429 A gimlet eyed rat steeped in classical learning. **1956** *Observer* 18 Mar. 15/2 'All for Love, or the World Ill Lost' is the message that issues from Shakespeare's gimlet-eyed indictment of Helen and Cressida.

gimlety (gi·mlĕti), *a.* Also **gimletty.** [f. Gimlet *sb.*¹ + -y¹.] Of eyes: piercing like a gimlet.
1899 *Daily News* 3 Jan. 5/6 Small, gimlety eyes. **1900** *Ibid.* 3 May 7/2, I can see the black, beady, gimletty, oblique eyes of his opponent glitter. **1919** J. Buchan *39 Steps* 13 Small gimlety blue eyes. **1955** *English Digest* June 44/2 Corporal of Horse Pomfret..subjected us to a dispassionate scrutiny. His gimlety eye fell on me.

gimme (gi·mi). Also **gi'me.** Colloq. contraction of *give me* (occas. of *give it to me*).
1883 [see Idioticon]. **1884** 'Mark Twain' *Huck. Finn* v. 44 Don't gimme no sass. **1888** 'R. Boldrewood' *Robbery under Arms* xi. 131 Now gimme some grub, for I've had nothing since sunrise. **1892** Kipling *Many Invent.* (1893) 163 Gimme a light—gimme a light! **1918** H. G. Wells *Joan & Peter* viii. 230 'Gimme my cane, someone,' said Mr. Mainwearing. **1923** Wodehouse *Adv. Sally* xv. 185 'Will you have a whisky and soda, Uncle Donald?'..'Yes,' said his relative... 'Gimme.' **1958** E. Nel *Mr. Churchill's Secretary* ii. 30 If the Prime Minister said 'Gimme Pug' I must fetch General Ismay. *Ibid.*, If he merely stretched out his hand and said 'Gimme' I must place in his hand whatever he needed.
Hence as *sb.*, acquisitiveness; greed; a desire for gifts; freq. *pl.*, in *to have* or *get the gimmes*. Also *attrib.* *slang*.
1927 Wodehouse *Meet Mr. Mulliner* v. 166 Boy, I use bank-notes for summer underclothing, and I don't care how bad you've got the gimme's if only you'll sign on the dotted line. **1928** *Sunday Dispatch* 2 Sept. 7/3 The vamp ..is what the Americans call a 'Gimme Girl'. **1963** C. Morris *I bought Newspaper* ii. 19 One could only write him off as a victim of our acquisitive, thrusting, philosophy of get and 'gimme'.

gimmick (gi·mik). orig. *U.S. slang.* [Origin unknown, but see quot. 1936.] A gadget; *spec.* a contrivance for dishonestly regulating a gambling game, or an article used in a conjuring trick; now usu. a tricky or ingenious device, gadget, idea, etc., esp. one adopted for the purpose of attracting attention or publicity.
1926 Maines & Grant *Wise-Crack Dict.* 8/2 *Gimmick*, device used for making a fair game crooked. **1926** *Amer. Speech* II. 62/1 Every snipe endeavors to impress the poor swabbos with his talk of gillguys, gadgetts, and gimmicks. [**1936** *Words* Nov. 12/2 The word *gimac* means 'a gadget'. It is an anagram of the word *magic*, and is used by magicians the same way as others use the word 'thing-a-ma-bob'.] **1948** J. Thurber *Beast in Me* (1949) 77 The wisecrack and the gag, the leg pull and the hotfoot, the gimmick and the switcheroo. **1951** *Manch. Guardian Weekly* 5 July 3/1 Washington has suspected that a political 'gimmick' might be wrapped up in the Malik offer. **1952** M. McCarthy *Groves of Academe* (1953) iv. 70 The perfect college they hinted at might exist on paper but it would never attract students, for it would have no selling-point, no gimmick, as they said in advertising. **1954** *Economist* 21 Aug. 602/1 Completely new 'gimmicks', such as the aerosol can, which is packed under pressure and blows its contents out as required, have opened up new fields for the tin-plate container. **1957** *Observer* 8 Sept. 9/5 Diamond..is unalarmed by Tory *élan*. His publicity gimmick, a small red diamond, is to be seen all over the city, sometimes stuck up in very curious places. **1958** I. Brown *Words in our Time* 58 Many comedians have their gimmicks, either as catch-phrase, theme-song, or bit of 'business', which they exploit in most of their appearances. **1959** 'P. Quentin' *Shadow of Guilt* xvii. 163 This was his new gimmick, his new device for making himself seem absolutely harmless. **1960** V. Nabokov *Invitation to Beheading* xii. 122 There are..all sorts of marvellous gimmicks. I remember, for instance when I was a child, there were objects called '*nonnons*' that were popular. **1969** *Daily Tel.* 1 Mar. 16/3 There is a growing tendency to use sport as a publicity gimmick for business. **1970** *New Society* 5 Mar. 385/2 It's a market that..boomed briefly in the mid-sixties with gimmicks like paper dresses.
b. *attrib.* and *Comb.*
1953 *Economist* 30 May 595/2 Some of Professor Triffin's colleagues at the conference were innately suspicious of what they called 'gimmick solutions' of this kind. **1958** *Spectator* 14 Feb. 197/1 A weekly devoted to one blind purpose—the promotion of a single, gimmick-geared pattern of love and marriage. **1961** *New Left Rev.* Jan.–Feb. 34/2 The speculators and gimmick-mongers. **1967** *Punch* 29 Mar. 441/2, I am inviting fellow revolutionaries to *refuse* to buy all goods with attached gimmick offers.
Hence **gi·mmick** *v. trans.*, to provide with a gimmick; to alter or tamper with; hence *gimmicked* ppl. adj.; **gi·mmickless** *a.*, devoid of or free from a gimmick or gimmicks; **gi·mmick(e)ry**, the use of gimmicks; gimmicks collectively; an abundance of gimmicks; **gi·mmicky** *a.*, employing or characterized by gimmicks; designed to attract attention or publicity.
1952 *Astounding Sci. Fiction* Feb. 55, I could gimmick your machine. **1952** *N.Y. Times* 9 Sept. 2/2 Senator Johnson's subcommittee also had attacked what it called excessive 'gimmickery and gadgetry' on today's warplanes. *Ibid.* 26 Oct. VI. 49/1 The Kennedy tea parties are a unique addition to the campaign gimmickry of one of the most spirited and critical Senatorial contests of the year. **1957** *Times* 2 Dec. 7 (Advt.), What I thought I'd do was write a gimmicky ad with a screwy drawing, and people would be bound to read it. **1958** *Observer* 28 Apr. 15/1 A Shakespeare play can be ripped apart by the twin steel claws of naturalism and gimmickry. **1959** *News Chron.* 18 Nov. 6/7 Eisenhower's own favourite speech writer, the master ghost of those gimmicked years. **1959** C. MacInnes *Absolute Beginners* 55 I'm not sure I care for that gimmicky girl. **1962** *Times* 9 Mar. 13/4 To declare oneself gimmickless is in these times to assert straightforward procedure. **1963** *Punch* 16 Oct. 576/2 A good, straight, gimmickless lecture on painting. **1964** *New Statesman* 6 Mar. 354/3 US policy in Vietnam has been a compote of gimmickry, half-truth and self-deception. **1966** I. Asimov *Fantastic Voyage* xiii. 151 He knows the ship. He designed it. He can best gimmick its controls. **1970** D. Bagley *Running Blind* ix. 202, I wasn't stupid enough to search Slade's luggage... He would have gimmicked it so that he could tell at a glance whether a suitcase had been opened. **1970** *Amat. Photographer* 11 Mar. 27/2, I don't therefore go along with those who condemn elaboration in cameras and projectors as futile gimmickry.

gimp (gimp), *sb.*³ *slang.* [Origin uncertain.] Courage, 'guts'; 'stuffing' (Stuffing *vbl. sb.* 2 d).
1901 *Munsey's Mag.* XXIV. 567/2 Sort of took the gimp out of you, didn't it? **1906** E. Dyson *Fact'ry 'Ands* xiii. 164 Ther wearer wouldn't have sufficient gimp t' get up off er tack. **1962** J. Potts *Evil Wish* xii. 159 She didn't even have the gimp to make the break herself.

gimp (gimp), *sb.*⁴ *slang* (orig. *U.S.*). [Origin uncertain; perh. a corruption of Gammy *a.* 3.] A cripple; a lame person or leg; a limp. Also as *vb.*, to limp, hobble (1961 in Webster). So **gi·mpy** *sb.*, a cripple; *adj.*, lame, crippled.
1925 *Flynn's* 31 Jan. 306/2 *Gimp*, a lame leg. **1925** G. H. Mullin *Adv. Scholar Tramp* xiii. 194 On the Road a lame man is a gimpy. **1929** *New Yorker* 9 Feb. 38/1 Eat-'em-up Jack McManus would never hit a cripple. He'd just kick a gimp in the good leg and leave him lay. **1931** G. Irwin *Amer. Tramp & Underworld Slang* 85 *Gimpy*, lame; crippled. **1931** D. Runyon *Guys & Dolls* (1932) viii. 160 She walks with a gimp in one leg, which is why she is called Madame la Gimp. **1965** J. Philips *Twisted People* iv. 77 Now I know you've got a gimpy leg. **1969** P. Craig *Gate of Ivory* (1970) xii. 161, I gimped back on deck.

gim-peg, var. *gem-peg* (Gem *sb.* 8 d).
1853 [see Gem *sb.* 8 d]. **1882** *Encycl. Brit.* XIV. 299/2 A very important substitute for the gim-peg-socket..is the dial.

gin, *sb.*¹ Add: **1. b.** A drink or glass of gin.
1922 Joyce *Ulysses* 236 A small gin, sir. **1938** C. Morgan *Flashing Stream* I. i. 54 Time for a gin before they come. **1938** G. Greene *Brighton Rock* I. i. 5 He only felt his loneliness after his third gin. *Ibid.* III. i. 98 I'll have a gin.
c. *Ellipt.* Gin rummy. *colloq.*
1956 E. Ambler *Night-Comers* viii. 175, I was able to persuade her to..begin another game of gin. **1959** M. Dolinsky *There is no Silence* i. 5, I was trying to salvage an incredible gin hand.
2. a. *gin-and-bitters* (earlier examples), *gin-and-orange*, *gin-and-tonic*; also *gin-and-French* (see *French B. 3).

a **1833** J. T. Smith *Bk. for Rainy Day* (1845) 168 She had taken her morning dose of gin and bitters. **1835** Dickens *Sk. Boz* (1836) 2nd Ser. 148 A couple of ladies.., having imbibed the contents of various 'three-outs' of gin and bitters. **1950** J. Cannan *Murder Included* i. 5 He himself took a sip of gin-and-orange. **1935** S. Box in *Best One-Act Plays of 1935* 56 Waiter! That will be two whiskies, and a gin and tonic. **1949** J. B. Priestley *Delight* 29 Just gin and tonic and some potato crisps.
b. gin and fog *colloq.*, a hoarse or broken-down voice; also *attrib.*; **gin and it** (or It), gin and Italian vermouth; **gin-and-Jag**(uar), colloq. phr. used *attrib.* to denote upper-middle-class people or areas; **gin-berry**, a juniper-berry; **gin-crawl** (see *crawl *sb.*¹ b); **gin-house** = Gin-shop; **gin rummy** orig. *U.S.*, 'a form of rummy in which a player who has cards that count no more than ten may "knock" in an effort to win the number of points by which his opponent's unmatched cards exceed his own' (D.A.); **gin-soaked** *a.*, soaked in gin; given to drinking large quantities of gin.
1907 *Westm. Gaz.* 18 May 7/1, I am badly bored by the man who, having an inclination to murder somebody, walks up and down the stage dissecting his inclination in what is known, I believe, as a 'gin and fog' voice. **1947** C. Witting *Let X be Murderer* xiii. 133 His voice was of the hoarse variety known as gin-and-fog. **1932** B. Worsley-Gough *Public Affaires* xiv. 257 As I was sipping my gin-and-it before lunch. **1933** *Punch* 15 Mar. 289/1 Soothed by many a gin and It. **1960** K. Amis *Take Girl like You* xix. 229 Her lighter and chocolates and gin-and-its with two cherries on sticks. **1969** *Sunday Tel.* 16 Mar. 3/3 (*headline*) The 'gin and Jag' rebels. *Ibid.*, The working-class boy is a dedicated and motivated student. It is students from the gin-and-Jaguar belt who often lack any sense of what a university is for. **1839** Z. Leonard *Adv.* (1904) 170 On the South side, where grows a kind of Juniper or Gin shrub, bearing a berry tasting similar to gin. Here we passed the night without any thing to eat except these gin berries. **1960** *Spectator* 18 Nov. 795/3 People..will turn and rend you..if you say you've put a couple of juniper berries in the sauce for the pork chops. Why can't we change juniper..to gin-berry? **1848** J. H. Newman *Let.* (1962) XII. 301 The owner of the Gin House has put off till Monday next to decide whether he will accept £90. **1941** W. S. Maugham in *Writer's Notebook* (1949) 317 After a good dinner..she suggests a game of gin-rummy. **1942** O. Nash *Good Intentions* 49, I love you more than gin rummy is a bore. **1965** M. Spark *Mandelbaum Gate* vii. 253 She found her playing gin rummy with Ruth Gardnor. **1899** *Daily News* 15 Apr. 2/1 Lurid accounts of gin-soaked Africa. **1908** *Daily Chron.* 20 Aug. 4/4 The gin-soaked grandmother.

gin (dʒin), *v.*³ *colloq.* [f. Gin *sb.*²] *intr.* To drink gin or other intoxicating liquor; to become drunk; usu. with *up*. Hence **gi·nned** (up) *a.*, intoxicated.
1894 *Midwinter Appeal* (San Francisco) 17 Feb. 4/5 As for jags, he held that he can gin up when he likes. **1900** W. F. Drannan *31 Yrs. on Plains* 121 This man Shewman got pretty well ginned up. **1905** H. A. Vachell *Hill* iii. 53 They're pretty well ginned-up, I can tell you. **1924** *T.P.'s & Cassell's Weekly* 6 Sept. 631/1 He danced with a good many girls who had whisky breaths. One girl clung to him..and whispered, 'Hold me up, kid; I'm ginned.' **1964** P. M. Hubbard *Picture of Millie* x. 103 He always drinks a good deal... I have seen him more or less ginned-up.

ginestra (dʒine·strā). [It., = broom.] = Broom *sb.* 1.
1899 *Daily News* 25 Feb. 5/1 The glorious ginestra, or mountain broom. **1921** *Q. Rev.* Jan. 52 Thou, Fragrant Ginestra, joyful in the wild, Scatterest thy solitary tufts around.

ginger, *sb.* Add: **6. a.** *ginger biscuit*, *-cookie.*
1845 E. Acton *Mod. Cookery* (ed. 2) xxiii. 519 Cheap ginger biscuits. **1969** D. Gray *Murder on Honeymoon* xvi. 98 A plate piled with ginger biscuits. **1880** *Harper's Mag.* Mar. 576/1 Aunty'll give you ginger-cookey this very minute! **1930** J. Dos Passos *42nd Parallel* 445 He started to think about the smell of gingercookies.
d. *attrib.* or as *adj.* Applied to a group, person, etc., which provides the 'ginger', spirit, or stimulus of a party or movement; esp. in *ginger group*.
1923 G. D. H. Cole *Workshop Organiz.* 37 A 'ginger' organization within the Trade Union movement. **1927** *Punch* 9 Mar. 269/1 The 'ginger group' of Conservative social reformers. **1944** H. Croome *You've gone Astray* ii. 18 This little ginger group campaigning for better houses. **1958** *Manch. Guardian* 1 Aug. 6/7 (*headline*) Ginger grouper. *Ibid.*, One of a small ginger group of Left-wing Catholics. **1965** *Economist* 29 May 1039/1 The Ministry of Public Building and Works was reorganised into a ginger-ministry designed to co-ordinate Britain's industrial building systems. **1970** *New Society* 5 Feb. 210/2 The appearance of ginger groups to fight specific proposals, is not necessarily a bad thing—particularly if the established bodies aren't prepared to fight.
7. ginger-cake, a cake flavoured with ginger; gingerbread in the form of a cake; **ginger lily** = *garland-flower* (*b*); **ginger-nob** *slang*, a red-haired person; also, the head of such a person.
1771 R. Drummond *Let.* 23 Feb. in *Corr. Garrick* (1831) I. 414, I have sent you the receipt for the ginger-cakes. **1824** J. Pickering *Jrnl.* 25 Dec. in *Emigration* (1830) ii. 11, I was presented..with ginger-cake, and a cake with raisins in it, which is their 'Christmas cake'. **1843** W. T. Thompson *Major Jones' Courtship* 43 I'm jest as good for old Miss Stallionses consent as a thrip is for a ginger cake. **1936** Lucas & Hume *Au Petit Cordon Bleu* 172 Ginger cake (A 2-lb. cake). **1900** L. H. Bailey *Cycl. Amer. Hort.* II. 717/2 Hedychium...Ginger Lily. **1926** M. Leinster *Dew on Leaf* 8 A sickly odour of ginger-lilies. **1969** 'J. Morris' *Fever Grass* xxi. 189 The thick clumps of ginger lilies. **1907** J. Masefield *Tarpaulin Muster* iv. 68 It wouldn't give me no pleasure..to have that ginger-nob in my chest. **1959** I. & P. Opie *Lore & Lang. Schoolch.* ix. 170 Red heads attract a barrage of nicknames:.. gingernob, [etc.].

ginger-beer. Add: **b.** *ginger-beer plant*, a mixture of a yeast (*Saccharomyces pyriformis*) and a bacterium (*Bacillus vermiformis*) used to ferment sugar solution to make ginger-beer.
1892 H. M. Ward (*title*) The ginger-beer plant and the organisms. **1953** J. Ramsbottom *Mushrooms & Toadstools* xviii. 213 The old English Ginger-Beer plant is a globular gelatinous white mass usually about the size of a pea and is used for fermenting sugary fluids.
2. Rhyming slang for 'queer': a homosexual; also as *adj.* Also *ellipt.* as *ginger*.
1959 E. Mannin *Blue-Eyed Boy* I. i. 14 There were blokes..that were definitely ginger-beer—that never had a fancy for the opposite sex, but plenty for their own. **1964** C. Rougvie *Medal from Pamplona* vii. 91 'There are too many ginger beers in the place.' 'Well, if you don't like queers..what are you doing here with that fellow?' **1968** A. Williams *Brotherhood* i. 38 'Unless you prefer ginger.' 'Ginger?' 'Beer, dear.'..'You ever meet an Aussie who was queer?'

gingerbeery, *a.* Delete *nonce-wd.* and add earlier and later examples.
1852 Dickens *Househ. Words* 31 July 454/1 A brisk, ginger-beery sort of speech. **1948** J. Betjeman *Coll. Poems* (1958) 151 And ginger-beery surf hissed 'Christabel!'

gingerish (dʒi·ndʒəriʃ), *a.* [f. Ginger *a.*¹ + -ish.] Somewhat ginger in colour.
1910 W. de la Mare *Three Mulla-Mulgars* 85 In this tree grows a small round, gingerish nut. **1959** *Listener* 23 July 129/1 A gingerish moustache.

gingivostomatitis (dʒi:ndʒivo,stǫmătəi·tis). *Path.* [f. Gingiv(itis + -o + Stomatitis.] Gingivitis combined with stomatitis.
1935 Prinz & Greenbaum *Dis. Mouth* vii. 102 The disturbances of the oral mucosa may be conveniently divided into gingivitis, stomatitis and gingivo-stomatitis. **1964** M. Hynes *Med. Bacteriol.* (ed. 8) xxv. 396 Primary infection with the herpes virus may be symptomless, but in childhood usually causes an acute gingivostomatitis.

gingko. Substitute for entry:
ginkgo (gi·ŋkgo, gi·ŋko). Also **gingko, gingo, ginko.** [Jap., f. Chinese *yinhsing* silver apricot.] The maidenhair tree (*Ginkgo biloba*) native to China and Japan and cultivated elsewhere, with wedge-shaped leaves and yellow flowers, the only living species of the order Ginkgoales which flourished in the Mesozoic era. Also *attrib.*, as *ginkgo-nut*, *-tree*. (Add earlier and later examples.)
[**1727** J. G. Scheuchzer tr. *Kæmpfer's Hist. Japan* ix. 116 Another sort of nuts, call'd Ginau..grow very plentifully..on a fine tall tree.] **1773** [see *maidenhair-tree* s.v. Maidenhair 6]. **1912** A. Conan Doyle *Lost World* x. 166 One huge gingko tree, topping all the others, shot its great limbs and maidenhair foliage over the fort which we had constructed. **1955** *Sci. News Let.* 26 Mar. 207/2 There are a lot of young ginkgos growing in Battery Park, New York. China is the native home of the ginkgo tree. **1957** R. Carrington *Mermaids & Mastodons* xiii. 175 Ginkgos are exceptionally resistant to fire.

gink (giŋk). *slang* (orig. *U.S.*). [Of obscure origin.] A fellow; a man. (Freq. pejorative.)
1910 *Sat. Even. Post* 22 Oct. 12/3, I don't believe that all these ginks have got coin enough to support one good game. **1919** Wodehouse *Damsel in Distress* ii, I'm certain this gink is giving her a raw deal. **1927** *Sunday Express* 6 Feb. 4 One of America's noblemen, a great pirate, a gink with nerve. **1930** G. Mitchell *Myst. Butcher's Shop* xiv. 164 When I 'ears silly ginks in this 'ere bar talking about 'im doin' this 'ere murder, I says to 'em [etc.]. **1957** L. P. Hartley *Hireling* 134 Hughie was a heel or a gink or anything you liked to call him. **1961** *Times* 19 Dec. 7/7 There are ginks running these schools and interfering with small boys. **1970** A. Draper *Swansong for Rare Bird* viii. 71 George wasn't the most talkative gink alive.

ginkgo: see *gingko.

ginnel, var. Gennel.

ginny, *a.*² Add to def.: resembling, addicted to, or characterized by gin. (Add further examples.)
1926 S. T. Warner *Lolly Willowes* 39 A hot ginny churchyard smell. **1948** G. Greene *Heart of Matter* III. III. i. 294 His ginny breath. **1964** R. Braddon *Year Angry Rabbit* vi. 58 Suddenly he felt very tired, and very reluctant to endure the ginny *bonhomie* of a wardroom.

gin-sling. (Earlier and later examples.)
1790 J. Baxter *Jrnl.* 19 Feb. in *Amer. Speech* (1965) XL. 199 Drank a gin sling with him at fat Simmon's. **1837, 1843** [see *cocktail 3 a]. **1938** G. Greene *Brighton Rock* v. i. 192 They use their own eggs in the gin slings. **1943** N. Coward *Middle East Diary* 15 Aug. (1944) 49 Sipping Gin-Slings and cocktails.

ginzo (gi·nzo). *U.S. slang.* Also **guinzo.** [perh. ad. *Guinea 1 b.] In contemptuous use: a person of Italian extraction; also as *adj.*
1931 *Amer. Mercury* Dec. 413/1 He's a Roumanian or some kinda guinzo. **1964** W. Markfield *To Early Grave* (1965) vii. 125, I have a boss, a ginzo—though he speaks a great Jewish. **1967** H. Harrison *Technicolor Time Machine* (1968) v. 51 I'm *Americano*, but I got ginzo relatives all around the Bay of Naples.

Gioconda (dʒiǫkǫ·ndă). [It., fem. of *giocondo* (see Jocund *a.*).] The name of a painting (in full La Gioconda; also known as the *Mona Lisa*) by Leonardo da Vinci (1452–1519), used allusively attrib. to describe an enigmatic smile resembling that of the woman in this painting.
1921 A. Huxley in *Eng. Review* XXXIII. 89 Miss Spence was smiling too: her Gioconda smile, he had once called it... Miss Spence..always tried to live up to the Leonardo standard. **1964** T. Helmore *Affair at Quala* iv. 113 She looked at me and gave me a Gioconda smile. **1967** L. Peters *Riot '71* 110 Smith smiled again, putting a slightly mysterious, Gioconda quality into it this time.

‖ **giocoso** (dʒiǫkō·so). *Mus.* [It. = merry.] A direction denoting that a composition is to be executed merrily. Also *transf.*
1828 T. Busby *Mus. Man.* 81 Giocoso, in a gay and joyous style. **1960** *Times* 25 July 3/1 It grew steadily more sinister, further from the *giocoso* drama of its title. **1962** *Times* 16 Feb. 15/5 The *giocoso* element [in *Don Giovanni*] was quite excessively overshadowed.

Giorgi (dʒǫ·ɹdʒi). [The name of Giovanni Giorgi (1871–1950), Italian physicist, who first proposed the system in *Atti dell'Assoc. elettrotecn. ital.* (1901) V. 402–18.] *Giorgi system*: a system of units based on the metre, the kilogramme, the second, and one other (electrical) unit; now *spec.* a system in which the electrical unit is the ampere.
1905 *Jrnl. Inst. Electr. Engin.* XXXIV. 179 The Giorgi system is a rational one (in the Heaviside sense). **1933** *Bull. Nat. Research Council* No. 93, 103 Like the Giorgi system, it could be adopted either unrationalized or rationalized. **1934** *Nature* 21 Apr. 597 (*heading*) The Giorgi (M.K.S.Ω) system of units. **1964** *Internat. System (SI) Units (B.S.I.)* 6 This system, called the MKSA or Giorgi System, was adopted by the International Electrotechnical Commission in 1950. **1969** *Nature* 8 Nov. 581/2 Physically all four of E, H, D and B act as 'force vectors'. Thus a dimensional difference between the members of each pair, as in the Giorgi systems, results in distortion.

Giorgionesque (dʒǫɹdʒǫne·sk), *a.* [f. *Giorgione* + -esque.] Resembling the style of the Italian painter *Giorgione* Barbarelli (1478–1510).
1906 *Westm. Gaz.* 19 Dec. 10/3 Mr. Lane's beautiful Giorgionesque half-length portrait of a young man. **1929** *Daily Tel.* 8 Jan. 8/5 The rich, almost Giorgionesque modelling, of the Diana. **1956** K. Clark *Nude* iv. 119 The most beautiful of Titian's Giorgionesque nudes is in the picture known as Sacred and Profane Love. *Ibid.* vii. 271 One of them..looks unbelievably Giorgionesque.

gip, var. *gyp *sb.*³ or *v.*

gipper, var. *jipper.

gippo² (dʒi·po). *slang* (chiefly *Services'*). Also **gippy, gyp(p)o, gypoo.** [Var. *jipper, *jippo.] (See quots.)
1914 *Times* 23 Dec. 3/2 You take your mess-tin, [and] empty out..the greasy 'gyppo' (that means stew). **1916** *Daily Mail* 1 Nov. 4/4 'Gipoo' (gravy or grease). **1925** Fraser & Gibbons *Soldier & Sailor Words* 105 *Gippo* (also *gypo*), gravy. Bacon fat. Butter. Any greasy sauce.—*e.g.*, 'Is there any Gippo?'

gippy (dʒi·pi). *slang.* Freq. with capital initial. Also **gippo, gypo, gyppie, gyppo, gyppy.** [f. Gip(sy *sb.* + -y⁶, infl. by Egyptian *a.*] **1.** An Egyptian, esp. a native Egyptian soldier. Also *attrib.* or as *adj.*
a **1889** *World* (Barrère & Leland), Colonel Kitchener will probably stick to his original intention of having only *gippies* (as they call the Egyptian soldiers here) at Suakim. **1915** V. Horsley in S. Paget *Sir V. Horsley* (1919) III. ii. 308 The Gippy himself has his points. **1919** *Athenæum* 8 Aug. 727/2 'Gypos' for Egyptians. **1925** Fraser & Gibbons *Soldier & Sailor Words* 105 The 'Gyppie' Army was the common colloquial term for the native force organized by Sir Evelyn Wood and Lord Kitchener. *Ibid.*, *Gippy bint*, an Egyptian girl. **1929** C. Mackenzie *Gallipoli Mem.* iii. 26 A captain of the Gyppy Army with a pleasant oblong face. **1955** E. Waugh *Officers & Gentlemen* 208 'What's to stop him coming round the other side?' asked Tommy. 'According to plan—the Gyppos,' said the Brigadier.
b. An Egyptian cigarette.

1920 *Punch* 7 Apr. 270 I'll smoke with pleasure if they're Gyppies. Can't *stand* gaspers. 1926 W. S. MAUGHAM *Constant Wife* (1927) II. 122 When you once get the taste for them, you prefer them to gippies.

c. *gippy tummy*: diarrhœa suffered by visitors to hot countries.

1943 A. MOOREHEAD *Year of Battle* i. 22 Few set foot in Egypt without contracting 'Gyppy Tummy' which is a mild stomach disorder lasting usually a couple of days. It recurs at irregular intervals and it makes you feel terrible. 1961 G. EGMONT *Art of Egmontese* iv. 71 Always take..whatever is your favourite antidote to gippy tummy when you go abroad.

2. A gipsy. Also *attrib*. or as *adj*.

1902 E. G. HAYDEN *From Thatched Cott.* xi. 95 Dally-oh if Jim ha'n't..left the place fur tramps an' gippoes to walk in. 1912 *English Rev.* Oct. 476 He could get some one else, perhaps a half gypo, or a lad, to take my place. 1913 *Chums* 13 Dec. 253/2 'What's your name, sonny?'.. 'No idea,' answered the gipsy with a soft laugh. 'I'm generally called Gippy in England. I've got other names in France, and Spain, and Italy.' 1924 KIPLING *Debits & Credits* (1926) 126 There's no sayin' what Gippos know. *Ibid.* 127 What the Gippo girl 'ad told 'er. 1953 DYLAN THOMAS *Under Milk Wood* (1954) 21 Ducking under the gippo's clothespegs.

gipsy, gypsy, *sb.* Add: **1. b.** (Examples.)

1800 W. WHITER (*title*) Etymologicon Magnum..with illustrations drawn from various languages:..Sanscrit, Gipsey, Coptic, etc. 1841 BORROW *Zincali* I. ii. 53 We have found this beautiful metaphor both in Gypsy and Spanish. 1875 SMART & CROFTON *Dial. Eng. Gypsies* 276 Tell me, old fellow, what the sun is in Gypsy. *Ibid.*, I see you know plenty of Gypsy, and I dare say you know more words than any of us. 1930 J. SAMPSON *Wind on Heath* x. 277 (*heading*) One Use Of Gypsy.

2. c. *U.S. slang.* (See quots.)

1942 BERREY & VAN DEN BARK *Amer. Thes. Slang* § 766/6 *Gypsy*, an out-of-town truck with no terminal in its destination city. 1954 *Ibid.* (ed. 2) § 765/5 *Gipsy*, an independent truckman. *Ibid.* § 766/6 *Gipsy*, an independently owned truck. 1967 *Boston Sunday Globe* 23 Apr. A. 36/1 The primary violators among truck drivers are the so-called 'gypsies' who operate independently.

5. (Earlier example of *gipsy party*.)

1816 JANE AUSTEN *Emma* III. vi. 87 There is to be no form or parade—a sort of gipsy party.

6. gipsy('s) pig, the hedgehog; **gipsy('s) pork,** the flesh of the hedgehog; **gipsy's warning,** a cryptic or sinister warning.

1928 *Sunday Disp.* 2 Sept. 3/3 'Gypsy Pork.' Hedgehogs are succulent this month. September is the month when the 'gypsies' pig' is plump and tender. 1918–19 *T. Eaton & Co. Catal.* 369/2 Vocal records...The Gipsy's Warning (Tenor). 1923 W. J. LOCKE *Moordius & Co.* viii. 113 Bid him beware of Peter Moordius...The gipsy's second warning. 1928 D. L. SAYERS *Unpleasantness at Bellona Club* xi. 126 'What price the gipsy's warning now?' said Lord Peter. 1941 'P. WENTWORTH' *Danger Point* (1942) xvi. 100, I expect Robson's got you taped...And that being so, suppose you listen to the gypsy's warning. 1967 A. CHRISTIE *Endless Night* xiii. 112 'You'll have to fend for yourself.' 'Cut out the gipsy's warning, Santonix,' I said.

giraffe. Add: **5.** Special Comb.: **giraffe acacia, giraffe tree** *S. Afr.*, names occas. used for the camel-thorn tree.

1896 H. A. BRYDEN *Tales of S. Afr.* 44 Groves of giraffe acacia (*kameel doorn*). 1815 A. PLUMPTRE tr. *Lichtenstein's Trav. S. Afr.* II. xlix. 288 A tall and wide spreading giraffe tree, the *acacia giraffæ* of Wildenow.

giraffid (dʒiræ·fid), *a.* and *sb.* Read: [f. GIRAFFE + -ID³.] **A.** *adj.* Belonging to the family Giraffidæ; like or resembling a giraffe. **B.** *sb.* A member of the family Giraffidæ. So **gira·ffine, gira·ffoid** *a.* and *sb.*

1889 in *Cent. Dict.* 1891 FLOWER & LYDEKKER *Introd. Stud. Mammals* ix. 332 A gradual diminution..can be traced from the more Giraffoid to the more Bovoid forms. 1901 *Proc. Zool. Soc.* II. 280 [The skull] showed itself to be that of a Giraffine animal, and not that of a Bovine. 1902 *Rep. Brit. Assoc. Adv. Sci.* 623 The African Okapi.. is a giraffine, horned in the male. 1945 G. G. SIMPSON *Princ. Class. Mammals* III. 266/2 The giraffid resemblance seems to be the result of ..common ancestry. *Ibid.* 268/2 The giraffids..are a distinctive and fairly compact family unit. 1965 M. H. DAY *Guide to Fossil Man* III. 120 The fossil mammalian fauna..includes..several bovids and giraffids. 1968 C. A. SPINAGE *Bk. of Giraffe* i. 15 The earliest giraffid appeared in the Miocene. *Ibid.* ii. 28 It [*sc.* an animal in a rock painting] is walking with a giraffid gait.

girba (gə̄·ɹbă). Also **gerba(h), gurba.** [ad. Arab. *ḳirba* waterskin.] A water-vessel made of leather.

1790 J. BRUCE *Trav.* IV. VIII. iv. 334 A girba is an ox's skin squared, and the edges sewed together very artificially by a double seam. 1821 G. F. LYON *Narr. Trav. N. Afr.* 35 Our road was through very difficult passes in the mountains, where we found some rain water, with which we filled our gerbas, or water-skins, with a sufficient supply for three days. 1864 J. A. GRANT *Walk across Africa* 418 We had twenty Seedees, and each required to have two goatskins, or 'girba', to carry his drinking-water. 1866 S. W. BAKER *Albert Nyanza* I. iv. 150 This is the unexceptional rule in African travelling—'fill your girbas before starting'. 1925 W. J. H. KING *Myst. Libyan Desert* viii. 97 A *gurba*, raised slightly off the ground in this way, so that the air can circulate round it, keeps the water much cooler.

girder (gə̄·ɹdəɹ), *v.* [f. GIRDER¹.] *trans.* To support or strengthen with or as with a girder; freq. *fig.* So **gi·rdered** *ppl. a.*

1934 DYLAN THOMAS *18 Poems* 32 Strong As motor muscle on the drill, driving Through vision and the girdered nerve. 1938 L. MACNEICE *Mod. Poetry* i. 18 There is a chance for poets..to girder it [*sc. élan vital*] with a structure supplied partly by reason. 1956 D. BARNHAM *One Man's Window* 19 The hangar is a vast girdered cavern. 1969 *Observer* 12 Jan. 27/1 A vision in which titanic masonry enveloped mankind like the girdered belly of a whale.

girdle, *sb.*¹ Add: **1. c.** *girdle of chastity* = *chastity belt; *girdle of Venus*: see VENUS¹ 9.

1802 tr. *Pallas's Trav. S. Prov. Russ. Empire* I. 399 Their girls..are provided with laced stays, or a broad girdle made of untanned leather: this singular coat of mail is among the common people tightly sewed round the waist, but in the higher classes it is fastened with silver hooks, so that they are obliged to wear it till their wedding-night...Besides the girdle of chastity above mentioned there is another circumstance which contributes to preserve the elegant shape of the girls: they are sparingly nourished.

d. = CORSET 2; *spec.* a corset, usu. elasticated, that does not extend above the waist. orig. *U.S.*

1925 *Eaton's News Weekly* 24 Oct. 19 Brassiere of rayon jersey silk and girdle of mercerized cotton and silk brocade with panels of elastic. 1928 *Delineator* Mar. 109 (Advt.), Girdle. 1932 *Daily Mail* 3 Oct. 10/1 A two-way stretch elastic girdle. 1942 D. POWELL *Time to be Born* (1943) xi. 266 Go out and get yourself a girdle. 1957 H. ROOSENBURG *Walls came tumbling Down* 14 The girdle certainly didn't support her kidneys; it flapped..loosely around her. 1963 *Punch* 20 Mar. 426/3 His elevation from girdle ads to the glossies.

4. g. (Earlier U.S. example.)

1825 J. LORAIN *Pract. Husb.* 63 All of them eventually die, provided the girdle be carefully cut through the sap into the heart-wood of the tree.

6. girdle-hanger (see HANGER² 4 b).

1921 *Brit. Mus. Return* 66 Anglo-Saxon iron girdle-hanger from Cliffe, near Rochester. 1923 C. FOX *Archaeol. Cambr. Region* vi. 271 Girdle hangers. The simplest forms are a close copy in bronze of the housewife's keys of iron (a Roman type), the possession of which they doubtless symbolized.

girdle, *v.* Add: **1. b.** To travel round. Cf. GIRDLE *sb.*¹ 3 b.

1901 F. H. BURNETT *Making of Marchioness* II. vi. 180 Agatha Norman, at present joyously girdling the globe with her bridegroom.

girdled, *ppl. a.* (Further examples.)

1888 [see GIRDLE *v.* 1]. 1940 E. S. GARDNER *Case of Silent Partner* xii. 196 Her tightly girdled, snappily dressed figure. 1961 A. SMITH *East Enders* vii. 119 Women with plucked eyebrows and men with girdled waists.

girl, *sb.* Add: **2. a.** *old girl* (earlier and later examples); *spec.* one's mother or wife; also, a former pupil of a girls' school or college; also *attrib.*

1826 T. CREEVEY *Let.* 22 Aug. in J. Gore *Creevey* (1948) xvii. 266 The old girl has a jointure of £5,000 a year. 1853 DICKENS *Bleak Ho.* xxvii. 272 You know me. It's my old girl that advises. *Ibid.* 273 'Old girl,' says Mr. Bagnet, 'give him my opinion.' 1875 C. M. YONGE *Let.* 6 Apr. in C. Coleridge *C. M. Yonge* (1903) 262 Fifty-four mothers had tea in the school... So many are my own old girls. 1898 J. D. BRAYSHAW *Slum Silhouettes* 220 He lets aht that Liz an' 'er ole gal was going ter the Crystal Palace. 1947 D. M. DAVIN *Gorse blooms Pale* 55 'Poor old girl,' he said as he leg-roped her [*sc.* a cow]. 1954 E. JENKINS *Tortoise & Hare* ix. 89 It was Old Girls' Week-end at her school. 1954 G. SMITH *Flaw in Crystal* x. 95 One hears these things on the old-girl network. 1967 E. LEMARCHAND *Death of Old Girl* ix. 109 She said she might still be held up at some Old Girls' supper.

c. (Earlier and later examples.) Also, one's wife.

1648 HERRICK *Hesper.* 24 Some ask'd how Pearls did grow, and where? Then spoke I to my Girle, To part her lips, and shew'd them there The Quarelets of Pearl. 1772 J. WEDGWOOD *Let.* 4 Oct. (1965) 137 Your good Lady is really recovering her health..though more slowly than we could wish, which is exactly the case of my poor Girl. 1887, etc. [see *BEST *a.* 2 c]. 1899 W. BESANT *Orange Girl* I. v. 59, I drew my girl closer and kissed her. 1912 T. DREISER *Financier* v. 54 Before leaving to call on his girl, Marjorie Stafford. 1917 *Punch* 15 Aug. 125/2 And when the War is over, some knight or belted earl, What's survived from killin' Germans, will take 'er for 'is girl. 1940 L. MACNEICE *Last Ditch* 23 Life in a day: he took his girl to the ballet. 1952 M. R. RINEHART *Pool* xxviii. 249 He even had a girl, although he said he wouldn't marry her until he was cleared of the murder charge.

†**e.** A black woman. *U.S. colloq. Obs.*

1835 J. H. INGRAHAM *South-West* II. 242 They always address them [*sc.* the slaves] as 'boy' and 'girl', to all under forty years of age. 1879 A. W. TOURGÉE *Fool's Errand* x. 42 You must remember that all colored women are 'girls'.

f. Colloq. phr. (*all*) *girls together*, on terms of close friendship with another girl or girls; also *attrib.* (hyphened).

1931 R. FERGUSON *Brontës went to Woolworth's* xii. 141 It would be terrible if she wanted to be all-girls-together with me about him. 1946 'S. RUSSELL' *To Bed with Grand Music* viii. 101 She seemed more than willing to re-establish a girls-together relationship with Deborah. 1959 J. BRAINE *Vodi* xiv. 185 That would-be refined, all-girls-together voice. 1961 A. CHRISTIE *Pale Horse* xii. 128, I got her softened up... Girls-together stuff.

g. *les girls*, girls collectively; *spec.* chorus girls.

1938 S. BECKETT *Murphy* ix. 193 The sceptic rut that places the objects of its curiosity on the level of Les Girls. 1955 E. WAUGH *Officers & Gentlemen* 178 He had come to the bar for stimulus, for a spot of pleasantry with 'les girls'. 1967 J. PORTER *Chinks in Curtain* ix. 89, I haven't seen hide or hair of him. He's probably still shacked up with les girls.

h. (*the*) *girl next door*, the girl in a conventional romance; a trusting, sweet, and faithful but usu. unimaginative young woman.

1961 *Sunday Express* 28 May 18/5 June Thorburn..has been trying..to break away from her usual 'girl next door' film roles. 1962 C. N. PARKINSON *In-Laws & Outlaws* 37 It is the itch to marry the girl next door which is the mark of those predestined for merely average success. 1963 *Listener* 24 Jan. 165/1 She had all the physical equipment of the vamp, but the spirit of the girl next door. 1968 *Times Educ. Suppl.* 23 Feb. 602/2 Diana Quirk's Ophelia was very much the girl-next-door.

4. a. *girl-bride, -clerk, -soldier*; **c.** *girl-crazy, -shy* adjs. Also **girl friend, girl-friend,** a female friend; *spec.* a sweetheart; a man's favourite female companion; **girl guide**: see *GUIDE *sb.* 2 d; **girl scout** *U.S.*, a girl guide.

1847 C. BRONTË *J. Eyre* II. ix. 220 Young Mrs. Rochester—Fairfax Rochester's girl-bride. 1935 *Burlington Mag.* Mar. 143/2 Quite a young man by the side of his girl-bride. 1901 *Daily Chron.* 29 May 3/6 No redeeming feature of girl-clerk labour. 1902 *Ibid.* 1 Sept. 3/6 The market is vastly overflooded with incapable girl-clerks, who can just manipulate a machine, at 10s. or 12s. a week. 1908 *Westm. Gaz.* 17 June 9/3 A girl-clerk does not expect to obtain an engagement without the necessary knowledge of shorthand and typing. 1930 I. GERSHWIN (*title*) Girl crazy. 1948 G. VIDAL *City & Pillar* (1949) x. 251 He was girl crazy, too, I guess. 1859 *Harper's Mag.* Aug. 337/1 A demure little widow, much more gay and girlish than any of her girl-friends when she chose to forsake her rôle. 1892 F. J. FURNIVALL *Hoccleve's Minor Works* p. iv (*dedication*), To the memory of Teena Rochfort Smith my much-respected and deeply-regretted girl-friend. 1896 Girl-friend [in Dict.]. 1907 A. WOOLLCOTT *Let.* 22 June (1946) 11 My best girl friend is the daughter of the big Grace Church in Utica. 1926 H. FIELDS (*title*) The girl friend. 1928 R. MACAULAY *Keeping up Appearances* xi. 118 Daphne seems quite to have become Raymond's girl friend again. 1962 'C. E. MAINE' *Darkest of Nights* v. 67 Mom made up her mind to tolerate his girl friends years ago. 1909 *Daily Chron.* 27 July 4/7 The girl-scout has arrived...This writer saw six of them..on Saturday—neat blue serge skirts, straw hats, haversacks, and poles. 1959 *Listener* 1 Oct. 542/3 A small contribution from Henry Miller and the Girl Scouts. 1925 T. DREISER *Amer. Trag.* (1926) I. xi. 71 He, at once girl-hungry and girl-shy, held himself nervously aloof. 1895 'MARK TWAIN' in *Harper's Mag.* July 232 To see and listen to the wonderful girl-soldier [*sc.* Joan of Arc]. 1944 BLUNDEN *Shells by Stream* 34 Girl-soldiers hasten.

girl, *v.*¹ Add: **2.** *intr.* To consort with women. Hence **gi·rling** *vbl. sb.*, usu. in phr. *to go (a-)girling.*

1787 *Sessions Papers* 10 Jan. 325/2 The maid said two men were missing, and the others said, G—d d—n them, they are gone a girling,..but we will pay for them. 1922 JOYCE *Ulysses* 537 No girl would when I went girling. 1931 L. STEFFENS *Autobiogr.* I. I. i. 8 Cowboys..used to come shouting on bucking bunches of bronchos into town to mix with the teamsters, miners, and steamboat men in the drinking, gambling, girling, fighting, of those days.

girldom (gə̄·ɹldəm). [See -DOM.] The domain or world of girls; girls collectively.

1864 D. W. THOMPSON *Day Dreams of Schoolmaster* x. 111 Frittering away the time and intellects of all subsequent generations of girldom. 1896 J. ASHBY-STERRY *Tale of Thames* ii. 20 Because..you choose to quarrel with a girl..there is no reason whatever that you should declare war against entire girldom. 1955 *Time* 29 Aug. 57 His house..was ringed all day and half the night by gazing girldom. 1968 *Newsweek* 12 Aug. 80 While he is playing a specific Mr. Singer, she is playing all young girldom.

girl Friday (gə̄·ɹl frai·di). Also **Girl Friday.** [After the man *Friday* in Defoe's *Robinson Crusoe*: cf. MAN *sb.*¹ 10 a.] A resourceful young woman assistant (to a man).

1940 *Time* 10 June 66/3 Irna is assisted in her work by two secretaries and a pair of literary Girl Fridays. 1940 *Times* 29 July 6/7 (Advt.), Cary Grant, Rosalind Russell in His Girl Friday. 1951 C. W. MILLS *White Collar* p. x, There are a dozen kinds of social engineers and mechanical technicians, a multitude of girl Fridays. 1958 *Woman* 5 July 19/1 What's it like to be Girl Friday to a Very Important Person? 1967 *Times* 5 Apr. 25/6 (Advt.), Secretary/Girl Friday required for principal of professional firm in Mayfair. 1971 *Petticoat* 17 July 38/1 They could be earning several pounds more, as Girl Fridays, personal assistants or private secretaries to rising executives.

girlie (gə̄·ɹli), *a.* *colloq.* [f. GIRL *sb.* + -IE; cf. GIRLIE; GIRLY *a.*] Denoting a publication, entertainment, etc., featuring young women, esp. scantily clothed or in the nude.

1942 H. FELSEN *He's in Submarines Now* 65 Being rather partial to girlie pictures. 1950 *N.Y. Times* 9 July II. 1/1 Mr. Todd..has produced a colossal spectacle and girlie show. 1952 B. MALAMUD *Natural* 104 It's a night

club with a nice girlie show. 1953 *Books* Mar. 38/2 The only purpose of the 'girlie' magazine..was 'pictorial prostitution'. 1959 *Encounter* Jan. 94/1 The picture magazines known to the trade as 'girlie books'. 1969 V. GIELGUD *Necessary End* v. 44 The barman, completely absorbed in the pages of a 'girlie' magazine. 1970 G. GREER *Female Eunuch* 35 Buttock fetishism is comparatively rare in our culture, although Kenneth Tynan did write a connoisseur article for a girlie mag..not so long ago. 1970 C. WHITMAN *Death out of Focus* vii. 92 All I printed for Forley were the usual girlie pictures.

girl-less (gə̄·ɹl,lės), *a.* [GIRL *sb.* + -LESS.] Without or devoid of a girl or girls.

1903 *Daily Chron.* 15 Apr. 5/1 If 'Harper's Weekly' may be trusted, the girlless telephone will soon be as familiar as the horseless carriage or the boneless sardine. 1929 J. MASEFIELD *Hawbucks* 147 It's a very girl-less neighbourhood.

girly-girly (gə̄·ɹligə̄·ɹli), *a.* and *sb.* *colloq.* [f. GIRLY *a.*] **A.** *adj.* Girlish in an exaggerated or affected manner; effeminate.

Girly-girly was the name of a music-hall song *c* 1900.

1883 'MARK TWAIN' *Life on Mississippi* xlvi. 467 The very feature that keeps it [*sc.* Mardi Gras] alive in the South—girly-girly romance. 1891 *Outing* (U.S.) Dec. 246/1 They despised dancing as 'girly-girly' and silly. 1900 H. LAWSON *On Track* 73 She'd..make a sweet little girly-girly dive for her chair. 1927 *Daily Express* 20 June 8 Typical girly-girly 'romp' stuff.

B. *sb.* A girl; a little girl.

1888 *Puck* 17 Oct. 123/2 The rusticator sighs, With a strange look in his eyes: 'Oh, where's my girly-girly of the Summer?' 1899 G. M. COHAN *Dear Little Girly, Girly*, Most ev'ry fellow tells about a little girly girly.. with teeth so white and pearly.

giro² (dʒaiə·ro). [G., a. It. *giro* circulation (of money).] A system whereby credits are transferred between banks, post offices, etc.; *spec.* a system operated by the British Post Office for the banking and transfer of money. Freq. *attrib.*

1896 P. DES ESSARS in *Hist. Banking* III. IX. 384 Germany has a slight advantage arising from the fact that in that country cheques and transfers (giro accounts) are more in use than in France and Belgium. 1896 R. VAN DER BORGHT *Ibid.* IV. III. i. 199 It will be perceived that the Amsterdam Bank was originally a bank of deposit and 'giro', in the primary sense of the word. 1907 *Westm. Gaz.* 17 Sept. 10/2 Invoices, note headings, &c., bearing in the corner the words 'Reichsbank Giro-konto', indicate to all that accounts due to that firm may be paid to their credit by cash or transfer (i.e., Giro) at any branch of the Reichsbank. 1930 *Economist* 27 Sept. 562/1 Other suggestions are concerned with the increased use of cheques, post-office banking facilities, giro transfers, clearings, etc. 1959 *Rep. Comm. Working Monetary System* xi. 331 in *Parl. Papers 1958–9* (Cmnd. 827) XVII. 389 The method of operation of a 'giro' operated by the post office is very briefly as follows. 1961 *Daily Tel.* 10 May 15/8 To provide a simple method for transferring payments a 'Giro' system, similar to that used in some continental countries, may soon be introduced by the Post Office, Mr Bevins, Postmaster-General, said yesterday. 1963 *Economist* 29 June 1375/1 On the Continent nearly all letter heads and invoices carry a giro number. 1968 *Listener* 22 Aug. 255/1 The Post Office Giro..will probably open in the first week of October. Basically, the Giro banks money for you, which you can draw out for yourself, or which you can instruct the Giro to transfer to other people.

Gironde. (Earlier and later examples.)

1795 H. M. WILLIAMS *Lett. Politics of France* I. v. 126 This calumny..formed the basis of the accusation which was framed against the Gironde. 1943 J. B. SIRICH *Revolutionary Committees* ii. 15 There were, however, still other enemies in Paris which the Gironde had to fear—the committees of surveillance.

Girondin (dʒirǫ·ndin), *sb.* (*a.*). [Fr. (see GIRONDIST).] = GIRONDIST *sb.* (*a.*).

1837 CARLYLE *Fr. Rev.* III. III. i. 160 As for the Girondin Formula, of a respectable Republic for the Middle Classes, all manner of Aristocracies being now sufficiently demolished, there seems little reason to expect that the business will stop there. *Ibid.* 166 The Girondins have left Memoirs, which are too often little other than long-drawn Interjections, of *Woe is me* and *Cursed be ye.* 1890 [see next]. 1965 M. J. SYDENHAM *Fr. Rev.* vii. 185 Thus the trial fulfilled its purpose, and the mythical 'Girondin party' was finally imposed on history. 1968 P. O'BRIAN tr. *Fay's Louis XVI* III. ix. 375 The Girondins, more eloquent and greedy than intelligent, wanted to compel the king to give them back their ministerial posts.

Girondism (dʒirǫ·ndiz'm). [ad. F. *Girondisme*, f. *Gironde* (see GIRONDIST) + *-isme*, -ISM.] The doctrine or practice of the Girondists.

1837 CARLYLE *Fr. Rev.* III. III. ii. 173 Bourdeaux demands a reign of Law and Respectability, meaning Girondism, with emphasis. 1890 E. B. BAX *Story Fr. Rev.* xv. 66 All the leading Girondins, including the twelve forming the Commission, were placed under arrest... Thus perished Girondism.

Girondist, *sb.* (*a.*). (Earlier and later examples.)

1795 H. M. WILLIAMS *Lett. Politics of France* I. iv. 68 But we need not here recur to conspiracies either of Jacobins or Girondists to discern the motives of Dumourier's conduct. 1943 J. B. SIRICH *Revolutionary Committees* ii. 14 The minds of the Girondists were not far from the enemies of the interior, the demagogues, the anarchists.

girth, *sb.*¹ Add: **8.** *girth-groove*; *girth-high* adj.

1923 C. FOX *Archaeol. Cambr. Region* iii. 92 Cineraries of hard paste with burnished girth-grooves. 1950 *Oxoniensia* XV. 48 A typical Oxford-style tripod-pitcher, with tubular spout and wavy applied vertical strips over rather irregular girth grooves. 1908 *Daily Chron.* 29 Aug. 7/2 They rode girth-high through the grass. 1936 R. CAMPBELL *Mithraic Emblems* 166 Girth-high, the poppies and the daisies To brush the belly of my mule.

gismo (gi·zmo). *U.S. slang.* Also **gizmo.** [Origin unknown.] A gadget, gimmick, 'thingumajig' (see quots.).

1943 *Time* 19 July 69/1 Gizmo—a term of universal significance, capable of meaning 'gadget', 'stuff', 'thing', 'whozis' or almost anything else the speaker wants it to. 1949 R. A. HEINLEIN *Red Planet* iii. 26 Now what is this gismo? When you brought it aboard I thought it was a volley ball. 1954 C. WILLIAMS *Touch of Death* (1955) i. 7 She was working the strap of the halter gizmo up between her arm and her side. 1967 *Observer* 21 May 21/2 To help him 'shoot' [photographs], he used..a 'gismo'—a small cardboard tube which he presses against the camera lens. 1970 *New Yorker* 1 Aug. 26/3 Every gismo that made use of a clothes hanger will be demonstrated by its inventor.

git (git). *slang.* [var. GET *sb.*¹] In contemptuous use: a worthless person.

1946 *Penguin New Writing* XXVIII. 171 Chalky! You idle git! 1960 H. PINTER *Caretaker* I. 9 Who was this git to come up and give me orders? 1967 *Listener* 3 Aug. 136/3 That bald-headed, moon-faced, four-eyed git Garnett gristling on about Harold Wilson. 1967 *Observer* 24 Sept. 36/6 The girl scarcely turned her head: 'Shutup yerself yer senseless git!'

Gitane (ʒitan). [Fr. *gitane* gipsy woman.] A cigarette of a proprietary French brand. Also *attrib.*

1933 *Trade Marks Jrnl.* 6 Sept. 1095 *Gitanes*. Cigarettes. Service d'Exploitation Industrielle des Tabacs... Paris. 1937 *Tobacco World Ann.* 1 (Advt.), Cigarettes: Elegantes, Gauloises, Celtique, Gitanes. 1961 E. HYAMS tr. *Crespi's Cigarette* i. 8 A certain way of offering a packet of cigarettes to a departmental head—filter-tip Gitanes Bleues, or American cigarettes..may enable a man to raise his prices by from twenty to a hundred per cent. 1967 V. CANNING *Python Project* viii. 157 Blowing a cloud of Gitane smoke, he went on, 'Your Prime Minister is not a wealthy man, is he?' 1968 N. FLEMING *Counter Paradise* iv. 58 'Here, have a Gitane,' he said, tossing the pack over to Jake. 1969 S. COULTER *Embassy* xxi. 237 A thick-set man smoking a yellow-paper Gitane cigarette.

giulio, var. JULIO.

1909 *Westm. Gaz.* 17 Mar. 5/1 A..purse containing twenty-five giulii.

Giuoco Piano (dʒuǭ·ko pyā·no). *Chess.* [It., lit. 'plain game'.] (See quots.)

1813 J. H. SARRATT *Wks. Damiano, Ruy-Lopez & Salvio* 214 An opening, denominated by Salvio, and by all Italian Players, Giuoco Piano; that name being given to all openings in which *no Pawn is sacrificed* for the sake of an attacking position. 1860 'OXFORD AMATEUR' *Hand-bk. Chess* viii. 32 If each player then brings his King's Bishop to QB 4th, we have the Giuoco Piano—a deservedly favourite opening with many players. 1965 I. A. HOROWITZ *Chess Openings* v. 25 Together with some related openings, the Giuoco Piano was extremely popular for many years; but the gradual change of insight..tended more and more to a lasting initiative rather than a splurge of brief fireworks and has slowly replaced the Giuoco Piano and the King's Gambit with the Ruy Lopez and the Queen's Gambit.

give, *sb.*² (Earlier and later examples.)

1868 G. M. HOPKINS *Jrnl.* 15 Sept. (1959) 189 They look like the swaling or *give* of water in a river when you look across it. 1921 *Daily Colonist* (Victoria, B.C.) 11 Mar. 2/3 One member in our business structure alone lacks 'give' —labor. If there is no 'give' something is going to smash. 1970 *Times* 13 Mar. 13/1 They began levelling the pitch at 8 a.m. this morning and if the weather stays fine we should have a fair surface with some give in it.

give, *v.* Add: **B. 6. e.** *give me* ——: form of words used as a request by a telephone-user to be connected with a specified person, number, etc.

1887 J. M. W. YERRINGTON *H. K. Goodwin* 26, I heard him call, 'Give me police station.' 1908 *Sears, Roebuck Catal.* 199/1 Vocal Solos..Hello, Central: Give Me Heaven (sentimental). 1914 *Bell Telephone News* Mar. 19/1 Subscriber: 'Give me 1000.' Operator: 'What office, please?' 1928 HECHT & MACARTHUR *Front Page* II. 72 Endicott calling. Gimme a rewrite man.

9. b. Also *to give* (a person) *as good as one gets.*

1952 M. LASKI *Village* vii. 123 If there was only someone she could talk to who could give her as good as they got. 1956 A. L. ROWSE *Early Churchills* 408 She answered her grandmother back, gave as good as she got. 1967 A. CHRISTIE *Endless Night* xv. 131 We went on, hammer and tongs..and I pretty well gave her as good as I was getting. Told her she was a bossy, interfering female.

e. Used in negative contexts with various complements to indicate indifference or unconcern, as in *not to give a curse, damn*, etc.; = CARE *v.* 4 a (*a*). Cf. *DAMN *sb.* 2.

1672 [see BUTTON *sb.* 1 b]. 1763 [see CURSE *sb.* 2 ¶]. 1895, etc. [see *DAMN *sb.* 2]. 1900 *Westm. Gaz.* 6 Nov. 1/3 Conductors who don't give a curse for the public. 1922, 1960 [see *BUGGER *sb.* 2 d]. 1969 *Advertiser* (Adelaide) 12 May 5/4, I don't give a bucket of dandruff for a bloke's lingo.

11. b. *refl.* Of a woman: to devote (herself) completely *to* (a man); *spec.* to accept sexual intercourse.

1860 J. W. PALMER tr. *Michelet's Love* II. ix. 130 She aspires only to give herself to you, to belong to you still more entirely. 1895 HARDY *Jude* (1896) I. ii. 10 Under the hedge which divided the field from a distant plantation girls had given themselves to lovers. 1926 R. MACAULAY *Crewe Train* III. i. 263 Barbara..gives herself to a young fisherman in Cornwall. 1967 *Times* 6 Apr. 9/2 A youngish woman making a last bid for emotional adventure, gives herself to a lonely old man in an empty Jamaican hotel.

14. c. Colloq. phr. *to give* (*her, it*, etc.) *the gun*: to cause (a vehicle) to accelerate; to open the throttle of (an engine). orig. *U.S.*

1917 B. K. ADAMS *Let.* 10 Oct. (1918) 35 He..saw that he wasn't going to even reach the aerodrome, so he gave her the gun, but the engine wouldn't pick up. 1919 *Red Cross Mag.* Feb. 30/2 He 'gave her the gun'—speeded up the engine—but the altimeter told him instantly that he was falling fast. 1945 G. CASEY in *Coast to Coast 1944* 6 He gave it the gun and went ahead. 1968 'G. BAGBY' *Corpse Candle* iv. 47 She slid behind the wheel, gave her hearse the gun, swung it around.

e. To punish (a person) for (doing something), freq. with reference back to what the other person has just said. *slang.*

1906 E. DYSON *Fact'ry 'Ands* ii. 23 'I'll give yeh whisperin' an' tisperin',' cried Annie. *a* 1930 D. H. LAWRENCE *Phoenix II* (1968) 136 Hark at her clicking the flower-pots, shifting the plants. He'd give her shift the plants! He'd show her!

f. To play music, esp. jazz, excitingly or enthusiastically; also with *out. slang* (orig. *U.S.*).

1936 *Esquire* June 92 The singer with the outfit can do with his or her voice just what the soloist can do with his instrument, he can *give*. 1952 B. ULANOV *Hist. Jazz in Amer.* (1958) 351 *Give* or *give out*, swing parlance for 'let yourself go'. 1955 SHAPIRO & HENTOFF *Hear me talkin' to Ya* x. 141 We would give out with such tunes as *Tiger Rag*. 1958 *Woman's Own* 19 Feb. 22/1 You feel that you're in a real jam session with everybody giving, the joint jumping.

16. c. To tell (a person); to offer for acceptance; esp. to tell or offer (a person) something incredible or unacceptable. Also *absol.* as *imp.*, speak! tell me! *colloq.* (orig. *U.S.*).

1883 J. HAY *Bread-Winners* xviii. 275 'Why, what are you givin' me now?' 'I'm a-givin' you truth and friendship.' 1889 'MARK TWAIN' *Yankee* 6 What are you giving me?..Get along back to your circus, or I'll report you. 1956 E. POUND tr. *Sophocles' Women of Trachis* 48 Thazza good tough start. Give. 1960 N. HILLIARD *Maori Girl* 123 He drew down the corners of his mouth. 'Don't give me that.' 1968 'P. HOBSON' *Titty's Dead* xvii. 165 'Come on. Give.' 'That ruddy policeman went digging things up and he found out I'd written my own testimonials.'

18. b. (Earlier example in *Cricket*.)

1816 W. LAMBERT *Instr. & Rules Cricket* 23 Such Umpire..is authorized to *give him out*, for thus impeding the progress of the Ball towards the Wicket.

30. d. To allow (a person) a specified period of time in which to do something; to predict a certain period remaining for (a person's life, a marriage, etc.). *colloq.*

1835 DICKENS *Let.* 30 Dec. (1965) I. 114, I give Cruickshank 'till Saturday:—I hope we shall have something to look at by that time. 1881 STEVENSON *Virg. Puerisque* 181 By all means begin your folio; even if the doctor does not give you a year.., make one brave push and see what can be accomplished in a week. 1913 F. L. BARCLAY *Broken Halo* xviii. 217 'How long do you give me, Sir James?' she asked, in a very quiet voice. 1962 WODEHOUSE *Service with Smile* vii. 107 'Well, all right,' said Lord Ickenham, rising. 'I can give her five minutes.' 1972 *Observer* 20 Feb. 8/5 The wiseacres..gave Charles Chaplin and Oona O'Neill's marriage six months. It's lasted decades.

37. d. *to give of oneself*: to devote oneself unselfishly; to give one's time and energy to some specified thing.

1926 G. HUNTING *Vicarion* xiii. 222 She had come here to him that she might find a way to give of herself, her love, her loyalty, in the time of his need. 1961 R. & C. WINSTON tr. *Mann's Genesis of Novel* xi. 118, I gave too much of myself. 1964 *Bull. Amer. Library Assoc.* Jan. 49/2 I'm not sure I want to give of myself that way. *Ibid.* 52/2 People will still rise out of their fast-paced lives and give unstintingly of themselves when they are shown a worthy project.

38. Esp. in phr. *to give* (*one*) *furiously to think* [tr. F. *donner furieusement à penser*], to set (a person) thinking very hard or seriously; to give (one) much food for thought; to puzzle.

1910 W. J. LOCKE *Simon* i. 8 This gives one furiously to think. 1920 A. CHRISTIE *Mysterious Affair at Styles* xiii. 280 This attitude of his gave me furiously to think, and I was slowly forced to the conclusion that Alfred Inglethorp wanted to be arrested. 1926 H. W. FOWLER *Mod. Eng. Usage* 210/1 If writers knew..how furious is the thinking that 'give furiously to think' stirs in the average Englishman, they would leave such paltry borrowings alone for ever. 1936 'J. TEY' *Shilling for*

Candles iv. 32 That Jammy Hopkins should stay without moving for more than three consecutive minutes argued that he was being given furiously to think.

39. d. *to give and take*, (*a*) to exchange repartee, blows, etc.; (*b*) to make mutual allowances, concessions, or compromises. Cf. GIVE AND TAKE *sb.*

1519 W. HORMAN *Vulgaria* f. 63ᵛ A man muste somtyme gyve and somtyme take. 1594 SHAKS. *Rich. III* v. iii. 6 We must both giue and take my louing Lord. *a* 1661 FULLER *Worthies* (1662) III. 96 The King [*sc.* Henry VIII], who in this kind would give and not take, being no Good Fellow in tart Repartees, was..highly offended. 1853 T. C. HALIBURTON *Sam Slick's Wise Saws* II. xii. 300 Give and take, live and let live, that's the word. You can't do without me, for you hante got no pilot, and I can't do without you, for I want your cash. 1963 *Higher Educ.* xv. 225 in *Parl. Papers 1962–3* (Cmnd. 2154) XI. 639 The habit of informal conference and a disposition to give-and-take.

e. *give or take*: to subtract or add (a period of time, a quantity, a sum of money, etc.): an indication that the estimate is to be considered as approximate. *colloq.*

1958 'P. BRYANT' *Two Hours to Doom* 11 Estimate Lakenheath twelve thirty hours. Give or take a few minutes. 1960 'R. SIMONS' *Frame for Murder* v. 59 The account..[has] stayed more or less the same, give or take a few pounds. 1962 N. FREELING *Love in Amsterdam* i. 15 'What time would that have been..?' 'Don't know; about a quarter to ten maybe, give or take. Who's dead?' he joked. 1967 M. WADDELL *Otley Pursued* xii. 109 It was 6.42 by my Petticoat Lane watch, give or take ten minutes or so.

f. *I('ll) give you that*: I admit that; I concede the truth of what you say. *colloq.*

1917 [see *bone-head* s.v. *BONE *sb.* 17]. 1966 *Daily Mail* 7 Nov. 4/5 He..savoured the flavour with great deliberation. 'Sort of salty mate, but tasty, I'll give you that.'

40. d. (*c*) (Later examples.)

1894 'A. ST. AUBYN' *Orchard Damerel* II. ix. 188 The old stones in the aisles were 'giving', as the sexton used to say in his homely way of describing mildew and damp. 1961 *Countryman* LVIII. III. 468 'The walls are giving' does not mean that they are collapsing, but that they are moist.

f. *what gives?*: what is happening? (freq. as a question or merely as a form of greeting); so *what gives with* (someone or something)? = what is happening to?; what is (he, etc.) doing? (Cf. G. *was gibt's?*) *colloq.* (orig. *U.S.*).

1940 J. O'HARA *Pal Joey* 133 What gives, I asked her. 1952 S. KAUFFMANN *Philanderer* (1953) iv. 56 'What gives?' 'I don't know, he didn't tell me,' said Rose. 1953 F. ROBB *Sea Hunters* vi. 80 George, whistle those lubbers again and ask them what gives. 1963 J. N. HARRIS *Weird World Wes Beattie* (1964) viii. 101 What gives with this cottage anyway? 1969 *Private Eye* 28 Mar. 14 What gives with this sheilah?

54. give away. a. (Usu. in pa. pple.) *given away with a pound of tea*: (of an article, usually something of little value) given, or as if given, free with a pound of tea. Esp. in derisive use.

1890 *Punch* 28 June 306/2 But Shopman Salisbury, why should he stand And advertise goods of his master J. B. As 'Given away with a Pound of Tea'? 1937 D. L. SAYERS *Busman's Honeymoon* xix. 314 They would not have it if it were given away with a pound of tea. 1958 *Times* 4 Dec. 13/4 It used to fall to the lot of some hapless object to be 'given away with a pound of tea'. 1966 'L. LANE' *ABZ of Scouse* 40 *Given away wid a pound of tea*, said of something considered tawdry and worthless such as a very cheap engagement-ring.

d. For *slang* read orig. *U.S. slang*, and add later examples. See also SHOW *sb.*¹ 16.

1914 G. B. SHAW *Pygmalion* (1916) I. 114 They want to drop Kentish Town; but they give themselves away every time they open their mouths. 1922 D. H. LAWRENCE *England, my England* 132 He'd burn your letters for fear they'd give him away.

f. Latterly *U.S.*

1816 U. BROWN *Jrnl.* in *Maryland Hist. Mag.* (1916) XI. 232, I was Just going to pay the Chain carriers & dismiss them, James gave away, Although full 6 feet high & well made. 1833 H. BARNARD *Ibid.* (1918) XIII. 305 He spoke about an hour and a half, and gave away on account of a sudden dizziness in his head and failure of voice. 1903 R. T. ELY *Stud. Evol. Industr. Soc.* 42 The cattle-raising frontier has given away continually to higher stages.

h. Esp. *to give it away.* To abandon, give up, stop. *Austral. slang.*

1949 L. GLASSOP *Lucky Palmer* iii. 37 It's about time you mugs woke up to yourself. You're not in the race to get in without a ticket. Why don't you give the game away? 1950 'N. SHUTE' *Town like Alice* v. 148 'I've changed my mind,' he said. 'I'm going to give it away... I won't be writing any letter.' 1955 J. CLEARY *Justin Bayard* v. 72 When I spoke to you.., you didn't say you'd given the idea away. 1961 P. WHITE *Riders in Chariot* ix. 260 He was at that time driving a truck for a firewood contractor, though he was thinking of giving it away and starting a line in poultry manure. 1968 D. IRELAND *Chantic Bird* iii. 28, I only hit him a few times, then I gave it away for the night.

59. give in. b. Also, to yield *to* a person.

1873 M. COLLINS *Squire Silchester's Whim* I. x. 138 You advise me to give in to Louisa on this point?

61. give on. b. = *CONTANGO *v.*

1928 [see *CONTANGO *v.*]. 1928 *Morning Post* 19 Nov., He will approach a fellow-member in the House, and say that he wishes to give on 50 shares.

give and take, *sb.* Add: **2.** (Earlier examples.)

1778 F. BURNEY *Evelina* I. xxv. 207 Give and Take is fair in all nations.

attrib. *a* 1817 JANE AUSTEN *Persuasion* (1818) III. xi. 232 So unlike the usual style of give-and-take invitations, and dinners of formality and display.

give-away (gi·vǎwē¹). *colloq.* (orig. *U.S.*). Also **give away, giveaway.** [f. the vbl. phr. *to give away* (see GIVE *v.* 54).] **1.** The act of giving something away; a thing or things that are given away; a free gift; *spec.* (*a*) a game in which the object is to lose points or to force one's opponent to make captures, etc.; (*b*) a price so low that the goods so sold are virtually 'given away'; (*c*) a radio or television programme or the like in which prizes are given to participants for performing some specified task. Freq. *attrib.*

1872 *Newton Kansan* 19 Dec. 2/2 We are decidedly opposed to the give-away game. 1889 'MARK TWAIN' *Yankee* xlii. 486 The stock was for sale at a giveaway. 1899 J. D. CHAMPLIN *Young Folks' Cycl. Games* (ed. 2) 181 *Give-away chess*...The Give-away game differs from the ordinary one. 1905 *Westm. Gaz.* 13 Sept. 8/1 They are advertising their stock at give-away prices. 1916 R. K. WOOD *Tourist's Northwest* xiii. 361 Alert Bay and other Indian villages of the British Columbia coast are occasionally enlivened by Potlatch or Give-away festivals, at which the Indian host bestows his worldly goods upon his invited amid formal dancing. 1934 M. H. WESEEN *Dict. Amer. Slang* (1935) XII. 167 *Give away*, free samples and prizes. 1940 *Time* 22 June 56/2 *Information Please*, opposite *Pot o' Gold*, in self-defence instituted a giveaway on its own high intellectual plane—sets of Encyclopaedia Britannica. 1948 *Ibid.* 4 Oct. 33/2 Plugged as a giveaway show..*Split the Atom* was even endowed with a sponsor. 1948 *Tuscaloosa* (Ala.) *News* 11 Nov. 4/6 He denies that 'audience participation' shows—he dislikes the term 'giveaways'—are only nuisance programs that attract morons. 1952 *Manch. Guardian Weekly* 13 Mar. 4/4 The state pays out..'giveaway grants' to Negroes who.. deserve a university education. 1953 *Ibid.* 11 June 4/3 Where a considerable choice of viewing exists..the bulk of viewers would immediately transfer their allegiance.. to a..big 'give-away' show. 1958 *Punch* 8 Oct. 457/3 Railwaymen loading cereal packets at Stretford found a live 3·7 anti-aircraft shell in the wagon. They say this is taking give-away too far. 1960 *Guardian* 16 Mar. 8/5 They have their 'give-aways', ranging from a set of books to a motor-car. 1962 *Rep. Comm. Broadc. 1960* 58 in *Parl. Papers 1961–2* (Cmnd. 1753) IX. 259 The appeal of the 'give-away', of prizes of values quite disproportionate to the skill or knowledge involved in the contest. 1969 *Daily Tel.* 20 Oct. 1/1 The feeling among.. MPs is that the Prime Minister will wait for the electoral benefits of a 'give-away' Budget before calling an Election.

2. An inadvertent betrayal or revelation of oneself, of plans, the truth, etc.; freq. in *dead give-away*, a complete betrayal; also, a person or thing that causes such a betrayal or revelation.

1882 W. A. BAILLIE-GROHMAN *Camps in Rockies* 14 The dead 'give away'..removed the shadowy blemish from his character. 1882 G. W. PECK *Peck's Sunshine* 83 A piece of that dog as big as a finger would ruin a butcher. It would be a dead give away. 1897 C. M. FLANDRAU *Harvard Episodes* 224, I gave my own name, but Dilly didn't; he had one all ready that went with the initials on his underclothes, so it wouldn't be a give-away. 1901 *Westm. Gaz.* 28 Sept. 2/1 'When I was at Poona in '76—' 'My dear Colonel,' purred Reginald, 'fancy admitting such a thing! Such a give-away for one's age!' 1904 *Ibid.* 19 May 3/2 When Mr. Chamberlain went on to declare that the Government's amendment and his own had the same idea, the give-away was complete. 1924 GALSWORTHY *White Monkey* II. ix, She also realised the give-away. 1924 G. OVERTON *Cargoes for Crusoes* 156 The patient has come to the conclusion that he is a dead giveaway, that everybody sees in him what the psychoanalyst sees. 1952 [see *ANALYSE *v.* 4 b]. 1959 'P. QUENTIN' *Shadow of Guilt* xi. 96 Her expression was a dead give-away. 1970 *Daily Tel.* 11 Aug. 11/2 An ancient parquet-floored hall which even the stealthiest house breaker couldn't negotiate without a salvo of giveaway creaks.

given, *ppl. a.* Add: **1. b.** (Earlier and later examples.)

1827 J. F. COOPER *Red Rover* I. 19 The youth, whose christian or 'given' name, as it is even now generally termed in New-England [etc.]. 1936 *S.P.E. Tract* XLVII. 244 Thoughtful parents often provide Smith with a second surname, usually as his second 'given' name, as John Howard Smith. 1954 *Manch. Guardian Weekly* 11 Nov. 2 His given names recall the majesty of the great judge.

4. Also *given-out.*

1868 DICKENS in *All Year Round* 19 Dec. 63/2 It come to her through two hands... The second hand took the risk of the given-out work. 1920 J. M. HUNTER *Trail Drivers of Texas* I. 211 The Indians had just as soon have these..given-out cattle as any.

B. *absol.* as *sb.* What is given; the known facts, situation, etc.

1879 W. JAMES *Coll. Ess. & Rev.* (1920) 132 If the philosopher fails to find a satisfactory formula of exorcism for his datum, the only thing he can do is to..assume the Given as his necessary ultimate. 1883 F. H. BRADLEY *Princ. Logic* I. i. 32 There is no practical interest in anything but the given, and what does not interest is not anything at all. 1965 *Case Conference* Mar. 280/2 The past cannot be seen as a fixed 'given', its meaning is imposed upon it by the stand that we take in relation to it. 1970 E. LEACH *Lévi-Strauss* iii. 45 'The English language'..from the point of view of any individual speaker..is a 'given', it is not something he creates for himself.

Givetian (ʒivī·ʃən, -ēi̯ʃən), *a.* *Geol.* [ad. F. *Givétien* (J. Gosselet 1879, in *Ann. de la Soc. géol. du Nord* VI. 130), f. *Givet*, name of a commune in north-eastern France: see -IAN. The Fr. spelling was formerly in English use.] Name of the upper one of the two stages that constitute the Middle Devonian in Europe; of or pertaining to this stage or the geological period during which it was deposited. Also *absol.*

1882 A. GEIKIE *Textbk. Geol.* VI. 702/1 Givetien. The great limestone of the middle Devonian series, well seen at Givet. 1902 A. J. JUKES-BROWN *Student's Handbk. Stratigr. Geol.* x. 176 [In the Belgian Ardennes] the upper stage of the Middle Devonian is a massive limestone 1300 feet thick, and is well seen at Givet on the Meuse, whence it is known as the Givétien [ed. 2, Givetian]. 1931 GREGORY & BARRETT *Gen. Stratigr.* vi. 96 In central Russia the Eifelian is also absent, but the Givetian is represented by marls and limestones. 1967 TEICHERT & YOCHELSON *Essays in Paleont. & Stratigr.* 184, 15 genera from the Eifelian and only ten genera from the Givetian Stage. 1967 D. H. RAYNER *Stratigr. Brit. Isles* v. 161 Near Padstow the Middle Devonian is represented by grey slates with Eifelian and Givetian goniatites.

gi·ve-up. *U.S.* [f. the vbl. phr. *to give up* (see GIVE *v.* 64).] **1.** The act of giving up; submission, relinquishment. *colloq.*

1895 *Harper's Mag.* Nov. 887/1 There was no give-up to those English. 1942 BERREY & VAN DEN BARK *Amer. Thes. Slang* § 306/2 Submission; surrender,..*give-up.*

2. *Stock Exchange.* (See quots.)

1934 WEBSTER, *Give-up*, a [stock-exchange] transaction in which the broker reveals the name of his principal, who is under obligation to complete the transaction. 1963 B. E. SHULTZ *Securities Market* (ed. 2) xvi. 302 A 'floor give-up' in which specialists and two-dollar brokers give up the names of the firms for which they are acting. 1964 HAZARD & CHRISTIE *Investment Business* xxiii. 368 There is no reason to countenance give-ups in the over-the-counter market. 1969 *Times* 5 May (Suppl.) p. v/5 Many of them depended for their very existence on commission-splitting, or the 'give-up' as it was known informally.

give-way sign. [Cf. GIVE *v.* 49 b.] A traffic sign ordering the road-user to yield the right of way to traffic on the road he is joining.

1963 *Rep. Committee Traffic Signs (Min. Transport)* 12 In recommending this *Give way* sign we are deliberately departing from the Protocol not only by adding words but in advocating that the sign should be made mandatory. 1969 *Daily Tel.* 15 Aug. 17/5 Roundabouts and 'give way' signs offer a far more fluid method of traffic control than lights. 1970 *Times* 11 Mar. 14/2 He ignored a 'give way' sign and white lines on a side road.

givey, givy (gi·vi), *a.* *dial.* or *colloq.* [f. GIVE *v.* 40.] Inclined to yield or give way; pliable, springy, soft; moist (see also quots.).

1829 *Virginia Lit. Museum* I. 457 *Givy*, 'muggy'. The weather is said to be givy, when there is much moisture in the atmosphere. 1859 BARTLETT *Dict. Amer.* (ed. 2), *Givy*, a term applied to tobacco leaves, in a certain condition of their preparation for market. 1887 PARISH & SHAW *Dict. Kentish Dial.* 64 The ground is said to be *givey* when the frost breaks up and the roads become soft and rotten. 1965 'S. HARVESTER' *Assassins Road* xiii. 129 He longed to sprawl on a bed. Any bed provided it was givey.

giving, *vbl. sb.* Add: **5.** *attrib.*, as **giving-set,** an apparatus for giving blood transfusions or the like.

1955 D. F. E. NASH *Princ. Surgical Nursing* II. ix. 138 All giving-sets must of course be supplied in a sterile container. 1961 *Lancet* 2 Sept. 537/1 The recent appearance of new sterile-packed, disposable giving-sets. *Ibid.* 16 Sept. 662/1 Any general practitioner undertaking domiciliary midwifery should..have at hand..a transfusion giving set.

gizmo, var. *GISMO.

‖ **gjetost** (yē·tost). Also **gjedeost.** [Norwegian, f. *gjet, geit* goat + *ost* cheese.] A Norwegian cheese made from goat's milk.

1908 DOANE & LAWSON *Varieties of Cheese* 27 Hvid gjedeost. This is a goat's-milk cheese made in Norway.. only for local consumption. 1923 A. WARD *Encycl. Food* 92 *Gjedeost*, a sweet, rich, strong-flavored, brown, Norwegian, goat's milk cheese, moist but firm enough to slice well. 1956 A. L. SIMON *Cheeses of World* 124 *Mysost* or whey cheese was first made from goat's milk and named *Gietost* (*gjet* means goat). 1960 *House & Garden* Apr. 111/2 A cheese called Gjetost which..has been produced in Norway for over 1,000 years.

glacé, *a.* Add: **1.** (Earlier and later examples.) Also *absol.* = glacé leather.

1847 E. GRAY *Let.* 5 May in W. James *Order of Release* (1947) iii. 31, I see a great many cloaks of pale glacé silk. 1900 *Queen* 28 Apr. (Advt.), Patent 'Lantry' shoe.. in Glacé. Buckles extra. 1936 N. COWARD *To-night at 8.30* II. 38 A navy blue school dress with a glacé red leather waist belt. 1955 *Times* 11 July 13 *Glacé* kid..is also a full chrome upper leather.

glacial (glēⁱ·ʃial, -ʃăl), *sb.* [f. the adj.] A glacial epoch or period.

1935 *Discovery* Nov. 317/2 Hardly one of the species present in England today could possibly have persisted here from the Tertiary Period through the glacials. 1957 G. E. HUTCHINSON *Treat. Limnol.* I. i. 49 High lake levels in equatorial Africa represent pluvial periods corresponding to the glacials.

†**glacialin(e** (glēⁱ·ʃiălin). *Obs.* [f. GLACIAL *a.* + -IN¹.] A food preservative composed of boric acid and glycerine. In full *glacialin(e salt.*

1876 W. F. GRIER *Brit. Pat.* 3107, My said Invention.. relates to a new or improved compound for preserving meat, fish, poultry, game, eggs, butter, milk... This liquid preservative I call 'glacialine', and its special value consists in the large proportion of hydric borate obtained in a solved or liquid state. 1883 *Encycl. Brit.* XVI. 304/2 Boracic acid and borax are also employed by dairymen, the former being known as glacialine salt. 1904 *Westm. Gaz.* 7 Jan. 7/1 Various preparations of boric acid are in use, the most common of which are glacialin salt and boro-glycerine.

glacier. Add: **2. a.** *glacier-river.* **c.** **glacier breeze** (see quot.); **glacier burst,** the sudden release of water impounded by a glacier; **glacier snout** = SNOUT *sb.*¹ 4 c; **glacier tongue** (see quot. 1956).

1930 *Meteorol. Gloss.* (ed. 2) 86 *Glacier breeze*, a cold breeze, blowing down the course of a glacier, which owes its origin to the cooling of the air in contact with the ice. 1904 *Nature* 29 Sept. 541/1 The importance of glacier-bursts in shaping the topography of glaciated areas. 1950 T. LONGSTAFF *This my Voyage* ix. 178 We saw a 'glacier-burst' from a small hanging glacier..a crashing roar of falling stones followed by a great rush of black, muddy water down the hillside. 1914 D. H. LAWRENCE *Prussian Officer* 16 The whity-green glacier-river twisted through its pale shoals. 1936 *Discovery* May 139/2 Another excellent river..[for fishing] is the glacier river Blandá. 1933 *Ibid.* Jan. 28/1 Nor is the correlation of glacier-snout movement with Brückner cycles by any means a fact. 1930 *Ibid.* Mar. 91/2 Glacier-tongues provided the starting point. 1956 ARMSTRONG & ROBERTS *Illustr. Ice Gloss.* 6 *Glacier tongue*, an extension of a glacier, projecting seaward, and usually afloat.

‖ **glacière** (gla·siēr). [Fr., = ice-house.] An ice-cave.

1865 G. F. BROWNE *Ice-Caves* iv. 46 The upper glacière of the Pré de S. Livres. 1879 *Encycl. Brit.* X. 270/1 Cases..occur where ice forms in caverns (*glacières*) even in warm and comparatively low districts.

glacieret (glæsiəre·t, glēⁱ·s-, glēⁱ·ʃ-). [f. GLACIER + -ET.] A small glacier; applied by Leconte to a mass of ice revealed after an extended period of dry weather has caused the wastage of overlying névé in the Sierra Nevada.

1875 J. LECONTE in *Amer. Jrnl. Sci.* X. 138 The thin moving ice-fields, the glacierets which still linger among the highest peaks and shadiest hollows of the Sierra. 1904 G. K. GILBERT *Glaciers & Glaciation* i. 11 Small alpine glaciers are sometimes called *glacierets*, or, if visible high on the sides of mountain valleys, *hanging* glaciers. 1922 WRIGHT & PRIESTLEY *Glaciology* 109 A snow-drift glacier, or glacieret. 1951 G. TAYLOR *Geogr. 20th Cent.* 614 *Glacierets*, small glaciers, may develop from snow-drifts.

glacierization. Restrict *nonce-wd.* to sense in Dict. and add: **2.** The covering of land by an ice-sheet; the state of being so covered.

1922 WRIGHT & PRIESTLEY *Glaciology* v. 134 'Glacierisation'—the inundation of land by ice. 1954 W. D. THORNBURY *Princ. Geomorphol.* vi. 149 Pleistocene glacierization caused many of the valleys beyond the margins of the ice sheets..to receive quantities of glacial outwash beyond the transporting capacities of the streams in them. 1968 R. W. FAIRBRIDGE *Encycl. Geomorphol.* 472/2 The continuity of the present ice age is suggested by the continued ice coverage (glacierization) of Antarctica.

glacifluvial (glēⁱʃi-, glēⁱsiflū·viăl), *a.* [f. L. *glaci-ēs* ice + FLUVIAL *a.*] = *fluvio-glacial adj.

1937 *Proc. Prehist. Soc.* III. 154 The erratics found in the latter must be derived, either directly or indirectly by means of glacifluvial transport, from the deposits of the North Sea Drift. 1946 F. E. ZEUNER *Dating Past* III. v. 111 Two superimposed glacial deposits (such as boulder clays or bottom moraines; or glacifluvial or meltwater gravels).

glacio- (glēⁱ·si,o, glēⁱ·ʃi,o), used as the combining form of L. *glaciēs* ice, as in **gla:cio-eu·stasy, -eu·statism,** changes in the sea-level caused by the waxing and waning of ice-sheets; hence **gla:cio-eusta·tic** adj.; **gla:cio-flu·vial** *a.* = *fluvio-glacial adj.; **gla:ciolacu·strine** *a.*, of or pertaining to a lake or lakes deriving water from the melting of ice.

1962 *New Scientist* 2 Aug. 243/1 Given a few decades or millenia of warmer summers, and the glaciers melt; the hydrologic balance swings to the positive side and the ocean level rises. This is the phenomenon of 'glacio-eustasy'. 1963 D. W. & E. E. HUMPHRIES tr. *Termier's Erosion & Sedimentation* vi. 152 The glacio-eustatic retreat of the seas at the time of the first glaciation. 1935 H. BAULIG *Changing Sea Level* 4 This kind of phenomena, for which the name glacio-eustatism would seem appropriate. 1954 Glacio-eustatism [see *EUSTATIC *a.*]. 1903 *Amer. Geol.* May 285 The glacio-fluvial deposits of the Wisconsin epoch. 1935 *Discovery* Mar. 78/1 Glacial and glacio-fluvial streams. 1910 *Encycl. Brit.* XII. 59/1 Professor Chamberlin and other American geologists have recognized the following stages in the glaciation of North America:..The Glacio-lacustrine substage. 1963 D. W. & E. E. HUMPHRIES tr. *Termier's Erosion & Sedimentation* vii. 162 The attribution of ancient deposits of similar type to the same cause, thus identifying them as glacio-lacustrine in origin.

glaciological (glēⁱsi-, glēⁱʃiǫlǫ·dʒikăl), *a.* [f. GLACIOLOGY + -ICAL.] Concerned with glaciology.

1889 in *Cent. Dict.* 1933 *Discovery* Mar. 89/1 Glaciological observations will be made at the ice-cap station. 1955 *Times* 13 July 4/4 One of the oversea expeditions will be in northern Norway engaged on geological, glaciological, biological, and survey work. 1970 *Nature* 28 Feb. 793/1 Because of the greater depth of erosion, glaciological knowledge based on Quaternary information may be enhanced.

glaciologist. (Further examples.)

1937 H. W. TILMAN *Ascent of Nanda Devi* i. 2 To other learned men, glaciologists, ethnologists or geographers, the Himalaya are a fruitful source of debate. 1958 *New Scientist* 13 Nov. 1245/2 Glaciologists in all parts of the world are beginning to watch the ice [of glaciers] with much interest. 1970 *Daily Tel.* 3 Sept. 13/7 Our glaciologists, working to a depth of 10,000 feet on the Antarctic ice cap, will be able to show how that cap has been polluted during the many centuries past.

glaciology. (Earlier and later examples.)

1889 in *Cent. Dict.* 1921 H. G. PONTING *Gt. White South* 138 Taylor could tell us more in an hour about Physiography and Glaciology than we could have absorbed in months of reading on such matters. 1971 *Nature* 12 Feb. 508/3 A synoptic view of the properties of ice for those engaged in research in glaciology, cloud physics and other related fields.

glacis. Add: **2.** Also *fig.*

1955 *Times* 3 Aug. 9/2 This view of east Germany as part of Russia's defensive glacis has never been dropped. 1960 *Observer* 7 Feb. 7/6 The countries of Eastern Europe ..form the glacis between the Soviet Union and the West.

b. In full *glacis plate.* A sloping armour-plate protecting an opening, etc., in a ship.

1889 J. J. WELCH *Text Bk. Naval Archit.* xiv. 149 These openings are generally further protected by working sloped armour plates—called *glacis plates*—around them about 3" thick. 1933 *Jane's Fighting Ships* 520 Armour (Krupp): 2" deck (amidships), ¾" deck (ends), 3½" glacis to engine room hatches.

‖ **glaçon** (glasoñ). Also **glacon.** [Fr.] A medium-sized ice-floe (see quots.).

1933 *Geogr. Jrnl.* LXXXI. 60 Drift-ice was pretty thick and we were constantly under helm to avoid the 'pans', as the seamen and fishermen call them, or 'glacons', which, I believe is the scientific term. 1963 D. W. & E. E. HUMPHRIES tr. *Termier's Erosion & Sedimentation* iv. 85 '*Glacons*', smaller than 650 feet [in diameter].

glad, *sb.*¹ Restrict †*Obs.* to sense in Dict. and add: **2.** *Ellipt.* for *glad eye* (see *GLAD *a.* 4 d).

1927 W. E. COLLINSON *Contemp. Eng.* 93 To give the glad-eye.., often abbreviated to the glad. 1956 A. WILSON *Anglo-Saxon Attitudes* I. iii. 46 Like a lot of old birds giving the glad in the Circus, or the York Road, Waterloo, more likely.

glad (glæd), *sb.*² Colloq. abbrev. of GLADIOLUS. Usu. *pl.*

1923 *Bull. Canad. Gladiolus Soc.* I. 3 We shall have a membership of at least five hundred (500) active 'Glad' enthusiasts. *Ibid.* 43 Cutting of Glads calls for a little discrimination. *Ibid.* 44 The wonderful Shows..have placed the 'Glad' in the very forefront of flower culture. 1934 S. ROBERTSON *Devel. Mod. Eng.* (1936) v. 120 Sometimes the problem of rival plural forms is resolved by retaining both..so it is with foci and focuses, gladioli and gladioluses. The last instance is further complicated by varying accentuation..hence some who have occasion to use the word dodge the issue altogether and say *glads.* 1949 [see *CUKE]. 1969 'J. FRASER' *Cockpit of Roses* xx. 151, I think them's my irises..and I'm certain those are my glads because that's my knot on the raffia.

glad, *a.* Add: **4. d.** *the glad eye*: a look or movement of the eyes designed to attract a person of the opposite sex; hence **glad-eye** *v. trans.*, to give (someone) the glad eye. *colloq.*

1911 *Punch* 22 Nov. 382/2 The lover, the elderly philanderer, the girl with the glad eye. 1913 'I. HAY' *Happy-go-Lucky* ii, Miss Welwyn..from the safe harbourage of her mother's arms, was endeavouring to administer to him what is technically known, I believe, as The Glad Eye. 1923 W. L. GEORGE *Hail, Columbia!* iv. 119, I have never seen an American girl give to a man in the street what the English call the 'glad eye'. 1935 A. J. CRONIN *Stars look Down* I. xvi. 150 Purves..'glad-eyeing' Hetty, trying 'to get off with her'. 1939 A. HUXLEY *After Many a Summer* II. i. 189, I *do* see her giving the glad eye to Pete.

e. *glad hand* (freq. used somewhat ironically): (the) hand of welcome; a cordial handshake or greeting; a welcome; esp. in phr. *to give* (someone) *the glad hand*; also *attrib.*; hence **glad-hand** *v. trans.*, to greet cordially, to welcome, to please; also *intr.*; **glad-handed** *ppl. a.*; **glad-hander,** one who gives people the glad hand; one who acts cordially towards everybody; **glad-handing** *vbl. sb.* and *ppl. a.* *colloq.* (orig. *U.S.*).

1895 W. C. GORE in *Inlander* Dec. 116 *Give the glad hand*, to welcome. 1896 ADE *Artie* i. 4 She meets me at the door, puts out the glad hand and says: 'Hang up your lid and come into the game.' 1903 —— *People You Know* 175 Every time he started down town he would have to zigzag so as to cover both sides of the street and glad-hand all of his Acquaintances. 1908 J. M. SULLIVAN *Criminal Slang* 11 *Glad hand*, handshake without any real friendship behind it. 1911 H. S. HARRISON *Queed* ii. 14 Mr. Bylash didn't go out to give her the glad hand, and welcome her into our humble *coturee*. 1929 C. E. MERRIAM *Chicago* 275 One type is the good fellow,..the glad hander, whose chief reliance is the cultivation of the personal friendship of individuals and the acquaintance with all sorts of groups and societies. 1930 *Time & Tide* 11 Oct. 1256 He makes a habit of cutting out glad-hand phrases and smooth generalities. 1939 MICHIE & RYHLICK *Dixie Demagogues* x. 207 Happy was stumping the hinterlands, glad-handing, back-slapping, and singing. 1945 AUDEN *Sea & Mirror* ii. 55 Such are the alternative routes, the facile glad-handed highway or the virtuous averted track. 1957 M. GAIR *Sapphires on Wednesday* ix. 114 He was probably quite a kindly old geezer..but a professional glad-hander. 1958 *Economist* 8 Nov. 497/2 General de Gaulle has stood aloof from the backslapping and glad handing. 1959 *New Statesman* 17 Jan. 62/2 Crude economic reasons do not explain why Mikoyan should have been given the glad hand. 1960 *Guardian* 8 Dec. 9/7 The glad-handing undertaker loves his job because 'it makes you feel closer to people'.

f. *glad rags* (occas. *glad clothes*): (one's) best clothes; very smart or fancy clothes; *spec.* formal evening dress. *colloq.* (orig. *U.S.*).

1902 'D. DIX' *Fables of Elite* 72 All the Females Assembled in their Glad Rags and proceeded to go through their particular Stunts for his Benefit. 1905 *Daily Chron.* 11 Jan. 4/5 Only when starvation stares him in the face will he relinquish his 'glad clothes', as the cowboys call them. 1907 N. MUNRO *Daft Days* vii. 60 'There's a big garden-party to be at it—' 'My! I guess I'll put on my best glad rags.' 1916 J. FARNOL *Definite Object* iii, 'Are you comin' wi' me, sir?' inquired Spike... 'I surely am.' 'But—but not in them glad rags?' and Spike pointed to Mr Ravenslee's exquisitely tailored garments. 1922 H. B. HERMON-HODGE *Up against it in Nigeria* v. 76 We all turned out in our glad rags to join in the procession. 1931 E. LINKLATER *Juan in Amer.* II. xii. 143 All dolled up in glad rags,.. They call you Glad Rag Doll!

8. *glad-eyed* adj.

1878 J. G. WHITTIER *Vision of Echard* 59 The freedom of the unshamed wind Among the glad-eyed flowers.

glade, *sb.*² Add: **4. glade mallow** *U.S.*, a tall herb, *Napæa dioica*, of the family Malvaceæ.

1849 SPRAGUE & GRAY *Genera Plants U.S.* II. 55 Glade Mallow. Flowers diœcious. 1963 H. A. GLEASON *Illustr. Flora Northeastern U.S.* II. 529 Napaea L. Glade Mallow... A monotypic genus.

Gladstonism (glæ·dstǫniz'm). [f. the name of W. E. *Gladstone* + -ISM.] The principles or policies of Gladstone. So **Gla·dstonite** = GLADSTONIAN *sb.* 1.

1873 CARLYLE *Let.* 15 Nov. (1904) II. 300 Stephen has evidently got intertwisted with Gladstonism. 1887 SWINBURNE *Let.* Apr. (1962) V. 184 Feeling deeply that the cause of Gladstonism can only be served by truth. *Ibid.* 186 A convinced and conscientious Gladstonite. *c* 1890 J. GOTT *Lett.* (1918) 110 My host is a lawyer—M.P. Gladstonite.

glaistig (gla·stik). Also **glaishrig, glaisrig, glastick, -ig.** [Gael.] A supposed she-devil or hag in the shape of a goat; also, a beautiful female fairy, usu. seen at the bank of a stream.

1903 *Q. Rev.* July 24 Fairies, brownies, kelpies and glasticks. 1925 *Contemp. Rev.* Oct. 537 The water-kelpies, the water-bulls, the mermaids, the glastigs. 1926 T. HANNAN *Beautiful Isle of Mull* xiii. 83 The green glaistig is always represented with masses of long yellow hair. 1959 K. M. BRIGGS *Anat. Puck* 189 The Glaistig is a female fairy, generally half-woman, half-goat, but sometimes described as a little, stout woman, clothed in green.

glam (glæm), colloq. abbrev. of (*a*) GLAMOROUS *a.*; (*b*) *GLAMORIZE *v.*; (*c*) GLAMOUR *sb.* Hence **glammed-up** *a.*, glamorized, 'tarted up'; **glammy** *a.*, glamorous.

1936 in *Amer. Speech* XI. 192/2 The glammiest of the glamour girls—Tallulah Bankhead. 1937 *Motion Picture* June 17/1 Loretta Young—does that girl know how to glam! 1957 J. OSBORNE *Look Back in Anger* III. i. 85 Get yourself glammed up, and we'll hit the town. 1961 PARTRIDGE *Dict. Slang* Suppl. 1109/1 *Glam*, glamour: film-world hangers-on: since ca. 1940. 1963 'M. CORRIGAN' *Why do Women..?* iv. 28 Have you any 'glam' night spots near by? 1964 C. DALE *Other People* v. 107 She was..wearing eye-shadow and a great deal of lipstick. 'You're looking very glam,' he said. 1966 J. CAIRD *Perturbing Spirit* vii. 72 We dress up for our dinner. We're not usually so glam. 1968 L. BERG *Risinghill* 45 In Risinghill Street itself..there are no glammed-up places.

glamorize, glamourize (glæ·mǒrəiz), *v.* orig. *U.S.* Also **-ise.** [f. GLAMOUR *sb.* + -IZE.]

trans. To make glamorous or attractive. Hence **glamo(u)riza·tion**; **gla·mo(u)rized** *ppl. a.*
1936 *Silver Screen* Dec. 5 They [*sc.* film stars] are so glamorized and vaporized and made to appear in print as somebody they aren't at all. **1938** *News of World* 12 June 7/2 Is this the way to glamorise 'a godsend'..? Shouldn't your advisers be aware that..his employers are..mightily concerned in finding him tough, he-man roles? **1941** *Amer. Speech* XVI. 98 Although the advertising writer does a creditable job of glamourizing colors and fabrics, he reserves his best efforts for the finished products. **1942** 'M. Innes' *Daffodil Affair* I. ii. 13 The glamorized advertisements, the pulsing sexy music. **1952** *Scrutiny* XIX. I. 39 A startling and quite unacceptable palliation, indeed glamorization, of acts which are by the book's own canons grievous sins. **1953** A. W. Fielding *Stronghold* 28 An outlaw whom the passage of time has glamourized into a figure..improbably noble. **1966** *Listener* 1 Sept. 300/2 What about the glamorization of crime as it is alleged to take place on certain mass media, particularly television? **1967** W. Soyinka *Kongi's Harvest* 24 The State will adopt towards him and to all similar institutions the policy of glamourised fossilism.

glamorous, *a.* Add: (Further examples.) Cf. next.
1938 D. Smith *Dear Octopus* I. 22 *Cynthia.* I work for her. *Belle.* As a mannequin? *Cynthia.* Nothing so youthful and glamorous. **1942** Partridge *Usage & Abusage* 144/2 *Glamorous* for *romantic* or (of a scene, a night, etc.) *lovely* or (of a woman) *beautiful and attractive* or (of a way of life) *exciting* or *adventurous* or (of a love-affair or a flirtation) *sex-filled* or *amorous* belongs to the advertising of films—and should be treated as the dubious privilege of boss-driven copywriters. **1946** F. Wakeman *Hucksters* (1947) xiv. 195 Her innocence was wonderful. In his world of hep, glamorous dames, he'd forgotten about the jeune fille. **1967** *Times* 8 July 11/4 Recently I visited a West End store and told the glamorous lady assistant that I wanted a bottle of scent.

glamour, *sb.* Add: **2. b.** Charm; attractiveness; physical allure, esp. feminine beauty; freq. *attrib.* (see sense *3). *colloq.* (orig. *U.S.*).
1937 *Daily Tel.* 26 Oct. 8/5 His hero is just another Glamour Guy. **1947** Partridge *Usage & Abusage* 353/1 A girl or a gigolo may possess glamour; and it makes no matter whether the girl is glamorous in her own right or by the catch-guinea arts of her dressmaker or her cinematographic producer. **1951** M. McLuhan *Mech. Bride* 94/1 The trade motto 'Bodies by Fisher'..insists on the close relation of motorcar glamour to sex. **1958** *Woman* 22 Feb. 8/2 Kitchen needs glamour badly; looks too antiseptic due to all-over white painted walls.
3. glamour boy, a young man who possesses glamour; *spec.* (*slang*) a member of the R.A.F.; **glamour girl,** a glamorous young woman; a 'pin-up girl'; **glamour puss** *slang*, a glamorous person.
1939 M. Brinig *Anne Minton's Life* (1940) 91 People do him a great injustice by calling him a glamour boy. **1941** *New Statesman* 30 Aug. 218/3 *Glamour boys*, R.A.F., especially flying crews. **1942** T. Rattigan *Flare Path* I. 105 I'm going to pour it on with a bucket. If I can't look like the screen's great lover, I can at least smell like a glamour boy. **1954** M. Croft *Spare Rod* I. v. 36 Those are your two glamour boys. You can't afford to leave them alone in the classroom or you'll have half the girls in as well. **1935** *Mademoiselle* Dec. 66/1 Horrible grimacing pictures of the Glamour Girl. **1936** [see *glam]. **1940** *Illustr. London News* CXCVI. 464/2 There emerged the new glamour-girl, as one must call her nowadays, as thin and slender as a flake of silver leaf, as blanched as an almond, as 'platinum' as a wedding-ring. **1956** Glamour girl [see *answer *sb.* 6 b]. **1952** N. Streatfeild *Aunt Clara* 95 Though Aunt Bess is an angel, nobody could say she was a glamour-puss. **1959** C. MacInnes *Absolute Beginners* 54 'Now listen, glamour puss,' I said, flicking his bottom with my towel. **1968** *Times* 6 Jan. 17/6 As every good Butterfly must, she develops visibly and vocally from glamourpuss to grand tragic heroine.

glamour, *v.* Add: **b.** To glamorize. Const. *up. colloq.* (orig. *U.S.*).
1951 L. Hobson *Celebrity* (1953) xv. 243 It's lost some of the subtlety of the book..and they've glamoured it up a little too much. **1958** B. Ruck *Third Love Lucky* ix. 58 Where was his sweet, sporting, glowing, skating partner with such pretty natural ways? All glamoured up and spoiled.

glance, *sb.*¹ Add: **1. b.** (Earlier example.)
1883 *Cricket* 19 Apr. 39/1 Leg glances being his favourite stroke.

glance, *v.*¹ Add: **5.** (Illustrations of the use with various preps. and advs.)
1816 Scott *Old Mort.* xxxv, A horseman..gave a letter. Claverhouse glanced it over. **1831** T. L. Peacock *Crotchet Castle* xvi. 252 Her eye glanced on something which made her change colour. **1843** W. H. Ainsworth *Windsor Castle* I. iii, The duke..was glancing rather wistfully at them. **1862** G. Macdonald *D. Elginbrod* II. xviii, Every now and then glancing up at her from her work. **1865** M. C. Harris *Christine* (1866) xxx. 181 Mrs Sherman and the party..glanced off their cards and chatted. *Ibid.* 187 She glanced up at the clock. *Ibid.* xxxiii. 208 He would..open her books, and glance through them for some trace of her in them. **1881** C. E. L. Riddell *Sen. Partner* I. v. 96 He did not glance round as the manager entered. *Ibid.* III. vi. 139 'No,' she agreed, glancing nervously around her. **1881** J. Fothergill *Kith & Kin* I. ii. 22 She..glanced for a moment into his face. **1894** A. Conan Doyle *Mem. Sherlock Holmes* 147 Glancing very keenly across at me. **1902** R. Machray *Night Side of London* v. 93 You indulge in pleasant little dreams, or glance away from what may become a tragedy. **1907** *Smart Set* Jan. 54 Her brother-in-law glanced after her. *a* **1953** E. O'Neill *Long Day's Journey* (1956) II. i. 57 As she talks, she glances everywhere except at any of their faces. *Ibid.* ii. 70 He glances away, ignoring her question. *Ibid.* IV. 132 He glances up at the chandelier disapprovingly. *Ibid.* 152 She glances around vaguely.
8. c. *Cricket.* To deflect (the ball) with the glance-stroke (see Glance *sb.*¹ 1 b). Also *absol.*, and with the bowler as obj.
1898 K. S. Ranjitsinhji *With Stoddart's Team* (ed. 4) iii. 50 He seemed able to 'drive',..or 'glance'..with equal skill and success. **1899** *Captain* I. 593/1 You had better practise slipping and glancing the ball. **1899** *Daily News* 22 July 4/2 Men were then less apt to 'glance and glide', like The Brook, and K. S. Ranjitsinhji. **1928** *Daily Express* 19 Dec. 3/2 White..glancing Ironmonger for three. **1963** A. Ross *Australia 63* vii. 129 Simpson glanced Statham's fourth ball and Smith, diving, caught it.

gland¹. Add: **4.** *Hist.* An acorn-shaped ball of lead, used as a missile.
1852 Burn *Naval & Milit. Dict.* 183 *Gland*,..leaden ball for a sling, in the form of an acorn, having frequently devices engraved upon it. **1927** J. Mothersole *In Roman Scot.* xviii. 252 These leaden sling-bullets (called 'glandes' because they were shaped like acorns) were only used up to the close of the first century.

gland². Add: **3.** *gland-pit*, *-tube.*
1906 *Academy* 6 Jan. 8/2 The pre-orbital gland-pit which has been traced in the skulls of Hipparion. **1881** *Trans. Obstetr. Soc. Lond.* XXII. 48 The lumen of the gland-tubes. **1902** *Encycl. Brit.* XXV. 394/1 The forms of the several kinds of glands depend on the degree of branching of these gland tubes.

glareal (glēᵊ·riăl), *a. Bot.* [f. L. *glarea* gravel + -al.] Growing on dry exposed soils. (Cf. Glareous *a.*¹ b.)
1847 H. C. Watson *Cybele Britannica* I. 66 *Glareal*, plants of dry exposed ground, chiefly on gravel or sand. **1930** G. C. Druce *Flora Northants.* 278 *Brachypodium pinnatum*... Native. Glareal.

glareless, *a.* Delete *rare* and add later examples.
1941 R. A. Heinlein *Menace from Earth* (1966) 58 A steady glareless light filled the room from no apparent source. **1962** H. C. Weston *Sight, Light & Work* (ed. 2) v. 149 This 'indirect' method of artificial lighting is sometimes claimed to be the best on the grounds that it is 'glareless and shadowless'.

glarney (glā·ɹni). [? Corruption of Glassy.] A glass marble.
1953 *Eng. Digest* June 93 The girls play skipping... The boys favour..'glarneys'. **1961** Partridge *Dict. Slang Suppl.* 1109/1 *Glarney*.., a corruption of *glassy*. **1969** E. Wilkins *Rose-Garden Game* 15 Seeing my bowls of taws and glarnies, some people would ask what they were *for*.

glass, *sb.*¹ Add: **15. c.** *glass-green* (later examples), *-grey.*
1912 E. Pound *Ripostes* 16 Out through the glass-green fields. **1939** T. S. Eliot *Old Possum's Pract. Cats* 40 He gives one flash of his glass-green eyes. **1910** W. de la Mare *Three Mulla-Mulgars* vi. 81 Whose eyes were pink, rather than glass-grey. **1920** A. Huxley *Leda* 74 The glass-grey silver of rivers.
d. *glass-cased* (cf. Glass case), *-clad*, *-distilled*, *-doored*, *-fronted*, *-jewelled*, *-lidded*, *-lined*, *-sided*, *-topped*, *-walled.*
1901 *Westm. Gaz.* 30 Dec. 2/1 The black marble glass-cased clock. **1902** *Ibid.* 8 Feb. 2/1 The glass-cased eatables. **1961** *Listener* 28 Sept. 464/1 He would say that his glass-clad buildings bring man in contact with nature. **1956** *Nature* 10 Mar. 471/2 Dissolved in glass-distilled water. **1924** H. Crane *Let.* 23 Sept. (1965) 190 Books.. in the glass-doored bookcase. **1902** *Westm. Gaz.* 11 Aug. 9/1 The great glass-fronted, gilded coach. **1960** C. Day Lewis *Buried Day* v. 90 Theological works in glass-fronted bookcases. **1916** H. G. Wells *Mr. Britling* I. ii. 55 Fastened with a large green glass-jewelled brooch. **1947** C. Morgan *Judge's Story* xxx. 199 The girl sitting.. at a glass-lidded table. *a* **1877** Knight *Dict. Mech.* II. 980/2 *Glass-lined pipe*, one in which the metal of the outer pipe is protected from corrosion by any liquids. **1960** *Farmer & Stockbreeder* 16 Feb. 72/1 There was the bottom unloading mechanism as with the glass-lined silo. **1901** 'L. Malet' *Hist. R. Calmady* v. viii. 443 The glass-sided hearse. **1905** *Westm. Gaz.* 15 May 10/2 One of the compartments of the glass-topped case. **1966** A. La Bern *Goodbye Piccadilly* I. 10 Glass-topped tables furnished with ashtrays too heavy to steal. **1959** N. Mailer *Advts. for Myself* (1961) 397 The institutional world, the monumental world, the world of skyscrapers and glass-walled banks. **1963** B. Fozard *Instrumentation Nucl. Reactors* v. 57 In the latter the envelope is most commonly of glass (glass-walled tube).
16. glass brick (see quot. 1909); **glass disease** (see quot. 1937); **glass-furnace** (substitute quot. 1671 for quot. *a* 1704 in Dict. and add later U.S. example); **glass-paper** (earlier U.S. example); **glass silk** (see quot. *a* 1884); **glass silkworm,** the cylinder on which glass silk is spun; **glass slipper** [mistranslation of Fr. *pantoufle en vair* fur slipper, mistaken for *verre* glass], a slipper made of glass, esp. the one lost by Cinderella in the fairy-tale; **glass wool** (earlier and later examples). Also *glass fibre.
1909 *Chem. Abstr.* III. 1210 Glass Brick: A New Building Material... Description of a hollow glass brick. In use the brick is laid up in the usual way and the inside filled with concrete, forming a monolithic concrete wall with a glass surface. **1938** *Archit. Rev.* LXXXIII. 205/2 (*caption*) The office counter, showing the glass-brick wall which is used as a display background. **1902** *Lancet* 25 Oct. 1143/1 Lately it has been found that a peculiar 'glass disease' has broken out amongst the windows of York Cathedral. **1937** *Burlington Mag.* Nov. 218/1 There is..[a] sort of decay, proceeding from within, which is due to too large a proportion of alkali in the composition of the glass itself. This 'glass disease' (as it is often called) may, and generally does, show itself within a short time after making, and takes the form at first of an interior network of very fine cracks. **1671** Locke *Hum. Und.* (*MS. draft*) (1931) xxxix. 93 The glowing heat of a glass furnace. **1880** *Harper's Mag.* Dec. 63 Since..Pittsburgh's first glass furnace in 1796 this industry has found in that city..congenial soil. **1815** *Niles' Weekly Reg.* IX. 94/2 Glass paper [was manufactured]. *a* **1884** Knight *Dict. Mech.* Suppl. 405/1 *Glass silk*, a product obtained by winding fine threads of glass in fusion on rapidly rotating and heated cylinders. **1937** *Archit. Rev.* LXXXII. 120 Glass silk has been used for the sound insulation of internal partitions. **1899** *Jrnl. Soc. Arts* XLVIII. 64/1 Should they [*sc.* bubbles] pass into the glass silk-worms, the continuity of the thread would be broken. **1729** tr. *Perrault's Histories* 73 (*title*) Cinderilla: or, The little Glass Slipper. **1819** M. Wilmot *Let.* 21 Dec. (1935) 46 *Cinderella* in her glass slippers and Fairy gifted finery was *dull* to the brilliancy of every creature. **1969** 'J. Munro' *Innocent Bystanders* iv. 64 If Cinderella had lost her glass slipper in New York, Loomis had said, her foot would have been in it all the time. **1879** *English Mechanic* 30 May 282/3 The patentee, therefore, proposes to employ..glass wool. **1962** A. Nisbett *Technique of Sound Studio* 239 Soft absorbers (glass wool, fibre board, etc.)..are poor absorbers at low frequencies.

glass, *v.* Add: **4. c.** To view or look for (something) with the aid of field-glasses.
1935 E. Hemingway *Green Hills Africa* (1936) II. vii. 166 We glassed the country. **1952** *Blackw. Mag.* Feb. 100/1 It was going to be a fine day with adequate light to 'glass' birds at a distance. **1964** F. O'Rourke *Mule for Marquesa* (1967) iv. 53 Fardan glassed the south, then the north where pursuit would come first, if it came.
7. *to glass off.* (Surfing.) Of the sea: to become smooth and transparent.
1967 J. Severson *Great Surfing* Gloss., *Glass-off*—when the wind dies (usually in the afternoon), causing the water to become very smooth or glassy slick. **1968** W. Warwick *Surfriding in N.Z.* 22/1 When the waves glass off in this way..surfriding takes on a truly aesthetic appearance not often seen.

glassed, *ppl. a.* **1. b.** (Further examples.)
1955 E. Bowen *World of Love* xi. 222 A glassed-in bar. **1969** J. Ross *Dead at First Hand* xvi. 152 Above the main doors a glassed-in bulb pushed out a sickly nimbus of light.

glass fibre. a. An individual filament of glass, of any length.
1824 *Edin. Philos. Jrnl.* X. 39, I brought to within half an inch of each other, two conductors, and I united them by a very fine glass thread. One..was connected with an electrical machine, and the other communicated with the ground. In this manner, the light appeared to pass continuously along the glass fibre, which..formed a fine and brilliant line of light. **1949** R. N. Haward *Strength of Plastics & Glass* vii. 228 Many of the outstanding mechanical properties of the fibre-glass laminates appear to be largely determined by the properties of the glass fibres. **1954** R. H. Sonneborn *Fiberglas Reinforced Plastics* p. v, *Fiberglas is the trademark..of Owens-Corning Fiberglas Corporation for a variety of products made of or with glass fibers. **1963** H. R. Clauser *Encycl. Engin. Materials* 569/1 Glass fibers for reinforcement are available in several forms. Individual fibers have a diameter of from 0·00025 to 0·0006 in.
b. (Also **glass-fibre, glassfibre.**) Collectively, glass in the form of such filaments, as used for manufacturing purposes or the like; any material made from such filaments, as a textile woven from them or a plastic containing them as reinforcement: = *fibreglass.
1882 Caulfeild & Saward *Dict. Needlework* 223/1 Ecclesiastical decorative fabrics composed of glass fibre. **1913** *Chem. Abstr.* VII. 3212 Porous articles of glass fiber. **1947** W. J. Brown *Fabric Reinforced Plastics* ii. 23 Glass fibre cloth is not suitable for high-pressure laminating processes because of a tendency to crush the fibres. **1955** *Times* 6 July 13/1 Structural plastics consist of a reinforcing agent, such as glass fibre, bonded with a synthetic resin. **1958** C. A. Redfarn *Guide to Plastics* (ed. 2) iii. 51 Glass fibre is of great interest as the reinforcing agent in plastics designed for use under highly stressed conditions. **1959** *Listener* 5 Mar. 435/1 For curtains, glass fibre has several advantages. **1961** *Times* 7 Aug. 2/2 A remarkable performance was given by the small glass-fibre boat Winkle. **1967** *New Scientist* 25 May 448/3 The glassfibre has been tried on other parts like the compressor casing. **1968** *Daily Tel.* 30 Oct. 25/6 The glass-fibre tanks will have an initial storage capacity of 400,000 gallons. **1970** *Commercial Motor* 25 Sept. 121 The roof is of translucent glass fibre.

glass-house. Add: **2. c.** A military prison or guard-room (see quot. 1925). Also *attrib.* *slang.*
1925 Fraser & Gibbons *Soldier & Sailor Words* 105 *Glass house*, detention barracks. An old Army expression, originally particularly applied to the detention barracks for the Aldershot Command at Woking. Later used of any place of military detention. Guard-room, Cells, etc.

1945 *Penguin New Writing* XXVI. 49 But surely they didn't give you the glasshouse for going to Dunkirk? *Ibid.*, The glasshouse man continued to recount his experiences in detention. **1946** *News Chron.* 2 Mar. 4/4 Glasshouse rioters' final stand... Some of the men at Northallerton barracks..holding out on the roof. **1947** J. Bertram *Shadow of War* 202 Someone with a lengthy 'crime sheet'—perhaps..a notorious frequenter of the glasshouse.

gla·ssichord. *U.S. Obs.* exc. *Hist.* Also **glassy-chord.** [f. Glass *sb.*¹ 4 + *-chord* as in *harpsichord*.] = Harmonica 1 a.
1835 J. H. Ingraham *South-West* I. ii. 24 The musical rippling of the eddies—like a glassichord, rapidly run over by light fingers. **1886** *Harper's Mag.* July 286/1 Turning upon his perch, he [*sc.* the cricket] brings to view his 'glassichord', or shrilling organ. **1954** *Grove's Dict. Mus.* (ed. 5) I. 205/2 Franklin's invention..was also popularly known as the 'glassy-chord'.

glassine (glɑ·sīn). [Glass *sb.*¹ + -ine⁵.] A glossy transparent paper. Also *attrib.* or as *adj.*
1916 W. W. Young *Story of Cigarette* vi. 98 Each carton is wrapped by hand in moisture-proof and germ-proof glassine paper and firmly sealed. **1925** *Pulp & Paper Mag.* 6 Aug. 881 The manufacture of glassine paper requires great care, exceptionally clean raw materials, experienced labor, and specially designed machinery... Glassine is a paper which is made from very clean pulps, and which must be highly transparent. **1931** F. Hurst *Back St.* lviii. 507 Her dry mouth open, with a bubble, as if of glassine paper, spanning it. **1957** P. Mansfield *Final Exposure* xvi. 242 Enclosed in a glassine sheath, there lay a negative. *Ibid.* 244 The negative had been carefully released from the glassine envelope. **1970** S. Ellin *Man from Nowhere* xxxii. 155 He had stored Thoren's envelope and scrap paper in glassine packets.

glassy, *a.* Add: **B.** *sb.*¹ *Surfing.* (See quot. 1967.)
1962 T. Masters *Surfing made Easy* 64 Glassy, smooth water, no wind, if the surf is up then the best of surfing conditions. **1965** J. Pollard *Surfrider* ii. 20 Last week the surf was a 'glassy', no wind and the waves smooth. **1967** J. Severson *Great Surfing* Gloss., *Glassy*, an extremely smooth surface or wave, usually giving off a glasslike reflection.

glassy (glɑ·si), *sb.*² Also **glassey, glassie.** [f. Glass *sb.*¹ + -y⁶.] A glass marble.
1887 T. Darlington *Folk-Speech S. Cheshire* 203 *Glassey*, a marble or 'taw' made of glass of various colours. **1915** D. H. Lawrence *Rainbow* viii. 200 Then suddenly she ceased to hear, having caught sight of a glassie rolled into a corner. **1933** J. Thorpe *Happy Days* i. 29 The owner of a 'glassy', which was decorated internally with beautifully twisted colours, placed it in the centre of the road. **1952** [see *agate *sb.* 1 b].

glastick, -ig, varr. *glaistig.

Glaswegian (glæswī·dʒiăn, glæz-), *sb.* and *a.* [f. *Glasgow*, after Galwegian, Norwegian.] **A.** *sb.* A native or inhabitant of Glasgow. **B.** *adj.* Belonging to or characteristic of Glasgow or its inhabitants.
1818 Scott *Rob Roy* xxii, The Glaswegian took him by the hand. **1884** *Illustr. Lond. News* 12 Jan. 27/1 Any of the ten words..in addition to the Glaswegian four. **1923** *Glasgow Herald* 15 Nov. 8 'Glaswegian' is both ugly and absurd... Let us in the name of etymology and common sense be Glasgovians. **1945** *Archit. Rev.* XCVII. 122/1 The fertility of these people will not be maintained when they become Glaswegians or Londoners. **1971** *Times* 26 Nov. 12/8 One of the tarts becomes Glaswegian. **1971** T. J. Honeyman *Art & Audacity* 247 In Pollok House there will be an abundance of riches for the art lover and scholar, and a magnetic attraction for Glaswegians and their visitors.

glazed, *ppl. a.* Add: **3. d.** *Metallurgy.* Having a smooth shining surface or fracture owing to a high silicon content (in the case of pig-iron) or fusion of the surface (in the case of blister steel).
1874 W. H. Greenwood *Man. Metallurgy* I. ix. 160 Aired and glazed bars are unfit for melting into best cast-steel, and require reconverting. **1884** —— *Steel & Iron* iv. 55 The excess of fuel employed when first blowing in a furnace often results in the metal first tapped being more highly siliceous than that produced in subsequent workings, and..the siliceous pig known as glazed or blazed pig often results. **1923** Harbord & Hall *Metall. Steel* (ed. 7) I. x. 245 Sometimes..the pots become too highly heated, with the result that the surface of the bars is slightly fused, and these are known as 'glazed bars'. **1967** A. K. Osborne *Encycl. Iron & Steel Industry* (ed. 2) 182 *Glazed pig*, pig iron containing a very high silicon content, e.g. 5%, which gives a very fine grained lustrous fracture.

glazier. Add: **2. b.** In colloq. phrases addressed to a person who is obstructing one's view, etc. (see quots.).
1738 Swift *Pol. Conversat.* I. 13 *Lady Smart*... You stand in your own Light... *Ld. Sparkish.* I'm sure he sits in mine: Prythee, *Tom*, sit a little farther: I believe your Father was no Glasier. **1891** J. M. Dixon *Dict. Idiomatic Eng. Phrases* 132 *Is your father a glazier?*—a vulgar expression, signifying, 'Do you suppose that I can see through you?' It is used when a person in front of you obstructs your view. **1910** P. W. Joyce *English as we speak it in Ireland* viii. 113 'Your father was a bad glazier': said to a person who is standing in one's light.

gleba (glī·bă). *Bot.* [mod.L. use of *glēba*, *glæba* clod, lump, Glebe.] The fleshy, spore-bearing tissue of certain kinds of fungi, the Gasteromycetes and the Tuberales.
1847 J. Lindley *Elem. Bot.* (ed. 5) p. xliii/1 Gleba, Glebula.—The peridium or the fleshy part of certain Fungals. **1887** H. E. F. Garnsey tr. *De Bary's Compar. Morphol. & Biol. Fungi* v. 309 They form together a mass of tissue which is distinguished from the adjoining tissue by its chambered structure and by the formation of spores and is known as the *gleba*. **1950** E. A. Bessey *Morphol. & Taxon. Fungi* xv. 531 The gleba consists of a more or less fleshy mycelial growth containing usually numerous hymenium-lined cavities (hymenial cavities) but only one in a few cases. **1968** J. H. Burnett *Fund. Mycol.* v. 119 One or more hymenium-lined cavities develop from potentially sporogenous tissue, the gleba.

gleditschia (gledi·tʃiă). *Bot.* Also **gleditsia, gleditzia.** [mod.L. (C. Linnæus *Genera Plantarum* (ed. 2, 1742) 480), f. the name of J. G. *Gleditsch* (1714–1786), a German botanist + -ia¹.] An ornamental tree of the genus so called, belonging to the family Leguminosæ, and including the honey locust tree of North America.
1770 R. Weston *Univ. Botanist* I. 116 Gleditsia of Java. **1838** J. C. Loudon *Arboretum et Fruticetum Britannicum* II. 650 The three-thorned gleditschia, or honey locust, in favourable situations in its native country, attains the height of 70 ft. or 80 ft. **1855** A. M. Murray *Let.* 7 May (1856) II. xxv. 157 A stream..bordered by.. gleditzias, cedars, sweet and black gum-trees. **1905** C. S. Sargent *Man. Trees N. Amer.* 556 Gleditsia is confined to eastern North America..south-western Asia, China, Japan, and west tropical Africa. **1914** W. J. Bean *Trees & Shrubs Hardy in Brit. Isles* I. 594 No Gleditschia has any beauty of blossom, the flowers being small, green, and borne in racemes a few inches long. **1957** M. Hadfield *Brit. Trees* 360 Locust is a name given in North America both to trees of the genus *Robinia* and to the rather similar *Gleditsia*.

glee, *sb.* Add: **5.** *glee-book*, *-singer* (earlier example), *-singing*; **glee-club** (examples); also *transf.*
1862 O. W. Norton *Army Lett.* (1903) 124, I would rather see two or three pretty girls and a glee-book.. than the whole of them. **1814** R. Clark (*title*) The words of the most favourite pieces, performed at the Glee Club, the Catch Club, and other public Societies. **1879** J. R. Planché *Extrav.* I. 46 Come to the Glee-club, at the Cat and Swallow. **1879** *Encycl. Brit.* X. 677/1 The most famous amongst these—The Glee Club—was founded in 1783, and at first used to meet at the house of Mr R. Smith, in St Paul's Churchyard. **1934** H. Nicolson *Let.* 7 Nov. (1966) 187 He would ask the glee-club to sing that song. **1971** *Times* 30 Jan. 13/1 Labour members will neither win back the confidence of the country nor prevent the passage of the Industrial Relations Bill by turning themselves into a political glee club. **1811** Jane Austen *Let.* 18 Apr. (1952) 269 There is to be some very good Music, 5 professionals, 3 of them Glee singers. **1854** E. Twisleton *Let.* 29 June (1928) xi. 213 Mrs. Adderley's concert was all glee-singing, English music and very pretty. **1912** W. Owen *Let.* 22 Sept. (1967) 162 Putting together..a Glee-Singing Band.

glei, var. *gley.

‖ **Gleichschaltung** (gləi·χʸʃaltuŋ). [G.] Also with lower-case initial. The standardization in authoritarian states of political, economic, and cultural institutions. Also *transf.*
Other parts of speech related to the G. word (e.g. *gleichschalten* v. trans., to impose *Gleichschaltung* upon; *gleichgeschaltet* ppl. a.) are also occas. found in English contexts.
1933 *Week-End Rev.* 12 Aug. 149/2 The only test can be whether or not the efforts to *Gleichschalten* Austria continue. **1934** *Surv. Internat. Affairs 1933* II. i. 140 The new masters of the Reich had succeeded in carrying out their revolutionary programme of *Gleichschaltung* far enough to have changed the whole face of German life. **1935** S. Lewis *It can't happen Here* xv. 150 His ripe experience in the *Gleichschaltung* of Mississippi. **1937** Auden in Auden & MacNeice *Lett. fr. Iceland* v. 57 You would, hearing honest Oswald's call, Be gleichgeschaltet in the Albert Hall. **1940** *Economist* 31 Aug. 280/2 Holland under German control... Slowly the *Gleichschaltung* takes place. Many trade unions and similar organizations are unified. **1941** Koestler *Scum* 209 It looked as if France were already *gleichgeschaltet* before we even knew the armistice terms. **1950** *Ann. Reg. 1949* 264 The heavy hand of Soviet *gleichschaltung* descended on art, literature, and even music. **1964** D. Daiches *Crit. Hist. Eng. Lit.* iv. 85 One cannot but welcome this failure of critical *Gleichschaltung*. **1969** J. Mander *Static Soc.* viii. 254 There was no effective resistance..to Peron's gleichschaltung of the CGT.

glen¹. Add: **b.** Also *attrib.*, designating check fabrics, esp. *Glenurquhart check.
1923 *Daily Mail* 13 Feb. 1 (Advt.), Smartly tailor made Glen Check Tweed. *Ibid.*, The new glen and plumage checks. **1967** T. Wells *What should you know of Dying?* x. 115 A double-breasted glen plaid suit was his uniform of the day. **1968** J. Ironside *Fashion Alphabet* 218 *Glen check*. 'Glen' describes many check fabrics originally Scottish in design. It is similar to Shepherd's check but can be in any combination of colours and has a large overcheck.

glengarry. (Earlier example.)
1841 in *Hist. Rec. Queen's Own Cameron Highlanders* (1909) II. 280 When the glengarry bonnet is used for evening dress, it is to be made of fine cloth, [etc.].

Glenurquhart, Glen Urquhart (glenɜ·ɹkǝt). [f. *Glenurquhart* in Inverness-shire.] A Scottish district check (see quot. 1956). Also *attrib.*
1923 *Daily Mail* 23 Apr. 8 Checked materials are not in fashion, with perhaps the exception of brown Glen-urquharts—of which the nicest carry a red line or over-check. **1944** C. Mackenzie *North Wind of Love* I. 10 This remark came from a lanky young man..in plus-fours of a modest Glen Urquhart tweed. **1956** M. B. Picken *Fashion Dict.* 255/1 (*caption*) Glenurquhart Plaid. Squares of small woven checks alternated with squares of larger checks, in one or two muted colors with white. **1959** M. Steen *Tower* I. vii. 93 A dapper little character in Glenurquhart tweed. **1967** R. Wilkinson *Pressure Men* x. 95 She wore a green glenurquhart check two-piece suit.

gletcher, var. Gletscher.
1932 Auden *Orators* III. 104 Or be ducked in a gletcher, as they ought to be. **1938** L. MacNeice *Earth Compels* 32 Riding the sullen landscape far from friends Through the jungle of lava..Fording the gletcher.

gley (glēⁱ). *Soil Science.* Also **glei.** [Ukrainian, = sticky bluish clay (see quot. 1963); cogn. w. Clay *sb.*] A blue-grey soil or soil layer in which iron and manganese compounds are reduced through being waterlogged; also, such a soil mottled with brownish oxidized patches as a result of periods of relative dryness. Also *gley horizon*, *soil*, etc.
1927 C. F. Marbut tr. *Glinka's Great Soil Groups* 36 In the last two groups the 'Glei' horizon, developed under the influence of ascending moisture, is often present. **1953** H. L. Edlin *Forester's Handbk.* v. 77 Gleys..are soils that develop where drainage is impeded, so that there is a fluctuating water-table or saturated layer in the soil structure. **1963** Birron & Cole tr. *Vilenskii's Soil Sci.* xxi. 308 The words 'glei' and 'gleification' were derived from the popular Ukrainian and introduced into scientific terminology in 1905 by G. N. Visotskii. **1969** *Jrnl. Soil Sci.* XX. 207 Failure to distinguish between sulphide-containing and sulphide-free gley soils causes considerable confusion.
Hence **gley·ed** *ppl. a.*, turned into a gley; **gley·ing, gleiing** *vbl. sb.*, **gleiza·tion** (glēⁱ-), the formation of a gley.
1934 *Forestry* VIII. 30 The gleyed horizon is usually of a greenish or bluish colour. **1938** H. G. Byers et al. in *Yearbk. Agric.* (U.S.) IV. 975 The bluish or greenish waterlogged horizons are sometimes called glei or gley, and the process by which they are formed is sometimes called gleization. **1949** *Jrnl. Soil Sci.* I. 205 It is not clear when it was first realized that gleying is caused by a microbial reduction of ferric compounds. *Ibid.*, A laboratory study of the gleying process. **1949** J. S. Joffe *Pedology* (ed. 2) xi. 430 Morphologically, the B horizon has also been affected by the gleiing process. **1956** C. D. Pigott in D. L. Linton *Sheffield* 80 Where water-logging occurs..the soils are gleyed and in many places the ground flora is almost exclusively dominated by *Allium ursinum*. **1971** *Nature* 1 Jan. 45/1 Marked peatiness of the soil and gleying are only found toward the upper limit of lower Montane forest.

glia. Substitute for def.: = Neuroglia; freq. *attrib.*, as *glia-tissue*; **glia cell,** any of the different kinds of cell in neuroglia. (Add earlier and later examples.)
1886 W. R. Gowers *Man. Dis. Nerv. Syst.* I. III. 107 Fine fibres..form a network. At their intersections are peculiar cells consisting of a nucleus and small cell body ('glia-cells', 'cells of Deiters'). **1890** Billings *Nat. Med. Dict.* I. 586/1 *Glia*, neuroglia. *G.-cells*,..small flattened cells found in the spinal cord. **1899** *Allbutt's Syst. Med.* VII. 172 Investigations of Weigert..seem to establish that new formations of glia may be either cellular or fibrous in structure. *Ibid.* 171 A subsequent proliferation of the embryonal epiblastic elements, or a thickening of the normal glia tissue. **1908** *Practitioner* Oct. 562 By a new staining method, which stains the neuro-fibril networks but leaves the glia totally unstained, [they] have demonstrated..that the Golgi pericellular nets are glial and not nervous in origin. **1964** J. Z. Young *Model of Brain* iv. 60 Glia cells are said to show electrical phenomena when suitably stimulated. **1964** *Jrnl. Neurophysiol.* XXVII. 291 No special mechanism providing for electrical interaction between neurons and glia could be detected.

glial (gləi·ăl), *a.* [f. prec. + -al.] Of or pertaining to glia.
1888 B. Bramwell *Intracranial Tumours* ix. 224 The tumour is found to be composed of small round or oval cells, and of extremely delicate fibres (glial threads). **1908** [see prec.]. **1910** Allbutt & Rolleston *Syst. Med.* (ed. 2) VII. 856 The normal glial structure of the spinal cord. **1954** S. Duke-Elder *Parsons' Dis. Eye* (ed. 12) xxiii. 321 These sheaths are separated by glial tissue and a neurilemma is absent. **1966** *New Statesman* 9 Sept. 349/1 The so-called glial cells, for example, which greatly outnumber the nerve cells, turn out to be more than the mere padding they were thought to be.

glide, *sb.* Add: **1. b.** spec. *Cricket.* A stroke by which the ball is deflected towards long leg by the turned blade of the bat; = Glance *sb.*¹ 1 b. In full *glide stroke.*

1888 Steel & Lyttelton *Cricket* ii. 67 Fig. 10 shows W. G. Grace attempting the glide... This is a stroke in which W. G. Grace excels. **1897** K. S. Ranjitsinhji in *Encycl. Sport* I. 218/1 There is no more effective stroke on the leg side than the 'glide' or 'glance'. **1911** C. B. Fry in P. F. Warner *Bk. Cricket* xiii. 217 His [*sc.* Ranjitsinhji's] so-called 'glide' strokes. **1955** *Times* 9 July 4/7 Never was beautifully caught at the wicket off a thin leg glide.
c. A step in certain ballroom dances; a gliding type of dance.
1889 *Cent. Dict.*, *Glide*,..in dancing, a peculiar waltz-step performed in a smooth and sliding manner. **1926** Whiteman & McBride *Jazz* xi. 224 'Avalon'..was one of the greatest fox trots of the late 'glide' period. *Ibid.* 230 The original 'glide two-step' fox trot of the 'Japanese Sandman' period. **1939** *Britannica Bk. of Year* 197/1 The 'Palais Glide', another popular dance from England, made some headway in America.
d. *Aeronaut.* The act of gliding; a flight accomplished by gliding.
1902 *Encycl. Brit.* XXV. 103/1 He made over 2000 glides safely, using gravity as a motive power. **1909** A. Berget *Conquest of Air* II. iii. 175 This descending glide. **1916** H. Barber *Aeroplane Speaks* iv. 53 The Pilot is satisfied that he is now sufficiently high to secure..a long enough glide to earth to enable him to choose and reach a good landing-place. **1940** L. Barringer *Flight without Power* v. 87 On windy days.. [illegible] glide. **1955** Welch & Irving *Soaring Pilot* vii. 135 The sort of approach which is often seen on aerodromes—a good deal of air-braking early on followed by a long flat glide—is useless for getting into small fields. **1971** *Sailplane & Gliding* XXII. 364/1 Many hours later [illegible] final glide to receive a tumultuous welcome.
4*. *Cryst.* Plastic deformation of a crystal in which there is a movement of one atomic plane over another, resulting in the lateral displacement of part of the lattice.
1934 *Nature* 16 June 922/1 Glide commences in a single crystal when the shear stress on the glide plane, and in the glide direction, reaches a certain value. **1952** *Jrnl. Iron & Steel Inst.* CLXXI. 225/1 [illegible] in single crystal form is, after a few per cent. glide, harder than cadmium. **1954** E. O. Hall *Twinning* ii. 31 The areas where glide occurs then appear as steps on the surface of the crystal. **1960** *Metallurgia* Mar. 125/1 He demonstrates glide, partial dislocations..and a number of other imperfections in structure. **1970** Kelly & Groves *Crystallogr.* vi. 169 At low temperature crystals yield plastically by a process called glide.
5. glide bomb, a bomb fitted with aerofoils that enable it to glide towards its target when released from an aircraft; hence as *v. intr.*, to drop glide bombs; **glide-direction,** a direction in a glide-plane in which glide can occur; **glide path,** the line of descent followed by a landing aircraft; *spec.* one indicated to the pilot by radar, etc., from the ground; **glide-plane** *Cryst.*, a plane in a crystal in which glide occurs; also, a symmetry element of a space-lattice such that reflection in the plane followed by a translation parallel to it produces a lattice congruent with the original; **glide-sound,** in *Phonetics*, the sound of a glide; **glide-twinning** *Cryst.*, the formation of a twin by the gliding of adjacent layers of a crystal lattice over one another; so **glide-twin.**
1943 *Newsweek* 8 Mar. 24 A divebomber pilot must be able to glide bomb in certain circumstances. **1943** *Time* 25 Oct. 23/1 The airmen knew that 1,800 fighters equipped with cannon, machine guns, some with glide bombs, are concentrated between Denmark and Belgium. **1954** K. W. Gatland *Devel. Guided Missile* (ed. 2) v. 135 The Petrel..rocket-propelled 'glide bomb', was capable of a short undersurface run, the wings and rocket motors breaking off as the missile entered the water close to the target. **1933** W. H. & W. L. Bragg *Crystalline State* I. vii. 198 The relative movement occurs along a definite crystallographic axis lying in the plane, the 'glide direction'. **1934** Glide direction [see 4* above]. **1936** *Electr. Commun.* XV. 196/1 The experimentally tested glide path [illegible] process..can be utilized as, for example, in Switzerland, where the glide path is followed down within a few meters from the ground. **1958** *Jrnl. R. Aeronaut. Soc.* XLII. 747 Such essential elements as runway localisers, glide path and markers are analysed. *Ibid.* 490 There is a vertical glide path indicator. **1966** *Guardian* 28 Dec. 1/5 Apollo 8 had to aim at a 'keyhole' entrance to the earth, an imaginary corridor only 35 miles wide. This is the so-called 'glidepath' they had to shoot at as they entered the upper atmosphere. **1970** *Times* 8 Apr. 10/3 A lower approach, much more like coming down the glide path of an airport, should create fewer troubles from a dust storm raised by the rocket motors. **1895** N. S. Maskelyne *Crystallogr.* i. 7 The glide-planes, in the case of deformed crystals, are..planes along which disruption can be easily effected. **1946** *Nature* 21 Sept. 395/1 Space groups, rotation-axes, glide-planes. **1953** E. S. Hills *Elem. Struct. Geol.* 137 The relatively high ductility of metals is due to the non-directional nature of the metallic bond.., which permits ready re-establishment of cohesion across glide planes and crystal boundaries. **1911** *Encycl. Brit.* XXI. 465/2 Acoustically speaking,..voiceless stops are pure glide-sounds, the stop itself being inaudible. **1933** Bloomfield *Lang.* vi. 98 The intervening non-distinctive glide-sounds that are produced while the vocal organs change their position. **1958** W. A. Wooster *Text-bk. Crystal Physics* ii. 32 The indices of the crystallographic twin..and those of the glide-twin. **1951** N. F. M. Henry et al. *Interpr. X-Ray Diffraction Photogr.* i. 17/2 In certain substances showing the special type of homogeneous deformation called glide twinning the amount of relative displacement is absolutely fixed for a particular glide in a given substance. *Ibid.*, In order to define a glide twin completely, it is necessary to specify (i) the glide plane, (ii) the glide direction, and (iii) the amount of glide. **1957** *Encycl. Brit.* VI. 828/2 Plasticity [of a crystal] is sometimes associated with glide-twinning, a process in which there is a sudden switching of the atoms to a second stable position related in a definite geometrical way to the first.

glide, *v.* Add: **1. c.** *spec.* Of an aeroplane: to fly without engine power. Also *trans.*, to traverse in a glider.
1910 C. C. Turner *Aerial Navig.* xx. 252 The best means of becoming proficient in flying is first to learn to glide. **1916** H. Barber *Aeroplane Speaks* iv. 53 The Aeroplane with noiseless engine glides over the boundary of the Aerodrome. **1934** *Times* 23 June 15/4 The claim that he was the first man to 'glide' the Channel. **1940** L. M. Barringer *Flight without Power* ii. 15 The very efficient high performance gliders can glide a long way without losing much height. **1958** D. Piggott *Gliding* ii. 12 Launching signals and procedure are more or less standard..wherever you glide in England.
9. *Cryst. intr.* Of particles in a crystal: to move, be displaced. Also of a crystal: to undergo glide. Cf. Glide *sb.* 4*.
1895 N. S. Maskelyne *Crystallogr.* i. 7 The particles.. aligned on all lines parallel to the edge..glided into new positions in the crystal-block. **1924** A. E. H. Tutton *Nat. Hist. Crystals* xviii. 273 Many of the softer crystallised substances develop the property of permitting one layer to glide over another by gentle side pressure with a knife blade. **1938** W. A. Wooster *Text-bk. Crystal Physics* ii. 29 The crystal glides on the (011) plane in the [011] direction. **1970** Kelly & Groves *Crystallogr.* vi. 169 Sapphire crystals can be made to glide at room temperature under a pressure of 25,000 atm.
10. *Cricket. intr.* To make the glide stroke (see prec. 1 b).
1899 *Daily News* 22 July 4/2 Men were then less apt to 'glance and glide', like The Brook, and K. S. Ranjitsinhji. **1932** T. B. Casson *Century of Roundels* 17 Ranjitsinhji, when he glides, Stands at the crease in posture cringy.

glideless (glai·dlės), *a. Phonetics.* [-less.] Unaccompanied by a glide (Glide *sb.* 4).
1900 Sweet *Primer Phon.* 65 Combinations of stops and vowel like consonants..are glideless in English.

glider. Add: **1. c.** (See quot. 1940.)
1926–7 *Army & Navy Stores Catal.* 192/3 Unbreakable glass gliders... Makes furniture easily movable. **1940** *Chambers's Techn. Dict.* 379/1 *Gliders*, dome-shaped metal pieces fastened to the feet of chairs, settees, etc., in place of casters.
d. A runner (Runner 13 b) for a curtain.
1957 *Archit. Rev.* CXXII. 356/3 A concealed curtain track..called 'Silent Gliss' and manufactured by Silent Gliss Ltd. will be shown; nylon gliders to which curtains may be attached direct or by means of hooks, runners or eyes, run in aluminium runners. **1961** *Lancet* 5 Aug. 297/1 Typical cubicle equipment, comprising aluminium tube and curtain-rail with nylon gliders.
2. *Aeronaut.* **a.** An engineless aeroplane.
1897 *Aeronaut. Ann.* 144, I made my first trials with a soaring machine in the summer of '95... It is exceedingly difficult to make a glider with one surface only which will sail properly. **1903** *Daily Chron.* 29 Oct. 6/7 The longitudinal stability of aerial gliders. **1909** H. G. Wells *Tono-Bungay* III. i. 208, I was developing a glider into a 'flyer'. **1933** *Daily Tel.* 6 Jan. 12/4 [He] had a leg and an arm broken when his glider crashed. **1950** *Engineering* 20 Feb. 158/1 The glider..had the advantage of being able to land with great accuracy. **1965** 'Sergeant' & Watson *Gliding* IX. iv. 50 The upward flow is called an 'anabatic' wind and it is occasionally strong enough to support a glider in flight.
b. One who glides or is an expert in gliding.
1910 C. C. Turner *Aerial Navig.* xx. 253 It will be well to take the experiences of the great gliders in proper order, beginning with Lilienthal. **1960** C. H. Gibbs-Smith *Aeroplane* 97 As to the tower-jumpers of history, there is naturally no telling who was the first to deliver himself from the vertical sufficiently to earn the title of first 'flier' or 'glider' in history.
3. (See quot. 1939.) *U.S.*
1932 *Hadley Co., Springfield, Mass., Advt.* July, Junior Glider Swing. **1939** Webster *Add.*, *Glider*, a porch swing with cushioned seat or couch suspended by links from an upright framework. **1967** P. Highsmith *Tremor of Forgery* xxi. 223 Adams's terrace faced the gulf and had a glider..and a table and chairs.
4. *attrib.* and *Comb.*, as **glider air-train** (see quot.); **glider bomb** = *glide bomb* (s.v. *Glide *sb.* 5); **glider-borne** *a.*, of troops, etc.; carried by glider; **glider train,** = **glider air-train*; **glider-tug,** a powered aircraft that tows gliders.
1935 C. G. Burge *Compl. Bk. Aviation* 377/1 *Glider 'air train'*, an 'air train' consists of an aeroplane attached to which are one or more gliders in tow. **1943** *Times* 20 Jan. 4/4 The enemy pressed home determined attacks on the convoy with glider bombs. **1945** W. Ley *Rockets* 272 There is no technological similarity between the radio-controlled glider bomb and the jet-propelled flying bomb. **1956** *Amer. Air Mail* Oct. 27/2 Existing equipment could drop glider-borne troops. **1946** G. Gibson *Enemy Coast Ahead* xii. 157 Combined use of parachute troops, glider-borne troops, anti-flak fighters and bombers. **1940** *Aeronautics* Nov. 44/3 Flight tests have shown that more than one glider can be towed by an aeroplane, so that a glider train is not out of question. **1944** *Jane's All World's Aircraft 1943–44* 432/2 The principal glider-tug..is capable of towing up to three gliders at a time.

gliding, *vbl. sb.* Add: **1. b.** *Cryst.* = *Glide *sb.* 4*; also, the lateral movement of particles that occurs in glide. So *gliding-plane* = **glide-plane.*
1878 *Mineral. Mag.* Dec. 82 In every crystal, in addition to..the planes along which slipping accompanied by a rotation of certain of the molecules occurs (gliding-planes), there is a..set of 'structure-planes'. **1938** W. A. Wooster *Text-bk. Crystal Physics* vii. 273 Gliding in crystals is restricted to certain crystallographically defined planes and directions. **1942** M. P. Billings *Struct. Geol.* iv. 26 Gliding is of two types, translation-gliding and twin-gliding. **1959** *Chambers's Encycl.* X. 771/1 Plastic deformation occurs as a result of the gliding of these planes over one another.
2. *Aeronaut.* Flight which is not dependent on engine power; the action of flying in this way. Freq. *attrib.*, as *gliding angle*, *contest*, *flight*, *school*; **gliding-boat,** a hydroplane; **gliding machine** = *Glider 2 a; **gliding path** = **glide path.*
1896 O. Chanute *Diary* 23 June in *Wright Papers* (1953) I. 645 It is a good beginning, but not a soaring machine and little is to be expected from it. **1898** *Sci. Amer. Suppl.* 22 Jan. 18490/2 It was a great deal more difficult to control any gliding machine on the ground than when the operator was in the air. **1902** *Encycl. Brit.* XXV. 103/1 Gliding flight is a matter requiring..practice. **1904** *Trans. Inst. Naval Archit.* XLVI. 245 Mr. Wilbur Wright found that [illegible] the gliding machine. **1906** *Sci. Amer.* 30 Sept. 211/1 The term gliding machine (meaning soaring machine) should be dropped from the nomenclature of aeronautics, as it is liable to be confused with the hydroplane or gliding boat, which is also a gliding machine. **1910** *Daily Chron.* 12 Mar. 3/3 To establish a gliding school at Hockley, in Essex. **1922** *Daily Mail* (Continental ed.) 19 Oct., The third day of the gliding contest at Itford Hill, near Lewes, Sussex. **1929** *Westm. Gaz.* 23 Oct., 'Gliding' is no longer an adequate name for the new form of flight which has been discovered. **1930** *Proc. Inst. Radio Engin.* May 833 A special radiobeam..is so oriented..as to define the proper gliding path which will permit..landings. **1950** *Gloss. Aeronaut. Terms* (B.S.I.) i. 8 *Gliding angle*, the angle between the flight path in a glide and the horizontal. **1960** C. H. Gibbs-Smith *Aeroplane* 197 The historical gliding experiments of Lilienthal.

glim, *sb.* Add: **4.** *glim lamp*, *light.*
1942 W. Simpson *One of our Pilots is Safe* ii. 40 At dusk the control officer for the night arrived and arranged a path of small electric glim lamps in the form of a capital L. **1944** [see *bat *sb.*¹ 3 f].

glimmer, *v.* Add: **2. d.** *to go glimmering*: to die away, die out, vanish, 'peter out'. *U.S. slang.*
1941 *Memphis Appeal-Avalanche* 8 May 4/1 The union depot project appears to have gone glimmering down the vale of things that were. **1966** W. M. Raine *B. O'Connor* 28 My reputation has gone glimmering. **1968** *La Junta* (Colo.) *Tribune-Democrat* 1 June 1/8 Plans for a new building..went glimmering today as School District 11 was turned down on its application for Lenham funds for this purpose.

glimmery, *a.*¹ Add: **c.** Glimmering. Also *adv.*
1906 Kipling *Puck of Pook's Hill* 36 A loose glimmery gown of chain-mail. **1907** *Daily Chron.* 4 Mar. 6/4 The marble bust..glimmery white. **1927** *Blackw. Mag.* Mar. 308/1 A great white wall of snow and rock, all glimmery in the rising sun, rose above us. **1953** H. E. Bates *Thinking & Exper.* ii. 47 When I see the ice, I..do not infer the coldness from the glimmery look.

glint (glint), *sb.*² *Geol.* [ad. Sw. *klint* cliff: cogn. w. Clint *sb.*] An escarpment of almost horizontal strata formed as a result of denudation of the adjacent lower rocks. Also *attrib.*, as **glint lake,** a lake formed by glacial excavation at a glint; **glint line,** the line followed by a glint, esp. that on the edge of the Baltic shield.
1906 H. Sollas tr. *Suess's Face of Earth* II. 65 In Russia the feature formed by these basset edges is known as the *glint*, and we shall in future employ this term [illegible] when they are due, not to fracture, but to denudation. [illegible] The glint line of the Baltic shield runs..through the gulf of Finland, which also has the position of a glint lake. **1939** *Geogr. Jrnl.* XCIII. 234 The large glint lakes bordering the Canadian Shield. **1959** Wooldridge & Morgan *Outl. Geomorphol.* (ed. 2) vi. 101 The south-eastern limit in Scandinavia [of the Caledonian ranges] is the famous 'glint-line', along which the folded terrain adjoins the Baltic or Fenno-Scandian block, which functioned as the hinterland in the original folding.

glioblastoma (glai,oblæstō·mă). *Path.* [f. Glio(ma + -blast + *-oma.] = *spongioblastoma. **glioblastoma multiforme,** a fast-growing malignant glioma of the cerebrum in

which there are cells of many kinds, most of them resembling spongioblasts.
1926 Bailey & Cushing *Class. Tumors of Glioma Group* v. 95 One might..designate the..[bipolar and unipolar spongioblasts] as 'bipolar and unipolar glioblastomas'... One would then employ glioblastoma multiforme and unipolare in the place of *spongioblastoma multiforme* and *spongioblastoma unipolare.* **1956** *Lancet* 22 Sept. [illegible] months after the patient's first admission to hospital a cystic glioblastoma multiforme was demonstrated at operation. **1970** Passmore & Robson *Compan. Med. Studies* II. xxviii. 42 The more rapidly growing and pleomorphic of these tumours [*sc.* gliomata] are often called glioblastomata.

gliomatosis (glai:omātōu·sis). *Path.* [ad. G. *gliomatose* (F. Schultze 1882, in *Arch. f. path. Anat. und Physiol. und f. klin. Med.* LXXXVII. 535), f. mod.L. *glioma-*, Glioma + -osis.] A diffuse proliferation of glia cells associated with or arising from a glioma; gliosis.
1886 W. R. Gowers *Man. Dis. Nerv. Syst.* I. 436 The gliomatosis may be regarded as a persistent, slightly modified, embryonal tissue, and these exuberant masses of tissue bear considerable resemblance in structure to glioma, so that the condition has been called 'gliomatosis'. **1899** *Allbutt's Syst. Med.* VII. 48 From syringomyelia and central gliomatosis it may be impossible to distinguish the later results of hæmatomyelia. **1924** R. Muir *Text-bk. Path.* xvi. 624 Intermediate varieties are met with, between a distinct tumour and a more diffuse overgrowth of glia [illegible] gliomatosis... Gliomatosis occurs also in relation to the central canal of the spinal cord in connection with syringomyelia. **1947** W. R. Brain *Dis. Nerv. Syst.* (ed. 3) iii. 285 Sometimes a glioma seems to arise diffusely, as in so-called 'gliomatosis cerebri', or from multiple centres at the same time. **1969** Russell & Rubinstein *Path. Tumours Nerv. Syst.* (ed. 3) ii. 33 Diffuse gliomatosis of the pia-arachnoid of the spinal cord was recorded by Harbitz (1932, case 1), the glioma apparently being of astrocytic character.

gliosis (glaiōu·sis). *Path.* [ad. G. *gliose* (F. Schultze 1882, in *Arch. f. path. Anat. und Physiol. und f. klin. Med.* LXXXVII. 535), f. Gli(a (see -osis).] Proliferation of glia cells, esp. of the fibrillar processes of astrocytes.
1890 *Brain* XIII. 299 Inflammation confined to a special tissue, as the neuroglia, has then the particular characteristics of gliomatous proliferation...Schültze would naturally call this special kind of inflammation *gliose* or *gliomatose.* **1892** W. Osler *Princ. Med.* 849 The condition is now regarded as a gliosis, a development of embryonal neuroglial tissue in which hæmorrhage or degeneration takes place with the formation of cavities. **1899** *Allbutt's Syst. Med.* VII. 560 He [*sc.* Weigert] wholly discards the hypothesis of syringomyelia to which the name spinal gliosis has been applied. **1934** H. W. C. Vines *Green's Man. Path.* (ed. 15) xxvi. 809 If any destructive lesion of the brain tissue should occur, the astrocytes proliferate, and repair of the lesion occurs by gliosis comparable to the reparative fibrosis in other tissues. **1962** S. Duke-Elder *Parsons' Dis. Eye* (ed. 14) xxxiv. 545 In syringomyelia cavities form around which secondary gliosis develops in the cervical and upper dorsal cord. **1966** Wright & Symmers *Systemic Path.* II. xxxiv. 1143/2 Gliosis is not always pathological. It may be found to some extent in normal brains in such places as the olivary nuclei..and the central canal of the spinal cord.

gliotoxin (glai,otǫ·ksin). *Biochem.* [f. mod.L. *Glio-cladium*, name of a genus of fungi (f. Gr. γλία glue + κλάδ-ος shoot, branch) + Toxin.] An indole derivative, $C_{13}H_{14}N_2O_4S_2$, which has antibiotic activity against many fungi and bacteria and is produced by the growth of some fungi, esp. various species of *Penicillium* and *Aspergillus.*
1941 R. Weindling in *Phytopathology* XXXI. 991 The principal purpose of this article is to present data on factors determining production, stability, and activity of the crystalline *Gliocladium* toxin (gliotoxin)... The term 'gliotoxin' has been suggested by J. H. Johnson. **1944** *Nature* 25 Nov. 669/2 Strains of *Trichoderma viride* produce gliotoxin, and..it is extremely probable that the fungus used by Weindling was not *G[liocladium] fimbriatum*, but was a *Trichoderma* as he originally supposed. **1951** Pratt & Dufrenoy *Antibiotics* (ed. 2) xiv. 286 Some antibiotics (actidione and gliotoxin, for example).. are too toxic for systemic chemotherapy in man. **1970** *Nature* 18 July 300/2 *T. viride* may retard the development of other fungi by producing the antibiotic gliotoxin or viridin.

glisk, *sb.* **1.** (Earlier and later examples.)
a **1692** A. Pitcairne *Assembly* (1722) 18 The Malignants, whom they knew by the first Glisk of their Face. **1922** J. Buchan *Huntingtower* vii. 136, I took a walk and got a glisk of the House and I liked the look of it.

gliss., gliss (glis), abbrev. of next.
1926 *Melody Maker* Mar. 31 The aforementioned mute [illegible] the necessary 'gliss' which I have marked by means of slurs. **1931** G. Jacob *Orchestral Technique* vii. 76 [illegible] gliss [illegible] two notes [illegible]. **1966** *Crescendo* Jan. 6/2 After a trombone gliss there's a comic Eastern bit.

glissando (glisæ·ndo). *Mus.* Pl. **glissandi, -dos.** [Italianized form of Fr. *glissant*, gerund and pres. pple. of *glisser* to slide.] A slurring or sliding effect produced by a musical instrument; = Portamento.
1873 H. C. Banister *Music* 240 *Glissando*, gliding; applied to the playing of several notes successively, on the Pianoforte, by the same finger, sliding from one to the other. **1876** Stainer & Barrett *Dict. Mus. Terms* (1898) 202/2 *Glissando*.., (1) playing a rapid passage in pianoforte music, by sliding the tips of the fingers along the keys instead of striking each note with a separate finger. (2) A rapid slur in violin playing. **1924** *Living Age* 18 Oct. 170/1 The trombone whose glissandos are becoming a favorite. *Ibid.*, A whole special technique of the violin..employing..the very slowest of glissandos. *Ibid.* 172/2 A combination of glissando and vibrato techniques. **1926** Whiteman & McBride *Jazz* ix. 202 The glissando is one of the chief embellishments of jazz. **1931** G. Jacob *Orchestral Technique* viii. 76 Harp glissandos are frequently used. **1939** *Scrutiny* VII. IV. 398 The vocal descriptions (by means of leaps of major sevenths, quarter tones, shouts, croaks and glissandi) of hail and lightning in *Soliloquy* are open to the same criticism. **1958** J. Aldam in P. Gammond *Duke Ellington* III. 200 One of the greatest New Orleans clarinettists..he [*sc.* Barney Bigard] plays with a warm, vocalized tone, full of inflections and long glissandi. **1970** R. Smith Brindle *Contemp. Percussion* vi. 32 Sometimes glissandos are written out in full, but it is much more usual to indicate the first and last notes played, and connect them by a straight or wavy line.

‖ **glissé** (glise). *Ballet.* [F., pa. pple. of *glisser* to slide.] A sliding step in which the flat of the foot is often used; also *pas glissé.* Cf. Glissade 2.
1913 C. d'Albert *Dancing* 78 *Glissades, Glissés.* These steps are used both in stage and ballroom movements. The foot can glide in any direction. By the term *pas glissé* is understood a step sideways and a gliding of the other foot. Temps glissé is made by raising on both feet, fall back on one foot and glide the other in any direction. **1953** E. Taylor *Sleeping Beauty* vi. 116 She then did a little glissé.

glitch (glitʃ). [Etym. unknown.] **a.** *slang.* A surge of current or a spurious electrical signal (see quots.); also, in extended use, a sudden short-lived irregularity in behaviour. **b.** *Astronauts' slang.* A hitch or snag; a malfunction.
1962 J. Glenn in *Into Orbit* 86 Another term we adopted to describe some of our problems was 'glitch'. Literally, a glitch is a spike or change in voltage in an electrical circuit which takes place when the circuit suddenly has a new load put on it... A glitch..is such a minute change in voltage that no fuse could protect against it. *Ibid.* 245 *Glitch*, a momentary change in voltage in an electrical circuit; (*slang*—a hitch). **1962** R. F. Graf *Mod. Dict. Electronics* (1963) 124 *Glitch*, low-frequency interference in a television picture. It is seen as a narrow bar moving vertically. **1969** *Product Engineering* 27 Jan. 15/3 It generated digital transients that caused the abort guidance to send false signals. Phillips said it took an inordinately long time to find this glitch. **1969** *Funk & Wagnalls Dict. Electronics* 70 *Glitch*, a stray current or signal, usually one that interferes in some way with the functioning of a system. **1969** *Daily Tel.* 15 Nov. 1/3 [Apollo moon flight] Thinking back to when we had our big glitch, I remember seeing it get light outside the window after we were in the clouds, and I'm pretty sure we got hit by lightning. **1971** *Nature: Physical Sci.* 14 June 146/2 Two pulsars have exhibited a sudden increase in frequency (glitch), after which the usual slowing down has been resumed.

glitterer (gli·tǝrǝɹ). [f. Glitter *v.* + -er¹.] One who or that which glitters.
1823 Byron *Juan* XIV. lvii, Fondly the wheeling fireflies flew around her, Those little glitterers of the London night. **1945** G. Millar *Maquis* i. 2 At the Berkeley, among the glitterers that were so unlike the major.

glitterwax (gli·tǝɹwæks). [f. Glitter *v.* + Wax *sb.*¹] A kind of coloured modelling-wax.
1944 *Horizon* May 307 They give their children glitterwax or plasticine to play with. **1967** M. Drabble *Jerusalem the Golden* vi. 149 Clelia, who used as a child to have a passion for modelling things out of an obsolete substance known as Glitterwax.

gloaming. Add: **2. gloaming sight,** a front sight specially adapted for evening shooting.
1895 *Army & Navy Co-op. Soc. Price List* 15 Sept. 925/1 New gloaming sight for guns and rifles. **1907** *Yesterday's Shopping* (1969) 639 Gloaming sights for rifles... Specially adapted for evening sport.

gloat, *sb.*¹ Restrict †*Obs. rare* to sense in Dict., and add: **2.** An act of gloating; a look, feeling, or expression of triumphant satisfaction.
1899 Kipling *Stalky & Co.* 13 They spun wildly on their heels, jodelling after the accepted manner of a 'gloat'. **1921** Galsworthy *To Let* II. iii, To watch the gloat in his eye. **1927** *Bulletin* 7 Oct. 17/2 Preparing to enjoy a satisfying gloat over some of his critics. **1930** 'Hay' & King-Hall *Middle Watch* xx, She now stood regarding the Admiral's wife with a certain deliberate.. insolence of expression. One might almost call it a gloat. **1970** A. Hunter *Gently with Innocents* viii. 106 He was having a gloat over his gold..picking it up, gloating over it.

glob (glǫb). *slang.* [? Blend of Blob *sb.* and Gob *sb.*¹] A mass or lump of some liquid or semi-liquid substance.
1900 H. Lawson *Over Sliprails* 44 The pup curled like a glob of mud on the sand in the moonlight. **1957** J. Masters *Far, Far the Mountain Peak* 96 Very large, mysterious snowflakes, wet and solid, drifted like globs of ice cream into their eyes. **1962** *New Scientist* 5 Apr. 807/2 Throughout the long coasting time, the fuel has been free of the pull of gravity. It is probably floating around the half-empty fuel tank in globs. **1965** E. Tunis *Colonial Craftsmen* vi. 138 The blower gathered a parison, a 'glob' of hot glass, from the pot by dipping the tip of a six-foot blowing iron (tube) into the mass.

global, *a.* Restrict *rare* to sense in Dict. and add: **2.** [After Fr. *global.*] Pertaining to or embracing the totality of a number of items, categories, etc.; comprehensive, all-inclusive, unified; total; *spec.* pertaining to or involving the whole world; world-wide; universal.
1892 *Harper's Mag.* Sept. 492/2 M. de Vogüé loves travel: he goes to the East and to the West for colors and ideas; his interests are as wide as the universe; his ambition, to use a word of his own, is to be 'global'. **1927** *Contemp. Rev.* Aug. 241 The essence of the American proposal therefore was its 'global' criterion. **1928** *Times* 1 Oct. 14/1 The proposal for a readjustment of tonnage proportions within the global limits originally proposed by the United States. *Ibid.*, The original French proposal was for global tonnage. **1928** *John o' London's* 24 Nov. 252/1 Adding figures of commerce and foreign investments..so as to show to-day's global contacts. **1943** *Air Force* Feb. 22 (*title*) Guides for global war. *Ibid.*, In this global war they [*sc.* maps] are vital to airmen. **1943** *Ann. Reg. 1942* 283 The hard lesson of modern global warfare. **1944** *Amer. Speech* XIX. 137 Its extremely healthy global attitude ('Linguistic isolationism..will be regarded as..outmoded and ridiculous..'). **1946** J. S. Huxley *Unesco* i. 8 A scientific world humanism, global in extent and evolutionary in background. **1948** *Ann. Reg. 1947* 14 The global sum of £300 million looked like the result of bargaining with the Treasury. **1951** *Sun* (Baltimore) 7 Jan. 2/6 American 'global bombers'—giant B-36's which can carry an atom bomb 10,000 miles. **1952** *Brit. Jrnl. Psychol.* May 101 There are..other and more satisfactory objective methods for investigating.. the temperamental and general behavioural traits... A 'global' picture can be obtained from the use of such techniques as Rorschach or T.A.T. **1957** *Economist* 12 Oct. (Suppl.) 17, 10 days of global cruising. **1959** P. H. Spaak *Why Nato?* iv. 30 The allies could only meet the global challenge of the Soviet Union with a global retort. **1968** McLuhan & Fiore (*title*) War and peace in the global village. **1970** *Sci. Jrnl.* Apr. 52/3 The meteorological global telecommunications system required poses a second major problem in the development of an effective system of global numerical weather prediction.
Hence **glo·balism,** internationalism; **globaliza·tion,** the act of globalizing; **glo·balize** *v. trans.*, to render global; so *globalized* ppl. adj.
1959 *Economist* 4 Apr. 65/1 Italy's 'globalised quota' for imports of cars has been increased. **1961** Webster, Globalism... Globalization. **1962** *Spectator* 5 Oct. 495 Globalisation is, indeed, a staggering concept. **1962** *Sunday Times* 28 Jan. 12/2 Our own comparatively timid intentions towards globalising the Common Market. **1965** *Economist* 23 Jan. 316/1 Between globalism and isolationism there is extensive middle ground.

globally (glōᵘ·bāli), *adv.* [f. prec. + -ly².] In a global manner; to a global extent; comprehensively; universally; throughout the world.
1930 *N.Y. Times* 9 Nov. 1 The 'minor technical question' of whether to limit naval officers and men globally or by categories. **1957** *Economist* 5 Oct. 31 To-day, more than ever before, exporters must plan *globally*, especially in view of the growing economic importance of the world's new markets. **1958** *Engineering* 21 Feb. 236/2 Globally, 50 to 100 times this amount may be produced annually by the end of the century. **1958** *New Biol.* XXVII. 53 Two world wars and facilities for increasingly rapid travel and communication have induced the more perceptive to think globally—that is to say to consider the total situation whether in politics or science. **1971** *Nature* 7 May 17/1 There is some evidence that lead from car exhausts is building up globally.

globe, *sb.* Add: **10. a.** *globe-girdling* adj.
1875 *Temple Bar* June 255 A somewhat old-fashioned house, not working any of your globe-girdling speculative gigantic operations. **1943** *Amer. Mag.* Mar. 98/1 A network of globe-girdling airways.
b. globe-lamp (earlier examples); **globe-trot** *v. intr.* [back-formation from *globe-trotter*, *-trotting*], to go globe-trotting; also *trans.*; **globe-trotter** (earlier example).
1788 B. Franklin *Autobiogr.* (1905) 381 The globe lamps we were at first supply'd with from London. **1825** H. Wilson *Mem.* IV. 67 This room..was lighted by large, ground-glass, French globe-lamps, suspended from the cieling. **1883** *Manch. Guardian* 6 June 6/1 He drives from London to York..and is amusing; he globetrots and is not amusing. **1883** Ld. R. Gower *My Reminisc.* II. 180 Nothing nowadays is easier and safer than to 'globe-trot' round the world. **1928** *Observer* 8 Apr. 5/7 You are never instructed, never globe-trotted. **1929** C. Day Lewis *Transitional Poem* I. 16 To..globe-trot with the wind. **1970** *Oxf. Univ. Press Record* xv. 3 Bill Waterfield ..globe-trotted for several years after reading Botany at University College. **1875** E. K. Laird (*title*) Rambles of a globe-trotter in Australia, Japan, China, Java, India and Cashmere.

globin. Add: Now more widely: any of various colourless basic polypeptides that are the protein components of hæmoglobin, myoglobin, and related molecules (the non-protein part in each case being hæm). (Add further examples.)

1898 *Jrnl. Chem. Soc.* LXXIV. I. 719 Nearly pure globin is obtained. This..must be distinguished from Preyer's 'globin' obtained from methæmoglobin. *Ibid.* 720 The blood of the goose, however, yields a globin which possesses different properties. 1925 [see *HÆM]. 1930 *Jrnl. Gen. Physiol.* XIII. 475 The denatured globin may be largely converted into a soluble, apparently native form which can combine with heme to form hemoglobin. 1957 Fox & Foster *Introd. Protein Chem.* xviii. 348 The isoelectric points of the globins, which are very rich in histidine, are usually on the alkaline side of 7·0. 1962 E. G. Young in Florkin & Stotz *Comprehensive Biochem.* VIII. i. 5 Globin has been the most carefully studied of the histones. It was first prepared from crystalline hæmoglobin by Preyer in 1868. 1970 R. W. McGilvery *Biochem.* ii. 13 Each peptide chain of globin has a sequence of amino acids that makes certain three-dimensional arrangements more likely than others, and the most likely of all is one that, in the presence of heme, causes it to fold around the heme.

globosite (glōu·bǒusait). *Min.* [ad. G. *globosit* (Breithaupt 1865, in *Berg- und hüttenmännische Zeitung* XXIV. 321/2), f. L. *globōs-us* GLOBOSE: see -ITE¹.] An ill-defined phosphate of ferric iron found in small globular concretions.

1868 J. D. Dana *Syst. Min.* (ed. 5) v. 584 Globosite. This name is given by Breithaupt..to a mineral occurring at the Arme Hilfe mine near Hirschberg. 1949 *Proc. Acad. Nat. Sci. Philadelphia* LXXVII. 8 Globosite was described by Breithaupt from the Arme Hilfe mine Hirschberg, where it occurred as wax-yellow to yellowish-gray globular concretions... No reliance can be placed in the analysis... It is probably identical with strengite.

globular, *a.* Add: **1. b.** Of protein: having a relatively compact molecular structure showing considerable folding.

1935 *Biochem. Jrnl.* XXIX. 2353 Edestin is thus a typical 'globular' protein. 1954, 1968 [see *FIBROUS *a.* 2 d]. 1970 R. W. McGilvery *Biochem.* ii. 9 Hemoglobin is a representative of the globular proteins—those with a compact structure that are usually quite soluble in the water phase of tissues.

3. *globular cluster*, a spherical cluster (see *CLUSTER *sb.* 3 c) of stars; also *ellipt.* globular; *globular lightning* = FIRE-BALL 1.

1859, 1928 [see *CLUSTER *sb.* 3 c]. 1959 *Listener* 2 July 14/2 Altogether, about 100 globulars are known. 1961 *Ibid.* 7 Dec. 973/1 Globular clusters are of particular importance in modern astronomy, since studies of them have provided a great deal of information about the shape and structure of the Galaxy. *Ibid.*, From Britain only the great globular in Hercules—Messier 13—is distinctly visible to the naked eye. 1964 Goodchild & Tweney *Technol. & Sci. Dict.* 259/1 *Globular lightning*, an electrical phenomenon whose nature is not understood: an appearance resembling a globe of fire, sometimes visible for a number of seconds, is produced. 1937 *Jrnl. R. Aeronaut. Soc.* XLI. 739 A Junkers Ju. 86..was struck by globular lightning.

glochidium (glǫki·diǝm). Pl. -idia (-i·diă). [mod.L.: see GLOCHIDIATE.] **1.** The larva of a pond-mussel of the family Unionidæ. Also *attrib.*

1875 Huxley & Martin *Elem. Biol.* 110 There is a single adductor muscle and a rudimentary foot, from which one or two long structureless filaments, representing the byssus of the sea-mussel, proceed. These byssal filaments become entangled with one another and tend to keep the 'Glochidia' in their places. 1888 Rolleston & Jackson *Forms Anim. Life* 137 This Glochidium is eventually set free from the parent. 1930 G. R. De Beer *Embryol. & Evol.* 44 The Glochidium larva of Unio. 1961 P. Street *Shell Life on Seashore* v. 78 Many fresh-water fish serve as hosts to these fresh-water glochidia.

2. *Bot.* (See quot. 1900.) (Also **glochid**.)

1882 S. H. Vines tr. *Sachs' Text-bk. Bot.* (ed. 2) II. 454 In some species (of *Azolla*) these massulæ have their surfaces covered with hair-like appendages, barbed at their free ends (Glochidia). 1888 Bennett & Murray *Handbk. Cryptogamic Bot.* i. 30 In some species, the surface of these masses is covered with hair-like appendages, barbed at the apex, the *glochidia*. 1900 B. D. Jackson *Gloss. Bot. Terms*, *Glochid*, *glochidium*,..(1) a barbed hair or bristle; (2) a similar structure on the massulæ of certain Cryptogams which act as organs of attachment to a macrospore. 1905 D. H. Campbell *Struct. & Devel. Mosses & Ferns* (ed. 2) xi. 417 Upon the outside of the massulæ are formed the curious anchor-like 'glochidia'. 1938 G. M. Smith *Cryptogamic Bot.* II. ix. 360 Massulae of species belonging to the section *Rhizosperma* lack glochidia. 1965 Bell & Coombe tr. *Strasburger's Textbk. Bot.* iii. 582 Arising from the surface of each massula are a number of small, barbed hooks or glochidia.

glockenspiel (glǫ·kĕnʃpil, -spīl). [G., lit. 'bell-play'.] **1.** (See quots.)

1825 J. F. Danneley *Dict. Music* s.v. *Carillon*, *Glockenspiel*, in Italian, *campanetta*, an organ-stop, composed of bells, comprising the upper half of the clavier. 1898 Stainer & Barrett *Dict. Mus. Terms* (ed. 2) 202/2 *Glockenspiel*,..an organ stop of two ranks.

2. A musical instrument consisting of a series of small bells or metal bars which are struck with a hammer, or by levers acted upon by a keyboard.

1833–4 Carlyle *Sart. Res.* (1836) I. vii. 46 Some few, of musical turn, have a whole chime of bells (*Glockenspiel*) fastened there [*sc.* on their girdles]. 1880 Grove *Dict. Mus.* II. 7/1 The instruments of percussion which give definite notes..are the Kettle Drums.., Glockenspiel (bells used in military bands and occasionally with orchestra), and the Harmonica. 1889 G. B. Shaw *London Music 1888–89* (1937) 193 In the waltz from Die Meistersinger I heard no glockenspiel. 1896 *Wood* 2 July 253/3 The glockenspiel proper consists of a set of eight or more clock bells, mounted on a central spindle, which is inserted in a wooden handle. 1957 Auden & Kallman *Magic Flute* 59 Though Papageno, one is sad to feel, Prefers the juke-box to the glockenspiel. 1970 R. Smith Brindle *Contemp. Percussion* vii. 53 By reason of the brief duration, limited volume and relative indefiniteness of its sounds, the glockenspiel is rarely used to contribute in a decisive way to 'large' orchestral climaxes.

‖ **gloire** (glwār). In full *la gloire*. [Fr.] Glory; *spec.* the national glory and prestige of France.

1854 E. Twisleton *Let.* 28 Nov. (1928) xiii. 252 If *la gloire* forsakes the French eagles, they will say..that [*sc.* Louis Napoleon] has driven away all their best generals. 1861 Mrs. Beeton *Bk. Housek. Managem.* 902 M. Curmer..tells us..that French beef is far superior to that of England. This is mere vaunting on the part of our neighbours, who seem to want *la gloire* in everything. 1867 'Ouida' *Under Two Flags* II. i. 11 She had..but two deities, 'la Gloire' and 'la France'. 1963 *Times* 7 Feb. 9/1 After the debacle with the High Military Tribunal, the present spectacle, both in and out of the court-room, hardly adds to the *gloire* of French justice. 1965 A. Horne *Fall of Paris* iii. 31 Had Louis Napoleon succeeded in his pursuit of '*la gloire*', the dynasty might have been assured a much longer life in France. 1969 *Listener* 16 Oct. 518/2 They [*sc.* the French] care only..for a slightly fake sense of their own importance, as Frenchmen, for that umbrella concept, *la gloire*.

Gloire de Dijon (glwār dǝ diʒon). *Bot.* [Fr., lit. 'glory of Dijon'.] A yellow hybrid tea rose. Also *attrib.*

1854 C. Bryant *Rose Amateur's Guide* (ed. 5) 119 A new variety exhibited last year in Paris, raised at Dijon, and called Gloire de Dijon, is a great acquisition. 1866 G. Meredith *Let.* 6 Sept. (1970) I. 286, I want you to see my Marie blooming. She is a full and perfect rose. You know the Gloire de Dijon? 1869 S. R. Hole *Book about Roses* ix. 132 All the Roses in this list, except Gloire de Bordeaux, Gloire de Dijon, and Jaune Desprez, are of the Hybrid Perpetual family. 1885 C. M. Yonge *Two Sides of Shield* I. v. 72 The old house hung over with..a Gloire de Dijon rose. 1928 Galsworthy *Swan Song* i. viii. 102 He took up a rose and sniffed at it deeply. So many different kinds now—he had lost track! In his young days one could tell them—La France, Maréchal Niel, and Gloire de Dijon—nothing else to speak of. 1928 D. H. Lawrence *Lady Chatterley* xiv. 255 She still lay naked and faintly golden like a Gloire de Dijon rose on the bed. 1969 *Jrnl. R. Hort. Soc.* XCIV. 241 'Gloire de Dijon' over a century old, is also of the tea rose persuasion.

glom (glǫm), *v. U.S. slang.* Also **glahm**. [Var. GLAUM *v.*] *trans.* To steal; to grab, snatch. Also *intr.*, usu. const. *on to*. Hence **glo·mming** *vbl. sb.*

1907 J. London *Road* (1914) 182 We..discovered that our hands were gloved. 'Where'd ye glahm 'em?' I asked. 'Out of an engine-cab,' he answered. 1914 Jackson & Hellyer *Vocab. Criminal Slang* 38 *Glom*, to grab; to snatch; to take; implying violence. Example: 'Glom this short and drop off two blocks below.' 1926 G. H. Mullin *Adv. Scholar Tramp* xii. 160, I learnt that stealing clothes from a clothes-line is expressed in Hoboland by the hilarious phrase, 'Glomming the grape-vine'. 1926 *Flynn's* 16 Jan. 638/1 'Course, th' rule is, glom while th' glommin's good. 1930 *Ibid.* 25 Jan. 344/1 In his hip-pocket, where even the lowest kind of lush-worker would have no difficulty in glomming it. 1960 J. Philips *Whisper Town* (1961) ix. v. 102 You think we ought to go out to the school and glom on to that gun? 1962 — *Dead Ending* (1963) i. 7 In the process of glomming onto that property in Venezuela some people got badly burned. 1969 C. Armstrong *Seven Seats to Moon* xiii. 126 Trust Lily Eden, though, to glom on to a customer.

glomerular, *a.* Add examples of the sense: of or pertaining to the glomeruli (of the kidneys).

1885 *Jrnl. Physiol.* VI. 385 To determine that..circulation through the glomeruli and glomerular activity are wholly stopped Nussbaum employed two methods. 1917 A. R. Cushny *Secretion of Urine* iv. 44 The blood pressure in the glomerular capillaries suffices for filtration. 1962 *Gray's Anat.* (ed. 33) 1501 The glomerular capsule is the blind, expanded end of the renal tubule, indented for the reception of the glomerulus. 1966 Wright & Symmers *Systemic Path.* I. xxiv. 733/2 The glomerular changes in the chronic stage of type 1 glomerulonephritis. 1970 Passmore & Robson *Compan. Med. Stud.* II. xi. 2/1 The rate of glomerular filtration is raised by agents which increase renal blood flow, cardiac output or blood volume.

glome:rulosclero·sis. *Path.* [f. GLOMERULAR *a.* + -O + SCLEROSIS.] Fibrosis or hyalinization of the glomeruli of the kidneys.

1936 Kimmelstiel & Wilson in *Amer. Jrnl. Path.* XII. 85 We have..contrasted the features of this special group [of cases] with those of true intracapillary and extracapillary glomerulonephritis... The term intercapillary glomerulosclerosis has been applied to the group of cases under discussion. 1964 S. Duke-Elder *Parsons' Dis. Eye* (ed. 14) xxii. 328 The fully developed ocular picture with micro-aneurysms is often associated with evidences of glomerulosclerosis in the kidney. 1966 R. H. Heptinstall *Path. Kidney* iii. 59 Congenital glomerulosclerosis is almost always a purely incidental finding.

glomus (glǭ·mǝs). Pl. **glo·mera**, (erron.) **glomi**. [L., = ball of thread.]

†**1.** *Bot.* = GLOMERULE 1 a. *Obs.*

1832 J. Lindley *Introd. Bot.* 111 Suppose the flowers of a simple umbel to be deprived of their pedicels, and to be seated on a receptacle or enlarged axis, we have a *capitulum* or head, named *glomus* by some, *glomerulus* by others. 1856 J. S. Henslow *Dict. Bot. Terms* 81 *Glomus*, synonyme for capitulum.

2. *Anat.* A small body consisting of a plexus of blood-vessels and associated tissue. **a.** Applied, chiefly in mod.L. phrases, to various specific structures in the body.

1839–47 R. B. Todd *Cycl. Anat.* III. 635/2 The former are most numerous at that part which was called by the Wenzels *glomus*, where the choroid plexus turns up from the inferior cornu into the horizontal portion of the lateral ventricle [of the brain]. 1897 W. N. Parker tr. *Wiedersheim's Compar. Anat. Vertebr.* (ed. 2) xi. 341 These vessels become coiled to form a mass which is known as the *glomus*. 1903 J. S. Ferguson *Normal Histol.* xxii. 458 The Carotid Gland. This body..is closely connected in its relation to the blood vessels and nerves, is also known as the glomus caroticum. *Ibid.* 459 The Coccygeal Gland.. It is richly supplied with broad capillaries or sinusoids and hence is also known as the glomus coccygeum. 1958 G. R. de Beer *Vertebr. Zool.* xii. 189 Capillaries grow out from the dorsal aorta forming the glomus. 1949 A. S. Romer *Vertebr. Body* xiii. 435 The tails of many mammals (such as dog, cat, rat) show small tissue masses, termed caudal glomeruli, which contain retia mirabilia. 1968 W. J. Hamilton *Textbk. Human Anat.* 315 A chemoreceptor is also present in this region [*sc.* the aortic arch] and constitutes the aortic body, or glomus aorticum. 1962 W. H. Hollinshead *Textbk. Anat.* xxiii. 823/2 Behind the upper end of the common carotid..is a flattened, somewhat ovoid body, the carotid body (glomus caroticum). This is a highly vascular epithelial body that contains special nerve fibers. 1964 J. Mason tr. *Baer's Compar. Anat. Vertebr.* 155 This latter area of the blood vessels forms the glomus or glomerulus, where filtration of the blood occurs. 1963 R. P. Morehead *Human Path.* xxxiv. 1381/1 Tumors of the glomus jugularis are rare..they arise in the middle ear.

b. [named by P. Masson 1924, in *Lyon chirurgical* XXI. 257.] One of the large number of structures in the skin of the digits in which a special arrangement of muscle and nerve tissue encloses an arterio-venous anastomosis by-passing the capillaries.

1938 *Physiol. Rev.* XVIII. 334 Masson was dissatisfied with the name 'arteriovenous anastomosis' because of the confusion with the traumatic and other unusual connections between arteries and veins... Since those in the finger often occur in the form of a group which makes a sort of ball, or skein, and since they have a morphology similar to those found in the glomus coccygeum, he adopted the term 'glomus'... Consequently he then termed them 'glomus tumors' or 'glomic tumors'. 1950 A. W. Ham *Histology* xxii. 365/1 In the skin of the fingers and toes, and in certain other sites, arteriovenous anastomoses are present in little bodies called glomi that are disposed deep in the dermis. 1962 *Gray's Anat.* (ed. 33) 727 The glomera play an important part in regulating body temperature. 1963 R. P. Morehead *Human Path.* xxii. 580 A glomus tumor is a benign mass of the cutaneous glomus consisting of epithelium-lined spaces surrounded by glomus cells.

gloom, *v.*¹ **1.** Add to def.: To be gloomy.

1913 'A. Garve' *Very Quiet Place* ix. 59 'What did you do?' 'Gloomed for a week—then started to write it again.' 1968 H. Franklin *Crash* vi. 77, I sat and gloomed in the hotel lounge.

glop (glǫp), *sb.*¹ *U.S. slang.* [Cf. GLOP *v.*²] A liquid or viscous substance or mixture; *spec.* inferior or unappetising food.

1945 *N.Y. Times Mag.* 4 Nov. 12/1 The somewhat pleasanter words, mostly referring to food, such as 'glop', meaning any food mixture. 1952 Berrey & Van den Bark *Amer. Thes. Slang* (ed. 2) §912/2 Inferior food,..*glop* [index has 'glop']. 1962 J. Potts *Evil Wish* xii. 88 A cheap, soiled cosmetic case crammed with little bottles of glop. 1965 *New Yorker* 15 May 41 (Advt.), This glop can choke an Army whirlybird.. The blobs on the left are drops of helicopter fuel—after it has been contaminated by dust, grit and dirt.

gloria. Add: **3.** The French name for coffee with brandy or rum.

1845 E. Acton *Mod. Cookery* xxvii. 547 Burnt coffee. (In France vulgarly called *Gloria*.) Make some coffee.. sweeten it..pour brandy on the top..set fire to it..and drink the gloria quite hot. 1934 S. Beckett *More Pricks than Kicks* 119 Reader, a gloria is a coffee laced with brandy.

4. A closely-woven fabric of silk and wool or cotton, etc., used for dresses, the covering of umbrellas, etc.; also, a dress of this fabric.

1895 *Montgomery Ward Catal.* 4/2 Black Silk Warp Sublime or Gloria Cloth..much used for dresses, blouse waists and underskirts. 1897 *Sears, Roebuck Catal.* 248/1 Men's Fine Black Gloria Silk Umbrella. 1908 *Daily Chron.* 3 Aug. 4/5 Men abroad carry a grey or tan gloria sunshade. 1908 L. M. Montgomery *Anne of Green Gables* xxv. 280 Anne took the dress and looked at it in reverent silence. Oh, how pretty it was—a lovely brown gloria with all the gloss of silk. 1917 — *Anne's House of Dreams* (1926) i. 8, I remember the first pretty dress I ever had—the brown gloria Matthew gave me. 1921 *Daily Colonist* (Victoria, B.C.) 18 Oct. 19/6 (Advt.), Umbrellas at All Prices... Strong durable covers in mercerized cotton, gloria and laventine. 1970 T. S. Crawford *Hist. Umbrella* vii. 142 [In the 1860s] the

silk mixture known as Gloria was also introduced as a covering.

glorious, *a.* Add: **3. a.** *spec.* As an epithet of: (*a*) the 'First of June', the date of a sea battle between the British and French in 1794, ending in victory for the British; (*b*) the 'Twelfth' of August [TWELFTH 1 c]; (*c*) (*U.S.*) the 'Fourth' (of July).

1794 (*title*) Songs, Duets, Choruses, &c. In a New and Appropriate Entertainment, Called The Glorious First of June. 1827 [see FOURTH *a.* 2 b]. 1854 W. G. Simms *Southward Ho!* xvi. 253 Our dinner on the glorious Fourth was worthy of the occasion. 1895 E. G. Mackenzie *In Greenland* xvii. 125 The glorious twelfth! Despite critical politics, the continuous fine weather and the certainty of sport has again filled all the saloon carriages bound northward. 1899 W. L. Clowes *Royal Navy* IV. xxxv. 230 (*heading*) The Glorious First of June. 1948 *Miami* (Okla.) *Daily News-Record* 4 July 1/1 Where are Miami's glorious fourths of yesteryear? 1966 *Chambers's Encycl.* V. 660/1 The Glorious First of June..the name given to a sea battle of the French revolutionary war. 1971 D. Avery *Guardian* xxx. 468 With the 'glorious twelfth' only three days off politicians were deserting Westminster for the grouse moors.

glory, *sb.* Add: **4. c.** Also in form *glory be!*

1893 [in Dict.]. 1954 Koestler *Invis. Writing* xxxvi. 392 For, glory be, man is a stubborn creature. 1968 B. Turner *Sex Trap* xii. 120 'Your hours are numbered.' Grange's eyes sparkled...'Glory be!'

9. c. *spec.* An anthelion or fog-bow.

1823 [in Dict.]. 1842 W. Howitt *Rur. & Dom. Life Germany* 432 If the fog is dry, you see only the shadow; but if moist, round your neighbour; if very damp, only yourself, surrounded by a rainbow-coloured glory. 1884 P. G. Tait *Light* § 167 It seems possible that glories may be due to a cause somewhat analogous to that which produces the spurious rainbows. 1910 C. Turner *Aerial Navig.* 141 Halos round the sun and moon, rainbows, 'glories', or 'aureoles'—the coloured rings seen round the shadow cast by the balloon on the clouds. 1956 C. F. Brooks in *Why the Weather?* x. 122 Sometimes there is a double glory, that of larger diameter containing the full range of colors of the rainbow, from violet to red. 1963 G. G. Bowden tr. *Schöpfer's Young Specialist looks at Weather* vi. 95 The airman is also familiar with the 'glory', as he often sees the system of coloured rings surrounding the shadow of his aircraft cast by the sun on a layer of cloud below.

11. glory-box *Austral.* and *N.Z.*, a box in which a woman stores clothes, etc., in preparation for marriage. Cf. GLORY-HOLE 1 b.

1949 D. M. Davin *Roads from Home* ii. iv. 129 Seeing the neatly folded linen there, the carefully worked doilies and table spreads, the unworn silk nightdress, she knew she had blundered on Moira's hoard, her glory-box. 1963 *Wanganui Herald* (N.Z.) 9 July 16/2 Sideboard (as new), 2 wardrobes, 4 fireside chairs, tin trunk, light oak glory-box. 1971 *Sunday Australian* 8 Aug. 26/8 The traditional glory-box was based on the 'kit'—a list of what was required, and would be used on all the linen items in a girl's glory box.

glory-hole. Add: **1. b.** (Earlier example.)

1825 M. Wilmot *Let.* 20 Nov. (1935) 227 A sort of play room or glory hole for poor Wilmot to keep his rubbish.

c. *Naut. colloq.* Any of various compartments of a ship, as (*a*) the lazaretto; (*b*) one or more rooms between or below decks used as sleeping-quarters for stewards. Also *fig.*

1839 W. H. Leigh *Reconnoitering Voy.* ii. 16 The other one was for the ship's speculum, and had nothing in his 'glory hole' worth notice. 1897 E. G. Constantine *Marine Engin.* ix. 108 In the 'glory-hole', which had to be passed through to get into the inner trunk gland near the crank pit. 1898 *Pearson's Mag.* V. 213/2 They went through every part of her, from the sodden shaft-tunnel, to the glory-hole where the stewards live. 1927 G. Bradford *Glossary Sea Terms* 74/1 *Glory hole*, a name sometimes given to the firemen's quarters of a steamer; sometimes applied to the lazarette. 1942 R. Bedford in *Austral. Short Stories* (1951) 95 'I'm..waitin' till I can sneak out on the boat deck to sleep—the glory hole being so hot.'

3. A large cavernous opening or pit into a mine; an open quarry. Hence as *v. intr.*, to carry on surface mining. *N. Amer.*

1902 *Geol. Survey Canada Summary Report 1902* 64A, The larger properties..are now mining by large quarries or 'Glory holes', from which the material will be loaded directly into ore cars. 1916 *Daily Colonist* (Victoria, B.C.) 21 July 3/3 The Mother Lode mine has been working for some sixteen years, partly as a glory hole and partly by shaft and drift. 1926 *Ibid.* [illegible] the glory hole of the property..and there is considerable of the high grade ore. [illegible] W. N. Burns *Tombstone* 381 In this tunnel beneath the town the Grand Central mine 'gloryholed', taking out $840,000. 1943 *Copper Camp* 29 [illegible] turned out over a million dollars' worth of high grade copper. 1966 *McGraw-Hill Encycl. Sci. & Technol.* VIII. 481/1 Glory-holing is an underhand stoping method commonly used in the dry district to develop funnel-shaped excavations in large ore bodies.

4. *Army slang.* (See quots.)

1925 Fraser & Gibbons *Soldier & Sailor Words* 105 *Glory hole*, a colloquial expression for any small billet or dug-out. 1950 Partridge *Here, There & Everywhere* 67 The small dug-outs are *cubby-holes*..and the smallish are *glory-holes*.

gloss, *sb.*² Add: **3. b. gloss-meter**, a photometric device for measuring the gloss of surfaces; **gloss paint**, paint that contains varnish and gives a glossy finish; also **gloss enamel**.

1908 *Sears, Roebuck Catal.* 74/1 Seroco gloss enamel—a high gloss interior paint. 1926–7 *Army & Navy Stores Catal.* 297 Jackson's gloss black enamel. 1930 *Jrnl. Optical Soc. Amer.* XX. 24 The entire range of objective gloss..may be covered. The essentials of the gloss-meter are shown in Fig. 1. 1961 J. H. Goodier *Dict. Painting & Decorating* 118 The gloss meter works on the principle of directing a beam of light on to a painted panel and allowing the reflected beam to excite a galvanometer. 1964 *New Scientist* 10 May 299/1 Though the instrument has been developed specifically for paper, the principle of the glossmeter can be adapted for measurements of the gloss of ball bearings and other industrial items with curved surfaces whose gloss may be a criterion of their excellence. 1926–7 *Army & Navy Stores Catal.* 299/1 Hard gloss finishing paint. 1933 *Drugs, Oils & Paints* Apr. 135/1 Portions of both the one and two-coat finishes are then given a final coat of gloss paint. 1960 *House & Garden* June 72/2 Gloss paints seem to get better and better in quality.

glossematic (glǫsimæ·tik), *sb. pl.* and *a.* *Linguistics.* [f. *GLOSSEM(E + -atics, after Gr. words like φώνημα, φωνηματικ- and θέμα, θεματικ- and the names of sciences like *mathematics* (see -IC 2); perh. influenced by med.L. *glōssēmaticus* adj., though its sense is not related.] **A.** *sb. pl.* const. as *sing.* A theory of structural linguistics introduced by the Danish scholar Louis Hjelmslev (born 1899) in 1936, and concerned esp. with developing an abstract theory of the distribution of minimal forms (glossemes) and their mutual relationships.

1936 Hjelmslev & Uldall in *Humanistisk Samfund Skrifter* (Aarhus) I. (*title*) Synopsis of an outline of Glossematics. 1939 L. Hjelmslev in *Proc. Third Internat. Congr. Phonetic Sci.* 272 This whole deductive theory of phonematics and cenematics, established by Mr. Uldall and myself under the common name of *glossematics*, bases the definitions of forms on their function among themselves. 1956 J. Whatmough *Language* vii. 115 According to the theory of the system (the pattern of mutual relationships of linguistic elements); of the *norm* (i.e. a set of rules based on the system and describing the limits of variation for each element); and *usage* (a set of rules based on the norm and describing the limits of variation tolerated in a given speech-community at a given time)... The comprehensive theory is designated *glossematics*. 1964 *Language* XL. 231 'Form' and 'substance' is a basic dichotomy which glossematics makes in both expression and content. 1965 M. Heppell tr. *Ivič's Trends in Linguistics* xv. 178 Glossematics is concerned with the systematic comparison of the structures of existing languages with the basic structures of all semiotic systems.

B. *adj.* Of, relating to, or characteristic of glossematics.

1952 A. Cohen *Phonemes of English* 8 The glossematic approach..aims at investigating various codes of communication, not merely speech, from the structural point of view. 1962 *Amer. Speech* XXXVII. 63 Application of glossematic phonological analysis to the Asele dialect of Swedish. 1965 M. Heppell tr. *Ivič's Trends in Linguistics* xv. 183 The beginner in glossematic studies is held up at every step by terminological difficulties.

Hence **glossemati·cian**, **glosse·matist**, an expert in, or student of, glossematics.

1937 *Trans. Philol. Soc.* 137 Phoneticians, phonologists, and the very recent glossematicians were prone to emphasize their differences. 1950 *Archivum Linguisticum* II. ii. 180 Here Lotz is in agreement with the glossematicians. 1953 C. E. Bazell *Linguistic Form* 9 Glossematists would distinguish two morphemes. 1963 J. Lyons *Structural Semantics* iii. 48 The Glossematicians' dependence on *translation* is especially clear in Uldall.

glosseme (glǫ·sīm). *Linguistics.* [ad. Gr. γλώσσημα, f. γλῶσσα GLOSS *sb.*¹] Any feature in a language (whether of form, stress, order, or the like) that carries meaning and cannot be analysed into smaller meaningful units.

1926 Bloomfield in *Language* II. 161 Whatever has meaning is a glosseme. 1933 — *Language* ix. 264 Just as a *phoneme* is the smallest unit of structural analysis, being a bundle of *distinctive features*, so a term is required for the smallest unit of *significance*; for this the term *glosseme*..has been introduced.

glosso-. Add: **glo:sso-kinæsthetic** *a.*, relating to control of the movement of the tongue and speech organs; **glo:sso-labio-laryngeal** *a.*, relating to the tongue, lips, and larynx; **glo:sso-pa·latine** *a.* = PALATO-GLOSSAL *a.*; **glo:sso·pharyngeal** *sb.*, the glosso-pharyngeal nerve.

1939 L. H. Gray *Found. Lang.* 89 The glossokinæsthetic (see *CHEIROKINÆSTHESIA). 1908 *Practitioner* June 758 All the usual phenomena of a glosso-labio-laryngeal paralysis. 1888 *Encycl. Brit.* XXIII. 79/2 The glosso-palatine arch. 1875 Huxley & Martin *Elem. Biol.* 182 The trunk of the glossopharyngeal.

glossophagine (glǫsǫ·fădʒǝin), *a.* [f. mod.L. *Glossophaga* (f. Gr. γλῶσσα tongue + φαγεῖν to eat) + -INE¹.] Belonging to or characteristic of the Glossophaginæ, a subfamily of South American bats which have long tongues used for obtaining food. Also as *sb.*

1884 T. Gill in J. S. Kingsley *Stand. Nat. Hist.* V. 173 Glossophagines have the muzzle long and narrow..and the tongue is much more elongated. 1899 *Cent. Dict.*, Glossophagine, adj. 1913 *Ann. & Mag. Nat. Hist.* 8th Ser. XII. 271 Other Glossophagine bats have light bases to the fur. 1964 E. P. Walker et al. *Mammals of World* I. 283/1 *Glossophaga* is one of the long-snouted, extensible-tongued glossophagine group of bats.

glossopteris (glǫsǫ·ptĕris). [a. F. *Glossopteris* (A. Brongniart 1822, in *Mém. Mus. d'Hist. Nat. Paris* VIII. 232), f. Gr. γλωσσο-, γλῶσσα tongue + πτερίς fern.] A fossil fern belonging to the genus so named. Also *attrib.*

1883 *Encycl. Brit.* XX. 173/1 The northern coal-fields [in Queensland] display the *Glossopteris*, [etc.]. 1897 A. C. Seward in *Science Progress* VI. 178 The Glossopteris Flora; an extinct flora of a Southern Hemisphere Continent. 1905 E. A. N. Arber *Catal. Fossil Plants Glossopteris Flora Brit. Mus.—N.H.* p. xviii, This flora is especially characterised by the frequent occurrence of *Glossopteris*, a fern-like plant not only of wide distribution but often found in..extreme abundance. 1912 J. W. Gregory *Making of Earth* III. x. 184 Glossopteris was a fern or fern-like plant with large blunt leaves, each of which has a prominent midrib. 1956 *Nature* 11 Feb. 337/2 The sorus-like body borne by a *Glossopteris*.

glossy, *a.* Add: **a.** *spec.* Designating photographic or printing paper that is smooth and shiny; hence denoting a magazine, etc., printed on such paper, or something that is characteristic of the type of material which is published in such magazines.

1897 C. M. Hepworth *Animated Photogr.* p. xv (Advt.), Citos glossy P.O.P. Citos matte P.O.P. 1930 *Post Office Electr. Engin. Jrnl.* July 101/1 The stock of printing papers, which all are 'Glossy'. 1934 R. Macaulay *Going Abroad* xii. 88 One of those smart, glossy, illustrated magazines. 1936 *Archit. Rev.* LXXX. 135/1 Salzburg, full of Americans in green corduroy shorts reading their glossy magazines. 1955 M. Gilbert *Sky High* viii. 117 She had made a pile of the glossy weeklies, and was looking at the latest copy of *Country Life*. 1958 *Times Lit. Suppl.* 7 Nov. 636/4 Like *Harper's* it is 'semi-glossy' and determinedly middle-brow, half way between the stringently intellectual and specialized *Kenyon*, *Sewanee*, *Hudson* and *Partisan Reviews* and the glossy and sophisticated *Ladies' Home Journal*, *Harper's Bazaar* and *Vogue*. 1967 *Observer* 1 Oct. 9/1 The ignorant and the downright stupid, whose only idea of what to expect is derived from glossy travel brochures.

glossy (glǫ·si), *sb. colloq.* (orig. *U.S.*). [f. the adj.] **a.** A photograph with a glossy surface. Also *fig.*

1931 J. E. Bakeless *Mag. Making* 303 Photographs for half-tone reproduction should be 'glossies', that is, prints with a high gloss finish. 1946 T. Godsey *Free Lance Photogr.* iii. 45 Most of the photos produced by a free lance are glossies. 1967 *Boston* (Mass.) *Herald* 8 May 25/1 Poor Robert, I thought, it is still true what another critic once wrote of him: he is 'an eight by 10 glossy'.

b. A 'glossy' magazine or the like (see *GLOSSY *a.*); also, a copy of such a magazine; material printed in such magazines.

1945 G. Endore *Methinks the Lady* (1947) ii. 25 She wrote most everything: pulp, glossy, copy, one-act plays. 1952 A. Wilson *Hemlock & After* I. iii. 55 The slightly too smart appearance, which the world of women's glossies had imposed upon her. 1955 J. Cannan *Long Shadows* iv. 65 Running your scissors through the poor girl's picture and then leaving the damned glossy about for anyone to see. 1960 *Punch* 17 Feb. 267/2 Give me the glossiest glossy and I mightn't glance at it. 1970 *Daily Tel.* 31 Mar. 13/2 The smart new decoration colours for the Seventies as decreed by the American glossies.

c. *Cinemat.* A film depicting fashionable life.

1960 *Guardian* 19 Nov. 4/7 One of those large-scale Hollywood 'glossies'. 1961 *Ibid.* 18 Mar. 5/3 Both [films] are large-scale glossies.

glottal, *a.* Add: *glottal catch* or *stop*, a sound produced by the sudden opening or shutting of the glottis with an emission of breath or voice.

1877 [in Dict.]. 1888 H. Sweet *Hist. Eng. Sounds* 1 The Glottal stop is produced by a sudden shutting or opening of the glottis, as in a cough. 1911 W. H. Van der Smissen *Harrap's Mod. German Gram.* p. xiii, The utterance of every German initial vowel, unless wholly unstressed, begins with the 'glottal stop'. 1964 R. A. Hall *Introd. Ling.* I. viii. 42 A complete stoppage of the breath-stream by the vocal cords is called a *glottal stop* or *glottal catch* (such as we make between the two *oh's* of 'Oh-oh!' when said in surprise or reproof).

glottalic (glǫtæ·lik), *a. Phonetics.* [f. GLOTTAL *a.* + -IC.] Relating to the glottis and its total or partial closure.

1942 Bloch & Trager *Outl. Ling. Analysis* 31 [Stops] with inner closure at the glottis, glottalic. 1959 *Archivum Linguisticum* XI. 1. 6 The source and direction of the air-stream: pulmonic, glottalic, velaric. 1964 R. H. Robins *Gen. Ling.* iii. 103 Sharpness when the glottalic pressure is released.

glottalize (glǫ·tălǝiz), *v. Phonetics.* [f. GLOTTAL *a.* + -IZE.] *trans.* To articulate with total or partial closure of the glottis. So **glo:ttaliza·tion**; **glo·ttalized** *ppl. a.*; **glo·ttalizing** *ppl. a.*

1916 E. Sapir in *Geol. Survey Canada Mem.* 90 83 One of the most striking American examples of phonetic accord..is the occurrence in a considerable number of West Coast linguistic groups..of..glottalized ('fortis') stops. 1921 — *Lang.* ix. 199 A 'glottalized' series of stopped consonants. 1933 Bloomfield *Lang.* vi. 101 Some languages have glottalized spirants (preceded, accompanied, or followed by a glottal stop). *Ibid.* 102 Some American languages have a whole series of laterals, with differences of position, glottalization, or nasalization. 1949 E. A. Nida *Morphology* (ed. 2) iii. 63 The final vowel ..in turn glottalizes the preceding vowel. 1953 W. J. Entwistle *Aspects of Lang.* iv. 136 The original *p t k* series is amplified in some languages, such as Quechua, by aspiration and glottalization. 1965 *Canad. Jrnl. Linguistics* Spring 179 A glottalizing form. *Ibid.* Fall 35 The laryngeal pair of aspiration or breathiness and glottalization.

glottochronology (glǫ:tokrǫnǫ·lŏdʒi). *Linguistics.* [f. GLOTTO- (see GLOSSO-) + CHRONOLOGY.] The application of statistics to vocabulary to determine the degree of relationship between two or more languages and the chronology of their splitting off from a common ancestor. Hence **glo:ttochronolo·gi-c(al** *adjs.*

1953 J. B. Carroll *Study of Lang.* ii. 62 Glottochronology, a technique for inferring, from statistical comparisons of language systems, estimates not only of their probable familial relationships but also of the dates when the branches of a language family broke off from one another and from the parent language. 1955 *Internat. Jrnl. Amer. Ling.* XXI. 116 There has been considerable interest in the application of statistical techniques to the study of language vocabulary, variously called glottochronology and lexico-statistical dating. *Ibid.* 118 An entirely different glottochronologic study. 1957 J. H. Greenberg *Ess. in Linguistics* iv. 54 Recently glottochronological methods have been used. 1961 *New Scientist* 13 July 77/3 The application of lexicostatistics (glottochronology) to African languages. 1964 R. A. Hall *Introd. Ling.* lxvi. 381 The glottochronological method is based on a count, not of structural, but of lexical items.

glove, *sb.* Add: **1. f.** *to handle without gloves* (earlier U.S. example); so *to handle with gloves off*, etc.; *to take the gloves off*: to 'set to' in earnest; to use no mercy. (Cf. sense 2 in Dict.)

1827 A. Sherwood *Gaz. Georgia* 94 Marion County has been handled without gloves. 1828 *Richmond Enquirer* 20 May 3/4 (Th.), The Baltimore Republican handles Mr. C[lay] with gloves off. 1838 J. C. Neal *Charcoal Sks.* 217 I'll give you a touch of natur' without no gloves on. 1922 *S.P.E. Tract* xi. 15 Dead metaphors lately noticed:..The flower of our manhood. Taking off the gloves. 1928 *Daily Express* 21 Mar. 1/1 It is time, Mr. Mayor, that we took the gloves off and showed the County of London Electric Supply Company that we will fight. 1931 *Times Lit. Suppl.* 10 Sept. 669/4 Like Lewis Carroll, who was too polite to 'take the gloves off' in his assault on 'Hiawatha'. 1934 J. E. Neale *Queen Eliz.* xii. 199 At still greater length, in caustic tones, and with many shrewd hits, Mary answered. The gloves were off.

5. a. *glove-case*, *-shop*.

1874 'S. Coolidge' *What Katy did at School* xi. 206 Glove-cases, of quilted silk, delicately scented. 1926–7 *Army & Navy Stores Catal.* 814/1 Glove case. Pigskin. 1813 Jane Austen *Let.* 20 May (1932) II. 78, I went into it rather because it was near than because it looked at all like a glove shop. 1859 G. A. Sala *Tw. round Clock* 157 Jewellers, French glove shops, perfumery, and point lace shops.

6. glove box, (*a*) a box for holding gloves; (*b*) = *glove compartment*; (*c*) a closed chamber into which a pair of gloves project from openings in the side, enabling radioactive or other material to be handled while isolated from the operator; **glove compartment**, a recess in the dashboard of a motor car for small articles such as gloves, etc.; **glove-fit**, something that fits like a glove; **glove-fitting** *a.*, that 'fits like a glove' (cf. sense 1 f in Dict.); **glove puppet**, a puppet consisting of a dress, a head and hands, made to fit on the hand like a glove; **glove stitch** (see quot. 1964).

1852 *Harper's Mag.* July 185/2 We..beheld our glove-box enriched with half-a-dozen pair of snowy French sevens! 1858 Glove-box [see sense 5 a in Dict.]. 1897 *Sears, Roebuck Catal.* 351/1 Silk Plush Glove Box, lined with fine puffed satin, silvered catch and hinges. 1946 'P. Quentin' *Puzzle for Fiends* xxv. 239, I sat for a minute in the front seat...I looked in the glove box. 1950 Glasstone *Sourcebk. Atomic Energy* xviii. 521/2 Glove boxes are also often used for experimental work with emitters of soft, i.e., low energy and short range, beta particles. 1959 *New Scientist* 23 Apr. 919/1 The glove box..is in essence a sealed plastic-windowed container kept at a pressure slightly below atmospheric. Manipulations are carried out through plastic gloves. 1971 *Daily Tel.* (Colour Suppl.) 4 June 39/1 This little hole..which they actually call a glove-box. 1939 R. Chandler *Big Sleep* xvii. 138, I went to the car and got a pair of handcuffs out of the glove compartment. 1959 I. Fleming *Goldfinger* xii. 173 He took a small pair of binoculars out of the glove compartment. 1910 C. E. Montague *Hind let Loose* viii. 145 The putting of glove-fits on Brumby and Pinn was too near the craftsman's heart to be quite given up. 1967 *Listener* 8 June 760/1 The rather odd and amateur scansion of the lyrics is a glove-fit to the melody. 1868 *Englishwoman's Dom. Mag.* Jan. 17 (*heading*) Thomson's Glove-fitting Corset. *Ibid.* 18/1 The name Glove-Fitting is certainly well chosen. 1920 D. H. Lawrence *Lost Girl* i. 17 The slim, glove-fitting Princess Robe. 1937 W. S. Lanchester *Hand Puppets & String Puppets* 10 Glove Puppets. This is the simplest form of puppetry. *Ibid.* 15 Some glove puppet workers use spring clothes pegs for hands. 1969 E. H. Pinto *Treen* 213 Hand or glove puppets are the simplest type and, for the travelling showman, have the great advantage of taking up little space. 1964 *McCall's Sewing* ii. 29/1 *Glove stitch*, decorative top-stitching made by taking the same size stitch on both sides of the work.

Glover² (glʌ·vəɹ). [The name of John *Glover* (1817–1902), English plumber and chemical engineer.] *Glover* (†*Glover's*) *tower*: in the chamber process for the manufacture of sulphuric acid, the tower in which 'nitrous vitriol' from the Gay-Lussac tower is denitrated and sulphuric acid from the chambers is concentrated by being passed down through packing against an upward flow of hot sulphur dioxide and air, which undergo cooling on the way to the chambers; also *ellipt.*; *Glover acid*: the sulphuric acid formed at the bottom of the Glover tower, of a higher concentration than the chamber acid.

1871 *Jrnl. Chem. Soc.* XXIV. 1100 A detailed description, with drawings, of 'Glover's towers', as used on the Tyne. 1873 *Chem. News* 21 Mar. 135/2 There are two methods at present in use on the Tyne for the denitration of the nitro-sulphuric acid; the Glover towers, and denitration by steam. 1873 J. Glover in *Ibid.* 28 Mar. 152/2 With a properly proportioned and worked Glover's tower none of the drawbacks..mentioned by him exists. 1896 Blount & Bloxam *Chem. for Engineers & Manuf.* II. i. 13 The saturated acid ('nitrous vitriol') running out at the foot of the Gay-Lussac is pumped..to the top of the Glover. *Ibid.* 15 For alkali making by the Leblanc process, Glover acid is suitable. 1902 *Times* 6 May 10/3 Though the first one was not built till 1859, and the system not adopted in Lancashire till nearly ten years later, the Glover towers are now almost universal in sulphuric acid manufactories. 1936 A. M. Fairlie *Sulfuric Acid Manuf.* viii. 159 Most lead Glover towers of modern design have brick walls laid up in 'acid-proof' cement. 1962 J. A. Kent *Riegel's Industr. Chem.* (ed. 6) iv. 71 It is the modern practice to pass all the chamber acid through the Glover and thus concentrate it to 60° Bé without expense for heat. The Glover acid is cooled in double-walled lead receiving tanks.

glow, *sb.* Add: **4. glow-discharge** (earlier and later examples); also, the luminous electrical discharge in a gas-filled tube; **glow-light**, a glowing light; *spec.* a glow-lamp; **glow plug**, an electrically heated plug used to ignite the gas in a gas turbine or rocket engine.

1844 H. M. Noad *Lect. Electr.* (ed. 2) ii. 45 *Glow discharge*, when a fine point is used to produce disruptive discharge from a positively charged conductor, the brush gives place to a quiet phosphorescent continuous glow. 1937 *Discovery* Feb. 54/1 The glow discharge from the first electrode. 1963 B. Fozard *Instrumentation Nucl. Reactors* ii. 25 Any further increase in applied voltage is likely to initiate a continuous glow discharge which may cause irreparable damage to the tube. 1891 *Cassell's Family Mag.* Oct. 700/2 Ten shillings is paid for an equivalent glow-light in the same time. 1913 *Pop. Mag.* 1 May 85/2 The *Potomac's* captain ordered even the tiny glow light illuminating the compass..to be covered. 1934 T. S. Eliot *Rock* ii. 84 Glow-worm glow-light on a grassblade. 1947 *Jrnl. Brit. Interplanetary Soc.* VI. 106 A spark or glow plug projecting directly into the chamber can be used. 1961 *Guardian* 3 Apr. 4/1 When you press the switch, an electric glow-plug heats up and ignites the oil vapour in the burner.

glow, *v.*¹ Add: **2. c.** To pass *to* or *into* a glowing colour.

1888 Mrs. H. Ward *R. Elsmere* xi, The stretches of purple heather, glowing into scarlet under the touch of the sun. 1906 E. Phillpotts *Portreeve* II. i, The breath of ocean made visible..glowed into red gold as the sun descended.

5. Also with predic. adj.

1855 Kingsley *Westw. Ho!* i, Turning first deadly pale, and then glowing red. 1907 *Smart Set* Jan. 126/2 With cheeks glowing red.

glucagon (glⁱū·kăgǫn). *Physiol.* [f. *GLUC(O- + Gr. ἀγών AGON or ἀγών, pres. pple. of ἄγειν to lead, bring.] A crystalline polypeptide hormone formed in the pancreas which stimulates glycogenolysis.

1923 Kimball & Murlin in *Jrnl. Biol. Chem.* LVIII. 342 This hyperglycemic substance has been given the name glucagon and its properties will be described more fully in a later paper. 1961 *Times* 22 Dec. 5/7 Insulin reduces the blood-sugar level and glucagon raises it. 1967 *Martindale's Extra Pharmacopœia* (ed. 25) 673/1 Glucagon has also been employed in terminating insulin coma. 1968 *New Scientist* 29 Feb. 481/1 Glucagon, a hormone produced in and excreted by the pancreas together with insulin, was recently successfully synthesized.

glucase (glⁱū·kēⁱz, -s). *Biochem.* [ad. G. *glukase* (R. Géduld 1892, in *Jahresber. Fortschr. Lehre Gährungs-Organismen 1891* II. 250), f. *GLUCO- + *-ASE.] = MALTASE.

1894 *Jrnl. Chem. Soc.* LXVI. I. 258 Glucase which occurs in maize slowly hydrolyses dextrin. 1910 [see *AMYLASE]. 1950 A. Gottschalk in Sumner & Myrbäck *Enzymes* I. I. xv. 553 He [*sc.* Géduld] found the enzyme in the grains of many cereals and called it 'glucase'.

gluco- (glⁱū·ko), before a vowel also **gluc-**, used as comb. form of Gr. γλυκ-ύς sweet (see GLYCO- and note s.v. GLUCIC *a.*), and of GLUCOSE in the designation of (*a*) substances containing, related to, obtained from, or producing glucose, or affecting its metabolism; (*b*) processes affecting the metabolism of glucose. Also, in some terms not now widely used, = *GLYCO-.

glu:co-asco·rbic *a.*, in *gluco-ascorbic acid*, a compound, $C_7H_{10}O_7$, related to ascorbic acid; **glucoco·rticoid** *Biochem.*, any of the steroid hormones produced in the adrenal cortex which are esp. concerned in carbohydrate metabolism; **glucohe·ptose** [a. G. *glucoheptose* (E. Fischer 1890, in *Ber. d. Deut. Chem. Ges.* XXIII. 934)], any heptose derived from glucose; **gluco·lysis** *Biochem.*, the metabolic breakdown of glucose to pyruvic acid or lactic acid; glycolysis in which glucose is the starting point; **glu:coneoge·nesis** *Biochem.*, the metabolic formation of glucose from non-carbohydrate precursors, esp. from proteins; hence **glu:coneogene·tic**, **-ge·nic** *adjs.*; †**glucopro·teid**, = *glycoprotein*; **glucopro·tein** [ad. F. *glucoprotéine* (P. Schutzenberger 1879, in *Ann. de Chim. et de Phys.* 5th Ser. XVI. 365)], †(*a*) any of various amino acids or mixtures of amino acids to which were ascribed the empirical formula $C_nH_{2n}N_2O_4$, where *n* is between 6 and 12 (*obs.*); (*b*) = *glycoprotein*; **glu:cosaccha·ric** *a.*, in *glucosaccharic acid*, the saccharic acid obtained by the oxidation of glucose; **gluco·sazone** [see *OSAZONE], the osazone derived from glucose, as well as from fructose and mannose; **glu·cosone** [ad. G. *glucoson* (E. Fischer 1889, in *Ber. d. Deut. Chem. Ges.* XXII. 87): see *OSONE], the osone derived from glucose, as well as from fructose and mannose; **glucova·ni·llin** [a. G. *glucovanillin* (Haarmann & Reimer 1883, in *Deutsches Reich Pat.* 279,992)], a glucoside, $C_{14}H_{18}O_8$, of vanillin which has been isolated from several plants.

1934 D. K. Baird et al. in *Jrnl. Chem. Soc.* 62 We have also described improved methods for the preparation of gluco-ascorbic acid (3-keto-*d*-glucoheptonofuranolactone). 1967 Cantarow & Schepartz *Biochem.* (ed. 4) xxv. 667 Although scurvy-like manifestations are produced in mice and rats..by glucoascorbic acid, this condition is not cured by administration of ascorbic acid. 1950 E. H. Venning in E. S. Gordon *Symposium Steroid Hormones* 98 Certain compounds..will cause an increase in liver glycogen. These substances will be referred to as glucocorticoids, as they appear to be intimately associated with the regulation of carbohydrate metabolism. 1970 R. W. McGilvery *Biochem.* xxiii. 565 The most active of the glucocorticoids in humans is cortisol, or hydrocortisone. 1890 *Jrnl. Chem. Soc.* LVIII. I. 599 Glucoheptose..crystallises from water in beautiful tables which melt at about 190° with decomposition. 1963 K. Mayer tr. *Staněk's Monosaccharides* ii. 47 Fischer defined the heptose obtained from D-glucose by the cyanohydrin synthesis as D-glucoheptose. 1932 Glucolysis [see *FRUCTOLYSIS]. 1943 *Thorpe's Dict. Appl. Chem.* (ed. 4) VI. 74/1 'Glucolysis' has been the name given to the phenomenon of glucose breakdown to lactic acid. 1912 *Jrnl. Biol. Chem.* XII. 511 (*heading*) The chemistry of gluconeogenesis. I. The quantitative conversion of propionic acid into glucose. 1968 Passmore & Robson *Compan. Med. Studies* I. xxxi. 4/2 If the dietary carbohydrate intake is insufficient, gluconeogenesis helps to replenish the stores of liver glycogen. 1961 Webster, Gluconeogenetic. 1969 Pontremoli & Grazi in Florkin & Stotz *Comprehensive Biochem.* XVII. iv. 184 At one time, the reductive carboxylation of pyruvic acid..had been assigned a key role in the gluconeogenetic process. 1954 A. White et al. *Princ. Biochem.* xvii. 441 Glucogenic materials include many of the amino acids. 1894 *Jrnl. Chem. Soc.* LXVI. I. 310 A knowledge of glucoproteids (proteids from which sugar is obtainable by hydrolytic agents) is therefore of practical importance. 1900 *Jrnl. Chem. Soc.* LXXVIII. I. 478 Galactosamine.. was obtained as a decomposition product of the glucoproteid of the albuminous gland of the frog. 1879 *Jrnl. Chem. Soc.* XXXVI. 545 For these latter bodies, the name glucoproteins has been adopted, on account of their sweet taste. 1906 *Jrnl. Chem. Soc.* XC. I. 776 His [*sc.* Lepierre's] so-called 'α-glucoproteins' are mixtures of monoamino-acids. 1911 J. A. Mandel tr. *Hammarsten's Text-bk. Physiol. Chem.* (ed. 6) iii. 163 (*heading*) Glycoproteins (glucoproteins). 1963 E. G. Young in Florkin & Stotz *Comprehensive Biochem.* VII. i. 10 The class was initially known as the gluco- or glycoproteins. 1941 *Brit. Chem. Abstr.* A. II. 350 (*heading*) Grignard synthesis of glucosaccharic acid from *l*-arabinose. 1964 N. G. Clark *Mod. Org. Chem.* xvi. 316 Somewhat more vigorous oxidation employing aqueous nitric acid brings about the additional oxidation of the primary alcohol group, resulting in a tetrahydroxy-dicarboxylic acid, or saccharic acid; glucose gives glucosaccharic acid. 1886 *Jrnl. Chem. Soc.* I. 934 Isoglucosamine..is obtained by treating.. phenylglucosazone..with..zinc-dust and..acetic acid. It is kept at a temperature of 50°, and vigorously shaken

until the glucosazone is entirely dissolved. 1964 N. G. Clark *Mod. Org. Chem.* xvi. 319 Additional reactions give rise to glucosazone. 1889 *Jrnl. Chem. Soc.* LVI. 484 The author [*sc.* E. Fischer] proposes 'osone' as a general term for compounds of the type of oxyglucose, and throughout the paper refers to this compound as glucosone. 1956 *Science* 27 July 171/3 Crude extracts of *Iridophycus* were also capable of forming glucosone and galactosone from the corresponding sugars. 1884 *Jrnl. Chem. Soc.* XLVI. 1343 When coniferin is oxidised with aqueous chromic acid, it is converted into glucovanillin. 1966 *McGraw-Hill Encycl. Sci. & Technol.* VI. 224/2 Vanillin is the aromatic principle of commercial vanilla extract, and is released from the glucoside (glucovanillin) by enzymatic hydrolysis during curing of vanilla beans.

gluconate (glⁱū·kǫnēⁱt). *Chem.* [f. *GLUCONIC (ACID + -ATE⁴.] A salt or ester of gluconic acid; *calcium gluconate*, an odourless, tasteless, water-soluble, crystalline or granular substance, used therapeutically as a source of calcium for the body.

1884 *Jrnl. Chem. Soc.* XLVI. 424 Calcium gluconate, $(C_6H_{11}O_7)_2Ca + H_2O$, crystallises in microscopic needles. 1915 *Jrnl. Amer. Chem. Soc.* XXXVII. 355 The new, normal β-lactone has been obtained both from free *d*-gluconic acid and from ethyl *d*-gluconate. 1938 *Thorpe's Dict. Appl. Chem.* (ed. 4) II. 297/1 Calcium gluconate, being soluble, practically tasteless, and non-irritating to tissues, is being increasingly used in medicine. 1963 Florkin & Stotz *Comprehensive Biochem.* V. ii. 73 Ferrous gluconate..is used in the treatment of iron-deficiency anaemia, and calcium gluconate can be given in cases of calcium deficiency.

gluconic acid (glⁱukǫ·nik). *Chem.* [tr. G. *gluconsäure* gluconic acid (Hlasiwetz & Habermann 1870, in *Ann. d. Chem. und Pharm.* CLV. 125), f. *GLUC(O- + *-ONIC.] The acid derived from glucose by oxidation of the aldehyde group to a carboxyl group.

1871 *Jrnl. Chem. Soc.* XXIV. 547 Hlasiwetz..considers gluconic acid to be bibasic. 1929 *Industr. & Engin. Chem.* XXI. 1198 (*heading*) Semi-plant production of gluconic acid by mold fermentation. 1963 E. Percival in Florkin & Stotz *Comprehensive Biochem.* V. ii. 73 Gluconic acid is important in the pharmaceutical, food, animal feeding stuffs and general chemical industries. 1970 J. F. R. Kuck in C. N. Graymore *Biochem. Eye* iii. 213 The role of gluconic acid, found in the human lens by van Heyningen (1958), is unknown.

glucosamine (glⁱūkōᵘ·zămīn, -sămīn). *Chem.* [f. *GLUCO- after *GLYCOSAMINE.] Any amino-sugar derived from glucose by substitution of an amino group for a hydroxyl group; *spec.* 2-amino-2-deoxy-D-glucose. Cf. *GLYCOSAMINE.

1884 *Jrnl. Chem. Soc.* XLVI. 724 G. Ledderhose gave the name glucosamine hydrochloride to a crystalline substance, $C_6H_{13}NO_5$, HCl, obtained by boiling chitin with concentrated hydrochloric acid. 1956 [see *GALACTOSAMINE]. 1957 W. Pigman *Carbohydrates* viii. 465 The *N*-methyl derivative of L-glucosamine (the enantiomorph of the common D-glucosamine) has been isolated from the degradation product of streptomycin. 1959 H. S. Goldberg *Antibiotics* ii. 140 The third moiety in the hydrolyzates of kanamycin B yields a ninhydrin-positive reducing test, but it is not 6-glucosamine present in kanamycin itself. 1967 Cantarow & Schepartz *Biochem.* (ed. 4) ii. 21 Glucosamine, also called 'chitosamine', is the basic unit in the polysaccharide, chitin, which forms part of the integument of arthropods. 1969 [see *GALACTOSAMINE].

glucosan (glⁱū·kŏsæn). *Chem.* [ad. F. *glucosane* (A. Gélis 1860, in *Compt. Rend.* LI. 333), f. GLUCOSE.] **a.** Any of several isomeric anhydrides, $C_6H_{10}O_5$, of glucose.

1862 H. Watts tr. *Gmelin's Hand-bk. Chem.* XV. 329 Glucose, dried between 100° and 110°, gives off water at 170°, becoming coloured at the same time, and is converted into glucosan. 1950 G. M. Dyson *Man. Org. Chem.* x. 805 An enquiry into..the structure of the remaining product led to the recognition of a number of isomeric glucosans, for it appears that glucose can lose one molecule of water in almost every conceivable way. 1963 E. Percival in Florkin & Stotz *Comprehensive Biochem.* V. i. 54 The constitution of β-glucosan..was established by reaction with periodate.

b. Any polysaccharide composed chiefly of glucose residues.

1925 *Jrnl. Chem. Soc.* CXXVIII. I. 233 Brigl's observation..that Pictet's supposed method of preparation of unimolecular α-glucosan..yields polymeric glucosans is confirmed. 1948 R. J. McIlroy *Chem. Polysaccharides* iv. 37 In addition to the above mannan, a glucosan has been isolated from bakers' yeast. It has molecular weight 6,500, and consists of glucopyranose units. 1966 P. McDonald et al. *Anim. Nutrition* ii. 17 Cellulose is a glucosan and is the most abundant plant constituent.

glucose. Delete def. and add: [First formed as F. *glucose* (Dumas 1838, in *Compt. Rend.* VII. 109).] **1. a.** One of the aldo-hexoses, $CHO(CHOH)_4CH_2OH$, obtainable as dextro- and lævo-rotatory enantiomorphs and as a racemic mixture; *spec.* the dextro-rotatory form (D-*glucose*, *dextrose*, *grape-sugar*), a white or colourless sweet-tasting crystalline solid which is the commonest sugar in nature; it is important in the metabolism of almost all organisms and occurs naturally in the free state, as in sweet fruit juices and in the blood, and in combination, as in glucosides, in some oligosaccharides (such as sucrose), and in some polysaccharides (such as cellulose, starch, and glycogen, all of which are composed wholly of glucose residues). (Later examples.)

In chemical terms glucose is the aldo-hexose in which the hydroxyl groups attached to the second, fourth, and fifth carbon atoms can be regarded as being on the same side of the molecule when this is represented as a linear structure.

1938 M. L. Wolfrom in H. Gilman *Org. Chem.* II. xvi. 1400 The central compound of the carbohydrates is *d*-glucose. 1951 A. Grollman *Pharmacol. & Therapeutics* xxvi. 572 In diabetic animals and patients, the injection of insulin is followed by..the disappearance of glucose and acetone bodies from the urine. 1959 *Chambers's Encycl.* III. 92/2 Although fructose is laevo-rotatory, it is configurationally related to dextrorotatory and not to laevorotatory glucose. 1965 Bell & Coombe tr. *Strasburger's Textbk. Bot.* (new ed.) I. 19 Soluble carbohydrates, especially the disaccharides sucrose and maltose and the monosaccharides glucose and fructose, are the most frequent components of the cell sap. 1968 Passmore & Robson *Compan. Med. Stud.* I. ix. 13/1 There is a continual uptake of glucose from the blood by many organs, notably the brain, which probably cannot utilize other sources of energy.

†**b.** Any of the hexoses, $C_6H_{12}O_6$, (or more broadly, any monosaccharide), including dextrose along with fructose, galactose, etc. *Obs.*

[1859 M. Berthelot in *Ann. de Chim. et de Phys.* 3me Sér. LV. 270 Le nom de *glucose*, appliqué jadis au sucre de raisin seulement, désigne aujourd'hui toute une série de sucres distincts;..séchés à 110 degrés, ils sont isomères et répondent à la formule C¹²H¹²O¹². Tels sont le glucose de raisin, le glucose de malt, le glucose de fruits lévogyre, le glucose lactique.] 1866 H. E. Roscoe *Less. Elem. Chem.* xxxvii. 325 Dilute acids convert lactose into a peculiar glucose, called galactose. 1887 *Encycl. Brit.* XXII. 623/1 'Sugar' is now a collective term for two chemical genera named saccharoses (all $C_{12}H_{22}O_{11}$) and glucoses (all $C_6H_{12}O_6$). 1889 G. M'Gowan tr. *Bernthsen's Org. Chem.* 285 The glucoses result..from the carbohydrates of the cane sugar and starch groups by the taking up of water. 1911 *Encycl. Brit.* XXVI. 33/1 It is convenient to divide the sugars into two main groups: monosaccharoses (formerly glucoses) and disaccharoses (formerly saccharoses).

2. A non-crystallizable colourless or yellowish brown syrup containing glucose (dextrose), maltose, dextrin, and water and obtained by the incomplete hydrolysis of starch. Also called *glucose syrup*, *starch syrup*, etc.

1852 T. R. Betton tr. *Regnault's Elem. Chem.* II. iv. 488 Glucose is found in commerce under three different forms: syrup of fecula, glucose in mass, and granulated glucose... Glucose, in syrup or in bulk, is used in the manufacture of beer and alcohol. 1885 E. R. Southby *Syst. Handbk. Pract. Brewing* (ed. 2) xv. 222 There is a liquid form of glucose known amongst other names as glucose syrup, which when well refined is a very good brewing material. 1906 Thresh & Porter *Preservatives in Food* xxxi. 396 It was proved at Quarter Sessions that glucose was composed of 40 per cent. dextrose, 40 per cent. dextrine, and 20 per cent. water,..and that glucose was not injurious to health, but was used to prevent the marmalade crystallizing. 1950 G. H. Rauch *Jam Manuf.* iii. 50 The addition of glucose gives jam a brighter appearance. 1963 Triebold & Aurand *Food Composition & Anal.* v. 211 Some of the syrup sold as commercial glucose is colored by the addition of other syrups and materials.

3. *attrib.* and *Comb.*, as *glucose ferment*, *syrup* (see *2), *tablet*; **glucose phosphate**, any of the phosphoric acid esters of glucose; esp. either of two monophosphates, $C_6H_{11}O_5(O{\cdot}PO_3H_2)$, important in carbohydrate metabolism in animals and plants: (*a*) *glucose-6-phosphate*, the compound formed by the reaction of glucose with adenosine triphosphate in the first stage in the metabolism of glucose; (*b*) *glucose-1-phosphate*, the compound formed from glucose-6-phosphate in the synthesis of polysaccharides; **glucose tolerance test**, a test for abnormalities in carbohydrate metabolism, esp. diabetes, in which glucose is given orally after a period of fasting and the blood sugar measured at intervals afterwards, the results being depicted graphically as a *glucose tolerance curve*.

1885 Glucose ferment [in Dict.]. 1912 E. F. Armstrong *Simple Carbohydrates & Glucosides* (ed. 2) 145 (*heading*) References to glucose phosphates. 1938 A. L. Raymond in H. Gilman *Org. Chem.* II. xvii. 1479 In this manner there have been synthesized the Robison ester..as well as glucose-1-, 3-, 4-, and 5-phosphates, leaving only the 2-phosphate unknown in the glucose series. 1963 H. Burn *Drugs, Med. & Man* (ed. 2) xiv. 146 One of the first steps in using glucose within the cell is its conversion into a glucose-phosphate combination. 1964 A. White et al. *Princ. Biochem.* (ed. 3) xx. 366 In the cells of liver, intestine, and kidney only, glucose 6-phosphate may be hydrolyzed, releasing glucose to the environment. 1965 Bell & Coombe tr. *Strasburger's Textbk. Bot.* (new ed.) II. 266 The formation of..starch in the living cell appears to begin not from α-glucose itself but from the energy-rich glucose-1-phosphate, which..is probably the actual substance produced in photosynthesis. 1891 Glucose syrup [in Dict.]. 1959 J. Braine *Vodi* xiii. 172 He started carrying digestive tablets and glucose tablets. 1923 *Arch. Internal Med.* XXXI. 248 It would be interesting to hear from someone doing frequent glucose tolerance curves how low the reaction may be. *Ibid.* 258 Ohler has reported several cases in which glucose tolerance tests were done three times at intervals of about one year. 1966 *Lancet* 24 Dec. 1377/2 The findings depended on..the reliability of the glucose-tolerance test in athero-sclerotic patients.

glucoside. For 'glucose' read 'a sugar' and add: Formerly also **glycoside** (now distinguished in sense). Now usu. restricted to mean: a *GLYCOSIDE which on hydrolysis gives glucose. (Additional examples.)

The earliest known example is as G. *glucosid* (W. Mayer 1854, in *Ann. d. Chem. u. Pharm.* XCII. 125), but it is not clear whether he was the first to use the word; the F. word *glucosamide* was coined in 1852 with the same meaning (A. Laurent in *Ann. de Chim. et de Phys.* XXXVI. 333). For examples of the form *glycoside* see *GLYCOSIDE.

1857 W. A. Miller *Elem. Chem.* III. vii. 511 (*heading*) Glucosides, or compounds which contain glucose. 1889 G. M'Gowan tr. *Bernthsen's Org. Chem.* 512 As glucosides are designated a series of vegetable substances which are so broken up by alkalies, acids, or ferments, that one of the products of this decomposition is a glucose, usually grape sugar. 1910 *Encycl. Brit.* XII. 142/1 Although glucose is the commonest sugar present in glucosides, many are known which yield rhamnose or iso-dulcite; these may be termed pentosides. 1930 [see *GLYCOSIDE 2]. 1932 I. D. Garard *Introd. Org. Chem.* x. 150 The most important glucosides are natural products. The sugar is generally but not always glucose. 1957 *Encycl. Brit.* X. 444/1 The complex sugars (polysaccharides) may..be regarded as glucosides since they consist of ethers formed from two or more simple sugars. 1967 I. L. Finar *Org. Chem.* (ed. 5) I. xviii. 486 The hemiacetol form (lactol) of a sugar [can] react with a molecule of an alcohol to form the acetal derivative, which is known under the generic name of glycoside; those of glucose are known as glucosides; of fructose, fructosides, etc.

Hence **glucosi·dal**, **-ic** *adjs.*, characteristic of a glucoside or glucosides; also, glycosidic; *glucosidic bond* or *linkage* = *glycosidic linkage*.

1927 *Jrnl. Biol. Chem.* LXXIV. 787 Certain substances of glucosidic nature..exhibit a characteristic cardiac or so called digitalis action. 1929 R. A. Gortner *Outl. Biochem.* xxiii. 513 The linkage between the glucuronic acid and the benzoic acid is not a glucosidal linkage, for the aldehyde group is still free. 1954 A. White et al. *Princ. Biochem.* xvii. 393 Amylose is a linear polysaccharide comprising glucose units linked to each other in a repeating fashion by an α-1,4-glucosidic bond. 1963 R. R. A. Higham *Handbk. Papermaking* ii. 33 The glucosidic chain linkages constituting the cellulose formation.

glucuronate (glⁱukiūᵊ·ronēⁱt). *Chem.* [f. next + -ATE⁴.] A salt or ester of glucuronic acid.

1904 J. A. Mandel tr. *Hammarsten's Text-bk. Physiol. Chem.* (ed. 4) xv. 522 According to the body with which they are conjugated, the glucuronates show different behavior. 1966 G. J. Dutton *Glucuronic Acid* iii. 261 A growing weight of evidence..suggests that administration of glucuronate..assists the excretion, or decreases the toxicity, of certain compounds.

glucuronic (glⁱūkiurǫ·nik), *a.* *Chem.* [f. *GLUCO- after *GLYCURONIC *a.*] *glucuronic acid*: the uronic acid derived from glucose; cf. *GLYCURONIC *a.*

1909 *Cent. Dict. Suppl.*, *Glucuronic*, same as glycuronic. 1911 *Encycl. Brit.* XXIII. 969/2 Sodium amalgam reduces it [*sc.* saccharic acid] to glucuronic acid, $C_6H_{10}O_7$ or $OHC[CH{\cdot}OH]_4CO_2H$. 1955 *Sci. News Let.* 2 July 6/3 The aspirin apparently acts by combining, or conjugating, with glucuronic acid. 1966 G. J. Dutton *Glucuronic Acid* p. xv, The compound now known as glucuronic acid was first described as a curious sugar acid occurring in the urine conjugated with certain drugs. This led to it being christened 'glykuronsaüre' by Schmiedeberg & Meyer.

glucuronide (glⁱukiūᵊ·ronəid). *Chem.* [f. prec. + *-IDE.] Any glycosidic compound formed by glucuronic acid.

1934 *Chem. Abstr.* XXVIII. 184 A specific enzyme may..be assumed for the cleavage of glucuronides. 1948 W. Pigman *Chem. Carbohydrates* xi. 509 Many phenols and alcohols when injected into animals are detoxified by conjugation as glucuronides. 1967 Cantarow & Schepartz *Biochem.* (ed. 4) ii. 21 Glucuronic acid is found in the urine, conjugated (as a glucuronide) to foreign substances.

glucuronoside (glⁱūkiurǫ·nosəid). *Chem.* [f. as prec. + *-OSIDE.] = prec.

1934 E. F. & K. F. Armstrong *Carbohydrates* (ed. 5) vi. 67 Menthol glucuronate is hydrolysed by emulsin, and therefore is a β-glucuronoside. 1963 E. Percival in Florkin & Stotz *Comprehensive Biochem.* V. ii. 74 A molecule of water is eliminated in the glucuronoside formation.

glue, *sb.* Add: **6. glue-sniffer**, a person who inhales the fumes of plastic cement for their narcotic effects; so **glue-sniff** *v. intr.*, **glue-sniffing** *vbl. sb.*

1963 *New Society* 20 June 14/1 A new threat to teen-age stability..is glue-sniffing, the deliberate inhaling of the fumes from plastic cement. *Ibid.* 14/2 The glue sniffers..

frequently fall asleep in class. **1968** *Daily Colonist* (Victoria, B.C.) 6 Oct. 21/1 That particular glue-sniffer told police his story in Los Angeles, but it could just as easily have been in any police department in Greater Victoria. **1971** E. E. Landy *Underground Dict.* 90 *Glue sniff* v., inhale model-airplane glue...Glue sniffing is regarded as an adolescent type of drug abuse, but it causes organic brain damage, and it can cause liver, kidney and bone-marrow damage. Eventually most glue sniffers outgrow glue and go to more adult-type drugs such as marijuana, [etc.]. **1971** J. Drummond *Farewell Party* xxv. 131 One of those red brick boxes..with a nice view of the glue factory...A glue-sniffer could get high there..just by holding his breath.

glue, *v.* **3. b.** Delete †*Obs.* and add later example.
1885 *Spons' Mechanics' Own Bk.* 131 The wood glues well.

glued, *ppl. a.* Add: **2. glued-on,** affixed by means of glue; *fig.* of literary devices, effects, etc.: carelessly superimposed and not integrated with the body of the work; **glued-up** *fig.*, applied to a medley of scenes or incidents with little apparent connection or unity.
1906 *Westm. Gaz.* 15 Feb. 2/3 What the Americans call a 'glued-up' or nailed-up drama. *Ibid.* 16 Sept. 3/2 To avoid auxiliary complications and eschew 'glued-on' comic relief. **1969** *Jane's Freight Containers 1968–69* 482/2 Covered on both sides with glued-on aluminium alloy panels.

glueily (glū·ili), *adv.* Also **gluily.** [f. Gluey *a.* + -ly².] In a gluey manner.
1925 F. M. Ford *No More Parades* i. 37 His very thick soles moved gluily and came up after suction. **1928** A. Huxley *Point Counter Point* ii. 31 The great Pongileoni glueily kissed his flute.

glue-pot. Add: **2.** A patch of wet or muddy ground in which one 'sticks'. *colloq.*
1892 *Daily News* (Morris), The Bishop of Manchester.. assures us that no one can possibly understand the difficulties and the troubles of a Colonial..clergyman until he has..struggled through what they used to call 'glue-pots'. **1907** *Daily Chron.* 18 July 7/2 The veriest 'glue-pot' of a wicket. **1916** J. B. Cooper *Coo-oo-ee* x. 137 Was it surprising that in a short time the 'glue pot' no longer bogged the jinker? **1963** *Times* 14 Jan. 3/2 If Cardiff is not a gluepot these two should be able to launch some fine attacks.

‖ **glühwein** (glū·vain). Also **gluhwein.** [G.] Mulled wine.
1898 *Elizabeth & her German Garden* (1900) 158 Waiting for the New Year, and sipping *Glühwein*...It was hot, and sweet, and rather nasty. **1929** E Hemingway *Farewell to Arms* xxxix. 232 We..drank hot red wine with spices and lemon in it. They called it *glühwein.* **1959** *Listener* 15 Jan. 134/3 The hot bath and the *Glühwein.* **1967** 'G. Carr' *Lewker in Tirol* vi. 82 A glass of *gluhwein*, red wine sugared and heated.

glulam (glⁱū·læm, glū·læm). [f. Glu(e *sb.* + Lam(ination.] A building material consisting of laminations of timber glued together.
1953 *Civil Engineering* Feb. 87 (Advt.), The glulam girders which carry the load are formed of kiln dried timber, bonded together with waterproof glue which is as strong and permanent as the wood. **1961** *Engineering* 29 Sept. 411/3 Rectangular glulam sections.

glutæo-, gluteo- (glⁱuti·o), combining form of Glutæus, = pertaining to the glutæal region (and some other part of the body), as *glutæo-femoral* adj.
1890 Billings *Med. Dict.*, *Gluteo-femoral crease*, gluteal fold. **1900** Dorland *Med. Dict.*, *Gluteofemoral*, pertaining to the buttock and thigh. **1962** *Gray's Anat.* (ed. 33) 690 A second [bursa] is found between the tendon of the muscle and that of the vastus lateralis (gluteofemoral bursa).

glutamate (glⁱū·tămēⁱt). *Chem.* [f. next + -ate⁴.] A salt or ester of glutamic acid; *esp.* the monosodium salt, widely used to flavour food.
1876 *Jrnl. Chem. Soc.* I. 906 Glutimide..obtained by heating ammonium glutamate. **1929** *Industr. & Engin. Chem.* Oct. 984/2 In China and Japan monosodium glutamate is manufactured on a commercial scale and consumed as a condiment. **1943** *Thorpe's Dict. Appl. Chem.* (ed. 4) VI. 34/1 The glutamates..have a characteristic meat-like flavour. **1960** *Spectator* 16 Sept. 417 That curious prickly after-taste which appears to be characteristic of every foodstuff in which monosodium glutamate figures. **1970** C. N. Graymore *Biochem. Eye* x. 670 This work immediately suggested that glutamate might play some special role in nervous tissue. **1971** *Times Lit. Suppl.* 12 Feb. 189/2 Flour..was said to be adulterated by things that make our ubiquitous monosodium glutamate sound positively delicious.

glutamic, *a.* Delete entry (in Dict. s.v. Glut-) and substitute: (glⁱutæ·mik). *Biochem.* [tr. G. *glutaminsäure* glutamic acid (H. Ritthausen 1866, in *Jrnl. f. prakt. Chem.* XCIX. 6), f. Glut- + Amic *a.*] *glutamic acid*: an amino-acid, $HOOC{\cdot}CH_2CH_2CH(NH_2){\cdot}COOH$, widely distributed in nature and a normal constituent of proteins.
1871 *Jrnl. Chem. Soc.* XXIV. 721 Aspartic acid, $C_4H_7NO_4$, and glutamic acid, $C_5H_9NO_4$. **1921** [see *glutathione]. **1946** *Nature* 9 Nov. 676/1 (*heading*) Action of penicillin in preventing the assimilation of glutamic acid by *Staphylococcus aureus.* **1970** C. N. Graymore *Biochem. Eye* x. 670 The role of glutamic acid in the transport of potassium in nervous tissues.

glutamyl (glⁱū·tămail, -il). *Chem.* [f. prec. + -yl.] Either of the univalent radicals derived from glutamic acid by the loss of one or other of the hydroxyl groups.
1909 *Jrnl. Chem. Soc.* XCVI. i. 369 The latter substance is converted by aqueous ammonia into glycylglutamyldiglycine, $C_{11}H_{18}O_7N_4$. **1956** *Nature* 25 Feb. 377/1 Secondary non-enzymic breakdown of γ-glutamyl peptides. **1965** *Nomencl. Org. Chem.* (*I.U.P.A.C.*) C. 120 α-Glutamyl: $HOOC{-}CH_2{-}CH_2{-}CH(NH_2){-}CO{-}$. γ-Glutamyl: $-CO{-}CH_2{-}CH_2{-}CH(NH_2){-}COOH$. **1968** A. White et al. *Princ. Biochem.* (ed. 4) v. 108 Glutamyl derivatives may be of the γ type as well as the α variety.

glutathione (glⁱūtăþai·ōⁿn). *Biochem.* [f. *Gluta(mic *a.* (see quot. 1921).] A tripeptide of glutamic acid, cysteine, and glycine, metabolically important, esp. as a coenzyme, and perhaps present in all cells.
1921 F. G. Hopkins in *Biochem. Jrnl.* XV. 297 Provisionally, for easy reference, the name *Glutathione* will perhaps be admissible. It leaves a link with the historic *Philothion*, has the same termination as *Peptone*, which has long served as a name for the simpler peptides, and is a sufficient reminder that the dipeptide contains glutamic acid linked to a sulphur compound. **1925** *Glasgow Herald* 5 Dec. 4 The oxygen-transporting and oxygen-liberating power of glutathione. **1949** H. W. Florey et al. *Antibiotics* II. xxi. 812 Glutathione and thiothreonine antagonized the anti-bacterial action of penicillin. **1969** *Nature* 11 Oct. 117/1 The tripeptide glutathione is an almost universal constituent of functioning biological systems.

glutelin (glⁱū·tělin). *Biochem.* [f. Glute(n + *-lin*, perh. after Globulin.] Any of a group of simple proteins all characteristically soluble in dilute acids and alkalis but insoluble in neutral saline solutions and found esp. in seeds.
1908 *Jrnl. Biol. Chem.* IV. p. xlix, *Glutelins*, simple proteins insoluble in all neutral solvents but readily soluble in very dilute acids and alkalies. **1963** E. G. Young in Florkin & Stotz *Comprehensive Biochem.* VII. i. 7 By modern criteria of purity the glutelins are heterogeneous and poorly characterized.

glutenin (glⁱū·těnin). *Biochem.* [f. Gluten + -in¹.] A glutelin found esp. in wheat.
1893 Osborne & Vorhees in *Amer. Chem. Jrnl.* XV. 457 It would be desirable to return to Taddei's original name and in future call this proteid zymon. Unfortunately this name is derived from the Greek word ζύμη, a ferment, and..this name is undesirable. As this proteid is especially characteristic of gluten, it seems appropriate to call it *glutenin*, a name suggested to us by Professor S. W. Johnson. **1963** E. G. Young in Florkin & Stotz *Comprehensive Biochem.* VII. i. 8 Those proteins of the group which have been isolated and studied are glutenin of wheat..and avenin of oats.

glutethimide (glⁱute·þimaid). *Pharm.* [f. *α-ethyl-α-phenyl*glut*ar*imide, the systematic chemical name for glutethimide, by rearrangement of some of its elements.] A colourless, odourless crystalline compound with hypnotic properties.
1955 *Unlisted Drugs* Jan. 4/2 Doriden (C-11, 511, glutethimide). **1966** *New Scientist* 1 Dec. 530/2 Glutethimide, a widely-sold sleeping tablet. **1967** *Martindale's Extra Pharmacopoeia* (ed. 25) 304/1 It has been suggested that, since glutethimide is chemically related to thalidomide, it may also have teratogenic effects.

glutton, *sb.* and *a.* Add: **A.** *sb.* **2. a.** Also const. *for.*
1895 Kipling *Day's Work* (1898) 197 He's honest, and a glutton for work. **1971** C. Egleton *Last Post for Partisan* iii. 29 This bug may not look up to much but the Volks is a glutton for punishment. **1971** E. Lemarchand *Death on Doomsday* ix. 131 Glutton for work, aren't you?

glycæmia (glaisī·miă). *Physiol.* Also **glycemia.** [f. Glyc(o- + Gr. αἷμα blood: see -ia¹.] The presence of sugar in the blood; the blood-sugar level. (Rare, except in translations.) Cf. *glycohaemia* (s.v. Glyco-). Hence **glycae·mic** *a.*
1901 *Index-Catal. Surg.-Gen. U.S.* 2nd Ser. VI. 266/1 (*heading*) Glycaemia. *See* Blood (Sugar in); Diabetes (Blood in). **1913** *Jrnl. Chem. Soc.* CIV. i. 923 The injection of adrenaline can induce perturbations in the glycæmia of an animal. **1923** *Archives Internal Med.* XXXI. 242 The term blood sugar curve is here applied to any glycemic reaction obtained in a tolerance test. **1936** Glycæmic [see *glycotropic* s.v. *glyco-]. **1946** *Nature* 26 Oct. 589/1 Both groups of dogs show the known glycæmic response to the alloxan. **1966** *Canad. Jrnl. Physiol. & Pharmacol.* XLIV. 615 (*heading*) Glycemia and consecutive passive transfers of an anaphylactoid reaction inducing factor in rats.

glyceraldehyde (gli:sĕræ·ldĭhaid). *Chem.* Also †**-aldehyd.** [f. Glycer(ic *a.* + Aldehyde.] A colourless, sweet-tasting, crystalline compound, $CHO{\cdot}CHOH{\cdot}CH_2OH$, which is the simplest optically active sugar.
1882 *Jrnl. Chem. Soc.* XLII. 1308 The oxidation of the glyceraldehyd to tartronic acid. **1899** *Ibid.* LXXVI. i. 110 It [*sc.* acraldehyde dibromide] apparently gives rise to glyceraldehyde when left in contact with water. **1951** I. L. Finar *Org. Chem.* I. xviii. 358 Glyceraldehyde has now been chosen as the standard configuration in sugar chemistry. **1965** *New Scientist* 17 June 761/2 The patients ..were also given glyceraldehyde, a normal intermediate product of glucose metabolism.

glyceria (glisiᵊ·riă). [mod.L. (R. Brown *Prodromus Floræ Novæ Hollandiæ* (1810) 179), f. Gr. γλυκερός sweet, in reference to the leaves and roots of some species.] A plant belonging to a widespread genus of aquatic grasses so named, of the family Gramineæ. Cf. Sweet-grass.
1829 J. C. Loudon *Encycl. Plants* 62 *Glyceria*..is found in stagnant water, and its long narrow leaves float on the surface. **1847** [see Manna¹ 9]. **1934** A. Arber *Gramineae* xvi. 376 At the periphery of the circular tufts of *Spartina*, the turf of *Glyceria* is, little by little, raised up and forced outwards. **1964** D. H. N. Spence in J. H. Burnett *Veget. Scotl.* ix. 352 *Glyceria* is a typical fen dominant in East Anglia.

glycerine, glycerin. Add: **4. glycerine tear,** a drop of glycerine used in theatrical make-up to simulate a tear (Tear *sb.*¹).
[**1926** Maines & Grant *Wise-Crack Dict.* 7/1 *Cork the glycerine*, stop the movie tears.] **1934** C. Lambert *Music Ho!* iv. 270 Even synthetic sentiment, the musical equivalent of glycerine tears, is harder to achieve than abstraction. **1969** *Listener* 13 Nov. 669/1 In a scene where Genevieve Page is producing the required tears with frightening realism, Wilder instructs the make-up man to add to them. 'Glycerine tears..great big Hollywood false tears...'

glycero-. Add: **gly:ceroki·nase** *Biochem.*, an enzyme found chiefly in the liver and adipose tissue which catalyses the reaction of glycerol and adenosine triphosphate to glycerophosphate and adenosine diphosphate.
1954 Bublitz & Kennedy in *Jrnl. Biol. Chem.* CCXI. 951 The name *glycerokinase* seems to be appropriate and convenient for this enzyme, although the purified enzyme also phosphorylates dihydroxyacetone and glyceraldehyde. **1956** *Nature* 7 Jan. 22/1 The isolation and partial purification of a rat liver enzyme, glycerokinase. **1962** A. Pirie *Lens Metabolism Rel. Cataract* 433 The relative activities of glycerokinase and α-glycerophosphate dehydrogenase in rabbit tissues.

glycerose (gli·sĕrōᵘz, -ōᵘs). *Chem.* [a. G. *glycerose* (Fischer & Tafel 1888, in *Ber. d. Deut. Chem. Ges.* XXI. 2636), f. Glycer(ine, -ol + -ose².] **a.** A mixture of glyceraldehyde and dihydroxyacetone, two isomers obtained by the oxidation of glycerol. **b.** = *glyceraldehyde.
1888 *Jrnl. Chem. Soc.* LIV. 1264 For the preparation of glycerose, the lead glyceroxide is spread in thin layers [etc.]. **1937** M. Steel *Biol. & Clin. Chem.* ii. 27 The simplest carbohydrate to possess an asymmetric carbon atom is glycerose. **1967** I. L. Finar *Org. Chem.* (ed. 5) I. xi. 287 Bromine water, sodium hypobromite and Fenton's reagent..oxidise glycerol to a mixture of glyceraldehyde (predominantly) and dihydroxyacetone; this mixture is known as glycerose.

glyceryl. Add: *glyceryl trinitrate* = Nitroglycerine.
1892 *Jrnl. Chem. Soc.* LXII. 692 Glyceryl trinitrate reduces alkaline copper tartrate. **1963** *Brit. Pharm. Codex 1963* 342 Glyceryl trinitrate solution causes the relaxation of involuntary muscle. **1968** J. H. Burn *Lect. Notes Pharmacol.* (ed. 9) 34 Glyceryl trinitrate..is used by those who suffer from 'effort angina'.

glycin. Delete entry and substitute: (glai·sin). *Photogr.* [f. *glycin(e.] A crystalline derivative of glycine, used as a developer; *p*-hydroxyphenylaminoacetic acid.
1893 [in Dict.]. **1961** A. L. M. Sowerby *Dict. Photogr.* (ed. 19) 355 In buying glycin from a pharmacist, make sure he is not supplying glycine (aminoacetic acid) which ..has no developing properties at all. **1963** John & Field *Textbk. Photogr. Chem.* v. 67 Glycin finds some use in fine-grain developers for negatives.

glycine (glai·sīn). *Biochem.* [ad. G. *glycin* (J. Berzelius 1848, in *Jahres-Ber. über d. Fortschritte d. Chem. und Min.* XXVII. 654), f. Gr. γλυκ-ύς sweet + -ine⁵.] A colourless sweet-tasting crystalline compound, NH_2CH_2COOH, which is the simplest amino-acid and a general constituent of proteins; = Glycocoll, *amino-acetic acid.
1851 G. E. Day tr. *Lehmann's Physiol. Chem.* I. 152 Glycine has not yet been found in an isolated state in the animal body. **1885** [in Dict. s.v. Glycin]. **1957** *New Biol.* XXIII. 79 In one experiment, radioactive glycine, valine and lysine were injected simultaneously

into a lactating goat, and casein, a milk protein, was isolated from milk collected 4 hours later. **1965** A. Meister *Biochem. Amino Acids* (ed. 2) I. i. 11 In 1820, Braconnot obtained glycine from a sulfuric acid hydrolyzate of gelatin...Subsequently Braconnot's 'sugar of gelatin' was named glycocoll and later glycine. **1969** *Times* 25 Aug. 4/8 In the genetic code three guanylic acid units are..the code for glycine.

glyco-. Add: In mod. chemical nomenclature, usu. used to refer to sugars generally, in contradistinction to *gluco-: so *glycose, glycoside*, etc. (cf. *glucose, glucoside*); **glycoli·pid** *Biochem.*, any substance which is a combination of both a carbohydrate (or carbohydrate derivative), esp. a sugar, and a lipid; **gly·cophyte** *Ecology*, a plant whose growth is inhibited by saline soil (cf. *halophyte* s.v. *halo-); hence **glycophy·tic** *a.*; **glycopro·tein** (also †**-proteid**) *Biochem.*, any of a group of proteins with one or more usually relatively short side chains composed generally of a mixture of carbohydrates or carbohydrate derivatives; **glycotro·pic** *a. Biochem.*, antagonistic to insulin.
1940 *Jrnl. Biol. Chem.* CXXXII. 178 The cerebroside.. belongs to a new class of glycolipids which probably differ from previously described cerebrosides only because they contain glucose in place of galactose. **1970** C. N. Graymore *Biochem. Eye* vii. 507 Cerebrosides and gangliosides may also be classified as glycolipids, both containing hexoses. **1947** R. F. Daubenmire *Plants & Environment* ii. 61 Because salts so evidently interfere with the absorption of water by glykophytes, saline soils have long been considered 'physiologically dry'..for these plants. **1964** V. J. Chapman *Coastal Veget.* iv. 89 Other evidence of elevation is the existence of obvious invasion by glycophytic species. *Ibid.* 96 Glycophytes such as White clover. **1969** Salisbury & Ross *Plant Physiol.* xxix. 692/1 Ecologists further classify those plants that are sensitive to relatively high concentrations of salt in the surrounding liquid medium..as glycophytes. **1898** *Jrnl. Physiol.* XXIII. 177 It is probable that every tissue containing proteids will be found to contain some admixture of glycoproteids. **1902** *Encycl. Brit.* XXXI. 724/1 Glycoproteids—a number of proteids, on treatment with acid, yield various sugar-like bodies. **1908** *Jrnl. Biol. Chem.* IV. p. l, *Glycoproteins.* Compounds of the protein molecule with a substance or substances containing a carbohydrate group other than a nucleic acid. **1945** *Adv. Protein Chem.* II. 250 The group of proteins which contain less than 4 per cent hexosamine, classified as glycoproteins, embraces many proteins listed as albumins and globulins in the accepted classification of proteins. **1968** A. White et al. *Princ. Biochem.* (ed. 4) vi. 118 The distinction between glyco- and mucoproteins is arbitrary, and intermediate types exist. **1970** R. W. McGilvery *Biochem.* xxiv. 585 The carbohydrate chains on glycoproteins are oligosaccharides. **1936** F. G. Young in *Lancet* 8 Aug. 301/2 The pituitary substance responsible for the rapid development of an exaggerated glycæmic response to hypoglycæmia will be called the 'glycotropic factor' without prejudice to the question of its existence as a separate entity. *Ibid.*, Glycotropic extracts. **1955** E. B. Astwood in Pincus & Thimann *Hormones* III. vii. 256 A prolactin preparation had some glycotropic activity in the dog.

glycogenesis. Add: **b.** The formation of glycogen in the animal body from sugar. So **glycogenic** *a.*, in this sense.
1864 W. B. Carpenter *Princ. Human Physiol.* (ed. 6) ix. 387 It appears that another purpose is fulfilled by the Liver—the production of an amyloid substance termed Glycogen; and we shall now proceed to consider the chief facts..in reference to this so-called 'Glycogenic function of the Liver'. **1897** *Lippincott's Med. Dict.* 434/2 *Glycogenesis*, the formation of sugar, or of glycogen. **1900** Dorland *Med. Dict.* 282/1 *Glycogenesis*, the production of sugar or of glycogen. *Ibid.*, *Glycogenic*, pertaining to glycogenesis or to glycogen. **1913** *Proc. Roy. Soc.* B. LXXXVI. 560 The liver will be found to contain only traces of glycogen. This effect is due to an inhibition of the glycogenic function of the liver. **1926** J. J. R. Macleod *Carbohydrate Metabolism & Insulin* xi. 159 In the strict sense, the term glycogenic should be restricted to the formation of glycogen from protein (and fat) and its breakdown to sugar, rather than to the polymerisation of sugar itself into that substance. **1970** W. V. Thorpe et al. *Biochem.* (ed. 9) xix. 251 The reactions involved in muscle glycogenesis are the same as those in liver glycogenesis.

glycogenolysis (glai:kodʒĕnǫ·lisis). *Phys.* [f. Glycogen + -o + *-lysis.] The hydrolytic breakdown of glycogen to glucose. Hence **gly:cogenoly·tic** *a.*, of, pertaining to, or causing glycogenolysis.
1909 *Jrnl. Biol. Chem.* VI. p. xl, It was found that glycogenolysis is quite slow immediately after death but afterwards becomes very rapid. It therefore takes some time after death for the glycogenolytic process to become active. **1929** *Proc. Soc. Exper. Biol. & Med.* XXVI. 490 The glycogenolytic properties of insulin. **1968** A. White et al. *Princ. Biochem.* (ed. 4) xix. 443 Glycogenolysis (glycogen breakdown) is hydrolytic in the intestine.

glycolysis (glaikǫ·lisis). *Biochem.* [f. Glyco- + *-lysis.] An important metabolic process, the initial pathway in cell respiration, in which sugar, sugar phosphates, or precursors of these are broken down by enzymic reactions releasing energy and yielding esp. pyruvic acid or lactic acid in animal tissues; by extension, used of similar processes in other organisms yielding different products, such as ethyl alcohol in fermentation. (Orig. used of the disappearance of sugar from drawn blood.)
1892 *Jrnl. Chem. Soc.* LXII. i. 89 (*heading*) Hematic glycolysis. Estimation of glycogen in the blood. **1895** *Ibid.* LXVIII. ii. 361 Glycolysis..is not a vital process, but the blood corpuscles contain substances which possess the power in question. **1937** Halliburton & McDowall *Handbk. Physiol. & Biochem.* (ed. 35) xl. 591 The blood is kept under oil to prevent loss of CO_2. Oxalate is added to prevent clotting and fluoride to prevent glycolysis. **1959** *New Scientist* 29 Jan. 218/1 On death the blood supply stops and the tissues become depleted in oxygen and fermentation (or anærobic glycolysis) sets in. **1968** A. White et al. *Princ. Biochem.* (ed. 4) xviii. 395 Meyerhof prepared soluble extracts of muscle that catalyzed glycolysis and later demonstrated that, except for the final steps, glycolysis and alcoholic fermentation are essentially similar. **1968** Miall & Sharp *New Dict. Chem.* (ed. 4) 270/1 The metabolic breakdown of carbohydrates..may occur in the presence of oxygen ('aerobic glycolysis') or in its absence ('anaerobic glycolysis').
Hence (as a back-formation) **gly·colyse** *v. trans.*, to metabolize (a substance) in glycolysis.
1951 F. Dickens in Sumner & Myrbäck *Enzymes* II. i. lxiii. 625 Allowing the somewhat vague expression 'glycolysis' to be replaced by the more specific indications of the sugar glycolyzed, if this is known. **1962** R. van Heyningen in A. Pirie *Lens Metabolism Rel. Cataract* 399 Most of the glucose used by the lens is glycolysed.

glycosamine (glaikōᵘ·zămīn, -sāmīn). *Chem.* [ad. G. *glycosamin* (G. Ledderhose 1876, in *Ber. d. Deut. Chem. Ges.* IX. 1200), f. Glycos(e + Amine.] **a.** = *glucosamine. **b.** Any amino-sugar in which the amino group replaces a hydroxyl group other than that attached to the first carbon atom.
1877 *Jrnl. Chem. Soc.* I. 64 (*heading*) Glycosamine hydrochloride... When purified chitin is heated..with pure strong hydrochloric acid, it dissolves..forming a blackish-brown solution. **1948** W. Pigman *Chem. Carbohydrates* ix. 375 An important additional group of nitrogenous derivatives [of the carbohydrates] is the amino sugars, among which are the glycosamines. **1963** *Biol. Abstr.* XLII. i. 720/2 A method of colorimetric determination of d-glycosamine has been developed by the authors. **1967** Cantarow & Schepartz *Biochem.* (ed. 4) ii. 21 The second type of amino sugar is the glycamine or glycosamine, in which alcoholic hydroxyl groups in the sugar molecule are replaced by amino groups.

glycose. Add: **2.** [cf. Glyco- in Dict. and Suppl.] Any of the monosaccharides.
1938 M. L. Wolfrom in H. Gilman *Org. Chem.* II. xvi. 1418 The methylglycosides (glycose referring to any sugar) and their acetates. **1957** W. Pigman *Carbohydrates* i. 3 A glycose is any monosaccharide.

glycoside (glai·kosaid). *Chem.* [f. Glyco- after Glucoside.] †**1.** Used as a variant spelling of *glucoside* before the two words were distinguished in sense (see sense 2 and *glucoside). *Obs.*
1878 *Chem. News* 2 Aug. 60/1 The glycosides and colouring matters of quercitron and of Persian berries are very near to each other. **1890** Billings *Med. Dict.* I. 592/1 *Glycoside*, glucoside.
2. Any of a class of compounds, many of which are present in plants, which on hydrolysis give a sugar and one or more other compounds, one of which is usu. an alcohol or a phenol.
1930 *Jrnl. Amer. Chem. Soc.* LII. 1681 The word glycoside is used in the generic sense to include glucoside-like derivatives of any of the aldoses or ketoses; the term glucoside is limited to the glycosides of the particular sugar glucose. **1951** R. J. McIlroy *Plant Glycosides* i. 1 Glycosides are derivatives of the cyclic forms of sugars in which the reducing or potential aldehyde group of the sugar is substituted by condensation with an alcohol or a phenol to form a hemi-acetal. **1960** S. F. Dyke *Carbohydrates* vii. 112 Many plant pigments responsible for the color of flowers are glycosides. **1961** *New Scientist* 7 Dec. 639/3 The active principles of digitalis are glycosides (such as Digitoxin and Digoxin). **1965** Bell & Coombe tr. *Strasburger's Textbk. Bot.* (new ed.) i. 21 The enzyme emulsin decomposes amygdalin, the glycoside of bitter almonds, into glucose, benzaldehyde and hydrocyanic acid. **1968** A. White et al. *Princ. Biochem.* (ed. 4) iii. 41 All glycosides, including oligo- and polysaccharides, can be hydrolyzed with varying degree of difficulty.
Hence **glycosi·dal, -ic** *adjs.*, of the nature of or characteristic of a glycoside or glycosides; *glycosidic bond* or *linkage*, a bond between the hydroxyl group attached to the first carbon atom of a cyclic monosaccharide and a hydroxyl group of another compound; **glycosi·dically** *adv.*
1925 *Jrnl. Amer. Chem. Soc.* XLVII. 269 He adopted Tollens' suggestion that the union between the monosaccharides in the structures of the compound sugars is of the semi-acetal ring form and this idea of the 'glycosidic union' has proved to be correct. **1931** E. F. & K. F. Armstrong *Glycosides* ii. 10 Glycosidic compounds. **1938** R. A. Gortner *Outl. Biochem.* (ed. 2) xxiv. 619 The linkage between the glucuronic acid and the benzoic acid is not a glycosidal [ed. 1 (1929), glucosidal] linkage, for the aldehyde group is still free. **1943** *Thorpe's Dict. Appl. Chem.* (ed. 4) VI. 83/1 The majority of polysaccharides have linkages of the glycosidic type. **1948** W. Pigman *Chem. Carbohydrates* i. 13 The uronic acid components of the polyuronides are likewise joined glycosidically by 1,4′-linkages. **1953** Fruton & Simmonds *Gen. Biochem.* xvii. 380 The glycosidic linkage occupies a place in carbohydrate chemistry analogous to the role of the peptide bond in protein chemistry. **1964** N. G. Clark *Mod. Org. Chem.* xvi. 331 This sugar comprises two D-glucose units linked glycosidically. **1965** Bell & Coombes tr. *Strasburger's Textbk. Bot.* (new ed.) i. 21 Tannins are mixtures of quite different aromatic substances, partly glycosidic in nature.

glycuronic (glaikiurǫ·nik), *a. Chem.* [tr. G. *glykuronsäure* glycuronic (glucuronic) acid (Schmiedeberg & Meyer 1879, in *Zeitschr. f. physiol. Chem.* III. 433), f. Glyc(o- + *uronic *a.*] *glycuronic acid*: = *glucuronic acid.
1882 *Jrnl. Chem. Soc.* XLII. 952 Urochloralic acid..is decomposed, on boiling with dilute acids, into trichlorethyl alcohol and glycuronic acid. **1949** H. W. Florey et al. *Antibiotics* II. xxxvii. 1219 There was a marked increase in the excretion of glycuronic acid in rabbits given penicillin. **1968** J. H. Burn *Lect. Notes Pharmacol.* (ed. 9) 53 In the body it [*sc.* chloral] is first reduced to an alcohol, and then combined with glycuronic acid, to be excreted in the urine as urochloralic acid.

glycyl (glai·sil, glai·sail). *Chem.* [a. G. *glycyl* (Fischer & Fourneau 1901, in *Ber. d. Deut. Chem. Ges.* XXXIV. 2869), f. *glyc(ine + -yl.] The univalent radical $NH_2{\cdot}CH_2{\cdot}CO{-}$ derived from glycine.
1901 *Jrnl. Chem. Soc.* LXXX. i. 675 This paper [of Fischer & Fourneau] deals with molecular anhydrides of amino-acids.., and in it the name glycyl is used for the radicle $NH_2{\cdot}CH_2{\cdot}CO$. *Ibid.*, Glycylglycine hydrochloride. **1924** T. B. Robertson *Princ. Biochem.* (ed. 2) vii. 155 Chloracetyl chloride for the introduction of glycyl. **1968** A. White et al. *Princ. Biochem.* (ed. 4) v. 108 The synthesis of a dipeptide, glycyl-L-proline.

glycyphyllin (glai-, glisifi·lin). *Chem.* [f. mod.L. *glycyphyll-a*, epithet of a species of *Smilax*, f. Gr. γλυκύ-ς sweet + φύλλ-ον leaf + -in¹.] A rhamnoside, known only from *Smilax glycyphylla*, having phloretin as the aglycone.
1881 Wright & Rennie in *Jrnl. Chem. Soc.* XXXIX. 240 Were its identity and distinctiveness established, the term glycyphyllin might perhaps be applied to it. **1966** A. H. Williams in T. Swain *Comp. Phytochem.* xvii. 304 In the monocotyledons we have the isolated example of glycyphyllin in some samples of one species of *Smilax* in the Liliaceae.

glycyrrhetic (glisiri·tik, -re·tik), *a. Chem.* [f. next + -ic.] *glycyrrhetic acid*: a triterpene, the aglycone of glycyrrhizic acid.
1907 *Jrnl. Chem. Soc.* XCII. i. 545 When it [*sc.* glycyrrhizic acid] is boiled for five hours with 3% sulphuric acid, air being excluded, glycyrrhetic acid, $C_{32}H_{46}O_7$.., m.p. 210° (hitherto known as 'glycyrrhetin'), is precipitated. **1966** *McGraw-Hill Encycl. Sci. & Technol.* XIV. 112/1 The ammonium salt of glycyrrhetic acid is the taste principle of licorice. Therapeutically, glycyrrhetic acid is reported to be the active ingredient of licorice having desoxycortone-like activity.

glycyrrhetin (glisiri·tin, -re·tin). *Chem.* Also **glycyrretin.** [ad. F. *glycyrretin* (E. von Gorup-Besanez 1861, in *Ann. d. Chem. und Pharm.* CXVIII. 242), f. Glycyrr(hizin + *-etin.] = prec.
1864 H. Watts *Dict. Chem.* II. 919 *Glycyrretin*,..a substance obtained, together with glucose, by boiling glycyrrhizin with dilute acids. **1907** [see *glycyrrhetic a.]. **1943** *Thorpe's Dict. Appl. Chem.* (ed. 4) VI. 98/1 Glycyrrhetin, glycyrrhetic acid, $C_{32}H_{44}O_7$.

glycyrrhizic (glisirai·zik), *a. Chem.* [f. Glycyrrhiz(in + -ic.] *glycyrrhizic acid*: a glycoside obtained from liquorice root.
1879 *Jrnl. Chem. Soc.* XXXVI. 1040 Glycyrrhizic acid prepared by the action of sulphuretted hydrogen on the lead salt, resembles dried albumin in appearance. **1907** *Ibid.* XCII. i. 545 Glycyrrhizin consists of the potassium and calcium salts of glycyrrhizic acid and was isolated from liquorice. **1969** J. H. Thompson in J. A. Bevan *Essent. Pharmacol.* xxix. 303 Carbenoxolone sodium..is a pentacyclic triterpene prepared from glycyrrhizic acid and licorice glycoside.

glyoxal. Delete ref. to the former use of *glyoxylic* (cf. *glyoxylic *a.*). Add earlier example.
1857 H. Debus in *Phil. Mag.* 4th Ser. XIII. 40 This residue consisted almost entirely of the aldehyde of glyoxylic acid; I propose to call it Glyoxal, $C_4H_2O_4$.

glyoxalase (glai,ǫ·ksălēⁱz, -s). *Biochem.* [f. prec. + *-ase.] Either of two enzymes which together convert methylglyoxal to lactic acid in the presence of glutathione; orig. considered a single enzyme.
1913 Dakin & Dudley in *Jrnl. Biol. Chem.* XIV. 423 The catalyst..effects the rapid conversion of methyl glyoxal into lactic acid... We propose that it be named

'glyoxalase'. **1951** *Ibid.* CXC. 685 The first step consists in a condensation reaction between methylglyoxal and glutathione catalyzed by an enzyme referred to in this paper as glyoxalase I. This condensation product, in a second step which is catalyzed by glyoxalase II, is then broken down into glutathione and lactic acid. **1961** *Biochem. Jrnl.* LXXIX. 481/1 Ophthalmic acid..was a competitive inhibitor in the reaction catalysed by glyoxalase I.

glyoxaline (glai,ǫ·ksălīn). *Chem.* [f. Glyoxal + -ine⁵.] = *imidazole.
1858 H. Debus in *Chem. Gaz.* XVI. 353 The other [organic base] in solution, to which in this paper the name of glyoxaline will be applied. **1943** *Thorpe's Dict. Appl. Chem.* (ed. 4) VI. 99/1 Glyoxalines are weak bases. **1946** A. A. Morton *Chem. Heterocyclic Compounds* xiii. 363 Glyoxaline is an empirical name, used sometimes for imidazoles, because the parent member of the series was first derived from glyoxal and ammonia. **1968** I. L. Finar *Org. Chem.* (ed. 4) II. xii. 532 Imidazole (iminazole, glyoxaline) is isomeric with pyrazole.

glyoxime (glai,ǫ·ksim, -im). *Chem.* Also **glyoxim.** [Blend of Glyoxal and Oxime.] **a.** The dioxime, $HON{:}CH{\cdot}CH{:}NOH$, of glyoxal. **b.** Any of the dioximes of the α-diketones.
1895 *Jrnl. Chem. Soc.* LXVIII. i. 36 Configuration of certain glyoximes. **1906** Muir & Morley *Watts' Dict. Chem.* II. 642/2 *Glyoxim*... Formed by the action of hydroxylamine on glyoxal. **1951** A. W. Johnson et al. in E. H. Rodd *Chem. Carbon Compounds* I a. xi. 728 When glyoxal, of which the monoxime is not known, pyruvic aldehyde and the α-diketones are treated with hydroxylamine, the α-dioximes or glyoximes are formed. *Ibid.*, Glyoxime, $CH({:}NOH){\cdot}CH({:}NOH)$, m.p. 178°, is readily soluble in hot water.

glyoxylate (glai,ǫ·ksĭlēⁱt). *Chem.* [f. next + -ate⁴.] A salt or ester of glyoxylic acid.
1856 H. Debus in *Phil. Mag.* 4th Ser. XII. 362 The filtrate yields..a crystallization of a new compound of lime, which I propose to call glyoxylate of lime. **1969** M. F. Utter in J. M. Lowenstein *Citric Acid Cycle* v. 254 These two reactions form the basis of the glyoxylate cycle..which can convert acetyl-CoA to oxalacetate.

glyoxylic (glai,ǫksi·lik), *a. Chem.* [f. Glyox(al + -yl + -ic.] *glyoxylic acid*: the simplest aldehydic acid, obtained as a thick syrupy liquid, $CHO{\cdot}COOH$, with a suffocating smell or as a hydrated crystalline solid, $CH(OH)_2{\cdot}COOH$; it is an important metabolic intermediate and occurs in animal and plant tissues and fluids, esp. in unripe fruit. Also called *glyoxalic acid.*
1856 H. Debus in *Phil. Mag.* 4th Ser. XII. 363 Oxalic acid separates the lime completely as oxalate of lime, and liberates the glyoxylic acid. **1903** Walker & Mott tr. *Holleman's Text-bk. Org. Chem.* 290 Glyoxylic acid contains one molecule of water, which cannot be separated from the acid or its salts without their undergoing decomposition. **1966** *McGraw-Hill Encycl. Sci. & Technol.* VI. 222/1 During metabolic degradation, glycine is deaminated to glyoxylic acid, which is oxidized to formate and carbon dioxide.

glyptal (gli·ptăl). Also †**glyptol.** [Proprietary name, perh. f. Gly(cerol, -yl + *phthalic*, -ate.] Any of the alkyd resins, esp. those formed from glycerol and phthalic anhydride or phthalic acid. Also *glyptal resin.*
1915 *Bull. U.S. Bureau Standards* XI. 385 The glyptol is an artificial resin furnished by the research department of the General Electric Co. *Ibid.* 406 Glyptol is a synthetic compound which has the appearance and many of the properties of amber. **1923** C. Ellis *Synthetic Resins* xvii. 293 Thus far the commercial applications of glyptal resins are of a minor character. **1926** *Trade Marks Jrnl.* 5 May 1033 *Glyptal.* Varnishes and Lacquers, included in Class 1. British Thomson-Houston Company. **1937** R. S. Morrell et al. *Synthetic Resins* ix. 205 The glyptal resins are formed by condensation of polybasic acids with polyhydric alcohols. **1943** [see *alkyd]. **1967** I. L. Finar *Org. Chem.* (ed. 5) I. xi. 286 With dibasic acids, glycerol forms condensation polymers known as alkyd resins, the commonest of which is glyptal, formed by heating glycerol with phthalic anhydride.

gm. Abbrev. of Gramme, Gram².
1875 J. D. Everett *Illustr. C.G.S. Syst. Units* p. ix, The abbreviation *gm.* is used for *gramme* or *grammes.* **1885** [see *g* (s.v. *G III. f)]. **1962** A. R. W. Hayes *Revision Physics* 12 Density of hydrogen = 0·09 gm/litre.

G-man (dʒī·,mæn). Also **G man.** [Probably f. the initial letter of *government* but perh. in Ireland an arbitrary use of the letter 'G'.] **a.** A political detective in Ireland. **b.** In the U.S., a special agent of the American Federal Bureau of Investigation.
1917 M. Skinnider *Doing Bit for Ireland* 177 One morning I was informed there was a 'G-man', as we call government detectives, waiting down-stairs to see me. **1922** Joyce *Ulysses* 160 Jack Power could a tale unfold: father a G man. *Ibid.* 294 The bloody old lunatic is gone round to Green street to look for a G. man. **1926** P. Béaslaí *Michael Collins* I. v. 107 The political detectives of the 'G' division of the Dublin Police came like a flock of carrion crows to pick out 'suspects'... Anyone who had seen that sight may be pardoned if he felt little compunction at the subsequent shooting of these same 'G' men. **1930** F. D. Pasley *Al Capone* (1931) i. 33 He offered a G man (Government agent) ten gran' to forget it. **1935** *Punch* 3 July 6/2 For G-men, as they are called.., form the new licensed killers organised by the Federal Government to override all State laws in the suppression of gangster crime and the capture of every Public Enemy. **1946** Wodehouse *Joy in Morning* iii. 24 He was still giving me that searching look, like a G-man hobnobbing with a suspect. **1958** *Spectator* 3 Oct. 447/1 Mr. Hoover has built up..an immense personal reputation..by creating a legend of the G-man that has largely transferred the romantic interest from the robbers to the cops.

gmina (mi·nă). Pl. **gminy.** [Polish.] A local division of the Polish administrative organization.
1885 *Encycl. Brit.* XIX. 311/1 The 'primary cell' of the administrative organization of Poland is the *gmina. Ibid.* 311/2 Justice is represented by the *gmina* tribunals. **1905** *Daily Chron.* 19 Sept. 5/4 Voters are divided into three classes. 1.—Landed proprietors. 2.—Members of gmina or cantons who own three to twenty acres. 3.—Urban electors. **1959** *Chambers's Encycl.* XI. 8 For purposes of local government the provinces are divided into districts ..which are sub-divided in turn into sub-districts (*gminy*).

gnamma hole (næ·mă hōᵘl). *Austral.* [Aboriginal.] A natural hole in rock, containing water; a native well.
The same word as *namma.
1903 J. Marshall *Battling for Gold* 14 The gnamma holes were full of water. **1927** M. Terry *Through Land of Promise* xii. 154 He found two gnamma holes (or rock holes) each holding about 40 gallons of water. **1950** C. Good *Yarns of Yilgarn* 9 A native well (or gnamma hole) in the rocks.

gnat¹. Add: **3.** *gnat-like* adj. and adv.; **gnat's** (or **gnats'**) **piss** *slang*, a very weak beverage; a drink of poor quality.
1905 *Westm. Gaz.* 1 Feb. 2/3 Drifts gnat-like to the flame. **1915** E. R. Lankester *Divers. Naturalist* 216 In early September, golf links..swarm with a large gnat-like fly of reddish-brown body. **1950** W. de la Mare *Inward Compan.* 37 A gnatlike wail. **1959** I. & P. Opie *Lore & Lang. Schoolch.* ix. 164 Weak tea may be 'gnat's piss'. **1963** B. S. Johnson *Travelling People* v. 110 Where'd you get this gnatspiss from, Maurie?.. I can get you gnatspiss as good as this gnatspiss for sixteen bob a bottle. **1966** 'L. Lane' *ABZ of Scouse* 41 *Gnats' piss*, cider, near beer, weak tea or any drink.

gnathal, *a.* (Later examples.)
1935 R. E. Snodgrass *Princ. Insect Morphol.* vi. 100 The embryonic gnathal segments constitute a distinct gnathal section of the trunk. **1964** R. W. & J. M. Fox *Introd. Compar. Ent.* v. 138 It is interesting to find that the Diplopoda, despite the difficulty in homologizing their gnathal segments, also have four pairs of glands.

gnathion (nē·þi,ǫn). *Anat.* [f. Gr. γνάθ-ος jaw + *-ion².] In the cranium, the lowest point of the inferior border of the mandible in the mid-line; the menton.
1888 *Catal. Marsupialia Brit. Mus.* p. viii, Basal length.—From the 'basion', or lower front edge of the *foramen magnum*, to the 'gnathion'. **1920** H. H. Wilder *Lab. Man. Anthropometry* i. 44 *Gnathion* (gn). The lowest median point on the lower border of the mandible. **1951** *N. & Q. Anthrop.* (ed. 6) i. 12 The distance between the nasion and the gnathion, which is the lowest and most anterior point on the lower border of the lower jaw. **1968** D. H. Enlow *Human Face* xii. 281/1 The subjective location of gonion and menton (or gnathion), another method of forming the mandibular plane, tends to be just that.

gnatho- (nē·þo). Combining form of Gr. γνάθος jaw, as in **gnathoce·phalon** (see quot. 1935); hence **gnathocepha·lic** *a.*; **gnathochila·rium**, a mouth-part found in Diplopoda; **gnathodynamo·meter** (see quot. 1969); **gnathoso·ma, gna·thosome** (see quot. 1964); hence **gnathoso·mal** *a.*; **gnathothe·ca**, the horny covering of the lower beak of a bird.
1935 R. E. Snodgrass *Princ. Insect Morphol.* vi. 102 The gnathocephalon is the region of the insect head that supports the mandibles, the maxillae, and the labium. **1964** R. M. & J. W. Fox *Introd. Compar. Ent.* ii. 42 The gnathocephalon bears the mouthparts. *Ibid.*, The gnathocephalic ganglia frequently are fused into a single mass. **1887** *Ann. New York Acad. Sci.* June 39 This genus [*sc. Nannolene*]..approaches *Julomorpha*, especially in the construction of the gnathochilarium. **1964** R. M. & J. W. Fox *Introd. Compar. Ent.* iii. 85 Diplopods have an accessory feeding structure located behind and between the mandibles, a highly modified affair called the gnathochilarium. **1959** *New Scientist* 2 Apr. 736/1 Many devices [for measuring biting power] have been produced.., with the titles of oral force-meters or gnathodynamometers. **1969** *Gloss. Terms Dentistry* (*B.S.I.*) 75 *Gnathodynamometer*, an instrument for measuring the force exerted in closing the jaws. **1948** *Smithsonian Misc. Coll.* CX. x. 61 The distinctive feature relating to the feeding apparatus of the Acarina is the presence of a discrete head structure carrying the mouth parts, known as the *capitulum, capitellum*, or *gnathosoma.* **1959** T. E. Hughes *Mites* x. 135 The gnathosoma is movable on the rest of the body and is well sclerotized... The gnathosomal roof consists of the fused tergites of the segments concerned. **1964** R. M. & J. W. Fox *Introd. Compar. Ent.* iii. 81 The large, complex order Acarida is characterized by the presence of a gnathosome (capitulum) formed from anterior segments and having the appearance of a little head bearing the mouthparts. **1889** *Cent. Dict.*, Gnathotheca. **1890** E. Coues *Handbk. Field & Gen. Ornithol.* ii. iii. 151 The entire covering of both jaws together is called *rhamphotheca*..of the under, *gnathotheca*..but these terms are not much used. **1959** Van Tyne & Berger *Fund. Ornith.* ii. 31 The oral cavity is bounded by the horny coverings of the upper (*rhinotheca*) and lower (*gnathotheca*) mandibles.

gnathobase (nē·þobēⁱs). *Zool.* [f. *gnatho- + Base *sb.*¹] A process developed from the lower part of a segmented limb in certain arthropods, modified to bite or crush food. Hence **gnathoba·sic** *a.*
1881 E. R. Lankester in *Q. Jrnl. Micr. Sci.* Apr. 348 Of the six endites the proximal is somewhat isolated and pushed towards the middle line... It is a jaw process, and may be spoken of as the 'gnathobase'. **1902** —— in *Encycl. Brit.* XXV. 695/1 The five pairs of appendages of the post-oral somites of the head or prosoma thus constituted all primitively carry gnathobasic projections on their coxal joints, which act as hemignaths; in the more specialized forms the mandibular gnathobases cease to develop. **1904** W. T. Calman in *Ann. & Mag. Nat. Hist.* Feb. 155 The double series of epipodial lamellæ, the segmentation of the thoracic limbs, the double gnathobasic lobes of the first pair. **1932** Borradaile & Potts *Invertebrata* x. 274 Some of the evidence suggests an archetype [arthropod] with a nine-segmented axis bearing on the median side of the first segment a biting process (gnathobase). **1964** *New Scientist* 13 Aug. 365/2 The mandibles themselves may either derive from the basal part of a segmented limb (the gnathobase) or from a whole jointed limb.

gnathostome (nē·þostōᵘm). [a. G. *gnathostome* (E. Haeckel *Anthropogenie* (1874) xviii. 432), f. *gnatho- + Stoma.] A member of the Gnathostomata, a superclass of vertebrates with fully developed upper and lower jaws. Also as *adj.*, of or pertaining to this group. Also **gnathosto·m(at)ous** *a.*
1889 *Cent. Dict.*, Gnathostomatous. **1905** J. McCabe tr. *Haeckel's Evol. Man* II. xxi. 560 All the vertebrates or gnathostomes, from the fishes to man, descend from a common, extinct, fish-like ancestor. **1908** W. H. Gaskell *Orig. Vertebr.* i. 60 From such gnathostomatous fishes the cyclostomes were supposed to have descended. **1909** E. S. Goodrich in R. Lankester *Treat. Zool.* IX. 92 The Gnathostomes might be divided into the Ichthyomorpha (Pisces) and the Tetrapoda (Amphibia, Reptilia, Aves, Mammalia). **1933** A. S. Romer *Vertebr. Paleont.* iii. 35 All possess jaws, structures indicating so great and significant an advance that these forms..are often spoken of as gnathostomes ('jaw-mouthed'). **1936** Neal & Rand *Compar. Anat.* viii. 305 Gnathostome fishes have an immovable tongue. **1954** G. C. Kent *Compar. Anat. Vertebr.* iii. 48 Agnathons lack true jaws..; all other vertebrates, because they exhibit jaws, are gnathostomes. **1968** R. Zangerl tr. *Peyer's Compar. Odontology* 30 The placoderms are all more or less clearly gnathostomous.

gnatty (næ·ti), *a.* [f. Gnat¹ + -y¹.] Resembling a gnat; infested with gnats.
1846 J. J. Hooper *Adv. Simon Suggs* ii. 28 You.. gnatty, flop-eared varmint! **1909** *Amat. Photographer* 14 Sept. 252/2 They can venture into the 'gnattiest' woods without suffering any assault.

gnaur, var. Knar.
1869 M. T. Masters *Veget. Teratol.* 158 The huge gnaurs and burrs met with occasionally on some trees often produce great quantities..of roots. **1900** B. D. Jackson *Gloss. Bot. Terms*, *Gnaurs*, burrs or knotty excrescences on tree-trunks or roots, probably from clusters of adventitious buds. **1903** F. W. Burbidge *Let. to W. T. Thiselton-Dyer* 20 June (MS.), I beg to hand you a 'gnaur' or swollen, arrested branch of a Tulip tree.

gnocchi (nǫ·ki, ‖nyǫ·ki), *sb. pl.* [It., pl. of *gnocco*, f. *nocchio* a knot in wood.] Small dumplings made with flour, semolina, potato, etc., cooked and served with cheese or seasoning.
1891 A. B. Marshall *Larger Cook. Bk. of Extra Recipes* 541 Gnocchi à la Française. *Ibid.* 542 Stand the tin containing the Gnocchi in a tin with boiling water. **1927** M. Diver *But Yesterday* xii. 142 He no longer damned lunch when he found himself eating gnocchi flavoured and fried to perfection. **1963** *Punch* 12 June 869/3 People gave me spaghetti and gnocchi and pannini.

gnome¹. Add: Specifically with reference to Old English verse.
1914 B. C. Williams *Gnomic Poetry Anglo-Saxon* 66 The *dol bið* gnome and its analogues have been noticed as occurring in Christian passages. **1935** A. C. Bartlett *Larger Rhet. Patterns Anglo-Saxon Poetry* 73 One..does find..most gnomes in *Beowulf, Wanderer, Widsith*, and most homiletic passages in the 'Christian' poems. **1948** S. O. Andrew *Postscript on Beowulf* viii. 93 The gnome serves to connect the idea of evil endured in the sentence preceding it with that of deliverance..in the sentence following it.

gnome². Add: **1. b.** A statue or figure of a gnome, esp. one placed in a garden. Cf. *garden gnome* (*garden *sb.* 6).
1938 D. Kincaid *Brit. Social Life in India* xii. 288 An imitation pergola and a coloured plaster gnome or two. **1955** E. Coxhead *Figure in Mist* ii. 65 'She sculpts too, did you know?' Agnes said. 'No, how unspeakable. You mean those gnomes one sees about?' **1969** V. C. Clinton-Baddeley *Only Matter of Time* 18 In the centre

of the grass was a lily pond. On the edge a gnome sat fishing. **1971** J. Betjeman in Betjeman & Vaisey *Vict. & Edw. Oxford* p. vii, They are now acres of detached and semi-detached houses, blocks of flats and housing estates, gnomes, birdbaths and shopping arcades.
c. *colloq.* An international financier or banker, *spec.* one who is Swiss; esp in phr. *the gnomes of Zurich.*
1964 *New Statesman* 27 Nov. 822/3 The gnomes of Zurich and their related goblins in the more politically involved capitals. **1968** P. Einzig *Leads & Lags* p. vii, Sterling was devalued in 1949 and again in 1967, not because of speculation by the 'gnomes' of Zurich. **1969** *Listener* 12 June 828/2 One doesn't have to posit gnome couriers flying in from Zurich with weekly instructions.

gnomic, *a.* Add: Specifically with reference to Old English verse.
1826 W. D. Conybeare in J. J. Conybeare *Illustr. Anglo-Saxon Poetry* 205 Another gnomic poem..concludes by tracing the origin of discord to the homicide of Cain. **1842** B. Thorpe *Codex Exoniensis* p. viii, The 'Gnomic Verses' are, as their title imports, a string of proverbial sentences. **1871** H. Sweet in W. C. Hazlitt *Warton's Hist. Eng. Poetry* II. 18 The curious poem, *Salomon and Saturn*, consists also of a variety of gnomic sentences, mixed..with a variety of other matter. **1914** B. C. Williams *Gnomic Poetry Anglo-Saxon* 6 Anglo-Saxon verse is gnomic so far as the presence of gnomic lines here and there adds sententiousness. **1948** K. Malone in A. C. Baugh *Lit. Hist. Eng.* i. i. iii. 26 One might have expected to find end-stopping used a good deal in the gnomic verses.

gnoscopine (nǫ·skōpīn, -in). *Chem.* [irreg. f. Gr. γι-γνώσκ-ειν to know + Op(ium + -ine⁵.] Either of two isomeric alkaloids, $C_{22}H_{23}NO_7$, designated α- and β-gnoscopine, of which α-gnoscopine is found in opium and β-gnoscopine obtained synthetically.
1878 T. & H. Smith in *Pharm. Jrnl. & Trans.* 3rd Ser. IX. 82/1 In the mother liquors from the purification of narceine we have now repeatedly met with a crystalline body, which..was..ascertained to be a hitherto unknown alkaloid, and which we have named Gnoscopine. **1914** Hope & Robinson in *Jrnl. Chem. Soc.* CV. ii. 2087 It is proposed to call the substance melting at 180° β-gnoscopine, reserving the name α-gnoscopine for the compound previously known as gnoscopine or *r*-narcotine, which melts at 229°. **1951** A. Grollman *Pharmacol. & Therapeutics* iv. 88 The remaining [opium] alkaloids (laudanosine,..codamine, gnoscopine, etc.) occur in too small quantity to have any influence on the action of the crude drug. **1954** Stanek & Manske in Manske & Holmes *Alkaloids* IV. xxxii. 181 Both of the gnoscopines were convertible into narceine, indicating that the only differences were stereochemical ones, and β-gnoscopine was partly isomerized to α-gnoscopine.

gnosiological (nōᵘzi,ǫlǫ·dʒikăl), *a.* Also **gnoseological.** [f. Gnosiology + -ical.] Of or pertaining to gnosiology.
1928 in *Funk's Stand. Dict.* **1938** G. Reavey tr. *Berdyaev's Solitude & Society* ii. i. 38 We have also to elucidate the relationship between the pure subject of cognition, the *gnosiological* subject, and the Ego, the knowing subject. **1963** V. Nabokov *Gift* iv. 235 In eliminating metaphysical dualism he fell into gnoseological dualism.

gnotobiology (nōᵘ:tobai,ǫ·lŏdʒi). *Biol.* [f. as next + Biology.] = *gnotobiotics.
1963 T. D. Luckey *Germfree Life & Gnotobiology* i. 29 Many quantitative biological relationships..need simple and exact description... A system which describes the quantitative relationships presently known in gnotobiology..is recommended. **1970** *Sci. Jrnl.* Apr. 46/2 A highly specialized segment of the ultraclean technology is gnotobiology, the raising of germ free animals, largely for research purposes.

gnotobiotic (nōᵘ:tobai,ǫ·tik), *a. Biol.* [f. Gr. γνωτ-ός known (f. γιγνώσκειν to know) + -o + βιωτικ-ός pertaining to life, f. βιωτός worth living, f. βιοῦν to live (βίος life).] Of an organism (esp. a higher animal) or its environment: artificially rendered devoid of bacteria and other organisms which would normally be present as parasites, commensals, symbionts, etc., or having only a few known organisms of this kind present.
1949 J. A. Reyniers et al. in *Lobund Rep.* Feb. 160 [After] the process of freeing an organism from its contaminants..the organism becomes gnotobiotic. **1964** F. Pohl in *Galaxy* Oct. 176/2 They had a gnotobiotic atmosphere, a little rich in oxygen, a little more humid than the ambient air. **1969** *Nature* 29 Nov. 846/1 Bacterial interactions..in gnotobiotic flies (grown from germ free eggs and fed a known mixture of bacteria). **1970** *New Scientist* 15 Jan. 103/1 Gnotobiotic animals..provide a highly sensitive system for detecting small numbers of micro-organisms.
Also **gnotobio·sis** [Gr. βίωσις way of life], the state of being gnotobiotic; **gnotobio·tics** [-ic 2], the field of study concerned with rearing and using gnotobiotic organisms.
1949 J. A. Reyniers et al. in *Lobund Rep.* Feb. 160 Gnotobiotics..would be the field of investigation concerned with growing living things by themselves or in an association with other completely known kinds of organisms. *Ibid.*, The state of 'germ-freeness' or being without contaminants is gnotobiosis. **1953** *Parasitology* XLII. 260/1 As defined by Reyniers et al., *gnotobiosis* is synonymous with *axenity* only, even though *gnotobiotics* is proposed for the science of *both* axenic and synxenic cultivation; this seems inconsistent, and I feel that the term *gnotobiosis* should be used in the broader sense. *Ibid.*, The literature on gnotobiotics of animal organisms. **1970** *New Scientist* 15 Jan. 101/1 Gnotobiotics includes the study of both 'germ-free' animals and animals whose microbial flora can be completely specified.

go, *sb.* Add: **2. b.** Vigorous activity; hard work; esp. in phr. *it's all go. colloq.*
1965 *New Statesman* 14 May 777/1 Believe me, it's all go with these tycoons, mate. Life's just one frenetic whirl of soigné secretaries and sex-mad air 'ostesses. **1967** *Boston Sunday Herald* 14 May ii. 16/6 All the time it's go in Florida.
4. Add to def.: a single spell of doing something, esp. in phr. *at one go.* Also *to have a go*: to make an attempt; to act resourcefully or with initiative; *spec.* (*a*) in *Cricket*, to hit out recklessly in an attempt to make runs; (*b*) to take independent or single-handed action against criminals.
1825 J. Neal *Bro. Jonathan* I. 8 He would rivet his large eyes, for half an hour at a 'go', upon some part of the wall. **1894** N. Gale *Cricket Songs* 25 Toss him down a slow, you see, He's sure to have a go, you see! **1912** W. Elmhirst *Freshman's Diary* 28 Apr. (1969) 67 We paddled most of the way up..& then Beach had a go at punting. Coming back I had a go for some time. **1933** *Punch* 1 Feb. 126/1 Whether my voice would work the charm, Frankly, I didn't know; But as it couldn't do much harm I thought I'd have a go. **1963** *Times* 28 Feb. 3/6 After several overs, during which he seemed to command the bowling, he announced to Reid, who was his partner, that he had 'had it' and was going to have a go. **1965** *Times* 1 Jan. 4/6 Mr. Bacon was asked what advice he would give to members of the public who saw a gunman carrying out a raid. '..if you can have a go then have a go,' he replied. *Ibid.*, It is completely wrong to encourage people to 'have a go'..particularly when the criminals they are..having a go at are trigger-happy hooligans. **1966** *Economist* 6 Aug. 534/1 There is a cryptic passage about legal protection for citizens who fight against crime (Russians who 'have a go'?). **1971** *Daily Tel.* 4 Mar. 15/4 A cashier who tried to 'have a go' was hit on the head with a gun butt.
b. A contest or battle; a fight, esp. a prize-fight or boxing-match; an argument. *colloq.*
1890 *Texas Siftings* 1 Nov. 7/3 Cost me five dollars the other day to see the tamest kind of a go. There wasn't a knockdown in ten rounds. **1893** K. Mackay *Out Back* (ed. 2) ii. i. 153 Douglas and Wrixon dearly loved a 'go' between horses. **1902** W. Satchell *Land of Lost* xxvi. 243 There was the little go we 'ad on the gumfield—'e 'ad all the best of that. **1959** I. & P. Opie *Lore & Lang. Schoolch.* x. 197 The Liverpudlian says 'Come on I'll 'ave you a go'..or 'I'll have you a scrap'.
c. An attack or bout of some sickness or ailment.
1890 Ld. Lugard *Diary* 3 Jan. (1959) I. 62 Afraid shall have a go of fever, for which this place is notorious. **1920** *Blackw. Mag.* May 614/2 A bad go of snow-blindness had driven me in to Leh. **1928** Galsworthy *Swan Song* v. 45 No second go of measles, Jon.
d. (*a*) *fair go*: equitable treatment; a fair chance, a 'square deal'; *spec.* a fair fight; also without *fair*; also *int.* and as an expression of enquiry; *to give it a go*: to give it a try; to make an attempt. Chiefly *Austral.* and *N.Z. colloq.*
1911 L. Stone *Jonah* ii. vi. 215 The spinner handed his stake of five shillings to the boxer, who cried 'Fair go!' The spinner placed the two pennies face down, and then ..the coins flew twenty feet into the air. **1918** *Chrons. N.Z.E.F.* 11 Oct. 130/1 An interesting competition. Give it a go! *Ibid.* 22 Nov. 264/2 It's a bit tight, but we'll give it a go. **1919** W. H. Downing *Digger Dial.* 22 *Fair go*, equitable treatment; a fair field and no favor. **1937** J. A. Lee *Civilian into Soldier* i. 25 'I'm not getting a fair go.' Guy tried to explain the injustice. **1941** Baker *Dict. Austral. Slang* 27 *A fair go*, a fight. *Ibid.* 28 *Fair go!*, be reasonable! Give him (it, etc.) a chance! **1944** W. E. Harney *Taboo* (ed. 3) 125 So he said, 'All right, old man, you sing rain and that boy can have Rosie. Fair go.' **1947** 'A. P. Gaskell' *Big Game* 22 No, fair go, Bomb. The Adjutant wants you. **1959** H. P. Tritton *Time means Tucker* vii. 55/1 Dutchy..asked him to give us a fair go. **1961** B. Crump *Hang on a Minute Mate* xiii. 149 Fair go? asked Jack, scenting a yarn in Sam's tone. **1961** P. White *Riders in Chariot* xi. 346 A man stands a better chance of a fair go if he's got a mate. **1963** *Times* 31 Jan. 4/1 Had England been left to score 300 or more at something like 70 an hour as was possible and 'given it a go', the odds would have been heavily against them. **1969** *Advertiser* (Adelaide) 12 May 5/4 Stop whingeing and give a bloke a go, mates. **1969** *Coast to Coast 1967–68* 23 He braced himself, but staggered under their onslaught and came down on his knees. 'Fair go!' 'Poor Daddy,' she murmured, settling down beside him. **1971** *N.Z. Listener* 22 Mar. 13/1 Finger's out, honest. I'm giving it a go, Dad.
7. Delete *rare* exc. *U.S.* and add earlier and later examples; esp. in phr. *to make a go of it.*
1876 Sothern in J. F. Daly *Life A. Daly* (1917) 222, I learn yr. new piece is a 'great go'—so it's quite on the cards you can do as well without me. **1896** [see B.P. (a) s.v. B III]. *a* **1911** D. G. Phillips *Susan Lenox* (1917) I. xv. 254 He had told the company that Susan was sure to make a go; and after she had made a go, he announced the beginning of a season of triumph. **1933** *Punch* 19 Apr. 442/3 There seemed every reason for Tilly and Stevan to make a go of it. **1942** E. Paul *Narrow St.* xxix. 261 The Amances had been unable to make a go of the hotel, and Monsieur Henri had been obliged to take it over. **1959** G. Freeman *Jack would be Gent.* viii. 168 You have to make a go of marriage, you have to work to make a marriage a success.
b. A bargain, an agreement, a 'deal'; usu. in phr. *it's a go. colloq.*
1878 B. Harte *Man on Beach* 61 'Then it's a go?'.. 'It's a go.' **1908** *Magnet* I. 1, The Remove don't like you now, but we'll stick together, and bring them round. Is it a go? **1936** Wodehouse *Laughing Gas* i. 15 'Then say no more,' I said. 'It's a go.'
8. a. Also applied to a person: no good; not a success. See also No-go.
1829 T. Creevey *Let.* 20 Oct. in J. Gore *Creevey's Life & Times* (1934) xiv. 311 According to the Earl and myself, Fanny Kemble is *no go.* **1912** J. London *Let.* 19 Nov. (1966) 368 No; I told Eliza, George is no go.

go (gōᵘ), *sb.*² Also **Go, Goh, I-go.** [Jap. *go* small stone.] A Japanese board game of territorial possession.
1890 B. H. Chamberlain *Things Japanese* 136 *Go*, sometimes, but with little appropriateness, termed 'checkers' by European writers, is the most popular of the in-door pastimes of the Japanese. **1911** H. F. Cheshire *Goh or Wei Chi* 15 In the East proficiency in Goh is one of the best recommendations to high places. **1958** *Listener* 13 Nov. 786/2 The game that does seem to me to be superior to chess, in that it has both depth and simplicity, is the Japanese game of Go. **1960** R. C. Bell *Board & Table Games* iii. 96 This Japanese game [*sc.* Gobang]..is a poor relation of the intellectual I-go.

go, *v.* Add: **A. 8. b.** For *Obs.* read *Obs. exc. dial.*, and add later examples.
1883 'Mark Twain' *Life on Mississippi* xliv. 450 The unpolished [Southerners] often use 'went' for 'gone'... 'He had n't ought to have went.' **1884** —— *Huck. Finn* xxviii. 287 I'll tell Miss Susan to..say you've went away for a few hours. **1886** F. T. Elworthy *W. Somerset Word-Bk.* 825 You never didn ought to a-went; for— You ought not to have gone. **1890** *Dialect Notes* I. 7 Mr. Emerson mentioned the principal parts *go, went, went* (*have went* for *have gone*).
B. 7. Now usu. in phr. (*to be*) (a specified period) *gone*: to have been pregnant for that amount of time.
1747 [in Dict.]. **1748** Smollett *Rod. Rand.* II. xlvi. 107, I am now four months gone with child by him. **1900** *Daily Express* 19 June 4/5 Vincent was turned out into Commissioner-street with Mrs. Vincent seven months gone. **1931** W. Holtby *Poor Caroline* iv. 133 Brought her to the Home, four months gone, and won't be fifteen till next March. **1935** N. Mitchison *We have been Warned* iv. 407 My mother found she was six months gone. **1955** [see *catch *v.* 9 b].
19. b. *to go big*: to be a big success, have a large sale. Cf. *87 h. *slang* (orig. *U.S.*).
1903 Ade *People you Know* 88, I done that with a Piece called 'A Boiled Dinner', and it always went big. **1930** *Publishers' Circular* 22 Feb. 186 We have reason to believe that *The Miracle of Peille*..will go big. **1936** Wodehouse *Laughing Gas* xxi. 231 The nosegay didn't seem to go very big. I was not feeling strong enough to pick it up, but I shoved it forward with my foot.
c. *spec.* Of a route, etc., in Mountaineering: to be usable.
1883 W. S. Green *High Alps of N.Z.* xiv. 232 The route to the northern ridges of the peak..would not 'go'. **1937** H. W. Tilman *Ascent of Nanda Devi* vi. 52, I wanted to have a look at the difficult part of the gorge to see if the route would still 'go'. **1956** C. Evans *On Climbing* vii. 112 It is important, as on rocks, to cultivate an eye for what will 'go' and what will not, so that no time is wasted.
d. To be acceptable or permitted; esp. in phr. *anything* (or *everything*) *goes. colloq.*
1879 G. Meredith *Egoist* II. i. 17 Everything goes on the stage, since it's only the laugh we want on the brink of the action. **1887** *Lantern* (New Orleans) 19 Feb. 5/1 As everything goes..I'll pass it by with never a word to say. **1891** *Harper's Mag.* Dec. 104/2 Any other night goes, but not this night. **1892** Kipling *Many Invent.* (1893) 168 Remember, everything goes in the States, from a trouser-button to a double-eagle. **1921** *Ladies' Home Jrnl.* June 80/3 One of the few real 'movie' fortunes has been made by a man who..has constantly exploited the vicious theory that 'anything goes in fun'. **1934** Cole Porter (*show- and song-title*) Anything goes. **1958** M. Allingham *Hide my Eyes* 191 Anything goes if it's done by someone you're fond of. **1959** *Listener* 15 Jan. 113/1 In a brawl, they're deadly. Anything goes—spurs, hobble chains, the lot. **1960** *Sunday Express* 12 June 14/3 In the evening anything goes..from Baby Doll frills to matador pants and a billowing overskirt.
e. To be accepted or carried into effect; to have authority or effectiveness; to be valid; to be obeyed without question; esp. in phr. *what I say goes. colloq.* (orig. *U.S.*).
1891 *Century Mag.* Dec. 190/1 'Can you fix it?'..'I guess what I say to Jim goes,' she said. **1901** Merwin & Webster *Calumet 'K'* 62 My only order was, 'Clear the road—and be damn quick about it.' What I said went. **1931** *Punch* 18 Mar. 286/3 An emergency Cabinet of five.. with supreme emergency powers..subject of course to full democratic control. But what I say goes. **1949** G. B. Shaw *Buoyant Billions* iii. 31 He makes so much money that whatever he says, goes.
21. e. In fig. phr. *let it go at that*: let that account, estimate, conclusion, etc., be accepted; let us say no more about it.
1881 J. C. Harris *Uncle Remus* xxx. 111 Hit wuz Miss Molly Cottontail, en I speck we better let it go at dat. **1898** E. N. Westcott *David Harum* 306 'Very well,' said John, 'we will let it go at that.' **1917** J. Farnol *Definite Object* ii. 19 Eleven will do as well as any other time; let it go at that. *Ibid.* xxxvi. 312 Somebody tried to kill me, but somebody didn't kill me; here

I am, getting stronger every day, so we'll let it go at that. 1955 J. D. Salinger *Franny & Zooey* (1962) 13 I'm no Freudian Man or anything like that, but certain things you can't just pass over as capital-F Freudian and let them go at that.

f. *go well.* A form of address, of African vernacular origin, used at parting.

1948 A. Paton *Cry, Beloved Country* I. ii. 14 That was good of you. Go well, small one. 1951 P. Abrahams *Wild Conquest* II. ii. 298 'Where do you journey?' 'To the land of the Basuto.' 'Go well, my friend.' 'Go well, white man.' 1961 H. Stanton *Go Well stay Well* 5 'Go well.' 'Stay well.' Sesuto: 'Tsamea pila.' 'Sala pila.' These words constitute the customary expressions of good-will when African friends are parting. The friend who stays says to the friend who goes, 'Go well', conveying the thought, 'May God protect you on your journey.'

22. *from the word Go*: for *U.S. colloq.* read *colloq.* (orig. *U.S.*), and add earlier and later examples.

1834 D. Crockett *Narr. Life* 59, I was plaguy well pleased with her from the word go. 1934 H. G. Wells *Exper. Autobiogr.* I. v. 211, I lost him from the word Go. 1937 —— *Star Begotten* vii. 135 People will hate them from the word Go! 1963 *Times* 24 Jan. 6/6 It was wrong from the word 'go' to put in a limitation such as 60.

b. To play jazz or similar music excitingly or uninhibitedly; to 'swing'. Freq. *imp.*

1926 *Melody Maker* Jan. 19 Atta-boy, let's go! 1935 *Vanity Fair* Nov. 71 Hot artists or bands that can put across their licks successfully..can 'go'. 1958 *Punch* 27 Aug. 270/1 In a kind of trance I began to cry rhythmically 'Go man go!'

24. a. (Further examples.)

1764 H. Walpole *Let.* 10 May (1904) VI. 62 There was a whole-length of Sir Henry Sidney, which I should have liked, but it went for fifteen guineas. 1968 M. Torrie *Your Secret Friend* v. 54 What did Little Monkshood go for? Do you know? 1971 *Radio Times* 30 Oct.–5 Nov. 36/3 Sunday tv.. Going for a Song. Customers and connoisseurs explore the world of *Antiques*.

27. b. Also, of a batsman: to be 'out'.

1862 *Baily's Mag.* Aug. 90 Hornby..was the first to go. 1904 P. F. Warner *How we recovered Ashes* vii. 105 Three of the greatest batsmen in the world gone for nine runs on a perfect wicket!

30. d. Colloq. phr. *where do we go from here* (or *there*)? = what shall we do next? what happens next?

1922 Wodehouse *Clicking of Cuthbert* ix. 222 'I've had three ginger ales,' observed the boy. 'Where do we go from here?' 1937 A. Christie *Death on Nile* II. xxvi. 255 Well, sir, where do we go from here? 1952 R. Vickers *Sole Survivor* 45 He was as bewildered as I was myself... I asked: 'Where do we go from *here*, Gramshaw?' 1954 M. Hutton in *Best One-Act Plays of 1952–53* 57 Right. We've got her name. Where do we go from there?

31. f. *to go places*: to go to various places; to travel; *spec.* (*fig.*) to be successful, to 'make the grade'; to make progress. *colloq.* (orig. *U.S.*).

1925 S. Lewis *M. Arrowsmith* xx. 232 The habit of social ease, of dressing, of going places without nervous anticipation. 1932 *Lit. Digest* 30 Jan. 36/2 All Russians love to 'go places'. 1933 A. Waugh *Wheels within Wheels* xi. 222 You want to be able to go places and do things. *Ibid.* vii. 137 He hasn't been places yet. 1934 M. Weseen *Dict. Amer. Slang* x. 142 *Go places*, to be successful; to get ahead. 1944 L. A. G. Strong *Director* 254 They were jealous because she'd made the grade... She was going places. 1947 N. Cardus *Autobiogr.* 250 This for me, is to live and to 'go places'. 1958 'A. Gilbert' *Death against Clock* 85 We're going places right away.

g. To urinate or defecate; to go to the lavatory; freq. *to go somewhere*. *colloq.*

1926 E. Bowen *Ann Lee's* 71 Do you want to go somewhere? Have you got a pain? 1935 *Time* 24 June 45/1, I took off all my clothes but my drawers and—well—I had to go. 1959 W. Golding *Free Fall* ii. 61 But I been three times... I can't pee any more! 1962 J. Wain *Strike Father Dead* 138 He's in the men's room. He's been wanting to go all evening, but as long as you were playing he didn't want to miss a note. 1967 F. Mullally *Prizewinner* i. 19 Excuse me..I've got to go somewhere.

32. a. For 'Now *arch.* and *dial.*' read: 'Now *colloq.* and *U.S.*', and add later examples; *to go get*: to go and get; to reach; to work hard or ambitiously (cf. *go-getter).

1893 [see *chase *v.*[1] 7 c]. 1912 *Pedagogical Seminary* XIX. 96 A rebuke to pride with the notion of 'get out',.. 'Go jump in the lake.' 1922 Joyce *Ulysses* 750, I dont like a man you have to climb up to go get at. 1928 F. Hurst *President is Born* viii. 113 They're the go-getters without your brains to go-get! 1929 Galsworthy *Roof* iii. 45 *Froba.* Heu! You are go-getters. *Bryn.* What is a go-getter? *Froba.* Kind of an early bird—go gets the worm. 1968 *Globe & Mail* (Toronto) 17 Feb. 35 (Advt.), Now there is polish as well as boyish personal appeal. Go see. 1968 *Encounter* Sept. 22/1 Go hit your head against the wall. 1969 A. Glyn *Dragon Variation* vii. 216 Let's go get ourselves a drink.

34. c. (Further examples.)

1847 C. M. Yonge *Scenes & Characters* vi. 57, I told my brother I did not think Mr. Devereux would go for to say such a thing. 1886 H. Baumann *Londinismen* 65/1 Don't you go for to think thet 'cos my 'air is grey, I ain't got any strength left me. 1895 'M. E. Francis' *Daughter of Soil* iii. 29 O' course, I wouldn't go for to say sich a thing o' th' Squire.

35. e. *spec.* To eat or drink (something specified); esp. in phr. *I, you*, etc., *could go* (a drink, etc.). (Cf. quot. *c* 1882 in Dict.) *colloq.*

1929 F. Bowen *Sea Slang* 58 *To go*, to eat. 1944 F. I. Cooze *Kiwis in Pacific* 12 God, how I could go a nice, long cool beer. 1948 D. Ballantyne *Cunninghams* (1963) I. iv. 25, I could go a good feed of eels just now.

f. To yield, produce (a certain amount). *U.S.*

1816 U. Brown *Jrnl.* 23 July in *Maryland Hist. Mag.* (1915) X. 369 None [of the fields] that I saw will go 15 Bushels to the Acre. 1868 *Iowa State Agric. Soc. Rep. 1867* 160 Corn.—On well-manured land..will go seventy-five bushels. 1869 'Mark Twain' *Innoc. Abr.* 57 An altar with facings of solid silver—at least they call it so and I think myself it would go a couple of hundred to the ton (to speak after the fashion of the silver miners).

36. Also *to go* (an amount) *better*: to raise the bet by (so much); *to go* (a person) (*one*) *better*: to outbid or outdo (someone). *U.S.*

1845 S. F. Smith *Theatr. Apprent.* 148 My adversary went the dollar, and five better. 1864 W. B. Dick *Amer. Hoyle* (1866) 174 B. puts down a dime. C. says, 'I'll go a dime better,' and he puts down two dimes. 1874 *Congress. Rec.* 17 Feb. 1556/1 The Senator from North Carolina..proposes an increase of 46 millions of banking circulation. The Senator from Pennsylvania..goes him better, and proposes an unlimited increase. 1889 K. Munroe *Golden Days* 66 [I was] going him one better every time for a fifty vara lot. 1890 [see See *v.* 13—second *transf.* example]. 1953 W. Stevens *Let.* 21 Dec. (1967) 805 The weather was constantly going me one better. 1960 W. H. Whyte *Organization Man* 34 Some European critics of America have gone them one better.

b. With personal object: to take on (a person) in a wager, game, or offer. *U.S. colloq.*

a 1846 W. T. Porter *Quarter Race Kentucky* 179, I knows Bill well, and I'll go you an independent on his beating yon feller bad. 1876 [see Dict., 36]. *a* 1889 *Spirit of Times* (Farmer), I goes you five dollars, this time. 1902 S. E. White *Blazed Trail* xxiv. 161 'Surely you won't refuse to be my guest here..!' 'Wallace,' said Thorpe, 'I'll go you.' 1909 'O. Henry' *Options* (1916) 38 'Believe I'll go you,' he said, brightening. 'I'll accept the invitation gladly.'

c. *Cards.* To make a call or declaration of; to accept (a specified number of cards).

1876 [see Napoleon 4]. *c* 1884, 1898 [see Nap *sb.*[2] 2 b]. 1932 *Sunday Express* 6 Mar. 13/2 Cries of 'I'll stick' and 'I'll go four' heard by police..led to..a summons for permitting gaming. 1964 N. Squire *Bidding at Bridge* ii. 25 The principle of the rebid of 1 NT..may cause you to go a trick too high. *Ibid.* iii. 30 Four No-trumps over an opening bid of One No-trump means: 'Go six if you are maximum!'

42. Esp. in phr. *to go to show*: to tend to show; to indicate; to serve as evidence *that*. Freq. *absol.*, to indicate or prove something implied.

1842 Emerson *Jrnl.* May in *Sel. Writings* (1965) 111 Then this new molecular philosophy goes to show that there are astronomical interspaces betwixt atom and atom. 1889 [in Dict.]. 1889 *Advent Rev. & Sabbath Herald* 16 July 450/3 Events..which go to show, with peculiar eloquence, to what extent the Romish policy has succeeded. 1925 *New Yorker* 25 July (Advt. inside front cover), Which just goes to show. Wherever you go in Paris you will find The New Yorker. 1937 D. L. Sayers *Busman's Honeymoon* 17 It does rather go to show, doesn't it, that there is something not quite right about the Wimseys. 1962 *Listener* 7 June 1000/1 Dramatic cases of child neglect only go to show, and wild generalizations are based upon them. 1966 A. Thomas *Judge* i. 16 It seemed unlikely that our paths would cross... Which, as they say, only goes to show. 1969 'A. Gilbert' *Missing from Home* viii. 134 'It just goes to show, doesn't it, you can't be too careful.' 'Yes,' agreed Mr. Crook hollowly, 'it just goes to show.'

44. a. (Further examples.) Also with *sb.* complement: to become, use, or adopt the characteristics of (something specified); *to go all ——*: see *all C. 2 c; *to go bush*: see *bush *sb.*[1] 9 e; *to go missing*: to get lost; *to go native*: to turn to or relapse into savagery or heathenism; also *transf.* (cf. *Fanti b); *to go —— on* (someone): to adopt a particular mode of behaviour towards or affecting (that person); *to go public*: to become a public company.

1901 Kipling *Kim* vii. 177 Kim did not sweep the board with his reminiscences; for St. Xavier's looks down on boys who 'go native altogether'. 1928 L. P. Greene *Red Idol* 253, I do all the heavy work... That's the only thing that's saved me from going native altogether. 1929 J. P. McEvoy *Hollywood Girl* (1930) vii. 102, I thought he was the one who had gone Hollywood. Pictures in the paper, guest of this and that. It's a wonder you notice me, I told him. 1930 *Times Lit. Suppl.* 4 Sept. 694 Religious ceremonies which suggest to him that the new religion of Christianity [in S. America] has, after the fashion of new religions, gone native. 1953 *Economist* 24 Oct. 273/2 (*headline*) BBC goes VHF. 1958 F. Norman *Bang to Rights* 56 The snout had gone missing. 1960 E. W. Hildick *Boy at Window* iii. 29 The best white mouse I ever had! Nearly human! And now it's gone native! 1963 *Times* 28 Jan. 2/7 Tentative plans are that on reaching a total of 3,000 acres a private limited company should be formed, and that at 5,000 acres it should 'go public'. 1963 *Sat. Even. Post* 15 June 4/2 Don't go too 'arty' on us. 1965 *Listener* 16 Dec. 983/1 Within twelve years 2,500,000—one in seven—had gone missing. 1966 [see *city page* s.v. *city 9]. 1966 *New Society* 17 Mar. 25/3 (*headline*) Amis goes serious on us. 1966 *New Yorker* 1 Oct. 183 (Advt.), Then I write my letter accusing them of going establishment. 1970 *Sci. Jrnl.* Aug. 3/3 It is not the fact that we are going nuclear that should give us pause. 1971 *Guardian* 6 July 7/6 A warning about the affairs of V and G was given as early as 1962, the year after the company went public.

46. c. *to go it alone*: to act by oneself, without support or assistance; hence *go-it-alone* adj., characterized by independent action. See also *bald-headed *a.* b, Strong *adv.* 1 c.

1842 *Spirit of Times* 9 Apr. 67/1 George pulled..to the head of the next quarter stretch, where his jockey looking at what he had in 'hand', determined to 'go it alone'. 1855 *Knickerbocker* Apr. 335 A ball through his frontal bone Laid him flat on his back on the hard-fought ground, And left Captain Davis to go it alone. 1907 J. London *Iron Heel* xv. 193 He broached a scheme to have the Locomotive Engineers make terms with the railroads and to 'go it alone' so far as the rest of the labour unions were concerned. 1956 A. L. Rowse *Early Churchills* 225 Holland could not 'go it alone'. 1961 *Daily Tel.* 27 Apr. 19/5 A 'go-it-alone' British space programme would be a waste of scarce scientific manpower and resources. 1964 M. Gowing *Britain & Atomic Energy* vi. 196 In England..a 'go it alone' programme was being drawn up. 1966 *Rep. Comm. Inquiry Univ. Oxf.* I. 28 The appropriate question must be posed: not 'how far can any college go it alone?' but 'how far can the colleges.. co-operate to form the policies of Oxford?'

48. f. Add to def.: Usu. const. *in*; *spec.* extremely insane, drunk, or evil. See also Far *adv.* 3 b. (Add further examples.)

1616 T. Roe *Jrnl.* 25 Oct. (1899) II. 303 The king returned at Euening, hauing beene ouer night farr gone in wyne. 1771 Smollett *Humph. Cl.* II. 49, I will follow you to the world's end, if you don't think me too far gone to be out of confinement. 1815 H. C. Robinson *Diary* 26 Feb. (1967) 39 Charles Lamb..[took] too much wine. He was not so far gone as to be outrageous..but he was flurried in his manner. 1849 C. M. Yonge in *Magazine for Young* June 188 Poor Ned, he must have been far gone, indeed, to let his children go without Baptism! 1939 F. Thompson *Lark Rise* i. 8 They were not too far gone in poverty to neglect such means as they had to that end. 1947 J. Harvey *Gothic England* ii. 62 The latter [*sc.* painting] is now too far gone for detailed stylistic analysis.

58. go for —. a. (Earlier and later examples.)

1595, 1660–1 [see For *prep.* 12 a]. 1883 'Mark Twain' *Life on Mississippi* xxiv. 269 They break camp and go for the woods.

b. (Later examples.)

1923 [see *bird dog* s.v. *bird *sb.* 9]. 1937 D. L. Sayers *Busman's Honeymoon* vi. 110 Leave everything as it is. Crutchley, you'd better go for the police.

d. Also, to be enamoured of or enthusiastic about; to care for, like, or prefer; to choose, accept, or support.

1830 W. L. Garrison in *Life W. L. Garrison* (1885) I. 201 In politics..I go for the people—the whole people. 1835 P. Hone *Diary* 13 Nov. (1889) I. 172 Daniel Webster's claim is incomparably stronger than that of either of the other candidates. He is entitled to the people's votes... I go, therefore, for Webster. 1839 *Congress. Globe* 6 Dec. 25 No, sir, I go for the laws and the Constitution, whether they define the qualifications of the voter, or prescribe the manner in which this right shall be exercised. 1930 *Amer. Mercury* Dec. 456/1, I go for that gee. He's a righto. 1936 *Studies in Eng.* (Univ Texas) XVI. 44 *Go for*..signifies..to become enthusiastic over something or somebody. 1940 Harrisson & Madge *War begins at Home* ix. 241 Comedy songs that are anti-Hitler the public are at first inclined to go for. 1950 A. Baron *There's no Home* 65, I could go for you in a big way, kid. 1952 M. R. Rinehart *Pool* xxi. 196 She doesn't go for whiskers. 1962 *Listener* 8 Nov. 759/1, I myself don't go at all for that heartiness, that matey stuff.. which figures so largely in mountaineering books. 1962 K. Orvis *Damned & Destroyed* i. 12 The people will never go for that guff.

f. To go for the purpose of becoming; to become; to act as.

1741, etc. [see For *prep.* 8 b]. 1957 A. Wilson *Bit off Map* 1 I'm well made all right. I could go for a model if I wanted.

g. To be valid or applicable for (a person); usu. in phr. *that goes for me*, etc. = that applies to me; that is my opinion. orig. *U.S.*

1923 C. E. Mulford *Black Buttes* iii. 36 In case nobody ever told you to go to hell before, I'm tellin' you now. That goes for the town an' everybody in it. 1925 T. Dreiser *Amer. Trag.* (1926) I. II. xxx. 360 She thinks I don't like her, and that's right, I don't... And that goes for that little Cranston show-off, too. 1936 Wodehouse *Laughing Gas* v. 63, I don't care if Pittsburgh chokes. And that goes for Cincinnati, too. 1941 *Punch* 2 July p. v/2 (Advt.), If you require *anything* that can possibly be connected with wire or wire-ropes (and this goes for manila and hemp ropes and canvas, too), British Ropes Ltd. can supply you. 1959 *Listener* 5 Mar. 431/2 That goes for all his music written..in the last twenty years and it applies to his lighter style of writing as well.

h. *to go for broke*: to make strenuous efforts; to go 'all out'. *U.S. slang.*

1951 *Amer. Speech* XXVI. 26 *Go for broke* means to bend every effort, to 'shoot the works'. 1963 *Guardian* 5 June 6 If he were to go for broke on behalf of the Negroes..the President would endanger the moral reform cause. 1968 *Ibid.* 19 Feb. 1/7 The enemy is 'going all out—..he is going for broke'.

i. To be in one's favour; to be favourable or advantageous to; esp. in phr. *to have* (something) *going for one*. *colloq.*

1967 *Melody Maker* 29 July 6/7 But his swing is ridiculous. He's got everything going for him. 1968 *Listener* 5 Dec. 749/2 Mr and Mrs White Rhodesia have quite enough going for them to make do with a few minor self-denials. 1970 *New Yorker* 10 Oct. 174/2 She has a lot going for her. Her serve..comes in hard and fast.

60. go off —. c. To cease to like; to begin to dislike.

1934 E. Bowen *Cat Jumps* 232 That's where I went right off him. In fact, he's not a nice man. 1965 M. Spark *Mandelbaum Gate* iii. 76, I simply don't feel anything for him any more. In fact, I've gone off him. 1969 *Listener* 24 July 127/1 Then word came down that the editor had gone off the idea and was averse to naming individual pubs.

61. go on —. b. Delete †*Obs.* and add later examples; usu. in *pres. pple.*; freq., to approach (a specified age). Cf. sense 84 k in Dict.

1876 T. E. Brown *Doctor* 17 The only child,..And just about goin' on twenty-one. 1880 'Mark Twain' *Tramp Abroad* xx. 193 Been here going on two years. 1924 C. C. O'Connor *Case of Galileo* viii. 45 He..appealed to his judges to consider his age (he was going on seventy). 1959 O. Hammerstein II (*song-title*) Sixteen going on seventeen.

e. Substitute for def.: To care for, concern oneself about; usu. in negative contexts, esp. in phr. *not to go much on* (something). *colloq.* (orig. *U.S.*). (Add earlier and later examples.)

1824 in N. E. Eliason *Tarheel Talk* (1956) 274 The people goes more on making nice cloth hear than they did thare. 1940 F. Sargeson *Man & Wife* (1944) 80, I don't go much on putting people away, I said. 1941 *Coast to Coast 1941* 197 They didn't go much on any of the chaps. 1958 *Daily Express* 17 Feb. 3/6 Waterloo was fascinating. But I didn't go much on the old armour. 1960 'N. Shute' *Trustee from Toolroom* 12 Jo says she wants to live in Tahiti, but I don't go much on that, myself. 1963 P. Willmott *Evolution of Community* vii. 73 This estate is low class. I don't go on the other people here myself.

g. To use (something) as evidence or as a starting-point. *colloq.*

1947 K. Tennant *Lost Haven* xix. 318 'Course I've got nothing to go on, but I wouldn't be a bit surprised if the johns wasn't on to us.

62. go over —. a. *to go over the top*: see *top *sb.*[1] 3 d.

63. go through —. f. (Earlier and later U.S. examples.) Also, to search (a person).

1861 *Calif. Police Gaz.* (San Francisco) 31 Mar. 2/4 Upon 'going through him', over $2,000 was found upon his person. 1865 T. W. Knox *Camp-fire & Cotton-field* 421 Not being privileged to 'go through' me as they had anticipated, the gentlemanly guerrillas went through the overseer. They took his money, his hat, his pantaloons, and his saddle. 1931 'D. Stiff' *Milk & Honey Route* 205 Train crews also *go through* the hobos. 1945 R. W. Service *Ploughman* 194 The girls were 'going through' a drunken sailor.

g. To wear out, make holes in; to use up.

1959 H. Pinter *Birthday Party* I. 2 He goes through his socks terrible. 1966 *Which?* Jan. 25/1 No fewer than 80 (of 118) reported water pump failure, and these members have between them gone through 103 pumps.

63*. go to —. Colloq. phr. *to go to it*: to go ahead; to get to work; to 'get cracking'; freq. *imp.*

1735–6 S. Legge *Alphabet of Kenticisms* (1876) 30 *Going to 't*, i.e. going to do it; as, 'do this or that;' the answer is—'I am going to 't.' 1917 Megrue & Hackett *It pays to Advertise* I. 40 Go to it. 1932 Wodehouse *Louder & Funnier* 12 Stoke up and go to it. 1952 C. Day Lewis tr. *Virgil's Aeneid* IX. 188 How they went to it then!

66. go upon —. f. = sense *61 g above.

1909 F. Barclay *Rosary* xv. 154 You see, this gave me something to go upon.

71. go ahead. (Earlier and later examples.)

1831 *American* (Harrodsburg, Ky.) 25 Mar. 2/5 We say to our Clay friends, 'go ahead.' 1839 Mill in *Westm. Rev.* XXXII. 508 The man..who 'goes a-head' with a policy adapted for uniting the Reformers, will find all things prepared for him. 1840 [see Ahead *adv.* 5]. 1845 *Punch* 8 Mar. 116/1 We should still go a-head, as this moment we do. 1966 *Listener* 23 June 904/2 In the meantime Sixtus was authorized to go ahead.

72*. go around. a. = *go round* (sense 88 g). *U.S.*

1883 [see Around *adv.* 4]. 1965 *Word Study* Feb. 4/1 There are simply not enough words to go around for things.

b. To go here and there; to wander about; *spec.* to be regularly in company *with* (someone, esp. a sweetheart) (= *go with*, sense 67 a).

[1896 in *Eng. Dial. Dict.* s.v. *around*, A seed em gangen aroond.] 1959 J. Braine *Vodi* xii. 162 Once he started going around with her there were more withdrawals than deposits in his Post Office savings book.

74. go back. d. (Earlier example.)

1859 G. W. Matsell *Vocabulum* 38 He won't go back on the cove; he is staunch.

f. (Earlier example.)

1789 A. Young *Jrnl.* 17 Aug. in *Travels* (1792) I. 165 The family of Polignac claim an origin of great antiquity; they have pretensions that go back, I forget whether to Hector or Achilles.

†g. *Bridge.* To redouble. *U.S. Obs.*

1907 R. F. Foster *Bridge* 16 If either the eldest hand or the pone doubles, it is the privilege of the player who named the trump to double him again, the usual expression being; 'I go back'. *Ibid.* 60 When you go over, never forget the possibility of their going back. 1920 —— *Auction made Easy* 111 *Going back*, redoubling.

78. go down. c. (Earlier examples in *Cricket*.)

1749 in H. T. Waghorn *Cricket Scores* (1899) 42 They.. had two wickets to go down. 1788 *World* 25 Aug. 3/2 Hampshire won, with 5 Wickets to go down.

h. To deteriorate; to decline in health or prosperity; to collapse or die. Also, to be subject to or suffer *with* (a specified illness or disease).

1857 [see Down *adv.* 17 a]. 1892 M. E. Freeman *Jane Field* 10 Well, I hope Lois ain't goin' down. I heard she looked dreadful. 1911 A. Bennett *Hilda Lessways* II. ii. 153 Calder Street's going down—it's getting more and more of a slum. 1934 W. Saroyan *Daring Young Man* (1935) 11 Hope he hasn't gone down; he deserved to live. 1953 'N. Shute' *In Wet* 4, I went down with a severe attack of malaria. 1968 K. Weatherly *Roo Shooter* 41 The air remained as dry as ever. On some of the stations the cattle were going down; all the earth tanks were dry.

i. To go away from a university or college; *spec.* to leave it permanently.

1861 J. A. Symonds *Let.* 4 Mar. (1967) I. 279 Another plan..is that I should go down myself next Term—take a Grace Term. 1883 [see Down *adv.* 2]. 1914 C. Mackenzie *Sinister St.* II. III. xii. 738 Guy Hazlewood had gone down and was away in Macedonia, trying to fulfil a Balliol precept to mix yourself up in the affairs of other nations or your own as much as possible. 1955 *Times* 12 May 14/3 From the time he went down from Cambridge until war broke out in 1914. 1965 J. Fleming *Nothing is Number* II. iii. 67 'I haven't seen him since the summer term. He's gone down.' 'Gone down?' 'Left Oxford.'

j. To be sent to prison. *slang.*

1906 Russell & Rigby *Making of Criminal* vi. 76 'Going down', as it is termed, for seven or fourteen days. *Ibid.*, The same youth will 'go down' time after time, and become more reckless and indifferent with every repetition. 1920 E. Wallace *Daffodil Mystery* iii. 29 Twice Sam had gone down for a short term, and once for a long term of imprisonment. 1945 M. Allingham *Coroner's Pidgin* xvii. 142 He went down for eighteen months and is now in Italy pulling his weight, I believe. He's a crook, but not a traitor.

k. *Bridge.* To fail to fulfil one's contract.

1918 E. Bergholt *Royal Auction Bridge* (ed. 2) 57 If he calls Four Hearts, as is probable, Z. and A. pass, but Y. doubles; and AB. are bound to go down. 1933 A. G. Macdonell *England, their England* vi. 78 [He] had gone down 650 points above the line whereas he ought to have made two no-trumps. 1964 N. Squire *Bidding at Bridge* ii. 24 You may go down quite often in game contracts.

l. In a card-game: to put one's cards on the table; to reveal one's cards.

1934 *Neuphilologische Mitteilungen* XXXV. 131 *To go down* 'to put one's hand down (as dummy)'. 1964 A. Wykes *Gambling* vii. 176 A player may declare his hand ('go down') when the unmatched cards in his hand count 10 or less.

m. To happen. *slang* (orig. *U.S.*).

1946 Mezzrow & Wolfe *Really Blues* 374 *Go down*, happen. 1956 'B. Holiday' *Lady sings Blues* (1958) xxi. 171 If they'd known about that they might never have let him off. Or they might have. In view of what went down later, who can say? 1970 *It* 12–25 Feb. 4/4 If everyone was aware of what went down in these organisations perhaps there would be enough response to keep them from petrifying and dying.

80. go in. c. (Earlier example.)

1718 *Weekly-Jrnl.* 6 Sept. 543/1 Three of their Men made an Elopement, and got off the Ground without going in.

81. go in *for*. b. (Earlier example.)

1845 *Punch* 6 Sept. 108/2 Having heard that there was a better thing than the Newfoundland Judgeship to be had in the neighbourhood, he determined to 'go in for it'.

83. go off. i. (Earlier examples.)

1749 Gray *Let.* 8 Aug. (1935) I. 324, I know not how, it has gone off again, & we have heard no more lately about it. 1813 Jane Austen *Pride & Prej.* II. ii. 18 It seems likely to have been a desirable match... I am sorry it went off.

k. (Earlier example.)

1775 E. A. Linley *Let.* 9 Mar. in T. Moore *Life of R. B. Sheridan* (1825) 102 In my life, I never saw any thing go off with such uncommon applause.

m. To experience a sexual orgasm. *slang.*

1928 D. H. Lawrence *Lady Chatterley* v. 62 You couldn't go off at the same time as a man, could you? *Ibid.* xiv. 244 They always make you go off when you're *not* in the only place you should be, when you go off. 1949 H. Miller *Sexus* (1969) ii. 59 Bango! I went off like a whale.

n. *to go off at* (a person): to reprimand angrily; to abuse. *colloq.*

1941 Baker *Dict. Austral. Slang* 31 *Go off at*, to reprove, express anger towards someone. 1948 D. W. Ballantyne *Cunninghams* (1963) II. xi. 175 He had to get away home else his father would go off at him.

84. go on. b. *to be going on with*: to start with; for the present; usu. in phr. *enough to be going on with*.

1967 *Listener* 19 Jan. 105/2 There are other semi-concealed social problems.., but to be going on with Cathy is quite enough to prove [etc.]. 1968 *Ibid.* 29 Aug. 260/3 Milton seems to have thought the English climate was a punishment for sin. In which case we have surely been punished enough to be going on with.

e. Add to def.: to happen, occur.

1873 C. M. Yonge *Pillars of House* I. iii. 55 How long has this been going on? 1886 F. R. Stockton *Hundredth Man* v, People would stop to look into Vatoldi's to see what was going on. 1955 *Times* 30 June 8/3 The governing council has in a sense admitted that it did not know all that went on.

g. Also, to talk excessively or tiresomely *about* (a subject); to discourse *ad nauseam*.

1921 M. Arlen *Romantic Lady* 235 For God's sake don't go on about it, Iris, else I won't be able to bear it at all. 1921 H. Crane *Let.* 21 Nov. (1965) 70, I don't need to go on about it, as you have probably seen it. 1960 L. Cooper *Certain Compass* 61 She doesn't go on about things, one of the many reasons why I like her. 1969 *Listener* 24 July 114/2 How much of what I have been so tediously going on about here is reflected in the programme itself?

i. (Earlier examples.)

1769 S. Barry *Let.* 27 Oct. in *Corr. Garrick* (1831) I. 370 The managers..expected she would continue to go on in the pageant as long as she was able. 1839 Dickens *Nich. Nick.* xiv. 123 The daughter of a theatrical fireman, who 'went on' in the pantomime. 1861 *Times* 25 May 9/4 A change of bowling was tried, Atkinson going on at Hodgson's end.

j. (Examples.) Also expressing surprise or incredulity.

1886 F. T. Elworthy *W. Somerset Word-Bk.* 295 In a quarrel either of the parties themselves, or a third, who wishes to stop it, says '*go on*'! and means 'be quiet'. 1894 Somerville & 'Ross' *Real Charlotte* II. xix. 67 Ah, go on, Mr. Dysart Why, I see the white water, and the black rocks, and all! 1940 L. A. G. Strong *Sun on Water* 59 'Ah, go on.' 'Well, we added it up and..it came out incontrovertibly..two years, one month, and five days.' 'Eighteen months, if it's a day.' 1960 [see *get *v.* 54 b].

m. To go on to another engagement.

1903 Mrs. H. Ward *Lady Rose's Daughter* i. 17 'Now, you are going on,' said Lady Henry... 'Freddie says I must,' said the other. 1934 P. Bottome *Private Worlds* iii. 29 Charles noticed that Myra was elaborately dressed. ..She must be going on somewhere else afterwards. 1961 F. & R. Lockridge *Murder has Points* (1962) ix. 98 Faith Constable had had to 'go on' from the party and had, presumably, gone on. 1969 D. Rutherford *Gilt-Edged Cockpit* v. 79 'Would you like to go on somewhere else?' Patrick had asked when coffee and a Calvados had been disposed of.

85. go out. c. Also, to leave one's country for the battle front.

1917 S. McKenna *Sonia* viii, 'Is David going out?.. What's he in?' 'The Midland Fusiliers.' 1928 Blunden *Undertones of War* i. 1, I read the notice that I was under orders for France... Berry, a subaltern of my set,..might pipe to me, 'Hi, Blunden, we're going out: have a drink;' I could not dance. 1936 S. Sassoon *Sherston's Progress* II. i. 98 She had hoped and prayed that I might get a home-service job; but now she just accepted the fact that I'd got to go out again.

m. (Earlier example.)

1816 Jane Austen *Emma* III. ix. 154 He had long made up his mind to Jane Fairfax's going out as a governess.

o. Also, to be broadcast.

1968 *Listener* 26 Sept. 420/1 Yorkshire's *Gazette*, a drama series about the lives and hard times on a weekly paper, goes out in London at 11 p.m. and cannot therefore hope for an audience. 1969 *Ibid.* 20 Feb. 241/2 In the discussion of abortion that went out on 13 January.. Philippa Foot raised the essential question: when does a human being begin? 1971 R. Busby *Deadlock* xiv. 210 'Cut—for Christ's sake, cut. This is all bloody going out.' ..The screens went blank.

q. (Earlier examples.) Also, of a side: to be dismissed from batting.

1735 in H. T. Waghorn *Cricket Scores* (1899) 10 London headed the county 42 notches before they went out. 1854 J. Pycroft *Cricket Field* (ed. 2) xi. 247 Had not an easy catch been missed, The Eleven of All England would have gone out for a run apiece.

r. (Earlier examples.) Esp. in phr. *my heart goes out*.

[1779 F. Burney *Diary* Feb. (1842) I. 184 [Johnson *loq.*] I love Burney: my heart goes out to meet him!] 1842 E. B. Browning *Eliz. Barrett to Miss Mitford* 22 Aug. (1954) 130 My heart goes out to you in full devotion. 1860 Geo. Eliot *Let.* 27 Aug. in J. W. Cross *Life* (1886) II. 228 My heart goes out with venerating gratitude to that mild face. 1881 *Atlantic Monthly* May 594/2 His own sympathy went out fully to cases of individual suffering.

t. To go hunting.

1877 Trollope *American Senator* I. i. 7 He..is not much thought of..except by those who go out with the hounds. 1932 R. Lehmann *Invitation to Waltz* III. v. 209 'Were you out today?' 'Oh yes.'.. 'Um. Going out on Saturday—?'..'Oh, I thought you meant... I misunderstood. I don't—as a matter of fact, I don't really hunt.' 1967 R. Rendell *New Lease of Death* ii. 24 He's very conscious of his status..since he bought Forby Hall. Goes out with the Pomfret hounds and all that.

u. To faint, lose consciousness, collapse; esp. in phr. *to go out like a light*. *U.S. colloq.*

1934 J. M. Cain *Postman always rings Twice* ix. 81 A doctor was working on my arm. I went out again as soon as I saw it. It was running blood. 1934 R. Chandler in *Black Mask* Oct. 15/1 Something swished and I went out like a light.

87. go over. a. *spec.* To go 'over the top' (see *top *sb.*[1] 3 d).

1919 Galsworthy *Saint's Progress* II. i. § 3 Calculating exactly where he meant to put foot and hand for the going over. 1931 W. V. Tilsley *Other Ranks* 49 His only comment on 'going over' had been: 'Well, Dick, Brettle's going to have his chance.'

f. Of a bill, motion, etc.: to be postponed for consideration. *U.S.*

1894 'Mark Twain' in *Century Mag.* Feb. 556 According to the by-laws it must go over to the next regular meeting for action. 1900 *Congress. Rec.* 3 Jan. 632/2, I ask that it may go over until to-morrow, so that we can have an opportunity to see it. *Ibid.*, The resolution goes over under the rule.

†g. *Bridge.* To double. *U.S. Obs.*

1902 J. B. Elwell *Bridge* 111 *Going over*... The effect of 'over', 'over', etc., is that the value of each trick point is doubled, quadrupled, etc. 1907 R. F. Foster *Bridge*

60 If you are reasonably certain of the odd trick, you should go over in order to make it more valuable. *Ibid.* [see *74 g]. 1920 —— *Auction made Easy* 111 *Going over*, obsolete for doubling.

h. Of a play, speech, etc.: to be successful in its appeal to the audience; to 'get across', 'get over'. Also *gen.* to have a vogue or success; to be well received; esp. in phr. *to go over big* (cf. *19 b). orig. *U.S. slang.*

1923 H. C. Witwer *Fighting Blood* ix. 281 These synthetic actors..are going over big with their parents. 1923 H. Crane *Let.* 18 Feb. (1965) 126 One of my campaigns that not only 'went over big' but is theoretically a good piece of work. 1924 *Ibid.* 3 Feb. 174 My copy on the cheese book went over without any changes. 1927 *Daily Express* 29 Aug. 5 'Little Bits of Love' went over with its accustomed success. *Ibid.* 23 Nov. 13 'Escape' has gone over with a great bang. 1930 *New Statesman* 1 Nov. 115/2 Experienced novelists know that a good trial always goes over big. 1936 D. Powell *Turn, Magic Wheel* i. 39 As this went over big he forgave Dennis. 1961 J. B. Priestley *Saturn over Water* xiv. 193 Dr Magorious had rated a whole column report, and clearly was going over big in Melbourne.

88. go round. g. (Earlier U.S. example.)

1869 P. T. Barnum *Struggles & Triumphs* xvi. 259 When it [*sc.* a turkey] was carved, there was not enough of it to 'go round'.

h. *Theatr.* Of a member of an audience: to go behind the scenes at a theatre; esp. to visit an actor, etc., backstage after a performance; (in quot. 1866, to go into the audience from backstage).

1866 'Old Stager' *Stage Reminisc.* i. 7 The scene-shifters..had thought it would be worth while seeing Kean's reception from the front, and had gone round for that purpose. 1900 T. E. Pemberton *Kendals* xi. 326 After the piece was over, I went 'round' to offer her my congratulations. 1952 A. Christie *Mrs. McGinty's Dead* xvii. 128 The play itself she had enjoyed, but the ordeal of 'going round afterwards' was fraught with..terrors. 1966 D. Blakelock *Eleanor* iii. 29 I'm not fond of 'going round' to see people.

89. go through. c. Of a deal: to be completed.

1889 *Kansas Times & Star* 8 Apr., Kump intimates the deal won't go through.

d. To desert, decamp, abscond; also, to desist, give up. *Austral. slang.*

1943 Baker *Dict. Austral. Slang* 35 *Go through*, to desert from a northern base to the south. War slang. 1944 J. Devanny *By Tropic Sea & Jungle* xi. 89 'Cutting cane's not living: it's death in life... You're going through,' I told myself. 'You're going through!' 1949 J. Cleary *You can't see round Corners* I. viii. 63 The corporal looked up and there was a faint stirring in the ranks... 'He's gone through. Went last night.'

92. go together. d. To keep company as lovers; to 'go steady'; to be courting or engaged.

1899 M. E. Wilkins *Jamesons* iii. 77 People began to say that Harry Liscom and the eldest Jameson girl were going together. 1958 E. Dundy *Dud Avocado* I. ix. 154 Now that Jim and I were seen around..in public, it became obvious..that we were 'going together'.

94. go up. a. *spec.* To go to a university or college; also, to enter *for* an examination.

1861 [in Dict.]. 1875 L. Troubridge in *Life amongst Troubridges* (1966) 23 'Ernest has passed *first of all!*'... Fancy, first of all the sixty-eight boys who went up. 1879 C. M. Yonge *Magnum Bonum* I. xxi. 426, I mean to go up for a scholarship next year. 1885 A. Edwardes *Girton Girl* III. xiii. 230 Very likely I may go up to Girton as a bye-term man in January. 1955 *Times* 9 May 13/1 He was educated at Sywell House School, Llandudno, whence he went up to Oxford.

b. Add further examples. Also, to be killed (or †hanged); to die; to be done for; esp. in phr. *to be gone up*. *colloq.*

1825 J. Neal *Bro. Jonathan* III. 233 Whose narrow escape, when his brother spy 'went up', he said, was quite a 'muricle'. 1867 W. H. Dixon *New America* I. xi. 132 Gone up, in the slang of Denver, means gone up a tree—that is to say, a cotton tree... In plain English, the man is said to have been *hung*. 1888 P. H. Sheridan *Mem.* I. 86 He remarked, 'Well, I fear that they are gone up,' a phrase used..to mean that they had been killed. 1907 J. Masefield *Tarpaulin Muster* 101 To your prayers, boys. We're gone up. *Ibid.* 209 You're gone up, my son.

g. To 'ascend' (see Ascend *v.* 9) in ancestry.

1872 A. de Vere *Leg. St. Patrick* p. xiii, Her genealogies went up to the first parent. 1930 J. L. Myres *Who were the Greeks?* 308 Hecataeus' genealogy went back sixteen generations, and then 'went up to a god'.

h. To explode, burn, be consumed. Also *fig.*

1932 A. J. Worrall *Eng. Idioms* 77 A bomb is dropped on a model house which goes up in flames. 1933 *Punch* 3 May 498/1 Civilisation Would probably go up in smoke If unsophisticated folk like you and me..Employed companionation. 1940 'N. Blake' *Malice in Wonderland* I. vii. 96 He fairly went up in smoke about the camp. 1950 A. P. Herbert *Independent Member* 167 The shattered 'skid' astern of a mine-sweeper that meant a mine gone up. 1971 *Daily Tel.* (Colour Suppl.) 5 Nov. 29/2 He left the sea, having had what amounted to a nervous breakdown, 'always thinking of the other ships that went up, the bombings and suchlike'.

VIII. go-anywhere *a.*, that can go anywhere; **go-ashore,** (*a*) also, designating clothes worn or articles used by sailors when ashore; (*c*) (examples); **go-as-you-please(ness** *sb.*, want of regularity or order; freedom; **go-at-it** *a.*, full of dash; **go-fever,** a feverish restlessness or longing for movement; **go-stop** *a.*, designating an economic system or situation characterized by alternating periods of inflation and deflation; cf. *stop-go *a.*

1963 Bird & Hutton-Stott *Veteran Motor Car* 27 The sturdy, no-nonsense, go-anywhere machine. 1966 G. N. Leech *Eng. in Advertising* xv. 139 The go-anywhere Electrolux 16 refrigerator. [1834 E. Markham *N.Z. or Recollections of It* 39 The Natives now begin to use Iron pots, known to them by the name of 'Go on Shores' from the circumstance of Boats going up and down the Rivers, coming to an Anchor and going on Shore to Cook.] 1839 J. D. Lang *N.Z. in 1839* ii. 34 Bought a large tract of eligible land..for two check shirts and an iron-pot, or go-ashore, as it is called by the natives. 1840 R. H. Dana *Bef. Mast* xvi. 140 Go-ashore jackets and trousers [were] got out and brushed. 1849 W. T. Power *Sketches in N.Z.* xviii. 160 Engaged in the superintendence of a Maori oven, or a huge gipsy-looking cauldron, called a 'go-ashore'. 1879 C. L. Innes *Canterbury Sketches* iii. 23 There was another pot used, called by the euphonious name of a 'Go ashore', which used to hang by a chain over the fire, this was used for boiling. 1933 J. Masefield *Bird of Dawning* 246 All his go-ashore things were there, laid up with camphor in tissue paper, including his tall hat. 1953 *Landfall* Sept. 173 Edwards had a whare,.. A Brown Bess, a go-ashore. 1889 *Judge* 20 Apr. 21/2 A Western Go-As-You-Please. 1920 H. G. Wells *Mr. Britling* I. iv. § 6. 117 Explaining..to this American.. how excellent was the backwardness of Essex and English go-as-you-please. 1935 B. Russell *In Praise of Idleness* 120 An element of free growth, of go-as-you-please and untrained natural living, is essential if men are not to become misshapen monsters. 1971 R. Roberts *Classic Slum* viii. 117 In later years, after cinema had begun to outstrip live entertainment..our theatre, like many others, tried 'go as you please' competitions on Friday evenings when local amateurs..trod the boards. 1927 *Observer* 1 May 15/3 The impromptu quality of the entertainment —its untidy go-as-you-please-ness. 1904 *Daily Chron.* 27 June 7/2 They are a versatile, well-balanced, go-at-it side, and they field as if they like it. 1891 Kipling *Light that Failed* 125 He's as restless as a swallow in autumn... He has the beginnings of the go-fever upon him. 1900 J. Buchan *Half-hearted* vi. 78 There comes a thing called the go-fever, which is not amenable to reason. 1907 N. Munro *Daft Days* xxiv, I'd not been twenty minutes in her society before I found out she had the go-fever pretty bad. 1966 *Word Study* Dec. 2/1 'Everyone I think sort of gets go-fever,' Colonel Glenn was heard to remark at his news conference. 1964 *Punch* 21 Oct. 591/2 The damage done by 'go-stop' economics. 1965 *New Statesman* 5 Nov. 680/2 The natural 'go-stop' cycle of the capitalist economy.

go (gōu), *a.* *colloq.* (orig. *U.S.*). [f. the vb.] **1.** Functioning properly; ready and prepared; esp. of devices in space-craft. So **go-no-go** *a.* (see quot. 1959).

A 'go' gauge (quot. 1951) is one which must pass through (or be passed through by) the piece, a 'no(t)-go' gauge one which must not, thus ensuring that the dimensions are within acceptable limits.

1951 *Engineering* 21 Sept. 369/2 It is necessary to decide on the tolerance on the shaft, the tolerance on the bush, and the relationship between the two. From the production point of view, go and not-go gauges must be selected for the shaft and the bush. 1959 W. A. Heflin *Aerospace Gloss.* 43/2 *Go-no-go a.*, of a rocket missile launch: so controlled at the end of the countdown as to permit an instantaneous change in decision on whether to launch or not to launch. 1961 *Words of Space Age* (*Newsweek*) 12 The fuel system is go. 1962 *Guardian* 22 Feb. 8/6 Colonel Glenn reported.. 'I am Go; all systems are Go.' 1962 *Flight International* LXXXII. 164/2 One of the biggest single contributions which could be made on jet aircraft is a Go–No Go indicator. 1962 A. Shepard in *Into Orbit* 102 Inside the Mercury Control Centre itself, all the lights were green. All conditions were 'Go'. 1962 *Sunday Times* 4 Mar. 31/8 A..better University.. creatively responding to the new challenges, and good luck indeed to them, their system too is no doubt Go. 1963 [see *abort *v.* 1 c]. 1969 J. Gardner *Founder Member* ii. 26 Fuel is go...Oxygen is go.

2. Fashionable, modern, progressive, 'go-ahead', 'with it'. Cf. *go-go *a.*

1962 in *Amer. Speech* XXXVII. 286 What is a Junior? ..A size, of course, not a matter of age! It stands for 'Go' in the fashion world. 1963 *Time* 4 Jan. 36/3 Beatniks, whose heavy black turtleneck sweaters had never looked particularly go with white tennis socks. 1964 *Punch* 8 July 43/1, I am not a go person.

go-ahead (gōu·ăhed), *sb.* [f. Go *v.* 71 and Go-ahead *a.*] **1.** The action or the spirit of going forward; progress; ambition, energy, or initiative. *U.S. colloq.*

1840 C. F. Hoffman *Greyslaer* I. iii. 32 Sarting! he does make a clean go-ahead of it. But when did he come up here to mix in our doings? 1844 G. W. Kendall *Narr. Santa Fé Exped.* II. xix. 362 There is a little 'go-ahead' in a spirited, showy, well-trained Mexican horse. *a* 1861 T. Winthrop *Canoe & Saddle* (1883) xii. 251 It racks my heart to know that I must still demand much go-ahead from you. 1885 W. D. Howells *Rise Silas Lapham* xi, Some of the fellows that had the most go-ahead were fellows that hadn't ever had much more to do than girls before the war broke out. 1942 Berrey & Van den Bark *Amer. Thes. Slang* § 240/3 Initiative; enterprise; 'push',.. go-ahead.

†2. An advance agent of a touring theatrical company, circus, etc. *Obs.*

1861 *All Year Round* 16 Nov. 184/2 'The agent in advance', or *go-ahead*, as he is now called,..accompanied by a bill-sticker, starts off in advance of the troupe.

3. A signal, order, or permission to proceed. Also *attrib.* *colloq.*

1941 *B.B.C. Gloss. Broadcasting Terms* 13 *Go-ahead signal*, spoken instruction from a control room to the personnel at a programme source, indicating that all is ready for a programme to begin. 1942 Berrey & Van den Bark *Amer. Thes. Slang* § 226/2 *The go-ahead*, the command or permission to do. 1948 *Sierra Club Bull.* (San Francisco) May 2/2 It must receive Congressional approval..for its go-ahead on construction. 1953 *Encounter* Oct. 14/1 The party had given the go-ahead signal after months of discreet silence. 1958 *Times* 6 June 12/3 Given the go-ahead, the system is likely to be in operation within two years. 1967 *Listener* 30 Mar. 429/1 B.E.A., which now has the backing of four major airlines for its project, is very confident that it will get the go-ahead.

goal, *sb.* Add: **2. b.** *spec.* in *Psychol.* An end or result towards which behaviour is consciously or unconsciously directed. Freq. *attrib.* and *Comb.*

1917 Glueck & Lind tr. *Adler's Neurotic Constitution* (1921) p. viii, The entire picture of the neurosis as well as all its symptoms are influenced by..an imaginary fictitious goal... The potency of this 'goal idea' is revealed to us by the trend and evaluation of the pathological phenomena. 1934 *Mind* XLIII. 111 Not all action is due to goal-seeking propensities. 1940 R. S. Woodworth *Psychol.* (ed. 12) ii. 42 The reader becomes set or adjusted for the situation portrayed in the story and for the 'goal', the outcome of the story. 1945 E. S. Russell *Directiveness of Organic Activities* i. 8, I shall use, with due caution, and without any implication of conscious purpose on the part of the organism or other organic agent, the concepts of goal or completion and of biological end. *Ibid.* iv. 81 It is not implied by the use of the word 'goal' that the agents concerned are conscious of it before it is reached; their action is directive, but not purposive. *Ibid.* v. 110 Goal-directed activity is limited by conditions, but is not determined by them. 1949 Koestler *Insight & Outlook* xv. 214 All cultural achievements appear as ersatz formations for goal-inhibited sexuality. 1951 R. Firth *Elem. Social Organiz.* vii. 220 This action is goal-oriented. 1952 T. Parsons *Social System* 8 The goal-directedness of action. 1953 *Brit. Jrnl. Psychol.* Aug. 203 When this state is absent or disturbed the system becomes active and..continues to be active until that state is established or restored. We may say of such a system that its activity is goal directed. 1956 J. Klein *Study of Groups* 141 'Instrumental', 'conative', 'goal-oriented' behaviour is concerned with decision-making. 1964 Gould & Kolb *Dict. Soc. Sci.* 290/1 *Goal* in psychology and in some social psychology has come increasingly to denote an end result of an act or series of acts whether or not it can be said to be intended by the organism acting. 1965 P. Caws *Philos. Sci.* xl. 310 The deliberate and conscious production of novelty, that is,..what we ordinarily call purposive or goal-seeking behaviour. 1965 *Philos.* XL. 348 Central-state behaviourism..hopes to account for intentionality in terms of neuro-physiology and goal-directed systems. 1968 *Canad. Jrnl. Linguistics* XIII. 126 He found..that a tendency towards hypercorrect forms in the lower middle class seems rooted in a profound linguistic insecurity—evidence that linguistic behavior is highly normative or goal-directed.

6. *goal-getter, -getting, -scorer, -shooter*; **goal-kick,** (*a*) in association football (see quot. 1960[1]); (*b*) in rugby football, an attempt to kick a goal; (*c*) = **goal-kicker*; **goal-kicker,** one who scores a goal or makes a goal-kick; so *goal-kicking*; **goal-mouth,** the space between the goal-posts and under the cross-bar in association football, hockey, etc.; also, the area near the goal; **goal-net,** the net behind the goal-mouth; **goal-tender** *N. Amer.*, a goal-keeper in ice-hockey; so *goal-tending* vbl. sb.

1887 M. Shearman *Athletics & Football* II. iv. 345 The doctor had the reputation of being the best centre and goal-getter in Scotland. 1910 *Westm. Gaz.* 21 Jan. 12/1 The goal-getters for the winners were [etc.]. 1932 Auden *Orators* III. 102 Goal-getter, holer-in-one. 1904 *Daily Chron.* 29 July 4/5 His goal-getting stroke. 1904 *Ibid.* 21 Nov. 8/3 Goal-getting did not..enter largely into the argument. 1891 *Football: its Laws, Rules, & Definition of Terms* 15 Two linesmen shall..decide when the ball is out of play, and which side is entitled to the corner flag-kick, goal-kick, or throw in. 1897 *Encycl. Sport* I. 429/1 *Goal-kick.* (R.) The attempt to kick a goal from a try. (A.) The kick-out when the ball has gone behind. It must be made from a spot within 6 yards of the nearest goal-post. 1960 J. R. Witty in Fabian & Green *Assoc. Football* I. II. iii. 169 The ball may have gone over the goal-line, but not into goal. If it were last played by an attacking player, the re-start is a *goal kick*. Here the ball is placed somewhere within the goal area (the 6 yards area) on that side where it passed out of play, and it must be kicked directly, by a defender, somewhere into the field of play beyond the penalty area (18 yards) marking, and it is not in play until then. 1960 T. McLean *Kings of Rugby* xi. 160 Henderson..was a goalkick of excellence. 1965 D. Law *Tackle Soccer this Way* v. 44 In most league games today..modern goalkeepers prefer the short goal-kick just out of the box to a defender,..to the long goal-kick downfield. 1909 *Bendigonian* 24 Aug. 21/2 The goal-kickers were: [etc.]. 1961 *Times* 5 Jan. 3/5 Du Preez probably owes his selection to his recent success as a goal-kicker. 1963 *Times* 29 May 3/7 Hosen..had a singularly unhappy day in both goalkicking, in which he failed with half a dozen attempts, and also in fielding. 1882 in Charles-Edwards & Richardson *They saw It Happen* (1958) 300 The ball..was headed back from the goal mouth by Suter. 1908 *Westm. Gaz.* 7 Oct. 4/2 The yawning goal-mouth was not fed with that elusive leather sphere. 1970 *Times* 8 June 5/3 Charlton looking younger than ever swept into the Brazilian goalmouth. 1897 *Windsor Mag.* Dec. 22/1 Goal-nets..would have had to be pretty strong. 1909 *Daily Chron.* 20 Mar. 8/5 He

heads the list of West Ham goal scorers. 1929 *Evening News* 18 Nov. 13/4 The League's leading goalscorer. 1961 *Times* 20 Jan. 17/4 The embracing and mobbing of the goal-scorer is entirely unnecessary. 1959 J. Fingleton *Four Chukkas to Australia* 155 The soccer field where the players fall upon the neck of the goal-shooter. 1909 Webster Goal tender. 1926 T. K. Fisher *Ice Hockey* ii. 12 A good goal-tender should be an alert and agile athlete. 1967 *Boston Sunday Herald* 26 Mar. 18/4 Charlie Driscoll, whose son Charlie Jr., was a great hockey goal-tender at Boys' High. 1968 *Globe & Mail* (Toronto) 5 Feb. 17/2 Ed Giacomin's shutout goaltending..ended Montreal's National Hockey League winning streak.

goal (gōul), *v.* *Rugby Football.* [f. the sb.] *trans.* To convert (a try) into a goal.

1922 *Weekly Dispatch* 29 Oct. 10 Bennett..intercepted a pass and scored a try, which Tebbutt goaled. 1954 J. B. G. Thomas *On Tour* 138 Saxton made a magnificent blind-side burst which put Sherratt over for a second try, which Cook goaled.

goalie (gōu·li). Also **-ee.** [f. Goal *sb.* + -ie.] A goal-keeper.

1921 *Oxf. Mag.* 28 Jan. 158/2 C. V. Hill..journeyed to Henley..much to the discomfiture of the opposing goalee. 1926 *Spectator* 24 Apr. 753/1 One can imagine an indignant Cockney ejaculating as he regarded the rotundity of the Red goalee. 1957 *Encycl. Brit.* XII. 42A/1 The positions on a six-man [ice] hockey team are goal tender (more commonly called goalies in the United States and goaler in Canada), [etc.]. 1966 *Reader's Digest* Mar. 64/2 The players [at ice hockey] (three forwards, two defencemen, one goalie)..are chosen for their ability to skate swiftly, [etc.]. 1967 J. Potter *Foul Play* vii. 84 Most English goalies were prize examples of British phlegm, but Basil outdid the continental keepers in panache.

goalless (gōu·l,lĕs), *a.* [See -less.] **1.** Purposeless; having no destination.

a 1886 E. Dickinson *Compl. Poems* (1960) 230 A goalless Road. 1906 *Daily Chron.* 23 Oct. 4/4 He has never taken a solitary, goalless walk. 1923 W. de la Mare *Riddle* 277 One late afternoon, in my goal-less wandering.

2. *Association Football.* Without a goal scored by either side.

1904 *Daily Chron.* 18 Nov. 1/7 At Southall Oxford City played a goalless draw with the local team. 1928 *Daily Tel.* 28 Aug. 9/5 The game..ended in a goalless draw. 1971 *New Society* 1 July 24/3 Pools companies had to re-appraise their points system because of the number of goal-less draws.

Goan (gōu·ăn), *a.* and *sb.* [See -an.] = next.

1927 J. Furtado (*title*) A Goan Fiddler. 1927 E. Gosse *Ibid.* Pref. p. vi, He was..not happy among the Goans. 1961 *Daily Tel.* 8 Dec. 1 There is no solution to the Goan problem except the Portuguese walking out. 1966 C. Sweeney *Scurrying Bush* viii. 119 An Indian or Goan at the other end began asking me seemingly irrelevant questions. 1971 *Illustr. Weekly India* 25 Apr. 9/1 Goans and Parsis were the pioneers but now the [modelling] profession has full representation.

Goanese (gōuănī·z), *a.* and *sb.* [f. Goa[1] + -ese, after *Japanese*, etc.] **A.** *adj.* Of or belonging to Goa. **B.** *sb.* A native of Goa; the Goanese people.

1851 R. F. Burton *Goa* v. 91 The Goanese smoke all day, ladies as well as gentlemen. *Ibid.* 93 A Goanese noble. 1909 *Chambers's Jrnl.* Aug. 523/1 A batch of Goanese tailors' shops. 1920 *Ibid.* May 299/2, I was obliged to read the burial service in Latin out of a prayer-book lent me by a Goanese cook. 1961 *Daily Tel.* 20 Dec. 8 Did the Goanese really yearn for 'liberation'? 1970 *Times* 10 Feb. 17/2 (Advt.), A rare late 17th century Goanese cabinet.

goanna (go,æ·nă). *Austral.* Also formerly **gohanna.** [Corruption of Iguana; cf. Guana.] Any of various large monitor lizards of the genus *Varanus*.

1891, 1898 [see Guana 2]. 1900 H. Lawson *Over Sliprails* 108 The four little Australians..ate underdone kangaroo..with an occasional treat of oak grubs and gohanna. 1903 *Blackw. Mag.* Apr. 468/2 My tongue was like a gohanna's back. 1923 *Chambers's Jrnl.* Aug. 527/1 Snakes, goannas (guanas), scorpions, and centipedes are numerous. 1946 I. L. Idriess *In Crocodile Land* ix. 60 The goanna burrows into the nest. 1965 A. Keast *Window to Bushland* 102 Goannas are the giants of the lizard tribe.

goat. Add: **2. d.** Short for Goatskin.

1895 *Montgomery Ward Catal.* 290/2 Men's goat driving gloves. 1927 J. S. Hewitt-Bates *Bookbinding for Schools* 13 Goat or Morocco. *Ibid.* 14 Persians..may be made either from goat or sheep. 1927 *Longman's Class. Cat. Educ. Works* 12 Hand grained goat, gilt edges.

3. b. Add to def.: to play the fool; to behave in an irresponsible manner. Also, *to act the goat*.

1879 H. Hartigan *Stray Leaves from Mil. Man's Note Bk.* i, Don't be actin' the goat. 1887 Kipling *From Sea to Sea* (1900) I. xiv. 162 You'll find some o' the youngsters play the goat a good deal when they come out o' stable. 1888 —— *Under Deodars* (1890) 91 Generally, as he explained, 'playing the giddy garden goat all round'. 1924 Galsworthy *White Monkey* I. v, It's playing the goat for no earthly reason. 1929 W. P. Ridge *Affect. Regards* 61 Haven't I got enough trouble without you acting the goat in this fashion?

c. *to get* (a person's) *goat*: to make (him) angry; to annoy or irritate. *slang* (orig. *U.S.*).

1910 J. London *Let.* 2 Aug. (1966) 316 Honestly, I believe I've got Samuels' *goat*! He's afraid to come back. 1912 C. Mathewson *Pitching in a Pinch* ii. 28 Lobert.. stopped at third with a mocking smile on his face which would have gotten the late Job's goat. 1914 *Sat. Even. Post* 4 Apr. 10/3 It got my goat—that and the cold and that light in all the dark. 1919 H. Jenkins *John Dene of Toronto* (1920) iv. 70 There are some things in this country that get my goat. 1922 *Weekly Westm. Gaz.* 27 May 8/1 What gets my goat is the assumption that the misty subject is necessarily more artistic than the sharp and regular one. 1924 Galsworthy *White Monkey* II. i, That had got the chairman's goat!—Got his goat? What expressions they used nowadays! 1929 J. B. Priestley *Good Companions* III. i. 474 Now this is what gets my goat, and you can't blame me. 1960 B. Keaton *Wonderf. World of Slapstick* (1967) i. 22 What got my goat was that when I finally did get knocked off..it was due to an accident outside the theatre.

d. A fool; a dupe. *colloq.*

1916 C. J. Dennis *Songs of Sentimental Bloke* 39 The drarmer's writ be Shakespeare, years ago, About a barmy goat called Romeo. 1947 K. Tennant *Lost Haven* i. 20 You old goat! Shut up and get out before I slam you out. *Ibid.* xxi. 365 'Don't be a goat.' Silly young fools, all three of them. 1949 Partridge *Dict. Underworld* 296/1 Goat,..a dupe; swindler's victim. 1971 *Inside Kenya Today* Mar. 37/2 'I must discipline these idiots,' Omolo said to himself...'I must beat them today, goats!'

4. b. goat and bee, used *attrib.* to designate a type of Chelsea porcelain; also, applied to silverware bearing the same pattern; **goat-foot** [after Gr. αἰγιπόδης, αἰγίπους], a faun or satyr; the god Pan (cf. also *goat-feet* in Dict.); also *attrib.*; **goat hair,** the hair or skin of the goat.

1931 E. Wenham *Domestic Silver* v. 98 Various curious shapes were adapted to these small jugs, one being the so-called 'goat and bee'. This is supposed to have been designed by Nicholas Sprimont, a silversmith, and.. examples in silver are extremely rare. 1957 Mankowicz & Haggar *Encycl. Eng. Pott. & Porc.* 97/1 '*Goat and Bee' jug*, a jug decorated with goats and a bee, incised in the base with the word Chelsea, a triangle, and a date, generally 1745. 1878 Wilde *Ravenna* v. 10 Some goat-foot Pan. 1898 G. Meredith *Odes Fr. Hist.* 6 To veil an evil leer, And bid a goatfoot trip it like a fay. 1906 *Daily Chron.* 13 Aug. 4/4 It was the hour of Pan. I could almost think I saw the goat-foot playing his pipes by the brook. 1912 R. Brooke *Poems* (1918) 54 To glimpse a Naiad's reedy head, Or hear the Goat-foot piping low. 1925 E. Sitwell *Troy Park* 9 The goat-foot satyr waves. 1895 *Montgomery Ward Catal.* 102/3 Infants' white goat hair brushes, fine and soft. 1960 *Farmer & Stockbreeder* (Suppl.) 29 Mar. 4/1 A multilayer goat-hair fleece. 1967 J. Rathbone *Diamonds Bid* iii. 27, I have plenty of kilims and goat-hair rugs.

goatee. Also **goaty.** (Earlier U.S. examples.)

1844 Lee & Frost *Ten Yrs. Oregon* viii. 102 A few individuals..leave what is called, by some of their politer neighbors, a 'goaty' under the chin. 1847 J. K. Paulding *Noble Exile* 107 A queer little amber-colored goatee.

go-away (gōu,ăwē·i). (Stress variable.) Also **go-way.** [Imitative of its cry *kway-kway* (cf. Bechuana *maquaai*).] In full *go-away bird*. A South African touraco of the genus *Corythaixoides*.

1881 E. E. Frewer tr. *Holub's Seven Yrs. S. Afr.* I. 289 A great grey lory, that from its cry is called the 'go-away' by the English, whilst by the Boers it is known as the 'grote Mausevogel'. 1896 H. L. Tangye *In New S. Afr.* II. vii. 401, I continue my walk along the kopje top, succeeding in obtaining specimens of the 'Go away' bird. 1897 J. P. Fitzpatrick *Outspan* 55 There is a sort of bastard cockatoo in those parts which is commonly known as the 'Go way' bird, on account of its cry, which closely resembles these words. 1906 *Chambers's Jrnl.* Mar. 214/2 A pair of gray crested parrots, or 'go-away birds', as they are called. 1940 [see *alarm-bird* s.v. *alarm *sb.* 13]. 1951 R. Campbell *Light on Dark Horse* v. 93 The 'Go 'way Bird' or grey lory. 1958 M. Spark *Go-away Bird* 74 All over the Colony it was possible to hear the subtle voice of the grey-crested lourie, commonly known as the go-away bird by its call, 'go 'way, go 'way'. 1964 A. L. Thomson *New Dict. Birds* 843/2 The characteristic call of *Corythaixoides* spp. is a nasal 'gwaaa', hence the name of 'go-away-bird'.

gob, *sb.*[1] Add: **1. c.** *Glass-making.* A lump of molten glass that is used to make a single bottle, jar, etc.

1925 Hodkin & Cousen *Textbk. Glass Technol.* xxxii. 427 This plunger, moving vertically with a reciprocating motion, serves to form and deliver gobs of suitable size and shape. 1941 C. J. Phillips *Glass* ix. 191 As each section passes a feeding station served by a gob feeder, a gob of glass is dropped into the press mold and pressed into a parison. 1967 *Times Rev. Industry* July 47/2 (Advt.), The shear blades cut off a lump of this molten glass (bottlemakers call it a 'gob') from which each bottle is made.

gob, *sb.*[2] Add: **b. gob-stick,** (*c*) *slang* a clarinet; **gob-stopper** *slang*, a large, hard, freq. spherical sweet for sucking.

1936 *Amer. Mercury* May x/2 *Gob stick*, a clarinet. 1938 Dylan Thomas *Let.* 16 May (1966) 195 The double-bed is a swing-band with coffin, oompah, slush-pump, gob-stick. 1959 'F. Newton' *Jazz Scene* 292 The player as such calls his instrument a clarinet..; *qua* hipster he may call it..the obs. *gobstick*. 1928 W. de la Mare *Come Hither* (ed. 2) 769 *Gob-stoppers* and *toffee*—are these not 'good' names for goodies? 1959 D. Barton *Loving Cup* 89 The sky looks like one of those orange gobstoppers we used to buy in the tuckshop. 1971 *Times* 6 Jan. 12/5 A large ball of green leaves is formed in the cheek which makes *qat* chewers looks [*sic*] as though they are perpetually sucking gobstoppers.

gob (gǫb), *sb.*[5] *slang* (orig. *U.S.*). [Cf. *gobby.] An American sailor or ordinary seaman. Also *attrib.*

1915 *Recruiters' Bulletin* (U.S.) Dec. 34/2 Even though us 'gobs' were on the 'Conny' standing by, Why, we all had a finger in the pie. 1923 R. D. Paine *Comrades of Rolling Ocean* iii. 44, I was feeding the coal to her when most of those..amateur gobs were seasick and useless. *Ibid.* xiv. 241 Here is one Yankee gob..that will be there on the first call. 1925 J. Gregory *Bab of Backwoods* xvi. 193 Step lively, you gobs! 1944 T. Rattigan *While Sun Shines* I. 187 Can you beat that—an earl being a gob. 1951 *Landfall* V. III. 193 He wore..a little white gob hat.

gob (gǫb), *v.*[3] [f. Gob *sb.*[1]] *trans.* and *intr.* To spit. Also with *up*.

1881 A. B. & S. Evans *Leicestershire Words* 161 *Gob*, to spit out; expectorate. 1933 L. A. G. Strong *Sea Wall* II. xvi. 223 Come on, man, gob it up. 1953 Dylan Thomas *Under Milk Wood* (1954) 50 And they thank God, and gob at a gull for luck.

gobbet, *sb.* Add: **1. d.** A piece of a literary or musical work removed from its context; *spec.* an extract from a text set for translation or comment.

1912 *Punch* 6 Mar. 173/1 He'll gorge you with gobbets of Homer. 1930 D. L. Sayers *Strong Poison* 160 Playing the most ghastly tripe, sandwiched in with snacks of Mendelssohn and torn-off gobbets of the 'Unfinished'.

gobbledygook (gǫ·b'ldigu:k, -gŭ:k). orig. *U.S.* Also **gobbledegook.** [Prob. repr. a turkey-cock's gobble.] Official, professional, or pretentious verbiage or jargon.

1944 *Amer. N. & Q.* Apr. 9/1 Gobbledygook talk: Maury Maverick's name for the long high-sounding words of Washington's red-tape language. 1944 M. Maverick in *N.Y. Times Mag.* 21 May 11/1 Just before Pearl Harbor, I..got my baptism under 'gobbledygook'..its definition: talk or writing which is long, pompous, vague, involved, usually with Latinized words. It is also talk or writing which is merely long. 1945 *Tuscaloosa* (Ala.) *News* 7 Aug. 4/3 The explanations sound like gobbledegook to me. 1947 *Time* 7 July 7/1 The Veterans Administration translated its bureaucratic gobbledygook. 1950 'S. Ransome' *Deadly Miss Ashley* ii. 16 It now seemed a tricked-up system of gobbledegook wherein justice had foundered. 1951 Wodehouse *Old Reliable* vii. 94 You insult my intelligence by trying to put gobbledy-gook like that over me. 1959 M. Dolinsky *There is no Silence* i. 5, I had been subjected to too much psychiatric gobbledygook. 1968 M. Black *Labyrinth of Lang.* vi. 124 Jargon (or 'gobbledygook', to use the more expressive term).

gobby (gǫ·bi). *slang.* [Said to be f. dial. and vulgar *gob* expectoration + -y[6].] (See quot. 1929.)

1890 in Farmer & Henley *Slang* III. 168/1 When a meeting takes place the men indulge in a protracted yarn and a draw of the pipe. The session involves a considerable amount of expectoration all round, whereby our friends come to be known as gobbies. 1904 Kipling *Traffics & Discoveries* 126 Think o' her Number One chasin' the mobilised gobbies round the lower deck flats. 1929 F. Bowen *Sea Slang* 58 *Gobby*, in the British Navy, a coastguard, or in the old days a quarterdeckman. In the American, any bluejacket.

Gobelin. Add: **1.** Also *ellipt.* (Earlier examples.)

1788 A. Young *Travels* 10 Oct. (1792) 98 Four fine tablets of the Gobelin tapestry. 1802 C. Wilmot *Let.* 31 Jan. (1920) 44, I tell you of the 'Gobelins tapestry' which we went to see. *Ibid.* 19 June 72 The apartments were hung over with fine Gobelins. 1817 M. Edgeworth *Harrington* I. xv. 399 The gobelin tapestry of the back drawing-room.

†goblin[2] (gǫ·blin). *Obs. slang.* Also **o'goblin.** [Shortened form of *Jemmy O'Goblin.] A sovereign; twenty shillings.

1887 W. E. Henley *Poems* (1908) II. 231 Your merry goblins soon stravag. 1909 Wodehouse *Swoop* II. ii. 68 Come now, your Grand Grace, is it a deal? Four hundred and fifty chinking o'Goblins a week for one hall a night. 1925 —— *Carry on, Jeeves* iv. 99 Five hundred o'goblins a year.

goblinesque (gǫ:blinĕ·sk), *a.* [f. Goblin + -esque.] = Goblinish *a.*

1916 *Proc. Mus. Assoc. 1915–16* 106 The weird, goblinesque effect of the bass-clarinet accompanied by the celesta will be remembered. 1934 *Times Lit. Suppl.* 1 Feb. 65/3 The goblinesque figure of 'old Wicked-shifts', Henry Brougham.

gobo (gōu·bōu). orig. *U.S.* [Origin unknown.] (See quots.)

1930 *Sel. Gloss. Motion Pict. Techn.* (*Acad. Motion Pict., Hollywood*) 15/2 *Gobo*, portable wall covered with sound-absorbing material. 1936 *Words* Oct. 6/2 A 'gobo' is a small black screen used to deflect light. 1970 *T.V. Times* (Austral.) 1 Apr. 8/3 A gobo is anything that goes between, e.g., the light and the set.

God, *sb.* Add: **1. e.** (Further examples.)

GOD-ALMIGHTY

1761 BOSWELL *Let.* (1857) 17 Dec. 383 It is Captain Andrew! it is! it is! Ye gods, he seizes! he opens! he reads! **1867** C. WILMOT *Let.* 15 May *Rust. Jrnls.* (1934) ii. 243 Oh! ye Gods! How you are to be envied & every Mortal alive. **1871** L. M. ALCOTT *Little Men* ii. 27 But out of school,—Ye gods and little fishes! how Tommy did carouse! **1909** H. G. WELLS *Ann Veronica* i. 9 'Ye gods!' she said at last. 'What a place!' **1927** W. E. COLLINSON *Contemp. Eng.* 26 We used harmless expletives like.. Ye Gods and little fishes. **1964** W. MARKFIELD *To Early Grave* (1965) xi. 187 He cried to himself 'Ye Gods!' and 'Whoosh!'

f. *god from* (or *out of*) *the* (or *a*) *machine* = *DEUS EX MACHINA.

1868 TROLLOPE *Phineas Finn* (1869) I. xxxi. 257 A gallant young member of that House..had appeared upon the spot at the nick of time;—'As a god out of a machine,' said Mr. Daubeny, interrupting him. **1888** KIPLING *Soldiers Three* 1 (*title of story*) The god from the machine. **1910** CHESTERTON *G. B. Shaw* 116 Shaw.. disliked the god from the machine—because he was from a machine. **1959** *Listener* 26 Nov. 911/2 The heads of government of the Great Powers are not gods from the machine. **1970** N. FISHER *Walk at Steady Pace* III. 157 If the God from the Machine was to solve my troubles he was by far the most convincing candidate.

8. a. *God bless*: also used *ellipt.* for 'God bless you' as a wish for God's blessing on a person or as an expression of goodwill, esp. at parting.

1964 P. M. HUBBARD *Picture of Millie* iv. 42 She took the drink..'I don't know what to say,' she said. 'Would "God bless" do?' **1964** 'J. ROFFMAN' *Likely to Die* vi. 69 'That would mean that I'd arrive here soon after half-past five.' She smiled. 'That will be fine. God bless, Albert.' **1966** E. PETERS' *Piper on Mountain* i. 10 Now good night, and God bless! Don't stay up too late! **1967** 'M. HUNTER' *Gambling Horses Disaster* xvi. 109 Try and forget me, David, God bless.

d. *God forbid* Rhyming slang (see quots.). Cf. *GAWD-FORBID.

1909 J. R. WARE *Passing Eng.* 144/2 *God-forbids*, kids —a cynical mode of describing children by poor men who dread a long family. **1960** J. FRANKLYN *Dict. Rhyming Slang* 70/1 *God forbids*, (1) Kids (child or children), (2) Yids (Jews), (3) *lid* (hat).

16. a. God-box, (*a*) *slang* a church or other place of worship; (*b*) (with lower-case initial) (see quot. 1923); **god-shelf,** a shelf-like shrine of white wood holding the sacred images in a Shinto household.

1923 OGDEN & RICHARDS *Meaning of Meaning* ii. 37 The priests in whom gods were supposed to dwell [a belief which induced the Cantonese to apply the term 'god-boxes' to such favoured personages]—are amongst the victims of this logophobia. **1928** GALSWORTHY *Swan Song* III. xii. 305 This great box—God-box the Americans would call it—had been made centuries before the world became industrialised. **1962** *New Statesman* 25 May 768/2 A ring-along God-box that will get your bird up with the flat-bottomed latitudinarians. **1876** F. V. DICKINS tr. *Chiushingura* 171 One of these O-kami-no-are ought to find a place upon every domestic *Kami-dana*, or god-shelf—a small model of a Shinto temple to be found in almost every house, labelled with the names of various deities. **1903** S. L. GULICK *Evol. Japanese* xv. 291 There is hardly a house in Japan but has some.. of these charms, either nailed on the front door or placed on the god-shelf. **1905** H. SLADEN *Playing Game* xii. They are never too poor to have a little light burning on the god-shelf.

b. *God- (or god-) consciousness, -idea.*

1894 G. M. GRANT *Relig. World* ii. 29 The God-consciousness of Israel expanded under the leadership of a long succession of prophets and psalmists. **1914** F. B. WILSON (*title*) The man of to-morrow. Human evolution impelling man onward to God-consciousness. **1968** W. JAMES *Coll. Ess. & Rev.* (1920) 429, I am now using the God-idea merely as an example, not to discuss as to its truth or error. **1910** E. S. AMES *Psychol. Relig. Experience* 319 The God-idea is a teleological idea. **1924** W. B. SELBIE *Psychol. Relig.* 239 The origin of the god-idea varies with different peoples. **1949** *Horizon* Mar. 225 However primary and archetypical the God-idea may be.

c. God's (own) country (see *COUNTRY 2 b); **God's-eye-view,** a view as seen by God; **God's (own) gift,** a godsend; **God's quantity** *colloq.*, a large amount; an abundance.

1920 God's-eye view [see *EYE VIEW]. **1936** A. HUXLEY *Eyeless in Gaza* vii. 85 One has made a habit of not feeling anything very strongly; it's easy, therefore, to take the God's-eye view of things. **1970** *Guardian* 14 May 9/6 Frank Tuohy's..short stories..are mostly studies in suburban isolation,..the God's-eye view with God on the stage. **1938** 'E. QUEEN' *Four of Hearts* (1939) I. i. 9 [*heading*] God's gift to Hollywood. **1941** M. ALLINGHAM *Traitor's Purse* xx. 228 Amanda was God's own gift to anyone in a hole. **1953** H. CLEVELY *Public Enemy* xxi. 151 It may do him a bit of good to find out he isn't God's gift to women walking the earth. **1911** C. E. W. BEAN '*Dreadnought' of Darling* xxviii. 242 There was God's quantity of fish. **1922** JOYCE *Ulysses* 287 Big foxy thief ..lifted any God's quantity of tea and sugar.

17. b. (Further examples.)

1929 W. B. YEATS *Winding Stair* 11 And God-appointed Berkeley that proved all things a dream. **1956** C. S. LEWIS *Let.* (1966) 269 The test of music or religion..is always the same—do they make one..more God-centred..and *less self-centred*? **1845** WHITTIER in *Amer. Liberty Almanac 1846* 26 And the solemn priest to Moloch, on each God-deserted shrine. **1863** TENNYSON *Milton* in *Cornh. Mag.* Dec. 707 God-gifted organ-voice of England. **1907** 'D. DONOVAN' *Gold-spinner* I. i 'God-gifted and beautiful' was Helga Arnold! **1927** W. B. YEATS *October Blast* 22 These God-hated children. **1865** MILL *Comte* 30 Hippocrates..could say with impunity,

speaking of what were called the god-inflicted diseases, that to his mind they were neither more nor less god-inflicted than all others. **1936** L. MACNEICE tr. *Aeschylus' Agamemnon* 52 From whence these rushing and God-inflicted Profitless pains?

God-almighty. Add: **B.** *adv.* Extremely, excessively.

1906 E. DYSON *Fact'ry 'Ands* vii. 85 It's that go'-mighty funny I can't laugh, 'n' I'm dyin' to. **1923** D. H. LAWRENCE *Kangaroo* viii. 164 Why should people..be so God-almighty puffed up?

God-awful (gǫ·dǫ·ful), *a. slang* (orig. *U.S.*). Also **God awful, Godawful.** [f. GOD *sb.* + AWFUL *a.*] Terrible; extremely unpleasant. (In quot. 1878 the sense is 'impressively large'.)

1878 J. H. BEADLE *Western Wilds* xxxvii. 612 Put thirty acres..into wheat, and went to work with a hurrah in 1874 to make a God-awful crop. **1897** C. M. FLANDRAU *Harvard Episodes* 68 Ellis is such a God awful rest. **1930** W. S. MAUGHAM *Breadwinner* II. 124 Your affairs are in a god-awful mess. **1946** 'S. RUSSELL' *To Bed with Grand Music* I. 14 Listen to the most godawful programmes on the radio. **1958** R. GRAVES in *Times Lit. Suppl.* 15 Aug. p. xiv The credible and vivid story that any context (red-brick, yellow-brick, or otherwise God-awful) offers. **1959** P. MCCUTCHAN *Storm South* iv. 63, I heard the most God-awful racket above my head.

go·dbush. (Also gad-.) A Jamaican name for several plants of the mistletoe family, Loranthaceæ.

1851 P. H. GOSSE *Nat. Sojourn Jamaica* 131 But what interests me most of all is to discover a flourishing Mistletoe, or God-bush, as the negroes call it. **1864** A. H. R. GRISEBACH *Flora of Brit. West Indian Islands* 784/1 God-bush: *Arceuthobium gracile*. **1924** FAWCETT & RENDLE *Flora Jamaica* III. 88 The members of this family [*sc.* Loranthaceæ] are generally known by the popular names..Godbush and Mistletoe. **1961** F. G. CASSIDY *Jamaica Talk* xvi. 380 There is *god-bush* too, the mistletoe.

God-damn(-me. Add: **1.** (Later examples.) Also quasi-*adj.*, accursed, damnable. So **God-damned** *a.*

1869 *St. James's Mag.* IV. 226 It was the fashion in France, on the stage, to represent the Englishman as habitually saying, 'Goddam'. **1898** W. P. DRURY *Tadpole of Archangel* 149 Before I could say 'Goddam',..they would be fired. **1918** *Wine, Women & War* (1926) 29 If and when I do get home, I'll be the same God damned crab I was when I went away. **1921** J. DOS PASSOS *Three Soldiers* (1932) I. vi. Yer right not to go with any of the girls in this goddam town. **1928** *Queen* 7 Nov. 6/3 Mr. Frederick Lonsdale gave his usual, 'No, God damn it' —..in talking to the actors. **1936** P. G. FOSSE *Land Called Naomi* 22 A half-drawn man's face has but ..'ad'l 'God-damn-me's Vermin. **1929** ..Nabakov 'Loud' 'Nor Almost enough of the nation a Wil..with the guilt 'Goddam' ..[illegible] **1945** F. L. WOLFE ..It's all so goddam ..my God damned engine. **1952** S. KAUFFMANN *Philanderer* (1953) 36 Stop being such a goddam fool! **1958** A. WILSON *Middle Age Mrs. Eliot* II. iv. 265 That goddam fool! **1963** J. B. WARD *Wind of Change* ..it goddam and.. ..'Jesus, the God-damned fool's taking'. **1964** *New Statesman* 6 Sept. 330/3 And I'll ponder upon the son-of-a-gun. Who turned my goddamned thesis down. **1970** W. C. WOODS *Killing Zone* (1971) ii. 23 Now you men knock off the goddam chatter in there and listen up.

Gödel (gō·dəl). The name of Kurt *Gödel* (born 1906), Austrian mathematician, used *attrib.* and in the possessive to designate his metamathematical theorems and related techniques and constituents; as *Gödel number, numbering*; *Gödel's proof*; **Gödel('s) theorem,** the demonstration (first published in *Monatshefte f. Math. u. Physik* (1931) XXXVIII) that in logic and in mathematics there must be true formulas which are neither provable nor disprovable, thus making mathematics essentially incomplete, and also the corollary that the consistency of such a system as arithmetic cannot be proved within that system. Also **Gödelian** (gōdīˑliăn), *a.*

1933 H. BLACK *Nature of Math.* 170 [*heading*] Note on Gödel's Theorem. **1942** *Mind* LI. 259 (*title*) Gödelian sentences: a non-numerical approach. **1952** S. C. KLEENE *Introd. Metamath.* viii. 204 We designate the first of these theorems, which entails the other as corollary, as 'Gödel's theorem'. *Ibid.* 206 We call this a Gödel numbering, and the correlated number of a formal object its Gödel number. **1956** E. H. HUTTEN *Lang. Mod. Physics* ii. 34 Gödel's theorem, in which Gödel showed that, in certain logical systems ..shall be conformed to..the incomplete Gödel..theorem in mathematics. **1959** NAGEL & NEWMAN (*title*) Gödel's proof. **1963** W. M. KNEALE *Devel. Logic* vii. 476 This discovery is closely akin to Gödel's theorem on the incompletability of formal arithmetic. **1964** *Philos.* XXXIX. 196 Machines could never be selfconscious in the way that human minds are, and..Gödel's theorem illustrates this. **1965** R. L. WILDER *Introd. Found. Math.* (ed. 2) xi. 275 In some of these systems, analogues of the Gödel theorems have been shown to hold.

‖ **godemiche** (godəmiʃ). [Fr.] = *DILDO[1].

1879–80 *Pearl* (1970) 265 'A godemiche? What's that for?'..'Oh! Oh! We ladies use it to excite ourselves.'

1886 R. F. BURTON *Arab. Nts.* X. 239 Of the penis succedaneus, that imitation of the Arbor vitæ,..which..the French [call] godemiché. **1887** L. C. SMITHERS tr. *Forberg's Man. Class. Erotol.* vi. 148 There are expounders..who..have imagined that Bassa misused women by introducing into their vaginas a leathern contrivance, an olisbos, a godemiche. **1966** L. COHEN *Beautiful Losers* II. 168, I also have in this trunk a number of artificial phalli (used by women), Vaginal Vibrators, the Rin-No-Tam and Godemiche or Dildo.

godet (godeˑ, goˑdet). Restrict †*Obs.* to the sense in Dict., and add:

2. A triangular piece of stuff inserted in a dress, glove, etc. Also *attrib.*, as *godet pleat, skirt.*

1872 *Young Englishwoman* Dec. 646/1 A basque plaited behind in three godet pleats. **1896** *Strand Mag.* July Advt. p. xiv, Costume..consisting of wide Godet Skirt. **1923** *Daily Mail* 10 Apr. 14 Black 'godets' in white kid (or white in black). **1925** *Brit. Weekly* 15 Oct. 59/1 Some of the smartest models have a full centre-piece instead of side godets. **1926** *Queen* 17 Feb. Advt. p. vii, The simple bodice has the new long sleeves and the full godet skirt is finished with binds. **1928** *Daily Mail* 31 July 1/2 The skirt has full godet each side. **1960** CUNNINGTON & BEARD *Dict. Eng. Costume* 96/1 *Godet pleat*, a hollow tubular pleat, narrow above and expanding below. **1960** E. W. MONCRIEFF *Artificial Fibres* viii. 100 As the filaments emerge from the jet,..they are partially..hardened by the bath, and thence guided led to an eye at the surface of the bath, and round the bottom godet rollers. **1957** *Textile Terms & Defs.* (ed. 3) 52 *Godet*, a pulley, usually having one flange, round which threads pass in order to regulate their speed during the extrusion of certain man-made fibres. **1960** J. G. COOK *Handbk. Textile Fibres* (ed. 2) 174 On leaving the Godet wheel, the fibres pass around a second wheel which is moving faster than the first.

3. A driven roller or wheel around which filaments of certain man-made fibres are drawn during their manufacture. Also *attrib.*

1927 T. WOODHOUSE *Artificial Silk* 41 The group [of filaments] is then passed behind a guide-wire, partly round a roller, usually termed a godet, and often made of glass..the godet roller takes the filaments..

go-devil (gōu·dev'l, gō·devil). orig. and chiefly *U.S.* [f. GO *v.* + DEVIL *sb.*] A name for various contrivances used in farming, logging, drilling for oil, etc. (see quots.).

1835 *Knickerbocker* Apr. 273 Led on by what they call in school-sports, a go-devil, prancing about in high horns, and a spear on the end of his tail. **1854** C. L. FLEISCHMANN *Wegweiser* 173 In Indiana und Illinois bedient man sich zum Zudecken der Maiskörner einer Art Hacke, welche unter dem Namen 'Go-On-Devil' bekannt ist. **1889** *Harper's Mag.* June 14/2 The graceful 'go-devil' rake,.. gathering up the hay with all the ease of a lady's carpet-sweeper. **1886** *St. Nicholas* Nov. 48/1 A queer looking, pointed piece of iron, called the 'go-devil', is dropped down the well, and [strikes]..a cap on the top of the torpedo. **1889** *Cent. Dict.*, *Go-devil*,..a movable-jointed contrivance..introduced into a pipe-line for the purpose of freeing it from obstructions... A rough sled used for holding one end of a log in hauling it out of the woods, etc. **1896** B. REDWOOD *Petroleum* I. 275 To explode the charge, an iron weight, known as a *go-devil*, was dropped into the well, and, striking the disc, exploded the cap and fired the torpedo. Now, however, a miniature torpedo known as a *go-devil squib*..is almost invariably employed. *Ibid.* II. 473 To remove obstructions in the pipes,..an automatic rotary scraper is forced through... The scraper is known as a 'go-devil'. **1931** *Wallers (Okla.) Herald* 14 Feb. 8/4 Farm Implements (Advt.), ..Sleighs, 'Yankee Jumpers' and Go Devils. **1937** D. LUTES *Home Grown* 64 Old Man Covell came over to borrow a go-devil with which to split a log. **1950** *Publ. Amer. Dial. Soc.* XXX. 53 *Go-devil*,..a U-shaped rig for sledding logs. **1964** *Listener* xxxxv. 41 *Go-devil*, V-shaped ditch-cleaner. **1959** *Times* 13 Apr. 14/2 A piece of equipment called a 'go-devil' is inserted every 24 hours at one end of the pipe and it emerges some hours later at the other end. **1969** *New Scientist* 20 Apr. 163/2 'Go-devils' have been used to scour and clean out oil, gas and water pipes... The go-devil, a sort of torpedo with rubber washers, scraper vanes or wire brushes mounted on it, is forced through the pipe under air or water pressure. **1962** *Amer. Speech* XXXVII. 268 A rather confusing situation exists with regard to *go-devil* in Colorado. One Colorado informant reported that the word as a generic term for 'all kinds of contraptions'... It may, in eastern Colorado, refer to a cultivator, but several times it clearly means 'buck rake'. In central and western Colorado it is much more likely to refer to a Y-shaped ditch cleaner. **1962** *Lebende Sprachen* VII. 8/1 Pipelines are regularly cleaned by a bristling metal contraption known as a 'go-devil'. This device scrapes out the clinging sludge which collects inside the pipe.

godfatherly, *a.* Delete † and add later examples. Also *transf.*

1948 *Observer* 29 Jan. 17/2 That 'brighter cricket' which Lord Hawke, on behalf of Yorkshire, promises for the coming season. This taking of godfatherly vows for a county team is a picturesque departure, which, we may hope, will have no anti-climax. **1958** WODEHOUSE *Cocktail Time* xv. 128 Lord Ickenham patted his arm in a godfatherly manner.

Go·dfea:ringly, *adv.* In a God-fearing way.

1899 J. H. RIGG *Oxford High Anglicanism* (ed. 2) 404 If, from point to point, the wise and equitable thing is Godfearingly carried out.

Go·d-forsa:ken, *ppl. a.* Of persons: depraved, profligate, abandoned. Of places: desolate, dismal, dreary. Hence **Go:d-forsa·kenly** *adv.*, **Go:d-forsa·kenness.**

1856, etc. [see GOD *sb.* 17 b]. **1860** G. DU MAURIER *Let.* (1951) 23 Tom, nurse & foster thine aversion towards the Godforsaken city in which thy lines are cast now. **1903** *Westm. Gaz.* 11 Feb. 12/1 Of course, it is not of the same date as Brive. But it has the God-forsakenness, the *misère*, the penetrating sadness, so essentially French charm. **1913** D. H. LAWRENCE *Let.* 19 May (1932) 123 Some of the reviews have been so God-forsakenly stupid. **1929** W. P. KER *Art of Poetry* 60 You come with Milton.. to the Paradise of Fools in a dry, parched, and god-forsaken land on the outside of the fixed stars. **1959** 'M. CRONIN' *Dead & Done With* vi. 61 You wouldn't know any place in this God-forsaken spot?

Godfrey[1] (gǫ·dfri). *Med.* [f. the name of Thomas *Godfrey* of Hunsdon, Hertfordshire, fl. early 18th c.] In full, *Godfrey's cordial.* (See quot. 1961.)

1742 *Applebee's Original Weekly Jrnl.* 17 Feb. 2298/1 To all Retailers and Others. The General Cordial formerly sold by Mr. Thomas Godfrey of Hunsdown in Hertfordshire, deceased, is now prepared, according to a Receipt written by his own Hand, and..is now sold by me Thomas Humphreys of Ware in the said County, Surgeon. **1786** *Act 26 Geo. III* c. 79 Schedule..Containing the names by which many medicinal preparations..are known and distinguished....Godfrey's Cordial. **1845** L. PLAYFAIR *Rep. Large Towns Lancs.* 125 We want more Godfrey for children, for it does not produce convulsions in our children, like some of the other Godfreys. **1856** C. M. YONGE *Daisy Chain* II. xx. 563 'What have you heard?' 'Godfrey's Cordial, sir,' quietly suggested the nurse. **1899** A. S. TAYLOR *Poisons* (ed. 3) IV. 563 Godfrey's Cordial.. This is chiefly a mixture of infusion of sassafras, treacle, and tincture of opium. **1929** ..[illegible]. **1937** C. H. LA WALL *Four Thousand Years' Pharmacy* xi. 373 Godfrey's Cordial was a popular household remedy in Great Britain in 1722, when an advertisement of it appeared claiming the transfer of the formula to John Humphreys, Apothecary, by the estate of Thomas Godfrey, deceased. **1961** *Brit. Med. Dict.* 622/2 *Godfrey's cordial*, a sweetened and flavoured tincture of opium, a preparation rarely prescribed in modern medicine.

Godfrey[2] (gǫ·dfri). *U.S. slang.* A euphemistic assimilation of *God* to the name *Godfrey*, used as an exclamation of strong feeling or surprise (cf. GOD *sb.* 7 a); also in *phr. by guess and by Godfrey*: see *GUESS *sb.* 1.

1904 *Dialect Notes* II. 425 *Godfrey!* An ejaculation. Also *godfrey mighty*, and *godfrey Lijah*. **1907** *Ibid.* III. 196 *Godfrey mighty*, softened form of God Almighty. **1909** *Ibid.* 411 *Godfrey*, used as a mild oath in the speech of the young. **1926** W. CHURCHILL *Coniston* 274 'Godfrey!' exclaimed Ephraim. **1909** J. C. LINCOLN *Keziah Coffin* vii. I'd bet those ..and taken that old hooker of Zach's..Godfrey in it. **1962** ..Hat! **1946** B. M. BOWER *Phantom Herd* v. 71 Why my godfrey, man, the stuff's all punch. **1930** S. HENRY *Conquering Plains* 136 And there are the Government troops at Riley and Harker —by Godfrey!—if it comes to that. **1944** W. FAULKNER *Go Down, Moses* 15 They hadn't even cast the dogs yet when Uncle Buck roared, 'Gone away! I godfrey, he broke cover then!'

Godism (gǫ·diz'm). [See -ISM.] A derisive term for belief in God. Also **Godite** (gǫ·dait), one who believes in God.

1861 J. M. ROBERTSON *Mod. Humanists* 18 Call his [sc. Carlyle's] creed 'Godism,' and you limit the confusion of words by separately labelling his confusion of thought. *Ibid.* 37 At the mere sound of that word [*sc.* atheism], the Godite skies have been raked. **1909** *Lit. Guide* 1 Aug. 114/1 *The Contemporary Review* permits Mr. G. K. Chesterton to make faces over Meredith's grave, and claim him as a believer in godism.

godless, *a.* Add: **b.** *godless florin* (earlier and later examples).

1870 [see *PACELESS *a.* 1 a]. **1966** P. A. RAYNER *Coin Collecting for Amateurs* v. 84 The florins of 1849 are known as 'godless' florins, because they omit the words in Queen Victoria's title 'Dei Gratia'.

c. *the Godless* [Russ. *bezbózhnik*]: the title of a union (and its press organs) in Soviet Russia, having for its primary object the suppression of religion.

1930 OLESCHUK in *The Anti-Religious* No. 10, The Union of the Godless makes every effort in order to transform the International Union of the Godless into an active staff of class conflict,..and to make the fight with religion in the West a factor in the class struggle of the proletariat with the bourgeoisie. **1930** *Slavonic Rev.* Mar. 518 The Union of the Godless is a semi-State institution... The Union organises special 'cells of the Godless' in factories, villages and units of the Red Army. **1967** G. VON STACKELBERG in Fletcher & Strover *Relig. & Search for New Ideals in U.S.S.R.* 95 Publication of atheist propaganda was..revived in 1947 when the Association of Militant Godless was dissolved and the Society for Dissemination of Political and Scientific Knowledge established.

God-man. Add: Also *transf.*

1899 C. J. C. HYNE *Further Adv. Capt. Kettle* ii. 34 'He God-man. Lib for here on gin-palaver.' 'Trading missionary, is he?' **1924** E. WALLACE *Room 13* 56 You couldn't undo what that old God-man did this morning.

Godwinian (gǫdwi·niăn), *a.* and *sb.* [f. the name of William *Godwin* (1756–1836) + -IAN.] **A.** *adj.* Pertaining to or characteristic of Godwin's views on politics, social reform, etc. **B.** *sb.* An adherent of Godwin's views. So **Go·dwinism,** Godwin's doctrine or ideas; **Go·dwinist,** an adherent of Godwinism.

1800 C. WILMOT *Let.* 14 Dec. (1920) 15 But I begin to smoke a little of the Visionary on the Godwinean System of living beyond the term of Man. **1812** SHELLEY *Let.* 26 Nov. (1964) I. 334 Jealousy has no place in my bosom; I am indeed at times very much inclined to think the Godwinian plan is best. **1891** L. STEPHEN *Hours in Library* (new ed.) III. 79 The Godwinism, indeed, is strongest in the crude poetry of *Queen Mab*. **1923** H. N. BRAILSFORD *Shelley, Godwin* 216 The ideal man of the Godwinian conception, who lives by reason. **1937** 'C. CAUDWELL' *Illusion & Reality* 97 A Godwinist, Shelley believed that man is naturally good. **1946** BLUNDEN *Shelley* 125 She was a child, a Godwinian. **1951** *Essays in Criticism* I. i. 17 The Godwinian Urging the gallows for the subject citizen. *Ibid.* 28 The aftermath of Godwinism. *Ibid.* 31 *Guilt and Sorrow*, that Godwinian plaint of man's inhumanity to man.

God-wottery, Godwottery (gǫdwǫ·tari). Also with lower-case initial. [f. *God wot* (cf. GOD *sb.* 10) in the line 'A garden is a lovesome thing, God wot!' in T. E. Brown's poem *My Garden* (1876) + -ERY.] An affected or over-elaborate style of gardening or attitude towards gardens (see quots.); also (in quot. 1939), archaic language.

1939 N. LOFTS *Colin Lowrie* (Author's note), I have written this so-called historical novel in so-called modern language... I am foolish enough to believe that people.. will appreciate this lack of 'God-wottery'. **1952** *Archit. Rev.* CXI. 59 The God-wottery..in that anaemic kind of 'Trajan' which societies concerned with what they generally call 'amenities' seem to think particularly relevant. *Ibid.* CXII. 235 He pays fast and loose with the average Englishman's sentimental leanings towards God-wottery, drawing his clichés from the Cotswolds where, it is widely believed, lie the typical English villages and the homeland of the picturesque. **1960** A. BURGESS' *Right to Answer* i. 7 Who shall describe..those semi-detacheds with the pebble-dash..., the tiny gates which you could step over, the god-wottery in the toy gardens? **1969** *Guardian* 18 Aug. 7/1 'Godwottery', the sentimental preconception of what a garden should be, results in a very strange collection of elements... Cotswold stone retaining walls; 'vaguely Spanish wrought iron gates; 'crazy' paving, nowadays often coloured yellow, green, and pink; plainly irregular ponds, now usually of pale blue fibreglass, fed by streams of impossible source; gnomes, fairies, and animals, usually plastic.

go-easy, *a.* [f. the vbl. phr. *to go easy*: see GO *v.* 2 b and EASY *adv.* 4 a.] Easy-going; characterized by leisurely ease.

1877 *Rep. Vermont Dairym. Assoc.* VIII. 22 The many serious drawbacks which the 'go easy' dairymen of Vermont are compelled to encounter. **1936** R. C. K. ENSOR *England 1870–1914* ii. 12 The purchase system kept officering as an aristocratic hobby.. and a trade for professional men. If it became the latter, it might menace our go-easy oligarchic liberties. **1940** S. O'CASEY *Star turns Red* I. 8 The workers..with their lightning strikes, stay-in-strikes, sit-down strikes, sit-down strikes, go-easy strikes. **1959** *Times* 19 Oct. 17/3 It was impossible to goggle-dive with-out seeing a fish.

goer. Add: **1. c.** Also, one who behaves in a lively, persevering, or profligate manner; also, a successful man; an expert.

c 1810 W. HICKEY *Mem.* (1960) v. 78 The whole party, male and female, were of the description yclept 'hard goers'. This did not alarm me, for..I could keep way with the best of them at fair drinking. **1811** [see *ATTICK sb. 1 a]. **1857** G. A. LAWRENCE *Guy Livingstone* xx. 185 She was always deeply engaged, and generally with one of the 'fast' set. **1888** MRS. A. GUNN *We of Never-Never* i. 5 By George! she's a regular goer. **1946** F. SARGESON *That Summer* 38 Miss Briggs is a goer anyhow. You want to see her on a wet day. **1959** *Observer* 1 Mar. 10/2 Princie is a 'goer', nerveless and brave, say the other people who know her. **1966** P. WILLMOTT *Adolescent Boys E. London* iii. 51 'She was a right banger,' said a 17 year old of one girl. 'A banger's a goer—a girl who'll go with anyone.'

d. One who frequently or regularly attends a specified place, type of entertainment, etc.; usu. with defining word prefixed: see CHURCH-GOER, *cinema-goer, *concert-goer, *film-goer, theatre-goer.

gofer. (Earlier examples of form *gaufre*.)

c 1845 C. BRONTË *Professor* (1857) II. xxi. 109 Having eaten an unlimited quantity of 'gaufres'. **1853** — *Villette* II. viii. 142 Regaled with *gaufres* and *vin blanc*.

go-getter (gōu·ge:təɹ). *colloq.* (orig. *U.S.*) [f. *to go get* (see *GO *v.* 32 a) + -ER[1].] An active, enterprising, pushing person. Also *attrib.*

1921 S. FORD *Inez & Trilby May* xii. 216 Think you're one of these go-getters, do you? **1922** P. B. KYNE (*title*) The Go-Getter. A story that tells you how to be one. **1926** F. SWINNERTON *Summer Storm* v. i. 197 The go-getter despises the non-go-getter; but never as much as the non-go-getter despises the go-getter. **1928, 1929** [see *GO *v.* 32 a]. **1932** J. CARY *Aissa Saved* 159 MacEwen did not write in the vein of a go-getter business expert seeking out inefficiency. **1959** M. CUMBERLAND *Murmurs in Rue Morgue* vii. 48 He was a go-getter, an *arriviste*,.. a bull charging at competitive life.

So **go·-ge:tting** *ppl. a.*, enterprising, pushing; also as *sb.*, the behaviour of a go-getter; enterprise, ambition. Hence **go-ge·ttingness.**

1921 J. G. FREDERICK *Great Game of Business* p. v, The true forward-looking 'go-getting' American business point of view. **1924** R. CUMMINS *Sky-High Corral* 25 He was one of them flyin' son-of-a-guns an' they say he was a go-gettin' fool. **1928** *Daily Express* 27 June 10/7 Such jobs generally call for..a go-getting attitude to life that the public school boy does not possess. *Ibid.* 4 July 13/6 All of which has been substantiated by my faith in the 'pep' and 'go-gettingness' of the American reporters. **1932** F. E. BAILY *Mr. Miller* 109 The students herded to hear breezy young instructors exhale the new gospel of 'go-getting'. **1946** R. KNOX *Epistles & Gospels* 93 There are some critics who maintain that when St Paul uses this particular word (it is really 'go-getting' rather than 'covetousness'), he is thinking of the ninth commandment, not of the tenth. **1959** *Punch* 21 Oct. 340/2 My future as a go-getting reporter was bleak indeed.

‖ **gogga** (xo·xa). *S. Afr.* Also †**goggo** (dim. *goggatje*, now *goggatjie*). [ad. Hott. *xoxon* insects collectively.] An insect (of any type).

1905 J. DU PLESSIS *Thousand Miles in Heart of Africa* iii. 34 This country must be 'gogga' land, it simply swarms with insects. **1909** *East London Dispatch* 8 Jan. 5 We have heard South Africa described as a land of goggas. **1911** *Ibid.* 27 Nov. 6 Another old, well-grown tree..is infested with those abominable goggatjes which have ..gnawed their way into the tree... **1926** E. LEWIS *Mantis* II. vi. 103 He said he'd find me lots of beautiful goggas to make pictures of much better than that. **1927** W. PLOMER *I Speak of Africa* 243 *Mrs. White*. Another huge black beast! What is it? *White*. It looks like a gogga. **1960** A. G. MCRAE *Hill called Grazing* vii. 62 There were other goggas in Italy.

goggle, *sb.* Add: **4. a.** Esp. with reference to their use by motor-cyclists (and formerly motorists) and underwater divers.

1903 A. H. JEKINSON *Three Men on Bummel* xiii, Their goggle-covered eyes, their necks tied up in comforters. **1903** G. B. SHAW *Man & Superman* II. 55 He cares for nothing but tearing along in a leather coat and goggles, at sixty miles an hour. **1904** A. B. F. YOUNG *Compl. Motorist* (ed. 2) 316 Goggles are, unhappily, almost a necessity when travelling at any but the lowest speeds. **1908** *Motor Cycle* 12 Feb. 132 A new goggle constructed after the principle of the four-glass goggle. **1908** E. J. BANFIELD *Confessions of Beachcomber* I. iv. 152 All were wearing swimming goggles which enable them when diving to distinguish objects at a considerable range. **1957** T. GUNN *Sense of Movement* II In goggles, donned impersonality.

6. *attrib.* and *Comb.*, as **goggle-box** *slang*, a television set; **goggle-dive,** an underwater dive made by a person wearing goggles; hence as *v. intr.*; also *goggle-diver, -diving.*

1959 *Guardian* 9 Nov. 5/5 Switch the goggle-box on at 10 a.m. **1960** *Times* 20 Oct. 1/1 Mr. Wilson was..so good at television appearances, that he had convinced himself that he, single-handed, could win elections with the help of the goggle box. **1963** J. V. CORNELL *Silent World* 5 Two years of goggle-diving passed before I met Dumas. *Ibid.* 7 In the goggle-diving era Dumas made a lightweight belt at Le Brusq that he could spear two hundred and twenty pounds of fish in two hours. *Ibid.* 16 The merou, virtually unknown in the Provençal markets until goggle divers went down and speared them. **1958** *Sunday Times* 19 Oct. 17/3 It was impossible to goggle dive without seeing a fish.

goggle, *v.*[1] **1. a.** Delete 'Now *rare*' and add later examples.

1938 G. GREENE *Brighton Rock* II. i. 67 She goggled hopelessly at the Boy. **1960** I. CROSS *Backward Sex* 105 It was too unbelievable of me to have expected that I would be well past goggling at her body when we were such friends. **1965** *Listener* 24 June 949/1 The conformity-poster freaks..but better things to do than goggle into the dim past.

go·ggled, *ppl. a.*[2] [f. GOGGLE *sb.* + -ED[2].] Equipped with or wearing goggles.

1903 *Westm. Gaz.* 22 Aug. 2/3 These ghastly goggled motor guys. **1908** *Daily Chron.* 21 July 1/6 The swimmer's head, goggled and capped, emerging from the water. **1916** H. G. WELLS *Tono-Bungay* II. ii. 137 A short figure, ..hugely goggled,..and surmounted by a table-land of motoring cap. **1919** N.Y. *Times* 3 Feb. A goggled aviator making ready for a flight. **1945** J. R. ULLMAN *White Tower* 425 He shifted up the goggled eyes.

goggle-eyed, *a.* Add: (Later examples.) Also *fig.*

1919 W. H. DOWNING *Digger Dial.* 26 *Goggle-eyed*, dazed. **1942** E. PAUL *Narrow St.* xvi. 128 The way in which he rode a bicycle as he pushed on and pushed around by his womenfolk left Parisians goggle-eyed. **1957** *Times Lit. Suppl.* 20 Dec. 771/1 We are all goggle-eyed, like a tele-viewer, at the automated production line.

goggly, *a.* **1.** Delete †*Obs.* and add later examples.

1923 WODEHOUSE *Inimitable Jeeves* ix. 93 He was a thin, tall chappie with..pale-blue goggly eyes which made him look like one of the rarer kinds of fish. **1969** *Daily Tel.* 2 Oct. 22/3 The great goggly eyes are unforgettable and endearing.

go-go (gōu·gōu), *a.* [Reduplication of Go *sb.* (sense 2) or *v.* (cf. sense *22 b), influenced perh. by *GO *a.*] **a.** Fashionable, 'swinging', 'fabulous', unrestrained; (of funds on the stock exchange) speculative. Cf. *GO *a.* 2. **b.** *spec.* Of a dancer or a dance, the music, etc., at a discothèque, strip club, etc.: full of verve, excitement, and movement (often deliberately erotic). Also as *v. intr.*, to dance in this manner. Also **go-go(-go)** *sb.*, continual movement, hustle and bustle.

1962 V. PACKARD *Pyramid Climbers* (1963) xiii. 156 Most executives of promise have a built-in go-go-go. **1964** *Punch* 8 July 38/1 It's fab..and withitly gogo. **1965** *N.Y. Times Book Rev.* 31 Jan. 42/4 (Advt.), The gorgeous Go-Go girls are your escorts to the discotheque that swings with the latest in dance crazes. **1966** H. NIELSEN *After Midnight* (1967) xii. 149 The room exploded into a wild go-go beat. **1966** *T.V. Times* (Austral.) 4 May 6/4 In clubs they Go-Go in cages on elevated platforms, under red lights. **1967** *Boston Herald* 1 Apr. 17/6 Brash, young Gypsy Joe Harris, squirming and twisting like a go-go dancer. *Ibid.* 8 May 21/4 Spring has come and it will be 'Go-Go' at the Cambridge Boat Club's annual spring regatta. *Ibid.*, Five girls..have volunteered to be Go-Go Regatta Girls. **1968** *Economist* 3 Aug. 69/2 Tesco.. became an early favourite for London's equivalent of New York's go-go funds. **1968** *Guardian* 23 Dec. 1/3 The main point is that it [*sc.* journey of spacecraft to the moon] is all go-go-go. **1968** 'O. MILLS' *Sundry Fell Designs* ii. 24 It'd take someone with a bit of go-go-go to take on swarming up one of those pylons with a banner. **1968** O. WYND *Sumatra Seven Zero* ii. 18 He..stared out..at a Post Office Tower erected by the go-go Britain boys. **1969** *Sunday Times* 9 Feb. 32/4 Only seven of the big 'go-go' funds managed to out-perform the market as a whole. **1969** *Winnipeg Free Press Weekly* 2 Aug. 18/3 Everybody is 'on the go' in this go-go generation. **1970** *Daily Tel.* 14 Feb. 12 Lurid invitations to see the topless go-go girls..and the pornographic peep-shows.

‖ **gogo** (gogo). Also **go-go.** [Fr. *à gogo.*] Only in the phr. *à gogo*, in abundance, galore, no end of.

1965 *Economist* 25 Dec. 1416/2 A lot more has been added, even before *après ski* life *à gogo* begins. **1966** *New Yorker* 24 Sept. 52 This is really nothing but Leninism à go-go! **1967** *Listener* 6 Apr. 468/3 (*heading*) Franglais à gogo. **1967** *Observer* 7 May 21/4 At Expo every nation is petrochemical a-go-go.

gohanna, var. *GOANNA.

going, *vbl. sb.* Add: **4.** (Earlier and later examples.) Also, a line or route, considered as difficult or easy to follow; advance or progress as helped or hindered by the nature of the ground; *heavy going*: something difficult to negotiate; slow or difficult progress; freq. *fig.*

1848 BARTLETT *Dict. Amer.* 159 The going is good since the road was repaired. **1901** 'LINESMAN' *Words by Eyewitness* v. 101 A narrow path just above the waterline, overhung with bushes in parts, formed the 'going'. **1925** E. F. NORTON *Fight for Everest, 1924* 114 We made very poor going, descending at a very much slower pace than we had made two years before. **1930** J. B. PRIESTLEY *Angel Pavement* iii. 124 He found such books too heavy going and preferred a detective story. **1935** *Economist* 5 Oct. 647/1 The 'going' was then still good. For the immediate future during the last quarter of the year, the 'going' in the new capital market seems likely to be heavy and uneven. **1936** *Discovery* May 142/1 The next stage, up the North Ridge, is not very difficult technically but is, nevertheless, heavy going. **1958** *Listener* 18 Sept. 433/3 A book that is not only full of interest but is completely without heavy going.

b. Colloq. phr. *while the going is good*: while the conditions are favourable; freq. *to go while the going is good.*

1916 H. L. MENCKEN *Let.* 10 July (1961) 85 You would be a maniac not to go out for all that money while the going is good. **1927** H. WADDELL *Wandering Scholars* ii. 48 Warned in time, the two..had gone while the going was good. **1958** HAYWARD & HARARI tr. *Pasternak's Dr. Zhivago* I. vii. 199 She..made off with her, home to their village while the going was good.

5. a. *going-away*: used *attrib.* to designate: (*a*) clothes worn by a bride when she departs for her honeymoon; (*b*) a savings club in which members build up holiday funds by small part-payments.

1884 [in Dict.]. **1910** H. G. WELLS *Mr. Polly* vi. 175 The heroine had stood at the altar in 'a modest going-away dress'. **1912** A. BENNETT *Matador* 135 William Henry began grimly to pay his subscriptions to the next year's Going Away Club. **1928** *Observer* 1 July 21/4 The amount of money which has been disbursed by the 'Going-away clubs' is as large as ever. **1959** D. EDEN *Sleeping Bride* iv. 30 This is my going-away suit.

b. (Later examples.)

1866 A. D. WHITNEY *L. Goldthwaite* iv, They watched the long, golden going-down of the sun. **1917** W. B. YEATS *Wild Swans at Coole* 15 From going-down of the sun.

6. going-to-press *a.*, designating the latest items of news in a newspaper or the like.

1906 *Westm. Gaz.* 24 Apr. 7/3 The following are going-to-press Stock Exchange prices.

going, *ppl. a.* Add: *going concern* (further examples); also, a flourishing business; a profitable enterprise.

1930 *Economist* 26 Apr. 938/2 It is..unlikely that Europe will recover her pre-war economic position unless and until Russia becomes a 'going concern' again. **1932** N. HODGINS *Some Canadian Essays* 111 If a religion is a going concern, in the sense of helping a man to face life and death honestly, it has already proved its substantial truth. **1939** *Brit. Jrnl. Psychol.* Jan. 221 Stragglers will return once they see the church is a 'going concern'. **1955** T. WILLIAMS *Orpheus Descending* II. i. 62, I got a going concern in this mercantile store.

going over, going-over (gōu·iŋōu·vəɹ). [See GOING *vbl. sb.* 5 a and cf. GO *v.* 62.] †**1.** A passage over a stream. *U.S. Obs.*

1662 in *Early Records of Providence, R.I.* (1893) III. 17 The high way which Leadeth from the goeing over att the River. **1782** *Southampton* (N.Y.) *Records* (1878) III. 292 The path from the going over at the River head up to the Great Pond.

2. A scolding, a talking-to, a dressing-down. *U.S. colloq.*

1872 *Chicago Tribune* 23 Oct. 4/2 The Cincinnati *Commercial* gives these male Mrs. Grundys a 'going over' in an article well worth reading. **1884** 'MARK TWAIN' *Huck. Finn* iii, I got a good going-over in the morning from old Miss Watson on account of my clothes. **1937** D. RUNYON *More than Somewhat* viii. 155 Many a time I hear Bookie Bob giving her a going-over about something or other, and generally it is about the price of something she orders to eat. **1969** E. BLISHEN *This Right Soft Lot* II. i. 130 Sir, don't give me a going over—but this desk's too small for me. Honest!

b. An inspection or examination; an overhaul; attention to or work on something. *colloq.* (orig. *U.S.*).

1919 H. L. WILSON *Ma Pettengill* x. 286 She wanted to give these here accounts a thorough going-over while the sensation lasted. **1931** *Kansas City Star* 18 July, The lawn mower could profit by a complete 'going over'. **1950** 'J. TEY' *To love & be Wise* ix. 108 Having given the room a quick going-over, he now went over it in detail. **1958** B. HAMILTON *Too Much of Water* i. 9 His first serious going-over of the scores of the..Bruckner symphonies. **1960** H. PINTER *Caretaker* II. 48 How do you think the place is looking? I gave it a good going over.

c. A beating; a thrashing. *slang* (orig. *U.S.*).

1942 BERREY & VAN DEN BARK *Amer. Thes. Slang* § 510/2 'Third degree'..*going-over*. **1948** W. G. SMITH *Last of Conquerors* (1949) IV. 193 'Let's give him a little going over,' one of the M.P.s suggested. **1963** T. & P. MORRIS *Pentonville* xi. 240 Most serious of all are the premeditated 'goings over' of individuals by small groups of men who are the bodyguard of a gang leader. **1970** A. ROSS *Manchester Thing* 81 'Got a going over, did you?' 'Not much, I got a going over. Want to see the bruises?'

†**3.** *Bridge.* Doubling. *U.S. Obs.*

1902, etc. [see *GO *v.* 87 g]. **1904** *To-Day* 5 Oct. 286/2 If any player double out of turn here, there is no penalty, but over the water the declaring player has the right to say whether the doubling, or 'going over', shall stand or not.

goitred, *a.* (Earlier and later examples.)

1851 E. RUSKIN *Let.* 24 Aug. in M. Lutyens *Effie in Venice* (1965) II. 183 The number of Cretins and Goitred persons is perfectly dreadful. **1923** *Blackw. Mag.* Aug. 152/1 A gentle frog-like croak proceeded from the goitred throat.

goitrigenous (goitri·dʒinəs), *a. Med.* Also **goitro·genous.** [f. GOITR(E + -I-, -O + *-GENOUS.] = *GOITROGENIC *a.*

1917 R. MCCARRISON *Thyroid Gland* III. i. 88 When children are subjected to goitrigenous influences for the first time they are considerably more susceptible to the disease than are adults. **1928** *Bull. Johns Hopkins Hosp.* XLIII. 281 Exposure to goitrogenous influences. **1937** J. H. MEANS *Thyroid & its Diseases* iv. 104 It seems likely..that certain so-called goitrigenous principles.. depend for their action upon a reduction in the responsiveness of tissue cells to thyroid hormone. *Ibid.* viii. 175 The goitrigenous properties of cabbage.

goitrogenic (goitrodʒe·nik), *a. Med.* Also **goiteroge·nic.** [f. GOITR(E (or the variant *goiter*) + -O + *-GENIC.] Causing goitre. Hence **goi·trogen,** a goitrogenic substance; **goitroge·nesis,** the production or development of goitre.

1929 *Proc. Soc. Exper. Biol. & Med.* XXVI. 823 Although cabbage is an excellent food for rabbits it is evident that it contains a powerful goiterogenic agent. **1931** *Indian Jrnl. Med. Res.* XVIII. 1311 It is now definitely known that a *positive* goitrogenic agent exists in cabbage. **1947** *Endocrinology* XL. 348 The ability of a solution containing elemental iodine..to prevent the formation of a goiter is not due to a decreased intake of goiterogen. **1950** R. H. WILLIAMS *Textbk. Endocrinol.* iii. 103 Goitrogenesis due to cabbage apparently is related to its cyanide content. **1961** *Lancet* 30 Sept. 742/1 The spontaneous regression of iodide goitres when the goitrogen is recognised and withdrawn makes surgery unnecessary. **1966** WRIGHT & SYMMERS *Systemic Path.* II. xxxi. 1099/1 Goitres can also be caused by the administration of chemical goitrogens, or antithyroid drugs. *Ibid.*, The goitrogenic effect of thiocyanates and perchlorates.

go-kart (gōu·kāɹt). [Commercial adaptation of GO-CART.] (See quot. 1963.) Also *attrib.* Cf. *KART.

1959 *Times* 17 Sept. 5/7 The R.A.C. are prepared in principle to accept the control in Britain of go-kart racing—the new miniature car racing popular in the United States. **1959** *Daily Tel.* 26 Oct. 15/4 (*caption*) Drivers negotiating straw-bale obstacles while competing ..yesterday in a Go-Kart meeting, the first in England. **1960** *Sunday Express* 11 Sept. 5/3 The Go-karts, which were bought for Prince Charles and Princess Anne..are capable of about 15 miles an hour. **1963** *Times* 12 Feb. 15/3 The name 'go-kart' was given to a miniature racing car, which consisted of a bare skeleton or chassis of small size mounted on four wheels and powered by a light two-stroke internal combustion engine, with the driver's seat a few inches from ground level. **1971** *Flying* Apr. 47/2 Tallman has a wooden leg, but he got it in a freak accident on a go-kart.

gold[1]. Add: **1. b.** *gold medal*: a golden medal, usu. awarded as the first prize in a contest, esp. in the Olympic Games. Also *ellipt.* as *gold.*

1908 T. A. COOK *Olympic Games* i. 16 For all these gold, silver, and bronze prize medals have been allotted. **1960** *Times* 5 Sept. 4/6 Hill..was representing a Germany united for Olympic purposes, which won three gold and a silver. **1968** *Guardian* 22 Oct. 1/1 In the equestrian event, the gold was taken by the team.

3. a. *heart of gold*: also, a kind or noble heart.

1858 LYTTON *What will he do with It?* VII. i. 208 If, with gentle blood, youth, good looks, and a heart of gold, that fortune does not allow him to aspire to any girl whose hand he covets, I can double it. **1923** WODEHOUSE *Inimitable Jeeves* xi. 123 While she may have had a heart of gold, the thing you noticed about her first was that she had a tooth of gold. **1936** G. GREENE *Journey without Maps* III. iii. 244 He had a heart of gold under that repressive exterior. **1971** *Times* 12 Dec. 19/4 Tarts invariably turn out to have a heart of gold.

8. b. *old gold* (earlier and later examples); also as *sb.*

1879 M. E. BRADDON *Vixen* III. xi. 313 Curtains..in dirty yellow, or, in upholsterer's language, 'old gold'. **1880** *Drapers' Jrnl.* 27 May 5/2 Edged with a narrow plissé of old gold. **1935** N. L. MCCLUNG *Clearing in West* xxix. 227 Mother had a silk eiderdown on her bed, old gold on one side and flowered with red geraniums on the other. **1966** J. B. PRIESTLEY *Salt is Leaving* ii. 25 The general effect suggested a palace made of milk chocolate. The old-gold lighting did nothing to spoil this effect.

c. With reference to the use of gold for coinage and as a standard of value, as *gold currency, standard, value.* Also designating a money of account: reckoned at its full undepreciated value according to a gold standard.

1776 ADAM SMITH *W. N.* II. ii, The..nominal sum of the gold and silver currency of the country. **1831** *Deb. Congress U.S.* 22 Feb. p. cl/2 The present rate (of our gold standard) was the result of information clearly incorrect. **1832** *Mechanics' Mag.* 11 Feb. 344/1 Notwithstanding 'the gold standard' is not the United States the modern *El Dorado*? **1857** GEO. ELIOT *Let.* 23 Dec. (1954) II. 414 The opponents of the 'gold standard' do not make it sufficiently clear..that they presuppose..some guarantee such as government security, as a basis for confidence. **1868** J. LAING *Theory of Business* (ed. 2) iv. 65 Had India possessed a gold currency. **1898** T. H. FARRER *Stud. Currency* p. xviii, The relation between the Gold value of the legal Rupee and the Gold value of Silver. **1911** *Encycl. Brit.* XVIII. 704/2 The 'gold-exchange standard' system, in which the ordinary currency is of a metal coined only by the state, and so limited as to keep it in a prescribed value ratio to another metal (gold) which.. acts as the standard of value. **1914** *Rep. R. Comm. Indian Finance* 14 in *Parl. Papers* (Cd. 7236) XX. 709 The system adumbrated by the Committee of 1898, viz., a gold standard based on a gold currency in active circulation such as the system in the United Kingdom is commonly held to be. *Ibid.* 15 The Indian currency system based on what is now known as the gold exchange standard. **1917** W. F. SPALDING *East. Exch. Currency* ii. 13 The exchange value of the Rupee was satisfactorily maintained between the gold points. **1923** *Westm. Gaz.* 22 Aug., If we take the franc of to-day as having one-quarter of the purchasing power of the pre-war gold franc. **1926** *Encycl. Brit. Suppl.* I. 776/2 In the final phase people reckoned in gold marks and stipulated for payment in paper marks at the exchange of the day... This competition of the gold mark as a money of account was the final undoing of the paper mark. **1933** B. ELLINGER *This Money Business* i. 9 This state of affairs lasted until September 18th of 1931, when we went off the gold standard. **1948** G. CROWTHER *Outl. Money* (ed. 2) ix. 279 Nowadays, virtually every currency in the world consists (apart from subsidiary..coins) entirely of paper. When this paper money is made by law freely interchangeable with gold at a fixed ratio the currency is on the gold standard. **1970** *Encycl. Brit.* X. 547/2 Gold standard, a monetary system in which the standard unit is a fixed weight of gold or is kept at the value of a fixed weight of gold.

9. b. *gold-bearing* (earlier and later examples), *-producing*; *gold-miner.*

1799 M. PARK *Trav.* (ed. 2) 304 Were the gold-bearing streams to be traced to their fountains. **1894** *Pop. Sci. Monthly* June 174 The northern rivers and creeks have gold-bearing sand. **1853** *Visit to Australia* iii. 82 Unless the great body of gold miners should themselves be seized with a fit of reflection, we can see but little hope. **1860** H. GREELEY *Overland Journey* 142 Experienced gold-miners from Georgia, California, and even Australia. **1934** *Times Lit. Suppl.* 4 Jan. 2/3 The Lydenburg gold-miners (mostly British), and the Boers of that district also, were contemptuous of Pretoria. **1960** G. BLAINEY *Mines in Spinifex* viii. 70 He wrote a letter..to an uncle who was a gold miner in Victoria. **1852** MILL *Pol. Econ.* (ed. 3) II. III. ix. § 2. 26 A pound weight of gold will, in the gold-producing countries, ultimately tend to exchange

for as much of every other commodity, as is produced at a cost equal to its own. **1943** WYNDHAM LEWIS *Let.* 26 Jan. (1963) 343 This [*sc.* Canada] is after Africa the second largest gold-producing centre on the planet. **1970** *Encycl. Brit.* X. 536A/1 The major gold-producing areas are in British Columbia, Ontario..and the Northwest Territories.

c. *gold-embroidered* (later example).

1954 L. MACNEICE *Autumn Sequel* 61 Gold-embroidered brocade.

d. *gold-bronze, -brown.*

1909 *Westm. Gaz.* 23 Oct. 7 Her gold-bronze hair was dressed low on her neck. *a* **1930** D. H. LAWRENCE *Last Poems* (1932) 204 Gold-bronze flutters beat through the thick upper air. **1881** WILDE *Poems* 154 Dame Jeannette had not that gold-brown hair. **1944** BLUNDEN *Shells by Stream* 39 And coloured, like a country grange, Gold-browns and bluebell grays.

e. *gold-backed, -chained, -rimmed, -stopped.*

1874 HOTTEN *Slang Dict.* 179 *Goldbacked uns*, body lice. **1963** C. R. COWELL et al. *Inlays, Crowns & Bridges* vii. 66 A very close bite may necessitate a gold-backed crown. **1905** *Daily Chron.* 27 Feb. 4/7 This first of Count Benckendorff's predecessors..was received..by eighty gold-chained City merchants. **1900** 'S. GRAND' *Babs* (1901) xxxix, Mr. Jellybond Tinney adjusted his gold-rimmed pince-nez. **1965** A. NICOL *Truly Married Woman* 17 He unfolded his gold-rimmed spectacles. **1915** E. CORRI *30 Yrs. Boxing Ref.* vii. 149 The good-natured black giant.. smiled and laughed all the time, revealing his rows of glittering, gold-stopped teeth. **1948** C. DAY LEWIS *Otterbury Incident* 5 He grinned a lot, showing his bad teeth and a gold-stopped one.

10. gold bloc, the bloc of countries having a gold standard; **gold blocking** (see quot. 1960); **gold braid,** a collective slang name for naval officers or senior prison warders; **gold bullion standard** *Economics* (see quots. and cf. sense 8 c above); **gold certificate** *U.S.*, a certificate or note certifying that gold to the amount stated on the face of the certificate has been deposited and is available for redeeming it; **gold clause** *U.S.* (see quots.); **gold-copper ore,** ore yielding both gold and copper; **gold-digging** (earlier and later examples); also *fig.* (cf. *GOLD-DIGGER 2) and as *ppl. adj.*; **gold-dredge(r,** a dredger by means of which gold is dug up from river-beds, etc.; so *gold-dredging*; **gold-fever** (earlier U.S. example); **gold-film (glass)** (see quot. 1958); **gold-lip,** a yellow-edged oyster shell, used as money in parts of Melanesia; **gold-pan** = PAN *sb.*[1] 2 e; **gold point** (see quot. 1925); **gold-quartz,** quartz containing gold; freq. *attrib.*; **gold reserve,** the reserve [RESERVE *sb.* 1 b] of gold coin or bullion held by a central authority, bank, etc.; **gold salt,** a salt of gold; in *Pharm.*, esp. any of several compounds containing gold and sulphur that have been used in the treatment of rheumatoid arthritis and lupus erythematosus; so **gold therapy,** treatment with sodium aurothiomalate or other compounds containing gold; **gold-tipped** *a.*, having a gold tip; *spec.* of a cigarette having a band of gilded paper at one end; **gold-washed** *a.*, thinly plated with gold.

1935 *Economist* 12 Jan. 57/2 The figures for the gold *bloc* countries reveal a noticeable contrast between the movement of French wholesale prices on the one hand and of Italian and German prices on the other. **1970** *Encycl. Brit.* X. 548/2 A small group of continental countries led by France continued the struggle to maintain convertibility at the old price until 1936. This gold bloc collapsed because the depreciation of sterling and the dollar meant that the exports of the gold bloc countries were at a competitive disadvantage in world markets. **1902** *Ibid.* XXVI. 302/2 In the matter of gold blocking there must be great care exercised in the matter of the heat of the block. **1960** G. A. GLAISTER *Gloss. Bk.* 155 *Gold blocking*, the stamping of a design on a book cover, using a heated die or block in a press, and gold leaf. **1933** G. INGRAM *Stir* (ed. 2) iii. 52 The chief warder... All the screws get their orders through him. The other gold braid you 'ave to be careful of are the principals, those blokes with the gold braid on the flaps of their caps. **1945** 'TACKLINE' *Holiday Sailor* xi. 113 There'd be more gold-braid than that, buzzing around like a fly in a strange gash-basket. **1933** B. ELLINGER *This Money Business* i. 6 After the Act of 1925 it was only gold bullion which could be obtained in exchange for notes, and therefore it was known as the 'gold bullion standard' instead of the vaguer term 'gold standard'. **1948** G. CROWTHER *Outl. Money* (ed. 2) ix. 280 When gold coins do not circulate, but the Central Bank is nevertheless under legal obligation to buy and sell gold in exchange for currency at a fixed price..it is known as the 'gold bullion standard'. **1864** *Santa Fe Weekly New Mexican* 27 May 2/3 The boys in New York..join[ed] in the procession of gold certificate buyers. **1935** *Economist* 23 Feb. 415/1 The changes in the item 'gold certificates' have given some suggestion that the fund has been used in greater volume than has been expressed in gold movements to date. **1935** *Ibid.* 19 Jan. 116/1 The clause..was quite definite, calling for payment in United States gold coin of the weight and fineness existing at the time of the issue of the bond. The holding, and..receipt, of gold coin was made illegal last year and Congress passed a resolution invalidating all 'gold clause' contracts. **1965** J. L. HANSON *Dict. Econ.* 201/1 A 'gold clause', which offers lenders the security of repayment in terms of the gold equivalent of the currency at the time the loan was floated. **1906** *Daily Chron.* 26 Jan. 4/5 A large deposit of gold-copper ore. **1908** *Westm. Gaz.* 1 July 6/3 There are piles of gold-copper ore. **1802** *Monthly Mag.* XIV. 93/1 (*title*) Gold-digging ants, and the griffins of the ancients. **1831** *Boston Transcript* 27 May 2/3 Gold digging, combined with slave holding, may always keep those states a whole century behind their neighbors. **1927** *Cleveland Press* 29 Jan., The charge of gold-digging is one of the major counts in Browning's [divorce] case. **1928** *Daily Tel.* 18 Sept. 6/5 She feels the consequences of a 'gold-digging' mother's love affairs without entirely understanding them. *Ibid.* 16 Oct. 10/7 She will go to his house, and (in the American phrase) do a little gold-digging—without, you understand, giving anything in return. **1936** I. L. IDRIESS *Cattle King* xiii. 121 So started the first ration store on the gold-diggings at Tibooburra. **1958** *Times* 2 Oct. 3/1 To his gold-digging mistress the publisher..is an ageing man foolishly pretending to be younger than he is. [**1945** *Jefferson Co. Republican* (Golden, Colo.) 31 Oct. 2/1 The troops visited a large gold mining dredge.] **1948** P. JOHNSTON *Lost & Living Cities Calif. Gold Rush* 25/1 Land that has once been mined by a gold dredge is utterly ruined for all time. **1901** *Daily Colonist* (Victoria, B.C.) 1 Oct. 10/4 The new North Thompson River gold dredger, in a fraction over four days' actual gold dredging operations, secured 32 ounces of gold. **1909** *Daily Chron.* 23 Jan. 1/1 Houses have been swept away [in the Transvaal] and a gold-dredger has been sunk. **1892** W. E. SWANTON *Notes on N.Z.* iii. 159 The Gillespie Beach Gold Dredging Company['s]..object is to dredge the sea at the mouth of the Grey River and bring up the sand and mud, which contain large quantities of gold-dust washed down by the stream. **1959** A. MCLINTOCK *Descr. Atlas N.Z.* p. xvi, The unsightly piles of gravel or 'tailings'..a legacy from the gold-dredging boom of 60 years ago. **1847** *Californian* (San Francisco) 22 May 4/1 After the gold fever was over, these ships were very successful in taking oil in the bay. **1958** *Chambers's Techn. Dict.* (ed. 2) 983/1 *Gold-film glass*, glass incorporating a thin gold film which can be electrically heated for demisting and de-icing. **1962** *Engineering* 2 Feb. 179/1 Heated gold-film windows for the flight deck of the de Havilland Trident aircraft. **1908** E. J. BANFIELD *Confessions of Beachcomber* II. i. 267 When such a prize as a gold-lip shell was found, it was used to the last possible fragment. **1929** A. B. LEWIS *Melanesian Shell Money* 11 In south-western New Britain a [pearl] shell of the golden yellow variety ('gold-lip') was specially valued. **1957** M. WEST *Kundu* ii. 29 Time was when a man's wealth was measured by the number of his pigs or by his store of gold-lip shell. **1873** J. MILLER *Life amongst Modocs* 57 [The miners].. washed their hands and faces in the gold-pan that stood by the door. **1901** *Daily Colonist* (Victoria, B.C.) 22 Oct. 1/5 The party of six men had but one gold pan. **1882** *Peel City Guardian* 2 Dec., The New York exchange has kept hovering at only a little above the gold point. **1925** S. E. THOMAS *Elem. Econ.* xxix. 461 We find that the rates at which one currency will exchange for another fluctuate between two limits on each side of the Mint Par, marking the points at which it becomes more profitable to send or to receive gold rather than to send or receive a credit instrument. These theoretical limits are known as the *gold points.* **1930** J. M. KEYNES *Treat. Money* II. 320 The degree of separation of the gold points is a vital factor in the problem of managing a country's currency. **1850** *Calif. Courier* (San Francisco) 18 July 2/3 If gold quartz rock comes in, in this way, we will begin to believe that..we are wrong. **1908** *Chambers's Jrnl.* Sept. 637/1 The discovery of some very rich gold-quartz veins. **1870** *N.Y. Herald* 8 July 4/3 [The national banks'] interests.. lie in a paper circulation only, and in not being compelled to hold a gold reserve. **1879** A. J. WILSON *Banking Reform* vii. 189 More stringent provisions for maintaining an adequate gold reserve would be required here than seems to have been thought necessary in the American Union. **1967** *Times* 5 July 1/8 Britain's gold and convertible currency reserves fell by £43m. last month. **1849** *Q. Jrnl. Chem. Soc. Lond.* I. 169 The crystals are a mixture of the gold salts of cyaniline and aniline. **1907** G. S. NEWTH *Text-bk. Inorg. Chem.* (ed. 12) III. v. 568 Most metals, when placed in a solution of a gold salt, precipitate the gold. **1937** *Lancet* 9 Oct. 840/1 The cause of the jaundice is the temporary toxic effect of the gold salts upon the liver cells. **1970** *Guardian* 20 Feb. 6/7 People being treated with gold salt for rheumatoid arthritis have no cause for concern. **1937** *Lancet* 9 Oct. 840/2 Nervous complications are perhaps the rarest seen in gold therapy. **1970** A. & E. F. GROLLMAN *Pharmacol. & Therapeutics* (ed. 7) xlii. 926/1 It would thus appear that [for rheumatoid arthritis] conservative therapy..is the preferred method of treatment and that the hazards of gold therapy do not support its use. **1882** 'SHWAY YOE' *Burman* I. xxiv. 311 The gold-tipped ox-goad..is covered with jewels. **1904** A. E. W. MASON *Truants* x. 92 Lighting a gold-tipped cigarette he..began to talk. **1951** R. SENHOUSE tr. *Colette's Chéri* 174 Stub-ends of gold-tipped cigarettes. **1872** E. EGGLESTON *End of World* ix. 65 Pewter watch-seals, gold-washed.

b. gold-moh(u)r, also **gul mohur,** the Indian name for *Poinciana regia* (or *Delonix regia*), an ornamental tree bearing scarlet flowers, which has been introduced into several tropical countries from Madagascar; cf. FLAMBOYANT *sb.*, *peacock-flower.*

1874 E. LEAR *Indian Jrnl.* June (1953) viii. 147 The wonderful orange blossomed gold-mohr tree is now a sight here as every compound has one or two in it, besides lots of other flowers. **1901** A. PERRIN *East of Suez* 83 The clambering bougainvillea and gold-mohur blossoms. **1953** N. L. BOR *Man. Indian Forest Bot.* 71 The flowers of the *Gul Mohur* are described as having four scarlet petals, while the fifth, or standard, is variously variegated with gold and vermilion.

goldarn (gǫ:ldā·ɹn), *v.*, *a.*, and *adv. U.S. slang.* Also **gol darn, goldurn,** etc. [Euphemistic substitute for GOD-DAMN.] = DAMN *v.*, *DAMN *a.* and *adv.* Cf. DARN, DARNATION, DARNED. So **go:lda·rned** *ppl. a.*, damned.

1832 *New England Mag.* III. 380 We have..'*Gaul darn you*' for G— D— you. **1849** *Picayune* (New Orleans) 6 May 2/6 I'll be gaul-durned ef I deu. **1856** F. L. OLMSTED *Journey Slave States* 312 Seems to me them gol-durned lazy niggers aint a goin' to come over arter you now. **1870** B. HARTE *Sensation Novels* iv. 87 Dog-gone. *Note*, A euphemism common with the men of the West, and equal to the English 'O drat it', or 'Gol darn'. **1885** *Harper's Mag.* Aug. 397/2 Hannah-Maria-Jemimy! goldarn an' blue blazes! *Ibid.* 400/2 You're a goldarn liar, Balaam. **1907** N. MUNRO *Daft Days* xxxii, Every gol-darned idiot in England. **1910** R. W. SERVICE *Ballads of Cheechako* 50 Goldarn his eyes. **1920** S. LEWIS *Main St.* 431 Cy Bogart had the 'flu, but of course he was too gol-darn mean to die of it. **1948** *Reader's Digest* Mar. 128 Another great story ruined by a goldurned eyewitness. **1964** MRS. L. B. JOHNSON *White House Diary* 17 July (1970) 183 Well, I'll be gol-durned—the salesman that sold it to me said it was Harry Truman.

Goldbach's conjecture. *Math.* The hypothesis put forward by the German mathematician C. Goldbach (1690–1764) in 1742 (in a letter to a fellow mathematician, Euler) that every even number greater than 2 can be represented as the sum of two primes. Also *Goldbach's hypothesis, theorem*, etc.

1919 L. E. DICKSON *Hist. Theory of Numbers* I. xviii. 422 G. Cantor verified Goldbach's theorem up to 1000. **1941** COURANT & ROBBINS *What is Math.?* 31 Until recently, a proof of Goldbach's conjecture seemed completely inaccessible. **1949** W. KNEALE *Probability* § 18. 79 The attitude of mathematicians..can..be expressed by.. 'Goldbach's conjecture looks like a theorem, but it may conceivably be false'. **1960** P. SUPPES *Axiomatic Set Theory* vii. 217 Goldbach's Hypothesis..is false for ordinal numbers. **1965** PHILLIPS & WILLIAMS *Inorg. Chem.* I. i. 25 There still exist a number of very simple problems which have yet to find solutions. These include the four-colour problem, Goldbach's conjecture, and Fermat's last theorem.

gold brick, gold-brick, *sb.* [f. GOLD[1] + BRICK *sb.*[1] 4.] **1.** A brick-shaped piece of gold.

1853 *San Francisco Sun* 7 June 2/2 (*heading*) Gold brick. **1877** R. W. RAYMOND *Statistics of Mining* VIII. 354 Individuals are constantly carrying out bags of gold and gold bricks and some silver bricks. **1899** *Monthly South Dakotan* I. 196 [He] found in a water hole a gunny sack in which was a gold brick.

2. A brick that appears to be made of gold; hence, something having only a surface appearance of value; a fraud; esp. in phr. *to sell* (someone) *a gold brick*, to swindle. Also *attrib. slang* (orig. *U.S.*).

1881 *National Police Gaz.* (U.S.) 24 Dec. 10/2 (*heading*) The Gold Brick Trick. *Ibid.*, The gold brick swindle is an old one but it crops up constantly... The bar, or brick as it is called,..is really of base metal. One corner, however, is of gold. **1887** *Chambers's Jrnl.* 1 Oct. 637/1 A 'gold-brick swindler'. **1889** *Kansas Times & Star* 30 Nov., What's new down at the old union depot shack? Any eruptions, gold bricks being sold or important arrivals? **1901** S. E. WHITE *Westerners* xiii. 94 Bunco men can clean him out in a gambling joint, but who ever heard of their selling him a gold brick? **1911** B. WASHINGTON *My Larger Educ.* 292 In many cases, the diploma that the student carries home at the conclusion of his course is nothing less than a gold brick. **1915** WODEHOUSE *Something Fresh* vi, Preventing Fate from working off on us any of those gold bricks, coins with strings attached, and unhatched chickens at which Ardent Youth snatches. **1923** H. C. BAILEY *Mr. Fortune's Practice* iii. 82 'He is said to be negotiating deals in Russian mining properties.' 'Sounds like a gold brick.' **1947** *Chicago Daily News* 16 May 18/5 It used to be the city slicker who sold gold bricks to the hick from the country.

3. A shirker; a lazy person (see also quots.). *U.S. slang.*

1914 *Dialect Notes* IV. 107 *Gold-brick, n.*, applied to army lieutenants appointed from civil life. 'The gold-bricks are overbearing.' **1926** L. NASON *Chevrons* viii. 275, I think you're a goldbrick... You don't look as if you were wounded the slightest bit. **1929** *Papers Mich. Acad. Sci.* X. 296/2 *Gold brick* (Am.), I, an unattractive girl; II, a shirker; one who tries to avoid work, or to get an easy job. **1943** *Reader's Digest* Oct. 97 The wise guy always complains when there is work to do. Sometimes [in the army] he is called a Gold Brick. **1958** J. STEINBECK *Once there was War* (1959) p. xviii, In the ranks, billeted with the stinking, cheating, foul-mouthed goldbricks, there were true heroes.

Hence **gold-brick** *v.*, (*a*) *trans.* to cheat, swindle, defraud; (*b*) *intr.* to shirk, to have an easy time; **gold-bricker,** (*a*) a shirker; (*b*) a swindler; **gold-bricking** *vbl. sb.* and *ppl. a. slang* (orig. and chiefly *U.S.*).

1902 H. L. WILSON *Spenders* xxviii. 328 He'll be gold-bricked if he wears 'em [*sc.* his whiskers] scrambled that way around this place. **1914** *Munsey's Mag.* Jan. 738/1 Well, look out they don't gold-brick you, sonny. **1926** L. NASON *Chevrons* ii. 39 You were somewhere in hospital goldbricking your time away. **1928** J. B. WHARTON *Squad* vi. 208 If there's any one here sick enough to stay, I'll see the Lieutenant... But no gold brickin' now, mind! **1932** *Blue Valley Farmer* (Okla. City) 4 Feb. 8/4 No country worth fighting for was ever built on slacker's devotion or gold-bricker's patriotism. **1937** *Amer. Speech* XII. 74/2 *Gold bricker*, duty-shunner, lazy soldier or enrollee. **1952** M. MCCARTHY *Groves of Academe* (1953) iv. 67 Students with applied art or science majors tended to gold-brick on their reading courses. **1958** J. K. GALBRAITH *Affluent Society* vii. 63 The typical business executive..

would endanger his chance for advancement if he were suspected of goldbricking because of his resentment over his taxes. *Ibid.* viii. 91 We have feather-bedding unions and gold-bricking workmen and slothful supernumeraries everywhere. **1968** *Courier-Mail* (Brisbane) 18 May 3/7 As in any large outfit, there is obvious goldbricking.

go·ld-di:g, *v. slang* (orig. *U.S.*). [Back-formation f. *GOLD-DIGGER 2.] *trans.* To behave as a gold-digger (*GOLD-DIGGER 2) towards (a man); to extract money from.

1926 S. Lewis *Mantrap* xii. 148, I gold-dig you for all the money I can get. **1931** A. Powell *Afternoon Men* xiii. 141 Of course he was intended by nature to be gold-dug and as a compensation given a temperament that enjoys the process. **1947** J. Steinbeck *Wayward Bus* 206 I'll bet she just gold-dug Eddie.

go·ld-di:gger. [Gold[1].] **1.** One who digs for gold. Also *fig.*

1830 *Cherokee Phoenix* (New Echota, Ga.) 24 Mar. 3/3 There are tippling shops on every hill where these gold diggers are collected. **1850** [see Gold[1] 10]. **1855** R. Henning *Let.* 29 Mar. (1966) 25 Dressed-up in a blue gold-digger's shirt over his clothes. **1948** O. Weston *Mother Lode Album* 28 Savage and his party of Indian gold diggers encamped under a big oak. **1954** Koestler *Invis. Writing* xix. 218, I had been done in..as thoroughly and completely as a choir-boy in a gold-digger town.

2. A girl or woman who attaches herself to a man merely for gain. *slang* (orig. *U.S.*).

1920 B. Mantle in *Best Plays of 1919–20* 360 'Jerry' Lamar is one of a band of pretty little salamanders known to Broadway as 'gold diggers', because they 'dig' for the gold of their gentlemen friends and spend it being good to their mothers and their pet dogs. **1928** *Sunday Dispatch* 19 Aug. 20 The professional gold-digger is generally a girl of good family who finds she can supplement her allowance by going out with, say, half-a-dozen men. **1934** G. B. Shaw *On Rocks* ii. 263 All I can get out of her is that she is not a gold digger, and wouldnt be seen at a wedding with a lousy viscount. **1959** J. Braine *Vodi* xii. 162 It was expensive; that appealed to Lois. Not that she was a gold-digger; but once he started going around with her there were more withdrawals than deposits in his Post Office savings book.

go·ld-du:sty, *a.* [f. Gold dust + -y[1].] Resembling or covered with gold dust.

a **1907** F. Thompson *Shelley* in *Dublin Rev.* (1908) July 37 He is gold-dusty with tumbling amidst the stars. **1931** R. Lehmann *Let. to Sister* 12 The infinitesimal flicks of gold-dusty midge-swarms.

golden, *a.* Add: **1. b.** *to kill the goose that lays the golden eggs*: see *goose *sb.* 1 d.

2. *Golden State* (earlier and later examples).

1847 *Congress. Rec.* 7 May, App. 246/2 From the hills of the Golden State we will send..cattle of every breed. **1967** *Boston Sunday Globe* 21 May A5/2 If Mayor Yorty gets the Senate nomination, Lyndon Johnson may have to write off the Golden State in 1968.

5. c. Also = *golden section* (see next sense).

1917 D. W. Thompson *On Growth & Form* xiv. 643 This celebrated series, which..is closely connected with the *Sectio aurea* or Golden Mean, is commonly called the Fibonacci series. **1929** [see *Fibonacci]. **1945** *Burlington Mag.* Jan. 7/1 Various people have tried to prove..that the 'golden mean', in which Italian artists believed, was applied here by Franciscus Florentinus.

d. *golden section*, (the proportion resulting from) the division of a straight line into two parts so that the ratio of the whole to the larger part is the same as the ratio of the larger part to the smaller, viz. $\frac{1}{2}(\sqrt{5}+1)$, or 1·61803...; 'extreme and mean ratio'.

This celebrated proportion has been known since the 4th century B.C., and occurs in Euclid (II. 11, VI. 30). Of the several names it has received, *golden section* (or its equivalent in other languages) is now the usual one, but it seems not to have been used before the 19th century.

[**1835** M. Ohm *Die reine Elementar-Mathematik* (ed. 2) II. 194 Diese Zertheilung einer beliebigen Linie *r* in 2 solche Theile, nennt man wohl auch den goldenen Schnitt. **1844** J. Helmes in *Arch. Math. und Physik* IV. 15 (*heading*) Eine..Auflösung der sectio aurea. **1849** A. Wigand (*title*) Der allgemeine goldene Schnitt und sein Zusammenhang mit der harmonischen Teilung.] **1875** *Encycl. Brit.* I. 220/1 Zeising..asserts that the most pleasing division of a line, say in a cross, is the golden section. **1898** [see Section *sb.* 3 e]. **1914** T. A. Cook *Curves of Life* xx. 426 In Sandro Botticelli's Venus..the line containing the figure from the top of the head to the soles of the feet is divided, at the navel, into the exact proportions given by that ancient formula the 'Golden Section'. **1931** T. L. Heath *Man. Gr. Math.* viii. 181 Proclus speaks of theorems which Plato 'originated regarding the "section"', and, if this 'section' was what came to be called the 'golden section',..Plato may well have had this case in mind. **1951** G. Sarton in *Isis* XLII. 47/2 As far as we know now, the expression divine proportion was introduced by Pacioli in 1509 [in his treatise, *Divina proportione*], and the expression golden section by Martin Ohm in 1835. It is almost certain that they had predecessors [*printed* predecessors]. Who were they? **1953** 'M. Innes' *Christmas at Candleshoe* i. 5 It so happens that in the picture-space we are contemplating, the one line cuts the other in a ratio which artists call golden section. **1955** T. Dantzig *Bequest of Greeks* v. 59 Just as specious are the contentions that certain proportions in human anatomy conform to the golden-section rule. **1963** H. Read *Contrary Experience* iv. viii. 345 Certain correspondences are easily established—the prevalence, for example, in art and in both organic and inorganic matter of the proportion known as the Golden Section. **1970** [see *Fibonacci]. **1970** *Oxf. Compan. Art* 489/1 Statistical experiments are said to have shown that people involuntarily give preference to proportions that approximate to the Golden Section.

Similarly *golden rectangle*, a rectangle in which the ratio of the longer side to the shorter is the golden section.

1953 *Scripta Math.* XIX. 136 Since the diagonal of a pentagon is τ times its side, this rectangle is a golden rectangle, whose sides are in the ratio τ:1. *Ibid.* 138 The golden rectangle can be dissected into two pieces: a square and a smaller golden rectangle. **1967** W. J. Reichmann *Spell of Math.* v. 90 The sides of the rectangle are thus in golden section, and the rectangle may therefore be called a golden rectangle... If we draw the largest square possible within this rectangle..then the next remaining rectangle will also be golden.

9. a. *golden-brown* (earlier example).

1845 E. Acton *Mod. Cookery* (ed. 2) ii. 60 Fry them a clear golden brown in plenty of boiling lard.

c. *golden-girdled, -hearted.*

1848 J. R. Lowell *New Year's Eve, 1844* in *Uncoll. Poems* (1950) 52 Blithely as the golden-girdled bee Sinks into the sleepy poppy's cup of flame. **1907** *Dublin Rev.* Jan. 30 She is the golden-hearted rose that held our perfect joy. **1960** *Times* 30 Jan. 11/2 We have..never until now seen him in other than golden-hearted parts.

10. a. golden boy, a popular or successful boy or man; similarly *golden girl*; **golden cat,** either of two honey-coloured wild cats, *Felis aurata* or *F. temmincki*, found in west Africa and south-east Asia; **Golden Chersonese** [tr. Gr. Χρυσῆ Χερσόνησος], the Malay Peninsula; **golden girl** (see *golden boy*); **golden handshake,** a gratuity given as compensation for dismissal or compulsory retirement; also *transf.*; **golden hoof,** said of the action of sheep or other animals trampling the ground beneath them, consolidating and improving its texture; **golden retriever,** a retriever with a thick golden-coloured coat; **golden sherry,** a type of sweet sherry.

1937 C. Odets (*title*) The golden boy. *Ibid.* III. i. 196 He walks down the street respected—the golden boy! **1964** 'J. Welcome' *Hard to Handle* viii. 91 Poor dear Richard... What a change from being the golden boy of English racing. **1965** P. Moyes *Johnny under Ground* xx. 250 *That* would have been the end of Beau Guest, the young chevalier, the golden boy. **1971** *Sunday Times* (Johannesburg) 28 Mar. 24/1 Ever since he assumed the 'golden boy' mantle, Richards has studiously avoided local opposition. [**1867** *Proc. Zool. Soc.* 815 An adult specimen of the Golden Tiger-cat of Sumatra (*Felis aurata*, Temm.) received June 19th.] **1883** D. G. Elliot *Monogr. Felidæ* tab. xvi. (*caption*) Temminck's Golden Cat. **1954** G. Durrell *Bafut Beagles* ix. 154 The Golden Cat, one of the smaller, but one of the most beautiful, members of the cat family. **1667** Golden Chersonese [see Chersonese]. **1883** I. L. Bird (*title*) The Golden Chersonese and the way thither. **1896** R. Le Gallienne (*title*) The quest of the golden girl. **1966** 'W. Cooper' *Mem. New Man* III. iv. 245 Alice liked him, but I'm bound to admit that she didn't see him as such a golden boy as I did. Nor..did she see Roz as such a golden girl. **1960** *Economist* 9 Apr. 179/2 There is little public sympathy for the tycoon who retires with a golden handshake to the hobby farm. **1960** *Times* 8 July 13/2 On the financial side, Cyprus receives its golden handshake of over £14 m. **1969** 'B. Graeme' *Blind Date* i. 13, I knew there would be no future in the Forces... So I came out, and have been living on the golden handshake ever since. **1927** *Daily Tel.* 15 Nov. 11/7 Much of the country [*sc.* Denmark] has land of the typical sheep and barley class, but it is successfully farmed without the aid of the golden hoof. **1941** M. Graham *Soil & Sense* iii. 47 It is presumably this essential difference between crumb structure and floury structure that has made the 'golden hoof' of the sheep important in consolidating land in which there is too little clay. **1946** L. D. Stamp *Britain's Struct.* xi. 99 On light soils..they [*sc.* cattle] pack together the particles of soil by the treading action of the 'golden hoof'. **1957** E. J. Russell *World of Soil* vii. 171 So great was the benefit to the succeeding crop and to the condition of the soil that farmers regularly spoke of the 'golden hoof' as the best amendment for light soils. **1919** T. Marples *Show Dogs* (ed. 2) xv. 73 The following is the Golden Retriever Club's standard of points for Golden Retrievers. **1959** R. Collier *City that wouldn't Die* v. 66 Then he set off on patrol like a country squire inspecting his coverts—tweeds, walking-stick, golden retriever Punch trotting at his side. **1840** Thackeray in *Fraser's Mag.* July 93/1 A bottle of fine old golden sherry..for 1s. 9d. **1854** Golden sherry [see *brown sherry* s.v. *Brown *a.* 7]. **1957** *Encycl. Brit.* XX. 500/2 A 'sweet', 'cream' or 'golden' sherry [contains] not over 7% [grape sugar].

b. golden bell = *forsythia; **golden-cup oak** = *golden oak* (b); **golden oak** *U.S.* (*a*) the false foxglove, *Aureolaria virginica*; (*b*) the canyon live-oak, *Quercus chrysolepis*; (*c*) a light-coloured finish used on furniture; **golden pert** (earlier examples); **golden-rayed lily,** *Lilium auratum*; **golden-seal** (earlier examples); **golden shower** = *pudding-pipe tree* (Pudding *sb.* 11 c); **golden-top** *U.S.*, a grass, *Lamarckia aurea*; **golden willow** = *golden osier*.

1893 W. Robinson *Eng. Flower Garden* (ed. 3) II. 417/1 Forsythia (Golden Bell). **1968** N. Taylor *Guide to Garden Shrubs & Trees* v. 341 The golden bells are of the easiest culture in any ordinary garden soil. **1897** B. B. Sudworth *Nomencl. Arborescent Flora U.S.* 164 *Quercus chrysolepis*,.. Golden-cup Oak (Cal.). **1830** C. S. Rafinesque *Med. Flora U.S.* II. 223 *Gerardia Quercifolia*, Mx. Golden Oak. Specific of the Sioux for the bite of rattle snakes, used also for the tooth ache. **1899** *Chicago Daily News* 30 May 14/3 Solid golden oak finish handsome dressers. **1909** *Cent. Dict.* Suppl. s.v. *Live-oak*, An evergreen oak of the Pacific coast,.. also called golden oak, maul oak, and Valparaiso oak. **1913** Britton & Brown *Illustr. Flora Northern U.S.* (ed. 2) III. 208 *Dasystoma virginica*.. In dry or moist woods, Maine to Minnesota, south to Florida and Illinois. Golden-oak. July–Sept. **1921** A. F. Hall *Handbk. Yosemite Nat. Park* 127 The Transition Zone..is characterized by the yellow pine, Douglas spruce, golden oak, black oak, and incense cedar. **1928** F. N. Hart *Bellamy Trial* i. 2 Nine rows of the golden-oak seats packed with grimly triumphant humanity. **1969** *New Yorker* 25 Oct. 58/1 His wife was black and blue as a new stovepipe, but their children and grandchildren are best described as 'golden oak'. **1784** Cutler in *Mem. Amer. Acad. Arts & Sci.* (1785) I. 403 *Veronica*...Goldenpert. **1821** W. P. C. Barton *Flora N. Amer.* I. 71 *Gratiola aurea.* Golden pert. [**1862** *Curtis's Bot. Mag.* LXXXVIII. 5338 (*caption*) Golden-striped Lily.] **1870** W. Robinson *Wild Garden* II. 118 Golden-rayed Lily. *Lilium auratum.* **1880** H. J. Elwes *Monogr. Genus Lilium* Pl. 15, I believe the Golden-rayed Lily can be grown to greater perfection in pots than in the open ground. **1908** *Pall Mall Gaz.* 20 Apr. 3/2 The golden-rayed lily, be it never so gorgeous. **1970** M. Templeton tr. *Feldmaier's Lilies* i. 7 Japan is the home of some of the finest lilies of all, such as *L[ilium] auratum*, the Golden-Rayed Lily of Japan. **1855** *Trans. Mich. Agric. Soc.* VI. 179 We have the sarsaparilla, ginsing, goldenseal, sweet cicily. **1881** S. P. McLean *Cape Cod Folks* ii. 38 The golden seal..was served in a diluted state with milk and sugar and taken as a beverage. **1914** L. H. Bailey *Standard Cycl. Hort.* II. 680/2 [*Cassia*] *Fistula*, Linn. Pudding-Pipe Tree. Golden Shower. **1934** R. Campbell *Broken Record* 82 Palms, golden shower, grenadillas. **1953** D. Lessing *Five* i. 14 The other neighbour was a house whose walls were invisible under a mass of golden shower—thick yellow clusters, like smoky honey, dripped from roof to ground. **1909** *Cent. Dict.* Suppl., *Goldentop*, an ornamental grass..introduced from the Mediterranean region into southern California. **1934** A. Arber *Gramineae* ix. 186 In Goldentop, however, the tendency to reduced fertility is stronger than in Dog's-tail-grass. **1959** P. A. Munz *California Flora* 1495 *Lamarckia* Moench. Goldentop. **1861** *Trans. Ill. Agric. Soc.* IV. 447 The Golden Willow has been a favorite with me. **1866** [see Willow *sb.* 2 b].

c. golden berry (see quot. 1951).

1951 *Good Housek. Home Encycl.* 385/1 Tinned cape gooseberries are imported from South Africa (usually under the name of golden berries). **1958** *Times* 6 Dec. 1/4 (Advt.), One tin each:—..Goldenberries,..Strawberries.

d. golden cuckoo, an African cuckoo belonging to one of the races of *Chrysococcyx cupreus*; cf. *emerald cuckoo* (s.v. *Emerald 5 d).

[**1811** W. J. Burchell *Jrnl.* 29 Dec. in *Trav. S. Afr.* (1822) I. xix. 502 The Green-and-gold Cuckoo was found in abundance.] **1827** G. Thompson *Trav. & Adv. S. Afr.* I. i. 6 His chief occupation was the stuffing of birds for sale, especially that very beautiful and much-prized species called the Golden Cuckoo. **1876** H. Brooks *Natal* iv. 136 It [*sc.* the emerald cuckoo] is in all probability nearly allied, if not identical, with the golden cuckoo..of the Cape. **1937** *Discovery* XVIII. 265/2 Two more new races—a new Scops owl and a new golden cuckoo—have been added to the Arabian list. **1953** Bannerman *Birds W. & Equat. Afr.* I. 583 This bird [*sc.* the yellow-throated cuckoo] is about the size of Klaas' Cuckoo but has not the brilliance in the plumage of the other Golden Cuckoos.

golden age. Add: Hence euphemistically **golden-a·ger** *U.S.*, an old person.

1961 *Front Page Detective* 25 June 68 A discriminating clientele of golden-agers. **1963** in *Post & Times-Star* (Cincinnati) 9 Oct. 23. **1970** *New Yorker* 15 Aug. 57/1 There are no euphemisms in Dutch for being old—no 'senior citizen', no 'golden-ager'. **1970** H. Waugh *Finish me Off* (1971) 135 Frank bought himself a drink in the bar..while watching the golden agers gossip in the lounge area.

goldfielder (gōu·ldfīldəɹ). [f. Gold-field + -er[1].] One who works in a gold-field; a gold-miner.

1903 *Westm. Gaz.* 28 Jan. 9/1 All the goldfielders were not mere adventurers. **1933** *Bulletin* (Sydney) 20 Sept. 9/2 Goldfielders swept Westralia into the Commonwealth.

gold-fish. Add: Also **goldfish. b. gold-fish bowl,** (*a*) a bowl, usually a glass globe, in which gold-fish are kept (cf. Globe *sb.* 6 b); (*b*) something resembling such a bowl; *spec.* a place or situation affording no privacy (see also quot. 1942).

[**1904** 'Saki' *Reginald* 115, I might have been a gold-fish in a glass bowl for all the privacy I got.] **1935** M. Sullivan *Our Times* VI. viii. 150 The situation was the more trying because it was as plain to the newspapers as to Harding and Daugherty—their relations for several months were carried on in a goldfish bowl. **1941** J. P. Marquand *H. M. Pulham, Esq.* xxxi. 304 She was..telling Mrs. Jones..not to let the maid pick up the broken pieces of the goldfish bowl. **1942** Berrey & Van den Bark *Amer. Thes. Slang* § 466/11 *Gold-fish bowl*, the room in which the third degree is administered. **1951** *Sat. Even. Post* 22 Dec. 44/4 The towermen in their elevated air-conditioned goldfish bowls. **1958** R. Stow *To Islands* viii. 158 It's certainly not going to be easy for us, the white people. Living in a goldfish bowl is the last thing we'd do for fun. **1965** *Listener* 20 May 730/1 The White House has always been a gold-fish bowl, but under Presidents Kennedy and Johnson it has become an educational theatre as well. **1966** J. Dos Passos *Best Times* (1968) iv. 148 They served excellent beer in glasses the size of goldfish bowls. **1969** *Guardian* 26 Mar. 2/2 President Nixon..disclosed his belief that progress towards a Vietnam settlement would come only through private talks... Serious negotiations could not take place in a goldfish bowl, he said.

goldilocks. 2. Delete †*Obs.* and add later examples. Also as a person's name.

1889 M. Wright in A. Lang *Blue Fairy Bk.* 193 The Story of Pretty Goldilocks. **1930** *Observer* 25 May 15 But why is Mr. Max Montesole in Cassio such a quaint goldilocks?

go·lding[2]. [f. the surname: see quot. 1798.] A kind of hop. Also *golding hop, vine.*

1798 W. Marshall *Rur. Econ. S. Counties* I. III. xxvi. 183 The 'golding' has, of late years, been in high repute. It is a sub-variety..of the Canterbury; which was raised by a man still living (1790) Mr. Golding, of the Malling quarter of the district [of Maidstone]. **1810** —— *Review W. Eng.* 378 There are two [varieties of hops]..in more particular esteem, both with the planter and merchant; the Golding Vine..and the Mathon White. **1900** C. Salter tr. *Gross's Hops* 39 The Goldings are the best class of English red hops. **1902** *Times* 5 Sept. 2/5 The golding hops are reported to be developing slowly. **1959** A. Cronk *Eng. Hops Gloss.* 15 *Golding's*,..the name given to a category embracing the choicest English varieties. *Ibid.*, *Golding variety*, one of a category of varieties, having affinity to Golding's, but somewhat inferior.

goldite (gōu·ldəit). *U.S.* [f. Gold[1] + -ite[1]. Cf. Silverite.] An advocate of a gold standard.

1878 *Nation* 21 Feb. 126/1 They were 'goldites'. **1886** *Congress. Rec.* 24 June 6090/2 It does not frighten me to see Goldites voting to preserve greenbacks. **1904** *Westm. Gaz.* 27 Oct. 2/3 In some States there were only two sets of electors, these being practically Silverites and Goldites.

goldless, *a.* Delete *rare* and add later examples.

1896 W. Raleigh *Some Authors* (1923) 146 He belongs, by right of kinship, to the 'threadbare, goldless genealogy' of those who indulge themselves with that most costly dish—speech for its own sake. **1900** *The King* 4 Aug. 135 A goldless gold region.

gold-mine. Also **gold mine.** (Add further *fig.* examples, esp. in the sense 'a source of abundant income or profit'.)

1882 'Thormanby' *Famous Racing Men* 81 Mendicant ..was destined to prove a 'gold mine', for ten years afterwards she brought her owner £80,000 through her famous son. **1888** J. A. Froude *Eng. in W. Indies* v. 56 The island was a gold mine to the Attorney-General. **1894** *Vermont Agric. Rep.* XIV. 101 We have a gold mine in our Morgan stock if we only continue to breed and develop them. **1943** E. O'Neill *Moon for Misbegotten* (1952) 65 This farm has suddenly become a gold mine. **1967** *Listener* 22 June 829/2 These two volumes will certainly be a gold mine for future unofficial biographers. **1971** *Sunday Times* 28 Nov. 40/8 Even as he was hailing the 'gold-mine' he had struck in starting the Mail he was venting anxiety lest advertising should threaten his position as 'sole possessor'.

‖ **goldwasser** (gǫ·ldvasəɹ). [G., lit. 'gold water'.] A liqueur originally made at Danzig (see quot. 1958).

[**1848** Mrs. Gaskell *Mary Barton* II. xiv. 180 He produced a square bottle of smuggled spirits, labelled 'Golden Wasser', from a corner cupboard.] **1920** G. Saintsbury *Notes on Cellar-Bk.* ix. 140, I hope that the rather unsettled fate of Dantzic will not dry up the fount of Goldwasser, which pleased sight and smell and taste alike. **1945** C. S. Forester *Commodore* xviii. 198 Bavarian beer, Swedish schnapps, Danzig goldwasser—he had drunk of them all. **1958** A. L. Simon *Dict. Wines* 83/2 *Goldwasser*..is a highly rectified spirit, flavoured with aniseed, cinnamon, and a number of herbs and spices.. Its outstanding feature is that it contains a large number of very small pieces of gold leaf or yellow and gold-like pieces.

go·ldwater. = prec.

1877 E. S. Dallas *Kettner's Bk. of Table* 152 Dantzic.. has produced the liqueur which is the most beautiful of all to look at—gold-water, a bright colourless liquid like water, with chips of gold-leaf floating in it. **1965** R. Manheim tr. *Grass's Dog Years* I. 52 A little glass of goldwater.

Goldwaterism (gōu·ldwǭtəriz'm). [f. the name *Goldwater* + -ism.] Rigid conservatism as represented by the views and policies of the American Republican politician Barry Morris *Goldwater* (b. 1909); adherence to or support for such conservatism. Hence **Go·ldwaterite,** an adherent of Goldwater or Goldwaterism. Also *transf.*

1960 *Nation* 18 June 531/3 How deep is Goldwaterism? **1963** *Sunday Times* 24 Nov. 3/3 He [*sc.* Kennedy] planned to take sharp aim at Goldwaterism. **1964** *Listener* 25 June 1018/1 He..tried in public to belittle the difference between the moderates and the Goldwaterites. **1964** *Ibid.* 24 Sept. 471/3 Mr Calder-Marshall seems to be a Goldwaterite where modern art is concerned. His 'vulgar economic realities' are the cloak for a conspiracy-theory, and he is disappointed that I don't subscribe to it. **1965** *Spectator* 22 Jan. 102/1 Goldwaterism—that is to say, far-out conservatism in general—has been the butt of Broadway humour for so long. **1970** *Times* 9 May 2/1 Let me suggest..that we solve our..British problems in a British way, not by the importation of Goldwaterism.

Goldwynism (gōu·ldwiniz'm). [f. the name *Goldwyn* + -ism.] A witticism uttered by or typical of Samuel G. Goldwyn, American film producer (1882–1974), esp. one that revolves round a contradiction, a colourful image, etc.

1937 *Sat. Evening Post* 8 May 82/2 'It sounds like Sam Goldwyn.' Chaplin said, 'We'll pin it on Sam,' and he repeated it until it became a world-famous Goldwynism. *Ibid.*, A kind of Goldwyn scholarship has grown up; there are specialists who can detect fake Goldwynisms as an archeologist can spot a phony Greek vase. **1938** I. Goldberg *Wonder of Words* xii. 245 In a phrase that has become humorously popular, and that has been attributed to the moving-picture producer Sam Goldwyn, renowned for his so-called Goldwynisms, the ideas of exclusion and inclusion have been telescoped. 'Include me out!' he is said to have exclaimed. **1959** *Globe & Mail* (Toronto) 24 Sept. 34/1 A Goldwynism..is a slight twisting of a word or a slight misstatement that makes an ordinary remark sound wittily off-beat. **1967** *Sunday Mail* (Brisbane) 13 Aug. 30 He [*sc.* S. Goldwyn] gave a new word to the English language: Goldwynism, meaning those delightful malapropisms such as: 'in two words—impossible', 'verbal contracts are not worth the paper they're written on', and 'I'll give you a definite maybe'. **1969** *Daily Tel.* 5 Nov. 17/3 He will be remembered as the master of innumerable Goldwynisms, many of which, he says, were invented by comedians and pinned on him.

goldy, *a.* Delete *Obs.* exc. *dial.* and add later examples. Also, gold-like; resembling gold in colour or sheen.

a **1861** T. Winthrop *John Brent* (1883) iii. 27 'I don't think that quartz looks quite so goldy as it did at a distance' said he. **1904** R. Bridges *Demeter* 81 Now maids playfully dance o'er enamel'd meadows, And with goldy blossom deck forehead and bosom. **1920** Galsworthy *Awakening* 35 There were silver threads in her dark goldy hair. **1969** 'J. Morris' *Fever Grass* v. 57 The room was illuminated with a soft, goldy-bronze glow.

golem (gōu·lĕm, goi·lĕm). Also **Golem.** [ad. Yiddish *goylem*, f. Heb. *gōlem* shapeless mass.] In Jewish legend, a human figure made of clay, etc., and supernaturally brought to life; in extended use, an automaton, a robot.

1897 H. Iliowizi *In Pale* 171 The Baal-Shem smiled when he was told that the official version of the *Golem's* work proclaimed, that 'the accident was caused by a red bolt of lightning, which killed the people and destroyed the building'. **1925** H. Schneiderman tr. *Bloch's Golem* 67 They formed out of clay the figure of a person,.. And the Golem lay before them. **1928** *Funk's Stand. Dict.*, *Golem*, a homunculus or figure made to represent a man: said to have been made and endowed with life by Reb Löw of Prague in the middle ages; hence, any one who acts like an automaton. **1942** B. Berenson *One Year's Reading* 12 Feb. (1960) 24 What a belief that the great masses are pails into which you can pour any kind of slop.., and make them act like your golems! **1958** *Times* 5 Dec. 16/6 The ungainly bronze golems that stand around the Hanover Gallery. **1964** N. Wiener *God & Golem* (1965) (front jacket-flap), The ability of machines to learn, and their potential ability to reproduce themselves, lead to the question: 'Can we say that God is to Golem as man is to machine?' (In Jewish legend *Golem* is an embryo Adam, shapeless and not fully created, hence an automaton). *Ibid.* v. 55 Rabbi Löw of Prague, who claimed that his incantations blew breath of life into the Golem of clay. **1969** *Listener* 10 July 33/1 So let us forget about robots as serfs, which is the way they were originally proposed in Capek's *RUR* (*robotnik*, in Czech, means a serf). Such robots are essentially in the 'Golem' image and have no further interest except as ingenious dolls for grown-ups.

golf, *sb.* Add: **b.** *golf bag, cap, match*; **golf ball,** (*b*) a colloquial name given to a spherical ball in certain kinds of electric typewriter on which all the type is mounted and which is caused to move to present the required symbol to the paper; **golf cart,** (*a*) a trolley for carrying golf clubs; (*b*) a motorized cart for transporting golfers and their equipment; **golf-club** (examples of sense 'a society for playing golf'); **golf-croquet** (see quot. 1960); **golf-drive,** a drive (Drive *sb.* 1 d) in golf; also, a similar stroke in Cricket; **golf-links** (examples); **golf shot,** a shot in golf; **golf-widow,** a woman whose husband spends much of his spare time playing golf.

1895 *Army & Navy Co-op. Soc. Price List* 1446 The New Golf Bag. Made same style as a cricket bag and large enough to take clubs, sling, balls, etc. **1921** *Daily Colonist* (Victoria, B.C.) 12 Oct. 16/3 (Advt.), English golf bag and clubs, also violin, for sale. **1966** *Gloss. Automated Typesetting* (ed. 2) 87 The IBM 72 electric typewriter characterized by its stationary platen and the concentration of all type characters on a single, interchangeable globe-shaped unit called a typing element..; sometimes referred to as the 'golf ball' typewriter. **1969** *Computers & Humanities* Sept. IV. I. 76 This arrangement allows more flexibility than a line printer, since one can change the type 'golfball' but it is very, very slow. **1970** *British Printer* Dec. 73/2 Hard copy is produced by the IBM 'golfball'. **1970** A. Cameron et al. *Computers & O.E. Concordances* 32 There is a machine which will read the product of a selectric typewriter with a special golfball. **1897** *Sears, Roebuck Catal.* 235/3 Men's Fancy Golf Caps at 21c. **1938** Golf cap [see *balding *a.*]. **1951** *Golfers' Year* II. 136 (Advt.), No other golf cart has the following unique features. **1963** *Golf World* Jan. 42/2 An American-style golf cart designed to carry both you and your clubs might seem of little use on Britain's courses. **1964** Mrs. L. B. Johnson *White House Diary* 17 July (1970) 183 We had two swift and pleasant hours—and then..into the hangar where the white golf cart waited for us. **1971** 'D. Halliday' *Dolly & Doctor Bird* iii. 28 He had already hired an electric golf-cart, a sorry sight. **1834** in R. Clark *Golf* (1875) 79 The Royal and Ancient Golf Club of St. Andrews. **1890** H. G. Hutchinson *Golf* 325 The Royal Liverpool Golf Club. **1931** T. S. Eliot *Triumphal March*, Those are the golf club Captains, these the Scouts. **1965** A. S. Graham *Golf Club* x. 64 Every golf club has its distinct golfing types. **1920** W. Deeping *Second Youth* xv, To play them at golf-croquet. **1960** E. P. C. Cotter *Tackle Croquet This Way* i. 13 A game of Golf Croquet. This is a game in which the hoops are treated as 'holes', as in Golf. The balls play in sequence..and the first ball to run the hoop wins the hole and then all proceed to the next hoop. **1909** *Westm. Gaz.* 16 Feb. 12/2 In such a stroke as a golf-drive the arm that reaches its fullest extension first is almost certain to be the dominating factor in regulating impact. **1913** *Daily Mail* 7 July 9/1 A plucky forcing batsman, rather partial to the on 'golf drive'. [**1877** J. Blackwood *Let.* 27 Mar. in Geo. Eliot *Lett.* (1956) VI. 357, I am going..to North Berwick.. It is a pretty country and there is a Golfing Links.] **1891** H. G. Hutchinson (*title*) Famous golf links. **1919** Wodehouse *Damsel in Distress* xv. 174 He seemed to spend all his spare time frolicking with the man on the golf-links. **1857** J. Blackwood *Let.* 30 Apr. in Geo. Eliot *Lett.* (1954) II. 324 If you saw me starting for a Golf match you would think my tastes..simple, if not even childish. **1926** Wodehouse *Heart of Goof* ii. 60, I am playing a very important golf-match this morning. **1971** 'H. Howard' *Murder One* xiii. 157 He wanted to get away..because he had a golf match at two. **1903** H. D. G. Leveson-Gower in H. G. Hutchinson *Cricket* xi. 352 'You want the golf shot?'..He went to the wicket and made ninety. **1928** M. H. Weseen *Crowell's Dict. Eng. Gram.* 274 *Golf widow*, humorous colloquial name for a woman whose husband spends much time playing golf and little time at home. **1965** *Punch* 19 May 725/1 At the last census, there were more stamp-widows than golf-widows.

go·lfdom. [f. Golf *sb.* + -dom.] The realm of golf.

1902 in W. W. Tulloch *Tom Morris* (1907) 290 Tom Morris, King of Golfdom. **1926** *Contemp. Rev.* Nov. 679 Mr. Wodehouse..pictures..type after type of the creatures that have beset golfdom. **1946** *Time* 24 June 51 In the U.S. Open, the steadiest wrists in golfdom get to shaking. **1971** *N.Y. Times* 17 June 51 This 71-year-old championship, most prestigious in golfdom.

golfer. Add: **2.** A cardigan.

1911 *Alfred Weeks' Sales Catal.*, Our Stock of Golfers will be disposed of at one price only..2/11¾. **1965** *Punch* 6 Jan. p. xii/1 Lambswool golfers. **1969** *Vogue* Nov. 56/1 A cashmere golfer with an interesting reversed rib texture.

Golgi (gǫ·ldʒi). *Anat.* The name of Camillo *Golgi* (1844–1926), Italian anatomist, used *attrib.* and in the possessive to designate various microscopical methods introduced by him, and various types of cell, cell organelle, etc., discovered by or named after him.

a. *Golgi('s) method, technique*, any of various staining methods employing silver salts or osmium tetroxide; so *Golgi stain*.

1885 *Jrnl. R. Microsc. Soc.* 2nd Ser. V. 904 Golgi's methods for staining nerve-elements black are based on the action of nitrate of silver and perchloride of mercury following the use of bichromate of potash. **1891** *Jrnl. Anat. & Physiol.* XXV. 448 By the Golgi method all the cells are rarely stained. **1892** *Bull. Johns Hopkins Hosp.* III. 26 (*heading*) The Golgi silver stain with the central nervous sytem, and its results. **1910** Allbutt & Rolleston *Syst. Med.* (ed. 2) VII. 853 The cavity is usually surrounded by a zone of thick neuroglial tissue which..is deeply stained by the Weigert haematoxylin stain, and by the Golgi stains. **1968** G. C. Hirsch in McGee-Russell & Ross *Cell Struct. & Interpret.* xxxi. 396 This method revealed the same cell-constituents as those blackened by the classical Golgi techniques.

b. *Golgi('s) cell*, any of various nerve cells, esp. *Golgi('s) type I* and *type II cell*, respectively nerve cells with long and with short axons.

1892 *Bull. Johns Hopkins Hosp.* III. 28/2 Cells of the posterior horns, whose nerve processes become richly branched and correspond to Golgi's II type. **1932** W. Penfield *Cytol. Cell. Path. Nerv. Syst.* II. ix. 433 Golgi's epithelial cells (Bergmann cells). *Ibid.* 435 These feathered cells are probably of the same order as Golgi's epithelial cells and if completely stained would prove to have subpial expansions like the Golgi cells. **1961** T. L. Peele *Neuroanat. Basis Clin. Neurol.* (ed. 2) i. 4/1 Structurally, nerve cells have been classed according to the length of their axons, as Golgi type I (long) and Golgi type II (short).

c. *Golgi corpuscle, (tendon) organ, tendon spindle*; *Golgi-Mazzoni corpuscle, organ* (see quots.).

1897 *Proc. R. Soc.* LXI. 248 The terminal arborisation which the nerve-fibres finally make is as a rule small as compared with the end-arborisations of ordinary Kühne-Ruffini 'spindles' or the Golgi 'tendon-organs'. **1900** *Jrnl. Compar. Neurol.* X. 165 The Golgi tendon spindles were found by him in the tendons of the eye-muscles of cattle, swine, dogs, cats, rabbits and men. *Ibid.* 174 Where a single branched nerve or several independent nerves enter the granular substance..the resemblance to the Pacinian corpuscles is lost and it is to these especially that the name 'Golgi-Mazzoni organs' is sometimes applied. **1950** R. Wyburn-Mason *Trophic Nerves* viii. 105 The sensory endings in deep tissues are of four main types:—...(2) The Golgi corpuscles found in tendons and supplied by myelinated afferent nerves. They are stimulated by tension. **1964** E. G. Walsh *Physiol. Nerv. Syst.* (ed. 2) ii. 54 Golgi tendon organs seem to bear a special relationship to the cerebellum.

d. *Golgi apparatus, body, complex, net, network*, etc., a cytoplasmic cell organelle of complex structure that is now believed to be involved in secretion. Also used to designate components of this structure, as *Golgi element, material, region, rod, sac, substance, vesicle*, etc.

1916 A. M. Pappenheimer in *Anat. Rec.* XI. 110 None of these terms appear to be entirely satisfactory. I shall, therefore, refer to the structures simply as the Golgi apparatus. **1924** Hogben & Winton *Introd. Rec. Adv. Compar. Physiol.* 197 In the cytoplasm are present granular bodies, of which two sorts are commonly distinguished, namely, the mitochondria and Golgi rods. **1925** E. B. Wilson *Cell* (ed. 3) 50 In many cases the Golgi 'net' is built up from originally separate bodies—lamelliform, rod-like, banana-shaped or the like... These bodies are variously designated as 'batonettes', 'dictyosomes', or Golgi-bodies. **1926** L. W. Sharp *Introd. Cytol.* (ed. 2) vii. 126 For many years the Golgi material has been chiefly the concern of students of animal tissues. **1946** *Nature* 24 Aug. 274/1 It has come to be realized that the Golgi bodies are to be found in most, if not all, living animal cells. **1949** *Q. Jrnl. Micros. Sci.* XC. 293 (*title*) Further remarks on the Golgi element. **1952** *Sci. News* XXIV. 23 The much debated reticular structure known as the Golgi apparatus. **1967** L. T. Threadgold *Ultrastruct. Anim. Cell* v. 164 The Golgi complex, which was such a controversial structure when it could only be observed with the light microscope, has emerged from electron microscopy with a distinct and characteristic fine structure which makes it readily recognizable in all cells. **1970** Ambrose & Easty *Cell Biol.* v. 164 The Golgi region was first observed in certain nerve cells.

Goliath: Add: **2. b.** The African giant heron, *Ardea goliath*. In full *Goliath heron*.

1860 *Ibis* II. 220 *Ardea goliath*, Temm. Goliath Heron. **1906** W. L. Sclater *Birds S. Afr.* IV. 55 Goliath Heron. *Ibid.* 56 The Goliath is found throughout the greater part of Africa. **1970** *Animals* July 127/1 Huge goliath herons stand like statues, suddenly darting to life to spear a fish or a frog.

c. A giant frog, *Cyclorana goliath*, found in S. Cameroon. In full *Goliath frog*.

[**1914** E. G. Boulenger *Reptiles & Batrachians* II. 208 The Giant Frog, *Rana goliath*, of Cameroon..was discovered only quite recently.] **1931** H. W. Parker in W. P. Pycraft *Standard Nat. Hist.* XII. 501 We may mention the..largest of all frogs, the giant Goliath-frog (*R[ana] goliath*) of the Cameroons, which is as big as a medium-sized terrier and can eat a full-grown rat. **1961** D. M. Cochran *Living Amphibians of World* 159/1 It seems very unlikely that a successful cross could ever be made between the goliath frog and any frog from the northern hemisphere.

3. A kind of powerful travelling crane. Also *attrib.*

1888 *Lockwood's Dict. Mech. Engin.* s.v., Goliaths, like travellers, are worked both by hand and by steam power. **1892** E. Marks *Notes Constr. Cranes* 107 A Goliath supplied to the Government of New South Wales. *Ibid.* 108 The Goliath crane, known also as the Wellington crane. **1892** F. Colyer *Lifting & Pressing Machinery* (ed. 2) 164 'Goliath' Travellers. **1928** *Daily Express* 5 Dec. 13 Floating cranes and land cranes, goliaths and derricks. **1970** *Railway Modeller* May 155/1 Three closely spaced sidings are spanned with a Goliath crane.

golliwog (gǫ·liwǫg). Also **gollywog, Golliwogg.** [perh. suggested by Golly *int.* (=God) and Polliwog, pollywog (*dial.* and *U.S.*, = tadpole).] A name invented for a black-faced grotesquely dressed (male) doll with a shock of fuzzy hair. Also shortened **go·lly, Golly.** Also *attrib.* and *Comb.*

1895 B. Upton (*title*) The adventures of two Dutch Dolls—*and* a 'Golliwogg'. **1904** *Woman's Life* 7 May 227/1 One of her hobbies is the collecting of those quaint, but unbeautiful, dolls known as Gollywogs. **1907** *Westm. Gaz.* 28 May 10/2 A clever golliwogg dance received the enthusiastic applause it deserved. **1922** Joyce *Ulysses* 347 Madcap Ciss with her golliwog curls. **1925** E. F. Norton *Fight for Everest, 1924* 104 Every now and then I was struck afresh with the absurdly 'gollywog' appearance of the party. **1934** *Neuphilologische Mitteilungen* XXXV. 130 Abbreviated forms such as *golly* 'gollywog'. **1956** G. Durrell *Drunken Forest* vi. 126 We came to the police-station,..shaded by a large, golliwog-headed palm tree. **1969** I. Kemp *Brit. G.I. in Vietnam* viii. 161 A short, slightly-built negro whose appearance reminded me of one of those golliwogs that decorate the labels of Robertson's marmalade jars.

gollop (gǫ·lǫ̆p), *v. dial.* and *colloq.* Also **gollup.** [Perhaps an extended form of Gulp *v.*, influenced by Gobble *v.*[1] Cf. next.] *trans.* To swallow greedily or hastily. Freq. with *down* or *up*. Also *fig.* Hence **go·lloping** *vbl. sb.*

1823 Egan *Grose's Dict. Vulg. T.*, *Gollup up*, to drink down quickly. **1851** T. Sternberg *Dial. & Folk-lore Northants* 42 *Gollop* (southern), to swallow greedily. **1854** A. E. Baker *Gloss. Northants Words* 283 Gollop it down. **1927** *Brit. Weekly* 14 July 326/4 The Ancient Order of Frothblowers. The 1st Qualification for Membership is that he or she 'should gollop their beer with a zest', three gulps to the pint and 14 to the gallon. **1928** Wodehouse in *Strand Mag.* July 10/1 He was aware that the golluping noises [of two boys eating oysters] at his side had ceased. *Ibid.* Dec. 532/1 'Don't gollup your food, Harold,' said the second burglar. **1930** A. Kennedy *Orra Boughs* i. 4 His conclusion, his attempt at constructive criticism, had been almost an afterthought. 'All things considered' (Ah, how he had golloped Chesterton) 'Burns does have a claim to immortality.' **1941** L. A. G. Strong *Bay* 14 'In sips, now.'..'Don't go gollop it down.' **1960** *Spectator* 21 Oct. 595 The licentious Allied soldiery golloped them up. **1969** C. Geeson *Northumberland & Durham Word Bk.* 96 *Gollop.* In general dialectal use in Scotland and England. To gulp one's food. 'Dinna gollop your broth like that!' **1970** D. M. Davin *Not Here, Not Now* III. vii. 210 Mind if I have one before Martin gollops the lot?

gollop (gǫ·lǫ̆p), *sb. dial.* and *colloq.* [Cf. prec.] A greedy or hasty gulp (in quot. 1912, a gulping sound). Also *fig.*

1912 W. Deeping *Sincerity* v. 43 Wolfe uncorked it, and turned it upside down. The water went 'gollop, gollop', and splashed the stones at Mr. Turrell's feet. **1933** F. E. Baume *Half Caste* 76 They washed down the lot with great gollops of tea. **1933** J. Masefield *Bird of Dawning* 152 We've all had a good gollop of fortune, being alive now. **1941** L. A. G. Strong *Bay* 14 If I took it all in a gollop, I never had to give it another thought. **1946** B. Fergusson *Wild Green Earth* I. ii. 45 The mule rolled over and over down the hillside until he lay, breathing great gollops of air, helpless until his load was taken off him.

golly, *int.* (Earlier example.)

1775 G. White *Jrnls.* 21 Dec. (1931) 117 *Gouleins* (Gothic) salutatio: hence perhaps our word Golly, a sort of jolly kind of oath, or asseveration much in use among our carters, & lowest people.

gollywog, var. *golliwog.

goluptious, *a.* Add: Also **galumptious, goloptious.**

1903 J. Vaizey *Pixie O'Shaughnessy* vi. 64 She gives you the most galumptious teas, and..you can eat as much as you like. **1926** Fowler *Mod. Eng. Usage* 164/1 Goloptious. **1935** W. de la Mare *Early One Morning* 79 'Dreams' so galumptious and prolonged.

gom[4] (gǫm). *Ireland.* Also **gaum.** [Cf. Ir. *gamal* stupid-looking person.] A poor silly fellow.

1834 S. Lover *Leg. Irel.* 2nd Ser. 241 Do you think me sitch a *gom*, all out, as to put me off wid four pence ha'pny. *a* **1849** J. Keegan *Leg. & Poems* (1907) 114 'By gor, you are no gom,' said Biddy. **1870** P. Kennedy *Fireside Stories Ireland* 29 (E.D.D.), You were a gaum before you went to travel, and you are a gaum after it. **1894** E. McNulty *Misther O'Ryan* xiv. 150 But, shure, I cudn't kape quiet an' see him standin' there purtindin' to be a gaum. **1928** D. Byrne *Destiny Bay* i. § 2. 11 You big, thick-footed, herring-fed Southern gom! **1970** M. Kenyon *100,000 Welcomes* iv. 24 Corrigan's a gom. This one has breeding.

‖ **gompa** (gǫ·mpă). [a. Tibetan *gön-pa, gŏm-pa* (Jäschke *dgón-pa*), a solitary place, a hermitage.] A Tibetan temple or monastery.

[**1863** E. Schlagintweit *Buddhism in Tibet* xiii. 179 The monasteries, in Tibetan Gonpa, 'a solitary place', are mostly at some little distance from the villages. **1895** L. A. Waddell *Buddhism of Tibet* xi. 255 The monastery is named in Tibetan *Gön-pa*, vulgarly *Gŏm-pa*, or 'a solitary place' or hermitage; and most monasteries are situated..at least some distance off from villages.] **1902** [see *chorten]. **1933** *Times Lit. Suppl.* 28 Sept. 644/3 Vivid descriptions of the..Lamas..of their *gompa* temples. **1939** M. Pallis *Peaks & Lamas* xiii. 147 Time enough to sort out the remaining stores and attend a service at the *Gompa* (monastery) to which we had been specially bidden. **1960** 'S. Harvester' *Chinese Hammer* iv. 46 A *gompa*, one of those religious establishments whose walls housed more than a fifth of Tibetan males.

gompaauw, gompauw (gǫ·mpāu, ‖χǫ·mpōu). *S. Afr.* [f. Afrikaans and Du. *gom* gum + Du. *paauw, pauw* peacock (cf. Paauw).] The giant or kori bustard, *Ardeotis kori*.

1838 J. E. Alexander *Exped. Disc. Int. Africa* II. viii. 141 Here a gum pauw, or bustard, which subsists partly on gum, was shot. **1905** W. L. Sclater in *Science in S. Afr.* 143 The Bustards (*Otidæ*) are represented by no less than twelve species, ranging from the large *Gom Paauw* (*Otis kori*), the male of which sometimes weighs as much as 40 lbs. to the smaller Knorhaan (*Otis afra*) about the same size as a partridge. **1936** *Blackw. Mag.* Mar. 296/2 An occasional gom-paauw, the great bustard of South Africa, largest of all the family. **1948** L. G. Green *To River's End* iv. 40 Both the gom paauw and the smaller duin paauw feed on the gum from the kameeldoorn trees. **1962** Mackworth-Praed & Grant *Birds S. Third Afr.* 262 Kori bustard or gom paauw: *Ardeotis kori.*

gomuti (gomū·ti). Also **gomuta.** [ad. Mal. *gĕmuti* sugar-palm fibre.] A palm tree, *Arenga pinnata*, native to Malaya and the East Indies, and cultivated elsewhere, particularly in India; also, the fibre obtained from this tree.

1811 W. Marsden *Hist. Sumatra* (ed. 3) 88 This palm, named in Sumatra *anau*, and by the eastern Malays *gomuto*, is the borassus gomutus of Loureiro. *Ibid.*, The *anau* palm produces also (besides a little sago) the remarkable substance called *iju* or *gomuto*, exactly resembling coarse black horse-hair. **1836** Loudon *Encycl. Plants* (ed. 2) Suppl. 1290 *Gomutus*,..a palm, growing to the height of 40 ft. **1886** Yule & Burnell *Hobson-Jobson* 295/1 *Gomuti*, a substance resembling horsehair, and forming excellent cordage.., sometimes improperly called coir, which is produced by a palm growing in the Archipelago, *Arenga saccharifera*. **1889** G. S. Boulger *Uses of Plants* vi. 168 *Arenga saccharifera*, Labill., Gomuti Palm. Ramenta, or Vegetable bristles, suggested as a covering for telegraph cables. **1908** G. Watt *Commercial Prod. India* 91 At the base of the petiole is found a beautiful black horsehair-like fibre known as the *Ejú* or *Gomuta* Fibre. **1926** E. Blatter *Palms Brit. India* IV. 357 Like the true Sago Palm..the Gomuti Palm affords a medullary substance, from which a meal is prepared. **1953** N. L. Bor *Man. Indian Forest Bot.* 351 At the base of the petioles is found a beautiful black fibre, exactly like horse-hair, known as Gomuta fibre, which is very strong and can be used under water.

-gon, *suffix.* Add: Also with numeral prefixed, as *16-gon.*

1898 T. J. McCormack tr. *Schubert's Math. Ess.* 125 He constructed an inscribed 16-gon. **1964** L. F. Tóth *Regular Figures* vi. 205 The vertices of a regular 12-gon.

gonad. Now usu. pronounced (gōu·næd). Substitute for def.: Any organ in an animal (as a testis or an ovary) that produces gametes. (Further examples.)

1900 J. S. Kingsley *Text Bk. Vertebr. Zool.* 125 The ovaries are those gonads which are to give rise to the eggs or ova. **1927** Haldane & Huxley *Animal Biol.* ix. 191 The changes which take place in body and mind at puberty are also under the control of the endocrine system. The immediate control is exerted by the gonad itself. **1957** Richards & Dawes *Imms's Textbk. Ent.* (ed. 9) I. 193 The hermaphrodite reproductive system is not unlike the usual female condition but the gonad.. produces both spermatozoa and eggs. **1957** *Times* 3 Sept. 9/2 The report of a study group on the effect of radiation on human heredity says that in every exposure the X-ray beam should..be so directed that a minimum of radiation reaches the gonads. **1962** D. J. B. Ashley *Human Intersex* iii. 34 The gonads, testes and ovaries, are the primary sex organs which house the definitive male and female germ cells which differentiate to form the gametes. **1964** Parker & Haswell *Text-bk. Zool.* (ed. 7) II. 318 The gonads are of great size in the sexually mature fish.

Hence **gonadal, gonadial** (gonē̆i·dăl, -diăl), **gonadic** (gonæ·dik) *adjs.*, of or relating to the gonads; **gonade·ctomy** [*-ectomy], excision of a gonad.

1902 A. Sedgwick in *Encycl. Brit.* XXVIII. 138/1 Formation..of gonadial sacs. **1914** Geddes & Thomson *Sex* iv. 88 The gonadial glands of internal secretion. **1925** F. A. E. Crew *Animal Genetics* 194 The operation of gonadectomy (extirpation of the sex-gland). *Ibid.* 198 The animals which were merely castrated failed to develop the secondary gonadic characters which they would presumably have shown otherwise. **1934** Webster, Gonadal. **1946** *Nature* 20 July 95/2 Trichloracetic acid filtrates of serum were prepared during an investigation of the effects of gonadal hormones on the mineral metabolism of the immature pullet. **1963** *Lancet* 5 Jan. 3/2 There is no evidence in favour of transplants of gonadal tissue rejuvenating the senescent human.

gonadotrophic (gōu:nădotrōu·fik), **-tropic** (-trōu·pik, -trǫ·pik), *a. Physiol.* [f. Gonad + -o + *-trophic, *-tropic.] Regulating the activity of the gonads; of or pertaining to gonadotrophins.

1931 Wiesner & Marshall in *Q. Jrnl. Exper. Physiol.* XXI. 147 (*title*) The Gonadotropic Hormones (ρ-Factors). *Ibid.* 148 It has been shown that the gonadotropic substances can produce haemorrhagic follicles and *corpora atretica*. **1936** *Lancet* 8 Aug. 309/2 It has afforded an opportunity of investigating the gonadotropic hormone production under the exceptional circumstances of pituitary hypofunction during gestation. **1938** *Encycl. Brit. Bk. of Yr.* 227/2 The adrenotrophic, thyrotrophic, and gonadotrophic hormones. **1957** *Science News* XLIV. 85 The gonadotrophic hormones secreted by the pituitary gland stimulate equally both testis and ovary. **1968** R. H. Williams *Textbk. Endocrinol.* (ed. 4) ii. 32/1 A provisional separation of two types of gonadotropic cells has been made in the mouse by Barnes. **1970** *Sci. Jrnl.* June 46/3 The pituitary produces at least two gonadotrophic principles which in their turn stimulate the secretion of two quite separate hormones from the ovary—oestrogen and progesterone.

So **go:nadotro·phin, -tro·pin,** any of several gonadotrophic hormones (as follicle-stimulating hormone) that originate in the pituitary or the placenta.

1937 *Endocrinology* July 489 (*title*) Excretion of gonadotropin by normal human males after the ingestion and injection of extracts of pregnancy urine. *Ibid.* 493 Biological and chemical differences may be presented as evidence for the existence of two or more gonadotropins, representing distinct products of internal secretion. **1957** *Encycl. Brit.* VIII. 434/1 Two gonadotrophins are secreted by the anterior pituitary, follicle-stimulating hormone (FSH) and luteinizing, or interstitial cell-stimulating, hormone (LH). **1964** H. H. Cole *Gonadotropins* p. iii, Until the mode of action of gonadotropins becomes known, it is unlikely that uniformity of spelling will be achieved. Arguments based on etymology are persuasive for the use of either suffix. In this volume, the 'h' has been dropped, arbitrarily. **1966** *New Scientist* 15 Dec. 618/3 Gonadotrophins are the pituitary hormones which regulate the activity of the sex glands—the ovaries in the female, and the testicles in the male. **1968** *Daily Tel.* 12 Nov. 1/6 Mrs. Pennington..has been taking the fertility drug gonadotrophin. **1970** *Sci. Jrnl.* June 48/2 Demonstration of the presence of chorionic gonadotrophin is also the basis of nearly all pregnancy tests.

Gond (gǫnd), *sb.* and *a.* [Hind., f. Skr. *goṇḍa* fleshy navel, person having this, Gond.] **A.** *sb.* **a.** A member of a Dravidian people, many of them jungle-dwellers, of central India. **b.** = *Gondi. **B.** *adj.* Of or pertaining to this people or their language.

The native name is *koitor*.

1810 Moxon in G. Smith *S. Hislop* (1888) 38 The Gonds..live mostly in the hills and jungles. **1854** S. Hislop *Ibid.* 167 We found the inhabitants to consist of Gonds and Dheds. **1855** *Ibid.* 169 The Gond Raja of Dewagad. **1856** R. Caldwell *Gram. Dravidian Langs.* 8 The Gōnd or Goand. *Ibid.* 9 The people by whom the Gond and the Ku are spoken. **1867** W. D. Whitney *Lang. & Study Lang.* ix. 327 The wild Gonds and Khonds of the hilly country of Gondwana. **1914** W. H. Rivers *Kinship & Soc. Organis.* 26 The earliest reference to the cross-cousin marriage which I have been able to discover is among the Gonds of Central India. **1924** *Blackw. Mag.* Apr. 545/1 Several Gond women washing clothes. **1925** *Ibid.* Jan. 64/1 The Gond nowadays is becoming civilised and Hinduised. **1936** *Times Lit. Suppl.* 4 Jan. 18/2 And so among these Gond songs there are many which seem to be quite inconsequential. **1961** P. Spear *India* vii. 89 The Chandel Rajputs are thought to be of Gond descent. *Ibid.* xiii. 157 Tribes like the Bhils, the Gonds, or the Santhals carried on their forest and hunting craft. **1971** *Illustr. Weekly India* 25 Apr. 42/1 Bastar is a land of Marias, Murias and Gonds.

Gondal (gǫ·ndăl), *sb.* and *a.* **A.** *sb.* The name of an imaginary island invented by Emily and Anne Brontë; also, an inhabitant of this island. Also **Gondaland. B.** *adj.* Of or pertaining to this island. Hence **Go·ndalan, Gonda·lian** (-dē̆i·l-) *adjs.*

1834 E. & A. Brontë *Diary* 24 Nov. in M. Spark *Brontë Lett.* (1966) 18 The Gondals are discovering the interior of Gaaldine. **1841** E. Brontë *Diary* 30 July in C. W. Hatfield *Poems E. J. Brontë* (1941) 167 The Gondolians [*sic*] are at present in a threatening state, but there is no open rupture as yet. **1908** C. Shorter *Brontës Life & Lett.* I. x. 215 The *Gondaland Chronicles*, to which reference is made, must remain a mystery for us. **1911** J. M. Dembleby *Key to Brontë Works* i. 17 Emily Brontë..was writing, on July 30th, 1845, that she, Emily, was..engaged upon and intended to continue some puerile compositions called *The Gondal Chronicles*, which she spoke of as 'delighting' her and Anne. **1912** M. Sinclair *Three Brontës* 193 There are no recognisable references to the Gondal poems. **1932** C. Morgan in H. J. & H. Massingham *Great Victorians* 70 The opening of 'The Prisoner', telling how narrator and jailer visit a dungeon and speak with a female prisoner confined there, is evidently fictitious, probably Gondal. **1941** F. E. Ratchford *Brontës' Web of Childhood* xiii. 252 Gondalan Vikings. **1941** *N. & Q.* CLXXXI. 182/2 Those [*sc.* poems of Emily Brontë] which were prepared for publication had Gondalian names and references carefully expunged. **1966** M. Spark *Brontë Lett.* 17 The Gondal manuscripts were not preserved, but the love..and excesses of the exotic people of Gondal moved in the poetry of Emily and Anne to the time of their latest compositions.

Gondi (gǫ·ndi). [Hind., f. *Gond.] The native (Dravidian) language of the Gonds.

The native name is *koiyān*.

1855 S. Hislop in G. Smith *S. Hislop* (1888) 228 Inquired about the Gondi. **1875** R. Caldwell *Gram. Dravidian Langs.* (ed. 2) 513 A translation of the Gospels of St. Matthew and St. Mark into Gôndi by the Rev. J. Dawson. **1888** G. Smith *S. Hislop* iii. 62 Gondi was the tongue of the aborigines in the uplands. **1967** D. S. Parlett *Short Dict. Lang.* 52 Gondi (Dravidian), 2 m speakers scattered over Gondwana region, now giving way to Hindi. **1970** *Encycl. Brit.* X. 568/2 The majority of Gonds speak various, and in part mutually unintelligible, dialects of Gondi, an unwritten Dravidian language.

gondola. Add: **4.** *U.S.* (Earlier and later examples.)

1871 Schele de Vere *Americanisms* (1872) 480 *Gondola*...The use of the word for a peculiarly shaped railroad-car is not unknown in England. **1906** *Westm. Gaz.* 31 Dec. 7/3 The Central News New York correspondent says..When the driver pulled up a heavy 'gondola', or low goods wagon, broke loose from the adjacent siding. **1907** *Cosmopolitan* July 390 A flat-car.. is known amongst the fraternity as a 'gondola', with the second syllable emphasized and pronounced long. **1914** *Sat. Even. Post* 4 Apr. 10/1 At the end of the stretch of wheels was the hangout, barricaded on one side by a corrugated iron windshield..and on the other by a derelict gondola freight car. **1922** H. Titus *Timber* iii. 37 Two Indians were loading pulp wood into a gondola on the siding. **1932** E. Wilson *Devil take Hindmost* xvii. 176 They left..on a freight-train..travelling in a low roofless car known as a 'gondola': the gondola was about two-thirds full of gravel.

b. An elongated car attached to the under side of a dirigible balloon or airship. (G., Du. *gondel*.) Also *transf.*, applied to a structure that resembles such a car in its purpose and mode of attachment.

[**1881** W. D. Hay *300 Years Hence* xi. 301 You step into an aërial gondola,..and are at once borne upwards.] **1896** *Strand Mag.* July 78/2 The finished balloon..is 97 ft. in all from the cap to the bottom of the basket or gondola. **1908** *World's Work* XVI. 10804/2 He suspended his gondolas close underneath the hulk. **1914** *Evening News* 15 Oct. 1/6 The distance between the gondola carrying the engines and the body of the airship has been very much reduced. **1930** G. G. Jackson *World's Aeroplanes* 17 Each ship was given four propellers..and these were driven by exceptionally powerful motors, divided between two cars, or gondolas, as they came to be called... The gondolas resembled the Italian craft, from which they took their name, in general design, but they were, of course, much larger. **1953** J. Ramsbottom *Mushrooms & Toadstools* viii. 89 When the balloon 'Explorer II' made its ascent to 72,395 feet..in 1935, spores of seven moulds were enclosed in small quartz tubes fastened to the outside of the gondola. **1955** M. Savill tr. *Galland's First & Last* xxiv. 202 Two additional 20-mm. cannons were mounted below the wings. These 'gondolas' or 'bath-tubs', as we called them, [etc.]. **1959** *Sunday Times* 5 Apr. 7/8 With a float containing petrol to provide buoyancy, and a pressure-proof 'gondola' underneath. **1962** J. Glenn in *Into Orbit* 70 The 'Wheel', as we called it, combined many of the control problems we would encounter in real space flight. It consists of a pill-shaped gondola.. suspended at the end of a 50-foot arm.

c. The car attached to a ski-lift. Also *gondola lift*.

1957 in *Amer. Speech* (1963) XXXVIII. 205 *Gondola, gondola lift*, a number of smaller enclosed cabins seating two or four persons each. The gondolas are stored at the base station poma lift style and are clamped onto the moving cable after loading. **1958** M. West *Second Victory* vii. 97 The Gondelbahn, the long aerial cable that swung the shining gondolas up the slope to the summit of the Grauglockner. **1964** *Harper's Bazaar* Nov. 140/1 *New this season*: gondola lift. **1970** *Times* 18 Dec. 20/2 Vail offers superb skiing... A new development..adds a second gondola.

d. An island counter used in self-service shops for the display of merchandise.

1942 *Super Market* Nov. 53 End of gondola displays may be constructed by using disc-shaped cardboard in graduated sizes. One edge of the cardboard should be cut straight and even in order to rest against the gondola. **1960** *Guardian* 16 May 3/5 Enough space is left round the 'gondolas' (the movable counters) to let shoppers take short cuts. **1962** E. Godfrey *Retail Selling & Organiz.* ii. 19 Other types of counter in common use are refrigerated runs..and gondolas—a self-service central fixture.

5. gondola wag(g)on = *gondola car*.

1918 *Chambers's Jrnl.* July 459/2 The comparatively new 'gondola' wagon serves that purpose admirably.

gondole (gǫndōu·l), *v.* [Back-formation from Gondola.] *intr.* To travel in a gondola. Also *trans.*, to propel (a gondola).

1874 W. James *Let.* 13 Feb. in R. B. Perry *Tht. & Char. W. James* (1935) I. 354 Three days have I gondoled and picture-gazed, under a cloudless sky. **1928** *Daily Express* 6 July 13/6 Sir Robert Witt..used to 'gondole' his own gondola when he went to Venice. **1956** R. Macaulay *Towers of Trebizond* 281 The one who had lent us this palazzo had left his servants in it—a cook and a gondolier—so we gondoled everywhere and that was well.

gondolier. Add: Also as *vb. trans.*, to carry in a gondola. Hence **gondolie·ring** *vbl. sb.*

1889 W. S. Gilbert *Gondoliers* II. 32 Take it altogether, is it—Better fun than gondoliering? **1936** F. Clune *Roaming round Darling* xxiv. 254 Albert gondoliered us across the stream. **1965** *Listener* 15 Apr. 560/2 Gondoliering runs in families.

Gondwana (gǫndwā·nă). *Geol.* Also **Gondwána.** [Name of a region in central north India, f. Skr. *goṇḍavana*, f. *goṇḍa* *Gond + *vana* forest.] **a.** The name given orig. to an extensive series of rocks in India, chiefly sandstone and shales of fluviatile origin, ranging in age from the Upper Carboniferous to the Jurassic or the Cretaceous; later used to designate similar rock systems with the same characteristic fossil flora found in other countries. Also = *Gondwanaland. **b.** Hence as quasi-*adj.*: of, pertaining to, or characteristic of the Gondwana systems or the period during which they were formed. *Gondwanaland*: see next.

1873 H. F. Blanford *Rud. Phys. Geogr. Indian Sch.* x. 116 An enormous series of deposits in Central India and Bengal, apparently of fresh water origin, for which the name 'Gondwána series' has been proposed by Mr. Medlicott. **1876** O. Feistmantel in *Rec. Geol. Surv. India* IX. 28 The best known, because almost the only fossiliferous, rock-series in the peninsula area of India, is that usually spoken of collectively as the plant-bearing series. This is an awkward designation; and I will at once adopt instead the name Gondwána series or system, to be understood in the same wide sense as when we speak of the Jurassic or Silurian series or system. The name was proposed some years ago by Mr. Medlicott, and has since been more or less current on the survey. **1880** *Ibid.* XIII. 190 (*heading*) The correlation of the Gondwána flora with other flora. **1895** *Ibid.* XXVIII. 117 The discovery of Gondwana plants in Argentina. **1909** *Q. Jrnl. Geol. Soc.* LXV. p. ii, Expedition to the Falkland Islands... Fossils, principally leaves of *Glossopteris*, occur in many places, and it is evident that the whole southern part of East Falkland..belongs to the Gondwana System. **1936** J. C. Brown *India's Min. Wealth* (ed. 2) i. 11 Over 98 per cent of the coal is mined from the Lower Gondwana rocks of the peninsula. **1943** M. S. Krishnan *Geol. India & Burma* (1949) x. 242 The name Gondwana was introduced by H. B. Medlicott in 1872 in a manuscript report. **1956** W. J. Arkell *Jurassic Geol. World* xiv. 384 It is necessary to return to the threefold division of the Gondwana system... According to this classification the Lower Gondwanas are Upper Palaeozoic, the Middle are Triassic and the Upper are Cretaceous. **1961** L. C. King in A. E. M. Nairn *Descr. Palaeoclimatol.* 309 By virtue of its size Gondwanaland must have extended beyond any simple climatic girdle... The huge landspread of Gondwana, indeed, does exhibit palaeoclimatic zoning appropriate to two or three of these girdles. **1965** *New Scientist* 14 Oct. 96/2 Any reconstruction of the Gondwana supercontinent results in several Texas-sized missing pieces to this jigsaw puzzle. **1968** R. W. Fairbridge *Encycl. Geomorphol.* 484/1 The Gondwana landscape was broken up by the fragmentation of Gondwanaland.

Also *absol.* or as *sb.*, short for *Gondwana bed, formation, period, rock*, etc.

1880 *Rec. Geol. Surv. India* XIII. 90 The fluviatile deposits of the upper Gondwánas. **1895** *Ibid.* XXVIII. 117 A patch of Gondwanas..contained several rather fair coal-seams. **1933** *Q. Jrnl. Geol. Soc.* LXXXIX. p. lxxxiii, These latter beds..are typical 'Gondwanas', corresponding with the upper part of the Santa Catherina system in Brazil, the Stormberg of South Africa, and the Rajmahals of India. **1953** D. N. Wadia *Geol. India* (ed. 3) ix. 188 The Middle Gondwana was also the epoch of most extensive land-conditions in India. During the Upper Gondwana, epeiric seas began to encroach on its borders from the north-west and south-east. **1956** [see above].

Gondwanaland (gǫndwā·nălænd). *Geol.* Also **Gondwana (Gondwána) land.** [a. G. *Gondwána-Land* (E. Suess *Antlitz d. Erde* (1885) I. xii. 768), f. prec. (see quot. 1904 below) + Land *sb.*] A vast continental area or supercontinent thought to have once existed in the southern hemisphere and to have broken up in Mesozoic or late Palæozoic times forming Arabia, Africa, South America, Antarctica, Australia, and the peninsula of India. Also, these land masses collectively as they exist today.

1896 W. T. Blanford in *Rec. Geol. Surv. India* XXIX. 52 The southern continent, of which India formed part, and which is widely known by an Indian name, the Gondwána land of Suess. **1904** H. Sollas tr. *Suess's Face of Earth* I. 596 We call this mass Gondwána-Land, after the ancient Gondwána flora which is common to all its parts. **1909** *Q. Jrnl. Geol. Soc.* LXV. p. ii, Expedition to the Falkland Islands... A claystone, containing blocks, apparently of glacial origin,.. undoubtedly corresponds with the moraines from other parts of Gondwanaland. **1927** Peake & Fleure *Apes & Men* 18 The movement of Africa northwards is said to be a part of a general breaking up of the ancient southern continent known to science as Gondwanaland. **1936** J. C. Brown *India's Min. Wealth* (ed. 2) i. 13 The coalfields of Bengal, Bihar and Orissa..are isolated fragments of the once continuous Gondwana land. **1965** A. Holmes *Princ. Physical Geol.* (ed. 2) xxx. 1190 The peripheral ring of Gondwanaland is interrupted not only by the South Atlantic, but also by the Indian Ocean. **1970** *Nature* 10 Jan. 142/2 Estimates of the time of break-up of Gondwanaland range from the Permian to the Cretaceous.

gone, *ppl. a.* Add: **1.** *gone goose* or *gosling*: a person or thing that is beyond all hope; a 'gone coon'; a 'dead duck'. *colloq.* (orig. *U.S.*).

1830 *Massachusetts Spy* 7 July (Th.), You are a gone goose, friend. *a* **1865** Smyth *Sailor's Word-Bk.* (1867) 343 *Gone-goose*, a ship deserted or given up in despair (*in extremis*). **1886** M. Thompson *Banker of Bankersville* (1887) xix. 285 If they do git 'im he's a gone goslin'. **1931** D. Runyon *Guys & Dolls* (1932) 59 But I catch pneumonia, and it looks as if maybe I am a gone gosling. **1949** *Sat. Even. Post* 19 Mar. 108/3 Two minutes more of her and I'm a gone goose. **1958** J. & W. Hawkins *Death Watch* (1959) 88 If my luck won't hold..I'm a gone goose anyway.

3. b. With adj. complement, as *gone-soft, -wild*. (Cf. Go *v.* 44 a.)

1925 A. S. M. Hutchinson *One Increasing Purpose* III. xv, Not a fit man..but a gone-soft and nerve-wracked man. **1960** *Times* 13 Jan. 13/6 The mammalian predators of the rabbit: fox, badger, stoat, weasel, tame and gone-wild cats.

4. Very inspired or excited; 'out of this world'; extremely satisfying; excellent; esp. in phr. *real gone*. *slang* (orig. *U.S. jazz musicians*').

1946 Mezzrow & Wolfe *Really Blues* 374/1 *Gone*, out of this world, superlative. *Gone with it*, really inspired, completely in control of the situation. **1948** *Jazz Jrnl.* Aug. 2 (*heading*) Real gone gal—the history of Nellie Lutcher. **1957** J. Kerouac *On Road* (1958) I. xi. 60, I have found the gonest little girl in the world. **1958** *Sunday Mail* (Glasgow) 10 Feb. 11 *Gone*—the best, in the top rung, the coolest. **1959** *Spectator* 1 May 613/2 Snapping his fingers in a very gone fashion to the beat. **1959** *News Chron.* 14 Oct. 8/6 The jazz-loving 'hep-cat' who claims that the music 'sends' him until he is 'gone'. **1967** L. J. Brown *Cat who ate Danish Modern* xvi. 141 This is a real gone pad..it's what the clients expect.

5. Used *absol.* as *sb.* Those who are dead.

1908 *Daily Chron.* 13 May 3/3 Unconscious imitations of Browning and others of the great and the gone. **1914** Hardy *Satires of Circumstance* 52 The speakers, sundry phantoms of the gone.

goner. (See under Gone *ppl. a.*) (Add earlier U.S. and later examples.) Also, something which is doomed or ended.

1850 'Dow, Jr.' in *Sunday Mercury* (N.Y.) 6 Jan. 2/6 Last Monday..the old year was not quite a goner. **1854** M. J. Holmes *Tempest & Sunshine* v. 211 I'd soon give you up as a goner. **1930** 'E. Bramah' *Little Flutter* xiii. 153 If it failed he was—if one may be permitted the word in the excitement of the moment—a 'goner'. **1933** *Boys' Mag.* XLVII. 124/2 When I found the car burnt out I thought you were a 'goner'. **1945** Auden *For Time Being* 12 Rome will be a goner.

gong^2. Add: (Further examples.) Also *Chinese gong*, a type of gong used in orchestras to give special effects. (See Tom-tom *sb.* 1 b.)

1806 T. Busby *Dict. Mus.* (ed. 2) *Gong*, a Chinese instrument of the pulsatile kind. *Ibid.*, The *Gong* is never introduced, except to give a national cast to the music in which it is employed, or to awaken surprise, and rouse the attention of the auditors. **1888** Stainer & Barrett *Dict. Mus. Terms* (ed. 3) 435/2 *Tom-tom*,.. a Chinese gong. **1900** Grove *Dict. Mus.* (ed. 2) IV. 56/2 *Tam-tam*, the French term for the gong in the orchestra. **1961** A. Baines *Mus. Instruments* xiv. 341 The tam-tam or gong, from Eastern Asia, has been used in the orchestra to assist a climax, with its ominous note to suggest sadness or despair. **1962** *Listener* 22 Nov. 885/3 A vibraphone, a zylophone, and four Chinese gongs. **1968** *Observer* 14 Jan. 4/7 We don't make gongs very much. *Ibid.*, The traditional home of the gong is the Far East.

c. *gong-drum, -hammer* (examples), *music; gong-like, -tormented* adjs.

1926–7 *Army & Navy Stores Catal.* 1086 Gong drum jazz outfit...Comprising:—17-in. Gong Drum, 10-in. Side Drum and Sticks, [etc.]. **1954** *Grove's Dict. Mus.* (ed. 5) II. 773/1 A single-headed type of bass drum is the 'gong drum' or 'gong bass drum'...Introduced for theatre use to save space...it was adopted towards the end of the 19th century...and is still in use...to-day. **1889** *Cent. Dict.*, Gong hammer. **1906** *Westm. Gaz.* 4 Sept. 8/2 The lever which governs the escapement of the alarum makes a noise sufficient almost to wake a light sleeper without the aid of the gong-hammer striking. **1924** A. D. Sedgwick *Little French Girl* II. v. 140 Now and then she emitted a loud gong-like laugh. **1969** *Australian* 7 June 16/6 Indonesian gong music and singing is the most accessible Asian music for Western ears. **1932** W. B. Yeats *Words for Music* 2 That dolphin-torn, that gong-tormented sea.

2. *slang.* A medal or decoration (see quot. 1925).

1925 Fraser & Gibbons *Soldier & Sailor Words* 106 *A gong*, a medal. (An old Army term suggested by the shape.) **1942** 'B. J. Ellan' *Spitfire!* xv. 80 Wilf, G— and F/Sgt. S— had all been awarded 'Gongs' (medals to you!) after Dunkirk. **1944** *Lancet* 9 Sept. 359/1 To balance my civilian contemporaries' achievements of the past four years, I have acquired a wife and family, some expensive tastes, the '1939–43 gong'. **1954** G. Smith *Flaw in Crystal* 144 He'd been invalided out...There he was...unadorned among all the wings and pips and gongs. **1958** M. Dickens *Man Overboard* iii. 35 Other people came out of the war with Mentions and worthwhile gongs that tacked letters after their names. **1959** [see *clanger].

b. A warning bell on a police car.

1938 F. D. Sharpe *Sharpe of Flying Squad* xxvii. 277 When they spotted the police car on their trail they opened their car out and pretended that they couldn't hear the gong,...but they...were overhauled and pinched.

3. [? A different word.] A narcotic drug. Also **go·nger**, opium; an opium pipe; **gongeri·ne**, an opium pipe. *U.S. slang.*

1914 Jackson & Hellyer *Vocab. Criminal Slang* 38 *Gonger*. Current amongst smokers and drug fiends. An opium pipe. Also used in the diminutive form 'gongerine'. **1915** G. Bronson-Howard *God's Man* vii. i. 393 Come, lie 'round and join in the fun; With the aid of 'the gong'. **1933** *Amer. Speech* VIII. 27/2 *Hitting the gong*,...kicking the gong around. **1938** *Ibid.* XIII. 185/1 *Gonger*, any opium derivatives. **1952** J. Steinbeck *East of Eden* 198 Let the gong alone for a couple of weeks. **1955** *U.S. Senate Hearings* (1956) VIII. 4162 *Beat the gong*,...to smoke opium. *Ibid.*, *Gong beater*, one who smokes opium.

gong (gɒŋ), *v.* [f. Gong^2.] **1.** To sound a gong; to make a gong-like sound; to summon (a person) with a gong.

1903 H. G. Wells in *Strand Mag.* Apr. 426/1 He has just gonged, no doubt to order another buttered teacake! **1959** J. Wain *Travelling Woman* 28 The vase, which was a metal one, gonged on the floor, and the flowers fell messily at his feet. **1959** D. Barton *Loving Cup* 167, I gong them into meals on the dot.

2. Of traffic police: to call upon (a driver) to stop by ringing a powerful 'gong'. Also *intr.* (Cf. *Gong^2 2 b.)

1934 in Webster. **1935** *Times* 9 Oct. 9/3 'If Major Gwynne had passed you a little farther down the road you would not have gonged him then because that part is not restricted?'—'No.' **1936** *Times* 12 Mar. 8/5 He...was approaching a stationary car outside the Royal Oak when he gonged, slowed down at a pedestrian crossing, and [etc.]. **1966** T. Wisdom *High-Performance Driving* xvi. 137 He will then have to 'gong' you into the side on a busy trunk road.

gongora (gɒŋgŏră). [mod.L. (Ruiz & Pavon *Floræ Peruvianæ et Chilensis Prodromus* (1794) 117), f. the name of Don Antonio Caballero y Góngora (fl. 1782), Viceroy of New Granada.] A plant or flower of the genus of tropical American orchids so named.

1827 W. J. Hooker *Exotic Flora* III. tab. 178 (*heading*) Dark-flowered Gongora. **1910** C. H. Curtis *Orchids for Everyone* 188 The Gongoras have a quaintness that is attractive, but the species are of little value except as curiosities. **1951** *Dict. Gardening* (R. Hort. Soc.) II. 907/2 Gongoras often develop numerous erect, aerial roots.

gongorism. Add: Also **gongori·stic** *a.*

1925 *Times Lit. Suppl.* 26 Mar. 224/3 The suggestion,...of omitting various passages...which are likely to be too 'gongoristic' for contemporary taste, merits favourable consideration. **1944** *Downside Rev.* 182 We may in dealing with this question of Gracián's style and his supposed condemnation of gongoristic effects draw the attention of the reader to the richness of the paragraph on Gongora himself in the *Criticon*.

gonimoblast (gɒ·nimoblæst). *Bot.* [f. Gr. γόνιμος productive + -blast.] In the red algæ: see quot. 1898.

1898 H. C. Porter tr. *Strasburger's Text-bk. Bot.* 337 The fertilised egg does not become converted directly into an oospore, but, as a result of fertilisation, numerous branching filaments termed gonimoblasts grow out from the sides of the ventral portion of the carpogonium. **1945** F. E. Fritsch *Struct. & Reprod. Algae* II. 413 In most Florideae, however, the gonimoblasts do not arise directly from the zygote. *Ibid.* 599 In the life-cycle of Nemalionales the zygote constitutes the only diploid stage, while there are two haploid phases, the ordinary seaweed and the gonimoblast-threads originating from the zygote.

goniometer. Add: **2.** An apparatus for determining the direction from which detected radio waves are coming, without the need for a rotating aerial, consisting essentially of a rotatable detector coil coupled to two fixed primary coils at right angles to one another, each of these being in turn connected to one of a pair of directional aerials also at right angles. Also (and orig.) called a *radio-goniometer.

1921 B. Leggett *Wireless Telegr.* viii. 233 The station was located by the Germans, doubtless by means of first a goniometer and then balloon observation, and for several days it was heavily shelled. **1932** F. E. Terman *Radio Engin.* xvi. 591 When it is desirable to avoid rotating a loop antenna...it is possible to obtain the effect of rotation by using two loop antennas at right angles to each other and combining the outputs in a goniometer. **1961** H. Jasik *Antenna Engin. Handbk.* xxviii. 21 Since dipoles are usually employed at frequencies above 50 Mc, a capacity goniometer is frequently used to avoid the necessity of shielding an inductive goniometer.

gonion (gōu·niɒn). *Anat.* [a. F. *gonion* (P. Broca 1875, in *Bull. de la Soc. d'Anthrop. de Paris* X. 362), f. Gr. γωνία angle + *-ion^2.] The outermost point on the angle of the lower jaw on each side.

1878 R. Bartley tr. *Topinard's Anthrop.* ii. ii. 235 *Gonion*, the region of the angle of the lower jaw. **1968** [see *gnathion].

gonioscope (gōu·nioskōup). [f. Gr. γωνί-α angle: see -scope.] An instrument for observing the angle of the anterior chamber of the eye, usu. consisting of a contact lens to which is attached some kind of optical system. Hence **gonioscopic** *a.*; **gonio·scopy**, the art or science of using the gonioscope; inspection of the angle of the anterior chamber.

1923 A. Duane *Fuchs's Text-Bk. Ophthalmol.* xviii. 372 If the observer, standing to one side, holds the ophthalmoscopic mirror almost at right angles to the patient's line of sight, he can see deposits and adhesions in the angle of the anterior chamber, which ordinarily are invisible because hidden by the limbus (Salzmann). Such an examination is called gonioscopy. **1925** M. U. Troncoso in *Amer. Jrnl. Ophthal.* VIII. 433/2 Hoping to obtain a clear, unobstructed view of the angle [of the anterior chamber of the eye] in all directions and a sufficient magnification, I have devised and perfected a new instrument, the gonioscope, which is a microscope and a periscope combined. *Ibid.*, The term 'Gonioscopy' suggested by the author in 1921, has been adopted. *Ibid.* 445/2 A second gonioscopic examination two weeks afterwards detected a total peripheral synechia all around the limbus. **1951** H. S. Sugar *Glaucomas* ii. 27 By applying a prismatic contact lens (gonioscopic lens) to the eye, we are able to see the structures of the chamber angle. **1961** *Lancet* 22 July 166/2 The implant should not be inserted until the eye has recovered from this operation and gonioscopy shows that the desired effect has been achieved. **1961** *Ibid.* 26 Aug. 467/1 Optical methods of examination [of the eye] (chiefly the slit-lamp microscope and the gonioscope).

gonk (gɒŋk). [Arbitrary formation.] An egg-shaped doll. Also *attrib.* and *Comb.*

1964 *Spectator* 29 May 726/1 Those neckless dolls called—I think—gonks, which witless adults are said to give to other adults. **1964** *Daily Mail* 2 Sept. 4/3 Gonks...are those nasty, expensive, fat balls of felt and rag that are squatting all over our houses and toy shops. **1964** *Daily Tel.* 11 Sept. 17/3 The principal of a technical college said...'...We had one with what I believe is a "gonk" cut. His ears were invisible and you could just see his eyes and nose peeping out from under shoulder-length hair.' **1968** A. Diment *Gt. Spy Race* iii. 44 The double bed...was covered by a large tribe of Gonk dolls, paperbacks and grimy LP sleeves. **1969** A. E. Lindop *Sight Unseen* xiv. 120 Her hair had degenerated into a gonk style.

gonna (gɒ·nə), colloq. (esp. U.S.) or vulgar pronunciation of *going to* (see Go *v.* 47 b).

[Cf. the earlier Sc. *ganna, gaunna*: see Eng. Dial. Dict. s.v. *Go*, quots. 1806, etc.]

1913 C. E. Mulford *Coming of Cassidy* ix. 149 Yo're gonna get a good lickin'. **1929** E. W. Springs *Above Bright Blue Sky* 136, 5684 has a busted cylinder. Gonna put a new motor in it. **1952** A. Baron *With Hope, Farewell* 56 Put 'em all in clover, that's what I'm gonna do. **1967** M. Shulman *Kill 3* ii. iv. 81 I'm gonna keep on yelling tell you let me out.

gono-. Add: **gonocho·rism** *Biol.* [ad. G. *gonochorismus* (E. Haeckel *Gen. Morphol.* (1866) II. i. 60)], = Diœcism; hence **gonocho·rist**, a diœcious organism; **gonocho·ric, go:nochori·stic** *adjs.*; **gonoco·ccal** *a.*, of, pertaining to, or caused by gonococci; **go·nocœl, -cœle** (-sīl) *Zool.* [Gr. κοῖλον cavity], a body cavity that contains or gives rise to gonads; **go·nocyte** *Biol.* [-cyte], any cell which may potentially undergo meiosis and produce gametes; also, any of the gametes so produced; **go·noduct** *Zool.* [cf. Gonaduct], in some invertebrates, a duct from a gonad to the exterior through which gametes are discharged; **go·nomere** *Biol.* [a. G. *gonomer* (V. Häcker 1902, in *Jenaische Zeitschr. f. Naturwiss.* XXXVII. ii. 312): see *-mere], any nucleus which, after plasmogamy, coexists with and does not fuse with another nucleus in the same cell; hence **gonome·ric** *a.*; **gono·mery**, a condition characterized by the presence of gonomeres; **go·noplasm** *Mycology* [see Plasm], in some genera of the fungal order Peronosporales, that portion of the contents of the antheridium which passes down the fertilization tube; **go·nopore** *Zool.*, a genital pore; **go·notome** *Zool.* [ad. G. *gonotom* (J. W. van Wijhe 1889, in *Arch. f. mikrosk. Anat.* XXXIII. 466): see *-tome], a block of tissue within a somite destined to form a gonad; also, any somite that contains a gonad.

1876 E. R. Lankester tr. *Haeckel's Hist. Creation* I. viii. 175 (*heading*) Distinction of sexes, or gonochorism. [*Ibid.*, Every organic individual, as a non-hermaphrodite (Gonochoristus), produces within itself only one of two generative substances.] **1904** J. McCabe tr. *Haeckel's Wonders of Life* 255 When the ovum and the sperm-cell...are formed in two different individuals (male and female), we call them monosexual, or gonochorists. *Ibid.* 258 The oyster is usually gonochoristic, but sometimes hermaphroditic. *Ibid.* 259 Such structures...have clearly been developed from gonochoristic structures in lower forms. **1950** *Biol. Abstr.* XXIV. 2706/1 Spp. with present haploid hermaphroditism might have been derived from gonoch[o]ric ancestors. **1951** *Ibid.* XXV. 3378/2 *L[ysmata] seticaudata* is a protandric hermaphrodite whose sexual periodicity places it physiologically among the gonochoric conditions found in most malacostraceous Crustacea. **1963** E. Mayr *Animal Species & Evol.* xi. 316 In many marine invertebrates certain geographic races are gonochoristic, while others are hermaphrodite. **1965** G. Bacchi *Sex Determination* i. 21 Unisexuality, dioecism or gonochorism indicate that male and female sex organs occur in different individuals, plants or animals. The three terms are almost perfectly equivalent although the last one is mostly used by zoologists. *Ibid.* iv. 64 Significant genetical work on sex in hermaphrodites and in labile gonochorists is still very scarce. *Ibid.* vii. 156 Labile gonochoric individuals of the *Bonellia* type. **1900** J. O. Symes *Bacteriol. Every-day Pract.* 39 Now and then the gonococcal infection may run an extremely mild course, showing only a slight muco-purulent discharge. **1923** *Daily Mail* 15 June 9 Severe forms of streptococcal, staphylococcal, and gonococcal infections. **1970** *Nature* 25 July 383/1 A few gonococcal isolates have proved resistant to rabbit antiserum. **1893** *Funk's Stand. Dict.* I, Gonocœle. **1900** E. A. Minchin in E. R. Lankester *Treat. Zool.* II. ii. 35 Each protocœlom is in its nature a gonocœl (Goodrich), that is to say a cœlomic pouch, the epithelial walls of which produce ova or sperm or both. **1940** Parker & Haswell *Text-bk. Zool.* (ed. 6) I. ix. 631 The cœlome consists of the pericardium and the gonocœle. **1967** P. A. Meglitsch *Invert. Zool.* ix. 310 The gonocoel theory visualizes the coelom as arising from the lumen of the gonads. **1900** E. A. Minchin in E. R. Lankester *Treat. Zool.* II. iii. 60 In sponges generally two classes of tokocytes can be distinguished: first, sexual cells or gonocytes, the mother cells of ova and spermatozoa of the normal type. **1904** J. McCabe tr. *Haeckel's Wonders of Life* 254 The two copulating sexual cells (gonocyta). **1956** *Nature* 21 Jan. 142/1 The conditions which determine the differentiation of indifferent gonocytes in hermaphrodite glands...constitute a major problem in the biology of sexuality. **1893** *Funk's Stand. Dict.* I, Gonoduct. **1900** E. A. Minchin in E. R. Lankester *Treat. Zool.* II. ii. 36 The cœlomoducts belonging to gonocœls may be called 'gonoducts' (Lankester). **1951** L. H. Hyman *Invertebrates* II. ix. 50 Coelomate animals may...lack gonoducts and use the nephridia for the discharge of sex cells. **1903** *Bot. Gaz.* June 443 The nuclear stages in which the idiomeres (partial nuclei) and gonomeres (double nuclei) appear are closely related. **1920** W. E. Agar *Cytol.* 78 In the germ-track...evidences of gonomery can be found at a much later stage of development than in the somatic cells. **1925** E. B. Wilson *Cell* (ed. 3) 433 Gonomeric grouping. **1969** Brown & Bertke *Textbk. Cytol.* xxii. 531 This condition of gonomery approaches the dikaryotic phase in Basidiomycetes. **1887** H. E. F. Garnsey tr. *De Bary's Fungi* 495/2 *Gonoplasm*, in Peronosporeae: portion of protoplasm of antheridium which passes through the fertilisation-tube and coalesces with the oosphere. **1952** F. L. Wynd tr. *Gäumann's Fungi* 66 Later the sexually functional cytoplasm accumulates in the central portion, forming the gonoplasm. **1897** Parker & Haswell *Textbk. Zool.* I. vi. 276 In the female [round-worm] the reproductive aperture or gonopore is separate from the anus, and is situated on the ventral surface. **1951** L. H. Hyman *Invertebrates* II. ix. 50 Male and female gonopores in hermaphroditic species may be separate,...or there may be a common gonopore. **1900** J. S. Kingsley *Text Bk. Vert. Zool.* i. 103 Whether we have metamerically repeated gonotomes, is as yet a disputed question. **1912** —— *Compar. Anat. Vert.* 319 At one time it was thought that the anlage of the gonad was segmental in character and 'gonotomes', comparable to nephrotomes and myotomes, were described. **1969** A. J. Grove et al. *Anim. Biol.* (ed. 8) xvi. 382 [Amphioxus.] For a time each gonotome remains connected to its own somite by a short stalk, but eventually the stalk is severed and in this way a series of young gonads is formed.

go-no-go: see *go *a.*

gonotocont (gɒnǫ·tokɒnt). *Biol.* Also **gonotokont.** [ad. G. *gonotokont* (J. P. Lotsy 1904, in *Flora* XCIII. 70), f. Gono- + Gr. τοκωντ-, τοκῶν pres. pple. of τοκᾶν to be near delivery.] Any cell that may undergo meiosis; also, any organ containing such a cell.

[**1905** *Amer. Naturalist* XXXIX. 494 Lotsy...proposes the term 'Gonotokonten'...for the mother-cells which inaugurate reduction phenomena.] **1909** A. C. Seward *Darwin & Mod. Sci.* vi. 105 In the nuclei of all those cells which we may group together as gonotokonts (i.e. cells concerned in reproduction) there are fewer chromosomes than in the adjacent body-cells (somatic cells). **1928** C. W. Dodge tr. *Gäumann's Compar. Morphol. Fungi* i. 1 The product of fertilization is called a zygote as long as it remains unicellular; it develops into a diplont which forms gonotoconts (organs in which meiosis occurs). **1965** J. Wilkinson tr. *Langeron's Outl. Mycol.* (ed. 2) ix. 375 The diplont produces gonotoconts or spore mother cells.

gony. 2. (Earlier and later examples.)

1839 *Knickerbocker* XIII. 386 May the 'Goneys' eat me, if he [*sc.* the whale] dodges us this time. **1957** W. L. McAtee *Folk Names Canad. Birds* 3 Black-footed Albatross. *Gony, gony bird, goony* (B.C.); goony bird (North Pacific). These names appear to have the significance of booby or dullard and probably are applied to any albatross. **1966** R. Ardrey *Territorial Imperative* (1967) iv. 148 The albatross...became known to all the American fleet as the gooney bird.

goo (gū). *slang* (orig. *U.S.*). [Of obscure origin but possibly a shortening of Burgoo.] A viscid or sticky substance. Also *fig.*, sickly sentiment, gush.

1911 E. Ferber *Dawn O'Hara* iii. 31 You mean to tell me that you woke me...to make me drink that goo?...I'll bet it's another egg-nogg. **1922** Joyce *Ulysses* 305 Bloom putting in his old goo. **1944** J. H. Fullarton *Troop Target* 43 It's all over blood and goo. **1949** S. Gibbons *Matchmaker* xiv. 163 He...began to measure and mince vegetable scraps and scoop out grey, gritty, oily goo from a large tin. **1951** M. McLuhan *Mech. Bride* 29/2 A crude morality which puts the best-seller goo to shame. **1959** *Times Lit. Suppl.* 20 Mar. 156/3 He writes about subjects which, in less skilled hands, have so often and so embarrassingly degenerated into a mess of gush and goo. **1959** *Punch* 17 June 815/1 *Lonely-hearts*...angered me very much by finally dissolving in a bath of sweet goo after being an excellent film for three-quarters of its length. **1960** *News Chron.* 16 Feb. 6/6 Barbara goes into the dressing-room to slap on the old goo. **1969** *Daily Tel.* (Colour Suppl.) 14 Mar. 50/3 *Crêpes Suzette*...precooked in the kitchen and reheated in the buttery goo.

goober. (Earlier examples.)

1833 *Louisville Publ. Advt.* 7 Nov., A few bags Gouber Pea, or Ground Pea [for sale]. **1834** *Cherokee Phoenix* (New Echota, Ga.) 24 May 3/4 But he so seam I frade of he, I guess he steal my goober. **1848** *Rep. U.S. Comm. Patents 1847* 190 The ground pea of the south, or as it is sometimes called, the gouber or pindar pea. **1871** Schele de Vere *Americanisms* (1872) 57 The peanuts or earth-nuts known in North Carolina and the adjoining States as Goober peas, so that during the late Civil War a conscript from the so-called 'piney woods' of that State was apt to be nick-named a Goober.

good, *a.*, etc. Add: **A.** *adj.* **1. d.** In mod. use freq. as an intensive, or as an indication of adequacy, with *money*.

1889 *Kansas Times & Star* 22 May, Mr. Hammerslough is putting a lot of good money into it. **1896** Ade *Artie* xi. 98 They say he makes good money. **1915** Conrad *Victory* iii. iii. 193 Father was earning good money. **1928** E. Wallace *Again Three Just Men* x. 212 But for his inherent meanness, he would have gladly paid good money to be rid of her. **1966** 'M. Renault' *Mask of Apollo* x. 173 We made good money, and lingered in the pleasant places.

f. *good opera, radio, television, theatre*: said of an entertainment that is effective, or well suited to a specified medium.

1926 C. Morley (*title*) Good theatre. *Ibid.* 216 Don't be sore on them laughs, brother. Every one o' them screams is a meal ticket. That's what I call good theayter. **1928** *Observer* 11 Mar. 15/2 This is an excellent example of what we mean when we say that a play is 'good theatre'. **1928** *Daily Tel.* 15 May 14/5 The play...is, to use the modern phrase, 'very good theatre'. **1958** *Listener* 24 July 138/1 This was not only good television: it also made sense of getting the two men there in the first place. *Ibid.* 140/1 It made very good radio. **1962** *Ibid.* 7 June 1009/3 It may be that this sincere, searching composer has developed an abstract theory in this case into good theatre. Whether it is good opera is another matter. **1971** *Times* 18 Nov. 15/2 There is the temptation for producers to seek what they regard as 'good television', which can mean a search for drama as such and for conflict in the studio no matter how unnecessary it may be for the presentation of a case.

g. *that's a good one* (or *'un*): used ironically to characterize a statement that is incredibly mendacious or absurdly exaggerated. (Cf. 'I like that'.) *slang.*

1813 J. K. Paulding *J. Bull & Br. Jonathan* v. 26 Now this was a good one, for every body knew [etc.]. **1869** *Punch* 30 Jan. 44 Medical-Attendance, Two-an'-Six! Well, that's a good 'un! Why, I attended on 'im. **1914** *Concise Oxf. Dict.* Addenda s.v., *That's a good 'un* (slang), what a lie. **1920** C. Sandburg *Smoke & Steel* 45 That's a good one.

h. In the colloq. U.S. phrase *to look* or *listen good* = to look or sound promising.

1911 R. D. Saunders *Col. Todhunter* iii. 43 It looks good to me, suh. **1916** H. L. Wilson *Somewhere in Red Gap* vi. 252 That listens good to her till she finds she has to give fifty-two dollars for the deck first. **1923** R. D. Paine *Comrades of Rolling Ocean* xii. 215 'Thanks, you look good to us,' yelled Judson. **1932** *Kansas City Star* 1 Mar., It listened good.

2. c. (Further example.) Also *your* (*his*) *good lady, your* (*his*) wife (see Lady *sb.* 7). *your good self* (or *selves*), a commercial form of polite address or reference.

1860 G. H. Lewes *Let.* 6 Mar. in Geo. Eliot *Lett.* (1954) III. 269 'My good lady' (style choisi!)...is reddening her eyes...over the foolish sorrows of two foolish young persons of her imaginary acquaintance. **1923** *Daily Mail* 20 Feb. 3 It is more than probable our next orders will be placed with your good-selves. **1931** *Ch. Times* 8 May 569/3 The kindness and consideration shown by your good self to me. **1957** Partridge *English gone Wrong* i. 17 Then there are such monstrosities as your good self or selves; of recent date; [etc.]. **1967** *Daily Tel.* 30 Aug. 7, I have always understood that the words 'goodself' and 'goodselves' related to credit-worthiness.

d. *good neighbours*: delete † and add later example (cf. also *good neighbour); *good people* (later examples).

1951 C. S. Lewis *Let.* (1966) 234 The desolate coast on which it stands is haunted by 'the good people'. There is also a ghost but...the faeries are a more serious danger. **1959** K. M. Briggs *Anat. Puck* 192 People of Peace...is the Highland name for the fairies, corresponding to the Lowland 'Good Neighbours'. **1966** G. E. Evans *Pattern under Plough* xiii. 128 The countryman in Ireland poured out a little of his draught as a compliment to the 'good people' or the fairies.

e. *good Samaritan*: see Samaritan b.

3. a. (Examples of *good-quality* used *attrib.*)

1897 *Sears, Roebuck Catal.* 289/1 A Good Quality White Marseilles Bed Spread. **1936** *Discovery* Dec. 375/1 A good quality silk. **1956** *Nature* 11 Feb. 289/1 A common method [of improving food intake] has been to give a concentrate feed containing about 10 per cent of good-quality protein. **1962** P. Strevens *Papers in Lang.* (1965) xii. 146 Equipment now exists for making good-quality recordings.

b. (Later examples.)

1872 Geo. Eliot *Middlem.* I. i. i. 2 The Brooke connections, though not exactly aristocratic, were unquestionably 'good'. **1892** G. & W. Grossmith *Diary of Nobody* xxiii. 287 Lupin...has taken furnished apartments at Bayswater...Lupin says one never loses by a good address. **1942** D. Powell *Time to be Born* (1943) x. 227 West Thirteenth?...Are you sure that's a good neighbourhood? **1956** J. D. Carr *P. Butler for Defence* vi. 61 Monnie wanted Jim...to be elected to a 'good' club and join a 'good' golf-club where he might find useful contacts.

c. *to feel good*: to feel oneself to be in good spirits or health. *U.S. colloq.*

1839 F. Marryat *Diary Amer.* 1st Ser. II. 224, I have heard a lady say, 'I don't feel *at all good*, this morning.' **1854** H. C. Kimball *Address* 17 Sept. in *Jrnl. Discourses* (1855) II. 224/1 You will see how good we will make the transient residents feel. **1883** [see *frolicky *a.*]. **1888** *Texas Siftings* 15 Sept. (Farmer), The saloons are going Saturday afternoons, and the men feel pretty good before they come abroad. **1904** *N.Y. Even. Post* 23 June 3 The Captain himself said, 'I feel good', but he did not look well. **1958** *Times Lit. Suppl.* 1 Aug. 438/5 [In Amer. novel] Even Bunny Muldoon, the golf professional, is kept on not because he teaches anyone golf but 'because he makes people feel good'.

4. c. Also (chiefly Austral. and N.Z.) *good-oh!, good-o!, goodo!*, etc., when the same words are used as *adjs.* = *good* (sense 4) and *advbs.* = 'well'.

1916 'Anzac' *On Anzac Trail* 34 The rain came down good-oh. **1917** *Chrons. N.Z.E.F.* 31 Oct. 133 Belting her about goodo. **1926** I. M. Peacocke *His Kid Brother* x. 158 'Goodo!' said Dal. **1928** A. Waugh *Last Chukka* 85 'Good-oh!' she [an Australian] said, 'that'll be bonzer!' **1930** V. Palmer *Passage* ii. x. 184 This is goodo! **1934** T. Wood *Cobbers* 210 You'll be good-O there. **1944** in *Coast to Coast 1943* 122 'How's young Jim getting along?'...'Good-o,' she said. 'I've had a letter only this morning.' **1946** F. Sargeson *That Summer* 118 Yes, good-oh, I said, and thanks very much. **1958** D. Niland *Call me when Cross turns Over* v. 136 It sounds good-o the way he tells it. **1964** *Navy News* Dec. 1/2 What I really thought were goodo were those films on Religion and Science with that chap Erwin A. Moon.

d. *good for you* (*him*, etc.)!: a colloquial expression of approval of something said or done by the person addressed or spoken of.

1861 in M. W. Disher *Cowells in Amer.* (1934) 297 Sam's share $43.33.—Good for him. **1870** L. M. Alcott *Old-fash. Girl* (1874) iii. 35 Good for you, Polly! **1896** C. M. Yonge *Release* ii. xvi. 234 'Old Sukey Shrimper,...has orders to call every Friday!' 'Good for Sukey,' quietly observed Mr. Darpent. **1904** H. James *Golden Bowl* I. x. 193 'Good for *you*!' Maggie smiled. **1922** D. H. Lawrence *England, my England* 142 'And did she take it in?' he asked. 'As much as she took anything else.' He stood grinning fixedly. Then he broke into a short laugh. 'Good for *her*!' he exclaimed cryptically. **1925** 'D. Yates' *As Other Men Are* 111 'Good for you,' she said. 'You've put it uncommonly well.' **1949** E. H. W. Meyerstein *Let.* 8 Jan. (1959) 355 Typical ex-Service husband with his reiterated 'good-for-you'. **1956** A. L. Rowse *Early Churchills* 74 Good for Sir Winston, we may say.

e. *good on you* (*him*, etc.)! = prec. Chiefly *Austral.* and *N.Z.*

1914 A. N. Lyons *Simple Simon* i. v. 84 'What ho!' she exclaimed. 'You've biffed him. Good on you, my lad!' **1944** L. Glassop *We were Rats* ii. xi. 65 'What a bloke believes is his own business.' 'Good on you, mate,' said somebody behind me. 'Give it to him.' **1949** H. Wadman *Life Sentence* i. i. 8 Good on you. Of course, I oughtn't to care twopence about your opinion, but it's nice to know you agree with me. **1953** K. Tennant *Joyful Condemned* xxvii. 268 Vance laughed thunderously. 'Good on you, Rose.' **1959** *N.Z. Listener* 21 Aug. 8/4 'Good on you!' said Dad, smacking my new leg approvingly, 'that's the spirit.' **1961** P. White *Riders in Chariot* xiii. 456 'Goodonya, mate!' called the heartier of the females.

f. *good man!*: an exclamation of approbation.

1887 *Murray's Mag.* Jan. 97 She held out her hand. 'Good man! that is what I call a friend!' **1913** F. L. Barclay *Broken Halo* ii. 21 Good man! The very thing! **1933** A. G. Macdonell *England, their England* vi. 82 You can keep your mouth shut. Good man. **1955** L. P. Hartley *Perfect Woman* ix. 85 Oh, so you did go?...Good man.

5. b. *good deed*: spec., an act of service for others. (Adapted from the Scout Law 3.)

1928 R. Knox *Footsteps at Lock* v. 42 This...was the spot where the boy scouts were encamped;...fourteen good deeds were registered. **1951** J. G. Fennessy *Sonnet in Bottle* viii. ii. 253 You've done your Good Deed for the Day, visiting the sick. **1965** J. Potter *Death in Office* xiii. 127 He was wearing a frank open expression, like a Boy Scout anxious to do his good deed.

6. b. *good grief!* (see *grief *sb.* 8 a); *good iron!* (chiefly *Austral.*); also as adj. phr.

1895 *Bulletin* (Sydney) 9 Feb. 15/4 Oh, she's good iron, is my little clinah; She's my cobber an' I'm 'er bloke. **1899** *Ibid.* 22 Apr., *Ringer* and *good iron* are both derived from the game of quoits...*Good iron* corresponds to *good ball* at cricket. **1909** J. Masefield *Tragedy of Nan* ii. 41 Good iron! A old chanti-cleer. Balm in Gilead, as the saying is. **1936** M. Franklin *All that Swagger* xi. 100 Good iron! I don't rob little boys. **1965** J. S. Gunn *Terminol. Shearing Indust.*, Pt. ii. s.v. *Quoits*. This term...'good iron' for the shed's best shearer.

9. Also of adults (converging with sense 5 a). Phr. *as good as gold* (further examples).

1932 N. Royde-Smith *Incredible Tale* 100 Here we are as good as gold. **1951** M. McLuhan *Mech. Bride* 118/2 Coke ads concentrate on the 'good girl' image. *Ibid.*, The 'good girl' is the nineteenth-century stock model which has long been merged with the mother image. **1955** L. P. Hartley *Perfect Woman* xxiv. 219 'Good girl,' he commented. **1958** Hayward & Harari tr. *Pasternak's Dr. Zhivago* i. vii. 212 As soon as a few were bumped off by way of example, all the others became as good as gold.

b. *be good!; if you can't be good, be careful!*, as jocular exhortations to good behaviour, esp. at parting.

1908 S. E. White *Riverman* iii. 29 Well, good-bye, boys...Be good! **1908** 'O. Henry' *Gentle Grafter* 119 I'll drop in on you...and we'll have dinner together. Be good. **1911** *MacLean's Mag.* Mar. 96/2 Well, old man, if you can't be good, be careful. **1929** J. B. Priestley *Good Companions* i. v. 185 I'm off then! Be good! **1951** T. Rattigan *Who is Sylvia?* iii. 267 Good night, ladies. Be good.

10. d. Delete '(Now regarded as an Americanism.)' and see note s.v. Time *sb.* 6.

13. *a good press*, a favourable reception in newspapers and journals.

1908 *Times Lit. Suppl.* 99/1 Mr. Leaf...has not had a good press lately. **1915** *Truth* 25 Aug. 301/2, I suppose he knew that was the sort of thing that would ensure a good press for him when he got back to Berlin. **1928** *Observer* 22 Jan. 14/6 The new Measure has not, upon the whole, such a 'good Press' as that which the House of Commons rejected in December. **1928** *Sunday Dispatch* 8 July 22/7 A considerable time has passed since a Scotch boxer received such a good Press in the South.

14. c. *a good question*, one that requires careful consideration before answering; one that is very difficult or awkward to answer.

1918 F. von Hügel *Let.* 11 Dec. (1927) 259 What is the precise meaning of Thekla's insistence upon religion as primarily an is-ness, not an ought-ness? A good question. **1945** C. S. Lewis *That Hideous Strength* ix. 235 'That's a very good question,' said MacPhee. **1960** G. W. Target *Teachers* 64 'I'd write something right across their arse.'...'Wouldn't do any good.' 'What would?' said Bert. 'That's a good question, Bert.'

16. b. *good for*: (*a*) (examples); (*c*) capable of producing; valid for, etc. orig. *U.S.*

1865 Trollope in *Fortn. Rev.* 1 Oct. 419 The porter...had taken his luggage eagerly, knowing that Mr. Belton was always good for sixpence. **1873** J. H. Beadle *Undevel. West* xviii. 337 From thirty to forty tons of ore...good for an average profit of a hundred and fifty dollars per ton. **1901** Merwin & Webster *Calumet 'K'* vi. 104 'How's it coming out?' he asked. 'Do we know how much we're good for?' **1903** *N.Y. Tribune* 20 Sept., A 50-cent combination ticket good for every amusement on the island. *a* **1916** H. James *Middle Years* (1917) i. 12 In possession of a return ticket 'good', as we say, for a longer interval than I could then dream about. **1967** *Listener* 14 Sept. 326/1, I look at a person and I say well, he's good for £5 or £6 for one watch, which cost me 30 shillings.

19. freq. in *a good way* (dial. and U.S. *ways*), a considerable distance. Also *transf.* of time.

1551, 1662 [see Way *sb.*^{1} 8]. **1588, 1594** [Way *sb.*^{1} 23 c]. **1809** M. L. Weems *Life of Marion* xiii. 116 Yes, by jing, does he live a good way up!...a matter of seventy miles. **1862** O. W. Norton *Army Lett.* (1903) 125 That day may be a good way off but still I do not get homesick in the least. **1864** T. L. Nichols *40 Yrs. Amer. Life* I. 250 It's a good way, and you will be out late.

22. make good. i. To succeed; to achieve success; to satisfy expectations; to fulfil a promise or obligation. orig. *U.S.*

1901 Merwin & Webster *Calumet 'K'* ii. 20 It'll play the devil with us if we can't make good. **1908** G. H. Lorimer *J. Spurlock* v. 89, I need work and I need it quick. Give me a show and I'll make good. **1908** *Daily Chron.* 25 Feb. 6/7, I made good, as the Yankees say, with my songs. **1909** H. G. Wells *Tono-Bungay* iii. i. 214 They couldn't for a moment 'make good' if the quarter of what they guarantee was demanded of them. **1910** W. M. Raine *B. O'Connor* 55 All I ask of you is to make good. **1914** G. Atherton *Perch of Devil* i. 58 Ability and talent make good as always. **1927** *Daily Tel.* 7 Mar. 2 The board consider that the company will now make good. **1946** E. B. Thompson *Amer. Daughter* 225 You go on and make good. **1958** *House & Garden* Mar. 111 (Advt.), To record certain extracts from the Schweppshire Roll, wherein are recorded the names of Schweppshire Lads who have Made Good.

j. *Poker.* (See quots.)

1882 *Poker; how to play It* 8 When all who wish to play have gone in, the person putting up the ante...can play like the others who have gone in, by 'making good',—that is putting up in addition to the ante as much more as will make him equal in stake to the rest. **1895** 'Templar' *Poker Man.* 4 If he determines to play on, he 'makes good', as the expression is; that is, he adds to his ante as much as will make his total stake equal to that of each of the other players. **1904** R. F. Foster *Pract. Poker* 232 *Make Good.*—Adding enough to the blind or straddle to make it equal to the ante. **1929** Arnold & Johnston *Poker* 150 Make good. To add sufficient to an ante or bet to meet a raise. **1950** L. H. Dawson *Hoyle's Games Modernized* i. 122 B...does not fancy his chance of improving as worth another yellow, so refuses to 'make good', and retires.

24. Proverbial phr. *too good to be true.*

[**1578** G. Whetstone *Promos & Cassandra* B3, I thought thy talke was too sweete to be true.] **1580** T. Lupton (*title*) Siuquila. Too good to be true. **1594** J. Lyly *Mother Bombie* iv. ii, in *Wks.* (1902) III. 208 It was too good to be true. *a* **1691** J. Flavel *Method of Grace* (1699) vii. 133 They thought it was too good to be true. **1849** Geo. Eliot *Let.* May (1954) I. 282 There is a sort of blasphemy in that proverbial phrase 'too good to be true'. **1908** W. S. Churchill *My African Journey* v. 90 It *is* too good to be true. One can hardly believe that such an attractive spot can be cursed with malignant attributes. **1961** New English Bible *Lk.* xxiv. 41 They were still unconvinced, still wondering, for it seemed too good to be true.

B. *adv.* **a.** Well, properly. Now chiefly *U.S.*

1834 D. Crockett *Narr. Life* xii. 86, I...shot him [*sc.* a bear] the third time, which killed him good. **1838** C. Gilman *Recoll. Southern Matron* 32 We will behave. We will behave good. **1840** *Southern Lit. Messenger* VI. 386/1 She used to tap her with it on the hands, when she behaved bad, or did not say her lesson good. **1859** Bartlett *Dict. Amer.* (ed. 2) s.v., English travellers have repeatedly noticed the adverbial use of this word. 'He cannot read good.' 'It does not shoot good.' **1865** in S. E. Morison *Oxf. Hist. U.S.* (1927) II. 318 Columbia!—pretty much all burned; and burned *good*. **1885** W. L. Alden *Adv. Jimmy Brown* 90 The bee...lit on Tom's hand and stung him good. **1901** S. E. White *Westerners* xv. 113 He'd have trimmed th' little cuss good. **1904** W. N. Harben *Georgians* 119, I stayed all day an' looked about good before I traded. **1946** K. Tennant *Lost Haven* (1947) vi. 89 We're doing pretty good. **1962** J. Ludwig in R. Weaver *Canad. Short Stories* 2nd Ser. (1968) 258 She stunk up that ritzy dress shop good. **1971** *Observer* (Colour Suppl.) 21 Nov. 65/1 If he makes it [*sc.* steel] good, it rolls good. [Steelworks in Cumberland.]

d. *good and*, as an intensive. *colloq.* (orig. *U.S.*).

[**1834** C. A. Davis *Lett. J. Downing* 6 Don't forgit my face, and the Gineral's face; and let the likenesses be good and natural.] **1885** W. L. Alden *Adv. Jimmy Brown* 88 So I got out the needle, and jammed it into his leg with both hands, so that it would go in good and deep. **1889** *Kansas Times & Star* 18 Mar., The shamrock doubtless will be wet often, and the tyrannical lion's tail twisted good and plenty. **1892** Kipling *Barrack-r. Ballads* 43 We met them good and large. **1896** Ade *Artie* xvi. 146, I was good and sore. **1901** Merwin & Webster *Calumet 'K'* i. 14 We got the letter the same day the red-headed man came here. His hair was good and red. **1904** J. London *Let.* 17 Nov. (1966) 165 The lawyers...waded into me good and hard for the cash. **1923** R. D. Paine *Comrades of Rolling Ocean* iv. 57 I'll roll out there when I get good and ready. **1926** *Publishers' Weekly* 15 May 1593 That made me good and provoked. **1928** *Daily Express* 2 Feb. 9/2 Colonel Ernest Cassell Maxwell...said 'She went through it...good and proper, and I am sorry for her.' **1931** Galsworthy *Maid in Waiting* iii. 18 Castro got it good and strong this morning. **1954** *Encounter* Nov. 16/2 The American Machiavelli is tethered good and fast to the public base of Communism. **1969** B. Knox *Tallyman* ii. 22 [It] can wait until we're good and ready.

C. *sb.* **5. b.** Also, *all to the good*, generally advantageous.

1943 E. Glasgow *Let.* 25 June (1958) 323 This is all to the good, I think. **1955** D. Murray *Species Revalued* p. xi, It is a kind of revolt against the stark materialism of the present day; this is all to the good. **1962** P. Gregory *Like Tigress at Bay* v. 49 The fact that you're wealthy and high up in society may attract certain people, and that's all to the good.

g. Also of things, *a bit of no good*, quite a lot of harm.

1958 E. A. Robertson *Justice of Heart* iii. 34 They've frittered the money away, and done themselves a bit of no-good, all round. **1958** L. A. G. Strong *Treason in Egg* x. 183 The pair of them were up to quite a bit of no good.

8. (Later examples.) Now used occas. in *sing.*

1936 *Q. Jrnl. Econ.* May 436 All that follows will hold true of any storable good, like cotton, wool, rubber, tobacco, wheat, coffee, sugar, oil, copper, or tin; but the theory will be expounded in terms of only one of these, namely cotton, because it is easier to deal with a particular case. **1958** *Times Lit. Suppl.* 15 Aug. p. xxiii/1 As a steady, cheap, business-making consumer good,...the book is out. **1964** Gould & Kolb *Dict. Social Sci.* 533/1 Complexities arise from many sources. For example, the existence of stocks of goods which might have to be reduced in some amount before additional resources were guided to the favoured good were ignored. **1968** *Times* 2 Jan. 18/5 Knowledge of the legal definition is important, for, under the Sale of Goods Act, 1893, a special set of rules applies to agreements involving goods, which, in legal language, are 'movable and tangible pieces of property'.

b. (Later examples.)

1895 [see Piece *sb.* 3 d]. **1955** Lord Chancellor in *Hansard, Lords* 8 Dec. 1241, I should ask the noble and learned Earl for his considered construction of the phrase 'a nice piece of goods'.

c. *goods box, guard, lift, set, train* (earlier example), *yard*.

1850 L. V. Loomis *Jrnl. Birmingham Emigr. Co.* (1928) 131 As we...made our way to the center [of Sacramento], we would see an old pork barrell, or an old goods box. **1880** 'Mark Twain' *Tramp Abr.* xix. 166 A little bit of a goods-box of a barn. **1890** *Chambers's Jrnl.* 21 June 385/1 Passenger guards are men of experience, and many of them have had to work as brakesmen and goods-guards many years before they are appointed to a passenger train. **1902** *Daily Chron.* 5 Feb. 3/4 Every shunter, and...every goods guard. **1959** M. Taylor *Railways as Career* iii. 47 The goods guard spends almost all his working time alone in his brake-van. **1909** *Westm. Gaz.* 1 Apr. 8/3 He got into the goods lift with some fifteen other men. **1927** W. E. Collinson *Contemp. Eng.* 8 Goods sets i.e. a set of goods trucks. **1861** J. A. Symonds *Let.* 26 Aug. (1967) I. 308 If sent by passenger train they wd reach me by Saturday...If sent by goods train they wd take longer. **1891** *Leisure Hour* 194/1 The goods-yard porter is not the least important of the railway workmen. **1900** *Westm. Gaz.* 20 Jan. 8/3 The dreary goods-yard which does duty for the entraining station of war-bound troops. **1959** M. Taylor *Railways as Career* iv. 64 In bigger goods yards there will be several trains being made up.

9. b. (See quot.)

1953 *Word for Word* (Whitbread & Co.) 21/1 *Goods*, the name often used by the brewer to describe the crushed malt grains in the mash tun.

D. *Comb.* **1. a.** *good-class.*

1901 *Daily Chron.* 2 Sept. 8/2 Good-class rudd have also been secured in this river. **1909** *Westm. Gaz.* 13 Jan. 12/2 A club...which has a number of good-class players. **1925** E. F. Norton *Fight for Everest, 1924* 39 He must be a good-class man of some intelligence. **1960** *Farmer & Stockbreeder* 29 Mar. 81/1 (*caption*) A typical good-class female of the breed.

b. *good-intentioned, -mannered.*

Some of these combinations have parallel forms with *well*: e.g. *well-conditioned, -intentioned, -mannered*.

1905 H. G. Wells *Kipps* ii. i. 165 That sinister passion for pedagogy to which the Good-Intentioned are so fatally liable. **1907** *Daily Chron.* 6 Mar. 3/2 It is all very nice and sentimental, and good intentioned. **1906** *Macm. Mag.* July 695 The librarian, a functionary whom he desired good-looking, good-natured, good-mannered, and ready of speech.

c. *good-going* (suggested by *easy-going*), *-living, -selling.*

1927 J. Adams *Errors in School* iii. 69 If we find that three metals that we have tested all expand when heated, we jump to the easy-going conclusion that all metals expand when heated—which is a good-going error. **1903** *Westm. Gaz.* 2 Nov. 9/2 The city had a population of ten thousand, all good-living people. **1909** *Ibid.* 1 Mar. 1/3 A most respectable and good-living man. **1908** *Daily Chron.* 10 Jan. 3/3 Ordinarily 'Edwin Drood' is one of the least 'good-selling' novels of Charles Dickens.

3. good-bad *a.*, of someone or something good but of an inferior class; also in appositive use = good and bad; **good-enough** *a.* (earlier Sc. and U.S. examples); **good fairy** = *fairy godmother*.

1899 *Chambers's Jrnl.* 23 Sept. 674/1 Smugglers in the good-bad old times pursued what they euphemistically called the 'fair trade'. **1921** E. Sapir *Lang.* 99 It is difficult to get into the frame of mind that recognizes that any particular thing may be both good and bad...Still more difficult to realize that the good-bad or black-white categories may not apply at all. **1933** A. Thirkell *High Rising* ii. 41 'Good bad books?' 'Yes. Not very good books,...but good of a second-rate kind.' **1940** G. Orwell *Inside Whale* 134 Good bad books like *Raffles*. **1949** M. Mead *Male & Female* xvii. 346 A frequent theme of modern movies is the 'good-bad' girl. **1966** J. Laver *Victoriana* i. 14 Many of the products of the age must, one fears, be consigned to the category of what has been half humorously called 'Good Bad Art'. **1570** R. Bannatyne *Mem.* 33 This was a guid aneuch obligatioune that the castle shuld be thair friend. **1647** *Stirling Chart. & Corr.* 485 The daik [= dyke] is in good aniogh order. **1856** *Congress. Globe* 53 You will have victims who can answer as 'good enough Morgans' at least until after the election. **1807** M. Wilmot *Let.* 27 Dec. in *Russ. Jrnls.* (1934) 313, I lit upon her sentiments on religion...being pearls & rubys dropping from the Good Fairy's lips. **1855** Mrs. Gaskell *Let.* ?5 June (1966) 866 What will our heroine do in such a dilemma? When lo and behold the good fairy steps in. **1876** L. Troubridge *Jrnl.* 6 May in *Life amongst Troubridges* (1966) xi. 143 If only some good fairy would present us with a ten pound note each. **1920** Galsworthy *In Chancery* ii. ii. 140 Forgiving and forgetting, and becoming the good fairy of her future. **1966** *Guardian* 16 Dec. 4/7 The familiar figure of Mrs Anne Kerr, Labour MP...dressed all in white like the Good Fairy.

goo:d-loo·ker. Chiefly *U.S.* [Cf. Good-looking *a.*, Good looks.] One who has good looks.

1894 *Harper's Mag.* Mar. 498 She's a good-looker...although they say she's gone off a little lately. **1902** O. Wister *Virginian* ii. 26 'She's a good-looker.' Hm! Yes, the kind of good looks I'd sooner see in another man's wife than mine. **1920** *Blackw. Mag.* June 786/2 He was a 'goodlooker', tall, blond, gentle-spoken. **1928** *Daily Tel.* 29 May 14/2 Felstead [a horse]...is quite a good looker and is well bred. **1929** W. P. Ridge *Affect. Regards* 121 There's a lady coming along...A good looker, if ever there was one. **1966** 'J. Hackston' *Father clears Out* 189 She [*sc.* a cow] was a...good-looker, too.

good luck. Add: As a salutation or wish. Also further *attrib.* examples.

1805 E. Cavanagh *Let.* 4 Oct. in M. Wilmot *Russ. Jrnls.* (1934) ii. 187 'Well to be sure,' sais I. 'Russia! & good luck to you, you are a comical place!' **1829** G. Griffin *Collegians* II. xvii. 33 Good *look* to them both, wherever they are. ?**1861** Mrs. Gaskell *Let.* 28 Feb. (1966) 643 Is he really 'after' Sophy? If so, good luck for her. **1898** J. D. Brayshaw *Slum Silhouettes* 1 'E ain't come in fer a bloomin' forchin 'ave 'e? Jolly good luck to 'im if 'e 'as! **1942** M. McCarthy *Company She Keeps* (1943) v. 130 He was a mascot, a good-luck piece. **1958** 'J. Bell' *Seeing Eye* x. 115 'He has kept the Bert Lewis drawing...' 'Good luck to him!' **1958** *Photoplay* Oct. 54/2 We have received some good-luck telegrams and letters. **1963** P. Willmott *Evolution of Community* ix. 99 People seem to be glad if someone else gets something. They don't grudge it. They say, 'Good luck to them.'

good-naturedly, *adv.* (Earlier example.)

1765 C. Brietzcke *Diary* 18 Feb. in *N. & Q.* (1963) 109/2 He good-naturedly promised me he would.

good neighbour. [See Good-neighbourhood.] In U.S. politics, a neighbouring country, esp. in Latin America, with which the U.S. has good relations. Also (with hyphen) *attrib.* Also in extended use.

1928 H. C. Hoover *Mem.* (1952) II. 213 We have a desire to maintain not only the cordial relations of governments with each other but the relations of good neighbors. **1933** F. D. Roosevelt (Inaugural Address), In the field of world policy I would dedicate this Nation to the policy of the good neighbor. **1937** *Ann. Reg. 1936* 293 The keynote of the conference was struck by Mr. Roosevelt as the policy of 'the Good Neighbour', and the President could assure the Latin-American countries that the United States...now interpreted the role of 'Good Neighbour' as one forbidding forcible intervention in the affairs of a neighbouring State. *Ibid.*, The President visited Quebec...The visit was intended to stress the 'Good Neighbour' policy. **1944** *Ann. Reg. 1943* 346 The 'good neighbour' policy of the U.S.A. to the Latin American States. **1963** *Listener* 28 Mar. 542/2 General de Gaulle may calculate that the Soviet Union may want to become a good neighbour in Europe again. **1967** *Ibid.* 1 June 721/1 The shelving of a good neighbour policy in favour of the Dulles mania.

goodness. 5. (Additional examples.)

1806 M. Wilmot *Let.* 13 Dec. in *Russ. Jrnls.* (1934) 274, I wish to Goodness all my Books were this moment in the Music Room. **1875** Trollope *Way we live Now* II. lxi. 68 He 'wished to goodness' that he had dined at his club.

good-o, good-oh: see *good *a.* 4 c.

goods (gudz), *sb. pl.* Chiefly or orig. *U.S. colloq.* [See Good *sb.* 8.]

1. Phr. **a.** *to deliver the goods*: to supply the objects contracted for; hence, to perform the contract undertaken; to do what one has undertaken to do; to supply what has been promised or is expected; hence, to come up to requirements or expectations.

[**1781** J. Burgoyne *Ld. Manor* iii. iii, *Sir John.* Hussy! how came you by all that money? *Peggy.* Perfectly honestly—I sold my mistress and myself for it—it is not necessary to deliver the goods, for his honour is provided with a mistress.] **1879** *Congress. Rec.* 4 Apr. 236/1 There are men in the North who walk around...saying; 'See me,...I will take you to victory.' They cannot deliver the goods. **1901** Merwin & Webster *Calumet 'K'* xi. 198, I told him that...when we paid blackmail it would be to some fellow who'd deliver the goods. **1904** F. Lynde *Grafters* viii. 120 Of the three justices, one of them was elected on our ticket; another is a personal friend of Judge MacFarlane. The goods will be delivered. **1909** H. G. Wells *War in Air* iv. § 5 As yet he was only in the beginning of the adventure. He had still to deliver the goods and draw the cash. **1911** R. W. Chambers *Common Law* ii. 45 'She certainly is a looker,' nodded Annan. 'She can deliver the cultivated goods, too.' **1919** *Economist* 12 July 44/1 To win a few votes at an election a Cabinet Minister declares it to be the policy of the Government to nationalise the railways. He is duly elected, and those who are in favour of nationalisation ask him and his colleagues to deliver the goods. **1922** Dk. Devonshire in *Hansard, Lords* 4 Dec. 233, I am convinced that the Irish Government intend...to deliver the goods...in the true spirit of the Act. **1936** 'F. Beeding' *Eight Crooked Trenches* vi. 93 The goods, as they said in England, would this time be delivered. **1946** W. S. Maugham *Then & Now* xxxvi. 207 She knew quite well that the chain was the price he was paying her to arrange things for him, and when she didn't deliver the goods surely the least she could do would have been to return the purchase price. **1968** *Listener* 18 July 88/1 This body has sometimes offered help to coloured workers when they were on strike, but it has never delivered the goods.

b. *the goods*: what is supplied or provided; what is expected or required (for a purpose expressed or implied); the real thing; the genuine article.

1812 *Norfolk* (Va.) *Herald* 29 May 314 Federalists call the troops now raising 'a standing army'. They are mistaken in the goods. **1880** A. A. Hayes *New Colorado* (1881) vii. 103 When the mariner heard an expert, who was chipping away at the wall with a little hammer, remark, 'That's good goods,' this purist stopped both ears. **1904** *Cosmopolitan* May 122 'I'll agree to make it 25 [dollars] at the end of 60 days if you are the goods,' said the editor. *a* **1910** 'O. Henry' *Rolling Stones* (1916) 200 Take it from me—he's got the goods. **1912** C. Mathewson *Pitching in Pinch* ii. 33 Now O'Toole is all right if he has the pitching goods. **1915** Wodehouse *Psmith Journalist* xv. 109 You are, if I may say so, the goods. You are, beyond a doubt, supremely the stuff. **1918** E. M. Roberts *Flying Fighter* 35 Some of the road pickets would want to see our identification papers as dispatch riders, and being unable to produce the goods we were often turned back. **1919** T. K. Holmes *Man fr. Tall Timber* xvi. 196 Believe me! this Gypsy is all the goods and then some. **1931** *Times Lit. Suppl.* 28 May 426/2 But to the 'General'...President Bonilla and the Honduras were 'the goods' in 1911. **1956** A. Wilson *Anglo-Saxon Att.* ii. iii. 369 He *was* the most awful old fraud himself, you know. Oh, not as an historian, you always said he was the goods.

2. *the goods*: the stolen articles found in the possession of a thief; unmistakable evidence or proof positive of guilt; chiefly in phr., e.g. *to catch with the goods*.

1900 Ade *More Fables* 94, I insist that he is an American traveling Incog. I suspect that I have Caught him with the Goods. **1911** *N.Y. Even. Post* 15 June (Th.), 'We've got you...now, and you're going to yield the stolen goods.' The goods in question were the office of Commissioner of Jurors, [etc.]. **1919** *Detective Story Mag.* Nov. 50 Detective Craddock had informed Thubway Tham that, sooner or later, he was going to 'catch him with the goods'. **1923** R. D. Paine *Comrades of Rolling Ocean* xiv. 245 You have caught me with the goods, Wyman. It was my way of getting a slant on you.

b. *to have* (or *have got*) *the goods on*: to have the advantage of or superiority over; to have knowledge or information giving one a hold over (another).

1913 A. Bennett *Regent* i. v. 134 You got the goods on her. And she deserved it. **1920** Wodehouse *Coming of Bill* ii. xiv. 239 I'd been in a ring-seat and had the goods on him same as if I'd taken a snap-shot. **1924** W. M. Raine *Troubled Waters* xxii. 233 They had the goods on us. We were going to hang—every one of us. **1928** *Observer* 15 July 18/2 'Well, the Old Country sure has the goods on everyone else,' said one of them [*sc.* Canadian teachers]. **1933** D. L. Sayers *Murder must Advertise* v. 90 We got the goods on that couple you helped us to arrest the other night. **1952** M. McCarthy *Groves of Academe* (1953) xiii. 263 He had a sudden inkling that they would have liked to get the goods on Mulcahy. **1967** *Boston Sunday Herald* 7 May (Show Guide) 17/2 The senator stubbed his toe just once, and Overbury has the goods on him. When the time comes to peddle them...knocking off Burden Day is easy.

3. *ellipt.* = *goods train* (Good *sb.* 8 c).

1855 *Punch* XXIX. 163/2 Each player is furnished with a small railway train, such as an 'express', a 'stopping', a 'goods'. **1868** 'Journeyman Engineer' *Great Unwashed* ii. 198 After the 'through goods', perhaps, comes the fiery wi-sh-sh of the night mail north. **1939** *Punch* 20 Sept. 310/2, I said couldn't they go by goods. **1959** A. Wood *Engine-driver & Signalman* iv. 35 (*caption*) Express goods braked by engine and brake-van...Branch goods.

good thing. Add: **a.** (Further examples.) **c.** Also, luxuries in general. **d.** A course of action, etc., that is commendable, desirable, etc. (cf. Good *a.* 4).

a. 1888 H. James *Reverberator* ii. 35 He had a genius for happy speculation, the quick, unerring instinct of a 'good thing'. *a* **1889** J. Albery *Dram. Works* (1939) I. 191 I'll put her on to a good thing. **1889** E. Sampson *Tales of Fancy* 31, I was so fast that I had been kept at these 'stables' until some 'good thing' should offer. **1898** J. D. Brayshaw *Slum Silhouettes* 100 As luck would have it, I managed to put the old man on to a good thing. **1913** *Field* 4 Jan. 26/1 Master at Arms is not a big horse, and with 12 st 7 lb in the saddle he hardly looked like a good thing for the Sunbury Steeplechase.

c. 1821 P. Egan *Life in London* i. iv. 58 Tom was *born* to be a happy fellow, if the enjoyment of the 'good things' of this world could have made him so.

d. 1897 *Sears, Roebuck Catal.* 198/2 This Shoe we have called 'A Good Thing', because the name signifies just what we believe the shoe is. **1930** Sellar & Yeatman *1066 & All That* i. 3 The Roman Conquest was...a *Good Thing*, since the Britons were only natives at that time. *Ibid.* xxi. 29 Simon de Montfort, though only a Frenchman, was...a Good Thing. *Ibid.* xlvii. 90 Fox said in the House of Commons that the French Revolution was a Good Thing. **1942** E. Waugh *Put out More Flags* iii. 213 He...had a liking for books; he thought them a Good Thing. **1960** *Observer* 17 Apr. 21/6 The general assumption that the French influence on England has been 'a good thing'. **1960** *Guardian* 26 Aug. 8/6 We all believe that group counselling...[is] a 'good thing'.

2. *too much of a good thing*: an act, behaviour, etc., spoilt by its excess.

1600 Shaks. *As You like It* iv. i. 123 Why then, can one desire too much of a good thing. **1809** [see Too *adv.* 5 b]. **1852** Mrs. Gaskell *Let.* 4 Sept. (1966) 197 We went at ½ p. 9, & did not get out till ¼ to 4, which was too much of a good thing. **1895** S. Crane *Red Badge of Courage* vi. 66 This is too much of a good thing! Why can't somebody send us supports? **1932** A. Huxley in D. H. Lawrence *Lett.* p. xvi, The great work of art and the monument more perennial than brass are, in their very perfection and everlastingness, inhuman—too much of a good thing. **1954** J. R. R. Tolkien *Fellowship of Ring* i. i. 29 Some...shook their heads and thought this was too much of a good thing.

good-time, *a.* [Cf. Good *a.* 3 c, 10 d.] Of a person: recklessly pursuing pleasure; esp. in *a good-time girl*. Also **good-timer**, one of this character.

1928 *Publishers' Weekly* 9 June 2393 Gerry Harris was 'a good time girl', who sought men only as playmates. **1928** A. Huxley *Point Counter Point* xxvi. 444 The boozers,...the business men, the Good-Timers. **1941** O. Sitwell in *Open Door* 82 The plane was crowded. A few 'good-timers' were still there...loyal to their favourite beat: Paris for Whitsun. **1943** *John Bull* 20 Nov. 6/2 Contribution to the controversy about war-time morals came from Capt. Cunningham Reid, who told...the pitiful story of Mary, the 'good-time girl'. **1948** 'J. Tey' *Franchise Affair* viii. 82 Bert deserved better out of life than a good-time wife. **1957** Wodehouse *Over Seventy* xvii. 160 Once a combination of Santa Claus and Good-Time Charlie, Hollywood has become a Scrooge. **1958** *Times Lit. Suppl.* 18 July 414/1 This is...a most eloquent protest against the present-day values of...casual good-timers. **1959** *Listener* 5 Mar. 428/2 The murder of a local good-time girl. **1970** *Globe & Mail* (Toronto) 25 Sept. B2/4 So much for the view, strongly fostered by the Minister of Finance, that convention-goers are good-time Charlies.

goodwill. Add: **2.** *attrib.* (Later examples.)

1906 *Westm. Gaz.* 20 Feb. 3/2 A large number of 'goodwill' workers are...enlisted. **1935** *Economist* 18 May 1125/2 A lesser American naval squadron is making a 'goodwill tour' in Japanese waters. **1936** *Daily Mirror* 22 Mar. 9/1 Welsh children will broadcast their 'Goodwill Day' peace message to the World on May 18th. **1947** *Ann. Reg. 1946* 222 In July the British Labour Party sent a 'goodwill' mission to Moscow. **1961** *Listener* 21 Dec. 1081/2 President Kennedy arrives in Caracas...at start of a 'goodwill' tour of Latin America.

3. b. (Later example.)

1816 Jane Austen *Emma* II. xiv. 265 Augusta, I believe, with her own good will, would never stir beyond the park paling.

Goodwood (gu·dwud). A race-meeting held near *Goodwood* Park, Sussex; freq. *attrib.*, as *Goodwood cup, races, week.*

1839 W. H. Mason *Goodwood* 182 (*heading*) Goodwood races. Were established in 1802. **1840** J. C. Whyte *Hist. Brit. Turf* II. xi. 498 (*heading*) The Goodwood cup.—Betting: Even on Harkaway, 2 to 1 against Deception. **1875** Trollope *Way we live Now* I. xxxv. 219 Mr. Melmotte had bought Pickering Park...There were rumours that it was to be made ready for the Goodwood week. **1896** J. Kent *Rec. & Reminisc. Goodwood* x. 96 Improvements have been made to meet the requirements of the most enjoyable race-meeting in England, until it has attained the greatest pre-eminence, and is justly known as 'glorious Goodwood'. **1922** E. Wallace *Flying Fifty-Five* x. 59 Where are you going to stay during the Goodwood week? **1951** B. W. R. Curling *Brit. Racecourses* 84 The present Duke of Richmond is more actively interested in motor-racing than in horse-racing, but Goodwood week does not suffer on that account. **1956** *Radio Times* 27 July 38/2 (*heading*) Goodwood. *Ibid.*, The Goodwood Cup. For three-year-olds and upwards over two miles and five furlongs.

goody, *sb.*^{1} **2.** (Later examples.)

1893 W. K. Post *Harvard Stories* 79 There are many individuals that make up the university population of Cambridge—unofficial members. There are the...goodies. **1902** J. Corbin *American at Oxford* 12 The scout is in effect a porter, 'goody', and eating-club waiter rolled into one.

goody, *sb.*^{3} (Further examples.)

1853 Mrs. Gaskell *Cranford* v. 75 The 'mother dear' probably answered her boy in the form of cakes and 'goody', for there were none of her letters among this set. **1890** C. M. Yonge *More Bywords* 137 People thought they had come through Lady Bountiful's goody-box. **1931** E. Wilson *Axel's Castle* vii. 253 She is an American woman of the old sort, she who cares for the handmade goodies and who scorns the factory-made foods.

goody, *a.* and *sb.*^{4} Add: **A.** *adj.* **2.** (Additional example.) *goody-two-shoes* (see Two IV. 2).

1871 *Monthly Packet* Christmas No. 103 Two girls who had stopped...to see if there were anything new among the...goody-books.

B. *sb.*^{4} (Further examples.) Now usu. in *colloq.* use opp. *baddy.

1873 C. M. Yonge *Pillars of House* IV. xxxix. 147 She is the most thorough Goody I ever came across. **1901** *Contemp. Rev.* Mar. 436 This goody ought to moderate the rancour of his tongue. **1951, 1958** [see *baddy].

goody (gu·di), *int.* Chiefly *U.S.* Also **goody goody.** [f. Good *a.* + -y^6. Cf. *lordy *int.*] A childish exclamation denoting delight, satisfaction, or surprise.

1796 'A. Barton' *Disappointment* (ed. 2) ii. iii, Oh! goodee, goodee, oh! we shall see presently. **1853** B. F. Taylor *Jan. & June* (1871) 125 Port's tongue [being] busy the while with...'may I go?' and 'goody! goody!' to a provisional affirmative. **1886** Baumann *Londinismen* 67/1 *My goody*, goodness gracious! **1890** *Harper's Mag.* Mar. 608/1 You're coming home with us?...Yes? Oh, goody! You'll come? **1898** P. L. Ford *Hon. Peter Stirling* 244 'That makes five,' said Peter. 'Oh, goody!' said Leonore, 'I mean,' she said, correcting herself, 'that that is very kind of you.' **1949** *Landfall* III. i. 57 His mother was out, goody. **1953** H. Miller *Plexus* (1963) x. 352, I see Halvah and Baklava too. Goody goody! **1959** I. & P. Opie *Lore & Lang. Schoolch.* ix. 161 Cries of jubilation include: Wow! Whacko! Goody gumdrops! Lovely grub! and By gog jolly custard! **1960** J. Grant *Come Again, Nurse* x. 54 'Just in time,' said the Registrar jovially. 'Goody goody gum drops.' He walked over to the coffee pot and helped himself. **1967** N. Freeling *Strike Out* 16 Buttered toast, and cherry cake, as well as Marmite. Goody, goody gumdrops.

goodyera (gudyī^ə^·rā, gu·dyərā). [mod.L. (R. Brown in W. Aiton *Hortus Kewensis* (ed. 2, 1813) V. 197), f. the name of John Goodyer (1592–1664), an English botanist.] A plant or flower of the genus of small terrestrial orchids so named.

1817 *Bot. Cabinet* I. (*index*), Goodyera pubescens—Downy Goodyera. **1825** *Curtis's Bot. Mag.* LII. 2540 (*heading*) Smaller Pubescent Goodyera. **1868** B. S. Williams *Orchid-Grower's Man.* (ed. 3) 142 The dark foliage of the *Goodyeras*. **1921** *Glasgow Herald* 16 July 4 That interesting little orchid called goodyera. **1963** W. Blunt *Of Flowers & Village* 104 The seeds of the creeping goodyera, a little British orchid, are so light that I could send you 14 million of them by post for threepence.

goody-good, *a.* Add: (Later examples.) Also as *sb.*

1904 *Daily Chron.* 12 Apr. 4/4 One is the bad; the other is the goody-good. *Ibid.*, They must be on their guard...against cultivating the goody-good. **1922** Joyce *Ulysses* 418 Baddybad Stephen lead astray goodygood Malachi. **1929** D. H. Lawrence *Pansies* 128 So our goody-good men betray us.

goody-goody, *a.* (and *sb.*). (Further examples.)

1873 *Punch* 4 Jan. 4/2 There are goody goody books; there are also baddy baddy books. *Ibid.* 11 Jan. 17/2 The three other Goody-goodies were *Messieurs Cabanet*, [etc.]. **1922** C. E. Montague *Disenchantment* viii. 122 A man who did not care to use so sound a means to his ends was thought to be a goody-goody ass. **1949** R. Graves *Seven Days in New Crete* 12 The atmosphere...would be described as goody-goody, a word that conveys a reproach of complacency and indifference to the sufferings of the rest of the world. **1959** I. & P. Opie *Lore & Lang. Schoolch.* x. 191 One who makes up to a teacher...is a 'goody-goody', a 'namby-pamby'.

gooey, *a. slang* (orig. *U.S.*). [f. *goo + -ey = -y^1.] Of a viscid or sticky nature; *fig.*, sentimental, mawkish. Also as *sb.*, sticky food; *fig.*, a man of weak character.

1906 E. Dyson *Fact'ry 'Ands* xiii. 165 Half ther 'ouses in town 'ud give er gooey iv that sort er billet rather 'n' take on er lad with er tizzy's worth iv grit in him. **1919** *Quill* Feb. 16 We may briefly state that there are three principal schools of dunking; viz., in hot liquids, in cold liquids, and in gooey substances. **1923** H. L. Foster *Beachcomber in Orient* i. 9 She...extracted a gooey substance from a tube of tinfoil and smeared it with a stick upon the bowls. **1928** *Papers Mich. Acad. Sci. & Arts* X. 296 *Gooey*, hash. **1930** Wodehouse *Very Good, Jeeves* v. 121 Plenty of chocolates with that goo-ey, slithery stuff in the middle. **1935** N. Mitchison *We have been Warned* ii. 208 You'd look all goo-ey and jumpy about your Dione and give the show away. **1944** L. A. G. Strong *All fall Down* 102 You ought to see those letters. All intense and gooey and squarmy. **1948** R. Knox *Mass in Slow Motion* i. 3 What you mean by a dance is the wireless in the hall playing revolting stuff and you lounging round in pairs and feeling all gooey. **1959** *Manch. Guardian* 24 June 6/2, I never see you without something gooey in your hands or your mouths. **1971** *News of World* 21 Nov. 4/4 'Mmmmm, lovely,' she sighed as she tucked into calorie-laden hors d'oeuvres, fattening spaghetti, and an enormous plateful of gooey chocolate gateau.

goof (gūf), *sb. slang.* [App. a use of dial. *goof*, Goff^2.] **1.** A silly, stupid, or 'daft' person.

1916 *Sat. Even. Post* 19 Feb. 37/2 It ain't the same show, you goof!...They change the bill every day. **1918** L. E. Ruggles *Navy Explained* 113 To cope with the situation some goof ashore made a salt water soap. **1925** C. R. Cooper *Lions 'n Tigers* iv. 99 The most idiotic, dunce-like goof that ever struggled about on four legs. **1930** 'Hay' & King-Hall *Middle Watch* xviii, Have you stopped to think what is happening to that poor old goof in the day-cabin, right now?

b. A mistake, esp. in an entertainment; a gaffe. Also *goof-up*.

1955 *Springfield* (Mass.) *Union* 30 May 21 He was convinced there was no 'goof' by the government at all in the polio vaccine distribution program. **1959** *Britannica Bk. of Year* 490/1 *Goof*, a mistake made in a show. **1956** *TV Guide* 13–19 Oct. 4 Randolph Churchill..has told friends his embarrassment is assuaged by past goof-ups among English men of letters. **1960** *Guardian* 15 July 19/5 His teleprompter promptly went on the blink. [illegible] **1960** M. Phillips in *Analog Science Fact & Fiction* Nov. 112/2 Every one of them came up to me to prove that the goof-ups in his particular department weren't his fault. **1970** *Daily Tel.* 11 Feb. 17/5, I believe they have made a goof.

2. *attrib.* and *Comb.* **goof ball**, (*a*) (a tablet of) any of various drugs, *spec.* marijuana; a barbiturate tablet or drug; (*b*) = sense 1 a above; **goof pill** = *goof ball (*a*).

1938 *Amer. Speech* XIII. 185/1 *Goof-ball*, marijuana. [illegible] **1957** J. Kerouac *On Road* (1958) 148 She took tea, goofballs, benny. **1959** *Encounter* XII. ii. 33 San Francisco, which 'is already a haven for wandering psychotics and goofballs of every description'. **1962** K. Orvis *Damned & Destroyed* v. 54 You're the granddaddy of all goofball windbags. **1963** *New Scientist* 28 Nov. 534/1 The barbiturates, are known as 'goof balls'. [illegible] **1968** H. L. Mencken *Amer. Lang. Suppl.* II. 682 A sodium pentobarbital capsule is a goof-pill. **1960** *Guardian* 21 Oct. 9/6 'Goof' pills was the term he used to refer to barbiturates.

goof (gūf), *v. slang.* [f. the sb.] **1.** *intr.* **a.** To dawdle, to spend time idly or foolishly; to 'skive'; to gawp; to let one's attention wander. Sometimes const. *off*.

1932 J. T. Farrell *Studs Lonigan* (1936) I. i. 5 To get even with..the Hunkie janitor, because he always ran them off the grass when they goofed on their way home from school. **1940** *Times* 23 July 2/4 Among other points of advice were:—Go quickly to your shelter or refuge room; suppress your curiosity and don't 'goof'. **1952** B. Ulanov *Hist. Jazz in Amer.* 351 *Goof* or *goof off*, to wander in attention, to fail to discharge one's responsibility. **1956** F. Castle *Violent Hours* (1966) ii. 21 Have you been goofing off? **1958** E. Dundy *Dud Avocado* I. vii. 117, I shouldn't have goofed off like that. Should have stayed on with you. **1959** J. Winton *We joined Navy* 64 'It's safer to take them [*sc.* caps] off and hold them in your hands while you're goofing.' 'Goofing?' 'Watching the flying. Anyone who watches the flying is known as a goofer. Where you're standing now is a goofing position.' **1962** J. Baldwin *Another Country* I. ii. 113, I used to like to just..go to the movies by myself or just read or just goof. **1963** C. D. Simak *They walked like Men* vii. 41 I'd work over the weekend, getting out the columns, to make up for goofing off. **1968** *New Yorker* 28 Dec. 25 If you ever feel like goofing off sometime, I'll be glad to keep the old ball game going and fill in for you here.

b. To blunder, to make a mistake. Occas. const. *off*.

1941 *Amer. Speech* XVI. 166/1 *Goofs off*, makes a mistake. **1954** *Time* 8 Nov. 42 *Goof*, make a mistake. **1958** J. D. MacDonald *Executioners* (1959) iii. 37, I goofed and I've got no apologies. **1966** *Word Study* Feb. 5/2 If a student goofs and says the tone is melancholy or sadness. **1970** W. Smith *Gold Mine* xiii. 38 'What the hell you goofing off—' The words choked off in his throat. **1971** *Daily Tel.* 2 Sept. 4/6 The Census Bureau has admitted that it 'goofed' when it wrote it off as a ghost town.

c. Const. *up*. To take a drug or drugs. Cf. *goof *sb.* 2.

1962 K. Orvis *Damned & Destroyed* xxi. 155, I hope they goof-up in the dispensary.

2. *trans.* To take a stupefying dose of. Also *fig.* Freq. in pa. pple., const. *up*.

1944 *Amer. Speech* XIX. 104 There is some allusion in sailors' language to the use of drugs...Gassed up..and goofed up are cognate and self-explanatory. **1957** R. A. Heinlein *Door into Summer* (1960) ii. 25, I was as goofed up about Belle as is possible for a man to be. **1970** *Guardian* 7 Aug. 11/8 Thousands of youths openly.. 'goofed' amphetamines. **1970** *Lebende Sprachen* XV. 104/1 Goofed up. Durch Barbiturate berauscht.

3. *trans.* To bungle, mess *up* (something or someone).

1960 M. Phillips in *Analog Science Fact & Fiction* Nov. 22/2 What could be anybody's purpose in goofing up a bunch of calculators the way they had? **1969** *Life* 4 Apr. 65 Now, it's hard to goof up pictures.

Hence **goo·fer**[1], one who goofs (in various senses).

1925 *College Humor* Aug. 97/2 They had me fixed up to loop with that terrible egg Buster Slaton who is a nephew of the two old Slaton goofers. **1941** *Daily Mail* 31 July, *Goofers* is the term applied to people who ignore orders to seek shelter during raids, but stand out in the streets gaping up at the bombers. **1959** [see sense 1 above]. **1961** E. Brown *Wings on my Sleeve* 98 The shortness of our take-off run astonished all the goofers on the island.

goofer[2], **goopher** (gū·fəɪ). [Of Afr. origin.] A witch doctor; a curse, spell, or conjuration; *goofer dust*: a powder used in conjuration.

1887 C. W. Chesnutt in *Atlantic Monthly* Aug. 257/1 Aun' Peggy say dat bein' ez Henry didn' know 'bout de goopher..she reckon she mought be able fer ter take de goopher off'n him. **1926** N. N. Puckett *Folk Beliefs Southern Negro* iii. 215 Some hoodoos burn a kind of powder called 'goopher dust', which represents the person being hoodooed, who is perhaps miles away at the time. This causes the conjured individual to lose his personality and to become sick or insane. **1932** B. De Voto *Mark Twain's Amer.* iii. 69 That the residue has not vanished is attested by the appearance of goofers [illegible]. **1935** Z. N. Hurston *Mules & Men* II. v. 281 [illegible] goofer dust as it is often called. **1941** *Sat. Even. Post* 10 Sept. 56/3 No goofer dust..could kill an enemy. **1967** W. C. Handy *Father of Blues* x. 142 Goofer dust all about—I'll fix him!

goofiness (gū·finės). *slang.* [f. *Goofy *a.* + -ness.] The state of being 'goofy'; stupidity.

[illegible] Evidence of goofiness is easily found: the more you dig into yourself, the more you can find. [illegible] **1934** Wodehouse *Thank You, Jeeves* vii. 94 It was just plain, straight goofiness, and I can quite understand now why Sir Roderick told father that you ought to be under restraint. **1939** C. Morley *Kitty Foyle* 22 In spite of the general goofiness of our home doings I like to think of the little backyard.

goofus (gū·fəs). *Mus.* [Arbitrary formation?] (See quot. 1952.)

1928 *Melody Maker* Feb. 142 (Advt.), With the coming of Rollini to the Savoy Hotel, the Goofus has leapt into popularity. It is an original instrument with two chromatic octaves, and plays both melody and accompaniment. **1935** *Vanity Fair* Nov. 71/2 New effects and new instruments (the 'goofus') have been invented. **1952** *Conc. Oxf. Dict. Mus.* (1960) 239/2 *Goofus*, an instrument introduced in the 1920s, or thereabouts. It looks like a saxophone, but has 25 finger-holes, each with its own reed, and is thus capable of producing chords.

goofy (gū·fi), *a. slang.* [f. *Goof *sb.* + -y[1].] **1.** Stupid, silly, daft.

1921 *Collier's* 19 Feb. 6/3 'How d'ye like California?' 'Fried?' says Knockout with a goofy grin. **1923** Wodehouse *Inimitable Jeeves* xvii. 232 He was lying back in an arm-chair with his mouth open and a sort of goofy expression in his eyes. **1928** *Daily Express* 26 Sept. 13 'I have gone completely goofy over Mr. Robey', writes one Montreal critic. **1931** *Observer* 11 Oct. 15 The 'sap' gone 'goofy'. **1932** E. Wilson *Devil take Hindmost* xxi. 214 They describe it as a 'hell-hole' where 'you get goofy from the heat'. **1938** E. Bowen *Death of Heart* II. ii. 169 Your dear little goofy face. **1951** S. Kaye-Smith *Mrs Gailey* 25 Two women..so utterly unlike her friends—one snooty and the other goofy. **1958** [see *Animation 6].

2. *Surf-riding.* **goofy foot, footer, surfer**, one who rides a surfboard with the right foot forward instead of the left.

1962 *Austral. Women's Weekly* 24 Oct. Suppl. 3/2 *Goofy foot*, a very good [surf] rider who reverses the usual stance by standing with the right foot in front of left. **1966** *Surfer* VII. No. 4. 53 Fred Hemmings displayed spectacular foot work at Yokohama, riding left foot forward and then switching to goofy foot. **1965** J. Pollard *Surfrider* ii. 19 I'm sorry, I didn't notice you were a 'goofy footer'—a board man who rides with the right instead of the left foot forward. **1967** *International Surfing* III. 16 [illegible] is a power surfer. **1968** *Surfer Mag.* Jan. 48/1, I realised goofy-footer [illegible] would have some problems. **1968** W. Warwick *Surfriding in N.Z.* 11 This turn is performed by right foot forward goofy surfers, when they turn to the left. **1970** *Surf'ye* (N.Z.) 13/1 Allan Byrne is one of the top goofy-foots in the world.

goog (gug, gʊg). *Austral. slang.* [Origin unknown.] An egg. Phr. *full as a goog*, drunk.

1941 Baker *Dict. Austral. Slang* 32 *Goog*, an egg. [illegible] say *full as a goog* [illegible] plastered or tanked. [illegible] *Goog*, an egg (a word formed perhaps on the sense of *gog*, in *googog*, a gooseberry). U.S. slang has *goog*, spectacles—in all these cases roundness is implied.

google (gū·g'l), *v. Cricket.* [Back-formation from *Googly *sb.*] *intr.* Of the ball: to have a 'googly' break and swerve. Of the bowler: to bowl a googly or googlies; also (*trans.*), to give a googly break to (a ball). Hence **goo·gler**, a googly bowler.

1907 *Badminton Mag.* Sept. 289 The googlies that do not google. **1909** *Westm. Gaz.* 1 July 7/4 Mr. Lockhart, having 'googled' for two years from the 'nursery' end. **1923** *Daily Mail* 9 July 11 In R. H. Bettington they have a googler who might triumph over the best of wickets. **1928** *Daily Tel.* 12 June 19/2 Constantine..was out to a semi-yorker, which also 'googled'. **1930** *Ibid.* 25 Apr. 8/5 Grimmett..can spin the ball and google it.

googly (gū·gli), *sb. Cricket.* Also **goglie, google.** [Origin unknown.] A ball which breaks from the off, though bowled with apparent leg-break action.

1903 C. B. Fry in P. F. Warner *How We recovered Ashes* (1904) 19 You must persuade that Bosanquet of yours to practise..those funny 'googlies' of his. **1904** P. F. Warner *How We recovered Ashes* 106 Bosanquet..can bowl as badly as anyone in the world, but, when he gets a length, those slow 'googlies', as the Australian papers call them, are apt to paralyse the greatest players. **1909** P. A. Vaile in *Westm. Gaz.* 17 Sept. 14/2 The 'googly' is merely the American service at lawn-tennis introduced into cricket. **1920** [see *Bosie]. **1924** N. Cardus *Days in Sun* 48 Hirst cultivated the swerve and Bosanquet the 'googly'. **1930** [see *Bosie]. **1954** J. H. Fingleton *Ashes crown Year* 46 Australians call it bosie after Bosanquet; Englishmen call it the googly, or googly. **1955** [see *Chinaman 4].

fig. and *transf.* **1916** *Anzac Book* 128 You could reach it in three bomb-throws, if the last of the three happened to be a 'googly' and swerved in from the off. **1941** Baker *Dict. Austral. Slang* 32 *Googly*, an awkward question which is calculated to upset. [illegible] **1947** J. Brown *Say Word* 60 Australian airmen called a bomb both a bosey and a googly during the war.

b. *attrib.* or as *adj.*, esp. in *googly bowler, bowling*.

1909 *Westm. Gaz.* 8 June 16/1 The discovery of so capable a 'googlie' bowler as Mr. Lockhart. *Ibid.* 12 Aug. 3/2 Googlie bowling is very wearisome work both to the fingers and the right side. *Ibid.* 17 Sept. 14/2 One 'googly'-man does not necessarily win Tests. **1921** P. F. Warner *Bk. Cricket* iii. 68 Mr. Bosanquet has been called the 'Googly King'. **1921** A. W. Myers *20 Yrs. Lawn Tennis* 9 Fifteen years ago, Brookes mainly employed a 'googly' service. [illegible] **1971** *Sunday Express* (Johannesburg) 28 Mar. 22/3 Kerry O'Keefe, 21-year-old leg break and googly bowler whom the Australian regard as the new Bill O'Reilly, has agreed to join Somerset.

googly (gū·gli), *a.* Also **-ey.** [Cf. *goo-goo *a.*] **1.** Of eyes: large, round, and staring. Hence **goog·ly-eyed** *a.*

1901 'H. McHugh' *Down Line* 35 Is id to my face you go behind my back to make googley-googley eyes. **1926** *Spectator* 11 Aug. 287/2 A gollywog hugging in its arms a googly-eyed [illegible]. **1927** *Daily Mirror* 15 Dec. 16/2 Others with movable googly eyes in small wooden heads. **1958** *New Yorker* 15 June 119 (Advt.), Strange, googly-eyed goldfish. **1959** I. & P. Opie *Lore & Lang. Schoolch.* [illegible]

2. Disposed to love-making, 'spoony'.

1919 W. Deeping *Roper's Row* ix. § 1. 107 She ascribed Mr. George's googly, amorous interest to fatherliness. **1932** C. Williams *Greater Trumps* v. 83 And father would say, 'Really, Sybil!' without being googly.

googol (gū·gɒl). [Arbitrary: see quot. 1940.] A fanciful name (not in formal use) for ten raised to the hundredth power (10^{100}). Also **googolplex** [cf. *plex* in *multiplex, complex*], a name for ten raised to the power of a googol.

1940 Kasner & Newman *Math. & Imagination* i. 23 The name 'googol' was invented by a child (Dr. Kasner's nine-year-old nephew) who was asked to think up a name for a very big number, namely, 1 with a hundred zeros after it... At the same time that he suggested 'googol' he gave a name for a still larger number: 'Googolplex'. *Ibid.* 23 A googol is 10^{100}; a googolplex is 10 to the googol power. [illegible] **1964** [illegible] *Excurs. Number Theory* ix. 111 The googol..can be written out in full in about two lines of print.

goo-goo (gū·gū), *a. slang.* [Sometimes connected with Goggle *v.*] **a.** Of the eyes or glances: amorous, 'spoony'. Also *sb.*, an amorous glance, a 'glad eye'.

1900 Goodwin & Hildrup *He used to play on the Oboe*, She'd make goo-goo eyes at the bandsmen above. **1901** 'H. McHugh' *John Henry* 13 'It is awfully nice of you to ask me to see Bernhardt,' says The Real Thing, throwing a goo-goo at me that settles everything. *Ibid.* 76 He'll turn such a warm pair of goo-goo eyes on her that somebody will have to..yell for the fire department. [illegible] **1906** *Westm. Gaz.* 22 Sept. 5/3 I don't mind her giving goo-goo [illegible]. **1924** C. Hamilton *Prisoners of Hope* 101 [illegible] goo-goo girl. **1959** J. Thurber *Years with Ross* ix. 158 There was so much spooning and goo-goo eyes.

goo-goo (gū·gū), *int.* [Echoic.] An imitative representation of baby talk. Hence **goo-goo** *v. intr.*, to talk in the manner of a baby; also *trans.*

1863 *Harper's Mag.* Nov. 402/1 Yes, you pet, goo goo. **1884** Mark Twain *Huck. Finn* xxiii. 295 The duke..just went a goo-gooing around, happy and satisfied. **1907** E. Dyson *Fact'ry 'Ands* xiv. 184 Baby's a baddy baddy 'icky bubb-bubb to goo-goo the wicked man! **1922** *Ladies' Home Jrnl.* Feb. 176 Goo! Goo!—is all I can say just now. I'm happy, healthy and comfy. [illegible]

gook (gūk, guk). *slang* (orig. and chiefly *U.S.*). [Origin unknown.] Used as a term of contempt: a foreigner; *spec.* a coloured inhabitant of (south-)east Asia or elsewhere. Also *attrib.* or as *adj.*

1935 *Amer. Speech* X. 79/1 *Gook*, anyone who speaks Spanish, particularly a Filipino. **1947** *N.Y. Herald Tribune* 2 Apr. 28/6 The American troops..don't like the Koreans—whom they prefer to call 'Gooks'—and, in the main, they don't like Korea. **1951** D. Cusack *Say no to Death* xix. 109 The fur coat she wore must

have cost her black-marketeer husband the best part of a thousand. He had seen ones like it in Tokyo when the Gooks were selling them for what they could get. **1953** *New Yorker* 7 Mar. 23/1 You'll notice it's not a gook car. **1959** R. Kirkbride *Tamiko* iii. 17 Ivan looked at the.. Jap... 'You get back to work, gook,' he said. **1959** N. Mailer *Advts. for Myself* (1961) 132 Miguel, he said a lot, but I just can't follow that Gook talk. **1968** *Guardian* 23 Feb. 11/3 The Gooks [*sc.* Viet Cong] hit from bunkers and the Marines had to carry half the company back. **1969** *Sunday Mail Mag.* (Brisbane) 6 July 4/2 This is a gook grave.

gooly (gū·li). [App. of Indian origin; cf. Hindustani (Yates) *goli*, a bullet, ball, pill, and see R. L. Turner *Compar. Dict. Indo-Aryan Lang.* s.v. *guḍá-*[1] and *gōla-*[1].] **1.** Usu. *pl.* The testicles. *slang.*

1937 Partridge *Dict. Slang* 343/1 Goolies. **1966** 'L. Lane' *ABZ of Scouse* 43 'Ow'd yer like a kick in ther goolies? **1967** B. Norman *Matter of Mandrake* ix. 68 Strapped naked to a bottom-less chair, having his goolies whacked with a carpet-beater. **1971** *Guardian* 27 Sept. 10/4 To get a good performance out of them [*sc.* actors].. it is sometimes necessary to kick them in the goolies.

2. A stone, pebble. *Australian slang.*

1941 Baker *Dict. Austral. Slang* 32 *Gooly*, a stone or pebble.

goombah, goombay, varr. Gumby.

1774 E. Long *Hist. Jamaica* II. 423 The goombah, another of their musical instruments, is a hollow block of wood, covered with sheepskin stripped of its hair. **1901** *United Free Ch. Mission Record* Jan. 20/2 It was surely the beating of the Goombah drum. **1960** *Harper's Bazaar* Oct. 131 The night was filled with hi-fi—no goombay calypso. **1963** *New Yorker* 15 June 19 (Advt.), Cosmopolitan Nassau. Goombay goes till the wee hours. **1968** *Globe & Mail* (Toronto) 17 Feb. 33 Buses wait at St. George dock to carry students to old Fort St. Catherine, where goombay dancers and limboists provide exciting entertainment.

goon (gūn). *slang.* [Perhaps a shortened form of dial. *gooney* (Gony 1) 'a booby, a simpleton'; but more immediately from the name of a subhuman creature called Alice the *Goon* in a popular cartoon series by E. C. Segar (1894–1938), American cartoonist.] **1.** A stolid, dull, or stupid person. orig. *U.S.*

1921 F. L. Allen in *Harper's Mag.* Dec. 121/1 (*title*) The Goon and his Style. *Ibid.* 121/2 A goon is a person with a heavy touch as distinguished from a jigger, who has a light touch. While jiggers look on life with a genial eye, goons take a more stolid and literal view. **1938** *Life* 14 Nov. 6/3 The word 'Goon' was first popularized by college students who used it to mean any stupid person. Labor union lingo has given it a second meaning: a tough or thug. Rival unions and factions speak of another's 'Goon Squads'. **1938** R. Chandler *Trouble is my Business* (1950) 80 Some goon here plays chess. You? **1940** R. Stout *Over My Dead Body* iv. 57 You may be a couple of goons.. But I'm asking you a damn straight question. **1942** D. Powell *Time to be Born* (1943) vii. 175 You sit there gawping at him like some little goon. **1945** Partridge *Dict. R.A.F. Slang* 30 *Goon*, a fool, very stupid fellow; a gaper. **1951** 'J. Wyndham' *Day of Triffids* viii. 154 The goon started to argue. **1957** R. M. Wardle *Oliver Goldsmith* i. 3 It was Goldsmith's misfortune that he was a jigger fallen among goons. *Ibid.* 5 William Cooke was a goon. Apparently it never entered his head that a man—especially an Irishman—might have preferred, for the joke's sake, to use a word which didn't make sense. **1959** S. Clark *Puma's Claw* xii. 135 There, you goon. You'll bump into them if you don't watch out.

2. A person hired (esp. by racketeers) to terrorize workers; a thug. orig. *U.S.*

1938 *Amer. Speech* XIII. 178 In the Pacific Northwest we hear the word *goon* on every hand. Locally a goon is a member of a labor-union's beef-squad; that is, a person of imposing physique and inferior moral and mental qualities who can be depended on to cow and frighten recalcitrant union-members. **1938** [see sense 1 above]. **1940** *Chicago Tribune* 28 Jan. 1/8 'Goon' is a term applied to hired sluggers used in labor troubles... A typical goon murder was the recent killing in Chicago of..a garage-man involved in a union dispute. **1959** [see *fink *sb.*[2]]. **1969** *New Yorker* 20 Dec. 88/3 Many so-called 'goons'—civilian terrorists, sometimes dressed in Constabulary uniforms—took over in many provinces. **1971** *Blitz* (Bombay) 6 Mar. 13/4 Attempts on his life by goons allegedly employed by the Calcutta police authorities. **1971** *It* 2–16 June 5/1 Heath orders Habershon of Barnet CID to 'turn London over'. And he does exactly that.. with 500 goons and a score of specially trained dogs.

3. A nickname given by British and U.S. prisoners of war to their German guards in the war of 1939–45. Also *transf.* So *goon-up!*, a warning cry.

1945 G. Morgan *Only Ghosts can Live* 140, I think it was an Australian who first called the Germans 'Goons'. *Ibid.*, The cry, 'Goon up!' remained in many camps a warning of the approach of the Detaining Power. **1948** *Amer. Speech* XXIII. 218 The word for German was invariably *goon* among airmen, American and British alike. ..It was also used as the adjective part of a compound noun,..e.g. a *goon-box* was one of the guard towers along the fences around the camp. **1952** E. F. Davies *Illyrian Venture* xi. 218 At midnight the Goon *Postern* (German sentry) was astounded to see a large naked form flying three times round the courtyard in a temperature of 25 degrees of frost. **1962** V. Nabokov *Pale Fire* 55 Morning finds us marching to the wall Under the direction of some goon Political, some uniformed baboon.

4. (With capital initial.) Any one of the members of the cast of a popular British radio comedy series, *The Goon Show*, noted for its crazy and absurd brand of humour.

1951 *Radio Times* 25 May 18/2 (*heading*) Crazy people. Radio's Own Crazy Gang 'The Goons'. *Ibid.* 18/3 Spike Milligan..has compiled the 'Goon Show' material. **1951** [see below s.v. *goonery]. **1958** *Observer* 18 May 14/4 The department of jokes, only a shadow of its former self without the Goons. **1958** *Listener* 25 Sept. 481/2 Lovers of the Goon Shows. **1962** A. Nisbett *Technique Sound Studio* x. 173 Just about the ultimate in this line of effects was reached when in a 'Goon Show' the script demanded 'a 16-ton, 1½-horsepower, 6-litre, brassbound electric racing organ fitted with a cardboard warhead'. **1970** *Sun* 28 July 3/5 The Goons are returning to radio—after 10 years in cold storage. **1971** S. Milligan in D. Nathan *Laughtermakers* ii. 49 Prisoners of war called their German guards goons but I got it from Popeye. There was a creature called the Goon which had nothing in the face at all except hair... I liked the word and we called it *The Goons*.

5. *attrib.* and *Comb.*, as (sense 2) *goon squad, tactics*; (sense 3) *goon-baiting* vbl. sb.; **goon box**, a guard tower at a P.O.W. camp; (sense 4) *Goon-like, -type* adjs. Also **goonskin** (see quot. 1943).

1962 *Times* 12 Oct. 15/4 'Goon-baiting', which was the favourite occupation of the prisoners. **1948** Goon-box [see sense 3 above]. **1956** A. Crawley *Escape from Germany* 31 Watch towers..covering the interior of the camp. To the prisoners they were known as 'goon boxes'. **1963** *Times* 21 Feb. 16/7 Mr Horst Buchholz, whose mixture of American and Indian accents is given an even more goon-like quality by the over-used attempt to inject local colour with the tag-phrase 'isn't it?' **1943** C. H. Ward-Jackson *Piece of Cake* 32 *Goonskin*, observer's flying suit and parachute harness made in one piece. **1957** Rawnsley & Wright *Night Fighter* 37 The fur-lined leather trousers and jacket known to us as 'Goon Skins'. **1937** *Nation* 4 Sept. 239/2 The goon squad, as it is commonly called, consists of at least twenty picked thugs and ex-convicts. **1951** E. Paul *Springtime in Paris* xi. 202 The Existentialists allege that the Communists send goon squads to the district, as a part of their anti-American, anti-tourist campaign. **1962** K. Orvis *Damned & Destroyed* ix. 62 I'm lucky the goon-squad haven't back-handed me into a lane and kicked me to death. **1967** N. Mailer *Cannibals & Christians* I. 32 He talked of..'strong-arm and goon tactics'. **1967** *Listener* 10 Aug. 187/3 And *I'm Sorry, I'll Read That Again*, a happy reminder of happy Goon-type days.

Hence **goo·nery**, a foolish or absurd kind of humour typical of a Goon (sense 4); **goo·nish** *a.*, of, pertaining to, or resembling a goon (senses 1 and 4).

1921 F. L. Allen in *Harper's Mag.* Dec. 122/1 A goonish style is one that reads as if it were the work of a goon. It is thick and heavy. **1951** *Picture Post* 16 June 34 Four young comics—Michael Bentine, Harry Secombe, Spike Milligan and Peter Sellers—have at last got together in a radio programme. In 'Crazy People' they put across their favoured kind of humour. This they call 'goonery'... General opinion was that if you like crazy, pun-dizzy, logic-smashing comedy that doesn't despise your intelligence, you'll like the Goons. **1958** *Vogue* Jan. 49 George Devine and Joan Plowright.. made this extraordinary play a startlingly moving piece of theatre and gave the dialogue..a quality of abstract goonery. **1959** *Listener* 12 Mar. 485/2 An hour..which had some elements of Goonery. **1959** *Sunday Times* 29 Mar. 17/5 Photographs by Inge Morath that have that air of solemnity suppressing a giggle that has become known, in this country at least, as 'goonish'. **1960** *Design* July 23/1 If you can force yourself past the first few sentences the Goonish mixture is oddly funny. **1968** *Times* 27 Feb. 3/8 This is a victory for Spike Milligan and goonery.

‖ **goonda, goondah** (gū·nda). [a. Hind. *guṇḍā* rascal.] (See quots.)

1926 *Glasgow Herald* 27 Apr. 14 A general round-up of goondahs, or roughs, took place this morning [in Calcutta]. **1931** *Daily Tel.* 21 May 11/5 They [*sc.* the Cawnpore massacres] started through the importation of 'goondas' (hired rowdies). **1960** 'S. Harvester' *Chinese Hammer* xxv. 202 Bloated millionaires and lean *goonda* thugs. **1969** *Times of India* 30 July 7/2 The Lt.-Governor had denied having stated that anti-social elements or goondas enjoyed the patronage of politicians. **1970** *Guardian* 14 Apr. 11/6 Since the parties let their goondas loose in the streets to plunder and terrorise, business has been redeploying itself in safer parts of India.

goondie (gū·ndi). *Austral.* Also **gundy.** [Aboriginal.] = Gunyah.

1890 'R. Boldrewood' *Col. Reformer* xvii. 204 There were a dozen 'goondies' to be visited. **1908** Mrs. A. Gunn *We of Never-Never* xiv. 186 The man..rode out of the gundy camp, and out of our lives. **1966** Baker *Austral. Lang.* (ed. 2) vi. 80 We have taken the words..gunyah, goondie or gundie,..from the Aborigines and used them to denote huts or shelters.

goop (gūp). *slang* (orig. *U.S.*). [Arbitrary formation; cf. *goof *sb.*] A stupid or fatuous person. So **goo·py** *a.* fatuous, esp. fatuously amorous; stupid.

[**1900** G. Burgess (*title*) Goops, and how to be them. Manual of manners for polite infants, inculcating many juvenile virtues, both by precept and example.] **1914** *Dialect Notes* IV. 107 (Kansas) *Goop*, a boor. **1918** *Story-Teller* Feb. 695/2 You rabbit-faced 'goop! **1925** N. Venner *Imperfect Impostor* i, Go on, it's a bargain. I'll be the poor goop of this piece. **1926** *Chambers's Jrnl.* Dec. 770/1 The baggage always affects strangers like that... Makes 'em frightfully goopy till they discover her for the cockatrice she is. **1929** 'A. Berkeley' *Wychford Poisoning Case* xx. 250 Oh, ass, dolt, fool, goop and *mutt!* **1946** Wodehouse *Joy in Morning* i. 8 My first emotion.. had been a gentle pity for the unfortunate goop slated to step up the aisle with her. **1955** *Sci. Amer.* Apr. 84/1 Americans are shocked when they go abroad and discover whole groups of people behaving like goops—eating with their fingers, making noises and talking while eating. **1959** I. & P. Opie *Lore & Lang. Schoolch.* x. 179 A person who is 'wanting in the upper storey' is:..dotty, goofy or goopy. **1966** *Punch* 21 Dec. 908/1, I am very jealous of my position as chairman of *Juke Box Jury*,.. and I don't believe one can be a placid smiling goop all the time.

goopher, var. *goofer[2].

goorie, goory (gū^ə·ri). *N.Z. slang.* Also **goori.** [Corruption of Maori *kuri*.] A (mongrel) dog. Hence as a term of abuse.

1937 in Partridge *Dict. Slang* 343/2. **1942** R. Finlayson *Sweet Beulah Land* 57 You can tell it's no more'n a goorie fight to him. **1959** M. Gee in C. K. Stead *N.Z. Short Stories* (1966) 272 It's over to you, you goori. **1970** *N.Z. Listener* 12 Oct. 13/1 'Are you going to marry her?' I said. 'Why should I? Let go of me, you goorie,' he said.

goose, *sb.* Add: **1. d.** *gone goose*: see *gone *ppl. a.* 1; *to cook one's goose* (earlier and later examples); *to kill the goose that laid* or *lays the golden eggs*, to destroy a source of one's wealth by one's own heedless action; to sacrifice future advantage to the greed of the moment; also used allusively.

[**1484** Caxton *Esope* (1967) 190 This fable sayth of a man whiche had a goos that leyd euery day an egge of gold.] **1589** Lyly *Pappe w. Hatchet* III. 404 A man.. had a goose, which euerie daie laid him a golden egge; hee..kild his goose, thinking to haue a mine of golde in her bellie, and finding nothing but dung..wisht his goose aliue. **1845** J. R. Planché *Golden Fleece* I. 7 To save my bacon I must cook his goose! **1862** G. Dodd *Where do We get It?* ii. 103 The natives adopted a reckless way of cutting down the trees in order to obtain the sap; but they are now gradually accustoming themselves to a more economical method—they preserve the 'goose that lays the golden eggs'. **1887** W. E. Norris *Major & Minor* v, If Brian had only known how immensely he had risen in her respect by the not very extraordinary display of talent and ability which he had just made, he would doubtless have hastened to kill the goose that laid the golden eggs by playing classical compositions till he wearied her. **1917** Galsworthy *Five Tales* (1918) 77 You're getting a thousand a year out of my fees. Mistake to kill the goose that lays the golden eggs. I'll make it twelve hundred. **1921** T. R. St.-Johnston *Islanders of Pacific* 295 Even an insouciant native hesitates to kill the goose that lays his 'golden eggs', for the tapping of the crown is generally fatal to the palm-tree. **1923** D. H. Lawrence *Birds, Beasts & Flowers* 207 Is that you, American Eagle? Or are you the goose that lays the golden egg? **1930** A. E. Housman *Let.* 21 Mar. (1971) 293 On the one hand I must thank and congratulate you, but on the other you have cooked your own goose. **1935** T. S. Eliot *Murder in Cath.* i. 25 Leave well alone, Or your goose may be cooked and eaten to the bone. **1946** W. S. Maugham *Then & Now* xii. 71 'I can count on your discretion, Messer Niccolo? My life would be short if it were discovered that I have told you what I have.' 'I know. But I am not one to kill the goose that lays the golden eggs.' **1965** *Melody Maker* 25 Sept. 20 Let's hope that promoters have learned from past experience and don't kill the geese that lay the golden pop eggs.

7. a. *goose-breast.*

1891 W. Morris in Mackail *Life W. Morris* (1899) II. 261 Goose-breast colour. **1904** *Daily Chron.* 19 Mar. 8/5 Smoked goose-breasts. **1963** A. Simon *Conc. Encycl. Gastron.* vii. 565 Minced Goosebreast.

8. goose bumps *N. Amer.* = Goose-flesh 2; **goose dinner** (see *goose match*); **goose drownder** *U.S. dial.* (see quot. 1969); **goose-eye**, a pattern used in weaving; **goose-fair** (later example); **goose game** *Cricket*, very cautious play adopted by a batsman; **goose-gamer**, one who plays the goose game (*these terms no longer current*); **goose liver** = *foie gras; **goose man** *N.Z.*, one who operates a goose saw; **goose match** *Cricket* (see quots.); **goose pimples** = Goose-flesh 2; **goose saw** *N.Z.* (see quot. 1957); also *ellipt.*; **goose-shot** (later U.S. examples); **goose-trap**, also *U.S.*, a swindle; **goose-yoke** *U.S.*, a yoke to hamper the movements of a goose.

1933 C. Miller *Lamb in Bosom* xi. 148 She rubbed down the skin of her arms and legs where goose-bumps stood on every pore as though it were cold weather. **1968** *Publ. Amer. Dial. Soc.* XLIX. 17 *Goose bumps*..seems to be replacing both *goose flesh* and *goose pimples*. **1970** *Washington Post* 30 Sept. D3/1, I no longer get goose-bumps before a game. **1929–33** in Wentworth *Amer. Dial. Dict.*, Goose drownder. **1969** *Daily Progress* (Charlottesville, Va.) 22 Aug. 4/6 Other two-word names for a heavy rain..are: bresh- or brush-mover, bridge-lifter, goose drownder, gully-washer, sand-packer, toad-strangler, and trash-mover. **1957** Simpson & Weir *Weaver's Craft* (ed. 8) xii. 151 The patterns most generally used for tweeds are: 1. Twill... 4. Bird eye twill. 5. Goose eye. **1960** G. Lewis *Handbk. Crafts* 121 For this purpose [*sc.* variety] twill, goose-eye and rosepath are all excellent. ..Goose-eye with its diamond-shaped pattern is perhaps best employed in a rug made in one colour on a different-coloured warp. **1970** *Daily Mail* 3 Oct. 7/1 In Notting-

ham, stalls at the famous Goose Fair were overturned by a gale. **1899** J. C. Snaith *Willow the King* xiv. 224 Don't play the goose game. Hard slogging's the sort o' thing for Grace. **1928** *Daily Tel.* 26 June 17/1 Jupp took four wickets for 37 runs. The batsmen would not go to fetch him, and nearly all of them are free players by inclination and habit. They are not good goose-gamers. **1860** Dickens in *All Year Round* 7 Apr. 560/1, I set him up in business in the goose-liver line. **1958** *Catal. County Stores Taunton* June 20 Goose Liver Purée with Truffles.. a glass 3/9. **1967** L. Deighton *London Dossier* 55 Sandwiches..with unusual fillings like game pâté, rillette, and goose liver. **1943** J. A. W. Bennett in *Amer. Speech* XVIII. 85 The timber trade..has supplied a wide variety of occupational terms..*goose man* ('drag' and 'goose' are various types of saw; cf. U.S. *drag-saw*). **1957** *N.Z. Timber Jrnl.* July 49/1 *Gooseman*, the operator of a goose saw. **1885** P. M. Thornton *Harrow School* xiv. 339 The Goose Match is the last game of cricket played in the year at Harrow. A goose dinner follows. **1905** H. A. Vachell *Hill* ii. 27 The Goose Match, the last cricket-match of the year, played between the Eleven and the Old Boys, on the nearest Saturday to Michaelmas Day. **1970** *Sunday Tel.* 27 Sept. 29/8 He refers to the rained off Goose Match at Harrow, cancelled, he believes, for the first time in 165 years. **1889** *Cent. Dict.*, *Goose-pimples*, the pimples of goose-flesh. **1914** *Dialect Notes* IV. 155 Don't stay in bathing so long that you're all goose-pimples when you come out. **1956** L. Sieveking in *Plays of Year* XV. 255 'E ain't arf got a wicked fyce, 'asn't 'e. Makes one come out in goose-pimples to look at 'im. **1959** *Times* 5 Nov. 14/6 A make-up man kept dodging forward to cover up goose-pimples on the bare shoulders of the two parasolled and picture-hatted belles. **1943** J. A. W. Bennett in *Amer. Speech* XVIII. 85 [In New Zealand] 'drag' and 'goose' are various types of saw; cf. U.S. *drag-saw*. **1950** *Landfall* IV. 125 The planer..spits out faced boards for the tailer-out to stack by the goose saw. **1957** *Brit. Commonw. Forest Terminol.* II. 163 *Pendulum* [*saw*], a crosscut circular saw mounted on a swinging arm and moved down to and across the timber to be sawn. Syn. *Swing saw, Swinging crosscut saw, Goose saw* (N.Z.). **1789** *Amer. Museum* V. 580 A major..received a wound in the cheek with a goose shot. **1898** P. L. Ford *Tattle-Tales Cupid* 51 It passeth human intelligence how Freddy could inspire any sort of feeling except an intense longing for a gun loaded with goose-shot. **1799** *Aurora* (Philad.) 31 Jan. (Th.), The gulls and goose-traps that have been sported for some time past all come from the shop in which the Washington Lottery wheels remain undrawn, and where a new goose-trap, the Amuskeag canal, was some time since hammered out. **1842** C. M. Kirkland *Forest Life* I. 120 A variety store, offering for sale every possible article of merchandize, from lace gloves to goose-yokes, ox-chains, [etc.]. **1863** 'E. Kirke' *Southern Friends* iii. 48 One half of it [*sc.* a building] was sparsely occupied with..fishhooks, log chains, goose yokes, etc. **1879** B. F. Taylor *Summer-Savory* xvii. 138 And you find it, the variety store of a hundred years ago, where needles and crowbars, goose yokes and finger-rings, liquorice-stick and leather are to be had for cash or 'dicker'. **1896** J. C. Harris *Sister Jane* 2/3 Go show your grandmother how to make a goose-yoke.

goose, *v.* Add: **2.** (Earlier example.)

1838 *Actors by Daylight* 31 Mar. 35 In every scene, O! think of me! And may they goose thee, when you die!

5. *slang.* To poke, tickle, etc., (a person) in a sensitive part, esp. the genital or anal regions; sometimes, more specifically, = *fuck *v.* 1.

1879–80 *Pearl* (1970) 257, I don't like to see vulgar girls in the town Pull their clothes up, and stand to be goosed for a crown. **1881** F. Griffin in J. R. Ackerley *My Father & Myself* (1968) xvii. 200 As soon as..I had learned the goose-step, I had learned to be goosed. **1906** *Dialect Notes* III. 138 *Goose*, to create nervous excitement in a person by pointing a finger at him or by touching or tickling him and making a peculiar whistle. **1932** J. Farrell *Studs Lonigan* (1936) 106 Paulie slapped Denny's face. Denny bawled... Paulie goosed him. Denny squirmed. **1943** M. Shulman *Barefoot Boy with Cheek* x. 99 As she was bending over her work-table.., a playful lab assistant goosed her. **1959** W. Burroughs *Naked Lunch* 82 Boys..goose each other at the peep show. **1960** I. Wallach *Absence of Cello* (1961) 109 Elliot..lightly kissed the top of her head. It would be vulgar to say that she leaped as though goosed, but truth can survive anything including vulgarity. **1965** G. McInnes *Road to Gundagai* vi. 111 He used to..urge them up the rope with a little skilful goosing. **1967** Partridge *Dict. Slang* Suppl. 1152/2 *Goose*, the predominant post-World War II meaning is 'to jab a finger *in ano*, in order to surprise or annoy'.

6. *slang.* Only in passive: to be finished, ruined.

1928 *Sunday Dispatch* 5 Aug. 3/2 We were just about goosed with nothing to think about when our football news supply began and put new life into us. **1959** 'J. Welcome' *Stop at Nothing* viii. 127 If I've guessed wrong and Jason has found out right, then we're goosed.

gooseberry. Add: **4.** Also applied jocularly to inferior or spurious brands of champagne.

1850 Thackeray *Pendennis* iv, Pen could not but respect Foker's connoisseurship as he pronounced the champagne to be condemned gooseberry. **1893** K. Deighton *Lamb's Ess. Elia* 130 Whether used literally of gooseberry wine, or of champagne, inferior brands of which wine are often spoken of contemptuously as 'gooseberry' [etc.].

6. b. *gooseberry bush*: used allusively in reference to the explanation of child-birth sometimes given in answer to a child's question.

1944 Brahms & Simon *Titania has Mother* xiii. 146 Fairy Peaseblossom..had never thought she would find herself hankering after one of Simple Simon's curious questions, but now she found she simply couldn't wait for the next one—even if it should be 'but why a gooseberry bush' again? **1952** V. Wilkins *King Reluctant* I. iii. 47 When girls come home and tell their fond relations that they have just found a baby..under a gooseberry bush, you know what the world says, don't you? **1956** B. Goolden *At Foot of Hills* x. 234 Perhaps she's one of the gooseberry bush brigade and is horrified by the precocity of the modern young. Or is it just because she loathes babies? **1964** G. L. Cohen *What's Wrong with Hospitals?* iv. 69 Middle-class mothers are an anxious lot; they have no precedent on child-rearing which hasn't been kicked into limbo along with gooseberry bushes. **1969** *Guardian* 28 Oct. 11/5 Many children said they were glad to *know* what happened, and not be fobbed off with a lot of gooseberry bushes.

7. a. *gooseberry jam, jelly, sauce, -tart* (earlier U.S. example), *-wine* (later examples).

1846 'A Lady' *Jewish Man.* viii. 165 Strawberry jam... By this recipe also are made raspberry, currant, gooseberry, apricot, and other jams. **1861** Mrs. Beeton *Bk. Househ. Managem.* 778 (*heading*) Gooseberry jam. *Ibid.* 779 (*heading*) Gooseberry jelly. **1845** E. Acton *Mod. Cookery* iv. 137 (*heading*) Gooseberry sauce for mackerel. **1785** A. Ellicott in *Life & Lett.* (1908) 44 Our waiters are now preparing some Goose-Berry Tart. **1849** Gooseberry wine [see Marigold 1 a]. **1971** *Times* 10 Nov. (Wine Suppl.) 3/6 Skilfully made gooseberry wine can be successfully passed off as champagne to the uninitiated.

goose-egg. (Earlier and later examples of *transf.* sense.)

1867 H. Chadwick *Scrapbooks* XI, The Buckeyes in this innings were treated to a goose-egg. **1878** C. Hallock *Amer. Club List & Sportsman's Gloss.* p. v, *Goose-egg*, a miss, so-called, in rifle-shooting at a range. **1931** *Amer. Speech* VII. 22 Goose egg usually means 'no tip'.

goose-flesh, gooseflesh. Add: Hence (nonce-wds.) **goo·seflesh** *v. intr.*, to experience 'goose-flesh'; **goo·se-fleshed** *ppl. a.* = *gooseflesby* adj.; **gooseflesby** *a.* (later examples in sense 'exhibiting "goose-flesh"').

1904 *Daily Chron.* 2 Aug. 7/1 In the life of every married man comes a moment when, with a goose-fleshy shudder, he realises that the face opposite to him at the breakfast-table is the face that will always be there. **1924** L. Abercrombie *Theory of Poetry* i. 18, I gooseflesh one day, but another day I do not, at the same passage. **1932** Kipling *Limits & Renewals* 47 A Fear leaped out of the goose-fleshed streets of London between the icy shop-fronts. **1934** A. Christie *Murder on Orient Express* I. vi. 59 This is where I'm supposed to go all goose-fleshy down the back. **1955** *Times* 13 July 10/5 Bathing under regularly warm skies spoils anyone, who is not a keen swimmer, for the chilly dips in and out, with the skin blue and goose-fleshy, which the hardy English will take. **1971** O. Norton *Corpse-Bird Cries* iv. 80 There was a lot of stuff about whether her skin was goose-fleshy or not.

goo·se-girl. [After G. *gänsemagd*.] A girl employed to tend geese.

1826 E. Taylor tr. *Grimm's German Popular Stories* II. 1 The Goose-girl. **1906** *Westm. Gaz.* 28 July 12/3 Mr. Randal Charlton has written a fairy-tale, and his heroine is called Mave. Of course she ought to be a Princess, but she is rather of the goose-girl type. *Ibid.* 24 Nov. 7/1 Like all Bohemian geese-girls, she carried a long branch, which served as a wand to drive her flock home. **1954** Wilkins & Kaiser tr. *Musil's Man without Qualities* II. cv. 245 Of course, an Arnheim might permit himself to marry even a goose-girl if he wished.

goosegog (gu·zgɒg). A widespread local form for *gooseberry*. (Cf. dial. *goose-gob, -bob*.)

1823 E. Moor *Suffolk Words*, *Gusegog*, a gooseberry—also gew-gog. *a* **1825** R. Forby *Voc. E. Anglia* (1830) *Goose-gog*, a gooseberry; particularly when ripe. **1864** Mrs. H. Wood *William Allair* iv, Green goosegogs. **1916** E. V. Lucas *Vermilion Box* xcix, Gertie's growing goosegogs for the Ghurkas. **1945** [see *goog]. **1959** B. Kops *Hamlet of Stepney Green* I. 13 Redcurrants and black-currants and golden goosegogs.

goose-neck. Add: **1.** (Later example: see quot.)

1961 F. H. Burgess *Dict. Sailing* 103 *Goose neck*, a ventilating cowl bent over to keep out water.

2. (Further examples.)

1929 Wilson & Webb *Mod. Gramophones* 125 Immediately after the war a new type of tone-arm with a 'trombone goose-neck'..was introduced. **1957** *N.Z. Timber Jrnl.* July 49/1 *Gooseneck*, (1) a strong wood bar connecting two logging trucks; (2) a curved iron at the bottom of a slide to check the speed of moving logs. **1958** A. D. Merriman *Dict. Metallurgy* 113/2 *Goose-neck*, a bent air-duct or penstock that is fire-brick lined and joins the bustle-pipe to the belly pipe of a blast furnace. It conveys the hot blast to the tuyeres.

4. *attrib.* and *Comb.*, as *goose-neck binder, handle, lamp*.

1940 E. C. Studholme *Te Waimate* 155 The Woods was probably the first string binder tried here. It was known as the 'goose-neck' binder because it had a long circular arm resembling the neck of a goose, which threw the sheaves away. **1957** Mankowitz & Haggar *Eng. Pott. & Porc.* 23/2 Vases with goose-neck handles were peculiar to the [Bilston] factory. **1929** M. Lief *Hangover* iii. 31 Books and magazines lay piled up on a library table alongside his typewriter and a gooseneck lamp. **1969** J. Ross *Dead at First Hand* ii. 20 The pool of light cast by the gooseneck lamp on his desk.

goo·se-necked, *a.* Shaped like the neck of a goose.

1907 *Yesterday's Shopping* (1969) 999/2 Goose-necked Putter... Fitted with hickory shaft—each 5/3. **1910** *Westm. Gaz.* 22 Apr. 6/3 Goose-necked and other daffodils. **1968** L. Durrell *Tunc* v. 247 The tall, goose-necked alabaster lamps with red parchment shades.

goose-step. Add: A balance step, practised esp. by various armies in marching on ceremonial parades, in which the legs are alternately advanced without bending the knees.

1916 H. G. Wells *Mr. Britling* II. i. 204 The small boys had discovered the goose-step. *Ibid.*, They tried it themselves, and then set out upon a goose-step propaganda. **1922** C. E. Montague *Disenchantment* xiii. 182 Doing the Prussianist goose-step by way of *pas de triomphe*. **1941** 'G. Orwell' *Lion & Unicorn* I. ii. 21 The goose-step..is one of the most horrible sights in the world. **1971** *Daily Tel.* 16 Oct. 3/6 As columns approached the Shah they broke into a goose-step.

goose-step, *v.* Add: (Further examples.) Also occas. *trans.* Hence **goo·se-stepping** *vbl. sb.* and *ppl. a.*

1935 H. R. L. Sheppard *We say No* viii. 97 Hitler's goose-stepping conscripts. **1938** *Reader's Digest* Jan. 110/1 Chilean citizens of German blood are already goose-stepped by Nazi missionaries. **1941** Koestler *Scum of Earth* 79 A helpless, bleeding prey to be stamped on by the boots of the goose-stepping conqueror. **1951** M. McLuhan *Mech. Bride* 122/2 The remorseless goose-stepping of the glamorous female warrior. **1964** C. Chaplin *Autobiogr.* xxv. 433 Greed has poisoned men's souls.., has goose-stepped us into misery and bloodshed. **1967** *Listener* 16 Mar. 370/3 The Nazi jackboot goose-stepping on the boulevards got his goat. **1970** *Daily Tel.* 16 June 11/7 Women as well as men goose-step around the stage to the tat-tat of drums. *Ibid.* 10 Dec. 4/5 President Nyerere reviewed goose-stepping troops of Tanzania's army yesterday.

goo·se-ste:pper. One who practises the goose-step (used contemptuously of supposed dupes of militarism).

1923 H. L. Mencken *Prejudices* 3rd Ser. 10 The most timorous, sniveling, poltroonish, ignominious mob of serfs and goose-steppers ever gathered under one flag. *Ibid.* 49 The first made them almost incapable of soldierly thought and conduct; the second converted them into cringing goose-steppers. **1928** *Daily Express* 15 Oct. 9 An old topical print showing 'Big Willie' and 'Little Willie' inspecting a parade of goosesteppers.

goo·se-walk. *Draughts.* (See quots.)

1892 J. Lees *Compl. Guide to Draughts* 60 (*Single Corner*) 27–24 here loses, and forms what is known as the 'Old Farmer', or the 'Goose Walk'. **1929** *Encycl. Brit.* VII. 621/1 The trap which White falls into in the above game when playing 27–24 at the 12th move is of very common occurrence, and is known by the quaint designation of the 'Goose Walk'.

goose-winged, *a.* Add: (Earlier and later examples.) Also **goo·se-wing** *v.*, to set the sails of a vessel in a 'goose-wing' fashion (see sb. 2, esp. quot. 1867); in fore-and-aft rig, when running before the wind, to boom out two working sails, one on either side.

1866 J. MacGregor *Thousand Miles in Rob Roy Canoe* (ed. 2) vi. 108 And the white sails swell towards you, goosewinged, before a flowing breeze. **1896** Kipling *Seven Seas* 61 And the *Northern Light* stood out again, goose-winged to open sea. **1920** *Blackw. Mag.* Mar. 320/1 An ability to keep clear of shifting sandbanks was deemed of more value than the correct way to 'goose-wing' a tattered topsail. **1961** F. H. Burgess *Dict. Sailing* 103 *Goose-wing*, sail with mainsail set one side and foresail set the other side, so that one will not blanket the other.

goosy, *a.* Add: **3.** [Cf. *goose *v.* 5.] Ticklish, nervously excited, touchy.

1906 *Dialect Notes* III. 138 *Goosy*, used of a person who is susceptible to nervous excitement when a finger is pointed at him, or when he is hardly touched or tickled. **1932** J. T. Farrell *Studs Lonigan* (1936) I. ii. 22 Davey goosed Hennessey. Hennessey was goosey anyway, and he jumped.

‖ **gopak** (gōᵘ·pæk). Also **hopak** (hōᵘ·pæk). [Russ. *gopák*, f. Ukrainian *hopák*.] A lively Ukrainian dance in ²⁄₄ time.

1929 G. Gould *Democritus* 80 They..see the same Ukrainians that the two men have previously seen in Moscow, dancing the gopak. **1943** K. Ambrose *Ballet-Lover's Pocket-Bk.* 21 Turns such as these are well suited to the virility of the Russian *Gopak*. **1958** *Economist* 25 Oct. 303/1 For some time ahead he [*sc.* Mr Gaitskell] could make Mr Bevan dance the *gopak* if he wanted to. **1959** *Collins' Mus. Encycl.* 324/1 *Hopak*, a Russian folk-dance in a lively 2–4 time, occasionally used by Russian composers. **1962** *Observer* 20 May 21/4, I was brought before her and told to dance a 'gopak' and a 'lesginka'.

gopher, *sb.*[1] Add: **1. b.** A native or inhabitant of Arkansas or Minnesota.

1845 in C. Cist *Cincinnati Misc.* 240 The inhabitants of..Arkansas [are called] Gophers. **1872** *Harper's Mag.* Jan. 317/2 The various nicknames given to the States and people of this republic..Minnesota, Gophers. **1873** J. H. Beadle *Undevel. West* xxxiii. 706 In May, 1859, I first became a 'gopher',—practical Western title of the Minnesotians. **1949** *Amer. Speech* XXIV. 26.

3. b. A native or inhabitant of Florida.

1869 *Overland Monthly* III. 129 On account of the great

number of gophers in that State,..a Floridian is called a 'Gopher'.

4. (Earlier example; also, the bull-snake.)

1837 J. L. Williams *Territory of Florida* 68 The Bull Snake..is sometimes called the Gopher snake. **1853** Baird & Girard *Catal. N. Amer. Reptiles* I. 165 Gopher Snake (*Georgia Couperi*).

6. *gopher-burrow*; **gopher man** (examples; also *gopher*, ellipt.); **Gopher State**, a nickname for Minnesota.

1850 L. H. Garrard *Wah-To-Yah* (1927) x. 138 The animals..stumbled more than once in the numerous gopher burrows. **1903** A. Adams *Log of Cowboy* iv. 42 Officer's horse suddenly struck a Gopher burrow with his front feet. **1901** 'J. Flynt' *World of Graft* 220 Gopher-men, safe-blowers. **1926** J. Black *You can't Win* ii. 12 Famous 'gopher men' who tunneled under banks like gophers and carried away their plunder after months of dangerous endeavor. **1928** M. C. Sharpe *Chicago May* 287 *Gopher*, one who tunnels to steal. **1880** J. M. Farrar *5 Yrs. Minnesota* 166 Gophers are here such a pest to the farmer that Minnesota has been called the 'Gopher State'. **1963** R. I. McDavid in H. L. Mencken *Amer. Lang.* x. 695 Minnesota..popularly..is the *Gopher State*,..and the football representatives of its state university are known as the *Golden Gophers*.

gopher, *v.* Add: **2.** (Later N. Amer. examples.) So **go·phering** *vbl. sb.*

1905 R. E. Beach *Pardners* i. 22 We kept gophering around till March, in hopes. **1910** *Sat. Even. Post* 13 Aug. 4/1 Promising mines did 'gophering' or development work by contract. **1916** *Daily Colonist* (Victoria, B.C.) 19 July 3/3 The numerous dumps along the hillside indicate how thoroughly the gulch has been gophered in the search for gold. **1927** C. M. Russell *Trails plowed Under* 129 This old boy is a prospector and goes gopherin' 'round the hills, hopin' he'll find something. **1966** *McGraw-Hill Encycl. Sci. & Technol.* VIII. 481/1 Gophering, or coyoting, refers to small-scale mining utilizing small, irregular excavations in ground that usually stands without support.

‖ **gopi** (gōᵘ·pi). *India.* Also **Gopi.** [Skr.] (See quot. 1962.)

1880 *Encycl. Brit.* XI. 845/2 The mystic dwelling on the amours of Krishna and the *Gopis* of Braj has been often compared to the mystical interpretation of the *Song of Solomon*. **1893** Kipling *Day's Work* (1898) 33 The young herd [*sc.* herdsman], the darling of the Gopis,..Krishna the Well-beloved. **1933** S. Sitwell *Canons of Giant Art* 121 They are the gopis, milkmaids of the herd, Dancing-girls of paradise, the park's fine pleasures. **1962** Brahmacharini Usha *Ramakrishna-Vedanta Wordbk.* 33 *Gopi*, a milkmaid of Brindavan. The gopis were companions and devotees of Sri Krishna. **1968** *Jrnl. Mus. Acad. Madras* XXXIX. 21 Krishna's sport of going on boat with the Gopis.

gopura (gōᵘ·pura). *India.* Also **-am.** [Skr. *gōpura* city gate, f. *gō* eye + *pura* city.] The great pyramidal tower over the entrance-gate to the precinct of a temple (in S. India).

1862 C. R. Markham *Trav. Peru & India* 408 The *gopurams* or towers of the great pagoda. **1895** R. W. Frazer *Silent Gods* 24 The iron-studded massive gate beneath the many storied gopura. **1967** Singha & Massey *Indian Dances* ii. 41 A permanent visual record of all these karanas adorns the four gopurams or gateways of the great temple at Chidambaram. **1969** *Indian Mus. Jrnl.* V. Suppl., Arunagiri sat under the northern gopura of the temple and did penance for a long time.

goral. Add: Also **goorul.** (Additional examples.)

1864 J. A. Grant *Walk across Africa* iii. 35 Goorul or chamois of the Himalayas. **1925** G. Burrard *Big Game Hunting* 101, I have always liked gooral meat. **1932** *Discovery* Nov. 344/2 On the wooded crags [in N. Burma] 2,000 feet above the river, there were *gooral*, but it needed dogs to drive them out from cover.

gor-amity (gɒrămǝi·ti). Also **garamity.** In representations of Negro speech, representing *God almighty*.

1834 T. Wentworth *West India Sketch Bk.* II. 16 Da kow no hab no tail, Gor-a-mity brush fry [= if a cow has no tail, God Almighty brushes the flies]. **1835** *Tough Yarns* 158 Gor Amighty send rain,—Gor Amighty send sun: but Gor Amighty send poor nigger too. **1837** *United Service Jrnl.* June 209 Goramity, whar for dem nigger makee me wait? *Ibid.* 211 Tank Garamity,..me hab de felicity to find my massa. **1891** *Strand Mag.* Oct. 77/1 I'm the gorramighty of the *Biddy McDougal*.

goramy, var. *gourami.

gorblimey (gɒɪblǝi·mi). Also **gaw-, -blime, -blimy.** [Cf. *blimey *int.*] Vulgar corruption of the imprecation *God blind me!* See also quots. 1919, 1925.

1896 A. Morrison *Child of Jago* i. 16 Gawblimy, not what? **1909** J. R. Ware *Passing Eng.*, *Gorblimy* (about 1875). A gutter phrase. **1911** L. Stone *Jonah* I. ix. 105 'Gorblimey! A knock-out!'..Stinky, with a haphazard blow, had given Chook the dreaded knock-out. **1914** T. A. Baggs *Back from Front* xix. 92 Gor blimey, 'ow are ye, then, old townie? **1915** A. D. Gillespie *Let.* 27 Mar. (1916) 74 Most of the infantry now wear the soft 'Gor'bli'me' hat which looks horrid, but does not give such a mark as the flat-topped 'Brodrick'. **1918** W. J. Locke *Rough Road* v. 51 'Gorblime!' said Chipmunk, 'that's the first I 'eard of it.' **1919** *War Terms* in *Athenæum* 8 Aug. 729/1 'Gor-blimey', a soft service cap. **1925** Fraser & Gibbons *Soldier & Sailor Words*, *Gorblimey*...An exclamation or adjective of emphasis. A 'Gorblimey' was the common colloquial term for an unwired, floppy, field-service cap worn by a certain type of subaltern in defiance of the Dress Regulations. Lines from a song, popular before the War, ran:—'He wears Gorblimey trousers An a little Gorblimey 'at.' **1956** J. Masters *Bugles & Tiger* 49 A tweed gorblimey cap worn well forward on the head. **1958** *Oxf. Mag.* 27 Feb. 326/1 The British and American tendency is to emphasise the Gorblimey aspect of history, the feelings of the ordinary man on the spot at the time. **1962** *Listener* 31 May 967/1 She offered a gorblimey cheerfulness.

gorbuscha (gɒɪbu·ʃă). Also **garbusche.** [ad. Russ. *gorbúsha*, f. *gorb* hump, humpback.] The humpback salmon, *Oncorhynchus gorbuscha*.

1792 T. Pennant *Introd. Arctic Zool.* p. ccvii, The Gorbuscha, or Hunch-back, arrives at the same time with the last [*sc.* the keta salmon]. **1884** D. S. Jordan in G. B. Goode *Nat. Hist. Aquatic Animals* III. 476 This species is known to the Russians still..by the name of 'Gorbuscha'. **1891** *Fur, Fin & Feather* Mar. 151 The salmon most plentiful in the Alaska waters is known as the humpback or garbusche. **1960** *Guardian* 15 Nov. 6/7 Soviet scientists had been transporting the eggs of gorbuscha from Sakhalin Island..to the..rivers of the Kola Peninsula.

Gordon (gǭ·ɪdǫn). [f. the name of Alexander Gordon, 4th Duke of Gordon (1743–1827), who promoted the breed.] In full *Gordon setter*. A black and tan setter (cf. Setter *sb.*[1] 11), used as a gun dog.

1865 *Field* 7 Oct. 255/2 (*heading*) The Black-tan Gordon setter. *Ibid.*, Several of the very best Gordons have even a white foot or feet. **1867** 'Ouida' *Under Two Flags* I. ii. 27 Three or four Gordon setters, an Alpine mastiff, and two wiry Skyes dashed at their chains. **1907** B. M. Croker *Company's Servant* xxviii. 278 The dog with the yellow face proved to be the Gordon setter belonging to Major McPherson. **1931** *Times Lit. Suppl.* 26 Nov. 939/2 Three setters (English, Irish and Gordon). **1968** E. H. Hart *Encycl. Dog Breeds* 165 The Gordon Setter. The fourth Duke of Gordon..brought into prominence the black and tan setter of Scotland, which eventually, as a breed, adopted his name. *Ibid.*, Gordons have beauty and brains.

goree, *sb.*[2], **gori** (gɒri·). Also **gore.** [ad. Chinyanja *goḷi*.] A forked stick used by the Arabs to fasten slaves together by their necks. Also *gori-stick*.

1865 D. & C. Livingstone *Narr. Exped. Zambesi* v. 137 'Goree', or Slave-stick. **1891** *Life & Work* May [4] This boy..worked in a gore-stick for two years. **1899** A. Werner *Captain of Locusts* 243 Uledi and certain men came down..from the Unango country, bringing with them people tied in gori-sticks. **1921** W. P. Livingstone *Laws of Livingstonia* III. ix. 213 The slaves.. fastened with chains or the gori stick. **1925** *Blackw. Mag.* Aug. 276/2 The young ones already had their necks in the *goree*.

gorge, *v.* Add: **4. b.** *intr.* Of ice: to become fixed so as to form an obstruction. *U.S.*

1852 *Knickerbocker* XL. 157 After an hour's plunging through the ice, which had accumulated in such masses as almost to 'gorge', we came to where it lay. **1873** J. H. Beadle *Undevel. West* xxxiv. 738 The ice..gorged against a bluff bank in a short bend of the stream, and dammed the water.

Gorgonzola (gǫɪgǫnzōᵘ·lă). A type of blue cheese, usu. made from cow's milk, orig. produced at Gorgonzola, a village near Milan.

1878 [see *Camembert]. **1885** [see *Edam]. **1896** Long & Benson *Cheese* 49 The Gorgonzola process. **1910** *Practitioner* Feb. 211 It has become quite an easy matter to make Gorgonzola, Camembert, Pont l'évêque, Gruyère, anything you like, in Nottinghamshire. **1955** *Times* 10 May 12/4 Gorgonzola, the Italian blue veined cheese, on the other hand, is dependent on no unique caves; the process of manufacture is copied elsewhere and Danish cheesemakers send their own brands of Gorgonzola to this country. **1971** *Sunday Times* (Colour Suppl.) 28 Mar. 39/4 *Gorgonzola*, one of Europe's three great blue cheeses, it is naturally veined, creamy, and made of cow's milk.

gorgy (gǭ·ɪdʒi), *a.* [f. Gorge *sb.*[1] + -y[1].] Full of gorges or ravines.

1891 C. E. Douglas *Diary* 25 Feb. in J. Pascoe *Mr. Explorer Douglas* (1957) iv. 130 I'll have to bridge a few Creeks & Torrents going up to the Glacier, some very ugley gorgy looking gaps are looming up at the head of the flat. **1927** *Blackw. Mag.* Oct. 468/1 As soon as Pyramid Island is reached, the country becomes broken & gorgy [in Tasmania]. **1960** B. Crump *Good Keen Man* 135 We struck a few more small waterfalls and a lot of steep, gorgy places where the slippery rocks made travel slow and difficult.

gori: see *goree *sb.*[2]

gorilloid (gǫri·loid), *sb.* and *a.* [f. Gorilla + -oid.] **A.** *sb.* An animal resembling a gorilla. **B.** *adj.* Resembling or characteristic of a gorilla.

1946 F. Weidenreich *Apes, Giants & Man* i. 24 (*caption*) Gorilloids. **1949** W. E. Le Gros Clark *Hist. Primates* 58 Large apes of gorilloid proportions. **1957** L. Eiseley *Immense Journey* 89 Some have..massive gorilloid crests atop their skulls.

gorm (gǭɪm), *v.*[1] Also **gawm.** A vulgar substitute for '(God) damn'.

1849 Dickens *Dav. Copp.* xxi. 220 Gorm the t'other one. **1850** *Ibid.* lxiii. 618 If a ship's cook..didn't make offers fur to marry Missis Gummidge, I'm gormed. **1853** *Punch* 19 May 230/2 I'm gormed if there was more than six of one and half-a-dozen of the other. **1905** E. Phillpotts *Secret Woman* I. iv, An' coming for to count 'em.. be gormed if I didn't find but three! **1907** *Daily Chron.* 9 Mar. 8/5 'Well, I'm gawmed!' exclaimed Mr. Bungard, unable to suppress his surprise. **1910** J. Farnol *Broad Highway* II. xviii, I'll be gormed if it ain't a'most onnat'ral!

gorm (gǭɪm), *v.*[2]: = Gaum *v.*[3]

1928 E. W. Hendy *Lure of Bird Watching* xi. 196 The youngsters sat gorming on its edge, while she stuffed gobbets of bread..into their insatiable maws.

gorm (gǭɪm), *v.*[3]: = Gaum *v.*[2] Also with *up*.

1962 *Listener* 1 Nov. 732/3, I have always hoped against hope that one day Mr Bucknell would..get gormed up in a glue-pot.

gorm (gǭɪm), *sb.* Also **gawm.** [cf. next and Gaumless *a.*] **a.** An undiscerning person, a fool. **b.** (See quot. 1921.) **c.** (See quot. 1933.)

1912 J. Stephens *Crock of Gold* v. xiv. 228 'I must have forgotten about the other one...' 'You gawm!' gritted the sergeant. **1921** F. B. Young *Black Diamond* ix. 106 He asked Abner if he could spare him a gorm. Abner had not the least idea what a gorm was, but the voice explained that it meant a little..bit of tobacco—'for to chew'. **1933** *Amer. Speech* VIII. 49/2 *Gorm*, a mess, a muddle, a poor job. **1936** 'N. Blake' *Thou Shell of Death* ii. 33 There's a gorm of a girl comes up from the village every morning. **1959** I. & P. Opie *Lore & Lang. Schoolch.* x. 181 A brainless gorm. **1969** H. Carvic *Miss Seeton draws Line* ix. 171 'It's all finished and it's all wunners.' She smiled with pride. 'Isn't it, you great gorm?'

gormless (gǭ·ɪmlės), *a.* orig. *dial.* Now the usual spelling of Gaumless *a.* and in general (i.e. non-dialectal) use. Hence **go·rmlessness**, the quality of being gormless.

1883 in B. Webb *My Apprenticeship* (1926) iii. 161 Parliament is such a far-off thing, that [they]..say that it is 'gormless meddling with it'. **1925** J. Agate *Contemporary Theatre 1924* 44 Cordelia is a 'gumph' or, as we say in Lancashire, 'gormless'. **1932** L. Golding *Magnolia St.* II. ii. 304 She just went on pulling the [beer] handle and in a moment..the floor was swilling. 'Mother!' cried little Nellie sharply. 'You *are* gormless!' **1934** A. Ransome *Coot Club* xix. 233 Ye'd better. He'll ferget the salt else, the gormless old lummocks! **1940** *Illustr. London News* CXCVII. 164/2 He looks at once genial and 'gormless'. **1951** 'E. Crispin' *Long Divorce* viii. 84 Pen's been running after a gormless little twerp of a foreign schoolmaster. **1958** *Times* 12 Nov. 3/3 Mr. Danny Ross suffers terrific indignities as the youth and makes his gormlessness not only amusing but genuinely likeable. **1959** *Punch* 17 June 813/2 As a result of this unorthodox treatment Quince's gormless Thespians emerge not as rough comics but as decent, sensible men. **1970** *Daily Tel.* 4 Sept. 11/2 That other skilled writer in the genre..depended on queers or gormlessness for two West End successes.

gormy, var. Gaumy *a.*

In quots., said of painting: coarsely executed, dauby. **1907** W. De Morgan *Alice-for-Short* xxxii. 331 'What's his work like though, reely?' 'Footy stuff. Gormy colour. No drawin'!' **1919** —— *Old Madhouse* 306, I wish she wouldn't trot out her gormy daubs and ask my honest opinion of them.

go-round (gōᵘ·raund). *U.S.* [Go *v.* 88.] The act of going round or something that goes round; *spec.* (*a*) a merry-go-round; (*b*) a fight; a beating; an argument; (*c*) an experience or attempt, esp. an unpleasant one.

1886 *Harper's Mag.* July 172/2 The rink and the go-round opposite the hotel were in full tilt. *Ibid.* 174/1 The rink band opposite..grinding out its go-round music. **1891** M. E. Wilkins *New Eng. Nun* 382 If I was to say what I thought, it would be something pretty plain. All this go-round——. **1898** E. N. Westcott *David Harum* ii. 13, I had to give him one more go-round..an' after that I didn't have no more trouble with him. *Ibid.* xvi. 142 Me an' him had a little go-round to-day. **1938** M. K. Rawlings *Yearling* xx. 248 Let's knock her [*sc.* a panther] out and leave the dogs have a go-round. *Ibid.* xxiii. 286, I had me a good go-round with them jessies. We cain't see it the same way about killin' 'em. **1968** R. Macdonald *Instant Enemy* vi. 34 She's a nice girl in serious trouble. This seems to be her second go-round with sex. The first go-round made her suicidal.

gorp, *dial.* var. Gaup, gawp *v. intr.*, to gape. Hence **go·rping** *ppl. a.*

1915 D. H. Lawrence *Rainbow* 19 There he sat stubbornly in his corner at the Red Lion, smoking and musing and occasionally lifting his beer-pot, and saying nothing, for all the world like a gorping farm-labourer. *Ibid.* v. 121 The little crowd at the gate gorps and stretches. **1930** R. Campbell *Adamastor* 43 Funnelled with roaring mouths that gorp like cod. **1961** J. I. M. Stewart *Man who won Pools* 81 The budgerigars in the windows..must be gorping at them. **1969** —— *Cucumber Sandwiches* 39 He would bring down mediums and other psychically well-accredited persons to gorp and gape at that lake.

gorry (gɒ·ri), *int.* [Substituted for God in oaths or exclamations; cf. Golly *int.*] In (*by*) *gorry* = (*by*) *God*.

1854 THOREAU *Walden* 162 Then, by gorry, your mind must be there. 1931 G. STERN *Meaning & Change of Meaning* 331 The well-known and well-worn [distortions] *blooming* for *bloody, dash, darn,*.. *Golly, Gorry,*.. and many others, which need not be translated. 1961 R. PARK *Hole in Hill* (1962) xii. 101 'Another dead-end, darn it.' 'By gorry, it isn't,' said Tom jubilantly.

gorse. Add: **3. a.** *gorse-knife, -slasher*: used for clearing land of gorse.

1907 'G. B. LANCASTER' *Tracks we Tread* vi. 80 Randal kicked aside his gorse-knife—he had been cutting brush in the gully beyond. 1938 'R. HYDE' *Nor Years Condemn* xiv. 253 We..were given grubbers and gorse-slashers.

Gorsedd (gǫ·ɹseð). [W., = throne, tribunal, session.] A meeting of Welsh bards and druids; esp. the assembly which meets each day during a certain period as a preliminary to the eisteddfod. Also *attrib.*

1794 E. JONES *Relicks Welsh Bards* (ed. 2) 60 Likewise, we held a Gorsedd, Tribunal-meeting, or Supreme Congress of the Bards of the Isle of Britain, according to the ancient form of a Druidical Assembly. 1876 *Programme Wrexham Eisteddvod* 6 The Gorsedd will be held every morning..at 9 o'clock, in a field in Grosvenor Road. 1908 *Daily Chron.* 9 June 5/7 The Eisteddfod..will be duly proclaimed..by the Arch-Druid, who will be assisted by the Gorsedd officials and bards. 1909 *Ibid.* 10 June 4/4 The Gorsedd ceremonial takes place within a circle marked out by twelve massive unhewn stones placed a few feet apart. 1928 *Observer* 15 July 11/3 The early morning sessions of the Gorsedd, with the bards in their multi-coloured robes. 1955 *Times* 3 Aug. 3/2 Early this morning the Gorsedd in their cloaking robes, green for the initiates, blue for the bards, and white for the druids, met in the sunlight within the circle of 12 stones to hold the traditional eisteddfod ceremony. 1959 *Manch. Guardian* 5 Aug. 3/1 The Gorsedd ceremony. *Ibid.*, The Gorsedd stones. *Ibid.*, The Gorsedd procession.

gosain (gosəi·n). *India.* Also (formerly) **gosaing, gosine, gossein.** [Hindi, etc. *gosāīn* (Skr. *gosvāmin* 'lord of cows', f. *go* Cow *sb.*[1]).] A Hindu who professes a life of religious mendicancy.

1774 G. BOGLE in C. R. Markham *Tibet* (1876) 124 The Gosains, the trading pilgrims of India, resort hither in great numbers. 1793 W. HODGES *Trav.* 112 A Gosine, or Hindoo Religious. 1813 J. FORBES *Orient. Mem.* II. 9 This village belongs exclusively to the Gosaings, or Senassees, a caste of religious Hindoo mendicants. 1826 W. B. HOCKLEY *Pandurang Hari* I. 97 About five miles from the city I reached the hut of a Gossein. *Note*, A religious mendicant. 1923 *19th Cent.* Jan. 107 A certain *gosain* or Hindu devotee. 1957 *Encycl. Brit.* XXII. 939/2 Chaitanya himself, as well as his immediate disciples, have come to be regarded as complete or partial incarnations of..Krishna himself; and their modern successors, the Gosains, share to the fullest extent in the devout attentions of the worshippers.

Goschens (gǭu·ʃənz), *sb. pl.* (*Temporary.*) A colloquial name for consols after their conversion from 3 to 2¾ per cent by G. J. *Goschen* (Chancellor of the Exchequer) in 1888 (later to 2½).

1889 *Man of World* 29 June (Farmer), The nickname Goschens is going out of fashion. The new 2¾ stock is now called by the old name. 1891 *Punch* 4 Apr. 161/1 Securities yielding a larger return than 2¾ Goschens. 1903 *Westm. Gaz.* 15 Aug. 7/1 For various reasons, chief of which is the multitude of better-paying investments, the public are not buying Goschens.

gosh. Add: Also *my gosh!* So **go·shawful** *a.* = *GOD-AWFUL *a.*; *go·sh-a:wfulness*; **go·sh-darned** *a.*, God-damned.

1906 H. GREEN *At Actors' Boarding House* 169 I'd a been free from that De Shine a long time ago if it wasn't fur the notoriety, an' me so gosh darned sensitive! 1909 H. L. HINDLEY *Gentleman from Hayville* 36 That gosh-awful long-winded, Vermont heestorical society meeting. 1920 F. SCOTT FITZGERALD *This Side Paradise* I. iv. 143 '*The Life of St. Teresa*,' read Alec aloud. 'Oh, my gosh!' 1928 *Daily Express* 9 July 9 Stripped of that language the result is what the flapper critic would call 'goshawful'. 1934 WODEHOUSE *Right Ho, Jeeves* iv. 39 That subtle gosh-awfulness which renders..my Aunt Agatha the curse of the Home Counties. 1942 M. LOWRY *Let.* (1967) 44 The Goshdarned thing will put you on the spot. 1967 J. FLEMING *No Bones about It* iv. 80 Oh, my gosh! We'll have to have meals in the dining-room. 1967 LORD REITH in *Listener* 30 Nov. 688/3 Blood pouring down my new tunic, my gosh.

go-slow (gǭuslǭu·). [f. Go *v.* + SLOW *adv.*] A form of industrial protest in which employees work at a deliberately slow pace. Also *attrib.* Hence **go-slo·wer,** one who works in this manner. Cf. *CA'CANNY.

1930 *Times* 27 Mar. 19/5 The receipts..showed large decreases..due to the recent 'go-slow' movement amongst the men. 1937 *Daily Herald* 11 Feb. 3/1 Drivers on the Morden–Edgware tube and the Bakerloo line had been adopting a 'go-slow' policy, because it is alleged, they resented being reprimanded by inspectors for speeding on bends. 1955 *Times* 12 Aug. 8/4 The men's representatives undertook that there would be no further 'go slows' or stoppages and that every man would work his full shift. 1963 *Listener* 10 Jan. 76/1 A 'go slow' at power stations. 1963 *Times* 12 Jan. 9/5, I would ask potential strikers or 'go-slowers' not to add to the hardship of old people, for whom each winter is a recurring agony.

gospel, *sb.* Add: **8. a.** *gospel ministry, music, shout.* **b.** *gospel singer.*

1721 *Jrnls. Ho. Repr. Mass.* III. 187 The first settled and Ordained Minister..that Shall Live and Dye in the Work of a Gospel Ministry. 1831 J. M. PECK *Guide Emigrants* 255 The object is, 'to educate pious, indigent young men for the gospel ministry'. 1955 KEEPNEWS & GRAUER *Pict. Hist. Jazz* i. 3 The spirituals and gospel music that were the Negro's own interpretation of the white man's religion. 1958 'F. NEWTON' in P. Gammond *Decca Bk. Jazz* v. 73 The great gospel singers..are the legitimate heirs of Ma Rainey and Bessie Smith, and their rise in the late 1930s and 1940s is one of the most cheering phenomena for the blues-lover. 1959 'F. NEWTON' *Jazz Scene* iv. 54 A local piano style, based on ragtime and Appalachian gospel shouts. 1971 *Ink* 12 June 19/1 In an attempt to give *bigness* to their white gospel music, they tried everything from accordion to horns.

9. gospel-sharp, a Western U.S. term for a Christian minister of religion; **gospel song,** a song characterized by its fervour or evangelistic message; **gospel-truth** (*b*), (earlier and later examples).

1872 'MARK TWAIN' *Roughing It* xlvii. 333 'What we want is a gospel-sharp. See?' 'A what?' 'Gospel-sharp. Parson.' 1897 A. H. LEWIS *Wolfville* 50 I've took the trouble to bring a gospel-sharp over from Tucson to do the marryin'. 1902 C. MORRIS *Stage Confidences* 224 There's another gospel-sharp out on the edge of town. 1905 *Methodist Rev.* LXXXVII. 704 The attitude to take toward the sort of tune..variously denominated, 'gospel song', 'spiritual song', 'pennyroyal', has cost the Commission a good deal of vexation of spirit. 1959 'F. NEWTON' *Jazz Scene* ii. 32 Good examples of contemporary negro gospel song groups. 1964 L. NKOSI *Rhythm of Violence* 41 Next time they came we were singing gospel songs. 1833 *Jurist* Feb. 5 It has been believed, nay, clung to as Gospel truth. 1911 BRERETON & ROTHWELL tr. *Bergson's Laughter* i. 41 'A red nose is a painted nose', 'A negro is a white man in disguise', are also absurd to the reason which rationalises; but they are gospel truths to pure imagination.

gospeller. 1. Delete †*Obs.* and add later example.

1933 V. McNABB *Nazareth or Social Chaos* iv. 18 This phrase of S. John is all the more striking because he more than the other three gospellers is insistent upon the Will of God.

Gosplan (gǫ·splæn). [Russ. *gosplán*, abbrev. of *Gos*udárstvennyĭ *plán*ovyĭ komitét (sovéta ministrov) SSSR, State Planning Committee (of the Council of Ministers) of the U.S.S.R.] An organization formed in 1921 to draw up plans for the development of the national economy of the U.S.S.R. (The governments of the constituent republics also have Gosplans.)

1926 *Encycl. Brit.* III. 580/1 He [*sc.* Sokolnikov] was..appointed deputy president of the 'Gosplan', or state-planning commission. 1930 W. H. CHAMBERLIN *Soviet Russia* vi. 136 The main stumblingblock in the path of the Gosplan forecasters. 1938 *Ann. Reg.* 1937 196 The Sovnarkom ordered the Gosplan to finish the schedule for the third five-year period. 1959 *Chambers's Encycl.* IV. 69/1 The state planning commission (Gosplan) set up in February 1921 and given formal status in August 1923. 1962 *Economist* 16 June 1078/2 There is no sign as yet of a budding Gosplan in the Soviet block.

goss[3] (gǫs). *slang.* [Origin unknown.] *to give, get goss*: to give, receive punishment.

1840 *Daily Picayune* (New Orleans) 29 July 2/4 Six victims to report this morning—..offences trivial—. Some of them got *gos*, and some got nothing. 1847 J. S. ROBB *Streaks of Squatter Life* 75 —Gin him goss without sweeten'—. 1864 HOTTEN *Slang Dict.* (ed. 3) 146 *To give a man goss*, to requite for an injury, to beat, or kill him. 1914 A. A. GRACE *Tale of Timber Town* xxxviii. 173 That's right, ole man. Give 'em goss.

Goss (gǫs). The name of W. H. *Goss* (1833–1906), of Stoke-on-Trent, used *attrib.* to designate a kind of armorial china orig. manufactured by him.

[1878 L. JEWITT *Ceramic Art Gt. Brit.* II. v. 224 The works of Mr. William Henry Goss were commenced in 1858 for the production of Parian, ivory-porcelain [etc.]. *Ibid.* 541 Goss's china and parian.] 1906 *Daily Chron.* 31 Aug. 1/7 The Goss china and porcelain. 1926 M. LEINSTER *Dew on Leaf* II. iii. 176 The jingling piano, the goss china, the gummed shells. 1936 G. GREENE *Gun for Sale* iii. 105 Dark curtains and Gosse [*sic*] china. 1969 H. SANDON *Brit. Pott. & Porc.* 112 The title Goss China is nowadays used to describe the huge quantities of holiday mementos, usually on a thin parian body, with the name and crest of the town on it, the shape often in the form of ancient Greek vessels.

gossip, *sb.* Add: **6.** (sense 4) *gossip-exchange, -shop* (later example), *-writing*; *gossip-greedy* adj.; **gossip column,** the column or columns of a newspaper or magazine in which gossip about persons or social events is printed; so *gossip-columnist, -writer.*

1859 J. BLACKWOOD *Let.* 18 Apr. in *Geo. Eliot's Lett.* (1954) III. 52 We had better set a paragraph afloat...Can you do it in the Athenæum or any of the other weekly literary gossip columns? 1897 G. MEREDITH *Let.* 20 Dec. (1970) III. 1285, I comply, but love your Gossip Column least. 1930 J. B. PRIESTLEY *Angel Pavement* ii. 57 Mr. Smeeth..next tried the gossip columns. 1933 H. NICOLSON *Diary* 29 Jan. (1966) 135 A gossip-column writer in a Chicago newspaper. 1938 L. MACNEICE *Earth Compels* 59 It's no go the gossip column. 1962 *Times* 7 June 17/3 Rich gossip-column meat. 1945 KOESTLER *Twilight Bar* 8 Glowworm, poet and gossip-columnist. 1959 *Times Lit. Suppl.* 4 Sept. 505/2 His friends are song-writers, gossip-columnists, [etc.]. 1963 *Times* 25 May 9/7 Everyone will know through the sewing-shop gossip-exchange. 1904 *Daily Chron.* 12 Dec. 4/4 Another of those scandal books of which the credulous, gossip-greedy public have..been treated to a singular crop. 1965 M. MORSE *Unattached* v. 153 The Hartford Swimming Club, which he described as 'virtually a gossip shop'. 1931 *Punch* 4 Nov. 499/2 Mr. Punch's Crown of Buttered Buns for Deserving Gossip-Writers goes this week to the creator of the following. 1961 *Ann. Reg.* 1960 6 One pleasing feature of the engagement was that the gossip writers of the press..had entirely failed to forecast the event. 1936 A. HUXLEY *Olive Tree* 117 Nowhere else in Europe is gossip-writing a highly paid and creditable profession; nowhere else would such a headline as 'Peer's Cousin in Car Smash' be..imaginable.

gossip, *v.* Add: **4.** Also with *around, over.*

1880 [in Dict.]. 1951 AUDEN *Nones* (1952) 5 Pawed-at and gossiped-over By the promiscuous crowd. 1956 W. NEIL tr. *Keller's Bible as Hist.* VII. xxxi. 294 Two unusual dreams..were gossiped around the whole of the ancient orient.

gossipy, *a.* (Additional example.)

1923 F. M. FORD *Let.* 20 Jan. (1965) 147 Suppose you write me a nice gossipy letter.

gossypol (gǫ·sipǫl). *Chem.* [a. G. *gossypol* (L. Marchlewski 1899, in *Jrnl. f. prakt. Chem.* LX. 84), f. mod.L. *Gossyp-ium* (see GOSSYPINE *a.*) + -OL.] A toxic phenolic crystalline compound, $C_{30}H_{30}O_8$, that is an important constituent of cotton-seed.

1899 *Brit. Chem. Abstr.* LXXVI. I. 821 (*heading*) Gossypol, a Constituent of Cotton-seeds... When the phenolic constituents of cotton-seed oil are purified..a crystalline product is obtained...This substance, to which the name of *gossypol* is given, has a composition corresponding..with that required for the formula $C_{18}H_{14}O_4$. 1937 *Jrnl. Amer. Chem. Soc.* LIX. 1725/1 A red isomeric form of gossypol has been isolated recently by Podolskaja. This material melts at 184–185° and gives the reactions of ordinary yellow gossypol...Merely grinding the red form in a mortar converts it to the yellow. 1955 *Sci. News Let.* 24 Sept. 196/2 The poison glands, scattered throughout the cotton plant, especially in the seeds, produce a toxic pigment called gossypol. Because of the presence of gossypol, feeding of cottonseed meal was long restricted to cattle and sheep, unaffected by the poison. 1961 L. F. & M. FIESER *Adv. Org. Chem.* xxii. 767 Gossypol..was shown by R. Adams (1938) to be a symmetrically substituted 2,2'-di-(1-naphthol) derivative.

goster, gosther. (Earlier and later examples.)

1734 LADY B. GERMAIN in *9th Rep. Hist. MSS. Comm.* App. iii. (1904) I. 157 Upon which I could not hold, but fell a laughing as loud as Horace could 'goster' himself. 1836 T. POWER *Impress. Amer.* I. 376 Not another word could we coax out of him: he was, however, quite willing and able to make it up in good Irish, and much did I regret not being able to have a 'goster' with him. 1914 JOYCE *Dubliners* 155 He was leaning on the counter..having a deep goster with Alderman Cowley. 1941 L. A. G. STRONG *Bay* 181 Uncle John did not write letters. He would explain that a good gosther and a drink were better than a cold bit of paper.

gotch. Add: **b.** *gotch-eared* adj.

1905 C. DRISCOLL *Girl of La Gloria* iii. 21 The *maveriquer*..usually rode a gotch-eared Mexican pony. *a* 1910 'O. HENRY' *Strictly Business* (1917) ix. 91 You are a concentrated, effete, unconditional, short-sleeved, gotch-eared Miss Sally Walker.

gotcha, gotcher (gǫ·tʃə), a representation of the colloq. or vulgar pronunciation of (*I have*) *got you* (see esp. GET *v.* 21 a).

1932 E. WALLACE *When Gangs came to London* xxii. 197 The 'plane was nearing the centre of Cavendish Square, when it suddenly heeled over. Its tail went down and it fell with a crash in the centre of the garden which occupied the middle of the square. 'Gotcher!' It was Jiggs' triumphant voice. 1966 H. WAUGH *Pure Poison* (1967) xviii. 112 'Give her background a once-over on your way to Springfield...You might try for a record of her blood type first. She claims it's O but she doesn't carry any card.' Wilks sighed. 'I gotcha.'

Goth. Add: **2. b.** = GOTHICIST.

[*c* 1812 in W. Thornbury *Haunted London* (1880) App. 465 The Modern Goth.] 1837 *Q. Rev.* LVIII. 66 That preterpluperfect Goth, Mr. S. Pugin.., regrets the mistake of Sir Christopher Wren in the construction of St. Paul's. 1849 T. W. ALLIES *Jrnl. in France* 142 The cathedral [of Milan] itself..is quite indefensible in the eyes of a thorough-going Goth. 1857 (*title*) A Word to the Goths. 1866 FR. PIUS *Life Fr. Ignatius* xii. 276 The famous Goth [*sc.* Pugin]. 1907 R. A. CRAM *Gothic Quest* 134 Bentley, in some ways perhaps the greatest of all the new Goths of England.

Gothic, *a.* Add: **1. b.** = MOZARABIC.

1867 tr. *Guéranger's Life St. Cecilia* xviii. 164 The Gothic Church of Spain, whose Liturgy was compiled by St. Leander, Archbishop of Seville. 1874 *Month* Feb. 223 The old Gothic or Mozarabic rite. 1911 E. B. O'REILLY *Heroic Spain* 235 The Christians who were under Moorish rule..kept to the old Gothic ritual.

3. b. Also *transf.* (of the wing of an aeroplane).

1959 J. L. NAYLER *Dict. Aeronaut. Engin.* 121 *Gothic wing*, a wing whose plan form is like a Gothic window. 1961 *Flight* LXXX. 966/2 The Super Caravelle wing is of gothic delta plan form with considerable leading edge camber.

d. Gothic Revival = REVIVAL 1 d. Also *attrib.* So *Gothic Revivalist.*

1869 C. L. EASTLAKE *Hints Household Taste* (ed. 2) i. 32 The earliest promoters of the Gothic revival appreciated the superficial effect of such features...The glories of the 'fretted vault' were not unfrequently imitated in lath and plaster. 1934 A. HUXLEY *Beyond Mexique Bay* 114 The Gothic revival in England was a product of the Oxford Movement. 1950 *Oxoniensia* XV. 118 Jackson witnessed the evolution of the Gothic Revival. *Ibid.*, They, the Gothic Revivalists, had got the old dead style on its legs and propped it up, but they could not make it walk. 1958 R. LIDDELL *Morea* III. ii. 243 A Gothic revival school building was a relic of the British protectorate. 1963 H. READ *Contrary Experience* III. ix. 276 The Gothic Revival was almost a spent force when Ruskin began to publish *The Stones of Venice* in 1851.

6. In combination with an adjective formed on a proper name: Gothic and; Gothic in connection with; as *Gothic-Finnish, -Sarmatian, -Scandinavian.*

1928 C. DAWSON *Age of Gods* iv. 84 The Gothic-Sarmatian kingdom from the Crimea to the lower Danube. 1931 A. SENN in *Jrnl. Eng. & Germ. Philol.* XXX. 143 (*title*) A contribution to Gothic-Finnish relations. 1965 *Language* XLI. 36 The theory of a Gothic-Scandinavian linguistic community distinct from the West Germanic languages.

B. *sb.* **b.** Delete *nonce-use* (quot. 1825) and add later examples.

1915 *Irish Eccl. Record* July 50 Some of these new Gothics were designed by a priest. 1916 *Ibid.* Sept. 209 The fine old walls..gave place to the Gothics.

Gothicism. 2. (Later example.)

1953 *Archit. Rev.* CXIII. 123/3 The Gothic motifs of Wollaton rightly registered as Gothicism, that is revival and not survival.

Gothiglacial (gǫþiglē̇i·ʃiăl, -ʃăl, -glē̇i·siăl), *a. Geol.* Also **Gotiglacial** (gǫti-). [f. L. *Gothi-a* country of the Goths + GLACIAL *a.*] Epithet of the second of the three divisions or 'sub-epochs' of the Late Glacial epoch in north-western Europe, when the ice-sheet of the last glaciation retreated from the tip of the Scandinavian peninsula across southern Sweden. Also *absol.* Cf. *DANIGLACIAL, *FINIGLACIAL *adjs.*

1910 G. DE GEER in *Geologiska Föreningens i Stockholm Förhandlingar* XXXII. 1146 It seems appropriate to designate as *gothiglacial* that part of the late glacial epoch during which the ice-border receded through Götaland—the old Gothia—from Skåne to the great moraines, crossing its northernmost provinces, Dalsland, Västergötland, and Östergötland. 1912 —— in *Compt. Rend. XI Congr. Géol. Internat.* I. 249 Already now I consider it possible to state that we are on the safe side assuming the whole goti-glacial sub-epoch, or the time of the ice-recession from the central parts of Skåne past the old Gotia to the Fennoskandian moraines, almost to amount to, but probably not to exceed 3000 years. 1923, 1927 [see *DANIGLACIAL *a.*]. 1957 J. K. CHARLESWORTH *Quaternary Era* II. 1062 The Gotiglacial tundras of south Sweden. 1960 L. D. STAMP *Britain's Struct.* (ed. 5) xiv. 170 The Gothi-Glacial was the period of the retreat of the ice across southern Sweden and lasted from about 15,000 B.C. to 9,000 B.C.

Gothonic (gǫþǫ·nik), *a.* and *sb.* Also †**Gotthonic.** [ad. L. *Gothōn-es* Goths + -IC.] A name introduced by Dr. Gudmund Schütte to include all early Germanic and Scandinavian peoples. **A.** *adj.* Of or belonging to the primitive Germanic stock. **B.** *sb.* The common language of this stock.

1912 G. SCHÜTTE in *Soc. Advancem. Scand. Study, Proc. Ser.* Dec. 93 To avoid ambiguity I have adopted the classical form 'Guttones, Gothones', and from it formed the adjective 'Gotthonic'. 1922 O. JESPERSEN *Lang.* 42 What is now ordinarily called Germanic and which is in this work called Gothonic. *Ibid.* 195 There is a wide gulf between Keltic and Gothonic. *Ibid.*, The oldest Gothonic languages. 1929 J. YOUNG tr. *Schütte's Our Forefathers* p. x, I call our forefathers 'Gothonic nations' so as to avoid the ambiguity attending all the synonyms now in use, both 'Goths', 'Teutons' and 'Germanic nations'. *Ibid.* vi. 152 Long words formed by derivation are at hand even in the oldest recorded Gothonic. 1948 PARTRIDGE *World of Words* (ed. 2) 77 The Teutonic (or Germanic or, better, Gothonic) languages.

Gotiglacial, var. *GOTHIGLACIAL *a.*

Gotlandian, Gothlandian (gǫt-, gǫþlæ·ndiăn), *a. Geol.* [ad. F. *gothlandien* (A. de Lapparent *Traité de Géol.* (ed. 3, 1893) II. II. ii. 748), f. *Got-, Gothland*, name of an island in the Baltic + -IAN.] Of or pertaining to the geological system between the Ordovician and the Devonian or to the geological period during which it was deposited. Also *absol.* Cf. SILURIAN *a.* and *sb.*[1] 2 in Dict. and Suppl.

1909 *Cent. Dict.* Suppl. I, *Gothlandian*, noting the uppermost division of the Silurian system of Murchison. 1914 *Proc. Geol. Assoc.* XXV. 213 In northern Gotland reef-formations play a predominant part at certain horizons of both lower and upper Gotlandian age. 1955 G. G. WOODFORD tr. *Gignoux's Stratigr. Geol.* iii. 87 At the beginning of the Gothlandian, the shaly facies seem to have invaded several domains. *Ibid.* 99 At the present time, many British geologists prefer not to use the term Gothlandian, which they deem not precise enough. They replace it by that of Silurian, used in a restricted sense.

go-to-meeting, *a.* and *sb. colloq.* (orig. *U.S.*). [See Go *v.* VIII.] **A.** *adj.* **a.** Suitable for use on Sundays or at church; *spec.* of clothes: suitable for wearing to church. Cf. *Sunday-go-to-meeting* (SUNDAY *sb.* 1 c). †**b.** Of people: church-going (*obs.*).

1790 R. TYLER *Contrast* III. i, All my tunes [are] go to meeting tunes. 1800 T. G. FESSENDEN *Orig. Poems* (1806) 115 Each scrapes, huzzas, and kicks and bounces, Waves high her go-to-meeting cap. 1835, etc. [see Go *v.* VIII]. 1853 *Harper's Mag.* VII. 562/2 He was looked up to with emulation..by the 'go-to-meeting' young folk of the town. 1868 G. G. CHANNING *Early Recoll. Newport, R.I.* 83, I was quite a go-to-meeting lad. 1933 O. JESPERSEN *Essentials Eng. Gram.* vii. 74 A *run-away* match, *go-to-meeting* clothes, etc. 1937 C. M. WILSON *Aroostook* 41 A final rinsing for Sunday shirts and go-to-meeting dresses.

†**B.** *sb.* A go-to-meeting garment; usu. *pl. Obs.*

[1831 *Boston Transcript* 12 Dec. 1/1 They tossed on their 'Sunday-go-to-meetings', and crossed into Jersey.] 1841 *Chicago Morning Democrat* 26 Feb. 2/1 Their Servant's a regular 'Miss Nancy with her go-to-meetings on'. 1878 R. T. COOKE *Happy Dodd* 70 She didn't never have more'n one caliker gown to her name, an' an old alipacky for a go-to-meeting. 1881 —— *Somebody's Neighbors* 265 He was gone and his shirt an' go-to-meetin's too.

‖**gotra** (gǭu·trā). [Skr. *gotrá.*] (See quot. 1877.)

1877 M. MONIER-WILLIAMS *Hinduism* xi. 160 All tribes of Brāhmins are divided into *Gotras* or groups of families, according as they are supposed to be derived from one or other of the seven mythical sages. 1902 *Encycl. Brit.* XXIX. 283/2 Members of the three higher castes [of Hindus] are forbidden to marry a woman of the same *gotra* as themselves. Literally a *gotra* means a cattle-yard. 1938 *Nature* 19 Feb. 310/1 The disappearance of totems in favour of gotras of Puranic heroes. 1971 *Illustr. Weekly India* 4 Apr. 9/2 The whole tribe is divided into a certain number of *gotras*, after the name of their patriarchs.

gotta (gǫ·tə), a representation of the colloq. or vulgar pronunciation of (*have*) *got a* or (*have*) *got to* (see GET *v.* 24).

1924 J. BUCHAN *Three Hostages* xviii. 263 He..went forward as if to take her arm. 'You gotta come along,' I heard. 1929 E. W. SPRINGS *Above Bright Blue Sky* 176 'That oughta be easy,' the sergeant suggested. 'You gotta couple of new pilots that could crash an ice wagon on a prairie.' 1932 T. S. ELIOT *Sweeney Agonistes* 29 But I've gotta use words when I talk to you. 1955 M. HASTINGS *Cork & Serpent* xiv. 210, I gotta little present for you. 1968 A. DIMENT *Gt. Spy Race* iii. 35 Sorry, can't stay, gotta rush.

gotten, *ppl. a.* Add: **3. gotten-up** = *got-up* (GOT *ppl. a.* b). *U.S.*

1931 O. NASH *Hard Lines* 47 In a tastily gotten-up flat.

‖**Götterdämmerung** (gōtəɹde·mĕruŋ). [G., lit. twilight of the gods.] Used *fig.* to denote the complete downfall of a régime, institution, etc.

Popularized by Wagner's use of it as the title of the last opera of the Ring cycle.

1909 C. F. G. MASTERMAN *Condition of Eng.* x. 293 Mr. Wells requires for his *Götterdämmerung* no fresh influx of barbarian hordes. 1930 G. B. SHAW *Perfect Wagnerite* 169 What would he [*sc.* Wagner] have said had he lived to see 1917 in Russia and 1918 in Germany..such a Götterdämmerung. 1934 *Mind* XLIII. 388 If..Dr Inge should chance upon a plain unsophisticated Christian, and try to scare him with his Thermodynamic *Götterdämmerung*, would he not be told that the Heavenly Jerusalem was incomparably preferable to the Palestinian, in which there were no abiding habitations? 1949 KOESTLER *Promise & Fulfilment* xv. 170 The Götterdämmerung of British rule in the Holy Land was not in the Wagnerian, but rather in the Dostoievskian style. 1959 *Listener* 1 Oct. 525/1 A totalitarian régime must either prevail absolutely or else it heads straight for some kind of Götterdämmerung.

Gouda (gau·dă). In full *Gouda cheese*. A flat round cheese orig. made at Gouda in Holland.

1885, 1890, 1902 [see *EDAM]. 1902 J. T. LAW *Grocer's Man.* (ed. 2) 326/1 Derbyshire Cheese... One variety is shaped like Gouda cheese, and known as 'Derby Goudas'. *Ibid.* 429/1 Factory Goudas are usually made of skimmed milk in Friesland. 1955 J. G. DAVIS *Dict. Dairying* (ed. 2) 195 *Gouda cheese.* This is a Dutch cheese made from the whole milk of cows. 1971 *Sunday Times* (Colour Suppl.) 28 Mar. 39/1 Gouda, which is made of whole milk, is less soapy [than Edam] and improves remarkably with age when the flesh which is straw yellow in immaturity deepens to a pale mahogany.

Goudy (gau·di). The name of F. W. *Goudy* (1865–1947), American typographer, used *attrib.* to designate type-faces designed and made by him.

1933 D. L. SAYERS *Murder must Advertise* viii. 132 Re-set in Goudy Bold. 1960 G. A. GLAISTER *Gloss. Bk.* 158/1 Among the more important of the hundred or so types he designed are Village, 1903,.. the old-face Goudy family, 1921. 1962 D. B. UPDIKE *Printing Types* (ed. 3) II. xxiii. 237 For lines set in capital letters on covers and in title-pages, the Goudy Old Style roman capitals are good.

gouge, *v.* Add: **5.** *Mining.* (See quots. 1964, 1971.) Also more generally, to dig for opal (cf. *GOUGING *vbl. sb.*).

1931 M. S. BUCHANAN *Prospecting for Opal in Australia* 8 Gouge your drive, viz., push a cut under the roof searching after the seam of potch. 1936 A. RUSSELL *Gone Nomad* i. 7 In chasing my rainbow I have..delved for gold; 'gouged' for opal; fossicked for diamonds. 1958 M. D. BERRINGTON *Stones of Fire* 27 We'll gouge..to start with; and when we strike something we can drive properly. 1964 A. NELSON *Dict. Mining* 202 *Gouging*, working only the rich pockets of ore and leaving the low-grade or marginal ore unmined. 1971 J. S. GUNN *Opal Terminol.* 21 *Gouge*, to cut carefully under the roof searching for a seam of potch [*i.e.* worthless opaliferous material] so that full-scale cutting of the drive can begin.

gouger. Add: **c.** *Austral.* An opal-digger. Also, less commonly, a miner seeking other minerals, e.g. copper. Cf. **opal-gouger*.

1898 *Barrier Weekly Post* 17 Sept. 13 The reason assigned for the absence of the 'gouger' from this hill is the extreme hardness of the ground not altogether the absence of opal. 1936 I. L. IDRIESS *Cattle King* xiii. 112 The silver gougers relied for their supplies on bullock-teams from Menindee. 1936 A. RUSSELL *Gone Nomad* vii. 58 Immature opal..was only too often..the usual portion of the gouger. 1937 *Discovery* June 185/2 All their worldly wealth was their copper gouger's outfit. 1948 V. PALMER *Golconda* iii. 19 Leave that piddling hand-to-mouth work to the gougers! Real mining was a long-range business. 1960 *Times Rev. Industry* Mar. 96/3 Because the firm can handle only 650 tons of ore a year from North Queensland copper-gougers—miner-prospectors who rely on established mining companies to treat their ore—the gougers have invited Japanese mining companies to erect a treatment plant. 1968 *Sunday Mail* (Brisbane) 4 Aug. 30/8 Copper gougers were warned last week against southern speculators moving in on Mt. Isa.

gouging, *vbl. sb.* (Further examples.)

1902 *Chambers's Jrnl.* Mar. 175/1 In the 'back blocks' of New South Wales opal is abundant and 'gouging'—the term given to opal-mining—is the chief pursuit of every man on the western side of the Darling River. 1971 J. S. GUNN *Opal Terminol.* 21 *Gouging-pick*, short-handled pick, chisel nosed at both ends of the head, which can be used in a confined space.

gouging, *ppl. a.* [f. GOUGE *v.* + -ING[2].] That practises gouging.

1796 *Gazette of U.S.* (N.Y.) 10 May (Th.), Brave Abraham..Dar'd tell them all..That gouging Gunn had challenged him. 1825 J. K. PAULDING *J. Bull in Amer.* i. 2, I also fully believed that the people were a bundling, gouging, drinking, spitting impious race.

goujon (gu·dʒən). [Louisiana Fr., f. Fr.: see GUDGEON *sb.*[1]] = *BASHAW 3, *Pylodictis olivaris.*

1883 *Nat. Mus. Bull.* No. 27, 491 *Leptops olivaris*.. Mud Cat;..Goujon. 1888 [see *BASHAW 3]. 1964 H. T. WALDEN *Familiar Freshwater Fishes Amer.* xiv. 201 The flathead catfish has acquired several local names, such as mud catfish, goujon, shovelhead, and yellow cat.

goulash (gū·lɑʃ, -læʃ). Also **goulasch, gulyas,** etc. [Hung. *gulyás*(*hús*), f. *gulyás* herdsman + *hús* meat.] **1.** A stew or ragout of meat and vegetables highly seasoned. Often called *Hungarian ragout.*

1866 CROWN PRINCESS OF PRUSSIA *Let.* 11 Sept. in R. Fulford *Your Dear Letter* (1971) 97, I have all their favourite dishes cooked..for them—goulash for the Hungarians, and polenta and macaroni for the Italians. 1900 F. B. & W. H. WORKMAN *Ice World Himálaya* 193 Irish stews and *goulasches*, composed of animal odds and ends. 1906 *Westm. Gaz.* 27 Sept. 5/1 The national Hungarian dishes of Gulyas, or ragout, and Paprikahuhn. 1915 WODEHOUSE *Something Fresh* viii, In his hour of affliction it soothed him to read of Hungarian Goulash and Escalloped Brains. 1929 E. LINKLATER *Poet's Pub* i, Sucking-pig, celery, and goulasch. 1958 W. BICKEL tr. *Hering's Dict. Cookery* 426 *Goulash*, a brown beef stew of Hungarian origin, prepared with beef, veal, lamb or pork or a mixture of all with plenty of sliced onions, paprika and other ingredients. 1963 R. CARRIER *Great Dishes of World* 160/1 Hungarian veal gulyas.

2. *Contract Bridge.* A re-deal of unshuffled cards after the hands have been thrown in without bidding.

1927 in E. V. Shepard *Correct Contract Bridge* (1930) 245 When all four players pass, no bid having been made, and the players desire to play a Goulash, the cards shall be redealt by the same dealer...No shuffling of any kind permitted. 1930 *Ibid.* 129 Each player arranges all the cards of a suit in sequence before a goulash deal. 1930 M. KERWIN *How to bid at Contract Bridge* xv. 47 In goulashes lies the greatest gambling element of Contract. 1959 REESE & DORMER *Bridge Players' Dict.* 108 For a goulash hand the cards..are dealt in a manner likely to

produce freakish distribution. 1964 *Official Encycl. Bridge* 222/1 *Goulash*, a deal in which the cards are not shuffled, and are dealt five to each player for two circuits, and finally three to each player.

‖**goum** (gūm). Also **Goum.** [Fr., ad. Arab. *gūm*, dial. var. of *ḳaum* band, troop.] **1. a.** A group of North African tribesmen. **b.** A contingent of North African soldiers (in French service). **2.** Such a tribesman or soldier. So **goumier** (*-ie*), a North African serving in a goum.

1845 *Times* 5 Nov. 5/1 Bou Maza left his dead on the field of battle, and retreated with his goum, composed of about 300 cavalry, without counting a great number of infantry. 1846 *Times* 10 Mar. 6/3 Captain Ducrot, with his Goums and a squadron of Spahis, was the first to overtake the fugitives. 1859 F. C. L. WRAXALL tr. *Mem. Robert-Houdin* II. vii. 185 The strange aspect the capital [of Algeria] assumed on the arrival of the *goums* of the Tell and the South. 1903 *Daily Chron.* 25 Feb. 9/2 Far more interesting,..was the fantasia of the Goums...The five hundred horsemen assembled at the top of the course. 1907 *Westm. Gaz.* 28 Aug. 5/1 The goumiers are paid for their services, and..pillaging is a crime which is severely punished. 1908 *Ibid.* 17 Mar. 5/1 The first section of the new *goum* from Algeria. 1946 E. LINKLATER *Private Angelo* x. 114 A wild and huge array of tribesmen from the Atlas mountains..known as Goums. 1955 *Times* 3 May 8/2 A band of 60 Algerian rebels on Sunday night surprised a detachment of 21 goumiers, or native auxiliary troops, in a forest guard post. 1961 B. FERGUSSON *Watery Maze* xii. 315 General Magnan, the French Commander, whose force comprised the 9th Colonial Division, 2,000 Moroccan *goums* and some French commandos. 1961 R. MAUGHAM *Slaves of Timbuktu* 207 In the back were two Goumiers attached to the Commandant's staff and an African servant. 1970 I. ORIGO *Images & Shadows* ix. 242 The Moroccan troops with the Fifth Army, the Goums, who had ransacked everything.

gou·ra[2]. Also **gowra.** = GORAH.

1790 tr. *Le Vaillant's Trav.* II. 104 The *goura* is shaped like the bow of a savage Hottentot, it is of the same size. 1801 J. BARROW *Trav. S. Afr.* I. 149 This instrument was called the *gowra*. 1902 *Jrnl. Anthropol. Inst.* XXXII. 156 The *goura*, a stringed-wind musical instrument of the Bushmen and Hottentots. 1905 *Rep. Brit. Assoc.* 529 The *goura* of the Bushmen..a bow-like instrument having a piece of flattened quill interposed between one end of the string and its attachment to the bow.

gourami (gūə·rāmi, gūərā·mi). Also **goramy, gouramy.** [ad. Mal. *gurāmī.*] A large fresh-water food fish, *Osphronemus goramy*, of the family Anabantidæ, native to south-east Asia; also, any of various smaller members of the family, freq. kept in aquaria.

1878 F. DAY *Fishes of India* I. 372 At the end of a month numerous fry appear, over which the old gouramies keep watch many days. 1879 *Encycl. Brit.* X. 781/1 Goramy, or Gouramy. 1880 A. GÜNTHER *Introd. Study Fishes* xiv. 185 In the first third of the present century, the Javanese Goramy was acclimatised in Mauritius and Guiana. *Ibid.* II. 517 To this genus belongs the celebrated 'Gourami' (*Osphromenus* [sic] *olfax*), reputed to be one of the best flavoured Freshwater-fishes in the East-Indian Archipelago. 1931 J. R. NORMAN *Hist. Fishes* iv. 81 Some or all of the rays may be drawn out into lengthy filaments, as in the Dwarf Cod-fish (*Bregmaceros*) and Gourami (*Osphronemus*). 1952 H. R. AXELROD *Tropical Fish as Hobby* iii. 55 The *Colisa lalia*, or Dwarf Gourami, is..the smallest of the Gouramis. 1961 E. S. HERALD *Living Fishes of World* 244/1 Although many members of this family have had the term 'gourami' included in their common names, the true gourami, *Osphronemus goramy*, is the only species to which the name should apply.

gourbi (gūə·ɹbi). Also 8 **gurbie.** [ad. Algerian Arab. *gurbi.*] A tent, or poor dwelling-place, in Northern Africa.

1738 T. SHAW *Trav.* 288 These *Gurbies* are generally raised either with Hurdles, daubed over with Mud, or else they are built out of the Materials of some adjacent Ruins, or with square Cakes of Clay baked in the Sun. 1854 J. R. MORELL *Algeria* 262 The mountains they were now traversing are intersected by very deep and beautiful valleys, up the steep slopes of which were clustered numerous *gourbies*, or huts forming villages. 1895 F. B. & W. H. WORKMAN *Alger. Mem.* 113 In the Province of Constantine the people..live in tents or gourbis made of white or light-coloured canvas. 1908 *Westm. Gaz.* 20 Jan. 2/3, I have watched for thee From my lone gourbi. 1960 *Spectator* 14 Oct. 561 To be a refugee from Algeria means that..you live in a leaky 'gourbi'. 1963 *Economist* 29 June 1352/3 A number [of Algerians]..must remain..in the 'gourbis', their floorless, mosquito-ridden mud or wattle hovels.

gourd[1]. Add: **7.** *gourd-like* adj. and adv.; **gourd-seed corn, maize** *U.S.*, a variety of Indian corn; **gourd tree** (earlier U.S. example).

1911 E. M. CLOWES *On Wallaby* ii. 30 The odd impression that all the lodging-houses had sprung up, gourd-like, to their present proportions the very night after the lease had been signed. 1927 PEAKE & FLEURE *Priests & Kings* 149 The gourd-like form of the earliest Moravian wares seems to indicate an eastern origin. 1952 A. G. L. HELLYER *Sanders' Encycl. Gardening* (ed. 22) 288 *Luffa*,.. stove climbing annual, bearing curious gourd-like fruits. 1780 W. DUNBAR *Diary* 27 May in E. O. Rowland *Life W. Dunbar* (1930) 73 Planted white Corn & goard Seed Corn. 1831 J. M. PECK *Guide Emigrants* II. 38 The species of corn called the gourd seed. 1835 *Knickerbocker* VI. 173 The rich scenery of forty acres of most luxuriant gourd-seed corn. 1872 E. EGGLESTON *End of World* viii. 60 The relative merits of 'gourd-seed' and 'flint corn'. 1827 *Western Monthly Rev.* I. 313 Gourd-seed maize as high as the waist. 1854 R. GLISAN *Jrnl. Army Life* (1874) xii. 161 The palm, mango, bread-tree, gourd-tree, [etc.].

gourmand, *a.* and *sb.* **B.** *sb.* **1.** Delete †*Obs.* and add later examples.

1848 THACKERAY *Van. F.* vi. 49 Jos, that fat *gourmand*, drank up the whole contents of the bowl. 1939 JOYCE *Finnegans Wake* (1964) 235 Our cousin gourmand, Percy, the pup.

2. (Example of fem. use.)

1853 C. BRONTË *Villette* I. x. 182 Fifine was a frank gourmande; any body could win her heart through her palate.

gourmet. Add: **b.** *attrib.* and quasi-*adj.*

1904 *Westm. Gaz.* 21 Jan. 8/1 The public in the matter of jokes is gourmand rather than gourmet. 1908 *Athenæum* 11 Apr. 447/3 Few can hope to rival the gourmet-author [of *The Gourmet's Guide to Europe*] in the extent of their experiences. 1967 K. GILES *Death in Diamonds* vi. 111 It was a wonderful, gourmet meal. 1971 M. McCARTHY *Birds of America* 72 Gift shops selling 'gourmet' food.

‖**gousblom** (χǭu·sblǫm). *S. Afr.* Pl. **gou·s-blomme.** [Afrikaans, ad. Du. *goudsbloem*, marigold.] A plant with showy flowers belonging to any of various species of *Arctotis, Gazania,* and some other related genera of the family Compositæ.

1822 W. J. BURCHELL *Trav. S. Afr.* I. xi. 229 The term *Goudsbloem*, like too many of the colonial names, is applied gratuitously to various plants, fancied to have a resemblance to the *Marygold*. Different species of *Arctotis* have generally been pointed out to me for it, and sometimes a kind of *Cotula*. 1902 H. J. DUCKITT *Hilda's Diary Cape Housekeeper* 190 In September the ground is literally carpeted with endless varieties of gazenias—local name *Gousbloom*. 1950 *Cape Argus* 5 Aug. 7/5 Namaqualand's most famous flower is the large, brilliant yellow gousblom... Smaller varieties of gousblom are seen in many colours. The arctotis, for example, is a more delicate flower. 1957 *Cape Times* 31 July 1/3 Every piece of uncultivated ground splashed with orange, white and yellow of the gousblomme. 1971 *Ibid.* 28 Aug. 1/7 The display of cinerarias, forget-me-nots, daisies, vygies, gousblomme and kalkoentjies would not be a mass of colour till mid-September.

gout, *sb.*[1] Add: **7.** *gout-ridden* adj.

1901 A. E. W. MASON *Clementina* xix. 235 [He] told the poor gout-ridden man that the Princess..had put up at the 'Cervo' Inn. 1961 *Times* 31 July 14/3 The peppery, gout-ridden Don Lope.

‖**goûter** (gūte). [Fr., f. *goûter* to taste.] A light afternoon repast; five-o'clock tea.

1792 *Wynne Diaries* 16 July (1935) I. 158 When you go anywhere you are shure to find a tea and a good gouté prepared. 1794 *Ibid.* 28 Apr. 149 We went to a Gouter at Thiergarten. *c* 1854 C. BRONTË *Professor* (1857) I. viii. 136 My 'gouter' (a meal which answers to our English 'tea'). 1905 *Westm. Gaz.* 28 Sept. 2/1 At five o'clock we return home to a sort of afternoon tea, or goûter,..wines, cakes, and fruit being served, as well as tea. 1958 R. GODDEN *Greengage Summer* iv. 37 'There wasn't any tea,' said Vicky; we had not learned about the French children's goûter yet.

governess, *sb.* Add: **2. b.** (Additional examples.)

1850 C. M. YONGE *Langley School* xix. 172 There is governess gone to Mr. Howard's. 1939 F. THOMPSON *Lark Rise* xi. 204 There was no assistant mistress; Governess taught all the classes simultaneously.

4. *attrib.* and *Comb.*, as *governess-instructor, -tongue, -trade*; *governess-like, -made, -moulded, -taught, -trained* adjs.; **Governess-General** (cf. 3), the wife of a Governor-General.

1878 *Porcupine* 16 Nov. 516/1 The brothers of the Governess-General arrived at the station. 1909 *Westm. Gaz.* 5 July 5/2 The governess-instructor stood at the right and the nursery-governess at the left. 1928 *Manch. Guardian Weekly* 10 Aug. 114/1 Some of them—and this applies chiefly to..the calfas and elderly governess-like slaves—are gradually selling off the treasures that their sojourn in the Palace brought to them. 1934 H. G. WELLS *Exper. Autobiogr.* II. ix. 769, I have already said that Tyrell's mind was governess-made. *Ibid.* 771 Another of these governess-moulded minds. *Ibid.* 772, I declare that the greatest present dangers to the human race are those governess-trained brains which apparently monopolize the Foreign Offices of the world. 1906 *Westm. Gaz.* 28 Aug. 2/1 That governess-taught conception of the classics. 1892 HARDY *Well-Beloved* (1897) I. ii. 20 To drown the local ballads by songs purchased at the Budmouth fashionable music-sellers', and the local vocabulary by a governess-tongue of no country at all. 1816 JANE AUSTEN *Emma* II. xvii. 324, I was not thinking of the slave-trade..; governess-trade, I assure you, was all that I had in view.

governess, *v.* **2.** Delete (*nonce-use.*) and add earlier example.

1847 C. M. YONGE *Scenes & Characters* xiii. 162 She has been governessed and crammed till she is half sick of all reading!

governessy, *a.* (Earlier and later examples.)

1872 *Spectator* 30 Mar. 406/1 Mary Garth has none of her mother's didactic precision, none of her *governessy* qualities—if one may be permitted to coin a word. 1879 C. M. YONGE *Magnum Bonum* II. xxv. 498 [She] has broken out of all her primmy governessy crust. 1963 *Times* 24 May 15/4 It is an ungrateful task, but her governessy approach does not make the best of it.

governing, *ppl. a.* (Later examples.)

1901 *Westm. Gaz.* 20 Apr. 2/3 He [*sc.* Whitty] originated the term 'governing classes'. 1903 G. B. SHAW *Man & Superman* p. xxiv, We know now that there is no hereditary 'governing class' any more than a hereditary hooliganism. 1938 *New English Weekly* 21 July 276/1 It may be that behind the apparent split in governing-class opinion there is a conflict of financial interests.

government. Add: **9. government house** (earlier examples); **government surplus,** unused equipment sold by the government through retailers. Also objective, as *government-making*; instrumental, as *government-appointed, -backed, -certificated, -controlled, -directed, -financed, -held, -issued, -imposed, -made, -mounted, -organized, -run, -sponsored, -supported, -worked* adjs.

1909 *Westm. Gaz.* 10 June 4/1 The Government-appointed Air Committee stepped in. 1963 *Rep. Comm. Inquiry Decimal Curr.* i. 1 in *Parl. Papers* 1962–3 (Cmnd. 2145) XI. 195 Following reports in 1959 by Government-appointed committees, Australia will adopt a decimal currency in 1966. 1951 A. L. ROWSE *England of Elizabeth* iv. 116 The Merchant Adventurers, the great government-backed body of exporters. 1909 *Westm. Gaz.* 11 Oct. 4/1 Four..Government-certificated schoolmistresses answered an advertisement. 1964 M. GOWING *Brit. & Atomic Energy* xii. 329 Everyone agreed that a Government-controlled Experimental Establishment should be created. 1959 G. D. MITCHELL *Sociol.* ix. 148 In industry or in government-directed organisations. 1964 *Ann. Reg.* 1963 189 To this end the U.S.A. on 6 February imposed a prohibition on U.S. government-financed cargoes to foreign merchant ships engaged in trade with Cuba. 1965 H. KAHN *On Escalation* iii. 60 A government-held village. 1802 A. ELLICOTT *Jrnl.* 43 The Governor..consented that we should meet at the government house. 1803 in *Ann. 8th Congress* 2 Sess. 1507 A Government house, stables, and garden. *a* 1828 D. WORDSWORTH *Jrnl. Tour Continent* 11 Sept. in *Jrnls.* (1941) II. 269 The bathing-house, a square building, is as conspicuous as a Government House. *a* 1830 W. HICKEY *Mem.* (1960) viii. 124, I..posted away to the Government House and made my request in person. 1960 *Farmer & Stockbreeder* 15 Mar. 66/1 The reward for this patriotic effort has been a Government-imposed cut of £9 million in the guarantees. 1964 A. WYKES *Gambling* 335 The latter are government-issued bonds. 1907 *Westm. Gaz.* 7 Nov. 1/3, I could not smoke the filthy Government-made stuff that French people call tobacco. 1905 *Ibid.* 30 Dec. 2/2 Government-making is a great sport for the public. 1907 *Daily Chron.* 1 Apr. 4/6 A special Government-mounted messenger was carrying the document along a country road. 1946 *Nature* 19 Oct. 529/2 What stands out from these discussions..is..a distrust..of Government-organised or -sponsored research. 1964 A. WYKES *Gambling* x. 241 Britain's government-run lottery..is commonly called 'Premium Bonds'. 1946 *Nature* 13 July 39/2 In its opinion, a government-sponsored public corporation financed by the Exchequer is, in general, the most suitable agency. 1956 *Ibid.* 18 Feb. 312/2 Graduates who take up government or government-supported jobs related to national defence. 1952 'E. C. R. LORAC' *Murder in Mill-Race* xv. 151 Grey blankets, (good 'government surplus'). 1962 'O. MILLS' *Headlines make Murder* xiii. 146 He kept a short-wave Government-surplus radio. 1900 *Daily News* 20 Apr. 3/2 Government-worked mines.

governor. Add:

4. *spec.* One who is in charge of a prison.

1807 SOUTHEY *Lett. fr. Eng.* II. xxxix. 162 Of late years the office of jailor has become of considerable importance, and ennobled by the title of Governor. 1864 C. M. YONGE *Trial* II. vii. 137 Mr. Ernescliffe sent in his card at the governor's house... They were told that the prisoner they wished to see was at work. 1897 O. WILDE *Let.* 1 Apr. (1962) 514 Of the many, many things for which I have to thank the Governor there is none for which I am more grateful than for his permission to write fully to A[lfred] D[ouglas]. 1911 *Encycl. Brit.* XXII. 366/1 The convict..exercises and goes to chapel..in the society of others, but holds no communication with them; his only intercourse with his fellow-creatures is when he is visited by the governor, chaplain, schoolmaster or trade instructor. 1965 T. PARKER *Plough Boy* III. i. 196 Then the governor come in..rabbiting..about 'respite' or 'reprieve' or something like that.

7. c. (Earlier examples.)

1844 DICKENS *M. Chuzz.* xxiii. 284 'My youngest died last week.' 'I'm sorry for it, governor, with all my heart,' said Mark. 1852 *Punch* 2 Oct. 152/1 (*caption*) I say, Guvner, give us a hist with this 'ere bilin' o' greens!

8. (Later examples.)

1918 E. M. ROBERTS *Flying Fighter* 63 By means of a governor, the speed of the lorries was being limited to fifteen miles an hour. But we discovered that if the ball in the governor was held up we could get as much as twenty-five miles out of the motor. 1940 'GUN BUSTER' *Return via Dunkirk* I. iv. 40 They do forty miles per hour, and can do eighty if you take the governor off. 1961 *Lancet* 2 Sept. 549/2 We might consider obliging the police to seal a governor set at 50 m.p.h. into the carburation of everything on wheels.

9. (Earlier example.)

1856 KINGSLEY *Let.* May in *A. Locke* (1876) p. lvii, Fished all the morning... Killed eight on 'March brown' and 'governor', by drowning the flies.

governorate (gʌ·vəɹnŏrĕt). [f. GOVERNOR + -ATE[1], after *consulate*, etc.] **a.** A province or

portion of country ruled by a governor, esp. in the Ottoman empire and subsequently in Egypt. **b.** The residence of a governor. **c.** The office of governor.

1899 *Standard* 8 Apr. 4/2 Suakim will form a separate Governorate. 1900 *Pilot* 1 Sept. 263/1 Samos, the freest of Turkish governorates. 1926 *Blackw. Mag.* Mar. 412/1 Kena..is the Mudirieh, or Governorate town, of the province which embraces Luxor. 1930 CHESTERTON *Four Felons* 25 The gardens round the Governorate. 1961 B. FERGUSSON *Watery Maze* xvi. 402 The block of flats behind the Governorate [at Port Said]. 1962 *Economist* 17 Nov. 673/1 Another..is in the running for one or two important governorates.

govey, govy (gʌ·vi), hypocoristic [-Y[4]] forms of *governess.*

1903 *Westm. Gaz.* 28 Mar. 2/2 I'm trying to avoid the govy difficulty by sending my girls to classes. 1904 H. O. STURGIS *Belchamber* xx. 272, I am..just your old govey that you loved when you were a little boy.

gow (gau). [Shortening and adaptation of Chinese *yao-kao* (Mathews), opium, f. *yao* drug + *kao* an oily, fatty substance, esp. an unguent.] A drug; *spec.* opium. Hence **gow·ster,** a drug addict.

1922 *Dialect Notes* V. v. 182 Terms for opium. *Gow*, a Chinese word which..is meaningless unless allied with another Chinese word. 1926 J. BLACK *You can't Win* xii. 159 You're in with what gow I've got. 1933 *Amer. Speech* VIII. II. 27/1 When one has contracted the [drug] habit..he is..hitting the gow. 1942 BERREY & VAN DEN BARK *Amer. Thes. Slang* § 509.19 *Gowster*,..esp. an opium addict.

gowai, gowhai, varr. KOWHAI.

go-way (bird): see *GO-AWAY.

gown, *sb.* **2.** Delete †*Obs.* and add later examples.

1907 *Yesterday's Shopping* (1969) 758 Dressing Jackets, Tea Jackets, and Gowns...Graceful Gown. In Nuns' Veiling, accordion pleated, trimmed white embroidery. 1969 E. McGIRR *Entry of Death* iii. 51 She wears thick Viyella pyjamas...And her gown...Brushed nylon, a kind of housecoat really.

gownless (gau·nlės), *a.* [See -LESS.] Not provided with or not wearing a gown.

1895 HARDY *Jude* II. vii. 144 Two devil-may-care young men who proved to be gownless undergraduates. 1901 'L. MALET' *Sir R. Calmady* IV. i, The picture of those reportedly gownless backs had depressed him abominably.

goy (goi). Pl. **goyim** (goi·yim), **goys.** Fem. **goy·ah.** [Heb. *gōy* people, nation, pl. *gōyim.*] A Jewish designation of a non-Jew, a Gentile. Hence **goy·ish** (also **goyisch,** etc.) *a.*, resembling, in the manner of, having the characteristics of, a goy.

Quot. 1835 may not belong here.

1835 [see GUY *sb.*[3] 3 c]. 1841 BORROW *Zincali* I. x. 159 The children of Roma and those of Israel..are..dispersed among the Gentiles, whom they hate and despise, under the names of Busnees and Goyim. 1892 I. ZANGWILL *Childr. Ghetto* I. vi. 158 The *Goyah*, which is literally heathen female, did everything required on the Sabbath. 1919 F. HURST *Humoresque* 205 A *goy* play-actor. *Ibid.*, She reproaches me with having brought about this *goy* mix-up! 1931 'D. STIFF' *Milk & Honey Route* vi. 64 Jewish hobos are so scarce that when business gets low the agencies will sometimes help the 'goys' or gentile hobos. 1940 E. POUND *Cantos* lii. 11 Paying for a big jews' vendetta on goyim. 1950 *Penguin New Writing* XL. 35 He would join the others in speaking derisively of the Dropper and his girl as a couple of Goys. 1959 M. LEVIN *Eva* 32 My father..had spoken of the Austrians as..among the best of the *goyim*. 1959 H. SLESAR *Grey Flannel Shroud* (1960) viii. 125 My mother..never really approved of Sword's Point [a district]; it was a little too *goyisher*..for her. 1960 *Commentary* Aug. 131 We got pros for the Yiddish parts and for the English roles I'm lining up all my *goyishe* friends. 1963 *Guardian* 23 Dec. 4/5 It would certainly have gone against their Jewish grain..to follow so *goyish* a custom. 1964 S. BELLOW *Herzog* (1965) 28 A Christian lady came... Beside his bed, the goyische lady sat in her long skirts.

Goyaesque (goiyă,e·sk), *a.* Also **Goye·sque.** [f. the name of Francisco de *Goya* y Lucientes (1746–1828), Spanish painter + -ESQUE.] Having the character of the pictures by Goya.

1934 R. CAMPBELL *Broken Record* 108 Squalid Goyesque groups of hags in conversation. 1939 *Burlington Mag.* Apr. 154/2 Not..Goyaesque in spirit. 1959 *Listener* 30 Apr. 774/2 A Goyesque ruthlessness. 1961 *Times* 24 May 16/6 A Goyaesque print of a massacred family.

goyle. (Later examples.)

1902 A. CONAN DOYLE *Hound of Baskervilles* ii. 24 Three of them..rode forward down the goyal. 1963 *Times* 26 Apr. 16/6 On my left tumbles the brook at the bottom of a deep gully—what we call in Devon a goyal—deep enough to take a horse and cart.

Gozitan (gǭu·zităn), *sb.* and *a.* [f. *Goz*(*o*, name of an island near Malta + *-itan* as in NEAPOLITAN.] **A.** *sb.* An inhabitant of Gozo. **B.** *adj.* Of or pertaining to Gozo.

1870 *Malta & Gozo Vade Mecum* 1871 83 Gozo is under the rule of the Governor of Malta... The Gozitans return 1 member to Council. 1958 SITWELL & JONES *Malta* vi. 69 Plenty of the Gozitan boats of peculiar design. 1962 *Times* 10 Feb. 11/6 A diminutive motor-vessel squeezes in..and this is the event of the Gozitan day. 1963 *Guardian* 17 May 11/4 The intricate carving at which the Gozitans excel.

grab, *sb.*[2] Add: **1. b.** *up for grabs*: open to offer; easily obtainable. *slang* (chiefly *U.S.*).

1945 L. SHELLY *Jive Talk Dict.* 35 *Up for grabs*, easy to make gal. 1954 BERREY & VAN DEN BARK *Amer. Thes. Slang* (ed. 2) § 356/7 *Up for grabs*, easy to pick up. 1967 *Boston Globe* 5 Apr. 51/5 Right now every position is up for grabs. Every player is going to get a shot. 1968 *Globe & Mail* (Toronto) 3 Feb. B3/3 At the time that General Bakeries made its bid, Mr. Conrad said the Ogilvie companies appeared to be 'up for grabs'. 1971 *Financial Mail* (Johannesburg) 26 Feb. 675/2 So the hotel reservations set-up looks up for grabs.

3. (Further examples.)

1919 B. BOOTH in H. Begbie *Life Wm. Booth* (1920) ii. 24 My father was a Grab, a Get. 1958 *Amer. Speech* XXXIII. 225 Less frequently used among nonmusicians ..are..*shamus, fuzz, grab* (all meaning policeman).

4. (Earlier and later examples.)

1865 A. M. EATON *Diary* 14 Oct. in *West. Pennsylvania Hist. Mag.* (1935) XVIII. 205 We saw a 'grab' an instrument designed to get out tools stuck in any well. 1955 *Times* 19 May 7/6 Cranes and mechanical grabs heave up loads of what the council call 'reluctant London clay'.

5. (Later example.) Cf. also *ANIMAL C. 1.

1941 H. G. WELLS *You can't be too Careful* II. i. 46 They had a pack of real cards..and gambled with them at *Beat your Neighbour out of doors* and Grab.

b. *Chess.* Applied to a particular class of problems: in full *grab theme.* (See quot. 1913.)

1909 A. C. WHITE *Knights & Bishops* p. iii. 1913 —— *Sam Loyd* 357 The Grab Theme...The theme includes, in its broadest sense, all problems where a Black piece is captured on two or more squares...In its narrower sense it is limited to the problems where a particular Black piece is captured on two or more squares by a single White piece or by two White pieces of the same kind. *Ibid.* 359 The Grab by the concerted action of the two White Knights is extremely pretty. 1943 B. HARLEY *Mate in Three Moves* 52 The grab theme..is essentially a brutal and monotonous business. 1963 M. LIPTON et al. *Chess Probl.* 259 A doubling of the grab theme, in which a black unit is captured on a number of different squares, leading to zugzwang.

6. *grab-digger, dredger*; **grab-bag** (earlier and later examples); **grab bucket** = *BUCKET *sb.*[1] 3 b; **grab-game** (earlier example); **grab handle,** a handle fitted in a motor car to assist passengers entering or alighting, or to steady them when the car is moving; **grab-hook** (later examples); *pl.* (*Naut. slang*) fingers; **grab rail, strap,** a rail or strap inside a motor vehicle for standing passengers to hold.

1855 M. M. THOMSON *Doesticks* xvi. 135 [A] young woman wanted me to invest in the 'grab bag'; [I] gave half a dollar, and fished in. 1960 *20th Cent.* May 460 Sociology is the great intellectual grab-bag of our times. 1967 *Canadian Ann. Rev.* 1966 6 It met strong resistance from those who supported the traditional formula of a grab-bag of promises. 1967 C. O. SKINNER *Madame Sarah* xii. 269 Sarah would hold out a grab bag from which each child drew a number. 1885 *Marine Engin.* 1 Aug. 139/2 (*heading*) Recent applications for patents..7333 C. W. Hunt. Dredging or grab buckets. 1936 *Oxoniensia* I. 81 Without their ready co-operation and their keenness to save everything possible from the very teeth of their 'grab'-digger, we should not have been able to record and preserve such a wealth of detail in advance of the gravel-digging. 1909 *Man. Seamanship* (*H.M.S.O.*) II. xiii. 224 In hard ground it would be necessary to loosen the earth ..or dig a hole with a grab dredger. 1846 R. B. SAGE *Scenes Rocky Mts.* xxxii. 282 Provided you won't attempt the grab game on us. 1959 *Observer* 1 Mar. 21/6 The.. walnut screen rail with grab handle. 1961 *Times* 28 Mar. 4/6 Large 'grab' handles of a flexible material fitted to the roof above each of the four doors are useful when passengers are entering or leaving. 1970 *Motoring Which?* Apr. 50/1 Armrests were only fair, but the passenger had a good grab handle. 1889 *Cent. Dict.*, *Grab-hook*, in angling, a hook made by fixing four large fish-hooks in a piece of lead. 1905 *Terms Forestry & Logging* 38 *Grab hook*, a hook having a narrow throat, adapted to grasp any link of a chain. 1946 J. IRVING *Royal Navalese* 86 *Grabhooks*, fingers. 1955 E. K. WENLOCK *Kitchin's Road Transport Law* (ed. 10) 94 A clear height of 5 ft 10 in excluding any grab rail or strap shall extend over all space intended for use by standing passengers. 1963 *Times* 21 May 5/6 Sensible points include a padded grab-rail.

grab, *v.* Add: **1.** Phr. *to grab hold of* (cf. HOLD *sb.*[1] 2).

1894 A. CONAN DOYLE *Mem. Sherlock Holmes* 123 The fellows evidently grabbed hold of anything they could get. 1900 BAYLY & BRISCOE *Chrons. Country Cricket Club* (1908) x. 111, I was walking outside the cricket grounds and you grabbed hold of me.

2. b. To arrest the attention of (a person); to make an impression on.

1966 *Gramophone Popular Record Catal.* (Artist Section) Dec. 190/2 *Sinatra, Nancy*... *How Does That Grab You?* Not the lovin' kind; Shadow of your smile;..How does that grab you darlin'?; Bang, bang. 1968 *Canadian Mag.* 15 June 27/2 D'you think *that* will grab them? 1970 *Daily Tel.* (Colour Suppl.) 18 Sept. 46/4, I suppose, as my daughter puts it, 'life is whatever grabs you'. 1971 *Post* (Cape ed.) 9 May 9/5 Elton John is big but if his music doesn't grab you then it just doesn't grab you.

3. b. Of the brakes of a motor vehicle: to act harshly or jerkily. So **grabbing** *vbl. sb.*

1919 FRASER & JONES *Motor Vehicles* xxii. 235 Cone clutches are usually faced with leather..the leather becomes hard and dry resulting in 'grabbing'. *Ibid.* xxxi. 328 If the brakes grab or screech a few drops of castor oil..may stop the trouble. 1959 *Motor Man.* (ed. 36) iv. 71 Quite elaborate springing arrangements may be built into the clutch disc to avoid 'grabbing' when the friction surfaces are just on the point of gripping. 1962 *Which?* (Suppl.) July 90/1 The brakes 'grabbed' very badly, because the friction pad assemblies on the front wheels had moved in the caliper units.

grabber. (Earlier and later examples.)

1849 T. MILLER in G. W. M. Reynolds *Myst. of London* V. lxxvi. 243/1 Instead of making you either a..buzzman, grabber, or sneak, we..makes a cracksman of you at once. 1920 R. MACAULAY *Potterism* I. iii. 37 He thinks Johnny's a profiteer, too, because of the ribbons and things... It's in the blood. We're grabbers. 1961 NEW ENGLISH BIBLE *I Cor.* v. 10, I was not, of course, referring to pagans who lead loose lives or are grabbers and swindlers or idolaters. *Ibid.* vi. 10 No thieves or grabbers or drunkards..will possess the kingdom of God.

grabbit (boat) hook. = *grab-hook.*

1936 J. GRIERSON *High Failure* v. 92, I made fast, to the first small buoy which I could see, by means of the Grabbit boathook and my special gear for mooring-up single-handed. 1948 *Shell Aviation News* cxxv. 20/3 The pilot using a short boathook fitted with a grabbit hook engages the strop on the buoy.

grabby. Substitute for def.: *slang.* A Service (esp. Naval) term for a foot-soldier. Add later examples.

1916 *Chambers's Jrnl.* Jan. 11/1 Being about five miles out from the shore, we knew that the local 'grabbies' would not see us. 1919 'ETIENNE' *Strange Tales from the Fleet* 123 'Do you mean the grabby's dinner-party we gave?' inquired the..first lieutenant.

grabby (græ·bi), *a. colloq.* [f. GRAB *v.* + -Y[1].] Having a tendency to grab; greedy, grasping.

1910 A. ROSS in *Penguin Dict. Mod. Quots.* (1971) 196/1 I don't believe in publishers who wish to butter their bannocks on both sides while they'll hardly allow an author to smell treacle. I consider they are too grabby together and like Methodists they love to keep the Sabbath and everything else they can lay their hands on. 1924 E. FERBER *So Big* xiv. 245 'What's the matter with her hands?'..'They're brown, and awfully thin and sort of—grabby.' 1942 BERREY & VAN DEN BARK *Amer. Thes. Slang* 279 *Avaricious; Greedy*, Grabby, a hog for,..on the make, piggy. 1953 E. HAHN *J. Brooke* vi. 99 James was not particularly patriotic, nor grabby on behalf of Britain. 1970 J. H. VANCE *Deadly Isles* xi. 76 Nice when they don't steal my copra. Some people are pretty grabby.

graben (grā·bĕn). *Geol.* Pl. **grabens, graben** (after G. pl. *gräben*). [a. G. *graben* ditch, introduced in its Geol. sense by E. Suess (*Antlitz d. Erde* (1883) I. i. iii. 166).] A depression in the earth's surface bounded by faults; a rift valley.

1896 J. W. GREGORY *Great Rift Valley* xii. 220 This left a great open Rift Valley (or, to use Prof. Suess's term, a 'Graben'). 1910 *Encycl. Brit.* X. 598/1 Crust-blocks resembling 'graben' and 'horsts'. 1913 *Bull. Geol. Soc. Amer.* XXIV. 201 The broad valley, or graben, of the middle Rhine is inclosed by maturely dissected scarps. 1968 R. W. FAIRBRIDGE *Encycl. Geomorphol.* 485/1 The association of volcanism with grabens is well known. 1970 *Nature* 18 Apr. 247/1 The whole region is a large graben which has been filled with up to 3 km of evaporites.

grace, *sb.* Add: **2. b.** (Later example.)

1837 DICKENS *Let.* 8 May (1965) I. 258 A young and lovely girl, who has been the grace and ornament of our home for the whole time of our marriage, died here yesterday.

6. g. *grace and favour house, residence,* etc.: accommodation held 'by grace and favour' of the Crown, the Government, or other owner.

1909 *Daily Chron.* 7 May 1/2 Someone to whom the King may grant a 'Grace and Favour' residence will find himself bound to pay for..work which he had not ordered. 1953 'M. INNES' *Christmas at Candleshoe* i. 14 Benison is a Grace and Favour house, the patronage of which is vested in two or three powerful persons in the City. 1956 MONCREIFFE & POTTINGER *Blood Royal* 54 Others contain 'grace and favour' apartments, for rewarding distinguished servants of the country in their retirement. 1967 *Whitaker's Almanack* 652/1 The Minister of Public Buildings and Works reported in Parliament..that, of 140 grace and favour residences, [etc.].

21. b. grace-note *transf.*

1927 *Blackw. Mag.* Dec. 827/1 A lady was holding forth, a born narrator, recklessly lavish of grace-notes and embroidery. 1957 J. BRAINE *Room at Top* xx. 177 The extra refinement, the grace-note, was Jack's waving away of my offer to buy the drinks. 1960 *Guardian* 30 Sept. 12/5 Its architecture, for all its English grace-notes, is fundamentally international.

grace, *v.* **3.** Delete †*Obs.* and add later example.

1961 F. J. RIPLEY *Last Gospel* xii. 114 God may have graced them more than he has graced us.

graceless, *a.* Add: **1. a.** *graceless florin* (see quot. 1870).

1870 H. W. HENFREY *Guide to Study Eng. Coins* II. 137 The usual letters D.G., for Dei Gratia, were omitted...

This raised a storm of remonstrance against this coinage [of 1849], which at once received the name of the 'godless' or 'graceless florin'. **1897** [see GODLESS *a.*].

gracilis (græ·silis). *Anat.* [L., = slender.] Also *musculus* (*adductor*) *gracilis*: a superficial muscle on the medial side of the thigh, passing from the hip-bone to the tibia, and acting as an adductor of the hips and a flexor and medial rotator of the leg.

1615 H. CROOKE *Descr. Body Man* xxxvii. 804 The second muscle of the Legge which is also called *Gracilis* or the slender muscle. **1685** S. COLLINS *Syst. Anat.* I. xviii. 111 The *Gracilis*..is inserted with a round Tendon into the inside of the *Os Tibiæ*. **1727** CHAMBERS *Cycl.* I. 176/1 *Gracilis*, in anatomy, a muscle of the leg, thus called from its slender shape. **1858** H. GRAY *Anat.* 281 The *Gracilis* is the most superficial muscle on the inner side of the thigh. **1964** *Cunningham's Textbk. Anat.* (ed. 10) 353/1 The gracilis arises by a tendon, short, thin, and wide, from the lower half of the edge of the pubic symphysis.

gracility. Add: **2.** *fig.* Of literary style: unornamented simplicity.

1900 T. W. WRATISLAW *A. C. Swinburne* 146 The dexterous verses *To a Cat* are in a lighter tune than usual, recalling the gracility of Hugo. **1906** *Academy* 1 Sept. 201/2 Their spontaneity, their *gracility*, to borrow a word from the Latin, is perfect.

gracing (grēⁱ·siŋ), *vbl. sb.*² *slang.* Also **greycing.** Contracted form of *greyhound racing* (see *GREYHOUND 4).

1928 *Star* 8 June, Gracing at Wimbledon. **1928** *Daily Express* 20 July 17/5 Greycing...Programmes..for to-night's greyhound racing meetings. **1935** E. C. ASH *Bk. Greyhound* viii. 141 Greyhound Racing, or 'Gracing', as it is sometimes termed, started in 1926.

gracious, *a.* Add: **2. c.** *gracious living*: an elegant way of life, esp. with reference to the proprieties and niceties in standards of house-keeping. Occas. ironical. Hence *gracious liver.*

[**1932** Q. D. LEAVIS *Fiction & Reading Public* 319 Cf. the following advertisement taken at random from one of the luxurious women's magazines: Those who golf at St. Andrews..shop in the Rue de la Paix..those who live graciously, are fastidious in their choice of ships.] **1937** *New Yorker* 16 Jan. 24/3 Many a shoe wholesaler has learned more about what constitutes a Pattern of Gracious Living. **1945** *Palestine Post* 26 Oct. 7/6 American magazines..are all geared to some super-glamourized Gracious Living in relation to the opinion of the neighbours. **1951** J. CANNAN *And All I Learned* ix. 154, I don't demand luxury..but I do like gracious living. **1953** K. AMIS *Lucky Jim* xiv. 145 It should be possible for the right man to stop, or at least hinder, her from being a refined gracious-liver and arty-rubbish-talker. **1958** *Observer* 13 Apr. 9/3 One of the main problems the Russians are wrestling with to-day is how to achieve 'gracious living'. **1959** 'M. DERBY' *Tigress* iii. 129 The suburban vulgarity of gracious living. **1969** M. PUGH *Last Place Left* xxiii. 178 In a booming bistro..we sat hard against a gracious liver who called the bread the club of death.

grad¹**,** abbrev. of GRADUATE *sb.* 1; also of UNDERGRADUATE *sb.* 1.

1871 L. H. BAGG *4 Years at Yale* 45 *Grad*, abbreviation for graduate. **1893** W. K. POST *Harvard Stories* p. ix, Reverend grads., from the tales I have heard ye tell, I opine that the undergraduate is still the same. **1920** F. SCOTT FITZGERALD *This Side of Paradise* (1921) iv. 143 Amory 'ran it out' at a great rate, bringing the most eccentric character to dinner, wild-eyed grad students, preceptors with strange theories. **1950** M. MARPLES *University Slang* 184 The provincial universities..have even produced analogous abbreviations [to undergrad], such as grad = graduate (Durham, 1934+) and post-grad = post-graduate (Leeds, 1940). **1957** W. H. WHYTE *Organization Man* xxi. 269 Only 27·3 per cent of high school grads aged twenty-five to thirty-four..were inter-state migrants. **1958** *Times* 3 Nov. 6/1 Town and gown in Cambridge have..been inclined to settle their differences through fists..and once again conflict is causing some concern, this time under the heading of 'grads' versus 'teds'. **1969** *Daily Tel.* 10 Mar. 1/4 Cambridge police were last night looking for three youths who beat up Rajeer Dhavan, 22, undergraduate son of the Indian High Commissioner, in the latest episode of 'grad-bashing'.

grad² (græd). [prob. f. GRAD(E *sb.*] = GRADE *sb.* 1 b.

1909 in WEBSTER. **1953** *Amer. Jrnl. Physics* XXI. 294 Proposed symbols for physical units...Grad (=10⁻² rt. angle). **1959** G. & R. C. JAMES *Math. Dict.* (ed. 2) 176/2 *Grad*, one-hundredth part of a right-angle in the centesimal system of measuring angles, also called a grade or degree.

grad³**,** abbrev. of *GRADIENT *sb.* 5, used chiefly in symbolic contexts.

1909 J. G. COFFIN *Vector Analysis* v. 102 The vector.. is called the gradient or the slope of *V*..and is written grad *V* or slope *V*. *Ibid.* 103 It remains to show that the operator ∇ applied to *V* gives the grad. **1957** R. S. LONGHURST *Geom. & Physical Optics* xix. 393 If ϕ is a scalar point function, grad ϕ is a vector. **1962** T. M. APOSTOL *Calculus* II. iv. 170 The notation grad *f* is also used instead of ∇f.

grade, *sb.* Add: **1. b.** (Later example.)

1968 *Times* 24 June 25/5 May I make a plea that the makers of theodolites cease producing circles marked in the sexagesimal system and use the metric system—that is, divide a right angle into 100 grades?

4. c. A class at school in relation to advancement. (Cf. GRADED *ppl. a.* 2.) *N. Amer.*

1835 *Southern Lit. Messenger* I. 275 In the first and second grades boys and girls are schooled together. **1852** *Indiana Hist. Soc. Publ.* III. 619 Qualified to teach in any of the grades, from the primary to the grammar school. **1903** A. B. HART *Actual Govt. Amer. Conditions* 543 Many cities have public kindergartens, which take children of 4 to 5 years of age in hand and teach them simple beginnings. The next division is usually the primary, extending over 3 to 6 years, followed by about 4 years of the grammar school; these two systems taken together are often called simply 'the grades'. **1904** O. H. LANG in *Forum* (U.S.) Oct. 268 Polite conduct and usages might advantageously be taught in the grades and in the high school. **1906** C. A. MCMURRY (*title*) Course of study in the eight grades. **1909** *Springfield* (Mass.) *Weekly Republican* 18 Feb. 2 In the British isles the classes are designated standards and these are supposed to correspond to our grades. **1934** W. SAROYAN *Daring Young Man* (1935) 75 A little girl named Maxine, in the third grade. **1967** *Atlantic Monthly* Apr. 102/2 My plan after being passed out of Grade VIII, King Edward School, was to go with my best girlfriend. **1968** *Globe & Mail* (Toronto) 3 Feb. 8/4 Ontario..will have public French education from Grade 1 through university. **1970** *Publ. Amer. Dial. Soc. 1968* L. 50 Attended Memphis public schools. Completed fifth grade.

d. A mark (usu. alphabetical) indicating an assessment of the year's work, examination papers, etc., of a student.

1886 *Ann. Rep. Pres. Harvard 1885–6* 72 The marks received by a student in the several studies of his college course have until now been combined to determine the grade of his degree. **1889** *Harvard Faculty Rec.* 15 Oct., Any member of the graduating class who has attained Grade C or a higher grade in eighteen courses [etc.]. **1963** BARNARD & LAUWERYS *Handbk. Brit. Educ. Terms* 101 *Grading*, a method of placing pupils in groups according to their academic attainment in examinations or tests. Frequently five grades, named A, B, C, D, E, are used. **1964** MRS. L. B. JOHNSON *White House Diary* 10 June (1970) 163 Her grades were excellent and she planned to go to the University of Southwestern Louisiana. **1966** *Rep. Comm. Inquiry Univ. Oxf.* II. 68 Where more than three A levels in different subjects had been taken, the rating is based on the three best grades. *Ibid.* 463 Questionnaire for Schools Survey...What had been their performance in 'A' Level examinations (subjects attempted and grades achieved).

5. a. (Earlier U.S. examples.)

1807 J. MARSHALL *Life of Washington* V. 213 To talents of the highest grade, he united a patient industry. **1816** U. BROWN *Jrnl.* in *Maryland Hist. Mag.* (1915) X. 274 Jack Oaks of the meanest grade. **1818** H. B. FEARON *Sk. Amer.* 30 Neither trades are (to use an Americanism) of the first grade. *Ibid.* 191 Because their neighbour's face was (to use their own phrase) a grade darker than their own.

c. *grade A, Grade A*: of the highest grade in value, of the best or highest quality; hence *colloq.* (occas. ironically), extremely good, first-rate.

1911 *Proc. 5th Ann. Conf. Amer. Assoc. Medical Milk Commissions* 100 On or soon after the first of January 1912, there will be three grades of milk, A, B, and C. Grade A—for infants, children, and adults certified (or equivalent such as guaranteed)...Grade B—for adults only, Pasteurized...Grade C—for cooking purposes only. **1920** *Collier's* 20 Mar. 35/3 All you could see was her little peaches and grade A cream face stickin' out over the top. **1935** N. MITCHISON *We have been Warned* iv. 385 She'd always insisted on Grade A milk. **1936** A. THIRKELL *August Folly* i. 10 A very fine herd of cows which supply Grade A milk, at prices fixed by the Milk Marketing Board. **1942** O. NASH *Good Intentions* 45 Give me a grade-A May day. **1948** H. R. DAVIDSON *Prod. Market. Pigs* iv. 51 The 'basic price' was the price for a Grade C pig in each class. Bonuses above and reductions below the basic prices applied to Grades A and B and to D and E. **1968** R. STOUT *Father Hunt* (1969) xiv. 161 If the fingerprints didn't match we were left with a Grade A mess.

7. (Additional examples.)

1851 J. F. W. JOHNSTON *Notes N. Amer.* I. vi. 164 Most of the stock are *grades*, as they are called, or crosses of the pure Devon bull with the older stock of the country. **1852** *Trans. Mich. Agric. Soc.* III. 142 A few full blood Saxons; the rest are a grade sheep. *Ibid.* 182 My cattle are yet grades, and am getting rid of them as fast as possible. *Ibid.* 184 One good grade sow. **1857** *Ibid.* VIII. 707 The greatest show, however, for number and size, was among the grades and natives. **1882** *Rep. Maine Board Agric.* XXVI. 168 In every dairy of six cows I would keep not less than two Jerseys, or their grades. **1939** [see *AFRIKANER 2]. **1945** *New England Homestead* 13 Oct. 9/2 If you mess around with breedin' purebreds a while longer you'll be so much in love that you'll never go back t' keepin' grades. **1949** *Caribbean Q.* I. ii. 36 A small herd of grade red poll cattle was maintained. **1970** *Daily Nation* (Nairobi) 16 Jan. 17/4 Cattle rearing is not encouraging but grade cattle has [*sic*] been introduced in some areas.

8. b. A group of animals at a similar level of development, but not necessarily having a common genetic origin. Cf. *CLADE *sb.*²

1957 J. S. HUXLEY in *Nature* 7 Sept. 455/2 Stasigenesis results in the formation of delimitable and persistent anagenetic units, or grades. **1958** —— in *Uppsala Univ. Årsskrift* vi. 27 The best general term for such anagenetic units would seem to be *grade*. **1961** G. G. SIMPSON *Princ. Animal Taxon.* iv. 126 That grade is..now known to have been reached independently by a considerable number of different clades. **1967** R. E. BLACKWELDER *Taxon.* xiii. 220 It is not uncommon to interpolate such levels as Branch, Grade, and Series between Subkingdom and Phyla.

10. b. (Earlier U.S. examples.)

1811 *Deb. Congress U.S.* (1853) 8 May 2171/1 Each grade of the [Cumberland] road to be perfectly levelled. **1850** R. GLISAN *Jrnl. Army Life* (1874) ii. 12 The cars are moved on these heavy grades by stationary engines. **1861** *Harper's Mag.* Jan. 147/1 The descent of the 'grade' was the next rough feature in our day's journey...The trail on the grade was slippery with sleet, and walking upon it was out of the question.

d. *Physical Geogr.* The condition of a river in which, after initial down-cutting of its bed, further down-cutting is balanced by aggradation; a state of equilibrium between the erosion of material from a river-bed and the deposition of fresh sediment; also, the profile of a river or part of one throughout which this condition exists; *at grade*, in this condition.

In quots. 1876, 1894 the sense is closer to 'slope' (GRADE *sb.* 10).

[**1876** G. K. GILBERT in *Amer. Jrnl. Sci.* CXII. 100 In this way a stream, which has a supply of debris equal to its capacity, tends to build up the gentler slopes of its bed and cut away the steeper. It tends to establish a single, uniform grade. **1894** W. M. DAVIS in *Jrnl. Geol.* II. 78 In certain cases, it seems to be possible for a stream to cut down its profile to a gentler grade in its early adolescence than is suitable to later adolescence and maturity.] **1902** —— in *Ibid.* X. 89 The use of 'grade', in the sense here advocated, was almost reached by Gilbert. *Ibid.* 92 The development of grade depends on the spontaneous adjustment of the capacity of a river to do work, and the quantity of work to be done by the river. **1934** C. R. LONGWELL et al. *Outl. Physical Geol.* iii. 42 In places where there is an approximate balance between erosion and deposition, the stream is said to be graded or at grade. **1936** *Geogr. Jrnl.* LXXXVII. 25 At least twice the river cut down to grade and had begun to widen its channel when rejuvenation forced it to trench further. **1942** O. D. VON ENGELN *Geomorphol.* viii. 134 Maintenance of grade consists of a continuous, infinite series of adjustments between volume, slope, and sediment supply. **1963** D. W. & E. E. HUMPHRIES tr. *Termier's Erosion & Sedimentation* v. 106 The profile is actually a composite curve, showing breaks at each important confluence and irregularities caused by the nature of the rocks which it crosses. It is a uniform slope varying regularly, a 'grade'. **1970** O. T. JONES in G. H. Dury *Rivers & River Terraces* iii. 75 Between Carmarthen and Fanog the Towy is graded to present sea-level, and the tributaries which enter it are at grade with the existing river.

e. *to make the grade*: to reach the proper standard, to be successful. orig. *U.S.*

1912 J. SANDILANDS *West. Canadian Dict.* 20 Make the grade, make the running. **1921** S. FORD *Inez & Trilby May* ix. 168 It's Gwendolyn that's got to do the hustling. Three days! I doubt if she can make the grade. **1922** *Collier's* 7 Oct. 5/1, I don't think he can make the grade. **1930** *Publishers' Weekly* 5 July 29/1 Can the seasonal bookshop make the grade, and under what conditions? **1932** A. D. POWELL *Venusberg* xxiii. 235 How will it feel when the Recording Angel calls your bluff for the last time? How many of us will make the grade? **1958** *Listener* 9 Oct. 568/3 A would-be thief who cannot make the grade. **1958** *Times* 24 Oct. 14/7 Word spread quickly round the village: my cottage had made the grade.

11. (Earlier examples of *at grade.*)

1849 *Acts & Resolves Prov. Mass. Bay* 126 Number of public ways crossed at grade, Number of railroads crossed at grade. **1869** *Congress. Globe* 24 Mar. 252/3 Does not the junction road..cross the streets..at grade? ..No, sir; it crosses them above grade in nearly every instance.

12. grade-crossing (earlier and Canad. examples); **grade school** (later examples); **grade teacher** *N. Amer.*, a teacher in a grade school.

1890 *Boston Jrnl.* 26 Aug. 4/2 (*caption*) Want a Grade Crossing. **1906** *Daily Colonist* (Victoria, B.C.) 27 Jan. 2/4 (*heading*) Fatal Grade Crossing Accident. **1967** *Boston Globe* 5 Apr. 51/6 The Celtics have been the greatest since Muhammad Ali was in grade school. **1970** *New Yorker* 14 Nov. 209/1 College, high-school, and grade-school teaching is unsatisfactory almost everywhere. **1906** *Springfield* (Mass.) *Weekly Republican* 12 Apr. 6 The grade teachers attempting little..except setting the class to sing music already familiar. **1909** G. STRATTON-PORTER *Girl of Limberlost* ix. 165 She gathered..all sorts of natural history specimens and sold them to the grade teachers.

grade, *v.*² Add: **5.** (Earlier U.S. example.)

1873 *Rep. Vermont Board Agric.* II. 92 He..thought he should improve the color of his butter by grading his herd with Jersey blood.

8. *intr.* To pass imperceptibly from one grade *into* another. Also with *down, off, over, toward, up*. Also, *to grade up*: to take rank with a higher grade or class (cf. 5); *to grade up with*: to compare with, to be like.

1892 R. D. SALISBURY in *Ann. Rep. Geol. Surv. New Jersey 1891* 74 While the drumlin type is fairly distinct, drumlins grade into hills which are not drumlins. **1903** *Harper's Mag.* July 186 The type graded downward into the lower invertebrates. **1907** 'O. HENRY' *Heart of West* i. 5 When a man marries a queen he ought to grade up with her. **1921** *Brit. Mus. Return* 157 Rock-crystal grading into quartzite, from..Minas Geraes, Brazil. **1922** C. E. MULFORD *Black Buttes* xiv. 227 Strikes me funny, though, the way they [*sc.* the new cattle] grade up. **1925** ODELL in E. F. Norton *Fight for Everest, 1924* 292 This granite..frequently graded off into pegmatite. **1927** HALDANE & HUXLEY *Anim. Biol.* ix. 198 The activity grades down from one pole of the egg to the other. **1943**

J. S. HUXLEY *Evol. Ethics* iv. 26 Ethics grades over into the prescriptions of totem and tabu. **1952** P. ABRAHAMS *Path of Thunder* II. ii. 91 If the whites and the black people were equal, if there were no colour bar, if a black man could go to Parliament and had all the same rights as a white man, and the Coloured people wanted to grade toward the whites, then it would be all right. **1971** *Nature* 1 Jan. 16/1 Basalt, up to 2 km thick, grades downwards into gabbro and metamorphic rocks.

9. To read and mark (a student's paper) with a grade.

1931 H. F. PRINGLE *Theodore Roosevelt* I. iii. 37 Lodge ..graded their papers with undue severity. **1948** *Christian Sci. Monitor* 22 Apr. 4/3 The robot prof. is a machine that automatically grades homework and examination papers, at the rate of 10 a minute.

-grade, an adj. suffix repr. L. *-gradus* stepping, walking (f. stem of *gradus* step, *gradi* to step, walk), occurring in a few L. compounds, e.g. *retrōgradus* RETROGRADE, *tardigradus* TARDIGRADE. Hence in mod.L. scientific formations, as *digitigradus* DIGITIGRADE, *plantigradus* PLANTIGRADE.

gradeability (grēⁱdăbi·lĭti). [GRADE *sb.* + ABILITY.] The ability of a vehicle to climb a gradient at an efficient speed.

1952 *Automobilist* Jan. 10 Hill climbing. Per cent of gradeability. Synthetic rubber tires 4·3%...Winterized mud-snow tires 7·6%...Tire chains 23·5%. **1959** *Chambers's Encycl.* IX. 572/1 Even though a vehicle may have a relatively poor top gear 'gradeability' it must still be able to climb all ordinary road gradients with its full load on low gear. **1967** *Jane's Surface Skimmer Systems 1967–68* 99/1 Characteristics..Gradeability, forward slope 60%. Gradeability, side slope 30%.

graded, *ppl. a.* Add: **4.** (Earlier U.S. examples.)

1876 H. T. WILLIAMS *Pacific Tourist* 185/2 The immense range fenced in at this point is occupied by a select herd of graded stock. **1879** *U.S. Dept. Agric. Spec. Rep., No. 12. Invest. Dis. Swine* 187 The graded calves of this county have this year suffered severely by a disease called black-leg.

5. *Physical Geogr.* Of a river or its profile: at grade (see *GRADE *sb.* 10 d); having attained grade.

1894 W. M. DAVIS in *Jrnl. Geol.* II. 77 Mr. Gilbert has recently suggested to me that a stream in this condition of balance between degrading and aggrading might be called a graded stream; and its slope, a graded slope. **1934** C. R. LONGWELL et al. *Outl. Physical Geol.* iii. 42 When a part of a main stream reaches grade, the local tributaries soon become graded with respect to it. **1944** A. HOLMES *Princ. Physical Geol.* x. 154 When the profile is developed so that it everywhere provides the necessary minimum gradient, it is called a graded profile or a profile of equilibrium. **1946** L. D. STAMP *Britain's Struct.* v. 47 A river tends to reach a state of equilibrium and its longitudinal profile will form a smooth curve from source to mouth. When it reaches this stage a river is said to be graded. **1970** [see *GRADE *sb.* 10 d].

grader. 2. (Later examples.)

1953 R. J. C. ATKINSON *Field Archaeol.* (ed. 2) ii. 58 For stripping large areas a bulldozer or grader is to be preferred. **1963** A. LUBBOCK *Austral. Roundabout* 11 The roads are kept up by a 'grader', a kind of steamroller with a wide scraper attached in front. **1963** *Field Archaeol.* (Ordnance Survey) (ed. 4) 14 When a grader is removing topsoil as a preliminary to the extension of a chalk pit it may reveal ditches, post-holes, beam slots, rubbish pits [etc.]. **1971** *Timber Trades Jrnl.* 21 Aug. 26/2 No grading machine is in operation at present in Sweden, but the major sawmillers are actively studying the potential offered by machine graders.

Gradgrind (græ·dgraind). Name of the mill-owner in Dickens's *Hard Times* (1854), 'a man of facts and calculations', used allusively for: one who is hard and cold, and solely interested in facts. Hence **Gra·dgrinding, Gra·dgrindery.**

1855 *Putnam's Mag.* Jan. 76/2 There have not been wanting travelled Gradgrinds to assure us that the song from his lips was a humbug and a sham. **1871** E. EGGLESTON *Hoosier Schoolmaster* (1872) v. 37 You, my Gradgrind friend, you think me sentimental. **1920** *Glasgow Herald* 14 Aug. 5 A..metropolis of ant-like industry and social Gradgrindery. **1924** *Ibid.* 18 Apr. 8 A nation of Gradgrinds, immersed in work and money-getting, and denying themselves and their employees any opportunity of recreation. **1925** *Public Opinion* 7 Aug. 122/1 The gradgrinding system. **1927** *United Free Ch. Mission Rec.* Sept. 379/2 These self-appointed Gradgrinds seem to imagine that their actions are pleasing to the Almighty. **1958** R. WILLIAMS *Culture & Society* I. v. 94 Public commissions, Blue Books, Parliamentary legislation—all these, in the world of *Hard Times*—are Gradgrindery. **1968** *Listener* 25 July 124/1 Sidney and Beatrice [Webb] instructed Asquith, Balfour, Churchill... Gradgrinds they may have been—it is Beatrice's own self-description—but one cannot deny them greatness. **1971** *Where?* Oct. 308/1 The interest motif can be just as bewildering as Gradgrindery, and it should be put into reverse; namely, it is the teacher's task to make interesting *that which is relevant*.

gradient, *a.* and *sb.* Add: **B.** *sb.* **1. b.** (Examples.)

1845 *Rep. Brit. Assoc. 1844* II. 96 It was necessary that that railway should present long and very steep gradients. **1915** R. B. HOLT *Tramway Track Constr. & Maintenance* ix. 114 The wear on the rails on all parts of the gradient, both on the up track and on the down track, is exceedingly irregular. **1971** *Homes & Gardens* Aug. 90/1 The train.. could be heard puffing like an old man, 'Chuff, chuff, chuff', as it travelled up the gradient approaching the cutting. **1971** *Daily Tel.* (Colour Suppl.) 27 Aug. 12/3 Snow drove into our faces and on the steep gradients where skis had to be removed we stumbled in deep powder.

2. In wider use: a continuous increase or decrease in the magnitude of any quantity or property along a line from one point to another; also, the rate of this change, expressed as the change in magnitude per unit change in distance.

1886 J. A. FLEMING *Short Lect. Electr. Artisans* vii. 122 Along the lead there is a regular fall or gradient of [electrical] pressure. **1892** W. PEDDIE *Man. Physics* ix. 132 The rate of variation of density per unit of length is *r*...The quantity *r* is generally called the 'concentration-gradient'. **1898** *Proc. R. Soc.* LXIII. 364 The kathode fall is constant for all pressures and currents whilst the potential gradient along the rest of the tube is variable. **1902** POYNTING & THOMSON *Prop. Matter* xviii. 205 The ratio of the stress to the velocity gradient is called the viscosity of the fluid. **1910** *Encycl. Brit.* V. 891/2 This outflow of heat necessitates a rise of temperature with increase of depth. The corresponding gradient is of the order of 1°C. in 100 ft. **1948** GLASSTONE *Textbk. Physical Chem.* (ed. 2) iv. 260 The gradient is actually negative, that is the concentration decreases from left to right. **1957** *Encycl. Brit.* X. 681/2 He proposed to measure the rate of change or gradients in the gravitational field. **1962** A. R. W. HAYES *Revision Physics* 98 We must measure..the uniform temperature gradient along the bar—found from readings of thermometers placed in mercury..in holes bored in the specimen. **1970** *Nature* 19 Dec. 1225/1 There is a gradient of dormancy within the spikelet, the larger proximal seed being less dormant than the smaller distal seed, while a much smaller third seed..is extremely dormant.

b. *spec.* in *Embryol.*, such an increase or decrease, along an axis of an organism or a part, in the potential for developing into an organ or in a related bodily process.

1911 [see **axial gradient*]. **1915** C. M. CHILD *Individuality in Organisms* iii. 65 Gradients in rate of cell division, size of cells, condition or amount of protoplasm in the cells, rate of growth, and rate and sequence of differentiation are very characteristic features of both animal and plant development. Such gradients are definitely related to the axes of the individual or its parts, and are..expressions of axial metabolic gradients. **1924** BELLAMY & CHILD in *Proc. R. Soc.* B. XCVI. 141 In a protoplasm of specific hereditary constitution, such a gradient is adequate as the initiating factor in the axial differentiation characteristic of that species. **1927** [see *FIELD *sb.* 17 c]. **1953** J. S. HUXLEY *Evol. in Action* i. 29 Gradients exist in the developing organism—gradients in metabolism, growth-potential, and other factors. Genes altering the shape and intensity of such gradients will affect a number of parts simultaneously. **1957** [see *FIELD *sb.* 17 c]. **1970** F. CRICK in *Nature* 31 Jan. 420/1 It is an old idea that 'gradients' are involved in embryological development...Many of the gradients to which Child referred seem more likely, in retrospect, to be the results of development rather than its cause. An outsider to embryology has the impression that in recent years gradients have become a dirty word. *Ibid.* 422/1 If this approach serves to make the idea of diffusion gradients respectable to embryologists it will have served its purpose.

4. The degree of steepness of a graph at any point, measured by the tangent of the angle between the horizontal axis and either the line (if straight) or the tangent to the curve; (see also quot. 1937).

1897 H. LAMB *Elem. Course Infinitesimal Calculus* ii. 67 It is convenient to have a name for the property of a curve which is measured by the derived function. We shall use the term 'gradient' in this sense. **1937** E. J. MCSHANE tr. *Courant's Diff. & Integral Calculus* (ed. 2) I. xi. 90 The slope or gradient of the curve is given by tan α, and hence the term *gradient* is occasionally used for the derivative of the function represented by the curve. **1942** C. E. K. MEES *Theory Photogr. Process* xix. 702 The *D*, log *E* curve continues..into the region of decreasing exposure with constantly decreasing gradient. **1958** A. BARTON *Introd. Coordinate Geom.* v. 64 A line whose gradient is zero is parallel to the *x*-axis; as the gradient increases the line gets steeper. *Ibid.* 66 Two lines are..perpendicular if the product of their gradients is −1.

5. *Math.* A vector function whose components along the co-ordinate axes are the partial derivatives with respect to the corresponding variables of a given scalar function; it is denoted by ∇f (see *DEL) or by grad *f*, where *f* is the scalar function.

1901 E. B. WILSON *Vector Analysis* iii. 138 The vector sum which is the resultant rate of increase of *V* is denoted by ∇V...The terms gradient and slope of *V* are..used for ∇V. **1936** E. J. MCSHANE tr. *Courant's Diff. & Integral Calculus* II. iii. 89 The direction of the gradient is the direction in which the function increases most rapidly. **1966** *McGraw-Hill Encycl. Sci. & Technol.* II. 413/1 There are three differentiation processes that are of conceptual value in the study of vectors: the gradient of a scalar, the divergence of a vector, and the curl of a vector.

6. *attrib.* and *Comb.*, as **gradient wind** *Meteorol.*, the (hypothetical) wind whose direction is that of the geostrophic wind but whose speed is calculated by allowing for the effect on the geostrophic wind of the centrifugal force that results from its curved path.

1908 E. GOLD *Barometric Gradient & Wind Force* 24 We can construct a scale..which shall give..the Beaufort number corresponding to the theoretical gradient wind for straight isobars for any pressure distribution. **1928** [see *GEOSTROPHIC *a.*]. **1966** *McGraw-Hill Encycl. Sci. & Technol.* VI. 244/2 The gradient wind is a good approximation to the actual wind and is often superior to the geostrophic wind, particularly when the flow is strongly curved in the cyclonic sense.

grading, *vbl. sb.* Add: **2. c.** The placing of school teachers or pupils in groups according to ability and other qualifications.

1903 A. B. HART *Actual Govt. Amer. Conditions* 543 The number of children is great enough to allow complete grading. **1923** *N.Z. Educ. Gaz.* 3 Apr. 39/1 Revised grading-marks will reach all primary-school teachers this month. **1923** *Ibid.* 1 Oct. 157/1 The grading of primary-school teachers. **1958** S. ASHTON-WARNER *Spinster* 90, I have tried.., giving far more to my work than many a crack Infant Mistress in town, dancing upward on the grading list. **1963** [see *GRADE *sb.* 4 d].

gradiometer (grēⁱdiǫ·mĭtəɹ). [f. GRADI(ENT *sb.*: see -METER.] **a.** Any of various surveying instruments used for setting out gradients or for measuring the gradient of a slope (see quots.).

1899 W. G. BLIGH *Notes on Instruments Engin. Field-Work* v. 89 The ordinary type of level is always at a disadvantage when the ground is sloping. To overcome this difficulty..a variation has been made..which permits the telescope to be inclined. One of these is Stanley's Patent Gradiometer, an excellent combination of the level and clinometer. **1940** *Chambers's Techn. Dict.* 384/2 *Gradiometer*, an instrument for setting out long uniform gradients; it consists of a level that may be elevated or depressed, by known amounts, by means of a vertical tangent screw. **1960** H. L. MICHAEL in K. B. Woods et al. *Highway Engin. Handbk.* ii. 11 Devices called 'gradiometers' or 'inclinometers', essentially bubble tubes graduated to read either percentage of grade or degree of incline, have been constructed to obtain grade information.

b. An instrument for measuring the gradient of a field, esp. the horizontal gradient of the earth's gravitational or magnetic field.

1929 *Mining Mag.* XL. 272 (*heading*) The gravity gradiometer. *Ibid.* 277 On account of its small size and its rapid speed of operation the gradiometer is admirably adapted for use in confined spaces such as underground mine galleries. **1934** *Trans. Amer. Inst. Mining & Metall. Engineers* CX. 383 At right angles to the magnetic meridian..a positive gradient implies that the vertical component of the magnetic field is increasing to the east. In the magnetic meridian, the effective component is that in the magnetic vertical plane through the center of the gradiometer. **1957** *Encycl. Brit.* X. 681/2 In a design by Shaw and E. Lancaster-Jones [of an instrument for the measurement of gravity gradients] the beam system is made irresponsive to 'curvature' effects and is called a gravity gradiometer. **1971** *Sci. Amer.* Aug. 70/2 During the time when the moon was inside this region of very steady magnetic fields we sent commands to the instrument that enabled it to function as a gradiometer.

gradocol (græ·dokǫl). Also **Gradocol.** [See quot. 1931.] Used *attrib.* in *gradocol membrane*: a membrane made by the controlled evaporation of a collodion solution, having a high uniformity of pore size and used as a filter, esp. for viruses.

1931 W. J. ELFORD in *Jrnl. Path. & Bacteriol.* XXVIII. 507, I propose to refer to these new membranes as gradocol membranes, since they are products of a graded coagulation of collodion. **1953** RHODES & VAN ROOYEN *Textbk. Virol.* (ed. 2) I. ii. 12 Suspensions of virus particles are filtered through a series of 'gradocol' membranes of varying degrees of porosity, and the filtrates tested for the presence of virus. **1964** WILSON & MILES *Princ. Bacteriol. & Immunity* (ed. 5) I. xl. 1166 In estimating the size of virus particles ultrafiltration through Gradocol membranes is widely used.

gradometer (grădǫ·mĭtəɹ). [f. GRAD(E *sb.*: see -METER.] Any instrument for measuring the gradient of a slope or the deviation from the horizontal.

1901 J. H. BULLARD *U.S. Pat.* 685,569, This invention relates to gradometers, and has for its object the construction of an instrument of this class adapted especially to use on self-propelled vehicles...The usual curved glass tube..is provided with graduation-marks, indicating the per cent. of inclination of the device relative to the horizen [*sic*]. **1904** GOODCHILD & TWENEY *Technol. & Sci. Dict.* 266/1 *Gradometer*, an instrument for measuring the angle of dip of a rock formation or mineral deposit. **1924** F. CALDWELL *U.S. Pat.* 1,492,156, Improvement in Gradometers..relates to the visible indication of the grade up or down on which a vehicle..is travelling or standing, on which an aircraft is travelling, or the extent to which a ship is rolling or pitching.

gradualism. (Later examples.)

1931 *Time & Tide* 5 Sept. Suppl. 19/2 The reading of the facts now most popular in the Labour Movement will inevitably strengthen enormously the 'dictatorship of the Proletariat' solution... 'This,' remarked one of the younger Labour M.P.'s the other day, 'is the end of gradualism.' **1959** *Times* 16 Mar. 9 Gradualism is to continue to be Mr. Gomulka's agricultural watchword [in Poland]. **1963** *Economist* 15 June 1134/1 Gradualism is in disfavour with increasing numbers of Negroes. **1965** *Listener* 3 June 812/2 We were thinking in terms of the politics of democracy and gradualism, of separate African states, each different. **1969** *Times* 6 Jan. 9/2 The best hope for Ulster..is..to allow the gradualism of Captain O'Neill's ministry to go forward, without direct intervention from Westminster.

gradualist. (Later examples.) Also **gra·dualist, graduali·stic** *adjs.*, of or pertaining to gradualism.

1926 *Glasgow Herald* 25 Jan. 10 The duty of the Left Wing would be to fight with all its power against evolutionary and gradualistic theories. **1931** *Economist* 17 Oct. 698/2 It [*sc.* Labour] now sees no chance of securing these ends by merely gradualist methods within the general structure of the capitalist system. **1931** *Time & Tide* 7 Nov. 1273/1 And this most effective body of Labour is gradualist, not revolutionary. **1945** *Chicago Daily News* 4 Oct. 12/7 On a higher level of thought, we've labeled ourselves 'gradualists'; and stilled any uneasy twinge of conscience by assuming that everything will wash out in the fullness of time. **1947** KOESTLER in *Partisan Rev.* XIV. II. 142 You know that we are a reformist, gradualist movement. **1953** K. BRITTON *J. S. Mill* v. 170 Such a solution..allows us to regard inductive inference as inconclusive and gradualistic. **1960** *20th Cent.* Nov. 391 The fighters of the country are not generally the gradualists. **1961** *Listener* 21 Dec. 1059/2 We cannot sit back and *hope* that..Africa will retain our gradualist and liberal traditions.

graduate, *a.* and *sb.* Add: **B.** *sb.* **1.** (Earlier and later examples of the Amer. application to a schoolboy.)

1773 J. ADAMS *Diary* in *Works* II. 321 Their academy [in Phila.] emits from nine to fourteen graduates annually. **1952** *Manch. Guardian Weekly* 18 Sept. 13/2 To an astonishing degree Groton graduates have made names for themselves in public life.

4. (Earlier and later examples.)

1871 L. H. BAGG *4 Years at Yale* 112 Delta Phi has also four alumni associations, or 'graduate chapters'. **1880** *Harvard Catal.* 190 (*heading*) Graduate department. *Ibid.*, Any Graduate course which is taken by less than three students may be withdrawn at the option of the Instructor. **1880** *Harper's Mag.* July 251/2 The solution of the difficulty lies in..putting the extra studies in the graduate courses. **1893** *Bryn Mawr Program* 34 The most distinguished place among graduate students will be held by the Fellows. **1926** *Encycl. Brit.* II. 318/2 The period under review [1909–26] was marked by constant developments in the graduate schools of the university [*sc.* Harvard]. **1951** M. MCLUHAN *Mech. Bride* 43/2 The Harvard methods and pursuits differ little from those of any other graduate school. **1958** *Times Lit. Suppl.* 10 Oct. 578/2 Just another piece of graduate-student exhibitionism.

graduate, *v.* Add: **3.** To complete a high school course and receive a diploma. *U.S.*

1882 I. M. RITTENHOUSE *Jrnl.* in *Maud* (1939) 77 The very minute that she found out she was too far behind the class to graduate she stopped school. **1935** H. NICOLSON *Dwight Morrow* i. 14 Dwight was..able to graduate from High School at the premature age of fourteen.

graduation. Add: **6.** Also *attrib.*

1901 *Daily News* 2 Mar. 4/7 There are only seven signatures of Milton known, the first occurring in the Graduation Book of Cambridge, 1628–9. **1903** *N.Y. Times* 7 Oct. 6 The annual graduation exercises of the schoolship St. Mary's were held last night on board the ship. **1906** M. E. W. FREEMAN *By Light of Soul* 217 Maria dressed herself in her graduation gown.

Græco-, Greco-. Add: **b.** Græco-Roman *a.*, of a style of wrestling, resembling that used by the ancient Greeks and Romans, in which attacks are directed only at the upper part of the body.

1901 J. W. MCWHINNIE *Mod. Wrestling* 7 Græco-Roman wrestling, as now practised all over the civilised world, is understood to be a last-century continental revival of the athletic contests which figured prominently in the festivities of ancient Greece and Rome. **1934** [see *ALL-IN 2]. **1968** G. KENT *Pict. Hist. Wrestling* viii. 175 Alberg laid claim to the Greco-Roman championship of the world.

graft, *sb.*¹ **3.** (Earlier and later examples.)

1871 J. WOODMAN *Notes Transpl. or Engraft. Skin* 19, I took three small pieces of skin from her own arm and engrafted them...On the fourth day I removed the plaster, and with it one of the grafts. **1913** *Interstate Med. Jrnl.* XX. 573 Only the peripheral part of thyroid grafts persisted, the central part becoming necrotic... Regeneration was complete at the end of three months, and the graft differed from a normal gland by an infiltration of connective tissue. **1970** *Oxford Times* 16 Jan. 1/1 This is the first transplant operation to be carried out at the hospital. The other kidney graft in Oxford was at the Radcliffe Infirmary two years ago. **1970** *Nature* 28 Feb. 851 Liver grafts may be less exacting in their genetical requirements than those of kidney.

graft, *sb.*⁴ **a.** (Earlier and later examples.)

1853 J. ROCHFORT *Adv. Surveyor N.Z.* v. 47, I could make more money by 'hard graft', as they call labour in the colonies. **1933** *Bulletin* (Sydney) 15 Nov. 20/3 Another three miles' tramping, and four hours' hard graft. **1968** *Times* 27 June 25/1 This view is that salvation..is to be won by long, hard graft by industrial management. **1971** *Observer* 14 Mar. 7/7 They're too busy turning down 14 per cent pay offers to fuss about the three-quarters of a million out of graft.

graft (graft), *sb.*⁵ *colloq.* (orig. *U.S.*). [Origin uncertain. Perhaps a use of GRAFT *sb.*⁴ 'work' (cf. *job*); but some authorities connect it with GRAFT *sb.*¹ with the notion of 'excrescence'.] The obtaining of profit or advantage by dishonest or shady means; the means by which such gains are made, esp. bribery, blackmail, or the abuse of a position of power or influence; the profits so obtained.

1865 *Nat. Police Gaz.* (N.Y.) 8 July 1/3 'Twas handy that we were so related, as, when about a 'graft', or 'doing stur', both sisters could keep each other company. **1886** W. NEWTON *Secrets Tramp Life Revealed* 14 This 'Guide' cannot work this 'graft' alone, for he has to have a good supply for stock, a bag of 'snide' or base coins. **1889** in J. B. Thoburn *Hist. Oklahoma* (1916) I. xxix. 407 The enterprising individual sold water at so much a drink until he was ousted from his profitable graft. **1896** ADE *Artie* i. 3 To the church show—the charity graft. **1901** 'J. FLYNT' *World of Graft* ii. 12 Chi ain't no free soup kitchen. The City Hall people want their graft just as much as I [*sc.* a criminal] do. **1903** H. HAPGOOD *Autobiogr. Thief* (1904) ii. 34 In those days..Moll-buzzing, as well as picking pockets in general, was an easy and lucrative graft. *Ibid.* x. 205 The boy had a much better chance to learn the graft than I had when a kid, for my father was an honest man. *Ibid.* 222, I was too sleepy those days to go out of town much on the graft. **1903** *Daily Chron.* 21 Oct., A Chicago paper has the headline, 'Labour revolts at paying graft'. *Ibid.* 3 Nov. 5/3 Are you ready to support a government of law against a government of 'graft', an administration of the city's resources in the interest of the public and of the public treasury, against their dissipation for the benefit of a favoured few? **1904** *Ibid.* 17 Feb. 3/5 Glimpses are also given of the shady side of American politics, where 'graft' is only another word for plunder. **1904** *Athenæum* 30 Apr. 560 In New York 'graft' is thieves' patter for stealing. **1905** *Daily Chron.* 13 Sept. 5/2 It is now the turn of the War Department to start a campaign against 'graft' among the officials of the army. **1909** *Westm. Gaz.* 13 Oct. 5/4 Showing how Tammany robs the city, bribes the judges, protects criminals, and generally carries on the game of graft, an Americanism for bribery, corruption, and illicit commission. **1915** *Lit. Digest* (N.Y.) 21 Aug. 340/1 There is no evidence that Minister Sullivan received any money or participated in any way in what has come to be known as 'graft'. **1928** *Sat. Even. Post* 4 Feb. 35/1 'We had a slick graft' he told me, 'We was taking about two hundred smackers a week.' **1945** C. S. LEWIS *Hideous Strength* xi. 301 Here was a world of plot within plot, crossing and double crossing, of lies and graft and stabbing in the back. **1970** *Daily Tel.* 13 Jan. 2/4 Victims in a wave of graft, corruption and fear were making regular payments for protection.

b. *attrib.* and *Comb.*

1905 *Daily Chron.* 26 May 5/2 He..says he is tired of 'graft' politics and the sale of city franchises to monopolies [in Philadelphia]. **1908** *Ibid.* 24 Dec. 1/6 During the hearing of the latest 'graft' scandal here [*sc.* in Pittsburgh] evidence was given that sixty members of the City Council received 45,000 dollars as bribe money. **1908** 'O. HENRY' *Gentle Grafter* 48 I'd like, myself, to hedge a bet or two in the graft game. **1910** *Westm. Gaz.* 4 Apr. 7/2 He expressed himself as overwhelmed with..shame at the stain which had been revealed by the graft exposures upon the fame of the city [*sc.* Pittsburgh]. **1955** D. W. MAURER in *Publ. Amer. Dial. Soc.* xxiv. 150 Is there already a well-established, well-oiled graft-machine set up to process all *fixed* cases?

graft, *v.*³ (Earlier and later examples.)

1859 HOTTEN *Slang Dict.* 47 *Graft*, to go to work. **1936** J. CURTIS *Gilt Kid* ii. 19 'Where did you graft in Wandsworth?' 'Cleaner.' **1958** *Times* 18 Oct. 3/3 But his 90 minutes in the middle were worth many nets, and he was quite imperturbable in the way he grafted along [at cricket]. **1966** A. PRIOR *Operators* xvi. 246 The great mass of mugs were law-abiding..doing as they were told, working, grafting.

graft (graft), *v.*⁴ *colloq.* (orig. *U.S.*). [f. *GRAFT *sb.*⁵] *intr.* To practise 'graft'; to make money by shady or dishonest means.

1859 *Nat. Police Gaz.* (N.Y.) 14 May 3/4 Liz Thompson and her husband..do not intend going out to 'graft' until the summer season sets in, when they are going to Newport, Saratoga, and other fashionable watering resorts, at which game she made out so good last season. **1863** *Illustr. London News* 13 June 658/3, I am progressing wonderfully, and I expect Poll and Bob will be able to go out with me and graft (pick pockets) in a few days. **1895** *McClure's Mag.* Feb. 247/2 He had been 'grafting' with a 'mob' of pickpockets at county fairs. **1903** H. HAPGOOD *Autobiogr. Thief* (1904) ii. 48, I know some thieves who, although they have grafted for twenty-five years, have not yet 'done time'. **1905** D. G. PHILLIPS *Plum Tree* 61, I don't see how those in politics that don't graft, as they call it, are any better than those that do. Would they get office if they didn't help on the jobs of the grafters? **1960** *Observer* 25 Dec. 7/6 Anybody who had neglected to have a nice tickle during the late autumn would be out grafting for all he was worth. **1967** J. MORGAN *Involved* 27 They used to graft together..they pulled one or two big capers.

Hence **gra·fting** *vbl. sb.*³ and *ppl. a.*

1859 MATSELL *Vocabulum* 39 *Grafting*, working; helping another to steal. **1901** 'J. FLYNT' *World of Graft* 78 They make their living, such as it is, by grafting. **1904** *Treasury* Oct. 8/2 We excel other countries in the phenomenal corruption of our city Governments and Legislatures. The evil system of 'grafting', so called, extends everywhere. **1912** F. J. HASKIN *Amer. Govt.* 71 Large business houses felt the loss from the petty grafting of stamps by office boys. **1921** *Glasgow Herald* 13 June 9 The efforts of professional and grafting Irish agitators. **1960** *Observer* 25 Dec. 7/7 Christmas Day was not likely to be a big grafting day for various reasons.

grafter³ (gra·ftəɹ). *colloq.* (orig. *U.S.*). [f. *GRAFT *sb.*⁵ or *v.*⁴ + -ER¹.] **1.** One who makes money by shady or dishonest means; a thief; a swindler.

1899 'J. FLYNT' *Tramping with Tramps* 394 *Grafter*, a pickpocket. **1900** *Daily Express* 22 June 5/3 American thieves, hotel robbers, 'bunco steerers', 'grafters', 'con' men, and 'crooks' of every degree. **1901** 'H. MCHUGH' *John Henry* 49 No more swell Sandwich Salons for me, where the grafters want to butt in all the while. **1903** *Westm. Gaz.* 16 July 10/1 Camp followers, adventurers, and a weird aggregation of grafters. **1916** S. LEACOCK *Ess. & Lit. Stud.* 205 All the world loves a grafter—at least a genial and ingenious grafter—a Robin Hood who plunders an abbot to feed a beggar, an Alfred Jingle, a Scapin, a Raffles. **1926** E. WALLACE *Yellow Snake* ix. 76 You're not exactly wrapping up your words now... You've called me a grafter and a crook. **1967** J. MORGAN *Involved* 82 She's a straight bird..not a grafter.

2. One who practises 'graft', esp. in public life; a politician, official, etc., who misuses his position in order to reap dishonest gain or advantage.

1896 *Columbus* (Ohio) *Dispatch* 20 Apr. 3/8 Most of the 'grafters' have left the town, and not many of them will remain here. **1904** F. LYNDE *Grafters* xxiv. 298 You were to crush the grafters in this railroad struggle—show them up. **1904** *Daily Chron.* 28 Mar. 3/5 Kuropatkin shares with Prince Khilkoff, the Minister of Railways, the distinction of being one of the few high officials who are not called grafters. **1908** *Ibid.* 16 Nov. 1/7 Many of his 'grafter' friends will be tried for attempted murder. **1914** G. ATHERTON *Perch of Devil* I. 58 And although she has her pestilential politicians, her grafters.., yet ability and talent make good as always. **1935** A. J. CRONIN *Stars look Down* III. vi. 526 They've always been a set of grafters down there; local government has been one long sweet laugh. **1958** S. ELLIN *Eighth Circle* (1959) II. viii. 95 And what happens to somebody who..isn't a born grafter? They make him one, that's what!

grafter⁴ (gra·ftəɹ). *colloq.* [f. GRAFT *v.*³ + -ER¹.] One who works; a (hard) worker.

1900 H. LAWSON *On Track* 41 'What are we to do now?' enquired Andy, who was the hardest grafter, but altogether helpless, hopeless, and useless in a crisis like this. **1906** E. DYSON *Fact'ry 'Ands* vi. 62 Several of the piece-workers, confirmed grafters, toiled with swift fingers. **1913** W. K. HARRIS *Outback in Austral.* 24 Shared and shared alike, the loafer with the grafter. **1927** *Glasgow Herald* 19 Mar. 9 The visiting eight are heavy and expert in the tight, and to meet them we have a pack which contains at least three players who are by no means 'grafters'. **1959** *Times* 24 Jan. 3/7 He is a grafter rather than a fluent striker, with little back-lift, plenty of concentration, and a willingness to use his feet.

grafting, *vbl. sb.*¹ Add: **3. b.** (Earlier example.)

1888 H. MORTEN *Sk. Hosp. Life* 66 We must try grafting on that boy..who was burnt. You don't mind parting with a few portions of your epidermis, I suppose?

d. (Earlier example.)

1841 LADY WILTON *Art Needlework* (ed. 3) xx. 317 There is..darning—grafting—and patching.

grafting, *vbl. sb.*³: see *GRAFT *v.*⁴

graftonite (gra·ftǫnəit). *Min.* [f. *Grafton*, name of a village in New Hampshire, U.S.A.: see -ITE¹.] A pink to brown crystalline phosphate of calcium and bivalent iron and manganese, $(Fe,Mn,Ca)_3(PO_4)_2$, usually found associated with triphylite.

1900 S. L. PENFIELD in *Amer. Jrnl. Sci.* CLIX. 20 (*title*) On graftonite, a new mineral from Grafton, New Hampshire, and its intergrowth with triphylite. **1968** *Amer. Mineralogist* LIII. 742 Graftonite is a phosphate mineral of wide occurrence.

Graham¹ (grēⁱ·əm). Also **graham.** [See GRAHAMISM.] Used *attrib.* to designate unbolted wheaten flour, and bread or biscuit prepared from this. Also *absol.* = Graham bread.

1834 *Knickerbocker* IV. 305 'Hail!' said I, 'thou pure, unadulterated substitute—Graham bread!' **1834** W. L. GARRISON in *Life* (1885) I. 428 If they are Grahamites we have a fine spring of water in our cellar, and plenty of Graham flour upstairs. **1859** *Amer. Agriculturist* May 150/2 (*caption*) Graham biscuits or bread. **1874** *Rep. Vermont Board Agric.* II. 509 Large enough to hold the flour, the meal, the rye, the graham, butter. **1882** 'M. HARLAND' *Eve's Daughters* 443 Eat lightly—dry bread or biscuit, Graham crackers—anything that is easy of digestion. **1896** *Godey's Mag.* Feb. 207/1 When Mr. Dives turns from his recondite menu to nibble at wheaten grits and graham bread. **1947** J. STEINBECK *Wayward Bus* 220 It's got graham crackers instead of crust. **1957** G. CARSON *Cornflake Crusade* (1959) iv. 53 They shared their beans, boiled rice, graham bread and puddings. **1958** L. C. PRITCHETT *Cabin at Medicine Springs* (1959) xiv. 118 Ma often..made gruel out of coarse graham flour.

Graham² (grēⁱ·əm). [The name of Thomas *Graham* (1805–69), Scottish chemist.] *Graham's law*: that the rates of diffusion and of effusion of a gas are inversely proportional to the square root of its density.

1845 TODD & BOWMAN *Physiol. Anat. & Physiol. Man* II. xxix. 409 Oxygen being the lighter gas, a larger quantity is required to replace the carbonic acid; 81 parts of the latter will require 95 of the former to replace it according to Graham's law, that the diffusion volume of different gases varies inversely as the square root of the density. **1946** GLASSTONE *Elem. Physical Chem.* iv. 86 The results of Graham's law of diffusion have been used to determine gas densities by measuring the time required for a definite volume of the gas to effuse through a small hole in a thin metal plate.

Grahamite (= a follower of Graham). (Earlier and later examples.)

1834 [see *GRAHAM¹]. **1839** C. F. BRIGGS *Adv. H. Franco* II. iii. 27 'Have you got the dyspepsia?' asked Mrs. D. 'Ain't you a Grahamite?' asked Mr. D. **1864** B. TAYLOR *H. Thurston* iii. 45 They were Grahamites for a year or two—lived on bran bread and turnips, boiled wheat and dried apples. **1957** G. CARSON *Cornflake Crusade* iv. 49 The Grahamite was required not to drink water with his meals.

grain, *sb.*¹ Add: **7. e.** Any of the irregularly shaped discrete particles or crystals in a rock or a metal, usu. but not necessarily small.

(i) *Petrol.* **1813** R. BAKEWELL *Introd. Geol.* ii. 25 *Granitic*, composed of grains or crystals closely united without a cement. **1836** *Edin. New Philos. Jrnl.* IX. 268 It [*sc.* the granite] is not throughout pure, but is occasionally mixed with the gneiss,..or its ingredients, especially felspar, are disseminated in grains or crystals. **1882** A. GEIKIE *Text-bk. Geol.* II. ii. 140 Pitchstone...Examined microscopically, it is found to consist of glass in which are diffused..angular or irregular grains, or more definitely formed crystals. **1939** A. JOHANNSEN *Descr. Petrogr.* (ed. 2) I. iii. 31 When the constituents [of the rock] are..from walnut to cocoanut size, it is very coarse-grained. Occasionally rocks are of still larger grain; certain pegmatites, for example, have grains of several cubic meters or more in size. **1970** *Encycl. Brit.* X. 163/1 The high-silica rocks are generally light coloured and their excess of silica is expressed in abundant grains of quartz.

(ii) *Metallurgy.* **1899** EWING & ROSENHAIN in *Phil. Trans. R. Soc.* A. CXCIII. 355 When the polished surface of a metal, such as gold or iron, is lightly etched, and is then examined by means of normally reflected..light, the surface appears divided up into a number of areas separated by more or less polygonal boundaries. These areas are the sections of the crystalline grains which constitute the mass of the metal. **1923** GLAZEBROOK *Dict. Appl. Physics* V. 392/2 A section cut parallel to the direction of extension shows that the metal still consists of an aggregate of grains, but instead of a system of equi-axed crystals we now find grains elongated in the direction of extension. **1953** *Science News* XXIX. 36 The atoms [in a metal] are..aligned in small regions called grains; one grain containing rows of atoms lying at an angle to the rows in the neighbouring grains. **1965** W. A. TILLER in R. W. Cahn *Physical Metall.* ix. 431 The main volume of the ingot generally consists of a zone of long columnar grains and a zone of equiaxed grains.

f. *Aeronaut.* A piece of solid propellant of the shape and size used in a rocket engine.

1952 K. W. GATLAND *Devel. Guided Missile* 125 Tiny Tim...Solid-propellent (4×40 lb solventless extruded ballistite grains, discharge through 24 nozzles). **1954** *Ibid.* (ed. 2) i. 34 The smokeless propellants used in modern rocket projectiles..may be produced in the form of tubes, or grains, of any desired length and thickness. **1962** F. I. ORDWAY et al. *Basic Astronautics* x. 417 Early solid rocket grains were ignited at one end and burned 'cigarette fashion' along the chamber. **1966** *McGraw-Hill Encycl. Sci. & Technol.* XI. 606/2 In some rockets there is more than one grain inside the same combustion chamber.

12. (Later example.)

1968 *Gloss. Terms Offset Lithogr. Printing* (*B.S.I.*) 21 *Grain*, a roughened state of a lithographic surface which assists the retention of moisture and control of the image.

13. d. Short for *grain-leather* (see sense 19 in Dict.).

1895 *Montgomery Ward Catal.* 516/1 A shoe..made from the finest grade of English imported grain...This grain consists of a very fine selection of calfskin, finished on the grain side. **1897** *Sears, Roebuck Catal.* 193/3 Ladies' Bright Grain Button Shoe. **1930** *Daily Tel.* 1 Dec. 15/1 Washable Grain Cape Gloves. **1949** D. WOODROFFE *Stand. Handbk. Industr. Leathers* i. 16 The grain is dressed for handbags, luggage, men's dress belts.

15. d. In paper.

1922 *Paper Trade Jrnl.* 15 June 50/2 *Grain*...In paper the direction of manufacture on the machine, as 'across the grain'. **1924** *Ibid.* 7 Feb. 56/2 *Grain direction*, the direction in which the fibers flow on a papermaking machine. **1949** *Manual of Style* (*Chicago Univ. Press*) (ed. 11) 251 Paper resists bending and folding against the grain. For this reason printers take care to make sure that the grain will run vertically in the completed book, in order that..the book pages will lie flat when the book is opened. **1961** J. P. CASEY *Pulp & Paper* (ed. 2) III. xvii. 1258 The grain of paper must be taken into account in measuring all physical properties.

18. a. *grain-barge, -barn, -bin* (U.S. example), *-boat, boundary, -box, -chamber, -land, -market, -mill* (example), *-room, -ship, -size, -trade.*

1902 S. E. WHITE *Blazed Trail* xxix. 204 They were locked through after some delay on account of the grain barges from Duluth. **1844** *Knickerbocker* XXIII. 439 Let us drop into the 'grain-barn'. **1879** B. F. TAYLOR *Summer-Savory* xiii. 112 The pulpit, with the architecture of a grain-bin and two stories high. **1891** KIPLING *Light that Failed* 122 An Odessa grain-boat. **1920** *Jrnl. Inst. Metals* XXIII. 462 This is rather deep etching, but the grain boundaries were revealed better than by light etching. **1957** D. MCLEAN *Grain Boundaries in Metals* i. 1 A grain boundary in a piece of metal is the boundary separating two crystals (or 'grains') that differ either in crystallographic orientation, composition, or dimensions of the crystal lattice, or in two or all of these properties. **1872** *Rep. Vermont Board Agric.* I. 312 A convenient grain-box and root-cellar are great aids. **1887** H. H. JACKSON *Between Whiles* i. 26 A winding staircase outside led to what had been the grain-chamber. **1868** *Rep. Iowa Agric. Soc. 1867* 158 Near the river a portion [of the corn crop is] sold to grain-dealers where it finds a market at St. Louis. **1817** S. R. BROWN *Western Gaz.* 84 It proves to be excellent grain land. **1938** *Times Lit. Suppl.* 19 Mar. 181/1 A distinction between grainland, paying tax or rent in kind, and vine-, orchard- and garden-land, paying money-dues. **1871** SCHELE DE VERE *Americanisms* (1872) 481 *Grain* is used in America as *corn* is in England,..the papers quote therefore daily an account of the Grain Market. **1870** J. K. MEDBERY *Men & Myst. Wall St.* 335 All our great grain-merchants..do the same. **1867** H. LATHAM *Black & White* 27 Ellicott.. dammed up the Patapsco..and built grain-mills there. **1873** J. G. BEADLE *Undevel. West* xxv. 524 The other officials and employes were..in charge of [the] grain room. **1928** L. P. SMITH *Words & Idioms* 12 The basket..hoisted by the Egyptian grainships as an ensign. **1935** *Discovery* Feb. 61/2 These photographs give a good idea of life on a grain ship. **1912** SEXTON & PRIMROSE *Outl. Metallurgy Iron & Steel* (ed. 2) xliv. 545 The grain-size is reduced to very small dimensions by each pass through the rolls, or by each blow of the hammer. **1956** W. EDWARDS in D. L. Linton *Sheffield* 13 Inferior thickness and grain-size of sandstones. **1958** F. E. ZEUNER *Dating Past* (ed. 4) 21 Varves..composed of sand below (grain-size chiefly 1·0–0·1 mm.) and silty clay in their upper portion (grain-size under 0·1 mm., chiefly 0·1–0·01 mm.). **1966** D. G. BRANDON *Mod. Techniques Metallogr.* 246 Etched grain boundaries are commonly 1 μm or so in width at the low magnifications usually used in grain-size analysis. **1889** *Kansas Times & Star* 22 June, The new Missouri inspection law..will seriously hurt the grain trade here.

b. *grain-carrier, -crusher, -dealer, -farmer, -grower, -huller, -scourer; grain-grinding, -growing* vbl. sbs.; *grain-carrying, -cutting, -growing* (earlier U.S. example) ppl. adjs.

1908 E. NOBLE (*title*) The grain carriers. **1901** *Daily Chron.* 19 Aug. 5/7 Twenty-nine grain-carrying ships, chartered for European ports. **1850** *Rep. U.S. Comm. Patents, Agric. 1849* 113 Their newly invented horse-powers..their seed sowers and grain-crushers..do much to expedite the labors of the farm. **1850** *Mary Wedlake's Priced List Farming Implements* 16 (*heading*) A General Grain Crusher, Crushing the smallest Pulse and the largest: viz., Lentils and Beans. **1838** H. W. ELLSWORTH *Valley Upper Wabash* v. 47, I have a plan..to introduce the mowing and grain-cutting machine into this state. **1840** C. MATHEWS *Politicians* 5 To the invading grain-dealer, the voracious statesman sends a furious inspector. **1804** J. BARROW *Trav. S. Afr.* II. vi. 386 The population of this colony may be reduced into four classes...3. Grain-farmers. **1959** *Cape Times* 2 July 2/8 Grain farmers are now anxiously looking out for rain for their young crops. **1845** F. DOUGLASS *Life* (1846) 12 The black-smithing, cartwrighting, coopering..and grain-grinding, were all performed by the slaves. **1863** D. G. MITCHELL *My Farm* 131 A professed grain-grower. **1927** PEAKE & FLEURE *Peasants & Potters* 22 Grain is a more sustaining diet than whelks and limpets, and the grain-growers had more time and more energy to improve the amenities of their surroundings. **1963** *Times* 13 Mar. 10/7 Stalin accused some of the 'respected grain-growers' of staging a sit-down strike and leaving workers and the Red Army without bread. **1813** *Niles' Weekly Register* IV. 385/2 The public vigilance and scorn must aid the legal authorities; and so it will, in the grain-growing states. **1858** J. A. WARDER *Hedges & Evergreens* I. x. 144 Just as..grain-growing prevails, we find the fences are legally considered inclosures for the cattle, or barriers against them. **1868** *Rep. Iowa Agric. Soc. 1867* 148 The enemies to grain-growing are numerous. **1850** *Rep. U.S. Comm. Patents 1849* 302 What I claim..is covering..grain hullers with vulcanized India rubber. **1868** *Rep. Iowa Agric. Soc. 1867* 420 We are not behind the rest of the world in inventive skill, for we have invented..Kent's grain scourer.

19. grain-cradle (examples); **grain-elevator** (see ELEVATOR 3 a, b); **grain-gold** (*a*) (later U.S. example); **grain growth**, an increase in the average grain size of a metal; **grain-moth** (earlier and later U.S. examples); **grain-oriented** *a.*, (of steel) having had the grains oriented predominantly in one direction in order to modify the magnetic properties; **grain roll**, an iron roller made by casting the metal in sand.

1824 *18th Congress 1 Sess.* H.R. Doc. No. 25, 7 Improvement in grain cradle [patented March 24, 1823, by] Isaac Babcock. **1845** *Cultivator* New Ser. III. 17 My method is to..cut with a grain cradle previous to the first frost. **1897** *Sears, Roebuck Catal.* 50/3 Morgan Grain Cradle, 4 fingers, grape vine pattern, wood brace, ring fastening, silver steel scythe. **1923** J. H. COOK *50 Yrs. Old Frontier* 3 Sturdy sons of the forest, they could swing the scythe or the grain-cradle from sunup to sundown. **1852** L. B. MACKINNON *Atlantic & Transatlantic Sketches* I. 57 To accelerate the introduction of the cargo, a grain-elevator was employed. This novel machine pumped the grain from barges,..in a continuous stream into the ship's hold...It was..accurately measured in the operation. **1873** 'MARK TWAIN' & WARNER *Gilded Age* xxii. 203 Pictures of wharves, crowded with steam boats, and of huge grain elevators on the bank. **1905** *Macm. Mag.* Nov. 47 The wheat..is warehoused ready for shipment in grain-elevators, which are large rectangular buildings of great height, consisting of vertical bins, some of which are a hundred feet in depth. **1926** *Daily Colonist* (Victoria, B.C.) 3 Jan. 1/2 Chief of the commercial buildings contemplated is the proposed new grain elevator of the Panama–Pacific Grain Terminals Company. **1967** *Canadian Ann. Rev. 1966* 283 Among the buildings deferred were a large grain elevator in Prince Rupert and several proposed government buildings in Ottawa. **1850** W. COLTON *Deck & Port* xiv. 397 Each has a bag of grain-gold in his hand, which he must double or lose. **1928** *Jrnl. Iron & Steel Inst.* CXVII. 920 Grain growth started at the surfaces of the samples by small grains of iron absorbing the diffusing metal, and extended by migration of grain boundaries in the direction of the diffusing force. **1842** T. W. HARRIS *Insects Injur. Veget.* 363 The European grain-moth (*Tinea granella*), in its perfected state, is a winged insect. **1932** METCALF & FLINT *Fund. Insect Life* viii. 273 Among the most destructive and best-known species [of Gelechiidæ] are the pink bollworm.., the Angoumois grain moth. **1967** S. O. NELSON in Kilgore & Doutt *Pest Control* iii. 107 The Angoumois grain moth and the lesser grain borer were more resistant to control by infrared treatment. **1951** *Trans. Amer. Inst. Electr. Engineers* LXX. 840/1 The successful application of grain oriented strip steel to turbine generators, marks an important new step in the field of power generation. **1960** *Times* 11 Feb. 17/3 A modern plant for the integrated and continuous processing of ultra-low-loss grain-oriented electrical quality sheet and strip. **1904** HARBORD & HALL *Metallurgy of Steel* xvi. 291 Such rolls cost from 50 to 100 per cent. more than those cast in sand, and known as 'grain rolls'. **1932** E. GREGORY *Metallurgy* i. 18 Small amounts of chromium also bring about considerable grain refinement, which is desirable in ordinary 'grain' rolls used for roughing purposes.

grain, *sb.*²: see *GRANE *sb.*²

grain, *v.*¹ **1.** Delete †*Obs.* and add later example.

1924 *Glasgow Herald* 12 Nov. 16 Arable ground would grain even faster than it does if the dole-fed masses of the great cities, such as Buenos Aires and Sydney, were to be induced to lend a hand at the plough.

2. c. Also of syrup.

1865 *Trans. Ill. Agric. Soc.* V. 566 The yield of stalk was enormous but the sirup made from it was quite dark, and refused to grain. **1893** R. WELLS *Toffy & Sweets* 7 When lump or crystallised sugars are boiled to the heat..of 250 degrees, the sugar is liable to grain, and to turn out a solid mass on the slab. **1906** *Daily Chron.* 31 Oct. 8/5 When the syrup has boiled for fifteen minutes add the chestnuts and stir the mixture until it 'grains' and turns white.

8. *trans.* To feed with grain. *U.S.*

1852 H. MELVILLE *Pierre* 40 No one grained his steeds, but himself. **1874** *Rep. Vermont Board Agric.* II. 406 Older sheep should be grained the first of the season, after which they may do without till the first of March. **1949** *Sat. Even. Post* 9 Apr. 132/4 We throwed our drive into a pole-fence pasture, grained Blaze and Blackie's *grullo*, then went up to the main house.

grain, *v.*² Restrict †*Obs.* to sense in Dict. and add: **2.** *trans.* To spear (fish) with a grains.

1892 STEVENSON & OSBOURNE *Wrecker* xii. 196 Something struck me right through the forearm and stuck there. I put my other hand up, and, by George, it was the grain; the beasts had speared me like a porpoise. 'Cap'n!' I cried...'They've grained me.'

graine (grēⁱn). Also **grain.** [Fr.] The eggs († or an egg) of the silkworm; cf. SEED *sb.* 5 a.

1835 URE *Philos. Manuf.* II. vi. 230 The eggs or grains of the silk-worm are covered with a liquid, which glues them to the piece of cloth or paper on which the female is made to lay them. **1887** *Encycl. Brit.* XXII. 58/2 The eggs of the silkworm, called *graine*, are hatched out by artificial heat. *Ibid.* 59/2 The sources of healthy graine became fewer and fewer. **1887** *Colonial & Indian Exhib., Rep. Col. Sect.* 337. **1930** R. CUTHILL tr. *Schober's Silk & Silk Industry* i. 14 Silkworm eggs (graine) are bluish violet granules. *Ibid.* 31 Persia itself does not produce good graine. **1941** C. J. HUBER in *Matthews' Textile Fibers* (ed. 5) xvii. 680 The production of industrial eggs by cocoon raisers consisted of..43 per cent grain method where the eggs are distributed in 10-, 20-, and 30-grain unit receptacles.

grainering (grēⁱ·nəriŋ), *vbl. sb.* [f. GRAINER¹ + -ING¹.] The preparation of hides with a grainer or bate.

1857 *Encycl. Brit.* XIII. 307/1 Tanning is preceded by what is called *abating* or *grainering*. **1882** *Ibid.* XIV. 386/1 The skins are washed in the dash wheel, and undergo a process of *bating* or *grainering*. **1897** C. T. DAVIS *Manuf. Leather* (ed. 2) 153 The operation of immersing hides and skins intended for the manufacture of pliable leathers, in an alkaline solution consisting of the dung of chickens, pigeons, dogs, or in bran water..is termed either 'bating', 'abating', 'grainering', 'reducing', 'drenching', or 'puring'.

‖ **graineur** (grenör). [F. *graineur*, also *greneur*. Cf. *GRAINE.] A producer of silkworm eggs.

1913 J. H. LONGFORD *Evol. New Japan* vii. 87 French and Italian graineurs, while eager to buy her raw silk, were still more eager to buy the eggs of her healthy silkworms in order to replenish their own exhausted stock. **1919** R. C. RAWLLEY *Econ. Silk Ind.* v. 80 In the province of Var there are 'graineurs' or seed-producers.

graininess (grēⁱ·ninės). [f. GRAINY *a.* + -NESS.] The quality of being grainy or granular; granularity. Also *fig.*

1921 *Chem. Abstr.* XV. 476 Graininess in photographic deposits. *Ibid.*, The word 'graininess' is used to denote the inhomogeneity of the deposit due to aggregations of particles. **1923** A. E. CONRADY et al. *Photogr. as a Sci. Implement* iv. 199 *Graininess or granulation.* The 'grain' of the developed image so far considered is the size of the silver particle. *Ibid.*, Corresponding to every stage of magnification..a certain degree of granulation or 'graininess' will exist. **1956** E. M. HUTTEN *Lang. Mod. Physics* iii. 96 The concept of probability, and statistics, is used to 'smooth out' the graininess in the particle picture. **1966** *New Statesman* 25 Feb. 269/1 Like so many television originals, his play's a Victorian genre-painting disguised under contemporary black-and-white graininess.

graining, *vbl. sb.*[1] Add: **1.** (Additional examples.)
1951 R. Mayer *Artist's Hand-bk.* xii. 379 *Graining.* The grain is imparted to the stone by grinding its surface with flint, sand, or other abrasive. **1961** T. Landau *Encycl. Librarianship* (ed. 2) 154/1 *Graining*, preparation of the surface of metal lithographic plates by grinding them with a muller and sand or mechanically, by pebbles and abrasive.
3. *graining block, board* (later example), *gouge, machine* (later example).
1846 R. B. Sage *Scenes Rocky Mts.* xxxiii. 288 Near this is his 'graining block', planted aslope, for the ease of his operative in preparing his skins for the finishing process in the art of dressing. **1960** G. A. Glaister *Gloss. Bk.* 158/2 *Graining boards*, boards or metal plates used by the binder to produce a diced effect on covers. The boards have a pattern in relief of parallel lines running diagonally. **1875** T. Seaton *Fret Cutting* 141 The details of the hair and curls must now be worked out with fine hollow gouges and graining gouges. **1959** R. Hostettler et al. *Techn. Terms Printing Industry* (ed. 3) 114/1 Graining machine for offset plates.

grainy, *a.* Add: **1.** (Additional examples.)
1940 A. L. M. Sowerby *Wall's Dict. Photogr.* (ed. 15) 350 A negative is said to be 'grainy' when an enlargement from it shows the structure of the image. **1947** J. Steinbeck *Wayward Bus* 33 He was tired and his skin felt grainy. **1961** G. Millerson *Technique Telev. Production* iii. 42 Pictures will be indistinct, smeary, lifeless, and scintillating with the grainy effect of picture-noise. **1967** *Times* 27 Dec. 11/1 The coarsely grainy photography which not very long ago was a sign of spontaneity and originality. **1970** *Nature* 5 Sept. 1064/1 The very dark and grainy appearance of many of the photographs.
b. Of a voice or sound: rough, gritty.
1963 W. K. Rose in *Lett. Wyndham Lewis* p. xxi, The everyday tone of Lewis's voice—grainy, insistent. **1969** *Listener* 20 Mar. 398/2 Jack Bruce's bass-guitar work..on the live tracks has a wonderfully grainy, growling sound.

Gram[3] (græm). Also **gram.** The name of H. C. J. *Gram* (1853–1938), Danish physician, used *attrib.* and in the possessive to designate his method of staining bacteria and the iodine solution he employed in this method, as *Gram('s) method, solution, stain* (so *Gram-stained, -staining* ppl. adjs.). Hence **Gram-positive, -negative** adjs., respectively staining, or not staining, by Gram's method.
1884 *Brit. Med. Jrnl.* 6 Sept. 487/2 Gram's method gives good results with many bacteria. **1886** E. M. Crookshank *Introd. Pract. Bacteriol.* 246 (index) Gram's solution. **1902** R. T. Hewlett *Man. Bacteriol.* (ed. 2) iii. 91 By this method the ordinary Gram-staining organisms are stained. **1903** W. D. Frost *Lab. Guide Elem. Bacteriol.* (ed. 3) i. 60 (*heading*) Gram's stain. **1907** *Practitioner* Aug. 277 The Boas-Oppler bacillus is Gram-positive, whereas the normal bacillary flora of the large intestine is mainly Gram-negative. **1908** Park & Williams *Pathogenic Micro-organisms* (ed. 3) xii. 133 A Gram-stained smear may show all Gram-negative or all Gram-positive bacteria. **1949** H. W. Florey et al. *Antibiotics* II. xv. 632 The active agent inhibited the growth of certain gram positive pathogenic organisms and gram negative cocci. **1961** *Lancet* 29 July 228/1 To avoid errors due to contamination, Gram-stained films of the growth were examined. **1963** H. Burn *Drugs, Med. & Man* (ed. 2) xx. 202 Most of the bacteria which stain by the gram stain..are sensitive to the action of penicillin. **1964** M. Hynes *Med. Bacteriol.* (ed. 8) v. 44 All [bacteria] may be placed into one of two broad groups according to whether they stain by Gram's technique. *Ibid.* v. 45 The organisms are first stained with methyl-violet or gentian-violet and then treated with iodine as a mordant; Gram-positive bacteria then resist decolourization by alcohol.

gram[4] (græm), colloq. abbrev. of (*a*) Telegram or Cablegram; also **'gram**; (*b*) Gramophone.
1891 'F. Leslie' *Let.* 24 Aug. in W. T. Vincent *Recoll. F. Leslie* (1893) II. xxv. 140, I wired you date of production and result, and sincerely hope the 'grams reached you safely. **1928** *Sunday Express* 19 Aug. 1 Grams: 'Mouldboard, London.' Phones: 1615 1616 East. **1960** Wodehouse *Jeeves in Offing* iii. 37 What's a guffin?..That's what she calls you in her latest 'gram. **1964** —— *Frozen Assets* iii. 51, I cabled the New York lawyers asking if..there was some small legacy coming my way, and back comes this gram informing me that I cop the lot.
1959 *New Statesman* 26 Dec. 904/3 The thing he wanted to buy most in the world was a gram and lots of jazz records. **1970** *Guardian* 24 Dec. 9/3 There was Edmundo Ros and his Cuban band on the gram.

grama, gramma. (Earlier examples.)
1828 A. Wetmore *Diary* 28 July in *U.S. Senate 22nd Congress 1st Sess. Doc.* 90 (1832) 39 Our mules have been recently much benefitted by the *gramme grass*, the best pasturage between the Atlantic and Pacific Ocean. **1844** J. Gregg *Commerce of Prairies* I. 160 A highly nutritious grass called *grama.*

grama[2] (grā·mă). [Skr. *grāma.*] A scale used in Indian music.
1807 J. D. Paterson in *Asiatick Researches* IX. 446 (*title*) On the Grāmas or Musical Scales of the Hindus. *Ibid.* 457 The scale is denominated *Grāma* (literally village) because there is in it the assemblage of all the notes. **1891** C. R. Day *Mus. & Musical Instr. S. India* ii. 15 The srutis are differently arranged in grāmas, or scales, three in number. **1913** E. Clements *Introd. Study Indian Mus.* i. 2 First came the *Grāmas*, which may be regarded as collections of notes definitely related to one another by musical intervals. **1954** *Grove's Dict. Mus.* (ed. 5) IV. 456/2 *Grāma* means 'village' and thus 'scale' or 'collection of notes'. **1968** *Indian Mus. Jrnl.* V. 49 The whole framework of *grāma* and its adjuncts was evolved on the basis of the actual musical practice that had crystallised as *Jāti-s.*

‖**Gramdan** (grāmdā·n). Also **gram-dan, gramdan.** [Hindi, f. *grām-a* village + *dān* gift.] In India, (a movement for) the free gift of a village for the benefit of the community. Cf. *Bhoodan.
1957 *Economist* 28 Sept. 1037/1 The most specifically Indian political innovation since Mahatma Gandhi's satyagraha was bhoodan, the gift of land. Acharya Vinoba Bhave, its founder, has now extended it to gramdan, the gift of villages. **1958** [see *Bhoodan]. **1959** Hough & Madhava Das *Co-op. Movement in India* (ed. 4) 425 The problem of meeting the credit needs of the gramdan villages in Koraput. **1969** *Times* 13 Oct. (Indian Suppl.) p. vi/7 The movement is at present in its second stage: that of Gramdan... So far all that is being done is to collect 'declarations of intent' on prescribed forms, a village being declared *Gramdan* when at least 75 per cent of its population has signed the form...Only after 80 per cent of the villages in a state have been brought under Gramdan in this manner, would the implementing of the declared intentions be taken up. **1971** *Catholic Worker* Feb. 7/2 We see in a gramdan village how a meeting house, a nursery, a village store, a school, a milk cooperative can be started through voluntary cooperation.

gramicidin (græmisəi·din). *Chem.* [f. *Gram[3] + -i- + -cid(e + -in[1].] Any of several polypeptides produced by *Bacillus brevis* which have antibiotic properties, esp. against Gram-positive bacteria, and are used externally to treat local infections; orig. considered a single compound.
1940 Hotchkiss & Dubos in *Jrnl. Biol. Chem.* CXXXII. 791 Three crystalline preparations highly bactericidal for Gram-positive microorganisms have been obtained... The third substance, which we have named gramicidin, is concentrated by repeatedly recovering the fraction which remains soluble in alcohol. **1967** Hunter & Schwartz in Gottlieb & Shaw *Antibiotics* I. 642 Countercurrent distribution techniques resolved gramicidin into four groups of polypeptides now specifically designated as gramicidins A, B, C, D. **1969** *New Scientist* 10 July 65/2 The antibiotics which are produced include.. gramicidin and tyrocidin.

grammar, *sb.* Add: **5. c.** The name of a class in certain Jesuit schools or colleges.
1629 J. Wadsworth *Pilgr.* iii. 13 Father Lacy, the Reader of Poetry, and Master of the Syntax. Father Henry Bentley and Father Iohn Compton of Grammer. **1667** in *Cath. Rec. Soc. Publ.* III. 63 He was newly entered into grammer. *Ibid.* 69 Answering to hard and intrecat questions publickly in our Refectory even the first week he was in grammer. **1773** in *Mem. Stonyhurst Coll.* (1881) 22 The former Master of Poetry, the latter of Grammar, at Bruges. **1837** J. C. Fisher in *Ushaw Mag.* (1904) Dec. 262–3. **1904** *Ibid.* June 201 On Tuesday, May 17th, Syntax played Grammar.
d. Short for *grammar school. Also *attrib.*
1950 J. Cannan *Murder Included* iv. 62 He won a scholarship to Harborough Grammar, but his father wouldn't let him take it up. **1959** I. & P. Opie *Lore & Lang. Schoolch.* xvi. 356 The home-work toilers are called 'Grammar grubs'. *Ibid.*, At Caistor, in Lincolnshire, the Moderns chant: Grammar fleas, [etc.]. **1964** A. Prior *Z Cars Again* xv. 147 A 'girl in the Grammar' meant much to a family in that neighbourhood. **1965** *Listener* 22 July 125/1 'Grammar grubs,' the secondary schoolboys shouted at us, and we passed by, noses lifted, precociously dignified.
6. a. (Further examples.)
1958 *Listener* 18 Sept. 441/2 Reizenstein's dissonances do not make one 'sit up' in the way Haydn's do if we attend to his musical grammar. **1963** *Times* 5 Mar. 15/1 The grammar of the film was established.
7. *grammar-ridden* adj.
1906 *Westm. Gaz.* 28 Aug. 1/3 Opportunities for experiment are not often forthcoming in our much-examined and grammar-ridden schools. **1935** R. Paget *This English* 7 English..is much less grammar-ridden than most other languages.

grammarian. Add: **2. b.** A member of the class named 'Grammar' in certain Jesuit schools or colleges.
1705 in *Ushaw Mag.* (1903) Dec. 298. **1837** J. C. Fisher *Ibid.* (1904) Dec. 250 In the cyphering school with the Grammarians and High Fig[ures]. **1904** *Ibid.* Mar. 98.

grammar school. 1. Add to def.: Since the Education Act of 1944, any secondary school with a 'liberal' curriculum including languages, history, literature, and the sciences, as distinct from technical or modern schools.
1963 Barnard & Lauwerys *Handbk. Brit. Educ. Terms* 101 *Grammar school*,... The term nowadays is used for a secondary school with an academic curriculum, particularly suited for preparing pupils for entry to the universities or professions.

grammatical, *a.* Add: **1. b.** *Logic.* Of or relating to the mere arrangement of words in the sentence or proposition, in contrast to its logical structure. So esp. *grammatical form, subject.* (Opp. *logical form*, etc.)
1874 W. S. Jevons *Princ. Sci.* vi. 137 Another..difficulty is to decide when a change is merely grammatical and when it involves a real logical transformation. Between *a table of wood* and *a wooden table* there is no logical difference. **1883** F. H. Bradley *Princ. Logic* I. i. § 17 It is false that the grammatical subject is the reality of which the predicate is held true, yet in every judgment there must be a subject. *Ibid.* III. I. ii. 394 But this differs from the result given by Professor Jevons in nothing except grammatical form. **1903** B. Russell *Princ. Math.* iv. 48 The question is: what logical difference is expressed by the difference of grammatical form? **1910** —— in Whitehead & Russell *Principia Mathematica* I. Introd. iii. 66 The proposition must be capable of being so analysed that what was the grammatical subject shall have disappeared. **1933** L. S. Stebbing *Mod. Introd. Logic* (ed. 2) ix. 153 The point that is of importance is to distinguish the grammatical subject of a sentence from the logical subject of the proposition expressed by the sentence. **1951** A. Flew *Ess. on Logic & Lang.* 7 It would be absurd, but it would also be easy, to be misled by the grammatical similarity of 'It goes on to London' to 'It goes on to Infinity'. **1959** P. F. Strawson *Individuals* II. v. 148 Grammatical classifications do not unequivocally or clearly declare their own logical rationale.
c. *Philol.* *grammatical change* [tr. G. *grammatischer wechsel*]: the system of contrasting consonants found in the strong verb in Germanic languages, exemplifying Verner's Law.
1926 *Language* II. 177 Another article by Braune..on what he calls 'The Grammatical Change in the Inflection of the German Verb'. **1934** Priebsch & Collinson *German Lang.* II. i. 95 Grammatical change (*grammatischer Wechsel*), which is limited in Gothic to a few cases..is well preserved in O.H.G. **1963** J. T. Waterman *Perspectives in Linguistics* 49 Grimm..had been especially impressed by the curious interplay of stop and spirant in the morphology of the strong verb, applying the term 'grammatical change' to this phenomenon.
d. Of languages: having relatively greater structural resources, and relying less on lexical richness.
1937 J. Orr tr. *Iordan's Introd. Romance Ling.* 287 Where there is greater solidarity between the semantic and the formal associations, the signs appear less arbitrary, and these he [sc. Saussure] calls grammatical languages. **1959** W. Baskin tr. *Saussure's Course in Gen. Linguistics* II. vi. 133 We might say that languages in which there is least motivation are more *lexicological*, and those in which it is greatest are more *grammatical.* **1962** S. Ullmann *Semantics* iv. 105 It was one of Saussure's most important discoveries that the proportion of transparent and opaque words varies characteristically...'Grammatical' languages..favour the transparent type.
2. *grammatical meaning* (later examples). Also *grammatical category, feature, form, word* (see quots.).
1891 H. A. Strong et al. *Hist. Lang.* xx. 343 The grammatical categories of substantive, adjective, and verb correspond to the logical categories of substance, quality, and..occurrence. **1924** O. Jespersen *Philos. Gram.* ii. 44 The paradigmatic arrangement is not one of grammatical form. *Ibid.* iii. 54 The grammatical category of number evidently corresponds to the distinction found in the outside world between 'one' and 'more than one'. **1925** P. Radin tr. *Vendryes' Lang.* ii. 90 To the concepts expressed by means of morphemes, we give the name grammatical categories. Thus, gender, number, person, tense and mood, interrogation and negation,..etc., are grammatical categories in languages where these concepts are expressed by special morphemes. **1933** Bloomfield *Lang.* x. 166 A simple feature of grammatical arrangement is a *grammatical feature* or *taxeme*...The utterance *Run!*, for example, contains two grammatical features (taxemes), namely, the modulation..and the selective feature. *Ibid.*, The smallest meaningful units of grammatical form may be spoken of as *tagmemes.* *Ibid.* 169 Some pitch-scheme.., in English at any rate, lends it a grammatical meaning such as 'statement', 'yes-or-no question', 'supplement-question', or 'exclamation'. *Ibid.*, The grammatical forms of a language can be grouped into three great classes. **1958** C. F. Hockett *Course in Mod. Ling.* xxvii. 231 A *specific grammatical category* is an element in a system or a class in a classification: English plural, Spanish masculine, Latin accusative. **1963** *Listener* 3 Jan. 21/2 The feeling of climax comes..from increasing syntactical concentration. More merely grammatical words—'see', 'are', 'was', 'is'—are omitted. **1964** M. Joos *Eng. Verb* iv. 81 The necessity of distinguishing between lexical meaning and grammatical meaning. **1966** G. N. Leech *Eng. in Advertising* ix. 88 'Grammatical words' such as prepositions, pronouns, and auxiliary verbs. **1968** M. Black *Labyrinth of Lang.* iv. 85 The totality of rules that determine the correct use of a given word or expression may be said to specify the grammatical form of that word or expression.

grammaticality (grămæ:tikæ·lĭti). *Linguistics.* [f. prec. + -ity.] The quality of being grammatical (sense 3).
1961 A. A. Hill in *Word* XVII. 1 (*title*) Grammaticality. **1965** *Language* XLI. 403 Within such a framework, deviance as well as grammaticality can tentatively be made explicit. **1968** J. Lyons *Introd. Theoret. Ling.* ix. 422 Grammaticality is that part of the acceptability of utterances which can be accounted for in terms of the rules.

grammaticalize (grămæ·tikăləiz), *v.* *Linguistics.* [f. Grammatical *a.* + -ize.] To express by means of the grammatical structure; to adopt as a grammatical requirement. Usu. in *pass.*

1937 J. Orr tr. *Iordan's Introd. Romance Ling.* 337 A similar order [sc. inversion in interrog. sentences] is observable in other languages, but has not become 'grammaticalized' as in French. **1961** *Brno Studies* III. 10 The position of the word within the sentence context is grammaticalized to a much higher degree in analytical than in synthetic languages. **1968** J. Lyons *Introd. Theoret. Ling.* ix. 438 We cannot assume..that such notions..will necessarily be 'grammaticalized', rather than 'lexicalized', in the structure of any particular language.
Hence **gramma:ticaliza·tion**, the action or process of grammaticalizing.
1955 *Archivum Linguisticum* VII. 1. 28 The affectivity of reprise has been weakened by grammaticalization. **1965** *Language* XLI. 107 The creation of oppositions through grammaticalization.

gramme. Delete 'Troy' and add: Later redefined as a unit of mass equal to 1/1000 of a *kilogramme, although it is still used as a unit of force equal to the *gramme force. (The spelling *gram* is usual in the U.S. and is preferred by some writers elsewhere.) (Further examples.)
1877 *Rep. Brit. Assoc. 1876* II. 32 In the system already adopted by the British Association Committee on Dynamical and Electrical Units..the Centimetre, the Gram, and the Second were taken as the units of length, of mass, and of time. [*Note*] The spelling Gram, instead of Gramme, for the English word is adopted..in accordance with the spelling put forward in the Metric Weights and Measures Act, 1864, which legalizes the use of the Metric System. **1892** *Proc. Amer. Assoc. Adv. Sci. 1891* 176 Rules for the orthography and pronunciation of chemical terms...Gramme. **1894** J. Parker *Thermodynamics* 3 The weight of a gramme has no definite value unless we specify the place where the weight is to be found, because the weight of a given mass is not quite the same in all parts of the world...The weight of a gramme at Paris is 980·868 dynes. **1911** Kaye & Laby *Tables Physical & Chem. Constants* 3 Mass. Unit—the gramme, 1/1000 of the International Prototype Kilogramme. **1954** *Amer. Jrnl. Physics* XXII. 298/1 Weighing in grams..is extremely familiar even though few calculations are carried out in this system [sc. the metric gravitational system], conversion usually being made to the cgs system. **1958** *Van Nostrand's Sci. Encycl.* (ed. 3) 1744/2 The abbreviation gf is used to indicate gram (force), the unit of force, which is defined as the weight of a one-gram mass under the action of a gravitational acceleration of 980·665 cm/sec². **1966** J. R. Partington *Gen. & Inorg. Chem.* (ed. 4) i. 7, 1 mol. wt. of a gas at s.t.p...in grams occupies 22·4 lit. **1967** *Units & Standards of Measurement: Mechanics* (H.M.S.O.) (ed. 4) 10 These densities..are expressed in terms of grammes per millilitre. **1970** *Spec. Univ. Syst. desig. Linear Density Textiles* (B.S.I.) 5 The linear density in 'tex' expresses the mass in grammes of one kilometre of yarn.
b. **gramme-atom**, the quantity of an element having a mass in grammes numerically equal to its atomic weight; **gramme calorie** = *calorie b; **gramme force**, a unit of force equal to the weight of a mass of 1 gramme, esp. under standard gravity; also called *gramme* and *gramme weight*; **gramme-ion**, the quantity of an ionic substance having a mass in grammes numerically equal to the atomic weight of the ion or the sum of the atomic weights of the constituent atoms; **gramme-molecule**, the quantity of a substance having a mass in grammes numerically equal to its molecular weight; so **gramme-molecular** adj.; **gramme-rad**, a unit of the energy absorbed by any quantity of a substance when irradiated with ionizing radiation, equal to 100 ergs; **gramme weight** = *gramme force.
1899 *Jrnl. Chem. Soc.* LXXVI. II. 587 The magnetic susceptibility of a number of the elements,..has been determined. The coefficient of susceptibility for each element, when divided by the number of gram-atoms per litre, gives the atomic magnetism. **1938** R. W. Lawson tr. *Hevesy & Paneth's Man. Radioactivity* (ed. 2) i. 7 A divalent ion requires 96,500 coulombs to deposit half a gram-atom. **1902** *Encycl. Brit.* XXXIII. 280/1 Small calorie or gramme calorie. **1951** *Engineering* 29 June 792/2 Expressed in the absolute units of the C.G.S. system,..he had found that a gram-calorie was equal to 4·181 Joules, and the mean gram-calorie corresponded to 4·187 Joules. **1966** Kaye & Laby *Tables of Physical & Chem. Constants* (ed. 13) 12 It is often convenient to use submultiples and multiples of these units, e.g., gramme-force (gf), ounce-force (ozf). **1898** *Jrnl. Chem. Soc.* LXXIV. II. 210 The contraction which occurs when a substance is dissolved in water is proportional to the concentration of the solution..the contraction is about 13·5 c.c...for every dissolved gram ion of an electrolyte. **1902** J. McCrae tr. *Arrhenius' Text-Bk. Electrochem.* i. 9 One gram-ion of chlorine signifies 35·45 grams of chlorine in the ionic condition (Cl̄). **1906** A. Smith *Introd. Gen. Inorg. Chem.* xii. 199 It is called, therefore, the gram-molecular volume (G.M.V.) or the molar volume...It may be defined as that volume which contains one mole (gram-molecular weight) of any gas at 0° and 760 mm. **1931** E. C. Miller *Plant Physiol.* ii. 51 A solution made up after this manner is termed a 'molar',..'gram-molecular', or 'molecular' normal solution. **1894** G. S. Newth *Text-bk. Inorg. Chem.* I. vii. 56 The number of grammes of a substance, equal to the number which represents its molecular weight, is spoken of as the gramme-molecule. **1958** W. K. Mansfield *Elem. Nucl. Physics* i. 2 It follows that a gramme-molecule of any substance contains the same number of molecules. **1954** *Brit. Jrnl. Radiol.* XXVII. 243/2 Integral absorbed dose is the integration of the energy absorbed throughout a given region of interest. The unit is the gramme-rad. 1 gramme-rad = 100 ergs. **1963** Quimby & Feitelberg *Radioactive Isotopes in Med. & Biol.* viii. 120 The maximum permissible dose to a normal individual in 3 months is 3 rads of radiation to the whole body...For a 70 kg man this would be an integral dose of 210,000 gram-rads. **1871** *Phil. Mag.* 4th Ser. XLII. 375 The value of T' is about ·074 of a gramme weight per centimetre. **1894** J. Parker *Thermo-dynamics* 3 The value of this arbitrary unit of pressure is found to be 1033·279 gramme-weights, or 1,013,510 dynes, per square centimetre. **1960** *Amer. Jrnl. Physics* XXVIII. 480/1 The weight of a body is a force measured in dynes or gram weight.

Gramme[2] (græm). The name of the Belgian electrician Zénobe Théophile *Gramme* (1826–1901), used *attrib.* to designate a form of dynamo armature introduced by him in 1870 (*Gramme ring, armature, winding*, etc.).
1884 F. Krohn tr. *Glaser de Cew's Magn. & Dyn.-Electr. Mach.* 255 The inductive actions in the coils of a Gramme ring. *Ibid.*, The ring of the Gramme machine. **1893** Hawkins & Wallis *Dynamo* 115 The first or 'Ring' method..is also frequently called the 'Gramme' winding. *Ibid.* 193 In the Gramme-wound ring armature the number of loops and the number of inductors are identical.

gramophile (græ·mŏfəil). [f. Gramo(phone + -phile.] An enthusiast of the gramophone and gramophone records.
1922 C. Mackenzie in *Daily Tel.* 2 Sept. 5/3, I wish Stevenson could have had a gramophone in Samoa...I feel sure that he would have become a 'gramophile'. **1926** *Gramophone* IV. 294/2 The price of the instrument.. (£120) is such that the number of gramophiles who could afford to possess one must be strictly limited. **1927** *Observer* 11 Dec. 22/3 Records of Ravel's Quartet in F.. that have received the composer's imprimatur. They will..please the increasing number of gramophiles who look to the [National Gramophonic] society for work of particular excellence. **1932** *Gramophone* July 60 (*heading*) Borodin for the gramophile. **1958** *Times* 19 Apr. 7/3 The ordinary gramophile is beginning to demand his high fidelity as a packaged whole.

gramophone. Substitute the following def. and add earlier and later examples. An instrument for the reproduction of recorded sound, similar in principle to the phonograph but using, instead of a drum, a flat disc containing a spiral groove; a stylus is allowed to rest in the groove as the disc is rotated on a turntable, and the vibrations communicated to the stylus by the irregularities in the groove are transformed into sound vibrations. (In the U.S., *phonograph* is the generic name for such an instrument.) In its modern form, with an electric motor, electronic amplification, and one or more loud-speakers, it is now more commonly termed a 'record player'.
The earliest gramophones were also used to cut records, using blank discs.
1887 *Pat. Off. Gaz.* 8 Nov. 620/2 Gramophone [patented by] Emile Berliner, Washington, D.C. **1899** E. Wharton *Greater Inclination* 59 Her voice..was like a voice reproduced by a gramophone: the real woman seemed far away. **1906** *Daily Chron.* 26 Nov. 4/7 What's wearing me to skin and bone? My neighbour's grinding gramophone. **1913** B. Clements-Henry *Gramophones & Phonographs* 5 The disc machine is known as the 'gramophone', and the cylinder machine as the 'phonograph'. **1917** A. Waugh *Loom of Youth* II. i. 89 Every night closed with a feed in Mansell's big study, while the gramophone strummed out rag-time choruses. **1921** P. A. Scholes *Learning to Listen* p. xv, The accumulated results of their already large experience of the educational use of the Gramophone. **1922** O. Mitchell *Talking Machine Industry* iv. 31 The gramophone, or disc machine, has..practically ousted the older invention from the English market. **1923** T. S. Eliot *Waste Land* III. 256 She smoothes her hair with automatic hand, And puts a record on the gramophone. **1928** Galsworthy *Swan Song* II. iv. 140 'What is the most pitiable sight in the world?'..'Oh! I think—a rich man listening to a bad gramophone.' **1952** Godfrey & Amos *Sound Recording & Reproduction* i. 33 The chief advantage of the gramophone over the phonograph as claimed by Berliner was this comparatively simple method of producing copies. **1955** *Times* 3 May 6/2, 10s. notes were found stuffed in the horn of a gramophone in his bedroom. **1967** *Times Rev. Industry* Mar. 41/3 The Gramophone Co. was refused the right to the trade mark 'Gramophone' in 1910. **1971** *Daily Tel.* 4 Nov. 12/4 Antoine speaks posthumously to the backbiting assembly through the horn of a 1913 gramophone.
2. *attrib.* and *Comb.*, as *gramophone bank, needle, pick-up, recital, rights*; **gramophone-cut**, the form of record-cutting in which the recording stylus vibrates parallel to the surface of the disc; **gramophone record**, a flat disc on which sound has been recorded for reproduction by a gramophone, the recording taking the form of a spiral groove starting at the periphery.
1941 *B.B.C. Gloss. Broadc. Terms* 14 *Gramophone bank*, group of turn-tables and other equipment for playing gramophone records or disc recordings. **1913** B. Clements-Henry *Gramophones & Phonographs* 4 The so-called 'gramophone-cut'. **1907** *Yesterday's Shopping* (1969) 1038/1 Only genuine Gramophone Needles should be used to play Gramophone Records. **1960** J. Rae *Custard Boys* I. v. 51 When he spoke it sounded as though a gramophone-needle was scratching across worn grooves in his throat. **1929** Wilson & Webb *Mod. Gramophones* x. 230 The change-over from radio to gramophone pick-up is effected by means of a switch. **1962** A. Nisbett *Technique Sound Studio* 247 *Crystal microphone or gramophone pick-up.* This generates a signal by means of a crystal bimorph. **1913** G. F. Rowell *Hints about Gramophone* 11 A gramophone recital. **1888** *English Mechanic* 24 Aug. 558/2 (*heading*) The preparation of grammophone and telephone records. **1918–19** T. Eaton & Co. *Catal. Fall & Winter* 369/2 The Gramophone Record Needles here listed..are made of good hard steel. **1924** P. A. Scholes *1st Bk. Gramophone Rec.* p. vii, Many of the best 'tunes'..await their enjoyment in the form of Gramophone Records. **1941** *Libr. Assoc. Rec.* Aug. 149/1 The report..records that Herefordshire County Library is one of the few possessors of a Gramophone Record Library in the country. **1947** *Ibid.* Sept. 224 (*title*) A gramophone record library service. *Ibid.* 224/1 The gramophone record library..comes at a time when there is a growing demand for music by the best artists. **1964** P. J. Guy *Disc Recording & Reproduction* v. 68 Up till a few years after the last world war, nearly all gramophone records were recorded at 78 r.p.m. with coarse grooves..and would accommodate only 4–4½ minutes of programme on a 12-in. record. **1971** *Radio Times* 16/23 Dec. 26/4 8.0 am. News...8.5 Aubade. Gramophone records of music by Parry, Delius, Elgar, and Grainger. **1921** A. E. Housman *Let.* 20 Mar. (1971) 184, I do not want revenue from gramophone and mechanical rights.

gra·mophone, *v.* [f. the sb.] *trans.* To reproduce by means of the gramophone; to record for the gramophone. Also *fig.* and *absol.* Hence **gra·mophoned** *ppl. a.*
1908 *Daily Chron.* 8 Apr. 3/7 The Tories are canvassing, 'gramophoning', &c. **1926** G. B. Shaw *Let.* 14 Sept. (1960) 104 Albert Coates has gone to London for 10 days to gramophone the Ninth Symphony. **1927** *Sunday Express* 28 Aug. 5/2 America's noise was gramophoned everywhere. **1928** *Manch. Guardian Weekly* 10 Aug. 113/4 The perils of a syndicated or grammophoned press. **1931** *Times Lit. Suppl.* 17 Dec. 1024/3 If it be true that the novel..is to be dramatized, filmed, gramophoned and everything else, one can only remark that this is another example of the incalculability of public taste. **1935** N. Mitchison *We have been Warned* IV. 463 Bubbling voices and laughter, penetrated by gramophoned jazz.

gramophonic (græmŏfǫ·nik), *a.* [-ic.] Of, pertaining to, or of the nature of the gramophone or gramophone records.
1905 *Westm. Gaz.* 2 Sept. 9/1 She has what I call a 'gramophonic mind'—that is, she assimilates other people's ideas and then rolls them off as if they were her own. **1915** *Morning Post* 4 Mar. 4/4 A gramophonic reproduction of a tune which was being played in Berlin. **1921** P. A. Scholes *Learning to Listen* p. xiv, The illustrated volume to which this one is, frankly, a gramophonic companion. **1927** *Observer* 24 Apr. 4/3 The National Gramophonic Society..issues the first of its orchestral records. **1946** *Penguin Music Mag.* I. 122 That a margin of extra time should be allowed—to be filled up if necessary by the time-honoured gramophonic stop-gap. **1965** *Listener* 3 June 837/2 Recording Wagner's *Götterdämmerung*...The biggest achievement in gramophonic history.
Hence **gramopho·nically** *adv.*, in a gramophonic manner; by or on a gramophone.
1911 W. J. Locke *Clementina Wing* xx, A stupendous woolly lamb..which, on something being done to its anatomy, opened its mouth and gramophonically chanted the 'Jewel Song' from *Faust.* **1913** *Christian* 3 Jan. 10/1 To find..the daily Press..giving voice to facts instead of gramophonically repeating the sentiments uttered by huge vested interests. **1924** P. A. Scholes *1st Bk. Gramophone Rec.* 30 The only piece of our great Purcell gramophonically available was the song *Nymphs and Shepherds.* **1948** *Scrutiny* XV. 337 The now gramophonically celebrated Concertgebouw Orchestra.

gramophonist (græmǫ·fŏnist, græ·mŏfōunist). [-ist.] One who uses or operates a gramophone; a 'gramophile'. (Not in current use.)
1907 *Daily Chron.* 12 Dec. 7/4 The gramophonist will redistribute the pearls of wisdom which have fallen from the lips of great Unionist statesmen to crowds of admiring villagers. **1923** *Weekly Dispatch* 4 Feb. 2, I must modestly disclaim being a passionate gramophonist (if that is the word). **1924** P. A. Scholes *1st Bk. Gramophone Rec.* 123 The description..will..be thought sufficient by many gramophonists. **1934** C. Stone in H. C. Bryson *Gramophone Record* p. xii, The true gramophonist will prefer to melt an old worn-out favourite and shape it into a flower-bowl..rather than to drop it into the dustbin.

gramophony (græmǫ·fŏni, græ·mŏfōuni). [f. Gramophone, after *telephony.*] The art of the gramophone; gramophonic reproduction. (Not in current use.)
1925 P. A. Scholes *2nd Bk. Gramophone Rec.* p. xix, It is one of the regrettable features of gramophony to-day that the songs of Schubert..are not to be obtained. **1927** *Daily Mirror* 10 Dec. 12/3 An impression is rapidly gaining ground that the whole future of gramophony rests with the electrical 'pick-up'.

gramp (græmp). [Abbrev. of Grandpapa.] A colloq. or dial. substitute for *grandfather.*
1898 in *Eng. Dial. Dict.* **1912** A. N. Lyons *Clara* xxvii. 297 'You see,' explained the maiden, 'we be going to Gramp's.' **1959** E. Pound *Thrones* xcvi. 4 And a wolf acting as guide Till it thought gramp looked too hungry.

1966 'L. Lane' *ABZ of Scouse* II. 25 That ther kid's ther dead spit of 'is gramp.

gran[2] (græn). A childish or familiar shortening of *granny* or *grandmother.*
1863 Dickens in *All Year Round* 3 Dec. 11/2 And now dear Gran let me kneel down here where I have been used to say my prayers. **1890** E. Lytton *Let.* 16 Apr. in E. Lutyens *Blessed Girl* (1953) iii. 29 Mother had a telegram last night to say that Gran is dead. **1908** *Westm. Gaz.* 8 Feb. 12/3 People say it was just because Gran lived so vividly that she flickered out like a candle come midsummer. *Ibid.*, The lady who received the Grans. **1960** R. Daniel *Death by Drowning* iii. 40 By the time she gets back to 'Mum' and 'Gran' she'll be wet through. **1960** *Sunday Express* 10 July 15/5 There is Gran Tatlock, tired and crusty after bringing up a family of 11.

grana, pl. *granum.

Granadine (græ·nădin), *a.* [ad. Sp. *granadino* pertaining to Granada.] Of or pertaining to Granada, a province of southern Spain (formerly a Moorish kingdom).
1865 H. O'Shea *Guide to Spain* 183/1 Alpine peaks..shine forth in the African sun like a mass of sparkling mother-of-pearl, to use the metaphoric definition of the Granadine Arab poets. **1959** *Times* 5 Aug. 9/6 But his rule marked the beginning of two centuries of Granadine grandeur. **1961** *John o' London's* 9 Feb. 137/3 The differences between Castilian, Sevillean, and Granadine solutions of similar architectural problems. **1962** Y. Malkiel in Householder & Saporta *Probl. Lexicogr.* 12 Granadine Arabic. **1963** *Times* 16 Feb. 4/3 He describes his play as a 'Granadine' poem of the twentieth century divided into different gardens, with scenes for singing and for dancing.

grand, *a.* and *sb.* Add: **A.** *adj.* **8.** (Further examples); *Grand Army* = *Grande Armée (earlier and later examples); *grand manner* [after F. *grande manière*], the style of plastic art suited for noble subjects and vast design; also *attrib.* and *transf.*; *grand passion* = *grande passion.
1775 T. Campbell *Diary* 27 Apr. (1947) 87 Revisited the exhibition of the Royal Academy & am confirmed in my opinion of the grand manner of Barrys Venus lamenting over Adonis. **1827** Scott *Napoleon* V. viii. 190 This formidable assemblage of troops, laying aside the appellation of the Army of England, was hereafter [sc. from 1805] distinguished by that of the Grand Army. **1850** Thackeray *Pendennis* II. v. 47 My father was a General of the Grand Army. *Ibid.* xxii. 210, I..am the person who eight years ago had a grand passion. **1850** C. M. Yonge (*title*) Kenneth, or the Rearguard of the Grand Army. **1905** A. Bennett *Tales of Five Towns* I. 170 He seemed to sink luxuriously into this grand passion of hers. **1905** *Daily Chron.* 28 Dec. 3/1 The Royal Academy tradition of the Grand Manner in painting. *Ibid.*, Sculpture groups, illustrating a classical theme, built up on Grand Manner traditions. **1925** F. F. Potter in *Teacher's World* July Extra No., The boys of Manchester Grammar School were engaged in a project in the grand manner, when they drained, levelled, and turfed their playing fields. **1929** A. Huxley *Do what you Will* 138 Where there are no psychological or external restraints, the Grand Passion does not come into existence. **1957** E. H. Gombrich *Story of Art* xxv. 381 David and his school cultivated the Grand Manner. **1962** 'H. Lourie' *Question of Abortion* ix. 76 We amuse each other...But it doesn't necessarily mean a grand passion or marriage. **1971** *Guardian* 14 Jan. 2/4 La Madeleine..began as a neo-Greek temple dedicated to the glory of Napoleon's Grand Army.
9. d. *grand-scale a.* (cf. Scale *sb.*[3] 13 b).
1959 *Listener* 29 Jan. 217/3 Today these works [of Wagner] are truly 'popular', the latest addition to the unquestioned grand-scale masterpieces. *a* **1963** L. MacNeice *Astrol.* (1964) i. 20 These [eclipses] (and comets) are still assumed to portend grand-scale happenings. **1964** I. L. Horowitz *New Sociol.* 22 A theoretical option to narrow empiricism and grand-scale rationalism.
10. e. With *not*, etc. Of persons: indisposed, unwell.
1934 N. Marsh *A Man lay Dead* viii. 145 'May I see your patient, Doctor Young?'..'She's not so grand,' he said doubtfully. **1951** E. Coxhead *One Green Bottle* x. 257 You don't look too grand...Are you sure you're all right? **1960** H. Pinter *Room* 98, I told him you hadn't been too grand.
12. Grand Canyon *Geol.* [name of the gorge of the Colorado River in Arizona, U.S.A.], a rock formation laid down during the pre-Cambrian era in the south-western U.S.A.; hence, in local use to designate the geological period during which these rocks were formed; **grand chain**: see *chain *sb.* 5 f; **Grand cordon, Cross** (see Cordon *sb.* 6, Cross *sb.* 19); **grand cru**, wine of a superior quality (cf. *cru); **grand fir**, *Abies grandis*, a large fir native to the west coast of North America; **Grand Fleet,** (*a*) the 18th-cent. fleet based at Spithead; (*b*) the name during the war-period 1914–16 for the British Battle Fleet operating in the North Sea; **Grand National**, the great steeplechase established in 1839 and run annually at Aintree, Liverpool, in the first week of the flat-racing season; **grand prize,** anglicization of *Grand Prix 2; **grand quarter** *Her.*, see quots. and cf. *grand-quartering*; **Grand Remonstrance** (see Remonstrance *sb.* 3 b); **grand slam** (see Slam *sb.*[2] 2 b).
1876 J. W. Powell *Rep. Geol. Eastern Portion Uinta Mount.* 70 The Grand Cañon Group rests unconformably upon the crystalline schists... Fossils have been found at the base of the Grand Cañon series... Red Creek Quartzite and Grand Cañon schists..are believed to be Eozoic. **1925** J. Joly *Surface-Hist. Earth* viii. 131 The third (the Killarney or Grand Canyon) closing pre-Cambrian time. **1958** R. C. Moore *Introd. Hist. Geol.* (ed. 2) iv. 61/2 Collectively, they are called the Grand Canyon Sequence of rocks. **1905** A. L. Simon *Hist. Champagne Trade England* ix. 119 There are three groups of famous growths in Champagne...These..may be called the Grands Crûs of Champagne. **1932** E. Hemingway *Death in Afternoon* i. 11 The Grand crus of Medoc. **1951** N. Mitford *Blessing* II. ix. 230 'Quite an honourable wine,' said Sigi, 'but not grand cru.' **1965** A. Sichel *Penguin Bk. Wines* 132 *Grand cru* or *grand vin.* Any wine bearing this description on the label must contain at least 11° of alcohol from the natural sugar of the grapes before the addition of extra sugar at the time of fermentation. **1897** G. B. Sudworth *Nomencl. Arborescent Flora U.S.* 54 *Abies grandis*...Grand or Oregon White Fir. **1913** S. B. Elliott *Important Timber Trees U.S.* II. 192 The important Western Firs are Grand Fir (*Abies grandis*), White Fir (*Abies concolor*), [etc.]...It is proposed by Mr. Sudworth to discard the appellation 'white' and adopt that of 'grand' in its place, making it Grand Fir. **1969** *Northwest* (Sunday Oregonian Mag.) 14 Dec. 21/3 While our familiar Douglas fir is still the most popular species..the..Grand firs are very close seconds especially in container plants. **1745** R. Goadby *Life & Adv. B. M. Carew* 119 Impressed Me.., who were all put on board the Winchester,..and carried to the grand Fleet then lying at Spithead. **1914** J. R. Jellicoe in *Times* 16 Sept. 8/6 The officers and men of the Grand Fleet beg that you will convey to their comrades of the British Army their intense admiration for the magnificent fight they have made. **1919** —— *Grand Fleet 1914–16* iii. 34 The Grand Fleet may be said to have come into being only at the outbreak of the War, when it was so christened. **1922** *Encycl. Brit.* XXXII. 294/1 Up to the outbreak of war, Rosyth was regarded as the principal base and headquarters for the Grand Fleet. **1966** A. J. Marder *From Dreadnought to Scapa Flow* III. vi. 207 The moral ascendancy of the Grand Fleet over the High Seas Fleet remained and, if anything, was stronger. **1839** *Sporting Rev.* Mar. 164 The Grand National Steeple Chase. **1866** *Field* 10 Mar. 199/3 Why should he be a favourite for the Grand National? **1886** Earl of Suffolk & Berks. *Racing & Steeple-chasing* 352 Tom Olliver won three Grand Nationals. **1894** J. D. Astley *Fifty Yrs. Life* II. 281, I hoped to be able to pick out the winner of the Grand National when the weights appeared. **1967** *Observer* Suppl. 1 Oct. 36/1 The Prime Minister is on the verge of his first political success: he is about to save the Grand National. **1866** *Lond. Gaz.* 26 June 3645/2 Paris Universal Exhibition of 1867...17 grand prizes, each of the value of 2000f. **1880** *Rep. Paris Univ. Exhib. 1878* II. 365, 100 Grand Prizes and exceptional awards in money. **1869** J. E. Cussans *Handbk. Heraldry* ii. 45 If one or more of these quarters should be subdivided into other like divisions, it is said to be Quarterly-quartered; and the quarter thus quartered is called a Grand quarter. **1896** J. Woodward *Her.* II. 102 It may happen that one of the heiresses whose arms are to be quartered, herself bore a quartered coat, in this case the quarter appropriated to her contains her whole bearings..and..is called a Grand-quarter. **1969** Franklyn & Tanner *Encycl. Dict. Heraldry* 271/1 In Scotland, quarters are limited to four, but each may be a grand quarter, i.e. a quarter which is itself quarterly of four sub-quarters.
B. *sb.* **7.** [G. (a. Fr.) *grand*, formerly *grando* (see quot. 1893).] In the game of skat, a bid to play with only the four matadores (knaves) as trumps. Varieties of this are called *solo grand, gucki grand, tournee grand, open grand* (or *grand ouvert*).
[**1893** 'L. Hoffmann' tr. *Hertefeld's Skat* 7 In *Grando*, the only trumps are the four knaves. *Ibid.* 46 The basis value is in *Grando Tourné*, 12; in *Grando Solo*, 16; and in *Grando Ouvert*, 24.] **1906** R. F. Foster *Skat Manual* 121 This he can do by declaring a Grand. **1957** *Encycl. Brit.* XX. 727/2 The games in which only knaves are trumps are called grand (or grando). *Ibid.*, Solo grand is played without use of the skat. In open grand the player's entire hand is exposed before the opening lead, and the player contracts to win every trick.
8. *slang* (orig. *U.S.*). A thousand dollars. Also occas., a thousand pounds. (The *sing.* form is often used for the *pl.*)
1921 *Collier's* 26 Mar. 24/2 'A hundred and fifty grands!' I breathed. 'You're cuckoo.' *Ibid.* 27 Aug. 4/3 '*I lose twenty-five thousand dollars!*'..Twenty-five grand! **1924** G. C. Henderson *Keys to Crookdom* 406 Grand. $1,000. Called 'a grand'. **1931** E. Linklater *Juan in America* IV. x. 359 D'you think I'd pay a hundred grand for protection if it wasn't worth it? **1932** *Amer. Mercury* Jan. 16, I don't know how much it is, but I suppose around ten, twelve, fifteen grand. **1946** *People* 7 Apr. 2/3, I stepped out with the spree-bent suckers..into this..world..where the black market boys..gamble in 'grands' (£1,000). **1958** *Times* 25 Feb. 5/3 He wanted his 'whack of the grand'. **1967** *Sunday Tel.* 23 Apr. 6/1 One 26-year-old [criminal]..insisted that he picked up a regular £1,000 a week working with a professional gang. 'Honest, a grand or a couple of grand isn't really big stakes in my game.'

‖**grand battement** (grañ batəmañ). [Fr.] A ballet exercise in which both legs are kept straight and one leg is raised into the air and brought down again, with the function of loosening the hip joints.
1830 [see *battement]. **1888** S. D. Headlam *Theory Theatr. Dancing* x. 59 Grands battements enable a dancer to turn her legs completely outwards, and give much facility to the motions of her thighs. **1933** C. W. Beaumont *Primer Class. Ballet* 34 *Grands battements*, which means *large beatings*, are used to loose the hip-joints and to turn out the legs from the hips. **1952** Kersley & Sinclair *Dict. Ballet Terms* 27 *Grands battements* are taken next. The student throws one leg high to front, side or back, the knee straight and the foot pointed, keeping the rest of the body as still and as upright as possible.

‖**grand coup** (grañ ku). [Fr.: see Coup *sb.*[2]] **1.** A great and important stroke or hit; a bold and successful effort.
1813 Byron *Let. to Moore* 22 Aug. in Moore *Life* (1832) II. 234, I hope you are going on with your *grand coup*—pray do—or that damned Lucien Buonaparte will beat us all. **1856** C. F. Adams in *Life & Wks. J. Adams* I. 352 Justly was it denominated by one who had spent his life in the diplomatic service, a '*grand coup*'. **1883** *Standard* 17 Sept. 5/2 (Stanford), [The police] then make a *grand coup* all at once.
2. *Whist* and *Bridge.* The getting rid of a superfluous trump by ruffing a winning card from the opposite hand.
1874 'Cavendish' *Whist* (ed. 10) 130 Sometimes..a player has a trump too many. To get rid of this trump..is to play the grand coup. **1939** N. de V. Hart *Bridge Players' Bedside Bk.* xxxvii. 116 This ruffing of a winner to use up an embarrassing trump constituted the Grand Coup. **1952** I. Macleod *Bridge* xiv. 169 *Grand Coup*...It is a simple exercise in trump reduction and the fact that you trump winners alone distinguishes it from its humbler cousin the trump coup. **1964** *Official Encycl. Bridge* 222/2 *Grand coup*, a play by which declarer deliberately shortens his trump holding by ruffing a winner in order to achieve a finessing position over an adverse trump holding in an end position.

gran(d-)daddy. Add: (*a*) *fig.* examples. Also used *colloq.* with *of* to designate something large, notable, etc., of its kind. Cf. *father *sb.* 1 h.
1956 *Amer. Speech* XXXI. 255 This [sc. a spoonerism by a radio announcer] is known as the 'grandaddy' of all bloopers. **1960** *Woman's Own* 23 July 54/4 That young fellow is going to have the granddaddy of all headaches for about a week. **1961** M. Beadle *These Ruins are Inhabited* (1963) xi. 156 The granddaddy of all electrical storms dumped a cloudburst. **1965** R. & D. Morris *Men & Snakes* vi. 132 The mythical Serpent of Midgard has been described as the grand-daddy of all sea-serpents. **1968** *New Yorker* 18 May 60 I've had some granddaddies of hangups.
Also **gran(d-)daddy-longlegs** = Daddy-long-legs b.
1875 'Mark Twain' *Sk. New & Old* 130 The learned and aged Lord Grand-Daddy-Longlegs..had been sitting in deep study, with his slender limbs crossed. **1890** W. D. Howells *Boy's Town* 201 You must not kill a granddaddy-long-legs, or a lady-bug, it was bad luck. **1907** *Dialect Notes* III. 188 Don't kill that granddaddy-long-legs; he won't hurt you.

grand-daughter. Add: **d.** *attrib.* **grand-daughter clock,** a clock similar to, but smaller than, a grandfather clock; also *ellipt.*
1940 *Chambers's Techn. Dict.* 385/2 *Granddaughter-clock*, a long-case clock about 3½ ft. high, with movement usually provided with lever escapement; a grandfather-clock in miniature. **1962** E. Bruton *Dict. Clocks* 83 *Grand-daughter clock*, modern clock of grandfather style standing under about 4 ft. 6 in. high. **1968** R. Thomson *Antique Amer. Clocks & Watches* iii. 80 The New England tall clock reached the pinnacle of development about 1810...On the right is a miniature (or 'granddaughter') in the same style.

‖**grande amoureuse** (grañdamurøz). [Fr.] A woman of passion; a woman skilled in the arts of love.
1925 A. Huxley *Those Barren Leaves* v. i. 341 She found herself..playing the part of the *grande amoureuse.* **1937** *Times Lit. Suppl.* 30 Oct. 796/3 She [sc. Pepita] became a very popular dancer at other European capitals and, obviously, a *grande amoureuse.* **1966** G. Greene *Comedians* I. iii. 82 She could have played the *grande amoureuse* with the English tourist. **1967** *Times* 23 Oct. 6/4 Kedrova [a Russian actress] is yet another version of the grande amoureuse—..who weeps for her Cherry Orchard—..and for her worthless lover.

‖**Grande Armée** (grañdarme). [Fr.] Napoleon's main army, from the campaign of 1805 to that of 1814.
1844 W. Siborne *Hist. War France & Belgium* I. ii. 20 The renowned '*grande armée*' of Imperial France. **1896** G. A. Henty *Through Russian Snows* vii. 142 All believed that the Grande Armée would be invincible. **1942** C. Oman *Britain against Napoleon* x. 302 Her army could not compare in numbers with the *Grande Armée.* **1958** *Listener* 18 Sept. 441/2 A son of an officer of the Grande Armée. **1969** *Times* 31 May 17/6 When the Emperor got back to Paris in December, 1812, he organized a new *Grande Armée.*

‖**grande dame** (grañd dam). [Fr.] A great lady; a lady of rank and dignified bearing. Also as *attrib.* phr.
1744 H. Walpole *Let.* 8 May (1857) I. 301, I have seen her but once, and found her..*très grande dame.* **1775** A. Storer *Let.* 29 July in *Hist. MSS. Comm. 15th Rep. App. VI.* (1897) 281 And little Ursula a finer lady, from being at Bath, than ever, there being no such *grande dame* there as herself. *c* **1845** C. Brontë *Professor* (1857) viii. 140 A licence of manners..such as our venerable grand-dames would recoil from. **1855** J. S. Mill *Let.* 15 Jan. in F. A. Hayek *J. S. Mill & Harriet Taylor* (1951) 215 O those grandes dames how all vestige of the very conception of strength or spirit has gone out of them. **1862** Thackeray *Philip* i, She had been a great beauty, and was a perfect *grande dame* always. **1865** 'Ouida' *Strathmore* I. ii. 35 Or you've made love to some grande dame because it answered a political purpose. **1908** *Daily Chron.* 21 Feb. 3/2 Mrs. Saker, who is an accomplished player of 'grand-dame' parts. **1936** *Burlington Mag.* Apr. p. xv/2 A Florentine *grande dame* seated in a canopied carriage. **1969** *New Yorker* 11 Oct. 43/1 A *grande dame* caught in the middle of dressing.

‖**grande épreuve** (grañdeprœv). [Fr., lit. 'great test'.] An international motor-race.
1955 *Autocar* 10 June 802/1 Francorchamps, the setting for the second of the European *grandes épreuves*, contrasts with Monaco. **1958** *Times* 4 Aug. 8/5 In 1956, his first season as a member of the Ferrari team, he won two of the *grandes épreuves*—the French and Belgian Grands Prix. **1959** *Times* 1 June 16/1 For him there is the added satisfaction of winning his first *grande épreuve.*

‖**grande passion** (grañd pasyoñ). [Fr.] An overmastering love for another person; an engrossing love affair.
1823 Byron *Juan* xii. lxxvii, And if in fact she takes to a *grande passion*, It is a very serious thing indeed. **1847** C. Brontë *Jane Eyre* xv. 278 A French opera-dancer..towards whom he had once cherished what he called a 'grande passion'. **1854** 'C. Bede' *Verdant Green* II. xi. 95 Despite the hindrance which the *grande passion* is supposed to bring to the student. **1865** 'Ouida' *Strathmore* I. ii. 30 The perpetual gallantries, the never-ending, ever-changing *grandes passions*..that were characteristic of the Sabreur. **1938** *Times Lit. Suppl.* 24 Sept. 612/2 So many Lives have been written without any mention of the *grande passion.* **1960** J. Bayley *Characters of Love* ii. 84 Troilo—after several affairs—is now embarked on a *grande passion.*

‖**grande sonnerie** (grañd sonəri). [Fr., = great striking, chiming.] (See quot. 1960.)
1938 H. Cescinsky *Old English Master Clockmakers* iii. 13 There are..quarter-strike, chiming, 'Grand Sonnerie', and musical clocks. **1960** H. Hayward *Antique Coll.* 133/2 *Grand sonnerie*, a system of clock striking whereby the hour and the quarter are struck at each quarter. The earliest known example is the movement..made by Tompion about 1676–80. **1963** *Times* 12 Jan. 10/6 A late eighteenth-century *grande sonnerie* mantel clock.

‖**grande tenue** (grañd tənü). [Fr.] Full dress; *esp.* full military costume. Also *fig.* (See also s.v. *en *prep.*)
1849 C. Brontë *Shirley* II. v. 132 Peter Augustus came up, all in 'grande tenue', gloved and scented. **1865** 'Ouida' *Strathmore* I. x. 171 The ease of the dressing-gown after the restraint of the *grande tenue.* **1880** in *Manchester Stage 1880–1900* (1900) 145 Mme. Bernhardt shows the possession of the *grande tenue* necessary for classic tragedy. **1886** *Athenæum* 24 Apr. 560/2 The little soldier..is almost a caricature; the *grande tenue* is so ridiculous as to lose the charm which belongs to grotesqueness. **1952** C. Day Lewis *Grand Manner* 23 Lost, too, are the rhetoric of pulpit and parliament, the ceremonious style of letter-writing, the high-flown leading article—all that *grande tenue* in everyday communication which, though not directly the source of a poetic grand manner, seems so often to be its concomitant.

grandfather. Add: **1. b.** (Additional examples.)
1961 P. Barry *Unwillingly to School* xvi. 221 There's been the grandfather of a mix-up somewhere. **1971** *Daily Tel.* 18 Dec. 10 Sir Julian Huxley, whom Peter Scott yesterday described as the 'grandfather' of the World Wildlife Fund.
4. b. Short for *grandfather clock* (see sense 5 in Dict. and Suppl.).
1894 F. J. Britten *Former Clock & Watchmakers* 175 Some of these primitive 'grandfathers' were exceedingly narrow in the waist. **1906** *Bazaar, Exch. & Mart* Suppl. 3 Oct. 1302/3 Wanted, old grandfathers, brass arched dials. **1909** E. Rickert *Beggar in Heart* 1 The house is so still I can hear grandfather ticking on the stairs with the door shut. **1967** B. Palmer *Treasury Amer. Clocks* 1 Grandfathers were the first type of clocks to be made in the Colonies.
5. grandfather clause *U.S. colloq.*, a clause in the constitutions of some Southern states, exempting from suffrage restrictions the descendants of men who voted before the Civil War; also *transf.*; **grandfather's clock** (examples); also **grandfather clock** (now the usual name).
1900 *Congress. Rec.* 22 Jan. 1033/1 The grandfather clause will not avail those citizens who..are unable to pay their poll tax. **1903** *N.Y. Evening Post* 22 Dec. 6 It is proposed [in Maryland] to remodel the Constitution so as to exclude colored men from voting. The usual Southern method is followed, 'grandfather clause' and all. **1906** W. H. Fleming *Slavery* 44 This proviso was popularly known as the 'Grandfather clause'. **1948** *Georgia Hist. Q.* Mar. 1 In 1898, Louisiana wrote the notorious 'Grandfather Clause' into its constitution. [**1876** H. C. Work *Grandfather's Clock*, My grandfather's clock was too large for the shelf, So it stood ninety years on the floor.] **1892** J. F. Kendal *Hist. Watches* 153 Longcase or 'Grandfathers'' clocks. **1910** *Encycl. Brit.* VI. 552/2 The long or 'grandfather' clock dates from about the fourth quarter of the 17th century. **1922** O. Mitchell *Talking Machine Industry* vi. 68 Others use weight-driven mechanism[s] like those of the old-fashioned grandfather's clocks. **1928** J. E. Haswell *Horology* 74 A typical grandfather clock. **1952** H. Gordon *Antiques* ii. 36 A particular development associated with the [Restoration] period was the first appearance of the 'Grandfather' clock. **1970** G. Savage *Dict. Antiques* 191 *Grandfather clock*, a comparatively recent term for the long-case clock.

grandfer, granfer (græ·nfəɹ), dial. shortening of Grandfather.
1874 Hardy *Far fr. Mad. Crowd* viii, Ah, he's his grandfer's own grandson. **1889** A. Gissing *Both of this Parish* I. i. 21 'E be a good girl to look after your old granfer. **1930** *Daily Express* 23 May 10/3 What is good enough for granfer is good enough for us, look'ee.

‖**grand feu** (grañ fø). [Fr., lit. great fire.] (See quot. 1960.)
1863 W. Chaffers *Marks Pott. & Porc.* 164 When the mark is not indented on the paste, or baked with the porcelain when at its greatest heat (*au grand feu*), it gives no guarantee for its genuineness. **1906** S. W. Bushell *Chinese Art* II. 37 Under-glaze copper red (*rouge de cuivre*) of the *grand feu.* **1959** G. Savage *Antique Coll. Handbk.* 227 Tin-enamelled wares are usually decorated in colours applied directly at the same time as the glaze (high-temperature colours, or colours of the *grand feu*). **1960** R. G. Haggar *Conc. Encycl. Cont. Pott. & Porc.* 210/1 *Grand feu*, a term introduced by Brongniart for the hottest kiln, used for firing the glaze of hard-paste porcelain. *Grand feu* colours are those which require a high temperature to fire them.

Grand Guignol (grañ ginyol). [Fr. (= Great Punch), the name of a theatre in Paris.] A dramatic entertainment in which short pieces of a sensational or horrific kind are played successively. Also *transf.* Hence **Grand Guignole·sque, Grand Gui·gnolish** adjs.; **Grand Gui·gnolism.**
1908 *Sat. Rev.* 397/2 To act well in the Grand-Guignol postulates..as much art as in any other theatre. *Ibid.* 398/1 The most typically Grand-Guignolesque thing in the programme. **1920** H. B. Irving in M. Level *Crises* p. iii, M. Level has given literary expression of a high order to the compact horrors of the Grand Guignol. **1922** *Westm. Gaz.* 1 Dec., Violent eccentricity is on the decrease; I have seen Salons of a much more Grand Guignol character than this one. **1923** *Daily Mail* 12 June 8 'The Drums of Oude'..is perhaps unequalled as a one-act thriller outside frank Grand Guignolism. **1928** *Daily Tel.* 27 Mar. 6 'The Unknown', a morbid Grand Guignolesque conception rendered tolerable..by the actor's arresting personality. **1958** *Listener* 7 Aug. 211/2 Nor did any Grand Guignolish nonsense spoil his story which was full of genuine horror. **1961** A. Wilson *Old Men at Zoo* iv. 214 Did you see this grand guignol of Leacock's yesterday morning?..The destruction of the lynx had a morbid effect on the spectators. **1967** *Times* 3 Oct. 11/5 Many more [secrets] perished with the death of the sinister man who managed Stalin's *Grand Guignol* in the last years. **1969** *Observer* 12 Jan. 26/5 The film winds, with brilliant clarity, through a maze of shadowy emotions to a splendidly Grand-Guignolesque ending.

grandiflora (græ:ndiflǭ·ră), *a.* [a. mod.L. *grandiflora* a specific epithet often used in the names of large-flowered plants, f. L. *grandi-s* great + Flora.] Bearing large flowers.
1901 *Vegetables & Flowers from Seeds* (Sutton & Sons) 274 The single [Petunia] flowers are of the fine Grandiflora type. **1933** *Jrnl. R. Hort. Soc.* LVIII. 393 *Phlox Drummondii nana compacta*..are similar to the well-known grandiflora type. **1938** *Harper's Mag.* Nov. 668/2 We, the unholy innocents, study the bulb catalogue and order one dozen paper-white Grandiflora Narcissus. **1952** V. Gollancz *My Dear Timothy* 279 Grandiflora geraniums. **1960** *Times* 9 Jan. 9/1 For those who like larger flowered petunias..the 'grandiflora' types are excellent. **1970** C. Lloyd *Well-Tempered Garden* iii. 273 Would a devotee of wild gladioli and of the dwarf or small-flowered hybrids ever desert them in preference for the heavy, showy spikes of the Grandiflora types?

‖**grand jeté** (grañ ʒəte). [Fr.: see *jeté.] A ballet-step (cf. *jeté and see quot. 1957).
1930 Craske & Beaumont *Theory & Pract. Allegro Class. Ballet* 37 (*heading*) Grand jeté en attitude. **1941** C. W. Beaumont *3rd Primer Class. Ballet* ii. 16 *Grand jeté*, as its name implies, is a step of big elevation, executed *en tournant en diagonale.* **1948** A. Chujoy tr. *Vaganova's Basic Princ. Class. Ballet* vii. 76 For execution on the stage, grand jeté demands an entirely different approach than the small jeté. **1957** G. B. L. Wilson *Dict. Ballet* 154 *Grand jeté*,..a leap from one foot to the other, finishing in any required position. **1960** *Spectator* 4 Nov. 691 Held in the air, momentarily frozen in a *grand jeté.* **1962** *Ibid.* 2 Mar. 272 His sensational *grand jeté* from the Kirov ballet naturally helped to establish him. **1962** *Times* 27 Nov. 14/7 The generous *grands jetés.*

‖**grand mal** (grañ mal). [Fr., = the great sickness.] General convulsive epilepsy with loss of consciousness; epilepsis gravior. Sometimes artificially induced as part of electro-shock treatment. (Cf. Petit mal.)
[**1842** J. Jackson *Inquiry into Nature & Causes of Epilepsy* 10 The peculiar giddiness or momentary state of half consciousness, to which epileptics are so liable, is indeed an imperfect seizure, and differs only in degree and duration from the state called catalepsy; it is termed by the French 'vertige epileptique' or 'petit mal', in contradistinction to the perfect seizure, or 'grand mal'.] **1879** *Encycl. Brit.* VIII. 479/2 There are two well-marked varieties of the epileptic seizure...To these the terms *epilepsia gravior* and *epilepsia mitior*, *le grand mal* and *le petit mal*, are usually applied. **1948** *Jrnl. Nerv. & Mental Dis.* CVIII. 296 No patient was treated with petit mal seizures, and a grand mal response was achieved with each single treatment in each patient. **1960** M. Spark *Bachelors* i. 10 The patient might learn to exercise some control during the *petit-mal* stage to stand him in good stead during the *grand-mal* convulsions. **1965** J. Pollitt *Depression & its Treatment* iv. 50 Electro-convulsive treatment..is still the most rapid and effective treatment. It consists of the passage of a brief alternating electric current across the anterior temporal regions of the head to induce a grand mal convulsion.

grandmamma. Add: (Earlier example.) Also **grandmama, grandma** (examples).
1749 J. Cleland *Mem. Woman of Pleasure* I. 140 The maid..added, that she was sure this usage of her sweet young master, would be the death of his grandmamma. **1836** Dickens *Let.* 30 July (1965) I. 159 Perhaps you will fill up the blank, and place 'Grandmama' and 'Grandpapa' at the top. **1867** W. James *Let.* 26 Sept. (1920) I. 105, I remember days passed in Grandma's old house in Albany. **1871** L. M. Alcott *Little Men* iii. 30 Mrs. Bhaer with Daisy and her own two boys drove into town, to pay the weekly visit to Grandma. **1939** Joyce *Finnegans Wake* 580 To make grim grandma grunt and grin again.

Grand Marnier (grañ marnye). [Fr.; *grand* great, fine, and *Marnier*-Lapostolle, name of manufacturer.] A French cognac-based liqueur.
1905 *Daily Chron.* 28 Apr. 6/7 A publican who supplied Benedictine to several customers who asked for Grand Marnier was fined £5. **1910** *Encycl. Brit.* VII. 636/1 A lighter variety of Curaçoa, made with fine brandy, is known as 'Grand Marnier'. **1924** Galsworthy *White Monkey* II. ii, He went to a cupboard and returned with two small glasses of a brown fluid. 'Have a Grand Marnier?' **1958** A. L. Simon *Dict. Wines* 85/1 *Grand Marnier*, popular French liqueur with an orange flavour and a Cognac brandy basis. **1962** *Cocktails* (Cunard Line) Grand Marnier 2/6.

grand master. Add: **3*.** *Chess.* (Freq. written as one word.) Orig. the title accorded to a player who won first prize in a great international tournament (opp. *Master* in national tournaments). Now often used more generally of an unusually skilled player. Also *attrib.* and *transf.* Hence **grandma·sterly** *a.*
1927 *Brit. Chess Mag.* XLVII. 61 Yates made a characteristic recovery and added to his growing bag the scalps of two 'grand masters'. **1936** *Chess Rev.* (N.Y.) Jan. 17 (*title*) A Gallery of Grandmasters. **1936** B. Harley *Chess & its Stars* 121 The ultra-caution which is merely a symptom of her progress to the grand-master class of chess. **1937** *Chess* III. 116 The victory was achieved by the only one of the Grand Masters who had not received a retainer: Keres! **1946** *Brit. Chess Mag.* LXVI. 11. 50 Grandmaster Lilienthal..insistently advises us to study the Slav defence. **1952** *Sci. Amer.* May 30 To take an apt title from the terminology of chess players, Sherrington was a grand-master of science. **1955** *N.Y. Times* 20 Feb. 2/5 United States chess players..will oppose the famous Soviet grandmasters who have maintained world supremacy ever since their series of victories during 1954. **1965** *New Statesman* 9 Apr. 590/1 Here..is a case of..truly grandmasterly chess. **1967** *Times* 11 Feb. 12/5 What constitutes a grandmaster at chess?..The term itself is supposed to have been invented by the Tsar to describe the competitors in the great international tournament at St. Petersburg in 1914. *Ibid.* 14 Nov. 6/3 He [sc. a Russian grandmaster since 1939] attributed much of the deterioration in chess to changes in the rules of the World Chess Federation, made in the 1950s. Abandoning the procedure under which the title of grandmaster could be won only by taking first prize in a tournament, the federation liberalized the rules to permit the title to be conferred upon any player winning 55 per cent of the points in a tournament.

‖**Grand Monarque** (grañ monark). [Fr.] The title given to Louis XIV. Also *transf.*, a supreme and absolute ruler.
1699 M. Lister *Journey to Paris* 25 There are an infinite number of Busto's of the Grand Monarque every where put up by the Common People. **1716** Addison *Freeholder* No. 10 ¶6 His Governours of Towns and Provinces, who form'd themselves upon the Example of their *Grand Monarque*, practised Rapine, Violence, Extortion. **1842** Barham *Ingol. Leg.* 2nd Ser. 22 'Twas the *Grand Monarque's* birthday. **1856** Bagehot *Coll. Works* (1965) I. 379 How gravely admirable to see the *grand monarque* shaved, and dressed, and powdered. **1911** *Encycl. Brit.* XVII. 42/2 Louis was singularly well fitted by his physical and intellectual gifts for the rôle of *Grand Monarque* and he played it to perfection. **1933** D. Ogg *Louis XIV* 250 This fact may temper enthusiasm for the cult of the *Grand Monarque.* **1967** D. W. Brogan in L. Norton tr. *Mem. St.-Simon* I. p. xix, Saint-Simon shows us the King with..his 'courage never to submit or yield' which enabled him..to be in 1715 still *le grand monarque.*

‖**grand monde** (grañ mǫ̃nd, mǫ·nd). [Fr., = great world.] The beau-monde, the 'upper ten'.
1704 Swift *Tale Tub* ii. 57 A Sect arose, whose Tenents [*sic*] obtained and spread very far, especially in the *Grand Monde*, and among every Body of good Fashion. **1725** —— *Let.* 29 Sept. in *Corr. Pope* (1956) II. 324, I am now returning to the noble Scene of Dublin in the Grande [*sic*] Monde. **1823** Byron *Juan* xiv. xlii, She was a fine and somewhat full-blown blonde,..For several winters

in the grand, *grande* [sic] *monde*. **1892** W. James *Let.* 29 Apr. (1920) I. 319 Strange to say, altho' practically bed-ridden for years, her mental atmosphere..was altogether that of the *grand monde*. **1925** D. H. Lawrence *St. Mawr* 13 Mrs. Witt..always expected to find the real *beau monde* and the real *grand monde* somewhere or other.

grandmother, *sb.* Add: **1. b.** *your grandmother!*: said of something with which one disagrees. (Cf. *granny 1 c.)
1870 'Mark Twain' *Screamers* (1871) 50 'Shake the tree—' 'Shake your grandmother! Turnips don't grow on trees!' **1874** Trollope *Phineas Redux* II. xiii. 111 'Did you see her?' said Ned...'See your grandmother.' **1909** Galsworthy *Joy* I, *Mrs. Hope.* You'll just attend to what I say and look into that mine! *Colonel.* Look into your grandmother! **1911** G. B. Shaw *Getting Married* 268 *Lesbia.* I hate..sentimental people. *Mrs. George.* Oh, sentimental your grandmother! **1934** E. Waugh *Handful of Dust* ii. 22 'I think she [*sc.* a horse] put in a short step.' 'Short step my grandmother.'
c. (Additional example.)
1959 *Woman* 24 Oct. 12/1 My mother and Laura Simmonds, who had lived in each other's pockets since the age of five, had the very grandmother of a row.
d. *Grandmother's (Foot)steps*, name of a children's game in which one player stands with his back turned to the rest and the others try to approach him in a stealthy manner and touch his back without his seeing them move. The person in front is allowed to turn round often and without warning and any player caught moving is sent back to the starting-line.
1937 'N. Blake' *There's Trouble Brewing* i. 30 The children's game called 'Grandmother's Steps'. **1945** E. Waugh *Brideshead Revisited* 122 She took hold of her subject in a feminine, flirtatious way... She played 'grandmother's steps' with it, getting nearer the real point imperceptibly while one's back was turned, standing rooted when she was observed. **1956** L. McIntosh *Oxford Folly* 204 We were playing grandmother's footsteps. **1966** J. Betjeman *High & Low* 37 A game of Grandmother's Steps on the vicarage grass.
4. grandmother clock, a clock resembling a grandfather clock, but with a smaller case. Also *ellipt.*
1922 H. S. Barrett *A.B.C. Hist. Antique Eng. Furnit.* 86 By Grandmother clocks I refer to clocks not exceeding about 5 ft. to 6 ft. in height. **1927** *Daily Tel.* 6 Dec. 3/7 A grandmother clock with brass dial. **1930** *Aberdeen Press & Jrnl.* 31 May 8/3 The gift to ex-Constable Jamie was a beautiful grandmother clock. **1931** H. Sutcliffe *Persons Unknown* i. 11 The quieter pendulum-swing of the slim clocks known as 'grandmothers'. **1965** E. Tunis *Colonial Craftsmen* vi. 145 In addition to the taller clocks, there were also 'grandmother' clocks about five feet high, controlled by shorter, half-second pendulums. **1967** B. Palmer *Treasury Amer. Clocks* 23 Grandmother clocks are diminutive editions of grandfather clocks, standing four feet or less in height.

gra·ndmother, *v.* [f. the sb.] *trans.* and *intr.* To take care of as a grandmother; to behave in a grandmotherly way (towards); to be the grandmother of. Hence **gra·ndmothering** *vbl. sb.* (Cf. Grandmotherly *a.*)
1901 Kipling *Kim* xv. 396 When one cannot dance in the festival one must e'en look out of the window, and grandmothering takes all a woman's time. **1903** H. James *Better Sort* 42 Do you mean by his idea his proposal that I should grandmother his wife? **1923** *Chambers's Jrnl.* Feb. 102/1 A frail little lady who had grandmothered a hefty brood of men. **1929** *Daily Express* 8 Jan. 8/7 Political power has naturally passed into the hands of people with a passion for grandmothering. *Ibid.* 12 Jan. 8/7, I refuse to believe that Britons are so excessively unmanly as to be driven to this state of grandmothering. **1966** E. H. Jones *Margery Fry* xv. 203 Agnes was grandmothering two schoolboy evacuees.

Grand Old Party. *U.S. politics.* [See Grand *a.* 10 d.]
†**1.** The Democratic party. *Obs.*
Used without capitals and perhaps not specific.
1879 *Congress. Record* 11 June 1913/1 We are for national politics now. We come back to the grand old party of the North. **1888** *Ibid.* 10 May 3981/1, I am glad that I am a member of that grand old party that assures a better trade to our people, larger wages &c.
2. The Republican party. Now usu. in abbreviated form *G.O.P.*
1876 *Cincinnati Comm.* in *Harper's Weekly* (1884) 576/3 Grand Old Party. **1884** *N.Y. Tribune* 15 Oct. 4/5 'The G.O.P. doomed,' shouted the *Boston Post*...The Grand Old Party is in condition to inquire [etc.]. **1888** *Congress. Record* 1 May 3598/1 Old Farmer: Is this Democratic doings or Republican doings? Collector: O, it is the doings of the G.O.P.,—the grand old party,—the Republican party. **1898** *Ibid.* 7 Jan. 444/1 Has the Grand Old Party (G.O.P.) so called never been corrupt? Has it never had dishonest men in it? **1904** *N.Y. Even. Post* 25 Aug. 6 A close examination of Republican speeches fails to reveal an instance in which the Democracy is portrayed as on a parity with the Grand Old Party. **1960** *Detroit News* 16 Aug. 18B/1 The Nixon influence..has sought to capitalize on the new appeal of the GOP. **1964** Knebel & Bailey *Convention* vi. 88 Right now, this has the makings of the closest G.O.P. nomination contest since the Eisenhower-Taft fight in 1952.

grand opera. [See Grand *a.* 8 b.] Serious opera without spoken dialogue. (Cf. quot. 1879 s.v. Grand *a.* 8 b; the term has not become obsolete.)
1803 *Lett. Miss Riversdale* II. 317 A few nights of the *Grand Opera*..will bring you over to my opinion. **1879** [see Opera 3]. **1954** *Grove's Dict. Mus.* (ed. 5) III. 757/1 Grand Opera, a term with a definite meaning in French, unlike the English 'grand opera', which..is useless for classification, since it is loosely used to mean serious as distinct from comic opera. **1967** *Listener* 5 Oct. 443/1 She [*sc.* Melba]..bulldozed her initially reluctant countrymen into a fervid admiration for Grand Opera. **1967** *Times* 7 Oct. 12/8 After 12 years in opera—both light and grand—soprano Joyce Blackham is to appear in her first West End musical.

grandpapa. Also **gra·ndpappy, gra·ndpop** (Pappy *sb.*¹, *pop *sb.*⁶).
1890 *Texas Siftings* 22 Nov. 10/1 'Grandpop,' said he, 'did you know that there was flying fish?' **1919** *Dialect Notes* V. 39 Gran'-pappy. **1931** H. Crane *Let.* 10 Jan. (1965) 363 Spinoza (Einstein's grandpop) furnishes plenty of discipline. **1936** F. Clune *Roaming round Darling* xiii. 107 Grandpop betrayed them, and at daybreak the police attacked. **1938** M. K. Rawlings *Yearling* x. 92 Now Grandpappy, I kin see you settin' there on your stoop. **1946** M. Lowry *Let.* 26 Dec. (1967) 134 Just like the old days of my grandpappy. **1948** D. Ballantyne *Cunninghams* (1963) i. xviii. 96 'You don't say, grandpop,' Joy said. **1953** 'N. Blake' *Dreadful Hollow* iv. 51 Have a glass of port, won't you? It's rather delish. Grand-pop laid it down. **1959** [see *break *sb.*¹ 8 k].

grandparental (grændpărε·ntăl), *a.* [See -al.] Of, belonging to, or derived from a grandparent or grandparents.
1903 *Lancet* 18 Apr. 108/1 Cases in which the taint was grandparental. **1929** *19th Cent.* Dec. 810 A collateral relative at the level of the grandparental or great-grandparental generation. **1930** R. A. Fisher *Genet. Theory Nat. Selection* 198 The grandparental effect..is subject to somewhat large sampling errors. **1958** *New Biology* XXVI. Pl. 3(b), At the base of the parent corm is the remains of the connexion to the 'grandparental' corm.

grandpop: see *grandpapa.

Grand Prix (gran pri). [Fr., = great or chief prize.] **1. a.** (In full *Grand Prix de Paris.*) An international horse race for three-year-olds established in 1863 and run annually in June at Longchamps, Paris. **b.** Any of certain motor-races held annually in different countries and governed by international rules.
1863 *Times* 3 June 9/4 The great event which had been looked forward to with such deep interest..both in England and France,..resulted in the victory of an English horse...The Grand Prix de Paris was looked forward to by the partisans of Lord Clifden, Saccharometer, and the Ranger. **1866** *Field* 10 Mar. 199/3 Some of our English friends, too, are travelled gentlemen, who go to the Grand Prix. **1899** [see *Ascot]. **1902** *Encycl. Brit.* XXIX. 336/2 The Grand Prix de Paris, an international race for three-year-olds, run at Longchamps... The first Grand Prix fell to an English horse. **1908** *Westm. Gaz.* 2 Apr. 4/3 The drivers have now definitely been selected for the team of three Austin cars entered for competition in the Grand Prix race. **1922** *Encycl. Brit.* XXXI. 1003/2 In 1911 began a new series of races for the *Grand Prix* of the Automobile Club of France. **1923** W. J. Locke *Moordius & Co.* viii. 99 Already, after the Grand Prix, the ultra fashionables were leaving Paris. **1931** *Times* 5 June 4/6 To-day's practice for the Irish Grand Prix Race for motor-cars. **1958** [see *grande épreuve]. **1959** *Chambers's Encycl.* IX. 559/1 In 1906 the French organized the first Grand Prix on a circuit near Le Mans, and by making it open to all, instead of to three cars only from each country, set the pattern for all future races. **1960** *Times* 28 Sept. 16/5 The cars to be run..will comprise three Grand Prix cars. **1966** *Times* 14 Feb. 6/3 Grand prix racing under the new 3-litre Formula 1 gets into its stride this year.
2. *gen.* The highest prize awarded for products in some particular line at an exhibition.
Also in anglicized form, *grand prize*: see *grand *a.* 12.
1880 *Rep. Paris Univ. Exhib. 1878* II. 425. **1916** *Oxf. Univ. Press Gen. Cat.*, The Twenty-three Grands Prix Awarded to the Oxford University Press At International Exhibitions 1889–1911.

grand siècle (gran syɛkl). [Fr., = great century or age.] The reign of Louis XIV, the classical or 'Augustan' age in France. Also *transf.*
1838 *Westm. Rev.* XXXI. 7 Many a Liberal or Republican critic will stand up stiffly..for the *grand siècle*. **1912** L. Strachey *Landmarks Fr. Lit.* v. 133 While the masterpieces of the *Grand Siècle* served no ulterior purpose.. those of the eighteenth century were works of propaganda. **1937** *Burlington Mag.* Dec. 294/2 An essay in praise of the *grand siècle*. **1963** *House & Garden* Mar. 97/1 The hundred years from 1811 to 1911 deserve to be called the *Grand Siècle* of wine.

grandstand, grand-stand. [See Grand *a.* 12.] **1.** The principal stand for spectators at a race-course, football ground, etc.
1834 *N.Y. Sporting Mag.* Nov. 169/1 At the Grand Stand Chapple let him go, and he won by a length in a canter. **1841, 1872** [see Grand *a.* 12]. **1961** F. C. Avis *Sportsman's Gloss.* 28/1 *Grand Stand*, that part of the football ground generally containing the business offices, dressing rooms, press box, reserved and special seats, etc.
2. *attrib.* Intended to impress the spectators in the grandstand; as good as if viewed from a grandstand. So *grandstand finish*, a close and exciting finish to a sporting contest; *grandstand play* (U.S.), a way of playing a game with an eye to the applause of the spectators in the grandstand.
1888 M. J. Kelly *Play Ball* viii. 39 It's little things of this sort which makes [*sic*] the 'grand stand player'. They make impossible catches, and when they get the ball they roll all over the field. **1893** W. K. Post *Harvard Stories* 308 They all hold on to something or clasp their knees tightly—to faint or fall over would be a grand-stand play. **1904** *Utica* (N.Y.) *Observer* 23 June 6 The ultimatum to the Sultan of Morocco, 'Perdicaris alive or Raisuli dead' was a good one. But telegraphing it to the National Convention at Chicago made it look very much like a grand stand play. **1906** S. Ford *Shorty McCabe* (1908) xiii. 276, I makes a grandstand finish, and then has the nerve to face the audience and do a matinée bend. **1942** E. Paul *Narrow St.* xx. 163 Then [André] Breton made another grandstand play. **1958** *Spectator* 1 Aug. 172/2 Dumaine's shrewdness and his grandstand view of French post-war history. **1958** F. C. Avis *Boxing Ref. Dict.* 49 *Grand stand finish*, very vigorous exchanges of blows in the final round of a boxing contest. **1967** *Bucks Examiner* 6 Oct. 2/1 He..had a grandstand view of Sir Francis [Chichester] and his yacht.
Hence **gra·ndstand** *v. intr.*, to perform with an eye to the applause of the spectators in the grandstand. Also **gra·ndstander**, (*a*) one who occupies a seat in a grandstand; (*b*) one who 'grandstands'.
1891 *Sporting Times* (N.Y.) 23 May 3/3 During the four New York games there were never less than 2,200 people at a game, and 50 per cent of the patrons here were 'grand-standers'. **1900** *Cincinnati Enquirer* 23 June 1/9 [Kentucky will go for McKinley] if Teddy can only be secured to do some 'Grand Standing'. **1914** S. Blythe *Fakers* 163 That old grandstander, Rollins, is making a good deal of a row over the franchise matter. **1920** C. H. Stagg *High Speed* vii. 121 A car like that, and in the hands of a grand-stander! **1927** K. Nicholson *Barker* 132, I ain't grand-standin' an' I'm tellin' you what I'm goin' to do. **1935** M. M. Atwater *Murder in Midsummer* xiii. 120 'Sure, he's grandstanding,' said Matter. **1935** E. S. Gardner *Case of Howling Dog* xv. 151 The public will think you're simply grandstanding for the purpose of getting a big fee out of the trial. **1948** *Sat. Even. Post* 24 July 21/1 Editorial blasts..have described the general in many unflattering terms—namely: a blunderer; a grandstander; a bull in a china shop; a trouble causer. **1970** B. Knox *Children of Mist* v. 114 Adam Jennings loves a chance to grandstand. This was made to order.

grand tour. b. (Further *transf.* examples.)
1970 *Sci. Jrnl.* May 9/3 The 'grand tour of the planets' (the opportunity at the end of this decade to send a single rocket around Jupiter, Saturn and the outer planets, using the gravitational pull of each planet it passes to send it on to the next one). **1971** *Daily Tel.* 2 Dec. 6 (Advt.), The complete story of man's conquest of the moon—glimpses of the future and the fantastic 'grand tour' of the planets.

grane *sb.*², var. of Grain *sb.*² 5 b.
1899 F. T. Bullen *Idylls of Sea* xvii. 136 A few good lines and hooks, and a set of granes. **1951** R. Campbell *Light on Dark Horse* xx. 285, I went on fishing, with my spare grane (fish-spear).

granfer: see *grandfer.

granita (grænī·tă). Pl. **granite** (grænī·te). [It. (cf. Granite).] **a.** = Granite 2. **b.** An iced drink.
1869 'Mark Twain' *Innoc. Abr.* xxiii. 174 People at small tables [in Venice]..smoking and taking *granita* (a first cousin to ice-cream). **1963** *Times* 14 Jan. 11/5 No account of Italian sweets would be complete without mention of the ices, and of these the simplest and perhaps the best are the *granite*, sorbets or water ices. **1968** R. Collin *Locust on Wind* iii. 25 A tall glass of *granite*, crushed ice in fresh lemon juice. **1970** Simon & Howe *Dict. Gastron.* 205/1 In Italy *granita* heralds the first sign of summer.

granite. Add: **1. b.** *fig.* Applied to 'stony', hard-headed, or hard-hearted persons. Often *attrib.* and *Comb.* (cf. *granite-like* in 3 b).
1839 J. R. Lowell *Ye Yankees* 3 in *Uncoll. Poems* (1950), Stern granite hearted ones! **1902** *Daily Chron.* 12 Nov. 5/6 That granite-headed ship-owner, Sir Alfred L. Jones. **1908** *Ibid.* 7 Nov. 4/4 His countenance expressed neither the sweetness and tenderness of the saint nor the granite severity of the prophet. **1916** *Punch* 14 June 398/2, I stole a look at Hercules over my shoulder, but he was granite. **1920** C. Jerdan *Scott. Clerical Stories* viii. 164 Strong-boned, granite-headed, and endowed with all kinds of vitality.

granitization (grænitəizē·ʃən). *Geol.* [f. Granit(e + -ization.] The process by which granitic rocks are formed from other rocks; formerly also, the injection of granite into a rock. Also **gra·nitized, gra·nitizing** *ppl. adjs.*
1893 A. Geikie *Text-bk. Geol.* (ed. 3) 579 Round some bosses of granite the adjacent rocks are injected or impregnated with such an abundance of minute threads or veins of granite substance..that they are said to be 'granitized'. *Ibid.* 604 This impregnation or granitization has been strongly insisted upon by M. Michel Lévy. **1944** *Proc. Geol. Assoc.* LV. 78 The view..that the granitising material must have varied with the rocks undergoing granitisation. **1963** D. W. & E. E. Humphries tr.

Termier's Erosion & Sedimentation p. vii, Many mechanical, physical and chemical phenomena resulting from internal geodynamic processes, such as folds, metamorphism and granitization, are only known by their effects. **1965** G. J. Williams *Econ. Geol. N.Z.* i. 1/1 This gneiss probably represents an intensely metamorphosed and granitized complex of ancient sediments.

granny, grannie. Add: **1. c.** *your granny!*: used as an exclamation suggesting derision or disbelief. (Cf. *grandmother *sb.* 1 b.)
1838 J. C. Neal *Charcoal Sks.* 35 'War!' ejaculated the party; 'oh, your granny!' **1858** Geo. Eliot *Scenes Cler. Life* I. 65 'Well,' suggested John,..'you should wet the bottom of the *duree* a bit, to hold it from slippin'.' 'Wet your granny!' returned the cook. **1863** *Ladies' Repository* XXIII. 482/2 'Repose, your granny,' answered Addie, who, when vexed never stopped for elegant phrases. **1876** 'Mark Twain' *Tom Sawyer* xxv. 193 'Do they hop?' 'Hop—your granny! No.' **1939** 'F. O'Brien' *At Swim-two-Birds* i. 13, I open several books every day, I answered. You open your granny, said my uncle.
6. *granny-armchair, collarette, dress, glasses, gown, hat, print, skirt*; **granny knot** = *granny's knot*; **granny's knot** (earlier example); **granny's nightcap**, *dial.* name for various plants.
1946 Koestler *Thieves in Night* 146 There were mattresses, saucepans, a cuckoo-clock, a granny-armchair, a bicycle and even a bird-cage. **1880** *Queen* 10 July (Advt.), The 'Granny' shade hat,..The 'Granny' collarette. **1909** *Westm. Gaz.* 3 Apr. 15/2 The best dressmakers are preparing what they call 'Grannie' dresses. **1966** *Punch* 29 June 946/1 The social columns continue to keep us informed in their eager girlish way about where the 'in' crowd are currently surging in their Granny dresses, Stalin caps, sawn-off skirtlets and mini-beards. **1968** J. Hudson *Case of Need* iii. i. 171 She had a flower painted on her cheek, and large, blue-tinted granny glasses. **1970** *Guardian* 22 June 1/1 A mop-headed youth with granny glasses. **1965** J. Hart *File for Death* iv. 31 The girl, dressed in a long granny gown and wearing a bright pink hairnet. **1967** *Boston Sunday Globe* 23 Apr. A 28/3 These favorites will be offered along with the teens' 'granny hats'. **1853** Thoreau *Jrnl.* 25 July in *Writings* (1906) XI. 335, I had been all the while tying what is called a granny's knot. **1905** *Eng. Dial. Dict.* Suppl., Granny-knot. **1932** Kipling *Limits & Renewals* 196 They'd been busy since light unpickin' the wire granny-knots this so-called Noo Navy had tied 'em in with. **1959** *Listener* 7 May 808/2 The bag..that took half an hour to rope up with eighteen feet of clothes-line and about forty granny-knots. **1961** *Sunday Express* 28 May 15/2 Small black-on-grey 'Granny' prints. **1966** H. Yoxall *Fashion of Life* viii. 76 In the autumn of 1965..girls in Los Angeles started a fad for wearing 'granny' skirts, down to their heels. **1863** *Phytologist* New Ser. VI. 416 The *Caltha palustris* is called 'May-blobs' by the children who gather the flower in the meadows near Warwick; they also call the Wood Anemone (*A. nemorosa*) 'Granny's Nightcap'. **1892** C. M. Yonge *Old Woman's Outlook* 119 The odd red-calyxed *Geum rivale*, called by the village children Granny's nightcaps. **1893** Dartnell & Goddard *Gloss. Wilts.* 69 *Granny* (or *Granny's*) *Nightcap*, (1) *Anemone nemorosa*,..Wood Anemone..(2) *Aquilegia vulgaris*,..Common Columbine ..(3) *Convolvulus sepium*,..Great Bindweed..(4) *Convolvulus arvensis*,..Field Bindweed.

Granny Smith. [f. the name of Maria Ann *Smith* ('Granny Smith'), d. 1870.] An Australasian variety of apple, which is bright green and suitable for eating raw or cooked.
1895 *Agric. Gaz. N.S. Wales* VI. 900, I think that..Granny Smith's Seedling..would be worth a trial. *Ibid.*, If it is desirable to go in more for the late home market, then I would include..Granny Smith's Seedling. **1932** *Discovery* July 220/2 Australasia has Granny Smith, Cleopatra, Jonathan, Sturmer, Pippin etc., which excel. **1933** *Bulletin* (Sydney) 11 Jan. 12 Seventy years ago there dwelt in the isolated hamlet of Eastwood..an old lady named Smith..the original source of the 'Granny Smith' apple. **1943** *Gen* 10 Apr. 60/2 The stubby 'Granny Smith' apple tree on the lawn. **1960** W. A. Beatty *Treas. Austral. Folk Tales* 186 The Granny Smith is unique among apples inasmuch that there is none its equal for cooking purposes, it is a splendid eating apple, and when stored keeps better than any other variety.

grano² (græ·no), abbrev. *granolithic *sb.*
1940 *Archit. Rev.* LXXXVIII. 40 The floor of the stoep is finished in shiny black grano. **1951** *Ibid.* CIX. 290 A wood roof with corrugated asbestos covering, wood and grano floors. **1958** J. S. Scott *Dict. Civ. Engin.* 165 *Granolithic* or *grano*, a screed of cement, sand, and granite chippings floated over concrete floors to give a smooth, hard-wearing surface about 1½ in. thick.

granodiorite (grænodəi·ŏrəit). *Petrogr.* [f. *grano-*, taken as combining form of *granite* (cf. Granophyre) + Diorite.] Any of the plutonic igneous rocks intermediate in composition between true granite and quartz-diorite, which are similar in appearance to granite but are usu. darker in colour; *esp.* one in which there is at least twice as much plagioclase as alkali feldspar.
1893 W. Lindgren in *Amer. Jrnl. Sci.* 3rd Ser. XLVI. 203 On the Survey maps of the Gold Belt of the Sierra Nevada, of which district Mr G. F. Becker is geologist in charge, it has therefore been determined to indicate this rock as granodiorite, which term it is hoped will find general acceptance. **1932** A. Johannsen *Descr. Petrogr.* II. 323 Granodiorites occur as intrusive bodies, in the same manner as granites, and often form huge masses. **1961** *Bull. Board Celtic Stud.* XIX. 167 A magnificent implement [*sc.* a Neolithic axe-head] made of an augite granodiorite, almost certainly of Pembrokeshire origin. **1965** G. J. Williams *Econ. Geol. N.Z.* xiv. 216/1 The central granite consists of types ranging from pink, generally porphyritic, alkali granites to the more common white to grey biotite calc-alkali granite, adamellite or granodiorite.

granolithic, *a.* Add: Also as *sb.*, = granolithic concrete.
1881 P. Stuart *Brit. Pat.* 610 4 The granit and cement is mixed...This composition, which I call 'granolithic', is spread on the concrete so as to embed the iron rods. **1909** *Pall Mall Gaz.* 12 Apr. 2/2 An inscription in brass letters set in granolithic. **1956** Davies & Petty *Building Elem.* ix. 282 Special abrasive aggregates can be added to the granolithic to provide a non-slip surface.

grant, *sb.*¹ Add: **6.** *grant-earning* (later examples; so *grant-earner*), *-giving* adjs.
1909 *Daily Chron.* 8 Sept. 6/3 Boys who will reach this age [14] between now and next Easter are inferior grant-earners as compared with boys who can put in a full year's attendance. **1900** H. G. Wells *Love & Mr. Lewisham* xxv. 231 Some grant-earning grammar-school. **1960** *Farmer & Stockbreeder* 16 Feb. 146/1 (Advt.), In Britain, Canadian Douglas fir plywood has been designated a grant-earning material. **1963** *Higher Educ.* (Cmnd. 2154) xiv. 213 The increased burden on the local grant-giving authorities. **1964** J. S. Huxley *Ess. Humanist* 267 The great grant-giving..agencies such as the U.N. [etc.].

grant, *v.* Add: **4. e.** In *pa. pple.* as a polite rejoinder to an apology.
1902 Kipling *Traffics & Discoveries* (1904) 238 'Granted—granted as soon as asked,' he said, unbending. 'I *did* think it a shade odd at the time.' **1924** —— *Debits & Credits* (1926) 311 '..I beg your pardon...' 'Granted.' **1926** R. Macaulay *Crewe Train* ii. v. 103 When others craved their pardon for stepping on their toes, their reply was, 'Granted.' **1951** E. Coxhead *One Green Bottle* v. 115 'Pardon?' said Cathy, momentarily bewildered; whereat Mr. Derwent..replied: 'Oh-er, granted.' **1967** 'H. Calvin' *Nice Friendly Town* vi. 87 She yawned a great yawn and said, 'Sorry.' 'Granted,' I said.

gra·nt-aided, *a.* [Cf. *grant-in-aid.] Of a school or other institution: assisted by an allowance from government or official funds. Hence, by back-formation, **grant-aid** *v. trans.*, to afford financial support of this kind to (a school, etc.). Also **grant-aid** *sb.*, short for *grant-in-aid; **grant-aiding** *a.*
1927 Carr-Saunders & Jones *Soc. Struct. Eng. & Wales* 120 Grant-aided secondary schools. **1930** *Times Educ. Suppl.* 18 Oct. 436/4 One-quarter of the grant-aided schools were now co-educational. **1933** *Planning* I. xii. 8 In road transport the adoption after the war of a national system of classified numbering and a policy of rapid grant-aided development opened the way to a national policy for road transport. **1943** *Our Towns. A Close-up* iii. 96 The Board [of Education] should..ensure, by special grant-aid if necessary, the observance of a minimum standard of hygiene. **1944** H. C. Dent *Educ. Act 1944* 71 A point of interest is that for the first time the Minister is empowered to grant-aid educational research. *Ibid.* 72 Special agreement schools will, under Clause 102, be grant-aided in accordance with the special agreement in respect of each school. **1959** *20th Cent.* Nov. 345 Independent Television..is no less concerned with programmes on current affairs than the grant-aided corporation. **1962** *Economist* 13 Oct. 148/2 The function ..will be largely that of a grant-aiding body.

Granth (grʌnt). Also **Grantha, †Grooht, Grunth.** [Hindi *granth* book, code, a. Skr. *grantha* tying, knot, literary composition, book, text, f. *granth, grath* to tie.] The sacred scriptures of the Sikhs.
1798 G. Forster *Journey f. Bengal to England* I. xi. 255 A book, entitled the *Grunth*,..is the only typical object which the Sicques have admitted into their places of worship. **1837** G. R. Siddons in *Jrnl. Asiatic Soc. Bengal* VI. i. 480 The Granthas, or sacred books, of the Dadupanthi Sect. **1838** E. Eden *Let.* 10 Dec. in *Up Country* (1866) II. i. 7 There was a large collection of priests, sitting in a circle, with the 'Grooht', their holy book, in the centre. **1839** H. Bevan *30 Yrs. India* II. i. 7 This compilation is called *Granth*, and is held at least in as high veneration by the Sikhs as the Bible by Christians. **1852** J. M. Honigberger *35 Yrs. in East* I. 99 The Gooroos, or priests of the Sikhs, did the same, from their holy scripture called Grunthsaheb. **1877** *Good Words* 699 Nanuk was rather the bard, who handed down in his Grunth the verses of Kubeer and other fakeers. **1877** E. Trumpp tr. (*title*) The Ādi Granth or Holy Scriptures of the Sikhs. **1901** *N. Amer. Rev.* Feb. 303 The Sikh still..visits the temple to listen to the reading of the Granth. **1927** *Daily Tel.* 22 Feb. 11/5 Three hundred Sikhs, carrying the Granth Sahib, or Holy Book,..were attacked by infuriated Moslems. **1934** H. H. Gowen *Hist. Religion* xxiii. 347 The fifth Guru, in 1601, compiled the sacred book known as the *Adi Granth* which is now reverenced almost as a god. *Ibid.*, When the tenth Guru died the succession was regarded as closed and the *Adi Granth* remained the sole authority. **1961** P. Spear *India* xxii. 249 The fifth Guru, Arjun, compiled the Adigranth, or first scripture of the community, containing many devotional poems which are now continuously recited by devotees in the Golden Temple. *Ibid.* Guru Gobind lost his sons in battle; after his death in 1708 the guruship lapsed and its place was taken by the Sikh scriptures or Granth Sahib as the sole object of worship.

Grantha (grʌ·ntă). Also **Grantham.** [Skr. *grantha* (see prec.).] A South Indian alphabet used by the Tamil Brahmans for the Sanskrit transcripts of their sacred books.
1874 A. C. Burnell *Elem. S. Indian Palæogr.* ii. 34 The name 'Grantha' by which the E. coast variety has been known for some centuries indicates that it was merely used for 'books' or literary purposes. **1883** I. Taylor *Alphabet* II. x. 334 Across the whole breadth of India two alphabetic types are superimposed—one was a cursive script..; the other being a literary alphabet, represented in the north by the Devanagari, in the south by the Grantha. **1959** V. Cronin *Pearl to India* vi. 86 One discovery Nobili soon made: Grantham was not, as Europeans thought, a name for Sanskrit, but merely the angular script evolved by the Tamils to write Sanskrit. **1962** D. Diringer *Writing* vi. 147 South-Indian..Grantha.

grant-in-aid. [Grant *sb.*¹ 3 b.] A sum granted, esp. by a government or institution, to a school, institution, scholar, etc. Also *attrib.*
1881, 1899 [see Grant *sb.*¹ 3 b]. **1890** G. B. Shaw *Let.* 26 Sept. (1965) 266 That the London Society shall make a grant-in-aid of £15. **1906** *Daily Chron.* 21 May 7/4 Mr. Hart Davies suggests that in future there should be two kinds of schools—'State' schools and 'grant-in-aid' schools. **1941** *Ann. Reg. 1940* 260 The grant-in-aid from the British Government..was almost doubled. **1946** *Nature* 10 Aug. 208/2 It has been decided to adopt a scheme of senior research fellowships and grants-in-aid to operate over the next two years. **1955** *Times* 30 June 17/3 The grant-in-aid of $8,000,000 in goods and services under the U.S.A./Jordan pact of June, 1954, has done much to encourage the realization of these various plans. **1959** *N. & Q.* Dec. 448, I am indebted to the University of Alabama Research Committee for a grant-in-aid which has enabled me to carry on research at Harvard University. **1971** *Nature* 12 Feb. 444/3 For the financial year 1970–71 the budget of the centre is £975,000, of which £589,000 is grant-in-aid.

‖ **gran turismo** (græn turi·zmo). [It., lit. great touring.] A touring-car (see quot. 1967). Abbrev. *G.T.* (see *G III. f).
1960 *Times* 28 Sept. 16/5 Sports and *gran turismo* cars. **1961** I. Fleming *Thunderball* ix. 90 He had..a white convertible Lancia Gran Turismo. **1963** *Observer* 24 Nov. 37/3 What the Italians call the Gran Turismo..is a high-performance car with good luggage space, very comfortable seats for two and..occasional seats. **1967** E. Rudinger *Consumer's Car Glossary* (ed. 2) 52 *GT*, short for *gran turismo*, or grand touring. Originally a car designed particularly for covering long distances at high speeds, now sometimes used to describe a sporty version of an ordinary car.

granular, *a.* Add: **1. b.** *granular pearlite* Metallurgy, a constituent of carbon steels produced by the disintegration of plates of pearlite into ferrite and spheres of cementite; divorced pearlite.
1910 C. H. Desch *Metallogr.* xvii. 372 The cementite forms granular masses or oval globules, surrounded by areas of ferrite. This granular pearlite may be of all degrees of coarseness. **1958** A. D. Merriman *Dict. Metallurgy* 115/1 *Granular pearlite*, a structure produced when pearlite steels are annealed for long periods at temperatures below, but approximating to, the lower critical point.

granulo-. Add: Also in wider use, as in *granulocyte, *granulometric *a.*

granulocyte (græ·niŭlosəit). *Med.* [f. Granulo- + -cyte.] Any cell that contains or is destined to contain (conspicuous) granules in its cytoplasm; *spec.* any of the mature granular leucocytes (comprising neutrophils, eosinophils, and basophils) or the immature precursors of these cells.
1906 G. A. Buckmaster *Morphol. Normal & Path. Blood* iv. 87 According to Pappenheim, all the members of the three groups of erythroblasts, lymphocytes, and granulocytes originate from an ancestral lymphocyte. **1951** Whitby & Hynes *Med. Bacteriol.* (ed. 5) v. 61 The body falls a ready prey to infections of all sorts when the bone-marrow for any reason fails to produce granulocytes. **1960** *McGraw-Hill Encycl. Sci. & Technol.* VII. 485/2 Of the three kinds of white cells found in normal blood, the granulocytes, the lymphocytes, and the monocytes, it is the granulocytes that are largely affected in most forms of leukopenia. **1968** Passmore & Robson *Compan. Med. Stud.* I. xxvi. 12/1 The time taken for maturation of a granulocyte in the bone marrow prior to its release into the circulation is believed to be of the order of 3–4 days. **1971** T. L. Lentz *Cell Fine Structure* 282 (*heading*) Endometrial granulocyte. *Ibid.*, The granulocyte has been implicated in the secretion of the peptide hormone relaxin.
Hence **granulocy·tic** *a.*
1908 *Practitioner* Feb. 244 A form of lymphatic (*i.e.*, non-granulocytic) leucocythaemia occurs, in which the parent cells have given rise to an excessive formation of Türk's 'stimulation' cells. **1962** *Lancet* 27 Jan. 209/1 We thus came to the tentative conclusion that in the marrow of our standard guineapig..the small lymphocyte functioned as a stem cell for both the red cell and the granulocytic series.

granulocytopenia (græ:niŭlosəitopī·niă). *Med.* [f. prec. + Gr. πενία poverty.] A condition characterized by an abnormally small number of granulocytes in the blood.

1931 H. N. Harkins in *Arch. Internal Med.* XLVII. 408 In this article the terms agranulocytosis and agranulocytic angina are used interchangeably for the Schultz syndrome specifically, and the general term granulocytopenia is used to include atypical and related cases. **1951** A. Grollman *Pharmacol. & Therapeutics* i. 27 One of the most feared reactions observed in drug idiosyncrasies is granulocytopenia which may progress to agranulocytosis. **1961** *Lancet* 22 July 208 Patients with hepatitis were separated into two groups...In the treated group there was..1 episode of mild granulocytopenia lasting two weeks.

granulocytosis (græ:niŭlosəitōᵘ·sis). *Med.* [f. as prec. + -osis.] The presence of an abnormally large number of granulocytes in the blood.
1937 Kracke & Garver *Dis. Blood* i. ii. 23 *Granulocytosis*, the accumulation of an unusual number of granulocytes in the blood stream. **1940** R. L. Haden *Princ. Hematol.* (ed. 2) ix. 161 Granulocytosis (increase in polymorphonuclear cells) may be classified as follows. **1947** *Radiology* XLIX. 301/2 This leukocytosis has been found ..to be predominantly a granulocytosis in the rabbit and the chicken. **1961** *Lancet* 12 Aug. 377/2 In all patients there was a granulocytosis.

granuloma. Add: Also with qualifying L. adj., as *granuloma inguinale*, a chronic, probably venereal infection produced by the bacterium *Calymmatobacterium* (*Donovania*) *granulomatis* and characterized by creeping granulomatous lesions in the inguinal and genital regions.
1918 *Jrnl. Med. Res.* Jan. 427 Granuloma inguinale is endemic in certain tropical and subtropical countries. **1957** *Encycl. Brit.* XXIII. 48/1 In all cases of suspected *granuloma inguinale*, syphilis should be thoroughly ruled out.

granulomatosis (græ:niŭlōᵘmătōᵘ·sis). *Med.* [f. *granulomat-a*, pl. of Granuloma + -osis.] Any condition characterized by multiple granulomas.
1911 Dorland *Med. Dict.* (ed. 6) 360/1 Granulomatosis. **1966** Wright & Symmers *Systemic Path.* I. xxiv. 726/1 Some of these are cases of Wegener's granulomatosis and show the necrotizing granulomatous foci in the nose..and lung..that are characteristic of this condition.

granulometric (græ:niŭlome·trik), *a.* [ad. F. *granulométrique* (R. Feret 1892, in *Ann. d. Ponts et Chaussées* IV. 15): see Granulo- and Metric *a.*¹ and *sb.*] Pertaining to the distribution of grain sizes in a sample of sand, etc.
1905 L. C. Sabin *Cement & Concrete* xi. 163 Thus, all of the sands tested had the same 'granulometric' composition. **1964** *Economist* 28 Mar. 1254/1 Granulometric measurements of the alluvial deposits. **1970** R. C. Selley *Anc. Sedim. Environm.* i. 5 The granulometric analysis of ancient sediments is a declining art, to the delight of laboratory technicians.

granum (grā·-, grēⁱ·nǒm). *Bot.* Usually as pl. **gra·na.** [a. G. *granum* (A. Meyer *Das Chlorophyllkorn in chem. morphol. und biol. Beziehung* (1883) ii. 24), a. L. *grānum* grain.] Any of the discs which are arranged in stacks in the chloroplast, which are formed of membranes and in which the chlorophyll is incorporated.
1894 S. H. Vines *Student's Text-bk. Bot.* I. i. 100 The chlorophyll appears to exist in an oily solution, and to be confined to the fibrillar portions of the plastid, in the form of droplets (*grana*). **1898** H. C. Porter tr. *Strasburger's Text-bk. Bot.* i. i. 57 The fundamental substance of the chlorophyll bodies is itself colourless, but contains numerous coloured drops, which are termed Grana. These consist of an oleaginous substance, which holds various pigments in solution. **1966** R. B. Park in Vernon & Seely *Chlorophylls* ix. 286 The chlorophyll fluorescence is seen to reside primarily in the grana stacks. **1967** Kirk & Tilney-Bassett *Plastids* i. 30 Most of the thylakoids in a granum extend beyond the edge of the granum.

grape, *sb.*¹ Add: **1.** *the grapes are sour*: now usu. *sour grapes* (see Sour *a.* 9 c).
1760 A. Murphy *Way to keep Him* i. 5 You'd be glad to have me!—But sour Grapes, my Dear. **1876** 'Mark Twain' *Tom Sawyer* vi. 62 Another boy said 'Sour grapes!' **1923** A. Huxley *Antic Hay* xiii. 190 The concert had begun. 'Never mind,' said Gumbril. 'We shall get in in time for the minuetto. It's then that the fun really begins.' 'Sour grapes,' said Emily, putting her ear to the door. 'It sounds to me simply too lovely.' **1958** C. P. Snow *Conscience of Rich* xxxv. 261, I have never been able to understand the fascination which makes my brother Philip and others wish to spend their entire lives in this neighbourhood. I once said as much to Hannah, and she replied that it was sour grapes on my part.
8. a. *grape-arbour, -cutting, -rot, -seed* (earlier U.S. example). **b.** *grape-grower, -picker, -treader*. **e.** *grape-bloom* adj.
1810 F. Cuming *Sk. Tour W. Country* 167 He has opened a little publick garden behind his house, which he calls Vaux-hall. It has a most luxuriant grape arbour, and two or three summer houses! **1898** M. Deland *Old Chester Tales* 107 They were sitting in the grape-arbor with a little table between them. **1907** *Chicago Evening Post* 4 May 15 (Advt.), Fine summer home on good lake; grape arbors; garden. **1926** M. Leinster *Dew on Leaf* 214 Grape-bloom darkness. **1950** D. Gascoyne *Vagrant* 27 Rockets released to-night rush up to rape the grape-bloom sky. **1768** Washington *Diaries* (1925) I. 267 Planted out Grape Cuttings. **1848** *Rep. U.S. Comm. Patents 1847* 469 The landlord to furnish the dwelling house..fruit trees and grape cuttings at his own expense. **1862** *Rep. U.S. Comm. Patents 1861: Agric.* 525 At present there are very few practical grape growers who ever saw such a document. **1889** *Harper's Mag.* Jan. 261/2 The better class of laboring agriculturists, grape-growers, and stock-raisers. **1970** Grape-grower [see *grape-picker below]. **1923** R. Herrick *Homely Lilla* 68 She found herself laughing freely with the grape-pickers. **1970** *Times* 14 July 8/1 The poverty-ridden grape pickers, most of them semi-literate Mexican-Americans, have brought the wealthy grape growers to the negotiating table. **1863** *Horticulturist* Sept. 287/2 Grape Rot and Mildew. **1886** *Harper's Mag.* June 44/1 When mildew and grape-rot first appear. **1786** Washington *Diaries* (1925) III. 36 Tho' the ground was nearly prepared for my grape Seeds, I could not sow them on acct. of the Weather. **1889** Wilde *Birthday of Little Princess* in *Paris Illustré* 30 Mar. 207/2 At vintage time came the grape-treaders. **1931** *Times Lit. Suppl.* 29 Jan. 68/4 The proud young grape-treader defying her rustic lover.
9. grape-berry moth (example); **grape hyacinth** (later examples); **grape-nuts**, the trade name for a breakfast cereal, a patent preparation of maize or wheat in a crisp granular form; **grape-scissors** (earlier example).
1871 *Trans. Ill. Agric. Soc.* VIII. 158 The *Penthina Vitivorana*, or Grape-berry moth. **1882** Grape hyacinth [see Hyacinth 2 b]. **1897** [see *baby's breath* (*baby *sb.* B. 2)]. **1950** G. Brenan *Face of Spain* vi. 146, I picked a blue grape hyacinth, the only flower growing there among the rushes. **1898** *Off. Gaz. U.S. Pat. Off.: Trade Marks* 14 June 1657/1 Cereal food for human consumption. Postum Cereal Company, Limited, Battle Creek, Mich... Grape-Nuts...Used since December 1, 1897. **1902** *Granta* 3 May 286/1 It will be a pleasant puzzle for your readers to discover where the Grape-Nuts come in. **1903** *Army & Navy Co-op. Soc. Price List* Sept. 11/2 Grape Nuts..per packet, about 2 lb. 0/7. **1905** Chesterton *Heretics* 136 There is more simplicity in the man who eats caviar on impulse than in the man who eats grape-nuts on principle. **1927** T. E. Lawrence *Let.* 8 Sept. (1938) 536 You [*sc.* E. M. Forster] called your novel-book 'a saucerful of last week's grapenuts'. **1956** *Trade Marks Jrnl.* 18 Apr. 243/1 Grape-Nuts...General Foods Corporation..City of White Plains, State of New York,..Manufacturers. **1861** Mrs. Beeton *Bk. Househ. Managem.* 802 Grape-scissors, a melon-knife and fork, and nutcrackers, should always be put on the table, if there are dishes of fruit requiring them.

grapefruit (grēⁱ·pfrūt). Also **grape-fruit.** [f. Grape *sb.*¹ + Fruit *sb.*; so called because it grows in clusters.] For def. (s.v. Grape *sb.*¹ 9) read: The globular fruit of *Citrus paradisi*, having a yellow skin and pale yellow (occas. pink), juicy, acid pulp. Also *attrib.* and *Comb.*, as *grapefruit cocktail, juice, knife, marmalade*.
1814 J. Lunan *Hortus Jamaicensis* II. 171 The shaddock was originally regarded by Linneus as only a variety of the orange...There is a variety known by the name of *grape-fruit*, on account of its resemblance in flavour to the grape; this fruit is not near so large as the shaddock. **1859, 1885** [in Dict.]. **1904** *Daily Chron.* 4 May 10/5 The grape-fruit, which is gradually growing in popularity in England. **1926–7** *Army & Navy Stores Catal.* 581/3 Grape Fruit Knife..Grape Fruit Spoon..Grape Fruit Fork. **1911** 'M. Ronald' *Century Cook Bk.* Suppl. 600 Grape-fruit marmalade. **1930** J. Cannan *No Walls of Jasper* vii. 125 She bought some grape fruit for dinner. **1934** R. Stout *Fer-de-Lance* v. 67 I'll have grapefruit juice. **1935** M. Morphy *Recipes of all Nations* 598 (*heading*) Crab and grapefruit cocktail. *c* **1938** *Fortnum & Mason Catal.* 65/2 Knives..Grape Fruit, English—each 3/-. **1959** J. Braine *Vodi* vii. 108 'Have a drink with me. Gin-and-orange?'.. 'Thanks very much, I'll have a grapefruit juice if you don't mind.' **1967** K. Giles *Death & Mr Prettyman* viii. 145 Breakfast in bed..kidneys and her grapefruit marmalade. **1967** *Listener* 12 Jan. 53/1 Modern hotels (complete with..grapefruit cocktails, and roast lamb).

grape-vine. 2. a. Add to def.: Now in general use to indicate the route by which a rumour or a piece of information (often of a secret or private nature) is passed.
1891 *Century Mag.* Mar. 713/2 The 'grape-vine' spoke to us of little else. **1934** J. T. Farrell *Studs Lonigan* (1936) ii. xv. 337 Down there at that express company they find out about everything a guy does. They got the best grapevine in the world. **1948** *Daily Tel.* 3 Sept. 4/5 The guerrillas know the jungle, and they have an almost incredible 'grapevine' which gets information from one State to another with uncanny speed. **1955** *Times* 11 June 9/6 Of the younger men, the Moscow grape-vine reported that Mr. Shepilov, editor of *Pravda*, was coming forward to strengthen the party theoreticians. **1962** K. Orvis *Damned & Destroyed* vii. 54 Hurrying a message through the grapevine to the trafficker. **1970** *New Yorker* 3 Oct. 100/2 The art-world grapevine buzzed with rumors.
b. (Example.)
1968 T. Clayton *Handbk. Wrestling Terms & Holds* 103 (*caption*) Double grapevine with arm tieup.
c. (Later examples.)
1903 *Westm. Gaz.* 30 Dec. 3/2 The expert dons his skates and glides..off with the air of one to whom threes, grape-vines,..and other mysterious figures are..familiar. **1952** Visc. Templewood in *Skating with T. D. Richardson* xiii. 104 In days long ago it was the fashion to perform a series of two-footed movements called grapevines.
3. *attrib.*, as *grape-vine journalist, method, moth, rumour, telegraph* (see 2 a), *telegraphic* adj., *weevil, wire*.
1959 *Economist* 4 Apr. 20/2 The grapevine journalists have been listening; up to Thursday morning the grapevine still had not yielded any word. **1942** *Daily Tel.* 31 Aug. 3/4 Each province is presumed to have a leader.. whose instructions are transmitted by 'grapevine' methods from one person to another. **1950** *N.Z. Jrnl. Agric.* Oct. 326/1 The caterpillars of the grapevine moth, which found its way to New Zealand from Australia in recent years, depredate vines by devouring the foliage and damaging the developing grapes. **1941** Auden *New Year Let.* iii. 61 The careless victor never knew Their grapevine rumour would grow true. **1889** Farmer *Americanisms* s.v., During the Civil War exciting news of battles not fought and victories not won were said to be received by grape-vine telegraph. **1936** J. G. Brandon *Pawnshop Murder* iii. 26 I'll see what I can get over the 'grapevine' telegraph. **1951** *John o' London's* 17 Aug. 494/2 First with the news was..the little old man who cleans our windows... He is our grapevine telegraph. **1953** A. W. Fielding *Stronghold* i. i. 5, I had long ago ceased to wonder at the workings of their grape-vine telegraph. **1864** in *Southern Hist. Soc. Papers* (1876) I. 437 Many 'grape-vine' telegraphic reports are afloat in camp. **1950** *N.Z. Jrnl. Agric.* Oct. 326/1 The grapevine weevil can cause considerable damage by eating the leaves, shoots, and bunches. **1907** J. L. Given *Making Newspaper* xiii. 230 Not often does a telegraph editor..manufacture news, or, in the vernacular, employ the 'grapevine wire'.

grapey, var. Grapy.
1962 *Harper's Bazaar* Aug. 33 The plummy, grapey tones of the harvest. **1968** *Daily Tel.* 17 Dec. 13 There is nothing astringent about it [*sc.* a variety of Madeira]—it is always 'grapey'.

graph, *sb.*¹ Add: **1.** (Further examples.) In abstract terms: A finite, non-empty set of elements together with a set (empty or non-empty) of unordered pairs of these elements.
1931 *Proc. Nat. Acad. Sci.* XVII. 122 Suppose we assign to each vertex of a graph a color in such a way that each pair of vertices joined by an arc are of different colors. **1966** *McGraw-Hill Encycl. Sci. & Technol.* VI. 255/2 If points represent people and lines their interrelationships, then a graph may be used to depict the structure of a social group. **1967** H. C. Sneyd tr. *Kaufmann's Graphs* i. 10 The map of all the roads in a country..forms a graph. **1969** F. Harary *Proof Techniques in Graph Theory* i. 1 By far the most celebrated problem concerning graphs is the Four Color Conjecture. **1971** W. L. Price *Graphs & Networks* i. 1 Many problems of sequencing and scheduling can be looked upon as problems in graph and network theory. *Ibid.* ii. 7 If the set *A* is empty and the graph consists only of nodes, the result is called a null graph.
2. (Further examples.) In wider use: A line or curve representing the variation of one quantity with another, each quantity being measured along one of a pair of axes at right angles.
1926 *Encycl. Brit.* III. 643 (*caption*) Graph showing consumption per head of tea and sugar in the United Kingdom in each year 1864–1913. **1951** Roberts & Miller *Heat & Thermodynamics* (ed. 4) v. 155 The graph of entropy as a function of magnetic temperature is obtained by plotting this temperature against the calculated change of entropy. **1962** A. R. W. Hayes *Revision Physics* i. 27 For a wire subjected to a stretching force four points on the load-extension graph can be distinguished. **1963** L. Nobles in H. M. Burlage et al. *Physical & Techn. Pharm.* ii. 63 The plotting of a graph will establish quickly the meaning of a set of experimental data. **1971** *Nature* 24 Sept. 271/1 These data are easier to comprehend in the form of a graph (Fig. 1).

graph (græf, grɑf), *sb.*³ *Philol.* [ad. Gr. γραφή writing; cf. Digraph.] A visual symbol representing a phoneme or a segment or feature of speech; esp., a letter, or one of its occurrent forms, or a combination of letters.
1933 Bloomfield *Lang.* xvii. 294 For the writers, the *gh* was now a mere silent graph, indicative only of vowel-quantity. **1953** *Medium Ævum* XXII. 14 The Exeter Book has one example of a clear t-graph where a 3 would normally be expected. **1958** C. F. Hockett *Course in Mod. Ling.* lxii. 544 In Egypt certain graphs very early came to represent syllables rather than morphemes. **1968** *English Studies* XLIX. 513 The graphs *t* and *d* in post-stress, intervocalic position can be ambiguous. **1970** A. Cameron et al. *Computers & O.E. Concordances* 49 In the final printout, a simple graphic accommodation was employed to designate..the Old English graphs, ash, eth, and thorn.

graph, *v.*¹ Add: (Later examples.) Also *fig.*
1951 M. McLuhan *Mech. Bride* 79/1 When producers want to know what the public wants, they graph it as curves. **1964** *Listener* 20 Aug. 280/1 The book graphs and annotates the hostility between generations..and this is shaming to read. **1966** I. Jefferies *House Surgeon* iii. 31 Anybody graph the survival rates against the day of admission?

graphematic (græfĭmæ·tik), *a.* *Linguistics.* [f. *graphem(e + -atic.] = *graphemic *a.*
1956 A. McIntosh in *Trans. Philol. Soc.* 42 The complexity..of a full graphematic analysis of a text. **1959** *Brno Studies* I. 15 The graphematic difference. **1969** *Computers & Humanities* IV. 131 The graphematic oppositions are being grouped as to their positions and analyzed statistically.

grapheme (græ·fīm). *Linguistics.* [f. *graph *sb.*³ + *-eme; cf. *morpheme.] The class of letters and other visual symbols that represent a phoneme or cluster of phonemes, as e.g. the grapheme ⟨f⟩ consists of the *allographs *f, ff, F, Ff, gh, ph*, and *Ph* which represent the phoneme /f/ in *fun, huffy, Fingal, Ffoulkes, cough, graph*, and *Philip* respectively; so, in a given writing system of a given language, a feature of written expression that cannot be analysed into smaller meaningful units.

1937 R. H. Stetson in *Mélanges Ling. et Philol. offerts à J. van Ginnekin* 353 The unit of writing may be called the grapheme. **1951**, etc. [see *allograph¹]. **1955** H. A. Gleason *Introd. Descr. Ling.* xxi. 302 A writing system consists of a set of graphemes plus certain characteristic features of their use. **1958** A. A. Hill *Introd. Ling. Struct.* 443 The grapheme for /+/ was normally the blank space left between words. **1964** R. A. Hall *Introductory Linguistics* xliv. 265 In the Greek alphabet, the grapheme ⟨σ⟩ 'sigma', corresponding to our ⟨s⟩, has the two allographs ς (occurring only at the end of a word, before space) and σ (used elsewhere).

graphemic (grăfī·mik), *a.* and *sb.* *Linguistics.* [f. *graphem(e + -ic.] **A.** *adj.* Of or relating to graphemes. **B.** *sb. pl.* **graphemics.** The study of systems of written symbols (letters, etc.) in their relation to spoken languages.
1951 Stockwell & Barritt (*title*) Some Old English graphemic-phonemic correspondences. **1951** E. Pulgram in *Word* VII. 19 It is precisely that parallelism of phonemics and graphemics which renders feasible a phonemic transcription. **1953** J. B. Carroll *Stud. of Lang.* ii. 13 This branch of linguistics has been termed graphemics by some linguists, and graphonomy by others. **1958** A. A. Hill *Introd. Ling. Struct.* 442 Graphemic symbols are enclosed in angle marks (⟨⟩). **1964** R. A. Hall *Introductory Linguistics* xliv. (*heading*) Graphemics.
Hence **graphe·mically** *adv.*
1956 *Trans. Philol. Soc.* 50 Graphemically irrelevant variations in the shape of *þ*. **1964** *Language* XL. 167 One-syllable words, graphemically defined, have the same part-of-speech assignments.

graphic, *a.* Add: **2.** *graphic arts*: also, the techniques of production and design involved in printing and publishing; *graphic design*: graphics (sense B. 2 below); so *graphic designer*.
1949 *Sci. Amer.* Nov. 29 The new Graphic Arts Research Foundation, whose work is supported by 139 newspapers, printing firms and other interested groups. **1956** H. Williamson *Methods Bk. Design* Pref., Principles of graphic design which apply to all kinds of book production. **1967** Graphic design [see sense B.3 below]. **1967** Karch & Buber *Offset Processes* ii. 9 Of the many methods and procedures used in the graphic arts, letterpress and offset-lithography account for 66·1% in monetary value. **1971** Perry & Aldridge *Penguin Bk. Comics* (ed. 2) Bastard title p., Alan Aldridge is one of the most original and creative graphic designers... His unique style of illustration has appeared in countless magazines, newspapers, record sleeves, posters.
B. *sb.* For sense 6 in Dict. substitute: **1.** *pl.* The technical use of diagrams and figures as an aid to mathematical calculation or to engineering or architectural design.
1889, 1898 [in Dict., sense 6]. **1912** D. A. Low (*title*) Practical geometry and graphics. **1929** W. Abbott *Pract. Geom. & Engin. Graphics* 3 In Part I considerable space has been devoted to Engineering Graphics, particularly to the applications of graphical integration. **1960** *McGraw-Hill Encycl. Sci. & Technol.* IV. 615/1 Engineering graphics includes preparation of drawings that show the shape of objects.
b. *pl.* The production of diagrams, patterns, etc., by means of a computer.
1966 *Sci. Amer.* Sept. 184 (*caption*) Perspective drawing of a vehicle designed for reentry from space was produced by a computer-graphics program. **1969** *Computers & Humanities* IV. 66 Production of Visual Material by Computer. The formal name of this category is computer graphics, but you might prefer calling it computer art. **1970** *Ibid.* 172 By concentrating on a single graphics language, each student will have the opportunity to prepare a computer-generated film sequence.
2. *pl.* Design and decoration that involves typographic elements; the production of pictures, diagrams, etc., in association with text.
1960 *Design* Sept. 79/1 The type of credits designed by Saul Bass are an isolated example of good graphics for movies. **1965** *Amer. N. & Q.* Apr. 122/1 A graphics exhibition centering on the production of a recent book published by the famous Gehenna Press. **1967** M. McLuhan *Medium is Massage* (back-flap of dust-cover), Quentin Fiore is a distinguished graphics designer and artist. **1971** *Guardian* 12 Oct. 8/2 Graphics, textile design, ceramics..and many other design courses.
3. An example of the graphic arts or of graphic design; also, a diagram, pattern, picture, etc., produced by means of a computer.
1961 in Webster. **1967** *Listener* 2 Mar. 296/3 There is an exhibition of Polish graphics. These are not prints but examples of graphic design...Typography is used as an essential element in the design rather than as adjunct to photography and illustration. **1969** *Computers & Humanities* IV. 66 Since the first computers used typewriters and punched cards, the first graphics were illustrations using clever combinations of typing characters, or simple arrays of holes. **1970** *Guardian* 17 June 12/6 At present neither multiples nor graphics (that is prints of all kinds) sell in really mass editions.

graphite. Add: (Later examples.)
1896 *Trans. Amer. Soc. Mech. Engineers* XVII. 106, I have..advised the introduction of graphite of fine grades, and specially purified, wherever..great loads are to be carried on heavy and slow-moving machinery. **1963** Gregory & Pitt in F. J. Pearson *Nuclear Power Technol.* ix. 228 The extract properties of the graphite cannot be predicted and all batches have to be tested for their neutron absorption properties.
b. *attrib.* and *Comb.*, as *graphite moderator, pile, reactor*; *graphite-moderated* adj.
1945 H. D. Smyth *Gen. Acct. Devel. Atomic Energy Mil. Purposes* 63 The prospects for a graphite pile with helium cooling looked promising. *Ibid.* 68 The lattices.. consisted of lumps of uranium imbedded in the graphite moderator. *Ibid.* 79 In a typical graphite-moderated pile a neutron that has escaped from the uranium into the graphite travels on the average about 2·5 cm between collisions. **1950** *Amer. Speech* XXV. 24 Reactors may be named according to fuel (uranium reactor), moderator (graphite reactor), [etc.]. **1958** *Times Rev. Industry* Aug. 7/1 Britain will depend mainly on the gas-cooled, graphite-moderated type reactor, using natural uranium. **1962** *Newnes Conc. Encycl. Nuclear Energy* 511/1 With gas-cooled reactors the graphite moderator normally acts as a guide for the fuel elements.
Hence **graphited** (græ·fəitĕd) *a.*, supplied with graphite as a lubricant; also (of a lubricant), having had graphite added to it.
1922 *Autocar* 10 Nov. 997 Graphited oilless bushes are freely used. **1939** J. I. Clower *Lubricants* v. 65 Perhaps the chief advantage of graphited oils and greases is their ability to form films of graphite on the bearing surfaces of mechanical devices. **1955** A. F. Brewer *Basic Lubrication Pract.* iv. 61 Bronze and cadmium-base graphited alloys involve somewhat lower maximum operating temperature ranges.

graphitization (græ:fitəizēⁱ·ʃən). [f. Graphit(e + -ization.] **1.** A process in which an allotrope of carbon becomes wholly or partly converted into graphite or becomes more graphitic in nature; also, the graphitic character of the resulting carbon.
1899 E. G. Acheson in *Jrnl. Franklin Inst.* 3rd Ser. CXLVII. 484 One [form of the product] consists of.. articles made out of amorphous carbon,..which will afterwards be heated in an electric furnace and converted ..into graphite. I have been..using the furnaces of the Carborundum Company to produce the graphitization. **1951** *Proc. R. Soc.* A. CCIX. 215 Graphitization [of the coal] does not occur because the crystallites are in random orientation and the pore structure remains intact. **1956** *Nature* 4 Feb. 239/1 Short periods of oxidation of the coal ..decreased the graphitization of the carbon. **1958** *Sci. News* L. 118 The graphite obtained from the Dayton, Ohio, meteorite..is like the graphite obtained by graphitization of diamond. **1970** *New Scientist* 5 Feb. 253/2 They may not reach the same degree of graphitization as in the absence of oxidation..but they can still be reasonably graphitic.
2. A kind of corrosion in which grey cast iron becomes soft and porous, the metallic iron constituents being converted to corrosion products whilst the graphite is left intact; graphitic softening.
1916 *Yorks. Post* 4 Apr. 8/1 An instance [of internal disintegration] may be taken from the 'graphitisation' of iron pipes, a form of decay to which cast iron is liable when buried in the ground. **1941** *Nature* 22 Mar. 343/2 Cast iron behaved well in air, but when submerged was very liable to 'graphitization', the outer form being retained but the interior being converted into a spongy mass without strength. **1967** Fontana & Greene *Corrosion Engin.* iii. 71 White cast iron has essentially no free carbon and is not subject to graphitization.
3. The formation of graphite from combined carbon in a ferrous alloy.
1919 *Jrnl. Iron & Steel Inst.* XCIX. 569 Graphite is always to be seen in an ingot which was slowly cooled, even when there is practically no silicon, which was formerly supposed to be necessary in graphitisation. **1951** W. B. Shannon in W. E. Benbow *Steels in Mod. Industry* xiii. 275 Of the carbide forming elements, chromium is particularly effective in inhibiting graphitization. **1961** *Metals Handbk.* (ed. 8) I. 587/3 Graphitization lowers the room-temperature strength somewhat by removing the strengthening effects of finely dispersed carbides.

graphitize (græ·fitəiz), *v.* [f. as prec. + -ize.] **1.** *trans.* **a.** To convert (an allotrope of carbon) into graphite, to make more graphitic.
1899 E. G. Acheson in *Jrnl. Franklin Inst.* 3rd Ser. CXLVII. 485, I have also graphitized some tons of carbon plates, to be used in making dynamo and motor brushes. **1912** W. L. Landis in Rogers & Aubert *Industr. Chem.* xii. 276 After such a run the granulated coke forming the heating core has been graphitized. **1952** *Mineral. Mag.* XXIX. 805 Attempts were also made to graphitize diamonds in a carbon arc.
absol. **1910** *Encycl. Brit.* XII. 365/1 In graphitizing *en masse* large lumps of anthracite are treated in the electric furnace.
b. To cause the combined carbon in (a ferrous alloy) to form graphite; to convert (a carbon-containing constituent of a ferrous alloy) into graphite and something else.
1919 *Jrnl. Iron & Steel Inst.* XCIX. 582 Carbon-iron alloys within the range 2·0 to 4·3 per cent. carbon, when..slowly cooled past the eutectic freezing temperature 1130° C., were partially graphitised. **1934** *Chem. & Ind.* 14 Dec. 1052/2 With the aid of silicon, which breaks down or graphitizes the iron carbide.., the bulk of the carbide can be converted to iron and graphite. **1937** Epstein & Sisco *Alloys Iron & Carbon* II. x. 369 All the remaining austenite changes to ferrite and cementite; the remaining cementite has, therefore, still to be graphitized. **1960** Laing & Rolfe *Man. Foundry Pract.* (ed. 3) x. 366 In the manufacture of whiteheart..the white iron is not only graphitised, but the carbon is also very largely removed.
2. *intr.* **a.** Of a ferrous alloy: to undergo a change characterized by the conversion of the combined carbon into graphite. **b.** To change into graphite.
1919 *Bull. Amer. Inst. Mining & Metall. Engineers* July 1068 The gray iron..begins to graphitize further at temperatures lower than those at which there is initial graphitization of the white iron. **1920** *Jrnl. Iron & Steel Inst.* CII. 293 White pig iron always graphitises when annealed at a temperature above 1000°. **1961** *Metals Handbk.* (ed. 8) I. 587/3 Many of these steels do not graphitize completely. **1968** C. G. Kuper *Introd. Theory Superconductivity* i. 11 Diamond at room temperature shows no tendency to graphitize. **1970** *New Scientist* 5 Feb. 252/2 The non-graphitizing type [of carbon] can be induced to graphitize by the use of a catalyst.
Hence **gra·phitizable** *a.*, convertible into graphite; so **gra:phitizabi·lity**; **gra·phitizing** *ppl. a.*, that graphitizes or brings about graphitization; also as *vbl. sb.*
1904 *Electrochem. Ind.* II. 492/2 When carbon is raised to the graphitizing temperature its properties will depend largely on the kind of amorphous carbon used. **1920** *Jrnl. Iron & Steel Inst.* CII. 288 This process continues..until the temperature falls below the graphitising range. **1930** *Engineering* 11 July 58/1 Work carried out on the influence of nickel on cast iron has.. demonstrated that nickel is, like silicon, a graphitising agent. **1951** *Proc. R. Soc.* A. CCIX. 198 Those non-graphitic carbons which, on heating to temperatures between 1700 and 3000° C, form graphitic carbons will be called graphitizing carbons. *Ibid.* 203 At a certain temperature..there is formed, in a non-graphitizing carbon, a certain proportion of graphitic or graphitizable carbon. *Ibid.* 207 The graphitizability of a carbon is apparently related to its colloidal or fine-pore structure. **1970** *New Scientist* 5 Feb. 252/2 With increasing temperatures the structure and properties of graphitizing carbons approach those of graphite. *Ibid.* 253/1 A synthetic fibre of the graphitizable type.

graphitized (græ·fitəizd), *ppl. a.* [f. prec. + -ed¹.] That has undergone graphitization; that has been graphitized.
1899 *Jrnl. Franklin Inst.* CXLVII. 484 The life or efficiency of these graphitized electrodes is many times that of the same electrodes ungraphitized. **1930** *Jrnl. Iron & Steel Inst.* CXXI. 355 Values for pure graphite and the graphitised silicon alloy. **1951** *Proc. R. Soc.* A. CCIX. 203 The small amount of graphitic carbon which is formed in certain non-graphitizing carbons is more highly graphitized than is a true graphitizing carbon heated under the same conditions. **1963** H. R. Copson in LaQue & Copson *Corrosion Resistance of Metals & Alloys* (ed. 2) i. 15 Superficially the casting appears unattacked, but the graphitized layer is soft and weak.

graphitizer (græ·fitəizəɹ). [f. as prec. + -er¹.] A substance that is added to molten metal to promote the formation of graphite rather than a carbide.
1940 *S.A.E. Jrnl.* Aug. 28/3 In a summary of the effect of the various elements of cast iron on its microstructure when other variables are considered constant, it is brought out..that nickel acts as a graphitizer, promoting machinability. **1959** H. F. Taylor et al. *Foundry Engin.* xii. 319 Grain refiners and graphitizers probably promote nucleation in the melt by introducing 'foreign' nuclei into the liquid metal.

graphologist. (Later examples.)
1932 C. Morgan *Fountain* ii. v. 62 Lewis did not examine the handwriting as a graphologist would have examined it. **1961** *Evening Standard* 2 Aug. 12/5 Please write in longhand (we are..graphologists).

graphology. Add: **1.** (Later example.)
1960 M. Spark *Bachelors* i. 11 Ronald got a job in a small museum of graphology in the City...To Ronald's museum came criminologists from abroad, people wishing to identify the dates of manuscripts, or the handwriting attached to documents of doubt.
4. *Linguistics.* The study of written and printed symbols and of writing systems (see quot. 1964).
1961 A. McIntosh in *Archivum Linguisticum* XIII. 107, I have used the word 'graphology' in a sense which is intended to answer, in the realm of written language, to that of 'phonology' in the realm of spoken language. **1962** D. Diringer *Writing* 21 Graphology, 'the science of writing', is more concerned with the subject from the biological and psychological points of view than in terms of its history. **1964** M. A. K. Halliday et al. *Linguistic Sci.* ii. 50 Graphology..includes orthography, punctuation, and anything else that is concerned with showing how a language uses its graphic resources to carry its grammatical and lexical patterns.
Hence also (sense 4) **grapholo·gical** *a.*, **grapholo·gically** *adv.*
1964 M. A. K. Halliday et al. *Linguistic Sci.* ii. 50 The 'graphological' level of language, whatever the nature of the script, is characterized by its own distinctive pattern, just as is the level of phonology. **1964** A. McIntosh in *Archivum Linguisticum* XVI. 79 These two graphs should at least be recognised as graphologically (even if not graphemically) contrastive.

graphy (græ·fi). *Philol.* [ad. Fr. *graphie* system of writing.] A graphic symbol representing a phoneme; = *GRAPH *sb.*^{3}

1955 *Trans. Philol. Soc.* 138 The characteristic insular graphies of the later twelfth century *boef, beof, bøf* do not deserve to be dismissed as Anglo-Normanisms. 1962 DAVIS & WRENN *English & Medieval Stud.* 56 Whereas original long *e* and *o* are not infrequently represented by the graphies *ee* and *oo*, original short *e* and *o* are never so represented. 1968 *Medium Ævum* XXXVII. 87 The same graphy appears in the same manuscript in the form *beniscun*. 1970 *Jrnl. Eng. Place-Name Soc.* II. 16 The graphy *gu* for [g] before *e, i* arose in French after initial *gw* had been reduced to *g* at some time before the late twelfth century.

grappa (græ·pă). [It.] A brandy distilled from the skins, pips, and stalks of the grapes after they have been pressed for wine-making. (Cf. MARC.)

1893 in *Funk's Stand. Dict.* 1921 *Chambers's Jrnl.* 6 Aug. 575/1 We drew our bombs and our *grappa*, and soon we were whirling..to the attack. 1929 E. HEMINGWAY *Farewell to Arms* iv. 25 He poured two glasses and we touched them...The grappa was very strong. 1947 DYLAN THOMAS *Let.* 20 May (1966) 307 He was..drinking, by the gallon, grappa. 1960 C. DAY LEWIS *Buried Day* i. 25 Auden and I were sitting by the Grand Canal soaking up grappa. 1971 M. MCCARTHY *Birds of America* 247 He accepted a *grappa* on the house.

grappier (græ·piəɹ, ‖ grapie). [Fr., f. *grappe* (as in *grappes de la chaux*).] One of the hard lumps of unslaked material sometimes left in hydraulic lime after it has been slaked; **grappier cement**, cement made by grinding grappiers to a powder.

1897 *Jrnl. Soc. Chem. Ind.* XVI. 889/1 The hardest burned portions, called 'Grappiers', of the celebrated Teil hydraulic lime. 1905 E. C. ECKEL *Cements* 185 Grappier cements are made by grinding finely the lumps of unburned and overburned material which remain when a hydraulic lime is slaked. 1922 A. P MILLS *Materials of Construction* (ed. 2) I. iv. 30 As a rule all of the grappiers are finely ground under millstones and a certain proportion is added to the lime...The ground grappiers are also separately marketed as a special cement known as grappier cement. 1970 F. M. LEA *Chem. Cement & Concrete* (ed. 3) ii. 12 Such sintered lumps are sometimes separated and form, after grinding, the French 'grappier' cement, a product closely akin to the natural cements.

grapple, *sb.* Add: **4. b.** 'A tool with spring jaws which are closed by striking the fish' (Knight *Dict. Mech.* Suppl. 1884).

1872 *Game Laws Maine* in *Fur, Fin & Feather* (1872) 162 No person shall be allowed to take or catch any pickerel with spears, hooks or grapples.

grapsoid, *a.* (Later examples.)

1934 *Nature* 29 Sept. 500/1 Species of cancroid, grapsoid and spider crabs in the Gulf of California. 1941 [see *AUTOTOMY].

grass, *sb.*^{1} Add: **1. c.** Phr. *between grass and hay* (see quots.).

1743 ELLIS *Mod. Husb.* III. I. 78 April and September are reckoned the worst Months to make Butter in, because then the Season is between Grass and Hay. 1848 in *Amer. Speech* (1935) X. 40 *Betwixt hay & grass*, between Boyhood & Manhood. 1891 H. C. BUNNER *Zadoc Pine* 17 He..got a couple of eggs cooked for his private supper...The eggs were, as he told Mr. Bryan, 'kinder 'twixt grass and hay'.

e. Marijuana, used as a drug. *slang* (orig. *U.S.*).

1943 *Time* 19 July 54 Marijuana may be called..grass. 1945 L. SHELLY *Jive Talk Dict.* 25 *Grass reefers*, marijuana cigarettes. 1967 *Boston Sunday Herald* 26 Mar. IV. 1/1 According to one Federal Narcotics Bureau agent, California 'is flooded with marijuana', which is better known by the increasing numbers who smoke it as 'pot', 'grass' and 'Mary J.'. 1968 A. DIMENT *Gt. Spy Race* vi. 88 Pure Grass cigarettes, at two dollars a pack and none of your watering down with tobacco. 1969 *Win* 15 May 31/2, I consider grass and mescaline to be extremely important and inherent parts of this social revolution.

4. b. Also, the young shoots of the carnation.

1820 T. HOGG *Pract. Treat. Culture of Carnation* 48 The propagation by piping..ought to commence as soon as the shoots or grass is [*sic*] ready. 1836 N. PATERSON *Manse Garden* (1860) 189 The young shoots [of carnations] near the ground which do not run to flower are denominated grass. *Ibid.* 190 Pipings (as the grass shoots taken off and stuck in the ground are called)..will take root. 1925 W. WATSON *Gardener's Assistant* (ed. 4) V. 22/1 The 'grass', or young growths produced at the base of the plant, form the layers.

9. *esp.* Ground covered with grass closely mown and rolled, forming a lawn in a public or private garden. Phr. *keep off the grass*: a notice frequently posted in a park or garden to which the public are admitted; also used *fig.* as a warning not to take liberties, encroach, or interfere.

[1846 *Punch* 12 Sept. 113/2 If from the gravel pathway hard He turn to tread the verdant sward,..What bids his happy dream to pass?—'Get off the Grass! Get off the Grass!'] 1850 *Punch* 5 Oct. 144/1 The public, who are here and there 'requested to keep off the grass'. 1897 W. S. MAUGHAM *Liza of Lambeth* v. 59 'Na then,' she said, 'keep off the grass!' [i.e. don't take liberties with me]. 1904 *Daily Chron.* 27 Oct. 4/7 'Now, then, some girl can tell me about grass. What is grass?' The protagonist of the class..gave the definition. 'Please, it's what you got to keep off of.' 1925 P. GIBBS *Unchanging Quest* xxi. 156 Of course you wouldn't be left alone to do what you like under some forms of government. Not entirely under ours, as you'll find if you don't keep off the grass, old lad. 1953 K. AMIS *Lucky Jim* xx. 211 What I want to say to you is, keep off the grass, that's all. You're causing nothing but trouble by behaving as you are.

d. *Electr.* A fuzzy appearance along the time base-line of a cathode-ray tube display due to random, fluctuating deflections caused by electrical noise.

1947 *Amer. Speech* XXII. 154 The radar-man..speaks of 'losing the pip in the grass'. *Ibid.*, *Grass*,..electronic haze at the bottom of the screen. 1957 RAWNSLEY & WRIGHT *Night Fighter* iv. 58 'What's all that stuff?'.., 'Grass,..it's like the background noise of a wireless set.' 1961 PARTRIDGE *Dict. Slang* Suppl. 1116/2 *Grass* was the normal 'picture' seen on certain types of radar cathode-ray tube, as distinct from the signals produced by aircraft, etc. It looked like waving grass. 1969 *Sunday Mail Mag.* (Brisbane) 16 Feb. 1/3 'Grass'—useless noise from outer space.

11*. *slang.* A police informer. (See also *GRASSER^2, *GRASSHOPPER 1 c.)

1932 A. GARDNER *Tinker's Kitchen* 283 *Grass*, an informer. 1936 J. CURTIS *Gilt Kid* ii. 22 Tell you the details and then you'll do the gaff on your jack..or else turn grass. 1954 'N. BLAKE' *Whisper in Gloom* ii. 31 He was a nark, nose, snout, grass, squeaker, or whatever coarse word is current for it. 1955 P. WILDEBLOOD *Against Law* 105 'What are grasses?' I asked. 'Informers. Short for "grasshoppers", which is rhyming slang for "shoppers", meaning people who go to the cop-shop and squeal on their friends.' 1961 *Guardian* 6 Dec. 4/5 Throughout the gaols..the word 'grass' is an abbreviation for grass-snake, which..means informer.

12. a. *grass-blade* (later examples), *-heath, -pollen, -ranch, -seed* (later examples), *-spire* (later example), *-tuft, -veld* (S. Afr.). **b.** *grass-catcher, -mower; grass-clipping.* **c.** *grass-fed* (later examples).

1934 T. S. ELIOT *Rock* ii. 84 Glow-worm glowlight on a grassblade. 1949 E. POUND *Pisan Cantos* lxxxiii. 124 When the mind swings by a grass-blade an ant's forefoot shall save you. 1895 *Montgomery Ward Catal.* 392/3 Grass Catcher..has new device for attaching to mower. 1971 *CGA Ann. Price List* 39/2 Motor Mowers..(4-stroke) complete with grass catcher. 1954 J. R. R. TOLKIEN *Fellowship of Ring* ii. 73 He lifted the astonished Sam, shears, grass-clippings and all, right through the window. 1966 G. W. TURNER *Eng. Lang. Austral. & N.Z.* iv. 68 *Lawn mowings* in Beatrix Potter's *The Tale of the Flopsy Bunnies* was rendered *grass clippings*. 1880 *Vermont Agric. Rep.* VI. 26 Time was when the butchers of Brighton claimed that they could distinguish between the grass-fed beef fattened in this valley, and that from other sections, by its superior quality. 1892 A. C. GUNTER *Miss Dividends* (1893) 213 What's champagne muscle to grass-fed muscle, you dainty cut of New York. 1960 *Times* 1 Oct. 7/6 Grass-fed cattle. 1936 *Discovery* Jan. 24/2 The unique 'grass-heath' of fescue and bent grasses. 1964 V. J. CHAPMAN *Coastal Veget.* vi. 148 Many of them can be regarded as 'grass-heath' species. 1894 *Country Gentlemen's Catal.* 240A 'Stamford' Grass Mower. 1913 W. J. LOCKE *Stella Maris* xii. 165 The grass-mower driven over the grass. 1921 *Amer. Jrnl. Bot.* VIII. 473 Liefmann..found 2,500,000 grains of grass pollen in one square meter. 1957 *Granta* 9 Mar. 19/2 The girl sneezed with the grass-pollen. 1905 *Westm. Gaz.* 20 May 4/1 One of the main objects of the Bill was to get rid of the agricultural slums by splitting up the grass-ranches. 1880 *Vermont Agric. Rep.* VI. 32, I cannot recommend the use of oats in connection with grass seed. 1965 *R. Tucker & Sons Catal.* (Autumn) 48 We have considerable experience of Lawn Grass Seed...Many types of grass seed sold are quite unsuitable for making good lawns. 1867 'T. LACKLAND' *Homespun* I. 99 The busy spiders..had spun slenderest ropes of very gossamer, and swung them across from one grass spire to another. 1891 KIPLING *Light that Failed* i. 10 Maisie was picking grass-tufts. 1909 *Westm. Gaz.* 11 Dec. 16/1 Only recovering his feet after much floundering in one of the sandy hollows which occur between the grass-tufts. 1933 R. TUVE *Seasons & Months* iv. 165 Shepherds tending sheep on green grass-tufts. [1844 J. BACKHOUSE *Narr. S. Afr.* 115 The Hottentots..could obtain from one to two rixdollars a day in the Gras *Veld*, grass field.] 1958 *Cape Argus* 1 Nov. (Mag. Section) 11/1 We have exchanged..grassveld for karoo. 1971 *Nature* 6 Aug. 374/3 This animal originally occurred extensively in..the central and southern Cape grassveld.

13. grass-box, the receptacle on a lawn-mower into which the cut grass is projected; **grass carp,** a vegetarian fish, *Ctenopharyngodon idella*, native to Asia, sometimes introduced elsewhere to control aquatic vegetation; **grass court,** a grass Lawn Tennis court; **grass-cutter,** (*b*) = *grass-hand* (*a*); (*c*) *slang* (see quots.); (*d*) = *cane-rat* (*CANE *sb.*^{1} 10); **grass-eating** *a.* = *grass-feeding* adj.; **grass fire,** a fire that destroys an area of grass; **grass-frog,** the common frog, *Rana temporaria*; **grass grub,** a New Zealand grass-eating grub, the larva of a cockchafer beetle, *Odontria zealandica*; **grass hockey** *Canad.*, hockey played on grass, as opp. ice hockey; **grass-hook,** a sickle for cutting grass; **grass-line** (*a*) *N.Z.*, the line or level on a mountain above which no grass grows; (*b*) = *grass rope*; **grass-parakeet** (earlier example); **grass-parrot,** a small brightly-coloured Australian parrot of the genus *Neophema* or *Psephotus*; **grass scythe,** a scythe for mowing grass; **grass-seeder** *N.Z.*, a person who gathers grass-seed for sale; also *attrib.*; hence **grass-seeding** *vbl. sb.*, the act of gathering grass-seed; **grass sickness,** an equine disease, usually fatal, which can occur when a horse is put on to certain pastures; **grass skirt,** a skirt made from long grass and leaves secured to a waistband, orig. worn by the hula dancers of some Pacific islands; **grass staggers** = *GRASS TETANY; **grass verge,** a strip of grass at the side of a garden path or road; **grass-way** = *grass-siding*; **grass-wrack** (later examples); **grass wren,** any of several small Australian birds of the genus *Amytornis*.

1894 *Country Gentlemen's Catal.* 289 The Grass Box can be placed either behind or in front of the cutters. 1885 D. J. MACGOWAN in *Bull. U.S. Fish Comm.* V. 240 Some [Chinese minnows] fatten on grass, and are called 'grass carp'. 1964 *Listener* 17 Sept. 429/1 Grass carp have been keeping Chinese rivers clear of weed for millions of years, and they have been cultivated for food in Chinese fishponds for the past 2,000 years. 1971 *Nature* 15 Jan. 154/1 The Asiatic grass carp, *Ctenopharyngodon idella*, a gross browser on aquatic vegetation. 1883 *Lawn Tennis for 1883* 98 One of the most important accessories to a grass court..is a good lawn mower. 1930 W. S. MAUGHAM *Cakes & Ale* iv. 48 She's got quite a good grass court and she does one very well. 1853 C. M. SMITH *Working-Man's Way in World* ii. 20 My father was a grass-cutter for twenty years on the *Morning* ——. 1918 *Independent* 6 July 11 This plane with clipped wings which keep it no more than six feet above the ground is variously nicknamed by the aviators 'grass-cutter', 'creeper', 'two-spot'. 1930 BROPHY & PARTRIDGE *Songs & Slang 1914–18* 128 *Grass-cutters*, small bombs dropped by aeroplanes on camps and bivouacs behind the lines, bursting on hitting the ground and scattering shrapnel pellets at a low level. 1944 *Word Study* Apr. 4/2 *Grass Cutter*. This is what pilots call the A-20A Attack Plane because it flies so low. 1946 G. S. CANSDALE *Animals W. Afr.* 62 Next in size is the Cutting Grass, Grass Cutter or Cane Rat. 1961 *Times* 12 May 18/7 We had also acquired a dried grass-cutter, a sort of bush-rat. 1888 *Amer. Naturalist* XXII. 260 The Grass-Eating Thrips. 1904 W. H. HUDSON *Green Mansions* ix. 117, I have found you..and your grass-eating dogs as well! 1946 *Nature* 28 Dec. 928/2 The grass-eating, short-tailed, or meadow vole. 1882 W. R. LUDLOW *Zululand* xxi. 182 As we approached the Umslatoos, we came in sight of an immense grass fire. 1891 R. WALLACE *Rural Econ. Austral. & N.Z.* xxii. 296 It must not be forgotten that a bush-fire or a grass-fire at a wrong season is one of the greatest causes of loss..to the stock-owner in Australia. 1927 W. PLOMER *I speak of Africa* i. 13 Thirty miles away a grass-fire gave the air a bluish tinge. 1967 *Evening News* 2 Nov. 7/1 (*heading*) Clamp-down on grass-fires [on railway embankments]. 1901 H. GADOW *Amphibia & Reptiles* 253 The habits of the Grass-frog are essentially terrestrial. 1931 H. W. PARKER in W. P. Pycraft *Standard Nat. Hist.* XII. 498 The Common English Grass-frog may be taken as an example of the normal structure and commoner habits of the whole group. 1910 *N.Z. Jrnl. Agric.* 15 Aug. 223 No means are yet known for the thorough control of the New Zealand grass-grub. 1946 *Nature* 21 Dec. 920/1 The Entomology Division has focussed its attention on the grass-grub, the major insect pest affecting pastures [in N.Z.]. 1969 *N.Z. News* 9 Apr. 7/1 Grass grub and porina caterpillar, numbered among the most devastating of New Zealand pasture pests, face a three-pronged attack. 1921 *Daily Colonist* (Victoria, B.C.) 2 Apr. 1/1 The Victoria Grass Hockey Club is holding a practice this afternoon. 1964 *Maclean's Mag.* 16 Nov. 81/2, I..do solemnly swear never to waste company time arguing..or raving about..lawn bowling and grass hockey. 1812 *Niles' Weekly Reg.* II. 131/1 The purveyor of public supplies advertises for..1000 grass hooks. 1858 J. A. WARDER *Hedges & Evergreens* 97 Using for the purpose [of pruning] a strong knife about two feet long, or a common grass-hook. 1969 *Sears, Roebuck Catal.* 1079 Craftsman Grass Hook. Traditional scythe type, heat-treated, 12½-in. blade adjusts to two cutting angles. 1892 C. E. DOUGLAS in J. Pascoe *Mr. Explorer Douglas* (1957) 176 It is very common all along the ranges close to the grass line. 1909 *Man. Seamanship* II x. 180 A grass line attached to a breaker should be towed astern in case..the boat missed the towing ship. 1941 J. MASEFIELD *Nine Days Wonder* 25 Various devices were tried for heaving off strings of boats together on messengers of grass-line. 1959 *Tararua* XIII. 45 *Scrubline* for the upper limit of the scrub seems to be confined to New Zealand as does the corresponding *grassline*. 1840 J. GOULD in *Proc. Zool. Soc.* VIII. 147 Those [birds] now exhibited were three new species of small Grass Parrakeets. 1913 G. M. MATHEWS *List Birds Austral.* 138 Neophema pulchella. Red-shouldered Grass-Parrot. 1936 A. RUSSELL *Gone Nomad* vi. 44 'It's out there!' he continued pointing to a flock of budgerigars flashing across the plain. 'Them grass parrots are makin' in fer it.' 1966 EASTMAN & HUNT *Parrots of Australia* 153 Characteristics of Neophema Group...Known as Grass Parrots because they are strictly ground-feeders on grass and herbaceous seeds. 1573 TUSSER *Husb.* (1878) 37 A brush sithe and grasse sithe. 1787 WASHINGTON *Diaries* (1925) III. 243 Called on my return at French's where I had begun with grass Scythes (a cradle having been found not to answer). 1908 *Sears, Roebuck Catal.* 522 (*caption*) Double rib extra grade all steel grass scythe. 1922 C. G. TURNER *Happy Wanderer* 211 The official issue of clothing is a wide-brimmed 'grass-seeder' straw hat. 1936 C. R. ALLEN *Poor Scholar* xix. 202 The grass-seeder would entrust his cheque to the squatter. *Ibid.*, Grass-seeding is conducive to a princely thirst. *Ibid.* 204 Grass-seeding is rather characteristic of these parts [Banks Peninsula]. 1920 *Glasgow Herald* 9 Sept. 9 The discovery of a bipolar organism in grass-sickness. 1923 *Daily Mail* 18 June 7 The disease in horses known as

grass sickness which first appeared in Forfarshire in the summer of 1909...The principal symptoms are paralysis of the palate and gullet, causing inability to swallow. 1955 GAIGER & DAVIES *Vet. Path. & Bacteriol.* (ed. 4) xxix. 555 It would appear, therefore, as if little or no immunity is left by an attack of grass sickness. [1898 J. R. MUSICK *Hawaii* vi. 79 Kalakaua's hula girls sometimes danced nude but the usual costume is a skirt made of grass, coming to the knees.] 1937 C. GESSLER *Hawaii* xxix. 334 The grass skirt introduced in Kalakaua's time survives in Hawaii mainly for sale to tourists. 1938 R. FINLAYSON *Brown Man's Burden* 68 She loved to dance the hulas in that bright-coloured grass skirt. 1949 M. MEAD *Male & Female* ii. 27 They [*sc.* anthropologists] are accused of having..put on a grass skirt or a loin-cloth. 1970 *Observer* (Colour Suppl.) 15 Feb. 20/2 Cruise passengers [are] very much in so far as the multitude of small shops selling dress materials and grass (plastic grass) skirts go. 1883 [see STAGGER *sb.*^{1} 2]. 1889 [see LOCO^1]. 1960 *Farmer & Stockbreeder* 29 Mar. 99/1 A disorder of cattle and sheep which is now more widespread than is generally realized has been christened with a host of names. The commonest is 'grass staggers', because it generally occurs in grazing animals and those which are affected tend to stagger. 1969 *Times* 13 Oct. 14/6 No accurate estimate has been made of the toll that grass staggers, or hypomagnesaemia, takes of beef cattle. 1824 LOUDON *Encycl. Gard.* (ed. 2) § 1338 Verge-shears..are chiefly used for trimming the sides of box-edgings and grass-verges. 1930 *Morning Post* 12 June 12/5 He was within four feet of the grass verge and was unable to avoid the approaching motor cycle. 1937 *Sunday Times* 10 Jan., Nearly all the new roads have broad central 'reserves' and broad grass verges on each side. 1963 *Radio Times* 14 Mar. 5/1 Even if there was a grass verge or something handy to wipe a toe-cap you would have to lie face down and drag yourself along. 1927 *Sunday Express* 17 July 17/5 Sometimes the road was so bad that, dodging between the trees, they left it for the flat grass-way beside it. 1838 AUDUBON *Ornith. Biogr.* IV. 10 Its subsistence..is chiefly derived from the grass-wrack or Eel-grass, *Zostera marina*. 1961 R. W. BUTCHER *New Illustr. Brit. Flora* II. 595 The Common Grass-wrack or Eelgrass is a stout to slender, green, marine plant with compressed, keeled, much-branched stems. 1898 MORRIS *Austral. Eng.* 519/1 *Grass W[ren]*...Called by Gould the *Textile Wren*. 1934 A. RUSSELL *Tramp-Royal in Wild Australia* xvii. 103 The grass wren he called the 'jump-along'. 1965 *Austral. Encycl.* IV. 363 All grass-wrens, though able to fly fairly well, move mainly on foot, and with remarkable speed.

grass, *v.* Add: **4. b.** *spec.* in Rugby and Australian National Football.

1937 in PARTRIDGE *Dict. Slang.* 1963 *Times* 13 June 13/3 In Rugby football no try could be simply converted; the oval had to be propelled through the uprights and players were not tackled but grassed. 1968 EAGLESON & MCKIE *Terminology Austral. Nat. Football* II. 13 *Grass*, knock a player to the ground by a hip or a shoulder bump.

e. *Cricket.* To drop (a catch).

1960 E. W. SWANTON *W. Indies Revisited* v. 124 Illingworth had a very sharp, low c and b chance from Sobers. He grassed it.

8. *trans.* and *intr.* To betray (someone); to inform the police about (someone). (Cf. *GRASS *sb.*^{1} 11*.) *slang.*

1936 J. CURTIS *Gilt Kid* xxvii. 269 Anyhow it was a dirty trick grassing his pals. 1938 G. GREENE *Brighton Rock* III. ii. 118, I wouldn't grass, Spicer said, unless I had to. 1958 F. NORMAN *Bang to Rights* 86 What is more he didn't grass any one else. 1963 *Daily Tel.* 28 Feb. 15/7 (*heading*) 'Grassing' discussed at Co-op murder trial. *Ibid.*, The underworld code dealing with criminals in prison who 'grass' or inform on colleagues was discussed. 1965 J. PORTER *Dover Two* vii. 77 It won't come out! Not unless you start grassing.

grassed, *ppl. a.* Add: **1.** Also *grassed-down*.

1960 *Times* 23 Jan. 11/4 The rest of the grassed-down area. 1967 *Heretaunga Plains* (Ministry of Works, N.Z.) 49/1 A crop rotation programme is normally observed and this provides for grassed-down periods (for grass seed and stock fattening) which improve soil fertility.

grasser^2 (grɑ·səɹ). [f. *GRASS *v.* 8 + -ER^1.] = *GRASS *sb.*^{1} 11*.

1950 P. TEMPEST *Lag's Lexicon* 97 *Grasser*. One who gives information. A 'squealer' or 'squeaker'. The origin derives from rhyming slang: grasshopper—copper; a 'grass' or 'grasser' tells the 'copper' or policeman. 1966 'L. LANE' *ABZ of Scouse* 46 *Grasser*, a police informant; a stool-pigeon. 1968 'P. ALDING' *Circle of Danger* vii. 57 Five minutes alone with you and he'll be babbling like a grasser. 1968 R. JEFFRIES *Traitor's Crime* iii. 31 'How reliable was the original information?' 'As reliable as any information is from a grasser.'

‖**grasserie** (gras'ri). [Fr., f. *gras* fat.] A virus disease of silkworms, characterized by yellowing of the skin and liquefaction of the internal tissues.

1836 F. G. COMSTOCK *Pract. Treat. Culture Silk* II. 40 (*heading*) The grasserie.—The period at which the Worms are most subject to this disease, is before the second moulting, and in the third and fourth ages. 1879 C. U. RILEY *Silkworm* 14 Silkworms..may become yellow, limp, and die of a malady called *grasserie* or jaundice. 1923 *Nature* 24 Mar. 411/2 Flacherie is of less importance in mulberry worms, while grasserie is stated to cause loss to all species. 1943 W. BORAH *Silk Raising in Colonial Mexico* vi. 64 Normally,..grasserie is a moderate scourge which does not result in epidemics. *Ibid.* 65 Even when confined to single silk houses, however, grasserie..can reduce or ruin many a crop.

grasshopper. Add: **1. b.** (Later examples.)

1935 *N. & Q.* CLXIX, 365 Use of animal names as epithets to man...*Grasshopper*, undependable; mental effort inconstant and purpose shifting. 1968 *Listener* 27 June 827/3 My natural inclination was to watch the screen for a moment, then look away, then look back again. This, after I began to study my own behaviour, turned out to be the way I watched not just the news but all television. I was the grasshopper viewer; and, I suspect, but one of millions of grasshopper viewers.

c. *Rhyming slang* = COPPER *sb.*^{4} Also ellipt. *grass*. (Cf. also quot. 1955 s.v. *GRASS *sb.*^{1} 11*.)

1893 in FARMER & HENLEY *Dict. Slang.* 1907 *Daily Chron.* 1 Apr. 4/4 The criminal classes always speak of policemen as 'grasshoppers'. 1938 F. D. SHARPE *Sharpe of Flying Squad* 330 *Grasshopper*, a policeman. 1939 H. HODGE *Cab, Sir?* xv. 222 A policeman is the usual cockney 'Grass' (copper, grass-hopper). 1950 [see *GRASSER^2].

d. (See quot. 1956.) *U.S. slang.*

1942 *Pop. Sci.* Jan. 63 Grasshoppers...that's what the army calls its new odd-job planes. 1956 W. A. HEFLIN *U.S.A.F. Dict.* 234 *Grasshopper*, any small, light, cabin monoplane used for observation, liaison, or training.

3. (Later example.)

1845 G. DODD *Brit. Manuf.* IV. 160 To give the technical terms applied to all these little pieces would be of no use; for after saying that the key acts on the grass-hopper..we have done but little towards explaining the particular construction and action of each.

5. grasshopper mouse, either of two species of North American mice belonging to the genus *Onychomys*; **grasshopper year,** a year when the crops are destroyed by grasshoppers.

1904 W. T. HORNADAY *Amer. Nat. Hist.* vii. 91/1 The Grasshopper Mouse..strongly resembles the white-footed mouse. 1964 E. P. WALKER et al. *Mammals of World* II. 779/2 Grasshopper mice live in practically any shelter they can find at ground level. 1880 *Scribner's Monthly* July 458/1 Then came 1875 and 1876, which were 'grass-hopper years' when no crops of consequence were raised in the whole state. 1949 *Chicago Tribune* 5 June 1/3 He predicted it will be the worst grasshopper year since 1936.

Hence **gra·sshoppery** *a.* = *grass-hopperish* adj.

1921 'K. MANSFIELD' *Let.* 21 July (1928) II. 121 If only one could make some small grasshoppery sound of praise. 1926 *Brit. Weekly* 2 Sept. 456/4 She announces the result in her little grasshoppery voice.

gra·sshopper, *v.* [f. the sb.] *intr.* To live, or jump, like a grasshopper. Also with *over*.

1873 J. H. BEADLE *Undevel. West* xxxii. 704 When they spar [=push (a steam-boat) with poles] thus on both sides, they are said to 'grasshopper over'. 1956 'A. GILBERT' *Riddle of Lady* i. 12 [We] enjoy ourselves..by working. We shouldn't get any fun out of grasshoppering. 1966 D. VARADAY *Gara-Yaka's Domain* iii. 32 All others were ignored, even one that grass-hoppered directly into her path. 1966 J. FOWLES *Magus* xxxvii. 230 My mind travelled up to the Bonnards, and grasshoppered from them to Alison.

grassiness. Delete *rare*$^{-0}$ and add examples.

1787 WASHINGTON *Diaries* (1925) III. 199 The harrow was ordered to preceed it once, or as many times as the ground from the clodiness, or grassiness of it, should appear to need it. 1924 C. MACKENZIE *Heavenly Ladder* x. 138 The infinitely various earthy odours and the cool grassiness of the place allayed his feverish regrets.

grass land, grassland. orig. *U.S.* [GRASS *sb.*^{1}] Land producing grass; pasture or grazing land. Also *attrib.*

1682 in C. W. Manwaring *Digest Early Connecticut Probate Rec.* III. 114 In any corn feild or grass land. 1697 *Connecticut Col. Rec.* IV. 212 About one acre of grasse land in the said meddowe. 1765, 1846 [see GRASS *sb.*^{1} 12]. 1807 J. BARLOW *Columb.* v. 300 The conscious flocks..spread thro the grassland. 1903 W. R. FISHER tr. *Schimper's Plant-Geogr.* II. i. 174 A good grassland climate is..composed of the following elements. 1920 *Nature* 27 May 408/2 It is too often the case that grassland is left to take care of itself, and that no steps are taken for its improvement. 1954 M. BERESFORD *Lost Villages* v. 148 From 1445 to 1491 he had been near Warwick at the heart of the grasslands. 1959 A. MCLINTOCK *Descr. Atlas N.Z.* 41 During the 1920s farmers adopted the idea of intensive grassland farming. 1971 *Sci. Amer.* July 86/1 The Serengeti National Park is an area of 5,600 square miles of grassland and open woodland.

Grassmann (grɑ·smăn). The name of Hermann *Grassmann* (1809–77), German mathematician and polymath, used in the possessive to designate his discoveries, esp. *Grassmann's Law* (Philol.): see quot. 1960 and Dict. s.v. LAW *sb.*^{1} 17 c.

1891 MAX MÜLLER *Science of Lang.* II. v. 270 The following are the principal words in which, according to Grassmann's Law, double media in Gothic can be accounted for: Sk. *Gardh*..Sk. *Dabh* [etc.]. 1935 G. K. ZIPF *Psycho-Biol. Lang.* iii. 81 We view in turn the conditions under which Grassmann's Law operated in Sanskrit and Greek. 1939 L. H. GRAY *Foundations of Lang.* 70 A noteworthy form of incontiguous dissimilation is seen in the operation of Grassmann's Law. 1960 S. POTTER *Lang. in Mod. World* x. 152 According to Grassmann's Law of the dissimilation of aspirates, the speakers of these ancient languages could not tolerate the aspirated plosives *ph th kh* at the beginning of two successive syllables. Wherever such a sequence of sounds developed, they just de-aspirated one of them, generally the first. *Ibid.*, Grassmann's law cannot fail to warm the heart of every true-born philologist.

grass root. [GRASS *sb.*^{1}] Usu. in pl. *fig.* **a.** The fundamental level; the source or origin. Some of the later examples are influenced by sense b.

1901 KIPLING *Kim* xiv. 371 Not till I came to Shamlegh could I meditate upon the Course of Things, or trace the running grass-roots of Evil. 1907 *Putnam's Monthly* July 483/1 Carter had taken out wonderful grass-root values from the first. 1947 H. READ (*title*) The grass roots of art. 1952 *Times* 22 Mar. 7/4 Pushkin and Bakunin were only two of many famous men who were..given an opportunity to study at leisure what would now be called the 'grass-roots' of Russian civilization. 1960 *Times* 13 Jan. 21/7 Thus the jobs themselves generate a range of industrial development at the grass roots of the economy. 1963 *Listener* 21 Feb. 323/2 The modernization of agriculture requires the most skilful combination of central direction and grass-roots co-operation for success. 1966 *Oxf. Univ. Gaz.* 23 Dec. 438/2 If one goes back, to use a dull cliché, to the grass roots of academic life,..then it seems to me the Franks proposals follow logically. 1969 *Oz* May 32/3 Some of the attitudes and values to be found inherent in the music at its grass roots level.

b. *Politics.* Used spec. to describe the rank-and-file of the electorate or of a political party. Also *attrib.* orig. *U.S.*

1912 *McClure's Mag.* July 324/1 From the Roosevelt standpoint, especially, it was a campaign from the 'grass roots up'. The voter was the thing. 1935 *Nation* 19 June 697/2 'No crisis so grave has confronted our people since the Civil War, Mr. Lowden told the grassroots convention at Springfield. 1948 *Times Lit. Suppl.* 13 Mar. 143/4 The self-governing congregation is a unique element in English 'grass-root' democracy. 1955 *Times* 12 Aug. 9/6 These are the complaints at grass-root level; in more sophisticated circles the N.L.M. politicians talk of disregard of minority rights and incipient dictatorship. 1964 *Listener* 30 April 746/2 The rather narrow oligarchic regime of President Ngo Dinh Diem [in South Viet-Nam]..seems to be losing contact with the 'grass roots'. 1966 *New Statesman* 25 Mar. 410/1 The grassroot Tory still prefers to touch his forelock and reverence his 'betters'.

Hence **gra·ss-rooted** *ppl. a.*, **gra·ss rooter.**

1935 *Harper's Mag.* Sept. 484/1 'We believe,' the embattled Republican Grass Rooters resolved, [etc.]. *Ibid.* 489/2 Do we hear the Grass Rooters, the financial interests, and the industrialists raising their voices in protest against this usurpation? 1947 *Chicago Times* 28 June 13/4 Other straw polls in other states indicate that Republican 'Grass rooters' quite generally feel the same way. 1957 D. DAVIE *Winter Talent* 21 Grass-rooted goodness and a joy unmixed Parch unbaptized inside a droughty head.

grass tetany. [ad. G. *grastetanie* (Sjollema & Seekles 1930, in *Biochem. Zeitschr.* CCXXIX. 358).] A disease in cattle caused by magnesium deficiency, occurring esp. when there is a change of diet such as from indoor feeding to outdoor grazing.

1933 *Jrnl. Biol. Chem.* May 636 It is interesting that Sjollema and coworkers have studied in a comprehensive manner a bovine disease which they term grass tetany. 1955 J. G. DAVIS *Dict. Dairying* (ed. 2) 1132 The term 'Grass Tetany' is frequently used to designate a disorder of dairy cows which may occur a week or so after the cows are first admitted to grass in springtime. 1960 *Farmer & Stockbreeder* 29 Mar. 78/1 It is important to ensure that animals have the necessary amount of magnesium in their diet to avoid the danger of grass tetany. 1967 *Times* 28 Dec. 7/3, I remember about 15 years or so ago when I first started spraying my pastures with magnesium sulphate as a preventive against grass tetany.

grass widow. 2. (Earlier examples.)

1846 J. J. HOOPER *Taking Census* ii. 183 John Green's sister, (the grass widder, as lives with 'em,) she goes for her battling bench. 1853 E. CLACEY *Lady's Visit Gold Diggings Austral.* 255 The absence of so many of 'the lords of creation' in pursuit of what they value..more than all the women in the world—nuggets. The wives thus left in town to deplore their husbands' infatuation, are termed 'grass-widows'—a mining expression.

Hence **gra·ss-widowed** *ppl. a.*, temporarily living apart from one's husband; **grass-widower** (earlier U.S. example); **grass-wi·dowerhood,** the condition of a grass-widower; **grass-widowhood** (earlier example).

1862 *Rocky Mt. News* (Denver) 14 June (Th.), David is a bachelor again, or rather a 'grass-widower'. 1870 W. B. DONNE *Euripides* ix. 186 The real Helen..passed the score of years between the visit of Paris to Sparta and the fall of that city in a respectable grass-widowhood. 1926 W. J. LOCKE *Stories Near & Far* 112 She could never resolve the problem whether she would have been happier or unhappier in a grass-widowed state. 1930 HARWOOD & BROWNE *Cynara* 34 We're celebrating his grass-widowerhood.

grassy, *a.* **3.** Add to def.: Also, resembling grass in smell.

1931 H. CRANE *Let.* 11 Sept. (1965) 379 A marvelous stillness and grassy perfume pervade the district. 1957 G. E. HUTCHINSON *Treat. Limnol.* I. xvii. 901 Blue-green algae are described as having a grassy odor.

grate, *sb.*^{1} Add: **9. grate-fire,** a fire in an open grate.

1907 *Daily Chron.* 30 Nov. 4/4 When other reformers insist on our abolishing grate fires altogether. 1909 E. BANKS *Myst. Frances Farrington* 87 Your utterly useless, but expensive, grate-fire.

grate, *v.*^{1} Add: **8. c.** *trans.* To utter (words) in a harsh tone.

1921 GALSWORTHY *To Let* II. v. 165 Gradman grated: 'Rather extreme at your age, sir; you lose control.' 1969

J. ROSS *Dead at First Hand* i. 8 'I'm a gambler, Rogers,' he grated.

graticule. Add: **2.** A transparent plate or cell bearing a grid, cross-wire, or scale, designed to be used with an optical instrument or cathode-ray oscilloscope for the purpose of positioning, measuring, or counting objects in the field of view; the scale, grid, etc., on such a plate. Hence **gra·ticuled** *ppl. a.*, fitted with a graticule.

1914 *Handbk. Artill. Instrum.* 42 In front of the eye-piece is fixed..a diaphragm, with spider's web graticules attached to it. 1919 *Trans. Opt. Soc.* XX. 277 Generally the graticules are on glass and it is usual to refer to the complete discs or plates with the measuring scales or marks on them, as 'graticules'. *Ibid.* 286 Graticuled binoculars are not used much for peace purposes. 1920 *Nature* CV. 563/1 Such motion being observed by a plate micrometer or 'graticule' in the observing telescope. 1922 *Encycl. Brit.* XXXII. 243/1 Graticuled binoculars. 1924 L. C. MARTIN *Opt. Meas. Instr.* 27 The use of lines engraved on a glass (a graticule) is finding an increasing favour. 1952 M. TRIPP *Faith is Windsock* ii. 34 By means of a line-of-flight marker, a controllable circular graticule and a movable compass ring, the relative position of the aircraft and a ground feature could be ascertained. 1966 *McGraw-Hill Encycl. Sci. & Technol.* IX. 430/1 Practically all laboratory oscilloscopes have calibrated horizontal sweeps so that time interval measurements may be read directly from a graticule over the cathode-ray tube screen. 1970 E. M. SLAYTER *Optical Methods in Biol.* xii. 282 When exact measurements are required, however, an ocular can be used in which a ruled scale (graticule) is incorporated. The graticule is placed at the first focal plane of the 'eye lens' of the ocular. 1971 *Physics Bull.* July 398/2 A graduation line is centred in the microscope eyepiece graticule.

gratin. Add to def.: *spec.* the light crust on the surface of such dishes, now usu. formed by a sprinkling of breadcrumbs or grated cheese browned in the oven or under the grill. Hence phr. *au gratin* (o gratæ̃): cooked in this way; also, as *sb.*

1806 J. SIMPSON *Compl. Syst. Cookery* 139 Crayfish au gratin. 1844 THACKERAY in *New Monthly Mag.* July 418 Eels, salmon, lobsters, either *au gratin* or in cutlets. 1846 A. SOYER *Gastronomic Regenerator* 112 Sole au gratin. *Note*, In France we have silver dishes on purpose for *au gratins*, in which they are dressed and served to table, the gratin adhering to the bottom of the dish. 1889 A. B. MARSHALL *Cookery Bk.* 219 Maccaroni au Gratin. *a* 1936 KIPLING *Something of Myself* (1937) v. 144 The diamond's tip bubbled like cauliflower *au gratin*. 1964 *Good Housek. New Cooking* v. 57 *Egg and Spinach Au Gratin*...Sprinkle on the remaining cheese and grill until golden.

b. *attrib.*

1889 A. B. MARSHALL *Cookery Bk.* 17 Gratin Sauce (for Fish). 1901 C. H. SENN *New Cent. Cook. Bk.* 599 Butter a silver-plated gratin-dish. 1902 *Daily Chron.* 15 Feb. 8/4 Drain and turn into a gratin dish,..cover with fine breadcrumbs. 1965 *House & Garden* Dec. 84/3 A *gratin* dish often does contain cheese, but it is not a vital ingredient.

c. *transf.* The 'upper crust' of society.

1934 E. F. THOMAS *Portrait of Widow* iii. 28 Did you chance to know the Duchesse de Mallincourt in Paris? She always received the *gratin* of the American colony. 1959 *Sunday Times* 17 May 15/4 She belonged to the Edwardo-Georgian *gratin*. 1967 S. PAKENHAM *Sixty Miles from England* xiv. 185 Madeleine Lemaire had one of the most famous Paris salons, where all but the very highest *gratin* of the Franch nobility congregated.

Hence **gra·tinate** *v.* [after F. *gratiner*: see -ATE^3], to cook (food) *au gratin*.

1902 in WEBSTER Suppl.

grating, *vbl. sb.*^{2} Add: **5.** *grating constant, photograph, space, spectrometer.*

1926 R. W. LAWSON tr. *Hevesy & Paneth's Man. Radioactivity* iv. 44 The grating space of a Rowland grating is about 10^{-4} cm. 1938 *Ibid.* (ed. 2) iv. 49 The 'grating constant' is here the distance between two adjacent lattice planes in the crystal. 1941 *Nature* 24 May 643/2 The spectrum was obtained from a discharge tube of the type described by Pearse and Gaydon, and grating photographs were taken (dispersions 2·6 and 1·9 A./mm.) in the region λλ 6400–4900. 1963 G. TROUP *Masers & Lasers* (ed. 2) viii. 149 The fluorescent light emitted from the ends and from the sides was examined, using a grating spectrometer capable of resolving the R_1 and R_2 lines.

grattage (gratāʒ). *Med.* [Fr., = scraping, f. *gratter* to scrape, scratch (see GRATE *v.*^{1}).] The scraping or scrubbing of a surface (as the conjunctiva in cases of trachoma) to remove granulations.

1890 in BILLINGS *Med. Dict.* 1895 S. STEPHENSON *Epidemic Ophthalmia* 143 Scarification of the conjunctiva, combined in various ways with 'brassage' and 'grattage', has been recommended. 1911 A. DARIER in W. L. Pyle *Internat. Syst. Ophthalmic Pract.* xvii. 217 Hypertrophic blepharitis..is best treated surgically by..longitudinal incision and grattage of the entire palpebral border. 1949 M. WIENER *Surg. Eye* (ed. 2) xiii. 200 The upper lid is grasped with the Darier grattage forceps, the male blade engaging the conjunctival surface about 4 mm. away from the lid margin.

gratters, colloq. (school and university) for *congratulations*: see *-ER^6. (Cf. *CONGRATTERS.)

1903 D. COKE *Sandford of Merton* xiii. 98 'Gratter [*sic*], Sandford,' he said, 'on your rowing to-day.' 1906 —— *Bending of Twig* xii, Gratters, Marsh, on being monitor. 1919 W. DEEPING *Second Youth* xiii, Right-ho! Good luck and gratters!

‖**grattoir** (gratwār). *Archæol.* [Fr., f. *gratter* to scrape, scratch.] A flint scraping tool in which the working edge is at the end of the blade or flake and lies across its long axis; an 'end-scraper'. (Cf. SCRAPER 4 e.)

1872 [see SCRAPER 4 e]. 1887 *Amer. Antiquarian* IX. 341, 4000 grattoirs, blades, knives and saws. 1915 W. J. SOLLAS *Anc. Hunters* (ed. 2) 298 The grattoirs or end scrapers are generally short and rough. *Ibid.* 485 The characteristic keeled grattoir. 1926 *Guide Antiquities Stone Age* (Brit. Mus.) (ed. 3) 130 A good specimen of the double end-scraper, consisting of a flintflake rounded at both ends on one face only by use as a plane (*grattoir*) is here illustrated.

graunch (grǭntʃ), *v.* *dial.* and *N.Z.* [Onomatopœic: cf. GRANCH *v.*] *intr.* To make a crunching or grinding sound; *trans.*, to cause to make such a sound; hence, to damage (a mechanism of some kind). Also **grau·nching** *vbl. sb.*; **grau·nchy** *a.*, difficult, testing.

1881 A. B. EVANS *Leics. Words* 163 *Graunch*, var. of 'crunch' and 'scrunch', to crush or grind with a noise; crash. 'I'm sure it freezes, for I heard the ice *graunching* under the wheels of the carriage.' 1954 *Dominion* (Wellington, N.Z.) 1 July, As far as I know 'to graunch' means to damage an engine, instrument, machine, etc., by using wrong tools and/or repair methods, and a 'graunch-artist' is 'a person who does that'. *Ibid.* 9 July, The first time I heard the word [graunch] was some time in '39 or '40, and it was used by an English airman. To the best of my knowledge it originated in the R.A.F. and was pronounced 'garraunch' to reproduce the sound of a plane as it crashed and slid...Later it became a one-syllable word and refers to any metal torn or damaged by force. 1957 *Evening Post* (Wellington) 17 Apr. 8 (*headline*) Graunch!—Bang goes more than the door. 1959 D. BEATY *Cone of Silence* xxviii. 288 'Have you tried this new take-off technique?' 'Yes sir..we're in for a hell of a graunching.' 1964 *Observer* (Colour Suppl.) 11 Oct. 42/1 Many people 'graunch' their gears. 1965 *N.Z. Listener* 27 Aug. 9/1 John Pascoe himself knew that editing an encyclopaedia would be a lengthy project. 'When I started I knew it was a graunchy job. But then I've always liked long graunchy jobs. It's tied in with my youthful experiences of long distance running...I rather like long, slow patient plodding.' 1968 *Dominion Sunday Times* (Wellington) 10 Apr. 2/3 They said they could hear the ship 'graunching' on the rock.

grave, *sb.*^{1} Add: **1. d.** *to dig the grave of*: to cause the ruin, downfall, end of (a person or thing).

1934 F. SCOTT FITZGERALD *Let.* 8 Dec. (1963) 397 Of course any *apologia* is necessarily a whine to some extent; a man digs his own grave and should, presumably, lie in it. 1963 *Listener* 31 Jan. 207/2 The delegation called for the convening of a conference next month to 'dig the grave' of the Federation.

5. a. *grave-field, -group.*

1868 G. STEPHENS *Runic Monuments* 1026/1 Grave-fields. 1937 *Jrnl. R. Anthrop. Inst.* 233 To point out to me the sight of the grave-field. 1963 *Times Lit. Suppl.* 18 Jan. 44/2 The Viking character of the Gnezdovo grave-field. 1937 *Jrnl. R. Anthrop. Inst.* 232 Nothing is said as to the original composition of the grave-groups.

6. grave-cloth, delete †*Obs.* and add earlier and later examples; **grave-furniture** = *grave-goods*; **grave-post** (earlier U.S. example); **grave-trap** (earlier and later examples); also *fig.*

1646 in C. W. Manwaring *Digest Early Conn. Probate Rec.* (1904) I. 16, 1 graue cloath 3s. 1925 V. WOOLF *Common Reader* 35 The Prior of Bromholm sent word that the grave-cloth was in tatters. 1937 *Discovery* 152/1 The excavation of the churchyard produced virtually nothing in the way of grave furniture. 1939 G. CLARK *Archæol. & Society* iii. 55 Any archaeologist digging in England would give his head to find grave-furniture in anything approaching such a state of preservation as that in the young Pharaoh's tomb. 1840 *Southern Lit. Messenger* VI. 191/1 When an Indian dies, it is his family or surname, that is put on his grave-post, or *adjedatigwon*. 1844 J. R. PLANCHÉ *Drama at Home* I. 8 I'll propose her [*sc.* Ophelia] to be resident directress, with a bed in the grave trap. 1859 E. FITZBALL *35 Yrs. Dram. Author's Life* II. 211 On one side, was the grave trap made use of in 'Hamlet'. 1919 M. BEER *Hist. Brit. Socialism* I. II. viii. 251 He was then firmly convinced that England..was tottering to the brink of the grave-trap in which exhausted nations disappear from the scene of history.

grave-digger. Add: **3.** *fig.* (See quots.)

1887 F. GALE *Game of Cricket* 193 Let me say a word to captains..about the gravediggers...They grow despondent for lack of encouragement, and there is a sameness in constantly bringing up the rear. 1896 *Daily Tel.* 21 Aug. 3/2 Too much 'route-marching, pipe-claying, and starching' tends to dulness and apathy, whilst it leads the British soldier, when off duty,..to make too free an acquaintance with the 'grave-digger' [*sc.* strong drink], as it is termed in India. 1934 LEWIS *Lang. Cricket* 111 *Grave diggers*, the last batsmen of a side in the order of going in to bat. 1949 I. DEUTSCHER *Stalin* 94 Labelled by Lenin 'the liquidators', the grave-diggers of the party. 1957 C. HUNT *Guide to Communist Jargon* l. 163 According to Marx, the proletariat is brought into existence by the capitalist system, of which it is destined to be the 'grave-digger'.

gravel, *sb.* Add: **8. a.** *gravel bar*; **b.** *gravel-bottomed* adj.; **c.** *gravel-strewn* adj.

1821 T. NUTTALL *Jrnl. Trav. Arkansa* vii. 136 Four miles above Dardennes commences the first gravel-bar, accompanied by very rapid water. 1968 R. M. PATTERSON *Finlay's River* 20 We followed the trail through the water-lilies and slipped over the last gravel bar into the little stream that calls itself the Crooked River. 1864 J. A. GRANT *Walk across Africa* 38 Clear, gravel-bottomed river M'gazee. 1960 *Times* 13 Feb. 9/4 Clean gravel-bottomed reaches..will also yield good quality roach. 1927 W. G. KENDREW *Clim. Cont.* (ed. 2) 241 The rivers, whose beds, dry, wide, and gravel-strewn in summer, often become filled in a few hours in winter by swollen torrents.

9. gravel court, a lawn-tennis court with a gravel surface; **gravel-crusher, -crushing** *ppl. a.*, *slang* (see quots.); **gravel culture,** a hydroponic method of plant cultivation, using beds of gravel supplied with nutrient solutions; **gravel eye, eyed** adj. (see quot. 1951); **gravel-throated** *a.* = **gravel-voiced*; **gravel voice,** a thick, husky voice; so **gravel-voiced** *a.*

1890 C. G. HEATHCOTE *Lawn Tennis* xv. 294 Gravel courts, though at first sight attractive, have many serious defects. 1934 T. S. ELIOT *Rock* i. 30 In the land of lobelias and tennis flannels..The nettle shall flourish on the gravel court. 1889 BARRÈRE & LELAND *Dict. Slang Gravel-crusher* (military), a soldier compelled to tramp about a square at defaulter's drill. 1901 *Daily News* 9 Jan. 5/2 The 'gravel-crushers' (as the dismounted service is generically known). 1918 E. S. FARROW *Dict. Mil. Terms, Gravel crushers*, a slang expression equivalent to doughboy or infantry soldier, and the French *fiflot*. 1948 PARTRIDGE *Dict. Forces' Slang* 86 *Gravel-crusher*, a Drill Instructor; a P.T. Instructor. (Mostly Air Force.) 1900 *Kynoch Jrnl.* Feb.–Mar. 63/2 Cyclists..act more in conjunction with and as the eyes of their gravel-crushing comrades. [1936 WITHROW & BIEBEL in *Jrnl. Agric. Res.* LIII. 697 Fine gravel..appears to be the most suitable for this type of culture.] 1940 A. LAURIE *Soilless Culture Simplified* viii. 136 The advantage often claimed for sand or gravel culture—that of increased production—can easily be overstressed. 1942 LAURIE & RIES *Floriculture* v. 104 If flat-bottomed concrete benches have already been built, they can be converted to gravel culture. 1966 *New Scientist* 23 June 784/3 Gravel culture may also open possibilities for basic food production in malnourished countries. 1855 *Poultry Chron.* III. 9/1 The Suabian Pigeons..have generally a turned crown, gravel eye, and clean feet. 1879 L. WRIGHT *Pract. Pigeon Keeper* 101 An altogether red, gravel, or orange eye is a decided fault. 1951 E. HAEDY *A–Z Pigeon Guide* 71 *Gravel-eyed*, white or pearl eye with red mixed in it. [1942 BERREY & VAN DEN BARK *Amer. Thes. Slang* § 188.7 *Gravel-throat*, a granular enunciation. *Ibid.* § 422.9 *Gravel-throat*, one with a husky granular voice.] 1955 *Jazzbook 1955* 45 Joseph 'De De' Pierce, gravel-throated vocalist. 1962 K. ORVIS *Damned & Destroyed* viii. 157 A gravel-throated switchboard cop. 1947 *Time* 24 Nov. 29 The very sound of Earl Long's gravel voice. 1958 *Times* 5 July 4/1 The umpire's voice echoed the gravel voice on the radio which tells us what 'the next object will be' in a parlour game. 1947 *Time* 29 Dec. 15 Gravel-voiced Joe Curran, president of the National Maritime Union. 1952 *Time* 2 June 46/2 When *Collier's* hired gravel-voiced Louis Ruppel as editor three years ago, it knew it was buying a whirl-wind. 1963 *Daily Tel.* 25 Nov. 1/2 Dallas's detective chief, Capt. Will Fritz, a gravel-voiced Texan who sports a white cowboy hat.

gravel, *v.* Add: **6.** *intr.* = DUST *v.*^{1} 3 b.

1870 D. P. BLAINE *Encycl. Rur. Sports* § 2618 Where they [*sc.* partridges] bask at noontide, and where they preen, scratch, and gravel.

7. (See quot.)

1902 C. J. CORNISH *Naturalist on Thames* 216 In winter the eelman goes 'gravelling', that is, scooping up gravel from the bottom to deepen any part of the channel.

gravelled, *ppl. a.* Add: **2. a.** (Later examples.)

1865 DICKENS *Mut. Fr.* II. IV. xii. 262 He never did know such a move,..he never had been so gravelled. 1967 *Listener* 7 Dec. 763/2 Sir Francis Chichester, temporarily gravelled for an admired man of the past, appealed to the studio audience for names.

e. Of a voice: cf. **gravel voice*.

1958 *Listener* 26 June 1071/2 A gravelled voice with superb powers of timing.

gravelly, *a.* Add: **5.** *transf.* Of a voice (cf. **gravel voice*).

1944 *Harper's Mag.* June 38/1 The tired, imperturbable man aboard the carrier flagship called his orders over the gravelly-voiced TBS (talk between ships). 1958 *Listener* 7 Aug. 203/1 Sunderland's gravelly voice insinuating that they were all a dead loss. 1959 'F. NEWTON' *Jazz Scene* vii. 135 The gravelly and expressive voice of the great Louis Armstrong. 1971 *Daily Tel.* 13 Oct. 15/3 William Costello..provided the gravelly voice of Popeye the sailor in film cartoons.

Gravenstein (grā·vənstəin). [The German name for Graasten, a village in Denmark formerly in Schleswig-Holstein, Germany.] A variety of dessert apple, which has large fruit with yellow, red-streaked skin.

1821 *Trans. Hort. Soc. Lond.* IV. 216 Mr. John Wilmot sent specimens of the Gravenstein Apple, the produce of

a tree imported from Holland. 1841 J. LINDLEY *Pomologia Britannica* III. 98 What the English call the Gravenstein is an Apple of great merit. 1860 R. HOGG *Fruit Man.* 10 Gravenstein...Large, round, flattened at the ends, and angular... A very valuable apple. 1908 L. M. MONTGOMERY *Anne of Green Gables* xxix. 325 Mr. Harmon Andrews took second prize for Gravenstein apples. 1969 *Oxf. Bk. Food Plants* 52 '*Gravenstein*' is often considered the best flavoured apple in Northern Germany and Denmark. It is a fairly large fruit, its strong sweetness balanced by a marked acidity and a penetrating scent.

Graves (grāv). [Fr. (pl.), a name for gravelly sandy parts of the Bordeaux country.] A wine produced in the Graves district of France.

1605 P. ERONDELL *French Garden* xi. sig. L7^v What wine will it please you to drink? Claret wine, Graue wine,..Greeke wine? 1630 J. TAYLOR (Water P.) *Wks.* Fff 4/1 The French Frontiniacke, Claret, Red nor White, Graues nor High-Country could our hearts delight. 1833 C. REDDING *Hist. Mod. Wines* 148 Macau produces seven or eight hundred tuns of red Graves. 1908 E. & A. VIZETELLY *Wines of France* 81 Cérons..produces a fine white Graves. 1935 *Punch* 30 Jan. 131/1 Add one bottle champagne cider and one bottle Sauterne or Graves, and stir again. 1971 *Times* 10 Nov. (Wines Suppl.) p. vi (Advt.), White Bordeaux...Chateau Olivier 1967 Graves 86p.

Gravesian (grēi·vziăn), *a.* [f. the name of R. R. *Graves* (b. 1895) + -IAN.] Resembling in matter, style, or quality the work or manner of Robert Ranke Graves, writer, poet, and scholar.

1961 *Times* 7 Jan. 9/5 Sir Cedric Hardwicke reads..and adopts exactly the approved Gravesian tone. 1968 *Listener* 6 June 736/3 There is an almost Gravesian myth now discovered by Burgess in the Kell deposit.

Gravette (grăve·t). *Archæol.* [f. *La Gravette*, name of a site in the Dordogne.] Name of a long, narrow knife-like flint of Upper Palæolithic date, having a sharp cutting-edge and blunted back. Usually *Gravette point*.

1911 W. J. SOLLAS *Ancient Hunters* viii. 218 A new form of implement came into use. This is a pointed flake, carefully retouched all along the cutting edge; it..is known as the type of La Gravette. *Ibid.*, (*caption*) The La Gravette Point. 1920 QUIGGIN & HADDON *Keane's Man Past & Pr.* 12 Finely worked knife-like blades (Châtelperron point, Gravette point). 1921 R. A. S. MACALISTER *Text-bk. Europ. Archaeol.* I. 368 The Gravette point..is a flake having one edge treated with secondary chipping and the other left sharp and untouched. *Ibid.* 369 In some Gravette knives the chipped edge is not straight, but has a kind of hump. 1932 *Antiquity* VI. 364 A number of points of gravette character..combine with the burins to suggest an upper Palaeolithic facies. 1959 *Chambers's Encycl.* V. 448 Straight knife-blades with blunted back (La Gravette blade), steep scrapers, end-of-blade scrapers, burins or chisels, usually with straight working-end, have been found. 1963 R. CARRINGTON *Million Years of Man* xii. 147 One of the uses of their *gravettes* was to decorate the ivory of the slaughtered mammoths.

Gravettian (grăve·tiăn), *a.* and *sb.* *Archæol.* [See *GRAVETTE + -IAN.] **A.** *adj.* Of, characteristic of, or designating the Upper Palæolithic culture represented by remains found at La Gravette, France. **B.** *sb.* **a.** The Gravettian culture. **b.** A man or woman of this culture.

1938 D. A. E. GARROD in *Proc. Prehist. Soc.* IV. 23, I would suggest for these two very closely related levels the names of Lower Gravettian and Upper Gravettian respectively. *Ibid.*, The theory of an eastern centre of dispersion for the Gravettian. 1940 V. G. CHILDE *Prehist. Communities* i. 13 Gravettians whose kinsmen and ancestors had specialized in hunting big game. *Ibid.*, Gravettian tools. 1946 F. E. ZEUNER *Dating Past* ix. 289 From the beginning of LGI_2 onwards..the archaeological succession is clear. It is a surprisingly rapid one. While the middle Aurignacian may have lasted into this time, upper Aurignacian or Gravettian dominates, both in the west ('western' Gravettian) and in the east (Predmost or 'eastern' Gravettian). 1947 J. & C. HAWKES *Prehist. Brit.* (ed. 2) i. 17 The next stray bands to arrive were Gravettians, a people with a rather more northerly range than their predecessors, who in south Russia and elsewhere had used their human wits to prey successfully upon the formidable mammoth herds. 1952 V. G. CHILDE *New Light Most Anc. East* (ed. 4) ii. 19 Even in Africa Minor great heaps of small shells in the deserts and rock shelters yield blade industries..Caspian and Oranian—comparable to the Châtelperronian and Gravettian of Europe. 1952 *Proc. Prehist. Soc.* XVIII. 105 Examples of the Upper Aurignacian, with backed 'Gravettian' points, often microlithic in their size and delicacy, are found in the Ardèche and Gardon caves. 1953 A. L. KROEBER *Anthrop. Today* 173/2 It is possible that further excavations in the plains region of southern Russia will reveal that the Gravettian was derived from this source. 1967 *Listener* 30 Mar. 426/1 Among the Upper Paleolithic tools which appear first in these two Greek caves are narrow flint blades which have been intentionally blunted down one side, and they are said to belong to a culture called the Gravettian.

graveyard. Add: (Earlier U.S. examples.)

1773 P. V. FITHIAN *Jrnl.* (1900) 74 He meant it for a Satire upon the neglect of the people in suffering their Grave-Yard to lie common. 1806 M. L. WEEMS *Lett.* (1929) II. 344 Constantly walking over the grave yard of Foreigners. 1822 J. F. COOPER *Spy* (1831) xiv. 168 The grave-yard was an enclosure on the grounds of Mr. Wharton.

transf. (Additional examples.)

1843 *Quincy* (Ill.) *Whig* 7 Jan. 2/6 The iron steamer Valley Forge has been sunk at the 'Grave Yard' below St. Louis. 1879 *Chambers's Jrnl.* 18 Oct. 663/2 Those whose biographies lie in the sub-editorial desk—'grave-yard' this compartment is grimly called. 1933 P. A. EADDY *Hull Down* vii. 157 She had gone ashore on what was known as the 'Graveyard', on the northern side of the entrance to the Kaipara River. 1953 R. GRAVES *Poems* 30 Must the book end, as you would end it, With testamentary appendices And graveyard indices? 1965 *New Statesman* 7 May 715/2, I survived part of adolescence in the despair of Dundee, that industrial graveyard of the Thirties. 1969 *Daily Tel.* 8 Feb. 1 The M1 motorway was a graveyard of cars abandoned across all six lanes.

2. *attrib.* and *Comb.*, as *graveyard cough, test*; *graveyard-minded* adj.; **graveyard shift, watch** (see quots.).

1873 J. H. BEADLE *Undevel. West* 33, I was shaken by an ominous graveyard cough. 1890 E. DOWSON *Let.* 11 Feb. (1967) 137, I have a graveyard cough of the most alarming dimensions. 1948 I. BROWN *No Idle Words* 71 No doubt Housman did overplay the lads, lightfoot or grave-yard-minded. 1907 *Collier's* 26 Jan. 14/1 From the saloons came the clink of the chips. For it was the 'grave-yard gamblers' shift...The small hours of the morning..are theirs. 1908 *Sat. Even. Post* 7 Nov. 27/2 A month later he and his fellows went on 'graveyard' shift. 'Graveyard' is the interval between twelve, midnight, and eight in the morning. 1965 'E. MCBAIN' *Doll* (1966) ii. 22 The afternoon shift is from four p.m. to midnight. And the graveyard shift is midnight to eight a.m. 1957 *Brit. Commonw. Forest Terminol.* II. 85 *Graveyard test*, a test conducted out of doors on pieces of timber in contact with the ground to determine their durability. 1927 G. BRADFORD *Gloss. Sea Terms* 76/1 *Graveyard watch*, the middle watch or 12 to 4 a.m., because of the number of disasters that occur at this time. 1928 *Papers Mich. Acad. Sci. & Arts* X. 297 Graveyard watch, the watch from midnight till 4 A.M., so called on account of the silence throughout the ship. 1933 J. H. MCCULLOCH *Million Miles in Sail* vii. 125 Without this arrangement..the 'graveyard watch'—or..the middle watch..would fall to the lot of the same men each night.

gravicembalo (grævi͵tʃe·mbălo), It. corruption of *CLAVICEMBALO.

1858 SIMMONDS *Dict. Trade*, Gravecembalo. 1899 KELLETT & NAYLOR tr. *Bie's Hist. Pianoforte* 133 When Italy decided..to replace the Gravicembalo by the pianoforte. 1958 A. JACOBS *New Dict. Mus.* 151 *Gravicembalo*, harpsichord.

gravimeter. Add: **2.** Any instrument designed to measure the variation in the force of gravity from one place to another; usu. applied to those which involve determining the force exerted on a suspended mass.

1932 *Physics* II. 124 (*caption*) Diagram of gravimeter. *Ibid.* 128 (*caption*) The gravimeter in the field. 1940 *Geogr. Jrnl.* XCV. 135 An exhaustive test of all modern gravimeters should be made..to compare the Lejay astatic pendulum, the new Nörgaard instrument, the gravimeters depending on the properties of helical springs..and finally the Nörgaard and Halck patterns of hydrostatic gravimeter. 1955 *Sci. Amer.* Sept. 165/1 The pendulum is still the standard for absolute values of gravity, but nowadays the instrument most commonly used is the gravimeter, a supersensitive version of the ordinary spring scale. The earth's pull is measured by the amount of stretching of a thin wire of silica or invar (a nickel-steel alloy) on which a small weight hangs. 1957 [see *gravity meter* s.v. *GRAVITY 8 b].

gravimetric, *a.* (Later examples in sense 2 of *GRAVIMETER.)

1935 *Geogr. Jrnl.* LXXXV. 199 Conditions of gravimetric survey. 1955 *Sci. Amer.* Sept. 170/3 The fundamental idea of the gravimetric method of mapping, and of the present world-wide gravity measuring program, is that the undulations of the geoid and its tilt at every place can be computed from the observed gravity anomalies.

‖**gravitas** (græ·vitɑs). [See GRAVITY.] = GRAVITY 3.

1924 *Manch. Guardian Weekly* 10 Oct. 313 He never sheds a certain Roman *gravitas*. 1958 *Spectator* 30 May 698/2 A certain *gravitas* in the atmosphere of the Scottish universities. 1961 *Times* 2 Aug. 11/6 Its leading articles, and even its news coverage, will have a superb Victorian *gravitas*. 1961 *Listener* 30 Nov. 901/2 As for the Prime Minister [of Nigeria], to see him at close quarters is to recognize the true *gravitas* of the statesman. 1969 *Ibid.* 20 Mar. 376/2 *Gravitas*, the heavy tread of moral earnestness, becomes a bore if it is not accompanied by the light step of intelligence.

gravitation. Add: **4. gravitation constant** = **gravitational constant*; **gravitation stamp** = *gravity stamp* (see *GRAVITY 8 b); †**gravitation unit** = **gravitational unit* (obs.).

1889 *Cent. Dict.* s.v. *constant*, The gravitation constant is about 0·0000000658 of a c.g.s. unit. 1894 *Nature* 2 Aug. 330/2 This G, the gravitation constant, or as I prefer to call it..the Newtonian Constant of Gravitation, has nothing to do with that other quantity generally written *g*, which represents the attraction of the earth's surface. 1910 *Encycl. Brit.* XII. 385/1 The law of gravitation states that two masses M_1 and M_2, distant *d* from each other, are pulled together each with a force $G.M_1M_2/d^2$, where G is a constant for all kinds of matter—the gravitation constant. 1894 T. K. ROSE *Metallurgy Gold* 99 Californian 'gravitation' stamps are in general use..for crushing gold ores. 1914 W. GOWLAND *Metallurgy Non-Ferrous Metals* 200 A modern heavy gravitation stamp. 1872 *Rep. Brit. Assoc. 1871* II. 29 The object of the paper was to urge the necessity of giving names to *absolute* units of force and energy, that is, units not varying with locality, like the gravitation units vulgarly employed (pound, foot-pound, &c.), but defined by reference to specified units of length, mass, and time. 1885 A. MACFARLANE *Physical Arithmetic* xxxiii. 192 Work is also measured in terms of gravitation units, by taking the corresponding gravitation unit of force instead of the absolute unit of force.

gravitational, *a.* Add: *gravitational constant*: in classical physics, the constant of proportionality in the equation relating the strength of the gravitational attraction between two bodies to their mass and their separation, equal to approx. $6{\cdot}67 \times 10^{-11}$ N. m.^{2} kg.$^{-2}$ ($6{\cdot}67 \times 10^{-8}$ dyne cm.^{2} gm.$^{-2}$); symbol *G*; *gravitational mass*: the mass of a body as measured by the force exerted on it by a gravitational field; cf. **inertial mass*; *gravitational potential*: the potential of a gravitational field, the gradient of which at any point is equal in magnitude and direction to the field at that point; *gravitational system* (*of units*): a system of units based on a fundamental unit of weight rather than a unit of mass; *gravitational unit*: a unit the value of which depends upon the value of *g*, the acceleration due to gravity; formerly called *gravitation unit*; *gravitational water*: the water in saturated soil which can drain away under the influence of gravity.

1904 *Science* XIX. 928/2 The adoption of this unit of mass involves a change in the numerical value of the gravitational constant. 1966 *McGraw-Hill Encycl. Sci. & Technol.* VI. 264 Determining the gravitational constant by a suitable experiment is therefore equivalent to 'weighing the Earth'. 1918 A. S. EDDINGTON *Rep. Relativity Theory Gravitation* i. 11 In astronomy..the masses of heavenly bodies are measured by their gravitational effects; naturally we cannot legitimately apply (7.8) to gravitational mass without a full discussion of the law of gravitation. 1955 *Sci. Amer.* June 32/2 The theory of general relativity was constructed on the basis of a physical observation of the equivalence of inertial and gravitational mass under certain simple circumstances. 1928 NEWMAN & SEARLE *Gen. Prop. Matter* iii. 59 To move this unit mass from one point to another in the gravitational field would require an expenditure of work against the attraction. The amount of this work..is called the difference in gravitational potential at the points. 1938 W. M. SMART *Stellar Dynamics* ii. 32 The motion of an individual star at any instant will be determined by the gravitational potential of the system. 1971 *Sci. Amer.* May 69/1 Light emitted from such an object would have to overcome an immense gravitational potential and would be red-shifted just as it is in quasars. 1888 *Proc. Physical Soc.* X. 41 To find J in the metre-kilogram-second gravitational system, when it is given in the British gravitational system. 1899 W. WATSON *Text-bk. Physics* I. x. 85 Such a system of units is called a gravitational system, and it is this system which is almost exclusively used by engineers. 1966 KAYE & LABY *Tables of Physical & Chem. Constants* (ed. 13) 6 It is convenient to distinguish the basic units of force in gravitational systems from the basic units of mass in other systems by using the symbols kgf for the kilogramme-force and lbf for the pound-force. 1896 NICHOLS & FRANKLIN *Elem. Physics* I. iii. 31 (*heading*) Gravitational units of force. 1923 H. MOORE *Textbk. Interm. Physics* viii. 67 The English gravitational unit of power is one foot-pound per second. 1954 *Amer. Jrnl. Physics* XXII. 293/2 The gravitational unit of force..is related to the weight of the standard object. [1897 L. J. BRIGGS in *U.S. Bur. of Soils Bull.* No. 10, 6 The water contained in a soil may be considered to be of three kinds—gravitation water, capillary water, and hydroscopic water.] 1907 C. W. BURKETT *Soils* iv. 40 Three kinds of water are present in soils: gravitational water, capillary water, and hydroscopic water. Immediately after a rain, gravitational water, or that which will move under the influence of gravity, is present in the soil. 1917 MOSIER & GUSTAFSON *Soil Physics & Management* xvii. 217 The movement of gravitational water downward through the soil by the force of gravity is called percolation. 1960 TEUSCHER & ADLER *Soil & its Fertility* vii. 85 Gravitational water is of next to no importance for plant growth.

b. *gravitational wave*: (*a*) = **gravity wave*; (*b*) a periodic variation in gravitational field-strength which is propagated through space (the existence of such waves is not yet established with certainty).

1899 W. WATSON *Text-bk. Physics* III. i. 345 The large waves seen on the surface of the sea are well-known examples of gravitational waves. 1941 B. HAURWITZ *Dynamic Meteorol.* xiv. 276 (*heading*) Wave motion at the free surface of a single layer. Gravitational waves. 1963 *Meteorol. Gloss.* (*Met. Office*) 120 *Gravity wave*, a type of wave, also referred to as a gravitational wave, in which the controlling forces are gravity and buoyancy. 1906 E. E. FOURNIER D'ALBE *Electron Theory* x. 191 If we had any means of creating or destroying matter, we could produce gravitational waves of infinite wave-length. 1922 A. S. EDDINGTON in *Proc. R. Soc.* A. CII. 278 The chief point of practical interest in the problem of the spinning rod is the question whether its energy is gradually carried away by the gravitational waves which are created, so that the rotation would slow down. 1937 EINSTEIN & ROSEN in *Jrnl. Franklin Inst.* CCXXIII. 43 The approximate method of integration of the gravitational equations of the general relativity theory leads to the existence of gravitational waves. 1970 *Sci. Amer.* 58/3 Einstein's general theory of relativity predicts that gravitational waves should be produced by matter that is accelerating with respect to the observer. 1971 *Nature* 15 Jan. 185/2 Weber has..presented a plausible case for believing that his detectors may indeed have monitored frequent and sharply pulsed gravitational waves, which seem to be emanating from the centre of our galaxy.

gravitationally, *adv.* (Further examples.)

1935 *Discovery* Dec. 352/2 Meteorites are gravitationally drawn towards the Earth. 1965 J. D. North *Meas. of Universe* 417 Using Whittaker's and McCrea's idea of gravitational mass..and restricting his attention to what he termed a 'gravitationally steady-state' in which σ is held to a constant value, McVittie found [etc.]. 1968 R. A. Lyttleton *Myst. Solar Syst.* ii. 77 The requisite pressure arises gravitationally.

graviton (græ·vitǫn). *Physics.* [f. Gravit(ation + *-on.] A hypothetical sub-atomic particle thought of as propagating the action of gravitational force.

1942 *Chem. Abstr.* XXXVI. 6074 Plane waves of a particle with spin 2 (graviton). 1960 *New Scientist* 2 June 1433/2 To produce gravitons, one would need rapidly varying gravitational fields. Such fields are familiar in astronomy and accompany double stars. However..it will be a very long time before (astronomical) gravitons can be detected even if these exist. 1971 *Sci. Amer.* May 22/2 This fact, together with the small value of the gravitational constant, means that only one graviton (the quantum of gravitational radiation) would be emitted by an atom or a molecule for every 10^{43} photons (light quanta) emitted.

gravity. Add: **5.** *absol.* A force equal to the accelerating force of gravity; abbrev. *g.*

1945 *Jrnl. Exper. Zool.* C. 398 A sample of lobster-serum was..subjected to ultracentrifugation in a small air-turbine..machine at approximately 120,000 gravities. 1949 W. Ley *Conquest of Space* (1950) 44 If we assume the average acceleration of a moonship amounts to 4 *g* (gravities), which is something we are sure the pilot can stand.

8. b. gravity anomaly, the difference between the observed acceleration due to gravity at a point on the surface of the earth (or another planet) and a value derived either from calculations of the geoid or from observations at some reference point; **gravity balance,** a type of torsion balance formerly used to measure the variation in the force of gravity from one place to another; **gravity-collapse structure** (see quot. 1961); **gravity conveyor,** a conveyor in which material slides, rolls, or falls under its own weight, the rate of descent being determined by friction; also, a conveyor in which the material is contained in freely suspended buckets kept upright by gravity; **gravity dam,** a dam that resists the pressure of the water by its weight; **gravity die-casting,** die-casting in which the metal is poured into the mould rather than forced in under pressure; a casting so made; **gravity meter** = *gravimeter 2; **gravity stamp,** a machine for crushing ore in which a heavy weight is repeatedly raised by a revolving cam and allowed to drop on the ore; **gravity survey** (see quot. 1923); **gravity tank,** a fuel container from which the petrol is fed by gravity to the engine; **gravity water system** (see quot. 1940); **gravity wave,** a wave on the surface of a liquid in which the dominant force is gravity rather than surface tension; also, a wave in the atmosphere propagated because of gravity; **gravity wind** (see quot. 1959).

1912 Hayford & Bowie *Effect of Topography & Isostatic Compensation upon Intensity of Gravity* 112 A study was made to see if a possible relation could be discovered between the gravity anomalies..and the geoid contours. 1924 H. Jeffreys *Earth* ix. 121 It is seen from an examination of the formula for the gravity anomaly that only the second term arises from the attraction of the mountain itself. 1959 B. F. Howell *Introd. Geophysics* xxi. 326 Positive rather than negative gravity anomalies would be expected in the regions into which more mass is sliding. 1969 Z. Kopal *Moon* xiii. 202 Local concentration of denser material would accelerate overflying spacecraft; and, conversely, negative gravity anomalies would slow the motion down. 1890 Threlfall & Pollock in *Phil. Trans. R. Soc.* A. CXCIII. 215 (*title*) On a quartz thread gravity balance. *Ibid.* 216 Our..attempts to construct a gravity balance began in September, 1889. 1923 Glazebrook *Dict. Appl. Physics* III. 403/1 In Threlfall and Pollock's gravity balance a quartz thread is mounted horizontally and is attached, at one end to a spring which takes up variations of tension, and at the other end to an axle which can be rotated, in line with the thread. 1936 Harrison & Falcon in *Q. Jrnl. Geol. Soc.* XCII. 91 (*title*) Gravity collapse structures and mountain ranges, as exemplified in south-western Iran. 1961 J. Challinor *Dict. Geol.* 95/2 *Gravity-collapse structures*, structures, in stratified rocks, produced on the limbs of (simple) folds as a result of collapse under the force of gravity. 1910 *Encycl. Brit.* VII. 56/1 The gravity or tilting bucket conveyor can be used as a combined elevator and conveyor. 1966 *McGraw-Hill Encycl. Sci. & Technol.* I.1. 447/2 The most economical means for lowering articles and materials is by gravity conveyors. 1940 *Chambers's Techn. Dict.* 388/1 Gravity dam. 1957 *Encycl. Brit.* VII. (*caption*, facing p. 8), Solid masonry gravity dams. 1971 N. Smith *Hist. Dams* ii. 33 A gravity dam is in general a straight wall of masonry or earth which resists the applied water-pressure because of its sheer weight. 1940 *Chambers's Techn. Dict.* 388/1 *Gravity diecasting*, a process by means of which castings of various alloys are made in steel or cast-iron moulds, the molten metal being poured by hand. 1960 *Metallurgia* LXI. 65/2 Although there are many methods of producing castings in aluminium alloys, only three are of major importance, namely sand casting, gravity die casting, and pressure die casting. 1964 S. Crawford *Basic Engin. Processes* (1969) xi. 246 Gravity die castings are generally superior in structure and strength to both sand castings and pressure die castings. 1941 J. M. Bruckshaw in *Proc. Physical Soc.* LIII. 449 The present interest in gravity meters arises from two..considerations. *Ibid.* 452 It is by no means an easy matter to produce a gravity meter..which..is robust and sufficiently transportable to be employed in the field. 1955 *Sci. Amer.* Sept. 164/1 The only way we can trace its girth and plot its distances accurately is to travel over its surface with a gravity meter, measuring the tiny differences in gravity from point to point as a guide to the ups and downs of the globe's undulating shape. 1957 *Encycl. Brit.* X. 677/1 The gravity meter or gravimeter is simply a spring balance comprising a constant mass supported by a spring system, the changes in elongation of which may be read with precision. 1903 R. H. Richards *Ore Dressing* I. v. 144 Gravity stamps are lifted by cams and drop by their own weight. 1965 E. J. Pryor *Mineral Processing* (ed. 3) iv. 60 The gravity stamp..is a fixed-path machine. It is obsolescent, its place being taken by the rod mill. 1923 Glazebrook *Dict. Appl. Physics* III. 398/1 The primary object of a gravity survey is to obtain values of the force and direction of gravity at various points of the sea-level surface. 1959 *New Scientist* 5 Feb. 274/2 The Americans are now planning an airborne seismic and gravity survey of the central section of the supposed *graben*, in Marie Byrd Land. 1917 'Contact' *Airman's Outings* 225 A small gravity tank for his machine, to be used when the pressure tank is ventilated by a bullet. 1934 V. M. Yeates *Winged Victory* 48 A small reserve supply [of fuel] in the gravity tank. 1940 *Chambers's Techn. Dict.* 388/1 *Gravity water system*, a system in which flow occurs under the natural pressure due to gravity. 1960 R. Davies *Voice from Attic* 99 The same unpleasing charts of the sexual organs (the male, like plans for a gravity water system; the female like the skull of an elk, with vastly branching antlers). 1877 *Proc. Lond. Math. Soc.* IX. 22 The following particular cases..are here tabulated for convenience:—$V \propto \lambda, U = 0$, Reynold's disconnected pendulums. $V \propto \lambda^{\frac{1}{2}}, U = \frac{1}{2}V$, Deep-water gravity waves. $V \propto \lambda^0$, $U = V$, Aërial waves, &c. $V = \lambda^{-\frac{1}{2}}, U = \frac{3}{2}V$, Capillary water waves. $V \propto \lambda^{-1}$, $U = 2V$, Flexural waves. 1930 N. Shaw *Man. Meteorol.* III. i. 30 The genesis of gravity-waves in air is not generally understood. 1957 *Jrnl. Marine Res.* XVI. 107 At wind velocities between 16 and 20 knots it is common for capillary waves having the same velocity as the gravity waves to ride just at the beginning of the crest of the gravity waves. 1971 *Nature: Physical Sci.* 29 Mar. 99/2 The excitation of atmospheric gravity waves by a nuclear test in the atmosphere. 1928 N. Shaw *Man. Meteorol.* II. 255 Katabatic winds...Such winds are gravity-winds which pay no attention to isobars until they get into the open where they have time to adjust themselves to the requirements of the earth's rotation. 1959 R. E. Huschke *Gloss. Meteorol.* 259 *Gravity wind*, a wind..directed down the slope of an incline and caused by greater air density near the slope than at the same levels some distance horizontally from the slope.

gra·vity-fed, *a.* [See Feed *v.* 7, 8 c.] Supplied with material by the action of gravity; utilizing a gravity feed. Hence (as a back-formation) **gra·vity-feed** *v. trans.*, to supply (material) in this way.

1908 *Westm. Gaz.* 20 Feb. 4/2 The gravity-fed oil-pump. 1936 *Archit. Rev.* Dec. p. lxii/3 The hopper is so designed that the fuel is gravity-fed into the back of the fire space. 1962 *Times* 26 Mar. 5/3 Automatic gravity-fed boilers. 1971 *New Scientist* 18 Mar. 626/2 The normal pre-1950s reservoir makes use of upland topography to capture a volume of water and gravity-feed it to the towns.

gra·vity(-)feed, *sb.* [Feed *sb.* 5. Cf. prec.] A supply system that makes use of gravity to maintain the flow of material; the supplying of material in this way. Also *attrib.*

1914 *Programme Isle of Man Tourist Trophy Races* 9/1 Either gravity or pressure feed may be used. 1925 A. W. Judge *Carburettors* 142 The three systems of fuel supply used on motor vehicles are as follows: (1) The Gravity Feed. (2) Pressure Feed. (3) Vacuum Feed. 1934 *Aircraft Engin.* Nov. 286/1 The simplest form of petrol system is one employing direct gravity feed to the engine. 1961 *Listener* 12 Oct. 583/2 A more automatic type is the gravity-feed boiler, generally using a special grade of small anthracite. 1964 S. Crawford *Basic Engin. Processes* (1969) xi. 246 Molten metal at the correct temperature is poured from a ladle under gravity feed into a split metal die.

gravy. Add: **2. c.** *Theatrical slang.* (See quot. 1952.)

1864 P. Paterson *Glimpses Real Life* x. 107 The farce ..was gone through with equal rapidity—of course, all the 'points' were carefully given, 'cartfuls of beefsteaks and bucketsful of gravy' especially. 1952 W. Granville *Dict. Theatr. Terms* 92 *Gravy*, easy laughs from a friendly audience. (2) Good lines, or business, in a farce or comedy.

d. Money easily acquired; an unearned or unexpected bonus; a tip. Hence *to ride (board) the gravy train* (or *boat*), to obtain easy financial success. *slang* (orig. *U.S.*).

1910 *Sat. Even. Post* 30 July 13/1 Stick him for all you can. You're a hard worker, and you mustn't let somebody else git the gravy. 1927 *Amer. Speech* II. 276 *Gravy train*, sinecure. 1932 Wodehouse *Hot Water* i. 30 'Sixty thousand dollars' worth, at least.' Mr. Slattery was impressed. Sixty grand, he agreed, was pretty good gravy. 1933 *Amer. Speech* Feb. 32/1 *Ride a gravy train*, to continue to receive more than one's deserts. 1934 J. O'Hara *Appointment in Samarra* (1935) 11 If you sell two Cadillacs a month, you make expenses, and anything over that is so much gravy. 1939 E. S. Gardner *D.A. draws Circle* (1940) xv. 269 We started riding the gravy train like I said. 1942 D. Powell *Time to be Born* (1943) viii. 180 An apartment which he was able to furnish almost completely with 'gravy'—sofas, mattresses, gadgets pressed on him by earnest manufacturers in hopes of public mention of their products. 1948 Menjou & Musselman *It Took Nine Tailors* 141 Once you get on the Hollywood gravy boat, it is no trick to make money; the trick is to keep it. 1952 M. McCarthy *Groves of Academe* (1953) x. 197 There was a moment in the spring when the whole Jocelyn sideshow seemed to be boarding the gravy train, on to fatter triumphs of platitude and mediocrity. 1960 *Guardian* 17 Oct. 16/3 In 1960, says the cool Bob Kennedy, if his brother gets 55 [electoral votes]..'anything else will be gravy'. 1967 *Sun* 20 Apr. 5/8 The tip is called the gravy, and when there is a mix-up at the table and two diners leave separate tips, it becomes double gravy. 1968 *Globe & Mail* (Toronto) 17 Feb. 8/1 In the past 10 years, the Manitoba Government has reaped about $8-million from the Downs (more than $1-million last year). This revenue is almost pure gravy.

3. *gravy-boat* (examples), *-salt.*

1895 *Montgomery Ward Catal.* Index, Gravy Ladles and Boats. 1916 *Daily Colonist* (Victoria, B.C.) 7 July 7/1 (Advt.), Covered Gravy Boats with Plate. 1970 Simon & Howe *Dict. Gastron.* 206 *Gravy boat*, the name commonly given to a low, boatshaped container used to dispense gravies, sauces etc. 1907 *Yesterday's Shopping* (1969) 33/1 Gravy Salt..pkt., 0/1. 1958 J. Cannan *And be a Villain* iii. 68, I went down to the shop for your gravy salt.

grazet(t. Delete †*Obs.* and add later *U.S.* and *poet.* examples.

1745 in J. F. Watson *Ann. Philad.* (1830) 179 (Advt.), Turkettees, grassetts, [etc.]. 1894 A. M. Earle *Costume Colonial Times* 120 Grazzets, a dress-stuff appearing in lists in the [Boston] *News Letter* from 1712 to 1768, often specified as 'changeable grazzets'. 1933 E. Sitwell *Five Variations* 2 To make more glittering the gowns of these Smooth nymphs that wear a grazet petticoat.

‖ **grazioso** (grātsiō·so), *adv.* and *a.* *Mus.* [It. = gracious, graceful.] A direction denoting that a composition is to be played in a graceful manner.

1806 T. Busby *Dict. Mus.* (ed. 2), *Grazioso*,..a term implying that the movement to which it is prefixed is to be performed in a smooth, flowing, and graceful style. 1935 D. F. Tovey *Ess. Mus. Analysis* II. lxi. 138 The seventh variation (grazioso 6/8) is the crowning point of new melody and new lusciousness. 1965 *Listener* 24 June 953/1 During an early scene with Moses, Aaron's *grazioso* line..is first accompanied by counter-lines on the divided cellos.

grease, *sb.* Add: **2. d.** Later *gen.* examples not necessarily implying rancidity or inferiority.

1919 *Athenæum* 8 Aug. 727/2 When 'gyppo' or 'grease' was asked for at mealtimes, gravy or butter (?) was meant. 1928 *Papers Mich. Acad. Sci. & Arts* X. 297 *Grease*, butter. 1953 J. Masefield *Conway* (ed. 2) III. 165 As to the food..we had many names for it;..grease for butter.

4. *in the grease* (earlier example).

1863 S. Butler *1st Yr. in Canterbury Settlement* x. 160 If you wash [the sheep]..you should do it thoroughly.. otherwise you had better shear in the grease i.e. not wash. Wool in the grease weighs about one third heavier.

5. b. (Later example.)

1959 N. Mailer *Advts. for Myself* (1961) 123 You should have seen the grease job I gave to Carter. I'm dumb, but man, he's dumber.

6. *grease-free, -laden* adjs.; **grease-ball,** (*a*) a medicinal ball of grease for giving to a horse; (*b*) *U.S. slang*, a derogatory term for a foreigner, esp. applied to one of Mediterranean or Latin American origin; **grease-band** (see quot. 1953); hence **grease-band** *v. trans.*; **grease boil** *N.Z.*, a boil caused by contact with the grease in sheep's wool; **grease-cap** = *greaser 1 d; **grease-gun** (see quot. 1963); **grease monkey** *slang*, a mechanic; **grease-paint** (later *attrib.* examples); **grease-pan** (see quot. 1960); **grease-proof** *a.*, impermeable to grease; **grease-spot** *slang* (earlier example); also (*b*) a spot of grease used in photometry; so *grease-spot photometer*; **grease-tight** *a.* = **grease-proof* adj.

1926–7 *Army & Navy Stores Catal.* p. xlv/5 Grease Balls. 1934 *Amer. Ballads & Folk Songs* (1957) 559 Our grease-ball is a goddam dirty bum. 1958 S. Ellin *Eighth Circle* (1959) II. xvii. 176 A certain Mr. García—some greaseball who runs a lunch stand. 1969 I. Kemp *Brit. G.I. in Vietnam* xii. 193 The gunner was..Rick Francese; tormented with such unflattering names as 'Wop' and 'Greaseball' and mercilessly teased..about his Sicilian extraction, he was at the same time universally liked and enormously respected. 1900 W. D. Drury *Bk. Gardening* 1093 Grease-bands put round the trees..will prevent the females from ascending. *Ibid.* 1100 As a preventive measure all orchard trees should be grease-

banded in autumn, using Willesden or similar grease-proof paper. 1930 J. Coutts et al. *Compl. Bk. Gardening* 704 The stake supporting the tree must also be grease-banded. 1953 *Brit. Commonw. Forest Terminol.* I. 71 *Grease band*, a band of sticky or greasy material applied to a stem, as a barrier to insects. 1965 *Listener* 7 Oct. 555/3 In really big fruit trees, you can get control of, say, caterpillar, by grease-banding now. 1926 J. Devanny *Butcher Shop* vii. 73 Hone Reki is pretty bad with grease boils. 1956 G. Bowen *Wool Away!* (ed. 2) iv. 52 When a shearer does get grease boils he should rest with the legs up. 1924 A. W. Judge *Mod. Motor Cars* I. 245 Screw-down grease caps are generally provided, and a turn should be given every time the car is used for more than a few miles. 1908 *Practitioner* Sept. 488 Cyllin obstetrical lubricant..has the advantage that it is grease-free. 1917 *Catal. T. Eaton* Spring & Summer 282/1 in *Shopper's View of Canada's Past* (1969) 194 Ford Grease and Oil Gun, made specially to fill rear axle housing with grease. 1923 *Daily Mail* 11 May 12, I liked very much the accessibility of all the greasers on the car. Every one can easily be reached with a grease gun without any acrobatic feats. 1963 R. F. Webb *Motorists' Dict.* 117 *Grease gun*, a tool or device designed to pump grease into the required place under high pressures. 1890 A. Conan Doyle *Firm of Girdlestone* (1926) 238 Grease-laden hold. 1928 L. Gravatt *Pioneers of Air* 251 All the way down the line we find them from skilled draftsmen in a polished office to the 'grease monkeys' with blackened faces and smeary over-alls. 1946 V. Tempest *Near Sun* viii. 63 Flight mechanics and fitters, known in the Battle of Britain as 'grease monkeys' but generally called in the Royal Air Force 'Erks'. 1959 *Times Lit. Suppl.* 27 Nov. 700/3 In Australia he was impressed by the 'grease-monkey' at Broken Hill who could afford to run a racing stable. 1928 H. Crane *Let.* 27 Apr. (1965) 324 Hawaii.. the Pollyanna greasepaint pinkpoodle paradise. 1944 L. MacNeice *Christopher Columbus* 16 The grease-paint voice will stick out all the more when there is no real grease-paint to look at. 1958 *Listener* 9 Oct. 578/1 'The Vortex', faithfully performed in 1920 greasepaint voices, is a very poor play. 1936 *Burlington Mag.* July 26/2 Candlesticks with candle-holders and grease-pans. 1960 H. Hayward *Antique Coll.* 134/1 *Grease-pan*, the circular dish beneath the nozzle of a candlestick into which grease from the burning candle might drip. 1895 *Montgomery Ward Catal.* 573/1 Waxed Butter Paper, grease proof. 1900 Grease-proof [see *grease-band* above]. 1910 *Daily Chron.* 8 Mar. 3/3 Each loaf..is packed and sealed in a paper wrapper, which is grease-proof and germ-proof. 1940 L. A. G. Strong *Sun on Water* 201 Secreted about her were various parcels of food, and it was the stiff greaseproof paper in which they were wrapped that gave off the cracklings. 1829 in *Amer. Speech* (1965) XL. 129, I hit a man..dere was nothing left, Sept a little grease spot. 1882 *Encycl. Brit.* XIV. 584/1 Bunsen has recently suggested the very simple expedient of making a grease-spot on white paper for photometric purposes. When the paper is equally illuminated from both sides, the grease-spot cannot be seen except by very close inspection. ..The amounts of light are as the squares of the distances of the sources from this point [*sc.* the grease-spot]. 1911 R. S. Clay *Treat. Pract. Light* xviii. 388 The grease-spot photometer is perhaps the most sensitive to scattered light. 1923 Glazebrook *Dict. Appl. Physics* IV. 416/2 As a convenient example, the simple form of the Bunsen grease-spot photometer may be described. 1925 A. W. Judge *Mech. Car* 163 The nipples are..grease-tight and dust-excluding.

grease, *v.* Add: **4. b.** *to grease the fat pig* (or *sow*) (fig.): to give to those who do not lack.

1566 J. Heywood *Dialogue containing Proverbs* (ed. 2) I. xi. Ei What should we (quoth I) grease the fat sow in thars. 1670 J. Ray *Coll. Eng. Prov.* 178 To grease a fat sow on the A... 1791 'P. Pindar' *Works* (1794) I. 287 'And then why vore?' the peepel rail:—'To grease a vat ould pig in the tail—*Old Weymouth* o' Long Leat.' 1923 *Daily Mail* 7 Aug. 8/6 These instances of extravagance, which seem to recall the old saying of greasing the fat pig.

greaser. Add: **1. b.** *spec.* An engineer on a ship.

1888 L. A. Smith *Music of Waters* 38 And who do you think was second greaser? 1890 [in Dict.]. 1906 *Daily Chron.* 20 Mar. 3/6 Every A.B. signing on for a British ship must prove that he has served already at least three years as an ordinary seaman on a British vessel, and that every fireman, greaser, &c., has served two years as trimmer, &c. 1909 J. R. Ware *Passing Eng.* 147/1 *Greaser* (Navy, 1860–82), a scornful way of describing naval engineers. 1942 *Penguin New Writing* XV. 9 Seamen don't salute bloody greasers.

c. (See quot. 1906.)

1906 L. Claremont *Gem-Cutter's Craft* 100 The heavy material among which are the diamonds..is..passed through a machine called the 'greaser', which consists of a shaking table made of five shallow steps..coated with a thick layer of grease, and the diamonds adhere to the grease while the remainder o the gravel is washed away. 1962 F. C. Phillips *Smith's Gemstones* (ed. 13) xx. 276 In this ingenious machine, the 'jigger' or 'greaser' as it is commonly termed, the concentrates are washed over a series of galvanized-iron trays, which are covered with a thick coat of grease.

d. A device for lubricating the parts of a motor vehicle with grease.

1908 *Westm. Gaz.* 17 Nov. 5/2 Greasers are fitted everywhere necessary. 1923 [see *grease-gun* s.v. *grease *sb.* 6]. 1924 A. W. Judge *Mod. Motor Cars* I. 245 The shackle pins..should be provided either with screw-down greasers or dust-proof oil cups.

2. b. An objectionable person; a sycophant. (Cf. Grease *sb.* 5 b.)

1900 J. S. Farmer *Public School Word-Book* 100 *Greaser*,..a cad. 1911 D. Coke *Wilson's* vii. 74 Dick could not help wondering..what possible breach of House discipline could brand a fellow as a beastly greazer. 1958 *Spectator* 7 Feb. 163/1 The dismissive contempt the little greaser had so richly earned. 1959 I. & P. Opie *Lore & Lang. Schoolch.* x. 191 *Grease* is a very popular word in school just now, and seems to be short for 'greaser' —i.e. sycophant, soft-soaper, teacher's pet.

c. A term applied, orig. in California, later elsewhere, to long-haired youths who, as members of a group or gang, ride about habitually on motor-cycles (in California, in 'hot rods').

1964 *San Francisco Examiner* 26 Jan. I. 14 The boy distinguished..'the Surfers'..from less-favored individuals called 'Greasers', who indulge in such gauche activities as drag-racing, putting grease on their hair, smoking marijuana and dancing the Twist. 1965 *Sun* 8 June 7/7 You can call rockers Greasers if you like... Greasers just means they have to put a lot of work into bikes. 1969 *Guardian* 23 Sept. 5/1 The London Street Commune..intends forming a 'working coalition between greasers, skinheads, and beats'. 1970 *Daily Mirror* 31 Mar. 24/4, 200 teenagers—both skin-heads and 'greasers', who are long-haired youths—fought for two hours with bricks and chains. 1971 *Daily Tel.* (Colour Suppl.) 3 Sept. 41 The surfers used to refer to the hotrodders neither as hotrodders nor as Hair Boys, but as 'greasers'. 1971 *It* 7–21 Oct. 21/3 From the very start he seemed bound for Angeldom, leading a young pack of greasers down Leyton Baths on a regular basis and battling with the Mods.

greasy, *a.* Add: **9. greasy spoon (restaurant)** *slang* (orig. *U.S.*), a cheap and inferior eating-house.

1925 *Writer's Monthly* June 486/2 *Greasy spoon*, a low-class restaurant. 1951 *Time* 31 Dec. 29/2 They [*sc.* the Marx brothers]..ate in coffee pots and greasy spoons. 1966 C. Himes *Heat's On* iii. 25 A room behind a greasy-spoon restaurant. *Ibid.*, The cook came from the greasy spoon. 1968 L. Deighton *Only when I Larf* viii. 110 Bob said he was hungry and wanted to pull up at every greasy spoon we passed.

great, *a.*, *adv.*, and *sb.* Add: **A.** *adj.* **6.** *great big* (later examples).

1841 Lytton *Night & Morning* II. xi. 58, I send you a great big sum of 20*l*. 1961 L. P. Hartley *Two for River* 94 It was a great big thing, the size of a small haystack.

b. See also Primer *sb.*[1] 3 a.

11. d. Later examples of use with *power.*

1863 [see Power *sb.*[1] 6 b]. 1876 Carlyle *Let.* Nov. in H. J. Nicoll *Carlyle* (1885) ix. 231 The thing to be desired is concord between the three Great Powers. 1955 *Times* 11 May 10/3 Diplomatists remark, however, that Moscow invariably speaks of a meeting of the Great Powers rather than the four Powers. 1965 *New Statesman* 30 Apr. 686/3 An agreement with Russia which would restore Germany's great-power status.

12. c. (Further examples.) Also *Great-Scott* v. intr.

1889 J. K. Jerome *Three Men in Boat* vi. 81 Great Cæsar! man,..you don't mean to say you have covered over carved oak with blue wall-paper? 1902 'Mark Twain' in *Harper's Weekly* 6 Dec. 9/1 'Ger-reat Scott!' ejaculated the Major... The secretary said, wonderingly: 'Why, what are you Great-Scotting about, Major?' 1947 *Mind* LVI. 356 Lord Russell mentions an exciting moment in 1894...He suddenly had a flash of apparent insight.. exclaiming: 'Great Scott, the ontological argument is sound!' 1963 Wodehouse *Stiff Upper Lip, Jeeves* i. 10 'You don't mean that old crumb was there?' I said, Great-Scott-ing.

h. *great (white) father, chief*, etc. (also with capital initials). A term of address orig. used by American Indians to refer to the President of the United States; now usu. *transf.* and ironical. Cf. *chief *sb.* 6 b.

1808 Z. M. Pike *Sources Mississ.* 5, I spoke to them [*sc.* Indians] to the following purport: 'That their great father, the president of the United States, [etc.]'. 1832 F. Trollope *Dom. Manners* I. xx. 314 All the chiefs who..have come to negociate with their great father, as they call the President. 1881 *Harper's Mag.* Apr. 671/1 Spotted Tail has been to the Great Father's house so often that he has learned to tell lies and deceive people. 1936 *Time* 25 May 11/1 The Indians came bearing gifts, a blanket for the Great White Father [*sc.* President Roosevelt], a ring for the Great White Mother. 1937 *Discovery* Sept. 276/1 The 'Proclamation of the Great White King' which granted this clan 'complete protection from the interference of strangers in this reserved area which the White King had given'. 1960 *Ontario Legislature Debates* 6 Dec. 247/2 And when they had it arranged, the great white father blows into town and gives the people a party. 1963 *Amer. Speech* XXXVIII. 272 The disparaging use of the term *Great White Father* for the superintendent, an unpopular authoritarian figure, appears to be limited to the staff, only half of whom are Indian. 1966 *Listener* 15 Dec. 895/1 To call a discussion influenced by such low-grade remarks as these..'a brisk intellectual exercise' can only suggest how little the great white scientist expects of the natives. 1966 'A. York' *Eliminator* iv. 73 She acknowledges you as the great white chief.

16. a. *a great one for* (cf. *one 15 c).

1927 C. Asquith *Black Cap* 100 Now Mrs. Mingle, unlike Hetty, had been a great one for reading. 1934 T. S. Eliot *Rock* i. 13 There's David. One o' them fancy lads—a good soldier and fond o' the ladies—but a great one for 'is church. 1957 J. Kirkup *Only Child* ix. 127 Isa was a great one for the proprieties.

b. (Later examples.) Also as *int.*

1967 *Listener* 12 Oct. 465 Great! I see a great headline. 1968 *Ibid.* 26 Sept. 423/2 Gary's mum's bread pudding is *great.* 1969 D. Francis *Enquiry* v. 66 'We're going down to the yard.' 'Great,' said Roberta...'I'll come too.'

20. Great British Public, a jocular, usu. ironic, way of referring to the British people; cf. *G.B.P.* (s.v. *G III. f); **Great Cham,** a nickname applied to Samuel Johnson; **Great Dane** (see Dane 2); **(Great) Deliverer,** a title given to King William III; **great game,** (*a*) golf; (*b*) spying; **great insertion,** the section of St. Luke's Gospel, ix. 51–xviii. 14, which is independent of St. Mark; **Great Lakes** (see *lake *sb.*[4]); **great omission,** St. Mark vi. 45–viii. 26, which is omitted in St. Luke; **great red spot** *Astr.*, an oval feature in the outer gas of the planet Jupiter, occas. red but now usu. pink in colour; **great thought,** a maxim, an apophthegm; now freq. used ironically; **great tradition,** a phrase employed by F. R. Leavis (b. 1895) to denote the corpus of great English fiction; **Great War,** (*a*) the French Revolutionary and Napoleonic Wars, 1793–1815; (*b*) the war which began on 28 July 1914 with hostilities between Austria-Hungary and Serbia, and ultimately involved the majority of the nations of the world; it was suspended by armistice 11 Nov. 1918; **Great White Way,** Broadway in New York City, in reference to the brilliant street illumination.

1866 J. Blackwood *Let.* 26 June in *Geo. Eliot's Lett.* (1955) IV. 279 The 'Great British Public' is not yet sufficiently aware of the book [*sc.* *Felix Holt*] to affect the sale. 1924 J. Doe in M. Joseph *Journalism for Profit* xii. 210 The next business in hand is getting on the right side of the Great British Public. 1927 T. E. Lawrence *Let.* 30 Dec. (1938) IV. 562 The soul of the great British public will be turned with rage at its surfeit of my rareness and virtuosity. 1932 Q. D. Leavis *Fiction & Reading Public* I. iii. 38 The major achievements of contemporary novelists appear to be unknown even by name to that part of the community journalists call the Great British Public. 1968 'L. Marshall' *Blood on Blotter* xvi. 108 The Great British Public would sooner eat dry bread and processed cheese..in a house they thought was tops, than enjoy a good meal in a place that didn't have any reputation. 1759 Great Cham [see Cham]. 1958 P. Kemp *No Colours or Crest* vii. 146, I would jerk awake to find myself declaiming one of the magisterial pronouncements of the Great Cham. 1886 C. M. Yonge *Chantry House* I. x. 92 There was a dissenting chapel, old enough to be overgrown with ivy.., erected by the Nonconformists in the reign of the Great Deliverer. 1898 S. Weyman *Shrewsbury* xi. 101 'The Deliverer', as the Whig party still love to call him, landed at Torbay. 1969 *Listener* 9 Oct. 475/2 A gilded equestrian statue of William III. Erected in 1734..and inscribed 'To the Great Deliverer'. 1866 J. Blackwood *Let.* 26 Apr. in *Geo. Eliot's Lett.* (1955) IV. 245 The old golf ball maker's shop is associated in my mind..with elevating talks about 'the great game'. 1901 Kipling *Kim* vii. 183 When he comes to the Great Game he must go alone—alone, and at peril of his head. 1961 *Guardian* 17 Mar. 10/7 Some John Buchan hero, busily playing the Great Game for Queen and Country. 1964 'J. Welcome' *Hard to Handle* ii. 28 Originally the Secret Service..was entrusted to amateurs..who played 'the great game', as they romantically called it, amongst themselves. 1911 J. V. Bartlet in *Stud. Synoptic Problem* 336 The part of Luke's Gospel prior to the Great Insertion. 1927 A. H. McNeile *Introd. N.T.* 26 The next non-Marcan block, [Luke] ix. 51–xviii. 14, containing more than 30 per cent. of the Third Gospel, is often called the 'Great Insertion'. 1911 J. C. Hawkins in *Stud. Synoptic Problem* 61 This well deserves its usual name of St. Luke's 'great omission'. 1927 A. H. McNeile *Introd. N.T.* 26 Whether intentionally or not he [*sc.* St. Luke] omits [Mark] vi. 45–viii. 26, which is sometimes called the 'Great Omission'. 1881 *Monthly Not. R. Astr. Soc.* XLI. 44 The great red spot..has maintained a constancy of appearance and regularity of motion. 1936 F. Reh *Astron. for Layman* xviii. 243 Occasionally a more or less fixed spot appears. Such a spot is the so-called 'great red spot'..still faintly visible in good photographs. 1967 P. Moore *Amat. Astron. Gloss.* 61 Of particular interest is the *Great Red Spot*, which can be traced on drawings made as long ago as 1631. It became prominent in 1878, and is still conspicuous, though it has been known to vanish temporarily. It is decidedly pinkish in colour; its exact nature is uncertain, though presumably it is a solid or semi-solid body floating in Jupiter's outer gas. 1821 Hazlitt *Table-T* xi. 253 Great thoughts reduced to practice become great acts. 1913 C. Mackenzie *Sinister St.* I. I. iv. 48 A calendar of Great Thoughts was roughly divested of ninety-eight great thoughts at once, in order that..a correct announcement should celebrate the ninth of April. a 1916 'Saki' *Coll. Short Stories* (1930) 407 A William the Conqueror calendar..with a quotation of one of his greatest thoughts for every day in the year. 1925 A. Huxley *Along Road* I. 67 There is nothing more dismal than a 'Great Thought' enunciated by an author who has not himself the elements of greatness. 1969 K. Giles *Death cracks Bottle* iii. 23 'How nice!' said Noni, with what passed with her for wit. 'Any more great thoughts on offer?' 1948 F. R. Leavis *Great Tradition* i. 7 By 'great tradition' I mean the tradition to which what is great in English fiction belongs. 1969 *Guardian* 21 Aug. 8/3 The inheritors of Leavis's Great Tradition..mutter about pretentious, jumped-up, pulp writers. 1970 *Ibid.* 16 Sept. 10/3 You can't go on writing Leavisite criticism when you've reviewed everything that relates to the great tradition. 1971 *Human World* Nov. 88 The Great Tradition from Jane Austen to Conrad is that of the fine individual consciousness, typically considered under the aspect of conscience, implying self-knowledge and self-transcendence. 1887 R. D. Blackmore (*title*) Springhaven. A tale of the Great War. 1911 J. W. Fortescue (*title*) British Statesmen of the Great War 1793–1814. 1914 *Maclean's Mag.* Oct. 53/1 Some wars name themselves...This is the Great War. 1915 Grahame-White & Harper (*title*) Aircraft in the Great War. 1921 A. Huxley *Crome Yellow* xxii. 238 Luther was reality—like the Great War. 1922 S. Weyman

Ovington's Bank i. 7 By that coach had come eleven years before, the news of the abdication of the Corsican and the close of the Great War. Laurelled and flagged, it had thrilled the town a year afterwards with the tidings of Waterloo. 1923 [see *gallery-play* s.v. *gallery *sb.* 12 b]. 1925 D. H. Lawrence *Refl. Death Porcupine* 61 Even the great war does not alter our civilization one iota, in its total nature. 1937 [see *atrocity 5]. 1969 A. Toynbee *Experiences* II. iii. 221 'The Great War'—as, until the fall of France, the British continued to call the First World War in order to avoid admitting to themselves that they were now again engaged in a war of the same magnitude. 1901 A. B. Paine (*title*) The great white way. 1908 G. V. Hobart *Go to It* 22 Eight weeks since we left Chicago, three shows to the bad, and still a thousand miles from the Great White Way. 1936 H. Miller *Black Spring* 248 The Great White Way is blazing with spark-plugs. 1967 *Act One Scene Two* II. Nov./Dec. 12/3 All we can do is to keep our bit of the Oxford Playhouse staked out worthily along the Great White Way.

21. b. *great-granduncle* (later example). Also *great-grandfatherly, -grandparental* adjs.

1903 *Daily Chron.* 2 Jan. 5/2 Is 1903 to revert to the great-grandfatherly ways of 1803? 1922 Joyce *Ulysses* 123 His granduncle or his greatgranduncle. 1929 *19th Cent.* Dec. 810 A collateral relative at the level of the grandparental or great-grandparental generation.

c. great(-great), an ancestor or descendant of 'great (-great)' degree. *colloq.*

1905 Mrs. H. Ward *Marriage W. Ashe* I. ii. 33 'We—you and I—are a little bit cousins too, aren't we?'.. 'Was our "great-great" the same person?' he said, laughing. 1907 A. Quiller-Couch *Major Vigoureux* xxii, Your grandfathers and grandmothers, and right back into the greats and great-greats. 1926 W. R. Inge *Lay Thoughts* 181 Its great-great-grand-offspring.

C. *sb.* **1. d.** *a great*, a large part or amount. *no great*, not a great deal, nothing great; *adv.* not much. *U.S. colloq.*

1724 *Essex Inst. Hist. Coll.* XXXVI. 337 Mackey's sloop sunk at Boston, & spoild a great of our English goods. 1854 'O. Optic' *In Doors & Out* (1876) 186 I've got consider'ble, but I don't care no great about sellin' it. 1885 A. Gray *Lett.* (1893) 772 No great to see, except a spick and span new Hotel. 1890 *Harper's Mag.* Apr. 715/1, I wa'n't no great of a boy, an' let little things wear on me. *Ibid.* Dec. 146/2, I hadn't been round no great in New York, an' there ain't no general store there.

7. Delete †*Obs.* and add later examples.

1912 E. Pound in *Poetry* I. i. 7 You also, our first great, Had tried all ways. 1947 R. de Toledano *Frontiers of Jazz* xvi. 176 The passing of another one-time great. 1959 *Encounter* Aug. 32/1 Their admiration for coloured greats like Tusdie and Maria really *meant* something to them. 1963 J. Walsh *Shroud* (1964) viii. 73 Statues and paintings of the greats of French science and literature. 1965 *Listener* 17 June 894/1 Tilden was an all-time great with no technical weakness anywhere. 1970 *Times* 13 July 3/1 El Vino has been the haunt of Fleet Street's greats and not-so-greats for nearly a century.

greater, *a.* Add: **1. c.** Applied to and descriptive of towns (to indicate the inclusion of adjoining boroughs, suburbs, etc.) and countries (with dependencies, etc.).

1616 T. Coryat *Traveller for English Wits* 12 *Traiecto Tigride*, I entred Armenia the greater: After that, Media the lower. *Ibid.* 19 My whole perambulation of this Asia the greater. 1868 Greater Britain [see Britain *sb.* 1]. 1882 *Encycl. Brit.* IV. 834/1 The London police district, or 'Greater London', is divided into two police jurisdictions, that of the metropolitan police,..and that of the City police. 1898 *Appletons' Dict. of N.Y.* (title-page), First 'Greater New York' Edition. *Ibid.* p. iii, The present edition of this book is the first to deal with the territory of the 'Greater New York' as constituting one municipality. 1939 *Ann. Reg. 1938* 190 After the Anschluss with Austria, Czechoslovakia was surrounded on three sides by Greater Germany. 1955 *Times* 2 Aug. 6/5 Motorists living in the Greater London area were on the roads early in the morning bound for the seaside or the country. 1958 *Listener* 21 Aug. 274/2 Apart from Greater London and Greater Birmingham all the conurbations were either increasing their population at a rate far lower than the nation as a whole or (in the case of Greater Manchester) were even losing population.

greatest. Add: **A.** *adj.* **3.** (Later examples in *slang* use.)

1946 *Jazzways* I. 104/1 'Duke's the greatest' is certainly the easiest cliché tossed around swing circles. 1954 *Time* 8 Nov. 70 A term of high approbation in the swing era.. is 'the greatest'. 1964 *Globe & Mail* (Toronto) 30 Nov., Joe Kapp and Willie Fleming swung in a jubilant jig, and Kapp hollered, 'Baby, you're the greatest.' 1970 *New Yorker* 17 Oct. 39/1 Everything about this broad spelled class,..she was the greatest.

Great Spirit. [tr. Ojibway *kitchi manitou*: see Manitou.] The supreme spirit or deity recognized by the North American Indians.

1703 tr. *Lahontan's New Voy. N. Amer.* II. 19 All the Savages are convinced..that Man was not made by chance, and that he's the Work of a Being superior in Wisdom and Knowledge, which they call the Great Spirit, or the Master of Life, and which they Adore in the most abstracted and spiritual manner. 1805 J. Sibley in *Deb. Congr. U.S.* (1852) 9th Congr. 2nd Sess. 1076 The Great Spirit placed on an eminence, near this lake, one family of Caddoques. 1812 *Niles' Weekly Reg.* II. 81/1 We return thanks to the Great Spirit for the many favors he has bestowed upon us. 1836 [see Spirit *sb.* 4]. 1855 Longf. *Hiaw.* i. 96 Listen to the words of warning, From the lips of the Great Spirit. 1869 C. L. Brace *New West* xi. 140 Their idea of a Great Spirit is undoubtedly a reflex of Christian teachings. 1873 J. Miller *Life amongst Modocs* (1876) xix. 264 The Indians said the Great Spirit made this mountain [*sc.* Shasta] first of all. 1958 *Native Voice* (Sp. ed.) 48/1 'Woe! Woe! My children,' cried the wise Chief. 'Do you not know that those who harm one of the Great Spirit's creatures will suffer in a like way.'

Grecian, *a.* Add: **1. b. Grecian coil** (see quot. 1966); **Grecian curve** = *Grecian bend*; **Grecian knot** (example); **Grecian nose,** one that is straight and continues the line of the forehead; **Grecian plait,** an elaborate plait of hair made from about thirteen strands; **Grecian slipper,** a shop name for a soft slipper cut low at the side; **Grecian splice** *Naut.* (see quot. 1883).

1874 C. M. Yonge *Lady Hester* v. 115 Her black hair in the Grecian coil we used to wear. 1966 J. S. Cox *Illustr. Dict. Hairdressing* 66 *Grecian coil*, a kind of spiral roll by which the front hair on either side of a middle parting is brought forward onto the forehead and from the forehead rolled spirally towards the ears. 1846 J. G. Saxe *Progress* (1847) 18 'She stoops to conquer' in a 'Grecian curve'. 1931 *Queen* 25 Feb. 38/1 Little twists of hair, Grecian knots and thick, soft plaits that clip on the back of the head..can be obtained very cheaply. 1830 T. E. Hook *Maxwell* viii, A beautiful girl,..—Italian eyes—Grecian nose. 1873 L. Troubridge *Life amongst Troubridges* (1966) ii. 8 She has..a very good long, straight Grecian nose. c 1845 C. Brontë *Professor* (1857) II. xix. 58 Her plenteous brown hair arranged in smooth bands on her temples, and in a large Grecian plait behind. 1851 *Ladies' Compan. & Monthly Mag.* July 266/2 The 'Grecian' plait, and the 'basket' or 'chain plait', and the 'cable plait', are repeated again and again in various forms. 1968 J. Ironside *Fashion Alphabet* 188 *Plait..Grecian*, a more complicated version of the braid. 1926–7 *Army & Navy Stores Catal.* 654/1 Red, blue or brown leather Grecian Slippers, with heels. 1953 J. W. Johnson in J. H. Thornton *Textbk. Footwear Manuf.* ix. 520 The Grecian slipper usually has a folded or bagged edge and orthodox methods of closing are employed. 1970 *Harrods Christmas Catal.* 15/3 Grecian slipper with shiny patent stripes. 87/-..Man's Shop, ground floor. 1883 *Man. Seamanship for Boys* 124 A Grecian Splice [description follows]. *Ibid.* 125 There is also another way to make a Grecian splice, by making all the yarns into foxes, leaving no heart...This splice is also used for tailing a smaller to a larger size rope, when it has to travel through a block. 1944 C. W. Ashley *Bk. Knots* 434/2 (*heading*) The Grecian Shroud Splice.

greedy-guts. (Later examples.)

1950 A. Wilson *Such Darling Dodos* 210 'Greedy guts,' he said smiling, as she took a second helping of watercress. 1959 *Elizabethan* Apr. 10/1 'Here, steady on with the sugar, greedy guts, you've taken it all,' Friday shouted.

greegree. (Later examples.)

1931 R. Aldington *Colonel's Daughter* v. 287 He shared with her the common but essentially mystic passion for gadgets—the white man's mechanical grigris. 1935 G. Gorer *Africa Dances* I. iii. 47 The great majority of both Catholics and Protestants believe in the efficacy of amulets or grisgris. 1956 M. Stearns *Story of Jazz* (1957) v. 48 Even today, voodoo drugstores in New Orleans are doing a profitable business in *gris-gris* or magic charms.

Greek, *sb.* Add: **10. Greekess** (later example).

1902 J. London *Let.* 3 Mar. (1966) 277 Charmian sends her love to Greek and Greekess and all the Crowd.

Greek, *a.* Add: **4. Greek chorus,** used *transf.*, in comparisons, etc., to indicate the wise, sympathetic comments or open wailing of the chorus in Attic tragedies (see Chorus 1); **Greek god,** (*a*) used *transf.* and in comparisons to denote a paragon of male physical beauty; (*b*) a short hairstyle with curls close to and all over the head; **Greek lace** = *Greek point.*

1863 F. Nightingale in C. Woodham-Smith *F. Nightingale* (1951) xvii. 291 She [*sc.* Queen Victoria] always reminds me of the woman in the Greek chorus.. wailing out her inexpressible despair. 1893 E. F. Benson *Dodo* I. iii. 56 He had no desire to interrupt this rapid monologue of Dodo's. He was quite content to play the part of the Greek chorus. 1913 J. Vaizey *College Girl* xxiii. 319 Hannah, as Greek Chorus, interposed moral remarks. 1947 A. Huxley *Let.* 9 Mar. (1969) 568, I visualize him as a kind of philosophical recluse, who comments upon all that is happening from the vantage ground of eternity, and acts as a kind of Greek chorus. 1962 J. Braine *Life at Top* v. 78 'You show him, Killer. You show him.' It was like a Greek chorus. a 1910 W. F. Butler *Autobiogr.* (1911) iii. 42, I often look now as soldiers pass and marvel what has become of those old Greek gods. 1955 E. Coxhead *Figure in Mist* iv. 106 It was irritating that Greek gods should drop off the bus and suddenly reduce Matthew to insignificance. 1968 J. Ironside *Fashion Alphabet* 193 *Greek god*, introduced in 1967, a close haircut with flat curls all over the head as seen in statues of Apollo. 1968 *Guardian* 7 Oct. 7/1 They will give you Greek god treatment, a tight mass of tiny curls. 1969 G. Sims *Sand Dollar* v. 72 He emerged from the sea looking like a Greek god, his hair bleached nearly white. 1970 O. John *Diamond Dress* viii. 90 Don't stand there like some Greek god! You've got about three minutes in which to get dressed. 1865 F. B. Palliser *Hist. Lace* v. 74 The greater part of the conventionally termed Greek lace is really the Venetian reticella; the designs are of geometric fashion, and often of Oriental character. 1880 L. Higgin *Handbk. Embroidery* 63 Design for quilt...To introduce squares of Greek or guipure lace. 1960 H. Hayward *Antique Coll.* 134/1 'Greek lace', a name given to drawn and cut work embroidery, often combined with geometrical needlepoint or pillow lace.

green, *a.* and *sb.* Add: **A.** *adj.* **1. i.** Used of lights and signals on the road, railway, etc., to indicate that the traffic is free to proceed (see also quot. 1970). Hence fig. phr. (*to give*) *the green light*, (to give) permission to proceed on a course of action; also *green-light* as *v. trans.* Cf. *green *sb.* 7 e.

1883 *Encycl. Brit.* XX. 238 *Railway Signals*... At night the place of semaphores or disks is supplied by large and powerful lamps with reflectors... In general practice, two lights..are shown,—red and green. 1937 T. Rattigan *French without Tears* III. i. 126 We had a bottle of wine and got pretty gay, and all the time she was giving me the old green light. 1954 Wodehouse *Jeeves & Feudal Spirit* xxii. 216 Carry on, old sport. You have the green light. 1968 *Highway Code* 44 *Traffic light signals*... *Green* means you may *go on* if the way is clear...*Green arrow* means that you may go in the direction shown by the arrow. 1968 F. Mullally *Munich Involvement* ii. 15 'Anything else I can do for you?' Her smile green-lighted the innuendo. 1968 *Times* 19 Feb. 3/7 The Greek revolutionary rulers have warned King Constantine not to attempt to return to his throne until the Papadopoulos Government gives him the green light. 1968 *Guardian* 11 Apr. 1/2 A more selective system of Customs questioning of passengers..is provided for in the finance bill published yesterday... Those who have anything to declare will go through a 'red' channel...Those who have nothing to declare will pass through a 'green' channel. 1968 *Listener* 27 June 837/2 But Professor Barnard decided to take a chance and have a go—and as so often happens, his example triggered off attempts by a whole host of imitators who had been impatiently waiting for the green light. It now seems sadly obvious that the light was not green. 1970 *Daily Tel.* 16 Dec. 3/2 He calmly drove through the 'green' channel, indicating he had nothing to declare.

j. Used to denote the colour of absinthe, esp. in *green peril*; also *ellipt.*

1889 E. Dowson *Let.* 18 Feb. (1967) 36, I have not felt myself since my generous allowance of the potent green on Thursday & was right 'orf' the fascinating fluid yesterday. 1905 *Westm. Gaz.* 4 May 12/1 Some statistics ..of the growth of the absinthe habit in France seem to justify the alarmists who speak of the beverage as 'the green peril'. 1908 *Daily Chron.* 21 May 1/5 This taxing of the 'green peril' will no doubt be popular.

k. Phr. *to have green fingers* (or *a green thumb*), to be unusually successful in making plants grow; also *transf.*; hence *green-fingered* adj.

1934 R. Arkell (*title*) Green fingers. 1943 R. Carson *Bride saw Red* xvii. 203 Plants don't grow for me, but my wife's got a green thumb. 1943 S. Cloete *Congo Song* xvi. 150 Some men have green fingers. Plants like them. They can make things grow because they love them. 1946 *Nature* 28 Dec. 941/2 The arts of budding and grafting can only be fully acquired by observing the green-fingered dexterity of the experienced propagator. 1949 *Eng. Studies* I. 2 Yet we are such born meddlers, and so convinced..that we have 'green fingers' of the spirit to give help and guidance in these difficult years. 1962 R. Page *Educ. Gardener* 16 To have 'green fingers' or a 'green thumb' is an old expression which describes the art of communicating the subtle energies of love to prosper a living plant. 1962 *Listener* 19 July 89/1 Every kind of briar, of bush rose, of rare bulb, and flowering tree flourished under her green thumb. 1963 [see *agin *prep.*]. 1966 *Lancet* 31 Dec. 1461/2 Trees like this.. would soon be produced by hybridisation and plant-hormones under the green-fingered genius of him and his helpers. 1969 *Daily Tel.* 26 Apr. 6/6 'Success with money is often accidental,' she sighed. 'One needs "green fingers" to make it grow.'

2. *green field(s)*: used *attrib.* to denote building away from existing developments.

1962 *Economist* 24 Mar. 1150/1 Continental shipyards.. had the advantage of being rebuilt *de novo*, often on 'green field' sites, after wartime destruction. 1968 *Times* 9 Nov. 2/3 It would not be a 'green fields' new town but an addition to an existing development, or in an open space.

b. *green road*: a permanent farm-road giving access to fields, etc.; an untarred road (see also quot. 1938); *green way* (later non-dial. examples).

1900 *Eng. Dial. Dict.*, Green-road. 1938 *Oxoniensia* III. 9 These limits embrace all the country between..the Northamptonshire forest-land on the one hand, and, on the other, all the land immediately visible or accessible from the 'green road' leading from the eastern counties to the Wiltshire Downs. 1943 *N. & Q.* CLXXV. 203 Green-road. Farm-road to fields. Kent. 1968 *Guardian* 11 July 3/6 The estate will..be broken into four units, each separated from the others by wide spaces, which the planners call 'greenway lungs'. 1970 *Times* 26 Aug. 10/3 A plan for turning two railway lines..into greenways for walkers, horseriders, and cyclists..has been sent to Somerset and Devon county councils.

8. c. (Further examples.)

1876 *Coursing Calendar* 5 Mr Vyner's puppy appearing to run very green to commence with. 1965 *Observer* (Colour Suppl.) 30 May 34 *Running green*, a horse that gets upset by the crowd and the noise and other unfamiliar surroundings.

d. So phr. *to be not as green as one is cabbage-looking*: to be less of a fool than might be assumed.

1898, 1922 [see *cabbage *sb.*[1] 1 d]. 1931 E. Raymond *Mary Leith* IV. ii. 318, I ain't the sort to be taken in by his gaff. I may look a fool..but I'm not as green as I'm cabbage-looking. 1969 J. N. Smith *Is he Dead, Miss ffinch?* xviii. 117 Women don't have a sense of direction. ..But I'm not as green as I'm cabbage-looking. 1970 J. Fleming *Young Man, I think you're Dying* i. 16 Mind you, she's not as green as she's cabbage-looking, she never leaves you alone in the room.

11. a. *green-capped, -faced, -shaded, -shadowed, -turfed* adjs.

1922 Joyce *Ulysses* 499 A greencapped dark lantern. 1963 G. McInnes *Road to Gundagai* x. 154, I saw..some green-capped boys marching. 1916 H. G. Wells *Mr. Britling* I. v. 138 Green-faced and pitiful under an anaesthetic. 1909 *Westm. Gaz.* 15 May 6/2 A green-shaded city nestling oasis-like in its arena. 1854 'G. Greenwood' *Haps & Mishaps* 19 Every hill & green-shadowed vale..spoke to my heart. 1955 P. Larkin *Less Deceived* 36 Green-shadowed people sit, or walk in rings. 1851 H. Melville *Whale* II. xii. 92 His green-turfed, flowery Nile. 1933 W. de la Mare *Fleeting* 101 Meek harebell hung her head Over the green-turfed chalk.

12. Green Beret, the nickname for a member of the British, and later the American, Army Commandos; **green butter,** a savoury butter (see also quot. 1938); **green card,** an international insurance document required by motorists taking their cars abroad; **green cross,** designating a poison gas shell marked with a green cross, or its contents; **green curtain** *Theatr.* (see quot. 1961); **green fingers** *pl.*, **-fingered** *a.* (see *green *a.* 1 k); **green flash** (see quot. 1925); **green goods** *pl.*, (*b*) vegetables and fruit, greengroceries; **Green Jackets** *pl.*, a name applied to the King's Royal Rifle Corps and the Rifle Brigade, from the dark green colour of their uniforms; **Green Line,** a service of green express coaches in London and the Home Counties; also *attrib.*, as *Green Line coach*; **Green Linnets** *pl.* (see quots.); **Green Mountain State,** the state of Vermont, U.S.A.; **green paper** (see quot. 1969); **green ray** = **green flash*; **green salad,** a salad made from one or more ingredients, esp. lettuce, chicory, cucumber, watercress, etc., freq. served with French dressing; **Green Striper** (see quot. 1948); **green tea** (see Tea *sb.* 1 b); **green thumb** (see *green *a.* 1 k).

1949 H. St. G. Saunders (*title*) The Green Beret. The story of the Commandos. 1955 *Mountaineer* (Fort Carson, Colo.) 2 Dec., Twelve of the green beret Special Forces troops will jump in the next test scheduled today. 1962 *Army* June 34/2 Before each exercise starts, the Green Berets—Special Forces guerrilla leaders—have the advantage of being established in the area first. 1964 C. B. Colby *Special Forces* 18 In the top photo a 'Green Beret' takes to the air over the cornice of a hotel. 1968 *Listener* 22 Aug. 229/2 The Green Berets, as the Special Forces are nicknamed, are guerrilla teams who live in the bush and work with bigger detachments of South Vietnamese. 1970 A. Sinclair *Guevara* iii. 40 Che['s]..manual..even serves as a text-book for the Green Berets and other North American counterinsurgency special forces. 1970 *Sunday Times* 22 Nov. 23/6 Southampton [football team], as notorious as the Green Berets for their policy of search and destroy. 1889 A. B. Marshall *Cookery Bk.* ii. 38 Montpellier or green butter. 1938 *Thorpe's Dict. Appl. Chem.* (ed. 4) II. 187/2 'Green butters' (*i.e.* vegetable tallows which may be coloured artificially to resemble Borneo tallow). 1951 *Good Housek. Home Encycl.* 495/1 *Green butter*, a savoury spread used for biscuits, sandwiches, etc. 1965 *Sunday Times* (Colour Suppl.) 5 Sept. 96/2 To make Green Butter; cream butter with garlic.. chopped parsley..lemon juice, salt and pepper. 1959 *Motor Man.* (ed. 36) xi. 263 The motorist must get from his insurers a 'Green Card', which confirms that he has Third Party insurance cover. 1963 *Daily Tel.* 29 Nov. 1/4 Green cards were introduced 10 years ago. 1969 J. Leasor *Week of Love* v. 90 I've owned the car outside for years. I've all the papers...Green card. Log book. 1918 E. S. Farrow *Dict. Mil. Terms*, *Green cross shell*, very dangerous asphyxiating shell, first used by the Germans, filled with diphosgene or phosgene. 1928 *Daily Express* 22 May 1/2 An immense steel flask of phosgene, the notorious Green Cross poison gas employed by Germany with such deadly effect during the war. 1931 Brophy & Partridge *Songs & Slang 1914–18* (ed. 3) 314 *Green cross shell*, an enemy gas-shell of an emetic and lachrymatory nature. 1805 *Wynne Diaries* 15 June (1940) III. 172 The Ballets have in general been curtailed..but this evening the Green Curtain..dropped at twenty minutes after Eleven. 1826 *Life & Times Frederick Reynolds* I. i. 16, I was much surprised by seeing a person put his head through the hole in the green curtain. 1859 J. R. Planché *Love & Fortune* 31 Then amidst your applause may the green curtain fall. 1961 Bowman & Ball *Theatre Lang.* 162 *Green curtain*, a heavy outer curtain, traditional from the time of the Restoration, but now outmoded, serving variously as an act drop, fire curtain, etc. 1912 *Nature* 6 June 351/2 The well-known phenomenon of the green flash at sunset. 1925 R. Clements *Gipsy of Horn* 125 For the first time I saw the 'Green Flash', as it is called. Just as the sun is about to sink below the horizon a flash of vivid green seems to leap from it. It only lasts a second and is gone. 1931 *Discovery* Apr. 112/2 The Green Flash is an interesting meteorological phenomenon. 1950 *Caribbean Q.* II. II. 40 In the West Indies..we have the finest opportunities for observing the phenomenon of the Green Flash..the bright green-blue light flashed from the sun at the moment of sunrise and at sunset. 1963 Green flash [see **green ray* below]. 1888 *Boston Transcript* (Farmer), Get a good melon, and if you can't tell for yourself by that intuition which is the best guide in such matters, then trust to your green goods grocer's judgement. 1920 E. Bok *Autobiogr.* (1921) 99 A market dealer in green goods. 1824 in Sir. H. Smith *Autobiogr.* (1901) I. 3 'Well, I will make you a Rifleman, a green jacket,' says the General. 1927 *Observer* 1 May 19 The Duke [of Connaught] loves the Green Jackets best of all in spite of his other military associations. 1970 *Times* 24 Nov. 3/6 Men of the Green Jackets created a very favourable impression of their verbal dexterity. 1932 *Times* 6 Feb. 14/3 Green Line Coaches, Limited, announce that arrangements have been made for the acquisition of the Skylark Motor Coach Company..and the London to Ongar service of Associated Coaches... Licences held by the above companies to be regranted to Green Line Coaches. 1935 A. E. W. Mason *They wouldn't be Chessmen* v. 67 He..was almost run over by a Green Line charabanc. 1940 Auden *Another Time* 77 Went by Green Line bus. 1961 J. Stroud *Touch & Go* ii. 21 'How do we get there?' 'Train, I suppose. Or Green Line.' 1968 *Times* 22 Aug. 3 (*heading*) New towns are proposed within the Green Line range. 1870 Brewer *Dict. Phrase & Fable* 365/2 *Green Linnets*, the 39th Foot, so called from the colour of their facings. 1901 'Linesman' *Words by Eyewitness* (1902) 191 Dorsets and Middlesex (famous old corps, with famous old *sobriquets*, 'Green Linnets' and 'Die-Hards'). 1925 Fraser & Gibbons *Soldier & Sailor Words* s.v. *Nicknames, Green Linnets, The*: The Dorsetshire Regiment. Through the 1st Battalion, as the 39th Foot. From the green facings. 1838 *N.Y. Advert. & Express* 7 Feb. 3/4 A Mr. Fletcher of Vermont, the only Administration member from the Green Mountain States [*sic*]. 1948 *Vermont Q.* July 74 It is of much historic interest to the Green Mountain State. 1967 K. Robinson in *Hansard Commons* 6 Nov. 644, I wish to make a statement concerning the administrative structure of the medical and related services for which I am responsible. I shall..set out my views, probably in the form of a Green Paper, as a basis for public discussion and wide consultation. 1969 *Times* 1 Mar. 8/8 Green Papers.. originated with the D[epartment] [of] E[conomic] A[ffairs] in 1967. *Ibid.*, Mr. Michael Stewart..defined a Green Paper as 'a statement by the Government not of policy already determined but of propositions put before the whole nation for discussion'. 1970 *Daily Tel.* 10 Apr. 2/1 Doctors are divided over the second Green Paper on the future structure of the National Health Service. *Ibid.*, The Green Paper, put forward for discussion, proposes the scrapping of the present hospital boards and committees. 1906 W. Marriott *Hints to Meteorol. Observers* (ed. 6) 66/2 *Green ray*, a flash of greenish-blue light seen, when the sun's disc appears or disappears, in a sunrise or sunset on a clear horizon. 1963 *Meteorol. Gloss.* (*Met. Office*) (ed. 4) 120 On still rarer occasions a 'green flash' or 'green ray', also lasting a few seconds, shoots above the horizon from the upper limb. 1971 *Islander* (Victoria, B.C.) 7 Mar. 2/2 It was in March I saw the rarely seen green ray. Twice..when the sun had almost disappeared the very top of its disc seemed to shoot upward, sudden and brilliant, in an emerald flame. 1891 A. B. Marshall *Larger Cookery Bk.* x. 416 *Green Salad à la Bretonne*...Take two hearts of well-washed and dried crisp lettuce. 1939 *Vogue's Cookery Bk.* 69 Fennel is another herb that adds distinction to an ordinary green salad. 1969 P. Highsmith *Tremor of Forgery* xi. 108 They had scrambled eggs with fried salami and a green salad. 1948 Partridge *Dict. Forces' Slang 1939–45* 87 *Green Striper*, an officer in the Special Branch of the R.N.V.R. who wears an emerald-green stripe between the gold lace on his sleeves. 1951 A. H. Cherry *Yankee R.N.* 417 The third was Lieut. Archie Pitt, *Starling*'s green-striper Asdic specialist.

b. green dolphin, an aphid, *Acyrthosiphon pisum*, that attacks peas and other Leguminosæ; **green heron** (earlier examples); **green mamba,** a venomous African snake, *Dendraspis angusticeps* or *D. viridis*; **green monkey,** the West African race of the grass monkey, *Cercopithecus æthiops*; formerly used for several other monkeys with greenish fur; **green pigeon,** a pigeon of the genus *Treron*, which is widely distributed in Africa south of the Sahara and southern Asia; **green racer** *U.S.*, a popular name for several snakes belonging to the genera *Coluber* and *Masticophis.*

1850 *Rep. U.S. Comm. Patents, Agric. 1849* 339 These plants are often smothered with lice, or green-dolphin, as they are termed. 1876 G. B. Buckton *Monogr. Brit. Aphides* I. 134 *Siphonophora pisi*...Vulgariter. Green Dolphin. 1926 F. V. Theobald *Plant Lice Gt. Brit.* I. 132 This common aphid is usually spoken of as the Green Pea Louse or Green Dolphin. 1785 T. Pennant *Arctic Zool.* II. 447 Green Heron... *Ardea virescens.* Lin. Syst. 238... Inhabits from New York to South Carolina. 1855 *Knickerbocker* XLVI. 222 Night-herons, snowy-herons, green-herons, and little-herons construct their nests so closely together that four or five hundred of them may be counted upon twenty or thirty cedars. 1882 C. C. Hopley *Snakes* ix. 154 Another African snake, the 'Green Mamba', has such very bad manners that it not only hisses, but spits and darts at you. 1912 F. W. Fitzsimons *Snakes S. Afr.* (ed. 2) vi. 196 Green Mambas are always found in the forests, clumps of tangled creeper-covered bush, and wooded valleys. 1969 J. L. Cloudsley-Thompson *Zool. Tropical Afr.* iv. 74 Although longer, heavier and more terrestrial than the forest-dwelling green mamba..it [*sc.* the black mamba] is not confined to the ground. [1727 Green monkey (see Monkey *sb.* 1).] 1840 *Cuvier's Anim. Kingd.* 57 Several of these smaller kinds are very common in Guinea. Allied to them are the larger green Monkeys. 1866 *Proc. Zool. Soc.* 80 The animals were undoubtedly referable to the common Green Monkey (*Cercopithecus callitrichus*, Geoffr.) of Western Africa. 1965 D. Morris *Mammals* 150 The most important [grass monkeys] are: the West African form, known as the Green Monkey, with a very dark face and a stronger greenish tinge in its fur; the East African form [etc.]. 1967 *New Scientist* 9 Nov. 330/2 Green monkey fever has baffled microbiologists throughout the world. 1832 J. Gould *Century of Birds* Tab. LVIII, The present as well as the preceding species, together with several others, are known to the natives of India by the general name of the Green Pigeon. 1884 Layard & Sharpe *Birds S. Afr.* 557 This species [*sc.* *Treron calva*] is easily distinguished from the other Green Pigeons of South Africa by its grey tail. 1928 H. Whistler *Popular Handbk. Indian Birds* 297 The Green Pigeon is found almost throughout India, Burma and Ceylon, and farther east. 1952 Mackworth-Praed & Grant *Birds E. & N.E. Afr.* I. 486 The normal voice of all Green Pigeons is most curious, it is a sort of clucking whistling yap. 1967 D. Goodwin *Pigeons & Doves of World* 300 In both the wedge-tailed and pin-tailed green pigeons there are some morphological differences. 1870 *Amer. Naturalist* III. 124 Green Racer (*Boscanion vetustus*). I saw one dead specimen of this snake along Hell Gate River. 1957 A. H. & A. A. Wright *Handbk. Snakes U.S. & Canada* I. 464 Green racer...*Masticophis taeniatus schotti.*

c. green rose, *Rosa chinensis viridiflora.*

1911 E. Willmott *Genus Rosa* I. 80 The curious Green Rose belongs to this section [*sc.* *chinensis*]. It is in no way beautiful, but is remarkable from having all its floral organs transformed into leaves. 1932 'J. Hill' *Curious Gardener* viii. 126 The 'Green Rose', which appeared about seventy years ago in America..is sometimes listed in catalogues. 1962 G. S. Thomas *Shrub Roses of Today* x. 122 [*Rosa*] *chinensis viridiflora* (*R. monstrosa*). The 'Green Rose'.

d. green gold. Delete ?*Obs.* and add later examples.

1935 W. G. Hardy *Father Abraham* III. iii. 265 They opened the chests and yellow gold blinded his eyes and massy bars of silver and green gold. 1951 J. R. Partington *Gen. & Inorg. Chem.* (ed. 2) xiii. 353 *Electrum* is a native alloy of gold and 15–45 p.c. of silver; green gold contains 10 p.c. of silver: these alloys were called *asem* in ancient Egypt.

B. *sb.* **2. c.** (Later examples.)

1936 F. Clune *Roaming round Darling* xxii. 219 The governor, however, hadn't any green in his eye, so in despair Andy smuggled a letter to Dr John Dunmore Lang. 1966 O. Norton *School of Liars* vi. 106 'You don't have to believe all you hear.'..He leant forward and pulled down his lower eyelid. 'See any green, Mrs Sumner?'

7. b. = *green tea.*

1728 [see Tea *sb.* 1 a]. 1835 Dickens in *Bell's Life* 4 Oct. 1/1 Two ounces of seven and sixpenny green. 1838 —— *O. Twist* III. ii. in *Bentley's Misc.* IV. 115 Half a pound of seven and sixpenny green, so precious strong.. it'll go nigh to blow the lid of the teapot off. 1903 *Westm. Gaz.* 4 Feb. 4/2, 4,000,000 lb. of black tea had been taken off the market last year by being converted into greens.

c. *fig.* A greenhorn, simpleton. (Cf. A. 8 d.)

Cf. *Verdant Green*, the name of the hero in the title of the story (1853–6) of Oxford university life by 'Cuthbert Bede'.

1825 *Spirit of Pub. Jrnls. 1823* (ed. 2) 63 It appears that George Charteris..had been 'doing' the *green*, and taking in the 'deep ones', quite in the gull-catching style, for a considerable period. 1837 Dickens *O. Twist* xviii, 'Well, well,' said the Dodger...'That hasn't got anything to do with young Green here.' 1840 G. Thompson *Newgate Calendar* 280, I then with my comrade stole from a green twelve shirts..and some stockings. 1841 *Southern Lit. Messenger* VII. 54/2, I knifed a flat-boat Hoozier—took his lucre—Went up the country—rifled twenty *greens.*

d. Also *pl.* Money. *slang* (orig. *U.S.*). Cf. Greenback *sb.* 1 and *long green* (*long *a.*[1] 18).

1925 *Flynn's Mag.* in Partridge *Dict. Underworld* (1968) 307 *Greens*, paper currency; bank notes. 1942 Berrey & Van den Bark *Amer. Thes. Slang* §559/5 *Paper money*..the green, green boys,..greens, the long green. 1951 E. Paul *Springtime in Paris* vi. 125 'A banker!.. Nobody comes on this scene wearin' any green,' said another taller Negro. 1961 *New Statesman* 21 July 81/2 The hours proved to be from four to half eleven in the morning and the greens would amount to eight ten a week. 1962 L. Deighton *Ipcress File* vi. 44 At five shillings a dose that's a lot of green. 1968 *Scottish Daily Mail* 3 Jan. 6 What had been 'dough' in the 20's and became 'readies' and 'greens' in the 50's turned up again as 'bread'. 1971 'R. Crawford' *Badger's Daughter* I. ii. 24 When finally we did lay our mitts on a nice pile of green, Arthur simply knuckled under to luxury.

e. A green light, lamp, flare, etc., meant as a signal. Cf. *green *a.* 1 i.

1953 R. Chisholm *Cover of Darkness* ii. 30 At last I was given a 'green', but the dim pattern of aerodrome lights made little sense by this time. 1962 J. Braine *Life at Top* ii. 31 The car in front of me stalled and I missed the green. 1962 J. Glenn in *Into Orbit* 213 You have a green. You look good on attitude. 1963 *Amer. Speech* XXXVIII. 119 *In the green*, slang expression meaning that all instruments show safe readings. Many of the instruments have colored markings on them: red for danger, yellow for caution, and green for safe ranges. 1963 'W. Haggard' *High Wire* xi. 116 Overhead there was the unmistakable clatter of a helicopter, then another. Somebody fired a green.

f. Marijuana of poor quality. *slang* (orig. *U.S.*).

1957 J. Kerouac *On Road* (1958) 184 He got hold of some bad green, as it's called in the trade—green, uncured marijuana. 1969 R. R. Lingeman *Drugs from A to Z* 86 *Green*, a 'grade' of marijuana sometimes of relatively low potency because it is low in resin content. 1969 *Observer* 12 Jan. 30/3 Everybody was talking pot last week (or Mary Jane, tea, grass, weed, hay, boo, gage, green or technically *cannabis sativa*).

11. f. Sexual activity, esp. intercourse. *slang.*

1889 Barrère & Leland *Dict. Slang* I. 429 'To have one's *greens*', to have sexual intercourse. 1893 Farmer & Henley *Slang* III. 206/1 *To have, get,* or *give one's greens*, to enjoy, procure, or confer the sexual favour. Said indifferently of both sexes. 1963 C. Connolly *Bond strikes Camp* 3 Section A make a study of the kind of greens the big shots go in for. Sometimes we know more about what these people are like between the sheets than they do themselves. 1963 L. Meynell *Virgin Luck* vii. 164 Mr. Cahill..is an adult male with healthy

GREENBACK

instincts. He wants his greens regularly. **1967** G. Greene *May we borrow your Husband?* 27 Why not go after the girl?.. She's not getting what I believe is vulgarly called her greens.

14. b. Phr. *on the green*: on the stage (see *Greengage 2).

1940 *N. & Q.* 29 June 462/2 'On the Green' is perfectly good rhyming slang for 'On the Greengage' = on the stage. As such, it is familiar to every touring actor, stage-door keeper and stage-hand of over forty, and is in constant use to-day. **1957** *Times Lit. Suppl.* 6 Dec. 742/1 If a modern producer asks his stage-manager to summon down a man from the flies, we might well hear the cry: 'Bill, come down on the green a minute.'

17. (in reference to a golf-links) *green(s) committee, -fee, -keeping, -man, -putter, -record.*

1896 *Rules of St. Andrews* in J. Kerr *Golf-bk. E. Lothian* App. p. xxii, When the Green Committee consider it necessary, a telegraph board shall be used to give the numbers for starting. **1909** *Westm. Gaz.* 5 May 12/4 The green committee did not consider that golf was a game likely to be benefited by inclusion in any programme of the Olympic Sports. **1926** Wodehouse *Heart of Goof* vii. 230 The Greens Committee..have altered the Mossy Heath course. **1909** *Westm. Gaz.* 20 Oct. 12/2 The committee suggested the charge of a green fee of one shilling a round on each player. **1962** *Punch* 21 Nov. 746/2 The economy-size golfer only needs a nearby course and a green-fee. **1971** 'D. Halliday' *Dolly & Doctor Bird* ii. 22 Greens fees eight dollars, power cart ten dollars, balls eighteen dollars fifty the dozen. And a complete set of clubs and bag...Four hundred dollars? **1907** *Westm. Gaz.* 18 Oct. 3/1 All that has gone past him, like the scientific green-keeping. **1961** *Technology* Oct. 257 Courses are also held in horticulture and greenkeeping. **1905** *Westm. Gaz.* 10 Feb. 3/1 The green-men use various liquids to bring the worms to the top, where they may be swept away and destroyed. **1928** *Daily Express* 3 Jan. 9/2 After the snow had fallen greenmen tried to clear it away by flooding the course with hosepipes. **1881** Green putter [see Putter *sb.*[2] 2 a]. **1902** *Westm. Gaz.* 28 July 4/1 Had Herd not been badly bunkered at the fourteenth hole he would probably have beaten the green record. **1908** *Ibid.* 22 June 9/4 At the age of sixteen he.. had won a scratch medal and broken a green-record.

greenback, *sb.* Add: **1.** (Later example.)

1966 *New Yorker* 22 Oct. 164 We observe him on his way to Mexico with a suitcase full of green-backs.

3. (Later examples.)

1953 R. Fuller *2nd Curtain* iv. 61 The quart bottle of beer, the Penguin greenback, incongruous modernities. **1960** E. Morgan in M. Barry *Television Playwright* 357 Harry re-enters with the book, a green-back Penguin.

4. A deficiency disease of tomatoes, shown by the calyx end of the fruits failing to ripen.

1926 *Ann. Appl. Biol.* XIII. 338 Exposure of the fruits to excessive sunlight causes a type of blotchiness known as 'green back' if the nitrogen and potash supply is inadequate. **1943** T. Wallace *Diagnosis Mineral Deficiencies in Plants* iv. 32 Tomato plants have a high requirement for potassium. When the element is deficient..fruits often fail to ripen evenly, and may show sharply defined, green patches near the stalks (green-back). **1961** *Amat. Gardening* 21 Oct. Suppl. 25/1 Green-back, a common tomato disorder in which the fruits do not ripen completely but remain green near the stalk.

5. *Surfing slang.* = *Greenie.

1965 H. A. Klein *Surfing* iii. 58 Australian surfers use the picturesque word 'greenbacks' for a swell as it peaks higher and higher but has not yet begun to spill 'white water' down its front face. **1965** J. Pollard *Surf-rider* ii. 20 Just out a little further are the 'green-backs', the unbroken waves. **1970** [see *Greenie].

green baize. (Earlier examples.)

1801 C. Wilmot *Diary* 29 Nov. (1920) 4 Even after trunks, Pocket Books, Writing Cases, Green baize bags, &c., were quietly deliver'd in. **1812** *Dramatic Censor 1811* col. 362 'Tis bekase I'm a poet you see, That I kiver my head with green baize.

green-baized, *a.* (Earlier example.)

1818 Keats *Let.* 7 July (1958) I. 319 A nice carpeted Room with Sofa hair bottomed chairs and green-baized mehogany.

green belt. [*Belt *sb.*[1] 5 a.] An officially designated belt of open countryside in which all development is severely restricted, usu. enclosing a built-up area and designed to check its further growth.

1932 S. Freese *Ten-Year Plan* 8 This had once been a 'green belt of country' as spoken of so glibly by the town-planning 'experts' of to-day. **1935** *Planning* 26 Feb. 3 Mr. Herbert Morrison's offer of a sum of £2 millions to neighbouring local authorities in order to create a green belt round London represents a belated assumption by the L.C.C. of the wider responsibilities already recognised last century by the City Corporation. **1937** *Daily Herald* 15 Jan., (*headline*) Green belt maps televised. **1937** *Discovery* Nov. 328/2 To draw the attention of the cities and towns of England to the need for safeguarding the country in their immediate vicinity by means of the establishment of green belts. **1945** P. Abercrombie *Greater London Plan 1944* ii. 29 One of the first attempts to prevent the expansion of London was made by Queen Elizabeth..on 7th July, 1580, in a proclamation from Nonsuch Palace near Ewell in Surrey (now interestingly enough one of the properties secured as part of the Green Belt). **1958** *Listener* 19 June 1022/3 Parking facilities in the 'green belt' areas. **1961** E. A. Powdrill *Vocab. Land Planning* iii. 34 Green belts have at least three specific functions: to prevent the unrestricted sprawl of large built-up areas; to provide areas of unspoilt country-side wherein people from the towns can find recreation and enjoyment; and to provide for the continued use of farmland, woodland, heath and common land. **1972** *Daily Tel.* 1 Jan. 2/1 Mr Page, Minister for Local Government and Development, is working on a scheme for extending green-belt policy so that it becomes..a more general protection of the countryside.

Greeneian (grī·niăn), *a.* Also **Greene-ian, Greenian.** [f. the name of the English writer Graham *Greene* (b. 1904) + -ian.] Typical of, or resembling the style, matter, or quality of the works of Graham Greene.

1961 *Encounter* Apr. 70 The idiom is often satisfyingly Greeneian—there is an ample provision of those new proverbs of hell. **1963** *Punch* 26 June 938/1 'Under the Garden'..begins in the cold Greene-ian manner. **1968** *Listener* 10 Oct. 475/1 A palpable difference between Greene and Amis is that everything Greenian is available in the first of his 'Catholic' novels..while there is something approaching an emergent moral philosophy in Amis.

Greeneland (grī·nlænd). [f. the name *Greene* (see prec.) + Land *sb.*; with a pun on Greenland.] A term used to describe the world of depressed seediness reputedly typical of the setting and characters of the novels of Graham Greene.

1940 A. C. Marshall in *Horizon* May 369 'The seedy level!' That is the location of Greeneland. **1966** D. Lodge *Graham Greene* 3 Over Clapham Common solitary walkers move with bowed heads through the slanting rain...in West Africa the laterite roads turn a fragile pink at sundown, then are swallowed by darkness. These are some characteristic scenes from a country of the mind known internationally as Greeneland. **1970** G. Greene *Brighton Rock* p. x, Some critics have referred to a strange violent 'seedy' region of the mind..which they call Greeneland, and I have sometimes wondered whether they go round the world blinkered. 'This is Indo-China,' I want to explain, 'this is Mexico, this is Sierra Leone carefully and accurately described. I have been a newspaper correspondent as well as a novelist.'

Greener[2] (grī·nəɹ). Applied to rifles, cartridges, etc., made by, or according to the designs of, W. *Greener* (1806–1869), gunsmith and author, or his son, W. W. Greener, gunsmith and author.

1895 *Montgomery Ward Catal.* 450/1 Facile Princeps Greener.. fine Damascus barrels...No better shooting gun at any price. **1901** *Kynoch Jrnl.* Apr.–May 93/2 Diagrams obtained with a Greener ·310 rifle and orthoptic sights. **1902** *Ibid.* Apr.–May 84/1 Of these cartridges the Greener sharpshooter is probably the most popular. **1918** E. S. Farrow *Dict. Mil. Terms* 271 *Greener bullet*, one of the first bullets made to take the grooves by the expanding action of the powder. It had a conical pewter wedge which was driven into a cavity in the base of the bullet by the powder and forced the outer walls of the bullet into the grooves. **1921** *Outward Bound* June 45/1 Tucking away his valuable Greener under his arm. **1949** H. Wadman *Life Sentence* 21 One of the beaters gave me a gun—it was a Greener.

greenery-yallery (grī·nəri yæ·ləri), *a. colloq.* [f. Green *a.* + *yaller*, vulgar pronunc. of Yellow *a.*, with ending *-ery* repeated to produce a jingling compound.] Of, pertaining to, or affecting the colours green and yellow, in accordance with the style or fashion of the aesthetic movement of the late nineteenth century; hence, typical of this movement; affected.

1880 W. S. Gilbert *Patience* ii. 37 A greenery-yallery, Grosvenor Gallery, Foot-in-the-grave young man. **1886** 'S. Coolidge' *What Katy did Next* (1887) iii. 59 She had chosen a 'greenery, yallery' paper for her walls. **1896** *Amer. Bookman* III. 131/1 The greenery-yallery school would have found no disciple in him [*sc.* Dickens]. **1896** Mrs. H. Ward *Sir G. Tressady* 194 You needn't wear greenery-yallery gowns, you know. **1945** E. Waugh *Brideshead Revisited* ii. 49 Nothing greenery-yallery about her. So gay, so correct, so unaffected. **1956** C. Willock *Death at Flight* ii. 18 Two articles, including the one about the boy artist, dealt with the more greenery-yallery end of the Arts.

greenfeed (grī·nfīd). Chiefly *Austral.* and *N.Z.* [f. Green *a.* + Feed *sb.*] Forage grown to be fed fresh to livestock. Also *attrib.*

1898 *Bulletin* (Sydney) 30 Apr. 31/4 A large hairy dairyman began to tell how many pounds of butter his little 'yaller' heifer..yielded in a week.. For ten minutes nothing could be heard but 'pounds', 'quarts',..'separators', 'green feed', and 'Ayrshire'. **1916** *N.Z. Jrnl. Agric.* 20 Sept. 190 The wise dairyman will provide sufficient soilage (green-feed-producing crops) and silage to make good all possible shortage of pastures in summer or winter. **1928** *Bulletin* (Sydney) 12 Jan. 31/2 Although green feed is essential for poultry, it should be given in moderation and regularly. **1947** E. A. McCourt *Music at Close* 51 By the end of May the spring rush was over and all the crops in except late oats and greenfeed. **1950** *N.Z. Jrnl. Agric.* July 37/2 Maize for greenfeed is sown at the rate of 2 bushels per acre. **1960** *Farmer & Stockbreeder* 23 Feb. 61/3 They claim that it gives a greater tonnage of green-feed per acre than any other variety. **1967** *Heretaunga Plains* (Ministry of Works, N.Z.) 45/2 Many farmers grow lucerne, chou moellier, and greenfeed ryecorn.

greengage. (Earlier example.)

1724 R. Bradley *New Improvements of Planting & Gardening* (ed. 4) 222 Queen Mother; Green Gage, Violet.

2. *Rhyming slang.* (*a*) the stage (cf. *Green *sb.* 14 b); (*b*) *pl.* wages.

1931 *Even. Standard* 19 Aug. 10/1 We speak of..the stage as the 'green-gage'. **1931** 'G. Orwell' *Coll. Essays* (1968) I. 71 Greengages, meaning wages. **1932** P. P. *Rhyming Slang* ii. 20 *Greengages*,..Wages. **1964** *Guardian* 2 Mar. 7/6 The money? Greengages we call it, greengages—wages. You'll be surprised. In a lot of places it's a fiver a night.

greenhead[2]. Add: **4.** One of several biting flies of the family Tabanidæ, esp. *Tabanus nigrovittatus.*

1837 A. Wetmore *Gaz. Missouri* 65 In the early settlement of the country, the value of the prairies was underrated by a knowledge of the mischievous power vested in the greenheads, or prairie fly. **1838** E. Flagg *Far West* II. 107 (Th.), [My horse was] severely troubled by that terrible insect, so notorious all over the West, the large green-bottle prairie fly, called the 'greenhead'. **1888** J. Kirkland *McVeys* 18 What can be the matter with that horse?..Is he crazy? Pretty near, I guess, with a greenhead on him somewhere where he can't get at it. **1951** E. Paul *Springtime in Paris* v. 113 He twitched like a steer which green-heads were biting.

5. An Australian ant, *Ectatomma metallicum.*

1907 W. W. Froggatt *Austral. Insects* 93 The 'Greenhead', *Ectatomma metallicum*, is a common ant of medium size that lives in small communities under stones or logs. **1926** *Glasgow Herald* 19 Dec. 6 The greenhead is the most vicious attacker of man. **1970** E. F. Riek in *Insects of Australia* (C.S.I.R.O.) xxxvii. 910/2 One specimen is recorded from the nest of *Ectatomma metallicum*, the greenhead ant.

green hide, greenhide: see Green *a.* 9 c.

greenhouse. Add: **1. b.** **greenhouse effect** *Meteorol.*, the phenomenon whereby the surface and the lower atmosphere of a planet are maintained at a relatively high temperature owing to the greater transparency of the atmosphere to visible radiation from the sun than to infra-red radiation from the planet.

1937 G. T. Trewartha *Introd. Weather & Climate* i. 25 The so-called greenhouse effect of the atmosphere. **1962** F. I. Ordway et al. *Basic Astronautics* iii. 78 On Mars the greenhouse effect is not expected to be very significant. **1968** *Observer* 17 Nov. 9/5 Carbon dioxide..is responsible for a 'greenhouse' effect which allows in heat from the sun but prevents it escaping back into space.

3. *Aeronaut. slang.* (See quots.)

1941 *Life* 24 Mar. 85/1 In the slang of the Royal Air Force man, the cockpit of his plane is the 'pulpit' or 'office', the glass covering over it the 'greenhouse'. **1942** *Gen* 1 Sept. 14/1 A fighter pilot..pulls the 'greenhouse'—cockpit cover—over him. **1944** *Word Study* Apr. 4/2 *Greenhouse.* This is what pilots call the glass cockpit covering over observation and similar planes. **1947** Auden *Age of Anxiety* (1948) i. 18 'Why have They killed me?' wondered Bert, our Greenhouse gunner.

greenie (grī·ni). *Surfing slang.* [f. Green *a.* + -ie.] A large wave before it breaks.

1962 *Austral. Women's Weekly* (Suppl.) 24 Oct. 3/2 *Greenie*, a big wave before it breaks into white foam. **1970** *Studies in English* (Univ. of Cape Town) I. 27 A wave can be a mere *green back*, or *greenie*, but more often it is a *roller*, *grinder*, *pounder* or *dumper*.

greening, *vbl. sb.* Add: **2. c.** In sense of Green *v.*[1] 3.

1879 *Boy's Own Paper* 18 Jan. 3/2 Leave to others the shamming, The 'greening' and 'cramming', In fun and in earnest, be true, boys! **1907** *Daily Chron.* 29 Jan. 3/3 The method of 'greening' is the schoolboy trick of making a statement which is a fabrication and getting another boy to believe it.

greenness. Add: **2. f.** Of horses: want of training. (See Green *a.* 8 c.)

1900 *Daily Mail* 30 Apr. 4/3 They were due to the greenness of many of the horses, to the fact that they were not acclimatised. **1927** *Ibid.* 30 June 13 The failure of Mrs. Whitburn's colt at Ascot may have been due to greenness.

greensome (grī·nsəm), *a.* [See -some[1].] = Greeny *a.* 2.

1901 *Westm. Gaz.* 19 Oct. 2/1 The goat..browsed.. quite close to a suit, a shirt, and other man's belongings that were strewn on the bank of a greensome bywater. **1908** *Ibid.* 15 June 2/3 Good-day, good-day! bright hazel there, What makes you so greensome?

Greenwell (grī·nwĕl). [f. the name of William *Greenwell* (1820–1918), archæologist and angler.] In full *Greenwell's glory*: A trout fly designed by the Revd. W. Greenwell. Also, a salmon fly of his invention.

1879 W. Henderson *Life Angler* xxi. 237, 104 fish [trout], chiefly captured with the 'Greenwell's Glory' fly. *Ibid.* xxii. 251 'The Greenwell' fly did most execution [of salmon]. **1899** E. Grey *Fly Fishing* viii. 216 In wet fly fishing for trout I am content with March-browns, Greenwell's glories, [etc.]. **1920** H. Maxwell in F. Francis *Bk. Angling* 177 The Greenwell salmon fly, a favourite on the Tweed. **1928** W. F. R. Reynolds *With Fly Only* 65 Perhaps an iron blue or two, and a few Greenwells. **1961** A. C. Williams *Dict. Trout Flies* (ed. 3) ii. 197 There can be no fly the name of which is better known to the general public than the great Greenwell's Glory which is as it should be, for a more generally useful wet fly has never been made.

Greenwich Village (gre·nidʒ vi·lědʒ). The name of a quarter of New York, used *attrib.* to denote the Bohemian outlook and way of life typical of many of its inhabitants. So **Greenwich Villager.**

1924 H. Crane *Let.* 12 Jan. (1965) ii. 170, I wish you could meet some of my friends, who are not the kind of 'Greenwich Villagers' that you may have been thinking they were. **1931** *Ibid.* 22 June iv. 376 Peggy (Baird) will be here [*sc.* in Mexico] in a few days—I'd rather.. she didn't step into a truly Greenich Village scene. **1962** *Listener* 4 Oct. 529/2 It was opposed to..socialists, capitalists, Greenwich Villagers, pedants, and Prohibition. **1968** *Ibid.* 3 Oct. 434/3 Muggeridge impressed me as very much a Bloomsbury type, or what in New York we would perhaps call a Greenwich Village type.

gree·n-winged, *a.* [Green *a.* 11 a.] Having green wings: in spec. names or descriptions of animals.

1637 T. Morton *New Eng. Canaan* 68 Teales, there are of two sorts, greene winged, and blew winged. **1849** C. Lanman *Lett. Alleghany Mts.* ix. 73 The Cherokees relate that there once existed among these mountains a very large bird, which resembled in appearance the green winged hornet. **1861, 1895** [see Green *a.* 11 a].

greeting, *vbl. sb.*[1] Add: **b. greeting(s)-card,** a card sent to relatives and friends at Christmas (or another festival); **greeting stamp** (see quot.); **greeting(s) telegram,** a coloured and illustrated telegram form conveying congratulations for weddings, birthdays, etc.

1898 *Weekly Budget* (Christmas Suppl.) 24 Dec. 14/5 (*heading*) Yule-tide parcel..containing:..greeting cards, [etc.]. **1907** *Westm. Gaz.* 10 Dec. 12/1 The sending of the Royal greeting-cards is also a very big business. **1936** *Discovery* Dec. 366/2 Attractive greeting cards for Christmas use. **1951** *Catal. Exhibits, S. Bank Exhib. Festival of Britain* 49/2 Greetings Cards. **1967** F. Mullally *Prizewinner* iv. 57 She had been behind the counter, serving a woman with greetings cards. **1936** *Discovery* Dec. 378/2 Christmas seals, which are little greeting stamps used for sticking on the flaps of Christmas letters and on parcels. **1937** *Rep. Proc. 14th Conf. ASLIB* 76 The design of a new 'greetings' telegram. **1952** M. Steen *Phoenix Rising* ii. 51 The floor was littered with the gilt envelopes of 'greeting' telegrams. **1967** V. Canning *Python Project* vi. 110 The British Post Office..would deliver it as a greetings telegram with a border of fluffy rabbits, song birds and nosegays.

greeze, greaze (grīz). *Westminster School slang.* [Orig. unknown.] (See quots.)

1881 W. A. Peck in C. E. Pascoe *Everyday Life in our Public Schools* 115 Few..will forget how they pressed through the surrounding 'greeze' (*Anglice* crowd), fought their way up to the notice-board. **1884** F. H. Forshall *Westm. School* i. 14 In my first week I learnt the meaning of 'the greaze'. **1927** *Weekly Times* 10 Mar. 267/2 The historical scramble for the pancake, or the 'Pancake Greeze', as it is called by the boys, took place. **1934** *Times Educ. Suppl.* 17 Feb. 1/3 Two pancakes had to be tossed at Westminster School on Tuesday for the annual 'greeze'.

gregarine, *a.* and *sb.* **A.** *adj.* (Examples.)

1903 E. A. Minchin in E. R. Lankester *Treat. Zool.* i. ii. 152 The first notice of a gregarine parasite. *Ibid.* 161 Gregarine spores. **1968** R. D. Manwell *Introd. Protozool.* (ed. 2) xxiii. 463 Some features of gregarine life cycles have already been briefly considered.

gregarization (gregărəizē·ʃən). [f. mod.L. *gregaria*, used to describe a locust during the gregarious phase of its life + -ization.] The swarming of locusts.

1939 J. S. Kennedy in *Trans. R. Ent. Soc.* LXXXIX. 531 The 3rd stage in the outbreak process is termed 'Gregarisation', that is, the development of typically gregarious behaviour as a result of intensive aggregation. **1947** *New Biol.* III. 22 The whole process of gregarisation is a matter of interplay between the insect and the environment. **1970** *Sci. Jrnl.* Jan. 65/1 Phase change plays a central part in the population changes of locusts, but exactly what conditions produce gregarization and swarms is not known.

grège (grēʒ, grɛ̄ʒ), *a.* and *sb.* Also **greige.** [F. *grège* raw (silk).] (Of) a colour between beige and grey.

1926 *Dry Goods Economist* 18 Sept. 13 Vera Dexter designed this sports dress of Belding's Crêpe in grège and beige. **1927** *Weekly Dispatch* 6 Nov. 10 The newest colours [for stockings]..are a deep peach-beige, light tan, and a soft grège shade something between a fawn and a grey. **1928** *Daily Express* 3 Apr. 5/4 Different shades of beige and greige are most in demand. **1931** *Times* 21 May 19/4 A gown of greige satin. **1949** *Brit. Colour Council Dict. Colours* III. 9/2 The colour names of Ecru, Beige and Grège..mean exactly the same thing—the colour of the condition of cloth in its raw, unbleached state. **1960** *Harper's Bazaar* Oct. 22 Wool-belted in Bronze, Ebony, Cypress, Graphite, Greige, Roan and African Violet.

greisen. Add: Hence **grei·sening** *vbl. sb.*, **greiseniza·tion,** the pneumatolytic conversion of granite or similar rocks into greisen.

1907 Reid & Flett *Geol. Land's End District* iv. 58 There is not much evidence to show whether the tourmalinisation preceded, accompanied or followed the greisening. **1920** A. Holmes *Nomencl. Petrol.* 114 The process whereby igneous emanations transform granite into greisen is called greisening or greisenisation. **1934** *Q. Jrnl. Geol. Soc.* XC. 348 Greisenization is always regarded as a typical example of pneumatolysis. **1962** W. T. Huang *Petrol.* iv. 97 Three main types of alteration primarily due to the action of volatiles or pneumatolysis are connected with the intrusion of granitic plutons. They are tourmalinization, greisening, and kaolinization. **1970** *Mineral. Abstr.* XXI. 39/2 The author discusses the chemical changes involved in greisenization.

Grelling (gre·liŋ). The name of Kurt *Grelling*, an early 20th-cent. German logician, used attrib., absol., in the possessive, and coupled with that of L. *Nelson*, to designate the paradox of heterologicality, which they discovered and published in 1907; as *Grelling antinomy*; *the Grelling*; *Grelling's paradox*, etc.; *Grelling-Nelson antinomy*, etc. (The paradox is sometimes referred to as *Weyl's contradiction*, by a mistaken attribution.)

1937 A. Smeaton tr. *Carnap's Logical Syntax of Lang.* § 60a. 211 (*heading*) Grelling's antinomy. **1955** A. N. Prior *Formal Logic* iii. iii. 287 A particularly clear example is..Grelling's paradox about 'heterological'. **1960** P. Suppes *Axiomatic Set Theory* i. 10 The third.. semantical paradox..is the Grelling-Nelson paradox..of heterologicality. **1962** W. & M. Kneale *Devel. Logic* xi. 656 Grelling's paradox has to do with a newly invented word. **1963** G. T. Kneebone *Math. Logic* iv. 128 The Grelling-Nelson antinomy (1908). **1966** L. J. Cohen *Diversity of Meaning* (ed. 2) § 26. 214 Similar antinomies, like the Grelling, may arise. **1966** W. V. Quine *Ways of Paradox* i. 6 Thus viewed, Grelling's paradox seems unequivocally falsidical.

gremlin (gre·mlin). orig. *R.A.F. slang.* [Orig. unknown, but probably formed by analogy with Goblin.] **1.** A mischievous sprite imagined as the cause of mishaps to aircraft; later, an embodiment of mischance in other activities.

1941 C. Graves *Thin Blue Line* ix. 123 As he flew round, he wished that his instructor had never told him about the Little People—a mythological bunch of good and bad fairies originally invented by the Royal Naval Air Service in the Great War... Those awful little people, the Gremlins, who run up and down the wing with scissors going 'snip, snap, snip' made him sweat. **1942** *Observer* 8 Nov. 6/5 Behaviour of..machines couldn't always be explained by..laws of aerodynamics. And so, lacking a Devil, the young fliers..invented a whole hierarchy of devils. They called them Gremlins, 'on account of they were the goblins which came out of Fremlin beer bottles'. They were the genii loci of the R.A.F. messes in India and the Middle East, where Fremlin's beer bottles were plentiful. **1943** C. Graves *Seven Pilots* 89 'Any Gremlin trouble last night?' Wing Commander Prall asked Paddy... 'None, thank goodness,' Paddy replied. **1944** *Times* 12 Aug. 2/1 The King said that on his way back from Italy they thought they heard a gremlin in the royal aeroplane. **1944** *Amer. Speech* XIX. 280 Gremlins are mythical creatures who are supposed to cause trouble such as engine failure in aeroplanes, a curious piece of whimsy-whamsy in an activity so severely practical as flying. Now the gremlin seems to be extending its sphere of operations, so that the term can be applied to almost anything that inexplicably goes wrong in human affairs. **1955** M. Gilbert *Sky High* xi. 163 'I hope you had a comfortable night,' she said. 'Not a single gremlin,' said Tom. **1958** *Times* 20 Feb. 5/2 A gremlin in the organ that had decided to add a noisy C sharp to the beautiful soft D major chords. **1959** *New Statesman* 2 May 611/3 Unfortunately the misprint gremlin has raised its ugly head in my letter to you on 18 April. **1969** *Daily Tel.* 15 Aug. 4/3 Gremlins were at work on this column yesterday and a 7 p.c. stake of the Furness Withy equity appeared in print as 27 p.c.

2. *Surfing.* **a.** A young surfer. **b.** A trouble-maker who frequents the beaches but does not surf. *slang.*

1961 *Life* 1 Sept. 51 If [a beginner's] surfing career goes well, he will be a 'gremlin' in a few years. **1965** J. M. Kelly *Surf & Sea* 286 Gremmie or gremlin, a young inexperienced surf enthusiast to whom nothing counts in life but surf-board and water. Also, an uncomplimentary term applied to those who hang around the beaches for reasons other than surfing. **1967** *Internat. Surfing* III. iii. 23 There is really a lot of talent running around these days in the form of young gremlins. **1967** J. Severson *Great Surfing* (Gloss.), *Gremlin*, a trouble-maker; one whose misdeeds cause surfing beaches to be lost and gives surfing a bad name.

gremmie (gre·mi). *Surfing slang.* Also **gremmy.** [Shortened form of *Gremlin.] = *Gremlin 2. Also *attrib.*

1962 T. Masters *Surfing made Easy* 64 *Gremmies*, young poor mannered surfers. **1965** *N.Z. Listener* 17 Dec. 4/5 Surfers..are..antagonistic towards learners ('gremmies'). **1967** *Internat. Surfing* III. iii. 13 About girl surfers, they are great to surf with until you are driving left and some gremmy girl takes off in front of you and goes straight off, forcing you to straighten out. **1968** *Surfer Mag.* Jan. 35/1 He worked all morning with several beach gremmies piling 12-foot sections of plywood and rocks into a small reef on the wet sand.

‖ **grenache** (grənaʃ). [Fr., the name of a sweet wine grape.] A sweet wine, usually drunk with dessert, produced in the Languedoc-Roussillon region of France.

1851 C. Redding *Hist. Mod. Wines* vi. 151 Good Grenache wine is made in the communes of Banyuls sur Mer, Collioure, Port Vendres, and some in the canton of Rivesaltes. **1946** G. Millar *Horned Pigeon* xxii. 372 There was good *grenache* for an *apéritif*. **1958** A. L. Simon *Dict. Wines* 86 *Grenache*, a sweetish red or tawny dessert wine from the Pyrénées (Roussillon) made chiefly in the *Banyuls* district from Grenache grapes. **1963** L. Durrell in *Holiday* Jan. 74/3 A *Blanc de Blanc* to accompany the *pâté* and the fish, and then a *grenache*.

grenade, *sb.*[1] Add: **3.** *grenade-launcher, -thrower.*

1959 C. Ogburn *Marauders* (1960) iv. 125 The Japanese ..redoubled the fire of their grenade launchers. **1969** I. Kemp *Brit. G.I. in Vietnam* iv. 83 There was a good deal of shouting from the garrison, several of whom started to fire grenade launchers into the night. **1920** *Glasgow Herald* 14 Apr. 9 For one moment Commander-in-Chief and grenade-thrower stood hand in hand looking long and confidently into each other's eyes.

grenade, *v.* Delete *nonce-wd.* and add later examples.

1947 Dylan Thomas *Let.* 29 May (1966) 309 Our little spankers make so much noise I cannot work anywhere near them. God grenade them. **1959** C. Ogburn *Marauders* (1960) iv. 120 The sound of anything stirring was to be instantly grenaded.

Grenadian (grĕnē·diăn), *a.* and *sb.* Also **Grenadan.** [f. *Grenada*, an island in the W. Indies + -ian.] **A.** *adj.* Of or pertaining to Grenada. **B.** *sb.* An inhabitant of Grenada.

1923 A. Aspinall *Pocket Guide to West Indies* (ed. 4) vi. 142 Grenadians..have been acquiring land. **1950** *Caribbean Q.* II. ii. 21 He is the product of urbanization, and, be he Grenadian, or Hindu or Trinidad creole, he has rapidly taken on characteristics of his own. **1962** *Listener* 5 July 9/2 The Grenadans have also a French tradition of desperate political remedies. **1967** V. S. Naipaul *Mimic Men* i. v. 82 We had a Grenadian woman cleaner. **1969** G. Sims *Sand Dollar* iv. 55 Pleasure before business. That's a good old Grenadian motto. **1969** 'L. Drummond' *Man with Tiny Head* x. 124 Some Grenadans tried talking to her in French, but she could not understand their debased slaves' argot, nor they her Parisian.

grenadier[2]. Add: **1. c.** *transf.* A person with the build of a grenadier.

1805 E. Cavanagh *Let.* 20 Aug. in Londonderry & Hyde *Russ. Jrnls. M. & C. Wilmot* (1934) ii. 182 In walk'd a Grenadier of a Man with a silver Tray. **1879** R. L. Stevenson *Trav. Donkey* 105, I found two other guests. One was a country parish priest... He was a grenadier in person with the hale colour and circular wrinkles of a peasant. **1929** W. Faulkner *Sartoris* (1954) 30 That straight, grenadier's back of hers. **1949** M. Laski *Little Boy Lost* iii. 57 No child of mine could ever have reminded this grenadier of her Isidor.

grenadin (gre·nădin). [Fr.: see Grenadine[1].] A French variety of carnation (see quot. 1910).

1904 R. P. Brotherston *Bk. Carnation* 18 The French Grenadin, truly a biennial, both single and double, with small flowers of a sweet scent, and extraordinarily floriferous. **1910** T. W. Sanders *Carnations* 45 Grenadin Carnations. A Continental strain of dwarf habit, growing about 1 ft. high, and bearing scarlet or white flowers. **1928** L. J. Cook *Carnation Culture* (ed. 5) viii. 69 What is known as the Grenadin Carnation is a race possessing a dwarf, stiff habit of growth.

grenadine[1]. (Earlier examples.)

1852 E. Twisleton *Let.* 15 July (1928) ii. 30, I..went in my purple Grenadine..and my lace bonnet. **1861** Mrs. Gaskell *Let.* 23 May (1966) 652 Silk barège scarf shawls, 35s—grenadine *shawls* ditto.

grenadine[3] (grenădi·n). [ad. F. (*sirop de*) *grenadine*, f. *grenade* Grenade *sb.*[1]] A syrup made from pomegranates (or other fruit).

1896 F. B. & W. H. Workman *Algerian Mem.* 43 We found syrup of grenadine, or pomegranate with water, a delicious drink. **1905** W. J. Locke *Morals Marcus Ordeyne* xii. 148, I..restore her to beatitude with grenadine syrup and soda-water. **1906** *Daily Chron.* 21 July 8/5 One of them, prepared from ripe gooseberries, is particularly delicious. It is called grenadine. **1913** C. Mackenzie *Sinister St.* I. i. ix. 130 Michael thought grenadine sucrée was just as nice as it looked. **1958** R. Godden *Greengage Summer* ix. 106 Mademoiselle Zizi had joined us for drinks; we had the pink grenadine sirop.

Grenfell (gre·nfĕl). [Name of Sir Wilfred Thomason *Grenfell* (1865–1940), English medical missionary who in 1893 founded the Labrador Medical Mission.] Used *attrib.* in *Grenfell cloth*, the proprietary name for a tough, closely woven, windproof cotton fabric. Also *Grenfell clothing.*

1926 *Trade Marks Jrnl.* 3 Nov. 2471/2 *Grenfell Cloth*.. Cotton Piece Goods made in England. **1933** W. Grenfell *40 Yrs. Labrador* x. 126 The first thing in clothing is the material—impervious to wind and water, light in weight, inexpensive, tough, smooth on the outside so as to shed snow... Just such a material can be ordered through our offices..and the English firm which invented it calls it 'Grenfell cloth'. **1933** G. D. Abraham *Mod. Mountaineering* x. 186 Closely woven, wind-proof cloth is best, such as 'Grenfell' cloth, which is made in Manchester. **1947** G. G. Denny *Fabrics* (ed. 6) 34 *Grenfell cloth*, closely woven, water-repellent, windproof fabric made of

Egyptian cotton... Uses: sports clothes, shower curtains, upholstery, drapery. **1955** 'G. Carr' *Corpse at Camp Two* i. 20 His Grenfell clothing and so forth would be turned over to you.

Grenzbegriff (gre·ntsbĕgrif). *Philos.* [G., f. *grenze* limit, boundary + *begriff* concept.] In Kantian philosophy, a concept which shows the limitation of sense-experience; a limiting concept (also *limit-concept, limitative conception*); also, loosely, a conception of an unattained ideal.

[**1787** Kant *Kritik der reinen Vernunft* (ed. 2) 310 Der Begriff eines Noumenon ist also bloss ein Grenzbegriff.] **1893** P. Carus *Primer of Philos.* 98 The straight line..must be taken as a *Grenzbegriff*, i.e., a conception which denotes the utmost limit to be reached by a certain operation. **1896** W. James *Will to Believe* (1897) 16 Objective evidence is never triumphantly there; it is a mere aspiration or *Grenzbegriff*, marking the infinitely remote ideal of our thinking life. **1908** —— *Meaning of Truth* (1909) xii. 239 The notion of an *absolute* reality inevitably arises as a *grenzbegriff*. **1941** *Mind* L. 309 Kant's 'thing in itself', understood either as a transcendent real or as a Grenzbegriff, could be only a prolific source of error.

Grepo (gre·po). [G. *gre(nz)po(lizei* frontier police.] An East Berlin border guard.

1964 L. Deighton *Funeral in Berlin* v. 31 The Grepo nodded. **1966** R. Thomas *Spy in Vodka* (1967) xiii. 136 'We run to an apartment building directly in front of the wall.' 'What are the Vopos and Grepos doing all this time?' **1967** J. Gardner *Madrigal* iv. 74 The Vopos and Grepos..were all young men,..each one washed from childhood into the Communist pattern.

‖ **grès** (grę). [Fr.] Stoneware. *grès de Flandres*, Cologne ware.

1872 Lady C. Schreiber *Jrnl.* (1911) I. 153 Perez.. showed us a grès de Flandres jug. **1874** *Ibid.* 244 The collection..chiefly consists of Grès and Delft. **1882** *Hamilton Sale Catal.* No. 950 A Pilgrim's Bottle, of old grès-de-Flandres stone ware. **1902** *Encycl. Brit.* XXXI. 875/1 The influence of the Japanese potters on the production of *grès* in France. **1962** D. M. Billington *Technique Pottery* 206 *Grès de Flandres*, the salt-glazed stoneware of the Low Countries.

Gresham's law: see Law *sb.*[1] 17 c (*d*).

Gretchen (gre·tʃĕn). [The name of the ignorant girl seduced by Faust in Goethe's play.] A girl resembling Gretchen; a German girl or woman. Also used *attrib.* to denote typically German hairstyles, etc., as *Gretchen braid, plait*, etc.

1890 *Amer. Mail Order Fashions* (1961) 11 Linen Lawn Gretchen Apron. **1905** *Daily Chron.* 16 Jan. 8/5 The English hairdressers are rejoicing..in the new Gretchen plait. **1913** D. H. Lawrence *Sons & Lovers* xi. 284 'Then why shouldn't we belong to each other altogether?' ..'I know it's a lot to ask,' he said; 'but there's not much risk for you really—not in the Gretchen way.' **1917** —— in *Seven Arts* July 283 No, but really, Marta, you're not going to wait any more—really! It's stupid for you to play Gretchen—your eyes are much too green. **1940** 'Gun Buster' *Return via Dunkirk* ii. x. 162 With that flaxen hair and those blue eyes they reminded one of a couple of typical Saxon Gretchens. **1952** Koestler *Arrow in Blue* x. 85 Seducing rustic Gretchens who were either the innkeeper's or the miller's daughter. **1966** J. S. Cox *Illustr. Dict. Hairdressing* 67/2 *Gretchen style*, a hair style usually worn by young girls, in which the hair, 6″ to 10″ long, is worn in tightly plaited stiff plaits hanging over or just behind the ears. **1970** *Sunday Times* 8 Feb. 61/4 Gretchen braids wound round small chignoned heads.

Gretna Green (gre·tnă grī·n). The name of a village in Dumfriesshire just across the Border, used *attrib.* in reference to the custom by which runaway couples from England have been married there according to Scots law, without the parental consent required in England for those who have not attained their majority. Also ellipt. as *Gretna.*

Such marriages first became common after Lord Hardwicke's Act of 1753, prior to which it was possible to contract a valid marriage in England clandestinely. Since then lack of parental consent, while not invalidating an English marriage, has meant that it is virtually impossible to get the ceremony performed. The convenience for elopers of a Scottish marriage has been reduced by successive Acts of Parliament, notably by the Marriage (Scotland) Act of 1939, prior to which a valid marriage could be contracted in Scotland merely by a declaration of consent by the two parties before a witness (traditionally the village blacksmith).

[**1791** C. Stuart *Gretna Green* ii. 19 Never think of dying a maid, for that would be setting a bad example, and Gretna Green might then go a begging.] **1813** [see Blacksmith]. **1823** Galt *Entail* III. x. 95 Whomsoever ye hear giving me the wyte o' ony sic Gretna Green job, I redde ye put your foot on the spark, and not let it singe my character. **1852** *Househ. Words* V. 199/2 It was my impression Gretna marriages were quite matters of the past. *Ibid.* 201/1 Old Colthard..was the first regular Gretna Green parson. **1863** *Chambers's Encycl.* V. 106/2 The Gretna Green marriages may yet be resorted to by English parties, provided the intended husband comply with this requisite [*sc.* 21 days' residence in Scotland]. **1904** *Daily Chron.* 17 Feb. 7/1 Caroline, Marchioness of Queensberry..was the heroine of a genuine Gretna Green marriage. *Ibid.*, The period of the Gretna Green weddings ..extended over almost exactly a century. **1959** *Chambers's Encycl.* VI. 590/2 Gretna marriages still attracted the sentimental until the Scottish irregular marriage..was abolished in 1939. **1968** D. J. Steel et al. *Nat. Index Parish Registers* I. 317 Information on the history of Gretna marriages.

Greuze (grōz). The name of J. B. *Greuze* (1725–1805), French painter, used in *Greuze-like* adj., resembling a figure in his pictures. Also *Greuze-ish.*

1860 O. W. Holmes *Elsie Venner* xxi. 232 A lax,.. Greuze-ish looking blonde. **1905** *Westm. Gaz.* 13 May 5/2 'This is my first dance!' said a young and very pretty girl with a Greuze-like face. *Ibid.*, We did not know whether that remark applied to the Greuze-like young girl. **1948** L. Spitzer *Linguistics & Lit. Hist.* 178 Her Greuze-like décolleté.

Grévy's zebra (gre·viz zi·brā, ze·brā). [tr. of *Equus grevyi* (A. Milne-Edwards 1882, in *La Nature* 3 June 13/1), f. the name of F. P. J. Grévy (1807–91), president of the French Republic.] A member of a species of zebra, *Equus grevyi*, found in southern Ethiopia and Somalia.

[**1882** *Proc. Zool. Soc.* 721 Mr. Sclater exhibited some photographs of the new Zebra, lately named *Equus grevyi* by M. A. Milne-Edwards.] **1891** W. H. Flower *Horse* 196/2 (*index*) Zebra, Grevy's. **1894** *Proc. Zool. Soc.* 320 Grévy's Zebra was, I think, first shot in Somaliland by Colonel Paget. **1910** R. E. Drake-Brockman *Mammals of Somaliland* 105 Grevy's zebra seems to prefer undulating rocky bush country to any other. **1931** *Times Educ. Suppl.* 7 Nov. 11 This species is known as Grevy's zebra, and is taller and slimmer than the true zebra. **1957** R. Campbell *Portugal* 46 The shimmering skin of a Grevy's zebra. **1965** D. Morris *Mammals* 353 Grévy's Zebra is the largest and most narrowly striped of the zebras.

grex (greks). *Biol.* [a. L. *grex* flock.] A clump of myxamœbæ formed during a phase of the life cycle of cellular slime moulds, the Acrasina or Acrasiomycetes. Also *attrib.*

1962 B. M. Shaffer in M. Abercrombie *Adv. Morphogenesis* II. 112 In the second phase, the aggregate has a fairly definite shape, whatever its origin, and reactions that are foreign to those of the component cells: *it* has become the individual. As no existing words connoting 'many-in-oneness' seem to be available, the name *grex*—the classical product of ad-gregation—is put forward for it. *Ibid.* 162 If the grex cells move like stream cells, we must again consider to what extent they are guided by acrasin. **1971** *Nature* 4 June 329/2 Although the aggregate can differentiate directly into a fruiting body it usually migrates as a coordinated irritable whole in the form of a tapered cylinder called the grex.

grey, gray, *a.* and *sb.* Add: **A.** *adj.* **5. b.** *spec.* Of a person: dull, anonymous, 'faceless'. Cf. *sb.* *5 c.

1967 [see *sb.* *5 c]. **1969** *Times* 8 Aug. 3/3 The identity of these grey men of politics should be revealed. **1969** *Sunday Times* (Colour Suppl.) 9 Nov. 80/3 Oh no, it's no myth. When he leaves university he is often very, very grey. **1969** *Gandalf's Garden* iv. 22/2 What about the Universities? (Surely ranking among Britain's most proficient and prolific in the churning out of grey people.)

c. Used in various collocations in the place of *black* to indicate a less extreme form of the activity, object, etc. See also *grey market, grey-out* (*8 below).

1966 *Punch* 14 Dec. 900/2 *A Hero of Modern Industry*.. could, I suppose, be described as a grey rather than a black comedy, on the grounds that no one actually gets killed. **1967** *Economist* 29 Apr. 487/3 The grey list.. cannot strictly be defined as a banned list, but the Ministry of Health has drawn up for doctors' guidance, ten pages of drugs..that it would prefer them not to prescribe. **1970** *Guardian* 26 Mar. 26/4 As for the over-prescribing doctors, it was not just the black sheep we had to worry about. It was the grey sheep too.

6. e. Of a person: white-skinned. *U.S. Negro slang.* Cf. *sb.* *5 b.

1962 *Times Lit. Suppl.* 21 Sept. 706/3 The 'grey boys', as Mr. Simmons calls white men. **1965** E. Lacy *Moment of Untruth* iv. 69 Funny thing with grey chicks.. They're always so sure their white skin is the sexiest ever. *Ibid.* vii. 112, I appreciate your socking that lousy grey dick.

7. b. *grey-cheeked, -crowned.*

1893 B. Torrey *Footpath-Way* 94 The evergreens immediately about the house were full of gray-cheeked thrushes. **1852** S. F. Baird in H. Stansbury *Exped. Valley Gt. Salt Lake* 317 (*heading*) Leucosticte Tephrocotis,..Gray-crowned Finch. **1945** Baker *Austral. Lang.* xii. 211 The Grey-crowned Babbler is known also as apostle-bird, [etc.].

8. grey area (see quot. 1968); **grey box,** an Australian tree, esp. *Eucalyptus moluccana*; **grey cells** *colloq.* = *grey matter*; **grey cloth,** unbleached cloth; **grey eminence** = *Éminence grise; **grey goods** (see quots.); **grey leaf** = *grey speck*; **grey market,** a less extreme form of black market; **grey oak,** any of several American oaks, esp. *Quercus coccinea* or *Q. borealis*; **grey oil,** olive oil and lanolin containing mercury; **grey-out** *Aeronaut.*, a less severe form of black-out; so **grey-out** *v. intr.*, to suffer a 'grey-out'; **greying-out** *vbl. sb.*; **grey pine,** either of two North American pines, *Pinus banksiana* or *P. sabiniana*; **grey powder** (earlier example); **grey scale** (see quot. 1959); also *attrib.*; **grey speck,** a disease in oats caused by a manganese deficiency and characterized by specks on the leaf-blades; **grey wethers** (earlier example); **grey willow,** one of several willow trees or bushes with grey foliage or down, esp. *Salix sericea* or *S. cinerea*; **grey wood** (see quot. 1934).

1963 *Times* 22 May 9/3 Within urban areas Mr. Wates said there were vast 'grey' areas—he mentioned Fulham, Battersea and Islington—which could not be classified as slums but were in need of rebuilding. **1968** *Guardian* 18 Mar. 8/1 The 'grey areas'—the term now often used for parts of the country which are in poor or declining economic health, but not so dramatically stricken..[as] the development areas. **1970** *Daily Tel.* 9 Apr. 21/4 Local authorities in administrative counties which include development areas and intermediate ('grey') areas will be able to borrow up to 50 p.c. of their capital finance needs from the Public Works Loans Board in 1970–1971. **1879** F. von Mueller *Eucalyptographia* i, The tree [*sc. E. goniocalyx*] passes among the woodmen as Blue and White Gum-tree, in the other case as Grey or Bastard Box. **1884** A. Nilson *Timber Trees N.S.W.* 136 *Grey box*, Eucalyptus saligna. Myrtaceæ. **1944** F. Clune *Red Heart* 51 Colson noted the grey box and whitewood trees became sparser. **1969** *Age* (Melbourne) 24 May 24/5 The estate is natural bushland crowded with grey box, [etc.]. **1920** A. Christie *Mysterious Affair at Styles* x. 226 'This affair must all be unravelled from within.' He tapped his forehead. 'These little grey cells. It is "up to them"—as you say over here!' **1960** Wodehouse *Jeeves in Offing* iii. 38 You can't hold down an editorial post on an important London weekly paper without being fairly well fixed with the little grey cells. **1930** *Aberdeen Press & Jrnl.* 1 Apr. 8/4 Plain greycloth—that is, unbleached cloth, or cloth dyed in the piece. **1959** *Listener* 9 July 46/2 The weavers manufacture, from the yarn, cloth in an unfinished state, known as grey cloth. **1968** Grey cloth [see *grey goods below]. **1941** A. Huxley (*title*) Grey Eminence. **1945** R. Hargreaves *Enemy at Gate* 151 Bismarck's 'grey eminence', the enigmatic Holstein. **1956** F. Swinnerton *Background with Chorus* 78 The parts played by men not visibly figuring in 'movements'—the grey eminences whose unpublished efforts give new ideas currency and new authors, or old authors, great places in literature. **1965** *Listener* 21 Oct. 614/1 That grey eminence of British communism Mr Palme Dutt. **1954** *Textile Terms & Defs.* 21 *Grey goods*, woven or knitted fabrics as they leave the loom or knitting machine, i.e. before any bleaching, dyeing, or finishing treatment has been given to them. **1968** J. Ironside *Fashion Alphabet* 231 *Grey cloth or goods*, woven fabric before it is dyed or printed. **1928** F. T. Brooks *Plant Dis.* ii. 15 On some alkaline soils oats cannot be grown profitably on account of a disease known as Grey-leaf. **1946** *Life* 1 Apr. 32/1 In the Rocky Mountain region there is a gray market in toilet paper. **1952** *Economist* 15 Nov. 451/1 My own experience is that the effect of the grey or black market is grossly exaggerated by the Ministry's estimate of 50 per cent of home production. **1963** *Listener* 14 Feb. 275/2 In the countryside there is a gigantic grey market where the collective farms get together to defeat the demands of the state [*sc.* Czechoslovakia]. **1969** *Times* 16 July 9/6 It was suggested that instead of the grey market where a few doctors deliberately overprescribed [heroin] and addicts sold off their surplus supplies there would be a..black market. **1697** *Rec. Early Hist. Boston* (1880) VI. 8 From thence to another wallnut tree and so straight to a gray oak. **1813** H. Muhlenberg *Catal. Plant.* 87 Upland willow-oak, or gray oak. **1832** D. J. Browne *Sylva Amer.* 261 The Gray Oak is found farther north than any other species in America. **1908** N. L. Britton *N. Amer. Trees* 290 Gray Oak—*Quercus borealis.* **1908** *Practitioner* Sept. 467 The use of grey-oil in subcutaneous injections. **1945** *Reader's Digest* May 43/2 Formerly a pilot had only two chances:..or go in all the way and turn sharply, incurring enough G to grayout or blackout. **1946** A. M. Low *How Secrets Work* 153 The first noticeable effect is a 'grey-out' due to a reduced pressure on the capillaries at the back of the eye. **1956** J. E. Johnson *Wing Leader* iii. 44 The Spit protests and shudders, and when the blood drains from your eyes you 'grey-out'. **1962** W. Schirra in *Into Orbit* 46 When a man is pulling heavy Gs in the normal sitting position, the blood drains down from his head, and he can have a grey-out or a black-out. **1961** *New Scientist* 25 May 457/3 At slightly lower *g*, the individual experiences greying-out: there is an overall dimming of vision and loss of peripheral vision. **1810** F. A. Michaux *Arbres* I. 16 *Pinus rupestris*.. Grey pine.., dénomination donnée..en Canada. **1832** D. J. Browne *Sylva Amer.* 240 In Nova Scotia and the state of Maine, where it is rare, it is called Scrub Pine, and in Canada, Gray Pine. **1908** N. L. Britton *N. Amer. Trees* 37 This tree [*sc. Pinus sabiniana*], also called Gray pine..occurs locally in the foothill region of western California. **1923** L. H. Bailey *Cultivated Evergreens* iii. 103 The gray pine, *Pinus Banksiana*, is found farther northward than any other American pine. **1956** B. R. Morton *Native Trees of Canada* (ed. 5) 16 Banksian pine [is also called] grey pine. **1842** C. Ridley *Let.* 4 Dec. (1958) ix. 111 Baby is not quite well. I consulted Sir John Fife, who gave him grey powder. **1940** *Chambers's Techn. Dict.* 389/2 Grey scale. **1957** V. J. Kehoe *Technique Film & T.V. Make-Up* i. 23 (*caption*) The Gray Scale. Diagrammatic representation of gradation from white to black, the in-between shades being tones of gray. **1959** W. S. Sharps *Dict. Cinematogr.* 101/1 *Grey scale*, a graduated range of tones extending from white to black, with intermediate greys consisting of intermixed white and black. **1961** G. Millerson *Technique Telev. Production* iii. 45 (*caption*) Sample paint-cards and materials are compared on camera with a standard grey-scale chart. **1968** *Brit. Med. Bull.* XXIV. 262/2 The grey scale is a graded series of densities from transparent (white) to black. **1928** *Jrnl. Dept. Agric. S. Austral.* XXXI. vii. 696 (*title*) Grey speck (manganese deficiency) disease of oats... For many years the Grey Speck disease of oats has been known to occur on certain alkaline soils in Holland, Norway, Sweden, Denmark, and Germany. *Ibid.* 704 In the pots treated with nitrate of soda and cyanamide the Grey Speck symptoms were worst. **1929** *Ann. Appl. Biol.* XVI. 497 It does not seem to have been recognised that the symptoms of the Grey Speck disease are essentially the symptoms of manganese deficiency. **1947** *Sci. News* V. 88 Manganese is essential for healthy plant life. Its deficiency in soil..leads to plant diseases such as grey speck of oats. **1794** J. B. S. Morritt *Let.* 13 Nov. (1914) vi. 143, I saw a hill covered with stones and grey-wethers, but could not make out many of the foundations he mentions. **1813** H. Muhlenberg *Catal. Plant.* 91 Gray willow. **1861** *Trans. Ill. Agric. Soc.* IV. 448 Mr. Overman has the Pennsylvania Gray Willow—the most rapid growing variety he has ever known. **1930** J. E. Kirkwood *N. Rocky Mt. Trees & Shrubs* 88 The Gray Willow, so called in allusion to its covering of long gray hairs, is variously understood by various authors. **1951** *Dict. Gardening* (R. Hort. Soc.) IV. 1852/1 *S[alix] cinerea.* Grey Willow...almost wholly covered with grey down. **1933** *Times Educ. Suppl.* 15 Apr. 115/3 The former is faced with Indian silver greywood. **1934** *Archit. Rev.* LXXVI. 70/1 Sycamore has also been used in decorative work and furniture, dyed and known as 'greywood' or steamed and known as 'weathered' sycamore. **1952** H. H. Saylor *Dict. Archit.* (1963) 79 *Greywood*,..used extensively in England for decorative woodwork and veneer.

b. grey-fin, a variety of trout found in the Tweed; **grey-fish,** (*b*) *U.S.* the common dog-fish; **grey nurse** (later examples); **grey snipe** (example); **grey wolf,** the common wolf, *Canis lupus*; also *fig.* (see quot. 1904).

1847 T. Stoddart *Angler's Comp.* 210 The grey-fin or bull-trout smolt. **1923** W. A. Herdman *Founders Oceanogr.* 310 People..are prejudiced against 'dog-fish', so the [United States] Bureau [of Fisheries] altered the name of the latter to 'gray-fish'. **1917** *Chambers's Jrnl.* Sept. 589/2 The gray nurse, like the white shark, is noted for its daring ferocity. **1969** *Man* (Austral.) Mar. 12/2 There's even been grey nurse hooked here, and a blue marlin that had been hunted over from the deep-sea beds. **1870** *Game Laws Penn.* in *Fur, Fin & Feather* (1872) 120 No person shall kill, capture, take..any gray snipe. **1814** Lewis & Clark *Hist. Exped. Missouri* (1815) I. 206 We caught..a large gray wolf. **1831** J. M. Peck *Guide Emigrants* 162 The large grey wolf, or *canis lupus* of Linneus, is not very plenty. **1880** *Harper's Mag.* July 171/1 You will find bear and the great gray wolf..in the wilderness. **1904** *Grand Rapids Even. Press* 8 June 4 In plain words, a gray wolf, in Chicago phraseology, is a professional grafter. **1936** D. McCowan *Anim. Canad. Rockies* iii. 30 The grey wolf, also known as timber wolf and formerly as buffalo wolf, is grey in colour with a liberal sprinkling of black and brown in the coat. **1964** E. P. Walker et al. *Mammals of World* II. 1152/1 Two species of wolves are recognized: *C. lupus*, the gray or timber wolf, and *C. niger*, the red wolf.

c. Also in the names of other substances, as **grey (cast) iron,** cast iron that has a grey fracture and contains most of its carbon in the form of flakes of graphite; **grey tin,** a grey powdery allotrope of tin, to which ordinary (white) tin is converted at low temperatures; so **grey modification.**

1665 D. Dudley *Metallum Martis* 47 The Gray Iron.. is most fined, and more sufficient to make Bar-Iron with, and tough Iron to make Ordnance, or any Cast Vessels, being it is..more malliable and tough, then the other two sorts. **1802** *Phil. Mag.* XII. 28 By increasing the dose of carbon you increase the fusibility, and it passes at length into the state of gray cast iron. **1824** T. Tredgold *Pract. Ess. Strength of Cast Iron* (ed. 2) i. 7 Gray cast-iron is used for artillery, and is sometimes called gun-metal. **1880** *Encycl. Brit.* XIII. 281/2 A coarse-grained crystalline structure results, the product being then termed grey cast iron. **1911** *Daily Colonist* (Victoria, B.C.) 28 Apr. 1/6 Fire..destroyed the blacksmith and forging shop, the machine shops and the Grey iron and brass founders. **1955** *Times* 15 July 2/6 Qualified Metallurgist preferably with experience with malleable and grey iron. **1965** *Economist* 23 Oct. p. xxii/2 Even a quick look at the processes of grey-iron castings—the type of iron used for the bulk of auto castings—shows up the fundamental differences. **1967** *Times Rev. Industry* Mar. 76/1 High production costs amounting to over £98 per ton of grey cast iron. [**1879** *Jrnl. Chem. Soc.* XXXVI. 888 The grey modification of tin was found to be electronegative towards ordinary tin.] **1886** *Ibid.* L. 124 The author concludes that only the following allotropic modifications may be regarded as different:—Grey tin..; rhombic tin..; and tetragonal tin. **1911** *Encycl. Brit.* XXIX. 996/2 When exposed for sufficient time to very low temperatures (to −39°C for 14 hours), tin becomes so brittle that it falls into a grey powder, termed the *grey modification.* **1933** *Jrnl. R. Aeronaut. Soc.* XXXVII. 540 Tin shows a maximum ductility at about 100°C., and becomes increasingly brittle at lower temperatures, until, at 18°C., the change to its allotropic form, 'grey tin', commences. **1950** N. V. Sidgwick *Chem. Elements* I. 552 The change from white to grey tin at low temperatures is very slow. **1965** Phillips & Williams *Inorg. Chem.* I. iii. 67 The transition of diamond to graphite, which involves a profound change of structure, requires a total energy change per atom of as little as 0·02 eV as does the change of white to grey tin.

B. *sb.* **1. b.** Also phr. *in the grey.* (Cf. *grey cloth, *grey goods in sense A. 8.)

1860 S. Jubb *Hist. Shoddy-Trade* 40 Short Ends were sold to the merchants..in the grey. **1929** *Times* 7 Feb. 9/3 There were thousands of piece goods coming into this country 'in the grey' to be dyed here.

c. The uniform of the Southern troops in the American Civil War. *U.S.*

1863 *Rebellion Rec.* V. ii. 72 An Irishman of the Seventeenth New-York came up to the General,..driving three prisoners in gray before him. **1866** J. C. Gregg *Life in Army* xviii. 172 They were refused admission..on the ground of their uniform; when if they had been dressed in rebel gray..no doubt they would have met with a warm welcome. **1886** F. C. Baylor *On Both Sides* II. viii. 429 A military society composed of men who had worn the gray. **1948** *Realty & Building* 15 May 11/2 Colonel John was a Johnny Reb who delighted in telling of the exploits of the boys in gray.

d. *pl.* Grey flannel trousers. Also (colloq.) *greyers* [see *-er[6]].

1900 J. S. Farmer *Public School Word-Bk.* 102 *Greyers*, (Harrow) Grey flannel trousers. **1923** *Daily Mail* 27 June 8 The jacket and waistcoat may be worn with 'greyers' or white flannel trousers, thus providing all the sports clothes the average man needs. **1932** A. J. Worrall *Eng. Idioms* 22 I'll wear my greys for tennis to-day as the ground is rather slippery. **1948** D. W. Ballantyne *Cunninghams* (1963) ii. xxiii. 226 A couple of..boys, wearing sports coats, greys, and open-necked white shirts.

4. *fig.* (Later examples.)

a **1907** F. J. Thompson *Works* (1913) III. 218 The Englishman says, 'Black's black—*furieusement* black; and white's white—*furieusement* white.' De Quincey saw many blacks, many whites, multitudinous greys. **1923** J. Hergesheimer *Bright Shawl* 21 He saw..the good on one side facing the bad on the other. There was no mingling of the ranks, no grey; simply, conveniently, black and white. **1948** *Amer. Sociol. Rev.* June 284 To disclose the blacks and imply the whites of a quality scale, the middle grays being largely lost.

5. b. A white-skinned person. *U.S. Negro slang.* Cf. *adj.* *6 e.

1960 Wentworth & Flexner *Dict. Amer. Slang* 228 *Gray*, (derog.) a white person. **1963** in *Amer. Speech* (1964) XXXIX. 60 Beyond the ears of the greys. **1965** O. Harrington in J. H. Clarke *Harlem* 90 The year was 1936, a bad year in most everybody's book. Ellis the cabdriver used to say that even the grays downtown were having it rough. **1966** *Publ. Amer. Dial. Soc.* 1964 xlii. 31 *Caucasian*..gray.

c. A dull, anonymous, or 'faceless' person. Cf. *adj.* *5 b.

1967 *Observer* 17 Dec. 25/1 Let them..remember that many of the 'grey' citizens have been working for forty years or more for the forgiveness of enemies... The 'greys' would appreciate the co-operation of the bright ones in such activities. **1969** *Gandalf's Garden* vi. 11/1 *Greys*, those people whose minds have ossified and who have lost the wonder of living and the will to learn what it's all about. **1969** *It* 4–17 July 3/3 The whole scene is under the thumbs of the greys. Pretty well everywhere today the dead men, the square men and the greys are running things, calling the tune.

greycing, var. *Gracing *vbl. sb.*[2]

greyers, *sb. pl.*: see *Grey *sb.* 1 d.

greyhead. Add: **2. b.** A male sperm-whale, *Physeter catodon*, with grey markings on its head; cf. *Grey-headed *a.* 4.

a **1889** C. M. Scammon (Cent. D.). **1908** *Westm. Gaz.* 28 Dec. 2/1 The right whale and the grayhead are gone.

grey-headed, *a.* Add: **4.** Of a male sperm whale; cf. *Greyhead 2 b.

1839 T. Beale *Nat. Hist. Sperm Whale* i. 31 Old 'bulls', as full-grown males are called by whalers, have generally a portion of grey on the nose immediately above the fore-part of the upper-jaw, and they are then said to be 'grey-headed'. **1874** C. M. Scammon *Marine Mammals N.W. Coast N. Amer.* viii. 75 Sperm Whale... The oldest males are frequently well-marked with gray about the nose, or upper portion of the head, and when this is indicated, they are called 'gray-headed'.

greyhound. Add: **4.** *greyhound fan, kennel, owner, race, racecourse, track*; **greyhound racing,** a sport in which a dummy hare propelled mechanically round a set track is pursued by greyhounds. Cf. *Gracing *vbl. sb.*[2]

1946 R. Genders *Mod. Greyhound Racing* xiii. 134 One of the famous Wireless Rally/Erin Green litter so well known to greyhound fans, he has been one of the finest sprinters this country has ever seen. *Ibid.* vi. 58 The breeder or keeper of greyhound kennels. **1927** 'Leveret' *Greyhound Racing* iii. 12, I have been asked by several would-be greyhound owners to find them dogs for track-racing purposes. **1926** *Manch. Guardian* 19 July 5/6 (Advt.), Greyhound races. Belle Vue racecourse, Manchester. *Ibid.* 23 July 11/4 The Greyhound Racecourse, Kirkmanshulme Lane, Gorton, applied for an occasional licence to sell intoxicating drink. *Ibid.* 26 July 9/4 Something of a novelty has been added to Manchester's odd collection of diversions by the opening of the new greyhound racing track. **1933** E. C. Ash *Bk. Greyhound* (ed. 2) xii. 305 In the year 1876, two events took place,.. one a swimming race between a man and a dog, which the dog won, and the other a Greyhound race after a mechanical hare. **1928** R. Knox *Footsteps at Lock* viii. 79 The conversation turned..on greyhound racing...One very old gentleman had to be convinced..that it was the hare, not the hounds, which worked by electricity. **1949** H. E. Clarke *Mod. Greyhound* ii. 16 In the early days of greyhound racing it was customary to bolster up the antipathy to the coursing animal by the argument that no dog from the coursing field had ever won a track classic. **1935** J. Agate *More First Nights* (1937) 141 This category includes racecourses, greyhound and dog tracks, [etc.]. **1946** R. Genders *Mod. Greyhound Racing* iii. 27 There has been little change in the construction of greyhound tracks since the first ones were opened. **1970** *Encycl. Brit.* VII. 555/1 All major tracks are members of the American Greyhound Track Operators association.

grey squirrel. [Grey *a.* 1 and 8 b.] A common squirrel of the United States (*Sciurus carolinensis*), which was introduced into Europe in the late 19th century.

1674 J. Josselyn *Account Two Voy. New Eng.* 86 There are three sorts, the mouse squirril, the gray squirril, and the flying-squirril. **1781** T. Pennant *Hist. Quadrupeds* II. 410 Grey Squirrel...*Sciurus cinereus* L. Inhabits the woods of North America, Peru, and Chili; are very numerous in North America; do incredible damage to the plantations of Mayz. **1804** in *Maryland Hist. Mag.* (1909) IV. 9 Squirrels in this neighborhood are of a deep black color..less in size than the grey squirrels of Maryland. **1831** J. J. Audubon *Ornith. Biogr.* I. 247 The Grey Squirrel..migrates in prodigious numbers. **1831** J. M. Peck *Guide Emigrants* 164 The grey and fox squirrels often do mischief in the corn fields. **1850** N. Kingsley *Diary* (1914) 105 In the afternoon went out with a gun for the first time, grey squirrels are quite plenty but too much water to hunt them. **1935** *Discovery* June 169/1 In North Leicestershire red and grey squirrels have been noticed inhabiting the same wood without coming into conflict, while in Northamptonshire the red squirrels had decreased before the few greys arrived. **1964** H. N. Southern *Handbk. Brit. Mammals* ii. 275 Whereas in America Grey Squirrel is prized game animal, in British Isles regarded as forest pest. **1971** *Daily Tel.* 1 Dec. 8/6 By far the worst crime of the grey squirrel was its attacks on hardwood trees by stripping off the bark.

greywacke. Delete 'now almost *obs.* Rarely *pl.*' and add later examples.

1895 A. Harker *Petrol.* xvi. 193 The old term *greywacke* (Ger. Grauwacke) has been revived for a complex rock with grains of quartz, felspar, and other minerals and rocks united by a cement usually siliceous. **1939** *Amer. Jrnl. Sci.* CCXXXVII. 29 This is the direction of a marked change in the angle of slope of the hillside..which is inferred to mark the vertical or highly inclined outer margin of the granite at its contact with Silurian shales and greywackes. **1950** *N.Z. Jrnl. Agric.* Jan. 16/3 Sandy silts, silts, clays, and greywacke shingle have been deposited by the Waipawa and Tukituki Rivers. **1963** Krumbein & Sloss *Stratigr. & Sedimentation* (ed. 2) v. 171 Many Precambrian and early Paleozoic graywackes are partially metamorphosed. **1970** *Nature* 3 Jan. 57/2 They resemble 'gouge-channels' occurring in graded greywacke beds.

grid. Add: **1. b.** (Earlier and later examples.)

1884 F. Krohn tr. *Glaser de Cew's Magn.- & Dyn.-Electr. Mach.* 122 The red lead paste becomes very firmly gripped by the surrounding grid-work. **1964** G. Smith *Storage Batteries* ii. 13 A grid consists of an outer frame with take-off lug and a central mesh or lattice of vertical and horizontal ribs. **1968** R. H. Bacon *Car* x. 136 If the battery is old, or subjected to repeated complete discharge and charge, the paste may fall away from the grid.

4. (Examples.)

1927 *Stage* 27 Oct. 17/6 Ask him [*sc.* the young actor] questions about the grid, the apron,..and he will be at sea. **1933** P. Godfrey *Back-Stage* iv. 43 Seventy feet above the stage-level is the grid from which the scenery is flown. **1959** A. S. Gillette *Stage Scenery* vii. 171 Hoisting cables run from the winches to dollies that may be shifted to any position on the grid.

5. *Electr.* **a.** Orig., a wire gauze or helix or a perforated plate between the filament and the anode of a thermionic valve, forming a third electrode to which a voltage can be applied to control or modulate the flow of electrons from the filament. Later extended to any electrode in a valve, gas discharge tube, or similar device, having one or more apertures for the passage of electrons or ions; *spec.* **control grid,** one that serves as a control electrode. Cf. **screen grid*, **suppressor grid.*

1907 *Electrical World* 30 Nov. 1034/1 In his talk, Dr. De Forest stated that many types of receivers are suitable for wireless Telephone work...However, the Audion receiver, especially in its latest form in which the antenna is joined to an isolated grid interposed between a tantalum filament and a platinum wing, gives perfect 'articulation'. **1919** R. Stanley *Text Bk. Wireless Telegr.* (new ed.) II. xi. 208 The output is controlled by the control grid, which is between the filament and the heavier grid anode beyond. **1923** E. W. Marchant *Radio Telegr.* iv. 52 When a grid is introduced into the valve, it may be used, not merely for rectifying signals.., but..for amplifying or increasing the strength of a signal. **1947** D. G. Fink *Radar Engin.* xi. 538 The potential on the control electrode (grid) controls the brightness of the spot. **1950** J. C. Slater *Microwave Electronics* xi. 279 For a relatively short accelerator, there may be few enough grids so that a considerable fraction of the ion beam penetrates all the grids. **1950** Rider & Uslan *Encycl. Cathode-Ray Oscill.* v. 86/1 Surrounding the cathode [of the electron gun] is a cylinder *G* which has a baffle containing a tiny aperture at its center... The baffle with its aperture is the actual control grid. **1959** *Chambers's Encycl.* XIV. 241/1 The two control grids in a frequency changer are usually the first and third from the cathode, the other grids being screens, accelerators and suppressors.

b. Freq. *attrib.* and *Comb.* (usu. in the sense *control grid*), as *grid battery, condenser, potential, voltage*; **grid bias,** a steady, usu. negative, voltage applied to the control grid of a valve on which the signal voltage is superimposed; hence, the steady potential difference between the control grid and the cathode in the absence of a signal; **grid circuit**, the circuit connected between the grid and the cathode of a valve; **grid control**, control of the anode current, esp. the discharge in a gas discharge tube, by means of the grid voltage; so **grid-controlled** *a.*; **grid current**, the current flowing to the grid from outside the valve; **grid-dip meter** or **oscillator**, a calibrated valve oscillator used as a frequency meter, resonance of the tank circuit of the oscillator with the resonant circuit under test being indicated by a drop in the grid current of the valve; **grid leak (resistance)**, a high resistance connected between the grid and the cathode of a valve by which any excess charge on the grid can escape; **grid modulation**, modulation in which the modulating signal is applied to the grid of a valve in which the carrier is present.

1924 *Harmsworth's Wireless Encycl.* II. 1053/2 The grid battery plays an important part in the operation of power-amplifying valves. 1926 S. O. Pearson *Dict. Wireless Terms*, Grid bias. 1928 *Morning Post* 26 Jan. 13/4 When there are two stages of low frequency amplification a grid bias of negative 1½ volts is ample. 1928 *Listener* 20 June 800/3 There was endless trouble with grid-biases, valves, condensers, speakers. [illegible] *Discovery* Feb. 54/1 A simple spherical electrode is connected with the grid circuit of a valve oscillator. [illegible] *Electrician* LXXVI. [illegible] The audio frequency beats are then rectified by the grid and the grid condenser [illegible]. 1940 *Chambers's Techn. Dict.* 390/1 *Grid condenser*, a condenser connected between the grid and the cathode of a thermionic valve. [illegible] L. B. Turner *Wireless Telegr.* [illegible] (heading) Grid control of current [illegible] in a thyratron. [illegible] *Junction Transistors* [illegible] devices show promise as solid state 'thyratrons'. They have the advantage over grid-controlled gas-tube rectifiers that they can be turned off more readily. [illegible] *Radio Rev.* Dec. 14 The circuit will be called the grid circuit, and the current flowing in it the grid current. 1948 J. W. Gray in Valley & Wallman *Vacuum Tube Amplifiers* xi. 419 Positive grid current, caused by electron flowing to the grid from the cathode, is much greater and much less erratic than the negative grid current. [illegible] *Physics Bull.* Feb. [illegible] Conventional electrometer tubes have a minimum guaranteed grid current in the region of 10^{-14} A, somewhat large for a number of modern requirements. 1959 K. Henney *Radio Engin. Handbk.* (ed. 5) xiv. 11 Grid-dip meters. [illegible] The grid current of an oscillator will 'dip' slightly as the wavemeter is tuned through the operating frequency. [illegible] It is possible to measure the input and output impedances [illegible] of the transistor [illegible] in such a case with the aid of a signal generator and a grid-dip meter. 1929 *Radio Rev.* Nov. 82/1 R is a large resistance which acts as a grid leak, and should be chosen so as to keep the grids at a voltage found advantageous by trial. 1966 *McGraw-Hill Encycl. Sci. & Technol.* II. 164/1 There is always a small grid current flowing through the grid-leak resistor when cathode or fixed bias is used. 1942 F. H. Haynes *Amateur's Bk. Wireless Circuits* (ed. 2) 21 In another system of grid modulation the microphone secondary is connected between the grid itself and the L.T. minus, with the grid condenser in the lead to the grid coil and this time not fitted with a leak. 1967 R. L. Snader *Electronic Communication* (ed. 2) xvii. 541 Low audio power requirement is one of the advantages of grid modulation. 1926 *Electrician* LXXVI. 800/1 A point will finally be reached where the grid potential is sufficiently reduced to allow the wing current to flow. 1928 *Wireless World* June 14/ Up to the saturation currents for the gas space within the bulb, the change in plate current is approximately proportional to the change of grid bias.

6. A network of lines, esp. two series of regularly spaced lines crossing one another at right angles; *spec.* one provided on a map as a means of specifying the location of places and objects. Also *grid finder, navigation, reference, sheet, system; grid-like* adj.

1918 in *Geogr. Jrnl.* (1919) LIII. 33 Doubtless the German was amused at the conservative Briton, who at first preferred to use a 'grid' of squares 1000 yards a side. 1922 *Encycl. Brit.* XXXII. 623/1 For the use of large scale maps in trench warfare..it must be possible to read off at sight the coördinates of any desired point from a 'grid' or network of lines printed on the map... For ease and accuracy of reference the 'grid' should be in squares. 1924 *Times Trade & Engin. Suppl.* 29 Nov. 241 Simplicity of reference is ensured by the use of a novel transparent grid sheet, ruled in squares and numbered to correspond with the grid numbers given in the Gazetteer. The use of this grid sheet enables the most obscure place to be found on the map in a moment. 1925 Close & Winterbotham *Text Bk. Topogr. Surveying* (ed. 3) 135 Some form of reference grid was found necessary by all combatants in the late war. *Ibid.* 136 To overprint his available maps with the appropriate grid. 1930 G. R. de Beer *Embryol. & Evol.* iv. 29 The comparison of one adult form with another can be made very instructive by inscribing the shape of one form on a grid-system of

Cartesian co-ordinates. *Ibid.* 30 By a harmonious transformation of the grid, the skull of Hyracotherium can be distorted and made to resemble that of the horse. 1931 *Times Lit. Suppl.* 10 Sept. 674/4 This grid, which has the appearance of a transparent chessboard, can be made to appear at the same height as any object in the stereoscope picture. 1932 *Jrnl. R. Aeronaut. Soc.* XXXVI. 518 This template, often known as the grid finder, is placed on the photograph so that the line representing the apparent horizon coincides with the apparent horizon on the photograph. 1944 *Geogr. Rev.* XXXIV. 436 (title) Grid Navigation by Samuel Herrick. *Ibid.*, The introduction of a rectangular grid system and what is hereafter described as 'grid variation' makes it possible to navigate a straight line on any conformal..chart. 1946 R. J. C. Atkinson *Field Archaeol.* ii. 26 The intersecting grid system..enables a much larger area to be searched in a given time. 1950 *Mind* LIX. 302 The description..both suggests a kind of permanent grid-like world framework and denies it. 1954 M. Beresford *Lost Villages* vii. 270 Grid-reference to the 1-inch Ordnance maps. 1958 *Listener* 13 Nov. 779/1 The first [major development in polar navigation] is the substitution of a rectangular 'grid' for the conventional graticule of parallels of latitude and meridians of longitude which converge at the Poles. This simple technique, which has been extensively developed since the war, is generally called 'grid navigation'. 1963 E. S. Wood *Collins Field Guide Archaeol.* ii. 306 A card index should be kept; one card for each find or site, showing details, location, with grid reference. 1966 H. Williamson *Methods Bk. Design* (ed. 2) xiii. 211 The photographer places in the camera..a glass screen ruled with a rectangular grid of fine diagonal lines. 1971 D. Potter *Brit. Elec. Stamps* xiv. 142 This tissue is first exposed behind a fine grid of horizontal and vertical lines.

b. *Building.* (See quot. 1935.) Also *grid plan*.

1935 *Archit. Rev.* LXXVII. 65 The plan of the building conforms to a grid composed of squares of the building's proportion, measuring 8 ft. by 8 ft. 10 ins. 1964 J. S. Scott *Dict. Building* 153 *Grid plan*, a plan in which setting-out lines called grid lines coincide with the most important walls and other building components. Prefabricated buildings are usually designed to fit a grid plan.

c. *Motor Racing.* A pattern of lines painted on the track at the starting-point to indicate the position in which the cars are to line up (see also quot. 1971).

1956 A. W. May *Formula 3* x. 124 Imagine the feelings of young Jeremy Fry when..the engine stalled. 1957 ... the white car was last on the grid. 1957 S. Moss *In Track of Speed* xiii. 160 The front row of the starting grid were Fangio and myself on Mercedes and Ascari on a Lancia. 1966 *Publ. Amer. Dial. Soc.* 1964 xlii. 5 *Grid*, area behind the starting line, usually painted in checked pattern. 1970 *Times* 8 May 9/2 The controversial qualifying races..for deciding the grid positions. [illegible] 1971 *Observer* 1 Aug. 19/3 The grid means the order in which the cars start at the start—perhaps three cars on the first row of the grid, two on the second, and so on, the positions being allotted according to the best time they are recorded in practice laps.

d. *Archit.* (See quots.) Also *grid layout, plan, system.* Cf. *Gridiron *sb.* 3 d.

1954 *Ann. Reg.* 1953 104 The Punjab's new capital at Chandigarh..was built on a grid system which separated fast traffic from slow and divided the city into 25 rectangular sectors. 1958 *Listener* 20 Nov. 826/1 The town is laid out on a grid plan. *Ibid.*, In spite of the imposed colonial grid, the landscaping tradition of the old country reasserts itself successfully. 1959 *Chambers's Encycl.* XIII. 765/2 Whenever land is approximately flat there has been a tendency to use the grid layout, by which a town is divided into 'blocks' of approximately equal size by streets at right-angles to each other. 1959 *Listener* 26 Mar. 545/1 Le Corbusier..is indeed any architect today must accept, the rectangularities, the grid, the engineering forms, the big areas of glass that, for most people, are modern architecture. 1969 *Ibid.* 24 Jan. 98/1 Some of the factors blighting American cities (their ruthless grid-plan expansion, [etc.]).

e. *Phonetics.* 'A diagrammatic representation of approximate tongue-positions of average English vowels compared with those of cardinal vowels' (Daniel Jones).

1961 *Amer. Speech* XXXVI. 219 Daniel Jones's cardinal vowel grid is proposed as a more useful method of specifying vowel sounds in terms of auditory reference points. 1964 E. Palmer tr. *Martinet's Elem. Gen. Ling.* iii. 66 We can..attempt to represent schematically the positions of the system, by arranging in a 'grid' the sets of phonemes characterized by the same relevant feature.

7. A bicycle. *slang.*

1922 D. H. Lawrence *England, my England* 111 Oh, well! I wheel the grid, do I? 1943 *Coast to Coast* 1942 71 'I'll walk and wheel the bike, and if my dad's home he can drive out in the car to meet me.' 'Gosh, no!' you said. 'Here, you go on, on my grid, an' I'll do the walking.'

8. A network of high-voltage transmission lines and connections that supply electricity from a number of generating stations to various distribution centres in a country or a region, so that no consumer is dependent on a single station.

1926 *Public Opinion* 3 Apr. 331/2 The electrical 'grid' is absolutely necessary in the future just as our railway network has been in the past. 1930 *Times Financ. Rev.* 11 Feb. p. xxxiii/1 With transforming stations at the points most suitable for enabling the pool of electricity provided by the 'Grid' to be tapped for distribution throughout the area. 1955 *Times* 5 July p. ii/3 Water for the mill will be drawn from the Tarawera River at

the rate of 10,000,000 gallons a day, while hydro-electric power will be supplied by the national grid. 1955 *Oxf. Jun. Encycl.* VIII. 191 Engineers in charge of the grid system meet this problem of changing demand by running the most efficient power stations continuously and by switching in the less efficient smaller stations as the demand increases. 1971 *Sci. Amer.* Sept. 47/3 Electric power can be transmitted as needed over grids of continental size.

b. Used of any network that serves a similar purpose for other services.

1943 *Times* 8 Dec. 5/5 Should not our leading water engineers be called upon by the Ministry of Reconstruction to report and prepare comprehensive plans for a national grid? 1960 *Economist* 15 Oct. 271/3 Total gasification (linked with a national high-pressure grid supplying industry direct and local systems with lean gas for enrichment to town gas), rather than carbonisation, offers the main opportunity. 1970 *Sci. Jrnl.* Aug. 82/1 [In Holland] a little over 80 per cent of the 13 million inhabitants..are connected to the gasgrid.

9. A strong open framework of iron fixed to the back of a motor car to hold luggage.

1928 *Evening News* 24 July 4/1 It can be carried on either luggage grid or running-board. 1938 *Daily Mail* 25 July 11/1 All straps, loops, etc., which are necessary to fix it to the grid or the car.

10. The field on which American football is played; hence loosely, American football. Also *attrib.* Cf. *Gridiron *sb.* 3 e.

1928 *Chicago Tribune* 13 Dec. 25/8 (*headline*) Law..to lead Irish on Grid in 1929. 1968 *Globe & Mail* (Toronto) 13 Jan. 17/4 It was a tragic start to the grid season for the aging Packers, who showed their grid obituaries had been written prematurely. 1970 *New Yorker* 1 Oct. 34/1 You grid fans are just going to have to buckle down.. Anybody who can't straighten out a plum old two-league, six-division distribution..probably isn't in shape.

gridded (gri·dèd), *pa. pple.* and *ppl. a.* [f. prec. + -ed².] Covered with, forming, or containing a grid (in various senses).

1966 *Blackw. Mag.* Dec. 774/2 A white screen, gridded and lettered, covered the wall. 1940 *Illustr. London News* CXCVII. 202 While fighters fly to the convoy's defence, women of the W.A.A.F. are moving counters and models of ships and other symbols with 'croupiers' rakes' to record all the information from various sources upon a gridded map. 1964 *Archit. Rev.* CV. 113/2 The pyramidal museum of the plastic arts is stratified from industrial art at the base, to a scattering of rare masterworks at the summit; each floor is gridded, the co-ordinates expressing respectively periods and schools. 1968 *Oxf. Univ. Gaz.* 10 Mar. 771/1 Other..finds in Italy included gridded layouts of roads. 1963 B. Fozard *Instrumentation Nucl. Reactors* ii. 18 The arrangement of a gridded chamber is shown. 1971 *Physics Bull.* June 346/1 Many experimenters choose to plot their data on the standard gridded graph paper.

griddle, *sb.* Add: **2.** Add to def.: also used for cooking grills, etc. Also *attrib.*

1960 *Economist* 29 Dec. 1295/1 The latest [British Railway] idea is the 'griddle car'..with a 'chef-conductor' grilling steaks and poaching eggs to order. 1963 *Daily Tel.* 8 Jan. 14/1 Meals cooked on the griddle in a matter of minutes may be taken back to seats in other parts of the train or eaten in the Griddle Car itself. 1968 *Times* 2 May 2/3 British Railways are to try out two new types of refreshment car, known respectively as a Griddle Buffet and a Lounge Buffet.

4. *griddle-hot* adj.

1970 M. Renault *Mask of Apollo* xv. 252 The streets were griddle-hot and dusty.

gridiron, *sb.* Add: **3.** Earlier examples of sense 'the United States flag'.

1824 *Niles' Reg.* 11 Sept. 31/2 The masts from which they flew, went over the side, while Hull's four 'gridirons' floated in the air triumphant. 1866 E. A. Pollard *Southern Hist. War* II. 103 'It was,' says a Charleston paper, 'the identical "gridiron" carried from Fort Sumter in 1861.' 1871 Schele de Vere *Americanisms* (1872) 258 Sailors laugh at it good-naturedly, and seeing in it..the Stars and Stripes] humorously, say ['Gridiron'].

d. *Archit.* = *Grid 6 d. Usu. *attrib.*

1883 Lo. R. Gower *My Reminisc.* II. xxvii. 204 The formal and gridiron-like plan of the streets of this city. 1910 *Catal. Cities Exhib.* in P. Geddes *Cities in Evolution* (ed. 2) (1949) 167 The cities of the United States, with their gridiron plan..very monotonous. 1938 *Oceanianist* III. 85 Anything approaching a spider's web at the nucleus of an English town appears to be extremely rare, and apart from a few towns where the Roman street-plan influenced later development a proper gridiron system is uncommon in England. 1961 L. Mumford *City in History* xiv. 424 In the gridiron plan, as applied in the commercial city, no section or precinct was suitably planned for its specific function. *Ibid.* 425 The extension of the speculative gridiron and the public transportation system were the two main activities that gave dominance to capitalist forms in the growing cities of the nineteenth century. 1969 *Geography* LIV. 200 The grid-iron town plan of Ashburton.

e. = *Grid 10. Also *attrib.*

1896 [in Dict., 3 a]. 1900 *Dial. Notes* II. 39 *Gridiron*, foot-ball field. 1937 L. C. Douglas *Forgive us our Trespasses* vii. 126 He was gleefully welcomed by young Assistant Coach Roberts and the awkward assembly of prospective gridiron heroes. 1960 T. McLean *Kings of Rugby* 61 American, or gridiron, football. 1968 C. Drummond *Death & Leaping Ladies* i. 11 You can't just walk into a team like you can, say, in gridiron or soccer.

gridiron, *v.* Add: (Earlier and later examples.) Also *fig.*

1832 *Reg. Deb. Congress* 4 Apr. 2390 With this revenue we could gridiron our States with railroads. 1901 *Daily Chron.* 5 Apr. 5/2 He won the right to gridiron its hills with an electric system. 1914 *Chambers's Jrnl.* June 415/2 The Park is gridironed with its own railway system. 1923 R. Kipling *Limits & Renewals* 110 An open square, near by, eased the pressure before long. Here the Patrol broke into fours, and gridironed it, saluting the images of the Gods at each corner.

Hence gri·dironing *vbl. sb.*, *N.Z.* (see quot. 1910); gri·dironer *N.Z.*, one who practises gridironing.

1898 Morris *Austral Eng.* 176/1 *Gridironing*, a term used in the province of Canterbury, New Zealand. A man purchased land in the shape of a gridiron, knowing that nobody would take the intermediate strips, which later he could purchase at his leisure. 1910 La Rossignol & Stewart *State Socialism in N.Z.* iii. 37 'Gridironing' consisted in buying a series of 20-acre sections so arranged as to leave 19 acres unbought between each two sections bought; and as no one could buy less than 20 acres without going to auction, the alternating 19-acre sections were left to be occupied by the runholder. 1941 Baker *N.Z. Slang* iii. 26 Land was purchased in strips,..so that the intervening land was rendered useless to another prospective settler and might be bought at the gridironer's leisure.

grief, *sb.* Add: **8. a.** *good* (or *great*) *grief!*, an exclamation indicating surprise, alarm, etc.; *to come to grief* (further example).

1850 Thackeray *Pendennis* II. xxvii. 364, I knew that your father had come to—to grief. You don't think it was—it was for your connexion I married you? 1900 in *Eng. Dial. Dict.*, Good grief. 1924 *Dial. Notes* V. 268 Great grief. 1937 R. Chandler *Lady in Lake* (1944) 217 'Good grief,' De Spain said. 'He's up there right now.' 1950 'N. Culotta' *They're Weird Mob* (1958) 180 'Do I ask your father for his daughter's hand?' 'Good grief, no!' 1965 'A. Gilbert' *Death takes Wife* xvi. 246 'Wonder if he expected this?' said the policeman grimly. And he called Crook at his home address. 'Good grief,' said Crook, when he heard. 1969 N. Marsh *Clutch of Constables* iv. 66 'I'm sorry.' 'Good grief, I'd forgotten that gang!'

9. *c. grief-ridden, -stricken* adjs. Also **grief therapy** (see quot.).

1891 Swinburne *Poems* (1904) II. 47 In that heart-stricken grief-ridden time. 1905 *Daily Chron.* 8 Feb. 3/5 The husband was sent for, but he was so grief-stricken that he had to be removed. 1963 J. Pollitt *Depression & its Treatment* ii. 22 Similarly, in human affairs, bereavement occurring during a depressive illness is not followed by the normal process of forgetting, the patient remaining grief-stricken. 1963 J. Mitford *Amer. Way Death* ii. 58 'Grief therapy' is..commonly used by funeral men to describe the mental and emotional solace which, they claim, is achieved for the bereaved family as a result of being able to 'view' the embalmed and restored deceased.

grievance. Add: **6.** *grievance-monger*.

1828 *Toronto Publ. Lib. MS. B* 104 41 Like William Lyon Mackenzie, he was a confirmed grievance-monger. 1890 [in Dict.]. 1966 *Economist* 12 Nov. 647/2 Mrs Gandhi's government and party can hardly be considered blameless. What they have to do to meet the real grievances that the grievance-mongers exploit is, simply, to observe.

grievous, *a.* Add: **3. b.** *grievous bodily harm*, a legal term denoting a serious injury (see quot. 1959). Also *transf.*

1861 *Act* 24 & 25 *Vict.* c. 100 § 20 Whosoever shall unlawfully and maliciously wound or inflict any grievous bodily Harm upon any other Person, either with or without any Weapon or Instrument, shall be guilty of a Misdemeanor. 1958 [see G.B.H., *G III. 5]. 1959 Jowitt *Dict. Eng. Law* I. 525/2 *Grievous bodily harm*, any injury which is merely technical or trivial, as distinguished from the more serious injury termed 'actual bodily harm', or the still more serious injury termed 'grievous bodily harm'. 1968 *Listener* 11 July 62/1 The spectator whose heart lifts at the sight of Clark Graebner committing grievous bodily harm on a tennis ball is a fortunate man indeed.

griff (grif), *sb.*⁵ *slang.* [Shortened form of *Griffin⁴.] A 'tip' or news; reliable information.

1891 N. Gould *Double Event* (1892) iv. 22 'He's a Newmarketer, he is. Not up here for nothing, you bet. He's got the straight griff for something.'..He is got the 'straightest griff' he ever had in his life, and that was to mount *Dorsia* for Sydney. 1946 *Finish! Pt Valley Campaign* 30 The officer who broke off a stern argument..to inquire, 'Have you got the latest griff' from a passing 'Tommy'. 1948 A. Baron *From City, from Plough* vi. 49 'Is that the griff, Higgsy?'.. 'Have I ever given you a wrong tip?' 1961 Wodehouse *Service with Smile* v. 73 You can't come telling a man to go pinching pigs without giving him the griff about why he's doing it and who for. 1967 M. Procter *Exercise Hoodwink* xvii. 119 Somebody in the centre of things had 'given the griff' whenever there had been any to give. 1968 *Listener* 8 Feb. 167/1 He was to give us the approved line, the authorised version, the holy griff on what we were actually up to in Suez. 1968 J. Wainwright *Edge of Extinction* 104 The informant was saying: 'It's griff, guv. The real thing.'

griffe (grif). *Archit.* [Fr.: see Griff *sb.*²] A claw-shaped ornament carved at the angle of the square base of a column; a spur.

1875 *Encycl. Brit.* II. 465/1 *Griffe*, a French term for an ornament at the angles of the base of early pillars, for which we have no proper equivalent. It first consisted of a single leaf, which became more elaborate, and was, no doubt, the origin of the foliated bases. 1901 R. Sturgis *Dict. Archit.* II. 323/1 The griffe..is often used

for elaborate ornamentation, being carved into vegetable or even animal form. 1905 F. Bond *Gothic Archit.* 455 Griffes are found in Roman work in a similar position... Pliny describes griffes of lizards and frogs. 1952 H. H. Saylor *Dict. Archit.* (1963) 79 *Griffe*, an ornamental form, somewhat like a claw, extending from the torus of a column base and lying on the four corners of the plinth.

griffin⁴. *slang.* [Origin unascertained.] A tip (in betting, etc.); a signal, hint. Cf. *Griff *sb.*⁵

1889 *Cassell's Sat. Jrnl.* Jan. 305/3 Plank yourself at the corner to give the griffin (signal) if you hear or see owt. 1891 J. Newman *Scamping Tricks* 95 When he wanted to give the chaps in the office the straight griffin, he used to say, 'Nelson's my guide.' 1912 A. N. Lyons *Clara* xiv. 265 'This is the Straight Griffin, Fred,' said Mr. Coenza: 'the absolute straight Tip.' 1962 F. Shaw et al. *Lern Yerself Scouse* 63 Let's give de fellers de griffin.

griffon³. Add: **Brussels griffon** (also **Griffon Belge, Bruxelles** or **Bruxellois**), a small griffon of European origin, with a short nose, flat face, prominent eyes and reddish-brown hair.

1898 *Dog Owners' Ann.* 63 [heading] Dog Griffon..The Griffon Bruxellois Club. 1900 C. H. Lane *All about Dogs* xiv. 269 *Griffons Bruxelles*, one of the most recent of all the breeds suitable as pets and companions and which has been very much 'boomed' the last few years. 1904 H. Compton *20th Cent. Dog* I. 119 In the Griffon Bruxellois you have the Captain Kettle of dogdom. 1906 [see Pomeranian]. 1919 [see *Affen-pinscher] The Brussels griffon. 1922 [see *Affenpinscher]. 1945 C. L. B. Hubbard *Observer's Bk. Dogs* 100 *Griffon Belge*, a variety of Griffon Bruxellois. 1966 —— *Dogs in Brit.* 293 Also known as the Brussels Griffon, the breed [*sc.* Griffon Bruxellois] is one of the smallest of terriers. 1969 B. Vesey-Fitzgerald *About Dogs* iv. 29 I believe that somewhere in the make-up of the Griffon are rough-haired Terriers: and that that is the Griffon Bruxellois.

grift (grift), *sb.*¹ *U.S. slang.* [Perh. corruption of *Graft *sb.*⁵] = *Graft *sb.*⁵

1914 Jackson & Hellyer *Vocab. Crim. Slang* 39 *Grift*..*graft*. 1920 D. Hammett *Dain Curse* (1930) iii. 21 'How's the literary grift go?' I asked. He looked at me sharply. 1930 W. Faulkner *Sanctuary* xvi. 131 'If you'll just promise to get the kid a good newspaper grift when he gets big enough to make change,' he said. 1940 R. Chandler *Farewell, My Lovely* xxiii. 157 Hell, I thought he sold papers. 1958 'E. Warner' The right protection behind him. 1961 ... that's a small-time racket. A peanut grift. 1963 J. P. Johnson *Man who sold Eiffel Tower* (1962) iv. 51 A good confidence man..has to have what is known as 'grift sense'.

grift (grift), *v.* *U.S. slang.* [f. prec.] = *Graft *v.*⁵ So **grifting** *vbl. sb.* and *ppl. a.*

1915 G. Bronson-Howard *God's Man* iv. iv. 263 Grifting ain't what it used to be. Fourteenth Street's got protection down to a system—a regular underworld tariff on larceny. 1926 *Flynn's* 31 Jan., *Grift*, to steal; to graft. 1940 *Amer. Speech* XV. 119/2 *Grift*, to work any gyp concession included in the grift. 1956 *Publ. Amer. Dial. Soc.* xxiv. 124 Tim Baird gifted on to a grifting graft with a buzzer on his wheel. 1965 ... he's got what she was not with *It* (1965) ii. 30 She had enough..to grift with a buzzer on her wheel. 1966 ... grifters, bonnie princes of con with tension plans. 1969 ... How long you been grifting?

grifter (gri·ftəɹ). *U.S. slang.* [f. *Grift *sb.*² + -er¹.] = *Grafter².

1915 G. Bronson-Howard *God's Man* iii. iii. 213 Every time I pick up one of these here magazines..I just naturally seem to encounter a lot of junk..whenever I read about grifters or guns—always this 'master crackman' stuff,..but when it comes to beating an ordinary box I could kick my way into it any stocking feet. 1927 L. Hughes *Fine Clothes to Jew*.. 1929 Y. Lustgarten... 1927 K. Nicholson *Barker* 149 *Grifter*, grafter. 1930 J. Lait *Put Homer* i. 7 A card or dice cheater is a 'grifter'. 1955 *Publ. Amer. Dial. Soc.* xxiv. 98 A grifter denotes a pickpocket or sneak thief. The term..has..in recent years so generalized that it now includes all sorts of confidence men. 1965 N. O'Connor *Jack London* i. 18 He lived off the horoscope trade until the World Fair of 1893 suggested a move to Chicago, as it did to thousands of other..grifters.

griggish (gri·giʃ), *a.* [f. Grig *sb.*¹ 5 + -ish.] Merry.

1879 G. M. Hopkins *Let.* 12 Feb. (1956) 153 The most griggish little piece [of music] the more we clapped it. 1882 —— *Let.* 20 Dec. (1955) 167, I seem to be in a griggish mood; it must be because holidays have begun.

Grignard (gri·nyɑ̈). *Chem.* The name of F. A. V. *Grignard* (1871–1935), French chemist, used *attrib.* and in the possessive to designate organometallic compounds of magnesium represented by the formula RMgX (where R is an organic radical and X a halogen), and the organic syntheses performed using such compounds; so *Grignard('s) compound, reagent; Grignard('s) reaction.*

[1902 *Jrnl. Chem. Soc.* LXXXII. i. 70 The following compounds have been obtained by Grignard's methods.] 1904 *Ibid.* LXXXV. 1666 (*heading*) The Grignard reaction applied to the esters of hydroxy-acids. *Ibid.* LXXXVI. i. 38 (*heading*) Nitric oxide and Grignard's reagent. 1936 *Discovery* May 162/2 Active organic substances (like organo-metallic compounds and the Grignard reagents) which are used as reagents, and the function of anionoid and kationoid reagents in ionic reactions are treated. 1964 N. G. Clark *Mod. Org. Chem.* v. 69 When dissolved in

dry ether, and treated with dry magnesium metal, alkyl halides combine to give alkylmagnesium halides or 'Grignard compounds'. 1966 Phillips & Williams *Inorg. Chem.* II. xxxi. 331 Grignard reagents are also of quite general applicability in substitution reactions of metal halides.

Grignolino (grinyoli·no). Also **Grignoli.** [It.] **a.** The name of an Italian wine grape grown in Piedmont. **b.** The red wine made from this grape.

1894 J. L. W. Thudichum *Treatise on Wines* xxv. 333 A number of red wines are made from the Grignoli grape, and named Grignolino. 1935 Schoonmaker & Marvel *Compl. Wine Bk.* vi. 115 The vast bulk of Italian wine is still..insufficiently standardized (out of a hundred bottles of Grignolino..two doesn't taste alike) and many are unintentionally sparkling and the others still). 1958 A. L. Simon *Dict. Wines* 86 *Grignolino*, a red wine of good distinction but none the less greatly prized locally; it is made in various parts of Piedmont from Grignolino grapes. 1966 H. Johnson *Wine* 208 There is some better and higher-strength wine made of the Grignolino, but it is hard to find. 1970 *Guardian* 6 Mar. 9/5 Half a bottle of Grignolino, a *pétillant* red wine from Piedmont.

gri-gri¹ (gri,gri). Var. Groo-Groo.

1871 Kingsley *At Last* (1896) vii. 138 This [*Desmoncus*] furnishes the gri-gri palm.

gri-gri², var. Greegree (in Dict. and Suppl.).

‖ **grihastha** (grihæ·spa). Also gr(h)astha, grihast. [Skr. *grastha*, married Brahman householder.] A Brahman in the second stage of life, which carries with it certain social obligations, as the duty to marry and have children.

1871 E. Balfour *Cycl. India* (ed. 2) II. 409 *Grihastha*, a hindu married householder. 1876 *Encycl. Brit.* IV. 204/2 The pious Brâhman..was enjoined to pass through a succession of four orders or stages of life. In the first he was a student..In the second he was a householder (*grihastha*, or householder. 1894 *19th Cent.* Aug. 281 One [story] that gained some credence with the grihastha was, that three natives..saw headless beings. 1949 R. B. Pandey *Hindu Saṁskāras* viii. 264 One who longs for imperishable heaven and happiness in this world, should uphold the Grastha-Āśrama. 1964 *Religious Hinduism* x. 117 [Fidelity] to the ritual and social duties of the caste are the qualities of the *grihastha*. 1969 *Indian Mus. Jrnl.* V. 79 [Music and dance] were the essential equipment of the training of a monk, a prince, a grihastha.

grike (graik). Also **gryke**. [north. dial.: see *Eng. Dial. Dict.*] A crack or slit in rock, a ravine in a hill-side; *spec.* in *Geol.*, a fissure between clints (*Clint *sb.* 1 b).

1781–1883 in *Eng. Dial. Dict.* 1902 Ld. Avebury *Scenery of Eng.* 437 Bare surface of Carboniferous Limestone, near Shap, showing 'grikes' or widened joints. 1919 *Glasgow Herald* 26 Dec. 8 Rock which lay about the 'grike' in the crag face. 1925 [see *Clint *sb.* 1 b]. 1938 A. E. Trueman *Scenery England & Wales* xi. 160 Wide, level stretches of bare light-coloured limestone, with the joint planes enlarged by the solvent action of rain so as to form fissures (grykes) between clints or grikes) trench the surface. 1952, 1963 [see *Clint *sb.* 1 b].

grill, *sb.*⁴ Add: **4.** **grill-room**, also more generally, an informal restaurant. (Later examples.)

1910 *Bradshaw's Railway Guide* Apr. 1027 Midland Hotel, Bradford..popular grill room and restaurant. 1931 J. Betjeman *Mount Zion* 19, I wish you'd seen the tea we had in the Grill Room at the Cri. 1957 *Encycl. Brit.* XX. 229/2 Modern grill rooms are an even later offshoot of the hotels and restaurants..to the grill-room's and demand for dress.

grill, *sb.*⁴ Add: **1. b.** In modern use: a gas burner (on a gas cooker), or a hot plate or a set of elements (on an electric cooker), which directs radiant heat downwards. Also *attrib.*

1907 *Yesterday's Shopping* (1969) 218/2 Gas stoves..The Talbot, hot plate and grill, fitted with 1 Reversible Grill Burner. 1908 *Installation News* II. 413 (*heading*) The..fitted on the grill, so that cooking simultaneously with same..cooking on the grill..to the supply lines that three of these connections, hall, three quarters, or the whole of the coils fitted are placed in circuit, and the degree of heat can thus be regulated. 1952 *Harmsworth's Househ. Encycl.* II. 1094/2 Everything cooked under the whole surface of the grill, and resting on the griller tray, comes as a white enamelled tray. 1928 *Daily Mail Cookery Bk.* 113 In making toast under the grill, be careful to see that the gas-burner..under the grill, griddle-pan with the grill-pan in position is underneath. 1951 *Good Housek. Home Encycl.* 75/2 A solid rectangular plate called the grill boiler. 1961 *Which?* Oct. 248/1 A recent trend in cooker design has been to move the grills to eye level.

2. *to put on the grill*: to subject (a prisoner) to 'third degree' treatment. *U.S.* (Cf. next.)

1928 *Daily Express* 25 May 10/5 Mr. John Brown..is arrested on a murder charge. He does not sound frank, so he is 'put on the grill'.

grill, *v.*² Add: **1. d.** To subject to severe questioning.

1894 G. Meredith *Let.* 30 Nov. (1970) III. 1178 Henry Parkman promised she would refresh me with an account of her last visit to you. Not a sign of her since. She comes to-day and she shall be grilled. 1928 A. G. Hays *Let Freedom Ring* 289 The three men were grilled about their movements on the day of the..attempted hold-up. 1932 E. Wallace *When Gangs came to London* ii. 25 Some day I'll be grilling you, big boy, up at police headquarters. 1938 G. Heyer *Blunt Instrument* vii. 139

Why on earth did your Superintendent go and grill the poor girl? 1970 *Radio Times* 8 Oct. 11/1 Listeners will be able to 'grill' leading public figures over the air when *It's Your Line*, a new-style 'live' current affairs programme begins.

Hence **grilling** *vbl. sb.* (further examples); **grill room** [pun on sense of *grill-room* s.v. Grill *sb.*³ 4], a room in a police station where suspects are questioned.

1915 Kipling in *Nash's & Pall Mall Mag.* Oct. 137/1 'They had a court-martial on me.'..'We *did* give you rather a grilling.' 1930 'E. Queen' *French Powder Mystery* xii. 86 Welles is on his way here—now we'll have arrests, interviews, grillings, reporters. 1931 O. K. Fraenkel *Sacco-Vanzetti Case* 16 The defendants had upon arrest been subjected to heavy grilling regarding their radical beliefs. 1950 G. Greene *Third Man* xv. 129 You were brought here for a grilling almost as soon as you got back into the Inner City. 1958 M. Procter *Man in Ambush* iii. 21 The 'grill room' was a place to make a suspect yearn desperately to see God's sunshine. 1967 —— *Rogue Running* x. 64 The two men went along to the 'grill room'...A C.I.D. clerk sat at the corner table.

grille, grill, *sb.* Add:

1. b. *spec.* Such a structure fixed in the body of a motor vehicle in front of the radiator, which it protects without preventing the flow of air over it. Freq. as **radiator grill(e)*.

1930 *Motor* 7 Oct. 434 New radiator styles are noticeable everywhere...Many use a grille of wire or a pierced metal screen to place in front of the honeycomb. *Ibid.* 437 The polished grille now fitted to the Panhard-Levassor radiator. 1938 Newton & Steeds *Motor Vehicle* (ed. 2) xv. 260 The plated external casing is merely a dummy, and in the great majority of cars takes the form of an ornamental grille or stone guard. 1962 *New Scientist* 22 Feb. 439 (Advt.), The wide-span grille shows the new car in an expansive mood. 1970 *Times* 4 June 18/7 Many car components..are now plastic moulding,..the..radiator grille, for example. 1971 *Daily Tel.* (Colour Suppl.) 22 Oct. 50/1 Fluorescent orange discs..were mounted on the front of a dark blue Mercedes, one each side of the grille.

5*. A rectangular pattern of small dots impressed on some issues of postage stamps (see quot. 1962). Also *attrib.* Also **grilled** *ppl. a.*, **grilling** *vbl. sb.*

1887 J. K. Tiffany *Hist. Postage Stamps U.S.A.* 170 All these values were issued with a grille, of which there are several sizes. 1896 *New Eng. Mag.* Jan. 566/1 The grilling of the stamps was continued until 1872, when a new ink was used that could not be removed from the paper without injuring the stamps. 1916 F. J. Melville *Postage Stamps in Making* 19 This grille embossing was applied to stamps of the United States between 1867 and 1873. *Ibid.*, The variety known as 'grilled all over'. 1929 K. B. Stiles *Stamps* v. 78 In the year 1867 there was invented a metal roll with points, and these made grill impressions on certain stamps of the United States. *Ibid.* 81 To count the grill points, examine the reverse side of the stamp. *Ibid.*, Only one other country has ever issued grilled stamps..Peru. 1962 K. F. Chapman *Stamp Collecting* 137 *Grille*, a pattern of tiny square dots impressed into the paper of several United States issues with a view to hindering the removal of cancellations. The grille broke the fibres of the paper and permitted the cancelling ink to penetrate the paper instead of remaining on the surface.

‖ **grillo** (gri·lo). [It. and Sp.: see Grylle.] A cricket.

1845 R. Ford *Hand-bk. Trav. Spain* I. 520 The Spaniards, like the ancients, delight in the *Grillo*. 1929 C. Mackenzie *Gallipoli Mem.* xi. 189 A wonderful race it was with..the grillos shrilling far and wide across that moon-drenched island. 1949 E. Pound *Pisan Cantos* lxxviii. 67 Be welcome, O cricket my grillo, but you must not sing after taps.

Grimaldi (grimæ·ldi). The name of the caves in Liguria, Italy, where the skeletons of a type of Upper Palæolithic man were found by Emile Rivière in 1872, used *attrib.* as in *Grimaldi skull*. Hence **Grima·ldian** *a.*, pertaining to or characteristic of Grimaldi man or of this culture.

1902 *Daily Chron.* 8 May 3/3 The Grimaldi Skulls. Dr. Verneau proposes to call the skulls by the name of Grimaldi, and I presume they will henceforth be so designated. 1932 *Antiquity* June 193 Various names, more particularly Creswellian and Grimaldian, have been given to this 'developed' Aurignacian [culture]. 1948 [see *Creswellian a.]. 1958 F. E. Zeuner *Dating Past* (ed. 4) 301 In the layers above, upper Aurignacian stone tools become abundant, and the bone tools rarer and atypical. For this Italian facies of the Upper Palaeolithic, Vaufrey adapted Rellini's term, Grimaldian.

Grimm's law: see Law *sb.*¹ 17 c (*c*).

‖ **grimoire** (grimwār). [Fr.; altered f. *grammaire* Grammar.] A magician's manual for invoking demons, etc.

1849 W. H. Ainsworth *Lancashire Witches* III. iii. v. 201 A witch with a Bible! It should be a grimoire. 1891 A. E. Waite *Occult Sciences* 55 The most noticeable feature of the Grimoires..is their utter futility. 1902 *Daily Chron.* 9 July 3/2 A monstrous brood of 'grimoires' and 'clavicles'. 1923 *Blackw. Mag.* Aug. 153/2 The curé's argument almost legitimised the *grimoire* as a commentary on the Bible. 1926 *Chambers's Jrnl.* July 460/1 The *grimoires* of the Middle Ages. 1945 M. Summers *Witch-craft & Black Magic* v. 133 *The Grimoire of Pope Honorius* is pretended to be found in manuscript as early as the thirteenth century. 1967 *Punch* 6 Dec. 854/1 One could fill a library with grimoires, ghost stories, studies of occult phenomena.

grimpen (gri·mpən). [Etym. uncertain.] ?A marshy area.

1902 A. Conan Doyle *Hound of Baskervilles* vii. 153 Life has become like that great Grimpen Mire, with little green patches everywhere into which one may sink and with no guide to point the track. 1940 T. S. Eliot *East Coker* ii. 10 In a dark wood, in a bramble, On the edge of a grimpen, where is no secure foothold. 1968 W. S. Baring-Gould *Annotated Sherlock Holmes* II. xxxvi. 47 As is well known, Watson's 'Great Grimpen Mire' is *Grimspound Bog*, three miles to the north and west of Widecombe-in-the-Moor.

Grimston (gri·mstən). The name of several towns and villages in eastern England (*spec.* Hanging Grimston, Yorks.), used *attrib.* to designate archæological remains found there, as *Grimston ware*. Also **Grimston hybrid** (see quot. 1960).

1954 S. Piggott *Neolithic Cultures* iv. 114 Fine carinated bowls typified by those from the Hanging Grimston long barrow..which may conveniently be called *Grimston ware*. 1960 P. H. Reaney *Orig. Eng. Place-Names* vii. 170 'Grimston hybrid' is a convenient term for place-names consisting of a Scandinavian personal-name compounded with OE *-tūn*, of which Grimston is particularly common. It derives from ON *Grímr*, ODa *Grim*.

Grimthorpe (gri·mþǫɹp), *v.* [f. the name of Sir Edmund Beckett, first Lord *Grimthorpe* (1816–1905), whose restoration of St. Albans Cathedral, completed in 1904, aroused fierce criticism and controversy.] *trans.* To restore (an ancient building) with lavish expenditure rather than skill and fine taste.

1890 *Antiquary* Jan. 34 To this a keen and well-known Yorkshire ecclesiologist replied: 'Heaven forbid! the building might be grimthorped!' 1892 *Athenæum* 23 July 138/2 St. Albans and other great national fabrics that have been 'Grimthorped'. 1900 *Ibid.* 28 July 129/2 This is indeed grimthorping with a vengeance. 1909 *Daily Chron.* 9 July 6/6 The parish church, which despite of vigorous 'grimthorping' still shows a trace of its old Norman architecture.

grimy, *a.* Add: **b.** *fig.* Unpleasant, mean. (Cf. quot. 1848 in Dict.)

1833 Dickens *Let.* ? 15 Apr. (1965) I. 19 The machinery is finished,..the Orchestra complete—and the manager *grimy*. 1834 —— *Let.* 3 Sept. (1965) I. 41 The grimy monotony of my every-day life. 1946 Mezzrow & Wolfe *Really Blues* 374 *Grimy*, sour, unpleasant.

grin, *v.*¹ Add: **1. f.** Of a coat of paint: to show *through* (an upper coat). Also used of other surfaces that exhibit gaps.

1854 A. E. Baker *Gloss. Northants. Words* I. 291 You must undo your work, the stitches grin so. 1901 in *N. & Q.* 9th Ser. VIII. 225/2 The priming coat grins through the paint of the sashes. 1916 R. A. Freeman *Exploits of Danby Croker* vii. 145 When you have drilled the holes, you must put a drop of walnut stain in each, or else they 'grin'. 1966 *Times* 25 Apr. 13/1 Tufting can produce a wide range of fabric qualities,..though cheaper ones have a tendency to 'grin'.., that is, to show the backing through the all-too-sparse pile.

3. *to grin and bear it* (earlier examples).

c 1810 W. Hickey *Mem.* (1960) viii. 132 Vexed at..the childishness of his behaviour, I answered, 'I recommend you to grin and bear it'—an expression used by sailors after a long continuance of bad weather. 1813 E. S. Barrett *Heroine* III. xxxix. 151, I heard a sudden disturbance below; his lordship crying out 'Oh, what shall I do?' and Jerry bidding him 'grin and bear it'. 1859 Geo. Eliot in M. Porter *Ann. Publishing House* (1898) III. ii. 51 'Adam Bede' flourishes, so I grins and bears it!

grind, *sb.*¹ Add: **1. c.** *Cambridge.* (See quots. 1889 and 1950.)

1889 Barrère & Leland *Dict. Slang* I. 431/1 The ferry-boats at Chesterton, wound across by a winch and chain, 'to go over in the grind'. 1901 *Cambr. Rev.* 14 Nov. 76/1 Trinity Hall..had 50 yards to spare at the Red Grind. 1906 *Westm. Gaz.* 11 June 8/2 Besides Charon's, two other ferries—grinds, with chain from bank to bank. 1950 M. Marples *University Slang* 103 A chain-ferry at Chesterton near Cambridge was also known as the Grind, but it is not clear whether from its connection with walking or rowing, or from the effort required to operate it: later every ferry came to be so called.

d. The size of the particles of a powder, e.g. ground coffee.

1922 W. H. Ukers *All about Coffee* xxv. 401/1 A progressive coffee-packing house may have..a pulverizer for making a really fine grind. *Ibid.* 402/1 A mixture of a very fine with a coarse grind gives the best results in the cup. 1958 *Listener* 13 Feb. 299/1, I buy the grind [of coffee] I find suits my method of coffee-making best. *Ibid.* 299/2 The coarse grind [of coffee] gives the best results. 1959 *Gloss. Packaging Terms* (B.S.I.) 34 *Grind*, the size of the particles of which moulding powder is composed, generally graded according to the portions retained by different mesh sizes.

3. b. (Earlier and later examples.)

1893 W. K. Post *Harvard Stories* 11 Come now, old grind, do take a day off. 1908 R. L. Dunn *W. H. Taft* 210 He was keen to learn and if he had not been so lusty outside of the house, he would have been called a grind. 1951 S. Lewis *World so Wide* viii. 90 He told himself that, with this conceited grind, there was no merit in even a boarding-house courtesy.

4. *slang.* (An act of) sexual intercourse. Also *on the grind*.

1893 Farmer & Henley *Slang* III. 216/1 *On the grind*, said of incontinent persons of both sexes. Also of prostitutes. *c* 1912 D. H. Lawrence *Love among Haystacks* (1930) 26 'A good grind, eh?' said the tramp, nodding after the Fraülein. The men only half-understood him. 1969 J. Wainwright *Big Tickle* 112 A grind with a cheap scrubber?

5. The action of rotating the hips in a dance or the like. Cf. *Bump *sb.*¹ 1 f. *slang* (orig. *U.S.*).

1946, 1964 [see *Bump *sb.*¹ 1 f].

grind, *v.*¹ Add: **5. b.** Also *to grind in*: to smooth the surface of (a machine part) by moving it to and fro against the surface with which it is to fit or mate; *esp.* to make (a valve in a cylinder of an internal combustion engine) fit smoothly and tightly into its seat by rotating it to and fro against the seat with a suitable abrasive paste; occas. *to grind* (a valve) *in* or *into* or *on to* (its seat).

1888 [in Dict.]. 1895 *Boy's Own Paper* XVII. 350/3 To make the valves fit tight you should grind them in their seating with a little fine emery and oil. 1903 R. J. Mecredy *Dict. Motoring* 281 New valves should also be ground on to their seatings. 1905 H. J. Spooner *Motors & Motoring* 19 Grinding in valves is an operation that had better be left to the trained mechanic. 1916 J. E. Homans *Automobile Handbk.* xvii. 184 It is necessary in grinding a valve into its seat to place a ball of cotton waste..into the port leading to the combustion space. 1924 A. W. Judge *Mod. Motor Cars* III. 297 It is usually necessary to grind in the valves whenever the cylinders are decarbonised. 1928 —— *Car Maintenance* 43 After replacing a ground-in valve. 1935 Jelley & Harrison *De Luxe Ford Handbk.* xii. 86 It is absolutely essential that each valve is ground into and assembled into the seat from which it was removed. *Ibid.* 87 It is always bad practice to grind in a badly pitted valve. 1950 A. W. Judge *Motor Vehicle Engine Servicing* iv. 55 The valve face..can readily be restored to its original condition by grinding with an abrasive paste on to its seating. 1962 'S. Abbey' *Motor-Car Maintenance* iv. 55 Sooner or later..it will be necessary to remove the cylinder head from the engine to allow..the valves to be ground-in on their seatings. 1971 B. C. MacDonald *Ford Cortina Repairs* i. 25 The next step is to grind-in the valves on their seats in the cylinder head.

9. Also with adv.

1924 Galsworthy *White Monkey* I. iii, A taxicab ground up.

10*. *intr.* and *trans.* To copulate (with). Hence **grinding** *vbl. sb. slang.*

[1598 Florio *Worlde of Wordes* 210/2 *Macinio*, the grinding or greest. Also taken for carnall copulation.] 1647 *Ladies Parliament* sig. C2, Digbies Lady takes it ill, that her Lord grinds not at her mill. 1811 *Lex. Balatronicum*, *To grind*, to have carnal knowledge of a woman. 1879–80 *Pearl* (1970) 258 A married man grinding another man's wife. 1928 D. H. Lawrence *Lady Chatterley* xiv. 243 She had to work the thing herself, grind her own coffee. 1966 I. Jefferies *House Surgeon* ix. 162 Rob, what do you think about grinding..? I know it's time-wasting but it's so difficult to do without it.

grindability (graindăbi·liti). [f. Grind *v.*¹ + Ability.] The extent to which a material is readily ground or pulverized; susceptibility to grinding.

1932 *Trans. Amer. Soc. Mech. Engineers* LIV. Fuels & Steam Power 37/1 If accurate knowledge of the grindability and abrasive qualities of the coal could be determined..it would be of considerable help to the coal buyer. 1950 J. H. Perry *Chem. Engineers' Handbk.* (ed. 3) xvi. 1114/2 In the case of grinding to No. 48 sieve fineness, values of 0·304 and 0·26 are reported as the grindabilities for petroleum coke and graphite, respectively. 1958 *Engineering* 21 Mar. 375/2 Investigations in the field of coke testing, coal 'grindability', and petrographical work. 1965 G. J. Williams *Econ. Geol. N.Z.* xvi. 245/1 Where possible the species of diatoms are named in this section as the shape affects the grindability of the material.

grinder. Add: **5. b.** = Grind *sb.*¹ 3 b.

1852 J. C. Patteson *Let.* Aug. in C. M. Yonge *Life J. C. Patteson* (1874) I. iv. 116 The difficulty is great enough to discourage any but a real 'grinder' at such work. 1942 Berrey & Van den Bark *Amer. Thes. Slang* § 825.10 *Diligent student*.., grinder, grindstone, grub.

8. b. *U.S. slang.* (See quots.)

1954 Webster Add., *Grinder*, a large sandwich made of two slabs of bread cut lengthwise from the loaf and containing ham, salami, or other meat, usually cheese, and pickle, tomato and lettuce, or other appetizers. 1967 *Amer. Speech* XLII. 287 *Grinder*,..one explanation offered..is that the consumer must be able to grind the diverse ingredients in his mouth.

8*. *Radio.* An atmospheric disturbance of relatively long duration heard as a rumbling sound and probably caused by lightning.

1922, 1936 [see *Click *sb.*¹ 1 b]. 1940 *Chambers's Techn. Dict.* 390/2 *Grinder*, a type of atmospheric disturbance.., best characterised by its name.

gringo. (Earlier, later, and *attrib.* examples.)

1849 J. W. Audubon *Western Jrnl.* (1906) 13 June 100 We were hooted and shouted at as we passed through, and called 'Gringoes'. 1871 *Republican Rev.* (Albuquerque, N.M.) 14 Jan. 2/2 Three Mexicans from Socorro..calling

her a *gringo* bitch, finally threw her on the body of her husband. 1876 *Congress. Record* 30 June 4310/1 Cortina has never failed to rouse the hatred of the Mexican population against the 'gringos'. 1927 W. Cather *Death comes for Archbishop* v. i. 135 Any European, except a Spaniard, was regarded as a gringo. 1933 A. Huxley *Let.* 24 Mar. (1969) 369 Annoying foreigners and especially white Gringoes is a national sport in Honduras. 1962 N. Maxwell *Witch-Doctor's Apprentice* iii. 22 When he revived, he explained that cameras are known to be gringo machines emitting 'electricity' which hypnotizes the victim and robs him of his will. 1964 *Daily Tel.* 11 Jan. 1/1 Mobs conducting a 'gringo hunt' roamed the streets and looted homes of Americans living in Panama. 1969 J. Mander *Static Soc.* iii. 95 It is unadorned *gringo* imperialism.

grip, *sb.*¹ Add: **2. a.** Also *to get* (or *take*) *a grip on* (oneself), *to get to grips with* (something).

1895 *Harper's* Nov. 962/1 My dear boy, get a grip on yourself... I won't bite you. 1929 W. Faulkner *Sound & Fury* 174 My throat wouldn't quit trying to laugh, like retching after your stomach is empty. 'Whoa, now,' Anse said. 'Get a grip on yourself.' 1947 *Sci. News* IV. 7 They [*sc.* readers] have to translate his article into understandable language before they can get to grips with its actual subject matter. 1950 R. Ackland in *Plays of Year 1949* 611 Don't be such a foolish woman... Sit down and take a grip on yourself. 1955 *Times* 25 July 5/4 What we have now agreed makes it possible to get to grips with the twin problems of the unity of Germany and the security of Europe. 1967 S. Beckett *No's Knife* 52 Come now, come now, he said, get a grip on yourself, be a man.

b. *to lose one's grip* (cf. Lose *v.*¹ 3 d).

1875 J. Miller *First Fam'lies of Sierras* (1876) 246 Lost my 'grip'.., didn't have any 'snap' any more. 1894 'Mark Twain' *Pudd'nhead Wilson* xx, Come, cheer up, old man; there's no use in losing your grip. 1968 *Times Lit. Suppl.* 8 Feb. 122/5 His work after the war shows a steady decline.., until he seems to have lost his grip altogether.

5. b. A hair-grip.

1960 C. Storr *Marianne & Mark* xi, 144, I want two cards of grips and a set of rollers.

7. (Later examples.)

1961 A. Berkman *Singers' Gloss. Show Business* 26 *Grip*, stage hand, especially one who works on the stage floor. 1965 J. von Sternberg *Fun Chinese Laundry* (1966) viii. 191 Grip and Property Man..$100. 1967 H. Harrison *Technicolor Time Machine* (1968) ix. 92 One of the grips brought out a baby spot and plugged it in for light.

8. (Earlier and later examples in sense 'gripsack'.) (No longer restricted to the U.S.)

1879 *Chicago Tribune* 7 Mar. 9/5 At Cherokee I stepped from the train, took my 'grip', and began in earnest the life of a pilgrim. 1926 *Daily Colonist* (Victoria, B.C.) 14 July 5/6 Experienced travellers in all countries always take a bottle of ENO in their grip to offset changes of water and diet. 1928 W. Gillette *Astound. Crime Torrington Rd.* v. 282 'Want anything from the hotel—toilet articles—clothing—tobacco?' 'Thanks—I've got 'em outside in a grip.' 1960 J. Betjeman *Summoned by Bells* vii. 66 Clutching a leather grip Containing things for the first night of term. 1965 G. McInnes *Road to Gundagai* xii. 207, I toted my grip all the way back.

8*. (See quot. 1916.) *Austral.*

1906 E. Dyson *Fact'ry 'Ands* xviii. 243, I had t' do it 'r resign me grip on ther spot. 1916 C. J. Dennis *Songs Sentimental Bloke* 123 *Grip*, occupation, employment. 1941 Baker *Dict. Austral. Slang* 32 *Grip*, a job, regular employment.

9. *grip-bag* = Gripsack.

1958 *Listener* 17 July 107/2 Take, if you can, an extra grip-bag—a canvas one. 1963 T. Parker *Unknown Citizen* i. 22 In one hand he carried a blue grip-bag, like those sometimes used by airline passengers.

grip, *v.*¹ Add: **1. d.** To place (one's hands) so that they hold each other or an object in a grip.

1907 *Smart Set* Jan. 32/2 She fell back in the chair and gripped her hands round the arms of it. *Ibid.* Feb. 24/1 He gripped his hands together and put the doubt behind him. 1910 E. M. Albanesi *For Love of Anne Lambart* 112 Anne's two cold hands gripped themselves together.

gripe, *sb.*¹ Add: **6. c.** A complaint; a grumble. *slang* (orig. *U.S.*).

1934 J. Farrell *Studs Lonigan* (1936) III. xv. 341 The water was just right..and he took rhythmic strokes...It was like losing all the gripes that had been piling up within him. 1949 *Harper's* Jan. 61/1, I want to clear my desk of various matters, mostly gripes. 1954 *Chem. & Engin. News* 8 Mar., As a standard bearer in the cause of accurate nomenclature, you may be interested in one of my pet gripes. 1959 J. Thurber *Years with Ross* v. 76 Ingersoll was the main target of his gripes. 1964 S. M. Miller in I. L. Horowitz *New Sociol.* 306 The 'gripes' of low-income neighborhood..are political issues.

9. *gripe water* (examples).

1728 E. Smith *Compleat Housewife* (ed. 2) 231 *To make Gripe-water.* Take..Penny royal,..Coriander-seeds, Aniseeds, sweet Fennel-seeds, Carraway-seeds; bruise them all, and..sprinkle on them a quart of Brandy..distil it off..drink it warm, and go to bed. 1926 *Daily Colonist* (Victoria, B.C.) 3 July 4/6 (Advt.), The baby screamed day and night, having been slightly poisoned. I tried Woodward's Gripe Water when all else failed. 1953 E. Taylor *Sleeping Beauty* xiii. 198 The difficulties of obtaining the right brands of gripe-water, groats, or rusks.

gripe, *v.*¹ Add: **7.** (Further U.S. examples.)

1905 R. E. Beach *Pardners* (1912) i. 29. It gripes me to hear a man cry. 1941 J. M. Cain *Mildred Pierce* (1943) 88 What's griping him is that he can't do anything for the kids.

9*. *intr.* To complain, 'grouse'. So **griping** *vbl. sb. slang* (orig. *U.S.*).

1932 *Amer. Speech* June 332 *Gripe*, to complain. 1934 F. Scott Fitzgerald *Tender is Night* II. i. 153 In some moods he griped at his own reasoning: Could I help it. 1940 *New Yorker* 21 Sept. 37 He got good and sore and griped. 1945 E. Ford *Larry Scott* v. 51 I've already told him that the newspaper game is a lousy business, so you can save your griping for somebody else. 1947 D. M. Davin *Gorse blooms Pale* 199 Old Snow was griping away about his girl turning him down. 1959 *Times Lit. Suppl.* 20 Nov. 678/1 Let us get the griping over quickly. 1963 *Time* 30 Aug. 18/2 Ike..griped publicly: 'There are too many of these generals who have all sorts of ideas.' 1967 *Boston Traveler* 27/2 People are always griping about kids hanging around and being at the wrong places at the wrong time.

griper. Add: **7.** One who complains. *U.S. slang.* Cf. *Gripe *v.*¹ 9*.

1937 *Amer. Speech* XII. 315/2 Radio lacks a word to describe the disgruntled listener...Griper is not precise enough. 1938 *Ibid.* XIII. 315/2 *Griper*, a customer that always finds fault.

griphite (gri·fait). *Min.* [f. Gr. γρῖφ-ος riddle (see quot. 1891) + -ite¹.] A basic phosphate of manganese, sodium, calcium, iron, and aluminium occurring as dark brown or black masses.

1891 W. P. Headden in *Amer. Jrnl. Sci.* CXLI. 416, I would propose to call this new phosphate Griphite, from γρῖφος *puzzle*, in allusion to its unusual and somewhat enigmatical composition. 1942 *Amer. Mineral.* XXVII. 452 Griphite is cubic and its structure is similar to that of garnets.

grippe. Add: Hence ‖ **grippé** *a.* = Gripped *a.*

1890 W. Bateson *Let.* 16 Jan. in B. Bateson *W. Bateson* (1928) 40, I am glad to hear that you were so slightly '*grippé*' after all. 1933 J. Joyce *Let.* 18 Oct. (1966) III. 288, I..have to put it off, being slightly *grippé*.

gripper. Add: **4.** *gripper edge.*

1932 W. Atkins *Art & Practice Printing* II. iv. 49 Turn the Cylinder forward until the gripper edge is in a position convenient for undoing the three screws. 1940 *Chambers's Techn. Dict.* 391/1 *Gripper edge*, the edge of a sheet of paper caught by the grippers when it is fed into a cylinder machine.

gripping, *ppl. a.* Hence **gri·ppingly** *adv.*

1934 in Webster. 1961 *John o' London's* 14 Dec. 664/1 It was grippingly well acted. 1969 *Daily Tel.* 5 Nov. 13/5 Nearly all the work's varied utterances were similarly well characterised by the Group, the clarinet solo..grippingly so if with a certain unauthorised licence.

grippy, *a.* Add: **3.** Capable of holding the attention and interest of a spectator, reader, etc.

1921 *Pall Mall & Globe* 15 Nov. 2/3 With a little cutting-down, 'Wat Tyler' would make an entirely good and grippy little play.

gripsack. (Earlier and later examples.)

1877 R. J. Burdette *Rise & Fall of Mustache* 59 If I was going to Europe, I would just rush into the house, put on a clean shirt, grab up my gripsack, and fly. 1909 W. J. Locke *Septimus* xiv. 213 Gives instant relief for blistered feet. Every mountaineer should carry it in his gripsack. 1911 *Daily Colonist* (Victoria, B.C.) 29 Apr. 3/1 This most refreshing, invigorating and purifying of Fruit Salts should be in..the gripsack of every tourist. 1926 Chesterton *Incredulity of Father Brown* i. 4 His grip-sack held in a steely grip. 1949 E. Pound *Pisan Cantos* lxxvii. 49 Preceded by a crash, i.e. by a Huge gripsack or satchel.

Griqua (grī·kwā), *sb.* and *a.* [Hottentot.] **A.** *sb.* One of a coloured people of mixed Hottentot and European descent inhabiting chiefly East and West Griqualand in the Cape Province of South Africa; = *Bastard *sb.* 1 b. **B.** *adj.* Of or pertaining to this people. Cf. *Coloured *ppl. a.* 2 b.

1731 G. Medley tr. *Kolb's Pres. State Cape Good-Hope* I. vi. 78 Next to the Damaquas, lie the Gauros or Gauriquas. 1793 tr. *Thunberg's Trav.* I. 308 Gauriquas land extends more to the north-eastward: this is a very fine country, and abounds in grass. 1815 J. Campbell *Trav. S. Afr.* (ed. 2) xiii. 162 The people were employed by one of the Griqua captains to watch his cattle. *Ibid.* 163 We partook together of the Lord's Supper, at which were present, Griquas, Hottentots, Dutch, English and Scotch. *Ibid.* xxvii. 349 The people in this part, being a mixed race, went by the name of Bastards; but having represented to the principal persons the offensiveness of the word to an English or Dutch ear, they resolved to assume some other name. On consulting among themselves, they found the majority were descended from a person of the name of Griqua, and they resolved hereafter to be called Griquas. 1822 —— *Trav. S. Afr. 2nd Journey* I. xiii. 130, I marked the Hottentot, Caffre, Bushmen, Griqua, Namaqua, and Matchappee countries. 1833 J. Philip *Let.* May in D. J. Kotzé *Lett Amer. Missionaries* (1950) 30 Our missionary stations among the Griquas. 1839 F. Owen *Diary* (1926) 149 Several Boers reside in the Griqua country. *Ibid.*, The Griqua or Hottentot language. 1848 H. H. Methuen *Life in Wilderness* iii. 72 Thence we proceeded to the Griqua village of Ramah, and rejoined our waggons there. 1850 [see Laager *sb.*]. 1857 Livingstone *Missionary Trav. S. Afr.* v. 104 On crossing the Orange River we come into independent territory inhabited by Griquas and Bechuanas. By Griquas is meant any mixed race sprung from natives and Europeans. 1866 [see *Bastard *sb.* 1 b]. 1897 [see Hottentot 2]. 1905 G. W. Stow *Native Races S. Afr.* xvii. 316 Among the old Hottentot tribes..was a clan..variously called Chariguriqua and Grigriqua... It was from this tribe the modern Griquas derived their name. 1958 L. van der Post *Lost World of Kalahari* i. 24 Griqua robbers. 1970 *Cape Times* 28 Oct. 2/9 In it he tells how, during the second British occupation of the Cape, there lives a Griqua-Hottentot, by name Nossie.

Hence **Gri·qualander**, an inhabitant of E. or W. Griqualand; a Griqua.

1897 A. J. Butler tr. *Ratzel's Hist. Mankind* II. iii. 295 The term 'Griqualanders'..has become usual since the Griquas have..had a country of their own allotted them.

grisaille. Add: Also *fig.*

1937 R. Storrs *Orientations* ii. 38 As I look on my lines they seem, compared with what he was, the mere *grisaille* cartoon of some heroic Victorian radiance. 1965 *Listener* 4 Mar. 333/1 Schopenhauer was the first to admit his indebtedness to the French moralists, and the *grissaille* of his prose is relieved..by quotations not only from the French but [etc.].

b. (Later examples.)

1899 R. Glazier *Man. Hist. Ornament* 93 Painted Enamels were introduced and Limoges became the centre of this art, called late Limoges or Grisaille Enamel. 1968 *Medium Ævum* XXXVII. 63 One may not greatly admire..the 105 large grisaille pictures made by Jean Tavernier to illustrate the Brussels manuscript.

griseofulvin (gri:ziofʊ·lvin). *Chem.* [f. mod. L. *griseofulv-um* (f. Griseo- + L. *fulvus* reddish yellow), epithet of a species of *Penicillium*: see -in¹.] An anti-fungal antibiotic, $C_{17}H_{17}O_6Cl$, produced by several species of mould of the genus *Penicillium* and used esp. in the treatment of ringworm.

1939 A. E. Oxford et al. in *Biochem. Jrnl.* XXXIII. 240 The purpose of the present communication is to record observations on a hitherto undescribed chlorine-containing metabolic product of *P. griseo-fulvum* which has been isolated from the mycelium of this mould and for which the name griseofulvin is proposed. 1960 *N.Y. Times* 14 Aug. 77/3 The first few cases of improvement with griseofulvin therapy. 1968 J. H. Burn *Lect. Notes Pharmacol.* (ed. 9) 126 Ring worm (Tinea capitis) responds well to oral treatment with antibiotic Griseofulvin (also used for ringworm of glabrous skin which does not respond to fungicidal lotions or salves).

‖ **grissino** (grisī·no). Pl. **grissini.** [It.] A crisp bread made in long slender sticks.

1851 Lytton *My Novel* I. iv. xvii. 383 Crisp *grissins*,..they make so pleasant a noise between one's teeth. 1906 S. Beaty-Pownall *'Queen' Cookery Bks.* (ed. 2) xi. iii. 56 Abroad,..grissini are very popular. 1967 H. Porter in *Coast to Coast 1965–6* 171 The School Matron and the Housekeeper had got crisper and leaner than *grissini*. 1969 A. Laski *Dominant Fifth* i. 27 There [were]..sticks of celery and carrot and grissini to dip.

grist, *sb.*¹ **2. d.** (Earlier and later examples.)

1832 J. K. Paulding *Westward Ho!* I. 77 There has been a mighty grist of rain lately up above. 1881 S. P. McLean *Cape Cod Folks* xviii. 295 'Grists on 'em, this year!' he said. 'Heaps!' Aunt Patty responded. 1906 *Springfield* (Mass.) *Weekly Republ.* 8 Feb. 9 A good-sized grist of matters was presented in the House last week under suspension of the rules.

grit, *sb.*¹ Add: **6.** *grit-band, -bed; grit-tempered* adj.; **grit-blasting** *vbl. sb.*, the use of a stream of abrasive particles directed at a surface to clean it and roughen it; hence (as a back-formation) **grit-blast** *v. trans.* and *absol.*; **grit-cell** = *stone-cell* (Stone *sb.* 20), Sclereid.

1881 *Geol. Mag.* Nov. 488 The beds are traversed by a coarse cleavage dipping south, which usually ignores the hard grit bands. 1890 *Jrnl. Geol. Soc.* XLVI. 496 The red slates with hard grit beds south of Goodrington beach support a Triassic outlier. 1953 Arkell & Tomkeieff *Eng. Rock Terms* 53/1 The original sense of 'gravelly' seems to be preserved in the local term Grit Bed used for the Melbourn Rock in the Middle Chalk of Kent and Sussex, which is a band of limestone with a hard nodular structure. 1962 *Flight International* LXXXII. 201/1 The finished spar would be grit-blasted,..principally in order to remove small amounts of metal under closely controlled conditions. 1964 *Financ. Times* 8 Sept. 11/4 The operator can safely grit-blast without having to dress in protective clothing. 1936 *Metals Handbk.* (Amer. Soc. Metals) 841 (*heading*) Cleaning metals by blast cleaning process (also known as sand blasting, grit, and shot blasting). 1947 *Metallurgia* Apr. 279/2 Such a cleaning process as grit blasting lowers the corrosion resistance of the alloys. 1963 *Times* 22 Apr. p. v/1 To make sure the zinc sticks the steel must be roughened by gritblasting immediately beforehand. 1900 B. D. Jackson *Gloss. Bot. Terms* 115/2 *Grit-cell*, a sclerotic cell, as in the flesh of pears. 1960 W. B. R. Laidlaw *Guide Brit. Hardwoods* 103 Fruit pear-shaped; with grit-cells. 1939 V. G. Childe *Dawn Europ. Civilisation* (ed. 3) xii. 214 The 'classical' beaker..made of relatively fine grit-tempered ware.

grit, *sb.*² **2.** (Later examples.)

1938 M. K. Rawlings *Yearling* iii. 27 Breakfast was on the table...There were grits and gravy, hot cakes, and buttermilk. 1961 *Encounter* XVI. 20 Other eaters..were forking up eggs and grits. 1969 *New Yorker* 10 May 32/1 We stopped for scrambled eggs and grits in a little town in Alabama.

grit, *v.* Add: **4.** (Earlier examples.)
1797 Jefferson *Writings* (1903) I. 416 Mr. Adams,..gritting his teeth, said [etc.]. 1840 *Southern Lit. Messenger* VI. 735/2 The duellist gritted his teeth as he cocked the gun a second time.
b. To utter with gritting of the teeth.
1900 *Daily News* 11 Oct. 3/1 The Boers have gritted it between their teeth, mingling it with sullen curses—for to the veldtsman the name [of J. Chamberlain] is as the name of a thing accursed. 1910 C. E. Mulford *Hopalong Cassidy* i. 11 'I'll kill you some day, you whelp,' he gritted.
Hence **gri·tting** *vbl. sb.*
1823 *Massachusetts Spy* 30 Apr. (Th.), The harmony arising from the filing of a saw, or the gritting of teeth. *a* 1849 Poe *Tales & Sketches* (1852) 255, I could have sworn that it was the gritting of this vagabond's teeth. 1901 *Daily News* 13 Mar. 5/2 The gritting and corroding of tubes. 1908 *Westm. Gaz.* 14 Dec. 11/2 The Borough Council includes no less a sum than £2,000 for snow removal and the gritting of the roads. 1963 *Guardian* 21 Jan. 1/1 Todmorden local authority has a plough and gritting lorry out. 1963 *Times* 4 Feb. 6/6 Snow ploughs and gritting gangs worked throughout yesterday to clear 50,000 miles of snow-covered trunk roads. 1968 *Times* 26 Mar. 2/2 A big new road-gritting machine.

grittily (gri·tili), *adv.* [f. Gritty *a.*¹ + -ly².] In a gritty manner; with a gritty sound.
1911 D. H. Lawrence *White Peacock* II. i, He marched me..into the sanded passage of the little inn... As we tramped grittily down the passage, [etc.]. 1953 *New Statesman* 17 Oct. 462 One can see what a successful story Mr. Peake could have made out of it had he..made his 'natural' characters grittily real.

grizzle, *sb.*² Add: **2.** A bout of grumbling or sulking; a peevish mood; a fretful effusion.
1900 in *Eng. Dial. Dict.* 1908 A. E. Lyons *Arthur's* I. ii. 12 At first I thought she was goin' to do a grizzle, but she turned round quick, with a kind of smile. 1923 U. L. Silberrad *Lett. J. Armiter* xi. § 2, Opportunity to write a good long letter to make up for the short grizzle of last month. 1930 H. M. Tomlinson *All our Yesterdays* I. viii. 85 Out he went without saying good-bye.., which would have given any girl the grizzles.

grizzle (gri·z'l), *v.*³ *local.* [Origin unknown.] *trans.* and *intr.* To fry, frizzle, over-cook.
1900 in *Eng. Dial. Dict.* 1913 H. S. Walpole *Fortitude* I. viii. 98 Poor old man..nobody loves him..to hell with the lot of 'em..let 'em grizzle in hell fire. 1941 J. Cary *House of Children* 202 Four or five men surrounded the table,..eating and calling out to the girl not to grizzle the bacon.

grizzler (gri·zləɹ). *dial.* or *colloq.* [f. Grizzle *v.*² + -er¹.] One who frets or grumbles.
1900 *Eng. Dial. Dict.* s.v. *Grizzle v.*¹ 2, If ever there was a grizzler she's the one. 1910 H. G. Wells *Mr. Polly* vi. 201 'I don't think I *could* abide a grizzler,' said Uncle Pentstemon. 1924 'R. Dehan' *Pipers of Market Place* xvi. 153 She who scorned the women who cried as 'wimmickers' and 'grizzlers'.

grizzly, *a.* and *sb.*¹ Add: **A.** *adj.* **b. grizzly bear**: also, the name of an American dance in which the hug and walk of a bear are imitated.
1912 *Lit. Digest* 30 Mar. 656/2 Steps that would have ruled him off any cotillion floor in New York in these days of the ban on the grizzly-bear and kindred dances. 1928 F. Scott Fitzgerald in *Sat. Even. Post* 29 Sept. 118 They moved around the room, locked in the convulsive grip of the grizzly bear. 1959 M. T. Williams *Art of Jazz* (1960) xii. 209 The halls where they had so recently danced the Grizzly Bear and the Buzzard Lope to the music of the jazz bands.

groaner. Add: **b.** A whistling buoy. *local U.S.*
1903 G. S. Wasson in *Century* Aug. 538/1 'These here plaguy bell-b'ys an' groaners is a ter'ble old nuisance, you' exclaimed Cap'n Roundturn. 1941 T. S. Eliot *Dry Salvages* 5 Groaner: a whistling buoy. [*Ibid.* 1. 8 The heaving groaner Rounded homewards.] 1947 C. D. White *Handbk. Sailing* xxii. 236 Effective at night in fog or low visibility, whistling buoys do not sound like a whistle but like a groan, and are called 'groaners' by seafaring men.

grocer. Add: **2. c.** *grocer's paper* = *grocery-paper*.
1863 Rossetti *Let.* Jan. (1965) II. 392, I shall have it printed..on blue grocer's-paper. 1925 E., O. & S. Sitwell *Poor Young People* 59 And tied in a packet of blue grocer's paper.
d. *grocer's Graves, port, sherry, wine*, cheap wine bought at a grocer's shop.
1931 W. Holtby *Poor Caroline* i. 21 The cultivation of a palate cannot be achieved on grocer's port. 1936 'N. Blake' *Thou Shell of Death* i. 22 His idea of dinner is boiled mutton and grocer's Graves. *Ibid.* iii. 48 I've..met him..at a dinner at Christ Church..you know the sort of grocer's port they dish out there. 1940 Graves & Hodge *Long Week-End* xxiii. 399 Ordinary cheap grocer's wine. 1958 'W. Haggard' *Slow Burner* iv. 72 Dublin glass..and grocers' sherry at best. 1960 J. Morris *Hired to Kill* III. xvi. 141 To that it bears about as much resemblance as does a bottle of the finest vintage claret to grocer's wine.

grocery. Add: **4. b.** (Earlier and later examples.)
1806 *Balance* (Hudson, N.Y.) 28 Jan. 31 (Th.), There are 174 licensed groceries in the city of Albany. 1830 *Jeffersonian* (Albany, N.Y.) 30 June (Th.), Wilson told the Sheriff to take the jury to a grocery, that he might treat them. 1872 E. Eggleston *Hoosier Schoolmaster* (1872) xi. 99, I must a took a little too much at Welch's grocery. 1923 *Dial. Notes* V. 240 *Grocery*, a grocery store where liquor was sold. 1946 G. Foreman *Last Trek Indians* 201 The Missouri state line, all along which were placed the so-called 'groceries', which were nothing more than grog-shops.
5. *grocery keeper* (U.S.).
1830 *Indiana Ho. Repres. Jrnl.* 274 An act to be passed repealing the present law granting license to grocery keepers. 1872 'P. V. Nasby' *Struggles* 112 The grocery-keepers are intimatin' that before long I may expect to hev my pay fer licker. 1888 J. Bryce *Amer. Commw.* II. III. xiii. 134 Grocery keepers, liquor dealers, insurance, vendors of patents.

groceteria (grōusĕtī°riă). *N. Amer.* Also **grocetería.** [f. Grocery, after *Cafeteria.] A grocery store in which customers serve themselves and pay the cashier as they leave.
1913 *Official Gazette* (U.S. Pat. Off.) 25 Mar. 1036/1 Lutey Bros., Butte, Montana. Filed Mar. 14, 1912. Groceteria. 1916 *Illustr. World* (Chicago) Jan. 655/1 In Pomona, California, a small grocery store has suddenly sprung into prominence..by adopting the 'wait-on-yourself' plan of cafeteria. A 'Groceteria' if you please. 1920 see *Cootery. 1921 *Daily Colonist* (Victoria, B.C.) 24 Mar. 20/4 (Advt.), Groceteria Daily Bulletin. All Day Specials. 1921 *Dial. Notes* V. 112 *Groceteria*, n., [grō-ser-te'-ja; grō-ser-te'-ri-a; grō-se-te'-ri-a]. Pronunciation varies. First pronunciation dominant in California; but second commoner here and there in West. 1925 *Canad. Labor Advocate* 18 Sept. 8/1 Kirkham's Groceteria Ltd. 1935 *Words* Jan. 11/2 Some years ago in Oregon a self-service grocery opened under the sign 'Groceteria', evidently under the impression that the suffix '-teria' means 'serve yourself'. 1962 *Guardian* 30 Aug. 6/6 An American 'groceteria' [*sic*], a sort of super super-market.

Groenendael (grō·nendāl). [f. the name of the town in Belgium where the breed was developed.] A black, smooth-coated Belgian sheep-dog.
1929 *Kennel Gaz.* Aug. 706/2 (*heading*) Any other breed or variety..not classified.. Groenendaels. 1945 C. L. B. Hubbard *Observer's Bk. Dogs* 80 The Groenendael Sheepdog was recognized as a pure breed by the Belgian Kennel Club in 1891. 1964 E. F. Dalziel tr. *Schneider-Leyer's Dogs of World* xii. 132 The Groenendael and the Tervuren are popular in Belgium as watch and companion dogs. 1971 *Times* 28 June 24/6 (Advt.), Pedigree Groenendael (Belgian Shepherd) puppies.

grog, *sb.* Add: **1. c.** *Austral.* and *N.Z. colloq.* Alcoholic liquor, including beer.
1949 *New Statesman* 23 Nov. 375/1 A 'Wowser' seeks to interfere with or limit the pleasure of others.. Common types are the Grog Wowser, Sheila Wowser, Sunday Wowser, Cine-wowser. 1948 D. W. Ballantyne in *Landfall* II. 110 Taking the old man's car, and there'll be some grog. 1955 'N. Shute' *Requiem for Wren* i. 9 The man was always on the grog, 'n your Dad gave them the sack. 1966 G. W. Turner *Eng. Lang. in Austral. & N.Z.* vi. 129 *Grog* is still used as a general term for drink, including, or even especially beer.
4. †**grog-den** *N.Z.*; †**grog-hole** *U.S.*; †**grog-shanty** *U.S., Austral.*, and *N.Z.* = Groggery. *Obs.* exc. *Hist.*
1840 *N.Z. Jrnl.* I. xvi. 198/2 Where a year ago only three grog-dens warned the passer-by..not to land on the beach, a neat row of wooden houses..extends along the line at high water. 1848 *Knickerbocker* XXVII. 311 He was busy about the village, penetrating every grog-hole and gambling-shop. 1875 *Scribner's Monthly* X. 533 Groggeries, billiard saloons..were well patronized. 1869 *Auckland Punch* 165/1..reached the grog-shanty in safety. 1888 C. D. Warner *On Horseback* (1896) 47 The woods were full of 'grog-shanties'. 1963 A. Lubbock *Austral. Roundabout* 77 It consists of a weatherboard, tin-roofed grog-shanty. 1966 *Telegraph* (Brisbane) 7 Dec. 10/3 As Australia developed, pubs and drinking habits changed. The grog shanties were replaced by comfortable inns.

groggery. (Earlier U.S. examples.)
1822 J. A. Quitman *Let.* 21 in J. F. H. Claiborne *Life of Quitman* (1860) 71 Consisting of warehouses, low taverns, groggeries, dens of prostitution, and gaming-houses. 1835 J. H. Ingraham *South West* II. 190 Wretched looking dwellings, occupied as 'groggeries' by free negroes.

groggily (grǫ·gili), *adv.* [-ly².] In a groggy manner; shakily.
1897 Kipling *Capt. Cour.* ii. 31 He stepped, rather groggily..to the cabin steps. 1927 *Daily Express* 2 May 11/3 The bull staggers groggily like a pugilist who has received a severe body blow. 1936 P. Fleming *News from Tartary* I. iii. 196 Groggily, but full of hope..the little caravan picked its way through the marsh. 1958 J. Courage in C. K. Stead *N.Z. Short Stories* (1966) 33, I had..gone groggily down into..the main lounge. 1972 R. Perry *Fall Guy* v. 89 Groggily I scambled to my feet.

grogging (grǫ·giŋ), *vbl. sb.* [-ing¹.] The process of extracting spirits from an empty cask by soaking the interior with hot water (see Grog *v.* 2).
1901 *Daily News* 5 Mar. 7/5 [A wine and spirit merchant] was fined £5 and costs for 'grogging'. 1902 *Daily Chron.* 18 Oct. 8/6 Many casks were found on the defendant's premises undergoing the process of grogging.

‖ **grognard** (grǫnyār). [Fr., lit. 'grumbler'. 'Nom donné aux soldats de la vieille garde sous le premier empire, et, en général, à un vieux soldat, le plus souvent en un sens favorable' (Littré).] A soldier of Napoleon's Old Guard. Also *transf.*, a veteran soldier.
1912 J. H. Rose *Personality of Napoleon* iii. 66 He [*sc.* Napoleon] loved to talk with his Old Guard, asking them how long and where they had served... He it was who nicknamed them *les grognards*. 1945 G. B. Shaw *Geneva* (1946) 79 They did not receive him as the grognards of the Bourbon army received Napoleon on his return from Elba. 1959 *Spectator* 10 Aug. 218/3 No wonder some of the grognards of the Soviet old guard mumble that the Czechs have always had it too easy.

groin, *sb.*¹ Add: **3*.** A ring. *slang.*
1931 W. F. Brown in *Police Jrnl.* IV. 500 *Henry.* 'Did he get any sparkle?' *Georgie.* 'Yes, a couple of kettles,..a lovely groin and a prop.' 1936 J. Curtis *Gilt Kid* v. 57 There was one [woman] with three groins on her fingers.

groise (groiz). *Public School slang.* Also **groize.** [Perh. altered form of Grease *sb.* 5 b.] **a.** A hard worker, a swot; one who curries favour. **b.** Hard work. So **groise** *v. intr.*, to work hard, to swot; to curry favour; **groi·ser, groi·sing** *vbl. sb.*
1913 A. Lunn *Harrovians* ii. 37 'Are you a scholar?' 'Yes, in history.' 'Oh, then you'll probably be in our form,' said Cadby. 'I suppose you're a horrid little groise!' *Ibid.* 108 We all have to groise a lot harder than we used. *Ibid.* 197 The whole time-table's altered... They try and make groise more amusing... You can choose your own line of sweat much more. *Ibid.* 211 'What's the old dear groising on to?' 'It's in a very bad way, I'm afraid. Why, groising is in a fair way to becoming fashionable.' 'Oh rot! I say, you fool! Ye telling me next that a man's keener on getting some bally scholarship than on getting his "fez".' 1936 I. Hay *Housemaster* 115 The School looked forward to it [*sc.* the Regatta] with strained anticipation from the first day of term,—oarsmen, cricketers, and groisers alike. 1959 M. Marples *Public School Slang* 130 *Groise*, to curry favour (Cheltenham, 1928+). *Ibid.*, *Groise* (Uppingham, 1950+), one who is over-efficient—e.g. a corps groize is one who tries to gain favour by his efficiency in the O.T.C.

Grolier (grǫ·lye, grōu·liəɹ). Name of a famous French book-collector, Jean Grolier de Servin, Vicomte d'Aiguisy (1479–1565), used *attrib.* to designate the interlacing geometrical designs which adorn the bindings of his books. Also *absol.*, a Grolier binding. So **Grolier Club** (see quot. 1960). Hence **Grolieresque** *a.* (also *absol.*).
1857 *Catal. Lib. T. Williams* 9 Extremely rare, beautiful copies in Grolier binding, red morocco. *Ibid.* 118 Green morocco, with insides most richly tooled in the Grolier style. 1880 J. W. Zaehnsdorf *Art of Bookbinding* p. xix, A centre block of Grolier. *Ibid.* 124 Groliers. 1882 H. B. Wheatley *Bookbinding for Amateurs* xxi. 173 The pattern generally used was that of interlaced strapwork.. Books so bound are often called 'Groliers'. *Ibid.* 175 Fig. 150 shows a side of an ancient book in the Grolier style. 1889 B. Quaritch *Coll. Facsim. Bookbinding* 15 The design is usually called Grolier that of the true Grolieresque. 1893 *Q. Rev.* July 196 The geometrical strapwork patterns known as 'Grolier' were introduced into England in the time of Edward VI. *Ibid.* 199 Another small volume in a Grolier binding. 1896 B. Matthews *Book Bindings* v. i. 294 The Grolier Club is a gathering of those who love books for their external beauty. 1928 E. P. Goldschmidt *Gothic Bookbindings* I. 102 The earliest Groliers. 1936 *Times Lit. Suppl.* 13 June 503 The estimated geometric Grolieresque style was elaborated into the richer and more exuberant design which is described by the phrase *à la fanfare*. 1952 J. W. Carter *ABC for Bk. Collectors* 94 The Grolier Club of New York has published many valuable exhibition catalogues and other bibliographical reference books. 1960 G. A. Glaister *Gloss. Bk.* 160/2 *Grolier Club*, a New York club for bibliophiles, founded in 1884. 1968 *Times Lit. Suppl.* 1 Feb. 116/3 The Grolier Club's *Gazette*, which has of late carried useful handlists of minor and major exhibitions. 1969 *Bodl. Libr. Rec.* VIII. 141 One (lot 18) is described as a 'Grolier' binding, while a number of others are described as 'Grolier style'.

groom, *v.* Add: **2. b.** *fig.* To prepare as a political candidate; in extended use, to prepare or coach for a career, a sporting contest, etc. orig. *U.S.*
1887 *Courier-Jrnl.* (Louisville, Ky.) 3 May 4/5, I learn that Sam Hill, of Hartford, is being groomed for the temporary chairmanship of the Convention. 1903 J. Hawthorne *Hawthorne & his Circle* 264 Grover Cleveland was being groomed for his first Presidential term. 1922 Wodehouse *Clicking of Cuthbert* v. 115 A man whom the committee were grooming for the amateur championship. 1955 *Times* 15 June 12/3 He did not agree that Professor Dent..had groomed him (the witness) to become president in order to keep out an 'Iron Curtain' delegate. 1957 *Listener* 19 Sept. 416/1 Committing the same mistake as Bismarck in not grooming his successor. 1959 *Times* 26 Aug. 4/1 Swetman has been groomed to succeed him [*sc.* Evans] in the Test matches. 1964 C. Chaplin *Autobiogr.* xxv. 435, I was surprised that Mr. Hoover should remember, because at the time he had seemed intensely preoccupied with grooming himself for the White House. 1966 *Globe & Mail* (Toronto) 3 Feb. 25/2 The Music Canada School in Montreal, which grooms pop musicians.

groove, *sb.* Add: **2. b.** (Earlier and later examples.)
1866 W. Davidge *Footlight Flashes* xv. 230 The scenery is pushed back as far as it will go in the slides, or grooves, so called. 1966 *Amer. N. & Q.* Sept. 13/2 Some promptbooks of the eighteenth century and most of those of the nineteenth show at the head of each scene..the number or numbers of the grooves in which the wings and shutters..are to stand.
c. The spiral cut in a gramophone record (earlier, in a phonograph cylinder) which forms the path for the needle.
1902 *Encycl. Brit.* XXXI. 679/1 In the first phonograph a spiral groove was cut on a brass drum fixed on a horizontal screw. *Ibid.*, The sharp ridge of the needle ran in the middle of the spiral groove when the cylinder was rotated. *Ibid.* 680/1 The grooves on the cylinder are 1/8 of an inch apart. 1913 B. Brown *Talking Machines* ix. 194 Examining an ordinary record we find the spiral grooves of very fine pitch somewhere about 100 to the inch. 1956, etc. [see *fine-groove* (Fine *a.* D. 3)]. 1957 *Records & Recording* Nov. 20/1 It is these grooves which must be tracked with absolute accuracy by the pick-up needle. 1958, 1962 [see *centre groove* (*Centre sb.*)].
4. b. Phr. *in the* (or *a*) *groove* (cf. *2 c above = *Groovy *a.* 3. Hence *groove* is used to mean: a style of playing jazz or similar music, esp. one that is 'swinging' or good; a time when jazz is played well; more widely, one's predilection or favourite style, = *bag *sb.* 1 d; something excellent or very satisfying. *slang* (orig. *U.S.*).
1932 *Melody Maker* Oct. 808/1 Having such a wonderful time which puts me in a groove. 1933 *Fortune* Aug. 90/2 The jazz musicians gave no grandstand performances; they simply got a great burn from playing in the groove. 1935 *Hot News* Sept. 17/1 The Boswells are not in the hot groove. 1936 *Rhythm* Apr. 27/2 His first chorus in the latter is really in the right groove, but he loses it completely in the next one. 1940 *Swing* Nov. 27 *Travelin'* has a sax-unison melody somewhat in the Tuxedo groove. 1946 B. Treadwell *Big Book of Swing* 124/2 *In the groove*, everything going O.K. 1954 *Jive Jungle* 32 The all night 'grooves' began. 1957 M. Mezzrow in S. Traill *Concerning Jazz* 18 What we had played was so good I doubted if we could even get in that same groove again. 1958 G. Lea *Somewhere there's Music* iv. 35 Romance? No, bruz, that's not my groove. 1959 N. Mailer *Advts. for Myself* (1961) 296 If you as a cat are way out too, and we are in the same groove.., why then you say simply, 'I dig.' 1962 R. Manheim tr. *Grass's Tin Drum* III. 518 We made music, played ourselves into the groove. 1966 *Melody Maker* 15 Oct. 19 The rhythm team..developed a very propulsive rhythmic groove. 1967 *Ibid.* 16 Dec. 8 This is what makes the Indian one such a groove for me.
6. groove cast *Geol.*, a ridge on the lower surface of a layer of sandstone corresponding to a groove on underlying mudstone; **groove-like** *a.*, wanting in novelty or originality; **groove-locating unit,** a device that indicates the position of a stylus on a record as the record is played.
1948 R. R. Shrock *Sequence in Layered Rocks* iv. 163 These interesting and puzzling ridges, here designated groove casts, seem to represent sand fillings (casts) of rectilinear, V-shaped and U-shaped grooves existing in the upper few millimeters of the bottom sediment on which sand was deposited. 1963 Krumbein & Sloss *Stratigr. & Sedim.* (ed. 2) iv. 130 The upper right illustration shows groove casts, elongated parallel grooves or scratches oriented in the direction of current flow. 1902 *Daily Chron.* 5 July 8/3 There are commendations without end waiting for the linen frock that displays just a little originality, so groove-like is the manifestation of that material as a general rule. 1908 *Westm. Gaz.* 4 June 10/3, I think the modern tendency is too groove-like. Once make a success as a Cockney or a love-sick maiden, and a Cockney or a love-sick maiden you will be to the end of time. 1941 *B.B.C. Gloss. Broadc. Terms* 14 *Groove-locating unit*, device forming part of a reproducing desk, and consisting of an arm carrying a pick-up, together with a pointer and scale to indicate the position of the needle on the record. 1962 A. Nisbett *Technique Sound Studio* viii. 145 BBC studios are equipped with record players which have optical groove-locating units. On these a mirror is fixed beneath the pivot and throws the image of a scale on to a ground-glass screen.

groove, *v.* Add: **4. b.** *fig.* To settle or be settled *into* (or *in*) a routine of work, habit, etc. Also with *down*.
1866 J. Conington *Let.* 28 June in *Misc. Writ.* (1872) I. p. lvi, I am grooving down into work here. 1879 Froude *Cæsar* ii. 10 Morality thus engrained in the national character and grooved into habits of action creates strength, as nothing else creates it. 1922 A. S. M. Hutchinson *This Freedom* I. v, She found Anna grooved in the business of helping her mother in the house.
5. *intr.* To play jazz or similar music with 'swing'; to be 'in the groove' (see *Groove *sb.* 4 b); to dance or listen *to* such music with great pleasure; hence, to make good progress or co-operate; to get on well *with* someone; to make love. Also *trans.*, to play (music) swingingly; to give pleasure to (a person). *slang* (orig. *U.S.*).
1935 *Vanity Fair* Nov. 38/1 That's the third date we've grooved half a dozen schmaltzy tunes. 1937 in *Amer. Speech* XII. 182/1 Men who can lay on sugar or groove it. 1945 'Dizzy' Gillespie (*title of tune*) Groovin' high. 1959 *Esquire* Nov. 70/1 To groove someone means to provide them with enjoyment. Example: Her singing grooved me. 1966 *Melody Maker* 31 Dec. 11/5 (*record title*) Benny Golson 'Groovin' with Golson'. 1967 *Observer* (Colour Suppl.) 4 Dec. 28 Groove, make good progress, co-operate. 1967 *Melody Maker* 16 Dec. 10/7 The rhythm section..grooves along in true Basie manner. 1968 *Listener* 5 Sept. 306/3 The radio Peel is quiet, self-mocking, sardonic and scornful of the 'let's move and groove to this latest gas group from Croydon' school of presentation. 1970 *New Yorker* 14 Mar. 34/2 Sad Arthur put away his boots and helmet..to stay in Nutley and groove with the fair Lambie. 1970 *Observer* 14 May 42/6 We're trying to get humanity to transcend its cultural limitations and groove with it.

grooved, *ppl. a.* Add: Also in *Archæol.*
1936 *Proc. Prehist. Soc.* II. 197 Significant associations are few, but of these two..suggest that grooved-ware was contemporary with Neolithic B pottery. 1939 V. G. Childe *Dawn Europ. Civilisation* (ed. 3) 338 Grooved—with broad incisions, not vertically round-bottomed. 1967 *Antiquaries Jrnl.* XLVII. 169 The Grooved Ware obtained from Durrington is closely allied to that from Woodhenge.

grooving, *vbl. sb.*¹ Add: **3.** *grooving saw.*
a 1877 Knight *Dict. Mech.* III. 2033/1 List of sawing-machines invented and manufactured by him [*sc.* Gen. Sir Samuel Bentham]..previous to 1800... Double-grooving saws. 1882 R. Grimshaw *Suppl. to Grimshaw on Saws* 155 Fig. 322 shows a form of sectional grooving saw in which the action is gradual throughout the width of the cut. 1913 *Saw in History* (Henry Disston & Sons) iii. 35 Grooving saws, as their name indicates, are designed for cutting grooves of various widths and depths. 1940 *Chambers's Techn. Dict.* 391/1 *Grooving saw*, a circular saw used for cutting grooves. 1964 W. L. Goodman *Hist. Woodworking Tools* 155 (*caption*) Grooving Saws, dated 1771 and 1803.

groovy, *a.* Add: **1.** (Later example.)
1966 *New Statesman* 29 Apr. 623/2 The flat tops..are richly textured to resemble pieces of groovy mud.
3. Playing, or capable of playing, jazz or similar music brilliantly or easily; 'swinging'; appreciative of such music, 'hep', sophisticated; hence as a general term of commendation: excellent, very good. Cf. *Groove *sb.* 4 b. *slang* (orig. *U.S.*).
1937 *Amer. Speech* XII. 46/2 *Groovey*, name applied to state of mind which is conducive to good playing. 1944 *Sat. Even. Post* 13 May 89/2 A boy or girl who is really 'groovy' is 'skate wacky' or a 'skate bug'. 1946 Mezzrow & Wolfe *Really Blues* 52 When he was groovy..he'd begin to play the blues on a beat-up guitar. 1948 *Cosmopolitan* Dec. 163/1 'I pitched a no-hit game last summer,' said Georgie. 'Hey, groovy,' said Sally. 1951 W. Morum *Gabriel* II. vii. 225 The boys have a groovy number they want to put across. *Ibid.* viii. 243 It's damned silly to say that. Just because I was extemporising Bach—feeling a bit groovy. 1958 *Spectator* 11 July 67/2 That was a good record..cool and groovy. 1959 *Observer* 1 Nov. 7/7 To-morrow I'll tell him to go to hell, and what's so groovy is, he will. 1968 *Listener* 5 Sept. 307/1 There are a lot of guys going round with groovy hair-styles.

groper. Add: **4.** A jocular appellation for a West Australian. So **Gro·perland,** West Australia; **Gro·perlander,** a West Australian. Cf. *sand groper* (Sand *sb.*² 10).
1924 Lawrence & Skinner *Boy in Bush* iii. 39 Western Australia is full of old prisoners, black fellers, and white ones too. The whites, born here, is called 'gropers', if you take me, sir. 1926 J. Doone *Timely Tips for New Austral.*, *Groper*, a West Australian. 1941 Baker *Dict. Austral. Slang* 33 *Groperland*, Western Australia. 1945 —— *Austral. Lang.* x. 186 Western Australians.., groperlanders. 1949 *Geogr. Mag.* Feb. 373 *Groper*, a Western Australian.

groper². Read: In Australia and N.Z., with pronunciation (grōu·pəɹ), the more usual form of Grouper 2 a, b.
Further examples of the form *groper* are provided s.v. Grouper 2 a, b.
1878 P. Thomson in *Trans. N.Z. Inst.* XI. 383 The Hapuku, or Groper, was in pretty regular supply. 1908 E. J. Banfield *Confess. Beachcomber* I. iv. 152 Blacks..are..fearful of the monstrous groper (*Promicrops itaiara*), which lying inert among the coral blocks and boulders of the Barrier Reef, bolts anything. 1947 I. L. Idriess *Isles of Despair* xxxv. 234 'What was it?' Talalulu laughed. 'Groper,' she answered. 'Old man devil of the sea. You would hardly have made one mouthful for him, Gi'Om.' She was right. For a giant groper is known to have bitten a man in half. 1953 *Landfall* VII. 174 Where they sailed the groper fled. 1957 A. W. Parrott *Sea Angler's Fishes N.Z.* 69 The Groper or Hapuka is well known throughout New Zealand. 1962 L. Wedlick *Fishing in Australia* iv. 159 The blue groper is occasionally caught on offshore reefs. 1965 *Austral. Encycl.* IV. 396/1 *Groper*, a name applied in Australia to several distinct types of fish, including the southern gropers (*Achoerodus*), which are discussed under wrasses; the Queensland groper (*Promicrops lanceolatus*) and various species of *Epinephelus*..and the New Zealand groper (*Polyprionum oxygeneios*).

grorudite (grōu·rudəit). *Petrogr.* [a. G. *grorudite* (W. C. Brögger 1890, in *Zeitschr. f. Krystallogr.* XVI. 1. 66), f. *Grorud*, name of a locality now included in the north-eastern part of the city of Oslo, Norway + -ite¹.] (See quot. 1960.)
1896 J. F. Kemp *Handbk. Rocks* 140 *Grorudite*, Brögger's name for a porphyritic dike rock from Grorud, near Christiania, Norway. 1926 H. H. Read et al. *Geol. Strath Oykell* 72 The grorudites of the Loch Ailsh mass carry much less ægirine than the type grorudites of Norway. 1960 *Gloss. Geol.* (Amer. Geol. Inst.) (ed. 2) 131/1 *Grorudite*, a hypabyssal rock with trachytoid texture containing phenocrysts of alkalic feldspar and aegirine and much quartz.

gros (grō), *a.* [Fr. (see Gross *a.*).] Occurring in various French designations, as **gros bleu,** a dark blue used to paint china; **gros Michel,** the West Indian banana; **gros point,** (*a*) **(de Venise)** a type of lace worked in bold relief, originally from Venice; (*b*) any of a variety of stitches employed in canvas embroidery; **gros sel,** coarse salt (in quot. 1917 *fig.*).
1870 Lady C. Schreiber *Jrnl.* 22 Feb. (1911) I. 74 Sèvres cups and saucers, gros bleu, with gold decoration. 1882 *Hamilton Sale Catal.* No. 495 A Gros-bleu and Gold Sevres Coffee-cup and Saucer. 1885 *Encycl. Brit.* XIX. 638/1 The chief colours [of Sèvres porcelain] are *gros bleu*, a very dark blue; *bleu du roi*, [etc.]. 1961 *Connoisseur New Guide Ant. Eng. Pott., Porc., & Glass* 57 The ground colours of Sèvres were also imitated, such as the 'gros bleu' (called 'mazarine' blue). 1970 *Times* 10 Mar. 7/2 (Advt.), Gros-bleu and blue-scale wares with floral and bird decoration. 1913 W. Fawcett *Banana* i. 16 An improvement on the ordinary fruit occurred in Martinique, and eighty years ago M. Jean François Pouyat..was..sufficiently alive to its importance to introduce it into Jamaica. This variety..is known now as the Jamaican or Gros Michel banana. 1927 *Observer* 17 Apr. 7/1 The Gros Michel..which comes from Jamaica. 1951 *New Biol.* XI. 67 But, when it is a question of really large scale enterprise, *the* banana is the Gros Michel banana, sometimes known as the Jamaica banana. 1962 *Queensland Fruit & Veg. News* 11 Oct. 338/1 Bodles Altafort has good qualities, which are superior to those of Lacatan and Gros Michel. [1390 [see Point *sb.*¹ B. 5].] 1865 F. B. Palliser *Hist. Lace* iv. 47 This is our Rose (raised) Venice point, the Gros Point de Venise, the Punto a rilievo, so highly prized and so extensively used for albs, collerettes, berthes, and costly decoration. 1882 *Encycl. Brit.* XIV. 186 Works of bold design done in relief are called 'gros point de Venise'. 1911 *Ibid.* XVI. (*caption*, facing p. 40), Jabot of needlepoint lace worked partly in relief, and usually known as 'gros point de Venise'. 1934 M. Thomas *Dict. Embroidery Stitches* 110 *Gros point*, a term sometimes used when referring to certain Canvas Stitches worked over two or more horizontal threads of the canvas, such as the variations of Cross Stitch, Florentine Stitch and the Gobelin Stitches. 1958 *Observer* 20 July 9/1 What appears as gros-point embroidery worked with the greatest skill in the shape of birds, trees and flowers, proves to be a design woven into the silk fabric. 1962 *Victoria & Albert Mus. Internat. Art Treas. Exhib.* 9/1 A George I needlework winged armchair with upholstered adjustable back and loose cushion seat covered with floral gros-point needlework. 1968 *Times* 8 Feb. 8/8 Since beginning rehearsals..she has almost completed a chair cover in *gros point*. [1849 Thackeray *Pendennis* I. xxxiv. 340 He instantly related a funny story, seasoned with what the French call *gros sel*.] 1917 G. Saintsbury *Hist. Fr. Novel* I. viii. 169 There is no want of salt, though there is no (or very little) *gros sel* in the *Astrée*. 1935 M. Morphy *Recipes of all Nations* 30 Sprinkle it with salt—the French always use 'gros sel', a very coarse salt, which gives an excellent flavour. 1958 *Spectator* 15 Aug. 222/2 The Maldon Crystal Salt Company's product is apparently the only *gros sel* sold in this country and is a 99·617 per cent. pure sea-water evaporation from the Blackwater estuary. 1967 *Sunday Times* (Colour Suppl.) 5 Mar. 42/1 That first purchase of *gros sel* for cooking.

grosgrain (grōu·grēⁱn, ‖ grogræ̃). [Fr., = coarse grain.] Any of various corded fabrics. Hence **grosgrained** *a.* Also *fig.*
1869 [see Grogram 1]. 1897 *Sears, Roebuck Catal.* 408 Silk Watch Fobs... The Ribbons are of Fine Quality Black Gros Grained Silk. 1905 *Westm. Gaz.* 5 Aug. 10/2 Grosgrain is far better with alpaca than any shiny make of taffeta. 1927 *Daily Express* 14 Mar. 5 Two toned grosgrained ribbon. 1930 *Times* 17 Mar. 15/6 On the black grosgrain ribbon hat is a jewelled pin. 1932 F. L. Wright *Autobiogr.* III. 309 Every flat plane grosgrained like the sahuato itself. 1969 *Sears Catal.* Spring/Summer 18 Widely-brimmed Hat. Braid straw with grosgrain ribbon streamers.

gross, *a.* and *sb.*⁴ Add: **A.** *adj.* **6. a.** *gross reproduction rate*: a reproduction rate representing the average number of girls born to each woman of a population when deaths before the end of the child-bearing period are neglected, calculated from the average fertility rates of each age-group during the period considered.
1928 R. R. Kuczynski *Balance Births & Deaths* I. ii. 25, 1·958 would then be what we may call the gross reproduction rate of Sweden in 1891–1900. 1945 *New Biol.* I. 42 A gross reproduction rate sustained at a level below 1·0 signifies that no reduction of mortality could save the community from eventual extinction. 1970 W. D. Borrie *Growth & Control World Popul.* vi. 146 In terms of the gross reproduction rate there seems little danger of population decline among European peoples.

c. *gross national product*, the total monetary value of all goods and services provided in a country during one year.
1947 J. F. Dewhurst *America's Needs & Resources* I. iv. 52/1 Gross national product represents the gross value of all the goods and services produced by business enterprises, including farmers, professional persons and other self-employed individuals. 1962 *Listener* 17 May 836/1 Economic growth is a purely materialistic concept. The rate of growth is usually defined as the rate of increase in gross national product or in gross national product per head, that is productivity, gross national product itself being the sum of all goods and services for which money changes hands. 1969 *Times* 12 Feb. 9/1 Although the gross national product (g.n.p.) may be increasing at about 4 per cent a year, it has to be shared out among a population growing at 2·5 per cent.
B. *sb.* **4. c.** Delete †*Obs.* and add later U.S. examples.
1930 F. Scott Fitzgerald in *Sat. Even. Post* 18 Jan. 109/1 A hit at the New Strand, a hit at the Prince of Wales, and the weekly grosses pouring in. 1969 Gish & Pinchot *Lillian Gish* xii. 157 *The Birth of a Nation* has become the all-time money maker in film history. There have been so many black-market prints in circulation that no one will ever know its true gross.

gross, *v.* Add: **4.** (Additional examples.)
1936 Wodehouse *Laughing Gas* iv. 52 My last picture but one grossed twenty-two thousand there on the week. 1970 *Observer* 19 Apr. 9/3 The..brokerage business brought him far more prestige than cash; specifically, it grossed something like $60,000 a year, but the firm's overhead was high.
b. With *up*: to count, add as part of the total; to treat (a payment) as if it were a larger taxable amount of the same net value. Hence **grossed** *ppl. a.*; **gro·ssing** *vbl. sb.*
1931 *Economist* 28 Feb. 456/2 This [tax] is charged..on the 'grossed-up' dividend. 1954 *Times* 24 Feb. 12/4 When discussing M.P.s' emoluments we ought to gross-up the expectations of pension rights. 1959 *Economist* 11 Apr. 146/1 For that rare bird paying surtax at the top rate, the 'grossed up' yield on savings certificates earning £4 4s. od. tax free has fallen by over a third. 1969 *Times* 25 Jan. 2/3 Some estimates had been made by means of plain 'grossing-up'—a very loose way of computing.

‖ **grossdeutsch** (grōu·sdoitʃ), *a.* [G.] Pan-Germanic; referring to a Greater Germany, including Austria.
1946 A. J. P. Taylor *Course of Germ. Hist.* 8 The interplay of *Kleindeutsch* and *Grossdeutsch* programmes. 1959 *Listener* 21 May 902/1 The Germans as a whole reverted to *grossdeutsch* ambitions for the absorption of Austria and expansion at the expense of the Slav nations to their east. 1963 *Ibid.* 14 Mar. 469/1 Schuschnigg himself emerges as all too honest, but as so drugged by his *grossdeutsch* mysticism as to be politically quite incompetent.

grossen, *v.* Delete *rare*⁻¹ and add later example.
1947 *Penguin New Writing* XXIX. 10 This face..had now become expanded and grossened.

grosser¹ (grōu·səɹ). [f. Gross *v.* + -er¹.] A film, etc., that 'grosses' or brings in a sum of money: usually qualified by a defining word (*big*, etc.).
1959 N. Mailer *Advts. for Myself* (1961) 335 We've had approximately the same number of big box-office grossers and box-office duds. 1964 C. Chaplin *Autobiogr.* xxv. 432 *The Great Dictator*..turned out to be the biggest grosser of all my pictures up to that time. 1966 *Listener* 15 Dec. 889/3 The gross last year of *Goldfinger* was £15,000,000. When the film is re-released it will almost certainly become the biggest grosser in the history of the cinema.

‖ **grosso modo** (grǫ·so mōu·do), *adv. phr.* [It.] Roughly, approximately.
1952 G. Sarton *Hist. Sci.* I. xvi. 405 There is a general agreement *grosso modo* on the following basis. 1961 *Amer. Speech* XXXVI. 219 Change of Formant 2 frequency of vowels in disyllables corresponds *grosso modo* to the change in tongue-alveolar distance. 1964 G. Germani in I. L. Horowitz *New Sociol.* 396 The first phase of disposability and mobilization may be followed by a second one of integration, which may, *grosso modo*, occur in different ways.

grot³. **1.** (Later examples.)
1901 'M. Franklin' *My Brilliant Career* xxii. 192 She and Harold would rest while I did the honours of the fern grots to my companion. 1958 R. Liddell *Morea* II. i. 45 On the plain there is a spring of excellent water, which comes here underground from the grot of Hermes.

grot⁴ (grǫt). Abbrev. of Grotesque *sb.* 3.
1961 *SIA Jrnl.* Sept. 6/2 Typography..is dominated by a particular sort of sans-serif letter, the 'grotesque'... It is useful to examine this dictatorship of the grot. 1964 L. Deighton *Funeral in Berlin* i. 11 He sipped..tea from an antique Meissen cup; around mine it said 'British Railways S.R.' in brown grot letters.

grotesque, *sb.* **1. b.** The Italian form **grottesco** (pl. **grotteschi**) is sometimes used.
1926 A. Huxley *Essays New & Old* 180 Very pretty little *grotteschi* in the Pompeian manner. 1934 *Burlington Mag.* Apr. 199/1 A certain type of symmetrical *grotteschi* design. 1958 'M. Innes' *Long Farewell* 11, I like *grotteschi* on my walls. All these little nudes like amorous shrimps.

Grotian (grōu·ʃiăn), *a.* [See -an.] Of or pertaining to the Dutch lawyer, statesman, and theologian Hugo *Grotius* (1583–1645), who founded the modern science of international law and propounded the 'governmental' view of the Atonement. Also as *sb.*, an adherent of the tenets or policies of Grotius; **Gro·tianism,** the views or teaching of Grotius.
1864 W. G. T. Shedd *Hist. Chr. Doctr.* II. 366 The Grotian soteriology. 1920 *Expositor* May 372 Jonathan Edwards the younger, went over to Grotianism, bag and baggage. *Ibid.* 373 Johnson had enunciated views of the doctrine [of the Atonement] which one can only characterize as strongly Grotian. 1934 J. Laird *Hobbes* III. vii. 255 Boëcler, a Grotian, was said to have counted Hobbes a great intellect who had gone astray by neglecting the classics.

grotto. Add: Hence **grottoe·sque** *a.*, resembling a grotto. Also *absol.*
Quot. 1863 illustrates a nonce *absol.* use.
1863 J. C. Atkinson *Stanton Grange* 107 A massive piece of rock..supplying them with a kind of natural grotto-esque back to the recess. 1881 J. P. Briscoe *Old Notts.* 123 Only a groto-esque [*sic*] summer house to the Castle when it was inhabited. 1955 S. Spender *Making of Poem* 101 Gilded Romanticism is atmosphere in the Ballad-makers, fancy in Shakespeare and the other Elizabethans, desperation in the Jacobeans, the grottoesque in Pope.

grotty (grǫ·ti), *a. slang.* [Shortened form of Grotesque *a.* + -y¹.] Unpleasant, dirty, nasty, ugly, etc.: a general term of disapproval.
1964 J. Burke *Hard Day's Night* iv. 88 'I wouldn't be seen dead in them. They're dead grotty.' Marshall stared. 'Grotty?' 'Yeah—grotesque.' 1964 *Daily Mail* 18 Aug. 1/6, I felt dead grotty. 1964 *Sunday Times* 25 Oct. 22/5 A charming touchline companion called the [hockey] match 'grotty' which seemingly means disappointing. 1966 M. Hastings *Cork on Telly* ii. 29 I've been checking my tapes...They sound pretty grotty when they're scrambled. 1967 *House & Garden* Apr. 59/3 The house was a ruin—dirty, really grotty, but with obvious potential. 1970 *Times* 19 June 10/4 'I don't like the grotty old pub,' says Miss McCormick.

grouch (grautʃ), *sb.* orig. *U.S.* [Var. of Grutch *sb.*] **1.** Grumbling; a complaint or grumble; a grumbling, sulky mood; a fit of ill temper or sulkiness.
1895 W. C. Gore in *Inlander* Nov. 66 *Grouch*,..a fit of ill-humor. 1900 'J. Flynt' *Notes Itinerant Policeman* 160 They began to get a grouch on against the gay-cats that kep' comin' to their camps. 1903 *Harper's Bazaar* Oct. 947 No woman who comes down to her breakfast table with what her son frankly calls a 'grouch on' is grouchy to herself alone. 1905 *N.Y. Even. Post* 2 Sept. 4 We need still another breakfast food..one that will take away matutinal grouch. 1906 B. von Hutten *What became of Pam* I. ii, You are not going to spoil my birthday with one of your grouches. 1913 C. E. Mulford *Coming of Cassidy* iii. 61 Longhorn..felt savage elation at this opportunity to unload quite a cargo of accumulated grouches of various kinds and sizes. 1921 *Chambers's Jrnl.* Jan. 37/1 He was known there as the Englishman whom Providence had a grouch against. 1940 'G. Orwell' in *World Review* (1950) June 33 Part of his grouch was that he had tried to join the Air Force.. and always been put off. 1966 *Listener* 8 Sept. 355/3 The students were not demonstrating on behalf of a fashionable grouch.
2. A grumbler.
1900 *Yale Fun* 25. 1916 *Recruiter's Bulletin* (U.S.) Apr. 12/2 Why, even the old-fashioned grouch, The biggest in the land, With 29 years' service will On payday shake your hand. 1919 H. L. Wilson *Ma Pettengill* ii. 61 A fifty-five-year-old male grouch..who had been snarling at everyone that came near him ever since the train left New York. 1923 *Public Opinion* 12 Oct. 361/1 Whether I thought myself a Gentleman or a Grouch..is no part of this Parable. 1957 *Listener* 10 Oct. 581/1, I am probably a humourless old grouch. 1966 D. Varaday *Gara-Yaka's Domain* x. 112 Had I been a gipsy palmist I might have paid more attention to the old grouch's lines of life and fate.
3. *Comb.* **grouch-bag** *U.S. slang*, a hidden pocket or a (draw-string) purse carried in a concealed manner for the hiding and storing of money, etc.; also, money saved and kept hidden.
1908 K. McGaffey *Sorrows of Show Girl* xiii. 152, I have met gentlemen who threw the lid of their grouch bag in the gutter and didn't care if they ever found it again. 1914 Jackson & Hellyer *Vocab. Criminal Slang* 39 *Grouch bag*, a place, as a pocket or receptacle, for concealing money or valuables; a reserve fund held in secret to the exclusion of fraternists. 1956 N. Algren *Walk on Wild Side* (1957) i. 17 Circus roustabouts..took their money out of grouch-bags, pouches drawn by string, like tobacco pouches. 1960 B. Keaton *Wonderf. World of Slapstick* (1967) 24 Like every other woman in vaudeville she carried the act's cash in a grouch bag. This was a chamois purse which hung under her dress from a string round her neck. 1969 *Telegraph* (Brisbane) 6 Dec. 2/2 Groucho..earned his nickname in poker games because he always carried his money in a 'grouch bag'.

grouch: see *kurus.

grouch (grautʃ), *v.* orig. *U.S.* [var. of Grutch *v.*] *intr.* To grumble. Also with words spoken as object.
1916 C. J. Dennis *Moods of Ginger Mick* 133 *Grouch*, to mope; to grumble. 1925 H. L. Foster *Trop. Tramp Tourists* 137 The tourists..all came back to the train at a painfully slow walk,..and grouched all the way home. 1926 J. Black *You can't Win* viii. 90 'Everything's all right now, ain't it?' 'Oh, sure,' he grouched. 'Everything's all right—just like Denmark.' 1931 H. de V. Stacpoole *Pacific Gold* II. i. 99 'Now what *she* grouchin' about?' asked the Captain.

grouchy (grau·tʃi), *a.* orig. *U.S.* [f. *Grouch *sb.* or *v.* + -y¹.] Grumbly, ill-tempered.
1895 W. C. Gore in *Inlander* Nov. 66 *Grouchy*, gloomily irritable; cross. 1902 *Daily Chron.* 25 Jan. 7/2 Thus we may learn which of them, in the opinion of his fellows, is..the slouchiest, the biggest fusser, the 'grouchiest'. 1903 [see *Grouch *sb.* 1]. *a* 1910 'O. Henry' *Trimmed Lamp* (1916) 212 What's the matter, Andy, you are so solemn and grouchy to-night? 1915 *Times* 12 Apr. 10/1 The Germans are still pretty grouchy over their beating. Both sides begin firing on the slightest alarm. 1929 C. H. Smith *Bridge of Life* i. 5 My maternal grandfather..was a grouchy, crusty old fellow. 1932 H. J. Massingham *Wold without End* iii. 77 Father Hiems rallied from his swoon and was grouchy with snow-showers. 1942 E. Paul *Narrow St.* xii. 90 Maggie, the unspeakable terrier beloved by the grouchy Madame Marie at the Caveau, took every advantage of her mistress's indulgence.
Hence **grou·chily** *adv.*, **grou·chiness.**
1906 H. Green *At Actors' Boarding House* 314 Ned was grouchily chopping ice and shooing flies away. 1923 R. D. Paine *Comrades of Rolling Ocean* iii. 41 'They are all new to me,' grouchily replied Judson. 1925 C. E. Mulford *Cottonwood Gulch* vi. 87 The second bar-tender, whose grouchiness was due to lack of proper sleep. 1930 *Time & Tide* 25 Apr. 529/2 He..had decided—a trifle grouchily—to stay on and make the best of a bad job.

ground, *sb.* Add: **1. d.** *Theol.* [repr. G. *grund* as used by 14th-c. mystics, notably Eckhart and Tauler.] (*a*) The divine essence or centre of the individual soul, in which mystic union lies. (*b*) Godhead as the source of all that is.
a 1400 *Book of Privy Counselling* (1944) 144/19 God, þi grounde & þi purete of spirit. 1865 J. H. Stirling *Secret of Hegel* I. II. i. 235 Being is posited as Existence, and the Mediating agency of this Being as the Ground. 1899 W. R. Inge *Chr. Mysticism* i. 7 The curious doctrine which we find in the mystics of the Middle Ages, that there is at 'the apex of the mind' a spark which is consubstantial with the uncreated ground of the Deity. 1911 E. Underhill *Mysticism* iii. 64 The point of contact between man's life and the divine life..is called the Ground of the Soul, the foundation or basal stuff whence springs all spiritual life. 1945 A. Huxley *Time must have Stop* viii. 92 There was the ultimate all-embracing field—the Brahma of Sankara, the One of Plotinus, the Ground of Eckhart and Boehme. *Ibid.* xxx. 289 There is a Godhead or Ground, which is the unmanifested principle of all manifestation... The Ground is transcendent and immanent. 1945 —— *Perennial Philos.* (1946) ii. 29 The divine Ground of all existence is a spiritual Absolute, ineffable in terms of discursive thought, but..susceptible of being directly experienced. 1949 P. Tillich *Shaking of Foundations* vi. 47 The God Whom he cannot flee is the Ground of his being. 1950 W. R. Trask tr. J. Bernhart in *Theologia Germanica* 95 What Plotinus had called the 'kentron' (center) of the soul, and Richard of St. Victor the 'height and inwardness of the spirit'..; what Bernard of Clairvaux called the 'point of the mind' (*acies mentis*), or again the 'spark' (*scintilla*) is now given a variety of German names and is indefatigably discussed and speculated upon. It is called the 'soul's essence' (*Wesen*), 'soul' (*Gemut*),..'ground' (*Grund*)... These concepts are intended to designate that which God and man must have in common if a contact is to be established between them... The thing in which they [*sc.* individual mystics] agree is always the idea of the mystic function of the 'ground' of the soul. 1961 J. Walsh tr. *Dame Julian's Rev. Divine Love* lxii. 168 God..is the Ground; he is the Substance. 1963 J. A. T. Robinson *Honest to God* iii. 45 (*heading*) The Ground of our Being.
2. b. Also *to take ground* (i.e. without article).
1893 Stevenson *Catriona* Summary p. viii, The *Covenant* took ground and sank off the coast of Mull. 1893 'Q' *Delect. Duchy* 295 Miss..Lear heard her brother's boat take ground on the narrow beach.
4. b. The floor.
1847 Webster, *Ground*, a floor or pavement. 1900 *Eng. Dial. Dict.*, Put the baby an the ground and let 'er craal. 1921 E. O'Neill *Emperor Jones* (1925) i. 7 (*stage-direction*) Woman (seeing the uselessness of struggling, gives way to frantic terror, and sinks to the ground). 1937 A. Christie *Murder in Mews* i. 14 We..forced the door open. Mrs. Allen was lying in a heap on the ground shot through the head. 1939 Joyce *Finnegans Wake* 452 Pricking up ears to my phono on the ground and picking up airs from th'other over th'ether.
8. b. *down to the ground* (earlier example); *from the ground up* (colloq., orig. *U.S.*), completely, entirely; 'down to the ground'; *to get off the ground*, to make a successful start; *on the ground*, in situ, on the spot.
1867 [see Down *adv.* 30]. 1895 *Congress. Record* 6 Feb., App. 207/1 There never has been a time that a democratic administration has been American from the ground up. 1910 W. M. Raine *B. O'Connor* 52 We suited each other from the ground up. 1960 *Guardian* 25 Nov. 15/1 On-the-ground investigations. 1961 *New Statesman* 28 July 129/3 Intended as a half-way point of the Festival, at which audience and platform might fruitfully interact, it never got off the ground. 1963 *Listener* 10 Jan. 59/2

There is no longer any good reason why the young..American writer should undergo a European apprenticeship unless it be to satisfy his curiosity or to watch the operations of another literature on the ground. 1969 *Ibid.* 3 Apr. 469/1 It soon became evident..that the history of contemporary music required reconsidering from the ground up. 1969 *Guardian* 4 July 5/5 If thefts continue, the future plans for the Crewe to Glasgow [railway] line can never really get off the ground.
d. (Additional examples.)
1938 B. Lunn in 'H. Kingsmill' *Eng. Genius* 205 The Presbyterian divines were maddened by answers which cut the ground from institutional religion. 1962 *Christian Cent.* 18 July 886/2 In short, *Veterum Sapientia* has actually succeeded in cutting the ground from under the feet of the exponents of a living liturgy.
f. *to go to ground*: also said of a dog. Also in other phrases, and *fig.* (of a person), to withdraw from public notice and live quietly or 'lie low'. Cf. Run *v.* 42 e.
1860 G. D. Prentice *Prenticeana* 175 A Party of our friends..chased a fox thirty-six hours. They actually 'ran the thing into the ground'. 1871 Mrs. Stowe *My Wife & I* ix. 93 Show me up the weak points of those reformers; raise a laugh at those temperance men,—those religionists, who, like all us poor human trash, are running religion, and morals, and progress into the ground. 1900 *Daily News* 23 Oct. 6/2 The British infantrymen watched the race for shelter, their sporting spirit rising.. above all racial hatred, and hailing with a 'gone to ground' whoop the final disappearance of the gun. 1905 Loder-Symonds & Crowdy *Hist. Old Berks Hunt* xv. 292 Hatford. Gorse, where they soon marked him [the fox] to ground. 1920 A. C. Smith *Dog* 18 Strictly speaking..Airedales and bull terriers should not be classified among the terriers, both being much too big to go to ground. 1925 *Times* 7 Jan. 5/6 Sticking to their fox, the pace continued good to Chesterton, where he was marked to ground. 1930 'Sapper' *Finger of Fate* 265 It so happens that on occasions members of the fraternity [*sc.* snakes] go to ground in the bunches of fruit as they lie stacked beside the railway line. 1931 *Our Dogs* 23 Oct. 292/2 Working Terrier Dog..goes to ground to fox or badger, and stays. 1964 *Ann. Reg. 1963* 326 The four men 'went to ground', probably in Johannesburg. 1968 K. Weatherly *Roo Shooter* 39 When they found where a fox had been caught they would track it, sometimes for miles, and shoot it, but often the fox would go to ground and another trap was lost. 1968 *Times* 11 May 4/6 They are looking for a suburban villa where they can go to ground.
9. a. (Later example.)
1883 R. W. Dixon *Mano* III. iii. 123 The truest gentleman that is on ground.
11. b. *to gain ground* (later examples); *to make* (*up*) *ground*, to make progress; *to break* (*new*) *ground*, to make progress in a new direction (see Break *v.* 44 c).
1895 *Brewer's Dict. Phr. & Fable* 557/2 *To break ground*, to be the first to commence a project, etc.; to take the first step in an undertaking. 1906 H. C. Wyld *Hist. Study Mother Tongue* v. 94 Those tendencies..which are peculiar to the individual, and which are not shared by the community, will not gain ground, but will be eliminated. 1921 *Granta* 30 Nov., The local side again and again made ground galore with long kicks down wind. 1928 *Nation & Athenæum* 7 Jan. 537/2 Montesquieu..had been the first to break the new ground. 1931 F. L. Allen *Only Yesterday* ix. 229 Theodore Dreiser, Willa Cather,..the Imagists and exponents of free verse had been breaking new ground since before the war. 1932 *Sunday Express* 3 July 22/7 Udaipur is gradually making up ground on the colts in Butters' stable. 1954 G. D. H. Cole *Hist. Socialist Thought* II. xiii. 362 Communities breaking new ground were in constant danger of becoming the prey of fraudulent financiers and bankers. 1954 A. S. C. Ross in *Neuphilologische Mitteilungen* LV. 45 *Posh* 'smart' is essentially non-U, but recently, it has gained ground among schoolboys of all classes. 1966 *Listener* 10 Mar. 345/2 I've had to break new ground in all directions in order to say them.
d. (Additional example.)
1933 H. L. Ickes *Diary* 12 Sept. (1953) I. 88 At eleven o'clock we had a meeting of the Public Works Board and we covered a great deal of ground.
14. b. (Earlier and later examples.)
1718 in G. B. Buckley *Fresh Light on 18th Cent. Cricket* (1935) 2 Three of their men made an elopement and got off the ground without going in. 1733 in H. T. Waghorn *Cricket Scores* (1899) 6 There will be a line around the ground as usual, within which none but the gamesters are to be allowed. 1880 J. Lillywhite *Cricketer's Ann.* 49 The ground staff for 1879 at Lord's consisted of [the twenty-two professional players named]. 1955 T. H. Pear *English Social Diff.* xi. 260 Boys of outstanding promise [at cricket] can receive coaching if they take jobs on a club ground-staff.
17. a. *ground clearance, fabric, -level* (examples), *-nester*; *ground-based, -feeding* (later example) adjs.
1958 *Chambers's Techn. Dict.* Suppl. 1018/1 Ground-based..radio duct. 1965 *Punch* 7 Apr. 506/2 A report just published by the National Academy of Sciences (counterpart of the Royal Society) surveys the requirements of 'ground-based' astronomy over the next ten years. 1959 *Motor Man.* (ed. 36) 4 The chassis frame cannot be lowered beyond a certain point without endangering the ground clearance of the car. 1938 *Burlington Mag.* Sept. 115/1 The loosely-woven linen ground-fabric is entirely covered..with rich ornament. 1967 E. Short *Embroidery & Fabric Collage* i. 19 These all entail cutting away part of the ground fabric. 1938 *Brit. Birds* XXXII. 222 An attempt to measure the frequency of association in the same fields of the more conspicuous ground-feeding birds was made. 1910 *Daily Chron.* 8 Mar. 1/6 The Parisian 'star', returning to her room at midnight, sees a strange object approaching her ground-level windows from the garden. 1959 *Times* 22 Aug. 9/4 Being nearer to eye-level can be better appreciated than when at ground-level in the border or rockery. 1875 *Field & Forest* I. 10 It was a very neat structure, and looked to me as though the owner was habitually a 'ground-nester'. 1964 A. L. Thomson *New Dict. Birds* 524/1 Other non-passerines may be broadly divided into ground-nesters, hole-nesters, and the builders of simple nests in trees.
b. *ground-basis, -fact, -form* (earlier and later examples), *-principle* (earlier example), *-quality, -sense, -tone* (later example).
1920 T. P. Nunn *Education* 156 The function of the self-regarding sentiment is to exercise control over the 'objective' sentiments that form the ground-basis of the self. 1905 *Spectator* 11 Mar. 353/1 The underlying ground-fact of Russia, the inadequacy of her food-supply. 1847 J. D. Morell *Hist. View Philos.* (ed. 2) I. i. 118 The native construction of the intellectual faculty..contains all those ground-forms of the understanding, by which knowledge from experience can be assimilated. 1938 J. R. Carpenter *Ecol. Gloss.*, *Ground form*, elementary form, as distinguished from growth form. 1847 J. D. Morell *Hist. View Philos.* (ed. 2) I. 3 The primary efforts of reason to get at the ground principles of human knowledge were naturally weak and imperfect. 1897 Hardy *Well-Beloved* II. iii, Avice..had yet possessed a ground-quality absent from her rivals. *Ibid.* III. vii, Pierston heard a voice below, the accents of a woman. They had a ground quality of familiarity, a superficial articulation of strangeness. 1909 E. B. Titchener *Text-bk. Psychol.* I. 116 The sense of smell..is also a ground-sense:..our own disregard of smell sensations is largely due to our assumption of the upright position. 1909 *Westm. Gaz.* 8 Mar. 5/2 Her sash will repeat the ground-tone of her dress.
d. In Aviation, as *ground alert, attack, boost, control* (landing, etc., by instrument direction from the ground; so *ground control(led) approach* (abbrev. *G.C.A.*), *ground controller*); *ground crew, cushion, defence, effect, engineer, loop* (so as vb.; also *ground looping* vbl. sb.); *ground marker, mechanic, organization, position, resonance, school, speed, staff, stunt, support, troops, wallah*; *ground-based* adj.; *ground-straf(f)ing* vbl. sb. (so, as a back-formation, *ground-strafe* v. trans., *-strafer*); *ground-to-air, ground-to-ground*, used esp. as attrib. phrs.
1965 H. Kahn *On Escalation* 294 A ground-alert bomber. 1917 'Contact' *Airman's Outings* I. vii. 203 We shall see a great extension of ground attacks by air cavalry. 1954 *Economist* 11 Sept. 11/1 In some countries ..special aircraft for ground attack duties only have been developed. 1960 *Times* 23 Feb. 5/3 In the semi-active homing system a ground-based radar illuminates the target. 1965 H. Kahn *On Escalation* ix. 171 Ground-based missiles. 1930 *Flight* 14 Mar. 305/2 Supercharging was the expression used to denote restoring the ground h.p. at some height, while ground boost was used to get increased power at ground level. 1933 *Jrnl. R. Aeronaut. Soc.* XXXVII. 31 How would he suggest the air control and ground control should be organised there? 1945 *Newsweek* 20 Aug. 42/3 GCA (Ground Control Approach) which permits blind landings of planes through overcast. 1969 *Listener* 1 May 596/1 He explained to ground control what was happening and ordered the crew to bail out. 1945 *Amer. Speech* XX. 309/2 GCA, Ground Controlled Approach, Ground radar landing system. 1959 R. Collier *City that wouldn't Die* vii. 105 His navigator.. was in minute-by-minute touch with the new G.C.I. (Ground Controlled Interception). 1958 *Listener* 21 Aug. 259/1 It is the ground controllers' job to see that collisions do not happen. 1970 *Daily Tel.* 15 Apr. 1/7 Ground controllers..decided against a speed-up plan that would have brought Apollo 13 down by tomorrow night. 1934 *Sci. Amer.* Feb. 83 (*caption*) The ground crew guiding the ship [*sc.* airship] into the hangar. 1940 *Flight* 7 Nov. *b*/2 The efficient devotion of ground crews and excellence of material is responsible for the trouble-free journeys made each night. 1952 *Ann. Reg. 1951* 109 The Government announced that pilots and ground crews had won 154 U.S. decorations. 1949 *Jrnl. R. Aeronaut. Soc.* LIII. 317/1 For take-off at altitude, a strong 'ground cushion' is an advantage. 1952 R. Sherbrooke-Walker *Khaki & Blue* i. 3 The problem became acute when 'Ground Defence' came along in the early days. 1956 *N.Y. Times* 8 Jan. p. X 41/5 The issue revolves about a phenomenon peculiar to helicopters known as ground cushion effect. What it means is this: Up to ten or twelve feet off the ground, or over water, a helicopter receives added buoyancy by the packed mass of air churned downward from the overhead rotors. 1935 *Jrnl. R. Aeronaut. Soc.* XXXIX. 277 The machine was a high wing monoplane and one would not expect much 'ground effect'. 1938 *Aeronaut. Res. Committee Rep. & Mem. No. 1865* 1 General formulae for..corrections to ground effect have been obtained for wings of any span. 1959 *Observer* 18 Jan. 15/4 The principle on which the Hovercraft is built has a good deal in common with an infuriating phenomenon known to pilots as 'ground effect', which occurs when certain types of aircraft come in to land. Their closeness to the ground creates something akin to a pad of air on which the aircraft floats tantalisingly a foot or so above the runway instead of sinking to the ground. 1966 *Electronics* 17 Oct. 131 It would be carried on anything from a hydrofoil craft and a ground-effect machine, which rides over water and land on a cushion of air, to helicopters and pilotless drone aircraft. 1920 *Flight* 11 Nov. 1182/2 Under present arrangements, a machine is supposed to receive an inspection by a ground engineer. 1928 *Daily Mail* 9 Aug. 7/1 Ground engineer of the London Aero Club. 1932 D. Garnett *Rabbit in Air* i. 7 The ground engineer gave us pamphlets about the Flying Club. 1958 'N. Shute' *Rainbow & Rose* i. 2 He had a ground engineer. 1921 *Flight* 5 May 315/2 The aeroplane has a tendency to ground loop in landing or in taxying in a strong wind. 1928 *Daily Mail* 7 May 6/4 Ground Loops.—Touching the ground and rising again. 1959 F. D. Adams *Aeronaut. Dict.* 86/1 *Ground loop*, 1. a violent, whirling turn of an airplane while moving on the ground... 2. loosely, a nose-over. 1937 *Jrnl. R. Aeronaut. Soc.* XLI. 829 The author investigates some of the causes of so-called 'ground looping', i.e., the instability observed with certain machines when landing in a cross wind and which causes them to swing violently after touching ground. 1950 *Gloss. Aeronaut. Terms* (*B.S.I.*) I. 11 *Ground looping*, an uncontrollable violent turn of an aircraft while taxying, alighting or taking-off. 1944 *Times* 26 Apr. 4/1 A number of crews were bombing from a clear sky while others at the same moment were aiming at the ground markers through the clouds. 1935 C. Day Lewis *Time to Dance* 57 But those hands have been always The ground mechanics of our wide-wing pride. 1920 *Proc. Air Conference, London* 11 A problem of considerable difficulty is the ground organisation for night flying. 1933 *Discovery* Dec. 367/1 All these difficulties could..be got over if there was a good ground organization and aircraft..to operate at such heights with multi-engines. 1951 *Gloss. Aeronaut. Terms* (*B.S.I.*) III. 7 *Ground position*, the position on the earth vertically below an aircraft. 1940 *Jrnl. Aeronaut. Sci.* Aug. 449/2 A series of high-speed motion pictures were taken of a turning rotor which was made to oscillate in ground resonance. 1959 F. D. Adams *Aeronaut. Dict.* 86/1 *Ground resonance*, a self-excited, mechanical, potentially destructive vibration of a rotary-wing aircraft in operation on the ground or surface, involving a coupling between the motion of the rotor blades and the motion of the supporting structure or the motion of the aircraft as a whole on its landing gear. 1924 Webster Addenda, *Ground school*, a school giving courses in aërodynamics, map-making, photography, etc., for aviators. 1917 *Blackw. Mag.* May 805/2 Our ground speed was now a good deal greater than if we had travelled directly west. 1924 Webster Addenda, *Ground speed*, the horizontal component of the velocity of an aircraft relative to the earth. 1928 C. F. S. Gamble *N. Sea Air Station* i, If there be no wind..a very much greater 'ground speed' has to be attained..before the machine will gain enough flying speed for the wings to take some load off the floats, and eventually enable the machine to 'take-off' from the surface of the sea. 1967 F. G. Mercer *Applied Aviation Sci.* viii. 48 An airplane flying eastward at a true airspeed ..of 120 mph in still air, will have a ground speed exactly the same—120 mph. 1933 P. Fleming *Brazilian Adv.* iv. 35 He..joined the ground staff of an aerodrome there. 1957 R. W. Zandvoort et al. *Wartime English* 8 *Aircrew*, ..a collective term for flying personnel, as opp. to ground staff. 1943 C. H. Ward-Jackson *Piece of Cake* 35 *Ground-strafe*, to attack ground objectives. 1938 *Flight* 15 Sept. 222/2 The ground-strafer's weapon should be..the light bomb. 1928 C. F. S. Gamble *N. Sea Air Station* iv, 'Ground-straffing' by low-flying machines. 1934 V. M. Yeates *Winged Victory* I. iii. 31 Unfortunately they were good machines for ground-strafing. They could dive straight down on anything, and when a few feet off the ground, go straight up again. 1941 *Times Weekly* 5 Feb. 8 In Libya..air activity was mainly confined to protective fighter patrols for our advancing troops and ground-strafing of the retreating enemy between Derna and Barce. 1943 Hunt & Pringle *Service Slang* 38 *Ground-strafing*, low-flying attack on transport or trenches; careless driving by servicemen. 1963 S. Douglas *Years of Combat* viii. 194 Ground strafing..had been a somewhat haphazard sort of business. 1917 'Contact' *Airman's Outings* I. vii. 196 Fighting squadrons soon caught the craze for ground stunts and carried it well beyond the lines. 1953 *Aero Digest* Oct. 37/3 An entire system of enemy destruction..includes a four-jet delta-wing supersonic bomber including its..reconnaissance, training, ground support and logistics systems. 1956 W. A. Heflin *U.S.A.F. Dict.* 238/2 *Ground support*, 1. either close air support or general air support. 2. the maintenance and care of flying equipment by the ground echelon. 1962 J. Glenn in *Into Orbit* 11, I flew..jets for sixty-three ground-support missions. 1967 *Technology Week* 20 Feb. 41/3 Design openings include development of..ground support equipment. 1920 *Flight* 10 June 624/1 Radio-Telephony..*Procedure*.—Ground to air and *vice versa*. The following illustrates the procedure adopted for work between a ground and an air station. 1945 *Aeroplane* 30 Nov. 619/2 An interesting point about the German development of air-to-air and ground-to-air weapons was the way in which 'pukka' aircraft designers were brought in to design missiles. 1951 *Gloss. Aeronaut. Terms* (*B.S.I.*) III. 18 *Ground-to-air communication*, one-way communication from ground stations to aircraft. 1958 *Observer* 15 June 18/6 The Russians can be assumed to be building ground-to-air missiles that are every bit as good. 1920 *Flight* 10 June 624/2 Radio Telephony..*Procedure*.—Ground to ground. The following abbreviated procedure to be used between ground stations A and B. 1943 L. Cheshire *Bomber Pilot* 18 It was ground-to-ground tracer, firing east and so probably British. 1958 *Times* 11 Sept. 4/5 The huge Corporal ground-to-ground guided missile. 1941 *Flight* 6 Feb. 107/2 A Rhodesian squadron has been giving support to the ground troops. 1925 Fraser & Gibbons *Soldier & Sailor Words* 112 *Ground wallah*, an Air Force term for a member of the R.A.F. whose duties were concerned with administrative, or office and aerodrome technical work.
18. ground and lofty, applied to acrobatic feats or performers on the ground and on a rope, etc.; also *transf.*; **ground ball** *Cricket* and *Baseball* = Grounder 3 c; **ground-bass,** also *fig.*, an undercurrent; **ground-bowler** (see quot. 1934); **ground-breaking** *a.* (cf. Break *v.* 44 c *fig.*); **ground chain** *Naut.* (see quot.); **ground cloth,** (*a*) (see quot. 1919); (*b*) = *ground-sheet*; **ground-colour,** (*b*) (later examples); **ground cover,** the plants covering the surface of the earth, esp., in horticulture, plants whose low, spreading habit of growth smothers weeds; so *ground-coverer*; **ground detector** *Electr.*, any instrument which is used to detect an accidental connection to

earth in a circuit; **ground force** *Mil.* = LAND FORCE; also *attrib.*; **ground(-)frost,** a frost on the surface of the ground, or in the upper layer of the soil (see also quot. 1963); **ground-keeper,** (*a*) *Cricket* = GROUND-MAN; (*b*) a root vegetable accidentally left in the ground during harvesting; **ground level** *Physics* = **ground state*; (see also *17 a above); **ground noise,** in sound reproduction, noise that is introduced by the recording medium (e.g. needle hiss on a gramophone record); **ground pin** (later U.S. example); **ground return** *U.S.* = **earth-return* (a); **ground rule,** (*a*) *Sport,* a rule devised for a particular ground; (*b*) a basic principle; **ground-sheet,** a waterproof sheet for spreading on the ground as a protection against damp; **ground-sluice** *sb.* and *v.* (earlier examples); so **ground-sluicing** *vbl. sb.*; **ground-space,** the area of ground occupied by a structure; **ground state** *Physics* [tr. G. *grundzustand*, lit. 'fundamental state'], the stationary state of lowest energy of a quantized system (as an atom or molecule); **ground station** *Radio,* a complex of buildings where radio and radar equipment is used in connection with aeronautical and aerospace projects; **ground-stroke** *Lawn Tennis,* a stroke played near the ground, after the ball has hit the court; **ground-tier,** (*a*) (earlier example); **ground wave,** the radio wave that passes from a transmitter to a receiver other than by reflection from the ionosphere, comprising one or more of the direct wave, the ground-reflected wave, and the surface-wave; also *attrib.*; **ground wire,** (*b*) *U.S.*, an earth wire, i.e. a wire that is connected to earth, either directly or through another earthed conductor (the usual sense); **ground zero,** that part of the ground situated immediately under an exploding bomb, esp. an atomic one.

1796 *Gazette of U.S.* (Philad.) 19 Nov. (Advt.) (Th.), Ground and Lofty Tumbling [at the Pantheon, Philadelphia]. **1840** *Southern Lit. Messenger* VI. 386/1 He is an adept in the art of walking on his head, turning somersets, and ground and lofty tumbling generally. **1843** T. WEED *Let.* 19 July (1866) 108 A strolling company of 'Ground and Lofty' Tumblers. **1898** J. HOLLINGSHEAD *Gaiety Chron.* viii. 330 The 'talented Cocksure family' in their celebrated ground and lofty performance. **1907** M. C. HARRIS *Tents of Wickedness* III. iv. 260 He has resigned his parish, left the ministry and bought a seat on the Stock Exchange. Isn't that ground and lofty tumbling? **1839** *Bell's Life* 13 Oct., It was for the umpire at the bowler's end to decide whether it was a 'ground' ball. **1851** J. PYCROFT *Cricket Field* vii. 99 The toss, the tice, the half volley, the long hop, and ground balls. **1948** *P.C.C. Chron.* (Pasadena, Calif.) 7 May 4/5 Salter hit a hard ground ball to shortstop Bill Davis. **1955** A. L. ROWSE *Expansion Eliz. Eng.* ii. 76 Their correspondence, their reports on the condition of their dioceses, always come back to this ground-bass. **1963** *Guardian* 20 Feb. 7/1 With rising unemployment, scarcer money, and a mutinous groundbass already discernible about the new rates householders will soon be called upon to pay, this could be a vintage year for bailiffs. **1968** *Listener* 23 May 657/3 These highlights apart, there is a ground bass of beatings, shootings and torture, all weltering in blood. **1874** *Baily's Mag.* June 225 The club has an efficient staff of ground bowlers. **1891** W. G. GRACE *Cricket* xi. 314 So pleased were the authorities of the M.C.C. with his [*sc.* W. Gunn's] first display at Lord's, that they made him the offer of a place as one of the ground-bowlers. **1934** W. J. LEWIS *Lang. Cricket* 113 *Ground-bowler,* a professional bowler attached to a club as a member of its 'ground staff'. **1907** W. JAMES *Let.* 6 Oct. (1920) II. 299, I am going to settle down to the composition of another small book, more original and ground-breaking than anything I have yet put forth (!). **1965** *Language* XLI. 138 She has had to do groundbreaking work in establishing the phonological..correspondences between a number of foreign languages and Russian in the 16th and 17th centuries. **1883** *Man. Seamanship for Boys* 195 Q. What is ground chain? A. A piece of small chain shackled to the anchor shackle,..of sufficient length to come through the hawse pipe when the anchor is high enough for catting. **1919** W. B. FARADAY *Gloss. Aeronaut. Terms* 54 *Ground cloth,* a floor covering, usually of canvas, placed on the ground under an envelope to protect it from damage. **1931** T. S. STRIBLING *Forge* xxiv. 204 The oilcloths were for raincoats or for ground cloths, or they could be propped up with sticks for tents. **1904** W. H. HUDSON *Green Mansions* vi. 72 A coral snake.. its ground colour a brilliant vermilion. **1912** W. R. OGILVIE-GRANT *Catal. Birds' Eggs Brit. Mus.* V. 105 The ground-colour varies from greyish-white to very pale greyish-green. **1970** H. E. SMITH *Bantams* iii. 20 Females ..should be chosen for clear ground colour. **1900** L. H. BAILEY *Cycl. Amer. Hort.* II. 629/2 In suitable soil they [*sc.* Gaultherias] are apt to form a handsome, evergreen ground-cover. **1906** *Westm. Gaz.* 31 May 4/2 As soon as the ground-cover was gone, the soil lost its moisture. **1946** *Nature* 13 July 71/1 Sparse ground-cover of herbs and dense undergrowth of hawthorn, bird-cherry, etc. **1970** G. S. THOMAS *Plants for Ground-Cover* p. xvii, Herbaceous plants..were limited to the stalwart clump-formers and the ground-coverers. *Ibid.* i. 3 Ground-cover can be of any height in nature or in the garden. *Ibid.* 4 The use of ground-cover plants may be the epitome of natural gardening. **1904** SWENSON & FRANKENFIELD *Testing Electro-Magn. Mach.* I. 36 Figure Q shows a General Electric Ground Detector. **1958** *Van Nostrand's Sci. Encycl.* (ed. 3) 768/1 Lamp type ground detectors are used to a considerable extent on low-voltage circuits because they are reliable and cheap. **1929** F. GIBBONS *Red Napoleon* ix. 222 The ground forces were exposed constantly day and night to bomb and gas attacks from above. **1951** *Ann. Reg.* 1950 322 Our people's ground forces must be strengthened continuously so that they can defeat any aggression. **1959** *N.Z. Listener* 17 Apr. 6/3 The reduction in the ground forces of the United States and Britain impairs their ability to fight limited wars. **1965** H. KAHN *On Escalation* vi. 128 Tactical nuclear weapons were a relatively small part of the NATO ground-force structure. **1900** *Daily News* 12 Oct. 5/1 Towards night, the thermometer fell briskly, and it seemed probable that a sharp ground frost would occur. **1958** HAYWARD & HARARI tr. *Pasternak's Dr. Zhivago* II. viii. 249 Probably there's still ground frost in the mornings. **1963** *Meteorol. Gloss.* (Met. Office) (ed. 4) 122 From 1906 to 1960, inclusive, the [British] Meteorological office practice was to record a 'ground frost' when the grass minimum thermometer reached 30°F or below... From 1 January 1961..no statistics have referred to 'ground frost'. The use of the term 'ground frost' in forecasts signifies a grass minimum temperature below 0°C (32°F). **1876** *Haygarth's Cricket Scores* v. 16 (G. H. Wright was engaged) on the Bramhall Ground, at Sheffield, where he still remains as groundkeeper. **1938** *Nature* 17 Sept. 530/1 Future policy..should aim at..the suppression of ground-keepers, always a fertile source of virus infection [in potatoes]. **1961** *New Scientist* 30 Mar. 795/3 The mild winter has also meant that more beet crowns, 'groundkeepers' (the beet missed by the lifting machines), and common weeds infected by the virus will have lasted through the winter. **1923** H. L. BROSE tr. *Sommerfeld's Atomic Struct. & Spectral Lines* vi. 325 In our diagram the absorption lines would have to be represented by arrows that start out from the natural or ground level and are directed upwards. **1953** L. H. ALLER *Astrophysics* 25 The excitation potential in ev is the potential through which a bombarding electron must drop in order to acquire sufficient energy to excite an atom from the ground level to the level in question. **1929** *Trans. Soc. Motion Picture Engin.* XIII. 53 *Ground noise,* any noise due to foreign matter or imperfections in or on the film arising during manufacture, processing or handling; does not include amplifier or photoelectric cell noises. **1938** *Motion Picture Sound Engin.* (Acad. Motion Pict. Arts & Sci.) iv. 44 Such a system..will differ from the ideal in several respects, one of which is the introduction of 'ground noise' during the recording process. **1942** *Electronic Engin.* XIV. 709 The fundamental failing of the disk is the fact that reproduction of the full frequency range recorded involves the production of atrocious scratch...We must not regard this ground noise as a defect altogether beyond improvement. **1843** J. C. FRÉMONT *Exped.* (1845) 54 Our lodge had been planted, and, on account of the heat, the ground pins had been taken out, and the lower part slightly raised. **1893** E. J. HOUSTON *Electr. Transmiss. Intell.* i. 9 The line wire or conductor may form what is technically known as a ground-return circuit. *Ibid.* 10 This is..called a ground-return, because the ground acts as the return conductor. **1968** FINK & CARROLL *Stand. Handbk. Electr. Engin.* (ed. 10) xiv. 3 In d-c transmission ground return can be used as one conductor. This means that each separately insulated transmission conductor, together with the ground-return path, forms a separate electric circuit. **1890** H. C. PALMER *Stories of Base Ball Field* 70 In the olden days there was a ground rule which only allowed two bases for a hit over this fence. **1953** *Manch. Guardian Weekly* 27 Aug. 1 Ground rules of American press-conferences. **1965** G. MCINNES *Road to Gundagai* vi. 104 His ground-rules were so elastic. **1967** M. MCLUHAN *Medium is Massage* 68 The groundrules, pervasive structure, and overall patterns of environments elude easy perception. **1907** *Daily Chron.* 15 Mar. 9/5 Groundsheet (waterproof). **1928** *Daily Express* 12 May 5/5 Some of the motorists were having tea in the cars, and some had spread ground-sheets and cushions on the ground. **1959** S. CLARK *Puma's Claw* v. 65 We could not afford heavy luxuries like..groundsheets. **1970** *Which?* May 132/1 The other half of the outer tent (the living room) does not have a groundsheet. **1857** *Hutchings' Mag.* July 8/1 Among the more important operations connected with gold mining upon an extensive scale, is 'ground sluicing'. **1860** *Harper's Mag.* Apr. 612/2 Ground-sluicing accomplishes the same result..with the chance of obtaining from the upper earth some gold, which..would be lost by the first plan. **1862** J. G. WALKER *Jrnl. Voy. N.Z.* (1863) [1] Jan., We..watched a man ground-sluicing. **1865** V. PYKE *Goldfields Report* in *Appendices to Jrnls. House of Reps. N.Z.* C. IVA. 10 Of the various methods of working, that of ground sluicing is most universally adopted—a ground sluice being nothing more than a rectangular drain cut for a depth of about a foot into the surface soil. **1874** A. BATHGATE *Colonial Experiences* viii. 91 In Otago, the principal kind of mining is ground-sluicing..similar to the method of working for tin in Cornwall. **1866** *Chambers's Jrnl.* 18 Aug. 521/2 The houses..are large.., with very little accommodation, considering the ground-space they occupy. **1908** *Westm. Gaz.* 28 Mar. 9/2 All citizens are users or consumers of air, water, ground-space. **1926** H. L. BROSE tr. *Sommerfeld's Three Lect. Atomic Physics* ii. 35 To every electron in its ground state there belongs an original momentum $s = \frac{1}{2}$. **1946** *Nature* 26 Oct. 593/2 To evaluate *D* for a diatomic gas, it is necessary to examine the vibrational energy-levels of the ground-state. **1963** B. FOZARD *Instrumentation Nuclear Reactors* ii. 13 This is duly emitted as radiant energy as the excited electrons return to the ground state. **1919** *Radio Rev.* Dec. 105 On aeroplanes, the problem of the reception of wireless signals is complicated...There is always so much noise that signals must be of much greater intensity than at ground stations in order to be of any use. **1966** *Electronics* 14 Nov. 58 The ground station, which was publicly introduced Nov. 10 in Melbourne, Fla., is also behind schedule and running 40% over cost expectations. **1895** H. W. W. WILBERFORCE *Lawn Tennis* 51 There are two ways in which a ground-stroke may be taken, namely, at the top of the bound, and again quite late, when the ball is near the ground. **1970** *Times* 5 June 14/6 His services and ground strokes were hit to a good length. **1820** W. SCORESBY *Acc. Arctic Reg.* II. 305 The hold of the ship must be cleared of its superstructure of casks, until the 'ground-tier', or lowest stratum of casks, is exposed. **1927** E. V. APPLETON in *Wireless World* 5 Jan. 3/2 There is a very real difference between a ground wave and an atmospheric wave if we consider the magnetic force in the wave as well as the electric wave. For a horizontally travelling ground wave, such as that which travels direct from transmitter to receiver, [etc.]. **1941** K. HENNEY *Radio Engin. Handbk.* (ed. 3) xv. 514 The signal..may have been propagated either by the ground wave, which travels along the earth's surface, or by the sky wave. **1943** F. E. TERMAN *Radio Engineers' Handbk.* x. 675 The surface wave..represents the whole of the ground wave when both transmitting and receiving antennas are located at the surface of the earth. **1965** *BBC Handbk.* 115 The signals which carry domestic broadcasting programmes are usually designed to be received by ground-wave on medium and long waves. *Ibid.*, Ground-wave propagation of short waves is not feasible over long distances. **1922** J. C. WRIGHT *Automotive Repair* II. 208 If the ground wire is disconnected the generator will build up an excessive pressure within itself. **1966** *McGraw-Hill Encycl. Sci. & Technol.* XIV. 54/1 Lines built where severe thunderstorms are prevalent are equipped with overhead ground wires..for intercepting the lightning stroke and leading it to ground at the nearest tower. **1946** *N.Y. Times* 7 July E10/1 The intense heat of the blast started fires as far as 3,500 feet from 'ground zero'. **1955** *Bull. Atomic Sci.* Sept. 255/1 There was no noticeable contamination even at ground zero at Hiroshima.

b. ground-bee, a bee that nests in the ground; **ground-hornet** (earlier U.S. example); **ground-mouse** *U.S.*, a field mouse of the genus *Reithrodontomys*; **ground rattler** *U.S.* (see quot. 1931); **ground-robin**: add def. and examples: any of several small American buntings, esp. the most common one, *Pipilo erythrophthalmus*; **ground-sparrow** *U.S.* (examples); **ground-spider** (earlier example); **ground woodpecker,** a member of the family Picidæ that lives on the ground, esp. *Colaptes campestris* of South America and *Geocolaptes olivaceus* of South Africa.

1849 C. BRONTË *Shirley* xxv, The nest..of some ground-bees, which had burrowed in the turf under an old cherry-tree. **1890** LD. LUGARD *Diary* 17 Apr. (1959) I. 195 Found ground bees' nest, and tried to dig it out, but it was some 16 inches down. **1822** Z. HAWLEY *Tour* 95 (Th.), A nest of ground hornets, concealed under the logway. **1839** J. BUEL *Farmer's Comp.* 99 Moles or ground-mice cannot penetrate and find a shelter. **1883** *Harper's Mag.* Aug. 462/2 A storm of expletives that must have startled the ground-mice and the birds. **1908** *Daily Chron.* 29 Aug. 7/5 The red dog..jumped a ground-rattler. **1931** R. L. DITMARS *Snakes of World* x. 105 The Pygmy Rattlesnake or 'Ground' Rattler, *Sistrurus miliarius,* seldom more than sixteen to twenty inches long, has a rattle so minute it would be unnoticed outside a distance of about eight feet. **1794** *Philos. Soc. Trans.* IV. 110 This bird was the chewink, or ground robin. **1844** J. E. DE KAY *Zool. N.Y.* II. 172 (*heading*) The Chewink or Ground Robin. Pipilo Erythrophthalmus. *Ibid.*, This beautiful and unobtrusive little species is..known..under the name of Ground Robin. **1955** *Sci. News Let.* 23 Apr. 271 The towhee is a bird of many aliases. 'Ground robin' is a popular name, and justified by his deceptively robin-like appearance. **1874** B. F. TAYLOR *World on Wheels* II. vii. 249 The ground-sparrows build in its margins. **1882** *Vermont Agric. Rep.* VII. 67 The blue bird, cat bird, wren and ground sparrows are acknowledged beneficial. **1867** *Amer. Naturalist* Oct. 410 Some of the ground spiders carry their eggs in a sack attached to the tip of their abdomen. **1867** E. L. LAYARD *Birds S. Afr.* 238 The Sub-Family, Colaptinæ, or Ground Woodpeckers, have the bill broad at the base. **1958** E. T. GILLIARD *Living Birds of World* 258/2 Tunnels of the Ground Woodpecker are often several feet deep. **1964** A. L. THOMSON *New Dict. Birds* 896/2 The 'ground woodpeckers' prefer stumps or rotten branches where many insects (especially ants) have their favourite haunts.

c. ground flower, a low-growing wild flower; **ground-hemlock** (earlier example); **ground laurel** (earlier and later examples).

1829 A. EATON *Man. Bot.* (ed. 5) 336 *Polygala polygama,* ground flower. **1902** *Daily Chron.* 1 Apr. 2/1 At Easter there are but the ground-flowers. **1928** D. COTTRELL *Singing Gold* I. vi. 50 The cup-like tufts of the little white ground-flower, with the sweetest scent in all Australia. **1807** F. PURSH *Jrnl. Bot. Excursion* 20 July (1869) 64 Here I found..taxus braccata or procumbens, calld Ground Hemlock. **1814** J. BIGELOW *Florula Bostoniensis* 101 Ground laurel..grows in woods. **1867** *Amer. Naturalist* May 154 In the books, this plant is known as the 'Epigea repens', but otherwise as the Trailing Arbutus, May Flower, and Ground Laurel. **1932** P. A. RYDBERG *Flora Prairies & Plains Cent. N. Amer.* 615 Epigaea L. Trailing Arbutus, Ground Laurel.

ground, *v.* **9.** Delete ? *Obs.* and add later examples.

1908 H. G. WELLS *War in Air* viii. 252 Not one third.. succeeded in getting back to the mother airship. The rest were either smashed up or grounded. **1968** EAGLESON & MCKIE *Terminol. Austral. Nat. Football* II. 13 *Ground,* a variant for bring down, recorded by two informants.

10. (Additional examples.)

1773 J. DUNCOMBE *Surry Triumphant* xlvii, in R. Freeman *Kentish Poets* (1821) II, Ere the foe could ground his bat, His ardour Lewis quell'd. **1879** *Boy's Own Paper* 18 Jan. 2/3 The players [at Rugby football] are scrambling to their feet, and waiting for the ball to be 'grounded'. **1899** *Westm. Gaz.* 10 Nov. 2/1 The Rules Committee have decided that..the [golf] club may be grounded.

c. (Earlier *intr.* and later examples.) Chiefly *U.S.*

1881 *Operator* 15 June 218/1 Never, if possible, ground within a hundred feet of any place where an electric light wire is grounded. **1882** T. D. LOCKWOOD *Pract. Inf. Telephonists* 28 It is not well to ground on a plate of one metal at one end and another metal at the other. **1967** *Technology Week* 20 Feb. 35/1 A resistance inserted in the 'V's' effectively 'grounds' the antennas.

d. (Later example.)

1908 H. G. WELLS *War in Air* viii. 258 The *Zeppelin*.. circled down and grounded in Prospect Park, in order to land the wounded.

e. To keep 'on the ground', prevent (an air-craft, pilot, etc.) from flying.

1931 *Vanity Fair* Nov. 78/1 A *flying ticket* is the pilot's license, which he holds until he is *grounded* by having the license cancelled. *a* **1940** F. SCOTT FITZGERALD *Last Tycoon* (1949) i. 6 The assistant pilot is always in uniform. ..I want to find out if we're going to be grounded in Nashville. **1940** *War Weekly* 2 Feb. 464/2 With catapults at an aerodrome, damage done by the enemy would not 'ground' all the aeroplanes in the hangars. **1944** *Flight* 20 July 61 That's young Begley. He's being grounded to-morrow! **1970** *Daily Tel.* 28 Apr. 1/8 Three..jets were grounded by mechanical faults and an accident.

ground, *ppl. a.* Add: **1.** (Additional examples.)

1781 in D. Davis *Hist. Shopping* (1966) x. Pl. 11 Rice whole. Do. ground. **1846** 'A LADY' *Jewish Man.* i. 2 Mix a little potatoe-flour, ground rice, or pounded vermicelli, in a little water. **1905** H. G. WELLS *Kipps* I. i. 15 Toke and cold ground-rice puddin' with plums. **1947** 'G. ORWELL' *Shooting Elephant* (1950) 165 Cold ground-rice pudding.

b. *ground-down*: also *fig.*, exhausted.

1911 J. MASEFIELD *Everlasting Mercy* 49 The ground-down starving man. **1946** S. SPENDER *European Witness* (1947) 32 The German soldiers now have the soulless ground-down expression as in carved-wood faces of Slav peasants.

4. **ground wood** (see quot. 1937).

1885 G. F. GREEN in Rattray & Mill *Forestry & Forest Products* xviii. 473 Ground wood was first used for paper-making about the year 1846, when it was manufactured by Keller. **1937** E. J. LABARRE *Dict. Paper* 119/1 *Ground wood,* wood pulp produced by grinding wood; another term for mechanical wood. **1955** *Times* 5 July p. ii/2 Some will be cut into lengths for the groundwood mill.

grounded, *ppl. a.*[1] Add: **7.** *Electr.* Electrically connected with the ground, either directly or through another conductor; earthed. Chiefly *U.S.*

1889 [see sense 6 in Dict.]. **1913** *Wireless World* I. 99/2 Grounded conductors, such as gas, steam, and water pipes. **1938** L. F. BLUME *Transformer Engin.* vii. 198 Grounded neutral power systems generally connect to ground through the neutral of a step-up delta-Y bank, located at the generating station. **1949** *Electronics* Aug. 120/2 The grounded-emitter arrangement [of a transistor circuit]..is analogous to a grounded-cathode tube circuit. **1971** *Sci. Amer.* Aug. 107/1 The charging current is measured with a microammeter in series with the grounded side of the coil.

8. Of an aeroplane, pilot, etc.: unable, or not allowed, to fly. Also *transf.* of sportsmen, etc., suspended, disqualified.

1939 in *Amer. Speech* (1955) XXX. 286 When a jockey is suspended or disqualified, he is said to be 'grounded'. **1942** T. RATTIGAN *Flare Path* II. ii. 140 And on my confidential report they'd put—grounded. Lack of moral fibre. **1942** *Amer. Speech* XVII. 103/2 *Grounded,* license revoked [of a truck-driver]. **1954** in *Amer. Speech* (1955) XXX. 286 'We're all grounded.' It took only another question or two to discover that for these youths 'grounded' meant not having the use of the family car. **1967** *Sunday Times* (Colour Suppl.) 10 Sept. 45/4 *Grounded,* stranded for lack of petrol. Borrowed from the R.A.F.

grounder. 3. c. Add examples (in cricket, baseball, and football).

1849 *Boy's Own Bk.* 69 Grounders and home tosses. **1862** J. PYCROFT *Cricket Tutor* 8 The old bat used to be heavy at the point—very requisite for picking up a Grounder. **1867** H. CHADWICK *Base Ball Player's Bk. Reference* 137 *A grounder,* a ball hit along the ground, either on a line or on a series of bounds. **1927** *Daily Express* 27 May 13/7 Brown opened the scoring, receiving a fine pass from Dean and sending in a beautiful grounder from eighteen yards. **1932** J. T. FARRELL *Studs Lonigan* (1936) viii. 617 Studs watched the infield practice, the grounders slapped hard, cutting over the dirt, the ball snapped around from player to player. **1970** *Washington Post* 30 Sept. D1/7 A fielder's-choice grounder by Dave Nelson knocked in Mike Epstein.

ground floor. Add: **b.** For (*U.S.*) read (orig. *U.S.*) and add further examples of this phrase and similar uses.

1872 T. TALMAGE *Abom. Mod. Society* 118 A select number go in on the 'ground floor'. **1901** MERWIN & WEBSTER *Calumet 'K'* xi. 211 Well then, we'll have to let you in on the ground floor. **1904** 'O. HENRY' *Cabbages & Kings* xii. 206 Says he's heard of the boom along this coast, and wants to get in on the ground floor. **1909** E. S. BARNETT *Dragnet* 12 To take advantage of this ground floor proposition Alexander and Company will have to give up its identity, and be a branch. **1916** J. BUCHAN *Greenmantle* xii. 159 The promoters are keeping it to themselves. They aren't taking in more than they can help on the ground-floor. **1930** W. S. MAUGHAM *Cakes & Ale* xiv. 167 It was out of the question then for Mrs. Barton Trafford to get in on the ground floor. She could only buy in the open market. **1939** J. B. PRIESTLEY *Johnson over Jordan* 57 Still plenty of good things if you know where to find 'em and get in on the ground floor. **1945** R. HARGREAVES *Enemy at Gate* 82 Spain bestirred herself to get in on the nearest thing she could find in the way of a ground floor, by declaring war on England in the June of [1779]. **1946** K. TENNANT *Lost Haven* (1947) xix. 319 It'd be a big thing, Alec... You're in on this, in on the ground floor. **1958** *Spectator* 7 Feb. 169/1 It was thus comparatively easy to get in on the ground floor of the occasional project. **1966** A. LOOS *Girl like I* (1967) iii. 68 My problem was that, without realizing it, I was in on the ground floor of a sex revolution: the twentieth century's breakdown of romantic love between the sexes. **1968** *Globe & Mail* (Toronto) 3 Feb. 48/5 (Advt.), Surgical sales representative... Outstanding future—ground floor opportunity. **1970** *Observer* 1 Mar. 31/6 I'd get in on the ground floor and see it now if I were you.

c. The lower deck of a bus. *slang.*

1936 *Daily Herald* 5 Aug. 8/4 Here is a short list of busmen's slang phrases:..*ground floor,* inside.

ground-hog. Add: **2.** (Earlier examples.)

1784 J. FILSON *Kentucke* 28 Nor are the animals common to other parts wanting, such as foxes,..ground-hogs, pole-cats, and oppossums. **1789** *Maryland Jrnl.* 13 Nov. (Advt.) (Th.), A Monack, or Ground-Hog, presented [to Peale's Museum] by Mr. Johnston. **1807** P. GASS *Jrnl.* ii. 25 Two of our hunters went out and killed an animal called a prarow, about the size of a ground hog.

3. A slang designation for various workmen whose occupation keeps them on the ground. Also *spec.* a caisson worker. *U.S.*

1926 *Amer. Speech* I. 651/2 *Ground hogs,* men who work in compressed air at caisson work. **1931** *Ibid.* VII. 48 'Tie men' who camp far out from the main camps are called 'Ground Hogs' or 'Pole Cats'. **1931** G. IRWIN *Amer. Tramp & Underworld Slang* 92 *Ground hog,* a railroad brakeman. **1934** WEBSTER, *Ground hog,* a man who works under air pressure in caisson sinking or shield tunnelling. Called also sand hog. **1948** H. L. MENCKEN *Amer. Lang.* (Second Suppl.) xi. 713 A brakeman is a.. ground-hog. *Ibid.* 718 A lineman's helper..who never leaves the ground is a..groundhog. **1960** WENTWORTH & FLEXNER *Dict. Amer. Slang, Ground-hog,* a pilot who does not like to fly. Early aviator use... A railroad brakeman.

4. *Comb.* **ground-hog case** *U.S.*, a desperate or urgent affair; **Ground-hog Day** *N. Amer.*, (see quots. 1871, 1948).

1885 C. A. SIRINGO *Texas Cow Boy* 125 Dangerous to cross. But the wagons being over made it a ground hog case. **1897** KIPLING *Capt. Cour.* x. 234 'Ground-hog case,' said the Galway man. 'Badly lighted port, Danny.' **1927** *Amer. Speech* II. 356/1 [W. Virginia dial.], *Ground hog case,* something that has to be settled immediately. **1871** SCHELE DE VERE *Americanisms* 369 Candlemas is known as Ground-hog Day, for on that day the ground-hog comes annually out of his hole, after a long winter nap, to look for his shadow. If he perceives it, he retires again to his burrow, which he does not leave for six weeks —weeks necessarily of stormy weather. But if he does not see his shadow, for lack of sunshine, he stays out of his hole till he can, and the weather is sure to become mild and pleasant. **1948** A. L. RAND *Mammals Eastern Rockies* 128 Groundhog day is February 2. Current belief has it that if on this day the groundhog sees its shadow, there will be 6 weeks more of winter; if it does not see its shadow, winter will be soon gone. **1965** *Kingston* (Ont.) *Whig-Standard* 2 Feb. 11/1 Today is groundhog day.

groundling. Add: **2. d.** A person on the ground, as opposed to an airman, passenger in an aircraft, etc.

1940 *Manch. Guardian Weekly* 25 Oct. 304/1 Of the R.A.F.'s courage and skill it would be almost presumptuous for any mere groundling to speak. **1966** M. R. D. FOOT *SOE in France* iv. 81 The groundlings assumed, wrongly, that the aircraft had accidentally dropped a couple of small bombs. **1969** *Daily Tel.* 13 Sept. 10 An occasional crash [of an aeroplane] may be expected, possibly involving the death of 500 passengers and an unpredictable number of groundlings. *Ibid.* 15 Nov. 22/6 The giant Saturn 5 rocket..sent the lunarnauts soaring up into a blue sky hidden from groundlings by heavy clouds.

ground-pea. [GROUND *sb.* 18 c.] = GROUND-NUT 1 and 2.

1769 [see GROUND-NUT 2]. **1796** B. HAWKINS *Let.* 2 Dec. in *Georgia Hist. Soc. Coll.* (1916) IX. 21 They made beans, ground peas, cymblins. **1823** E. JAMES *Long's Exped. Rocky Mts.* I. 200 The squaws..are often necessitated to dig the pomme de terre..and to scratch the groundpea. **1854** *Florida Plant. Rec.* 115, I have picked 80 bushels of ground peas. **1893** *Jrnl. Amer. Folklore* V. 95 *Arachis hypogæa,* ground-peas. *Ibid.* VI. 140 *Apios tuberosa,* ground-pea. **1946** *Publ. Amer. Dial. Soc.* v. 24 *Ground peas,* peanuts.

groundsman. Now the form in general use = GROUND-MAN b.

groundwork. Add: **5.** *Cricket.* = *ground-fielding* (GROUND *sb.* 18).

1898 K. S. RANJITSINHJI *With Stoddart's Team* (ed. 4) iv. 79 The catching of the team was far better than the ground work. **1905** *Westm. Gaz.* 19 Sept. 3/1 The ground-work was always splendid, and the throwing a joy to behold.

6. *Boxing* and *Judo.* Physical exercises performed by a sportsman whilst lying on the ground.

1906 MIYAKE & TANI *Game of Ju-Jitsu* (ed. 2) vi. 40 Ground-work cannot be described as systematically as standing work... Ju-jitsu ground-work is merely the proper use of the legs. **1948** D. S. LISTER *Boxing* 24 Gymnasium training..consists of..top work, ground work and, of course, boxing. **1954** E. DOMINY *Teach Yourself Judo* vii. 77 Groundwork ability is seldom related to the ability of the judoka with the standing up side of the sport.

group, *sb.* Add: **1. d.** A set of letters used in coding.

1911 *Encycl. Brit.* XXV. 72/2 Each word or 'group' sent by the Morse code must be 'answered' before the sender passes on to another...All cipher 'groups' are repeated *en bloc.* **1916** T. E. LAWRENCE *Lett.* (1938) 211, I don't suppose my wire of conditions in Feisul's camp will get through for some days yet, as it is a long one of about 400 groups. **1966** M. R. D. FOOT *SOE in France* ix. 241 A suspicious-minded staff officer noticed both coded telegrams had the same number of groups.

2. c. *spec.* A group of hits made by a series of shots fired at a target; = **shot group.*

1911 *Encycl. Brit.* XXVI. 419/2 The position of his shot group with reference to the bull's-eye does not matter; if his group is comprised within a 6 or 12-inch ring (at 100 yards range) he is passed on to more advanced practices at service targets. **1913** A. G. FULTON *Notes on Rifle Shooting* 27 If the rifle gradually or suddenly.. changes to a lower or higher group, there is no doubt that some adjustment is necessary to the bands or fore-ends. **1958** J. A. BARLOW *Elem. Rifle Shooting* (ed. 5) i. 4 When the firer is satisfied that the rifle is maintaining a good group under these conditions, a more severe test must be applied, since some rifles shoot very nicely when cold, but tend to throw a more scattered group when hot.

3. a. (Earlier and later examples.)

1782 W. COWPER *Poems* I. 6 But where, good Sir, do you confine your kings? There—said his guide, the groupe is full in view. **1807** [see *FAMILY *sb.* 9 a]. **1935** HUXLEY & HADDON *We Europeans* iii. 104 It follows that practically all human groups are of decidedly mixed origin. **1970** E. D. CHAPPLE *Culture & Biological Man* xv. 307 During the struggle, life crises occur and, afterward, rites of passage..are necessary mediators to re-establish the equilibrium of the group.

c. Specific senses in *Chem.*: (i) In qualitative analysis, a set of ions or radicals the presence of at least one of which in a presumed mixture of compounds can be ascertained by a single test specific for that set, provided that a standard procedure is followed; *esp.* any such set of the more usual cations which are precipitated from solution together in the course of analysis.

1843 J. L. BULLOCK tr. *Fresenius's Elem. Instr. Chem. Anal.* I. iii. 77 To teach the relation of the various bodies to reagents, it is usual..to treat of the substances individually..and to point out their characteristic reactions. I have, however,..deemed it more judicious..to collect into groups those substances which are in many respects analogous. *Ibid.* 82 The solutions of the salts of the alkaline earths are..not precipitated by sulphuretted hydrogen,..but alkaline carbonates and phosphates do precipitate them. This relation distinguishes the oxides of the second group [*sc.* barytes, strontian, lime, magnesia] from those of the first [*sc.* potash, soda, ammonia]. *Ibid.* II. ii. 257 If, therefore, we add to our solution.. Hydrochloric acid, we remove from the solution the metallic oxides of the first section of the fifth group. **1869** HARCOURT & MADAN *Exercises Pract. Chem.* 250 The chlorides, iodides, nitrates, &c. are also classed together in one group, as having the common property of not being precipitated from aqueous solutions by barium chloride. **1888** A. H. SEXTON *Outl. Qual. Anal.* II. 24 The metals.. are classified for analytical purposes into six groups—not according to their natural affinities, but according to their behaviour with certain reagents. The number of divisions adopted, and the exact lines of demarcation vary. **1938** *Thorpe's Dict. Appl. Chem.* (ed. 4) II. 551/1 The filtrate from the foregoing sulphides should be examined thoroughly as to the complete precipitation of all metals of analytical Group II. **1957** H. HOLNESS *Adv. Inorg. Qual. Anal.* 35 The case of platinum appearing in this Group of elements is of some interest for it is only precipitated here when selenium is present as well.

(ii) One of the sets of elements which show some similarity in chemical and physical properties and which are commonly represented in a column of the periodic table.

Nine groups are now recognized, numbered O–VIII, which represent the eight groups of Mendeleev's revised periodic table (1871) with the addition of the noble gases. Groups I–VII are each divided into two sub-groups in which the similarity in properties is more marked.

[**1863** J. A. R. NEWLANDS in *Chem. News* 7 Feb. 71 Many chemists..have..pointed out some very interesting relations between the equivalents of bodies belonging to the same natural family or group. *Ibid.*, Group I. Metals of the alkalies:—Lithium, 7; sodium, 23; potassium, 39; rubidium, 85; cæsium, 123; thallium, 204.. Group II. Metals of the alkaline earths:—Magnesium, 12; calcium, 20; strontium, 43·8; barium, 68·5... Group VII. —Nitrogen, 14; phosphorus, 31; arsenic, 75; osmium, 99·6; antimony, 120·3; bismuth, 213... Group XI.— Mercury, 100; lead, 103·7; silver, 108.] **1871** H. E. ROSCOE *Less. Elem. Chem.* (new ed.) xxv. 278 We have the carbon group, the nitrogen group,..that of the alkaline metals, and that of the metals of the alkaline earths. **1922** J. W. MELLOR *Inorg. & Theoret. Chem.* I. vi. 258 The valency of the elements shows a peculiar relation, for the maximum valency rises from 1 to 8 in passing along a given series from the first to the last group. **1950** N. V. SIDGWICK *Chem. Elements* I. 11 Hydrogen..was one of the chief problems of the original Periodic Table, since it has close affinities both with the alkali metals of Group I and with the halogens of Group VII. **1952** T. MOELLER *Inorg. Chem.* xix. 859 The sulfides of the Group IIb elements are water insoluble. **1959** R. A. SMITH *Semiconductors*

iii. 61 (*heading*) Energy levels of group III or group V impurities in group IV semiconductors. **1963** J. HICKS *Comprel. Chem.* xiii. 247 Elements in the same group usually have the same valency.

(iii) Any combination of atoms (usually composed of more than one element) which, being recognizable in a number of compounds and persisting through a series of chemical changes, is regarded as a distinct entity.

1869 HARCOURT & MADAN *Exercises in Pract. Chem.* I. v. 140 In many cases a group of two or more elements appears to be more easily detached from a substance than the individual elements; the substance, like a crystal, cleaving more easily in some directions than in others. **1894** G. S. NEWTH *Text-bk. Inorg. Chem.* iv. 22 It is sometimes necessary to represent the presence in a molecule of certain groups of atoms, groups which seem to hold together, and often to function as a single atom. **1962** D. H. CALAM in A. Pirie *Lens Metabolism Rel. Cataract* 439 At the low pH employed, only strongly acidic groups remain charged, most of the carboxyl groups are unionized.

d. One of the constituent bodies of the (*Oxford*) *Group(s) Movement* (or *Oxford Group*), a religious revivalist movement brought from America to England in 1921 by Frank Buchman, characterized by the 'sharing' of personal problems by groups. Hence **Grou·per, Grou·pist,** a member of the movement; **Grou·pism,** the principles of the movement; **grou·py** *a.* (used disparagingly). Cf. *BUCHMANISM, **Moral Rearmament.*

[**1923** H. BEGBIE *Life Changers* iv. 101 A group soon formed in his room of men who really longed for spiritual life...F.B. [*sc.* Frank Buchman]..sent him over to Oxford..to begin there a similar work of personal religion.] **1928** *Isis* 16 May 1/1 Attendance at several of them [*sc.* meetings] is a preliminary step to admission to a 'group'—a gathering of perhaps four or five friends... Here..souls are laid bare by hysterical confession, and with a fervour which no longer pretends to be religious the tenets of the doctrine are discussed. *Ibid.* 25/1 Buchmanism, to give the Group its popular name, bases the whole of Christianity on four points—Honesty, Unselfishness, Purity and Love. *Ibid.* 24 May 34/1 In three of the Societies, no group meetings had been held. **1931** *A Group Speaks* iii. 100 What I owe to the Cambridge Group can never be estimated. **1932** A. J. RUSSELL *For Sinners Only* viii. 109 The well-dressed Doctor of Divinity..founder of the Oxford Group. *Ibid.* 116 A kindly and fruitful worker in the Oxford Group Movement. *Ibid.* 118 The Thursday night Group meeting. **1933** H. H. HENSON *Oxford Groups* 7 Religion—unless all the masters of the spiritual life throughout Christian history are mistaken—is not quite so cheerful a matter as these gay, almost uproarious teams of missioning Groupists affirm. *Ibid.* 9 A religious system like Groupism, which appears to be mainly adapted to the needs of adolescents, can hardly have much staying power. *Ibid.* 15, I have read as many publications of the Groups Movement as I could. *Ibid.* 24 The movement commonly designated, though rather misleadingly, 'The Oxford Group Movement', of which the founder and director is an American Lutheran minister named Frank Buchman. **1934** R. MACAULAY *Going Abroad* v. 49 The next thing will be.. you'll go groupy. *Ibid.* xx. 160 The eyes of the Oxford Groupers brightened. *Ibid.* xxiv. 201 A cheerful company —the four Groupers. *Ibid.* xxix. 252 There is a book the Groupists all seem to recommend and like called *Inspired Children.* **1937** W. H. AUDEN in Auden & MacNeice *Lett. fr. Iceland* xiii. 203, I am no Grouper, I will never share With any prig. **1937** F. T. JESSE *Act of God* xiii. 163 'Oxford Groupers?' said Erskine vaguely. 'Oh yes. They're harmless people, I believe, though rather bright and jolly.' **1939** *Times* 7 Mar. 15/5 Again, not by any means all of our youth are represented by those who frequent mass meetings, 'go all groupy', and discuss with great emotion topics on which, by the nature of things, they can be but very inadequately informed. **1940** GRAVES & HODGE *Long Week-end* xii. 205 One of the Groupist songs. **1959** *Chambers's Encycl.* VI. 604/1 The organization in Great Britain was incorporated in 1939 as the Oxford Group... Its object is to 'change lives' by the power of Christ. *Ibid.*, The Cambridge Group Movement, no longer in existence, was a fellowship of young methodists which in the 1930s adopted many of the principles and methods of the Buchmanite movement in a less extreme form. **1961** [see *BUCHMANISM].

e. An air-fleet or division of an air-fleet; *spec.* one in the Royal Air Force.

1922 *Times* 15 Feb. 14/5 The Officer Commanding Irak Group is directly responsible to the Air Ministry for the command and administration of the Air Force units located in Irak. **1939** *War Illustr.* 29 Dec. 538/2 A Squadron in the R.A.F...is the basic tactical unit, a number of Squadrons forming a Wing, so many Wings a Group, and so to the Command. **1942** T. RATTIGAN *Flare Path* I. 114 Been a bit of a muck-up at Group. *Maudie.* What's Group? *Dusty.* Group headquarters. Where the orders come from. **1959** *Chambers's Encycl.* I. 189/1 The main formations of the R.A.F. are the flight, the squadron, the wing and the group.

f. An ensemble of popular musicians; = **pop group.*

1958 in *Conc. Oxf. Dict.* Add., Skiffle group. **1964** *Gramophone Pop. Record Catal.* (Artist Section) Dec. 44/1 *Barron-Knights, The*... Call up the groups. Medley. **1967** *Melody Maker* 27 May 9 Groups who are going to give us action. **1967** *Listener* 20 July 80/1 It is an eye to detail..which keeps the group on its toes.

4. b. In *Stratigraphy*: formerly, any of various categories into which rocks were classified and which would now correspond to the following modern geological time units: (i) Period (in some systems also Epoch); (ii) Age; (iii) Era. (iv) In modern use: a stratigraphic unit consisting of two or more formations.

(i) **1830** H. T. DE LA BECHE *Geol. Notes* p. xxxv, The superior stratified or fossiliferous rocks are divided into groups. *Ibid.* p. xxxvi, Group 3. (Supercretaceous)... Group 4. (Cretaceous). **1838** C. LYELL *Elem. Geol.* xiii. 281, I proposed to give short technical names to these four groups [*sc.* Eocene, Miocene, Older Pliocene, Newer Pliocene], or the periods to which they respectively belonged. **1893** *Jrnl. Geol.* I. 187 De la Beche..carries out the system more completely, calling the first, or superior order, Supercretaceous group, and applying the terms Cretaceous, Oölitic and Red sandstone to three groups into which he divides the second order, and giving the third the name Carboniferous group.

(ii) **1883** [see sense 4]. **1912** A. J. JUKES-BROWNE *Student's Handbk. Stratigr. Geol.* (ed. 2) i. 11 Systems are divided into sections or formations... These sections are again divided into stages or groups.., and these again are often divisible into zones.

(iii) **1883** [see sense 4]. **1898** *Jrnl. Geol.* VI. 353 The terms, Group, System, Series, Stage, and the correlative time-divisions, Era, Period, Epoch, Age, are to my mind very satisfactory. **1927** LAKE & RASTALL *Text-bk. Geol.* (ed. 4) xvi. 299 In this way the rocks of the earth's crust have been divided into four great divisions, which are often known as groups. The groups are subdivided into systems, the systems into series, and the series into stages.

(iv) **1933** *Bull. Geol. Soc. Amer.* XLIV. 429 Article 2. The following divisions or units of rocks are recognized:.. (3) Group, a local or provincial subdivision of a system, based on lithologic features. It is usually less than a standard series and contains two or more formations. **1963** KRUMBEIN & SLOSS *Stratigr. & Sedim.* (ed. 2) ii. 33 Two or more successive formations, related by lithology or by position with reference to unconformities, may be assembled as a group.

5. (Later examples.) Now more generally defined as: A set of elements, together with a single-valued associative binary operation, which is closed with respect to the operation and contains an inverse for each element and an identity element. (Except for the property of being closed these properties were formerly tacitly assumed.) [The sense is due to E. Galois, who used F. *groupe* (*Bull. d. Sciences math., astron.,* etc. (1830) XIII. 435).]

1907 M. BÔCHER *Introd. Higher Algebra* vi. 82 The positive and negative integers with zero form a group if the rule of combination is addition... These same elements, however, do not form a group if the rule of combination is multiplication,..since zero has no reciprocal. **1941** BIRKHOFF & MACLANE *Surv. Mod. Algebra* vi. 130 A group whose operation satisfies the commutative law is called a 'commutative' or 'Abelian' group. **1958** SYKES & BELL tr. *Landau & Lifshitz's Quantum Mech.* xii. 317 The set of all symmetry transformations for a given body is called its symmetry transformation group (or simply its symmetry group). **1965** PATTERSON & RUTHERFORD *Elem. Abstract Algebra* ii. 31 Any element of a group *S* can be expressed in one and only one way in the form $a \wedge x$, where *a* is a fixed element of the group. **1969** A. P. CRACKNELL *Crystals* ii. 65 All the operations of this type, combinations of a point-group operation and a lattice translation are said to make up a space group, and it is really a space group which is needed to describe the symmetry of a real crystal.

b. group theory, a branch of algebra concerned with the properties of groups and their applications, esp. in mathematical physics (e.g. to the investigation of symmetries in physical systems).

1898 *Proc. Calif. Acad. Sci.* (ser. 3: Maths-Physics) I. iv. 29 In the application of group theory to problems of geometry and analysis, simple groups play the fundamental rôle. **1919** F. CAJORI *Hist. Math.* (ed. 2) 335 A line-geometry and kinematics are elaborated, partly by the use of group theory, which are carried over to non-Euclidean spaces. **1965** J. C. DAVIS *Adv. Physical Chem.* iii. 75 The student of quantum mechanics and spectroscopy will find the structure of group theory beautiful and its utility exciting.

6. *group-burial, -name, -photograph, -portrait*; **Group Areas Act** (see quots.); **group captain,** a rank in the Royal Air Force equivalent to colonel in the army; **group dialect,** distinctive language used by members of a group sharing the same occupation or interests; so **group language**; **group-flashing,** the repeated emission of a set group or pattern of flashes by a lighthouse; also *attrib.*; **group genitive** (see quot. 1957); **group language** (see **group dialect* above); **group (life) assurance** or **insurance** (see quot. 1927); **group-marriage,** a primitive form of familial relationship hypothesized by some anthropologists in which certain groups within a tribe were considered husbands and wives; also *transf.*; **group practice,** a medical practice in which several doctors are associated; **group velocity,** the speed at which the energy of a wave or wave-group travels: so called because if the sinusoidal components of a wave-group do not differ greatly in frequency, the group-velocity is the speed at which the group as a whole travels; **group-verb** (see quots.); **group-wise** *adv.*, (performed) by groups; **group-word** (see quot.); **group work** (see quots.).

1952 L. MARQUARD *Peoples & Policies S. Afr.* vii. 152 In 1950 the Group Areas Act was passed empowering the Government to declare any area a group area for Coloured, European, African, or Asian. **1957** *Encycl. Brit.* XXII. 426/2 The Group Areas act of 1950 was one of a series of acts which sought to implement Smuts's Natives Land act of 1913, which checked the penetration of natives into European areas and vice versa. **1932** *Times* 12 Feb. 11/1 The significance of this unique group-burial must remain a mystery. **1920** Group captain [see *air officer,* *AIR *sb.*[1] B. III. 3]. **1922** *Man. Seamanship* I. i. 11 Marks of Rank.. R.A.F... Group Captain. R.N... Captain. **1923** *Daily Mail* 13 Feb. 9 Group-captain in the Air Force. **1934** PRIEBSCH & COLLINSON *German Lang.* II. v. 260 We propose to use the term 'group dialects' as the equivalents of the German *Sondersprachen* and French *Langues spéciales.* The word contrasts with 'regional dialects'. **1959** I. & P. OPIE *Lore & Lang. Schoolch.* viii. 152 'Pax' is group dialect not regional dialect. **1891** A. G. FINDLAY *Lighthouses of World* Add. facing p. 32, Its [*sc.* electric light's] range, definition, and, where a distinctive character is employed as *Group-Flashing,* its unmistakable superiority to all other modes of Illumination, pronounce its excellence and pre-eminence. **1911** *Encycl. Brit.* XVI. 629/1 The 12 lens panels are arranged in groups of two, thus producing a group flashing light. **1958** R. DE KERCHOVE *Internat. Maritime Dict.* (ed. 2) 310/2 *Group flashing light,* a light showing at regular intervals a group of two or more flashes. **1894** O. JESPERSEN *Progress in Lang.* viii. 306 It will not be easy to lay down fully definite and comprehensive rules for determining in which cases the group genitive is allowable and in which the *s* has to be affixed to each member. **1927** *Rev. Eng. Studies* Oct. 438 There are other practical problems of syntax... One might instance the Split Infinitive, and the modern development of the group genitive. **1957** R. W. ZANDVOORT *Handbk. Eng. Gram.* II. 91 The genitive-suffix is always added to the last element of a word-group: *the Prince of Wales's birthday, Miss Mansfield's wedding.* This is known as the group-genitive. **1934** PRIEBSCH & COLLINSON *German Lang.* II. v. 262 Like all group languages student slang has a number of different designations for the people and things most familiar in student life. **1964** C. BARBER *Present-Day Eng.* iii. 68 The importance of a special group-language in promoting feelings of cohesion can easily be seen. **1913** *Craftsman* Sept. 652/1 A new and progressive form of life insurance has recently been developed, which is of interest to every employer of labor and every employee. 'Group insurance' it is called, and by its means an employer is able to insure the lives of those who work for him at a much less cost than would be possible if each employee were to become individually insured. **1920** H. DOUGHARTY *Pension, Endowment, Life Assurance* xi. 51 Group Life Assurance, introduced by an American Life Office some four years ago, is a non-contributory system of Staff Life Assurance. *Ibid.* 52 The following features of Group Assurance are common to practically all Group Policies—1. The insurance is effected by the employer on the lives of his employees. **1927** B. C. HOSKINS *Insurance Lex.* 105 Group Life Assurance.—Under this contract groups or numbers of lives are assured instead of individuals. **1928** *Daily Mail* 25 July 19/5 Group Insurance came into existence only seventeen years ago and it took seven years for all companies to accumulate as much Group Life Assurance as this one contract. **1964** G. L. COHEN *What's Wrong with Hospitals?* v. 103 Nor do large firms relish the idea of highly paid men frittering hours away: hence the marked spread of professional and industrial group insurance. **1880** Group marriage [see MARRIAGE 1 d]. **1899** A. H. KEANE *Man Past & Present* v. 153 Here it is necessary to distinguish carefully between *class*-marriages and the so-called 'communal' or 'group' marriages. **1906** *Westm. Gaz.* 13 Aug. 10/1 Dr. J. W. Gregory, in his 'Dead Heart of Australia'..includes them [*sc.* the Aborigines] in the same race-group as ourselves..and reminds us that their system of 'group marriage' was prevalent in Britain at the time of Julius Cæsar's invasion. **1921** E. WESTERMARCK *Hist. Human Marriage* (ed. 5) I. vii. 268 Even if it were worth while inventing such a society in the interest of the cherished institution of group marriage, it would be impossible to find mothers who were equally ignorant of their children as the children were of their mothers. *Ibid.* III. xxxi. 260 Nor can the hypothesis that it [*sc.* the classificatory system] is an indication of group-marriage or sexual communism be accepted as even probably correct. **1937** R. H. LOWIE *Hist. Ethnol. Theory* viii. 88 Morgan had suggested a stage of group-marriage. **1957** AUDEN & KALLMAN *Magic Flute* 60 Whether they live in air-borne nylon cubes, Practise group-marriage or are fed through tubes. **1902** *Folk-Lore* XIII. 386 It appears to me that group-names may, originally, have been imposed *from without.* **1935** *Discovery* Sept. 253/1 *Arya* has been used..as a religious group-name, to distinguish the worshippers of the gods of the Brahmans from the worshippers of certain other Indian deities. **1968** *Times* 13 Mar. 11/5 Surnames are group names which subdivide the Johns and Jameses into..Smiths,..&c. **1918** W. OWEN *Let.* c 9 June (1967) 558 Preserve me from..plush chairs, group-photographs, flowers under glass-shades. **1935** J. JOYCE *Let.* 18 Jan. (1966) III. 342 Send me a group photograph like the last lovely one we had taken. **1937** H. READ *Art & Society* iv. 171 There had grown up in Holland..a custom of commissioning group-portraits—just as we still commission group-photographs of..wedding parties—and Rembrandt was not above undertaking such a commission. **1942** *Lancet* 19 Sept. 345/1 Group practice was defined here as referring to general practitioner practice only. **1947** *Jrnl. Amer. Med. Assoc.* 6 Dec. 904/1 (*heading*) Medical group practice in the United States. **1954** S. TAYLOR *Good General Practice* iv. 90 It is useful to begin by testing the hypothesis that group practice is inherently better than individual practice. *Ibid.* 93 True group practice is comparatively rare. **1958** *New Statesman* 18 Oct. 530/2 The development of health centres and group practice has been much too slow, and the education of general practitioners barely touches on..social medicine and thus offers no training for over two-thirds of their future work. **1961** *Brit. Med. Dict.* 1152/1 *Group practice,* the co-operation of several medical practitioners, usually in partnership, for the diagnosis of and treatment of patients. Frequently, one partner specializes in medicine,

another in surgery, etc. **1968** *Guardian* 6 May 7/2 There are strong incentives to form group practices. **1887** *Proc. Inst. Mech. Engin.* Aug. 426 Calculate out the result from the law that the group-velocity is half the wave-velocity —the velocity of a group of waves at sea is half the velocity of the individual waves. **1948** MOTT & SNEDDON *Wave Mech. & its Applications* i. 24 The group and wave velocities are equal only if the wave velocity is independent of frequency (as for light in a vacuum). **1892** H. SWEET *New Eng. Gram.* I. 138 We may regard *pass-by* and *run-across* in such constructions as group-verbs, logically equivalent to such simple transitive verbs as *pass* and *cross* in *he passed the house, he crossed the road.* **1924** H. E. PALMER *Gram. Spoken Eng.* II. 169 Some adverbs are so intimately associated with verbs that such combinations may be considered as group-verbs. **1963** F. T. VISSER *Hist. Syntax* I. iv. 407 Prepositional object after such group-verbs as *break in upon, look out for.* **1901** E. A. ROSS *Social Control* 29 The fittest to survive when the competition is man-wise, may be eliminated when the competition is group-wise. **1953** C. E. BAZELL *Linguistic Form* vi. 73 (Type *queen of France's son*). This abnormality led to the whole group (e.g. *queen of France*) being taken as a word, albeit of a special kind ('group-word'). **1941** J. S. HUXLEY *Uniqueness of Man* xi. 233 We may distinguish such [joint] work from true group work, using the term group in the sense of a body of people pooling their different knowledges and skills to cope with quantitatively differentiated problems. **1958** *Times* 6 Sept. 10/2 Another attractive group-work is the huge 'Wise Man' (Hornsey High School). **1959** *Gloss. Terms Work Study* (*B.S.I.*) 24 *Group work,* work done by a number of workers in close association, each worker contributing towards a completed unit of production. **1970** *New Society* 5 Mar. 401/1 Social group work is a tool to solve external problems and is based on activity, discussion and again action.

b. *Social Sciences.* In many *Combs.* in which *group* means 'a group of people, esp. a social group or community', as *group-action, -analysis, -attitude, -behaviour, -centred* adj., *-characteristic, -conflict, -consciousness, -counselling, -counsellor, -discussion, dynamics, -experience, -feeling, -interest, -life, -living, -loyalty, -mate, -membership, -memory, -mentality, -migration, -mind, -norm, -organization, -person, -personality, psychiatry, -psychologist, psychology, psychotherapy, -sentiment, -solidarity, -spirit, status, survival, theory, therapy, -think, -thinking, -type, -will.*

1931 *Times Lit. Suppl.* 20 Aug. 635/1 The work of Mr. Edmund Selous on group-action in bird flocks recently reviewed by us. **1943** *Mind* LII. 228 The term 'I' could be said to be a sociological term, because it designates a class of group-actions. **1927** D. H. LAWRENCE *Let.* 3 Aug. (1932) 687, I must come and be present at your group-analysis work one day. **1943** H. READ *Educ. through Art* vi. 198 Therefore, for the Freudian method of individual analysis, Dr. Burrow substitutes a method of group analysis, by which he does not mean an analysis of the group, but a 'phyletic principle of observation', that is to say, group activities which involve the group's analysis of any one of its component individuals. **1937** C. M. ARENSBERG *Irish Countryman* 116 For we are dealing.. with group-attitudes reciprocally held. **1920** E. D. MARTIN *Behavior of Crowds* iv. 79 We would sit in chapel and hear a wrathful president denounce our group behavior as 'boorishness and hoodlumism'. **1927** G. A. DE LAGUNA *Speech* xiv. 261 The adaptation of group-behaviour to the situation. **1959** *Times Lit. Suppl.* 27 Mar. 181/2 It is to cover..biological reproduction, dynamic psychology, group-behaviour patterns, and so on. **1951** *Jrnl. Abnormal & Soc. Psychol.* XLVI. 521/1 This study is concerned with the effects of two contrasting group-leadership techniques—group-centered..and leader-centered. **1958** W. J. H. SPROTT *Human Groups* ix. 144 'Group-centred' ones [*sc.* groups], in which interaction was encouraged. **1944** KOESTLER in *Horizon* Mar. 162 Historically, it is..the 'aspiration to independent thinking' which provides the only valid group-characteristic of the intelligentsia. **1954** H. GIBBS *Background to Bitterness* 8 Nevertheless, the country contains many who fear that future group-conflicts or racial conflict may occur within the next decade. **1915** A. HUXLEY *Let.* Dec. (1969) 87 All the men who are running the *Palatine,* who are infused with the Palatinate group-consciousness, which is a good group-consciousness, are good men. **1951** R. FIRTH *Elem. Social Organiz.* iii. 115 In their own social and ceremonial life they display a strong group consciousness. **1957** H. READ *Tenth Muse* xxxi. 277 This quality may sometimes be due to some kind of collective intuition —the working of several minds to a common conception; the spontaneous overflow of a group consciousness. **1950** HAHN & MACLEAN *Gen. Clin. Counseling* i. 11 To talk of 'group counseling' and to imply that it is similar in structure and outcomes to one-to-one clinical counseling is as silly as talking about 'group courtship'. **1959** *News Chron.* 4 Dec. 7/5 Describing the experiment, known as 'group counselling', as 'the most important thing to happen in penology for a very long time', the report stresses that the discussions were spontaneous and wide. **1961** 'C. H. ROLPH' *Common Sense about Crime* ix. 144 In prisons and Borstals what is now called Group Counselling is now being developed... The technique in its modern form..began at San Quentin prison in California. *Ibid.* 145 Selected members of the..staff at San Quentin.. were doing the work of group counsellors. **1912** C. H. COOLEY *Social Organiz.* (ed. 2) iii. 24 As regards play, I might..multiply illustrations of the universality and spontaneity of the group discussion and coöperation to which it gives rise. **1963** A. HERON *Towards Quaker View of Sex* v. 44 They may..feel it necessary to initiate group discussion on sexual matters. **1953** A. K. C. OTTAWAY *Educ. & Society* 13 Group dynamics is an excellent example of a subject which overlaps with sociology and psychology. **1959** G. D. MITCHELL *Sociol.* i. 23 This modern micro-sociological development, often called group dynamics. **1941** *Mind* L. 395 The members participating in a family meal round the same table share, each in his own way, in the group-experience of eating together. **1945** A. L. ROWSE *Eng. Spirit* 46 Patriotism is a form of group-feeling. **1904** Group-interest [see *COMMUNITARY *a.*]. **1931** A. L. ROWSE *Politics & Younger Generation* 232 The basic factors in society are the impersonal *blocs* of group-interest. **1902** *Amer. Jrnl. Sociol.* VIII. II. 189 In the case of the dyad and triad configurations, we had to do with that inner group-life, with all its differences, syntheses, and antitheses. **1913** J. N. FIGGIS *Churches in Mod. State* 226 Facts so tremendous as the complex group-life which is to most of us more than the State. **1941** *Mind* L. 396 The group-life of the rural parish. **1951** *Essays in Criticism* I. 1 The literary problem must not be divorced from the problems of group-living, in the widest sense, that lie behind it. **1958** HAYWARD & HARARI tr. *Pasternak's Dr. Zhivago* i. 18 It is always a sign of mediocrity in people when they herd together, whether their group loyalty is to Solovyev or to Kant or Marx. **1927** *Mod. Philol.* XXV. 213 Successive actions of his group-mates (parents, etc.).. 'condition' him to the social habits. **1949** M. MEAD *Male & Female* x. 205 He will very likely live longer than his more active group-mates. **1950** T. H. MARSHALL *Citizenship & Social Class* 105 My consciousness of age..does not take the form of a feeling of group-membership. **1923** J. S. HUXLEY *Ess. Biologist* i. 52 What may be called the 'group-memory'—the power of storing and rendering knowledge available. **1965** *Listener* 30 Sept. 491/1 Group memory..is no more than the transmittal to many people of the memory of one man or a few men, repeated many times over. **1920** B. RUSSELL *Pract. & Theory Bolshevism* I. i. 19 The group-mentality that Communism requires. **1960** P. H. REANEY *Orig. Eng. Place-Names* 103 This evidence of somewhat extensive group migration is a matter of some importance to the historian and student of dialect. *a* **1899** D. G. BRINTON *Basis of Social Relations* (1902) I. ii. 28 The actual existence of the group-mind can no more be denied than the constant inter-relation between it and the individual mind. **1960** *Times Lit. Suppl.* 20 May 323/4 We know that men may lose their individual identities..by music and rhetoric, and that (as the careers of Dr. Goebbels and Senator McCarthy have shown) this state of group-mind can be stretched out over long periods. **1936** M. SHERIF *Psychol. Soc. Norms* vi. 105 If the leader changes his norm after the group norm is settled he may cease thereupon to be followed. **1958** R. K. MERTON *Soc. Theory* 317 Departures from the strict definitions of group-norms. **1913** L. T. HOBHOUSE *Development & Purpose* x. 186 Group-organisation becomes a system of peace and, on the whole, co-operation. **1940** H. READ *Annals of Innocence & Experience* III. 210 Such a society..would give the individual the greatest degree of liberty consistent with a group organization. **1915** E. BARKER *Polit. Thought in England* 175 Permanent groups are themselves persons, group-persons, with a group-will of their own. **1934** —— tr. *Gierke's Natural Law & Theory of Soc.* I. III. i. 81 In his [*sc.* Hobbes's] theory of corporations, as in his theory of the State, the central conception is that of the unity of group-personality. **1942** L. B. NAMIER *Conflicts* 91 As members of a group-personality most people enjoy greater freedom from moral scruples and inhibitions. **1944** *Horizon* Jan. 79 Dr. Maxwell Jones, who is experimenting in group-psychiatry. **1969** E. MCGIRR *Entry of Death* vi. 114 An interest in group psychiatry. **1933** H. G. WELLS *Shape of Things to Come* III. § 5. 291 The new economists on the one hand and the group psychologists on the other. **1920** W. MCDOUGALL *Group Mind* 8 Group Psychology has, first, to establish the general principles of group life. **1949** KOESTLER *Insight & Outlook* xiii. 194 Its demonstrable connections with the phenomena of projection,..group psychology, and so on. **1938** P. SCHILDER *Psychotherapy* x. 197, I shall try here to give a detailed description of a method of group psychotherapy which I have employed..in the out-patient department of Bellevue Hospital. **1948** *Sci. News* VIII. 106 Group psychotherapy..implies that a group of patients is treated together rather than individually. **1970** *New Society* 5 Mar. 401/1 Group psychotherapy is based on '"free floating discussion"', the equivalent in the group of "free association" in the one-to-one psycho-analytic process'. **1935** HUXLEY & HADDON *We Europeans* i. 12 For the moment we will refer to the sentiment which animates tribal and national units alike, by the non-committal phrase 'group sentiment'. **1927** G. A. DE LAGUNA *Speech* xviii. 318 The importance of the celebration of tribal deeds in fostering group-solidarity. **1951** M. MCLUHAN *Mech. Bride* 10/1 A very able person may often choose to freeze or anesthetize large areas of his mind and experience for the sake of..the pleasures of group solidarity. **1920** W. MCDOUGALL *Group Mind* 63 The group spirit, the idea of the group with the sentiment of devotion to the group developed in the minds of all its members. **1934** R. BENEDICT *Patterns of Culture* (1935) iv. 103 It [*sc.* initiation] makes the children valuable by giving them group status. **1946** *Mind* LV. 45 It is quite possible that the community will develop a philosophical view that serves the end of group survival. For example, this is what appears to be happening very rapidly in South Africa among the Afrikaans-speaking community ('the Dutch'). **1916** C. C. J. WEBB (*title*) Group theories of religion and the individual. **1943** S. R. SLAVSON (*title*) An introduction to group therapy. **1948** *Sci. News* VIII. 107 The first recorded use of group therapy was the experiment of Dr. J. H. Pratt in Boston [in 1905]. **1957** A. HUXLEY *Let.* 18 Nov. (1969) 830 Group therapy.. I personally wd try, if I got into a state of psychological distress. **1970** *Daily Tel.* 18 May 4/8 Group therapy is to be tried as a means of curbing bad driving in New York State under a three-year experimental plan to be started on June 1. **1959** *Sunday Times* 22 Nov. 9/6 The group-think that is one of Chelsea's strongest assets, is also a source of weakness. **1969** D. E. WESTLAKE *Up your Banners* (1970) xxxvii. 266 If ever I'd seen a document that was the result of group-think, that was it. **1923** J. MACCURDY *Probl. Dynamic Psychol.* xxi. 324 This rationalization tendency points to the fact that we dislike the thought of not forming our own opinions and that there is some antagonism between individual, intellectual activity and acceptance of 'group thinking'. **1933** 'R. WEST' *St. Augustine* i. 18 This led inevitably to comic fatuities of the sort that Gibbon loved to mock, and to the depreciation of thought by the hasty and facile processes inevitable in group-thinking. **1903** E. A. ROSS in *Amer. Jrnl. Sociol.* VIII. 762 The hundred interlacing groups into which men combine, are the proper subject of study. This..conception..excuses us from showing.. how a group-type or a group-will arises. **1939** V. A. DEMANT *Religious Prospect* iii. 66 Politics rapidly loses its character as the corporate attempt to embody a Natural Law and becomes increasingly a conflict of bare group-wills.

group, *v.* Add: **3. b.** *Med.* To assign to a particular blood group; to determine the blood group of.

1936 *Brit. Med. Jrnl.* 28 Mar. 651/2 The patient's blood should in any case be grouped, so that an immediate transfusion may be given later if necessary. **1968** PASSMORE & ROBSON *Compan. Med. Stud.* I. xxvi. 19/2 With improved laboratory techniques and increased care in grouping and cross matching the blood of both patient and donor, reactions are now uncommon.

4. *intr.* Of shots from a fire-arm: to cluster about a point on the target; to form a compact shot group. Also, of a fire-arm or the firer: to fire shots which do this (*trans.* and *intr.*).

1900 W. W. GREENER *Sharpshooting* v. 84 Having learned so to shoot that all shots group closely, proceed by altering the sights to get the groups centrally placed upon the bull's-eye. **1911** *Encycl. Brit.* XXVI. 419/2 The use of the bull's-eye to-day is to teach the soldier to shoot uniformly, that is, to 'group' his shots closely. **1913** A. G. FULTON *Notes on Rifle Shooting* 28 It is often difficult to account for some beginners grouping right away and others proving almost hopeless. **1958** J. A. BARLOW *Elem. Rifle Shooting* (ed. 5) i. 2 The one and only essential in any rifle for competition purposes is that it should group within a certain predetermined maximum.

grouper: see *GROPER[2], *GROUP *sb.* 3 d.

groupie (grū·pi). **1.** *R.A.F. slang.* Short for *group captain* (*GROUP *sb.* 6).

1943 in C. H. WARD-JACKSON *Piece of Cake* 35. **1952** M. TRIPP *Faith is Windsock* iv. 74 'Oh sergeant, don't leave us for a moment,' said Groupie acidly. **1966** *Listener* 21 Apr. 589/3 There is a horrible Cyclops of a groupie. **1968** I. LAMBOT *Queen dies First* xii. 84 Groupie's a devil for the girls.

2. Also **groupy.** [*GROUP *sb.* 3 f.] (See quots. 1967 and 1970.)

1967 *Sun* 22 Feb. 6/6 (Teenage terms) *Groupy,* a girl who follows the pop groups. **1969** *Examiner* (San Francisco) 29 Jan. 39 'The Groupies'—girls who chase boys in the rock groups—are now getting so way out in their adulation that the whole mess warrants a federal investigation. **1969** *Private Eye* 10 Oct. 2/1 (Advt.), Read the sensational story, in her own words, of Katie a nineteen-year-old Groupie, as she 'pulls' from pop group to pop group. **1970** *Times* 15 Sept. 5/3 His defence described the sisters as 'groupies', girls who deliberately provoke sexual relations with pop stars.

grouping, *vbl. sb.* Add: (Further examples.)

1902 *Encycl. Brit.* XXX. 256/1 One of the most useful distinctions for lighthouses consists in the grouping of two or more flashes separated by short intervals of darkness, the group being succeeded by a longer eclipse. **1931** *Economist* 15 Aug. 315/2 This shows that the position of the railways, when 'grouping' was achieved in 1923, was particularly difficult. **1958** J. A. BARLOW *Elem. Rifle Shooting* (ed. 5) i. 2 Provided it [*sc.* a rifle] satisfies grouping requirements, anything else about it that may not be quite up to standard can be altered or rectified quite easily. *Ibid.* 5, I..tested it [*sc.* a rifle]..and found that its grouping powers had more or less vanished. **1959** *Times Lit. Suppl.* 17 July 423/1 Our new international, continental groupings clash. Inside the same religion, political philosophy, or system of thought there are bitter clashes.

b. = *blood grouping* (*BLOOD *sb.* 19).

1907 *Jrnl. Infectious Dis.* IV. 298 The exceptions to Landsteiner's grouping are given in heavy type. *Ibid.* 299 There is no striking alteration in the agglutinative grouping of the blood of persons suffering with various diseases. **1921** *Jrnl. Immunol.* VI. 363 In 1908 while working on the relation of the isoagglutination to blood transfusion, I noticed that the groupings were hereditary. **1957** *Encycl. Brit.* XI. 495/2 In the ABO system, which is very important in blood transfusion, there are four classes of individuals, O, A, B and AB, the grouping being based on certain agglutination reactions. **1968** 'C. AIRD' *Henrietta Who?* xvi. 152 'I've done a grouping.' 'Yes, doctor?' 'The girl's group O.'

Groupism, Groupist: see *GROUP *sb.* 3 d.

groupment (grū·pmĕnt). [f. GROUP *v.* + -MENT.] **a.** A group. **b.** The action of placing in groups.

1887 A. M. BROWN *Contrib. Animal Alkaloids* 87 It corresponds to two molecules of kreatinine, plus the CNH groupment. **1933** *Burlington Mag.* Mar. 144/1 A student needs these for purposes of groupment and comparison. **1953** *Britannica Bk. of Year* 639/1 *Groupment,* a national [U.S.] military unit of 1,300 men.

groupy: see *GROUP *sb.* 3 d; *GROUPIE 2.

grouse, *sb.*[1] Add: **3.** *grouse-butt, -pie, -shooter* (earlier example); *grouse-driving, -shooting* (earlier example) vbl. sbs.; **grouse-berry** *U.S.*, the blueberry, *Vaccinium scoparium*; cf. DEER-BERRY.

1804 A. F. M. WILLICH *Domestic Encycl.* III. 150/2 It is called Canadian Gaultheria, or Mountain Tea, Grouse-berry, Deer-berry, Ground-ivy. 1932 P. A. RYDBERG *Flora Prairies & Plains Cent. N. Amer.* 618 *V[accinium] scoparium*... Grouse-berry. 1904 *Westm. Gaz.* 26 Aug. 3/1 One of the unfailing topics that come up..for debate..is the mental attitude of the driven grouse towards the grouse-butt. The opinion of the good men of old, who were in their prime in the beginning of the grouse-driving, [etc.]. 1933 *Times Lit. Suppl.* 16 Mar. 175/4 The no less difficult art of grouse-driving on a Yorkshire moor. 1770 E. MONTAGU *Let.* 31 Aug. in *Private Corr. David Garrick* (1831) I. 395, I shall endeavour to send you a grouse pie. 1861 MRS. BEETON *Bk. Househ. Managem.* 516 (*heading*) Grouse pie. 1931 H. NICOLSON *Diary* 26 Sept. (1966) 92 We have..grouse-pie and mushrooms. 1832 *Chambers's Edinb. Jrnl.* I. 186/3 A neat little mountain inn, much frequented..by fishers and grouse-shooters. 1785 G. A. BELLAMY *Apology for Life* (ed. 2) I. 163 Grouse shooting claimed his attention in Autumn, and hunting in winter.

grouse (grɑus), *sb.*^{2} *slang.* [f. GROUSE *v.*^{2}] A grumble or complaint: a reason for grumbling.

1918 W. OWEN *Let.* 17 July (1967) 564 I'll make my grouse. Why didn't you come to Scarborough for a few days? *Ibid.* 15 Oct. 585 The Parcel came all right; my only grouse is that it was too big. 1923 *Pictorial Weekly* CXIX. 127/1 A Transfer 'Grouse'. 1927 *Daily Express* 27 Apr. 3/1, I cannot understand the point of view of the hanging committee...I have no grouse against them. I am not an Academician, but I do not agree with their choice. *Ibid.* 5 Oct. 3/4 The 'moan' of a Gun Room is the naval version of an Army 'grouse'.

grouse, *v.*^{2} Add: (Earlier example.) The pronunciation is now (grɑus).

1887 KIPLING *From Sea to Sea* (1899) II. 396 That's the only thing as 'ill make the Blue Lights stop grousin' and stiffin'...'Grousing' is sulking, and 'stiffin' is using unparliamentary language.

grouse (grɑus), *a. Austral.* and *N.Z. slang.* Formerly also **grouce.** [Origin unknown.] Excellent, very good.

[1938 PARTRIDGE *Dict. Slang* (ed. 2) Add. 1000/2 *The grouse.* (Only predicatively.) Very good: Australian cant: C.20. Origin?] 1941 BAKER *Dict. Austral. Slang* 33 *Grouce, grouse,* good, excellent. 1944 L. GLASSOP *We were Rats* I. i. 5 You know them two grouse sheilas we've got the meet on with tomorrer night? 1947 D. M. DAVIN *Gorse blooms Pale* 200 An Iti bint, a real grouse brush she was, with bonzer black eyes. 1963 *Australasian Post* 14 Mar. 51/2 'Real good, Knobs... How about you?' 'Extra grouse.' 1968 *Telegraph* (Brisbane) 1 May 24/3 (*heading*) Grouse slang guide.

grouser^{3} (grɑu·səɹ). [f. GROUSE *v.*^{2} + -ER^{1}.] One who grumbles or complains.

1885 J. B. PATTERSON *Life in Ranks* 120 Impossible to do anything at all entirely to the satisfaction of a certain class of individuals... This..body of men is commonly designated by their..comrades as the 'grousers'.

grout, *v.*^{2} Add: **1.** Also *transf.* and *fig.*

1907 C. E. RYDER *Bede Papers* xi. 170 Modern science ..with pickaxe and spade grouting at the immemorial roots of her tree of life. *a* 1941 V. WOOLF *Captain's Death Bed* (1950) 31 He went off..to grout for fossils. 1951 N. ANNAN *L. Stephen* iv. 136 Jowett..encouraged his undergraduates to grout among the pearls that he cast before them and select whichever they regarded valuable. 1956 J. MASTERS *Bugles & Tiger* 240 The tribesmen could have walked in anywhere while we grouted around in the soup [i.e. liquid mud] for our rifles and machine-guns.

grouting (grɑu·tiŋ), *vbl. sb.*^{2} [f. GROUT *v.*^{2} + -ING^{1}.] The action of GROUT *v.*^{2} (in quot. *fig.*).

1961 M. KELLY *Spoilt Kill* ii. 36 By a bit of discreet grouting in the cupboards..I've found it.

grow, *v.* Add: **6. a.** *to grow away,* to develop (well).

1933 *Jrnl. R. Hort. Soc.* LVIII. 99 When..the requisite number of shoots are growing away well. 1961 *Listener* 10 Aug. 222/2 All Talisman plants will be small and behind all other varieties, but they grow away to produce good crops without any difficulty.

13. e. To be sensible, mature; freq. *imp.*

1951 [see *CHRIS(S)AKE]. 1959 A. WESKER *Chicken Soup with Barley* in E. M. Browne *New Eng. Dramatists* II. ii. 212 Oh, grow up, Ronnie. You should know that by now. 1967 G. NORTH *Sgt. Cluff & Day of Reckoning* xx. 181 'The Abbot who shirked his obligations hasn't lived!' 'Grow up!' 1969 'A. GILBERT' *Missing from Home* v. 55 You're surely never on that old game. You want to grow up, Dad. 1971 D. DEVINE *Dead Trouble* ii. 17 That was Dorothy's constant refrain: 'Grow up, Nev!'..He'd show her who was immature.

14. f. *Cryst.* To bring about the formation of (a crystal); to cause (a crystal) to increase in size.

1911 [implied in *GROWN *ppl. a.* 1 b]. 1915 *Amer. Jrnl. Sci.* CLXXXIX. 567 We can, by proper treatment, grow crystals that are nearly symmetrical and complete in their parts. 1950 *Sci. News* XV. 55 We start with our spherical crystal and grow it into a larger crystal by depositing more atoms on it. 1971 *Sci. Amer.* May 130/2 Diamonds are best grown from a solution of carbon in a molten metal such as nickel.

g. *to grow on,* to keep (seedling plants) in suitable situations or conditions as they develop to maturity.

1947 R. P. FAULKNER *Commercial Hort.* xix. 134 They [sc. cinerarias] should then be transferred to cooler conditions, and may eventually be grown on in frames. 1971 L. N. & V. L. FLAWN *Gardening under Glass* xvii. 208 In March move [the Cape primroses] into 5 or 6 in. pots and grow on in a cool temperature.

growing, *vbl. sb.* Add: **5.** *growing-region*; **growing-on,** the cultivating of seedlings, the breeding of young chicks, etc., to maturity or full size; **growing pains,** also *fig.*; **growing point** (earlier examples); also *fig.*; **growing season,** the season when rainfall and temperature permit plants to grow; **growing stock** *Forestry,* the total quantity of trees in an area; **growing zone,** the region of an annelid worm in which growth or regeneration is initiated.

1960 *Farmer & Stockbreeder* 15 Mar. 150/2 A new popular system for growing-on from 8 to 10 weeks to laying stage. 1960 *Jrnl. R. Hort. Soc.* LXXXV. 89 Cost can also be reduced by purchasing small specimens and growing on. 1962 J. N. WINBURNE *Dict. Agric.* 354/2 *Growing-on-house,* a greenhouse used for growing potted plants from the small plant stage to maturity. 1810 Growing pains [see Dict.]. 1915 A. HUXLEY *Let.* Oct. (1969) 80 Germany seems to be like..a new growing country, swelled with its own pride, filled by its growing pains with an immense folie de grandeur. 1923 J. M. MURRY *Pencillings* 70 The struggles of a generation towards complete rationality..are growing pains. 1835 J. LINDLEY *Introd. Bot.* (ed. 2) I. ii. 56 They [sc. the leaf-buds] consist of scales imbricated over each other..and surrounding a minute cellular axis, or growing point. 1880 S. H. VINES tr. *Prantl's Elem. Textbk. Bot.* ii. 64 The growing end or apex of an organ, such as a root or a stem, is called the growing-point (*punctum vegetationis*). 1948 *Mind* LVII. 103, I shall..indicate what seem to me the growing points in his theory. 1959 *Listener* 15 Jan. 108/2 Muir..had truly perceived where lay the growing-point of poetry in our time. 1962 *Punch* 28 Feb. 343/2 Secondary Modern..is undoubtedly one of the 'growing points' in English education. 1927 HALDANE & HUXLEY *Animal Biol.* xii. 282 This growing-region often continues active throughout life. 1958 BROCKLEHURST & WARD *Gen. School Biol.* 146 (*caption*) Growing region of a root. 1924 W. S. JONES *Timbers* i. 5 The process [of tissue growth] continues throughout the growing season. 1957 G. E. HUTCHINSON *Treat. Limnol.* I. vii. 526 The incidence of small lakes may increase the growing season for crops. 1889 W. SCHLICH *Man. Forestry* I. I. i. 15 The capital employed in forestry consists principally of the soil and the growing stock of wood. 1967 D. R. JOHNSTON et al. *Forest Planning* xviii. 287 The level of the growing stock is clearly an important factor in thinning. 1921 *Phil. Trans. R. Soc.* B. CCXI. 145 There were behind this four segments with setæ, one without setæ and a growing zone. 1952 *Biol. Rev.* XXVII. 408 The posterior residual growing zone itself grows progressively more slowly.

growl, *sb.* Add: **3.** In Jazz, a deep rasping sound made on a wind instrument. Also *attrib.*

1935 *Hot News* Aug. 18/3 He used the non-pressure method common to most white players, producing a tone..with just a suggestion of a growl. 1946 R. BLESH *Shining Trumpets* (1949) II. xii. 280 Here, too, is the growl trumpet, an atmospheric part of the tissue paper jungle in which the band was then ensconced. 1959 'F. NEWTON' *Jazz Scene* vii. 122 Bubber Miley (1902–32) pioneered the systematic use of the mute and the 'growl'. 1961 *Times* 20 May 11/4 The art of 'growl' trumpet playing has declined so much over the years.

growl, *v.*^{2} Add: **1. c.** (Additional examples.)

1958 *Spectator* 15 Aug. 225/2 The bulldozers of the New Towns growl nearer. 1970 *Observer* (Colour Suppl.) 15 Feb. 24/1 The big jets of Qantas and BOAC growl in and out daily on their way round the world.

3. *intr.* Of a wind instrument: to make a low, rasping sound. Of a musician: to make such a sound on an instrument. Also *trans.*, to play (music) in a growling manner.

1935 N. E. WILLIAMS *His Hi de Highness of Ho de Ho* 35 Even white musicians will say 'growl it' to a trumpet player when they are asking him to play it 'lowdown' or 'dirty'. 1935 *Hot News* Apr. 19/1 He has the dirtiest tones imaginable, using his hand in front of the bell to produce a unique growling effect. *Ibid.* Aug. 6/2 Only a trumpeter who can growl well is necessary. 1935 *Metronome* Nov. 25/3 A trumpet..growls really effectively for a change. 1955 DUKE ELLINGTON in Shapiro & Hentoff *Hear Me Talkin' to Ya* xii. 195 Everybody told him he'd have to use a plunger and growl all night long.

growler. Add: **5.** A compact mass of floating ice, smaller than a 'bergy bit', but of such a size that waves do not quite break over its top; a small iceberg.

1912 *Standard* 4 May 7/7 He thought the distinction between icebergs and growlers was that the growler was an iceberg with very little protruding above the water. 1922 JOYCE *Ulysses* 609, I seen icebergs plenty, growlers. 1934 *Geogr. Jrnl.* LXXXIII. 393 The faces [of the glaciers] break away in small pieces making brash and growlers common around the islands. 1958 *New Scientist* 10 July 359/1 As it weathers away, a berg becomes known as a 'bergy bit'.., and finally a 'growler' (when almost awash).

6. *Electr.* An electromagnet with two poles designed to receive an armature, used for testing the windings for short circuits.

If a short circuit exists in the armature coils its presence is indicated by a 'growling' noise.

1922 J. C. WRIGHT *Automotive Repair* II. 73 For making tests showing the general conditions of an armature, an instrument commonly known as a 'growler' is desirable. 1951 H. COTTON in P. Kemp *Electr. Engin.* II. 30 The growler is applied to the armature as shown..and a thin steel rule held to the teeth at a distance equal to the coil pitch away from the centre line. 1961 K. WILKINSON *Armatures & Field Coils* v. 104 The growler is a very simple and extremely valuable piece of test equipment.

7. *dial.* and *slang.* A dog. (In quot. 1822 *fig.*)

1822 'REAL PADDY' *Real Life Ireland* xii. 264 All the blackguards backed the growler. 1959 I. & P. OPIE *Lore & Lang. Schoolch.* ix. 154 Barker or growler.

growly (grɑu·li), *a.* [f. GROWL *sb.* or *v.*^{2} + -Y^{1}.] Resembling a growl.

1920 GALSWORTHY *In Chancery* I. vii, Val, uttering a growly sound, followed her. 1954 *Time* 25 Oct. 74 Terry plays some pretty, growly trumpet.

grown, *ppl. a.* Add: **1. b.** Of a crystal: produced by growing (*GROW *v.* 14 f).

1911 *Encycl. Brit.* VII. 584/1 The pitted and cavernous faces of artificially grown crystals of sodium chloride and of bismuth are..a result of rapid growth. 1962 SIMPSON & RICHARDS *Junction Transistors* iii. 46 The manufacture of *n-p-n* grown-junction transistors. 1966 *McGraw-Hill Encycl. Sci. & Technol.* VII. 318/2 There are several variations of the grown-junction technique. The simplest consists of successively adding different types of impurities to the melt from which the semiconductor crystal is being grown.

grown-up, *ppl. a.* and *sb.* Add: **A. 2.** Befitting, suitable to, characteristic of, an adult; sensible, worthwhile.

1852 C. M. YONGE *Two Guardians* vi. 93 As to books, all the real good grown-up ones are down in Mr. Lyddell's library where no one can get at them. 1907 E. WHARTON *Fruit of Tree* xxv. 382 Gravely measuring Cicely's milk in to a 'grown-up' teacup. 1958 *Sunday Express* 22 June 15/4 This film somehow manages to be memorable, moving, and outstandingly grown-up. 1960 *Times* 17 May 3/7 It is also capable of undertaking long journeys at 'grown-up' average speeds. 1964 V. NABOKOV *Defence* x. 164 It was impossible to express his recollections in words—there simply were no grown-up words for his childish impressions.

Hence **grown-up-dom; grown-up-ness** (later examples).

1871 G. M. HOPKINS *Let.* 25 Apr. (1938) 41 There was such a youngladyship and grownupdom about the address. 1900 *Westm. Gaz.* 7 July 2/1 They will say a conventional 'Thank you'..; the manners of youth being..superior to those of grown-up-dom. 1925 'R. CROMPTON' *Still—William* v. 85 He seemed to them to be free of all the drawbacks that usually accompany the state of grown-upness. 1944 'G. ORWELL' *Crit. Ess.* (1951) 154 The grown-upness, the lack of surprise or denunciation, the pity and irony with which the story is told, show the advantage, when one is handling a theme of this kind, of being a European. 1969 *Daily Tel.* 2 May 17 Most parents really want their children to grow up and become successful adults. But many find it hard to accept the onset of grown-upness soon enough.

growth^{1}. Add: **1. a.** (Additional examples.)

1848 tr. *Regnault's Elem. Treat. Crystallogr.* 59 Sometimes the crystal assumes the form shewn... The growth of the crystal, perpendicularly to the horizontal faces, has been almost nothing. 1911 *Encycl. Brit.* VII. 583/2 When crystals are aggregated together, and so interfere with each other's growth, special structures..often result. 1962 F. C. PHILLIPS *Smith's Gemstones* (ed. 13) ii. 39 Natural crystals frequently depart from..ideally symmetrical growth and show instead a variable degree of distortion.

c. *spec.* in *Economics.* See also sense *5 (*growth area, industry,* etc.).

1952 W. W. ROSTOW *Process Econ. Growth* (1953) iv. 81 Growth is defined as a relation between the rates of increase in capital and the working force,..and in population,..such that *per capita* output (not necessarily consumption) is rising. 1965 H. WILSON in *Oxf. Times* 3 Dec. 16/2, I am now fighting a losing battle on another word I dislike—growth—which had a certain medical and agricultural connotation. 'Economic growth involves more purposeful work than leaving it to nature.'

d. A crop or yield as used in a classification of (esp. the best) vineyards to indicate the quality of the wine produced there. Cf. *CRU.

1707 [see CLARET *sb.*^{1} (*a.*) 1 b]. 1715 [in Dict., sense 4]. 1851 C. REDDING *Hist. Mod. Wines* (ed. 3) vi. 170 The first growths of Medoc are scarcely ever sent to England in a perfect state. 1862, etc. [see *CRU]. 1920 G. SAINTSBURY *Notes on Cellar-bk.* iv. 53 It was customary..to lay down..hogsheads of the best reputed first or second growths. 1964 *Harper's Bazaar* Dec. 60/2 A claret of a bourgeois growth and a moderate price. 1965 A. SICHEL *Penguin Bk. Wines* 135 The sixty-two best red wines were classified in 1855 by an official committee into five 'growths' or '*crus*', known collectively as the Classed Growths, or *Crus Classés*. 1970 *Times* 18 Apr. p. vii/2 First it showed how exaggerated is the importance which some attach to the words 'Great Classed Growth'.

3. (Later examples: cf. *GROW *v.* 14 f.)

1950 *Sci. News* XV. 56 A typical example of its use is..for the growth of metal crystals. 1962 SIMPSON & RICHARDS *Junction Transistors* iii. 40 Crystal growth by zone melting. *Ibid.*, Floating-zone crystal growth. The

floating-zone refining process may also be used for the production of single crystals.

5. *growth-control, -direction, -hormone, -measurer, -phase, -policy, -rate, -ratio, -regulation, -target; growth-controlling, -inducing, -influencing, -inhibiting, -making, -promoting, -regulating, -retarding, -seeking, -stimulating* ppl. adjs.; **growth area,** an area designated for economic growth; **growth company,** a company (sense 7) that has expanded, or is likely to expand, more than the average; **growth curve,** a line drawn so as to represent growth diagrammatically by showing how some quantity like size, weight, or numbers varies with time; **growth factor** *Biol.*, any substance required by an organism in minute amounts in order to maintain its growth; **growth-form** (see quot. 1960); **growth-gradient** *Biol.*, a continuous variation in the rate of growth along an axis of an organism, limb, etc.; **growth industry,** an industry which has been, or is in process of, developing at a faster rate than other industries; **growth leader,** an investment stock with much past and potential growth; **growth-man,** a person who advocates a policy of economic growth; hence **growthmanship,** such a policy; **growth point,** (*a*) = *growth area*; (*b*) = *growing point*; **growth potential,** (*a*) potentiality of growth in living matter; (*b*) potentiality of economic growth; **growth regulator,** an organic substance, such as a hormone, vitamin, or auxin, which regulates growth; **growth ring,** the layer of wood added to a tree during one growing season; cf. *annual ring* (s.v. *ANNUAL *a.* and *sb.* A. 1 c; RING *sb.*^{1} 7 c); **growth stock** (see quot. 1965); **growth zone,** (*a*) = *growing-region*; (*b*) = *growth area*; (*c*) = *growing zone.*

1963 *Daily Tel.* 5 Dec. 14/2 The conception of 'growth areas' is new. 1966 *Listener* 2 June 788/2 The Humber estuary is already one of the most significant economic growth areas in the country. 1959 T. E. & D. L. BABSON *Investing for Successful Future* xii. 157 Listed..are a few outstanding growth companies, with the length of their business lives. 1969 *Times* 8 Jan. 23/2 It has been the growth company par excellence. 1964 *New Statesman* 20 Mar. 468/3 Calves..injected with growth-control substances. 1922 *Times Lit. Suppl.* 27 Apr. 279/1 A growth-controlling principle which he named tethelin. 1916 *Jrnl. Biol. Chem.* XXIV. 368 In the growth curves the dots represent the actually determined weights. 1946 *Nature* 5 Oct. 462/1 Fischer describes the sigmoid growth-curve of the culture, its limiting size, shape-regulation, and power of true reconstitutive regeneration. 1916 *Ibid.* 1 June 290/2 The word 'tropism', first used to indicate the growth-direction of plant-members under the influence of some stimulus. 1928 *Biochem. Jrnl.* XXII. 790 There is, therefore, no justification for calling vitamin B_2 a 'growth factor' in contradistinction to vitamin B_1. 1937 Growth factor [see *BIOTIN]. 1964 WAGNER & FOLKERS *Vitamins & Coenzymes* i. 5 The antiberiberi factor, designated vitamin B, was later found to contain both a heat labile antiberiberi factor and a thermostable growth factor. 1887 H. E. F. GARNSEY tr. *De Bary's Fungi* i. 2 Both [forms] are growth-forms (Wuchsformen) comparable with those growth-forms in the higher plants which are known as the tree, shrub and herb. 1909 GROOM & BALFOUR tr. *Warming's Oecol. Plants* ii. 4 In 1884, Warming, having in view the North-European Spermophyta, gave a general survey of growth-forms which he arranged in fourteen chief groups with many sub-groups, based upon morphological and biological characters. 1960 N. POLUNIN *Introd. Plant Geogr.* iii. 92 The 'life-form' or 'growth-form' of a plant is the form which its vegetative body produces as a result of all the life-processes, including those that are affected by the environment within the plant's life-time and are not heritable. 1929 *Nature* 15 June 910/1 (*heading*) Growth-gradients and the axial relations of the body. 1932 J. S. HUXLEY *Probl. Rel. Growth* iii. 80 It was found that there existed within the limb what we may call a growth-gradient. 1956 C. H. WADDINGTON *Princ. Embryol.* xiii. 292 It appears probable that there is a single continuous growth gradient with its high point towards the feet, falling off as one goes higher up the body. 1924 T. B. ROBERTSON *Princ. Biochem.* (ed. 2) xx. 579 Such catalyzers being distributed by the circulation and operating as growth-hormones. 1953 Growth hormone [see *ACROMEGALY]. 1969 *Times* 25 Mar. 12/7 When growth hormone is injected into young rabbits the fleas are induced to copulate. 1961 L. MARTIN *Clin. Endocrinol.* (ed. 3) i. 5 Insulin itself, when combined with a high carbohydrate diet, has also been shown to have some growth-inducing effect. 1957 *Economist* 7 Sept. 778/2 The pattern of growth in Britain's biggest growth industry..is not a balanced one. 1961 *Times* 30 May 13/6 Publishing is judged to be a 'growth' industry. 1966 *Listener* 15 Sept. 397/3 Only one laird preferred to keep his moor for private use...The grouse shooting agent's view was rather different—his is a growth industry. 1918 *Nature* 21 Feb. 484/2 The growth-influencing substance 'Tethelin', which Prof. Robertson had isolated from the anterior lobe of the pituitary body. 1922 *Jrnl. Exper. Med.* XXXV. 647 (*heading*) Heat and growth-inhibiting action of serum. 1946 *Nature* 10 Aug. 200/2 The antibiotic substances present in these extracts had a marked growth-inhibiting effect both on litter-decomposing and mycorrhiza-forming fungi. 1967 *N.Y. Times* (Internat. Ed.) 11–12 Feb. 9/6 The growth leaders with phenomenal rises are now well known stocks to all investors. 1922 *Chambers's Jrnl.* Dec. 875/1 The natives..chew it from boyhood, and attach great importance to it as a growth-making agent. 1961 *Britannica Bk. Year* 537/2 From commerce and organization generally came growthman, one favouring a policy of expansion, and growthmanship, the fixing of specific economic goals. 1964 *New Statesman* 3 Apr. 539/1 Political growthmen. 1966 *Ibid.* 18 Nov. 748/2 It is well-fed public opinion which presents the growthman with his chief problem. 1960 *Times* 18 Aug. 9/2 Both parties advocate a higher rate of economic growth (though not long ago Mr. Nixon was deriding 'growthmanship'). 1961 C. CLARK (*title*) Growthmanship. 1967 *Economist* 26 Aug. 733/1 'Growthmanship'..has dominated economic policy-making throughout Europe since the end of the second world war. 1924 J. A. THOMSON *Science Old & New* xxxvi. 206 Growth-measurers (auxanometers). 1915 P. GEDDES *Cities in Evolution* xvii. 361 Our synoptic vision of the city, for each and all of its growth-phases, thus ranges through region to homes, and back again. 1932 J. S. HUXLEY *Probl. Rel. Growth* i. 8 The weights of chela and rest-of-body..over the earlier and longer growth-phase. 1963 *Economist* 11 May 520/2 Turning many more people to new houses at new or refreshed growth points. 1968 *Listener* 4 Jan. 8/3 His energy and intellectual passion place him at the growth-point of English study to-day. 1960 *Encounter* Oct. 11/2 Growth-policies in relatively young economies like China. 1932 J. S. HUXLEY *Probl. Rel. Growth* i. 40 The distribution of growth-potential in different regions. 1958 *Listener* 13 Nov. 792/2 He regards any repression of demand below the growth potential as a development that would aggravate a rise in the cost of living [etc.]. 1968 *Globe & Mail* (Toronto) 17 Feb. B6 (Advt.), Growth potential is strong due to the enthusiastic management support this position enjoys. 1914 *Jrnl. Biol. Chem.* XIX. 248 The growth promoting substance was transferred from the butter fat to the olive oil by the procedure described. 1926 J. S. HUXLEY *Ess. Pop. Sci.* viii. 90 More growth-promoting proteins. 1966 *Economist* 11 June 1206/1 Strategy for growth—specifically the question of growth-promoting industries—is treated by Perroux (France). 1927 *Biol. Abstr.* I. 547/2 A quantitative analysis of the growth-rate of the chick embryo. 1930 R. A. FISHER *Genet. Theory Nat. Selection* 45 The vital statistics of an organism in relation to its environment provide a means of determining a measure of the relative growth-rate of the population. 1932 J. S. HUXLEY *Probl. Rel. Growth* i. 3 The relative growth-rates of..various parts [of the body]. 1970 P. R. & A. H. EHRLICH *Population, Resources, Environment* x. 243 The population of Europe is growing at considerably less than 1 percent per year...Only Albania, Rumania, and Iceland have growth rates that exceed 1·2 percent. 1924 J. S. HUXLEY in *Nature* 20 Dec. 895 Constant differential growth-ratios and their significance. 1929 *Biol. Bull.* LVII. 176 Growth regulating substances in echinoderm larvæ. 1948 *New Biol.* V. 47 'Growth regulating substances'..will inhibit, increase, or otherwise alter..the subsequent growth of the plants. 1956 *Nature* 4 Feb. 201/1 Pigments, weed-killers, insecticides and plant growth-regulating substances. 1927 HALDANE & HUXLEY *Animal Biol.* viii. 165 Sometimes growth-regulation breaks down, and the cells of some part of the body grow too quickly, causing a tumour. 1936 G. S. AVERY tr. *Jensen's Growth Hormones in Plants* i. 3 Plant-growth substances, i.e., *Wuchsstoffe*, have been referred to by various workers as growth hormones, growth regulators, growth enzymes, phytohormones, and auxins. 1954 *Plant Physiol.* XXIX. 307 The term *growth regulator*, as recommended here, includes the auxins, but it is broader, and it encompasses other regulators that may modify growth. 1969 SALISBURY & ROSS *Plant Physiol.* xx. 444/2 Synthetic growth regulators often mimic the effects of naturally occurring growth regulators (including hormones) or interfere with their actions. 1930 *Biol. Abstr.* IV. 351/1 It is concluded that a growth-retarding substance has not been found. 1932 J. S. HUXLEY *Probl. Rel. Growth* vi. 188 A much greater growth-retarding effect. 1907 D. P. PENHALLOW *Man. N. Amer. Gymnosperms* ii. 24 In proceeding to a study of the transverse section the first feature to which attention is naturally directed is the growth ring. 1937 *Discovery* Apr. 98/1 The largest of the Sequoias of California, estimated by growth-rings to exceed 3,500 years of life. 1960 N. POLUNIN *Introd. Plant Geogr.* i. 13 Where cells of different sizes are produced at different seasons, annual 'growth-rings' are formed which may easily be seen in most timbers. 1966 *Times* 28 Feb. (Canada Suppl.) p. iv/1 Growth-seeking industries. 1914 *Jrnl. Biol. Chem.* XIX. 246 (*heading*) Growth-stimulating substance in butter fat. 1957 *Economist* 9 Feb. 490/2 Undistributed income..has been transmuted into capital...Hence the investor's emphasis on 'growth' stocks. 1965 J. L. HANSON *Dict. Econ.* 209/1 '*Growth*' *stock*, a stock or share which can be expected to appreciate in value in the future, the policy of the directors of the company being to plough back each year a considerable portion of the profits for the purposes of expansion. 1969 *Times* 13 Jan. 11/3 This is a growth stock yielding only about 2 per cent, and no investment adviser would advocate putting that proportion of their total worth into that kind of investment. 1962 *Listener* 8 Mar. 400/2 The question of labour costs cannot be excluded from any realistic discussion of a specific industrial growth-target. 1927 HALDANE & HUXLEY *Animal Biol.* xii. 282 A growth-zone near the hind-end. 1963 *Daily Tel.* 15 Nov. 27/1 The three main centres of expansion in the growth zone. 1967 P. A. MEGLITSCH *Invert. Zool.* xiv. 636/1 During metamorphosis, this part of the [annelid] larva grows very rapidly; it may be termed the growth zone.

grub, *sb.* Add: **5.** (Earlier examples.)

1870 *Baily's Mag. Sports* Aug. 355 Nor did we ever before see a species of underhand bowling, known to schoolboys as 'grubs', made use of in the crack match of the year. 1882 *Australians in England* 92 The latter bowls left-handed 'grubs' of the most pronounced description.

6. a. Colloq. phrs.: *grub up!*, the food is ready; time to eat!; *lovely grub,* good food; also *transf.*

1953 'R. GORDON' *Doctor at Sea* xiv. 165 All the way down—*Miss, Miss, Miss.* Lovely grub! 1957 E. TAYLOR *Angel* v. i. 220 We're here, madam. Grub up! 1959 I. & P. OPIE *Lore & Lang. Schoolch.* ix. 161 Cries of jubilation include.. Lovely grub!

7. (Earlier examples.)

1788 WASHINGTON *Diary* 20 Feb. (1925) III. 306 At Muddy hole they were..taking up grubs in the Winter fallow of No. 4. 1825 J. LORAIN *Pract. Husb.* 334 The Yankee farmer first chops the fallen timber, then scalps off the grubs level with the ground. 1839 C. M. KIRKLAND *New Home* xx. 133 Grubs are, in western parlance, the gnarled roots of small trees and shrubs.

8. *grub wagon*; **grub-kick** *Rugby Football,* = *GRUBBER 6; hence as *v. intr.*; **grub-screw,** a screw without a head but with a recess at one end to receive a screwdriver or key; **grub-stake** (earlier and later examples; also *transf.*); **grub-stake** *v.* (earlier, later, and *transf.* examples); **grub-staker,** a prospector who is supplied with a grub-stake, or the supplier of a grub-stake; **grub-worm** (later examples).

1951 *Rugger* Dec. 5 He uses the grub kick so much favoured by Northern Union players. 1959 *Times* 6 Mar. 16/2 Elliot and Patterson engineered a try at the posts for the latter from a grub kick by Horrocks-Taylor. 1960 *Times* 4 Jan. 4/3 Tolson grub-kicked ahead. 1903 *Jrnl. Inst. Electr. Engin.* XXXII. 384 It is..becoming quite common..to fasten the armature connection to the commutator segment by means of steel grub screws. 1930 *Engineering* 26 Dec. 815/2 The terminals..are fixed by a grub screw. 1863 H. EDGAR *Jrnl.* 27 May in *Montana Hist. Soc. Contrib.* (1900) III. 138 'A grub stake is what we are after' was our watchword all day, and it is one hundred and fifty dollars in good dust. 1879 *Chicago Tribune* 15 May 9/6 Judge Pendery, a former Congressman, politician, and lawyer, has been grubstaking a party of miners who were digging a shaft down near the base of the hill. 1919 W. A. FRASER *Bulldog Carney* 38 I'm goin' to grubstake you,..leave you rations for three days. 1932 *Atlantic Monthly* Mar. 322/1 The farmer realizes the.. plight of the out-of-work who..is left without a grub-stake between himself and hunger. 1937 PARTRIDGE *Dict. Slang* 358/1 *Grub-stake,* to give (an author) money to keep him going while he writes a book. 1947 L. HASTINGS *Dragons are Extra* i. 30 Grub-staked by some optimistic colleague in town, they could be found at every wayside pub. 1947 V. PALMER *Golconda* ii. 10 We'd been out for six months on a grubstake from the store here, looking for mica. 1957 *Times* 12 Nov. (Canada Suppl.) p. xi/4 The prospector can no longer look forward to making his fortune with a 'grubstake' of a few hundred dollars. 1959 *Economist* 21 Mar. 1073/2 For over seven years, from 1951, soon after the Korean outbreak, until mid-1958, the federal government in effect 'grubstaked' owners of promising deposits of strategic minerals. 1965 G. MCINNES *Road to Gundagai* x. 176 A..dentist whose son Ian..had been grubstaked to Wembley *hors concours*. 1968 R. M. PATTERSON *Finlay's River* 96 On the Ingenika itself three prospectors were making at least a good grubstake: they finally took out seventy ounces of gold. 1880 A. A. HAYES *New Colorado* (1881) vii. 107 Here does the whilom grub-staker and present millionaire purchase his corner lot. 1897 A. C. HARRIS *Alaska & Klondike Gold Fields* 444 A grub-staker is a man who wants someone to stake him with grub, and 'grub' is Klondike for beans, bacon and tea. 1898 J. GRINNELL *Gold Hunting in Alaska* (1901) xii. 45 We may have to foot it home just like..prodigal sons who have wasted their substance and that of our grub-stakers in 'riotous living'. 1926 *Glasgow Herald* 8 Mar. 10 But the grub-staker who is financed by some wealthy man and will share his discoveries with his backer, is in the fortunate position of having enough money to avoid the dangerous trail. 1969 *Daily Colonist* (Victoria, B.C.) 27 Nov. 9/4 The telegram recalled the chamber had recommended..that the prospector and grubstaker tax exemption be maintained in its present form. 1891 *Fur, Fin & Feather* Mar. 150 They take with them a grub wagon and ten saddle horses, and expect to be absent four months. 1920 J. M. HUNTER *Trail Drivers of Texas* 69 We were two miles from the grub wagon. 1849 C. BRONTË *Shirley* v, My grub-worm is always a straitened, struggling, careworn tradesman. 1966 *Publ. Amer. Dial. Soc.* XLV. 14 *Grubworm,* the larva of the Green June Beetle.

grubber. 2. Delete *local* and add further examples.

1911 *Encycl. Brit.* VII. 618/2 *Cultivator,* also called scuffler, scarifier or grubber. 1940 *Chambers's Techn. Dict.* 392/2 *Grubber,* a heavy type of cultivator in which the teeth are set rigidly in a frame. 1950 *N.Z. Jrnl. Agric.* Feb. 271/2 The aim should be to secure a seed-bed with a rubbly surface and with the fine soil worked to the bottom through frequent use of grubber and harrows.

4. (Earlier example.)

1838 'P. PRY' *Oddities London Life* I. 235 She chucks ony von tater at me, and a bit of meat vot aint of no use to stitch a heavy grubber as I am.

b. *slang.* (See quot. 1940.) **c.** Food.

1940 M. MARPLES *Publ. School Slang* 91 *Grubber,* tuck-shop..tuck-box. 1959 I. & P. OPIE *Lore & Lang. Schoolch.* ix. 163 Food in general is referred to as 'bait'.., 'grub', or 'grubber'.

5. *Cricket.* = GRUB *sb.* 5.

1837 *Bell's Life* 15 Oct. 4/1 The Catapulta..was capable of..giving a home toss or a grubber. 1924 LAWRENCE & SKINNER *Boy in Bush* 116 Ross..sent down a sulky grubber. 1963 *Times* 23 May 4/1 Constable was undermined by a 'grubber'.

6. In full *grubber kick.* In *Rugby Football,* a forward kick of the ball along the ground. Hence **grubber-kick** *v. intr.*, to make a grubber kick.

1950 *Adam* (Sydney) Feb. 29 If it wasn't Sullivan's boot winning kicking duels, it was Parkin's grubber-kicks. 1956 V. JENKINS *Lions Rampant* xiii. 199 This time Ulyate put through a grubber kick from fly-half

which Cameron failed to gather. 1958 *N.Z. Listener* 18 July 6/4 He tried grubber kicking. All right, if they're going to grubber kick there's only one thing to do. 1960 *Sunday Times* 27 Nov. 20/1 From a set scrum, Lockyer fed Kirkpatrick, going right, and he put a grubber kick between Rogers and D. Bebb.

grubbery. 2. b. (Earlier example.)

1823 'J. BEE' *Slang* 91 *A grubbery*, a cook's-shop.

‖ **Grübelsucht** (grü·bəlzūχt). *Psychiatry.* [Ger., f. *grübeln* to brood + *sucht* mania.] A form of obsession in which even the simplest facts are compulsively queried.

1876 *Mind* I. 413 Under the name *Grübelsucht*, Griesinger..described three examples of a peculiar mental condition, characterised by continuous uncontrollable questionings as to the origin and causes of things great and small. 1885 W. JAMES in *Mind* X. 37 It would practically be a case of *Grübelsucht*, if a ruffian were assaulting and drubbing my body, to spend much time in subtle speculation. 1909 W. F. ROBERTSON tr. *Tanzi's Textbk. Mental Dis.* v. 150 Some cannot cease from putting to themselves metaphysical and insoluble problems. This constitutes interrogative obsession, the obsession of the why, or *Grübelsucht*.

grudge, *sb.* Add: **6. grudge fight,** a fight based on personal antipathy; also *fig.*

1927 *Daily Express* 19 July 3/4 It is a 'grudge fight',.. and it is clear that Stalin is trying to expel Trotsky.. from the Central Committee of the Communist Party. *Ibid.* 22 Sept. 2/5 It may be that the interchange of letters between the Dempsey and Tunney camps has convinced the sporting public that this will be a 'grudge fight'. 1964 J. HALE (*title*) The grudge fight.

gruelling, *ppl. a.* (Earlier example.)

1852 J. F. BATEMAN *Aquatic Notes* v. 70 Their rivals [in a boat-race]..were pulling a very quick and 'gruelling' stroke.

gruenlingite, var. *GRÜNLINGITE.

grumble, *sb.* Add: **3.** *slang.* [Shortened from *grumble and grunt,* rhyming slang for *cunt.*] = *CRUMPET 4 c.

1962 E. BROCK *Little White God* v. 68 There's this copper..and he puts away a local tea-leaf. And this tea-leaf's old woman's a fair bit of grumble. 1966 *Melody Maker* 30 July 8/5 American visitors are invariably delighted by references to birds, scrubbers, grumble.

grumblesome (grʌ·mb'lsəm), *a.* [f. GRUMBLE *sb.* + -SOME *suffix*^{1}.] Grumbling, complaining.

1925 *Brit. Weekly* 9 July 333/1 It doesn't lose its temper when you lose the way: it doesn't get tired and grumblesome. 1940 BLUNDEN *Poems 1930–40* 193 The gold-chained bee..seemed inclined, though grumblesome, To pass the time of day. 1950 *Publ. Amer. Dial. Soc.* XIV. 33 *Grumblesome,* irritable, complaining.

grumbling, *ppl. a.* (Later examples.)

1935 J. C. THOMSON *Appendicitis* 21 If he [the surgeon] is to keep his death-rate low..he must operate in many normal conditions, but he must have a good excuse... Hence we have the 'grumbling appendix'. 1966 *Lancet* 10 Dec. 1308/2 If interaction is responsible for, say, a 'grumbling' appendix, the removal of the appendix will also remove the symptoms.

grumly, *adv.* Delete *rare*$^{-0}$ and add examples.

1827 J. F. COOPER *Red Rover* iii, 'Any fool knowd it,' returned Scipio grumly. 1854 H. H. RILEY *Puddleford* 92 (Th.), Mr. Bird very grumly said he'd better hold on. 1880 G. W. CABLE *Grandissimes* vi. 41 'Um-hum,' he said grumly.

grummet^{2}, **grommet.** Add: **1. e.** A washer used to insulate electric conductors passing through a hole in a conducting material.

1942 *Electronic Engin.* XV. 303 The power cord should have been threaded through that grommet first. 1959 *B.S.I. News* June 10/2 Tiny grommets for aircraft instruments.

f. A stiffener used inside a Service cap.

1953 J. MASEFIELD *Conway* (ed. 2) III. 164 Next term, arriving back with no grommet in my cap as an 'old hand', and promptly being told to put it back. 1956 W. A. HEFLIN *U.S.A.F. Dict.* 236/2 *Grommet,* a ring-like device of rubber, roll cloth, or metal used inside the top of the service cap to keep it tightly stretched.

grump, *sb.* Add: **2.** (Earlier examples.)

1844 C. RIDLEY *Let.* 6 Oct. in *Life & Lett.* (1958) xiv. 173 Wells has ceased to be under the dominion of the grumps. 1868 L. M. ALCOTT *Little Women* (1871) iv. 41 Hannah had the grumps, for being up late. Never did suit her.

3. A gruff, grumpy, or ill-humoured person. *colloq.*

1900 *Eng. Dial. Dict.*, *Grump* (pl.), a surly person. 1947 P. BENNETT *Varmints* 225 The grumps who secretly did a little charity. 1959 I. & P. OPIE *Lore & Lang. Schoolch.* x. 186 In Fife 'grump' and 'peenger' apply more to people who are snivelling and fretful than openly sobbing. 1959 T. GRIFFITH *Waist-high Culture* (1960) iii. 36, I called on an affectionate grump known throughout the journalism department as 'Pa' Kennedy. 1970 *New Yorker* 12 Sept. 39/1 Oh, you are the world's worst grump.

grunion (grʌ·nyən). [Prob. ad. Sp. *gruñón* grunter.] A small Californian marine fish, *Leuresthes tenuis*, which comes ashore to spawn.

1917 *Overland Monthly* Dec. 529/1 Of all the peculiar fish of which the southern coast of California boasts, the Grunyon is the most peculiar, on account of its phenomenal manner of spawning. 1919 *Fish Bulletin* (Calif. Fish & Game Commission) 15 July 3 At Long Beach no name other than 'grunion' is ever heard, although one gleans from scientific works such names as 'silver-sides' and 'little-smelt'. 1927 *Glasgow Herald* 17 Sept. 4 A very striking rhythm is exhibited by a Californian smelt, called the Grunion. *Ibid.*, This is a very adaptive rhythm that works out profitably for the race of grunions. 1932 M. MILLER *I cover Waterfront* 112 The spring tides, which accompany the full and the dark of the moon, are the time-tables by which the grunion runs can be predicted. 1965 A. J. MCCLANE *Standard Fishing Encycl.* 167/1 The grunion come completely out of the water to spawn in moist beach sand at which time they are caught by hand.

grünlingite (grü·nliŋgəit). *Min.* Also gruenlingite. [ad. G. *grünlingit* (Muthmann & Schröder 1897, in *Zeitschr. f. Kryst.* XXIX. 145), f. the name of F. *Grünling,* formerly keeper of the mineral collection of the Univ. of Munich: see -ITE^{1}.] A sulphide and telluride of bismuth, similar to or identical with joseite, which occurs in steel-grey lamellar masses.

1898 *Jrnl. Chem. Soc.* LXXIV. II. 78 A tellurium mineral from Cumberland... This new mineral is named grünlingite. 1941 *Amer. Mineral.* XXVI. 200 The pattern is identical with that given by joseite.., 'grünlingite'.. from Carrock Fells, Cumberland (type locality), and 'oruetite'. 1944 C. PALACHE et al. *Dana's Syst. Min.* (ed. 7) I. 165 The relation of gruenlingite to joseite is uncertain. 1950 *Mineral. Abstr.* XI. 112 Results..for joseite, grünlingite, a new mineral Bi_4TeS_4, and seleniferous cosalite, all from Rézbánya [in Rumania].

grunt, *sb.* Add: **2. b.** *U.S. slang.* An infantry soldier.

1969 I. KEMP *Brit. G.I. in Vietnam* v. 106 The sound of..engines, among the most welcome of all music to the average infantryman—or 'grunt', as we were impolitely called—in Vietnam. 1970 *Times* 28 May 7/5 These luckless victims of the American military machine are known as 'grunts', a name said to be derived from their way of complaining as they trudge along the jungle trails.

Grunth, var. *GRANTH.

gruntled (grʌ·nt'ld), *ppl. a.* [Back-formation f. DISGRUNTLED *a.*] Pleased, satisfied, contented.

1938 WODEHOUSE *Code of Woosters* i. 9 He spoke with a certain what-is-it in his voice, and I could see that, if not actually disgruntled, he was far from being gruntled. 1962 C. ROHAN *Delinquents* 76 Come on, Brownie darling, be gruntled. 1966 *New Statesman* 11 Nov. 693/2 An action against a barrister for negligence..would open the door to every disgruntled client. Now gruntled clients are rare in the criminal courts. 1967 E. MCGIRR *Hearse with Horses* i. 17 The Agency has a nice file of gruntled exes who have found their talents in a great variety of jobs.

‖ **gruppetto** (grupe·to). *Mus.* Also **grupetto.** Pl. **gruppetti, gruppettos.** [It., dim. of *gruppo* turn.] = TURN *sb.* 5.

1842 J. F. WARNER *Univ. Dict. Mus. Terms* p. xliv/2 *Groppo, Gruppo, Gruppetto.*—These terms..are rather indefinitely employed in music to denote, in general, every species of musical ornament which consists of several small notes. 1950 G. B. SHAW *How to Become Mus. Critic* (1960) 330 Where we fall short is in *roulades,* shakes, and *gruppettos,* which many of our singers simply cannot sing at all. 1958 *Times* 3 Nov. 14/4 Small *grupetti* were sometimes taken in one bow, but mostly it was the single stroke to the single note. 1959 *Listener* 29 Jan. 228/1 She sang the Queen of Night's aria..her *gruppetti* being ..'bang on'.

grutch, *sb.* **3.** (Later U.S. example.)

1898 A. NICHOLAS *Idyl of Wabash* 36 There'd been some old grutch atween him an' Bill.

Gruyère. (Earlier examples.)

[1775 T. BLAIKIE *Diary Scotch Gardener* 7 June (1931) 39 This is the Mountaine that furneshes a vast quantity of cheese to France under the name of fromage du Gruyere.] 1802 W. DYOTT *Diary* 28 June (1907) Got excellent trout and good Gruyère cheese and also good wine. 1822 L. SIMOND *Trav. Switzerland* I. 22 Great quantities of cheese are made here in imitation of Gruyère cheese. 1827 B. DISRAELI *Vivian Grey* VI. vi. 204 The Prince..contented himself for the present with assisting his Gruyère with one of the very fine-looking cucumbers.

G string. 1. *Mus.* [G 2, STRING *sb.* 3.] (See quot. 1876.)

1831 T. L. PEACOCK in *Examiner* 12 June 374/1 The real magic is..the surpassing beauty of the effect. It is the same with his [*sc.* Paganini's] performance on the single string (the fourth, or G string), as with his performance on four. 1876 STAINER & BARRETT *Dict. Mus. Terms, G string,* the name of the first string on the double bass, the third on the violoncello, viola, and guitar, and the fourth on the violin. 1884 W. HOE *Dict. Fiddle* s.v. *Strings,* To preserve strings in good condition they should be kept in a tin box...This prevents the G string from shrinking in the heat of the summer. 1961 A. BAINES *Mus. Instruments* vi. 110 Originally all four strings were of gut, but early in the eighteenth century some players began to use a G string wound with silver or copper wire to improve its speaking properties. 1964 *Conc. Oxf. Dict. Mus.* 242/2 *G string* (violin), the lowest string, with a rich tone.

2. (Usu. hyphenated.) Also **gee-string. a.** A loin-cloth or breech-cloth, or the string supporting this, worn by American Indians, etc.

1878 J. H. BEADLE *Western Wilds* xvi. 249 Around each boy's waist is the tight 'geestring', from which a single strip of cloth runs between the limbs from front to back. 1885 *Rep. Indian Affairs* 179 The more industrious, intelligent, and independent [Indians]..dress partially in civilized clothing and invariably stick to the blanket, leggins, and gee-string. 1891 *Harper's Mag.* Dec. 36/2 Some of the boys wore only 'G-strings' (as, for some reason, the breech-clout is commonly called on the prairie). 1907 S. E. WHITE *Arizona Nights* I. ix. 163 These Yuma Indians..were a peaceful, fine-looking lot, without a thing on but a gee-string. 1913 C. E. MULFORD *Coming of Cassidy* iv. 67 'It's a G-string an' a medicine-bag..' cried Dad from the harness-shop.

b. A similar piece of material worn by show-girls, strip-tease artists, etc.

1936 J. DOS PASSOS *Big Money* 488 One of the girls.. wiggled her geestring at him. 1939 A. HUXLEY *After Many a Summer* I. xii. 168 It had happened, but quite unwillingly..at Congo Club with nothing on but a G-string and some talcum powder. 1951 I. SHAW *Troubled Air* xxii. 377 What I know about politics you could put in a chorus girl's g-string and it wouldn't raise a lump. 1960 V. H. GIELGUD *To Bed at Noon* I. ix. 61 Her costume consisted of two tinsel stars and a G-string.

G-suit, g-suit (dʒi·siūt). [f. *g as a symbol for the acceleration due to gravity.] A garment designed to enable a person to withstand high accelerations; also, a similar garment used in surgical operations. Cf. *ANTI-G *a.*

1945 *Aeroplane* 12 Jan. 44/2 Many experiments..have been conducted..and success has now been accomplished with the United States Berger G-suit, used by operational fighter pilots. 1952 A. COPPEL in Bleiler & Dikty *Best Sci. Fiction Stories* 4th Ser. (1955) 92 There was pain in the vice-like grip of the g-suit as it banged hard against his flesh. 1956 L. MALLAN *Men, Rockets & Space* 295 *G-suit,* a protective costume designed so that expanding air bladders increase pressure against the body as acceleration increases. 1956 *Jrnl. Amer. Med. Assoc.* 22 Sept. 275/2 The increased venous filling that the g-suit creates decreases the possibility of air embolism during operations with the patient in this position. 1962 J. GLENN et al. *Into Orbit* 244 *G Suit* is a garment worn by pilots which contains inflatable bags positioned over various parts of the body. When the pilot is subjected to abnormally high G forces, the bags inflate automatically and prevent his blood from draining from one area of his body and pooling in another.

‖ **guacamole** (gwakāmōᵘ·li). [Amer. Sp. *guacamole*, ad. Nahuatl *ahuacamolli*, f. *ahuacatl* avocado + *molli* sauce.] A Mexican dish made from avocado pears mixed with onions, tomatoes, chili peppers, and seasoning.

1920 W. POPENOE *Man. Tropical & Subtropical Fruits* ii. 23 In the tropics, the fruit is added to soups at the time of serving; mashed with onions and lemon juice to form the delectable *guacamole* of Cuba and Mexico. 1943 I. S. ROMBAUER *Joy of Cooking* (rev. ed.) 38/1 (*heading*) Avocado Spread (Guacamole). 1956 C. SPRY *Cook Bk.* i. 13 There is one particularly delicious cocktail savoury, Guacamole, which one serves in this country only when avocado pears can be bought inexpensively. 1966 *Listener* 4 Aug. 164/2 A piquant *guacamole*, a paste of avocado pear seasoned with tomatoes, onions, coriander, and.. chilli peppers. 1969 E. STEWART *Heads* (1970) 18 A bowl of tortillas and chilli with guacamole. 1970 *Good Housek. Cooking for Company* III. 121 *Guacamole,* a Mexican dish that's good to start a meal.

guaco. Add: **2.** The Rocky Mountain bee plant, *Cleome serrulata*; also, the pigment made from it and used by the Pueblo Indians for decorating pottery. *U.S.*

1844 J. GREGG *Commerce Prairies* I. 278 This kind of crockery..is often fancifully painted with colored earths and the juice of a plant called *guaco*, which brightens by burning. 1936 *Burlington Mag.* Nov. 241/2 The use of 'guaco' to bind the glaze. 1956 A. O. SHEPARD *Ceramics for Archeologist* i. 33 Before firing, the paint, which is called *guaco*, is a dark olive-green.

guaco, var. *HUACO.

Guadalupe (gwa:dālū·p, gwa:dālū·pe). The name of a Mexican island off the coast of California, used *attrib.* in the names of plants found wild there, as **Guadalupe palm,** *Erythea edulis,* an ornamental palm bearing edible pulpy fruit.

1895 F. FRANCESCHI *Santa Barbara Exotic Flora* 11 More plentiful still and represented by large specimens, is Erythea Edulis, the Guadalupe Palm. 1969 T. H. EVERETT *Living Trees of World* 77/1 Among those [species] that exceed 20 feet in height [is]..the Guadalupe palm.

guaiac. Add: **b. guaiac test,** a test for the absence of blood from the urine or fæces involving guaiac as a reagent.

1894 GOULD *Illustr. Dict. Med.* 1479/1 Guaiac-test. Blood. 1904 C. H. BEDFORD *Clin. Handbk. Urine Anal.* (ed. 2) 72 The Guaiac test is an exceedingly delicate one. 1953 J. C. TODD et al. *Clin. Diagnosis* (ed. 12) iv. 326 The benzidine test..is similar to the guaiac test and has the same fallacies, but is distinctly more sensitive.

guaiaretic (gwəiˌāre·tik), *a. Chem.* [f. GUAIA(C + Gr. ῥητ-ίνη resin + -IC, after G. *guajakharzsäure* guaiac resin acid (H. Hlasiwetz 1861, in *Ann. d. Chem. und Pharm.* CXIX. 267).] *guaiaretic acid:* a white compound, $C_{20}H_{24}O_4$, found in gum guaiacum. So **guaia·retate,** a salt of this acid.

1866 H. WATTS tr. *Gmelin's Hand-bk. Chem.* XVII. 243 Guaiaretic acid dissolves very slightly in cold or warm ammonia-water. 1892 *Photogr. Ann.* II. 73 Guaiaretic Acid,..when freshly prepared, is soluble in alcohol, ether, [etc.]. *Ibid.*, Addition of small quantities of colouring matters..to the guaiaretic acid or guaiaretates. 1964 *Jrnl. Chem. Soc.* 4011 Fractionation of the resin, by solvents, pyrolysis, etc., resulted in numerous products. Outstanding among these is the dihydric phenol, guaiaretic acid, readily isolated as its sparingly soluble sodium salt.

‖ **guajira** (gwāhiᵊ·rā). Also **guajiro.** [Cuban Sp.] The music of a Cuban peasant dance whose rhythm shifts from 6/8 to 3/4 time.

1923 *Blackw. Mag.* Nov. 711/1 A traveller could sing soleares and guajiras which I could accompany. 1947 A. EINSTEIN *Mus. Romantic Era* xvii. 331 Pieces characteristic of the South, with musical impulses from the slums of New Orleans, from Cuba and Brazil..habaneras, tangos, *guajiros*. 1956 [see *GUARACHA].

‖ **guajiro** (gwāhiᵊ·ro). [Cuban-Sp. *guajiro* rustic, rural.] A Cuban agricultural worker.

1901 *U.S. Bur. of Statistics* (*Treasury Dept.*) *Monthly Summary of Commerce & Finance* Oct. 1352 The 'estancieros', as well as those who are engaged in stock raising, are known in the country under the name of 'guajiros'. 1937 E. HEMINGWAY *To have & have Not* III. x. 164 We want to end the slavery of the *guajiros*, you know, the peasants. 1961 *New Left Rev.* May–June 6/2 Consumer goods have been priced down drastically to place them within reach of the poorest *guajiro*. 1965 H. JOHNSON *Bay of Pigs* 18 From Fidel Castro to the humblest *guajiro* toiling in the fields, everyone talked about it.

Guam (gwām). [The name of the largest island of the Marianas.] Phr. *to clear out, sail for Guam:* to sail 'for some port unknown'.

1881 *Times* 23 June 4/5 Having..docked the steamer.. they cleared out 'for Guam'. 1898 *Westm. Gaz.* 7 Dec. 12/1 Sailed for Guam.

guanajuatite (gwānāhwā·təit). *Min.* [f. *Guanajuat-o,* the name of a city and state in central Mexico + -ITE^{1}: see quot. 1877.] A bluish-grey selenide of bismuth, Bi_2Se_3, often containing a high proportion of sulphur.

1877 *Amer. Jrnl. Sci. & Arts* CXIII. 319 It seems, however, that the same mineral was at first named *guanajuatite* by Fernandez, who described it in full in the Guanajuato paper, 'La República' for July 13th, 1873. The latter name should consequently be accepted instead of the name frenzelite. 1967 *Mineral. Abstr.* XVIII. 281/2 The rare bismuth selenide, guanajuatite, is recorded associated with native gold.

guanase (gwā·nēiz, -s). *Biochem.* [a. G. *guanase* (Jones & Partridge 1904, in *Hoppe-Seyler's Zeitschr. f. physiol. Chem.* XLII. 347), f. GUAN(INE + *-ASE.] An enzyme that hydrolyses guanine to xanthine and ammonia; guanine aminohydrolase, guanine deaminase.

1904 *Jrnl. Chem. Soc.* LXXXVI. I. 838 Fresh pancreas, when subjected to autolysis, does not yield adenine or guanine, and guanine which is added is transformed into xanthine within 3 days. The enzyme present in pancreas which induces this change is termed Guanase. 1967 *Biochem. Jrnl.* CIV. 634/1 Guanase (guanine aminohydrolase, EC 3.5.4.3) was of special interest as guanine is metabolically a precursor of xanthine and is thus indirectly a source of uric acid.

guanay (gwā·ne). [ad. Amer. Sp. *guanae*, prob. from Quechua.] The Peruvian cormorant, *Phalacrocorax bougainvillei.*

1860 [see GUANO *sb.* 3]. 1925 R. C. MURPHY *Bird Islands of Peru* iv. 73 The range of the guanay is restricted to coastwise waters along the arid, western shores of South America. 1968 B. NELSON *Galapagos* xvi. 272 Guanays are heavy birds, none too well endowed for getting off the ground and landing again. 1971 *Islander* (Victoria, B.C.) 25 July 3/1 The guanay cormorant of South America has poured millions into the Peruvian economy.

Guanche (gwā·ntʃe). Also **Guancho.** [Sp.] One of the aboriginal inhabitants of the Canary Islands, who were absorbed by the Spanish on their conquest of the islands in the fifteenth century. Also, the language of the Guanches.

1599 T. NICOLS in Hakluyt *Voy.* (ed. 2) II. II. 5 These people were called Guanches, naturally they spake another language cleane contrary to the Canarians. 1797 *Encycl. Brit.* IV. 81/1 The inhabitants are chiefly Spaniards; though there are some of the first people remaining, whom they call Guanches, who are somewhat civilized by their intercourse with the Spaniards. 1814 H. M. WILLIAMS tr. *Humboldt's Trav.* I. 117 Santa Cruz, the Annaza of the Guanches, is a neat town, with a population of 8000 souls. 1876 *Encycl. Brit.* IV. 796/1 Many of the Guanches fell in opposing the Spanish invasion. 1884 tr. *De Candolle's Orig. Cultiv. Plants* 320 It is not known whether the Guanchos (the Berber people of the Canaries) knew the bean. 1890 J. G. FRAZER *Golden Bough* I. i. 19 In times of drought the Guanches of Teneriffe led their sheep to sacred ground. 1939 L. H. GRAY *Foundations of Lang.* 366 To these [dialects] must be added the *Guanche* of the Canary Islands, which became extinct in the seventeenth century.

guanethidine (gwāne·þidīn). *Pharm.* [f. GUANIDINE, by insertion of ETH(YL.] A drug, 2-(octahydro-1-azocinyl)-ethyl guanidine, used to reduce blood pressure.

1959 *Jrnl. Amer. Med. Assoc.* 11 July 1265/1 The former has been given the generic name guanethidine by the manufacturer. 1959 *Lancet* 12 Dec. 1048/2 The hypotensive action of guanethidine appears to be due to selective blockade of the sympathetic nervous system. 1970 *Jrnl. Pharmacol. & Exper. Therap.* CLXXIII. 166/1 Reserpine and guanethidine prevented the glandular enlargement induced by papain or amputation.

guango (gwæ·ŋgo). [a. Amer. Sp. *guango*, prob. from Quechua.] The Jamaican name for an ornamental tree, *Samanea saman,* of the family Leguminosæ, the pods of which are used as cattle fodder; also called *rain-tree* (s.v. *RAIN *sb.*^{1}) and *SAMAN.

1886 E. D. M. HOOPER *Rep. Forests of Jamaica* 31 Guango, Pithecolobium (Inga) saman. 1920 FAWCETT & RENDLE *Flora of Jamaica* IV. 152 Guango... The wood is hard and ornamental, but cross-grained and difficult to saw. 1956 J. HEARNE *Stranger at Gate* vii. 55 Under the guango tree Tom and Joshua..were saddling the ponies. 1963 ROBERTSON & GOODING *Bot. for Caribbean* xxiii. 183 Guango, Saman or Rain Tree—*Enterolobium saman.* This is a large, spreading tree, a native of Central and South America, naturalised or cultivated throughout the West Indies. 1971 H. D. CARBERRY in J. Figueroa *Caribbean Voices* I. 25 Also there are the days when the leaves fade from off guango trees.

guanidino- (gwā:nidi·no). *Chem.*, combining form of GUANIDINE. So *guanidino acid, group,* etc.

1910 *Jrnl. Chem. Soc.* XCVIII. I. 825 The guanidinobutyric acid which crystallises out is purified by conversion into the hydrochloride. 1951 *Jrnl. Amer. Chem. Soc.* LXXXIII. 1870/1 The basic nature of the guanidino group indicated that it might also be a specific site for water absorption. 1952 E. H. RODD *Chem. Carbon Compounds* I. xiv. 932 Guanidino acids. This group includes the biologically important compounds creatine and arginine. 1965 P. A. S. SMITH *Chem. Open-Chain Org. Nitrogen Compounds* I. vi. 277 The prefix 'guanidino', for $H_2NC(=NH)NH$—, is used in cases where naming the substituent on guanidine would be awkward or would misplace the desired emphasis; e.g. guanidinoacetic acid, $H_2NC(=NH)NHCH_2COOH$.

guanido- (gwā·nido). *Chem.*, = prec. So *guanido group.*

1874 *Jrnl. Chem. Soc.* XXVII. 906 Guanido-dibenzoic acid. 1951 *Chem. Abstr.* XLV. 3433 (*heading*) Participation of the guanido group of arginine in the gelling of gelatin. 1967 CANTAROW & SCHEPARTZ *Biochem.* (ed. 4) xxi. 579 In the first reaction, the amidine group of arginine is transferred to glycine, forming guanidoacetate (glycocyamine) and ornithine.

guanin. Add: Now spelt **guanine** (-īn). (Later examples.)

1952 *New Biol.* XIII. 75 This same silvery sheen, due to the deposition of crystals of guanine in the skin, also appears in the rainbow trout of Michigan streams. 1966 *New Scientist* 24 Nov. 424/1 These are the four main genetic 'code letters' or bases adenine, guanine, thymine and cytosine.

guanophore (gwā·nŏfōəɹ). *Zool.* [a. G. *guanophore* (W. J. Schmidt 1912, in *Zeitschr. f. wissenschaftl. Zool.* CI. 188), f. GUAN(INE + -O + -PHORE.] A chromatophore containing crystals of guanine, found in the skin of fishes and reptiles; cf. *IRIDOCYTE, *LEUCOPHORE.

1924 L. T. HOGBEN *Pigment. Effector Syst.* 24 In the skin of the Chameleon there are present, immediately below the epidermis, cells charged with yellow pigment.. described as guanophores. 1960 FOX & VEVERS *Nature of Animal Colours* x. 150 When guanophores are overlain by a reddish carotenoid pigment a golden hue results. 1963 M. FINGERMAN *Control of Chromatophores* i. 4 Cells that contain guanine or guanine-like substances are called guanophores.

guanosine (gwā·nŏsīn). *Chem.* Formerly also -in. [ad. G. *guanosin* (Levene & Jacobs 1909, in *Ber. d. Deut. Chem. Ges.* XLII. 2470), f. GUANINE, by insertion of -OS(E^{2}.] A nucleoside, $C_{10}H_{13}N_5O_5$, which yields guanine and ribose on hydrolysis and is itself a hydrolysis product of ribonucleic acid.

1909 *Chem. Abstr.* III. 2562 Guaninepentoside, $C_{10}H_{13}O_5N_5.2H_2O$, termed 'guanosin' by the authors. 1968 A. WHITE et al. *Princ. Biochem.* (ed. 4) xlix. 1035 Guanosine is directly utilized for pteridine synthesis.

guanylic (gwāni·lik), *a. Chem.* [tr. G. *guanylsäure* guanylic acid (I. Bang 1898, in *Hoppe-Seyler's Zeitschr. f. physiol. Chem.* XXVI. 136), f. GUAN(INE + -YL + -IC.] *guanylic acid:* a nucleotide, $C_{10}H_{14}N_5O_8P$, which is a constituent of ribonucleic acid; guanosine monophosphate.

1899 *Jrnl. Chem. Soc.* LXXVI. I. 180 Nine-tenths of the nitrogen originally present in guanylic acid is obtained, after hydrolysis, in the form of guanine. 1969 *Times* 25 Aug. 4/8 In the genetic code three guanylic acid units are..the code for glycine.

guar (gwāɹ). [a. Hindi *guar.*] **1.** An Indian plant, *Cyamopsis tetragonoloba* (*psoralioides*), of the family Leguminosæ, which tolerates drought and is grown as a vegetable, fodder crop, and green manure, and as a source of guar gum.

[1832 W. CAREY *Roxburgh's Flora Indica* III. 317 Goor-chikurkai..is an annual esculent plant, much cultivated in the gardens of the natives on the coast of Coromandel.] 1882 DUTHIE & FULLER *Field & Garden Crops* II. 24 Guár is grown in these Provinces for two very different purposes,—as a vegetable for human consumption, and as a pulse for horses and cattle. 1908 G. WATT *Commercial Prod. India* 449 Guar is specially suitable as a green manure or green fodder crop. 1948 *Chem. Industries* Jan. 60/3 As a consequence of the demand for a gum such as may be obtained from guar, General Mills undertook large scale agricultural production of the plant. 1965 R. R. AGARWAL *Soil Fertility in India* vii. 157 Cowpea (*Vigna catjang*), guar (*Cyamopsis psoralioides*).. are examples of other leguminous plants that are used as green manures.

2. Commercial guar gum.

1945 *Paper Mill News* 8 Dec. 52/1 It is intended to explore the possibilities of tub sizing with converted guar. 1952 J. P. CASEY *Pulp & Paper* I. vii. 406 When added dry at the beater, guar acts as a strong cementing agent which greatly increases the strength of the paper. 1966 *Kirk-Othmer Encycl. Chem. Technol.* (ed. 2) X. 750 Guar solutions can be gelled by the addition of small amounts of borax.

3. *attrib.*, as *guar seed*; **guar flour,** commercial guar gum; **guar gum,** a fine greyish-white powder obtained by grinding the endosperm of guar seeds and used in the food, paper, and other industries; also, the polysaccharide that is the main constituent of this.

1945 *Sat. Even. Post* 30 June (Back cover), Science has found that guar seeds contain a gum which has..dozens of important uses. 1946 *Progress thru Research* (General Mills, Inc.) Fall 6/2 (*caption*) This experimental mill.. produces guar flour for development research and commercial evaluation. 1950 *Jrnl. Amer. Chem. Soc.* LXXII. 4634/1 Guar gum extracted from the endosperm of guar seed is a galactomannan polysaccharide in which the ratio of D-mannose to D-galactose is 2:1. 1959 SMITH & MONTGOMERY *Chem. Plant Gums* xii. 325 The polysaccharide in guar flour, as the crude, powdered gum is sometimes called, contains 84 per cent of D-mannose and 16 per cent D-galactose...The crude guar gum..may contain 1 to 5 per cent protein. 1969 *Encycl. Polymer Sci. & Technol.* XI. 404 Guar seeds have been the source of large quantities of industrial gum.

guaracha. (Later example.) Also, the music to which it is danced.

1956 M. STEARNS *Story of Jazz* (1957) iii. 26 Of the native Cuban dances, the Habanera, Guajira, Punto, and Guaracha contain strong Spanish elements. 1958 E. BORNEMAN in P. Gammond *Decca Bk. Jazz* xxi. 270 The music of the Machito band..boiled down to two types: (1) Afros, rumbas, guarachas, congas, sones and boleros, with jazz trumpets, jazz tenor and alto-saxophones improvising against the Cuban rhythms; (2) jazz numbers of the sort..written for Chick Webb. 1962 J. STEWART tr. *Cousseau's Death Miss Cunningham* I. 58 To the raucous sound of a *guaracha*, Indian women.. were dancing.

guarache, var. *HUARACHE.

Guarani (gwā·rānī). Also **Guarany.** [Sp.] **1.** One of the main divisions of Tupi-Guarani, a family of South American Indian languages; also, a speaker of one of these languages. Also *attrib.* or as *adj.* Cf. *TUPI-GUARANI.

1797 *Encycl. Brit.* XIII. 730/1 The Guaranies resolved to quit their settlements in the neighbourhood of the Portuguese, and to remove into the province of Paraguay. 1840 *Penny Cycl.* XVII. 226/1 The Guaranis compose the bulk of the population [of Paraguay]. 1875 *Encycl. Brit.* I. 702/1 The supposition..that the Guarani tribes are the remnants of a once powerful and united people, is scarcely admissible. 1885 *Ibid.* XVIII. 244/2 The inhabitants of Paraguay are mainly Guaranis or half-breeds with a strong proportion of Guarani blood. 1900 *Fortn. Rev.* Jan. 79 There the women speak nothing but the Guarany language, and the men use only Guarany in talking to them, whereas among themselves the men always speak Spanish. 1947 *Whitaker's Almanack* 939/1 Most of the inhabitants are bi-lingual, speaking Guarani (the language of the extinct tribe of Indian inhabitants at the time of the Spanish occupation) and Spanish. 1959

Times 1 June (Lat. Amer. Suppl.) p. ii/2 Guaraní is still the language of many people.

2. (With lower-case initial.) The currency unit of Paraguay, consisting of 100 centimos, introduced in 1943.

1943 *Economist* 6 Nov. 624/2 The guarani will replace the 'peso fuerte' at the rate of 1 guarani for 100 pesos. **1947** *Whitaker's Almanack* 939/1 From November 4, 1943, a new currency unit (*Guarani*, plural *Guaranies*) symbolised by a crossed G, replaced the *Peso Fuerte* at the rate of 1 *Guarani*=100 *Pesos*, and 1·75 *Guaranies*=1 Gold *Peso*.

guaranteed, *ppl. a.* (Further examples.)

1931 *Economist* 24 Jan. 163/1 The remaining proposals involve the cancellation of the present agreements as to the guaranteed day and guaranteed week, overtime on a daily basis, [etc.]. **1939** *Monthly Labor Rev.* (U.S.) Apr. 914 Guaranteed annual wage encouraged by Fair Labor Standards Act. Interpretation..of the Fair Labor Standards Act, 1938..has opened the way for growth of the policy of guaranteed annual wages in industry. **1941** Alley & Hall *Farmer in N.Z.* iv. 113 The Primary Produce Marketing Act 1936..originated the principle of the guaranteed price, which may be summarised as the principle of a state guarantee of a definite price each season to dairy farmers for their butter and cheese and the taking over by the state of the responsibility for marketing these products overseas. **1958** *N.Z. News* 29 July 2/2 The Dairy Board flatly rejected the Government's proposed guaranteed price for butterfat which reduced the return to the farmer by 12 per cent. and upwards. **1964** *Lebende Sprachen* IX. 100/2 *Guaranteed bonds* are backed by the guaranty of another company. **1970** *Times* 11 Mar. 14/4 A memorandum of March, 1948, headed 'Guaranteed week', stated that where workmen regularly worked, say, a six-day week, they should be entitled to a six-day guarantee.

guard, *sb.* Add: **3. b.** *Cricket.* (Earlier example.)

1843 'Wykehamist' *Pract. Hints Cricket* 4 Let the player first obtain guard..for the middle stump.

7. *spec.* (chiefly *U.S.*), a warder in a prison or other place of detention.

1889 in *Cent. Dict.* **1907** J. London *White Fang* 299 During Jim Hall's third term in prison he encountered a guard that was almost as great a beast as he. **1936** *S.P.E. Tract* XLV. 190 Why has American practice decided in favour of..guard rather than warder? **1952** *Manch. Guardian Weekly* 15 May 4 The guard knew they would do it and retreated. **1968** *Times* 11 Mar. 4/2 There were rumours that some of the men had threatened to decapitate a guard every 40 minutes.

b. (Earlier examples with reference to a railway train.)

1842 Dickens *Amer. Notes* I. iv. 147 The conductor or check-taker, or guard,..wears no uniform. **1863** G. Meredith *Let.* 27 Apr. (1970) I. 200 On Tuesday next, my little recovered rosy man will journey down, alone, by that train.. The Guard will be bribed to keep eye on him.

d. Also in *Basketball*, either of the two players who are chiefly responsible for the marking of opposing forwards.

1905 *Official Basket Ball Rules* 30 The position of the guard is the most difficult and unsatisfactory place in the team... He is expected to prevent his opponent from throwing a goal, and that without making a foul himself. **1929** *Encycl. Brit.* III. 182/2 The players are divided into three groups, forwards, centres and guards. **1961** J. S. Salak *Dict. Amer. Sports* 207 *Guard*, a defensive player in basketball and football.

9. b. (Earlier and later examples of *Old Guard*.)

1850 H. Melville *White Jacket* I. iii. 10 Hearty old members of the Old Guard; grim sea grenadiers. **1852** J. A. Quitman *Let.* 20 Feb. in J. F. H. Claiborne *Life of Quitman* (1860) II. xvii. 164 What shall we of the strict state-rights school, what shall the 'Old Guard' do? **1902** A. Bennett *Anna of Five Towns* ii. 28 Tellwright belonged by birth to the Old Guard of Methodism; there was in his family a tradition of holy valour for the pure doctrine. **1954** N. Coward *Future Indefinite* iv. 183 The old guard however..sang some of the vintage Maori songs. **1965** N. Gulbenkian *Pantaraxia* viii. 151, I am a member of the Old Guard. 'Leave my gloves?..Not on your life! I was sent home [from court] twenty-five years ago for not having them.' **1968** *Listener* 25 July 102/1 There is..another reason for the angry panic among the Old Guard in Warsaw.

10. *spec.* in *Card-playing.* (See quots.)

1885 R. A. Proctor *How to play Whist* 241 A third best card generally requires *two* guards. **1932** H. Phillips *Week-End Probl. Bk.* 299 He will lead through South's guard in Hearts. **1936** E. Culbertson *Contract Bridge Complete* xxxviii. 423 A card which accompanies a higher card of the same suit is called a *guard*. **1960** *Times* 7 Dec. 16/7, I was sure that my little partner had a guard in the suit.

16. j. (Earlier example.)

1829 T. Flint *George Mason* 140 Madam Mason was..sitting on the guards of the boat.

p. A protector worn on various parts of the body by cricketers or other sportsmen.

1889 in *Cent. Dict.* **1906** A. E. Knight *Compl. Cricketer* ii. 49 Many players and most wicket-keepers wear additional guards for the more delicate parts of the body. **1939–40** *Army & Navy Stores Catal.* 811/3 Football shin guards. *Ibid.* 812/3 The 'Stopem' hockey leg-guard.

18. guard band (*a*) *Telecommunications*, one of the frequency bands on either side of a communication band which serve to protect it from interference from adjacent communication bands; (*b*) a strip separating neighbouring recording tracks on magnetic tape; **guard book,** (*a*) (further examples); **guard-chain,** (*a*) (earlier examples); **guard-changing** *vbl. sb.*, the action of changing the guard, esp. at St. James's and Buckingham Palace; **guard cradle** (see quot. 1924); **guard dog**: delete † and add later examples in sense 'a watch-dog'; **guard hair** = Overhair; **guard-lock** (earlier U.S. example); **guard net** = **guard cradle*; **guard-ring,** (*a*) (example); **guard's van,** the railway coach or compartment occupied by a guard; **guard tent** (U.S. example); **guard tube,** (*b*) a cylindrical conductor surrounding part of a wire anode (in an ionization chamber or proportional counter) which modifies the shape of the electric field and makes the sensitive volume of the chamber more clearly defined; (*c*) one of a series of ionization chambers placed so as to surround a weakly radioactive source and the chambers used to detect its radiation, and connected so as to prevent radiation from other sources from being counted (see quot. 1964); **guard wire** (see quot. 1941).

1956 W. C. Michels *Internat. Dict. Physics* 79/1 Interference guard bands. **1958** H. G. M. Spratt *Magn. Tape Recording* viii. 219 The centre 0·05 in. wide strip is left as a guard band and reduces the possibilities of cross-talk. **1960** J. L. Bernstein *Video Tape Recording* v. 91 When a signal track slips off the playback gap so that ⅛ of the gap length is covered by a guard band, the noise level increases by 3 db. **1960** M. S. Kiver *F-M Simplified* (ed. 3) i. 8 Each station..is assigned a bandwidth of 200 kc. Of this 200-kc bandwidth, 150 (±75) kc is to be employed for the modulation and the remaining 50 (±25) kc is to function as a guardband. **1839** J. Romilly *Diary* 8 May (1967) 169 Pasted docum[ts] in Guard Book all the Ev[s]. **1903** *Daily Chron.* 18 Nov. 8/6 Book Sewers wanted, used to guardbooks and scrap-books. **1907** *Westm. Gaz.* 9 Dec. 7/2 He also produced the 'transfer guard-book'. **1933** D. L. Sayers *Murder must Advertise* i. 18 He propped up the Dairyfields guard-book before him, and fell to studying his predecessor's masterpieces on the subject of Green Pastures Margarine. **1961** T. Landau *Encycl. Librarianship* (ed. 2) 158/2 *Guard book catalogue*, a catalogue in book form (bound or loose-leaf), the entries being..made on slips and pasted on blank pages. **1832** F. Trollope *Dom. Manners* II. xxx. 211 [His] toilet was equally studied; eye-glass, guard-chain, nothing was omitted. **1837** Dickens *Pickw.* liv. 591 Very little watches and very large guard chains. **1904** *Daily Chron.* 30 May 8/2 The minor formalities of guard-changing are completed. **1907** *Westm. Gaz.* 12 Nov. 9/1 The ceremony of guard-changing in the quadrangle. **1924** S. R. Roget *Dict. Electr. Terms* 102/1 *Guard cradle or net*, an earthed network of wires placed under overhead conductors at points where danger might be produced by their falling. **1940** *Chambers's Techn. Dict.* 393/1 *Guard cradle*, a network of wires serving the same purpose as a guard wire. Also called guard net. **1967** 'M. Hunter' *Cambridgeshire Disaster* v. 31 The double chainlink fencing and the guard dogs. **1971** B. Callison *Plague of Sailors* 7 Ancillary boards stated..: Unrestricted guard dog operation do not pass this point. **1913** J. W. Jones *Furs Farming in Canada* iv. 96 Mammals which have a short, fine, soft coat of fur through which grows hair, usually of greater length, variously called overfur, water-fur, guard-hair, are known as fur-bearers. **1921** *Daily Colonist* (Victoria, B.C.) 9 Oct. 19/4 The fur of the muskrat ..is very much like the beaver's fur..having two kinds —the long, coarse guard hairs and the soft undercoat. **1930** M. de la Roche *Finch's Fortune* xxiii. 303 It was growing into a rugged animal of good girth, the glossy black of its pelt shading to blue-black, the silver bands on the guard hairs bright as polished metal. **1966** *Times* 25 Apr. 13/1 Simulated..furs..now..have..'guard-hair'. **1815** *Niles' Weekly Reg.* IX. 44/2 The boat, having entered the guard lock, went through the new canal. **1924, 1940** Guard net [see **guard cradle*]. **1817** M. Edgeworth *Harrington* I. xvii. 462 She never wore Sir Josseline's ring, without putting on..another,..which she called her guard ring, a ring which being tighter than Sir Josseline's, kept it safe on her finger. **1881** [see *van boy* (Van *sb.*[3] 3)]. **1916** W. Owen *Let.* 29 Dec. (1967) 417 Train was wickedly late. I travelled in Guard's van. **1848** E. Bryant *California* iv. 52 They were greatly alarmed when brought to the guard tent, expecting immediate punishment. **1951** *Rev. Sci. Instr.* XXII. 38/1 The best possible insulation of the collecting electrode is required, and..the use of earthed guard tubes to protect the wire is generally a necessity. **1958** H. Etherington *Nuclear Engin. Handbk.* v. 14 End effects will involve regions of non-uniform *A* and of uncertain limits of the counting volume unless 'guard tubes' or field-shaping features are employed. **1963** B. Fozard *Instrumentation Nuclear Reactors* v. 60 Counting work of this kind requires the use of anticoincidence circuits and guard tubes. **1964** R. D. Rusk *Introd. Atomic & Nucl. Physics* (ed. 2) xiii. 308 The surrounding ring of guard tubes is connected in anticoincidence with the counting tubes that are near the source. A cosmic ray passing through one of the counting tubes would not be counted because it would first have to pass through one of the anticoincident guard tubes. **1902** *Jrnl. Inst. Electr. Engin.* XXXI. 91 Probably the most efficient safeguard at present is *earthed* guard wire put up in accordance with Post Office regulations. **1920** R. E. Neale *Whittaker's Electr. Engin. Pocket-bk.* (ed. 4) 579 Earthed guard wires..must be erected, at a minimum height of 2′ above the trolley wires, where telephone and telegraph wires cross the latter. **1941** S. R. Roget *Dict. Electr. Terms* (ed. 4) 156/2 *Guard wire*, an earthed wire, above an overhead line, to catch other wires which may fall accidentally across the line, and to ensure their being earthed before coming into contact with the line to be protected.

guard, *v.* Add: **12.** *Cricket.* To defend, protect, or cover (the wicket). (Cf. Guard *sb.* 3 b.)

c **1742** J. Love *Cricket* iii. 20 Now the two mightiest of the fainting Host..With pow'rful Skill, their threat'ned Wickets guard. **1851** J. Pycroft *Cricket Field* vii. 117 Yet few, very few, ever play as upright as they might play, and that even to guard their three stumps.

guarded, *ppl. a.* **1. d.** (Earlier examples.)

1742 Hoyle *Whist* iv. 34 Do not throw away any of that Suit, even to keep a King or Queen guarded. **1821** J. Burney *Essay on Game of Whist* 20 If you have the queen in sequence, guarded with a small card.

guardee (gā˞ıdī·). [f. Guard *sb.* 8 + -ee[2].] A familiar name for a guardsman, esp. as a type of smartness and elegance. Also *attrib.* or as quasi-*adj.*

1904 Kipling *Traffics & Discov.* 281 That old chap.. is an ex-Guardee. That's why he's saluting in slow time. **1931** Brophy & Partridge *Songs & Slang 1914–18* (ed. 3) 213 *Guardsman's wriggle*, exaggerated salute affected by the Guards. Variants, 'tickling his ear', 'Guardee wriggle'. **1932** 'A. Bridge' *Peking Picnic* xv. 184 Music, of which Touchy, for all his rather Guardee appearance, was passionately fond. **1937** *Oxford Mag.* 6 May 565 Blow the trumpet! Guardee tramp it! **1939** J. Cary *Mr. Johnson* 151 He hollows his back like a guardee.

‖**guardia civil** (gwardi·a þivi·l). Pl. **guardias civiles.** [Sp., = civil guard.] A force formed in Spain in 1844 to take over police duties from the military, and chiefly responsible for public order and safety. Also *ellipt.*

1846 R. Ford *Gatherings from Spain* xiii. 149 The alcalde, the muleteer, the guardia, and all who have shared in the booty. *Ibid.* xvi. 188 The mounted guards, who are now stationed in towns, and by whom the roads are regularly patrolled..are called *guardias civiles*. **1946** G. Millar *Horned Pigeon* xxiii. 416 They had found no sign of the *Guardia Civil*. **1955** J. Thomas *No Banners* iii. 26 On the Pyrenees Franco's *guardias civiles* blocked the roads into Spain. **1966** M. Catto *Bird on Wing* x. 153 The *Guardia Civil* shouted again, raising his rifle.

guardo (gā˞ıdo). *U.S.* [Arbitrarily f. Guardship + -o, simulating Sp. words.] A receiving-ship for enlisted men who are to be drafted to sea-going vessels. Chiefly *attrib.* and *Comb.*

1834 W. N. Glascock *Naval Sketch-bk.* 2nd Ser. I. 232 Why, bless ye, some o' your *guardo*-chaps come over you now with the pride o' their pratin' schoolmaster. *a* **1846** J. A. Gardner *Recoll.* (1906) 116 A droll old guardo midshipman. **1850** H. Melville *White Jacket* II. xxvi. 168 Familiarity with the *guardo-moves*..of a frigate. **1909** *Cent. Dict.* Suppl., *Guardo-move*, a trick played upon a landsman on a receiving-ship.

guardy (gā˞·ıdi). *colloq.* Also **guardie.** Shortened form of Guardian.

1833 J. Romilly *Diary* 15 Jan. (1967) 26 Box from Manchester containing wedding cake from Williams (Ackers' guardian): suppose Guardy's daughter has entrapped the wealthy heir. **1858** A. Mayhew *Paved with Gold* I. i. 33 When he and the other little ones were allowed..to go out for a walk..he would call every well dressed man that passed 'a guardy'. **1924** Galsworthy *Old English* I. i, You *are* an awful boy! Guardy, he really is too awful.

Guarnerius (gwaɪnīə·riɒs). Also **Joseph Guarnerius; Guarn(i)eri.** The name of a family of famous Italian violin- and violoncello-makers of Cremona of the seventeenth and eighteenth centuries, used to designate a violin or violoncello made by a member of this family. (Cf. *Joseph.)

1866 J. Pearce *Violins & Violin Makers* 130 Fifty years ago, a Guarnerius of the best time might have been bought for £50 that will now command £500. **1875** G. Hart *Violin* 96 A 'Joseph Guarnerius' for one hundred and fifty guineas! **1898** H. R. Haweis *Old Violins* 217 It was no more a Guarnerius than a Strad. [**1902** *Grove's Dict. Mus.* IV. 284/2 A full-sized Stradivari or Giuseppe Guarnieri.] **1905** O. Racster *Chats on Violoncellos* (1908) iv. 164 Another violoncello by the redoubtable Andreas ..was found..six months after the sale of Mr Gillot's Guarnerius. **1923** *Weekly Dispatch* 18 Feb. 2 A new violin alongside his old Guarnerius. **1971** *Daily Mail* 19 Jan. 6/4 A Del Gesu Guarnerius made in 1743.

guarri (gwa·ri). *S. Afr.* Also **ghwarrie, guarry, guerrie, gwarri(e,** quarri. [Kaffir *um Gwali*.] Any one of several trees or shrubs of the genus *Euclea*, esp. *E. undulata* or *E. lanceolata*; the fruit of these trees or shrubs. Also *attrib.*

1789 W. Paterson *Narr. Journeys Country of Hottentots* 43 The ground is covered with shrubs about four feet high, called by the natives Guerrie, a species of Royena. **1793** tr. *Thunberg's Trav.* I. 202 The berries of the Guarri bush (*Euclea undulata*) had a sweet taste, and were eaten by the Hottentots. **1801** J. Barrow *Trav. S. Afr.* I. iii. 149 The *guerrie bosch*, apparently a species of rhus. **1843** J. C. Chase *Cape of Good Hope* II. 152 The wild fruits, indigenous to the country, are also incredibly numerous. ..Among these are the wild grape (*Vitis Capensis*),.. quarri (*Euclea undulata*) [etc.]. **1866** Lindley & Moore *Treas. Bot.* s.v. *Euclea*, The fruit is globular, fleshy, and juicy, sometimes as large as a cherry. Those of many of the species, known by the colonists as Guarry, are eaten.

1897 S. J. Du Toit *Rhodesia* 32 Various kinds of sweet grass and small shrubs, varied with very good large bush and trees, as 'quarri', &c. **1913** Pettman *Africanderisms*, *Guarri honey*, a very pale honey in much esteem in the Riversdale District, obtained by the bees from *Royena sp. Guarri tea*, this is made by the Hottentots from the leaves of *Euclea lanceolata*. **1939** E. N. Marais *My Friends the Baboons* iv. 42 Near the sleeping-place there was, in a corner of the rocks, a large gwarrie-bush. **1966** E. Palmer *Plains of Camdeboo* xvii. 271 They made vinegar from the little fruits of the gwarri tree.

Guatemalan (gwătĭmā·lăn, gwā·t-), *sb.* and *a.* Also **Guatema·lian** and (no longer current) **Guatelma·lan.** [f. *Guatemala* + -an.] **A.** *sb.* A native or inhabitant of Guatemala, the most northern republic of Central America, bordering on Mexico. **B.** *adj.* Of or pertaining to Guatemala. Also **Guatema·ltec(an)** [Sp. *guatemaltecano*].

1823 *Gentl. Mag.* XCIII. II. 636/2 The kingdom of Guatemala..has since separated, and the Guatemalans have established themselves as a sovereign State. **1831** J. Bell *Syst. Geog.* V. 617 The Guatelmalans declared themselves independent. **1877** Boddam-Whetham *Across Cent. Amer.* 19 The kind-hearted Guatemaltecans have a custom of driving out to the plain to meet expected friends. **1880** *Encycl. Brit.* XI. 240/1 Singing-birds are commonly kept in the Guatemalian houses. **1888** *New Princeton Rev.* May 356 Zaldivar transmitted a series of despatches misrepresenting the situation, and appealing for protection against Guatemalan tyranny. **1902** *Encycl. Brit.* XXIX. 151/1 A Guatemalan company. **1920** *Glasgow Herald* 7 Apr. 10 Ayutla, on the Guatemalian border. **1927** *Blackw. Mag.* Aug. 173/1 This map, the work of a well-known Guatemalan engineer. **1932** *Times Lit. Suppl.* 24 Nov. 869/3 Harassed by Guatemaltec Jacks-in-office. **1972** *Buenos Aires Herald* 3 Feb. 7/6 The British troops issue was put to the juridical committee by the Guatemalan delegate.

guayule (gwai,ū·li, hwai,ū·li). [a. Amer. Sp. *guayule*, f. Nahuatl *cuauhuli*.] A silver-leaved shrub, *Parthenium argentatum*, of the family Compositæ, native to northern Mexico and adjacent parts of Texas, formerly cultivated as the source of a type of rubber; also, the rubber produced from the plant.

1906 *Bull. Imper. Inst.* IV. 114 The Guayule rubber of Mexico. *Ibid.*, The Guayule plant. *Ibid.* 115 The Guayule is found at altitudes ranging from 3,000 to 5,600 feet. **1926** P. C. Standley *Trees & Shrubs of Mexico* v. 1520 The guayule rubber plant is one of the most important members of the Mexican flora. **1939** J. Steinbeck *Grapes of Wrath* ix. 102 Remember a fella wanted us to put in that rubber bush they call guayule. **1944** *Bot. Gaz.* CVI. 198/1 The experiments indicate that substances unfavorable to the growth of guayule plants emanate from the roots of actively growing plants of this species [*sc.* guayule]. **1964** F. O'Rourke *Mule for Marquesa* 72 The distant green of deceptively deadly ocotillo savannahs, filled with the scorched grey of candelilla and guayule. **1969** A. S. Craig *Dict. Rubber Technol.* 75 Intensive research, started in 1942, has not succeeded in making guayule competitive with *Hevea* rubber, and since 1962 there has been no commercial production.

guazu (gwăsū·, gwā·zū). [a. Guarani *guaçú* deer.] The South American marsh-deer, *Blastoceros dichotomus*. Also **guazapuco, guazu-bira, guazu-pita,** etc. (formerly regarded as comprising more than one species; see quots.).

1837 *Penny Cycl.* VIII. 361/2 The Guazapuco deer, *Mazama paludosa* of Smith, *Cervus paludosus* of Desmarest and Lichtenstein. *Ibid.*, The Guazu-bira, *Cervus nemorivagus* of Lichtenstein.. Locality the same as that of the Guazu-pita. **1871** C. Kingsley *At Last* x, A little brown guazu-pita fawn. **1838** W. P. Hunter tr. *Azara's Nat. Hist. Quadrupeds of Paraguay* II. 126 The Guaranese denominate all deer *Guazu*. *Ibid.* 130 Guazupucu signifies *large deer*... We caught..two female of the guazupucu deer. *Ibid.* 145 The fourth species is universally called Guazubira; and its skin is held in high repute, for gloves. *Ibid.*, It was stated that the guazupita is met with in New Spain. **1850** *Proc. Zool. Soc.* XVIII. 237 Blastoceros paludosus. The Guazu-puco... Inhabits the Brazils. **1898** R. Lydekker *Deer of all Lands* xii. 282 (*heading*) The Guazu Group—Sub-Genus Blastoceros. *Ibid.* 286 He [*sc.* Illiger] refers to the gouzoupoucou of Azara, which is the present species.

guazuti (gwăsūtī·, gwazū·tī). [a. Guarani *guaçuti* white deer.] The brown and white South American pampas deer, *Ozotoceros* (or *Blastoceros*) *bezoarticus*.

1837 *Penny Cycl.* VIII. 361/2 The Guazuti of D'Azara. **1838** W. P. Hunter tr. *Azara's Nat. Hist. Quadrupeds of Paraguay* II. 135 Guazuti, the name given to this species by the Guaranese, signifies *white deer*..others call it guazay. **1850** *Proc. Zool. Soc.* XVIII. 237 Blastoceros campestris. The Mazame or Guazuti. **1898** R. Lydekker *Deer of all Lands* xii. 291 In certain parts of Brazil this deer [*sc.* the pampas deer] is known as the guazuti.

gubbins, *sb. pl.* **1.** Delete †*Obs.* and add later examples, passing into senses: trash; anything of little value; a gadget, thingummy.

1918 P. MacGill *Glenmornan* v. 106 That gubbin iv land was at one time nothin' but a bare rock. **1925** Fraser & Gibbons *Soldier & Sailor Words* 112 *Gubbins*, mere stuff. Trash. Anything of no value—*e.g.*, 'That's only gubbins, all rot!' Also, personal effects—*e.g.*, 'See after my gubbins, will you?' **1944** *Amer. Speech* XIX. 280 A *gubbins* is used to describe almost any part of the equipment of a plane, with about the same meaning as *gadget*. **1958** I. Brown *Words our Time* 60 You can save more petrol by how you drive than with the gubbinses now floating around. **1965** *Sunday Times* (Colour Suppl.) 5 Dec. 16 Many machines flying have a vast illicit complement of rivets, nails, nuts, bolts, torches, pliers and half-eaten sandwiches.. One of the modern test pilot's less enviable jobs is to fly new aircraft upside down and try to catch the gubbins as it hurtles past his face. **1968** *New Scientist* 3 Oct. 8/2 Behind that again is the engine and propeller, the fuel tank and various bits of 'gubbins'.

2. b. *colloq.* A fool, a duffer.

1916 E. F. Benson *David Blaize* vii. 124 'Silly gubbins,' she said. **1955** 'E. C. R. Lorac' *Ask a Policeman* ii. 25 If we only get these old gubbinses out I could let the rooms proper. **1957** [see *Batter *sb.*[4]].

Gubbio (gu·bio). The name of a city in northern Italy, used *attrib.* to designate majolica made there in the sixteenth century, particularly a ruby-lustred majolica made by Giorgio Andreoli.

1857 J. Marryat *Hist. Pott. & Porcelain* (ed. 2) ii. 24 The Gubbio ware unites in some degree the qualities of the mezza majolica and fine majolica; its prismatic glaze has never been surpassed. **1885** *Encycl. Brit.* XIX. 626/2 The early Pesaro and Gubbio ware. *Ibid.* 627/2 The pictured wares of Urbino sometimes have the Gubbio lustre colours. **1942** *Burlington Mag.* June 156/2 Several pieces of Italian Maiolica,..a Gubbio dish by Maestro Giorgio, signed and dated 1522.

gubble (gɒ·b'l), *v.* [Echoic.] *intr.* To make the sound rendered by 'gub': a verb formed to imitate an inarticulate sound.

1904 H. F. Day *Kin o'Ktaadn* 88 Like molasses gubblin' out of a bung-hole. **1929** A. Mason *Swansea Dan* 199 The ocean plunked and gubbled as if in a nightmare. **1944** L. A. G. Strong *All fall Down* 22 If you'd have the goodness to let me finish, instead of praying and gubbling.

guddle, *v.*[2] Add: **1.** (Later examples.)

1951 F. S. Chapman *Mem. Mountaineer* I. i. 10 About this time I learned the art of poaching... I learned to 'guddle' trout. **1955** 'M. Innes' *Man from Sea* vi. 66 You and a friend had been guddling Lord Urquhart's trout.

2. (Later example.)

1966 J. Caird *Perturbing Spirit* xix. 204 He went about on his own, guddling in the burns, bird's-nesting.

guddler (gɒ·dləɹ). Chiefly *Sc.* [f. Guddle *v.*[2] + -er[1].] One who guddles for fish.

1880 in Jamieson *Dict. Sc. Lang.* **1919** G. Rae *'Tween Clyde & Tweed* 72, I close my een, and a' is seen, The burn ed, and the guddler's play. **1928** *Daily Express* 4 Aug. 4/6 The guddler has his chance. **1940** J. Buchan *Memory Hold-the-Door* 304, I was an expert 'guddler', scooping up trout from below stones in the channel.

gudmundite (gu·dmundəit). *Min.* [ad. G. *gudmundit* (K. Johannson 1928, in *Zeitschr. f. Krist.* LXVIII. 87), f. *Gudmundstorp*, name of a locality near Sala, Västmanland, Sweden + -ite[1].] A rare greyish-white antimonide-sulphide of iron, FeSbS, related to arsenopyrite.

1928 *Chem. Abstr.* XXII. 4411 Gudmundite is a new mineral belonging to the marcasite group. **1959** *Mineral. Mag.* XXXII. 133 The apparent rarity of gudmundite may be due solely to the failure of many geologists to distinguish it from arsenopyrite.

guemal (gwē·măl). Also **gemul, guemul, huemal, huemul.** [Amer. Sp., ad. native name.] A small Andean deer of the species *Hippocamelus bicolor* or *H. antisensis*, having the antlers simply forked.

1808 R. Alsop tr. *Molina's Geogr. Nat. & Civil Hist. Chili* I. iv. 226 The *guemul*, or *huemul* (equus bisulcus) is an animal which I have classed with the horse, although it ought to form a separate genus in consequence of its hoofs being divided. **1850** *Proc. Zool. Soc.* XVIII. 236 The female Gemul in the British Museum and in Lord Derby's Museum at Knowsley is considerably larger, and has the legs thicker, than the Siberian Ahu. **1875** *Proc. Zool. Soc.* 44 The Huemul is found in the Andes on either side from Magellan to near Santiago. **1898** R. Lydekker *Deer of all Lands* xii. 297 Ancient travellers in Chili brought reports of a cloven-hoofed animal known to the natives as guemal, or huemal. **1966** *Punch* 25 May 775/3 Chile should also be all right with its *huemal*, an elusive furciferine deer.

guéridon (ge·ridǫn, ‖geridoṅ). [Fr.] A small ornamental table or stand, usually ornately carved.

1853 C. Brontë *Villette* xiii. 329 In one corner appeared a guéridon with a marble top, and upon it a work-box. **1938** *Burlington Mag.* May 227/1 Candlesticks..could be placed on 'guéridons', nowadays usually called 'torchères'. **1955** *Times* 7 July 5/7 £340 for a pair of early Georgian carved gilt and painted wood *guéridon*, resting on carved tripods terminating in claw-and-ball feet (4 ft. high). **1967** M. Ward *Du Barry Inheritance* xii. 184 The Empress Josephine..particularly cherished the Sèvres porcelain guéridon.

Guernsey. Add: **1.** Guernsey frock, shirt (earlier examples).

1832 *Chambers's Edin. Jrnl.* I. 307/2 Before any of his associates had entered the cabin, he had secreted his prize under his Guernsey frock. **1835** Dickens *Sk. Boz* II. 184 Two or three fellows in great rough trousers and Guernsey shirts.

2. a. (Earlier examples.) Also *Austral.*, a football jersey. So *to get* (or *draw*) *a guernsey*, to be selected (for a team); also *transf.*, to be successful.

1839 *Bell's Life* 16 June 3/5 The Etonian crew were dressed in white guernseys, with pale blue facings; rosette, sky blue. **1845** *Ainsworth's Mag.* VII. 499 An amateur boatman..in a Fez cap, striped Guernsey. **1963** *Times* 5 June 5/4 The flying gold guernsies of the Australians lit the gloomy arena with their explosive action. **1966** Baker *Austral. Lang.* (ed. 2) xix. 392 *Get a guernsey* (succeed or be selected, from football). *Ibid.* xxi. 428 Success: get (or draw) a guernsey. **1969** *Australian* 24 May 17/3 Goold..wears either a hacking jacket or a football guernsey with easy grace.

Guesdist (gē·dist). [f. name of Jules *Guesde* (1845–1922), French political leader + -ist.] A follower of the principles of revolutionary Marxism advocated by Guesde. So **Gue·sdism,** the policy or principles of the Guesdists.

1886 'Alb' *Living Paris & France* 337 The Marxists or Guesdists form the next considerable revolutionary party. **1901** *Daily Chron.* 25 May 5/4 The only true Socialists, the Guesdists. **1927** H. W. Laidler *Hist. Socialist Thought* xxii. 357 Guesdist's Revolutionary Program... The party of Guesde emphasized its revolutionary and Marxian character. **1940** D. W. Brogan *Devel. Mod. France* vi. iii. 295 Blanquism, Broussism, Anarchism..all divided with Guesdism the allegiance of the French workers. *Ibid.*, Electorally, the Possibilists naturally did better than the less practical Guesdists. **1966** Collins & Mitchell tr. *Braunthal's Hist. of International* xvi. 197 The Guesdists, at loggerheads with the Possibilists from the beginning, boycotted the congresses called by their rivals.

guess, *sb.* Add: **1.** *by guess and by God* (or *Godfrey*) (slang, orig. Naval): (to steer) at hazard without a set course or without the guidance of landmarks; *my guess is* or *it is my guess*: I am tolerably sure; *to miss one's guess* (U.S.), to be wrong in one's assumption; *you have another guess coming*: you are mistaken; *your guess is as good as mine*: a phrase used to indicate uncertainty about facts or circumstances or about the outcome of a set of events; *anybody's guess* (see *Anybody 3); *anyone's guess* (see *Any 8 c).

1909 [see *Godfrey[2]]. **1931** W. G. Carr (*title*) By Guess and by God. **1931** R. Graves *Poems 1926–30* 41 He lurches here and there by guess And God and hope and hopelessness. **1957** E. Whate *Press on Regardless* i. 10 We drove by guess and by God, and the rest of that journey is best left to the obscurity which shrouded it at the time.

1916 T. Wolfe *Let.* Sept. (1958) 4, I hope I will do well in all my studies and my guess is I'll have to 'bone' on math. **1930** J. C. Ransom *God without Thunder* xi. 231 It is my guess, in brief, that the scientists, with their Tables No. 2, hope to have, in the first place, a means of predicting Tables No. 1. [**1939** A. Derleth *Let.* 29 Dec. in *N. & Q.* (1965) July 268/2 My guess, naturally, is the latter—but I would value a letter from you setting down your own convictions.] **1958** I. Murdoch *Bell* xx. 246 My guess is that once you start to fight you'll know you can't stay with Paul. **1967** *Observer* 30 Apr. 29/3 Will bloomers and shorts catch on? My guess is that they will. **1921** *Sat. Rev.* 21 May 413/2 Governor Miller, who is responsible for the new legislation, seems to have missed his guess, if he means business. **1936** M. de la Roche *Whiteoak Harvest* xviii. 199 You're wot I calls a well-plucked 'un... *And*, w'en you're older, an overdose of sex appeal, or I miss my guess. **1965** F. L. Utley in Bessinger & Creed *Medieval & Linguistic Stud.* 308 If I don't go down into history as the martyr President I miss my guess. **1969** *New Yorker* 12 Apr. 121/1 It is not certain that there are any moonquakes, though Dr. Latham says he will miss his guess if the passive seismic experiment records none.

1935 *Punch* 3 July 12/2 If you think I am fool enough to be hoodwinked.., you have another guess coming. **1939** C. Day Lewis *Child of Misfortune* III. vi. 337 If you think that's your doing..you've got another guess coming.

1939 L. Baird *Waste Heritage* xxii. 310 Your guess is as good as mine. **1943** N. Balchin *Small Back Room* xii. 135 'The question is, what will happen then?'..'Your guess is as good as mine.' **1951** M. Kennedy *Lucy Carmichael* II. v. 130 'Who's going to get in?'..'Your guess is as good as mine.' **1967** M. Procter *Exercise Hoodwink* vi. 47 'Did he give you the impression of being honest.' Evans allowed himself a shrug. 'Your guess is as good as mine, sir.'

guess, *v.* Add: **6.** (Examples illustrating wider distribution.)

1916 A. Huxley *Let.* 21 Jan. (1969) 89 Say, kiddo, guess you're between the Devil and the W.C. **1927** M. de la Roche *Jalna* xv. 165, I guess I've just the right amount of brains for that. **1938** R. D. Finlayson *Brown Man's Burden* 46 Py korry she cost the Pakeha thousands and thousands of pounds I guess. **1946** K. Tennant *Lost Haven* (1968) I. 19, I guess it was a bandicoot, or else an ole swamp wallaby that died in the lonely scrub. **1959** M. Callaghan in R. Weaver *Canad. Short Stories* 2nd Ser. (1968) 5 There's no harm in you talking to Mother if you want to, I guess.

7. b. In phr. *to keep* (a person) *guessing*: to keep in a state of uncertainty. *colloq.* (orig. *U.S.*).

1896 Ade *Artie* v. 41 When your old college chum gets that letter it'll keep him guessin' where to begin on it. **1905** *Springfield* (Mass.) *Weekly Republ.* 24 Nov. 1 More doubt is now thrown upon the question whether he will go to Washington at all. The governor seems determined to keep us guessing. **1924** A. C. MacLaren *Cricket Old & New* xii. 116 A. G. Steel..putting the ball where he liked and keeping the batsman guessing all the time. [**1927** M. A. Noble *Those 'Ashes'* 175 Mailey..varied his pace splendidly and had everyone guessing.] **1930** H. Zink *City Bosses in U.S.* 23 Murphy proceeded with considerable caution, sometimes withdrawing from a position, sometimes forcing it, and altogether keeping his opponents guessing what he would do next. **1955** L. P. Hartley *Perfect Woman* xiii. 124 She always tries to keep me guessing.

guessable, *a.* (Earlier example.)

1828 Disraeli *Voyage Capt. Popanilla* iii. 16 My plots are not of that extremely guessable nature.

guessed-at, *ppl. a.* (Later examples.)

1917 J. Masefield *Lollingdon Downs* 61 The discoverer scans..the guessed-at satellite. **1930** Auden *Poems* 26 Areas..Whose guessed at wonders would be worth alleging.

guessing, *vbl. sb.* Add: *guessing game*, a parlour game in which much of the playing consists of guessing. Also *transf.*

1889 W. C. Brownell *French Traits* iii. 98 The French intelligence seems to have almost no frivolous side. The different varieties of mental arithmetic, guessing-games, puzzles, puns, spiritualism, theosophy, fanaticisms, have no attractions for it. **1936** E. S. Gardner *Case of Stuttering Bishop* (1937) viii. 122 If you want to waste your time playing guessing games with me, that's your funeral. **1938** D. Kincaid *Brit. Social Life in India* ix. 254 After dinner..there would be guessing games and competitions. **1941** E. Bowen *Look at all those Roses* 125 Juvenile parties..when guessing games could be played. **1955** I. Fleming *Moonraker* xiv. 139, I can't spend all the morning playing guessing games. **1966** 'R. Standish' *Widow Hack* xiv. 153, I played a little guessing game in an effort to divine in advance what Madeleine's next move would be. **1969** I. & P. Opie *Children's Games* ii. 108 Popular at the beginning of the century, was the guessing game 'Dead Men, Dark Scenery'.

guesstimate (ge·stimĕt), *sb.* orig. *U.S.* Also **guestimate.** [f. Guess *sb.* + Es)timate *sb.*] An estimate which is based on both guess-work and reasoning.

1936 *N.Y. Times* 22 Dec., 'Guesstimates' is the word frequently used by the statisticians and population experts. **1943** *N.Y. Times* 19 July 4/6 Many Americans.. think the Axis..will be defeated in less than two years. Intangibles..might make this popular 'guestimate' close to correct. **1948** *Jrnl. R. Aeronaut. Soc.* LII. 614/2 The former is a firm figure which arises in fact. The latter is an arbitrary 'guesstimate'. **1957** R. Watson-Watt *Three Steps to Victory* liii. 316 This mixture of estimates and guestimates comes out better than we deserve, for the Opana clock was very unlikely to agree with whatever clocks and watches were consulted in the aircraft and in Pearl Harbor. **1967** *New Scientist* 24 Aug. 373/2 Faced with engineering problems of this magnitude, £20 million is likely to prove as elusive a 'guesstimate' as the original figure for Concord. **1970** *Daily Tel.* 1 Oct. 19 (Advt.), £1000 tax free clear profit... This is proved performance —not an optimistic guesstimate.

Hence **gue·sstimate** (-ĕit) *v. trans.* and *intr.*, **guesstima·tion, gue·sstimator.**

1942 Berrey & Van den Bark *Amer. Thes. Slang* § 179.5 *Guesstimate*, blend of 'guess' and 'estimate'. **1943** in *Amer. Speech* (1944) XIX. 149/2 At no time in the past has so much depended upon the correct guesstimation of the tactical requirements of the future. **1948** *Elks Mag.* Oct. 11 One conservative guesstimator predicts 375,000 [television] sets in the country by January 1. **1950** *Jrnl. R. Aeronaut. Soc.* LIV. 175/1 50,000 lb...he had arrived at by a bit of 'guesstimating'. **1969** *Daily Tel.* (Colour Suppl.) 31 Jan. 32/3 The experts can 'guesstimate' how much gas is there [*sc.* under the North Sea], but there is no sure way of telling.

guest, *sb.* Add: **1. c.** Phr. *be my guest*: you are welcome to (something); do as you wish.

1955 *Amer. Speech* XXX. 302 *Be my guest*, used when someone asks for something: 'Join me' or 'Go right ahead'. **1962** D. Mayo *Island of Sin* xiii. 118 Rum? Scotch? Bourbon? Be my guest. **1965** *New Statesman* 14 May 760/3 'Did you say something, man?' the face asked. I took a deep breath..and then I address the face. 'Not a thing, brother,' I reply, 'not a thing. Be my guest.' **1967** J. Gardner *Madrigal* i. 18 Mostyn gestured towards the telephone. 'Be my guest,' said the Chief. **1968** D. Francis *Forfeit* iv. 55 'Goodnight, honey.' 'Thanks for everything.' 'Be my guest.' **1969** B. Malamud *Pictures Fidelman* 159 'Do you mind if I get on top?' she asked. 'It's hard to breathe since I had my illness.' 'Be my guest.'

3. b. An occasional performer, one not belonging to the regular company in a theatre, etc.

1900 W. A. Ellis tr. *Glasenapp's Life R. Wagner* I. vii. 259 She..appeared four times as 'guest' at the theatre.., playing Preziosa, Maria Stuart, [etc.]. **1901** *Athenæum* 27 Apr. 539/2 In addition to the artists of the Hofoper, 'guests' from Frankfort, Wiesbaden,..and Vienna will take part in the performances.

6. a. *guest-ranch, register.* **b.** (sense *3 b) *guest actor, actress, artist* (so as v. intr.), *conductor* (so, as a back-formation, *guest-conduct* vb.), *critic, producer, soloist, speaker, star* (so as v. intr.). Also **guest-line** = Guest-rope; **guest-night** (earlier and later examples); **guest-size** *a.*, of a size (usually smaller than the 'regular' size) suitable for a guest; **guest-towel,** a small hand-towel intended for visitors' use.

1961 Bowman & Ball *Theatre Lang.* 164 *Guest actor*, a distinguished actor playing temporarily with a repertory or stock company. **1910** *Busy Man's Mag.* Jan. 26 (*caption*) Miss George has abandoned her regular season to become 'guest actress' at the New Theatre. **1934** A. Haskell *Balletomania* ix. 183 She appeared for him again many times, but only as a guest-artist during the London seasons. **1942** Berrey & Van den Bark *Amer. Thes. Slang* § 619.7 *Guest-artist*, to appear as a guest performer. **1959** *Listener* 19 Mar. 512/3 A guest artist from the Royal Swedish Ballet. **1945** *Time* 16 Apr. 58/3 He will guest-conduct the Minneapolis Symphony. **1933** *Sat. Even. Post* 7 Jan. 15/3 (*caption*) William Mengelberg, who has served as guest conductor for many of the leading philharmonic orchestras. **1968** *London Symphony Orch. Programme Bk.* 7 Mar. 3 At this period he began to make appearances as a guest conductor with most of the important orchestras in Europe. **1925** *New Yorker* 14 Mar. 20/2 Ernest Newman, guest critic of the *Evening Post*, packs up his troubles and returns to London. **1926** *Blackw. Mag.* May 624/2 As the *Carmania* came down she took the force of wind and sea off the boat so that those in her were able to grab the guest-lines and hold the boat against the ship's sides. **1885** A. Edwards *Girton Girl* I. xiv. 283 The usual guest-night at mess. **1968** *Times* 11 Dec. 13/8 Guest night in the officers' mess and the bellicosity of sergeants. **1958** *Times* 7 Oct. 5/5 The Belgrade Theatre director's hope to have guest-producers working in Coventry. **1932** *Bull. Arizona Agric. Exper. Station* CXLI. 2 Unclassified. Land owned and operated by the State or a public agency, guest ranches, private estates, forest products. **1968** *Punch* 31 Jan. 174/3 A ritzy guest-ranch overlooking Santa Barbara. **1926** E. Hemingway *Torrents of Spring* (1933) 58, I..demanded to see the guest-register. **1928** *Daily Mail* 25 July 16/6 The sample box containing two guest-size tablets of Erasmic. **1942** 'R. Crompton' *William carries On* i. 38 Lemon soap. Guest size. **1951** in M. McLuhan *Mech. Bride* 47/2 We'll send you a guest-size tube absolutely free. **1962** *Melody Maker* 7 July 10/4 The full..bands have already agreed to appear for no fee, as have guest soloists Terry Lightfoot..and Mike Cotton. **1959** *Listener* 26 Mar. 569/1 The debate..at which Professor A. J. Ayer was the guest speaker. **1942** Berrey & Van den Bark *Amer. Thes. Slang* § 619.7 *Guest-star*, to appear as a guest performer. **1951** L. Hobson *Celebrity* (1953) xiii. 206 Kitterley had begun to guest-star on just about every radio programme on the air. **1957** Webster, Guest artist or guest star. **1921** *Daily Colonist* (Victoria, B.C.) 6 Apr. 4/5 (Advt.), Huckaback Guest Towels in a very fine quality; offered at a snap price. **1922** S. Lewis *Babbitt* i. 6 He wiped his face on the guest-towel.., a pansy-embroidered trifle. **1967** 'K. O'Hara' *Unknown Man* vi. 55 Giving..attention..to the match of James's guest-towels and the serving of James's dinners.

guest, *v.* Add: **3.** *intr.* To appear as a guest or as a guest artist, etc. orig. *U.S.*

1936 H. L. Mencken *Amer. Lang.* (ed. 4) xi. ii. 586 *To guest*, to appear as a guest. **1942** Berrey & Van den Bark *Amer. Thes. Slang* § 593.18 *Guest*, to appear as a guest artist. **1943** in *Amer. Speech* (1944) XIX. 102 Many stars guest on the program. **1959** *Spectator* 3 July 9/1 Stars from one company 'guesting' with the other. **1965** *Melody Maker* 10 July 10/4 Ian Carr guested with the Roy Budd Trio..on Saturday.

guest house. Add: **3.** A house for the reception of paying guests.

1925 *Daily Tel.* 13 May 19/4 For Sale, Guest House, South London. **1953** E. Taylor *Sleeping Beauty* i. 11 That house on the cliff is a guest-house, is it?

guestimate, var. *Guesstimate.

gufa (gu·fā). Also **gufar.** [Arab., f. *quffa* basket.] A round boat, made from straw and palm branches, found in Mesopotamia since ancient times.

1914 R. Meinertzhagen *Army Diary* 30 Jan. (1960) 65 Fortunately we were surrounded by gufas into which we quickly transferred our belongings and ourselves. **1919** *Chambers's Jrnl.* Jan. 23/2 Moored behind the *mahailas* are two strings of the up-river *gufars*. **1925** *Countries of World* IV. 2729/2 The 'gufas', the round cauldron-shaped coracles of the Tigris and Euphrates are the oldest river-craft in existence. **1951** A. Christie *They came to Baghdad* xii. 111 A *gufa*, the primitive craft of the Tigris.

guff[2]. **2.** For ? *U.S.* read orig. *U.S.* and add later examples.

a **1911** D. G. Phillips *Susan Lenox* (1917) II. vi. 161 Don't listen to their guff about wanting to see you again. ..There's nothing in it. **1930** G. B. Shaw *Apple Cart* II. 78 You cant run them [*sc.* power stations] on patriotic songs..and guff and bugaboo. **1936** 'F. O'Connor' *Bones of Contention* 187, I never thought..that my Tom had such ould guff in him. **1959** *Encounter* Aug. 64/1 All that guff in *Look Back in Anger* about how there are no causes left to die for. **1966** J. Dos Passos *Best of Times* (1968) iv. 144 Whether it was politics or literary work or painting he [*sc.* Hemingway] would take the guff out of a situation with a wellplaced fourletter word. **1966** *Crescendo* Jan. 32/1 The sleeve-notes give us a lot of guff about getting with it and so on and tell us nothing constructive.

guga (gū·gā). *Sc.* Also **goug.** [Gael.] A young gannet.

c **1840** N. Mackenzie in J. B. Mackenzie *Episode Life Rev. N. Mackenzie at St. Kilda* (1911) iii. 47 When it [*sc.* the young gannet] is about a month old it is called 'guga', and looks like a young white lamb. **1897** H. A. Macpherson *Hist. Fowling* xxii. 207 Before the young Solan-Geese, which they call *Goug*, fly off, they are larger than the mothers, and excessively fat. **1938** *Brit. Birds* XXXI. 292 A party of men..undertake the unpleasant voyage to Sula Sgeir and stay there a few days, taking all the gugas, or nestling Gannets, they can. **1953** *Scotsman* 13 Aug., Eight Ness men..are spending 14 days on the uninhabited rocky islet of Sulisgeir,..killing and salting the young of the gannet known locally as the 'guga'.

‖**gugelhupf** (gū·g'lhupf). Also **guglhupf, kugelhupf.** [Ger. dial.] A light Austrian cake baked in a ring-shaped mould. Also *attrib.*

1886 Kugelhupf [see *Centre *v.* 4]. **1936** M. Morphy *Recipes of All Nations* 316 *Wiener Gugelhupf*... This cake is also made in Alsace and Germany, but the Viennese claim it as one of their national cakes. **1954** G. Beer *Austrian Cooking* 163 A *Guglhupf* really does taste better when baked in its proper 'setting'. *Ibid.*, On the Continent *Guglhupf* forms are generally available in copper. **1965** M. Stewart *Airs above Ground* xi. 141 You wouldn't say no to some really nice *Gugelhupf*, would you? **1967** *House & Garden* Apr. 85/3 A *gugelhupf* mould (a deep ring mould for making a traditional Austrian sponge-like cake in).

Guggenheim (gū·gĕnhəim, gu·g-). [f. the John Simon *Guggenheim* Memorial Foundation, established in 1925 by Simon Guggenheim (1867–1941), American senator.] Designating a grant for advanced study and research provided by the Guggenheim Foundation. Also *ellipt.*

1930 H. Crane *Let.* 21 Nov. (1965) 358, I have applied for a Guggenheim Scholarship (which would give me a year's study and creative freedom abroad). **1931** *Ibid.* 379 My father's will left modest provisions for me, but they are as good as an annual Guggenheim, anyway. **1934** H. Miller *Tropic of Cancer* (1948) 220 Artists who had a little money, Guggenheim prize men, etc. **1956** O. Nash *Good Intentions* 74 He had won a Nobel award and a Pulitzer Prize, A Guggenheim and a leg on the Davis Cup. **1958** *Times Lit. Suppl.* 24 Jan. 37/4 Has not 'He's on his fourth Guggenheim' become part of the life story of many a quondam rebel or of his nominal successors?

Guianese (giăni·z), *a.* and *sb.* [f. *Guiana* + -ese.] **A.** *adj.* Of or pertaining to Guiana, a tropical region in north-eastern South America. **B.** *sb.* An inhabitant of the region of Guiana; also in collective sense. So **Guianan** (giā·năn), *a.*; **Guia·nian** *a.* and *sb.* (See also *Guyanese *a.* and *sb.*)

1600 Raleigh *Discoverie of Guiana* in Hakluyt *Voy.* III. 636 Those Guianians..are marvellous great drunkards. *Ibid.* 641 The poore Guianian, betrayed on all sides, was delivered to the campe-master of Berreo. **1869** *Month* Feb. 191 The chief varieties of native Guianese now seen are the Arawaks, Acawoios, Waraus, and the Caribs. **1879** J. G. Wood in C. Waterton *Wand. S. Amer.* (ed. 6) 478 The European and Guianan types of the human frame. *Ibid.* 479 The Guianan type of female beauty. **1881** C. D. Dance (*title*) Chapters from a Guianese log-book. **1883** H. V. P. Bronkhurst *Colony of Brit. Guyana* p. vii, My numerous English and West Indian and Guyanian friends. **1888** Rodway & Watt *Chron. Hist. Guiana* i. 11 Martinez sought for and obtained permission to depart, a number of Guianians being deputed to conduct him to the Orinoco. **1896** J. Rodway *West Indies & Spanish Main* 44 These were Guianians, who had never before seen a white man. **1954** G. Durrell *3 Singles to Adv.* ii. 43 A Guianian sand reef may be of great botanical and zoological interest, but it is a most exhausting place to cross in a hurry. **1960** *Guardian* 9 Apr. 2/1 Guianese leaders complain of threats at London conference. **1964** M. Dickson *World Elsewhere* iv. 147 The Guianese are naturally hospitable. *Ibid.* 148 The Guianese staff at the school.

‖**guichet** (giʃe). [Fr.] A wicket, grating, or hatch, *spec.* one through which tickets are issued.

1839 Thackeray in *Corsair* (N.Y.) 26 Oct. 521/2 The porter closes the *guichet* behind us. **1848** H. Greville *Diary* (1883) 280 Hundreds of prisoners have been shut up. In the Church of L'Assomption..they are fed through a guichet. **1907** *Westm. Gaz.* 22 July 3/1 The *Bigliettaio*, ducking his gold-braided cap through the guichet, triumphantly gave us a ticket. **1926** 'C. Barry' *Detective's Holiday* 131 The post-mistress herself came to the guichet. **1931** H. G. Wells *Work, Wealth & Happiness of Mankind* (1932) ix. 381 Within are the familiar bank desks and counters with their brass rails and guichets. **1961** *Times* 6 Nov. 11/4 Not everybody is a *guichet* user. Some there are so insulated from the bustle of life that they would not go to a theatre or a football match unless someone provided them with a ticket in advance.

Guickwar, Guicowar, varr. *Gaekwar.

guidance. Add: **1. a.** spec. in *Astronautics.*

1962 F. I. Ordway et al. *Basic Astronautics* p. vii, High-thrust space carrier vehicles with accurate guidance systems. **1969** *Sunday Times* 13 July 13/2 *Guidance system*, the system which measures and evaluates flight information, correlates this with target data, converts the results into the conditions necessary to achieve the desired flight path and communicates this data in the form of commands to the flight control system. **1970** *Guardian* 18 Apr. 1/1 Guidance control officers reported ..that it had gone well, putting the spacecraft within a hundredth of the 6.5 re-entry angle.

c. With reference to problems and advice relating to education, marriage, careers, etc. Also *attrib.*

1927, etc. [see *child guidance*, *Child *sb.* 22]. **1935**, etc. [see *marriage guidance*, *Marriage 8]. **1947** Erickson & Smith *Organ. & Admin. Guidance Services* viii. 207 The administrator should recognize the necessity for beginning the guidance program at a point consistent with its existing stage of development. The nature and amount of guidance training present among staff members and the facilities that the school has for proceeding with the extension and refinement of guidance services must be considered. **1951** M. McLuhan *Mech. Bride* 42/1 Vocational-guidance investigation has turned up the curious fact that executives..do show an aptitude for words. **1957** *Encycl. Brit.* XII. 295/2 The primary problem of vocational guidance is, given a person, to find the most suitable occupation. **1967** *Medium Ævum* XXXVI. 144 Benoit and Guido..characterized him [*sc.* Calkas] as the guidance counsellor. **1968** *Lancet* 9 Mar. 518/2, 41% [of medical students] acknowledged that they had been given guidance about sexual difficulties unrelated to marriage.

guide, *sb.* Add: **2. d.** (Usu. with capital initial.) A girl aged between about 10 and 16 who is a member of the Girl Guides Association, an organization of girls, established in 1910, corresponding to the Boy Scouts. In full (formerly) *Girl Guide*. Also *attrib.* and *Comb.*, as *Guide camp, -mistress*, etc.

1909 *Boy Scouts' Headquarters' Gaz.* Nov. 12/1 (*heading*) The scheme for 'Girl Guides'. *Ibid.* 12/2 Each Guide must be able to describe the..flower..and must be able to draw its outline. *Ibid.* 13/1 Where it is desired to start 'Girl Guides' it would be best for ladies interested to form a Committee. **1916** [see *Brownie[1] 2]. **1916** *Daily Colonist* (Victoria, B.C.) 23 July 4/5 While he was in that part of the grounds he happened to notice Guide-mistress Miss Leighton in the crowd. He greeted her again, telling her how well he thought the Girl Guides had marched. **1924** A. D. Sedgwick *Little French Girl* I. v, Alix heard of a Women's Institute, of Boy Scouts and Girl Guides. **1932** H. Nicolson *Public Faces* viii. 213 She entered, swinging her hips a little with what Shorland had once called her girl-guide gait. **1941** J. S. Huxley *Uniqueness of Man* xii. 253 A girl-guide organizer. **1962** *Guardian* 21 Mar. 6/7 The elder children will..go to Guide camps. **1969** *Policy, Organisation & Rules of Girl Guides Assoc.* (ed. 33) 40 A Brownie Guide, Guide, or Ranger Guide may belong to only one Unit but may be attached to another. **1971** *Guider* Oct. 363 May I please make a further appeal..for all Guiders to make sure that their Brownies, Guides and Rangers know all about their respective section magazines.

Hence **Guider,** an adult leader in the Girl Guide movement (in full formerly *Girl Guider*; now *Brownie Guider, Guide Guider,* or *Ranger Guider*); **Guide's honour,** the oath taken by a Girl Guide, used as a protestation of honour and sincerity; also *transf.* in jocular use; **girl-guidey, -guidish** *adjs.*, pertaining to, or typical of, a Girl Guide; **guiding** *vbl. sb.*, the activities practised by Girl Guides (cf. Scouting *vbl. sb.*[1] 1 b).

1912 A. Baden-Powell *Handbk. Girl Guides* i. 38 A Guide's Honour is to be trusted. If a Guide says 'On my honour it is so', that means that it *is* so just as if she had taken a most solemn oath. **1924** A. Kindersley *Guiding Bk.* 7 It is these things, then, that poets have always sung and writers celebrated, and it is to them once more that we have looked..to help with this true explanation of Guiding. **1931** E. M. R. Burgess *Girl Guide Bk. Ideas* 3 It is an inspiring thought that throughout the British Empire there are Guides and Guiders like ourselves. **1944** *Times* 5 May 7/2 Guiders, Rangers, Guides and Brownies continued to render valuable war service. **1953** K. Tennant *Joyful Condemned* xv. 135 If there was a fat, thumping falsehood to be propped up, 'Guide's Honour' was always promptly evoked. **1959** Guide's honour [see *Brownie[1] 2]. **1960** J. Betjeman *Summoned by Bells* iv. 37 She came, a woman of the open air, Swarthy and in Girl-Guide-y sort of clothes. **1962** A. Lejeune *Duel in Shadows* vii. 96 'You must keep this to yourself.' 'My dear man,' she protested. 'Professional ethics. Guide's honour.' **1963** *Listener* 10 Jan. 96/3 If occasionally girl-guidish, she has a sharp eye. **1963** I. Fleming *On H.M. Secret Service* x. 106 The girls all seemed to share a certain basic, girl-guidish simplicity. **1966** *Listener* 20 Oct. 571/3 They see Scouting (and Guiding) as a unique medium for this 'social education'. **1968** *Even. Standard* 23 Oct. 15 (*heading*) Former Girl Guider jailed for contempt. **1969** *Policy, Organisation & Rules of Girl Guides Assoc.* (ed. 33) 42 The Guide Guider and the Assistant Guide Guider are the adult leaders of the Company. *Ibid.*, All Units..are known officially as Ranger Guide Service Units, their members as Ranger Guides, and their Guiders as Ranger Guiders.

3. c. *Spiritualism.* = *Control *sb.* 4 b.

1856 *Spiritual Herald* July 191 Annie passed into the trance, and said: I can see my guide now, my own guide. **1885** *Cent. Mag.* XXX. 381/2, I can't seem to do anything in these days, now that I no longer have a Guide. **1957** J. S. Huxley *Relig. without Revelation* i. 19 The so-called 'controls' or 'guides' of mediums... It is considered *de rigueur* for a professional medium to be under control by a spirit guide.

10*. *Electr.* A linear structure (as a pair of wires or a waveguide) or a surface along or over which an electromagnetic wave is propagated and to which it is confined. Now usu. short for *Waveguide.

1893 O. Heaviside *Electromagnetic Theory* I. iv. 399 When waves are left to themselves in ether without the presence of conductors, they expand and dissipate themselves... To prevent this we require conducting guides or leads..as a pair of parallel wires. **1902** *Encycl. Brit.* XXXIII. 215/1 The effects of the resistance of the guides are very complicated in general. **1936** *Bell Syst. Techn. Jrnl.* XV. 284 Extremely high-frequency waves may be transmitted from one point to another, through specially constructed wave guides. The guide..may be a hollow copper pipe. **1947** W. H. Watson *Physical Princ. Wave Guide Transmission* p. v, A dielectric cylinder may also be used to guide electromagnetic waves of sufficiently high frequency and is usually called a dielectric guide. **1962** Corson & Lorrain *Introd. Electromagnetic Waves* xii. 424 The guide wave length..is longer than the free space wave length.

12. *guide-face, -frame, -framing, -groove.*

1900 P. N. Hasluck *Mod. Eng. Handy-bk.* 59 The guide-faces may be got up with a file. **1901** P. Marshall *Metal Working Tools* 44 A rising and falling guide-frame. **1936** Guide-frame [see *all-welded* adj. (*All- III. 13)]. **1900** *Engineering Mag.* XIX. 781/1 The Guide-Framing of Gasholders. **1903** *Brit. & Col. Printer* 19 Nov. 12/2 The ends slide in parallel guide-grooves.

13. guide-board (earlier U.S. example); **guide card** (see quot. 1923); **guide coat** (see quot. 1953); **guide dog,** a dog trained to lead the blind; **guide fossil** (see quots.); **guide letter** (see quot. 1960); **guide-line,** (*a*) a line used as a guide, a guiding line; also *fig.*; (*b*) = **guide-rope* (*c*); **guide mill** (later example); **guide number** *Photogr.* (see quot. 1962); **guide-rope,** (*c*) *Aeronautics*, a long rope hung from a balloon or small airship so as to trail along the ground and to preserve altitude automatically by the drag of the rope without loss of ballast or gas; also, one of a number of ropes used to steady an airship before flight; hence **guide-rope** *v. intr.*, to use a guide-rope; **guide vane** *Aeronaut.* (see quot. 1962); **guide-way** (earlier and later examples); **guide-wheel,** a wheel used to guide a moving structure or vehicle.

1810 M. V. H. Dwight *Journey to Ohio* 27 Oct. (1912) 13 Soon after we cross'd the mountain, we took a wrong road, owing to the neglect of those whose duty it is to erect guide boards. **1923** H. A. Maddox *Dict. Stationery* 35 *Guide cards*, a segment of the card index and vertical filing system. Guide cards, or guiders, are stiff manilla cards, usually in distinctive colourings and provided with projecting tabs..and the purpose of the cards is to act as dividers, or indicators. **1969** R. L. Collison *Indexes & Indexing* (ed. 3) i. 27 In the case of 5 in. by 3 in. cards, suitably printed guide cards of this description can be purchased,..and there are also larger sets of guide cards which can break up the sequence into any number of parts. **1930** *Motor Body Building* May 105/1 *Guide coat*, the staining coat put on top of the filling as a guide to rubbing down. **1937** *Times* 13 Apr. p. xiii/1 The various stages.., therefore, are as follows:—First coat of primer, or foundation coat; first, second, third, and fourth coat of oil filler; guide coat for rubbing down; oil filler, [etc.]. **1953** *Gloss. Paint Terms* (*B.S.I.*) 8 *Guide coat*, a very thin coat of loosely bound paint applied over a continuous coating of surfacer or filler, prior to rubbing down... It serves as a guide to the operator in producing a smooth surface. **1932** *Proc. World Conference on Work for Blind 1931* 183 (*caption*) Guide dogs for the blind. **1944** D. Hartwell *Dogs against Darkness* 7 The modern guide dog provided by the Guide Dogs for the Blind Association is of the very greatest use. *Ibid.* 161 The majority of Guide Dogs in England are now types of British sheep-dogs. **1960** *News Chron.* 13 Jan. 5/6 Margaret Barber, 21, blind since she was six, was promised a guide dog. **1909** *Cent. Dict.* Suppl., *Guide-fossil*, a fossil species regarded as specially characteristic of a given geological formation, horizon, or fauna. **1914** Schuchert & Barrell in *Amer. Jrnl. Sci.* 4th Ser. XXXVIII. 6 In the chronologic correlation of the stratified rocks most dependence is put upon a few species, known as 'guide fossils', together with the collateral evidence of associated forms. These guide fossils may be represented by many or few individuals. **1957** Dunbar & Rodgers *Princ. Stratigr.* xvi. 279/1 A good guide fossil should have relatively wide geographic distribution and limited stratigraphic range. **1961** J. Challinor *Dict. Geol.* 97/1 *Guide fossil*, a species (or genus) of fossil useful as a guide to stratigraphical horizon or to the conditions under which the organism lived. **1931** A. F. Johnson *Decorative Initial Letters* 6 Zainer's first outline initials were probably intended as guide-letters for the rubricator. **1960** G. A. Glaister *Gloss. Bk.* 162 *Guide letters*, small letters inserted in the otherwise blank space left for an illuminated, historiated, or rubricated capital to be executed by hand after a work had been printed: a feature of early printed books and still earlier manuscripts. **1785** A. Ellicott in C. V. Mathews *A. Ellicott* (1908) 44 My Brother Joseph at Present runs the guide Line for the choppers. **1846** *Illustr. London News* 30 May 356/1 The difficulty,..of getting a fulcrum.. without which all attempts at steering an aëronaut machine must be failures, Mr. Green meets by the adoption of the guide-line, as he calls a rope about 1000 feet long. **1867** *All Year Round* XVIII. 452/1 In hauling in his guide-line, lest it should entangle itself with a factory chimney. **1881** Guide-line [see Guide *sb.* 12]. **1931** *Times* 27 Aug. 13/6 The coast..serves as a guide-line to their destinations. **1948** A. Toynbee *Civilization on Trial* 4 The national history of England itself as the principal guide-line. **1952** A. Grimble *Pattern of Islands* 161 Tekirei and Mautake first drew guide-lines on my arms with stretched strings, which they dipped in their tattooing dye. **1963** *Listener* 7 Mar. 404/2 Today, in an age of religious doubt, the moral guide-lines are less obvious than they were. **1964** I. L. Horowitz *New Sociol.* 37 The past offers guidelines (not mandates) to the present and future. **1904** J. W. Hall in Harbord & Hall *Metallurgy of Steel* xvi. 311 The term 'guide mill' is reserved for mills in which the bar could not be properly entered with the workmen's tongs, and must have a guide to hold the bar on edge when it is being entered. **1948** A. L. M. Sowerby *Dict. Photogr.* (ed. 17) 317 s.v. *Flash bulbs*, The figures given are guide-numbers; by dividing these by the distance..between object and flash the stop to be used is given directly. **1959** L. A. Mannheim *Successful Flash Photogr.* 44 Guide numbers vary with film speed. **1962** M. L. Haselgrove *Photographer's Dict.* 120 Manufacturers of flash bulbs specify a guide number for each bulb to be used in conjunction with a given shutter speed and given film. This..simplifies the matter of determining what aperture to use to obtain the correct exposure for any given subject at a known distance from the flash bulb. **1838** M. Mason *Aeronautica* 23 An incident connected with the use of this guide-rope. **1848** *Chambers's Jrnl.* 6 May 301/2 Mr. Green as a substitute, uses a long rope, called the 'guide-rope'. **1897** *Strand Mag.* XIII. 227/2 One rope (the guide-rope) is securely tied to this crow-bar, and then thrown on the cliff. **1903** *Ibid.* 27 June 476/2 A life-saving kite. The kite carries a guide-rope. **1904** *Pall Mall Mag.* XXXII. 20 One can guide-rope in the centre of Paris. **1905** *Spectator* 11 Mar. 371/1, I hopped over the trees of the Bois..and guide-roped down the Avenue des Champs Elysées to my door at the corner of the Rue Washington. **1928** *Daily Express* 12 Oct. 2/2 Four hundred sinewy fists released their hold on the guide-ropes that still leashed the mammoth [Zeppelin] to earth. **1941** *Illustr. London News* 28 June 827/1 (*caption*) At Langley Field, Virginia, U.S.A., where these powerful guide-vanes direct the air in pressure wind-tunnels. **1962** *Gloss. Aeronaut. Terms* (*B.S.I.*) iv. 15 *Guide vanes*, a cascade of fixed vanes which guide the fluid stream round the bends in the passages of a wind tunnel. **1876** J. S. Ingram *Centenn. Exposition* ix. 270 Both the upper and lower guide-ways were adjustable for keeping the saw in line whenever required by the settling of floors. **1966** *Electronics* 14 Nov. 16 Seifert will analyze such problems as propulsion, guideway, communications and control. **1969** *Times* 16 Apr. 11/6 The dual mode vehicle..is driven normally at either end of its journey, but..automation takes over once it runs on the 'automated guideway'. *Ibid.*, The automated guideway system has been under study..for some 18 months. **1840** *Mag. Sci.* 29 Aug. 173/2 Have it [*sc.* a tricycle] upon three wheels, the first of them a guide wheel. **1847** *Patent Jrnl.* III. 471/1 The next feature is the provision of safety guide-wheels [on a locomotive engine]. **1885** *Marine Engineer* 1 July 91/2 The guide-wheel supports the 'bag' of the bucket chain. **1909** *Westm. Gaz.* 22 July 5/2 The pilot must be careful to hold the guide-wheel so as to maintain this position.

guide, *v.* Add: **1.** Also *refl.*

1865 M. C. Harris *Christine* iii, Moving cautiously upon the ice..he..lay down, guiding himself by his hands alone. **1888** Mrs. H. Ward *Robt. Elsmere* I. viii. 231 And refusing all help, she guided herself out of the room. **1907** *Smart Set* Mar. 128 He guided himself cautiously with his left arm stretched out against the object of quest.

d. *Electr.* To serve as a guide for (an electromagnetic wave); so *guided wave*. Cf. *Guide *sb.* 10*.

1893 O. Heaviside *Electromagnetic Theory* I. iv. 368 If we abolish the fictitious magnetic conductivity throughout the medium traversed by the waves we should, to have distortionless transmission, also abolish the electric conductivity. This is only to be attained by using wires of no resistance to guide the waves through a non-conducting medium. **1902** *Encycl. Brit.* XXXIII. 214/2 The use of conductors will now be seen partly. They serve to *guide* a wave along from place to place without loss, and with a limited amount of energy. **1925** A. M. Morse *Radio: Beam & Broadcast* iii. 66 [He] enunciated the theory that a 'Heaviside Layer' was unnecessary, since all radio waves were guided by the earth, which was the real conductor; just as was the wire, in guided-wave telephony or telegraphy. **1936** *Bell Syst. Techn. Jrnl.* XV. 285 It is possible to reduce the size of the guiding structure for a given frequency by the use of a suitable dielectric. *Ibid.* 312 In the ordinary type of guided-wave systems..the guiding conductors form two sides of a circuit in which equal and opposite currents flow. **1947** [see *Guide *sb.* 10*]. **1962** Corson & Lorrain *Introd. Electromagnetic Waves* xii. 429 In the process of guiding electromagnetic waves conductors waste part of the wave energy in the form of Joule losses. This is because the guided waves always induce electric currents in the conductors.

2. b. *refl.*

1816 Scott *Old Mort.* xxxix, He..formed the resolution of guiding himself by the circumstances in which he might discover the object of his quest. **1874** J. W. Draper *Hist. Conflict Relig. & Sci.* 136 He guides himself by past as well as by present impressions.

guide-book. Add to def.: Also *fig.* (Earlier and later examples.)

1814 J. Mayne *Jrnl.* 25 Oct. (1909) x. 164 At Mr. Vasi's we purchased his guide-book, which is considered the best of its kind. *a* **1846** B. R. Haydon *Autobiogr.* (1927) ix. 135, I mean my Life, if possible, to be a guide book to youth. **1866** J. Macgregor *Thousand Miles in Rob Roy Canoe* i. 8 In his guide-book route,..from town to town. **1935** T. S. Eliot *Murder in Cathedral* ii. 86 The sanctity shall not depart from it..though sightseers come with guide-books looking over it.

guide-booky (gəi·dbu:ki), *a.* [f. Guide-book + -y[1].] Resembling or characteristic of (that of) a guide-book; having the style of a guide-book. Also **gui·de-boo:kishly** *adv.*

1878 A. Mackennal in D. Macfadyen *Life & Lett.* (1905) 72 This talk is a little guide-booky. **1883** W. H. Russell *Hesperothen* I. 165 There are, I am told, nearly one hundred churches—to be guide-bookishly accurate, ninety-six—in Cleveland. **1893** *Academy* 25 Mar. 251/3 The guide-booky lines on p. 222 of the same volume.

guided (gəi·dèd), *a.* [f. Guide *sb.* + -ed[1].] Accompanied by a guide, having a guide in charge.

1909 *Daily Chron.* 6 Sept. 4/6 On a guided ascent. **1969** *New Yorker* 30 Aug. 30/1 There is a very good museum.. There are guided tours. **1970** *Ibid.* 12 Sept. 31/1 We were taken to see it [*sc.* Lenin's body] as part of a guided tour.

b. Of weapons: operating by remote control or as directed by equipment carried in the weapon, as *guided bomb, missile, rocket, weapon.*

1945 *Newsweek* 27 Aug. 25/1 The 'guided missile'—a rocket projectile that can be aimed accurately over great distances. **1948** J. Steinbeck *Russ. Jrnl.* (1949) i. 3 Experiments with atomic weapons and guided missiles. **1952** *Ann. Reg. 1951* 96 United Kingdom experts attended tests of rockets and guided weapons at the Woomera range. **1954** *Economist* 11 Sept. 12/3 Guided, containing electronic control mechanisms, designed to seek out the target. Capable of altering course during flight. **1954** K. W. Gatland *Devel. Guided Missile* (ed. 2) v. 130 A guided bomb of this type need have no inherent propulsion unit. It will leave the bomber in the normal way.. The task of the bomb-aimer will then be to guide the weapon by remote radio-control. **1954** *Amer. Rev.* **1955** i. 81 In the field of armament research.. work on guided rockets had been intensified. **1955** *Times* 30 June 7/4 Probably few people know that the first guided missile was designed and made in Great Britain. It was the invention of Mr. A. M. Low, developed by him in the course of the war of 1914–18. **1956** *Nature* 7 Apr. 658/1 The first guided-weapons trial ship should be in operation next summer. **1958** *Daily Mail* 30 June 2/6 The guided-missile ship.. will be in Greek or Dodecanese ports. **1958** 'N. Blake' *Penknife in my Heart* i. 20 Swift and feared as a volley of guided missiles out of control. **1965** M. Spark *Mandelbaum Gate* i. 12 The State of Israel had that day sent up its first guided rocket.

guidee (gəidī·). [f. Guide *v.* + -ee[1].] One who is guided.

1923 *Chambers's Jrnl.* Apr. 250/2 A 'guide'.. who only started duties after the 'guidee' had made himself acquainted with his guider. **1934** J. B. Priestley *Eng. Journey* 220, I am probably too conceited and dogmatic a person myself to make a good guidee. To be at ease, I ought to be doing the guiding.

guide-post. (Earlier and later examples.)

1761 L. Sterne *Tr. Shandy* III. xxiii. 90 Wherever.. three several roads meet.. set up a guide-post in the centre of them. **1940** [see *audience 7 d]. **1967** *Times Rev. Industry* Feb. 16/1 Our prices and incomes board is an attempt to apply the experience of the USA in using guide posts to influence increases in wages and prices.

guidguid (gwi·d,gwid). Also hued-hued, huet-huet. [Probably echoic: cf. Guit-guit.] A ground-dwelling South American bird, *Pteroptochos tarnii*, belonging to the tapaculo group.

1845 [see Barking *ppl. a.*[1] 2 b]. **1912** Brabourne & Chubb *Birds S. Amer.* I. 183 Hylactes.. tarnii.. Tarn's Barking-bird. Hued-hued; Guid-guid. **1932** C. E. Hellmayr *Birds of Chile* 215 The 'Hued-hued' is a common bird in southern Chile.. and ranges across the Andes to the extreme western section of Rio Negro and Chubut. **1964** A. L. Thomson *New Dict. Birds* 806/1 The Huet-huet *Pteroptochos tarnii* has the upper parts and forenck very dark brown.

guiding, *ppl. a.* (Further examples.)

1856 Bagehot *Coll. Works* (1965) I. 362 When we had learned the great landmarks, understood the guiding-stars, we might amuse ourselves with small points. **1960** A. O. J. Cockshut *Imagination of Charles Dickens* x. 141 Here Dickens's sense of the superiority of life to fact, which is the guiding star of the novel, up to this point, seems ironically to have deserted him. **1962** *Daily Tel.* 11 Jan. 1/1 'Guiding light' on pay rises. *Ibid.*, A 'guiding light' figure, about 2½ per cent., which unions and employers would be expected to accept.

c. *guiding telescope*: a visual telescope fixed rigidly to a photographic telescope so that the latter may be 'guided' manually to follow the course of a star, etc., kept under observation during an exposure.

1867 A. H. Miles *Concise Knowl. Astron.* ii. xv. 195 In the guiding telescope are two spider threads at right angles to each other, and it is by constantly keeping the image of a star at the intersection of these 'wires' that the operator ensures the images remaining in a constant position upon the sensitive plate. **1945** Dimitroff & Baker *Telescopes & Accessories* viii. 252 It is not advisable to use as a guiding telescope one whose aperture is less than one half of that used in the photographic telescope. **1959** T. Rackham *Astron. Photogr. at Telescope* iv. 68 It has long been the custom to mount a short focal length wide aperture camera on a larger telescope of greater focal length and, using the latter as a guiding telescope, make long exposures at the night sky. **1966** *Listener* 21 Jan. 169/1 The 50-inch reflector does not possess a separate guiding telescope.

Guignet's green (gi·nʸez grīn). [f. the name of C. E. *Guignet* (b. 1829), French chemist, who discovered and patented it in 1859.] A green pigment, consisting chiefly or wholly of hydrated chromic oxide, which gives a bright, transparent, very fast colour and is used esp. in high-quality paints and printing inks; = Viridian.

1862 C. O'Neill *Dict. Calico Printing* 52/2 A pigment green, which is an oxide of chromium,.. known as Guignets' [*sic*] green, is obtained by making an intimate mixture of about three parts of boracic acid with one part of bichromate. **1876** *Encycl. Brit.* IV. 51/1 A very beautiful pigment, now much used in calico-printing, under the name of Guignet's green, is a borate of chromium. **1920** F. A. Mason tr. *von Georgievics's Text-bk. Dye Chem.* (ed. 2) 541 Frequently Guignet's Green itself is termed Permanent Green when used for chromolithography, bank-note printing, etc., though not in calico printing. **1967** *Introd. Paint Technol.* (Oil & Colour Chemists' Assoc.) (ed. 2) v. 93 The hydrated oxide (Viridian or Guignet's Green) is bright but transparent.

Guignol (ginʸol). [see *Grand Guignol.] **a.** = *Grand Guignol. **b.** A Punch and Judy show. Hence **guignole·sque** *a.* and *sb.*

1882 G. Meredith *Let.* 8 Sept. (1970) II. 671, I think you would have been impressed by the Guignol, or rather by an episode in the career of Polichinelle. **1919** G. B. Shaw *Great Catherine* Pref. 114 They were great sentimental comedians, these Peters, Elizabeths and Catherines who played their Tsarships as eccentric character parts.. Catherine kept this vast Guignol Theatre open for nearly half a century. **1925** J. M. Murry *Pencillings* 205 Molière's attitude to life 'reminds him too often of the attitude of *Punch*'—not the famous figure of Guignol, I imagine, but our own comic weekly. **1935** *Punch* 17 Apr. 440/1, I think the truth is that here are three clever portraits woven into a legend of strong Guignol with the ends left loose. **1953** J. Y. Cousteau *Silent World* 26 Inspired by the camera he enacted fantasies. He leaned down dramatically and clutched a reef starfish to his breast. We filmed 'Guy's roll' of this underwater guignol. **1957** *Listener* 18 July 101/1 The Roman ghost in a Senecan shroud stalked through the guignolesque sensationalism of the minor Jacobeans. *Ibid.*, I Oct. 515/1 Guignol, the French Punch and Judy shows. **1964** C. Wildman tr. E. Ionesco in *Listener* 2 Dec. 858/2 *Ionesco*: I was in the gardens of the Tuileries or the Luxembourg, and I saw the *Guignol*. *Wildman*: Punch and Judy! They are violent too! *Ionesco*: Indeed. After that, I've always sought the guignolesque, in other words, an extremely schematic, simplified, elemental and caricatural type of theatre... A kind of theatre very close to Jarry's. In *Ubu-Ros*, Jarry wrote a play which is guignolesque.

guild. Add: **4. guild-socialism,** an economic system by which the profits, resources, and methods of each industry are to be controlled by a council of its members, on the model of medieval guilds; so **guild socialist.**

1912 *New Age* 10 Oct. 560/2 Unless we can prove the practicability of Guild Socialism, and so attract the practical man, we admit that we are preparing for a moral catastrophe. **1913** C. Booth *Industr. Unrest* 16 The Guild Socialists in England are young middle ground between Syndicalist and Socialist. *Ibid.* 21 Syndicalism, Guild Socialism, and State Socialism hold no terms with each other. **1915** D. H. Lawrence *Let.* 27 Dec. (1932) 300 That is why we have to get something like Guild-Socialism in the long run. **1919** G. D. H. Cole *Guild Socialism* (1920) 4 The desire of the Guild Socialist is.. to convert the Socialist Movement as well as the Trade Union Movement to its point of view. *Ibid.* 5, I do not pretend.. that Guild Socialism is the right way for all the peoples of the world to tackle their economic problems. **1930** *Times Lit. Suppl.* 26 June 525/3 Back to the guild! is.. the reformer's cry, back to the land,.. back to the Middle Ages and guild-socialism. **1957** V. A. Demant *Religion & Decl. Capitalism* ii. 54 The condemnation of proletarianism was taken up later by the Guild Socialists who numbered many churchmen and who unsuccessfully sought to divert British labour from state collectivism to the recovery of authentic artisan status through industrial corporations.

guild-hall. Add: (spelt *Guildhall*), *spec.* the hall of the Corporation of the City of London, used for ceremonial meetings, state banquets, etc.

1598 Stow *Surv.* 217 William Hariot Draper Mayor 1481, gaue 40 pound to the making of two lauers in the said Guildhall. **1797** *Encycl. Brit.* X. 243/1 The lord-mayor elect.. is soon after presented to the lord-chancellor.. ; and on the 9th of November following is sworn into the office of mayor at Guildhall. **1841** H. Thomas *City of London* I. 423 At the northern extremity of King Street is Guildhall yard, the north side of which is occupied by the principal front of the Guildhall or common hall of the corporation of London. **1965** C. Trent *Greater London* xvi. 264 The rebuilding of the City is far from complete but the skyline has changed out of all recognition since 1945.. The Guildhall has been rebuilt, many new fine blocks of offices have appeared.

‖ **guillaume** (giyōm). [Fr., a use of the proper name (= William).] A rabbet-plane.

1889 *Spons' Mech. Own Bk.* 378 The ends are.. worked to the given shape with an iron guillaume. *Ibid.*, The checks are worked out with fillester and guillaume planes. **1906** W. L. Goodman *Hist. Woodworking Tools* 106 Félibien's plate gives several types of *guillaume* or rabbet plane.

guillotine, *sb.* Add: **3. a.** (Earlier example.)

1850 N. Hawthorne *Scarlet Letter* 56 Keeping up the metaphor of the political guillotine, [etc.].

4. (sense 3 b) *guillotine closure, motion, resolution, time.*

1909 *Westm. Gaz.* 14 May 2/2 Let it be understood.. that the guillotine closure will not be used, however prolonged the sittings may be. **1927** *Daily Express* 10 May 2/7 To introduce the guillotine or kangaroo method of closure. **1946** *Ann. Reg. 1945* 94 The guillotine motion in Standing Committee should take the form of naming the date by which the Bill should be reported. **1958** *Ann. Reg. 1957* 7 A 'guillotine' motion for its [*sc.* the Bill's] acceleration was moved and carried. **1927** *Daily Tel.* 10 May 12/3 The Government will bring in a 'guillotine' resolution. **1968** *Globe & Mail* (Toronto) 1 Feb. 3/4 *(heading)* Guillotine time only days away as broadcasting debate goes on.

b. guillotine shears, a form of shearing machine having a stationary lower blade and used chiefly for cutting metal sheet and strip.

1884 W. H. Greenwood *Steel & Iron* xvi. 348 A form of powerful guillotine shears. **1967** *Times Rev. Industry* Feb. 96/2 The Swedish company's range of hydraulic press brakes and guillotine shears.

guilt, *sb.* Add: **5. d.** (Later examples.)

1932 *Brit. Jrnl. Psychol.* Apr. 377 The author considers that all individuals nourish large numbers of fads, fancies, hypothetical and unreal 'guilts'. **1966** *Listener* 10 Mar. 341/2 If he [*sc.* a child] does take something that he knows is not his property, and if you detect him at it you will often find guilt written all over his face.

6. b. Phr. *guilt by association* (see quot. 1964).

1952 *Brit. Jrnl. Sociol.* III. 336 Guilt by association. **1959** The doctrine of guilt by association is abhorrent enough in the criminal and deportation fields without being extended into the relation between lawyer and client. **1960** *Listener* 3 Mar. 415/1 He introduced those concepts of guilt by association and guilt by intention which have always been a feature of political trials and disputes in Russia. **1964** Gould & Kolb *Dict. Soc. Sci.* 298/2 Another form of legal usage comprehends the idea of guilt by association or guilt attaching to an individual through his relation or connection with a group of persons who are charged with a violation of a legally prohibited type of behaviour.

7. *guilt-feeling, -sense; guilt-free, -haunted, -laden, -ridden, -stricken* adjs.; **guilt-complex** (see *complex *sb.* 3), a mental obsession with the idea of having done wrong.

1927 Herrmann & Galbraith *Textbk. Psychiatry* ix. 198 Guilt complexes.. also lead to compensating phantasies. **1929** A. Adler *Probl. Neurosis* vi. 84 The complicated state of self-accusation and repentance at the same time, which we call a guilt complex, which is always a superiority-striving on the useless side of life. **1950** Koestler *Lotus & Robot* 276 Auden's *mea culpa*.. might serve as a motto for the Western guilt complex towards Asia. **1951** —— *Age of Longing* I. i. 25 We can only achieve a constructive attitude if we rid ourselves of fallacious guilt-feelings about the past. **1963** A. Heron *Towards Quaker View of Sex* iii. 18 Without the special guilt-feelings which her upbringing has so often laid on the girl, man and maid are.. on equal terms. **1960** Koestler *Lotus & Robot* ii. viii. 198 The permitted, and therefore guilt-free, pleasures of eating and chase pleasures. **1953** S. Spender *Shelley* 11 His friendship with Hogg now entered a new and perhaps guilt-haunted phase. **1944** *Mind* LIII. 247 The repressed guilt-laden hatred, originally felt towards the mother for checking a certain impulse. **1951** S. Spender *World within World* 31, I was intensely self-conscious, guilt-laden, undisciplined, curious, inspired and lazy. **1960** Koestler *Lotus & Robot* 280, I accepted it as a truism, as most guilt-ridden Westerners do. **1959** *Times Lit. Suppl.* 26 Jan. 57/1 Unless the distinction is clearly made and recognized between objective and subjective guilt (between 'guilt' and 'guilt-sense'), the significance of the former may all too easily escape the attention of the analyst. **1930** E. M. Aveling tr. *Marx's Secret Diplomatic Hist. 18th Cent.* v. 74 Whether we consider her [*sc.* Russia's] power, or the more vision of the guilt-stricken consciences of the European peoples—the question remains the same.

‖ **guimauve** (gimōv). [Fr.] The marshmallow, *Althæa officinalis*; also, the medicinal preparation made from its root.

1812 J. Stokes *Bot. Mat. Med.* III. 531 Marsh Mallow.. Guimauve. **1857** A. Henfrey *Elem. Bot.* II. 245 The ordinary properties of this order [*sc.* Malvaceæ] depend on the abundance of a bland mucilage, especially in the roots, as in the Marsh Mallow (*Althæa*). **1866** Lindley & Moore *Treas. Bot.* I. 48/2 The roots of marshmallow are much used, especially in France, under the name of Guimauve, to form demulcent drinks. **1870** [see *pastil]. **1890** Bentley & Trimen *Medicinal Plants* I. 35 A favourite preparation in France is the *pâte de guimauve*.] **1895** *Army & Navy Co-op. Soc. Price List* 15 Sept. 702 Guimauve Pastilles. **1950** R. C. Wren *Potter's Cycl. Bot. Drugs* (ed. 6) 228 Marshmallow... Syn.—Mallards, Guimauve, Schloss Tea.

guimpe (gæmp, gimp). Also guimp. [Fr.: cf. Gimp *sb.*[1], *Wimple *sb.*] A chemisette coming high up the throat; an under-blouse designed to be worn with a low-necked frock.

1850 *Ann. Reg.* 1840 ... Guimpe bouillonné [see *entredeux]. **1880** *Cent. Dict.* ... **1909** *Westm. Gaz.* 15 Feb. 15/3 One of the high corselet bodices permit of nothing more than a small guimpe with sleeves. **1914** 3 Jan. 18/5 The corsage is finished with a guimpe of white spotted net. **1931** F. Hurst *Back St.* iii. xxxi. 276 Remaking last year's blouse into this year's guimp. **1957** M. B. Picken *Fashion Dict.* 155/2 *Guimpe*,.. 1. Short blouse, often with sleeves. Usually worn with a pinafore type of dress... 2. Chemisette or yoke with high, standing collar. Worn about

1910 to fill in neck-line when low-necked dresses were introduced.

Guinea. Add: **1. Guinea negro** or **nigger**; also *ellipt.*

1748 in *Amer. Speech* (1952) XXVII. 283 Run-Away, a likely well-made Guiney Negro Man, named Toney. **1789** S. Low *Politician Outwitted* III. i. He talks as crooked as a Guinea niger. **1823** J. F. Cooper *Pioneers* xxx, One of them Guineas down in the kitchen there. **1856** P. Cartwright *Autobiog.* 192 [De Vere], If he don't get his soul converted God will damn him as quick as he would a Guinea Negro. **1861** J. Kemble *Resid. in Georgia* 29 Called ... **1883** J. C. Harris *Nights with Uncle Remus* ... 'Dey's Guinea niggers ... **1923** J. L. Williams *Ole Plantation* Pref., I remember hearing the old plantation negroes before the war speak of one as a 'Gullah nigger' and another as a 'Guinea nigger'.

b. A derogatory term for an immigrant of Italian or Spanish origin, or one of similar appearance. Also **ginny, guinny.** *U.S. slang.*

1896 *Dial. Notes* I. 418 Guinea, an Italian. **1897** F. Moss *Amer. Metropolis* III. viii. 113 Boys and girls together, we would sing and waltz. While the 'Ginnies' played the organ on the sidewalks of New York. **1910** Dutchman who lands in New York has the address of some.. cheap hotel. **1927** J. Barbican *Confess. Rumrunner* ix. 94 Tell that squint-eyed guinea to throw the mud-book overboard. **1932** D. Runyon *Guys & Dolls* x. 221 A bomb such as these Guineas like to chuck at people they do not like, especially Guineas from Chicago. **1934** J. O'Hara *Appointment in Samarra* (1935) ii. 53 Tony Moracchi.. who up to that time had been known only as a tough little guinny, was matched to fight a preliminary bout at McGovern's Hall.

5. guinea-gold, (*c*) *adj.*, of the colour of a gold guinea.

1908 M. K. Rawlings *Yearling* xi. 110 The river shone in the late sunlight, Guinea-gold, like the bright flowers. **1966** E. Palmer *Plains of Camdeboo* xiii. 260 They [*sc.* cobras] must just have sloughed their old skins, for they shone as if fresh minted. Often they are brown or yellow, sometimes a guinea-gold or the handsome red gold.

Guinean (gi·niăn), *a.* and *sb.* Also 6 **Guyneane.** [f. Guinea + -an.] **A.** *adj.* Of or pertaining to Guinea. **B.** *sb.* An inhabitant of Guinea.

1589 J. Hawkins *Voy. Coast of Guinea* in Hakluyt *Voy.* 525 Their foode.. poaster.. with foode for kine. **1697** Towrson... Gvinean ... and out ... Guineans ... midmost ... **1849** *Encycl. Brit.* III. 758/2 Beginning on the West Coast, where the Libyan Sub-region stops, we have another Subregion, the 'Guinean', comprising the seaboard from Sierra Leone to somewhere about Angola. **1893** —— *Dict. Birds* 335 This 'Guinean' Province occupies what is commonly called the 'West Coast' of Africa. **1965** *Daily Tel.* 24 Feb. 16/1 The teacher, Miss Svetlana Ushakova, 20, offended the ambassador by fraternising with Guineans. **1965** *Guardian* 11 Mar. 9/1 All Guinean francs held abroad were yesterday declared valueless, to combat currency trafficking and the circulation of counterfeit money. **1970** *Times* 6 May 10/6 She wears a striking brown and gold Guinean warrior's hat.

guinea-pig. Add: **1. b.** A person or thing used like a guinea-pig as the subject of an experiment. Hence **guinea-pig** *v. intr.*, to act as the subject of an experiment; also *trans.*, to use (a person) thus.

1913 G. B. Shaw *Quintessence of Ibsenism Now Completed* 135 The.. folly which sees in the child nothing more than the vivisector sees in a guinea pig: something to experiment on with a view to rearranging the world. **1920** U. Sinclair *Brass Check* xxii. 202 Say to yourself that Upton Sinclair is a guinea-pig. **1923** H. G. Wells *Men Like Gods* 180 And may I ask.. the nature of this treatment of yours, these experiments of which we are to be the—guinea pigs, so to speak? Is it to be anything in the nature of a vaccination? **1955** *Times* 15 June 5/3 The volunteers.. are also guinea-pigs, because it is hoped that decisions arising out of these cases will save a great amount of public money and trouble, and provide a guide in law and principle which may go a long way in aiding other cases. *Ibid.* 20 June 4/4 We.. might like to spend a 'holiday' at Salisbury, guinea-pigging for the Common Cold Research Unit. **1955** R. Bannister *First Four Minutes* 129 In some of my experiments I used other athletes as guinea-pigs. **1961** *Time* (Atlantic ed.) 3 Mar. 19/The astronauts] were guinea-pigged into hot chambers. **1969** *Times* 24 May 13/6 Something of a guinea-pig, the ship has demonstrated its value. **1970** *Daily Tel.* 13 Oct. 19/5 'You can't detect any difference unless you know it's synthetic material,' one guinea-pig commented yesterday.

3. a. (Later *attrib.* examples.)

1929 *Daily Chron.* 18 Jan. 4/6 Greedy manufacturers.. have already laid their plans for large commercial trusts with their 'experts' and Members of Parliament, purchased Press and guinea-pig peers. **1930** *Economist* 9 Aug. 284/2 A large number of the 'bubble' company boards of the 1928 boom were packed with guinea-pig directors.

c. An evacuee or billetee in the war of 1939–45. *colloq.*

1939 *Daily Dispatch* 10 Oct., We are known here [West of England] as the *guinea-pigs*.. and we are being treated locally like those tailless rodents. **1940** *Manch. Evening News* 16 July, I toured a part of Manchester in which every other house seems to be teeming with *guinea-pigs*, the nickname for the men [*sc.* soldiers] because the billeting allowances are about a guinea. **1941** *New Statesman* 30 Aug. 218/3 Guinea Pig—Evacuated civil servant.

Guinness (gi·nĕs). [Family name.] A brand of stout manufactured by the firm of Guinness of Dublin; a bottle or glass of this. Also *Comb.*, as *Guinness-coloured adj.*

1836 Dickens *Sk. Boz* I. 190 A large hamper of Guinness's stout. *Ibid.* 191 Taking.. a draught of Guinness. **1842** Barham *Ingol. Leg.* 2nd Ser. 79 With a three-corner'd Sandwich, and *soupçon* of 'Guinness's'. **1897** *Sat. Rev.* 2 Jan. 2/1 An Irishman drinking his Guinness with uncharacteristic quietude in a London 'pub'. **1922** G. Saintsbury *Scrap Bk.* viii. 93 You can't see the stones for the Guinness-coloured foam. **1928** H. Jenkins *Stiffsons* iv. 118 Oh! for some fish and chips and a bottle of Guinness. **1949** [see *black velvet*, *black *a.* 19].

guira (gwai·rā). [Amer. Sp., f. Tupi *guira* bird.] In full *guira cuckoo*. A non-parasitic cuckoo, *Guira guira*, found in eastern tropical South America.

1866 *Proc. Zool. Soc.* 79 Guira Cuckoo (*Guira piririgua*). One of these birds during the early part of the summer dropped or laid an egg on the ground in the aviary. **1867** *Ibid.* 385 When alarmed it [*sc. Ophisthocomus cristatus*] played the same part as the Guira.. [ed.] the crest of the Guira. **1889** P. L. Sclater *Argentine Ornith.* II. 37 The Guira Cuckoo is about sixteen inches long, has red eyes and blue feet, and an orange-red beak. **1904** *Westm. Gaz.* 13 May 12/1 The guira has two bright red eyes. **1913** ... network of chalky matter. **1957** M. H. Mitchell *Observ. Birds S.E. Brazil* 93 Guira Cuckoos are odd-looking birds, with their shaggy heads, very long tails.. and baleful yellow eyes. *Ibid.* 94 The voice of the Guira is louder and the calls more varied.

‖ **guiro** (gwi·ro). Also † guira. [Sp., gourd.] A musical instrument consisting of a gourd with a serrated surface, the rubbing of which produces a rasping sound.

1898 J. E. Haskins in *Amer. Anthropologist* XI. 344 An instrument of this class which has seemed to be confined to the West Indies.. is called the *guira*, and is made of a gourd varying in size in different instruments... In the hand of a native *guira* player a wonderful rhythmic sound comes from this dried.. shell. **1914** F. Morris *Catal. Crosby Brown Coll. Mus. Instr.* II. ii. 183 In.. Cuba the *maruga*.. is rubbed with pieces of wire. The *guiro*, still another form, is made of a long-necked gourd with lines cut in the surface while the gourd is still green. The scraper is a small stick. **1934** S. R. Nelson *All about Jazz* vii. 166 The Guiro is a native vegetable which is scraped with a fork of two prongs, or, alternatively, struck with a pair of xylophone sticks. **1944** W. Apel *Harvard Dict. Mus.* 566/2 Primitive instruments such as.. the guiro (a serrated gourd scraped with a stick), have been used by Prokofiev.. and Stravinsky. **1961** A. Baines *Mus. Instruments* i. 27 Scrapers have survived into modern times, for instance in the folk music of Venezuela, where the *guiro* or *charrasca* is made from a bull's horn, or in the rumba bands of Cuba and the fashionable Western world which use a scraping gourd, also known as *guiro*. **1966** *Crescendo* Dec. 27/1 The rest of the front-line thump tambourines, click claves, scratch guiros or shake maracas.

guiser. Add: Hence **gui·ser** *v. intr.*, to act as a mummer, to go masquerading; cf. Guisard *v.*

1907 D. H. Lawrence in *Notts. Guardian* 7 Dec. 17/2 Hey, we're going to the Mill guysering. **1932** J. M. E. Saxby *Trad. Lore* 77 On Hallowe'en the Grüliks went a-guisaring.

guitar, *sb.* Add: **b.** *guitar-case, -picker, -picking*; **guitar-fish,** a ray belonging to the family Rhinobatidæ.

1657 J. Verney *Let.* 13 Jan. in M. M. Verney *Mem.* (1894) III. x. 365 You left your gittarre Case att Bremers. **1834** Dickens *Sk. Boz* (1836) I. 152 Parasols, guitar-cases; and parcels of all imaginable shapes. **1848** *Knickerbocker* XVIII. 225 Mr. Thwackit.. took his guitar-case in one hand and his double barreled fowling-piece in the other. **1959** *Sears, Roebuck Catal.* Spring & Summer 889/1 Popular-priced guitar cases. **1905** D. S. Jordan *Guide to Study of Fishes* I. xxx. 550 The Rhinobatidæ (guitar-fishes) are long-bodied, shovel-nosed rays, with strong tails... The numerous species abound in all warm seas. **1963** D. G. Stead *Sharks & Rays Austral. Seas* xv. 138 We come to a division of rays, almost all of which would be classed as Shovelnoses by Australian fishermen coming in contact with them, and known in America and elsewhere as Guitar-fishes. **1951** E. Paul *Springtime in Paris* xi. 203 For many years, while the French were learning about jazz, no Frenchman except possibly Django Reinhardt, the guitar picker, could play it. **1959** 'F. Newton' *Jazz Scene* v. 78 The 'urban blues singers' circuits'.. served by the guitar-pickers and itinerant piano players. **1956** M. Stearns *Story Jazz* (1957) xv. 169 The guitar-picking colleagues of Huddie Ledbetter.

guiver, var. *gyver.

Gujarati (gūdʒārā·ti), *sb.* and *a.* Also **Gujarathi, Gujerati, Gurjarati, Guzarat(h)i, Guzatta, Guzerattee,** etc. [Hind., f. *Gujarāt* (Skr. *Gurjarā*).] **A.** *sb.* A native or inhabitant of Gujarat, a state in Western India; a speaker of the Gujarati language; this language. **B.** *adj.* Of or pertaining to the state of Gujarat, its inhabitants, or its language.

c **1607** in Birdwood & Foster *1st Let. Bk. of E. India Co. 1600–19* (1893) 85 And alsoe the Guzaratte do saile in the Portugalls shipps in euery parte of the East Indies. **1615** T. Roe *Jrnl.* 24 Aug. (1899) I. 31, 12 of his priuat Guard, hiered Guzaratts, some with Turkish bowes, some with Pistolls. *Ibid.* 37 Besides the Cloth he and his people weare are stuffes and Callicoes of Suratt, which Guzeratts bring. **1808** R. Drummond *(title)* Illustrations of the grammatical parts of the Guzerattee, Mahratta, and English languages. **1835** J. Bird tr. *Ali Mohammed Khan's Pol. & Stat. Hist. Gujarát* iii. x. 413 They advanced to battle.. Muzaffar.. was accompanied by a great crowd of Gujaráts and Rajputs. **1838** *Penny Cycl.* XII. 227/2 *Guzerati*, or Guzerati, spoken in Gujarat, and along the coast as far as Bombay. **1925** A. Huxley *Let.* 4 Oct. (1969) 254 Somebody recited a poem in Gujarati.. chanting lugubriously. **1927** *United Free Church Miss. Rec.* May 220/1 The programme was a mixture of English, Marathi, and Gujerathi things. **1931** *Times Lit. Suppl.* 3 Dec. 983/1 An adept in the Gujarati language. **1936** *Ibid.* 18 Jan. 57/4 The Gujarati writers became inspired by Scott, Byron, Mill, and especially Shakespeare. **1937** L. Bromfield *Rains Came* i. iv. 26 That's what is the matter with Gandhi—besides being a Gujerati by race and a Bunya by caste. **1964** E. Huxley *Back Street New Worlds* vii. 76 There's been a lively demand for teachers with a knowledge of Urdu, Gujerati, Punjabi, Hindi and other Asian tongues. **1969** *Sunday Standard* (Bombay) 3 Aug. (Magazine Section) 5/3 The modern Gujarati stage is about two decades old. **1970** H. R. F. Keating *Inspector Ghote breaks Egg* iv. 41 His appearance.. marked him not as a native of the place but as a Gujarati.

gulch, *sb.*[2] Add: **1.** (Earlier and later examples.)

1832 A. Earle *Narr. Res. N.Z.* 1 May (1966) 220 From the Peak.. to the sea shore, the earth is cut into gullies.. The settlers call these ravines gulches. **1964** F. O'Rourke *Mule for Marquesa* (1967) x. 124 The canyon.. became.. a mere gulch with fifty-foot walls that debouched in-gloriously into the gap.

2. (Earlier examples.) Also *gulch-man, -miner.*

1859 *San Francisco Bulletin* 5 May 1/2 The gulch diggings are.. paying well for the labor expended. **1876** R. W. Raymond *Statistics of Mines* VIII. 203 The production of gulch-gold on and below Spanish Bar has amounted.. to \$80,000. **1869** A. K. McClure *Rocky Mtns.* 210 (Th.), The unfortunate politician is 'corraled' by the mountaineers, the gulchmen, or the settlers. **1866** *Beadle's Monthly* Oct. 279/1 The diggings yielded very richly.. but, like most gulch mines, were soon exhausted. **1867** J. F. Meline *2000 Miles on Horseback* 63 The gulch miner has been here in all his pristine strength and glory. **1867** *Harper's Mag.* June 11/2 We.. passed over a creek which had been brought by an artificial ditch for gulch-mining purposes. **1876** R. W. Raymond *Statistics of Mines* VIII. 186 The gold comes from the gulch-washings in Indian district near the Eagle Mine.

gull, *sb.*[1] Add: **c. gull wing,** (*a*) an aeroplane wing of which the short inner section slants upwards from the fuselage, and the longer outer section is approximately horizontal; (*b*) of a car door (used *attrib.*): opening upwards from the body of the car.

1932 *Aviation* Jan. 39/3 The machine is a wire-braced monoplane of the gull-wing type designed primarily to afford the widest possible angle of vision for both pilot and observer. **1934** *Flight* 15 Feb. 156 The 'gull' wings of the earlier type have been dispensed with. **1957** *Motor Body* Apr. 8/1 This was a streamlined design with.. the gull wing or butterfly wing door. **1966** N. Freeling *King of Rainy Country* 122 We had a new sports car—you remember the first three-hundred SL—with the gullwing doors. **1967** C. H. Barnes *Shorts Aircraft* 421 The new craft was a very elegant and efficient gull-wing cantilever monoplane. **1969** *Daily Tel.* (Colour Suppl.) 17 Oct. 71/3 The gull-wing door is likely to be seen more widely in future as a new breed of low-slung sports two-seaters is developed.

Gullah (gʊ·lă). *U.S.* Also **Golla, Goolah.** [Conjectured to be either a shortening of *Angola*, or from a Liberian group of tribes known as Golas.] Used *attrib.* or *absol.* to designate Negroes living on the sea-islands and tide-water coastline of South Carolina and Georgia, and the dialect spoken by them.

1739 *South-Carolina Gaz.* 12 May 2/2 Run away a short well set Negro Man, named Golla Harry. **1822** *Account Late Intended Insurrection among Blacks of this City* (Corporation of Charleston) 22 These disclosures.. were obtained from Harry Haig (whose confession and subsequent testimony went to implicate a corps of Gullah or Angola negroes). **1835** W. G. Simms *Partisan* 224 To their arts the Gullah and the Ebo negroes.. added their spells and magic in no stinted quantities. **1838** *Southern Lit. Messenger* IV. 641/1 The etymology of all which terms.. is quite as untraceable as that of any terms in the Goolah negro dialect. **1896** J. G. Williams *Ole Plantation* Pref. p. v, The older ones of that set of negroes.. speak as pure Gullah as their grandfathers... They seem to have been scarcely affected in their low-country Gullah speech [etc.]. **1908** *S. Atlantic Q.* Oct. 339 The vocabulary of our *Gullah* patois is.. five times as great. *Ibid.*, To some *Gullah* remains a closed book. **1931** E. Linklater *Juan in Amer.* ii. vii. 111 They spoke mostly in the argot of Naples, with here and there a little Czech, Finnish, Gullah, and Yiddish. **1957** W. C. Handy *Father of Blues* viii. 116 A humorous Negro custom that could be traced to the Gullahs and from them all the way back to Africa. **1969** *Times* 19 July 9/5 Gullah is a form of creolized English taken to the United States by slaves from West Africa,.. rather than a regular dialect of black American English.

gully, *sb.*[1] Add: **2. c.** In extended meaning (see quots. 1871 and 1966). *Austral.* and *N.Z.*

1840 F. Mathew *Founding of N.Z.* (1940) ii. 48 Crossing a deep and broken gully, the sides of which are so

precipitous that [etc.]. **1840** *N.Z. Jrnl.* XIX. 245/1 The timber grows principally in the gullies between the hills and mountains. **1846** C. Rowcroft *Bushranger of Van Diemen's Land* I. xi. 109 By this time they had descended into a deep and narrow gulley. **1856** *Richmond-Atkinson Papers* I. v. 218 Colson asked Hinde £300 for his 37½ acres of gulley. **1862** H. C. Kendall *Poems* 17 The gums in the gully stand gloomy and stark. **1871** C. L. Money *Knocking about in N.Z.* i. 9 'Gully' means nothing more than a strip of ground lying between two hills, and having a 'creek' flowing down its centre. **1875** Wood & Lapham *Waiting for Mail* 16 The terrible blasts that rushed down the narrow gully, as if through a funnel. **1908** E. J. Banfield *Confessions of Beachcomber* I. i. 16 *Pandanus aquaticus* marks the courses and curves of some of the gullies. **1911** C. E. W. Bean *'Dreadnought' of Darling* xxvii. 230 For six months they had to live in this little gully—as barren as a stone quarry and not unlike it to look at. **1930** L. G. D. Acland *Early Canterbury Runs* iv. 66 Harper's Homestead was on the property which belonged to Mrs. Dunlop, but further up the gully than the present house. **1938** R. D. Finlayson *Brown Man's Burden* 63 They were winding along the side of a deep gully. **1966** G. W. Turner *Eng. Lang. Austral. & N.Z.* iii. 57 The word *valley* has fallen from use in Australia and is not very common in New Zealand except in a few place-names. It is replaced by *gully*.

d. *Cricket.* The fielding position between point and the slips; the fieldsman in this position. Also (Austral.) *gully-slip.*

1920 G. L. Jessop in P. F. Warner *Cricket* 165 The position which has been favoured in recent years by fast bowlers, whenever signs of the wicket bumping have been apparent, namely, 'the gully'. **1921** P. F. Warner *My Cricketing Life* xii. 229 He [*sc.* A. O. Jones] was, indeed, quite exceptional as a fieldsman in any position, but especially in the slips and at short third man, or in the 'gully', as it is called nowadays. **1927** *Observer* 19 June 25/1 Macaulay fell to a dazzling left-hand catch high up in the gully by Kidd. **1954** J. H. Fingleton *Ashes crown Year* 43 Morris.. found himself caught high at gully-slip. **1955** *Times* 12 July 12/4 Ralph was indeed unlucky, for Arnold with good intentions of not wasting time, cut him straight into gully's hands, when exit followed entry. *Ibid.* 15 July 3/3 At 252 Insole was caught in the gully slashing at Titmus. **1970** [see *edge *v.*[1] 7].

4. (sense 2) *gully-bottom*; (sense 3) *gully-emptier, -grating*; **gully erosion,** the erosion of soil by rain-water forming channels; **gully-washer** *U.S.*, a heavy downpour.

1917 J. Masefield *Old Front Line* 49 He.. tunnelled long living rooms, both above and below the gully-bottom. **1965** F. Sargeson *Mem. Peon* iv. 58 Finally you reached the gully-bottom. **1929** *Even. News* 18 Nov. 16/4 Following a collision between an L.C.C. tramcar and a Fulham Borough Council gulley-emptier in the Fulham Palace-road, S.W., to-day, passengers in the tramcar received a showerbath from the contents of the gulley-emptier, which poured into the tramcar. **1928** *Proc. 1st Internat. Congr. Soil Sci.* VI. 755 That phase of normal gully erosion which gradually cuts out V-shaped ravines usually not excessively deep. **1937** E. J. Russell *Soil Conditions & Plant Growth* (ed. 7) viii. 579 The erosion takes two forms: sheet erosion, which goes on slowly and evenly over a large area, and gulley erosion, which is more localised and washes out the soil into great gulleys or ravines. **1905** *Daily Chron.* 7 Aug. 6/5 Volumes of steam issued through the gully grating. **1903** J. Fox *Little Shepherd* iv. 59 Send us, not a gentle sizzle-sozzle, but a sod-soaker, a gully-washer. Give us a tide, O Lord! **1961** *Amer. Speech* XXXVI. 153 An old farmer looked at the threatening sky.. and said, 'It's goin' to be a gully washer and a chunk floater.' **1969** *Daily Progress* (Charlottesville, Va.) 22 Aug. 4/6 Other two-word names for a heavy rain [are].. gully-washer, [etc.].

gully, *v.* Add: Hence **gu·llying** *vbl. sb.*

1928 *Proc. 1st Internat. Congr. Soil Sci.* VI. 755 The same general class of wide-spread carving-out of the soil material or deep gullying that characterizes the devastation of Memphis. **1958** *New Biol.* XXV. 51 The flooding in Devon was associated with gullying and soil erosion and much of upland Britain is scarred with gullies similar to those formed so dramatically in 1953. **1963** D. W. & E. E. Humphries tr. *Termier's Erosion & Sedimentation* i. 20 In contrast, the equatorial climate which is constantly hot and humid tends to remove soil by gullying.

gully-gully (gʊ·ligʊli). Also **gulli-gulli.** [Of unknown origin.] A conjuror's catch-word. So **gully-gully man,** in Egypt, esp. at Port Said, a conjuror who works with live chickens.

1930 E. Waugh *Labels* iii. 69 [At Port Said] a conjuror came and did tricks for us with live chickens. These are called 'gully-gully men' because of their chatter. They are the worst possible conjurors but excellent comedians. They squat on the ground, making odd clucking noises in their throats and smiling happily, and proceed with the minimum of deception to pop things in and out of their voluminous sleeves. *Ibid.* 70 A little Arab girl.. who had taught herself to imitate them perfectly.. used not to bother about the conjuring at all, but would scramble from table to table in the cafés, saying, 'Gully-gully', and taking a chicken in and out of a little cloth bag. **1955** A. Ross *Australia 55* 24 The decks were deserted, the gulli-gulli men departed. **1960** *Spectator* 15 July 95/3 The 'gully-gully' men ply between Port Said and Suez, boarding the ships and giving performances on deck at advertised times. **1962** D. J. Enright *Addictions* iv. 67 *(title)* I was a gulli-gulli man's chicken. **1969** J. R. D. Tata in *Assoc. Chambers Comm. & Industr. India Golden Jubilee Commem. Vol.* 74 A belief that all you had to do to create.. large industry was to produce the money, order the plant, build the sheds, advertise for posts and 'presto, gully-gully!', you had production and sales following automatically.

gulose (giū·lōᵘz, -s). *Chem.* [a. G. *gulose* (Fischer & Piloty 1891, in *Ber. d. Deut. Chem. Ges.* XXIV. 521), f. Glucose by omission and transposition of letters.] An artificial hexose sugar, $C_6H_{12}O_6$, which is stereoisomeric with glucose.

1891 *Jrnl. Chem. Soc.* LX. I. 677 The authors.. call.. the corresponding sugar gulose. **1963** K. Mayer tr. *Staněk's Monosaccharides* ii. 31 D-Glucose (6) and L-gulose (10) form the same dicarboxylic acid.

gulyas, var. *goulash.

gum, *sb.*[1] Add: **5.** *gum-chewed* adj.; **gum-digger** *Austral.* and *N.Z. slang*, a dentist; so **gum-digging** *vbl. sb.*; **gum-ridge,** the ridge of gum behind the upper teeth; **gum-ring,** a child's teething-ring; **gum-shield** *Boxing* (see quot. 1954); **gum-tickler** (earlier U.S. example); **gumwork** *Dentistry* (see quot. 1969).

1922 Joyce *Ulysses* 364 Chap in the Burton today spitting back gumchewed gristle. **1941** Baker *Dict. Austral. Slang* 33 *Gumdigger*, a dentist. **1932** A. S. Bruce *Early Days of Canterbury* xii. 135 Purdie the Dentist.. was.. among the leading practitioners in the somewhat primitive days of the art of 'gum digging'. **1938** I. Goldberg *Wonder of Words* ix. 179 A consonant is called palato-alveolar because it is made by the palate and the gums... This term is applied, among other things, to the teeth sockets, or the gum-ridge. **1965** W. S. Allen *Vox Latina* i. 13 An alveolar articulation (in which the tongue makes contact with the gum-ridge behind the upper teeth rather than with the teeth themselves). **1856** F. S. Cozzens *Sparrowgr. Papers* x. 138 It.. sat up rigidly in its mother's lap, twirling its thumbs and cutting its teeth without a gum-ring. **1954** F. C. Avis *Boxing Ref. Dict.* 50 *Gum shield*, a soft pad worn in the mouth by boxers to protect their teeth and gums during a contest. **1959** *Times* 27 Aug. 3/7 In the eighth [round] Erskine's gumshield went skidding across the canvas. **1963** *Times* 7 Feb. 3/6 He took a hammering and had his gum-shield knocked loose. **1810** J. Lambert *Trav. N. Amer.* (1813) II. 299 A gum-tickler is a gill of spirits, generally rum, taken fasting. **1881** P. H. Austen *Harris's Princ. & Pract. Dentistry* (ed. 10) iv. xiii. 633 It [*sc.* platinum] is also the only metal used in a remarkably beautiful style of work known as the Continuous Gum Work. **1940** J. Osborne *Dental Mechanics* viii. 88 The absence of the anterior gumwork has a serious effect upon the retention of the denture in the mouth. **1969** *Gloss. Terms Dentistry (B.S.I.)* 84 *Gumwork*, that part of the denture which replaces lost natural gum and alveolar process.

gum, *sb.*[2] Add: **1. d.** (Examples.)

1842 *Spirit of Times* (Phila.) 11 Apr. (Th.), [She] asked me if I didn't want A piece of gum to chaw. **1915** J. Webster *Dear Enemy* (1916) 273 A painted yellow-haired thing who chewed gum like a cow. **1936** R. E. Sherwood *Idiot's Delight* I. 36 You've got to hoard your gum here in Europe.

f. The viscid or waxy substance which surrounds the filaments of silk in its natural state.

1774 in *Trans. Amer. Philos. Soc.* II. (1786) 352 The substance which forms the silk, is in their stomach, which is very long; wound up as it were on two spindles and surrounded with a gum, commonly yellow, sometimes white, not often greenish. **1825** J. Nicholson *Oper. Mech.* 399 The silk being now spun, is put into a boiler filled with hot water, into which is put a small quantity of soap, in order to divest the silk of its gum. **1835** Ure *Philos. Manuf.* vi. 248 Marabout... Being white as it comes from the worm, it takes the purest and most delicate shades of colour at once, without the discharge of its gum. **1887** *Encycl. Brit.* XXII. 61/2 The natural gum of the cocoons which holds the filaments together. *Ibid.* 64/2 It has long been the practice to dye some dark silks 'in the gum'. **1959** *Chambers's Encycl.* XII. 555/1 After the first boiling the silk is hydro-extracted which removes the dirt and the bulk of the gum.

g. A mixture, of which gelatine is a main ingredient, from which a hard sweetmeat is made in a mould; a sweetmeat made of this.

1827 W. A. Jarrin *Italian Confect.* (ed. 3) xxvi. 220 Pastilles, Mille-Fleurs are made with fine gum paste, of different colours. **1868** L. M. Alcott *Little Women* vii. 101 Mr. Davis.. had succeeded in banishing gum. **1894** E. Skuse *Compl. Confect.* 103 There is a quantity of goods sold as French, American, German, &c., gums, all more or less a mixture of the genuine article, with gelatine, farina, &c. **1921** *Dict. Occup. Terms* (1927) § 434 *Starch hand* (male),.. *gum boiler*; weighs on scales, or, measures by means of measuring glass, ingredients for gums, mainly gelatine and butter;.. pours the mixture, known as 'gum' or 'boil', when boiled, into trays or moulds. **1950** C. T. Williams *Chocolate & Confect.* xiv. 172 Excessive stirring in any form of gum, pastille or jelly is to be discouraged, since granulation readily occurs. **1962** *Which?* Sept. 283/1 Rowntree's Fruit Gums (tube).

h. ellipt. for Kauri *gum.*

1839 J. D. Lang *N.Z. in 1839* 59 This gum has recently been sold in some quantity.. to the Americans who manufacture it into varnish. **1887** *Col. & Indian Exhib., Rep. Col. Sect.* 287 The ordinary gum of commerce is the semi-fossilized turpentine of the [Kauri pine] tree. **1906** *Macm. Mag.* Apr. 478 Not having caught on to the feel of the gum.

i. The substance whose presence causes a ropy condition in wine.

1888 *Encycl. Brit.* XXIV. 603/2 It sometimes happens that wine becomes viscous and forms threads when poured from the bottle. This mischief, which is caused by the development of a foreign ferment, can be cured by the judicious addition of a solution of tannin, which precipitates the 'gum'.

j. A non-volatile solid or semi-solid substance apt to be deposited by some petroleum products when stored for long periods or heated, and formed by the oxidation of certain of their constituents; it varies in nature from a soft, sticky mass to a hard, resinous layer.

1922 *Rep. Investigations U.S. Bur. Mines* No. 2394. 4 A study of the gums that develop spontaneously when cracked gasoline is stored. **1926** *Industr. & Engin. Chem.* Nov. 1198/2 Gum formation in gasoline claimed public notice when cracked gasoline began to be widely marketed. **1935** Nash & Howes *Princ. Motor Fuel Prep. & Applic.* II. xiii. 105 The formation of gum in motor fuels.. can lead to the seizure of inlet valves in their guides. **1944** M. Van Winkle *Aviation Gasoline Manuf.* vii. 206 The addition of certain gum inhibitors to all grades of aviation gasoline is permitted. **1967** W. A. Gruse *Motor Fuels* iii. 68 One type of instability, troublesome twenty-five years ago, and still occurring occasionally when storage conditions are bad or when a very unstable stock is employed, is the development of gum content in gasoline.

3. a. *Chagual gum* (see quot. 1880); *Sonora gum*, resin obtained from the creosote-bush; **gum accroides** = *accaroid; **gum benjamin** (see Benjamin[1] 1); **gum-dammar** (see Dammar); **gum-kino** (see Kino[1] 1).

1830 Gum kino [see Kino[1] 1]. **1851** *Illustr. Catal. Gt. Exhib.* IV. I. 997 Gum kino, from the blue gum-tree, the stringy bark, and other *Eucalypti*. **1851–9** Gum benjamin [see Benjamin[1] 1]. **1868** Watts *Dict. Chem.*, *Sonora gum*, a kind of lac produced by the puncture of a coccus in *Mimosa cerifera*, a tree growing in Mexico. **1869** A. R. Wallace *Malay Archip.* II. 8 Tortoise-shell, rattan gum-dammer, and other valuable products. **1880** *Encycl. Brit.* XI. 275/2 Chagual gum, a new variety brought from St. Iago de Chili, resembles gum senegal. **1908** W. Schlich *Man. Forestry* (ed. 3) V. 730 Gum-kino, a bright red, astringent gum-resin.. from India and Ceylon. **1909** Webster, Gum accroides. **1922** Joyce *Ulysses* 502 Pellets of new bread with fennygreek and gumbenjamin. **1930** *Discovery* Aug. 260/2 A coating of red ochre and then.. another of gum damar boiled in oil. **1937** Gum accroides [see *accaroid].

5. With various other defining epithets. (For *red, rough, rusty, sugar, swamp, water, white gum*, see the first word.)

1833 C. Sturt *Southern Australia* II. viii. 236 Eucalypti were the general timber on the ranges; one species.. resembling strongly the black-butted gum, was remarkable for a scent peculiar to its bark. **1846** J. L. Stokes *Discov. Australia* II. iv. 132 York gum... Abundant in York—on good soil. *Ibid.* xii. 387 The trees, which grew only in the valleys, were small kinds of banksia, wattles, and drooping gums. **1847** F. W. L. Leichhardt *Jrnl. Overland Exped. Austral.* 6 The prevailing timber trees are Bastard box,.. and the Flooded Gum. *Ibid.* 283 On the small flats, the apple-gum grew. **1848** T. L. Mitchell *Trop. Australia* 107 A small group of trees of the yellow gum, a species of eucalyptus growing only on the poor sandy soil near Botany Bay. **1852** Mrs. C. Meredith *My Home in Tasmania* I. xi. 169 A kind of *Eucalyptus*, with long drooping leaves, called the 'Weeping Gum', is the most elegant of the family. **1887** *Col. & Indian Exhib., Rep. Col. Sect.* 420 Other noble trees, as the Blue, White, Red, Swamp, Water-rooted and Manna-drooping Gums. **1889** Cider Gum [see Sugar *sb.* 5]. **1893** *Australasian* 5 Aug. 252/4 The bark of the salmon gum approaches in colour to a rich golden brown. **1893** *Sydney Morning Herald* 19 Aug. 7/1 Here are no straight and lofty trees, but sprawling cinnamon gums. **1930** Billis & Kenyon *Pastures New* viii. 123 The trees were very pretty, being a kind of weeping gum. **1947** I. L. Idriess *Isles of Despair* vii. 44 Horn Island, with its stunted gums and cabbage-tree palms.

8. a. *gum-leaf* (earlier and later examples), *-log* (earlier example), *-tip.*

1874 Trollope *Harry Heathcote* i. 2 When the gum leaves crackle.. before Christmas, there won't be a blade of grass by the end of February. **1966** 'J. Hackston' *Father clears Out* 182 They might have chewed the moss off old fences and, at a pinch, say, taken on the gum-leaves. **1836** D. Crockett *Exploits & Adv. Texas* 82 (Th.), A chap just about as rough hewn as if he had been cut out of a gum log with a broad axe. **1955** *Sci. News Let.* 12 Mar. 168/2 In 1952 it was discovered the koalas had eaten nearly all their food supplies and were in danger of starving to death. The public rushed to the rescue with carloads of gum-tips.

9. gum-bichromate *Photogr.*, designating a process of printing on paper coated with a mixture of pigment, gum-arabic, and potassium bichromate; also designating a print so produced; **gum-boiler,** one who boils certain sweetmeat mixtures; **gum-boots**: for '*U.S.*' read 'orig. *U.S.*' (earlier and later examples); hence **gu·m-booted** *a.*, wearing gum-boots; **gum-bucket** *Naval slang*, a smoker's pipe; **gum-chewer,** one who chews chewing-gum; so **gum-chewing** *vbl. sb.* and *ppl. a.*; **gum-digger** (earlier and later examples); **gum-digging** (earlier N.Z. examples); **gum-drop** (earlier example); **gum-field** (earlier N.Z. example); **gum-game** *U.S. slang*, a trick or dodge; **gum-hole** *N.Z.*, the hole a gum-digger sinks; **gum-land** *N.Z.*, land where kauri gum is found; **gum-nut** *Austral.*, the hardened flower-cup of *Eucalyptus gummifera*; **gum-platinum** *Photogr.* (see quots.);

gum-shake (see Shake *sb.*[1] 9); **gum-shoe** (earlier example); (*b*) *fig.* (*U.S. colloq.*), used *attrib.* to describe something done stealthily; hence, a detective; hence **gum-shoe** *v. intr.* (*colloq.*, orig. *U.S.*), to move or act with stealth as if wearing gum-shoes; **gum silk,** silk from which the natural gum has not been removed; **gum-spear** *N.Z.*, a spear used in probing soil for kauri gum; **gum-sucker,** (*b*) a fool or simpleton; **gum-worker** *Photogr.*, one who makes prints by any of the processes, such as the gum-bichromate process, in which gum-arabic is used.

1897 Wall *Dict. Photogr.* (ed. 7) 117 The gum-bichromate or photo-aquatint process. **1900** *Daily News* 1 Oct. 7/4 A striking profile done in red by the gum-bichromate process. **1919** *Brit. Jrnl. Photogr. Alm.* 252 In the gum-bichromate process, also termed aquatint or photo-aquatint. **1962** M. R. Haselgrove *Photographer's Dict.* 124 *Gum bichromate prints*. These are made by a nearly obsolete process, but are luckily still to be seen in our major exhibitions. **1921** Gum boiler [see *1 g]. **1930** Blunden *De Bello Germanico* 10 The unshaven, clay-cased, and gum-booted one. **1960** *News Chron.* 27 Apr. 1/7 Gum-booted searchers waded in the shallows. **1850** E. Christman *One Man's Gold* (1930) 119, I put on my long gum boots and waded through the water. **1917** 'Contact' *Airman's Outings* 220 A bus lands and taxies to a shed. From it descends the Squadron Commander, who, with gum-boots and a warm coat over his pyjamas, has been 'trying the air'. **1921** *Daily Colonist* (Victoria, B.C.) 28 Oct. 20/3 (Advt.), Men's pure gum boots, with red soles, hip length \$6.95. **1962** *Lancet* 15 Dec. 1272/1 Twice a day I put on my gumboots to go out to feed [chickens]. **1970** *Times* 10 Feb. 1/7 More than 200 policemen in gumboots.. drained ponds and dragged the stream. **1893** *Funk's Stand. Dict.*, *Gum bucket*, a tobacco pipe. **1917** 'Taffrail' *Sub* viii. § 2. 204 'The Bloke' was an inveterate smoker. I never remember seeing him off duty without a 'gum-bucket', as he called it, in his mouth. **1919** 'Etienne' *Strange Tales from Fleet* 143 Mr Smith, revived by the cocoa and soothed by the pipe, known as the 'gum-bucket' to his pals. **1850** S. Judd *R. Edney* 158 There are the Gum-chewers,—all backlotters; and vulgar. **1938** I. Kuhn *Assigned to Adventure* v. 51 His conviction that the gum-chewers relish stories about the upper classes. **1889** *Sunday Opinion* (Pueblo, Colo.) 14 July 4/5 The careful observer can not fail to note the growing prevalency of gum chewing. **1907** *Daily Chron.* 29 July 5/2 The gum-chewing habit. **1960** D. Storey *This Sporting Life* I. ii. 31 The other gum-chewing player. **1961** *Encounter* Apr. 24/1 Local hot-rodders and their gum-chewing molls. **1967** *Coast to Coast 1965–6* 101 He became the victim of their gum-chewing.. inattention. *c* **1858** in *Richmond-Atkinson Papers* (1960) I. 437 Of the kauri gum diggers 9/10ths are furnished by this tribe. **1871** *Evening Post* (Wellington, N.Z.) 4 Jan. 2/3 A gum-digger named Denis McManus, has been burned to death at Riverhead, Auckland. **1921** *N.Z. Jrnl. Agric.* 20 May 260 To break in these lands one must first face the legacy left by the gum-digger. **1965** G. J. Williams *Econ. Geol. N.Z.* ii. 14/1 Coal had been discovered by gum-diggers near Kawakawa in Northland in 1861. **1871** *McLean Papers* XXXVII. 72 (MS.), Paora Toki.. is gone gum-digging up the Thames. **1879** J. Grey *His Island Home* iv. 34 A great many of the natives abandoned their kaingas and went gum-digging, when that article brought a high price. **1860** *North-West* (Port Townsend, Wash.) 5 July 3/3 Candies, gum drops, mottoes. **1880** W. Senior *Travel & Trout in Antipodes* ii. ii. 182 The gum-fields.. indicate that the fine forests have disappeared at an alarming rate. **1840** in *Amer. Speech* (1941) XVI. 299 I've come the gum game over you. **1872** E. Eggleston *Hoosier Schoolmaster* xiv. 118 Now, looky here... You don't come no gum games over me. **1885** *Lisbon* (Dak.) *Star* 18 Sept., They tried the gum-game on me down in Pennsylvania.. but I came out ahead. **1882** W. D. Hay *Brighter Britain!* II. i. 17 Rich gum-holes here and there. **1900** *N.Z. Illustr. Mag.* III. 205/1 Each man sinks his own gumhole where he strikes the first gum. **1882** W. D. Hay *Brighter Britain!* II. i. 16 On our farm and in the surrounding bush, though these are distinctly not gum-lands. **1900** *N.Z. Illustr. Mag.* III. 203/1 Here, then, on this gumland is where the old Kauri forests grew. **1950** *N.Z. Jrnl. Agric.* June 535/1 Gumland soils vary in type from peat, on some of the flats, through light to medium clay loams, to heavy clay soils. **1936** F. D. Davison *Children of Dark People* x. 143 They'd tell the gum-nuts and they would drop to the ground and tell the grass. **1965** *Austral. Encycl.* III. 406/1 When stamens fall and the young seeds [of *Eucalyptus gummifera*] are fertilized, the flower-cup hardens into a woody capsule ('gum-nut') which opens.. to shed the seed. **1965** M. Shadbolt *Among Cinders* xx. 189 Gum-nuts. There might be a few scattered round. **1918** *Photo-Miniature* XV. Mar. (Gloss.), *Gum-platinum process*—of first making a light print on platinum paper, then coating the print with sensitive gum mixture and reprinting from the same negative. **1919** *Brit. Jrnl. Photogr. Alm.* 252 *Gum-platinum* is a compound process, in which a pale print on platinum paper is coated with the sensitive gum mixture, and a second (pigment) image produced by re-printing under the same negative. **1887** *Col. & Indian Exhib., Rep. Col. Sect.* 427 These are all large trees..—some are a little liable to gum-shakes. **1863** P. S. Davis *Young Parson* 12 A little boy who wore his father's gum shoes in dry weather. **1904** 'H. McHugh' *I'm from Missouri* iii. 44 It seemed that every voter in the community quit work and gum-shoed after the Two Candidates. **1904** *Omaha Bee* 24 Oct. 4 No gumshoe democratic campaign in Nebraska. **1906** A. H. Lewis *Confess. Detective* 198 You're d'gum-shoe guy I was waitin' fer... It was Inspector Val tells me to lay for you. **1907** *Springfield* (Mass.) *Weekly Republ.* 4 Apr. 2 He.. was forced to accomplish his ends by main strength rather than by gum-shoe methods. **1908** J. M. Sullivan *Crim. Slang* 11 *Gumshoe worker*, a private detective; a spotter. **1913** H. A. Franck *Zone Policeman 88* 159 But the 'gum shoe' naturally cannot twirl a police club. **1927** D. Hammett in *Black Mask* Nov. 21/2 He.. looked me up and down, growled: 'So you're a lousy gum-shoe.' **1930** J. K. Winkler *J. Pierpont Morgan* 264 That eminent political gum-shoe artist. **1930** *Publishers' Weekly* 8 Feb. 718 Under the present obscene book law the vice crusader.. goes gum-shoeing around from one bookseller to another. **1951** J. Cannan *And All I Learned* ix. 140 A seventeenth-century President of Trinity who gum-shoed round the college gardens carrying a whip. **1954** 'N. Blake' *Whisper in Gloom* iv. 58 What *were* you doing, gum-shoeing around Mr. Gray's premises? **1962** *Spectator* 1 June 728/3 Ross MacDonald.. is a thoroughly competent detective-story writer, specialising in West Coast private-eye stories. In *The Wycherly Woman*.. his gumshoe hero Archer searches for a missing student. **1885** *Encycl. Brit.* (1887) XXII. 64/1 Sugar is known to have been used for adulterating and loading gum silk for a very long time. **1873** J. E. Tinne *Wonderland of Antipodes* 54, I saw them at work with their gum-spears. **1888** P. W. Barlow *Kaipara* xix. 147 A gum-digger's outfit.. consists of a spade, a gum-spear and a piece of sacking... The gum-spear is a four-sided rod of steel, about four feet long, and pointed at one end. **1906** *Macm. Mag.* Apr. 478 A green hand of a gummy, that hadn't quite got the trick of it yet, went poking around that very cabbage-tree with his gum-spear. **1936** W. Lawson *When Cobb and Co. was King* xii. 223 Some men.. called them 'gumsuckers', and a few other things. **1941** Baker *Dict. Austral. Slang* 33 Gumsucker. **1908** *Westm. Gaz.* 3 Oct. 14/2 Some photographers now classify themselves as gum-workers, oil-workers, and so forth.

gum, *sb.*[6] Add: (Earlier and later examples.) Also **gummie, gummy.**

c **1815** J. K. Paulding *Bucktails* (1847) ii. 34 By gum, that's jist what I want you to tell me, I swow. **1827** T. Hood *Works* (1862) I. 311 But Hunks still ask'd to see the tooth, And swore by gum! he had not drawn it. **1845** S. Judd *Margaret* I. xvi. 139 'Gummy!' retorted the woman. 'He has been a talkin' about me.' **1932** W. L. Graff *Lang.* 283 The name of God is avoided and gives way to such substitutes as Jove,.. gory, gummie. **1970** *Private Eye* 22 May 16 By gum, it must be visiting day up at hall.

gum (gʊm), *sb.*[7] *Mining* (orig. *Sc.*). [Origin uncertain; perh. related to Culm[1], Coom *sb.*[1]] Coal dust, fine coal; now esp. that produced by a coal-cutting machine.

1790–1925 in *Scottish Nat. Dict.* **1883** Gresley *Gloss. Coal-M.* 125 *Gum*, free-burning small slack or duff. **1938** *Colliery Engin.* Mar. 81/2 The amount of gum left in the bottom of the cut never exceeded ½ in. in thickness. **1964** V. Shiffer tr. *Shevyakov's Mining of Mineral Deposits* xi. 236 To prevent the formation of gum and dust, the combine is furnished with a spraying device.

b. *Comb.*, as *gum-flinger, -loader, -thrower.*

1940 *Trans. Inst. Mining Engin.* May 53 Mechanical devices which trap the broken material brought out by the cutter-chain.. have been given such names as jud cleaners,.. gummers, and gum loaders. **1956** E. Mason *Deputy's Man.* II. xxxiv. 501 The same machine can be equipped with a 'gum flinger'. **1960** Shepherd & Withers *Mech. Cutting & Loading of Coal* v. 75 'Gum-throwers'.. take up the cuttings from the chain and eject them into the goaf.

So **gu·mmer**[3], a man or a machine that clears away the fine coal and small bits of debris, etc., from under a coal-cutting machine.

1921 *Dict. Occup. Terms* (1927) § 042 *Gummer*,.. rakes small coal or stone by means of long-handled flat shovel.. out of groove cut by coal cutting machine, and loads into tubs or throws back into goaf. **1940** *Trans. Inst. Mining Engin.* May 54 Except in special cases, undercutting coal-cutters will be fitted with gummers as standard practice in the future. **1959** *New Scientist* 23 July 102/3 Beneath the cutting head is a 'gummer'—a paddle-bladed scraper which removes all the small coal from under the machine. **1959** G. D. Mitchell *Sociol.* iii. viii. 136 Four gummers, who clean out the undercut.

gum, *v.*[1] Add: **5. b.** (Examples.) Also with *up.*

1929 *Motor World* 24 May 328/2 The valves of car engines have always exhibited a tendency to 'gum-up' under certain conditions. **1931** *Carnegie Scholarship Mem.* XX. 96 The most unsatisfactory material in this respect was copper, which drew rather badly with all the lubricants tried, and seemed to 'gum up' in the die very readily.

c. *trans. fig.* To interfere with the smooth running of (something); to spoil, wreck. Chiefly with *up.* Freq. in phr. *to gum the game, to gum (up) the works.* orig. *U.S.*

1901 *Yale Fun* 27 (Weingarten), The plot that was gummed. **1911** L. J. Vance *Cynthia* 174 You've just about gummed things up good and plenty, that's what you've done. **1915** *Dial. Notes* IV. 222 *Gum the game*, delay the game. 'Jack's tactics were to gum the game.' **1920** Wodehouse *Coming of Bill* i. v. 60 It sure would get my goat.. to have the old man gum the game for them. **1932** —— *Hot Water* xi. 181 When it comes to you horning into this joint and aiming to gum the works for me.. well, that's something else again. **1932** Kipling *Limits & Renewals* 281 The main point, as I read it, is that it makes one—not so much think—Research is gummed up with thinking—as imagine a bit. **1936** J. Dos Passos *Big Money* 263, I hope it wasn't me gummed the game. **1938** G. Heyer *Blunt Instrument* ix. 174 Helen's getting mixed up in it gummed up the works. **1948** 'M. Westmacott' *Rose & Yew Tree* xxii. 183 She hasn't gummed up the works after all. What a relief that will be to her. **1964** *Listener* 8 Oct. 548/2 Their Land Commission—far from providing more and cheaper houses, would.. gum up house-building.

6. (Earlier example.)

1840 *Frankfort* (Ky.) *Commonwealth* 20 Oct., You are always right as a book and nobody can gum you. In short, you are O.K.

gum, *v.*[2] (Earlier example.)

1777 W. Dunbar in E. O. Rowland *Life W. Dunbar* (1930) 41 Begun to gum one of our old saws, having unfortunately broke one of the new ones by the fall of a Log.

Gumban (gʊ·mbăn), *a.* [f. Kikuyu *Gumba*, name of a race of diminutive people believed to have been former inhabitants of Kikuyu country.] Of or pertaining to a late Stone Age culture of East Africa. Also *absol.*

1931 L. S. B. Leakey *Stone Age Cultures Kenya Colony* xi. 243 In attempting to discuss the relation of the Gumban to cultures in other parts of the world.. it should be noted here that the name of this culture has been derived from a Kikuyu term which is applied to-day primarily to a semi-mythical race of small people... In using the word for this culture, I am not referring to this, its primary meaning, but to its present more usual meaning, that is in reference to any people who preceded the Kikuyu. From the moment that I first started excavations, old Kikuyus who visited the camp used to talk of my 'search for the Gumba', and when thinking of a name for this comparatively recent culture the word Gumban seemed to be not inappropriate. **1931** *Times Lit. Suppl.* 20 Aug. 627/2 Coming to the Kenya cultures after the Aurignacian, Mr. Leakey believes that in one of the latest, the Gumban, we may have evidence 'of the contact of a late Stone Age people with one of the early civilizations of the world'. **1957** G. Clark *Archaeol. & Society* (ed. 3) v. 166 Faience beads.. were found with a burial at Nakuru in Kenya associated with the so-called Gumban A culture of sub-Neolithic type. **1959** *Chambers's Encycl.* I. 143/2 The best known of the East African neolithic cultures are the Gumban and Njoroan of Kenya described by Leakey. The Gumban is characterized by stone bowls, pestles and mortars, saddle querns and pottery with rolled-cord decoration and lugs.

gümbelite (gū·mbĕləit). *Min.* [ad. G. *gümbelit* (F. von Kobell 1870, in *Sitzungsber. d. königl. bayer. Akad. d. Wissensch.* I. 294), f. the name of C. W. von *Gümbel* (1823–1898), German geologist: see -ite[1].] A magnesium-containing variety of hydromuscovite.

1871 *Jrnl. Chem. Soc.* XXIV. 325 *(heading)* Gümbelite, a new mineral. **1962** W. A. Deer et al. *Rock-forming Minerals* III. 15 A fibrous magnesium-bearing hydromuscovite called gümbelite was investigated by Aruja (1944). Its formula is approximately

$$(K, H_3O)_2(Al_3Mg)Si_6Al_2O_{20}(OH)_4$$

and it crystallizes in the ZM_1 polymorph.

gumbo. Add: **1. a.** (Earlier example.)

1845 *Bangor* (Maine) *Mercury*, At St. Peter's [Ill.] there is a large commerce carried on between the whites and redskins, for heads and whiskey, in exchange for skins and gumbo.

b. (Earlier and later examples.)

1805 J. F. Watson in *Amer. Pioneer* (1843) II. 233 Shrimps are much eaten here; also a dish called *gumbo*. This last is made of every eatable substance, and especially of those shrimps which can be caught at any time. **1813** *Cramer's Alman. 1814* (Pittsburgh) 55 Beyond our usual dinner [we had].. gumbo soup. **1957** H. Croome *Forgotten Place* 95 [An] affable American brigadier who told her that the hot and tasty stuff they were eating was gumbo. **1970** *New Yorker* 10 Jan. 62/1 (Advt.), You'll want to try—Crawfish Gumbo—a kind of spicy bouillabaisse.

3. (Earlier example.)

1838 E. Flagg *Far West* II. 36 A spirited colloquy ensued in the patois of these old hamlets—a species of gumbo-French.

gumbo-limbo (gʊ·mbo li·mbo). [f. Gumbo + second element of unknown origin.] A gum-yielding timber tree, *Bursera simaruba*, found in Florida, Central America, and the West Indies.

1837 J. L. Williams *Terr. of Florida* 98 Gum Elemi.—Called by the inhabitants, Gumbo-limbo, is a large spreading tree, with a smooth brown bark. **1890** H. M. Field *Bright Skies* v. 70 Perhaps some old savage.. has been punished for his cruelty by being turned into that gumbo-limbo tree. **1908** N. L. Britton *N. Amer. Trees* 584 Gumbo Limbo, known also as Bitter-wood, Paradise tree, and on the Bahama islands as Ash, inhabits southern Florida, the Bahamas and Jamaica. **1924** J. A. Thomson *Science Old & New* v. 27 The gumbolimbo, which exudes fragrant balsam when it is wounded. **1964** Gleason & Cronquist *Nat. Geogr. Plants* xxii. 340 *(caption)* Gumbo limbo in a hammock on Big Pine Key, Florida.

gumbotil (gʊ·mbotil). *Geol.* [f. Gumbo + Til(l *sb.*[2])] A leached grey clay, very sticky when wet and very hard when dry, found extensively in Iowa and neighbouring states.

1916 G. F. Kay in *Science* 3 Nov. 638/1 It is now proposed that the somewhat indefinite term 'gumbo' be no longer used for these super-drift clays, but that the name 'gumbotil' be used. **1957** J. K. Charlesworth *Quaternary Era* II. xxxvii. 967 The gumbotils have often been regarded as loesses.

gumlah (gʊ·mla). *India.* Also **gumla.** [Hind. *gamlā*.] An earthenware water-jar.

1834 H. Caunter *Oriental Ann.* ix. 114, I had seen her as she occasionally repaired to the river to replenish her gumlahs. **1837** T. Bacon *First Impr. Hindostan* II. xii.

415, I caught sight of the large *gumla* (earthern jar, containing about eighteen gallons) filled with water. **1892** H. D. INGLIS *Tent Life* 285 Encasing his head in a great wide-mouthed *gumla*, or earthen pot.

gummauve, var. *GUIMAUVE.
1900 in *N. & Q.* (1959) CCIV. 418/2 Medicated and other Lozenges, Jujubes, Pastilles, Drops and Gummauves.

gummed, *ppl. a.* **b.** (Later examples.)
1937 E. J. LABARRE *Dict. Paper* 154/2 Gummed paper, also termed adhesive (U.S.A.), is any kind of paper gummed on one or both sides. **1962** F. T. DAY *Introd. to Paper* 116 *Gummed tape*, Kraft paper in brown or colours, glued and dried.

gummer^{2} (gʌ·məɹ). [f. GUM *v.*^{1} + -ER^{1}.] One who gums (in various technical uses).
1881 *Instr. Census Clerks* (1885) 82 Envelope Maker:.. Gummer, Cementer. **1902** *Daily Chron.* 24 July 9/4 Pattern-Card Mounters, Cutters and Gummers wanted. **1905** *Ibid.* 11 Aug. 10/6 Envelope Hand Gummers, also fountain gummers. **1921** *Dict. Occup. Terms* (1927) § 549 *Gummer* (bookbinder's); applies gum to back of books,.. also gums edge of illustration pages, etc. for insertion in books. **1960** M. SPARK *Ballad of Peckham Rye* i. 11 Raymond Lowther, packer, Lucille Potter, gummer.

gummer^{3}: see *GUM *sb.*^{7}

Gummidge (gʌ·midʒ). The name of a peevish, self-pitying, and pessimistic widow in Dickens's novel 'David Copperfield' (1850), used to describe a person of such a nature, or their complaints. Also *attrib.* Hence **gu·mmidge** *v. intr.*; **gu·mmidging** *ppl. a.*; **gu·mmagy, Gu·mmidgey** *adjs.*
1873 A. HELPS *Animals & their Masters* viii. 191 There is a great deal of what I call 'Gummidge' talked in the world, and very unreasonable talk it is. **1889** A. LANG in *Longman's Mag.* May 111 One might do very well in the Transvaal if one did not 'gummidge'. **1889** BARRÈRE & LELAND *Dict. Slang* I. 437 To be *gummagy*, to be of a snarling, scolding disposition. **1895** *Westm. Gaz.* 14 Jan. 7/1 What good is there in merely Gummidgeing on like this? **1916** *Daily Colonist* (Victoria, B.C.) 20 July 9/7 (*heading*) A Nation of Gummidges. **1918** *Best Short Stories 1917* 209 At twenty-seven Jo had been the dutiful, hard-working son..of a widowed and gummidging mother. **1927** *Sunday Express* 1 May 12/3 The Gummidge chorus is never silent. **1928** D. H. LAWRENCE *Let.* 9 Mar. (1962) II. 1042 It's true I've felt rather Gummidgey and 'low' and disheartened these last two years.

gumming, *vbl. sb.* Add: **2.** (Earlier and later examples.)
1703 tr. *H. van Oosten's Dutch Gardener* v. xii. 284 The common Ailments that the Orange-Trees are subject to in these Countries, are, Gumming, growing Yellow,.. and Changing. **1931** in *New Biol.* (1953) XV. 38 Malformations of nuts, such as occur in gumming..were found to be physiological diseases.
b. The deposition of gum by petroleum products (see *GUM *sb.*^{2} 1 j); the clogging of an engine as a result of this. Also *gumming-up*.
1915 *Jrnl. Inst. Petroleum Technologists* I. 179 Any oil which was to be used satisfactorily in a motor-car must naturally have the terpene products removed, otherwise gumming took place, and the valves became clogged. **1927** *Daily Express* 15 July 9 The total absence of 'gumming-up' of piston rings. **1935** *Nature* 14 Sept. 413/1 Even the schoolboy and his sister talk glibly of anti-knock and gumming. **1942** A. W. JUDGE *Automobile Eng.* (ed. 4) vii. 307 Important claims in regard to non-oxidising, gumming and sludging tendencies have been made for such oils. **1961** G. CLAXTON *Benzoles* xiv. 705 The extensive use of cracked petrols for blending purposes..has drawn attention to the important problem of the gumming of internal-combustion engines when run on fuels containing unsaturateds.

gumming (gʌ·miŋ), *vbl. sb.*^{2} *Mining.* [f. *GUM *sb.*^{7} + -ING^{1}.] **a.** The clearing away of the fine coal and debris from the groove or undercut made by a coal-cutting machine. Also *attrib.*
1934 WEBSTER, Gumming spade. **1938** *Colliery Engin.* Mar. 81/1 The operation of 'gumming' or 'scuffling' out the kerf in the track of the coalcutter is at best a brutal and ineffective business. **1958** A. NELSON *Methods of Working* xiii. 156 Coal-cutters are now provided with efficient automatic gumming devices. **1960** J. SINCLAIR *Winning Coal* vi. 167 Gummers are provided as a substitute for a man with a gumming shovel.
b. *pl.* = *GUM *sb.*^{7}
1938 *Colliery Engin.* Sept. 293/2 The gummings brought out by the cutter chain are received in this spiral and delivered on the goaf side of the machine. **1956** E. MASON *Deputy's Man.* I. xviii. 254 The smalls produced by the picks are termed kirvings, gummings or cuttings. **1958** A. NELSON *Methods of Working* xiii. 158 Various appliances have been designed to trap the gummings from the cutter-chain and in this way the dust clouds are considerably reduced. **1959** G. D. MITCHELL *Sociol.* III. viii. 136 The first shift is concerned with..clearing out the 'gummings'.

gummosis (gʌmōu·sis). *Bot.* [mod.L., f. L. *gumm-i* GUM *sb.*^{2} + -OSIS.] The pathological production and exudation of gum by trees; *spec.* a disease of fruit trees marked by this process. (Cf. GUMMING *vbl. sb.* 2).
1882 *Gardeners' Chron.* 23 Sept. 396/2 The bacterial slime spreads, together with the sap-current, through the xylem, or woody part of the vascular bundles, and consequently infects the whole plant...The gum-disease (gummosis) of our stone-fruit trees might be explained in like manner. **1887** H. M. WARD tr. *Sachs's Lect. Physiol. Plants* xi. 180 Such gum-reservoirs are distinguished from the cases of proper Gummosis, where large groups of tissue in the older organs are changed into basorin and other kinds of gum. **1911** *Ann. Bot.* XXV. 108 Gummosis is a fairly common phenomenon in the vegetable kingdom. **1951** *Dict. Gardening* (R. Hort. Soc.) II. 936/2 In gummosis the tree is in ill health and even small shoots show the exudation of gum. **1970** W. H. SMITH *Tree Path.* xiii. 129 Both canker types are characterized by exudation or gummosis (leakage of fairly viscous sap) during wet weather in the spring.

gummy, *a.*^{1} Add: **3. d.** Used as a derogatory epithet of varying application.
1907 [see *BACK- A. 3]. **1909** J. R. WARE *Passing Eng.* 148/2 *Gummy composer*, old and insipid. **1928** S. VINES *Humours Unreconciled* xi. 143 'Oh, for the wings——' Mendelssohn is pretty gummy; how'd that do? **1936** E. M. FORSTER *Abinger Harvest* 139 He [*sc.* T. E. Lawrence] called the style gummy.
5. (sense 3 a) *gummy-eyed* adj.
1911 R. BROOKE *Poems* 33 Oft she weeps, gummy-eyed and impotent. **1960** C. DAY LEWIS *Buried Day* viii. 162 Stumbling out of the cinema, gummy-eyed from an afternoon's session with Greta Garbo.

gummy (gʌ·mi), *a.*^{3} [f. GUM *sb.*^{1} + -Y^{1}.] Toothless. Hence **gu·mmily** *adv.*
1906 E. DYSON *Fact'ry 'Ands* iii. 29 Sarah was a large.. woman..with..a gummy smile. **1947** 'A. P. GASKELL' *Big Game* 94 An old crone came out..smiling gummily. **1970** *Daily Tel.* 15 July 14 So great is the demand for false teeth among our gummy tots that in many cases they may have to wait for a year or so before being fitted up.

gummy (gʌ·mi), *sb.*^{1} [f. GUM *sb.*^{1} + -Y^{6}.] **1.** The Australian name for *Mustelus antarcticus*, a small shark found in the Pacific Ocean. In full *gummy shark*.
1893 *Funk's Stand. Dict.*, *Gummy*, a galeoid shark. **1898** MORRIS *Austral Eng.* 185/2 *Gummy*, name given to a shark of Victorian and Tasmanian waters, *Mustelus antarcticus* Günth., and called *Hound*..in New South Wales, Victoria and New Zealand. The word *Gummy* is said to come from the small numerous teeth, arranged like a pavement, so different from the sharp erect teeth of most other sharks. **1933** *Bulletin* (Sydney) 31 May 17/1 All sharks are said to be excellent eating, but the usual variety retailed as snapper or Murray cod are the small crustacean-eaters known as 'gummies' in the trade. **1963** D. G. STEAD *Sharks & Rays Austral. Seas* x. 108 The name of Gummy is applied to it because of the rhombic, pavement-like teeth, which, at a passing glance into the fish's mouth, give the impression of toothless gums. **1965** G. MCINNES *Road to Gundagai* iii. 47 All we caught was a gummy shark.
2. A sheep that has lost, or is losing, its teeth. *Austral.* and *N.Z.*
1910 H. JACKSON in *Golden Fleece* 92, I have seen scores of ewes, not 'gummies' either, cutting fleeces of pocket handkerchief dimensions. **1933** L. ACLAND in *Press* (Christchurch, N.Z.) 21 Oct. 15/7 *Gummy*, an old sheep that has lost its teeth. **1933** *Bulletin* (Sydney) 6 Sept. 24/1 One young breeding ewe is worth three old 'gummies'. **1965** J. S. GUNN *Terminol. Shearing Indust.* i. 27 When the sheep's teeth are beginning to decline it is called a 'faulty mouth' or 'failing mouth'. This is the stage before 'broken mouth', after which it becomes a 'gummy'.

gummy (gʌ·mi), *sb.*^{2} *N.Z. colloq.* [f. *GUM *sb.*^{2} 1 h + -Y^{6}.] A gum-digger.
1906 *Macm. Mag.* Apr. 476 We soon were giving those gummies a hand to get some tucker ready.

gump, *sb.*^{1} Add: Also gumph.
1866–1930 in *Sc. Nat. Dict.* **1920** R. MACAULAY *Potterism* VI. i. 209, I never know whether to believe Clare; she's such a gumph. **1945** 'A. GILBERT' *Don't open Door* xv. 124 She might do her best to attract attention—any girl who wasn't a complete gumph would.

gump (gʌ·mp), *sb.*^{2} *U.S. slang.* [Perh. the same word as *sb.*^{1}] A chicken.
1914 JACKSON & HELLYER *Vocab. Crim. Slang* 39 *Gump*,..a chicken; a fowl. **1926** J. BLACK *You can't Win* vi. 64 'I've got a gump in my bindle.'..He unrolled his blankets and produced a live chicken, big and fat. **1931** 'D. STIFF' *Milk & Honey Route* 193 While 'Slippery Slim' and 'Bashful Tim' Croaked gumps for our menu. **1934** *Amer. Ballads & Folk Songs* (1960) i. 24 Not even a shack to beg for a lump, Or a hen-house to frisk for a single gump. **1960** WENTWORTH & FLEXNER *Dict. Amer. Slang* 234/2 *Gump*, a chicken; esp. a live chicken that can be stolen to provide a meal...Because chickens are considered among the most foolish of creatures.

gump (gʌmp), *sb.*^{3} Also gumph. Abbrev. of GUMPTION 1.
1920 *Chambers's Jrnl.* 13 Mar. 233/1 If they weren't so sure of getting me, they would have had the gump to wait a few days till the court adjourns. **1946** 'S. RUSSELL' *To Bed with Grand Music* iii. 44 Any impertinent stranger who had the gumph to ask her out. **1968** K. BIRD *Smash a Glass Image* i. 10 A phoney who hasn't the gump to feed and exercise properly.

gumption. Add: **1.** Also, initiative, enterprise, 'drive'. (Further examples.)
1825 J. T. BROCKETT *Gloss. N. Country Words* 87 *Gumshon, gumption*, common sense, combined with energy. **1844** J. SLICK *High Life N.Y.* II. 134 Sez she, 'Jonathan, what *du* you mean?' My heart riz, it was the only sign of gumption she had made for a hull day. **1862** C. F. BROWNE *A. Ward his Book* 213, I like..your enterprise, gumpshun, [etc.]. **1903** M. E. FREEMAN *Six Trees* 193 Ain't you got any gumption, no snap at all? **1941** J. STREET *In my Father's House* ix. 170 Me and Papa and Mama kissed Teenie and then she just stood there looking at Woodie. Papa said, 'Buss her one, boy. Ain't you got no gumption?'

gumptious, *a.* Add: Also, clever, vain, self-important [by association with BUMPTIOUS]. (Examples.)
[**1823** E. MOOR *Suffolk Words* 158 *Gumshus*, quarrelsome, offensive, obstinate.] **1853** LYTTON *My Novel* I. IV. xii. 320 When I say that sum un is gumptious, I mean..sum un who does not think small beer of hisself. **1860** F. & J. GREENWOOD *Under Cloud* I. iii. 68 You are a gumptious old ooman, you are! Tarrible smaart! **1875** W. D. PARISH *Dict. Sussex Dial.* (1957) 52 *Gumptious*, smart. **1877** BARTLETT *Dict. Americanisms* 271 *Gumptious*, one who has a good opinion of himself; a 'knowing one'.

gum-tree. 2. *to be up a gum-tree*: delete *U.S.* and substitute for def.: to be in great difficulties. Cf. TREE *sb.* 7. (Examples.) *'possum up a gum-tree* (earlier example).
1829 in *Amer. Speech* (1965) XL. 131 Dere's possum up de gum tree. **1926** D. L. SAYERS *Clouds of Witness* iii. 86 S'long as they're in his trousers pocket you're up a gum-tree. **1927** *Daily Express* 15 July 1/4 Captain Lancaster cannot obtain any information, and, as he says, he is 'up a gum tree'. **1959** *Encounter* Nov. 60/2 Until somebody solves the problem of an English idiom we're going to be up a gum-tree.

Gumza, var. *GAMZA.

gun, *sb.* Add: **3. b.** A pistol or revolver. orig. *U.S.*
1744 A. HAMILTON *Itin.* (1907) 150 'Then surely you had needs ride with guns' (meaning my pistols). **1851** R. GLISAN *Jrnl. Army Life* (1874) 80 He might..not fire unless his gun has a revolving chamber with more than one load. **1890** *Harper's Mag.* Dec. 160/2 That six-shooter you gave Pete was such a pretty gun I couldn't resist when Pete offered to swap. **1902** C. HYNE *Mr. Horrocks Purser* 56 Then he made a great fuss and pulled out a gun. **1948** *This Week Mag.* 9 Oct. 22/2 Police believe that if more people carried guns, murders and suicides would zoom. **1971** *Daily Tel.* 26 Oct. 1 The dockers had been unloading a cargo of 72 tea chests containing pistols brought from Rotterdam...The discovery of the guns led to an immediate alert.
c. Any of various devices for discharging missiles or substances through a tube, as by the expansive force of compressed air; usually with defining word, as AIR-GUN, *blow-gun*, **Flit gun*, **grease-gun*, POP-GUN 1, SPRING-GUN 2 (which see).
1895 *Montgomery Ward Catal.* 261/1 The best Insect Powder Gun in the market in which to use insect powder. **1930** *Engineering* 31 Jan. 126/1 The Webb concrete gun has been used by the city's day labour gangs in lining operations. **1937** *Times* 13 Apr. p. iii/3 As many as 3,000 gallons of cellulose preparations are mixed each week, so that 1,000 car bodies can receive colour sprayed from 120 'guns'. **1938** F. D. SHARPE *Sharpe of Flying Squad* xxi. 227 The drug used was in a liquid form and one of the gang possessed a 'gun' loaded with it. He sprayed this dope at the favourites [at horse-racing]. **1968** *Times* 29 Apr. 2/7 The 'gun' is a new way of giving injections without puncturing the skin. It uses a fine but very powerful jet to penetrate the skin. *Ibid.* 27 May 25/2 The company has developed a new tear gas gun.
d. *spec.* in *Athletics.* The starting pistol; hence, the start of a race.
1900 [see NEEDLE *sb.* 12 b]. **1925** T. E. JONES *Track & Field* 18 Keep the mind concentrated on the gun. **1959** *Times* 23 Apr. 16/6 Smith..took the lead from the gun.
e. A hypodermic syringe used by drug addicts. *U.S. slang.*
1904 *San Francisco Chron. Suppl.* 30 Oct. 4/1, I.. reached out my hand for my master, the little syringe, called the 'gun', which always lay ready at my bedside for the early morning 'shot'. **1923** N. ANDERSON *Hobo* vii. 102 One type of dope fiend is the Junkie. He uses a 'gun' or needle to inject morphine or heroin. **1926** MAINES & GRANT *Wise-Crack Dict.* 8 *Gun-toter*, user of a hypodermic needle. **1933** *Amer. Speech* VIII. 27/2 The hypodermic needle and its accessories used for the injection of narcotics are called the gun or artillery. **1955** *U.S. Senate Hearings* (1956) VIII. 4164 *Gun, dropper*, a syringe.
f. = *electron gun* (*ELECTRON^{2} 2 b). Also *attrib.*, as *gun electrode*.
1933 *Electronics* Dec. 333/1 We shall now consider the gun. **1953** AMOS & BIRKINSHAW *Telev. Engin.* I. iv. 52 In cathode-ray tube guns the beam strikes the gun electrodes and releases secondary electrons from them. **1961** G. MILLERSON *Technique Telev. Production* ii. 19 A small gun in the camera-tube generates a continuous beam of electrical particles (electrons). **1971** *Physics Bull.* Oct. 590/1 It offered the advantages of..a colour tube with a single gun.
5. a. (Earlier and later examples.)
1818 KEATS *Let.* [29?] Dec. (1958) II. 18, I went.. shooting on the heath...There were as many guns..as Birds. **1822** VISCOUNTESS ANSON *Let.* 5 Nov. in *Creevey*

Papers (1903) II. ii. 52, 780 head of game were killed by 10 guns. **1970** *Courier-Mail* (Brisbane) 13 Aug. 4/9 The price of being a 'gun'—the name for the shooter—is almost prohibitively high.
c. *pl.* = gunnery-lieutenant. *Naval slang.*
1916 'TAFFRAIL' *Carry On* 25 The first lieutenant..is 'Jimmy the One'; the gunnery and torpedo lieutenants, the 'Gunnery Jack' and 'Torpedo Jack' respectively, but, to their messmates in the wardroom, these three officers, with the officer borne for navigation duties, are usually 'Number One', 'Guns', 'Torps' and 'Pilot'. **1925** FRASER & GIBBONS *Soldier & Sailor Words*, *Gunnery Jack* (also *Guns*), the Gunnery Lieutenant on board ship. **1962** W. GRANVILLE *Dict. Sailors' Slang* 57 *Guns*, wardroom nickname and vocative for the gunnery officer.
d. = *GUN-MAN 1. *U.S.*
1931 C. W. WILLEMSE in *Detective Fiction Weekly* 15 Aug. 123/1 Hey, cap, there's a 'gun' outside. Wants to see you. **1958** R. CHANDLER *Playback* xxiii. 182 Goble was beaten up..tonight—by a hired gun named Richard Harvest. **1965** T. CAPOTE *In Cold Blood* (1966) iv. 275 He was always talking about..making his living as a hired gun.
6. d. *to carry* (or *hold*) (*big*) *guns*: to be in a position of strength or power; *to have* (or *carry*) *the guns for*: to have the ability for (something).
[**1732** T. FULLER *Gnomologia* no. 1824, He carries too big a Gun for me; I must not engage him.] **1867** G. MEREDITH *Let.* 13 Dec. (1970) I. 364 We carry big but immoveable guns, and the work you can supply will be heartily acceptable. **1887** S. BUTLER *Notebks.* (1912) xvi. 256 This gentleman had a decided manner and carried quite as many guns as the two barristers. **1930** *Times* 25 Mar. 17/3 The Chancellor—whose..concern is to make the two ends of his Budget meet—necessarily carries the biggest guns. **1939** A. POWELL *What's become of Waring* iv. 106 But do you really think I carry the guns?..I shouldn't like to think that I was not going to do him justice. **1961** I. MURDOCH *Severed Head* xii. 104 'Why she should have followed it up beats me.' 'You didn't ask her?'..'Of course not! As I told you, she carries too many guns.' **1961** *Times* 8 Nov. 18/7 Miss Catherine Lacy has not the vocal guns for the part of Clytemnestra. **1963** *Times* 26 Feb. 3/5 It was Rangers and Celtic who held the biggest guns.
e. *to beat* (or *jump*) *the gun*: in Athletics, to make a false start; hence *fig.*, to act before the permitted or agreed time.
[**1905** S. CROWTHER *Rowing & Track Athletics* 302 False starts were rarely penalized..and so shiftless were the starters and officials that 'beating the pistol' was one of the tricks which less sportsmanlike runners constantly practised.] **1933** C. LITTLEFIELD *Track & Field Athletics* 31 Do not learn how to try to beat the gun. **1936** WODEHOUSE *Laughing Gas* xxii. 239 Acting swiftly, I did a backwards leap of about five feet six. It was the manœuvre which is known in America as beating the gun. **1942** BERREY & VAN DEN BARK *Amer. Thes. Slang* § 59 *Jump the gun*, to make a false start. **1951** *Economist* 24 Nov. 1258/1 Col. Hanley, judge-advocate of the Eighth Army in Korea, first jumped the gun with statistics. **1955** R. BANNISTER *First Four Minutes* 20 It seemed so unnecessary to beat the gun in a race that would last for 3¾ minutes. **1958** *Economist* 1 Nov. 391/1 The Prime Minister has jumped the gun by announcing that it will take the form of government advances to building societies. **1960** *Guardian* 7 Nov. 8/4 Both candidates jumped the traditional gun of Labour Day.
f. Used in *pl.* and contrasted with *butter* to describe a government policy in which the necessity for military expansion is weighed against the importance of social and economic development.
1936 *Times* 18 Jan. 12/3 Speaking on Germany's rearmament Dr. Goebbels said:—We can well do without butter, but not without guns, because butter could not help us if we were to be attacked one day. Some people say there is a world conscience which is the League of Nations,..but I prefer to rely on guns. **1937** *Daily Herald* 15 Jan. 2/5 A scheme to dissuade Hitler from his 'guns rather than butter' policy. **1938** 'G. ORWELL' *Let.* 26 May in *Coll. Essays* (1968) I. 331 In every country..the supposed necessity to prepare for war is being systematically used to prevent every kind of social advance. It goes without saying that this happens in the Fascist countries, but 'guns before butter' also rules in the democracies. **1968** *Sat. Rev.* 23 Nov. 32/1 The incredible American economy has such unprecedented wealth that it can afford both guns and butter. **1968** *Guardian* 4 Dec. 8/2 The wars in the Yemen and against Israel have added economic depression to endemic poverty. Is it the beginning of a 'guns or butter' argument in Egypt?
g. *at gunpoint*: threatened by a gun.
1958 *Globe & Mail* (Toronto) 4 June 1/5 Baskar was charged after two men robbed cab driver Benjamin Katz..of $25 at gunpoint. **1962** *Times* 3 May 17/3 Three escaping criminals..board a lightship and order the crew at gunpoint to help them reach shore.
h. *to give the gun* (see *GIVE *v.* 14 c).
7. a. Also *big gun*.
1886 [see GATLING]. **1900** J. RALPH *Towards Pretoria* xii. 153 Our big guns were painted like the ruddy earth, and our Maxims were wrapped in canvas great-coats. **1915** A. D. GILLESPIE *Let.* 11 July (1916) 231 Some big guns were firing on our right.., making a horrid noise.
b. (Further examples.)
1834 *Knickerbocker* III. 439 The big guns of the nation are there. **1958** *Times Lit. Suppl.* 14 Feb. 83/1 We must now brace ourselves to receive broadsides from the great guns of science and technology. **1966** B. KIMENYE *Kalasanda Revisited* 41 Mrs. Lutaya's set absolutely refused to accept this high-handed ruling, preferring to remain large fish in their own small pond, rather than compete with the big guns of Gumbi and Male villages.
c. Also *to blow guns*.
1833 A. CONSTABLE *Let.* 15 Feb. in *J. Constable's Corr.* (1962) 273 It rains every night & the wind has blown guns. **1920** 'K. MANSFIELD' *Let.* 7 Oct. (1928) II. 50 It's blowing guns to-day.
d. (Earlier U.S. example.)
1884 'MARK TWAIN' *Huck. Finn* xiii. 115 Great guns! is *he* her uncle?
e. Phr. *to go great guns*: to have a run of success, to advance (rapidly) towards success.
1913 *Field* 3 May 849/3 A moment later Louvois shot out, passed Sanquhar and Fairy King, and going great guns..beat the favourite by a head. **1958** *Times* 26 Sept. 19/2 Local partisanship will no doubt run high between Bristol, who are going great guns so far, and their old.. rivals Gloucester. **1971** *Times* 16 Jan. 16/8 Arsenal, going great guns in their functional, efficient way, must see the league title within their sights.
13*. In full *gun shearer*. An expert sheep-shearer. *Austral.* and *N.Z.*
1898 *Bulletin* (Sydney) 17 Dec., *Gun*..generally speaking a man who can shear over 200 a day. **1947** P. NEWTON *Wayleggo* (1949) iii. 39 There I saw some of our greatest gun (fast) shearers in action. **1952** J. CLEARY *Sundowners* iii. 121 A 'gun' shearer, a crack man, was always welcome in a team. **1956** G. BOWEN *Wool Away!* (ed. 2) iii. 24 While perhaps this may be all right for odd 'guns' it is not a good practice for the majority of shearers or learners. **1970** *Tel.* (Brisbane) 18 Feb. 5/1 (*heading*) 'Gun' shearer only 11.
13.** *Surfing slang.* A large heavy surf-board used for riding big waves.
1963 *Pix* 28 Sept. 62/1 *Big Gun*, big surfboard for heavy surfs. **1965** FARRELLY & MCGREGOR *This Surfing Life* vi. 69, I haven't a gun board myself. For the Australian surf that I call big..the board I use is just a long hot-dog board. **1969** *Surfer* IX. vi. 57 Aipa rides the first wave, a long green wall, accelerating his gun to tremendous velocity across the face of the wave. **1970** *Surf '70* (N.Z.) 44/2 While in Hawaii I had two boards. They were an 8 ft 9 in 'hot-dog' and a 9 ft 6 in tracker type gun.
14. a. *gun-battery, -belt, -butt, -cupboard, -detachment, factory* (earlier U.S. examples), *-flash, -licence* (later examples), *-line, -park* (later examples), *-position, -rack* (later examples), *-ship* (earlier U.S. example), *-shop, -team, -trial.*
1816 H. CLARKE *Hist. War* I. 319/2 The mortar and gun-batteries of the enemy. **1918** E. S. FARROW *Dict. Mil. Terms*, *Gun Battery*, a defense constructed of earth faced with green sods or fascines, sometimes of gabions filled with earth. **1965** *Times Lit. Suppl.* 25 Nov. 1048 A..cop who wears his gunbelt in bed. **1891** KIPLING *Light that Failed* ii, To drag down the slayer till he could be knocked on the head by some avenging gun-butt. **1932** W. FAULKNER *Light in August* (1933) xi. 228 A hand more apt for a rope or a gunbutt..than a pen. **1892** W. W. GREENER *Breech-Loader* 180 If..a dust-proof gun-cupboard, it will last longer. **1860** *Man. Artill. Exerc.* II. 22 The medium 12-pounder requires two gun detachments. **1918** E. S. FARROW *Dict. Mil. Terms* s.v., The cannoneers assigned to the service of a single gun, formed in double rank, constitute a gun detachment. **1780** in *Cal. Virginia St. Papers* I. 372, The warrant for Six thousand pounds on account of the Gun Factory. **1812** *Niles' Reg.* III. 60/2 Messrs. Coggswell and Hosford are erecting a gun factory in Albany. **1950** J. BUSSELL *Puppets & I* iv. 85 Gun flashes (made with wickless cigarette-lighters) denote the start of a battle. **1957** M. K. JOSEPH *I'll soldier no More* (1958) ix. 170 Over to the left, gunflashes lit the sky. **1921** *Daily Colonist* (Victoria, B.C.) 1 Oct. 9/4 For not having his gun licence on his person, W. Whitta was fined $10 and costs. **1965** A. NICOL *Truly Married Woman* 92 He had not bothered to apply for a gun licence. **1945** *Diamond Track* (Army Board, N.Z.) 25/2 The defensive power of an anti-tank gun-line. **1968** *Sunday Truth* (Brisbane) 11 Aug. 2/1 HMAS Hobart has won an American 'fleet citation' for action..while on gunline duty off the Vietnam coast. **1940** 'GUN BUSTER' *Return via Dunkirk* I. ii. 27 Guides dashed off to meet the column and lead it to the gun-parks and vehicle-parks already selected. **1943** *Roof over Britain* 25 The guns are spaced around the sides of the gun park, with the command post at the centre. **1901** 'LINESMAN' *Words by Eyewitness* (1902) 73 From the gun-position one could look down upon line upon line of trenches. **1838** J. MCDONALD *Biogr. Sk. N. Massie* 38 His gun-rack was examined, and there hung his rifle and his pouch in their usual place. **1969** *Daily Colonist* (Victoria, B.C.) 19 Dec. 9/7 (Advt.), Wooden Gun Rack. A favorite gift for the hunter in the family. **1841** L. M. CHILD *Lett. New York* viii. 59 You probably recollect that he built a large gun-ship for the Turkish Sultan. **1865** *Atlantic Monthly* XV. 717 The better class of workmen had gone..to private gun-shops in the North. **1893** M. BEERBOHM *Let.* 13 Aug. (1964) 47 The window of a gun-shop. **1904** *Illustr. London News* CXCVII. 22 (*caption*) Ranks of guns—some of them of the largest calibres—in this British gun-shop betoken good supplies of these naval weapons. **1908** *Westm. Gaz.* 30 Oct. 3/2 Horses.. capable of drawing weight at the pace required in a gun-team. **1898** *Engineering Mag.* XVI. 112/1 Krupp's gun-trial grounds.
b. *gun-pulling, -toter, -toting.* **c.** *gun-battle; gun-armed* adj.
1938 *19th Cent.* Feb. 198 Germany had but few submarines, and of these not many were gun-armed. **1945** *Everybody's Digest* Aug. 89 A gun battle used to bring a puncher out 'a-smokin''. **1967** *Listener* 13 Apr. 486/1 In Aden, British troops and extremists fight gun battle. **1909** 'O. HENRY' *Roads of Destiny* xvi. 271 The bystanders asserted that it was met by the most beautiful exhibition of lightning gun-pulling ever witnessed in the Southwest. **1925** O. P. WHITE *Them was Days* 120 This opened up the field for the renegade white man..the gun-toter, [etc.]. **1948** *Sat. Rev.* 28 Aug. 37/1 His steps were the measured pace of a gun toter. **1912** I. COBB *Back Home* 293, I reckon none of you young fellows..can remember when this wasn't a gun-toting country down here? **1969** *Listener* 23 Jan. 103/3 One can imagine the sight of the gun-toting..in all the Westerns, joined together as one fusillade.
15. gun-barrel (further examples of *attrib.* and *transf.* use); **gun-bright** (see quot.); **gun bus** [*BUS *sb.*^{2} 2 b] (see quot. 1925); **gun-camera** (see quot. 1948); **gun captain,** the captain of the crew of a ship's gun; **gun-case** (earlier examples); **gun club,** the name of a fabric design, freq. used in tweeds, consisting of large checks superimposed over small ones; **gun-dog,** a dog trained to accompany the 'guns'; **gun-fight** *U.S. colloq.*, a fight with revolvers, a shooting affray; **gun-fighter,** one who frequently participates in gun-fights; also *fig.*; **gun-fire** (earlier examples); *spec.* rapid firing in which each gun acts independently and fires as rapidly as it can be loaded; also *fig.*; (*b*) *Army slang*, an early morning cup of tea served out to troops before going on first parade; **gun-fodder** = **cannon-fodder*; **gun-howitzer, gun-how** (see quot. 1942); **gun-lance,** see LANCE *sb.*^{1} 2; **gun-layer,** one who aims or lays a gun; hence *gun-laying* vbl. sb.; **gun microphone,** a moving-coil microphone having a number of parallel tubes of different length in front of the diaphragm to increase its directional property; **gun moll** *U.S. slang*, a female thief (cf. sense 13); an armed woman; **gun-pit,** (*c*) in a fighting aeroplane, the compartment for a gun and gunner; **gun-plane,** a fighting aeroplane armed with a gun or guns; **gun-pointer** = **gun-layer*; **gun-power,** the number and strength of guns available in any given place or circumstances; **gun-range,** (*a*) the range of a gun's fire; (*b*) a place where gun-firing is practised; **gun-reach** = **gun-range* (*a*); **gun-rest,** also, a wall-fixture for portable fire-arms, a gun-rack; (**helicopter**) **gunship,** a heavily armed helicopter; **gun-sight** (see SIGHT *sb.*^{1} 14 b); hence **gun-sighting** *vbl. sb.*; **gun-site,** an emplacement, usually fortified, for guns; **gun-sling,** also, a sling for carrying a portable fire-arm; **gun-slinger** = *GUN-MAN 1; hence **gun-slinging** *vbl. sb.*; **gun turret** (see quot. 1959); **gun-well,** in a submarine, the sunk compartment for a gun; **gun-worker,** one who works in a gun-foundry.
1864 S. HIBBERD *Rose Bk.* 245 Gun-Barrel Budding. *Ibid.* 246 Bud it there at once just under one of the leaf-rings, 'gun-barrel' fashion. **1904** *Westm. Gaz.* 19 Jan. 10/1 The recent gun-barrel fight in Birmingham. **1907** 'ARTIFEX' & 'OPIFEX' *Causes of Decay in Brit. Industry* ii. 25 Gun-barrel welding is one of the handicrafts lost to Birmingham.., whilst it is thriving in Belgium. **1961** C. H. DOUGLAS-TODD *Pop. Whippet* 53 If she has no spring of ribs, a gun-barrel front and so on..do *not* regard her as a foundation brood bitch. **1970** *New Yorker* 22 Aug. 67/1 The children's kind of blindness was identified as tunnel, or gun-barrel, vision—a constriction of the visual fields. **1918** E. S. FARROW *Dict. Mil. Terms*, *Gun-bright*, Dutch rush (*equisetum hyemale*) much used in scouring gun barrels. **1919** BLACKBURN & NEWBY *All about Aircraft* 63 The Vickers' 'Gun Bus'..having a Gnome engine. **1925** FRASER & GIBBONS *Soldier & Sailor Words* 112 *Gun-bus*, Air Force slang for a gun-carrying aeroplane. Specifically applied to the first Vickers' 'pusher' machine, the first aeroplane specially built to carry a machine-gun. **1970** R. JOHNSTON *Black Camels* xii. 183 Out on the flanks, four of Kassim's gun buses were standing by. **1921** *Flight* XIII. 414/1 Dealing with the gun-camera, he said airmen were trained to aim and 'fire' with the gun, and the camera, which was attached, showed what part of their opponent they were actually on when they 'fired'. **1948** A. L. M. SOWERBY *Dict. Photogr.* (ed. 17) 356 *Gun camera*, a camera attached to a gun, usually in a fighter aeroplane, and operated when the trigger of the gun is pulled. These cameras were introduced during the war of 1914–18 for use in training fighter-pilots. **1901** *Westm. Gaz.* 27 June 8/1 The gun captain and layer. **1690** S. NILES *Indian Wars in Mass. Hist. Soc. Coll.* (4th Ser.) V. 275 We took two guns,.. gun-cases and four canoes. **1839** C. SINCLAIR *Holiday House* xv. 333, I observed a gun-case in the saloon. **1848** MRS. GASKELL *Mary Barton* II. v. 70 Having abstracted the paper, and bullets, &c., she saw a woollen gun-case, made of that sort of striped horse-cloth you must have seen a thousand times appropriated to such a purpose. **1939** M. B. PICKEN *Lang. Fashion* 25/1 *Gun club check*, check design used frequently in tweeds, consisting of large check over smaller one. **1967** *Guardian* 7 Sept. 4/3 Pattern and colour in trousers are 'in', Tattersalls, gunclubs, dice checks, overchecks and stripes to blend with jackets. **1744** W. ELLIS *Mod. Husb.* VI. I. 184 The Dog..which, you said, was a Gun-dog and Setter. **1908** *Westm. Gaz.* 5 Mar. 9/3 The most handsome of all English gundogs. **1959** *Elizabethan* July 12/2 The *haute-école* of gundog training demands an intricate relationship of understanding between man and dog. **1898** *McClure's Mag.* Feb. 380 You don't mean there is going to be a gun-fight? **1907** S. E. WHITE *Arizona Nights* II. ii. 252 I'll go yore little old gunfight to a finish. **1961** K. REISZ *Technique Film Editing* (ed. 9) ii. 75 The gun-fight is simply presented in to-and-fro reaction shots. **1894** *Midwinter Appeal* (San Francisco) 27 Jan. 2/3 The gun-fighters rushed up with cocked revolvers and ordered him to halt. **1910** J. HART *Vigilante Girl* xxvii. 374 This man Hawke is a gun-fighter, and as cool and courageous as Tower can be. **1950**

Manch. Guardian Weekly 17 Aug. 7/3 All 'Westerns' are.. strict observers of a moral and social code—..But 'The Gunfighter' goes much farther in moral lecturing. **1964** D. F. DOWD in I. L. Horowitz *New Sociol.* 59 To become ..an intellectual gunfighter. **1801** MARIA LADY NUGENT 30 Oct. in *Jrnl. Voyage..Jamaica 1801–1805* (1839) I. ii. 83 Up at gun-fire. **1814** T. E. HOOK *Let.* 24 Mar. in A. Mathews *Mem. Charles Mathews* (1838) II. xii. 269 Always up by gun-fire, five o'clock. **1870** *Daily News* 13 Oct. 5/5 This same shell disturbed a hare, which..scampered across the battlefield right in a line with the gun fire. **1912** S. E. BURROW *Friend or Foe* x. 125 In the prayer-room they gathered at noon day by day for their 'Gun-fire', and around the Word had the most helpful fellowship. **1916** 'BOYD CABLE' *Action Front* 210 The gunners.. will tell you how they stretched themselves to the call for 'gun-fire'. **1919** *War Slang* in *Athenæum* 18 July 632/2 'Gun fire' for early morning tea. **1926** *Times* 1 Jan. 13/3 After a sharp exchange of gunfire the massive tanks of the new property legislation have rolled over the last ditch. **1928** *Daily Mail* 31 July 13/1 A typical day in the life of a Territorial in camp..is as follows: 6 a.m. Réveillé. 6.30 'Gunfire' (morning tea and biscuits), [etc.]. **1940** 'GUN BUSTER' *Return via Dunkirk* II. xvii. 201 'Dawn just breaking, sir,' he affirmed, shoving into my hand a mug of hot 'gunfire'. **1951** M. MCLUHAN *Mech. Bride* 137/2 One has only to listen to the tense gunfire delivery of radio sports announcers to understand this. **1900** *Westm. Gaz.* 9 July 2/1 Exceedingly useful in the capacity of gun-fodder and stop-gap. **1925** P. GIBBS *Unchanging Quest* xxvii. 207 From historic houses..these boys of ours came as gun-fodder. **1941** KOESTLER *Scum* ix. 47 To fight against its enemies at home, instead of serving as gun-fodder for their purposes. **1940** *Illustr. London News* 20 Jan. 75/1 This drawing shows a gun-howitzer—a weapon unknown in the World War, but of increasing importance in recent years—in the development of which British artillerymen have played a leading part. *Ibid.* 16 Mar. 345 (*caption*) Here we give some photographs of the new 25-pdr. 'gun-how.'—the outstanding artillery novelty of the war. **1942** J. T. GORMAN *Mod. Weapons War* iv. 80 Guns and howitzers, as separate weapons, have been largely superseded by a single, all-purposes 'Gun-How', combining the long range of guns with a howitzer's greater weight of fire. **1901** *Westm. Gaz.* 27 June 8/1 Acting Bombardier Mullen, the gun captain and layer, had a truly marvellous escape. **1906** *Daily Chron.* 13 Aug. 5/7 While carrying out gunlayers' tests with the six-inch guns. **1938** C. DAY LEWIS *Overtures to Death* 47 Brisk at their intricate batteries the German gun-layers go About death's business. **1909** *Westm. Gaz.* 26 July 7/2 (*heading*) Remarkable gun-laying tests. [**1941** W. ABBOT *Handbk. Broadcasting* (ed. 2) i. 8 Two interesting microphones are the machine-gun and the parabolic. The machine-gun accessory.. consists of a series of tubes strapped together through which sound is conveyed to a dynamic microphone which fits into the end.] **1962** A. NISBETT *Technique Sound Studio* i. 23 A 'gun' microphone..is sufficiently directional to pinpoint surfaces which cause echoes in concert halls. **1967** *Punch* 25 Jan. 132/2 The camera-team.. trained their directional gun-microphones on guilty couples. **1908** J. M. SULLIVAN *Criminal Slang* 2 *A gun-moll*, a woman thief. **1910** *National Police Gaz.* (U.S.) 31 Dec. 3/1 When the professional woman thief, who is known to the denizens of the underworld as a gun moll is arrested and taken back to the office, she is searched thoroughly. **1928** M. C. SHARPE *Chicago May* 286 Gun Molls, women who steal from men in the street, or carry guns. **1949** KOESTLER *Promise & Fulfilment* II. v. 279 Fierce-looking Yemenite gun-molls, Sephardi beauties. **1928** C. F. S. GAMBLE *N. Sea Air Station* xii, The German officer..standing in the after gun-pit. **1915** *Times* 4 Oct. 8/4 Our gunplanes carried out during the night a bombardment of the German lines. **1915** W. E. DOMMETT *Aeroplanes & Airships* vi. 75 What has latterly been described as a battleplane or gunplane..does not yet exist in very great numbers. **1904** *Collier's* 16 July 15 As the breech-blocks close with a snap the gun-pointer bends over his sights. **1918** *Chambers's Jrnl.* Dec. 839/1 This time the gun-pointer, having overcome his pardonable excitement, aimed true. **1890** G. S. CLARKE *Fortification* xiii. 176 The actual gun power of the broadside iron-clads. **1928** *Daily Tel.* 11 Sept. 12/4 A division today lacks the tank-power and the gun-power necessary for it to strike as a whole. **1940** W. S. CHURCHILL *Into Battle* (1941) 244 None of the British ships..was..affected in gun-power or mobility. **1852** tr. *Görgei's My Life in Hung.* I. 398 At the distance of three or four gun-ranges from the Monostor. *Ibid.* II. xix. 182 They were..far out of gun-range of our trenches. **1856** Gun-range [see GUN *sb.* 14 a]. **1904** *Daily Chron.* 21 Nov. 5/2 The gun-range at Brassact, near Antwerp. **1954** W. FAULKNER *Fable* (1955) 283 As soon as they can get us up in gun-range again. **1825** C. WATERTON *Wand. S. Amer.* 118 Almost out of gun reach. **1918** W. BEEBE *Jungle Peace* xi, Within gun-reach in front of me. **1925** A. S. M. HUTCHINSON *One Increasing Purpose* III. xv, Pike-rests.. Not gun-rests; they are too far apart for that. **1968** *Times* 3 Feb. 8/3 Helicopter 'gunships' armed with machine-guns accounted for most of the toll. **1969** I. KEMP *Brit. G.I. in Vietnam* iii. 63, I saw two Huey gun-ships—assault helicopters—swooping down towards us...I listened gratefully to the whoosh of its [*sc.* the leading gun-ship's] 2.75 inch rockets and the stutter of its M60 machine guns. **1969** *Australian* 7 June 2/7 Other RAAF gunships remained overhead until the crew were lifted out. **1867** Gun-sight [see GUN *sb.* 14 a]. **1908** *Westm. Gaz.* 17 Sept. 5/1 It was discovered that all the gunsights in the ship had been removed. **1941** C. MORGAN *Empty Room* i. 10 'Bomb-sights and the Paramounts.' 'The what?' 'That's what I call the fighter gunsights.' **1905** *Daily Chron.* 5 Apr. 8/5 Gun-sighting platforms. **1943** *Hutchinson's Pict. Hist. War* 17 Feb.–11 May 43 (*caption*) A Bofors anti-aircraft gun manned by men of the U.S. Army at a gun-site situated on the coast of Algeria. **1944** *Times* 3 Feb. 6/1 At the beginning of this year the American gunners took over a gunsite on London's outskirts. **1812** *Niles' Reg.* II. 131/1 The purveyor of public supplies advertises for..25000 gun slings. **1907** *Yesterday's Shopping* (1969) 653/2 Gun and Rifle Slings. Webbing—3/9. **1953** in Wentworth & Flexner *Dict. Amer. Slang* (1960) 236/1 The gun-slinger will spend ..his life behind bars. **1960** *Spectator* 4 Mar. 321 Yet another brutalised gun-slinger. **1967** *Boston Sunday Herald* 7 May *Show Guide* 2/4 (*caption*) The gunslinger.. comes to town, cigar between teeth, his prowess with a gun for sale. **1944** R. F. ADAMS *Western Words* (1945) 70/1 *Gun slinging*, slang for the act of shooting. **1958** *Church Times* 12 Sept. 3/1 The EOKA boycott is resented even more than the EOKA gun-slinging, for it affects every single citizen. **1916** *Flying* (Aero Club of America) Jan. 820/1 The Sturtevant Battleplane is a biplane of tractor type built with remarkable simplicity and..attention to efficiency. There are many novel features, including the steel construction, the placing of gun turrets on either side of the central body. **1919** A. KLEMIN *Text-bk. Aeronaut. Engin.* 175 Pilot forward machine gun firing through propeller. Passenger in rear with circular gun-turret. **1935** *Jrnl. R. Aeronaut. Soc.* XXXIX. 988 In the case of rotatable gun turrets for aircraft, it is proposed to provide a removable top so as to allow the gunner to escape in case of emergency. **1959** J. L. NAYLER *Dict. Aeronaut. Engin.* 125 *Gun turret*, a gun position in an air-craft under the control of an air-gunner...Modern gun turrets are power-operated, equipped with gyro gun-sights and often radar ranged and fired automatically. **1915** *Illustr. London News* CXLVI. 234/1 The deck of a German submarine with the hatch of the gun-well open. **1905** *Spectator* 4 Mar. 311/2 A meeting of gun-workers.. held at Birmingham on Monday.

gun, *v.* Add: **1. c.** *Stock Exchange.* (See quot.)
1870 J. K. MEDBERY *Men & Myst. Wall St.* 136 Gunning a stock, is to use every art to produce a 'break', when it is known that a certain house is heavily supplied, and would be unable to resist an attack.
d. To shoot (a person). Also with *up*. *U.S. colloq.*
1898 H. S. CANFIELD *Maid of Frontier* 83 I'll gun you if you do that again. **1916** H. L. WILSON *Somewhere in Red Gap* i. 35 Wilfred went pasty, indeed, thinking his host was going to gun him. **1923** L. J. VANCE *Baroque* xxvii. 178 If you don't want to have your Fiancy gunned up without notice by some wild-eyed wop. **1934** R. CHANDLER in *Black Mask* Oct. 36/1 Canales had no motive to gun Lou, unless it got back the money.
e. *Forestry.* (See quot. 1957.)
1905 *Terms Forestry & Logging* 39 *Gun*, to aim a tree in felling it. In the case of very large, brittle trees, such as redwood, a sighting device (gunning stick) is used. **1957** *Brit. Commonw. Forest Terminol.* II. 88 *Gun*, in felling, to select the direction in which a tree should fall.
2. *to gun for*: also, to go after or in search of; to seek to attack, harm, or destroy (someone).
1893 W. K. POST *Harvard Stories* 188 That bull Mick Shreedy is gunning for me just at present. **1903** *N.Y. Times* 29 Sept. 1 Others talked of mysterious influences that had been 'gunning' for financiers of prominence. **1922** *Daily Mail* 5 Dec. 9 Observing that the Company's statement is not a denial of the assertion that it is 'gunning' for the Mesopotamian oilfields claimed by the heirs of Abdul Hamid. **1930** 'E. QUEEN' *French Powder Mystery* xix. 171 Mr. Trask has been gunning for Bernice [with a view to marriage] for over a year. **1936** WODEHOUSE *Laughing Gas* xviii. 198 Nice little bit of luck, finding her like that...Matter of fact, I wasn't gunning for her at all, really. I came to get that notebook. **1950** G. GREENE *Third Man* iii. 31 I'm gunning..for Colonel Callaghan. **1955** *Times* 16 June 12/2 You found when you came back from Oslo that for other reasons the Communist Party was 'gunning' for Mr. Frankel? **1958** *Observer* 10 Aug. 3/2 Last week American commentators were gunning for Mr. Dulles ('too busy, too tired, too discouraged, too stale,' said Walter Lippmann..). **1960** C. DAY LEWIS *Buried Day* ix. 204, I felt that 'They' were gunning for me again.
3. *trans.* To look at closely, to examine.
1819 J. H. VAUX *Mem.* II. 179 To *gun* anything is to look at or examine it. **1859** G. W. MATSELL *Vocabulum* 39/2 The copper gunned me as if he was fly to my mug. **1946** MEZZROW & WOLFE *Really Blues* xii. 223 They wanted to be..alert and keen-sighted.., gunning everything.
4. = *to give the gun* (*GIVE *v.* 14 c). Hence **gunned** *ppl. a.*, **gunning** *vbl. sb.*
1930 *Amer. Speech* V. 290 *Gun the motor*, accelerate the motor [of an aircraft]. **1941** N. ALLEY *I Witness* 308 We gunned into an easy takeoff. **1943** R. CHANDLER *Lady in Lake* (1944) xxxiv. 180 Degarmo let the clutch in and gunned the motor and hit forty in the first block and a half. **1965** G. MCINNES *Road to Gundagai* xiii. 224 Dad's favourite manoeuvre..was to..gun the bike roaring down the front path. *Ibid.* xv. 274 A car door crashing shut and a terrific gunning of the motor. **1967** 'J. CROSS' *To Hell for Half-a-Crown* i. 13 The car went by, with the heavy roar of the gunned motor. **1968** P. DURST *Badge of Infamy* xvi. 171 He gunned the Volkswagen and fell in behind.

guna, *sb.* Add: **2.** In the Sankhya philosophy of India, any one of the three dominating principles of nature.
1862 VIJNANA BHIKSHU *Sankhya-Sâra* 6 S'ankara's own definition is in these words: 'The reflecting, that the *gunas*,—goodness, passion, and darkness,—are objects of my perception, [etc.]'. **1875** M. MONIER-WILLIAMS *Indian Wisdom* 67 The dominance of one or other of the three Gunas, goodness, passion, or darkness. **1962** BRAHMACHARINI USHA *Ramakrishna-Vedanta Wordbk.* 34 Man's mood and character vary according to the predominating guna. **1968** *Jrnl. Mus. Acad. Madras* XXXIX. 25 An exposition of Rama's *Gunas* as depicted by Tyagaraja.

gunboat. Add: **1. b.** *gunboat diplomacy*, diplomacy supported by the use, or threatened use, of military force.
1927 *U.S. Naval Inst. Proc.* Feb. 234 It has been said that the days of 'gunboat diplomacy' in China are over. **1960** *Daily Tel.* 2 Sept. 12/2 'Gun-boat' diplomacy has long and properly disappeared. **1961** *Ibid.* 6 July 1 The Iraqi delegate called the British action in Kuwait 'gun-boat diplomacy at its worst'. **1969** *Times* 3 Mar. 23/4 (Advt.), There's nothing to regret about the passing of gunboat diplomacy.
3. Usu. in *pl.* Large shoes or feet. *U.S. slang.*
1870 D. MACRAE *Americans at Home* I. 68 Most of the people wear rubbers over their boots—gunboats as they sometimes call them from their size. **1919** *Dialect Notes* V. 69 *Gunboats*, the feet. 'Keep your *gunboats* out of my way.' **1932** J. FARRELL *Studs Lonigan* (1936) I. ii. 22 Reardon nodded as he shifted his weight from the right to the left gunboat. **1939** *Amer. Speech* XIV. 90 *Gun boats*, big shoes. **1951** in Wentworth & Flexner *Dict. Amer. Slang* (1960) 235 He brought some of the [size 14 EE] gunboats with him from the states, but they wore out.

gundy, var. *GOONDIE.

gung ho (gʌŋ hōu). Also **kung-hou.** [Chin. *kung* work + *ho* together.] A slogan adopted in the war of 1939–1945 by the United States Marines under General E. Carlson (1896–1947); hence as *adj.*: enthusiastic, eager, zealous.
1942 *Times Mag.* (New York) 8 Nov. 13/4 Borrowing an idea from China, Carlson frequently has what he calls 'kung-hou' meetings...Problems are threshed out and orders explained. **1943** *Life* 20 Sept. 58, I [*sc.* E. Carlson] told them of the motto of the Chinese Co-operatives, *Gung Ho.* It means Work Together...My motto caught on and they began to call themselves the *Gung Ho* Battalion. **1959** *She* May 21/3 All would be *gung ho* (in favour) of declaring him a *fungus* (musty character). **1967** R. M. STERN *Kessler Legacy* (1968) xvi. 143 In those days he was very gung ho for National Socialism and the pan-Germanic grandeur it was going to produce. **1968** M. LOCKWOOD *Accessory* (1969) ii. 59 I've always thought of you as being—I don't know, full steam ahead about life. Not gung-ho exactly, but—well always ready to manage and organize things. **1969** I. KEMP *British G.I. in Vietnam* iii. 49 He..was one of the most 'gung-ho' (exceptionally keen to be personally involved in combat) characters I ever met. **1970** *Times* 28 May 7/5 Today's grunts are noticeably different from those who filled the enthusiastic 'Gung-ho' units of a few years back.

gunibri, gunibry, varr. *GIMBRI.

gunite (gʌ·nəit). *Build.* Also gunnite.. [f. GUN *sb.* + -ITE^{1} 4 a.] A mixture of cement, sand, and water applied through a hose. Also *attrib.*
1914 C. WEBER in *Jrnl. Western Soc. Engineers* Mar. 278 The product of the Cement Gun is called 'Gunite' in order to distinguish it from the ordinary cement stucco or plaster work, and for the reason that the material develops qualities which put it in a class all by itself. Gunite is usually a cement mortar composed of about one part of Portland cement and three parts of coarse, sharp sand passing through a ⅜ in. screen. **1930** *Engineering* 21 Mar. 368/3 The upstream face of the domes was covered with a layer of gunite, 1½ in. thick, reinforced with wire mesh. **1962** *Ibid.* 12 Jan. 52/1 The arduous conditions imposed on the hosing of cement guns has led to the development of an improved form of gunite hose. **1970** *Telegraph* (Brisbane) 30 Jan. 18/2 Swimming pools..built to the highest engineering standards of concrete and steel, using the latest gunnite spray methods.

gunk (gʌŋk). orig. *U.S.* **1. a.** [Proprietary term.] (See quots. 1948 and 1970.)
1932 *Official Gaz.* (U.S. Pat. Off.) 23 Aug. 864/1 A. F. Curran Co.,..Gunk. For Liquid Soaps and Liquid Cleaners for Hard Surfaced Materials or Articles. **1948** MENCKEN *Amer. Lang.* Suppl. I. v. 347 *Gunk*, a self-emulsifying colloidal detergent solvent...Launched by the Curran Corporation, Malden, Mass., in 1932. **1969** J. LEASOR *They don't make them like that any More* i. 14, I..poured Gunk all over the engine to freshen it up. **1970** *UK Trade Names* 221/1 Gunk—degreasing compound—Bennett (Hyde) Ltd, Boston Mills, Hyde, Cheshire.
b. Any of a variety of viscous or liquid substances. *slang.*
1949 in Wentworth & Flexner *Dict. Amer. Slang* (1960) 235/2 You can actually see the gook and gunk that drains out with your [motor] oil. **1950** R. STARNES *Another Mug for Bier* (1952) xx. 82, I smeared additional gunk on my eye. **1958** C. C. ADAMS *Space Flight* iii. 84 Work has also been carried on with jelly and related propellants, sometimes affectionately known as 'gunk'. **1958** I. ASIMOV *Whiff of Death* (1968) v. 46 Foster referred to the useless materials prepared by side-reactions in the course of a synthesis as 'gunk'. **1959** *N.Y. Times* 30 May 2/1 Its openings have been washed with a black preservative 'gunk'. **1965** *Guardian* 19 Aug. 6/5 'Gunk' is..a crude rubber which he moulds on to costumes. **1966** C. HENRY *Third Twin* v. 66 Too much eye gunk and lipstick—that sort of girl.
2. A derogatory term for a person. *slang.*
1964 *Publ. Amer. Dial. Soc.* XLII. 30 Chinaman, Jap.. gunk. **1968** P. MARLOWE *Hire me Hearse* vi. 88 A couple of gunks who used to be bouncers at the 'Golden Pagoda'. **1968** J. D. MACDONALD *Pale Grey for Guilt* (1969) vi. 74 It was a drag to listen big-eyed to that tired gunk and say Oh and Ahh.
3. *Comb.* **gunk-hole** *colloq.*, a small cove or fishing hamlet.
1908 G. S. WASSON *Home from Sea* x. 317, I rec'lect well hearing of her call it nothing only a dead-and-alive little two-cent gunk-hole of a place. **1927** G. BRADFORD *Gloss. Sea Terms* 78/1 *Gunk hole*, a small narrow channel

dangerous to navigate owing to current and to numerous rocks and ledges. **1958** K. M. WELLS *Cruising Georgian Bay* 94 And then you may cruise back to your hidden gunk hole near the inlet entrance and enjoy peace.

gun-man, gunman. 1. Delete 'Now *rare*' and add further examples. Also *fig.*
1903 *N.Y. Sun* 23 Nov. 1 A notorious outlaw and one of the most expert gun men of the West. **1913** *Industrial Worker* (Spokane, Wash.) 3 July, Last evening one of the gunmen of the A.C.M. Co. came to the jail and tried to get a blacksmith out of here. **1921** *Daily Colonist* (Victoria, B.C.) 15 Oct. 4/7 The gunman who yesterday was foiled in his attempt to rob the Bank of Hamilton, robbed the Union Bank today. **1931** *Star* 8 May 6/2 Lord Hailsham is rapidly making a name for himself in the House of Lords as a gunman of debate. **1946** KOESTLER *Thieves in Night* xiv. 277 He seems to regard him as a kind of gunman-messiah. **1972** *Daily Tel.* 13 Jan. 1 The Army said five IRA gunmen were shot and injured during the two-and-a-half-hour battle.
Hence **gu·nmanship,** the status, performance, or practice of a gunman.
1925 L. O'FLAHERTY *Informer* vi. 78 We can imagine him perfecting himself in the arts of gunmanship. **1959** I. JEFFERIES *13 Days* ii. 29 We would practice academic gunmanship...The idea was to jerk the gun and fire it off like a rattle. **1965** *Economist* 20 Feb. 791/1 Just how serious was London's recent outbreak of criminal gunmanship?

gun-metal. Add: **2.** A colour resembling that of gun-metal, a dull bluish-grey.
1905 *Daily Chron.* 12 July 6/7 Gun-metal coloured taffeta. **1923** *Daily Mail* 8 Oct. 1/2 In Gun Metal, Shoe Grey and Black. **1931** *Daily Tel.* 21 May 6/3 A train of gun-metal chiffon. **1952** C. DAY LEWIS tr. *Virgil's Aeneid* II. 42 Its angrily-rearing head, its gun-metal neck.

Gunn (gʌn). [The name of John Battiscombe *Gunn* (b. 1928), physicist, who first observed the effect in 1963.] *Gunn effect*: a phenomenon observed in certain semiconductors (as gallium arsenide and indium phosphide), in which a (constant) electric field greater than some threshold value applied between opposite faces of a thin piece of the material results in an oscillatory electric current with a frequency in the microwave region. So *Gunn diode*: a semiconductor diode in which the Gunn effect is produced.
1964 H. KROEMER in *Proc. IEEE* LII. 1736/1 Gunn (1) has recently discovered a new kind of current oscillations at microwave frequencies, in *n*-type GaAs and InP...The purpose of this correspondence is to point out that most.. of the known properties of the Gunn effect can be explained, at least quantitatively, if [etc.]. **1966** *Ibid.* LIV. 1480/2 Oscillations covering 4 Gc/s to 31 Gc/s from a single Gunn diode. **1969** *New Scientist* 4 Dec. 511/2 A solid-state microwave source—a Gunn diode—is used as part of a low-cost electron spin resonance (ESR) experimental system. **1971** I. J. KAMPEL *Semiconductors* xix. 198 Devices such as the Gunn-effect diode considerably extend the range of the electromagnetic spectrum in which we can control and generate frequencies.

gunner. Add: **1. c.** *Master gunner*: delete †*Obs.* and for def. read: the chief gunner in charge of ordnance and ammunition, formerly *spec.* an officer under the Crown, the name still being retained as an honorary title conferred on distinguished soldiers; also, in more recent use, a warrant officer in the Royal Artillery who has charge of the stores and equipment in a fortification or other armed place.
1647 L. HAWARD *Charges Crown Rev.* 21 Master Gunner of England: Fee, *per diem* 2.s. 6.d. **1802** C. JAMES *Mil. Dict.*, *Master gunner*, in a ship of war, an officer appointed to take charge of the artillery and ammunition aboard, and to teach the men the exercise of the great guns. **1876** VOYLE & STEVENSON *Mil. Dict.* **1880** TENNYSON *Revenge* xi, Sink me the ship, Master Gunner. **1904** *Standard* 20 Dec. 5/3 Field Marshal Earl Roberts, V.C., K.G., has been appointed Master Gunner of St. James's Park. **1918** E. S. FARROW *Dict. Mil. Terms*, *Master Gunner*...In the United States, an enlisted specialist of the non-commissioned staff Coast Artillery Corps, who is assigned to duty as assistant to the Artillery Engineer in connection with the preparation of charts, maps, drawings, range tables, etc., in a coast defense command. **1966** *Chambers's Encycl.* VI. 667/1 Master-gunners are warrant-officers of artillery, generally placed in charge of one or more forts.
e. A member of an aircraft crew who operates a gun. Cf. *aerial gunner* s.v. *AERIAL *a.* 5, *air-gunner* s.v. *AIR *sb.*^{1} B. III. 3.
[**1910** C. C. TURNER *Aerial Navig.* xv. 211 Swift must be the aerial gunner's calculation.] **1918** E. MIDDLETON *Glorious Exploits of Air* (*caption*, facing p. 240) He dived to 500 feet, thus enabling his gunner, 1st A.M. Donald, to open heavy fire on them. **1919** BLACKBURN & NEWBY *All about Aircraft* 25 To be efficient a gunner must have a good eye. **1935** C. G. BURGE *Complete Bk. Aviation* 168/2 Many interesting developments have been designed to protect the gunner from the air pressure when flying at these high speeds. **1968** *Encycl. Brit.* X. 1037/1 To assist the gunner in manual aiming, a movable wind vane type of gun sight, calibrated and designed to compensate for the forward motion of the gunner's own aircraft, is often placed on the forward end of the gun barrel.
5. *horse gunner.* For def. read: A member of the Royal Horse Artillery.
1896, 1897 [see Dict.]. **1931** *Geogr. Jrnl.* LXXVIII. 121 Two other officers, one a Sapper and one a Gunner—I believe I ought to say one of the Royal Engineers, one of the Royal Artillery.

gunnera (gʌ·nĕrā, gʌnīə·rā). *Bot.* [mod.L. (Linnæus *Mantissa Plantarum* (1767) 16), f. the name of J. E. *Gunnerus* (1718–73), Norwegian bishop and botanist.] A plant of the genus so called, esp. *G. manicata*, cultivated for its large ornamental leaves; cf. *prickly rhubarb* (s.v. *PRICKLY *a.*).
1789 W. AITON *Hortus Kewensis* III. 304 Marsh Marygold-leav'd Gunnera. **1823** *Curtis's Bot. Mag.* L. 2376 (*heading*) Cape Gunnera. **1900** W. G. BAKER in W. D. Drury *Bk. Gardening* xx. 891 Conspicuous among waterside plants that are valued chiefly for their ornamental foliage are the giant Gunneras. **1930** *Times Lit. Suppl.* 18 Dec. 1079/2 We find bilberry, ling and woody nightshade included, but not guelder-rose nor gunnera. **1961** *Amateur Gardening* 23 Dec. 2 From the water garden we could take a stalk or two of the giant gunnera.

gunnery. Add: **6. gunnery jack** *Naval slang*, a gunnery-lieutenant.
1904 KIPLING *Traffics & Discov.* 53 The Gunner mops up a heathenish large detail for some hanky-panky in the magazines, an' led 'em off along with our Gunnery Jack. **1908** *Daily Chron.* 5 Oct. 4/6 A smarter man than Gunnery Jack is not to be found in the services. **1925** [see *GUN *sb.* 5 c].

gunning, *vbl. sb.* **3.** Delete †*Obs.* and add later example.
1917 H. MACFALL *Germany at Bay* (1918) 210 The enormous gunning and munitionment of the Germans made trench warfare a costly affair for such as had it not.
4. gunning-stick, a device used by lumbermen in guiding the falling of a tree.
1905 [see *GUN *v.* 1 e].

gunny. Add: **b.** *gunny-bag* (earlier examples), *-sack* (earlier and later examples), *-sacking.*
1764 *Account of Expences* 5 Sept. in J. Long *Select. Unpubl. Rec. Bengal* (1869) I. 384 Baskets, gunny bags, and dubbers. **1798** WELLINGTON *Suppl. Desp.* (1858) I. 101 Shot carried on the backs of bullocks in gunny bags. **1862** W. S. EBEY *Diary* 6 July in *Pacific Northwest Q.* (1943) XXXIV. 42 To look at the claims on the hill. They are on the rime rock & prospect very fine, as much as a dollar to the gunny sack. **1960** 'A. BURGESS' *Doctor is Sick* 166 He had brought with him a gunny sack which he emptied on to the floor—more fruit and vegetables. **1965** G. MCINNES *Road to Gundagai* ix. 142 Across the opening hung gunny-sacking.

gu·n-play. orig. *U.S.* [*GUN *sb.* 3 b.] The use of fire-arms; a shooting affray; skill in shooting.
1897 *Typogr. Jrnl.* Jan., The talk of 'gun-plays' at the Federation Convention..is without a germ of truth. **1899** ADE *Doc. Horne* vi. 57 So you made good with the gun-play did you. **1904** *N.Y. Even. Post* 7 Nov. 1 The feud has been the cause of several assaults and gunplays in Chinatown. **1913** C. E. MULFORD *Coming of Cassidy* ii. 30 I've heard so much about Bradley's gun-play that I'm some curious. **1928** *Sunday Dispatch* 22 July 9/3 Other charges of gunplay were made against the officer. **1956** *Times Lit. Suppl.* 9 Nov., Years of disorder and lightly controlled gun-play. **1958** *Listener* 9 Oct. 553/1 Perhaps gunplay and fisticuffs are as formal a part of the 'western' as blood and bodies are of the detective novel.

gunpowder. Add: **1. c.** (Later example.)
1791 J. IRELAND *Hogarth Illustr.* I. 226, I think it is probable that these gunpowder initials are merely the marks of a woman of the lowest rank, and most infamous description.
4. gunpowder weed *S. Afr.*, a name for *Silene gallica*, a plant having small black seeds.
1860 HARVEY & SONDER *Flora Capensis* I. 127 *Silene gallica*...This is the 'Gunpowder-weed' of the colonists; its black seeds resembling powder. **1897** EDMONDS & MARLOTH *Elem. Bot. S. Afr.* xvii. 139 *Silene*...Several species of Campion, also the so-called 'Gunpowder Weed' (*S. gallica*). **1926** J. B. DAVY *Man. Fl. Plants Transvaal* I. 149 S[ilene] gallica...Gunpowder weed...The black seeds resemble grains of gunpowder, hence the vern. [*sic*] name. **1946** M. WILMAN *Check List Fl. Plants Griqualand W.* 16 S[ilene] gallica..an increasingly frequent weed; September. *Gunpowder weed.*

gunsel (gʌ·nsəl). *U.S. slang.* Also gonsil, gunshel, gun(t)zel. [ad. Yiddish *genzel*, f. G. *gänslein* gosling, little goose.] **1.** A (naïve) youth; a tramp's young companion, male lover; a homosexual youth.
1914 JACKSON & HELLYER *Vocab. Criminal Slang* 40 *Gunshel*, current amongst yeggs chiefly. A boy; a youth; a neophyte of trampdom. **1923** N. ANDERSON *Hobo* vii. 101 *Gonsil*, youth not yet adopted by jocker. **1927** *Amer. Speech* II. 387/2 Men can be observed traveling with boys...The boy has many names,—punk, gazooney, guntzel and bronc. Punk, guntzel and gazooney are also used to refer to any sort of green lad. **1929** D. HAMMETT in *Black Mask* Nov. 43/1 Keep that gunsel away from me while you're making up your mind. **1931** G. IRWIN *Amer. Tramp & Underworld Slang* 88 *Gonsil*, a young tramp, not yet taken in hand and bent to his will by an older man. A boy. A passive male homosexual, usually a youth or younger man. Also gunsel. **1946** MENCKEN *Amer. Lang.* xi. 582 The tramp..carries a boy with him, to rustle food for him and serve him otherwise..the boy is a punk,..guntzel. *Ibid.* 584 Gonov, which means a thief to thieves, means a fool on the carnival lot, and the same meaning is given to guntzel, which means, in the jungles, the boy companion of a tramp.
2. [as if f. GUN *sb.*] An informer, a criminal, a gunman.
1950 H. E. GOLDIN *Dict. Amer. Underworld Lingo* 89/1 *Gunzel*,..(by extension) an informer, a weasel, an unscrupulous person. **1951** *New Yorker* 3 Mar. 26/3 Scores of hoodlums, gunsels, informers. **1959** G. FISHER *Hospitality for Murder* xix. 151 Bruce thought he'd never make a reliable gunsel. **1964** W. MARKFIELD *To Early Grave* (1965) vi. 109 After all, didn't Ben Gurion himself hand her a blank cheque, she should have what to hire a couple gunsels? **1965** J. WAINWRIGHT *Web of Silence* 34 The voice said: '..This is a gun—and it's silenced. Okay?'.. I tried to work out what sort of a voice the gunsel had.

gun-shy, *a.* Add: (Later examples.) **gun-shyness** (later example).
1938 C. DAY LEWIS *Overtures to Death* 59 He was gun-shy, Hating all quarrels. **1957** R. CAMPBELL *Portugal* 116 The gun-shy poets of the 'left wing'. **1960** C. DAY LEWIS *Buried Day* v. 91 What caused this gun-shyness, I have no idea.

guntz (gʌnts). *slang.* [Perh. f. Yiddish *gants* whole, f. G. *ganze* whole, entirety.] 'The whole lot, the whole way' (Partridge, *Dict. Slang* Suppl., 1961).
1958 F. NORMAN *Bang to Rights* III. 137 This time I went the guntz and blaged her for a grand. **1960** *News Chron.* 16 Feb. 6/3 This bird Barbara Windsor, who's had the real guntz from the audience on opening night—cheers, encores, the lot. **1967** J. MORGAN *Involved* 147 You don't want a pay-day, you boys are asking for the guntz.

gunyang (gʌ·nyæŋ). *Bot.* [Native name.] An Australian shrub, *Solanum laciniatum* or *S. vescum*, which bears edible orange fruits; cf. *kangaroo apple* (s.v. KANGAROO *sb.* 4 b).
1867 W. WOOLLS *Contrib. Flora Austral.* 125 The native name of this fruit is 'gunyang'. **1874** LINDLEY & MOORE *Treas. Bot.* Suppl., Gunyang, *Solanum vescum*, known by this name in Victoria, where the native tribes collect its fruits eagerly as an article of food. **1888** D. MACDONALD *Gum Boughs* 222 A dwarfed ti-tree, clinging low about the ground, like the gunyang or kangaroo apple. **1911** W. R. GUILFOYLE *Austral. Plants* 334 *Solanum aviculare* 'Kangaroo Apple'. 'Bulli Bulli', or 'Gunyang' (herbaceous shrub, 6 to 8 ft.). **1965** *Austral. Encycl.* VIII. 196/1 S[olanum] aviculare, ranging almost across the temperate part of the continent..is the kangaroo apple—called *gunyang* by former Victorian aborigines and *poroporo* or *kohoho* by the Maoris.

gunyio, var. GUNYAH.
1848 H. W. HAYGARTH *Recoll. Bush Life Austral.* x. 105 Comfortably sleeping in an adjacent 'gunyio', or camp. *Ibid.* xii. 132 Perhaps the most primitive boat in the world: like the 'gunyio's', or huts, of the aborigines, it is built in a few minutes.

Günz (gʏnts). *Geol.* The name of a tributary of the Danube in southern Germany, adopted by A. Penck (in Penck & Brückner *Die Alpen im Eiszeitalter* (1909) I. I. 110) and used *attrib.* to designate the first Pleistocene glaciation in the Alps. Also *absol.*
1910 *Zeitschr. f. Gletscherkunde* IV. 244 The two older, known as Günz and Mindel, are separated from each other by a long interval of deglaciation. *Ibid.* 251 The Günz drift is almost completely covered by later glacial deposits. **1923** A. L. KROEBER *Anthropol.* ii. 21 The period of the stratum is generally considered early Pleistocene, possibly approximately contemporary with the first or Günz glaciation of Europe—nearly a million years ago. **1970** I. CORNWALL *Ice Ages* iii. 62 The North American correlative of European Günz was generally supposed to be the Nebraskan glaciation.

gup, *sb.* Add: Also in general *colloq.* use with the sense: silly talk, blather, nonsense. (Further examples.)
1848 J. H. STOCQUELER *Oriental Interpreter* 94/2 *Gup*, or *Gup-shup*, the origin of *gossip*, to which, in India, it bears the closest possible affinity. **1868** F. MARRYAT (*title*) 'Gup'. Sketches of Anglo-Indian Life and Character. **1883** B. M. CROKER *Pretty Miss Neville* II. xvi. 23 Passing friends..related the local 'gup'. **1924** GALSWORTHY *White Monkey* II. i, The usual 'gup' over the dividend. **1927** *Punch* 30 Mar. 340/3 Need I give the jury any more of this gup? **1942** S. HOPE *Sea Breezes* 36 With little to do except drink sherbert and listen to the gup—gossip, which, incidentally, they didn't understand. **1962** 'W. HAGGARD' *Unquiet Sleep* v. 41 The gup in the business is that Hassertons have agreed.

guppy^{1} (gʌ·pi). [f. the name of R. J. L. *Guppy*, a Trinidad clergyman who sent the first recorded specimen to the British Museum, used as the specific epithet in *Gerardinus guppyi* (A. Günther *Catal. Fishes Brit. Mus.* (1866) VI. 353), the name used when the fish was first described.] A small fish, *Lebistes reticulatus*, originally from the West Indies, well suited for the aquarium. Cf. *MILLION 2 c.
1925 *Aquatic Life* Nov. 110/2 The following are live-bearing tropicals:..Guppy (Lebistes reticulatus). Males

small and brilliantly colored. **1927** A. E. Hodge *Tropical Aquarium-Fishes* v. 75 One of the best known of the Killifishes..is the Guppy, or Rainbow Fish (*Lebistes Reticulatus*) commonly known as 'Millions'. **1950** O. Nash *Family Reunion* (1951) 71 Seals have puppies, But guppies just have little guppies. **1962** *Listener* 22 Nov. 852/2, I got some guppies—hardy little rainbow-coloured fish.

guppy[2] (gʌ·pi). orig. *U.S.* [f. greater under-water propulsive power + -Y[6].] A submarine which has been streamlined and equipped with a schnorkel. Also *attrib.* Hence **guppy** *v. trans.*, to streamline (a submarine).

1948 *Sci. News Let.* 27 Mar. 200/3 Called the 'Guppy' program, the alterations involve streamlining the hulls of the fleet submarines by reducing the size of their superstructures and by removing deck guns and other topside appendages to cut down on under-water resistance. **1948** *News* (Birmingham, Alabama) 13 Oct. 12/6 However, the Tigrone was not streamlined into the high speed 'guppy' class with which the Navy is experimenting. **1949** *Jane's Fighting Ships 1949–50* 380 *Pickerell* [U.S. Navy] is of the new 'Guppy' (Greater Underwater Propulsive Power) design and is equipped with the latest devices. *Ibid.*, *Cutlass, Sea Leopard, Sirago* converted to 'guppy-schnorkel'. **1959** *Jane's Fighting Ships 1959–60* 44 The boats in this class which have not been 'guppied'.

Gupta (gu·ptă), *a.* and *sb.* [f. *Chandragupta*, name of the founder of the dynasty.] **A.** *adj.* Of or pertaining to a dynasty which ruled in north India from the fourth to the sixth century A.D. **B.** *sb.* A member of this dynasty. Hence **Gu·ptan** *a.*

[**1845** *Encycl. Metropolitana* XVI. 354/2 Four families of the Vaidyas were raised by the same Prince to the rank of Culînas; their family names are Sêna, Mallica, Datta, and Gupta.] **1871** E. Balfour *Cycl. India* (ed. 2) II. 440/2 The following is a table of the Gupta kings. *Ibid.*, 'Elder Guptas'. **1880** *Encycl. Brit.* XIII. 120/2 All the Gupta inscriptions are dated in the *Gupta-kâla*, the Gupta era, the epoch of which has long been and still remains a subject of dispute. *Ibid.*, After the Sâhs come the Guptas of Kanauj, a dynasty which must not be confounded with the Maurya dynasty. **1939** A. J. Toynbee *Study Hist.* VI. 193 A Guptan 'Indian Summer'..was followed ..by the blight of an irruption of Eurasian Nomads. **1947** Auden *Age of Anxiety* (1948) v. 103 The Ghuzz, the Guptas, the gloomy Krimchaks. **1963** *Times* 12 Feb. 12/4 A fifth century A.D. Gupta sandstone carving. **1969** *Sunday Statesman* (Calcutta) (Mag. Sect.) 27 July p. i/5 The most prolific period of Hindu art, from the Gupta age to the mediaeval times, saw the rapid development of sculptural ornamentation.

gur, var. Goor. (Examples.)

1886 [see Goòr]. **1934** M. L. Darling *Wisdom & Waste in Punjab Village* 26 Half a dozen figures were crushing cane... I stopped, and at once they uncovered a pan and brought me a handful of the warm gur beloved of horse and peasant. **1965** E. Linton *World in Grain of Sand* x. 173 The harvesting of sugar cane followed by its crude and simple processing into 'gur'—the round brown unrefined sugar cakes. **1969** *Femina* (Bombay) 26 Dec. 57/2 Add a pinch of camphor and if desired grated gur (jaggery).

gurah, var. *GURRAH[2].

gural, var. Goral.

1919 *Chambers's Jrnl.* 8 Nov. 777/1, I hoped to get a gural or two on my way. **1922** *Blackw. Mag.* May 558/2 They reminded me of the Himalayan gural.

Guran (gū·răn). One of a people of Kurdistan; also, the language of this people. Also as *adj.*

1882 *Encycl. Brit.* XIV. 157/2 The Gurán dialect.. which is spoken throughout Ardelán and Kirmánsháhán. *Ibid.* 158/1 The Guráns have for a long period abandoned nomadic habits. **1912** E. B. Soane *To Mesopotamia & Kurdistan* x. 218 There was even a special court language, the graceful and euphonious Guran dialect, an ancient Persian tongue. *Ibid.* 224 Those who took protection under the Guran tribe, and have become Guran in name, were the Qadir Mir Waisi..and the Gurgkaish. *Ibid.* xvi. 382 The Gurans ruled with great power from their old capital at Gahwara. **1948** A. Safrastian *Kurds & Kurdistan* vi. 89 The many tribes which bear the general denomination of Guran.

‖ **gurdwara** (gɔ̈ɪdwā·rā). [Punjabi *gurduārā*, f. Skr. *guru* teacher + *dvāra* door.] A Sikh temple.

1909 M. A. Macauliffe *Sikh Religion* I. i. 47 A temple at a place visited by a Guru is now called Gurdwāra. **1924** *Glasgow Herald* 15 Apr. 9 The [Sikh] community have numerous temples or shrines called gurdwaras. **1954** J. Masters *Bhowani Junction* II. xxvii. 228 This bare square temple, the Sikh gurdwara.

gurgeon (gɔ̈·ɪdʒən). [perh. related to Gurjun.] *gurgeon-stopper*, a hardwood tree, *Eugenia buxifolia*, found in Florida and the West Indies; cf. Stopper *sb.* 8.

1884 [see Stopper *sb.* 8]. **1908** N. L. Britton *N. Amer. Trees* 724 Spanish Stopper—*Eugenia buxifolia*... Also called Gurgeon stopper; this small tree or shrub of the West Indies enters our area in southern peninsular Florida and the Keys. **1921** C. S. Sargent *Man. Trees of N. Amer.* (ed. 2) 771 *Eugenia buxifolia* Willd. Gurgeon Stopper. Spanish Stopper.

gurgitate, *v.* Restrict †*Obs. rare*[−0] to sense in Dict. and add: **b.** = Ingurgitate *v.* 2 b. Also *fig.*

1907 A. Quiller-Couch *Major Vigoureux* xxiv. 242 The ledge looked full upon the Mermaid's Rock, and upon the heave of black water surging past it to gurgitate between the narrowing walls of rock. **1963** *Evening Standard* 12 Nov. 18/3 The memoranda ceaselessly gurgitated by government departments.

Gurian (gūᵊ·riăn), *sb.* and *a.* **A.** *sb.* One of a Caucasian race, inhabiting Tiflis, closely related to the Georgians. **B.** *adj.* Of or pertaining to this race, or the region in which they live.

1876 *Encycl. Brit.* V. 257/2 The Gurians, a small people occupying the strip of land between the Rion and the mountains on the south, which form the frontier between Russia and Turkey. **1882** R. Brown *Peoples of World* V. i. 22 Like all the Georgians, the Gurian is fond of music. **1906** *Daily Chron.* 9 Aug. 3/1 The Gurians are the ..most warlike..of the Georgians. **1932** W. E. D. Allen *Hist. Georgian People* v. 50 Many more were the sea-ports, towns and castles of the Colchian-Lazic basin..inland to Petra, a Roman stronghold under the Gurian hills. *Ibid.* xviii. 208 The debris of their army was cut to pieces in the Gurian forests. *Ibid.* 210 In Imereti Solomon the Great, after a disastrous defeat at the hands of the Turks and Gurians, had died recently in Kutais (23rd April, 1782). **1963** C. Toumanoff *Stud. Chr. Caucasian Hist.* 269 The Imeretian house of Mik'elaje (1850) and the Gurian house of T'avdgirije (1850) appear to be of Bagratid ..origin.

gurjun. Add: Also **gurjan.** (Later example.)

1957 J. H. Williams *Spotted Deer* xi. 143 The two main species of which we were in search were gurjan and padauk... In others [*sc.* areas] [were] evergreen broad-leaved gurjan forests.

gurk (gɔ̈ɪk), *v. colloq.* [Echoic.] To belch. Hence **gu·rking** *vbl. sb.* and *ppl. a.*

1923 in Manchon *Le Slang* 146. **1946** B. Marshall *George Brown's Schooldays* 34 Belching and gurking with ghoulish glee. **1948** C. Fry *Thor with Angels* 7 Spit some words at me instead, and gurk Away your grudge. **1954** G. Durrell *Bafut Beagles* iii. 56 The loudest and most awesome gurking noises I have heard from a squirrel. **1966** *New Statesman* 10 June 839/2 They grunted and gurked with an unconcern that amazed me.

gurk (gɔ̈ɪk), *sb. colloq.* [See prec.] A belch.

1932 N. Mitford *Christmas Pudding* x. 151 That's right, darling, have a good gurk. *Ibid.* 153 The silence was broken by a shattering gurk from Elspeth Paula. **1961** G. Durrell *Whispering Land* iii. 75 It was not an accidental gurk, the minute breach of good manners that we are all liable to at times. This was a premeditated, rich and prolonged belch, with all the fervour of the Orient in it. **1962** C. Middleton *Torse 3* ii. 38 A too hard sniff culminates in a sly gurk.

Gurkha (gūᵊ·ɪkă). Also **Ghoorka, Ghorkha, Ghurka, Goork(h)a.** A member of one of the dominant races of Nepal, of Hindu descent and Sanskritic speech, and especially famous for prowess in fighting. Also *attrib.*

1811 W. Kirkpatrick *Account Kingdom Nepaul* App. I. 339 The Goorkha envoy proposed, that, as they could not stop the circulation of the base coin with which they had been supplied, they should, at least, establish a just rate of exchange. **1822** (*title*) Military Sketches of the Goorka War, in India, in the years 1814, 1815, 1816. **1836** *Penny Cycl.* VI. 65/1 It was clearly established by the Ghorkhas..that the village of Butool had never rightly formed part of the dominions of Oude. **1848** [see *Bhutanese *sb.* and *a.*]. **1848** J. D. Hooker in L. Huxley *Life J. D. H.* (1918) I. 269 The Goorkha Sepoys are immense fellows, stout and brawny. **1875** *Encycl. Brit.* II. 590/1 The four Ghoorka regiments, enlisted from Nepaul. **1905** *New Internat. Encycl.* IX. 388 In physique the Gurkha is very short and stout. **1914** G. B. Shaw in *New Statesman* (Suppl.) 14 Nov., A Ghoorka's religious conscience is respected: an Englishman's is insulted and outraged. **1952** Morin & Smith tr. *Herzog's Annapurna* ii. 30 For many years he had served as an officer in Gurkha battalions, crack units of the British Army. **1964** *Ann. Reg. 1963* 358 Great concern was caused..by the..decision to reduce the size of the Brigade of Gurkhas.

Gurkhali (guᵊɪkā·li). **a.** *pl.* The Gurkhas. **b.** The language spoken by this people.

1884 *Encycl. Brit.* XVII. 341/2 The Ghôrkhâlis use Parbatiya, a modern dialect of Sanskrit. **1911** *Ibid.* XIX. 379/2 The Gurkhalis (Gurkhas or Ghurkhas) are descendants of the Brahmans and Rajputs who were driven out of Hindostan. *Ibid.*, The Gurkhalis and western tribes use Khas [language]. *Ibid.* XX. 453/1 Khas-kura is also called Gōrkhālī, or the language of the Gurkhas. **1953** J. Masters *Lotus & Wind* xii. 168 Jagbir would have guessed that the Muralevs could not recognize Gurkhali when they heard it. **1956** —— *Bugles & Tiger* 131 At evening in my tent I studied Gurkali with the Jemadar head clerk.

Gurmukhi (gūᵊ·ɪmuki). Also **Gurumukhi.** [Punjabi, f. Skr. *guru* teacher + *mukha* mouth.] The alphabet used for writing the Punjabi language; also, the language of the Punjab. Also *attrib.* or as *adj.*

1858 M. H. Court in *Hist. of Sikhs* p. iv, I..have added a reverse index in English to enable a word to be easily found, if required, in translating English into Gurmukhī. At the commencement of the grammar, I have given the alphabet in the Gurmukhī character. **1907** G. A. Grierson in *Imp. Gazetteer India* (ed. 3) I. 369 An improved, and legible, form of Landā is known as Gurmukhī. This was invented about three hundred years ago for writing the Sikh scriptures, and is now the character in ordinary use for printing. **1909** *Westm. Gaz.* 22 Feb. 7/3 A seditious letter has been sent from England, written in Gurumukhi, the language of the Sacred Scriptures of the Sikhs. **1937** *Scrutiny* Mar. 447 In the Punjab ..only 3 and ·4 per cent. of the candidates for Matriculation answered their history papers in Hindi and Gurmukhi respectively. **1948** D. Diringer *Alphabet* 378 The Gurmukhi script seems to be a polished form of the Landa character. **1971** *Illustr. Weekly India* 11 Apr. 37/1 At Lahore, some time ago, Punjabi in the Gurmukhi script was demanded and introduced as a part of the University syllabus.

gurrah[2] (gʌ·rā). Also **gurah.** [Hind. *ghaṛa*, Skr. *ghaṭa*.] An earthen jar.

1828 G. C. Mundy *Jrnl.* 22 Mar. in *Pen & Pencil Sketches* (1832) iii. 137 The Rajah sent..sixty gurrahs (earthen vessels holding a gallon) of sugar candy and sweetmeats. **1864** J. A. Grant *Walk across Africa* 41 With a gourd cup they drew water from their wells and filled it into earthen 'gurahs', similar to those in India. *Ibid.* 67 A few earthen jars, made like the Indian 'gurrah', for boiling vegetables or their stirabout. **1895** B. M. Croker *Village Tales* (1896) 153 He..had been flung in the dust, like a broken gurrah.

gurry[5] (gʌ·ri). [Etym. unknown.] *gurry sore*, a kind of boil.

1897 Kipling *Capt. Cour.* v. 106 The affliction of gurry-sores being the mark of the caste that claimed him. *Ibid.* 118 Uncle Salters had a gurry-sore on his palm. **1950** C. S. Forester *Mr. Midshipman Hornblower* 109 Boils—gurry sores—blains—all the plagues of Egypt.

gurtcher, var. *Gertcha.

1937 C. Day Lewis *Starting Point* 204 Gurtcher! If Voycey was to let 'imself go, Sid'd wake up in 'ospital.

guru. Now the usual spelling of Gooroo. Also in gen. or trivial use: an influential teacher; a mentor; a pundit. (Further examples.)

1826 W. B. Hockley *Pandurang Hàri* III. xii. 347 He was in want of a gùrù, a religious instructor, to his household. **1877** H. P. Blavatsky *Isis Unveiled* II. ii. 114 It is in the presence of his instructor, the guru..that he is suddenly placed face to face with the unknown *presence*. **1884** in *1st Rep. Comm. Soc. Psychical Research* (1885) 58 A letter..being opened it was found to contain a distinct order from my Guru that I should stop at Calcutta. **1940** H. G. Wells *Babes in Darkling Wood* III. i. 245, I ask you, Stella, as your teacher, as your Guru, so to speak, not to say a word more about it. **1949** Koestler *God that Failed* 40 My self-confidence as a *Guru* had gone. **1957** *Sunday Times* 12 May 9/3, I learnt to love French poetry from it, a schoolboy in a class of one, with my guru Mr de Solgé. **1967** *New Scientist* 14 Sept. 559/1 Marshall McLuhan is (or is about to be turned into) one of those gurus whom the United States hungers for more than most nations. **1968** *Times* 4 Oct. 11/5 In what was obviously intended to be a sneering reference, the Prime Minister in his facile glibness.. described Mr. Enoch Powell as 'the guru of Wolverhampton'. **1969** *Eve's Weekly* (Bombay) 20 Dec. 11/4 Hannah Sahney, the chief instructress for the Indian Airlines' hostesses, is a guru to many of her young trainees.

b. *Comb.* **guru jacket,** a high-necked jacket fastened at the front by a vertical row of many small buttons; also *ellipt.*

1966 *Vogue* Nov. 21/2 Most colourful man's shop, Testa's..newest here—the *guru* jacket in brilliant red, blue or white silk or wool buttoned up high Chinese style. **1968** *Guardian* 29 Aug. 5/4 The tailors began it all three years ago with the 'guru' jacket,.. the guru with its high neckband notched at the throat and its multiple buttons.

gusher. Add: **2.** (Later examples.) Also *fig.* and *transf.*

1930 *Times Lit. Suppl.* 24 July 608/2 Mr. Thomas Wolfe's novel..might be called a 'gusher'; for Mr. Wolfe's words come spouting up with all the force of a subterranean flood..breaking through the overlying strata of repression. **1934** Dylan Thomas *18 Poems* 30 Nor fenced, nor staked, the gushers of the sky Spout to the rod. **1936** I. L. Idriess *Cattle King* xxiv. 214 He dreamt of artesian bores. Out in the 'dry belt' a few both private and government 'gushers' had now been struck. **1959** C. Ogburn *Marauders* (1960) vii. 223 A PFC..awakened at first light by a gusher from a mule towering above him from which he was protected by the fortuitous interposition of a log. **1970** *Times* (Canada Suppl.) p. iv/3 Only a few weeks ago Imperial Oil of Canada said cautiously they had found a gusher with considerable promise, the first oil strike in the Canadian North-West Territories.

gushily (gʌ·ʃili), *adv.* [f. Gushy *a.* + -ly[2].] In a gushy manner, gushingly.

1921 W. De Morgan *Old Man's Youth* xxiii. 211 Our visitor gushily asked what the dear boy was saying.

gushiness (gʌ·ʃinės). [f. Gushy *a.* + -ness.] Violent or copious outflow; effusiveness.

1856 *Punch* 5 Jan. 7/1 We presume that it is a similar spirit of economy that has recently reduced one of the Trafalgar Fountains from the exuberant gushiness of a ginger-beer bottle, to the slow, uncertain trickling of a watering-pot. **1937** 'C. Caudwell' *Illusion & Reality* 240 We get the leaping and gushiness of 'spring' in poetry's use of it as a word for the idea 'season', but we do not get the fountain or the jump.

gushing, *vbl. sb.* **2.** (Earlier examples.)

1852 *Punch* 17 July 39/1 The Colonel vehemently condemned the bran new Bribery Act; and with his philanthropic gushings, the Colonel has good, benevolent cause for reprobation of the measure. **1884** W. James *Coll. Essays & Rev.* (1920) 262 The sentimentalist is so constructed that 'gushing' is his or her normal mode of expression.

gushing, *ppl. a.* **4. b.** (Earlier example.)

1859 E. Fitzball *35 Yrs. Dram. Author's Life* II. 77 The gushing young ladies..in possession of some of our theatres.

gushy, *a.* (Earlier examples.)

1845 *Punch* 12 Apr. 169/1 There is a gushy vivacity in the manner of telling the pleasant fact, that 'in the afternoon the Grand Duchess visited her daughter the Marchioness of Douglas'. **1880** 'Mark Twain' *Tramp Abroad* (App. D) 613 A mere rhythmical, gushy euphuism.

gusle (gu·slĕ). Also **gusla, gustlé, guszla, guzla.** [Serbian.] A musical instrument of the violin family found in the Balkans, usually having only a single string, and used chiefly to accompany and support the chanting of the epic poems of the southern Slavs.

1869 H. F. Tozer *Highl. Turkey* I. 265 The national instrument..is the guzla... In shape it is like an elongated pear cut in half, and it is something between a guitar and a violin. **1874** C. Engel *Descr. Catal. Mus. Instr. S. Kensington Museum* 204 This species of violin somewhat resembles the guszla which has been in use by most of the Slavonic tribes in Europe from an early period. **1880** *Grove's Dict. Mus.* I. 642 *Gusla*, a kind of rebab, a bow instrument with one string only, used in Illyria. **1922** D. H. Low *Ballads Marko Kraljević* p. xxxvi, He then makes ready his gusle which in shape bears a rough resemblance to a mandolin. **1930** *Contemp. Rev.* Feb. 225 The gustlé player is the Serbian equivalent of the Saxon harpist. **1941** 'R. West' *Black Lamb* (1942) I. 313 A male gipsy who sang and played the gusla. **1969** *Observer* 26 Jan. 32/5 This was the residence of Petar Petrovic Njegos II.., prince, crack shot, player of the *gusle* (the Montenegrin lute) and author.

So **gu·slar** (pl. *guslari*), one who plays the gusle; a singer of traditional epic poems.

1891 E. S. Hartland *Sci. Fairy Tales* i. 16 A viol having only one string accompanies the passages in verse..; and a similar instrument seems to be used.. among the orthodox Guslars of Bosnia and Herzegovina. **1908** *Daily Chron.* 29 July 4/4 A blind guslar, playing the gusla. **1922** D. H. Low *Ballads Marko Kraljević* p. xix, If the Turkish dominion had endured a century or two longer, the separate ballads of the Kossovo cycle chanted by the Serbian guslari, would have fused together. *Ibid.*, The bard or guslar is often blind. **1923** *Glasgow Herald* 13 June 7 The wandering minstrels or guslari. **1930** *Harvard Stud. Class. Philol.* XLI. 79 There is too little known about the making of the early poetry in hexameters for us to liken the Singers to the Serbian Guslars without more ado, or to make of Homer a Singer like any other.

gusli (gu·sli). [a. Russ. *gúsli.*] A Russian musical instrument resembling a zither.

1897 J. S. Shedlock tr. *Riemann's Dict. Mus.* 314/1 Gusli (*Gussel*), Russian stringed-instrument, a kind of zither. **1910** *Encycl. Brit.* XII. 733/2 Among the Russians, the gusli is..a kind of psaltery having five or more strings stretched across a flat, shallow sound-chest in the shape of a wing. **1959** *Listener* 9 Apr. 647/2 He [*sc.* Prokofiev] uses saxophones to imitate the sounds of the Russian *gusli*. **1961** A. Baines *Mus. Instruments* 206 A true psaltery of medieval vintage survives in Russia in certain forms of *gusli*, much used by ballad singers formerly.

guss (gʌs). *local.* (See quots.)

1883 Gresley *Gloss. Coal-m.*, *Guss* (Bristol Coal-field), a short piece of rope by which a boy draws a tram or sled in a pit. **1928** *Daily Tel.* 21 Aug. 13/2 A gentleman stated that he had seen in a newspaper an illustration of a 'guss', taken from a book dated 1842. **1928** *Times* 4 Oct. 11/4 The guss is a girdle fitted around the waist of boys who pull coal in 'putts' or sledges in the Somerset mines.

gussie[1] (gʌ·si). *Austral. slang.* [dim. of the name *Augustus.*] An effeminate man.

1901 M. Franklin *My Brilliant Career* xi. 94 I'm not a booby that will fall in love with every gussie I see. **1941** Baker *Dict. Austral. Slang* 33 *Gussie*, an effeminate or affected man.

gussy (gʌ·si), *v. slang.* [Cf. *gussie[1].] With *up*: to smarten up. Hence **gu·ssied** *ppl. a.*; **gu·ssy** *a.*

1940 M. Marples *Public School Slang* 126 Gussy: of over-dressed persons. **1952** in Wentworth & Flexner *Dict. Amer. Slang* (1960) 236 When I get all gussied up, somebody says, 'Pull in your pot!' **1959** M. Chamberlin *Dear Friends & Darling Romans* (1960) iv. 41 The French have gussied up their Riviera, so you can't see the foliage for the freaks. **1960** J. Kirkwood *There must be Pony* (1961) viii. 65 She was all gussied up..she looked very attractive. **1963** *Spectator* 1 Mar. 260 Dressy, not to say gussied-up. **1970** M. G. Eberhart *El Rancho Rio* (1971) iii. 23 'You're really all gussied up...Coast slang for dressed up,' she explained.

gust, *sb.*[1] Add: **3.** *attrib.* and *Comb.*, as (sense 1) *gust alleviator, effect, load, recorder, response, spectrum, tunnel*; *gust-flying* vbl. sb.

1919 A. Klemin *Text-bk. Aeronaut. Engin.* xix. 234 It is also clear from the above that the gust effects are most important, when the speed of the machine is lowest. **1947** *Shell Aviation News* cxii. 20/3 'Gust alleviators' as fitted to certain new British aircraft will provide some protection to passengers but even assuming these can relieve 50% of the gust effect, they will still only provide the same 'ride comfort' at 320 m.p.h. as one now gets in the D.C.3 at 160 m.p.h. **1922** *Flight* XIV. 659/2 At any rate, the mere trying should teach us quite a lot about air currents around hills, quite apart from the, as yet untouched, problem of real 'gust-flying', in which sudden changes in velocity of the wind itself are made use of. **1955** *Sci. News Let.* 1 Jan. 9/2 The accumulated effects of repeated, but mostly moderate, structural loads—normally gust loads—could sometimes cause failures of the primary structure before the airplane attains a reasonable service life. **1959** J. L. Nayler *Dict. Aeronaut. Engin.* 125 *Gust loads*, the loads on an aircraft structure due to gusts. **1955** *Sci. News Let.* 24 Sept. 197/1 Fifty gust recorders. **1965** *Economist* 20 Feb. 733/2 This aircraft now has to have very low gust response to keep the crew functioning. **1935** *Jrnl. R. Aeronaut. Soc.* XXXIX. 372 It seemed..that if the gust spectra were extended to include the short period changes, some very large ordinates would be added to the curves. **1939** *Ibid.* XLIII. 784 To assist in answering these questions an apparatus, known as the 'gust tunnel' has been developed, in which it is possible to determine experimentally the reaction of suitably scaled dynamic models in controlled artificial gusts.

gust, *v.*[2] Delete *rare*[−1] and add: Also without *up*, to blow in gusts. Also *fig.*

1899 J. M. Falkner *Moonfleet* xi, The wind came gusting round the corner. **1927** *Chambers's Jrnl.* 312/2 The wind was gusting. **1960** I. Shaw *Two Weeks in Another Town* v. 77 Jack..remembered what Delaney had looked like gusting into the dressing-room. **1963** *Times* 13 May 4/4 An exciting final, fought out in winds that gusted to more than 30 knots. **1964** N. Freeling *Double-Barrel* v. 146 An erratic wind..gusted at me from all quarters.

gustatory. Add: Hence **gu:stato·rial** *a.*

1937 F. M. Ford *Let.* 27 Sept. (1965) 284 For unless he will come to the rescue of our Western civilisations, in things gustatorial, what will become of us?

Gustavian (gʌstēⁱ·viăn), *a.* [f. the name *Gustavus*: see -ian.] Of or pertaining to the reign of any of the Swedish kings named Gustavus, *spec.* of Gustavus III (1771–1792) and Gustavus IV (1792–1809); used esp. with reference to the literature of the period.

1884 R. B. Anderson tr. *Horn's Lit. Scand. North* III. v. 356 The poets of the Gustavian period form two groups according to the prevalence, respectively, of the French and the national element. **1887** *Encycl. Brit.* XXII. 757/1 The excellent lyrical poet Frans Mikael Franzén..and a belated academician Johan David Valerius..fill up the space between the Gustavian period and the domination of romantic ideas from Germany. **1951** E. Bredsdorff et al. *Introd. Scandinavian Lit.* iv. 80 Like the Gustavian Classicists, she handled her technique skilfully and unobtrusively. **1961** A. Gustafson *Hist. Swedish Lit.* v. 135 Seldom if ever in literary history has a monarch's name more appropriately been attached to a literary period than those two or three decades in late eighteenth-century Swedish literature which it is customary to call 'The Gustavian Era'. *Ibid.* 145 Thorild and Lidner..proved to be anything but Gustavian in their literary tastes.

gustiness (gʌ·stinės). [f. Gusty *a.*[1] + -ness.] The condition or quality of being gusty.

1901 *Westm. Gaz.* 20 June 10/1 The everlasting gustiness for which the Maoriland capital is famed. **1920** *Times Lit. Suppl.* 9 Sept. 586/1 From the point of view of construction his stories are..irregular, but for sheer gustiness they are hard to equal. **1946** G. Millar *Horned Pigeon* ix. 128 Binns..thought all Italians were 'dirty people', and he could hate Garibaldi with fresh gustiness. **1957** G. E. Hutchinson *Treat. Limnol.* I. v. 275 The roughness of the water surface and gustiness of the wind.

gut, *sb.* Add: **1. b.** *to have* (a person's) *guts for garters* (a hyperbolical threat); *to hate* (a person's) *guts*: to dislike (a person) intensely; *to sweat* (or *work*) *one's guts out*: to work extremely hard.

a **1592** R. Greene *Sc. Hist. James Fourth* (1598) III. ii, Ile make garters of thy guttes, Thou villaine. **1601** B. Jonson *Cynthia's Rev.* IV. iii, Sir, I will garter my hose with your guttes. **1714** J. Walker *Sufferings Clergy* II. 341/2 He hoped to have the Parson's Guts to Garter his Hose with. **1918** *Wine, Women & War* (1926) 140 R— decided on different way, so did it all over again. Great boy, R—. Hate his guts! **1925** F. Scott Fitzgerald *Great Gatsby* (1926) i. 9 There were men at New Haven who had hated his guts. **1930** W. S. Maugham *Breadwinner* II. 101 God knows, it's been an uphill job, but I've done my best. I've just sweated my guts out. **1932** N. Coward *Words & Music* in *Play Parade* (1939) II. 111 We have to work our guts out... We have to hop and bustle. **1933** *Cornh. Mag.* Mar. 698 I'll 'ave yer guts fer garters. **1935** Auden & Isherwood *Dog beneath Skin* II. v, One o' these dys I'll 'ave 'is guts fer garters. **1936** N. Coward *To-Night at 8.30* II. 31 You know perfectly well I hate Freda's guts. **1937** 'G. Orwell' *England Your England* (1953) 191 It is brought home to you, at least while you are watching, that it is only because miners sweat their guts out that superior persons can remain superior. **1938** G. Greene *Brighton Rock* VII. viii. 338 He hates her guts. **1945** Wodehouse *Let.* 22 May in *Performing Flea* (1953) 126 The entire personnel of the cast sweat their guts out..and then the studio discovers that it doesn't own the rights to the novel. **1959** *Listener* 24 Sept. 495/3 Those who (to use a colloquial phrase that does justice to feelings, especially in war time) 'hated his guts'. **1967** *Guardian* 29 Dec. 6/3 Resentment in Service quarters is now focusing on Mr Healey... But those who are demanding his guts for garters are making a mistake.

d. *pl.* Energy, verve, staying power; courage, force of character. *colloq.*

Cf. dial. phr. *to have neither gut nor gall* (1887 in E.D.D.). **1893** Farmer *Slang* s.v., Put your guts into it.. = Row the very best you can. He (or it) has no guts in him (or it) = He (or it) is a common rotter. **1900** G. Swift *Somerley* 85 If you have what are, at Cambridge, vulgarly but expressively called 'guts'. **1924** W. M. Raine *Troubled Waters* ii. 22 It's about your size to send a skull-and-crossbones threat through the mail, but I notice you haven't the guts to sign it. **1924** R. Keable *Recompence* i. 9 Can't you dig me out a chap with some guts, who has learned to rough it? **1933** J. C. Powys *Glastonbury Romance* xxii. 713, I think, if you haven't the guts to act like a man in the matter, you ought to leave this girl alone. **1955** *Times* 30 Aug. 5/2 That policeman had plenty of guts. I have been informed that the policeman was not seriously hurt.

2. g. In machine sheep-shearing, a flexible shaft which conveys the power from an over-head source to the shearer's handpiece. *Austral.* and *N.Z.*

1956 G. Bowen *Wool Away!* (ed. 2) viii. 100 The correct length of a gut is shown when, with the long and short gut connected, they hang so that the short gut swings just clear of the floor. **1965** J. S. Gunn *Terminol. Shearing Industry* II. 35 The tube is a casing down which runs a flexible driving shaft known as the 'gut'.

h. *fig.* Used, chiefly *attrib.*, of an issue, question, etc.: basic, fundamental; also, of a reaction: instinctive and emotional rather than rational.

1964 *Economist* 17 Oct. 261/3 For Harold Wilson it was a carefully planned campaign:..the neo-Kennedyism combined with a concentration on gut issues. **1968** *Guardian* 26 Sept. 10/3 The three nights of rioting that followed his murder were an immediate gut reaction. **1969** *Times* 22 July p. ii/3 The moon programme..was a gut issue, as even the less enthusiastic realized. **1969** *Daily Tel.* 14 Nov. 5/2 When we [*sc.* the Americans] first went into space, we had no idea how much it was going to benefit the economy. We went in as a gut reaction to the Soviet challenge. **1970** *Win* 15 June 4/1 There are some gut questions the pacifist must face. *Ibid.*, Really, the questions are too gut for us. **1971** *Listener* 19 Aug. 223/2 Most people's reaction to the Oz trial and sentences has been what one might call a gut-reaction—whether of shock or satisfaction.

i. Phr. *to bust* (or *rupture*) *a gut*: to exert oneself, to make a great effort. *colloq.*

1912 *Dialect Notes* III. 572 *Bust a gut*,..to make a supreme effort. 'Just bust a gut now and see if we can't lift this log.' **1968** C. Drummond *Death & Leaping Ladies* i. 23 'I'll be back in twenty minutes.'..'Don't bust a gut,' advised Miss Winkelbaum, 'the hussies will be late.' **1970** J. Porter *Rather Common Sort of Crime* xiii. 154 If Mack'd been some fat, respectable, middle-aged old bastard, the cops'd've bust a gut nicking somebody fer croaking him. **1970** W. Smith *Gold Mine* xiii. 36 'Huh!' Popeye checked his watch. 'Two hours forty to get down, you don't reckon to rupture a gut do you?'

3. b. (Later example.)

1959 I. & P. Opie *Lore & Lang. Schoolch.* ix. 168 The unfortunate fat boy.. is known as..guts, [etc.].

5. b. Also, a street in Valletta, Malta.

1948 Partridge et al. *Dict. Forces' Slang* 89 *The Gut*, a notorious street in Malta. **1970** 'Zeno' *Grab* iii. 25 Strait Street, Valetta, better known as the Gut, the centre of Malta's red light district. **1970** M. Butterworth *Vanishing Act* x. 108 Mosta dome and the Città Notabile, the trashy souvenir shops of Kingsway..and the honky tonk dives of the Gut.

8. *gut-fat* (earlier U.S. example), *-substitute* (sense 4 d), *guts-box*; *gut-open* adj.; **gut-bucket** *slang* (orig. U.S.), a primitive, unsophisticated brand of jazz (see also quot. 1939); cf. *barrel-house 2; also *attrib.*; **gut-rot** *colloq.*, unwholesome or unpalatable liquor or food (cf. Rot-gut); **guts-ache** *colloq.* = Stomach-ache; **gut-shoot** *v. trans. slang*, to shoot in the guts; so **gut-shot** *sb.* and *ppl. a.*

1929 *N.Y. Age* 8 June 7/6 Using a mute, occasionally a small megaphone inserted at the bell of his trumpet, he [*sc.* Louis Armstrong] eschews the tin pail, hat, plunger and other devices of the 'gut bucket' player. **1934** *Melody Maker* 10 Feb. 5/1, I urge some good black-and-tan gutbucket records with Hawkins. **1938** *Manch. Guardian Weekly* 2 Sept. 188/3 Sometimes they play 'boogie-woogie'..sometimes 'gutbucket' (unrefined). **1939** Ramsey & Smith *Jazzmen* 12 From barrel-houses and honkey-tonks came many of the descriptive words which were applied to the music played in them; such as..'gut-bucket', referring originally to the bucket which caught drippings or 'gutterings' from the barrels, later to the unrestrained brand of music that was played by small bands in the dives. **1951** W. Morum *Gabriel* II. viii. 239 Jake's suggestion of another two hours session of dance music at a gut-bucket joint in Piccadilly. **1847** *Rep. U.S. Comm. Patents* (1848) 527 The slaughterers formerly got the gut fat for the whole of the labor thus described. **1935** L. MacNeice *Poems* 64 The town-dweller like a rabbit in a greengrocer's..Hangs by the heels gut-open against the fog. **1916** A. H. Macklin in A. Lansing *Endurance* (1959) v. vi, 217 'Gut Rot, 1916'..served only to turn most of us teetotallers for life. **1938** S. Beckett *Murphy* 83 The customer..was paying for his gutrot ten times what it cost to produce. **1941** Baker *Austral. Slang* 33 *Gutrot*, unhealthy-looking food or strong drink. **1965** F. Sargeson *Mem. Peon* ii. 32 The garish-looking sweet stuff he made his living from...'I make a dishonest living by trading in gutrot.' **1818** Keats *Let.* 13 July (1958) I. 324 Cant!..It is enough to give a spirit the guts-ache. **1934** Blunden *Mind's Eye* 145 Don't eat that, Jack; it'll give you the Guts'-Ache. **1946** K. Tennant *Lost Haven* (1947) vi. 89 'Ar,' Launce said contemptuously. 'You got a guts-ache or you're coming out in boils.' **1940** Dylan Thomas *Portrait of Artist* 133 He'd put his hand down in the guts-box and bring you out a rat with its neck broken clean as a match for the price of a glass of beer. **1939** 'G. Orwell' *Coming up for Air* II. iv. 87 Even now I could give you all the details about gut-substitute and gimp and Limerick hooks. **1878** C. Hallock *Amer. Club List & Sportsman's Gloss.* p. v, *Gut-shot*, hit in the belly; wounded, but not disabled. **1935** E. Hemingway *Green Hills Africa* (1936) II. iii. 103 I'm afraid I gut-shot him. *Ibid.* 114 It looked, now, like a gut shot or one through the paunch. **1960** B. Crump *Good Keen Man* 33 He came to the door and shouted into the frosty dark that he'd gut-shoot the first of us to come near the hut. *Ibid.* 50 Phillip was as slow as a gut-shot pig and started yelling as soon as he lost sight of me through the bush.

gut, *v.* Add: **2. a.** Now freq. used *pass.* and of destruction by fire.

1903 *Westm. Gaz.* 25 Aug. 8/2 The farmhouse and out-buildings were completely gutted. **1968** *Bucks Examiner* 3 May 1 (*heading*) Furniture factory gutted.

gutful (gʌ·tful). *dial.* and *slang.* Also **gutsful.** [f. Gut *sb.* + -ful.] = Belly-ful.

1900 *Eng. Dial. Dict.* II. 768/1 He's hungry when he's got a gutful (he is very stingy). **1932** N. Lindsay *Cautious Amorist* xvi. 226 Let me have some fun out of you to prove it's not a gutsful of grunts you are. **1960** I. Cross *Backward Sex* vi. 118 I've just had a gutful, and want to have hell scared out of her. **1965** M. Shadbolt *Among Cinders* ii. 13 He said he'd had a gutsful of the bush when he was younger. **1970** *Daily Tel.* (Colour Suppl.) 15 May 7/2 Lately, we have had a 'gutful' of the permissiveness that seems to tolerate violence against the police.

Gutian (gū·tiăn), *sb.* and *a.* **A.** *sb.* A member of a mountain people who lived *c* 2500–2000 B.C. in the area of the Tigris and the Euphrates. **B.** *adj.* Of or pertaining to this people.

[**1882** *Encycl. Brit.* XIV. 159/2 We now find that at the dawn of history the mountains overhanging Assyria were held by a people named *Gutû*, a title which signified 'a warrior'.] **1928** V. G. Childe *Most Anc. East* i. 19 The Gutians, the Kassites, the Chaldæans and the Persians. **1929** C. J. Gadd *Hist. & Monuments of Ur* iii. 99 Throughout all the years filled by the end of the Agade dynasty and the Gutian tyranny hardly anything is known concerning the fortunes of Ur. **1960** tr. *S. Moscati's Face of Ancient Orient* ii. 25 Then the Gutians, savage people from the East, spread destruction far and wide, and earn themselves the name of 'mountain dragons' (about 2150–2050 B.C.). **1971** W. W. Hallo in *Reallexikon der Assyriologie* 716/2 By the end of the Old Babylonian period, the term 'Gutian' was of little more than vague geographic or ethnic significance. *Ibid.* 719/1 The earliest contemporary references to Gutium..seem to place it, or persons designated as Gutians, in the mid-Euphrates area, in company with Amorites... In the Old Babylonian period, persons identified as Gutian, or bearing 'Gutian' names are attested throughout northern Mesopotamia and Assyria.

gutless (gʌ·tlės), *a.* Restrict †*Obs.* to sense in Dict. and add: **b.** Lacking in energy, courage, or determination. Also **gutsless.**

1915 E. Pound in *Lett. J. Joyce* (1966) II. 364 Gutless Oxford graduates. **1927** D. H. Lawrence *Let.* 19 May in E. & A. Brewster *D. H. Lawrence* (1934) 133 The rest [at an art exhibition] piffle... So old-fashioned and timid and gutsless. **1929** W. J. Locke *Ancestor Jorico* x, Dr. Chepstow firmly refused to allow the convalescent and..gutless Nicholas to venture a second time. **1930** G. M. Sanders *Burnt Man* xv. 175 He was thinking that the man before him was a poor gutsless creature after all, and that he had been foolish to suppose either of his sisters would ever be attracted to him. **1941** L. A. G. Strong *Bay* ix. 228 Now you see what a gutless poor worm I am. **1969** *Daily Tel.* 16 May 18 The housing situation in North Kensington has been created by a short-sighted and gutless national housing policy.

Hence **gu·tlessness,** lack of determination.

1936 R. Lehmann *Weather in Streets* II. 245 Isn't it [*sc.* psycho-analysis] an excuse for gutlessness?

Gutnish (gu·tniʃ). [ad. G. *gutnisch.*] **A.** *sb.* An East Norse dialect spoken on the island of Gotland. **B.** *adj.* Of or pertaining to the island of Gotland or its inhabitants.

1892 J. Wright *Primer Gothic Lang.* 1 *East Norse*, including Swedish, Gutnish, and Danish. **1908** J. & E. M. Wright *O.E. Gram.* 2 Old Norse..is sub-divided into two groups: (a) East Norse, including Swedish, Gutnish, and Danish; (b) West Norse. **1927** E. V. Gordon *Introd. O. Norse* 177 The Gutnish laws..state that after the 'bride-mass' is sung, 'the wedding shall be drunk for two days with all the folk'. **1934** Priebsch & Collinson *German Lang.* ii. 28 East Norse split up into Swedish, which now includes a peculiar dialect—Gutnish—spoken on the island of Gotland, and Danish.

guts (gʌts), *v. colloq.* [Cf. Gut *v.* 3.] To eat greedily, to gormandize. (See *E.D.D.*)

1903 S. Macplowter *Mrs. McCraw* 28 Ef ye hedna kep' us in sae lang, wi' yer haivers, A wid hae been hame tae keep her frae gutsin'. **1934** B. Penton *Landtakers* (1935) vii. 50 Did ye not guts up a dollop of that junbuck I drove off..last week. **1943** K. Tennant *Ride on, Stranger* iii. 24 'Gutsing again, Briscoe?' she reproved. **1956** B. W. Aldiss in *Pick of Today's Short Stories* 28 In the lunch hour, while I gutsed a bun in the background, we were sometimes quite crowded. **1967** D. Pinner *Ritual* xvii. 164 They gutsed the food without chewing.

gutser (gʌ·tsəɪ). *dial.*, *Austral.*, *N.Z.*, and *colloq.* Also **gutzer.** [f. *guts* (Gut *sb.* 3) + -er[1].] A heavy fall. Esp. in *fig.* phr. *come* (*fetch*, etc.) *a gutser*, come a cropper, make a mistake; *Air Force slang* (in concr. use), to crash.

1918 *Chrons. N.Z.E.F.* 21 June 221/1 The tenderfoot and Zambuk, Working madly in the trenches, Comes a 'gutzer' in those trenches, And is taught to take a rumble. **1919** *War Slang* in *Athenæum* 8 Aug. 728/1 A 'gutser' is the last straw. **1925** Fraser & Gibbons *Soldier & Sailor Words* 113 *To come a gutzer*, to 'crash' or fall badly. (Ordinarily an Air Force term with reference to an aeroplane.) Also used generally. To get into serious trouble, *e.g.*, 'He's before a court martial and looks like coming a gutzer.' 'Gutzer' is pre-war slang, and an old term among Scottish boys for falling flat on the water in diving, instead of making a clean header. **1933** *Bulletin* (Sydney) 6 Sept. 9/2, I work for the dole in a gutser gang, shovelling bits of coast..the down-and-out brigade. **1936** M. Franklin *All that Swagger* xl. 317 The banks will fetch old Robert a gutser one of these days. **1941** D. Masters *So Few* xxx. 335 The starboard Heinkel turned straight into..McKellar's stream of bullets and got what is known in the service as a 'gutser'. **1957** I. Cross *God Boy* (1958) xii. 98 The back wheel slid in some loose gravel. Down he went again in another beautiful gutzer. **1960** B. Crump *Good Keen Man* 111, I came one or two gutsers myself, and once, when I was carrying Flynn through a gorge below the forks, I got out of control.

gutsy (gʌ·tsi), *a.* Also **gutsey, gutzy.** [f. *guts* (Gut *sb.* 3) + -y.] **1.** Greedy, voracious. orig. *Sc.*

1803 T. Campbell *Let.* 2 Mar. in W. Beattie *Life & Lett. T. Campbell* (1849) I. 456 A John-Bull song..to the chorus of 'Gooseberry tarts'. Gutsy proud people—we sing songs in Scotland about her nature and love!..but here, there is no song but 'Roast Beef', or 'Gooseberry Tarts'! **1829** J. Wilson in *Blackwood's Edin. Mag.* XXV. 392 There's nae denying that maist o' them's gutsy. **1893** R. L. Stevenson *Catriona* xxx, Ye muckle, gutsey hash, here's a Scots boot to your English hurdies. **1893–4** R. O. Heslop *Northumberland Words* II. 350 *Gutsy, gutty*, gluttonous, fond of his belly. **1902** *Daily Chron.* 27 Mar. 3/1 An inspector having got from a class that a cow was a graminiverous and a tiger a carnivorous animal, went on to ask what an animal that eats everything, both grass and flesh, would be. 'A gutsy beast,' was the reply. **1946** B. Marshall *George Brown's Schooldays* 23 Half a rissole was quite enough for a gutzy young sod like him.

2. Tough, spirited, courageous; possessing or requiring 'guts' (*gut *sb.* 1 d).

1893 in Farmer & Henley *Slang* III. 237/2. **1936** *Amer. Speech* XI. 101/1 Gutsy folk do not care for much pussyfooting. **1937** S. O'Faolain in *John o' London's* 12 Mar. 984/3 *Kit Brandon* is the life-story of a woman gangster, a regular tornado, a passionate, lawless 'gutsy' young girl from the mountains. **1951** E. Coxhead *One Green Bottle* vii. 190 She took him up one of the horrid, gutsy little Cribin climbs. **1964** *Punch* 9 Sept. 372/1 The following gutsy alternative. **1970** *New Yorker* 3 Oct. 36/3 He would..sing out if a tune came along that was of his persuasion—something gutsy like 'Onward, Christian Soldiers'.

Hence **gu·tsily** *adv.*, gluttonously (1825 in Jamieson's *Sc. Dict.*; 1898 in *E.D.D.*); **gu·tsiness,** (*a*) greediness, gluttony (1825 in Jamieson; 1898 in *E.D.D.*); (*b*) energy, spirit; courage.

1893 Farmer & Henley *Slang* III. 237/2 Gutsiness. **1959** *Times* 1 Apr. 5/4 Israel audiences like and admire a play if it shows the qualities they most respect: directness..'gutsyness'. **1969** *Guardian* 7 July 9/4 His brilliance, his intellect, his gutsiness. **1970** *Ibid.* 31 Mar. 8/2 Tilson's work..has always been gutsily vulgar.

gutta[1]. Add: **3.** A drop-like marking on an insect's wing.

1826 Kirby & Spence *Introd. Ent.* IV. xlvi. 285 Gutta. ..A roundish dot, intermediate in size between an *atom* and a *macula*.. Linné..has employed the term *Gutta* for a white or yellow spot in a darker ground, and *Pustula* for a red spot in a black ground. We thought one term sufficient to express spots bigger than atoms. **1906** J. B. Smith *Explan. Terms Ent.* 59 *Gutta*, a light spot on a dark ground. **1913** N. K. Jardine *Dict. Ent.* 91 Gutta.—A light spot upon a light ground, *viz.*, white upon yellow; a roundish dot of colour, intermediate in size between an atom and a macula. **1937** J. R. de la Torre-Bueno *Gloss. Ent.* 121 *Gutta*, a roundish colored dot.

gutta[2]. Add: **1. b.** A gutta-percha golf ball, a gutty.

1857 H. B. Farnie *Golfer's Man.* 11 The first flight of 'Guttas' was hailed with a burst of joy financial by every-one except the old monopolists of the feather manufacture. **1858** *Chambers's Jrnl.* 4 Sept. 157/2 He seizes his sand-iron..and, with a skilful jerk behind his ball, frees it..and taking his long spoon—a wooden-headed club.. Jones drives forth his gutta from its not too favourably lying position. **1881** R. Forgan *Golfer's Handbk.* 5 The 'guttas' speedily became popular. *Ibid.* 36 The 'gutta' ball..was still far from perfect. **1905** A. S. Cunningham *Rambles in Scoonie & Wemyss* 99 He and his brother were credited with devising the hand hammering of balls, which led to the ultimate marking of guttas.

2. gutta-jelutong (or **joolatong**), a substitute for rubber obtained from any one of several apocynaceous trees of Malaysia and Indonesia of the genus *Dyera*, spec. *D. costulata*.

1904 *Electr. World & Engin.* 18 June 1150 (Cent. D. Suppl.), Importations of 'gutta-joolatong',.. which is used in certain industries as a substitute for india rubber. **1927** C. R. Toothaker *Commercial Raw Materials* 207 There are several substances similar to gutta-percha obtained from the milky juices of trees in the East Indies. Gutta-jelutong (jelatong), or pontianak,.. is the most important.

guttation (gʌtēⁱ·ʃən). *Bot.* [a. G. *guttation* (A. Burgerstein 1887, in *Verh. Zool.-bot. Ges. Wien* XXXVII. 692), f. L. *gutta* drop + -tion.] The exudation of liquid from a plant, esp. from hydathodes in the leaves or from fungal mycelia.

1889 in *Cent. Dict.* **1900** A. J. Ewart tr. *Pfeffer's Physiol. Plants* I. vi. 253 Burgerstein uses the special term 'Guttation' when the water escapes through water-pores. **1909** Groom & Balfour tr. *Warming's Oecology of Plants* xxix. 101 Not only in the tropical rain-forest, but also in the temperate countries, are there many plants, especially herbs, exhibiting the phenomenon of guttation. **1932** Fuller & Conard tr. *Braun-Blanquet's Plant Sociol.* v. 127 Guttation is especially common in warm, moist, tropical forests. **1966** F. M. Irvine tr. *Lundegårdh's Plant Physiol.* vi. 296 In some trees and in many herbs the active xylem stream usually leaks through the hydathodes or in other ways if transpiration is restricted; this is the guttation phenomenon. Guttation occurs in almost all plants.

gutter, *sb.*[1] Add: **1. c.** (Earlier example.)

1856 S. Davison *Let.* 13 Aug. in W. B. Clarke *Res. S. Gold Fields N.S.W.* (1860) iv. 50 Pebble-covered local gold in evenly-spread beds or linear troughs of leads or gutters.

8. *gutter-boy, -brat, -canal, -girl, -lout, -mongrel, -snippet, -sweeping*; **gutter-crawling** *vbl. sb.*, the action of driving a car, etc., slowly along a road close to the pavement and attempting to entice into it women, esp. prostitutes (cf. **kerb-crawling*); **gutter-man,** (*b*) (*U.S.*) *Logging*, one who removes under-brush, fallen trees, and other obstacles in making a gutter road; (*c*) one who cleans out the gutters of buildings; **gutter road,** the path or track followed in skidding logs (*Terms Forestry & Logging*, 1905); **gutter-splint,** a splint moulded to the shape of the limb; **gutter-way,** (*a*) = Gutter *sb.*[1] 2; (*b*) = *gutter waterway*.

1901 G. K. Chesterton *Defendant* 15 We rate the gutter-boys for their immorality. **1962** K. Orvis *Damned & Destroyed* xii. 80 He was a gutter-brat. **1946** Koestler *Thieves in Night* I. iii. 23 The gutter-canal that ran along the middle of the street as the serpent's inverted spine. **1948** Mencken *Amer. Lang.* Suppl. I. vi. 504 Gutter-crawling is practised by mashers who run close to the sidewalk, hoping to pick up light-headed girls. **1968** *Courier-Mail* (Brisbane) 30 Nov. 3/5 Thirty-five men were fined..on 'guttercrawling' charges... The charges, under an Act promulgated last December alleged the men 'loitered either to be accosted by a prostitute or for the purpose of inviting or soliciting any female to prostitute herself for pecuniary reward'. **1909** *Daily Chron.* 13 Nov. 3/1 A devoted priest, a noble gutter-girl,..—these are Mrs. Baillie Saunders's stock-in-trade. **1926** D. H. Lawrence *Plumed Serp.* i. 8 A real gutter-lout came to look at their courslips. **1904** *Dialect Notes* II. 397 *Gutterman*, a term used in logging camps. **1921** *Dict. Occup. Terms* (1927) § 970 *Roofman*; *gutterman*; sweeps roofs and removes dirt and other obstruction from guttering, rain pipes, etc., of large buildings. *a* **1930** D. H. Lawrence *Phoenix* (1936) 21 Not fit to be trusted with any dog but a gutter-mongrel. **1931** S. Beckett *Proust* 61 But he does not proceed pari passu with..the Parnassians to the ineffable gutter-snippets of François Coppée. **1919** W. Deeping *Second Youth* xxiv. 203 At his ease in a long cane chair, his left arm still in a gutter-splint. **1957** *Encycl. Brit.* IX. 577/2 Aluminium gutter splints are of value in some cases [of fracture] on account of their malleability and translucence to X-rays. **1954** W. Faulkner *Fable* (1955) 26 The African regiment recruited from the prison- and gutter-sweepings of Europe. **1908** *Daily Chron.* 19 Aug. 3/4 Choked gutter-ways, creepers stuffed with unsightly straw..induced him..to order that the sparrow nests should be removed from the walls and eaves of his house. **1923** *Man. Seamanship* II. 280 To drain water from the decks..scuppers are fitted... These are led from the gutter-ways.

b. (Earlier and later examples.)

1849 J. O. O'Connell *Recoll. Parl. Career* I. v. 104 Feargus O'Connor carried the election..he was everywhere and everything;—speechifier,..gutter-agent, mob leader. **1900** J. London *Let.* 1 Oct. (1966) 112 How different from the gutter attack of Robert Buchanan on Kipling and Besant! **1940** 'G. Orwell' *Inside Whale* 126 There is no clear reason why every adventure story should necessarily be mixed up with snobbishness and gutter patriotism. **1941** Auden *New Year Let.* i. 24 Conscious *Catullus*, who made all His gutter-language musical. **1953** *Manch. Guardian Weekly* 10 Dec. 4/4 Mr Truman, a graduate of the Pendergast school of gutter politics, called the Red issue a 'red herring'. **1955** J. Thomas *No Banners* xviii. 165 He found himself cursing under his breath, using foul Cockney gutter-slang that normally would have appalled him. **1957** 'P. Quentin' *Suspicious Circ.* viii. 87 One of..those terrible gutter magazines which make fortunes unearthing people's private lives. **1958** Wodehouse *Cocktail Time* v. 45 To..sell this information to the gutter press for what it will fetch. **1959** *Daily Tel.* 9 Apr. 22/4 One of the cheapest forms of gutter electioneering.

gutter, *sb.*[2] **1.** Add to def.: or animals. Also *fig.*

1851 C. Cist *Sk. Cincinnati* 280 Here the animal falls into the hands of the gutter who tears out the inside, stripping at the rate of three hogs to the minute. **1963** *Guardian* 18 Sept. 10/6 Is it possible that..there is a clever summary in the last few pages..making the document easy work for an accomplished 'gutter' which I feel sure Mr Wilson is?

gutter, *v.* Add: **5.** *transf.* (Earlier and later examples.)

1869 G. Meredith *Let.* 27 Dec. (1970) I. 409, I have turned Wendell Phillips like a drenching fireman's hose on a parson, and made him sputter and gutter and go to his wife to trim his wick. **1917** T. S. Eliot *Prufrock* 22 My self-possession flares up for a second... My self-possession gutters; we are really in the dark.

guttering, *vbl. sb.* **3. b.** (Later examples.)

1934 *Jrnl. R. Aeronaut. Soc.* XXXVIII. 326 This raises the working temperature of the former and may eventually cause 'blow past' of the gases when the valve should normally be closed, finally leading to valve burning or 'guttering'. **1948** *Shell Aviation News* cxix. 17/2 The condition of the valve face was good, and there were no signs of definite guttering or blow-by.

gutter-snipe. Add: **2. b.** (Earlier example.) Also *attrib.* and as *adj.*

c **1869** 'Mark Twain' *Sk. New & Old* (1875) 70 Unfurl yourselves under my banner, noble savages, illustrious guttersnipes. **1907** *Daily Chron.* 17 May 3/6 Harry Lauder, in his Glasgow gutter-snipe character. **1961** *Encounter* Mar. 72/1 The guttersnipe ecstasies of [Henry Miller's *Tropic of*] *Capricorn*.

c. (See quot. 1877.) *U.S.*

1857 Underhill & Thomson *Elephant Club* 57 He belongs to a class of beings in New York..known by the ornithological appelation of 'gutter-snipes'. **1870** J. K. Medbery *Men & Myst. Wall St.* 136 Gutter snipes. Curbstone brokers. **1877** Bartlett *Dict. Amer.* (ed. 4), *Gutter-snipes*, a Wall Street term for brokers who do business chiefly on the sidewalk or in the street, and who are not members of the Stock Exchange. **1942** Berrey & Van den Bark *Amer. Thes. Slang* § 563/4 *Guttersnipe*, one who operates in a curb market.

Guttman scale (gʌ·tmăn). *Psychol.* and *Sociol.* [f. the name of L. *Guttman* (b. 1916), American psychologist.] A type of scale used to measure and assess mental attitudes and properties (see quot. 1970).

1950 S. A. Stouffer et al. *Measurement & Prediction* IV. i. 5 The perfect Guttman scale is not likely to be found in practice, but satisfactory approximations to it are not so rare as some critics of the early papers describing it have implied. **1968** J. Whiting in Lindzey & Aronson *Handbk. Soc. Psychol.* (ed. 2) II. xvii. 714 Guttman scales have been used effectively in a number of cross-cultural studies. **1969** *Computers & Humanities* III. 144 [They] have used Guttman scale analysis to measure attitudes of congressmen and senators about roll call votes between 1830 and 1860. **1970** G. A. & A. G. Theodorson *Mod. Dict. Sociol.* 181 *Guttman scale*, a cumulative scale designed..to determine whether the items can be arranged along a single continuum of increasing intensity. When a series of items (attitudes) forms a Guttman scale, acceptance of any item on the scale indicates acceptance of all the items below it.

guttus (gʌ·tŏs). *Class. Archæol.* Pl. **-ti.** [L.] A narrow-necked cruet or oil-flask.

1842 A. Rich in W. Smith *Dict. Gr. & Roman Antiq.* 141/2 Oil..was dropped upon it from a small vessel called *guttus*, which had a narrow neck, so as to discharge its contents drop by drop, from whence the name is taken. **1883** J. W. Mollett *Illustr. Dict. Art & Archæol.* 163 *Guttus*,..a vessel with a very narrow neck and mouth, by means of which liquids could be poured out drop by drop; whence its name. **1921** *Brit. Mus. Return* 61 Guttus of early Italian red-figure style.

gutty, *sb.* Add to def.: In full *gutty ball.*

1881 A. Wardrop *Poems* 106 He's speel'd up tae the dresser heid, To get his gutty ba'. **1902** *Daily Chron.* 9 Dec. 9/3 He advised them to select the 'gutty' ball only, as the best ball for golf. **1924** C. J. H. Tolley *Mod. Golfer* 64 Thousands to-day would not be playing the game if the 'gutty' ball remained the official ball of to-day.

gutty, *a.* Add: **2.** *Jazz slang.* Earthy, primitive.

1939 in R. S. Gold *Jazz Lexicon* (1964) 133 Buster Bailey's reaction can be felt in the 'gutty' clarinet tones he uses. **1955** L. Feather *Encycl. Jazz* vii. 160 Higginbotham was..the supreme trombone stylist of the gutty, forceful school in the '30s. **1958** *Esquire* Feb. 35/2 You feel it in a beat, in jazzy..or a good gutty rock number. **1958** P. Gammond *Decca Bk. Jazz* vi. 80 The 1940 recordings..serve as good examples of Freeman's gutty, swinging tenor[-saxophone] style.

3. = *gutsy *a.* 2.

1953 in Wentworth & Flexner *Dict. Amer. Slang* 236/2 Here Come the Guttiest Guys of All. **1960** *News Chron.* 30 June 6/4, I can get gutty characters to play at last.

guv (gʌv). Also **guvner, guv'ner, guvnor, guv'nor** (gʌ·vnəɪ). Spellings representing vulgar or colloq. pronunciations of Governor.

1852 *Punch* 2 Oct. 152/1 (*caption*) Coster (to extremely genteel person). 'I say, Guvner, give us a hist with this 'ere bilin' o' greens!' **1886** [see Governor 7 c]. **1890** *Punch* 15 Mar. 123/1 Yes, you're right, Guv, this is dirty fun and dreary. **1895** *Ibid.* 17 Aug. 77/1 Guv wos in it all the while. **1959** I. & P. Opie *Lore & Lang. Schoolch.* xvii. 362 The headmaster may be known as 'The Gaffer', 'The Guv'nor', or 'The Boss'. **1960** *Observer* 24 Jan. 5/2 Sometimes the peterman finds his own jobs and acts as guvnor of his own team. *Ibid.*, The guvnor of a team of active criminals. **1968** *Listener* 15 Aug. 210/1 You can be sure that if somebody calls you 'mister' on the railways he doesn't like you. The term of endearment is 'guv'nor'.

guvacine (giū·văsīn). *Chem.* [ad. G. *guvacin* (E. Jahns 1891, in *Ber. d. Deut. Chem. Ges.* XXIV. 2615), f. Skr. *guvāka* betel-nut tree, areca + -ine[5].] A white crystalline alkaloid, $C_6H_9NO_2$, obtained from the areca nut.

1891 *Jrnl. Chem. Soc.* LX. II. 1520 The examination of larger quantities of material has led to the discovery of a fifth alkaloid in areca nut. This has been termed guvacine, from guvaca, the Indian name for the areca palm. **1946** *Jrnl. Amer. Chem. Soc.* LXVIII. 1053/1 The white solid remaining undissolved..was guvacine hydrochloride. **1970** R. K. Hill in S. W. Pelletier *Chem. Alkaloids* xiv. 411 The parent member of this family [*sc.* the areca alkaloids] is an amino acid guvacine, $C_6H_9O_2N$... Guvacine is a secondary amine, is optically inactive and nonresolvable.

guvner, guvnor: see *guv.

guy, *sb.*[1] Add: **3.** *guy-line*; **guy-rod,** a rod used in place of a guy-rope.

1903 A. Adams *Log of Cowboy* x. 152 We took a guy line from the wagon and snubbed it to a tree. **1928** *Daily Express* 7 Apr. 5/5 Drive in the remaining pegs and attach and true up the remaining guylines. **1903** *Electr. World & Engin.* 31 Oct. 711 (Cent. D. Suppl.), Guy-rods are 8 ft. × ⅝ in., provided with an eye at one end. **1910** N. Hawkins *Electr. Dict.*, *Guy Rod Bands*, bands by which a guy rod is attached to a telegraph pole.

guy, *sb.*[2] **1.** Delete 'This custom is now falling into disuse'. Add to def.: Also in full *Guy Fawkes.* (Later example.)

The figure is now rarely carried about in the streets, but is exhibited by children when collecting money for fireworks during the days preceding Nov. 5.

1861 C. M. Yonge *Young Step-Mother* xxix. 442 'There's Guy Fawkes,' cried Albinia, as a procession of scare-crows were borne on chairs amid thunders of acclamation. **1968** *Listener* 7 Nov. 606 I'm afraid Simon is rather burnt. It was so dark some of the children thought he was the guy. **1970** R. Garrett *Run Down* iv. 95 It was the first Saturday of November. There were dozens of Guy Fawkes of varying merit lined along each street.

b. *Guy Fawkes day* (*night*), 5 Nov., the anniversary of the 'Gunpowder plot'. Also *ellipt.*

1825 W. Hone *Every-day Bk.* 1430/2 'Guy Fawkes-day', or, as they as often call it, 'Pope-day', is a holiday, and.., on account of its festivous enjoyment, is the greatest holiday of the season. **1833** J. Romilly *Diary* 5 Nov. (1967) I. 41 Webb (the W. merchant's) long dull son dined in hall with me being Guy Fawkes day. **1913** C. Mackenzie *Sinister St.* I. I. iv. 63 Suddenly he heard the cry, 'Remember, remember the Fifth of November..,' and, almost before he had time to realize it was the dreaded Guy Fawkes, a band of..boys..held close to the window a nodding Guy. **1970** *Times* 4 Nov. 2/2 The reserves of stamina the [fire]men have to build up for Guy Fawkes night each year.

3. b. Also *to do a guy.*

1897 [see Guy *v.*[4]]. **1898** [in Dict.]. **1925** N. Venner *Imperfect Impostor* xiii. 221 He's just picked me up out of the road with a sprained ankle, or very near it, bandaged me up like a medical student, and brought me home. Then he wants to do a guy at the front door. **1954** N. Balchin *Last Recoll. Uncle Charles* iii. 43 It was..not so very surprising that he should 'do a guy', as they put it locally, with his employers' money.

d. (Earlier and later examples.)

The earliest examples may be influenced by sense 2.

1847 *Swell's Night Guide* 41, I can't tonight, for I am going to be seduced by a rich old Guy. **1863** C. Reade *Hard Cash* III. xiii. 270, I wouldn't speak to you in the street for fear of disgracing you; I am such a poor little guy to be addressing a gentleman like you. **1876** L. C. Barraud 15 Oct. in E. M. Barraud *Story of Family* (1967) xi. 133 The little children are such cures, and the nurses seem to go out with the master and mistress. The little boys look great guys. **1896** Ade *Artie* i. 3 You guys must think I'm a quitter. **1904** *Cincinnati Commercial Tribune* 29 June 4 Mr. Bryan is a hefty guy when it comes to Democratic conventions and the platforms thereof. **1928** D. H. Lawrence in *T.P.'s & Cassell's Weekly* 7 July 333/3, I says to Mother: *Show* me somebody happy, then! And she shows me some guy, or some bright young thing, and gets mad when I say: See the pretty monkey! **1931** R. Campbell *Georgiad* I. 16 These are the guys that find the world forlorn And wish (correctly) they had not been born. **1931** D. Runyon (*title*) Guys and Dolls. **1932** E. Wilson *Devil take Hindmost* xi. 114 The literary guys are taking public matters more seriously. **1951** Auden *Nones* (1952) 62 Thou shalt not be on friendly terms With guys in advertising firms. **1966** Wodehouse *Plum Pie* I. 32 All the other places.. had been full of guys and dolls standing bumper to bumper.

guy, *v.*[3] Add: **2.** (Earlier and later examples.) Also, to trifle with a theatrical part.

1854 A. C. Mowatt *Autobiogr. Actress* xv. 227 Good gracious! the audience will guy you!..'Guy me? What do you mean by *guy*?'..'Why, laugh at you, to be sure—and *chaff* you!' **1890** J. Jefferson *Autobiogr.* 219 With all this at stake, some wanton actor deliberately 'guys' his part and overturns the patient care of his comrade. **1895** W. Archer *Theatr. World of 1894* xliii. 287 Larking and guying on the stage are my abhorrence. **1898** J. Hollingshead *Gaiety Chron.* x. 414 The actors, I am bound to say, treated this curious fossil of dramatic protection with more than proper respect, and did not 'guy' the parts allotted to them. **1906** *Daily Chron.* 31 Oct. 5/5 We must make an end of that disgusting blunder of guying them [*sc.* arrested suffragettes] up in hideous prison uniforms. **1963** *Times* 27 May 8/2 Francoise Sagan, British phlegmatism and many other picturesque aspects of contemporary life in the 1960s are gently but tellingly guyed by Mr. Kohout. **1970** G. Greer *Female Eunuch* 328 Vociferous women are guyed in the press.

guy, *v.*[4] (Later examples.)

1938 F. D. Sharpe *Sharpe of Flying Squad* xxvi. 263 Jack, guy for your b— life. The Squad are here and they're pinching *everybody*. **1963** *Times* 15 Feb. 4/7 Hurry up, I have had to do a chap, we will have to guy out of here.

Guyanese (gəiănī·z), *a.* and *sb.* [f. *Guyana* + -ese.] **A.** *adj.* Of or pertaining to Guyana, formerly British Guiana. **B.** *sb.* A native or inhabitant of Guyana; also in collective sense. So **Guy·anan.** Cf. *Guianese.

1965 *New Statesman* 26 Nov. 814/1 If UDI had been proclaimed by Adenis or Guyanans, can anyone deny we would have sent in troops immediately? **1966** 'A. York' *Eliminator* vii. 131 The Guyanese police and us parted company. **1966** *Listener* 2 June 781/1 The uncertainties facing the Guyanese people as they acquire independence. *Ibid.* 781/2 Most Guyanese acquired a fairly uniform central culture. **1968** *Times* 10 Feb. 4/3 Five out of six Guyanese live in villages. **1969** *Daily Tel.* 11 Nov. 1/7 Sir David, first Guyanese Governor-General of Guyana, was leaving the West Indian Club. **1971** *Jamaican Weekly Gleaner* 3 Nov. 24/1 A Jamaican, a Trinidadian and two Guyanese women.

guyot (gī·yo). *Oceanogr.* [f. the name of the Swiss geographer A. H. *Guyot* (1807–84).] A flat-topped, submarine mountain; a kind of sea-mount.

1946 H. H. Hess in *Amer. Jrnl. Sci.* CCXLIV. 772 In discussing these submerged flat-topped peaks which rise from the normal ocean floor, the writer will henceforth call them 'guyots' after the 19th century geographer, Arnold Guyot. **1959** *New Scientist* 1 Jan. 14/3 These 'tablemounts' (or 'guyots', if one prefers to use the non-descriptive term of the discoverer, Prof. H. Hess of Princeton) must be deep-drowned ancient islands. **1962** *Ibid.* 12 Apr. 11/2 The deep flat-topped sea-mounts or guyots which are such a typical feature of the Pacific. **1970** *Sci. Amer.* Feb. 36 (*caption*) Mount Asmara.. resembles the numerous guyots, or submerged oceanic mountains, whose level summits are usually attributed to wave erosion.

guyver, var. *gyver.

guzla, var. *gusle.

guzzle, *sb.* Add: **5.** *attrib.*, as **guzzle-guts,** a glutton.

1959 I. & P. Opie *Lore & Lang. Schoolch.* ix. 167 They call him [*sc.* a greedy-guts]..guzzle-guts. **1968** *Gloss. Brit. Argot* (Paramount Pictures), *Guzzle-guts*, glutton.

guzzle, *v.* Add: **4.** *trans.* To seize by the throat, choke, throttle; to strangle, kill. (Cf. Guzzle *sb.* 4.) *slang* and *dial.*

1885 F. Gordon *Pyotshaw* 318 *Guzzle*, to choke violently. **1916** Joyce *Portrait of Artist* (1917) iv. 195 Duck him! Guzzle him now, Towser! **1931** D. Runyon *Guys & Dolls* (1932) 186 He will be safe from being guzzled by some of Black Mike's or Benny's guys. **1960** *Observer* 17 July 9/8 'Can't you..guzzle your friends, Hugh?'.. North of the Border it [*sc.* the word 'guzzle'] does not mean to stuff oneself: apparently it means to gag.

gwan, g'wan (gwān). *U.S.* and *Irish dial.* pronunciation of *go on* (esp. Go *v.* 84 j).

1906 C. E. Mulford in *Outing* Apr. 75/2 'Where'd he get you, Fat?' asked Hopalong. 'G'wan, don't get funny, son,' replied Skinny. **1924** *Dialect Notes* V. 269 *Gwan* (='go on'). **1968** 'N. Blake' *Private Wound* vii. 96 G'wan, Harry. Give him a kiss. He deserves it. **1970** M. Kenyon *100,000 Welcomes* xviii. 154 G'wan onto it! Anyone wantin' to interfere wid yous'd have to be hard up!

gwarri, var. *guarri.

gwely (gwe·li). [Welsh.] **a.** A social unit that was once traditional in Wales, consisting of four generations of one family in which the great-grandfather, the head of the group, had proprietary right over its landed property. **b.** The land held by the members of such a group.

1889 H. Lewis *Anc. Laws Wales* iv. 81 There was..a family or *gwely*, then a joint family sprung from or including three generations of ancestors. *Ibid.* vi. 167 Gavell is often used interchangeably with *wele*, which meant the holding of a family (gwely). **1895** F. Seebohm *Tribal Syst. Wales* i. 9 We shall see..that the *wele* or *gwely* was, strictly speaking, rather the family or kindred occupying the hamlet than the hamlet itself. **1900** Rhys & Brynmor-Jones *Welsh People* vi. 195 The rights of the chief of household..and the lands in the occupation of himself and other members of his household were termed his *gwely*. *Ibid.* 197 There was no re-division of the ancestor's *gwely* after the second cousins had divided it. **1907** H. M. Chadwick *Origin Eng. Nation* xii. 324 In some laws,..we find..evidence for a system similar to that of the Welsh gwely. **1951** *Trans. Anglesey Antiq. Soc. & Field Club* 35 The local unit was the 'gwely', a specific area of land held, semi-communally, by a group of relatives. *Ibid.* 36 After the Edwardian conquest, new factors reinforced those already at work... Even within the *gwely* itself, the individual tribesman of initiative could often buy up the portions of his less fortunate or less energetic kinsmen. **1968** *Encycl. Brit.* XXIII. 156/2 A movement toward more stationary settlement among freemen slowly spread..over the whole country, splitting the *maenor* into smaller township units (*trefs*)..and laying the foundations of the hamlets and clanlands (*gwelys*).

Gwentian (gwe·ntiăn), *sb.* and *a.* [f. the name *Gwent* + -IAN.] **A.** *sb.* **a.** An inhabitant of Gwent in Monmouthshire, historically a Welsh principality. **b.** The dialect of this region. **B.** *adj.* Of or pertaining to Gwent.

1831 S. LEWIS *Topogr. Dict. England* III. 318/2 The attempts of the Anglo-Saxon sovereigns to subjugate Wales were opposed by the Gwentians with extraordinary courage, insomuch that they do not appear to have been completely conquered during the Anglo-Saxon period. **1841** A. OWEN *Anc. Laws & Institutes of Wales* p. vi, The Laws and Institutes of Wales are here given in six parts: the three first of these are the Venedotian, the Dimetian, and the Gwentian Codes. *Ibid.* p. vii, The Gwentian Code..contains an account of territorial divisions peculiar to Gwent. **1901** *Daily Chron.* 16 Oct. 3/2 The Gwentian bowmen were famous. **1911** J. E. LLOYD *Hist. Wales* I. x. 342 The Gwentian Code..appears to be also a compilation, to be ascribed, perhaps, to Morgeneu and his son Cyfnerth, who are mentioned as the authors. **1913** J. M. JONES *Welsh Gram.* 8 Gwentian, the dialect of Gwent and Morgannwg, or South East Wales.

gwine (gwəin). Also **gwyn(e)**. Representing *dial.* pronunciation of GOING *pres. pple.* (Go *v.* A. 7).

1831 R. LOWER *Tom Cladpole's Jurney* XLII. 14 He said he must be gwyn. **1881** H. & C. R. SMITH *Isle of Wight Words* 50, I be gwine zoo vast as I can. **1882** 'MARK TWAIN' *Lett. to Publishers* 4 Mar. (1967) 152 I's gwyne to sen' you de stuff. **1896** *Longman's Mag.* Dec. 155 As I wur gwine up street, zo I looked in. **1908** *Sears, Roebuck Catal.* 200/1 Vocal Quartettes...I'se Gwine Back to Dixie (coon song). **1932** *Amer. Speech* VII. 178 By de time a'-wee was gwine tek de 'tart again, me foot get soak. **1950** *Publ. Amer. Dial. Soc.* XIV. 34 Whar you gwine? **1969** 'J. MORRIS' *Fever Grass* ii. 19 One day police gwine hol' you.

gy-. Add: Most scientific words of Greek etymology beginning with *gyn-* are now usually pronounced with (g) rather than (dʒ); in particular *gynæcology* and its derivatives are in general use with this pronunciation. *Gypsum*, *gyrate*, and *gyration* now always have the g 'soft', and *misogynist* rarely has it 'hard'.

gybe, *v.* **2.** Change first sentence of def. to read: To alter the course of a boat when the wind is aft so that her boom-sails gybe.

gybe, var. JIBE *v.*

1959 *Economist* 14 Feb. 612/1 A..fall in export deliveries..which gybes a little oddly with official trade figures.

gybing, *vbl. sb.* (Later example.)

1963 *Times* 31 May 5/1 Inexorably, Shadow sailed past to leeward, where there was still some wind and, though Wildfire did at last get moving again, Shadow was ahead at the gybing mark.

gym (dʒim). Colloq. abbreviation of GYMNASIUM or GYMNASTIC(S. Also *attrib.*, as *gym-dress*, *-slip*, *-tunic* (these also *spec.* of a schoolgirl's garment not primarily for use in the gymnasium); *gym-knickers*, *-mistress*, *-shoe*, etc.

1871 L. H. BAGG *Four Yrs. at Yale* i. 45 *Gym*, abbreviation for *gymnasium*. **1887** H. BAUMANN *Londinismen* 70/2 *Gym-shoes* (statt gymnastic-shoes) Schulsprache: Turnschuhe. **1888** *Boy's Own Paper* Summer No. 38/3 If you'll come round to the Gym. after, I'll show you. **1889** BARRÈRE & LELAND *Dict. Slang, Gymcad* (Royal Military Academy), a gymnasium instructor. **1891** H. S. HOLLAND in S. Paget *Mem.* (1921) 198 We had a superb exhibition in the gym. **1907** *Daily Chron.* 27 Nov. 3/3 The 'gym.' practices. **1922** JOYCE *Ulysses* 520 If you do tremble in anticipation of heel discipline to be inflicted in gym costume. **1926** F. M. FORD *Man could stand Up* I. i. 17 Five foot four in her gym shoes. **1930** *Cambridge Daily News* 25 Sept. 3/4 Gym slips should not be washed. **1932** D. C. MINTER *Mod. Needlecraft* 252 Gym Knickers.. 1¼ to 1½ yards blue or brown drill, nurse cloth or serge. **1936** *Daily Sketch* 9 Mar. 11/2 When a girl leaves school.. she throws away..her hockey sticks and her gym tunic. **1944** A. THIRKELL *Headmistress* iv. 79 Going to a physical physical culture college to learn to be gym and games mistresses. **1947** DYLAN THOMAS *Let.* 20 May (1966) 307 She thumped the hot Florence pavements..elbowing the droll Florentines from her gym-knickered way. **1949** 'G. ORWELL' *Nineteen Eighty-Four* I. 34 A youngish woman.. dressed in tunic and gym-shoes. **1949** D. SMITH *I capture Castle* I. iv. 41 My school gym-dress has a lot of life in it yet. **1961** M. SPARK *Prime of Miss Jean Brodie* ii. 24 Rose Stanley was pulling threads from the girdle of her gym tunic. **1967** H. W. SUTHERLAND *Magnie* iv. 64 That bloody gym dress with the black cotton stockings —and that old school hat. **1969** *Listener* 8 May 640/3 Yes, I've got a recollection of Barbara in a gym slip at Bradford Grammar School. **1970** D. CLARK *Deadly Pattern* vi. 145 One or two of the girls were married almost before they could get out of their gym knickers.

gymkhana. Add to def.: Now *spec.* a meeting at which horses and their riders take part in games and contests; also a competition designed to test driving skill.

1933 A. BLEWITT *Ponies & Children* viii. 120 Nowadays gymkhanas have become very much the fashion. **1955** *Times* 20 Aug. 3/1 Miss Valerie Engelmann is leading by five points from Miss Pat Moss in the gymkhana points competition which is proving a popular feature of this show. **1966** *Publ. Amer. Dial. Soc.* 1964 XLII. 5 *Gymkhana*, a tight, low speed sports car competition, in which only one car at a time runs a course. **1970** G. WHEATLEY *Let's start Riding* xxiv. 164 Gymkhanas can be great fun and there are many different events and games for which one can enter, if one has a fast, handy pony.

gymnadenia (dʒimnădī·niă). [mod.L. (see quot. 1813), f. Gr. γυμνός naked, bare + ἀδεν-, ἀδήν gland + -IA.] A terrestrial orchid of the genus so called, which includes *G. conopsea*, the fragrant orchid, native to Britain.

1813 R. BROWN in W. Aiton *Hortus Kewensis* (ed. 2) V. 191 Fragrant Gymnadenia. *Nat.* of Britain. *Fl.* June and July. **1856** W. A. BROMFIELD *Flora Vectensis* 479 Fragrant Gymnadenia..not very general. [**1905** LD. AVEBURY *Notes on Life Hist. Brit. Flowering Plants* 405 This genus [sc. *Habenaria*], in which Gymnadenia is generally included, hardly seems to me to be sufficiently distinct from Orchis.] **1926** G. C. DRUCE *Flora Bucks.* 334 It is somewhat remarkable that the *Gymnadenia* should flourish not only on bare chalk slopes but also in wet bogs. **1965** A. D. HAWKES *Encycl. Cult. Orchids* 234/1 Most of the Gymnadenias, because of their habitat in temperate areas, require rather cool temperatures at all times.

gymnastic, *a.* and *sb.* Add: **A. 1. d.** Of the initial letter of an illuminated manuscript: decorated with human figures, etc., which are portrayed climbing like gymnasts round the letter.

1945 F. WORMALD in *Archæologia* XCI. 127 These initials are composed of animals, monsters, and human figures, who clamber all over the framework of the letter as if using it as a kind of gymnastic appliance...In the Canterbury MSS. this gymnastic method is carried to extreme lengths. *Ibid.* 130 The Durham artists had some knowledge of the continental 'gymnastic' initial so popular at Canterbury. **1946** —— *Survival Anglo-Saxon Illumin.* 9 Another new type of English initial..is the so-called 'gymnastic' style, where animals and creatures clamber all over the frame of the initial rather in the manner of acrobats. This type was rare in England before the Conquest, but is found in the Durham MSS. and is an outstanding characteristic of Canterbury illumination of about the year 1100. **1952** D. T. RICE *Eng. Art 871–1100* vii. 216 The B of the Beatus page is in keeping with late Saxon developments, for in addition to the usual scroll-work, a number of little figures appear clambering amongst the stems; Wormald has aptly termed this the 'gymnastic' style. **1954** M. RICKERT *Painting in Britain: Middle Ages* iii. 66 The wholesale introduction of such human figures into the initial.. decorative, 'gymnastic' figures. **1970** *Anglo-Saxon Illumination in Oxf. Libr.* 7 Initial 'P'. Two dragons and a gymnastic figure make up the initial.

B. 2. b. (Later examples.)

1905 R. BROUGHTON *Waif's Progress* xxviii. 308 It seemed an impossible feat in mental gymnastics to.. wrench his thoughts away. **1957** C. L. WRENN *Word & Symbol* (1967) 193 Will the deliberate ordinariness of dramatic language of *The Confidential Clerk* come any nearer to bringing poetry to the people than the stirring mental and metrical gymnastics of Mr Christopher Fry? **1966** P. MOLONEY *Plea for Mersey* 25 These verbal gymnastics are heard in England elsewhere than in Liverpool.

4. (Later example.)

1907 'MARK TWAIN' in *N. Amer. Rev.* 15 May 4 When he had been teaching me twice a day for three weeks I introduced a new gymnastic—one that he had never seen before.

gymnemic (dʒimnī·mik, -ne·mik), *a. Chem.* [f. mod.L. *Gymnem-a*, name of an asclepiadaceous genus, f. Gr. γυμ-νός naked + νῆμ-α thread + -IC.] *gymnemic acid*: an impure substance which is obtained from the leaves of *Gymnema sylvestre* and is believed to be the cause of the temporary loss of the ability to taste sweetness and bitterness which results when leaves of this plant are chewed.

1887 D. HOOPER in *Nature* 14 Apr. 566/2 The body just described has the characteristics of an organic acid related in some particulars to chrysophanic acid, but having some distinctly peculiar reactions, and possessing the anti-saccharine property ascribed to the leaves. I propose to call it Gymnemic acid. **1958** *Jrnl. Indian Chem. Soc.* XXXV. 650 It is also not certain whether the amorphous substance, described as gymnemic acid ..by Hooper is a single substance or not. Further, Hooper's claim that gymnemic acid constituted the anti-saccharine principle of the leaves has been challenged.

gympie (gi·mpi). *Austral.* [Native name.] An evergreen shrub, *Laportea moroides*, which belongs to the nettle family, Urticaceæ, and has leaves covered with stinging hairs.

1895 A. MESTON *Geogr. Hist. Queensland* 55 Gympie. The Mary River blacks' name for the stinging tree. **1911** W. R. GUILFOYLE *Austral. Plants* 233 *Laportea moroides* 'Gympie Nettle Tree' or 'Mulberry Nettle Tree' (evergreen shrub, reputed poisonous and injurious to stock, 15 to 20 ft.). **1934** *Bulletin* (Sydney) 15 Aug. 21/2 Strangely enough, the weed is nearly always found near the gympie-gympie trees, and is easily identified by its narrow curled leaves on a pink stalk. **1963** W. V. MACFARLANE in Keegan & Macfarlane *Venomous & Poisonous Anim. Pacific* I. 31 In the eastern rain forests of northern New South Wales and Queensland, *L[aportea] gigas* becomes a tree 30–40 m high and the Gympie bush, *L. moroides*, grows only 6–8 m. **1965** *Austral. Encycl.* IV. 406/2 In 1868 the name [Nashville] was altered to Gympie, an aboriginal term for the stinging trees found in the district.

gynæ (gəi·ni). Also **gynie.** Colloq. abbreviation of GYNÆCOLOGICAL *a.* and GYNÆCOLOGY.

1942 BERREY & VAN DEN BARK *Amer. Thes. Slang* § 534/1 *Gynae*, gynaecology. **1960** J. GRANT *Come again, Nurse* xxiv. 158 Sister Judgeson on Gynie Theatre was a sweet middle-aged woman. **1962** G. BUTLER *Coffin in Oxford* ix. 123 'I'm over in Gynae now,' said the nurse. **1964** G. L. COHEN *What's Wrong with Hospitals?* iv. 75 'We didn't come across any horrors,' said Dr Duncum.. 'unless you count adolescent girls in gynae wards.'

gynæco-. See *GY- and add: **gynæcoid** *a.* (see *GYNE); **gynæcomasty,** in mod.L. form **gynæcoma·stia**; **gynæcoma·zia** [Gr. μαζός var. μαστός breast], gynæcomastia.

1881 *Index-Cat. Surg.-Gen. U.S.* II. 421/2 (*title*) On the male mammary glands, and on gynaecomastia. **1904** *Lancet* 26 Mar. 865/2 This was not so marked a case of gynæcomastia as in Case I. **1962** *Lancet* 29 Dec. 1358/1 Gynæcomastia in the male may be associated with an abnormality of œstrogen production or metabolism. **1850** J. BIRKETT *Dis. Breast* 254 Gynecomazia..signifies a development of the mammary organ in the male which more or less resembles the form of that of the female. **1923** W. H. EVANS *Dis. Breast* xliii. 473 The secretion of milk by the breasts of men is, however, by no means confined to those with Gynæcomazia.

gynandro-. Add: **gyna·ndromorph,** an individual which exhibits gynandromorphism; **gynandromo·rphic** *a.* = *gynandromorphous* adj.

1897 WEBSTER, Gynandromorph. **1903** *Bull. Amer. Mus. Nat. Hist.* XIX. 648 A single incipient colony.. was found to contain about twenty workers, a few worker cocoons, and a gynandromorph. **1968** M. W. STRICKBERGER *Genetics* xxi. 468 Gynandromorphs differ from intersexes in the sense that gynandromorphs are obviously mosaic. **1894** W. BATESON *Study of Variation* 68 Gynandromorphic insects, in which the characters of the whole or part of one side of the body, wings and antennæ, are male, while those of the other side are female. **1950** B. P. SONNENBLICK in M. Demerec *Biol. Drosophila* ii. 90 A wide assortment of gynandromorphic patterns can result.

gyne (dʒəin). *Ent.* [ad. Gr. γυνή woman.] The fertile female in a colony of social insects, esp. a queen ant. Hence **gy·næcoid** *a.* [see GYNÆCO- and -OID], showing some characteristics of this type of insect.

[**1905** tr. *E. Wasmann's Compar. Stud. Psychol. Ants* iv. 164 Intermediate forms between females and worker ants..female-like workers, which I have named pseudofemales (pseudogynes).] **1907** *Bull. Amer. Mus. Nat. Hist.* XXIII. 54 The *female* (gynē), or queen, is the more highly specialized sex among ants and is characterized, as a rule, by her large stature and the more uniform development of her organs. **1915** H. ST. J. K. DONISTHORPE *Brit. Ants* 38 The female (*gyne*), queen, or α-female, is the most highly specialized sex. **1907** *Bull. Amer. Mus. Nat. Hist.* XXIII. 24 Wasmann's 'gynæcoid workers', which are merely workers whose ovaries contain mature eggs. **1933** W. M. WHEELER *Colony-founding among Ants* 132 He cites the development of fertile gynaecoid workers and the rearing of substitution queens in ant and termite colonies that have lost their mother. **1953** *Q. Rev. Biol.* XXVIII. 145/1 *Ergatogyne.* Individuals falling along the allometric progression connecting the queen and worker castes, ranging from subapterous forms with queen-like alitrunks to slightly gynecoid workers.

gynee (gəi·ni). Also **gaini, ghinee.** [Hindi (related to *go* Cow *sb.*[1]).] One of a small variety of Indian cattle.

1829 J. SHIPP *Mem.* III. v. 132, I found that the said tiger had feasted on a more delicate morsel,—a nice little ghinee, a small cow. **1832** F. PARKS *Jrnl.* 1 Dec. in *Wand. Pilgrim* (1850) I. xxii. 251 We have become great farmers, having sown our crop of oats, and are building outhouses to receive some thirty-four dwarf cows and oxen (gynees), which are to be fed up for the table. **1873** H. BLOCHMANN tr. *Abul Fazl 'Allami's Ain i Akbari* I. 149 There is also a species of oxen, called *gainí*, small..but very beautiful.

gynie, var. *GYNÆ.

gyno-. See *GY- and add: **gynodiœcy** (-dəi,ī·si), gynodiœcism; **gynomonœcious** *a.* (earlier example); **gynomonœcism** (-mǫnī·siz'm), the condition of being gynomonœcious; **gynomonœcy** (-mǫnī·si), gynomonœcism; **gynopho·bia,** fear of women; **gy·nospore** = MEGASPORE.

1940 *Nature* 30 Mar. 486/2 Gynodiœcy should not be confused with unisexuality. **1956** *Evolution* X. 115 (*heading*) The genetics and evolution of gynodioecy. **1877** DARWIN *Diff. Forms Flowers* 12 Other species..bear on the same plant hermaphrodite and female flowers; and these might be called gyno-monœcious, if a name were desirable for them. **1897** Gynomonœcism [in Dict. s.v.

GYNO-]. **1963** DAVIS & HEYWOOD *Princ. Angiosperm Taxon.* xi. 369 Gynomonoecism where female and hermaphrodite flowers co-exist on the same plant. **1949** DARLINGTON & MATHER *Elem. Genetics* 394 *Gynomonoecy*, the condition of plant individuals, or of species having individuals, which bear both female (*i.e.* male sterile) and hermaphrodite flowers. **1886** O. W. HOLMES *Mortal Antipathy* xix. 231 If we give it a name, we shall have to apply the term *Gynophobia*, or Fear of Woman. **1886** *Academy* 16 Jan. 37 A man..has become afflicted with gynophobia. **1940** HINSIE & SHATZKY *Psychiatric Dict.* 245 *Gynophobia*, morbid fear of women. **1900** B. D. JACKSON *Gloss. Bot. Terms* 309/2 *Gynospore*.., formerly suggested for macrospore, that is, a Megaspore. **1937** *Sci. Proc. R. Dublin Soc.* XXI. 457 The development of several gynospores, supernumerary prothalli, lateral pollen tubes and lateral archegonia;..all seem to give the redwood an unique position. **1964** PRIESTLEY & SCOTT *Introd. Bot.* (ed. 4) xxxvi. 590 Alternative and more non-committal terms are indeed already proposed, viz...gynospore, or female spore, for the embryo sac.

gynogenesis (gəinodʒe·nesis). *Zool.* [f. GYNO- + -GENESIS.] Reproduction in which the development of the embryo occurs as a result of the penetration of an ovum by a sperm but does not involve the nuclear material of the latter.

1925 E. B. WILSON *Cell* (ed. 3) v. 460 In a nearly related phenomenon, which may be called gynogenesis, the sperm penetrates (and in some cases activates) the egg but otherwise takes no part in the processes of development. **1952** BOYD & HAMILTON in A. S. Parkes *Marshall's Physiol. Reprod.* (ed. 3) II. xiv. 42 Gynogenesis may occur to a limited degree in mammals as the result of irradiation of the semen. **1969** R. F. CHAPMAN *Insects* xix. 380 An unusual type of thelytoky, known as gynogenesis, occurs in the form *mobilis* of *Ptinus clavipes* (Coleoptera). The form *mobilis* exists only as triploid females which reproduce parthenogenetically, but the development of eggs is triggered by healthy sperm of *P. clavipes* or, less successfully, of *P. pusillus*.

Hence **gynogene·tic** *a.*

1925 E. B. WILSON *Cell* (ed. 3) v. 463 The conclusion seems probable..that even in such (presumably) diploid larvæ development is gynogenetic, the diploid number having been restored by a doubling of the maternal haploid group. **1964** L. B. RUSSELL in C. Pavan et al. *Mammalian Cytogenetics* 64 Ultraviolet irradiation of spermatozoa can destroy the paternally contributed chromatin without affecting the activating function of the sperm and thus lead to gynogenetic haploids.

gyp, *sb.*[1] Add: **3.** A fraudulent action; a swindle. Also as *adj.* Cf. *GYP *v.* orig. *U.S.*

1914 JACKSON & HELLYER *Vocab. Criminal Slang* 41 *Gyp*,..the act of short-changing; a defrauding by substitution; an action that belies a professed sincerity. **1930** 'A. ARMSTRONG' *Taxi* xvi. 220 The non-combine cabs..are referred to as 'gyp' taxis, which implies mingled piracy and deceit. **1967** *Boston Sunday Globe* 23 Apr. B61/2 Some are good, but gyps abound. Authorities report..phony practices.

gyp, *sb.*[2] (Earlier example.)

1878 C. HALLOCK *Amer. Club List & Sportsman's Gloss.* p. v, *Gyp*, the young female pup.

gyp (dʒip), *sb.*[3] *dial.* or *colloq.* Also **gip, jip.** [App. contraction of GEE-UP, which is used in dial. as sb.] *to give* (a person) *gyp*: to punish, thrash, treat roughly; to hurt, give pain.

1893 *Funk's Stand. Dict.*, *To give one gyp*, to make one smart for anything done. **1898** in B. KIRKBY *Lakeland Words*. **1902** *Eng. Dial. Dict.* s.v. *Jip*, Ah'll gi'e tha jip... 'Ah gav' it jip Ah can tell tha,' said of beating a carpet soundly with a stick in each hand. **1910** 'G. B. LANCASTER' *Jim of Ranges* xi. 212 'Jim Kyneton's a good man, and this is giving him particular gyp if I know anything of good men,' he [Jack West] said. **1915** 'BOYD CABLE' *Between Lines* 19 We'll give 'em gyp if they try it. **1917** P. MACGILL *Brown Brethren* xii. 170 A cramp in my guts! ..Gawd, it isn't 'arf giving me gyp! **1936** WODEHOUSE *Laughing Gas* iv. 46 If you knew what gyp those shoes were giving me that night. **1966** I. JEFFERIES *House Surgeon* viii. 155, I should think his tum is giving him gip.

gyp (dʒip), *v.* orig. *U.S.* Also **gip.** [Cf. *GYP *sb.*[1] 3.] To cheat, trick, swindle.

1889 in *Cent. Dict.* **1914** JACKSON & HELLYER *Vocab. Criminal Slang* 41 *Gyp*.., to flim-flam; to cheat by means of guile and manual dexterity...'Gyp this boob with a deuce.' **1925** F. SCOTT FITZGERALD *Great Gatsby* (1926) ii. 42 We had over twelve hundred dollars when we started, but we got gyped out of it all in two days. **1930** D. BYRNE *Golden Goat* xiii. 106 Dariano had gipped the Greek Government during the war of millions in contracts. **1932** J. DOS PASSOS *1919* 55 American dollars went pretty far if you knew enough not to let 'em gyp you. **1935** WODEHOUSE *Luck of Bodkins* xiv. 156 A suspicion was growing with him that..he had been gypped. **1962** *Punch* 28 Feb. 356/2 If he..thinks the conductor is trying to gyp him..he..need only look at the fares table. **1965** R. HOWARD tr. *S. de Beauvoir's Force of Circumstance* 658 Turning an incredulous gaze toward that young and credulous girl, I realize with stupor how much I was gypped.

gyppie, gyp(p)o, gyppy, varr. *GIPPY.

1916 *Anzac Book* 137 The erstwhile Land of Jacko did breed much 'flue' and 'pneu', and it did seem as though the plagues of the ancient Gyppos had descended upon them. *Ibid.* 138 And it came to pass that while they yet warmed their hands there was heard a mighty crash, and of the 'Gyppies' that remained were picked up seven stretchers full. **1928** *Papers Mich. Acad. Sci. & Arts* X. 297 *Gypos*, the Egyptian army; Egyptian laborers. **1942** C. BARRETT *On Wallaby* iv. 62 Marmalade was a luxury, which, on four-day old Gyppo bread, you spread as thinly as tinfoil. **1948** PARTRIDGE *Dict. Forces' Slang* 89 *Gyppy tummy*, a sharp griping pain, accompanied by diarrhœa, very common in Egypt. **1960** C. MACKENZIE *Greece in My Life* xiii. 190 Both of them besides the usual gyppy tummy had had bad colds. **1970** *New Scientist* 8 Jan. 47/1 The commonest ailment experienced by modern-day Marco Polos is an intestinal attack known as gyppy tummy..in the Middle East.

gypsophila (dʒipsǫ·filă). *Bot.* [mod.L. (Linnæus *Dissertatio Botanica, qua Nova Plantarum Genera* (1751) no. 1103), f. Gr. γύψος chalk + φίλος loving.] A member of the genus of herbs so called, belonging to the family Caryophyllaceæ and bearing small delicate flowers in panicles.

1771 R. WESTON *Universal Botanist* II. 373 Tall Siberian White Gypsophila. **1810** *Curtis's Bot. Mag.* XXXI. 1281 (*heading*) Trailing Gypsophila. **1909** *Daily Mirror* 13 Aug. 7/4 The graceful gypsophila (chalk plant). **1970** C. LLOYD *Well-Tempered Garden* ii. 63 Dianthus and gypsophilas should not be kept under close conditions a day longer than is necessary for rooting them.

gypsophilous (dʒipsǫ·filəs), *a. Ecol.* [See prec. and -OUS.] Of a plant: thriving on a soil rich in gypsum; formerly used in reference to limestone soils (cf. *CALCICOLOUS *a.*).

1902 F. E. CLEMENTS in *Botanische Jahrbücher* XXXI. LXX. 14 *Gypsophytia*,..limestone plant formations; gypsophyta, limestone plants; gypsophilus, dwelling on limestone. **1932** FULLER & CONARD tr. *Braun-Blanquet's Plant Sociol.* vi. 186 On the high Algerian plateau the gypsophilous plants form an upper zone of scanty vegetation bordering some schotts at Le Kreider. **1941** *Jrnl. Arnold Arboretum* XXII. 154 Several gypsophilous species confined to Texas and New Mexico have also been included. **1966** *Madroño* XVIII. 185 (*heading*) Two gypsophilous species of Muhlenbergia.

gyrator. Add: (Later example.)

1908 *Westm. Gaz.* 15 Jan. 3/3 'It's called a diable... He belongs to the best Diabolo Club, and he's never heard it called anything else.' Then he consulted the directions which had come with the box. The first of them began, 'Place the gyrator on the ground.'

2. *Electr.* A passive circuit element which has two pairs of terminals and introduces a phase shift of 180 degrees for one direction of propagation but no phase shift for the reverse direction.

1948 B. D. H. TELLEGEN in *Philips Res. Rep.* III. 87 Another such four-pole [network element], but violating the reciprocity relation, is described by $v_1 = -si_2$, $v_2 = si_1$... We shall denote such a four-pole by the name of ideal gyrator. **1952** *Bell Syst. Techn. Jrnl.* XXXI. 1 The microwave gyrator has been realized by making use of the Faraday rotation in pieces of ferrite placed in the waveguide. *Ibid.* 3 Network synthesis today is based upon the existence of four basic circuit elements: the capacitor, the resistor, the inductor, and the ideal transformer... The introduction of a fifth circuit element, the gyrator, would lead to considerably improved solutions for many network problems. **1957** *Jrnl. Brit. Interplan. Soc.* XVI. 241 The Hall gyrator..is a ferrite device using the Hall effect, which, used at high frequencies, can act as a radar isolator. **1959** H. W. KATZ *Solid State Magnetic & Dielectric Devices* iii. 115 A gyrator preceding a network terminal pair effectively interchanges voltage and current with a scale factor of *R* ohms. **1971** R. CLAY *Nonlinear Networks & Systems* ii. 55 If a capacitor is connected to a gyrator..the input port appears as an inductor.

gyratory, *a.* Add: **b.** Applied to a system of directing road traffic round a roundabout or through a system of one-way streets to avoid the need for one line of traffic to intersect another.

1909 *Westm. Gaz.* 7 Aug. 4/2 The gyratory principle, by which vehicles are directed into circular lines ingeniously devised to avoid intersection. **1926** *Rep. Comm. Police Metropolis, 1925* 16 in *Parl. Papers* (Cmd. 2660) XV. 239 Gyratory systems for the circulation of traffic, after years of discussion, reached the point of practical demonstration this year. **1928** *Observer* 5 Feb. 13/7 Now that every week dedicates a new bunch of streets to the Gyratory System. **1966** *Guardian* 8 Sept. 5/4 A new gyratory road system to ease traffic congestion..is to be built..at Stretford.

gyre, *sb.* **1.** (Later examples.)

1920 W. B. YEATS *Michael Robartes* 34 All our scientific, democratic, fact-accumulating, heterogeneous civilization belongs to the outward gyre. **1929** —— *Let.* (1954) 764, I believe I shall have a poetical re-birth for as I write about my cones and gyres all sorts of images come before me. **1928** —— *Coll. Poems* (1950) 217 O sages standing in God's holy fire..Come from the holy fire, perne in a gyre. **1930** R. CAMPBELL *Adamastor* 98 A serpent..With lifted crest and radiant gyre Revolving into wheels of fire. **1948** C. DAY LEWIS *Poems 1943–47* 64 Earth-souls doomed in their gyres to unwind Some tragic love-tangle. **1962** *Listener* 20 Dec. 1047/2 It is deeply satisfying both as riddle and as poem. The poet evokes an atmosphere of mystery within the frame of the eternal gyre.

gyre, *v.* **3.** (Later examples.)

1871 'L. CARROLL' *Through Looking-Glass* i. 21 'Twas brillig, and the slithy toves Did gyre and gimble in the wabe. **1920** W. B. YEATS *Demon & Beast* in *Coll. Poems* (1950) 210 To watch a white gull take A bit of bread thrown up into the air; Now gyring down and perning there. **1930** E. POUND *XXX Cantos* xxv. 114 Three lion cubs..which born at once began life and motion and to go gyring about their mother. **1951** S. SPENDER *World within World* v. 283 The bomber was gyring and diving.

gyrectomy (dʒəire·ktŏmi). *Surg.* [f. GYR(US + *-ECTOMY.] Surgical excision or resection of a gyrus of the brain.

1949 J. L. POOL et al. in F. A. Mettler *Selective Partial Ablation of Frontal Cortex* iv. 34 In 3 cases bilateral 'gyrectomy' was carried out; that is, removal of the superior frontal gyri..of the middle frontal gyri..and of the inferior frontal gyri. **1967** HORWITZ & RIZZOLI *Postoperative Complications in Neurosurg. Pract.* iv. 177 Destructive surgery on the cerebral hemispheres in the treatment of intractable pain has been attempted in a variety of ways to which a myriad of names has been applied (lobotomy, leucotomy, gyrectomy, topectomy, etc.).

gyrfalcon. Delete 'obs.'; this is now the usual spelling.

1884 [see GERFALCON]. **1910** *Encycl. Brit.* X. 139/1 Next to the typical falcons comes a group known as the 'great northern' falcons...Of these the most remarkable is the gyrfalcon. **1964** *Oxf. Bk. Birds* 46/2 Gyr falcon.

gyrinid (dʒiri·nid), *sb.* and *a. Ent.* [mod.L., f. *Gyrinidæ*, name of a family of Coleoptera.] **A.** *sb.* A member of the Gyrinidæ, a family of aquatic beetles; a whirligig beetle (see WHIRLIGIG *sb.* 4). **B.** *adj.* Of, pertaining to, or designating a beetle of this kind.

1925 *Pap. Michigan Acad. Sci.* V. 430 The first step that one must take in the study of gyrinid affinities is to divest the beetles of their adaptive characters. *Ibid.* 432 Nothing can be surer than that the gyrinids would never have developed in the absence of a surface habitat. **1943** *Brit. Jrnl. Psychol.* Sept. 26 Gyrinid beetles,.. maintain a constant general position [in flowing water] by means..of visual perception of the pattern of objects on the adjacent banks. **1955** BORROR & DELONG *Introd. Study Insects* xxii. 339 The gyrinids are oval black beetles that are commonly seen swimming in endless gyrations on the surface of ponds and quiet streams.

gyro-. Add: **gy·rocopter** [after *helicopter*], a kind of helicopter; now *spec.* a small, light, single-seater one (cf. *AUTOGIRO); **Gy·rodyne,** the proprietary name for a type of helicopter which has a horizontal rotor for providing lift and one or more propellers for providing forward thrust; **gy·rofrequency,** the frequency with which a charged particle spirals about the lines of force of a magnetic field through which it is passing; **gy·roplane** = *gyrocopter; **gyro·pter** (*disused*) [Gr. πτερόν wing], a rotary-wing aircraft; **gy·rotiller,** a type of cultivator in which the tines rotate about an axis.

1915 *Aeronautics* 9 June 385/1 The gyrocopter is a variation of the helicopter operated by a rotary motor. **1934** W. HOLTBY *Pavements at Anderby* (1937) I. 25 The gyrocopter sailed over the rim of the wold and sank silently into the deep meadow. **1962** *Britannica Bk. Yr. 1961* 545/2 Among new scientific gadgets there was the gyrocopter, a rotorcraft having both drive to rotors and a normal propeller, and usually a very light single-seater craft. **1965** *Kingston* (Ont.) *Whig-Standard* 21 Apr. 6/5 The club members hope to turn their gyrogliders into motorized gyrocopters. **1969** *Courier-Mail* (Brisbane) 1 Apr. 19/2 Gyrocopters are against the law in Australia. Nobody may fly them, even..in the remoteness of his own cattle station. **1946** *Flight* 14 Nov. 540/1 (*caption*) The Fairey FB-1 research Gyrodyne model which will be on view at the Paris Show. *Ibid.* 540/2 This is the first of a new class of rotary-winged aircraft known as Gyrodynes. **1948** *Jrnl. R. Aeronaut. Soc.* LII. 285/1 Delaying the stall by twisting of the blade, or by the use of some alternative configuration, such as the gyrodyne. **1958** LAMBERMONT & PIRIE *Helicopters & Autogyros of World* 172 After the end of World War II..Peter P. Papadakos founded a group of engineers into the Gyrodyne Company of America Inc. for the purpose of perfecting the 'gyrodyne' concept. **1941** 'R. STRANGER' *Dict. Radio & Telev. Terms* 101 Gyro-frequency. **1962** W. B. THOMPSON *Introd. Plasma Physics* i. 4 The character of the radiation depends on electron energy; if the energy is low the radiation occurs primarily at the gyrofrequency $\Omega = eB/mc$, where *e*, *m* = electron charge, mass, *c* = velocity of light and *B* = magnetic field strength. **1907** *Nature* 5 Dec. 107/1 The gyroplane of Messrs. Breguet..revives interest in the attempt to overcome gravity by vertical screw propellers. **1935** *Jrnl. R. Aeronaut. Soc.* XXXIX. 177 Aerodynamic analysis of the gyroplane rotating-wing system. **1969** *Daily Tel.* 10 Mar. 1/8 Mr. Ernest Brooks, 39, the garage owner who designed and built the Mosquito gyroplane, was fatally injured yesterday when it plunged 250 ft. to the ground. *Ibid.*, The tiny £1,000 gyroplane could cruise at 40 mph at 1000 ft and was capable of 80 mph. **1908** *Westm. Gaz.* 14 Aug. 10/2 A new machine, the Gyropter, or rotary-wing flying machine. **1912** *Yorks. Post* 13 Jan. (leader), Mr. Davidson proposes the building of what he calls a gyropter. **1938** *Encycl. Brit. Bk. of Yr.* 27/2 Combine harvesters, milking machines, gyrotillers, [etc.]. **1947** *Times* 8 Mar. 4/1 Every available piece of snow-shifting machinery had been pressed into service, including ploughs, tractors, bulldozers, blowers, gyrotillers, and disc harrows. **1962** J. N. WINBURNE *Dict. Agric.* 358/2 *Gyrotiller*, a machine used in England to pulverize subsurface soil below depths of ordinary plowing. **1970** G. E. EVANS *Where Beards wag All* ix. 106 The gyrotiller—an owd engine that come round and twizzle everything round.

gyro (dʒəi·ro). Colloq. abbreviation of (*a*) GYROSCOPE, (*b*) *gyro-compass* (see below).

1910 tr. *Anschütz Gyro Compass* 16 The ordinary mariner's compass merely points to a certain spot known as the magnetic pole, from the direction of which the true North can be deduced. The Gyro, however, avoids this operation by indicating the true North direct. **1914** H. CRABTREE *Spinning Tops & Gyroscopic Motion* (ed. 2) 77 If the position of the ends *N* and *S* of the gyro [gyro-compass] are reversed. **1922** *Encycl. Brit.* XXX. 733/2 In this model [of the gyro compass] three gyros are used in place of the single one of the earlier model. **1964** MRS. L. B. JOHNSON *White House Diary* 24 Mar. (1970) 99 A gyro in a suitcase—that's the machine that keeps the rocket on its proper course. **1966** *Electronics* 17 Oct. 38/3 In a strapdown system..the gyroscope bearings are connected to the vehicle without gimbals and the gyros roll with the vehicle.

b. *Comb.*, **gyro-bus, -cab,** names given in Science Fiction writing to gyroscopically controlled aircraft; **gyro-car,** (*a*) a mono-rail car or carriage which is balanced by means of gyroscopes driven at high speed in opposite directions; (*b*) a two-wheeled motor-car whose balance and steering are controlled by gyroscopes; **gyro-compass,** a form of gyroscope used as a compass, being continuously driven and specially mounted so that its axis remains parallel to that of the earth; **gyro-horizon** = **artificial horizon*; **gyro-pilot,** a gyro-compass used to steer a vessel or an aircraft without human agency; **gyro-sight,** a gun-sight fitted with a gyroscope; **gyro-stabilizer,** a gyroscopic device for maintaining the equilibrium of a vessel; hence **gyro-stabilized** *a.* (also of a compass).

1936 'J. BEYNON' *Planet Plane* v. 40 Machines of every kind from the dainty flipabout to the massive gyrobus.. started to float in from each quarter. **1956** 'W. TENN' *Time in Advance* (1963) 62 They climbed quickly into one of the many hovering gyrocabs. **1909** *Westm. Gaz.* 8 Nov. 8/3 The gyro-car, as Herr Scherl calls it, is to make a series of runs..in the Exhibition Hall at the Zoological Gardens. **1909** *Daily Chron.* 11 Nov. 1/6 We waited for the gyro-car to emerge from its shed. **1961** *Ford Bulletin* 14 Apr. 2/5 The Gyron is not the first two-wheeled gyro car. **1970** B. WALLIS in *Daily Tel.* (Colour Suppl.) 16 Oct. 46/2, I am convinced that the car of the future will be a gyroscopically-controlled two-wheeler, with one wheel in the front, the other behind...When my gyrocar cornered it would, irrespective of speed, heel over at the correct angle. **1910** tr. *Anschütz Gyro Compass* 10 An accurate means of determining direction by some 'nonmagnetic' appliance is of great importance; this is.. appreciated by many of the large navies, who are installing the Gyro Compass in their submarines. **1911** *Chambers's Jrnl.* 27 May 414/2 The magnetic compass is easily deranged...Its rival is the gyro-compass. **1913** J. H. POYNTING *Earth* 98 That very remarkable invention by the brothers Anschütz, termed the Gyro Compass. **1931** *Times Lit. Suppl.* 22 Jan. 51/3 A speck of dust might cause a gyro compass to 'wander'. **1966** *McGraw-Hill Encycl. Sci. & Technol.* VI. 305/2 Modern gyrocompasses..are now used as the prime navigational instrument on nearly every ship. **1939** *War Illustr.* 29 Dec. 539/2 At ten thousand feet, our gyro-horizon froze. **1942** Gyro horizon [see *GYROSCOPE]. **1923** *Westm. Gaz.* 24 Dec., Through the agency of an invention known as the gyro-pilot apparatus, the Cunard liner Laconia, which reached Liverpool yesterday, steered from New York a straight course without human aid. **1924** *Glasgow Herald* 8 Jan. 7 Gyro-Pilot for Cargo Boats. **1936** *Jrnl. R. Aeronaut. Soc.* XL. 555 Air sickness is practically non-existent on..air lines, which use scientific comfort conditioning and which also use gyropilots. **1941** F. H. JOSEPH *Lett. from Brit.* (1942) 55 Boulton-Paul [gun turret] with Gyro-Sight. Improved reflector sight with a rate gyro. **1946** Gyro-stabilised [see *GYROMAGNETIC *a.* 2]. **1959** F. D. ADAMS *Aeronaut. Dict.* 87/2 *Gyro-stabilized*, stabilized by a gyroscope, as in gyro-stabilized compass, gyro-stabilized drift meter. **1963** *New Yorker* 1 June 125 (Advt.), You'll stand at the rail, watch the infinite sea—so smooth to your gyro-stabilized ship. **1921** *Sci. Amer.* 18 June 482/3 The gyro-stabilizer damps out the rolling of a vessel. **1946** *Nature* 24 Aug. 250/1 The marine applications fall into three groups, the gyro compass, the gyro pilot and the gyro stabilizer. **1957** *Encycl. Brit.* XI. 51/1 By dealing with each wave increment individually and by exerting a small counteracting force against it at just the right moment, the gyrostabilizer quenches the force of each wave and never allows the vessel to build up a roll averaging more than three or four degrees.

gyromagnetic (dʒəi:romægne·tik), *a.* [See GYRO-.] **1.** *Physics.* Of or pertaining to the interdependence of the angular momentum of a spinning charged particle and its resulting magnetic moment; *gyromagnetic effect*: any effect arising from this interdependence, esp. the fact that a change in the angular momentum of a freely suspended magnetizable body will produce a change in its magnetic moment, and *vice versa*; *gyromagnetic ratio*: the ratio of the angular momentum of a spinning charged particle to its magnetic moment; also, the reciprocal of this, expressed either as a true ratio (symbol γ) or as a dimensionless quantity (symbol g').

1922 O. W. RICHARDSON in *Proc. R. Soc.* (1923) A. CII. 538 (*heading*) The magnitude of the gyromagnetic ratio. *Ibid.* 539 We are now practically driven..to the conclusion that the motions of the positively charged parts of the atoms cannot be disregarded in considering these gyromagnetic phenomena. **1922** *Phil. Trans. R. Soc.* A. CCXXIII. 257 (*heading*) On the Richardson gyromagnetic effect. **1930** S. J. BARNETT *Evidence on Elem. Magnet from Res. on Gyromagnetic Phenomena* 12 Rotating a rod about its axis at a speed of *N* revolutions per second magnetizes it exactly as it would be magnetized by applying to it an axial magnetic field with intensity $H = 2\pi sN$, where *s* is the ratio of the rotary momentum of the elementary magnet to its magnetic moment. This ratio is known as the gyromagnetic ratio, or magnetomechanical ratio. **1953** C. KITTEL *Introd. Solid State Physics* x. 165 Gyromagnetic experiments identify the magnetization in ferromagnetics as arising largely from the electron spin, rather than from the orbital moment. *Ibid.*, The magnetomechanical ratio is defined as the ratio of the magnetic moment to the angular momentum. [*Note*] This quantity..is sometimes called the gyromagnetic ratio, but strictly speaking the gyromagnetic ratio is the reciprocal of the magnetomechanical ratio. **1961** *Encycl. Dict. Physics* III. 551/1 Gyromagnetic phenomena are to be expected in paramagnetic as well as ferromagnetic materials. **1967** *Electronics* 6 Mar. 54/2 The NBS now uses the gyromagnetic ratio of a proton to determine the constancy of its standard voltage.

2. Applied to a type of compass in which a magnetic compass automatically corrects the gradual deviations of a directional gyroscope, which in turn provides the compass reading.

1946 WELLS & GLENNY in M. Davidson *Gyroscope* III. ii. 170 A gyro-magnetic compass..should not be confused with a gyro-stabilised magnetic compass. **1952** HITCHINS & MAY *Lodestone to Gyro-Compass* viii. 159 In the gyro-magnetic system a gyroscope is maintained in a nearly fixed direction by intermittent or continuous signals from some sort of magnetic compass. The gyroscope then operates the repeaters. **1957** *Encycl. Brit.* VI. 176/1 In the gyro-magnetic compass..the gyro-direction indicator and magnetic compass are combined in one instrument and the necessity for hand adjustment disappears.

gyroscope. Add: (Further examples.) Also used, in various modifications or in conjunction with other equipment, to provide a horizontal or vertical reference direction (as in the artificial horizon and the gyro-compass), to stabilize ships, mono-rail vehicles, etc., and to measure angular velocity and angular acceleration (as in the turn-and-bank indicator and other navigational devices).

1857 *Chambers's Jrnl.* 30 May 351/2 To overcome it [*sc.* rolling and pitching]..the professor invented a balanced frame, with free axes for rotation, somewhat on the principle of the gyroscope, which..remains uninfluenced by the..movements of the ship. **1871** *Cornh. Mag.* XXIV. 450 It is probable that the principle of the instrument called the gyroscope will be called into action to secure steadiness of motion [for aircraft]. **1891** *Cassell's Fam. Mag.* 187/1 A small gyroscope has also been fitted to the ship's sextants, so as to give an invariable line of sight. **1904** *Daily Chron.* 29 Feb. 4/5 The newest torpedoes, we are told, will have a speed of thirty knots for 3,000 yards, and, thanks to the gyroscope, almost perfect accuracy can be relied upon. **1907** *Standard* 23 Mar., The experiments showed conclusively that gyroscopes could be designed which would exercise a sensible steadying effect upon even the largest passenger steamers on service. **1942** B. A. SHIELDS *Princ. Flight* viii. 297 The gyro horizon, or the artificial horizon as it is often called, also depends on a gyroscope for its operation. **1962** F. I. ORDWAY et al. *Basic Astronautics* ix. 372 The gyroscope in conjunction with the accelerometer forms the heart of modern space vehicle guidance systems. **1966** BURGER & CORBET *Ship Stabilizers* ii. 24 The three types of 'three-frame' gyroscopes discussed are often called 'displacement' gyroscopes because they can measure angular displacement between the framework in which they are mounted and a fixed reference mark—the rotor axle.

b. *Comb.* **gyroscope-compass** = **gyro-compass*; **gyroscope top,** a spinning top on the gyroscope principle, which when spinning may be supported by one end of its horizontal axis.

1909 *Westm. Gaz.* 20 Apr. 9/3 The gyroscope-compass, an invention of Dr. Anschuetz-Kaempfe, of Kiel,..is based upon the familiar principle that a rapidly rotating body tends to keep in the same plane. **1880** *Encycl. Brit.* XI. 352/2 Perhaps the most common form of gyroscope is that which has been largely sold under the name of the gyroscope top. **1921–2** *T. Eaton & Co. Catal.* Fall & Winter 328 The Gyroscope Top..spins at either end or in the middle.

gyroscopic, *a.* Add: (Earlier and later examples.) Also *gyroscopic compass* = **gyrocompass*; *gyroscopic couple* (see quot. 1959[2]).

1871 *Cornh. Mag.* XXIV. 451 Since travelling-machines *must* travel swiftly, the gyroscopic portion of the machine may be made to support itself. **1907** *Standard* 23 Mar., Dr. Schlick's gyroscopic apparatus for steadying ships. **1920** T. W. CHALMERS (*title*) The gyroscopic compass. **1928** A. L. RAWLINGS (*title*) Theory of the gyroscopic compass and its deviations. **1937** *Jrnl. R. Aeronaut. Soc.* XLI. 2 The other dynamic force present in the spin, is the gyroscopic couple, due to the gyroscopic force of the propeller. **1948** A. C. HARDY *Mod. Marine Engin.* III. xvi. 149 The gyroscopic compass operates by the rotation of the earth and the pull of gravity. **1959** N. MAILER *Advts. for Myself* (1961) 183 The inner-directed man—the nineteenth-century businessman would be a prototype—is essentially self-directed, 'gyroscopic', production-minded, set for life from early childhood in pursuit of certain built-in goals. **1959** J. L. NAYLER *Dict. Aeronaut. Engin.* 126 *Gyroscopic couple*, the turning moment which resists any alteration of the inclination of the axis of rotation of a gyroscope. Gyroscopic couples are produced, for instance, by propellers when the plane of rotation is changed, e.g. as in a pull-out, or a spin. **1962** F. I. ORDWAY et al. *Basic Astronautics* v. 204 The probe was fitted with an orientation system that included optical and gyroscopic pickups.

gyroscopically (dʒəirǒskǫ·pikăli), *adv.* [f. GYROSCOPIC *a.*: see -ICALLY.] By or with a gyroscope; in a gyroscopic manner.

1910 R. M. NEILSON *Aeroplane Patents* II. 64 (*caption*) Herring's gyroscopically-controlled propellers. **1911** *Chambers's Jrnl.* 55/1 The success of the gyroscope in preventing the rolling of ships..would seem to augur well for the gyroscopically controlled aeroplane. **1970** [see *gyro-car* s.v. *GYRO b]. **1971** *Sunday Express* 24 Oct. 11 (Advt.), A low mass pick-up arm gyroscopically pivots in a concentric gimbal mount.

gyrostat. Add further examples and substitute for def.: A name given to various forms of gyroscope.

1902 *Encycl. Brit.* XXXII. 578/1 For the purpose of controlling the horizontal direction this boat [*sc.* a submarine] is fitted, in addition to a vertical rudder, with a 'gyrostat', the motion of which instrument is maintained by an electric motor. Previous to making a start the boat must be headed in the direction in which it is intended to move; when the gyrostat, once started, will..continue to rotate about an axis invariable in direction, and any departure from this will be detected, whether the boat be above the surface or below it. **1922** *Ibid.* XXX. 44/1 Gyrostats are also used in aircraft as azimuth indicators for experimental or test purposes; they may some day be used as part of a gyrostatic compass.

gyrus. (Earlier example.)

1842 DUNGLISON *Med. Lex.* (ed. 3) 339/2 *Gyrus*, anfractuosity, convolution.

gyttja (yi·tʃă). *Geol.* Also **gytje.** [Sw., = mud, ooze (adopted in its technical sense by H. A. von Post 1862, in *K. Sven. Vetenskaps-Akad. Handl.* I).] A sediment which is typically black, rich in organic matter, and deposited in productive lakes.

1887 *Encycl. Brit.* XXII. 740/1 In the shallow lakes and enclosed bays of the sea there began to be formed and still is in course of formation a deposit known by the name *gyttja*, characterized by the diatomaceous shells it contains. **1899** *Jrnl. Chem. Soc.* LXXVI. II. 39 At Sandefjord the *gytje* is greasy to the touch, and has the odour of hydrogen sulphide; it is used for mud baths. **1958** F. E. ZEUNER *Dating Past* (ed. 4) iii. 90 There is about one metre of blue lake marl below the polliniferous gyttja. **1970** *Nature* 19 Sept. 1235/1 At a few places, however, there are thin gyttja-like layers a few centimetres thick, which could have formed in pools over permafrost.

gyver (gəi·vəɹ). *slang.* Chiefly *Austral.* and *N.Z.* Also **givo, givor, guiver, guyver.** [Of unknown origin.] Affectation of speech or behaviour, esp. in phr. *to put on the gyver*. Also as *adj.*: smart, fashionable.

a **1866** VANCE *Chickaleary Cove* (Farmer & Henley), The stock around my squeeze of a guiver colour see. **1889** BARRÈRE & LELAND *Slang* I. 411/2 *Guiver* (theatrical), flattery, artfulness. **1897** D. MCK. WRIGHT *Station Ballads* (1945) 45 He'd put on 'gyver' enough to last a full-bred actor a week. **1899** W. T. GOODGE *Hits, Skits & Jingles* (1904) 18 And I ain't got the style and the guiver Of them bank clerks and students and sich. **1914** *Bulletin* (Sydney) 14 May, 'E 'ad a bit of guiver wif the secretary. **1916** J. B. COOPER *Coo-oo-ee* x. 142 I'm not takin' any of that sort of 'gyver', you know. **1918** *Chrons. N.Z.E.F.* 5 July 253/1 The 'givor' that kid puts on, you would think he was a captain at least. **1928** J. DEVANNY *Dawn Beloved* xviii. 179 Mrs. Rodda thinks herself too good for anybody else...To hear her and Mrs. Devoy talk! The guyver they put on! **1938** *Observer* 25 Sept. 9/2 Guyver: Make-believe, still used in Anglo-Jewish slang. It is Hebrew for pride, but has now come to mean pretence and is synonymous with..swank. **1948** V. PALMER *Golconda* xxxii. 272 Well, what's all this guyver about the going being tough? **1948** PARTRIDGE *Dict. Forces' Slang 1939–1945* 83 *Givo* or *gyvo*, an adjective which describes any sartorial eccentricity. A 'tiddley-suit' made by a shore-going tailor which has wider bell-bottoms than are permitted by the Service regulations. **1970** D. M. DAVIN *Not Here, Not Now* III. vii. 207, I wouldn't want you to get stuck-up and start putting on the gyver and forgetting your own.

H

H. Add: **I. 2.** *H girder, iron.* *H hinge*, a type of hinge which when open has the form of an H.

1726 in *Maryland Hist. Mag.* (1912) VII. 278 H hinges at 8s per pair. **1836** L. HEBERT *Engin. & Mech. Encycl.* I. 674 Another sort, called H..hinges, from their resemblance to those letters, are extensively employed for common purposes. **1888** *Lockwood's Dict. Mech. Engin.*, *H iron*, rolled wrought-iron bar whose section is that of the letter I. Used extensively for building up engineering structures. **1902** A. C. HARMSWORTH et al. *Motors* vi. 90 The roof of the house is strengthened at certain points by cross timbers which support two small H girders, and carry iron frames to which are attached pulley blocks. **1960** H. HAYWARD *Antique Coll.* 142/1 'H' *hinge*, like the cock's head hinge, an early external type of hinge in the form of the letter 'H' extensively used on cupboards of the 16th and 17th cent.

II. 3. b. Designation of a strong Fraunhofer line at 3969 Å, caused by calcium ions; † orig., (the position occupied by) the H and K lines as a pair. [Named by J. Fraunhofer 1817, in *Ann. d. Physik* LVI. 286.]

1823 tr. FRAUNHOFER in *Edin. Philos. Jrnl.* IX. 297 The two bands at H are of a very singular nature. **1865** *Phil. Trans.* CLIV. 149 A pair of strong lines..near the extreme refrangible end of the spectrum..may coincide with those of Fraunhofer's H. **1879, 1967** [see *K 3 c].

8. *Math., Physics.* *H* denotes the Hamiltonian function of classical mechanics or the Hamiltonian operator of quantum mechanics.

1835 W. R. HAMILTON in *Phil. Trans. R. Soc.* LXXV. 98 If then we introduce, for abridgement, the following expression H. **1935** PAULING & WILSON *Quantum Mech.* i. 16 Involving a function *H*..called the Hamiltonian function. **1965** W. HAUSER *Introd. Princ. Mech.* vi. 194 Whenever the Lagrangian is not explicitly a function of time, the function *H*..referred to as the *Hamiltonian* of the system, is a constant of the motion. *Ibid.* 195 This can be verified by considering the variation of *H*.

9. *Physics.* *h* denotes Planck's constant, the elementary quantum of action (M. Planck 1900, in *Verh. d. Deutsch. Physik. Ges.* II. 245). In more recent usage, the quantum of angular momentum $h/2\pi$ has been represented by $\hbar$.

1901 *Sci. Abstr.* IVA. 230, $\epsilon = h\nu$, where *h* is a constant. **1934** *Physical Rev.* XLVI. 925/2, $\hbar$. **1935** PAULING & WILSON *Quantum Mech.* ii. 25 The constant of proportionality, *h*, is a new constant of nature,..called *Planck's constant*... $h/2\pi$ [is] a natural unit or quantum of angular momentum. **1955** L. I. SCHIFF *Quantum Mech.* (ed. 2) i. 7 The product of the uncertainties of the..position and momentum components is at least of the order of magnitude of $\hbar$.

III. H., henry (*Electr.*), heroin; H, designating *horror* films; h., hot, as h. and c., hot and cold (water); H, hydrogen (bomb); so *H-bomb*, *-test*, etc.; H (on lead pencils) (examples); H. and D. (see quots. 1918, 1930); HB (on lead pencils) (examples); H.E., His Eminence, His (or Her) Excellency, high explosive; HF, H[F], H[F], Hf, healthy female(s); H.F., h.f., high frequency; *H-Hour*, the hour at which an operation is to begin; cf. *D-DAY; H.K., Hong Kong (in currency notation); HM, H[M], H[M], Hm, healthy male(s); H.M.C., Headmasters' Conference; H.M.G., His (or Her) Majesty's Government; H.M.I.(S.), His (or Her) Majesty's Inspector of Schools; H.M.S.O., His (or Her) Majesty's Stationery Office; H.N.C., Higher National Certificate; H.O., Hostilities Only (see quots.); H.P., high pressure, hire purchase, hybrid perpetual; *H.P. Sauce*, the proprietary name of a type of spiced brown sauce; H.Q., Headquarters; H.T., h.t., high tension; HUAC, House [of Representatives] Un-American Activities Committee; Hz, hertz (unit of frequency).

1910 N. HAWKINS *Electr. Dict.*, H. The symbol of induction. The Henry. h. An abbreviation for the henry, the practical unit of induction. **1926** *Clues* Nov. 161/2 *H*, heroin. **1929** *Sat. Even. Post* 13 Apr. 54/3 Heroin is referred to as H. **1933** C. DE LENOIR *Hundredth Man* iv. 61, I opened the packet of H. and took a generous sniff. **1962** K. ORVIS *Damned & Destroyed* v. 42 Suppose I.. ask you where to connect for H? **1938** *Ann. Reg. 1937* 339 In June, Lord Tyrell, President of the British Board of Film Censors, introduced a new film classification called 'H' to apply to horror films. Pictures so labelled will not be shown to children under 16, whether accompanied by an adult or not. **1958** *Times* 4 Aug. 10/2 'H' used to, and now 'X' does, among other things, stand for Horror, and the two new films this week..are rich in those qualities [etc.]. **1960** *Spectator* 1 July 20 *The Visit* is old-style *grand guignol* with a few modern H-certificate props. **1901** *Punch* 24 July 64/1 Bathroom (h. and c.). **1930** *Morning Post* 17 June 20/7, 4 bed-rooms (h. and c. in each). **1937** 'G. ORWELL' *Road to Wigan Pier* xiii. 250 The literary gent in his Tudor cottage with bathroom h. and c. **1950** *Hansard, Commons* 6 Nov. 114, I am not one who criticised the right hon. Gentleman..for drawing attention to the H-bomb. **1952** *Manch. Guardian Weekly* 20 Nov. 2/2 The first test model of the H-bomb will be followed shortly by even more violent versions. **1955** *Bull. Atomic Sci.* June 226/2 He asserts that H-war, terrible as it will be, is still better than Communist domination. **1957** *Observer* 8 Sept. 9/3 With bovine stolidity Western man carries on with his H-bomb tests. **1958** 'P. BRYANT' *Two Hours to Doom* 96 Their phoney ending of H-tests. **1959** *Listener* 16 July 88/2 United States H-bombers. **1968** *Times* 29 Oct. 7/3 The first Chinese H-bomb, tested in June last year, probably contained zinc in its construction. **1852** C. H. WEIGALL *Art of Figure Drawing* (Advts.) 14 H. Moderately hard (used for light sketching). HH. A degree harder (for outlines and fine Drawing). HHH. Very hard (for Architectural Drawing). **1895** *Army & Navy Co-op. Soc. Price List* 563/2 The 'Kohinoor', a high-class drawing Pencil..H, HH, HHH. **1926** R. MACAULAY *Crewe Train* II. xii. 211 She purchased a packet of notepaper and envelopes and a pencil (H.H.) at the post office. **1934** M. V. HUGHES *London Child of Seventies* xiv. 167 'I say, Molly, lend us your Scripture pencil,' Dym would say, for he knew that was an H, and good for his geometry figures. **1948** H. MISSINGHAM *Student's Guide Commerc. Art* II. 57 The degree markings on pencils are:..6H, 4H, 3H—Extra Hard. **1903** A. WATKINS *Photogr.* (ed. 2) 29 A box of one maker's plates marked H & D 100. **1918** *Photo-Miniature* XV. Mar. (Gloss.), H. & D. (Hurter and Driffield)—used as a prefix to numbers signifying the speed of plates measured by the system devised by these investigators. Fastest plates, H. & D. 400 to 500. **1930** *Sel. Gloss. Motion Pict. Techn.* (*Acad. Motion Pict., Hollywood*), *H and D curve*, the characteristic curve of a photographic emulsion. **1852** C. H. WEIGALL *Art of Figure Drawing* (Advts.) 14 H.B. Hard and Black (deeper shade than F). **1962** L. DEIGHTON *Ipcress File* xxv. 165, I took his wooden HB pencil. **1732** *Let.* 19 Feb. in *Calendar State Papers, Amer. & West Indies* (1939) 64 That H.E. and less than five of the Council could not hold Courts of Chancery. **1848** THACKERAY *Van. Fair* li. 453 H.E. Papoosh Pasha, the Turkish Ambassador (attended by Kibob Bey, dragoman of the mission). *Ibid.* lxiii. 571 H.E. Madame de Burst received once a week. **1946** KOESTLER *Thieves in Night* 198 He thought that H.E. went indeed a bit far in demonstrating his dislike of the Hebrew community. **1970** *Catholic Directory* 43 (*heading*) Ireland. The hierarchy at the present time... Armagh. H.E. Cardinal William Conway. **1901** *Daily Chron.* 27 June 3/2 To introduce a very much larger proportion of H.E. shell. **1915** D. O. BARNETT *Let.* 23 May 153 H.E. is the shell for attacking, because you blow the defenders out of their trenches. **1955** *Bull. Atomic Sci.* Feb. 55/3 The thoroughness of an atomic bombardment's 'area coverage' exceeds that of HE carpet bombing beyond comparison. **1880** H[F] [see *HM* below]. **1913** *Year-Bk. Wireless Telegr.* 289 To make the current-distribution over the cross-section more uniform for H.F. currents. **1923** *Popular Wireless* (Suppl.) 13 Oct. 10 'Plug-in' H.F. Transformers. **1924** *Exper. Wireless* Apr. 397/2 The atmospheric band frequencies which penetrate the H.F. filter. **1942** *Electronic Engin.* XV. 168 A hair-pin filament heated by h.f. current. **1967** *New Scientist* 25 May 455/3 All the frequencies in the HF band that are needed by the many transmitters at the station are produced by processing the 'master' frequency. **1918** in *Amer. Speech* (1944) XIX. 302 'Over the top' is now 'the jump off' and 'zero hour' has changed to 'H hour'. **1927** J. M. SAUNDERS *Wings* (1928) v. 210 The word went out that 'D' day was to be Sept. 12, and that 5 A.M. was to be 'H' hour. **1945** *Hutchinson's Pict. Hist. War 12 Apr.–26 Sept. 1944* 342 No fewer than 13 companies landed on the first tide, at H hour plus 25 minutes to be precise. **1952** *Hong Kong Trade Returns* Jan. 1 The unit of value is the Hong Kong dollar, the official rate of exchange being H.K. $16 = £1 sterling. **1964** *Asia Mag.* 12 July 24/1 According to official figures, on average nearly HK$1 million went into land and building every day in 1963. **1880** *Encycl. Brit.* XIII. 170/2 Table H[MF], comprising all the healthy lives, male and female, included in the observations... Table H[M], comprising the healthy male lives only... Table H[F], comprising the healthy female lives. **1898** *Westm. Gaz.* 14 Nov. 8/1 With its premiums and its reversionary bonuses, and its Hm tables and its surrender values. **1905** in *Rep. Headmasters' Conf. 1905* (1906) 97 The 1903 recommendations of the H.M.C. favoured solution (*b*) as more in harmony with the existing practice and principles of the Universities. **1966** *Rep. Comm. Inquiry Univ. Oxf.* II. 48 Among men from independent schools, the great majority are from H.M.C. schools. **1938** E. WAUGH *Scoop* II. xii. 137 We don't quite know what he's up to; whatever it is, it doesn't suit H.M.G.'s book. **1971** *Guardian* 28 July 11/1 Any deal with Ian Smith..would make it impossible for Labour men to support HMG on Europe. **1908** E. M. SNEYD-KYNNERSLEY (*title*) H.M.I.: Some passages in the life of one of H.M. Inspectors of Schools. **1963** S. MARSHALL *Exper. in Education* ii. 40 The modern H.M.I. sniffs the atmosphere of a school the moment he opens the door. **1905** F. H. COLLINS *Author & Printer* 164/1 *H.M.S.O.*, (His or Her Majesty's) Stationery Office. **1969** *Listener* 20 Feb. 255/3 *Britain: An Official Handbook*, published by HMSO two or three weeks ago at 32s. 6d. **1949** *Educ. in 1948* (Cmd. 7724) 44 (*caption*) H.N.C. **1967** *Times Rev. Industry* Aug. 70/3 The rising failure rate of the HNC candidates. **1942** PARTRIDGE *Dict. Abbrev.* 48/2 *H.O.*, Hostilities only; applied to a man that has joined for the duration of the war. (Naval.) **1961** B. FERGUSSON *Watery Maze* i. 55 H.O. ('Hostilities Only') ratings—men who had joined the Royal Navy for the duration of the war. **1902** *Encycl. Brit.* XXXII. 151 The H.-P. valves are worked by means of a simple lever from the L.-P. valve-rods. **1930** *Engineering* 9 May 599/1 Since both the H.P. and L.P. rotors have their own thrust blocks, the two are connected by a Wellmann-Bibby coupling, which permits of free axial expansion. **1945** *Daily Mirror* 6 Sept. 3/2 'Stop H.P. babies'... Mothers..have to pay for their babies on the hire-purchase system because of the high charges of maternity homes. **1958** *Spectator* 13 June 759/2 They cannot keep up the HP payments. **1959** *New Statesman* 17 Oct. 494/2 The artisan class lives in new houses and pays off the telly and the car on HP. **1869** S. R. HOLE *Bk. about Roses* xi. 174 Marguerite Dombrain, H.P...an early, reliable, vigorous, bright carmine Rose. **1893** W. ROBINSON *Eng. Flower Garden* (ed. 3) 644/2 Gabriel Luizet, General Jacqueminot and many other H.Ps do not usually bloom after the month of August. **1912** *Trade Marks Jrnl.* 22 May 768 HP..Sauce and pickles. Edwin Samson Moore, trading as 'The Midland Vinegar Company', 'The Trade Malt Vinegar Company', and as F. G. Garton & Co.,..Aston Cross,..Warwickshire, vinegar brewer and sauce and pickle manufacturer. **1926–7** *Army & Navy Stores Catal.* 37/1 Sauces..H.P.— -/9. **1940** J. BETJEMAN *Coll. Poems* (1958) 83, I pledge her in non-alcoholic wine And give the H.P. Sauce another shake. **1971** P. WORSTHORNE *Socialist Myth* viii. 171 He [*sc.* Mr. Harold Wilson] manages to retain a little of the working-class-lad-made-good appeal... The H.P.-Sauce style of leadership is a style, so long as it is abnormal. **1909** WEBSTER, H.Q. **1915** D. O. BARNETT *Let.* 18 Jan. 39 After reporting at Brigade H.Q. we went on to our regiments. **1958** M. SHAARA in 'E. Crispin' *Best SF Three* 23 To heck with the rest. We'll let HQ worry about that. **1931** *Daily Express* 18 Mar. 13/2 Fuller 100 volt H.T. and G.B. Battery. **1966** *Economist* 27 Aug. 817/1 HUAC, even in its very name, implies an inquisitiveness about individual political beliefs that is unconstitutional. **1968** *Listener* 31 Oct. 566/3 HUAC is trying to investigate how far the Chicago violence was the result of a communist conspiracy. **1958** *Quantities & Units Periodic Phenom.* (ISO Recommendation R 31) 5, 1 Hz is the frequency of a periodic phenomenon of which the periodic time is 1 s. **1966** *Wireless World* Sept. 50 (Advt.), In order to extend the flat response below 40 Hz the rear of the bass driver is loaded with an acoustic transmission line. **1969** *Sears Catal.* Spring/Summer 8 Frequency response of 35 to 18,000 Hz. Crossover is at 2500 Hz.

haanepoot, haanepot, varr. *HANEPOOT.

Haarlem (hā·ɹləm). The name of a town in Holland, used *attrib.* in **Haarlem blue,** a variety of blue containing alumina; **Haarlem oil,** 'a proprietary diuretic and stimulant oil' (Dorland 1900).

1885 *Spons' Mech. Own Bk.* 407 Slight differences in the manufacture [of Prussian Blue] cause considerable variation in tint and colour, which leads to the material being known by different names—such as 'Antwerp', 'Berlin', 'Haerlem', 'Chinese' Blue. **1885** *Encycl. Brit.* XVIII. 720/1 Barbados tar, Haarlem oil, Seneca oil, and American oil, all consisting wholly or in large part of crude petroleum, were sold by apothecaries for years before petroleum was obtained by boring. **1906** *Practitioner* Dec. 852 Methyloids.—These are an improved combination of methylene blue, Santal oil, copaiba, Haarlem oil, and cinnamon oil, and are supplied in capsoid form.

habanera (hæbănē·ră). Also **habanero.** [Sp., short for *danza habanera* Havanan dance, f. *Havana*, capital of Cuba.] A slow Cuban dance and song in ⅔ time. Also *attrib.*

1878 tr. *Bizet's Carmen* 1 'Love the Vagrant', the celebrated Habanera, with English words. **1887** F. C. GOOCH *Face to Face with Mexicans* viii. 289 The *danza* is the most distinctively national of all the dances, and bears a strong resemblance to the *Habanero*, as known in Cuba. **1926** WHITEMAN & MCBRIDE *Jazz* xi. 231 A fox trot was played in a rhythm exactly that of the Habanera or Tango, but much swifter in time. **1958** P. GAMMOND *Decca Bk. Jazz* xv. 176 He uses..often the minimum outline of a tango or habanera bass.

habara (hæ·bără). Also **habarah, habareh, habra, khabarah.** [Arab. *ḥabara*.] A woman's outdoor silk garment. Also *attrib.*

a **1817** J. L. BURCKHARDT *Trav. Arabia* (1829) I. 339 The women of Mekka and Djidda dress in Indian silk gowns, and very large blue striped trowsers..; over these they wear the wide gown called *habra*, of black silk stuff, used in Egypt and Syria. **1839** E. W. LANE tr. *Thousand & One Nights* (1859) I. 190 It [*sc.* the izár] is now generally made of white calico, but a similar covering of black silk for the married, and of white silk for the unmarried, is now worn by females of the higher and middle classes, and is called a 'ḥabarah'. **1851** *Illustr. Catal. Gt. Exhib.* v. II. 1410 Habara veil, in black silk, for females. **1923** *Sunday at Home* Oct. 51/2 A black Egyptian habareh and veil.

habdabs (hæ·b,dæ:bz). *slang.* Also **abdabs.** [Orig. obscure.] Nervous anxiety, the heebie-jeebies, esp. in phr. *to give* (a person) *the screaming habdabs*.

1946 *Penguin New Writing* XXVIII. 177 Come on, kid. This joint gives me the hab-dabs. **1962** *Spectator*

HABDALAH

8 June 761/3 *Treasure Island* gives pleasure and excitement to some and the screaming habdabs to others. 1963 *Ibid.* 19 July 72 A desperate tension which the slightest crisis will transform into the screaming abdabs once more. 1966 L. Davidson *Long Way to Shiloh* ii. 28 Uri's whimsy-shrouded secrecy, strenuously maintained throughout the journey, had already brought on a severe attack of the habdabs.

Habdalah (hævdā·lă). Also **Habdala, Havdal(l)ah, Hovdoloh.** [a. Heb. *habhdālāh* separation, division.] A Jewish religious ceremony celebrating the end of the Sabbath; a prayer said at this ceremony.
1733 tr. *B. Picart's Ceremonies & Relig. Customs* I. 62 The Festival concludes with the Ceremony which they call Habdala, as it is observed on the Sabbath. *Ibid.* 65 The Repetition of the Habdala. 1891 M. Friedländer *Jewish Relig.* ii. ii. 254 On Sabbath evening, after the close of the Sabbath, we recite the *Habdalah*, in which God is praised for the distinction made between Sabbath and the six week-days. 1892 I. Zangwill *Childr. Ghetto* (1893) i. xvi. 151 On Saturday night, immediately after *Havdalah*, Sugarman went to Mr. Belcovitch. *Ibid.* 409 *Havdalah*, ceremony separating conclusion of Sabbath or Festival from the subsequent days of toil. 1941 G. G. Scholem *Major Trends in Jewish Mysticism* ii. 67 An extremely interesting..magical text, the '*Havdalah* of Rabbi Akiba'. 1957 L. Stern *Midas Touch* iii. xx. 150 His other treasures..a silver spice box and two *Hovdoloh* cups. 1960 *Commentary* June 500/2 To observe the ceremonies of..the Sabbath meal, the blessing of children, and Havdallah. 1962 B. Abrahams tr. *Life Glückel of Hameln* iv. 102 At the close of Sabbath, while my husband was reciting the *Habdalah*.

haberdasher. Add: **b.** Formerly also a drink-seller (as a dealer in 'tape' = spirituous liquor).
1821 P. Egan *Life in London* ii. viii. 354 The Haberdasher is busily employed in measuring out tape for his customers. 1828 'W. T. Moncrieff' *Tom & Jerry* iii. v. 76 The haberdasher is the whistler, otherwise the spirit-merchant, Jerry—and tape the commodity he deals in. 1893 Farmer & Henley *Slang* III. 243/1 *Haberdasher*, (humorously) a publican.

haberdashery. Add: **1.** Also *fig.*
1773 G. Steevens *Let.* 8 Dec. in *Garrick's Corr.* (1831) I. 588 He might have made many discoveries of consequence to us who deal in the haberdashery of words. 1923 Kipling *Independence* 31 He may be festooned with the whole haberdashery of success.

habit, *sb.* Add: **9. e.** *spec.* in *Psychol.* An automatic, 'mechanical' reaction to a specific situation which usually has been acquired by learning and/or repetition.
1859 A. Bain *Emotions & Will* ix. 519 Some natures are distinguished by plasticity or the power of acquisition, and therefore realize more closely the saying that man is a bundle of habits. 1871 E. B. Tylor *Primitive Culture* I. i. 1 Custom, and any other capabilities and habits acquired by man as a member of society. 1890 W. James *Princ. Psychol.* I. iv. 104 The moment one tries to define what habit is, one is led to the fundamental properties of matter. The laws of Nature are nothing but the immutable habits which the different sorts of elementary matter follow in their actions and reactions upon each other. 1956 E. R. Hilgard *Theories of Learning* (ed. 2) i. 10 The stimulus-response theorist and the cognitive theorist come up with different answers to the question, What is learned? The answer of the former is 'habits'; the answer of the latter is 'cognitive structures'.
f. The practice of taking addictive drugs (see also quot. 1914). *colloq.* (orig. *U.S.*).
1887 in *Amer. Speech* (1948) XXIII. 246/2 May he continue to wage war against them [*sc.* Chinese opium dens] until the habit has been swept entirely out of existence. 1894 [see Opiumate]. 1914 Jackson & Hellyer *Vocab. Criminal Slang* 41 *Habit*, current amongst dope fiends. Necessity for opiates; a craving; the condition produced by habitual indulgence in drugs... Example: 'I must drop into the hotel donegan (lavatory) and fire (take a hypodermic injection), for I feel my habit coming on.' 1926 J. Black *You can't Win* xii. 161 The sufferings they would undergo when there was no more and the 'habit' came on. 1959 *Daily Mail* 17 Oct. 7/3 'Do you have the habit?' He knew she meant 'Do you take drugs?'
12. (sense 9, esp. *9 e) *habit-bound* (examples), *-breaker, -formation*; *habit-forming* vbl. sb. and adj., *-worn* adj.; **habit-memory,** one of the two kinds of memory first distinguished by H. Bergson, which consists of motor mechanisms or 'habits' fixed in the organism and which acts in response to an appropriate stimulus, e.g. when repeating a lesson learnt by heart; **habit-neurosis,** a neurosis caused by habit-bound behaviour; **habit pattern,** a pattern of behaviour created by habit; **habit-response,** a response induced by habit; **habit spasm** *Med.*, = Tic 1; **habit strength** (see quot. 1958); also called sHr; **habit-training,** the training of an infant or child in regular habits of behaviour, often specifically referring to hygiene, sleeping, and eating.
1863 J. G. Whittier *Poetical Wks.* (1874) 412/2 But what if, habit-bound, thy feet Shall lack the will to turn? 1922 W. B. Yeats *Trembling of Veil* 140 Old and habit-bound. 1932 *Brit. Jrnl. Psychol.* July 54 Box 7 [in set of boxes designed for intelligence testing] was introduced as a habit breaker. 1913 *Lancet* 27 Sept. 964/2 (*title*) A preliminary note on habit-formation in guinea-pigs. 1936 *Mind* XLV. 290 All habit-formation under the example, instruction, command, influence, of others is propaganda. 1961 *Lancet* 27 Aug. 485/1 Narcotics had not been used for fear of habit-formation. 1899 W. James *Talks to Teachers on Psychol.* p. viii, Maxims relative to habit-forming. 1913 A. E. Leach *Food Inspection & Analysis* (ed. 3) xxi. 955 (*heading*) Habit-forming drugs in beverages. 1958 J. Cannan *And be a Villain* vii. 151, I didn't take any [sleeping pills] last night because if you keep on they might become habit-forming. 1911 Paul & Palmer tr. *Bergson's Matter & Memory* ii. 99 (*marginal note of translators*) Automatism has a wide range, and representative memory is often superseded or masked by habit memory. 1912 *Mind* XXI. 226 M. Bergson, in contrasting these two forms of memory, makes the motor or habit-memory *too* mechanical. 1925 C. Fox *Educ. Psychol.* 140 Take two processes, habit-memory and image-memory respectively. 1907 W. James *Mem. & Stud.* (1911) x. 239 There seems no doubt that we are each and all of us to some extent victims of habit-neurosis. 1960 G. Sanders *Mem. Professional Cad* ii. iv. 133 Once a man has acquired this habit pattern it will be intolerable for him to stay home at night. 1964 *Word Study* Feb. 2/2 *Finalize* and *dollar-wise* were deep-seated habit patterns long before Webster's Third displayed them. 1960 B. Malinowski *Sex & Repress. Savage Soc.* 194 The zoologist deals with specific instinctive behaviour, the anthropologist with a culturally fashioned habit-response. 1888 W. R. Gowers *Man. Dis. Nervous Syst.* II. v. 586 Children often..present spasmodic movements such as winking, twitching the mouth, jerking the head..which the individuals are unable to control... This condition has been termed 'habit-chorea'..but..'habit-spasm' is, I think, a better name. 1940 S. A. K. Wilson *Neurology* II. 1629 In France the word 'tic' has been employed for centuries to denote a habitual, unpleasing gesture; but only within the last 30 years has it been current in English neurology, replacing the incorrect 'habit-spasm' or 'habit-chorea' of prior date. 1948 E. R. Hilgard *Theories of Learning* iv. 83 Habit strength increases when receptor and effector activities occur in close temporal contiguity. 1951 C. L. Hull *Essentials of Behavior* xiv. 57 We have presented evidence..to indicate the quantitative molar law according to which habit strength (sHr), primary motivation or drive (D), incentive motivation (K)..and the delay in reinforcement (J) respectively operate as functions in the determination of reaction potential. 1958 H. B. & A. C. English *Dict. Psychol. Terms* 235/2 *Habit strength* or *sHr*, (*C. Hull*) an inferred part of the organism that is determined by variation in four empirical determinants: number of reinforcements, amount of reinforcing agents, time between stimulation and response, time between response and reinforcement. 1959 Lambert & Fillenbaum in Saporta & Bastian *Psycholinguistics* (1961) vii. 455/1 An important paper by Pitres ..offered a different generalization in order to account for..the effect of aphasia on polyglots. In essence, Pitres' is a habit strength principle which states that the language or languages most used before the aphasic insult will be the first to recover. 1927 A. Gesell in C. W. Kimmins *Mental & Physical Welfare of Child* iii. 40 Wholesome habit training in infancy lays the foundation of mental health...Feeding, sleeping, bladder control, bowel control—these are not physical matters. They are 'mental'. 1939 E. R. Boyce *Infant School Activities* 240 Habit-training. The nursery school and class has been closely associated with training in good physical habits and with attention to the health of the child. 1960 I. Bennett *Del. & Neur. Childr.* v. 181 Difficulties in habit-training may be expected in both delinquent and neurotic children. Both types..are likely to express early difficulties in the form of sleep disturbances, feeding disturbances, or disturbances in bladder or bowel control. 1890 W. James *Princ. Psychol.* I. xvi. 655 These habit-worn paths of association are a clear rendering of what authors mean by 'predispositions', 'vestiges', 'traces', etc., left in the brain by past experience.

habitant. **B. 2.** (Earlier and later examples.)
1789 *Quebec Gaz.* 5 Feb. 4/1 My Brother Habitants will be..convinced of the expediency of the regulation. 1791 J. Long *Voy. & Trav. Indian Interpr.* 167 The Canadians are particularly fond of dancing, from the *seigneur* to the *habitant*. 1909 *Westm. Gaz.* 10 Apr. 6/2 From school Drummond became a clerk in a telegraph office at Bord-à-Plouffe, a little village on the Rivière des Prairies, where he was in the midst of habitants, lumbermen, and voyageurs. 1966 *Kingston* (Ont.) *Whig-Standard* 27 Aug. 4/3 As the old habitant joke had it, it's okay to t'row out de hank [*sc.* anchor], but suppose there's no rope on the hank?

habitat. Add: **2.** *Comb.* **habitat form,** the form developed by a race or organism in response to its habitat; **habitat group,** any group of species whose members favour a similar habitat.
1902 F. E. Clements in *Beiblatt zu den Botanischen Jahrbüchern* LXX. 17 A habitat form is the modified form of a species common to two or more formations produced by a particular formation, i.e. habitat, such as the alpine meadow habitat form of *Campanula rotundifolia*. 1916 B. D. Jackson *Gloss. Bot. Terms* (ed. 3) 169/1 *Habitat-form*, the impress given to the plant by the habitat. 1898 Pound & Clements *Phytogeogr. Nebraska* iv. 93 A habitat group is a group of species, which are subject to similar physical conditions, and frequent like habitats. 1959 E. F. Linssen *Beetles Brit. Is.* I. 57 This 'bionomic classification', as the method is called, is based on habitat-groups. 1962 *Conservationist* June–July 20/2 Four new life-size dioramas (natural habitat groups) of four types of fish areas in Rochester and Munroe County are featured in permanent exhibitions in the Hall of Natural Science at the Rochester Museum of Arts and Sciences.

habitation. Add: **5.** *Comb.* **habitation name,** a place-name in which at least one of the elements denotes an inhabited place; **habitation site** *Archæol.*, a site where there has been a settlement.
1936 *Oxf. Dict. Eng. Place-Names* p. xv, Near habitation-names stand names that originally denoted a pasture-ground or a shelter for the protection of animals, a cowhouse, a cattle-fold, etc. 1962 H. R. Loyn *Anglo-Saxon England* i. 9 There are more British habitation names in the region. 1925 *Antiquaries Jrnl.* Apr. 182 The author has more leisure to examine Meare, a similar habitation-site three miles distant. 1942 *Oxoniensia* VII. 106 The B-beaker and neolithic sherds were drift-sherds from one of the many habitation-sites..in the neighbourhood. 1962 H. R. Loyn *Anglo-Saxon England* i. 15 This is not to deny the possibility of continuity in habitation sites at places such as London or York. 1971 *World Archaeol.* III. 141 Some of the shelters may have been habitation sites in the past.

habitative, *a.* Delete *rare* and add earlier and later examples; now esp. in place-name studies.
1578 [see Possessive *a.* 2]. 1929 A. Mawer *Probl. Place-Name Study* iii. 124 By the time that *tun* had become the great habitative suffix there were probably few such groups left. 1961 K. Cameron *Eng. Place-Names* ii. 27 Place-names..can be divided into two main types, habitative and topographical. 1967 A. H. Smith *Place-Names of Westmorland* I. p. xxxviii, The older English settlements are represented by some 30 or more habitative names of parishes and villages.

habituation. **1.** Delete † *Obs.* and add: Esp. the formation of such habits as dependence on drugs. (Later examples.)
1929 *Jrnl. Pharmacol. & Exper. Therap.* XXXVI. 466 'Habituation' we interpret to mean a condition wherein one becomes accustomed to but not seriously dependent upon a drug. 1960 *Times* 21 Jan. 6/7 The two drugs specifically mentioned in this context are carbromal and bromvaletone. The committee state that cases of habituation arising from the widespread use of these by the public, largely without medical guidance, have been 'numerically very few, but individually serious'. 1962 *Jrnl. Amer. Med. Assoc.* 14 July 93/1 The inherent difficulty in establishing a uniform cut-off point where 'habituation' becomes 'addiction'. 1973 G. G. Nahas *Marihuana* v. 188 In Egypt 65% of the consumers of hashish declared they were unable to stop although they expressed a wish to discontinue their habituation.
2. esp. in *Psychol.* The diminishing of response to a frequently repeated stimulus. (Further examples.)
1895 *Amer. Jrnl. Psychol.* VII. 82 Gewöhnung, habituation. 1934 H. C. Warren *Dict. Psychol.* 120/1 *Habituation*, the process of becoming adapted to a given stimulus, situation, or general environment. [A pop. and rather loose term.] 1967 S. Arieti *Intrapsychic Self* iii. 47 Loss of awareness in the phenomenon of habituation is a parsimonious device used constantly by the nervous system. 1973 S. Rose *Conscious Brain* ix. 194 The habituation process is..both a behavioral and a physiological phenomenon.

habitudinal (hæbitiū·dinăl), *a.* [f. L. *habitūdo, -inis* Habitude + -al.] Of or pertaining to habit; habitual.
c 1380 Wyclif *Sel. Eng. Wks.* (1869) I. 78 But clerkes witen þat þer ben two manere of seyngis, þat ben personel seynge, and habitudinel seynge. 1925 *Glasgow Herald* 12 Apr. 12 Disease.., biologically considered, is of three great kinds—(a) constitutional, (b) occupational or habitudinal, and (c) parasitic and microbic. 1932 J. A. Thomson *Scientific Riddles* iv. 324 Deeply saturating changes, whether environmental, nutritional, or habitudinal.

haboob (habū·b). Also **haboub, habub,** etc. [Arab. *habūb* blowing furiously.] A violent and oppressive wind which blows at certain seasons in the Sudan, and which brings with it sand from the desert. Also *transf.*
1897 *Daily News* 2 Oct. 2/1 This was a real haboob—a tornado of sand and small stones. *Ibid.* 4/6 A soldier that had been swept into the river by the merciless haboob. 1928 *Blackw. Mag.* Feb. 259 May is the month of 'haboobs'. 1936 *Jrnl. R. Aeronaut. Soc.* XL. 91 Our captains pay the greatest respect to the 'haboobs' of the Sudan. 1959 R. E. Huschke *Gloss. Meteorol.* 268 *Haboob* (many variant spellings, including *habbub, habub, haboub, hubbob, hubbub*), a strong wind and sandstorm or duststorm in the northern and central Sudan, especially around Khartum, where the average number is about 24 a year. 1973 *Sci. Amer.* Jan. 46/3 The American haboobs are not so frequent as the Sudanese (two or three a year at Phoenix as compared with perhaps 24 a year at Khartoum).

habu (hā·bu). [Jap.: see quot. 1818.] A venomous pit-viper, *Trimeresurus flavoviridis*, native to the Ryukyu Islands and neighbouring areas.
[1818 B. Hall *Acc. Voy. Discovery to Great Loo-Choo Island* App., Vocabulary of English and Loo-Choo Words...Snake—Hāboo.] 1895 *Geogr. Jrnl.* V. 299 The poisonous Trimeresurus.., called *habu* by the natives, is 4 or 5 feet long by 2 inches in diameter, and is an object of universal fear. 1955 *Sci. News Let.* 15 Jan. 36/1 The habu and mamushi, native to certain islands of the Pacific and parts of the Asiatic mainland.

habutai (hā·butai). Also **-aye, -ae.** [Jap.] A fine soft Japanese silk.
[1822 F. Shoberl tr. *Titsingh's Illustr. Japan* 17 Tchouya followed, dressed in two robes of light blue, made of the stuff called *fabita*, with his hands tied behind him.] 1895 *Montgomery Ward Catal.* 12/3 Habutai Silk.. is a very soft, light weight of silk of Japanese make. 1902 *Encycl. Brit.* XXIX. 725/1 Pictures so elaborate and.. accurate as those produced by the *yuzen* process on silk crape or *habutaye*. 1921 *Daily Colonist* (Victoria, B.C.) 7 Apr. 13/6 Black Waists of heavy quality Habutai silk, with tucked fronts and neat collars. 1931 *Stud. Eng. Lit.* (Tokyo) XI. 515 *Habutae*, a kind of silk. This word is regularly used in newspaper advertisements in England. 1950 '*Mercury*' *Dict. Textile Terms* 254/2 A pure dye spotproof silk of the Habutai order, finished with a little lustre.

hacendado (ɑːsendā·do). Also **haciendado.** [Sp.] The owner of an hacienda.
1840 D. Turnbull *Trav. in West* 98 In the unexpected case of the confiscation of the rural property of a Hacendado, the civil judge of the district..is directed to proceed to the spot. 1862 Mayne Reid *Tiger Hunter* xix, To the haciendado he hired himself out a part of the year. 1897 *Blackw. Mag.* Nov. 685 The polity of the Mexican haciendados remains unchanged. 1897 *Outing* (U.S.) XXIX. 593/2 The plantation homes of the *haciendados*. 1920 *Glasgow Herald* 22 Sept. 8 The Chilean haciendado. 1934 A. Huxley *Beyond Mexique Bay* 284 These lands were once the property of rich *hacendados*. 1962 *Economist* 13 Oct. 157/3 The Constitutionalists [of Mexico] led by Carranza, a *hacendado*, and Villa, a former bandit.

hacienda. Add: Also *attrib.*
1860 *Ure's Dict. Arts* (ed. 5) III. 676 Working it on their own account, or, as it is termed, hacienda account. 1897 *Blackw. Mag.* Nov. 685/2 The pleasant picture of hacienda life in the land of the Aztecs. 1944 *Harper's Mag.* Aug. 201/2 The Rio Blanco textile works set up in Mexico..were designed to produce cloth for sale at hacienda stores.

hack, *sb.*[1] Add: **6.** An act of hacking; a hacking blow. Also *fig.*, now esp. (*U.S.*) a try, attempt.
1836 D. Crockett *Exploits & Adv. Texas* 79 Better take a hack by way of trying your luck at guessing. 1873–4 *Rep. Vermont Board Agric.* II. 238, I have a chance to have several hacks at the weeds before the crop is sown. 1898 M. Deland *Old Chester Tales* 244, I get more men in a saloon, that's why; and when the show's done I get a hack at 'em. 1969 *New Yorker* 12 Apr. 95/1 We go into the second order of testing,..which would give us a better hack, a better indication of what we are dealing with.

hack, *sb.*[3] Add: **7. b.** *hack-cab* (example). **c.** *hack writer* (earlier example); so *hack-writing*.
1851 *London at Table* i. 31, I..started in a hack cab for the scene of action. 1826 *Blackw. Mag.* XX. 296/2 You forget the effrontery of the hack-writer in the shamefacedness of the would-be gentleman. 1850 Kingsley *Alton Locke* II. ii. 14 My hack-writing was breaking down my moral sense, as it does that of most men. 1933 E. Pound *Let.* 24 Sept. (1971) 247 Teaching damn sight easier way of earning living than hackwriting.
8. (Earlier U.S. examples.)
1812 *Boston Gaz.* 10 Sept., Advt. (Th.), Hack Stand. 1835 in *Southern Lit. Messenger* IV. 197/1 My hack-driver..assured [me] that there was no other tavern in the city.

hack, *v.*[1] Add: **2. b.** Also in *Rugby Football.* Const. *over, up.*
1864 *Blackheathen* 9/1 'Hacking first man up'..remains at present quite a Rugby rule. 1887 M. Shearman *Athletics & Football* 395 No hacking, or hacking over, or tripping up shall be allowed under any circumstances. 1897 *Encycl. Sport* I. 404/1 Rugby Football... Not only was it legal to hack over the carrier of the ball, but also the first on side, and I have seen as many as four of the van brought to earth by this means. 1963 *Times* 24 Jan. 3/1 It had been agreed [when the laws of Rugby Union were drawn up] that hacking-over and tripping-up should not be permitted.
c. To embarrass, annoy; to disconcert, confuse. Freq. as **hacked** *ppl. a.* *U.S. slang* or *dial.*
1892 J. C. Harris *Uncle Remus & Friends* 349 When you once git 'em hacked dey er hacked fer good; dey des give right up en roll der eyes. 1908 *Dialect Notes* III. 318 We tried to hack the pitcher. 1917 *Ibid.* IV. 413 That joke hacks Steve to this day. 1969 *Rolling Stone* 28 June 19/1 The big word down there is *commercial*...I wouldn't be so hacked off about it if I didn't love country music.
d. To cope with, manage, accomplish; to tolerate, accept; to comprehend; freq. *to hack it. slang* (orig. *U.S.*).
1955 *Antioch Rev.* XV. 379, I can't hack something like stealing. 1968 *Maclean's Mag.* Dec. 29/1, I just couldn't hack teaching any more, it was as simple as that. 1970 *Globe Mag.* (Toronto) 26 Sept. 9/2 You know, they're shooting people at Kent State and we talk about amendments to the Warble Fly Act. I can't hack that; it drives me crazy. 1972 *Sunday Mirror* 16 Apr. 23/3 Now, suddenly and bewilderingly since President Nixon has ordered his legions home, the Arvin is a great little guy who *can* hack it. 1972 *Newsweek* 7 Aug. 18/2, I had proved to the world during my four years in the Senate..that I can hack it.

hackamore. Add: (Earlier and later examples.) Also, a headstall.
1850 W. R. Ryan *Upper & Lower California* I. 152 He overtook me, mounted on a well saddled horse, and leading another by the hackamore. 1926 D. Branch *Cowboy & his Interpreters* 39 But having the 'hackimore' rope fastened to my belt I held to him until help arrived. 1971 A. P. McInnes *Dunlevy* 86 Her only riding equipment was a rawhide hackamore already on the horse's head.

hackia (hæ·kiă). = Guaiacum 2.
1851 *Illustr. Catal. Gt. Exhib.* iv. i. 983/2 Hackia, lignum vitæ, transverse and vertical sections, from River Demerara. 1858 Simmonds *Dict. Trade*, *Hackia*, a wood ..used for mill cogs and shafts. 1969 S. M. Sadeek *Windswept & Other Stories* 36 The dark East Indian.. flicked a piece of paper before him with his hackia stick.

hackie (hæ·ki). *U.S. colloq.* Also **hacky.** [f. Hack *sb.*[3] 2.] A taxi-driver.
1937 *Daily Express* 10 Mar. 6/7 'Hackie' is taxi-driver. 1946 Mezzrow & Wolfe *Really Blues* (1957) viii. 114 Weaving..like an expert hackie in heavy traffic. 1959 'M. Nevill' *Sweet Night for Murder* xiii. 129 And now.. unearth some other blasted hacky that drove me there.

hacking, *vbl. sb.* Add: **1. c.** [After G. *hackung*; cf. F. *hachement*.] Massage with the edge of the hand.
1890 A. Kellgren *Ling's Syst. Man. Treatm.* 25 Tapotement means hacking or beating. 1893 A. S. Eccles *Sciatica* 64 Thorough rubbing, kneading, hacking, and passive movements are practised.
5. [Hack *v.*[3] 3.] **hacking coat, jacket,** a sports coat suitable for use when riding, often tailored in a tweed with vents at the side or at the back; **hacking length,** the length of a hacking jacket.
1948 'E. Crispin' *Buried for Pleasure* vi. 38 He wore jodhpurs, riding-boots, a violent check hacking coat, and a yellow tie. 1954 *New Yorker* 13 Nov. 168/2 A jacket with a center vent and four buttons, or a hacking coat with side vents and slanted flap pockets, is from $125 to $135. 1954 *Irish Digest* Nov. 40 Resplendent in cloth cap, tweed hacking jacket, [etc.].., Huston squires it happily among his beloved Irish. 1959 H. Hobson *Mission House Murder* ii. 11 My Harris tweed hacking-jacket. 1965 *Punch* 22 Dec. 933/3 Leo McKern's irascible Baron is a splendid comic creation, wrapped in a loud-checked hacking jacket (checks any larger and they wouldn't have fitted on to the material). 1966 *Guardian* 27 July 6/4 Jackets are hacking length.

hackle, *sb.*[2] Add: **1. b.** *Hairdressing.* (See quot. 1957.)
1903 A. M. Sutton *Boardwork* (ed. 2) i. 9 A 'card' or 'hackle', used for disentangling combings, smoothing and mixing hair, is a magnified comb composed of steel spikes or prongs. 1957 V. J. Kehoe *Technique Film & T.V. Make-Up* iii. 34 *Hackle*, a multi-spiked tool which is clamped to a bench and used for combing or carding skeins of hair.
3. Also, the feathers on the saddle of a cock.
1850 D. J. Browne *Amer. Poultry Yard* 22 The hackles of the lower part of the back. *Ibid.* 253 In capons..the hackle, the tail feathers, and the spurs grew to a much greater length than in cocks. 1970 H. E. Smith *Bantams* i. 9 Feathers towards the stern are correctly called 'saddle hackles'. *Ibid.* iii. 19 The colours of a Red Jungle Fowl.. male are neck hackle, golden; saddle hackle, orange.

hackle, *v.*[3] Add to def.: Also used of dressing the hair in wigmaking.
1931 G. A. Foan *Art & Craft of Hairdressing* i. 11/1 When dry the hair is ready for drawing off into roots and points. Taking each section separately the student should lightly hackle the extreme ends. 1966 J. S. Cox *Illustr. Dict. Hairdressing* 68/1 *Hackle*, to draw hair through a hackle to disentangle it.

hackling, *vbl. sb.*[1] *attrib.* Add: *hackling house.*
1849 E. Chamberlain *Indiana Gazetteer* (ed. 3) 132 A brick building, erected for a hackling house.

hackman. Add: (Earlier and later examples.) Also *Canada.*
1796 *Boston Directory*, passim. 1806 *Repertory* (Boston) 3 Oct. (Th.), Died, in this town, Mr. Daniel Henry, hackman. 1819 *N.Y. Gaz.* in *Massachusetts Spy* 16 June 3/1 The horses were stopt by the hackmen on the stand. 1898 H. E. Hamblen *Tom Benton's Luck* 56 The line of vociferous hackmen who formed a gauntlet across the exit from the railroad station. 1906 *Daily Colonist* (Victoria, B.C.) 9 Jan. 5/4 A hack case is being heard in the city police court, in which one of the local hackmen is being charged with overcharging.

hackmanite (hæ·kmănait). *Min.* [ad. Sw. *hackmanit* (L. H. Borgström 1901, in *Geol. Föreningens Stockholm Förh.* XXIII. 563), f. the name of Victor A. *Hackman* (1866–1941), Finnish geologist: see -ite[1].] A pink or reddish violet variety of sodalite which loses its colour when exposed to daylight but regains it in the dark.
1903 *Jrnl. Chem. Soc.* LXXXIV. ii. 304 Hackmanite, a new member of the sodalite group. 1941 *Amer. Mineralogist* XXVI. 441 The induced color change of hackmanite when exposed to ultra-violet light was first observed.. on hackmanite from Bancroft, Ontario, which changed from pink to a raspberry shade or deep violet on exposure to ultra-violet light. The induced color faded rapidly and nearly completely on exposure to strong light. 1970 *Physics Bull.* Nov. 487/1 Geologists had found that freshly cleared rocks of a naturally occurring form of sodalite, hackmanite, exhibited a pink colour.

hackney-carriage. (Earlier U.S. example.)
1796 *Mass. Acts & Laws* (1896) 62 The said Selectmen are hereby authorized to grant licences for such number of Hackney Coaches & Carriages..as they shall judge proper.

hackneydom (hæ·knidəm). [f. Hackney(ed *ppl. a.* 2 + -dom. Cf. Hackney *v.* 7.] A state of commonplaceness.
1897 G. B. Shaw *Our Theatres in Nineties* (1932) III. 235 The latest attempt to escape from hackneydom and cockneydom is the Chinatown play. 1959 *Times* 30 Nov. 14/3 Again this was a performance that shook the dust of hackneydom from the symphony.

hadada(h (hā·dadā). Also **hadadaw, haddada, hadeda, hadida.** [Onomatopœic from the bird's raucous call.] A large brown-green ibis, *Hagedashia hagedash.*
1801 J. Barrow *Trav. S. Afr.* I. iv. 264 The Egyptian black *ibis* (*niger*) and another species of *tantalus*, called by the farmers the *haddadas*, were procured at this place. 1846 tr. *Arbousset & Daumas's Narr. Tour Cape G.H.* 190 A large ibis of a brown lustre commonly called by onomatopy *addada*. 1862 J. S. Dobie *S. Afr. Jrnl.* 10 Oct. (1945) 40 Missed a hadadaw. 1907 *African Monthly* Oct. 445 Flocks of 'ha-di-da' grub silently and unconcernedly in close proximity to the camp. 1952 Mackworth-Praed & Grant *Birds E. & N.E. Afr.* I. 59 Key to the Adult Storks, Ibises...Back bronzy brown, no crest on nape, ridge of bill only red, bare skin confined to front of eye: Hadada *Hagedashia hagedash.* 1953 Bannerman *Birds W. & Equat. Afr.* I. 184 West African Hadada. *Hagedashia hagedash brevirostris*...The best guide to the Hadada's identity lies in its loud note—*hah-a-a-a*—uttered frequently, both in flight and on the ground. 1953 R. Campbell *Mamba's Precipice* v. 48 To the forest, on the left bank, flocks of hadadah ibis were flying back to roost. 1957 McLachlan & Liversidge *Roberts's Birds S. Afr.* 39 Hadeda...At a distance it appears dull olive-grey, but metallic reflections on back and wing-coverts may be seen. 1964 C. Willock *Enormous Zoo* vii. 120 Hadada ibises rise with the horrible complaining cry from which they get their title.

Hadassah (hădæ·sā). [Heb., = myrtle, name of the Biblical Esther (Esther 2:7).] An American Zionist women's organization, founded in 1912, which contributes to welfare work in Israel.
1913 H. Szold *Let.* 15 July in M. Lowenthal *H. Szold Life & Lett.* (1942) 79 If we can keep up this folder propaganda as the winter's work of Hadassah, we shall be doing admirably. 1952 S. Spender *Learning Laughter* vii. 100 In 40 years Hadassah has grown from being a small organization for training a few nurses, into a main channel for transfusing life-blood into a young nation. 1964 D. Greenburg in *Playboy* Sept. 170/2 Your family ..will expect you to be able to relate amusing stories which you have heard..at a meeting of the Hadassah. 1966 H. Kemelman *Saturday the Rabbi went Hungry* (1967) i. 8 It fell off and I sewed it on myself. You were at a Hadassah meeting. 1972 *Times* 16 Oct. 6/1 The letter ..was addressed to a member of Hadassah, a women's Zionist organisation.

hadda (hæ·dă), repr. colloq. pronunc. of *had to* or *had a.*
1945 A. Kober *Parm Me* 34, I hadda do a lotta talking to get her to come here to-night. 1967 'L. Egan' *Nameless Ones* xv. 192 Well, she hadda lot of trouble. *Ibid.* 193 She hadda give up an apartment, couldn't afford it. 1969 *Coast to Coast 1967–68* 4 There was that Alsatian they hadda shoot because it wouldn't leave a kid's grave and kept scrabbling at the loose dirt. 1971 'A. Blaisdell' *Practice to Deceive* x. 138 'Did you hit her?' 'Naw, ..I never hadda chance.'

haddada, hadeda, varr. *Hadada(h.

Hades. Add: **2. b.** Used trivially as a substitute for *hell* in imprecations, etc.
1912 A. Bennett *Matador* ii, What the hades are you waiting there for? 1917 [see *blind *sb.* 11]. 1942 T. Bailey *Pink Camellia* xxvii. 196 What in Hades is he doing here? 1972 G. Bell *Villains Galore* i. 4 Damn protocol to Hades!

hadida, var. *Hadada(h.

Hadith (hæ·diþ). Also **Hadis, Hadithah.** Pl. **Hadithat.** [a. Arab. *ḥadīṯ* a tradition.] The body of traditions relating to Mohammed, which now form a supplement to the Koran, called the Sunna.
a 1817 J. L. Burckhardt *Trav. Syria & Holy Land* (1822) 326 An Olema thinks he has attained the pinnacle of knowledge if he can recite all the Koran together with some thousand of Hadeath, or sentences of the Prophet. 1880 *Encycl. Brit.* XI. 367/2 Rejecting the *Hadith*, or traditional sayings of Mahomet. 1883 *Ibid.* XVI. 594/2 The traditions of Mohammed, or *Hadith*, the collective body of which constitutes the *Sunna*, or custom. 1922 *Blackw. Mag.* Mar. 375/2, I treasured this like a *hadis*, an authentic tradition of which I was the custodian. 1924 A. Guillaume *Trad. Islam* 15 The hadith literature

as we now have it provides us with apostolic precept and example covering the whole duty of man. *Ibid.* 150 Many of the hadith already cited will have shown the good sense, amiability, and liberality of the prophet. 1951 'N. Shute' *Round Bend* v. 137 Legacies are governed by *hadith*, based upon the Koran.

hadj. Add: Also **haj, hajj.** Also *transf.*
1910 *Encycl. Brit.* XII. 827/1 The word *hajj* is sometimes loosely used of any Mahommedan pilgrimage to a sacred place or shrine, and is also applied to the pilgrimages of Christians of the East to the Holy Sepulchre at Jerusalem. 1930 Kipling *Limits & Renewals* (1932) 217 He had forbidden music because it was a *haj.*

hadjeen, var. *Hygeen, Hajeen.

hadji, hajji. Substitute for def.: The title given to one who has made the greater pilgrimage (on the 8th to 10th day of the 12th month of the Muslim year) to Mecca.

Hadrianic (hēⁱdriæ·nik), *a.* [f. L. *Hadriān(us* + -ic.] Of or pertaining to the Roman emperor Hadrian (A.D. 76–138).
1886 W. P. Dickson tr. *Mommsen's Provinces of Roman Empire* I. v. 189 In the time of Diocletian we find the district between the two walls evacuated, but the Hadrianic wall occupied still as before. 1897 *Trans. Cumberland & Westmorland Antiquarian & Archæol. Soc.* XIV. 419 Hadrianic inscriptions occur at Chesterholm and Greatchesters. 1911 *Ibid.* New Ser. XI. 390 Vindolana and Magna, although occupied in Hadrianic times as Wall stations, differ from the others in Northumberland. 1921 *Jrnl. Roman Stud.* XI. 51 The idea of a single camp extending from sea to sea is a bold flight of imagination... It became..a cardinal feature in the 'Hadrianic Theory' of the nineteenth century. 1935 *Burlington Mag.* Feb. 97/1 Hadrianic art was certainly Greek in its nature. 1962 *Guardian* 11 July 5/7 The rampart and ditch of a small Hadrianic fort. 1972 *Daily Tel.* 5 Dec. 12/5 Coins and Antiquities gave £6,800 for a Hadrianic/Early Antonine period Roman marble figure called Paris.

hadrome (hæ·drōᵘm). *Bot.* [ad. G. *hadrom* (G. Haberlandt *Physiologische Pflanzenanatomie* (1884) vii. v. 265), f. Gr. ἁδρ-ός thick, bulky + *-ome.] The conducting tissue of the xylem, excluding fibres.
1898 H. C. Porter tr. *Strasburger's Text-bk. Bot.* 102 Other terms often used to designate the vascular bundles are folio-vascular bundles and mestome. The vascular portion is also termed the xylem or hadrome. 1914 M. Drummond tr. *Haberlandt's Physiol. Plant Anat.* vii. 347 The water-conducting vessels and tracheides constitute..the resistant hadrome portion [of the conducting strand]... The xylem includes the hadrome with its associated wood-fibres...Where..no wood-fibres are developed, xylem is the exact equivalent of hadrome. 1965 K. Esau *Plant Anat.* (ed. 2) xii. 272 The parallel term for the xylem is *hadrom*, which refers to the conducting part of the xylem..excluding the fibers.
Hence **hadroce·ntric** *a.*, having the hadrome surrounded by the leptome; **ha·dromal** [a. G. *hadromal* (F. Czapek 1899, in *Zeitschr. f. physiol. Chem.* XXVII. 163], a hydrolysis product of lignin; *para*-coniferyl aldehyde, $C_6H_3(OH)(OCH_3)CH{:}CHCHO$; **ha:dromyco·sis,** a fungal disease of plants in which the xylem is the part most affected.
1899 *Jrnl. Chem. Soc.* LXXVI. i. 560 A substance termed hadromal has been isolated from different woody tissues; it has the properties of a phenol and of an aldehyde. 1900 B. D. Jackson *Gloss. Bot. Terms* 310/1 *Hadrocentric*, having the hadrome in the centre surrounded by the leptome. 1914 M. Drummond tr. *Haberlandt's Physiol. Plant Anat.* vii. 349 If the hadrome is central and the leptome peripheral, the bundle may be termed *hadrocentric*. 1916 G. H. Pethybridge in *Sci. Proc. Roy. Dublin Soc.* XV. 87 The fungus mycelium is, at any rate in the early stages of the disease, confined to the wood vessels.. and I, therefore, suggest the word 'hadromycosis' for use in this connexion. 1917 *Nature* 22 Feb. 500/2 Plants suffering from the choking of their vessels [by fungi] (hadromycosis). 1928 *Chem. Abstr.* XXII. 399 Czapek's hadromal.., which he considers responsible for the characteristic red color given by wood tissues with phloroglucinol-HCl. 1931 *Hilgardia* V. 197 (*title*) Verticillium hadromycosis. 1952 F. E. Brauns *Chem. Lignin* iv. 37 Hadromal occurs in wood as a coniferyl aldehyde cellulose ester. 1971 G. C. Ainsworth *Ainsworth & Bisby's Dict. Fungi* (ed. 6) 254 Hadromycosis, a disease of plants in which the pathogen is confined to the xylem, e.g. *Verticillium* wilt of potato and tomato.

hadron (hæ·drɒn). *Physics.* [f. Gr. ἁδρ-ός thick, bulky + *-on[1]; first used in Russian, with the spelling *adron*.] Any strongly interacting sub-atomic particle. Hence **hadro·nic** *a.*
1962 L. B. Okun' in *Proc. Internat. Conf. High-Energy Physics* 845/2 In this report I shall call strongly interacting particles hadrons, and the corresponding decays hadronic. 1966 *New Scientist* 26 May 500/1 The particles, so-called baryons and mesons, collectively called hadrons. 1968 *Sci. Jrnl.* Nov. 34/2 Weak decays of hadrons—which are nucleons (neutrons and protons) and their heavy partners. 1969 *Physics Bull.* Jan. 32/1 In high energy hadron collisions, the fraction of the total cross section going into two-body final states decreases rapidly with energy.

Haeckelian (heki·liăn), *a.* [f. the name of E. H. *Haeckel* (1834–1919), German biologist: see -ian.] Of or pertaining to the opinions of Haeckel; also as *sb.*, a believer in Haeckel's theories. So **Haeckelism** (he·kəliz'm), **-i·smus,** the opinions and theories of Haeckel.
1894 *Natural Sci.* Mar. 162 We are well content to cease from controversy, to let Calcareous sponges, the Gastrula, and..Haeckelismus take care of themselves. 1897 *Ibid.* Jan. 31 The typical form of the Haeckelian genealogical tree. 1899 E. J. Chapman *Drama of Two Lives* 88 Thus, hæckelism's wondrous gleam Makes clear, to all, how all arose. 1930 G. R. de Beer *Embryol. & Evol.* xv. 102 There is then no recapitulation in the Haeckelian sense of accelerated repetition of adult stages. 1971 *Nature* 11 June 400/2 The German Monist League..was neither a 'scientific' nor a 'political' body but rather one devoted to Haeckelian naturalism.

hæm, heme (hīm). [Back-formation from Hæm(oglobin.] **a.** A chelation compound, $C_{34}H_{32}O_4N_4Fe$, of ferrous ion and protoporphyrin, obtained on reduction of hæmatin: the red-coloured non-protein constituent of hæmoglobin.
1925 Anson & Mirsky in *Jrnl. Physiol.* LX. 50 Haemoglobin is a conjugated protein consisting of globin and a non-protein part, containing pyrrol nuclei and iron, which we shall call haem. 1939 *Jrnl. Biol. Chem.* CXXXI. 661 Coryell..has applied the Pauling equation..to the oxidation of hemoglobin..and has shown that interaction among the four heme groups will account for the occurrence of an *n* which is not integral. 1956 *Nature* 11 Feb. 275/1 The angular variation of the *g* values enables an accurate determination to be made of the orientation of the hæm and porphyrin planes with respect to the external crystalline axes. *Ibid.* 275/2 Detailed information on the orientation of the hæm planes can be combined with X-ray measurements to calculate the polypeptide chain directions. 1970 R. W. McGilvery *Biochem.* ii. 13 The biological function of hemoglobin is therefore derived from both the heme and the peptides... Each of the four peptide chains has its own heme.
b. Any of various compounds of a ferrous or ferric ion and a porphyrin, present in biological pigments.
1948 *Biochem. Jrnl.* XLII. p. xlvii/2 The phase separation of haems has been applied to ox heart muscle. 1962 Rimington & Kennedy in Florkin & Mason *Comprehensive Biochem.* IV. xii. 563 By far the most important metalloporphyrins..are the iron complexes or hemes. In these nature has exploited the valency change from the ferric to ferrous state and vice versa to establish an electron transport system connecting the intracellular dehydrogenases with atmospheric oxygen. 1966 K. Okunuki in Florkin & Stotz *Comprehensive Biochem.* XIV. 233 Four types of haem have so far been known to occur in cytochromes...These are haem *a*, protohaem, haem *c* and so-called haem a_2.

hæmachromatosis, hem-, varr. *Hæmochromatosis.

hæmagglutinate, hem- (hiːm,ăglū·tineⁱt), *v.* [f. Gr. αἷμ-α blood + Agglutinate *v.*] *trans.* To cause (red blood cells) to agglutinate. So **hæ:magglu·tinating** *ppl. a.*
1921 *Jrnl. Immunol.* VI. 423 In..experiments with normal and 'immune' hemagglutinating sera we have used the slide method. 1922 *Jrnl. Amer. Med. Assoc.* 11 Nov. 1684/1 (*title*) The hemagglutinating fraction of human serums. *Ibid.* 1685/1 The hemagglutinating property is contained in the pseudoglobin fraction. 1936 *Nature* 26 Sept. 554/1 The haemagglutinating substance is present in the cotyledons of the seed of the runner bean. 1956 *Ibid.* 4 Feb. 234/2 Under the conditions of assay, five hæmagglutinating units were regularly inhibited by a minimum of 10–20µ gm. of preparation *IL-1*. 1961 *Lancet* 23 Sept. 718/1 Only 6 strains were found to hæmagglutinate the red cells of more than 5 species.
Also **hæ:magglutina·tion,** the action or process of hæmagglutinating.
1907 *Jrnl. Med. Res.* XVII. 323 Repeated controls have convinced me that the presence and degree of hemagglutination may be detected quite as accurately macroscopically as microscopically. 1919 *Jrnl. Immunol.* IV. 284 The use of hypertonic solutions of sodium chlorid are of no practical value in preventing hemagglutination in complement fixation tests. 1946 *Nature* 27 July 119/1 Vaccinia hæmagglutination is inhibited by appropriate immune sera either of animal or human origin. 1949 *Poultry Sci.* XXVIII. 622 (*title*) A plate hemagglutination-inhibition test for Newcastle disease antibodies in avian and human serums. *Ibid.* 622/1 Virus isolation and neutralization tests are carried out in embryonating chicken eggs; the hemagglutination-inhibition test is done in test tubes. 1969 *New Scientist* 30 Jan. 171/1 The basis of a simple laboratory test—the haemagglutination test—which is widely used in the study of influenza.

hæmagglutinin, hem- (hiːm,ăglū·tinin). [f. Gr. αἷμ-α + *Agglutinin.] A substance that causes agglutination of red blood cells.
1904 *Amer. Jrnl. Med. Sci.* CXXVIII. 669 (*title*) Concerning haemagglutinins of bacterial origin and their relation to hyaline thrombi and liver necroses. *Ibid.* 670 The observations of Hueter..are of interest in connection with our recently acquired knowledge of bacterial haemagglutinins. 1946 *Nature* 27 July 119/1 Work on the hæmagglutinin of vaccinia virus was initiated by an observation by Burnet in October 1941 that a chorioallantoic membrane emulsion agglutinated fowl cells to a low titre. 1969 *New Scientist* 23 Jan. 171/1 The presence of haemagglutinin on the virus surface enables the influenza virus to adhere firmly to the surface of chicken erythrocytes.

hæmangioma, hem- (hiːmændʒi,ōᵘ·mă). *Path.* Pl. **-ata, -as.** [f. Gr. αἷμ-α blood + *Angioma.] (See quot. 1900.)
1890 in Billings *Nat. Med. Dict.* 1900 Dorland *Med. Dict.* 291/2 *Hemangioma*, angioma containing blood-vessels, but not lymph-vessels; true angioma. 1913 C. P. White *Path. Growth Tumours* vi. 86 Angeiomata..are formed in connection with blood vessels (haemangeioma) or with lymphatic vessels (lymphangeioma). 1961 *Lancet* 22 July 211/2 (*table*) Hæmangiomata. *Ibid.* 26 Aug. 492/2 Patients with coarctation of the abdominal aorta..or large hæmangiomas. 1962 *Ibid.* 6 Jan. 46/1 A capillary hæmangioma of the upper lips, nose, and frontal area seems to be highly pathognomonic of this syndrome. 1970 Passmore & Robson *Compan. Med. Stud.* II. xxx. 16 It is difficult to draw a dividing line between fibrous xanthomata..and the sclerosing haemangiomata.
Hence **hæ:mangiomato·sis,** a condition characterized by the presence of many hæmangiomata.
1912 *Jrnl. Amer. Med. Assoc.* 27 Apr. 1311/1 (*title*) General hemangiomatosis of placenta. 1913 Dorland *Med. Dict.* (ed. 7) 415/2 Hemangiomatosis, a condition in which multiple hemangiomata are developed. 1970 *Gut* XI. 515 Radiotherapy has been reported..to be of some benefit in..hepatic haemangiomatosis.

hæmanthus, hem- (hīmæ·nþɒ̆s). *Bot.* [mod.L. (Linnæus *Hortus Cliffortianus* (1737) 127), f. Gr. αἷμ-α blood + ἄνθος flower.] A bulbous plant of the genus so called, belonging to the family Amaryllidaceæ, native to southern and tropical Africa, and bearing umbels of red, pink, or white flowers.
1771 R. Weston *Universal Botanist* II. 375 Hæmanthus, Blood flower, or African Tulip... Spotted-stalked Guinea Hæmanthus. 1834 *Curtis's Bot. Mag.* LXI. 3373 (*heading*) Hairy, Pink Hæmanthus. 1885 T. Baines *Greenhouse & Stove Plants* 199/1 Hæmanthus are increased like Amaryllis by offsets which the strong bulbs produce. 1961 *Amateur Gardening* 14 Oct. 29/1 Bulbs of haemanthus can be potted now.

hæmarthrosis, hem- (hīm,arþrōᵘ·sis). *Path.* Pl. **-oses.** [f. Gr. αἷμ-α + ἄρθρο-ν joint + -osis.] Hæmorrhage into a joint.
1883 *Brit. Med. Jrnl.* 22 Sept. 561/2, I diagnosed the case as one of hæmarthrosis. 1891 C. W. M. Moullin *Surg.* iii. vi. 613 In cases..in which the hæmorrhage is often considerable and the swelling immediate, it may be almost pure blood (hæmarthrosis). 1908 *Practitioner* Apr. 521 Other cases are given..of the association of fatal hæmorrhages from the bowels together with the hæmarthrosis. 1962 *Lancet* 27 Jan. 174/1 Their bleeding is similar to that seen in mild hæmophilia; they have had hæmarthroses, deep intramuscular hæmorrhage, and hæmaturia.

hæmato-, hemato-. Add: **hæ·matocri:t** [Gr. κριτ-ής judge], a centrifuge used to estimate the volume occupied by the red blood cells in a sample of blood; the value obtained, expressed as a percentage of the volume of the sample; also earlier **hæmatokrit**; †**hæ·matoge:n** [a. G. *hämatogen* (G. Bunge 1885, in *Zeitschr. f. physiol. Chem.* IX. 56)], a yellow powder obtained from egg yolk and supposed to be the precursor of hæmoglobin (*Obs.*); **hæmatolytic** *a.* (earlier example); **hæ:matomye·lia** [Gr. μυελ-ός marrow + -ia[1]], hæmorrhage into the substance of the spinal cord; **hæ:matopo·rphyrin** [a. G. *haematoporphyrin* (F. Hoppe-Seyler *Med.-chem. Untersuch.* (1871) IV. liii. 533): see *Porphyrin], a dark violet porphyrin compound, $C_{34}H_{38}O_6N_4$, obtained by the action of concentrated acids on hæm or its derivatives; **hæ:matosa·lpinx** [Salpinx 2] (see quot. 1890).
1894 *Med. News* 29 Sept. 348/2 (*heading*) A modification of Hedin's hematokrit. *Ibid.* 350/2 States of comparative health..seem to be of the least importance of all the data necessary for the present status of the hematokrit. 1946 *Nature* 31 Aug. 304/1 These patients had,..because of the low hæmatocrit, a significantly reduced blood volume. 1958 *Immunology* I. 206 Blood samples were centrifuged in Wintrobe haematocrit tubes. 1966 *Lancet* 24 Dec. 1381/2 Chamberlain and Millard (1963) reduced the haematocrit and red-blood-cell volume (R.C.V.) in their patients with polycythaemia by means of oxygen. 1890 L. C. Wooldridge tr. *Bunge's Text-bk. Physiol. & Pathol. Chem.* vi. 102 The iron is more firmly fixed in the nuclein of the yolk of egg than in the albuminates of iron. ..The nuclein which contained iron..is doubtless the precursor of hæmoglobin, for there is no considerable quantity of any other compound of iron in the yolk. I have therefore proposed that this compound should receive the name hæmatogen (blood-former). 1934 J. F. McClendon *Man. Biochem.* 112 Bunge..supposed that mammals are born with a store of iron. That led him to look for iron in the eggs of birds, and he found an iron compound which he called hematogen. 1875 R. Fowler *Med. Vocab.* (ed. 2) 222/1 Hæmatolytic...1. Accompanied with the escape of blood from distended capillaries. 2. Applied adj. and subs. to medicines, said

to, by long continued use, impoverish the blood. 1881 *Brit. Med. Jrnl.* 28 May 852/2 A case of haematomyelia in a man aged 19. 1940 H. G. Wells *Babes in Darkling Wood* iii. iii. 275 He was equal to hæmatomyelia, a sort of temporary stroke just at the back of the head. 1970 *Archiv. für Toxikol.* XXVI. 56 The apparent sudden onset, the lack of progression,..and the relatively advanced age of manifestation all point to hematomyelia as the cause of the lesion, which in turn follows as a consequence of the increased bleeding tendency associated with benzene poisoning. 1885 *Jrnl. Physiol.* VI. 27 The filtrate was reddish and shewed a spectrum which is that of acid haematoporphyrin. 1902 *Encycl. Brit.* XXXI. 726/2 By mineral acids the iron may be removed, leaving a purplish pigment, Hæmatoporphyrin, which has no power of taking up or giving off oxygen. 1928 J. Pryde *Recent Adv. Biochem.* (ed. 2) x. 315 It would seem that neither hæmatoporphyrin nor mesoporphyrin is formed in the human body. *Ibid.*, Hæmatoporphyrin has a very powerful light-sensitising action. 1955 *Sci. News Let.* 9 Apr. 240/3 Cancer tissue can be made to glow a bright red under ultraviolet light when a powder called hematoporphyrin is introduced intravenously before surgery. 1955 *Endeavour* XIV. 126/2 Haematoporphyrin..is accordingly described as 1,3,5,8-tetramethyl-2,4-di-(α-oxyethyl)-porphin-6,7-dipropionic acid. 1884 *Lancet* 2 Feb. 207/2 Haematosalpinx... Tumours were discovered to the left and right of the uterus. These..proved to be the tubes, full of tar-like blood and firmly adherent. 1890 Billings *Nat. Med. Dict.* 614/1 *Hæmatosalpinx*, collection of blood in the Fallopian tube. 1923 J. M. M. Kerr et al. *Combined Text-bk. Obstetr. & Gynæcol.* xli. 612 The fluid..may extend to the uterine cavity, forming a hæmatometra, and in the most extreme cases it may distend the Fallopian tubes, forming hæmatosalpinges. 1972 C. J. Dewhurst *Integrated Obstetr. & Gynaecol. Postgrad.* i. 12/2 Haematosalpinx is most uncommon except in cases of very long-standing [imperforate membrane], or in association with retention of blood in a fragment of upper vagina.

hæmatology. Add: Also **hem-.** **hæmatological** *a.* (examples). So also **hæmatolo·gic** *a.*; **hæmatolo·gically** *adv.*; **hæmato·logist,** one who specializes in hæmatology.
1904 *Lancet* 25 June 1790/2 The next method was demonstrated by Stengel some ten years ago and has since rapidly gained favour amongst hæmatologists, who frequently re-discover it. 1939 *Jrnl. Clin. Invest.* XVIII. 543/2 The serum or plasma iron fluctuations which occur in hematologically equilibrated subjects. 1946 *Nature* 6 July 24/2 We would like to take this opportunity of expressing our thanks to Dr. R. A. Kekwick for advising us on the hæmatological technique. *Ibid.* 21 Sept. 412/1 An International Hematology and *Rh* Conference will be held in Dallas, Texas, on November 15. 1947 *Radiology* XLIX. 286/2 The hematologic constituents of the peripheral blood were the most sensitive indicators of radiation effect. 1956 A. H. Compton *Atomic Quest* 333 Nuclear chemists,..metallurgists, hematologists, and meteorologists. 1965 *Math. in Biol. & Med.* (*Med. Res. Council*) iii. 90 (*title*) Digital computer as aid to differential diagnosis: use in hematologic diseases.

hæmerythrin, hem- (hīːmeri·þrin). *Biochem.* [f. Gr. αἷμ-α blood + Erythrin.] A red respiratory pigment in the blood of certain invertebrates.
1903 *Jrnl. Chem. Soc.* LXXXIV. ii. 741 Haemerythrin, the pink colouring matter in the blood of *Sipunculus* and a few other worms, is contained in the blood corpuscles. 1950 *Sci. News* XV. 103 A..rare red respiratory pigment is hæmerythrin possessed by certain marine animals. 1963 R. P. Dales *Annelids* iii. 70 Amongst other peculiar features they have enucleate corpuscles containing the respiratory pigment haemerythrin, unique in the Annelida, and found elsewhere only in the sipunculids.

hæmiglobin, hem- (hīmi,glōᵘ·bin, hīmai-). *Biochem.* [ad. G. *hämiglobin* (Kiese & Kaeske 1942, in *Biochem. Zeitschr.* CCCXII. 122), f. *hämoglobin* haemoglobin, by alteration.] = Methæmoglobin.
1944 *Chem. Abstr.* XXXVIII. 1537 In expts. dealing with the mechanism of chlorate poisoning, H. and J. [Heubner and Jung] found that hemiglobin (hitherto termed methemoglobin)..served to accelerate its own formation. 1965 *Clin. Chim. Acta* XI. 571 (*heading*) The formation of haemiglobin using nitrites. 1966 *Biol. Abstr.* XLVII. 5186/1 The rate of hemiglobin formation was measured in human blood during drying in air.

hæmo-, hemo-. Add: **hæ:mochromato·sis** (erron. **hæma-**) *Path.* [see Chromato- and -osis] = *bronze diabetes; **hæ:mochro·mogen** [Chromogen], a product obtained from hæmoglobin by hydrolysis; **hæ:moconcentra·tion** (see quot. 1949); **hæ:modia·lysis** *Med.* = *Dialysis 5 b; hence **hæ:modi·alyser,** an artificial kidney; **hæ:modyna·mic** *a.*, of or belonging to hæmodynamics; **hæ·mogram** [-gram], a systematic description of a patient's blood cells; **hæ:mogre·garine** [ad. mod.L. generic name *Hæmogregarina* (B. Danilewsky 1885, in *Archiv für mikroskopische Anatomie* XXIV. 589], a member of a group of coccidian parasites which infest the blood of vertebrates and are transmitted by invertebrates; **hæ:mopoie·sis,** var. Hæmatopoiesis; **hæmopoietic** *a.* (later examples); **hæ:mopoie·tin, -ine** [ad. F. *hémopoïétine* (Carnot & Deflandre 1906, in *Compt. Rend.* CXLIII. 386)], = *Erythropoietin; **hæ:mosi·derin** [a. G. *hämosiderin* (E. Neumann 1888, in *Arch. f. Path. Anat. u. Physiol.* CXI. 27), f. Gr. σίδηρ-ος iron: see -in[1]], a brownish-yellow granular iron-protein substance used to store iron in the body; **hæ:mosidero·sis** [Siderosis], accumulation of hæmosiderin in body tissues; **hæmo·stasis** [cf. Stasis], stoppage of the flow of blood; **hæ·mostat** [cf. -stat], †**a.** (see quot. 1900). *Obs.*; **b.** an instrument for retarding hæmorrhage.
1899 *Brit. Med. Jrnl.* 9 Dec. 1595/1 In the general haemochromatosis associated with cirrhosis of the liver, the pigment is the haemosiderin and has an ochre yellow colour, which gives to the organs..a most remarkable and characteristic appearance. 1907 *Practitioner* Aug. 214 Haemochromatosis is a rare disease; the pigmentation is often, but not invariably, associated with glycosuria and cirrhosis of the liver. 1932 *Sunday Pictorial* 17 Jan. 6/4 The cause of death was hæmachromatosis, an extremely rare disease of metabolism. 1964 L. Martin *Clin. Endocrinol.* (ed. 4) v. 179 In hæmochromatosis the pigmentation is primarily a slaty-grey colour and there is hepatic enlargement with glycosuria. 1885 *Jrnl. Physiol.* VI. 28, I have seen in the lobule of the liver of a pigeon..in one part haemochromogen, in another biliverdin. 1957 *New Biol.* XXIV. 65 Another possible route for haemoglobin loss is the following In the gut there is found a compound known as a haemochromogen, which is related to haemoglobin. *Ibid.*, In the laboratory haemochromogen is a breakdown product of haemoglobin. 1940 *Acta Med. Scand.* CIII. 548 We shall be able to find hyperglobulinemia without any displacement of the ratio of albumin to globulin in those cases where there is a hemo-concentration on account of desiccation. 1947 *Radiology* XLIX. 302/2 These dogs also showed a terminal hemoconcentration. 1949 *New Gould Med. Dict.* 453/1 *Hemoconcentration*, an increase in the concentration of blood cells resulting from the loss of plasma or water from the blood stream; anhydremia. 1964 L. Martin *Clin. Endocrinol.* (ed. 4) v. 170 Shock is a complex syndrome manifested by hæmoconcentration. 1959 Kupfer & Rosenak in *Jrnl. Laboratory & Clin. Med.* Nov. 746 (*title*) A new parallel tube continuous hemodialyzer. 1963 *Lancet* 12 Jan. 82/2 The dialysing area of the particular hæmodialyser employed is given, together with the urea clearance achieved by its use. 1947 *Q. Cumulative Index Medicus* XLII. 1186/1 Attempted therapy of anuria by intraperitoneal hemodialysis. 1962 *Lancet* 19 May 1055/1 Hæmodialysis is now commonly applied in acute renal failure. 1968 Hæmodialysis [see *dialysis 5 b]. 1907 *Practitioner* Aug. 217 Although fully recognising the importance of the diastolic pressure, when working at hæmodynamic problems,..I did not consider it essential. 1961 *Lancet* 12 Aug. 331/1 (*title*) Hæmodynamic effects of guanethidine. 1929 R. B. H. Gradwohl tr. *Schilling's Blood Picture* 17 With the aid of..simple measures the 'hemogram' is constructed; by its brevity and capacity to express many things it constitutes the basis for the practical usage of the blood picture. 1961 *Lancet* 9 Sept. 568/1 Other studies showed a normal hæmogram and urine analysis. 1908 *Practitioner* Feb. 226 (*heading*) Piroplasmosis, hæmogregarines and Leishman-Donovan body. 1961 C. H. Pope *Giant Snakes* (1962) 189 No one knows just what hæmogregarines do to their reptile hosts. This technical name for such parasites derives from the fact that they live in red blood cells. 1900 Dorland *Med. Dict.* 296/1 Hemopoiesis. 1948 *Amer. Jrnl. Med. Sci.* CCXV. 411/1 We have noted that a number of chemical substances stimulated hemopoiesis in persons with Addisonian pernicious anemia...One of these, thymine.., is a pyrimidine base. 1964 D. Nichols in *Oceanogr. & Marine Biol.* II. 398 Most of the vessels are composed of large, loosely-packed connective tissue cells with scattered regions of haemopoiesis. 1947 *Radiology* XLIX. 291/2 These studies indicate a sensitivity of the hemopoietic system of man. 1956 *Nature* 10 Mar. 452/1 Adult mice irradiated with an expectedly lethal dose of X-rays could recover if grafted or injected with hæmopoietic tissue from a normal mouse. 1926 *Chem. Abstr.* XX. 1839 Hemopoëtin, a substance which appears in the serum of organisms exposed to reduced pressure and has a marked stimulating effect on the bone marrow resulting in increased regenerative capacity of the blood. 1932 Wilkinson & Klein in *Lancet* 2 Apr. 721/1 Hæmopoietin may be identified with or allied to Castle's 'intrinsic factor' of normal human gastric juice. *Ibid.* 721/2 This 'enzyme' acting on the proteins in a normal diet may produce a substance which is stored as the active principle in liver until it is required for haemopoietic regeneration... It is proposed temporarily to term this substance in hog's stomach 'haemopoietin'. 1960 *Blood* XVI. 1407 Up to the present the only reproducible sources of hemopoietine are plasma and urine of animals made severely hypoxic. 1970 Krantz & Jacobson *Erythropoietin* i. 4 The plasma factor that increased erythropoiesis had been termed hemopoietine by Carnot and Deflandre; however, as work proceeded, it appeared to be involved exclusively in red cell production.., and erythropoietin became the adopted name. 1896 F. W. Mott in T. C. Allbutt *Syst. Med.* I. i. 196 When blood corpuscles undergo destruction, as in large extravasations of blood, two substances may be formed—(*a*) *Haemosiderin* and (*b*) *Haematoidin*... Haemosiderin may also be found in the renal epithelium. 1964 S. Duke-Elder *Parsons' Dis. Eye* (ed. 14), xvi. 221 A brownish ring, probably due to hæmosiderin, may form in the epithelium encircling the cone (Fleischer's ring). 1972 Balcerzak & Wheby in C. E. Mengel et al. *Hematol.* ii. 41 At physiologic levels of tissue iron, slightly more ferretin iron is present than hemosiderin iron. Hemosiderin predominates when excess iron develops. 1909 *Cent. Dict.* Suppl., Hemosiderosis. 1942 M. M. Wintrobe *Clin. Hematol.* x. 435 Enlargement of the liver with hemosiderosis has been noted in a number of instances [of acute hemolytic anemia]. 1963 H. Burn *Drugs, Med. & Man* (ed. 2) xvi. 159 Among the Bantu in Africa, who use cooking utensils of iron, some of the iron of the pan gets into the food, so that the intake of iron is very high. The absorption of iron continues, and the amount of iron in the liver and other tissues becomes very large. The condition is known as haemosiderosis. 1971 Leavell & Thorup *Fund. Clin. Hematol.* (ed. 3) v. 149 Post-transfusional hemosiderosis is an important development in some patients with chronic bone marrow failure or hemolytic anemia who require frequent blood transfusions. Usually the iron is stored in the reticuloendothelial cells. 1843 *Maryland Med. & Surg. Jrnl.* III. 265 (*heading*) On hæmostasis, and the physical phenomenon of circulation. 1848 Dunglison *Med. Lex.* (ed. 7) 411/1 Hæmostasis. 1907 *Practitioner* Aug. 302 Simple serum contains all the coagulating ferments necessary for haemostasis. 1914 *Brit. Med. Jrnl.* 4 July 8/2 (*heading*) Note on haemostasis by application of living tissue. *Ibid.*, I found that a muscle haemostasis would resist as much as 60 to 80 mm. Hg blood pressure. 1962 *Lancet* 27 Jan. 177/1 It is interesting to speculate whether the control could be so low that defective hæmostasis would result despite a normal total prothrombin content. 1900 Dorland *Med. Dict.* 296/2 *Hemostat*, a proprietary remedy for nose-bleed, containing tannin, quinin sulphate, lard, and benzoic acid: used externally. 1904 F. P. Foster *Appleton's Med. Dict.* 1033/2 *Hæmostat.* 1. A hæmostatic forceps or other appliance. 1929 F. A. Pottle *Stretchers* (1930) 110 The assistant mops it up with a gauze sponge, discovers the point where the blood vessel is severed, and the surgeon clips it with a haemostat, another variety of pincers with handles like manicure scissors. 1969 Troup & Schwartz in S. I. Schwartz *Princ. Surg.* iii. 106/1 The finger has the advantage of being the least traumatic vascular hemostat.

hæmoglobin. In def. for 'a solid substance.. globulin' read 'a protein which is resolvable into hæm and globin'. Add later examples. Also *attrib.* and *Comb.*
1907 *Yesterday's Shopping* (1969) 510/1 Hæmoglobin Tablets. 1950 *Sci. News* XV. 96 But these lake-dwelling *Daphnia*, if deprived of abundant oxygen in the laboratory, also become pink with newly-formed hæmoglobin in their blood. Thus they have the capacity of hæmoglobin synthesis when stimulated by lack of oxygen, although they do not profit by this gift in nature. *Ibid.* 103 Hæmoglobin is unusual among proteins in having a coloured part of its molecule, a coloured part with characteristic absorption bands in its spectrum which can be measured. 1956 *Nature* 17 Mar. 524/1 There was a drop in the hæmoglobin-level from 81 to 70 per cent when tested by the Sahli method. 1956 *New Biol.* XXI. 55 Sickle-cell haemoglobin is the best known of the abnormal haemoglobins in man, but several other types can be distinguished by electrophoretic and solubility tests. These are known as haemoglobin types C, D, E, G, H, I, J, K, and M. 1957 *Ibid.* XXIV. 61 The mean of the twenty values on the scale gives the 'haemoglobin index' of the population. 1963 R. P. Dales *Annelids* iv. 81 Haemoglobin-containing corpuscles are distributed in the coelomic fluid. 1964 G. H. Haggis et al. *Introd. Molecular Biol.* v. 115 In the red blood cell it is the only structure observed between the haemoglobin-laden cytoplasm and the blood plasma. 1968 H. Harris *Nucleus & Cytoplasm* vi. 118 The failure of high concentrations of actinomycin D to inhibit decisive events in the differentiation of colonial myxamoebae, of pancreatic cells in the mouse embryo, and of haemoglobin-forming cells in the chick embryo. 1968 *Times* 17 May (*heading*) Structure of haemoglobin solved.
Hence **hæmoglobinometry** (later example); **hæ:moglobino·pathy,** any condition in which the quality of the hæmoglobin in the blood is defective.
1961 *B.S.I. News* Dec. 27 (*heading*) Sealed glass cells for photometric hæmoglobinometry. 1957 A. W. Woodruff et al. in *Brit. Med. Jrnl.* 25 May 1235/1 (*heading*) Terminology of the hereditary haemoglobinopathies with haemoglobin variants. *Ibid.* 1235/2 The term haemoglobinopathy should be used to denote a condition in which the production of normal adult haemoglobin..is partly or wholly suppressed and it is partly or wholly replaced by one or more haemoglobin variants. 1962 *Lancet* 12 May 1006/2 As the hæmoglobinopathies grow in importance, a monograph taking stock of what we know of thalassæmia is welcome. 1966 *Ibid.* 31 Dec. 1435/1 The situation with respect to diabetes reminds one of attempts to analyse the hæmoglobinopathies before chemical techniques were available for the identification of hæmoglobins to discriminate between possible genotypes.

hæmolymph. (Later examples.)
1964 O. Kinne in *Oceanogr. & Marine Biol.* II. 302 Autoradiographs of *Asellus aquaticus* indicate that some 20 to 30% of the total body Na is located outside the haemolymph. 1968 H. Harris *Nucleus & Cytoplasm* iv. 79 Some authors believe that the proteins of the salivary secretion are not synthesized in the gland, but are simply extracted from the haemolymph.

hæmolysis, hem- (hīmɒ·lisis). *Med.* [f. Hæmo- + *-lysis.] The dissolution or lysis of red blood cells with the consequent liberation of their hæmoglobin.
1890 F. Taylor *Man. Pract. Med.* 663 The immediate cause of the anæmia is the destruction of red corpuscles in the blood (hæmolysis). 1892 Osler *Princ. Med.* 725 Increased hæmolysis and dissolution of the hæmoglobin in the blood-serum. 1901 *Jrnl. Chem. Soc.* LXXX. ii. 325 Hæmolysis produced by Solanine. 1906 *Practitioner* Nov. 591 The jaundice of the newly born..is dependent upon changes, probably toxic in character, with excessive hæmolysis. 1947 *Radiology* XLIX. 307/2 Increased red

cell hemolysis is indicated by elevated excretion of fecal urobilinogen and urinary bilirubin. **1966** *Lancet* 24 Dec. 1382/1 Pyridium also causes hæmolysis.

Hence **hæmo·lysate**, any preparation obtained from hæmolysed blood; **hæ·molyse, -lyze** *v. trans.*, to lyse (red blood cells); also *intr.* (of red blood cells or a preparation of them) to undergo hæmolysis; **hæ·molysed, -lyzed, hæmo·lysing, -lyzing** *ppl. adjs.*; **hæmolysin** (hīmǫ·lisin, hīmǫlai·sin) [see *LYSIN], any substance which causes hæmolysis; **hæmolytic** *a.* (in Dict. s.v. HÆMO-) (examples); **hæmolytically** *adv.*

1893 *Funk's Stand. Dict.*, Hemolytic. **1897** *Allbutt's Syst. Med.* II. 104 Pointing to a hæmolytic as well as a simple hæmorrhage origin for the anæmia. **1900** *Proc. Roy. Soc. Med.* LXVI. 435 Certain blood poisons, viz., the hæmolysins,..exercise a solvent action only on such red blood corpuscles as are able to unite chemically with them. **1901** *Lancet* 14 Dec. 1661/1 Since the discovery of tetanolysin by Ehrlich a series of hæmolysins have been described. **1901** *Trans. Path. Soc. London* XLII. 212 A substance is present in the serum which dissolves or hæmolyses the blood-corpuscles of the rabbit *in vitro*. *Ibid.*, In great every serum that acts hæmolytically on a number of different kinds of erythrocytes possesses a corresponding number of immune bodies and of complements. **1902** *Jrnl. Chem. Soc.* LXXXII. ii. 464 Hæmolysis of Bacillus Megatherium... In cultures of *B. megatherium* a specific lysin occurs which hæmolyses the corpuscles of guinea-pig, monkey, and man. **1903** *Ibid.* LXXXIV. ii. 443 Influence of Cold on the Action of some Hæmolytic Agents. **1908** *Practitioner* Feb. 249 To yield substances which have similar hæmolysing properties to the hæmolytic agent found in tape-worms. **1911** *Jrnl. Amer. Med. Assoc.* 23 Dec. 2050/2 The antireceptor should be used in twice the strength sufficient to hæmolyze the corpuscles in from fifteen to twenty minutes. *Ibid.*, The delay in hemolysis with tuberculous serums is striking in contrast to the promptness with which the controls hemolyze. **1916** *Jrnl. Immunol.* I. 37 The hemolyzed cells do not give up an effective hemolysin. **1920** *Nature* 13 May 347/2 The anti-coagulating and hæmolysing action of sodium nucleate. **1946** *Ibid.* 24 Aug. 269/2 It was found possible to rear first instar bugs to the adult stage by feeding them on defibrinated hæmolysed blood through a mouse skin membrane. **1952** *Q. Jrnl. Exper. Physiol.* XXXVII. 163 The methæmoglobin (MHb) formation which occurs spontaneously in hæmolysates of red blood cells occurs much faster when these have been treated so as to remove the posthæmolytic residue. **1957** *Times* 5 Sept. 15/4 Dr. Coombs, whose laboratory test for the diagnosis of hæmolytic disease of the new-born infant is in worldwide use. **1962** *Lancet* 8 Dec. 1184/2 The hæmolysate of unfractionated whole-blood cells obtained from the same subject was diluted in the same way. **1967** *Jrnl. Gen. Microbiol.* XLVII. 153 Two hæmolysins may be produced by *Escherichia coli*. **1968** *Sci. Jrnl.* Nov. 65/2 The cells are completely disrupted—hæmolysed. *Ibid.* 65/3 The cells will hæmolyse when subsequently exposed to some mild form of stress. **1972** *Science* 2 June 1030/2 After 72 hours, the tissue culture media were removed and assayed for hemolytically active C4 and C2.

hæmophilia. Add: **hæmophiliac** (-fi·liæk) *a.*, affected with hæmophilia; also as *sb.*, a person so affected; **hæmophilic**, also as *sb.*, a hæmophiliac.

1896 *Lancet* 18 Jan. 153/2 An arrest of severe hæmophiliac bleeding from the gums was obtained by an application of calcium phosphate. **1897** *Boston Med. & Surg. Jrnl.* 11 Mar. 227/1 In hemophiliacs, leeching, extraction of the teeth and circumcision are very hazardous operations. **1897** *Lippincott's Med. Dict.* 454/1 *Hemophilic*,... 2. A person affected with hemophilia. **1935** Whitby & Britton *Disorders of Blood* xiv. 272 On Mendelian principles a female may be a true hæmophilic if she is the daughter of a hæmophilia-transmitting woman and a hæmophilic male. **1936** *Discovery* Dec. 388/1 A preparation from egg-white, which reduces the clotting time of blood, provides new hope for hæmophiliacs. **1938** *New Statesman* 2 July 7/2 Between thirty-five and seventy hæmophiliacs are alive in Greater London to-day. **1946** *Nature* 28 Sept. 447/1 We have been able to study the effect, in some hæmophilic patients, of a product containing 82 per cent fibrinogen. **1962** *Lancet* 27 Jan. 194/1 A pharmacist who is a hæmophiliac had noted that by taking bioperidin (flavonoid) he could ward off hæmorrhagic episodes. **1966** Dunlop & Alstead *Textbk. Med. Treatm.* (ed. 10) 496 In centres with specialist facilities, a supply of this plasma specifically for use in hæmophilics serves a useful purpose. **1967** M. M. Wintrobe *Clin. Hematol.* (ed. 6) xviii. 937/1 Karyotype analysis has been carried out in several of the hemophiliac women and only in 2 instances has the karyotype been abnormal.

hæmorrhoid¹, hemorrhoid. Add: Hence **hæmorrhoide·ctomy**, the surgical removal of hæmorrhoids.

1917 V. C. David in *Surg. Clinics Chicago* I. 543 (*title*) Local anesthesia for hemorrhoidectomy. *Ibid.* 552 Infiltration anesthesia with novocain offers a safe and technically simple method for hemorrhoidectomy. **1949** M. Lowry *Let.* Oct. (1967) 182 I'm glad you're better now after your operation—the condition is much better for a hæmorrhoidectomy with a Catholic institution sounds sadistic. **1967** S. Taylor et al. *Short Textbk. Surg.* xxiii. 319 In third degree piles..., hæmorrhoidectomy is indicated.

hæmostatic. A. *adj.* (Earlier example.)

1834 *Lancet* 8 Mar. 889/2, I have resolved upon giving such a view of it [*sc.* torsion of arteries] as will connect it with the other hæmostatic processes now in use in surgery.

haenapod, var. *HANEPOOT.

haeremai (hā·ərəmāi, *anglicized* hoiᵊ·rēmai). *New Zealand.* Also **haere mai, haire mai, horomai.** [Maori, lit. = come hither.] A Maori term of welcome.

1769 J. Banks *Jrnl.* 12 Nov. (1962) I. 432 As soon as they [*sc.* the Maoris] came near enough they wav'd and calld *horomai* and set down in the bushes near the beach (a sure mark of their good intentions). **1832** H. Williams *Jrnl.* in H. Carleton *Life* (1874) I. 112 They were very glad to see us, and gave us the usual welcome, '*haere mai! haere mai!*' **1845** E. J. Wakefield *Adv. N.Z.* I. 249 No shouts of *haeremai*, so universal a welcome to the stranger, were to be heard. **1883** F. S. Renwick *Betrayed* 34 (Morris), Haire mai ho! 'tis the welcome song Rings far on the summer air. **1938** R. D. Finlayson *Brown Man's Burden* 9 As the visitors splashed across the ford, that time-honoured cry of welcome broke from every throat. 'Haere mai!' **1943** N. Marsh *Colour Scheme* iii. 55 The Maori people..would like me to greet him with a cordial *haeremai*.

‖ **haff** (hæf). [G., f. (M)LG. *haf* sea, corresp. to ON. *haf*, OE. *hæf* sea.] A shallow fresh-water lagoon found at a river mouth, esp. one of those on the Baltic coast.

1859 S. O. Beeton *Dict. Univ. Information* 582/2 *Haff, haf*, an extensive bay or gulf of Pomerania... 2. Of East Prussia... 3. A very extensive bay of the Baltic. **1875** *Encycl. Brit.* III. 294/1 The shore of the Baltic is generally low. Along the southern coast it is for the most part sandy... Where streams come down, there are often fresh-water lakes termed *haffs*, which are separated from the sea by narrow spits called *nehrungs*. **1879** *Ibid.* X. 447/1 The 'haffs' or lagoons on the Baltic. **1933** *Discovery* June 203/1 He writes from experience of fishing in general—on Exmoor, in a Prussian haff, in Cornwall, on Scottish waters, and in Ireland.

haffle (hæ·f'l), *v. dial.* [cf. Du. (local) *haffelen* (of a suckling baby) to pull and push at the breast; (of women) to talk a lot, argue.] *intr.* To speak in a hesitant or stammering manner; to prevaricate, shilly-shally. Cf. *CAFFLE *v.*

1790 Grose *Provincial Gloss.* (ed. 2) *Haffle*, to prevaricate. **1825** J. T. Brockett *Gloss. N. Country Words* 88 *Haffle*, to waver, to speak unintelligibly. **1869** R. B. Peacock *Gloss. Lonsdale* 39/1 *Haffle*, to stammer, to prevaricate, to falter. **1902** in *E.D.D.* s.v., [Nottingham] The doctor, he haffled and caffled, he didn't rightly know what war wrong wi' her himself. **1913** [see *CAFFLE *v.*]. **1913** D. H. Lawrence *Let.* 3 Mar. (1962) I. 191 The Nottingham people are still haffling and caffling about the children.

hafiz (hā·fiz). Also 7 **hafis**, 9 **hafeez.** [Pers., f. Arab. *ḥāfiẓ* watch, guard.] A Muslim who knows the Koran by heart.

1662 J. Davies tr. *Olearius' Voy. Ambass.* 314 [The] Turbants..of their Priests, and particularly, of the Hafis, are white. **1819** T. Hope *Anastasius* (1820) I. x. 192 Who, to obtain the epithet of hafeez, had learnt his whole koran by heart unto the last stop. **1927** *Blackw. Mag.* May 574/2 A hafiz chanted the Koran for the rest of her soul. **1965** *Encycl. Islam* (new ed.) III. 55/2 Ḥāfiẓ..no doubt in youth..earned the right to use the title *ḥāfiẓ* (Kur'ān-memorizer), which became his pen-name.

hafnium (hæ·fniəm). *Chem.* [f. *Hafnia* (f. Da. *Havn* harbour (see Haven *sb.*), orig. name of Copenhagen (Da. *København*)), mod.L. name of Copenhagen: see -ium.] A metallic element with a silver lustre usually found associated with zirconium, which it closely resembles chemically, and used in nuclear reactor control rods. Symbol Hf; atomic number 72. Earlier called *CELTIUM.

1923 Coster & Hevesy in *Nature* 20 Jan. 79/2 For the new element we propose the name Hafnium (Hafniae = Copenhagen). **1955** *Sci. Amer.* Oct. 35 In its ores zirconium is invariably accompanied by hafnium, which absorbs neutrons all too readily. **1957** *Bull. Amer. Physical Soc.* II. 269/1 Hafnium's thermionic efficiency in terms of grams evaporated per unit electron emission is slightly greater than that of Th metal. **1967** W. H. Kohl *Handbk. Materials & Techniques for Vacuum Devices* xii. 329/2 The high neutron absorption of hafnium, its excellent corrosion resistance in high-temperature water, and its adequate strength at reactor operation temperatures make this metal suited as a control material.

hafod (ha·vǭd). [W., = summer dwelling.] In Wales = Shiel.

[**1781** T. Pennant *Tour in Wales* II. 161 This mountainous tract scarcely yields any corn. Its produce is cattle and sheep, which, during summer, keep very high in the mountains, followed by their owners,..who reside.. in *Havodtys*, or summer dairy-houses.] **1952** *Proc. Prehist. Soc.* XVIII. 74 The evidence therefore points to summer pastures, the older pound being used as a corral for cattle, while the lowland farmer set up his hafod within or just outside the wall. **1958** *Rep. R. Comm. Common Land* 274 in *Parl. Papers* 1957–8 (Cmnd. 462) X. 1 *Hafod*,..the upland pastures in Wales to which transhumance took place in the summer months. **1963** *Times* 19 Apr. 6/1 Reminiscent of the earliest dwellings in Wales, the 'hafod' or summer home in the mountains, these will be available this season at a peak rental of 12 guineas a week.

haft, *sb.*¹ Add: **1. c.** *Bot.* Of an iris: the narrow part, or claw, at the base of the petal.

1924 W. R. Dykes *Handbk. Garden Irises* i. 2 An Iris flower consists usually of three outer segments called falls and of three inner segments called standards... The lower part of both the falls and the standards is usually called the haft. **1948** G. Anley *Irises* 113 *Haft*, the narrowed portion at the base of a perianth segment.

haft, *sb.*² See also *HEFT *sb.*²

hafta (hæ·ftə), repr. colloq. pronunc. of *have to* (see Have *v.* 7 c). *Chiefly N. Amer.*

1941 B. Schulberg *What makes Sammy Run?* v. 80 That's a boner.. I'll hafta remember that one. **1948** A. Kober *Parm Me* 58, I see... You don't hafta explain. **1952** E. Wilson *Equations of Love* 275 'I don't hafta marry the Aldridge girls,' he said urgently. **1968** N. Benchley *Welcome to Xanadu* vi. 133 You'll hafta carry him.

haftara, haftarot(h): see *HAPHTARAH.

hag, *sb.*¹ Add: **6.** *hag-like*, also *adj.*

1824 J. Morier *Adv. Hajji Baba* xiii. 128 There was also..an old woman of a hag-like and decrepit appearance.

Haganah (hăgānā·). Also **Hagana.** [ad. Heb. *hāgannāh* defence.] A group of Jewish settlers in Palestine who, as an underground defence force, played a leading part in the creation of the state of Israel in 1948.

1923 *Daily Mail* 29 Jan. 6 He knows more about the 'Haganah', the Zionist Self-Defence force, than the authorities in Palestine like. **1949** Koestler *Promise & Fulfilment* 96 Specially picked anti-terrorist Haganah squads... **1950** *Guardian* 26 Aug. 5/3 The Hagana was transformed from an underground guerilla force into a regular army. **1973** *Jewish Chron.* 19 Jan. 12/4 The Haganah [Jewish self-defence] movement..ultimately became Israel's army.

hagden, hagdown. Add: Also **hagdel, hagdon.**

1832 W. D. Williamson *Hist. State Maine* I. 150 The Hagdel [is] of a dark brown colour, about as large as a Murr, though its feathers are longer. **1954** Fisher & Lockley *Sea-Birds* i. 26 The Tristan great shearwater..also probably reaches its greatest abundance on the North American coast,..where it is known as the 'hagdon'. **1959** Bannerman *Birds Brit. Isles* VIII. 141 Wynne-Edwards reminds us that it [*sc.* the sooty shearwater] is known to the fishermen as the hagdown or black hagdon.

hagfish, hag-fish. (In Dict. s.v. Hag *sb.*¹ 5.) Add later examples.

1931 J. R. Norman *Hist. Fishes* iii. 47 The related Hagfish (*Myxine*) possesses still more singular habits, and bores right into the fishes it attacks. **1967** *Oceanogr. & Marine Biol.* V. 291 Aqua-lung diving is beginning to provide exact data about the natural habitats of such animals, for example, the hagfish, *Myxine glutinosa*. **1968** *Times* 19 Dec. 4/8 Lampreys, like hagfish, are surviving members of the jawless fishes, the first group of vertebrates to evolve.

haggadah. Add: **2.** The Jewish ritual for the first two nights of the Passover. Also the book containing the text of the service.

1733 tr. *B. Picart's Ceremonies & Relig. Customs* I. 61 Then each of them holding a Glass of Wine in his Hand, says the Hagada. [**1887** Jacobs & Wolf *Catal. Anglo-Jew. Hist. Exhib.* 194 Haggadah Pesach, or Liturgy of the Passover.] **1891** M. Friedlander *Jewish Relig.* II. iv. 379 The first two evenings of Passover are called '*seder*-evenings', and the book which contains this Service is generally called *Haggadah*. **1896** W. H. Greenburg *Haggadah* 6 Upon the first cup one says the benediction... Upon the second cup one recites the Haggada. **1904** *Daily Chron.* 30 Mar. 7/5 Perhaps the whole genius of the celebration of the Passover may be summed up in the words of the Haggadah: 'In every generation each Israelite shall behold himself as though he had been delivered from Egypt.' **1904** *Jewish Encycl.* VI. 142/2 The opinion of Friedmann, that special books containing the Passover service existed in Talmudic times, is based on a judgment of Raba in favor of a man who claimed a Haggadah..from an estate under the plea that he had lent it to the deceased. **1922** Joyce *Ulysses* 208 An ancient haggadah book. **1972** *Publishers Weekly* 7 Feb. 16 (Advt.), We have created what we feel is the most unusual Passover Haggadah... The Haggadah with both English and Hebrew text.

haggadically (hægədæ·dikăli), *adv.* [f. Haggadical *a.* + -ly².] As in the haggadah.

1920 Oesterley & Box *Lit. Rabbinical Judaism* 78 The Scriptural lesson..is haggadically developed.

haggis. Add: **1. d.** A mixture, hodge-podge; a mess.

1899 *Daily News* 13 Sept. 7/6 They cheerfully go through the curious haggis of social and philanthropic duties served up to them each week. **1928** W. J. Archbold (*title*) Bengal haggis. **1929** H. Marwick *Orkney Norn* 66/1 He'll just mak a haggis o' the job.

haggy, *a.*¹ [f. Hag *sb.*¹ + -y¹.] Of or pertaining to a hag.

The sense of quot. 1654 is uncertain: it may belong to Hag *sb.*¹ 1 or 2.

1654 M. Stevenson *Occasions Off-spring* 83 Didst

thou devise This haggy look, to be thought weather wise? **1964** S. Bellow *Herzog* (1965) 159 That bitch, Madeleine, whose face looks either beautiful or haggy.

haggy, *a.*² Chiefly *Sc.* [f. Hag *sb.*⁴ + -y¹.] Boggy and full of holes.

1794 *Scots Mag.* Oct. 624/1 The night was neither warm not [*sic*] dry, The road was rough and haggy. **1881** D. Thomson *Musings among Heather* 62 He thocht he had yet tae cross, A haggy, benty, splashy moss. **1959** D. D. C. P. Mould *Peter's Boat* vii. 113 This country of bare peat cut with haggy trenches.

haham (hā·hăm). Also **hakam**; *Yiddish* **chochem** (khǭ·khĕm), **cacham, chacham, -em.** [ad. Heb. *ḥăkām* wise, wise man.] One learned in Jewish law; a wise man, savant; *spec.* a Jewish rabbi among Sephardic Jews.

1676 L. Addison *Present State of Jews* (ed. 2) xxvi. 216 In the first rank march the Chachams or Priests. **1733** tr. *B. Picart's Ceremonies & Relig. Customs* I. 46 A Man who hath made the Oral Law his principal Study, he is looked upon by the Generality amongst them as a Doctor, and is therefore called Cacham, or Wise Man. **1892** I. Zangwill *Childr. Ghetto* II. xix. 103 The *Gemorah* says ve muz be vise, *chocham*. **1894** —— *King of Schnorrers* 106 The Haham himself, the Sage or Chief Rabbi of the [Sephardic] congregation. **1901** *Daily Chron.* 23 Nov. 3/2 The vice-presidents include...Mrs. Gaster, wife of the Haham—the spiritual head of the Spanish and Portuguese [Jewish] congregation. **1960** *Jewish Chron.* 8 Apr. 16/1 The Haham or such other person as may for the time being constitute the Ecclesiastical Authority of the Spanish and Portuguese Jews of Great Britain. **1967** D. T. Kauffman *Dict. Relig. Terms* 221/2 *Haham*, wise one. Chief rabbi in sephardim communities, and among Palestinian Jews in talmudic times. **1968** L. Rosten *Joys of Yiddish* 63 A proud young *chachem* told his grandmother that he was going to become a doctor of philosophy. **1973** *Jewish Chron.* 18 May 8/3 The memorial service was organised by the Haham.

hahnium (hā·niəm). *Chem.* [f. the name of Otto *Hahn* (1879–1968), German radio-chemist + -ium.] An artificially produced radioactive element, atomic number 105. Symbol Ha.

1970 A. Ghiorso et al. in *Physical Rev. Let.* XXIV. 1503/1 In honor of the late Otto Hahn we respectfully suggest that this new element be given the name hahnium. **1971** *Nature* 26 Feb. 607/2 The present multi-detector shuttle apparatus is quite a complicated instrument.. and its value as a research tool has been proved by the quality of the nuclear spectroscopic data obtained for.. hahnium.

Haida (hai·dă), *a.* and *sb.* Also **Haidah, Hydah.** [Native word meaning 'people'.] **A.** *adj.* Of or pertaining to a North American Indian people living on the Queen Charlotte Islands, British Columbia, and on Prince of Wales Island, Alaska. **B.** *sb.* **a.** A member of this people; also in collective sense. **b.** The language of this people.

1841 *Jrnl. R. Geogr. Soc.* XI. 219 The Haidah tribes of the Northern Family inhabit Queen Charlotte's Island. *Ibid.*, Since the sea-otter has been destroyed, the Haidahs have become poor. **1862** F. Poole *Diary* 5 Aug. in *Queen Charlotte Islands* (1871) vi. 73 Two Hydah chiefs and four of their women. *Ibid.* 75 He reciprocated by initiating me into the mysteries of the Hydah tongue. **1869** *Mainland Guardian* (New Westminster, B.C.) 30 Oct. 3/2 We bought a large Hydah canoe for $50, and hired ten siwashes (nine Hydahs and one bog-will Indian), for $10 a month. **1890** J. G. Frazer *Golden Bough* I. i. 26 When a Haida Indian wishes to obtain a fair wind, he..shoots a raven. **1914** W. H. Rivers *Kinship & Soc. Organisation* ii. 54 The only people among whom it has been recorded are the Haidahs of Queen Charlotte Island. **1921** E. Sapir *Lang.* iii. 56 Haida, the Indian language spoken in the Queen Charlotte Islands. **1959** E. Tunis *Indians* 136/1 In the north the Tlingit and the Haida were related to the Dené. *Ibid.* 136/2 A Haida house. **1969** *Times* 22 Sept. 14/2 A Haida amulet from the Queen Charlotte Islands has a bilaterally symmetrical design.

haikal (hai·kăl). [Coptic.] The central chapel of three forming the sanctuary of a Coptic church. Also *attrib.* in **haikal screen**, a screen, often elaborately carved or decorated, which separates the haikal from the body of the church.

1884 A. J. Butler *Anc. Coptic Churches* I. i. 28 The screen of the haikal, instead of aligning with that of the side chapels, projects out three or four feet into the choir. **1902** *Encycl. Brit.* XXVII. 238/1 The central division is called the haikal or sanctuary... Haikal screen and choir screen are often sumptuously carved and inlaid. **1935** D. Attwater *Catholic Eastern Churches* vi. 139 A church of the Coptic rite has a distinctive arrangement... Within the triple-domed sanctuary (*haikal*) are three altars... On the *haikal*-screen are a few pictures. **1961** O. Meinardus *Monks & Monasteries Egyptian Deserts* 302 Part of the haikal screen of the old church can still be seen.

‖ **haiku** (hai·ku). Also **haikai, hokku.** [Jap.] A form of Japanese verse, developed in the mid-16th century, usually consisting of 17 syllables and originally of jesting character; an English imitation of one.

The *hokku* was originally the opening hemistich of a linked series of *haiku* poems, but is now synonymous with *haiku* and *haikai*. An earlier meaning of *haikai*, an abbreviation of the phr. *haikai no renga* ('jesting linked-verse'), was a succession of *haiku* linked together to form one poem.

1899 W. G. Aston *Hist. Jap. Lit.* iv. 289 In the sixteenth century a kind of poem known as Haikai, which consists of seventeen syllables only, made its appearance. **1899** *Trans. Asiatic Soc. Japan* XXVII. iv. p. xiv, The *hokku* must be an exceedingly compact bit of word and thought skill to be worth anything—as literature. **1902** *Ibid.* XXX. ii. 243 The poets of Japan have produced thousands of these microscopic compositions... Their native name is *Hokku* (also *Haiku* and *Haikai*), which, in default of a better equivalent, I venture to translate by 'Epigram', using that term..as denoting any little piece of verse that expresses a delicate or ingenious thought. **1904** *Westm. Gaz.* 19 Apr. 10/1 The perfect haikai is a Lilliputian lyric of but three unrhymed lines of five, seven, and five syllables respectively—seventeen in all—in which is deftly caught a thought-flash or swift impression... An example..is the following: The west wind whispered And touched the eyelids of Spring: Her eyes, Primroses. **1957** C. Brooke-Rose *Lang. Love* 47 Her translations of *haiku* were elegant. **1969** *Radio Times* 15 May 9/1 A sequence of twenty-one sonnets and two haiku on the first American landing in Japan in the mid-nineteenth century.

Haileybury (hēⁱ·libəri). The name of a school (Haileybury College) in Hertfordshire, orig. owned by the East India Company, used to designate the system of providing civil servants, or the civil servants themselves, for service in India.

1864 in F. C. Danvers et al. *Mem. Old Haileybury Coll.* (1894) 95, I trust the new men will be found to furnish persons qualified to sustain the character of the Service.. [and] also worthily to fill those high posts of trust.. which we now see so happily filled by Haileybury civilians of the old school. **1902** *Encycl. Brit.* XXIX. 451/2 Towards the latter years of the 19th century the last of the old Haileybury civilians, who entered the service as nominees of the East India Company's directors under the system abolished in 1857, were leaving India. **1931** L. S. S. O'Malley *Ind. Civil Service* 241 A system of pass examinations, such as the Haileybury entrance examination. **1931** *Times Lit. Suppl.* 18 June 474/3 The modern Civilian is the descendant of the Haileybury students of the early nineteenth century. Whatever the merits or demerits of the Haileybury system, it at least 'led to a tradition of service handed down from generation to generation'.

hair, *sb.* Add: **8. b.** *in one's hair*: (*c*) being a nuisance or encumbrance, in one's way; usu. with *get* and *have*; so *out of one's hair*: out of one's way, not encumbering (see sense 8 r below). orig. *U.S.*

1851 *Oregon Statesman* (Oregon City) 30 Sept. 1/2, I shall depend on your honor..that you won't tell on me, cause if you did, I should have Hetty Gawkins in my hair in no time. **1880** 'Mark Twain' *Tramp Abroad* I. xx. 193 What you learn here, you've got to know..or else you'll have one of these..spectacled..old professors in your hair. **1935** S. Lewis *It can't happen Here* xiii. 123 Maybe there'll be a few Communist cells around here now, when Fascism begins to get into people's hair. **1936** 'J. Tey' *Shilling for Candles* x. 115 She got in my hair until I couldn't bear it another day. **1945** M. Lowry *Let.* (1967) 49 We had them in our hair all summer. **1951** C. Fry *Sleep of Prisoners* 4 You know what Absalom Said to the tree? 'You're getting in my hair.' **1957** R. Watson-Watt *Three Steps to Victory* 255 His endurance of a bunch of untidy civilians constantly 'in his hair'.

l. *to put up one's hair* (examples); *let down her hair*, also *fig.*; and of both men and women, *to let* (*take*) *one's* (*back*) *hair down*, to throw off reserve, to become confidential.

1662 [see Put *v.*¹ 53 a]. **1850** G. H. Lewes in *Leader* 7 Dec. 882/3, I am well aware that a little ranting and 'letting down the back hair' would have 'told' upon the audience with more noisy effect. **1921** W. de la Mare *Mem. Midget* iii. 15 On my seventeenth birthday I put up my hair, and was confirmed. **1925** N. Coward *Vortex* ii. 66 Helen and I have just had a grand heart-to-heart talk; we've undone our back hair. **1933** Wodehouse *Heavy Weather* vii. 116 You needn't be coy, Beach... No reporters present. We can take our hair down and tell each other our right names. **1951** Auden *Nones* (1952) 31 To let their hair down and be frank about The world. **1959** *Listener* 15 Oct. 608/1 Mr. Fredric Warburg has reminded us of this in a volume of autobiography..in which he lets down his hair. **1967** *Guardian* 3 Jan. 2/7 Lively young thing, I recall—but she'll have put her hair up by now. **1967** C. Fremlin *Prisoner's Base* ix. 67 After you'd gone, Mother—he really let his back hair down. I was right, you know—he *has* been in prison. **1967** B. Wootton *In World I never Made* i. 36 Before it became customary for women of all ages to wear their hair short, one of the marks of entering upon adult status was to put one's hair 'up'.

p. *to get* (a person) *by the short hairs* (formerly *to get* (a person, etc.) *where the hair is short*): to have complete control over.

1872 G. P. Burnham *Mem. U.S. Secret Service* 207 You've got me where the ha'r is short! What a cursed fool I have been. **1880** 'Mark Twain' *Tramp Abroad* I. xx. 184, I had to tackle this miserable language... I've got it where the hair 's short, I think. **1888** Kipling *Wee Willie Winkie* 67 Then they'll rush in, and then we've got 'em by the short hairs! **1928** *Blackw. Mag.* Feb. 150/1 Those Chinhwan really did seem to have got the rest of the world by the short hairs. **1930** Sayers & Eustace *Docts. in Case* i. 25 She's evidently got her husband by the short hairs.

q. *to make one's hair curl*: see *CURL *v.*¹ 4.

r. *out of one's hair*: opp. *in one's hair* (sense *8 b (*c*)).

1902 Kipling in *Sat. Even. Post* 6 Dec. 2/3 Get out o' my back-hair! **1949** 'J. Tey' *Brat Farrar* x. 81 They wouldn't bother to look for him. They would be too relieved to have him out of their hair. **1959** J. Masters *Fandango Rock* 173 He wouldn't want to interfere with her big moment, and he'd even managed to keep Peggy out of her hair. **1967** *Boston Sunday Globe* 23 Apr. 18/2 Two vice presidents of the First Pennsylvania Banking and Trust Co., the city's largest and most respected, said the bank paid Karafin and an associate $12,000 a year 'to keep him out of our hair'. **1971** Wodehouse *Much Obliged, Jeeves* xvi. 177 He wanted to get Florence out of his hair without actually telling her to look elsewhere for a mate.

s. *to lose one's hair* (or *to get one's hair off*): to lose one's temper.

c **1920** D. H. Lawrence *Phoenix II* (1968) 120 'Nay—nay,' said Lewis testily. 'Don't get your hair off, Mrs. Goddard.' **1931** T. R. G. Lyell *Slang* 356 *To lose one's hair*, to lose one's temper. 'Last night Jones quite lost his hair and made an awful fool of himself.' **1938** E. Bowen *Death of Heart* iii. ii. 343 *This* is what one gets for being so nicely nonchalant, for saving people's faces, for not losing one's hair.

9. a. *hair-chain, -combing, craft, crêpe, -fashion, fetishism, fetishist, -shaft, -work* (examples).

1895 *Montgomery Ward Catal.* 158 Hair Chains made to order. Send us the hair and we will braid. **1907** N. Munro *Daft Days* xxx, The lockets are large and strong, and hair-chains much abound. **1940** G. Greene *Power & Glory* ii. iv. 178 There was a hair-slide,..and a ball of hair-combings. **1962** *John o' London's* 4 Jan. 14/1 The haircraft women..used to tour the Swedish countryside..selling their products. **1957** V. J. Kehoe *Technique Film & T.V. Make-Up* xv. 203 *Hair crepe*..may be human hair (Caucasian, Chinese or Indian), yak or a combination of any or all of these types. **1944** Koestler in *Horizon* Mar. 162 There are..certain typical attitudes to life including clothing, hair-fashion, drink and food. **1951** C. Berg *Unconscious Significance of Hair* vii. 65 The universality of hair fetishism may be brought into relief by this short instance of its negative aspect: A young woman patient of mine, who had become completely bald,..had an indescribable horror of her predicament being seen or.. suspected by anyone. *Ibid.* 61 The hair fetishist loves the women's hair but frequently has the impulse to despoil or 'castrate' it. **1954** Koestler *Invisible Writing* xxiv. 284 The hair-fetishists who loiter in tube-stations with scissors in their pockets. **1906** *Practitioner* Nov. 692 Complete removal..of the hair-shaft, together with the root-sheath or papilla. **1924** *Chambers's Jrnl.* 668/2 The process flattens the almost spherical hairshaft and causes it to lean inwards. **1790** *Columbian Centinel* 13 Oct. 36/4 The Artists' ability in Painting and Hair-Work may be seen. **1959** *Times* 7 Mar. 9/4 Hairwork jewelry was already popular in the late seventeenth century.

b. *hair appointment, -clasp, -clip, conditioner, -cream, -dye* (examples), *-grip, lotion, -oil* (earlier examples), *-ornament, preparation, -slide* (Slide *sb.* 6), *-spray, -tonic, -wash* (earlier and later examples).

1938 D. du Maurier *Rebecca* xxiv. 402 Mrs. de Winter had a hair appointment from twelve until one thirty. **1894** A. M. Earle *Costume Colonial Times* 121 *Hair-clasps.* These ornaments for the hair—clasps to hold up the braided back hair—were advertised for sale in the New York newspapers and in the *Connecticut Courant* of January, 1791, and were worn until a simpler form of hair-dressing appeared about the year 1800. **1957** J. Frame *Owls do Cry* 55 Hairclips have been taken from them. **1951** *Catal. Exhibits, S. Bank Exhib., Festiv. Brit.* 63/2 Hair Conditioner. **1926–7** *Army & Navy Stores Catal.* 491/1 Hair Cream, for fixing the hair. **1843** *Ainsworth's Mag.* III. 554 Invent a new hair-dye expressly to accommodate his wife. **1933** W. S. Maugham *Sheppey* i. 20, I don't believe there's another man in the business could 'ave sold Mr Bolton a bottle of 'air-dye. **1896** *Woman's Life* III. 462/2 Hair-Bow (Fitted with New Safety Hair-Grip). **1938** 'J. Bell' *Port of London Murders* vi. 91 The jug..contained..half a bootlace and two rusty hair grips. **1955** *Sci. News Let.* 5 Mar. 150/3 Hair-grips and kirbi-grips are known in America as bobby pins. **1906** T. D. Lister *Chavasse's Advice to Mother* (ed. 16) iii. 370 Avoid grease, pomatum, hair lotions, and all abominations of that kind. **1962** N. Marsh *Hand in Glove* vii. 221 Mr. Period's bedroom smelt of hair lotion. **1810** E. Weeton *Let.* 25 Feb. in *Jrnl. of Governess* (1969) I. 233 A small phial of hair oil. **1853** Mrs. Gaskell *Cranford* xii. 174 The delusive lady was off upon..the merits of cosmetics and hair oils in general. **1895** *Montgomery Ward Catal.* 183 Real Tortoise Shell Hair Ornaments. **1967** H. Porter in *Coast to Coast 1965–6* 178 They were not discussing the weather or hair-ornaments. **1897** *Sears, Roebuck Catal.* 779/5 Hair Preparation. **1909** in A. Adburgham *Shops & Shopping* (1964) xiii. 273 A handsome sales-room where are sold..Hair Preparations, ..Hair nets, etc. **1895** *Army & Navy Co-op. Soc. Price List* 15 Sept. (Index), Hair Slides. **1927** *Glasgow Herald* 6 Oct. 11 Her hair-slide was found some distance from the body. **1968** J. Ironside *Fashion Alphabet* 166 A hair slide..is a clip for keeping the hair in place. **1959** *Sears, Roebuck Catal.* Spring & Summer 280/4 Glow Hair Spray. **1966** *Vogue* Nov. 81 Creamy Skin Perfume..and a hair spray. **1967** W. Pine *Protectors* i. 10 He smelt the scent of her hair-spray. **1895** *Montgomery Ward Catal.* Index, Hair Tonic. **1897** Hair tonic [see Tonic *sb.* 1]. **1938** Auden & Isherwood *On Frontier* i. i. 28 Surely he's the man who does the hair-tonic advertisements? **1869** D. G. Rossetti *Let.* (1965) II. 707 Certainly a hair-wash would be the unkindest cut of all to bring against

the Absalom of modern poetry. **1938** H. Nicholson *Let.* 18 May (1966) 342 My hairwash comes from Floris.

c. *hair-clipper, -curler* (examples), *-cutter* (later example), *-dryer, -remover, -straightener, -waver*; *hair-colouring, -conditioning, -cutting* (earlier examples), *-doing, -drying, -lifting, -straightening* vbl. sbs. and/or ppl. adjs.

1895 *Montgomery Ward Catal.* 444/1 The very best hair clipper in the market. **1930** *Daily Express* 6 Nov. 19/3 A display of the latest type of electrical hair-clippers. **1972** G. Durrell *Catch me a Colobus* iii. 58 The next thing was carefully to shave the area... This was done with an electric hair-clipper. **1959** *Punch* 3 June 752/1 Hair-colouring (modern usage for hair-dyeing) has become part of a woman's normal routine. **1966** J. S. Cox *Illustr. Dict. Hairdressing* 70/2 *Hair-conditioning*, external treatment designed to improve the condition of the hair by means of lotions, creams, massage and the application of steam to the head and hair. **1753** in E. Singleton *Social N.Y. under Georges* (1902) 176 Hair-curler and peruke-maker from London. **1872** *Rep. Comm. Patents 1870* (U.S.) II. 779/1 Hair-Curler... [A] combination, with a curling-iron tube [etc.]. **1929** *Bookman* May 270/1 A woman's steel hair curlers. **1936** *Discovery* Aug. 250/2 A long ivory rod with a pomegranate finial is probably a hair curler. **1889** *Monthly Packet* Christmas 102, I suppose—there—ain't no hair-cutters up in Heaven? **1832** *Chambers's Edin. Jrnl.* I. 60/2 The announcement 'Hair-cutting rooms' in the window. **1850** Dickens *Dav. Copp.* vii. 77 My recollectons of..canings, rulerings, hair-cuttings, rainy Sundays. **1875** C. M. Yonge *My Young Alcides* I. vii. 232 In the midst of my hair-doing.. Viola's running in to me. **1895** *Army & Navy Co-op. Soc. Price List* 15 Sept. 180/1 The Princess Patent Hair Dryer and Burnisher. **1909** *Installation News* III. 7 This Hair-Dryer works..by means of a small..electric fan. **1909** *Westm. Gaz.* 13 Oct. 7/4 One ounce of hair, which she was drawing through the hair-dryer in her hands. **1961** *Times* 26 Apr. 25/4 Domestic appliances such as..hair-driers. **1902** M. Barnes-Grundy *Thames Camp* viii. 159 You dive into the sparkling river..., forgetting all about hair-drying. **1906** *Chambers's Jrnl.* 30 June 495/2 In my lady's room may be found electrically heated curling-irons and an ingenious hair-drying machine. **1889** 'Mark Twain' *Connecticut Yankee* 354, I flung out a hair-lifting, soul-scorching thirteen-jointed insult. **1907** *Yesterday's Shopping* (1969) 532/3 Hair Remover. **1951** M. McLuhan *Mech. Bride* 60/2 Hair removers..are backed by long-standing national advertising campaigns. **1898** *Today* 5 Nov. 18/1 'The Hair Straightener Company manufactures an instrument that will at once remove the curl from the most stubborn hair. ..' It would be a waste of money..to advertise its wares in a climate like ours, where the moisture of the atmosphere does more hair straightening than is conducive to feminine happiness. **1966** J. S. Cox *Illustr. Dict. Hairdressing* 74/2 *Hair-straightener*, (1) A preparation to straighten frizzy or over-curly hair. (2) An implement that straightens frizzy hair. **1966** *B.B.C. Handbk.* 25 Could we broadcast something about a new hair straightening cream? **1892** *Queen* 27 Feb. in L. de Vries *Vict. Advts.* (1968) 42/1 Automatic hair waver and curler.. Price 2s. 6d. **1895** *Army & Navy Co-op Soc. Price List.* 15 Sept. 180/2 Hair Wavers (Patent). **1966** J. S. Cox *Illustr. Dict. Hairdressing* 75/1 *Hair-waver*, (1) An implement such as waving irons by the use of which hair can be waved. (2) Any apparatus such as a permanent waving machine which heats the hair wound on curlers during the permanent waving process. (3) A person who waves hair.

d. *hair-bottomed* adj. **e.** *hair-stripe.*

1818 Keats *Let.* 5 July (1958) I. 319 Hair bottomed chairs. **1920** *Blackw. Mag.* Aug. 161/2 They would, I understand, be described by tailors as 'fine cachemire with hair-stripe suitable for gents' morning wear'.

10. hair bag, (*a*) a bag made of hair or of very thin thread; (*b*) a bag in which human hair is kept; (*c*) (see quot. 1966); **hair-colour**, (*b*) the colour of a person's hair; **hair-cord** (examples); (*b*) a cord made of human hair; **hair crack** *Metallurgy* = *HAIR-LINE 7; **hair-cut, haircut**, (*a*) an act of cutting the hair by a hairdresser; (*b*) the shape or style in which the hair is cut; (*c*) a customer for a hair-cut; **hair-mattress**, a mattress stuffed with hair; **hair-net** (see Net *sb.*¹ 3); so **hair-netted** *a.*; **hair-pencil** (earlier and later examples); **hair-piece**, a length of false hair used to augment the natural hair; **hair-point** *Bot.*, an extension of the nerve at the top of some moss leaves, forming a fine tip; **hair-raising** *a.*, capable of causing the hair to 'stand on end' through fear or excitement; so **hair-raiser**; **hair-raisingly** *adv.*; **hair-restorer** (earlier example); **hair-seal** (earlier examples); **hair-slip**, a place on a green hide where the grain has decayed causing the hair to slip; so **hair-slipped** *a.*, marked with decayed places; **hair-spring**, (*b*) of a trap (perhaps Spring *sb.*³ rather than *sb.*¹); **hair-style**, a particular way of dressing the hair; hence **hair-styling** *vbl. sb.*; **hair-stylist; hair-tidy**, a tidy [Tidy *sb.* c] for hair-combings; **hair-trim** [Trim *sb.* 3 d]; **hair-tuft** (see quots.).

1712 Mortimer *Husb.* II. 2 Haws put in a Hair-bag, and soaked in Water all Winter..will come up the first Year. **1723** J. Nott *Cook's & Confectioner's Dict.* § 136 To make cider..stamp your Apples, press them in a Hair Bag. **1747** H. Glasse *Art of Cookery* vi. 65 Strain it through a coarse Hair-bag..then strain it through a Hair-sieve. **1824** J. Morier *Adv. Hajji Baba* xvii. 188 The different operations of rubbing with the hand, and of the friction with the hair bag. **1911** R. G. Anderson in *4th Rep. Wellcome Tropical Res. Lab.* B. 253 By Blood-brotherhood is meant a mutual coalition... The rite.. consists in incising the other's forehead..drinking the outflow of blood, smearing an adjacent lock of hair in its residue, and cutting this off to keep..in a neatly woven hair bag as a charm. **1966** J. S. Cox *Illustr. Dict. Hairdressing* 69/2 *Hair-bag*,..a bag to hold the queue of a bag-wig. **1885** J. Beddoe *Races of Brit.* xiii. 144 The division of hair-colours..into red, fair, brown, dark, and black. **1906** *Jrnl. Anthropol. Inst.* XXXVI. 325 Such statistics as those..of eye colour, hair colour, as in many anthropological works. **1972** *Woman* 22 Jan. 17 Do you know that British women spend a staggering £10 million a year on changing their hair colour? **1866** in A. Adburgham *Shops & Shopping* (1964) xii. 133, 1 White hair cord dressing jacket. **1899** T. Watts-Dunton *Aylwin* ii. 46 'This is her hair,' he said, taking the hair-cord between his fingers and kissing it. **1920** L. Harmuth *Dict. Textiles* (ed. 2), *Haircord*, English dress muslin made with thick warp cords. **1923** *Weekly Dispatch* 18 Feb. 12 (Advt.), Useful Shirt in White Haircord Voile. **1951** *Good Housek. Home Encycl.* 34/1 For a..hair cord carpet, herring-bone the raw edges on the underside. **1960** *Textile Terms & Defs.* (ed. 4) 76 *Haircord carpet*, a hair carpet produced by weaving over unbladed wires. **1896** *Trans. Inst. Naval Archit.* XXXVII. 215 A 10 in. steel shaft..had shown fine hair cracks on the surface near the propeller. **1925** *Jrnl. Iron & Steel Inst.* CXI. 113 A defect known as *snow-flakes* or *flakes* (America), *hair-cracks* or *hair-lines* (Great Britain), *Flocken* (Germany), and *cassures ligneuses* (France), has received much attention among manufacturers and inspectors of alloy steel forgings. **1959** J. H. Thornley *Foundation Design & Pract.* xiii. 103/2 The test for hair cracks consists merely of cleaning the pile and washing or immersing it in a tinted fluid. **1899** *Westm. Gaz.* 5 Dec. 8/1 The trade in Pretoria was kept very busy for about ten days giving the burghers a commando hair-cut. **1900** *Ibid.* 22 Jan. 2/3 He won mainly on his promise that he would reform the city barber into charging two dollars fifty cents for a hair-cut. **1904** *Daily Chron.* 8 Apr. 4/7 The barbers of Bethlehem, Pa., have raised the price of haircuts from sevenpence to tenpence. **1923** *Glasgow Herald* 10 Feb. 8/8 Commenting upon how few of his customers in recent days had been 'haircuts', he remarked... 'The change of the moon always brings more haircuts out.' **1924** R. Macaulay *Orphan Island* xviii. § 2. 241 They were interrupted by Mr. Albert Edward Smith, who had come for a shave and a hair-cut. **1836** in *Mass. Hist. Soc. Proc.* (1892) 2nd Ser. VII. 276 Mine an upper berth; a hair mattress to lie on. **1863** J. Nash *Brit. Pat.* 2681, I..take an ordinary wool, hair, or other mattress, and fasten it..to the top of the spring frame. **1931** *Times* 16 Mar. 2/7 Box-springs, hair mattresses. **1865** M. Eyre *Lady's Walks* xv. 185 Quilts, mittens, hair-nets, and other articles knitted, of Pyrenean wool. **1873** *Young Englishwoman* Jan. 38/2 Hair nets..may either be worked with coloured silk or, if intended for night wear, with white cotton. **1958** C. Fremlin *Hours before Dawn* ii. 22 The wisps of untidy grey hair protruding from her hairnet. **1950** J. Cannan *Murder Included* ii. 21 Elizabeth Hudson,..high-collared, hair-netted. **1951** E. Coxhead *One Green Bottle* vi. 164 The hair-netted lady at the next table. **1763** *Gentl. Mag.* XXXIII. 83/2 Let the spots be gently rubbed with a hair pencil. **1965** *Listener* 26 Aug. 316/2 Modelling, carving and engraving, prowess with the inked hair-pencil: these complete the specifically aesthetic skills of the Chinese. **1939** *Time* 25 Dec. 2/3 He wears a toupee (hairpiece or divot in Hollywood) for cinema and most public appearances. **1957** V. J. Kehoe *Technique Film & T.V. Make-Up* vi. 78 Incidentally, a wig covers not only the hair line, but the entire area of the hair on the head, while hairpieces are used to supplement the natural hair growth. **1969** *Times* 20 Mar. 27/1 The man sitting next to you may be wearing a hairpiece. **1818** Hooker & Taylor *Muscologia Brit.* 25 *P[olytrichum] juniperinum*... Except in the want of the hair-points to the leaves..we can find no essential difference between this and the preceding species [*sc. P[olytrichum] piliferum*]. **1893** H. G. Jameson *Illustr. Guide Brit. Mosses* 10 The single nerve [of the leaf] may either cease below..or in..the apex, or may run out beyond it so as to be *excurrent*, forming a *mucro*.., *cusp*.., or *hair-point*. **1966** *Oxf. Bk. Flowerless Plants* 74/2 In the form [of *Grimmia apocarpa*] shown here the leaves are tinged reddish-brown and have white 'hair points'. **1897** *Westm. Gaz.* 16 Nov. 3/2 The writer being put on his mettle merely to throw in what an American has felicitously called 'hair-raisers' by the way. **1900** *Daily News* 24 Apr. 7/5 The hair-raising, long, steep descent of Box Hill. **1902** *Daily Chron.* 16 Sept. 3/4 Marvellous yarns of hair-raising perils. **1928** *Daily Express* 17 Aug. 9/1 There were a few hair-raising mishaps, but nobody was hurt. **1957** *Times* 30 Aug. 8/6 A runaway 70-ton Army transporter..careered downhill into the village of Carlton, Notts., to-day at 60 m.p.h. after its brakes had failed.. 'It was a hair-raising experience..' said Driver Lee. **1960** *Times* 15 Feb. 15/2 England had left it hair-raisingly late, but it was enough. **1873** *Young Englishwoman* Aug. 414/1 Helena has heard 'hair restorers' so much condemned. **1824** *Pettigrew Papers Shipping & Commercial List* 31 July (D.A.E.), About 500 hair Seal Skins, ..were sold by auction. **1846** R. B. Sage *Scenes Rocky Mts.* vi. 56 A hair-seal cap and a frock-coat. **1903** L. A. Flemming *Pract. Tanning* 265 Grading and Classification of Green Calf-Skins... Second, regular No. 1. .. Scores are allowed in this grade, but there must be no holes, hair slips or other bad imperfections. *Ibid.*, Third, good No. 2. This term designates those skins that are slightly hair slipped. **1707** Mortimer *Husb.* xi. 244 With the small Stick (gently put into the hole to stop the knot of the Hair-spring,..) place it in the Earth in the Moles passage. **1854** Thoreau *Walden* 37 With consummate skill he has set his trap with a hair springe.. and then..got his own leg into it. **1913** *Vanity Fair* (N.Y.) Oct. 91/1 Mme. Fried is prepared to show all of the latest..Hair Styles. **1944** M. Laski *Love on Supertax* ix. 84 She longed for some new perfume, a new hair-style. **1963** V. Nabokov *Gift* ii. 139 He had a remarkable hair style that was also somehow indecent. **1936** *Harper's Bazaar* Mar. 88/4 They specialise particularly in 'Hair Styling', which of course means designing coiffures to fit the individual, as well as carrying them out. **1960** *Guardian* 19 Apr. 2/5 The Princess had led the world's hair-styling fashion. She had been 'an ambassadress' of British hair-styling. **1935** 'Madame Louise' *Mod. Hair Cutting & Styling* 15 A hair stylist is a hairdresser who has the artistic ability to suggest and create a new hair fashion. **1950** 'P. Quentin' *Follower* i. 6 One of her clients at Maurice's, where she worked as a hair-stylist. **1907** *Yesterday's Shopping* (1969) 1164 Silver-mounted Hair Tidy. **1918** 'K. Mansfield' *Prelude* 11 She.. found nothing except a hair-tidy with a heart painted on it. **1935** *Punch* 15 May 592/1 The whole affair of the rejection of Miss Rinse's beaded hair-tidy from our Institute Exhibition has been most unfortunate. **1960** B. Snook *Eng. Hist. Embroidery* 116 It suddenly became genteel to embroider..hair tidies, pin-cushions, spectacle and comb cases and what-nots. **1957** S. Beckett *All that Fall* 30 Hairtrims and shaves. **1905** E. Phillpotts *Secret Woman* i. ii. 21 The hair-tufts of his eyebrows had been tawny, but they were now turning. **1923** G. A. Gaskell *Dict. Sacred Lang.* 335/2 Hair tuft between the eyebrows of Buddha: An emblem of spiritual truth within the soul.

hair, *v.* Add: **4.** *trans.* To fit hairs to (a violin-bow).

1898 H. R. Haweis *Old Violins* 116 Most violinists prefer to pay a small sum and get their bows haired.

hairbrush. Add: **2.** Used (chiefly *attrib.*) of a kind of hand-grenade.

1916 J. N. Hall *Kitchener's Mob* ix. 132 Ten or a dozen varieties of bombs were in use... The 'hairbrush', the 'lemon bomb', the 'cricket ball', and the 'policeman's truncheon' were the most important of these, all of them so-called because of their resemblance to the articles for which they were named. **1919** J. Masefield *Battle of Somme* 9 Just before they got us we found some hairbrush bombs. **1923** Kipling *Irish Guards in Great War* I. 75 The 'stick' hand-grenade of the hair-brush type. **1925** Fraser & Gibbons *Soldier & Sailor Words* 113 *Hair-brush grenade*, the name for a type of hand-grenade used in the early part of the War, with a handle, the shape of which suggested a lady's hair-brush.

hair-do (hēᵊ·ɹdū). Also **hairdo.** [f. Hair *sb.* + Do *sb.*¹ 2.] **a.** A way or style of dressing the hair (orig. *U.S.*). **b.** A cutting and setting of the hair.

[**1875** J. G. Holland *Sevenoaks* 299 To do the bride's hair and act as the..supervisor of her dress.] **1932** E. Ferber *They brought their Women* (1933) 24 Why don't we wait until I get the new hair-do. **1935** *Mademoiselle* Oct. 42 He shows you how to achieve an evening look with the same hair-do, by drawing up the side curls high above the ears and clipping them with stones. **1941** *Illustrated* 6 Sept. 14/2 (*caption*) Judy tries a new hair-do. **1942** *Chicago Sunday Tribune* 5 July 1. 12/2 Spurts of rain inspired feminine picnickers eager to protect holiday hair-dos. **1946** 'S. Russell' *To Bed with Grand Music* iv. 55 Nail-varnish and perfumes and hair-do's. **1959** *Daily Tel.* 31 Dec. 9/7 The 'hair-do' took an hour and a half. **1971** *Petticoat* 17 July 2/2 Why do girls spend pounds on clothes, makeup and hairdos?

hairdressing. Add: (Later *attrib.* examples.) Also, a dressing (see Dressing *vbl. sb.* 4) for the hair.

1856 W. Ferguson *Amer. by River & Rail* 67, I.. resigned myself into the hands of one of the assistants in 'Phalon's Hair-dressing Saloon'. **1876** J. S. Ingram *Centenn. Exposition* 701 It contained..ladies' hair-dressing rooms. **1892** *York Co. Hist. Rev.* 60 William Schreiber's Fashionable Shaving and Hair-Dressing Parlors. **1907** *Yesterday's Shopping* (1969) 532/1 Paraffin Hair Stimulant... As a hair dressing it is unrivalled. **1908** *Westm. Gaz.* 29 Feb. 3/2 The hair-dressing sketch shows the Greek silhouette..inspired by Greek ideas. **1909** in A. Adburgham *Shops & Shopping* (1964) xxiii. 273 (*heading*) Ladies Hairdressing Courts. **1966** J. S. Cox *Illustr. Dict. Hairdressing* 71/1 *Hair-dressing*, a liquid preparation of a creamy consistency used to anoint the hair of the head to improve its condition, impart a sheen, facilitate its dressing or hold it in position. *Ibid.*, *Hair-dressing academy*, in the 18th cent. in Paris these academies were described as follows: 'The academy for teaching the art of female hair-dressing'.

haire mai, var. *HAEREMAI.

hairily (hēᵊ·rili), *adv.* [f. Hairy *a.* + -ly².] With hair or hairiness.

1925 W. Deeping *Sorrell & Son* xxx, A lone, grim, anthropoid creature, hairily grotesque.

hair-line. Add: **4.** = *hair-cord* (Hair *sb.* 10).

1862 *Catal. Internat. Exhibit., Brit.* II. No. 4104, Claret, drab, grey, and fancy hairlines. **1950** '*Mercury*' *Dict. Textile Terms* 256/1 An imitation hairline fabric, woven from woollen warp and worsted weft.

5. *Typogr.* The thin stroke in a letter form (as distinguished from the stem and the serifs). Also *attrib.*

1896 T. L. De Vinne in J. Moxon *Mech. Exerc. Printing* 415 No defined width is made for the thin-stroke, which is now called the hair-line. **1932** *Paper & Print* Dec. 326/2 Finely cut serifs, not the hair lines of Bodoni, but cut to a point, are characteristic of some of the latest types. **1970** W. P. Jaspert et al. *Encycl. Type Faces* (ed. 4) p. x, All book types show some variation of thick and thin; in the fifteenth century it was slight, and gradually became more pronounced, until it reached the extreme in the nineteenth century when they became called hair lines. **1972** P. Gaskell *New Introd. Bibliogr.* 29 Didot's

first neo-classic type did not show marked contrast, but later developments of the form, by Didot himself and by Bodoni in Italy, resulted by 1800 in faces of great contrast combined with vertical stress and unbracketed, hair-line serifs.

6. The limit-line of the hair on the head.

1922 S. Lewis *Babbitt* i. 8 A tremendous forehead, arching up two inches beyond the former hair-line. **1936** L. C. Douglas *White Banners* x. 225 The forward curve of the hair-line on the temples. **1959** A. Salkey *Quality of Violence* viii. 128 The rope round his neck was cutting into the hair-line at the back of his head.

7. *Metallurgy.* In full *hair-line crack*: see quot. 1949.

1923 J. A. Jones *Woolwich Res. Dept. Rep. no. 55* 51 The occurrence of hair-line cracks at one end of the forgings suggests that trouble might be experienced. **1925** [see *hair crack* s.v. *HAIR *sb.* 10]. **1949** R. T. Rolfe *Dict. Metallogr.* (ed. 2) 121 *Hair-line cracks* (or *hair c[racks]*), (1) very fine short cracks occurring in the interior of some steel forgings which have not been allowed to cool sufficiently slowly from the working temperature... (2) The term is also applied to any fine cracks which may occur in metals and alloys. **1962** G. R. Bashforth *Manuf. Iron & Steel* IV. ii. 31 When once this [hydrogen-rich] constituent has been formed, its breakdown at low temperatures must result in the formation of hairline cracks, but hairline cracks will not be formed if the breakdown..is brought about at higher temperatures. **1968** *Times* 28 Aug. 21/6 The South of England Electricity Board has had to take its newest.. power station..out of commission because of a discovery.. of hair-line cracks in welding.

8. In various technical uses: see quots.

1935 *Burlington Mag.* Sept. 109/2 The hair-line sprays with delicate gold leaves. **1955** *Sci. Amer.* May 124/1 Its operation resembles that of a slide rule. You first position the hairline of the slider over the caret between the first four balls. **1960** *Times* 25 Oct. 15/5 He has recovered from a hairline fracture of the wrist. **1961** T. Landau *Encycl. Librarianship* (ed. 2) 160/1 *Hair-line rule*, a fine line of varying length used for division of text matter. **1962** *Gloss. Terms Glass Ind.* (*B.S.I.*) § 72 *Hair line*, fine cord on the surface of glass. **1967** *Gloss. Paper/Ink Terms for Letterpress Printing* (*B.S.I.*) 12 *Hair lines*, fine filaments of foreign matter which forms barriers preventing the felting of many of the fibres and often leading to web breaks.

9. *fig.* A very thin dividing line. Also *attrib.*

1940 F. Scott Fitzgerald *Let.* 21 Sept. (1964) 124, I don't know how this job is going... Things depend on such hairlines here. **1959** *New Statesman* 29 Aug. 235/3 It is this hair-line compromise that Dr Stockwood has now challenged..in his statements, though not in his action, at Carshalton. **1962** *Times* 22 Mar. 3/3 It looked a hairline decision indeed.

hairpin. Add: **2.** A jocular word for: a person. Also, a thin person. *slang* (orig. *U.S.*).

1879 R. Grant *Little Tin Gods* 8 That is the kind of a hairpin that he is! **1884** E. W. Nye *Baled Hay* 103 That's the kind of hair pin he is. He never works. **1910** W. M. Raine *B. O'Connor* 214 Collins ain't that kind of a hairpin. **1959** I. & P. Opie *Lore & Lang. Schoolch.* ix. 168 The names [for thin people]..are merely descriptive, as..hairpin, [etc.].

3. In full *hairpin bend, corner*, etc. A sharp bend in a road or course likened to a hairpin in form.

1906 *Daily Chron.* 15 June 6/5 The length and steepness of the rise complicated by a double-hairpin corner. **1906** *Amer. Mag.* LXIII. 176/1 At hairpin turn, perhaps the worst of all, where the course doubles itself, were 500 machines and at least 10,000 people. **1912** *Motor* 23 Apr. Suppl. 2 As we dropped down the gentle 'hairpins' into Voreppe. **1912** G. B. Shaw *Let.* 9 Aug. (1952) 33 A journey of 800 miles includes crawls up endless hairpin zigzags. *Ibid.* 17 Sept. 148 Reversing at impossible hairpin corners. **1914** *Auto-Motor Jrnl.* 410/2 It is not unlike the higher part of Birdlip Hill, starting with a long and steep gradient and ending with a sharp hair-pin bend just below the summit. **1923** *Motor Cycle* 13 Sept. 373/1 There is only one acute hair-pin bend, but there are several corners. *Ibid.* 374/1 A rather tricky and loose-surfaced left-hand hairpin was the principal feature of Chinnor Hill. **1957** P. Kemp *Mine were of Trouble* iii. 36 Its sudden ascents and declivities, its blind curves and hairpin bends flanked by unguarded precipices. **1971** D. O'Connor *Eye of Eagle* iii. 20, I..was driving up the hairpins that mark the beginning of the Simplon Pass.

4. Other *attrib.* uses.

1887 Bury & Hillier *Cycling* xiii. 344 In many tangent wheels very fine spokes are used, and their most delicate point is at the bend in the hub, where the double spoke is bent 'hair-pin' fashion. **1895** *Montgomery Ward Catal.* 195 Fancy Chased Hairpin Box. **1906** *Westm. Gaz.* 29 Dec. 15/2 Joined to the ninon with coarse hairpin stitch. **1917** W. Owen *Let.* 10 Jan. (1967) 426 Celluloid hair-pin box from Boots. **1932** D. C. Minter *Mod. Needlecraft* 104 (*heading*) Hairpin Work. *Ibid.* 104/1 Hairpin-work D'oyley.

hair-splitting, *vbl. sb.* (Earlier example.)

1826 *Blackw. Edin. Mag.* XX. 854/2 A sort of game at hair-splitting.

hair-trigger. Add: *fig.* (Earlier U.S. example.)

1876 'Mark Twain' *Tom Sawyer* 226 The inmates were asleep, but it was a sleep that was set on a hair-trigger.

b. (Earlier and later examples.)

1834 M. Edgeworth *Tour Connemara* (1950) 4, I had been much amused by my father's account of Dick Martin—'Hairtrigger Dick'. **1948** C. Day Lewis *Poems 1943–1947* 60 The quick-set ears, the hair-trigger nerves. **1968** *Observer* 1 Sept. 8/7 A hair-trigger laugh.

Hence **hair-triggered** *a.* (earlier and later examples); also **hair-triggerish, -triggery** *adjs.*

1806 *Balance* 7 Jan. (Th.), I know not whether hair-triggered pistols are in use in Penn. **1937** L. C. Douglas *Forgive our Trespasses* xv. 304 Intolerant, irascible, pig-headed, hair-triggered. **1945** *Daily Express* 4 June 2/7 The A.A. gunners..had become so hair-triggered that no Allied pilot dare come anywhere near. **1928** Galsworthy *Swan Song* ii. vi. 162 'I admit,' said Michael, unhappily, 'it's all hair-triggerish.' **1937** L. C. Douglas *Forgive our Trespasses* ii. 41 He is always so autocratic and hair-triggery.

hairworm. Substitute for def.: An aquatic, nematomorph worm of the order Gordioidea, which, in its larval stages, is a parasite of insects, worms, or fishes. (Later examples.)

1897 [see *horsehair snake* (*HORSEHAIR c)]. **1930** E. C. Faust *Human Helminthology* xxix. 483 (*heading*) Gordiacea or 'hairworms'. **1951** G. Lapage *Parasitic Animals* vi. 175 The larvae of the hairworms may also develop in freshwater worms or fish. **1963** J. B. Goodey *T. Goodey's Soil & Freshwater Nematodes* (ed. 2) 522 The hairworms or gordiids..occur in freshwater ponds, ditches, streams, lakes, etc.

hairy, *a.* Add: **1. c.** In names of animals (further examples): **hairy armadillo**, an edentate mammal (*Chaetophractus villosus*) found in Argentina; **hairy frog**, a West African frog (*Trichobatrachus robustus*), the male of which shows filaments of skin on sides and thighs during the breeding season; **hairy woodpecker** *U.S.*, a common woodpecker (*Dendrocopus auduboni* or *D. villosus*) of the eastern parts of North America.

1840 E. Blyth tr. *Cuvier's Animal Kingdom* 125 The Pichiy of Azzara, and an allied species, the Hairy Armadillo (*Tatou velu*, Az.), resemble the Encoubert. **1892** W. H. Hudson *Naturalist in La Plata* i. 17 The fourth..is the hairy armadillo, with habits which are in strange contrast to those of its perishing congeners, and which seem to mock many hard-and-fast rules concerning animal life. **1956** G. Durrell *Drunken Forest* ii. 47 The hairy armadillo is the vulture of the Argentine pampa. **1925** *Jrnl. Morphol. & Physiol.* XL. 342 The occurrence in a frog of long, hair-like processes covering the sides of the body and part of the thighs with a thick growth has excited the curiosity of biologists since the first discovery of this 'hairy frog' nearly twenty-five years ago. **1960** H. W. Parker tr. *Mertens's World of Amphibians and Reptiles* viii. 138 The male of the large, West African, Hairy Frog..shows hairlike proliferations of the skin 10–15 mm. (0·4–0·6 of an inch) long, at the breeding season; they are completely absent in the female. **1731** M. Catesby *Nat. Hist. Carolina* I. 19 *Picus medius, quasi villosus.* The Hairy Woodpecker, weighs two ounces. **1808** A. Wilson *Amer. Ornith.* I. 150 [The] Hairy Woodpecker..is another of our resident birds,..a haunter of orchards, and lover of apple trees, an eager hunter of insects. **1839** J. J. Audubon *Ornith. Biogr.* V. 164 The Hairy Woodpecker, *P[icus] villosus*, is a constant resident in our maritime and inland districts. **1880** *Harper's Mag.* Oct. 672/2 *Picus auduboni* is not now recognized as a valid species, but only as a local variety of the hairy woodpecker. **1896** Hairy woodpecker [in Dict.]. **1956** L. W. Wing *Nat. Hist. Birds* ii. 26 The Downy and Hairy Woodpeckers of North America bear striking resemblances to each other.

d. *hairy at* (*about, in, round*) *the heel*(*s*) (*fetlocks*): deficient in breeding. So *hairy-heeled* adj., and simple *hairy*, in the same sense. *slang.* Cf. Hair *sb.* 8 d.

1890 R. L. Stevenson *Mem. & Portr.* 100 That hairy man of business knew his errand well. **1899** A. Conan Doyle *Duet* 212, I couldn't stand that chap at any price. A bit too hairy in the fetlocks for my taste. **1905** H. A. Vachell *Hill* xii, The Rev. Septimus scowled also, because he had always maintained that any Harrovian could accept defeat like a gentleman... 'I always said he was hairy at the heel.' **1906** *Macm. Mag.* Nov. 9, I would join you and cry *Viva Pio Nono!* with the hairiest. **1922** J. Buchan *Huntingtower* xi. 213, I can't say I ever liked him... Bit hairy about the heels. **1927** *Blackw. Mag.* Oct. 488/2 He took refuge in..the display..of an honest but slightly hairy heel. **1928** *Observer* 22 Jan. 10/7 (Advt.), There is an Atlantic [locomotive] over there..—a bit hairy about the heel.., but quite sporting on gradients. **1930** A. E. W. Mason *Dean's Elbow* xi, What would those people say..if they knew? Hairy-heeled, eh? **1962** N. Marsh *Hand in Glove* ii. 47, I always say that when people start fussing about family and all that, it's because they're a bit hairy round the heels themselves.

e. Excited, angry, 'out of temper'.

1914 J. Joyce *Dubliners* 54 She doesn't know my name. I was too hairy to tell her that. **1927** W. E. Collinson *Contemp. Eng.* 116 He got shirty or hairy.

f. *hairy ape*: a person of a low mental or social type.

[**1922** E. O'Neill (*title*) The Hairy Ape.] **1931** *Times Lit. Suppl.* 1 Oct. 750/3 The submerged tenth, the hairy apes of society.

g. In various *fig.* and slang senses: difficult (quot. 1848); out-of-date, passé; frightening, hair-raising; crude, clumsy, rough, erratic.

Some examples belong equally under sense 1 b. *transf.*

1848 A. H. Clough *Bothie* ix. 53 He..never once had brushed up his *hairy* Aldrich. **1914** D. O. Barnett *Let.* 25 Nov. (1915) 13, I..threw a hairy salute! *Ibid.* 2 Dec. 16 It's top-hole fun, with four hairy captains teaching us things. **1934** H. G. Wells *Exper. Autobiogr.* II. ix. 783 They were not throwing themselves into their parts as the hairy young Italians they were aping would have done. **1946** B. Marshall *George Brown's Schooldays* 7 There you go again using great long hairy words. **1950** Wentworth & Flexner *Dict. Amer. Slang* 239/1 *Hairy*, old, already known, passé; usu. said of a joke or story. **1962** D. Slayton in *Into Orbit* 22 If you happen to be pulling a lot of Gs..it might get a little hairy trying to manipulate the controls with all the finesse you'd need. **1966** 'W. Cooper' *Mem. New Man* iii. iv. 239 The problem was of the kind that Mike described in his up-to-date slang as 'hairy', meaning complex in surface detail and involving more parameters than anybody would want to cope with simultaneously. In a word, messy. **1966** J. Miles in T. Wisdom *High-Performance Driving* v. 45 You can go just as quickly if you brake and accelerate smoothly... If it's hairy its bad. **1966** *Surfer* VII. iv. 48 One of the fastest, hairiest waves I've ever ridden. **1967** *Autocar* 5 Oct. 24/1 This Healey had all the works racing mods which brought the engine power up to 210 b.h.p. and turned what is a hairy and perhaps a slightly clumsy road car into a Ferrari-beating racer. **1968** *Listener* 20 June 816/1 Khe Sanh wasn't too bad. They had good bunkers there, but most places since Tet have been pretty hairy. **1968** *Sun* 12 Nov. 8/5 *Hairy*: a fast driver is a hairy driver. **1968** H. C. Rae *Few Small Bones* i. iv. 39 'Were you ever at one of his parties?'..'Just one. .. It wasn't my style really. In fact it was pretty hairy... Too many jumped-up gentry.' **1969** J. Morris *Fever Grass* xvii. 154 Things may be rough now, baby, but they could get really hairy if you try to cross me. **1971** *New Yorker* 21 Aug. 39 And do you, Elizabeth, take this man, John, to have and to hold, to love and to cherish, until the going gets hairy? **1972** *Times* 14 Oct. 6/2 Lord Snowdon said during a break for an orange juice: 'I was a bit frightened. Some bends are a bit hairy.'

4. *hairy-arsed, -chested, -nosed* adjs.

1885 *Encycl. Brit.* XIX. 518/2 The Hairy-nosed Porcupine *H[ystrix] leucura.* **1911** *Ibid.* XXVIII. 782/1 In the hairy-nosed wombat (*P[hascolomys] latifrons*) of Southern Australia the fur is smooth and silky. **1937** C. W. Ferguson *Fifty Million Brothers* vii. 96 What is this thing that has..drawn the admiring gaze alike of dictators and hairy-chested novelists? **1944** A. Russell *Bush Ways* xliv. 192, I found there were numberless burrows of the hairy-nosed wombat. **1960** G. Sanders *Mem. Professional Cad* i. v. 44 Meanwhile the man will applaud in a manner that he feels will demonstrate a fine balance between hairy-chested virility and sensitive intellectuality. **1964** L. MacNeice *Astrol.* ii. 62 Saturn makes you hairy-chested. **1965** J. S. Gunn *Terminol. Shearing Industry* i. 31 *Hairy-arsed learner*, a man who has probably shorn hundreds of thousands of sheep..but whose skill has been much reduced by age or infirmity. **1967** Partridge *Dict. Slang* 1164/1 *Hairy-arsed*, no longer young... Mature and hirsute and virile.

B. *sb.* A heavy artillery horse, so called from its hairy fetlocks. *Army slang.*

1899 A. Conan Doyle *Duet* 215 The hairies—trooper's chargers, you know. **1924** *Blackw. Mag.* Mar. 365/2 We had the bar placed as high as possible and put the old 'hairies' as hard at the jumps as they could travel. **1930** *Even. Standard* 15 Feb. 15/1 Whipping up the lumbering hairies to a desperate canter. **1959** *Times* 31 Dec. 10/7 No longer should I be dependent on 'hairies' hired from the local cavalry regiment.

Haitian (hēⁱ·tiăn, hai·tiăn, -ʃăn), *a.* and *sb.* Also 9 **Haytian.** [f. *Haiti* + -an.] Of or pertaining to the island of Haiti or Hispaniola in the West Indies, or to the Republic of Haiti situated in the western part of that island. Also as *sb.*, a native or inhabitant of Haiti, or of the Republic of Haiti.

1805 M. Rainsford *Hist. Acct. Black Empire of Hayti* 448 Where is that Haytian so vile, Haytian so unworthy of his regeneration, who thinks he has not fulfilled the decrees of the Eternal, by exterminating these blood-thirsty tygers? **1811** *Gentl. Mag.* Sept. 275/1 After the ceremony, their Haytian Majesties..received the sacrament. **1828** J. Franklin *Pres. State Hayti* 6 The partial eulogists of the Haytians go to the length of asserting that they have arrived at a high degree of moral improvement. *Ibid.*, Instances of intelligence have been discovered in the Haytian citizen. **1863** *Chambers's Encycl.* V. 274/1 The inhabitants of the eastern or Spanish portion of Hayti, rising against their Haytian oppressors. **1880** *Encycl. Brit.* XI. 546/1 Haytian Republic. **1964** E. A. Nida *Towards Sci. Transl.* viii. 181 For Haiti, the Scriptures are published in Haitian Creole, which is essentially a colloquial dialect of French. **1966** G. Greene *Comedians* i. iii. 75 The walls were hung with pictures by Haitian artists. *Ibid.* 83 The little houses the Haitians constructed for their dead. **1972** *Daily Tel.* (Colour Suppl.) 10 Nov. 23/2 The Haitian government is encouraging the hotel boom with generous concessions.

haka (hā·kā). *N.Z.* [Maori.] A Maori ceremonial posture dance accompanied by chanting; one danced by members of a sports team, etc.

1832 H. Williams *Jrnl.* 13 Jan. in H. Carleton *Life H. Williams* (1874) I. 113 They now prepared for their *haka*, or dance. **1845** E. J. Wakefield *Adv. N.Z.* I. 98 A *haka* was now performed by about one hundred and fifty men and women. **1872** A. Domett *Ranolf* xv. vi. 19 The *hāka*-dances where she shone supreme. **1907** *Macm. Mag.* Sept. 855 The Maoris are a people with grand manners, and the *haka* is an amusement wherewith they beguile for their guests the long dark evening hours. **1920** *Glasgow Herald* 30 Apr. 9 At the close of the oration the warriors performed the 'Haka' war dance. **1934** *Bulletin* (Sydney) 15 Aug. 11/4 The haka is kept going by Maoriland footballers and hockey players. **1938** R. D. Finlayson *Brown Man's Burden* 13 Afterwards there were pois and there were hakas, and there was sure to be a dance. **1957** *N.Z. Listener* 22 Nov.

4/2 One common group..of Maori words has come right over into New Zealand English—whare, haka and mana (e.g.) have all acquired (when used in English) overtones and extra meanings that were not in the original Maori. They are true New Zealandisms. 1963 *Evening Post* (Wellington) 12 Oct., 'Kamate! Kamate!' the fierce Maori haka rang out today as the 1963–64 All Blacks for Britain had their first haka practice.

hakam, var. *HAHAM.

‖ **hakama** (ha·kămă). Also **hakkama**. [Jap.] Loose trousers with many folds in the front, worn in Japan.

1859 A. Steinmetz *Japan & her People* i. iii. 252 A very peculiar sort of trousers called *hakkama*, which may be called an immensely full-plaited petticoat sewed up between the legs. 1871 A. B. Mitford *Tales Old Japan* II. 284 The *hakama*, or loose trousers worn by the Samurai. 1893 A. M. Bacon *Jap. Interior* vii. 119 The Japanese costume of purple *hakama*, or kilt-plaited divided skirt, which forms the uniform of the little school-girls. 1963 'G. Black' *Dragon for Christmas* iv. 67 Mr. Kishimura opened the door wearing heavy grey-black silk robes, with the *hakama* over-garment.

‖ **Hakenkreuz, hakenkreuz** (ha·kənkrɔits). [G.] The Nazi swastika. Also *attrib.*

1931 *Times* 23 Dec. 7/4 A large Nazi Hakenkreuz flag, 'which can be seen for miles', flies from the tallest chimney. 1935 C. Isherwood *Mr. Norris changes Trains* xi. 165 Hitler's negotiations with the Right had broken down; the Hakenkreuz was even flirting mildly with the Hammer and Sickle. 1966 [illegible] 115 His eyes fell on a large hakenkreuz. To come upon the Nazi insignia so unexpectedly made Bronsky feel slightly sick. 1972 *Oxford Times* 28 July 9 Perhaps he [sc. Hitler] hoped the Hakenkreuz would bring bad luck to his enemies.

haker (hēi·kəɹ). [f. Hake *sb.*[1] + -er[1].] A fisherman or a fishing-boat engaged in catching hake.

1880 *Harper's Mag.* Aug. 340/2 The man who fished for hake, and also his boat, was a 'haker'.

Hakka (hæ·kă). [Chinese.] A member of a people now dwelling in parts of southern China, especially in the province of Kwangtung or Canton, and in Taiwan, Hong Kong, etc.; also the dialect spoken by this people. Also *attrib.* or as *adj.*

1867 *N. & Q. on China & Japan* I. 66/2 It is a common saying among the Hakkas, that a Punti may study Hakka for many years, and yet not be able to speak it correctly. 1878 H. A. Giles *Gloss. of Reference* 56 Hakkas, strangers. A race said to have migrated from the North of China (Kiangsu or Shantung) to the Kuang-tung province at the time of the Yuan dynasty [illegible]. 1879 *Encycl. Brit.* IX. 416/1 Hakkas from the vicinity of Swatow. 1881 J. D. Ball *Easy Sentences Hakka Dial.* p. i, While engaged in my studies in the Hakka dialect I put the sentences in Giles' *Handbook of the Swatow Dialect* into Hakka. 1921 *Outward Bound* July 17/1 He converses freely in Hakka, Cantonese, Mandarin [illegible]. 1926 *Blackw. Mag.* Nov. 618/1 The Hakkas, as woodcutters and hunters, had already penetrated into the foothills. 1964 *Asia Mag.* 30 Aug. 5/1 (*caption*) [illegible] a Hakka woman [illegible]. 1968 [illegible] xviii. 196 The old man at the helm..looked like a Hakka to me. They are fishermen, and nobody's fools. 1971 K. Hopkins *Hong Kong* 335 Cantonese is very much the predominant language but there are minorities who speak Hakka [illegible]. 1972 *Nat. Geographic* Oct. 550/1 (*caption*) [illegible] identifies a Hakka farm-woman. A distinctive group of immigrant Chinese, Hakkas till hillsides in the New Territories.

Halafian (hālă·fiăn), *a.* *Archæol.* [f. Tell *Halaf* in north-eastern Syria + -ian.] Denoting the chalcolithic culture which existed in northern Syria and Iraq, characterized by polychrome pottery, evidence of which was first discovered at Tell Halaf.

1937 G. E. Wright *Pottery of Palestine* ii. 30 Painted motifs used by the Palestinian potters are common in the region of the Halaf pottery... When it comes to the farms, however, the Ghassulian has very little in common with the Halafian. *Ibid.* 31 A type of small bowl with gently rounded sides seems also to be popular in the Halafian culture. 1947 *Iraq* IX. 45 It is evident that there must have been an extensive prehistoric settlement of the T. Halaf period. This Halafian culture has been found throughout the whole of Northern Syria from Ugarit in the west across the Upper Euphrates valley as far as the district of Nineveh on the Tigris. 1950 G. Daniel *100 Yrs. Archæol.* vi. 214 The Tell Halaf culture or the Halafian as it was called, could be accurately dated.

halal (hālā·l), *v.* Also **hallal**. [f. Arab. *ḥalāl* lawful.] *trans.* To kill (an animal) in the manner prescribed by Muslim law. Hence **halal** *sb.*, lawful food; also *attrib.* and as *adj.*

1855 R. F. Burton *Personal Narr. Pilgrimage to El-Medinah* I. xiii. 377 To '*halál*' is to kill an animal according to Moslem rites: a word is wanted to express the act, and we cannot do better than to borrow it from the people to whom the practice belongs. 1858 W. Jesse tr. *Ferrier's Hist. Afghans* xxi. 289 They will not eat meat unless it is *halal* (lawful), that is, the animal must have its face turned towards Mecca, and its throat cut in a particular part of the neck. 1879 R. A. Sterndale *Seonee* 454 *Hálál*. Slaughtered according to religious law. 1879 F. T. Pollok *Sport in Brit. Burmah* I. 142 My mahouts, when they have got down to halal, or cut the throat of a [illegible]. *Ibid.* 179 The mahouts would not cut the stag as it had not been hallaled, so my Madrass servants and the Burmese had it all to themselves. 1883 E. H. Aitken *Tribes on my Frontier* 167 To allow Peer Khan to make it halal, by cutting its throat in the name of Allah [illegible]. 1896 *Daily News* 1 June 5/5 The special 'Halal' meal, ordained for the [illegible], will be provided. 1910 *Scribner's Mag.* Apr. 404/1 Whenever possible the game being hal-lalled in orthodox fashion by the Mahometans. 1933 J. Masters *Deceivers* xvii. 193 So the two black goats died, one in the Mohammedan manner, the halal, and one in the Hindu manner, its head struck off at a single blow. 1966 *Guardian* 9 Dec. 8/6 Nuts, spices, curries, meat (avowedly halal at that!). 1970 *Listener* 9 July 44/2 You can get kosher meat next to halal meat.

halation. Add: **b.** A similar effect in television (see quots.).

1937 A. T. Witts *Television Cycl.* 56 *Halation*, the reflection of image rays by the back of a screen or film. Such reflection produces a blurring of the image as viewed by the observer, or is reproduced on the film. 1940 [see *blooming *vbl. sb.*[1] 3]. 1958 *Chambers's Techn. Dict.* Suppl. 983/2 *Halation*, bright annular area around the spot on cathode-ray tube phosphor spot, arising from internal reflection within the glass support.

halawi (hălā·wi). Also 9 **khalaweh**. [Arab.] A kind of sweetmeat; = *halva, *hulwa.

1836 E. W. Lane *Acct. Manners & Customs Mod. Egyptians* II. 14 The seller of a kind of sweetmeat (*halawee*), composed of treacle fried with some other ingredients. 1921 T. E. Lawrence *Lett.* (1938) 129 About halawi—I hope you remember the particular sticky sweetmeat of which Thompson & myself used to eat pounds last year. 1923 —— *Home Lett.* (1954) 248 Eat that halawi in my house.

halch, *v.* Add: **3. b.** *Cotton-spinning*, etc. (see quots.). Also **halch-band**.

1892 J. Nasmith *Cotton Spinning* 286 As yarn is always wound off a cop by drawing it upwards, any such condition of the cop more results in a number of coils being drawn off simultaneously in an entangled condition. In this case the cop is said to be 'halched', and a good deal of waste is produced when the unwinding takes place. 1921 *N. & Q.* 12th Ser. VIII. 81/2 'Halsh'..This word is in every-day use in various ways. So far as the cotton trade goes it refers to the band of coloured 'tie yarn' that encircles the 'knot', in addition to the ordinary tie yarn that holds each lea in the knot separately. This is called the 'halsh-band', and when the band is tied the knot is said to be 'halshed'.. The 'halsh' is also—in the case of a necktie in the form of a bow, for example—that part in the centre that runs in a vertical or slightly oblique direction, embracing the whole bow.. Sadlers also use the word, and possibly it is known in the woollen and worsted industries.

haldu (ha·ldu). [Hindi.] A tree, *Adina cordifolia*, of the family Rubiaceæ, found in Burma, India, and Thailand; also its yellowish hardwood timber.

[1881 J. S. Gamble *Man. Indian Timbers* 220 Adina cordifolia.. Vern. *Haldu*.] 1920 *Nature* CV. 693/1 *Haldu* (*Adina cordifolia*) is a bright canary-coloured wood [illegible]. 1934 *Jrnl. R. Aeronaut. Soc.* XXXVIII. 51 Other timbers used for carving on account of their even structure are [illegible] and the Indian wood known as Haldu (*Adina cordifolia*). 1958 [illegible] (1963) 17/2 [illegible] haldu (*Adina cordifolia*) and sandalwood (*Santalum album*), the texture is fine giving a smooth finish.

hale, *a.* Add: **3.** Often in phr. *hale and hearty*.

1860 *Leisure Hour* 174/2 As hale and hearty..as ever. 1863 *Good Words* IV. 495/2 He..was hale and hearty at the age of upwards of a hundred years old. 1899 *Captain* I. 249/1 Dr. Grace is close on fifty-one, hale and hearty. 1908 T. Hardy *Dynasts* I. i. i, 4 We be the King's men, hale and hearty, Marching to meet one Buonaparty. 1928 A. B. Callow *Food & Health* 7 In the past many people have lived perfectly hale and hearty lives without having any clear ideas about the science of nutrition.

half, *sb.* Add: **6. d.** Also in Rugby Union, Rugby League, etc. Cf. *fly-half* (*fly *sb.*[1] 8), *scrum-half.

1896 B. F. Robinson *Rugby Football* v. 90 The half 'working' the scrummage should rarely try to run himself. 1906 Gallaher & Stead *Complete Rugby Footballer* iv. 64 A new line of attack and defence between the halves and the three-quarters. 1951 *Men's Hockey* (Know the Game Series) 17/2 At long corners..the halves go out behind them. 1960 E. S. & W. J. Higham *High Speed Rugby* vii. 55 Give me a good pair of halves and I will give you a side.

f. *Golf.* A hole or point which is halved.

1881 [see *dead *a.* 21 b]. 1908 J. Braid *Advanced Golf* 213 Halves ought rarely to be agreed upon unless the balls are so close to the hole that it is not possible for the putts to be missed. 1931 *Daily Tel.* 22 May 18/1 He..secured the necessary 5 for the half and the match. 1959 *Times* 29 May 5/1 Sewell got a brave half at the 18th.

g. In Old English prosody = *half-line* (*half- II. n).

1892 S. A. Brooke *Hist. Early Eng. Lit.* I. p. x, The Anglo-Saxon line is divided into two halves by a pause. The first half has two 'measures'. 1940 J. R. R. Tolkien in J. R. C. Hall tr. *Beowulf* p. xxvii, The Old English line was composed of two opposed word-groups or 'halves'. Each half was an example, or variation, of one of six basic patterns.

h. A half-holiday.

1900 P. Traherne *Diary* Nov. (1918) i. 28 By getting up this entertainment on a 'half' when there was nothing else to do I found myself launched into about six rows. 1934 [illegible] He will acquire prep-school slang (words such as..'half' holiday).

i. Ten shillings, half of £1. *slang.*

[illegible] 1936 [illegible] 1958 G. Greene *Brighton Rock* [illegible]

j. A fare or ticket at a reduced (usu. half) rate. Also *quasi-adv.*

1935 *Punch* 30 Jan. 137 You two can't go half. You're over age. 1962 [illegible] Two adults and three halves, please.

7. c. Esp. in phr. *too clever by half*: trying too hard to be clever.

1858 [in Dict.]. 1889 W. Westall *Birch Dene* II. vi. 89 'Nobody can deny as he's clever.' 'Ay, too clever by half.' 1944 L. MacNeice *Christopher Columbus* 8 Constant ingenuity, often leads to an appearance of being too clever by half. 1961 *Listener* 2 Nov. 717/1 A bad one [sc. documentary], whether dull through laziness, or self-conscious or pretentious or too-clever-by-half, can be a real catastrophe.

g. *and a half* (following a sb.): and more; of an exceptional kind. Cf. Heart *sb.* 39 b. *colloq.*

1626, etc. [see Heart *sb.* 39 b]. 1832 J. K. Paulding *Westward Ho!* II. i. 7 Bushfield, too, was here in all his glory, and was not only a whole team, but a team and a half, good measure, as he affirmed. 1837 Dickens *Pickw.* xlvi. 501, I rayther think the gov'ner vants to have a word and a half with you. 1921 T. E. Lawrence *Lett.* (1938) 118 Last night was paradise and a half. 1925 *Strand Mag.* LIII. 606/2 Golly! He took a toss and a half! 1950 'M. M. Kaye' *House of Shade* x. 118 Roaring Rory must have been a hell-raiser and a half in his day.

i. *the half of it*: a significant or more important part of something. Usu. in negative contexts.

1932 Wodehouse *Hot Water* i. 27 It makes me sick. And that's not the half of it... She told me I've got to be American Ambassador to France. 1949 [illegible] *Minute for Murder* ii. 2 'We've not seen the half of it yet,' said the Messenger darkly. 'The half of what?' 'You mark my words, sir. When peace comes, as you might say real peace, there'll be chay-os in this country.' 1966 M. Brewer *Man against Fear* xi. 117 'You haven't heard the half of it yet.' I went on to tell him about the Carver Street ambush. 1971 M. Babson *Cover-up Story* x. 109 'How awful,' she said... I nodded, without telling her she didn't know the half of it.

half, *a.* Add: **1. a.** *half sheet* (earlier examples). See also *half-sheet* s.v. *Half-* II. n in Dict. and Suppl.

1733 J. Nott *Cook's & Confect. Dict.* 261 Lay every Cutlet on a half Sheet of Paper. 1852 E. Ruskin *Let.* 27 Apr. in M. Lutyens *Effie in Venice* (1965) II. 296, I was very grateful for..Papa's half sheet about the Ruskins.

b. (Later examples.)

1862 [see Battle *sb.* 4]. 1864 Trollope *Can you forgive Her?* I. xxxviii. 298 If you are half the woman that I take you to be, you will understand this. 1867 —— *Claverings* I. iii. 34 Though the lord might be only half a man, Julia walked out from the church every inch a countess. 1936 *Discovery* Dec. 397/1 The author has undoubtedly the gift of winning the confidence of his African hosts, which is half the battle. 1944 *Living off Land* i. 15 The correct mental approach is half the battle.

c. Also *half a bar* (see *bar *sb.*[1] 3 c); *half a bull* [Bull *sb.*[1] 7], half a crown (*slang*).

1812 J. H. Vaux *Flash Dict.*, *Half a bull*, half a crown.

half, *adv.* Add: **1. c.** (Later examples.)

1910 W. de la Mare *Three Mulla-Mulgars* 63 Half-hiding his face in his jacket. 1928 D. H. Lawrence *New Poems* 13 One side shadow, half in sight, Half-hiding the pavement-run.

3. *not half* (further examples): 'not half a bad fellow' = a good fellow; 'not half long enough' = not nearly long enough; also (*slang*), extremely, violently, as 'he didn't half swear'.

1851 Melville *Moby Dick* II. xxxi. 215, I don't half like that chap, Stubb. 1924 H. Ashton *First from Front* xiv. 99 It wasn't half all right, I tell you. 1929 V. Woolf *Night & Day* xv. 195 Not half so easy as people think.. It wouldn't be half bad. 1920 Galsworthy *Foundations* in. in Plays (1929) 498 Talk of your sacrifices in the war—they put you on your guard. The hops 'd not half be sized so mot it. Rations—not 'arf! 1953 L. P. Hartley *Go-Between* xix. 165 And we didn't half enjoy your cake. 1959 [illegible] I don't half want to. 1962 Parker & Allerton *Courage of his Convictions* iii. 143 It doesn't half nark them.

half-. Add: **I. 1. a.** in the predicate. (Further examples.)

1844 J. R. Lowell *Poems* 12 A youth half-smiling,

1846 'A Lady' *Jewish Manual* ii. v. 226 Dresses made high are..unbecoming; they should either be cut close up to the throat or low. 1849 Thoreau *Week on Concord* 399 They are half forgotten ere we have learned the language. 1858 Bagehot *Coll. Works* (1965) II. 55 Half-crazed as she [illegible] is described to be. 1881 'Mark Twain' *Prince & Paup.* 162 He was half-minded to resign. 1893 —— *Man corrupted Hadleyburg* (1900) 69 The station-master..became pleasant and even half-apologetic. 1904 W. de la Mare *Henry Brocken* vi. [illegible] 1920 J. Morley *Cromwell* (ed. 2) iii. 54 [illegible] 1929 D. H. Lawrence *Birds, Beasts & Flowers* 41 [illegible] 1936 *Mind* XLV. 252 What were the views concealed or half-concealed, expressed or half-expressed? 1937 *Brit. Birds* XXX. 340 We have two records of single adults spending one day in the nest when their young were half-fledged. 1961 R. W. Brunskill? *Eng. Farmhouse & Cottage* iv. ii. 196 Roofs gabled and half-hipped.

b. as *attrib.* (Further examples.)

c 1827 J. S. Mill in *Adelphi* (1924) I. 689 A half-cultivated person is likely to be gaudy, affected, and meretricious ornament. 1843 —— *Syst. Logic* II. vi. ii. 345 It is in those steps of the reasoning which are made in this tacit and half-conscious, or even wholly unconscious manner, that the error oftenest lurks. 1851 Melville *Moby Dick* I. xvii. 132 Their half crazy conceits on these subjects. [illegible] 1864 [illegible] a half formed intention of writing a History of the French Revolution. 1864 Thoreau *Walden* 180 [illegible] 1894 [illegible] 1901 [illegible] 1913 [illegible] 1922 [illegible] 1924 [illegible] 1959 *Daily Chron.* 6 Feb. 4/6 Is he really a half-saved personage? 1908 *Westm. Gaz.* 17 Dec. 3/2 Some trivial gossip in the half-lit hall. *Ibid.* 29 July 3/4 The forces of Free-Trade may be confidently reckoned on to square the half-measure of the proviso of Tariff Reform. 1920 W. de la Mare *Three Mulla-Mulgars* 120 Her half-blind whitening eyes. 1923 D. H. Lawrence *Birds, Beasts & Flowers* 37 Wavering men of all the nations, half-fledged fearsome of his thoughts, conscious of the highest of the far-flung. 1927 [illegible] 1930 [illegible] 1935 [illegible] 1936 [illegible] 1950 [illegible] 1959 [illegible]

2. *half-apologetically, -consciously, -jokingly.*

1932 J. London [illegible] 1933 [illegible] 1944 [illegible] 1949 [illegible] 1957 [illegible] 1966 [illegible]

3. *half-believe* (examples), *-laugh.* Also **half-inch** *v.* (*Rhyming slang*), to 'pinch', to steal; so **half-inching** *vbl. sb.*; **half-lap** [Lap *sb.*[3] 2 b] *v.* (see quot. 1825); **half-pay** *v.*

1825 Nicholson *Oper. Mech.* 653 The reason for making the joints half-lapped, or scarfed, (etc.). 1844 H. Stephens *Bk. Farm* II. 296 They are half-lapped in pairs at the centre. 1849 Thoreau *Week on Concord* 152, I half believed that I should get above it. 1924 J. Murry *Voyage* viii. 152 'I can't help it,' she said, half-laughing at her own confession. 1925 Fraser & Gibbons *Soldier & Sailor Words* 134 *Half-inch*, to steal. 1928 W. Clewes? *Journey into Spring* ii. 30 We can half-inch those [illegible]. 1930 [illegible] 1959 *Listener* [illegible]

4. **half-cut** *a.* (*slang*), half-drunk; **half-shaved** *a.* (*Obs. U.S. slang*), drunk.

1893 Farmer & Henley *Slang* III. 250/1 *Half-cut*, half-drunk. 1971 *Radio Times* 18 Nov. 80 Inebriation..is the sport of all ranks. How many executives can work reasonably effectively unless they are half-cut? 1818 M. L. Weems *Lett.* (1929) III. 225 One night, getting half shaved, he was easily over-persuaded (a common curse of whiskey) to try his luck at All Fours. a 1852 F. M. Whitcher *Widow Bedott Papers* (1856) xxviii. 354 I've seen that man half shaved on cider afore breakfast in the mornin'.

II. In attributive relation to a sb.

a. In names of Coins, Measures, etc. **half-bottle, -caser** [*caser[1]], **-litre**; **half-pint** (earlier and later *attrib.* uses); also *ellipt.*, a half-pint of beer; (see also *half-pint* under sense n below).

1877 E. S. Dallas *Kettner's Bk. of Table* 287 Where are these bottles and half-bottles of Madeira to be found? 1927 E. Hemingway *Men without Women* (1928) 172 The American lady bought..a half bottle of Evian water. 1950 T. S. Eliot *Cocktail Party* i. ii. 51, I found some champagne—Only a half-bottle, to be sure. 1907 H. Lawson in *Austral. Short Stories* (1951) [illegible] 1921 W. J. Locke *Mountebank* ii. [illegible] half-litre glass of beer [illegible]. 1967 'G. Carr' *Lewker in Tirol* iv. 56 He..ordered a half-litre of wine. 1611 J. Donne in Coryat's *Crudities* Pref. verse, [illegible] 1728 E. Smith *Compleat Housewife* (ed. 2) 151 Pour it into half-pint Basons. 1899 R. Whiteing *No. 5 John St.* xi. 103, I fell upon roast pork, and a foaming half-pint. 1937 *Discovery* Sept. 277/1 Fill a half-pint [illegible]. 1966 E. McGirr? [illegible]

i. In *Games.* **half-back** (examples in other games, e.g. American Football, Rugby Football, Australian Rules); **half-blast** *Golf*, a shot which is played with half the force of a 'blast' (explosive shot); **half-blue**, the 'colours' (see Blue *sb.* 9) awarded to a player representing his university in inter-university contests as second choice to a 'full blue', or to any chosen representative in sports or games not recognized by the Blues Committee as sufficiently important for the award of a 'full blue'; also, a competitor who has gained this award; **half-brassy shot** *Golf*, a brassy shot played with a half swing; **half-captain**, in women's colleges in Oxford, one who has attained a certain degree of proficiency in the management of a boat; so **half-captaincy**; **half-colour**, a badge showing that a stage of proficiency half-way towards getting one's colours has been reached (see *colour *sb.*[1] 6 c); **half-court** *Tennis* and *Rackets*, half the court divided by a line (the half-court line) parallel with the side lines; **half-fifteen, -forty** *Tennis* (see quot.); **half-forward, half-forward flanker** *Austral.* (see quot. 1968); so *half-forward flank*; **half-hit** *Cricket* (read.), a half-way hit, the ball falling short of the distance it would have travelled if properly hit; **half-iron shot** *Golf*, an iron shot played with a half swing; a **half nelson** *Wrestling*, a hold in which one arm is thrust under the corresponding arm of the opponent and the hand placed on the back of his neck; also *fig.* in phr. *to get a half-nelson on*, to hold in a crippling position, gain a complete hold over; hence as *v. trans.* with the sense (of quot.); **half-pass** (see quot. ...); **half-pin** *Chess*, that position in which a defending man lies between an attacking piece and the defended king and in the line of attack of the attacking piece, but has liberty to move along the line of attack; also, that position in which two defending men lie between the attacking piece and the king so that if either moves the other man becomes pinned; so **half-pinned** *a.*; **half-pinner**, a half-pin problem; **half-shot** *Golf*, a stroke made with a half swing, intended to carry less far than the full shot; **half-stroke** *Golf* = *half-shot*; **half swing** *Golf*, a swing of half the usual amount of distance; **half-thirty** *Tennis* (see quot. s.v. *half-fifteen*); **half-topped** *a.* *Golf*, topped; **half-volley** *Cricket* (earlier and later examples), also, a ball which pitches just in front of a fieldsman; (*b*) *Lawn Tennis*, a stroke made when the ball has just rebounded from the ground; **half-volley** *v.*, to hit a half-volley; hence **half-volleyer**; **half-volley** *vb.*

1887 *Century Mag.* Oct. 892 Behind the quarter-back, and covering the two sides of the field, are the 'half-backs', the cavalry of the team. 1906 Gallaher & Stead *Complete Rugby Footballer* v. 71 The half back as we know him in New Zealand is the donkey man of the team. 1959 N. Mailer *Advts. for Myself* (1961) 53 He paused, crouched into a halfback's position, waiting for the ball. 1965 *Sun-Herald* (Sydney) 4 July 51 Sydney's half-back flanker Bob Sterling won the President's trophy. 1965 *Advertiser* (Adelaide) 17 July 25 Its half-forward line.. was over-run by the powerful South Adelaide half-back line. 1968 Eagleson & McKie *Terminol. Austral. Nat. Football* ii. 74 There are three half-backs in each team: left, centre, and right half-back. 1969 *Australian* 24 May 39/2 Ian Bremner, Norm Bussell and Peter Chilton constitute a tight-checking half-back line for Hawthorn. 1928 *Weekly Dispatch* 24 June 21/6 He played a superb 'half-blast' out of a trap to lay the ball one foot from the cup. 1908 *Westm. Gaz.* 29 July 10/4 The half-blue for billiards. 1909 *Ibid.* 26 Feb. 12/2 For some time players of lacrosse at Oxford have been urging the Blues Committee to grant them the Half-Blue. 1963 *Times* 5 Feb. 3/5 Mr. J. J. Grinalds (Brasenose), an old half-blue, came back into the defence for the first time this season. 1903 *Westm. Gaz.* 28 Aug. 5/1 The half-brassy shot for... 1928 *Daily Express* 7 May 5/2 She may not go on the river unless she is accompanied by a half-captain or is one herself. Success in the 'half-colour' test... 1929 *Evening News* 28 Nov. 13/5 The players who appears in future Novic international trials will... 1908 *Ibid.* ... 1886 *Encycl. Brit.* XXIII. 152/2 A space bounded by the net, the side line, the half-court line, and the service line. 1895 H. W. W. Wilberforce *Lawn Tennis* 62 The half-court-line divide the court into two equal parts... 1898 *Encycl. Brit.* XXIII. 182/2 Half-fifteen is one stroke given at the beginning of the second and every subsequent alternate game of a set... 1963 *Courier-Mail* (Brisbane) 21 Nov. 7/5 A typical commentary from Australian Rules, 'a long drop-kick from the half-forward flank'. 1968 Eagleson & McKie *Terminology Austral. Nat. Football* ii. 74 *Half-forward*, a player occupying a half-forward position, the half-forward flankers... 1969 *Australian* 24 May 39/2 Bremner will probably be given the task of watching Geelong's mercurial Ken Newland, who will start on a half-forward flank. 1888 *Hall Inc.* (Ibid.) 1858 A. G. Steel & Lyttelton *Cricket* iii. 212 Extra cover-point...may be..placed for half-hits made on the on-side..... 1893 *Black & White Budget* 30 Nov. 35/1 The half-nelson... 1896 W. A. Bettesworth *Walkers of Southgate*... 1900 ... 1905 H. G. Hutchinson *Golf* (ed. 5) 44 The attitude..for the half-iron stroke. 1906 *Westm. Gaz.* 10 Nov. 15/2 ... 1908 *Ibid.* ... 1916 ... 1925 ... 1932 ... 1945 ... 1948 ... 1960 ... 1962 ... 1966 *New Statesman* 15 July 74/2 He knew..that Sir Alec Douglas-Home had to be..half-nelsoned and eagerly followed his ... 1880 *Illustr.* ... 1880 ... 1886 ... 1891 ... 1892 ... 1893 ... 1898 ... 1905 ... 1906 ... 1909 ... 1931 ... 1935 ... 1960 ... 1843 'A Wykehamist' *Pract. Hints Cricket* 12 All balls pitching between the first line and the crease..are technically termed half vollies. *Ibid.* 17 A leg half-volley..may be.. 'dropped into' now and then. 1851 J. Pycroft *Cricket Field* viii. 168 Every one knows the difficulty of making a good half-volley hit off a slow ball. *Ibid.* x. 189 If a bowler has half volleys returned to him, by stretching

and stooping after them, he gets out of his swing. 1867 G. H. Selkirk *Guide Cricket Ground* v. 83 The half volley..requires practice to ensure its being picked up properly. 1870 *London Society* Nov. 425/1 The mode of playing a ball so well known at tennis, not quite half-volley. 1897 *Encycl. Sport* I. 621/2 *Half-volley*, a stroke made the moment the ball leaves the ground. 1960 *Times* 21 June 16/1 Taylor half-volleyed a return into the net. 1963 A. Ross *Australia 63* iii. 76 He played at nothing he did not have to, leaving Davidson to flash the odd half-volley through the covers. 1969 *New Yorker* 14 June 45/1 Players call it a half-volley drop shot. Ashe reaches down, lightly touches the rising ball, and sends it on a slow, sharply angled flight toward the net. *Ibid.* 45/3 Ashe, moving up, is again confronted with the need to half-volley. 1912 *Daily News* 11 July 2 A famous half-volleyer.

l. *half-bath, -bathroom, -blind, -door* (later example), *-hoop* [cf. Hoop *sb.*[1] 7], *-tester* (examples).

1879 A. von Harlingen in A. H. Buck *Treat. Hygiene & Public Health* I. 373 The half-bath, in which the bather sits in a tub filled with water to the depth of from ten to twelve inches,..is adapted to invalids. 1953 R. Chandler *Long Good-Bye* xxvii. 171 There was a half-bath off the study. 1959 *News Chron.* 19 June 4/3 A half-bathroom ..has a shower, wash-basin and lavatory, but no bath. 1763 Boswell *London Jrnl.* 22 July (1951) 382, I was disturbed by the light..at the earliest dawn, as the windows have only half-blinds. 1936 J. Tickell *See how they Run* ii. 19 The sort of horse's face that looks out over the half-door of a loose-box. 1882 *Times* 31 Jan. 16/6 (Advt.), Single-stone, half-hoop, and cluster rings. 1892 Kipling *Lett. of Travel* (1920) 94 His wife..wears half-hoop diamond rings. 1902 *Daily Chron.* 14 June 10/4 The hair..is surmounted by a half-hoop diadem encrusted with precious stones. 1928 R. Hall *Well of Loneliness* xxi. 198 'I made your mother's engagement ring for him; a large half-hoop of very fine diamonds.' 1960 R. Collier *House called Memory* x. 150 Uncle William Henry bought Dora a half hoop of diamonds as an engagement ring. 1803 T. Sheraton *Cabinet Dict.* 44 As to the particular management of beds, and the articles required in mounting them, together with their various classes; these..will most conveniently come under their respective names, as..Half-tester,..&c. 1859 *Blackw. Mag.* Aug. 229/1 We approached the bed and examined it—a half-tester, such as is commonly found in attics devoted to servants. 1960 H. Hayward *Antique Coll.* 279/2 At this period [sc. late 17th cent.] a new type of bedstead without footposts was introduced, known as a 'half-tester'.

m. Further examples: *half-believer* (earlier example), *-consciousness, -dark, -darkness, -dream, -education, -hint, -humour, -laugh, -lengthening, -lie, -literate, -look* (examples), *-mind, -power, -savage, -spacing* (on a typewriter), *-whisper* (earlier example).

1814 Jane Austen *Mansf. Park* II. vii. 148 A certain half-look attending the..expression of his hope. 1816 —— *Emma* II. xvii. 320 Much that passed between them was in a half-whisper. 1836 J. S. Mill in *Westm. Rev.* XXV. 5 See how incapable half-savages are of co-operation. 1838 —— *Ibid.* XXVIII. 454 The sceptics and half-believers of the story. —— *Ibid.* XXXI. 484 Almost all rich veins of original and striking speculation have been opened by systematic half-minds. 1862 G. Borrow *Wild Wales* II. xxxii. 370 'In truth I am,' said she, with a half laugh. 1881 'Mark Twain' *Prince & Paup.* 208 His senses struggled to a half-consciousness. 1895 Kipling *Day's Work* (1898) 344 Leading him on to see, more by half-hints than by any direct word, how boys and men are all of a piece. 1898 *Pearson's Mag.* May 539/1 With a half-power steamer which had only one man all told upon her decks. 1900 *Daily News* 18 Aug. 6/1 How she did it she didn't know, she said, in a half-humour manner. 1904 W. de la Mare *Henry Brocken* 13 The half-dream [which] weariness brings. 1904 W. B. Yeats *Tables of Law* 9 The formalisms of half-education. 1905 *Westm. Gaz.* 25 Mar. 11/2 The inaccuracies do not matter very much unless they are so gross as to shock the great half-literate. 1926 Fowler *Mod. Eng. Usage* 399/2 The uneasy half-literates who like to prove that they can spell. 1927 A. Clarke *Son of Learning* ii. 44 In the half-darkness his cowled figure suggests demonic possession. 1928 D. H. Lawrence *Lady Chatterley* xvi. 278 She could still see on Connie's face.. the half-dream of passion. a 1930 —— *Last Poems* (1932) 257 A half-lie causes the immediate contradiction of the half-lie. 1934 Blunden *Choice or Chance* 40 Possessions too,—part fungus and part flower,—Forced on him their half-power. 1934 E. Pound *Eleven New Cantos* xxxix. 45 From star up to the half-dark. 1937 *Mind* XLVI. 101 Postulating a sort of extra, separate half-mind—entelechy—like Driesch. 1938 *Times Lit. Suppl.* 8 Oct. 638/3 The curious delicate half-humour which smiled at his own hypersensitiveness. 1941 T. S. Eliot *Dry Salvages* ii. 11 The backward half-look Over the shoulder, towards the primitive terror. 1948 —— *Notes Def. Culture* 105 But what is important is to remember that 'half-education' is a modern phenomenon. 1949 Koestler *God that Failed* 31 Once or twice she spoke on the telephone to comrades of hers—always in half-words and half-hints. 1953 K. Jackson *Lang. & Hist. Early Brit.* 342 Half-lengthening of penultimates could not have arisen until after the accent-shift. 1959 E. Pulgram *Introd. Spectrogr. Speech* viii. 64 The half-power points, whose power is..proportional to the square of the amplitude. 1961 *Imperial Type Faces* 10 A half-spacing device lends itself to display work. 1962 *Which?* Dec. 359/2 If you leave out a letter, it should be possible, after rubbing out the word, to fit in the extra letter by using half-spacing.

n. **half-Abo** *Austral.*, a person having one Aboriginal parent (cf. *Abo *a.* and *sb.*); **half-adder** [*Adder[1] 3], a unit in an electronic computer (see quot. 1962); **half-arsed, -ass, -assed** *adjs. slang* (orig. *U.S.*), ineffectual, inadequate, mediocre; stupid, inexperienced; **half-belt** (see quot. 1957); **half-bull**, (*b*) [Bull *sb.*[1] 7] *slang*, a half-crown; **half-cell** *Electr.* (see quot. 1943); **half-centre** (see quot.); **half-commission** *attrib.*, working for or based on half commission; **half-compression** *attrib.*, designating a device for lessening the compression of the explosive mixture in an internal-combustion engine; **half-day**, half a working day (cf. *day *sb.* 8 d); **half-dress** (see quot. 1960); **half-duck** = Half-bird; **half-evergreen** *a.*, of a plant that is evergreen in a mild climate; also as *sb.*; **half-foot** (see quot. 1880); **half-frame**, (*a*) *pl.* reading-spectacles consisting of only the lower half of the frames and lenses; also in sing. *attrib.*; (*b*) (*Photogr.*) half the standard 35 mm. picture size; **half-gerund** (see quot. 1924); **half-hose** (see Hose *sb.* 1 a γ); **half inferior** *a.* *Bot.* (see quots.); **half-integer**, any member of the set of numbers obtained by dividing the odd integers by two; hence **half-integral** *a.*; **half-landing**, a landing half-way up a flight of stairs; **half-lap** (see Lap *sb.*[3] 2 b); **half-lattice girder**, one consisting of a single system or row of triangles; **half-lift** [*lift *sb.*[2] 5 f], a medium-stressed lift in O.E. verse; **half-line** *Pros.* [Line *sb.*[2] 23 e], half of a line of verse, used esp. of O.E. and related verse; **half-pass** (see quot. 1948); **half-period**, the *half-life of a substance; see also quot. 1904; **half-pint** [cf. sense II. a] *fig.*, a small or insignificant person; also *attrib.*; **half-plane** *Math.* (see quot. 1959); **half-plate** *Photogr.* (see Plate *sb.* 5 c), also *attrib.*; **half-portion**, a half of a portion; *fig.*, a small or insignificant person; **half-race** *Bot.* (see quots.); **half-rhyme**, an imperfect or near rhyme; hence **half-rhymed** *ppl. a.*; **half-ripper, half-rip saw**, a finer-toothed ripping saw (see Ripper 2, Rip-saw *sb.*); **half-roll** *Aeronaut.*, a manœuvre in which the aircraft turns through 180° about the longitudinal axis; hence **half-roll** *v. intr.*; **half-secret dovetail** (see quot.); **half-sheet**, (*b*) (see quots.); **half-sibling**, each of two or more individuals having one parent in common; **half-slip**, the lower half of a slip (see Slip *sb.*[3] 4 c); a petticoat; **half-sole** *v.* (earlier U.S. example); **half-speed shaft**, the cam shaft of a four-stroke cycle internal-combustion engine, which rotates at half the speed of the crank shaft; **half-stress** *Pros.*, a secondary stress; **half-stuff** (later example); **half-term**, a period approximately half-way through a school or other term, often made the occasion of a holiday; freq. *attrib.* as in *half-term holiday*; **half-thickness** *Physics* = *half-value thickness*; **half-turning bolt** (see quot.); **half-uncial**, writing which combines the characters of uncial and cursive; semi-uncial; also *attrib.*; **half-valve** *Mus.* (see quot. 1955); hence **half-valved** *ppl. a.*; **half-valving** *vbl. sb.*; **half-verse** *Pros.*, = *half-line* (above); **half-virgin** = *demi-vierge; **half-watt** *a.* *Electr.*, applied to a gas-filled incandescent lamp consuming approximately a half-watt per candle-power; **half-wave**, one-half of a complete wave, esp. of electricity or radiation; freq. *attrib.*, utilizing only alternate halves of a sequence of waves, as *half-wave rectification* (*Chambers's Techn. Dict.* 1940); half a wave-length long, as *half-wave antenna, dipole*, etc.; **half-white** = Half-breed 2; **half-word, halfword**, a group of consecutive bits, which can be handled as a unit, occupying half a word storage unit of a computer; **half-world**, the demi-monde (earlier and later examples); also *transf.*

1945 C. Mann in B. James *Austral. Short Stories* (1963) 72 Little black half-abo. piccaninnies. 1954 *Electronic Engin.* XXVI. 288 Numbers always enter the accumulator loop via the two half-adders and are therefore automatically added to the previous contents. 1962 *Gloss. Terms Autom. Data Processing* (*B.S.I.*) 60 *Half-adder*, a logic element with two outputs and two inputs to which may be applied signals representing a digit or a number and a single addend or carry digit. 1961 A. West *Trend is Up* ix. 386 You don't know what it is to worry about what half-arsed thing your own son is going to pull on you next. 1972 *Observer* 24 Sept. 35/2 The sort of half-arsed dottiness they dish out in West End comedies. 1959 N. Mailer *Advts. for Myself* (1961) 399 He spent years hobnobbing with gentlemanly shits and half-ass operators. 1932 *Amer. Speech* VII. 333 *Half-assed*, mediocre; insignificant. 1955 W. Gaddis *Recognitions* I. v. 183 A half-assed critic..thinks he has to make you unhappy before you'll take him seriously. 1957 M. B. Picken *Fashion Dict.* 19 *Half b[elt]*, belt which extends only half-way around body; especially one across back section of garment, as on sports jackets or coats. 1960 *Farmer & Stockbreeder* 9 Feb. Suppl. 5/1 A..jacket with a half-belt at the back. 1789 Half bull [see *bob *sb.*[8]]. 1852 Halfbull [see Bull *sb.*[1] 7]. 1906 E. Dyson *Fact'ry 'Ands* xvi. 216 I've gotter get that 'arf-bull 'r somethin' dangerous may set in. 1940 *Chambers's Techn. Dict.* 399/1 Half-cell. 1943 *Gloss. Terms Electr. Engin.* (*B.S.I.*) v. 93 *Half cell*, of an electrolytic cell: an electrode and that part of the electrolyte with which it is in contact. 1888 *Lockwood's Dict. Mech. Engin.*, *Half-centre*, half-centre is sometimes used to denote the position of the crank-pin of an engine when midway between the two dead centres or dead points. 1909 *Westm. Gaz.* 16 Feb. 7/4 He became a half-commission man with a firm of stockbrokers. 1927 *Sunday Express* 13 Mar. 2 A half-commission stockbroker. 1931 *Times* 16 Mar. 18/1 The Half Commission Practice. 1901 *Motor-Car World* II. 317/1 To facilitate starting the engine a half-compression device is fitted which operates on the exhaust valve through the medium of a second or subsidiary cam attached to the main cam working the exhaust. 1907 *Westm. Gaz.* 11 Nov. 7/2 The simple half-compression gear. 1791 F. Burney *Jrnl.* 31 July (1972) I. 2 One precious half Day I was indulged with my kindest Mr. Lock. 1876 Lady C. Schreiber *Jrnl.* (1911) I. 477 We found that we could not execute all our little shoppings.., because the Saturday is but a half-day. 1932 *Discovery* Mar. 71/2 Now the minimum wage [for farm workers] is 30s. per week of five shorter days and a half day. 1973 *Daily Tel.* 8 Feb. 1/1 The Shadow Cabinet decided to give up a half-day of its Parliamentary time to complete the debate in the Commons. 1788 E. Sheridan *Jrnl.* (1960) 138 Great coats made very open before to shew the peticoat—in Undress—half dress, Night gown and peticoat with fine muslin Aprons—full dress I have not seen. c 1810 W. Hickey *Mem.* (1918) II. xx. 261 A tailor named Knill..advised my having a dark green with gold binding..and for half dress a Bon de Paris with gold frogs. 1815 *Belle Assemblée* Sept. Pl., Autumnal Walking Dress...The head dress must be either a half-dress cap, or a white satin gipsy, or Wellington hat. 1850 *Ladies' Gaz. Fashion* Aug. 255/1 Plain *mousseline de soie*.. begins to be a good deal seen in half-dress. 1960 Cunnington & Beard *Dict. Eng. Costume* 100/2 *Half dress*, late 18th and 19th c's...The costume worn at day functions and at informal evening ones. 1892 W. J. Gordon *Our Country's Birds* 10 Local and Popular Names...Half Duck. 1903 J. A. Hamilton *MS. in Red Box* 329 Good sport among the half-duck and mussel-duck which abounded at Tudworth. 1934 Webster, Half-evergreen. 1952 A. G. L. Hellyer *Sanders's Encycl. Gardening* (ed. 22) 131 Glaucophylla, half-evergreen. 1814 J. Sinclair *Agric. Scot.* App. ii. 396 *Half foot*, is another method of occupying a farm, equally barbarous in itself, and adverse to improvement. It is not so prevalent in the Highlands, as in some of the Western Isles. 1873 *Trans. Highland Soc.* 298 *Out* or *led* farms like the *metayers* of France, or the *half-foot* tenants of the Hebrides. 1880 W. Skene *Celtic Scot.* III. 370 A kind of tenancy called half-foot, where the possessor of the farm furnished the land and seed corn,..the produce being divided. 1961 *Colour Photogr.* vi. 249/3 The normal 20-exposure cassette gives 40 half-frame exposures and the 36-exposure cassette gives 72 so that each colour transparency would cost less than 6d. 1967 A. Flowers in L. Deighton *London Dossier* 170 Buy one of the half-frame cameras such as the Olympus Pen D2 or the Canon Dial. 1968 J. Hudson *Case of Need* i. i. 10 'Call me,' Sanderson said, peering over his half frames. 1968 L. Deighton *Only when I Larf* ii. 17 Strange half-frame spectacles that he peered over abstractedly. 1898 H. Sweet *New Eng. Gram.* ii. 121 The absence of a distinction between common case and genitive in the plural often makes it impossible in the spoken language to distinguish between gerund and half-gerund, as in *to prevent the ladies leaving us, I generally ordered the table to be removed*.., where the..alteration of *ladies* into *ladies'* would make *leaving* into a full gerund. *Ibid.*, There seems little doubt that the colloquial half-gerunds in such causal constructions as *she caught cold sitting on the damp grass*..have arisen through dropping a preposition. 1924 H. E. Palmer *Gram. Spoken Eng.* 168 In certain constructions the ing form has a function intermediate between that of the present participle and the gerund. Sweet suggests for such cases the term *half-gerund*. 1900 B. D. Jackson *Gloss. Bot. Terms* 118/1 *Half inferior*, used of an ovary when the stamens are perigynous. 1940 *Chambers's Techn. Dict.* 399/1 *Half inferior*, said of a flower in which the receptacle forms a cup which is adherent to the base of the ovary and partly up its side. 1928 *Proc. Physical Soc.* XL. 332 The number *m* can take only certain discrete values (either integers or half-integers). 1971 P. Hlawiczka *Introd. Quantum Electronics* xvi. 247 The quantum numbers j_1 and j_2 may be either integers or half integers. 1930 Ruark & Urey *Atoms, Molecules & Quanta* vii. 187 Many authors use $j+\frac{1}{2}$ instead, when *j* is half-integral. 1968 M. S. Livingston *Particle Physics* ii. 24 Nuclei with an odd number of protons plus electrons, each with half-integral spin, should result in half-integral nuclear angular momenta. 1910 *Daily Chron.* 1 Feb. 1/1, I saw the proprietor, ..perched on the half-landing of the stairs. 1965 'T. Hinde' *Games of Chance* i. i. 16 The black and lemon half-landing bathroom. a 1877 Knight *Dict. Mech.* II. 1049/2 *Half-lattice girder*, a form of girder..consisting of horizontal upper and lower bars, and a series of diagonal bars, sloping alternately in opposite directions, and dividing the space between the bars into a series of triangles. 1894 H. Sweet *Anglo-Saxon Reader* (ed. 7) p. xc, To make up for the want of an accompanying dip, an extra medium-stressed half-lift is made obligatory. 1967 C. L. Wrenn *Study O.E. Lit.* 38 Five basic combinations of stress or lift of the voice, half or secondary stress or half-lift, and unstressed syllables. 1864 H. Morley *Eng. Writers* I. 251 The most important is a heroic poem.. extending..to 6357 of the short Anglo-Saxon lines, or half lines, as they are usually printed. 1900 A. S. Cook *Christ of Cynewulf* 70 Half-line space. 1910 F. Tupper *Riddles of Exeter Bk.* 220 The half-line is of the A-type..

common in the *Riddles*. 1927 E. V. Gordon *Introd. Old Norse* 293 In ON. tradition the unit verse was not the long line, but the half-line, which was called a *vísa* or line. 1964 *English Studies* XLV. 38 The patronymic epithet is separated from its antecedent by at least a full half-line. 1965 *Ibid.* XLVI. 420 In Germanic the basic principle is stress, together with a division into half-lines. 1929 *Man. Horsemastership, Equitation & Driving* ii. 115 The bending lesson includes..the 'half passage' or 'half pass'... In all lateral movements the forehand must slightly precede the hind quarters. 1948 E. Schmit-Jensen *Equestrian Olympic Games* App. 95 At the Half Pass the horse moves on two tracks... The outside legs pass and cross in front of the inside legs... The legs on the side to which the horse is moving are the inside legs; those on the opposite side the outside legs. 1904 Goodchild & Tweney *Technol. & Sci. Dict.* 277/2 The area included in this curve is the first half period element. *Ibid.*, The effect of the whole wave can be expressed in terms of these half period components. 1905 *Nature* 13 Apr. 574/1 Different samples gave for the half-period of decay from 52 to 55 seconds. 1942 J. D. Stranathan *'Particles' of Mod. Physics* xi. 448 One product of the nuclear disruption was a Ba isotope having a half period of 86 minutes. 1926 Maines & Grant *Wise-Crack Dict.* 9/1 *Half pint*, stunted individual. 1929 W. R. Burnett *Little Caesar* iv. ii. 117 'Here's the half-pint,' said Killer Pepi, pushing Joe Sansone forward. 1930 G. C. Myers *Mod. Parent* ix. 165 Here are types of remarks which some parents..will stoop to make: 'What do you think of our half-pint?' 1938 Wodehouse *Summer Moonshine* xvii. 200 That wonder girl in whose half-pint person were combined all the lovely qualities of woman of which he had so often dreamed. 1943 C. H. Ward-Jackson *Piece of Cake* 35 *Half-pint hero*, swaggerer. 1948 'N. Shute' *No Highway* iii. 82 The little half-pint size, with thick glasses? 1891 Half-plane [see Orthomorphic *a.* 2]. 1927 H. G. Forder *Found. Euclidean Geom.* iii. 68 The regions into which a line *a* separates a plane *ω* in which it lies are the 'half-planes from *a* in *ω*'. 1959 G. & R. C. James *Math. Dict.* 182/1 *Half-plane*, the part of a plane which lies on one side of a line in the plane. 1968 P. A. P. Moran *Introd. Probability Theory* x. 473 In the cases considered the domain is a rectangle, an infinite strip, a half-plane, or infinite quadrant. 1877 *Design & Work* III. 451/1 Half-plate portrait lens. 1892 *Photogr. Ann.* II. 58 On your slide you require to get all the view on the half-plate negative. 1903 A. Watkins *Photogr.* (ed. 2) 13 Half-plate is the favourite amateur size. 1907 F. H. Burnett *Shuttle* xxxviii. 379 Adroit manipulation of 'portions' and 'half portions' ..enabled them to add variety to their bill of fare. 1919 Wodehouse *Coming of Bill* (1920) i. v. 59 He certainly is a kind o' half-portion, ma'am! 1967 —— *Company for Henry* vii. 130 Even when calling her a squirt and a half-portion he had thought of her as a comely squirt and a half-portion with plenty of sex appeal. 1967 L. Deighton *Expensive Place* ii. 14 [He] did not like orders for..*charcuterie* as a main dish or half-portions of anything. 1906 R. H. Lock *Variation, Heredity & Evolution* v. 141 In the case of a half-race a small percentage only of seedlings is found to produce plants which show the racial character. *Ibid.* 145 A half-race might have been defined as a strain in which the character of the complete race is usually latent, and only rarely appears. 1928 B. D. Jackson *Gloss. Bot. Terms* (ed. 4) 169/2 *Half-race*, a form intermediate between a species and a variety of it, producing but few seedlings of the racial character, the majority reverting to the specific type. 1830 B. Thorpe tr. *Rask's Gram. Anglo-Saxon Tongue* 139 Line-Rime is when two syllables, in the same line of verse, have their vowels and the consonants following them alike, which is called perfect rime (consonances), or unlike vowels, and only the following consonants the same, which is called half rime (assonances). 1860 G. P. Marsh *Lect. Eng. Lang.* xxv. 553 In Icelandic poetry,..imperfect rhyme is regularly employed, and..is called skothending,..which we may conveniently translate by *half-rhyme*. *Ibid.* 560 Although half-rhyme may be said to be peculiar to Icelandic poetry, ..yet there are examples of the employment of both full and imperfect line-rhymes in modern English. 1873–4 G. M. Hopkins *Note-Bks.* (1937) 243 In Icelandic verse an opposite kind of alliteration (skothending) is made use of, namely ending with the same consonant but after a different vowel, as *'bad' 'led', 'find' 'band', 'sin' 'run'* (from Marsh, who calls it *half-rhyme*). 1886 J. M. D. Meiklejohn *Eng. Lang.* ii. 186 The English language is very poor in rhymes, when compared with Italian or German. Accordingly, half-rhymes are admissible..: sun/gone, love/move, allow/bestow, etc. 1936 M. Roberts *Faber Bk. Mod. Verse* 28 In Owen's war poetry, the half-rhymes almost invariably fall from a vowel of high pitch to one of low pitch. 1960 D. S. R. Welland *Wilfred Owen* vi. 115 Only in 'Arms and the Boy', 'Wild with All Regrets', and 'Strange Meeting' does he [sc. Wilfred Owen] write in half-rhymed couplets throughout. 1841 *Penny Cycl.* XX. 476/2 The ripping-saw, half-ripper, hand-saw..are saws for the use of one person. 1846, 1875 Half-rip saw [see Rip-saw *sb.*]. 1926 J. M. Grider *War Birds* 206, I half rolled on top of him, he half rolled too. 1934 V. M. Yeates *Winged Victory* i. iii. 29 The same with the half-roll. Nothing would half-roll like a Camel. A twitch of the stick and flick of the rudder and you were on your back. 1904 Goodchild & Tweney *Technol. & Sci. Dict.* 277/2 *Half secret dovetail*, a dovetail of the form used in a drawer front; it is concealed in a front view, but visible in the side of the drawer when drawn out. 1683–4 J. Moxon *Mech. Exerc. Printing* (1962) 227 In Half-sheets, all the Pages belonging to the White Paper and Reteration are Imposed in one Chase, and are plac'd, as you see by the Drafts..of Half-sheet Forms. 1888 C. T. Jacobi *Printers' Vocab.* 56 Book-work is sometimes printed in 'half-sheet' fashion. When thus printed there are two copies on one sheet. 1914 T. L. De Vinne *Mod. Book Composition* ix. 336 It is called half-sheet because this larger sheet must be cut in halves before either half can be folded. 1952 E. J. Labarre *Dict. Paper* (ed. 2) 122/1 *Half-sheet*, printing a whole sheet of a book so that all the pages of a signature are in one forme. 1964 F. Bowers *Bibliogr. & Textual Crit.* iv. ii. 109 The work-and-turn method of printing by half-sheet imposition. 1903 *Biometrika* Nov. 391 The high values, however, found for half-siblings in the case of the thoroughbreds. 1938 *Jrnl. R. Anthrop. Inst.* 213 The mother allows the children to invite their half-siblings. 1959 B. Wootton *Social Sci. & Social Pathol.* iii. 85 Some authors state whether deceased, half- or step-siblings are included, or whether the delinquent himself is counted in the total of family members. 1952 H. Waugh *Last seen Wearing* (1953) 31 'What about her..undergarments?' 'Half-slip, pants, and bra.' 1957 M. B. Picken *Fashion Dict.* 313/1 *Half slip*, any of the various types of slips starting at the waistline. 1795 J. & E. Pettigrew *Let.* 5 Apr. (MS.) (D.A.E.), I have not got my shoes halfsoaled yet, as shewmakers are very scarce. 1902 A. C. Harmsworth et al. *Motors* viii. 152 A crank, operated by a connecting rod from the half-speed shaft on the engine. 1905 Goodchild & Tweney *Technol. & Sci. Dict.* s.v. *Motor Cycles*, The half speed shaft, rotating at one half the speed of the crank shaft. 1938 A. Campbell *Battle of Brunanburh* 24 Graz..regarding *butu* as having a half-stress on the second syllable. 1961 *Rev. Eng. Stud.* XII. 345 Long and short syllables must be distinguished in scansion, when they bear either a strong stress or a half-stress. 1912 *Chambers's Jrnl.* Oct. 671/2 The pulp—at this stage commonly called half-stuff—is fed into beating-engines. 1888 *Boy's Own Paper* Summer No. 16/2 At half-term it was Hoskyn's custom to write letters to all the parents with reports of their sons' progress. 1944 L. A. G. Strong *All fall Down* 55 It's half term, as even you must have realised. 1950 Hodgkinson & Muir in R. M. Scrimgeour *North London Collegiate Sch. 1850–1950* v. 101 Miss Drummond spent a half-term morning in her Sussex cottage trying to see whether, if girls spent the whole day at Canons, they could be supplied with adequate teaching there. 1950 Glasstone *Sourcebk. Atomic Energy* vii. 170/2 The mass half-thickness, i.e., the actual..half-thickness multiplied by the density, is almost independent of the material absorbing the gamma rays. 1958 W. K. Mansfield *Elem. Nucl. Physics* v. 45 To compare the penetrating power of the γ-rays with that of α and β-rays, it is necessary to estimate the half-thickness. 1960 J. N. Gregory *World of Radioisotopes* i. 9 The shorter the wavelength of the γ-radiation the greater the half-thickness. a 1877 Knight *Dict. Mech.* II. 1050 *Half-turning bolt*, one with a thread occupying one half of its cylindrical surface. 1885 *Encycl. Brit.* XVIII. 153/2 Examples of half-uncial writing. 1897 H. W. Johnston *Lat. MSS.* 70 Half-Uncials are derived from the uncials and represent the last efforts of the book hand to differentiate itself from the improved business hand of the time... It is also called the Roman Uncial and Pre-Caroline Minuscule. 1912 E. M. Thompson *Introd. Gr. & Lat. Palaeogr.* 305 It is the Half-uncial hand which we find employed as far back as the fifth century as a literary hand in the production of formally written MSS. 1926 E. A. Lowe in Crump & Jacob *Legacy of Middle Ages* 209 Before developing a minuscule Irish calligraphers had created a majuscule, the Irish half-uncial as it is styled, of which the Book of Kells, a work of unsurpassed skill and artistry, is the most eminent example. 1946 Mezzrow & Wolfe *Really Blues* i. 12 Yellow's half-valve inflections and slurs. 1955 L. Feather *Encycl. Jazz* vii. 292 *Boy meets Horn*, a showpiece displaying the novel style and tone he popularized using 'half-valve' effect (a squeezed tonal sound obtained by depressing the valve halfway). 1958 S. Dance in P. Gammond *Decca Bk. Jazz* xxiii. 297 Original creations like *Boy meets horn* with its stifled, half-valved statements. 1956 S. Traill *Play that Music* viii. 87 Some players achieve a sort of 'blue note' effect by 'half-valving', which would mean, in our particular example, pressing the middle valve (for a correct E♭) half down. The distortion of tone which this produces is not very attractive, and I would discourage 'half-valving' altogether. 1711 Half Verse [see Hemistich]. 1876 H. Sweet *Anglo-Saxon Reader* p. xcviii, There is often only one alliterative letter in the first half verse. 1907 F. A. Blackburn *Exodus & Daniel* p. x, Uncorrected errors are few, though occasional omissions occur, generally of a half-verse. 1938 A. Campbell *Battle of Brunanburh* 16 It will be convenient to group its half-verses in the 'five types' of Sievers. 1946 Koestler *Thieves in Night* 165 As obscene and shocking to me as a petting party with a half-virgin. 1965 N. Freeling *Criminal Conversation* ii. xv. 164, I imagined, being full of valuable premedical catch-phrases, that she was 'half-virgin' and therefore despicable. 1913 *Lighting Jrnl.* I. 207/1 (*heading*) A half watt lamp!..The types which it is expected to first develop..operate at an efficiency of half a watt per candlepower. 1915 *Nature* XCVI. 407/1 With electricity generated in modern power-houses, and ordinary metal filament lamps, 750,000 candle-power-hours are generated per ton of coal, compared with 260,000 C.P. per ton of coal when gas and modern gas mantles are used. The extended use of so-called 'half-watt' lamps will soon double this 750,000. 1932 N. Royde-Smith *Incredible Tale* 57 The white glare of the half-watt lamp hanging from the studio roof. 1904 Goodchild & Tweney *Technol. & Sci. Dict.*, *Half wave plate*, a plate of doubly refracting crystal, capable of splitting up a plane polarised ray into two portions, one of which is retarded half a wave length with respect to the other. 1912 *Motor Man.* (ed. 14) ii. 33 When the platinum contacts at the end of the armature touch, one-half of every complete wave flows into the accumulator, and when the contacts separate, the reverse wave of the current is interrupted at the zero or no-voltage line; thus only the half-waves of current flowing in the same direction are used. 1926 R. W. Hutchinson *Wireless* viii. 145 If the first half-wave is positive the grid will become positive. *Ibid.*, The next half-wave is negative and this still further lowers the potential of the grid. 1928 *Morning Post* 6 Feb. 3/4 A half wave rectifier. 1943 *Gloss. Terms Telecommunications* (*B.S.I.*) 67 *Half-wave dipole*, a straight aerial symmetrical in regard to its standing-wave current and usually approximately half a wavelength long. 1962 Simpson & Richards *Junction Transistors* xii. 276 The same is true of any amplifier, such as the Class AB, whose output is asymmetrical and consists of pulses that are larger than one half-wave. 1962 Corson & Lorrain *Introd. Electromagn. Fields* xiii. 446 After having mastered the electric dipole, we shall be able to study the radiation field of the half-wave antenna—the type commonly used for transmitting radio waves. 1866 'Mark Twain' *Lett. fr. Hawaii* (1967) 26 Foreigners and the better class of natives, and 'ha whites' in carriages. 1897 —— *Following Equator* 63, I asked after 'Billy' Ragsdale, interpreter to the Parliament in my time—a half-white. 1901 *Daily Colonist* (Victoria, B.C.) 23 Oct. 8/3 In this boat's crew..was Charlie Diamond... He was a Bonin island half-white and is well known to old time sealers. 1959 J. Jeenel *Programming for Digital Computers* ii. 60 One might choose a 12-digit word size for the calculator and have a word represent either one number or a pair of instructions. The single-address instructions would be represented by 'half words'. 1964 Halfword [see *byte]. 1970 O. Dopping *Computers & Data Processing* vi. 105 Some machines pack two instructions per cell, and there are different addresses for each half-word, e.g. even addresses for left half-words, odd addresses for right half-words. 1870 D. J. Kirwan *Palace & Hovel* xliii. 613 'Baby Hamilton' is another celebrity of the Half-World. Many stories are told about the recklessness of this girl. 1874 *Porcupine* 21 Feb. 742/1 Those moral magistrates who have so distinctly set their faces against Cremorne and other outdoor haunts of the 'half-world'. 1950 Half-world [see *folk-jazz* s.v. *folk 6]. 1972 *Times* 6 Apr. 7/5 Away from his chosen half-world, Munby's social life was passed in the first literary and artistic circles of his day.

half-and-half. Add: **2. b.** A half-breed or half-caste.

1827 J. F. Cooper *Prairie* iii. 58 The half-and-halfs, that one meets in these distant districts, are altogether more barbarous than the real savage. 1922 Joyce *Ulysses* 315 Pity about her... Or any other woman marries a half and half.

half-baked, *a.* Add: Hence as *sb.*, a half-baked person. *colloq.* and *dial.*

1866 J. Sleigh in *Reliquary* VI. 160 *Half-char* or *half-baked*, a foolish fellow. 1923 U. L. Silberrad *Lett. J. Armiter* ix. 192, I believe girls were better off really with the old lock-and-key, guard-the-girls sort than with these half-bakeds, who let 'em have their heads.

half-binding. (Earlier U.S. example.)

1821 M. L. Weems *Lett.* III. 325 [Books] in neat half binding, red backs, and corners.

half-blood. Add: **4.** *attrib.* Half-blooded.

1830 in *Wisconsin State Hist. Soc. Coll.* (1892) XII. 185 He is a half Blood St. Regis, with a half Blood Menomonie wife. 1835 C. F. Hoffman *Winter in West* I. 215 The driver was also accompanied on the box by a well made young half-blood Chippeway. 1837 H. Colman *Rep. Agric. Mass.* (1838) 52, I have slaughtered two half-blood heifers, which have weighed at four years old over 700 lbs. 1873 J. H. Beadle *Undevel. West* xxi. 406 A handsome half-blood daughter married to a white man. 1943 *Sun* (Baltimore) 4 Dec. 12/6 A further sprinkling of transactions is noted in fine and half-blood wool.

half-bred, *a.* Add: **1. b.** Of a sheep. *Austral.* and *N.Z.*

1891 R. Wallace *Rural Econ. Austral. & N.Z.* xviii. 265 The half-bred sheep, being the various crosses between Merino ewes and long-woolled rams. 1959 Baker *Drum* ii. 116 *Halfbred sheep*, orig. a sheep by a longwool ram from a merino ewe; now loosely applied to the type.

B. *sb.* A half-bred sheep. *Austral.* and *N.Z.*

1848 E. G. Wakefield *Let.* in *N.Z. Jrnl. Agric.* (1950) Feb. 136/3 The want of the more prolific and hardy constitution of the half-bred. 1891 R. Wallace *Rural Econ. Austral. & N.Z.* xviii. 265 The half-breds, or first crosses, are the most profitable sheep. 1930 L. G. D. Acland *Early Canterbury Runs* i. 7 Corriedales and Half-breds have displaced the Merinos except on the lightest and highest country. 1956 G. Bowen *Wool Away!* (ed. 2) xii. 134 Half-breds are recognised in New Zealand as a cross between Merinos and one of the English longwool breeds.

half-breed. Add: **1.** (Later example.)

1858 C. L. Flint *Milch Cows & Dairy Farming* (1860) ii. 69 Qualities which are, in a measure, artificial.. change not only with the breed of one species, but with the different individuals of the same breed, of the same half-breed, and often of the same family.

2. (Earlier U.S. example.)

1760 *Newport* (R.I.) *Mercury* 22 Apr. 2/1 On the 18th a Half-Breed, who is a Leader and Head Warrior.. came..to Fort Augusta.

b. *transf.* and *fig.*

1846 *Quincy* (Ill.) *Whig* 27 Jan. 2/4 All the Jacks in the county, consisting of T. H. Owen, John Harper, Backenstos, Bedell, and a few 'half breeds'. 1883 'Mark Twain' *Life on Mississippi* xliv. 402 This reminds me that a remark of a very peculiar nature was made here in my neighbourhood (in the North) a few days ago: 'He hadn't ought to have went.' How is that? Isn't that a good deal of a triumph? One knows the orders combined in this half-breed's architecture without inquiring: one parent Northern, the other Southern. 1952 A. Stevenson *Speeches* (1953) 179 I'm a half-breed myself. My father was from an old, staunch Democratic family, and he was a Presbyterian. My mother was from an equally old and staunch Republican family and she was a Unitarian.

4. (Earlier U.S. example.)

a 1762 S. Niles *Wars* in *Mass. Hist. Soc. Coll.* (1861) 4th Ser. V. ii. 538 One Molton, a half-breed fellow,.. seized the fellow that wounded Mr. Atkins.

half-cock, *sb.* Add: **3.** *attrib.*, as **half-cock shot** or **stroke** *Cricket*, a stroke begun as a forward stroke but checked half-way, the ball being allowed to hit the bat.

1888 R. H. Lyttelton in Steel & Lyttelton *Cricket* ii. 52 He may, after he has got forward and perceived his error, effect a compromise and perform what is sometimes called a 'half-cock stroke'. 1909 *Westm. Gaz.*

17 Apr. 16/2 He may occasionally use a half-cock stroke with the left leg well up to the bat. **1959** *Times* 29 May 4/4 Hallam..made a half-cock shot to be caught and bowled.

half-cocked, *ppl. a.* and *pa. pple.* [See HALF-COCK *v.*] **a.** Of a gun: at half-cock.
1809 [see COCKED *ppl. a.*[2] 2]. **1833** [see HALF-COCK *v.*].
b. Partly intoxicated. *dial.* and *slang.*
c **1830** T. WILSON *Pitman's Pay* (1843) 54 Half-cock'd and canty, hyem we gat. **1886** W. H. LONG *I.O.W. Dial.* 120 All on 'em was about half cocked. **1888** [see COCKED *ppl. a.*[2] 2]. **1910** *Dial. Notes* III. vi. 453 *Half-cocked,..* Half drunk.
c. *to go off half-cocked*: to speak or act prematurely. *U.S. colloq.*
1833 *Deb. Congress U.S.* 31 Jan. 1521 The gentleman from Maryland has gone off half cocked. **1877** J. HABBERTON *Jericho Road* xvi. 152 Just like you, always goin' off half-cocked. **1920** S. LEWIS *Main St.* 349 Well—I don't suppose I ought to have gone off half-cocked, and not jollied him along. **1940** E. POUND *Let.* 14 Mar. (1971) 340 No use my going off half-cocked on large subjects whereon I have not yet arrived at conclusion.
d. Incompletely prepared or realized. Also quasi-*advb.*
1946 M. SHULMAN *Zebra Derby* (1947) viii. 45 You know, we're not going into this thing half cocked. **1953** A. UPFIELD *Murder must Wait* xxi. 186 I've a half-cocked kind of idea.

half-crown. Add: (From 1970 no longer legal tender.)
Hence **half-crowner,** a person who pays a half-crown for a seat at a performance, etc.; a publication costing a half-crown.
1886 H. BAUMANN *Londinismen* 71/2 Half-crowner. **1890** G. B. SHAW *London Music 1888–89* (1938) 288 The half-crowners energetically cried 'Hear, hear'. **1893** FARMER & HENLEY *Slang* III. 250/1 *Half-crowner*, a publication costing 2s. 6d. **1959** *New Statesman* 25 July 106/2 One can watch the half-crowners filing..before the country-house displays of things properly designed and well made.

half-eagle. *U.S.* (Examples.)
1786 in *Amer. Museum* (1789) II. 182/2 There shall be two gold coins; one..equal to five dollars, to be stamped in like manner [to the eagle], and to be called a half-eagle. **1841** *Congress. Globe* 30 July 269/2 It was an open declaration of war upon the half eagles, the gold currency. ..This gold, in half eagles, was too good for us. **1852** *Knickerbocker* XL. 323 He was about to contribute a half eagle to the funds.

half-god. Delete †*Obs.* and add later examples.
1901 JOYCE *Day of Rabblement* 8 *The Adoration of the Magi*..shows what Mr. Yeats can do when he breaks with the half-gods. **1916** 'Æ' *National Being* ii. 14 The gods departed, the half-gods also. **1951** L. MACNEICE tr. *Goethe's Faust* I. 56 A half-god has dashed it asunder!

ha:lf-ha·rdy, *a.* [See HALF- 1 b and HARDY *a.* 4 b.] Of a plant: needing some protection from cold winter weather. Also as *sb.*
1824 LOUDON *Encycl. Gard.* 881 Half-hardy annual border-flowers. **1852** [see HARDEN *v.* 7]. **1862** [see HALF-1 b]. **1904** *Westm. Gaz.* 6 May 10/1 The fuchsias, the heliotrope, the geraniums..are half-hardies. **1960** *Guardian* 25 June 2/6 Most half-hardy plants are..a bit of a nuisance. *Ibid.*, The true alpine enthusiast will turn his nose up at some of these half-hardies. **1972** *Suttons Seeds (Catal.)* 18 Dahlias..Can be treated as an annual or as a half-hardy perennial. *Ibid.* 37 The annual phlox. One of the best half-hardy annuals.

half horse. *U.S.* [HALF *adv.* 2.] Formerly used in the phr. *half horse and half alligator* (see quots.). Usu. *attrib.*
[**1809** 'D. KNICKERBOCKER' *Hist. N.Y.* (1820) IV. ii. 360 The backwood-men of Kentucky are styled half man, half horse, and half alligator, by the settlers on the Mississippi.] **1828** J. HALL *Lett. fr. West* 47 Eight or ten of those 'half-horse and half-alligator' gentry, commonly called Ohio boatmen. **1847** T. B. THORPE *Big Bear of Arkansas* 14 The half horse and half alligator species of men, who are peculiar to 'Old Mississippi'. **1860** *Oregon Argus* 13 Oct. (Th.), These half horse and half alligator sort of politicians are becoming a stench in the nostrils of the American people. **1948** E. N. DICK *Dixie Frontier* 241 More than once a 'half-horse and half-alligator' possessed accurate information on politics and government.

half joe. *N. Amer.* [f. HALF- II. a + JOE *sb.*[1]] A Portuguese gold coin, worth 3,200 reis, formerly current in North America. (Cf. *Half Johannes*, s.v. JOHANNES.)
[**1772**: see JOE *sb.*[1]] **1775** in *New Hampshire Hist. Soc. Coll.* IX. 88 A Gentleman yesterday gave me a half Jo. **1809** in P. Horry *Life Marion* (1833) 29 He offered..a half joe a-piece for Marion and me to let the recruits go. **1955** E. POUND *Section: Rock-Drill* lxxxix. 53 Doubloons, guineas, half-Joes.

half-leg. *U.S.* Half the height of a man's leg. In phr. *half-leg deep, high.*
1752 J. HEMPSTEAD *Diary* (1901) 599 A great Snow knee deep..last night & this morning half Leg deep & more. **1825** J. NEAL *Bro. Jonathan* I. 112 Natty, makin' his way, through the bushes, half-leg deep. **1832** J. P. KENNEDY *Swallow Barn* II. i. 13 The snow was lying about half-leg deep. **1852** G. N. JONES *Florida Plant. Rec.* (1927) 65, I have a little corn half leg high. **1855** *Ibid.* 133 The spring branch newground [is] knee high, spring branch cut half leg.

half-length. 1. (Later examples.)
1812 *Theatrical Inquisitor* I. 21 *Resolved*... That the armorial bearings of the society be a half-length of the American critic, Mr. Snarler. **1911** *Encycl. Brit.* XXII. 128/2 The small half-length of young Martin van Nieumenhoven.

ha·lf-life. Also **half life. 1.** A life of half the full length; an unsatisfactory way of life. Also *attrib.*, denoting a size of painting half life-size.
1864 *Atlantic Monthly* XIII. 157 (*title*) A half-life and half a life. **1867** D. G. ROSSETTI *Let.* 25 July (1965) II. 624 He would prefer a half-life scale. **1963** *Times Lit. Suppl.* 31 May 393/4 The half-life which men were leading.
2. The time in which the quantity of a substance (or the number of similar objects) in a sample decreases by half, *spec.* (*a*) in *Physics* of a radioactive substance, (*b*) in *Med.* and *Biol.* of a substance accumulated in (an organ of) the body. Also *half-life period.*
1907 *Jrnl. Chem. Soc.* XCI. II. 1281 Rutherford and others have shown that whilst radium A, B, and C have a life-period of only a few hours, radium D has a half-life-period of forty years. **1954** H. SEMAT *Introd. Atomic & Nucl. Physics* (ed. 3) xi. 363 The value of the half-life of the neutron was calculated from a determination of the density of the neutron beam and the number of neutrons decaying per unit time per unit volume. **1955** *Metabolism* IV. 419 Two patients convalescing from recently active rheumatic fever..had hydrocortisone half-life values which were similar to those seen in the control subjects. **1955** *Sci. Amer.* Aug. 37/1 So the hazard of a given radio-isotope depends basically on a composite quantity called the 'biological half-life'—a measure of the duration of its activity within the body. **1964** *Times* 31 Mar. 11/4 The extension of the half-life concept, familiar in the field of radioactivity, to the break-down of pesticide residues is very welcome. **1971** MURPHY & MUSTARD in J.-M. Paulus *Platelet Kinetics* ii. 26 The slope of the regression curve divided into −0·30103..is the estimated platelet half-life. **1971** *Nature* 26 Nov. 233/3 The holes made in leaves by herbivorous insects, the half-life of leaves (four to nine months in most species) and a mass of interesting information about the roots and buttresses of tropical trees. **1972** *Ibid.* 22 Dec. 465/1 Because [244]Pu has an 82 m.y. half life, its presence today, 56 half lives after the formation of the Earth, is a most impressive accomplishment.

half-man. Add: **a, b.** (Later examples.) **c.** In Carlyle's use, one of two eminent men whose knowledge and attainments complement those of the other.
1832 CARLYLE in *Fraser's Mag.* VI. 413/2 They were the two half-men of their time: whoso should combine the intrepid Candour, and decisive scientific Clearness of Hume, with the Reverence, the Love, and devout Humility of Johnson, were the whole man of a new time. **1838** MILL in *Westm. Rev.* XXXI. 484 He [*sc.* Bentham] could be a systematic and accurately logical half-man; hunting half-truths to their consequences and practical applications, on a scale both of greatness and of minuteness not previously exemplified. **1932** R. KNOX *Broadcast Minds* vii. 160 We are introduced to a pack of half-men doing a maypole-dance round the Tree. **1941** V. WOOLF *Between Acts* 90 My child's not my child..I'm a half-man. **1953** R. P. ANSCHUTZ *Philos. J. S. Mill* iv. 61 As Carlyle had regarded Hume and Dr. Johnson as the two half-men of their time, so did Mill regard Bentham and Coleridge as 'the great seminal minds' who succeeded them. To Bentham it was given, Mill considered, to discern the truths with which existing doctrines and institutions were at variance; to Coleridge the neglected truths which lay *in* them. Since moreover it was in these terms also that the generality of early Victorians thought about politics, Mill and Carlyle in their turn came to be regarded as the two half-men of that age. **1954** P. H. JOHNSON *Impossible Marriage* 299 A lot of half-men who don't wash.

half-mast. Add: **b.** *transf.*
1940 L. A. G. STRONG *Sun on Water* 139 With his trousers at half-mast. **1966** *Listener* 10 Feb. 210/1 Its [*sc.* a dog's] tail at half-mast.

half-mile. *attrib.* [f. HALF- II. a.] Extending to, comprising or covering, half a mile. Hence **half-miler,** one who competes in a half-mile race; **half-miling,** the running of such a race.
1799 *Steele Papers* I. 176 The Purse is one hundred and fifty Dollars the first day 3 ½-mile heats. **1901** *Encycl. Sport* I. 48/2 The Amateur Championship meeting has..a Half Mile Race. **1934** T. O. BEACHCROFT *Young Man in Hurry* 49 'All out for the half-mile!'..Why did none of the other half-milers move? **1959** *Times* 18 May 3/3 Down the last back straight Valentin went like a half-miler. **1963** *Times* 27 May 5/6 British half-miling, which has been experiencing something of a renaissance this season, was further strengthened over the weekend.

half-moon, *sb.* Add: **2.** On a finger-nail.
1883 M. MORRIS *Bk. of Health* 912 The laminæ in the half moon, or lunula, near the root, are not supplied so abundantly with blood-vessels as those beneath the rest of the nail. **1914** JOYCE *Dubliners* 84 The half-moons of his nails were perfect. **1952** E. GRIERSON *Reputation for Song* xvii. 144 The nails trimmed short, the half-moons only crescents, almost swallowed by the encroaching skin.
6. half-moon spectacles (or **glasses, specs**), spectacles having lenses shaped like half-moons, used esp. for reading.
1952 B. MALAMUD *Natural* 73 Pop, wearing half-moon specs. **1969** 'G.' NORTH *Procrastination of Sergeant Cluff* xiv. 133 The Duty-constable's reading glasses, wire-framed, half-moon, perched on the tip of his nose. **1969** C. BOOKER *Neophiliacs* viii. 207 Lord Home..looked out over his half-moon spectacles and read the message. **1972** P. TOWNEND *Zoom!* xi. 190 He put on his half-moon spectacles and glanced over them.

halfpenny. Add: Also **ha'penny, hapenny.**
1. Since 1969 used with reference to the 'new penny'.
1969 *Guardian* 30 July 16/1 The halfpenny ceases to be legal tender on Friday [i.e. 1 August]. **1975** *Daily Tel.* 5 May 1/4 Bread prices go up today.., sending the standard loaf up by a halfpenny to 16p.
d. A halfpenny stamp.
1881 *Stamp-Collector's Ann* 9 The penny adhesive stamp of the new type..appeared on the 1st of January, and was followed by the halfpenny and three-halfpence on the 14th October. **1908** *Daily Chron.* 20 Feb. 4/7 Many people..think it necessary to fortify themselves with penny stamps. Others cram on a couple of halfpennies.
2. *not a halfpenny the worse; a bad halfpenny.* (Cf. PENNY.)
1603 S. HARSNET *Declar. Egreg. Popish Impostures* 17 Sylvester, Bonifacius, and some other Popes, haue beene errand deuill-coniurers, and yet theyr holinesse not an halfepeny the worse. **1819** J. H. VAUX *Mem.* II. 154 When a man has been upon any errand, or attempting any object which has proved unsuccessful or impracticable, he will say on his return, It's a bad halfpenny; meaning he has returned as he went. **1850** HAWTHORNE *Scarlet Let.* 22 It was not the first time, nor the second, that I had gone away—as it seemed, permanently—but yet returned, like the bad halfpenny. **1895** *Brewer's Dict. Phr. & Fable* (new ed.) 571/2 *I am come back again, like a bad ha'penny.* A facetious way of saying 'More free than welcome'. As a bad ha'penny is returned to its owner, so have I returned to you, and you cannot get rid of me.
3. b. A form of ear-mark on cattle and horses. *U.S.*
Cf. *halfpenny slit*, 'an ear-mark given to pigs or sheep' (*E.D.D.*).
1658 *Rec. East Hampton, N.Y.* (1887) I. 151 John Woodroff marked a horse colt with a hapenny under the left eare. **1666** *Early Rec. Portsmouth, R.I.* (1901) 266 A halfpeny from the route [of the ear]. **1667** *Ibid.* 269 A halfpeney out of the r[ight ear]. **1702** *Town Rec. Topsfield, Mass.* (1917) I. 124 A..horse..[with] a half penny cut out of the right Ear. **1845** *Early Rec. Portsmouth, R.I.* (1901) 387 The Ear Mark of the Creatures of David Baker is two half pennys before the near or left ear.
4. a. (Further examples.)
1729 E. SMITH *Compleat Housewife* (ed. 3) 86 Slice a halfpenny Loaf. **1747** H. GLASSE *Art of Cookery* xv. 140 Your Oven must be as hot as for Halfpenny Bread. **1762** BOSWELL *London Jrnl.* 21 Dec. (1950) 99 A halfpenny roll,..which I had bought at a baker's. **1903** G. B. SHAW *Man & Superman* II. 54 Here comes the New Man, demoralizing himself with a halfpenny paper as usual. **1908** *Chambers's Jrnl.* Jan. 62/1 It..marks as great advance upon the latter as does the incandescent gas-burner upon the halfpenny dip. **1909** H. G. WELLS *Tono-Bungay* III. iv. 423, I lost three pounds..at ha'penny nap and euchre. **1914** 'SAKI' *Beasts & Super-Beasts* 116 The office of one of the halfpenny dailies. *a* **1930** D. H. LAWRENCE *Last Poems* (1932) 169 Oh carcase with a board-school mind and a ha'penny newspaper intelligence.
b. (Later examples.)
1908 H. L. MENCKEN *Philos. Nietzsche* 284 Reich.. has attained the ha'penny celebrity he seems to crave in much the same manner. **1911** G. B. SHAW *Getting Married* 221 To lie down and let..every halfpenny journalist walk over us.

half-pie (hā·f,pəi), *a. N.Z. slang.* Also occas. half-pi. [perh. ad. Maori *pai* good.] Halfway towards, imperfect, mediocre. Also *absol.* or as *sb.*
c **1926** 'MIXER' *Transport Workers' Song Bk.* 11 There's no half-pie about this kid. **1938** R. FINLAYSON *Brown Man's Burden* 16 She would rather have a Maori who was a real man than a half-pie Pakeha who talked too much. *Ibid.* 78 A few straggling houses and a half-pie store. **1949** H. WADMAN *Life Sentence* I. i. 8 New Zealanders who go home on scholarships and come back half-pi Englishmen. **1952** R. FINLAYSON *Schooner came to Atia* x. 58 Half-pie commie and the gospel-thumpers together! **1955** *Landfall* IX. 274 He hadn't been a real officer, only a half-pie one.

half-round, *a.* and *sb.* Add: **A.** *adj. half-round bead, channel, chisel, gutter, screw.*
1888 *Lockwood's Dict. Mech. Engin.* (1918) 169 Half-round chisel. **1934** *Burlington Mag.* Nov. 209/2 Chinese details, such as the half-round beads ending in small volutes. **1940** *Chambers's Techn. Dict.* 122/2 *Button-headed screws*, screws having hemispherical heads, slotted for a screwdriver; known also as half-round screws. **1967** *Gloss. Sanitation Terms (B.S.I.)* 13 *Half-round channel*, a channel (def. 2) of semi-circular cross section. *Ibid.* 23 *Half round gutter*, an eaves gutter having a half round cross section.
B. *sb.* (Later example.)
1867 *Common Sense Cook Bk.* 38 Half-Round of Beef should be put into cold water.

half-shell. orig. *U.S.* Half an oyster shell. In phr. *on the half-shell*, served in this manner. Also *fig.*
1860 in *Amer. Speech* (1947) XXII. 203/1 Democrats fried,..roasted, or on the half shell. **1861** *Vanity Fair* 30 Mar. 148/1 Hard Shell,..and on the Half Shell Bapt-*ists*,..and all other *ists*. **1872** E. EGGLESTON *End of World* 155 The eggs..were not roasted on the half-shell. **1880** 'MARK TWAIN' *Tramp Abroad* xlix. 574 Blue points, on the half shell. **1940** A. SIMON *Conc. Encycl. Gastron.* II. 70/1 *Bluepoints* are one of the most popular varieties on the half shell. **1972** E. HARGREAVES *Fair Green Weed* iii. 49 They..ate..broiled lobster on the half-shell.

half-shot, *a. colloq.* (orig. *U.S.*). [f. HALF- 1; cf. SHOOT *v.* 32 d.] Half drunk.
1838 J. C. NEAL *Charcoal Sks.* 13 Moseying is only to be done when a gemman's half shot. **1943** N. MARSH *Colour Scheme* x. 187 'The chap was half-shot,' said Simon. 'They all say he smelt of booze.' **1948** J. M. CAIN *Moth* 64 Stuff for guys in college to gag about when they were half shot with beer.

half-time. Add: **1. d.** (See quot., also 3 below.)
1904 A. B. F. YOUNG *Compl. Motorist* iv. 118 All six valves are interchangeable and mechanically operated by rods worked from a cam shaft which is geared at half-time from the crank-axle.
e. Half the tempo of the performer; an accompaniment at half the tempo of the performer (see quot. 1961).
1938 D. BAKER *Young Man with Horn* iv. 265 They went into half-time together for a short coda. **1961** A. BERKMAN *Singers' Gloss. Show Business* 27 *Half-time*, while the singer continues singing the melody in the normal manner, the rhythmic complement is beat only half as fast as the original count.
f. *Chem.* (See quot.)
1957 *A.S.M.E. Gloss. Terms Nuclear Sci.* (1958) 72/2 *Half-time of exchange*, the time required for half the net realizable exchange of atoms in a chemical exchange reaction to take place.
3. half-time shaft = *half-speed shaft* (see *HALF- II. n).
1904 A. B. F. YOUNG *Compl. Motorist* iii. 49 The projecting part of a cam fixed on the half-time shaft. **1908** *Westm. Gaz.* 15 Oct. 4/2 The lubrication of the engine is carried out by a Dubrulle mechanical lubricator, fitted on the dash-board, and operated by an eccentric on the half-time shaft.

half-tone, *sb.* **2.** Delete 'used esp...*attrib.*' and read 'esp. in *Printing* and *Photogr.*, a photo-mechanical illustration printed from a block in which the tones are broken up into small or large dots by the interposition of a glass screen, ruled with fine cross-lines, between the camera and the object; this process. Also *attrib.*' Add earlier and later examples.
1867 G. W. SIMPSON *Photographs in Pigments* 51 The imperative condition upon which half-tone depends, the exposure of one side of the film to light..seemed to present an insuperable difficulty. **1911** *Encycl. Brit.* XIV. 325/1 Half-tone blocks..were used in the *Graphic* from 1884. **1937** E. J. LABARRE *Dict. Paper* 155/1 *Halftone paper*, a printing paper suitable for printing half-tone blocks. **1940** *Chambers's Techn. Dict.* 399/2 *Half-tone process*, a process of photographic reproduction in which the varying tones of the original are photographically translated into dots of uniform tone but varying size. **1958** *Times Lit. Suppl.* 17 Jan. 35/2 Such subjects as colour correction by masking, three-colour half-tone printing. **1959** *House & Garden* July 92/4 The publisher has had the sensible idea of backing his colour-plates..with half-tones in black-and-white. **1961** T. LANDAU *Encycl. Librarianship* (ed. 2) 160/2 *Half-tone screens*, transparent plates ruled diagonally with opaque lines at right angles to each other. **1967** KARCH & BUBER *Offset Processes* 541 *Halftone screen*, the ruled, plate glass dot-forming device used to translate continuous tones into halftones.

half-track. 1. [*TRACK *sb.*] A vehicle, usu. military, with wheels in front and traction chains in the rear; also *attrib.* Also *half-tracked* adj., of such a vehicle.
1927 *Daily Express* 7 Mar. 1/2 Though armoured cars are still tied to roads, the introduction of the half-tracked and the six-wheel carriage will..enable them to move over normal field land. **1935** *Sun* (Baltimore) 9 Feb. 2/3 This force would necessitate acquisition of 285 lightweight tanks..and 76 half-track cars. **1943–4** *Hutchinson's Pict. Hist. War* 27 Oct.–11 Apr. 236 Manned by American infantrymen, this half-track vehicle is stolidly ploughing its way through thick mud. **1945** *Times* 24 Mar. 4/1 Infantry with the task-force dropped off the half-tracks in which they travelled and flushed the woods. *Ibid.* 12 May 4/2 There were a few mechanical vehicles, including some of the standard half-tracked troop-carriers. **1955** *Ibid.* 6 June 5/1 Israel forces made use of machine-guns and half-tracked vehicles in an attack on Egyptian positions. **1973** *Jewish Chron.* 19 Jan. 1/1 (*caption*) An Israeli Army half-track vehicle carries civilians across a main square in Jerusalem as the capital lies under the heaviest fall of snow for 21 years.
2. Half the width of a magnetic tape.
1956 G. A. BRIGGS *High Fidelity* xi. 127 A full-width single track tape can always be played on a half-track machine. Half-tracks can, of course, be played with full-width heads only when the adjacent track is unrecorded. **1959** W. S. SHARPS *Dict. Cinematogr.* 101 *Half-track recorder*, a magnetic tape recorder using a recording head covering half the tape width, so that double the playing time is obtained with any given length of tape and speed. **1962** A. NISBETT *Technique Sound Studio* 273 The standard width for the recording of sound programmes is ¼-inch and this may be recorded full, half or quarter track.

half-value. A value of a physical property that is half an earlier value; used esp. *attrib.*, as *half-value period* = *HALF-LIFE 2; *half-value thickness* (see quot. 1938).
1903 *Phil. Mag.* V. 578 The activity of the radium emanation decays to half-value in four days. **1908** *Amer. Jrnl. Sci.* XXV. 506 The half-value period of radium is.. about 2000 years. **1910** *Nature* 6 Oct. 431/1 It is proposed that the term 'half-value period' should be used in all cases to represent the term required for a substance to be transformed to half its original value. **1922** *Times Lit. Suppl.* 18 May 318/2 The rate of decay of an element is measured by the 'half-value' period, which may vary from 10^{10} years to 10^{-11} of a second. **1938** R. W. LAWSON tr. *Hevesy & Paneth's Man. Radioactivity* (ed. 2) iii. 37 The intensity of the radiation is reduced to half its value in its passage through a layer of thickness 0·16 mm. Al... Accordingly, the 'half-value thickness' of aluminium for the β-rays from RaE has the value 0·16 mm. **1955** C. CROXSON in W. C. Newell *Casting of Steel* xii. 531 Some half-value thicknesses for steel are shown in the following table. **1962** F. I. ORDWAY et al. *Basic Astronautics* xii. 500 A useful concept in physical shielding is that of the *half value layer* (HVL) or the thickness of a specific material required to reduce a particular quantity of radiation by half. **1963** W. E. BURCHAM *Nucl. Physics* ii. 30 The interval..during which half the atoms disappear by decay (half-value period or half-life).

half-way, halfway. Add: **A.** *adv.* (Later examples.)
1926 J. BLACK *You can't Win* (1927) v. 60 She had cleaned up the room till it looked halfway decent. **1938** *Times* 25 May 17/3 He shows how much remains to be done before the standard of life of the West Indian labourer and of the West Indian peasant is raised to a level which can be regarded as halfway tolerable. **1942** *R.A.F. Jrnl.* 16 May 12 There is a drill which must be followed if the voyage is going to be halfway comfortable for everybody. **1960** H. PINTER *Caretaker* III. 77 He's nutty, he's half way gone.
B. *adj.* *half-way house* (earlier and later examples).
1694 W. BURNABY tr. *Petronius's Satyricon* 21 We could not reach Lycurgus's that Night, and therefore he brought us to a half-way House. **1901** S. E. WHITE *Westerners* xix. 164 Copper Creek had begun as a half-way house, and had ended as a camp. **1967** *Canad. Ann. Rev. 1966* 92 There was still need for such things as halfway group homes, preventive delinquency programs [etc.]. **1970** 'T. COE' *Wax Apple* (1973) iii. 24 Halfway houses are places for people returning to society but unable or unwilling to make the plunge all at once. There are halfway houses for ex-drug addicts, former convicts. **1973** *Canadian Antiques Collector* Jan.–Feb. 26/2 The buildings known as half-way houses..were the inns stationed halfway between the larger centres to accommodate weary travellers.
2. (Later examples.)
1903 *Daily Chron.* 3 Nov. 4/4 The halfway state between maidenhood and womanhood. **1909** *Ibid.* 31 Dec. 1/5, I am no halfway burglar, and I would kill a man for 2½d. **1962** *Guardian* 19 Nov. 16/1 These buildings are used by the LCC as 'halfway housing' for homeless families.
C. *sb.* **b.** The half-way line in a football field.
1960 V. JENKINS *Lions down Under* 202 New Zealand brought the ball back to half-way.

half-wit. Add: **2.** Also *attrib.*
1938 F. SCOTT FITZGERALD *Let.* Spring (1964) 29 We are setting it aside till we think of a way of half-witting halfwit Hayes and his Legion of Decency. **1960** *Times* 24 Sept. 10/7 Their half-wit uncle.

halide (hǣ·ləid). *Chem.* [f. HAL(OGEN + -IDE.] A binary compound formed from a halogen and a metal or radical. Also *attrib.*
1876 *Encycl. Brit.* V. 571/2 Acid Halides..may be regarded as the haloid ethers of acid radicles. **1927** N. V. SIDGWICK *Electronic Theory Valency* vi. 88 The halides of the elements show a similar differentiation into volatile non-salts and non-volatile salts. **1938** *Jrnl. R. Aeronaut. Soc.* XLII. 363 An attempt is made to explain the photolysis of silver halides in terms of the concepts of atomic physics. **1948** *Electronic Engin.* XX. 20 If a thin layer of alkali-halide crystals is bombarded with electrons, some electrons are displaced from the halides and secondaries are produced. **1957** *Jrnl. Chem. Soc.* 2071 (*title*) A new preparation of steroid halides.

Halifax (hæ·lifæks). [The name of a town in the West Riding of Yorkshire.] *Go to Halifax*: see GO *v.* 30 b. (Now regarded as a euphemism for *hell.*) Also in phr. *from Hell, Hull, and Halifax, Good Lord, deliver us* (see quot. *a* 1661).
In U.S., sometimes with supposed allusion to Halifax, Nova Scotia.
1630 J. TAYLOR *Works* 115/2 Now if my Whore or Thiefe play well their parts, Give them their due, applaud their good deserts. If ill, to Newgate hisse them, or Bridewell, To any place, Hull, Halifax or Hell. *a* **1661** T. FULLER *Worthies* (1662) Yorks. 189 From Hell, Hull, and Halifax, —— deliver us... This is part of the Beggars and Vagrants Letany... Halifax, is formidable unto them for the Law thereof, whereby Theeves taken..in the very Act of stealing of cloath, are instantly beheaded with an Engine, without any further Legal Proceedings. **1669** [see GO *v.* 30 b]. **1807** *Deb. Congress U.S.* (1852) 11 Dec. 1169 Instead of sending it [*sc.* a ship] where he wished it had gone, to Halifax, or to the bottom. **1875–** in *Eng. Dial. Dict.* s.v. **1876** *Congress. Rec.* 4 Aug. 5185/1 'Go to Halifax' was a substitute for a more impious, but not more opprobrious expression. **1882** *Ibid.* 13 July 6015/1 He told them..that he had no further use for them, and they could go home, ashore, or to Halifax. **1920** H. J. JENNINGS *Chestnuts & Small Beer* xiii. 140, I refused to admit that I had made a *faux pas*, and told my critics to go to Halifax.

haliotis. Add: Also *attrib.*
1845 E. J. WAKEFIELD *Adv. N. Zealand* I. 241 These hooks..take their name from the *haliotus*[sic]-shell, with pieces of which they are lined. **1931** *Oxf. Univ. Gaz.* 17 June 715/2 Toe-bone with carved face and *haliotis* eyes.

halitosis (hælitōu·sis). *Med.* [mod.L., f. L. *halitus* breath + -OSIS.] An abnormally odorous condition of the breath; foul breath.
1874 J. W. HOWE *Breath & Dis.* i. 20 Chronic poisoning from lead, arsenic, or mercury, may also be enumerated as a common cause of halitosis. **1928** *Punch* 17 Oct. 427/1, I shall become a mere mass of degenerate tissues, and flaccid muscles,..with probably a touch of halitosis and lethargy. **1934** S. ROBERTSON *Develop. Mod. Eng.* (1936) xi. 447 Thus *halitosis*, the happy discovery of purveyors of mouth-washes, has found a wider usefulness, and *bad breath* is taboo. **1957** *Times Lit. Suppl.* 12 July 426/4 The dismissed Mulcahy, sadly proud of his porous complexion, bad teeth and occasional morning halitosis.

hall[1]**.** Add: **7.** Also in *pl.* (occas. in *sing.*), abbrev. of *music-hall.*
1862 A. J. MUNBY *Diary* 29 Mar. in D. Hudson *Munby* (1972) 119 Socially speaking, the audience were a good deal higher than those I have seen in similar Halls at Islington & elsewhere. **1895** CHEVALIER & DALY *A. Chevalier* 115 He was one of the few actors of any prominence who had migrated to the halls. **1898** J. D. BRAYSHAW *Slum Silhouettes* 29 He've tried every small 'all in town, even the pubs wot gits up 'smokers'. **1905** MORTON & NEWTON (*title*) Sixty years' stage service, being a record of the life of Charles Morton, 'the Father of the Halls'. **1923** N. COWARD *Coll. Sk. & Lyrics* (1931) 75 Doing a turn of me own on the halls—very trying work—and so cosmopolitan. **1934** T. S. ELIOT *Rock* i. 25 Robey's on the 'alls; but this gentleman..used to hentertain the toffs. **1942** E. BLOM *Mus. in Eng.* x. 168 The 'halls' were the last places where anybody would have thought of going for the sake of music. **1967** *Listener* 8 June 760/2 The vital and self-confident art of the halls..a London Sound to set beside the Liverpool sort.
13. *hall-chair* (examples), *-porter* (examples); **hall-bedroom** *U.S.*, a small bedroom partitioned off at one end of a hall; **hall-bed roomer** *U.S.*, one who sleeps in this; **hall-boy,** a page-boy in a large house; a call-boy in the hall of a hotel or the like; **hall-man** (example); **hall-room** *U.S.*, a room at the end of and of the width of a hall; also *v. intr.*, to live in such a room; **hall stand,** a piece of furniture used to receive umbrellas, hats, coats, and brushes usually situated near the front door in the hall of the house; **hall-table** (def. and further examples), (*a*) a large, solid table belonging to the hall in a mansion; (*b*) a small table situated in the hall near the front door; **hall-tree** *U.S.*, a hall stand or hat rack.
1738 *New Hampsh. Probate Rec.* II. 280 Samuel Brewster shall have..ye Hall Bed Room. **1886** H. JAMES *Bostonians* II. xxi. 44 One of his rooms was directly above the street-door of the house; such a dormitory, when it is so exiguous, is called in the nomenclature of New York a 'hall bedroom'. **1922** A. BENNETT *Lilian* I. vi. 57 In New York it would have been termed a hall-bedroom. **1934** L. MUMFORD in W. Frank et al. *Amer. & Alfred Stieglitz* ii. 35 Lonely young men and women from the country..faced their first year in the city from hall bedrooms on the top-floor rear of unamiable boarding-houses. **1899** J. L. WILLIAMS *Stolen Story* 230 Like many an other lonely hall-bed roomer. **1884** *N.Y. Herald* 27 Oct. 2/2 Janitors and hall boys in attendance. **1885** C. M. YONGE *Nuttie's Father* II. xiii. 158 The hall boy, an alert young fellow, had already dashed down the steps. **1892** —— *That Stick* I. ii. 23 He had been hall boy to a duke, footman to a viscountess, valet to an earl. **1912** L. J. VANCE *Destroying Angel* xx, The hall-boys said you were busy on the telephone. **1864** DICKENS *Mut. Fr.* I. ii. 5 The Veneering establishment, from the hall-chairs..to the grand pianoforte. **1939** A. CLARKE *Sister Eucharia* i. 7 The stage is seen to be quite bare except for two high hall chairs. **1883** D. COOK *On Stage* I. ix. 204 There is no situation in the world where a man can better study his kind than the hall-porter's chair of a London theatre. **1934** *Punch* 5 Aug. 164/2 The hall-porter moved wearily across the floor to take my luggage. **1859** *Ladies' Repository* XIX. 466/2 The little hall-room is just large enough for the boys to sleep in. **1919** T. K. HOLMES *Man fr. Tall Timber* i. 3 'Shucks! why didnt you say H. Harvey Stafford?' interrupted the hall-man. **1886** S. W. MITCHELL *R. Blake* (1895) v. 39 Miss Darnell had for her own use a like space on the third floor, leaving to Miss Wynne a bed-chamber..known as a hall-room. **1906** 'O. HENRY' *Four Million* (1916) xiv. 139 The restaurant was next door to the old red brick in which she hall-roomed. *Ibid.* 140 Schulenberg was to send three meals per diem to Sarah's hall-room. **1882** *Times* 31 Jan. 16/6 (Advt.), A very fine carved oak bookcase, two cabinets, a ditto hall stand and table. **1897** R. KRON *Little Londoner* (1917) 28 Not far from the door, there are an umbrella-stand, a hat-rack, with several pegs on it, and a large looking-glass; if the three are combined, such a piece of furniture is called a hall-stand. **1911** *Daily Colonist* (Victoria, B.C.) 4 Apr. 19/1 (Advt.), Oak Hall Stand, Brussels Rugs. **1943** K. TENNANT *Ride on Stranger* iii. 23 By applying her eye to one of the coloured panes in the front door, she could make out the dim bulk of a hall-stand. **1952** J. GLOAG *Short Dict. Furnit.* 282 Hall stands were often made wholly of cast iron. **1869** L. M. ALCOTT *Little Women* II. xxiii. 327 He..never..expressed.. surprise at seeing the Professor's hat on the Marches' hall-table. **1902** H. JAMES *Wings of Dove* II. iv. 82, I shan't leave mine [*sc.* my letters] on the hall-table. *a* **1941** V. WOOLF *Haunted House* (1943) 44 She touched the letters on the hall table. **1891** *Harper's Mag.* June 79/1 One could distinguish..the hall tree, whereon Rhodes's hat swung in its place. **1900** E. E. PEAKE *Darlingtons* ix. 79 She..walked back to the sitting-room, stopping to touch up her hair before the glass in the hall-tree. **1954** J. STEINBECK *Sweet Thursday* xviii. 108 He busted two windows and run off with the deer-antler hall-tree.

Hall[2]**.** *Physics.* The name of Edwin H. Hall (1855–1938), American physicist, used attrib. to designate an effect discovered by him and various quantities associated therewith (see quot. 1958).
1902 Hall effect [see *electric potential* s.v. *ELECTRIC *a.* 2 b]. **1922** GLAZEBROOK *Dict. Appl. Physics* II. 468/2 The effect of a magnetic field upon the electrical conductivity of metals, first discovered by Lord Kelvin in 1856 and known as the Hall effect, is very marked in the case of bismuth wire or plates. **1940** *Chambers's Techn. Dict.* 400/1 *Hall effect*, a change in the distribution of current in a strip of metal, due to a magnetic field. **1958** *Van Nostrand's Sci. Encycl.* (ed. 3) 784/1 In 1879 Hall.. discovered that if a strip of gold-leaf, carrying an electric current longitudinally was placed in a magnetic field with the plane of the strip perpendicular to the direction of the field, points directly opposite each other on the edges of the strip acquired a difference of electric potential...The transverse electric potential gradient per unit magnetic field intensity per unit current density is called the 'Hall coefficient' for the metal in question... The Hall angle is the ratio of E_y..to the field E_x... The Hall mobility is the mobility of the electrons or holes in a semiconductor as measured by the Hall effect. **1965** PHILLIPS & WILLIAMS *Inorg. Chem.* I. vi. 195 Other methods which measure the mobility of the electrons, e.g. measurement of the Hall effect, also permit a distinction to be made between the two mechanisms of electron migration.

hallal, var. *HALAL *v.*

hallali (hæ·ləli). [Echoic.] A bugle call. Also *fig.*
1885 G. MEREDITH *Diana of Crossways* II. ii. 55 He knew enough to blow his huntsman's horn... His hallali rang high. **1898** —— *Reading of Life* 77 Right loud the bugle's hallali elate Rang forth. **1920** *Q. Rev.* July 13 The hallali was sounded in a famous letter to Lord Grey de Wilton, the candidate for Bath, in which Disraeli accused the Ministers of having for five years harassed every trade. **1923** W. J. LOCKE *Moordius & Co.* x. 140 On the stone of four the orchestra of French hunting horns blew the *hallali.*

Hallé (hæ·le). Applied *attrib.* to an orchestra, concerts, and other musical events which owe their inception to Charles Hallé (orig. Carl Halle) (1819–1895). Also used *absol.*
1852 MRS. GASKELL *Let.* 7 Dec. (1966) 217 We have 3 Hallé tickets for Thursday. *c* **1860** —— *Lett.* (1966) 614 Beautiful Hallé last night. **1936** R. C. K. ENSOR *England 1870–1914* x. 327 Outside the metropolis..the chief purveyor of orchestral concerts [was] the Hallé orchestra. **1954** *Grove's Dict. Music* (ed. 5) V. 542/2 In 1895 the Hallé concerts had become a national institution. *Ibid.* 544/1 Philip Godlee, the former Chairman of the Hallé. *Ibid.*, The Hallé Choir had also had its vicissitudes. *Ibid.*, Concerts outside the Hallé series. **1957** *Times Lit. Suppl.* 1 Nov. 652/3 There is something wrong with a system of public finance that rewards hard work and the enterprise to secure solvency with a cut in the next year's subvention, as was the Hallé's experience. **1968** *Guardian* 29 Mar. 10/6 Our North Midlands scout, driving hard to attend a Hallé concert in Hanley, had to pause at a garage for repairs..'I'm going to the Hallé in Hanley.'

hallelujah, *sb.*[1] Add: **1. c. hallelujah-lass** (examples).
1886 F. HUME *Myst. Hansom Cab* (1887) xvi. 109 It appears that she had been in the Army as a hallelujah lass. **1967** W. S. SMITH *London Heretics* III. v. 238 In 1879 ..more than 2,000 fellow citizens packed into Livingston Hall nightly to listen to the 'Hallelujah Lasses'.
2. A semi-Christian religion practised among the Carib-speaking peoples of Guyana.
1946 F. W. KENSWIL *Childr. Silence* vi. 13 Aboriginal Indians are a very religious set of people...Those of the hinterland of this colony [British Guiana] have their own religion which they call 'Hallilieujah'. This Hallilieujah is said to have been conceived, in the dim past, by an Indian whose name was Bi-chi-wung. **1955** *Times* 22 June 11/7 The most primitive religion which they still practise is a debased version of Christianity called Hallelujah... It..seems to date from some time in the nineteenth century.

hallelujah, *sb.*[2] (Further examples.)
1920 *Sunday at Home* June 569/2 Oxalis, the wood-sorrel, was known as hallelujah,..from its blossoming between Easter and Whitsuntide, when psalms were sung ending in the word hallelujah. **1923** *Times Lit. Suppl.* 3 May 293/3 How happy is the country polyonymosity that hails it also as sheep-sorrel,..hallelujah,..and God Almighty's bread-and-cheese?

hallful (hǭ·lful). [f. HALL *sb.* + -FUL 2.] As many or as much as will fill a hall.
1905 *Daily Chron.* 17 Feb. 3/6 The entertainment of a whole hallful of poor children. **1909** *Westm. Gaz.* 26 Sept. 6/3 What a hallful it was that received him with ringing, rousing, rollicking cheers!

halling (hæ·liŋ). [Norw., from *Hallingdal*, a valley in southern Norway.] A Norwegian country-dance in triple rhythm; also, the music for such a dance.
1866 PLESNER & RUGELEY-POWERS tr. *Björnson's Arne* ii. 13 Nils played; and the two gentlemen each gave a dollar for him, and then asked for the *halling.* **1924** *Glasgow Herald* 31 May 4 We have no space here to describe the wonderful Halling dance. **1947** A. EINSTEIN *Mus. Romantic Era* xvii. 320 Many of these dances show evidence of great age and uninterrupted tradition—the leaping dance (*springar*) in ¾ meter and the *halling* in duple meter.

hall-mark, *sb.* Delete def. and substitute: The official mark or stamp used by the three statutory Hall-marking Authorities in England (Birmingham, London, Sheffield) or by the one in Scotland (Edinburgh), in marking the standard of gold and silver articles assayed by them, without which articles of these metals may not legally be sold.

hall-mark, *v.* (Earlier example.)
1773 *Rep. Comm. Assay Offices* 25 He bought the Knives and Forks..and believed at that Time they had been Hall marked,..and refused selling them when he found they were not Hall marked.

hallo, *int.* Add: Used as a greeting, etc., on a telephone. Also, repeated, as a locution indicating surprise. Cf. *HELLO *int.* b, *HULLO *int.*
1932 *N. & Q.* 6 Aug. 105/2 The telephonic 'hallo'. **1942** A. CHRISTIE *Body in Library* ii. 24 Hallo, 'allo, 'allo, what's this? **1972** *Police Rev.* 10 Nov. 1473/2 You could have a walking, talking plastic Policeman..saying 'Allo, allo, allo.'

Hallstatt (ha·lʃtat). The name of a village in Upper Austria, the site of an ancient necropolis, where rich archæological finds have been made, used *attrib.* to denote a phase of the early Iron Age, and the type of civilization of that period. So **Hallsta·ttian** *a.*
1866 J. LUBBOCK *Addr. Primæval Antiquities Archæol. Inst.* 17 The significance of the absence of silver in the Hallstadt find is greatly increased when we see that in the true Iron age..silver was used. **1869** —— *Pre-Historic Times* (ed. 2) i. 24 M. Ramsauer..director of the salt mines at Hallstadt, near Salzburg, in Austria, has discovered an extensive cemetery belonging to this transitional period. *Ibid.* 25 Another interesting point in the Hallstadt bronze, as in that of the true Bronze Age, is the absence of silver, lead, and zinc. **1893** *Funk's Stand. Dict.* I. 812/3 Hallstattian. **1900** tr. *Deniker's Races of Man* 315 The so-called 'Hallstattian' period lasted in Central Europe, France, and Northern Italy from the tenth or ninth to the sixth century B.C. The Hallstattian civilisation flourished chiefly in Carinthia, Southern Germany, Switzerland, Bohemia, Silesia, Bosnia, the south-east of France, and Southern Italy. **1905** *Brit. Mus. Guide Antiq. Early Iron Age* 36 The drum-shaped brooch..is also not uncommon in the Hallstatt period. *Ibid.* 39 The Hallstatt stage of culture is well represented at Glasinatz. **1931** *Times Lit. Suppl.* 1 Oct. 753/3 Scanty traces of Hallstatt immigrants. **1958** *Ibid.* 19 Sept. 526/2 The general reader..will meet..the well-known cultures of Hallstatt and La Tène.

hallucinant (hæliū·sinănt), *sb.* and *a.* [f. HALLUCIN(ATE *v.* + -ANT[1].] **A.** *sb.* **a.** Someone who experiences hallucinations. **b.** A drug that induces hallucinations. **B.** *adj.* Producing or experiencing hallucinations.
1895 tr. *Nordau's Degeneration* I. iii. 32 The women devotees..did not merely believe that the hallucinant maiden had herself seen the vision, but all of them saw the Holy Virgin with their own eyes. **1910** A. W. VERRALL *Bacchants of Euripides* 72 In a hallucinant working by suggestion..such effects are perfectly natural. **1932** A. HUXLEY *Brave New World* iii. 63 Euphoric, narcotic, pleasantly hallucinant. **1964** *Punch* 30 Dec. 1013/2 Someone who knew her potions, her febrifuges and hallucinants.

hallucinate, *v.* **2.** Delete *Obs.* or *arch.* and add later examples. Also, to have a hallucination or hallucinations. Now chiefly *U.S.*
1847 WEBSTER, Hallucinate. **1930** C. SPEARMAN *Creative Mind* x. 135 A man hallucinated that the clothes of the girls 'flew off them'. **1958** E. DUNDY *Dud Avocado* III. vi. 270 My first thought was that I had gone stark raving mad..and that I was now hallucinating in a looney bin. **1964** A. CROSS *In last Analysis* iii. 31 Had such an idea crossed her mind, Kate would have decided that..she was 'hallucinating'. **1973** *Publishers Weekly* 19 Mar. 61/3 He describes her and is told, bluntly, that he is hallucinating.

hallucinating *ppl. a.* (later examples).
1903 E. WHARTON *Sanctuary* II. iv. 137 That hallucinating distinctness which belongs to the midnight vision. **1966** *New Statesman* 18 Feb. 233/2 Jennifer Dawson writes about the surface pain of living—with hallucinating effect.

hallucinatory, *a.* Add: Hence **hallucina·torily** *adv.*
1917 C. R. PAYNE tr. *Pfister's Psychoanalytic Method* xvii. 467 Then every time, out of hate and expiation, he changed hallucinatorily the feared one into the death's head. **1959** G. D. PAINTER *Marcel Proust* I. 178 They were irresistibly comic, and at the same time hallucinatorily accurate.

‖ **halluciné** (hæliū·sine). *rare.* [Fr.] A person who regularly suffers from hallucinations.
1886 *Buck's Hand-bk. Med. Sci.* III. 481/1 Nowadays the hallucinés hear voices through the telephone, and feel electric shocks. **1898** *Daily News* 20 Sept. 4/5 Genius goes somehow with what we call hysteria, most persons of genius being epileptic, or 'hallucinés'. **1909** W. JAMES *Let.* 19 Sept. in R. B. Perry *Thought & Char. W. James* (1935) II. lviii. 123, I strongly suspect Freud, with his dream-theory, of being a regular *halluciné.*

hallucinogen (hæliū·sinŏdʒen). [f. HALLUCIN(ATION + -O + -GEN.] A drug which causes hallucinations (see HALLUCINATION 2).
1954 A. HOFFER et al. in *Jrnl. Mental Sci.* C. 30 When the literature is examined to catalogue these hallucinatory substances, which for convenience we have called the hallucinogens, one is struck by their small number. **1954** A. HUXLEY *Doors of Perception* 6 Lysergic acid, an extremely potent hallucinogen derived from ergot. **1955** *Jrnl. Mental Sci.* CI. 318 There are so few hallucinogens known that we must study intensively the types of subgroups and molecules which cause hallucinations. **1958** *New Scientist* 28 Aug. 715/3 More recently..several hallucinogens have been observed which produce, in addition to other symptoms of schizophrenia, auditory hallucinations. **1969** *Times* 24 Jan. 6/6 The fly agaric fungus yields one of the most potent and anciently used hallucinogens, inducing a slight trembling followed by illusions.
So **hallu:cinoge·nic** *a.*, being or containing such a drug; causing hallucinations.
1952 *Jrnl. Mental Sci.* XCVIII. 311 There are many other hallucinogenic drugs, but none has either such striking properties or such a simple chemical constitution as mescaline. **1958** *Sci. News* XLVII. 36 Other lines of research..include the search for substances that antagonize the hallucinogenic drugs. **1959** *Times Lit. Suppl.* 27 Feb. 113/4 We believe our evidence points to a role for the hallucinogenic mushroom in the origins of the religious idea in primitive society. **1960** *Times* 31 Dec. 9/7 A liana of which the boiled juice has hallucinogenic effects. **1965** *Listener* 23 Sept. 465/1 It is a powerful mind-changer of the hallucinogenic variety. **1968** *Times* 13 Nov. 16/1 Many well known hallucinogenic drugs, such as LSD and mescaline, cause mice to exhibit the curious behaviour pattern of head twitching.

hallucinosis (hæliū:sinōu·sis). *Psychiatry.* [f. HALLUCIN(ATION + -OSIS.] A disorder of the nervous system associated particularly with alcoholism, marked by persistent hallucinations, commonly auditory, with little if any impairment of consciousness.
1905 S. PATON *Psychiatry* xi. 300 Acute alcoholic hallucinosis. **1908** *Practitioner* Jan. 9 The acute hallucinosis and paranoidal forms of alcoholic insanity. **1912** OSLER & MCCRAE *Princ. & Pract. Med.* (ed. 8) III. i. 399 Alcoholism... There is a condition termed *acute hallucinosis*, in which auditory hallucinations are marked..and the mental disturbances are fixed. Ideas of persecution are common. There are intermediate forms between this and the ordinary delirium tremens. **1914** *Amer. Jrnl. Insanity* LXX. II. 369 (*title*) Study of hallucinosis. **1962** HENDERSON & GILLESPIE *Text-bk. Psychiatry* xii. 291 Accompanied by..clear and orderly thinking, willing and acting; there is no real or genuine hallucinosis. **1964** LANDIS & METTLER *Varieties Psychopathol. Experience* xiv. 312 The horror, torture, and pain experienced during delirium tremens or alcoholic hallucinosis have been described many times.

hallway. For *U.S.* read orig. *U.S.* and add earlier and later examples.
1877 J. HABBERTON *Jericho Road* 173 It passed through the narrow hallway which separated the cell from the jailor's apartments. **1920** B. CRONIN *Timber Wolves* ii. 41 He stood hesitating at the entrance to the musty hallway. **1953** L. KUPER *Living in Towns* ii. 13 The front doors..give direct access to hallway and staircase. **1959** *Shropsh. Mag.* Mar. 9/2 He has a beautiful wooden chest in his hallway. **1971** *Real Estate Rev.* Fall 104/2 A woman who had been attacked in the common hallway of her apartment house.

halo, *sb.* Add: **1. e.** A style in women's hats (worn at the back of the head with the brim thus framing the face). Also *attrib.* and *Comb.*, as *halo-brim(med), hat, style*, etc.
1899 *Daily News* 22 Apr. 8/4 Some of the new models [*sc.* hats] are intended to be put on in the halo style. **1903** *Daily Chron.* 24 Oct. 8/4 The hat makes a halo in front. **1934** *Times* 22 June 17/4 A brown halo hat. **1935** *Times* 17 June 11/3 Felt halos are made with a velvet cap in front, and cost 3½ guineas. *Ibid.* 2 Oct. 17/4 A pale blue halo-brimmed hat. **1952** E. GRIERSON *Reputation for Song* xxvii. 244 Laura..wore a little black halo-hat to frame her pallid face.

f. A more or less circular bright or dark area formed in various photographic processes (see quots.).
1941 *Amer. Speech* XVI. 316/2 *Halo*, the effect obtained in portraiture when a strong back-light is used. **1961** G. MILLERSON *Technique Telev. Production* iii. 49 *Haloes* (throw-off). A black aureole surrounding an over-bright high-contrast area, and obliterating the nearby picture. **1967** KARCH & BUBER *Offset Processes* 541 *Halo*, a luminous circle or aura around the halftone dot.
4. halo blight, spot, either of two bacterial diseases caused by species of *Pseudomonas*, *P. coronafaciens* affecting oats, and *P. phaseolicola* affecting beans of the genus *Phaseolus*; characterized by brown spots surrounded by yellowish-green rings on leaves; **halo effect** *Psychol.*, the favourable bias in interviews, intelligence tests, and the like generated by an atmosphere of approbation; also *transf.*
1920 C. ELLIOTT in *Jrnl. Agric. Res.* XIX. 139 This 'halo-blight' is a disease which occurs to at least some extent each year throughout the oat-growing sections of the central and eastern States. **1930** W. H. BURKHOLDER in *Mem. Cornell Univ. Agric. Exper. Station* No. 127 37 The most striking symptom [of *Phytomonas medicaginis phaseolicola*] arises from a local infection, and is the spot to which Miss Hedges has applied the term *halo blight.* **1954** A. G. L. HELLYER *Encycl. Garden Work & Terms* 115 *Halo blight*, a disease of beans also sometimes known as halo spot. **1955** *Sci. Amer.* June 83 Bacterial blight of beans..embraces a group of diseases: common blight, fuscous blight and halo blight, each caused by a different bacterium. **1971** M. NOBLE in J. H. Western *Dis. Crop Plants* iii. 33 In New South Wales an inspection scheme for bean (*Phaseolus*) seed has been in operation..for control of halo blight. **1971** J. COLHOUN in *Ibid.* x. 211 Halo blight of oats..is not regarded as constituting an economic problem in the British Isles. [**1926** E. H. MAGSON in *Brit. Jrnl. Psychol. Mon. Suppl.* 9 89 It must be pointed out that it is quite unnecessary to employ a new technical term such as 'halo' or 'aura' to cover these cases.] **1938** *Brit. Jrnl. Psychol.* Jan. 285 Such general impressions, often called 'halo effects', have already been noted to affect the diagnosis of personal qualities. **1940** R. S. WOODWORTH *Psychol.* (ed. 12) v. 143 Another error [in rating intelligence tests] is known as the 'halo effect'. If an individual creates a favourable impression by his excellence in one trait, you are apt to rate him near the top in every trait. **1967** *Guardian* 20 Dec. 1/6 Mrs. Castle.. agreed that the new Act had a 'halo' effect in that it made drivers more careful. **1928** *Jrnl. Agric. Res.* 1 Mar. 428 Halo spot is known to occur in Georgia, Florida, and Connecticut.

halo- (hælo), combining form of Gr. ἅλς, ἁλός sea, salt, as in **ha·lobiont** *Ecol.*, an organism that lives in a saline habitat; hence **ha:lobio·ntic** *a.*; **ha:lobio·tic** *a. Ecol.*, living in the sea; **ha:lochro·mism** *Chem.* [ad. G. *halochromie* (Baeyer & Villiger in *Ber. d. Deut. Chem. Ges.* (1902) XXXV. 1190)], the property possessed by certain colourless or faintly coloured compounds of becoming brilliantly coloured in the presence of acids or of certain other compounds; **haloli·mnic** *a. Biol.*, living in fresh water but having an affinity with marine forms; **halomo·rphic** *a. Soil Sci.*, (of a soil) containing, or developed under the influence of, large quantities of salts other than calcium carbonate; **ha·lophyte** *Ecol.* [ad. mod. L. *halophyta* (J. F. Schouw *Grundtræk til en almindelig Plantegeographie* (1822) 138)], a plant which is adapted to grow in saline conditions; so **halophy·tic** *a.*, growing, or adapted to grow, in saline conditions; **halopla·nkton,** marine plankton; **ha·losere** *Ecol.* (see quot. 1930[1]); **ha·lowax** [HALO(GEN + WAX *sb.*[1]] (see quots.).
1928 K. E. CARPENTER *Life Inland Waters* ix. 228 Other halobionts are: all known species of *Ephydra* (Diptera)..and several species of *Ochthebius, Philydrus*, and *Paracymus* (Coleoptera). **1937** ALLEE & SCHMIDT *Hesse's Ecol. Animal Geogr.* xix. 370 'Halobionts' are limited to water of rather high salt content, and are more or less salt-tolerant stenohaline forms. **1928** K. E. CARPENTER *Life Inland Waters* ix. 228 Above this concentration, species rapidly diminish in numbers, and above 10 per cent. are only found the true 'halobiontic' forms, which rarely, or never, occur in waters other than saline. **1909** WEBSTER, Halobiotic. **1927** R. S. LULL *Org. Evol.* (rev. ed.) v. 70 Halobiotic or Marine Realm. **1902** *Rep. Brit. Assoc.* 119 Reference may be made to some recent work of v. Baeyer and Villiger on dibenzylidene acetone and triphenyl methane. They refer to the constitution of colourless substances which form highly coloured salts, and term the phenomenon *halochromism.* **1944** *Hackh's Chem. Dict.* (ed. 3) 395/1 *Halochromism*, the formation of colored salts from colorless organic bases by the addition of acids. **1952** K. VENKATARAMAN *Chem. Synthetic Dyes* I. viii. 326 In the phenomenon of halochromism, the neutral organic compounds, which become brightly colored on the addition of hydrogen ion, are colored to about the same depth and intensity by the addition of neutral substances such as boron trichloride or stannic chloride instead of hydrogen ion. **1898** J. E. S. MOORE in *Proc. R. Soc.* LXII. 453 They probably belong to the same quasi-marine, or what I shall in future call the Halolimnic group. **1903** —— *Tanganyika Probl.* vii. 141 The animals forming the invertebrate section of this peculiar group have an obviously marine aspect, and on that account I have spoken of them elsewhere as forming a halolimnic series in Lake Tanganyika—that is to say, they form a group of animals which, although living in a fresh-water lake, have at the same time the characters of animals that are typical of the sea. **1904** *Westm. Gaz.* 26 May 5/2 The shells of the halolimnic gasteropods. **1922** *Nature* 5 Jan. 28/1 The halolimnic forms..exhibit a marine-like appearance. **1938** *U.S. Dept. Agric. Yearbk.* 1169 *Halomorphic soils*, a suborder of intrazonal soils, the properties of which are determined by the presence of neutral or alkali salts, or both. **1968** R. W. FAIRBRIDGE *Encycl. Geomorphol.* 273/1 Other desert soils are intrazonal..and either contain appreciable amounts of calcium carbonate (the pedocal soils) or have relatively high concentrations of other soluble salts (the halomorphic soils). **1886** WEBSTER Add., Halophyte. **1894** F. W. OLIVER et al. tr. *Kerner's Nat. Hist. Plants* I. 74 Plants which only flourish abundantly on soils rich in alkaline salts are called halophytes. The same name has also been applied to plants which only thrive in sea-water. **1903** W. R. FISHER tr. *Schimper's Plant-Geogr.* 90 Halophytes can thrive on ordinary soil..without any addition of common salt. **1909** GROOM & BALFOUR tr. *Warming's Oecol. Plants* liv. 219 A halophyte..is one form of xerophyte. **1966** *New Scientist* 2 June 575/1 Because of the removal of water by transpiration or in the harvested crop, the concentration of salts in the system will rise so that even halophytes will suffer. **1895** G. HENSLOW *Orig. Plant Struct.* 83 Halophytic plants, and others yielding ethereal oils. **1950** *Engineering* 26 Mar. 610/3 While.. salt..is being washed out of the soil..the vegetation will still be halophytic. **1909** GROOM & BALFOUR tr. *Warming's Oecol. Plants* xxxviii. 160 The plankton of salt water may be subdivided into neritic and oceanic haloplankton. **1927** R. S. LULL *Org. Evol.* (rev. ed.) iii. 43 Marine or halo-plankton. **1929** WEAVER & CLEMENTS *Plant Ecol.* iv. 74 Hydroseres in saline areas are distinguished as (salt) haloseres. **1930** *Jrnl. Ecol.* XVIII. 201 *Halosere*, the sere commencing in saline water or upon saline soil. *Ibid.* 229 (*heading*) Communities developing within the halosere. **1964** K. A. KERSHAW *Quantitative & Dynamic Ecol.* iii. 39 Clements similarly termed the stages of salt marsh succession a halosere. **1922** *Halowax* (*Condensite Co. of America*) 3 Halowax for impregnating paper round electrical condensers. *Ibid.* 4 Halowax is a trade name for chloro-naphthalene substitution products, i.e., products in which chlorine atoms are substituted for those of hydrogen in the naphthalene. **1928** *Daily Express* 10 Jan. 3/7 The..Anti-Knock Compound..is a liquid consisting of tetra-ethyl lead, ethylene dibromide, halowax oil and red aniline dye. **1947** J. C. RICH *Materials & Methods of Sculpture* vi. 157 *Halowax* is a synthetic wax with a high melting point. It is a strong and hard material and imparts a milky opaqueness to a cool wax formula. **1963** R. F. WEBB *Motorist's Dict.* 121 *Halowax*, a type of oil, blended with a tetra-ethyl lead compound used to lubricate the working parts of some mechanical superchargers.

halogen. Add: In mod. use, any of the elements of group 7 of the periodic table, viz. fluorine, chlorine, bromine, iodine, and astatine. (Further examples.)
1867 C. L. BLOXAM *Chem.* 10 A salt-radical or halogen is a substance which forms an acid when combined with hydrogen. Examples.—Chlorine, which forms hydrochloric acid (HCl); Cyanogen (C_2N), which forms hydrocyanic acid (HC_2N). **1884** T. S. HUMPIDGE tr. *Kolbe's Short Text-bk. Inorg. Chem.* 96 The halogens or salt-producers. The four elements chlorine, bromine, iodine, and fluorine are grouped together under this name. **1959** *Nomencl. Inorg. Chem.* (I.U.P.A.C.) 6 The use of the collective names: halogens (F, Cl, Br, I, and At),..may be continued.
Hence **halogena·tion,** the introduction of an atom of a halogen into a molecule of a compound by addition or substitution.
1911 *Chem. Abstr.* V. 1436 (*heading*) Halogenation of benzene monohalides...The expts...were carried out at 55–75° without a diluent. **1965** PHILLIPS & WILLIAMS *Inorg. Chem.* II. xii. 463 Some of the interhalogens provide convenient solvents for halogenation reactions, particularly in view of their partial ionization.

haloid. Add: Now rare as adj. and superseded by *halide* as sb.

haloing, *ppl. a.* Add: Also *vbl. sb.*
1967 *Gloss. Paper/Ink Terms for Letterpress Printing* (*B.S.I.*) 9 *Haloing*, the appearance of vehicle from the ink round half-tone dots or characters.

halonate (hǣ·lonĕt), *a. Bot.* [f. mod.L. *halōn-*, HALO *sb.* + -ATE[2].] Surrounded by an outer ring.
1911 CROMBIE & SMITH *Brit. Lichens* II. 359 Halonate ..surrounded by an outer circle. **1921** A. L. SMITH *Handbk. Brit. Lichens* 102 Spores ellipsoid or oblong.. usually with a hyaline mucilaginous epispore (halonate). **1959** U. K. DUNCAN *Guide to Study of Lichens* 64 Th. Fries..cited the halonate character of the spores [of *Rhizocarpon*] as the distinguishing factor.

halophile, *sb.* Add: **2.** *Ecol.* An organism which grows in or can tolerate saline conditions.
1928 K. E. CARPENTER *Life Inland Waters* ix. 229 These 'halophiles' are true freshwater species endowed with powers of resistance. **1965** B. E. FREEMAN tr. *Vandel's Biospeleol.* xiii. 196 The Pogoninae are for the most part halophiles. **1966** T. D. BROCK *Princ. Microbial Ecol.* iii. 47 Halophiles are in ionic equilibrium with their environment.
B. *adj. Ecol.* Growing in or tolerating saline conditions.
1909 *Cent. Dict.* Suppl., Halophile. **1961** *Times Rev. Industry* June 82/1 A gradual increase of the halophile (or, better, salt-resisting) plants..at the fringe of the oasis and the great desert beyond.

halophilic (hæ:lofi·lik), *a. Ecol.* [f. HALOPHIL(OUS *a.* + -IC.] Growing in or tolerating saline conditions, halophilous.
1919 *Jrnl. Bacteriol.* IV. 177 (*title*) A preliminary report upon some halophilic bacteria. **1964** *Oceanogr. & Marine Biol.* II. 161 Few studies have been made of the role of detritus as a food for halophilic Crustacea. **1971** M. ALEXANDER *Microbial Ecol.* viii. 181 Some populations are obligately halophilic.

halosaurus (hæ:losǭ·rŏs). [mod.L. (see quot. 1863); see HALOSAURIAN.] A deep-water marine fish of the genus so called, characterized by an elongated body and long tapering tail. Hence **ha·losaur,** a member of the family Halosauridæ, which includes both fossil and living fishes; **ha:losau·roid** *a.*, of, pertaining to, or resembling a fish of this kind; also as *sb.*
[**1863** J. Y. JOHNSON in *Proc. Zool. Soc.* 406 *Halosaurus*, gen. nov. Body elongated, clothed with cycloid scales;..tail compressed and tapering to a point.] **1893** *Funk's Stand. Dict.*, Halosauroid, a. & n. **1897** *Proc. Zool. Soc.* 268 On *Echidnocephalus*, a Halosauroid Fish from the Upper Cretaceous Formation of Westphalia... Thanks to..new specimens of *Halosaurus* obtained by the 'Challenger' Expedition, it is now possible to demonstrate that..the strange Halosauroid type was already completely developed before the end of the Cretaceous period. **1904** G. A. BOULENGER in *Cambr. Nat. Hist.* VII. xxii. 622 The conformation of the pectoral arch has much in common with that of the Halosaurs. *Ibid.* 624 In *Halosaurus* the scales of the lateral line..are scarcely enlarged. **1957** R. CAMPBELL *Portugal* iv. 68 In the depth of Setúbal they were visited by many weird monsters,.. notably, the *halosaurus* with a long undulating tail. **1963** *New Scientist* 10 Jan. 80 The deeper-living benthic fishes, those that live at depths beyond 2,000 metres, are.. halosaurs (Halosauridae). **1971** *Nature* 2 Apr. 279/1 They include the halosaurs, which are known to develop from leptocephalus larvae, and the spiny eels or notacanths.

halothane (hæ·lǫþēin). [f. HALO(GEN + E)THANE.] A volatile liquid, $CF_3CHBrCl$, with a characteristic odour, used as a general anæsthetic.
1957 *Lancet* 8 June 1196/1 Dr Foster has done a grea service by his letter of June 1, if only by drawing attention to errors in the use of 'Fluothane' (halothane). *Ibid.* 7 Sept. 493 To date, cardiac arrest has been reported in five patients anæsthetised with halothane. **1961** *Brit. Med. Dict.* 653/1 *Halothane*, B.P. Commission approved name for 2-bromo-2-chloro-1:1:1-trifluoroethane, a volatile, non-explosive anaesthetic, more powerful than ether. **1971** *Nature* 1 Oct. 353/1 Cats anaesthetized with 50% nitrous oxide in oxygen and 1% halothane.

halt, *sb.*[1] Add: **b.** A small railway station without the ordinary accommodation or staff, at which only local trains normally stop.
1910 *Offic. Guide L.N.W.R.* (ed. 15) 410 Rail motor car halts at Wendlebury, Charlton, and Oddington. **1914** *Railway Mag.* Aug. 152/2 The provision of the 'halts' on the new line has been much appreciated locally. **1921** *Dict. Occup. Terms* (1927) § 706 *Halt attendant*; a porter who attends at roadside halt, where there is no proper station staff. **1973** *Country Life* 7 June 1612/4 A halt or unstaffed stopping place where trains called only if required.

halt, *v.*[2] Add: **1. b.** Also formerly used as a command in traffic regulations and on road signs. So *halt notice, sign.* Also *transf.*
1932 E. WALLACE *When Gangs came to London* xx. 175 He caught Terry's eye and abruptly changed the subject. When they were outside: 'What was that halt sign?' demanded Jiggs. **1935** *Highway Code* 21 (*caption*) Halt sign. **1958** A. WILSON *Middle Age of Mrs. Eliot* II. 153 David, recognizing the Grimm quotation as a halt sign, laughed too. **1965** D. M. DEVINE *His Own Appointed Day* I. ii. 17 Left, here. Then turn right at the halt sign. **1967** M. SUMMERTON *Memory of Darkness* ii. 21 He had driven through a halt sign, and collided with an on-coming tanker. **1969** *Times* 15 July 7/3 What are the advantages of the Give Way signs? Why cannot the Ministry go back to the old safer Halt notices?

halter, *sb.*[1] Add: **1. b.** A strap attached to the top of a backless bodice and looped round the neck; also, a bodice with such a strap or cut so as to give a similar effect. Hence *attrib.* and *Comb.*, as *halter neck(line), top.*
1935 *Mademoiselle* Aug. 1/2 (*caption*) Trunks with halter top. **1936** *New Yorker* 18 Jan. 50/2 When a dress terminates in a halter neck, they have an ingratiating habit of putting a little bolero jacket over it. **1939** M. B. PICKEN *Lang. Fashion* 102/3 Halter neckline.. introduced about 1933. Used in sports and evening clothes. **1948** N. MAILER *Naked & Dead* (1949) II. vii. 229 In the brothel the girls wear halters and trim panties with a tropical print. **1953** BERG *Dict. New Words* 91/1 *Halter*, a woman's bodice, held in place by straps around the neck and across the back, so as to leave the arms and the back free. **1958** J. D. MACDONALD *Executioners* (1959)

iv. 59 Nancy wore very short red shorts..and a yellow linen halter. **1959** *Vogue Pattern Book* June/July 23 A full, floating skirt and bare back halter top for a sun dress. **1971** *Vogue* Dec. 64/1 Black silk jersey halter-neck dress.. £70.

3. halter-break *v. U.S.* (earlier examples); **halter hitch** (see quot. 1944).

1837 *N.Y. Mirror* 28 Oct. 140/3 The moose has been frequently tamed, and unlike the common deer, can be halter-broken as easily as a horse. **1860** J. G. Holland *Miss Gilbert's Career* xix. 350 You want to halter-break 'em 'fore they're little and get 'em kind o' wonted to the feel of the harness. **1868** *Rep. Iowa Agric. Soc.* 1867 117 My colts are halter-broken as soon as foaled. **1944** C. W. Ashley *Bk. Knots* ii. 44 *Halter hitch.* Horses are hitched with this knot the world over. The end is stuck loosely through the loop, which is not tightened. The knot is easily slipped after removing the end from the loop. **1947** *Times Lit. Suppl.* 15 Nov. 594/4 When he was seven he was given a pony on condition that he mastered a halter hitch.

halteridium (hæltĕri·diəm). *Zool.* Pl. **-ia.** [mod.L. (A. Labbé 1894, in *Archives de Zoologie expérimentale et générale*, 3e série, II. 129), f. Gr. ἁλτήρ weight used in leaping.] A name for the gametocytes of the protozoan genus *Hæmoproteus*, which is parasitic in birds, used when the gametocytes were erroneously considered to be a separate genus.

1901 G. M. Sternberg in *Pop. Sci. Monthly* LVIII. 367 The mosquito..could not transmit the malarial parasite of man or another similar parasite of birds (halteridium). **1901** *Practitioner* Mar. 278 One of the malaria-like organisms of birds, namely, halteridium. **1962** J. D. Smyth *Introd. Animal Parasitol.* ix. 197 These gametocytes [of *Hæmoproteus columbæ*] are shaped like a curved sausage and encircle the nucleus in a halter-fashion, and are sometimes referred to as *Halteridia*, as they were at one time called by the generic name of *Halteridium*.

haluka: see *khalukah.

halutzim (hălū·tsim), *sb. pl.* Also **chalutzim, haluzim.** [Heb. *ḥălūṣ.*] Jewish pioneers entering Palestine in order to build up their future national home.

1921 *Daily Mail* 11 Apr. 6/5 The 'Haluzim' are the Jewish pioneers who are flocking into Palestine to help in building up the Jewish National Home. **1923** *Daily Mail* 18 July 7 Tel Aviv..is the headquarters of the Rutenberg Company and also of a number of co-operative building societies run by young Jewish halutzim pioneers. **1923** W. P. Livingstone *Galilee Doctor* 261 The halutzim, or 'pioneers'. **1971** D. Meiring *Wall of Glass* ix. 71 A few other Second Aliyah men from Russia, calling themselves Chalutzim, pioneers.

halva (halvā·, χ-). Also **halvah, halvas, halwa.** [ad. Turk. *helva*, mod. Gr. *halvas*, Arab. *halwā* *hulwa.] A sweetmeat made of sesame flour and honey.

1846 J. F. Romer *Pilgrimage* II. ii. 33 The vender of sweetmeats with a tempting array of *halva* and lumps of delight' set out upon a wooden tray borne upon his head. **1908** *Daily Chron.* 12 June 4/7 A Greek merchant in the East-end, who..would sell you..black olives, rose leaf jam, and halvas. *a* **1916** 'Saki' *Toys of Peace* (1919) 134 A tin of the best Smyrna halva. **1927** Kipling *Eyes of Asia* (1918) 80, I eat..halwa and puri. [illegible] *Halva*, a Turkish confection. **1963** H. Miller *Plexus* (1965) [illegible] Halvah and Baklava too. **1966** *Listener* 19 May 731/2 Halva cake, Many of the Greek cakes and pastries are beyond the scope of the home cook, but here is a simple one. [illegible] bought a handful of Hershey bars and Charms and sour balls and chocolate-covered halvah. **1973** *Sunday Express* 23 Sept. 7/1 From the first time he tasted halva as a child [illegible] the traditional Arabic sweetmeat made from honey.

halver¹. 2. Delete *Obs.* exc. *dial.* and add U.S. examples: used esp. in phr. *to go halvers* (*with*).

1865 'Mark Twain' *Celebr. Jumping Frog* (1867) 36 No man can say he ever see him go anything by halvers. **1873** J. H. Beadle *Undeveloped West.* ix. 93 'I'll go halvers with him,' shouts the conservative-looking chap. **1932** W. Faulkner *Light in August* xviii. 394 Byron Bunch, that weeded another man's laidby crop, without any halvers.

halwa, var. *halva.

ham, *sb.*¹ Add: **3. ham-curing, -sandwich; hamfatter** *U.S. slang*, an ineffective actor or performer; (also **hamfat**) a mediocre jazz musician; so **hamfat man,** etc.; **ham-fisted** *a.*, having large or clumsy hands, heavy-handed, awkward; bungling; hence **ham-fistedly** *adv.*, **ham-fistedness; ham-footed** *a.*, clumsy, awkward, stupid; **ham-handed** *a.*, = **ham-fisted*; hence **ham-handedly** *adv.*, **ham-handedness; ham loaf** orig. *U.S.*, a shaped mass of chopped cooked ham intended to be cut into slices.

1907 *Daily Chron.* 23 Oct. 4/4 Spinning, or bread-baking, or ham-curing. **1880** G. A. Sala *America Revisited* (1882) I. iv. 66 Every American who does not wish to be thought 'small potatoes' or a 'ham-fatter' or a 'corner loafer'. **1889** *Cent. Dict.*, *Hamfatter*,..a term of contempt for an actor of a low grade, as a negro minstrel. Said to be derived from an old-style negro song called 'The Ham-fat Man'. **1932** 'Spindrift' *Yankee Slang* 20 *Hamfatter*, loudly-dressed and loudly-decorated dude. **1938** *N.Y. Amsterdam News* 12 Mar. 17 The Harlem Hamfats grind out the tune. **1946** Mezzrow & Wolfe *Really the Blues* 58 A lot of beat up old hamfats..sang and played. **1959** S. B. Charters *Country Blues* 86 The singing of these little 'hamfat' bands never reached the artistic intensity of men like Blind Lemon. **1966** *New Yorker* 11 June 160/1 Most of the musicians playing in these clubs are old men. They're 'hamfat' musicians. In the old days, the rough musicians kept pieces of ham fat in their pockets to grease the slides of their trombones. **1928** *Daily Mail* 7 May 6/4 Ham Fisted.—Applied to pilots who are heavy on controls, or generally clumsy. **1928** *Sunday Express* 24 June 8/3 Two thousand lumber-jacks were in town, ham-fisted great fellows with hair on their chests and pine needles growing out of their ears. **1938** C. S. Forester *Ship of Line* xi God damn and blast all you ham-fisted swabs. **1942** H. Asken in Forbes & Allen *Ten Fighter Boys* 15 A dog-fight with a Hun very rarely entails a considered acrobatic movement as an evasive action. In fact, the more ham-fisted the movement, the better its effect. **1960** *Times* 20 Oct. 8/1 The play's basic idea implies a less ham-fisted humour than the authors can supply. **1964** *Punch* 2 Sept. 355/1 Some ham-fistedly insensitive moments. **1963** *Times* 16 Feb. 9/3 The campaign cannot be written off because of the ham-fistedness of its beginnings. **1960** E. S. & W. Higham *High Speed Rugby* 26 One 'ham-footed' forward makes a present of the ball to the other side. **1961** *Sunday Express* 7 May 14 Is he so thick-soled, ham-footed? **1918** W. A. Bishop *Winged Warfare* 30 First the instructor would tell me I was 'ham-handed'—that I gripped the controls too tightly with every muscle tense. **1928** *Punch* 4 Apr. 22/3 Second [Pilot]...I was getting ham-handed and mutton-fisted, flapping the old things every which way. **1930** C. Dixon [illegible] The pilot with sensitive hands is a better pilot than one with non-sensitive hands. The latter are bluntly called 'ham-handed'. **1952** E. Linklater *Mr. Byculla* [illegible] Are you trying to insult me, or is that your ham-handed idea of a compliment? **1946** *Times* 3 Dec. 8/3 There should be no ham-handed bulk purchasing of stuff which was not really wanted. **1968** *New Statesman* 12 Apr. 465/3 Much of the recipient's pleasure is taken away by the very ham-handed invitation. **1964** *Economist* 11 Apr. 168/1 The FMC has gone a bit hamhandedly about its job. **1928** O. Stewart *Aerobatics* 30 One of the main objectives is finesse in the development of good 'hands'. Ham-handedness is not often a gift of unkind fate; it is not necessarily incurable. **1963** *Economist* 8 June 1046/1 The Kennedy Administration has contributed its own moments of ham-handedness. **1902** *Encycl. Brit.* XXVI. 558/2 Hamburger steak with onions, veal loaf, ham loaf. **1927** *Daily Chron.* 23 Sept. 7/5 Veal loaf, ham loaf, beef loaf. *c* **1847** J. S. Coyne in M. R. Booth *Eng. Plays of 19th Cent.* (1971) IV. 186 We used to go together to Greenwich, with a paper of ham sandwiches in my basket. **1866** 'Mark Twain' *Lett. fr. Hawaii* (1967) 68 The Sandwich Islanders always squat on their hams, and who knows but they may be the old original ham-sandwiches? **1872** 'L. Carroll' *Through Looking-Glass* vii, I feed him with—with—with Ham-sandwiches and Hay. [illegible] ham-sandwich, and is ready for the 'En voiture, messieurs'. **1972** B. Everitt *Cold Front* xv. 145 The boy..sat between us, polishing off a gigantic ham sandwich.

4. Phr. **ham and beef** *Rhyming slang*, the chief warder in a prison; **ham and eggs,** a dish consisting of fried ham and eggs.

1941 J. Phelan *Murder by Numbers* iv. 46 There's the ham-and-beef and ticket-boo making rounds. **1962** *John o' London's* 29 Jan. 82/1 A chief warder or prison officer is known in rhyming slang as a *ham and beef*. **1837** W. H. Wills *Jrnl.* in *S. Hist. Assoc. Publ.* VI. 473 They gave me fryed ham and eggs and biscuit, bread & Coffee. **1838** Dickens *Let.* 1 Feb. (1965) I. 366 We have had for breakfast...ham and eggs. **1967** G. Drummond *Death at Furlong Post* iii. 31 'Get me ham and eggs,' he sent.

II. 5. [App. short for **hamfatter*.] An ineffective or over-emphatic actor, one who rants or overacts. *slang* (orig. *U.S.*).

1882 *Illustr. Sport. & Dram. News* 23 Dec. 355/2 'Banjo Hams' are held up to scorn. *Ibid.*, One writer proudly describes himself as 'no ham, but a classical banjo player'. **1902** S. A. Clapin *New Dict. Amer.* 220 *Ham*, in theatrical parlance, a third-rate actor or variety performer. **1911** *Hampton's Mag.* Aug. 178/1 It was the voice of what is known as a 'ham'. [illegible] a 'ham' actor. **1926** H. C. Witwer *Roughly Speaking* 223 Ham actors get a extra split week at a picture house if their fearful monologs put the ladies on the border. **1928** *Daily Express* 20 June 9/4 Sophie Tucker will, in all probability, appear in a revue next autumn...'You have never seen me in revue,' Sophie reminded me, 'I am a ham actor too, you know.' **1933** 'Hay & Armstrong' *Orders are Orders* ii. 51 'We'd better have Harvey..to double for him.' '*That old ham actor!*' **1936** Wodehouse *Laughing Gas* xviii. 200 Just one of these ham actors that's jealous of a fellow's screen genius. **1942** E. Wilson *Wound & Bow* i. 81 Dickens had a strain in him of the ham actor, and in the desperation of his later life, he gave in to the old ham and let him have his head. **1947** Marsh *Final Curtain* xii. 179 A squalid little ham actress. **1957** V. J. Kehoe *Technique Film & T.V. Make-Up* i. 15 The expression 'ham' actor originated from those performers who rubbed ham rind on their faces as a base for their colored powders when they could not afford the more expensive and less odoriferous oils. **1958** *Times* 16 Apr. 3/2 'He thought I was an old ham,' says Miss Seyler indulgently.

b. An inexpert or over-theatrical performance; ham acting. *slang*.

1942 R. Chandler *High Window* (1943) xxx. 195 Don't feed me the ham. I've been in pictures. I'm a connoisseur of ham. **1959** *Times Lit. Suppl.* 20 Feb. 95/3 Charles Dickens..saw Lemaître in his late period and was swept off his feet, but what he says might apply equally well to ham acting. In fact, it sounds suspiciously like ham. **1969** *Listener* 28 May 754/2 The mummer who thinks that all acting before his time was 'ham'.

6. An amateur telegraphist; now esp., an amateur radio operator. *slang* (orig. *U.S.*).

1919 C. H. Darling *Jargon Bk.* 17 *Ham*, a student telegraph operator. **1922** *Glasgow Herald* 13 Mar. 4 Any person who passes a test prescribed by the Government can obtain a licence to 'send' radio messages in the United States, and in popular parlance one who has qualified and taken this 'Radio Operator Amateur—First Grade' certificate is dubbed a 'ham'. **1928** *Collier's* 22 Sept. 26 The amateur radio 'hams' have the ends of the earth for neighbors. **1929** *Amer. Speech* IV. 288 At either end of a wire an unskillful operator is a 'lid', 'ham', 'bum' or 'plug'. **1936** *Daily Herald* 19 Sept. 7/5 (Advt.), Do you ever hear the 'hams'? It appears that 'hams' is American for amateur radio transmitters. Of course, the 'hams' use the short wavelengths. **1955** *Sci. News Let.* 19 Mar. 188/2 Now it will be easier for a blind person to qualify for a license as a radio 'ham'. **1957** *Oxford Mail* 9 Nov. 4/5 The Russians claimed that thousands of 'hams' throughout the world to send details to Radio Magazine, Moscow, of reception from their satellites. **1967** *New Scientist* 11 May 330/3 The army of radio 'hams', who reach out over fantastic distances with their single sideband transmitters and receivers, are about to be reinforced. **1973** D. Lees *Rape of Quiet Town* vi. 90 He'd heard the radio ham speaking into a microphone.

III. 7. [Partly from **ham-fisted, -handed* adjs.] An incompetent boxer or fighter. *U.S. slang.*

1888 *Missouri Republ.* 27 Mar. (Farmer), He is a good fighter but will allow the veriest ham to whip him. **1929** *Sat. Even. Post* 14 Dec. 144/3 They want me to slug with this big ham.

B. *attrib.* or *as adj.* **1.** Characteristic of or relating to a ham actor or an inexpert performer; self-consciously theatrical. *slang.*

1935 H. Williams *4 Yrs. Old Vic* xi. 186 Young players to-day are scared of being what they call 'ham', which I suppose is an abbreviation of what I used to be termed 'ham-bone'. **1938** *Evening Standard* 26 July 7/4 We hear a great deal about 'ham' acting nowadays. As far as I can judge, 'ham' acting is the habit of rolling sonorous speeches round the tongue and delivering them with extravagant relish to the gallery. **1944** Auden *Sea & Mirror* in *For Time Being* iii. 56 The animals tenor never quite able at his big moments to get right up nor the ham right back down. **1958** B. Nichols *Sweet & Twenties* xvi. 231 His conception of aristocracy was strangely out of date, and more than a little 'ham'. **1958** *Observer* 4 May 15/7 It is one of the most extraordinary exhibitions of ham acting I've ever seen.

2. [Partly from **ham-fisted, -handed* adjs.] Clumsy, ineffective, incompetent. *slang.*

1941 M. Allingham *Traitor's Purse* xii. 133 Campion's thin hands remained expressionless and Lugg's great ham-fisted did not stir. **1942** Forbes & Allen *Ten Fighter Boys* p. xv, What he obviously intended to do on overshooting me was to flick over and spin down, but being a little ham, he overdid the manoeuvre and came the right way up. *Ibid.* 84, I didn't stay to argue, but went bowling down in the hammiest manner possible. **1949** 'J. Tey' *Brat Farrar* xiv. 124 He was..reluctant to submit that tender mouth to the ham hands of a Westerner. **1963** *Times Lit. Suppl.* 1 Feb. 71/2 Nothing he hated more than 'ham' writing and 'prefabricated' characters.

ham (hæm), *v.*² *slang.* [f. Ham *sb.*¹] To act in a 'hammy' manner, to over-act. Freq. const. *up*. Hence **ha·mmed-up** *ppl. a.*; **ha·mming** *vbl. sb.* and *ppl. a.*

1933 Stanley & Maxfield *Voice* 268 Hamming. **1937** *Printers' Ink Monthly* Apr. 54 *Ham it*, overacts [sic] for emphasis—bluster. [illegible] What with toning my voice down to that kid's mewing, and then trying to balance that hamming delivery. **1946** *Daily Tel.* 18 Nov. 6/6 Thomas Mitchell, after a deal of recent hamming, is a convincing detective. **1955** A. Huxley *Genius & Goddess* 16 The performance was on the corny side; but it was a sympathetic part and, though he clearly loved to ham it up, Beulah was not merely a treasure. **1955** T. Sterling *Evil of Day* ix. 110 'Any actor would give twenty years of his life to play the part'. 'I thought if I told you what it was you'd ham it.' **1957** *Listener* 12 Sept. 403/1 Nor does he purvey anything of Wales as it is—rather the hammed-up version of Wales that the stupider sort of Englishman prefers. **1957** *Observer* 10 Nov. 19/2 The temptation of second-feature hamming. **1958** M. Dickens *Man Overboard* ii. 27 She was hamming it up a bit for him [illegible]. **1960** S. H. Courtier *Gently dust Corpse* iii. 38 Hamming it now, thought Bolton [illegible]. **1965** *Listener* 18 Nov. 795/1 Marie Bell..hams it up in a smugly self-conscious cameo portrayal. **1973** E. Page *Fortnight by Sea* xii. 132 A hammed-up impression of a military man.

hamada(h), varr. *hammada.

Hamadan (hā·mădān). The name of a town in north-west Iran, used *attrib.* or *ellipt.* to denote a kind of carpet or rug (see quot. 1960).

1901 J. K. Mumford *Oriental Rugs* xi. 199 There is little difficulty in distinguishing the Hamadan carpet from all other weavings. **1932** P. Selver tr. *Čapek's Tales from Two Pockets* 184 By Jove, if it isn't a Hamadan. ..and sometimes there are some fine Karavams and Kelims to be picked up. **1960** H. Hayward *Antique Coll.* 138/1 Hamadan is the marketing centre for Kurdish rugs and they are often known by this name. Heavy, long pile of great durability, rather coarse in stitch but well coloured, designs often embellished with animal figures. Hamadan rugs are frequently described as 'Persian'.

hamadryas. *hamadryas baboon* = Hamadryad 2 b.

1932 S. Zuckerman *Soc. Life Monkeys & Apes* xii. 200 The barks of the Chacma are almost indistinguishable from those of the Hamadryas baboon. **1967** *Listener* 6 Apr. 459/3 The gelada baboon and the hamadryas baboon.

hamal, var. Hammal (Later examples.)

1962 J. Fleming *When I grow Rich* iii. 44 Heavily disguised as hamals, or human mules. **1967** J. Rathbone *Diamonds Bid* xii. 105 Grey-clad hamals, porters who will carry anything anywhere.

hamamelis (hæmămī·lis). [mod.L. (J. F. Gronovius in Linnæus *Genera Plantarum* (ed. 2, 1742), a. Gr. ἀμαμηλίς medlar.] A shrub or small tree of the genus so called, which is native to North America and Eastern Asia, belonging to the family Hamamelidaceæ, and includes several species bearing yellow flowers late in winter before the leaves appear; a witch-hazel. Also, the extract made from the leaves and bark of *Hamamelis virginiana*. So **hamame·lin**, the dried extract.

1751 J. Bartram *Observ. Trav. Pensilvania to Canada* 19 July (1751) 35 Now we came to most excellent level ground..full of tall timber..and shrubs, as opulus, green maple, hornbeam, hamamelis. **1760** J. Lee *Introd. Bot.* 332 Witch Hazel, Hamamelis. **1890** Billings *Med. Dict.*, *Hamamelin*, name given to a dry powdered extract of hamamelis. [illegible] **1900** M. Thorn in W. D. Drury *Bk. Gardening* xi. 586 The Hamamelis here mentioned may be propagated by grafting in February or early March. [illegible]

Haman (hē¹·mæn). **1.** The name of the chief minister of Ahasuerus who was hanged on the gallows prepared for Mordecai, as related in the Book of Esther, used allusively (phr. *to hang as high as Haman*). So **Hama·nic** *a.*

[illegible] **1816** Scott *Old Mort.* xxxvi, The whig Captain Balfour..swore..that if the garrison was not given over in two days by daybreak, he would hang up the young lord, poor thing, as high as Haman. **1842** Barham *Ingol. Leg.* 2nd Ser. I'll hang him as high as Haman! **1881** H. Adler in *19th Cent.* Dec. 813 Prof. Goldwin Smith renews his onslaughts upon Jews and Judaism with an acerbity and virulence which I may venture to term Hamanic.

2. Haman's ears, formerly **Haman's fritters,** fritters or cakes eaten by the Jews at the festival of Purim.

1846 'A Lady' *Jewish Man.* vii. 123 Haman's Fritters. **1949** *Housewife* May 50/1 The pudding that we ate..on Purim..was appropriately called Haman's Ears—lovely crisp shape fried in oil, and then sprinkled with sugar and cinnamon. **1961** *Times* 23 Dec. 9/1 'Haman's ears': a cake eaten by Jewish people at the festival of Purim, to commemorate the downfall of Haman, their oppressor.

hamartia (hāmă·ɹtiă). [a. Gr. ἁμαρτία fault, failure, guilt.] The fault or error which entails the destruction of the tragic hero (with particular reference to Aristotle's *Poetics*).

[**1789** T. Twining *Aristotle's Treat. Poetry* 308 Dacier confounds himself and his readers in his note about *Thyestes*. He mistakes Aristotle's sense of ἁμαρτία.] **1895** S. H. Butcher *Aristotle's Theory Poetry & Fine Art* vii. 303 But with him [sc. Macbeth] the ἁμαρτία, the primal defect, is the taint of ambition. **1927** F. L. Lucas *Tragedy* iv. 105 If we seek the *hamartia* in more modern tragedy like Ibsen's, it becomes clearer than ever that an intellectual mistake is all that the term need mean. **1956** H. House *Aristotle's Poetics* vi. 94 All serious modern Aristotelian scholarship agrees..that 'hamartia' means an error which is derived from 'ignorance of some material fact or circumstance'. **1968** D. W. Lucas in *Aristotle's Poetics* 302 The essence of *hamartia* is ignorance combined with the absence of wicked intent.

hamartoma (hæmaɹtōu·mă). *Path.* [mod.L., ad. G. *hamartom* (E. Albrecht 1904, in *Verh. d. deutsch. path. Ges.* VII. 153), f. Gr. ἁμαρτ-άνω to go wrong: see *-oma.] A tumour-like mass resulting from the faulty growth or development of normal cells or tissue.

1904 *Index Med.* II. 142/2 (*index*) Hamartoma. **1909** J. G. Adami *Princ. Path.* I. iii. xxiii. 749 We must be governed here by our conception of the meaning of the word *angioma*...If we restrict it to mean a tumor, due to the independent growth of vessels, they [sc. other tumors having vessels as their main constituent] must be cast out of this class. To provide a class for them Albrecht has suggested the term *Hamartoma*. **1950** G. P. Wright *Introd. Path.* xxvii. 479 Striking examples of the vascular hamartomas may be seen in the simple, cavernous, or more complicated plexiform, overgrowths of blood vessels that form congenital nævi, birth-marks,.. or vascular warts. **1970** S. D. Kobernick tr. *Masson's Human Tumors* I. ii. 77 In one category, heterotopia consists of tissues which belong to the region where they are encountered, but have escaped the organization of this region and are found in excess: these are Hamartomas.

Hamathite (hē¹·măþəit). [f. *Hamath*, the biblical name for Hama in western Syria + -ite¹.] An inhabitant of the ancient Syrian city of Hamath; also, a script found in the Taurus mountains, now called 'Hittite'. Also *attrib.*

1611 Bible *Gen.* x. 18 The Aruadite, the Zemarite, and the Hamathite. **1880** *Encycl. Brit.* XII. 26/1 Professor Sayce's view that the Hittites were the authors of the Hamathite hieroglyphics. **1926** D. G. Hogarth *Kings of Hittites* [illegible] The Hamathite territory is..the nearest [illegible] evidence of occupation by [illegible] yet appeared. **1952** O. R. Gurney *Hittites* 3 The same script had been noticed by J. L. Burckhardt [illegible] rock-carving over a stream at Ivriz in the Taurus mountains. Davis had then called the script 'Hamathite'.

hambergite (hæ·mbəɹgəit). *Min.* [ad. G. *hambergit* (W. C. Brögger 1890, in *Zeitschr. f. Kryst. und Min.* XVI (Specieller Th.). 65), f. the name of A. *Hamberg* (1863–1933), Sw. mineralogist, who discovered it: see -ite¹.] A basic borate of beryllium, $Be_2(OH)BO_3$, occurring in colourless orthorhombic crystals that have strong double refraction.

1890 *Jrnl. Chem. Soc.* LVIII. ii. 1078 Hambergite, discovered by A. Hamberg in 1889, and named after him by the author, crystallises in the rhombic system. [illegible] xvi. 265 Hambergite..was originally found in southern Norway in material of non-gem quality, but a discovery has now been made of large colourless crystals in central Madagascar.

hambo (ha·mbo). [f. *Hambo*, name of a parish in Hälsingland, Sweden.] A Swedish folk dance in ¾ time.

1929 *Blackw. Mag.* Jan. 79/1 We had been dancing in the Nyboda barn—dancing polskas and hambos. *Ibid.* Feb. 196/2 We only play old Swedish tunes..only old peasant polskas, hambos, and waltzes.

ha·m-bone. 1. The bone in a ham of hog meat; also, such a bone with the meat attached. Also (in *pl.*) *transf.*

1855 W. G. Simms *Forayers* 340 He had..brought with him the ham-bone and several which he had so hastily appropriated. **1866** 'Mark Twain' in *Harper's Mag.* Dec. We have only left a lower end of a ham-bone, with some of the outer rind and skin on. **1908** J. London *Let.* 22 Dec. (1966) 275, I have to get down on my ham-bones and beg forgiveness. **1972** P. Dickinson *Lizard in Cup* vi. 97 He found a bone in a trash-can—a bit of ham-bone.

2. An inferior or amateur actor, esp. one who speaks in a spurious Negro accent; a mediocre musician. *U.S. slang.*

[illegible] *Lippo* v. 11 Hambones! I told you so. I could say that when I heard them [illegible] *No Wedding Bells for Her* [illegible] hambones means them as pays to act, which they can't do, and so saves the manager engaging deserving actors. **1942** Berrey & Van den Bark *Amer. Thes. Slang* § 583/38 *Hambone*, an unconvincing blackface dialectician. **1960** B. Keaton *Wonderful World of Slapstick* (1967) 25 Because I was also a born hambone, I ignored any bumps..I may have got at first on hearing audiences gasp.

3. A sextant. *Naval slang.*

1938 F. A. Worsley *First Voy. in Square-Rigged Ship* viii. 144 What officer have you got on that hambone, Johnnie? **1962** Granville *Dict. Sailor's Slang* 58/1 *Hambone*, sextant which is of much the same shape.

hambro, hambro-line, vars. Hamber-line.

1793 J. MacDonell *Diary* in C. M. Gates *Five Fur Traders* (1933) 75 The canoe line...consists of fine Hambro lines loosely twisted upon one another. **1867** [see Hamber-line]. **1961** F. H. Burgess *Dict. Sailing* 108 *Hambro, Hambroline*, a small line, made of three strands of hemp, hard laid, sometimes tarred, and used for lacings, seizings, etc.

Hamburger. Also **-burgher, †-bourger.** [G. *Hamburger* a native or inhabitant of Hamburg in Germany.] **1.** A native or inhabitant of Hamburg. Also *attrib.* or as *adj.*

1616 G. Carew *Let.* Jan. (1860) 81 The Kinge of Denmarke..purposeth to build a stronge fort vppon the River of Elbe, requiringe the Hambourgers to permitt 3,000 of his men to passe throughe there towne. **1653** W. Gee *Let.* 16 June in M. M. Verney *Memoirs* (1894) III. ii. 56 They had been fired upon by an English ship, and had pursued 'a Hamburger'. **1737** *London Mag.* Feb. 81/2 (*heading*) Of the City of Hamburgh, with several Observations on the Hamburghers, and other Germans. **1798** D. Wordsworth *Jrnl. Visit to Hamburgh* (1941) I. ii. 21 Hamburgher girls with white caps. **1856** Geo. Eliot in *Westm. Rev.* IX. 14 The fair Hamburgher acted in the spirit of Johnson's advice to Hannah More. **1932** L. Golding *Magnolia St.* ii. iv. 314 The stolid young Hamburgers who went out rowing on Sundays on the Alster. **1966** *Guardian* 5 Nov. 7/3 He had the welfare of the sleeping Hamburghers in mind.

2. (Now freq. with lower-case initial.) In full *Hamburger steak* = *Hamburg steak* (see Steak 2 c); also, a kind of sausage. Now, chopped beef, spiced and flavoured, formed into a cake and fried, often served between two halves of a toasted bun. So *hamburger bar*, etc. orig. *U.S.*

1889 *Walla Walla* (Wash.) *Union* 5 Jan. 2/4 You are asked if you will have 'porkchopbeefsteakhamandeggs-hamburgersteakorliverandbacon'. **1901** Ade *40 Mod. Fables* 285 After the kid had been carried out of the Ring looking like a Hamburger Steak. **1902** *Encycl. Brit.* XXVI. 558/2 Hamburger steak with onions. **1908** 'Yeslah' *Tenderfoot S. Calif.* xiv. 118 Out of date eggs, last year's hamburger and over ripe limburger. **1912** I. Cobb *Back Home* 147 A vendor..sold to the same customers..odorous hamburger and flat slabs..of striped ice-cream. **1920** *Chambers's Jrnl.* 348/2 Hamburger steaks and German pot roasts. **1929** E. Hemingway *Farewell to Arms* ii. xv. 106 They had the look of not too freshly ground hamburger steak. **1931** B. Stark *Touch & Go* iv. 58 A truck driver transporting two huge draft horses took us to Bryan, where he grandly 'set 'em up' to hamburgers and coffee. **1943** *R.A.F. Jrnl.* Aug. 10 She held a pair of..gloves in one hand and a hamburger sandwich in the other. **1950** *Manch. Guardian Weekly* 6 Apr. 3 After much pleasantry over the han·burgers and hot dogs. **1957** *London Mag.* Dec. 19 There is a pier with all the usual sideshows, hot dog, hamburger and ice-cream stands. **1960** *Times* 30 Jan. 11/7 The mounting number of hamburger bars suggests that it is here to stay.

Hamburg parsley: see Parsley 1, 2.

hame². Add: **b.** *hame-rein.*

1902 *Daily Chron.* 9 July 3/6 Why a tight hame rein should be used on so many builders' and other carts is a puzzle. **1908** *N. & Q.* 10th Ser. X. 106 At the foot of a hill leading from Blackrock, near Brighton, to Rottingdean is a board with the inscription: 'Please slacken hame-rein on going uphill.'

hamel (hā·məl, ‖ha·məl). *S. Afr.* [Afrikaans = Du. *hamel*, G. *hammel* castrated ram; cf. Hamble *v.*] A wether.

1835 A. Smith *Diary* 22 Apr. (1939) I. 376 If relatively rich, sometimes kill ten oxen and ten hamels. **1871** H. Dugmore *Remin. Albany Settler* 14 A dozen of startled hamels, just separated from a large flock, would be likely to try a driver's legs, and lungs too, in crossing it. [illegible] W. C. Scully *Kafir Stories* 28 Would your father have let me die rather than take a hamel from the flock of a rich Boer, who never counts his sheep? **1896** R. Wallace *Farm. Ind. Cape Col.* xvii. 346 Ram lambs are..'sorted' to make wethers or hamels. **1950** *Cape Times* 14 Dec. [illegible] At an auction sale..hamels fetched £6 11s.—the highest price for hamels in this district.

‖hamerkop (hā·məɹkɒp, ha·məɹkɒp). *S. Afr.* Also semi-anglicized **hammerkop** (hæ·-). [Afrikaans, f. *hamer* hammer + *kop* head.] = Hammer-head 4.

1834 A. Smith *Diary* 7 Dec. (1939) I. 163 A Basuto said the hammerkop gives rain. **1887** *Encycl. Brit.* XXII. 578/1 It [sc. *Scopus umbretta*]..is the 'Hammerkop' (Hammerhead) of the Cape colonists. **1890** [see Hammer-head 4]. **1895** *Funk's Stand. Dict.*, Hamerkop. [illegible] *Cape Argus* 7 Dec. 2 Hamerkops are remarkable fowls... The hamerkop's nest is a huge affair, generally in a tree. **1952** *Cape Times Week-end Mag.* 8 Nov. 6 A solitary hamerkop flies homeward every evening. **1964** C. Willock *Enormous Zoo* vii. 120 Hammerkops, the queer brown bird with the blunt head that gives it its name, pick over scraps in the rotting papyrus stems. [illegible] E. Palmer *Plains of Camdeboo* xi. 195 We never saw the hamerkop in numbers. **1971** *Country Life* 28 Oct. 1127/2 Hammercops, curious bulky brown birds with crests and thick bills, flew about in pairs.

‖hametz (hā·mets, χ-). Also **chametz, chometz,** etc. [Heb. *ḥāmēṣ*.] Leaven, or food that has been mixed with leaven, prohibited during the Passover.

1891 M. Friedländer *Jewish Relig.* 377 The head of the family..examines his residence thoroughly, and keeps the *chametz*, which he has found, in a safe place till the next morning. *Ibid.* 378 All the *chamets* that is left after the first meal on the 14th of *Nisan* must be removed. **1892** I. Zangwill *Childr. Ghetto* II. xxiv. 124 'Where is the Rabbi?' 'Up in the bedrooms gathering the *Chomets*..hunting with a candle for stray crumbs.' **1960** *Jewish Chron.* 8 Apr. 35/1 The Rev. S. Black..told the boys and girls of Pesach and its meaning. Of the search for chametz (leavened bread). **1960** *Commentary* June 499/2 Removing bread (but not other *hametz*) from the house on Passover. **1973** *Jewish Chron.* 9 Feb. 12/5 He is convinced that no one eats *hametz* during Passover.

Ham Hill stone, Ham stone, a Somerset stone, representative of the lower part of the Upper Lias, quarried in the Ham Hill quarry near Yeovil and used widely for building purposes in the area.

1889 H. B. Woodward in *Proc. Bath Nat. Hist. & Antiquarian Club* VI. 182 The celebrated building-stone of Ham Hill, near Yeovil..is not without geological interest. .. The Ham Hill stone is mainly composed of sand and comminuted shells. **1918** *Q. Jrnl. Geol. Soc.* LXXIV. 169 At Ham Hill..the portion of *moorei* date is exposed in the big quarry on the hill and the main mass of it is a 'freestone'—the celebrated Ham-Hill Building-Stone. In the big quarry the sequence is as follows: 1. Sand.. 2. 'Riddings'..3. Ham-Hill Stone. **1936** G. Pollett *Song for Sixpence* ix. 76 This grey Ham stone is most satisfying to the eye. **1961** *Countryman* LVIII. iii. 439 Sophisticated strangers are not yet trying to buy the lovely old Hamstone farmhouses.

Hamidian (hæmi·diăn), *a.* [f. the name of Abdul *Hamid* II + -ian.] Pertaining to or resembling the rule of Abdul Hamid, Sultan of Turkey from 1876 to 1909. Hence **Hami·dianism.**

1908 *Westm. Gaz.* 1 Aug. 2/3 Thirty years of wandering in the Hamidian wilderness. **1908** *Daily Chron.* 24 Oct. 4/4 The Hamidian rule. *Ibid.* 18 Dec. 4/4 A reversion to Hamidianism. **1930** *Times Lit. Suppl.* 4 Dec. 1047/3 His spy-system appears to have been Hamidian in its extent and efficiency.

Hamidieh (hæmi·die). [f. the name of Abdul *Hamid* II + *-ieh* adj. suffix.] A body of Kurdish cavalry formed by the Turks in 1891.

1898 H. A. G. Percy *Diary Asiatic Turkey* 83 Zekki,.. the reputed founder of the Hamidieh System. **1901** *Westm. Gaz.* 27 Aug. 2/2 The Hamidie Cavalry..defy the Porte by ignoring its commands. **1902** *Encycl. Brit.* XXV. 665/2 A tribal militia force (Hamidieh), consisting of 48 regiments, is formed somewhat on the lines of Cossacks.

Hamite, *sb.*¹ Add: Hence **Ha:miticiza·tion,** the action of becoming Hamitic; **Hami·ticized** *a.*, having become Hamitic; **Ha·mitoid** *a.*, resembling the Hamitic type.

1884 *Nature* 17 Apr. 581/1 These peoples should apparently be regarded rather as Negroes affected by Hamitic than as Hamites affected by Negro elements. In other words, they are Negroid rather than Hamitoid. **1911** H. H. Johnston *Opening up of Africa* iii. 91 The earlier and more elaborate of these works were inspired by Semites and executed by Hamiticized negroes. **1923** G. W. Murray *Eng.-Nubian Dict.* Introd., In the case of Nubian, the process of Hamiticization has gone so far that it has borrowed Hamitic personal-endings for its verb, Hamitic case-endings for its noun, and possesses a vocabulary largely Hamitic. **1936** *Discovery* June 171/1 The first great group of hamiticised Negroes, the Nilotes, constitute a well-defined physical type.

Hami·to-Semi·tic, *a.* Designating the language family including Hamitic and Semitic languages. Also as *sb.* Also **Hami·tic-Semi·tic** *a.* and *sb.*

1909 *Cent. Dict.* Suppl., *Hamito-Semitic*, relating to the peoples speaking Hamitic and Semitic languages which are considered members of one linguistic stock. **1936** *Science & Society* I. 25 He records similarities..between pronouns in Hamitic-Semitic. **1939** L. H. Gray *Foundations of Lang.* 357 Second in importance only to the Indo-European linguistic family comes the Hamito-Semitic group. **1964** R. H. Robins *Gen. Ling.* viii. 307 The Hamito-semitic family, represented by classical Arabic and the Arabic languages and dialects of the Middle East and North African coast.

Hamlet² (hæ·mlėt). The name of the prince of Denmark who is the hero of Shakespeare's play of this name, in allusive phr. *Hamlet without the Prince* (*of Denmark*): a performance without the chief actor or a proceeding without the central figure.

[**1775** *Morning Post* 21 Sept., *Lee Lewes* diverts them with the manner of their performing Hamlet in a company that he belonged to, when the hero who was to play the principal character had absconded with an inn-keeper's daughter; and that when he came forward to give out the play, he added, 'the part of Hamlet to be left out, for that night.'] **1818** Byron *Let.* 26 Aug. (1830) II. 445 My autobiographical essay would resemble the tragedy of Hamlet.., recited 'with the part of Hamlet left out by particular desire'. **1820** Lady Granville *Let.* 22 Aug. (1894) I. 161, I am not used to be newsmonger and perhaps I leave out Hamlet. **1825** Scott *Talisman* (1883) 5 The title of a 'Tale of the Crusaders' would resemble the playbill, which is said to have announced the tragedy of Hamlet, the character of the Prince of Denmark being left out. **1859** G. Meredith *Ordeal R. Feverel* I. vii. 109 'What have you been doing at home, Cousin Rady?' 'Playing Hamlet, in the absence of the Prince of Denmark.' **1902** *Daily Chron.* 22 Apr. 3/1 Of what avail is it to promise 'entirely new scenery' for 'Die Meistersinger', if the part of Hans Sachs is to be practically eliminated from the performance? And yet this 'Hamlet-without-the-Prince' method is consistently pursued season after season at Covent Garden. **1910** *Times Weekly* 17 June 452 The army without Kitchener is like Hamlet without the Prince of Denmark. **1918** L. Strachey *Emin. Victorians* 86 The Catholic Church without the absolute dominion of the Pope might resemble the play of Hamlet without the Prince of Denmark. **1967** J. Prescot *Case Counterfeit* viii. 96 Without Drax one can't do a thing. Hamlet without the Prince of Denmark, I guess. **1972** *Publishers Weekly* 3 Apr. 22/3 The article..in the March 6th *PW* was an attempt to stage Hamlet without the Dane.

Hence **Ha·mletish** *a.*; **Ha·mletism,** an attitude resembling that of Hamlet; **Ha·mletize** *v. rare*, to soliloquize or meditate after the manner of Hamlet.

1844 Hebbe & MacKay tr. *Sealsfield's Life in New World* 267 Halloo! Mr. Howard! Hamletizing? **1852** H. Melville *Pierre* vii. vi. 191 In this plaintive fable we find embodied the Hamletism of the antique world. **1854** 'G. Greenwood' *Haps & Mishaps in Europe* iii. 53 Herr Devrient is a handsome, Hamlet-ish man, with a melancholy refinement of voice. **1905** *Daily Chron.* 11 Apr. 4/7 Let us forget Hamletism and all its ills. **1920** D. H. Lawrence *Women in Love* xiv. 205 One shouldn't talk when one is tired and wretched.—One Hamletises, and it seems a lie. **1923** —— *Stud. Classic Amer. Lit.* ix. 180 So Dana sits and Hamletizes by the Pacific—chief actor in the play of his own existence. **1936** *Times Lit. Suppl.* 5 Sept. 711/2 Adams's madness is, indeed, a trifle Hamletish. **1945** W. Fowlie in *Mod. Reading* XII. 210 He is the one contemporary writer who has driven out from his nature all traces of hamletism, and yet he writes constantly about Hamlet. **1952** A. R. D. Fairburn *Strange Rendezvous* 25 He has played the gravedigger to many a Hamletish posture of my soul.

hammada (hæmā·dā). *Geol.* Also **hamada(h).** [f. Arab *ḥammāda*.] A flat rocky area of desert blown free of sand by the wind, typical of the Sahara.

1853 J. Richardson *Narr. Mission Cent. Afr.* II. iv. 60 Aghadez is situated on a hamadah, or lofty plateau of sandstone and granite formation. **1857** H. Barth *Trav. N. & Cent. Afr.* I. v. 133 Overweg and I had no time to lose in preparing for our journey over the hammáda, or plateau. **1886** *Encycl. Brit.* XXI. 149/2 Nearly all the rest of the Sahara consists..of undulating surfaces of rock (distinguished as *hammada*),..and regions of sandy dunes. **1934** W. Fitzgerald *Africa* i. ii. 60 Rocky wastes with the bare exposure of fissured rocks as dominant features of the scene, form the 'hamada' type of the Sahara. **1966** *McGraw-Hill Encycl. Sci. & Technol.* IV. 76/2 Ordinarily, a hammada is a bare rock surface composed of relatively flat-lying consolidated sedimentary rocks from which overlying softer sediments have been stripped, principally by wind erosion.

hammel, var. Hemel.

1812 J. Sinclair *Syst. Husb. Scot.* i. 21 Small open sheds, or what, in Berwickshire, are called *Hammels* or *Hemmels*, with separate straw-yards attached. **1851** H. Stephens *Bk. Farm* (1855) I. 242 The hammels [are occupied] by the two-year-olds, or such as are fattening for the butcher. **1893** W. Fream *Youatt's Compl. Grazier* (ed. 13) 171 The litter used in hammels is..less by one-third than that required for stall-feeding.

hammer, *sb.* Add: **6.** *hammer and sickle*: an emblem consisting of a crossed hammer and sickle, used as a symbol of the industrial worker and the peasant, e.g. on the national flag of the U.S.S.R.; hence used allusively of Soviet-type Communism.

1921 *Times* 20 Sept. 4/6 The subjects of the..designs [of Bolshevist postage stamps] are symbolical of Labour.. the 20 roubles a shield charged with the device of a hammer and sickle crossed. **1933** H. G. Wells *Shape of Things to Come* iii. § 11. 330 There was still no discord with Russia; there the blazon of the wings was put up side by side with the old hammer and sickle. **1935** E. Weekley *Something about Words* 27 A new ideal in literature and poetry, a kind of 'hammer and sickle' conception of artistic composition. **1937** H. G. Wells *Brynhild* v. 65 It might be possible to indicate whether the flavouring [of a book] were sexual, intellectual, left, right, or detective, by some variation in the general design, an obelisk, for example, the hammer and sickle, the swastika or what-not. **1958** *Listener* 5 June 928/2 An Algiers broadcast said the choice was 'between the Hammer and Sickle and the Cross of Lorraine'.

7. *hammer-boy, -shed*; **hammer-action,** (*a*) action of or as of a hammer; (*b*) those parts of a piano which compose and control the hammers; **hammer-axe** (later example); **hammer-block,** the steel face of a steam-hammer; **hammer-dress** *v.* (later examples); **hammer drill,** a percussion drill; **hammer-lock** *Wrestling*, a position in which a wrestler is held with one arm bent behind his back; also *fig.*; so **hammer-lock** *v. trans.*; **hammer-price** *Stock Exchange*, the price realized for shares (of a defaulter) closed at the hammer; **hammer-rifle,** a rifle fired by means of a hammer; **hammer-thrower** (see sense 1, note).

1885 *Encycl. Brit.* XIX. 71/2 An altered German harpsichord, the hammer action of which..may have been taken from Schroeter's diagram. *Ibid.* 72/1 In Frederici's upright grand action..the movement is practically identical with the hammer action of a German clock. **1906** *Westm. Gaz.* 22 Mar. 7/2 The explosion, which was probably caused by the hammer action of the water. **1927** Peake & Fleure *Priests & Kings* 165 Perforated hammer-axes..are said to have been found [at Tripolye]. **1861** W. Fairbairn *Iron* 121 The hammer-block is guided in its vertical descent by two planed guides or projections. **1881** *Instr. Census Clerks* (1885) 42 Forge and Hammer Boy. **1909** *Westm. Gaz.* 19 Aug. 9/4 There has been a considerable shortage of hammer boys in most of the mining districts. **1939** J. D. S. Pendlebury *Archaeol. Crete* iii. 98 The stones are invariably hammer-dressed, the saw not yet being used for masonry. **1940** *Chambers's Techn. Dict.* 401/1 *Hammer-dressed*, a term applied to stone surfaces left with a rough finish produced by the hammer. **1908** R. Peele *Compressed Air Plant for Mines* xx. 249 Numerous small air hammer drills..have come into favor in the past few years... The hammer drill strikes a light blow. **1922** *Encycl. Brit.* XXXI. 958/1 Machine drills underwent important changes during 1910–20, especially in the development of the 'hammer' drills... In the hammer drill, the bit is held stationary..and is struck a rapid succession of blows by the reciprocating piston-like hammer. **1897** *Pearson's Mag.* III. 638 Hammer lock and Nelson on the ground. **1905** *Daily Chron.* 21 Feb. 7/4 The very thought of being 'hammer-locked' should be enough to deter the most confirmed 'disorderly'. **1906** E. Dyson *Fact'ry 'Ands* vi. 72 Jest you take a 'ammerlock holt iv yerself, 'n' 'ave some dam consideration fer others. **1907** G. B. Shaw *Let.* 23 Sept. (1956) 107 Short of giving Phyllis a leading part, and thus giving you the hammer lock on him, I dont know what to do. **1944** *Infantry Jrnl.* (U.S.) June 25 He got his Jap in a hammerlock. **1965** *Economist* 4 Dec. 1072/2 These are fuzzy far-off dreams, considering the right wing's hammerlock on the Republican party today. **1900** *Westm. Gaz.* 4 June 7/1 He can have the stock closed at the hammer price. **1901** *Ibid.* 13 May 9/1 The actual dealings in the shares being between £6 and £8 per share and the hammer price £2. **1907** *Yesterday's Shopping* (1969) 634 Hammer rifles. **1920** G. Burrard *Notes on Sporting Rifles* 15 Hammerless ejectors are better than non-ejectors and hammer rifles. **1890** W. J. Gordon *Foundry* 13 The blast-furnaces that stand near the hammer-shed. **1899** *Daily News* 18 July 7/2 The hammer-throwers were out in the morning. **1968** *Listener* 11 July 49/2 There have been a number of marriages..between hammer-throwers and female discus-throwers.

hammer, *v.* Add: **1. a.** (Later examples.)

1907 F. H. Burnett *Shuttle* xxxviii. 379 Jem Belter, who 'hammered' a typewriter. **1959** M. Shadbolt *New Zealanders* 26 The Potoki boys hammered the piano and banged the drums.

2. e. To inflict heavy defeat(s) on, in war, games, etc.; to strike forcefully; to beat up. *colloq.*

1948 Partridge *Dict. Forces' Slang 1939–45* 90 *Hammer*, to shell severely. To inflict a heavy defeat on. **1959** *Times* 28 May 4/6 Smith hammered Slade for two fours and a six. **1973** *Times* 5 Jan. 17/5 Challenging the well-entrenched leaders in the United Kingdom car rental industry seems to hold no fears for Crook. He is hoping to hammer them on both quality and price. **1973** *Courier & Advertiser* (Dundee) 14 Feb. 5/3 He was severely injured about the face and his dentures were broken. He had no doubt that he had been 'hammered'.

3. b. Of a pipe: to make a knocking noise, as when a flow of liquid is suddenly stopped by turning a tap. (Cf. Water-hammer 2.)

1889 P. Hasluck *Model Engin. Handybk.* 108 The pump, owing to its not being filled properly at each stroke, will hammer very much.

hammerer. 1. Add: As a specific occupation.

1909 *Westm. Gaz.* 8 Feb. 3/1 The man was a 'hammerer' —i.e., a driver of rivets into boilers, &c. **1921** *Dict. Occup. Terms* (1927) § 278 *Hammerer*,..flattens saw blades,..by..striking any curved part with hammer.

hammer-head. 2. Delete † and add later example.

1947 R. Taylor *Bar Nothing Ranch* (1949) xvi. 151 The meanest old hammerheads under her tutelage became as cooing doves.

5. *hammer-head crane* = **hammer-headed crane.*

1910 *Encycl. Brit.* VII. 371/1 The Titan is portable and the hammer-head crane fixed. **1938** *Jane's Fighting Ships* 37 Fitting out berth equipped with giant 25 ton Hammer-head Crane.

hammer-headed, *a.* Add: **1. b.** *hammer-headed crane* (see quot. 1910).

1908 A. Tolhausen tr. *Böttcher's Cranes* ix. 492 (*title*) Hammer-headed crane of 150 tons, constructed by the Duisburger Maschinenbau-A.-G. **1910** *Encycl. Brit.* VII. 370/2 The so-called 'hammer-headed' crane consists of a steel braced tower, on which revolves a large horizontal double cantilever; the forward part of this cantilever or jib carries the lifting crab, and the jib is extended backwards in order to form a support for the machinery and counter-balance.

hammerkop: see *hamerkop.

hammerman. c. (Later example.)

1880 *Harper's Mag.* Dec. 59 The hammer-man, in a swinging seat, times the turning of his rod of steel to the quick stroke of the hammer.

hammily: see *hammy *a.*

hammock¹. Add: **4. hammock chair,** a folding reclining-chair with canvas support for the body, suitable for use in a sitting-room or garden; a deck-chair; **hammock-moth** (see quot.).

1881 *Graphic* 18 June in L. de Vries *Vict. Advts.* (1968) 127/1 The Yankee hammock chair..costs but 17s. 6d. complete. **1885** *Army & Navy Co-op. Soc. Price List* II. 1478 Portable Hammock Chairs. **1971** *Country Life* 1 Apr. (Suppl.) 44/2 (Advt.), Early 19th century hammock chair in mahogany upholstered in deep-buttoned Havana brown leather. **1899** *Cambr. Nat. Hist.* VI. 379 The Hammock-moth, *Perophora sanguinolenta*, of the centre of South America, the larva of which constructs its portable habitations out of its own excrement.

Hammond organ (hæ·mənd ǭ·ɹgăn). *Mus.* [See quot. 1960.] The proprietary name of an electric organ produced by the Hammond Organ Company, in which sounds are produced by generating and combining electric currents of suitable frequencies; applied also to similar instruments; also *ellipt.* as *Hammond*.

1935 *Electronics* VIII. 156 (*caption*) The Hammond electric organ. **1936** *Proc. Inst. Radio Engin.* XXIV. 1446 The amount of agreement and departure from this tempered scale of frequencies found in the Hammond organ have been listed in his [sc. Hammond's U.S.] patent No. 1,956,350. **1957** *Gramophone* Apr. 426/3, I confess that the Hammond organ is an instrument which I find hard to accept in the field of popular music. **1960** Gammond & Clayton *Guide Pop. Mus.* 100 *Hammond organ*, an electrical instrument invented by Laurens Hammond in 1929. The Hammond Instrument Company of Chicago was formed in this year and began marketing the instrument in 1935... In 1938 the Aeolian-Hammond Player-Organ was produced, an entirely mechanical instrument. **1970** *Melody Maker* 3 Oct. 15/1 Having squeezed every possible sound from his Hammonds, the Moog was the natural progression.

hammy, *a.* Add: Also, resembling ham.

1877 E. S. Dallas *Kettner's Bk. of Table* 413 It was a grand hit this—the introduction of the hammy taste.

2. Of, pertaining to, or characteristic of a ham actor or ham acting. *slang.*

1929 T. Wolfe *Look Homeward, Angel* (1930) xxvii. 367 With fat hammy sonority he welcomed them. **1933** Stanley & Maxfield *Voice* ii. iii. 80 In particular, the prolongation of the vowel sounds associated with the consonants m, n and l is cheap and 'hammy'. **1946** *Penguin New Writing* XXVIII. 182 Toni..put on a hammy deep-in-thought act. **1965** G. McInnes *Road to Gundagai* vi. 106 His part..was..hammy enough, but Carson managed to ham it up a good deal further. **1973** 'D. Jordan' *Nile Green* xxxv. 171 Condon raised an eyebrow in a hammy attempt to be supercilious.

Hence (sense 2) **ha·mmily** *adv.*

1942 *Time* 27 Apr. 61 The talented author..has told one of the most hackneyed of all sentimental yarns, and told it hammily. **1958** *Spectator* 7 Feb. 175/2 Rather hammily acted but extremely well sung. **1961** *John o' London's* 14 Sept. 307/1 The plot is hackneyed to the point of imbecility, the slapstick hammily archaic.

Hampshire (hæ·mpʃəɹ). **a.** The name of a county in the south of England, used (chiefly *attrib.*) to designate a breed of sheep; also *Hampshire Down*; also designating a breed of pig.

a **1661** T. Fuller *Worthies* (1662) Hants. 2 Hantshire Hoggs, are allowed by all for the best Bacon. **1813** C. Vancouver *Gen. View Agric. Hampshire* 371 The.. common Hampshire ewe will cost from 25s. to 40s. each. **1825** Loudon *Encycl. Agric.* 1123/2 The heath sheep, old Hampshire, or Wilts breeds. **1875** *Encycl. Brit.* I. 392/2 These sheep are now usually classed as Sussex Downs and Hampshire Downs, the former being the most refined type of the class.., and the latter..having a heavier fleece, stronger bone, and somewhat coarser and larger frame. *Ibid.* 400/2 The Berkshire and Hampshire hog seems originally to have been from the same stock, but by some early cross acquired the thicker carcase, prick-ears, shorter limbs, and earlier maturity of growth, by which they are characterised. **1886** C. Scott *Sheep-farming* 12 The Hampshire Down, though a larger sheep than the Southdown, does not mature so early. **1957** *Encycl. Brit.* XVII. 920/1 The Hampshire breed [of pig] originated in England and was later introduced into the United States... Hampshires possess good growing and fattening qualities. **1962** J. N. Winburne *Dict. Agric.* 361/1 *Hampshire swine*, an American, lard-type breed of black, white-belted swine. **1971** *Farmers Weekly* 19 Mar. 77/4 It was a risky step to take from the viewpoint of..Hampshire Down enthusiasts.

b. *Hampshire hog*: a colloq. or derogatory term for a native of Hampshire; also, a dish of boiled bacon and vegetables.

[**1622** Drayton *Polyolbion* ii. xxiii. 70 As Hamshire long for her, hath had the tearme of Hogs.] *c* **1720** *Vade Mecum for Malt-Worms* i. 50 Now to the Sign of Fish let's jog, There to find out a Hampshire Hog. **1861** C. M. Yonge *Stokesley Secret* i. 9 'You could not be more right if you were a Hampshire hog,' said Sam. **1937** J. Rayner *Shell Guide to Hampshire* 26/1 Hampshire Hog. You boil 4 to 5 lb. of bacon..in an iron saucepan, keep the extracted bacon hot on the hob, and put..cabbages..into the water. ..You can put potatoes in as well. *Ibid.* 30 There are three sorts of Hampshire hog, and they have given the county the subsidiary name of *Hoglandia*. One, the inhabitant of the county. Two, the less domestic animal from whose frequency the inhabitant gets his name...And three, the dish. **1944** in A. Wykes *Royal Hampshire Regiment* (1968) v. 104, I reckon us little lot of Hampshire Hogs have done well for his nibs Adolf in this invasion. **1963** C. Mackenzie *My Life & Times* II. 169 She was a Dorset woman, and both she and her husband had a profound contempt for what they called the Hampshire hogs with whom they were condemned to live.

c. *pl.* The Royal Hampshire Regiment.

1904 *Westm. Gaz.* 14 June 8/2 The Hampshires, who mustered ten officers and 484 men. **1968** A. Wykes *Royal Hampshire Regiment* v. 107 The Hampshires, with..the Dorsets and Devons, were the three battalions of infantry forming one of the spearheads that was to land on the Arromanches beach.

Hampstead Heath (hæ·mpsted hīþ). The name of a district in north London, (*a*) used in *Rhyming Slang* to designate the teeth; also *Hampsteads*; (*b*) *Hampstead Heath sailor* (see quot. 1889).

1887 *Referee* 6 Nov. 7/3 She'd a Grecian 'I suppose', And of 'Hampstead Heath' two rows In her 'sunny south' that glistened Like two pretty strings of pearls. **1889** Barrère & Leland *Dict. Slang* I. 444/2 *Hampstead Heath sailor*.., a term of ridicule—no sailor at all. **1932** *Daily Express* 25 Jan. 6/6 (*heading*) 'Hampsteads' and 'Yobs'. A common expression for the feet is 'plates o' meat' and for the teeth 'Hampstead Heath'... These become simply 'plates' for feet and 'Hampsteads' for teeth. **1962** R. Cook *Crust on its Uppers* (1964) ii. 23 The rot had set in something horrible with her hampsteads and scotches.

hamseen, var. Khamsin.

1923 F. S. Marvin *Sci. & Civiliz.* 29 Physical contrasts of seasonal and regional fertility are abrupt; solar heat contends with Nile water, sea breeze with scorching 'hamseen'.

Han¹ (hæn). Designating a Chinese dynasty (206 B.C.–220 A.D.) marked by the introduction of Buddhism, the extension of Chinese rule over Mongolia, the revival of letters, and increase of wealth and culture.

1736 R. Brookes tr. *Du Halde's Gen. Hist. China* I. 346 (*heading*) The Fifth Dynasty, called *Han*, which had twenty-five Emperors in the Space of 426 Years. **1837** *Penny Cycl.* VII. 81/1 About the year 201 B.C., the race of Tsin was succeeded by that of Hân, which filled one of the most celebrated periods of Chinese history. **1876** *Encycl. Brit.* V. 644/1 Lew Pang was then proclaimed emperor (206 B.C.) under the title of Kaou-te, and the new line was styled the Han dynasty. **1930** *Times Lit. Suppl.* 2 Oct. 774/3 The majority seem to reflect the Han style... Two famous homonymous generals who flourished under the Han dynasty. **1935** A. Toynbee *Study of Hist.* (ed. 2) II. 373 The Han Empire and the Kushan Empire marched with one another in Central Asia for at least a century. **1971** *Ashmolean Mus. Rep. of Visitors 1970* 51 Imperial seals of the 17th and 18th centuries as well as some early Han types.

han²: see Khan².

1903 *Westm. Gaz.* 10 Feb. 3/1 The *hans* are large, rambling inns, with a courtyard in the middle. **1920** *Q. Rev.* Apr. 395 Four hundred emaciated forms, the remnant of such convoys, are lying in one of the hans.

‖hanami (hanami). [Jap.] (See quots.).

1891 A. M. Bacon *Jap. Girls & Women* x. 295 The *hanami*, or picnic to famous places to view certain flowers as they bloom in their season. **1902** L. Hearn *Kottō* x. 97 All of us..should make up a party, and enjoy our *hanami* together. **1965** W. Swaan *Jap. Lantern* iv. 47 This collection is suitable for such festive occasions as *hanami* (flower- and more particularly cherry-blossom-viewing).

hanapoot, var. *hanepoot.

‖hanashika (hanaʃika). [Jap.] A professional story-teller.

1891 A. M. Bacon *Jap. Girls & Women* x. 294 Public halls, where professional story-tellers, the *hanashika*, night after night, relate long stories to crowded audiences. **1936** K. Nohara *True Face of Japan* v. 120 A *hanashika*, or story teller, is telling innocent stories.

hancockite (hæ·nkǫkəit). *Min.* [f. the name of E. P. *Hancock* (*c* 1834–1916), American artist and amateur mineralogist, who discovered it: see -ite¹.] A variety of epidote rich in strontium and lead.

1899 Penfield & Warren in *Amer. Jrnl. Sci.* CLVIII. 339 (*heading*) Hancockite. *Ibid.* 343 A considerable quantity of hancockite was taken from the mine at one time, and it is the most abundant of the new species described in this paper. It is named after Mr. E. P. Hancock of Burlington, N.J. **1900** *Jrnl. Chem. Soc.* LXXVIII. ii. 88 Hancockite. This occurs as brownish-red, cellular masses of minute, lath-shaped crystals, which are monoclinic. **1968** I. Kostov *Mineral.* 309 Hancockite contains Pb and Sr.

hand, *sb.* Add: **1. f.** *pl.* In Association Football, the illegal handling of the ball.

1894 Branscombe & 'Ross' *Morocco Bound* ii. 28 The statute demands A free kick for hands! **1897** [see *handling *vbl. sb.* 1 c]. **1967** *Assoc. Football* (Know the Game Series) 28 (*caption*) Area covered by 'Hands'.

5. Also as a symbol of acceptance of an invitation to dance.

1813 Jane Austen *Pride & Prej.* I. xviii. 208 When the dancing recommenced..and Darcy approached to claim her hand. *a* **1817** —— *Northanger Abbey* (1818) II. i. 15 After aspiring to my hand, there was nobody else in the room he could bear to think of.

7. b. (Later example.)

1965 *Listener* 3 June 835/3 A major document of the post-Symbolist movement in Spain, with English versions by eleven hands, the 'hands' including W. S. Merwin,.. and James Wright.

14. b. A member of a cricket eleven.

1731 in H. T. Waghorn *Cricket Scores* (1899) 4 The Duke's hands came in first. **1874** *Baily's Monthly Mag.* Dec. 155 Seven of the eleven..were new hands.

† c. A score in cricket. *Obs.*

1833 J. Nyren *Young Cricketer's Tutor* 104 He would often get long hands. **1836** *New Sporting Mag.* Oct. 361 [Which number] added to the byes they stole, and the wide balls bowled, sufficed to make a hands of eighty-six runs. **1875** *Baily's Monthly Mag.* Sept. 273 Let me see him make a good hand against good bowling.

15. (Later examples.) Esp. in present-day use in phr. *to give* (or *get*, etc.) *a big* (or *good*) *hand*: to give, etc., a large round of applause. orig. *U.S.*

1838 Dickens *Nickleby* xxix. 284 He has gone on night after night, never getting a hand and you getting a couple of rounds at least. **1849** *Theatrical Programme* 18 June 30 Buskin's part goes without a hand—Lady [illegible] carries off all the honours. **1883** G. B. Shaw *How to become Mus. Critic* (1960) 48 The dance-tunes, played by an indifferent band, went almost without a hand. **1886** *Lantern* (New Orleans) 6 Oct. 4/3 Their act always pulls a big hand. **1896** *Punch* 10 Oct. 180/2 Aeschylus..wrote tragedies in blank verse, but they are not now played at any London theatre. He would not get a 'hand' nowadays. **1922** C. Sandburg *Slabs of Sunburnt West* [illegible] It's a good act—we got a good hand. **1924** H. A. Vachell *Quinney's Adv.* 179 The second curtain fell without a 'hand'. **1927** Prince of Wales in *Even. News* 7 Oct. [illegible] They both do a great deal of hard work for the British Legion. It may be I am more the fellow who travels about and gets the hands. **1932** A. J. Worrall *Eng. Idioms* 40 He always gets a good hand when he appears in a London theatre. **1948** *Prairie Club Bull.* June 14 Three lusty cheers and a big hand for Charles, Our Square Dance Host! **1959** *Listener* 28 May 958/3 A [illegible] which earned what our Quiz compères insist on calling 'a big hand'.

18. c. A device shaped like a hand.

1830 M. Edgeworth *Let.* 6 Dec. (1971) 439 Mr. Turner ..had shewn me the bank of England and the famous machine-*hand* which weighs the guineas without assistance from mortal touch. **1873** *Young Englishwoman* Jan. 52/1 Will any one..tell her how to clean white..gloves? She possesses wooden hands for stretching them on. **1926–7** *Army & Navy Stores Catal.* 1008/2 Dairy utensils.. Scotch hands [for shaping butter].

21. d. (Earlier example.)

1756 P. Browne *Civil & Nat. Hist. Jamaica* ii. [illegible] I have sometimes seen a hand of ginger weigh near half a pound... The larger spreading roots are called Hands in Jamaica.

23. Also, the cards held at any stage of such a game as Poker.

1889 R. Guerndale *Poker Bk.* 25 To fill your hand, to improve it by the draw. **1913** 'A. B. Lougher' *Poker* [illegible] The next process is that of drawing to fill the hands.

d. *to declare one's hand* (fig.): to reveal one's circumstances or aims. (Cf. Declare *v.* ii. [illegible])

1922 D. H. Lawrence *England, my England* [illegible] Upstairs Fanny evaded all the thrusts made by her mother, and did not declare her hand.

29. in hand. k. *Billiards.* Of the cue-ball: having been retrieved by hand, after being pocketed, and having been placed on any selected spot within the D preliminary to the next stroke.

1860 'R. Crawley' *Handy Bk. Games* xiii. 106 Your ball being in hand, you must play for the hazard that shall bring the object-ball back to the opposite cushion. **1904** S. A. Mussabini *Mannock's Billiards* v. 228 The cue-ball is 'in hand' with the red ball, presenting a straightaway winning hazard into the right middle pocket.

l. *N.Z.* (See quots.)

1930 L. G. D. Acland *Early Canterbury Runs* i. [illegible] Until wire fences were introduced about 1862, all sheep on the plains were kept more or less in hand. The practice was for a shepherd to go round the boundary once or twice a day. **1933** —— in *Press* (Christchurch) 28 Oct. 15/7 Sheep are in hand when you have them in a mob near you.

33. c. *to eat* (or *feed*) *out of one's hand* (see *eat *v.* 3 d).

1958 Hayward & Harari tr. *Pasternak's Dr. Zhivago* i. vii. 212 'Well, have they had their tails twisted yet? Are they keeping quiet now?' 'The shopkeepers, you mean?..Feed out of your hand!'

44. c. *to give a hand*: to help a person. Also, *to lend a hand* (see Lend *v.*² 2 e).

1860 in A. F. Ridgway *Voices from Auckland* 71 His young wife..will readily give him a hand at the crosscut for a few hours. **1880** *Daily Tel.* 26 Nov., A policeman gave him a hand up. **1949** J. Routh in *Granta* Christmas ed., 'Here, let me give you fellows a hand,' I suggested.

45*. put (one's) **hand. a.** *to put one's hand*: to exert oneself, use one's energies; now always with *to*: to set about, undertake (a piece of work).

1388 Wyclif *Luke* ix. 62 No man that puttith his hoond to the plouȝ, and biholdynge bacward, is able to the rewme of God. **1439** in *Fenland N. & Q.* (1905) June 222 And yat..ye wole at yis tyme..putte youre handes and ese us by wey of lone of ye somme of C marc. *c* **1450** *St. Cuthbert* (Surtees) 6056 Þat to þi seruyce puttys þaire hands. **1535** Coverdale *Deut.* xii. 18 All yᵗ thou puttest thine hande vnto. **1631** J. Preston *Treat. Effect. Faith* 45 If God himselfe put not his hand to the worke no man is able to believe. **1633** J. Hall *Hard Texts* Zech. xi. 9, I will not put my hand to redresse it. **1879** M. J. Guest *Lect. Hist. Eng.* ix. 80 Whatever he put his hand to, he did it 'with all his might'.

b. *to put* (one's) *hand*(*s*) *on*, (†*in*, †*unto*): to lay hands on (see Lay *v.*¹ 21 c); †to do violence to (*Sc. Obs.*); to get hold of, seize (also *fig.*).

1535 Coverdale *Exod.* xxii. 8 ([He] shal sweare) that he hath not put his hande vnto his neghbours good. **1837** C. M. Goodridge *Voy. S. Seas* (1843) 44 [We] got in her with such articles as we could immediately put our hands on. **1842** J. H. Newman *Par. Serm.* VI. viii. 111 Perhaps..we can put our hand, as it were, on a time in our childhood [when, etc.]. **1972** L. Henderson *Cage until Tame* vii. 57 Right now he couldn't put his hands on a hundred quid.

53. hands off! Also *attrib.*

1902 *Daily Chron.* 23 Jan. 7/1 A protest must be made against the hands-off policy. **1908** *Times Lit. Suppl.*

3 Sept. 283/3 The hopelessly *doctrinaire* character of the old 'hands-off' individualism.

b. *Aeronaut.* Used as *adj.* and *adv.* in connection with an automatically controlled aircraft.

1932 *Flight* 20 May 443/1 By means of the adjustable tail plane the machine can be trimmed to fly 'hands off'. **1935** *Jrnl. R. Aeronaut. Soc.* XXXIX. 1041 Control of this airplane when operating on one engine was carefully studied with the result that it can be flown 'hands off'. **1959** *Times* 18 Sept. 7/3 Some 4,000 'hands off' landings have been made. **1966** *Electronics* 3 Oct. 134 On the last orbit, it guided Gemini 11 into a hands-off reentry and landing virtually down the stacks of the recovery ships. **1968** *New Scientist* 1 Feb. 251/1 Several companies have developed artificial stabilization systems which enable the helicopter to be flown 'hands-off'.

54. hands up! Also in *Curling* (see quot. 1897).

1873 J. Miller *Life amongst Modocs* 193 Hands up, gentlemen! **1897** *Encycl. Sport* I. 264/1 *Hands up*, the command of the Skip..to stop sweeping. **1910** *Encycl. Brit.* VII. 646/1 Curling has a language which contains many curious terms... *Hands up!* stop sweeping.

55. hand..fist. a. (Later examples.) Also, esp., *fig.* of the making of money.

1825 W. N. Glascock *Naval Sketch-Bk.* (1826) I. 26 The French..weathered our wake, coming up with us, 'hand over fist', in three divisions. **1833** S. Smith *Life Major J. Downing* (1834) 116 They..clawed the money off of his table, hand over fist. **1861** M. B. Chesnut *Diary* 8 Aug. (1949) 107 Fitzhugh Lee and Roony are being promoted hand over fist. **1888** 'R. Boldrewood' *Robbery under Arms* xxvii, We..made money hand over fist. **1901** *Daily Chron.* 27 Dec. 3/3 To use a phrase common to the Anglo-Saxon, they have been making money hand-over-fist. **1929** J. B. Priestley *Good Companions* II. vii. 445 She lost money hand-over-fist for weeks and weeks—and not a murmur—and now she is beginning to get a little back again. **1963** *Times* 12 June 13/4 It pays hand over fist always to take the secondary roads that run along the 'wrong' sides of rivers in preference to the main roads that run along their 'right' sides.

56. hand and foot (further example).

1955 L. P. Hartley *Perfect Woman* x. 96 He has everything he wants and servants who wait on him hand and foot.

59*. hands-across-the-sea, used *attrib.* of an act, etc., performed by one country as a gesture of friendship to an overseas country.

1899 *Westm. Gaz.* 1 Feb. 1/4 Mr. Tree has a new drama in his mind—an old-time hands-across-the-sea subject. **1955** *Sci. News Let.* 5 Feb. 88/3 The Missouri Commission is interested in the experiments, not only as a 'hands-across-the-sea gesture'..but because the Missouri bunny might also be susceptible to the dread disease, and advance knowledge would permit more effective preventive action. **1957** *Times Lit. Suppl.* 25 Oct. 635/2 What is this but empty rhetoric of the 'hands-across-the-sea' brand? **1971** C. Fick *Danziger Transcript* (1973) 88 It was supposed to be a hands-across-the-sea sort of thing.

60. h. *hands down*: with ease, with little or no effort; unconditionally, submissively; orig. in the racing phr. *to win hands down*, referring to the jockey dropping his hands, and so relaxing his hold on the reins, when victory appears certain.

1867 'Pips' *Lyrics & Lays* 155 There were good horses in those days, as he can well recall, But Barker upon Elepoo, hands down, shot by them all. **1882** *Moonshine* 3 June 265 (*caption*) Won!! 'Hands Down'. **1913** Mrs. H. Ward *Mating of Lydia* II. xii, That I should surrender, hands down, to a lot of trumpery complaints and grievances. **1920** W. B. Money *Humours of Parish* 126, I started off in the race in full nigger costume, and won hands down. **1958** *Times* 14 Aug. 9/7 Double this speed, however, and the submarine wins hands down.

61. a. *hand-gesture, -kiss, -movement, -rest.*

1930 R. Paget *Babel* ii. 56 The large number of ideas which cannot be symbolized directly by hand-gesture. **1962** *Listener* 1 Mar. 382/2 The energy and urgency of hand gesture, visible in his recent drawings, has extended itself on the larger scale of these paintings. **1861** C. Reade *Cloister & Hearth* ii. 32 A sweet little coaxing hand-kiss. **1958** F. Harris *My Life & Loves* V. i. 28 She sent me back with an imperious hand-kiss. **1924** R. M. Ogden tr. *Koffka's Growth of Mind* v. 254 During the hand-movements, the gaze is directed fixedly upon the object. **1904** Goodchild & Tweney *Technol. & Sci. Dict.* 279/1 *Hand rest*, the T-shaped support for supporting hand-turning tools in working at a lathe when a slide rest is not used. **1922** Joyce *Ulysses* 225 Two carfuls of tourists passed slowly, their women sitting fore, gripping frankly the handrests.

c. *hand-baggage, -camera* (earlier and later examples); so *hand-camerist*; *hand-lamp* (earlier and later examples), *-luggage, -microphone, -mike, -props.*

1902 *Daily Chron.* 27 Feb. 3/3 The Boer delegates have only brought with them hand-baggage. **1889** *Photogr. News* 15 Nov. 755/1 For the hand camera there is, I believe, a great and wide field of usefulness. **1910** *Chambers's Jrnl.* Sept. 588/1 Hand-cameras are made in a thousand patterns. **1892** *Photogr. Ann.* II. 52 It is this ungentlemanly abuse of the hand camera which brings the whole class of hand camerists into disrepute. **1897** C. M. Hepworth *Animated Photogr.* xiii. 96 Subjects which are suitably lighted and otherwise 'possible' for the hand camerist may be safely attempted with a cinematographic camera. **1831** Carlyle *Sart. Res.* I. x, Thou..wilt walk through thy world by the sunshine of what thou callest Truth, or even by the hand-lamp of what I call Attorney-Logic. **1836** *Mechanics' Mag.* XXV. 317 The fact is that no *unguarded* light should ever be permitted in any stable, warehouse, cellar, or bed chamber; cheap, convenient, and even elegant hand lamps and lanterns suited to these uses, are met with in abundance. **1940** *Chambers's Techn. Dict.* 401/2 *Hand-lamp*, a portable electric-light fitting suitable for carrying in the hand. Also called *inspection-lamp, portable-lamp.* **1965** 'Lauchmonen' *Old Thorn's Harvest* x. 134 They all sit down near to their handlamps. **1888** Ld. Macnaghten in *Law Rep. Ho. Lords* xiii. 55 Passengers take the lighter articles of luggage—or 'hand-luggage' as it is called,—in the carriage with them. **1908** *Daily Chron.* 8 Jan. 4/4 Glancing furtively at that terrible piece of hand-luggage, a New York Sunday newspaper. **1924** W. J. Locke *Coming of Amos* iv, Maxime possessed himself of her hand-luggage. **1970** *New Yorker* 16 May 41/3 'Any hand luggage?' says the clerk, peering over the top of the counter. **1968** *Radio Times* 20 June 58/1 For long I have been puzzled about the use of hand microphones. **1968** *Punch* 3 Jan. 29/1 He does know precisely what he's about when he clutches a handmike and sings. **1933** P. Godfrey *Back-Stage* iii. 34 'Props' is an abbreviation of 'stage properties', and therefore 'hand-props' are things handled by the actors, such as fans, snuff-boxes, etc.

d. *hand-brake* (later examples), *-brush, -carriage, -drum, -feed, -machine, -punch* (example; hence as vb.), *puppet, sewing-machine, -sled, -sledge* (earlier example), *-sleigh, -wheel* (examples).

1913 *Autocar Handbk.* (ed. 5) xiv. 216 In some Coventry Daimler cars the hand brake can be caused either to withdraw the clutch or not when it is applied. **1959** *Times* 15 Dec. 13/5 Consternation clutches at the heart while the driver clutches the handbrake. **1968** *Sun* 12 Nov. 8/5 *Handbrake turn*, the technique involving using the handbrake instead of reverse gear, for negotiating 'impossible' hairpins. **1747** H. Glasse *Art of Cookery* xvii. 150 Let them be scrubed clean with a Hand-Brush and Sand, and Fuller's Earth. **1904** J. Vaizey *More about Pixie* (1910) i. 7 She went down on her knees, and swept up the dust with a small hand-brush. **1745** Ellis *Mod. Husb.* VI. II. 14 This Farmer..carried his Wheat-sheaves into his Barn on a Sunday, by Hand-carriage. **1859** Dickens in *All Year Round* (1860) 11 Aug. 422/2 A hand-carriage, drawn by a man... I saw within it an old man. **1864** J. A. Grant *Walk across Africa* viii. 144 A band of hand-drums is near the sultan's hut, giving lighter dance-music for the amusement of the boys and girls. **1958** E. Borneman in P. Gammond *Decca Bk. Jazz* xxi. 275 A male leader and a small group..who accompanied themselves on..hand drums,..and gong-gong. **1940** *Chambers's Techn. Dict.* 401/1 *Hand feed*, the hand operation of the feed mechanism of a machine tool, as distinct from an automatic feed. **1962** *Gloss. Autom. Data Proc.* (*B.S.I.*) 91 *Hand-feed punch*, a key punch into which punched cards have to be fed manually one at a time. **1873** *Young Englishwoman* Mar. 131/1 A most useful..hand-machine..at the low price of 39s. **1927** T. Woodhouse *Artificial Silk* 81 Practically coincident with the hand knitting of jumpers and the like came the hand-machine knitting. **1962** *Gloss. Autom. Data Proc.* (*B.S.I.*) 91 *Hand punch*, a key punch into which punched cards have to be fed manually one at a time. *Ibid.* 96 *Hand punch*, a tape punch operated directly by hand. **1967** A. Battersby *Network Analysis* (ed. 2) xv. 264 Their decisions are hand-punched on to special cards. **1950** *Dryad Handicraft Catal.* 74 Hand puppets and string puppets. **1957** *Encycl. Brit.* XIV. 906/1 Hand puppets can be of wood, plaster,..or stuffed cloth. *a* **1877** Knight *Dict. Mech.* II. 1058/1 *Hand sewing-machine*, a form of sewing-machine in which the parts are pivoted jaws, operated in the manner of scissors. **1746** *Coll. New H. Hist. Soc.* IX. 141 [I] went to mill with a hand sled. **1780** in *Coll. Mass. Hist. Soc.* (1905) 7th Ser. V. 6 [They] hall their wood on hand sleds. **1843** *Knickerbocker* XXII. 294 The serjeant's hand-sled, piled with wood. **1877** *Rep. Vermont Board Agric.* IV. 92 Provided with a handsled, the boy would first roll on to it the back log. **1848** R. M. Ballantyne *Hudson's Bay* (1890) 83 The hand-sledge is a thin flat slip or plank of wood... Indians invariably use it when visiting their traps, for the purpose of dragging home the animals or game they may have caught. **1829** G. Head *Forest Scenes* 203 [The trees] had been..removed by means of small hand sleighs purposely prepared for them. **1836** C. P. Traill *Backwoods of Canada* 110 We were overtaken on our return by S— with a handsleigh, which is a sort of wheelbarrow, [etc.]. **1841** G. Powers *Hist. Sk. Coos* 70 A rude hand sleigh. **1936** D. McCowan *Anim. Canad. Rockies* xii. 103 An experienced trapper..delivered it on a hand sleigh. **1931** *Engineering* 9 Jan. 62/3 The roll adjusting gear is operated by a single handwheel, through an arrangement of steel bevel and spur gears. **1940** *Chambers's Techn. Dict.* 402/1 *Hand wheel*, a grooved pulley provided with a cranked handle and mounted on a universal form of vice, used for driving a lathe or other tool by hand. **1948** *Brit. Jrnl. Psychol.* Mar. 149 The pointer..moved steadily away from the line, and required a compensatory movement of the handwheel at a rate of 1 r.p.m. to keep it on the line.

62. a. *hand-holding* (hence as a back-formation *hand-hold* vb.), *-washing* (later example).

1963 *Movie* Apr. 12/1 Newsreel photographers were often forced to hand-hold their cameras. **1908** *Daily Chron.* 13 Mar. 4/6 Hand-holding ensures a rapid means of communication between the hand and the heart or brain. **1960** C. Day Lewis *Buried Day* v. 94 Hand-holdings and childish kisses. **1964** M. Hynes *Med. Bacteriol.* (ed. 8) ii. 22 Nevertheless the main sterilizing action of hand-washing is the mechanical removal of epithelial scales and surface bacteria.

b. *hand-done, -drawn* ppl. adjs.; *hand-feeding*; *hand-fired* ppl. adj. (so *hand-firing*); *hand-fisher*; *hand-flung, -held, -hewn* (ppl.) adjs.; *hand-knit* adj. (also as *sb.*), *-knitted* ppl. adj.; *hand-knitting*; *hand-milker, -milking*; *hand-operated, -set* ppl. adjs.; *hand-sew* vb. (so *hand-sewing, -sewn*); *hand-thrown, -tooled* ppl. adjs.; *hand-tufted* pa. pple. and ppl. adj.; *hand-washing* (see also 62 a); *hand-woven* (examples).

1907 N. Munro *Daft Days* xvii. 151 Another hand-done bill upon the counter. **1908** *Westm. Gaz.* 28 Dec. 5/2 Insets of hand-done crochet form one of the newest designs. **1907** *Yesterday's Shopping* (1969) 737/1 Hand-drawn linen. **1937** *Evening News* 15 Mar. 3/6 (Advt.), Hand-drawn top collar. **1957** Manvell & Huntley *Technique Film Music* iii. 167 Attempts at creating hand-drawn sound-tracks have been in progress since the coming of the sound film. *Ibid.*, He built up a sound-track of hand-drawn musical effects, mixing them occasionally with normal orchestral instruments to give variety. **1960** *Farmer & Stockbreeder* 29 Mar. 65/1 To the busy shepherd, hand-feeding may not seem worth while. **1968** *Gloss. Terms Offset Lithogr. Printing* (*B.S.I.*) 24 *Hand feeding*, the manual placing of single sheets of paper or other material to the machine lays. **1963** *Times* 23 May 7/3 Existing hand-fired plant still in fair condition. **1908** *Chambers's Jrnl.* Aug. 624/1 Owing to the fact that unscreened British coal is extensively employed in Hamburg hand-firing is more generally adopted. **1961** *Listener* 12 Oct. 583/2 'Hand-firing', or the fact that the fuel has to be brought to the boiler, is the drawback of all solid-fuel boilers. **1855** 'P. Paxton' *Captain Priest* 147 In the deeper places of such streams must the handfisher seek his prey. **1913** Kipling *Songs from Books* 113 But Tubal fashioned the hand-flung spears. **1923** Glazebrook *Dict. Appl. Physics* IV. 393/2 The hand-held camera implies 'exposures' of brief duration. **1969** *Daily Tel.* (Colour Suppl.) 10 Jan. 18 (*caption*) The photograph was taken by astronaut Anders with hand-held Hasselblad. **1969** *Jane's Freight Containers 1968–69* 584/1 All four motors are controlled from a single hand-held push-button control unit. **1938** M. K. Rawlings *Yearling* xiv. 143 The thick hand-hewn slabs of the shingled roof. **1969** *Computers & Humanities* III. 240 The work of Hagelman and Barnes is born into competition with one of the last of the hand-hewn concordances. **1920** F. Scott Fitzgerald *This Side of Paradise* (1921) I. ii. 63 The hand-knit, sleeveless jerseys were stylish. **1958** Handknit [see *bawneen]. **1959** *Manch. Guardian* 29 July 5/1 Woollens fall into two classes, the cashmere jerseys, sweaters, and cardigans..and the Scottish handknits. **1967** *Harper's Bazaar* Sept. 49 The neck, High-polo-ed, hand-knit. **1881** *Sylvia's Home Jrnl.* in A. Adburgham *Shops & Shopping* (1964) xvii. 189 The largest stock in the kingdom of hand-knitted socks and stockings. **1952** M. Laski *Village* viii. 135 A skirt and a hand-knitted jumper. **1902** *Daily Chron.* 20 Dec. 5/2 Another ancient industry is at its last gasp—viz., the hand-knitting of Kilmarnock bonnets. **1961** J. G. Davis *Dict. Dairying* (ed. 2) 740 The War of 1914–18, because of the acute shortage of hand milkers, gave an unexpected impetus to the adoption of machine milking. **1915** J. London *Let.* 26 Jan. (1966) 446 Get Timms' experience with hand-milking labor conditions. **1960** *Farmer & Stockbreeder* 8 Mar. 55/2 In the past when hand-milking was the rule. **1936** *Discovery* Nov. 350/1 The Electrotor meter has a hand-operated pump movement. **1908** *Daily Chron.* 23 Oct. 9/4 All hand-set [type] becomes unnecessary. **1938** *Times Lit. Suppl.* 26 Mar. 204/3 Hand-set in most attractive type. **1895** *Montgomery Ward Catal.* 318/3 Hame Strap, hand sewed. **1919** Barrie *Alice Sit-by-the-Fire* I. 21 You hand-sew them and stretch them over a tin cylinder. **1946** *Nature* 14 Dec. 868/1 The speed of expert hand-sewing, thirty stitches per minute, is slow and laborious compared with that of machine work. **1961** T. Landau *Encycl. Librarianship* (ed. 2) 161/1 *Hand sewing*, usually sewing through the fold by hand on the sewing frame, to suspended cords or tapes arranged across the back of a book. **1887** *Col. & Indian Exhib., Rep. Col. Sect.* 401 A handsome pair of men's hand-sewn Wellingtons. **1911** *Rep. on Labour & Social Conditions in Germany* III. vi–vii. 101, I could have my boots soled and heeled with this quality of leather, and hand-sewn for 3s. 5d. **1959** *Times* 7 Mar. 9/2 A pigskin-covered whip, tightly stretched and handsewn, makes a pleasing and serviceable article. **1909** *Westm. Gaz.* 22 Oct. 4/2 The barbed spear used..is a hand-thrown weapon. **1933** *Archit. Rev.* LXXIV. 38/1 (*caption*) These are hand-thrown pieces on the wheel. **1895** 'Mark Twain' in *North Amer. Rev.* July 2 An illustrated, gilt-edged, tree-calf, hand-tooled, seven-dollar Friendship's Offering. **1931** *Times Lit. Suppl.* 25 June p. vii/2 The Swiss hand-tooled bindings are disappointing. **1906** *Daily Chron.* 5 June 4/5 Killybegs carpets, which are hand-tufted by the peasants. **1922** Joyce *Ulysses* 697 Handtufted Axminster Carpet. **1962** *BSI News* Feb. 17/1 Methods for the determination of colour fastness of textiles to a number of agencies, for example, soda boiling, handwashing, [etc.]. **1880** L. Higgin *Handbk. Embroidery* 62 Neutral-tinted hand-woven linen. **1925** A. Huxley *Let.* 2 Nov. (1969) 258 We bought..twelve yards of hand woven material.

63. hand-bag, (*a*) a light travelling-bag, (*b*) a lady's bag for accessories; **hand-balancer,** an acrobat; **hand-bible** *slang* = Holy stone *sb.*; **hand block** (see Block *sb.* 7), a block used in printing textiles by hand; also *attrib.*; hence **hand-blocked** *a.* (cf. *block-printed* adj.); **hand-blown** *ppl. a.*, of glass blown by a craftsman; **hand-board** *U.S.*, a board in front of a preacher or speaker; **hand-fives,** the usual game of fives as distinguished from bat-fives (see Fives[2] 1); **hand-hole** (examples); **hand-jam** *v.* (*Mountaineering*), to wedge a hand in a crack as a handhold; hence as *sb.*; also **hand-jamming** *vbl. sb.*; **hand-jive** (see quot. 1961); hence **hand-jiving** *vbl. sb.*; **hand-laid** *ppl. a.* (cf. *laid paper*); **hand-letter** *v.* (see quots.); hence *hand-lettered* ppl. adj., *hand-lettering* vbl. sb.; **hand-light** (earlier example); **hand-pick** *v. trans.*, to pick by hand; also *fig.*; so **hand-picked** *ppl. a.*; **hand-piece, handpiece,** (*a*) the part of a dental drill that is held in the hand; (*b*) the part of a sheep-shearing machine that is held in the shearer's hand; **hand-plate,** (*a*) = *finger-plate*; (*b*) a small plate to pass over the surface of work to be tested; **hand-pollinate** *v. trans.*, to pollinate by hand; so *hand-pollination*; **hand-print,** the mark left by the impression of a hand; also (quot. 1886), a representation of a hand; **hand-reading,** palmistry; so *hand-reader*; **hand signal,** a manual indication by the driver of a motor vehicle, pedal cycle, etc., of his intention to stop, turn, etc.; **hand-stand,** an act in gymnastics in which the body is supported by the hands while the feet are in the air; also *attrib.*; **hand-towel,** a small towel for wiping the hands after washing; **hand traverse** *Mountaineering* (see quots. 1897, 1957).

1862 *Englishwoman's Domestic Mag.* July 143 Portable umbrellas..may easily be carried in the hand-bag. **1867** A. D. Whitney *L. Goldthwaite* ii. 32 Their hand-bags were hung up. **1880** Hand-bag [see Hand *sb.* 61 c]. **1896** G. B. Shaw *Let.* 9 Nov. (1965) 700, I want to buy a handbag for the journey. **1899** O. Wilde *Importance of being Earnest* I. 37 'Where did the charitable gentleman who had a first-class ticket for this seaside resort find you?..' 'In a handbag.'..'A handbag?' **1913** *Vanity Fair* Dec. 79/2 The latest novelty in hand bags. **1923** *Weekly Dispatch* 13 May 14 Crocodile Calf handbag. **1937** *Discovery* Dec. 374/2 Ladies' shoes and hand-bags. **1968** *New Society* 22 Aug. 266/1 Non-U handbag/U bag (the thing carried by women). **1927** *Daily Tel.* 30 Aug. 12/6 Masu, the Japanese hand-balancer and juggler. [*a* **1865** Smyth *Sailor's Word-Bk.* (1867) 98 *Bible*,..a squared piece of freestone to grind the deck with sand in cleaning it; a small holystone, so called from seamen using them kneeling.] **1908** O. Onions *Pedlar's Pack* 109 That was Ben, and i' the Dolphin, where all they knew o' Bibles was the hand-bibles they holystoned the decks wi'. **1839** Ure *Dict. Arts* I. 215 The hand blocks are made of sycamore or pear-tree wood, or of deal faced with these woods. **1936** *Archit. Rev.* LXXIX. 291/1 Any artist of decorative faculties can create patterns for printed cretonnes or linens,..provided he is sufficiently familiar with the advantages and drawbacks of roller-printing, screen-printing and handblock-printing. **1928** *Daily Express* 9 Jan. 5/2 The new..hand-blocked linens are a boon. **1931** *Ibid.* 21 Sept. 5/2 Pure silks, whose hand-blocked patterns have all been designed by famous..British artists. **1928** *Ibid.* 28 May 7/2 An exhibition of hand-blown and enamelled glass. **1958** *Archit. Rev.* CXXIV. 256 These fittings are made of thick handblown glass in white or dark grey-green with inner shades masking the bulbs. **1734** *Col. Rec. Georgia* III. 130 Part of the Twenty one Pieces of Mahogany, Ash, Sycamore, Ilex and Red Bay Timber the Growth of Georgia used in the Experiments for making Hand Boards &c. **1845** A. Wiley in *Indiana Mag. Hist.* (1927) XXIII. 165 Behold the..awkward man arise and place his chair before him for his pulpit and hand-board. **1857** P. Cartwright *Autobiogr.* xx. 203 They drove a stake down, and nailed a board to it,..and this was my hand board. **1905** *Westm. Gaz.* 17 Mar. 4/1 Our game of hand-fives is perhaps the closest approach we have to the central type of the games. *a* **1877** Knight *Dict. Mech.* II. 1055/1 *Hand-hole*, a small hole at or near the bottom of a boiler, for the insertion of the hand in cleaning, etc. It is closed by a hand-hole plate. **1967** *Gloss. Sanitation Terms* (*B.S.I.*) 24 *Handhole*, a small opening with an access cover to provide for the inspection, repair or cleaning of the inside of a vessel or pipe. **1948** H. C. Parker *Climbs on Gritstone* I. 37 The crack facing the Pinnacle is climbed, with the use of an awkward hand jam, to the second chockstone. **1937** *Mountaineering Jrnl.* V. 138/1 A crack in the right-hand corner of the rectangular grass platform was climbed with the aid of small holds and hand jamming. **1957** Clark & Pyatt *Mountaineering in Brit.* xvi. 237 A new method of hand-jamming, enabling a hold to be obtained in a cleft with far more flexibility and far less pain than the previously accepted method. **1958** M. Pugh *Wilderness of Monkeys* 33 There was no room for dancing but here and there couples were hand-jiving and one girl expressed ecstasy by pulling her hair down into her eyes. **1958** *Radio Times* 14 Feb. 5/1 The world of skiffle, rock 'n' roll, jazz, and the hand-jive. **1961** Partridge *Dict. Slang* Suppl. 1125 *Hand-jive*, system of rhythmic hand-movements in time to music where floor is too crowded to allow people present to jive (dance), esp. as in coffee bars. **1958** *Punch* 12 Feb. 288/3 Hand-jiving only employs the dancer from the waist up and affords a great saving on Espresso space, the calf muscles and shoe-leather. **1899** Hand-laid [see *deckle 2]. **1889** *Century Dict.*, *Hand-letter*, an impress on a book-cover by movable types from a hand-stamp. **1960** G. A. Glaister *Gloss. Bk.* 171/1 *Hand-letters*, brass letters, mounted in wooden handles, which are used by the finisher for lettering the title, etc., on the cover of a hand-bound book. **1907** N. Munro *Daft Days* xvii. 150 A large hand-lettered bill was in each window. **1969** 'I. Drummond' *Man with Tiny Head* xv. 171 A hand-lettered sign. **1967** Karch & Buber *Offset Processes* iv. 117 A few Filmotype lettering styles unlike the usual type faces appear to be special handlettering. **1824** M. R. Mitford *Our Village* 13 A melon bed!—fie! What a grand pompous name was that for three melon plants under a hand-light! **1831** *Sutherland Farm Rep.* 72 in *Brit. Husb.* (1840) III, A few boys and girls hand-pick the whole. **1881** *Chicago Times* 4 June, Good to choice mediums [*sc.* beans] were quotable at $2.25 & 2.40 per bu. for hand-picked. **1898** *Advance* (Chicago) 3 Mar. 282/1 [Loyola] face to face with individuals, hand-picking souls from the fire. **1907** *Westm. Gaz.* 12 Sept. 5/1 The most expensive of the ordinary coals—'large hand-picked coal'. **1918** *Times* 23 Jan. 6/3 True, there has been a widespread feeling that the Irish Convention was handpicked. **1925** J. Gregory *Bab of Backwoods* xii. 157 An able-bodied trio, hand-picked by William Badger. **1928** *Daily Express* 11 July 1/2 A handpicked 'National' Assembly for Parliament. **1949** I. Deutscher *Stalin* 353 Its members, though they had all been hand-picked by Stalin..differed on means and methods. **1959** *Manch. Guardian* 1 July 4/5 The director ..hand-picks his best students for design jobs in the potteries. **1889** C. A. Harris *Princ. & Pract. Dentistry* (ed. 12) 526 Its use will also keep the hand-piece in good condition. **1914** J. B. Parfitt in N. G. Bennett *Sci. & Pract. Dental Surg.* 326 Hand-pieces..are often liable to get very much soiled. **1949** F. Sargeson *I saw in my Dream* 115 The whirring of the hand-pieces. **1950** *N.Z. Jrnl. Agric.* May 463/3 Thorough scrubbing of the shearing board with disinfectant and the cleansing of handpieces are also worthwhile practices. **1960** *Farmer & Stockbreeder* 23 Feb. 67/2 Entrants for the final at the Royal Show..will be judged using their own handpieces. **1963** J. Osborne *Dental Mechanics* (ed. 5) x. 212 For these purposes..small rotary stones in a handpiece are advisable. **1900** H. Lawson *Over Sliprails* 49 Scraping old splashes of paint off the brass and hand-plate. **1918** *Nature* 15 Aug. 470/2 To hand-pollinate the flowers of a soft-shelled tree with pollen from a tree of similar character. **1954** A. G. L. Hellyer *Encycl. Garden Work* 130/2 As a result of the hand-pollination, seed may be formed. **1886** R. Tuck *Handbk. Biblical Difficulties* II. 324 This hand-print is made in order to avert the 'evil eye'. **1894** 'Mark Twain' in *Cent. Mag.* Feb. 553 The hand-print of one twin is the same as the hand-print of the fellow-twin. **1966** *New Scientist* 3 Mar. 539/3 The use of handprints has spread from the charge room to the clinic. Three paediatricians..have reported..that newborn babies with congenital defects frequently have abnormal palm and fingerprints. **1902** *Daily Chron.* 28 Nov. 6/3 S.S.,..'hand reader',..appeared..to answer a charge of pretending to tell fortunes by palmistry. **1867** A. R. Craig *Bk. of Hand* 31 In obedience to the stern dictates of the hand-reading art. **1960** C. Storr *Marianne & Mark* ii. 29 She asked if I wanted the cards or the crystal or a hand reading. **1922** *Collier's* 7 Jan. 8/1, I am asked by Collier's to suggest a simple, universal, and almost automatic system of hand signals for the automobile driver. **1934** *Amer. Speech* IX. 114/2 On the way to business [by car],..those who drive have to make allowances for..hand signals. **1960** E. H. Clements *Honey for Marshal* v. 94 He made the hand-signal impatiently..as the lorry still seemed shackled by its own indecision. **1965** Priestley & Wisdom *Good Driving* vi. 47 The Highway Code..lists the three hand signals. **1899** H. Butterworth *How To* iv. 15 (*heading*) Hand stand. **1909** W. Skarstrom *Gymnastic Kinesiology* 74 But in raising the legs and inverting the body to the 'Hand stand' there is at first a considerable bend at the hips and more flexion in the elbows than occurs in 'Free Front Rest'. **1946** 'J. Tey' *Miss Pym Disposes* vi. 111 When she goes out from thees [*sic*] plaace [*sic*] it will not matter any longer that she can do a handstand better than anyone else. **1951** *Swimming* (E.S.S.A.) v. 87 Before attempting handstand dives, you must be able to maintain a steady hand balance..grip the end of the board with the hands,..you can then throw up or press up into the armstand position. **1959** *Times* 17 Feb. 3/2 The little girl with upraised arms about to perform a handstand. **1598** Florio *Worlde of Wordes* 355/1 *Sciugatóio*, a hand-towell, a wiper, a rubbing cloth. **1778** in P. Ziegler *King William IV* (1971) ii. 26, 2 Dozen of Hand Towels. **1972** M. Kenyon *Shooting of Dan McGrew* iv. 32 In the..men's room he was drying his face on a hand-towel. **1897** O. G. Jones *Rock-Climbing* xvii. 268 We each in turn ventured on the *hand-traverse* from above... It is so named because the climber hangs by his hands,..and traverses across the face by sheer strength of his arms. **1935** D. Pilley *Climbing Days* i. 17 Above this the 'hand traverse' faced us. A crack..offers sloping and not very good holds to the hands..the slab gives a little friction to the knees, but not very much. **1957** Collomb *Dict. Mountaineering* 84 *Hand traverse*, a horizontal movement across a broad flake of rock, the body being supported entirely on the hands which grip the edge of the flake.

hand, *v.* Add: **4. a.** *spec.* To deliver or serve (food) at a meal. Also with passive force: to be served, to be delivered. Also with *round.*

1802 C. Wilmot *Let.* 3 Jan. (1920) 27 Cakes,..Lemonade, &c., continually handing about the Room. **1803** M. Wilmot *Let.* 13 May in *Russ. Jrnls.* (1934) I. 13 Soup was handed round. **1844** 'J. Slick' *High Life N.Y.* II. 250 The niggers..dodged about, fillin plates and a handin em round. **1851** *London at Table* II. 44 Don't omit to hand the vegetables and sauces. **1901** F. H. Burnett *Making of Marchioness* I. iv. 134 'I ought to go and help hand cake,' she said. **1945** M. Allingham *Coroner's Pidgin* i. 11 I'm going to 'and round at the reception. **1964** S. H. Nowell-Smith *Edwardian England* iv. 183 At smart tables, dishes were now handed by the servants —service *à la russe*, as it was called.

c. To give, convey: often with implication of palming-off or imposing. *U.S.*

1901 Merwin & Webster *Calumet 'K'* ii. 21, I told him he ought to give it to somebody else, and he handed me a lot of stuff about my experience. **1908** 'O. Henry' *Options* (1916) 30 I've had it handed to me in the neck, too. **1925** F. Lonsdale *Spring Cleaning* II You ought to have heard the stuff they have handed over to her about you! **1926** J. Black *You can't Win* vi. 75 You'll.. maybe get grabbed off a train and handed thirty days at Colorado Springs. **1970** *Morning Star* 17 Feb., The American Civil Liberties Union has condemned sentences for contempt handed down by Judge Julius Hoffman.

d. *to hand it to*: to acknowledge the superiority of; to congratulate; freq. in phr. *you have (got) to hand it to* (someone). orig. *U.S.*

c **1906** J. F. Kelly *Man with Grip* 14 You must hand it to the Jap. **1923** H. L. Foster *Beachcomber in Orient* xiv. 377, I do not like John [Chinaman]... But, to use the vernacular, you have to hand it to him. **1923** *Harper's Mag.* Apr. 558 You've got to hand it to that kid... He's stood everything and never squealed a yelp. Some young tough, believe me! **1926** G. D. H. & M. Cole *Blatchington Tangle* xli. 279 'I must hand it to you, sir,' the pseudo-American acknowledged. **1926** E. Wallace *Ringer* I. 11 The Ringer's clever. I hand it to him. **1927** 'A. Berkeley' *Mr. Priestley's Problem* ii. 30 'Guy, I hand it to you,' Laura was shrieking. **1965** *Listener* 30 Sept. 498/1 You've got to hand it to the Jerries, they know how to make cars. **1973** D. Jordan *Nile Green* xxi. 85, I had to hand it to him: he hadn't missed a trick.

hand-ax, -axe. Add: **b.** A prehistoric stone implement, *esp.* a bifacially worked cutting tool typical of certain Lower and Middle Palæolithic industries. (Cf. *coup de poing.)

1878 *Brooklyn Monthly* May 143/2 Another pattern have the groove extending partly round; others are wholly without a groove, and are of a pattern sometimes called hand axes. **1914** J. Geikie *Antiquity of Man in Europe* ii. 44 The *coup de poing* or hand-axe still occurs, but is rare, and would seem to have gone out of use in early Mousterian times. **1955** J. S. Weiner *Piltdown Forgery* xiii. 185 At Olorgesailie in Kenya Leakey found some good examples of these spherical stone balls with the Hand-axe culture. **1959** J. D. Clark *Prehist. S. Afr.* ii. 41 *Handaxe*, a heavy all-purpose tool, often pear-shaped and some 8–9 inches in length. Believed to have been used in the hand without hafting. **1972** *Times Lit. Suppl.* 31 Mar. 371/2 The chopper and biface core tools generally known as hand-axes.

hand-ball. Add: **4.** A game resembling fives.

1886 G. H. Benedict *Spaldings Hand Bk. Sporting Rules* 45 A game of hand ball shall consist of twenty-one aces, to be played with a ball about two inches in diameter. **1910** *Encycl. Brit.* X. 450/2 Handball, of ancient popularity in Ireland and much played in the United States, is practically identical with fives. **1957** *Ibid.* IX. 339/1 Certain forms of the games [of fives] in the United States, in Ireland and in some parts of the north of England are known as handball. *Ibid.* XI. 141/1 There is evidence that handball originated in Ireland about a thousand years ago. *Ibid.* 142/1 The game of Irish handball..was played on a hard clay floor, with one wall of stone,..against which the ball was struck. **1961** J. S. Salak *Dict. Amer. Sports* 215 *Handball*, a wall game in which a black rubber ball is struck with the hand against a wall, or walls, the ball being struck alternately by opposing players. **1968** *Globe & Mail* (Toronto) 17 Feb. 48/1 (Advt.), Regulation-size squash and handball courts.

handbook. Add: **c.** A betting-book; *handbook man*, a bookmaker. Also *hand-booking*, bookmaking. *U.S.*

1894 *Voice* 20 Sept., In every saloon which boasts a ticker are to be found men who will register a bet to any amount. These 'handbook' men are all [etc.]. **1903** *N.Y. Evening Post* 14 Sept., A case where an officer arrested a handbook man. **1904** *N.Y. Times* 13 June 1 The handbooking possibilities on the Derby. **1946** *Chicago Daily News* 26 June 14/2 The mob was..operating handbooks with full knowledge of your police department.

handclap. Add: (Later example.)

1962 A. Nisbett *Technique Sound Studio* ii. 33 You can use a hand-clap to give a rough guide to the reverberation.

b. Applause.

1907 *Daily Chron.* 1 Oct. 4/5 The curtain goes up..to some feeble handclaps from invited guests. **1908** *Westm. Gaz.* 13 Mar. 7/3 The chairman of the company.., with the directors, entered the room, a hand-clap greeting them.

c. *slow handclap*: slow applause expressing disapproval. Also (with hyphen) as vb. Hence *slow hand-clapping* vbl. sb.

1953 *Britannica Bk. of Year* 638/2 The freedom of usage seen in the formation of compounds also allows frequent changes in function, so that verbs, for example, may be formed from nouns. Examples are: to slow-handclap. **1955** Miller & Whitington *Cricket Typhoon* 215 Indeed, they jeered and slow-handclapped Cowdrey. **1958** F. C. Avis *Boxing Ref. Dict.* 103 *Slow handclap*, ironic applause expressed by means of a very slow clapping of the hands. **1959** *News Chron.* 13 July 4/6 Some cynical dons..were giving the slow hand-clap to the end of the procession. **1961** *Guardian* 3 Apr. 1/7 All the slow hand-clapping and hysterical cheering. **1966** *Listener* 20 Jan. 88/2 This destructive criticism was nothing compared with the..slow-handclaps and final booing.

Hence **hand-clapping** *vbl. sb.*

1838, 1888 [see Hand *sb.* 62 a]. **1948** B. G. M. Sundkler *Bantu Prophets S. Afr.* vi. 189 After the hymn..sung with gusto and more handclapping, the testimonies begin. **1959** I. & P. Opie *Lore & Lang. Schoolch.* vi. 94 Bangor children find that an innocent hand-clapping game goes neatly to the words.

handcraft. Delete †*Obs.* and add later examples. Hence **handcraft** *v. trans.*; **hand-crafted** *ppl. a.*

1933 *Catholic News* (Johannesburg) Mar. 15 (Advt.), Hand-craft tailored suits. **1965** *Punch* 3 Nov. 645/3 The stone, being hand-crafted and set, remained in place for more than a month before it was washed away in a rainstorm. **1967** *Listener* 5 Jan. 21/2 The bowls appear hand-crafted instead of factory-manufactured. **1968** C. Leader *Angry Darkness* iv. 39 The tiny shops were filled..with old silver and hand-crafted copper and brass. **1968** J. Arnold *Shell Book of Country Crafts* 59 Machine- and mass-production, together, have had a tremendous impact..on hand-crafts. **1971** *Daily Colonist* (Victoria, B.C.) 24 Oct. 4/2 All the fuel injection parts had to be handcrafted.

handcraftsman. Delete †*Obs.* and add later examples. Hence **handcraftsmanship.**

1923 *Daily Mail* 22 Jan. 6 (Advt.), A specialised handcraftsmanship places these brushes as a blessing [etc.]. **1933** *Archit. Rev.* LXXIII. 253/1 The subject of handcraftsmanship and mass production has been touched upon. **1954** B. Griffiths *Golden String* ix. 137

If a society is to be really human, there will always be a need for the hand-craftsman. **1961** *John o' London's* 7 Sept. 275/2, I think no one today, save a few incurable rustic-romantics and starry-eyed handcraftsmen, will imagine that a solution is to be found.

handedness (hæ·ndĕdnes). [f. Handed *a.* + -ness.] The tendency to, or the preference for, the use of either the right or the left hand. Also *transf.*

1921 *Trans. Utah Acad. Sci.* II. 59 (*title*) The problem of handedness. *Ibid.* 62 It is argued by some that a child's handedness is a result of imitating its parents. **1936** F. A. E. Crew in *Jrnl. Genetics* XXXIII. 67 It seemed to me necessary to examine each rat..for evidence of this right-hand and left-hand turning habit,... [68] In any examination of handedness the experimentation must not be complicated by the presence of an alternating light. **1937** *N. & Q.* 10 July 32/1 The First Series of an exhaustive scientific work on the subject of 'Handedness'. **1961** *Lancet* 12 Aug. 363/1 The left hemisphere is usually dominant for speech regardless of the handedness of the individual. **1962** *Listener* 10 May 814/3 Proteins..contain amino acids (all of one particular handedness) joined together. **1973** *Sci. Amer.* May 27/3 Background data, including..length of postpartum separation from siblings and handedness of the mother.

Handelian (hændī·liăn), *a.* and *sb.* [f. the name of Georg Friedrich *Handel*, originally *Händel* (1685–1759), German musician + -ian.] **A.** *adj.* Of, pertaining to, or characteristic of Handel, or his style of composition. **B.** *sb.* One who favours or imitates the style of Handel.

1770 *Priv. Lett. 1st Ld. Malmesbury* (1870) I. 205 Tenducci is amazingly improved; in his part the old Handelian songs were left out. **1788** F. Burney *Diary* (1842) IV. iv. 231 The concert was very Handelian. **1808** S. Wesley *Lett.* (1875) 9 This would nettle the Handelians devilishly. **1825** Lamb *Lett.* (1888) II. 132 My sister's cold is as obstinate as an old Handelian, whom a modern amateur is trying to convert to Mozart-ism. **1865** J. Hullah *Transition Per. Music* 244 It 'bears so genuine a Handelian impress'. **1885** *Athenæum* 28 Feb. 288/1 Revivals of Handelian oratorio. **1911** Lady Guthrie in R. L. Orr *Ld. Guthrie* (1923) 258 Those dusky simple souls interpreting the glorious Handelian strains. **1966** K. Amis *Anti-Death League* 349 The series of florid, rather Handelian amens swung to its close.

handful. Add: **5.** *slang.* A five years' prison sentence.

1930 J. Lait *Big House* i. 6 A five-year sentence is a 'handful'. **1953** M. Gilbert *Fear to Tread* ix. 118 He's had a two-stretch... He'll collect a handful next time. **1966** *New Society* 31 Mar. 22/2 Going up for a handful (receiving a sentence of five years' imprisonment).

hand-glass. 2. (Earlier examples.)

1788 G. White *Selborne* lxv. 304 The hail broke..all my garden-lights and hand-glasses. **1824** Loudon *Encycl. Gardening* (ed. 2) 287 The wrought-iron hand-glass is composed of solid iron sash-bars, and may therefore be formed of any shape or height.

handgun. Delete †*Obs.* exc. *Hist.* and add later *U.S.* examples. Hence **handgunner** (later example). Also **handgunning** *vbl. sb.*

1957 *Amer. Speech* XXXII. 190 It is inevitable..that colloquialisms of the handgunner overlap with those of the rifleman. *Ibid.* 191 The word handgun is very old... Since the 1930s its use as a generic term in place of pistol [etc.]... Recent use has made familiar, at least in writing, such derivatives as..handgun work,..applied to shooting the handgun at paper targets. *Ibid.*, Handgunning offers a great diversity of matches (slow, timed, and rapid fire). **1961** Webster, *Handgun*, a firearm held and fired with one hand. **1968** *Economist* 22 June 43/3 California's Department of Justice estimates that there are 3·5 million concealable hand-guns (pistols and revolvers) within the state.

handicap, *v.* Add: Hence **ha·ndicapped** *ppl. a.*, of persons, esp. children, physically or mentally defective. Also *absol.* as *sb.*

1915 L. D. Wald *House on Henry St.* 117 (*caption*) The Handicapped Child. **1919** *School & Society* 29 Aug. 256/2 There are, of course, other types of mentally handicapped children who should be sharply differentiated. **1942** *Q. Jrnl. Speech* Feb. 81/1 The child who is still babbling, lisping, stuttering..at thirty-six months of age is just as handicapped..as the child with a misshapen back. **1958** *Times Lit. Suppl.* 21 Nov. p. xiii/2 The approach is strictly a practical one and even extends to the needs of handicapped children. **1958** P. Townsend in N. Mackenzie et al. *Conviction* 118 The handicapped..still are treated too often as second-class citizens. **1959** *Housewife* June 33 Chronically sick or handicapped people.

handicapper. Add: **b.** A horse running in a handicap race.

1895 *Starting Price* 23 Mar. 1/3 *Barbary.*—A second-rate handicapper, but bad-tempered and disappointing. **1955** *Times* 25 Aug. 2/6 That fine handicapper Durante put up a splendid performance under 9 st. 7 lb. in coming away in the last two furlongs to beat Naval Patrol by five lengths.

handie-talkie (hæ·ndi,tǭ:ki). Also **Handie-Talkie, handy-talky.** [After *walkie-talkie.] Name of a light form of walkie-talkie two-way radio set, easily carried in the hand.

1942 *Nat. Geogr. Mag.* Nov. 680 Churchill..is holding a 'handie-talkie' radio used for conversation between ground points and planes in the air. **1943** *Time* 27 Sept. 83/2 (Advt.), He'll talk his way out..with his two-way 'handy-talkie'... The 'handy-talkie' is only one of many radio communication devices of our armed forces. **1960** *Times* 8 Mar. 15/3 The equipment is known as a Handie-Talkie and has been produced to replace the heavier walkie-talkie sets used during the last war. **1969** S. Greenlee *Spook who sat by Door* xiii. 113 Cops spoke busily into their car radios and handy-talkies.

handkerchief. Add: (Additional examples of sense 'kerchief worn about the neck'.) Phr. *to drop the handkerchief* (example); *to throw* (or *fling*) *the handkerchief* (earlier and later examples).

1749 H. Walpole *Let.* 20 July (1903) II. 396 Till all the juries of matrons have finished their inquest, one shall not care to make one's choice—I was going to say—*throw one's handkerchief*, but at present that term would be a little equivocal. **1764** —— *Let.* 5 June (1904) VI. 78 Lord Tavistock has flung his handkerchief to Lady Elizabeth Keppel. **1786** E. Sheridan *Jrnl.* 21–23 Jan. (1960) 79 The hankerchiefs are not so much puff'd out and there is now a very pretty sort of hankerchief much worn open at the neck and exactly made and trim'd like a Boy's shirt. **1825** H. Wilson *Mem.* II. 11 The system at White's Club..is..never to black ball any man, who ties a good knot in his handkerchief. **1870** *Brewer's Dict. Phr. & Fable* 384/1 '*The committee was at a loss to know whom to throw the handkerchief to*' (The *Times*)..the allusion is to the game called in Norfolk 'Stir up the dumplings', and by girls 'Kiss in the ring'. **1897** *Outing* (U.S.) Apr. 71/2, I was hoping that they would have an English May-pole dance, but instead they played 'drop the handkerchief', which Philip said, though not so ancient, was more fun. **1932** *Times Lit. Suppl.* 16 July 506/3 If he hesitate today whether he shall throw the handkerchief to Germany or Russia, does not such an embarrassment prove his power?

b. *handkerchief blouse, -case, -cloth, dress, -hat, -pin, pocket, sachet, table, -turban*, etc.; **handkerchief-head** (see quot. 1942).

In several of the collocations applied to parts of costume made up of squares resembling or suggesting handkerchiefs.

1790 E. Wynne *Diary* 17 Jan. (1935) I. ii. 30 Betzy bought..a handkerchief pin. **1880** L. Higgin *Handbk. Embroidery* 106 (Advt.), Handkerchief sachets, from £3. 3s. **1890** Ld. Lugard *Diary* 27 Mar. (1959) I. iv. 161, I..presented the women with a return present viz. a dhoti of 'handkerchief cloth'. **1893** 'Mark Twain' in *Cent. Mag.* Dec. 238/2 She took off her handkerchief-turban. **1895** *Montgomery Ward Catal.* Index, Handkerchief Boxes and Cases. **1896** E. Turner *Little Larrikin* x. 106 She had conceived the idea of making Ruffy a present of a handkerchief-sachet. **1899** *Daily News* 1 July 4/3 The revival of the handkerchief dress. **1900** *Westm. Gaz.* 9 Aug. 3/2 The handkerchief blouse. **1903** *Daily Chron.* 18 July 8/4 The sleeves are handkerchief ones. **1922** Joyce *Ulysses* 99 His inner handkerchief pocket. **1932** D. C. Minter *Mod. Needlecraft* 246/1 *Handkerchief case*..oblong folded to form pocket and flap. **1942** *Amer. Mercury* LV. 95 *Handkerchief-head*, sycophant type of Negro; also an Uncle Tom. **1950** A. Lomax *Mr. Jelly Roll* (1952) v. 231 This corny old handkerchief-head would assert that Count Basie did not know piano. **1956** S. Longstreet *Real Jazz* 147 A 'handkerchief-head' is an old-fashioned Negro who doesn't know his rights. **1960** *Encounter* XIV. II. 39 The Negro officer isn't a 'handkerchief head', an Uncle Tom. **1960** H. Hayward *Antique Coll.* 138/2 *Handkerchief table*, an American term for a single-leaf table with leaf and top triangular in shape. Closed, the table fits in a corner, opened it is a small square. **1963** *Sunday Express* 3 Nov. 19/1 A group of handkerchief hats in pastel glove leather. **1971** P. D. James *Shroud for Nightingale* v. 170 An embroidered handkerchief sachet with a dozen handkerchiefs carefully folded.

handky, var. *hanky[1].

handle, *sb.*[1] Add: **1. b.** *to fly off the handle* (later examples; no longer restricted to the U.S.); now usually = 'to lose one's temper'. Also, in same sense, *to go* (or *be*) *off the handle.*

1888 Kipling *Phantom Rickshaw* 2 Pansay went off the handle,..all that nonsense about ghosts developed. **1898** —— *Day's Work* 78 How are we to do our work if you fly off the handle that way? **1908** C. E. Mulford *Orphan* xxii. 271 He reckoned you would..get good and mad, fly off the handle, and raise h—l generally. **1915** A. Conan Doyle *Valley of Fear* I. v. 82 A kind of wave of jealousy would pass over him and he would be off the handle and saying the wildest things in a moment. **1932** Kipling *Limits & Renewals* 157 Jimmy went off the handle at once; and Nicol kept patting him on the back. **1958** *Times* 3 Nov. 11/7 Montgomery flew off the handle and told the Minister of Defence..that he must find out whether Bevin still stood by what he said. **1964** L. Nkosi *Rhythm of Violence* 64 Calm down, for God's sake! Everybody's flying off the handle. What's the matter with everybody?

c. *to the handle, up to the handle*: thoroughly, completely, up to the hilt. *U.S. colloq.*

1833 *Louisville Publ. Adv.* 9 May, He is determined to carry the contest 'to the handle'. **1835** A. B. Longstreet *Georgia Scenes* 234 We'll all go in for you here up to the handle. **1843** T. C. Haliburton *Attaché* 1st Ser. I. viii. 119 Give me your figgery-four, Squire, I'll go in up to the handle for you. **1855** *Knickerbocker* XLV. 435 (Th., s.v. *Up*), He was enjoying his trip 'up to the handle'. **1860** *Ibid.* LV. 415 He had for the last few years used a boy and dog as fencing material; he found it 'a good institution'; they did the thing up to the handle. **1877** J. Habberton *Jericho Road* xi. 101 If he isn't playin' possum right up to the handle, then he is a fool.

d. *to give, use, the long handle* (Cricket): to hit freely and continuously.

1888 Steel & Lyttleton *Cricket* ii. 77 Hold the bat nearer the top and give her the long handle. **1903** Warner in H. G. Hutchinson *Cricket* 71 As a rule the hitting or 'long-handle game', as it has been called, pays best under these circumstances [*sc.* on a sticky wicket]. **1928** *Daily Express* 20 Dec. 3/2 Ryder set about the bowlers unmercifully, using the long handle.

2. b. A small basket with a handle, in which soft fruit is packed for the market.

1900 *Daily Express* 30 June 5/5 French red currants reached 1s. 3d. a basket or 'handle'.

c. A measure of beer, approx. 1 pint. *N.Z.*

1938 R. D. Finlayson *Brown Man's Burden* 40 'A handle of beer,' Mr Puttle was saying easily to the barman. **1943** J. A. W. Bennett in *Amer. Speech* XVIII. 89 [In New Zealand] beer is dispensed in *handles* (in Australia, *pots*) or *half-handles*. **1947** 'A. P. Gaskell' *Big Game* 57 'He gets one handle every day,' said George [barman]. **1956** *N.Z. Listener* 8 June in J. Reid *Kiwi Laughs* (1961) 204 They still drink beer out of handles, sixpence a pop.

4. b. A person's name; a nickname. *slang* (orig. *U.S.*).

1870 J. C. Duval *Adv. Big-Foot Wallace* xxxviii. 236, I would rather be called 'Big-Foot Wallace' than 'Lying Wallace'... Such handles to my name would not be agreeable. **1927** *Dialect Notes* V. 449 Whut's yer handle? **1935** *Amer. Speech* X. 18/2 *Monicker*, a genuine name, as distinguished from an alias. Modern *handle*. **1964** D. Varaday *Gara-Yaka* xx. 180 One was Toothless Annie... She had come by her 'handle' when a hysterical grass-eater had kicked her teeth in. **1969** C. F. Burke *God is Beautiful, Man* (1970) 82 One night Jesus met a guy named Nicodemus. How's that for a handle?

5. handle-bar, a transverse bar, usually curved, with a handle at each end, connected with the driving- or steering-wheel of a cycle, by which the vehicle is guided by hand; *pl.* the right- and left-hand parts of which this is composed. Also *attrib.*, *spec.* of a (usually large) moustache of handle-bar shape.

1887 *Graphic* 3 Dec. 619 These machines..are fitted with adjustable handles and seat rod. Well finished in black enamel, with plated hubs, handle bars, &c. **1894** *Million* V. 377/1 Pick up your dress with your right hand, take hold of the handle bar with your left. **1898** *Science Siftings* XV. 170/1 Handle-bars in which the drop is greater than four inches below the seat are..dangerous. **1908** *Westm. Gaz.* 15 Apr. 10/2 A contributor to the *Motor Cycle* advises motor-cyclists to carry handle-bar mirrors on their machines. **1908** *Daily Chron.* 21 Nov. 9/4 One cannot effect this unobserved change when other than handle-bar control is fitted. **1909** *Captain* Aug. 448/2 Home-made Handlebar-grips. **1923** H. L. Wilson *Oh, Doctor!* xxiv. 333 [He] threw a confident leg across the saddle and worshipfully grasped the spreading handle bars. **1933** G. P. Jackson *White Spirituals* 65 An elderly man with handle-bar moustache. **1941** *Penguin New Writing* VIII. 17 One man was huge and swarthy, with a handlebar moustache. **1953** M. Dickens *No More Meadows* vii. 295 Chap with the handlebar moustache. Real Pilot-Officer Prune type. **1968** *Which?* Aug. 231/2 Turn the handlebars—the steering should not be too tight or too loose. *Ibid.* 232/1 If the brake levers are too far from the handlebar grips, a child with small hands might find it difficult to work them. **1972** J. Rossiter *Rope for General Dietz* ii. 23 We both had the enormous handlebar moustaches *de rigueur* in the RAF at that time.

handle (hæ·nd'l), *sb.*[2] [f. Handle *v.*[1] 1.] The feel of or sensation produced by goods, especially textiles, when handled.

1884 W. S. B. McLaren *Spinning* 19 It is not merely the coarseness or fineness of the fibre which guides him, but also the softness and kind 'handle', as it is called. **1898** *Daily News* 7 Mar. 2/1 For softness and beautiful handle they have no equal. **1927** T. Woodhouse *Artificial Silk* 2 The handle of many artificial silk articles of commerce compares favourably with that of most of the corresponding textures in the other branches of the textile industry. **1961** *Times* 26 Oct. p. iv, A new type of paper.. which had all the qualities—good surface and opacity, good colour and crispness of 'handle'. **1962** *Which?* Aug. 240/2 The blanket..felted considerably and had a harsh 'handle'. **1968** J. Ironside *Fashion Alphabet* 210 The filaments are 'crimped' to resemble wool, giving a warm, soft 'handle'. **1970** *Nature* 17 Oct. 212/1 The 'handle' and 'feel' of fabrics.

handle, *v.*[1] Add: **1. b.** (Later examples.)

1946 *Mod. Lang. Notes* LXI. 443 The use, in advertising, of the 'potential intransitive', in such examples as 'this car *operates, handles* smoothly'. **1958** *Times* 23 Sept. 14/2 Sceptre, handling excellently, rounded the mark to the sound of loud hootings. **1962** *Which? Car Suppl.* Oct. 142/1 It [*sc.* the car] handled very securely. **1972** *Country Life* 23 Mar. 703/3 The machine handles well, and I discovered no snags.

2. c. (Earlier example.)

1839 Ure *Dict. Arts* 764 They [*sc.* the hides]..are successively transferred into other pits with stronger ooze; all the while being daily *handled*, that is, moved up and down in the infusion.

d. *Cricket.* (See quots.)

1788 in H. T. Waghorn *Dawn of Cricket* (1906) 98 Their opponents were superior to them in handling [*sc.* fielding] the ball. **1797** in G. B. Buckley *Fresh Light on 18th Cent. Cricket* (1935) 184 Handling b. in play. **1841** *Manchester Chron.* 24 July, John Ogden..handled the ball well [i.e. as bowler].

handleability (hæ:nd'lăbi·lĭti). [f. Handleable *a.*: see -ity.] Ease of handling.

1947 N. Balchin *Aircraft Builders* v. 31 The problem

was one of combining performance with 'handleability' and toughness of condition. **1949** *Jrnl. R. Aeronaut. Soc.* LIII. 962/1 The general 'handleability' of an aeroplane must not be lost sight of in a welter of requirements which might lead to excessive stability. **1970** *Amat. Photographer* 22 Apr. 55/2 An indefinable quality of a camera is its 'handleability'.

handler. Add: **2.** One who shows the points of dogs at a trial, etc.

1897 [in Dict.]. **1931** *Our Dogs* 23 Oct. 296 Handlers and Breakers. Gun dogs wanted for training or boarding. **1959** *Times* 18 Sept. 7/5 Of the nine handlers who took part in the opening event—the hired shepherds' championship—only two failed to complete the course.

b. A police officer who is in charge of a trained dog.

1959 B. J. Farmer *Murder Next Year* xxi. 134, I know Sergeant Cristobel... He's handler for a trained Alsatian. **1962** *Times* 29 Dec. 6/7 Minivans..containing two highly trained police dogs with their handlers. **1971** B. Callison *Plague of Sailors* 10 They're not dogs, they're bloody werewolves. What d'you handlers do for leave, spend it in the bloody jungle?

c. *Boxing.* (See quot. 1961.)

1950 J. Dempsey *Championship Fighting* 9 His handlers threw in the towel. **1960** *Times* 1 Sept. 4/4 It took several seconds of rough first aid by his handlers before he was able to regain his seat. **1961** J. S. Salak *Dict. Amer. Sports* 216 *Handlers*, the chief second and assistant seconds of a boxer during a contest. **1973** *Sunday Express* (Trinidad & Tobago) 8 Apr. 29/2 (*caption*) One of his handlers..is near tears after Ali lost a 12 round non-title bout.

hand-line. Add: Hence **hand-line** *v.*, to fish with a hand-line; to pull in a fishing-line by hand.

1935 A. J. Cronin *Stars look Down* II. xix. 437 Old Macer..had to make the best of it by hand-lining off shore for whiting. **1969** *Islander* (Victoria, B.C.) 7 Sept. 6/1 Peter began commercial fishing, handlining for cod. **1972** *Shooting Times & Country Mag.* 24 June 15/3, I was at one stage handlining in like fury to keep in touch and then suddenly as the fish passed us, was slipping line out to the running fish.

handling, *vbl. sb.* Add: **1. c.** In games, the illegal touching of the ball.

1882 in Charles-Edwards & Richardson *They saw it Happen* (1958) 300 For a breach of rule, which forbids handling, a free kick was awarded against the Etonians. **1897** *Encycl. Sport* I. 429/1 *Handling*, or *Hands*, touching the ball with any part of the arm when in play. Only the goal-keeper can do so without a penalty. 'Hands' is given against the offender.

d. [Handle *v.*[1] 1 b.] The way in which a motor vehicle handles.

1962 *Which? Car Suppl.* Oct. 143/1 Its handling was very secure, but the car was badly affected by side winds. **1967** *Autocar* 28 Dec. 5/3 In general the handling of the car was satisfactory.

5. *attrib.* (Additional examples.)

1927 *Observer* 27 Mar. 28 In spite of the progress Rugby has made, the Association game at the Schools..has fully held its own with the handling code. **1949** *Archit. Rev.* CV. 218 The east bay includes..the polymer handling bay. **1954** *Economist* 11 Sept. 8/2 The central terminal area, which will hold the permanent control tower and the passenger handling buildings, is now being completed. **1955** *Times* 6 July 4/4 A long handling rush by Britain looked dangerous and from a loose scrum in the opposing twenty-five their backs gained possession for O'Reilly to cut in and score. **1962** *Which?* Dec. 367/1 Labour costs were covered, but there was a 'handling charge' of 2s. 6d. **1969** *Times* 13 Jan. 11/2 The increase in the handling margins would be nearer to 100 than 10 per cent.

hand-loom. Add: Hence **ha·nd-loomed** *a.*

1928 *Daily Express* 13 July 3/6 Hand-loomed leathers.. are a vogue in exclusive handbags. **1966** A. Adburgham *View of Fashion* 135 Hand-loomed skirts.

hand-made, *a.* Add: (Later examples.)

1959 Halas & Manvell *Technique Film Animation* vi. 69 The animated film is essentially a hand-made art. **1968** *Times* 31 Aug. 19/4 Some biographical facts (including the..item of her using the blue hand-made paper fabricated for Colette).

b. *fig.*

1936 *Punch* 7 Oct. 418/2 (*heading*) The hand-made short story. **1937** B. H. L. Hart *Europe in Arms* iv. 49 The fortified region..is garrisoned by units of varying composition which are 'hand-made' to suit the sector allotted to them. **1958** *Listener* 9 Oct. 549/2 By the very terms of his [*sc.* de Gaulle's] own newly adopted hand-made constitution, the President must not be anything but an arbitrator.

ha·nd-me-down, *sb.* and *a.* *dial.* and *colloq.* [f. the verbal phr. *to hand down* (see Hand *v.* 4 b).] **A.** *sb.* That which is handed down, as an heirloom, a second-hand garment, etc.; also, a ready-made garment. **B.** *adj.* Having been handed down or passed on; = Reach-me-down *a.* So *hand-me-down shop*, etc. Also *fig.*

1874 Hotten *Dict. Slang* 187 *Hand-me-downs*, second-hand clothes. **1882** G. W. Peck *Peck's Sunshine* 213 A hand bill for a Chicago hand-me-down clothing store. **1888** *New York World* 5 Mar. (Farmer), A twelve-dollar suit of hand-me-downs. **1889** *Sporting Times* 29 June (Farmer), Trousers..which all over proclaim themselves entitled to the epithet of hand-me-down. **1896** Ade *Artie* xviii. 70 They'll be workin' for some Reub that come into town wearin' hand-me-downs. **1897** *Congress. Rec.* 25 Mar. 274/1 These cheap-johns, ready-made, 'hand-me-down' statesmen. **1904** *Boston Herald* 15 Oct. 2 He wears a cheap suit of 'hand-me-down' clothing. **1909** *Daily Chron.* 2 July 7/4 He got it from a lady admirer..and he wanted me to 'ave it as a hand-me-down. **1914** Joyce *Dubliners* 150 His little old father kept the hand-me-down shop in Mary's Lane. **1925** S. Lewis *Martin Arrowsmith* viii. § 2 A dirty old office, with hand-me-down chairs and a lot of second-hand magazines. **1935** A. J. Cronin *Stars look Down* II. xiii. 375 A little hand-me-down factory. **1954** M. Mead *Growing up in New Guinea* 188 Their myths are dull hand-me-downs. **1960** *Economist* 31 Dec. 1382/1 Many large corporations are still flying converted bombers and hand-me-down transports, but these are being supplemented by the newer, smaller light models. **1966** *New Yorker* 5 Nov. 197 To dramatize this hand-me-down truth.

hand-off (hændǫ·f), *v.* *Rugby Football.* [f. Hand *v.* + Off *adv.*] *intr.* To push off an opponent with the hand. Also *trans.* Hence **ha·nd-off** *sb.*, the action of pushing off an opponent.

1897 *Encycl. Sport* I. 429 *Handing-off*, pushing off an opponent who endeavours to impede a player running with the ball. **1920** *Times* 8 Nov. 6/2 The wings ran well and were not afraid to 'hand-off'. **1922** *Daily Mail* 8 Dec. 12 A dangerous scoring wing with a powerful hand-off and an elusive swerve. **1923** W. J. A. Davies *Rugby Football* 135 Coates..ran with his head half turned to the right.. which gave one the impression that he was waiting and was anxious to hand-off some one. **1928** *Observer* 19 Feb. 27/1 [He] has a fine kick, with a strong hand-off. **1959** *Times* 21 Sept. 3/5 Gray, who used his hand-off effectively.

hand-out. [f. Hand *v.* + Out *adv.*]

1. *Tennis.* (See Hand *sb.* 63.)

2. a. That which is handed out; *spec.* (*a*) food or alms given to a beggar at the door; (*b*) a gift of money. orig. *U.S.*

1882 Sweet & Knox *Texas Siftings* 195 If I can't get a 'hand-out' for it I can at least expatiate on its merits. **1887** M. Roberts *Western Avernus* 71 'Bummers' is American for beggars, and a 'hand out' is a portion of food handed out to a bummer or a tramp at the door when he is not asked inside. **1896** *Dialect Notes* I. 418 *Hand-out*, clothes such as a tramp asks for. **1896** Ade *Artie* vi. 50, I see barrel-house boys goin' around for hand outs that was more on the level than you was. **1903** *Daily Chron.* 4 Apr. 5/2 The weekly hand-out for the butcher. **1904** 'O. Henry' *Trimmed Lamp* (1916) 32 Pretty soon I was in the free-bed line and doing oral fiction for hand-outs among the food bazaars. **1925** W. Cather *Professor's House* 195 He soon drank up all his wages. When Rapp picked him up there he was living on hand-outs. **1931** C. Massie *Confessions of Vagabond* vii. 74 Tramps will often travel a hundred miles to one particular spot where they are sure to get a 'hand-out'. **1946** Wodehouse *Joy in Morning* vi. 45, I can well imagine a man of conservative views recoiling from one which might come asking for handouts for the rest of its life. **1959** *Daily Tel.* 27 July 12/6 The report in yesterday's newspapers that Mr Nixon, on an early morning visit to a Moscow market, had tried to give a 100-rouble note as a 'hand-out' to a worker. **1968** *Globe & Mail* (Toronto) 3 Feb. B4/3 Poor countries realize part of the burden lies on them... They're no longer just looking for handouts.

b. *attrib.*; *spec.* providing light refreshments in a handy form.

1910 *Salt Lake Tribune* 27 Nov. 32/7 On the first floor 'hand-out' luncheons were being served. **1928** F. N. Hart *Bellamy Trial* viii. 277, I would take a good walk, get a bite to eat at one of the hand-out places in the vicinity of the station. **1960** *Farmer & Stockbreeder* 22 Mar. 79/3 The future of our industry is going to rely much more on capital grants than on hand-out grants.

3. Matter handed out to or by the newspaper press; more generally, matter handed out from any source to convey information, guidance, etc.

1927 *Amer. Speech* II. 242/1 To get pictures and 'hand-outs', that is, prepared statements given to the press by officials or other prominent persons. **1929** *Literary Digest* 12 Oct. 7/1 Mr. Shearer told..how he gave the newspaper men at Geneva 'hand-outs' to help them in preparing their despatches. **1929** *Sat. Even. Post* 7 Dec. 213/2 We have public-relations experts who do their stuff by means of propaganda in the press and hand-outs to the newspaper boys and girls. **1942** *Punch* 8 July 8/2 An N.C.O. distributes hand-outs in which we are warned that the information given on this course is going to be the most secret. **1942** *Gen* 15 Sept. 24/2 White feather hand-outs..to men and women not in uniform have now reached epidemic proportions. **1945** *Ann. Reg. 1944* 251 The Spanish official 'hand-out'..was a masterpiece of nebulous verbiage. **1951** *Manch. Guardian Weekly* 23 Nov. 2 Fakes a handout to several of his team. **1958** *Punch* 8 Jan. 84/2 B.O.A.C. hadn't given me a free briefcase full of handouts for nothing. **1959** 'H. Howard' *Deadline* viii. 66 A little loose-leaf notebook issued as a free handout by Hopalong Cassidy Enterprises. **1965** *Spectator* 22 Jan. 92/2 The handout is a (necessary) curse of modern political life. It shackles the speaker and bores his audience, but it delights the reporter. **1972** 'G. Black' *Bitter Tea* (1973) iv. 63 Not a handout likely to satisfy a newspaper reporter, Inspector.

hand over hand. Add: **c. hand-over-hand stroke,** a style of swimming in which each arm is alternately brought out of the water from behind and with a circular sweep returned to the water in front. Also as *adv. phr.*

1856 'Stonehenge' *Brit. Sports* 516/2 The Hand-over-Hand style is a very rapid mode of swimming. **1872** H. Gurr *Art of Swimming* 25 To Swim Hand-over-hand. **1904** R. Thomas *Swimming* 139 The hand-over-hand is the most ancient stroke, at all events that is recorded.

Hence **ha:nd-over-ha·nder.**

1924 R. Clements *Gipsy of Horn* vi. 104 Sending the royal yards aloft to a rattling hand-over-hander.

hand over head. Add: **2. b.** *Cricket.* Designating a style of bowling (see Overhand *a.* 2).

1899 A. Lang in *Daily News* 22 July 4/2 The modern hand-over-head style.

hand-press. (Later example.)

1967 E. Chambers *Photolitho-Offset* i. 3 The operation of printing consists, first, in damping the stone—with a wet sponge in hand-press printing or with a wet roller in power-press work.

handraulic (hændrǭ·lik), *a.* [f. Hand *sb.* + Hyd)raulic *a.*] Of something done by hand rather than by machine. Hence **handrau·lically** *adv.*

1948 Partridge *Dict. Forces' Slang* 90 *Handraulic power*,..with a pun on hydraulic. *Ibid.* 103 *Johnny Armstrong*, the elementary motive power known in the Navy as 'handraulic', used for 'pully-haully' work. **1962** *Times* 4 Aug. 8/6 The ease with which even a battleship could be set in motion 'handraulically' was once vividly demonstrated. *Ibid.* 8/7 'Rolling ship' is more than just a spectacular display of 'handraulic' power. It is a good way of getting refloated. **1963** *Flight International* LXXXIII. 291/3 There are two general approaches to automation of the complex organisation of a traffic control centre: either a complete system is designed and tested more or less in isolation and transferred as an entity from the experimental stage into the working, or the existing 'handraulic' system is improved step-by-step until a clearer picture of an ultimate requirement emerges.

handset (hæ·nd,set). Also **hand-set.** [Hand *sb.* + Set *sb.*[2]] A telephone transmitter and receiver combined in a single instrument.

[**1914** Smith & Campbell *Automatic Teleph.* vi. 131 The telephone instrument follows the general form which is so popular on the Continent, making large use of the combined transmitter and receiver, sometimes known as the hand microphone set.] **1930** *Electr. Commun.* VIII. 265/2 The tendency towards more comfortable and convenient apparatus has been evidenced..finally by the development of handsets, or as they are sometimes called, 'micro-telephones'. **1955** 'N. Shute' *Requiem for Wren* 283 They repeated it and booked the call, and I put down the handset. **1962** A. Nisbett *Technique Sound Studio* 176 For telephones a standard hand-set can be similarly adapted, so that the bell may be worked by a press-button. **1972** C. Drummond *Death at Bar* v. 127 There was a call box... He had to wait ten minutes while a young citizen..quacked into the handset.

handshake, *sb.* Add: **b.** A gift of money.

1960, etc. [see **golden handshake*]. **1968** *Times* 3 July 26/6 (*headline*) Dockers told of handshake. *Ibid.*, An offer of bigger 'golden handshakes'. **1971** *Financial Mail* (Johannesburg) 26 Feb. 649/2 (Advt.), However you invest with the United, you get a handsome handshake.

handshake (hæ·ndʃēik), *v.* [Back-formation from Hand-shaking.] *intr.* To shake hands. So **ha·ndsha:ker.**

1898 H. James *Two Magics* 8 We handshook and 'candlestuck', as somebody said, and went to bed. **1905** *Westm. Gaz.* 2 Nov. 12/1 As the line moves forward each hand-shaker is steadily pushed along. **1928** *Daily Express* 28 Aug. 8/3 Hearty handshakers. **1940** *Amer. Speech* XV. 211/2 Those who try to get promotions by pull with officers are called handshakers or suction kids. **1964** G. B. Schaller *Year of Gorilla* x. 240 The Belgians are the most confirmed handshakers I have ever met.

hand-shaking. (Later and *attrib.* examples.)

1964 Rose & Ziman *Camford Observed* vii. 129 *Hand-shaking*: at Oxford the Master of a College interviews every undergraduate at the end of each term and hears a report on his progress by his Tutor. **1965** *Listener* 16 Sept. 425/2 A hand-shaking tour of the Soviet Union.

handsome, *a.* Add: **6. b.** Used, sometimes ironically, to address, or as a designation of, a handsome person. *colloq.* (orig. *U.S.*).

1921 J. Dos Passos *Three Soldiers* (1922) VI. ii. 334 'Teach him how to salute,' the officer had said and Handsome had stepped up to him and hit him. **1940** S. Lewis *Bethel Merriday* i. 13 'Hya, Toots. Hya, handsome,' said her brother. **1945** E. Waugh *Brideshead Revisited* I. v. 104 Be a sport, handsome: no one's seen anything but you. **1963** 'H. Calvin' *It's Different Abroad* ii. 7 Laurent turned to him and sneered. 'Okay, handsome,' he said.

B. *adv.* Now usu. in phr. *handsome is as handsome does.*

1847 F. A. Durivage in W. T. Porter *Quarter Race in Kentucky* 48 She aint no Wenus, Sir,..but handsome is as handsome does. **1950** A. Christie *Murder is Announced* xiii. 139 'Such a handsome young man.'..'Handsome is as handsome does... Much too fond of poking fun at people.' **1963** *Times* 5 June 4/3 But handsome is as handsome does, and his job was done when he was caught off the last ball, when Kent still had a wicket to spare. **1970** G. Greer *Female Eunuch* 56 Men were slipping into relative anonymity and 'handsome is as handsome does'.

hands-up (hæ·ndz,ɒp), *v.* [f. the order *hands up!* (see Hand *sb.* 54).] *intr.* To put up the hands in token of surrender. Also *trans.*, to cause to surrender. So **ha·nds-up** *sb.*, the

HANDWRITE

action of putting up the hands (in quot. *attrib.*); **ha·nds-u:pper**, one who surrenders. Also **ha·nd-up** *sb.*, one who throws up his hands.

1901 *Contemp. Rev.* Mar. 327 A small patrol..went..to the farm of a 'hands upper', *i.e.*, one who had surrendered his arms. **1901** *Daily Chron.* 12 Nov. 5/4 They regard themselves as quite the aristocrats of the camp, and much superior to the 'hands-uppers', as they have delighted in calling the children of less obstinate patriots. **1901** 'Linesman' *Words by Eyewitness* 239 The refugee camps within the British lines, wherein dwell the hundreds of Dutchmen who have surrendered, or 'hands-upped'. **1902** *Westm. Gaz.* 20 Mar. 7/1 Trooper Long..was grabbed by the throat by a 'hands-up' prisoner, who threw down his rifle. **1902** *Appleton's Ann. Cycl.* 629/2 The Boers who had accepted British sovereignty at various times since the fall of Bloemfontein and Pretoria, contemptuously called 'handups' by the others. **1915** *Observer* 4 Apr. 7/2 We have now a case of 'hands-upping', the first in this war, by a whole unit of Germans. **1923** *Daily Mail* 9 Mar. 10 The Germans after 'hands-upping' Rumania proceeded literally to turn out their pockets. **1928** *Observer* 17 June 7 Those faint-hearted ones who are 'hands-uppers' in regard to aviation. **1929** J. Buchan *Courts of Morning* III. 307 They hands-upped like lambs. We've gotten a nice little bag—fourteen hundred and seventy-three combatant soldiers.

handwrite (hæ·nd,rəit), *sb.* *Sc.*, *Ir.*, and *U.S.* [f. Hand *sb.* + Write *sb.*[1] 5. Cf. Handwrit and *hand of writ, write* (Hand *sb.* 16 b).] Handwriting.

1483– in *Dict. Older Scot. Tongue.* **1617** in W. K. Tweedie *Sel. Biogr.* (1847) I. 95, I receaved a letter..whilk albeit it wanted a subscription, yet by the handwrite..I knew to be yours. **1638** S. Rutherford *Lett.* (1664) 14 His hand write, & his seal. **1688** in R. Wodrow *Hist. Suff. Ch. Scot.* (1722) II. 633 You..adhered to your preaching Book, and declared the same to be your own Hand-write. **1836** B. Tucker *Partisan Leader* (1861) 16 (Th.), He has got a paper in the captain's hand-write to show him the way. **1856** W. G. Simms *Eutaw* 429 (Th.), Thar's his name in handwrite!..Hyar's a boy that reads this hand-write. **1880** W. H. Patterson *Gloss. Antrim* 49 Whose hand write is that? **1907** N. Munro *Daft Days* xv, She knew she could never sustain the standard of hand-write, spelling, and information Bud had established in her first epistle. **1923** *Dialect Notes* V. 209, I know his hand write.

handwriting. Add: **2.** (Later *fig.* examples.)

1928 R. Fry in *S.P.E. Tract* XXXI. 331 *Ecriture.* Has a special sense in regard to painting, and refers to the rhythm of the handling of paint. *Handwriting* has hardly acquired this use, but perhaps might be adequate. **1959** *Sunday Times* 10 May 19/5 Certain designers possess an accuracy of taste and a precision of expression which produce an identifiable handwriting... Dior's was a great handwriting. **1959** *Observer* 20 Sept. 18/4 A style policy which is recognisably the 'handwriting of the store'. **1960** *Observer* 28 Feb. 5/3 Leonardo's left-handedness, his preoccupation with things observed..contribute to what Sir Kenneth [*sc.* Clark] calls his 'handwriting'. **1961** R. Seth *Anat. Spying* vi. 97 The individual's use of the morse-key is as distinctive as his handwriting—in fact, it is referred to as 'handwriting'.

3. *attrib.*, as **handwriting expert**, one who makes a study of handwriting in order to determine the authorship of disputed documents, to detect forgeries, etc.

1894 *Strand Mag.* VIII. 293/1 The methods employed by handwriting experts. **1897** *Westm. Gaz.* 2 Dec. 5/1 M. Bertillon, the famous handwriting expert, one of the witnesses at the Dreyfus trial. **1898** *Ibid.* 17 Jan. 7/2 The testimony of hand-writing 'experts'. **1967** G. B. Mair *Girl from Peking* iv. 52 Our handwriting experts say that they *could* have been written by the same person.

handy, *sb.* Add: **2.** A hand-bier.

1909 *Daily Chron.* 8 June 2/5 Hearses, Handys, Biers, &c. **1922** *Daily Mail* 4 Nov. 10 The charges for licences on motor-hearses and handies.

handy, *a.* Add: **2. b.** Phr. *handy by, to*: conveniently situated for. *dial.* and *U.S.* So *handy for* (general colloq.).

1825 J. Jennings *Observ. Dial. W. Eng.* 133 I've hir'd 'twar handy ta tha zea. **1893** 'O. Thanet' *Stories of Western Town* 136 It is customary in the Lossing Building to say, 'We are so handy to the cars.' **1934** H. G. Wells *Exper. Autobiogr.* II. viii. 602 A miscellany of people came and went there and to lodgings handy-by the smaller house at Dymchurch. **1968** *Globe & Mail* (Toronto) 17 Feb. 45 (Advt.), Not one street to cross to get to public school. Handy to shopping plaza. Beautiful view over park.

3. b. *handy dog* (see quots.). *N.Z.*

1933 L. G. D. Acland in *Press* (Christchurch) 25 Nov. 15/7 Some dogs will both head and huntaway and are called handy dogs. **1968** *N.Z. News* 28 Aug. 16/1 A pack usually has several 'huntaways', a 'heading' dog and a 'handy' dog... The handy dog is generally versatile at all jobs.

5. b. handy-billy (later example); **handy-man,** (*b*) a sailor; **handy-sized** *a.*, of a convenient or suitable size.

1933 J. Masefield *Bird of Dawning* 56 We handed up the handy billy, they made it fast to the davit-head. **1899** H. Begbie *Handy Man* iii, And the babe sleeps sound in her cot o' nights, and the trader may plot and plan, For under the stars on the rolling deep stands the vigilant Handy Man. **1900** *People* 1 Apr. (Ware), The handy man. High praise for the naval brigade. **1927** *Scots Observer* 19 Mar. 15/3 The need for a handy-sized and attractive edition of the Gospels. **1961** *Times* 7 Feb. 17/5 Few owners are building handy-sized tankers these days.

handy-dandy. **1. c.** (Later examples.)

1965 *New Statesman* 3 Dec. 879/3 Change places, handy-dandy and the tone..could almost be Peter Simple on Dr Castrumba. **1966** *Ibid.* 15 Apr. 539/3 Handy-dandy, which is the Underground Man and which is the complacent bourgeois?

hanepoot (hā·nəpɔ̄:t, -pū̆ə:t). *S. Afr.* Also **haanepoot, haanepot, haenapod, hanapoot,** and (corruptly) **honeypot.** [Afrikaans, f. Du. *haan* cock + *poot* foot.] **1.** The grape-variety Muscat of Alexandria, a table grape often also used for making wine and for raisins.

c **1798** Lady A. Barnard in A. W. C. Lindsay *Lives of Lindsays* (1849) III. 403 The Honipot grape.., a fleshy white grape, which is of the Muscatel nature and excellent. **1801** J. Barrow *Trav. S. Afr.* I. 65 A large white Persian grape, called here the haenapod. **1855** W. R. King *Campaigning in Kaffirland* (ed. 2) 190 The most deliciously flavoured grapes, one sort, called the 'honeypot',..of immense size. **1878** T. J. Lucas *Camp Life* 36 A fine fleshy well-favoured variety called hanne poot. **1887** *Colonial & Indian Exhib., Rep. Col. Sect.* 136 Raisins are made from the Haanepot grape. **1896** R. Wallace *Farm. Ind. Cape Col.* x. 202 Of grapes the Haanepoot..and the Barbarossa are considered the best for the British market. **1927** *Daily Express* 8 Apr. 5 The Cape grapes..either the gros Colmars or the white Hanapoots. **1945** *Cape Times* 30 Mar. 4/7 The hanepoots are golden-brown and sweet. **1946** *Ibid.* 11 Feb. 6/5 In 1938–39 it was easy to buy good hanepoots at six or eight lb. for a shilling. **1971** *Rand Daily Mail* 28 July 15 The Spanish variety was thought to be the forebear of the Hanepoot grapes of today.

2. A sweet white wine made from hanepoot grapes.

1804 R. Percival *Acct. Cape of Good Hope* xi. 188 The Hanepod made from a large white grape is very rich, but scarce and dear. **1952** C. L. Leipoldt *300 Yrs. Cape Wine* xv. 203 A more common Hanepoot wine is a golden coloured, fairly sweet wine, of which several kinds are on the market. **1966** C. De Bosdari *Wines of Cape* (ed. 3) vi. 66 There are also the Muscadels,..from Hanepoot..to Frontignac.

hang, *v.* Add: **1.** (Later examples.)

1896 R. Fry *Lett.* (1972) I. 168 Tonks's *Broadstairs* is a terrible thing to hang: it is so spotty and brilliant that it knocks the other things to pieces. **1967** *Listener* 2 Mar. 296/2 More rewarding, and better hung,..is the loan exhibition of graphics.

6. b. *fig.* To reduce to, or hold in, a state of indecision or inaction; *esp.* in phr. *to hang a jury*, to prevent (as a juryman) a jury from reaching a verdict (cf. *17 c). *U.S.*

1778 G. Washington *Let.* 15 June in *Writings* (1834) V. 405, I am hung in suspense. **1848** E. Bryant *California* xxvi. 291 The jury, after the case was referred to them, were what is called 'hung'; they could not agree. **1868** *Harper's Mag.* Mar. 542/2 The jury..returned with a verdict in favor of the *plaintiff!* On remonstrating with the Mexican why he did not 'hang' the jury, the lawyer asked him, 'Why did you bring in a verdict against yourself?' **1967** *Guardian* 3 Apr. 2/7 The lone juror who finally hangs the jury will not emerge,..unless at the start his view has some support.

7. b. To tie or hitch up (a horse). (Cf. 28 e.) Chiefly *U.S.*

1835 *Southern Lit. Messenger* I. 581 Having arrived at Blank, we *hung* our horses, as Virginians always do after riding them. **1843** 'R. Carlton' *New Purchase* xxvi, While *hanging* Dick to a gate post. **1900** H. Lawson *On Track* 30 He got down, wondering what was up, and hung his horse to the last post but one. **1916** J. B. Cooper *Coo-oo-ee!* xii. 164 You made the remark that 'you'd tie up your horse'—an Australian 'hangs' his horse to a fence.

10. Delete *arch.* and add further examples. Now usu. in phr. *to go hang*: to go and be hanged; to 'go to the devil'; to be dismissed or rejected; freq. *let (it, etc.) go hang.*

1921 R. Hichens *Spirit of Time* xii. 203 Hold on to the best in yourself and let all the rest go hang. **1937** J. Betjeman *Coll. Poems* (1958) 41 Other cars all go hang My little bus is enough for us. **1960** M. Sharp *Something Light* xix. 174 Louisa instantly resolved to let the room go hang. **1973** *Physics Bull.* June 345/3 It would even be proper for SRC to decide to support (say) only 20 postgraduate schools of chemistry and to let the rest go hang.

b. To be in desperate difficulties. *slang.*

1874 Hotten *Slang Dict.* 187 *Hanging*, in difficulties. A man who is in great straits, and who is, therefore, prepared to do anything desperate to retrieve his fortunes, is said, among sporting men, to be 'a man hanging', i.e. a man to whom any change must be for the better. **1889** Barrère & Leland *Dict. Slang* I. 446/2 *To hang* (popular and sporting), to be in a desperate state.

d. Of a horse: to veer towards one side.

1951 E. Rickman *Come racing with Me* ii. 16 Sarda II 'hanging' towards Native Heath..who won by a short head. **1958** J. Hislop *From Start to Finish* xi. 128 Courses such as Epsom and Lewes, where the ground slopes towards the rails and horses tend to hang that way. **1965** *Observer* (Colour Suppl.) 30 May 34 If he starts to hang before he jockey can generally straighten him up.

11. c. *Iron-founding.* = Scaffold *v.* 5.

1878 *Jrnl. Iron & Steel Inst.* XII. 202 When a furnace 'hangs' on one side, a more common occurrence with small old furnaces than with large modern ones, a system prevailed in some works of putting half a pig of lead in above the part that was fast. **1908** R. Forsythe *Blast Furnace* 242 When the stock becomes wedged so tightly that it can no longer descend, the furnace is said to 'hang'.

17. c. Of a jury: to fail to agree. (Cf. *6 b and *Hung *ppl. a.* 3.) *U.S.*

1859 Bartlett *Dict. Amer.*, *To hang*, to stick fast, come to a stand still; as, the jury hung, and 'the man got a new trial'. **1929** *Randolph Enterprise* (W.Va.) 24 Oct. 5/1 The jury hung up on the case and were discharged.

19. b. To slacken motion perceptibly; *spec.* in *Cricket* (see quots.) and *Baseball.* Occas. *trans.*

1838 *Bell's Life* 8 July 4/4 The dead state of the ground, which prevented the balls from working, and caused them to hang considerably. **1897** K. S. Ranjitsinhji *Jubilee Bk. Cricket* 77 The ball is made to hug the ground when it pitches, and to rise slowly afterwards, or 'hang', as it is called by cricketers... With some bowlers it either 'hangs' or more often comes fast off the pitch owing to something in their regular action. **1897** *Encycl. Sport* I. 246/1 A ball 'hangs' which rises unexpectedly slowly from the pitch. **1906** Jephson in H. G. Hutchinson *Cricket* 103 He ran up and delivered the ball, to all appearances, exactly similarly each time; but one found now that the ball was hanging in the air, now that it was on to one surprisingly soon. **1928** *Funk's Stand. Dict.* I. 1112/1 *Hang* (Sport), to slacken speed perceptibly and unexpectedly: said of a ball in flight in various games, and of a boat between strokes, in rowing. **1967** *Boston Globe* 5 Apr. 51/6 'It was a bad pitch,' Bennett admitted. 'I was trying to pitch low and instead I hung a high curve ball for him to hit.'

c. *to hang to* (see quot.).

1888 *Lockwood's Dict. Mech. Engin.* 172 *Hang to*, a term having several applications. A file hangs to its work when it cuts without slip. A saw hangs to, when it feels as though being drawn into the timber. A pattern hangs to the sand when it delivers with difficulty.

20. (Earlier and additional examples.) Also, esp. *U.S.*, *to hang around* (a person, place, etc.). So *hang-arounder.*

1830 *Corrector* (Sag Harbor, N.Y.) 26 June 1/3 What a number of young gentlemen you have in this city—hanging round the corners—standing in hotel doors. **1847** J. S. Robb *Streaks of Squatter Life* 133 Every time I come up from Lusiane, I found Jess hangin' round that gal. **1885** 'C. E. Craddock' *Prophet Gt. Smoky Mts.* 8, I hev seen that critter, that thar preacher, a-hangin' round you-uns house a powerful deal lately. **1897** S. T. Clover *Paul Travers' Adv.* 51, I guess I can fix you out if you hang around here, but keep shady. **1915** N. L. McClung *In Times like These* vi. 72 Although the polls are only open every three or four years, if women once get into the way of going to them, they will hang around there all the rest of the time. **1938** O. Nash *I'm Stranger Here Myself* 234 The hang-arounders' cheerful chirrups. **1939** I. Baird *Waste Heritage* xi. 136 He hung around the window then he stopped in the doorway and tried the door in case there was anyone inside the store could say whether Eddy had been hanging around. **1950** A. Lomax *Mr. Jelly Roll* 57 Buddy Bolden, the most powerful trumpet player I've ever heard..and the absolute favourite of all the hangarounders in the Garden District. **1970** G. F. Newman *Sir, You Bastard* viii. 244 He didn't hang around afterwards. **1973** *Melody Maker* 25 Aug. 27 In a front room in Shepherds Bush, however, plots are being hatched—and hang about, because I'm not going to bore you with yet another..yarn.

25. hang on. a. (Later examples.)

1899 G. B. Shaw *Let.* 20 Apr. (1931) 260 She is always hanging on by her eyebrows, whereas the German is comfortably seated in a solid, permanent, broadbottomed engagement. **1931** *Times Lit. Suppl.* 12 Mar. 193/1 Lacking roots in the soil of any particular country, Whistler had always to 'hang on by his eyebrows'. **1935** *Yachting* Dec. 82/3 *Hanging on by the eyelids*, the seaman's vivid description of his situation during a very heavy gale. **1958** *Listener* 21 Aug. 259/2 Each aircraft hangs on to the tail of the one directly in front.

c. To wait. Freq. in *imp.*, be patient, be reasonable!

1939 J. B. Priestley *Let People Sing* x. 262 I'd better hang on and have a word with her. **1941** Baker *Dict. Austral. Slang* 34 *Hang on!*, be reasonable! Not so fast. **1971** *Woman's Own* 27 Mar. 26/1 Hang on a minute...I'm coming with you.

d. Used in a telephone conversation in the sense of 'hold the line'.

1936 R. Lehmann *Weather in Streets* I. iv. 70 Hang on a moment...Mummy wants to speak to you. **1960** *Daily Tel.* 15 Aug. 17/5 Switchboard operators have been trained not to keep any caller 'hanging on'. **1969** S. Hyland *Top Bloody Secret* i. 37 'Shall I tell him you're coming?' 'Yes please. Tell him to hang on.'

e. *to hang on to* (something): to retain.

1871 Trollope *Eustace Diamonds* (1873) I. xvi. 220 It was manifest enough that she meant 'to hang on to them' [*sc.* the diamonds]. **1936** 'M. Innes' *Death at President's Lodging* ix. 166 He had in his possession certain valuable documents... Umpleby simply hung on to them. **1971** 'D. Halliday' *Dolly & Doctor Bird* ii. 16 The hotel wouldn't let her hang on to her room.

f. *to hang one on*: to deal (someone) a blow.

1908 K. McGaffey *Sorrows of Show-Girl* 200 Hauling off wifey hangs one on Alla's map. **1960** B. Crump *Good Keen Man* 44 I'd thought for a moment he was going to hang one on me. The idea..had got his goat all right. **1966** *Punch* 19 Jan. 69/1 There are moments when most of us have felt the keenest desire to hang one on the boss's chin and walk out.

g. Used in various technical senses (see quots.).

1963 *Amer. Speech* XXXVIII. 118 *Hang on*, to maintain a proper position for the receiver while it is coupled to the tanker's air refueling boom. 'Dingbat 27, can you hang on if I start a slow turn to the left?' **1967** *Gloss. Mining Terms* (*B.S.I.*) x. 7 *Clip on* or *hang on*, to attach a tub or tubs to a haulage rope by a clip or shackle.

26. b. Also, *to hang out to dry*: to suspend (wet washing) on a clothes-line in the open so that it can dry. Hence *transf.* in *Cricket*: *hang one's bat out to dry* (see quots.).

1893 [in Dict., in def. of *clothes-line*]. **1895** C. B. Fry in

Badminton Mag. Aug. 132 He [*sc.* the young player on hard wickets] gets into the habit of moving his right leg, leaving his bat hanging out to dry, and playing crooked. **1925** *Country Life* 25 July 142/1 In playing forward..never 'hang your bat out to dry' by not advancing your left foot to the pitch of the ball; if you do, you have neither power nor control.

c. hang out. (Later examples.) Also, of a job: to be available, to be found.

c **1926** 'Mixer' *Transport Workers' Song Book* 69 When there is a job hanging out. **1931** T. Lyell *Slang* 364, I hear you've got a job in Foster's factory. Where does it actually hang out? **1931** D. Runyon *Guys & Dolls* (1932) ii. 35 He cannot have a whole lot of sense, or he will not be hanging out with Handsome Jack. **1935** *Forres, Elgin & Nairn Gaz.* 6 Nov. 4/5 (*heading*) Later American word-imports... Phrases are very numerous:—Where do you hang out? **1936** Wodehouse *Laughing Gas* ii. 23 The head of the family has always hung out at the castle.

d. Also without *it*: to endure, hold out. Chiefly *Austral.* and *N.Z.*

1939 J. Dell *Nobody ordered Wolves* ii. 14 B. and P. offered her twelve thousand..but I told her to hang out and sure enough Bill sold her to M.B.G. for fourteen thousand flat. **1941** Baker *Dict. Austral. Slang* 34 *Hang out*, to endure: to delay (a matter). **1944** J. Fullarton *Troop Target* xi. 87 I've been pretty crook for the last hour. But I wanted to hang out till we saw a house. **1946** K. Tennant *Lost Haven* ix. 132 The old punt had broken down at last. He had been hoping against hope that it would hang out until the war ended, but the luck was against him.

e. Slang phr. *to let it all hang out*: to be uninhibited or relaxed; to be candidly truthful. orig. *U.S.*

1970 C. Major *Dict. Afro-Amer. Slang* 76 *Let it all hang out*, to be uninhibited, free. **1972** *National Observer* (N.Y.) 27 May 17/3 Give it expression, they say, 'Let it all hang out.' If it 'all hangs out', it is bound to do some good. **1972** *Village Voice* (N.Y.) 1 June 51/1 No names, of course, will be used; he doesn't expect everyone will be as willing as he is to let it all hang out.

28. hang up. a. Also *absol.* = to hang up the receiver of a telephone at the end of a conversation; *to hang up on*: to break off telephonic communication with.

1911 A. B. Smith *Mod. Amer. Teleph.* xxvi. 759 When the subscribers are through talking, they hang up their receivers. **1928** E. Wallace *Double* viii, 'Oh, Mr. Staines!..What a dull life yours must be!' And then she hung up on him, and left him feeling like a spanked child. **1928** F. N. Hart *Bellamy Trial* iii. 101 He'd hung up, I guess. Anyway he didn't answer. **1952** A. Baron *With Hope, Farewell* 103 He managed to say, 'Thank you,' and was about to hang up. **1960** *Daily Tel.* 15 Aug. 17/5 Several directors and secretaries of firms told me that they hung up within a minute if they could not get through. **1968** 'P. Barrington' *Accessory to Murder* vii. 125 Mrs. Lindley heard the click of the receiver and became indignant. He'd almost hung up on her.

b. *to hang up one's boots*, to give up playing a game; *to hang up the spoon*, to die; *to hang up one's hat* (further examples in various senses); *to hang up a record*, to set up a record.

1925 O. Jespersen *Mankind, Nation & Individ.* ix. 166 There are countless variants [for 'to die']..take an earth bath, hang up the spoon, snuff the candle, snuff it. **1930** *Publishers' Weekly* 15 Mar. 1508/2 A record sale was hung up..on Tuesday... Four hundred and ten copies of the book were sold in one hour. **1938** D. Runyon *Take it Easy* xv. 283 Professor D. says he has no doubt that under the old rule Nicely-Nicely will hang up a record that will endure through the ages. **1942** Berrey & Van den Bark *Amer. Thes. Slang* 44 *Reside*, hang up one's hat. *Ibid.* 132 *Die*, hang up one's hat. *Ibid.* 242 *Hang up one's hat*,..to be perfectly at ease, make oneself at home. **1949** F. Sargeson *I saw in my Dream* II. xiii. 113 Some said that..he'd have had more self-respect if he'd told the girl to go and hang her hat up somewhere else. **1963** *Times* 23 Jan. 3/4 Johnson, Miller, and Johnston hung up their boots soon afterwards and two years later Benaud began to build the side.

d. Also *to hang it up*, to chalk it up, to give credit. *slang.*

1841 *Swell's Night Guide* Gloss., *Hang it Up*, to go on Credit. **1874** 'Mark Twain' & Warner *Gilded Age* I. xiii. 172 The Colonel muttered something to the barkeeper about 'hanging it up'. **1942** S. H. Adams *Tambay Gold* xiv. 191 They hung me up for the parking fee.

e. (Earlier and later examples.)

1858 W. Kelly *Life in Victoria* (1860) 49 In Melbourne there are posts sunk in the ground almost opposite every door... Fastening your horse to one of these posts is called 'hanging him up'. **1966** 'J. Hackston' *Father clears Out* 118 On the Saturday many good hacks were hung up at the hotel.

f. Also, to suspend movement or action; to stop or stay.

1845 *Greenfield Fish. Rec., Chowan, N.C.* 6 May in N. E. Eliason *Tarheel Talk* (1956) 276 Made 2 hauls & hung up [for the fishing season]. **1854** *Congress. Globe* App. 108 (Th.), In reading the President's message,..he got befogged, and, in the language of the Kentucky boatman, 'hung up for the night'. **1874** E. Eggleston *Circuit Rider* xvi, You mou't get a place 'bout a mile furder on whar you could hang up for the night. **1895** *Dialect Notes* I. 372 A mower, when rain was coming on: 'I reckon we'll have to hang up for all day.'

g. *Cab-drivers' slang.* (See quots.)

1930 'A. Armstrong' *Taxi* xii. 164 'Hanging it up' is loitering past a theatre to snatch a fare away from the recognized rank. **1939** H. Hodge *Cab, Sir?* I. v. 50 Policemen in these outer districts are more easy-going than in the West End. So I chance 'hanging it up' as we call it, near the door, keeping my engine running in case the policeman looks too nasty.

hang, *sb.* Add: **1. b.** Also in *Cricket* (see *Hang *v.* 19 b).

1888 R. H. Lyttelton in Steel & Lyttelton *Cricket* ii. 48 Any break, hang, or rise that the bowler or the ground may impart to the ball must almost inevitably produce a bad stroke. **1897** K. S. Ranjitsinhji *Jubilee Bk. Cricket* 78 The ideal bowler..should do his best to acquire a command of off-break and leg-break, 'top' and 'hang'. **1901** [see *Bump *sb.*[1] 1 c].

2. a. Of a painting or work of art.

1959 *Listener* 5 Mar. 422/3 The Secretary of the Society, with no previous experience of the compromise necessary in rooms so unsuited to the display of very modern painting, has achieved a remarkably successful hang. **1964** *Guardian* 21 Apr. 9/1 At the great Tate Gallery exhibitions..the brilliance of the hang has invariably been cancelled out by the failure of..the lighting engineers.

3. For *U.S. colloq.* read orig. *U.S. colloq.* and add further examples.

1890 *Daily Chron.* 4 Apr. 7/2 He gets what some call 'the hang' of the place. **1895** R. Kipling in *Century Mag.* Dec. 271/1 I'm getting the hang of the geography of that place. **1918** *War Illustr.* 13 July 372/3 On the second day I had a 'flip' round the aerodrome to get the 'hang' of the country. **1931** H. G. Wells *Work, Wealth & Happiness of Mankind* (1932) 1 Never before has there been this need and desire to 'get the hang' of the world as one whole. **1957** *Listener* 17 Oct. 606/1 Children..in their desire to get the hang of their surroundings.

6. (*a*) *hang of a*: an Australian and N.Z. intensive phrase, variously spelt (*hangava, hanguva*, etc., and in altered forms, e.g. *hangashun*), used informally, sometimes with adverbial force, of something big, bad, vexatious, etc., of its kind. Also *like hang*, like hell. Cf. *Hellishing, Hellishun.

1941 Baker *N.Z. Slang* vi. 51 Expressions..in constant use by our youngsters..hangava, hangashun. **1943** J. A. W. Bennett in *Amer. Speech* XVIII. 90 The intensives hanguva, hangershun. **1945** F. Sargeson *When Wind Blows* ii. 14 They got down in a hang of a hurry. *Ibid.* iii. 16 All this was because Charlie was hang of a funny to be with. **1949** *Landfall* III. 145 Gosh, Dad's hangava crabby with you! **1950** B. Sutton-Smith *Our Street* ii. 33 It's a hang of a wet day. **1960** N. Hilliard *Maori Girl* 64 It hurts like hang.

hang-. Add: **hang-five, -ten** *Surfing* = *hanging five, ten* (*Hanging *ppl. a.* 6); so *to hang five, ten*; **hang-glider,** a suspension glider, controlled and stabilized by deliberate movements of the operator's body, which is suspended upright in the framework of the machine; hence *hang-gliding*; **hang-out** (further examples); **hang-up** *slang*, drawback, fault; difficulty, fixation, 'thing'; also *attrib.*

1962 D. Muirhead *Surfing in Hawaii* viii. 80 (*caption*) Film-maker Walt Phillips hangs five at Halewia, Hawaii. **1969** *Observer* 3 Aug. 35/1 He may 'nose ride', balancing on the very front of the board to achieve a 'hang five' or 'hang ten', with five or ten toes over the tip of the board. **1930** V. W. Pagé *Henley's ABC of Gliding & Sailflying* (1931) 202 On 'hang' gliders the lateral and longitudinal stability and the angle of attack are controlled by shifting the weight of the pilot. **1969** K. Munson *Pioneer Aircraft 1903–14* 125/2 The unpowered D.1 was launched from a 4-wheel trolley chassis for its first take-off in 1907 with Colonel Capper aboard. However, when it was damaged in a crash-landing further tests were abandoned in favour of the D.3 'hang-glider'. **1972** *Daily Tel.* 7 June 13/8 Considerable interest is being revived in hang gliders and at least four groups are now working on prototype flying wings... They rely on the pilot's leg-work to become airborne and span and performance is limited by the power of the human shoulder. **1973** *Daily Colonist* (Victoria, B.C.) 21 Aug. 16/3 She was flying a Black Hawk hang glider made of black plastic sheeting stretched on aircraft tubing and cable. It has a 38-foot wingspan and a 15-foot keel. **1972** *Daily Tel.* (Colour Suppl.) 13 Oct. 10/3 In America the sport of 'hang gliding'..is becoming extremely popular. **1973** *Daily Colonist* (Victoria, B.C.) 6 July 15/1 Hang gliding is the closest you can get to flying free. **1895** *Century Mag.* Oct. 943/1 In the afternoon some thirteen boys appeared at the 'hang-out'. **1896** *Atlantic Monthly* Jan. 67/2 His wanderings ended..oftener in some distracting vagabond's 'hang-out' in a neighboring city. *a* **1911** D. G. Phillips *Susan Lenox* (1917) II. x. 248 She avoided the tough places, the hang-outs of the gangs. **1951** R. Campbell *Light on Dark Horse* 244 But I had still another hang-out. That was the Harlequin Restaurant. **1957** *New Yorker* 5 Oct. 66/3 The most energetic dancer of her day, gyrating nightly at the Stork Club and other hangouts of the quality. **1960** *Guardian* 9 Nov. 8/6 Basement hangouts which have a vogue and then disappear. **1968** *Globe & Mail Mag.* (Toronto) 13 Jan. 7/3 It is 3 a.m. in a steam bath known as an after-midnight homosexual hangout. **1963** *Pix* 28 Sept. 62/2 *Hang ten*, ten toes over the nose of the board. **1967** *New Yorker* 25 Feb. 18/1 *The endless Summer*—A hang-ten documentary about surfing in various parts of the world. **1959** *Daily Colonist* (Victoria, B.C.) 16 Apr. 2/6 Man—Omnibus salutation extended to men, women, domestic animals—saves cool cat hangup of remembering names. **1967** A. Diment *Dolly Dolly Spy* ii. 27 Hash..is non-narcotic but whether..it leads you on to the hang up drugs I don't know. **1967** *Melody Maker* 27 May 10/6 All these hang-ups are eliminated by the truth—this is a great record. *Ibid.* 29 July 10/5 There are all sorts of hang-up noises going on in the background. **1967** *Ottawa Jrnl.* 31 May 39/2 Your husband's hang-up dates back to childhood and you must treat it as an illness, which it is most assuredly. **1967** *Crescendo* Dec. 27/4 He was always very kind and patient with me, as he was..with all people. A great degree of sensitivity to people's hang-ups, you know—he'd never put you on a spot. **1968** *Melody Maker* 22 June 2 The group didn't want the hang up of worrying about recording specific singles. **1968** *Observer* 22 Dec. 21/1 People have this hang-up about art. A woman will worry for days about spending money on a painting: is it a good investment, can she trust her own judgment? The same woman will spend $150 on a dress..without giving it a thought. **1973** *Black World* Apr. 17/1 Depressing piece about pushers, junkies, whores and their hang-ups.

hangar. Add, with pronunciation (hæ·ŋəɪ): **b.** A shed for the accommodation of aircraft or spacecraft.

1902 *Daily Chron.* 31 Oct. 5/3 Mr. Santos Dumont..will construct a hangar in the Bois de Boulogne. **1935** H. G. Wells *Things to Come* ix. 48 Inside an aeroplane hangar. **1962** A. Shepard in *Into Orbit* 97, I tried to avoid moving into Hangar S—our quarters at the Cape—for as long as I could. **1962** V. Grissom *Ibid.* 119 On 1 July the capsule was taken from the hangar to the launching pad to be mated to the Redstone.

hangarage (hæ·ŋərėdʒ). [Blend of *hangar* and *garage.*] Accommodation for aircraft in a hangar.

1932 *Flight* 10 Nov. 1044/1 When I thought I was almost beaten, I suddenly won, and hangarage was granted. **1961** R. Higham *Brit. Rigid Airship* vii. 121 This caused a tremendous demand for hangarage and other means of storing gear under cover. **1970** M. Kelly *Spinifex* iii. 54 Hangarage for twenty planes. Full servicing workshops.

hangashun, hangava, etc.: see *Hang *sb.* 6.

hangbird. (Earlier examples.)

1789 J. Morse *Amer. Geogr.* 59 Upwards of one hundred and thirty American Birds have been enumerated..[including the] Hangbird, Heron, Little white Heron. **1794** S. Williams *Hist. Vermont* 118 Hangbird, *Oriolus icterus.* **1824** Z. Thompson *Gazetteer Vermont* 18 The singing birds are the robin, thrush,..springbird, goldfinch and hangbird. **1831** J. Q. Adams *Mem.* (1876) VIII. 426 The oriole of Baltimore is the fiery hang-bird.

hang-down (hæ·ŋ,daun), *sb.* and *a.* [f. phrase *to hang down* (see Hang *v.* 8).] **A.** *sb.* That which hangs down, *spec.* in certain technical uses. **B.** *adj.* That hangs down.

1888 *Lockwood's Dict. Mech. Engin.* 172 *Hang down*, or *hanger*, a bearing suspended from a roof or beam for the journal of a shaft. **1904** Goodchild & Tweney *Technol. & Sci. Dict.* 279/2 *Hang down* (Eng.), a frame for suspending a bearing from a roof or beam.—(Foundry), the sling which supports heavy weights in the foundry; the upper ends of the rods of the sling are attached to the travelling crane. **1906** Kipling *Actions & Reactions* (1909) 212 As Guiseppe [*sic*] unshipped the working mechanism of the organ (it developed a hang-down leg) from its wheels. **1967** *Punch* 31 May 804/2 Vintage varieties of [cider] apples (with names such as Dabinett, Woodbine, Slack-Ma-Girdle, Hangdown, Sheep's Nose) can be used for no other purpose.

hanger[2]. Add: **2. g.** (See quot. 1905.)

1905 Calkins & Holden *Art Mod. Advertising* 352 Hangers are printed or lithographed cards of various shapes and sizes, to be hung up in a store. **1927** in *John Edwards Mem. Foundation Q.* (1969) V. iv. 144 Inclosed find a folder... Also we have a large hanger and dealers order blank.

4. e. A coat- or dress-hanger.

1873 *Young Englishwoman* Feb. 91/1 The two different kinds of hangers..will be found very advantageous for hanging up heavy articles of dress, as winter cloaks, etc. **1908** *Daily Chron.* 26 Feb. 8/5 Every coat and every skirt should have a hanger to itself. **1934** L. A. G. Strong *Corporal Tune* II. iv. 151 It does clothes no good to stay all folded up. The sooner they're out and on hangers the better. **1955** M. Allingham *Beckoning Lady* iii. 38 She was carrying a newly pressed dress on a hanger. **1970** *New Yorker* 28 Feb. 34/3 There are no hangers for suits which have always had hangers.

6. *Comb.* **hanger-back,** one who hangs back (see Hang *v.* 22); **hanger-board** (see quot. 1893).

1923 *Q. Register* May 583 He never..played the calculating hanger-back. **1962** *Times* 10 Aug. 9/4 Young novelists..must realize..that, however many people will unwisely refrain from reading their daring first novels, their own mothers, and even their own grandmothers, will not be among the hangers-back. **18..** *Electr. Rev.* (U.S.) XII. 8 (Cent. Dict.), Electrical connection between the conducting-wires and lamps must be made through a suitable hanger-board. **1893** T. O'C. Sloane *Stand. Electr. Dict.*, *Hanger board*, a board containing two terminals, a suspending hook, and a switch, so that an arc lamp can be introduced into a circuit thereby, or can be removed as desired.

hang-fire (hæ·ŋ,fəiəɹ). [f. phrase *to hang fire* (see Hang *v.* 6).] A delay in the explosion of the charge of a gun or of a blasting charge. Also *transf.*

1892 W. W. Greener *Breech-Loader* 170 Nothing is more tantalising to the sportsman than miss-fires; hang-fires, too, are a great nuisance. **1899** *Kynoch Jrnl.* Oct.–Nov. 6/2 Hang fires, soft shots, high pressures, and other defects. **1936** L. B. Lyon *Bright Feather Fading* 40 Dove's hang-fire height A long way falls. **1955** *Times* 16 May 12/5 There is..no hang-fire in a story which..lacks the long dramatic suspense of the year before.

hangi (hæ·ŋi). *N.Z.* [Maori.] A Maori earth-oven in which food is placed on heated stones.

1861 *Richmond-Atkinson Papers* I. 697 They had made a 'hangi' just before the front windows. **1882** W. D. Hay *Brighter Britain* II. iii. 153 Fish and meat were frequently roasted on the clear side of the fire..but the great national culinary institution was the earth-oven, the kopa or hangi. **1905** W. Satchell *Toll of Bush* xxx. 343 At least the recipients of the hangi should partake of his hospitality. **1905** W. Baucke *Where White Man Treads* 16 For in their season he [*sc.* the Maori] could supplement his dry fare..with the tender bulbous shoots of the tii (cabbage tree), at heart white and delicate, baked in a haangi. **1959** *Weekly News* (Auckland) 30 Dec. 42 Recently, 3404 guests sat down in relays to the wedding breakfast, cooked Maori style, in a hangi—a stone-lined pit—and served in woven flax. **1963** B. Pearson *Coal Flat* xxii. 372 Yesterday..we had the *hangi* ready.

hanging, *vbl. sb.* Add: **5. b.** *Iron-founding.* = Scaffolding *vbl. sb.* 2 a.

1878 *Jrnl. Iron & Steel Inst.* XII. 202 The modern system of putting the material round the in-wall and allowing it to roll to the centre, has diminished the heat at the in-wall of the furnace and greatly reduced the hanging and scaffolding. **1948** G. R. Bashforth *Manuf. Iron & Steel* I. x. 165 Hanging, which is sometimes referred to as wedging, is similar to scaffolding, but is due to carbon deposition.

8. *hanging day* (examples).

1795 tr. *Moritz's Travels* 60 Last Tuesday was (what is here called) hanging day... I only heard tolling at a distance the death-bell of the sacrifice to justice. **1806** *Balance* (Hudson, N.Y.) 11 Nov. 355 (Th.), Next Friday [the newspaper] promises to make its debut. Friday—that's hanging day—but no matter. **1857** D. G. Rossetti *Let.* June (1965) I. 325 *Friday* is the hanging day.

hanging, *ppl. a.* Add: **2. b.** *hanging gardens* (later examples).

1815 J. Fernie *Hist. Dunfermline* 16 On the sides or slopes of the mound, and at the back of the houses are hanging gardens. **1931** H. Crane *Let.* 21 Sept. (1965) 381 Dense tropical foliage and veritable hanging gardens. **1971** R. Russell tr. *Ahmad's Shore & Wave* i. 13 He had conjured up a picture of the hanging gardens of Malabar Hill in Bombay, overlooking the sea.

5. *hanging judge* (later examples). Also *transf.*

1929 J. B. Priestley *Good Companions* II. iv. 339 Your Bruddersfordian is a hanging judge of anything that costs money. **1937** 'G. Orwell' *Road to Wigan Pier* ix. 178 The worst criminal..is morally superior to a hanging judge. **1963** *Times* 9 Mar. 9/6 He became an advocate of reform and a hanging judge of the powers that be in politics, commerce and agriculture. **1972** M. Gee *In my Father's Den* 121 Price..is a combination public relations man and hanging judge.

6. hanging-block (see quot. *a* 1884); **hanging-bowl** *Archæol.*, name given to certain Celtic or Saxon bowls that were suspended from the roof; **hanging bridge,** a suspension-bridge; see also quot. *a* 1877; **hanging-compass** (see quot. *a* 1865); **hanging drop** *Biol.*, a drop of liquid suspended from a cover glass fitting on to a special transparent cell or microscope slide, by means of which living microorganisms or cells in the drop may be examined microscopically; usu. *attrib.*; **hanging five, ten** *Surfing*, used, freq. *attrib.*, with reference to the placing of all the toes of one foot (or of both feet) over the front edge of a surfboard; **hanging glacier** (see quot. 1940); **hanging glider** = *hang-glider* (see *Hang-); **hanging inden(ta)tion,** (*a*) *Printing* (see Indentation 3 and Indention 2); (*b*) *Librarianship* (see quot. 1941); **hanging lie** *Golf*, the position of a ball when it rests on ground sloping downwards in the direction of play; **hanging paragraph** = *hanging indent(at)ion*; **hanging pawn** *Chess*, one of two advanced pawns which are side by side with no pawns on the adjacent files that can support them; **hanging shelf,** a suspended shelf; **hanging steps** (see quot. 1904); **hanging ten,** see *hanging five* above; **hanging valley,** a valley which is abruptly cut across by the steep side of a larger valley or a sea-cliff; **hanging wall** (earlier and later examples); **hanging wardrobe,** (*a*) a wardrobe designed to accommodate clothes hanging at full length; (*b*) a row of hooks on which clothes may be hung.

a **1865** Smyth *Sailor's Word-bk.* (1867) 366 *Hanging-blocks*..are sometimes fitted with a long and short leg, and lash over the eyes of the topmast rigging; when under, they are made fast to a strap. *a* **1884** Knight *Dict. Mech.* Suppl. 436/2 *Hanging block*, a block through which the top-sail tye is rove, then through the tye-block on the yard, and the standing part made fast to the mast head. **1940** *Burlington Mag.* Dec. 180/2 The two bronze hanging-bowls (believed to be lamps) with enamelled escutcheons and mounts. **1956** I. S. Maxwell in D. L. Linton *Sheffield* 122 The presence of three hanging-bowls from this same area may perhaps indicate that Celtic art survived for a long time in this remote district. **1962** H. R. Loyn *Anglo-Saxon England* i. 14 In the case of..hanging-bowls, some of the richest work culturally of the whole settlement period may be attributed to Celtic craftsmen. **1815** *Niles' Weekly Register* IX. 92/1 The main post-road..crosses the Brandywine on a hanging bridge. *a* **1877** Knight *Dict. Mech.* II. 1060/2 *Hanging-bridge.* 1. A hollow, vertical partition depending from the bottom of a boiler and serving to deflect the flame... 2. *a.* A suspension bridge. *b.* A truss-frame bridge. *a* **1865** Smyth *Sailor's Word-bk.* (1867) 366 *Hanging-compass*, a compass so constructed as to hang with its face downwards. **1885** *Jrnl. R. Microsc. Soc.* V. 117 The 'hanging drop'..has some great..disadvantages. **1892** *Phil. Trans. R. Soc.* B. CLXXXIII. 130 Cultures in hanging drops, made in sterilised cells under the microscope. *Ibid.* 136, I..prepared a hanging drop culture of this. **1908** *Practitioner* Aug. 264 By observation of hanging-drop preparations from growth in glucose broth. **1970** Passmore & Robson *Compan. Med. Stud.* II. xviii. 13/1 The presence of flagella is usually inferred by observing..motility in hanging drop preparations of fluid cultures. **1963** *Sunday Mail Mag.* (Brisbane) 5 May 12/5 *Hanging five*, five toes over the nose of the board for maximum speed. **1965** P. L. Dixon *Compl. Bk. Surfing* vi. 78 Riding forward is a term used here to cover all sorts of nose-riding styles like hanging five and ten toes over. **1894** J. W. Gregory in *Q. Jrnl. Geol. Soc.* L. 515 The 'corrie' or 'hanging glaciers'. **1902** *Encycl. Brit.* XXXI. 23/1 Hanging glaciers (*i.e.*, glaciers perched on steep slopes) often discharge themselves over steep rock-faces, the snout breaking off at intervals. **1940** C. M. Rice *Dict. Geol. Terms* 168/1 *Hanging glacier*, a glacier of small size on so steep a slope that the ice breaks off and falls from its lower end. **1932** J. Manchot tr. *Kronfeld's On Gliding & Soaring* 254 'Hanging Glider' is the literal translation of the German 'Hängegleiter'. **1956** *Flight* LXIX. 270/1 This also was a 'hanging' glider. **1927** *Amer. Speech* II. 239/2 The hanging indention is built just the opposite of a paragraph. **1941** *A.L.A. Catalog Rules* (ed. 2) p. xxv, *Hanging indention*, a form of indention in which the first line begins at author indention and succeeding lines at title indention. **1961** T. Landau *Encycl. Librarianship* (ed. 2) 161/1 Hanging indentation. **1909** P. A. Vaile *Mod. Golf* pl. 96 The stance and address for a hanging lie. **1959** L. M. Harrod *Librarians' Gloss.* (ed. 2) 140 Hanging..paragraph. **1964** T. L. Kinsey *Audio-Typing & Electric Typewriters* vii. 66 New paragraph to be typed as a hanging paragraph. **1927** *Brit. Chess Mag.* XLVII. 269/2 The fact that Black has completed his development so early with no other disadvantage than being saddled with the 'hanging Pawns' goes to show that the nightmare..is ended. **1943** R. Fine *Ideas Chess Openings* iv. 131 The hanging Pawns need not be feared by White because of his excellent development. **1726** Swift *Gulliver* I. II. ii. 179 The Cradle was put into a small Drawer..and the Drawer placed upon a Hanging-shelf for fear of the Rats. **1825** J. Neal *Bro. Jonathan* I. 188 A hanging shelf..loaded with cheeses; ropes of onions; dried apples, [etc.]. **1881** S. P. McLean *Cape Cod Folks* ii. 31 In one dark recess I came into forcible contact with a hanging-shelf of pies. **1962** *Williamsburg Reproductions Catal.* 12 An approved reproduction of a hanging shelf of English design, about 1760. **1876** *Notes Building Construction* II. 108 Hanging steps are fixed at one end only. **1904** Goodchild & Tweney *Technol. & Sci. Dict.* 279/2 *Hanging steps*, stone steps having one end built into a wall. **1962** *Austral. Women's Weekly* Suppl. 24 Oct. 3/2 *Hanging ten*, a trick method of riding with toes tucked over the front of the surfboard. **1963** *Observer* 13 Oct. 15/4 The critical 'hanging ten' stance, in which the surfer speeds across the wave with his 10 toes actually hanging over the nose of the 10-ft. surfboard. **1900** W. M. Davis in *Proc. Boston Soc. Nat. Hist.* XXIX. 288 In the spring of 1899, I sent a brief note..to..Mr. G. K. Gilbert of Washington, telling him that all the lateral valleys seemed to be 'hung up' above the floors of the trunk valleys. His reply was long in coming..and he suggested that such laterals should be called 'hanging valleys'—a term I have since then adopted. He fully agreed that hanging valleys presented unanswerable testimony for strong glacial erosion. **1932** Auden *Orators* I. 22 Arguments from the other side of the lake on the formation of hanging valleys. **1938** *Sat. Rev. Lit.* 1 Jan. 17/3 We go..to hear a well-informed ranger explain how a glacier makes a U-shaped valley, leaving hanging valleys to dump waterfalls over the edge. **1952** H. W. Tilman *Nepal Himalaya* II. xii. 149 In a sort of hanging valley, where the slope eased off..we began searching for a camp site. **1963** D. W. & E. E. Humphries tr. *Termier's Erosion & Sedimentation* v. 126 It may have tributaries, but these are often 'hanging valleys' with waterfalls. **1968** R. W. Fairbridge *Encycl. Geomorphol.* 522 Interesting examples of hanging valleys may also be seen entering fjords, notably in Norway and New Zealand. Hanging valleys also occur sometimes along non-glaciated coasts where the rate of cliff retreat is higher than the adjustment potential of the smaller streams, e.g., in the chalk cliffs in the south of England. They are also to be seen along youthful fault scarps. **1778** W. Pryce *Mineralogia Cornubiensis* II. i. 79 When the Miners dig down.., then the roof, i.e. the upper, the hanging wall, or incumbent wall of the Lode or Fissure, is..over their heads. **1901** *Daily Colonist* (Victoria, B.C.) 31 Oct. 6/3 The quartz in the hanging wall here assayed $39.60 and that on the footwall $4.80. **1970** W. Smith *Gold Mine* iv. 9 Rod pondered the unfortunate choice of mining terminology that had named the roof of an excavation 'the hanging wall'. **1896** *Heal & Son Catal.* 169 The 'Eversfield' Suite..consisting of 2 ft. 9 in. Hanging Wardrobe, [etc.]. **1907** *Yesterday's Shopping* (1969) 278/2 Hanging Wardrobe, in oak, teak and mahogany—6 hooks. **1972** *Country Life* 25 May (Suppl.) 37/1 Full length, Hanging Wardrobe in the Chippendale style. It is 4 feet 3 inches wide, 7 feet 2 inches high and 21 inches deep.

hang-over, hangover (hæ·ŋōu:vəɪ). orig. *U.S.* [Hang *v.* 17.] **1.** A thing or person remaining or left over; a remainder or survival, an after-effect. (Later quots. influenced by sense 2.)

1894 *Outing* (U.S.) XXIV. 67/2 Then there are a few 'hang-overs' who have tried before, and two or three green candidates. **1920** C. Sandburg *Smoke & Steel* 153 A hangover of summer song. **1922** H. Crane *Let.* 23 June (1965) 77 Since I have been writing ads a certain amount of hangover work to be done evenings. **1930** L. Denny *Amer. conquers Brit.* 9 That easily inspired hatred of Germany remained as a hang-over in America long after it had been thrown over by the British. **1939** C. Day Lewis *Child of Misfortune* 136 At the beginning of his second University year, he was still suffering a little from the hang-over of public-school education. **1941** *Ann. Reg. 1940* 232 Owing to shortage of labour..as much as any hang-over from the Civil War, the..harvests were all unsatisfactory. **1958** *Economist* 20 Dec. 1054/2 There has been a slight move away from the previous invariable association of every increase in unemployment with a mental picture of lean and hungry men, in hangover from the grim thirties. **1959** *Times Rev. Industry* Dec. 54/3 Only just recovered from the ghastly hangover of that [buying] spree. **1963** *Times Lit. Suppl.* 8 Feb. 87/2 The bitter taste of the humanitarian hangover. **1973** *Daily Tel.* 19 Feb. 6/4 The oversized dormitories..are hang-overs from the old lunatic asylums.

2. The unpleasant after-effects of (esp. alcoholic) dissipation.

1904 'G. Wurdz' *Foolish Dict.*, *Brain*,..usually occupied by the Intellect Bros.,—Thoughts and Ideas—as an Intelligence Office, but sometimes sub-let to Jag, Hang-Over & Co. **1912** W. Irwin *Red Button* 93 This was the first time in his life that Tommy North had ever admitted a 'hangover'. **1935** D. L. Sayers *Gaudy Night* viii. 161 'How's Miss Cattermole?' 'Bad hang-over. As you might expect.' **1942** *New Statesman* 11 July 26/1 But the use of myths has a similar effect to the use of alcohol: an inevitable hang-over follows the original elation. **1957** *Listener* 18 July 105/2 Its [*sc.* coffee's] ability to quicken the spirits, and, above all, to remove the vestiges of those severe hang-overs which afflicted our hard-drinking forefathers. **1959** N. Mailer *Advts. for Myself* (1961) 220 It was the only good writing I ever did directly from a drug, even if I paid for it with a hangover beyond measure. **1962** K. Orvis *Damned & Destroyed* ix. 59 Her eyes were walled in panic, flaming with hangover pain.

3. *Electr.* (See quots.)

1940 *Chambers's Techn. Dict.* 402/1 *Hang-over*, the delay in restoration of speech-operated switches, as in the *Vodas*, to ensure the non-clipping of weak final consonants of words. **1943** *Gloss. Terms Telecommun.* (*B.S.I.*) 12 *Hangover time*, of an echo-suppressor, the interval of time that elapses between the instant when the operating signal ceases to be applied at the input terminals of the echo-suppressor and the instant when the suppression loss is reduced to 6 db. **1961** G. A. Briggs *A to Z in Audio* 95 A perfect loudspeaker would cease to vibrate immediately any applied signal is cut off. Failure to do so is mainly due to resonance, and the unwanted output is sometimes referred to by the unpleasant word hangover. Its effect is to colour the reproduction and spoil the transient response. The worst offender is often the cabinet. **1967** W. E. Pannett *Dict. Radio & Telev.* 125 *Hangover*, lack of 'attack' and extended decay in sound reproduction. It is most apparent with transients and is usually due to a resonance or insufficient damping in the system.

Hence **ha·ng-o:verish** *a.*, somewhat affected by a hang-over.

1936 'P. Quentin' *Puzzle for Fools* viii. 62, I felt a bit hang-overish, but that was nothing new.

hangul[1] (hɒ·ŋgŭl). Also **hungal.** [Kashmiri *hăngul.*] A deer, *Cervus cashmiriensis*, related to and perhaps a race of the red deer.

1858 A. L. Adams in *Proc. Zool. Soc.* XXVI. 529 *Cervus cashmeriensis*..*Barra Singa* and *Hanglu* of the Cashmerees. **1869** A. A. A. Kinloch *Large Game Shooting* 44 Cashmeerie hangul. **1898** R. Lydekker *Deer of all Lands* 83 The Hangul—*Cervus cashmirianus.* **1922** *Blackw. Mag.* Mar. 334/1 The hungal or Kashmir stag, found on the western side of Chamba. **1955** I. T. Sanderson *Living Mammals of World* 251/1 The real Red Deer...include..the Hangul of Kashmir. **1973** *Times* 20 Feb. (India Suppl.) p. xi/3 The lordly Hangul, the most magnificent of deer, also dwell here.

‖**hangul**[2] (ha·n,gul). Also **hankul.** [Korean, f. *Han* Korea + *kul* script, alphabet.] The Korean national phonetic alphabet (formerly called *Onmun).

1951 C. Osgood *Koreans & their Culture* xvi. 323 Books were printed in the native alphabet, önmun (or hangŭl), with the innovation of having the characters run horizontally from left to right instead of vertically. **1953** D. Portway *Korea* vii. 122 Sino-Korean words can be transcribed into hangul. **1966** S. McCune *Korea* xiii. 182 One of the benefits of the Korean phonetic writing system, *hangul*..has been that it could be quickly learned by those who speak Korean. **1972** P. M. Bartz *S. Korea* iv. 39/1 In its modern form, *Hangul* consists of 24 phonetic symbols and is considered one of the most ingenious writing systems ever devised. **1972** *Computers & Humanities* VI. 264 The entries are transcriptions of the Korean syllabary (hangul) used to annotate those Chinese characters.

‖**haniwa** (ha·niwa). [Jap.] A clay image or cylinder of a type anciently placed outside Japanese sepulchres.

1931 G. B. Sansom *Japan* I. i. 7 Outside the mounds, but evidently associated with them, are found clay figures (known as *haniwa*). **1960** B. Leach *Potter in Japan* vi. 136 Haniwa figures from A.D. 600. **1970** *Oxf. Compan. Art* 607/1 'Tomb figures' or *haniwa*, clay cylinders some of which were decorated with human or other figures. **1972** *Mainichi Daily News* (Japan) 7 Nov. 5/5 Shards of cylindrical Haniwa had been found at the site.

hanjee, var. Khanjee.

1920 *Cornhill Mag.* Oct. 438 The hanjee was taking down his shutters.

hank, *sb.* Add: **2.** (Later examples.)

1880 *Harper's Mag.* Sept. 534/2 The ceilings [were] hung with hanks of blue yarn. **1888** *Century Mag.* XXXVI.

768/2 These little silken 'hanks' were sometimes..prettily colored. **1957** *Vogue Knitting Bk.* L. 70/2 (Advt.), 3, 4 & 2-ply Super Botany Wool. oz. hanks 1/5. **1966** *Which?* Feb. 53/2 Most yarns are sold in balls nowadays; we tested only nine still sold in the hank.

4. c. *Wrestling.* In the Cumberland and Westmorland style, a throw made by putting the left leg between the legs of an opponent, catching his left leg, and leaning or pulling backwards. Also *back-hank.*

1870 W. Armstrong *Wrestliana* 44 Robinson lifted him up like a cat lifting a mouse, when, Plaskett immediately put in the hank. **1888** *Encycl. Brit.* XXIV. 690/2 Each man tries to throw his adversary by using the 'buttock',..the 'crossbuttock',..or the 'back-hank'. **1898** *Encycl. Sport* II. 547/2 The hank, when manipulated by an expert wrestler, becomes one of the hardest and most dangerous falls of all.

hank, *v.* Add: **1. b.** *Wrestling.* To throw (an opponent) by means of the hank (see *Hank *sb.* 4 c).

1881 *Sportsman's Year-Bk.* 314 The next fall resulted in favour of Pooley, who hanked his adversary. **1894** *Carlisle Patriot* 13 July 7/4 (Cumbld. Gloss. 1899), J— was hanked, S— trying the inside click.

hankul, var. *Hangul[2].

hanky[1] (hæ·ŋki). Also **handky, hankie.** Nursery and colloquial name for Handkerchief.

1895 J. Davidson *Earl Lavender* iv. 73 They..sighed, and looked up, and the schoolmaster's wife used her handky. **1902** *Westm. Gaz.* 17 Dec. 8/2 Lovely ladies' hankies in dainty lawn. **1924** E. Marsh tr. *La Fontaine's Fables* 52 Every occurrence was referred to her; Whether one lost a hanky or a lover. **1939** [see *Accessorize *v.*]. **1953** [see *fly-whisk* s.v. *Fly *sb.*[1] 11]. **1967** N. Freeling *Strike Out* 133 Janine was snuffling in a silly little hanky.

hanky[2]. = Hanky-panky. *rare.*

1924 Galsworthy *White Monkey* II. iv, On our floor, with Michael outside the door, one would know there couldn't be any hanky.

Hannibal (hæ·nibăl). The name of the famous Carthaginian general, who fought against Rome in the third century B.C. Hence, *allusively*, a great general. (Also, in Shakes., humorously confounded with Cannibal.) †*Hannibal eye*, a blind eye.

1585 T. Washington tr. *Nicholay's Navig.* Ded. ¶ iii b, Were it not that I feare the censure of some politike Hanniball. **1603** Shakes. *Meas. for M.* II. i. 187 Proue this, thou wicked Hanniball. *a* **1652** R. Brome *New Acad.* III. ii. in *Wks.* (1873) II. 58, I passe For a brisk youth, but for my Hannibal eye here.

Hence **Hannibalian** (hænibē·liăn), **Hannibalic** (-bæ·lik) *adjs.*, of, pertaining to, or characteristic of Hannibal.

1678 J. D. (*title*) The History of Appian..In Two Parts. The First consisting of the Punick..and Hannibalick Wars. **1862** W. P. Dickson tr. *Mommsen's Hist. Rome* II. III. vi. 189 Thus ended the second Punic, or as the Romans more correctly called it, the Hannibalic, war. **1880** *Encycl. Brit.* XI. 444/2 In the year 202 B.C. the Second Punic, or, more properly, the Hannibalian War was at an end. **1886** H. F. Lester *Under two Fig Trees* 135 And baby had registered a Hannibalic vow. **1934** A. Toynbee *Study Hist.* II. v. 162 The advance of the Roman frontier..was a direct..consequence of the Hannibalic War itself.

Hanover. Add: **2.** Phr. *what the Hanover*, an expression of irritation or impatience. *dial.* or *colloq.*

1902 *Eng. Dial. Dict.* III. 56/2 What the Hanover do I care about it? **1914** D. H. Lawrence *Prussian Officer* 223 'What the Hanover's got you?' asked Whiston. 'Nothing. Can't I get up?' **1915** —— *Rainbow* i. 27 He went home..wondering What the Hanover!

Hansen (hæ·nsən). *Med.* The name of G. H. A. *Hansen* (1841–1912), Norwegian physician, used *attrib.* and in the possessive to designate *Mycobacterium leprae*, the causative agent of leprosy, which he discovered, and occas. leprosy itself, as *Hansen('s) bacillus, Hansen's disease.*

1903 *Jrnl. Path. & Bacteriol.* VIII. 260 These bacilli had some resemblance in form to Hansen's bacilli. **1914** *Jrnl. R. Microsc. Soc.* 569 (*heading*) Behaviour of the Hansen bacillus in vitro. **1938** Dorland *Med. Dict.* (ed. 18) 430/1 *Hansen's disease*, leprosy. **1947** *Amer. Jrnl. Public Health* Mar. 313 (*title*) Education in Hansen's disease. **1955** *Sci. News Let.* 21 May 322/3 Changes in the proteins in the blood of patients with Hansen's disease (leprosy) were reported. **1970** A. W. Woodruff *Alimentary & Haematol. Aspects Trop. Dis.* i. 3 Hansen's bacilli will readily be found in the second [disease].

hant, ha'nt. Revived as a variant of Haunt *sb.* 5.

1933 M. Emmons in B. A. Botkin *Treas. S. Folklore* (1949) III. ii. 540 One never knows when the most sociable of cats may turn out to be a witch or a 'ha'nt'. **1934** B. A. Botkin in W. T. Couch *Culture in South* xxvi. 589 A Bible or a sharp object under the pillow will keep away both 'hants' and witches. **1935** *Scribner's Mag.* XCVII. 121/2 Old Joe's daid an' gone But his hant blows de hawn. **1943** W. C. Hendricks *Bundle of Troubles* 98 Then the wife told the hant who her husband is, and the hant begun at the start and told it all over agin. **1965** 'Malcolm X' *Autobiogr.* i. 20 It was spooky, with ghosts and spirituals and 'ha'nts' seeming to be in the very atmosphere when finally we all came out of the church.

‖**hantu** (hæ·ntu). [Malay.] An evil spirit, a ghost.

1821 J. Leyden tr. *Malay Annals* vi. 54 He saw a hantu, or spectre. **1839** T. J. Newbold *Straits of Malacca* II. xii. 191 The Hantu Ribut is the storm fiend that howls in the blast. **1900** W. W. Skeat *Malay Magic* iv. 101 *Hantu* and *sheitan* are generic terms for evil spirits, the former being the Malay term. *Ibid.* 103 The Hantu Kubor (Grave Demons) are the spirits of the dead, who are believed to prey upon the living. **1927** H. M. Tomlinson *Gallions Reach* xxx. 239 They tell me this land is full of hantus, things that ought not to be about; souls now stowed safely away in Gehenna. **1959** *Listener* 8 Oct. 579/3 The spirits, or *hantus*, as they were called in Malay, were not always of human beings. **1962** *Times* 30 June 10/6 There was the *hantu* that brings the lightning.

Hanukkah, var. *Chanukah, Chanukkah.

hanum: see *Khanum.

Hanuman (hɒnumā·n). Also **hoonoomaun, huniman,** etc. [Hind., Hindi *hanumān* (Skr. *hanumant*, f. *hanumat* large-jawed).] **1.** *Hindu Mythology.* Proper name of a monkey-chief; a semi-divine monkey-like creature, to whom extraordinary powers were attributed in legend.

1814 Southey in J. W. Robberds *Mem. W. Taylor* (1843) II. 427 For the last ten years..Buonaparte..was the God Hanuman—the monkeys, whom he commanded, did the mischief. **1883** *Trans. Asiatic Soc. Japan* XI. 270 For greater safety Rāmachandra..stationed Hanuman, the monkey-god..to guard the palace. **1886** Yule & Burnell *Hobson-Jobson* s.v. *Lungoor*, The monkey-god Hunimān. **1936** E. G. Boulenger *Apes & Monkeys* ix. 215 The uncanny revenge taken by priests..upon an Englishman, who..insulted the image of Hanuman in a wayside temple. **1965** P. C. Jay in I. DeVore *Primate Behavior* vii. 197 One kind of monkey, the langur, has also had an important part in the traditions and epics of India and is often referred to in its role as the monkey deity Hanuman.

2. (with small initial) An Indian monkey, *Presbytis entellus*, venerated by Hindus.

1843 E. Blyth in *Jrnl. Asiatic Soc. Bengal* XII. 174 The Hoonumans are strictly protected. *Ibid.*, I know of one locality where the whole numerous community of Bengal Hoonumans appears to consist of males only. **1867** T. Hutton in *Proc. Zool. Soc.* 944 The particular species of Monkey to which the name of Hoonoomaun now more especially and properly applies is known to naturalists as the *Semnopithecus entellus.* **1891** J. L. Kipling *Beast & Man in India* iii. 65 Of late years the tradesmen who form the bulk of the members of our municipalities have felt that there are too many Hanumāns abroad. **1897** *Q. Rev.* Oct. 395 No visitor to Hindostan..can have failed to see the sacred Monkey or Hanuman. **1936** E. G. Boulenger *Apes & Monkeys* ix. 215 The sacred langur, or hanuman..ranges over India from the Deccan northward to the south bank of the Ganges. In its *rôle* of a sacred animal, dedicated to the god Hanuman, it is unique.

Hanunóo (hā·nūnōᵘ). [Native name.] A member of a people inhabiting southern Mindoro in the Philippines; also, the language of this people.

1949 *Amer. Anthropologist* LI. 269 The Hanunóo raise their own cotton and indigo, and weave and dye skirts, blankets, and other fabrics with considerable skill. *Ibid.*, While visiting in dingy Hanunóo huts, large piles of bamboo poles, four to six feet long, were found. *Ibid.* 271 The ambáhan 'literary' language is either an archaic form of Hanunóo, or a rarer type known as *binúkid.* **1963** J. Lyons *Structural Semantics* iii. 38 Hanunóo color categories.

haole (hau·li, ha·ōli). *Hawaiian.* [Native word.] One who is not a native Hawaiian; a white man. Also *attrib.* or as *adj.*

[**1825** W. Ellis *Jrnl. Tour Hawaii* vii. 151 We had escaped, only because we were *haore*, (foreigners.) No Hawaiian..would have done so with impunity.] **1843** J. J. Jarves *Scenes Sandwich Islands* iii. 104 One brings vegetables, another fish..in short, any thing and every thing which they suppose the *haole*, (foreigner,) to want. **1866** 'Mark Twain' *Lett. fr. Hawaii* (1967) 161 But the thing was tabu..to foreigners—haoles. *Ibid.* 202 To the natives all whites are haoles—how-ries—that is, strangers, or more properly, foreigners. **1905** *Daily Chron.* 24 June 3/1 Stevenson 'fell in love' with the Polynesians,..and was consequently unjust to the *haoles*, or white people. *c* **1938** L. Mumford *Report on Honolulu* in *City Development* (1946) 75 A lasting link between their ancient ways and the less primitive life lived by the various haole groups that have followed. **1954** *Ellery Queen's Myst. Mag.* Oct. 8/2 The *haole*—white—characters are fiction. The Hawaiians..are authentic. **1970** *Language* XLVI. 981 Ever-increasing pressure for the use of English from 'haoles' (main-land whites) and others.

haoma (hōᵘ·mă). [Zend.] = Hom.

1890 *Ann. Rep. Smithsonian Inst.* 91 On the position of the Haoma in the Avesta of the Parsees. **1953** *Trans. Philol. Soc.* 22 Old Ind. *durósa-*, Avestan *dūraoša-*, the epithet of soma and haoma.

haori (hā·ori). [Jap.] A short loose coat worn in Japan.

1877 *Trans. Asiatic Soc. Japan* V. I. 8 A *haöri*,—the upper mantle worn by the military class. **1880** F. V. Dickins tr. *Chiushingura* (new ed.) iii. 16 Badge or device on the sleeves and back of the *haori* or mantle. **1896** L. Hearn *Kokoro* vi. 94 Haori, a sort of upper dress, worn by men as well as women. **1897** J. La Farge *Artist's Lett. from Japan* 274 Women under their umbrellas wore the graceful short overcoat they call *haori*, and tottered over the wet ground on high wooden pattens. **1907** *Daily Chron.* 15 May 3/5 The little ornament on his haori (the gown) was the family crest. **1922** J. Street *Mysterious Japan* ii. 19 One or two of them wore the graceful and dignified *hakama* and *haori*—the silk skirt and coat of formal native dress. **1970** J. Kirkup *Japan behind Fan* 125 There are even some small garments, jackets called *haori*, made from decorated paper. **1972** *National Geographic* CXLI. 692/2 In a quiet way several of the men were equally impressive in somber-hued kimonos complete with *haori*, the elegant outer jacket of dark silk.

‖**hapax legomenon** (hæ·pæks lėgɒ·mĕnɒn). Pl. **hapax legomena.** Also simply **hapax.** [Gr. ἅπαξ λεγόμενον (thing) once said.] A word or form of which only one instance is recorded in a literature or an author.

[**1654** J. Trapp *Minor Prophets* 605 'Tis ἅπαξ λεγόμενον read only here: and hence this variety of interpretations. **1801** W. Magee *Atonement & Sacrifice* 336 [The book of Job's] very great antiquity, and uncommon sublimity of elevation, which has occasioned a greater number of ἅπαξ λεγόμενα, and expressions difficult to be understood.] **1882** Farrar *Early Chr.* I. xi. 236 The number of the *hapax legomena* is remarkable, and some of them are full of picturesqueness. **1931** *English Studies* XIII. 124 An article that should certainly find a place in a miscellany in honour of the brave defender of Wulfila as a translator: Collitz on two hapax legomena in Wulfila's translation. **1956** J. Whatmough *Poetic, Scientific & other Forms of Discourse* ii. 37 The *hapax legomenon*, although statistically it hardly differs from a word of very low occurrence..is nevertheless anomalous, just like the scazon in Greek comedy. **1957** C. Brooke-Rose *Langs. Love* iv. 34 She saw herself go through the minutiæ of scansion, dialect forms, emendation, haplography, *hapax legomena* and *anacolutha* in Beowulf. **1962** *Amer. Speech* XXXVII. 54 He..rejects a Middle English *hapax* as a genuine idiom unless he can trace it back to an Old Norse..word.

ha'pence, var. *halfpence* s.v. Halfpenny.

ha'penny, hapenny, varr. Halfpenny.

haphazard, *sb.*, *a.*, and *adv.* Add: Hence **hapha·zardry.**

1932 V. Woolf *Common Reader* 2nd Ser. 63 But with all this haphazardry, the *Letters*..provide their own continuity. **1949** *Scrutiny* Sept. 196 *Antony and Cleopatra* has none of the haphazardries of *Pericles.* **1959** *Times Lit. Suppl.* 7 Aug. 459/1 A tape-recorder..that can reproduce in all their haphazardry the jumbled rhythms of modern conversation.

haphtarah. Add: Also **haftara, haphtara, hapht(h)orah.** Pl. **-rot(h).** (Further examples.)

1891 M. Friedländer *Jewish Relig.* II. 347 The lesson from the Prophets is called *haphtarah*, 'conclusion'. **1907** I. Zangwill *Ghetto Comedies* 141 You shall read the *Haphtorah* (prophetic section) next *Shabbos.* **1932** C. Roth *Hist. Marranos* xiii. 326 A translation of the Bible..contained a list of the *Haftarot* (Prophetical lessons). **1973** *Jewish Chron.* 2 Feb. 23/4 During the reading I sat with anticipation awaiting his rendering of the haftara.

haplo-. Add: **ha·plophase** *Biol.*, the phase in the life-cycle of an organism when the nuclei are haploid; **ha·plopore** *Zool.*, an isolated pore on the surface of the theca of certain cystoids (order Diploporita) in which the pores usu. occur in pairs; also, a thecal canal that ends in one of these pores.

1925 E. B. Wilson *Cell* (ed. 3) 1132 *Haplophase*, that phase of the life-history, particularly in the antithetic alternation of generations in plants, in which the nuclei are haploid. **1957** Fischer & Holton *Biol. & Control Smut Fungi* vii. 245 The point at which the haploid nuclei are reunited in conjugate association determines the duration of the haplophase. **1899** *Rep. Brit. Assoc. 1898* 917 The simple or irregular haplopores become connected in pairs (diplopores). **1962** D. Nichols *Echinoderms* xi. 138 The diploporite *Aristocystites*..is flask-shaped, with a theca composed of many irregularly arranged plates pierced by haplopores and diplopores.

b. *Genetics.* Used as a prefix to designate the presence of only one of a pair of homologous chromosomes.

1924 T. H. Morgan in E. V. Cowdray *Gen. Cytol.* xi. 720 The offspring are of two kinds, one kind normal with two chromosome IV's, the other kind with only one IV. These haplo-IV flies also give an interesting result if crossed out to a stock that carries a recessive factor in chromosome IV. **1932** Sinnott & Dunn *Princ. Genet.* (ed. 2) ix. 185 One whole IV chromosome is missing from the 'Diminished' flies; they have only one of the pair of small chromosomes and are thus known as haplo-IV individuals.

haploid (hæ·ploid), *a.* (and *sb.*). *Biol.* [a. G. *haploid* (E. Strasburger 1905, in *Jahrb. f. wissensch. Bot.* XLII. 62), f. Gr. ἁπλό-ος single: see *-ploid.] Having a single set of unpaired chromosomes, as in a gamete or germ-cell; made up of cells the nuclei of which contain such a set of chromosomes. (Distinguished from *DIPLOID, etc.) Also as *sb.*, a haploid individual. Hence **haploidy**, the state or condition of being haploid.

1908 W. H. Lang tr. *Strasburger's Text-Bk. Bot.* (ed. 3) I. i. 165 The organism with the single number of chromosomes may be termed the haploid, or haplophase, that with the double number the diploid, or diplophase generation. 1924 G. N. Calkins *Biol.* ix. 209 The chromatin of the nucleus collects in a thick fibrous mass on one side of the nucleus (synapsis stage) and from it emerge one-half as many chromosomes as are formed at ordinary vegetative divisions (in modern terminology this is called the haploid number). 1922 *Amer. Naturalist* LVI. 57 There is one striking difference between haploidy for X and haploidy for an autosome. 1925 *Nature* 10 Oct. 537/2 (*heading*) Haploidy in the male sawfly. 1926 T. H. Morgan *Theory Gene* x. 139 A cell with one set of chromosomes is said to be haploid, and an individual made up of such cells is..called, by extension, a haploid. 1960 *Nature* 17 Aug. 239/2 Since the basic haploid number of *Artemia salina* is known to be 21, the present race must be considered a decaploid with a slight augmentation of the number of 105 tetrads. 1957 C. P. Swanson *Cytol. & Cytogenetics* vii. 159 In plants, where haploidy has been more commonly observed as an abnormality than in animals, the haploid individuals can generally be characterized as being smaller than their diploid progenitors. 1968 J. A. Serra *Mod. Genetics* III. xx. 150 The course of meiosis is expected to be different in haploids derived from diploids than it is in haploids derived from polyploids.

haplont (hæ·plǫnt). *Biol.* [a. G. *haplont*, f. Haplo- + Gr. ὤν, ὀντ- being: see Onto-.] A sexual organism that is haploid at all stages of its life other than the zygote, which is diploid; an organism at a stage, or during the stages, in its life cycle at which it is haploid. So **haplo·ntic** *a.*, characteristic of, or having the characteristics of, a haplont.

1920 *Bot. Abstr.* V. 214 (*heading*) Biology and morphology of the male haplonts of some Oenotheras. 1925 E. B. Wilson *Cell* (ed. 3) vi. 492 The spores..receive the haploid number of chromosomes and develop without fertilization into a haploid, gamete-producing 'sexual' generation, known as the haplont (in plants the gametophyte) which intervenes between meiosis and the gamete-formation. 1929 *Hereditas* XIII. 311 I propose the terms haplontic and diplontic sterility. 1938 *Bot. Rev.* IV. 235 It is more probable that the above-described diplobiontic Ulvaceae and Cladophoraceae arose from a haplontic ancestor than from a primitive ancestral type with alternating unicellular haploid and diploid phases. 1951 M. O. P. Iyengar in G. M. Smith *Man. Phycol.* iii. 59 *Protosiphon* is another member of the Siphonales which is a haplont and not a diplont like the majority of the Siphonales. 1964 Priestley & Scott *Introd. Bot.* (ed. 4) xxiii. 499 If this is confirmed the nuclei in the *Vaucheria* thallus would be haploid and the life-cycle haplontic with the zygote (and zygospore) the only cell representing the diploid stage.

happen, *v.* Add: **1. a.** Said ominously of an accident or some serious thing (*spec.* death) happening to a person, with vague subject, *anything, something*.

1795 H. Nelson *Let.* 10 Mar. (1945) III. 18 A glorious death is to be envied; and if anything happens to me, recollect that death is a debt we must all pay. 1811 Princess Charlotte *Let.* 21 Oct. (1949) 9, I am going again on Monday at 1, unless anything should happen between this time & that. 1829 *Blackw. Mag.* June 715/1 In the event of 'any thing happening to his father', as the modern phrase for the termination of man's mortal career runs. 1862 *Cornhill Mag.* (1863) VIII. 574 Doctor says I shall not last long, so I don't think I shall be removed before anything happens. 1884 G. C. Davies *Peter Penniless* x, It isn't a night for any man to be left out in it anything has happened to him. 1886 Rider Haggard *K. Solomon's Mines* ii, I will..arrange that in the event of anything happening to me or to you, that your son shall be suitably provided for. 1965 N. Dunn *Talking to Women* 40 Lots of people they plan and they put this away for when they get old and that type of thing and then anything happens and who has it, their kids, don't they?

3. b. Used with varying degrees of intensity to support or imply an assertion, const. *inf.* Also used impersonally, in which case it is sometimes followed by a subordinate clause.

1933 F. Baldwin *Innocent Bystander* (1935) v. 95 She happens to be my only sister's child and I have an interest in her. 1937 S. Lewis in *Colophon* Feb. 240 *Main Street*, which is always put down as my first book, happens to have been my seventh. 1956 N. Coward *South Sea Bubble* ii. i. 65 You have got it in for her, haven't you? C. Certainly not. I just don't happen to like the way she goes on. 1957 B. & C. Evans *Dict. Contemp. Amer. Usage* 217/1 We happen to like her. *Ibid.*, It happens we like her. 1967 N. Innes *Appleby's Answer* vii. 74 You don't happen to have any cigarettes?

4. d. Also *happen along, around, back, by, over*. (Earlier and other U.S. examples.)

1749 in G. O. Seilhamer *Hist. Amer. Theatre* (1888) I. 29 Joseph Morris and I happened in at Peacock Bigger's and drank tea there. 1838 J. F. Cooper *Homeward Bound* viii. 112, I only happened in..to make a first call. 1845 C. M. Kirkland *Western Clearings* 116 He could hardly have 'happened in' at a more fortunate juncture. 1872 E. Eggleston *Hoosier Schoolmaster* xxxiii, Miss Nancy just happened over at Mrs. Thomson's humble home. 1882 A. Percy *Twice Outlawed* 101 Under-sheriff Knight, of Pepin county, happened along that way. 1893 R. D. Wiggin *Polly Oliver's Problem* (1894) ii. 20 A swarm of horrid insects might happen along and devour the plants. 1901 W. Churchill *Crisis* iii. ii. 366, I happened around at Colonel Carvel's this afternoon. 1930 R. Frost *Coll. Poems* 67, I go nowhere on purpose: I happen by. 1931 D. Runyon *Guys & Dolls* (1932) 261 There we are away over by the East River in the early morning, with no other taxis in sight, and a cop liable to happen along any minute. 1953 H. Miller *Plexus* (1963) viii. 228 We were literally without a cent when he happened along. 1970 C. F. Newman *Sir, You Bastard* viii. 146 [illegible] and to any of the girls who happened in before she fell asleep she explained that she was taking forty winks.

6. *intr.* To be successful; esp. in *it's all happening*, there is much activity or success. *slang.*

1949 A. Shaw *Vocab. Tin-Pan Alley* in *Music Libr. Assoc. Notes* Dec. 44/1 A song happens..when the preparatory work results in a successful bid for popularity. 1962 *Down Beat* 8 Nov. 38 It wouldn't [illegible] they were all striving to create..but it didn't really happen. 1966 *Crescendo* Mar. 2/3 'It's all happening' was one of the more tiresome items in the pseudo-hip phrase books of the recent past. It seldom meant much while it was in fashion, but now there has arisen a situation which it describes exactly. For it really is all happening for Stan Tracey this year. 1967 *Ibid.* Feb. 19/1 It's all been happening in the Village Vanguard lately. *Ibid.* Dec. 27/4 We could either go back to New York with a flop show, or try to stay over and make a band happen. 1971 *Melody Maker* 9 Oct. 18/5 The guitar solo didn't happen.

happenchance: see *HAPPENSTANCE.

happening, *vbl. sb.* Add: **2.** Also in *sing.*

1836 D. Crockett [illegible] xxviii, I could not find it in my heart to tell him of the happening. 1896 *Black & White* 27 June 824/2 [illegible] 1899 *Daily News* 5 Apr. 12 Such a happening would almost certainly have had much more serious results had it been a horsed carriage. 1907 *Westm. Gaz.* 30 Dec. 2/2 The Denshawi incident is not viewed by all Egyptians as [illegible]. 1970 E. E. Cummings *Let.* 12 Apr. (1969) 75 [illegible]

3. a. An improvised or spontaneous theatrical or pseudo-theatrical entertainment. Also extended use, any spontaneous or 'vital' display. *orig. U.S.*

1959 *Nation* (N.Y.) 24 Oct. 260/2 The first exhibition is not of painting but is an 'event' consisting of eighteen 'happenings' by Allan Kaprow. 1962 *Listener* 5 Apr. 604/1 It was with these 'happenings' that these artists first reached the public. 'Happenings', usually staged in downtown lofts, were performances [illegible]. 1963 *Guardian* 1 Mar. 8/7 The latest form of 'way out' entertainment in Washington, DC, is a 'Happening', a goofy party. 1963 *Observer* 15 Sept. 27/6 [illegible] 'Happening'. 1966 *Ibid.* 17 Apr. 11/4 [illegible] 1966 *Listener* 23 Mar. 339/2 The Japanese city is not a design that has been done badly: it's the negation of design, an urban happening with its own special vitality. 1970 *Daily Tel.* 29 Dec. 10 Tomorrow the 1,600 delegates will see a 'happening' called 'Thank God We're Normal' performed by 70 boys and girls from..comprehensive schools in London.

b. *Art.* (See quot. 1962.)

1962 *Listener* 5 Apr. 604/2 The room-sized collages called happenings or situations are works of art that actually stimulate the environment. *Ibid.* 605/1 In his happenings, he..uses cheap materials and found objects, lifted from the city's waste. 1965 *Times Lit. Suppl.* 25 Nov. 1042/2 Happenings may be seen as the logical extension of the collage principle.

happen-so (hæ·p'n sōᵘ). Chiefly *U.S.* [f. Happen *v.* + So *adv.*] A chance event.

1904 *Westm. Gaz.* 13 Sept. 1/3 Politics..may or may not be mixed up in it. It is just a happen-so if they are. 1938 M. K. Rawlings *Yearling* xiv. 130 Gittin' that buck was pure happen-so. 1941 [illegible] (1963) ix. 62 That Japanese girl—didn't have anything to do with us... It was just a happen so.

happenstance (hæ·p'nstæns). Chiefly *U.S.* [Amalgam of Happen(ing + Circum)stance *sb.*] A chance event; a coincidence. Occas. in altered form **happenchance**. Also *attrib.*

1897 *Outing* (U.S.) XXX. 557/1, I guess it was just a 'happen-stance'. 1911 *Dialect Notes* III. 544 *Happen-chance, happenstance*, happening, circumstance. Used facetiously. Blend-formations. 1937 *John o' London's* 12 Mar. 986/3 The buyer of *Human Psychology*..will be given the opportunity to defile his vocabulary with such terms as 'happenstance'. 1941 *Sat. Even. Post* 22 Mar. 24/3 Even if by happen-chance a hailstorm didn't come along and ruin the crop, there was always something to fight. 1946 M. C. Self *Horseman's Encycl.* 306 The introduction of polo to England was pure happen-chance. 1960 *Listener* 7 Jan. 17/2 They shrewdly refused to attribute this to happenstance. 1963 D. B. Hughes *Expendable Man* (1964) v. 149 The abortionist hadn't killed her, he would have no need of the happenstance tool. 1965 *Observer* (Colour Suppl.) 13 Apr. 27/2 Balloon races have a fixed beginning, all right, but their finishing line is a matter of happenstance. 1966 Ogilvy & Anderson *Excurs. Number Theory* x. 117 Again the last fraction, which we wish to discard, is a [it need not have been—that was just happenstance]. 1969 *Sci. Jrnl.* May 9/3 Berry could say little, beyond suggesting that the use of second sleeping pills by both Schweickart and Frank Borman, who suffered a vomiting attack on the *Apollo 8* flight, was probably a 'happenstance'. 1973 *Tablet* 17 Feb. 155/2 'Falsifiability' is duly recorded—not long after 'fall-out', with which some might think the word has more than happenchance..connection.

happi-coat (hæ·pikōᵘt). Also **happy-**, (ellipt.) **happi**. [Jap. *happi* a kind of coat.] A Japanese loose outer coat of various materials; a similar fashion garment.

1880 *Trans. Asiatic Soc. Japan* VIII. 344 Young men often wore a red Happi with large sleeves. 1890 B. H. Chamberlain *Things Japanese* 94 But jinrikisha-men wear the happi. 1921 [illegible] (Tokyo) XI. Oct. 515 *Happi-coat*, often misspelt 'happy' [illegible]. 1931 R. Crompton *William—the Rebel* vi. 121 [illegible] happi coat, William, she said. 1948 M. Allingham *More Work for Undertaker* (1949) vi. 70 She wore a gay little happi-coat. 1970 J. Kirkup *Japan behind Fan* 127 Here there was a crowd of men in happi coats. 1971 *Daily Tel.* 1 Nov. 11/1 They want to trot around in tuxedo trouser outfits, Japanese happi coats, or pedal-pusher pants. 1971 *Catal. Exhib. C. Beaton's Fashion* (V. & A. Mus.) 54/1 A beige suede 'happi coat'..worn with tapering suede trousers.

happify, *v.* Delete (Now unusual.) and substitute (Now *U.S.*).

1875 M. B. Eddy *Science & Health* vi. 315 To happify existence by constant intercourse with those adapted to elevate it, should be the motive of society. 1905 *Daily Chron.* 13 July 4/4 Her jargon about 'happifying existence' lays upon one [illegible] the hearts of men'. 1945 L. Shelly *Jive Talk Dict.* 12/1 Happify, to make happy. 1950 [illegible] 182/2 Keeping these happifying thoughts in mind, we are determined to maintain our integrity faithfully.

happily, *adv.* Add: **4.** *happily ever after*: see *HAPPY *a.* 7 b.

happy, *a.* Add: **2. c.** *happy land*, a prosperous, favourable, etc., land; *spec.*, heaven.

1787 S. Stennett in J. Rippon *Selection of Hymns* 584, I stand, and cast a wishful Eye, To Canaan's fair and happy Land. 1806 T. G. Fessenden *Democracy Unveiled* (ed. 3) I. 85 Such principles, alas, will flood Columbia's happy land with blood. 1843 C. H. Bateman *Children's Hymn-Bk.* 36 There is a happy land Far far away. 1853 [illegible] The old 'un will soon join the young 'un in the happy land. 1902 *Daily Chron.* 6 Feb. 5/2 During the great..strike..a rhyme went round beginning 'There is a happy land, Far, far away [etc.]'. 1948 [illegible] (*title*) Happy land. 1959 I. & P. Opie *Lore & Lang. Schoolch.* xvii. 365 There is a happy land Down in Duddley school Where Miss Macdonald stands, preaching like a fool.

3. (Earlier examples of *many happy returns.*) Also in certain familiar or conventional special collocations: *happy day*, wedding day; *happy days!*, a drinking toast; similarly, esp. in aviation circles, *happy landings!*; *happy ending*, an ending in a novel, play, etc., in which the characters acquire spouses, money, do not die, etc.; *happy event*, the birth of a baby; *happy pair*, an engaged or newly wedded couple; *happy release*, (esp.) death.

1697 Dryden *Alexander's Feast* 1 The Lovely Thais by his side, Sate like a blooming Eastern Bride. Happy, happy, happy Pair! 1702 C. Sedley (*title*) The happy pair. [illegible] Richardson *Pamela* (1740) II. 151 [illegible] 1789 G. Parker *Life's Painter* xiv. 115 (*title*) The happy pair. 1796 Lady Newdigate *Let.* 2 June in A. E. Newdigate-Newdegate *Cheverels* (1898) vi. 84 Many happy returns of the day to us De Love. [illegible] 1838 Dickens *Nickleby* (1839) xiv. 124 Many happy returns of the day, my dear. 1848 Mrs. Gaskell *Mary Barton* I. xi. 209 [illegible] 1850 Thackeray *Pendennis* II. xxxvii. 357 [illegible] 1856 Dickens in *Household Words* 19 Oct. 74/1 [illegible] happy release in every point of view. 1864 —— in *All Year Round* [illegible] 1913 G. B. Shaw *Quintessence of Ibsenism* (rev. ed.) 192 The substitution of a sentimental happy ending for the famous last scene. 1929 E. Bowen *Joining Charles* 122 [illegible] happy release. 1934 *Evening News* 25 July 4/5 Ronnie swallowed half the whisky.. 'Happy landings, Phyllis..dear!'.. The powder left his fingers, missed the glass. 1934 R. S. Lambert *For Filmgoers Only* 68 'Happy endings' are in much greater evidence on the screen than, for instance, in the play. 1934 E. Wharton *Backward Glance* vii. 147 The American public always wants..a tragedy with a happy ending. 1935 G. Greene *Basement Room* 106 'Your health, my dear. You look younger than ever.' 'Happy days,' Amy said. 1938 —— *Brighton Rock* v. v. 213 'When's the happy day?' Cubitt said and they all smiled. 1940 Partridge *Dict. Clichés* 100 *Happy event, a* or *the*, the birth of a child; esp. the first in a family: mostly lower-middle class: from ca. 1880. 1946 T. Rattigan *Winslow Boy* i. 29 Happy pair, I think, is the phrase that is eluding you. 1951 J. B. Priestley *Festival at Farbridge* II. i. 199 'Happy days!' cried Mobbs. 'Cheers!' said the Major gloomily. 1953 P. Frankau *Winged Horse* III. ii. 199 The glass lifted. 'Happy Landings,' Carey said. 1957 N. Frye *Anat. Criticism* 104 Most students of literature prefer to keep in the middle distance..run-of-the-mill Elizabethan sonnets and lyrics,..nineteenth-century happy-ending novels. 1960 [illegible] 'happy events' that occurred in 1959. 1963 [illegible] 1966 T. Walsh *Face of Enemy* [illegible] 'Happy days, old boy.' 1969 *Times* 20 Mar. 16/2 Aunt Juju, in her harping upon 'happy events'..knows more about life and death than Hedda.

4. b. Freq. with neg., as *not (at all), not entirely, not quite happy about* (or *with*), usually indicating substantial dissatisfaction. Cf. Not *adv.* 10 b.

1947 *People* 22 June 7/5 The receiving club were not at all happy about this. 1967 N. Freeling *Strike Out* [illegible] I dropped a monstrous clanger, letting anybody see I wasn't happy, but..I'm still not happy. 1971 *Guardian* 5 Dec. 11/2 She says, with some delicacy, that the studio, Paramount, was 'not happy with it' and failed to promote it.

c. *happy family*: (*a*) a conventional description of a harmonious family; also *fig.*; (*b*) (see Family *sb.* 2 b); (*c*) *Austral.*, a popular name of the grey-crowned babbler (*Struthidea cinerea*): also called *happy jack*.

1868 F. Harrison *Let.* 11 Nov. in *Geo. Eliot's Lett.* (1955) IV. 484 [illegible] 1901 [see *Apostle III]. 1927 T. E. Lawrence *Lett.* (1938) 539 The happy family is the squadron or flight, and the misery of discipline..is resident in depots and workshops. 1932 *Week-end Rev.* 30 July 119/1 [illegible] 'em like one big, happy family. 1939 *Times Lit. Suppl.* 14 Jan. 27/1 In spite of the title she has given to her book, Miss Stevenson's heroine comes from anything but a happy family. 1945 Baker *Austral. Lang.* xii. 211 The Grey-crowned Babbler is known..as the..happy family, happy jack. 1946 [illegible] 1958 N. W. Cayley *What Bird is That?* (ed. 2) 69 Also called Grey Jumper, Happy Family, Apostle-bird. 1965 *Austral. Encycl.* I. 385/1 Babblers..commonly known as catbirds (or cat-babies), chatterers, happy families, and apostle-birds.

d. *happy families*: a game played with a pack of special cards, each card depicting on its face a member of a tradesman's family of four; it is the aim of each player to make as many complete families as he can.

1881 *Cassell's Bk. In-Door Amusem.* 142 The well-known game of Happy Families is nothing but a variation of Spade the Gardener. 1928 C. Dane *First Blade* xxv. [illegible] 1954 E. Hyams *Stories & Cream* 50 'Play cards, pa!?'..The other man said, in a rather nasty way, 'Ah, 'appy families, I suppose?' 1955 J. Lehmann *Whispering Gallery* iv. 225 We were like a pack of cards for Happy Families.

5. a. *spec.* *happy warrior*, applied conventionally to an excellent soldier; also *fig.*

1806 [see Warrior II. 2]. 1915 D. O. Barnett *Let.* 1 July 200 The 'happy warrior' who did the deed is in my platoon, one Finlay, and his hair is red. 1924 F. D. Roosevelt in *N.Y. Times* 27 June 4/1 He [*sc.* Alfred E. Smith] is the 'Happy Warrior' of the political battlefield. 1959 *Listener* 12 Nov. 843/3 Ernest Jones was a happy warrior.

c. *happy medium* = *golden mean* (Golden *a.* 5 c).

1778 *English Mag.* Feb. 59/1 All extremes are ridiculous: the happy medium is to be aimed at. 1784 J. Priestley *Hist. Corrupt. Chr.* [illegible] ..he hit the happy medium between the popish doctrine of merit [illegible] of good works. 1901 Ade *Forty Modern Fables* 51 Moral: only one in a thousand ever strikes the happy medium. 1920 *Ladies' Home Jrnl.* Oct. 163/2 There is a happy medium. 1947 K. Tennant *Lost Haven* (1968) x. 164 Ain't there no bloody happy medium?

d. Exhibiting harmony or co-operation, esp. *happy ship*, a ship on which the crew work together harmoniously; also *transf.* of the conduct of any organization.

1905 *Westm. Gaz.* 9 Dec. 16/1 There never was a 'happy ship', and from captain to cabin-boy all were unhappy together. 1916 'Taffrail' *Pincher Martin* iii. 43 The *Belligerent* was essentially a happy ship. 1929 T. E. Lawrence *Lett.* (1938) [illegible] 1950 W. J. M. Mackenzie in G. F. M. Campion *Brit. Govt. since 1918* [illegible] a branch or a department may be a theoretical monstrosity and yet be a 'happy ship' [illegible]. 1958 *Times* 30 May 5/6 A visit to our Combined Training team recently gave the pleasant impression of a thoroughly 'happy ship' and, without undue confidence, that it will take a very good team indeed to beat them. 1958 P. Kemp *No Colours or Crest* iii. 26 *Fidelity* was not a happy ship. 1958 *Observer* 31 Aug. 19/3 The team was an undeniably happy one with immense spirit.

e. Of drugs: in certain colloquial phrases with the sense 'intended to produce or induce happiness', e.g. *happy dust*, cocaine; *happy pill*, a tranquillizer.

1922 E. Murphy *Black Candle* i. vii. 67 The boxes were found to contain cocaine, or 'happy-dust'. 1929 H. Miles tr. *Morand's Black Magic* i. i. 12 Cocaine (or 'happy-dust', as Congo said). 1937 E. St. V. Millay *Conversation at Midnight* iv. 114 Your head's So full of dope, so full of happy-dust, you're just a drug Addict. 1956 A. Huxley *Let.* 14 Mar. (1969) 791 The present mass consumption of 'Happy Pills', (Miltown-Equanil). 1964 E. Dundy *Old Man & Me* xviii. 176 Those heart-shaped 'happy-pills' of soft musty mauve, pale blue, or apple-green, with that faint incision down their middles. 1966 I. Asimov *Fantastic Voyage* ii. 11 You've got that tranquillizer gleam in your eye, doctor. I don't need any happy pills.

7. *happy-natured, -seeming*.

1920 D. H. Lawrence *Sea & Sardinia* 36 Old wood.. happy-seeming as iron never can be. 1924 M. A. Lowndes *Terriford Myst.* iii. 35 Yet she looked so happy-natured. 1946 E. Sitwell *Fanfare for Elizabeth* xv. 155 Katherine Parr, a happy-natured, placid woman. 1952 S. Spender *Learning Laughter* 103, I never noticed such a happy-seeming family.

b. Used in certain comparative or hyperbolical phrases, e.g. *(as) happy as the day (is long)*; *as happy as Larry* (see *Larry *sb.*[2]). Also, with reference to the happy endings of fairy tales, novels, etc., *happy* (also *happily*) *ever after(wards)*.

1768 Cowper *Let.* 9 Feb. (1904) II. 462 We will be as happy as the day is long. 1823 C. Lamb *Let.* 6 Jan. (1935) II. 361 May your grandsons be half, nay, a quarter as happy as you as the day is long! 1853 C. M. Yonge *Heir of Redclyffe* II. xii. 187 [illegible] 1858 Lytton *What will he Do?* IV. viii. iii. 61 And then they would live happy ever after. [illegible] (1956) IV. 168 He is as happy as the day is long [illegible]. 1873 L. Troubridge *Life amongst Troubridges* (1966) viii. 67 [illegible] 'happy-ever-after' endings [illegible] 1925 R. Hall *Saturday Life* xi. 122 [illegible] 1938 *Times Lit. Suppl.* 15 Jan. 4/3 [illegible] a happily-ever-after note. 1963 *Harper's Bazaar* May 11 [illegible] 1970 *Observer* [illegible] 'happy ever after' endings.

-happy. Used freely during and since the 1939–45 war as the second element in many combinations: **a.** In a dazed, nervous, or light-headed state as a result of excessive strain, e.g. by exposure to bombs (*bomb-happy*), anti-aircraft fire (*flak-happy*), the desert (*sand-happy*), etc. **b.** Acting in an irresponsible, obsessive, or precipitate manner, e.g. *gadget-happy* (= obsessed with the acquisition of gadgets), *trigger-happy* (= liable to shoot at anything at any time). Cf. also *slap-happy*. (Examples are entered under the first elements in this Supplement.)

happy-go-lucky, *adv.*, *a.* (*sb.*) Add: Hence **happy-go-luckiness**.

1893 Yonge & Coleridge *Strolling Players* xxii. 187 Her Irish happy-go-luckiness. 1928 *S.P.E. Tract* xxix. 264 The fertility and happy-go-luckiness of Elizabethan English. 1952 G. Raverat *Period Piece* iii. 40 The casual happy-go-luckiness..which was one of her most attractive qualities.

hapten (hæ·ptĕn). *Immunol.* Also **-ene**. [ad. G. *hapten* (K. Landsteiner 1921, in *Biochem. Zeitschr.* CXIX. 303), f. Gr. ἅπτειν to fasten.] A substance, usu. of low molecular weight, which cannot by itself elicit an antibody, but which can do so when combined with another substance, usu. a protein, the antibody thus produced being capable of reacting either with the free or the combined hapten.

1921 *Chem. Abstr.* XV. 3317 (*heading*) Heterogeneous antigens and haptenes. 1926 *Proc. Soc. Exper. Biol. & Med.* XXIII. 143 [illegible] 1928 *Jrnl. Immunol.* XV. 595 Specifically reacting non-antigenic substances—so-called haptens—play a great part in the constitution of the antigens of animal cells and bacteria. 1931 C. H. Browning in *Syst. Bacteriol.* (Med. Res. Council) VI. 206 These non-protein substances when isolated do not by themselves cause antibody production *in vivo*; therefore Landsteiner has classed them as 'haptens' in contrast to true antigens. 1969 *New Scientist* 12 June 575/1 Low molecular weight chemicals (haptens) which are themselves non-antigenic but become antigenic when conjugated to 'carrier' proteins.

Hence **hapte·nic** *a.*

1932 *Dorland's Med. Dict.* (ed. 16) 616/2 *Haptenic*, pertaining to or caused by haptens. 1966 *Lancet* 24 Dec. 1419/1 The capacity to manifest contact allergy to a haptenic substance is impaired in persons with leprosy.

hapteron (hæ·ptĕrǫn). Pl. **haptera**. [mod.L., badly f. Gr. ἅπτειν to fasten.] An organ of attachment by which certain aquatic plants or algæ fasten themselves to rocks.

1895 M. C. Potter tr. *Warming's Handbk. Syst. Bot.* i. 10 Hairs and organs of attachment (rhizoids and haptera), which biologically serve as roots, are developed. 1909 Groom & Balfour tr. *Warming's Oecol. Plants* lxii. 241 Lithophytes require haptera by which they can attach themselves to rock, unless the thallus itself adheres closely to this. 1967 C. D. Sculthorpe *Biol. Aquat. Vasc. Plants* v. 111 It creeps over the rocks to which it adheres by hairs or by exogenous projections known as haptera, which secrete a cement from their discoid tips.

haptic (hæ·ptik), *a.* (and *sb.*). [ad. Gr. ἁπτικ-ός able to come into contact with, f. ἅπτειν to fasten.] **a.** Of, pertaining to, or relating to the sense of touch or tactile sensations. **b.** Having a greater dependence on sensations of touch than on sight, esp. as a means of psychological orientation. Also *absol.*, a haptic person.

1890 in Billings *Med. Dict.* 1904 *Amer. Jrnl. Relig. Psychol.* May 379 [illegible] tactile or haptic sensations come next. 1939 *Mind* XLVIII. 360 There is the notion of pure 'touch', and there are 'kinaesthetic experiences', and we can have the one without the other; but when we speak of the world of 'touch', or 'tactile aesthetics', we are referring to the data provided by an intimate combination of them both [illegible]. *Ibid.* 364 How does Prof. Révész find out whether the blind have tactile aesthetic experiences? Does he treat haptics seriously in their own right? 1954 *Archit. Rev.* CXVI. 400/3 [illegible] 1964 *Listener* 30 July 156/2 [illegible] haptic types [illegible] visual imagery. 1966 *Publ. Amer. Dial. Soc.* 1964 XLI. 41 A complete record of the segmental, paralinguistic, kinesic, and haptic systems [illegible] 1971 [illegible] Scleral (sometimes called haptic) lenses [illegible] the size of a 5p piece, cover the whole front of the eye.

So **ha·ptical** *a.*, **ha·ptically** *adv.*; **haptics**, *Psychol.* and *Linguistics* [ad. G. *haptik* (M. Dessoir 1892, in *Arch. f. Physiol.* 242)], the study of touch and tactile sensations, esp. as a means of communication.

1895 E. B. Titchener in *Amer. Jrnl. Psychol.* VIII. 82 (*heading*) A psychological vocabulary. ..Haptik, haptics. 1899 *Amer. Jrnl. Psychol.* XI. 25 Haptical images, beside being vague and ill-defined, offer peculiar difficulties. 1904 G. S. Hall *Adolescence* II. ix. 7 Haptics is thus a paleopsychic field par excellence, and the exploration of this most extended of all senses involves a study of the entire dermal area. 1964 *Listener* 30 July 156/2 Those who are..essentially haptically minded—in other words, who primarily orientate themselves by means of touch, and their own bodily feelings and muscular sensations. 1966 *Publ. Amer. Dial. Soc.* 1964 XLII. 41 Haptics is that sub-system of nonlanguage communication which conveys meaning through physical contact. 1972 W. M. Austin in A. L. Davis *Culture, Class, & Lang. Variety* viii. 147 The highly social animals..greet each other haptically by briefly holding each other's muzzles in their mouths.

haptine (hæ·ptin, -īn). *Immunol.* [a. G. *haptine* (Ehrlich & Morgenroth 1900, in *Berlin. klin. Wochenschr.* 30 July 683/2), f. Gr. ἅπτειν to fasten: see -ine[5].] In Ehrlich's theory of immunization, a receptor detached from the parent-cell, circulating freely in the blood-stream, and acting as a protection against infection by combining with the foreign substance which would produce it. Cf. *haptophore *a.* (and *sb.*).

1900 E. F. Bashford tr. P. Ehrlich in *Proc. R. Soc.* LXVI. 448 [illegible] as haptines. 1902 *Brit. Med. Jrnl.* 22 Apr. 1017/1 [illegible] These side chains are spoken of as haptines (Ehrlich). 1931 Hewlett & McIntosh *Man. Bacteriol.* (ed. 9) v. 168 Ehrlich termed the diverse free receptors when liberated into the body fluids in various circumstances 'haptines'.

haptoglobin (hæptoglōᵘ·bin). *Biochem.* [ad. F. *haptoglobine* (Polonovski & Jayle 1940, in *Compt. Rend.* CCXI. 518), f. Gr. ἅπτειν to fasten + Hæmoglobin.] Any of several proteins of the α_2-globulin group that occur in blood serum and are able to combine with free hæmoglobin to form fairly stable complexes.

1941 in *Index Medicus* XXIX. 190/2. 1956 *Nature* 29 Sept. 695/1 Since differences in the haptoglobins of individuals determine their inherited serum groups, we suggest that the system comprising these groups be known as the 'haptoglobin system'. 1961 *Lancet* 23 Sept. 722/1 Normal sera contain three types of haptoglobins, which are now known to occur in three genetically controlled combinations. 1966 E. S. West et al. *Textbk. Biochem.* (ed. 4) xv. 573 Only the haptoglobins of types 1–1 and 2–2 have been shown to be homogeneous compounds... Type 2–1 haptoglobin has been found to be heterogeneous by ultracentrifugation.

haptophore (hæ·ptofōᵊɹ), *a.* (and *sb.*). *Immunol.* Also **-phor.** [a. G. *haptophor* (P. Ehrlich 1898, in *Deutsch. med. Wochenschr.* 22 Sept. 599/2), f. Gr. ἅπτειν to fasten + -o + -phore.] Applied, in Ehrlich's theory of immunization, to that group of atoms in the molecule of a toxin or other substance which enables it to combine with the corresponding receptors of a cell. Also *absol.* So **haptopho·ric**, **hapto·phorous** *adjs.*

1899 tr. P. Ehrlich in *Trans. Jenner Inst. Prevent. Med.* (2nd Ser.) 11, I have already..touched upon the question whether..the toxin molecule contains two independent groups, of which the one (the toxophore) conditions the toxicity, and the other (the haptophore) the combining property. *Ibid.* 12 With the help of the haptophore groups, the toxin molecule becomes 'anchored' to the cell. 1902 Vaughan & Novy *Cellular Toxins* (ed. 4) 182 Both the toxophil groups of the cell, and the cytophil groups of the toxin may be designated as haptophorous bodies. 1902 *Brit. Med. Jrnl.* 29 Mar. 785 The atom arrangement, or group in the toxin, which corresponds to the receptor, he [*sc.* Ehrlich] calls the 'haptophoric group'. 1904 *Ibid.* 10 Sept. 574 Although the toxophoric group may be similar, the haptophor is dissimilar. 1938 W. Bulloch *Hist. Bacteriol.* xi. 275 Of the two, the haptophore is the more stable. Ehrlich considered that the toxophoric atom group can deteriorate to a non-toxic state although the haptophoric group may at the same time remain unchanged. 1960 *New Biol.* XXXI. 104 It was very hard to conceive that 'haptophores' could pre-exist.

haptotropism (hæptotrōᵘ·piz'm, -trǫ·piz'm, hæptǫ·trŏpiz'm). *Bot.* [ad. G. *haptotropismus* (L. Errera 1884, in *Bot. Zeitung* 5 Sept. 564), f. Gr. ἅπτειν to fasten: see Tropism.] The phenomenon whereby plant organs, as the tendrils of climbing plants, exhibit tropic movements in response to the stimulus of touch. Hence **haptotro·pic** *a.*

1892 L. Errera in *Ann. Bot.* VI. 373 Thus, the geotropic, heliotropic, hydrotropic, haptotropic curvatures arise, which are familiar to vegetable physiologists. 1900 B. D. Jackson *Gloss. Bot. Terms* 118/2 *Haptotropism*, curvature induced in climbing plants by the stimulus of a rough surface. 1924 M. Skene *Biol. Flowering Plants* iv. 298 The response to contact stimulus is termed haptotropism. 1934 Webster, Haptotropic. 1953 Fritsch & Salisbury *Plant Form & Funct.* (rev. ed.) xxix. 263 Certain tropic growth-curvatures result from direct contact with a foreign body and are described as haptotropic. 1965 Bell & Coombe tr. *Strasburger's Textbk. Bot.* (new ed.) 375 Many plants..are sensitive to touch... This phenomenon is termed haptotropism (thigmotropism).

hapu (hā·pu). *N.Z.* Also (erron.) **harpu**. [Maori.] A clan, sub-tribe, or small community.

1843 E. Dieffenbach *Trav. N.Z.* II. iii. ix. 361/2 Hapu—tribe, family. 1857 C. F. Hursthouse *New Zealand* I. 162 The 70,000 semi-civilized natives now in New Zealand are divided into some dozen chief tribes, and into numerous sub-tribes and 'Harpu'. 1873 *Jrnls. Ho. Reps. N.Z.* III. App. G. vii. 87 (Morris), Were not all your hapu present when the money was paid? 1891 *Rep. Australas. Assoc. Adv. Sci.* III. G. 378 (Morris), Tribes or nations, each of which was divided into hapus, and the hapus into families. 1921 H. Guthrie-Smith *Tutira* viii. 52 In sympathy with this *hapu* or sub-tribe and its old-world ways. 1949 P. Buck *Coming of Maori* (1950) III. i. 333 To denote the groupings in English, the *iwi* has been termed tribe and the *hapu* a sub-tribe. 1960 N. Hilliard in C. K. Stead *N.Z. Short Stories* (1966) 241 He's East Coast, he don't know the *hapus* up our way.

hapuku, hapuka (hā·puku, -kā). *N.Z.* Also formerly **whapuku**, etc. [Maori *hapuku*.] A large marine food fish, *Polyprion oxygeneios*; = Cod *sb.*[3] 2 b and *groper[2].

1838 J. S. Polack *New Zealand* I. ix. 322 Some deep banks lie off the east coast, on which the *kanai*, or mullet, *wapuka*, or cod-fish, and..salmon abound. 1844 W. Wakefield in *N.Z. Co. Rep.* XXXVI. 31 Aug. 137 The habouka is taken in great quantities near the fishing town. c 1845 in C. F. Hursthouse *New Zealand* (1857) I. 217 We've..lowing herds on every side, Hapuka in every tide. 1855 R. Taylor *Te Ika a Maui* 411 *Hapuku*, or *whapuku* commonly called the cod, but a much richer fish in flavor. 1859 A. S. Thomson *Story N.Z.* I. 30 The Hapuku is the largest New Zealand salt-water fish. 1944 *Mod. Jun. Dict.* (Whitcombe & Tombs) (ed. 7) 193 *Hapuka, hapuku*, the Maori name for a large edible fish, also called 'groper'. Often mispronounced 'ha-pu'-ka'. 1960 Doogue & Moreland *N.Z. Sea Anglers' Guide* 209 *Groper* or *Hapuku*... Other names: *Polyprion oxygeneios*; whapuku (Maori)...The word 'hapuka' is a corruption of hapuku. 1966 *Encycl. N.Z.* I. 907/1 Hapuku or groper (*Polyprion oxygeneios*)..is a large, heavy, deep-sea fish closely related to the bass.

‖ **harai goshi** (hā·rai gǫ·ʃi). [Jap., f. *harai*, *harau* to sweep + *goshi, koshi* loin, waist.] A throw in Judo.

1941 M. Feldenkrais *Judo* vi. 114 Japanese experts are generally smaller than their foreign opponents, and still they find no difficulty in throwing them by *Harai-Goshi*, for example. 1954 E. Dominy *Teach yourself Judo* 190 *Harai Goshi*,..Sweeping Loin Throw. 1957 Takagaki & Sharp *Techniques of Judo* ii. iii. 31 Techniques such as *harai-goshi*. 1965 *New Statesman* 14 May 760/2 My son.. has applied the old Harai Goshi, a very effective throw in judo.

harambee (hăræ·mbi). [Swahili.] Pulling or working together; co-operation; the slogan of the Kanu government of Kenya at the time of independence. Also *attrib.*

1963 *Times* 13 Aug. 6/4 The farmers..joined him in the shout of 'Harambee'—the rallying call to the people of Kenya. *Ibid.* 12 Dec. (Kenya Suppl.) p. ii/7 They ran an effective Kanu government on the theme of *harambee* (working together) and at the pre-independence talks.. they broadly won..all their main objectives. 1969 *Reporter* (Nairobi) 16 May 40/1 The farmers of Nyanza Province..are giving an example to the rest of the country in launching *harambee* self-help schemes. 1971 *E. Afr. Standard* (Nairobi) 13 Apr. 2/4 The self-help work carried out by the people on a harambee basis was intended to make them join hands together in the development of a strong nation. *Ibid.*, Harambee groups should offer co-operation to one another in giving donations to fund-raising meetings.

harassing, *vbl. sb.* Add: Also *attrib.*, as **harassing agent**, a gas intended for use in harassing or incapacitating an enemy, a rioting crowd, etc., without being lethal; also *harassing gas*.

1968 R. Clarke *We All fall Down* iii. 40 No weapons of any kind are manufactured at Porton, although the harassing agent known as CS was developed there in the early 1950s. 1972 W. F. Biddle *Weapons Technology & Arms Control* xix. 286 The object of a harassing agent is to make it impossible for opposing troops either to stay in an area or to carry out their military duties. Harassing agents are not necessarily intended to be lethal. 1969 *New Scientist* 30 Jan. 219/2 The other major class of agents in use are the harassing gases CS, CN and DM. They are supposedly non-lethal.

harbour, *sb.*[1] Add: **3. b.** An airship shed or hangar.

1909 *Chambers's Jrnl.* Oct. 659/2 Work in connection with the other Zeppelin air-ships is so far advanced that as soon as the halls, or harbours, as they are called, are ready it will only be necessary to put the parts together. 1912 C. B. Hayward *Pract. Aeronaut.* 36 To the only two airship sheds or 'harbors' exceeding 400 feet in length.. no less than nine had been added [in France].

c. (See quot. 1948.)

1935 *Jrnl. R. United Service Inst.* Nov. 747 The aeroplane cannot hit a moving tank with a bomb, but when the tanks harbour, the aircraft will make every effort to locate and bomb them... The bombing of tanks in harbour will cause immediate dispersion. 1948 Partridge *Forces' Slang* 91 *Harbour*, halting place for the night for guns and tanks. Also a verb.

5. harbour seal *N. Amer.*, the common seal, *Phoca vitulina*, found along the shores of northern oceans; **harbour-side**; **harbour stow**, furling in a body (cf. Furling *vbl. sb.* 1); so **harbour-stowed** *a.*

1766 J. Banks *Diary* Oct. in A. M. Lysaght *Joseph Banks in Newfoundland & Labrador* (1971) ii. 145 They [*sc.* the fishermen] divide them [*sc.* the seals] into five sorts which they call Square Phipper Hooded Seal Heart or houke Bedlamer and harbour seal, which last stays in the Countrey all the year. 1832 J. McGregor *British America* I. iii. 107 The harbour seal (phoca vitulina)..does not seem to be migratory. 1958 A. W. Cameron *Canad. Mammals* 55 Apart from the grey seal, the harbour seal is the only member of the tribe that ordinarily spends the summer in southern Canada. 1964 E. P. Walker et al. *Mammals of World* II. 1302 (*caption*) Hair or harbor seals (*Phoca vitulina*). 1947 Crowther & Whiddington *Science at War* 180 Larger explosive charges can be used, and their effects registered by electrical recording on the harbour-side. 1962 *Daily Tel.* 11 Aug. 14/5 It was then decided to take Cowslip, still low in the water, to the nearby harbourside home of Mr. B. A. L. 1969 *Jane's Freight Containers 1968–69* 56/2 D & F Harborside Terminal. 1886 R. Brown *Spunyarn & Spindrift* xxv. 311 Every rope in its place and hauled taut, every sail neatly furled in a harbour-stow. 1924 R. Clements *Gipsy of Horn* v. 98 A 'harbour stow' we gave them, rolling the canvas into a neat skin as though it were covered with a jacket and passing the gaskets at regular intervals like seizings. 1924 J. Masefield *Sard Harker* 24 She was in lovely order; yards squared, harbour-stowed.

harbour, *v.* Add: **9. b.** Of tanks, military forces, etc.: to shelter; to halt for the night. Cf. senses 7 and 9 in Dict.

1935, 1948 [see *harbour *sb.*[1] 3 c]. 1956 W. Slim *Defeat into Victory* 498 That night our leading troops harboured two hundred and forty miles from Rangoon.

hard, *a.* (*sb.*). Add: **I. 1.** *hard egg* (earlier and later examples).

1607 E. Topsell *Hist. Foure-footed Beastes* 188 Prescribing him a diet; which is to drink water, and to eat hard Egs. 1962 L. Deighton *Ipcress File* xxi. 143, I was loaded with anchovy,..hard egg and salmon.

e. Of a lawn tennis court: made of asphalt or other hard material, as distinguished from a grass court.

1889 H. W. W. Wilberforce *Lawn Tennis* v. 19 Most people on a dry ground or a 'hard' court use brown leather or buck-skin shoes with thick, smooth, red rubber soles. 1909 *Westm. Gaz.* 30 Mar. 12/3 It is possible to place too much significance on hard-court results, grass conditions in England being so materially different. 1959 *Daily Tel.* 1 June 1/4 The French hard-court championships. 1973 G. Mitchell *Murder of Busy Lizzie* i. 10 A sunken garden, a hard tennis court, miniature golf.

f. Of silk: retaining its natural gum. Also applied to a worker in hard silk.

a 1877 Knight *Dict. Mech.* 2180/2. 1878 A. Barlow *Hist. Weaving* 395 Before the gum has been boiled off the silk it is said to be hard silk. 1921 *Dict. Occup. Terms* (1927) § 399 *Hard hand* (silk); general term for any worker engaged in treating silk while still hard, *i.e.*, before it is degummed.

g. Of porcelain: made of hard paste; *hard paste*: see Paste *sb.* 3 b, Porcelain 1 note.

1814 Rees *Cycl.* XXVIII. Dd 4/2 Porcelain made of the best proportions of these two substances..is called hard porcelain. 1832 G. R. Porter *Treat. Porcelain & Glass* iii. 43 This paste is not so cohesive or viscous as that which forms hard porcelain. 1848 H. R. Forster *Stowe Catal.* 6 The celebrated Porcelain of Dresden, or more properly, Meissen,..is the most choice..of German fabrication. The material is termed 'hard paste'. 1869 Lady C. Schreiber *Jrnl.* (1911) I. 44 A bird on a raised sort of foot, *possibly* hard paste English. 1879 Hard-paste [see Paste *sb.* 3 b]. 1881 *Harper's Mag.* Feb. 368/1 There are now hard or true porcelain manufactories in New York. 1885 *Encycl. Brit.* XIX. 642/1 Bristol porcelain is of interest as being the first hard natural porcelain made in England. 1909 *Chambers's Jrnl.* Nov. 751/1 China manufactured in the eighteenth century was of two kinds—namely, 'hard' paste or true porcelain, and 'soft' paste or artificial china. 1968 *Canad. Antiques Collector* Aug. 27/1 Hard-Paste porcelain is unaffected by the file where it is 'free from the glaze'. 1968 J. Arnold *Shell Bk. Country Crafts* 234 Hard porcelain, already glazed, is fired at 1300°–1400°C., causing complete fusion.

h. In many specific collocations, e.g. *hard brass, cheese, coke, cure, glaze, lights, mixture, pavior, pitch, solder, solderer, soldering, stock*.

Also *hard coal* = Anthracite; *hard rubber* = Ebonite, Vulcanite 2; *hard soap*, see Soap *sb.* 2.

1873 E. Spon *Workshop Rec.* 1st Ser. 10 Hard Brass, for Casting.—25 parts copper, 2 zinc, 4.5 tin. 1888 *Lockwood's Dict. Mech. Engin.*, *Hard Brass*.—(1) Brass which has not been annealed after drawing or rolling... (2) Hammered brass, and brass which contains a large proportion of tin. 1902 *Encycl. Brit.* XXVII. 355/1 A perfect Leicester is perhaps the most attractive of all the so-called 'hard' cheese. 1957 'K. Curragh' *Lady into Cook* 68 Cheese Soufflé... 4 tablespoons grated cheese (Parmesan or any hard cheese). 1846 *N.Y. Morning Express* 2 Oct. 4/2 Hard Coal is a little higher owing to the increased traffic and freight. 1960 *Gloss. Coal Terms* (*B.S.I.*) 8 *Hard coal*, all coal of higher rank than lignite. In the U.S.A. the term is restricted to anthracite. 1979 *Sci. Amer.* Jan. 28/3 Hard coal (anthracite and the various grades of bituminous coal) and soft coal (brown coal and lignite). 1888 *Lockwood's Dict. Mech. Engin.*, *Hard coke*, oven coke. 1907 Hard cure [see *cure *sb.*[1] 12]. 1909 *Westm. Gaz.* 9 Nov. 12/1 Fine Hard Cure Para Rubber. 1962 J. T. Marsh *Self-Smoothing Fabrics* vi. 80 Another feature of the treatment with triazones is the necessity for a 'hard cure', e.g. for 2 to 3 min. at 150 °C or 1 to 2 min. at 165° C. 1814 Rees *Cycl.* XXVIII. Ee 2/2 The hard and less fusible glaze of the hard porcelain, which is mostly feldspar. 1839 Ure *Dict. Arts* II. 1016 The hard glaze of pipeclay ware. 1930 *Sel. Gloss. Motion Pict. Techn.* (*Acad. Motion Pict., Hollywood*), *Hard lights*. (1) Arc lights. (2) Illumination from arcs, in general. Refers to the sharp shadows cast. 1909 *Practitioner* Feb. 266 The mixture of the cocculus with beer..was kept by brewers' druggists, and sold to brewers under the name of 'multum' or 'hard mixture'. 1904 Goodchild & Tweney *Technol. & Sci. Dict.* 280/1 *Hard paviors*.., malm bricks, over-burnt and slightly blemished in colour, used for paving, coping, etc. 1879 *Encycl. Brit.* X. 100/1 If the heat is forced, and the distillation [*sc.* of coal-tar] continued, a large amount of 'heavy' or 'dead oils' is obtained, and the mass left in the still is 'hard pitch'. 1860 Hard rubber [see Vulcanite 2]. 1846 *Pat. Jrnl.* 1 Aug. 174/2 Hard solder. Melt together two pounds of copper and one pound of tin. 1873 E. Spon *Workshop Rec.* 1st Ser. 364, 2 parts of good silver and 1 of ordinary brass pins, well melted, is a good, useful jewellers' hard solder. a 1877 Knight *Dict. Mech.* II. 1061/2 *Hard solder*, the solder used for uniting the more infusible metals. Spelter solder and silver solder are the two principal varieties. 1879 *Spon's Encycl. Industr. Arts* I. 324 Alloys employed for joining metals together are termed 'solders' and they are commonly divided into two classes: hard and soft solders. 1902 *Young Engineer* I. 104 The art of soldering may be divided into two distinct classes—soft soldering,..and hard soldering, in which the solders are composed of gold, silver, copper, zinc, or brass. 1921 *Dict. Occup. Terms* (1927) § 262 *Brazer, brazier*; *hard solderer*; joins together parts of steel, iron, brass or copper articles by brazing or hard soldering. 1836 *Penny Cycl.* V. 409/1 The bricks are now separated for sale; the hard sound stocks are the best, and are worth from 1*l.* 10*s.* to 2*l.* a thousand. 1879 *Notes Building Constr.* III. 105 Hard Stocks are overburnt bricks, sound, but considerably blemished in both form and colour. 1904 Goodchild & Tweney *Technol. & Sci. Dict.* 69/2 Those [bricks] which are less overburnt are termed 'hard stock', and are useful for many building purposes.

5. c. *hard of hearing* (later examples).

1950 *Lancet* 11 Nov. 532/2 Practical courses..on audiometry and hearing-aids, hard-of-hearing children, [etc.]. 1968 *Brit. Med. Bull.* XXIV. 256/2 There may be a real problem in distinguishing the hard of hearing from those with organic intellectual impairment or autism.

d. (*to do something*) *the hard way*: (to do it) by one's own unaided efforts, through bitter experience, or by the most difficult method.

1931 D. Runyon *Guys & Dolls* (1932) xiii. 276 'Charley,' he says, 'do you make it the hard way?' 1938 *Collier's* 2 Apr. 16 (*title*) The hard way. 1945 N. L. McClung *Stream runs Fast* ix. 80 You learned everything the hard way. 1951 'N. Shute' *Round Bend* 1, I came into aviation the hard way. 1954 M. Croft *Spare Rod* i. ii. 12 I'm starting you off the hard way. 1958 *Observer* 26 Jan. 15/8 Making a movie the hard way spiritually and intellectually is the thing that really matters. 1958 *Listener* 28 Aug. 292/1 Unqualified men who come up what used

to be called 'the hard way'. 1959 *Times Lit. Suppl.* 4 Dec. 701/4 The Japanese failed to recognize it in the 1930s, and learnt it the hard way. 1971 D. Lees *Rainbow Conspiracy* ix. 135 In the end I nearly found the reservoir the hard way.

6. b. *hard word*, used *dial.* in various senses, e.g. pass-word, abuse, scandal, marriage proposal, refusal. Phr. *to put the hard word on* (someone) *Austral.* and *N.Z. slang*, to ask for a favour or a loan, esp. to ask a woman for her favours.

1831 S. Lover *Legends & Stories of Ireland* 1st Ser. p. xxiii, *Hard word*, hint. 1843 W. Carleton *Traits & Stories Irish Peasantry* I. 78 So I gives Jack the hard word [*sc.* pass-word]. 1891 B. Stoker *Snake's Pass* xvi, He would send the hard word round the country about me and my leman. 1899 Somerville & 'Ross' *Exper. Irish R.M.* vi. 123, I had said what is called in Ireland 'the hard word' [*sc.* marriage proposal]. 1905 *Eng. Dial. Dict.* III. 63/1 Ah assed him for a shillin', an' he gev mi t'hard-word. 1919 W. H. Downing *Digger Dial.* 28 *Hard word*, an outrageous demand. (Put the hard word on.) 1927 J. Devanny *Old Savage* 144 He thinks she is putting the 'hard word' on him. 1943 *Coast to Coast 1942* 215 A tradesman notorious for putting the hard word on his typists. 1947 I. Douglas *Opportunity in Australia* 89 *Put the hard word on*, to cadge for a loan or favour. 1959 Baker *Drum* iv. 38 Establishing a suitable vantage point to 'put the hard word on' her. 1960 D. Lockwood *Fair Dinkum* ii. 11 He didn't put the hard word on me once, and my credit is still good. 1969 *N.Z. News* 28 May 1/1 (*headline*) Pilots put hard word on airline. 1970 *N.Z. Listener* 21 Dec. 8/3 'Don't you think hitching 's a little dangerous for females?' 'Well, some sheilas I know have had the hard word put on them.'

7. *hard case*: delete *U.S.* and add earlier and later examples; also *attrib.*, hardened, tough. In Australia and New Zealand, an amusing or eccentric but adventurous person, a 'character'; also called *hard doer* (see *doer 5), *hard shot, hard thing*. See also sense 11 below.

1836 W. T. Porter *Quarter Race Kentucky* (1854) 38 A 'hard case' called Emanuel Allen. 1842 *Life in West* 323 A canoe full of 'hard cases' (vagabonds) had passed up the river. 1920 *Punch* 7 Apr. 266/1 The hard-case mates a-bawlin'. 1928 *Sunday Express* 8 Jan. 4 With memories that go back to the days of 'cracker-hash', 'lobscouse', and hard-case, blue-nose, Nova Scotia mates. 1928 *Sunday Dispatch* 29 July 2 Half a dozen particularly hard-case units of the Flying Squad. a 1936 Kipling *Something of Myself* (1937) ii. 22 It [*sc.* a school]..had been made up..by drafts from Haileybury..and, I think, a percentage of 'hard cases' from other schools. 1896 H. Lawson *In Days when World was Wide* (1900) 197 Cause of half the fun that's started—'Hard-case' Dan—Isn't like a broken-hearted, Ruined man. 1896 —— *While Billy Boils* in *Wks.* (1948) 54 Steelman was a hard case... There was no shaking off Steelman. 1900 —— *Over Sliprails* in *Ibid.* 273 After dinner a humorous old hard case mysteriously took us aside. 1918 *Chrons. N.Z.E.F.* 7 June 204 Without a smoke he was a hard thing. 1938 'R. Hyde' *Nor Years Condemn* ix. 184 Fred's a hard shot. 1940 F. Sargeson *Man & Wife* (1944) 64 He was the hardest case bloke you ever came across. 1943 —— in *Penguin New Writing* XVII. 66 He looked a bit of a hard-shot. 1950 'A. P. Gaskell' in *Landfall* IV. 18 Cliff was a hard case.

b. Of facts: incapable of being denied or explained away, 'stubborn'.

1887 *Graphic* 29 Jan. 123/1 Hard Facts. 1906 *Daily Chron.* 11 Apr. 3/4 A few more hard-fact letters and less of this soft imagining might have made the body of the book as interesting as the appendix. 1929 Chesterton *Poet & Lunatic* iv, Thank God for hard stones; thank God for hard facts. 1956 A. H. Compton *Atomic Quest* 311 The hard fact is that war, like business, reduces to a question of gain versus cost. 1973 *Daily Tel.* 8 May 7/2 We have been unable to substantiate the allegations. We have few hard facts to go on but are continuing our investigations.

c. Of news or information: factual, real, objective, reliable, substantiated.

1938 E. Waugh *Scoop* ii. i. 117 There isn't any hard news. 1948 *Newsweek* 16 Aug. 51/1 The bulk of the broadcast time is given over to so-called 'hard news'—that is, straight newscasts of what is going on in the world and in the United States. 1956 A. J. Ayer *Probl. of Knowl.* iii. 92 Those who have sought to erect an edifice of knowledge on the basis of what Bertrand Russell..has called 'hard data', have commonly agreed that such data were yielded by sense-perception. 1958 *Listener* 16 Aug. 239/2 This would yield some interesting and some relatively 'hard' evidence. 1958 *Economist* 13 Sept. 815/2 Two hard items for the agenda. 1959 Duke of Bedford *Silver-Plated Spoon* ix. 185 Newspapers do not encourage telephone calls to the other side of the world unless they are in possession of pretty hard information. 1963 *Ann. Reg. 1962* 520 Upon receiving the first preliminary hard information of this nature..I directed that our surveillance be stepped up.

11. *hard case*: applied to a sailing-ship on which conditions are rough; *hard luck*: see Luck *sb.* 1; also *attrib.*; *hard-lying money*: corruption of *hard line money* (see Line *sb.*[2] 6); hence *hard lyer*; (joc.) *hard-liar*.

1920 *Blackw. Mag.* Mar. 322/2 The mate of a Yankee hard-case. 1924 R. Clements *Gipsy of Horn* vi. 111 He signed away as Third Mate on a hard-case Yankee barque. 1900 Ade *More Fables in Slang* (1902) 18 Her Hard-luck story. 1906 B. von Hutten *What became of Pam* ii. viii, Learning..something of..his hard-luck story. 1919 H. L. Wilson *Ma Pettengill* iv. 109 She said it [*sc.* the letter] would tell a new hard-luck tale for non-payment of a note. *Ibid.* 134 It was another hard-luck letter. 1959 P. H. Johnson *Humbler Creation* v. 32 The hard-luck stories of all parsons whose luck would never get any better. 1916 'Taffrail' *Pincher Martin* iv. 56 'They ain't so bad,' he murmured. 'You gits a tanner a day, 'ard lyers in 'em.' 1920 —— *H.M.S. Anonymous* xv. 291 What d'you think the government pay us one and six hard lyers for? Twenty seven extra bloomin' pounds per annum. 1916 —— *Pincher Martin* iv. 56 Men serving in destroyers receive sixpence a day extra pay. It is known as 'hard-lying money'. 1925 Fraser & Gibbons *Soldier & Sailor Words*, *Hard-lying money*, the extra allowance granted to officers and men for service in destroyers and torpedo boats, and as compensation for wear and tear of uniform and clothing, etc. Extended in the War to the crews of motor launches and other auxiliary small craft. (Abolished in 1923.) 1927 *Daily Express* 10 Oct. 3 Sometimes, in recompense for discomforts endured, the crews of drifters draw what is termed 'hard-lying money' (those who receive this are naturally known as 'hard liars').

12. c. (Later example.)

1968 *Times* 26 Aug. 7/2 The House of Commons today must take a hard look at British defence policy.

14. c. For *U.S.* read *orig. U.S.* and add earlier and later examples.

1789 F. Asbury *Jrnl.* (1821) II. 304 [A] drink made of one quart of hard cider, [etc.]. 1810 M. L. Weems *Lett.* (1929) III. 13 What could possibly have kept me from hard drink? 1840 *Congress. Globe* 13 Feb. 197/3 He had heard..the same arguments preached nine hundred and ninety-nine times over a barrel of hard cider. 1848 *Ibid.* 27 Apr. 688/2 They had charged him [*sc.* President Harrison] with drinking hard cider. 1857 *Spirit of Times* 3 Jan. 281/1 It was not infrequent, as late as the hard-cider campaign of 1840,..that [etc.]. 1861 H. W. Harper *Lett. from N.Z.* (1914) iv. 67 Order up some hard stuff to give them something to drink. 1946 F. Sargeson *That Summer* 35 They all started on hard stuff and went on to beer later. 1964 C. Willock *Enormous Zoo* ix. 169 With a hard drink in the hand the day lengthens and softens. 1965 O. A. Mendelsohn *Dict. Drink* 90 *Cider*,.. If fermented and therefore alcoholic, the term hard cider is frequently used.

d. Of oil (see Soft *a.* 23 d).

e. Of drugs: dangerous and habit-forming, addictive, e.g. heroin and cocaine.

1955 *Amer. Speech* XXX. 87 *Hard stuff*, opium. 1965 'Malcolm X' *Autobiogr.* vii. 110 As the pros did, I too would key myself to pull these jobs by my first use of hard dope. I began with..sniffing cocaine. 1967 *Listener* 10 Aug. 169/2 Nothing on earth would persuade me to try LSD or the hard drugs. 1967 *Times* 7 Oct. 5/3 The Court said that anyone supplying a child with a hard drug was doing a terrible deed which called for grave punishment. ..The Court refused an application..for leave to appeal against..conviction..of supplying a dangerous drug (six tablets of heroin). 1969 J. Gardner *Compl. State of Death* ix. 167 You start on pot and straight off you're mixing with people who've graduated to the hard stuff.

f. Of nuclear sites and structures (see quot. 1960). Also applied to nuclear missiles.

1958 R. D. Bowers in *Air Univ. Q. Rev.* X. 91 It would be useful to know how the cost of a hard base compares to that of a soft base if they have equal measures of merit (cost per surviving missile). 1960 *Amer. Speech* XXXV. 302 The adjective *hard* is now used in certain parts of the Air Force..to refer to the resistance to atomic explosions of airfields, missile launching pads, command posts, and other structures, their resistance coming from underground location or toughness of structure or both. 1962 *Listener* 29 Mar. 547/2 The American development of missiles such as Minuteman which can be fired from strongly protected pits in the ground—the so-called hard missile sites—as well as the Polaris submarines. 1965 H. Kahn *On Escalation* vii. 136 Let the reader assume.. that both the Soviet Union and the United States have 10,000 hard and dispersed missiles on each side.

15. (Earlier U.S. example.)

1838 D. Webster *Private Corr.* (1856) II. 37 Money is very hard, all along the coast, from here [i.e. Washington, D.C.] North.

16*. *Physics.* **a.** Of radiation: having great penetrating power.

1902 *Encycl. Brit.* XXVIII. 52/2 If the exhaustion of the bulb is carried further, so that there is a considerable increase in the potential difference between the cathode and anode and therefore in the velocity of the cathode rays, the Röntgen rays have much greater penetrating power and are often called 'hard rays'. 1938 R. W. Lawson tr. *Hevesy & Paneth's Man. Radioactivity* (ed. 2) iv. 49 The wave-length of the 'hard' γ-radiation is so small that even crystals are of no avail here as diffraction gratings. 1940 *Nature* 8 June 903/2 (*title*) Absorption of hard cosmic rays and mesotron decay. 1943 *Electronic Engin.* XVI. 54 Short wave, or hard, radiation penetrates more deeply into the tissues. 1955 *Gloss. Terms Radiology* (*B.S.I.*) 12 *Hard radiation*, a term used to describe qualitatively the more penetrating types of X-rays, beta rays and gamma rays.

b. Of a vacuum: complete or almost complete. Of a vacuum tube or thermionic valve: containing a high vacuum.

1899 W. Crookes in *Rep. Brit. Assoc. Adv. Sci. 1898* 23 Röntgen suggests a convenient phraseology; he calls a low vacuum tube, which does not emit the highly penetrating rays, a 'soft' tube, and a tube in which the exhaustion has been pushed to an extreme degree, in which the highly penetrating rays predominate, a 'hard' tube. 1919 R. Stanley *Textbk. Wireless Telegr.* (ed. 2) II. 22 With a hard vacuum none of these varying effects are present... The hard valve, made by Langmuir, was called by him a Pliotron. 1923 E. W. Marchant *Radio Telegr.* v. 61 *Hard Valves*.—The only way in which this can be achieved is by exhausting the air so completely from the bulb that there is nothing but a pure electron discharge. 1931 Duncan & Drew *Radio Telegr. & Telephony* (ed. 2) 214 The degree of vacuum in the tube would change and some tubes became *soft* (having less vacuum) while others became *hard* (having a higher vacuum, with little or no gas present). 1940 *Chambers's Techn. Dict.* 403/1 *Hard tube*, a high-vacuum discharge tube. 1943 *Gloss. Terms Telecomm.* (*B.S.I.*) 28 *Hard vacuum tube*, a vacuum tube evacuated to such a degree that its electrical characteristics are essentially unaffected by ionisation of included gas. 1953 *Electronic Engin.* XXV. 241 A blocking oscillator was first used as a hard-valve time-base in 1923. 1962 F. I. Ordway et al. *Basic Astronautics* xi. 449 The hard-vacuum condition of space.

18. b. *hard labour*: also *attrib.*

1905 *Daily Chron.* 10 July 5/2 Hard-labour convicts. *Ibid.* 6 Nov. 5/6 Hard-labour prisoners. 1908 *Ibid.* 7 Jan. 4/6.

20. *hard at it* (cf. Hard *a.* 19); *hard cases make bad law* (see quot. 1903); *to play hard to get* (cf. Play *v.* 34), to pretend to remain aloof, to act or behave as if unapproachable or uninterested; also *hard-to-get* attrib. phr., aloof, unapproachable.

1749 Fielding *Tom Jones* III. vii. v. 32 Pray who hath been the Occasion of putting her into those violent Passions? Nay, who hath actually put her into them? Was not you and she hard at it before I came into the Room? 1811 Jane Austen *Let.* 30 Apr. (1932) 278 By this time I suppose she is hard at it, governing away—poor creature! 1923 J. S. Huxley *Ess. Biologist* iii. 119 The two are hard at it, shaking their heads. 1964 C. Willock *Enormous Zoo* i. 13 The circular saws were soon hard at it. 1854 G. Hayes in W. S. Holdsworth *Hist. Eng. Law* (1926) IX. App. 423 Sur. B. [*i.e.* Baron Surrebutter] A hard case. But hard cases make bad law. 1903 V. S. Lean *Collectanea* III. 479 Hard cases make bad law, i.e. lead to legislation for exceptions. 1909 *Spectator* 22 May 809/1 'Hard cases make bad law', and also bad policy. 1945 A. Kober *Parm Me* 32 'I played "hard to get". Y' know,' he amplified, 'like I couldn't be bothered.' 1951 Wodehouse *Old Reliable* xi. 132 Why are you pulling this hard-to-get stuff on Joe? 1951 Auden *Nones* (1952) 18 But that Miss Number in the corner Playing hard to get. 1959 P. Capon *Amongst those Missing* 194 To be blunt, you sort of strike me as playing hard to get. 1959 J. Fleming *Miss Bones* xv. 168 Is she playing hard-to-get? he thought angrily. 1961 *Times* 26 Oct. 18/1 Miss Nancie Jackson's Millamant is a statuesque blonde, playing hard-to-get while keeping our sympathy. 1962 *Guardian* 8 Feb. 7/4 The wish to create a hard-to-get atmosphere for the coming negotiations.

21. *hard-backed, -based, -edged* (examples), *-faced* (examples), *-glazed, -leaved, -lipped, -nailed, -textured*.

1853 Mrs. Gaskell *Cranford* xvi. 312 Mr. Peter said he was tired of sitting upright against the hard-backed uneasy chairs. 1959 Hardbacked [see *hardback 2]. 1959 *Listener* 31 Dec. 1140/1 Within a few years it [*sc.* the American Strategic Air Command] will have at its disposal enough 'hard-based' missiles—that is, either housed in protected subterranean sites or mounted on mobile carriers—to make it impossible for any surprise attack to succeed at all. 1954 J. R. R. Tolkien *Two Towers* 22 Long slopes they climbed, dark, hard-edged against the sky. 1964 *English Studies* XLV. 28 Nor are registers hard-edged pigeon-holes. 1871 J. G. Whittier *Marguerite* in *Wks.* (1898), By her bed the hard-faced mistress sat. 1958 B. Nichols *Sweet & Twenties* 145 Harding was a hardfaced provincial politician. 1928 *Daily Express* 6 Oct. 11/7 A hard-glazed lacquer work upon tin-plate. 1926 J. Masefield *Odtaa* xiii. 223 Hi crackled through some hard-leaved scrub. 1849 Rossetti *Let.* 27 Sept. (1965) I. 60 Hunt reads Dumas, hard-lipped. 1961 *John o' London's* 6 July 57/2 One of his virtuoso essays in hard-lipped sensitivity. 1879 G. M. Hopkins *Sermons* (1959) 18 Christ has..made us deaf here,..with his hands hardnailed out and appealingly stretched on the cross. 1910 W. de la Mare *Three Mulla-Mulgars* xvii. 232 Long, hairy, hard-nailed toes. 1937 E. Sitwell *I live under Black Sun* 141 A sour hard-textured unripe plum.

22. a. hard fescue (see Fescue *sb.* 4); **hard-grass** (later examples); **hard wheat**, any wheat having a hard grain rich in gluten; also *attrib.* **b. hard bop** (see *bop *sb.*[2]); **hard-bread** (U.S. examples); **hard cheddar, cheese** *colloq.*, hard luck (see also sense 1 h above); **hard chine** [Chine *sb.*[3] 2] (see quot. 1961); so **hard-chined** *a.*; **hard copy** (see quots.); **hard core**, (*a*) (see Core *sb.*[1] 7 d); (*b*) an irreducible nucleus or residuum; also a stubborn or reactionary minority; something blatant or intractable; freq. *attrib.*; **hard cover** orig. *U.S.*, (of a book) a stiff binding case, chiefly (with hyphen) *attrib.*; also **hard-covered** *a.* = *hard-bound* (*hard *adv.* 8 d); **hard currency**, (*a*) in specie as opposed to paper currency, metallic currency (*Obs.*); (*b*) (see quot. 1949); **hard dot** *Printing* (see quot. 1968); **hard-edge**, a style of painting (see quot. 1962[1]); also freq. *attrib.*; **hard facing** *Metallurgy*, the application to the surface of a metal of a protective hard material resistant to wear, corrosion, etc.; hence **hard-faced** *a.*; **hard glass**, a borosilicate glass (cf. *pyrex); **hard hat**, (*a*) a hat made of hard or stiffened felt; a bowler hat; (*b*) a tin helmet; (*c*) a person who wears a tin helmet, *spec.* a construction worker; (*d*) a person who is reactionary or conservative; also *attrib.*; **hard-hitter** (**hat**) *Austral.* and *N.Z. colloq.*,

= Bowler³; **hard landing** *Astronaut.*, an uncontrolled landing in which the vehicle is destroyed; so **hard lander**; **hard-line** *attrib.*, adhering to a hard or firm policy without abatement or concession (cf. sense 12 c); so **hard-liner**; **hard-nosed** *a.*, (*a*) (see quot. 1889); (*b*) *U.S. slang*, obstinate, stubborn; **hard pad**, a form of distemper in dogs and sometimes other animals; **hard pear**: see *Hardpeer; **hard-rock**, (*a*) *attrib.* or as *adj.* (*N. Amer.*), 'experienced in underground work in hard massive formations;—said of a miner' (Webster 1934); (*b*) (*N. Amer. slang*) as noun phr., a hard, craggy person; also *attrib.*; (*c*) a type of strident music; **hard sell** orig. *U.S.*, aggressive salesmanship or advertising; also *attrib.*; **hard-sell** *v.*; **hard-selling**; **hard-sold** *ppl. a.*; **hard shoulder** (see quot. 1955); **hard-standing** (occas. **hard-stand**) (see quots. 1951, 1956); **hard top**, (*a*) a rigid or fixed roof of a motor-car (as opposed to one of soft material); also a car so fitted; (*b*) (see quot. 1957); **hard tube** (see sense 16* b); **hard twist**, a hard-spun yarn, a yarn with more than the usual amount of twist; **hard valve** (see sense 16* b); **hard waste** (see quot.).

1781 Heath in R. Putnam *Mem.* (1903) 187 The major is gone to the commissary to obtain some hard bread if possible. **1835** in J. B. Thoburn *Stand. Hist. Oklahoma* (1916) vi. 74 The ration of bread shall be one pound of wheat flour, Indian meal, or hard bread. **1857** W. Chandless *Visit Salt Lake* ii. 11 What we call rolls, in America are ycleped biscuits, and biscuits in their turn hard bread. **1905** G. E. Cole *Early Oregon* i. 12 Having no salt junk or hard bread left. **1931** 'N. Bell' *Life & Andrew Otway* 464 'He knew all about cutting throats. Seen a bit in his time.' 'Hard cheddar. I reckon a bullet'd be my mark if I wanted to pass in my checks.' *Ibid.* 465 No, I don't see how you can blame the bleedin' government. Hard hard cheddar on you but—. **1876**, etc. Hard cheese [see *Cheese *sb.*¹ 2 d]. **1913** C. Mackenzie *Sinister St.* I. ii. i. 149, I thought it hard cheese on her. **1973** J. I. M. Stewart *Mungo's Dream* xxii. 270 It was hard cheese on him coming up against another top-class specimen. **1912** *Motor Boat Man.* (ed. 5) v. 44 'Miranda IV'..has a single step... The hard chine or angular bilge is not an essential feature. **1951** *Engineering* 8 June 680/2 In general, the appraisal of the respective merits and demerits of round-bilge, hard-chine and stepped hulls is fair and temperate. **1961** F. H. Burgess *Dict. Sailing* 110 *Hard chine*, a feature of a boat in which the topsides and bottom meet at an angle instead of curving to a round bilge. **1967** *Jane's Surface Skimmer Systems 1967–68* 86/2 A small sports hydrofoil, the ST-1 has a typical hard-chine type hull of wooden construction. **1966** *Amer. Speech* Oct. 235 *Lightning*, a wood-planked, hard-chined boat made for racing. **1964** *Gloss. Automated Typesetting* (C.I.S.) 17 *Hard copy*, a printed (typewritten) record or copy of machine output. **1964** T. W. McRae *Impact of Computers on Accounting* vi. 165 An intermediate device, the computer, is necessary to translate the 'magnetic' records into a 'hard copy' format suitable for audit. **1936** *Nature* 12 Sept. 441/2 Possibly 200,000 would be practically unemployable on any ordinary basis—the 'hard core' as it is called. **1940** *Economist* 3 Feb. 193/2 One of the more encouraging developments of the last few months is a substantial loosening of what has hitherto been regarded as the 'hard core' of unemployment. **1951** J. Cornish *Provincials* 24 The party was acknowledged even by the hard-core cynics to be a credit to the Dunseith brothers. **1955** *Treatm. Brit. P.O.W.'s in Korea* (H.M.S.O.) 25 This camp..was the home of the hard core of Other Rank reactionaries—men who had distinguished themselves by their heroic resistance to all Chinese brutality. **1956** *Ann. Reg. 1955* 300 A hard core of Karens in both zones maintained resistance. **1958** *Daily Mail* 15 Aug. 2/1 More than 100 hardcore EOKA terrorists have been netted in the anti-terrorist operation. **1958** *Listener* 11 Dec. 982/1 In seven 'hard core' States, no Negro child attends a white school. **1959** F. van P. Bryan in Ernst & Schwartz *Censorship* (1964) xix. 130 A work of literature..stands on quite a different footing from hard core pornography. **1961** R. Kee *Refugee World* iii. 26 Weiss, the Frankfurt jeweller..was a 'hard-core' case in 1950. **1968** *Times* 2 Sept. 2/7 Heathrow airport..could, perhaps, be called the hard core of Britain's noise problem. **1973** *Times Lit. Suppl.* 20 Apr. 451/5 The leading modern writer of hard-core science fiction. **1949** in *Amer. Speech* (1952) XXVII. 148 Hard-cover reprints. *Ibid.*, Hard-cover reprint house. **1951** *Sat. Rev. Lit.* 12 May 31/1 Much of the reading involves books whose quality does not justify any particular consideration just because the pages are bound inside a hard cover. **1951** *Publishers' Weekly* 12 May 1949 Each issue will be a hard-cover book. **1957** *Economist* 5 Oct. 45/1 Critics of mass market publishing suspect that it spreads mediocrity, even to the hardcover business. **1960** *News Chron.* 10 Jan. 3/7 The Heinemann group..published..an expurgated version of 'Lady Chatterley'..in hard cover in this country. *Ibid.* 3/8, I would immediately bring out a hard-cover edition. **1968** *Times* 14 Sept. 21/4 The whole inquiry relates exclusively to hard-cover sales. **1952** *Amer. Speech* XXVII. 147 (*heading*) Books: hard-bound, hard-cased, hard-covered, limp-covered. **1851** J. H. Green *Twelve Days in Tombs* 33 The politicians were fiercely discussing the 'hard' and 'soft' currency question. **1940** *Economist* 6 Apr. 609/1 The phenomenon of a 'free' rate diverging from the 'official' rate occurs only in respect of those so-called 'hard' currencies for which official rates are fixed by the Bank of England (notably American dollars). **1948** *Hansard Commons* 29 Jan. 1247 As a hard-currency market, the United States is of great importance. **1948** *Ibid.* 9 Mar. 1003 Does the Chancellor of the Exchequer think it right to spend over £60 million a year in hard currency? **1949** *Times* 10 Sept., *Hard currency*, a term without precise meaning. In general, when used in relation to this country, it means the currency of any country with which this country has an adverse balance of payments in current transactions which has to be settled in gold or dollars...Hard currency is..a relative rather than an absolute term, reflecting as it does the relation between one currency and another. **1950** *Engineering* 10 Feb. 144/1 Vehicles for export to hard-currency countries. **1957** *Time* 2 Sept. 27/2 Canadians last week enjoyed the mixed blessing of having the world's hardest currency. **1969** *Listener* 20 Feb. 229/3 Arthritis would have helped me break into the hardest of hard-currency resort areas. **1961** Dugan & Pine in *Penrose Ann.* LV. 135 (*title*) The hard-dot positive in gravure. *Ibid.*, The need for a 'hard-dot' in the positive plates. **1968** *Gloss. Terms Offset Lithogr. Printing* (*B.S.I.*) 15 *Hard dot*, a half-tone dot with a high edge density. **1961** *Times* 6 June 16/5 The 'hard-edge' abstractions of recent years. **1962** *Guardian* 18 Jan. 7/7 During the last three years.. a new branch-line in British painting has opened up and been loosely labelled the 'hard-edge school'..The paintings tend to be precisely geometric, crisp..and totally withdrawn from any personal emotive content. *Ibid.* 12 July 6/4 Hard-edge is socially useful..but it is neither art nor painting. **1962** *Listener* 21 June 1080/2 Kelly is a 'hard-edge' abstractionist. He balances simple areas of sharply defined flat colour one against the other. **1965** *New Statesman* 30 Apr. 693/2 The exhibition effects an uneasy marriage between, on the one hand, hard-edge abstraction and stain painting, fields in which the Americans easily excel, and, on the other, optical art. **1960** *Jrnl. Iron & Steel Inst.* CXCIV. 269/3 Hard-faced rolls had a useful service life five times as long as the usual steel rolls. **1930** *Ibid.* CXXII. 542 Welded-on overlays or 'hard facings' have been applied to drill bits and other tools to combat abrasion. **1931** *Ibid.* CXXIV. 634 The author cites a number of applications of hard facing and indicates the economy of the process. **1904** S. P. Mulliken *Method Identification Pure Org. Compounds* I. 10 Prepare an ignition-tube 8–10 cm. in length from a piece of hard-glass combustion tubing. **1937** *Discovery* Nov. 360/2 Pyrex and other hard glasses. **1965** Phillips & Williams *Inorg. Chem.* I. xiv. 545 Hard glass or pyrex is made from mixed boron (30 per cent) and silicon oxides. **1961** R. W. Butcher *Brit. Flora* II. 978 *Parapholis* (*Lepturus*) *incurva*. The curved hard-grass is a small, tufted annual with many erect, curved stems 1–6 in. high. *Ibid.* 979 *Parapholis strigosa*. The sea hard-grass is a small, tufted annual with many slender, solitary, erect or bent stems 6–18 in. high. **1935** A. J. Pollock *Underworld Speaks* 51/1 *Hard hat*, a derby. **1945** Baker *Austral. Lang.* 181 The Australian equivalents of what the Englishman calls a bowler and the American a derby. Here are our contributions: boxer,..hard hat, [etc.]. **1953** *Collective Bargaining Agreement Giant Yellowknife Gold Mines* 3 Protective devices..shall be provided by the Company, but this shall not include personal necessities such as hard hats, hard toed boots and gloves. **1956** H. M. Newell *Dam* xvii. 84 Workmen were cocooned in clothing till he couldn't recognize them...Beneath their hard-hats, stocking caps or scarves or kerchiefs tied under their chins. **1963** A. Lubbock *Austral. Roundabout* 21, I was given a white tin hat, as it is a regulation that these safety helmets must be worn in the 'hard hat' areas of the mines. **1970** *Sunday Tel.* 14 June 12/7 All the dangerous implications of 'hard-hat' demonstrations by flag-waving workers who see the anti-war movement as a betrayal. **1970** *Sunday Mail* (Brisbane) 14 June 21/2 A 'Hard Hat' is a construction worker, but his helmet symbolises all those beefy blue-collar workers who have suddenly become the knuckleduster on the strong right arm of President Nixon's silent majority. **1972** D. E. Westlake *Cops & Robbers* (1973) v. 63 It was one of those huge office buildings being constructed there, and the hardhats kept steady working away at it. **1895** J. Roberts *Diary* 7/2, I had long ere this put on my own clothes..of the 'masher' type—white shirt, hard-hitter, tight trousers, etc. **1907** H. Lawson *Romance of Swag* (1948) 478 Jim sat in his shirt-sleeves, with his flat-brimmed, wire-bound, 'hard-hitter' hat on. **1924** H. T. Gibson *That Gibbie Galoot* i. 1, I didn't mind so much when my former mates sat on my hard-hitter hat. **1932** N. Scanlan *Pencarrow* xxiv. 242 Then hats. Of course they must have hard hitters..black bowler hats—hard hitters as they were usually called in the colony. **1962** F. I. Ordway et al. *Basic Astronautics* v. 176 (*caption*) Exterior view of Ranger lunar hard lander. **1958** *Times* 28 Mar. 10/3 The first [*sc.* landing on the moon] would be a simple shot, ending either in a 'hard' (uncontrolled) landing or a circling of the moon. **1967** *Technology Week* 23 Jan. 61/1 (Advt.), Grumman is engaged in major application research to develop vehicles for soft and hard landings on Mars. **1962** *Times Lit. Suppl.* 21 Dec. 992/5 The 'hard-line' periodical *Literatura i zhizn*, which has up to now dealt exclusively with the literature of the Russian Federation, will close down. **1964** *Economist* 18 Jan. 206/1 CIA's..reputation as a 'hard-line' agency. **1965** *New Statesman* 7 May 710/3 Among those whose toughness stiffened the rebels' will to resist were.. the 50 or 60 'hard line' communists. **1970** *Guardian* 28 Oct. 4/6 He has not concealed his hardline views, but has avoided coming into direct public confrontation with the party leadership. **1963** *Sat. Rev.* 25 May 22/2 The fact that war has now become an instrument of mutual suicide..has made no dent in the thinking of hardliners. **1966** *Observer* 17 Apr. 2/8 Vice-President Humphrey—widely regarded as a hard-liner on Asia. **1969** *Guardian* 5 Mar. 8/1 Hardliners of the Right and the Left will say..that there is no middle way in Latin-American politics. **1889** *Cent. Dict.*, *Hard-nosed*, in hunting, having little or no sense of smell: said of dogs. **1927** *Hollis St. Theatre Prog.* (Boston) 19 Sept. Gloss., *Hard nosed*, stubborn. **1949** *Penguin New Writing* XXXVIII. 49 And *there* there is a lock, too, with the saint who welched three times in charge of the keys, who can be trusted not to be too hard-nosed. **1965** *Economist* 3 Apr. 27/2 Washington has got pretty 'hard-nosed' about criticism of its Congo policy. **1971** *New Society* 7 Jan. 26/1 A whole host of prophetic ravers and hard-nosed technocrats. **1973** *Times* 8 Feb. 23/2 Dolly's hard-nosed business approach to publishers probably did not have universal support. **1948** Macintyre et al. in *Vet. Rec.* 28 Feb. 103/1 We now recognise a condition which we have tentatively called 'hard pad disease'. **1950** A. C. Smith *Dogs since 1900* 35 The disease should be more properly termed canine encephalitis...It became known as 'hard-pad' because a curious symptom.. is a hardening of the pads. **1958** *New Yorker* 22 Feb. 31/3 Even the fact that foxes are now carrying a disease called hardpad is insufficient reason for shooting a fox. **1923** 'B. M. Bower' *Parowan Bonanza* iv. 56 Tommy's an old, hard-rock man. **1926** *Amer. Speech* II. 87/1 The old hardrock miners (now nearly extinct) were either single jackers or double jackers. **1949** *Chicago Daily News* 9 Apr. 1/6 A machinist drilling with a crew of hardrock miners 75 feet below the surface. **1950** W. R. Bird *This is Nova Scotia* 19 One night..he ran up against a hardrock from Spencer's Island, and when the fracas ended he had been completely thrashed. **1962** K. Orvis *Damned & Destroyed* xvi. 119 A hard-rock by the name of Welch. **1965** *Globe & Mail* (Toronto) 6 Jan. 24/9 Page [is] a hard-rock defensive back. **1967** *Boston Sunday Herald* 26 Mar. 1. 11/2 'Sing-Out' shook Saunders Theater with its mesmeric hard-rock beat. **1969** *Rolling Stone* 28 June 38/2 Drummer, hard-rock, blues, heavy, wants dependable group with horns. **1952** *Business Week* 9 Aug. 40 A few months ago everyone had keyed himself up to the 'hard sell'. **1957** 'E. McBain' *Con Man* (1960) iii. 31 It's the hard sell and the soft sell, anywhere you go. **1958** D. Delman (*title*) The hard sell. **1959** *Times Lit. Suppl.* 13 Nov. 662/3 One does not see any examples..of what is called 'hard-sell' advertising. **1961** *Economist* 14 Jan. 114/2 The need for the 'hard sell' is evident. **1963** *Guardian* 16 Nov. 14/7 It is difficult to hard-sell the honest song the way they do the contemporary counterfeit. **1963** *New Scientist* 19 Sept. 613/1 The reader cannot miss the hard-sell line of advertising copy. **1966** *Ibid.* 21 July 125/1 Whatever one may think of the 'hard-sell' methods employed by big American corporations, they obviously work. **1960** *Economist* 8 Oct. 158/2 The current slump in sales has also prompted many firms to return to the 'hard-selling' practices of earlier recessions. *Ibid.* 172/1 The fuel-economy services offered by the coal and oil industries to sell their products may not prove serious competition for an independent organisation like Nifes, which may end up refereeing between them for hard-sold firms. **1955** *Times* 6 July 10/1 The motorways are to be constructed to modern standards, with hard shoulders of 9 ft. (land hardened and laid down for vehicles to get safely off the running lanes). **1959** *Ibid.* 4 Nov. 9/7 The hard shoulder, used by vehicles for emergency stops, has collapsed at one point under the weight of a heavy lorry. **1973** *Scotsman* 12 Jan. 11/4 Twenty-six of the accidents were on the inside lane or hard shoulder. **1944** *R.A.F. Jrnl.* Aug. 259 Our lorries, drawn up on the hardstanding ground, beside lines of gliders. **1951** *Gloss. Aeronaut. Terms* (*B.S.I.*) III. 22 *Hard standing*, a prepared hard surface for parking aircraft or heavy vehicles. **1956** W. A. Heflin *U.S.A.F. Dict.* 245/1 *Hardstand*, any paved, compacted, or otherwise specially-prepared surface or area set up either for parking an airplane or ground vehicle, or for storing supplies and equipment. *Ibid.* 245/2 *Hardstanding*, the facility provided by hardstands. **1957** *Times* 1 Mar. 9/6 These are certainly the aircraft which one sees on the hardstanding. **1960** *Guardian* 19 May 1/6 The rural council intends to provide hard-standing for caravans. **1968** *Bucks Examiner* 2 Aug. 2/7 Hardstanding for the parking of one coach or coaches. **1949** Frazee & Bedell *Automotive Fundamentals* 82 The convertible sedan features the same entrance and seating arrangements as are found in the regular hard-top sedan. **1951** *Amer. Automobile* Mar. 78/3 Plymouth has adopted entirely new designations... The coming hardtop will be called Belvidere. **1952** *Autocar* 12 Dec. 1663/1 The hardtop style of roof is now used on all bodies, whether they have two or four doors. **1956** *Sat. Rev.* 7 Apr. 40/2 The number of enclosed theatres now colloquially known as 'hard tops' declines. **1957** *Amer. Speech* XXXII. 239 The conventional movie house under a roof is now referred to as a hardtop. **1959** 'S. Ransome' *I'll die for You* iv. 43 Two Burt Fishers ..driving two identical Chevy hard-tops. **1967** *Guardian* 3 Oct. 5/3 The price of the hard top has gone up to £1,255. **1897** *Sears, Roebuck Catal.* 186/1 This suit is made from genuine Michigan hard twist cassimere. **1921** T. Woodhouse *Yarn Counts* vii. 87 There are..exceptional degrees of twist, some..of which might exceed the value of the so-called hard twist. **1963** A. J. Hall *Textile Sci.* iii. 114 In a similar manner much the same considerations apply in a multi-ply yarn if the two or more single yarns present are only slightly or highly twisted (often referred to as soft and hard twist) together. **1921** *Dict. Occup. Terms* (1927) § 362 *Breaker, hard waste; hard waste tenter*; feeds and operates machine which opens out hard cotton waste (waste from ring frames and from reeling and winding machines, cop bottoms and other thready waste) in preparation for re-spinning, or for use in manufacture of gun cotton. **1812** in H. Davy *Elem. Agric. Chem.* (1813) 133 Hard wheat always sells at a higher price in the market than soft wheat. *Ibid.*, The flour of hard wheat is in general superior to that made from soft. **1843** *Penny Cycl.* XXVII. 301/2 There are three principal varieties...These are the hard wheats, the soft wheats, and the Polish wheats. **1856** J. C. Morton *Cycl. Agric.* II. 1005 *T[riticum] vulgare durum*, hard African wheat. **1908** *Westm. Gaz.* 1 July 6/3 The hard-wheat lands of Canada. **1971** *Times* 23 Jan. 21/4 Britain should take commodities like oil, wine, hard wheat.

B. *sb.* **1. a.** Delete † *Obs.* and add later examples.

1795 A. Shirrefs *Sale Catal.* 3 (E.D.D.), A plain North-country bard, Who fain would cripple through the hard. **1808** Jamieson *Dict. Scot. Lang.*, *Hard*, difficulty, hardship. *To come through the hard*, to encounter difficulties, to experience adverse fortune. **1858** G. Roy *Generalship* vi. 101 The bits o' bairns run a great risk o' coming through the hard. **1902** *Westm. Gaz.* 10 July 10/1 He had 'come through the hards' himself. **1904** *Daily Chron.* 27 May 3/4 She is a lady who..has given her life to nursing, and has gone through its hards.

c. In various technical applications.

1855 *First Rep. Adulteration of Food* 2 in *Parl. Papers 1854–5* VIII. 221 Bread is adulterated with mashed

potatoes, alum, 'hards', and sometimes..with sulphate of copper. **1921** *Dict. Occup. Terms* (1927) § 238 *Sider, hard* (needles); burnishes sides of sewing machine needles in the hard, *i.e.*, after hardening on revolving bob with oil and emery. **1937** *Archit. Rev.* LXXXI. 269/2 Scrap lead may contain other metals, such as the tin in solder, resulting in what is known as 'hards'. **1956** F. S. Atkinson in D. L. Linton *Sheffield* 268 The 'hards' of the Barnsley seam and, to a lesser extent, of the Parkgate or Deep Hard, make an excellent locomotive coal. **1960** *Gloss. Coal Terms* (*B.S.I.*) 12 *Hards*, a commercial term for the larger sizes of dull hard coal, in contrast to 'brights'.

d. Also *hard-on* sb. (and as adj.). An erection of the penis. *slang.*

1893 Farmer & Henley *Slang* III. 269/1 *Hard-bit* (or *bit of hard*). 1. The *penis* in erection. *Ibid.* 270/1 *Hard-on* adj. phr., prick-proud. **1922** Joyce *Ulysses* 527 What, boys? That give you a hardon? **1937** Partridge *Dict. Slang* 375/2 *Have a hard up*, to have a priapism. **1966** N. Behn *Kremlin Let.* I. iv. 56 When he wasn't plundering I suppose he was raping. He was more like a dog with a hard-on than a man with a mission. **1967** A. Wilson *No Laughing Matter* III. 377 He pulled up her red woollen dress..but still no hard. **1971** B. W. Aldiss *Soldier Erect* 44 The bromide damped down desire—you really had to work to get a hard on, whereas before it always flipped up naturally. **1971** B. Thornberry tr. Hansen & Jensen's *Little Red School-Bk.* (ed. 2) 95 When boys get sexually excited, their prick goes stiff. This is called having an erection or 'getting a hard on'. **1972** *Screw* 12 June 10/2 Billy and I talked down our hardons and..went downstairs to load the truck.

8. Tobacco in a cake.

1865 T. Archer *Pauper, Thief & Convict* v. 83 Peaceable companions..smoking pipefuls of 'hard' which they cut from a flat cake with their clasp-knives. **1898** G. Bartram *White-headed Boy* iv. 102 Packages of shag tobacco, lumps of sweetened 'hard'. *Ibid.* 105 Lind me a hand..with this lump o' harrd. **1898** *Daily News* 24 Feb. 3/1 Mr. Atkins..pulled at his pipe until he floated off into dreamland on a whiff of 'hard'.

hard, *adv.* Add: **1. c.** Very, extremely. *U.S. colloq.*

1850 N. Kingsley *Diary* (1914) 97 Mr. Hopkins is hard sick. *a* **1910** 'O. Henry' *Trimmed Lamp* (1916) 16 He isn't a millionaire so hard that you could notice it, anyhow.

8. a. *hard-bit, hard-driven, hard-driving* (example), *hard-hitting* (earlier and later examples), *hard-hurled, hard-running, hard sought, hard-worked* (examples); **b.** *hard-lived, -looking, -pressing, -tried, -used* (example); **hard-hit**, severely stricken by misfortune, grief, or disaster; deeply in love; **hard-pushed**, in difficulties; **hard run** *U.S.*, in difficulties or want, esp. with regard to money; **hard-wearing**, able to stand a considerable amount of wear; **d.** *hard-pressed* (examples); **hard-bound** orig. *U.S.*, (of books) bound in boards; **hard-cased** *U.S.* = **hard-bound*; **e. hard-spun** (example); also *hard-twisted*.

1886 Kipling *Dep. Ditt.* (ed. 2) 108 What a hard-bit gang were we. **1946** *Publishers' Weekly* 5 Oct. 1971/1 Several publishers of hard-bound reprints offer new series or expanded ones. **1952** *Amer. Speech* XXVII. 148 The ubiquitous 'paper-back'..is undoubtedly the cause of a reversal in bookbinding nomenclature. Whereas the board-bound used to be the normal and expected kind of book it is now necessary to use the qualifying adjectives *hard-bound, hard-cased*, or *hard-covered* when one refers to any book not in paper covers. **1959** *Times Lit. Suppl.* 6 Nov. p. xxxviii/4 It is estimated that more than 233,000,000 copies were sold in 1957, as against 32,000,000 hard-bound adult trade books. **1951** *Publishers' Weekly* 2 June 2357 Using the conventional method, eight to ten hours is a fair estimate of the time required to build in each batch of hard-cased books. It takes that long for paste to set and hinges to be formed. **1902** 'Mark Twain' in *North Amer. Rev.* Dec. 762 The poor and the hard-driven. **1949** R. K. Merton *Social Theory* (1951) 17 A small, hard-driven group of professors. **1951** M. McLuhan *Mech. Bride* 157/1 The cowboy is as non-erotic as the hard-driving executive. **1860** 'Old Shekarry' *Hunting Grounds of Old World* i. 19, I feel sure he is hard hit. *Ibid.* 20 A bright crimson pool..showed that he was hard hit. **1884** G. C. Davies *Peter Penniless* xix. 145 Hard Hit. **1891** M. E. Braddon *Gerard* xxix, You've been hard hit. **1909** H. G. Wells *Ann Veronica* ix, She saw her aunt in tears, her father white-faced and hard hit. **1839** *Q. Rev.* LXIII. 25 Our hard-hitting Irish labourers. **1955** *Times* 16 July 5/3 His plea was for an immediately available joint military force of hard-hitting character. **1962** *Christian Cent.* 26 Sept. 1164/1 Hard-hitting new book and films to help you combat communism. **1876** G. M. Hopkins *Poems* (1918) 22 A released shower, let flash to the shine, not a lightning of fire hard-hurled. **1921** Galsworthy *To Let* II. i, A look of life hard-lived. **1884** 'Mark Twain' *Huck. Finn* 89 A couple of mighty hard-looking strangers. **1915** A. Conan Doyle *Valley of Fear* v. 79 They were a mighty hard-looking crowd. **1825** Mill *Speech* in *Autobiogr.* (1924) 282 The Lion, finding himself hard-pressed, called together the aristocracy of the forest. **1891** Hard-pressed [see Pressed *ppl. a.*¹]. **1961** *New Scientist* 16 Mar. 664/3 Hard-pressed managers and engineers can hope to read only a tiny fraction of it. **1938** *New Statesman* 20 Aug. 282/1 Mr. Lennox Robinson..said that..it was not fair to press the lecturer. But the S.J...was a hard-pressing man. **1950** D. Gascoyne *Vagrant* 43 And the heart's slowly dulled By the hard-pressing years. **1807** J. Barlow *Columb.* VII. 259 To aid her hard-pusht powers. **1834** [Asa Greene] *Perils of Pearl St.* 123 (Bartlett), We began to be hard pushed. Our credit, however was still fair. **1822** J. Fowler *Jrnl.* 22 June (1898) 163 We have left them all behind, and will be hard run for meat. **1834** *Deb. Congress U.S.* 10 Mar. 848 Men, I say, who, to use the mercantile phrase, are 'hard run' to make ends meet, and only wanting an honorable excuse to fail. **1845** *N.Y. Tribune* 1 Nov. (Bartlett), We knew the Tammany party were hard run; but we did not know it was reduced to the necessity of stealing the principles of Nativism. **1865** Rossetti *Let.* 27 July (1965) II. 562 I'm dreffle hard run for tin till the end of next week when I shall have some. **1939** C. Morley *Kitty Foyle* 324 Everybody there looked so hard-run it cheered me up. **1952** C. Day Lewis tr. *Virgil's Aeneid* v. 98 The hard-running waves off Malea. **1963** *Times* 29 May 3/5 Hard-running fairways and small, sometimes tricky greens. **1909** J. Jusserand *Lit. Hist. Eng. People* III. 162 His [*sc.* Shakespeare's] most wonderful inventions were not hard-sought finds. **1906** Goodchild & Tweney *Technol. & Sci. Dict.* 864/2 s.v. *Yarn*, The yarn is defined as soft spun, medium spun, hard spun, according to the amount of twist it has received. **1906** *Daily Chron.* 1 Oct. 3/2 Its purpose of helping the hard-tried bookseller. **1906** Goodchild & Tweney *Technol. & Sci. Dict.* 833/1 s.v. *Warp*, The term applied to the series of spun threads, usually stronger and harder twisted than the weft. **1962** J. T. Marsh *Self-Smoothing Fabrics* xi. 168 Cotton voiles with their hard-twisted yarns may be impregnated on a mangle whose bowls have been wrapped with a fine cloth. **1950** *Mind* LIX. 407 Difficulties which can in one sense of a hard-used word be called 'philosophical'. **1909** *Daily Chron.* 11 June 7/5 Everything possible to be done is achieved in the endeavour to make it hard-wearing. **1928** *Observer* 1 Apr. 13 [This] Lingerie is amazingly hard-wearing. **1894** 'Mark Twain' in *Century Mag.* Jan. 330/1 He was coarsely fed and hard worked. *a* **1930** D. H. Lawrence *Phoenix* (1936) 74 The busy, hard-worked-looking woman.

hard-and-fa·stness. The condition of being hard and fast; hard and fast character.

1903 *Daily Chron.* 5 Mar. 3/3 The 'hard-and-fastness' of experience. **1904** *Westm. Gaz.* 18 June 13/2 By denying the hard-and-fastness and asserting the strictly provisional character of the forms or categories.

Hardanger (hā·ɹdæŋəɹ). The name of a district in west Norway used *attrib.* or *absol.* in names of things connected with Hardanger, as *Hardanger cloth, embroidery, fiddle, violin.*

1883 J. M. Fleming *Old Violins & their Makers* v. 177 In the Exhibition of 1862, a specimen of their class of work—a Hardanger violin—was exhibited. **1900** Grove *Dict. Mus.* (ed. 2) IV. 663/1 The Halling..is accompanied on the Hardanger fiddle.., a violin strung with four stopped and four sympathetic strings. **1908** *Sears, Roebuck Catal.* 527/5 Hardinger [*sic*] Canvas. **1928** *Funk's Stand. Dict.*, *Hardanger*, ornamental needlework in the pattern of diamonds or squares, made at Hardanger. **1930** tr. *T. de Dillmont's Encycl. Needlework* 580 This border is a specimen of the Norwegian openwork known under the name of 'Hardanger' embroidery. **1957** M. B. Picken *Fashion Dict.* 159/1 *Hardanger cloth*, soft cotton cloth of excellent quality. *Ibid.*, *Hardanger embroidery*, heavy, symmetrical, Norwegian needlework done in elaborate diamond or square pattern. **1972** P. A. Whitney *Listen for Whisperer* vii. 120 What you heard was one of our Hardanger fiddles.

hardback. Add: **a.** (Later examples.)

1958 J. Carew *Wild Coast* ii. 22 Hector..watched a hardback beetle crawling up the wall...'Boy, if you kill all the hardbacks that come in here you will make a mess of my clean floor.' **1959** P. Capon *Amongst those Missing* 66 The insects..whirred and buzzed..and the noise made by the hardbacks..kept Harry's nerves on the stretch.

2. A book bound in stiff boards; cf. *Paperback(ed). Also *attrib.* So **hard-backed** *a.* (see also *Hard *a.* 21).

1954 *New Republic* 26 Apr. 18 (*heading*) New novels: hardbacks or paperbacks. **1957** *Harper's Mag.* Sept. 94/3 Is it not possible that he may come away reading nothing but paperback books, that he will have become attuned to never spending $4.50 on a hardback? **1958** *Economist* 8 Nov. Suppl. 1/1 The retailer's margin on paperbacks is just as profitable as on hardbacks... A hardback order may well be topped up with a couple of 'quality paperbacks'. **1959** *Times* 24 Nov. 6/4 Most 'respectable' American publishers respect the British publisher's hardbacked and paperbacked book rights. **1960** *Times* 3 Feb. 17/4 The big paperback publishers are not hardback publishers but specialists in what is virtually a new genre. **1970** G. Greer *Female Eunuch* 170 Love affairs whether in cheap 'romance' comic-papers or in hard-back novels.

hardboard. [Board *sb.* 3.] A stiff type of board made from wood-pulp fibre. Also *attrib.*

1929 C. J. West *Bibliogr. Pulp & Paper Making 1900–1928* 157 (*heading*) Preparation of hard boards. **1934** *House Building 1934–1936* xxvii. 264 (Advt.), Insulite is an ideal material for the lining of walls and ceilings... Insulite products include Hardboard and Boards for Roof Insulation. **1939** *Chem. Abstr.* 2675 A process of making dense hardboard sheet products, which consists in subjecting fibrous wood or woody lignocellulose material..to a heat treatment. **1959** *Times* 18 May 10/4 Composite hardboard..is displacing traditional panelled construction in many buildings. **1959** *Housewife* June 32 Many of the huts are divided..by hardboard partitions.

hard-boil, *v.* [Back-formation f. next.] *trans.* To boil (an egg) until hard-boiled. Also *transf.*

1895 'Mark Twain' in *Harper's* Nov. 886 No more time to decide it than it takes to hard-boil an egg. *a* **1930** D. H. Lawrence *Etruscan Places* (1932) 16 He [*sc.* a shepherd] is the faun escaping again out of the city precincts... You cannot hard-boil him. **1963** *Listener* 10 Jan. 103/3 Hard-boil the eggs. **1973** *Nature* 23 Mar. 258/2 Considerably less than 12 h is required, either at 91° or 86° C, to hard-boil an egg.

hard-boiled, *a.* [f. *to boil hard*, where *hard* is a predicative adj. Cf. Hard *adv.* 8.]

1. Of an egg: boiled till the white and yolk are solid.

1723 J. Nott *Cook's & Confect. Dict.* No. 21, Mince.. the Yolks of hard boil'd Eggs. **1747** H. Glasse *Cookery* (1784) 71 Chop two or three hard-boiled eggs fine. **1833** [see Hard *adv.* 8 d]. **1846** A. Soyer *Gastron. Regen.* 445 Prepare a border of hard-boiled eggs. **1968** C. Roden *Bk. Middle Eastern Food* 99 (*heading*) Fried hard-boiled eggs.

2. Of articles of clothing: stiff, hard. *U.S.*

1903 A. Adams *Log of Cowboy* ix. 58 That fellow in front of the drug store over there with the hard boiled hat on. **1919** S. Lewis *Free Air* 86 To Claire, traveling men were merely commercial persons in hard-boiled suits.

3. Hardened, callous; hard-headed, shrewd. orig. *U.S.* Hence, of measures, practical.

1886 'Mark Twain' *Speeches* (1923) 137 Hard-boiled, hide-bound grammar. **1915** in *Amer. Speech* (1937) 260 Hard boiled egg who wouldn't bid 90 on 100 aces. **1919** in F. A. Pottle *Stretchers* (1930) 354 We are too hard-boiled to make much of a demonstration. *Ibid.* 358 Two hardboiled Irish sergeants are terrorizing the barrack. **1926** *Publishers' Weekly* 10 July 120/1 Stone..being hard-boiled, waited a few days to notice any appreciable increase in sales. **1926** *Ladies' Home Jrnl.* 26 Aug., The hard-boiled cynic has a shell it [*sc.* satire] can never penetrate. **1928** *Weekly Dispatch* 3 June 10/3 From its obscure beginning down in the 'tough' section of New York, up through the 'hard-boiled' wards of the great city, into municipal politics and thence into the Governor's chair. **1929** A. Conan Doyle *Maracot Deep* vi. 153 The hard-boiled Scanlan actually fell down in a faint. **1931** Buck & Anthony *Bring 'Em Back Alive* 163 It is all a hard-boiled proposition of not treading on the other fellow's feet for fear he may rise up and poke his big toe in your eye. **1932** E. Wilson *Devil take Hindmost* viii. 80 That man of iron..a drastic-minded and hard-boiled Dane. *Ibid.* 82 There is a Detroit type which..has some of the energy and hard-boiled bluffness of the Chicagoans. **1934** *Archit. Rev.* LXXV. 116/2 Yet those old houses are safe and kind and probably better employed in modifying some hard-boiled business man's mentality. **1942** *Mind* LI. 274 It certainly is difficult to remain a stoic or a cynic, to be 'hard-boiled', for a long time. **1959** *Encounter* Sept. 62/2 The disregard of truth in favour of hard-boiled scientific ideals. **1968** *Times* 27 Sept. 2/3 Mr. Heath's hard-boiled image is beginning to crack.

Hence **hard-boi·ledly** *adv.*; **hard-boi·ledness.**

1933 H. J. Massingham *London Scene* iv. 76 No other quarter of London is so consciously, hard-boiledly, shamelessly middle-class [*sc.* as Kensington]. **1934** Webster, Hard-boiledness. **1936** *Times Lit. Suppl.* 28 Mar. 255/2 Pareto apprehended the essential 'hardboiledness' of politics by personal and bitter experience. **1939** A. Huxley *After many a Summer* I. vii. 86 He dreaded for her the influence of so much cynicism and hardboiledness.

ha·rd-burned, -burnt, *a.* [Hard *adv.* 8 d.] Made hard by intensified firing.

1851 C. Cist *Cincinnati* 214 Walls of hard-burnt brick. **1869** *Rep. Comm. Agric. U.S. 1868* 360 Hard-burned terra cotta pipes. **1893** Kate Sanborn *Truthf. Wom. S. California* 45 Half-cylindrical plates of hard-burnt clay.

harden, *v.* Add: **1. b.** *spec.* of metals.

1797 *Encycl. Brit.* VIII. 310/1 There are several ways of hardening iron and steel, as by hammering them, quenching them in cold water, &c. *a* **1877** Knight *Dict. Mech.* II. 1060/2 Iron is surface hardened by heating to a bright red, sprinkling with prussiate of potash, allowing to cool to a dull red, and cooling with water. **1957** *Encycl. Brit.* VI. 906/2 After the blades are forged or cut out they are hardened by heating in a suitable furnace to the correct temperature and then quenched.

7. *to harden off*: to inure (plants) to cold by gradually reducing the temperature of a hot-bed or forcing-house or by increasing the time of exposure to wind and sunlight.

1873 *Young Englishwoman* May 238/1 Everything which has been kept in the house during winter for summer planting, or raised in a frame..should be gradually hardened to endure the changes of life in the open air... This 'hardening off', as the gardeners term it [etc.]. **1905** *Terms Forestry & Logging* 13 *Harden off*, to prepare seedlings in the seedbed for transplanting by gradually exposing them to wind and sunlight. **1909** *Daily Chron.* 5 June 9/5 This cool treatment or 'hardening off' process. **1912** *Chambers's Jrnl.* Dec. 848/1 Plants raised in this frame require no hardening off. **1933** *Jrnl. R. Hort. Soc.* LVIII. 117 Such young plants are generally well hardened-off, and receive but little check when transferred to their new quarters. **1970** C. Lloyd *Well-Tempered Garden* ii. 54 When they [*sc.* the cuttings] have rooted, they can be ..returned to a close atmosphere but then gradually hardened off by the admission of more air.

b. To render (a nuclear missile or base) hard (see *Hard *a.* 14 f).

1958 R. D. Bowers in *Air Univ. Q. Rev.* X. 90 Another possibility..might be to harden our sites. *Ibid.* 92 Repeating the analyses and assigning various values to the parameters should provide a good feeling for the payoff in hardening missile sites. **1960** *Times* 11 Feb. 11/6 Though land-based missiles can be 'hardened' by burying them and surrounding them with concrete they are still vulnerable to..nuclear attack. **1972** *Sci. Amer.* June 15/3 Attempts to 'harden' such fixed missile-launchers (that is, to increase their resistance to the effects of nuclear explosions) are in the long run doomed to futility.

ha:rdenabi·lity. *Metallurgy.* [f. Harden *v.*: see -ity.] The extent to which a metal may be hardened (see also quot. 1954).

1932 *Jrnl. Iron & Steel Inst.* CXXVI. 609 (*heading*) Factors affecting the inherent hardenability of steel. **1950** J. H. Bateman *Materials of Construction* 411 The surface hardness and the distance from the surface of the hardening effect are measures of the hardenability of steel. **1951** *Engineering* 14 Dec. 764/2 He said that in some circumstances an alloy content was necessary in order to achieve the necessary hardenability of the steel. **1954** *Gloss. Terms Iron & Steel* (*B.S.I.*) 1. 16 *Hardenability*, the property which determines the depth and distribution of hardness after quenching under specified conditions. *Ibid.*, *Hardenability test*, a test to assess hardenability. A common example is the Jominy test.

hardened, *ppl. a.* Add: **1.** (Additional example.)

1730 [see Quenching *vbl. sb.* 1].

3. Rendered hard (see *Hard *a.* 14 f).

1960 *Aeroplane* XCIX. 588/2 In the case of Atlas, this hurried development has resulted in four different types of operational launch site—unprotected, semi-protected, semi-hardened and hardened—and immense cost has been a feature of the programme. **1962** *Listener* 5 Apr. 605/2 A relatively small number of 'hardened', invulnerable, I.C.B.M.s.

hardener. Add: **2.** That which hardens. **a.** *Photogr.* Any chemical used in the making of gelatine negatives to prevent the melting or frilling of the film in warm weather.

1909 *Cent. Dict.* Suppl., Hardener [Photogr.]. **1930** *Sel. Gloss. Motion Techn.* (*Acad. Motion Pict., Hollywood*), *Hardener*, solution used to harden photographic emulsion. **1948** A. L. M. Sowerby *Dict. Photogr.* (ed. 17) 362 A hardener containing 1½ per cent. of potash alum, used at this *p*H value, may be expected to raise the melting point of the gelatine.

b. In various technical applications (see quots.).

1903 *Westm. Gaz.* 30 Nov. 2/1 The hardening temperature for the steel called 'high carbon' is difficult to define; for the personal equation comes into play, and with different hardeners the variation in hardening temperature often reduces the quality of the steel. **1945** R. T. Rolfe *Dict. Metallogr.* 101 *Hardeners*, alloys prepared for the purpose of adding small quantities of additional elements to molten metals. **1951** *Gloss. Terms Plastics* (*B.S.I.*) 17 *Hardener*, a material used to promote the setting of certain types of synthetic resin. **1959** *Gloss. Packaging Terms* (*B.S.I.*) 13 *Hardener*, a chemical used to promote the setting of adhesive. **1966** A. W. Lewis *Gloss. Woodworking Terms* 43 *Hardener*, liquid used to speed up the setting of resin glue. **1967** E. Chambers *Photolitho-Offset* ix. 130 Potassium alum is recommended as a *hardener* in preference to chrome alum which, although more potent, loses its hardening power after short use and forms a sludge.

hardening, *vbl. sb.* and *ppl. a.* (Further examples: see Dict. s.v. Harden *v.*)

1877 *Encycl. Brit.* VI. 734/1 The hardening is accomplished by heating the blade to a cherry-red heat and suddenly quenching it in cold water. **1902** *Daily Chron.* 18 Jan. 5/4 The hardening of new-arrived drafts [of troops] is most noticeable. **1902** A. Bennett *Anna of Five Towns* viii. 176 The 'hardening-on' kiln, a minor oven where for twelve hours the oil is burnt out of the colour in decorated ware. **1908** *Westm. Gaz.* 15 Apr. 1/3 The inexorable and hardening passage of twenty years. *Ibid.* 21 Aug. 5/4 There has been a great hardening on the part of the merchants, who were formerly placing the stones [*sc.* diamonds] on the market for anything they could fetch. **1930** *Economist* 5 Apr. 758/2 The hardening of bill rates, which put a further reduction of the Bank Rate out of court for the time being. **1936** *Forestry* X. 124 By exposure to suitable, but not damagingly low, temperatures plants are rendered more resistant to frost; this is the process known as hardening, and takes place naturally during the autumn. **1940** *Economist* 6 Jan. 12/2 These difficulties have already had their effect a general hardening of prices. **1959** *Chambers's Encycl.* X. 686/2 Most fixing solutions also contain a tanning or hardening agent which unites with the gelatin of the emulsion layer, increases its melting-point and reduces its swelling in water. **1970** *New Yorker* 17 Oct. 171/1 These maneuvers have all added up to what one astute observer of the talks has described as 'a hardening of the arteries'.

harder (hā·ɹdəɹ). *S. Afr.* Also 8 **harter**, 20 **haarder.** [a. Afrikaans *harder*, Du., LG. *harder*, OE. *heardhara, heardra.*] Any of various species of the grey mullet family (Mugilidæ), of which *Liza ramada* and *M. cephalus* are well known.

1731 G. Medley tr. *Kolben's Pres. State Cape Good-Hope* II. 193 There is..about the Cape a Sort of Herrings the Cape-Europeans call Harters. **1838** D. Moodie tr. *Record* 13 We..caught and salted 400 large steenbrass, and about 2,000 harders. **1892** Simmonds *Dict. Trade* (new ed.), *Harder*, a kind of mullet about twelve inches long, caught near the coasts of the Cape colony. **1947** K. H. Barnard *Pict. Guide S. Afr. Fishes* 81 The family of Grey Mullets, called in South Africa Harders or Springers (*Mugilidae*) is economically very important. **1962** *Cape Argus* (Mag. Sect.) 11 Aug. 1/7 Bunches of harder, maasbanker and pilchards and many other kinds of fish. **1971** *Daily Dispatch* (East London, Cape Province) 8 Mar., There were hundreds of haarders (mullet) in the bay itself.

hardhack. (Earlier and later examples.)

1832 W. D. Williamson *Hist. State Maine* I. 116 The Hardhack, a barren bush, usually chooses poor cold ground for its residence and growth. **1880** *Harper's Mag.* Dec. 85 Them mulleins an' hardhacks in the buryin'-ground. **1968** E. R. Buckler *Ox Bells & Fireflies* vi. 95 The hardhacks, with roots like the roots of wisdom teeth, to be kept back from the edges of the cleared land.

hardhead¹, hard-head. Add: **1.** (Later example.)

1967 P. Jones *Fifth Defector* xiii. 190 I'd advise you to keep your mouth shut and let the hardheads handle it at embassy level.

b. A person not easily affected by alcohol.

1860 E. Cowell *Diary* 19 Mar. (1934) 41 Mr. Van Orden a very pleasant, but, to Sam, very dangerous companion being a great drinker, and one of the 'Hard Heads' whom drink does not seem to hurt.

Hardian: see *Hardyan *a.* and *sb.*

hardie, var. Hardy *sb.*

1957 R. Lister *Decorative Wrought Iron-Work* ii. 13 The square hole through the heel is called a *hardie hole*. *Ibid.* 228 *Hardie*, a small chisel, used in the anvil. **1964** [see *hot set* s.v. *Hot *a.* 12 c]. **1965** A. F. Shirley *Metalwork Techniques* vi. 84 Metal..should be heated where it is to be cut and this part placed on the hardie and hammered to form a vee cut.

hardly, *adv.* Add: **10.** *hardly-used* (earlier example), *-won.*

1866 Trollope *Belton Est.* II. vi. 158 The hardly-used groom had returned from his futile afternoon's inquiry. **1937** *Discovery* Aug. 240/2 The hardly-won natural gem. **1952** C. P. Blacker *Eugenics* 282 Each hardly-won improvement in human conditions.

hard metal. 1. Any of various alloys valued for their hardness.

1729, 1845 [see Metal *sb.* 5]. **1911** *Encycl. Brit.* XXI. 339/1 Hard metal (96 parts of tin, 8 of antimony and 2 of copper), a mixture very closely resembling..'Britannia metal'.

2. (See quot. 1967.)

1936 *Jrnl. Iron & Steel Inst.* CXXXIV. 66A The author discusses the application of hard metal alloys and of diamond substitutes (such as tungsten carbide) to the hard-facing of tools. **1960** *Ibid.* CXCIV. 532/1 The final section is devoted to a consideration of materials used for cutting tools, namely unalloyed tool steels, alloyed tool steels,..hard metals, i.e. sintered tungsten carbide and the like. **1967** A. K. Osborne *Encycl. Iron & Steel Industry* (ed. 2) 195/1 *Hard metals*, powdered carbides of tungsten, tantalum, or titanium, cemented into solid masses by mixing with powdered cobalt or nickel, then cold pressing and sintering. Used for cutting tools, wire drawing dies, and parts subjected to heavy wear and abrasion.

hardness. Add: **c.** *spec.* The degree of resistance of a mineral to abrasion or scratching.

1784 R. Kirwan *Elem. Mineral.* I. App. 171 (*heading*) Table of the comparative hardness of different Species of Stones, extracted chiefly from the Memoirs of Stockholm, for 1768. Mr. Quist, the author of this Memoir determined the hardness of most of the following stones. **1904** Goodchild & Tweney *Technol. & Sci. Dict.* 279/2 *Hardness*, the hardness of a mineral is determined by noting which of the standard minerals the specimen may be scratched by, and which of the ten it will scratch.

d. Of radiation (cf. *Hard *a.* 16* a).

1926 R. W. Lawson tr. *Hevesy & Paneth's Man. Radioactivity* IV. i. 45 The shorter the wave-length of the γ-rays, the greater is their penetrating power through matter, or the greater their 'hardness'.

hardometer (hāɹdǫ·mɪtəɹ). [f. Hard *a.* + -o + -meter.] An instrument for measuring the hardness of metals.

1934 H. O'Neill *Hardness of Metals* iii. 90 The 'Firth Hardometer'..is an excellent bench instrument. **1940** *Jrnl. R. Aeronaut. Soc.* XLIV. 845 The following chapter carries on the hardness tests with details of the Rothwell test, the Firth hardometer, the Firth-Brown variable load hardometer and notched bar impact testing. **1946** *Firth Brown Gloss. Metall. Terms* 26 *Firth hardometer*,... For testing harder materials a pyramid diamond indenter is recommended and can be supplied with the [Brinell] machine in addition to the hardened steel ball. **1958** *Oxf. Univ. Gaz.* 10 Mar. 772/1 Mr. Allen also completed the examination of about 40 Bronze Age implements of copper and bronze by hardometer.

hard-on: see *Hard *sb.* 1 d.

hard-pan. For *U.S.* read orig. *U.S.* Add: **1.** (Earlier and later examples.)

1817 T. Dwight *Trav. New Eng.* (1821) I. 374 What is here called hard pan, a very stiff loam, so closely combined, as wholly to prevent the water from passing through it. **1963** D. W. & E. E. Humphries tr. *Termier's Erosion & Sedimentation* 406 *Hardpan*, an English agricultural term (used mainly in the U.S.A., Africa and Australia) for a horizon in podsolic and lateritic soils hardened by precipitation and cementation. **1968** *New Scientist* 10 Oct. 79/3 The number of rice paddies under cultivation in some Far Eastern countries could be doubled using an asphalt 'hardpan'.

3. *attrib.* and *Comb.*

1870 J. K. Medbery *Men & Myst. Wall St.* 212 Hard pan is soon reached, and both old world and new are full of hard-pan capitalists. **1889** K. Munroe *Golden Days* xi. 122 To tell the honest hard-pan truth. **1907** R. W. Service *Songs of Sourdough* (1908) 77 When a man gits on his uppers in a hard-pan sort of town. **1928** *Bull. Amer. Soil Survey Assoc.* IX. 33 Immaturely developed soils may have a hardpan-like horizon.

hardpeer (hā·ɹtpiəɹ). *S. Afr.* Also **hardepeer**, and anglicized **hard pear.** [Afrikaans; f. Du. *hard* hard + *peer* pear.] A small tree of the Cape, *Olinia cymosa*, having hard wood; also applied to other trees (see quot. 1913).

1801 J. Barrow *Trav. S. Afr.* I. 340 Hard peer...Uses.. Sometimes in waggons. **1851** *Catal. Gt. Exhib.* IV. 1. 951 Pear (hard). **1874** Lindley & Moore *Treas. Bot.* Suppl. s.v. *Olinia*, The plant grows in rocky thickets and woods at the Cape, where it is known as Hardpeer. **1880** Hard Pear [see Pear *sb.* 3]. **1887** C. A. Moloney *Sk. Forestry W. Afr.* 354 Hardpeer of the Cape. Shrub 4 to 10 feet high. **1913** Pettman *Africanderisms*, *Hard Pear.* (1) In Natal this name is given to *Pleurostylia capensis*. (2) In the Cape Colony it is applied to *Strychnos Henningsii*. **1961** Palmer & Pitman *Trees S. Afr.* lii. 297 *Strychnos Henningsii*. *Hard pear*, hardepeer... This tree should not be confused with the 'hard pear' of the Knysna forests, which is *Olinia cymosa*, and which belongs to a different family.

hard scrabble (hāɹd skræ·b'l). *U.S. colloq.* [cf. Scrabble *v.* 4.] **1.** 'A place thought of as the acme of barrenness where a livelihood may be obtained only with great difficulty. Also attrib. Often as a proper name.' (*Dict. Americanisms.*)

1804 Lewis & Clark *Orig. Jrnls.* (1905) VII. 38 Got on our way at hard Scrable Perarie. **1904** *Pittsburgh Gaz.* 7 July 4 In the early days of my ministry..I was sent to take charge of a little hard-scrabble circuit. **1949** *Sat. Even. Post* 30 Apr. 22/3 She was the daughter of a hard-scrabble rancher. **1972** *Science* 26 May 891/1 The reservoir would back up along the creek and inundate 125 small, hardscrabble farms that lay along 24 miles of the stream's flood-plain.

2. 'A vigorous effort made under great stress.' (*Dict. Americanisms.*)

1812 *Salem Gaz.* 29 May 2/3 Presidential Hard Scrabble! **1851** H. Melville *Moby Dick* II. xxx. 205 While taking that hard-scrabble scramble upon the dead whale's back. **1854** S. Hale *Lett.* (1919) 7 By a well-organised hard-scrabble, Luc. and I get the breakfast things washed by nine o'clock. **1972** *Sat. Rev.* 26 Feb. 31/3 [Ulster] Catholics..enjoy the benefits of..an economy that may be hardscrabble but is still substantially more prosperous than the South's.

hardshell, *a.* and *sb.* Add: **A.** *adj.* **1.** (U.S. examples.) Also applied to the fruit of a nut-tree.

1798 *Spectator* (N.Y.) 7 Nov. 2/5 Hardshell almond trees. **1818** *Amer. Monthly Mag.* II. 296 The hard shell clam..is cooked by roasting. **1855** *Knickerbocker* XLVI. 222 'Hard-shell' clam-catchers. **1942** M. K. Rawlings *Cross Creek* xvii. 226 We have four turtles, the gopher;.. the hard-shell cooter; the soft-shell; and the alligator cooter.

2. (Earlier examples.)

1838 W. Y. Allen *Diary* 17 July in *S.W. Hist. Q.* (1914) XVII. 54 Was introduced to Daddy Spraggins, a Hardshell Baptist preacher. **1846** J. J. Hooper *Adv. Simon Suggs* i. 13 He lived with his father, and an old 'hard shell' Baptist preacher.

B. *sb.* **1.** Also *fig.*, a stubborn or unemotional person.

1858 *South. Cultivator* XVI. 187/2 We have, however, one or two specimens in our eye of the genus, *hard shell*, who still do as their *daddies* did. **1916** H. L. Wilson *Somewhere in Red Gap* iv. 135 A grouchy old hardshell with white hair and whiskers whirling about his head. **1919** T. K. Holmes *Man fr. Tall Timber* xiii. 156 I've ridden up here from Tall Timber Junction to get acquainted with you hardshells.

2. (Earlier and later examples.)

1845 *Knickerbocker* XXVI. 285 A 'Hard-Shell' recently turned a 'Soft-Shell' out of church. **1855** *Putnam's Monthly* V. 190 The claim of 'Hard-Shells', touching their familiarity with the Bible. **1872** E. Eggleston *Hoosier Schoolmaster* xii, Of course the Hardshells are prodigiously illiterate. **1908** *Dialect Notes* III. 319 *Hardshell*, a Primitive Baptist.

hardshelled, *a.* **1.** Having a hard shell; = Hardshell *a.* 1.

1611 [in Dict.]. **1782** J. H. St. J. de Crèvecœur *Lett. Amer. Farmer* iv. 135 The shores..abound with the soft-shelled, the hard-shelled, and the great sea clams. **1796** B. Hawkins *Lett.* (1916) 17 A grove of dwarf hard shelled hickory trees. **1839** C. F. Briggs *Adv. H. Franco* II. i. 2 Close by, was a negro opening hard-shelled clams. **1865** *Trans. Ill. Agric. Soc.* V. 408 Beetles, or, as they are sometimes called, 'hard-shelled bugs'. **1942** M. K. Rawlings *Cross Creek Cookery* xvii. 191 Why does a hard-shelled cooter lay a soft-shelled egg, and a soft-shelled cooter lay a hard-shelled egg?

2. *fig.* = Hardshell *a.* 2. Also, hardened, callous.

1842 J. S. Buckingham *Slave States Amer.* I. 197 The Baptists [in Macon, Georgia] are of the order called here 'Hardshelled Baptists'. **1872** W. Mathews *Getting on in World* xi. 153 There is no man so 'hard-shelled' that his soul cannot be reached by kindness. **1904** *N.Y. Herald* 23 Oct. 16 There are a good many hard shelled Bryan men who intend to vote for Roosevelt. **1909** R. A. Wason *Happy Hawkins* 108 It was a hard-shelled book. **1941** J. Stuart *Men of Mountains* 331 If you could see all of us Republicans, Democrats, Methodists, Forty-Gallon Baptists, Hard shelled Baptists,..shaking hands and asking the other how he is after the long night o' sleep. **1965** *Times Lit. Suppl.* 25 Nov. 1046/1 Various forms of hard-shelled Establishment versus dissent.

hard stone, ha·rdstone. a. Any type of hard stone. **b.** A precious or semi-precious stone.

The earliest examples are instances of the informal union of *hard* adj. and *stone*.

1568 [see Hard *a.* 1]. **1613** [see Stone *sb.* 2]. **1733** W. Townesend *Let.* 28 Aug. in *Archit. Rev.* (1945) XCVIII. 105/3 Eight feet below the top of ye hardstone plinth. **1905** *Daily Chron.* 11 July 5/7 The hard-stone works. **1921** *Dict. Occup. Terms* (1927) § 572 *Banker hand, hewer* or *mason*; *hard-stone banker mason*; cuts, and if necessary saws, blocks of stone at 'banker'. **1931** *Times* 16 Mar. 22/4 Chinese hard-stone carvings. **1935** *Burlington Mag.* June 299/1 (*title*) An exhibition of hardstone carvings. **1936** *Ibid.* Aug. 91/2 A few jades and hardstones. **1958** *Times* 2 Dec. 12/5 Chinese porcelain and hardstones realized £9,585 at Christie's yesterday.

hard-tack. Add: (Earlier and later examples.) Also, hard bread or biscuits generally. Also *fig.* and *attrib.*

1836 *Knickerbocker* VIII. 203 When I was the size of that monkey there, who knows how to do nothing but gnaw hard tack. **1888** *Century Mag.* XXXVI. 614/1 A little rabbit that kept..nibbling at our bread and hard-tack. **1899** T. Hall *Tales* 108 A meal of raw bacon, hard-tack and cold water. **1909** *Daily Chron.* 8 July 9/2 Of all the hard-tack breads..I have found..the small ringed bread of Siberia the most substantial. **1931** *Economist* 5 Dec. 105/1 It has paved the way..for the real hard-tack committee work on the thousand practical problems of the constitution builder. **1955** W. Foster-Harris *Look of Old West* ii. 56 Hardtack..was hard, unleavened bread, baked in cakes..about 3 inches square, decorated with what looked like nail punctures. **1960** *Economist* 15 Oct. 219/1 Some of those who raised left-wing political rather than hard-tack organisational questions being shouted down by genuine party workers.

hardtail (hā·ɹd,tēɪl). *U.S.* [f. Hard *a.* + Tail *sb.*¹] **a.** A marine fish, *Caranx crysos*, found in the western parts of the Atlantic Ocean.

1884 G. B. Goode *Fisheries U.S.* I. 324 The Jurel—Caranx Pisquetus. This fish, known about Pensacola as the 'Jurel', 'Cojinua', and 'Hard-tail'. **1902** Jordan & Evermann *Amer. Food & Game Fishes* 306 The runner, hard tail, or jurel, reaches a foot or more in length..and is a food-fish of considerable importance. **1968** *OECD Multiling. Dict. Fish* 139 *Caranx crysos*...Also called runner, hardtail, crevalle.

b. *slang.* A mule.

1917 A. G. Empey *Over Top* 294 *Hard tails*, mules. **1931** 'D. Stiff' *Milk & Honey Route* 207 *Hard tails*, mules, usually old ones. So named because they show little response to the skinner's whip. Young mules are shave-tails. **1966** *Publ. Amer. Dial. Soc.* XLVI. 26, I was driving an old pair of hard tails.

hard up. Add: Also *sb.* (see quots.).

1851 Mayhew *Lond. Labour* I. 3/2 The cigar-end finders, or 'hard-ups', as they are called. **1905** *Daily Chron.* 17 May 6/7 In tramp phrase they [*sc.* cigar and cigarette ends] are known as 'hard-ups', and are smoked along the road. **1933** 'G. Orwell' *Down & Out* xxxii. 236 Hard-up—tobacco made from cigarette ends. **1959** *Listener* 5 Mar. 406/1 We roll a couple of 'hard-ups' to smoke. Hard-ups are made of tobacco we collect from cigarette ends.

hard-upishness. (Earlier example.)

1859 D. G. Rossetti *Let.* 23 June (1965) I. 353 As for hardupishness..I have been literally penniless for two days.

hardware. Add: **1. b.** Weapons.

1865 L. N. Beaudry *Hist. Rec. 5th N.Y. Cavalry* (1868) 38 Capt. Hammond..charged upon the rebels.., crying as he flew forward, 'give them your hardware, boys!' **1885** *Daily News* 12 Nov. 5/4 The chances are that the authorities..may have had an eye on such kind of 'hardware' [*sc.* torpedoes]. **1914** Jackson & Hellyer *Vocab. Criminal Slang* 42 *Hardware*,..weapons; knives; razors; tools and paraphernalia used by safe-crackers [etc.]. **1929** *Amer. Speech* V. 59 His [*sc.* a Nebraska cowboy's] fire arms are frequently called his 'hardware'. **1955** *Bull. Atomic Sci.* Apr. 168/2 How much does our superiority in hardware contribute to our over-all security? **1966** *New Yorker* 29 Oct. 236 Oh, put your hands up, dear. He's got the hardware. **1967** *Observer* 11 June 11/2 The wholesale destruction this week of expensive hardware has brought home to some Arabs the folly of an extravagant armament policy.

c. The physical components of a system or device as opposed to the procedures required for its operation; opp. **software*.

1947 D. R. Hartree *Calculating Machines* 14 The ENIAC... I shall give a brief account of it, since it will make the later discussion more realistic if you have an idea of some 'hardware' and how it is used, and this is the equipment with which I am best acquainted. **1953** A. D. & K. H. V. Booth *Automatic Digital Calculators* xv. 169 The engineering difficulties encountered in this type of machine are great, and a considerable increase in the size and complexity of the 'hardware' seems inevitable. **1960** [see *Algorithmic *a.*]. **1960** *Times* 21 Mar. 13/6 Academician Sedov..knows the hazards of 'hardware' projected into space. **1960** *Times* (Computer Suppl.) 4 Oct. p. iii/7 Both punched card and computer 'hardware' will continue to develop very rapidly. **1962** J. Glenn in *Into Orbit* 6 The engineers and technicians had to start from scratch to develop the capsule and some of the other pieces of hardware which we are using. **1963** *Engineering* 23 Aug. 246/1 Hardware means the apparatus or machinery of computing, both the main instrument and its peripheral or ancillary equipment. **1964** *Daily Tel.* 14 Feb. 29/8 Engineers are leaving because..they cannot rely on..early development being continued to the 'hardware' stage. **1965** *New Scientist* 4 Nov. 331/1 All necessary hardware [for computers] exists today and is in production, but getting a comprehensive system off the ground requires the creation of a great deal of software. **1968** *Lebende Sprachen* XIII. 3/1 The practical requirements for the manned lunar landing program are being translated into hardware, experiments and programs. **1968** *Brit. Med. Bull.* XXIV. 198/2 Elaborate hardware is not a substitute for a thoroughgoing analysis of the whole area of medical vocabularies, procedures, information and systems for the care of patients. **1969** *Computers & Humanities* III. 139 Hardware refers to the physical apparatus while software describes the program languages that permit efficient use of the hardware.

d. Various slang and colloquial senses.

1839 *Spirit of Times* 1 June 153/3 He prepared to swallow his fifth invoice of 'hardware' [*sc.* whisky]. **1945** L. Shelly *Jive Talk Dict.* 12/2 *Hardware*, flashy jewelry. **1951** I. Shaw *Troubled Air* vii. 105 When the rating goes up, I buy my hardware [*sc.* jewellery] at Cartier's. **1963** *Amer. Speech* XXXVIII. 206 *Hardware*, slang for a medal or trophy won in a skiing competition. **1968** *Times* 9 Sept. 6/7 Prince William has some useful leisure hardware—like a private aircraft.

2. hardware paper, a make of durable wrapping-paper; **hardware store** orig. *U.S.*, an ironmongery shop.

1789 *Boston Directory*, Whitwell,..hardware store. **1886** *Harper's Mag.* June 48/1 Wrapping the stem [of the peach-tree]..with strong hardware or sheathing paper. **1964** M. Gallant in R. Weaver *Canad. Short Stories* 2nd Ser. (1968) 60 Across the Alps was the name of a hardware store and its address on the other side of Montreal. **1972** *P.O. Telephone Directory, London Yellow Pages Classified* (North) 152/1 Barker's Hardware Stores Ltd, 43 High St.

hardwood. Add: **1.** (Later examples of attrib. use.)

1840 *Knickerbocker* XXVIII. 337 Most unexceptional 'hardwood' land. **1869** *Trans. Ill. Agric. Soc.* VII. 578 Some elevated ridges..called technically, hard wood ridges..escaped wholly, or in part the effects of the fire. **1903** S. E. White *Forest* viii. 89 The trunks of the hard-wood forest. **1911** *Daily Colonist* (Victoria, B.C.) 15 Apr. 18/1 We mail on request an illustrated catalogue of plain and fancy hardwood floors with prices attached. **1968** E. R. Buckler *Ox Bells & Fireflies* xxi. 297, I stand on top of the hardwood hill.

b. *ellipt.* A hardwood tree.

1905 J. Nisbet *Forester* I. III. iv. 419 Pit-planting is the usual method of growing Oak and all other kinds of hard-woods. **1908** *Westm. Gaz.* 15 Aug. 15/3 The tender colours of the hardwoods bursting towards summer glory. **1972** *Country Life* 16 Mar. 653/3 The number of trees by species..together with some idea of their quality (in the case of hardwoods, veneer, good or.. low-grade mining and fencing timber).

hardy, *a.* **4. b.** (Earlier examples); *hardy annual* (earlier examples); *hardy perennial*, a herbaceous plant with a perennial rootstock; also *fig.*

1664 Evelyn *Kalendarium Hortense* in *Sylva* 59 Auriculas..need not be hous'd; it is a hardy Plant. **1783** T. Blaikie *Diary of Scotch Gardener* (1931) 187 A little way from St Germains..ther is a Curious Gentelman one Mr. Trochereau who has a curious collection of hardy exotick plants. **1813** [see Borecole]. **1831** *Athenæum* 5 Nov. 718 This, truly, is 'a hardy Annual'! **1852** R. Buist *Amer. Flower-Garden Directory* (ed. 5) 29 Hardy Annuals..are possessed of much beauty of hue. **1900** J. M. Abbott in W. D. Drury *Bk. Gardening* viii. 260 Hardy herbaceous perennials are a very popular set of plants. **1916** 'Taffrail' *Pincher Martin* xiv. 248 The subjects most often brought under discussion, however—the hardy perennials, so to speak—were: [etc.]. **1944** A. Huxley *Let.* 9 Apr. (1969) 502, I am very glad to hear the good news of *The Art of Seeing*. The book has all the appearance of a hardy perennial. **1967** C. O. Skinner *Madame Sarah* vii. 132 Thousands of playgoers travelled thousands of miles to sob over Marguerite Gauthier's departure from life..in that hardy perennial whose actual title is *La Dame aux Camélias*. **1967** C. Lloyd *Hardy Perennials* i. 9 The hardy perennial possesses every virtue that you could require of a plant, except for a permanently visible structure.

Hardyan, Hardian (hā·ɹdiăn), *a.* and *sb.* [f. the name of Thomas *Hardy* (1840–1928), novelist and poet + -ian.] Characteristic of the works of T. Hardy. Also *sb.*, an admirer or follower of Hardy. Similarly **Hardye·sque** *a.*

1910 R. Brooke *Let.* Jan. (1968) 216 That abysmal darkness..inspired me with thousands of Hardyesque short poems about people whose affairs went dismally wrong. **1927** H. Crane *Let.* 29 May (1965) 300 A footnote of Hardian doom. **1929** *Sat. Rev.* 24 Aug. 221/2 But in the dialogue of the rustics it shows a delicious Hardyesque sense of humour. **1931** *Times Lit. Suppl.* 28 May 423/1 Confirmed and receptive Hardians..will know how many ..approach it with preconceptions of various degrees of falseness. **1941** Blunden *Thomas Hardy* 271 It can be admitted by almost any Hardyan that the poor passages in his work are an offering to the wanton or the unsympathetic critic. **1944** —— *Cricket Country* 142 An unspoken Hardyan complaint. **1960** C. Day Lewis *Buried Day* vi. 111 The rich Dorset accents and the Hardy-esque names.

hare, *sb.* Add: **2.** *to make a hare of* (examples).

1830 W. Carleton *Traits & Stories* II. 111 What a hare Mat mad iv 'im;..and did not lave him a leg to stand upon. **1938** J. Cary *Castle Corner* x. 562 That fella thought he'd made a hare of me, but I knew one trick better.

3. b. *hare and hounds* (= *paper chase* s.v. Paper *sb.* 12): also *fig.*

1920 T. S. Eliot *Sacred Wood* 11 Coleridge is apt to take leave of the data of criticism, and arouse the suspicion that he has been diverted into a metaphysical hare-and-hounds. **1938** Partridge *World of Words* ix. 261 Well worth the hare-and-hound chase through the dictionary. **1963** *Daily Tel.* 5 Feb. 20/6 Throughout his speech he was constantly heckled and interrupted, but the scene cannot be described as 'Hare and hounds'.

6. hare-coursing: see Coursing *vbl. sb.*¹ 2; **hare-pocket**, a pocket in a shooting-coat, made of a size to hold a hare; **hare's fur** *Ceramics*, a brown or black glaze streaked with silvery white or yellow used on some Chinese pottery.

1840 D. P. Blaine *Encycl. Rural Sports* v. i. 562 The credit of the organisation of the sport of hare coursing.. [is] without all doubt the undisputed property of the English. **1972** *Times* 5 Feb. 1/2 The House of Commons ended its Friday sitting in uproar and confusion when the Hare Coursing (Abolition) Bill was talked out. **1925** G. Burrard *Big Game Hunting* 281 Two 'hare' or 'poacher' pockets will be found most useful on occasions. **1950** *Q. Jrnl. Forestry* XLIV. 60 The map should be made ..to fit into the forester's hare pocket. **1899** S. W. Bushell *Oriental Ceramic Art* xxvii. 724 The most highly appreciated ware at the tea-testing parties..was the dark-colored pottery of the province of Fuchien..the tea-bowls of which were known to Chinese virtuosos as 'hare's-fur bowls'. **1934** *Burlington Mag.* May 214/1 *Temmoku* tea bowls with the 'hare's fur' glaze... They have a blackish stone-ware body, and a thick, lustrous black glaze streaked with hair lines of brown and silver. **1959** G. Savage *Antique Coll. Handbk.* 52 The black glazed wares of Honan are of great interest, and tea-bowls with a variegated dark brown glaze, known as 'hare's fur', came from Chien-an in Fukien Province.

hare, *v.*² [f. Hare *sb.*] *intr.* †**a.** To double like a hare. **b.** To run or move with great speed. Also with *it*.

1893 Farmer & Henley *Slang*, *To hare it*, to retrace one's steps; to double back. **1908** D. Coke *House Prefect* xi. 141 He had heard..the order, 'Hare'! Now 'Hare'! is Seftonian for 'Run—and run jolly quick'! *a* **1914** J. E. Raphael *Mod. Rugby Football* (1918) 262 Receiving the ball well inside his own half-way, Palmer commenced to 'hare' for the touch-line. **1917** P. Gibbs *Battles of Somme* 173 There were other trenches ahead, and the men 'hared' off to these. **1923** Wodehouse *Inimitable Jeeves* xiv. 178, I..hared it rapidly to the spot. **1957** *Listener* 19 Dec. 1046/1 The producer..can't go haring about collecting the items. **1958** *Woman* 11 Jan. 47/1, I hared up to London, left my book with the publishers and went to my flat. **1963** *Times* 13 June 5/1 Boulter took over by the backstraight and went haring away past 660 yards in 1 min. 21.1 sec.

harebell. Add: **3.** *attrib.*, as *harebell blue*.

1909 *Daily Chron.* 15 June 7/5 The dress linens..have been prepared in many exquisite shades of leaf-green, pale primrose, hare-bell blue. **1925** T. Dreiser *Amer. Trag.* III. xvi. 647 His harebell eyes showing only cold..practical logic. **1940** C. Day Lewis *Poems in Wartime* 8 This hare-bell height of calm. *Ibid.* 9 The climb To a tremulous, hare-bell crest. **1969** R. Godden *In this House of Brede* (1970) i. 39 Dame Veronica of the wistful hare-bell blue eyes.

harefoot. Restrict ? *Obs.* to senses 2, 3, 4 in Dict. and add examples to sense 1. Also *hare's foot*.

1901 *Encycl. Sport* I. 329/2 *Harefoot*, a long, narrow foot, carried far forward. **1945** C. L. B. Hubbard *Observer's Bk. Dogs* 161 American standards permit hare-feet. **1952** R. Leighton *Compl. Bk. Dog* (ed. 6) v. 80 Faults: [in Boxer]..hare's feet.

5. Used adverbially: swiftly. *poet.*

1939 T. S. Eliot *Family Reunion* I. ii. 60 What of the terrified spirit Compelled..To rise toward the violent sun.. Harefoot over the moon?

Hare Krishna (hā·re kri·ʃnā). [f. Hindi *hare* O God! + *Kṛishṇā* name of an incarnation of the god Vishnu.] The title of a love-chant or mantra based on a name of the Hindu deity Vishnu; used esp. as an incantation by members of a religious cult in the U.S. and elsewhere; hence *attrib.* or *absol.* to designate this cult or its members.

1968 *New Yorker* 17 Aug. 36 Newspaper pictures of the poet [*sc.* Ginsberg] chanting '*Hare Krishna*' at one of Leary's sellout psychedelic celebrations. **1970** *Time* 3 Aug. 31 He [*sc.* David Hoyt] became a member of the Hare Krishna cult and custodian of the Radha Krishna temple. **1971** *Times* 5 July 3/6 The founder of the Hare Krishna Movement, his Divine Grace A. C. Bhakividanta, did not make his expected appearance. **1971** E. Larsen *Strange Sects & Cults* viii. 164 Small groups of smiling English youngsters..could be seen slowly dance-marching in file through London's West End..to the accompaniment of some tinkling Oriental instruments, chanting 'Hare Krishna' and offering passers-by literature on their 'Krishna Consciousness Movement'. **1972** G. V. Higgins *Friends of Eddie Coyle* xi. 68 Near the first subway kiosk the Hare Krishnas sang and danced, wearing saffron robes and tattered gray sweaters.

harem. Add: **2. b.** Applied *spec.* to the family units of various animals.

1898 D. S. Jordan *Fur Seals* I. 57 The average size of a harem..is about thirty females to a single bull. **1932** S. Zuckerman *Soc. Life Monkeys & Apes* xi. 178 Sokolowsky's account of sexual life in a chimpanzee harem is by far the best that has hitherto appeared. **1948** A. L. Rand *Mammals E. Rockies* 26 In the autumn..the bull elk come down from the mountains to gather their harems. **1955** L. Darling *Seals & Walruses* 24 There are an average of forty cows in a fur seal's harem. **1964** G. Durrell *Menagerie Manor* i. 32 The peacock..leading his vacant-eyed harem towards their roosting place.

4. harem dress, a dress with a harem hem; **harem hem,** a hem which draws in the material which then billows over it; **harem skirt,** a loose trouser-like skirt as worn in a harem, or an imitation of one; hence *harem-skirted* adj.

1911 *Sphere* 11 Mar. 219 The opinions of London and Paris over the harem skirt seem to be as divided as is the costume itself. **1927** *Delineator* June 21 *Chéruit* was the first to make taffeta frocks in great puffs with harem hems. **1952** *California Stylist* May, The harem dress..mushrooming fullness on a delicate stem. **1957** *Punch* 14 Aug. 191/1 His cocktail and evening dresses are harem skirted, and there was a bright yellow harem trouser-dress. **1958** J. Laver *Edwardian Promenade* vii. 163 Two symbolic kinds of skirt: the harem skirt and the hobbleskirt. **1958** *Vogue* Mid-Sept. Extra Issue 77/4 Electric blue chiffon softly wrapping a tiny waist and gently curving into a harem hem. **1966** *Sunday Times* (Colour Suppl.) 27 Feb. 32/4 Paul Poiret's harem dress..was inspired by Bakst's designs.

haremlik (hēə·rĕm-, hărī·mlik). [Turk., f. Harem + *-lik* place.] = Harem 1.

1920 *Blackw. Mag.* Nov. 661/2 High above their bench and entered from the Haremlik, is an iron grille or cage, in which the Padishah could overhear unseen his Ministers' deliberations. **1936** N. M. Penzer *Harēm* i. 16 Relations with European powers soon gave rise to the coining of a word that would embrace not only the *harēmlik* and the *serāmlik* but the entire Royal buildings as a whole... The word *seraglio* was chosen. **1941** *Archit. Rev.* XC. 101/2 The houses were divided vertically into two parts—the *Haremlik* or women's quarters, and the *Selamlik* for the men.

hare's-foot. Add: **2*.** A hare's foot used in applying rouge, etc., to the face.

c **1800** [see *burnt cork* s.v. *Burnt *ppl. a.* 7]. **1827** L. T. Rede *Road to Stage* 38 Burn a cork to powder, wet it with beer (which will fix the colouring matter), and apply it with a hare's-foot, or a cloth. **1835** Dickens *Sk. Boz* (1836) II. 206 The young lady with the liberal display of legs, who is kindly painting his face with a hare's foot. **1859** E. Winstanley *Shifting Scenes Theatr. Life* xxi. 200 There are pots of rouge, hare's feet, powder-boxes. **1877** 'Haresfoot & Rouge' *How to 'Make-up'* 13 Then with a hare's foot apply a colouring of Rouge to the cheeks. **1939–40** *Army & Navy Stores Catal.* 1097/1 Theatrical make-up...Hares' feet.

harewood (hēə·rwud). Also **hairwood, air-wood** (8 aire-); and simply 7 **ayer, ayre.** [ad. dial. G. *aehre, ehre*, or its apparent source Friulian, etc. *ayar, ayer, aire* :– Rom. **acre* = L. *acer* maple.] Stained sycamore wood, used by cabinet-makers.

1664 Evelyn *Sylva* i. x. 28 The Timber [of Maple] is far superiour to Beech for all uses of the Turner... Also for the lightness (under the name Ayer) imploy'd often by those who make Musical-instruments. **1676** T. Mace *Musick's Mon.* 49 Next, what Wood is Best for the Ribbs. The Air-wood is absolutely the Best. *Ibid.* 64 A Lute made of Ayre. **1723** *Evening Post* 30 May, He has..some fine Aire-wood for furnishing the Insides [of harpsichords]. **1843** Holtzapffel *Turning* etc. I. 107 A variety of sycamore, which is called harewood, is richer in figure and sometimes striped. **1873** E. Spon *Workshop Rec.* 1st Ser. 414/1 Have the veneers ready, which must be air-wood. *Ibid.* 423/2 Hair-wood. **1899** *Daily News* 22 June 8/7 A cabinet..of inlaid satin, hare, and other woods. **1901** *Westm. Gaz.* 9 May 1/3 On satin and harewood banded with rosewood. **1947** J. C. Rich *Materials & Methods of Sculpture* x. 296 English sycamore, or Harewood, is actually a variety of maple. It is a light-colored, fine-grained, figured wood and is generally available in the form of thin planks. **1968** *Times* 26 Nov. 13/6 A pair of George III satinwood and harewood commodes. **1973** *Country Life* 7 June 1587/1 A pembroke table in the silvery green of harewood.

Hargrave (hā·rgrēiv). The surname of Lawrence *Hargrave* (1850–1915), an Australian pioneer in Aeronautics, used *attrib.* to designate a cellular box-kite invented by him in 1894.

1900 R. S. Baker *Boy's Bk. Inventions* vi. 210 Those [tailless kites] of the Hargrave model giving the impression of a number of big pasteboard boxes with the bottoms knocked out. **1908** W. H. Story tr. *Hildebrandt's Airships* x. 117 (*caption*) Hargrave kite. *Ibid.* x. 121 Four Hargrave kites were used. **1945** Baker *Austral. Lang.* x. 200 Hargrave box kite.

haricot, *sb.* Add: **2.** Also *haricot blanc, pod, vert.*

1845 E. Acton *Mod. Cookery* xv. 328 The haricot blanc is the seed of a particular kind of French bean. **1877** E. S. Dallas *Kettner's Bk. Table* 242 (*heading*) Haricot pods, what the French call haricots verts; called also French beans. **1907** *Yesterday's Shopping* (1969) 45/1 Vegetables, preserved, in Tins..Haricots Verts, extra fins—tin, 0/7½. **1941** A. Simon *Conc. Encycl. Gastron.* III. 46/2 In the U.S.A...there is a variety of canned String Beans..marketed under the name of *Haricots verts.* **1966** *Harrod's Food News* Sept. 2/1 Haricot Verts—8 oz. 3/3. **1967** *Listener* 20 Apr. 533/3 Chefs who bite the *haricots verts* or spit on the icing sugar. **1970** *Sat. Rev.* 3 Oct. 44/3 *Cassoulet*..usually includes goose..sausages, and *haricots blancs* or white beans.

Harijan (hæ·ridʒăn). [a. Skr. *harijan(a)* person devoted to the god Vishnu, f. *Hari* Vishnu + *jana* person.] The name given by Gandhi to the Untouchables in India. Also as *adj.*

1931 M. K. Gandhi *Bleeding Wound* (1932) ix. 40 Only the other day a friend suggested to me that the word Harijana (man of God) be substituted for the word 'antyaja' (the 'lastborn') that is being used for 'untouchables'... I am delighted to adopt that word. **1932** *Ibid.* App. 186, I would like to assure my Harijan friends, as I would like henceforth to name them, that..I am wedded to the whole..agreement. **1958** G. Mikes *East is East* 162 But the untouchables—the *harijans*—are still not regarded as human beings. **1960** *Times* 23 May 13/7 Many more Harijan children are now attending the village schools. **1966** *New Statesman* 11 Feb. 190/1 Already the leader of the untouchables..has won a seat in the cabinet ..in recognition of the 70-odd Harijan votes cast against Desai. **1969** *Times* 13 Oct. (India Suppl.) p. vi/2 In Porbander, in spite of Gandhi's dream and work to eradicate untouchability and the Government's enlightened policy during the past 22 years, Harijans must still draw water from separate taps. **1973** *Times* 20 Aug. 12/1 Harijans, or 'Children of God', the name coined by Mahatma Gandhi—and now generally used—for the Untouchables.

hark, *v.* Add: **2. a.** Also const. *at.*

1887 G. M. Fenn *Dick o' the Fens* vii, Hark at him!.. young squire ar'n't going to eat any more bacon, 'cause it's cruel to kill the pigs. **1895** 'G. Mortimer' *Like Stars that Fall* xiii, 'You're so spry, I can't trust you for a grass widow.' 'Hark at him!' laughed Mrs. Larpenti.

4. d. *to hark after*: to go after, to follow.

1899 B. Tarkington *Gentleman fr. Indiana* vii. 109 Men were running around a corner of the court-house, and the women and children were harking after.

harka (hā·ɹkă). [a. Moroccan Arab. *ḥarka* military expedition, classical Arab. *ḥaraka* movement.] A body of Moroccan irregular troops.

1903 *Daily Chron.* 18 Sept. 4/6 The regions in the Sahara affected by the recent incursions of the Moorish harkas. **1909** *Ibid.* 26 July 1/5 The harka..consisted of about 15,000 Kabyles. **1925** *Blackw. Mag.* Nov. 624/1 [He] led out a harka to fight in the national cause.

harl, *sb.*[3] Add: **5.** A composition of lime and gravel or sand; roughcast.

1869 R. L. Stevenson *Let.* 18 June in *Scribner's Mag.* (1899) XXV. 42/1 The houses, white with harl. **1898** J. J. Burgess *Tang* ii. 23 The gable was white, for the 'harl' had been picked off in the spring. **1940** 'M. Innes' *Secret Vanguard* ix. 92 The walls of the sort of rough-cast which in Scotland is called harl.

Harlem (hā·ɹlĕm). The name of a predominantly Negro area in Manhattan, New York, used *subst.*, *attrib.* or in *Comb.* to designate a strongly swinging jazz style. Also as quasi-*adj.*

1934 C. Lambert *Music Ho!* iii. vii. 201 By jazz, of course, I mean the whole movement roughly designated as such, and not merely that section of it known as Afro-American, or more familiarly as 'Harlem'. *Ibid.* viii. 224 American jazz is either too Hollywood or too Harlem—it rarely suggests the dusty panorama of American life. **1946** R. Blesh *Shining Trumpets* xiii. 315 Harlem piano was, more than any other style, akin to the player piano, no doubt because many Harlem pianists made early player rolls. **1947** R. de Toledano *Frontiers of Jazz* p. ix, Harlem Jazz is..much more commercial, there's less going on, and therefore it's easier to understand. **1958** C. Fox in P. Gammond *Decca Bk. Jazz* vii. 87 A percussive, striding style of playing..still called ..'Harlem Piano'. **1959** M. T. Williams *Art of Jazz* (1960) iii. 17 The Harlem style of James P. Johnson and Fats Waller.

Harlemese (hāɹlĕmī·z). [f. prec. + -ese.] A regional type of speech used by the inhabitants of Harlem. Also *attrib.* or as *adj.*

1928 *Opportunity* VI. xi. 346/2 Dr. Fisher writes with equal authority and zest of..his Harlemese version of a sermon on the battle of Jericho, the technique of piano moving, [etc.] **1932** *Ibid.* X. x. 320/2 The escapades of these two roustabouts and the 'Harlemese' which they use so glibly. **1942** *Amer. Mercury* July 84 She offers a sketch of Harlem life couched in Harlemese. **1963** A. Bontemps *Amer. Negro Poetry* p. xvi, Hughes's art can be likened to that of Jelly Roll Morton and the other creators of jazz. His sources are street music. His language is Harlemese. **1972** *Listener* 27 Jan. 125/3 [They] assume we know Brooklynese..and Harlemese, and Bronxese. **1973** *Black World* Mar. 81 His [*sc.* Langston Hughes'] language has been appropriately called Harlemese: vibrant, rhythmic, direct, and racy.

Harlemite (hā·ɹlĕməit). [f. *Harlem + -ite[1].] A person born in or residing in the Harlem area in the city of New York.

1890 *Harlem Local Reporter* 1 Mar. 2/5 The average Harlemite is in a continuous swim of development and prosperity. **1896** *N.Y. Dramatic News* 4 July 6/3 With music nightly the place will be of great benefit to Harlemites. **1897** *Outing* (U.S.) XXX. 488/1 As a contrast to the scope and aim of the Harlemites,..the Riverside Wheelmen, of New York, may be quoted. **1926** A. Niles in W. C. Handy *Blues* 22 His satire on the prissy and citified Harlemite. **1950** A. Lomax *Mr. Jelly Roll* ii. 78 The passionate enthusiasm of a Harlemite for baseball. **1958** C. Fox in P. Gammond *Decca Bk. Jazz* vii. 86 The 'twenties found many Harlemites..throwing parlour-socials (or rent parties). **1971** *Black Scholar* June 18/1 The Harlemites appearing in the simple stories are..'like the knights and courtiers in *Le Morte d'Arthur*'.

harlequin, *sb.* Add: **6. harlequin bug** *Austral.*, either of two bugs with brightly-coloured markings, *Dindymus versicolor* or *Tectocoris diophthalmus*; **harlequin (Great) Dane,** a Great Dane having a black and white coat; **harlequin fish,** (*a*) *Rasbora heteromorpha*, a small cypriniform fish found in Thailand, Malaya, and Sumatra; (*b*) *Othos dentex*, the scarlet rock cod, a perciform fish found along the coasts of south and west Australia; **harlequin fly,** a fly of the genus *Chironomus*; **harlequin (eye)glasses, spectacles,** spectacles with the frame tilted upwards at the corners (named from their resemblance to a harlequin's mask); **harlequin opal** = Harlequin *sb.* 4; **harlequin smiler,** *Merogymnus eximius*, a small Australian perciform fish.

1945 K. C. McKeown *Austral. Insects* xv. 81 The commonest and best known species [of Pyrrhocoridæ] is the striking red and black insect popularly known as the Harlequin Bug (*Dindymus versicolor* Sch.), found throughout Australia, and attacking apples and other fruits. **1970** T. E. Woodward et al. in *Insects of Australia* (C.S.I.R.O.) xxvi. 450/1 *Tectocoris diophthalmus* (Thunb.), the 'harlequin bug' of Queensland, attacks the bolls of cotton, and is common on other malvaceous plants. **1800** Harlequin Dane [see Dane 2]. **1909** *Daily Chron.* 28 July 7/1 A black and white or harlequin Great Dane. **1948** 'Sigma' in B. Vesey-Fitzgerald *Bk. Dog* ii. 420 When smart men drove a well-appointed turnout, a Dalmatian or harlequin Great Dane was necessary to complete the picture. **1956** M. West *Gallows on Sand* x. 115 A school of harlequin fish flirted away from my descent, their tube-like bodies flashing blue and gold, their ugly faces smiling like a circus clown's. **1959** *Times* 3 Mar. 7/1 The 'guinea pigs' in these experiments are inch-long harlequin fish..from south-east Asia. **1900** Miall & Hammond (*title*) The structure and life history of the harlequin fly (*Chironomus*). **1956** *Nature* 17 Mar. 534/1 The family Chironomidae has scarcely been studied at all in New Zealand... Known colloquially as the 'Harlequin fly', they are, it seems, almost ubiquitous. **1945** 'L. Lewis' *Birthday Murder* (1951) i. 7 Her harlequin eyeglasses became crooked on her pointed face. **1961** Wodehouse *Service with Smile* ix. 130 She was regarding him austerely through her harlequin glasses. **1887** *Col. & Indian Exhib., Rep. Col. Sect.* 70 Some specimens are of a rare kind, known as 'Harlequin' opals. **1955** A. Ross *Australia 55* ix. 118 Sharks..whose presence causes.. coral trout and Harlequin Smilers suddenly to evaporate. **1964** T. C. Marshall *Fishes of Great Barrier Reef* 326 Harlequin Smiler *Merogymnus eximius.* **1940** *Optometric Weekly* 19 Dec. 1262 The Harlequin spectacle frame is protected by patents. **1962** J. Braine *Life at Top* xii. 154 Her harlequin spectacles didn't make her expression any less severe.

harlequin, *v.* Restrict *rare* to senses in Dict. and add: **c.** To colour, decorate with contrasting colours. So **ha·rlequined** *ppl. a.*

1941 'R. West' *Black Lamb* II. 304 A slope of long grass harlequined with flowers. **1959** *Housewife* June 5 The two colours daringly harlequinned. **1963** *Harper's Bazaar* July 44 Shetland jumper—white, harlequinned in different greys. **1965** D. Francis *For Kicks* iv. 49 She wore a black and white harlequined ski-ing jacket.

Harley (hā·ɹli). *Harley Street*: name of a street in London associated with eminent physicians and surgeons; hence used allusively for the specialists of the medical profession.

1830 *New Monthly Mag.* ii. 220 Harley-street was..in an uproar at these monstrous stipulations. **1905** G. B. Shaw *Shaw on Shakes.* (1962) 1 A regular company.. holding itself as exclusively above the casual barnstormer as a Harley Street consultant holds himself above a man with a sarsaparilla stall. **1958** B. Nichols *Sweet & Twenties* xi. 138 The sort of people who always contribute to symposiums—a bishop, a general..an 'eminent Harley Street surgeon'. **1961** L. Mumford *City in History* iv. 105 'Harley Street', 'Madison Avenue', 'State Street', are shorthand expressions not just for occupations, but for a whole way of life that they embody. **1972** M. Gilbert *Body of Girl* xvii. 156 Two Harley Street surgeons discussing a difficult case.

harm, *sb.* Add: **1.** Often in the set phrase 'to do more harm than good'. Cf. quot. 1875 in sense 1 in Dict.

1809 *Q. Rev.* May 305 The story should be suppressed altogether, as one which will do more harm than good. **1857** Dickens *Dorrit* ii. xxix. 723, I should have done you more harm than good, at first. **1914** G. B. Shaw *Misalliance* p. xxix, These rare cases actually do more harm than good.

harmattan. (Later examples.)

1906 F. B. Archer *Gambia Colony* i. 27 This excessive dryness is undoubtedly due to the severe 'harmattan' experienced in the locality. **1963** W. Soyinka *Lion & Jewel* 22 The dew-moistened leaves on a Harmattan morning.

harmonic, *a.* and *sb.* Add: **A.** *adj.* **1.** *harmonic telegraph* (earlier example). Also, *harmonic telegraphy.*

1878 *Telegraphic Jrnl.* VI. cxxxiii. 348/1 Gray's harmonic telegraph can now be seen in operation at the Paris Exhibition. **1902** *Westm. Gaz.* 8 Jan. 6/2 The extensive adoption of..harmonic telegraphy. **1925** *Telegr. & Teleph. Jrnl.* XI. cxxii. 152/2 Mr. Cromwell Varley, who seems to have been the first to get hold of the fundamental idea of harmonic telegraphy, of sending into the telegraph line a number of different frequencies of signalling current at the same time and sorting them out at other stations.

4. *harmonic minor mode* or *scale*: see quot. 1884. Also *harmonic series* = *harmonic scale.*

1884 Maitland in Grove *Dict. Mus.* IV. 666/2 Harmonic minor is the name applied to that version of the minor scale which contains the minor sixth together with the major seventh, and in which no alteration is made in ascending and descending. **1889** E. Prout *Harmony* (ed. 10) vii. § 171 This form is known as the Harmonic Minor Scale, the other two being called Melodic Minor Scales. **1910** *Encycl. Brit.* XIII. 1/1 The unisonous quality of octaves is easily explained when we examine the 'harmonic series' of upper partials.

5. a. *Math. harmonic average* = *harmonic mean*; *harmonic ratio* = *harmonic proportion*; *harmonic series* = *harmonic progression*; esp. the series $1+\frac{1}{2}+\frac{1}{3}+\frac{1}{4}+\ldots$

1866 Brande & Cox *Dict. Sci., Lit., & Art* II. 96/1 *Harmonic Progression or Series*, a series of numbers such that any three consecutive terms are in harmonic proportion. **1942** G. James *Math. Dict.* 199/1 *Harmonic ratio.* If the cross ratio of four points (or four lines) is equal to −1, it is called a harmonic ratio and the last two points are said to divide the first two harmonically. **1949** G. & R. C. James *Math. Dict.* 23/2 The harmonic average is the reciprocal of the arithmetic average of reciprocals of the observations. **1964** Crowder & McCuskey *Topics in Higher Analysis* iv. 193 Since $\sqrt{(n+1)}/n > 1/n$ for all $n > 0$, and

$$\sum_{n=1}^{\infty} 1/n$$

is the harmonic series that diverges,

$$\sum_{n=1}^{\infty} \sqrt{(n+1)}/n$$

also diverges.

b. *harmonic analyser* (example); *harmonic current*, an alternating current the variations of which, graphically represented, follow a harmonic curve.

1908 *R. Soc. Catal. Sci. Papers 1800–1900* I. *Pure Math.* 402/1 (*title*) Harmonic analyser. **1910** *Hawkins's Electr. Dict.* 193/1 Harmonic current.

c. *Electr.* Of or relating to harmonics (*Harmonic *sb.* 2 b), as *harmonic distortion*, non-linear distortion of a wave-form in which harmonics of the original frequencies are introduced into it; *harmonic generator*, a device that generates and combines harmonics of one or more sinusoidal oscillations to produce a complex wave-form; *harmonic interference*, interference caused by the reception of harmonics of a transmitted signal of some other frequency; *harmonic selective signalling* (see quot.).

1929 K. Henney *Princ. Radio* xvii. 450 We determined the percentage of harmonic distortion that occurred in an amplifier when it worked over a curved characteristic. **1930** *Terms & Def. Telegr. & Teleph.* (*B.S.I.*) 21 *Harmonic selective signalling*, signalling a number of stations on one circuit by means of alternating or pulsating currents of different frequencies, each individual station being tuned to one frequency only. A calling station can call any selected station independently of the others by employing the frequency particular to the selected station. **1930** *Proc. Inst. Radio Engin.* XVIII. 31 If a receiver with poorly designed selective circuits is subjected to relatively high local field intensities one of the radio-frequency tubes may be overloaded and may then function as a modulator or harmonic generator. *Ibid.*, Complaints of harmonic interference are, at times, received by the operators of broadcast stations which can be traced directly to deficiencies in the design of the receivers employed. **1931** *Trans. Amer. Inst. Electr. Engin.* L. 811/1 The vacuum-tube harmonic generators of present practise are fundamentally amplifiers operated under conditions of input voltage and grid bias. **1962** A. Nisbett *Technique Sound Studio* 249 Harmonic distortion is most easily caused by flattening of peaks in the waveform. *Ibid.*, 1% harmonic distortion is not usually noticeable.

B. *sb.* **2. b.** *Electr.* In an alternating circuit, a component current whose frequency is a multiple of the fundamental; also, a corresponding electro-magnetic oscillation.

1894 *Amer. Jrnl. Sci.* CXLVIII. 379 The presence of upper harmonics in an alternating current wave. *Ibid.* 383 For every harmonic of the inducing current we shall have a harmonic electromotive force of the same frequency in the resonant circuit. **1919** R. Stanley *Text-bk. Wireless Telegr.* (ed. 2) II. 164 When the fundamental oscillations in a circuit are accompanied by other subsidiary oscillations the latter are called harmonics. **1955** *Sci. Amer.* June 43/3 They act like radio transmitters, emitting radio waves at the critical frequency and at harmonics of this frequency.

harmonica. 1. c. (Examples.)

1895 *Montgomery Ward Catal.* 241/1 Concert harmonica, 10 double holes, 40 reeds, brass reed plates, celluloid covers, absolutely perfect in tone. **1938** S. G. Hedges *Hohner Harmonica Band Bk.* iii. 15 Harmonicas are made in several keys, the principal being G and C. **1966** L. M. Fox *Instruments Pop. Mus.* xiii. 84 Some harmonicas have a slider stop which can switch into play a second row of reeds tuned a semitone higher. **1973** *Advocate-News* (Barbados) 24 Feb. 3/6 (Advt.), Attention all musicians... Just arrived:—..Harmonica Holders.

harmonization. Add: **1. b.** Agreement in colour.

1897 R. Kearton *Nature & Camera* 252 Their wonderful harmonisation with the sand upon which they lay stretched. **1925** R. W. G. Hingston in E. F. Norton *Fight for Everest, 1924* 262 We are attracted by their example of harmonization, the pale grey colour of their fur blending well with the upland soil.

harmonize, *v.* Add: **3. d.** To form a harmonious combination *with.*

1852 *Art-Jrnl.* Apr. 117/3 If it is necessary that the colours of the different articles of dress should..harmonise with each other. **1862** *Englishwoman's Domestic Mag.* May 60/1 Flowers, and shells, and coloured fabrics that harmonise admirably with themselves and with the tropical scenery among which she lives. **1925** R. W. G. Hingston in E. F. Norton *Fight for Everest, 1924* 265 Then unexpectedly the bird alights, the crimson colour vanishes, the white spots disappear, and the bird again harmonizes with the hill. **1949** *Oxf. Jun. Encycl.* II. 72/1 Colour is used in one of the following ways: to break up the outline of the body, or to make it harmonize completely with the background, or to provide obliterative shading.

harmon mute (hā·ɹmǫn miūt). [perh. f. Harmonica.] A type of mute for a trumpet or trombone, also called *wa-wa mute.* Hence **harmon-muted** *a.*

1955 L. Feather *Encycl. Jazz* ii. 64 The sounds produced by both trumpet and trombone were changed by..a variety of mutes, including the straight, cup and Harmon mutes and the rubber plunger. **1966** *Crescendo* Dec. 29/2 A slow, moody piece, with flute and Harmon-muted trumpet. **1967** *Ibid.* Feb. 26/2 Pull out the tube on your harmon mute.

harmonogram (haɹmǫ·nŏgræm). [f. as Harmonograph: see -gram.] A figure or curve drawn by a harmonograph.

1902 *Pearson's Mag.* Apr. 445/1 Not only can an infinity of varying harmonograms be produced with one ratio of pendulum length, but the ratio can be altered. **1962** R. Quirk *Use of English* (*dust jacket*), The Harmonogram illustrated on this jacket is Crown Copyright, Science Museum, London.

harness, *sb.* Add: **2. e.** (See quot. 1940.) Also called *ignition harness.*

1938 R. Keen *Wireless Direction Finding* (ed. 3) xiii. 523 To reduce the capacity of the screening of the cables, and to avoid the bulk and inconvenience of large numbers of separately screened leads, the whole wiring system of an engine may be built into a unit known as 'screening harness'. Fig. 405 shows such a harness..for a Bristol 'Pegasus' radial engine. **1940** *Chambers's Techn. Dict.* 404/1 *Harness*, the entire system of screened ignition leads enclosed within their screening tubes to prevent electromagnetic radiation from affecting the radio-receiving equipment. **1956** W. A. Heflin *U.S.A.F. Dict.* 262/2 *Ignition harness*, a system or assembly of wires, together with any shielding or conduits inclosing them, for conducting electric current from the distributor to the spark plugs of an aircraft engine.

4. a. *double harness*, harness for two draught horses working side by side; *single harness*, harness for a draught horse working alone; *in harness*, side by side, together. Often *fig.*

1838 *Lexington Observer & Rep.* 2 June, We soon hitch'd traces to trot in double harness. **1846** R. Ford *Gatherings from Spain* viii. 88 Those who have a friend with whom they feel they can venture to go in double harness, had better do so. **1862** *Catal. Internat. Exhib., Brit.* II. No. 4732 Double and single harness, pads, collars, round reins, pole pieces. **1873** 'Mark Twain' & Warner *Gilded Age* 373 He and I are sworn brothers on that measure; we work in harness. **1901** 'M. Gray' *Four-Leaved Clover* i, And it's about time you went in double harness. I go better in single. To confess the solid truth, I was born an old maid. **1907** G. B. Shaw *John Bull's Other Island* i. 20 In the main it is by living with you and working in double harness with you that I have learnt to live in a real world and not in an imaginary one. **1937** D. L. Sayers *Busman's Honeymoon* vii. 152 It was her own feelings that didn't seem to be quite pulling in double harness with her intelligence. **1967** *Listener* 2 Feb. 177/3 Prokofiev enthusiasts will be delighted to see that Milstein has now recorded the two violin concertos in harness.

c. From their resemblance to the harness of a horse (see sense 4 a above): straps so arranged that they can be fitted for the protection of travellers in an aeroplane or car. Also used of straps fitted (*a*) on a dog, instead of a collar; (*b*) on a parachute; (*c*) in a perambulator; (*d*) round a child and held by an adult as a safety lead or leash. Also called *safety harness.*

1895 *Montgomery Ward Catal.* 484/2 Pug dog harness, black or russet leather. **1897** H. Dalziel *Brit. Dogs* (ed. 2) III. 43 A kind of dog-harness to mitigate the evils of.. choking by the collar. **1935** C. G. Burge *Compl. Bk. Aviation* 538/1 Safety belts and safety harness have been specially designed for use in aircraft... Harness must hold the wearer firmly in his seat against upward accelerations. **1939** *Sewing Machine & Pram Gaz., Buyers' Guide* Apr. 30 (*caption*) This firm have a very wide selection of reins and safety belts... A typical model of Safety harness is illustrated **1945** C. H. Ward-Jackson *Piece of Cake* (ed. 2) 38 *Harness*, strap holding one to one's seat in an aircraft. **1951** *Gloss. Aeronaut. Terms* (*B.S.I.*) iii. 13 *Harness*, an assembly of straps or cords worn by a parachutist or employed to suspend an inanimate load to which the parachute is attached. **1962** *Times* 23 Jan. 5/6 Every approved harness [in a motor-car] has a quick-release catch. **1962** *Which?* Jan. 8/1 The buckles of the three harnesses that survived the test crash intact were easy to release. **1963** *B.S.I. News* May 34 Safety harness for babies... The types dealt with will be suitable for attachment to perambulators, push chairs and high chairs. Provision will also be made for use of reins with the harness, when the child is able to walk. **1971** J. Philips *Escape a Killer* (1972) i. ii. 24 She unbuckled the dog's harness. **1972** P. Cleife *Slick & Dead* xviii. 233 Tripping the quick-release of my harness, I leapt from my seat.

5. b. Uniform, clothes. *harness bull, cop*, a policeman in uniform. *U.S. slang* (chiefly *criminals'*).

1891 'Mark Twain' *What is Man?* (1917) 225 At the Metropolitan in New York they sit in a glare, and wear their showiest harness. **1899** B. W. Green *Word-bk. Virginia Folk-Speech* 178 *Harness*,..clothing, dress garments. **1903** A. H. Lewis *Boss* 262 [The] Captain sends along a couple of his harness bulls from Mulberry Street. **1914** Jackson & Hellyer *Vocab. Criminal Slang* 42 *Harness*, general currency. A uniform...A 'harness bull' is the commonest form of the term's use. **1926** J. Black *You can't Win* iv. 31 The 'harness cop' who had been at the front door went back to his beat. *Ibid.* xii. 165 We're bang up against the city prison when a big, flat-footed, harness bull steps out an' yaffles us. **1930** E. H. Lavine *Third Degree* ii. 12 'Wise detectives', who dread going back into 'harness', or uniform,..sail along the lines of least resistance. **1931** 'D. Stiff' *Milk & Honey Route* iv. 45 Any harness bull can tell you where the municipal lodging house..is to be found. **1972** J. Godey *Three Worlds* (1973) iii. 32 The cops. From the chief on down to the harness bulls.

6. (Later U.S. example.)

1888 Eggleston in *Century Mag.* XXXVI. 529/2 When Barbara had tied a broken string in the 'harness' of the loom, she resumed her seat on the bench.

9. *harness-horse* (earlier example); **harness (horse) racing,** a race between horses harnessed to vehicles; also **harness race** (Webster, 1909).

1861 Walsh & Lupton *Horse* xv. 272 Hacks and Harness-horses demand nearly as much time and care to prepare them for their work. **1901** *World Almanac* 266 Harness racing. **1909** *Ibid.* 213 Harness horse racing. **1947** *Newsweek* 8 Sept. 71/1 Harness racing is doing very well in keeping up with the flashy bankrolls of the times. **1968** *Globe & Mail* (Toronto) 3 Feb. 36/2 About the only sure thing in harness racing is that Russ Miller..will come up with something special each year. *Ibid.* 17 Feb. 44 (Advt.), Nine harness races today. **1971** *Guardian* 9 June 6/5 Trotting or harness racing, which has become a major sporting attraction in Australia and the United States, is making a comeback in its place of origin, the Yorkshire dales and fells.

harness, *v.* Add: **3.** *fig.* (Later examples.) Now chiefly 'to utilize (a river, waterfall, natural forces, atomic energy) for motive power'.

1927 A. Christie *Big Four* xvii. 258, I believe that she has, to a certain extent, succeeded in liberating atomic energy and harnessing it to her purpose. **1935** *Discovery* Feb. 41/1 The business of harnessing cosmic rays, of forcing them to do the work of electricity, is proceeding apace. **1955** *Times* 19 May 3/6 This monster is, of course, the huge underwater vessel Nautilus..propelled by 'the dynamic force of the universe', which somehow he has succeeded in harnessing. **1965** *Listener* 3 June 823/1 This seems..to make sense: harnessing individual and group enthusiasm to enrich the region.

harnessed, *ppl. a.* Add: **4. harnessed antelope.** Also called *bushbuck.* (Earlier and later examples.)

c **1789** *Encycl. Brit.* IV. 149/1 The scipta or harnessed antelope..has straight horns nine inches long, pointing backwards. **1899** H. A. Bryden *Great & Small Game of Africa* 453 (*heading*) The bushbucks. Genus Tragelaphus. The harnessed antelopes, or bushbucks,..may be arranged as follows:— A... 1. The Bongo (*T. euryceros*). 2. The Nyala (*T. angasi*)... 3. West African Bushbuck (*T. gratus*). 4. Situtunga (*T. Spekei*)... 5. Lesser Bushbuck (*T. scriptus*). The last is represented by several local races. **1960** *Times* (Nigeria Suppl.) 29 Sept. p. xxi/4 The bushbuck, also called harnessed antelope from the pattern of white stripes on its coat, remains widespread.

harnser, dial. form of Heronsew.

haroses, haroset(h, varr. *Charoset(h.

harp, *sb.*[1] Add: **1.** *spec.* One used by Anglo-Saxon minstrels.

1767 Percy *Ess. Anc. Eng. Minstrels* 9 In the early times it was not unusual for a Minstrel to have a servant to carry his harp. **1807** S. Turner *Hist. Anglo-Saxons* (ed. 2) II. 407 Of the harp, Bede mentions, that in all festive companies it was handed round, that every one might sing in turn. **1898** S. A. Brooke *Eng. Lit. fr. Beginnings to Norman Conq.* iv. 82 We should place ourselves..in the hall..when the benches are filled..and hear the Shaper strike the harp to sing this heroic lay. **1903** L. F. Anderson *Anglo-Saxon Scop* 36 The harp was the instrument most used by the scop. **1942** J. C. Pope *Rhythm of Beowulf* 91 If the harp were keeping time, the voice might omit the first accent of a verse..without causing the slightest confusion. **1957** *Rev. Eng. Stud.* VIII. 7 The clear song of the bard is accompanied by the music of the harp.

e. *U.S. colloq.* Also **mouth harp** = Mouth-organ, harmonica. *colloq.* (orig. *U.S.*).

1887 *Scribner's Mag.* Oct. 481/1 She displayed a flimsy red silk handkerchief and a child's harp. **1903** Ade *In Babel* 40 I'd walked from Loueyville over to Terry Hut with a nigger that played the mouth harp. **1963** *Amer. Speech* XXXVIII. 246 *Harp* or *mouth harp* 'harmonica'. **1965** *Melody Maker* 10 July 12/6 For the best blues sound you have to..play the harp in a transposed manner.

f. An Irishman. *U.S. slang.*

1904 'No. 1500' *Life in Sing Sing* xiii. 249/1 *Harp*, an Irishman. **1926** T. Beer *Mauve Decade* iv. 162, I sewed up his head for a young Italamerican who had been trying to impress the haughty Harps on his street. **1936** J. Dos Passos *Big Money* 75 The foreman was a big loudmouthed harp.

harpacticid (hāɹpæ·ktisid). *Zool.* [f. Gr. ἁρπακτικ-ός rapacious + -id[3].] One of the family Harpacticidæ, tiny copepod crustacea. Also as *adj.*

1909 G. Smith in *Cambr. Nat. Hist.* IV. iii. 62 *Euterpe acutifrons*..exhibits the structure of a typical Harpacticid. **1932** *Discovery* Sept. 287/2 The Cyclopids and Harpacticids are, of course, represented in ponds and streams on the surface. **1957** *New Biol.* XXIV. 70 Yet other free-living copepods belong to the group called harpacticids. **1961** J. Green *Biol. Crustacea* iii. 45 Some crustaceans (e.g. harpacticid copepods) do not have a heart.

harpacticoid (hāɹpæ·ktikoid). *Zool.* [f. as prec. + -oid.] One of the order Harpacticoida, very small worm-like copepod crustacea. Also as *adj.*

1946 *Nature* 28 Dec. 935/1 Its [*sc. Bathynella*'s] habitat ..is analogous to that of the interstitial harpacticoid copepods. **1952** J. Clegg *Freshwater Life Brit. Isles* xii. 176 There are about forty-six species of Harpacticoids in Britain. **1956** *Nature* 11 Feb. 289/2 The sample contained ..one harpacticoid copepod and numerous nematodes. **1961** J. Green *Biol. Crustacea* v. 72 There may be one sac, as in many calanoids and harpacticoids, or two sacs as in most cyclopoids. **1970** *Nature* 24 Oct. 323/2 These fish use their gill rakers to strain off the minute animals such as harpacticoid copepods.

harper[1]. Add: **1.** *spec.* in the Anglo-Saxon period.

1767 Percy *Ess. Anc. Eng. Minstrels* 7 Much greater honours seem to have been heaped upon the northern scalds..than appear to have been paid to the minstrels and harpers of the Anglo-Saxons. **1883** Vigfusson & Powell *Corpus Poeticum Boreale* p. lv, In England an innovation appears, the harper who sits at the king's feet. **1912** W. M. Dixon *Eng. Epic & Heroic Poetry* 61 The braggart, the coward, the bard or harper, cunning with the glee-wood..all are there.

harpist. Add: *spec.* in the Anglo-Saxon period.

1898 S. A. Brooke *Eng. Lit. fr. Beginnings to Norman Conq.* ii. 41 As far as we can go back with certainty we find the Teutonic..harpists and singers. **1948** K. Malone in A. C. Baugh *Lit. Hist. England* i. 46 Whether Scilling was Widsith's harpist or a fellow scop..we cannot tell.

harpoon, *sb.* Add: **2. b.** *Med.* A trocar-like surgical instrument for removing small pieces of living tissue for examination.

1876 J. S. Bristowe *Theory & Pract. Med.* (1878) 719 The extraction by means of a suitable instrument (harpoon) of fragments of striped muscular tissue. **1897** *Allbutt's Syst. Med.* II. 1057 The harpoon designed..for this purpose produces an unsurgical wound.

‖ **harpuisbos** (harpō̆i·sbǫs). Also 9 **harpuisbosje** (-bǫsi) and (semi-anglicized) **harpuis, arpuse,** or **rapuis bush.** [Afrikaans, f. *harpuis* resin + *bos* bush.] An evergreen shrub belonging to the genus *Euryops*, esp. the resin-bush, *E. multifidus.*

[**1793** tr. *C. P. Thunberg's Trav. Europe, Afr., & Asia* I. 211 A species of *Coccus*, called *Harpuys*, that was found on the branches of trees, was said to prove mortal to sheep.] **1811** W. J. Burchell *Trav. S. Afr.* (1822) I. xii. 259 The inhabitants of this district, when in want of resin, use as a substitute, a gum which exudes from different species of shrubs; which they therefore call *Harpuis bosch* (Resin bush). **1815** A. Plumptre tr. *Lichtenstein's Trav.* II. 176 A shrub, which grows from two feet to three feet and a half high, called by the colonists *harpuisbosjes*, the rosin tree. **1846** H. H. Methuen *Life in Wilderness* 112 We again were in danger of being burnt; a sea of flame raging on one side of the road, and consuming the resinous *arpuse* bushes with a roaring noise, audible a long way off. [**1912** *East London Dispatch* 22 Aug. 5 (Pettman), Mr. Moffatt (Tarka) brought up the question of the noxious *rapuis* which had hitherto baffled their efforts.] **1926** O. Schreiner *From Man to Man* xi. 353 They burnt harpuis bushes on the lands at home. **1952** *Cape Argus* (Mag. Section) 1 Nov. 1/7 The yellow of the harpuisbos.

Harrian, var. *Hurrian *sb.* and *a.*

Harriet Lane (hæ·riĕt lēi·n). *slang* (chiefly *Naut.*). [f. the name of a famous murder victim; cf. *Fanny Adams.] Preserved meat, esp. Australian tinned meat.

1896 in Farmer & Henley *Slang* IV. 153/2. **1909** B. Lubbock *Deep Sea Warriors* ix. 117 I'd raythe eat this here than the Harriet Lane we get served out for our Sunday dinner. **1909** J. R. Ware *Passing Eng.* 150/2 *Harriet Lane*, Australian canned meat—because it had the appearance of chopped up meat; and Harriet Lane was chopped up by one Wainwright. **1916** 'Taffrail' *Carry On!* 28 'Fanny Adams' and 'Harriet Lane' were the names once given to the preserved meat issued to seamen. **1932** J. W. Harris *Days of Endeavour* viii. 134 A meal of pantiles and cold 'Harriet Lane' is provided. **1938** W. E. Dexter *Rope-Yarns* 30 On Sunday we were allowed 1 lb. of preserved meat, known as 'Harriet Lane', from the name of a woman who had disappeared.

Harris (hæ·ris). The name of the southern section of the island of Lewis with Harris in the Outer Hebrides, used (chiefly *attrib.*) to designate the hand-woven tweed produced by the inhabitants of this region. Also *ellipt.* (*Harris* is a proprietary term in relation to tweed manufactured in the island of Lewis with Harris.)

1892–3 *T. Eaton & Co. Catal.* Fall & Winter 11/2 The finest Harris tweed with detachable cape. **1894** *Strand Mag.* VIII. 661/2 My tailor tells me that Harris tweed cannot wear out. **1898** *Daily News* 5 July 2/4 The delightful 'Harris', 'Shetland', and 'Sutherland' tweeds that were being shown. **1924** Galsworthy *White Monkey* ii. ii. 131 Very elegant in smoke-grey Harris tweeds. **1939** 'N. Blake' *Smiler with Knife* ii. 33 Her trim, green Harris-tweed suit. **1949** H. Wadman *Life Sentence* 5 We have no tailors who know how to handle Harris. **1971** *Stornoway Gaz.* 21 Aug. 6/7 (Advt.), Gent's Harris heavy knit crew necks.

Harrogate (hæ·rogĕit). Name of a borough in the West Riding of Yorkshire used *attrib.* to designate (*a*) a medicinal water originating in Harrogate, (*b*) the proprietary name of a kind of toffee.

1771 Smollett *Humph. Cl.* I. 229 Harrigate-water, so celebrated for its efficacy in the scurvy..is supplied from a copious spring. **1867** Bloxam *Chem.* 45 The Harrowgate water is eminently sulphureous. *Ibid.* 654/1 (*index*) Harrogate water. **1890** in A. Davis *Package & Print* (1967) 64 (Advt.), Farrah's Original Harrogate Toffee. **1910** *Trade Marks Jrnl.* 28 Dec. 2069 Farrah's Harrogate Toffee... John Farrah, Limited..Harrogate; Toffee Manufacturers. **1931** W. Holtby *Poor Caroline* vii. 278 Eleanor came..bringing..the caramels and Harrogate toffee that Caroline loved. **1938** *Harrogate Spa Med. Jrnl.* Apr. 17 All the Harrogate waters are chemically unstable in contact with oxygen, indicating a measure of activity not common to most mineral springs. **1967** H. Johnson in C. Ray *Compleat Imbiber* IX. 141 Harrogate water is bottled to this day. **1967** E. S. Turner *Taking Cure* iii. 40 Knaresborough..was to become eclipsed in reputation by the near-by 'stinking' wells, the source of 'Harrogate water'.

Harrow, *sb.*[3] *Cricket.* The name of a public school at Harrow in Middlesex, used *attrib.* to designate: a bat of less than full size (also *ellipt.*); 'a stroke by which the ball is driven in the direction of mid-off' (Lewis); also, an ineffectual attempt at such a stroke; 'the position of the fieldsman placed for the Harrow drive' (Lewis).

1851 J. Pycroft *Cricket Field* ix. 171 'I beg your pardon, sir,' he..said.., 'but ain't you Harrow?'—'Then we shan't want a man down there,'..; 'stand for the "Harrow drive", between point and middle wicket.' **1877** C. Box *Eng. Game Cricket* 451 *Harrow drive.* Some persons define this phrase to mean a fluke in the slips, after an ineffectual attempt to play forward. **1887** J. Lillywhite *Cricketers' Ann.* (Advt.), 7 Youth's Cane-handled Bats. Harrow size. **1922** D. J. Knight *First Steps to Batting* i. 13 For a boy of 14 or 15 who has chosen a Harrow 2 lbs. 2¼ oz. should be satisfactory. **1958** *Times* 22 May 15/4 It was pure pantomime with Chinese cuts and Harrow drives flying off Lobb's bat. **1970** *Times* 6 Mar. 16/8 His one escape was from a 'Harrow' drive, off McKenzie, which narrowly missed his leg stump.

harrumph (hæ̆rʌ·mf). Chiefly *U.S.* Also **harrump.** [Imitative.] A guttural sound made by clearing the throat. Also *fig.* So as *vb.*, to make this sound; to speak in a rasping or guttural voice; to make a comment implying disapproval.

1936 'J. Tey' *Shilling for Candles* iv. 34 'Howdyudo. Harrump!' He cleared his throat. **1941** *Time* 16 June 29/2 This touching appeal evoked a harrumph from the *New York Times.* **1942** *N.Y. Herald Tribune Books* 11 Jan. 13 He seems to be a figure of fun, with his..fairly continuous harrumphing. **1943** *Time* 11 Oct. 86 The State Department harrumphed and other U.S. oil companies stood on their legal, unenforceable rights. **1957** *Ibid.* 2 Sept. 23/2 Harrumphed one official: 'The best way of putting it would be to say at this point we are tolerant of Dr. Nkrumah's actions.' **1965** G. McInnes *Road to Gundagai* ii. 33 The Captain harrumphed. **1967** *New Yorker* 25 Feb. 99 My goodness, Henry, you're much too young to be going *har-rumph, har-rumph* all the time! **1970** *Times Lit. Suppl.* 29 Jan. 113 Across the room he'd go, singing and harrumphing.

Harry, *sb.*[2] Add: **2. c.** *Flash Harry*: an ostentatious, loudly-dressed, and usually ill-mannered man; cf. Flash *a.*[3] Also *attrib.*

1960 J. Rae *Custard Boys* ii. xiii. 158 'They're just a lot of smart Alecs.' 'Flash Harrys,' suggested Peter. **1960** *Times* 31 Oct. 16/4 He registers emotional upset by a slightly raised eyebrow, and then briskly readjusts his flash-Harry tie. **1962** *Times* 22 May 15/4 Her flash-Harry boy-friend.

7*. *to box Harry*: see *box *v.*[2] 3 b.

8. b. In arbitrary appositive uses of which a few have emerged as set expressions, e.g. **Harry Flakers** *Naut. slang*, exhausted; **Harry Flatters** *Naut. slang*, (of the sea) calm; **Harry Freeman's** (also **Harry Frees**) chiefly *Naut. slang*, a gift; also as *adj.*, free; **Harry James** *slang*, nose.

1925 Fraser & Gibbons *Soldier & Sailor Words* 115 *It's Harry Freeman's*, a gift. Something gratis. (Navy.) **1929** F. C. Bowen *Sea Slang* 64 *Harry Frees*, the name given in the Grand Fleet to the very welcome fruit and vegetables sent up as gifts by the public. **1935** 'L. Luard' *Conquering Seas* xii. 139, I don't expect to supply cigarettes Harry Freemans. **1941** C. Graves *Life Line* 154 Fortunately, the sea has dropped and it is Harry Flatters. Harry Flatters means flat calm, and Harry is used as a predicate for almost any expression. **1946** *Lancet* 2 Feb. 177/1 Get in there, and strip off Harry Nuders. **1950** T. E. Lawrence *Mint* 32 Sort of thing the civvies in London pay fifty quid for, we get harry-freeman's. **1958** F. Norman *Bang to Rights* 36 Plenty of dust floating about in the air, which gets..up your Harry James. **1962** *John o' London's* 14 June 571/2 'Harry Flakers' to mean worn out after a party or heavy work. *Ibid.*, 'Harry Flatters' for a flat calm sea. **1962** P. Purser *Peregrination 22* xxii. 99 It's okay for our kind of thing but it would be Harry Grimmers for ordinary civilians. **1966** F. Shaw et al. *Lern Yerself Scouse* 58, *I wuz lookin fer some Arry Freeman's*, I was looking for something for nothing. *a* **1966** M. Allingham *Cargo of Eagles* (1968) viii. 103 Get me a Harry pinkers—a large one. **1969** *Guardian* 14 Mar. 10/5 It's derisory, old boy, they'll turn it down harry nem-conners. *Ibid.*, Harry shambles, old boy... In the old Imperial Aircraft days.. the engineer would bring the old kite down harry plonkers on the grass.

Harry Tate (hæ·ri tē̆i·t). [Stage-name of R. M. Hutchison (1872–1940), music-hall comedian.] Used *attrib.* or in the possessive to designate anything incompetent or disorderly. Also (by *Rhyming slang*), a state, usually of nervous excitement or irritability.

1925 Fraser & Gibbons *Soldier & Sailor Words* 115 *Harry Tate's Cavalry*, a nickname occasionally applied in jest to Yeomanry. *Ibid.*, *Harry Tate's Navy*, a nickname occasionally used in jest for the Royal Naval Volunteer Reserve. **1929** F. C. Bowen *Sea Slang* 64 *Harry Tate's Navy*, before the war a term of derision applied to the Royal Naval Volunteer Reserve generally, but dropped completely since they showed their value. During the War it was generally applied to the Auxiliary patrol, particularly the Motor Boat Reserve. **1932** 'P. P.' *Rhyming Slang* ii. 21 *Harry Tate*, state. **1935** *Brit. Jrnl. Psychol.* Jan. 359 Native courts have been established [in Uganda]... Their methods have been described as 'Harry Tate' procedure; but they are generally successful in arriving at the facts.

hartal (hā·ɹtæl, ‖hʊ·rtāl). *India.* [Hind. *hartāl* for *haṭṭāl* lit. 'locking of shops' (Skr. *haṭṭa* shop, *tālaka* lock, bolt).] Organized shutting of shops and cessation of business, to serve, usually, as a protest against government legislation or a political situation, or as an act of mourning.

1920 *Blackw. Mag.* Apr. 441/1 What I had seen there of the crowds at the Hartal..had made me nervous. **1921** *Q. Rev.* July 54 He proclaimed a universal 'hartal', or cessation from business, as a protest against the Rowlatt legislation. **1922** *Ibid.* Oct. 417 Gandhi was preparing a *Hartal* at Bombay. **1931** *Daily Tel.* 6 Jan. 9/5 The Moslems enforced a complete 'hartal' (day of mourning or strike). **1955** *Times* 18 Aug. 8/5 A hartal to be observed throughout India on Friday was called here to-day by the Goa liberation committee in memory of the Indians killed by Portuguese fire on Monday. **1958** J. V. Bondurant *Conquest of Violence* iii. 36 Non-cooperation may include strike, walk-out, *hartal*, and resignation of offices and titles. *Ibid.* iv. 119 The *hartal* is usually of short duration—a day or two. Shops close and work ceases. *Hartal* is also employed at times of national mourning. **1962** A. Carter *Direct Action* 8 At the beginning of the Rowlatt Act Satyagraha Gandhi called on his followers for a 'hartal'—a day of abstention from work and of fasting and prayer.

hartebeest. Add: Also in Afrikaans form **hartebees.** Also *attrib.*, as **hartebeest house,** hut, 'a frail structure of "wattle and daub", so called, apparently, because a similar primitive structure was often erected by the earlier hunters' (Pettman); also **hartebeeste house.**

[**1815** A. Plumptre tr. *Lichtenstein's Trav.* II. 95 Not far from this wretched cabin stood a somewhat more spacious, but very ruinous straw hut, of the sort which is here called *hartebeesthuisje.*] **1818** B. H. Latrobe *Jrnl.* 256 A hartebeest-house, being a roof, put upon a wall about two feet in height. **1863** W. C. Baldwin *Afr. Hunting* i. 16 What is called a hartebeest house, of very tall reeds, stuck close together in a kind of trench dug for them in bundles, and meeting over head. **1873** F. Boyle *To Cape for Diamonds* 242 A colony of Hottentot women had seized possession of our 'hartebeest's hut'. **1898** W. C. Scully *Vendetta of Desert* iv. 23 Uncle Diederick lived in a structure known in South Africa as a 'hartebeeste house'. **1959** *Cape Times* 27 Jan. 2/6 Rare kinds of buck, such as eland and red hartebees. **1961** L. van der Post *Heart of Hunter* iii. xiv. 178 Once, the story says, Mantis appeared to the children of the early race as a dead Hartebees.

harumfrodite (hĕə·rʊ·mfrŏdəit). *Jocular slang.* Also **harumphrodite, herumfrodite.** [After *hermaphrodite.*] = Hermaphrodite *sb.* and *a.*

So **harumphrodi·tic** *a.*, characteristic of an hermaphrodite.

1896 [see *giddy *a.* 3 b]. **1924** *Blackw. Mag.* Aug. 235/1 Take the young Jot from the plough and turn him to Pope's 'Rape of the Lock' or Shelley's 'Adonais', and you will make a comfortless harumphroditic amphibian of him. **1953** *Essays & Studies* VI. 103 A Welsh writer writes in Welsh..and those others who make articulate some two-thirds of the nation are, according to taste, 'lost to the English' or base harumphrodites.

Harveian (haɹvī·ăn, hā·ɹviăn), *a.* [f. the name of William *Harvey* (1578–1657), English physician, who discovered how the blood circulated + -an.] Pertaining or relating to, expounded by, or commemorating Harvey.

1755 T. Leman *Some Mem. Life & Writings Late Dr. Richard Mead* 27 In 1723, Dr. Mead was appointed to speak the anniversary Harveian oration, before the members of the college of physicians. **1837** *London Med. Gaz.* 8 July 565/1 The College of Physicians continues to show their classic taste in the Harveian Oration. **1839** J. Turner *Reg. Exper. Living Animals* 34 As a sceptic of the Hunterian and Harveian doctrines, I here take my stand. **1880** *Encycl. Brit.* XI. 505/1 Caspar Hoffmann,.. admitting the truth of the lesser circulation in the full Harveian sense. *Ibid.* 506/2 The Harveian Orations. **1903** *Lancet* 6 June 1608/1 The Harveian Lectures of the Harveian Society of London. **1928** *Daily Tel.* 15 May 14/3 MSS., books, pictures, and other objects of Harveian interest. **1972** *Times Lit. Suppl.* 7 Apr. 403/3 (Advt.), Qualified cataloguer (temporary) urgently required... Applications..to: Harveian Librarian, Royal College of Physicians of London.

harvest, *v.* Add: **1. c.** *trans.* To kill or remove (wild animals belonging to a local population) so as to provide food (or other useful product) or sport, or to reduce the population.

1947 *Biol. Abstr.* XXI. 1602/2, 14 tagged fish were recaptured later by anglers, suggesting that only a small % of the sauger crop is being harvested. **1948** *Jrnl. Wildlife Managem.* XII. 78/1 In 13 years of harvesting the surplus, 546 deer have been taken. **1960** *Biol. Abstr.* XXXV. 2529/1 *Aeromonas* caused heavy mortality of golden shiners..when these fish were harvested and moved to holding tanks. **1961** *Listener* 7 Sept. 348/2 Now 500 to 1,000 hippo are being harvested annually for food. **1970** *Daily Tel.* 30 Oct. 4/8 The tablets were made from the livers of seals harvested in Alaska in 1964. **1973** *Times* 10 Oct. 6/8 Shellfish in Italian waters can be harvested again after a month-long ban brought about by cholera.

d. To remove (cells) from a culture made *in vitro* or *in vivo*; to remove (cells, tissues, organs, or embryos) from an animal for experimental purposes.

1946 *Nature* 9 Nov. 677/2 Table 2 shows the general metabolic activities of normal cells compared with those of cells harvested from a culture grown for 90 min. in the presence of 10 units [of] penicillin per ml. medium. **1957** *Jrnl. Cellular & Compar. Physiol.* XLIX. 369 Various numbers of HeLa cells were added to duplicate Warburg flasks and oxygen consumption was measured for 68·5 hours. The results..indicated that..the rate of oxygen consumption was related linearly to number both of cells inoculated and cells harvested. **1960** *Biol. Abstr.* XXXV. 460/2 (*heading*) Effects of 2,4-dinitrophenol on endogenous respiration of yeast harvested during the first budding cycle. **1971** *Nature* 17 Dec. 385/3 The lymphoid organs are always harvested 24 h after the injection of labelled cells. **1972** *Ibid.* 24 Mar. 169/1 Macrophages were harvested from the peritoneal cavity 10 days after the second immunization. **1972** *Science* 5 May 519/1 Pregnant animals were killed 3 days later and the embryos were harvested.

harvester. Add: **3.** (Earlier and later examples.) Cf. **combine harvester.* **harvester-thresher,** a machine for both harvesting and threshing.

1848 *U.S. Pat.* 21 Nov., Harvester. **1851** C. Cist *Sk. Cincinnati* 161 Harvesters and mowing-machines. **1858** Simmonds *Dict. Trade, Harvester*, an American machine for cutting clover and timothy seed, &c. **1911** *Daily Colonist* (Victoria, B.C.) 5 Apr. 1/3 Large shipments of harvester machinery are arriving from overseas points. **1929** Newman & Blackaby (*title*) Report of the trials of the combined harvester-thresher in Wiltshire, 1928. **1950** *Engineering* 17 Nov. 391/3 Attempts to solve the potato-harvester problem. **1971** *Farmers Weekly* 19 Mar. 83/3 One man with the buckrake comfortably kept pace with a double-chop harvester.

harvesting, *vbl. sb.* **b.** (Earlier example.)

1836 *U.S. Pat.* 28 June, Harvesting machine.

Harvey, *sb.* Add: **2.** The name of Peter *Harvey* (see quot. 1959), English publican, in *Harvey's Sauce* (now a proprietary trade-mark). Also *Harvey Sauce* and *ellipt.*

1817 [see Ketchup]. **1856** Dickens in *Household Words* XIII. 555/2 The grocer's hot pickles, Harvey's Sauce, Doctor Kitchener's Zest. **1870** —— *E. Drood* xi. 80 A condiment of a profounder flavor than Harvey. **1876** *Trade Marks Jrnl.* 28 June 198 Harvey's Sauce for Fish, Game, Steaks etc. Prepared from the Original Receipt, only at E. Lazenby's Fish Sauce Warehouse. *Ibid.* 199 Harvey's Sauce—Caution: The admirers of this celebrated Sauce are..requested to observe that each bottle bears the well-known Label signed 'Elizabeth Lazenby'. **1905** E. Wharton *House of Mirth* i. xiv. 243 A bottle of Harvey sauce on the sideboard. **1959** *Tradition* (*E. Lazenby & Co.*) Jan. 2/1 The history of Elizabeth Lazenby can really be dated from 1760. In that year, a Mr. Peter Harvey, owner of an inn called the 'Black Dog' in Bedfont, Middlesex, invented a thin sauce known as 'Harvey's Sauce'. So good was it that many of his customers endeavoured to obtain the recipe, and one of them, a certain London grocer named M. Lazenby, offered to buy it, but Peter Harvey refused to part with his secret. Mr. Lazenby.. married Harvey's sister, Elizabeth. As a wedding present, Peter Harvey gave Elizabeth..the recipe for his famous sauce. **1967** A. Davis *Package & Print* 44 It is likely that by the end of the eighteenth century Burgess's essence of anchovies..and Lazenby's Harvey's sauce were on the market in labelled bottles not noticeably different from those in which they were to enter the twentieth.

harzburgite (hā·ɹtsbʊ̄ɹgəit). *Petrogr.* [ad. G. *harzburgit* (H. Rosenbusch *Mikrosk. Physiogr. d. Min. u. Gesteine* (ed. 2, 1887) II. i. 270), f. *Harzburg*, name of a town in Saxony: see -ite[1].] A rock of the peridotite group consisting basically of orthopyroxene and olivine.

1890 *Mineral. Mag.* IX. 41 The typical norite has 49·23 *per cent.* of silica,..the well-known 'schillerfels' or bastite-serpentine-rock (Harzburgite of Rosenbusch) 42·36. **1922** *Mineral. Abstr.* I. 396 The rocks of Mansjö Mtn. (61½°N., 15½°E.), Sweden, include crystalline limestone and gneiss with intruded amphibolites and later harzburgite and eulysite. **1924** *Trans. Geol. Soc. S. Afr.* XXVI. 14 The recognisable harzburgites exhibit fairly considerable differences in composition and grain. **1965** G. J. Williams *Econ. Geol. N.Z.* x. 149/1 Hutton (1942) identified chromite grains in dunite-serpentinite and harzburgite-serpentinite in a tributary of Gentle Annie Creek in western Otago. **1970** *Nature* 28 Mar. 1227/2 Most of the volcanics were basic, but during the upper Cretaceous ultrabasics (harzburgite, lherzolite, dunite and anorthosite) were intruded.

has-been, *sb.* Add: (Further examples.) Also **hasbeen.**

1853 B. F. Taylor *Jan. & June* (1871) 206 Dilapidated 'has-beens', and despised 'used-to-be's'. **1879** G. F. Jackson *Shropsh. Word-bk.* 180 'Er's a good owd 'as bin. **1904** *Philadelphia Even. Tel.* 9 Nov. 5 Parker and his party are among the has beens. **1905** B. Tarkington *In Arena* 3 I'll potter along trying to look knowing and secretive, like the rest of the has-beens. **1914** G. Atherton *Perch of Devil* i. 70 The obsolete notions that made most of our relations a sort of premature has-beens. **1929** *Psyche* Apr. 27 The age when physicians are divided into Has-beens and Neverwozzers. **1972** J. Wambaugh *Blue Knight* (1973) vii. 104 When I retire I'm just a has-been.

2. *pl.* Old times. *U.S.*

1904 W. H. Smith *Promoters* v. 91, I met old Bishop Slosher..and just for has-beens I took him to lunch with me.

hasenpfeffer (hā·zĕn(p)fefeɹ, hā·s-). [G.] A highly seasoned rabbit stew.

1892 *Sun's Guide to New York* 16 Pork and beans, *hassenpfeffer*, and corned beef and cabbage are equally palatable over there. **1909** L. Meier *Art of German Cooking* vii. 142/2 (*heading*) Hasenpfeffer. **1967** P. McGerr *Murder is Absurd* x. 127 There's a restaurant in Baltimore that has a real knack with Hasenpfeffer.

hash, *v.* Add: **2.** (Later examples.) Also *hash over.*

1920 F. S. Fitzgerald *This Side of Paradise* (1921) iv. 129 The things..they had hashed and rehashed for many a frugal conversational meal. **1931** Loeb & Schenker *Please stand By* i. iv. 45 But drop up anyway, I have something I want to hash over with you. **1950** *New Yorker* 16 Dec. 26/1 Asked him in to hash over a point or two. **1958** *Times Lit. Suppl.* 5 Sept. 493/3 It is the classic film formula, the sort of thing that dead-beat script-writers hash up for B. pictures.

Hence **hasher,** also *U.S. slang*, a waiter or waitress in a restaurant; **ha·shery** *U.S. slang*, a hash-house, a cheap eating-house.

1916 A. B. Dunn in *Editor* 11 Mar. 297/2 Hasher, meaning waitress. **1960** *Listener* 18 Aug. 250/2 When it came to making an impression on the 'hashers' in the railroad 'beaneries', the boomers really let themselves go... The 'hashers' were girls chosen for their looks. **1961** *Amer. Speech* XXXVI. 271 Somewhere on your run you will spend some time at a truck stop..while the hasher serves your diesel. **1870** *Alaska Times* (Sitka) 8 Jan. 1/3 Having lately opened a hashery, I send you this my rules and regulations. **1901** *Munsey's Mag.* XXIV. 568/2 The salary was ten times what she was getting at the hashery. **1957** J. Kerouac *On Road* (1958) 172 Her honest labors in the hashery.

hash, *sb.*[1] Add: **3. b.** (Earlier and later examples.)

1803 I. Cruikshank *Olympic Games* 16 June (*caption*), I think the first round will settle his hash. **1807** *Massachusetts Spy* 14 Oct. 4/1 This settles all the hash. **1809** T. G. Fessenden *Pills Poetical* 114 We therefore mean to make a dash, To settle fighting Europe's hash. **1822** [see Settle *v.* 21 b]. **1930** R. H. Mottram *Europa's Beast* xii. 292 He's settled my hash, right enough. **1933** C. St. J. Sprigg *Fatality in Fleet St.* x. 124 What are you going to do? Settle his hash and drop him overboard?

6. hash-house chiefly *U.S. colloq.*, a cheap eating-house, boarding house, etc.; also *attrib.*; **hash-joint** *U.S. slang* = **hash-house*; **hash-mark** *U.S. slang*, a military service stripe; **hash-slinger** *U.S. slang*, a waiter or waitress; **hash-up** *slang*, a hastily cooked meal; also *fig.*, something concocted afresh from existing material; a reworking.

1869 *Territorial Enterprise* (Virginia, Nev.) 21 Sept. 3/1 The Mayor proposes to double the tax on all 'hash houses'. **1875** *Scribner's Monthly* July 277/1 In the slang vernacular, an eating place is a 'hash house'. **1883** *Daily Tel.* 10 Jan. 5/4 (Farmer), Fifteen-cent restaurants, commonly called hash-houses. **1895** W. C. Gore in *Inlander* Dec. 116 *Hash-house*, boarding house. **1897** *Outing* (U.S.) XXX. 362/1 It has its swell hotels..and its 'hash-houses'. **1900** H. Lawson *On Track* in *Prose Wks.* (1948) 223 Fourpenny hash-houses (good beds, 6d.). **1903** A. H. Lewis *Boss* xx. 273 His is this deadfall on Barclay Street, with that hash-house keeper to give him th' dough for his checks. **1936** I. L. Idriess *Cattle King* xviii. 170 Sid Kidman found every hotel, every boarding-house, every hash-house, every room crowded. **1946** 'P. Quentin' *Puzzle for Fiends* (1947) xxiv. 171 You'll have to take that job in a hashhouse after all. **1960** N. Hilliard *Maori Girl* 221, I see the hotels and the hash-houses. **1895** W. C. Gore in *Inlander* Dec. 116 *Hash-joint*, boarding-house. **1930** J. Dos Passos *42nd Parallel* 101 Passing the same Chink hashjoint for the third time. **1909** *Man-o' Warsman* Dec. 24/1 First Sergeant John J. Maloney earned another hash-mark. **1935** G. & S. Lorimer *Heart Specialist* vi. 168 Slim and Shorty each had two gold stripes on their left sleeve that Slim called hash marks because they were service stripes and stood for the number of years of free food they'd had on Uncle Sam, he said. **1868** *Gold Hill News* (Nevada) 6 May, The nice young man of Washoe may or may not be some kind of a clerk, a hash-slinger, or a check-guerrilla. **1895** W. C. Gore in *Inlander* Dec. 116 *Hash-slinger*, table waiter. **1946** *Amer. Speech* XXI. 86 The cooks and 'hashslingers' of former years went off to war or to the shipyards. **1895** A. W. Pinero *Second Mrs. Tanqueray* ii. 71 Dreams are only a hash-up of one's day-thoughts. **1902** *N.Z. Illustr. Mag.* VI. 452 While sharing their 'hash-up', he had [determined] to pitch camp on the [gum] field. **1914** *Auto-motor Jrnl.* 816/2 The so-called 'motoring notes' in the daily press are a thinly disguised hash-up of the best stuff in the motor press proper. **1970** *Times* 28 Feb. p. iv, A style perilously close to certain Colour Supplement hash-ups and clearly aligned for Over-ground consumption.

hash, *sb.*[2] Colloq. abbrev. of Hashish.

1959 N. Mailer *Advts. for Myself* (1961) 245 Drawing upon hash, lush, Harlem, Spanish wife, Marxist culture [etc.]. **1967** *Listener* 31 Aug. 262/2 If people use dangerous machinery..while they are high on hash, the consequences may be..brash. **1968** A. Diment *Gt. Spy Race* viii. 143 What do you and Riordan do for a living—I thought he was just another part-time hash pusher? **1972** P. Dickinson *Lizard in Cup* x. 157 'It's morphine she's been on?' said Pibble. But Tony shook her head. 'Just grass. Hash.'

Hashimite (hæ·ʃiməit), *a.* and *sb.* Also **Hashemite.** [f. the name of *Hāšim*, great-grandfather of Mohammed + -ite[1].] **A.** *adj.* Of, pertaining to, or characteristic of the Hashimites. **B.** *sb.* A member of an Arabian princely family claiming descent from Hashim.

1697 H. Prideaux *True Nature of Imposture in Life Mahomet* 5 From him [*sc.* Hashem] the Kindred of Mahomet are called Hashemites. **1757** S. Ockley *Hist. Saracens* (ed. 3) I. 6 Hashem the great grandfather of Mahomet, whose descendants were from him called Hashemites. **1883** *Encycl. Brit.* XV. 672/1 Native princes claiming descent from the Prophet—the Hāshimite emirs of Mecca..—attained to great authority and aimed at independence. **1949** Koestler *Promise & Fulfilment* xvi. 177 Transjordan was and is as interested in a solution on these lines as the Jews, in view of the age-old rivalry between the Hashimite and Saudite dynasties. **1958** *Spectator* 7 Feb. 159/2 The Hashemites of Transjordan or Iraq. **1973** *Times* 22 Sept. 5/1 Despite a spiritual attachment to the guerrillas and a distaste for Hashemite rule, Jordan's Palestinian majority view any guerrilla return with far less enthusiasm than Arabs elsewhere.

Hashimoto (hæʃimōᵘ·tō). *Med.* The name of H. *Hashimoto* (1881–1934), Japanese surgeon, used in the possessive to designate struma lymphomatosa, a disease (described by him in 1912), usu. of women and probably of autoimmune origin, in which the thyroid is enlarged, usu. symmetrically, and infiltrated by lymphoid tissue, as *Hashimoto's disease, goitre, struma* (*lymphomatosa*), *thyroiditis.*

1935 *Arch. Surg.* XXXI. 424 No cervical adenitis is present in association with Hashimoto's struma. **1936** Stedman *Med. Dict.* (ed. 13) 475/2 *Hashimoto's disease*, struma lymphomatosa. **1937** J. H. Means *Thyroid & its Dis.* xix. 503 The histology of Hashimoto's goiter is that of extensive lymphoid infiltration. **1956** *Jrnl. Clin. Endocrinol.* XVI. 1570 (*title*) An unusual iodinated protein of the serum in Hashimoto's thyroiditis. **1962** *Times* 30 Nov. 19/4 Hashimoto's disease, in which there is a gradual destruction of the thyroid gland. **1968** *Brit. Med. Bull.* XXIV. 224/2 Shown are the observed frequencies of occurrence of signs, symptoms and results of laboratory tests in three diseases: Hashimoto's disease, simple goitre and thyroid cancer. **1972** Bastenie & Ermans *Thyroiditis & Thyroid Function* v. 110 The formal criteria of Hashimoto's goitre, namely a recently developed symmetrical and homogeneous goitre, the presence of very high thyroid antibody titres, and the diffuse lesions characteristic of the disease. *Ibid.* 118 The progress of untreated Hashimoto's thyroiditis is variable.

hashmagandy (hæ:ʃmăgæ·ndi). *Austral.* and *N.Z. slang.* Also **hash-me-gandy, hash magandy.** [f. Hash *sb.* 1.] A type of stew.

1919 W. H. Downing *Digger Dial.* 28 *Hashmagandy*, an

insipid and monotonous army dish. **1941** Baker *N.Z. Slang* vi. 54 Terms bequeathed to us by shearers and tramps and farmers..that appear to have originated this century..*hash-me-gandy*, station stew. **1945** —— *Austral. Lang.* 81 For stews our only original contributions appear to be *hash-me-gandy* and *mulliga stew*. **1951** L. G. D. Acland *Early Canterbury Runs* 381 *Hash-me-gandy*, station stew.

Hasid, -ic, -im, -ism: see *Chasid, Chassid.

haskinize (hæ·skinəiz), *v.* [f. the name of S. E. *Haskin*, the inventor of the process + -ize.] *trans.* To submit (green timber) to a process by which it becomes hard and durable through the application of heat of over 212°F. under a pressure of 200 pounds to the square inch. So **ha:skiniza·tion.**

1897 S. E. Haskin (*title*) Haskinizing—Vulcanizing—for the preservation of wood from decay. **1908** W. R. Fisher tr. *Gayer's Forest Utiliz.* (ed. 2) 509 The process is termed Haskinisation or Vulcanisation, and has given good results on the Manhattan Railway, New York.

Hasmonean (hæzmonī·ăn), *sb.* and *a.* Also Ash-, -æan, **Asmonean, -æan.** [f. mod.L. *Asmōnæus*, f. Ἀσαμωναῖος (Josephus) = *ḥašmōnāy*, name of the reputed grandfather of Mattathias.] **A.** *sb.* A member of a Jewish dynasty or family to which the Maccabees belonged. **B.** *adj.* Of or pertaining to this dynasty.

1620 Lodge tr. *Josephus* xiv. xxviii. 381 Thus ended the estate of the Asmoneans, after 120. and sixe yeeres. *Ibid. marg.*, The end of the Asmonean family. **1832** H. Cotton *Five Bks. Maccabees* 50 Ashmonæan princes. **1834** *Penny Cycl.* II. 485 Asmonæans. **1880** *Encycl. Brit.* XIII. 421/2 A certain priest Mattathias, of the family of the Hasmonæans. **1898** *Expositor* Apr. 273 The Hasmonean priestly dynasty. **1926** E. F. Scott *1st Age Christianity* i. 16 In virtue of his priestly descent the Asmonaean king could also hold the office of high-priest. **1956** A. Toynbee *Historian's Approach to Religion* x. 134 A short-lived Hasmonaean successor-state. **1973** *Sci. Amer.* Jan. 80/2 We would know nothing of the political fortunes of the Hasmonean dynasty, which followed the Maccabean revolt.

Hassid, -ic, -im, -ism: see *Chasid, Chassid.

hassle (hæ·s'l). *colloq.* (chiefly *N. Amer.*). Also **hassel.** [Eng. and U.S. dial.: see *E.D.D.* and Wentworth *Amer. Dial. Dict.*] A quarrel, argument, fuss; a difficulty, problem; trouble. Also as *vb.*, to quarrel, argue; to worry, harass.

1945 *Down Beat* 15 Feb. 1/5 Building bands is getting to be a habit with Freddie Slack. He broke up his last few after booking hassels. **1946** *Sat. Even. Post* 31 Aug. 72/2 'Hassle' is a gorgeously descriptive word which lately has won wide usage in show business. **1950** B. Shulberg *Disenchanted* (1951) ix. 102 She's actually a society girl.. who's had a hassel with her family and wants to prove she can get by on her own. **1957** J. Kerouac *On Road* (1958) 50 We'll both understand purely and without any hassle that we are simply stopping talking. **1959** F. W. Householder in Saporta & Bastian *Psycholinguistics* (1961) 18/1 The chief metaphysical bones hassled over in recent years concern such points as 'biuniqueness'. **1967** *Boston Sunday Globe* 23 Apr. 25/1 Now the zoning hassle has switched across the city to..where the Greek Orthodox Church is petitioning for a rezoning to allow a developer to erect..a $1.8 million office building complex. **1969** *Rolling Stone* 17 May 11/2 All others [*sc.* dancing clubs] had collapsed or been hassled to death. **1971** *Frendz* 21 May 16/1 The Edgar Broughton Band toured Germany earlier this year and were involved in some heavy hassles with the promoters of the various gigs.

‖ **hasta la vista** (a·sta la vi·sta). [Sp.] Good-bye, au revoir (used chiefly in Spanish contexts).

1935 C. Morley (*title*) Hasta la vista, or, A postcard from Peru. **1940** A. Huxley *Let.* 5 Jan. (1969) 449 Well, bless you both. Give our loves to the love-worthy. Hasta la vista. **1967** C. Armstrong *Gift Shop* ix. 76 Dorinda had bade him a gay *hasta la vista* in Copenhagen and gone off. **1967** K. Giles *Death in Diamonds* ix. 168 'Come and stay with us...' 'I hope to. Hasta la vista.'

haste, *sb.* Add: **5.** *to make haste.* **b.** *to make haste slowly*, after L. *festina lente* (Suet. *Aug.* 25).

1744 B. Franklin *Poor Richard* (1890) Apr. 146 Make haste slowly. **1831** *Deb. Congress U.S.* 4 Feb. 98 Thus far the committee have 'made haste slowly'. **1938** M. Teagle *Murders in Silk* iii. 22 Easy, son. Let's make haste slowly. Does Conner know where the knife came from?

c. *Cricket.* Of a ball: to come up from the pitch with increased speed.

1888 A. G. Steel in Steel and Lyttelton *Cricket* iii. 123 Every now and then one of their balls will, in cricket slang, 'make haste from the pitch'. **1904** P. F. Warner *How we recovered Ashes* ix. 177 The ball made haste off the pitch, kept a little low, and clean beat Duff. **1920** —— *Cricket Reminisc.* ii. 19 Australia, where the bowler who makes haste off the pitch is the most useful type.

hasten, *v.* Add: **2.** *to hasten slowly*: cf. *haste *sb.* 5 b.

1907 *Spectator* 12 Jan. 43 'Hasten slowly' is a very good motto in Imperial politics. **1958** *Oxford Mail* 14 Aug. 1/3 The Government is still hastening slowly on re-expansion.

hastener. Add: **1. a.** *Services' slang.* (See quot. 1946.)

1943 C. H. Ward-Jackson *Piece of Cake* 35 *Hastener*, a letter asking for a reply to a previous letter. **1946** J. Irving *Royal Navalese* 92 *Hastener*, a letter or a 'Minute' asking for a reply to some previous correspondence. **1955** *Times* 12 May 11/4 Those who were once temporary soldiers may recall how they used to send 'hasteners' for the stores they wanted.

hasty pudding. (Later U.S. examples.)

1879 B. F. Taylor *Summer-Savory* i. 7 Their green knapsacks are growing plump with rations of samp, hasty-pudding, and Indian bread. **1881** *Harper's Mag.* Jan. 227/1 Cod-fish balls for breakfast on Sunday morning, ..and fried hasty-pudding. **1948** *Newsweek* 5 Jan. 66/1 Cook in an iron pot; turn out on a dish and the result: hasty pudding.

hat, *sb.* Add: **3. a.** An office, position, occupation; esp. in phr. *to wear two hats*, to hold two appointments concurrently; *wearing one's* —— *hat*, in one's capacity as ——.

[**1869** S. R. Hole *Bk. about Roses* viii. 111, I never remember to have seen a scientific botanist and a successful practical florist under the same hat.] **1961** Webster *Hat*, an office symbolized by or as if by the wearing of a special hat. **1963** *Times* 25 Apr. 13/7 They..would perform that precarious feat known in the Whitehall idiom as wearing two hats. **1965** *Observer* 31 Oct. 21/4 Even when he is wearing his ecumenical hat he is reported to be speaking as Archbishop of Canterbury. **1966** *Rep. Comm. Inquiry Univ. Oxf.* I. 27 Members of the colleges have accustomed themselves 'to wear two hats' and to act as lecturers paid by the University and as fellows paid by their colleges. **1967** *Even. Standard* 29 Aug. 1/1 Wearing his new 'economic overlord' hat the Prime Minister summoned three key figures to Downing Street today. **1968** *Listener* 8 Feb. 177/2 Cecil Day-Lewis has two hats: one has laurel in it, the other is that of Nicholas Blake, who writes detective stories. **1972** *Village Voice* (N.Y.) 1 June 17/5, I wear two hats. Are you asking me this question as president of the Bartenders Union or as chairman of the ABC?

5. a. *to take off one's hat*: to doff or remove the hat, as a salute or sign of respect. Hence *hats off to..*, as a command or exhortation.

1856 *Punch* 5 Jan. 3/2 *Quaker*, a friend who..in the art of making inflammatory speeches, takes his hat off to no man. **1857** D. Livingstone *Miss. Trav.* viii. 272 This being the only hill we had seen since leaving the Bamangwato, we felt inclined to take off our hats to it. **1863** A. J. Munby *Diary* 7 Mar. in D. Hudson *Munby* (1972) 151 The populace..caught fire all at once. 'Hats off!' shouted the men: 'Here she is!' cried the women. **1881** *Harper's Mag.* Jan. 206/1 Over in Greenwood there is a stately monument, to the New York fireman,..before which I take off my hat. **1886** *Ibid.* June 45/2 We should take off our hats to them [*sc.* the 'lady-bugs'] and wish them godspeed. **1923** *Daily Mail* 22 Jan. 6, I say in all sincerity: 'Hats off to France!' **1947** 'P. Woodruff' *Wild Sweet Witch* iv. 106, I take off my hat to that boy. **1972** M. Farhi *Pleasure of your Death* vii. 198 'Hats off to them.' 'Yes, of course. Hats off to all the dead.'

b. (Earlier examples.)

1857 Kingsley *Two Y. Ago* I. v. 137 A little packet, containing not one five pound note, but four... The Mumpsimus men..had 'sent round the hat' for him. **1863** W. H. Goode *Outposts of Zion* xxi. 182 The hat for the collection was carried around by a hand disabled by a gunshot.

c. *bad hat* (later examples); *to be in a (the) hat* (later example); also *to make a hat of*, to make a mess of; *to hang one's hat*: to take up one's quarters (in a certain place); *to hang one's hat on*: to depend upon; *to talk through one's hat*: to make unsupported or 'wild' assertions, to talk nonsense; *to throw one's hat into the ring*: to take up a challenge; *under one's hat*: secret, sub rosa; *my hat!*, a trivial exclamation of surprise. See also *high hat.

1847 *Sporting Life* V. 224/2 Warren threw his hat in the ring. **1848** *Ibid.* 1 Jan. 237/1 Curtis..threw his hat into the ring. **1875** Trollope *Prime Minister* (1876) I. ii. xx. 335 Lopez can come in and hang up his hat whenever it pleases him. **1880** A. A. Hayes *New Colorado* (1881) viii. 118 Why that's my preacher. I hang my hat on him every time. **1885** C. M. Yonge *Nuttie's Father* I. xviii. 220 Nuttie..was taking in all these revelations with an open-eyed, silent horror... It was all under her hat, however, and the elder ladies never thought of her. **1888** *N.Y. World* 13 May 12/3 Dis is only a bluff dey're makin'—see! Dey're talkin' tru deir hats. **1899** Kipling *Stalky & Co.* 174 My Hat!..That's pretty average heroic. **1902** W. N. Harben *Abner Daniel* 81 All this talk about the devil makin' the bad an' the Lord the good is talk through a hat. **1904** 'O. Henry' *Cabbages & Kings* vi. 117 The governor man had a bit of English under his hat, and when the music was choked off he says: 'Ver-r-ree fine.' **1904** *N.Y. American* 18 July 2 If the Tammany leader expects to hang his hat inside Judge Parker's political headquarters, he must come here voluntarily. **1912** *Nation* 7 Mar. 226/1 When Mr. Roosevelt threw his hat into the ring the other day, he gave the signal for a contest the like of which has not been seen before in this country. **1914** C. Mackenzie *Sinister St.* II. iii. viii. 661 'My hat, what a frowst,' exclaimed Maurice. **1916** *Chambers's Jrnl.* May 302/2 Now, Joshua Billings, A.B., though officially a bad hat, was one of the best seamen in the ship. **1916** *Chums* 30 Sept. 37/2 I'm in no end of a hat, chauffeur. Can you give me a hand? **1917** W. J. Locke *Red Planet* xxiv. 'You dashed young idiot,' I cried, 'do you think I'm in the habit of talking through my hat?' **1923** Wodehouse *Inimitable Jeeves* xviii. 249 She kept it under her hat. She meant to spring it on me later, she said. **1925** J. Buchan *John Macnab* xv. 312 Palliser-Yeates lost at Glenraden..and now I've made a regular hat of things at Haripol. **1928** *Observer* 4 Mar. 11/2 Mr. Secretary Hoover has been forced to throw his hat into the ring for the Presidency, but he does not mean to follow it there. **1929** Galsworthy *Roof* iv, *Mr. B.* I suppose you think you never snore. *Mrs. B.* I know I don't. *Mr. B.* My hat! **1939** 'A. Bridge' *Four-Part Setting* 155 One couldn't just sit by and watch a person..make a complete hat of her life and herself and her character. **1945** M. Allingham *Coroner's Pidgin* xiv. 118 My hat! was it only last night? **1953** 'N. Shute' *In Wet* viii. 251 'Nothing about that in the papers, is there?' 'Not yet. Keep it under your hat.' **1956** 'A. Gilbert' *Death came Too* xiv. 152 'You,' suggested Frank, politely, 'are talking through your hat.' **1958** *Daily Mail* 6 Sept. 4/2 Some of them innocent hard-working people, others petty thieves and bad hats. **1963** N. Marsh *Dead Water* (1964) vi. 148 I'd be very grateful ..if you'd keep the whole affair under your hat.

Also, in contexts referring to the drawing of names from (or the putting of names into) a hat in selecting opponents in a competition, etc.; also with reference to the conjuring trick of producing a rabbit from a hat.

1929 *Evening News* 18 Nov. 13/2 Dagenham..will be amongst the distinguished clubs to go into the hat. **1958** *Listener* 18 Sept. 404/1 Mr. Dulles first pulled indirect aggression out of the hat in mid-July. **1963** A. Ross *Australia 63* v. 110 Simpson's five [wickets] were simply out of the hat. **1966** H. Waugh *Pure Poison* (1967) xv. 93 He picked Roger out of a hat as a victim? **1971** J. McClure *Steam Pig* ii. 26, I must say you've really pulled one out of the hat this time.

6. d. The creamy top of hatted kit.

1831 [see Hatted *ppl. a.* c]. **1946** *Farmhouse Fare* (new ed.) 124 Hatted Kit..can..be made without milking the cow into it, although direct milking puts a better 'hat' on the Kit. **1952** F. White *Good Eng. Food* iv. ii. 180 *Hatted Kit*...fresh good butter-milk, and a pint of milk hot from the cow. Mix well by jumbling... It will now firm, and gather a hat.

7. b. *dial.* A clump of trees.

1895 De Crespigny & Hutchinson *New Forest* 113 The term 'hat' is still in use for a little wood crowning a hill. **1895** G. Patterson in *Jrnl. Amer. Folk-Lore* VIII. 29 A hat of trees. **1936** C. R. Acton *Sport & Sportsmen of New Forest* ii. 43 A clump of trees is known as a 'Hat'; two examples being 'Crab Hat' and 'King's Hat'.

8. *hat-securer.*

1892 A. Conan Doyle in *Strand Mag.* III. 75/1 It was pierced in the brim for a hat-securer, but the elastic was missing.

9, 10. *hat-raising, -trimming*; **hat-check boy, girl** *U.S.*, a cloakroom attendant; **hat-guard** (examples); **hat leather** (see quot. 1888); **hat-pad,** a pad usually of velvet for wiping the dust off or smoothing the nap of a hat; **hat-rack,** (*a*) a rack to hold hats; (*b*) *slang*, a scraggy animal; (*c*) *slang*, the head; **hat-tip,** the circular piece of stuff used to line the crown of a hat; **hat-tree,** (*b*) *Austral.* (see quot.).

1917 *N.Y. Tribune* 19 June 8/4 How about the hat check boys? **1921** Wodehouse *Indiscretions of Archie* xiv. 159 He paid no attention to the hat-check boy. **1959** *Guardian* 22 Dec. 5/1 He found work as a hatcheck boy. **1920** Wodehouse *Jill the Reckless* (1922) xv. 223 When a burglar marries a hat-check girl, their offspring goes into the theatrical business automatically. **1938** *Times Lit. Suppl.* 3 Sept. 572/4 He..has included all the important information..even to..the name of the hat-check girl in the New York restaurant. **1899** *Catal.* in A. Adburgham *Shops & Shopping* (1964) xxii. 261 Hat guards. **1912** A. Bennett *Matador* 131 William Henry commanded her to buy a hat-guard. The hat-guard cost sixpence. **1888** *Lockwood's Dict. Mech. Engin.*, *Hat leather*, the leather ring packing used for hydraulic pistons. **1940** *Chambers's Techn. Dict.* 405/1 *Hat-leather packing*, an L-section leather ring, gripped between discs to form a piston, or similarly attached to the ram of a hydraulic machine to prevent leakage. **1902** W. W. Jacobs *Lady of Barge* 221 At the hall he paused, and busied himself with the clothes-brush and hat-pad. **1872** Hat-rack [in Dict., sense 8]. **1935** *Amer. Speech* X. 269/1 If he should try to hang his hat on the hatrack he will probably find that he will have to catch it first, for the hatrack will be found milling about in one of the pens. *Ibid.* 271/1 *Hatrack*, an old, thin cow, a nellie or canner. **1942** Berrey & Van den Bark *Amer. Thes. Slang* § 121/56 *Head*,..hat rack. **1957** R. Campbell *Portugal* v. 73 One trick is to deprive a hatrack of an old horse of water, and then give him a good lick of salt. **1964** L. Hairston in J. H. Clarke *Harlem* 286 If you spent half as much time tryin' to put something *inside* that worthless hat-rack as you did having your brains fryed. **1905** *Westm. Gaz.* 11 Aug. 7/1 At Trafalgar-square there was much hat-raising. **1908** *Ibid.* 30 Jan. 12/2 The hat-raising habit. **1921** *Dict. Occup. Terms* (1927) § 549 *Hat tip sizer*, prints hat tip with size before gold leaf or bronze dust is applied by hat tip printer. **1898** Morris *Austral Eng.*, *Hat-tree*, name given to a species of *Sterculia*, the Bottle-trees. **1895** *Montgomery Ward Catal.* Index, Hat trimmings. **1905** *Daily Chron.* 23 Dec. 6/5 The success of the hat-trimming competition.

hat, *v.* Add: **c.** *intr.* To work alone. (Cf. Hatter *sb.* 2.) *Austral.*

1891 *Age* 25 Nov. 6/7 (Morris), Two old miners have been..hatting for gold amongst the old alluvial gullies. **1900** H. Lawson *On Track* 88 And he 'hatted' and brooded over it till he went ratty.

‖ **hatamoto** (hætămōu·to). [Jap.] In the

Japanese feudal system, a vassal or member of the household troops of a Shogun.

1871 A. B. Mitford *Tales of Old Japan* I. 95 *Hatamoto.* This word means 'under the flag'. The Hatamotos were men who..rallied round the standard of the Shogun, or Tycoon, in war-time. **1899** L. Hearn *In Ghostly Japan* vi. 74 The *hatamoto* were samurai forming the special military force of the Shōgun. **1904** —— *Japan: Attempt at Interpretation* xii. 267 These two bodies of samurai formed the special military force of the Shōgun; the hatamoto being greater vassals with large incomes. **1968** J. W. Hall *Japan fr. Prehist. to Mod. Times* x. 166, 5,000 'bannermen' (*hatamoto*), who were privileged to come into the Shogun's presence.

hatch, *sb.*[1] Add: **3. d.** fig. *down the hatch*: a toasting or drinking phrase.

1931 *Amer. Mercury* Mar. 357/2 The boys didn't pester her to drink. 'Down the hatch!' they said. **1933** M. Lowry *Ultramarine* iii. 120 'Well, let's shoot a few whiskies down the hatch, and you'll see three,' I remarked fatuously. **1935** *Yachting* Dec. 32/2 'Down the hatch!' is a toast well known ashore. **1942** T. Rattigan *Flare Path* I. 110 That went down the old hatch pretty quick, didn't it? **1958** B. Hamilton *Too Much of Water* xii. 272 And so now, down the hatch, and let's..see what we can do with the pudding and souse. **1972** *House & Garden* Mar. 130/1 Unlike the professionals, who take a small sip ..and then spit it out..we, as amateurs, adopted the 'down the hatch' technique.

e. *Aeronaut.* An opening or door in an aeroplane or space capsule. See also quot. 1948.

1940 [see *escape hatch* s.v. *escape *sb.*[1] 8]. **1943** *Coastal Command* (Ministry of Information) ix. 89 At 1.55, after the rigger has reported that all hatches are closed, the klaxon sounds. The Catalina moves slowly at first... The take-off has begun. **1943** E. V. Rickenbacker *Seven came Through* i. 13, I helped Sergeant Alex pry open the bottom hatch in the tail and between us we dumped all that high-priority mail into the blue Pacific. **1948** Partridge *Dict. Forces' Slang* 92 *Hatch*, a bomb-hatch—the bomb-aimer's compartment, at the front of the kite, especially in 'Lanks' and 'Wimpeys'. **1956** W. A. Heflin *U.S.A.F. Dict.* 246/1 *Hatch*, a ship term sometimes applied to an opening or door in an aircraft, esp. one in the deck of an aircraft or in the top and bottom of the fuselage. **1962** D. Slayton in *Into Orbit* 26 We asked them to adapt the entry hatch and convert it into an exit, too. **1969** *Times* 23 May 1/2 The two craft are linked by a 3 ft. connecting tunnel, sealed at each end by a hatch.

hatch, *sb.*[2] Add: Also in phr. *hatches, matches, and dispatches* (occas. in *sing.*), a newspaper list of births, marriages, and deaths.

1878 J. Payn *By Proxy* I. xix. 217 First came the Births, Deaths, and Marriages... The female mind.. takes an interest in the 'Hatch, Match, and Despatch' of its fellow-creatures. **1880** *Times for Year 1980* 1/1 Hatches... Matches... Despatches. **1953** M. Steen *Anna Fitzalan* viii. 215 Dismissing reviews..Lin turned to what Mummy called Hatches, Matches and Despatches. **1959** F. King *So Hurt & Humiliated* 128 Glancing through the 'Hatches, Matches, Despatches' columns in *The Times* at breakfast, Emily suddenly interrupted my reading of the *Economist*. **1966** 'H. Howard' *Counterfeit* iii. 57 He might even be a registrar of births, marriages and deaths—the man who issued certificates for what people used to call Hatches, Matches and Despatches.

hatch, *v.*[1] Add: Hence **hatchabi·lity,** the condition or state of being likely to hatch, or able to produce eggs which will hatch.

1916 *Experiment Station Rec.* Feb. 178 The hatchability of eggs which are produced. **1950** *N.Z. Jrnl. Agric.* Jan. 14/1 Work is involved in keeping data about the hatchability of the eggs from each pen. **1956** *New Biol.* XXI. 116 There is evidence that the presence of earthworms in soil increases the hatchability of the cysts of the potato root eelworm. **1960** *Farmer & Stockbreeder* 9 Feb. 87 Greater egg production, better grading, increased hatchability.

hatchel, *sb.* See also *hetchel *sb.*

hatchel, *v.* Add: **1. b.** *transf.*

1845 [see *hetchel *v.*].

2. (Earlier U.S. example.)

1800 *Aurora* (Phila.) 20 Oct. (Th.), They have..hatchelled them with prosecutions, fines, and imprisonments.

hatchery. Add: (Later examples.) Also *fig.*

1932 A. Huxley *Brave New World* i. 1 Central London hatchery and conditioning centre. **1932** M. A. Jull *Poultry Breeding* xi. 347 Sanitary conditions at hatcheries must be approved by the hatchery inspector. Only eggs from approved hatchery flocks may be incubated. **1942** D. Mitrany in *Agenda* I. 305 The T.V.A. is itself a hatchery of public enterprise. **1952** *Oxf. Jun. Encycl.* VI. 360/2 Some poultry-farmers do not hatch eggs from their own birds but buy day-old chicks from 'hatcheries', which are places that do nothing but incubate eggs on a very large scale.

hatchet, *sb.* Add: **2.** See also Bury *v.* 2 a in Dict. and Suppl.

3. hatchet fish, a member of the family Gasteropelecidæ, South American flying characins which are often kept in aquaria, or one of the family Sternoptychidæ, deep-sea clupeiform fishes found in most of the oceans of the world; also *hatchet* ellipt.; **hatchet-job, -work** (see *hatchet-man 3).

1931 J. R. Norman *Hist. Fishes* xii. 231 (*caption*) Hatchet-fish (*Argyropelecus* sp.). **1959** P. Capon *Amongst those Missing* 196 Hatchet-fish skimmed the water. **1960** M. Burton *Under Sea* xi. 198 Another consumer of small prey is the 'hatchet' fish, so called because its body is flattened from side-to-side... For the most part hatchets are only a few inches long. **1962** K. F. Lagler et al. *Ichthyology* ii. 36 Family Sternoptychidae—deepsea hatchet fishes.

hatchet, *v.* Restrict *Obs.* to sense in Dict. and add: **2.** *transf.* To act as a hatchet-man against (someone), to do down.

1959 'B. O'Brien' *Operators & Things* (1960) i. 34 Even the Knoxes were willing to hatchet each other.

ha·tchet-man. [f. Hatchet *sb.* + Man *sb.*[1]] † **1.** A pioneer or axeman serving in a military unit. *U.S. Obs.*

1755 [see *hatchet-man* s.v. Hatchet *sb.* 3].

2. In the U.S., a hired Chinese assassin. Also *transf.*

1880 G. B. Densmore *Chinese in California* xii. 94 Some of them are called hatchet-men. They carry a hatchet with the handle cut off. **1888** *Boston Jrnl.* 3 May 1/2 The work of the hatchetmen among the enemies of the organization. **1913** J. London *Valley of Moon* iii. xx, Chan Chi, had been a hatchet-man of note, in the old fighting days of the San Francisco tongs. **1957** P. Frank *Seven Days to Never* iii. iii. 90 He was a hatchet man for the NKVD... He may have delivered Beria over to Malenkov and Krushchev.

3. (Now the usual sense.) A person, especially a journalist, employed to attack and destroy other people's reputations. So *hatchet job, work.* orig. *U.S.*

1944 *Time* 23 Oct. 20 Exuberant hatchet jobs were.. done on Foster Dulles because of his Wall Street connections. **1952** *Manch. Guardian Weekly* 3 Apr. 15/4 Republican hatchet-men. **1961** M. McCarthy *On Contrary* (1962) i. 87 The literary Communists..doing the hatchet work on artists' reputations. **1959** *Encounter* July 83/1 One has no difficulty in recognising the familiar tones of Dr. Leavis' hatchet-men when he is attacked. **1959** *Guardian* 13 Oct. 7/4 One critic..was the meanest son of a bitch that ever lived. His criticism was a hatchet job on every book. **1960** *News Chron.* 14 July 1/5 The Kennedy family went into action with a commando team of political hatchet-men. **1962** *Listener* 21 June 1089/1 It was difficult enough to sympathise with the hero once we'd seen him doing his hatchet work.

hatching, *vbl. sb.*[1] Add: **a.** Also, that which is hatched, a brood.

1905 *Kynoch Jrnl.* Apr.–June 108 The hatchings at the present time are quite up to the average of a good year.

hatchling (hæ·tʃliŋ). [f. Hatch *sb.*[2] + -ling[1].] A very young fish or bird, etc., usually artificially hatched and not old enough to take care of itself.

1899 *19th Cent.* Sept. 399 The ova hatched out *en route*, and the hatchlings died. **1899** *Field* 16 Sept. 496 This assertion may be verified by throwing some hatchlings into a tank where fish of all sizes are mixed together. It will be seen that the strangers are at once devoured. **1955** *Sci. Amer.* Oct. 98/3 It is curious that, although the young hatchling in the nest is in great hazard of its life, once it has begun to fly it is extremely unlikely to be lost during the remainder of the dependence period. **1957** *New Scientist* 24 Oct. 9 The female octopus..laid eggs..on 6 September. ..The first hatchlings appeared on 16 October.

hatch-out (hæ·tʃ,aut). [f. the verbal phr. *to hatch out* (see Hatch *v.*[1] 2).] The action of hatching out; also, the brood hatched out.

1898 *Westm. Gaz.* 13 May 4/1 It only depends upon climatic conditions to ensure a good hatch out. **1908** *Ibid.* 5 June 4/2 While the hatch-out is in progress the number of the fly is marvellous. *Ibid.* 14 Aug. 4/2 Partridges are more faithful to a fixed date for the hatch-out of their eggs than grouse.

hate, *sb.*[1] Add: **1. c.** In the war of 1914–18, a bombardment, a 'strafe'. *slang.*

A jocular use based upon the German 'Hymn of Hate', which was ridiculed in *Punch* 24 Feb. 1915, p. 150, in the legend of a drawing, 'Study of a Prussian household having its morning hate'.

[**1914** *Punch* 30 Dec. 530/1 Kaiser, what vigil will you keep tonight?..While your priesthood chants the Hymn of Hate, Like incense will you lift to God your breath?] **1915** D. O. Barnett *Lett.* 204 There are some unhealthy spots, 'Suicide Corner', 'Deadman's Alley' and others, where they drop shells regularly, trying to catch our transport at night. We call it the 'Evening Hate'. **1926** F. M. Ford *Man could stand up* ii. v. 174 There is not going to be a *strafe*. This is only a little extra Morning Hate. **1927** E. Thompson *These Men thy Friends* 112 He was watching a spasmodic 'hate' of some violence. **1968** D. Reeman *Pride & Anguish* x. 180 I'm going to turn in, Sub. I want a couple of hours before the night's 'hate' gets going.

d. Phr. *to have a hate on* or *against* (a person) (see quot. 1941).

1941 Baker *Dict. Austral. Slang* 38 *Have a hate against*, actively to dislike a person or thing. **1966** 'S. Woods' *Let's choose Executors* 62 Things have been perfectly horrid, ever since Mark started to have a hate against her. *Ibid.* 220 She seemed to have a complete hate on him.

2. *hate-maddened* adj.; **hate-love,** a conflicting emotion combining hate and love (cf. **love-hate*).

1915 J. C. Powys *Visions & Revisions* 244 This monstrous *hate-love*, caressing the bruises itself has made, and shooting forth a forked viper-tongue of cruelty from between the lips that kiss. **1962** *Listener* 5 July 11/2 He consciously contrasts his teaching with that of the object of his hate-love. **1921** R. Graves *Pier-Glass* 25 It beams on set jaw and hate-maddened eye. **1937** B. H. L. Hart *Europe in Arms* xxii. 284 To use force without limit and without calculation of cost may be instinctive in a hate-maddened mob, but it is the negation of statesmanship.

b. Used *attrib.* or as quasi-*adj.*: designed to stir up hate, e.g. *hate campaign*; marked or characterized by hate.

1916 *Daily Colonist* (Victoria, B.C.) 21 July 12/7 The official Cologne Gazette published the following excellent example of 'hate literature': 'Among those who are guilty of involving Europe in a bath of blood Lord Northcliffe is perhaps the guiltiest of all.' **1949** 'G. Orwell' *Nineteen Eighty-Four* i. 5 The economy drive in preparation for Hate Week. **1959** *Daily Tel.* 18 May 6/2 Hence, perhaps, the decision to revert to 'Western imperialism' as target of a fresh hate-campaign in Iraq. **1966** H. Waugh *Pure Poison* (1967) xii. 71 Have you or your wife ever received hate phone calls or hate messages before? **1969** *N.Y. Rev. Books* 16 Jan. 36/1 Mr Epstein reaches the heights..of absurdity by stating that the hate literature distributed in the Ocean Hill-Brownsville teacher mail boxes may have been fraudulent.

hateworthy (hē^i·t,wɒ̄:ɪðɪ), *a.* [f. Hate *sb.*[1] + Worthy *a.*] Worthy of hate, hateful.

1901 A. Symons *Poems* (1907) i. 180, I tremble lest a wrath so just avenge On him a mother so most hateworthy. **1924** *Public Opinion* 9 May 450/3 There is nothing sinister or hateworthy in Mrs. Carlyle's slowly and deliberately formed judgement.

hatha-yoga (ha·tăyōu·gă). [Skr., f. *haṭha* force, violence, forced meditation + Yoga.] A system of exercises and control of breathing forming part of the Hindu religious philosophy of yoga. So **hatha-yogi(n),** a devoteee of hatha-yoga.

1911 *Encycl. Brit.* XXVI. 791/1 The physical methods and spiritual exercises recommended by theosophists are those inculcated in the systems known in Hindu philosophy as Rājā Yoga in contradistinction to the Hatha Yoga system, which is most commonly to be met with in India, and in which the material aspects are given greater prominence. **1937** A. Huxley *Ends & Means* xiii. 234 The methods of Hatha Yoga, as they are called in India, are said to result in heightened mental and physical powers. *Ibid.* 247 It is possible for meditation to be practised by those who are neither extreme ascetics nor Hatha-Yogis. **1956** E. Wood *Yoga Dict.* 62 *Hatha-yoga*, a form of yoga which is concerned chiefly with the regulation of breathing, and secondarily with other bodily disciplines or training. **1956** A. Huxley *Adonis & Alphabet* 32 The training of the dervish or the hatha-yogin is a long laborious affair. **1963** *Times Lit. Suppl.* 11 Jan. 29/3 It may be observed that even Indian Buddha figures ..by no means express the strenuous constrictions of mediaeval Hindu Hatha-yoga. **1967** *Daily Tel.* 1 Feb. 13/5 Hatha yoga, he explained, deals with the mastery of thought and breath. 'If we control our breath we control our thought. When we control our thought we begin to understand the full meaning of life.'

hathi (hā·ti). *India.* Also **hotty, huttee,** etc. [Hind. *hāthī* (also Marathi, etc. *hattī*), f. Skr. *hastin* elephant, f. *hasta* elephant's trunk, hand.] An elephant. Also *attrib.*: **hathi tractor,** a kind of tractor used in the war of 1914–18.

[**1826** Leyden & Erskine tr. *Mem. Zehir-Ed-Din* 315 As for the animals peculiar to Hindustân, one is the elephant, the Hindustânis call it *Hathi*.] **1831** Tyerman & Bennet *Voy. & Trav.* II. 375 Our bearers suddenly set up the cry of 'Huttee! huttee!' **1838** in E. Eden *Up Country* (1866) I. 269 You are of course aware that we habitually call elephants Hotties. **1860** W. H. Russell *Diary India* I. 392 We came to the Ramgunga, a deep stream, which our elephant waded across... The hathi nearly floated his driver off his neck. **1890** Kipling *Barrack-r. Ballads* (1892) 51 An' the *hathis* pilin' teak. **1922** *Glasgow Herald* 11 Feb. 10 Golden also was the head of the 'hathi' selected for this honour. **1926** *Glasgow Herald* 27 Aug. 11 There was the Hathi tractor, constructed in the first place largely from German spare parts, captured during the war. The Hathi—its title is the Indian word for elephant—had besides the strength of the elephant qualities which that intelligent animal does not possess.

Hathor (hā·þǭɪ). Also **Athor.** [ad. Gr. Ἀθώρ, f. Egypt. *Ḥet-Ḥerh* the house above, or *Ḥet-Ḥeru* house of Horus.] The name of an Egyptian divinity, the goddess of love, often represented with the head or ears of a cow, used *attrib.* or *Comb.* to designate a type of column surmounted by a capital on which are carved one or more representations of the head of Hathor. So **Hathoric** (hăþǫ·rik), *a.* Hence **Hathore·sque** *a.*, in the style of a Hathor figure.

1786 tr. *C. E. Savary's Lett. on Egypt* II. xlviii. 351 Athor, or the night, in the opinion of the Egyptian priests, represented the darkness which enveloped the chaos before the creation. **1851** W. S. W. Vaux *Handbk. Antiq. B.M.* 355 The Venus of the Egyptians was called Athor, Hathor, or Athyr, and her name implied the abode of Horus. **1857** J. Gardner Wilkinson *Egyptians* 273 His hair is that of the milky way..his eyes, the symbolical eyes of Athor.

1896 W. M. F. Petrie *Koptos* i. 4/2 Below the scene is a frieze of *dad* signs alternating with figures, the lower parts of which are like the Isiac girdle tie, while above they have the human Hathor head, with cows' ears and horns... They seem as if they might be copies of some primitive Hathor idol. **1901** R. Sturgis *Dict. Archit.* I. 854/2 The columns are easily divisible into a few general types, such as the single and the clustered lotus-bud, the campaniform, the palm-capped, and the Hathor-headed. *Ibid.* II. 366/2 *Hathoric*, having to do with the Egyptian goddess Hathor. **1934** E. Pound *Eleven New Cantos* xxxix. 44 When Hathor was bound in that box afloat on the sea wave. **1960** *Times* 7 Mar. 8/3 A handsome jewelry box with an ivory inlay of Hathor beads. **1962** D. Harden *Phoenicians* xiii. 198 A peculiar multiple vase for offerings has..a Hathoresque head above a long-horned cow's head.

hatikvah (hati·kvā). [ad. Heb. *ha-tiḳwāh* the hope.] A national song, of which the words were written by N. H. Imber (1856–1909), adopted by the Zionist movement in 1907; since 1948 the Israeli national anthem.

1925 P. Guedalla *Napoleon & Palestine* 63 The proceedings concluded with the singing of the Hatikvah by some of the audience. **1932** L. Golding *Magnolia St.* i. viii. 130 The Jewish guests thought it would balance things nicely if they sang the Jewish anthem, *Hatikvah*. **1965** *Times Lit. Suppl.* 30 Dec. 1210/5 The Jewesses of Salonika singing the Hatikvah, the Jewish anthem. **1970** I. Sieff *Memoirs* vi. 111 We sang *Hatikvah*, and also 'God Save the King'.

hatless, *a.* Add: Hence **ha·tlessly** *adv.*; **ha·tlessness,** hatless condition.

1890 E. Dowson *Let.* 1 June (1967) 149 We sat & smoked for some hours hatlessly on the balcony. **1881** R. G. White *Eng. Without & Within* 271 The hatlessness, the shoelessness, the rags, and the dirt. **1902** *Westm. Gaz.* 17 Oct. 2/3 Hitherto hatlessness was only *de rigueur* in the stalls and the front row of the dress circle. **1924** *Glasgow Herald* 29 May 6 Many people, who advocate hatlessness during the summer months. **1933** E. A. Robertson *Ordinary Families* x. 229 He did not mind my looking out of place through hatlessness. **1960** *Guardian* 15 July 8/2 The prevailing..hatlessness of Frenchwomen.

hatter, *sb.* **2.** For def. read: One who lives or works alone, orig. a miner; a solitary bushman. *Austral.* and *N.Z.* Add earlier and additional examples.

1853 J. Rochfort *Adv. Surveyor* viii. 66 The Bendigo diggings are suitable for persons working singly... Such persons are humorously called 'hatters'. **1865** B. L. Farjeon *Shadows on Snow* ii. 76, I was working as a 'hatter'. **1889** E. Wakefield *N.Z. after 50 Yrs.* vi. 165 Miners who work alone are called 'hatters', one explanation of the term being that they frequently go mad from the solitude of their claim away in the bush, exemplifying the proverb 'As mad as a hatter'. **1903** 'S. Rudd' *Our New Selection* iv. 37 A weird, silent 'hatter' was there,.. the strange man who lived..away from everybody. **1914** J. M. Bell *Wilds of Maoriland* vi. 135 At times one comes across an old hatter [near Karamea]. **1924** H. T. Gibson *That Gibbie Galoot* xxvii. 124 The skipper [of a timber scow] I seldom saw, for he was a 'hatter' and kept to his cabin and keg. **1943** V. Palmer in *Coast to Coast 1942* 21 People on the mainland said that McGowan was a cranky old hatter who had gone off his head because his home was broken up and was now letting his mind rot in isolation. **1944** F. Clune *Red Heart* 66 The 'hatter' was mumbling to himself in the manner of lonely outback prowlers. **1966** *Southerly* XXVI. 108 Rueben McGrath was..a bush 'hatter', a loner.

Hattic (hæ·tik), *a.* Also **Kh-.** [f. Assyrian and Hittite *Ḫatti* + -ic.] Of or pertaining to the Hatti or their language, formerly regarded as conterminous with the Hittites, now as a section of them. Hence as *sb.*, their language. So **Ha·ttian** *sb.* and *a.* **a.** One of the race of Hatti or their language. **b.** = *Hattic* adj. above. **Ha·ttism,** the social and political system of the Hatti.

[**1874** *Trans. Soc. Bibl. Arch.* III. 245 The king of the Khati. **1880** Cheyne in *Encycl. Brit.* XII. 25/1 Hittites, a warlike and powerful nation... In the Egyptian inscriptions they are called the Khita or Kheta; in the Assyrian, the Khatti; in the Hebrew Scriptures, the Khittim.] **1924** D. G. Hogarth in *Cambr. Anc. Hist.* II. 253 Boghaz Keui..is the site of the Hattic capital. **1926** —— *Kings of Hittites* 3 The Hittite civilization of Hamath was but an outlier of 'Hattism', advanced southward along a trunk-road. *Ibid.* 8 Have any remains been revealed which manifestly are Cappadocian Hattic, wholly or in part? **1928** C. Dawson *Age of Gods* 302 The official language of the empire has been named by its discoverers Nashili or Kanesian; but since the ruling people have always been known as the Hittites, it seems better to retain the same name for their language and to describe the native Hattic tongue as Old Hittite. **1929** J. Garstang *Hittite Empire* ii. 39 The suggestion of language..would seem to indicate an original movement or series of movements from or affecting the Caucasian area, which at the same time peopled Armenia, northern Mesopotamia, and Elam, and won for the Hattians and other Hittite tribes a footing on the eastern mountains and plateau of Asia Minor. *Ibid.* 40 The Hattians themselves were an inland and not a seafaring people. **1933** E. H. Sturtevant *Compar. Gram. Hittite Lang.* i. 29 Fortunately our use of the biblical name Hittite leaves the ancient stem free for use in its original sense; we shall call the predecessor language Hattic. **1952** O. R. Gurney *Hittites* ii. 64 The original (Hattian?) form of the name. *Ibid.* 69 This conclusion agrees well with the linguistic evidence, according to which a group of Indo-European immigrants became dominant over an aboriginal race of 'Hattians'. **1958** *Archivum Linguisticum* X. ii. 82 Bilinguals whose native language was Hattic. **1963** *Times Lit. Suppl.* 1 Feb. 72/3 The fully prehistoric 'Hattian' period.

hat trick. Add: **2.** *Cricket.* (Earlier example.)

1877 J. Lillywhite *Cricketers' Compan.* 181 Having on one occasion taken six wickets in seven balls, thus performing the hat-trick successfully.

b. Hence *gen.*, a threefold feat in other sports or activities.

1909 *Daily Chron.* 12 Aug. 9/2 It is seldom that an apprentice does the 'hat trick', but the feat was accomplished by..an apprentice... His three successes were gained on Soldier.., Lady Carlton.., and Hawkweed. **1930** *Morning Post* 16 July 17/2 Wragg's mount..enabled his jockey to complete the so-called 'hat-trick'. **1931** *Statesman* (Calcutta) 5 Dec., British aircraft constructors are hoping that an official attempt will shortly be made on the world's height record, and the 'hat trick' accomplished by the annexation of all three of the records which really matter in aviation. **1958** *Economist* 13 Sept. 819/1 The Tories are excited because it looks as if they may flout all precedents and complete a hat-trick of wins. **1967** J. Potter *Foul Play* (1968) ix. 100 Apart from a hat trick by our centre forward it wasn't much of a game.

hatty (hæ·ti), *a.* [f. Hat *sb.* + -y[1].] Wearing showy hats; interested in wearing hats.

1909 H. G. Wells *Tono-Bungay* iii. ii. 294 They all sat about in the summer-house and in garden-chairs, and were very hatty and ruffley and sunshadey. **1959** *Star* 29 Jan. 4/2 'I'm not a hatty person really,' she confessed. 'They always take so long to put on when I'm rushing out shopping.'

hau (hau). *Bot.* Also **hau-tree.** The Hawaiian name for a tropical shrub or tree, *Hibiscus tiliaceus*, belonging to the family Malvaceæ.

1843 J. J. Jarves *Scenes & Scenery Sandwich Islands* iii. 117 Groves of dark-leaved *hau*. **1866** 'Mark Twain' *Lett. fr. Hawaii* (1967) 99 Large tracts were covered with large hau (how) bushes, whose sheltering foliage is so thick as to be almost impervious to rain. **1888** W. Hillebrand *Flora Hawaiian Islands* 49 A small freely branching tree... Occurs in all tropical countries and is abundant in all Pacific islands. Native name: 'Hau'. **1913** R. Brooke *Let.* 12 Oct. (1968) 518 I'm sitting under a busy 'Hau-Tree' (pronounced 'How'). **1915** W. A. Bryan *Nat. Hist. Hawaii* xv. 201 One of the most common, persistent and useful of the native trees.. is the hau. **1935** F. B. H. Brown *Flora S.E. Polynesia* III. 174 The native name [of *Hibiscus tiliaceus*] is..hau in the northern islands of the Marquesas,..and in Hawaii.

hauchecornite (haukĕkǭ·ɪnəit). *Min.* [ad. G. *hauchecornit* (R. Scheibe 1893, in *Jahrb. d. Preuss. geol. Landesanst. und Bergakademie zu Berlin, 1891* XII. 91), f. the name of Wilhelm *Hauchecorne* (1828–1900), German geologist: see -ite[1].] A bronze-coloured sulphide of nickel, bismuth, and antimony, $Ni_9(Bi,Sb)_2S_8$.

1893 *Jrnl. Chem. Soc.* LXIV. ii. 418 Hauchecornite is of a light, bronze-yellow colour. **1950** *Amer. Mineralogist* XXXV. 440 Study of a specimen..from the original locality, Friedrich mine, Hamm a. d. Sieg, Westphalia, confirms the individuality of hauchecornite. **1968** I. Kostov *Mineral.* 117 (*table*) Hauchecornite.

Hau Hau (hau·hau). *N.Z.* Also **Hauhau, Hau-hau, hau hau.** [Maori.] A follower of the Pai-Marire religion during the nineteenth-century Maori Wars. Also *attrib.* Hence **Hau·hauism.**

1865 *Richmond-Atkinson Papers* II. iii. 171 The excitement among the Hau-hau and other hostile natives was reviving. **1871** C. L. Money *Knocking about in N.Z.* x. 137 A large village..said to be a nest of Hau-haus. **1875** *Official Handbk. N.Z.* (ed. 2) 28/2 Many who eagerly adopted Hau-hauism at first, have since given it up. **1884** M. Martin *Our Maoris* xi. 169 Early in 1865 came the terrible news from the East Cape, of the Rev. Carl Volkner's murder by the fanatical Hauhaus there. [*Ibid.* 173 He proclaimed a new religion, though indeed it was a mixture of wild applications of Old Testament history with spells and incantations. A pole was set up in the pah, round which the people danced. They drew in their breaths all at once, somewhat in the way paviours used to do. This deep groan at the end of each sentence, 'Hau', gave a name to the fanatical movement which lasts to this day.] **1914** *Chambers's Jrnl.* Mar. 173/2 In religion he follows 'Hau-hauism', a strange intermingling of ideas, based largely on the Old Testament. **1930** J. Cowan in J. Reid *Kiwi Laughs* (1961) 97 They would have had his head to decorate the end of a Hauhau pole had they discovered the particular potato-pit in which he was hiding. **1949** P. Buck *Coming of Maori* (1950) iv. iii. 474 Possession was practised by the fanatical followers of the late post-European sect known as *hauhau*, when dancing around a pole termed the *niu*. **1959** M. Shadbolt *New Zealanders* 237 The great-grandfather was eaten in the latter stage of the Maori wars by the *Hau Hau*, that fanatic group which combined Christianity and cannibalism with apparent success.

hau-hau (hǭ,hǭ), *v.* [Echoic.] *intr.* To utter the cry of a hyena.

1924 *Other Lands* Jan. 43/1 A hyena went hau-hauing down the path between the tent and the garden.

haul, *v.* Add: **1. e.** To transport by cart or other conveyance; to cart, carry.

1741 *New Hampshire Probate Rec.* III. 43 Her fire wood from time to time shall be haul'd to Said house. **1787** [in Dict., sense 1 a]. **1814** H. M. Brackenridge *Views Louisiana* 141 They are sometimes employed in hauling lead from the mines. **1852** *Trans. Mich. Agric. Soc.* III. 179, I haul it [*sc.* manure] out in the fall, spread it and plow it in immediately after. **1880** *Harper's Mag.* Sept. 619/2 In winter I haul logs, and in summer I haul mealers. **1887** *Congress Rec.* 10 Jan. 484/1 There is not one-tenth part of the risk in hauling dressed beef that there is in hauling live animals. **1918** F. Hackett *Ireland* ii. 46 The more fish was caught..the less any one of them was worth. And when it came to salting them or hauling them, the same curse was in it. **1970** *Washington Post* 30 Sept. B13/4 The company sought a million-dollar contract with Ft. Rucker, Ala., to haul gasoline.

absol. **1871** R. L. Dashwood *Chiploquorgan* viii. 117 We had fifteen miles to haul along a lumber road to the mouth of Rocky brook. **1883** J. Hay *Bread-winners* vi. 96 You know Clinsty Fore, that hauls for the Safe Company? **1933** E. Merrick *True North* 338 We hauled across lots of yellow, slushy places.

f. *intr.* With *out, up.* Of bachelor seals: to come out of the water to rest on the hauling-grounds.

1869 *Overland Monthly* III. 39 To ascertain if any elephant-seal had 'hauled up' on the beach. **1894** Kipling *Jungle Bk.* 98 I've often thought we should be much happier if we hauled out at Otter Island. **1902** *Encycl. Brit.* XXXII. 488/1 The young males, or bachelors, haul out to rest and sleep on beaches adjacent to, but distinct from, the breeding-grounds. **1967** *Listener* 6 Apr. 459/1 Adult male seals hold territories on beaches where a population hauls out for breeding.

3. c. *to haul off* (chiefly *U.S.*), to withdraw or draw back a little before completing an action of any kind; *to haul out* (*U.S.*), to go out, depart.

1866 W. H. Jackson *Diary* 30 July in *Nebr. Hist. Mag.* (1932) XIII. 156 Hauled out before sunrise and corralled at the Springs by 9 o'clock. **1870** 'Mark Twain' in *Galaxy* Oct. 572/2 Suppose he should take deliberate aim and 'haul off' and fetch me with the butt-end of it [*sc.* a gun]? **1902** A. D. McFaul *Ike Glidden* xxv. 282 The train hauled out while the officer was taking him into custody. **1930** D. Runyon in *Collier's* 20 Dec. 32/3 Then Lily hauls off and gives me a big kiss right in the smush. **1961** M. McLuhan *Mech. Bride* 60/2 Looks like he's going to haul off and kiss her. **1960** Wodehouse *Jeeves in Offing* vi. 63, I shall have no alternative but to haul off and bop *him* one. *Ibid.* vii. 71 A cow that looked as if it were planning, next time it was milked, to haul off and let the milkmaid have it in the lower ribs.

haul, *sb.* Add: **1. c.** *spec.* The distance over which something is hauled, freq. in phr. *long* (or *short*) *haul*. Also *fig.*

1877 W. Rockefeller *Let.* 17 Oct. in *Philadelphia Inquirer* (1879) 8 Mar. 2/c We will endeavor to deliver the oil to you at points from which you will have short hauls. **1884** *Congress. Rec.* 18 June 5314/2 The farmer has to pay for short hauls just about what they ask him [etc.]... We must study the effect..of short hauls and long hauls. **1905** *Terms Forestry & Logging* 39 *Haul*, in logging, the distance and route over which teams must go between two given points, as between the yard or skid way and the landing. **1909** H. N. Casson *Life C. H. McCormick* 213 Today it is not the long haul of wheat, but the short haul, that is more expensive. **1936** L. C. Douglas *White Banners* vi. 123 Florid, peaches-and-cream blondes weren't intended for long hauls of worry. **1957** *Sunday Times* 13 Oct. 5/6 A new building for long-haul traffic is recommended. **1962** *Listener* 15 Feb. 307/1 Both points of view are necessary at different times, but the C.R.O. one is designed for the long haul. **1968** *Times* 1 Nov. 10/3 Improvements would be certainly possible; but he gave a warning: 'This is a long haul. You cannot suddenly change the existing systems.' **1973** *Daily Tel.* 15 Jan. 19/1 Everyone knows that you don't just buy investment trusts for the short haul.

haulabout (hǭ·lăbaut). *U.S.* [f. the verbal phrase *to haul about* (Haul *v.* 1).] A vessel, resembling a barge, used for coaling ships.

1903 *Trans. Inst. Naval Archit.* XLV. 221 These vessels, ..like the smaller barges, or 'coal haulabouts' as they are termed, have no means of propelling themselves.

haulage. Add: **4.** *haulage-engine.*

1909 *Daily Chron.* 16 Sept. 1/3 A haulage engine..was taking a load of bricks up the steep gradient. *Ibid.* 30 Sept. 1/5 The flood carried away the haulage engine at the entrance of the level.

haulier. Add: **2.** A firm or a person engaged in road transport.

1919 *Commercial Motor* 1 May 199/1 We do not find a motor haulier keeping his vehicles in the garage because his repair department informs him that this is the best way of reducing the cost of maintenance. **1951** *Oxf. Jun. Encycl.* IV. 277/1 Hauliers often have to transfer loads from one vehicle to another at the state border. **1959** *Times Rev. Industry* May 44/3 Hauliers are afraid to work out costs properly. **1967** *Listener* 20 Apr. 539/3 Interviews with drivers and hauliers.

hauling. Add: **b. hauling-ground,** a place where bachelor seals congregate, distinguished from the rookery or ground occupied by breeding seals.

1898 D. S. Jordan *Fur Seals* i. 36 Adjoining the breeding grounds and an essential part of each rookery are what are known as the 'hauling grounds' of the bachelors, frequented by the young males of the ages of 5 years and under. **1960** *Canad. Audubon* Jan.–Feb. 2/2 Localities

where sea lions come ashore for purposes other than breeding or pupping are called 'hauling grounds'.

haunch, *sb.*[1] Add: **3. b.** The side of a made-up road.

1937 [see *haunching 2].

4. c. The end of a tenon reduced in width. So **haunched** *a.*, (of a tenon) having its end reduced in width; **haunching,** a recess in a style for the end of a tenon.

1885 *Spons' Mech. Own Bk.* 276 The haunched tenon [is used] when the edge of the piece on which the tenon is formed is required to be flush with the end of the piece containing the mortice. **1904** A. C. Passmore *Handbk. Techn. Terms*, *Haunch*, the wide part left close to the root when part of a tenon is cut away. **1904** Goodchild & Tweney *Technol. & Sci. Dict.* 282/1 *Haunched tenon*, a tenon cut back in its width to allow for wedging. **1964** W. L. Goodman *Hist. Woodworking Tools* 53 The joints themselves are stub tenons, haunched and pinned in a very modern manner.

haunching. Add: **2.** (See quot.)

1937 *Times* 13 Apr. p. viii/2 In such cases (of excessive camber) the process known as 'haunching' should be carried out. The haunches or sides of the road are made up with stone,..and the whole road is then dressed with a new surface dressing.

haunk-haunk (hǭŋk,hǭŋk). [Echoic.] The cry of a hyena. Cf. *hau-hau *v.*

1895 B. M. Croker *Village Tales* (1896) 208 Another sound that made his heart beat very fast—the 'haunk-haunk' of a hyena.

haunt, *sb.* Add: **5.** (Earlier and later examples.) Also (occas.) in wider use. Cf. *hant, ha'nt.

1843 Winnemore & Reps *Cudjo's Wild Hunt* (song) 3 It am de hunt ob Cudjo dat nigger so bold. **1896** Mrs. Stowe *Oldtown Folks* vi. 80 But this 'ere's a regular haunt,..they both on 'em said..they'd seen a figger of a man. **1902** *Westm. Gaz.* 5 Feb. 2/1 This is the 'haunt' that troubles all our minds, and, especially, that comes forth..when the question is of peace by arrangement. **1952** W. R. Titterton in *Columba* Aug. 102/1 We had a haunt in our flat. Father Vincent came home with us that day, and blessed the place, and the haunt was no more.

haunted, *ppl. a.* Add: **3. b.** In wider use.

1906 *Daily Chron.* 23 Feb. 3/4 The beauty-haunted eyes of such painters as Gainsborough, Romney, Botticelli. **1906** Rider Haggard *Benita* vii, Staring at the white Benita and at her haunted eyes. **1908** *Westm. Gaz.* 17 Aug. 3/1 He paces the garden in this haunting, haunted fashion. **1910** A. C. Benson *Silent Isle* xv, You become aware that some exquisite haunted quality has slipped away from the later work.

c. *Comb.*, as *haunted-looking* adj.

1883 Ld. R. Gower *My Remin.* II. 26 A low, long, damp, haunted-looking gallery. **1918** Mrs. Belloc Lowndes *Out of the War?* xx. 257 Haunted-looking eyes.

Hausa (hau·să). Also **Haussa, Hous(s)a.** [Native name.] A widespread Negroid people of the Sudan and N. Nigeria, of the Bantu family with some Hamitic mixture; also, the language of this people, used, esp. in commerce, over much of W. Africa. Also *attrib.* or as *adj.*

1820 J. G. Jackson (*title*) Account of Timbuctoo and Housa. **1853** E. Norris (*title*) Dialogues and a small portion of the New Testament, in the English, Arabic, Haussa, and Bornu languages. **1879** *Encycl. Brit.* X. 59/1 An armed police force, recruited..from the Mahometan tribe of the Houssas. **1923** F. W. Taylor *Pract. Hausa Gram.* i. 9 Hausa occupies a position midway between the tone languages, such as Yoruba, and the stress languages, such as English. **1926** *Blackw. Mag.* Nov. 666/1 The troops were Hausas—I don't think we had begun to call them 'Waffs' then. **1957** M. Banton *W. Afr. City* viii. 158 Some blind Hausa beggars come from Nigeria to Freetown to sing in the streets. **1959** R. C. Abraham *Lang. Hausa People* i. 3 In Hausa..each word has its fixed tone, no matter what the emotional context. **1962** *Listener* 22 Feb. 335/1 A Hausaman, clad in a scarlet tarbush and flowing white robes. **1967** *Listener* 24 Aug. 230/1 There have been several uprisings already among Hausas and Yorubas trying to take revenge on Ibos. **1971** *E. Afr. Jrnl.* Mar. 8/1, I haven't worked and slaved..to have you give me a Hausa grandson.

‖ **hausfrau** (hau·sfrau). Also **house-frau.** [G.] A housewife. Also **hausvrow, huisvrouw** [after Du. *huisvrouw*].

1798 Lady A. Barnard *S. Afr. Cent. Ago* (1901) 157 This, as a careful haus-vrow, devolved on me. **1843** E. Hall *Diary* in O. A. Sherrard *Two Victorian Girls* (1966) ix. 92 [On my way] to do the haus-frau, dear Mr. Shore met me. **1848** *Wesleyan-Meth. Mag.* Aug. 886 You find the *huis-vrouw*, or 'mistress', seated at a small table. **1866** C. M. Yonge *Dove in Eagle's Nest* i. 29 Hausfrau Johanna adjured her father..to be a true guardian and protector to the child. **1873** —— *Pillars of House* II. xiii. 40 A simple painstaking businesslike man, who had married a German hausfrau. **1918** R. Wilson *Martin Schüler* xi. 120 My sister Bertha was charming: now she is a house-frau. **1925** 'E. Barrington' *Divine Lady* ii. xv. 213 Her Majesty Queen Charlotte, the prim German hausfrau. **1930** *Observer* 20 Apr. 8/4 The big German newspapers, cognisant of the power of the hausfrau. **1962** *Punch* 9 May 706/2 Women in West Germany appear to have taken a tremendous leap forward from *hausfrau* to high executive positions.

‖ **hausmaler** (hau·smālər). [G., = house-painter.] One who paints undecorated china in his own house or private workshop. Hence **hau·smalerei** (-əi), the painting and decorating of such china.

1935 *Burlington Mag.* June 271/1 The painting on this jug bears the signature of the well-known *Hausmaler* Abraham Helmhack. *Ibid.*, Johann Schaper, the first great master of *Hausmalerei* on faience. **1938** *Ibid.* Dec. p. xviii/2 Some interesting *Hausmalerei* from the same region. **1959** G. Savage *Antique Coll. Handbk.* 68 The work of outside decorators (or *Hausmaler*), who bought white porcelain and decorated it in their homes. **1959** *Times* 6 Jan. 16/7 A Meissen hausmaler bowl. **1971** *Times* 30 Nov. 24/3 (Advt.), An important German faience Hausmaler tankard.

haustellation (hǭstĕlē^i·ʃən). [f. Haustellate *a*: see -tion.] The action of sucking.

1901 *Practitioner* Mar. 278 If a mosquito be killed on the second day after haustellation.

haustrum (hǭ·strɒ̆m). *Anat.* Pl. **haustra.** [mod. use of L. *haustrum* bucket, scoop, f. *haurīre*, *haust-* to draw (water).] Each of the small sacs enclosed by folds in the colon. Hence **hau·stral** *a.*

[**1826** J. Lizars *Syst. Anat. Plates.* Descriptions of Plates. xi. 30 These bands purse the colon into these peculiar pouches or cells ([*Footnote*] Syn. Cellulæ seu haustra) so characteristic of this intestine.] **1889** A. Macalister *Text-bk. Human Anat.* § 257. 405 Its [the colon's] cavity is to some extent sacculated, the haustra, or pouches being separated by crescentic plicæ sigmoideæ. **1913** Dorland *Med. Dict.* (ed. 7), *Haustral*, pertaining to the haustra of the colon. **1936** *Med. Rec.* 1 Jan. 28/1 The loss of haustral markings with and without dilatation, was found much less frequently. **1959** R. D. Lockhart et al. *Anat. Human Body* 522 Compared with the smooth small intestine, the large is sacculated along its length, the sacculations or haustra bulging between three equidistant longitudinal bands.

Haut-Brion (obrī·oṅ). Also 7 **Hobriant, Ho Bryan,** 9 **Obryan.** [Fr., f. the name of an estate, Château *Haut-Brion*, in the commune of Pessac, near Bordeaux.] In full *Château Haut-Brion.* A variety of fine quality claret.

1663 Pepys *Diary* 10 Apr. (1893) III. 89 Here drank a sort of French wine, called Ho Bryan, that hath a good and most particular taste that I never met with. **1670** W. Hughes *Compl. Vineyard* (ed. 2) 65 A sort of Claret called Hobriant-wine, of a deep red colour. *a* **1700** [see Vignoble]. **1792** T. Jefferson *Let.* in A. Lichine *Encycl. Wines* (1967) 4/2 Bordeaux red wines. There are four crops of them more famous than the rest. These are Chateau-Margau, Tour de Segur, Hautbrion & De La Fitte. **1833** C. Redding *Hist. Mod. Wines* 145 The first growth of this noted commune [*i.e.* Pessac] is Château Haut Brion. **1845** *Encycl. Metrop.* XXV. 1287/2 The following are the principal wines, or those most celebrated in the different countries where the vine has long been cultivated:—France... Bordeaux. Lafitte, red... Haut Brion, ditto. **1851** C. Redding *Hist. Mod. Wines* (ed. 3) 174 In 1710, a wine called Obryan claret was sold in London at three shillings the bottle. **1888** *Encycl. Brit.* XXIV. 605/1 Château Haut-Brion, Pessac. **1935** *Punch* 28 Aug. 238/2 The Old Yquem which was not too sweet, And the Old Haut Brion so round and neat. **1959** W. James *Word-Bk. Wine* 93 Haut Brion for some odd reason has always been regarded by Englishmen with a peculiar tenderness.

‖ **haute Bohème** (ot bo^ɛ̄m). [Formed by M. Baring after *haute bourgeoisie.] A name applied to members of any fast or high-class Bohemian set.

1925 M. Baring *Cat's Cradle* i. xvii. 216 You see quite different people..people like the Svensens and all those musicians and archaeologists... *La Haute Bohème*... You never went near them before. **1939** O. Lancaster *Homes Sweet Homes* 46 The cult of æstheticism..was only accepted whole-heartedly by..the *haute Bohème* of the day. **1954** *Design & Industries Assoc. Yearbk.* 37 The intelligentsia is..becoming, in Maurice Baring's phrase, an Haute Boheme. **1964** M. Laski in S. Nowell-Smith *Edwardian England* iv. 196 A fast Bohemian set, what Maurice Baring had called the *haute Bohème*.

‖ **haute bourgeoisie** (ot buɾʒwazī). [Fr.] The French upper middle class; also extended to the upper middle class of other countries. Cf. Bourgeoisie in Dict. and Suppl.

1888 *Athenæum* 4 Aug. 153/2 The haute bourgeoisie and the humble shopkeeper, citizens by nature and condition, have interests as indivisible. **1901** G. B. Shaw *Sixteen Self-Sketches* (1949) ix. 51 On the down grade from the zenith of *haute bourgeoisie* and landed gentry. **1924** A. D. Sedgwick *Little French Girl* i. ii. 12 Alix was to tell her whether they were *petite noblesse* or *haute bourgeoisie*. **1934** A. Huxley *Beyond Mexique Bay* 7 A collection of the elderly *haute bourgeoisie*. **1940** F. Scott Fitzgerald *Let.* 14 Aug. (1964) 420 My father and Aunt Elise struggling to keep their children in the *haute bourgeoisie*. **1962** *Listener* 22 Feb. 350/1 A rich residential district for the prosperous Glasgow *haute bourgeoisie*. **1969** *N.Y. Rev. Books* 16 Jan. 34/3 The Business Council —which is about as close to an executive committee of the American *haute bourgeoisie* as one can get.

‖ **haute boutique** (ot būtīk). [Fr., f. *haute* (fem.) high + *boutique.] (See quot. 1969.)

1966 *Guardian* 25 July 6/2 Simonetta is said to be presenting a much bigger haute boutique collection this season. **1969** *Ibid.* 14 Jan. 7/6 The class that the French call *haute boutique*—midway between couture and ready-to-wear.

haute couture: see *couture.

‖ **haute cuisine** (ot kwizīn). [Fr., f. *haute* (fem.) high + Cuisine.] High-class (French) cooking.

1926 *Time* 5 July 12 In France, perhaps in France alone, the traditions of *la haute cuisine* survive from the days of the great gastronomes. **1928** S. Browne tr. *T. H. Van de Velde's Ideal Marriage* xv. 277 The most effective dish in the *haute cuisine* is supposed to be crayfish soup. **1930** A. Bennett *Imperial Palace* xxxvi. 246 *La Haute cuisine.* Not fifty people in the world were equipped by education and natural taste to comprehend it. **1935** *Time* 11 Mar. 22/1 If only English landladies spoke French, and if only their English cooks knew something of *haute cuisine*, Frenchmen with gold francs would be tempted across the Channel. **1951** E. David *French Country Cooking* 172 Their use in what was regarded as *Haute Cuisine* became ridiculously excessive, and no dish was considered really refined without a garnish of sliced truffle. **1959** *Listener* 30 Apr. 776/3 The miracles of *haute cuisine* often arrive from a kitchen where the scales are faulty! **1962** Auden *Dyer's Hand* 75 It is difficult to imagine a *haute cuisine* based on algae and chemically treated grass. **1966** *Observer* 25 Sept. 46/2 An egg-based haute cuisine dinner for 6.

‖ **haute école** (ot ekol). [Fr., = high school. Cf. School *sb.*[1] 3 d.] The more difficult feats of horsemanship. Also *attrib.* and *transf.* (esp. in *Mus.*).

1858 *Rarey's Art of Taming Horses* i. 5 The accomplished Colonel Greenwood, who was equally learned in the *manége* of the *Haute Ecole*, and skilled in the style of the English hunting-fields. **1864** G. A. Sala *Quite Alone* I. xi. 191 She was doing the haute école. **1889** G. B. Shaw *London Mus. 1888–89* (1937) 215 The haute-école acts of the prima donna and tenor. **1896** *Strand Mag.* Mar. 334/2 Five other 'artistes'—trapezists, *haute école*, and 'bare-back' ladies. **1931** *Times Lit. Suppl.* 26 Mar. 247/3 The excellent *haute-école* rider seen at the Kingsway Opera House. **1953** G. Brooke *Introd. Riding* vii. 69 Genuine 'Haute École', as maintained and demonstrated today. **1955** *Times* 15 Aug. 5/5 An energetic circus performer of the haute école. **1959** *Listener* 30 Apr. 776/2 Balakirev's *Islamey* is written in close imitation of Liszt's *haute école*. **1959** [see *gun-dog* (*gun *sb.* 15)]. **1960** *Times* 4 June 9/3 It is even more gratifying to hear the superbly groomed *haute école* pianism on the disc.

hautefeuillite (otfö·yəit). *Min.* [a. F. *hautefeuillite* (L. Michel 1893, in *Bull. de la Soc. française de Min.* XVI. 40), f. the name of P. G. *Hautefeuille* (1836–1902), French chemist.] A hydrous phosphate of magnesium and calcium that occurs in colourless crystals; possibly the same as bobierrite, the calcium being due to contamination with apatite.

1896 *Jrnl. Chem. Soc.* LXX. ii. 112 (*heading*) Hautefeuillite, a new mineral from Bamle, Norway. **1937** *Amer. Mineralogist* XXII. 337 Michel..introduces the name *hautefeuillite* for a bobierrite-like mineral which, according to his analysis, differs from bobierrite..in its optical orientation. However, the redetermined optical properties of bobierrite correspond to those of hautefeuillite as given by Michel. *Ibid.* 338 Until more evidence is produced ..it seems best to discard hautefeuillite as a mineral name. **1955** M. H. Hey *Index Min. Species* (ed. 2) 232 *Hautefeuillite*, $(Mg, Ca)_3PO_4.8H_2O$... Has been interpreted as a mixture of Bobierrite..and Apatite..but the H_2O is very high for this interpretation, and further study is desirable.

‖ **haute noblesse** (ot nobles). [Fr., f. *haute* (fem.) high + Noblesse.] The upper stratum of the aristocracy.

1787 W. Beckford *Let.* 8 Nov. in *Italy* (1834) II. xxx. 146 The famous tenor singer, who entertained us..with many private anecdotes of the *haute noblesse*. **1907** M. E. Braddon *Dead Love has Chains* ii. 23 She had friends among the *haute noblesse*..in the old St. Germain faubourg. **1934** A. Woollcott *While Rome Burns* 220 The panic among the English *haute noblesse* during the Oscar Wilde trial.

‖ **haute vulgarisation** (ot vülgarizasyoṅ; also freq. with (quasi-)anglicized pronunc. of second word). [Fr., f. *haute* (fem.) high + Vulgarization.] The popularization of abstruse or complex matters.

[**1943** *Mind* LII. 178 Mr. Skemp would refer us..to the *Timaeus* of which he regards the utterances of *Laws X* as only a 'popularised' version, in the nature of what the French call an *œuvre de haute vulgarisation*.] **1946** *Christian Sci. Monitor* 21 Dec. (Mag. Sect.) 10 Mr. Van Doren has written a work of '*haute vulgarization*'. **1958** *Times* 6 Dec. 7/7 This can only be achieved if some of the scholars themselves transform their knowledge into the *haute vulgarisation* of which Cicero himself, far more even than H. G. Wells, was the master. **1960** *Cambr. Rev.* 16 Jan. 232/3 Is the *haute-vulgarization* of archaeology, which we see happening.., a good thing? **1966** *Listener* 13 Oct. 544/2 Mrs. Thomson's edition is a model of *haute vulgarisation*. It is based on her father's..standard edition..but modernized and abridged. **1968** *Times* 9 Nov. 23/6 This book is an extraordinary mixture of

John o' London's 18 May 567/3 (*heading*) Have towel, will strip. **1961** *Sunday Times* 25 June 21/2 Have honours degree, will travel. **1965** *Harrods Xmas Catal.* 43/1 *Have iron, will travel*—featherweight iron weighs only 2¼ lbs. and travels..complete with universal adapters for use with any voltage..£4. 4. 0. **1966** *Listener* 16 June 889/1 *Have Gun Will Travel* was a much better western..than the ones they are making now. **1968** *Times* 29 Nov. p. vi/4 Have portable, will play. **1969** *Times* 14 July 5/5 The..scene has now gone one step further towards the American dream with the opening of Have Typewriter, Will Travel.

15. b. (Earlier example.)

1816 Scott *Old Mort.* in *Tales of Landlord* IV. vii. 125 He has you there, I think, my Lord Duke.

d. *to have on*: to puzzle or deceive intentionally; to chaff, tease; to hoax. orig. *dial.*

1867 J. T. Staton *Rays fro' Loominary* 117 It looks as if somebuddy wur havin me on. **1893** Farmer & Henley *Slang*, *To have on*, to secure a person's interest, attention, sympathy: generally with a view to deceiving him (or her). **1895** M. Mather *Lancs. Idylls* 46, I were nobbud hevin' her on a bit. **1928** *Daily Express* 31 Aug. 7 Speaking unjudicially and in ordinary language you are 'having him on'. **1951** L. P. Hartley *Travelling Grave* 52 'Of course,' said Dickie, when the boy had gone off with his *mancia*, whistling, 'he's having us on.'

16. b. *to have it out*: see Out *adv.* 7 b.

c. *to have it away*: to escape from prison or custody. *Criminals' slang.*

1958 F. Norman *Bang to Rights* 48 The P.O. who was in charge of the escort that was going to..make sure no one had it away. **1965** *New Statesman* 30 July 152/3 One thing broke the monotony of this dreary sentence and that was the occasion when a geezer, three peters away from me, had it away. **1969** T. Parker *Twisting Lane* 196 After I'd had it away three times, they decided it was no use bothering with me in these open places.

18. a. *to have it coming to one*: see *come *v.* 9 b.

21. The following phrases are also treated under the indicated words: *to h.* *everything, *h. a* *heart, *as* *luck *would h. it*, *what *h. you*.

24. d. (*I*) *have and* (*I*) *haven't*: a phrase indicating that a statement is true in some respects but not in others.

1858 Trollope *Dr. Thorne* II. xiv. 282 'Have you spoken to my niece about this, Sir Louis?' 'Well, I have, and yet I haven't; I haven't, and yet in a manner I have.' **1910** J. Buchan *Prester John* vi. 108 'Had the man any news?' I asked. 'He had and he hadn't.' **1933** A. Christie *Ld. Edgware Dies* iii. 31 'You have a problem for me—yes?'..'Well,..I have and I haven't.' **1967** 'L. Bruce' *Death of Commuter* v. 58 'You haven't got any suspicions about Mr. Parador's death, have you?'..'Well, I have and I haven't.'

26. ¶ (Later U.S. examples.)

1816 U. Brown *Jrnl.* in *Maryland Hist. Mag.* (1915) X. 282 If this forest had never have been fired it would have been a vast..Timbered country. **1869** *Trans. Ill. Agric. Soc.* VII. 444 If said hogs had, in style of Hanlon Brothers, have stood one on the other. **1911** J. F. Wilson *Land Claimers* i. 17 'If the fire hadn't have gone out,' he mused.

27. have-got: see *have *sb.* 2; **have-on** *slang* = Have *sb.* 3 (cf. sense *15 d).

1931 T. R. G. Lyell *Slang* 372 *Have* or *have on*, a swindle; a mild joke to deceive a person. **1967** *Listener* 16 Feb. 237/3 Puns, tropes, polyglot have-ons, batty new coinings.

have, *sb.* Add: **2.** Also, a nation or country that *has* or possesses; one of the wealthier nations. Also *attrib.* and (occas.) *have-got*. (Usu. opp. *have-not*.)

1919 J. L. Garvin *Econ. Found. Peace* xvi. 375 They contemplate a World-Federation when the international League of the Have-Nots has conquered all the Haves. **1937** E. Snow *Red Star over China* vi. iii. 227 The Reds..radically changed the situation for..all the 'have-not' elements. **1949** Koestler *Insight & Outlook* xvi. 227 The equalization of the steep gradients.. between have and have-not nations. **1955** *Bull. Atomic Sci.* Jan. 3/2 The proposal is for the atomic 'haves' to contribute crucial materials and at least some limited amount of technical information to the 'have nots'. **1959** *Times* 29 Sept. 18/5 Algerian oil is expected to change France from an oil 'have-not' to a 'have'. **1962** *Listener* 5 July 29/3 The greatest of the 'have' powers. *Ibid.* 30/1 The Soviet Union is a 'have' society that ought to be more generous. **1959** *Brno Studies* I. 70 The sharp distinction..between the 'have-gots' and the 'have-nots' was soon felt in politics. **1963** *Listener* 21 Mar. 487/1 Russia was becoming a 'have-got' power herself with a productive capacity second only to that of the United States. **1965** H. Kahn *On Escalation* xiii. 244 A 'have' nation might perceive a situation that threatened its possessions as a crisis. **1968** *Punch* 22 May 757/3 The country had a one-crop economy; the more cocoa it exported, the less the 'have' nations were willing to pay for it.

havelock. Add: *havelock cap*, a military cap provided with a havelock.

1880 *Harper's Mag.* Oct. 399 A poncho and havelock cap comprise the rubber clothing outfit.

haver, *v.* Add: **2.** Orig. *Sc. dial.* but now in general English use: to hesitate, to be slow in deciding.

1866 W. Gregor *Dial. Banffshire* 73 *Haiver*, to hesitate and make much ado about doing anything. **1955** J. Bayley *In Another Country* 75 It was a classic moment for polite havering, but the sensible girl did not haver: he was holding the front door open and she climbed in without more ado. **1957** *Times* 14 Nov. 13/3 No doubt the Government, in deciding to institute an inquiry.., might appear at first sight to have been havering and shifting their ground.

havoc, *sb.* Add: **2.** Also in weakened sense: confusion and disorder, disarray; *to play havoc* (examples); freq. const. *with*. The phrases *to work havoc*, *create havoc* are also common.

1812 M. E. Bicknell *Let.* 28 Oct. in *J. Constable's Corr.* (1964) II. 91 You perfectly well know, what terrible havoc it [*sc.* meeting often] makes with your time. **1900** J. Morley *Cromwell* I. 3 The thirst after broad classifications works havoc with truth. **1908** E. J. Banfield *Confessions of Beachcomber* iv. 129 Terrestrial storms work as much if not greater havoc in the shallow places of the sea as on the land. **1910** G. D. Abraham *Mountain Adv.* vi. 115 The hot sun, reflected off the snow, played havoc with his complexion. **1934** G. G. Coulton *H. W. Fowler* 156 He.. displayed..anxiety about the havoc made in the projected festivities. **1949** *Times Lit. Suppl.* 4 Nov. 715/2 History has played havoc with their hopes. **1961** J. E. Mansion *Harrap's French-Eng. Dict.* 705/2 The storm.. played havoc with the crops. **1961** Webster s.v. *havoc*, Several small children can create havoc in a house. **1964** *Times* 5 Sept. 9/5 Surely one can make up one's mind as to which [political party] would create less havoc if they came to power. **1965** A. Nicol *Truly Married Woman* 24, I have created enough havoc in one afternoon as it is. **1966** B. Kimenye *Kalasanda Revisited* 86 The fine, dust-like substance enveloped him in a cloud which played havoc with the delicate membranes of his eyes and nose. **1969** *Times* 25 Mar. 16/1 The noise and clatter of high-revving engines can play havoc with a driver's nerves. **1971** B. Patten *Irrelevant Song* 32 This creature singled out creates Havoc with intelligence.

haw, *int.*[2] and *sb.*[5] *dial.* and *U.S.* A call used to direct a horse or team to turn to the left.

1843 *Knickerbocker* XXI. 494 He admonishes them with his goad, and ejaculates, 'Haw'. **1843** 'R. Carlton' *New Purchase* xxvi. 239 Whoas, gees and haws. **1856** J. C. Morton *Cycl. Agric.* II. 723/2 Horses—terms used in directing—..Cheshire..To left. Haw. **1864** Webster s.v., *Haw, haw here*;—words used by teamsters in guiding their teams. **1930** *Amer. Speech* V. 419 *Haw*, direction given to oxen to turn to the left. **1972** *Even. Telegram* (St. John's, Nwfndl.) 24 June 14/3 'Gee' tells the dogs to take a right turn, and 'Haw' means left.

haw, *v.*[2] *U.S.* (but *Eng. dial.* in quot. 1911). [f. prec.] **a.** *intr.* Of a horse or team: to turn to the left. Also *fig.* (see quot. 1864).

1846 *Knickerbocker* XXVII. 119 The plough-boy has hardly energy to cry out..'Gee-haw, there, I tell you to haw, now.' **1861** *Trans. Ill. Agric. Soc.* IV. 99 They were required to plow lands of about fifteen rods in length, and 'haw' about. **1864** Webster s.v., *To haw and gee*, or *haw and gee about*, to go from one thing to another without good reason; to have no settled purpose; to be irresolute or unstable. (*Colloq.*) **1911** J. Masefield *Everlasting Mercy* 86 Now and then he seems to stoop To clear the coulter with the scoop, Or touch an ox to haw or gee.

b. *trans.* To direct (a horse, etc.) to turn to the left. Also *fig.*

1864 Webster s.v., *To haw and gee*, or *haw and gee about*, to lead this way and that at will; to lead by the nose; to master or control. (*Colloq.*) **1867** [see *gee *v.*[2] b].

Hence **hawing** *vbl. sb.*

1843 'R. Carlton' *New Purchase* xvi. 119 After performing wonders on the journey from Philadelphia to the West in hawing and geeing. **1867** [see *gee *v.*[2] b].

Hawaiian (hăwai·ăn), *a.* and *sb.* Also **Hawaian.** [f. *Hawaii* + -an.] **A.** *adj.* Of or pertaining to the island of Hawaii, or to the whole group of the Sandwich Islands in the North Pacific. **B.** *sb.* **1.** A native or inhabitant of Hawaii. **2.** The language of Hawaii, belonging to the Malayo-Polynesian group.

1825 W. Ellis *Jrnl. Tour Hawaii* 205 The account given this evening of the Hawaiian *hades*. **1859** [see *inter-island* s.v. Inter- 5]. **1864** W. D. Alexander *Hawaiian Gram.* 20 What would form a long sentence in English, in Hawaiian is generally broken up into several independent propositions. **1877** T. H. Streets *Nat. Hist. Hawaiian & Fanning Isl.* 8 The Fanning group, with the exception of the Hawaiian, were the only islands visited in the Pacific. **1877** L. H. Morgan *Anc. Society* iii. ii. 404 These are terms in Hawaiian for grandparent. **1893** *Funk's Stand. Dict.* I. 825/1 *Hawaiian*, a native..of Hawaii. **1913** R. Brooke *Coll. Poems* (1918) 28 And new stars burn into the ancient skies, Over the murmurous soft Hawaian sea. **1921** *Nature* 20 Jan. 673/1 Some broad features of Hawaiian petrology. **1929** C. H. Smith *Bridge of Life* ii. 29 White men, Chinese, Japanese, Hawaiians and many others slept in the common dormitory. **1936** *Discovery* June 198/2 He is bold enough to believe that the Hawaiian native practitioners can handle elemental spirits 'like tame animals'. **1957** P. Worsley *Trumpet shall Sound* i. 30 Sects..have also appeared amongst the.. Hawaiians. **1970** *Western Folklore* XXIX. 234 The slider occasionally covers several strings at once to produce different chords in the 'Hawaiian' style.

II. *Special Combs.* **Hawaiian goose,** a rare bird, *Branta sandvicensis*, formerly called the Sandwich Island goose; **Hawaiian guitar,** a type of guitar, usually held in a horizontal position, in which the pitch is obtained by placing a small metal bar on the strings and

The head and gullet are strengthened with steel plates and there is also a steel reinforcement laid on to the under-side of the tree from the head to the cantle.

r. The closed end of a cylinder of a pump or engine, esp. an internal-combustion engine; a cylinder-head or cylinder-cover.

1884, 1895 [see *cylinder head* s.v. *CYLINDER *sb.* 9 b]. **1904** A. B. F. YOUNG *Compl. Motorist* iv. 111 The cylinders consist of two separate parts. The body of the cylinder proper is a cast-iron liner... The head—containing the vertical valves and ignition-plug—is a separate casting. **1907** R. B. WHITMAN *Motor-Car Princ.* i. 5 While in the great majority of steam engines the steam acts first on one side of the piston and then on the other, in an automobile gasoline engine the pressure is exerted on only one side, the combustion of the mixture taking place between the piston and the closed end, or head, of the cylinder. **1965** P. H. SMITH *High-Speed Two-Stroke Petrol Engine* xiii. 258 On air-cooled engines, non-detachable heads are generally confined, in the case of iron castings, to the simplest and cheapest industrial engines.

s. Of a bicycle frame (see quot. 1904). Also *attrib.*

1887 BURY & HILLIER *Cycling* (1889) xiv. 321 Beneath the head and between the forks is placed the trouser guard. **1902** *Captain* VII. 82/1 That going from the head to the crank bracket is made duplex. **1904** GOODCHILD & TWENEY *Technol. & Sci. Dict.* 282/2 *Head* (Cycles), the socket or hollow tube through which the tube carrying the front fork runs. **1959** *Sears, Roebuck Catal.* Spring & Summer 1017/1 Universal Head Bearing Set... Fits all bicycles.

t. Of an explosive shell.

1898 [see *war-head* (WAR *sb.*[1] 11)]. **1899** *Kynoch Jrnl.* Oct.–Nov. 17/1 The head [of a shrapnel shell] is attached to the body by means of small rivets.

11. c. (Additional examples.)

1930 L. M. DAVIES in P. O'Connell *Science of To-Day* (1959) II. xii. 75 'Head' is a term applied to this rubble-drift where it masks an old raised beach. **1934** *Antiquity* VIII. 305 The angular deposit..corresponds to what is termed 'head' in Devon and Cornwall.

f. *Curling* and *Bowls.* (See quots. 1897.)

1828 *Kilmarnock Treat. Curling* (1883) 79 Head (probably a corruption of *heat,*) that portion of the game in which both parties play all their stones once. **1877** *Encycl. Brit.* VI. 713 [Curling Rules.] All matches to be of a certain number of heads. **1897** *Encycl. Sport* I. 129/1 (Bowls) *End*—One delivery of all the bowls upon the two sides, after which the jack is again 'set'. Also called Head. *Ibid.* 264/1 (Curling) *Head,* the portion of the game in which all the players have delivered their stones, and have counted the winning shot or shots. **1969** R. WELSH *Beginner's Guide Curling* xvi. 104 All matches shall be of a certain number of Heads, or Shots, or by Time as may be agreed on, or as fixed by the Umpire at the outset.

g. A device designed to convert variations in an electrical signal into variations in the motion of a stylus (in the making of a gramophone record) or *vice versa* (in the playing of one). Also, a device in which a small electromagnet is similarly used to produce or respond to variations in the magnetization of magnetic tape as the tape is moved past it. Freq. with defining word(s), as *cutting head, erase head, magnetic head* (see the defining words in Supplement).

1951 *Catal. Exhibits S. Bank Exhib., Festival of Brit.* 143/1 Gramophone pick-up with interchangeable heads. **1960** COOKE & MARKUS *Electronics & Nucleonics Dict.* 207/1 *Head.* 1. The photoelectric unit that converts the sound track on motion-picture film into corresponding audio signals in a motion-picture projector. 2. Cutter. 3. Magnetic head. **1962** L. DEIGHTON *Ipcress File* xxvii. 172 He laid a huge shiny L.P. on the..turntable and delicately applied the diamond head. **1962** A. NISBETT *Technique Sound Studio* 255 *Head,* transducer which converts electrical energy into magnetic or mechanical energy, or vice versa. Thus we have a tape recording and reproducing heads and disc cutter and pick-up heads. The electromagnet used for erasing tape is also called a head. **1963** *Which?* Jan. 8/2 The pick up is composed of two parts—an arm, and at the end of the arm..a head. The head contains the cartridge, and set into the cartridge is the stylus. **1964** *Honeywell Gloss. Data Proc.* 29/1 *Head,* a device that reads, records or erases information in a storage medium, e.g., a small electromagnet used to read, write or erase information on a magnetic drum or tape, or the set of perforating, reading or marking devices and block assembly used for punching, reading or printing on paper tape. **1964** F. L. WESTWATER *Electronic Computers* (ed. 2) iv. 68 Each track must have its read/write head.

13. b. The top of a book. Cf. HEADBAND 3.

1835 J. HANNETT *Bibliopegia* 26 The book is now taken between the hands and well beaten up at the back and head on a smooth board, to bring the sheets level and square. **1876** *Encycl. Brit.* IV. 43/1 The object of the binder in this operation is to make every page of uniform size, presenting a smooth and equal 'head', 'tail', and 'fore-edge'. **1930** *Godfrey's Catal.* No. 134. 26/1 Folio, old sheep (roughly repaired at head and heel).

c. A headline in a newspaper.

1911 H. S. HARRISON *Queed* xviii. 230 The Chronicle that afternoon shrieked it under a five-column head. **1915** J. WEBSTER *Patty & Priscilla* xi. 170 A life-size portrait of her..appeared in a New York evening paper, and scare-heads three inches high announced..that the champion athlete..was at death's door. **1962** *Amer. Speech* XXXVII. 200 A *Saturday Evening Post* editorial head: 'Good News.' **1967** *Guardian* 2 Nov. 8/7 'Ebullient Mr Brown hits out,' said the (changed) head on the last edition.

19. c. (*a*) *Phonetics.* The initial stressed element(s) in a sequence of sounds before the nucleus. (*b*) *Linguistics.* (See quot. 1964[1].)

1922 H. E. PALMER *Eng. Intonation* v. 17 Any syllable or syllables preceding the nucleus in the same Tone-Group is termed the 'Head' of the group. **1961** R. B. LONG *Sentence & its Parts* i. 20 Headed units are made up of (1) contained heads and (2) contained modifiers which attach to these heads. **1964** R. H. ROBINS *Gen. Ling.* vi. 236 The word or group sharing the syntactic functions of the whole of a subordinative construction is called the head, and the other components are subordinate. Thus in English adjective noun groups, the noun is head and the adjective subordinate. In adverb adjective groups the adjective is head and the adverb subordinate..; in *reasonably clever boys, boys* is head and *reasonably clever* is subordinate, and within this latter group *clever* is head and *reasonably* is subordinate. **1964** M. SCHUBIGER in D. Abercrombie et al. *Daniel Jones* 263, I am using the term head for the first stressed element pitched high. **1964** *Amer. Speech* XXXIX. 37 Nice city home... All the ten fine old stone houses. In these sequences the noun is more intimately tied to the head than is the adjective. *Ibid.* 38 A noun that has 'widespread' use as attributive to many different heads is an adjective.

21. d. A ship's latrine (in the bows). Often (in the U.K., usually) in *pl.* In the U.S. also used of W.C.s ashore.

1748 SMOLLETT *R. Random* I. xxviii. 253 The madman ..took an opportunity, while the centinel attended him at the head, to leap over-board. **1905** *Trans. Inst. Naval Archit.* XLVII. I. 29 The W.C.s for officers, and the seamen's head for the crew, are to be fitted where shown on the drawings. **1938** C. S. FORESTER *Ship of Line* 21 You'll clean out the heads of this ship every day. **1952** 'E. Box' *Death in Fifth Position* (1954) iv. 85, I saw Louis coming out of the head with a blond footman. **1957** *Partisan Rev.* 328 Yet it had happened and here I was, talking about algebra to a lot of boys who might..be popping off needles every time they went to the head. **1965** J. R. HETHERINGTON *Selina's Aunt* 48 The rating who cleans them is Captain of the Heads. **1972** *Daily Colonist* (Victoria, B.C.) 20 Jan. 28/2 The head, or shipboard bathroom, for women crew members is to be the same one that male crew members now use exclusively.

26. c. *Rugby Football.* In full *loose head*: in the front row of the scrummage the forward closest to the scrum half as he puts the ball into the scrummage; *to win the ball against the head,* to hook the ball notwithstanding the fact that the opposition front row holds the advantage by having a player in the loose-head position.

1917 in P. Jones *War Lett.* (1918) 259 We used to spend hours arguing over anything, from free-will to the 'loose-head'. **1959** *Times* 7 Sept. 16/2 A heel against the loose head was a prelude to Coventry's next try. **1960** *Times* 30 Nov. 3/6 He even managed to win the ball against the head in a five-yard scrummage.

35. on..head. e. *to do it* (*standing*) *on one's head*: to do it with ease. *slang.*

1896 G. B. SHAW *Our Theatres in Nineties* (1932) II. 227 Of course, Mr Waring does the thing on his head, so to speak; but how can I compliment an actor who has done what he has done on stuff like that? **1897** CONRAD *Nigger of Narcissus* i. 7 It's a 'omeward trip... Bad or good I can do it hall on my 'ed. **1922** A. A. MILNE *Red House Myst.* xvi, Right, old boy. Leave it to me. I can do this on my head. **1923** *Westm. Gaz.* 19 Mar., Mr. Wells, assuming the best Cockney accent, intimated that he could 'do it on 'is 'ead'. **1944** M. SHARP *Cluny Brown* xvi. 109 If there was one thing Betty could do on her head, it was handle a compliment. **1968** J. M. WHITE *Nightclimber* viii. 60 The climb he wanted me to attempt was a simple one. At Cambridge I could have done it standing on my head.

36. out of one's head. b. Out of one's mind. Chiefly *U.S.*

1825 J. NEAL *Bro. Jonathan* I. 267 You are out o' your head, I guess. **1878** H. JAMES *Fr. Poets & Novelists* 428 Pathelin pretends to be out of his head. **1902** C. E. JEFFERSON *Quiet Hints Preachers* xiii. 103 If they could not understand what was going on they..might think Christians out of their head.

38. b. *to go to* (some)one's *head*: (*a*) to intoxicate; cf. GO *v.* 37 b; (*b*) to make one vain or proud.

1912 A. LUNN *Oxf. Mountaineering Essays* ix. 233 The delight of watching distant hills..went to my head like wine. **1939** A. CHRISTIE *Ten Little Niggers* x. 143 He's played God Almighty for a good many months every year. That must go to a man's head eventually. **1942** —— *Body in Library* xii. 108 He settled a large sum of money on Frank... It went to Frank's head.

39. ¶ *head over ears* (later example).

1912 G. B. SHAW *Let.* 22 Dec. in *Times* (1968) 19 Oct. 19/5, I plunged in head over ears and..wrote off my 56 years.

39*. head in the air. *to go about with one's head in the air*: to assume a pose of superiority. Hence *head-in-air,* designating either one who is absent-minded and dreamy or one who is a snob or a 'highbrow', or the actions of such persons.

1848 *English Struwwelpeter* (ed. 4) 21 (*title*) The story of Johnny Head-in-Air. **1903** *Trawl* May 7 The Laureate crost over the lawn with the dreamy head-in-air gait that was known through five parishes round. **1906** RIDER HAGGARD *Benita* iv, On the ship I always thought him rather a head-in-air kind of swell, but he was a splendid fellow. **1942** J. PUDNEY *Dispersal Point* 24 Do not despair For Johnny-head-in-air; He sleeps as sound As Johnny underground.

41*. head and girth. See GIRTH *sb.*[1] 1 b.

43. head to head. Delete † *Obs. rare* and add later examples. Also *transf.*

1799 *Sporting Mag.* XIII. 311/1 The contest here commenced,..the horses never being more than a length asunder, and generally head to head. **1950** J. DEMPSEY *Championship Fighting* 63 Short range. That's the head-to-head slugging range. **1956** *Nature* 4 Feb. 206/2 There is the so-called head-to-head or head-to-tail arrangements. **1966** *B.B.C. Handbk.* 1966 14 Much of the time the BBC's competition is merely head-to-head. **1972** *Time* 17 Apr. 24/1 Until December the Gallup poll did not even pit McGovern in head-to-head polls with President Nixon and Wallace.

44. b. Also *fig.* (In quot. 1924 with contextual omission of *heels.*)

1834 D. CROCKETT *Narr. Life* i. 20, I soon found myself head over heels in love with this girl. **1924** GALSWORTHY *White Monkey* II. vi. 158 They were head over—the family feud stopped that [marriage].

c. *head over tip* = *head over heels.*

1824 P. EGAN *Boxiana* IV. 260 A first-rate *swell,* who was extremely eager to get on board, lost his footing, and went *head* over *tip* into the water. **1906** E. DYSON *Fact'ry 'Ands* xv. 202 Er stream iv water..sluices ther red-'eaded girl..'ead-over-tip down ther front stairs.

46*. heads and posts. Leather heads placed on posts for use in cavalry exercises.

1895 *Army & Navy Co-op. Soc. Price List* 15 Sept. 1456 Heads and Posts. For Military Tournaments. **1902** *Encycl. Brit.* XXVI. 156/1 Courses for jumping and 'heads and posts' exercise. **1907** *Yesterday's Shopping* (1969) 1017/2 Heads and Posts..for Military Tournaments.

47. head and shoulders. c. A portrait in which only the head and shoulders are shown. Freq. *attrib.*

1865 D. G. ROSSETTI *Let.* 30 July (1965) II. 562, I fear all I could undertake with prospect of bringing it to a conclusion without unreasonable delay would be a 'head and shoulders' portrait. **1897** H. B. WHEATLEY *Hist. Portraits* viii. 173 Stoop himself made an etching of this portrait... The picture is described as 'Head and shoulders'. **1902** *Daily Chron.* 7 July 3/5 A head and shoulders portrait of Lady Morshead in white dress and fichu. **1968** R. SAWKINS *Snow along Border* xv. 121 The picture was a head-and-shoulders portrait of a man of about thirty.

d. *head-and-shoulder target,* a target representing a head and shoulders.

1899 *Westm. Gaz.* 23 Nov. 2/2 A hostile force in entrenchments is represented by rows of 'head and shoulder' targets. **1901** *Ibid.* 11 Sept. 3/1 Not only were there the usual head-and-shoulder dummies, but there were several 'surprise' targets.

49*. get one's head down. To have a sleep.

1943 in Hunt & Pringle *Service Slang* 35. **1958** 'N. SHUTE' *Rainbow & Rose* 61 I'll have to get my head down for a bit, though, before going out again.

49. have one's head examined** (or need one's **head examining**). A jocular phrase suggesting that one is 'off one's head'.

1949 'J. TEY' *Brat Farrar* xxvi. 239 Of all the 'old soldier' tricks to fall for!..I ought to have my head examined. **1954** J. SYMONS *Narrowing Circle* xxxii. 145 Give me credit for a bit of sense...If I'd behaved in the way you suggest I should need my head examining. **1965** R. MCDOWELL *Hound's Tooth* (1967) vii. 69, I let Bowman persuade me to call in the state police...I should've had my head examined. **1966** J. BINGHAM *Double Agent* ii. 25 If you think these chaps know nuclear secrets..you want your head examined. **1972** *N.Y. Times* 3 Nov. 22/6 Anyone who votes for Nixon ought to have his head examined.

52*. open one's head. To speak. *U.S. slang.*

1849 *Neal's Sat. Gaz.* (Phila.) 17 Feb. 1/1 But don't you open yer head about it to no other individiwal—for I want to supprise the Wiggletown folks, and make 'em open ther eyes a leetle. **1885** H. JACKSON *Zeph* ii. 44 He never opens his head to nobody. **1895** *Century Mag.* Sept. 674/1 I'm glad you didn't open your head about it. **1898** M. DELAND *Old Chester Tales* 307 Jones said..that he hardly opened his head for the whole twenty-one miles.

53*. put a head on. 'To punch or assault another, and figuratively to silence, or shut up another' (Clapin). *U.S. slang.* (Cf. *1 d.)

1868 F. WHYMPER *Trav. Alaska* 283 One calls the other a 'regular dead beat!' at which he, in return, threatens to 'put a head on him!' **1869** *Overland Monthly* III. 63 The gentlemanly proprietor of the premises had kindly volunteered to 'put a head' on the man who fired the pistol. **1876** *Scribner's Monthly* Nov. 142/2 Threats, profanely emphasized, 'to put a head on' me! **1911** R. W. CHAMBERS *Common Law* ii. 46 Kelly will put a head on you!

61. (Further examples.)

1894 G. B. SHAW *Let.* 3 Dec. (1965) 467 You could, at your worst, talk the heads off most of them. **1931** H. CRANE *Let.* 11 Sept. (1965) 379 Yesterday we..worked our heads off digging into the side of a small hill. **1951** J. CORNISH *Provincials* 11 As term progressed, Saturdays and Sundays..we would sit in our den..talking our heads off. **1965** W. SOYINKA *Road* 17 The bishop sermonized his head off.

62*. *to put a pistol to* one's *head*: to commit suicide by shooting; *to put a pistol to* (someone's) *head*: to coerce (someone). Also *fig.*

1841 C. DICKENS *Let.* 26 Feb. (1969) II. 220 Put a penny pistol to Chapman's head, and demand the blocks of him. **1853** THACKERAY *Newcomes* (1854) I. ix. 91 I'm blowed if I don't put a pistol to my 'ead, and end it, Mrs. G. **1896** A. E. HOUSMAN *Shropshire Lad* xliv. 67 And early wise and brave in season [you] Put the pistol to your head.

63. a. *head-waiter* (earlier and later examples).

1805 G. COLMAN *John Bull* I. i. 17 *Dan.*..I be head-waiter and hostler:—only we never have no horses, nor customers. **1823** L. MINOR *Jrnl.* 29 Nov. in *Atlantic Monthly* (1870) XXVI. 171/1 It is a singular spectacle to see a man, who has occupied such high and varied stations, bustling about a tavern at once as landlord, barkeeper, and head waiter. **1971** *Good Food Guide* 484 Rich, male, business customers..will receive highly skilled attention from everybody, starting with the headwaiter and sommelier and going down to the busboy.

64. *head sea* (later example).

1886 R. BROWN *Spunyarn & Spindrift* ix. 167 A head-sea began to heave up.

65. a. *head-flannel, -wrap.*

1861 MRS. BEETON *Bk. Househ. Managem.* 1021 The infant..must not be exposed to strong light, or too much air; and in carrying it about the passages, stairs, &c., the nurse should always have its head-flannel on. **1880** Advt. in L. Higgin *Handbk. Embroidery* 106 Babies' Headflannels, from £1 3s. **1896** *Godey's Mag.* Feb. 202/1 Fastening..a sombre head-wrap over her..hair. **1905** *Daily Chron.* 15 May 3/3 A hood of this kind..will obviate any necessity for the wearing of those head-wraps.

b. *head-scratching* (also *fig.*), *-shaking* (additional examples), *-splitting*; *head-scratcher* (also *fig.*), *-shaker.* Cf. SHAKE *v.* 6 b.

1936 J. B. PRIESTLEY *They walk in City* 375 A pair of tweezers and a head-scratcher from Ur of the Chaldees. **1969** E. H. PINTO *Treen* ii. 26 The plain lignum vitae, English head scratcher..was typical of that 18th-century elegance which accepted public poking of wigs and itching heads as normal. **1971** *Daily Colonist* (Victoria, B.C.) 22 July 1/1 Now, the head-scratcher is how interest rates can be going up. **1926** J. S. HUXLEY *Ess. Biologist* v. 178 It was a hard nut for them, and there was much head-scratching. **1958** *Times* 22 Oct. 14/3 It is now that the Rugby football captains get down to their real head-scratching. **1973** *Times* 3 July 23/4 If it is cleared without reference, then there will be more head scratching. **1927** H. G. WELLS in *Sunday Express* 1 May 12/3 The Gummidge chorus is never silent; the thoughtful headshaker moping for a return to medievalism casts his daily shadow on every patch of sunshine. **1958** *Times Lit. Suppl.* 14 Mar. 140/3 Even the contemplative Henry James became a head-shaker. **1869** GEO. ELIOT *Lett.* (1955) V. 54 Best love to Sara—and some headshaking at her tendency to work too hard. **1961** *Times* 11 Mar. 3/3 Both new works caused widespread discussion and much head-shaking. **1903** W. J. LOCKE *Where Love Is* (1904) i. 3 Discussing the functions of art and other such head-splitting matters. **1953** W. STEVENS *Let.* 8 Dec. (1967) 804, I sat..listening to platitudes propounded as if they were head-splitting perceptions.

66. head arrangement *Jazz* (see quot. 1946); **head-ball** *Cricket,* a cunningly-bowled ball; so **head bowler, bowling; head-carry** *v.*, to carry (a load) on one's head; **head-cheese** (earlier and later examples); **head-clause** *Gram.*, the principal clause; **head-cover** *Mil.*, protection for the head, *spec.* a shield for protection against gun-fire; *Fortif.* (see quot. 1892); **head dip** *Surfing* (see quots.); **head doctor** *slang,* a psychiatrist; **head-feast,** a feast in celebration of successful head-hunting; **head-fold** (see quots.); **head-form,** (*a*) the form of the head, *spec.* in reference to the ratio of its breadth to its length; (*b*) the first word in a dictionary or glossary entry, lemma; **head girl** (see quot. 1963); also *transf.*; **headlamp,** one of the headlights of a (motor) vehicle; **head-load,** a load carried on the head; so **head-load** *v.*; **head-lock, -locking,** (*a*) (see quots.); (*b*) *Wrestling* (see quot. 1961); **head metal** *Founding,* the head of metal at the upper end of a cylindrical casting (see DEAD-HEAD 2 a); **head-reach** *v.* (later example); also *trans.*; **head-register** (see quots. and REGISTER *sb.*[1] 8 b); also *attrib.*; **head resistance,** resistance of a fluid to the movement of a body through it; **head-right** *U.S.*, an inheritable right to land, formerly granted by the state of Texas to the heads of immigrating families; **head-ring,** (*b*) (examples); **head-scarf, headscarf,** a scarf worn instead of a hat; hence **head-scarved** *a.*; **headset,** (*a*) a pair of earphones; (*b*) a combination of earphones and a microphone as worn by a telephone exchange operator; **head-shaking,** a display by certain birds at mating or egg-laying (see also sense 65 b in Dict. and Suppl.); so **head-shake** *v.*; **head-shrinker,** (*a*) a head-hunter who preserves and shrinks human heads; (*b*) *slang* (orig. *U.S.*) a psychiatrist; so (sense (*a*)) **head-shrinking** *vbl. sb.*; **head-shy** *a.*, of animals: afraid of having the head touched; so **head-shyness; head start,** an advantage at the beginning of a race; also *transf.*; *spec.* (with capitals) the name of an educational and welfare programme in the United States; **head-station** (further examples); also *N.Z.*; **head-stream, -tributary,** a head-water stream or tributary; **head-tie,** a head-band or scarf worn by women, esp. among peoples of African origin; **head-up** *Aeronautics,* used *attrib.* of a visual display system by which the pilot is able to read his instruments without averting his eyes from the aircraft's course; also, such a system in a motor vehicle; **head-wall** *Phys. Geogr.*, the steep slope at the head of a glaciated valley, esp. a cliff that rises abruptly from the floor of a cirque; **head-word,** (*b*) *Gram.* a word modified by another word or words; (*c*) = *HEAD *sb.* 19 c (*b*); = **head-form* (*b*).

1946 R. BLESH *Shining Trumpets* (1949) xi. 251 'Head' arrangement. *footnote.* A memorized, not written arrangement, that leaves ample room for improvisation. **1949** L. FEATHER *Inside Be-bop* iii. 21 He'd help to set up ideas for head arrangements. **1958** N. D. HINTON in *Publ. Amer. Dialect Soc.* XXX. 46 Head arrangement, a musical arrangement which is not written down and never has been, but is known by all the members of the ensemble. **1968** *Jazz Monthly* Feb. 21/1 The functional scoring of *John's idea, Shorty George* and *Cherokee* was the work of Jimmy Mundy,..the remainder probably being head arrangements. **1870** *Baily's Monthly Mag.* July 295 Alfred Shaw..is one of the few bowlers..qualified to attempt a 'head ball'. **1902** *Westm. Gaz.* 11 Aug. 7/1 Self-restraint that not even the most tempting 'head-balls' of Lockwood and of Rhodes could overcome. **1870** F. GALE in *New Sporting Mag.* New Ser. LX. 35, I must.. see another little square man..before I believe that a better head bowler ever lived than Lillywhite. **1867** G. H. SELKIRK *Guide Cricket Ground* ii. 28 Head Bowling. **1957** R. CAMPBELL *Portugal* 84 Trundling, head-carrying, or pedalling their various contraptions. **1968** *Times* (Pakistan Suppl.) 6 Apr. p. v/5 For eight or more hours a day, he head-carries his dish of earth from one spot to another and his pay is about three rupees, or roughly 4s. **1841** *Southern Lit. Messenger* VII. 39/2 The animal..may be traced in the stewed chine and souse, the head cheese and sausages. **1891** H. FREDERICK *Copperhead* (1894) 255 Reducing what remained of the [pig's] head into small bits, to be seasoned..and then fill other pans as head-cheese. **1942** C. MORLEY *Thorofare* (1943) xli. 159 Yes, Ma'am, over here we call it headcheese, but I remember my old gramp called it brawn. **1970** C. MAJOR *Dict. Afro-Amer. Slang* 65 *Headcheese,* various cheap grades of pork meat prepared and sold as lunch meat. **1928** H. POUTSMA *Gram. Late Mod. Eng.* (ed. 2) 38 A subordinate statement with modal *may* often stands with a head-clause containing *possible.* **1957** R. W. ZANDVOORT *Handbk. Eng. Gram.* I. iv. 62 The *perfect of experience*..is not unknown in other languages, at least in head-clauses, though an adjunct expressing repetition is usually added. **1890** LD. LUGARD *Diary* 26 Mar. (1959) I. iv. 160 Along the top [of the Stockade] a log for head cover, and line below it for firing. **1892** F. IRWIN *Fortification* (ed. 2) 37 Always place head-cover on wall when firing over the top. **1916** 'BOYD CABLE' *Action Front* 229 It's a good foot and a half I have of head-cover. **1923** KIPLING *Irish Guards in Gt. War* I. 24 The Battalion..next day quietly improved trenches and head-cover. **1962** *Austral. Women's Weekly* Suppl. 24 Oct. 3/2 *Head dip,* trick riding—putting head in and out of a wave while riding it. **1963** *Observer* 13 Oct. 15/5 The 'head dip', in which the rider bends double and dips his head into the wave at his feet. **1956** 'E. MCBAIN' *Cop Hater* (1958) xxi. 177 'What's the matter with this guy?' 'Nothing that a head doctor couldn't cure,' Byrnes said. **1959** M. DOLINSKY *There is no Silence* vii. 1:4, I was impressed in spite of my previous opinions about 'head doctors'. **1971** 'A. BLAISDELL' *Practice to Deceive* xiv. 210 Getting let loose by some damn-fool head doctor. **1882** H. DE WINDT *On Equator* 82 These 'Head Feasts' are general among the aboriginal tribes throughout the island of Borneo. **1890** BILLINGS *Med. Dict., Head-fold,* a fold of the blastoderm under the cephalic end of the embryo. **1893** A. M. MARSHALL *Vert. Embryol.* 226 The head of the embryo is lifted up above the yolk-sac by an anterior constriction or head fold. **1885** J. BEDDOE *Races of Britain* xiii. 259 The principal ethnical elements in Britain are too much alike in headform to yield their differences to an average constructed on but a few living heads. **1903** *Biometrika* II. 505 We are ignorant..of the characters of such a race, of its variability, for instance in head-, nose- or hair-form. **1927** PEAKE & FLEURE *Peasants & Potters* 128 There can be little doubt that profile and head-form have subtle..interrelations. **1935** *Proc. Prehist. Soc.* I. 4 The particulate inheritance of the several genes determining head-form. **1962** K. MALONE in Householder & Saporta *Probl. Lexicogr.* 112 Here such verbs are listed in full, and the irregular forms of each verb are entered against the head-form. **1846** C. M. YONGE in *Mag. for Young* 2 Sept. 196 At school..she went to.. the head girl. **1919** A. BRAZIL (*title*) The Head Girl at the Gables. **1941** *Brit. Jrnl. Psychol.* Jan. 194 The 'head-girls' [in a factory] wear an overall different in colour. **1963** BARNARD & LAUWERYS *Handbk. Brit. Educ. Terms* 104 *Head boy/girl,* a boy or girl appointed or elected as a leader of the other pupils, and who traditionally in public schools has considerable responsibilities for the maintenance of discipline acting through prefects. **1964** M. DRABBLE *Garrick Year* xii. 179, I was a prefect, but I wasn't head girl. **1885** KIPLING *Phantom 'Rickshaw* in *Quartette* 96 It [*sc.* a rickshaw] lay in readiness in the Mall, and..with a lighted head-lamp. **1912** *Motor Man.* (ed. 14) iii. 123 Paraffin Head-lamps. **1961** *Times* 25 Apr. 17/3 The first British production car to be fitted with two pairs of twin headlamps. **1972** *Country Life* 7 Dec. 1592/3 The brakes were effective..and wipers and headlamps very good. **1927** W. H. TODD *Tiger, Tiger!* 20 Carrying head-loads of sand. **1957** M. BANTON *W. Afr. City* iv. 67 Head-loading someone's baggage or helping to take goods out of a lorry. **1959** *Times* (Ghana Suppl.) 9 Nov. p. iv/1 Much of the cocoa starts its journey to the coast by being head-loaded along the bush tracks. **1971** J. SPENCER *Eng. Lang. W. Afr.* 29 *Head-load* (a term now surely becoming obsolescent with the advent of mechanical modes of transport). **1901** DORLAND *Med. Dict.* (ed. 2), *Head-lock,* the locking together of the chins in twin labor. **1905** F. R. TOOMBS *How to Wrestle* (1906) 107 Secure a head lock by putting your left forearm..on the defensive man's head. **1934** J. M. CAIN *Postman always rings Twice* x. 109 He's the only one in this town can throw the headlock on Sackett. **1957** *Encycl. Brit.* XXIII (*caption,* facing p. 804), Jim Browning..with a headlock on Danno O'Mahoney. **1961** J. S. SALAK *Dict. Amer. Sports* 220 *Headlock,* a hold in which the wrestler encircles the opponent's head with one or both arms. **1973** *Times* 16 Mar. 2/2 They dragged me out of the house and struck me on the head and legs, holding me in a headlock. **1890** BILLINGS *Med. Dict., Head-locking,* in twin labor, one child being born by the breech, its chin catches upon the chin of the second child presenting by the head. **1888** *Lockwood's Dict. Mech. Engin.* s.v., When the casting is removed from the mould the head metal is turned off, leaving the actual casting smooth and free from these foreign impurities. **1960** R. LISTER *Decorative Cast Ironwork* 229 *Head metal,* the metal in a feeding head. **1892** *Outing* (U.S.) Apr. 57/1 Soon she had head-reached them all, *Shadow* included, and showed to the front of the fleet. **1938** C. S. FORESTER *Ship of Line* 276, I want to hear instantly if they alter course, or if they headreach upon us. **1890** BILLINGS *Med. Dict., Head-register,* register in which the pitch is raised by shortening the vocal chords; second falsetto in females. **1909** H. KLEIN *Phono-Vocal Method* 37 The blending of the medium and head registers will be practised upon the same plan. **1966** H. L. SHORTO in C. E. Bazell *In Memory of J. R. Firth* 407 Henderson found that in Cambodian a sequence of head-register consonant and chest-register vowel..was the mark of a secondary pattern. **1889** J. J. WELCH *Text Bk. Naval Archit.* iii. 53 This [fluid] resistance is due:—..(2) To the opposition offered to the passage through the water..of projections such as the keel and bilge keels, and of the comparatively flat parts of the ship at the ends: this is known as direct or head resistance. **1891** *Railroad & Engin. Jrnl.* LXV. 465/1 The head or hull resistance will probably be found to be the chief element which will limit the possible speed of flying machines. **1909** *Westm. Gaz.* 1 July 4/1 Allowing a coefficient of ·3 for the pointed ends, the total head-resistance would be reduced to 3,324 lb. **1922** R. GLAZEBROOK *Dict. Appl. Physics* I. 717/1 A second form of eddy resistance developed by a ship, sometimes called 'head resistance', is due to such features as..web supports to the propeller shafts, thick stems, and stern posts. **1934** *Discovery* Dec. 352/2 Another cowl known as a Townend Ring..produces ..a reduction in head resistance. **1959** J. L. NAYLER *Dict. Aeronaut. Engin.* 128 *Head resistance,* a term used for the resistance, or drag at no yaw, of the front part of a projectile, the remainder of the drag being due to skin friction and base drag. **1703** in *Amer. Speech* (1961) XXXVI. 152 Head-right. **1799** in O. A. Rothert *Hist. Muhlenberg County* (1913) 45 Colonel William Campbell's head-right.. adjoining the lands [etc.]. **1837** *Laws of Texas* (1838) I. 266 So much of the vacant lands..to be surveyed and sectionized..as will be sufficient to satisfy all claims.. for scrip sold, soldiers' claims, and head rights. **1898** H. S. CANFIELD *Maid of Frontier* i. 13 He owned the headright of 160 acres on which his house was built. **1866** C. BARTER *Alone among Zulus* v. 51 When a [Zulu] soldier has attained a certain standing he receives the royal permission to marry, and adopt the head-ring as a mark of manhood. **1952** S. G. MILLIN *Burning Man* xxv. 227, I have worn the head-ring of a married man for three years. **1921** G. B. SHAW *Back to Methuselah* III. 101 (*stage direction*) A handsome negress is trying on a brilliant head scarf. **1955** G. FREEMAN *Liberty Man* I. i. 18 Maureen came out adjusting her headscarf. **1962** D. HARDEN *Phoenicians* vii. 103 A bearded priest with a head-scarf. **1960** M. SHARP *Something Light* II. x. 91 A head-scarved plumpish figure. **1970** *Guardian* 13 July 9/3 Head-scarfed refugees. **1921** *Telegr. & Teleph. Jrnl.* Dec. 46/2 Supervisor (speaking on head-set of bewildered learner). **1942** *Tee Emm* (Air Ministry) II. 84 The pilot..should see that his headset plug is firmly in. **1955** 'N. SHUTE' *Requiem for Wren* 64 A telephone headset strapped across his beret. **1957** *Spaceflight* I. 71/2 Each seat in the auditorium was fitted with a headset and switchbox, by means of which one could select the language one desired. **1970** N. ARMSTRONG et al. *First on Moon* iii. 63 They had to be checked out: the communications consoles, the technician headsets, the purge ventilators. **1930** J. S. HUXLEY *Bird-Watching & Bird Behaviour* iv. 70 It is doubtless very enjoyable to head-shake together. *Ibid.* v. 95 The head-shaking ceremony. **1959** BANNERMAN *Birds Brit. Isles* VIII. 217 The second important display is 'head-shaking'. The bird..waggles its head from side to side at another... Birds already paired together frequently head-shake together. **1926** G. M. DYOTT *On Trail of Unknown* xii. 173 (*heading*) The head shrinkers. *Ibid.* 190 In the process of head-shrinking, the hair retains its original length and..looks longer than it did on the original man. **1950** *Time* 27 Nov. 19 Anyone who had predicted that he would end up as the rootin'-tootin' idol of U.S. children would have been led instantly off to a headshrinker. **1957** A. MANEY in G. Oppenheimer *Passionate Playgoer* (1958) 381 Marcus Heiman, head of the United Booking Office, turned the play over to his psychiatrist. That headshrinker said a play glorifying a drunkard conflicted with public interest. **1958** *Spectator* 22 Aug. 241/2 The head-shrinkers would doubtless say it is the universal human search for a father-figure that is behind it. **1959** T. B. MORRIS *Death among Orchids* xiii. 121 There are still cannibals... But the worst of them are now..the head-shrinkers. **1968** *New Scientist* 8 Feb. 289/1 Dr. Louis West..may eventually be taking the caviare out of headshrinkers' mouths with his development of the robot psychiatrist. **1970** I. REED in A. Chapman *New Black Voices* (1972) 516 A vocabulary that calls things by their names: 'headshrinker' and 'egghead'. **1900** M. H. HAYES *Among Horses in Russia* iv. 82 They weren't head-shy, and liked their manes to be scratched. *Ibid.*, Every horseman knows that 'head-shyness' is one of the worst of vices. **1952** J. STEINBECK *East of Eden* 384 A few strokes on the nose will make a puppy head-shy. **1886** in *Amer. Speech* (1950) XXV. 34/1 Fifteen paces, head start. **1911** W. JAMES *Some Probl. Philos.* xi. 180 Owing to the tortoise's head-start, the tortoise's path is only a part of the path of Achilles. **1935** F. SCOTT FITZGERALD *Let.* (1964) 6 You and Peaches (who isn't selfish, I think) had a superficial head-start with prettiness. **1962** *Economist* 18 Aug. 620/2 The research data needed to give them a head-start in the market. **1965** Mrs. L. B. JOHNSON *White House*

Diary 3 Feb. (1970) 235 Sarge [Shriver] asked me if I would consider and sponsor the program Head Start. **1968** *Times* 9 Oct. 10/7 If we can give them a head start—..they're going to do as well as other children. **1970** *Washington Post* 30 Sept. 133/1 Volunteers are urgently needed for Head Start and day care programs throughout the Washington area. **1973** *Black World* Mar. 37 Effective or not, the Poverty Program, Headstart, compensatory education programs,..and so on were efforts toward establishing justice for Blacks. **1973** J. PATTINSON *Search Warrant* ii. 37 He decided to spend the night in Philadelphia... The other man had a head start anyway. **1862** R. HENNING *Let.* 19 Oct. (1966) 111 It is a most eligible spot for a head-station, and the two carpenters have already begun putting up the store. **1895** G. CHAMIER *South-Sea Siren* xi. 161 [The house] was used as the head-station for a sheep run. **1936** A. RUSSELL *Gone Nomad* iii. 14, I had left the head station at sunrise and ridden all day. **1817** S. R. BROWN *Western Gaz.* 8 Navigation..can be pursued up the Coose to one of its head streams. **1899** A. H. KEANE *Man Past & Present* 190 Northwards.. about the Irawadi head-streams. **1908** *Athenæum* 11 Apr. 456/3 From its head-streams in the glens. **1956** D. L. LINTON *Sheffield* p. xxiv, Turning southwards round the headstreams of the Don our boundary passes above the long Woodhead tunnel. **1857** M. GRIFFITH *Autobiogr. Female Slave* i. 17 One gave a yard of ribbon, another a half-paper of pins, a third presented a painted cotton head-tie. **1956** in Cassidy & Le Page *Dict. Jamaican Eng.* (1967) 222/2 [*Head-tie*] a head scarf. **1967** W. SOYINKA *Kongi's Harvest* 52 You see yourself How the courtesan is one hour escalating Her brocade head-tie. **1973** *Trinidad Guardian* 1 Feb. 11/1 She was wearing black shoes, a flowered dress and headtie. **1925** N. E. ODELL in E. F. Norton *Fight for Everest, 1924* iii. 292 A far-flung head tributary of the Dzakar Chu. **1960** *Times* 15 Dec. 4/3 Mr. Naish said that some 1,200 hours of laboratory 'flying' with the head-up device used in conjunction with the flight simulator had shown that accuracy in following flight directions was twice as good as with the normal instruments. **1968** *New Scientist* 8 Aug. 273/1 A new term has found its way into the cockpits of military aircraft in the past few years. It is 'head-up display'. It connotes a system whereby the pilot can see what tale his instruments are telling without taking his eyes off the scene ahead. *Ibid.* 273/2 Head-up flying is a bonus derived from the application of electronics in many new ways. **1972** *Drive* Spring 43/2 Road-speed reflected in the windscreen glass by the safety-first 'head up' display. **1972** *Times* 14 Sept. 31/1 At present only the drivers of a handful of experimental cars (and the pilots of supersonic aircraft) get the benefit of head-up displays... Smiths are working on head-up displays for family cars. **1904** *Jrnl. Geol.* XII. 570 The canyons, at their heads, were abnormally deep..and their head walls..stood as nearly upright, apparently, as scaling of the rock would permit. **1910** *Geogr. Jrnl.* XXXV. 154 Perhaps because of their small size these cliff glaciers have not developed cirques, though a Bergschrund parallels the generally straight headwall. **1954** W. D. THORNBURY *Princ. Geomorphol.* xv. 367 A cirque headwall may be as much as 2000 to 3000 feet high and is notably steep and free from talus at its base, even in empty cirques. **1898** SWEET *New Eng. Gram.* II. § 1759 Thus pre-adjunct or pre-adjective position means that the adjunct-word precedes its head-word. **1939** *English Studies* XXI. 71 It [*sc.* the genitive] was pinned down..to its head-word, first either in front or in post-position, eventually only in front position. **1940** *Ibid.* XXII. 88 *Headwords, attributes* and *adjuncts,* the terms representing their relative importance or ranks within the sentence, headwords coming first. **1957** S. POTTER *Mod. Ling.* v. 115 In the phrase *good men* there are two ranks: *men* is the head-word or primary and *good* is the attribute or secondary. **1957** R. W. ZANDVOORT *Handbk. Eng. Gram.* II. ii. 104 The headword need not be repeated if it occurs earlier or later in the sentence: She put her arm through her mother's. **1961** R. B. LONG *Sentence & its Parts* v. 121 But in true phrasal passives the auxiliaries set the time for actions whose semantic centers are the head-word participles. **1964** M. SCHUBIGER in D. Abercrombie et al. *Daniel Jones* 258 Restrictive relative clauses preceded by their head-words. **1966** *English Studies* LXVII. 211 Head-words..appear in their West-Saxon variant. **1967** *Listener* 26 Oct. 545/3 (Advt.), 97,000 headwords.

Head (hed). The name of Sir Henry *Head* (1861–1940), English neurologist, used *attrib.* and in the possessive with reference to his work on sensation, etc.

1908 *Jrnl. Nerv. & Mental Dis.* XXXV. 576 (*title*) On the hyperesthetic areas (Head's zones) in visceral disease. *Ibid.* 577 His experience would not corroborate Dr. Bloomfield's as to the great value of Head's test. *Ibid.*, Dr. McCarthy asked in what percentage of tuberculous cases Dr. Ludlum found the distinct Head sign. *Ibid.* 578 Head's lines cannot be elicited in chronic cases. *Ibid.*, That patient still shows Head's lines. **1910** *Practitioner* Jan. 119 Some of the headache of pyorrhœa alveolaris may be reflex (*cf.* Head's areas). **1913** DORLAND *Med. Dict.* (ed. 7) s.v. *Zone, Head's zones,* areas of cutaneous sensitiveness associated with diseases of the viscera: called also zones of hyperalgesia. **1941** *Brit. Jrnl. Psychol.* XXXII. I. 6 Physiological experiments upon the action of the vagus nerve in respiratory movements..[which were recorded] by the use of a strip of the diaphragm (known henceforth as 'Head's strip'). **1961** *Brain* LXXXIV. 530 He [*sc.* Head] devised the preparation which is known as the 'Head's diaphragm slip preparation'. *Ibid.* 533, I suggest the possibility that this gasping response in the newborn baby is in fact Head's Paradoxical Reflex. *Ibid.* 532 The Head Reflex is carried in small unmyelinated fibres.

head, *v.* Add: **9.** Also with *up.* Chiefly *N. Amer.*

1959 E. LIPSKY *Scientists* 178 Bronco and I feel you're the logical one to head up a committee. **1968** *Globe & Mail* (Toronto) 17 Feb. B7 (Advt.), Company..requires capable and professional person..to head up real estate department. **1971** *Daily Tel.* 21 Oct. 10 (Advt.), We need women who can head up the book department of several of our branches throughout England and Wales.

12. *intr.* Also *fig.*

1903 G. B. SHAW *Man & Superman* II. 56, I rather think Rhoda is heading for a row with Ann. **1922** H. CRANE *Let.* 7 Nov. (1965) 104 Matty's trans[lations] from Soupault in the last *Broom* are undoubtedly clever, but I don't see how he can rave so... Where is he headed for, anyway?

13. b. (Earlier example of *head off.*)

1841 J. F. COOPER *Deerslayer* v. 29 But 'head him off', as you say of the deer.

c. *N.Z.* (See quot. 1933.)

1933 L. G. D. ACLAND in *Press* (Christchurch, N.Z.) 28 Oct. 15/7 A dog goes round to the far side of a mob of sheep and stops them. This is called heading... The owner would also say 'I can head with him'. **1934** J. LILICO *Sheep Dog Mem.* 27 [The dogs] would head, lead, huntaway, force and back, though, of course, they were best at rouseabout work. **1947** P. NEWTON *Wayleggo* (1949) v. 52 This dog would 'head a nor'wester'.

head, colloq. abbrev. of *HEADLIGHT.

1959 I. JEFFERIES *13 Days* vi. 75 He..flashed his heads just as I got abreast. **1969** 'A. HALL' *Striker Portfolio* ix. 112 My undipped heads catching the Mercedes full across the screen. **1971** —— *Warsaw Document* xxii. 279 The patrol-car coming at us with the heads full on.

headache. Add: **1. b.** Phr. *to be no more use than* (or *as good as*) *a* (*sick*) *headache*: said of something quite useless. *colloq.*

1915 D. O. BARNETT *Lett.* 153 Shrapnel is for defenders, to stop an advance of infantry, but no more use against prepared positions than a sick headache. **1927** D. L. SAYERS *Unnatural Death* I. v. 50 That woman..was no more use than a headache—to use my brother's rather vigorous expression. **1931** W. HOLTBY *Poor Caroline* vi. 225 The Tona Perfecta's no more use to any company today than a sick headache. **1963** *Guardian* 3 Dec. 5/5 The car's contract of sale gives no undertaking or guarantee except the usual one—and that is as good as a sick headache.

c. A troublesome or annoying problem. *colloq.* (orig. *U.S.*).

1934 M. WESEEN *Dict. Amer. Slang* 347 *Headache,* anxiety; worry. **1937** *Punch* 1 Dec. 610/1 My headache is this—the Big Guy, my boss, won't go to the movies and see for himself what a newspaper girl can do. **1939** G. B. GILBERT *Forty Years a Country Preacher* 77 The new rectory was both beautiful and expensive, but it proved to be a great headache. **1942** 'H. HABE' *Thousand shall Fall* ix. 181, I asked him how we were expected to transport all these goods. He shrugged his shoulders. 'That's your headache!' **1945** *Times* 11 Jan. 2/4 Commander Bower continued: 'The biggest headache of all is undoubtedly Poland.' **1952** N. STREATFEILD *Aunt Clara* 251 They're my headache, not yours. **1968** *New Scientist* 25 Jan. 205/1 The single-celled organism *Euglena* is rather a headache for those who would like to divide living things neatly into plants and animals.

headachiness. (Earlier example.)

1862 GEO. ELIOT *Let.* 1 May (1956) IV. 28 In a moment of feverish headachiness I transgressed my own rule.

headachy, *a.* **1.** (Earlier examples.)

1795 LADY NEWDIGATE *Let.* Aug. in A. E. Newdigate-Newdegate *Cheverels* (1898) xi. 154 Want of sleep.. makes me feel Languid & headachy in a Morning. **1813** JANE AUSTEN *Let.* 29 Jan. (1952) 298, I was rather headachey that day & could not venture on anything sweet except jelly.

headage (he·dedʒ). [f. HEAD *sb.* 7 c + -AGE.] The number of animals; = HEAD *sb.* 7 c. Also *attrib.*

1957 *Liverpool Daily Post* 30 Jan. 5/9 Abbey-Cwm-Hir..Fox Destruction Society have decided to pay headage money on rabbits and carrion crows, as well as foxes, that are killed in the society's area. **1960** *Farmer & Stockbreeder* 5 Jan. 91/1 A headage price [for a steer] of 22,356 dollars. **1962** *Times* (Agric. Suppl.) 3 July p. ii/3 Winter reserves of fodder to carry a greatly increased headage of stock. **1972** *Guardian* 30 Dec. 4/7 'Headage' subsidies for sheep flocks are ruled out by the EEC commission.

headband. Add: **1. c.** The band connecting a pair of receivers or ear-phones.

1913 *Work* 17 May 145/3 Double Receivers, with adjustable head-band. **1962** A. NISBETT *Technique Sound Studio* 256 *Headphones,* a pair of electro-acoustic transducers..held to the ears by a headband.

head-block. Add: **4.** (See quot. 1905.)

1853 *Trans. Mich. Agric. Soc.* IV. 35 G. S. Snyder, Lancaster O. improved head block for setting logs on saw mills. **1905** *Terms Forestry & Logging* 39 *Head block,* the log placed under the front end of the skids in a skidway to raise them to the desired height.

head-chief. *U.S.* [HEAD *sb.* 63.] The paramount chief of an Indian tribe.

1806 J. ORDWAY in *Jrnls. Lewis & O.* (1916) 355 The head chief..informd us that the most of our horses and pack Saddles were Safe. **1837** R. M. BIRD *Nick of Woods* I. 236 From the head-chief to the commoner. **1881** *Harper's Mag.* Apr. 670/2 White Eagle, head-chief of the Poncas. **a 1918** G. STUART *40 Yrs. on Frontier* (1925) I. 96 The head chief proposed to meet the interpreter unarmed and talk with him.

hea·d-dresser. = HAIRDRESSER. Also, one who makes head-dresses (*Obs.*).

1727 DEFOE *Compl. Eng. Tradesman* II. II. v. 166 Now we see..the Millenary Trade separated into innumerable little Commode Shops, Head Dressers and such like People. **1859** TROLLOPE *Bertrams* ix. 81 No stray jagged ends would show themselves if by chance she removed her bonnet, nor did it even look as though it..required to be afresh puffed out by some head-dresser's mechanism. **1927** *Daily Tel.* 24 May 17/3 The 'head-dressers'' shops are nests of decoy birds.

header. Add: **4. c.** = *heading dog* (*HEADING *vbl. sb.* 4 b). *N.Z.*

1938 R. M. BURDON *High Country* x. 107 Dogs are usually kept in the proportion of three or four huntaways to one header now. **1958** *Landfall* XII. 17 Watching him work the new sheep dog, I see the taut thread of his whistle run from his mouth to the pricked quivering ears of the header. **1973** *Times* 13 Oct. 14/4 Early British sheep dogs were less versatile. They were divided into those which fetched the sheep and those which drove them away, as still occurs in New Zealand with header and hunt-away.

5. b. A top layer. *U.S.*

1867 *Trans. Ill. Agric. Soc.* VI. 641 Prime Pork—Shall be packed with a header of side cuts, the regular width, three half heads.

c. *Engin.* (See quot. 1940.)

1930 *Engineering* 25 July 121/1 They contain four headers, which are turned from solid mild-steel forgings. **1940** *Chambers's Techn. Dict.* 406/2 *Header,* a box or manifold supplying fluid to a number of tubes or passages, or connecting them in parallel. **1958** W. HRYNISZAK *Heat Exchangers* vii. 132 The function of the headers is either to distribute the gas over the matrix (inlet header), or to collect it from the matrix (outlet header) with the minimum possible loss of pressure not used for heat transfer purposes.

d. *Hedging.* (See quot.)

1941 *Archit. Rev.* LXXXIX. 85/2 The 'pleaches' too are rammed down, and when several yards are ready the two men work in the headers, which are slender and straight rods, twisted over and across, in and out of the stakes.

9. b. = HEADING *vbl. sb.* 11.

1877 R. W. RAYMOND *Statistics of Mines* 165 The header had reached..a length of 12,259 feet.

11. *Association Football.* A ball which is headed. Also, one who heads the ball (see HEAD *v.* 15).

1906 L. V. LODGE in B. O. Corbett *Ann. Corinthian Football Club* 188 A back must be not only a good kick and sound tackler, but at the same time an accurate header. *Ibid.* 189 A really high-class header, by a skilful movement of the neck.., can move the ball a surprising distance. **1927** *Daily Express* 20 Apr. 13/2 Trotter gave the home club the lead with a beautiful header, following a free kick well taken by Leach. **1955** *Times* 13 Aug. 4/3 Yashin brought off a wonderful one-handed save to push out Wilshaw's header, which looked a certain goal. **1969** *Times* 8 Oct. 13/3 It was Marsh who came closest to scoring, when Kelly, using his left arm like a scythe, turned a header over the bar.

12. *attrib.* and *Comb.*: **header-board,** a diving-board; **header brick** = HEADER 5; so **heading brick.**

1913 E. F. BENSON *Thorley Weir* i, The nude figure of a boy on the header-board in the act of springing from it into the water. **1897** F. C. MOORE *How to build Home* vii. 110 A 'header' brick is one laid in the wall so that only its end shows. **1901** J. BLACK *Illustr. Carp. & Build., Scaffolding* 24 Cavities in the brickwork obtained by leaving out 'header' bricks at proper intervals. *Ibid.* 26 A 'heading' brick.

heading, *vbl. sb.* Add: **4. b.** *N.Z.* Of a farm dog: see quot. 1933. Hence *heading dog.* Cf. *HEAD *v.* 13 c.

1913 A. I. CARR *Country Work & Life N.Z.* xix. 33 A new hand, if he intends to qualify for the work [as shepherd] is wise in investing in a good huntaway or a heading dog. **1933** L. G. D. ACLAND in *Press* (Christchurch, N.Z.) 28 Oct. 15/7 A dog goes round to the far side of a mob of sheep and stops them. This is called heading. Hence heading dog, one whose work this is. **1947** P. NEWTON *Wayleggo* (1949) 13 The heading dog is bred to run out silently, cast round sheep, and bring them back to his master. **1968** *N.Z. News* 28 Aug. 16/1 The heading dog brings sheep up to the shepherd and holds them at one spot. **1972** P. NEWTON *Sheep Thief* v. 40 He told her of his successes on the dog trial grounds. His old huntaway, Sam, was his particular pride, and he also had high hopes for his little heading dog, Smoke.

c. *Aeronaut.* (See quot. 1951.)

1935 T. C. LYON *Pract. Air Navigation* 29 *Compass heading,* the true course plus or minus variation and deviation, and *including* allowance for wind. **1951** *Gloss. Aeronaut. Terms* (*B.S.I.*) III. 7 *Heading,* the direction of the longitudinal axis of an aircraft defined by the angle it makes with a specified meridian. **1968** *New Scientist* 18 Apr. 133/1 The aircraft's heading is defined by the localizer's two overlapping beams.

7. (Earlier examples.)

1682 T. A. *Carolina* 6 With this [cedar] they make Heading for their Cask. **1752** J. MACSPARRAN *Amer. Dissected* (1753) 26 Barrel and Hogshead Staves and Heading.

headlight (he·dləit). orig. *U.S.* Also with hyphen. [HEAD *sb.* 66.] A powerful light carried on the front of a locomotive or on the mast-head of a vessel; each of two powerful lamps carried on the front of a motor vehicle. Also *attrib.* and *fig.*

1861 *Remin. Life Railroad Engineer* 124, I saw the glimmer of his head-light when he first turned the curve. **1862** in *U.S. Pat.* 35486. **1891** C. ROBERTS *Adrift Amer.* iii. 55 The great head-light which forms such a noticeable feature of American engines at night. **1904** A. B. F. YOUNG *Compl. Motorist* (ed. 2) xii. 257 On dark nights it is advisable, when driving in the country, to carry on such cars a single head-light of greater power... On very fast cars two of these head-lights should be carried. **1907** *Autocar* 28 Dec. 1067/2 The thief calmly took the large valuable head light off and disappeared. **1929** *Evening News* 18 Nov. 4/4 He had almost reached the high road when the headlights of an automobile swung round a corner. **1959** *Sears, Roebuck Catal.* Spring & Summer 1103/3 Chrome-plated headlight shields. **1963** *Times* 13 Mar. 10/6 It has transistorised ignition, a.c. electrical generator, automatic headlight dipper, [etc.]. **1971** *Daily Tel.* 11 Feb. 11/2 It is only when they commit some offence that they are caught in the headlights of history. **1973** 'M. INNES' *Appleby's Answer* IV. xx. 172 The sound of a motor engine..and the sudden appearance of wavering headlights.

head-line. Add: Now usu. **headline. 2. c.** *to make* or *hit the headlines*: to be given prominent notice in the newspapers.

1934 M. WESEEN *Dict. Amer. Slang* (1935) xvii. 257 *Hit the headlines,* to become famous; to gain notoriety. **1939** *War Illustr.* 21 Oct. 181 He [*sc.* Winston S. Churchill] 'hit the headlines' in 1899 with a dramatic escape from captivity in a Boer armoured train. **1944** F. CLUNE *Red Heart* 12 It was just another tragedy of the Outback, the sort of thing that doesn't make headlines in the City newspapers, but it wrings the hearts of people who really know. **1948** *Manch. Guardian Weekly* 29 Jan. 11 The publication..hit the headlines here last night. **1957** 'J. WYNDHAM' *Midwich Cuckoos* 12 Before that it [*sc.* the village] hit the headlines—well, anyway, the broadsheets—when Black Ned, a second-class highwayman, was shot on the steps of The Scythe and Stone Inn by Sweet Polly Parker. **1968** *Globe & Mail* (Toronto) 3 Feb. 35/1 Anybody who gets his picture on the cover of the magazine [*sc. Time*] immediately breaks a leg, or hits the headlines in embarrassing fashion about the date of publication.

d. *Broadcasting.* Usu. in *pl.* A summary of important news items, given at the beginning or end of a news bulletin.

1908 'O. HENRY' *Gentle Grafter* 39 He shows me a machine..with two things for your ears..I puts it on and listens. A female voice starts up reading headlines of murders, accidents and other political casualties. 'What you hear,' says the farmer, 'is a synopsis of to-day's news..wired in to our Rural News Bureau and served hot to subscribers.' **1934** *B.B.C. Year-Bk.* 82 They [*sc.* Topical Talks] were given five minutes every evening in which to deal with the 'head line' of the day, and were tacked on like a tail to the news bulletins. **1941** *B.B.C. Gloss. Broadc. Terms* 14 Headline News: Brief statements giving, within the space of not more than five minutes, salient news items without comment or background material. **1971** 'D. HALLIDAY' *Dolly & Doctor Bird* i. 6, I watched the news headlines on television.

3. b. A line fastening the head of a vessel to the shore.

1876 'MARK TWAIN' *Tom Sawyer* ii. 29 Get out that headline! **1877** J. HABBERTON *Jericho Road* i. 9 The headline was cast off as the pilot's bell rang. **1958** E. S. LAND *Winning War with Ships* 105, I 'went ashore with the headline', contacted the Italian Admiral, and cooled off somewhat.

4. A base-line in surveying.

1656 *Doc. & Rec. New Hampshire* I. 221 From the said head lyne we measured for the length..6 miles & a halfe. **1704** *New Hampshire Probate Rec.* I. 514 All my land.. up as far as the southern hills, viz. as far as to butt against Andrews head line.

5. *Palmistry.* A line of the head (cf. LINE *sb.*[2] 8 b).

1867 A. R. CRAIG *Bk. of Hand* xxiii. 189 If a line sets out from the head line, and rises straight to the mount of Jupiter, crosses it, and cuts the roots of the forefinger, it is excessive pride. **1894** 'MARK TWAIN' in *Century Mag.* Feb. 554/2 Wilson began to study Luigi's palm, tracing life lines, heart lines, head lines, and so on. [**1911** *Encycl. Brit.* XX. 650/1 s.v. *Palmistry.* A line starting above the head of the second metacarpal bone and crossing the hand to the middle of its ulnar border is the line of the head.] **1934** *Cassell's Mod. Encycl.* 733/1 The principal lines on the palm are named life line, head line, heart line, fate line, and line of Apollo.

6. *attrib.* and *Comb.*

1909 *Daily Chron.* 19 Aug. 3/1 His ingenuity is amazing, ..and not merely amazing in the headline sense of that ill-used word. **1933** *Amer. Speech* Dec. 6/2 'Headline English' has become almost a menace to standard English usage nowadays. **1958** *Times* 11 Aug. 11/3 She.. is usually to be seen..at Broadway first nights,..at headline parties. **1963** *Times* 11 June 17/4 The headline-maker, of course, has been the contract from United Airlines, of the United States, which has taken delivery of its 20-plane medium-range jet fleet from France. **1965** *Economist* 2 Oct. 57/3 He still uses that unfortunate American headline-verb 'to score' (meaning 'to attack').

head-line, *v.* Now usu. **headline.** Add later examples. Hence **hea·dlined** *ppl. a.*, furnished with a headline.

1891 [in Dict.]. **1912** *Out West* Apr. 237/2 A big headlined, illustrated..story, the pride of some reporter's heart. **1953** *Manch. Guardian Weekly* 12 Feb. 3 The 'New York Herald Tribune' headlined this story to-day. **1958** *Listener* 20 Nov. 811/2 The answer.. faithfully reported and perhaps headlined the next day in the local press. **1964** *Melody Maker* 28 Nov. 4/2 Nashville Teens, Kinks and Hullaballoos will be headlined in the 10-day Christmas holiday show..at Brooklyn's Fox Theatre.

headliner. Add: One whose name appears in a headline; a chief personage or performer. *U.S.*

1896 *N.Y. Dramatic News* 4 July 10/3 That clever pair..were the headliners..last week. **1905** *Daily Chron.* 11 Feb. 3/5 They..secure good 'head-liners' or 'stars' at paralysing salaries. **1907** *Chicago Tribune* 8 May 7 The 'Headliners' on the program will be James Whitcomb Riley, George Ade, etc. **1914** *Boston Herald* 23 June 8/4 The headliner at B. F. Keith's. **1966** R. STOUT *Death of Doxy* (1967) xii. 140 Julie's dressing room..was about six by eight, par for a headliner in a place with a four-dollar cover charge. **1970** *Globe & Mail* (Toronto) 26 Sept. 35/1 He is the centre on a checking line, a headliner on the penalty-killing unit.

2. One engaged in head-lining (HEAD *sb.* 66).

1963 *Times* 5 Mar. 13/1 A typical example were the head-liners, who fitted the roof linings in vehicles.

headline·se. [*-ESE.] The elliptical style of language characteristic of the headlines, esp. in popular newspapers.

1927 C. G. MARSHALL *Private Let.* (G. & C. Merriam Co. files) 12 Aug., In the headlines of general newspapers you see time after time such words as 'Probe', 'Quiz', 'Tilt', 'Pact', etc. In Newspaper offices such language is referred to as 'Headlinese'. We banned it from the headlines of The [United States] Daily. **1934** *Times Lit. Suppl.* 1 Feb. 66/3 They recognize that the difficulty nowadays is to keep the headline 'from shouting too loudly', and they cannot condemn too heartily mere 'headlinese'. **1935** H. STRAUMANN *Newspaper Headlines* 256 An essential feature of present-day headlinese is the typographical make-up. **1966** *Listener* 2 June 811/3 In headlinese you don't marry, you wed... You don't advance arguments against, you score.

head master, head-master. Add: Hence **headma·ster** *v.*, to act as headmaster; also *transf.*; **headma·sterly** *a.*

1940 J. H. JAGGER *English in Future* vi. 73 No novelist has *headmastered* his countrymen as frequently as Mr. H. G. Wells. **1959** R. FULLER *Ruined Boys* 154 The school..so brilliantly headmastered. **1964** *English Studies* XLV (Suppl.). 9 Further headmasterly duties took him to Klagenfurt. **1965** G. HOUSEHOLD *Olura* 175 A headmasterly grief that there should be nigger-lovers in the upper forms. **1967** *Guardian* 4 Oct. 1/4 A headmasterly chat.

head-note. Add: **3.** A note or comment inserted at the head of the text.

1863 D. G. ROSSETTI *Let.* (1965) II. 472 Alteration of the last part to suit the new plan of head-notes to Part II. *Ibid.* 479, I am glad you like the headnote plan. **1965** *Listener* 17 June 905/3 A headnote to each poem. **1972** *Times* 4 Mar. 12/7 Since performance details for individual [crossword] puzzles are likely to interest many more readers than they may distress, it is hoped that those in the latter category will agree to turn a blind eye to head-notes to puzzles.

head office. [f. HEAD *sb.* 63 + OFFICE *sb.* 8.] The principal, controlling office of a firm or organization, where the chief administration is carried out, policy decisions made, etc. Also *attrib.*

1869 *Bradshaw's Railway Man.* XXI. (Advt.), London & County Banking Company... Head Office. 21, Lombard Street. *Ibid.* 390 Head office accounts, local. **1933** A. G. MACDONELL *England, their England* vi. 70, I was warned by telephone, my Lord, from our Head Office. **1941** F. THOMPSON *Over to Candleford* xiv. 214 I'll speak to Head Office. **1952** A. BRIGGS *Hist. Birmingham* II. 54 In 1881 an export representative was sent out to Australia and New Zealand with his first head office in Sydney, and later on new offices were opened at Melbourne, Adelaide, and Brisbane. **1972** *Woman* 4 Mar. 10/2 If you ever feel you have been badly treated [by the bank], *complain*—first to the branch and, if that's no good, then wake up head office. **1972** C. WATSON *Broomsticks over Flaxborough* xi. 150 In this trade you get used to being buggered about a bit by head office.

head-on, *adv.* and *a.* orig. *U.S.* [HEAD *sb.* 21 b.] **A.** *adv.* (*hea·d-o·n*). With the head pointed directly towards or running full against or towards something. Also *fig.* **B.** *adj.* (*hea·d-on*). Of a collision: involving the direct meeting of the fronts of two vehicles in the same track, or of the head of a vehicle with an object. Also *fig.*

1840 R. H. DANA *2 Yrs. bef. Mast* ii. 15 The two vessels stood 'head on', bowing and curveting at each other. **1903** C. E. MERRIMAN *Lett. from Son* 33 It's a case of head-on collision with your pride. **1904** *N.Y. Evening Post* 11 May 2 A head-on collision between north and southbound passenger trains. **1907** *Westm. Gaz.* 16 Sept. 5/2 New York, September 16.—A head-on collision between two passenger trains. **1916** H. BARBER *Aeroplane Speaks* 46 Gliding just over the trees and head on to the wind. **1930** *Daily Express* 8 Sept. 1/3 When the vehicles collided head-on. **1932** *Economist* 20 Feb. 400/2 At the same time, a head-on collision between the chamber and the Senate is an unusual occurrence. **1952** *Manch. Guardian Weekly* 15 May 2 The General will here run head-on into the irony of the destiny he has chosen. **1954** J. STEINBECK *Sweet Thursday* 166 Despair and humour crashed head-on in Doc. **1957** *Economist* 9 Nov. 492/1 He is known to be working..to present foreign aid next year in a way that will meet head-on and in good time any indiscriminate renewal of the economy drive. **1958** E. L. MASCALL *Recovery of Unity* i. 4 It shows directly how a head-on theological conflict can arise from the unquestioned assumption of a common premiss. **1961** *Lancet* 5 Aug. 325/1 The Government had a head-on collision with the medical profession in 1956. **1973** *Listener* 14 June 790/3 The coming of the Renaissance..was a head-on collision with the medieval system.

headphone (he·dfōun). [f. HEAD *sb.* + *phone* of TELEPHONE (see *-PHONE).] = *ear-phone* (*a*) and (*c*), s.v. *EAR *sb.*[1] 16.

1914 *Work* 7 Mar. 506/1 Electrolite Head-phones. *Ibid.* 28 Mar. 576 We..consider our Headphones a perfect treat. **1926** C. SIDGWICK *Sack & Sugar* xxi. 250 At present they only had head-phones. **1926** *Proc. Musical Assoc.* 1926–7 19 Many seem to have the headphones permanently attached to themselves. **1929** *Strand Mag.* Aug. 152 Women whose head-phones appear so appropriate. **1965** *Listener* 20 May 743/2 These side plaits were gathered into nets..in something like our headphone style. **1970** J. EARL *Tuners & Amplifiers* iii. 76 Headphones..can nowadays give a subjective impression of quality, spaciousness and stereo effect equally as good as the best loudspeakers.

headquarter (he·dkwǭːɹtəɹ), *v.* Usu. in *pass.*: to be provided with headquarters.

1903 *Daily Mail* 3 Sept. 4/4 The off-scourings of civilisation which are headquartered in Constantinople. **1958** P. SCOTT *Mark of Warrior* III. 203 We can take it that Blake's headquartered on the ridge. **1963** *Time & Tide* 28 Mar. 24/2 The society is headquartered in Westminster. **1971** W. TUCKER *This Witch* (1972) v. 55 The refugee camps..are really terrorist camps as well. The Al Fatah are headquartered there.

head-quarters. Add: Also **headquarters. 2.** (Earlier and additional examples.)

1780 A. YOUNG *Tour in Ireland* 382 A good line in which to view these objects is..to make Dobbin's inn, at Ballyporeen, the head quarters, and view them from thence. **1809** J. M'MAHON in G. L. Wardle *Charges against Duke of York* 235 Although he has returned to town for the season as his head-quarters, he makes two or three days excursions from it as often as he can. **1834** G. CRABBE Jun. *Life G. Crabbe* ix. 270 Mr. Crabbe, in subsequent years, made Hampstead his head-quarters on his spring visits. **1836** DICKENS *Let.* ?22 Aug. (1965) I. 168, I shall be at head quarters by 12 Wednesday Noon. **1922** E. WALLACE *Crimson Circle* v. 34, I didn't think that head-quarters had much use for private men like you. *Ibid.* vi. 37 We view with consternation the seeming helplessness of police head-quarters to deal with this criminal gang. **1929** A. CHRISTIE *Partners in Crime* xi. 107 The local Inspector of Police had unemotionally arrested the second footman who proved to be a thief well known at headquarters.

head-rail[1]. Add: **3.** Usu. in *pl.* Teeth. *slang.*

1785 GROSE *Dict. Vulg. T., Head rails,* teeth; sea phrase. **1854** 'C. BEDE' *Verdant Green* II. iv. 31 Your head-rails were loosened, wasn't they? **1935** A. J. POLLOCK *Underworld Speaks* 52/1 *Head rails,* the teeth.

headroom. Restrict †*Sc. Obs.* to sense in Dict., and add: **2.** Room above the head; overhead space. Also *fig.*

1851 [see HEAD *sb.* 65 a]. **1876** *Encycl. Brit.* IV. 267/2 Sufficient light, and headroom of at least 48 inches, being provided. **1902** *Eng. Dial. Dict.* s.v. *Head* sb. II, *Head-room,* of ceilings, staircases, etc.: sufficient height overhead. **1908** *Installation News* II. 68/1 Arrangements are being made to provide 15 feet headroom. **1933** *Punch* 18 Oct. 421/1 In the two-decker Underground trains..the upper deck would have less than six feet of head-room. **1958** *Economist* 20 Dec. 1103/2 Greater 'headroom' to make..increases in passenger fares..was the main object of the application by the Transport Commission. **1960** *Farmer & Stockbreeder* 15 Mar. 157, 14 ft. clear headroom.

head-rope. Add: **4.** Also for other animals.

1936 P. FLEMING *News from Tartary* 328 The camels' headropes were fixed not to nose-pegs but to gaily decorated halters. **1957** P. KEMP *Mine were of Trouble* viii. 137 Hold on to the mules' head-ropes!

headstock. Add: **1. g.** The horizontal end members in the under-frame of a railway carriage or truck.

1888 *Lockwood's Dict. Mech. Engin., Headstock.* (2) The end timbers in the under frame of a railway truck. **1928** *Daily Express* 29 Dec. 9/4 The interval between headstocks of coaches should be as small..as practicable for necessary freedom of movement.

h. (See quot.[1])

1927 T. WOODHOUSE *Artificial Silk* 108 Occasionally both reeds are placed on what is termed a dividing head or headstock. *Ibid.* 112 Fig. 50 illustrates the delivery or headstock end with the weaver's beam partially filled with the sheet of sized and dried threads.

headward. Add: **C.** adj. *headward erosion*: erosion of a stream at its head, in such a way that the length of the stream is increased.

1916 H. F. CLELAND *Geol.* iv. 99 A valley is lengthened at its upper end and is cut back by the water which flows in at its head... This is called headward erosion. **1922** C. A. Cotton *Geomorphol. N.Z.* I. 72 They [*sc.* insequent streams] rapidly eat their way back into the interfluves by headward erosion. **1937** Wooldridge & Morgan *Physical Basis Geogr.* xii. 162 With the progress of time it [*sc.* down-cutting] proceeds backwards, *i.e.* headwards, from the lowest point. The grading of a stream course thus involves headward erosion. **1944** A. Holmes *Princ. Physical Geol.* II. x. 153 The torrent tract..evolves into the valley tract, and each gradually migrates inland as the source continues to recede by headward erosion. **1965** F. J. Monkhouse *Dict. Geogr.* 154/2 *Headward erosion*, the cutting back upstream of a valley above its original source by rainwash, gullying and spring-sapping. **1968** [see *dissection 6*].

head-water. Add: **1. b.** *ellipt.* = *head-water-mark.*
1908 *Westm. Gaz.* 29 Apr. 6/3 The Thames has risen rapidly during the night, and is now 2 ft. 6 in. above head-water at Windsor.

headway. Add: **6.** The interval of time or the distance between two consecutive trains, trams, buses, etc., running on the same route and in the same direction. orig. *U.S.*
1895 in *Funk's Standard Dict.* **1900** *Jrnl. Soc. Arts* 2 Mar. 315/1 The headway between the carriers is fixed say at ten or twenty seconds. **1930** *Oxford Times* 21 Mar. 17/4, I think Route 2 should have a six-minute headway between 8 a.m. and 10.30 a.m. *Ibid.*, Until the headway was closed to 15 minutes in the morning, and 12 minutes in the afternoon, serious inadequacy existed. **1970** *Courier-Mail* (Brisbane) 17 Dec. 1/3 The 30-minute headway between each bus reduced to a 50-minute headway. **1971** *Mod. Railways* May 193/3 Page 105 of the March *Modern Railways* indicates a new 17.11 Paddington to Bristol, first stop Didcot, thus providing a second 3 min headway to Didcot East Junction with the existing 17.15 to Worcester.

headwear (he·dwēəɹ). [f. Head *sb.* + Wear *sb.* 3.] = Head-gear 1.
1896 J. C. Harris *Sister Jane* 39 Gi' me my bonnet. It's all the head-wear I've got left. **1900** *Daily News* 27 Mar. 4/2 They were most easily classified by their headwear. Caps, bowlers, and felt hats were there in almost equal proportions. **1904** W. M. Gallichan *Fishing & Trav. Spain* vii. 76 He was much interested in my wife's hat. Such headwear had never been seen in Coria. **1937** *Discovery* Apr. 106/2 The head wear is the Kalpak, a tall cap, not unlike a busby in principle. **1939–40** *Army & Navy Stores Catal.* p. xxxviii/5 Headwear, Naval, Military and Tropical.

headwork. Add: **1.** (Earlier example.)
1837 Dickens *Pickw.* liv. 587 How the blazes you can stand the head-work you do, is a mystery to me.
b. The practice of carrying loads on the head.
1840 R. H. Dana *2 Yrs. bef. Mast* xiv, For we soon found that..'head-work' was the only system for California.
c. Skill in games and sports.
1851 J. Pycroft *Cricket Field* ii. 22 There is a deal of head-work in bowling. **1898** B. J. Angle in W. A. Morgan *'House' on Sport* 42 His quickness of foot and tricky head-work quickly demoralised the majority of his opponents. **1898** K. S. Ranjitsinhji *With Stoddart's Team* (ed. 3) iv. 84 The English bowling..aimed at steadiness, rather than head work and sting. **1958** F. C. Avis *Boxing Ref. Dict.* 52 *Headwork*, craftiness in boxing.
3. *pl.* **a.** Apparatus for controlling the flow of water in a river or canal. **b.** (See quot. 1905.)
1891 *Scribner's Mag.* X. 468 The river flowing between firm banks, could be permanently controlled by head-works of masonry. **1903** *Sci. Amer. Suppl.* 10 Jan. 22597/3 Headworks can be placed more easily along the banks of smaller streams, or dams built across their beds, raising and controlling the waters. **1905** *Terms Forestry & Logging* 40 *Headworks*, a platform or raft, with windlass or capstan, which is attached to the front of a log raft or boom of logs, for warping, kedging, or winding it through lakes and still water, by hand or horse power. **1963** *Times* 19 Apr. 14/6 Smoothly it flowed, the head-works of a carefully planned series of canals. **1971** N. Smith *Hist. Dams* i. 10 The headworks of the irrigation canal.

heady, *a.* Add: **2.** Also, that affects or turns the head; that turns one giddy.
1898 *Atlantic Monthly* Apr. 501/1 He would sit on a heady scaffold. **1924** A. I. I. Finch in G. I. Finch *Making of Mountaineer* xii. 177 Up and down we seemed to go, and once round a little natural balcony that hung out over space but proved not in the least heady.
c. Headachy (cf. *head *sb.* 1 d).
1872 Geo. Eliot *Let.* 4 June (1956) V. 277 George, being a little heady, and unable to occupy his mornings with writing, is going tomorrow to see the aquarium. **1934** *Air Rev.* Nov. 60, I was feeling a little 'heady' as..a good sea was running, and I was making an effort to eat some nice fat, greasy beef. **1955** E. Hillary *High Adventure* 69 In the morning I felt thick and heady and a sharp cough rasped my sore throat. **1965** 'J. Lymington' *Green Drift* iv. 63, I was feeling very heady and tired. *Ibid.* 64 'Do you still have this heady feeling?' 'Yes, I think it must be the heat. Thundery.'

heah. A representation of a *colloq.* pronunciation of Here *adv.* Freq. used in Black English.
1927 A. P. Randolph in A. Dundes *Mother Wit* (1973) 200 Bin heah too long. **1937** C. Himes *Black on Black* (1973) 139 A nickel victrola in the rear blared a husky, negroid bellow: 'Anybody heah wanna buy.' **1970** R. D. Abrahams *Positively Black* p. ix, Everybody 'round heah is talkin' 'bout 'ligion, gittin' happy an' shoutin'. **1973** *Black World* June 61 Yo ole man in heah?

healder (hī·ldəɹ). [f. Heald + -er¹.] An operative who draws the warp yarn through the eyes of a heald. So **hea·lding** *vbl. sb.*
1881 *Instr. Census Clerks* (1885) 64 Woollen cloth manufacture... Healder. **1888** R. Beaumont *Woollen & Worsted Cloth Manuf.* 138 After healding the work of loom mounting is comparatively simple, the only operation requiring attention being that of sleying.

health, *sb.* Add: **1. b.** Colloq. phr. *for one's health*, used esp. in neg. contexts or with negative implication, e.g. *to be not doing* (something) *for one's health*: to have a serious purpose in doing something, to be doing something for one's material advantage. orig. *U.S.*
1887 G. H. Devol *Forty Yrs. a Gambler* 133 We called it ours, for we had fitted it up just to suit us; and for fear someone would use it when we were out traveling for our health, we paid for it all the time. **1900** *Congress. Rec.* 5 Feb. 1520/2, I am not making this speech for fun, nor for my health, nor as an oratorical exercise. **1900** J. K. Jerome *Three Men on Bummel* iv. 96 What d'ye think I'm running this shop for—my health? **1909** *Westm. Gaz.* 5 July 2/2 We..doubt if it can really be said that Tariff Reformers are 'in it for their health' (to use a very expressive Americanism) or anybody else's. **1914** Wodehouse *Man Upstairs* 229 What is it that makes men do perilous deeds? Why does a man go over Niagara Falls in a barrel? Not for his health. **1927** G. K. Chesterton *Coll. Poems* 90 He will learn..Whether the Health Ministry Are in it for their health. **1944** L. A. G. Strong *Director* 31 I'm not in this job for my health, any more than you are. **1955** L. P. Hartley *Perfect Woman* ix. 86 You ought to have sold several hundredweight, at least, or several tons if you want to persuade the Inspector that you're not in business for your health.
8. a. *health-card, certificate.* **b.** *health-biscuit; health-screening, -seeker* (examples). **d. health camp** *N.Z.*, a camp open (for exercise, outdoor life, etc.) to children below the average in physique, etc.; **health centre** (cf. *centre *sb.* 6 a), a local headquarters of medical services, *spec.*, a local centre for a group practice; **health club**, an establishment where one can do exercises, have massage, etc.; **health farm** orig. *U.S.*, a place to which people resort in the hope of improving their health; **health food**, food chosen for its dietary or health-giving properties; **health insurance**, insurance against financial loss through illness; **health physics**, that branch of radiology which is concerned with the health of those working with radioactive material; **health salt**, freq. in *pl.*, name given to a number of salts, sold under various brand-names, obtained from or mixed with mineral water or other beverages; **health service**, name given generally or specifically to the aggregate of public (as opposed to private) medical facilities available to members of a community; **health visitor**, a specially trained nurse concerned with the welfare of sick or old people, expectant mothers, etc., in their homes.
1905 *Daily Chron.* 25 Apr. 4/5 An uneatable health biscuit..stood by his bedside. **1925** *N.Z. Educ. Gaz.* 1 May 68/1 The teachers decided to hold a 'health camp' for the twenty-six children. **1963** *Evening Post* (Wellington) 26 Oct., A woman doctor whose name will always be remembered in New Zealand as the pioneer of the health camp movement, Dr. Elizabeth Catherine Gunn, M.B.E., died in Wellington today. **1940** Auden *Another Time* 96 And his Health-card shows he was once in hospital but left it cured. **1970** T. Lilley *Projects Section* x. 123 You will not..have inoculations to bring your Health Card up to date. **1916** *Public Health Nurse Q.* (U.S.) Jan. 27 Historically the first Health Center started under that name was begun by the New York Health Committee in 1913. *Ibid.* 33 This Health Center..illustrates two fundamental principles, namely a definite area is selected for the field of operation; [etc.]. **1918** *Lancet* 29 June 922/2 With the removal of the medical officer of health from the jurisdiction of the borough council that official will need a new office in the town, with laboratories, museum, library, and lecture hall. This I call for want of a better title the future 'Health Centre' of the borough. **1934** T. S. Eliot *Rock* ii. 72 Libraries and health centres and milk for the children. **1968** *Brit. Med. Bull.* XXIV. 198/1 With the era of the medical team and health centre upon us, the necessity for adequate documentation already exists. **1972** *Times* 22 Feb. 3/1 Some 3,000 family doctors will be practising from more than 500 health centres by the end of 1974. **1938** F. G. Hobson *Med. Pract. Residential Schools* p. xv (*heading*) Organization of medical services. A. Health Certificates. B. Records. **1960** J. Betjeman *Summoned by Bells* vii. 66 My health certificate, photographs of home. **1961** *Economist* 27 May 872/1 The active gymnasia ('health clubs') are much of a type. **1962** *Which?* Oct. 303/1 These health clubs, which have been fashionable in the United States for many years, have appeared in this country over the last three. **1964** S. Bellow *Herzog* (1965) 84 Herzog had met him in the steam bath at Postl's Health Club. **1970** L. Sanders *Anderson Tapes* xliv. 121 We were in the steam room of that health club the Doc's got on West Forty-eighth Street. **1927** E. Hemingway *Men without Women* (1928) 115 Jack started training at Danny Hogan's health farm over in Jersey. **1928** Wodehouse *Money for Nothing* ix. 196 What if that health-farm was a mere blind for more dastardly work? **1966** G. B. Mair *Kisses from Satan* v. 55 He was going to enter one of the most fashionable health farms in the world. **1969** *Guardian* 18 Aug. 7/5 Their health farm..is the place where they go once a year ..and..lose 10 lb. or 15 lb. in two weeks. **1882** W. D. Howells *Mod. Instance* xxviii, I put the camp on a health-food basis. **1884** E. W. Nye *Baled Hay* 75, I have had occasion to thoroughly investigate the subject of so-called health food, such as gruels, beef tea. **1939** 'G. Orwell' *Coming up for Air* iv. v. 261 Health-food cranks. **1962** J. B. Priestley *Margin Released* i. iii. 26 Another shop, specialising in health foods, had a line in mashed dates and coconut. **1965** *Observer* 18 Apr. 45/2 Health food shops, where everything is free range, unsprayed, naturally fertilized. **1972** *New York* 8 May 49 *Health Food.* As used to describe stores, this term has taken on a generic meaning, encompassing everything from organic, natural, and specialized diet foods to whole wheat and other products mass-produced and refined. **1901** *Index-Catal. Library Surg.-General's Office U.S. Army* Ser. 2. VI. 849/1 (*heading*) Health-insurance. **1911** *Act 1 & 2 Geo. V* c. 55. 1 (*heading*) National Health insurance...All persons so insured..shall be entitled..to the benefits in respect of health insurance and prevention of sickness. **1916** *Machinery* July 1018/1 Booklet entitled 'Health Insurance'. **1941** J. S. Huxley *Uniqueness of Man* p. ix, Subsidized housing..health insurance..free education..are all symptoms of..change. **1946** R. S. Stone in *Proc. Amer. Philos. Soc.* XC. 13/2 The term 'Health-Physics' has been used..to define that field in which physical methods are used to determine the existence of hazards to the health of personnel. *Ibid.* 16/1 The instrument development problems that faced the Health-Physics section were numerous. **1961** *Engineering* 26 May 734/2 The syllabus will include lectures..with..instruction in health physics. **1900** *Confectioners' Union Hand-bk.* 169 Gums, jellies, lozenges..health salt, etc. **1921** D. H. Lawrence *Sea & Sardinia* 46 Like a health-salts..advertisement. **1962** A. Nisbett *Technique Sound Studio* x. 177 Health salts do fine for fizzy drinks. **1966** *New Scientist* 1 Dec. 499/2 Whether Britain goes ahead with an intensive programme of health screening is a decision for the politicians. **1968** *Daily Tel.* 22 Nov. 25 A £250,000 health-screening centre aided by computers in a new building..adjoining Harley Street. **1832** *Chambers's Edinb. Jrnl.* I. 113/2 When a health-seeker takes a walk, he keeps his coat wide open. **1953** Dylan Thomas *Under Milk Wood* (1954) 23 There is little to attract the hillclimber, the healthseeker, [etc.]. **1935** *Economist* 7 Sept. 456/2 Twenty-five years ago only the germ of our present health services existed, in the form of the old Poor Law and a rudimentary Old-Age Pension scheme. **1938** B. Webb *Let.* Jan. in K. Martin *Editor* (1968) iii. 73 The organisation of a public Health Service. **1958** *Times Lit. Suppl.* 12 Dec. 717/2 The Health Service and the welfare state...have brought appalling drabness into the doctor's life. **1901** Health visitor [see Visitor 1 c]. **1905** *Westm. Gaz.* 3 Apr. 12/2 An audience of health visitors, district visitors, charity organisation visitors. **1917** *New Witness* 28 June 202/1 The bare idea that a Health Visitor should attempt to force her way into a Frenchwoman's house would be regarded with horror. **1965** *Listener* 30 Sept. 483/1, I would think it most important by means of..the health visitor..to make assessments of the food which they require. **1970** G. Greer *Female Eunuch* 19 The revolutionary woman must know her enemies, the doctors, psychiatrists, health visitors.

healthy, *a.* Add: **2. b.** In ironical use.
1831 S. Smith *Life & Writings Major J. Downing* 149 Major Eaton, it won't be healthy for you to come on to these steps to-night. **1902** C. Hyne *Mr. Horrocks Purser* 251, I want to impress on them that they'll find it more healthy not to try for more. **1916** 'Boyd Cable' *Action Front* 179 'Additional artillery support would be useful a—a—a.' 'Sounds healthy, don't it?' said the sergeant reflectively.
4. *healthy-minded* adj. (earlier example); *healthy-mindedness.*
a **1882** H. James *Lit. Remains* (1885) 117 In a pluralistic philosophy the healthy-minded moralist will always feel himself at home. *Ibid.*, The feeling of *action*..makes us turn a deaf ear to the thought of *being*; and this deafness and insensibility may be said to form an integral part of what in popular phrase is known as 'healthy-mindedness'. **1906** *Westm. Gaz.* 25 Jan. 12/2 The spirit which animated Japan was chivalry and healthy-mindedness. **1931** *Times Lit. Suppl.* 19 Feb. 122/1 So well supplied is the Victorian era with names standing for stability, sanity and healthy-mindedness.

heap, *sb.* Add: **1. e.** Usually preceded by a defining word: a slovenly woman. *colloq.* (orig. *dial.*).
1806 A. Douglas *Poems* 125 She jaw'd them, misca'd them For clashin' claikin' haips. **1810** J. Cock *Simple Strains* II. 91 Foul fa' the sly bewitchin' heap Cou'd turn hersel' in ony shape. **1922** Joyce *Ulysses* 300 The fat heap he married is a nice old phenomenon with a back on her like a ball-eye. **1957** J. Frame *Owls do Cry* 106, I may be *forced* to [sell-out], if that lazy heap doesn't help me.
f. A battered old motor vehicle. *colloq.* (orig. *U.S.*).
1926 *Clues* Nov. 161/1 *Heap*, automobile. **1928** R. J. Tasker *Grimhaven* iii. 28 Once in a while some fellow who really did own a good car would come up to be topped, but, as a rule, I've noticed that kind never have much to say about their heaps. **1935** R. Chandler *Killer in Rain* (1964) 7, I got out of the Chrysler... I went back to my heap. **1951** J. Kerouac *On Road* (1958) 79 He gunned the heap to eighty. **1959** J. Braine *Vodi* xiv. 190 Bought two old heaps today. Just junk really, a '28

Chrysler and a '27 Essex. **1967** A. Hunter *Gently Continental* xi. 166 Stody too has driven away in his modest heap. **1969** C. F. Burke *God is Beautiful, Man* (1970) 56 You will be like a guy who paid no attention to his heap and it broke down in the traffic.
4. d. In the representation of the speech of North American Indians used adverbially and as quasi-adj.: very, very much, a great deal.
1832 W. Irving *Jrnl.* (1919) III. 180 'Look at these Delawares,' say the Osages, 'dey got short legs—no can run—must stand and fight a great heap.' **1848** *Blackw. Mag.* LXIII. 719 An Indian is always a 'heap' hungry or thirsty—loves a 'heap'—is a 'heap' brave—in fact, 'heap' is tantamount to very much. **1850** 'M. Tensas' *Louisiana 'Swamp Doctor'* 42 Whoop! whiskey lour! Injun big man, drunk heap. **1867** *Harper's Mag.* July 137/1 Disturb the game and you make the Indian 'heap big mad'. **1872** 'Mark Twain' *Roughing It* (1873) xxxix. 276 'Heap' is 'Injun-English' for 'very much'. **1902** —— in *Harper's Mag.* Jan. 270/2 Billy explained..'she heap much hungry'. **1958** B. Cerf *Shake well before Using* 17 President Coolidge posed later in the regalia of a heap-big chief. **1968** Mrs. L. B. Johnson *White House Diary* 21 June (1970) 688 His favorites among the presents were.. the gift wrappings, or maybe the rubber canoe that said 'Heap Big Indian Lyn'.
e. *a heap sight* (U.S. *dial.* and *colloq.*): see Heap *sb.* 4 c and Sight *sb.*¹ 2.
1874 E. Eggleston *Circuit Rider* i. 14 He 'lows there was a heap sight more corn. **1888** G. W. Cable *Bonaventure* 49 He's..a heap sight happier than us. **1906** *Smart Set* June 107/1, I care a heap sight too much for Ummy to let him go through what I know 's comin'. **1911** R. D. Saunders *Col. Todhunter* x. 152 You're a heap-sight smarter man than I gave you credit for bein'.
5. e. Also, *to knock all of a heap.*
1898 W. J. Locke *Idols* xiii, It knocked the prosecution all of a heap. **1928** *Manch. Guardian Weekly* 7 Sept. 183/3 Its owner's anxiety to knock the critics all of a heap.

heaping, *ppl. a.* *U.S.* [f. Heap *v.*] Of a spoonful: heaped. Also *fig.* mounting up.
1838 *Congress. Globe* June 470/2 App., The amount of money..is a very high and heaping price. **1868** L. M. Alcott *Lit. Women* xi, Amy..took a heaping spoonful, choked..and left the table precipitately. **1908** *Smart Set* June 25/1 Aunt Natica waddled off..to fetch Thorndyke a heaping portion of the *dulce.* **1965** C. D. Eby *Siege of Alcázar* (1966) xi. 221 He had just been served a heaping ration of rice and beans, a special treat.

hear, *v.* Add: **3. e.** *to like to hear oneself speak, talk* (and similar phrases): to be fond of talking; *to hear oneself think*: usu. in neg. contexts, not to be able to think because there is too much noise going on.
1592 Shakes. *Rom. & Jul.* ii. iv. 156 Pray what sawcie Merchant was this?..*Rom.*: A Gentleman Nurse that loves to heare himselfe talke. **1781** George III *Lett.* (1927) V. 304 Considering the great love modern Orators have of hearing themselves speak. **1920** R. Macaulay *Potterism* vi. iii. 226, I wish everyone would shut up, so that we could hear ourselves think. **1927** H. T. Lowe-Porter tr. *Mann's Magic Mountain* II. vii. 779 You won't be able to hear yourselves think. **1934** J. E. Mansion *Harrap's French & Eng. Dict.* 272/1 *Discoureur*... *C'est un grand d.*, he likes to hear himself talk. **1962** *New Yorker* 12 May 38/1 You have to wear earplugs to hear yourself think.
4. b. *to hear out*: also, to distinguish (the sounds of something heard).
1922 R. S. Woodworth *Psychol.* x. 230 By careful attention and training we can 'hear out' the separate overtones from the total blend.
7. c. *to hear to*, to listen to, to hear of. *U.S.*
1833 H. Barnard in *Maryland Hist. Mag.* (1918) XIII. 379, I made a move to depart—but they would not hear to that. **1869** Mrs. Stowe *Oldtown Folks* xx. 243 She has her own ways and doings, and she won't hear to reason. **1915** E. Poole *Harbor* 202 When I tried at last to turn our talk to our affairs at home, at first she would not hear to it.
11. *to hear from*: (also, pregnantly) to receive a reprimand from.
1907 *Munsey's Mag.* Dec. 307/1 If those louts up at the castle neglected to have dinner ready.., they would hear from him... If they didn't [spring at his word] they always heard from him.

hearing, *vbl. sb.* Add: **7. hearing aid**, a sound-amplifier for the hard of hearing.
1922 *Lancet* 11 Mar. 462/2 These electrical instruments should go far towards destroying the too general prejudice against the use of hearing aids. **1950** *Lancet* 11 Nov. 532/2 Practical courses..on audiometry and hearing-aids. **1951** *Consumer Reports* Jan. 13/1 Hearing aids and batteries are..supplied without charge. **1960** 'H. Carmichael' *Seeds of Hate* ii. 18 A grey-haired woman who wore a hearing aid. **1969** B. Patten *Notes to Hurrying Man* 57 Much later on in life I wear my hearing-aid.

hearse, *sb.* Add: *hearse-driver.*
1829 G. Griffin *Collegians* III. xxxiii. 60 ''Twill be a great funeral,' said the hearse-driver. **1841** J. S. Buckingham *Amer.* II. 322 The Whig authorities of New-Haven, have removed Mr. Willoughby..from the place of hearse-driver, and appointed another person in his place. **1851** H. Melville *Moby Dick* III. xiii. 90 The old hearse-driver, he must have been. **1922** Joyce *Ulysses* 590 Sober hearse-drivers a speciality.

hearsy (hɔ̄·ɹsi), *a.* [f. Hearse *sb.* 8 + -y¹.] Resembling or characteristic of a hearse; funereal.
1901 'A. Hope' *Tristram of Blent* v. 57 Mr. Gainsborough was obviously a man who would not waste his chance of a funeral;..it would need startling measures to keep him from a funeral. 'I hate hearsey people,' grumbled Harry as he threw the letter down. **1908** Hardy *Dynasts* III. vi. vi. 476 Full-clothed in black, with nodding hearsy plumes.

heart, *sb.* Add: **1. c.** A diseased or disordered heart: often with defining word; as *athletic heart*, simple hypertrophy of the heart with no disease of the valves; *fatty heart* (see Fatty *a.* 5); *smoker's heart* (see 1 b).
1862 W. H. Walshe *Pract. Treat. Dis. Heart* (ed. 3) II. 320 Patients..often express themselves, 'they have a heart', (the mildest form of cardiac paræsthesia). **1871** Da Costa in *Amer. Jrnl. Med. Sci.* LXI. 17, I noticed cases of a peculiar form of functional disorder of the heart, to which I gave the name of irritable heart. **1886** Fagge & Pye-Smith *Princ. Med.* II. 41 Rather more than a century ago Haller described the 'hairy' heart as occurring especially in bold and adventurous men. **1888** Smoker's heart [in Dict., sense 1 b]. **1902** *Daily Chron.* 3 Nov. 8/4 [He] has been forbidden to row again.. owing to his having developed 'a heart'. **1908** *Westm. Gaz.* 29 Oct. 14/1 [He] failed to qualify before the Medical Board of the police on the ground that he had an 'athletic heart'. **1929** E. Bowen *Joining Charles* 125 Cottesby the cow-herd, a greyish-faced man, had 'a heart'. **1965** 'W. Haggard' *Hard Sell* i. 4 He's got a heart, by the way, and I'm afraid this might finish him. **1971** *Current Slang* (Univ. S. Dakota) VI. 6 Slow down or you'll give me a heart.
10. d. *to have a heart* colloq., to be merciful. Freq. in *imp.*: come off it, be reasonable, show some pity!
1917 Wodehouse & Bolton (*play title*) Have a heart. **1928** *Observer* 1 Jan. 4 We only sigh for old delights, and in homely phrase beseech him..to 'have a heart'. **1936** P. Bottome *Level Crossing* xviii. 225 'Have a heart!' Nelly told her crossly. **1950** W. Stevens *Let.* 28 June (1967) 683 If you use the things..I shall have to go out and drown myself... Have a heart. **1967** J. B. Priestley *It's Old Country* xix. 209 'You haven't made any plans for him, have you?' 'How could I?..Have a heart!' **1970** *New Yorker* 12 Sept. 50/3 Spare us a reefer, beautiful. Have a heart.
24. b. *Hearts*, a card-game for three or four players, similar in principle to whist but without partners or a trump suit: the object of the game is to avoid taking a trick containing a Heart or the Queen of Spades.
1886 'The Major' (*title*) The game of hearts. Rules of the game. How to play hearts. **1907** *Yesterday's Shopping* (1969) 361/1 *Invitation Cards*... At Home, Progressive Hearts —— o'clock. R.S.V.P. **1930** W. S. Maugham *Writer's Notebk.* (1949) 231 In the evening the guests collect and play hearts for infinitesimal sums. **1943** 'C. Dickson' *She died a Lady* viii. 67 You don't call playing bridge or hearts on Saturday night a very Bohemian sort of life, do you? **1946** A. Christie *Hollow* viii. 75 Do you think Hearts or Bridge or Rummy? **1959** J. D. Salinger in *New Yorker* 6 June 101 At all card games, without exception—Go Fish, poker, cassino, hearts, old maid..—he was absolutely intolerable.
51. c. *heart-to-heart*: used to denote conversation, discussion, etc. of real frankness and sincerity; usually *attrib.* but also *absol.* as *sb.*
1867 *Mission Life* 1 Mar. 190 The visitation of an Australian Bishop..is a hand-to-hand and heart-to-heart visit to each Clergyman, and to his people with him. **1894** *Advance* 11 Oct., A kind of public religious 'orphanage', where no true heart-to-heart 'mothering'..was possible. **1902** A. H. Lewis *Wolfville Days* xi. 152 He don't own no real business to transact; he's out to have a heart-to-heart interview with the great Southwest. **1902** Kipling *Traffics & Discov.* (1904) 22 He began by a Lydia Pinkham heart-to-heart talk about my health. **1906** *Daily Chron.* 5 Mar. 6/4 A heart-to-heart discussion of the solar plexus and its part in the emotional economy of man. **1910** S. E. White *Rules of Game* v. xvi. 444 Let's have a heart-to-heart, and find out how we stand. **1918** E. M. Roberts *Flying Fighter* 201 After a heart-to-heart talk, I induced him to let me remain in the Flying Service. **1925** Wodehouse *Carry on, Jeeves* ix. 221 He and Jeeves had had a heart-to-heart chat in the kitchen. **1934** J. E. Neale *Queen Eliz.* ii. 31 Parry came back to have heart-to-heart talks with Mistress Ashley and to probe Elizabeth's mind. **1948** A. Waugh *Unclouded Summer* xv. 252, I have the girls up there in the evenings for 'heart-to-hearts'. **1951** L. MacNeice tr. *Goethe's Faust* 50 All this needs a little explaining And will keep till our next heart-to-heart. **1955** W. Gaddis *Recognitions* I. v. 180 Baby, I just make a few notes on them and write these heart-to-heart confessions.
55. a. (*a*) *heart attack, condition, failure* (examples), *rate, -shape* (earlier example), *-strain, -tube, -valve* (examples); (*b*) *heart-burst, -lift* (so *-lifter*), *-springs, -wound; heart-sinking* (earlier example).
1935 D. L. Sayers *Gaudy Night* xxi. 444 She's had rather a nasty heart-attack, but she's better now. **1845** P. J. Bailey *Festus* (ed. 2) 258 Like a horse Put to his heart-burst speed, sobbing up hill. **1896** A. Morrison *Child of Jago* xiii. 134 Dicky..had been afflicted to heart-burst by his father's dodging and running. **1946** *Mod. Lang. Notes* Nov. 442 Heart condition. **1958** *Listener* 13 Nov. 778/2 Before cleaning a car..be certain you haven't a heart condition. **1971** D. O'Connor *Eye of Eagle* viii. 53 He has a heart condition—nothing very serious. **1894** 'O. Henry' *Compl. Wks.* (1928) 797 'Read this,' he said, 'here is proof that Marie Cusheau died of heart failure.'
1906 *Lancet* 13 Jan. 96/2 Dr. C. Bolton..read a paper entitled 'The Treatment of Heart Failure in Diphtheria'. **1960** I. A. Stanton *Dict. for Med. Secretaries* 68/1 Occasionally heart failure denotes a sudden cessation of heart action, but generally it merely means insufficient circulation. **1893** 'Mark Twain' in *Cosmopolitan* Nov. 61/2 Oh, the heart-lift that was in those words! **1967** 'La Meri' *Sp. Dancing* (ed. 2) 7 Yet who can reflect in the immutable phrase the heartlift in watching emotion in motion? **1901** Kipling *Kim* x. 260 You will find one small silver amulet..a heart-lifter. **1959** *New Statesman* 25 Apr. 576/3 The heart-lifter that I chanced to hear was well up to her standard. **1936** *Discovery* 291/2 Adrenalin, by increasing the heart rate..facilitates the passage of the current. **1961** *Lancet* 22 July 190/1 An increase in heart-rate may also increase potassium efflux. **1842** W. Howitt *Rural & Dom. Life Germany* v. 62 The gingerbread was all made up into heart-shapes. **1743** D. Brainerd *Let.* 30 Apr. in J. Edwards *Life D. B.* (1765) 265 There seems to be little of the special workings of the divine Spirit among them yet; which gives me many a heart-sinking hour. **1903** B. Harraden *Kath. Frensham* xviii. 278 She, with ..perseverance, dug a hole in their frozen heart-springs. **1907** Kipling *Bk. of Words* (1928) 36 A people..whose heart-springs go down deep into the fabric. **1906** *Med. Ann.* 241 Heart-strain in growing boys. **1909** *Daily Chron.* 21 Aug. 6/2 Heartstrain and contraction of the joints. **1881** *Trans. Obstetr. Soc. Lond.* XXII. 78 An abnormal amount of tension on the primitive heart-tube. **1932** *Gray's Anat.* (ed. 25) 1437 Heart valves. **1963** *Daily Tel.* 21 Sept. 9/5 (*heading*) Heart valve operation. *Ibid.*, The limited number of heart valve replacement operations so far carried out. **1839** P. J. Bailey *Festus* 269 Her heart-wound. **1902** *Temple Bar* CXXVI. 111 It rained upon his bleeding heart-wound like balm. **1906** *Westm. Gaz.* 3 Aug. 10/2 The faint, fine smell of new-mown grass Stabs like a heart-wound as I pass.
b. *heart-holding, -shaking, -sickening, -swelling, -tearing, -warming, -wringing* adjs.
1897 J. L. Allen *Choir Invisible* xvi. 240 Universal fellowship with seeding grass and breeding herb and every heart-holding creature of the woods. **1913** E. F. Benson *Thorley Weir* iv, Things fairer and more heart-holding. **1907** *Tatler* 22 May 132/2 A heart-shaking tragedy. **1911** Kipling *Diversity of Creatures* (1917) 130 The heart-shaking jests of Decay. **1918** V. Woolf in *Times Lit. Suppl.* 31 Jan. 55/1 Effective and heart-shaking ghost stories. **1945** W. S. Churchill *Victory* (1946) 223 The decision..remained nevertheless a heart-shaking risk. **1820** *Edin. Monthly Rev.* Apr. 449 Can anything be more heart-sickening to such a philanthropist? **1902** *London Mag.* VIII. 432/2 It was heart-sickening, as his great form with its yellow skin and black stripes, as his blazing eyes, his flashing teeth and his outspread claws rose toward us through the air. **1814** Jane Austen *Mansf. Park* III. vi, Her happiness was of a quiet, deep, heart-swelling sort. **1884** W. James *Coll. Ess. & Rev.* (1920) 258 In listening to poetry..we are often surprised at the..heart-swelling and the lachrymal effusion that unexpectedly catch us. **1916** 'Boyd Cable' *Action Front* 149 Thirty-six solid hours of physical stress and heart-tearing strain. **1920** *Glasgow Herald* 21 Oct. 6 The latest phases of the heart-tearing Irish tragedy. **1899** *Daily News* 20 Apr. 5/7 They are a heart-warming cordial. **1966** *Times* (Austral. Suppl.) 28 Mar. p. viii/4 Perth..friendly enough to give..migrants..a heartwarming impression of their new country. **1932** H. Crane *Let.* 12 Apr. (1965) 408 The Mexican singer..is generally shrill but capable of heart-wringing vibrations.
c. *heart-wrung* adj. (earlier and later examples).
1791 Burns *Ae Fond Kiss* in *Wks.* (1871) 294 Deep in heart-wrung tears I'll pledge thee. **1948** C. Day Lewis *Poems 1943–47* 70 One heart-wrung phantom still.. Shadows my noontime still.
56. heart-balm, (*a*) something that soothes a person's emotions; (*b*) *U.S. slang*, alimony; **heart-block** [*block *sb.* 19 d] *Med.* (see quot. 1906); **heart brass**, a brass sepulchral tablet in which a heart is represented (see quot. 1912); **heart-hurry** *Med.*, tachycardia (see also quot. 1897); **heart-line** *Palmistry* = *line of the heart* (Line *sb.*² 8 b); **heart-lung** *attrib.*, involving or consisting of the heart and the lungs, esp. when removed together for physiological experimentation; *heart-lung machine*, a machine to which a patient's blood supply is connected during an operation and which by-passes and takes over the functions of the heart and the lungs; **heart-rot**, a disease which causes decay in the heart of a tree; also, a fungous disease of beetroots etc.; **hearts-and-flowers** orig. *U.S.*, undue sentimentality, cloying sweetness; also *attrib.*; **heart-talk**, a heart-to-heart talk; **heart-throb**, (*a*) lit. a pulsation of the heart; (*b*) *colloq.* (orig. *U.S.*) something or (esp.) someone that thrills the heart, a lover: freq. used of film stars and other entertainers; also *attrib.*; **heart transplant**, an operation in which a heart from one person is transplanted into the body of another; similarly of two animals; a heart so transplanted; also *attrib.* and *fig.*; **heart-worm**, a parasitic nematode worm which infests the hearts of some carnivores, or the disease caused by this worm; also *transf.*
1922 Joyce *Ulysses* 352 There were wounds that wanted healing with heartbalm. **1938** Wodehouse *Summer Moonshine* x. 126 This Miss Prudence Whittaker

issuing this T. P. Vanringham for breach of promise and heart balm. **1903** *Lancet* 22 Aug. 523/1 The jugular pulsations correspond to independent auricular contractions which are not propagated to the ventricles—a state of 'heart-block'. **1906** *Brit. Med. Jrnl.* 27 Oct. 1107/1 The term 'heart-block' is applied to that condition where the stimulus for contraction passing from auricle to ventricle, is stopped or 'blocked' on account of some defect in those muscle fibres. **1966** *Lancet* 31 Dec. 1441/1 Patients who had partial heart block while on P.G.I. therapy alone.. reverted to sinus rhythm. **1971** *Jrnl. Gen. Psychol.* Jan. 13 Magnesium sulphate was superior to sodium amytal and ether. Its main drawback was its tendency to produce heart block. **1907** H. W. Macklin *Brasses of England* viii. 205 The typical form of a heart brass is seen when this device is placed by itself in the midst of a monumental slab. **1912** J. S. M. Ward *Brasses* 80 Heart brasses proper fall into two main divisions: (*a*) plain, sometimes inscribed or with scrolls, (*b*) held by hands, usually coming out of a cloud. **1956** A. C. Bouquet *Church Brasses* vii. 114 There is a large heart brass at Melton Mowbray, Leicestershire. **1891** *Lancet* 18 July 118/2 (*title*) Paroxysmal heart hurry associated with visceral disorders. **1897** *Med. Times & Hosp. Gaz.* XXV. 33/2 By acceleration of the heart or 'heart-hurry', is meant a persistent increase of the pulse above eighty in a woman, well above seventy beats per minute in a man, and above ninety in a child. Heart-hurry is divided into two kinds; they are tachycardia and palpitation. **1893** Beerbohm *Let.* 14 Oct. (1964) 76 He has no heart-line on his right hand. **1894** 'Mark Twain' in *Century Mag.* Feb. 554/2 Wilson began to study Luigi's palm, tracing life lines, heart lines, head lines, and so on. **1956** N. D. Ford *Life in your Hands* v. 40 The Head and Heart lines join in forming one straight line... The Fate line begins well clear of the Heart line. **1912** *Jrnl. Physiol.* XLV. 213 The heart-lung preparation should serve therefore for investigations on the normal gaseous metabolism of the heart. *Ibid.* 214 The apparatus consisted of the heart-lung circulation apparatus as described by Knowlton and Starling, and of a respiration apparatus. **1925** *Ibid.* LX. 103 (*title*) A closed circuit heart lung preparation. **1945** *Amer. Jrnl. Physiol.* CXLIV. 191 No details of the experiments on the heart-lung preparation need be presented. *Ibid.*, The results of the thirty-two heart lung experiments can be summarized as follows. **1959** *Daily Tel.* 24 Apr. 13/3 In the party is Dr. Denis Melrose, inventor of the heart-lung machine which bears his name. This makes possible the by-passing of heart and lungs, and enables the operating surgeon to work on a heart which is bloodless, clear and stopped. **1961** *Lancet* 22 July 187/1 (*heading*) Variable atrial venting for the Melrose heart-lung machine. **1968** J. H. Burn *Lect. Notes Pharmacol.* (ed. 9) 37 Another way of demonstrating the action of ouabain on the ventricular contraction is in the heart-lung preparation of the dog. **1847** J. Brown *Forester* v. 193 That disease, now so prevalent among our larch plantations, generally termed the *heart-rot*—or, as some writers term it, *dry-rot*. **1882** *Encycl. Brit.* XIV. 311/2 A far more formidable enemy [of larches] is the disease known as the 'heart-rot'. **1909** *Cent. Dict.* Suppl. 571/2 *Heart-rot*..of beets. **1919** W. E. Hiley *Fungal Dis. Common Larch* v. 80 Heart-rot of trees is caused by fungi which grow saprophytically on the dead wood. **1945** *New Biol.* I. 52 Heart rot of swedes. **1955** Auden *Shield of Achilles* i. 19 An oak with heart-rot. **1968** *Gloss. Terms Timber Preservation* (*B.S.I.*) 10 *Heart rot*, a type of decay characteristically confined to the heart-wood. [**1908** A. Woollcott *Lett.* (1946) 13 Taking dinner with the mother of the girl I hope to marry some day, and she played 'Hearts and Flowers' for me.] **1942** Berrey & Van den Bark *Amer. Thes. Slang* § 265.1 *Sentimentality*, hearts and flowers. **1964** *Times* 16 Apr. 6/7 We are nearly betrayed into a hearts-and-flowers ending in domestic compromise. **1967** *Listener* 11 May 626/2 Hearts-and-flowers confrontations between..pop singer..and a girl friend. **1912** F. M. Hueffer *Panel* i. ii. 31, I want a regular—what you might call—heart-talk with Miss Delamere. **1839** P. J. Bailey *Festus* 62 We should count time by heart-throbs. **1846** [in Dict., sense 55 a]. **1908** *Modern Song Favorites: High Voices* 2 (*title*) Heart-Throbs. **1912** J. London *Let.* 19 Nov. (1966) 368 I've not much heart-throb left for my fellow beings. **1914** G. Burgess *Burgess Unabridged* 7 The 'jacket' of the 'latest' fiction.. tells of 'thrills' and 'heart-throbs'. **1926** *Atlantic Monthly* Mar. 390/1 Word has gone out to the writers..that the heart throb is what the reading world now pulsates to. **1928** J. P. McEvoy *Show Girl* (title-p.), Cast... Also.. the Heart-throb Poet. **1930** Wodehouse *Very Good, Jeeves* ix. 227 She has got that way..from a lifetime of writing heart-throb fiction for the masses. **1943** 'A. A. Fair' *Double or Quits* (1949) vii. 72 She's easy on the eyes, but she's a little too anxious to have it understood I'm her heart throb. **1958** G. Mitchell *Spotted Hemlock* ii. 16 He was quite a heart-throb, you know. **1959** D. Du Maurier *Breaking Point* 202 A heart-throb, a lover, someone with wide shoulders and no hips. **1966** *Listener* 23 June 911/2 Rudolph Valentino was the great heart-throb of the silent screen in the nineteen-twenties. **1952** *Surg. Forum 1951* 217 An arterial supply from the host was anastomosed to a pulmonary vein of the heart transplant and an outlet for the left ventricular output of the heart transplant was provided. **1960** *Ibid.* X. 103 Forty-eight puppy heart transplants are reported. **1963** *Surg. Gynecol. & Obstetrics* CXVII. 361/2 If a renal graft fails to function for several days after transplantation, the host can be supported by dialysis. A heart transplant at the present time enjoys no such privilege and must function vigorously immediately. **1967** *Times* 4 Dec. 1/7 The heart transplant operation, the first in the world, took Groote Schuur's surgical team five hours. **1968** *Guardian* 11 Sept. 1/5 Some of the gravest criticisms yet were yesterday levelled against the over-eagerness of heart-transplant surgeons to get hold of donors. **1973** *N.Y. Times Bk. Rev.* 21 Jan. 2 It is a real heart-transplant into English of the great Alexandrian love-poet and voluptuary. **1888** J. S. Stallybrass tr. J. Grimm's *Teut. Mythol.* IV. 1659 Stories of the heart-worm. *Ibid.* 1660 The miser's heart-worm. **1955** W. W. Denlinger *Compl. Boston* 94 Heart worms..in dogs are rare. **1957** *Encycl. Brit.* XVI. 207/2 *Dirofilaria immitis* (cause of heartworm in dogs). **1959** *Listener* 5 Nov. 796/1 The professional intimate, the confidential heart-worm with the hypodermic technique, is one of the horrors of television. **1965** E. J. L. Soulsby *Textbk. Vet. Clin. Path.* I. iv. 100 *Dirofilaria immitis* is the heartworm and is parasitic in the..dog, fox, wolf and various other wild carnivores.

hearth¹. Add: **1. b.** (Later example.)
1845 E. Acton *Mod. Cookery* vii. 191 The hot plates, or *hearths* with which the kitchens of good houses are always furnished.
4. hearth tidy, a pan for containing the ashes that fall from a fireplace.
1920 *Ironmonger* 18 Dec. 95 Saucepans, hearth tidies, curbs, plate racks.
b. hearth-bottom (examples).
1880 *Encycl. Brit.* XIII. 299/2 This is the hearth bottom, formerly made of one or more large slabs of sandstone. **1951** *Good Housek. Home Encycl.* (1956) 269/2 Sunk or hearth-bottom grates, in which the fuel rests on a bed of fire clay.

hearth-rug. Add: Also *attrib.* (*a*) fireside, domestic; (*b*) resembling a hearth-rug.
1901 *Daily Chron.* 31 July 7/2 Turning the hearthrug favourite into the streets is certainly better than shutting it up, slowly to starve in an unoccupied house. **1902** *Ibid.* 2 Sept. 5/2 The sparrow is far too 'fly' a bird for the hearth-rug-bred cat. **1909** *Westm. Gaz.* 20 July 5/1 The inelegant, cumbrous, and shaggy hearth-rug coats.

heartland (hā·ɹtlænd). [f. Heart *sb.* 17 + Land *sb.*] A (usually extensive) central region of homogeneous (geographical, political, industrial, etc.) character. Also *transf.*
1904 H. J. MacKinder in *Geogr. Jrnl.* XXIII. 434 But trans-continental railways are now transmuting the conditions of land-power, and nowhere can they have such effect as in the closed heart-land of Euro-Asia, in vast areas of which neither timber nor accessible stone was available for road-making. **1919** —— *Democratic Ideals & Reality* 96 Taken together, the regions of Arctic and Continental drainage measure nearly a half of Asia and a quarter of Europe, and form a great continuous patch in the north and centre of the continent..inaccessible to navigation from the ocean... Let us call this great region the Heartland of the Continent. **1947** *Landfall* I. 298 We are in the frontier West, the heartland of the American myth. **1949** 'G. Orwell' *Nineteen Eighty-Four* ii. 189 The territory which forms the heartland of each superstate always remains inviolate. **1959** A. J. Toynbee *Hellenism* 2 The matrix of the Indo-European languages, somewhere in the heartland of the Old World. **1966** *New Statesman* 13 May 674/3 Mr Heath..is right to make the attempt even if it means some ill-feeling in the Conservative heartlands. **1968** Powell & Wallis *House of Lords in Middle Ages* viii. 123 King John planned a concerted attack on the French heartland from east and west. **1972** *Observer* 30 July 9/8 Chobham Farm container depot and Midland Cold Storage nestling within a few hundred yards of each other, lie at the very heartland of British trade unionism.

heartwater (hā·ɹtwǭːtəɹ). *Vet.* [f. Heart *sb.* + Water *sb.*: so called from the characteristic accumulation of straw-coloured fluid in the pericardium of the heart.] A febrile disease of sheep, goats and cattle, caused by the virus *Rickettsia* (= *Cowdria*) *ruminantium*, transmitted by the bont tick (or other closely related ticks), and occurring in various parts of Africa, esp. South Africa, and in Madagascar.
1882 S. Heckford *Lady Trader in Transvaal* xiv. 134 The Nell family..swore it had died of what they call here 'heart-water'. **1896** R. Wallace *Farm. Ind. Cape Col.* xx. 380 Heart-water in sheep is another obscure disease of a specific nature, which seems to be unknown in other sheep countries. **1905** *Rep. Brit. Assoc.* 282 An old-time supposition that *A*[*mblyomma*] *hebræum* was associated with a disease called 'heartwater', which had practically put a stop to the farming of sheep and angora goats in several south-eastern districts. **1930** *Discovery* Aug. 265/2 Heartwater in sheep was..caused by the bite of a tick, *Amblyomma hebraeum*. **1949** *Cape Argus* 11 Aug. 3/6 They dealt with heartwater..in the Union.

heart-wood. Add: **2.** The Tasmanian ironwood, *Notelæa ligustrina.*
1889 J. H. Maiden *Useful Native Plants Austral.* 579 The heart-wood yields a very peculiar figure; it is a very fair substitute for Lignum-Vitæ. **1902** G. S. Boulger *Wood* ii. 221 *Notelæa ligustrina*...South-eastern Australasia. Known also as 'Heartwood' in Tasmania...exceedingly hard and close-grained.

hearty, *a.* and *sb.* Add: **C.** *sb.* **3.** At some English universities, used to denote an extrovert who enters heartily into college life and sports; an athletic (as distinguished from an æsthetic) man. Also in more general use (see quot. 1955).
1925 *Weekly Dispatch* 22 Nov. 9/2 The leaders in the sport ['debagging'] are a band of 'hearties' who hail mostly from Magdalen and 'The House'. **1928** in L. MacNeice *Strings are False* (1965) 274 Capell was knocked down in the Broad the other day by a hearty. **1930** *Times Lit. Suppl.* 24 July 610/4 At English universities undergraduates classify themselves into the mutually exclusive categories of 'aesthetes' and 'hearties'. **1934** C. Lambert *Music Ho!* iii. 216 The sheer anger aroused in 'hearties' of the Beachcomber order by such different manifestations of contemporary depression as jazz songs and the poetry of Eliot. **1955** *Times* 12 May 11/4 A hearty has come to mean an oppressively cheerful, muscular and back-slapping personage who is prostrating company. **1959** *News Chron.* 19 Aug. 4/3 There is no trace of the horse-play hearty in his make-up. **1964** C. Mackenzie *My Life & Times* III. iv. 130 To go back to the noise in Trinity quad on that Saturday evening in 1903. 'Oh, these hearties!' I said... From that moment, at first as a term for Trinity men and later more generally, 'hearties' became current.

heat, *sb.* Add: **11. a.** (Further examples.)
1958 *Listener* 30 Oct. 709/2 The heat is being pumped into utterly different quarrels. **1962** *Ibid.* 5 Apr. 587/1 His foreign minister..had set himself the task of taking the heat out of inter-Arab exchanges. **1964** *Ann. Reg. 1963* 100 This merely added more heat to the argument.
c. *U.S. slang.* A state of intoxication caused by alcohol or drugs, esp. in phr. *to have a heat on.*
1912 D. Lowrie *My Life in Prison* vii. 77 A few years ago this dump was full of dope. Every other man y'r met had a heat on, an' lots o' young kids what came here strong an' healthy went out with a habit. **1931** D. Runyon *Guys & Dolls* (1932) ii. 41 The party is going big along toward one o'clock when all of a sudden in comes Handsome Jack Maddigan with half a heat on, and in five minutes he is all over the joint, drinking everything that is offered him.
12. (Later examples.)
1944 'N. Shute' *Pastoral* ix. 206 He wants to get you both off the station on leave till the heat goes off. **1970** E. R. Johnson *God Keepers* (1971) xii. 132 There was a lot of merit in having the ranking man right where the heat was going to be.
b. *slang* (orig. *U.S.*), in various inter-connected senses, notably (*a*) a gun (? as an instrument of 'heat'); also *heater*; (*b*) in phr. *to turn on* (or *give*) *the heat*, to use a gun, hence *fig.*, *to turn the heat on* (someone), to apply pressure on; (*c*) involvement with or pursuit by the police; a police officer, the police.
1928 *Amer. Mercury* May 80/1 The greatest difficulty for such a mob was to avoid another's *heat*. *Ibid.*, It's not so much your own heat you got to watch, but you're apt to run into a bunch of hoosiers out looking for another outfit just hot from some caper. **1929** *Sat. Even. Post* 13 Apr. 54/3 A pistol may be a heat... A man shooting a gun is fogging... 'I fogged away with my heat until I pooped that dummy.' **1929** *Detective Fiction Weekly* 9 Nov. 651/2 Aw, put up your heaters. If you bump me you don't git anywheres. **1930** *Amer. Mercury* Dec. 456/1 Either take our beer or it's plenty of heat for yours. **1931** G. Irwin *Amer. Tramp & Underworld Slang* 96 *Heat*, the state of mind of the police or public following a crime or series of crimes, when the people are 'hot under the collar' or 'all heated up'. More lately, any trouble, as 'in hot water'. **1931** D. Runyon *Guys & Dolls* (1932) iii. 58 Maybe you remember John the Boss, and the heat which develops around and about when he is scragged in Detroit? **1932** W. R. Burnett *Silver Eagle* i. 7 'He don't even pack a heater.' 'Don't what?' 'He don't carry a gun.' **1934** H. N. Rose *Thesaurus of Slang* iii. 16/2 Cover One with a Gun (v. phr.): to turn on the heat. **1936** J. G. Brandon *Pawnshop Murder* xxv. 246 You planted yourself in a safe spot to give Lou the heat. **1936** H. Corey *Farewell, Mr. Gangster* xiv. 174 But the word went out that the government heat was on. The FBI was known to be relentless in its pursuit. *Ibid.* ii. 14 During the heat on the bank robbers the field agents almost lost the habit of sleep. **1937** E. H. Sutherland *Professional Thief* 238 *Heat*, danger in general; an investigation; a policeman. **1938** J. Curtis *They drive by Night* xix. 211 The bleeding heat's on here for me. **1939** R. Chandler *Big Sleep* xiv. 110 Then he leaned back..and held the Colt on his knee. 'Don't kid yourself I won't use this heat, if I have to.' **1944** W. R. Burnett *Nobody lives Forever* xvii. 137 Jim.. took out his gun..and..tossed it down a manhole-grating... 'I was hoping that I'd never have to use that heater.' **1957** *Listener* 24 Oct. 637/2 The moment seemed opportune to 'turn the heat' on Turkey. **1967** W. Murray *Sweet Ride* x. 168 He got busted last week and he don't take that too kindly. Guess he figured you was heat. **1969** *New Yorker* 19 July 20 Out the door comes this great big porcine member of the heat, all belts and bullets and pistols and keys. **1970** C. Major *Dict. Afro-Amer. Slang* 65 *Heat*, law-enforcement officer. **1972** Wodehouse *Pearls, Girls, & Monty Bodkin* xi. 178 And Dolly, drop the heater and leave that jewel case where it is, I don't want any unpleasantness.
14. a. *heat-capacity, -cloud, -flow, -haze, -insulation, -insulator, -isolation, -mist, -power, -radiator* (= Radiator b), *-retrogression, -shock, -test, -trap, -value*; also *heat-labile, -regulating, -resistant, -resisting, -sensitive, -stable* adjs. **b.** *heat-absorption, -evolution, -loss, -producer, -production, -storage.* **c.** *heat-crazed, -hazed, -killed, -misted, -set* adjs. (so *heat-setting* vbl. sb. and adj.); *heat-seal* vb. (so *heat-sealed, -sealing* ppl. adjs.); also with meaning 'against or from heat', as *heat-insulated, -isolated, -proof* adjs.
1902 *Encycl. Brit.* XXXIII. 280/1 Heat-evolution is reckoned as positive, heat-absorption as negative. *Ibid.* 279/2 The heat-capacity of the water. **1895** Kipling *2nd Jungle Bk.* 140 When the heat-cloud sucks the tempest. **1926** *Daily Colonist* (Victoria, B.C.) 23 July 1/4 In Jersey City, three heat-crazed dogs attacked two young boys. **1902** *Encycl. Brit.* XXXIII. 297/1 Heat-flow due to Conduction. **1925** J. Joly *Surface-Hist. Earth* vi. 104 Steady heat-flow to the surface. **1955** *Times* 11 July 2/6 A research group investigating problems of heat flow in supersonic aircraft. **1971** I. G. Gass et al. *Understanding*

Earth iii. 67/1 The 'heat flow'—the rate of escape of interior heat from the Earth's surface. **1899** *Daily News* 12 Jan. 6/2 The flat, endless continent, fading away in the heat-haze. **1901** 'Linesman' *Words by Eyewitness* (1902) 30 The ranks of little kopjes across the river slumbered in the heat-haze. **1960** C. Day Lewis *Buried Day* ii. 43 Summer lanes Whose sound quivers like heat-haze endlessly. **1913** E. F. Benson *Thorley Weir* iii, Over all lay a grey heat-hazed sky. **1963** A. J. Hall *Textile Sci.* v. 221 The pin or clip chains over the greater part of their travel run through a heat-insulated chamber. **1902** *Encycl. Brit.* XXXIII. 283/2 Expansion or compression under the condition of heat-insulation, represented by curves called Adiabatics. **1937** *Discovery* Feb. 35/1 Double walls of canvas enclosing an air space, which acted as a perfect heat insulator. **1902** *Encycl. Brit.* XXXIII. 288/1 If the system is heat-isolated. *Ibid.*, The difficulty of realizing experimentally the condition of heat-isolation. **1946** *Nature* 27 July 121/1 In heat-killed grain there was no change in nucleolar size. *Ibid.*, One vital heat-labile system in the earliest stages of the chain of activity preceding cell division. **1964** *Oceanogr. & Marine Biol.* II. 342 A hot-water extract (containing the substrate but not the heat-labile enzyme). **1899** Callendar & Barnes in *Rep. Brit. Assoc.* 626 The external heat loss is more regular and certain. **1902** *Encycl. Brit.* XXVI. 508/1 The heat-loss can be reduced to a minimum. **1935** *Archit. Rev.* LXXVIII. 129 A double window was evolved with central heating between the two glass lines to minimize the heat loss occasioned by the lavish use of glass. **1964** R. F. Ficchi *Electr. Interference* viii. 151 As ground current flows through the ground rod electrode, heat is generated that follows the well known I^2R heat-loss pattern. **1901** H. W. Wilson *With Flag to Pretoria* I. vi. 91 Indistinct lines of Boer entrenchments, flickering through the heat-mist. **1940** W. Empson *Gathering Storm* 48 The heat-mists that my vision hood Shudder precisely with the throng. **1894** M. Dyan *All in Man's Keeping* I. vi. 98 The deep heat-misted valley. **1905** *Westm. Gaz.* 26 Aug. 13/2 The practical science of heat-power-production. **1956** A. H. Compton *Atomic Quest* 52, 10,000 kilowatts of heat-power. **1905** *Daily Chron.* 14 July 4/4 Animal foods rich in fat..are heat-producers of the first order. **1884** *Jrnl. Nerv. & Mental Dis.* XI. 141 He believed the central nervous system to have an immediate influence on heat-production. **1927** Haldane & Huxley *Animal Biol.* iii. 87 If we put our man..into a calorimeter for a day and measure his heat-production. **1906** *Daily Colonist* (Victoria, B.C.) 6 Jan. 5/6 Get a 34-inch poker for your air-tight heater; they are nicely made and have the Alaska heat-proof handle. **1909** *Daily Chron.* 21 Jan. 4/7 Glasses treated in this manner become heat-proof, and may last for years. **1920** H. G. Wells *Outline Hist.* 21 This novel covering of feathers, this new heat-proof contrivance that life had chanced upon. **1867** *Trans. Ill. Agric. Soc.* VI. 53 Heat Radiator [exhibited]. **1904** Goodchild & Tweney *Technol. & Sci. Dict.* 283/2 *Heat radiator*, a device by which the cooling of the cylinder of a motor cycle or of the condenser of a car is promoted. **1897** *Allbutt's Syst. Med.* II. 26 The paralysis of the heat-regulating centres. **1904** *Daily Chron.* 29 Nov. 4/5 The heat-regulating mechanism of the body. **1964** L. Martin *Clinical Endocrinol.* (ed. 4) vii. 227 A varicocœle may also upset the heat-regulating mechanism and this is aggravated by a suspensory bandage. **1960** *Farmer & Stockbreeder* 22 Mar. Suppl. 11/3 The steel-reinforced, heat-resistant handle. **1961** Whitby & Hynes *Med. Bact.* (ed. 7) ii. 17 The ultimate test of a sterilizer is to show that live spores are killed. The spores must be carefully chosen—soil bacteria are often too heat-resistant for the purpose. **1899** *Daily News* 21 July 4/4 A heat-resisting alloy. **1934** *Archit. Rev.* LXXV. 24/2 With the removal of weight from partitions and external walls came a reduction in thickness of material, with a consequent loss of sound and heat-resisting qualities. **1962** *Gloss. Terms Glass Industry* (*B.S.I.*) 8 *Heat-resisting glass*, a glass able to withstand high thermal shock. **1880** S. Haughton *Phys. Geogr.* vi. 312 Periods of heat-retrogression (such as the glacial). **1961** *Lancet* 9 Sept. 592/1 This is heat-sealed across its width. **1964** *Discovery* Oct. 17/1 So impervious to water-vapour is the laminate, even along heat-sealed seams, that less than 0.012 grams per square metre can be leaked through samples every 24 hours. **1952** E. J. Labarre *Dict. Paper* (ed. 2) 125/2 Heat-sealing papers include several types of paper coated with wax, varnish..which will adhere when pressed together with heat. **1946** *Nature* 10 Aug. 194/1 Electronics have brought a contribution in the evaporation of solutions of heat-sensitive materials such as penicillin. **1964** N. G. Clark *Mod. Org. Chem.* i. 5 This may cause the decomposition of one or more of the heat-sensitive components. **1963** A. J. Hall *Textile Sci.* iii. 130 The yarn becomes bulky, with each filament having heat-set small loops closely but irregularly spaced. **1957** *Textile Terms & Defs.* (ed. 3) 88 (s.v. *Setting*), In order to ensure that the crimp is not readily removed..the fibre may be set to impart permanency of crimp, and the operation is known as heat-setting. **1962** J. T. Marsh *Self-Smoothing Fabrics* ii. 8 During the early investigations into the finishing of nylon fabrics, it was found that a heat-setting process had a stabilising effect. **1946** *Nature* 23 Nov. 763/1 The production in certain varieties of apples, of diploid pollen by heat-shock treatment of the pollen mother cells. **1956** *Ibid.* 4 Feb. 227/2 In *Drosophila*, heat-shock at an appropriate stage results in the development of the cross-veinless phenotype. **1946** *Ibid.* 23 Nov. 760/1 Heat-stable enzyme. **1964** M. Hynes *Med. Bacteriol.* (ed. 8) xxiii. 339 *R. prowazeki* and *R. mooseri*..are differentiated by specific heat-labile major antigens, but share a common heat-stable antigen. **1951** *Good Housek. Home Encycl.* 263/1 All heat-storage cookers have insulated hot-plate covers. **1901** *Kynoch Jrnl.* Feb.–Mar. 57/1 The Heat-Test of Nitro Explosives. **1906** W. De Morgan *Joseph Vance* xviii, I think of the *sole di marzo* blazing on the roses in that Tuscan heat-trap. **1887** *Chambers's Jrnl.* 24 Sept. 623/1 The exact heat-value of different kinds of liquid fuel. **1962** *Economist* 21 July 256/1 The main use of this gas should be to fuel power stations (at a 'heat value' parity price with coal or oil).
d. heat balance, the distribution of the flow of heat and other forms of energy into and out of a system in which there is no change in internal energy; also, an account or record of such a distribution, esp. as a means of evaluating the efficiency of boilers, etc.; **heat barrier** *Aeronaut.*, the limitation on the speed of aircraft, etc., due to heating by air friction; **heat bump**, a protuberance on the skin supposed to be due to heat; **heat-centre** *Physiol.*, any of several areas within the central nervous system which control the regulation of the body temperature; **heat coil** *Electr.*, a device fitted in a telephone exchange to protect the lines against small harmful currents; **heat cycle**, a cycle of operations or states in a heat engine; **heat-death** (see quot. 1930²); **heat-energy**, that form of energy which is manifested in heat; **heat equator** = *thermal equator* (see *equator 3 b); **heat exchanger**, a device used for the transference of heat from one medium to another; so **heat exchange, heat exchanging**; **heat filter**, any device that selectively removes heat radiation but permits the passage of light; **heat flash** (see quot. 1958); **heat-lightning** (earlier examples); **heat-pipe**, a closed, evacuated tube containing around its inner surface a wire mesh or other wick saturated with a working liquid, which through the capillary action of the wick and the higher vapour pressure of the liquid when heated makes possible the rapid conduction of heat away from a source; **heat-pump**, a heat-engine working in reverse (such as a refrigerator), in which work supplied to it is used to transfer heat from a colder to a hotter body; **heat-set, heat-setting ink** (see quots.); see also *heat-set* adj., *-setting* vbl. sb. and adj. sense 14 c above; **heat-shield** (see quots.); **heat-sink** (see quot. 1965²); **heat sponge**, a type of heat sink; **heat tinting** (see quot. 1958); **heat tonality, tone, toning** *Physical Chem.* [tr. G. *wärmetönung*], the sum of the heat produced in a chemical reaction and of the work done by the system, expressed in heat-units; the heat of reaction at constant volume (*disused*); **heat transfer**, the transfer of heat from one medium to another.
1898 B. Donkin *Heat Efficiency of Steam Boilers* xiv. 239/2 An approximate 'heat balance', or statement of the distribution of the heating value of the coal among the several items of heat utilised and heat lost, may be included in the report of a test. **1954** *Jrnl. Meteorol.* XI. 8/1 The heat balance between the surface of the earth and the atmosphere..involves a flux of latent heat and of sensible heat, in addition to the radiational items. **1971** *Nature* 25 June 540/1 Ecologists are therefore interested in ways of inferring the temperature of a leaf from a knowledge of its heat balance. **1953** H. Haber *Man in Space* 66 The designers of the Sky-rocket had to be on guard against not only the sonic barrier. With its high rate of speed their craft might run into an obstacle more serious than buffeting shock waves: the heat barrier. **1953** *Sci. Amer.* Dec. 80/1 This is the heat barrier: the heating of a plane by the friction and piling up of air on aircraft surfaces at supersonic speeds. **1954** *Times* 5 Mar. 11/5 They might well find that [the ultimate limits of manned aircraft] were very high and that in the same way as the sound barrier had been overcome the problems of the heat barrier would be solved also. **1957** [see *barrier *sb.* 4 b]. **1970** J. Chaplin *Wings & Space* 146/1 There is no way to break through the heat barrier as there is with sound. **1927** W. E. Collinson *Contemp. Eng.* 57 Spots, which, it is hoped, are heat-bumps. **1884** *Jrnl. Nerv. & Mental Dis.* XI. 141 Tscheschichin was the first to announce the existence of an inhibitory heat-centre in the nervous system. **1907** *Practitioner* June 771 The action of the heat-centres being sluggish. **1968** M. Monnier *Functions Nerv. Syst.* I. xv. 422 Successful protection from cold is possible through the central nervous co-ordination of several biophysical and chemical mechanisms. This is accomplished by the so-called heat center in the posterior hypothalamus. **1900** K. B. Miller *Amer. Telephone Pract.* (ed. 3) xxiii. 275 A device to afford protection against currents such as these [*sc.* sneak currents]..is termed a heat coil. **1971** *Gloss. Electrotechnical Power Terms* (*B.S.I.*) III. ii. 13 *Heat coil*, a thermal device to protect apparatus from damage by external currents. **1894** B. Donkin *Text-bk. Gas, Oil, & Air Engines* I. ii. 13 (*heading*) Heat 'cycles' and classification of gas engines... Engineers have agreed to designate as a 'cycle' the successive operations taking place in a heat motor. **1930** *Engineering* 8 Aug. 187/3 The following..trends were..observable:..design and operation on more efficient heat cycles [etc.]. **1930** J. Jeans *Mysterious Universe* i. 13 The second law of thermodynamics predicts that there can be but one end to the universe—a 'heat-death' in which the total energy of the universe is uniformly distributed, and all the substance of the universe is at the same temperature. **1959** J. Blish *Clash of Cymbals* iii. 73 Any cyclical theory of the universe, any continuous and eternal systole/diastole from monobloc to heat-death and back again. **1973** *Nature* 11 May 65/1 What lies ahead is, in Clausius's later term, 'a heat death'. **1876** P. G. Tait *Rec. Adv. Phys. Sci.* 138 We are led to speak of the availability of an amount of heat-energy. **1893** *Jrnl. Soc. Arts* 8 Sept. 897/1 The practically unavoidable waste of heat energy. **1902** *Encycl. Brit.* XXXIII. 283/2 The whole of its intrinsic heat energy might theoretically be recovered in the form of external work. **1915** *Chambers's Jrnl.* Jan. 43/1 For all the heat-energy wasted..the consumer has had to pay. **1968** R. A. Lyttleton *Mysteries Solar Syst.* ii. 77 The release exceeds the gentle loss of heat-energy arising from the very slow processes of conduction within the Earth. **1904** Goodchild & Tweney *Technol. & Sci. Dict.* 283/2 Heat Equator. **1911** M. I. Newbigin *Mod. Geogr.* iv. 87 Those regions of the earth which are directly beneath the vertical rays of the sun are heated most intensely... This belt of high temperature is called the heat equator. **1902** G. E. Davis *Handbk. Chem. Engin.* II. ii. 132 (*caption*) Diagrammatic sketch of heat-exchanging tanks. *Ibid.* 133 (*heading*) Heat exchangers... In no case would the cold water be heated to the temperature of the original hot water. **1908** *Sci. Abstr.* A. XI. 203 For snow the average total daily heat-exchange is 19 gm. cals. per cm.² **1915** *Chem. Abstr.* IX. 2332 (*heading*) Heat exchange apparatus wherein the one agent flows through one tube and the other agent flows through an annular chamber surrounding the said tube. **1924** R. Seligman *Brit. Pat.* 223,033, In some.. descriptions of heat exchanging or sterilising apparatus.. it has been proposed in order to obtain a tight jointing to groove and tongue the rims. *Ibid.*, The plates would be working in parallel and the heat exchange effected by counter current. **1947** *Science News* IV. 33 A heat exchanger, then, is merely a means whereby the heat which would normally be wasted is used for combustion. **1952** *Ibid.* XXV. 87 This is done by means of the heat exchanger, which by various means effects the transfer of heat from the gases leaving the turbine to the air entering the combustion chamber. **1958** *Engineering* 28 Feb. 284/1 As an aid in securing high thermal efficiency from gas-turbine plants, use is frequently made of a heat exchanger, whereby the turbine exhaust heat is used to preheat the combustion gas. **1959** *Listener* 29 Oct. 732/3 Twelve heat exchangers for the new Bradwell (Essex) nuclear power station. **1967** M. Chandler *Ceramics in Mod. World* v. 157 Where refractories are used to store and transfer heat, as in heat exchangers, the most important property required is high heat capacity. **1898** W. E. Woodbury *Encycl. Dict. Photogr.* 367, I have taken a powerful projection lantern and set it as near to the microscope as the intervening heat-filter will permit. **1962** *Which?* Mar. 68/2 The heat filter prevents much of the heat radiated from the lamp, from reaching the slide. **1958** *Chambers's Techn. Dict.* Suppl. 984/1 *Heat flash*, intense heat radiation from an elevated A or H bomb, detection of which, by heat-sensitive paint, gives the precise indication of ground-zero. **1961** 'C. E. Maine' *Man who owned World* x. 118 Central London was a wilderness of fused stone and leaning skeletal buildings, blackened and oxidised by nuclear heat flash. **1834** C. A. Davis *Lett. J. Downing* ii. 17 You may just as well try to paint a flash of heat-lightning in dog-days. **1849** Thoreau *Week Concord Riv.* 275 Friendship is..remembered like heat lightning in past summers. **1964** G. M. Grover et al. in *Jrnl. Applied Physics* XXXV. 1990/1 We will refer to devices of this general class, for brevity, as 'heat pipes'. *Ibid.* 1991/1 A liquid sodium heat pipe for operation at about 1100°K was constructed. **1969** *New Scientist* 19 June 641/1 A heat pipe is one of the major components of the most powerful and efficient radioisotope-heated power generator yet built. *Ibid.*, The advantage of the heat pipe is that the outside surface is at the same temperature along the whole of its length. **1894** J. A. Ewing *Steam-Engine* iv. 118 By a refrigerating machine or heat-pump is meant a machine which will carry heat from a cold to a hotter body. *Ibid.*, Any heat-engine will serve as a heat-pump if it be forced to trace its indicator diagram backwards. **1948** E. F. Obert *Thermodynamics* xiv. 520 The reversed heat-engine cycle is called a refrigerator (and, also a heat pump) when the evaporator is used for cooling purposes..; the same cycle is called a heat pump (but not a refrigerator) when the condenser is used for heating purposes. **1957** *Encycl. Brit.* XIX. 55/2 The heat-pump system..is a conventional refrigeration system where the heat rejected by the refrigerant at the condenser is utilized for heating during the winter while the evaporator absorbs heat from..any..low-grade heat source. **1966** *McGraw-Hill Encycl. Sci. & Technol.* VI. 369/1 Unless the price of electric energy is low..the heat pump cannot be justified solely as a heating device. However, if there is also need for comfort cooling..in the summer, the heat pump, to do both the cooling and heating, becomes attractive. **1941** *Inland Printer* Nov. 42/1 The new presses..would enable us to print the body of the magazine entirely with the improved heat-set inks. **1947** R. Burns *Printing Inks* v. 249 (*heading*) Heat-setting inks. The fresh prints are exposed to intense heat from gas flames or radiant surfaces for a very short period. **1963** Kenneison & Spilman *Dict. Printing* 91 *Heat-set inks*, printing inks manufactured in a special way to induce quicker drying... The vehicle of these inks is such that it vaporizes rapidly when the paper is heated after printing. **1957** W. E. Clason *Elsevier's Dict. Electronics* 226 *Heat shield*, a metallic surface surrounding a heat radiating element e.g. a hot cathode in order to reduce the radiation loss. **1962** J. Glenn et al. in *Into Orbit* 245 Heatshield, as used in Project Mercury missions ..consists of a coating of ablative material on the rounded base of the capsule which evaporates during re-entry and carries off much of the heat in the form of a gas. **1968** *Times* 16 Dec. 7/3 Reentry speed was slightly faster than expected for Apollo 8 and the heat shield on the spacecraft was charred to a depth of three-quarters of an inch. **1956** *Jrnl. Brit. Interplan. Soc.* XV. 302 The determination of optimum sink temperature is beyond the scope of this paper, particularly since thorough analysis of the entire radiation heat sink problem has been previously presented. **1957** W. E. Clason *Elsevier's Dict. Electronics* 226 *Heat sink*, used with power transistors to dissipate heat. **1959** *Listener* 28 May 930/1 By mounting the transistor on a relatively large piece of metal, which in turn is fixed to what is called a heat sink—something into which unwanted heat can be shot—the powers that transistors can handle..have been greatly increased. **1961** *Aeroplane* C. 372/2 For the Mach 2 aircraft the air supply from the main engines can be cooled by using the fuel as a heat sink. **1965** *New Scientist* 20 May 507/1 Satisfactory control of the rate and extent of cooling of the patient is obtained by regulating the temperature of the heat sink.

1965 W. H. Allen *Dict. Technical Terms for Aerospace Use* 132/2 *Heat sink*, (1) in thermodynamic theory, a means by which heat is stored, or is dissipated or transferred from the system under consideration; (2) a place toward which the heat moves in a system; (3) a material capable of absorbing heat; a device utilizing such a material and used as a thermal protection device on a spacecraft or reentry vehicle; (4) in nuclear propulsion, any thermodynamic device, such as a radiator or condenser, that is designed to absorb the excess heat energy of the working fluid. **1972** *Sci. Amer.* Mar. 118/2 All power transistors.. must be mounted on heat sinks that have large cooling fins. **1949** A. R. Weyl *Guided Missiles* ... **1958** A. G. Haley *Rocketry* iii. 57 The American Rocket Society...developed a 'heat sponge' motor, wherein blocks of aluminum absorbed large amounts of heat. **1910** C. H. Desch *Metallogr.* vii. 149 Stead has devised an electrical heater, by means of which the heat-tinting can be carried on on the stage of the microscope. **1958** A. D. Merriman *Dict. Metallogr.* 121/2 *Heat tinting*, a method of distinguishing and of identifying the micro-constituents of a polished surface of a metallographic specimen. The method is based on the fact that temper colours or heat tints...appear when oxidation begins on a polished surface that is being heated. **1895** C. S. Palmer tr. *Nernst's Theoret. Chem.* III. iv. 435 Instead of using the 'heat-toning' (heat tonality) to determine the ratio of distribution, one may employ...the changes in the volumes... [illegible] **1966** W. A. Heylin *Second Aerospace Gloss.* 60/2 *Heat transfer*, the transfer of heat within a substance or structure by radiation, conduction, or convection.

heated, *ppl. a.* Add: **1. b.** *heated term*, the hot season of the year. *U.S.*

1855 *N.Y. Herald* 26 Dec. 3/4 Our 'heated terms' are over, and we now begin to look out for the approach of 'the northers'. **1867** *Congress. Globe* 5 July 487/1, I think we could go on now during the heated term better than...during the cold season. **1873** J. H. Beadle *Undevel. West* 793 The average of the 'heated term', one day with another, is there recorded at eighty-four degrees. **1949** *Chicago Tribune* 11 Sept. 43/5 What a month ago appeared to be a trivial item of conversation during the heated term has become a raging topic among scientists.

heater. Add: **1. b.** *slang.* A gun (see *HEAT *sb.* 12 b).

2. a. (Earlier examples.) **c.** Also, a usual name for a domestic electric or gas fire.

1666 in *Essex Inst. Hist. Coll.* XXV. 147 It. boxe Iron & heaters. **1744** B. Franklin *Acc. Fire-Places* 27 You..may..warm the Flat-Irons, heat Heaters [etc.].

f. A triangular structure resembling in form the heater of a box-iron.

1797 J. A. Graham *Descr. Sk. Vermont* 119 There are two arches, with a pier in the center..with the addition of a heater, or triangular front. **1899** Dickinson & Prevost *Cumbld. Gloss.* 379 *Heater bit* is the triangular piece of ground, generally grass-grown, at the junction of three roads; so called because of resemblance to the iron heater in a box-iron.

g. A device used for the indirect heating of the cathode of a thermionic valve.

1940 *Chambers's Techn. Dict.* 407/2 *Heater*, the conductor carrying the current for heating an equipotential cathode, generally enclosed by the cathode. **1962** *Electronic Engin.* XVII. 454 Radio receivers and other electronic devices may have the valve heaters connected in series.

h. A device used for heating the interior of a motor car. Also *attrib.* and *Comb.*, as *heater-demister*, *-fan*.

1939–40 *Army & Navy Stores Catal.* 266/1 Car heater.. A robust and reliable heater. **1948** *Motor Man.* (ed. 33) xii. 232 (*heading*) Car heaters. The use of car heaters has spread in recent months, largely owing to the spur of the export trade. **1961** *Which?* (Reports on Cars) 14 Heater efficiency is measured and compared, and the results analysed for average interior temperatures and effective distribution of heat. **1962** *Ibid.* Oct. 310/2 Most modern cars can have a built-in heater-demister which blows warmed air into the car. **1969** S. Hyland *Top Bloody Secret* ii. 163 The [car] engine was silent, but the heater-fan was still humming.

3. heater-shaped (examples); **heater-piece** *U.S.* (examples).

1859 Bartlett *Dict. Amer.*, *Heater piece*, a gore or triangular piece of land, so called probably, from a flat iron, the form of which it resembles. **1863** D. G. Mitchell *My Farm* 243 Waal—kinder like to have a little 'heater' piece, the boys, you see, hoe it out in odd spells. **1847** C. Boutell *Monumental Brasses* 37 The shield is small, flat, and heater-shaped. **1927** A. C. Fryer in *Trans. Bristol & Glouc. Archæol. Soc.* XL. 41 A half angel vested in alb and holding a heater-shaped shield.

heath, *sb.* **4. a.** *heathland* (later examples).

1936 *Discovery* Jan. 25/1 Only about 50,000 acres of Breckland remain at the present moment as heathland. **1954** M. Beresford *Lost Villages* vi. 200 This was forest, scrub, or rough heathland. **1966** M. R. D. Foot *SOE in France* xii. 407 Bourgoin was too wily a fighter to be rounded up methodically in his heathland base.

heather. 3. a. *heather-honey* (earlier and later examples).

1826 *Blackw. Edin. Mag.* XX. 412/1 Heather-honey of this blessed year's produce. **1935** *Jrnl. Physical Chem.* XXXIX. 213 The term 'heather honey' is used to describe any honey derived largely from the nectar of *Calluna vulgaris*, *Erica cinerea*, and allied species. **1971** *Country Life* 28 Oct. 1101/2 The drawback to heather honey is that it is difficult to extract. **1973** *Harrod's Xmas Catal.* 59/3 'Double Scotch' Honey is a unique blend of Scottish heather honey and rare old malt whisky.

Heath Robinson (hīþ rǫ·binsǝn). [f. the name of the humorous artist W. *Heath Robinson* (1872–1944).] Used *attrib.* or *ellipt.* of any absurdly ingenious and impracticable device of the kind illustrated by this artist. Hence **Heath-Robinson·esque**, **Heath-Ro·binsonish** *adjs.*; **Heath Ro·binsonism.**

1917 'Contact' *Airman's Outings* i. 12 The movable mounting for the observer's gun in the rear cockpit... We classified it with the Christmas Tree, the Heath Robinson, the Jabberwock, the Ruddy Limit, and names unprintable. **1920** *Teleg. & Teleph. Jrnl.* June 180/2 The tour concluded at the principal machine shop. Here one is introduced to what on first acquaintance appears to be a Heath Robinson nightmare. **1931** D. L. Sayers *Five Red Herrings* xiii. 157 'Not very likely,' he mused; 'better, I think, for a Heath Robinson device.' **1934** *Discovery* Nov. 328/1 This 'Heath-Robinson' jumble of wooden sheds, sluices, and water troughs looks ridiculous, yet it works all right. **1951** R. Campbell *Light on Dark Horse* xii. 175 [illegible] **1960** L. Davidson *Night of Wenceslas* xiii. 207 It was surely a...Heath-Robinsonish way of passing valuable secrets. **1962** *New Scientist* 29 Mar. 760/2 [illegible] heathrobinsonism. **1963** *Times* 26 Feb. 9/5 [illegible] **1966** N. Freeling *Tsing-Boum* iv. 163 The English talked about things being 'Heath-Robinson' but this was quite tightly—he recalled the man from his own childhood, a caricaturist who filled his drawings with wonderful complicated mechanisms made out of every kind of rubbish-tip junk held together with knotted bits of string.

heating, *vbl. sb.* Add: **b.** *heating arrangement(s)*; **heating element** (see *ELEMENT *sb.* 4 c).

1873 *Leisure Hour* 18 Jan. 48/1 Cisterns in the upper parts of a house should be emptied, if the heating arrangement has been neglected. **1928** *Encycl. Brit.* XXVI. 310/2 [illegible] at any time in a modern laboratory with adequate heating arrangements. **1931** R. G. Collingwood *Roman Brit.* 54 The discovery of skeletons huddled inside the heating-arrangements beneath the floors.

heatronic (hītrǫ·nik), *a.* [f. Heat *sb.* + *electronic *a.*] (See quot. 1943.)

1943 *Plastics Engin.* Mar. 87/1 Heatronic molding is announced as a most significant advance in the art of molding plastics. This process is the result of...research involving the application of electronics to the molding of thermosetting materials. **1944** *Electronic Engin.* XVII. 40 Space is devoted to discussing the physical properties of heatronic mouldings.

heat treatment. a. The specialized application of heat to various substances to produce a desired metallurgical or physical condition, e.g. hardness, softness, toughness. Hence **heat-treat** *v.*; **heat-treatable** *a.*; **heat-treated** *ppl. a.*; **heat-treating** *vbl. sb.*

1895 H. M. Howe in *Trans. Amer. Inst. Mining Engin.* 1894 XXIV. 746, I...call your attention to two directions in which very important progress may be..hoped for..pyrometry and the heat-treatment of steel. *Ibid.* 747 In the hardening and tempering of tools, in the annealing of steel castings,...and in the manufacture of guns, projectiles, and armor, we have already extensive and careful if empirical heat-treatment. **1899** *Jrnl. Iron & Steel Inst.* LIV. 147 The simple heat treatment had converted the more or less brittle annealed material into tough and strong steel. **1904** F. W. Harbord *Metallurgy of Steel* xxxix, (*title*) Heat treatment of steel. **1906** *Chem. Abstr.* II. 1121 The Structure of Metals. Changes Produced by Working and by Heat Treatment. **1908** *Westm. Gaz.* 4 June 4/2 The wonderful effect of heat-treating [of steel]. *Ibid.*, With these alloying materials added to steel small gears can be made that, if properly heat-treated, they will be so tough and strong as to make it impossible to break out a tooth even with a sledge-hammer. **1946** *Nature* 27 July 120/2 Plants from heat-treated grain were..somewhat shorter in the straw. **1947** *Hansard, Commons* 4 Dec. 559 Mr. Lambert asked the Minister of Education the number of rural schools..in which the children are supplied with heat-treated milk. **1947** *Food & Drugs Statut. Rules & Orders* DCXII. (*heading*) The ice cream (heat treatment, etc.) regulations, 1947. *Ibid.*, The mixture shall be subjected to heat treatment as follows [etc.]. **1950** J. G. Davis *Dict. Dairying* 332 *Heat treatment of milk.* The term is a general one covering any of the ordinary commercial heating methods by which the keeping quality of milk is enhanced and pathogenic organisms destroyed. **1956** *Nature* 25 Feb. 360/1 He invented the heat-treatment which has for decades been applied as compulsory routine at every ginnery in the country. [illegible] The specimen [of an alloy] had been heat-treated. **1966** *Farmer & Stockbreeder* [illegible] All seed grown from heat-treated stocks. **1960** *Times* [illegible] Heat-treatable aluminium alloys. **1963** *B.S.I. News* Feb. 27/1 Fifteen different steels...were heat-treated by the National Engineering Laboratory. **1964** *Nature* 23 May [illegible] The heat treatment (57°C for 30 s in summer) disrupted the layer into irregular globules.

b. The therapeutic use of heat, esp. radiant and infra-red heat.

1934 J. E. Mansion *Harrap's Fr. & Eng. Dict.* 837/2 *Thermothérapie*, heat cure, heat treatment, thermotherapy. **1967** *Listener* 5 Oct. 427/2 He asked for..heat treatment, which my wife should apply herself.

heave, *sb.* Add: **1. c.** *Wrestling.* A chip performed by bringing the right arm round the opponent's right shoulder preparatory to a throw. *Cornwall heave*, a heave in which a wrestler places one hand in front and one behind his adversary, and falls with him.

1889 W. Armstrong *Wrestling* 224 The Heave, Cornwall and Devon. *Ibid.* 230 The principal Cornish and Devon chips are..the Back-heave, the Belly-heave, the Heaving-toe. **1898** *Encycl. Sport* I. 548/2 One way to the heave is to cross click your man and then ply the crossbuttock.

5. *Comb.*: **heave-gate** *local*, a gate which is opened by being lifted out of the sockets or mortises.

1736 Pegge *Kenticisms* (1876), *Heave-gate*, when the rails with the pales nailed to them, drop into mortises, and are lifted out of them, and put in again. **1876** [illegible] Surrey Provincialisms. [illegible] **1907** 'J. Halsham' *Lonewood Corner* 149, I perched myself on the heave-gate between the two fields. **1959** [illegible] Sussex heave-gates and small fences.

heave ho. Add: **b.** orig. *U.S. slang.* A snub or dismissal.

1944 D. Runyon *Runyon à la Carte* (1946) vi. 107 A most obnoxious character..tries to claim the deuce as a sleeper and gets the heave-o from Nathan Detroit. **1952** B. Cerf *Good for Laugh* (1953) 152 I'll bet that new girl..gave you the heave-ho. **1963** Wodehouse *Service with Smile* viii. 121 If I were you, I think I would reconsider this idea of giving Bill Bailey the old heave-ho. **1966** *New Yorker* 22 Jan. 30/2 Do we keep him on or give him the heave-ho? **1973** *Guardian* 20 Jan. 13/5 Mr Heath's prices and incomes package was given the old heave-ho by the TUC Economic Committee.

heave-ho *v.* (later example); also *v. trans.*, to heave or lift with force.

1964 *New Yorker* 18 Jan. 84 A groaning mass of men heave-ho'd the snow car up the ramp. **1968** *Listener* 5 Dec. 765 During a rehearsal of *Billy Budd*, a singer was asked why he was just lolling about in the wings, not heave-hoing with the rest of them. **1971** *Guardian Weekly* 7 Aug. 17 The [Congolese] women who, apparently 22 months pregnant, nonetheless are constantly hauling and heave-hoing on this packing case of merchandise or that basket full of provisions.

heaven, *sb.* **6. d.** Also, *Heavens above*, *alive!*; *Heavens to Betsy!* (U.S.).

1892 R. T. Cooke *Huckleberries fr. New England Hills* 173 'Heavens to Betsey!' gasped Josiah. **1895** W. Pinero in M. B. Booth *Eng. Plays of 19th Cent.* (1969) II. 275 [illegible] Heavens above! **1913** S. Rohmer *Mystery of Fu-Manchu* xx. 209 Yes yes—heavens above, the huge green eyes! **1924** *Dialect Notes* IV. 74 Heavens to Betsy! Common Exclamation among women. **1937** [illegible] 'Heavens alive, darling, but I haven't been up so near dawn for years.' **1958** Hayward & Harari tr. *Pasternak's Dr. Zhivago* II. viii. 441 But Heavens above! You misunderstood us. What are we talking about? **1968** 'E. V. Cunningham' *Cynthia* (1969) xi. 156 'Oh, heavens to Betsy,' [illegible] Lucille whispered.

e. *Heaven knows.* (*a*) Used to emphasize the truth of a statement. (*b*) Used to imply that something is unknown to the speaker, and probably also to others. Freq. with *what*, *where*, *who*. Cf. God 10.

[**1600** Shaks. *Macbeth* v. i. 52 Shee ha's spoke what shee should not, I am sure of that: Heaven knowes what shee ha's knowne.] **1711** [in Dict., sense 6 a]. **1805** Wordsworth *Prelude* xi. 124 Not in Utopia—subterranean fields,—Or some secreted island, Heaven knows where! **1872** Geo. Eliot *Middlemarch* IV. viii. lxxiv. 198 She invites clergymen and heaven-knows-who. *a* **1916** 'Saki' *Square Egg* (1924) 125 From privates in the Regular Army to Heaven-knows-what in some intermediate corps. **1936** *Delineator* CXXIX. 48/3 It was clearly apart from the spirituals..and heaven knows, was unlike any music that America had been playing before. **1967** *Listener* 26 Jan. 117/1 Heaven knows, there are old excuses for it. **1969** *Ibid.* 13 Mar. 351/2 Heaven knows, he'd been through this often enough in the past.

heavenly, *a.* Add: **4. b.** *colloq.* Excellent, particularly enjoyable.

1874 L. Troubridge *Life amongst Troubridges* (1966) ix. 88 We had a most heavenly bathe. **1931** R. Lehmann *Let. to Sister* 11 The heavenly mixing of paints and distempers. **1940** N. Mitford *Pigeon Pie* viii. 127 Sophia felt at once extremely dowdy. 'You are lucky,' she said, 'the way you always have such heavenly things.'

heavens, *adv. dial.* and *colloq.* (Earlier examples.)

1858 Dickens in *Househ. Words* Xmas no. 21/1 A shy company though its raining Heavens hard. **1870** —— *E. Drood* i. 2, I got Heavens-hard drunk for sixteen years afore I took to this.

hea:vier-than-ai·r, *attrib. phr. Aeronautics.* Designating a flying-machine whose weight is greater than the weight of the air which it displaces, and whose lift is not dependent on light gases; also applied to the use of such a machine or machines in flight.

[**1870** tr. *Marion's Wonderful Balloon Ascents* II. ix. 162 To form a 'Free Association for Aerial Navigation by means of Machines heavier than Air'. **1879** *Encycl. Brit.* IX. 309/2 Weight, however paradoxical it may appear, is necessary to flight. Everything which flies is vastly heavier than the air.] **1903** *Westm. Gaz.* 18 Sept. 9/3 The only example of the heavier-than-air machine. [**1904** *Chambers's Jrnl.* 1 Oct. 699/1 All who have sought to sail the skies divide themselves into the 'lighter-than-airites' and the 'heavier-than-airites'.] **1908** H. G. Wells *War in Air* viii. § 2 The most efficient heavier-than-air fliers. **1909** A. Berget *Conquest of Air* II. ii. 155 Many persons ask aviators why their 'heavier-than-air' apparatus is not provided with parachutes. **1909** *Flight* 19 June 356/1 Any heavier-than-air type of machine. **1909** *Daily Chron.* 9 Sept. 1/6 For a long time, Mr. Cody has practised heavier-than-air flying on Laffan's Plain. **1927** C. L. M. Brown *Conquest of Air* 21 When heavier-than-air flight was an accomplished reality. **1961** C. B. Smith *Testing Time* ii. 27 The War Office instructed O'Gorman to concentrate entirely on airships. They admitted that 'heavier-than-air dirigibles' might one day have military uses.

heavily, *adv.* Add: **6.** *Comb.*: often equivalent to parasynthetic comb. of the adj., as *heavily-booted*, having heavy boots.

1883 A. Thomas *Mod. Housewife* 131 Trodden by heavily-booted feet. **1883** Ld. R. Gower *My Remin.* II. 79 A large and heavily-veined nose. **1904** *Westm. Gaz.* 27 Aug. 6/2 This heavily-scented, image-laden atmosphere. **1905** *Daily Chron.* 20 Oct. 8/5 Those heavily-jetted waistbelts. **1906** *Ibid.* 11 June 5/2 A good deal of heavily-jewelled..speech. **1907** B. von Hutten *Halo* I. i, Very long, half-closed, heavily-lashed eyes. **1927** A. Conan Doyle *Case Bk. S. Holmes* xi. 293 A strong, heavily-moustached face and angry eyes. *c* **1909** D. H. Lawrence *Collier's Friday Night* (1934) i. 9 Then he drags his heavily-shod feet to the door on right. **1938** *Daily Tel.* 18 Jan. 6/3 As the heavily-policed funeral was marched down the street every window was flung open and red flowers showered down. **1961** *B.S.I. News* Feb. 6/1 Fast heavily-trafficked roads.

Heaviside (he·visəid). *Physics.* The name of O. *Heaviside* (1850–1925), English physicist, used attrib. to designate concepts proposed by him; esp. *Heaviside layer*, an ionized layer in the upper atmosphere able to reflect long radio waves (now usu. called *E layer of the ionosphere*); *Heaviside–Lorentz units* (H. A. Lorentz, 1853–1928, Dutch physicist), or *Heaviside rational units*, units of electric charge or magnetic pole defined in a certain manner which simplifies many formulæ; *Heaviside unit function*, a function which is zero when its argument is negative, and unity when its argument is positive; used esp. in *Electr. Communication.*

1911 *Encycl. Brit.* XXVII. 744/2 It will be seen..that the Heaviside rational units are all incommeasurable with the practical units. **1912** *Proc. R. Soc.* A. LXXXVII. 95 Both long and short waves are propagated through the lower and middle atmosphere in straight lines to great heights and reflected at the Heaviside layer. **1913** *Year-Bk. Wireless Telegr.* 395 The conductivity, and consequently the reflecting power, of the Heaviside layer depends greatly on the presence of local electromotive forces. **1926** R. W. Hutchinson *Wireless* vii. 131 It is clear that if the earth were surrounded by a spherical conductor some distance away from it the waves would travel between this conducting surface and the conducting surface of the earth, and they would follow the curvature of the earth... Such a conductor does exist and it is known as the Heaviside layer. **1937** *Discovery* Mar. 65/2 The well known Heaviside and Appleton layers which play such an important part in carrying broadcasting to long distances. **1940** *Chambers's Techn. Dict.* 407/2 *Heaviside unit function*, a step in which the change in amplitude is unity. **1955** O. Klein in W. Pauli *N. Bohr* 99 The units are the usual ones, $\hbar = c = 1$ and Heaviside–Lorentz units for the electromagnetic quantities. **1963** Jerrard & McNeill *Dict. Sci. Units* 14 The Heaviside–Lorentz units were the earliest rationalized units, they were proposed by Heaviside in 1883 and used by him in a classical paper on electrical theory published nine years later.

heavy, *a.*[1] (*sb.*) Add: **A. 1. c.** Also of timber: consisting of large trees. *U.S.*

1843 *Yale Lit. Mag.* VIII. 406 In this patch of 'heavy timber'. **1853** B. F. Taylor *Jan. & June* (1871) 252 [The storm] went crashing on, into the heavy timber.

2. c. Applied to elements whose specific gravity is relatively great; *heavy metal* (see also sense 6 b in Dict.), a metal of high specific gravity (see quot. 1955).

1864 *Jrnl. Chem. Soc.* XVII. 126 In support of the view that thallium is one of the heavy metals, the following reasons may be given. **1868** [in Dict., sense 2 a]. **1903** *Jrnl. Physiol.* XXIX. 165 Most of the heavy metals when injected directly into the circulation give rise to increased movements of plain muscle throughout the body. **1922** F. W. Aston *Isotopes* viii. 101 The nucleus of the atom of an ordinary element (not hydrogen)..is very small compared with the atom itself. Its dimensions can be roughly determined by actual experiment in the case of the heavy elements. **1936** *Discovery* Feb. 36/1 Heavy elements, such as gold, silver, and lead. **1936** R. P. Bell tr. *Bjerrum's Inorg. Chem.* 213 The metals fall naturally into two groups: the light metals with densities below four, and the heavy metals with densities above seven... The heavy metals have their electrons more firmly bound and are less electropositive than the light metals. **1946** *Monthly Notices R. Astron. Soc.* CVI. 357 Material at any point.. on the other side of the curve is composed almost entirely of heavy elements, the main mass of the elements in the latter case having atomic weight greater than 50. **1955** *Chem. & Engin. News* 2 May 1902/2 Karl F. Heumann wonders what is meant by 'heavy metal'. One authority says it is any metal having a specific gravity greater than 4.0. Another says it is sometimes applied to those of sp. gr. 5.0 or over... Has 'heavy metal' ever been officially defined? **1961** *Jrnl. Chem. Educ.* XXXVIII. 67/1 The present treatment will be restricted to the main process responsible for the heavy elements, of mass number *A* greater than 70. **1972** *Science* 14 Apr. 161 (*title*) Enrichment of heavy metals and organic compounds in the surface microlayer of Narragansett Bay, Rhode Island.

d. *Physics.* Of hydrogen: consisting of the isotope deuterium (which is of greater mass than protium, the normal isotope). Of a compound of hydrogen: having some or all of the hydrogen isotope present as deuterium. So **heavy water,** deuterium oxide, D_2O, or a mixture of this with ordinary water; **heavy-water-moderated** *a.*, of a nuclear reactor: employing heavy water as a moderator; **heavy water reactor,** a nuclear reactor in which the moderator is heavy water.

This usage is occas. extended to the isotopes of other elements to designate an isotope that is of greater mass than the normal isotope.

1933 *Nature* 22 Apr. 590/2 Heavy water freezes when surrounded by melting ice. **1933** *Jrnl. Chem. Physics* June 344/2 Let us make an estimate of the amount of the heavy hydrogen isotope in ordinary water. **1933** *Discovery* July 211/1 For the first time in history a chemical element has been divided into two completely different parts. A new 'heavy' hydrogen has been separated from the old. **1933** *Science* 29 Dec. 602/1 In fact, if there were only two waters, two ammonias, and so forth, the names 'light water', 'heavy water', 'light ammonia' and 'heavy ammonia' would be very satisfactory indeed. **1934** *Discovery* Jan. 1/1 There is one part only of heavy hydrogen to 35,000 parts of light hydrogen. **1934, 1935** [see *DEUTERIUM]. **1936** *Punch* 23 Sept. 337/1 'Heavy-water', the newly-discovered fluid, costs £120 a teaspoonful. **1937** *Discovery* Oct. 317/1 The value of heavy nitrogen for research in physiological chemistry is inestimable. **1938** *Encycl. Brit. Bk. of Year* 320/1 'Heavy water' or deuterium oxide is now manufactured commercially and is an article of commerce. **1938** R. W. Lawson tr. *Hevesy & Paneth's Man. Radioactivity* (ed. 2) xx. 187 The properties of 'heavy hydrogen' or 'deuterium'..differ..strongly from those of the much more abundant ordinary hydrogen. **1941** in M. Gowing *Britain & Atomic Energy* (1964) App. ii. 395 We know that Germany has taken a great deal of trouble to secure supplies of the substance known as heavy water. **1945** H. D. Smyth *Gen. Acct. Devel. Atomic Energy Mil. Purposes* i. 11 A frequently used 'beam' source of neutrons results from accelerated deuterons impinging on 'heavy water' ice. **1946** *Electronic Engin.* XVIII. 142 The deuteron..which is the nucleus of heavy hydrogen, or deuterium. **1947** Crowther & Whiddington *Science at War* 145 The slow neutrons produced by the uranium and heavy-water system would transmute many uranium atoms into the new element plutonium. **1955** *Ann. Reg. 1954* 393 Dimple (deuterium moderated pile, low energy) was Britain's first heavy water reactor. **1956** *Nature* 4 Feb. 205/2 Three enriched-uranium heavy-water-moderated..reactors are under construction. **1958** *Listener* 28 Aug. 294/2 Producing heavy water from electricity from the proposed Aswan High Dam. **1964** M. Gowing *Britain & Atomic Energy* ii. 73 Meanwhile doubts grew about the wisdom of pressing the Americans too hard about the heavy water project.

5. b. Esp. in phr. *heavy industry.* Hence *heavy-industrial* adj. Also *heavy chemicals*: see *CHEMICAL *a.* 6 b.

1932 *Times* 5 Jan. 11/2 The Central Committee of the Communist Party has decided to reorganize the Union Supreme Economic Council, which is to be styled Commissariat of Heavy Industries. **1938** *Archit. Rev.* LXXXIII. 117/2 The majority of 'heavy' industries are of the latter type. **1938** *Times* 17 Feb. 16/1 The comparatively high price of iron and steel in Japan (a result of the artificial character of heavy-industrial growth in a country with insufficient ore and unsuitable coal). **1944** J. S. Huxley *On Living in Rev.* I. iii. 6 The deliberate encouragement of heavy industry under a Five Year Plan, at the expense of all other kinds of enterprise which would have flourished in a *laisser-faire* economy, is the most clear-cut example. **1957** L. F. R. Williams *State of Israel* 44 The raw materials required by her expanding heavy and light industries.

6. Also applied to aerial bombs.

1917 'Contact' *Airman's Outings* 206 A line of narrow-nosed buses, with heavy bombs fitted under the lower planes, ready to leave for their objective.

7. Also of military aircraft, descriptive of a large type of bombing aeroplane.

1921 *Flight* XIII. 615/2 The S.E.5's made the first attack, and dropped about 40 25-lb Cooper bombs, with the object of..preparing the way for the heavy bombers. **1939** *War Illustr.* 18 Dec. 459 Described as modern heavy bombers, these 'planes of the Red Air Force are certainly larger than any standard bomber in the British Air Force.

9. b. *Golf.* Of a ball: lying in sand.

1886 H. Hutchinson *Hints Game Golf* 39 When lying 'heavy'..bear in mind that it is better to hit the ball with the iron than to miss it with a spoon.

13. Of an amatory relationship: intense, intensive; spec. *heavy petting*, non-coital physical contact between two people, involving sexual stimulation of the genitals.

1952 M. R. Rinehart *Pool* xii. 111 He has a sort of heavy date here with a girl called Janey. **1959** 'M. Neville' *Sweet Night for Murder* vii. 76 Duncan was making a very heavy pass at Cathy. **1960** 'M. Caine' *S Man* 126 What is called 'heavy petting' in which frank exploration of each other's bodies is permitted. **1968** M. Richler *Cocksure* xviii. 111 His thirteen-year-old daughter was the only girl in the fifth form to stop at.. *heavy* petting. **1972** *Daily Tel.* 29 Jan. 2/6 Heavy petting between boys and girls is not discouraged and intercourse is described in some detail.

14. b. *fig.*, esp. in phr. *to make heavy weather of*: to make (unnecessary) fuss or labour over.

1915 [see Weather *sb.* 2 c]. **1955** *Times* 21 July 8/5 The Geneva conference ran into heavy weather this morning, but made some ground later. **1957** *Listener* 24 Oct. 664/1 He makes rather heavy weather of the difference. **1960** V. Gielgud *To Bed at Noon* I. xii. 92 Aren't you making rather heavy weather out of nothing?

15. *heavy face* (*type*): see Face *sb.* 22.

1891 [see Face *sb.* 22]. **1898** J. Southward *Mod. Printing* I. xxii. 140 The first would be called a *light face*, and the second a *heavy face*.

17. b. Of a line in Old English verse: containing more than the normal number of stressed elements. Also, more generally, opp. to Light *a.*[1] 12.

1893 J. Lawrence *Chapt. Allit. Verse* 46 Verses with double alliteration are as a rule heavier than those with single. **1948** *Mod. Philol.* XLVI. 81 These heavy and extra-heavy verses, are the exceptions. **1958** A. J. Bliss *Metre of Beowulf* 8 There are also many verses which contain three stressed elements instead of the normal two: blæd | wide | sprang 18 b... All verses of this kind are here termed 'heavy' verses.

19. a. (Further examples.)

1962 *Which?* (Car Suppl.) Oct. 117/1 The Riley 4/72's steering was somewhat heavy and imprecise. *Ibid.* 119/2 The Ford Taunus foot brake was not too heavy at 30 mph but needed a great deal of pressure for gentle stops at 60 mph.

d. Of market conditions.

1831 [in Dict., sense 19 b]. **1843** *Times* 20 May 7/3 The English securities were heavy again to-day. **1935** *Economist* 2 Feb. 261/2 Japanese bonds were heavy. **1962** S. Strand *Marketing Dict.* 339 *Heavy market*, a market of declining prices.

e. *heavy going*: see *GOING *vbl. sb.* 4 a.

20. b. Of newspapers, journals, etc.: serious, addressed to the serious-minded.

1874 Trollope *Way we live Now* (1875) I. xxx. 187 Old Splinter,..who had written for the heavy quarterlies any time this last forty years, professed that he saw through the article. **1967** *Listener* 7 Dec. 743/1 The editors of the heavy dailies.

c. orig. in Jazz and popular music, used in various senses to designate something profound, serious, etc. *colloq.*

1937 B. Goodman *This Thing called Swing* 9 *Mugging heavy*: soft swing with a heavy beat. **1940** *Swing* July 17 Very fast semi-boogie blues in Gabriel with nasty, heavy off-beat drumming. **1958** Blesh & Janis *They all played Ragtime* vi. 117 *Victory Rag*, a 'heavy' number of great difficulty, went on the market in 1921. **1959** 'F. Newton' *Jazz Scene* xiv. 261 Jazz is not simply an ordinary music, light or heavy. **1969** *Rolling Stone* 28 June 38/4 Bass player wanted for heavy blues-rock band. **1969** *It* 4–17 July 10/2 The Rolling Stones..are well supported by such swingin' outfits as..the very heavy Third Ear Band. **1970** *Time* 17 Aug. 32 Marcuse is heavy stuff. **1971** *It* 2–16 June 2/1 The Bournemouth drug squad (reputed to be one of the heaviest squads in the country). **1972** *Last Whole Earth Catalog* (Portola Inst.) 30/1 Not heavy stuff about what is terrible or what should happen, but how to remake life and stay alive in the process. **1972** *Southerly* XXXII. 101 We talk about this and that and where's the heavy dope scene now.

21. (Earlier and additional examples.)

1814 Jane Austen *Mansf. Park* I. xv. 301 'Anhalt' is a heavy part. **1823** *Drama* IV. 209 Mr. Hillington takes the heavy line of business. *a* **1828** J. Bernard *Retrosp. Stage* (1830) I. 13 The Company consisted of a heavy man, who played the tyrants in tragedy. **1833** R. Dyer *9 Yrs. of Actor's Life* 237 This gentleman possesses natural requisite and acquired talents of the first order in heavy tragedy. **1858** H. J. Byron *Maid & Magpie* 35 Such a heavy villin. **1860** G. Vandenhoff *Dramatic Rem.* 176 There was no heavy lady for the Emilias and Lady Macbeths. **1870** T. A. Brown *Hist. Amer. Stage* 54/1 In California she played all lines of business, from walking ladies to heavy, and juvenile leading. **1884** [see Lead *sb.*[2] 7]. **1901** C. Morris *Life on Stage* 40 Then came the leading lady, the first old woman (who was sometimes the heavy woman). **1909** J. R. Ware *Passing Eng.* 151/2 *Heavy merchant*, man who plays the villain. **1941** *Picturegoer* 26 July 6/1 John [Barrymore] started off in heavy drama.

b. Also, ponderously dignified; stern, repressive, unbending: esp. *heavy father*, *heavy uncle*, which are also used as attrib. phrases (= sternly paternal or avuncular). orig. *Theatrical slang.*

1849 Thackeray *Pendennis* I. xxix. 281 Those parts in the drama, which we called the heavy fathers. **1853** 'C. Bede' *Verdant Green* v. 42 He took an affectionate farewell of his son, somewhat after the manner of the 'heavy fathers' of the stage. **1858** H. J. Byron *Maid & Magpie* 4 A Fine Specimen of the good old Heavy Father of Melodrama. **1864** H. Morley *Jrnl.* (1866) 339 A heavy father in broad farce. **1898** Rider Haggard *Doctor Therne* iii, Sir John..received me in his best 'heavy-father' manner. **1931** *Daily Express* 31 Jan. 15/2 The heavy uncle attitude. **1931** *Times Lit. Suppl.* 13 Aug. 613/4 The Venetian Pantalone becomes the Atellane 'heavy father' Pappus. **1956** 'M. Westmacott' *Burden* II. iii. 83 Really, Laura dear, you might be at least fifty. A heavy Victorian father rather than a sister.

22. c. *heavy man*: a criminal or law-breaker. *U.S. slang.*

1926 J. Black *You can't Win* xx. 302 It was the kind of safe that discouraged the 'heavy man' (safe breaker). **1963** H. L. Mencken *Amer. Lang.* (ed. 4) 730 *Heavy man*, one transporting narcotics.

24. b. *heavy-duty* (see Duty 6), used *attrib.*, of a machine, material, etc., designed to deal with heavy materials or to be suitable to stand up to hard wear. Also *transf.*

1914 *Engineering* 4 Dec. 670/2 (*caption*) Heavy-Duty Drilling-Machine. **1935** *Discovery* July 202/1 Fireproof materials can be substituted for practically every form of heavy duty cloth. **1958** *Listener* 20 Nov. 839/3 The heavy-duty rubber tyre. **1964** *English Studies* XLV. 426 Special attention has been given to 'heavy-duty' words such as *have*. **1969** *Computers & Humanities* III. 137 A stand-alone device which consists of a magnetic tape unit, keyboard, and heavy-duty selectric typewriter.

30. heavy bag, a punch-bag; **heavy franc,** name given to the new franc, equivalent to 100 old francs, introduced in France in 1960; **heavy mineral** (see quot. 1971); **heavy oil,** any oil of high specific gravity, orig. such an oil obtained from the distillation of coal-tar (cf. *dead oil* s.v. Dead *a.* D. 2); **heavy sugar** *U.S. slang*, 'big money' (see Sugar *sb.* 2 c); **heavy swell** (earlier example); **heavy-wooded pine,** the western yellow pine, *Pinus ponderosa.*

1950 J. Dempsey *Championship Fighting* 41 Canvas or leather 'dummy bag'—sometimes known as the 'heavy bag'. **1958** *Times* 29 Dec. 6/4 (*headline*) The 'Heavy' Franc. *Ibid.*, A new monetary unit is to be created [in France] worth 100 francs. It will be introduced gradually during the next 12 months. **1959** *Observer* 11 Oct. 3/8 The new 'heavy franc', which officially comes into use next January. **1893** A. Geikie *Text-bk. Geol.* (ed. 3) II. ii. 129 These heavy minerals constitute sometimes as much as 4 per cent of the Bagshot sand. **1939** *Proc. Prehist. Soc.* V. 109 A heavy-mineral analysis of a sample of the sand. **1971** I. G. Gass et al. *Understanding Earth* xiii. 166/2 Rather more satisfactory from the point of view of recognising parent rocks is the presence of a small proportion (often less than 1%) of what are known as 'heavy minerals'. These have a greater specific gravity than the common minerals quartz and feldspar (hence the name), and are separated by breaking up the rocks and floating off the lighter minerals in a heavy liquid (bromoform, S.G. 2.89 is commonly used). Assemblages of heavy minerals may be characteristic of certain groups of parent rocks. **1849** Heavy oil [see *dead oil* s.v. Dead *a.* D. 2]. **1913** V. B. Lewes *Oil Fuel* 129 The heavy tar oil, or 'creosote oil' forms a fairly good liquid fuel. The specific gravity is usually in the neighbourhood of 1.1, hence its name of 'heavy-oil', being heavier than water. *Ibid.* 180 Heavy oil engines. **1936** *Discovery* Feb. 37 Locomotives driven by heavy oil and electricity. **1926** Maines & Grant *Wise-Crack Dict.* 9/2 *Heavy sugar papa*, sweet old man with fat purse. **1928** *Flynn's* 4 Feb. 437/1 Johns with heavy sugar. **1819** Heavy swell [see Swell *sb.* 9 a]. **1836** P. & C. Lawson *Agriculturalist's Manual* 354 Pinus ponderosa—Heavy Wooded Pine... Introduced by Mr. Douglas from the west coast of North America in 1828. **1858** J. A. Warder *Hedges & Evergreens* II. 250 *Pinus ponderosa*, or Heavy-wooded Pine, has leaves from nine inches to a foot long. **1866** 'Senilis' *Pinaceæ* 125 *Pinus Ponderosa*: The Heavy-Wooded Pine. Introduced from North America nearly forty years ago. **1923** Dallimore & Jackson *Handbk. Coniferæ* 437 *Pinus ponderosa*, Douglas. Western Yellow Pine... Big Pine; Bull Pine; Heavy Pine; Heavy-wooded Pine.

31. *heavy-footed* (later example), *-framed*, *-jowled*, *-lidded* (example; so *heavy-liddedness*), *-scented*, *-set*; **heavy-faced,** having a heavy face (see Face *sb.* 22 and sense 15 above); **heavy-timbered,** (*a*) thickly furnished with growing trees; (*b*) large-limbed.

1917 F. S. Henry *Printing for School & Shop* vii. 90 Heavy-faced types are appropriate in printed matter for the iron and steel industry. **1957** T. Gunn *Sense of Movement* 13 Here is a room with heavy-footed chairs. **1898** *Daily News* 14 Mar. 7/2 A heavy-framed colt. *a* **1963** J. Fountain in B. James *Austral. Short Stories* (1963) 2nd Ser. 274 His heavy-framed bike loaded with packages. **1944** A. L. Rowse *Eng. Spirit* 88 That earlier Tudor type ..clean-shaven and heavy-jowled. **1919** V. Woolf *Night & Day* xxvii. 390 Camels slanted their heavy-lidded eyes at her. **1961** *New Yorker* 25 Feb. 129/1, I remember experiencing spells of heavy-liddedness during a fairly recent stage presentation of this talkfest. **1906** *Westm. Gaz.* 4 Jan. 2/1 The heavy-scented buds. **1908** *Ibid.* 7 Aug. 3/1 Never was such clover!..heavy-scented, rich, and generous. **1938** D. Runyon *Furthermore* vii. 130 He is a heavy-set guy. **1831** J. W. Peck *Guide for Emigrants* II. 40 The wide, level, and heavy timbered alluvions, are.. unhealthy. **1861** Whyte Melville *Market Harb.* 7 Deep-ribbed, heavy-timbered hounds. **1903** S. E. White *Conjuror's House* iv. 39 The fort itself, a medley of heavy-timbered stockades and square block-houses. **1908** *Westm. Gaz.* 25 Aug. 7/1 A dark, heavy-timbered wood.

B. *sb.* **1. b.** *the heavies*, the heavy artillery.

1908 *Daily Chron.* 6 Aug. 6/4 The excellent firing of the 4·7 guns by the First and Seconds (or, as they are more familiarly called, 'the Heavies'). **1916** 'Boyd Cable' *Action Front* 113 The Heavies as well as the Field guns were to bombard. **1918** E. M. Roberts *Flying Fighter* 57 Soon the field artillery and the 'heavies' woke up again.

c. A heavy bomber.

1943 *Time* 15 Nov. 26/2 Another co-ordinated series of punches..cost the Allies only ten heavies, two Marauders and five fighters. **1944** *Even. Standard* 16 Dec. 1/4 Heavy bombers could be heard massing for an attack... Wave after wave of 'heavies' went out. **1961** W. Vaughan-Thomas *Anzio* viii. 164 In cold blood the heavies may not have done vital damage to the Germans forming up on the ground.

d. *the heavies*, the serious newspapers, journals, etc. (see sense *A. 20 b).

1950 C. Woodham-Smith *F. Nightingale* 310 In 1857 great influence was exercised by 'the heavies'—the quarterlies and the reviews. **1962** *Guardian* 5 Dec. 7/4, I was reading the Sunday papers... I picked up one of the two Heavies. **1962** 'O. Mills' *Headlines make Murder* viii. 89 All three 'heavies', *The Times*, *Telegraph* and.. *Guardian*. **1971** *Author* LXXXII. 101 The popular press, thrown off balance and uncertain of its role, lost out to the heavies and the provincials.

2. b. Anything particularly large and weighty of its kind.

1897 *Daily News* 25 Nov. 5/1 Amongst the elephantine heavies is Mandarin, who killed a keeper during his last residence at Olympia. **1908** *Pall Mall Gaz.* 27 Mar. 12/3 Firms..which have specialised in the manufacture of 'heavies' [*sc.* motor vehicles]. **1935** *Amer. Speech* X. 271/1 *Heavies*, very heavy beef cattle, more than two years old. **1960** *Farmer & Stockbreeder* 15 Mar. 6/2 Of the uncertified beasts, lightweights were more plentiful than heavies. **1965** *Listener* 8 Apr. 537/1 The 'heavies' are on the march. By 1970 there will be 120 trucks for every 100 there are now on our inadequate roads. **1968** *Times* 25 Oct. 25/3 (*heading*) 60 mph for 'heavies' [*sc.* motor vehicles].

c. A heavyweight boxing-match or boxer.

1913 J. G. B. Lynch *Compl. Amat. Boxer* 221, I remember in the finals of the heavies at the All-India Championship of 1909 seeing Private Clohessy..take on Bombardier Wells. **1950** J. Dempsey *Championship Fighting* 16 Top-flight heavies like Frank Moran.

d. *pl. Racing.* Horses' work-shoes.

1930 *Times* 24 Mar. 4/2 It is almost impossible to tell by watching a horse walk in the parade ring whether he is plated, or whether he is carrying, to use a racing term, 'the heavies'.

e. A strongly built person, usu. of violent disposition.

1936 [see *COME *v.* 13 d]. **1962** R. Cook *Crust on its Uppers* i. 22 A good solid heavy like Chas to deal with the writ-servers. **1970** G. Greer *Female Eunuch* 194 Cherry is surrounded by threatening creatures, mostly the nightclub heavies. **1972** *Catholic Herald* 28 Jan. 2/5 Sit down, we want to talk to you... We are going out to get the 'heavies'. **1973** *Times* 12 July 4/1 Prostitutes were threatened with 'heavies' working for a man named Kenny Lynch.

5. esp. *Theatr.* Short for *heavy actor*, *villain*, etc. Cf. sense A. 21 in Dict. and Suppl.

1880 F. Belton *Random Recoll. Old Actor* viii. 132 Robertson (the celebrated author of 'Caste' 'School' etc. for 'second heavies'). **1906** S. Ford *Shorty McCabe* (1908) 70 So far it's as good as playin' leading heavy in 'The Shadows of a Great City'. **1928** *Observer* 22 July 15/2 The fun succumbed to a bucolic lethargy that was only partially shaken off by the retreat to Half Moon Street and the assault of the sentimental heavies. **1937** 'C. McCabe' *Face on Cutting-room Floor* vii. 53, I asked..who the man was... 'That's Vic's new heavy.' **1961** J. McCabe *Mr. Laurel & Mr. Hardy* (1962) ii. 49, I always played a 'heavy'—you know, the villain. *Ibid.* ii. 57 The villains in those days were always called 'heavies'. Their trademark was usually *heavy* eyebrows and moustache make-up. **1962** J. D. Salinger *Franny & Zooey* 143 I'm sick to death of being the heavy in everybody's life... They're as happy as pigs till I show up. I feel like those dismal bastards Seymour's beloved Chuang-tzu warned everybody against. **1966** *Listener* 15 Dec. 890/2 Two of the chief characters are avowed communists, and yet are not the heavies plotting to overthrow the free world. **1973** J. Wainwright *Pride of Pigs* i. 178 Two of the heavies dived for Tallboy.

6. *pl.* The heavy trades or industries (see Heavy *a.* 5); also, stocks or shares in such a trade or industry.

1900 *Westm. Gaz.* 26 Jan. 11/1 Hope for the Heavies. **1902** *Daily Chron.* 20 May 3/6 The ratio of working expenses for the past half-year on the North British railways was 49 per cent., compared with 65 per cent., or more, on the four 'heavies'. **1922** *Daily Tel.* 12 June 2/3 The prices of the other comparable 'heavies'—Great Western and London and North-Western—have also gone ahead.

heavy-handed, *a.* Add: **3. b.** Of a joke, humour, etc.: clumsy.

1910 A. C. Benson *Diary* 7 Dec. (1926) 199 Their jokes are very heavy-handed, and generally involve discomfort for the victim. **1934** *Amer. Speech* IX. 158/2 The work of heavy-handed humorists.

heavy-weight. Also **heavyweight. a.** Substitute for def.: A person or animal of more than the average weight; *spec.* a jockey, etc., of more than the average weight; a professional boxer weighing over 12 st. 7 lb., or *transf.*, a horse which carries more than the average weight.

1877 'Pugnus' *Hist. Prize Ring* II. 104 For such a heavy weight, Hooper had a particularly small foot, of which he was very proud. **1888** *Encycl. Brit.* XXIV. 691/1 Heavy weights [amateur] to be over 11 stone 4 lb. **1889** [see Feather-weight 3]. **1910** [see Welter weight 2]. **1917** *Sat. Rev.* 10 Nov. 373/2 Thanks to the muscle, he has become a champion heavy-weight. **1928** [see *CRUISER 4]. **1954** F. C. Avis *Boxing Dict.* 52 *Heavyweight*, a standard weight division for professional boxers weighing more than 12 st. 7 lb.; for amateurs, 12 st. 10 lb.

c. A work of large size or serious content. Cf. Heavy *a.*[1] 20 a (in Dict.), b (in Suppl.).

1928 Blunden *Undertones of War* x. 111 When I saw scattered about the porch and the doorstep,..a number of volumes,.. I could not but snatch up four or five... The heavyweight was..a treatise on Country Houses. **1963** *Punch* 3 Apr. 488/2, I have managed to convert my wisp of a *conte* into a heavyweight.

d. *attrib.* Also as *adj.*, particularly heavy of its kind.

1895 *Montgomery Ward Catal.* 283/1 Ladies' jersey knit ribbed vests, heavy weight Egyptian cotton. **1909** *Westm. Gaz.* 2 Mar. 5/2 In a variety of cloths, serges, and heavy-weight cashmere. **1931** *Bombay Chron.* 14 Oct. 1 Heavy-weight championship. **1934** G. B. Shaw *Too True to be Good* I. 40 Oh, sweetiest, why did you tell me that this heavyweight champion was a helpless invalid? **1958** *Times* 26 Sept. 6/3 Heavyweight American and Continental trucks and buses are not being exhibited. **1963** *Times Lit. Suppl.* 4 Jan. 16/5 Figgins's heavyweight version of Caslon's (1816) Sanserif in 1832. **1972** *Guardian* 31 Oct. 11/5 Underwear: Heavy-weight tights in nylon.

hebdomadarian (hebdǫmădēə·riăn). [f. Hebdomadary *sb.* and *a.* + -ian.] = Hebdomadary *sb.*

1898 W. St. Clair Baddeley *Cotteswold Shrine* 75 At a signal given by the hebdomadarian of the week they returned to the Cloister. **1949** M. Baldwin *I leap over Wall* vi. 139 The 'hebdomadarian' (such was the official title of the 'Great Week' keeper in the *Ceremonial*) was continually in evidence.

hebe[1]. Add: **2*.** *Bot.* [mod.L. (P. Commerson in A. L. Jussieu *Genera Plantarum* (1789) 105).] A member of a large genus of shrubs so called, mostly native to New Zealand, belonging to the family Scrophulariaceæ, and formerly included in the genus *Veronica.*

[**1921** F. W. Pennell in *Rhodora* XXIII. 2 The austral distribution, with its suggestion of genetic remoteness, emphasizes *Hebe's* claim to recognition as a genus. **1927** Cockayne & Allan in *Trans. N.Z. Inst.* LVII. 13 A species is usually transferred to *Hebe* only when..we are pretty well convinced it is valid.] **1961** *Amat. Gardening* 21 Oct. 5/1 The hebes, as the shrubby veronicas are now called. **1972** *Country Life* 23 Mar. 690/1 Our hebes have come through remarkably well: even the tenderest kinds like Simon Deleaux and Andersonii Variegata.

Hebe, hebe[2] (hīb). Also **Heeb.** [abbrev. of Hebrew *sb.* 1.] A derogatory term for a Jew.

1932 J. T. Farrell *Studs Lonigan* (1936) iv. 71 He should've been a nigger or a hebe instead of Irish. **1946** Mezzrow & Wolfe *Really the Blues* 374/2 *Heeb*, Jewish person. **1950** T. Sugrue in M. Hay *Foot of Pride* p. xx, He might go through the whole of his life without expressing more than a casual distaste for 'the Hebes'. **1953** E. F. Russell *Somewhere a Voice* (1965) 11 Lastly there was Sammy Finestone... A typical Hebe. **1972** *National Observer* (U.S.) 27 May 17/4 They will be followed close upon their heels by miserly Hebes, and cheating kikes.

hebe-. Add: **b. hebephrenia** (examples); **hebephreniac** *sb.* (examples); also **hebephre·nic** *a.* and *sb.*

1883 W. A. Hammond *Treat. Insanity* 556 Hebephrenia..is the term applied to the insanity of pubescence. **1948** *Brit. Jrnl. Psychol.* XXXIX. 89 Hebephrenia is featured by silliness, incongruity, mannerisms etc. **1956** C. P. Snow *Homecomings* 348 He lay on his back, his legs relaxed, like a figure on a tomb or one in a not disagreeable state of hebephrenia. **1884** *Jrnl. Nerv. & Mental Dis.* XI. 303 Imperative conceptions are relatively frequent among hebephreniacs. **1885** *Ibid.* XII. 516 (*heading*) Autopsy findings in a hebephreniac. **1908** *Practitioner* Jan. 12 The patient may gradually become imbecile and demented—the hebephrenic type. **1915** C. R. Payne tr. *Pfister's Psychoanalytic Method* 542 Dementia praecox (in catatonic, hebephrenic and paranoid forms). **1938** S. Beckett *Murphy* 168 A hebephrenic playing the piano intently. **1973** 'E. McBain' *Let's hear It* v. 59 He considered them [*sc.* the police] obsolete and essentially hebephrenic.

Heberden's nodes (he·bəɹdenz nōᵘdz). *Med.* [f. the name of William *Heberden*, Eng. physician (1710–1801), who described the condition.] Nodular enlargements of the terminal joints of the fingers due to osteoarthritis.

1889 D. Duckworth *Treat. Gout* iv. 72 (*heading*) Heberden's nodes in a case of gout—..patient has noticed a gradual increase in size of the joints of the fingers... Heberden's nodes, both hands. **1911** *Encycl. Brit.* XXIII 239/1 What are termed 'Heberden's nodes' are small hard knobs about the size of a pea frequently found upon the fingers near the terminal phalangeal joints... These nodes are..a manifestation of arthritis. **1961** R. D. Baker *Essent. Path.* xxi. 578 In the hands the hypertrophy of the bones about the distal interphalangeal joints is known as Heberden's nodes. **1972** H. L. Jaffe *Metabolic, Degenerative & Inflammatory Dis. Bones & Joints* xxv. 757 With advancing age, the incidence of Heberden's nodes increases.

hebetude. (Later examples.)

1918 E. Pound *Let.* 3 Apr. (1971) 133 There is *something* in his [*sc.* Jules Romains's] work. It is not the hebetude of a lignified cerebrum. **1955** W. Gaddis *Recognitions* II. vi. 564 The robe was too big. Nevertheless, the pattern was so conservative, and the material so fine, that this seemed rather a mark of luxuriance than some deliberate hebetude on the part of the giver.

hebra (he·brā). Also **chevra(h).** Pl. **hebras, hebroth, chevroth.** [Heb. *ḥeḇrāh*, association, society; group as small religious community.] (See quot. 1959.)

1880 *Jewish Chron.* 6 Feb. 9/2 The poorer classes prefer to belong to one of the numerous *Hebras* which.. abound in the East End. *Ibid.* 4/1 Many of these unattached Jews are not wealthy... They..have their *Hebras*, their customs and peculiarities. **1892** I. Zangwill *Childr. Ghetto* I. I. ii. 44 Even in the smallest *Chevrah* the high hat comes next in sanctity to the Scroll of the Law. **1918** C. G. Montefiore *Liberal Judaism & Hellenism* v. 259 This oligarchic tendency is counteracted by the little Synagogues and the *Chevras*. **1959** F. M. Wilson *They came as Strangers* III. iii. 195 Thousands of Jews were too poor to be seat-holders in the synagogue, so they formed *chevras*, which were partly Friendly Societies, and partly groups for public worship and the study of the Talmud. **1960** *Economist* 25 June 1331/1 Here and there are the little *chevras* or conventicles whose continued existence so much annoyed the big synagogues. **1960** L. P. Gartner *Jewish Immigrant* vii. 186 In the *hebra*..a Jew associated himself with fellow Jews..for..worship and study and conviviality.

Hebraist. Add: **4.** One who maintains that the New Testament was written in Greek that contained Hebrew idioms.

1859 E. Masson tr. *Winer's Gram. N.T. Diction* I. i. 25 Various..scholars (the Purists) perseveringly endeavoured to demonstrate that the style of the N.T. entirely reaches the standard of classical Greek purity..while others (the Hebraists) maintained..that it exhibits a..predominant Hebrew tincture. **1906** J. H. Moulton *Gram. N.T. Greek* (1908) 3 The Hebraist went absurdly far in recognising Semitic influence where none was really operative. **1907** [see Purist 2].

Hebrid (he·brid), *a.* [ad. *Hebrides*: see next.] = *Hebridean *a.* Also **He·bridal** *a.*

1748 Thomson *Cast. Indol.* I. xxx, A shepherd of the Hebrid Isles. **1841** W. Yarrell *Brit. Fishes* (ed. 2) II. 133 The Hebridal Smelt, *Osmerus Hebridicus*..is at once clearly distinguishable from our long-known and highly-esteemed favourite, the common Smelt. **1923** *United Free Ch. Mission Rec.* July 275 From many a Hebrid isle ..from plain manses and luxurious mansions these men and women have come.

Hebridean (hebridī·ăn, hēbri·dīăn), *a.* and *sb.* Also 6–9 **-ian.** [f. *Hebrides*, an alteration, said to have originated in an accidental misprint, of L. *Hebudes* (Pliny), Gr. *Ἐβουδαι* (Ptolemy).] **A.** *adj.* Of or pertaining to the Hebrides, a group of islands off the west coast of Scotland. **B.** *sb.* A native or inhabitant of the Hebrides.

1623 Cockeram, *Hebridean wave*, the Irish sea. **1632** Lithgow *Trav.* 494 The desperate courage of these awfull Hebridians. **1641** in J. Sylvester tr. *Du Bartas's Weeks* 331/1 *Hebridian Wave*, the Sea about the Isles Hiberides, to the North from Ireland. **1775** Johnson *West. Isl.* 221 Such intelligence the Hebridians probably receive from their transmarine correspondents. **1780** J. Johnstone (*title*) Anecdotes of Olave the Black, King of Man, and the Hebridian Princes of the Somerled Family. **1810** Scott *Fam. Lett.* 10 June (1894) I. 181, I intend to take the Hebridian character and scenery..for my subject. **1828** J. Tytler *Hist. Scot.* I. i. 29 Dugal and other Hebridean chiefs. **1833–4** J. Phillips *Geol.* in *Encycl. Metrop.* VI. 561/1 The Zetland Isles, which are in some measure to be viewed as a prolongation to the Hebridian group. *a* **1856** H. Miller *Cruise Betsey* (1858) vi. 90 The penetrating powers of a true Hebridean drizzle. **1887** H. B. Woodward *Geol. Eng. & Wales* (ed. 2) 38 The Lewisian (or Hebridean) group. **1897** R. H. Story *Apostolic Ministry Scot. Church* v. 154 The beehive cells, the remains of which may still be seen in Eilean Naomh and other Hebridean isles. **1911** *Encycl. Brit.* XXIV. 416/1 The oldest rocks of Scotland and of the British Isles are known..as Archean, and consist chiefly of gneiss (called..Lewisian and Hebridean). **1926** *Chambers's Jrnl.* Sept. 577/2 Our comely Hebridean. **1972** *Daily Tel.* 29 Jan. 16/5 The Hebridean processes of cloth-treatment are in danger of falling into oblivion.

hechima (hetʃi·mā). [Jap.] The sponge-gourd (see Sponge *sb.*[1] 13 c).

1883 *Trans. Asiatic Soc. Japan* XI. 13 (*heading*) Luffa petola, Ser. Hechima... Young fruit as food: and fibres of ripe fruit as brushes and sponges. **1889** J. J. Rein *Industries of Japan* i. 72 (*heading*) 21. *Luffa petola*, Ser., Jap. Hechima, Tô-guwa. The long cylindric fruit resembles a long straight cucumber.

heck (hek), *sb.* and *int. dial.* and *colloq.* Euphemistic alteration of *hell*. (Also *hecky* in dial. use.)

[**1865** [J. A. Ferguson] *Wot Aw seed ut th' Preston Eggsibishun* 88 (E.D.D.), Well, aw'll go to ecky, he cried. **1878** J. Almond *Bunch of Watercresses* 21 Where the hecky could he go to?] **1887** T. Darlington *Folk-sp. S. Cheshire* s.v., What the heck are yŏ up to? **1922** S. Lewis *Babbitt* xiv. 178 How it feels, by heck, to be up at five-thirty. **1925** *Blackw. Mag.* Oct. 545/1, I couldn't make out what in heck was going on. **1928** M. Walsh *While Rivers Run* ii. § 3 By heck! what a kick he must have in that right of his. **1930** *Daily Express* 23 May 10/3 Does the borough council care? By heck, it doesn't! **1932** J. T. Farrell *Studs Lonigan* (1936) iii. 68 He would have the heck of a time explaining his shiner to the old lady. **1933** *Punch* 11 Jan. 52/1 He insisted on St. Isinglas because he thought everything here was so well organised. The heck it is. **1936** M. H. Bradley *Five Minute Girl* xiii. 236 He had certainly played heck with that party. **1956** E. Pound tr. *Sophocles' Women of Trachis* 17 That fellow was lying, one time or the other, One heck of a messenger! **1957** I. Cross *God Boy* (1958) i. 12 Heck now, I started off with Dad talking to me..and here we are no further on. *Ibid.* xii. 95 People go all the way the heck over to France. **1966** *Guardian* 5 Feb. 6/4 Sometimes he sings for sheer fun and the heck of it. **1973** D. Westheimer *Going Public* i. 15 It's a heck of a responsibility.

heckelphone (he·kəlfōᵘn). Also **-phon.** [ad. G. *heckelphon*, f. name of Wilhelm *Heckel* (1856–1909), an instrument-maker of Biberich, after *saxophone*.] A baritone oboe.

1905 *Westm. Gaz.* 14 Dec. 8/2 A new instrument, called appropriately the 'Heckelphon', answered from the orchestra. **1914** C. Forsyth *Orchestration* 228 Strauss has made use of the Heckelphon in..*Salome*. **1940** G. Jacob *Orchestral Technique* (ed. 2) i. 3 Some very fully scored modern symphonic and operatic works employ quadruple woodwind, such exotics as the bass flute and heckelphone sometimes being added to the flute and oboe group respectively. **1966** *Guardian* 22 Apr. 12/5 A heckelphone (a kind of outsize 'baritone' oboe).

heckle, *sb.* Add: **4. b.** The action of heckling.

1905 *Westm. Gaz.* 29 Apr. 3/2 Our congratulations to.. Mr. Davies on his..successful heckle. **1944** H. van Zeller *Ezechiel* viii. 69 We again look for some sort of an apologia, an objection or two, a mild heckle, a question.

hecogenin (hekodʒe·nin). *Chem.* [f. mod.L. *Hec*(*htia*, name of a genus of plants + -o + *GENIN.] A steroid glycoside (see quot. 1965) occurring in various plants, as *Hechtia texensis* and *Agave* species, obtained commercially from sisal waste and used commercially as a precursor in the manufacture of cortisone and other steroid hormones.

1943 R. E. Marker et al. in *Jrnl. Amer. Chem. Soc.* LXV. 1199/2 We first isolated from *Hechtia texensis* (S. Wats.) a new steroidal sapogenin having the composition $C_{27}H_{42}O_4$..which we have named hecogenin. **1952** *Sci. News* XXV. 115 More than 100,000 tons of fibre are produced yearly from sisal,..and..approximately the same amount of waste material, containing 0.04 per cent to 0.1 per cent of hecogenin, are available. **1965** Pollock & Stevens *Dict. Org. Compounds* (ed. 4) III. 1809/1, 3β-Hydroxy-5α-spirostan-12-one (Hecogenin). **1972** *Materials & Technol.* V. xx. 752 Hecogenin is extracted from sisal waste.

hecte, var. *HEKTE.

hectic, *a.* Add: **4.** Stirring, exciting, disturbing; characterized by a state of feverish excitement or activity. *colloq.*

1904 Kipling *Traffics & Discov.* 210 Didn't I say we never met in pup-pup-puris naturalibus, if I may so put it, without a remarkably hectic day ahead of us? **1922** *Westm. Gaz.* 19 Aug., The hectic undulations of the mark. *Ibid.* 7 Nov., Those hectic inconsidered actions which kept the country in a state of crisis for some ten days. **1922** *Daily Mail* 21 Nov. 11 As additional excitements to the hectic finishes one horse was killed and the judge mistook the winner of the Leycester Nursery. **1925** Fraser & Gibbons *Soldier & Sailor Words* 117 *Hectic show*, an Air Force expression for flying very low. **1968** *Daily Mirror* 20 Aug. 9/2 Stretch bikini bottoms and loose towelling tops in hectic colours.

hectically. Add: **b.** With feverish activity. *colloq.*

1908 *Daily Chron.* 18 May 3/5 So many pictures in the Salons look as if they had been painted hectically a month before the exhibitions opened. **1972** *Daily Tel.* 15 Jan. 11 A hectically social couple of my acquaintance.

hectography (hektǫ·grăfi). [f. Hectograph *sb.*: see -graphy.] The use of the hectograph.

1889 J. H. Scrine *Mem. E. Thring* 31 Splice 3 pens together,..and execute the task..by fraudulent hectography.

hector, *v.* Add: Hence **he·ctoringly** *adv.*, in a hectoring manner.

1913 W. J. Locke *Stella Maris* xx. 251 He..questioned her further, almost hectoringly. **1963** *Punch* 27 Feb. 315/3 His adopting a hectoringly superior manner.

hedebo (he·debo). [ad. Dan. *hedebobroderi*, f. *hede* heath + *bo* dwelling + *broderi* embroidery.] (See quots.)

1932 D. C. Minter *Modern Needlecraft* 49/2 The drawing of threads..forms the basis of various interesting forms of embroidery. This network..is sometimes used as a contrast for parts of a richly embroidered surface. This kind of work is known as *Hedebo work*, and is usually executed almost entirely in white. **1957** M. B. Picken *Fashion Dict.* 171/1 *Hedebo embroidery*, Danish embroidery of cut and drawn work.

heder, var. *CHEDAR.

hedge, *sb.* Add: **5.** Also *Commercial*, *Financial*, and *transf.* (cf. *HEDGE *v.* 8 c).

1917 A. W. Atwood *Exchanges & Speculation* xiv. 196 The local elevator companies..place their hedges as soon as they begin to accumulate supplies of grain. **1955** *Times* 8 June 9/2 Your board are keenly aware of the need that the group should continue to build up, in the form of profitable interests elsewhere, 'hedges' against catastrophe in British Guiana. *Ibid.* 29 June 11/3 As for the hedge of going into television itself that may save the property but it will not save the Press. **1957** *Economist* 12 Oct. 152/2 In France, wool was probably taken into stock as a hedge against currency devaluation and the pressure of credit restriction is also at work there. **1958** *Punch* 19 Nov. 669/2 A good unit trust group..provides the best way of combining the safety of numbers with the promise of participating in the economic growth that lies before us and of providing a hedge against inflation. **1959** *Ibid.* 19 Aug. 54/2 The share of these companies would seem to be a perfect election hedge.

7. *hedge-cricket*, *-cutter*, *-cutting*: examples; *hedge-wren*.

1820 Keats *To Autumn* in *Lamia*, etc. 139 Hedge-crickets sing. **1881** *Encycl. Brit.* XII. 234/1 A new instrument for clipping hedges, Ridgway's hedge-cutter. **1960** *Farmer & Stockbreeder* 2 Feb. 80/1 There was a record entry of nearly 100 mechanical hedge-cutters. **1971** P. Gresswell *Environment* 125 The mechanical hedge-cutter clears the young saplings. **1899** *Westm. Gaz.* 22 Dec. 1/3 Hedge-cutting competitions have a..useful effect in checking the use of barbed wire. *Ibid.* 21 July 2/3 The swallow does not fear us, the hedge-wren does not flout us. **1907** *Academy* 9 Feb. 131/2 The hedge-wren..Is out in the open.

9. hedge-clause *U.S.*, a safeguarding clause in a contract; **hedge-fence**, a hedge serving as a fence; **hedge-grown** *a.* (see quots.); **hedge-hop** *v. colloq.*, to fly in an aircraft at low levels so as to suggest hopping over hedges; so **hedge-hopper, -hopping** *vbl. sb.* and *ppl. a.*; also *fig.*; **hedge selling** (see 5 above and *HEDGE *v.* 8 c); **hedge trimmer** (see Trimmer 3); **hedge trimming,** (*a*) (see Trimming *vbl. sb.* 1 b); (*b*) (see Trimming *vbl. sb.* 1 c).

1928 *Sat. Even. Post* 10 Mar. 185/2 In the Wall Street language..hedge clauses..signify that if the representations turn out to be wrong the banker shall not be held accountable. **1662** *Portsmouth* (R.I.) *Rec.* 396 The said fence..provided that it be a hedge fence. **1778** *Essex Inst. Hist. Coll.* LII. 13 Seeing this hedge fence, they might take it to be a breastwork thrown up to annoy them. **1826–44** Hedge fence [in Dict., sense 7 a]. **1820** Keats *Fancy* in *Lamia*, etc. 124 The first Hedge-grown primrose that hath burst. **1900** *Daily Express* 3 Aug. 2/7 Barley is not so good this year, for it has come up irregularly..; barley of this character is known with us as a 'hedge-grown crop'. **1926** *Nat. Geogr. Mag.* Jan. 18/2 Back he'd go 'upstairs' under a 200-foot ceiling, and hedge-hop along 20 miles or so, to the next emergency field. **1928** *Daily Express* 21 May 10/3 They can 'hedge-hop' with skill or fly to the greater heights with as much impunity as a man pilot. **1940** 'Gun Buster' *Return via Dunkirk* II. xi. 171 A German plane hedge-hopped right over us. **1962** L. Deighton *Ipcress File* vii. 49 The machine [*sc.* a helicopter]..hedge-hopped in 100 m.p.h. gallops towards the sea. **1940** H. E. Baughman *Aviation Dict.* 96/1 *Hedge-hopper airplane* (slang), any small, under-powered plane with enough power and lift to get off the ground a few feet for a brief period of time. **1957** R. W. Zandvoort et al. *Wartime Eng.* 127 *Hedge-hopper*, a low-flying aircraft. **1919** R. H. Reece *Night Bombing with Bedouins* 23 The British sport of 'hedge-hopping', i.e., flying close to the ground and 'zooming' up over trees. **1939** *War Illustr.* 11 Nov. 286 The German pilot's story of his eventful pursuit of a hedge-hopping English 'plane. **1955** *Amer. Speech* XXX. 72 Hedge-hopping..took on a special meaning in the 1952 presidential campaign—using an airplane in political campaigning. **1957** L. Durrell *Bitter Lemons* 100 My own ambitions were more hedge-hopping and my means forbade me to indulge in such delightful fantasies. **1920** J. Stephenson *Princ. Commercial Corr.* II. xiii. 151 Further liquidation and some hedge selling caused another decline. **1930** *Daily Express* 8 Sept. 2/7 A reaction occurred owing to liquidation, hedge selling, a bearish crop estimate. **1964** *Financial Times* 12 Mar. 2/7 Selling was stop loss, together with some hedge selling. **1870** *Trans. Ill. Agric. Soc.* VIII. 18 For want of opportunity there had been no test made of the Hedge Trimmer, entered by D. Oliver, of Galesburg, Ill. **1961** *Times* 26 May 16/7 Hedge-trimmers and verge-cutters are a common part of the roadside scene. **1859** *Trans. Ill. Agric. Soc.* III. 362, I believe that the men are..here present, who will live to see..hedge trimming &c., done by steam. **1960** *Farmer & Stockbreeder* 12 Jan. 63/2 Disposing of hedge-trimmings.

10. hedge-cactus *U.S.*, a cactus (*Cereus peruvianus*) grown as a hedge-plant; **hedge-nettle** (later example).

1883 J. H. Beadle *Western Wilds* xxxvi. 593 There is.. the hedge cactus, with which Mexicans fence their fields. **1869** J. G. Fuller *Flower Gatherers* 277 There are several other species of the Hedge-nettle, some of them without hairs.

hedge, *v.* Add: **8. c.** To insure against risk of loss by entering into contracts which balance one another. Also *trans.*, to operate (in a commodity) in this way.

1909 J. Fisher *Elimination of Risk* 12 An important method of shifting risks is 'hedging', whereby a dealer, for instance in transporting wheat, may be relieved of the risk of a change of price. **1917** A. W. Atwood *Exchanges & Speculation* xiv. 193 Hedging..consists in matching a purchase with a sale, or vice versa; in other words, it consists in making a purchase or sale for future delivery to offset and protect an actual merchandising transaction. *Ibid.* xiv. 197 It makes little difference to an elevator if wheat rises or falls fifty cents a bushel, provided its holdings have been hedged. **1957** *Times* 19 Dec. 16/1 We have drawn the attention of the stockholders to the difficulty in hedging our unsold stocks against a fall in cotton values.

hedge-bank. (Later examples.)

1900 *Daily News* 21 Sept. 3/2 Deeply laid roads and high, steep hedge banks. **1909** *Westm. Gaz.* 6 Mar. 16/3 A network of tiny banks..crossing and recrossing from hedge-bank to stack. **1937** *Discovery* Apr. 120/2 Feverfew, commonly found about hedgebanks.

hedgehog. Add: **4. c.** (Earlier example.)

1838 *Civil Engin. & Archit. Jrnl.* Dec. 391/1 (*title*) A machine called a hedgehog for removing mud etc. in rivers.

f. (Earlier and later examples.)

1723 J. Nott *Cook's & Confect. Dict.* 28 (*heading*) To make a Hedge-hog. *Ibid.*, Almonds.. Eggs..Cream... Butter..stirring; till it is stiff enough to be made in the Form of a Hedge-hog; then stick it full of blanch'd Almonds..like the Bristles of a Hedge-hog. **1960** *Good Housek. Cook. Bk.* 405/2 Hedgehog cake.

g. A fortified position 'bristling' with guns pointing in all directions.

1942 *Daily Tel.* 22 May 1/2 The German Infantry has been used to bring up tanks and, throughout the past winter, to holding strong points called 'hedgehogs'. **1943** *Times* 16 Dec. 4/2 The Germans fought with the utmost ferocity for their old 'hedgehog' position. **1951** *Time* 29 Sept. 18/1 Holdfast's strategists had developed their plan after studying German tactics in the long retreat from Stalingrad (in which the Germans first used the word 'hedgehog').

h. (See quot. 1947.)

1947 *Jane's Fighting Ships 1946–7* 6 Anti-Submarine Weapons..include the 'hedgehog', a salvo of 24 depth charges each containing 32 lb. of explosive fired ahead of a ship from a spigot mortar. **1968** D. McLachlan *Room 39* xiv. 329 The dangers of underwater hedgehogs to the Mulberry floating harbours [had not] been closely examined.

6. *hedgehog roller.*

1930 *Engineering* 14 Nov. 615/2 An elevator..delivered the clay into small hedgehog rollers.

7. b. **hedgehog converter, transformer** *Electr.*, a type of transformer (no longer used) with an open magnetic circuit, in which the ends of the iron wire core assume a bristling appearance; **hedgehog fish**, = *porcupine fish*; **hedgehog wheat**, a race of hardy dwarf wheats, grown in mountainous districts of Europe, having dense short ears and awned glumes.

1851 P. H. Gosse *Nat. Sojourn Jamaica* 244 Specimens of the Hedgehog-fish, or Sea Porcupine (*Diodon*), are frequently carried home by mariners. **1902** *Encycl. Brit.* XXVIII. 117/2 The wire..used..to form the core of his 'hedgehog' transformers. **1909** Webster, Hedgehog wheat. **1921** J. Percival *Wheat Plant* 307 Club, Dwarf, Cluster or Hedgehog Wheat, *Triticum compactum*.

hedging, *vbl. sb.* Add: **3.** (Later examples.)

1917 [see *Hedge v. 5]. **1940** *Economist* 11 May 863/1 Much of the apparent speculation taking place in markets, is, in fact, justifiable hedging either against receipts of sterling..or against the holding of sterling assets.. But over and above such hedging some outright speculation also is proceeding. **1954** *Ibid.* 22 May 642/2 The Liverpool market should..offer Lancashire a satisfactory hedging medium. **1958** *Spectator* 13 June 785/1 The tenacity of the 'inflation hedging' investor.

5. *hedging-glove* (later example).

1906 Kipling *Puck of Pook's Hill* 235, I was cheated.. over a pair of hedging-gloves.

hedgy, *a.* Add: **c.** Of behaviour (see Hedge *v.* 9).

1928 D. H. Lawrence *Lady Chatterley* iv. 38 Clifford was much more hedgy and nervous. **1955** 'C. H. Rolph' *Women of Streets* 154 Personality: Suspicious, hedgy, aggressively defensive.

hedonal (hĕ·dŏnăl). *Chem.* [a. G. *hedonal* (H. Dreser 1899, in *Verh. d. Ges. deutsch. Naturf. und Ärzte* II. 48), f. Gr. ἡδονή pleasure + -al.] A white crystalline compound, $C_6H_{13}O_2N$, that has been used as a hypnotic and an anæsthetic; methyl-propyl-carbinol urethane.

1900 *Brit. Med. Jrnl.* epit. 21 July 12/1 (*heading*) Hedonal. Schüler..publishes 21 cases in which this new hypnotic was used in Krafft-Ebing's clinic. Hedonal is methyl-propyl-carbinol-urethan, a white crystalline body. **1905** *Med. Ann.* 191 Hedonal has been used by Vargas in the treatment of chorea. **1927** *Observer* 27 Mar. 13/1 Drugs such as hedonal may be injected..to facilitate or produce surgical anaesthesia. **1971** McComish & Bodley *Anaesthesia for Neurol. Surg.* i. 15 The vogue for hedonal [as an anæsthetic] did not last long because the war in 1914 stopped supplies from Germany, and toxic effects later became apparent.

hedonic, *a.* and *sb.* Add: **A.** *adj.* In wider use, chiefly in *Psychol.*: of, pertaining to, or involving pleasurable or painful sensations or feelings, considered as affects. Spec. *hedonic tone*, the degree of pleasantness or unpleasantness associated with an experience or state, esp. considered as a single quantity that can range from extreme pleasure to extreme pain.

1900 G. F. Stout *Man. Psychol.* (ed. 2) I. i. 63 When we wish to say that pleasure or displeasure belongs to this or that mental process, we say that the process is pleasantly or unpleasantly toned. Hedonic tone is a generic term for pleasure and the reverse, considered as attributes of this or that mental process. *Ibid.*, Again his hedonic tone, mostly of an unpleasant kind. **1932** J. G. Beebe-Center *Psychol. Pleasantness & Unpleasantness* i. 6 In the present volume..the general algebraic variable, whose positive values correspond to pleasantness and whose negative values correspond to unpleasantness, will be called hedonic tone. **1940** *Jrnl. Exper. Psychol.* XXVI. 233 The oscillations of hedonic tone in his case are slight, and the tone rises continuously from the beginning, in spite of pain and fatigue. *Ibid.* 227 While S worked E took their tapping rate every minute..and at the same time called at stated intervals for a rating on a previously agreed hedonic scale. **1959** D. J. O'Connor *John Locke* 51 By pleasure and pain Locke..is referring to what the psychologists nowadays call the hedonic tone of our experiences which can be roughly measured on a scale ranging from very pleasant through mildly pleasant, neutral, mildly unpleasant to very unpleasant. **1966** P. T. Young *Motivation & Emotion* v. 153 The sign, intensity, and temporal changes of affective processes can be represented upon the hedonic continuum.

2. *Zool.* Of or pertaining to sexual activity; *hedonic gland*, any of various specialized glands found in many reptiles and amphibia that serve, apparently by secreting an attractive-smelling substance, to attract members of the opposite sex.

1901 H. Gadow *Amphibia & Reptiles* x. 443 All the recent Crocodilia possess two pairs of skin-glands, both secreting musk... The use of these strongly scented organs, which are possessed by both sexes, is obviously hedonic. *Ibid.* 658/2 (*index*) Hedonic glands (ἡδονή, lust). **1931** G. K. Noble *Biol. Amphibia* vi. 137 The secretions of the hedonic glands of newts and plethodontid salamanders have no recognizable odor and yet they seem to function in holding the attention of the female during courtship. **1962** H. M. Smith *Evol. Chordate Struct.* xiii. 324 A large number of integumentary glands of spotty distribution among vertebrates are of hedonic function.

So **hedo·nically** *adv.*

1951 H. A. Murray in Parsons & Shils *Toward Gen. Theory Action* iv. iii. 436 The kinds of events that are hedonically negative and the kinds that are hedonically positive. *Ibid.* 457 The aim of all needs is hedonically positive (in the imagination). **1961** P. T. Young *Motivation & Emotion* ii. 49 It is reasonable to assume that playful behavior is generally pleasant, hedonically positive.

hedonical, *a.* (Example.)

1897 B. Russell *Essay Foundations Geom.* iii. 158 They would leave Geometry in a position no better than that of the Hedonical Calculus, in which we depend on a purely subjective measure.

hedrumite (he·drŭmait). *Petrogr.* [a. G. *hedrumit* (W. C. Brögger 1890, in *Zeitschr. f. Kryst. und Min.* XVI. 40), f. *Hedrum*, the name of a village north of Larvik, Norway + -ite[1].] A hypabyssal porphyritic igneous rock having a trachytic texture and consisting essentially of a potash-feldspar with small amounts of pyroxene and usu. also nepheline.

1896 J. F. Kemp *Handbk. Rocks* 143 Hedrumite, a name proposed by Brögger for certain syenitic rocks that are poor or lacking in nepheline, but that have a trachytic texture. **1920** A. Holmes *Nomencl. Petrol.* 116 Hedrumite,..A trachytic variety of alkali-syenite containing subordinate nepheline. **1938** A. Johannsen *Descr. Petrogr. Igneous Rocks* IV. i. 25 Hedrumites..are essentially pulaskite-porphyries with a coarse trachytic texture. Brögger defined them as the mineralogical and chemical hypabyssal-trachytoid equivalents of the pulaskites. **1964** *Mineral. Abstr.* XVI. 574/1 Senonian hedrumites containing several earlier andesites and pyroclastics have been found..in Bulgaria.

hedychium (hĕdi·kiəm). [mod.L. (J. G. Koenig in A. J. Retzius *Observationes Botanicæ* (1783) III. 73), f. Gr. ἡδύς sweet + χιών snow, in allusion to the fragrant white flowers of one species.] A perennial herb of the genus so called, belonging to the family Zingiberaceæ, native to tropical Asia, and bearing showy flowers in a terminal spike; the garland-flower. Also, a fibre obtained from a species of this plant.

1822 *Curtis's Bot. Mag.* XLIX. 2300 Roots of this undescribed species of *Hedychium* were sent by Dr. Wallich of Calcutta, to our friend Mr. Kent. **1894** A. K. Nairne *Flowering Plants W. India* 339 *Hedychium*, with long and slender filament and broad lateral staminodes, is..a garden plant with pure white, fragrant flowers. **1920** Cross & Bevan *Text-bk. Paper-Making* (ed. 5) 171 *Hedychium coronarium*. Hedychium has lately come into prominence as a paper-making fibre. **1952** F. Kingdon-Ward *Plant Hunter in Manipur* 237 During the long summer..Hedychium, with several species and in many bright colours, comes into bloom.

Heeb, var. *Hebe, hebe[2].

heebie-jeebie(s (hī·bi,dʒī·bi(z). *slang* (orig. *U.S.*). Also **heebies, heeby-jeebies**, etc. A feeling of discomfort, apprehension, or depression; the 'jitters'; delirium tremens; also, formerly, a type of dance.

1923 W. De Beck in *N.Y. American* 26 Oct. 9/3 You dumb ox—why don't you get that stupid look offa your pan—you gimme the heeby jeebys! *Ibid.* 10 Nov. 10/1, 31,000 shares! Worthless stock of 'the Belgian Hair Tonic Company' wiped out! Every cent I had in the world... It gives me the heebie jeebies. **1924** H. C. Witwer in *Cosmopolitan* Oct. 114/2 That discovery gave my new found friend the hibby jibbys. **1926** Maines & Grant *Wise-Crack Dict.* 9/2 *Heebie-jeebies*, alcoholic shimmy. **1926** *Bulletin* 13 Dec. 5/5 The latest dance, the 'Heebie-Jeebies' is said to represent the incantations made by Red Indian witch doctors before a human sacrifice. **1927** *Punch* 2 Feb. 116/1 It is interesting to observe that in spite of artificial sunlight, television, winter sports and the heebie-jeebie there are still some stalwarts who stand by the old traditional amusements of the English people. **1927** *Weekly Dispatch* 1 May 8 The Heebie Jeebies is rich in haunch movements. *Ibid.* 3 May 3/7 A terrible girl in the next gallery, painted in the fearsome and fashionable 'pink and putty' manner, had given him what an American present might have called the 'heeby-jeebies'. *Ibid.* 17 Aug. 3/2 Does this work never give you the heeby-jeebies? Does it never depress you? **1929** R. C. Andrews *Ends of Earth* 89, I thought I had the 'hebe-jibies' and stepped up very gingerly. **1934** R. Nichols *Fisbo* 37 It would have given the downright heebee-jeebee To even the dullest of the dull *amœbæ*. **1959** J. Fleming *Miss Bones* viii. 94 You've given me the heeby jeebies... It'll be the end of me. **1971** R. Dentry *Encounter at Kharmel* iii. 57 That little creep..gives me the screaming heebies. **1972** Wodehouse *Pearls, Girls & Monty Bodkin* ix. 138 He was suffering from an ailment known to the medical profession as the heeby-jeebies, and anything having the appearance of a hitch in the programme might lead to a total collapse.

heel, *sb.*[1] Add: **1. e.** *heel of Achilles, Achilles' heel*: the only vulnerable spot (in allusion to the story of the dipping of Achilles in the river Styx: cf. *tendon of Achilles* s.v. Tendon a).

1810 Coleridge *Friend* 431 Ireland, that vulnerable heel of the British Achilles! **1864** Carlyle *Fredk. Gt.* IV. xvii. ii. 522 Hanover,..the Achilles'-heel to invulnerable England. **1897** G. B. Shaw *Let.* 2 July (1965) 777 Divorce is the Achilles heel of marriage. **1930** L. D. Bronshtein tr. *Trotsky's Life* xxv. 262 By his verbal artifices, he only discloses his own Achilles' heel. **1944** *Times* 19 June 5/6 Military observers have dubbed Viipuri the Achilles' heel of the Finnish defences. **1957** A. E. Coppard *It's Me, O Lord!* ii. 17 The three R's, the third of which..was..my Achilles heel. **1972** *Catholic Herald* 28 Jan. 1/5 It is this refusal to condemn which is the Achilles heel of contemporary Christian psychology.

f. *Horsemanship.* Management by the heel (in quot. the spurred heel).

1728 Chambers *Cycl.* s.v., This horse understands the Heels well.

3. a. In *Rugby Football*: a heeling of the ball from the scrummage. Cf. Heel *v.*[1] 5 b.

1937 *Times* 15 Feb. 5/3 A quick heel and the ball went through the hands of [etc.].

5. b. *high heel* (earlier and later examples).

1671 A. Wood *Life & Times* (1892) II. 226, 4*d* given to see a man at the King's Head 7 foot and an half high... He had a night gowne on, which made him seem taller, and high heels. **1950** G. Barker *News of World* 10 Heavy my heart walks ahead on the pavements With her high-heel shoe my martyrdom on stone.

7. b. (Further example.)

1933 L. G. D. Acland in *Press* (Christchurch) 28 Oct. 15/7 *Heel*, the corner of a shear blade, next the grip.

k. The lower part of the back of a book.

1930 *Godfrey's Catal.* No. 134. 26 Small piece gone from heel, and joint becoming tender.

l, m. (See quots.)

1880 E. D. Cope in *Amer. Naturalist* XIV. 836 Stages in the following modification of parts:—..(6) In the obliteration of the inner tubercle of the lower sectorial. (7) In the extinction of the heel of the same. **1888** *Lockwood's Dict. Mech. Engin.*, *Heel*, the thick or broad end of a wedge-shaped piece, the broad end of a railway switch for example. **1957** R. Lister *Decorative Wrought Ironwork* i. 12 The anvil's parts are known by special names... The part of the face and body that terminates in a thick wedge-shaped end is the *heel*.

11. (Further examples.) Also **down-at-heels** *attrib.*; **down-at-heeledness.**

1906 *Daily Chron.* 22 Dec. 3/2 A down-at-heels party hailed him as a countryman, and asked 'the lend of the loan of twopence'. **1919** C. Orr *Glorious Thing* iv. 37 The old down-at-heel slippers she kept for working. **1956** E. C. Hiscock *Around World in Wanderer III* vi. 68 To hear once more the shrill scream of pigs protesting their passage aboard some down-at-heel schooner. **1963** A. Lubbock *Austral. Roundabout* 31 The usual two-storey bush pub, rather scruffy and down-at-heel. **1924** Down-at-heeledness (see *daverdy *a.*).

13. Also as a word of command: *heel!*

1878 C. Hallock *Amer. Club List & Sportsman's Gloss.* p. vi, *Heel*, the order to dogs to come behind the gunner. **1923** D. L. Sayers *Whose Body?* ix. 200 The dog..barked. .. 'Heel,' said the man in velveteen, violently. The animal sidled up, ashamed. **1935** G. Heyer *Death in Stocks* ii. 13 She..was chiefly occupied in keeping back a powerful bull-terrier... 'Shut up, you fool!' commanded the girl. '*Heel!*' **1971** M. Tripp *Five Minutes with Stranger* i. vi. 64 She was saying 'Heel' in a voice that would have quelled a riot in hell.

14. c. Of motoring. Also as *vb.* (see quot. 1962). So **heeling-and-toeing** *vbl. sb.*

1937 O. Stewart *Learn to Drive* viii. 63 A method of gear changing..is that which employs heel-and-toe operation of clutch and accelerator pedals at the same time. **1962** *Which?* (Suppl.) July 96/2 If you want to, you can 'heel-and-toe'—work brake and accelerator at the same time. **1966** T. Wisdom *High-Performance Driving* viii. 73 Use of the 'heel-and-toe' technique.. reduces the time and distance taken to complete the slowing-down and gear-change operations. **1966** R. Maxwell in T. Wisdom *High-Performance Driving* viii. 72 Heeling and toeing..involves double-declutching into a lower gear while braking. **1973** 'J. Ashford' *Double Run* xiv. 114 With heel-and-toe braking and gear changing he flicked down through the gears.

16*. *dig in one's heels*: see *dig *v.* ii c.

22. (Earlier example.)

1751 Fielding *Amelia* III. ix. vii. 283 Instead..of attempting to follow her, he turned on his Heel, and addressed his Discourse to another Lady.

24. c. *to run heel.* Delete † and add later examples.

1923 *Times* 17 Jan. 5/5 The old Melbreak hounds will never run heel. **1946** M. C. Self *Horseman's Encycl.* 455 When hounds hit the line and run it backwards they are said to 'run heel'.

25. *heel-back, -chaser, -dance, -kicker*; *heel-clacking, -clicking* vbl. sbs. and ppl. adjs.; *heel-free* adj.

1936 *Times* 9 Jan. 4/1 A quick heel-back from a loose scrummage. **1938** Dylan Thomas *Let.* 1 June (1966) 199 It's the dog among the fairies..the wizard's heel-chaser. **1922** Joyce *Ulysses* 515 A firm heelclacking is heard. **1928** Blunden *Undertones of War* 155 Strutting with redoubled vanity and heel-clicking. **1970** R. Parkes *Death Mask* v. 64 The abrupt, heel-clicking return of Castilla. **1951** Koestler *Age of Longing* i. iv. 58 Loose, springy limbs which seemed specially designed for the Kaukasian heel-dance. **1948** B. Vesey-Fitzgerald *Bk. Dog* 223 Ten minutes a day for three days and most puppies will be 'heel-free' in the pen. **1926** D. H. Lawrence *Let.* 19 Jan. (1932) 647 Murry..wrote me impertinently..that I was a professional heel-kicker.

26. a. **heel-breast**, in a shoe, the inside edge of the heel, adjoining the waist; so **heel-breaster**, an operator who cuts heel-breasts; also, the tool used; **heel-breasting**, the cutting of heel-breasts; **heel-parer**, one who shapes and trims heel-blanks; **heel-scourer**, one who scours the surface of heels.

1921 *Dict. Occup. Terms* (1927) § 429 *Scourer*..designated according to parts upon which he works, *e.g.* bottom or naumkeag scourer, heel scourer, heel-breast scourer. **1905** *Westm. Gaz.* 30 Oct. 7/3 The same firm have several other novelties, including an automatic Louis heel-breaster. The uninitiated may like to know that 'heel-breasting' is the operation of bevelling out the curve on the inside edge of the heel to the familiar half-moon or other shape. **1921** *Dict. Occup. Terms* (1927) § 414 *Heel breaster*; cuts breast on front of heel square. **1881** *Instr. Census Clerks* (1885) 76 Heel Parer. **1904** *Daily Chron.* 11 June 8/6 Boot Trade.—Wanted good heel parers and heel scourers. **1921** Heel scourer [see *bottom-scourer* s.v. *bottom *sb.* 19].

c. **heel bug**, a harvest mite, *Trombicula autumnalis*, or the skin disease it causes in horses.

1920 *Vet. Rec.* 6 Nov. 218/1 Heel bug, or harvester, *Leptus autumnalis*, is an annual source of trouble to thoroughbreds in training. **1931** *Daily Tel.* 22 May 19/7 Lady Marjorie is suffering from lameness in the off-hind heel, due to variola, which is a similar complaint to heel-bug. **1950** W. E. Lyon *First Aid Hints Horse Owner* (rev. ed.) iv. 87 Heel Bug. As a rule only well-bred horses with thin skins are affected. The heels will be swollen and painful: lameness may also be present. **1954** P. Smythe *Jump for Joy* v. 83 She had contracted a heel bug disease. **1968** G. Lapage *Vet. Parasitol.* (ed. 2) xxxii. 771 *T[rombicula] autumnalis* may be the cause of 'heel-bug' of racehorses.

heel (hīl), *sb.*[3] *slang* (orig. *U.S.*). [Of doubtful origin though prob. f. Heel *sb.*[1] (cf. sense 3).] Among criminals: a double-crosser, a sneak-thief; more generally: a dishonourable or untrustworthy person, a rotter.

1914 Jackson & Hellyer *Vocab. Criminal Slang* 43 *Heel*,..An incompetent; an undesirable; an inefficient or pusillanimous pretender to sterling criminal qualifications. **1916** *Lit. Digest* 19 Aug. 425/1 She..is said to be running a respectable 'scatter' in Dayton, Ohio, for reformed pickpockets and 'heels' or 'pennyweighters', the argot for sneak-thieves and shoplifters. **1929** *Sat. Even. Post* 13 Apr. 54/4 If a crook becomes an informer, then he is a rat or a heel. **1932** J. T. Farrell *Studs Lonigan* xvi. 354 Studs watched him give the college handshake, thought what a heel O'Brien had turned into. **1949** 'J. Tey' *Brat Farrar* xvii. 157 Signing a paper didn't make him any more of a heel than he was being at the moment. **1949** R. Graves *Seven Days in New Crete* 67 She had not only treated me foully but managed at the same time to put me in the wrong and make me feel a thorough heel. **1957** L. P. Hartley *Hireling* 225 It doesn't matter how she feels, does it, when she's lost her fiancé—though he was a heel and she's well rid of him? **1958** *Times Lit. Suppl.* 26 Dec. 749/5 John Augustus Grimshawe was a heel about money and women.

heel, *v.*[1] Add: **2. b.** (Further example.)

1873 J. Miller *Life amongst Modocs* 301 This was his signal to 'heel' himself and come upon the ground.

4. Read: To follow at the heels of, chase by running or nipping at the heels; also *absol.* to follow at a person's heels. (Further examples.)

1940 E. C. Studholme *Te Waimate* (1954) xvi. 138 Two good dogs, one of which frightened the beasts by heeling them up (biting their heels) and the other by pulling their tails. **1947** R. B. Kelley *Sheep Dogs* (ed. 2) xv. 178 Dogs that heel when forcing can be made relatively harmless by removing their canine teeth. **1966** 'J. Hackston' *Father clears Out* 14 Our half-bred sheep-dog, which for years had poked about the kindly heels of callers' nags, got it in for the ribs. 'No dog's going to beat that horse.'

5. b. (Further example.) Also *trans.*

1930 R. Campbell *Poems* 11 See the fat nouns like porky forwards sprawl Into scrums that never heel the ball. **1936** *Times* 19 Jan. 4/3 In the earlier scrummages the Navy's forwards heeled the ball with commendable cleanness. **1955** *Times* 22 Aug. 9/4 One of the three [*sc.* abilities] is to heel out sufficiently cleanly to offer the halves a chance to open up the play.

7. *intr.* To run back on the scent, to run heel.

1898 *Daily News* 5 Oct. 6/6 One or two of the best hounds showed a disposition to heel—i.e.—go back on the line if they changed.

heel (hīl), *v.*[3] [A corruption of earlier Hele, heal *v.*[2]] With *in* = Hele *v.*[2] 2 (*a*). Hence **hee·ling-i·n** *vbl. sb.*

1857 *Rep. Comm. Patents 1856* (U.S.): *Agric.* 93 In nurseries, fruit-trees are often taken up and 'heeled in'. **1882** [see Hele *v.*[2] 2 (*a*)]. **1928** *Forestry* II. 54 A group of thirty young elms..which had been 'heeled-in' temporarily in the nursery. **1953** H. L. Edlin *Forester's Handbk.* iv. 59 The bundles [of plants] are then *heeled in*—that is, set in a trench..with their roots well covered. **1957** *N.Z. Timber Jrnl.* Aug. 59/1 *Heeling in*, placing plants temporarily in a shallow trench, the roots covered with soil to prevent loss of moisture before planting.

heelaman, -oman, varr. Hielaman.

1848 H. W. Haygarth *Recoll. Bush Life Austral.* x. 113 The heeloman is a sort of shield, made of the toughest wood procurable, about three feet in length, and six inches in breadth at the centre, whence it gradually tapers off to a point at either extremity.

heeled, *ppl. a.* **2.** For *U.S. slang* read *slang* (orig. *U.S.*). (Add earlier and later examples.)

1866 'Mark Twain' *Lett. fr. Hawaii* (1967) 86 In Virginia City, in former times, the insulted party..would lay his hand gently on his six-shooter and say, 'Are you heeled?' **1873** J. H. Beadle *Undevel. West* xii. 215 A man long out West a man must be, in the local phrase, 'well heeled'. **1915** A. Conan Doyle *Valley of Fear* II. ii. 153 'Halloa, mate!' said he. 'You seem heeled and ready.' **1928** 'I. Hay' *Poor Gentleman* xvii. 284 A scattered shot or two rang out—doubtless some of the defenders were 'heeled'. **1956** 'E. McBain' *Cop Hater* (1958) v. 47 'Were you heeled when they pulled you in?'..'We didn't even have a water pistol between us.'

b. Provided with money. Usu. preceded by *well*. *slang* (orig. *U.S.*).

1880 *Pacific Metropolis* (San Francisco) 12 June 8/4 His friends want him to go 'heeled' and so they've got up the biggest sort of a benefit for him..next Wednesday night. E. W. Brodhead *Bound in Shallows* 153, I ain't so well-heeled right now. **1958** J. Curtis *Gilt Kid* v. 51 He had done a gaff and was well heeled with dough. **1965** G. McInnes *Road to Gundagai* x. 176 Dr. Crapp was a prominent dentist..and was therefore obviously well heeled. **1968** *Daily Tel.* (Colour Suppl.) 19/1 Through the million and a quarter left by his grandfather they were heeled among a large family.

4. *heeled bet*: in card games (see quot.).

1923 L. H. Dawson *Hoyle's Games Modernized* 274 A 'heeled Bet' is said to be one in which the counters of the stake are placed diagonally across from one card to another signifying that the punter is playing both cards to win.

heeler. Add: **3. b.** (See quots.)

1929 F. Bowen *Sea Slang* 66 *Heeler*, a fast sailing ship. **1961** F. H. Burgess *Dict. Sailing* 114 *Heeler*, a sailing boat with a good, or clean, heel.

4. b. A working dog that urges animals, esp. cattle, onward by nipping at their heels. Also (quot. 1888), a nip on the heels. Also *blue heeler* (*blue *a.* 12 a). *Austral.*

1888 'R. Boldrewood' *Robbery under Arms* xii, He fetches him [*sc.* the horse] such a 'heeler' as gave him something else to think of for a few miles. **1928** 'Brent of Bin Bin' *Up the Country* 11 Bert's heelers and kangaroo dogs chased the packhorses ahead. **1940** F. D. Davison *Woman at Mill* iii. 214 As thick as the hair on a heeler's back. **1945** [see *Australian cattle-dog* s.v. *Australian B adj. b]. **1959** A. Upfield *Bony & Black Virgin* vii. 58 Bluey, the heeler dog, laid himself in the porch shade. 'J. Hackston' *Father clears Out* 116 Patting the Queensland heeler..that was..to be rechristened later on, 'lousy mongrel'. **1968** K. Weatherly *Roo Shooter* 139 He [*sc.* a dog] was a heeler, and it was his nature to attack from the rear.

5. (Later examples.)

1901 *Daily Chron.* 6 Nov. 6/2 The assurance of the Tammany 'Heelers' was less blatant than usual. **1933** H. G. Wells *Shape of Things to Come* iii. 311 The specialist demagogue, sustained by his gang and his heelers, his spies and secret police.

6. One who heels (cf. Heel *v.*[1] 5 b).

1898 Maclagan & Jeffery in W. A. Morgan '*House*' *on Sport* 157 An English [Rugby football] team is an amalgam of heelers, wheelers, pushers (scarce), and sprinters.

heeler[2]. *colloq.* [f. Heel *v.*[2] + -er[1].] A lurch to one side; also, a boat inclined to lurch.

1894 *Times* 6 Aug. 5/1 The wind came off in hard puffs. Each took a regular heeler as they crossed the mouth of the Medina. **1956** R. Clements *Stately Southerner* 106 The ship herself was a heeler.

heeling, *vbl. sb.*[1] Add: **1.** (Further example.)

1963 *Times* 14 Feb. 3/4 They were helped, it is true, by the quicker heeling, although perhaps hindered by their stand-off half's unwillingness to part with the ball until too late.

3. *heeling dog*, a heeler (*heeler 4 b).

1947 P. Newton *Wayleggo* 101 Jim had the severest heeling dog I have seen.

heel-piece. Add: **1. d.** *Shipbuilding.* An angle-bar joining the heels of a frame across the keel. **e.** *Electr.* The iron bar connecting the soft iron cores in an electro-magnet.

1904 A. C. Holms *Pract. Shipbuilding* I. 471 The frame heel pieces are usually fitted when the frames are screwed up ready for riveting. **1904** M. Kirkman *Telegr. & Telephone* 29 The magnet is constructed of a bar or heel piece of soft iron, into which are screwed two pencil-shaped pieces of iron which form the cores of the magnets.

heel-plate. Add: **2.** (Examples.)

1895 *Montgomery Ward Catal.* 526/1 Steel heel plates, for preventing boots and shoes from wearing off at the heels. **1909** *Daily Chron.* 27 Dec. 4/7 The accident was through a wicked heel-plate on my ski. **1959** *Sears, Roebuck Catal.* Spring & Summer 565/1 Home shoe repair outfit, includes ..6 pairs of heel plates.

3. A plate to support the heel of the boot in a metal skate; also, 'a slotted plate fixed on a boot-heel, to which a skate may be locked' (*Funk's Standard Dict.* 1893).

1890 D. Adams *Skating* 5 The heel plate..should be screwed on to the boot with ordinary screws. **1895** G. W. Meagher *Figure & Fancy Skating* 19 A perfect skate should be all of one piece. This is effected by welding the toe and heel plates on to the blade itself.

heel-tap, *sb.* Add: **1.** (Later example.)

1954 J. Steinbeck *Sweet Thursday* 136 Run up the street to Wildock's and get new heel-taps on these shoes.

2. (Later example.)

1933 C. St. J. Sprigg *Fatality in Fleet St.* v. 55 Wait, I have still a heel-tap. I must drink a toast.

heel-tap, *v.* Add: (Later example); also (*b*) to delay. So **heel-tapping** *vbl. sb.*

1909 *Westm. Gaz.* 15 May 2/3 He rivetted china, and clamped or heel-tapped boots. **1958** M. McMinnies *Visitors* 491 Twenty-four hours to git—and no heel-tapping. **1968** *Daily Colonist* (Victoria, B.C.) 19 Dec. 4/8 To put it bluntly the government is still heel-tapping and will continue to do so until and unless all those affected make themselves heard.

Heemraad (hī·mrat, hē·mrād). *Hist.* Also **-raad, -raat.** Pl. **-ra(a)den.** [Du., f. *heem* village, home + *raad* council.] A local petty court or council assisting the landdrost in South Africa and also formerly in Holland; also, a member of this council.

1801 J. Barrow *Trav. S. Afr.* i. 12 A civil magistrate called a *Landdrost*, who, with his *Heemraaden*, or a council of country burghers, is vested with powers to regulate the police of his district, [etc.]. **1833** W. W. Bird *State of Cape of Good Hope in 1822* ii. 23 An Englishman has been rarely called to the office of heemrad. **1876** *Encycl. Brit.* V. 47/2 Prior to 1827 there existed in the several districts of the colony an institution established by the Dutch called the Board of Landdrost and Heemraaden. **1888** Theal in J. P. Fitzpatrick *Transvaal* (1899) 20 The abolition in 1827 of the courts of landdrost and heemraden. **1900** *Westm. Gaz.* 19 Dec. 3/2 Lord Caledon, after our second annexation of the Cape, revived an old kind of elected Councils, named Heemraden, for the government of the inland districts. **1970** *S. Afr. Panorama* Feb. 45/2 In 1682 local administration was vested in a Heemraad.

heffalump (he·fălʌmp). A child's word for 'elephant'.

Now commonly in adult use.

1926 A. A. Milne *Winnie-the-Pooh* v. 66 He would go up very quietly to the Six Pine Trees now, peep very cautiously into the Trap, and see if there was a Heffalump there. **1928** —— *House at Pooh Corner* iii. 43 He guessed what had happened. He and Piglet had fallen into a Heffalump Trap. **1958** *Spectator* 22 Aug. 241/1 The Conservatives are not going to leap into the heffalump-trap in which their opponents..reside. **1959** *Manch. Guardian* 14 July 6/3 Hannibal's heffalumps can hardly have had any such protection against the weather.

Hefner (he·fnəɹ). *Physics.* The shortened name of F. F. von *Hefner*-Alteneck (1845–1904), German electrical engineer, used *attrib.* to designate esp. a lamp devised by him and formerly used as a photometric standard and the intensity of light obtained from it, as **Hefner candle** (see quot. 1943); **Hefner (amyl** or **amyl-acetate) lamp,** a lamp of standard

dimensions and with standard parts burning amyl acetate. Also *Hefner flame, kerze* (G., = candle), *standard, unit.* Also occas. in full *Hefner-Alteneck lamp* etc.

1891 *Jrnl. Soc. Chem. Industry* X. 685/2 A discussion on the subject of amyl acetate, the fuel of the Hefner lamp, has recently taken place. **1896** *Electrician* 2 Oct. 738/2 The standard of light will be either the Vernon-Harcourt or the Hefner-Amyl-Acetate. **1898** *Electr. Rev.* XLII. 795/1 As a result of the investigations of the German Reichsanstalt, the Hefner lamp alone fulfils all technical requirements. The light of the Hefner lamp is designated a Hefner candle. *Ibid.*, The Hefner candle was accepted as the international unit of light by the Electrical Congress at Geneva in 1896. **1901** *Phil. Trans. R. Soc.* A. CXCVI. 26 Knowing the value of the energy of the visible light of the Hefner standard, the heating effect of the rays can be deduced. *Ibid.* 37 The chief source of difficulty in the comparison is the difference in colour between the light from the Hefner lamp and a fluorescent screen. **1902** *Encycl. Brit.* XXX. 235/2 For accurate scientific purposes the best standard is the Hefner-Alteneck or amyl-acetate lamp. **1911** *Ibid.* XXI. 526/2 Various experimental investigations into the properties of the Hefner flame. **1914** S. E. Sheppard *Photo-Chem.* 23 Violle's unit was found by Lummer to equal 26 Hefner units. *Ibid.* 24 The light-unit 1 HK (Hefner-Kerze or Hefner candle) is taken as the mean of protracted observations on a Hefner lamp at the Physik. Techn. Reichsanstalt in Charlottenburg. **1917** G. Martin *Industr. & Manuf. Chem.* II. 357 The Hefner candle power is equal to about 0·9 British standard candle. **1943** *Gloss. Terms Electr. Engin.* (*B.S.I.*) 112 *Hefner candle*, a unit of luminous intensity equal to that of the Hefner lamp burning under specified conditions of atmospheric pressure and humidity. It is the official unit of luminous intensity in Germany and is accepted internationally as equivalent to 0·9 candle. **1963** Jerrard & McNeill *Dict. Sci. Units* 30 Germany, however, continued to use the Hefner candle (Hefnerkerze HK) unit which was derived from the Hefner lamp and had a luminous intensity of about 0·9 International candles.

heft, *sb.*[1] **1.** (Further examples.)

1864 'E. Kirke' *Down in Tennessee* viii. 107 I's six foot three,..weigh a hun'red an' eighty, kin whip twice my heft in Secesh, bars, or rattlesnakes. **1966** H. Roth *Button, Button* (1967) iv. 84 He was more on the lean side than supplied with heft. **1972** *Sci. Amer.* Dec. p. ii/2 Go ahead, pick it up. The heft tells you it's solid sterling silver.

‖ **heft** (heft), *sb.*[2] Pl. **hefte.** [G.] A number of sheets of paper fastened together to form a book; *spec.* a division of a serial work; a part of a serial publication, a fascicle.

1886 *Athenæum* 9 Oct. 464/1 This treatise forms the fifth *Heft* of the second volume. **1892** *Rev. Reviews* Jan. 58/1 There is another interesting article..in Heft 14 of the *Gartenlaube*.

heft, *sb.*[3] *local.* [Var. of Haft *sb.*[2] 2.] (The sheep in) a settled or accustomed pasture-ground.

1960 Williamson & Boyd *St. Kilda Summer* 84 The Hirta flock is divided into hefts, more or less discrete groups each restricted to its own particular range. **1961** *New Scientist* 9 Nov. 341/2 The natural unit in hill sheep farming is the heft—the group of sheep that habitually graze within the confines of a particular area of hill ground. **1971** *Country Life* 28 Oct. 1166/1 Anticipated difficulties from depriving the hefted sheep of their age-old hefts or heafs have not occurred.

heft, *v.*[1] Add: **1.** (Later examples.) Also *absol.*

1913 R. W. Service *Rhymes of Rolling Stone* 40 And here they must make the long *portage*, and the boys sweat in the sun; And they heft and pack, and they haul and track, and each must do his trick. **1932** W. Faulkner *Light in August* xiv. 308 He was hefting the bench leg. **1960** J. Maclaren-Ross *Until Day she Dies* ii. 36 'Can't see anybody,' I said, hefting the case.

3. (Further example.)

1893 C. M. Yonge *Treasures in Marshes* ii. 11, I do believe it is [gold]. Brass never would heft so much.

hefty, *a.* Add: **1.** Now in general colloq. use; also, large or significant in size. Also *advb.*, 'powerfully', exceedingly.

1871 *N.Y. Tribune* 21 Jan., He is, as a Yankee would say, a little hefty for the ideal lover. **1890** P. H. Emerson *Diary* 25 Nov. in *On Eng. Lagoons* (1892) xxii. 100 Rum night this, hefty weather, don't it blow and snow. **1898** Kipling *Land & Sea T.* (1923) 135 What are we going to do? It's hefty damp here. **1905** *Daily Chron.* 18 Sept. 8/2 When an American girl does that, you can guess there's something 'mighty hefty' weighing her down. **1908** *Daily Chron.* 1 July 3/3 Mr. Barnes of New York was hefty with the trigger. **1925** E. F. Norton *Fight for Everest, 1924* 39 The bucolic bumpkin with coarse features and slow brain fails no less than the 'hefty' giant. **1930** J. B. Priestley *Angel Pavement* ii. 85 It's a hefty commission all right. **1930** *Diary of Public School Girl* 32 Played in a game with Highlands. Got some good hefty bangs. **1958** *Oxf. Mag.* 22 May 461/1 Slighter than Hitchcock but heftier than Chaplin. **1959** *Manch. Guardian* 29 Jan. 5/5 You may protect yourself in respect of the very hefty bill for any medical treatment if someone falls ill..while you are abroad. **1972** *Sunday Times* 30 Jan. 63/5 On top of the hefty basic wage is a bonus system from the pool of tips. **1972** *Sunday Tel.* 6 Feb. 14/6 This hefty book, written in brisk journalese, brings Capone to life.

Hegelian. B. *sb.* (Earlier and later examples.)

1843 Mill *Logic* V. iii. 364 Whether in the Vedas, in the Platonists, or in the Hegelians, mysticism is neither more nor less than ascribing objective existence to the subjective creations of the mind's own faculties. **1891** [see Thing *sb.*[1] 14 e].

Hegelianism. Add: (Earlier example.)

1846 J. D. Morell *Hist. View Philos.* II. ii. v. 160 It is in the department of theology chiefly, that the great battle of Hegelianism has been, and is still being fought.

Hegelese (earlier example); **Hegelism** (further example). Also **Hege·lianizing** *ppl. a.* and *vbl. sb.*, rendering Hegelian; **He·gelizer** = Hegelian *sb.*

1879 W. James *Let.* 3 Sept. in R. B. Perry *Tht. & Char. of W. J.* (1935) II. 15 Poor Palmer has gone abroad to steep himself I suppose still more deeply in that priggish English Hegelism. **1886** *Mind* XI. 258 The chief point about the law of development is its containing what in Hegelese might be called *Aufgehobensein*. **1890** W. James *Princ. Psychol.* I. vi. 163 The Hegelizers amongst them will take high ground at once, and say that the glory and beauty of the psychic life is that in it all contradictions find their reconciliation. *Ibid.* xii. 464 A conception, according to the Hegelizers in philosophy, 'develops its own significance'. **1910** *Mind* XIX. 123 The Hegelianising of poetry. **1945** K. R. Popper *Open Society* II. xxii. 196 These Hegelianizing theories. **1970** A. MacIntyre *Marcuse* ii. 35 The counterpart to Marcuse's Hegelianizing of Marx is a total neglect of Engels.

hegemon (hi·dʒĭmǫn, he·-). [a. Gr. ἡγέμων leader.] A leading or paramount power.

1904 *Forum* Jan.–Mar. 347 (Cent. Dict. Suppl.), The *hegemon* of the western hemisphere is the United States. **1920** *Public Opinion* 2 July 6/1 Japan..asserting her ambition to become hegemon of a far East on which white influence shall be reduced to a vanishing point. **1936** W. J. Entwistle *Spanish Lang.* v. 135 He was king of León as others were sovereigns elsewhere in the Peninsula, but he alone was hegemon and the successor of those who had ruled all Spain from Toledo.

hegemonist (hĭdʒe·mŏnist). [f. Hegemon(y + -ist.] An advocate of hegemony. Also **hege·monizer.**

1898 *Pall Mall Gaz.* 12 Feb. 4/1 This Prince Kraft was also, it would seem, the earliest Prussian hegemonist, who has so far, and as such, revealed himself to us. **1921** *Pilgrim* Apr. 273 It does not..follow that the resistance of England to the previous hegemonisers would be condemned.

Hehner (hē·nəɹ). The name of the chemist Otto *Hehner* used attrib. in *Hehner number, value* (see quots.).

1909 Webster *Hehner value*, a number expressing the percentage of insoluble fatty acids in an oil or fat. **1913** G. Martin *Industr. & Manuf. Chem.* I. iv. 64 Hehner Value.—This test devised by Hehner indicates the percentage of insoluble fatty acids which can be separated from oil or fat... Thus butter usually has a Hehner value ranging from 85–88. **1915** *Chem. Abstr.* IX. 531 (*heading*) A new modification for making the Hehner number determination. **1918** T. H. Pope tr. *Villavecchia's App. Anal. Chem.* I. 382 Insoluble, Fixed Fatty Acid Number. (Hehner Number.)

Heian (hē·ān), *a.* [Jap.] Of or pertaining to a period in Japanese history from the late 8th to the late 12th century A.D.

1893 F. Brinkley tr. *Hist. Empire Japan* iii. 104 The people called the new capital 'Heian-kyo'... The interval ..from 794 to 1186 A.D.—is known in history as the 'Heian Epoch'. **1909** *Westm. Gaz.* 30 Dec. 9/1 It has been decided to arrange the pictures in strict historical sequence, commencing with examples of the Heian period. **1911** *Encycl. Brit.* XV. 259/1 All the pastimes of the Nara epoch were pursued with increased fervour and elaboration in the Heian (Kiōto) era. **1959** *Chambers's Encycl.* VIII. 41/2 (*title*) The Heian Period. *Ibid.*, A new capital was built on the site of the present city of Kyoto. It was given the auspicious name of Heian-kyō (the 'city of peace and tranquillity'). **1960** *Times* 1 Mar. 9/7 Professor Sakamoto..in Heian court robes. **1967** *Listener* 14 Dec. 791/2 In Japan quite a lot of Heian literature has been preserved. **1970** *Oxf. Compan. Art* 609 During the later Heian or Fujiwara period an easier, milder Buddhist sect prevailed. **1973** *Times Lit. Suppl.* 2 Mar. 237/2 The Heian period, perhaps the apogee of Japanese culture, was essentially peaceful.

heiau (hē·au). [Hawaiian.] A temple.

1825 W. Ellis *Jrnl. Tour Hawaii* 51 Tamehameha.. finished the heiau, dedicated it to his god of war. **1920** *Nature* 15 July 628/1 There are shorter notes on heathen prayers and the ceremonial erection of the *heiau* or god's house. **1954** J. Sheridan in J. Macdonald *Lethal Sex* (1962) 160 Sacred regions where ancient *heiaus* still stand.

Heidelberg (hai·dəlbɜ̄ɹg, ‖ hai·delbęrχ). The name of a German city used *attrib.* to denote a jaw found at Mauer near Heidelberg in 1907, or the type of prehistoric man (*Homo heidelbergensis*) indicated by this jaw.

1911 A. Keith in *Nature* 25 May 414/1 The Heidelberg mandible was found in 1907 embedded in the Mauer sand beds. **1912** W. L. H. Duckworth *Prehistoric Man* 10 The Heidelberg or Mauer jaw. **1927** R. S. Lull *Org. Evol.* xxxviii. 677 The Heidelberg man represents the oldest recorded European race, geologically speaking. **1964** K. P. Oakley *Frameworks for Dating Fossil Man* 108 He considered that the horizon of the Heidelberg jaw was at the base of the 30-m terrace which he equated with the Mindel-Riss Interglacial.

Heidsieck (hai·dsīk). [Name of the original producer and exporter.] The proprietary name of a brand of champagne.

1853 E. K. Kane *U.S. Grinnell Exped.* ix. 64 We tapped a bottle of Heidsiek.., and all hands spliced the main-brace. **1877** *Trade Marks Jrnl.* 29 Jan. 249/1 Charles Heidsieck. Charles Heidsieck, Reims (Marne), France; champagne merchant... Wine. **1890** Kipling *Life's Handicap* (1891) 163 The King's Peg..liqueur brandy for whisky, and Heidsieck for soda-water. **1904** A. Bennett *Great Man* xxv. 284 He was intoxicated..though not with the Heidsieck. **1920** G. Saintsbury *Notes on Cellar-Bk.* v. 70 The earliest pages of my book show Pommery itself and Heidsieck Monopole. **1956** A. L. Simon *Know your Wines* 82 Heidsieck & Co. are a firm of producers and shippers of Reims. It started as a man's name. **1959** *Trade Marks Jrnl.* 26 Aug. 731/1 Heidsieck.. Champagne wines. Kunkelmann et Compagnie.., Reims, France; wine shippers. **1965** O. A. Mendelsohn *Dict. Drink* 163 *Heidsieck*, old-established champagne of Reims, not to be confused with Piper-Heidsieck.

heifer. Add: **1. d.** A woman, a girl. *depreciatory slang.*

1835 A. B. Longstreet *Georgia Scenes* 143 He rushed into the Kitchen in a fury. 'You infernal heifer!' said he to Aunt Clory. **1853** T. C. Haliburton *Sam Slick's Wise Saws* II. xii. 282, I have half a mind to marry that heifer, tho' wives are bothersome critters when you have too many of them. **1940** M. Marples *Pub. Sch. Slang* 115 Charwomen were satirically known as heifers at Charterhouse. **1964** O. E. Middleton in C. K. Stead *N.Z. Short Stories* (1966) 201 Was that heifer of Blackie's the same one he had a fortnight ago? **1973** *Black World* Jan. 62/2 That heifer that been trying to get next to my man Lucky since the year one.

2. **heifer dust** *slang*, (*a*) nonsense (see also quot. 1927[2]); (*b*) (see quot. 1945); **heifer-paddock** *Austral. slang*, a girls' school.

1927 'J. Barbican' *Confess. Rum-Runner* xxiii. 256 Even if they do get pinched, they always have some heifer dust ready about laying a trap for a ship. **1927** *Dialect Notes* V. 449 *Heifer dust act, the*, an arrest and cross-questioning by the police. **1941** Baker *Dict. Austral. Slang* 35 *Heifer dust*, nonsense. **1945** —— *Austral. Lang.* vi. 123 Other Australianisms for girls or young women include:..a bit of heifer dust. **1885** Mrs. C. Praed *Sketches Austral. Life* ii. 50 Next year I shall look over a heifer-paddock in Sydney, and take my pick.

height, *sb.* Add: **1. c.** Of type: the distance from the foot to the face, called by printers *height to paper*.

1683–4 J. Moxon *Mech. Exerc. Printing* (1962) 157 If he finds that the edge of the Liner just touch..as well all the parts of his Proof-Letters as they do upon his old Letters, He concludes his Matrice is Sunk to a true Height against Paper. **1771** P. Luckombe *Hist. Printing* 243 They [*sc.* imperfections] are seldom exact to the prior sorts, but differ from them, sometimes in thickness, height to paper, or depth of Body. **1888** *Encycl. Brit.* XXIII. 698/2 The height of type varies slightly with different founders, the mean being 11/12 in. **1890** A. Oldfield *Pract. Man. Typogr.* xxii. 164 Each letter should be of exactly the same height to paper; the height of type being 11-12ths of an inch. **1900** H. Hart *Cent. Typogr.* 23 Five packets of types of the same face, but cast on a Pica body and Dutch 'height-to-paper', were found at the Oxford Press in 1898.

12. (Later example.)

1923 T. E. Lawrence *Lett.* (1938) 407 Knewstub.. thinks it's the height of John.

14. *height of land*, a watershed or ridge of high land dividing two river basins. *N. Amer.*

1725 in G. Sheldon *Hist. Deerfield* (1895) I. 559 They told us they we travel to the hight of land by black river. **1805–9** J. J. Henry *Camp. agst. Quebec* (1812) 36 On this lake, we obtained a full view of those hills which were then, and are now, called the 'Heighth of land'. **1860** H. Y. Hind *Narr. Canad. Red River Exped.* II. 225 The Vermilion Pass, which was traversed by Dr. Hector presents on the whole the greatest natural facilities for crossing the mountains without the aid of engineering work, as the rise to the height of land is gradual from both sides. **1875** *Encycl. Brit.* II. 201/1 In the north it [*sc.* the watershed] is found in a stretch of country, called the Height of Land, that lies between the White and the Green Mountains, and gives birth to the Connecticut and a number of smaller streams. **1902** *Ibid.* XXXI. 330/2 Beyond the Height-of-Land the Winnipeg and English rivers flow westward to Lake Winnipeg. **1918** H. Bindloss *Agatha's Fortune* xxv, It was hardly a range of hills, but rather what prospectors call a 'heighth' of land. **1930** G. L. Wood *Pacific Basin* 5 Behind the peninsula of California the height of land is a thousand miles from the sea.

19. For †*Obs.* read *Obs.* exc. in literary use.

1820 Lamb *Elia* (1823) 8 While he held you in converse, you felt strained to the height in the colloquy. **1871** Tennyson *Last Tournament* 658 For once—ev'n to the height—I honour'd him.

heik, var. *hike *v.* and *sb.*

‖ **heil** (hail), *int.* [G., = hail!] Used in the expression *Heil Hitler!* by the Germans during the Nazi regime. Also *transf.*, and as *sb.* So **heil** *v.*, to give the Nazi salute; **heiled** *ppl. a.* *Ski-heil!*: good skiing!

1927 E. Hemingway *Men without Women* (1928) 182 'Ski-heil!' said the innkeeper. 'Heil!' we said. **1937** *Nation* 31 July 114/2 The weekly scene of heiling. **1937** A. S. Neill *That Dreadful School* i. 20 In the absence of a government I herewith declare myself Dictator. Heil Neill! **1938** [see *click *v.*[1] 2 a]. **1939** *Journal* (Topeka, Kans.) 20 Apr. 1/7 Germany heils Hitler on fiftieth birthday. **1939** S. Spender tr. *Toller's Pastor Hall* i. 48 Extending your arm like a crane and yelling, 'Heil Hitler.' **1940** *Time* 15 Apr. 43/1 One of the most widely heiled of German sculptors. **1942** E. Waugh *Put out More Flags* iii. § 5. 234 An obnoxious young man..said 'Heil Mosley.'

1961 J. Heller *Catch-22* (1962) xxiii. 241 When the Germans marched into the city, I danced in the streets..and shouted, '*Heil* Hitler!' until my lungs were hoarse. **1965** Eva-Lis Wuorio *Z for Zaborra* vi. v. 175 The man..gave ..a sharp salute. 'Leader and blood brother,' he heiled.

‖ **Heilsgeschichte** (hai·lzgĕʃiːχʹtə). *Theol.* [G.] Sacred history, *spec.* the history of God's saving work among men; history seen as the working out of God's salvation. So **hei·lsgeschichtlich** *a.*

1938 C. H. Dodd *History & Gospel* v. 168 The whole of history is in the last resort sacred history, or *Heilsgeschichte*. **1952** G. E. Wright *God who Acts* v. 115 Biblical faith may be treated in such a way as to preserve its history-centred (or as the Germans call it, *heilsgeschichtlich*) nature. **1957** D. M. Baillie *Theol. Sacraments* ii. 69 It is bound up with the rediscovery that the Christian message is a *Heilsgeschichte*, a sacred story, running on from eternity through history to eternity again, with Christ as its central and determinative point. **1959** Guthrie & Hall tr. *Cullmann's Christology N.T.* i. 9 There can be no *Heilsgeschichte* without Christology; no Christology without a *Heilsgeschichte* which unfolds in time.

heimin (hē·min). [Jap., f. *hei* level (horizontal), common + *min* people.] In Japanese society of the feudal period, the common people, including the peasantry, craftsmen, and traders, as contrasted with the court aristocracy and samurai (the warrior class).

1875 *Trans. Asiatic Soc. Japan* (1884) III. ii. 104. **1891** A. M. Bacon *Jap. Girls & Women* ix. 228 The great hēimin class includes not only the peasants of Japan, but also the artisans and merchants. **1904** *Daily Chron.* 30 Mar. 4/5 Of those not Samurai, the heimin, or commoners, the peasantry ranked first. **1904** L. Hearn *Japan: Attempt at Interpretation* xii. 271 The Buddhist (like the Shintō) priests, though forming a class apart, ranked with the samurai, not with the heimin. **1951** D. H. James *Rise & Fall Jap. Empire* iii. 119 The profession of arms, previously the privilege of Samurai, was extended to heimin (commoners).

‖ **heimisch** (hai·miʃ), *a.* orig. *U.S.* Also **haimish, heimische, heimishe, heymish.** [a. G. *heimisch* domestic, homelike.] Homely; unpretentious. Hence **hei·mischness,** homeliness.

1964 S. Bellow *Herzog* (1965) 35 A politician..still found me good company, *heimisch*, and took me along to the races. **1968** L. Rosten *Joys of Yiddish* 148 *Haimish, haimisher,..haimisheh,*..informal, cozy,..having the friendly characteristics..that exist inside a happy home, ..unpretentious; putting on no airs... Jews put a high value on being *haimish*. **1970** L. M. Feinsilver *Taste of Yiddish* ii. 111 Heymish, homey, friendly, informal. *Ibid.* 279 This conveys the sense of good talk in congenial company—the 'Heymishness' of herd warmth. **1970** 'E. Queen' *Last Woman* i. 8, I own..a guest cottage... It's all terribly *heimisch*. **1973** *Times* 3 Feb. 13/4 Another word that may confuse is *heimishe*, which is Yiddish, and means roughly..'home cooking' or 'traditional' cooking. *Ibid.*, The heimishe cooking at Leslie's is only part of a large suburbanized menu.

‖ **Heimweh** (hai·mvēi). [G.] Home-sickness.

a **1721** Prior *Essay upon Opinion* in *Dialogues of Dead* (1907) 199 The Swiss are remarked to have a Distemper, which they call the Hemvie, a desire of going home, and where ever They are in Service they get leave to return to their Canton at least once in Some Years, and certainly desire to Dye there. **1756, 18..** [see Home-sickness]. **1845** R. Ford *Handbk. Trav. Spain* ii. 695 If debarred a hope of return, they [*sc.* the Asturians] pine from Nostalgia or Heimweh. **1850** Geo. Eliot *Let.* 9 Feb. (1954) I. 328, I have a little *Heimweh* 'as it regards' my friends. **1912** R. Brooke *Let.* 24 June (1968) 389 Writing about Grantchester gave me a bit of *Heimweh*. **1920** D. H. Lawrence *Let.* in C. Mackenzie *My Life & Times* (1966) V. 170 Heimweh or nostalgia there, for the North. **1971** *Guardian* 25 Nov. 13/4 The *Heimweh* of absences.

‖ **Heimwehr** (hai·mvēəɹ). [G., f. *heim* home + *wehr* defence.] Formerly, the German or the Austrian Home Defence Force. Also *attrib.*

1931 *Ann. Reg. 1930* 184 The usual collisions between Heimwehr and Socialists recommenced, in consequence of the revival of the Heimwehr policy of holding provocative marches. **1938** *Times Lit. Suppl.* 15 Jan. 34/4 A *Heimwehr* officer in the late twenties. **1939** *War Illustr.* 2 Dec. 384/2 Prince von Starhemberg, formerly leader of the Austrian Heimwehr, has been deprived of German citizenship. **1957** *Times Lit. Suppl.* 20 Dec. 771/4 He was later in charge of several *Heimwehr* newspapers—the organs of Austrian Fascism.

Heinesque (haine·sk), *a.* [f. the name of H. *Heine* (1799–1856), German poet + -esque.] Of, pertaining to, or resembling the style of Heine.

?**1892** E. Dowson *Let.* ?Jan. (1967) 222 Where would appreciation of the Heinesque style come in? **1899** *Academy* 24 June 677/1 A grimly tragic vignette of peasant life, with a Heinesque turn at the close. **1907** *Daily Chron.* 2 Aug. 3/3 His Heinesque moods are steeled through with a strong man's virility. **1915–16** *Musical Assoc. Proc.* 158 Beethoven is never sinister, never even Heinesque, if I may use the word.

Heinie (hai·ni). *N. Amer. slang.* Also **Heine, Hiney.** [f. the German Christian name *Heinrich*.] A German (soldier).

1904 'No. 1500' *Life in Sing Sing* xiii. 249/1 *Hiney*, a German. **1917** *Daily Chron.* 25 Aug. 1/7 The Canadians call their enemy Heine and not Fritz. **1918** *Daily Mirror* 12 Nov. 6/4 An Irish terrier of my acquaintance was perfectly certain that the maroons meant a visit from Heinie. **1925** Fraser & Gibbons *Soldier & Sailor Words*, *Heine* (or *Hiney*). **1929** E. W. Springs *Above Bright Blue Sky* 227 'There, you're all right,' the sergeant soothed him. 'Those yellow-bellied Heinies can't kill you. We got ten of 'em this morning.' **1931** [see *Dutchman 1 a]. **1961** *Listener* 20 Apr. 684/1 It's not the Russians we should be congratulating..but the Heinies. Sure, we got Von Braun, but the Russians grabbed all the rest of the German rocket guys.

heintzite (hai·ntsait). *Min.* [ad. G. *heintzit* (O. Luedecke 1890, in *Zeitschr. f. Kryst. und Min.* XVIII. 485), f. the name of W. H. *Heintz* (1817–80), German chemist: see -ite[1].] A borate of potassium and magnesium, at first also called *hintzeite and now regarded as the same as kaliborite.

1891 *Jrnl. Chem. Soc.* LX. i. 528 For the new borate, Milch proposes the name of *hintzeite*, after Professor Hintze, of Breslau; whilst Luedecke proposes that of *heintzite*, after Heintz, the discoverer of pinnoite. **1892** *Ibid.* LXII. ii. 791 The author [*sc.* O. Luedecke]..admits that the chemical composition of kaliborite and heintzite is practically identical, both containing about the same percentage of boric acid, potash, magnesia, and water. This does not, however, prove them to be the same mineral. Whilst kaliborite occurs in badly-formed, crystalline crusts on the surface of pinnoite, heintzite, which is found on the inside, forms well-made crystals with shining faces, and is readily cleavable in three directions. **1951** C. Palach et al. *Dana's Syst. Min.* (ed. 7) II. 368 Heintzite (and hintzeite, the two names being proposed simultaneously for the mineral) was later shown to be identical with kaliborite.

heir, *sb.* Add: **1. b.** *heir-designate*, one who has been designated as a person's heir.

1909 *Daily Chron.* 6 Sept. 3/3 Her relatives, heirs-designate of Charles Dorrien in the scrap of paper lying in his widow's writing-desk. **1961** B. Fergusson *Watery Maze* xii. 311 The meeting in Carthage on Christmas Day, with Churchill presiding, and Eisenhower and Wilson as his heir-designate.

heir, *v.* Add: **b.** *intr.* To inherit. *rare.*

1900 J. Hastings *Dict. of Bible* III. 270 The younger brother, instead of himself heiring, raises up heirs to the deceased.

Heisenberg (hai·zənbɜ̄ɹg). The name of Werner *Heisenberg* (b. 1901), German physicist, used esp. with reference to his matrix theory of quantum mechanics, and to the 'uncertainty principle' deduced by him in 1927.

1932 W. T. Stace *Theory of Knowl. & Existence* xiv. 381 Heisenberg's Principle of Indeterminacy..lays it down that an electron may have a determinate position or a determinate velocity, *but not both*. **1951** *Physical Rev.* LXXXII. 922/1 Note that these Schrödinger equations have been obtained from the Heisenberg picture. **1955** W. Pauli *Niels Bohr* 38 We..restrict ourselves to the discussion of the field operators (Heisenberg-representation). **1965** Phillips & Williams *Inorg. Chem.* I. i. 3 As this [*sc.* wave mechanics] provides a relatively simple method of representing atoms and molecules in pictorial terms, we use it in preference to such alternatives as Heisenberg's matrix mechanics which can be shown to be equivalent. **1968** *Peace News* 18 Oct. 6/3 A kind of Heisenberg effect which is far more serious than anything in the physical or biological sciences: the very act of observation distorts that which is being observed. **1968** J. J. C. Smart *Betw. Sci. & Philos.* 12 Some scientific results (Heisenberg's uncertainty principle, for example) do *not* bear on the problem of free will.

heist (haist). *slang* (orig. *U.S.*). [Repr. U.S. local pronunc. of Hoist *v.* and *sb.*] A hold-up, a robbery; also *attrib.* and *Comb.* Also as *vb.*, to hold up, rob, steal. So **hei·ster,** a robber, a hijacker; a shoplifter. Cf. *hist *v.* 2, *hoist *v.* 6 and *sb.* 5.

1927 *Dialect Notes* V. 449 *Heister*, n. (1) A nickname. Suggested etymon, Ger. 'heissen'. (2) A shoplifter. **1930** E. D. Sullivan *Chicago Surrenders* (1931) xiv. 229 Any such giant 'heist'. **1931** [see *hoist *v.* 6]. **1943** P. Cheyney *You can always Duck* xi. 170 If you think I'm gonna be heisted by a cheap thug like you, you made a mistake. **1947** S. J. Perelman *Westward Ha!* (1949) x. 123 His new ballpoint fountain pen..had been heisted by the attendants. **1953** 'S. Ransome' *Drag Dark* (1954) ii. 22 Any heister..would face a bit of a problem in moving his loot. **1955** D. W. Maurer in *Publ. Amer. Dial. Soc.* xxiv. 18 Very peaceful when he ain't on the heist. *Ibid.* 92 Thus a *heist mob* is one which brooks no interference and robs the victim willy-nilly. **1965** *Punch* 11 Aug. 199/2 Six years ago Jim Tempest was one of a bunch of tearaways heisting cars round the North Circular. **1967** 'D. Shannon' *Chance to Kill* (1968) i. 7 The pair of heist boys had been busy... Since ten days they had..hit four liquor stores, three small markets, two bars, and a drugstore, for a total take of around eighteen hundred bucks. **1968** 'E. Trevor' *Place for Wicked* ii. 22 A heist was when you took a motor with the idea of doing a repaint and flogging it with a bent log-book you'd got from a breaker.

‖ **hei-tiki** (hē·ti·ki). *New Zealand.* [Maori, f. *hei* to hang + *tiki* the first created being.] A greenstone neck-ornament worn by Maoris.

1835 W. Yate *Acc. N.Z.* (ed. 2) 151 The *hei-tiki* being taken off the neck, laid down..and then wept and sung over. **1843** E. Dieffenbach *Trav. N.Z.* II. iv. 55 Around the neck both sexes generally wear a figure cut out of jade. This they call E' Tiki: it has an enormous head, very large eyes, and monstrous and disproportionate arms and legs. **1880** *Encycl. Brit.* XIII. 540/1 The hideous breast ornament termed *hei tiki*. **1887** *Col. & Indian Exhib., Rep. Col. Sect.* 74 Heitikis or native deities. **1936** 'R. Hyde' *Check to your King* 153 Queer little amulets, some of which, the hei-tikis, were in the shape of the human embryo. **1949** P. Buck *Coming of Maori* (1950) ii. xii. 295 The term *tiki* was applied to the carved human figures set up at the gable end of important houses. When the ornaments in human form were made, they were also termed *tiki* but, to distinguish them from the larger wooden *tiki*, the nephrite ornaments were termed *hei tiki*, *hei* meaning to tie around the neck.

hejeen, var. *hygeen, hajeen.

hekistotherm (hĭki·stoþɜ̄ɹm). *Bot.* [ad. F. *hékistotherme* (A. de Candolle 1874, in *Archives des Sciences physiques et naturelles* L. 14), f. Gr. ἥκιστο-ς smallest + θέρμη heat.] A plant which can grow in very cold environments, as the arctic and antarctic lichens and mosses. Hence **hekistothe·rmic** *a.*

1875 J. H. Balfour *Man. Bot.* (ed. 5) 817 Hekistotherms, plants requiring a very small amount of heat, as arctic or antarctic plants. **1909** Groom & Balfour tr. *Warming's Oecology of Plants* 36 Hekistothermic: plants living beyond the limits of tree-growth, where the annual mean temperature sinks below 0° C. **1934** H. Gilbert-Carter tr. *Raunkiaer's Life Forms of Plants* ii. 6 A fourth group, Hecistotherms, comprises plants belonging to the cold regions. They have the lowest heat demand of all plants, will grow where the summer is short, and are able to endure a long and very cold winter. **1965** F. J. Monkhouse *Dict. Geogr.* 155/2 *Hekistotherm*, a plant such as reindeer moss or lichen, which can exist where the mean temperature of the warmest month is under 50°F.

hekte (he·ktĭ). Also **hecte.** [Gr. ἕκτη the sixth (of a stater), fem. (sc. μοῖρα part) of ἕκτος sixth.] A Greek silver coin.

1872 B. V. Head in *Synopsis Contents Brit. Museum* (Dept. Coins & Medals) 37 The work upon many of these staters and hektæ is exceedingly fine. **1906** G. F. Hill *Hist. Greek Coins* 17 A few specimens of an electrum *hekte*, or sixth of the stater. **1921** *Brit. Mus. Return* 79 An electrum hecte with the type of a crouching lion and a hemihecte with the type of a winged monster. **1933** C. Seltman *Greek Coins* vii. 113 The staters of Cyzicus, and her *hektai*, or 'sixths',..testify..to the large issues of these states.

HeLa (hī·lă). [f. the name, *Henrietta Lacks*, of the patient from whom the original tissue was taken (cf. *Obstetr. & Gynecol.* XXXVIII (1971). 945).] Designating a strain of human epithelial cells maintained in tissue culture and derived originally from tissue from a carcinoma of the cervix. Occas. *absol.*

1953 W. F. Scherer et al. in *Jrnl. Exper. Med.* XCVII. 695 This cellular strain, designated as strain HeLa by one of the authors (G. G[ey]) when he obtained it from an epidermoid carcinoma of the cervix has been maintained in continuous serial culture passage *in vitro* from February 8, 1951, until the present. *Ibid.* 705 Cultures of strain HeLa cells are capable of producing large quantities of poliomyelitis virus. **1959** *Laboratory Invest.* VIII. 278 Since the human cancer cell strain HeLa is in such wide use as an in vitro system for the study of cell processes, it appeared to be of value to study..aspects of DNA physiology in these cultures. *Ibid.* 283 The variability in DNA synthesis in HeLa is small in this medium. **1965** C. H. Andrewes *Common Cold* viii. 65 A few lines of cells..have been growing happily for years and these have been sent all round the world, so that the best-known lines, such as HeLa cells from a human cancer..are used in hundreds of laboratories.

Helanca (helæ·ŋkă). [Proprietary term.] (See quot. 1964.)

1944 *Trade Marks Jrnl.* 15 Nov. 544/2 Helanca. Textile yarns. Heberlein & Co. A.G. (a Joint Stock Company organized under the laws of Switzerland), Wattwil, Canton of St. Gallen, Switzerland. **1946** J. V. & S. L. Sherman *New Fibers* xii. 229 An interesting new type of viscose rayon, known as 'Helanca', was used extensively in Switzerland during the war as a substitute for wool. It is a yarn mechanically and chemically treated to produce a permanent wool-like effect. *Ibid.* 230 Helanca yarns are available in a wide variety of colors. **1959** *Times* 7 Dec. 13/2 Elasticised cloth (generally wool and Helanca) gives the greatest comfort as well as best and smartest fit. **1960** *Guardian* 26 Aug. 6/5 The brown and white Helanca bikini was lovely. **1964** *Which?* Sept. 286/1 *Helanca*, wide range of bulked and stretched yarns (made from a number of man-made fibres).

held, *ppl. a.* Add: Also with adverbs.

c **1611** Chapman *Iliad* xxiv. 275 With held vp hands. **1906** *Westm. Gaz.* 28 Dec. 7/2 A long string of 'held-up' cars.

‖ **Heldentenor** (he·ldĕntenōəɹ). [G.] A powerful tenor voice suited to the singing of heroic roles in opera; a person with such a voice.

1926 *Times* 18 May 6/3 Herr Melchior..has..the physical energy of voice and action which is the essential qualification of the *helden* tenor. **1931** *Gramophone* Suppl. Oct. 16/1 A *Heldentenor* whose career..has been almost exclusively restricted to the embodiment of Wagner's greatest heroes. **1947** N. Cardus *Autobiogr.* III. 234 A fearsome imitation of the latest Heldentenor at Covent Garden. **1962** J. B. Priestley *Margin Released* III. 175 The help he gave the *Heldentenor* Melchior. **1962** *Listener* 5 Apr. 627/1 Mr Thomas has the commanding quality of a real *Heldentenor*, not a pushed-up baritone as are many Wagnerian tenors. **1973** *Times* 20 Mar. 16/8 Mr Lauritz Melchior, whom English opera goers remember as the greatest Wagnerian *Heldentenor* of his generation between the wars, died in Santa Monica, California.

Helderberg (he·ldəɹbɜːg). *Geol.* The name of a range of hills in New York State, used *attrib.* to designate a group of strata found there and later also the lower division of the Lower Devonian in North America and the fossils, etc., typical of it. Hence **Helderbe·rgian** (-bɜːg-, -bəɹdʒ-), *a.*

1840 W. W. Mather in *Geol. Surv. State N.Y.* (Assembly No. 50) 212 The subjoined nomenclature of these rock groups is presented as local, and one of convenience merely... (3) The next in order is the Helderberg group, which is composed of various strata of common and hydraulic limestones.. intermediate with grits and shales. **1880** J. Dana *Man. Geol.* (ed. 3) III. ii. 236 The Lower Helderberg period.. is so named because its beds are well displayed in the Helderberg Mountains, south of Albany, beneath Devonian beds called the 'Upper Helderberg'. *Ibid.*, The Helderberg rocks outcrop also over a large area in western Ohio. **1906** Chamberlin & Salisbury *Geol.* II. 434 From this intermediate or transitional assemblage the Helderberg fauna seems to have taken its origin. *Ibid.* 455 The capulid shells which abound at some localities in the Helderbergian.. faunas. **1949** C. O. Dunbar *Hist. Geol.* x. 207 Two stages [of the American Devonian] are recognized, the Helderbergian and the Deerparkian, each with several formations. The Helderberg stage includes only limestone and calcareous shales.

helenium (helī·niəm). [mod.L., f. Gr. ἑλένιον, possibly commemorating Helen of Troy.] **1.** An early name for elecampane, the European herb *Inula helenium*, of the family Compositæ.

1608 Topsell *Serpents* 5 Helen.. planted the same there.. called ..after her owne Name Helenium, which the skilfull Herborists at this day affirme to grow in Pharus. **1777** R. Weston *Universal Botanist* II. 380 Many of the Varieties of the Helenium of Vaillant, are arranged under Aster and Inula. **1931** M. Grieve *Mod. Herbal* I. 279/1 Elecampane was known to the ancient writers on.. natural history.. *Inula*, the Latin classical name for the plant, is considered to be a corruption of the Greek word *Helenion*, which in its Latinized form *Helenium*, is also now applied to the same species.

2. [Adopted by Linnæus in *Hortus Cliffortianus* (1737) 418.] A plant of a large genus of North American annual or perennial herbs so called, belonging to the family Compositæ; a sneezeweed.

1789 W. T. Aiton *Hortus Kewensis* III. 227 (*heading*) Smooth Helenium. Nat[ive] of North America. **1900** J. M. Abbott in W. D. Drury *Bk. Gardening* vii. 272 Heleniums are valuable composite plants for back positions in mixed borders. **1961** *Amateur Gardening* 16 Sept. 7/2 Despite the epithet 'autumnale', many of the heleniums are summer-flowering plants.

helgramite: see Hellgrammite.

1878 C. Hallock *Amer. Club List & Sportsman's Gloss.* p. vi, *Helgramite*, the dobson. **1894** *Outing* (U.S.) XXIV. 228/2 Helgramites (purchasable at most any of the tackle stores during the bass season). **1957** *Times* 11 Oct. 12/6 A helgramite net—a flat wire tray that will catch any insects swept down by the current.

heli-, combining form, repr. the first element of *helicopter (cf. Gr. ἕλιξ), used (*a*) in the names of types of helicopters or aircraft resembling helicopters, as **helibus**, a helicopter with accommodation for a large number of passengers; (*b*) = 'helicopter', as **heliborne** *a.*, carried by helicopter; **heli-lift** *v.*, to transport by helicopter; **helipad**, a 'pad' or landing-ground for a helicopter; **helipod**, a 'pod' or container borne by a helicopter and carried e.g. to forward battle-areas for use as an operating theatre, workshop, etc. See also *Helidrome, *Heliport.

1949 *News-Age-Herald* (Birmingham, Ala.) 13 Nov. A18/3 There is much work.. to be done before the combination of the jetliner and the helibus can be fully utilized. **1956** *Britannica Bk. of Year* 493/1 *Heliear*, a combination of automobile and helicopter. **1960** *Times* 29 Oct. 4/5 A Rotodyne helicopter lifting the 'helipod', which contains two operating theatres. **1961** *New Scientist* 8 June 583/3 The principles of the helicopter and the ducted or flying platform are combined in a vehicle named 'Helipod' and designed by an American business man. **1962** *Aeroplane* CI. 822/2 There are now 487 established heliports or helipads in the U.S.A., Canada and Puerto Rico. **1966** *Guardian* 2 Apr. 14/2 Military requests for a hard 'helipad' that could be laid swiftly in jungles on mud, sand, grass, or dusty surfaces. **1966** *Atlantic Monthly* Oct. 14 Man for man, the U.S. troops may lack some of their enemies' jungle skill, but the rapid availability of firepower and heliborne mobility have tipped the scales decisively in their favor. **1966** *New Statesman* 14 Oct. 549/1 Heli-lifted Medivac Attends the WIA. **1968** *Guardian* 29 Feb. 5/5 Hovercraft and heli-borne troops would transport people and goods. **1968** *Courier-Mail* (Brisbane) 5 June 2/6 Because of the heliborne capability of the Government forces it was no longer possible for the Viet Cong to concentrate anywhere in company strength. **1969** I. Kemp *Brit. G.I. in Vietnam* xii. 197 A 'heliborne' assault in the jungle near Song Be mountain. **1970** *Guardian* 12 Sept. 11 The Israelis.. could heli-lift a half brigade to Dawsons Field. **1973** *Observer* (Colour Suppl.) 1 June 34/2 Units of the elite First Cav, the new heliborne division.

helianthemum (hīliæ·nþĭmĕm). *Bot.* [mod. L. (J. P. de Tournefort *Elemens de Botanique* (1694) I. 214), f. Gr. ἥλιο-ς sun + ἄνθεμον flower.] A plant of the very large, widely-distributed genus of evergreen shrubs or herbs so named, belonging to the family Cistaceæ; also called *rock-rose*, *sun-rose*, or *frost-weed*.

1822 Loudon *Encycl. Gard.* 1026/1 Helianthemum, sun-rose, and cistineæ.. grow in sandy loam and peat. **1827** R. Sweet *Cistineæ* 43 They were both sent from the Brazils.. as two distinct species, one marked *Helianthemum*, 19, the other 48. **1900** M. Thorn in W. D. Drury *Bk. Gardening* xi. 436 Helianthemums (Sun Roses).. are charming plants of dwarf habit. **1934** V. Rendall *Wild Flowers in Lit.* 67 The flower was called *Helianthemum*, 'sun-flower' in Greek, because the blossoms open out in sunshine. **1958** *Listener* 26 June 1058/3 Helianthemums are ideal plants for a new garden.

helical, *a.* Add: **helical gear, tube** (see quots.). Also *Comb.*, as *helical-cut* adj.

1888 *Lockwood's Dict. Mech. Engin.*, *Helical gear*, toothed gear in which the wheel-teeth instead of being at right angles with their faces are set at some other angle therewith. **1904** Goodchild & Tweney *Technol. & Sci. Dict.* s.v. *Cycles*, A form of tube known as 'helical', which is formed from a thin steel band or ribbon, wrapped into the form of a tube and brazed at the edges. **1908** *Westm. Gaz.* 25 Mar. 4/1 The love axle is chiefly noticeable for its helical-cut driving pinion. *Ibid.* 18 Nov. 5/1 Several of the wheels have helical-cut teeth. **1958** *Which?* I. III. 30 The new.. [drill] has.. helical gears, which give quieter running and longer life.

B. *sb.* = **helical gear*.

1923 *Lockwood's Dict. Mech. Engin.* (ed. 4) xv. *Helical gear cutting*, Double helicals are cut in the same way if the teeth are stepped and divided at the apex.

helically, *adv.* Add: Also *Comb.*

1908 *Westm. Gaz.* 14 Nov. 14/1 Helically-cut half-time gear-wheels to ensure silence. **1962** [see *Delay *sb.* 3].

helicity (hili·sĭti). [f. L. *helix, helic-em* (see Helix) + -ity.] **1.** *Physics.* The projection of the spin angular momentum of an elementary particle on the direction of its linear momentum.

1958 *Physical Rev.* CIX. 1017/1 Our result seems compatible with.. 100%.. negative helicity of the neutrinos emitted in orbital electron capture. **1961** R. M. S. Livingston *Particle Physics* vii. 141 Helicity must be defined relative to a particular coordinate system. For example, if an electron is emitted from a beta decay process with velocity v = 0·95c and is observed in the laboratory to have left-handed helicity, its helicity would be reversed if viewed from a coordinate system moving parallel to the particle with velocity v = 0·99c. **1970** *New Scientist* 19 Mar. 545/2 A particle has helicity + ½ if its spin is spinning counter-clockwise while approaching an observer, while if the observer sees the particle as spinning clockwise, it has helicity − ½.

2. *Biochem.* Helical character.

1966 *Biochem. & Biophys. Res. Communications* XIX. 231 We observe that the degree of helicity in native and denatured α-lactalbumin are the same. **1970** *Nature* 25 July 336/2 The side chains in a series of water-soluble glucamine derivatives, the helicity of which can be controlled.

helico- (he·liko), comb. form of Gr. ἕλιξ Helix, in names of chemical substances occurring in snails.

1914 A. Mandel tr. *Hammarsten & Hedin's Physiol. Chem.* (ed. 7) 174 Another phosphoglycoprotein is helicoprotein, obtained by Hammarsten from the glands of the snail Helix pomatia. **1917** *Jrnl. Chem. Soc.* CXII. I. 421 Helicorubin is thus closely related to hæmoglobin, and acts without doubt in the intestine of the snail as a respiratory pigment.

helicoid. **B.** *sb.* Delete †*Obs. rare* and add later example.

1959 *New Scientist* 5 Mar. 515/2 Inside the silencer cylinders there are two helicoids which have the effect of mincing the air flow passing through the cylinder.

helicon. Add: **2. b.** (Later example.)

1961 J. Richardson tr. *'Ilf' & 'Petrov's' Twelve Chairs* (1965) 257 The most powerful machine in the band was the helicon, encircled three times by a brass serpent.

3. In various techn. senses (see quots.).

1961 J. Bok et al. in *Proc. Internat. Conf. Semiconductor Physics* (Prague 1960) 225 Les temps d'atténuation des ondes basse fréquence.. que nous proposons d'appeler 'hélicon'. **1962** *Electronics* 9 Feb. 36/1 He'll talk about active homogeneous semiconductor devices.. including the Ecole [Normale Supérieure, Paris]'s helicon. Defined as devices without p-n junctions that amplify or oscillate by an action similar to that of a traveling-wave tube, these devices are expected to have a bright future. **1966** *New Scientist* 20 Jan. 155/3 The type of plasma-wave currently exciting most attention is called a helicon. It is transverse and circularly polarized.. and moves along the direction of a magnetic field. **1968** *Ibid.* 17 Oct. 13/2 One might also single out for their promise the intriguing 'helicon waves' which result from the interaction between electromagnetic radiation and a 'gas' of electrons within a solid.

helicopter (he·likǫptəɹ). [ad. F. *hélicoptère*, f. Gr. ἕλικο-, ἕλιξ spiral, Helix + πτερόν wing.] A flying machine sustained by one or more lifting screws revolving horizontally. Also *attrib.* and *Comb.* Also formerly in Fr. form.

[**1861** G. L. M. de Ponton *Brit. Pat.* 1599, The required ascensional motion is given to my aerostatical apparatus (which I intend denominating aeronef or helicoptère) by means of two or more superposed horizontal helixes combined together.] **1887** tr. *Verne's Clipper of Clouds* iv. We can look forward to such contrivances.. which we can call stropheores, helicopteres, orthopters.. by means of which man will become the master of space. **1908** O. & W. Wright in *Century Mag.* Sept. 641/2 Several years later we began building these helicopters for ourselves. **1909** *Westm. Gaz.* 28 Jan. 4/2 Mr. Howard Wright's new helicopter.. with which flying tests have been satisfactorily carried out, is now en route for Italy. *Ibid.* 2 Mar. 4/1 The Gobron engined Breguet helicopter aeroplane. **1921** *Glasgow Herald* 11 Nov. 6 Recently the Aero Club of France.. offered a prize.. to the first helicopter pilot in France to take a machine 25 metres up in the air. **1923** *Ibid.* 5 May 8 (*heading*) Helicopter Flight. **1927** C. L. M. Brown *Conquest of Air* ii. 29 The helicopter method of flight. **1958** *Jane's Fighting Ships 1957–8* 9 Official illustrations of the new.. helicopter carrier. **1959** *Daily Tel.* 24 Apr. 20/7 He was opening the helicopter station built in Battersea. *Ibid.*, It is intended primarily to be experimental and to assist in the development of helicopter services for London rather than as a permanent station. **1960** *Times* 11 Oct. 12/7 The men are normal infantry, not specialist helicopter-borne troops. **1963** *Economist* 30 Mar. 1213/1 Several helicopter-carriers and three anti-submarine hunter-killer air groups. **1967** *Courier-Mail* (Brisbane) 17 Mar. 4 The petrol drums.. were blasted alight by tracer bullets from R.A.A.F. helicopter gunships hovering 60 ft. above the treetops. **1973** *Guardian* 2 Feb. 3/2 Three helicopter-carriers had arrived in the Gulf of Tonkin to join the US naval fleet.

Hence **he·licopter**, **helicopter** *vb. trans.* and *intr.*, to fly with or as with a helicopter; to transport by helicopter; **helico·pterist** (now *rare*), one who uses a helicopter.

1911 *Daily Mail* 12 Feb. 7 M. Raoul de Pescara, the helicopterist. **1926** *Spectator* 10 Apr. 665/1 As albatross, helicoptering over the masthead, signalled the land. **1959** *Times* 23 Mar. 15 He might be helicoptered up to Camp David. **1961** *Aeroplane* CI. 121/2 The sequence is then: take-off.. and helicopter away. **1962** *New Scientist* 5 May 230/1 The rotor on its long spindle helicoptered up and out of its case into the air. **1963** *Sunday Times* 20 Oct. 24/3 They build it on deck, and then helicopt it ashore. **1968** *Radio Times* 10 Oct. 31/2 We joined her [*sc.* the aircraft carrier] off Singapore, helicoptering over the rubber forest.

helictite (heli·ktait). [f. Gr. ἑλικτός twisted: after Stalactite.] A distorted twig-like form of stalactite.

1882 H. C. Hovey *Celebrated Amer. Caverns* xi. 186 The term 'Helictite' has been suggested as appropriate to these contorted growths. **1941** *Illustr. London News* CXCIX. 106 (*caption*) The roof of the first grotto or chamber, festooned with stalactites of varied shapes and structure, some tree-like, termed helictites. **1954** W. D. Thornbury *Princ. Geomorphol.* xiii. 337 There is found in portions of some caves a form known as a helictite. It is so unusual in that its growth does not necessarily proceed along vertical lines. **1971** *Islander* (Victoria, B.C.) 24 Jan. 5/3 Helictites are the amazing coral-like stalactites which grow in every direction, seemingly to defy the law of gravity.

helidrome (he·lidrōum). [f. *heli- after *Aerodrome.] A landing-site for helicopters.

1947 *Daily Tel.* 1 June 4/6 Helidrome, rotorport, helistop and heliharbor are among recent verbal coinages] to signify a helicopter station. **1952** *Times* 10 Oct. 7/4 What.. is a 'helidrome'? So far as I can see it is either a place where marsh-meadows run.. or a place where [s]nails run.. But in either case the word ought to be 'helodrome'. **1953** *Listener* 25 June 1040/2 One can almost hear the sound of rockets being fired from the space ships moored at the South Bank helidrome to herald in the twenty-first century. **1966** *Punch* 6 July 8/2 The Director.. with a personal helidrome at Great Yarmouth.

helio[1]. Add: **b.** = *Heliograph *sb.* 4 b.

1886 Kipling *Departm. Ditties* (1904) 23 All honour unto Bangs, for n'er did Jones thereafter know By word or act official who read off that helio. **1901** 'Linesman' *Words by Eyewitness* (1902) 32 Then another helio, spelt out painfully by the frowning, staring signallers, 'Very hard pressed'.

helio[2]. [colloq. abbrev. of Heliotrope.] = Heliotrope 1 d.

1894 *T. Eaton & Co. Catal.* Spring & Summer 51/3 Pink, sky and helio, printed pique vest. **1922** *Daily Mail* 5 Nov. 5 Very exceptional value with Blue or Black or Helio border. **1928** *Ibid.* 31 July 1/3 Dainty.. Dressing Jackets in Pink, Apple and Helio.

helio-. Add: **helio-phyllite** *Min.* [ad. G. *heliophyllit* (G. Flink 1888, in *Öfversigt af kongl. Vetenskaps-Akad. Förh.* XLV. 575), f. Gr. φύλλον leaf: so called because of its colour and structure], a yellowish oxychloride of lead and arsenic, probably dimorphous with ecdemite; **heliopo·lar** *a.*, pertaining to the pole of the sun's rotation; **heliota·ctic** *a.*, responding to sunlight by movement; characterized by heliotaxis; **heliota·xis**, phototaxis due to the rays of the sun; **heliothe·rapy**, the treatment of disease by exposure to the sun's rays; **he:liozinco·graphy** (see quot. 1898).

1890 *Jrnl. Chem. Soc.* LVIII. I. 459 On specimens of the so-called rhodolithic from Pajsberg a yellow mineral has been observed.. The author proposes the name heliophyllite. **1968** I. Kostov *Mineral.* II. vii. 467 Ekdemite is tetragonal, dimorphous with heliophyllite which is orthorhombic pseudotetragonal. **1902** *Science* 7 Feb. 223/2 The vector diagram in heliopolar coordinates takes the form of a conical surface around the Sun. **1898** *Jrnl. R. Microsc. Soc.* 421 The progressive movements of the larvae of the gooseberry mite (*Bryobia ribis* Th.) are always positively heliotactic. **1904** *Biol. Bull.* VI. 253 The mating habits of these highly heliotactic males and.. wingless females. **1898** *Jrnl. R. Microsc. Soc.* 422 (*heading*) Heliotaxis of larval mites. **1890** Billings *Med. Dict.* I. 607/2 *Heliotherapy*, treatment of disease by exposure to sunlight. **1903** *Lancet* 11 July 104/1 The fundamental principles and practical applications of heliotherapy and phototherapy. **1921** *Glasgow Herald* 15 July 10 For the information of medical practitioners interested in heliotherapy. **1928** A. Huxley *Let.* 13 May (1969) 297 My medical uncle.. says that he has known TB of the intestine greatly benefited by heliotherapy. **1905** *Nature* 19 Nov. 60 Two new methods have now superseded photozincography; one of these, 'heliozincography', was worked out by the Ordnance Survey... The first method consists in reproduction direct on a sensitised zinc plate in contact with a reversed negative. **1969** H. S. L. Winterbotham *Kayto Maps* xix. 196 We have seen above how work from paper may go direct to the zinc by a process known as vandyking. To do the same with a negative and a sensitized sheet of zinc is known as heliozincography.

heliocentrically, *adv.* (Later example.)

1960 V. Nabokov *Bend Sinister* ii. 18 They of the solar side saw heliocentrically what you telurians [*sic*] saw geocentrically.

heliodon (hī·liodǫn). [f. Heli(o- + Gr. ὁδός way, path.] A mechanical apparatus for demonstrating the sun's apparent motion, used in astronomy or in architectural design.

1909 in Webster. **1931** Dufton & Beckett in *Jrnl. Sci. Instrum.* IX. 253 In the Heliodon, the apparatus which is the subject of the present paper, the representation of the motion of the sun is simplified. **1964** *New Scientist* 30 Jan. 288/1 Models of a building.. with an artificial Sun would be of immense value... Although individual machines were given such names as 'sunshine analysers' or 'solarometer', the generic name for them is 'heliodon'.

heliodor (hī·liodǫɹ). *Min.* Also **-dore**. [a. G. *heliodor*, f. Gr. ἥλιο-ς sun + δῶρον gift.] A richly coloured variety of golden beryl found in South West Africa.

1913 *Mineral. Mag.* XVI. 362 Heliodor... Trade-name for a golden beryl of gem-quality from German South-West Africa. **1933** L. G. Green *Lords of Last Frontier* (1953) xviii. 185 Heliodore is found only on the barren slopes of Rossing Mountain... Apparently it does not exist anywhere else in the world. German prospectors discovered the first deposit.. and the lovely greenish-yellow opalescent stones were set in the form of a cross. **1959** *Chambers's Encycl.* II. 281/2 The pink morganite, the golden heliodor and the green emerald are among the most attractive of coloured gems.

Heliogabalus (hī:liogæ·bălǝs). [Latinized f. *Elagabal*, Syro-Phoenician sun-god.] The adopted name of Varius Avitus Bassianus, Roman Emperor A.D. 218–222, famed for folly and profligacy, used allusively. Also **He:liogaba·lian** (-găbē·liăn) *a.*, resembling the character or tastes of Heliogabalus. **He:liogaba·lize** *v. intr.*, to act like Heliogabalus.

1589 Greene *Menaphon* (1880) 71 For his dissolute life he seemed another Heliogabalus. **1618** J. Taylor (Water P.) *Pennilesse Pilgr.* sig. F3ᵛ Had I beene a Sardanapalus, or a Heliogabalus, I thinke that.. the great Iurie of Gods and the Mountaines had tamed me. **1624** Burton *Anat. Mel.* (ed. 2) I. II. ii. 63 What Fagos, Epicures, Apitios, Heliogabales our times afford? **1859** *National Mag.* V. 242/1 In California.. the Celestials.. make a Heliogabalian [*sic*] feast of dried fish. **1895** W. S. Gilbert *Utopia Ltd.* I. 5 His Majesty is one of the most Heliogabalian profligates that ever disgraced an autocratic throne. **1930** D. H. Lawrence *Phoenix II* (1968) 492 Perhaps the mentality of a boy of fourteen.. is more wholesome than the mentality of the young cocktaily person.. whose mind has nothing to do but play with the toys of life... Heliogabalus [*sic*], indeed!

heliograph, *sb.* Add: **4. b.** A message sent by heliograph.

1899 *Pall Mall Mag.* Nov. 319, I have just received a heliograph that the Basutos have bands out through.. the valley.

heliolithic (hī:liǫli·þik), *a.* [f. Helio-, after *eolithic*, etc.] Designating a civilization characterized by megaliths and sun-worship.

1915 G. E. Smith *Migr. Early Culture* 4 The habit of megalithic building and sun-worship (a combination for which it is convenient to use Professor Brockwell's distinctive term 'heliolithic culture'). **1916** *Church Q. Rev.* Jan. 283 The world-wide migration of this 'heliolithic culture'. **1925** *Bull. John Rylands Libr.* IX. 402 There is no known heliolithic temple in the Thames Valley. **1929** W. Deeping *Roper's Row* xxxvi. But assuredly Ruth belonged to the old heliolithic people. She had something of the south in eyes, hair, temperament.

heliometer. Add: **1.** (Further example.) Also *attrib.*

1874 *Monthly Not. R. Astron. Soc.* XXXIV. 279 The planet Juno at the opposition of 1874 appears to be very favourably situated for a trial of this method.. and the Repsold Heliometer, with which the transit is to be observed, is a suitable instrument. **1893** D. Gill (*title*) Heliometer observations for the determination of stellar parallax made at the Royal Observatory, Cape of Good Hope. **1906** *Astrophysical Jrnl.* XXII. 103 The heliometer measures made in connection with transits of Venus in 1874 and 1882.

So **heliometry** (hīliǫ·mĕtri), the art or practice of using the heliometer.

helion (hī·liǫn). *Nuclear Physics.* [ad. F. *hélion* (G. Fournier 1930, in *Jrnl. de Physique et le Radium* I. 196), f. Heli(um + *-on[1].] **a.** The nucleus of the normal helium isotope (^{4}He), consisting of two protons and two neutrons; an alpha-particle. **b.** The nucleus of the helium isotope ^{3}He, consisting of two protons and one neutron.

1930 *Sci. Abstr.* A. XXXIII. 969 Equilibrium tends to be established between the number of helions (helium nuclei) in the nucleus and the number of free nuclear electrons. **1964** L. Pauling in *Nature* 4 Jan. 61/1, I suggest that the word helion should be used for the α-particle, the nucleus of the helium atom. **1965** *New Scientist* 14 Oct. 87/2 The helium nucleus, or helion, consisting of two neutrons and two protons. **1972** *Nature* 8 Dec. 325/1 In the case of deuterons, the low binding energy makes it easy to calculate the deuteron-nucleus potential from the separate neutron-nucleus and proton-nucleus optical potentials, but for the more tightly bound helions, tritons and α-particles this cannot yet be done with sufficient accuracy.

heliotrope. Also with pronunc. (he·liotrōᵘp).

heliotropin (hīliotrō·pin). *Chem.* Also **-ine**. [f. Heliotrop(e + -in[1].] A colourless crystalline compound, $C_8H_6O_3$, present in the heliotrope and other plants, that has a strong smell of heliotropes and is made synthetically for use in perfumes. Also called *piperonal*.

1881 *Chemist & Druggist* 15 Sept. 396/2 (*heading*) Heliotropine... Messrs. Schimmel & Co., of Leipsic, now prepare pure heliotropine in crystals... The heliotropine prepared by them is.. suitable for preparing perfumes. **1943** *Thorpe's Dict. Appl. Chem.* (ed. 4) VI. 191/1 Most of the artificial heliotrope perfumes contain modifiers of the pyranyl type, the basic ingredients being heliotropin, with vanillin, coumarin and dimethylhydroquinone. **1963** *New Scientist* 4 July 30/3 Two Japanese scientists have produced.. aromas for cosmetics and foodstuffs similar to the heliotropin.. obtained by chemical synthesis.

helipad, helipod: see *Heli-.

heliport (he·lipǭɹt). [f. *heli- after *Airport.] A landing-place for helicopters.

1948 *Amer. Aviation* 1 June 20 The helicopter is a marvelous vehicle. It just plops down anywhere. All that the city has to do is provide the right inside towns on small fenced-off portions of vacant lots which have been designated by the super-name of 'heliports'. **1950** *N.Y. Times* 28 Nov. 4/5 (*heading*) Construction of City's first aerial heliport began atop Port Authority's headquarters. **1958** *Economist* 5 Mar. 813/2 Plans for the first commercial heliport on Manhattan Island in New York City now await only the final approval of the city and port authorities. **1958** *Times* 15 May 16/5 A committee appointed by the Helicopter Association of Great Britain have come to the conclusion that there is an immediate need for a London 'heliport'. **1959** *Daily Tel.* 24 Apr. 20/7 'Guinea pig' heliport is opened. **1971** *Guardian* 17 Nov. 15/1 A public inquiry into a proposal for a heliport at Shadwell Basin on the Thames.. opens today.

helipterum (helĭ·ptĕrǝm). *Bot.* [mod.L. (A. P. de Candolle *Prodromus* (1837) VI. 211), f. Gr. ἥλιο-ς sun + πτερόν wing, in allusion to the feathery pappus.] A shrub or herb of the genus so called, belonging to the family Compositæ, native to South Africa and Australia, and including several plants described as everlastings or immortelles (q.v. in Dict.).

1862 *Curtis's Bot. Mag.* LXXXVIII. 5350 Major Sandford's Helipterum... Whether it be a true *Helipterum* or a *Helichrysum* I will not venture to say. **1886** G. Nicholson *Illustr. Dict. Gardening* II. 131/2 Helipterums may be raised from seeds. **1962** *Amateur Gardening* 7 Apr. 3/1 Catalogues may list the attractive little rhodanthes either separately or under helipterum, which is their proper name.

helium. Add: (Earlier and later examples.)

1872 W. Thomson in *Rep. Brit. Assoc.* p. xcix, Frankland and Lockyer find the yellow prominences to give a very decided bright line not far from D, but hitherto not identified with any terrestrial flame. It seems to indicate a new substance, which they propose to call Helium. **1958** *Sci. Amer.* Oct. 61/1 Helium is an ideal cooling gas because it has a high specific heat and does not capture neutrons, but it is very expensive and is not available in large quantities in Great Britain. **1958** *Chambers's Techn. Dict.* 984/2 *Helium*, proposed coolant for reactors because of negligible cross-section for neutrons.

b. *attrib.* and *Comb.*, as *helium atom*, *content*, etc.; **helium star**, a star which exhibits the helium lines in its spectrum.

1903 A. M. Clerke *Probl. Astrophysics* 93 A helium envelope surrounds the sun to a depth of five thousand miles. *Ibid.* 94 Now that the helium-spectrum has been unravelled. **1904** Goodchild & Tweney *Technol. & Sci. Dict.* 284/2 *Helium stars*, stars the spectra of which show prominently the lines of the gas helium. **1920** *Discovery* Apr. 111/1 Helium determinations never can provide data for more than a *minimum* estimate [of the age of a mineral]. All that the helium-ratio can tell us is that the age of the mineral to which it refers is greater than a certain minimum value. **1921** *Ibid.* Sept. 236/1 The B helium stars are on the crest of the evolutionary curve, at the meridian of stellar life. **1926** R. W. Lawson tr. *Hevesy & Paneth's Man. Radioactivity* xxvi. 216 (*heading*) Age determination from the helium content. **1938** *Ibid.* (ed. 2) xxv. 267/2 When the 'helium method' was first applied to minerals, it gave values for the age which were mostly only about one-half or one-third of the values found by the 'lead method'. **1940** *Chambers's Techn. Dict.* 409/1 *Helium diving bell*, a diving bell in which the nitrogen in the compressed-air is replaced by helium, thus reducing tendency to the bends. **1956** A. H. Compton *Atomic Quest* iii. 163 The only plans for a production plant.. were those of Moore's helium-cooled reactor. **1957** *Gloss. Terms Nucl. Sci.* (Nat. Res. Council U.S.) 7/1 s.v. *Age*, If the age is calculated from the relative number of atoms of a stable radiogenic end product and radioactive parent present, the method is designated by the name of the end product. Examples are the lead age and helium age of a uranium-containing and/or thorium-containing mineral. **1957** *Technology* Mar. 14/2 In the hydrogen bomb the energy generated is derived from the fusion of hydrogen atoms into helium atoms. **1958** F. E. Zeuner *Dating Past* (ed. 4) 329 Were there not certain difficulties connected with the gaseous nature of helium, age estimates could be carried out with the helium generated by radioactive substances just as well as with the lead. The amount of helium present is determined and compared with the amount of uranium (and thorium) contained in the mineral, in other words, the helium-ratio is determined. *Ibid.* 427 A sample of magnetite from Larder Lake District.. has yielded a helium-age of 2,400 million years. **1960** *Nature* 24 Sept. 1077/1 At the meeting of the International Committee of Weights and Measures held in Sevres, during September 29–October 3, 1958, the recommendation was adopted that the '1958 Helium-4 Vapour Pressure Scale of Temperatures'.. should be used as the international standard scale of temperature between 1° and 5·2°K.

helixin (hī·liksin). [f. Helix + -in[1].] **1.** *Chem.* A crystalline glucoside found in the seeds and leaves of ivy, *Hedera helix*, later called *hederin*. **2.** An antibiotic produced from species of *Streptomyces*, a genus of actinomycete fungi.

1894 C. E. Sohn *Dict. Active Princ. Plants* 56/2 (*heading*) Hedera helix (Ivy);.. Helixin G. (Hedera Glucoside). *Ibid.*, Fehling's solution not reduced till the Helixin has been boiled with acid. **1952** C. Leben et al. in *Mycologia* XLIV. 160 A name, helixin, is tentatively proposed for the antibiotic produced... The name was suggested by the helical nature of the spore chains produced by the antibiotic organism. *Ibid.* 167 Helixin appears to be an antibiotic potent primarily against fungi.

hell, *sb.* Add: **4. a.** (Further examples.)

1903 Kipling *Five Nations* 51 Yes, we shall be perfectly pleased with our work, And that is the perfectest Hell of it! **1944** *Living off Land* v. 103 A boggy area.. hell to plough through with a wheeled vehicle. **1955** N. Balchin *Way through Wood* viii. 111, I should think he'd be pretty average hell to live with. **1971** H. E. Bates *Blossoming World* xv. 176 If the times had been bad for writers in 1926,.. they were now hell.

d. *a hell of a* —: also, an exceedingly bad, great, loud, etc. Also *the hell of a* —. Cf. *Helluva.

1776 J. Leacock *Fall Brit. Tyranny* IV. vii, This is a hell of a council of war. **1778** in S. Curwen *Jrnl.* (1842) x. 207 After travelling in the heat of the season in a hell of a climate. **1806** M. L. Weems *Lett.* (1929) II. 354 I've had a hell of a time in your service. **1897** 'Mark Twain' *Following Equator* xxxi, It's a charming town, with a hell of a hotel... It's the worst hotel in Australia. **1910** R. W. Service *Ballads of Cheechako* 133 Lord! it's a hell of a night. *a* **1918** McCudden *Five Yrs. R.F.C.* (1919) 220 There was immediately a Hell of a yell. **1920** C. E. Mulford *J. Nelson* xii. 126 You must 'a' had one h—l of a time gettin' out. **1923** —— *Black Buttes* ii. 24 He was a hell of a trail-boss, an' he had a hell of an outfit, if you leave it to me! **1931** J. Betjeman *Mount Zion* 20 And we each had a couple of toy balloons and made the hell of a din. **1942** N. Coward *Blithe Spirit* I. ii. 34 Pedalling down the drive at the hell of a speed. **1944** *Living off Land* vii. 155 The Abo thinks himself a hell of a feller, the same as you do. **1947** 'N. Shute' *Chequer Board* i. 3, I had the hell of a headache. **1964** E. A. Nida *Toward Sci. Transl.* ix. 215 In Bassa, a language of Liberia, the English phrase *hell of a* has been borrowed through Pidgin

English as an attributive meaning 'tremendous, great, and important', so that a Bassa churchgoer can quite appropriately tell the pastor that his latest message was 'a helava sermon'. **1969** *New Yorker* 14 June 44/3 His forehand is a hell of a weapon.

e. Used in the genitive (esp. with *own*), or as *hells* quasi-adverbially, with intensive force.

1926 E. Hemingway *Fiesta* (1927) I. vii. 65 You've got hell's own drag with the concierge. **1962** 'J. Le Carré' *Murder of Quality* i. 10 He's entertaining every don... Hells extravagant. **1963** C. Bingham *Coronet among Weeds* ii. 25 They sit about and.. talk about their ancestors. Ancestors are hell's boring. **1968** 'M. Underwood' *Man who killed too Soon* ix. 84, I had a puncture. I had hell's own time changing the wheel.

7. b. Also designating similar receptacles for waste.

1872 *Saddl. Harn. & Carriage Builder's Gaz.* 1 Dec. 207/2 Each smith shop has what is termed the 'hell', and in cutting off a set of tires, if the farmer is not present, the largest half of the end cut off finds its way to the 'hell'. **1886** *Encycl. Brit.* XXI. 345/2 A useful adjunct to the many saw-mills, which produce more waste than can be consumed in raising the necessary steam, is the 'slab-burner' or 'hell'.

9. (Further examples.) *hell's bells!*: an expression of anger or annoyance. *what the hell!*, = what does it matter?, who cares? Also used in expressions of strong disagreement of the type 'Will I hell!' = 'I won't'.

1872 Geo. Eliot *Middlemarch* II. III. xxiii. 21 But, what the hell! the horse was a penny trumpet to that roarer of yours. **1893** *St. Louis Republic* 8 July 16/3 The dealer.. said: 'Say, Rick, do you know this gentleman? He's been playing mighty lucky.' Rickebaugh glanced at the great stack of chips.. and sarcastically remarked: 'Lucky h—!' **1902** R. H. Davis *Capt. Macklin* 295 'Then why in hell didn't you say so!' he roared. **1912** Z. Grey *Riders of Purple Sage* i. 10 To hell with your Mormon law! **1912** E. Hubbard *Age of Rubber* 11 Goshity gosh, helzbelz, there ain't no such animile. **1913** *Maclean's Mag.* Mar. 45/1 'To hell with 'em,' he grated. **1919** *Thrill Bk.* 1 May 13/1 '*Mister!*' I corrected him in his own tone... 'Mister, was it?' he rumbled... 'Mister—hell!' **1920** S. Lewis *Main Street* 352 Hell's bells,.. no harm in being polite. **1925** N. Coward *Fallen Angels* III. 83 *Maurice*: We are great friends—they confide in me. *Fred*: The hell they do! **1925** *New Yorker* 22 Aug. 5/1 Where the hell is my comb? **1927** D. L. Sayers *Unnatural Death* III. xxi. 244 Hell's bells. Here's somebody at the door. **1931** E. Linklater *Juan in Amer.* II. xvii. 183 'We'll all see her,' shouted the Snake's Hips. 'Will you hell!' said Rosy. **1931** G. B. Shaw *Too True to be Good* (1934) I. 46 *The Patient*.. Doesnt that tempt you? *The Nurse*. Tempt me hell! I'll see you further first. **1932** J. Maxwell (*title*) Hell's bells and mademoiselles. **1933** *Atlanta Jrnl.* 21 Jan., I've abandoned it completely. The hell with it. **1936** R. Lehmann *Weather in Streets* II. 235 As if she'd decided to say at last, 'Oh, what the hell! Let them rip.' **1942** N. Coward *Blithe Spirit* I. ii. 42 *Mrs. Bradman*: Ought we to pick it up or leave it where it is? *Dr. Bradman*: How the hell do I know? **1943** F. J. Bell *Condition Red* 38 Carrier, hell! It's a goddamn submarine! **1957** *New Yorker* 5 Oct. 37/2 'The hell with organization,' Todd said. **1958** *Engineering* 4 Apr. 424/3 If *we* can't do it how the hell can I expect Government to do it? *Ibid.* 425/1 Why the hell haven't we got a computer? **1959** M. Hastings *Hour-Glass to Eternity* II. i. 154 Hell's bells! You talk and I'll spill the beans. **1961** *John o' London's* 18 May 567/3 The hell with realism. **1962** P. Gregory *Like Tigress at Bay* vi. 70, I wish to hell I was out of it. **1962** *Sunday Express* 1 Apr. 19/5 Am I dressed for ease and comfort? Am I *hell*? **1968** *Landfall* XXII. 195 Why in hell didn't you get John to build it for you? **1968** A. MacLeod *Dam* xiv. 140 'Hell's teeth!' he swore furiously.

10. Phrases. **With another substantive.*

b. not a chance, hope, in hell: no possibility; also, *a snowball's chance in hell* and similar phrases.

1923 O. Onions *Peace in Our Time* iii. 37 'I rather fancied Lovelightly.' 'Lovelightly? Not a hope in Hell!' **1931** *Amer. Speech* VI. 435 As much chance as a snowball in hell. **1961** K. Vonnegut *Sirens of Titan* (1962) v. 129 The Army of Mars didn't have the chance of a snowball in hell. **1963** J. T. Story *Something for Nothing* iii. 79 'What are the chances of a job here, then?' Albert asked. 'For you—not a chance in hell.' She spoke matter-of-factly. **1966** J. Porter *Sour Cream* vii. 94 One telephone call from Melkin.. and Babak wouldn't have a snowball's chance in hell. **1972** *Listener* 27 Jan. 126/1 Poor Robert's empirical doubts don't stand a snowflake's chance in hell.

c. hell and (or **or**) **high water**: any great difficulty or obstacle.

1915 *Everybody's Mag.* June 69/2 He'll be one of us in spite of hell and high water. **1918** C. E. Mulford *Man fr. Bar-20* xii. 120 Logan found out that he was a *real* man, a *gun*-man, an' not scared of h—l an' high water. **1939** A. Keith *Land below Wind* I. ii. 26 'Let empires be built!'—and, come hell or high water, they build 'em. **1939** P. I. Wellman *Trampling Herd* viii. 93 'In spite of hell and high water'.. is a legacy of the cattle trail when the cowboys drove their horn-spiked masses of longhorns through high water at every river and continuous hell between. **1962** *Sunday Times* 12 Aug. 27/7 A superb instinct for working with the camera guided her to rough out a public image which, come hell or high water, she was not going to change.

d. hell for leather: at breakneck speed, orig. used with reference to riding on horseback; also (usu. with hyphens) *attrib.* or as *adj.*

1889 Kipling *Story of Gadsbys* (1891) 116 Here, Gaddy, take the note to Bingle and ride hell-for-leather. **1893** —— *Many Invent.* 47, I perceived a gunner-orf'cer in full rig'mentals perusin' down the road, hell-for-leather, wid his mouth open. **1915** D. O. Barnett *Lett.* 176 The little English plane went humming back, hell-for-leather. **1927** *Sunday Express* 10 July 4 A long line of stage coaches starting on a hell-for-leather race. **1928** H. W. Freeman *Joseph & His Brethren* vi. 48 Charging down hell for leather with your sabres all flashing in the sun. **1930** *Daily Express* 6 Sept. 8/7 That magnificent, hell-for-leather, boiling verse. **1963** *Times* 21 Feb. 3/2 Australia's plan was to make 90 during the afternoon, if they could, without losing too many wickets and to go hell for leather afterwards.

e. hell on wheels: someone or something regarded as resembling hell; also *attrib.* or quasi-*adj.*

1843 *Quincy* (Ill.) *Herald* 10 Mar. 1/4 Hell-upon-Wheels!.. the most appropriate name for that craft [*sc.* a steam-boat]. **1868** S. Bowles in F. L. Paxson *Hist. Amer. Frontier* (1924) lii. 497 'Hell on Wheels' was the appropriate name that Samuel Bowles of the Springfield *Republican* bestowed upon the town he visited in 1868. **1897** P. Warung *Tales of Old Regime* 50 To look up an' know heaven 's above, an' not the roof of a hell-on-wheels—oh, that'll be grand! **1945** Wyndham Lewis *Let.* 13 Mar. (1963) 381 We learn here that the 'Hell-on-wheels' outfit has reached the Elbe. Hooray! **1966** J. Pearl *Crucifixion P. McCabe* (1967) ii. 24 He's hell on wheels on Monday mornings. **1968** S. Challis *Death on Quiet Beach* xii. 174 You don't pull any imitation disease over the immigration doctors. Those guys are hell on wheels.

f. hell's delight: pandemonium.

1823 'J. Bee' *Slang* 95 Kicking up hell's delights. **1835** *Sessions Paper* Apr. 959 She said if I went out, she would kick up *hell's delight*. **1888** 'R. Boldrewood' *Robbery under Arms* II. xix. 287 If these fellows are half drunk they'll.. play hell's delight. **1918** W. J. Locke *Rough Road* xi. 131 Just listen to the hell's delight that's going on over yonder. **1958** L. A. G. Strong *Light above Lake* 26 Once let anything bad go wrong with them, and you'd hell's delight to mend it. **1961** M. Kelly *Spoilt Kill* iii. 134 There'll be hell's delight ringing round all the hotels till we find him.

***With a verb.*

g. to beat, blast, knock, etc., **hell out of** (a person): to pound heavily, thrash, 'beat up'; also *fig.* to achieve supremacy over.

1922 Joyce *Ulysses* 247 His old fellow welted hell out of him. **1925** [see *Gang *v.*[2] 2 b]. **1937** E. Ambler *Uncommon Danger* viii. 110 Are you going to be sensible or do I knock hell out of you first? **1945** C. Isherwood *Prater Violet* 95 If anybody says you're not, I'll help you beat the hell out of him. **1958** P. Scott *Mark of Warrior* II. 166 How did we interrogate Mr Baksh? Beat the hell out of him, I hope? **1965** *Times* 15 Apr. 12/1 Given the response of which our people are capable.. we shall be ready to knock hell out of you.

h. when, till, until hell freezes (over): advb. phr. indicating a date in the impossibly distant future, for ever.

1919 J. A. Fisher *Let.* 13 June in *Henry Bristow Ltd. Catal.* (1973) No. 203.9 Yours till hell freezes. **1931** *Amer. Speech* VI. 435 Till hell freezes over. **1949** *Romance Philol.* II. 105 We have the meaning 'forever' in 'I'll wait until Hell freezes over' and the meaning 'never' in 'I'll do it when Hell freezes over'. **1961** 'A. A. Fair' *Stop at Red Light* (1962) ii. 36 If their suspicions once get aroused, they'll investigate until hell freezes over. **1962** *Listener* 1 Nov. 704/1 'I am prepared,' Mr Stevenson rasped out, 'to wait for an answer till hell freezes over.' **1966** *Guardian* 17 Aug. 9/2 The Texan chairman.. declared that he would.. open the hearing today even if it meant going to gaol 'until hell freezes'.

i. to get hell: to be given hell, to be reprimanded, dressed down.

1938 E. Bowen *Death of Heart* III. iv. 382 'I was using the telephone in Miss Paullie's study, and she came in and caught me...' 'So then you got hell, I suppose.' **1951** Wodehouse *Old Reliable* ii. 32, I would only get hell from Emily Post.

j. to get the (or **to**) **hell out** (**of** a place): to make a hasty retreat.

a **1911** D. G. Phillips *Susan Lenox* (1917) II. x. 257 Get the hell out... I want to sleep. **1929** E. Linklater *Poet's Pub* vii. 83 Get to hell out of this, you accidental offspring of a Marine sentry. **1934** W. Saroyan *Daring Young Man* (1935) 262 Get the hell out of here, I reply quietly. **1944** 'Brahms' & 'Simon' *Titania has Mother* xiii. 147 'Get to hell out of here,' he roared. **1952** 'J. Tey' *Singing Sands* ix. 139 You want her to get the hell out of here. **1961** J. Heller *Catch-22* (1962) vi. 57 He.. felt that any who did not share this confidence he had placed in them could get the hell out. The only way they could get the hell out, though.. was by flying the extra ten missions. **1972** Wodehouse *Pearls, Girls, & Monty Bodkin* v. 67 You ought to be in bed. Get the hell out of here, Bodkin.

k. to give (a person) **hell**: to give him 'a bad time'.

1851 *Harper's Mag.* III. 461/1 Riley shouted, 'Forward and give them h—ll.' **1863** O. W. Norton *Army Lett.* (1903) 161 We have met the enemy and given them hell. *a* **1917** E. A. Mackintosh *War, the Liberator* (1918) 141 You swine, I'll give you hell for this. **1940** N. Marsh *Surfeit of Lampreys* (1941) xv. 235 Gabriel would give me hell and we would both get rather angry with each other.

l. hell to pay: great trouble, discord, pandemonium.

1807 Ld. Paget *Let.* 29 July in A. Paget *Paget Papers* (1896) II. 311 Did you not know.. that there has been hell to pay between the Dukes of York and Cumberland. **1811** [in Dict.]. **1956** Wallis & Blair *Thunder Above* (1959) iv. 42, I got you all in this mess... There'll be hell to pay.

m. to play hell (with): to upset, confuse; to alter for the worse; to make a fuss (cf. quots. 1832–4 and 1879 in Dict.).

1803 G. Colman *John Bull* III. ii. 39 I'll be good to the landlord, but I'll play hell with his wife! *a* **1911** D. G. Phillips *Susan Lenox* (1917) I. xiii. 218 Don't drink.. it'll play hell—excuse me—it'll spoil your looks. **1927** H. Crane *Let.* 19 Dec. (1965) 312 Port every night for dinner is playing hell with my waistline. **1937** T. Rattigan *French without Tears* I. 22 As a matter of fact it would rather amuse me to see you play hell with the Commander. **1959** *Listener* 4 June 979/1 Wingate and his Chindits would play hell with the Japanese communications. **1960** 'A. Burgess' *Right to Answer* xix. 201 E talks about gross neglect.. and e plays ell. **1960** L. Cooper *Accomplices* II. v. 119 The firm.. wanted delivery and were playing hell about it.

n. to raise hell: to create a disturbance; to cause great trouble. (The slogan 'Kansas should raise less corn and more hell' is attributed to Mrs Mary Ellen Lease (1853–1933) but proof is lacking. See *Kansas Quarterly* Fall 1969, 52–58.)

1896 *Emporia Weekly Gaz.* 20 Aug., We have decided to send three or four harpies out lecturing, telling the people that Kansas is raising hell and letting the corn go to weeds. *a* **1911** D. G. Phillips *Susan Lenox* (1917) II. viii. 214 What hell Jim will raise when he finds I spent the night working in this house. **1959** M. Scott *White Elephant* iii. 24 He would go home, ring Bert and 'raise hell'.

****Other phrases.*

o. (just) **for the hell of it**: out of (pure) devilry, (merely) for fun.

1934 'J. Spenser' *Limey breaks In* x. 166 Both of them were of the mischievous type that misbehaves 'just for the hell of it'. **1939** R. Chandler in *Sat. Even. Post* 14 Oct. 74/3, I wouldn't be telling you just for the hell of it. **1951** W. Stevens *Let.* 7 Sept. (1967) 726, I assume that he is merely doing it for the hell of it. **1959** P. McCutchan *Storm South* xiii. 197 The kind of bloke you'd expect to find taking passage in a venture like ours, just for the hell of it.

p. like hell: recklessly, desperately; extremely, very much: freq. as a mere intensive; also ironically, to indicate emphatic contradiction: not at all, on the contrary.

1855 [in Dict.]. **1892** Kipling *Lett. of Travel* (1920) 66 'Hit, old man?' 'Like hell,' he said. **1922** D. H. Lawrence *England, my England* 231 'And I shall miss thee, Jack.'.. 'Miss you like hell.' **1925** F. Lonsdale *Last of Mrs. Cheyney* I. 19 *Maria*: Enjoying the concert, Willie? *Willie*: Like hell! **1930** D. Hammett *Maltese Falcon* xx. 260 'You can't say that.' 'Like hell I can't,' Spade said. **1941** H. MacInnes *Above Suspicion* ix. 76 'I've quite enjoyed it here.' Like hell I have, she added under her breath.

q. merry hell: a disturbance, upheaval, great trouble; severe pain.

a **1911** D. G. Phillips *Susan Lenox* (1917) II. xi. 279 We don't dare let you off... They'd send down along the line, to have merry hell raised with us. **1922** S. Lewis *Babbitt* xix. 229 I've come to raise particular merry hell. *c* **1926** 'Mixer' *Transport Workers' Song Bk.* 13 We don't get drunk to fight the boss Or kick up merry hell. **1931** D. L. Sayers *Five Red Herrings* xxii. 248, I am supposed to have faked an alibi, suborned my friends and played merry hell generally. **1938** S. V. Benét *Thirteen O'Clock* IV. 268 If you think it was all romance.. you're wrong. A lot of it was merry hell. **1944** J. H. Fullarton *Troop Target* 86 This arm's giving me merry hell. **1961** B. Fergusson *Watery Maze* iv. 102 The Special Boat Squadron.. was to play merry hell in the Eastern Mediterranean during the next two years. **1963** M. Duggan in *Landfall* Mar. 9 Watching mum with a shoehorn wedging nines into sevens and suffering merry hell.

r. to hell and gone: used hyperbolically = 'a long way', 'for ever', etc.

1938 S. J. Perelman in *New Yorker* 15 Oct. 17/2 Zarah Trenwick just got blasted to hellangone in her tepee at the Gayboy. **1944** N. Mailer in *Cross-Section* 332 Picking up two-foot piles of plates and lugging them to hell and gone. **1957** M. Millar *Soft Talkers* 39 That's my business. I can contradict myself to hell and gone if I feel like it. **1972** R. Lockridge *Preach No More* ii. 24 Name's Manuel something... Lives to-hell-and-gone downtown.

11. a. *hell-brew, -pack, -pride, -queen, -rake, -rook, -spurge.*

1923 Wodehouse *Inimitable Jeeves* viii. 69 'Have some lemon squash,' I said... The hell-brew appeared to buck him up, for he resumed in a slightly more pally manner. **1935** *Discovery* Sept. 264/2 The dart poison.. is indeed a hell-brew. **1923** R. Graves *Whipperginny* 56 Twenty swans glide out With hell-packs loathlier yet to amaze the night. **1944** Blunden *Shells by Stream* 50 Passions armed with horror and hell-pride. **1918** D. H. Lawrence *New Poems* 63 Out of the hell-queen's cup, the heaven's pale wine. **1794** R. F. Greville *Diary* 28 Aug. (1930) 309 His M. order'd Me to call at Farmer Sherring's where I order'd two of those Broad Rakes called Hell Rakes. **1879** G. M. Hopkins *Poems* (1918) 44 The hell-rook ranks. **1849** D. G. Rossetti *Let.* 18 Oct. (1965) I. 74 Hell-spurge of geomaunt and teraphim.

b. *hell-buster, -raiser*; *hell-raising, -raking, -roaring, -tearing.* Also *hell-rake* vb.

1929 J. B. Priestley *Good Companions* III. iii. 2 They're all damned good, but the last two are real hell-busters. **1914** *Emporia* (Kans.) *Gaz.* 13 Jan., He is a.. rip-snorting hell-raiser. **1925** S. Lewis *Martin Arrowsmith* iii. 20 Young men technically known as 'hell-raisers' looked forward to his lectures on physiology. **1928** *Daily Express* 1 June 9 She dislikes the 'hell-raiser' that he likes to make the public believe he is. **1971** *Guardian* 24 July 10/6 The ex-hell raiser of the Bevanite group [*sc.* Michael Foot].. seemed to have settled comfortably into the new role of Left-wing Whip. **1922** S. Lewis *Babbitt* iii. 25 When it comes to.. a lot of hell-raising all the while.., it's too rich for my blood! **1936** W. Stevens *Let.* 27 Jan.

(1967) 307 Any form of hell raising is simply out. **1966** Wodehouse *Plum Pie* ix. 215 She's the hell-raising type, always apt to be starting something. **1915** J. E. Flecker *Old Ships* 5 The pirate Genoese Hell-raked them till they rolled Blood, water, fruit and corpses up the hold. **1960** *Spectator* 17 June 870 Chatter nostalgically but amusingly about the hellraking times at Balliol. **1920** J. Gregory *Man to Man* iv, Your hell-roaring old grand-dad. **1915** W. J. Locke *Jaffery* xxi, These hell-tearing fellows.

c. and **d.** *hell-black* (later example), *-purple*; *hell-mouthed*, *-plumed*.

1904 Swinburne *Poems* VI. 392 Till murder dawns Blood-red from hell-black treason's heart of hate. **1934** Dylan Thomas *Let.* 11 May (1966) 126 Today I complain again for a hell-mouthed mist is blowing. **1876** G. Meredith *Vittoria* iii. 35 Those hell-plumed Tyrolese. **1923** D. H. Lawrence *Birds, Beasts & Flowers* 57 At her white ankles Hell rearing its husband-splendid, serpent heads, Hell-purple, to get at her.

12. hell-box (examples); **hell-diver** *U.S.*, the dabchick; **hell-driver**, delete sense (*b*) and substitute: (*b*) one who drives a motor vehicle in a very fast or dare-devil manner; hence *hell-driving*; **hell-hole** (further examples); **hell's angel** (usu. in *pl.*), name given in the 1950s in the U.S. (later to similar people elsewhere) to a member of a group of lawless, usually leather-jacketed, motor-cyclists notorious for their disturbances of civil order in California (featured, but there called 'Black Rebels', in a 1954 film entitled 'The Wild One'); also earlier casual uses, e.g. (in *pl.*) as the name of a film about air-battles in the 1914–18 war (quot. 1930) and (in *sing.*) as the name of a Flying Fortress aeroplane (quot. 1943); **hell-ship**, a hell-afloat (cf. Hell *sb.* 4 a).

1889 Barrère & Leland *Dict. Slang* I. 458/1 *Hell-box*, the receptacle for bad, broken, or 'battered' letters, which are eventually melted down. **1909** 'Mark Twain' *Is Shakes. Dead?* vii. 73 If a man should.. say '..empty.. the imposing stone i ito the hell-box..' I should.. know that the writer was only a printer theoretically, not practically. **1839–40** Hell-diver [in Dict.]. **1940** E. T. Seton *Trail of Artist-Naturalist* 89, I traced them to the pied-bill grebe, or little helldiver. **1942** Berrey & Van den Bark *Amer. Thes. Slang* § 723/2 *Automobile racer*,.. hell driver. **1971** *Cape Times* 13 Feb. 7/6 The helldriver with the 99-to-1 chance of becoming the first Cape Times helldriving champion. *Ibid.*, One of the hottest nights of helldriving. **1866** J. C. Gregg *Life in Army* xxi. 184 All the whiskey shops, even down to the lowest hell-hole, adopt the decent name.. of a 'Coffee-house'. **1891** G. M. Hopkins *Let.* 15 June (1956) 63 Liverpool too, 'hellhole' though it is. **1945** J. B. Priestley *Three Men in New Suits* ii. 26 Go and drudge in some hell-hole of an office. **1930** (*film title*) Hell's Angels. **1943** *Examiner* (San Francisco) 8 Aug. 3/1 The ten-man crew which manned the flying fortress, 'Hell's Angel', on her thirty-three raids across the English channel. **1957** *Chronicle* (San Francisco) 3 June 5/1 It also attracted several hundred cyclists who are not American Motorcycle Association members, but who belong to such clubs as the Vampires, Scavengers and Hell's Angels group that rides in the Bay Area. **1957** *Call-Bulletin* (San Francisco) 9 Aug. 3/4 Police raiders cracked down last night on the Hell's Angels, a hard-riding motorcycle outfit whose members have had brushes with the law in the past. **1958** *Ibid.* 1 Aug. 11/1 The main wheel of the 'Hell's Angels' motorcycle club is a more or less conservative delivery man during the week, but Fridays he.. combs his hair into a wild mane, gets into black denims and motorcycle boots, plus a red-fringed vest with the club's death's-head emblem, fastens his pierced ear-lobe and dyes his blonde handlebar mustache with heliotrope pencil. **1967** G. Legman *Fake Revolt* 21 The hoodlum drug-addicts and homosexual motorcyclists.. who've had two 'Hell's Angels' movies already exposing (*read*: glorifying) them. **1968** *Listener* 22 Feb. 253/2 The Hell's Angels—those much-publicised outlaw motor-cyclists, noted for their wearing of leather jackets and Nazi insignia, whose hunting ground is mainly California. **1969** *Oz* Apr. 3/2 Hell's Angels were moved in on the 3rd floor as a protection against external aggression. **1971** *New Scientist* 11 Feb. 331/2 The Hell's Angels created rather than prevented disorder when Mick and the Stones were dispensing their magic. **1927** *Observer* 21 Aug. 15/2 Hell-ships they must be if three and four months' passages.. is anything from which to form an opinion. **1934** *Times Lit. Suppl.* 18 Oct. 713/2 The Dovenby Hall, a notorious hell-ship in her day. **1971** D. Niven *Moon's a Balloon* (1972) vii. 93 The troopship was a hell-ship of about 11,000 tons.

hell, *v.*[2] Add: Revived in various, chiefly *slang*, uses: **a.** to make into a hell. **b.** to give (a person) hell. **c.** to hurry, to go 'hell for leather', to 'fly' *around* (esp. in some activity disapproved of by the speaker).

1897 O. Wister *Lin McLean* (1898) 60 A man was liable to go sporting and helling around till he waked up. **1903** P. F. Rowland *New Nation* 34 The raging bushfires that hell the Australian plains. **1924** Kipling *Debits & Credits* (1926) 242 *That's* not his real trouble... I wonder what's really helling him. **1928** J. P. McEvoy *Show Girl* 166 You were in the show business and throwing your best years away helling around. **1929** W. Faulkner *Sound & Fury* 243 She had to come helling in there at twelve, worrying me about that letter. **1929** —— *Sartoris* (1933) II. 53 Men can't stand anything... Can't even stand helling around with no worry. **1959** E. Fenwick *Long Way Down* x. 83 Had his supper.. just like he always did. I missed *mine*, helling up here this way. **1960** F. Sullivan *Let.* 9 Jan. in *Groucho Lett.* (1967) 147 That oppressed and downtrodden share-cropper, Massa Nunnally Johnson, is hellin' around with Ava Gardner, making a picture. **1969** 'E. Lathen' *Come to Dust* xvii. 172 'If he did any helling around, it wasn't here,' the janitor continued.

Helladic, *a.* Add: **b.** *Archæol.* Denoting the Bronze Age cultures of Greece, lasting from about 2800 to 1200 B.C. Also *absol.*

1921 Wace in *Jrnl. Hellenic Studies* XLI. II. 260 Early Helladic pottery. *Ibid.* 265 With the Third Late Helladic Period Mycenae reached the zenith of its dominion and riches. **1925** V. G. Childe *Dawn Europ. Civilization* v. 74 The first metal-using culture therefore is called Early Helladic. **1927** Peake & Fleure *Priests & Kings* 113 The sub-periods range from Early Helladic I to Late Helladic III. **1940** *Antiquity* XIV. 244 Deposits earlier than the beginning of Late Helladic III, the fourteenth century B.C. **1971** J. L. Caskey in *Cambr. Anc. Hist.* (ed. 3) I. II. xxvi(*a*). 783 There has been a question whether material elements of the latest Stone Age culture persisted for a time alongside the earliest Helladic, since pottery of both types was found together.

hellandite (he·lăndəit). *Min.* [ad. G. *hellandit* (W. C. Brögger 1903, in *Nyt Mag. f. Naturvid.* XLI. 213), f. the name of A. *Helland* (1846–1918), Norwegian geologist: see -ite[1].] A hydrous borosilicate of the rare earths and calcium, samples of which also contain aluminium, manganese, and iron, perhaps as impurities, that is found in pegmatite veins in Norway.

1903 *Jrnl. Chem. Soc.* LXXXIV. II. 657 Both in crystalline form and chemical composition, hellandite is analogous to guarinite. **1924** J. W. Mellor *Inorg. & Theoret. Chem.* V. 512 Brownish-red hellandite from Lindvikskollan near Kragerö, Norway. **1964** *Mineral. Abstr.* XVI. 644/2 Hellandite is a boro-silicate with more than 10% B_2O_3 in addition to the known major constituents.

he·ll-bent, *a.* and *adv. colloq.* (orig. *U.S.*). [Hell *sb.* 11 d, Bent *ppl. a.* 3.] 'Fiendishly', doggedly, or recklessly determined (*on* or *upon* a certain course). Also *advb.* determinedly, recklessly.

1835 *Knickerbocker* VI. 12 A large encampment of savages,.. 'hell-bent on carnage'. **1840** *Pol. Song* (Cent. Dict.), Maine went Hell-bent For Governor Kent. **1904** *Boston Herald* 2 Aug. 6 The Populist Democrats are going 'hell-bent', as the old song says, for Roosevelt. **1910** W. M. Raine *B. O'Connor* ii. 21, I know your kind—hell-bent to spend what you cash in. **1910** C. E. Mulford *Hopalong Cassidy* xxviii. 184 As soon as we lick this aggregation of trouble-hunters, what's left will ride hell-bent for that valley. **1912** L. J. Vance *Destroying Angel* ix, Unless you're hell-bent upon sticking around here. **1918** C. E. Mulford *Man fr. Bar-20* xv. 152, I was hell-bent to get down here,.. an' now I'm hell-bent to get back again. **1926** B. Cronin *Red Dawson* vi, Shaw sending the coach hell-bent round the curve of Jumping Lead. **1935** A. Squire *Sing Sing Doctor* iii. 32 We'll always have people hell bent on doing what they want to. **1957** *Times* 27 Dec. 6/1 Sir Edmund Hillary's message.. went on to say: 'We are heading hellbent for the Pole, God willing and crevasses permitting.' **1967** *Spectator* 24 Nov. 633/1 This report has been widely used to sustain the charge that the French government was hell-bent on feeding speculation against the pound. **1968** *Times* 31 Oct. 11/3 It is now becoming.. clear that an intelligent plan may have to be drawn, according to which those elements hell-bent on.. embarrassing the School will have to be expelled from it. **1973** *Times* 31 July 1/3 A minority of Unionist Party members.. feel obliged to vote with those who are hell-bent on destroying the first democratically elected assembly the Ulster people have had since the dissolution of Stormont.

helleborine. (Later examples.)

1833 Smith & Sowerby *Eng. Bot.* 2775 Purple-leaved Helleborine.. was discovered in Worcestershire in 1807. **1900** *Daily Express* 22 June 2/7, I was surprised.. to find the large white helleborine.. flowering in plenty. **1951** V. S. Summerhayes *Wild Orchids Brit.* vii. 123 The helleborines have rather tall slender stems.

Hellenian. **B.** *sb.* (Later example.)

a **1846** B. R. Haydon *Autobiogr.* (1927) III. xvi. 308 Two complete subjects of combat,—viz. between the Centaurs and Lapithæ, and between the Amazons and Hellenians.

Hellenic, *a.* Add: (Additional examples.)

1936 A. Thirkell *August Folly* viii. 240 Everyone had a suggestion to make for spending the gift... Mrs. Tebben thought an Icelandic or Hellenic cruise. **1971** A. Christie *Nemesis* vii. 76 They had been on an Hellenic cruise last year and a tour of bulbs in Holland the year before.

2. *Typogr.* Designating a variety of Greek type.

1927 *Greek Printing Types* (Exhib. B.M.) 3 It is thus only natural that the Graeco-Latin founts should now make way for more directly Hellenic styles of type. *Ibid.* 21 Thucydides, Funeral Oration of Pericles (part). Printed with 12-point 'New Hellenic' type.

Hellenophile (he·lĕnŏfəil), *a.* and *sb.* Also **Hellenophil** (-fil). [f. Hellene + -o + -phil, -phile.] = Philhellene *a.* and *sb.* Also **Hellenophi·lic** *a.*

1897 E. A. Bartlett *Battlefields of Thessaly* iii. 60 A little splutter of Hellenophile and Radical agitation in this country. **1898** [see *Armeno-]. **1959** *Encounter* July 48/1 A few Greek and hellenophil critics acclaimed Kazantzakis's poem as a masterpiece. **1966** *Listener* 6 Jan. 34/3 Another Hellenophile's reluctant conversion to the spell of the old Turkey. **1970** *Nature* 21 Nov. 711/1 Travers reports the discovery of another such protein, which in the Hellenophilic tradition of Harvard he duly names psi.

heller[2] (he·ləɹ). *U.S. slang.* [f. *hell *v.*[2] + -er[1].] One who 'hells around'.

1895 W. C. Gore in *Inlander* Nov. 67 Heller, a remarkable person. 'He is a heller at foot-ball.' **1933** *Amer. Speech* VIII. I. 81/2 *Heller*, one who is unusually daring or aggressive, intensified usually as *a regular heller*. **1939** J. Steinbeck *Grapes of Wrath* viii. 107 Tom grinned affectionately at him. 'Ain't he a heller?' he said. **1959** *Listener* 17 Dec. 1086/1 Jack Harrick, the old hillbilly satyr or 'heller'.

hell-fire. Add: **1.** (Later examples.)

1915 W. S. Maugham *Of Human Bondage* xxviii. 122 He could go his way without the intolerable dread of hell-fire. **1972** P. M. Hubbard *Whisper in Glen* vi. 59 Abomination was one of the devalued words. No one had used it wholly seriously. Mrs Haskell did, so that there was the reek of hell-fire to it.

3. (Further examples.)

1952 R. Campbell *Lorca* 10 The hell-fire sermons of the local priest. **1957** J. S. Huxley *Relig. without Revelation* (new ed.) ii. 31 A Hell-fire revivalist preacher. **1957** *Economist* 21 Dec. 1046/3 He reacted violently.. against the hellfire faith which had laid waste his childhood. **1972** P. M. Hubbard *Whisper in Glen* xviii. 175 The old woman tried to purge her own indiscretions by giving her daughter a hell-fire upbringing.

hell-fired, *a.* **2.** (Further examples.)

1833 J. Neal *Down-Easters* I. 79 See what a hell fired noise it [*sc.* the watch] makes. **1972** J. S. Hall *Sayings from Old Smoky* 80 It was the hell-firedest wreck I've ever seen!

hell-gate. (Later examples.)

1892 Kipling *Barrack-Room Ballads* 195 He yearned to the flare of Hell-gate there as the light of his own hearth-stone. **1934** T. S. Eliot *Rock* i. 47 The Heart of Man.. Swinging between Hell Gate and Heaven Gate.

hellgrammite, helgramite. Add: (Earlier example.) Also *transf.*

1866 *Wilkes' Spirit of Times* 14 July 315/3 There is another bait for bass called *kill-devil*—a sort of indescribable Barnum-what-is-it thing... An old friend of mine denominated them hell gramites. **1935** O. Nash *Primrose Path* (1936) 122 This human hellgramite that I think we could all dispense with.

hellion, hellyon (he·lyən). *U.S. colloq.* [prob. variant of Hallion, hallyon, with assimilation to Hell *sb.*] A troublesome or disreputable person; a mischievous child.

1846 J. J. Hooper *Adv. Simon Suggs* i. 18 The 'oudacious' little hellions! **1857** *Jrnl. Discourses* V. 135/2 We are going to dig a cache.. and put all the whining men and women into it... We want to be released from such poor hellyons. **1896** J. C. Harris *Sister Jane* 136 If dey ever was a hellian he wuz one. **1906** 'O. Henry' *Four Million* 8 Jawn, did ye ever see a straighter-nosed gang of hellions in the days of your life? **1919** H. L. Wilson *Ma Pettengill* vii. 225 Three children that was known to be hellions. **1941** H. G. Wells *You can't be too Careful* III. xxii. 218 That vision of Swedenborg's where all the damned and blessed fly of their own accord to the particular places appointed for them, hellions of every sort to their hells and the blessed to their heavens. **1957** M. Mezzrow in S. Traill *Concerning Jazz* 26 Baby Dodds, our drummer in Nice, is a wonderful person, but in his younger days he was a real hellion, and would fight at the drop of a hat. **1973** J. Wainwright *Pride of Pigs* 83 They were young animals.. pure, down-to-earth hellions.

he·llishing, he·llishun. *slang* (chiefly *Austral.* and *N.Z.*). Used as intensive adj. or adv.: terrible; very. Cf. *hangashun* s.v. *Hang *sb.* 6.

1931 *Amer. Speech* VI. 434 He seems to be in a hellishin' hurry about something. This is a hellishin' fine time to tell us about it. **1941** Baker *Dict. Austral. Slang* 35 *Hellishun*, an intensive used widely by children. **1950** *Landfall* IV. 38 It's hellashin funny. **1967** K. Giles *Death & Mr. Prettyman* vii. 145 The journey was as hellishing as he remembered. **1968** E. McGirr *Lead-Lined Coffin* iii. 114, I don't know that anybody.. has any knowledge of how hellishing thorough we are.

hello. Add: **A.** *int.* (Further examples.) Also as a greeting.

1967 *Listener* 5 Oct. 427/2 'Hello,' I thought, 'Now she's overdoing it.' **1971** *Farmer & Stockbreeder* 23 Feb. 3/3 Next week.. we shall say hello again to most of you, and to 100,000 new readers as well.

b. Used as an answer to a telephone call.

1892 Kipling *Lett. of Travel* (1920) 94 A.. millionaire.. clawing wildly at the telephone... 'Hello!.. Yes. Who's there?' **1922** S. Lewis *Babbitt* iv. 41 On the telephone they said only: '..Oh, Hello, 343?' **1973** J. Wainwright *Pride of Pigs* 169 She.. picked up the receiver, waited for the S.T.D. pips to stop, said 'Hello?' and.. recognised her brother's voice.

B. *sb.* *Comb.* (Earlier and later examples.)

1889 'Mark Twain' *Connecticut Yankee* 176 The humblest.. hello-girl,.. could teach the highest duchess. **1928** *Daily Chron.* 4 Feb. (*headline*) Brave Hello Girls. **1971** *New Scientist* 17 June p. iv, That was the day we said Goodbye to the Hello girls.

helluva: used freq. to represent 'hell of a'. Cf. *hell *sb.* 4 d.

1910 C. E. Mulford *Hopalong Cassidy* xxvi. 176, I got money—helluva lot of money. **1934** E. Pound *Eleven*

New Cantos xxxv. 24 It must be one helluva country. **1959** 'H. Carmichael' *Stranglehold* ii. 25 Be a helluva thing if I've left it just too late. **1959** 'M. M. Kaye' *House of Shade* ii. 26 It looks like saving you a helluva headache. **1967** *Crescendo* Dec. 33/1 Although she may not be as good a jazzer as Humph, she's certainly one helluva lot prettier. **1968** *Times* 23 Oct. 10/8 It's very unfortunate looking like him: he must have a helluva life.

helly, *a.* Revived in literary example.
1934 Dylan Thomas *Let.* 11 May (1966) 127 Every doubt and misgiving that an hereditary..imagination, an hereditary thirst..are capable of conjuring up out of their helly deeps.

Helmholtz (he·lmhǫlts). *Physics.* The name of H. L. F. von *Helmholtz* (1821–1894), German scientist, used attributively with reference to various devices and theories invented by him. Also **Helmho·ltzian** *a.*
1890 W. James *Princ. Psychol.* II. xx. 170 The Helmholtzian theory is probably not the last word in the physiology of hearing. **1920** G. B. Shaw *How to become Mus. Critic* (1960) 311 The Helmholtzian chords of Scriabin. **1930** R. Paget *Human Speech* 8 The ocarina is, I believe, the only well-known wind instrument which operates on the principle of the 'Helmholtz' resonator. **1940** *Chambers's Techn. Dict.* 409/2 Helmholtz galvanometer, Helmholtz resonance, Helmholtz resonator. **1962** Corson & Lorrain *Introd. Electromagn. Fields* v. 215 A pair of Helmholtz coils consists of two identical circular current loops placed coaxially so as to obtain uniformity of the magnetic induction *B* in the region midway between the loops.

helmitol (he·lmitǫl). [a. G. *helmitol* (E. Eichengrün 1902, in *Pharm. Zeitschr.* XLVII. 866/2).] A proprietary name of a derivative of hexamethylenetetramine that has been used as a urinary antiseptic and in the treatment of rheumatism.
1903 *Jrnl. Chem. Soc.* LXXXIV. I. 195 Helmitol.. forms colourless crystals. **1908** *Practitioner* Jan. 64 The newer drugs, urotropine, hetraline, helmitol, are useless in gleet. **1933** A. J. Clark *Appl. Pharmacol.* (ed. 5) ii. 50 Hexamethylenetetramine anhydromethylene citrate (helmitol) yields formaldehyde with alkalis as well as with acids, but it is doubtful whether it yields formaldehyde in alkaline urines. **1967** *Martindale's Extra Pharmacopoeia* (ed. 25) II. 1524/1 Proprietary preparations of formamol were formerly marketed in Great Britain under the name helmitol (Bayer Products).

helophyte (he·lŏfəit). *Bot.* [mod. f. Gr. ἕλος marsh + φυτ-όν plant.] A marsh plant.
[**1902** F. E. Clements *Syst. Nomencl. Phytogeogr.* 6 Helium, a marsh formation; helia, a group of marsh formations; helophyta, marsh plants; helophilus, marsh-loving.] **1909** Groom & Balfour tr. *Warming's Oecology of Plants* xxxiv. 131 There is a group of plants, marsh plants (*helophytes*), which..develop their lower parts.. in water or at least in soaking soil, but have their assimilatory organs mainly adapted to existence in the air. **1913** *Jrnl. Ecol.* I. 17 Another division [of plants] is characterised by semi-aquatic dormant buds—helophytes and hydrophytes. The helophytes or marsh-plants do not include all so-called marsh species. **1926** Tansley & Chipp *Study of Vegetation* ii. 22 The marsh plants (*Helophytes*) and water-plants (*Hydrophytes*), whose [perennating] buds are situated at the bottom of the water or in the subjacent soil. **1960** N. Polunin *Introd. Plant Geogr.* xv. 505 Marsh-plants (*helophytes*..) and water-plants (*hydrophytes*) are much alike in their morphological and anatomical characteristics.

helotage. (Later examples.)
1957 K. A. Wittfogel *Oriental Despotism* 414 The helotage-based societies of ancient Greece. **1962** *Listener* 27 Sept. 478/3 Spartan helotage barely lasted into the second century B.C.

helotism. Add: **2.** *Biol.* [ad. Sw. *helotisme* (E. Warming *Plantesamfund* (1895) II. iv. 85), prob. after G. *helotenthum* (S. Schwendener *Die Algentypen der Flechtengonidien* (1869) 4).] A form of symbiosis in which one organism makes use of another as if it were a slave, by causing it to function to its own advantage; used *esp.* of the relationship of the fungus and alga in a lichen by those who regard it as neither mutualism nor parasitism.
1900 B. D. Jackson *Gloss. Bot. Terms* 119/2 *Helotism*, Warming's term for the relation of the symbionts in the Lichen thallus. **1909** Groom & Balfour tr. *Warming's Oecology of Plants* 85 The symbiosis between lichen-fungi and algae is obviously most correctly interpreted as helotism. **1932** Fuller & Conard tr. *Braun-Blanquet's Plant Sociol.* i. 6 The original individualities of fungus and alga are lost and merged in a new and more aggressive organism, so that the term helotism does not seem any more fortunate than the term mutual parasitism. **1962** C. J. Alexopoulos *Introd. Mycology* (ed. 2) xxii. 539 Botanists of a middle-of-the-road group..state that the fungus holds the alga imprisoned in a state of slavery, helotism being the word used, thus granting that the fungus has the upper hand.

help, *v.* Add: **1. c.** Also ellipt. *so help me*, and as a variant *so help me bob*. Cf. Swelp.
1821 P. Egan *Life in London* II. iii. 229 She tripped me up, my Lord, so help me bob, it is true. **1869** Trollope *Vicar of Bullhampton* (1870) iii. 20 Just go home to father's, sir; not a foot else, s' help me. **1936** L. C. Douglas *White Banners* iv. 69 That he would never again ..so help him..fritter away precious time.

5. a. Delete 'this is now *dial.* or *vulgar*' and read 'this is now a common *colloq.* form'.
1941 *Punch* 2 July 13/3 Sir Kingsley Wood..asked the House for another £1,000,000,000, to help pay for the next three months of war.
b. (Further examples.)
1936 *Punch* 1 Apr. 375/1, I suppose you two fellows wouldn't help me get the stuff into the coal-cellar, would you? **1940** *Ibid.* 5 June 620/3 The collection of waste-paper that's going to help us win the war. **1971** D. E. Westlake *I gave at Office* (1972) 121 None of the locals.. had any desire to help us off-load the plane.
6. d. To render assistance in dealing *with*.
1924 A. D. Sedgwick *Little French Girl* I. v, If he sat there..not helping with the water-cans, the baskets of flowers, the scissors, it was because he loved her and wanted to watch her. **1933** *Punch* 26 July 104/2 'Poetry. I believe people use it in exams to remember rules.. and things...' 'That wouldn't help with Greek and Latin,' said Charles.
e. Phr. *to help the police in* (or *with*) *their inquiries*: to be questioned by the police in connection with a crime, often regarded as having the implication of being the chief suspect; also *to help the police, to help with inquiries*.
1957 *Times* 3 Sept. 4/3 The police are anxious to trace an itinerant photographer... It is believed that he may be able to help them in their inquiries. **1970** *Guardian* 10 Nov. 20/4 A man was helping police last night after the body of Susan Young, aged 12..was found. **1971** 'J. Fraser' *Death in Pheasant's Eye* xxvi. 162 What's the deadline for arresting Stanley Robinson? At the moment he's 'helping with enquiries'. **1972** V. C. Clinton-Baddeley *To study Long Silence* ii. 72 In newspaper language..the man who is 'helping the police' is the chap who's being badgered into a confession. **1973** *Sunday Times* 14 Oct. 1/1 A 17-year-old girl..was found battered to death... Later, a man was helping police with their inquiries.
8. b. (Later examples.)
1919 V. Woolf *Night & Day* xxxi. 2 She behaves very oddly. She forgets to help the pudding. **1938** M. K. Rawlings *Yearling* i. 11 Ma Baxter sat at the table waiting for them, helping their plates.
c. help-yourself, used *attrib.* esp. of a restaurant or cafeteria where one serves oneself, or of the meal obtained there; also as *sb.* Also *transf.*
1894 M. Fry *Let.* in E. H. Jones *Margery Fry* (1966) v. 37 Then at 1 is lunch..a help-yourself meal and you get up and come away as soon as you've done without waiting for anyone else. **1923** [see *cafeteria]. **1955** T. H. Pear *Eng. Social Diff.* 183 The 'help yourself' cafeteria system. **1959** R. Postgate *Good Food Guide* 280 There is an à la carte menu and a 'help-yourself' lunch for the hasty at 6/6. **1961** *Times Lit. Suppl.* 13 Oct. 712/5 Up to a point he has invented the help-yourself novel. **1967** 'A. Gilbert' *Visitor* iii. 38, I got a cup of coffee and a sandwich at a help-yourself, and went back to the flat.
11. a. (Later example.)
1963 *Sunday Express* 3 Mar. 2/5 You are aware that the archdeacon swears that he never wears pyjamas?— I can't help that.
b. (Later examples.)
1928 *Manch. Guardian Weekly* 5 Oct. 263/4 If clairvoyants are to be attached to police stations they can hardly help but become officials. **1952** G. Sarton *Hist. Sci.* I. xxi. 526 They were brutally raped by conquistadors ..who could not help increase geographic knowledge.

help, *sb.* Add: **3. c.** (Further examples.)
1743 Ellis *Mod. Husb.* III. II. 2 Next to them [*sc.* hired servants] we should be provided with auxiliary Helps. **1807** C. W. Janson *Stranger in Amer.* 87, I am Mr —'s *help*. I'd have you know..that I am no *sarvant*. **1815** *Massachusetts Spy* 23 Aug. (Th.), Our lady and gentleman 'hired helps' do not understand who is meant when their master is inquired for. **1818** H. B. Fearon *Sk. Amer.* 80 Servants, let me here observe, are called 'helps'. If you call a servant by that name they leave you without notice. **1838** J. F. Cooper *Amer. Democrat* 122 Those who aid their masters in the toil may be deemed 'helps', but they who perform all the labor do not assist..but they do it themselves. **1899** *Westm. Gaz.* 4 Aug. 2/3 Judge: What is a 'help'? Plaintiff: Well, she's a cook-housemaid-barmaid. **1949** 'J. Tey' *Brat Farrar* xi. 84 Lana, their 'help'..'obliged' only because her 'boy friend' worked in the stables. **1971** [see *daily *a.* 1 b].
d. For *U.S.* read orig. *U.S.* Add examples.
1959 S. Gibbons *Pink Front Door* xviii. 221 The months when she had been without help had established Molly's unexpected visits as a habit. **1962** P. Mortimer *Pumpkin Eater* xi. 106 'We got help. I don't know why it's called help.' 'You mean servants?' 'We don't call them servants.'
7. *help-girl*; **help-mate** *Chess*, a type of chess problem in which Black is required to play so that White may give mate in a certain number of moves.
1863 Mrs. Gaskell *Cousin Phillis* in *Cornhill Mag.* Nov. 633 Betty..carried off the great dish to the kitchen, where an old man..and a help-girl, were awaiting their meal. **1897** Rowland & Rowland *Problem Art* 91 Another class of problems..in which *both players concur* in endeavouring to effect the speediest mate—..which we term Help-mate Problems. **1913** A. C. White *Sam Loyd* 31 Whether Loyd was the inventor of the Help-mate problem..I do not know. **1966** *New Statesman* 11 Nov. 718/3 'Help-mate'..Black moves first and helps White to mate in a given number of moves.

helpable. (Earlier example.)
1833 J. S. Mill *Lett.* (1910) I. 48, I believe I am the least *helpable* of mortals.

helped (helpt), *ppl. a.* [f. Help *v.* + -ed[1].] That has been helped, aided, or assisted. Also with advs., as *helped-out*.
1905 *Daily Chron.* 15 May 3/5 The dowagers of to-day, with their helped-out complexions and Venetian red hair. **1910** E. M. Albanesi *For Love of Anne Lambart* 104 But this is no helped success, this is real.

helter-skelter. Add: **C.** *sb.* **b.** (Also *helter-skelter lighthouse.*) A tower-like structure used in fun fairs and pleasure-grounds, with an external spiral passage for sliding down on a mat.
1906 *Westm. Gaz.* 1 June 10/2 The World's Manufacturing Company, examples of whose 'helter-skelter' lighthouses are at Earl's Court, Blackpool, Southport, and other places. **1907** *Daily Chron.* 4 Sept. 2/7 All sorts of fearsome things..from a helter-skelter..to a smashing saloon and a 'coker nut' alley. **1927** *Sunday Express* 5 June 4 They will travel on the merry-go-rounds... They will go down with the mats on the helter-skelter, and up with the swings. **1945** G. Millar *Maquis* i. 8, I began to feel that I was sitting on a mat on top of a helter-skelter. That somebody would give me a push and I would be away with..nobody at the bottom to pick me up. **1968** D. Braithwaite *Fairground Archit.* 24 Vertical features like the 'Big Wheel', 'Helter Skelter' and 'Chair-o-Planes'..are valuable advertising symbols.

helvellic (helve·lik), *a.* [tr. G. *helvellasäure* helvellic acid (Boehm & Külz 1885, in *Arch. f. exper. Path. u. Pharm.* XIX. 414), f. mod.L. *Helvella*, a genus of ascomycetous fungi: see -ic.] *helvellic acid*, a poisonous acid, $C_{12}H_{20}O_7$, present in some fungi of the sub-family Helvellæ.
1906 H. B. Shaw in Allbutt & Rolleston *Syst. Med.* (ed. 2) II. I. 871 Helvellic acid and extracts of the Helvella produce nausea, vomiting, hæmoglobinuria, icterus,.. and uræmia. **1947** F. A. & F. T. Wolf *Fungi* II. xv. 351 There occurs in *Helvella esculenta* a water-soluble, heat-labile, hemolytic principle that has been identified as helvellic acid, $C_{12}H_{20}O_7$. **1953** J. Ramsbottom *Mushrooms & Toadstools* vi. 54 *Gyromitra esculenta, Helvella crispa*..and several other Discomycetes contain helvellic acid, which has a strong dissolving action on the red corpuscles of the blood.

helvetium (helvī·ʃiɒm). *Chem.* [ad. G. *helvetium* (W. Minder 1940, in *Helvetica Physica Acta* XIII. 152), f. mod.L. *Helvet(ia* Switzerland + -ium.] Earlier name for *astatine. Cf. also *alabamine and *Anglo-Helvetium. (*Disused.*)
1940 *Times* 13 Aug. 3/2 The Swiss scientist, Dr. W. Minder, of Berne University, has now succeeded in isolating element '85', which, in honour of his Fatherland, he has named 'Helvetium'. He has proved that 'Helvetium' is produced in small quantity from the decomposition of actinium, which is radio-active.

helvolic (helvǫ·lik), *a. Biochem.* [f. mod.L. *helvola* yellowish, the name of the mutant variety of fungus, characterized by its buff colour, from which the acid was first isolated: see -ic.] *helvolic acid*, an antibiotic with the probable formula $C_{32}H_{42}O_8$ produced by some strains of the fungus *Aspergillus fumigatus*. Also called *fumigacin*.
1943 E. Chain et al. in *Brit. Jrnl. Exper. Path.* XXIV. 119 A new antibiotic has been isolated in the crystalline state from culture filtrates of *Aspergillus fumigatus*, mut. *helvola* Yuill. It has been named helvolic acid. **1944** *Jrnl. Bacteriol.* XLVII. 392 Helvolic acid is apparently identical with the fumigacin from which the gliotoxin fraction has been removed. **1949** H. W. Florey et al. *Antibiotics* i. 66 Helvolic acid..had some protective action against streptococcal infection in mice, but its other properties made it unsuitable for use in man. **1953** J. Ramsbottom *Mushrooms & Toadstools* xxiii. 289 Fumigacin (helvolic acid) produced by *Aspergillus fumigatus* has all the necessary qualities except that bacteria readily acquire resistance to it. **1965** P. K. C. Austwick in Raper & Fennell *Genus Aspergillus* vii. 106 *Aspergillus fumigatus* is known to produce three antibiotics that are toxic to experimental animals: fumigatin,..helvolic acid,..and gliotoxin.

helxine (helksəi·ni, helzəi·ni). *Bot.* [mod.L. (E. Requien 1825, in *Annales des Sciences Naturelles* V. 384), f. Gr. ἑλξίνη pellitory, a related plant.] *Soleirolia soleirolii*, a creeping, perennial herb of the family Urticaceæ, native to Corsica and Sardinia, formerly called by the generic name *Helxine*.
1873 Mrs. Hooker tr. *Le Maout & Decaisne's Gen. Syst. Bot.* II. 667 Flowers [of Urticaceæ]..very rarely solitary and axillary (*Helxine*). **1924** L. H. Bailey *Man. Cultivated Plants* 241 Helxine..making a matted moss-like covering. **1964** M. Fish *Ground Cover Plants* v. 44 If one happened to have a dark, dank courtyard which needed the softening influence of tender green, nothing could be more charming than helxine. **1969** O. Polunin *Flowers of Europe* 58 *Soleirolia soleirolii*..(*Helxine*).. Native of Mediterranean islands; naturalized in Western Europe.

hem, *sb.*[1] Add: **5.** *Comb.* **hem-line,** the outline of the hem, hence the height from the ground, of a woman's skirt.

1923 *T. Eaton & Co. Catal.* Spring & Summer 34 The modish uneven hemline. **1927** *Vanity Fair* Sept. 4/3 Liberty of opinion, in skirt lengths, hemlines, waistlines, ..was convincingly demonstrated. **1929** *Daily Tel.* 16 Jan. 8/4 It is admitted that the hem lines of 1928 changed the whole contour of the evening mode as they flounced and floated in uneven length to the ground. **1957** *New Yorker* 30 Mar. 98/3 This year, she puts a minute and completely absurd circular godet, only three inches deep, at the hemline in front—to allow for striding, they say. **1958** *T.V. Times* 10 Oct. 21/2 Other girls put up their hem-lines a couple of inches as Paris dictates. **1971** R. Garrett *Spiral* x. 94 She fidgeted with the hemline of her skirt.

he-man: see *he *pers. pron.* 8 a.

heme, var. *hæm.

hemera[2] (he·mĕră). *Palæont.* and *Geol.* Pl. **-æ.** [mod.L., f. Gr. ἡμέρα day.] A period of geological time in which any particular species was most abundant as represented in strata; an interval between times when two successive species were dominant. Hence **he·meral** *a.*
1893 S. S. Buckman in *Q. Jrnl. Geol. Soc.* XLIX. 481 For a palaeontological purpose... I propose the term 'hemera'... I wish to use it as the chronological indicator of the faunal sequence. **1898** —— *Ibid.* LIV. 443 The hemeral names are taken from the names of ammonites. *Ibid.*, The shortest geological time-division is a hemera: that is, the time during which a particular species..had dominant existence. *Ibid.* 448 The genera *Dactylioceras* and its allies..almost disappear with the close of the hemera *bifrontis*. **1902** —— in *Geol. Mag.* Dec. 555 A 'hemera' was designed to mark the time from..when one species or set of species becomes dominant to the time when another..does so. **1920** *Q. Jrnl. Geol. Soc.* LXXVI. 63 The range of strata which Mr. Richardson has divided into seven or at the most eight hemeræ cover, according to my more detailed subdivisions, thirty-two hemeræ, and should afford a good test as to whether this more detailed method of dating makes for greater precision. **1933** W. J. Arkell *Jurassic Syst. Gt. Brit.* i. 20 Nine years later, finding that there were still some who considered that a hemera was simply a subdivision of a zone, Buckman published a fresh explanation. *Ibid.* 27 In the course of an argument with Buckman over his insertion of an excessive number of hemeræ into the time-table of the Corallian rocks, I attempted to justify my view..that many of his hemeral indices lived side by side on the same sea-bed. **1966** D. T. Donovan *Stratigr.* vii. 160 Several other time terms have been proposed but have never achieved general recognition. Perhaps the most notorious is the *hemera*. *Ibid.* 161 Hemerae were much used by English Jurassic workers for about forty years, but hardly at all by anyone else.

hemerocallis. For etym. and def. read: [mod.L., ad. L. *hemerocalles*, Gr. ἡμεροκαλλές a kind of lily, f. Gr. ἡμέρα day + κάλλος beauty, adopted by Linnæus in his *Hortus Cliffortianus* (1737) 128 as the name of a genus.] A herbaceous perennial plant of the genus so called, belonging to the family Liliaceæ, mostly native to temperate, eastern Asia, and bearing corymbs of yellow or orange, trumpet-shaped, short-lived flowers; a Day-lily. (Later examples.)
1900 L. H. Bailey *Cycl. Amer. Hort.* II. 728/1 The flowers of Funkia are borne in racemes; of Hemerocallis in corymb-like panicles. **1938** F. Perry *Water Gardening* xvi. 271 Hemerocallis have some economic qualities, for in the Orient the flower petals are gathered. **1970** *Observer* 23 Aug. 25/4 In the Orient hemerocallis flowers are often used for food.

hemi-. Add: **hemi-a·cetal** *Chem.*, any of a class of compounds having the general formula R·CH(OH)(OR′), differing from an acetal in having an -OH group in place of one of the -OR groups; **he:miamblyo·pia** *Ophthalm.*, amblyopia of half of the field of vision; hence **he:miamblyo·pic,** one suffering from hemiamblyopia; **he:miangioca·rpic, -ca·rpous** *adjs.*, designating a fungus in which the hymenium is enclosed during the early part of its development; characteristic of such a fungus; **he:micole·ctomy** *Surg.*, excision of part of the colon, *esp.* of the right or left half; **hemidemisemiquaver** (later example); **hemihydrate** (examples); **hemihype·rtrophy,** unilateral or partial hypertrophy; **hemika·ryon** *Cytol.* [a. G. *hemikaryon* (T. Boveri 1905, in *Jenaische Zeitschr. f. Naturwiss.* XXXIX. 447), f. Gr. κάρυον: see Karyo-] (see quot.); so **he:mikaryo·tic** *a.*; **hemime·llit(h)ene, -ine** *Chem.* = *trimethylbenzine*; **hemiobole:** also **hemiobol;** **hemipa·rasite** *Bot.* [G. (F. Johow 1890, in *Verhandl. Deutsch. Wissensch. Ver. Santiago* II. II. 67)], a facultative parasite, e.g. certain fungi; also a plant which is partially parasitic, drawing water and mineral nutrients but not synthesized foods from its host, e.g. certain higher plants, as the mistletoe; hence **he:miparasi·tic** *a.*; **hemiparesis** (examples); **hemipe·nis** *Zool.*, one of the paired eversible copulatory organs in snakes and lizards; **hemisa·prophyte** *Bot.* [ad. G. *hemisaprophyt* (F. Johow 1889, in *Jahrb. f. wissensch. Bot.* XX. 479)], a facultative saprophyte, being alternatively either parasitic or autotrophic; hence **he:misaprophy·tic** *a.*
1893 *Jrnl. Chem. Soc.* LXIV. I. 563 (*heading*) Hemi-acetals derived from substituted chloranils. **1964** N. G. Clark *Mod. Org. Chem.* xiv. 273 Hemi-acetals are generally unstable intermediates in the formation of true acetals. **1890** W. James *Princ. Psychol.* I. ii. 44 According to Loeb, the defect is a dimness of vision ('hemiamblyopia') in which (however severe) the centres remain the best seeing portions of the retina. **1947** F. B. Walsh *Clin. Neuro-Ophthalm.* i. 56/1 Along with hemiamblyopia, hemiachromatopsia is evidence for early and partial involvement of the primary visual pathways. **1960** H.-L. Teuber et al. *Visual Field Defects* vi. 64 (*caption*) The hemiamblyopia was more disturbing to this patient than an outright hemianopia. **1933** *Mind* XLII. 386 In hemiamblyopics there is a tendency for a displacement, towards the sound side, of stimuli falling on the amblyopic area. **1902** *Encycl. Brit.* XXVIII. 558/2 Gymnocarpic and Hemiangiocarpic:—*Hymenomycetes.* **1900** B. D. Jackson *Gloss. Bot. Terms, Hemiangiocarpous*, when the hymenium of some Fungi is for some time covered with a membrane, the gonidiophore is so termed. **1902** *Encycl. Brit.* XXVIII. 561/1 The Discomycetes and their immediate allies are termed Hemiangiocarpous, because however much their fructifications are closed at first, they ultimately open and expose the layers of asci. **1928** C. W. Dodge tr. *Gäumann's Compar. Morphol. Fungi* xxv. 410 In the second, or hemiangiocarpous, stage.. the sporiferous tissue is differentiated from the tissue in the interior of the fundaments. **1967** M. E. Hale *Biol. Lichens* ii. 32 Henssen has recently traced consistent hemiangiocarpous development in a group of genera. **1926** R. J. E. Scott *Gould's Med. Dict.* 596/1 Hemicolectomy. **1963** *Lancet* 19 Jan. 133/1 A man, aged 53, had had right hemicolectomy performed eight months previously on account of carcinoma of the cæcum. **1959** D. Cooke *Lang. Mus.* iii. 133 A bass of rushing hemi-demi-semi-quavers. **1909** Webster, Hemihydrate. **1946** *Nature* 6 July 13/2 Calcium sulphate exists in three states of hydration, anhydrite, gypsum, and the lower hydrate, generally known as hemihydrate or plaster of Paris. **1965** *New Scientist* 18 Mar. 709/2 Gypsum is calcined to form the hemi-hydrate which sets after wetting to re-form gypsum. **1900** Dorland *Med. Dict.* 294/1 Hemihypertrophy. **1922** *Proc. R. Soc. Med.* XV Child. 51 Forty cases..of complete hemihypertrophy where all the structures on one side of the body were involved. **1964** S. Duke-Elder *Syst. Ophthalm.* III. xvii. 1028 Facial hemihypertrophy is a much rarer condition which may involve the whole of one side of the body. **1925** E. B. Wilson *Cell* (ed. 3) 1132 *Hemikaryon*, a nucleus containing the haploid number of chromosomes. *Ibid.* ix. 728 (*caption*) Karyoplasmic relation in embryos of the sea-urchin..stage of haploid (hemikaryotic) dwarf, from merogonic egg-fragment. **1889** Muir & Morley *Watts' Dict. Chem.* (rev. ed.) 671/1 *Hemimellithene*,..formed by distilling (*a*)-cuminic acid with lime. *Ibid.*, Hemimellithene may also be isolated from coal-tar oil. **1956** *Nature* 18 Feb. 301/2 The great variety of organic compounds contained in this latest issue [of a text-book] may be indicated by mentioning azelanitrile,..hemimellitine, [etc.]. **1921** *Brit. Mus. Return* 81 A gold hemiobol of the fourth century B.C. **1941** *Antiquity* XV. 302 Issues of tetrobols and hemiobols. **1891** *Jrnl. R. Microsc. Soc.* 70 Each of these classes, except the last, may be again divided into Holoparasites and Hemiparasites. **1900** B. D. Jackson *Gloss. Bot. Terms* 120/1 *Hemiparasite*, a facultative saprophyte, a parasite which can exist as a saprophyte. **1927** Hemiparasite [see *facultative *a.* 1 c]. **1960** N. Polunin *Introd. Plant Geogr.* xiv. 437 There are..two main synusiae in the tropical rain forest—the root-parasites growing on the ground..and the semi-parasites (often termed hemi-parasites) growing epiphytically on the trees. **1970** *Nature* 21 Mar. 1162/1 Annual hemiparasites characteristically form haustorial connexions (root grafts) with most of the plants that surround them. **1902** *Encycl. Brit.* XXV. 439/2 Loranthaceæ and Santalaceæ are chiefly hemiparasitic. **1970** *Nature* 21 Mar. 1161/2 The hemiparasitic flowering plants that live in annual grasslands are of special interest. **1893** *Funk's Stand. Dict.*, Hemiparesis. **1952** M. E. Florey *Clin. Appl. Antibiotics* ii. 37 One patient developed hemiparesis and prolonged convulsions. **1909** Webster, Hemipenis. **1913** G. A. Boulenger *Snakes of Europe* ix. 83 Each hemipenis is lodged in a cavity on each side of the base of the tail. **1965** R. & D. Morris *Men & Snakes* viii. 190 Male snakes possess two hemipenes. Only one hemipenis is inserted at a time and apparently males do not switch from one to the other during the same mating bout. **1895** *Ann. Bot.* IX. 337 The structure of the stele in absorbing rhizome-axes of hemi- and holo-saprophytes is frequently remarkably like that of a root. **1900** B. D. Jackson *Gloss. Bot. Terms* 120/2 *Hemisaprophyte*, a facultative parasite. **1909** Groom & Balfour tr. *Warming's Oecology of Plants* II. xxv. 90 *Hemisaprophytes* have the external appearance and structure of normal plants. **1927** Hemisaprophyte [see *facultative *a.* 1 c]. **1895** *Ann. Bot.* IX. 337, I have already shown that in a hemisaprophytic Orchid (*Corysanthes*) there is an atrophy in the leaves of the absorbing rhizome.

hemianopia (he:miănǭu·piă). *Ophthalm.* [mod. L., f. Hemi- + Gr. ἀν priv. + -ωπία sight.] = Hemianopsia. Hence **hemiano·pic** *a.*, of, pertaining to, or characterized by hemianopia; also as *sb.*, a person with hemianopia.
1882 *Ophthalmic Rev.* I. 253 The more exactly the cause of a homonymous hemianopia can be localised in the cortex of one occipital lobe, the more improbable becomes the theory of Charcot and Landolt. **1889** G. A. Berry *Dis. Eye* xi. 340 Hemianopia may be to the right or left, and partial or complete. *Ibid.* 341 Ferrier's experiments, which led him to localise the visual centres in the angular gyri, have given rise to the hemianopic symptoms. **1891** F. Taylor *Man. Pract. Med.* (ed. 2) 161 A transient hemianopia may occur in cerebral hæmorrhage. **1898** *Nature* 13 Jan. 255/1 Hallucinations connected with hemianopia. **1908** *Practitioner* Oct. 558 Hysterical hemianopia. **1943** *Mind* LII. 363 This seems ruled out by the experiments of Gelb and Goldstein on hemianopics, who saw a field of vision of the normal shape, though the receptive area had been severely damaged. **1964** S. Duke-Elder *Parsons' Dis. Eye* (ed. 14) iv. 37 A lesion in the optic tract will produce a hemianopic reaction involving both eyes. **1966** D. G. Cogan *Neurol. Visual Syst.* xiv. 265 The most common complaint of patients with hemianopia, aside from collision with objects on the blind side, is the difficulty with reading.

hemianopsia. Add: (Earlier and later examples.) Also **hemianopsy.**
1883 *Ophthalmic Rev.* II. 82 The above named recent works on hemianopsia and the decussation of the optic nerve contain much of interest. **1884** *Encycl. Brit.* XVII. 785/1 Hemianopsia means loss of one-half of the visual field. **1908** *Practitioner* Jan. 15 Attacks of aphasia, monoplegias, hemiplegias, word-blindness, and word-deafness, or hemianopsy. **1962** L. S. Sasieni *Optical Dispensing* viii. 200 Hemianopsia spectacles.

hemicellulose (hemise·liulōus). [a. G. *hemicellulose* (E. Schulze 1891, in *Ber. d. Deut. Chem. Ges.* XXIV. 2286), f. Hemi- + Cellulose *sb.*] Any of various non-cellulosic polysaccharides, of simpler composition than cellulose, that are major constituents of the cell walls of many plants and are characterized by undergoing hydrolysis by acids more readily than cellulose to give a variety of simple sugars and other carbohydrates and by being extractable with dilute alkaline solutions.
1891 *Jrnl. Chem. Soc.* LX. 1179 Those constituents of the cell which dissolve easily in dilute mineral acids, with formation of glucose, he [*sc.* E. Schulze] calls hemicelluloses. **1921** A. L. Smith *Lichens* v. 212 In *Cladonia rangiferina*..the cell-membranes of the hyphae contained, as hemicelluloses, pentosans in small quantities and galactan. **1948** *New Biol.* IV. 87 Interspersed in the meshes of the cellulose, which constitutes approximately 50 per cent of the weight of the dry wood, are a number of other substances which apparently serve to stiffen the framework, principally hemicelluloses..and the more complex carbohydrate lignin. **1963** R. R. A. Higham *Handbk. Papermaking* ii. 33 Hemi-celluloses are found in two main groups, i.e., those associated mainly with the plant cellulose and centred largely in the structure of the fibre and those associated with lignin, in the middle lamella and primary walls of fibres. *Ibid.*, The presence of hemi-celluloses in papermaking pulps is very important, as they assist in internal fibrillation of the fibres, but owing to their short chain structure they easily become broken down during digestion and are lost in the form of by-products.

hemicryptophyte (hemikri·ptŏfəit). *Bot.* [ad. Da. *hemikryptofyte* (C. Raunkiaer 1904, in *Bot. Tidsskrift* XXVI, p. xiv), f. Hemi- + Crypto-: see -phyte.] (See quot. 1932.) Also *attrib.* Hence **he:micryptophy·tic** *a.*
1913 *Jrnl. Ecol.* I. 24 The *Anemone nemorosa* facies of the beech-wood is geophytic since the 0·1 sq. m. readings show geophytes 82 per cent., hemicryptophytes 18 per cent. *Ibid.* 25 The contrast between spruce hemicryptophytic wood and beech geophytic wood. **1932** Fuller & Conard tr. *Braun-Blanquet's Plant Sociol.* xii. 291 Hemicryptophytes..plants with perennial shoots and buds close to the earth's surface. **1938** J. R. Carpenter *Ecol. Gloss.* 130 *Hemicryptophyte climate*, the climate of the greater part of the cold temperate zone. **1964** V. J. Chapman *Coastal Veget.* vi. 150 A hemicryptophyte flora.

hemicyclic, *a.* (Later example.)
1878 [see Acyclic *a.*].

he·mi-de:mi-se·mi: used as adj. and sometimes as a combining form in imitation of the use in *hemidemisemiquaver* (s.v. Hemi-).
1929 J. Laird *Idea of Value* ix. 302 Even believers in the 'unconscious'..seem commonly to hold that the 'unconscious' is at least hemi-demi-semi conscious. **1958** *Spectator* 3 Jan. 14/3 What is the point of ideological compromises and hemi-demi-semi pro-Soviet hedging? **1965** O. Barfield in J. Gibb *Light on C. S. Lewis* p. xiii, Hemi-demi-semitone of alteration in the pitch of his voice.

hemidesmus (hemide·smɒs). *Bot.* [mod.L. (R. Brown, 1809) f. Hemi- + Gr. δεσμός bond; so named in allusion to the incomplete coherence of the anthers with the stigma.] A small, swimming herb of the genus so named, belonging to the family Asclepiadaceæ, and native to India and Ceylon; *esp.* a plant of *Hemidesmus indicus*, the root of which is used as a substitute for sarsaparilla; also, a syrup prepared therefrom. Hence **hemide·smic** *a.*
1809 R. Brown in *Mem. Wernerian Nat. Hist. Soc.* I. 56 Hemidesmus..whose name is derived from the partial connection of the stamina, is composed of *Periploca Indica*, and two very nearly related unpublished species. **1844** *Pharm. Jrnl.* III. 239 The root of hemidesmus indicus has for some years been sold in this

country under the name of Smilax Aspera. **1880** Garrod & Baxter *Mat. Med.* 311 Hemidesmus Root... It.. contains a peculiar volatile, crystallizable substance, with acid properties: this has been called hemidesmic acid. **1898** *Revised Brit. Pharmacop.* 4 The retention of hemidesmus is noteworthy, and a tribute to the St. Bartholomew's school of physic. **1968** D. C. Gunawardena *Genera et Species Plantarum Zeylaniae* 122 Hemidesmus... Anther connective prolonged, covering over stigma.

hemimorphite. (In Dict. s.v. Hemimorphic *a.*) Add: Now generally used in place of earlier names such as *calamine*. (Examples.)
1868 J. D. Dana *Syst. Min.* (ed. 5) v. 409 Unfortunately, Brooke & Miller, in 1852, reversed Beudant's use of these names [*sc. calamine* and *smithsonite* for the silicate and carbonate of zinc], with no good reason; and in 1853, Kenngott, on account of the confusion of names, as he says, introduced for the silicate the new name *Hemimorphite*, and so added to the confusion. **1910** *Encycl. Brit.* VII. 553/1 Two hemihedral or hemimorphic crystals (e.g. of diamond or of hemimorphite) are often united in twinned positions. **1951** A. N. & H. Winchell *Elem. Optical Mineral.* (ed. 4) II. x. 452 Hemimorphite is found in veins with smithsonite and sphalerite.

hemin, var. Hæmin.
1955 *Sci. News Let.* 29 Oct. 274/1 Hemin, a chemical related to the red color chemical of blood, and protein. **1963** *New Scientist* 25 July 182 Biliverdin is formed by biological degradation of the red blood pigment hemin.

Hemingwayesque (he:miŋwĕi,e·sk), *a.* [f. the name of Ernest *Hemingway* (1898–1961), American novelist + -esque.] Characteristic of the works of E. Hemingway. So **He:mingwa·yan** *a.*; **He:mingwaye·se**; **He·mingwayish** *a.*
1942 H. Haycraft *Murder for Pleasure* viii. 171 Hemingwayesque courage and fatalism. **1957** J. Kerouac *On Road* (1958) vii. 41 Composing his latest Hemingwayan short story. **1959** *Times Lit. Suppl.* 4 Sept. 505/3 His father, a Hemingwayish figure. **1964** *Listener* 28 May 882/2 Plain unvarnished Hemingwayese looks decidedly mannered. **1970** *English Studies* LI. 491 In *Ernest Hemingway, A Life Story*..Carlos Baker has adopted his subject's deliberately simple declarative style. The opening and closing of the book are archly Hemingwayesque; some parts of the 640 pages that come between are simply Hemingwayese.

hemiopia. Add: So **hemio·pic** *a.*
1873 *Arch. Sci. & Pract. Med.* I. 293 (*title*) Hemiopic and sector-like defects in the field of vision. **1890** W. James *Princ. Psychol.* I. ii. 42 A hemiopic disturbance of vision is one in which neither retina is affected in its totality, but in which, for example, the left portion of *each* retina is blind, so that the animal sees nothing situated in space towards its right.

hemiplegic, *a.* Add: Also *sb.*, one who is affected by hemiplegia (see also quot. 1970).
1890 *Retrospect Med.* CII. 155 Convalescent hemiplegics. **1970** *New Scientist* 3 Sept. 473/1 It would enable amputees and hemi-plegics—persons paralyzed below the waist—to walk about under their own power.

hemispherectomy (he:misfere·ktŏmi). *Surg.* [f. Hemispher(e 5 + *-ectomy.] Excision of a cerebral hemisphere.
1950 *Jrnl. Neurol. & Psychiatry* XIII. 243/1 Over a period of five years..12 hemispherectomies were performed on patients suffering from infantile hemiplegia. *Ibid.* 263/1 Because of the ventricular distortion, hemispherectomy was carried out. **1956** E. L. Mascall *Christ. Theol. & Nat. Sci.* vi. 252 There are no *theoretical* problems for Christian faith..in the remarkable psychological effects of operations such as pre-frontal leucotomy, lobotomy and hemispherectomy.

hemistich. Add: *spec.* such a half-line or line in Old English verse.
1823 J. Bosworth *Elem. Anglo-Saxon Gram.* 246 The question, as to whether the two hemistichs shall be regarded as one or two lines, is evidently that of a writer or printer, not of a singer or reciter. **1857** C. Patmore in *North Brit. Rev.* XXVII. 148 Each hemistich contains two accented syllables. **1888** A. S. Cook *Judith* p. l, The line of poetry consists of two hemistichs, separated by the cæsura. **1925** M. D. Clubb *Christ & Satan* p. xv, The mark most consistently..employed is the metrical point, indicating the pause between hemistichs. **1970** *Jrnl. Eng. & Gmc. Philol.* LXIX. 86 One would not expect *þā* to provide the only alliteration in the second hemistich.

hemixis (hemi·ksis). *Biol.* Also **hemimixis.** [f. Hemi- + Gr. μίξις mixing, f. μ(ε)ίγνυμι to mix: see Mixo-.] In *Paramecium*, any of several types of change in the macronucleus, such as fission or elimination of chromatin, which take place without involving the micronucleus. So **hemi·ctic** *a.*, of, pertaining to, or characterized by hemixis.
1936 W. F. Diller in *Jrnl. Morphol.* LIX. 18 The attempt at interpreting the varied cytological conditions which P. aurelia exhibits under different circumstances has led to the discovery of a hitherto undescribed sexual reproductive process which corresponds to autogamy as it is known in other animals, and to a group of asexual phenomena which will be called 'hemixis'. *Ibid.* 36 It is proposed to group these reorganizations together.. and refer to them collectively by the term 'hemixis'. By this term is meant a series of autonomous changes which the macronucleus undergoes in vegetative life, exclusive of binary fission. *Ibid.*, No genetic relationships are necessarily implied between the varieties of hemictic phenomena. **1940** L. H. Hyman *Invertebrates* I. iii. 176 In hemimixis the macronucleus undergoes degenerative changes. **1953** R. Wichterman *Biol. Paramecium* iii. 90/2 Hemictic animals with the two micronuclei may show variable numbers and sizes of macronuclear fragments. **1961** Mackinnon & Hawes *Introd. Study Protozoa* iv. 292 A variety of nuclear changes may occur [in *Paramecium*] without the intervention of a micronucleus; they are collectively known as hemixis..and their significance is uncertain.

hemizygous (hemizəi·gəs), *a. Biol.* [f. Hemi- + *homo)zygous *a.*] Having a single unpaired allele at a particular genetic locus, as at all the loci in an XO pair of sex-chromosomes and some of the loci in an XY pair, rather than having two paired alleles, one on each homologous chromosome, as normally occurs in a diploid. So **hemizy·gote,** a hemizygous organism; **hemizygo·tic** *a.*, hemizygous; **hemizy·gously** *adv.*, in a manner characteristic of a hemizygote.
1921 W. A. Lippincott in *Amer. Naturalist* LV. 570, I should like to suggest the noun *hemizeuxis* (a half yoking) and the corresponding adjective *hemizygous* (half yoked). Should such a suggestion prove acceptable there would be the three adjective series: homozygous, heterozygous, and hemizygous, referring to the three possible conditions with respect to any single gene, namely, 'like mates', 'differing mates', and 'no mate'. **1935** H. J. Muller in *Jrnl. Genetics* XXX. 407 The terms 'hemizygous' and 'hemizygote' (Serebrovsky) refer to the condition of being haploid for a given gene when the genome as a whole is diploid; its usual usage is in connection with sex-linked genes in the sex in which they are haploid. *Ibid.*, The normal gene, although apparently quite dominant to white and to the other mutants, must in reality be incompletely so, and the heterozygotes, being like uncompensated hemizygotes, must have a definitely lower average survival rate. **1939** C. H. Waddington *Introd. Mod. Genetics* ii. 62 Muller has suggested that it would be better to speak of factors in haploid organisms such as male bees as hemizygous; and this word can also be used for factors (e.g. in an unpaired *X* chromosome) for which there is no allelomorph in a normal diploid. **1961** *Lancet* 16 Sept. 626/2 In a hemizygous XO female, a recessive gene could express itself as in a heterozygous XY male. **1965** J. A. Serra *Mod. Genetics* I. iv. 103 The genes for haemophilia and for colour-blindness have no allele in the Y chromosome, that is the males are hemizygotic for these factors. *Ibid.* 109 The yellow-black pair of alleles in cats producing either yellow or black hemizygous males but giving patched black and yellow or 'tortoise-shell' females when these are heterozygous. *Ibid.* 111 Such a character is transmitted from father to son and because there is no corresponding allele in the X chromosome, it manifests itself hemizygously.

hemlock, *sb.* Add: **2.** (Later examples.)
1927 M. de la Roche *Jalna* xviii. 213 She stood.. looking at the sombre shapes of the hemlocks. **1932** *Atlantic Monthly* Mar. 331/2 We come to that hidden glade, under the beeches, under the hemlocks.

hemlock (he·mlǫk), *v.* [f. the sb.] *trans.* To poison with hemlock. Hence **he·mlocked** *ppl. a.*
1846 Thackeray *Cornhill to Cairo* v. 66 Of the race of Englishmen who come wondering about the tomb of Socrates, do you think the majority would not have voted to hemlock him? **1908** L. Abercrombie *Interludes & Poems* 18 The slave Fate who serves Gods..fetched Skill'd poison,..and with this stew Hemlock'd the wine of Heaven. **1934** Dylan Thomas *Let.* 11 May (1966) 129 A twisted veil of evil..coils up from the pit to the top of the hemlocked world.

hemming, *vbl. sb.*[1] Add: Also *hemming-in*.
1905 *Fabian News* Mar. 14/2 It is enacted that 'he who receives relief must submit to the hemmings in of his personal liberty which the law imposes'.
b. *attrib.*
1858 *U.S. Patent 21,355* (*title*) Improvement in hemming guides for sewing machines. *a* **1865** in M. Johnson *Amer. Advertising* (1960), Patent binding folders and hemming guides on hand. **1932** D. C. Minter *Mod. Needlecraft* 196/2 *For Machining Hems.*—Screw on the hemming foot in place of the presser foot. **1972** *Canad. Antiques Collector* Mar.–Apr. 32/2 A hemming bird, ornate and beautifully made..brought over..one hundred and fifty years ago from Scotland.

hemp-agrimony. (Earlier and later examples.)
1760 J. Lee *Introd. Bot.* 304 Hemp agrimony, Eupatorium. **1908** *Westm. Gaz.* 31 Aug. 2/3 Nature had.. enriched the banks with..meadow-sweet and hemp-agrimony. **1971** *Country Life* 17 June 1520/1 The muzzy mauve hemp-agrimony (*Eupatorium purpureum*).

hen, *sb.* Add: **1. b.** *like a hen with one chick(en)*: indicating extreme solicitude or fussiness about a small matter; *(as) mad as a wet hen*: very angry; *(as) scarce* (occas. *rare*) *as hen's teeth* (orig. *U.S.*): very scarce.
1595 G. Delamothe *Treasure French Toung* 19 He is as busie as a henne that hath but one chicken. **1670** J. Ray *Coll. Eng. Prov.* 203 As busie as a hen with one chicken. **1732** T. Fuller *Gnomologia* 25 As busy as a Hen with one Chick. **1823** J. Doddridge *Logan* 42 Every body that was not ax'd was mad as a wet hen. **1854** A. E. Baker *Gloss. Northampt.* I. 320 'As busy as a hen with one chick.' Unnecessarily solicitous or active over trifles. **1858** in N. E. Eliason *Tarheel Talk* (1956) 276 As scarce as hen's teeth. **1863** 'E. Kirke' *Southern Friends* 250 [Horses are] scarcer than hen's teeth round here. **1881** A. Parker *Oxfordshire Words* (Suppl.) s.v., To be as busy as a hen with one chick is to make a great fuss over a little work. They also say 'as proud as a hen with one chick'. **1893** *Congress. Rec.* 2 Oct. 2044/1 North of Mason and Dixon's line, colored county officials are scarce as hen's teeth. **1904** E. F. Benson *Challoners* x, Karl is devoted to him, just like a beautiful old hen in spectacles with one chicken. **1907** *Amer. Mag.* Feb. 339/1 He just looked at me and then flounced out, mad as a wet hen. **1962** *Listener* 5 July 8/2 A good guide for Western leaders, among whom communiqués were 'as scarce as hens' teeth'. **1969** *Times* 12 June 3 (Advt.), Stoppages are as rare as hen's teeth.
4. (Earlier example.)
1747 H. Glasse *Art of Cookery* xxi. 164 Cock Lobster is known by the narrow back Part of the Tail... The Hen is soft, and the back of her Tail broader.
8. *hen-run, -yard*; *hen-brained, -headed, -housed, -toed* adjs.; **hen-and-egg,** used *attrib.* of the unresolvable problem of the 'first cause' (freq. in somewhat trivial contexts); cf. *chicken *sb.*[1] 6 b; **hen-cackle** *N.Z. slang* (see quots.); **hen-fruit** chiefly *U.S. slang*, eggs; also *hen's fruit*; **hen-party** (later example); **hen scratch** *U.S.*, 'a chicken feed made from grain for scattering in litter or on the ground to induce chickens to scratch' (*Dict. Americanisms*); **hen-scratch** *v.*, to scratch in the manner of a hen; also *transf.*; **hensure** *a.* [joc. formation after Cock-sure *a.*], = Cock-sure *a.* 5; so **he:nsu·reness.**
1931 A. L. Rowse *Politics & Younger Generation* 146 It is the old hen-and-egg argument, that there is no knowing which comes first. **1951** W. Empson *Struct. Complex Words* 436 But firstly, there is a typical hen-and-egg problem. *a* **1963** L. MacNeice *Astrology* (1964) ii. 39 The old hen-and-egg dilemma (did god or planet come first?). **1923** W. de la Mare *Riddle* 93 Poor hen-brained things, they came to be fed. **1965** A. Garner *Elidor* xix. 147 You landed us in enough trouble yesterday with your hen-brained ideas. **1939** J. Pascoe *Unclimbed N.Z.* ii. 33 All the lasting Alpine partnerships in Canterbury have been formed on the easy expeditions. 'Hen-cackles' we call these expeditions. **1941** Baker *N.Z. Slang* vi. 57 Hen-cackle..is applied by mountaineers to a mountain that is easy to climb. Doubtless there is a wider application of the term since a mere hen-cackle, a trifle, seems to have been the origin of the application. **1854** *Harper's Mag.* Jan. 280/2 A young lady is said to have asked a gentleman at the table of a hotel 'down East' to pass her the 'hen fruit'. She pointed to a plate of eggs. **1873** C. G. Leland *Egypt. Sketch-Bk.* 71 Their 'hen-fruit', as it is elegantly termed in America. **1887** *Boston Guide* (Farmer), If he confines his Hen Fruit to the vintage of '87. **1942** *Sunday Chron.* 1 Mar. 1/1 To him [*sc.* a ward-room steward] egg and bacon is 'hen's fruit and hog's body'. **1912** *Dialect Notes* III. 578 *Hen-headed*, brainless. 'That hen-headed cuss can't do anything you tell him.' **1938** A. H. Bill *Astrophel* iv. 76 Elizabeth, always short of money..., railed against the required outlay like a hen-headed housewife over a coal bill. **1960** *Farmer & Stockbreeder* 16 Feb. 152/1 A hen-housed average of 104·4 eggs per bird in 112 days. **1960** *Guardian* 15 Feb. 4/3 A hen-party can be a very pleasant, relaxing affair, particularly for the older woman. **1897** Hen-run [in Dict.]. **1929** J. B. Priestley *Good Companions* I. i. 20 He was now 'on his own' at Wabley, the proud proprietor of a large hen-run. **1887** V. Pyke *Hist. Early Gold Discoveries in Otago* 35 Peter was only *hen-scratching* on the edges of the creek. **1921** D. H. Lawrence *Sea & Sardinia* 267 It was a small, stony, hen-scratched place of poor people. **1931** *Daily News-Jrnl.* (Murfreesboro, Tenn.) 15 Apr. 4/2 Corn,..White oats,..Hen Scratch. **1957** V. J. Kehoe *Technique Film & T.V. Make-Up* vii. 86 Facial lining for old age and other character make-ups should not look like 'hen scratchings'. **1929** D. H. Lawrence *Assorted Articles* (1930) 72 There are the women who are cocksure, and the women who are hensure. *Ibid.* 76 The lovely henny surety, the hensureness which is the real bliss of every female, has been denied her. **1951** M. McLuhan *Mech. Bride* 64/1 The old age reserved for the hensure types. **1937** Partridge *Dict. Slang* 388 *Hen-toed*, with one's feet turned in as one walks. **1955** I. Peebles *Ashes* 36 He stands at the wicket rather hen-toed. **1816** M. L. Weems *Lett.* (1929) III. 166 Yr. Bible carts had been here as thick as weasels in a hen yard selling Bibles at nearly half price. **1876** *Scribner's Monthly* Apr. 813/2 The best places in which to look for Jacobean sideboards..are found to be the hen-yard. **1960** *Farmer & Stockbreeder* 2 Feb. 121/1 The breeders have been housed in the henyard previously used for layers.

hen and chickens. Add: **3.** The name of a children's game.
1894 A. B. Gomme *Trad. Games Eng. Scot. & Irel.* I. 201 Hen and Chicken... The game is played in the usual manner of 'Fox and Goose' games. One is chosen to be the Hen, and one to be the Fox. The rest are the Chickens. **1969** I. & P. Opie *Children's Games in Street & Playground* xi. 311 It was played..under the names 'Fox and Chickens'..and 'Hen and Chickens'.

hence, *sb. U.S.* [Hence *adv.* 3 b and 4 c.] **a.** The other world. **b.** The future.
1884 E. W. Nye *Baled Hay* 26 All-wool delaine that was worn by one who is now in the golden hence. **1904** F. Lynde *Grafters* xviii. 233 Now suppose you hint.. that more..developments may be safely predicted in the immediate hence.

hen-coop. Add: **b.** *attrib.*
1898 *Daily News* 28 Sept. 5/3 The huge hen-coop crinoline disfigured the women. **1937** E. Sitwell *I live under Black Sun* 296 There came a sound of crazy hen-coop laughter.

hendeca-. Add: **b.** *Organic Chem.* Occas. used in place of the synonymous and more usual prefix *undeca-* to denote the presence in a molecule of eleven carbon atoms, as in **he·ndecane,** undecane; **hendeco·ic acid,** undecoic acid.
1889 Muir & Morley *Watts' Dict. Chem.* II. 673/1 n-*Hendecane* $C_{11}H_{24}$... Formed by the action of HI and phosphorus at 230° upon hendecoic (undecylic) acid. **1943** *Thorpe's Dict. Appl. Chem.* (ed. 4) VI. 204/1 (*heading*) Hendecoic acids (undecoic acids), $C_{11}H_{22}O_2$. **1960** *Handbk. Chem. Soc. Authors* iii. 49 Greek numeral roots are used, except that 9 (alone or in combination) is rendered by the Latin nona (not the Greek ennea) and 11 by the Latin undeca (not the Greek hendeca).

hen-egg. (Additional examples.)
1900 *Daily News* 23 July 5/1 When they [*sc.* hail-stones] attain the size of hen-eggs..the matter is beyond a joke. **1961** *Ann. Reg.* 1960 508 The guaranteed prices for..hen eggs..were reduced.

heneicosane (henəi·kosēin). *Chem.* Also **henicosane.** [f. Gr. ἑν-, εἷς one + εἴκοσι twenty + -ane.] Any of the hydrocarbons of the paraffin series having twenty-one carbon atoms, *esp.* the unbranched isomer (*n-heneicosane*). Hence **heneicosa·nic, -cosano·ic, -coso·ic acid,** the saturated fatty acid, $C_{20}H_{41}COOH$, derived from *n*-heneicosane.
1889 Muir & Morley *Watts' Dict. Chem.* II. 674/1 *Henicosane* $C_{21}H_{44}$... Formed by reduction of the dichloride..of the ketone. **1915** *Jrnl. Chem. Soc.* CVII. 738 Heneicosoic acid, $CH_3{\cdot}[CH_2]_{19}{\cdot}CO_2H$, is readily soluble in ether. **1924** *Jrnl. Biol. Chem.* LIX. 920 n-Heneicosanic Acid..melted at 75–76°C. **1945** A. N. Sachanen *Chem. Const. Petroleum* iv. 232 The heneicosanes substituted in position 11 ($C_{10}H_{21}{\cdot}CHR{\cdot}C_{10}H_{21}$) have low melting points (below −7°C). **1951** Kirk & Othmer *Encycl. Chem. Technol.* VI. 257 Heneicosanoic acid, $C_{21}H_{42}O_2$, has been reported to occur in Japanese wax and earth-nut oil, but the evidence presented indicates that it is not found in natural fats and waxes.

henge[2] (hendʒ). [f. Stone)henge.] **1.** In particular reference to the name *Stonehenge*: something 'hanging' or in suspense. **2.** *Archæol.* A term (first applied by T. D. Kendrick) for classes of monuments more or less akin to the stone circle of Stonehenge. Also *attrib.*
1740 W. Stukeley *Stonehenge* ii. 8 Pendulous rocks are now called henges in Yorkshire, and I have been informed of another place there called Stonehenge, being natural rocks. So that I doubt not, Stonehenge in Saxon signifies the hanging stones. **1742** in Defoe *Tour Gt. Brit.* (ed. 3) I. v. 257 The present Name [*sc.* Stonehenge] is Saxon, tho' the Work is beyond all Comparison older, signifying a hanging Rod or Pole, *i.e.* a Gallows, from the hanging Parts, Architraves, or rather Imposts; and pendulous Rocks are still in Yorkshire called Henges. **1932** Kendrick & Hawkes *Archæol. in Eng. & Wales* vii. 83 (*heading*) The 'henge' monuments. **1936** *Proc. Prehist. Soc.* II. 1 A new monument of the 'henge' class. **1951** *Field Archæol.* (*Ordnance Survey*) (ed. 3) 17 The critical event in the modern study of 'henges' was the discovery of the site known as 'Woodhenge' from the air, two miles east-north-east of Stonehenge... The term henge monument..is an unsatisfactory term since, on strict etymological grounds, it should only be applied to sites which contain a 'hanging' element like the lintels of Stonehenge. **1967** *Antiquaries Jrnl.* XLVII. 166 The term 'henge' is applied to those monuments which enclose a circular or oval area by means of a bank and *internal* ditch and which possess one or two opposed entrances (Atkinson, Piggott, and Sandars, 1951, p. 82).

hen-hawk. (Earlier examples.)
1806 W. Clark in Lewis & Clark *Orig. Jrnls.* (1905) IV. 131, I have observed..a hawk of an intermediate size with a long tail and blewish coloured wings, remarkably swift in flight and very ferce. Sometimes called in the Un. States the hen Hawk. **1819** D. Thomas *Trav. W. Country* 210 The Hen-Hawk is not very numerous.

hen-house. Add: **b.** *fig.* A house or establishment inhabited chiefly by women.
1785 Grose *Dict. Vulg. T.*, *Hen house*, a house where the woman rules. **1931** W. Faulkner *Sanctuary* xxi. 193 She's got two daughters... I'm heading for the hen-house. **1963** P. Moyes *Murder à la Mode* ii. 29 'You know what this place is, don't you, sir?' The sergeant was plunged in gloom. 'Fashion magazine. Ruddy hen-house.'

henid (he·nid). *Philos.* [ad. G. *henide*, coined by Weininger on the basis of Gr. ἑν- one; cf. Henism.] In the philosophy of Otto Weininger (1880–1903): see quots. Hence **heni·dical** *a.*
1906 tr. *O. Weininger's Sex & Character* II. iii. 99, I propose for psychical data at this earliest stage of their existence the word Henid (from the Greek ἕν, because in them it is impossible to distinguish perception and sensation as two analytically separable factors, and because, therefore, there is no trace of duality in them)... The very idea of a henid forbids its description; it is merely a something. **1909** J. London *Martin Eden* xxxvii. 322 By some henidical process—henidical, by the way, is a favourite word of mine which nobody understands—by some henidical process you persuade yourself that you believe in the competitive system and the survival of the strong. **1914** —— *Let.* 10 Sept. (1966) 428 The word henid was coined by a crazy German philosopher... All persons possess henids. **1915** —— *Jacket* xiv. 160 'I'll—' he began explosively, proving, by his inability to conclude the remark, that he thought in henids. **1946** D. Abrahamsen *Mind & Death of Genius* 112 Weininger introduces a special name for the psychological data at the earliest stage, before clarification has begun. He calls the vague perception the henid. *Ibid.* 113 According to Weininger, the henid is the form of perception known to the lower types of organism. In mankind development from the henid to the completely differentiated form of perception and idea is possible.

Henle (he·nli). *Anat.* The name of F. G. J. *Henle* (see Henlean *a.*) used in the possessive and with *of* adjunct to designate numerous anatomical structures, as: **a.** *Henle's layer* (or *layer of Henle*), a single layer of cubical cells in the inner root sheath of the hair follicle, between Huxley's layer and the outer root sheath. Formerly also called *Henle's sheath* etc. (cf. sense c).
1853 Busk & Huxley tr. *Kölliker's Man. Human Histol.* I. 186 The outermost layer [of the inner root-sheath of the hair], which alone was formerly known, the inner root-sheath of Henle, is formed of elongated cells without nuclei. **1860** G. Buchanan tr. *Kölliker's Man. Human Microsc. Anat.* 109 (*caption*) A portion of the root of a dark hair..f. outer fenestrated layer (*Henle's* layer). **1892** H. E. Clark *Wilson's Anatomist's Vade Mecum* (ed. 11) 70 The inner root-sheath [of a hair] is again divisible into..an outer layer (Henle's sheath). **1970** T. S. & C. R. Leeson *Histol.* (ed. 2) xiii. 263/2 Henle's layer..is a single layer of flattened, clear cells which contain hyaline fibrils.
b. *Henle's loop* (or *loop of Henle*), a portion of a uriniferous tubule from where it passes from the cortex of the kidney into the medulla to where it returns into the cortex, also *spec.* the central part of this where it forms a distinct loop at its deepest point in the medulla. Also called *Henle's* (*looped*) *tube* or *tubule* etc.
1867 *Quain's Elem. Anat.* (ed. 7) II. 930 The tubes in question have been designated looped tubes of Henle. **1877** W. Turner *Introd. Human Anat.* II. xi. 785 Each intermediary tube..descends..into the medullary pyramid, where it turns on itself, forms a loop, known as the looped tube of Henle, and reenters the cortex as the ascending limb of the looped tube. **1885** [see Loop *sb.*[1] 4 a]. **1890** Billings *Med. Dict.* I. 631/2 *Henle's loop*, loop of Henle. *Ibid.*, *Henle's tubules*, looped tubules of Henle. **1970** T. S. & C. R. Leeson *Histol.* (ed. 2) xvi. 374/2 The loop of Henle consists of the straight part of the proximal tubule in the descending limb, a thin segment in descending and ascending limbs, and the straight part of the distal tubule in the ascending limb.
c. *Henle's sheath* (or *sheath of Henle*) (see quot. 1942). Cf. sense a.
[**1878** M. L. Ranvier *Leçons Histol. Syst. Nerveux* I. x. 159 Nous pouvons vous montrer sous un de ces microscopes des nerfs composés d'un seul tube nerveux qui possèdent une membrane enveloppante; j'appellerai cette membrane, gaine [= sheath] de Henle, du nom de l'auteur qui l'a découverte.] **1887** *Buck's Handbk. Med. Sci.* V. 145/1 The sheath in this form has been named the sheath of Henle (Ranvier). There is a space between the nerve-fibre and the sheath. **1890** Billings *Med. Dict.* I. 631/2 *Henle's sheath*, a continuation of the fibrous tissue of the perineurium. **1942** O. Larsell *Anat. Nervous Syst.* iv. 48 A delicate sheath of connective tissue fibers, continuous with the endoneurium of the nerve trunk, is intimately associated with and surrounds the neurolemma of most individual peripheral fibers. This is usually called the sheath of Henle.
d. *Henle's gland* (or *gland of Henle*), any of numerous gland-like structures in the conjunctiva of the eye-lid (see quot. 1933).
1890 Billings *Med. Dict.* I. 631/2 *Henle's glands*, short tubular glands said to be found on the palpebral conjunctiva. **1933** E. Wolff *Anat. Eye & Orbit* iii. 105 Henle's 'Glands' occur in the palpebral conjunctiva between the tarsal plates and the fornices. They are probably not true glands, but folds of mucous membrane cut transversely. **1961** S. Duke-Elder *Syst. Ophthalm.* II. iii. 115 The epithelial depressions between them were originally described by Henle (1866) as glands (the *glands of Henle*) and were termed conjunctival crypts by Dubreuil (1908).

Henley (he·nli). The name of a town on the Thames, in Oxfordshire, used alone or *attrib.* to designate the annual regatta held there since 1839. Hence **He·nleyite,** a Henley enthusiast.
1839 *Times* 6 June 3/5 Henley Regatta..takes place on Friday, the 14th. **1861** T. Hughes *Tom Brown at Oxf.* II. 218 You know the Leander are to be at Henley... There will be a splendid race for the cup. **1868** *Broadway* I. 104/2 Roland rowed his man—a Henley winner—down. **1887** [see Regatta 2 *attrib.*]. **1893** *Isis* 29 Apr. 130/2, I have heard nothing more about the Henley arrangements. *Ibid.* 3 June 207/2 The Oxford arrangements for Henley are now getting into rather a more settled condition. **1901** *Westm. Gaz.* 2 July 7/2 Henleyites..will be depressed to hear that the Meteorological Office forecasts..rainy and colder weather. **1902** *Encycl. Brit.* XXXII. 307/1 A motion was proposed at the October meeting of the Henley Stewards to exclude foreign crews from the regatta. **1959** *Chambers's Encycl.* VII. 13/2 Henley Royal Regatta (1851—formerly Henley Regatta, 1839).

henna. b. Add: (with reference to dyeing or staining with henna) *henna-dyeing*; *henna-coloured, -dyed, -haired, -tipped, -tressed* adjs.
1954 M. Mead *Growing up in New Guinea* 143 Henna-coloured betel juice. **1920** *Chambers's Jrnl.* May 299/1 He..showed a handful of his henna-dyed beard. **1906** *Daily Chron.* 27 June 6/4 From tight-lacing to henna-dyeing the frisky matron is a study in successful artifice. **1907** *Ibid.* 31 Dec. 3/1 His American trotting-car and his henna-haired wife. **1923** *Chambers's Jrnl.* 306/2 Their henna-tipped fingers are loaded with rings. **1939** R. Campbell *Flowering Rifle* I. 24 Amongst her Modern Southeys, henna-tressed.

hennaed (he·năd). [f. Henna + -ed[2].] Dyed or stained with henna.
1924 *Countries of World* II. 1145/1 A slim hand with hennaed nails. **1925** H. V. Morton *Heart of London* 53 Maud's hennaed hair. **1961** A. Wilson *Old Men at Zoo* vii. 309 Her dark hennaed hair.

Henoch (hī·noχ). The name of E. H. *Henoch* (1820–1910), German pædiatrician, used in the possessive and occas. *attrib.* to designate esp. purpura associated with abdominal symptoms, as *Henoch('s) purpura*; also used *attrib.*, in combination with the name of Schönlein (see *Schönlein), to designate purpura associated with both the abdominal symptoms of Henoch's purpura and the articular symptoms of Schönlein's purpura, as *Henoch–Schönlein* (or *Schönlein–Henoch*) *purpura, syndrome*.
1893 Dunglison *Dict. Med. Sci.* (ed. 21) 522/1 *Henoch's purpura*, purpura complicated with infection of the intestines. **1896** *Intercolonial Med. Jrnl. Australasia* I. 364 (*title*) A case of Henoch's purpura hæmorrhagica. **1897** *Brit. Med. Jrnl.* 18 Dec. 1800/1 Henoch's purpura. Dr. Dreschfeld made some observations on the symptoms, pathology, and treatment of this disease, cases of which, though not under that name, had been repeated, observed and described in English journals. **1908** *Practitioner* June 824 Gastro-intestinal crises are common with both Henoch's purpura and angioneurotic oedema. **1947** *Gastroenterology* IX. 610 My working diagnosis in this case was Henoch's Purpura. **1948** *Q. Jrnl. Med.* XLI. 95 (*title*) The Schönlein-Henoch syndrome (anaphylactoid purpura). **1952** *Arch. Dis. Childhood* XXVII. 480 (*title*) The Schoenlein-Henoch syndrome in childhood with particular reference to the occurrence of nephritis. **1959** *Brit. Med. Jrnl.* 19 Dec. 1385/2 (*heading*) Familial incidence of the Henoch-Schönlein syndrome. **1966** Dunlop & Alstead *Textbk. Med. Treatm.* (ed. 10) 502 (*heading*) Henoch-Schönlein purpura. *Ibid.*, In this condition,.. abdominal symptoms,..(Henoch purpura), and pain and swelling in joints (Schönlein purpura), may occur together. **1967** A. C. Allen *Skin* (ed. 2) xvii. 661/1 In Henoch's purpura..the prominent symptoms are recurrent abdominal pain..and colic, associated with moderate fever and with melena, tenesmus, and mucus in the stools. **1968** *Gastroenterology* LIV. 260 (*title*) Gastro-intestinal and roentgenological manifestations of Henoch-Schoenlein purpura.

henpeck, *sb.* Add: **1. b.** A husband so domineered.
1765 Garrick *Let.* 23 May in *Corr. Garrick* (1831) I. 185 More of the sneaking hen-peck, than of the tender enamoured husband.
hen-peckery. (Later examples.)
1869 *Harper's Mag.* Mar. 508/2 Husbands flee from hen-peckery, and wives desert bearish husbands. **1958** *Daily Mail* 15 July 3/4 Charmian Eyre..remains disarmingly human at the height of henpeckery.

hen-pecked, *ppl. a.* (Later examples.)
1923 D. H. Lawrence *Kangaroo* i. 4 A little red-faced man, rather beery and hen-pecked looking. **1930** G. B. Shaw *Apple Cart* Interlude 57 *Orinthia*: Why are you so afraid of your wife? You are the laughing stock of London, you poor henpecked darling. **1939** —— *In Good King Charles's Golden Days* I. 57 What! that henpecked booby! *Ibid.* 58 He may be henpecked: what married man is not?

Henrician, *sb.* Add: **3.** A supporter of the ecclesiastical policy of Henry VIII. Hence **Henricianism** (henri·ʃiăniz'm), the ecclesiastical policy of Henry VIII.
1828 E. Nares *Mem. Ld. Burghley* I. iv. 53 The king's supremacy, which, like a true *Henrician*, he was very careful to maintain. **1900** F. W. Maitland *Coll. Papers* (1911) III. 159 Calvin had spoken ill of Henricianism. **1903** —— in *Camb. Mod. Hist.* II. 555 A Reformed religion, or some northern version of Henricanism [*sic*]. **1946** A. L. Rowse *Use Hist.* 180 Yet we cannot be grateful enough to the (sometimes unattractive) Henricians and Elizabethans who pushed us through it.

Henri Deux (ańri dǿ). [Fr., = Henri II.] Designating the style of Renaissance architecture or art developed in France during the reign of Henri II, king of France 1547–59; *spec.* the purest style of the French Renaissance.
1863 W. Chaffers *Marks Pott. & Porc.* 89 The following is a list of all the pieces of Henri II. ware now known to be extant. **1873** *Young Englishwoman* May 234/2 The

Henri II. hat is very becoming to a young face. **1881** C. C. HARRISON *Woman's Handiwork* II. 104 Modern English potters have put within our reach reproductions of that exquisite (so-called 'Henri Deux') faïence bequeathed to the world by the lady Hélène de Hangest-Genlis—the ware of the Chateau d'Oiron. **1884** KNIGHT *Dict. Mech.* Suppl. s.v., Henri-Deux Ware (*Faïence d'Oiron*). **1960** R. G. HAGGAR *Conc. Encycl. Cont. Pott. & Porc.* 397 Saint-Porchaire earthenware *tazza* with inlaid decoration in coloured clays (so-called Henri Deux Ware), c. 1540.

Henrietta (henrie·tă). *Disused.* [Female name.] Designating a light-weight dress fabric, sometimes with a silk warp.

1851 *Illustr. Catal. Gt. Exhib.* III. 494/2 Henrietta cloths, with silk warp and worsted weft. **1862** *Illustr. Catal. Internat. Exhib., Industr. Dept., Brit. Div.* II. No. 4018 Paramatta, or Henrietta Cloth, twill. **1890** *Advt.* (Ann Arbor, Mich.) 1 Mar., We offer a 46-inch Black Silk Warp Henrietta. **1901** *Daily News* 23 Feb. 6/7 Henrietta cloths, which wear so well and drape so charmingly. **1908** *Sears, Roebuck Catal.* 933/1 Cotton henrietta cloth..having all the appearance and touch of an imported all wool henrietta.

hen-roost. Add: **b.** *fig.* A source of plunder: in allusion to a political speech referring to 'the robbing of hen-roosts'.

1909 *Westm. Gaz.* 16 Apr. 5/1 Mr. Lloyd George's now historic reference to 'hen-roosts'. **1928** *Britain's Industr. Future* (*Lib. Ind. Inq.*) v. xxix. § 2. 420 Apart from the public hen-roosts which Mr. Churchill has raided, it is impossible for an outsider to estimate what private hen-roosts inside the Treasury he has also helped himself to.

Henry[1] (he·nri). The name of Benjamin Tyler *Henry* (1821–98), American inventor, used attrib. to designate a breech-loading magazine rifle or parts thereof, subsequently used in the Martini-Henry rifle. Also *ellipt.*

1859 G. A. JACKSON *Diary* 25 Jan. in F. Hall *Hist. Colorado* (1890) II. 522 Packed up our things for the trip and got Oakes' Henry rifle for Phil. **1869** in *Frontier* (1929) IX. 157 One of them..lost one of our Henry carbines. **1880** *Encycl. Brit.* XI. 282/1 In the Henry action the barrel does not move, but is closed at the breech end by a sliding vertical block. *Ibid.* 282/2 The combination of the Martini breech action with the Henry barrel. *Ibid.* 283/2 Henry rifling. **1902** *Ibid.* XXXII. 242/1 In 1861 the Henry grooving for a cylindrical bullet, a modification of the Whitworth, first appeared. **1927** C. M. RUSSELL *Trails plowed Under* 71, I guess his weapon's a Henry. **1964** H. L. PETERSON *Encycl. Firearms* 180 (*caption*) Iron-frame Henry rifle and brass-frame Henry rifle.

Henry[2] (he·nri). The name of William *Henry* (1774–1836), English chemist, used in *Henry's law* (see quot. 1940).

1886 *Syd. Soc. Lex.*, H[enry]'s law. **1910** *Encycl. Brit.* XIII. 302/1 The conclusion he reached ('Henry's law') being that 'water takes up of gas condensed by one, two or more additional atmospheres, a quantity which, ordinarily compressed, would be equal to twice, thrice, &c. the volume absorbed under the common pressure of the atmosphere'. **1940** *Chambers's Techn. Dict.* 411/1 *Henry's Law.* The amount of a gas absorbed by a given volume of a liquid at a given temperature is directly proportional to the pressure of the gas. **1966** PHILLIPS & WILLIAMS *Inorg. Chem.* II. xix. 43 Oxygen dissolves as atoms (obeying Henry's law) in molten silver.

henry[3] (he·nri). *Physics.* The name of Joseph *Henry* (1797–1878), American physicist, used (pl. *henrys, henries*) to designate the practical unit of inductance, now incorporated in the International System of Units, i.e. the inductance of a circuit in which an electromotive force of one volt is produced by a current changing at the rate of one ampere per second. Abbrev. H or (*rare*) h. Also **he·nrymeter** (see quot. 1940).

1893 *Electrician* 29 Sept. 577/2 There was the proposal [at the International Electrical Congress] to christen the unit of self-induction as the *henry* in honour of Joseph Henry its discoverer. **1915** *Proc. IRE* III. 223 The transformer is made up of coils having an inductance of the order of a henry or more. **1929** K. HENNEY *Princ. Radio* iv. 62 The coils used in radio apparatus vary from inductances of the order of microhenries to very large ones having over 100 henries in inductance. **1940** *Chambers's Techn. Dict.* 411/1 *Henrymeter*, an obsolete apparatus for measuring inductance; in it an alternating current was passed through the inductance under test and a standard inductance in series, the voltage drop across the two being compared. **1947** *Jrnl. Inst. Electr. Engin.* XCIV. 342/1 From the 1st January the units employed [at the National Physical Laboratory] will be those derived from the centimetre, gramme and second, i.e. the so-called 'absolute' units. The effects of this change may be seen from the following table:..One international henry = 1·00049 'absolute' henrys. **1952** *Electronic Engin.* XXIV. 465 A 10 henry A.F. choke has been provided in the cathode circuit.

Henry Clay. The name of an American statesman (1777–1852) used to designate a type of cigar.

1867 in *Amer. Speech* (1965) XL. 130. **1884** *Harper's Mag.* Sept. 647/1 The dealer..asked him if he would 'like to 'ave a 'Enry Clay'. **1888** KIPLING *Departmental Ditties* (1890) 105 There's peace in a Laranaga, there's calm in a Henry Clay. **1893** *Harper's Mag.* Dec. 34/1 And bring some cigars—Henry Clays. *Ibid.*, My father was always a Henry Clay man. **1922** JOYCE *Ulysses* 243 He removed his large Henry Clay decisively. **1969** L. GROW tr. *Davidoff & Lambert's Connoisseur's Bk. Cigar* 71 Henry Clay (now planted in the Canary Islands) has changed its name; its owners have struck the new name of Don Miguel.

Hentenian (henti·niăn), *a.* [f. the name of John *Henten* or *Hentenius* (1499–1566), a theologian of the Dominican order at Louvain: see -IAN.] Of or pertaining to Henten, or to the editions of the Vulgate (Louvain 1547, often reprinted) prepared by him.

1902 H. J. WHITE in J. Hastings *Dict. Bible* IV. 880/2 The various Hentenian editions remained for some years as the standard text of the Roman Church, but were still private publications. **1930** S. ANGUS in *Internat. Stand. Bible Encycl.* V. 3061/2 Hentenian critical ed. (Louvain, 1547).

hentriacontane (he:ntraiăkǫ·ntēⁱn). *Chem.* [f. Gr. ἐν-, εἷς one + τριάκοντα thirty + -ANE.] A hydrocarbon of the paraffin series, $C_{31}H_{64}$, *esp.* the solid unbranched hydrocarbon $CH_3{\cdot}(CH_2)_{29}{\cdot}CH_3$ present in petroleum and many natural waxes.

1887 *Jrnl. Chem. Soc.* LII. I. 124 The most soluble portion of the extract melting..at 67°, is probably identical with normal hentriacontane, $C_{31}H_{64}$. **1950** J. BONNER *Plant Biochem.* xxiv. 366 Commercial candelilla wax (*Euphorbia* sp.) contains..50–60% of a paraffin, *n*-hentriacontane.

heortology (hī,ǫɹtǫ·lŏdʒi). [ad. G. *heortologie*, F. *héortologie*, f. Gr. ἑορτή feast: see -OLOGY.] The science or study of the origin, meaning, growth, and history of the religious feasts and seasons of the Christian year. Hence **heortolo·gical** *a.*, of or pertaining to heortology; **heorto·logist**, one who studies heortology.

1900 *Expositor* Nov. 348 We are to regard the statement of the calendars as the conjecture of a heortologist. **1901** J. R. HARRIS in *Soc. Hist. Theology* 31 Oct. 5 The Study of Christian Heortology... The problems that belong to the region of Christian Heortology. **1913** J. R. MCKEE (*title*) The Church's year, a handbook of heortology. **1918** E. BISHOP *Liturg. Hist.* 258 Recalling too how the recent heortologist Dr. Kellner considers that the mention of the feast in the Irish calendars does not prove the celebration of the feast.

hep (hep), *a. slang* (orig. *U.S.*). [Of unknown origin.] Well-informed, knowledgeable, 'wise to', up-to-date; smart, stylish. Hence as *sb.*, the state of being 'hep'. Also as *vb.*, to pep *up*; **hepped** *ppl. a.* (often with *up*); *to be hepped on*, to be enthusiastic about, 'bitten with'. Cf. *HEP-CAT, *HIP *a.*

1908 *Sat. Even. Post* 5 Dec. 17/1 What puzzles me is how you can find anybody left in the world who isn't hep. **1914** JACKSON & HELLYER *Vocab. Criminal Slang* 43 *Hep*,... Sapiency; understanding... Derived from the name of a fabulous detective who operated in Cincinnati. **1918** WODEHOUSE *Piccadilly Jim* xi. 118 'You see in me a confidant. I am hep.' 'You know—' 'Everything.' **1923** —— *Adv. Sally* xiii. 148 He was aware that women were seldom hep to the really important things in life. **1927** P. MARKS *Lord of Himself* 47 You're pleased because the top-notchers wanted me, but that doesn't make you think I'm a top-notcher. I'm just getting hep. There you have it. **1938** 'J. SPENSER' *Crime against Society* xxiv. 235 The coppers are hep and we've got to stage a cover-up. **1941** *Amer. Speech* XVI. 154/1 'Tis said that back in the 1890's Joe Hep ran a saloon in Chicago... Although he never quite understood what was going on, he thought he did... Hence his name entered the argot as an ironic appellation for anyone who thought he knew but didn't. The ironic sense has now largely disappeared..in..to get hep to. **1951** M. MCLUHAN *Mech. Bride* 68/2 His failure to be hep to success doctrines. **1956** D. KARP *All Honorable Men* 107 You know how hepped on the matter of wasting time the whole Board is. **1957** C. MACINNES *City of Spades* I. iii. 19 Where can I get a shirt like that?.. It's hep. Jumble style, but hep. **1958** N. D. HINTON in *Publ. Amer. Dial. Soc.* XXX. 40 In the swing period, 'hep' was widely used by musicians to mean 'in the know', 'possessed of good taste', or to indicate simple understanding... The boppers quickly changed the word to 'hip'. Use of 'hep' was then regarded as a sign that the speaker was not the right sort. **1958** *Punch* 27 Aug. 270/1 And when I stood up and he began to get hep I noticed that his shoes were cunningly low cut. **1959** 'M. M. KAYE' *House of Shade* xviii. 246 Are you, in the distressing jargon of the age, 'hep'? **1959** *Guardian* 16 Oct. 10/6 Columns of drug-hepped, ragged men. **1959** *News Chron.* 14 July 4/4 A slightly hepped-up version of the old deck chair and concert party formula. *Ibid.* 21 July 1/6 The pills were being taken..to give the addicts a form of 'hep'. **1959** T. GRIFFITH *Waist-high Culture* (1960) 238 A California chemist who is hep to every current allusion and is an asset to any party. **1960** *News Chron.* 6 July 3/1 Even some of the classics..have been hepped up to circus style. **1960** *Guardian* 12 Aug. 8/3 Not even its bitterest critics could accuse the Labour party of being 'hep'. **1962** *Listener* 6 Sept. 350/2 'I wasn't hepped on becoming a painter,' he [*sc.* Henry Miller] said. **1970** *Cape Times* 28 Oct. 1/8 Are you hep to what the Beatles are saying? **1972** J. L. DILLARD *Black English* iii. 119 It is, of course, a commonplace of the jazz language that *hep* is a white man's distortion of the more characteristically Negro *hip*.

hep (hep), *int.* [Said to be f. the initials of *H*ierosolyma *E*st *P*erdita; or, the cry of a goatherd.] Usu. *hep, hep!* The cry of those who persecuted Jews in the 19th century. Also *attrib.*

1839 *Penny Cycl.* XIII. 122/1 They [*sc.* the Jews] were massacred at the cry of 'Hep', 'Hep', the initials of the words 'Hierosolyma est perdita'. **1879** GEO. ELIOT *Impressions of Theophrastus Such* xviii. 313 (*heading*) The Modern Hep! Hep! Hep! **1930** D. PHILIPSON *Reform Movement in Judaism* (rev. ed.) i. 25 The *hep hep* cry resounded in the streets of..Frankfort and Würzburg. *Ibid.* vi. 108 A violent anti-Jewish literary campaign ensued..which culminated in the..disgraceful hep-hep outbreaks of the year 1819. **1971** *Encycl. Judaica* VII. 1227/1 Hamburg Jews were molested during the Hep! Hep! riots of 1819.

heparin (he·părin). *Biochem.* [f. HEPAR + -IN[1].] A sulphated polysaccharide present in various body tissues and organs, esp. the liver, lungs, and muscles, and used therapeutically as an anti-coagulant.

1918 HOWELL & HOLT in *Amer. Jrnl. Physiol.* XLVII. 328 A phosphatid, not previously described, which exists in various tissues but is found in greatest abundance in the liver. This phosphatid is designated as *heparin* to indicate its origin from liver... This substance was [previously] described under the name of antiprothrombin. **1933** *Jrnl. Biol. Chem.* CII. 435 Dog liver contains approximately twice as much heparin as does beef liver. **1946** *Nature* 24 Aug. 270/1 Even drastic blood changes brought about by the injection of dicoumarol or heparin had only a slight effect on the rate of growth of the bugs. **1968** A. WHITE et al. *Princ. Biochem.* (ed. 4) xxxi. 732 Many tissues of the body contain heparin since it specifically originates in the metachromatic granules of mast cells.

Hence **he:pariniza·tion**, the process of heparinizing; **he·parinize** *v. trans.*, to treat with heparin and thus reduce the clotting power of the blood; **he·parinized** *ppl. a.*

1940 J. E. R. MCDONAGH *Universe through Med.* 24 Snake venom has the same action in this condition as has calcium in making heparinised blood coagulate. **1943** *Science* 2 July 20/1 In all experiments in which heparinized whole blood or plasma was administered the lipemia was abolished. **1956** *New Eng. Jrnl. Med.* 29 Nov. 1027/2 General heparinization carries with it the risk of serious hemorrhage from bleeding lesions elsewhere in the body as well as at the operative site. **1959** *Lancet* 11 July 25/1 The risk of heparinising a patient who is receiving a coumarin drug. **1961** *Ibid.* 14 Oct. 858/1 The principle of regional heparinisation is to supply anticoagulated blood to the artificial kidney, whilst not interfering with the normal coagulation of blood in the patient.

hepat-. Add: **hepate·ctomy**, excision of (a part of) the liver; also **hepate·ctomized** *ppl. a.*; **hepato·ma**, a tumour of the liver (see quot. 1934).

1900 DORLAND *Med. Dict.* 296/2 Hepatectomy. **1910** *Practitioner* Mar. 383 Hepatectomy under these conditions does not appear to add to the gravity of the prognosis. **1946** *Nature* 31 Aug. 310/2 The so-called xanthorubin, a yellow compound present in the serum of hepatectomized dogs. **1905** H. D. ROLLESTON *Dis. Liver* 457 The term 'hepatoma' was suggested by Sabourin to describe the transitional stage between adenoma and carcinoma. **1912** *Ibid.* (ed. 2) 474 This condition [*sc.* primary carcinoma developing in a cirrhotic liver] was described..as Hepatoma by Rénon, Géraudel, and Monier-Vinard who insist that it is not a carcinoma. 'Hepatoma', also employed by Sabourin..is a confusing title. **1934** *Brit. Jrnl. Surg.* XXI. 684 The next advance of note was made by Yamagiura in 1911. He..made two simple divisions: (1) Hepatoma, i.e., carcinoma of hepatic cells; and (2) Cholangioma, i.e., carcinoma of bile-ducts. The term 'hepatoma' had previously been used by Sabourin in reference to a condition of nodular hyperplasia which in his opinion was a transitional stage between adenoma and carcinoma. Most modern writers, however, accept Yamagiura's interpretation, and use it as a term for primary carcinoma of the liver cells. **1971** *New Scientist* 17 June 668/3 From one strain of a mouse hepatoma they have a factor which they describe as being 'a heat stable molecule of low molecular weight'.

hepatic, *a.* and *sb.* Add: **B.** *sb.* (Later example.)

1908 *Chambers's Jrnl.* Sept. 671/2 An East Indian Aloes used to..be quoted in trade papers under the distinction of 'Hepatic'.

2. *Bot.* Usu. in *pl.* = HEPATICA.

1939 *Nature* 2 Sept. 416/2 The three smallest plants which have left recognizable fragments are a fungus and two liverworts or, as they are often called, hepatics, a group allied to the mosses but of simpler construction. **1964** V. J. CHAPMAN *Coastal Veget.* vi. 152 It is here also that some hepatics..can be found.

hepatico- (hĭpæ·tiko), combining form of HEPATIC, = HEPATO-.

For further examples see medical dicts.

1910 *Practitioner* Mar. 384 The hepatico-cystic confluence. *Ibid.* 385 Vautrin put a drain in the hepatic duct, thus making a hepaticostomy. **1933** *Med. Rec.* 18 Jan. 52/2 Hepaticocholangiogastrostomy..should be used only when other methods are impracticable.

hepatin (he·pătin). *Biochem.* Also (sense 1) -ine. [f. HEPAT- + -IN[1].] †**1.** = GLYCOGEN. *Obs.*

1858 F. W. PAVY in *Guy's Hosp. Rep.* IV. 316 [In calling it glucogenic] we are giving a name..to a substance which implies a purpose to which the facts..show it does not naturally administer in the living animal... I therefore propose to call it hepatine—a term which..cannot convey an erroneous impression..and which, nevertheless, is strictly pertinent. **1860** —— in *Phil. Trans. R. Soc.* CL. 608, I have made some analyses to show..how much sugar is formed for the hepatine that disappears. **1865** W. B. CARPENTER *Man. Physiol.* (ed. 4) I. iii. 108 The conversion of hepatine into sugar seems to be promoted by the presence of a 'ferment' not merely in the liver itself, but also in the blood circulating through it. *Ibid.* II. viii. 454 There is evidence that Hepatin may be formed in the Liver at the expense of Albuminous substances.

2. [a. G. *hepatin* (S. S. Zaleski 1886, in *Zeitschr. f. physiol. Chem.* X. 494)], an iron-containing protein reported to occur in liver.

1886 *Jrnl. Chem. Soc.* L. 1054 Iron..is found in all the morphological constituents of the liver tissue in organic combinations, both with albuminates and with nuclein. In the iron-nuclein group of compounds, one is present which gives the ordinary tests for iron in contradistinction to the others which do not; from this latter group one compound, hepatin, has been isolated. **1891** W. D. HALLIBURTON *Text-bk. Chem. Physiol. & Path.* xxv. 551 The quantity of iron in the blood-free liver was found to vary between wide limits, but it was constantly found in organic combinations in the liver-cells, especially with nuclein; and one of the iron-nuclein compounds named hepatin was isolated. **1914** G. M. NILES *Diagn. & Treatm. Digestive Dis.* xiv. 357 Other somewhat vaunted preparations are gasterin and hepatin, which are obtained from the gastric juice of dogs through gastric fistulas.

hepatitis. Delete ‖. Substitute for def.: Inflammation of the liver. (Later examples.)

1879 A. FLINT *Clinical Med.* III. 370 Diffuse, or parenchymatous hepatitis and yellow atrophy of the liver are considered as one affection. **1938** YATER & AULT in W. M. Yater *Fund. Internal Med.* IV. 375 The differential diagnosis of the various types of hepatitis brings into consideration the differentiation of the causes of jaundice. **1955** GAIGER & DAVIES *Vet. Path. & Bacteriol.* (ed. 4) xxxiii. 644 Apart from the specific forms of hepatitis met with in tuberculosis,..the two main forms of inflammation met with are suppurative hepatitis due to bacterial activity within the liver tissue and chronic interstitial hepatitis due to blood-borne toxins and other agents. **1959** *Chambers's Encycl.* VII. 23/2 Acute infective hepatitis is the newer name for a condition long known in medical practice as catarrhal jaundice. **1963** L. SCHIFF *Dis. Liver* (ed. 2) xii. 370/1 At least two forms of viral hepatitis are recognized: the naturally occurring type referred to as infectious hepatitis (catarrhal jaundice, infectious jaundice, etc.), and homologous serum hepatitis (transfusion jaundice, yellow fever vaccine jaundice, syringe jaundice, postarsphenamine jaundice, etc.). *Ibid.* xiv. 453 (*heading*) Toxic and drug-induced hepatitis.

hepato-. Add: **he:patoce·llular** *a.*, of or pertaining to hepatic cells; **hepatofla·vin** *Biochem.*, a substance first isolated from liver and later found to be the same as *RIBOFLAVIN; **hepato-lenticular degeneration** [tr. F. *dégénérescence hépato-lenticulaire* (H. C. Hall, 1921)], a progressive disease of the nervous system (see quot. 1955); Wilson's disease; **he:patomega·lia, -me·galy**, abnormal enlargement of the liver; **hepatoscopy** (later examples); hence **hepato·scopist**, one who practises hepatoscopy; **he:patosple:nomega·lia, -me·galy**, abnormal enlargement of the liver and spleen; **hepatoto·xic** *a.*, having a toxic effect on the liver; so **he:patotoxi·city**; **hepatoto·xin**, (*a*) any substance which has a toxic effect on the liver; (*b*) an antibody produced by injecting liver tissue into an animal.

1940 E. ROSENTHAL *Dis. Digestive Syst.* III. 212 Jaundice may be i. Mechanical ii. Functional (hepatocellular) iii. Hæmolytic. **1949** KANTOR & KASICH *Handbk. Digestive Dis.* (ed. 2) xvii. 415 The hepatocellular and the obstructive forms of jaundice make up more than 95 per cent of the cases encountered in clinical practice. **1962** *Lancet* 13 Jan. 67/1 Patients with hepatocellular disease are faced with the prospect of hepatic coma and death every time they have a haemorrhage. **1933** K. G. STERN in *Nature* 18 Nov. 784/1 The isolation in a crystalline state of the lyochrome from horse liver, to be designated as 'hepatoflavin', has been achieved. **1936** *Jrnl. Nutrition* XI. 75 Fractions prepared from liver extract which were rich in vitamin G (B_2) and from which the hepatoflavin had been removed..were very active in the cure of black tongue. **1943** M. E. REHFUSS *Indigestion* xxiv. 392 The terms lactoflavin, ovoflavin, hepatoflavin, and so on serve to indicate the source of the particular riboflavin under discussion. **1960** A. E. BENDER *Dict. Nutrition* 62/1 *Hepatoflavin*, name given to substance isolated from liver, shown later to be riboflavin. **1922** *Lancet* 29 Apr. 849/2 This is an account of the disease known as progressive lenticular degeneration, which Dr. Hall christens hepato-lenticular degeneration. **1925** *Brain* XLVIII. 332 It would be better to adopt Hall's terminology, and refer to the condition as hepato-lenticular degeneration... This title fails to indicate the peculiar type of the disease in the liver..but it indicates..that the lenticular disease is a sequel to liver damage. **1945** *Archives Internal Med.* LXXV. 151/1 Wilson in 1912 ..first clearly defined the condition now most widely known as hepatolenticular degeneration. **1955** S. SHERLOCK *Dis. Liver & Biliary Syst.* xiii. 339 Hepatolenticular degeneration is a rare disease of young people characterised by portal cirrhosis of the liver, bilateral softening and degeneration of the lenticular nuclei of the basal ganglia of the brain, and greenish-brown pigmented rings in the periphery of the cornea. **1893** DUNGLISON *Dict. Med. Sci.* (ed. 21) 524/1 Hepatomegalia. **1910** COLLINS & LIEBMANN tr. *Dieulafoy's Text-bk. Med.* II. VIII. i. 1916 He [*sc.* Glénard] found that hepatomegalia is the most frequent of the organic changes in diabetes. **1904** STEDMAN *Dunglison's Dict. Med. Sci.* (ed. 23) 527/1 Hepatomegaly. **1937** J. L. KANTOR *Synopsis Digestive Dis.* xxi. 230 Cancer of the liver must be differentiated from other conditions causing hepatomegaly. **1969** Hepatomegaly [see *hepatotoxicity* below]. **1947** AUDEN *Age of Anxiety* (1948) vi. 121 Peace was promised by the public hepatoscopists. **1928** C. DAWSON *Age of Gods* xiii. 307 The practice of Hepatoscopy or divination from the liver of the sacrificial victim, which reached Asia Minor from Mesopotamia, was carried by the Etruscans to Italy. **1957** *Encycl. Brit.* XI. 451/1 The theory underlying hepatoscopy consists of the belief (1) that the liver is the seat of life, or the soul of the animal; and (2) that the liver of the sacrificial animal..took on the same character as the soul of the god to whom it was offered. **1930** *Chem. Abstr.* XXIV. 886 In hepatosplenomegalia and hepatic cirrhosis, the urobilin increased in the urine and feces. **1939** STEDMAN *Med. Dict.* (ed. 14) 496/2 Hepatosplenomegaly. **1961** *Lancet* 19 Aug. 434/2 Physical examination revealed considerable hepatosplenomegaly. **1940** *Jrnl. Amer. Med. Assoc.* 28 Dec. 2264/2 (*heading*) Alleged hepatotoxic action of stilbestrol. **1961** *Lancet* 16 Sept. 623/1 Each of the drugs which has caused jaundice is a derivative of hydrazine, itself a potent hepatotoxic agent in laboratory animals. **1972** *Nature* 4 Feb. 279/1 Both compounds..have been reported to be hepatotoxic, carcinogenic, teratogenic and neurotoxic. **1952** *New Eng. Jrnl. Med.* 20 Nov. 797 (*heading*) The hepatotoxicity of intravenous aureomycin. **1969** *Nature* 19 Apr. 223/2 Increase in size of the liver (hepatomegaly) is not a reliable indication of hepatotoxicity. **1904** STEDMAN *Dunglison's Dict. Med. Sci.* (ed. 23) 527/1 *Hepatotoxin*, a cytotoxin having a specific action on the cells of the liver. **1909** J. G. ADAMI *Princ. Path.* I. III. viii. 489 Ciliated epithelium was shown to have its cytotoxin..., as have kidney cells (nephrotoxin), liver cells (hepatotoxin), pancreatic, adrenal, and, in fact, every form of animal cell that has been tested. **1929** *Chem. Abstr.* XXIII. 5509 Hepatotoxins, prepd. by immunizing rabbits with emulsions of rabbit and rat livers, were injected into rabbits and rats. **1951** A. GROLLMAN *Pharmacol. & Therapeutics* xxvii. 607 Because of their lipotropic action, choline and, to a lesser extent, methionine..have been used therapeutically in cirrhosis of the liver,..and as a prophylactic in poisoning by hepatotoxins. **1963** G. KLATSKIN in L. Schiff *Dis. Liver* (ed. 2) xiv. 453/1 Hepatotoxins, a heterogeneous group of naturally occurring and synthetic chemical agents, produce a variety of lesions in the liver that are classified as forms of toxic hepatitis.

Hepburn (he·pbɜ̄ɹn, he·bɜ̄ɹn). The name of J. C. *Hepburn* (1815–1911), Amer. physician and missionary, used *attrib.* in *Hepburn system*, a Romanized transcription of Japanese characters. So **Hepbu·rnian** *a.*

[**1867** J. C. HEPBURN (*title*) A Japanese and English dictionary.] **1937** *Mélanges Ling. et Phil. offerts à J. van Ginneken* 357 The Nipponsiki or Japanese system versus the Hepburnian system. *Ibid.* 358 It was this Hepburn system..which was adopted by the Romajikai. **1950** D. JONES *Phoneme* 38 The old (Hepburnian) Roman writing of Japanese. **1961** T. LANDAU *Encycl. Librarianship* (ed. 2) 369/1 These [Japanese] sounds are transcribed into Roman letters, either by the Hepburn system ..or by the Japanese system of 'New spelling'... The outside world sticks to the Hepburn system.

hep-cat (he·pkæt). *slang* (orig. *U.S.*). [f. *HEP *a.* + *CAT *sb.*[1] 2 c.] An addict of jazz, swing music, etc.; one who is 'hep'; = *HIPSTER[1].

1938 in *Amer. Speech* (1939) XIV. 140 *Hep cat*,..guy who knows what it's [*sc.* swing music is] all about. *a* **1940** F. SCOTT FITZGERALD *Last Tycoon* (1949) v. 127 Suddenly they were at work again—taking up this new theme in turn like hepcats in a swing band and going to town with it. **1955** *Sci. News Let.* 1 Oct. 221/2 This is not cool chatter between some young hep-cats in a smoke-filled jazz joint. **1957** C. MACINNES *City of Spades* II. xii. 184 You's the hep-cat what stole Mr. Vial's puss-cat! **1959** *Listener* 9 Apr. 646/2 The lament at a local hep-cat's funeral. **1959** *News Chron.* 14 Oct. 8/6 The jazz-loving 'hep-cat'. **1961** *Times* 20 May 5/1 Mr. Louis Armstrong and his fellow hepcats.

Hepialid (hĭpiæ·lid), *a.* and *sb.* *Zool.* Formerly also **epialid.** [ad. mod.L. *Hepialidæ*, f. *Hepialus* (J. C. Fabricius, *Systema Entomologiæ* (1775) 589), a. Gr. ἠπίολος moth; sometimes *Epialidæ, Epialus*, perhaps through confusion with Gr. ἠπίαλος nightmare: see -ID[3].] (Of or pertaining to) a moth of the family Hepialidæ, the ghost-moths or swifts.

1888 *Proc. Linnean Soc. N.S.W.* II. 1015 We have drawn up a description of the finely coloured Hepialid which was exhibited at the June meeting. **1895** *Funk's Stand. Dict.*, *Ghost-moth*, an epialid moth, especially *Epialus humuli*. *Ibid.*, *Swift*, n... 2. An epialid or ghost-moth. **1900** *Trans. Ent. Soc.* III. 411 (*title*) Life histories in the Hepialid group of Lepidoptera. **1931** *Ann. Appl. Biol.* XVIII. 54 The larva of a Hepialid moth, *Oncopera intricata*.

Hepplewhite (he·p'l,wəit). The name of George *Hepplewhite* (died 1786), who was succeeded by A. *Hepplewhite* and Co., used *attrib.* to designate an English style of furniture of the latter part of the eighteenth century, characterized by lightness, delicacy, and graceful curves, being an adaptation of current French styles.

1897 K. W. CLOUSTON *Chippendale Period Eng. Furnit.* vii. 177 The Hepplewhite commode has long been obsolete or transformed into the modern cabinet with mirror back. **1900** *Jrnl. Soc. Arts* 23 Mar. 380/1 Hepplewhite and Sheraton furniture should be studied by designers for *motifs*. **1901** *Connoisseur* Dec. 272/2 Six Hepplewhite arm-chairs. **1903** *Chambers's Jrnl.* 20 June 460/1 Six ball-and-claw Hepplewhite chairs. **1957** *Encycl. Brit.* XI. 453/1 The smaller Hepplewhite pieces are much prized by collectors.

hepster (he·pstəɹ). *slang* (orig. *U.S.*). Now *rare.* [f. *HEP *a.* + -STER.] = *HEP-CAT, *HIPSTER[1].

1938 in *Amer. Speech* (1939) XIV. 140/2 Cab Calloway's Cat-alogue, a 'hepster's' dictionary. **1948** [see *BOP *sb.*[3]]. **1958** *Spectator* 21 Nov. 702/1 Yet although jazz seems to have burst out of the locked treasure casket over which an egghead minority of hepsters crooned for so many years, it still remains a curiously unreal cult.

hepta-. Add: **he·ptachlor** *Chem.*, a chlorinated hydrocarbon, $C_{10}H_5Cl_7$, used as an insecticide; **hepta(i)co·sane** *Chem.* [Gr. εἴκοσι twenty: see -ANE], any of the hydrocarbons of the paraffin series having twenty-seven carbon atoms, *esp.* the unbranched isomer, which is present in tobacco oil and many natural waxes; **heptade·cane** *Chem.* [Gr. δέκα ten: see -ANE], any of the hydrocarbons of the paraffin series having seventeen carbon atoms, *esp.* the unbranched isomer; **he:ptahydra·ted** *a.* *Chem.*, containing seven molecules of water; **heptahy·dric** *a.* *Chem.*, containing seven hydroxyl groups; **he·ptastyle** *Archit.* [*-style*, a. Gr. στῦλος pillar], (a building or portico) having seven columns in front; **he·ptose** [a. G. *heptose* (E. Fischer 1890, in *Ber. d. Deut. Chem. Ges.* XXIII. 934): see -OSE[2]], any of a group of monosaccharides, $C_7H_{14}O_7$, present in some plants and as constituents of some bacterial polysaccharides.

1949 *Jrnl. Econ. Ent.* XLII. 328/1 Heptachlor, a close relative of chlordan, gave results superior to chlordan. **1961** *New Scientist* 6 July 9/2 British manufacturers of agricultural chemicals have agreed with the Government to restrict the use of three pesticides..aldrin, dieldrin and heptachlor. **1963** R. CARSON *Silent Spring* x. 140 Heptachlor, after a short period in the tissues of animals or plants or in the soil, assumes a considerably more toxic form known as heptachlor epoxide... The Food and Drug Administration took action which had the effect of banning any residues of heptachlor or its epoxide in food. **1968** M. PYKE *Food & Society* viii. 119 [They] found not only DDT and its breakdown products but BHC, heptachlor and dieldrin as well in the tissues of penguins. **1889** MUIR & MORLEY *Watts' Dict. Chem.* II. 675/1 *n*-Heptaicosane $C_{27}H_{56}$. **1901** *Jrnl. Chem. Soc.* LXXIX. I. 986 The results of the analysis..indicate that the second hydrocarbon is in all probability heptacosane, $C_{27}H_{56}$. **1961** L. F. & M. FIESER *Adv. Org. Chem.* iv. 99 Beeswax contains heptacosane ($C_{27}H_{56}$) and hentriacontane ($C_{31}H_{64}$). **1882** *Jrnl. Chem. Soc.* XLII. 1272 Heptadecane (from margaric acid, or from the ketone $C_{33}H_{66}O$ obtained from barium palmitate) crystallizes in thin hexagonal plates. **1895** G. LUNGE *Sulphuric Acid* (ed. 2) II. 22 The anhydrous salt..if protected from the air is changed into heptahydrated..salt. **1892** E. F. SMITH tr. *V. von Richter's Chem. Carbon Compounds* (ed. 2) 49 The heptahydric aldehydes, $C_7H_{14}O_7$, resemble the sugars in their behavior. **1968** J. A. MONICK *Alcohols* i. 442 Perseitol..is a heptahydric alcohol with a 7-carbon atom straight-chain that contains seven hydroxyl groups. **1843** *Civil Engin. & Archit. Jrnl.* VI. 167/1 The temple of the giants, at Agrigentum (heptastyle). **1909** A. MARQUAND *Greek Archit.* vi. 313 Another classification of temples notes merely the number of columns exhibited in the façade... The Theseion at Athens was hexastyle; the Temple at Thorikos, heptastyle. **1890** *Jrnl. Chem. Soc.* LVIII. I. 598 The author proposes to employ the terms pentose, heptose, octose, &c., for the sugars..and heptonic acid, octonic acid, &c., for the acids of the sugars, according to the number of carbon-atoms in the compounds. **1916** Heptose [see *BIOSE]. **1952** J. K. N. JONES in E. M. Rodd *Chem. Carbon Compounds* IB. xx. 1264 Heptose sugars may exist in ketose and aldose forms. **1960** *Adv. Carbohydrate Chem.* XV. 288 Whereas, in the Enteric group of bacteria, there is as yet no evidence for the presence of heptoses other than L-*glycero*-D-*manno*-heptose, this is not true of other Gram-negative groups.

heptane. Add: **heptano·ic acid**, œnanthic acid, $CH_3{\cdot}(CH_2)_5{\cdot}COOH$; **hepteno·ic acid**, any of several monocarboxylic acids, $C_6H_{11}COOH$, having one double bond.

1928 *Chem. Abstr.* XXII. 5909 (*index*) Heptanoic acid. See Enanthic acid. **1964** N. G. CLARK *Mod. Org. Chem.* xvi. 317 By heating the heptonic acid with hydriodic acid and red phosphorus, (*n*-)heptanoic acid is produced. **1889** MUIR & MORLEY *Watts' Dict. Chem.* II. 676/1 Heptenoic acid $C_7H_{12}O_2$. **1948** A. W. RALSTON *Fatty Acids* ii. 86 None of the heptenoic acids has been identified in the naturally occurring fats and oils.

heptarch (he·ptɑɹk), *a.* *Bot.* [f. Gr. ἑπτά seven + ἀρχή beginning, origin: cf. DIARCH, MONARCH, OCTARCH, POLYARCH, TETRARCH, TRIARCH *adjs.*] Arising from seven distinct

points of origin, as the xylem of the root of some plants.

1884 [see DECARCH *a.*]. **1914** M. DRUMMOND tr. *Haberlandt's Physiol. Plant Anat.* vii. 353 (*caption*) The heptarch radial bundle [*sc.* stele] of an adventitious root. **1951** MCLEAN & IVIMEY-COOK *Textbk. Theoretical Bot.* I. xx. 791 Dicotyledons usually have two..four..or five xylem groups, less frequently three (triarch) or seven (heptarch), and rarely more.

heptode (he·ptōᵘd). *Radio.* [f. HEPT(A- + *-ODE.] A valve with seven electrodes. Also *attrib.*

1932 *Post Office Electr. Engin. Jrnl.* XXIV. IV. 299/2 A complete electrode assembly of this type includes seven electrodes—hence the name 'Heptode' for the double-acting balanced thermionic valve. **1934** *Times* 11 Aug. 7/3 The days when one could speak simply of a screen-grid valve or a pentode are almost gone, for now there are in general use the double-diode-triode, the heptode, and many others with a double and even treble purpose. **1942** *Electronic Engin.* XIV. 629 The frequency changer follows normal practice..a separate triode with hexode or heptode comprising the local oscillator. **1943** *Gloss. Terms Telecomm.* (*B.S.I.*) 30 *Heptode*, a vacuum tube with seven electrodes normally comprising a hexode with an auxiliary anode between the first grid and the first screen grid or with a suppressor grid between the second screen grid and the (main) anode. **1945** *Electronic Engin.* XVII. 648 A circuit composing a heptode oscillator.

heptose: see *HEPTA-.

Heracleid. Add: **c.** A poem describing the exploits of Heracles.

1725 [see *Theseid* s.v. THESEAN *a.*]. **1904** T. R. GLOVER *Stud. Virgil* iii. 75 Poets who have composed a Herakleid, a Theseid, or other poems of the kind.

heracleum (heræ·kliəm, herăkli·əm). *Bot.* [mod.L. (C. Linnæus *Systema Naturæ* 1735), f. Gr. Ἡράκλεια, the plant named after Heracles.] A plant of a genus of large herbs of this name, belonging to the family Umbelliferæ and native to northern temperate regions; COW-PARSNIP or HOGWEED.

1787 W. WITHERING *Bot. Arrangem. Brit. Plants* (ed. 2) I. 287 Heracleum. **1824** J. E. SMITH *Eng. Flora* II. 101 Heracleum. Cow-parsnep. **1847** H. C. WATSON *Cybele Britannica* I. 451 This [*Angelica sylvestris*] and the Heracleum are the two most widely distributed species of their order. **1864** D. OLIVER *Less. Elem. Bot.* II. 175 (*caption*) Vertical section of flower of Common Heracleum. **1894** W. ROBINSON *Wild Garden* (ed. 4) xiii. 133 Such plants as Heracleum, Willow Herb, and many others..should be planted only in outlying positions. **1951** *Dict. Gardening* (R. Hort. Soc.) II. 986/1 The Heracleums are sometimes grown in shrubberies or the rougher parts of the pleasure grounds.

Heraclitean, *a.* Add: (Earlier and additional examples.) Also **Heracleitean.**

1791 W. ENFIELD *Hist. Philos.* I. 443 Plato himself, when he was young, learned the Heraclitean philosophy from Cratylus, and adopted that part which treated of the nature and motion of matter. **1875** [in Dict.]. **1955** J. K. BAXTER *Fire & Anvil* 65 A Heracleitean cosmos.

Heracliteanism: also **Heracleiteanism.**

1932 *Times Lit. Suppl.* 21 July 531/1 Professor Laird yet finds it in him to relax his comity when dealing with the epistemological Heracleiteanism of Gentile.

herald, *sb.* Add: **5. herald-snake,** the southern African snake, *Crotaphopeltis hotambœia hotambœia*, which has red or yellow lips and is also called the red-lipped snake.

1910 F. W. FITZSIMONS *Snakes S. Afr.* iii. 57/2 The Red-lipped or Herald Snake..is one of the best-known and most widespread snakes in Africa. **1947** J. STEVENSON-HAMILTON *Wild Life S. Afr.* xxxvi. 330 The red-lipped or herald snake (*Leptodira hotamboea*).—This is distinguished by its upper lip being of bright red colour; it carries large light-brown scales and a black headband. **1970** V. F. M. FITZSIMONS *Field Guide to Snakes S. Afr.* 118 *Herald* or *Red-lipped Snake*.., according to the prevailing colour on the upper lips, it is variously known as the White- or Yellow-lipped Snake.

Herat (herā·t). The name of a city in north-western Afghanistan, used to designate a kind of carpet and rug made there, and the small, close design of leaf and rosette patterns characteristic of such rugs. Also **Hera·ti.**

1917 in R. Storrs *Orientations* (1937) x. 261 Some fine old carpets, including a brilliant Herât. **1931** A. U. DILLEY *Oriental Rugs & Carpets* Pl. 51 (*caption*) Herat rugs... Herat borders. *Ibid.* iv. 118 Herat, applied both to the 'Ispahan' weaving of the sixteenth century and to the blue Herati-patterned rug of the nineteenth century, is, rightly or wrongly, the outstanding name in rugs. **1957** *Encycl. Brit.* XIX. 622/1 (*caption*) The Herati pattern. *Ibid.* 628C/2 The Ferraghan [rugs], with their so-called Herati pattern—an all-over, rather dense design with a light green border on a mordant dye that leaves the pattern in relief. **1960** H. HAYWARD *Antique Coll.* 141/2 *Herat carpets*, fine quality Persian carpets... Knot: Ghiordes or Sehna. **1967** 'W. HAGGARD' *Conspirators* ix. 99 There was..fine old mahogany furniture and a Herati rug. **1969** —— *Doubtful Disciple* xv. 166 The dealers call them Isfahans and most of the time they know they're Heratis. **1972** *Observer* 8 Oct. 14/1 (Advt.), A central panel leading to the reception panel shows the Herati pattern on a light red ground.

herb, *sb.* Add: **6.** *herb-lore, -master.*

1955 J. R. R. TOLKIEN *Return of King* 145 A chance of talking herb-lore with me. *Ibid.* 140, I will go and ask of the herb-master.

7. herb-doctor *local U.S.*, one who treats or cures ailments by means of herbs. So **herb-doctress.**

1854 THOREAU *Walden* 150 Hygeia, who was the daughter of that old herb-doctor Aesculapius. *a* **1864** HAWTHORNE *S. Felton* (1871) 113 [Aunt Keziah was] a mixture of an Indian squaw and herb doctress. **1881** *Harper's Mag.* July 305/2 The herb-doctor was not so fortunate as another practitioner of his own class who came to England some years ago. **1891** *Ibid.* Jan. 220/1, I would say that Mr. Pettingill has behaved very strangely —sending for a herb doctor.

herbaceous, *a.* Add: **4.** *herbaceous border*, a border filled with herbaceous perennial plants (so, as nonce-wd., *herbaceous borderer*); *herbaceous perennial*, a plant whose roots live for several years, although stem and leaves die down to the ground each year, after flowering.

[**1822** LOUDON *Encycl. Gard.* III. 993 (*heading*) Species and Varieties of..Herbaceous Border-flowers. **1868** D. THOMSON *Handy Bk. of Flower Garden* 6 There is enough in a border of hardy herbaceous plants..to gratify the keenest sensibility.] **1881** T. MOORE *Epitome of Gardening* vi. 190 The herbaceous border should be a distinct compartment, and not less than 10 ft. in width. **1883** F. MILES in W. Robinson *Eng. Flower Garden* p. xl/2 What cannot be done with an herbaceous border edge when that edge is the green grass? **1909** H. G. WELLS *Tono-Bungay* III. ii. § 1 An ardent rose grower and herbaceous borderer. **1931** C. ST. JOHN *Ellen Terry & Bernard Shaw* 132 Miss Audrey Campbell: an enthusiastic amateur actress, and a plant of 'vigorous habit' in Ellen Terry's herbaceous border of friends. **1968** R. HAY *Gardener's Round* 213 A friend, anxious to replant a fairly large herbaceous border, asked me to mark a catalogue for him. **1868** D. THOMSON *Handy Bk. Flower Garden* vi. 153 Herbaceous perennials are a class of plants distinct in their nature. **1871** W. ROBINSON *Hardy Flowers* III. 277 (*heading*) A choice selection of the very finest herbaceous perennials. **1959** C. H. POTTER *Perennials in Garden* i. 8 The term perennial..refers to hardy herbaceous perennials which tend to live year after year without replanting.

herbarium. Add: Also *attrib.*

1849 J. H. BALFOUR *Man. Bot.* 616 This [*sc.* the vasculum] should be of sufficient length to receive a plant of the full size of the herbarium paper. **1887** C. A. MOLONEY *Sk. Forestry W. Afr.* 319 He compared the available herbarium material of the two plants. **1898** B. TORREY in *Atlantic Monthly* Apr. 461/2 A comparison with herbarium specimens. **1962** D. B. O. SAVILE *Coll. & Care Bot. Specimens* i. 50 (*heading*) Herbarium sequence. The operation of the phanerogamic herbarium of the Plant Research Institute may serve as an example of herbarium management. *Ibid.* 52 Ideally the herbarium units should be built like library stack rooms.

Herbartian (həɹbā·ɹtiăn), *a.* and *sb.* [f. the name of J. F. *Herbart* (1776–1841), German philosopher + -IAN.] **A.** *adj.* Of or pertaining to Herbart, or to the system of psychology and teaching originated by him. **B.** *sb.* A disciple or follower of Herbart. Hence **Herba·rtianism**, the doctrines of Herbart.

1884 W. JAMES *Coll. Ess. & Rev.* (1920) 267 The Herbartian psychologists have tried to distinguish feelings due to the *form* in which ideas may be arranged. **1886** *Encycl. Brit.* XX. 41/1 The whole Herbartian psychology. *Ibid.* 62/2 This difference between a conflict of presentations to enter consciousness..and that opposition or incompatibility of presentations which is only possible when they are in consciousness has been strangely confused by the Herbartians. **1903** HAYWARD & THOMAS (*title*) The critics of Herbartianism. **1904** C. I. DODD (*title*) Introduction to the Herbartian principles of teaching. **1908** H. G. WELLS *New Worlds for Old* (1912) v. § 1. 91 This change in the circle of ideas (as the Herbartians put it) is the essence of the Socialist project. *Ibid.* x. § 1. 225 The majority of Socialists still fail to grasp completely the Herbartian truth. **1932** L. BLOOMFIELD in *Language* VIII. 225 Interpreting their methods in terms of Herbartian or Wundtian psychology. **1952** J. A. PASSMORE in G. F. Stout *God & Nature* p. xxix, The Herbartian ethnographic psychologists—Waitz, Lazarus, Steinthal—were at this time exerting a powerful influence upon him. **1971** *Language* XLVII. 980 The choice was between Herbartian and Wundtian doctrines.

herbicide (hɜ̄·ɹbisəid). [f. L. *herba* grass, green crops + -CIDE.] †**a.** Proprietary name for a preparation, prob. of sodium arsenite, used as a weed-killer. *Obs.* **b.** Any chemical agent that is toxic to some or all plants and is used to destroy unwanted vegetation. Hence **he·rbicidal** *a.*

1899 *Vermont Agric. Exper. Station Ann. Rep. 1898–99* 185 Carbolic acid is a valuable herbicide..but..the herbicidal action is of short duration. **1906** *U.S. Trademark 48,757* My [*sc.* W. J. Reade's] trade-mark consists of the coined word 'Herbicide'. The trade-mark has been continuously used in my business since 1894. **1915** *Chem. Abstr.* IX. 1973 (*caption*) Herbicides with arsenical base. **1942** W. W. ROBBINS et al. *Weed Control* xiii. 229 Studies of herbicides in soils show that the following factors determine the herbicidal effects of any chemical: (1) inherent toxicity of the chemical; (2) adsorption of the chemical by the soil; (3) decomposition..tending to reduce toxicity;..and (6) species tolerance. **1947** *New Biol.* II. 109 During recent years various organic compounds have been tested for their possible herbicidal value. **1954** *Sci. News* XXXI. 104 Weeds, therefore, could easily get life insurance were it not for the modern technique of chemical weed-killing, in particular by selective herbicides. **1959** *New Scientist* 5 Nov. 894/2 In abstracting and indexing the world's literature on weed control at Oxford [Department of Agriculture] we are constantly faced with difficulties caused by there being no agreement on what to call a herbicide. **1969** N. W. PIRIE *Food Resources* ii. 65 In Britain nearly all farms of more than 100 acres use herbicides on cereals. **1971** *Nature* 22 Jan. 224/1 Herbicidal attack appears to prevent the re-establishment of any new plant community..for at least six years.

Hercynian, *a.* Delete the phr. beginning 'esp.' and add: **1.** (Later examples.)

1890 J. G. FRAZER *Golden Bough* I. i. 56 Down to the first century before our era the Hercynian forest stretched eastward from the Rhine for a distance at once vast and unknown. **1935** W. G. EAST *Hist. Geogr. Europe* ii. 51 The Hercynian forest, as Caesar described it, could be crossed in nine days by a fast runner.

2. In *Geol.* used by different writers in various senses, with allusion to the Harz Mountains.

The word was first used in geology as the G. *hercynisch* (according to Suess by von Buch) in sense c, and was adopted by several writers, chiefly German and French, in sense a; in 1887 Bertrand (*Bull. de la Soc. géol. de France* XV. 438) used it in sense b to replace the *variscisch und armoricanisch* of Suess, and this has become the usual sense in English.

a. Designating one of the Devonian formations of the Harz Mountains; so *Hercynian fauna* (after Kayser, 1879), *gneiss* (after Gümbel, 1868).

1880 J. D. DANA *Man. Geol.* (ed. 3) III. i. 151 In Europe, the Archæan system has been distinctly recognized in.. Bavaria (Hercynian and Bojie Gneiss). **1885** Hercynian gneiss [in Dict.]. **1893** P. LAKE tr. *Kayser's Text Bk. Compar. Geol.* II. 101 First described by H. Römer and Giebel as Silurian, this Hercynian fauna of the Lower Harz has more recently been compared by E. Beyrich with the fauna of the Bohemian stages F,G, H of Barrande, and was afterwards described by Kayser..as Lower Devonian. **1895** J. D. DANA *Man. Geol.* (ed. 4) IV. ii. 570 Kayser concluded..that the Lower Helderberg formation of America was Hercynian, that is, lowest Devonian. **1906** CHAMBERLIN & SALISBURY *Geology* II. viii. 450 The Hercynian fauna which characterizes this stage of the Devonian in southern Europe has much in common with the Helderberg fauna of America. **1955** G. G. WOODFORD tr. *Gignoux's Stratigr. Geol.* iv. 128 The Hercynian facies of the Lower Devonian appears in the Kellerwald.

b. Of, pertaining to, or contemporaneous with the mountain-building movements that occurred in Europe in late Carboniferous and early Permian times, or the mountains then formed; hence, late Palæozoic; = *ARMORICAN *a.* 2.

1895 J. D. DANA *Man. Geol.* (ed. 4) IV. iii. 734 The 'Hercynian system' of Bertrand includes a long range of dislocated Devonian and Carboniferous rocks extending from Brittany to the Vosges and Ardennes, and beyond along the Black Forest, the Harz to Bohemia. **1926** *Publ. Carnegie Inst. Washington* No. 375. v. 69 The movements in late Stephanian and in Permian time can best be considered as the final effects of the Hercynian movement. **1935** E. B. BAILEY *Tectonic Ess.* i. 5 The Ardennes of Belgium and much of the Appalachians of America are alike members of the Hercynian System. *Ibid.* 8 Hercynian in Bertrand's sense has established itself apparently for all time in the literature of France, Switzerland, Britain, and America. **1948** R. L. SHERLOCK *Permo-Triassic Formations* iv. 40 Like the present Alps, the Hercynian Mountains were the result of the squeezing of the sediments deposited in a Mediterranean Sea, the Palaeozoic Tethys. **1967** D. H. RAYNER *Stratigr. Brit. Isles* i. 27 Extending down the centre of England..there are various Carboniferous outcrops whose structures are largely Hercynian in age but which follow older directions.

c. Applied (rarely in English) to those faults, folds, and other geological features in Europe with a predominantly north-western direction.

1904 H. B. SOLLAS tr. *Suess's Face of Earth* I. I. iii. 121 This..does not exclude the existence in Central Europe of two different directions, which have produced folds and mountain chains striking more to the north-east in the one case, in the other more to the north-west. The former is known as the direction of the Netherlands, the latter as the Hercynian direction. **1909** *Ibid.* IV. v. i. 2 The important point for L. von Buch, when he created the 'Hercynian system', was the (orographical) north-westerly direction, for Marcel Bertrand it was the tectonic age. *Ibid.* 33 These fractures..determine L. von Buch's 'Hercynian system', which embraces all the mountains striking to the north-west; many eminent German geologists still describe these long lines as Hercynian. **1935** E. B. BAILEY *Tectonic Ess.* i. 5 For von Buch, the Hercynian character was a west-north-westerly, or north-westerly, *trend*... Accordingly, among German-speaking tectonists, Hercynian is still commonly employed to group together a very important set of more or less north-westerly fractures and folds that characterize much of Czechoslovakia, Germany, and Scania.

herd, *sb.*[1] Add: **4. a. herd-testing** *vbl. sb.*, testing of the butterfat content of the milk from cows of a specified herd and their productivity; so **herd-tester**; also *herd test.*

1962 J. N. WINBURNE *Dict. Agric.* 374/2 *Herd test*, a

type of semiofficial testing for milk production in which the whole herd of cows of milking age are included. **1960** B. CRUMP *Good Keen Man* 94 He told me..all about one of his sisters who was training to be a herd-tester. **1966** G. W. TURNER *Eng. Lang. Austral. & N.Z.* vii. 149 A herd-tester is a man or woman who goes from farm to farm sampling milk to test cows for their productivity and the fat content of their milk. **1911** *Jrnl. Dept. Agric.* (N.Z.) 15 July 26 A striking case of the value of testing the individual members of a herd is reported from a district where a herd-testing association has been established. **1956** AMESS & JOHNSON *Dairying* (ed. 4) xv. 195 All herd-testing is now done by the New Zealand Herd Improvement Associations under direction from the Dairy Board.

b. *Psychol.* Denoting feelings, actions, thoughts, etc., common to a large company of people; esp. **herd instinct,** an instinctive tendency to think and act as one of a crowd. (Cf. sense 3.)

1908 W. TROTTER in *Sociol. Rev.* I. 227 (*title*) Herd instinct and its bearing on the psychology of civilised man. **1908** *Westm. Gaz.* 22 Aug. 16/1 The fundamental assumptions of the Liberal and the Conservative are hostile, and are the outcome of herd tradition. **1912** J. LONDON *Let.* 18 Jan. (1966) 359 There is a sort of herd psychology in this. **1914** G. B. SHAW *Androcles & Lion* (1916) 46 That herd instinct which makes men abhor all departures from custom. **1919** M. K. BRADBY *Psycho-anal.* 113 Not even abandonment to 'herd enthusiasm' could destroy the terrible loneliness of death. **1920** B. RUSSELL *Pract. & Theory Bolshevism* I. 125 The Marxian assumes that a man's 'herd' from the point of view of herd-instinct, is his class. **1922** *Brit. Jrnl. Psychol.* Oct. 118 The love complex, the religion complex, the herd complex and many others. **1923** J. S. HUXLEY *Ess. Biologist* vii. 275 The herd ideas..may be those of a nation or of a stratum within the nation. **1924** W. B. SELBIE *Psychol. Relig.* 148 Man is a gregarious animal, and even in his civilized condition never quite loses his herd consciousness. **1927** N. P. WILLIAMS *Ideas of Fall & Orig. Sin* p. xxix, We therefore identify the 'inherited infirmity' of theology with 'inherited weakness of herd-complex'. **1927** M. SADLEIR *Trollope* ii. 49 She was.. lacking in that pride of individuality which throws persons of a different type into automatic opposition to herd-bias. **1928** G. B. SHAW *Intell. Woman's Guide Socialism* v. 11 Our minds are mostly herd minds, with only a scrap of individual mind on top. **1932** Q. D. LEAVIS *Fiction & Reading Public* I. iii. 67 These writers are using the technique of Marie Corelli..to work upon and solidify herd prejudice. **1939** J. MASEFIELD *Live & Kicking Ned* 344 What saved us was the fact that Mimbo is an animal passion or herd-madness, which blinds each of the herd to all other things than the herd-enemy. **1942** R. A. KNOX *In Soft Garments* xxi. 162 There is such a thing as herd-morality. You notice it especially in matters like divorce, where social considerations necessarily apply. **1960** C. DAY LEWIS *Buried Day* ii. 43 At the age of eight or nine the herd instinct begins to operate.

herd, *v.*[2] **1. b.** Delete *rare* and add further examples. Also *transf.*, spec. (*U.S. slang*) to drive (a car, aircraft, etc.).

1895 KIPLING *Second Jungle Book* 67 Who is Man that we should care for him..? I have followed him all day— at noon—in the white sunlight. I herded him as the wolves herd buck. **1927** W. FAULKNER *Mosquitoes* 261 'Come on here, you men.' He named over his depleted watch and herded it forward. He herded it down to his cabin and nourished it with stimulants. **1954** —— *Fable* (1955) 100 He saw the two-seater..the two S.E.'s above and behind it, herding it down. **1955** *Sunday Times* 25 Sept. 3/3 (*heading*) Herd that beast!.. I append an up-to-the-minute glossary of motoring terms..*herd*, to drive a car. **1971** M. TAK *Truck Talk* 82 *Herd*, to drive a truck.

herdsman. Add: So **he·rdsmanship,** the performance or occupation of a herdsman.

1889 J. M. ROBERTSON *Christ & Krishna* xvi. 113 The theory of Krishna's herdsmanship being derived from the cloud-cows of the Vedas is new. **1960** *Farmer & Stockbreeder* 8 Mar. 120 (*caption*) Good herdsmanship, prompt veterinary attention and close liaison between the herdsman and the farm staff are essential.

here, *adv.* Add: **1. e.** *here today and gone tomorrow*: a catch-phrase indicating a constant change of events or someone (or something) remaining in a place for a short time.

1687 A. BEHN *Luckey Chance* iv. 47 Faith Sir, we are here to Day and gone to Morrow. **1776** H. NEWDIGATE *Let.* in A. E. Newdigate-Newdegate *Cheverels* (1898) i. 10 Going over for a day to Arbury to beat for Wood Cocks.. as they are birds of passage, here today and gone tomorrow. **1895** KIPLING *Day's Work* (1898) 172 Here today and gone tomorrow. Didn't come to stay for ever. **1944** W. S. MAUGHAM *Razor's Edge* IV. iv. 132 Even in the old days you could never count on his being where you expected him to be. He was here today and gone tomorrow. **1972** *Listener* 23 Nov. 690/2 We [*sc.* the TUC] put forward proposals for a general rise in pensions—not a 'Christmas Box' that is here today and gone tomorrow.

2. b. *here's hoping, how, looking* (*at you*), *luck*, formulas used in drinking healths.

1888 KIPLING *Soldiers Three* (1889) 51 He opened a bottle... 'Here's luck!' **1896**, etc. [see How *adv.* 18 in Dict. and Suppl.]. **1910** W. M. RAINE *B. O'Connor* iii. 40 'Here's hoping,' Bucky nodded gaily. 'I bet there will be a right lively wolf hunt.' **1933** M. LOWRY *Ultramarine* 164 Here's looking at you! **1938** E. HEMINGWAY *Fifth Column* (1939) 184 Here's looking at you... Here's how. **1973** E. LEMARCHAND *Let or Hindrance* xii. 142 Pollard..raised his glass to her. 'Here's hoping.'

5. b. *here's where*.., this is the point at which. *U.S. colloq.*

1923 R. D. PAINE *Comr. Rolling Ocean* xii. 203 'It makes me feel sick at my stomach', declared Briscoe. 'Here's where you feel sicker. Great Scott, look at that.' *Ibid.* xiv. 250 Here's where I slip it out to the old gink.

13. b. here we go again: we are off on the same undesirable course, project, etc., as before. Also used as *attrib. phr.*

1954 R. BISSELL *High Water* (1955) xix. 161 'Time to get up, Mister Duke.' 'Oh, hell,' I thought. 'Here we go again.' **1958** B. MALAMUD *Magic Barrel* (1960) 72 'Ah, there I can't help you,' said the portiere. 'I haven't got the key.' 'Here we go again,' Carl muttered. **1962** J. BALDWIN *Another Country* (1963) II. iii. 270 'Oh, shit,' he said, 'here we go again.' **1969** *Listener* 12 June 836/1 One of William Glock's most difficult tasks in planning each year's series of Promenade Concerts is to write that desperate here-we-go-again introduction to the prospectus.

B. as *sb.* Delete (*nonce-uses*) and add: (Instances of objective uses.) *here and now* (later examples); also as *advb. phr.*; so *here-and-nowness*. Also *here and there* sb. phr.

1816 M. E. BICKNELL *Let.* 26 Sept. in J. Constable *Corr.* (1964) II. 210, I know of no other plan but of my leaving here early on Monday. **1838** DICKENS *O. Twist* II. xxxiv. 269, I—I—ought to have left here before. **1839** —— *Let.* c 24 July (1965) I. 567, I dine in town tomorrow and shall leave *for here* at 10 at night. **1857** —— *Dorrit* II. xxxiv, You would rather not leave here till to-morrow morning. **1874** GEO. ELIOT in *Macm. Mag.* XXXVIII. July 177 Widening his consciousness from Here and Now to larger wholes. **1887** RIDER HAGGARD *A. Quatermain* xvi. 188 Here and now for thy dear sake I will forget my people and my father's house. **1922** W. S. MAUGHAM *On Chinese Screen* xliii. 172 Your thoughts travel through time and space, far from the Here and Now. **1933** P. GODFREY *Back-Stage* xiv. 176 Its romantic appeal is an escape from here and now into an artificial world of somewhere else or some time past. *a* **1943** R. G. COLLINGWOOD *Idea of Hist.* (1946) 248 The detail of the here-and-now as given him in perception. **1957** C. DAY LEWIS *Pegasus* 56 The truant here-and-there of the Stour. **1959** *Listener* 26 Feb. 385/1 Whether Condorcet or Huxley is right is an undecidable question for us here and now. **1959** *Times Lit. Suppl.* 11 Sept. 522/3 All combine to form a subtle escape-route from the unwelcome here-and-now of the twentieth century. **1961** I. MURDOCH *Severed Head* x. 85 She was all gentleness and filled with so genuine a concern to save me here and now from distress and anxiety. **1962** *Listener* 26 Apr. 717/2 Here and now Russia, while she does not want a war with us, is nevertheless our antagonist. **1963** *Times* 27 Feb. 13/1 They came, it seems, to a realization of the here-and-nowness of life. **1971** *New Scientist* 27 Apr. 263/1 Cocking's group have *their* minds firmly concentrated on the here-and-now.

hereditary, *a.* Add: **2. a.** (Further examples.)

1833 *Destructive* 22 June 166/3 (*heading*) Absurdity of 'hereditary wisdom'. **1899** W. JAMES *Talks to Teachers* xiv. 164 The foreign terms 'déséquilibré', 'hereditary degenerate', and 'psychopathic' subject, have arisen in response to the same need. **1928** B. D. JACKSON *Gloss. Bot. Terms* (ed. 4) 176/1 *Hereditary symbiosis*, the presence of mycobacteria in the tissues, including seeds. **1941** J. S. HUXLEY *Uniqueness of Man* ii. 48 The latter [*sc.* fraternal twins] will have hereditary outfits as different as those of members of the same family born at different times. **1967** MAHLOUDI & PIKIELNY in *Brain* XC. 672 With such divergence of opinion, it may be wise to drop the term paramyoclonus multiplex altogether. Instead we suggest the term *hereditary essential myoclonus*.

B. as *sb.* A hereditary ruler; in *pl.*, the House of Lords.

1836 *Radical* 13 Mar. 4/1 The debate, or rather debates in the House of Commons, on the question of justice to the Irish, are but a sad augury of its ever passing the 'Hereditaries' unemasculated. **1932** G. B. SHAW *Platform & Pulpit* (1962) 246 The hereditaries are brought up to exercise their personal power conventionally and leave the rest to their ministers.

Hereford (he·rifəɹd). The name of the county town of Herefordshire, a county in the west of England, used to designate a breed of cattle originating there, distinguished by their reddish-brown colour and white faces. Also **He·refordshire.**

1789 W. MARSHALL *Rural Econ. Glouc.* II. 226 The Herefordshire breed of cattle..may..be deemed the finest breed of cattle in this island... In general appearance the Herefordshire cattle resemble very much those of Sussex. **1805** J. LAWRENCE *Gen. Treat. Cattle* 46 Are we to conjecture, that the Herefords owe their bald face to the smoky white faces of the red cattle of Montgomery? *Ibid.* 48 The distinguishing qualities of Hereford oxen. **1805** J. DUNCUMB *Agric. Hereford* 118 The Leicestershire cow..will never breed a rival to the Herefordshire ox... There is an extraordinary difference between the weight of a Herefordshire cow and the ox bred from her. **1807** R. SOUTHEY *Lett. from Eng.* (1951) lxxi. 452 Here people will apply to..a Herefordshire bull, the same epithets of praise..which a sculptor would bestow upon the Venus de Medici. **1834** W. YOUATT *Cattle* iii. 31 The Herefordshire white-faced breed... The Hereford oxen are considerably larger than the North Devons... The old Herefords were brown or red-brown. **1836** *Penny Cycl.* VI. 379/2 The Herefordshire oxen are best suited to the rich pastures of their native county. **1844** H. STEPHENS *Bk. Farm* III. 1273 Another breed ..is the Hereford, which has long been famed for its excellent steers. **1862** J. C. MORTON *Farmer's Cal.* 578 The Herefords, another leading breed of cattle, characterized by red body and white or mottled face, come almost as early to maturity as the short-horn. **1875** *Encycl. Brit.* I. 388/1 The Hereford is the breed which in England contests most closely with the short-horns for the palm of excellence. **1889** R. WALLACE *Farm Live Stock* (ed. 2) vi. 74 *A Cross*, bred from a Galloway cow by a bull of one of the favourite or improved breeds, such as short-horn or Hereford, is usually an excellent butchers' animal... The Hereford crosses are..polled, with a white head and a red or dark grey body. **1931** *Times Lit. Suppl.* 20 Aug. 625/3 The fat Herefords that graze lazily in English meadows. **1960** *Farmer & Stockbreeder* 15 Mar. 95 A yardful of well-finished Hereford-cross beef. **1973** *Country Life* 8 Mar. 654 Cattle Societies... The Hereford herd book society.

herem, var. *CHEREM.

Herero (herēᵊro): One of a negroid people in South-West Africa, also called *Cattle Damara*; also their Bantu language, called *Otshi-Herero* by the Hereros themselves.

1862 W. H. I. BLEEK *Comp. Gram. S. Afr. Langs.* I. 8 The language of *Benguela*..is quite distinct..from the Herero species. *Ibid.* ii. 22 Kongo *bhobha* (to speak, talk) is Herero *pópa* (to warn, persuade). **1868** F. W. KOLBE (*title*) A brief statement of the discovery of the laws of the vowels in Herero. **1880** *Encycl. Brit.* XI. 731/2 *Hereroland*..so called from the native race known to the Namaqua as Herero and to the Cape colonists as Damara. **1904** *Westm. Gaz.* 30 Jan. 2/2 The Kaffirs are joining the Hereros in attacking the Germans. **1952** L. G. GREEN *Lords of Last Frontier* (1953) ii. 18 Midway in the sixteenth century those forceful Bantu, the Hereros, reached the end of a long migration, and settled in the cattle country. *Ibid.* xv. 148 Hereros are tall people, especially the men... Features are almost European, with well-shaped noses, high foreheads and oval faces. **1961** L. VAN DER POST *Heart of Hunter* I. iii. 64 The black man, the Herero, the Bastaards, had kraals and lands of their own. **1972** *Cape Times* 10 Mar. 1/1 Mr. John Garvey Muundjua, a Herero who is the South West Africa National Union's 'Secretary for External Affairs'. **1973** *Times* 8 Mar. 6/2 Herero tribal leaders have rejected an invitation from the Government to appoint representatives to sit on the advisory council.

heresy. Add: **4.** *heresy-hunter.*

1765 A. MACLAINE tr. *Mosheim's Eccl. Hist.* (1844) I. xiii. 344 This new set of heresy-hunters. **1831** CARLYLE in *Edin. Rev.* LIII. Mar. 168 Scenting out Infidelity with the nose of an ancient Heresy-hunter, though for opposite purposes. **1902** *Westm. Gaz.* 7 May 12/1 A proceeding quite in harmony with the usual methods of heresy-hunters. **1906** *Daily Chron.* 16 Oct. 3/3 The heresy-hunter made him his quarry.

herewith, *adv.* Add: **B.** *adj.* Accompanying.

1917 'CONTACT' *Airman's Outings* p. xii, It was a squadron that possessed..the herewith testimonial.

Hering (hē·riŋ, he·-). The name of Karl Ewald Konstantin *Hering* (1834–1918), German psychologist and physiologist, used *attrib.* or in the possessive to designate certain physiological effects observed, and principles enunciated, by him.

1891 M. FOSTER *Textbk. Physiol.* (ed. 5) IV. III. iii. 1232 Hering's theory attempts to reconcile..the various facts of colour vision with the supposition that we possess ..six fundamental sensations. **1902** *Encycl. Brit.* XXI. 749/2 Hering's theory proceeds on the assumption of chemical changes in the retina under the influence of light. *Ibid.*, Hering's theory accounts satisfactorily for the formation of coloured after-images. **1911** GOULD *Pocket Med. Dict.* (ed. 6) 415 *Hering's Law.* The distinctness or purity of any sensation or conception depends upon the proportion existing between their intensity and the sum total of the intensities of all simultaneous sensations and conceptions. **1934** H. C. WARREN *Dict. Psychol.* 123/2 *Hering after-image*, the first positive after-image, or after-sensation, which occurs following a brief light-stimulus. *Ibid.*, *Hering illusion*, an illusion, or distorted perception of visual form, observed when a number of lines radiating from a point are crossed by two parallel lines on opposite sides of the point and equidistant from it; the parallel lines appear to bow outward, i.e. away from the central point. *Ibid.*, *Hering theory of vision*, the theory..according to which visual sensations are due to three pairs of antagonistic processes in the optic system. **1971** *Jrnl. Gen. Psychol.* LXXXIV. 164 The Zöllner and Hering illusions are examples of phenomena in which the overall pattern dominates the geometry and leads to the perceptual distortion of straight lines into curved lines.

herit. Delete †*Obs.* and add *rare.* (Later examples.)

1876 'INDIAN CHAPLAIN' *Mahabuleshwar Hills* 40 Each hallow'd spirit Hath gone the land of life and love to herit. **1922** JOYCE *Ulysses* 401 To herit the tradition of a proper breeding.

hermeneut and related words: (Earlier and later examples.)

1678 R. BURTHOGGE *Organum Vetus & Novum* 70 Ratiocination Speculative, is either Euretick or Hermeneutick, Inventive or Interpretative. **1906** *Westm. Gaz.* 19 Jan. 4/2 More than could be expected from the most practical and most skilful hermeneutist. **1965** *Listener* 3 June 820/1 The word-event which once happened and in happening became the text..must become a word-event again through the interpretation of the 'hermeneut', who transposes it into the thought-mode of his own day. **1967** J. MACQUARRIE *God-Talk* vii. 148 We could say that history is the hermeneutic of historical existence, or even that physics..is the hermeneutic of nature. **1968** C. E. BRAATEN *History of Hermeneutics* vi. 138 Preaching today is the goal of exegesis and hermeneutical reflection. **1972** *Times Lit. Suppl.* 20 Oct. 1262/1 The quest for a hermeneutics.

Hermesianism. (Earlier example.)
1847 J. H. NEWMAN *Let.* 10 Jan. (1962) XII. 8, I dread Hermesianism.

hermetic, *a.* **2. a, b.** *fig.* and *transf.* Further examples, esp. in senses 'unaffected by external influences, recondite'.
1954 *Ann. Reg.* 1953 204 From the published texts [of the Spanish agreement with America] might be inferred the existence of certain secret clauses, guardedly described as 'hermetic'. 1965 *Listener* 2 Sept. 351/2 Poems ranging from the hermetic to the directly obscene. 1966 *Ibid.* 5 May 663/2 It is hermetic music and solely concerned with problems of formal design, structure, and instrumental texture. 1972 *Times Lit. Suppl.* 25 Aug. 984/3 The poetry which he had written between the wars was considered hermetic by some.
3. a. (Examples.)
a 1891 J. R. LOWELL *Old Eng. Dramatists* (1892) 17 They [*sc.* the Elizabethans] had the Hermetic gift of buckling wings to the feet of their verse. 1951 AUDEN *Nones* (1952) 60 In fake Hermetic uniforms. *Ibid.* 61 Keep well the Hermetic Decalogue.

hermetism (hɜ̄·ɹmĭtiz'm). [f. HERMET(IC *sb.* + -ISM.] Hermetic or theosophical philosophy; hermetics. So **hermeto·logist**, a hermetist.
1894 G. S. HALL in *Forum* Aug. 719 (Cent. D. Suppl.), Its teachings..make the doctrine of sin as vital as with the most ardent of the old hermetologists. 1898 H. C. BOLTON in *Smithsonian Rep. 1897* 213 Traditions of the kabala, the mysteries of hermetism, and the practice of transmutation. 1927 *Contemp. Rev.* July 59 M. Valéry's poetics have been accused of hermetism and of preciousness.

hermit, *sb.* Add: **3.** *spec.* of a sheep. Also *hermit sheep*. *Austral.* and *N.Z.*
1874 A. BATHGATE *Colonial Experiences* xv. 212 A sheep which has been badly tutued and recovers, loses its gregarious habits, and becomes what the shepherds call a 'hermit'. 1917 E. GLEN *Six Little N.Zers* vii. 95 They brought in a 'hermit' sheep which lived by itself, and had been overlooked in the last muster. 1933 L. G. D. ACLAND in *Press* (Christchurch, N.Z.) 28 Oct. 15/7 *Hermit*, a single sheep which for some reason takes to living by himself, away from the mob. 1966 G. W. TURNER *Eng. Lang. Austral. & N.Z.* viii. 165 Sheep that recovered [from eating tutu] sometimes became hermit sheep, losing their gregarious habits.

hermithood (hɜ̄·ɹmithud). [f. HERMIT *sb.*: see -HOOD.] The state or condition of a hermit.
1915 C. C. MARTINDALE *In God's Army* I. ii. 62 Vocations to hermithood are few. 1938 *Times Lit. Suppl.* 19 Nov. 745/2 Sentenced to a compulsory hermithood in a hideous African desert, the convict contrives to be brought back to Constantinople.

Hermitian (hɜɹmi·ʃăn), *a.* *Math.* Also herm-, -ean. [ad. F. *hermitien* sb. and adj. (L. Autonne 1902, in *Rendiconti d. Circolo matem.* XVI. 104), f. the name of C. *Hermite* (1822–1905), French mathematician: see -IAN.] Applied to a matrix in which pairs of elements symmetrically placed with respect to the principal diagonal are complex conjugates, and to other quantities (such as a self-adjoint operator or transformation, or a bilinear form) whose coefficients or elements form such a matrix; *Hermitian conjugate*, the complex conjugate of the transpose of a matrix.
1927 *Proc. R. Soc.* A. CXIII. 628 If *b* commutes with H, the new matrix..will be a diagonal matrix, and if in addition the elements of the matrices *b* and b^{-1} satisfy the condition that $b(\alpha'\alpha'')$ and $b^{-1}(\alpha''\alpha')$ are conjugate imaginaries, each matrix G will be Hermitian when the corresponding matrix *g* is Hermitian. 1935 P. A. M. DIRAC *Princ. Quantum Mech.* (ed. 2) ii. 29 These are just the relations required for the matrix formed by the α_{rs} to be what is called a Hermitian matrix. A linear operator that satisfies this condition may conveniently be called a Hermitian operator. 1940 E. T. BELL *Devel. Math.* xiv. 285 From Hermite's arithmetical theory of these forms..evolved the extensive theory of Hermitian forms and matrices, which after 1925 became familiar to physicists through the revised quantum theory. 1951 J. A. SCHOUTEN *Tensor Analysis for Physicists* x. 241 For co- or contravariant hermitian tensors the following theorem holds. 1961 POWELL & CRASEMANN *Quantum Mech.* ix. 295 Other matrices related to $A = (a_{ij})$, which occur frequently in the theory, are the transpose of *A*,.. and the Hermitian conjugate of *A*, $A^{\dagger} = (a^{*}_{ji})$. 1965 *Czech. Math. Jrnl.* XV. 37 Let *A*, *H* be *n*-by-*n* hermitian matrices. 1967 G. G. HALL *Appl. Group Theory* iii. 33 A less obvious example is that complex Hermitean 2 × 2 matrices form a four-dimensional real Euclidean space. 1969 T. F. JORDAN *Linear Operators Quantum Mech.* ii. 23 A Hermitian operator *A* is characterized by the equation $(\phi, A\psi) = (A\phi, \psi)$ or $(\phi, A\psi) = (\psi, A\phi)^{*}$ holding for all vectors ψ and ϕ. 1973 *Physics Bull.* July 449/2 The analogous results for hermitean matrices.

herniation (hɜɹniē·ʃən). *Path.* [f. HERNIA + -TION.] Protrusion as in a hernia.
1897 *Allbutt's Syst. Med.* III. 712 A series of herniations of the intestinal walls. 1962 *Lancet* 22 Dec. 1332/2 Powell and Johnstone have reported the case of a mother who had taken phenmetrazine during two of her pregnancies and produced both times an infant with a defect of the left diaphragm and herniation of abdominal organs into the thoracic cavity.

hernio-. Add: **herniorrha·phy** *Surg.* [*-RRHAPHY], the operation of repairing a hernia and suturing the opening.
1919 *Surg., Gynec. & Obstetr.* XXIX. 201 Treatment of the sac in herniorrhaphy... High ligation and amputation of the hernial sac is a point..on which most surgeons are agreed. 1962 *Lancet* 27 Jan. 221/1 This problem engaged our interest in 1959 when we reviewed 762 patients who had had herniorrhaphies performed in the previous ten years. 1963 *Ibid.* 19 Jan. 133/1 A man, aged 62, who had a large indirect inguinal hernia with an associated undescended testis, underwent right inguinal herniorrhaphy and orchidectomy.

hero, *sb.* Add: **5.** *hero-cult*, *-fantasy*, *-figure*, *-image*; **hero sandwich** *U.S. slang*, a very large sandwich; also *ellipt.* **hero.**
1945 KOESTLER *Yogi & Commissar* III. ii. 197 His famous speech on the 24th anniversary of the Soviet Revolution..which started the new hero-cult. 1960 C. DAY LEWIS *Buried Day* 22 Nor..did I indulge in the hero-fantasies of the only child. 1958 *Punch* 23 July 102/3 But surely his agent..built up an image of this untalented scallywag as a top-line genius and hero-figure for millions who little guessed? 1963 *Times Lit. Suppl.* 11 Jan. 29/3 The bodily posture adopted by Japanese hero-figures. 1951 M. MCLUHAN *Mech. Bride* 63/1 Li'l Abner himself is a cluster of the swarming hero images. 1955 *Sat. Even. Post* 1 Jan. 16 When he got back to Brooklyn, the first thing he asked for was an Italian hero sandwich.., two inches thick and eighteen inches long. 1957 *Britannica Bk. of Year* 512/1 *Hero-sandwich*, a sandwich made with a whole loaf of bread. 1959 R. CONDON *Manchurian Candidate* v. 91 She wiped her mouth with the back of her hand like a labourer who had just finished a hero sandwich and a bottle of beer. 1963 M. SCHERF *Death & Diplomat* (1964) v. 62 They have hot and cold heros next door. 1972 *New Yorker* 1 July 21/2 An office just above a hero-sandwich shop.

Herodotean (hĭrǫdǫ̆ti·ăn), *a.* [f. the name of *Herodotus* (Gr. Ἡρόδοτος), Greek historian of the fifth century B.C.: see -AN.] Of, pertaining to, characteristic of, or mentioned by Herodotus.
1856 K. H. DIGBY *Lover's Seat* II. xiii. 67 They are remarkably Herodotean in their style of operation. 1857 DUFFERIN *Lett. High Lat.* 54 The Herodotean work of Sturleson. 1872 W. MINTO *Man. Eng. Prose Lit.* II. vi. 413 A Herodotean knack of giving numerical measures of extent. 1881 *Athenæum* 2 Apr. 456/2 Some of the reports which he received he recounts with Herodotean incredulity. 1883 *Macm. Mag.* May 67/2 This is what I mean by calling him Herodotean. Nothing was too small nor too apparently remote from the main studies of his life to escape him or be without interest for him.

heroic, *a.* Add: **2.** *heroic age*: also *transf.*
1897 W. P. KER *Epic & Romance* i. 7 What the 'heroic age' of the modern nations really was, may be learned from what is left of their heroic literature, especially from three groups or classes,—the old Teutonic alliterative poems on native subjects; the French *Chansons de Geste*; and the Icelandic Sagas. 1912 H. M. CHADWICK *Heroic Age* v. 105 This carries us back..to what we may call the Russian Heroic Age. 1927 E. V. GORDON *Introd. Old Norse* p. xxix, The Germanic heroic age of the fourth to seventh centuries. 1928 W. W. LAWRENCE *Beowulf & Epic Trad.* 24 This was the usual procedure of a minstrel of the Heroic Age, who knew of all notable men about the circle of the seas. 1948 K. MALONE in *English Studies* XXIX. 170 All these passages serve to make our hero part and parcel of the heroic age of Germanic antiquity. 1965 K. SISAM *Struct. Beowulf* 7 The doors of Heorot opened into the Heroic Age.

heroin (he·ro,in; formerly also hĭrōᵘ·in). [a. G. *heroin*, f. Gr. ἥρως HERO; said to be so derived because of the inflation of the personality consequent upon taking the drug: see -INE⁵.] A white crystalline alkaloid prepared from morphine by acetylation, which is administered usu. in the form of its hydrochloride as a hypnotic and analgesic (though many countries prohibit this medical use), and which is illicitly used as a powerful and addictive drug producing intense euphoric sensations. Also *attrib.*
1898 *Lancet* 3 Dec. 1511/1 A new hypnotic, to which the name of 'heroin' has been given, has been tried in the medical clinic of Professor Gerhardt in Berlin. According to a communication made by Dr. Strube to the *Berliner Klinische Wochenschrift* it is a product of the di-acetic ester of morphia, and it was discovered by Professor Dreser, chief of the chemical department of the Elberfeld Farben Fabriken. 1908 *Practitioner* Apr. 436 Subcutaneous injections of morphia or heroin locally. 1910 *Ibid.* Apr. 542 A sedative may be prescribed... Heroin hydrochloride is the best drug for this purpose. 1920 A. B. BAXTER *Parts Men Play* xviii, She..took to opium cigarettes, and then to heroin. She disappeared one night. 1955 *Sci. News Let.* 2 Apr. 219/2 Morphine and heroin, for example, do not give normal persons the 'kick' and pleasant sensations they are supposed to give. 1962 K. ORVIS *Damned & Destroyed* iv. 30 One would be compelled to rush wildly off in search of heroin-relief. *Ibid.* xi. 76 A heroin-party. 1968 *Lancet* 7 Dec. 1239/2 The cost of treating heroin addiction is particularly great. 1968 *Even. News* 11 Dec. 13/1 The 'erratic and unsatisfactory life' of a 19-year-old heroin addict ended when he took a barbiturate overdose. 1969 *Times* 9 July 2/3 (*headline*) Heroin 'black market' in West End. *Ibid.* 2/4 Scotland Yard's Drug Squad..have been raiding premises..for a gang of heroin 'pushers'. 1972 *Listener* 27 Jan. 125/3 The Marseilles–New York heroin traffic.

heroine, *sb.* Add: **5.** *heroine-worship*, *-worshipper* (cf. HERO-WORSHIP).
1900 *Westm. Gaz.* 1 May 4/2 Mrs. Cock has succeeded in giving a true and striking portraiture, without falling into mere heroine-worship. 1916 A. BENNETT *Lion's Share* xxiii. 165 'Jenny!' Audrey protested, full of heroine-worship. 1943 BEERBOHM *Lytton Strachey* 11 He was not a hero-worshipper, or even a very gallant heroine-worshipper. 1970 R. RENDELL *Guilty Thing Surprised* ix. 111 It was a case of heroine worship on one side and a sort of flattered acceptance on the other.

herola (herōᵘ·lă). Also **hirola.** [Galla name.] A small, rare antelope, *Damaliscus hunteri*, native to Kenya and Somalia, and more frequently called Hunter's hartebeest.
1894 H. C. V. HUNTER in Sclater & Thomas *Bk. Antelopes* I. 57 Their Galla name is 'Herola', not 'Horonta'. 1915 *Chambers's Jrnl.* Nov. 700/2 Then follows the curious Hunter's hartebeest or herola, having horns which somehow suggest a connecting link with the fleet and supremely graceful impala. 1957 *Encycl. Brit.* II. 22/2 In *Damaliscus*, the bastard hartebeeste,..the horns arise directly from the head. The hirola or Hunter's hartebeeste (*D. hunteri*) has a white chevronlike marking on the forehead; it is found only on the north side of Tana valley in Jubaland, east Africa.

heron. Add: **2.** *heron pie*; *heron-billed*, *-built*, *-feathered*, *-haunted* adjs.
1932 W. B. YEATS *Words for Music* 12 The heron-billed pale Cattle-birds. 1851 H. MELVILLE *Moby Dick* III. xxii. 153 Tall, heron-built captains. 1935 C. DAY LEWIS *Time to Dance* 21 The heron-feathered sky. 1903 *Daily Chron.* 10 Dec. 3/2 It [*sc.* a mist]..hid all the heron-haunted flats and marshes. 1723 J. NOTT *Cook's & Confect. Dict.* 32 H, To make a Hern Pye. 1963 V. CRONIN *Comp. Guide Paris* iv. 67 Where else but in Paris would a king raise a restaurateur to the nobility simply because he enjoyed his heron-pies?

heroon (hĭrōᵘ·ǫn). *Gr. Antiq.* Pl. **-a.** Also **herōon, heroum.** [L. *heroum*, a. Gr. ἡρῷον, f. ἡρώιος of a hero, f. ἥρως HERO.] A sepulchral monument in the form of a small temple; orig. a temple or sanctuary dedicated to a hero, often over his supposed tomb.
1775 R. CHANDLER *Trav. Asia Minor* 111 Searching about, we found..an inscription, which has belonged to an heroum or sepulchre. 1820 T. S. HUGHES *Trav. Sicily* I. 298 The Heroa of Theseus and Pirithous. 1883 A. S. MURRAY *Hist. Greek Sculpture* II. 289 The herōon at Gjölbaschi in Lycia. 1904 T. R. GLOVER *Stud. Virgil* iv. 89 A herôon of Aeneas in Ambracia. 1950 *Antiquity* XXIV. 131 Dyggve himself has excavated such a Heroon at Kalydon.

hero-worship, *v.* Delete (*nonce-wd.*) and add examples. Also **he·ro-wo:rshipful** *a.*
1914 G. B. SHAW *Pygmalion* (1916) 195 The weak may not be admired and hero-worshipped; but they are by no means disliked and shunned. 1918 BEERBOHM *And even Now* (1920) 203 The hero-worshipful gaze. 1936 N. MARSH *Death in Ecstasy* vii. 86 Maurice hero-worshipped Father Garnette. 1948 'J. TEY' *Franchise Affair* xix. 224 The centre of an adoring family—secure, loved, hero-worshipped. 1960 C. DAY LEWIS *Buried Day* i. 21, I hero-worshipped him.

herpes. Add: **1.** Now recognized as a group of virus diseases, the chief of which are **herpes si·mplex,** ordinary or 'simple' herpes (as contrasted with *herpes zoster*), distinguished as *herpes facialis*, *genitalis*, *labialis*, etc., according to the part of the body affected, and caused by *Herpesvirus hominis*; **herpes zo·ster,** shingles, caused by *H. varicellæ*.
1807 MORRIS & KENDRICK *Edin. Med. & Physical Dict.* II, Herpes zoster. 1813 T. BATEMAN *Pract. Synopsis Cutaneous Dis.* vi. 233 The Herpes labialis..occurs most frequently in the course of diseases of the viscera. 1886 FAGGE & PYE-SMITH *Princ. & Pract. Med.* II. 670 His [*sc.* Willan's] species of herpes included (1) *Herpes zoster* or zona... (2) *Herpes circinatus*, which..is now classed with *Tineæ*. 1907 *Jrnl. Amer. Med. Assoc.* 2 Mar. 746/2 (*heading*) The nature of herpes simplex. *Ibid.* 750/1 Herpes zoster and herpes simplex—both facial and genital varieties—while not clinically identical, are closely related. *Ibid.* 749/2 Another form of simple herpes..is herpes genitalis. 1946 G. C. ANDREWS *Dis. Skin* (ed. 3) xx. 537 Certain types of herpes zoster have received descriptive terms, the more important of which is herpes zoster frontalis, which is an involvement of the area supplied by the upper branches of the trigeminal nerve. 1962 *Lancet* 13 Jan. 105/1, I have often been impressed by the frequent association of herpes zoster with stress and anxiety, especially in the elderly. 1967 A. C. ALLEN *Skin* (ed. 2) viii. 335/2 Herpes simplex (fever blisters, cold sores, herpes facialis, herpes labialis) is..a mild, although recurrent, eruption.
b. *herpes virus* (also as one word), any of a group of related viruses that includes those causing shingles and chicken-pox, esp. *Herpesvirus hominis*, the cause of herpes simplex.
1925 *Amer. Jrnl. Path.* I. 337 (*heading*) Studies on filterable viruses... Cultivation of herpes virus. *Ibid.*, We have used herpes virus and a note on the cultivation of this virus is given below. 1955 *Sci. Amer.* May 33/1

The benign but recurrent lesions known as fever blisters or cold sores, caused by the herpes virus. 1968 A. ROOK et al. *Textbk. Dermatol.* I. xxiv. 763/2 The particles of herpesvirus are first found in the nucleus and later appear in the cytoplasm from which the virus is gradually released with destruction of the cell.

Herr (hęr). Pl. **Herren.** [G., master, lord; Mr. Cf. HER, HERE *sb.*] The German equivalent of Mr.; a German gentleman.
1653 R. CREIGHTON *Let.* 28 Jan. in M. M. Verney *Mem. Verney Fam.* (1894) III. iii. 81 Herr Skatt hath undertaken the cure. 1828 T. CARLYLE *Let.* 16 Apr. (1887) 81 How is it that the Author of *Faust* and *Meister* can *tryste* himself with such characters, as 'Herr ——' (the simplest and stupidest man of his day, a Westmoreland Gerundgrinder and *Cleishbotham*)? 1849 THACKERAY *Pendennis* I. xiv. 124 Herr Garbage's lions and tigers had drawn for a little time, until one of the animals had bitten a piece out of the Herr's shoulder. 1879 C. M. YONGE *Magnum Bonum* (1882) xxi. 294 He..was a decided favourite with Fräulein Rosalie, who would do anything for her dear young Herr. 1940 W. S. CHURCHILL *Speech* 21 Oct. in *Second World War* (1949) II. xxvi. 452 But Herr Hitler is not thinking only of stealing other people's territories. 1954 'M. COLES' *Not for Export* iii. 34 Life was difficult for the poor in Berlin, as the *Herr* would understand. 1959 *Times* 30 Jan. 11/2 The possibility that Herr Alfred Krupp may be allowed to retain his main holdings in coal, iron, and steel. 1967 'G. CARR' *Lewker in Tirol* iii. 36 Your glass, *mein Herr*. *Ibid.* 53 It was in the corner of the big bar..the guides..meet their *Herren* there. 1971 W. LAQUEUR *Dict. Politics* 207 Brandt's attempts..incurred the wrath of Herr Ulbricht.

‖ **Herrenvolk** (hę·rənfǫlk, -fōᵘk). Also **h-.** [G., master-race.] The Nazi conception of the German people as born to mastery; also *transf.* as an appellation of other 'superior' groups. Also *attrib.* or as *adj.*
1940 *New Statesman* 17 Feb. 198/1 Show how German industry can enjoy prosperity without rearmament and no German worker will accept the starvation wages of a *Herrenvolk* pursuing the mirage of a closed imperialism. 1941 *Amer. Guardian* 15 Oct. 6/1 Hitler's race program was stolen from the White Trash Herrenvolk of Dixie. 1941 M. TREADGOLD *We couldn't leave Dinah* xii. 184 The representative of the Herrenvolk nodded. 1941 H. G. WELLS *You can't be too Careful* v. ii. 245 A people.. who were united in their dislike to the German *herrenvolk*. 1941 D. WILSON *Germany's 'New Order'* 22 A *Herrenvolk* to dominate other peoples. 1943 W. S. CHURCHILL *End of Beginning* 103 Forms of warfare which, according to the German view, should be the strict monopoly of the *Herrenvolk*. 1944 G. B. SHAW *Everybody's Political What's What?* x. 82 Nations each of which regards itself as The Chosen Race or *Herrenvolk* appointed by God to own and rule all the others. 1944 H. G. WELLS *'42 to '44* 187 The evil suggestion of inherently inferior peoples, predestined hewers of wood and drawers of the water for some imaginary *Herrenvolk*. 1947 *Penguin New Writing* XXXI. 157 He referred to them [*sc.* Sinhalese] in their presence as 'wogs'..and forced them to admit his herrenvolk creed. 1958 *Listener* 9 Oct. 566/2 The *Herrenvolk* ideas, against which so many of us..fought and suffered during the seven terrible years of the war. 1960 *Guardian* 28 June 10/5 A.. Portuguese..cursing an African..in that unmistakable way which those who know Africa cannot fail to recognise—the herrenvolk attitude. 1964 *Punch* 1 July 31/3 Brutality and *herrenvolk* arrogance.

herring. Add: **1. b.** *fat herring* (*FAT *a.* 2 f).
2. Phrase (local): *every herring should hang by its own head, gills, neck, tail.*
1639 J. CLARKE *Paroemiologia Anglo-Latina* 20 Every herring must hang by th'owne gill. 1672 W. WALKER *Paræmiologia Anglo-Latina* 23/1 Every herring must hang by its own neck; gill. 1694 T. D'URFEY *Comical Hist. D. Quixote* I. III. ii. 32 Let not the fault of the Ass be laid upon the Pack-Saddle, every Herring must hang by its own tail. 1721 J. KELLY *Compl. Coll. Scot. Prov.* 240 Let every Herring hing by its own Head. 1818, 1824 [in Dict.]. 1890 HALL CAINE *Bondman* II. ii, Adam, thinking as little of pride, said No, that every herring should hang by its own gills.
3. a. *herring-spawn* (examples).
1906 *Daily Colonist* (Victoria, B.C.) 11 Jan. 3/4 The practice of using herring spawn..has been the habit of the Indians all over British Columbia for many years. 1972 *Guardian* 1 Apr. 7/8 Herring spawn is relatively vulnerable and relatively few survive.
b. **herring choker** *slang*, (*a*) *Canad.* a nickname for a native or inhabitant of the Maritime Provinces; (*b*) *U.S.* a Scandinavian.
1899 *Yarmouth Telegram* (Nova Scotia) 20 Oct. 1/1, I am down among the 'herring chokers' and 'blue noses' for a few weeks. *Ibid.*, Happy the wearied globe-trotter and denizen of the 'herring chokers' of Nova Scotia. 1944 H. WENTWORTH *Amer. Dial. Dict.* 290/1 *Herring-choker*. 1. A Prince-Edward-Islander, or native of any of the Provinces 'down east'. 2. A Scandinavian. 1954 *Fundy Fisherman* (Black's Harbour, N.B.) 3 Mar. 4/4 These Herring Choker senators point out that Duplessis has already peddled a lot of horse power to Ontario and indeed is selling plenty to Premier Frost et al. right now. 1968 *Word Study* Dec. 3/2 The Roman Catholic is frequently called a *fish*, short for *fish-eater*... Similar is the term *herring-choker*, another name for a Scandinavian.

herring-bone, *sb.* Add: **1. b.** *pl.* Small cirrocumulus clouds (cf. *mackerel sky*).
1905 *Westm. Gaz.* 22 Apr. 4/2 The sky was sunny, but mottled in the north-west with 'herring-bones', which prophesied wind.
c. *pl.* The herring-bone-line hatching used to represent mountains on a map.
1900 E. T. FOWLER *Farringdons* v, 'I see. And please what are the mountain-ranges that you are drawing now?' 'These,' replied Elisabeth, covering her map with herring-bones, 'are your scruples.'
d. *Furniture.* (See quot. 1952.)
1937 *Burlington Mag.* July 45/2 All the ornament required is to be found in the finely figured wood, the craftsmanship displayed in the herringbone and cross-banded borders. 1952 J. GLOAG *Short Dict. Furniture* 285 *Herringbone*, patterns in the form of a herring bone were used on the banding of drop fronts on early 18th century walnut furniture. 1960 H. HAYWARD *Antique Coll.* 142/1 *Herring-bone banding*, a decorative veneer border much used on walnut furniture of the late 17th and early 18th cent. consisting of two narrow strips of veneer laid together diagonally forming a pattern resembling a herring-bone or feather [etc.].
e. *Skiing.* (See quot. 1924.) Also *attrib.*
1904 E. C. RICHARDSON *Ski-Running* 35 A short, steep bit of slope may be overcome by the methods illustrated in Figs. 12 and 13, but both are fatiguing,.. especially the 'herring bone'. 1924 *Tourist* Winter Sports No. 12/1 *Herring-bone*, a method of climbing without kick-turns which leaves tracks like the well-known stitch. 1972 M. YORKE *Silent Witness* iii. 58 [Her] muscles always screamed after..side-stepping or herring-bone climbing.
2. e. *herring-bone coralline* (see quots.).
1755 J. ELLIS *Ess. towards Nat. Hist. Corallines* 17 Herring-bone Coralline. This Coralline is often found sticking to Oysters as they are brought to the London Market, during the Winter-season. 1850 A. PRATT *Common Things Sea-side* 258 The Herring-bone Coralline ('Halecium halecinum') which grows on stones and shells, in the deep parts of the sea. 1915 E. R. LANKESTER *Divers. Nat.* 97 Very minute jelly-fish,..[which] originate as buds from small branching polyps, one kind of which is common on oyster-shells, and is called 'the herring-bone coralline'.
f. Designating a type of milking parlour in which the stalls are arranged in a herring-bone pattern.
1957 *Farmer & Stockbreeder* 9 Apr. 45/1 The yard and parlour system obviously attracts those changing over to dairying and the main interest centred on the..herring-bone parlour design. 1963 *Ibid.* 5 Mar. 33/2 (Advt.), Gascoigne milking systems. Herringbone parlour for batch milking. 1970 *Times* 9 Mar. 13/5 Another innovation costs £120 a milking stall, or £600 for this item alone in the five-point herringbone parlour.

herring-bone, *v.* Add: **3.** To make (a wall, floor, etc.) of herring-bone work. Hence **he·rring-boned** *ppl. a.*
1907 W. DE MORGAN *Alice-for-Short* ix, A 'erring-boned brick floor.
4. *Skiing.* (See *HERRING-BONE *sb.* 1 e.) Hence **herring-boning** *vbl. sb.*
1904 E. C. RICHARDSON *Ski-Running* 35 (*caption*) Herring-boning. 1914 *Queen* 17 Jan. 128/2 He climbed very quickly, looking uncommonly like a monkey up a stick as he herring-boned in jerks up the slope. 1919 *Century Mag.* May 43/1 'You're herring-boning!' shouted Dean. 'Don't herring-bone! We'll come to that later.' 1946 H. CROOME *Faithless Mirror* 55 Half a dozen boys herringboning up from the road. *Ibid.* 67 Half a dozen skiers herringboned up the trail towards them.

hers, *poss. pron.*¹: see *HIS *absolute poss. pron.* b.

Herschel. Add: **2.** Used *attrib.* or in the possessive to designate certain phenomena or principles discovered by or related to the work of Sir William or Sir John *Herschel* (see HERSCHELIAN *a.*), as *Herschel('s) condition*, a condition that must be satisfied by a system of lenses if spherical aberration is to be absent in the images of two neighbouring objects; *Herschel effect*, the partial destruction of the latent image on a photographic negative by the action of infra-red radiation; *Herschel's fringes*, spectral fringes observed in a prism standing on a plane mirror at the line of separation between the brighter field, due to incident light internally reflected by the prism, and the darker field, due to light externally reflected by the mirror, first observed by Sir William Herschel.
1890 T. PRESTON *Theory of Light* viii. 153 (*heading*) Herschel's fringes. 1909 *Jrnl. Chem. Soc.* XCVI. II. 141 Warnerke's modification of the Herschel Effect and the Preparation of the Substance of the Latent Image. 1910 J. P. C. SOUTHALL *Princ. & Methods Geom. Optics* xii. 394 It has..been suggested that the other requirement should be the so-called Herschel-Condition; that is, that the function $\phi_1 Z_1 + \phi_2 Z_2$ should vanish not only for the particular value of x_1 but also for object-points on the axis very near to the point M_1 to which the value x_1 belongs. 1932 HARDY & PERRIN *Princ. Optics* vi. 98 The condition that $n \sin \frac{1}{2}\theta/n' \sin \frac{1}{2}\theta' =$ a constant must be satisfied for all values of θ. This is commonly called the Herschel condition although it had not been formulated mathematically when Herschel constructed his objective. *Ibid.* xi. 233 This Herschel effect has a very important significance in connection with theories of the latent image but has not been found very useful as a technique for infrared photography. 1957 R. S. LONGHURST *Geom. & Physical Optics* xiv. 280 Herschel's condition is the condition that must be satisfied if the system is to form, simultaneously, a good image of a neighbouring axial object point.

Hertz (hɜ̄ɹts, ‖ hęrts). **1.** The name of H. R. *Hertz* (1857–1894), German physicist, used *attrib.* to denote apparatus used or invented, phenomena discovered, or concepts elaborated by him.
1890 *Nature* Feb. 368 By separating the coats of the jar as far as possible we get a typical Hertz vibrator. 1892 O. LODGE *Mod. Views Electr.* 361 Hertz waves can get through deal doors and stone walls. 1898 *Science Siftings* XIV. 117/2 A Hertz-wave 'detector'. 1898 *Westm. Gaz.* 12 May 9/2 Experiments in Hertz-wave space telegraphy. 1902 *Encycl. Brit.* XXXIII. 230/2 This aerial being used as a Hertz oscillator or radiator. 1962 CORSON & LORRAIN *Introd. Electromagn. Fields* xiii. 488 The Hertz vector *Π* is defined by the equation

$$\mathbf{A} = \epsilon\mu \frac{\partial \Pi}{\partial t}.$$

2. *Physics.* (Usu. *hertz.*) A unit of frequency, equal to one cycle per second. Abbrev. Hz.
1928 *B.B.C. Handbk.* 270/1 *Hertz*, a term sometimes used to designate frequency, meaning one cycle per second. 1934 *Electr. Engin.* (U.S.) LIII. 403/2 It was proposed by the Italian committee that the name *hertz* be given to the unit of frequency—one cycle per second... The *hertz* is used almost universally in German technical literature. 1967 *Electronics* 6 Mar. 325/1 Markus continues to use 'cycles per second' instead of 'hertz'. 1967 *New Scientist* 5 Oct. 6/2 The frequency band between about 10 megahertz and 20 gigahertz is technically suitable for space communication.

Hertzian (hɜ̄·ɹtsiăn), *a.* [f. the name *Hertz* (see prec.) + -IAN.] Of or pertaining to Hertz or to the phenomena discovered by him.
Hertzian telegraphy: wireless telegraphy. *Hertzian waves* (see WAVE *sb.* 5 a).
1890 A. E. BOSTWICK in *Appleton's Ann. Cycl.* 716/1 (Funk), Trouton..found that glass absorbs Hertzian vibrations with comparative rapidity. 1897 RUTHERFORD in *Phil. Trans. R. Soc.* A. CLXXXIX. 1 They were also found to be a sensitive means of detecting electrical radiation from Hertzian vibrators at long distances from the vibrator. *Ibid.* 8 A detector for electrical waves was devised, which proved to be a sensitive means of detecting Hertzian waves. 1898 *Daily News* 3 May 5/3 The General Post Office..is also carrying out experiments in wireless Hertzian telegraphy. 1900 *Jrnl. Soc. Arts* XLVIII. 788/1 The sensation created in 1897 by Mr. Marconi's application of Hertzian waves. 1900 J. J. FAHIE *Hist. Wireless Telegr.* 197 Sir William Crookes..was..the first to distinctly foresee the applicability of Hertzian waves to practical telegraphy. 1907 J. ERSKINE-MURRAY *Handbk. Wireless Telegr.* 36 The transmitter consisted of a Hertzian oscillator placed in the focal line of a parabolic mirror. 1908 J. A. FLEMING *Elem. Man. Radiotelegr.* 132 An arrangement of two rods.. with a spark gap in the centre constitutes the simplest form of linear radiator or Hertzian Oscillator for the production of damped electro-magnetic waves. 1914 R. STANLEY *Textbk. Wireless Telegr.* 91 Using a suitable length of spark gap the discharge of this Hertzian open circuit is oscillatory. 1922 A. N. WHITEHEAD *Princ. Relativ.* xiv. 117 Now consider the internal vibration of a molecule which radiates light of period *T*..as capable of being represented as the vibration of a variable electric Hertzian doublet with this period. 1940 *Chambers's Techn. Dict.* 412/2 *Hertzian waves*, electromagnetic waves having frequencies between zero and approximately ten thousand million cycles per second; especially those towards the upper limit of this range. 1971 *Gloss. Electrotechnical Power Terms* (*B.S.I.*) III. vii. 7 *Hertzian waves*, electromagnetic waves having frequencies below those of infra-red waves.

herumfrodite, var. *HARUMFRODITE.

Herzegovinian (hɜ̄ɹtsəgǭvi·niăn), *a.* and *sb.* Also **Herzo-.** [f. *Herzegovina*: see -IAN.] (Of or pertaining to) a native or inhabitant of Herzegovina, a region to the south of Bosnia, now forming part of Bosnia-Herzegovina, a constituent republic of Yugoslavia. So **Herz(e)govine·se** *a.* and *sb.*
1876 *Fraser's Mag.* May 541/2 The Herzgovinese insurgents. 1880 *Encycl. Brit.* XI. 775/2 Much of the old Slavonic customs..still holds among the Herzegovinian Mussulmans. *Ibid.*, The Herzegovinians are tall, broad, and darker..than the Bosnians. 1900 tr. *Deniker's Races of Man* 345 The southern [Slav] group.. comprises the Herzogovinians, Bosnians, Montenegrins. 1904 M. E. DURHAM *Through Lands of Serb* 6 The Herzegovinese migrated to Montenegro. 1918 R. J. KERNER *Jugo-Slav Movement* 4 The Herzegovinian Serbs of the Narenta. 1935 HUXLEY & HADDON *We Europeans* vii. 213 The Herzegovinian type has a broad head, and is rather tall and inclined to be fair.

Heshvan, var. *HESVAN.

Hesiodic (hĭsiǫ·dik), *a.* [f. the name of *Hesiod* (Gr. Ἡσίοδος), Greek poet of about the eighth century B.C. + -IC.] Of, pertaining to, or resembling the poetical style of Hesiod, or the school of poetry which followed him. Also **Hesio·dian** *a.*
1838 *Penny Cycl.* XII. 186/2 Ulrici considers..the story of Prometheus and that of the Five Ages as much altered from their original Hesiodic form. 1847 J. LEITCH tr. *Müller's Anc. Art* § 77 The Hesiodic bards

come down to about the 40th Ol[ympiad]. 1873 J. DAVIES *Hesiod & Theognis* i. 19 Under one or other of these heads it is easy to group the Hesiodic poems. *Ibid.* ii. 23 The Hesiodian rhapsodists. 1908 A. W. MAIR *Hesiod* p. xi, The Hesiodic epic is the antithesis of the Homeric. 1970 G. E. EVANS *Where Beards wag All* xx. 223 In literature there was little beyond a Hesiodic theogony or some Orphic hymns to exhibit.

hesitate, *v.* Add: **1. a.** Const. various preps.
a 1849 POE *Tales* (1884) I. 93 At the baptismal font I hesitated for a name. 1856 DE QUINCEY *Confess.* (1896) 216 This..surgeon saw no reason whatever in the simple practice of opium-eating for hesitating upon a life-insurance proposal. 1860 W. COLLINS *Woman in White* I. iv. 33, I hesitated about answering it. 1908 *Grand Mag.* Dec. 612/2 I'm only hesitating over the price.
d. *spec.* in *Dancing* (see quot. 1919).
1914 V. CASTLE *Mod. Dancing* iv. 72 In my opinion it is much better to hesitate when the music hesitates. 1919 E. SCOTT *All about Latest Dances* 87 The term 'hesitate', as regards the waltz we are now considering, implies merely that you pause, or move in what appears a faltering manner at a given juncture.
e. To move in an indecisive, faltering manner.
1908 H. G. WELLS *War in Air* v. 168 He hesitated towards the door of the cabin. 1917 T. S. ELIOT *Prufrock* 27 Regard that woman Who hesitates toward you in the light of the door.
3. b. With sentence as quasi-obj.
1816 SCOTT *Tales my Landlord* II. xii. 311 'I am not sure,' hesitated Edith. 1881 C. E. L. RIDDELL *Sen. Partner* xvi, 'It looks awful like the stuff last night,' hesitated Mr. McCullagh.

hesitation. Add: **2. b.** *hesitation-form*, a sound or form, e.g. (ə), used deliberately or accidentally when faltering or stammering in speech. So *hesitation-vowel*, etc. (Cf. *ER.)
1933 BLOOMFIELD *Lang.* xii. 186 When a speaker hesitates, English and some other languages offer special parenthetic *hesitation-forms*, as [əː] or [ɛ] in *Mr.—ah—Sniffen*. 1957 S. POTTER *Mod. Ling.* v. 122 He may even inflict indeterminate *hesitation-forms* [əː] or [ɑː] upon his suffering hearers. 1965 *Canad. Jrnl. Ling.* XI. I. 36 This is often called the 'hesitation vowel' although it is not a vowel qua vowel at all.
3. In full *hesitation waltz* or *valse*: a variety of waltz, characterized by the *hesitation step* (see quot. 1919). Also *fig.* So **hesitation-valse** (*nonce-wd.*) *v. intr.*
1914 V. CASTLE *Mod. Dancing* iv. 71 It is the Hesitation Waltz. 1918 J. M. GRIDER *Diary* 29 Jan. (1927) 71 It's a long jump from the Boston and the hesitation to the giant swing, but we had them all fox-trotting in no time. 1919 E. SCOTT *All about Latest Dances* 86 As you lift the left foot..from the floor, count a short one, two; and on three drag the other foot along the floor, bringing it over in front of the first foot, which is by this time on the floor... That is the hesitation step. 1920 A. M. CREE *Ball-Room Dancing* 41 The Hesitation Valse is a variety of the true valse that can very easily be performed once the valse is known. 1924 'L. MALET' *Dogs of Want* v. § 2 For over four hours a mixed multitude..had one-stepped, two-stepped, hesitation-valsed, and fox-trotted. 1927 *Daily Express* 31 Oct. 11/6 The foxtrot, tango, and the hesitation. 1959 *Chambers's Encycl.* IV. 364/1 During the first world war.. the occasional waltzes..were danced with a slight hesitation movement. 1972 R. MAYNE *Europeans* v. 136 Britain's slow hesitation waltz with European unity.

Hesped (he·sped). [Heb.] A funeral oration pronounced over the dead at a Jewish memorial service.
1650 E. CHILMEAD tr. *Leon Modena's Hist. Rites of Jews* v. ix. 242 If he were a Rabbine that is dead, or a person of quality, they then have Sermons, and Funeral Orations, which they call..*Hesped*, made for him. 1892 I. ZANGWILL *Childr. Ghetto* I. I. xiv. 295 Moses Ansell was at *Shool*, listening to a *Hesped* or funeral oration. 1970 L. M. FEINSILVER *Taste of Yiddish* 131 *Hesped*, eulogy.

Hesperian. B. *sb.* **2.** (U.S. examples.)
1869 *Rep. U.S. Commissioner Agric. 1868* 314 The family of skippers, Hesperians, are rather small, thick-bodied butterflies. 1881 *Papilio* I. 132 (*title*) Two new Hesperians.

Hesperid. Add: **2.** *Ent.* (Also **Hesperiid.**) One of the family Hesperidæ or Hesperiidæ of lepidopterous insects; a Hesperian butterfly; a skipper. Also *attrib.*
1889 in *Cent. Dict.* 1912 *Entomologist* XLV. 5 (*title*) A Supplementary Note on Hesperiid Classification. 1930–1 *Proc. Ent. Soc.* V. 88 Similar observations had been made on African Hesperiids. 1931 *Oxf. Univ. Gaz.* 17 June 704/1 The Hesperid butterfly *Parnara zelleri cinhara*. 1957 RICHARDS & DAVIES *Imms's Gen. Textbk. Ent.* (ed. 9) III. 560 Its larvae and pupae exhibit definite Hesperiid characters.

Hessian, *a.* and *sb.*¹ **B.** *sb.* **1.** (Earlier examples.)
1729 A. IRWIN *Let.* 1 Feb. in *15th Rep. Hist. MSS. Commission* App. VI. (1897) 56 The House..voted 30,000 English and 12,000 Hessians, which is the same number I think we had last year. 1776 *Jrnls. Continental Congress* (1906) V. 640 The Hessians, and other foreigners, employed by the King of Great Britain. 1846 *Knickerbocker* XXVII. 559 A Montreal editor [noted] the demise at that city of an old Hessian who was in Burgoyne's army when he surrendered.

hessonite (he·sǒnəit). *Min.* Also **essonite.** [ad. F. *essonit* (R. J. Haüy *Traité Pierres Préc.* (1817) 51), f. Gr. ἥσσων less + -ITE¹: so called because it is less hard and heavy than some minerals, such as hyacinth, which it resembles.] A variety of garnet containing calcium and aluminium. Also called *cinnamon-stone*, *grossularite*.
1820, 1884 [see ESSONITE]. 1892 E. S. DANA *Dana's Syst. Min.* (ed. 6) 440 Cinnamon-stone, or essonite (more properly hessonite), included a cinnamon-colored variety from Ceylon, there called hyacinth. 1895 *Amer. Jrnl. Sci.* L. 128 (*heading*) Apatite and hessonite in a pegmatite from Canaan, Ct. 1933 *Mineral. Abstr.* V. 281 The other [belt], from Ramona to Jacumba, is in granodiorite and the gem minerals in the pegmatites are mainly hessonite, beryl, and topaz. 1959 C. S. HURLBUT *Dana's Man. Min.* (ed. 17) v. 403 *Grossularite* (*Essonite, Cinnamon Stone*). Often contains ferrous iron replacing calcium and ferric iron replacing aluminium.

Hesvan (he·svăn). Also **Chesvan, Heshvan,** etc. [Heb. *ḥešwān*, f. earlier *marḥešwān* (recorded in the Mishnah, 2nd cent. B.C.), ad. Akkadian *Araḫ samna* eighth month.] The eighth month of the Jewish ecclesiastical year and the second month of the civil year, corresponding to parts of October and November; also formerly called Bûl.
1833 H. NICOLAS *Chronol. Hist.* 167 Marchesvan, Chesvan, or Bul 29 or 30 [Days]. 1838 E. H. LINDO *Jewish Cal.* 5 The perfect [year] has 355 days, and is when the months of Hesvan and Kislev have each 30 days. 1876 *Encycl. Brit.* IV. 678/1 The signs + and − are respectively annexed to Hesvan and Kislev to indicate that the former of these months may sometimes require to have one day more, and the latter sometimes one day less, than the number of days shown in the table. 1903 M. JOSEPH *Judaism as Creed & Life* 310 The ordinary Jewish year consists of twelve months... Cheshvan or Marcheshvan (November). 1927 G. F. MOORE *Judaism* II. 67 If by the seventeenth of Marḥeshvan the autumn rains had not begun, the religious heads of the community began to fast in a mitigated fashion. 1956 PEARL & BROOKES *Guide Jewish Knowledge* 2 (*heading*) Cheshvan (Oct.–Nov.).

het, *ppl. a.* **1.** Add to def.: also *transf.*, and for 'Now *dial.*' read 'Orig. *dial.* and *U.S.* Also with *up. colloq.*'
[1886 S. W. MITCHELL *R. Blake* ii. 17, I don't het up easy.] 1894 KIPLING *Day's Work* (1898) 46 You look consider'ble het up. Guess you'd better..cool off a piece. 1902 G. H. LORIMER *Lett. Merchant* v. 59 But you mustn't get yourself all 'het up' before you take the plunge. 1930 W. DE LA MARE *On Edge* 244 You must have been overdoing it. You look quite het up. 1932 C. WILLIAMS *Greater Trumps* x. 182 Yours is a remarkable family, Henry; you get all het up over your hobbies. 1945 L. A. G. STRONG *Othello's Occupation* ii. 43 The poor chap gets terrifically het up. 1967 *Listener* 20 Apr. 518/3 One thing that I think endears him to the normal young intellectual, is that he can get tremendously het-up about a cause.

hetærio: see *ETÆRIO.

hetærolite (hetiᵊ·rolait). *Min.* [f. Gr. ἑταῖρο-ς companion (see quot. 1877) + -LITE.] A black oxide of zinc and manganese, $ZnMn_2O_4$, isostructural with hausmannite and found in New Jersey associated with franklinite and with chalcophanite.
The mineral of quot. 1913 was later regarded as a separate species and named *hydrohetærolite*.
1877 G. E. MOORE in *Amer. Jrnl. Sci. & Arts* XIV. 423 From its invariable association with and close genetic relation to chalcophanite, I propose for the species the name *Hetærolite*. 1913 *Amer. Jrnl. Sci.* XXXV. 601 The hetærolite from Leadville was found..about 700 feet below the surface. 1928 [see *hydrohetærolite* s.v. *HYDRO-]. 1942 *Amer. Mineralogist* XXVII. 50 It was..proposed by Palache to adopt Moore's name hetaerolite for the anhydrous mineral, analogous to hausmannite,..since this was the sense of the original definition. The actual material of Moore..was separated under the name hydrohetaerolite as a distinct species. The hydrous mineral from Leadville..also is to be classed as hydrohetaerolite.

hetchel, *sb. dial.* and *U.S.* = HATCHEL *sb.*
1869 MRS. STOWE *Oldtown Folks* xlii. 530 She don' know no more 'bout religion than an old hetchel. 1900 J. SHELTON *Salt-box House* xvii. 143 Mops were made of corn-husks bound to a handle, the husks having been drawn through a hetchel which shredded them.

hetchel, *v. dial.* and *U.S.* = HATCHEL *v.*
1845 S. JUDD *Margaret* I. xiii. 100 The clouds hung low, and their floating skirts seemed to be pierced and hetchelled by the trees. 1878 *Harper's Mag.* LVII. 576 She'll hetchel the old woman mortally, I be afraid. 1897 [see HATCHEL *v.* 2]. 1906 C. H. PARKHURST *Lower than Angels* 18 Mercilessly hetchelled by some prosecuting attorney.

hetero, colloq. abbrev. of HETEROSEXUAL *a.* and *sb.*
1933 E. A. ROBERTSON *Ordinary Families* xiii. 272 The odd thing about me is that..I should be so purely 'hetero' in spite of lack of opportunity. 1957 J. OSBORNE *Look Back in Anger* II. i. 50 I'm tired of being hetero. 1960 I. JEFFERIES *Dignity & Purity* ii. 28 Is that really a fact about Proust being a secret hetero? 1968 J. R. ACKERLEY *My Father & Myself* xii. 117 Almost the first mischievous question he shot at me was 'Are you homo or hetero?' 1970 *Sunday Times* 6 Sept. 24 Talking.. about homosexual life and letters..I meant to ask him.. why homos found it so much easier to find sleeping partners than heteros.

hetero-. Add: **he:terarom·a·tic** *a.* = *hetero-aromatic* adj.; **heterauxesis,** now applied to animals as well as plants, with a more specialized meaning (see quot. 1941); **he:tero-agglu:tinabi·lity,** the ability to undergo heteroagglutination; **he:tero-agglutina·tion,** agglutination of cells due to the action of a hetero-agglutinin; so **he:tero-agglu·tinating** *vbl. sb.*; **he:tero-agglu·tinative** *a.*, producing heteroagglutination; **he:tero-agglu·tinin,** an agglutinin that causes agglutination of foreign cells, esp. red blood cells of another group or from an animal of another species; **he:teroa·lbumose** *Biochem.*, an albumose insoluble in water but soluble in solutions of sodium chloride; **he:teroaroma·tic** *a. Chem.*, heterocyclic and aromatic; also as *sb.*; **he·tero-atom** *Chem.*, an atom in the ring of a cyclic compound other than a carbon atom (also as two words); so **he:tero-ato·mic** *a.*; **he:tero-a·xial** *a.* [a. G. *heteroaxial* (V. Goldschmidt *Index d. Krystallformen d. Mineralien* (1891) III. 136)], having a structure based on two axes or sets of axes; *spec.* of a geological feature: having an external symmetry that does not correspond with the symmetry of the individual components of the fabric; **heteroblastic** *a.*, (*b*) *Bot.*, (characterized by) having a marked difference between the immature and adult forms; (*c*) *Petrol.*, composed of grains of two or more distinct sizes; opp. *homœoblastic*; **he:terobla·stically** *adv.*, in a heteroblastic manner; **he·terocharge,** the charge on an electret that is polarized in the opposite direction to that of the original polarizing field; hence **he·tero-charged** *a.*; **he:terochlamy·deous** *a. Bot.* [Gr. χλαμύς, χλαμυδ- cloak], having a perianth in which the calyx and corolla are of a different colour or texture; **he·terochrome** *a.* = *HETEROCHROMATIC *a.* (sense 1); **he:terochro·mia** (-krōᵘ·miă) *Med.*, a difference in colour between two organs (esp. the eyes), or between different parts of the same organ, that are usually the same colour; so **he:terochro·mic** *a.*; **he:terochro·mosome** *Cytol.*, a modified or differentiated chromosome, *esp.* a sex-chromosome; **heterochthonous** (-ǫ·kþǒnəs) *a. Path.* [after AUTOCHTHONOUS *a.*], originating in or derived from another organism; **he·teroclin, -cline** *sb. Min.* [ad. G. *heteroklin* (A. Breithaupt 1840, in *Ann. Physik und Chem.* XLIX. 205), f. Gr. ἑτεροκλινής leaning to one side] = MARCELINE²; **he:teroco̅e·lous** *a. Zool.* [Gr. κοῖλος hollow], applied to vertebrae in which the articular facets are saddle-shaped, as in certain birds; **heterocyst** (earlier example); **he:terocy·stous** (-si·stəs) *a. Biol.*, containing heterocysts; **heterode·smic** *a. Chem.* [Gr. δεσμός bond], containing chemical bonds of more than one type; **heterodu·plex** *a. Biochem.*, containing or consisting of polynucleotide strands derived from two different parent molecules; also as *sb.*, a heteroduplex molecule; **he:terodyna·mic, -dy·namous** *adjs. Ent.* [ad. F. *hétérodyname* (E. Roubaud 1922, in *Bull. Biol. de la France et de la Belg.* LVI. 471)], (of an insect, its life cycle, etc.) characterized by having a continuous succession of generations only during the favourable part of the year; **heterœ·cious** *a.* [first formed as G. *heterōcisch* (A. de Bary 1866, in *Monatsber. d. K. Preuss. Akad. d. Wissensch. zu Berlin 1865* 32)]; **he:tero-immu·ne** *a.*, immune to the cells or cell-products of an animal of a different species, or producing such immunity; **he:tero-inocula·tion,** inoculation from another organism; **he:tero-ino·culable** *a.*; **he·terojunction** *Electronics* [*JUNCTION 2 b], an area of contact between different semiconducting materials; **he:terokine·sis** *Cytol.*, the division of a cell into cells having dissimilar hereditary tendencies; opp. *homœokinesis*; **he:terole·cithal** (-le·sipăl) *a. Embryol.* [Gr. λέκιθος yolk of an egg], (of an egg cell) having the yolk unevenly distributed in the cytoplasm; **he:terone·reid** *a.*, of, pertaining to, or of the character of a heteronereis; also as *sb.*, a heteronereis; **he:terone·reis** *Zool.*, a dimorphic sexual form of certain worms of the genus *Nereis*, so called because originally regarded as a distinct genus; also *attrib.*; **he:teropho·ria** *Ophthalm.*, a latent tendency to squint; hence **heteropho·ric** *a.*; **heteroph-tha·lmic** *a.*, characterized by heterophthalmy; **he:teropolymeriza·tion** *Chem.* [a. G. *heteropolymerisation* (T. Wagner-Jauregg 1930, in *Ber. d. Deut. Chem. Ges.* LXIII. 3213)], a reaction in which a polymer is formed from two or more different molecules, esp. such a reaction when one of the monomers will not polymerize by itself; so **he:teropo·lymer,** a polymer so formed; **he:teropolysa·ccharide** *Chem.*, any polysaccharide composed of molecules of two or more different monosaccharides; **he:teropro·teose** *Biochem.*, any of a class of proteoses that are insoluble in water but soluble in dilute salt solutions and are formed during gastric digestion; **he:terosceda·stic** *a. Statistics* [Gr. σκεδαστ-ός capable of being scattered (σκεδάννυμι I scatter)], of unequal scatter or variation; having different variances; so **he:teroscedasti·city; heterosome,** (*b*) *Cytol.*, a sex chromosome; **heterospeci·fic** *a.*, (*a*) said of blood or serum of different blood groups; *heterospecific pregnancy*, one in which the red blood cells of the fœtus would be agglutinated by the serum of the mother; (*b*) derived from an organism of a different species; **heterospo·ric** *a.* = *heterosporous* adj.; **hetero·spory** *Bot.*, the condition of being heterosporous; **he:terosugge·stion** *Psychol.*, suggestion from another person, contrasted with *AUTO-SUGGESTION; **he:terosylla·bic** *a. Philol.*, belonging to a different syllable (opp. to *tautosyllabic*); **he:terosy·llis** *Zool.*, a breeding form of worms of the family Syllidæ; **he:terote·lic** *a.*, having or being an external end or purpose; **he:terotha·llic** *a. Biol.*, (of a fungus) having an incompatibility system by which only genetically different strains can undergo nuclear fusion during sexual reproduction; so **he:terotha·llism, -tha·lly,** the condition of being heterothallic; **he·terotope** *Physics* and *Chem.* [Gr. τόπος place], each of two or more elements the atoms of which have different atomic numbers and so occupy different positions in the periodic table; **he·terotype, heteroty·pic, heteroty·pical** *adjs. Cytol.* [ad. G. *heterotypisch* (W. Flemming 1887, in *Arch. f. mikrosk. Anat.* XXIX. 400)], designating the first division of meiosis; **hetero-xa·nthine** *Biochem.* [ad. G. *heteroxanthin* (G. Salomen 1885, in *Ber. d. Deut. Chem. Ges.* XVIII. 3407)], 7-methylxanthine, $C_6H_6O_2N_4$, a purine sometimes found in human urine; **heterozoœcium** (-zōᵘ,ī·siǒm) *Zool.* [Gr. ζῷον animal + οἰκία house] = *heterozooid*; so **heterozoœ·cial** *a.*; **he·terozooid** (-zōᵘ,oid) *Zool.*, any reduced or modified form of the typical bryozoan zooid, found chiefly in the class Gymnolæmata.
1960 *Tetrahedron Lett.* XXIII. 8 (*heading*) New heteraromatic compounds containing two boron atoms. 1940 NEEDHAM & LERNER in *Nature* 9 Nov. 618/1 We welcome a suggestion..by Dr. Arthur L. Peck..that for relative *growth*, in contradistinction to relative *proportions*, the word *heterauxesis* should be used... It is true that the terms auxesis, heterauxesis,..etc., were formerly employed in plant physiology, but they have long been obsolete there. 1941 *Ibid.* 23 Aug. 225/1 *Heterauxesis*, the relation of the growth-rate of a part of a developing organism (whether morphological or chemical) to the growth-rate of the whole or of another part; a comparison between organisms of the same group but of different ages and hence sizes. 1964 *Biol. Abstr.* XLV. 946/1 Hetero-agglutinability of goat erythrocytes by zebu serum. 1913 *Jrnl. Exper. Zool.* XIV. 564 The iso-agglutinating action was noted as undiminished, whereas the hetero-agglutinating action was entirely lost. *Ibid.* 561 (*heading*) Hetero-agglutination and the question of specificity: reactions between Nereis and Arbacia. 1949 *Biol. Abstr.* XXIII. 156/2 Lecithin suppresses the auto-agglutination as well as heteroagglutination of rabbit erythrocytes by bovine plasma. 1938 BELDING & MARSTON *Textbk. Med. Bacteriol.* ix. 451 The demonstration of the M and N antigens in human cells requires the production of heteroagglutinative immune sera by the injection of human cells into some experimental animal. 1906 DORLAND *Med. Dict.* (ed. 4) 332/1 Hetero-agglutinin. 1913 *Jrnl. Exper. Zool.* XIV. 564 The egg-extract contained two agglutinating substances at least, namely: An iso-agglutinin and a hetero-agglutinin. 1945 *Biol. Bull.* LXXXIX. 193 Lobster-serum contains at least ten hetero-agglutinins for sperm or blood cells of various animals. 1956 *Nature* 18 Feb. 329/2 The injections caused no in-

crease in the titre of natural heteroagglutinin against human red cells. 1884 KÜHNE & CHITTENDEN in *Amer. Chem. Jrnl.* VI. 33 We name..No. II. Heteroalbumose. *Ibid.* 103 Nothing characterises heteroalbumose more than its alteration by boiling and the properties of the coagulum thus formed. 1908 J. R. BRADFORD in Allbutt & Rolleston *Syst. Med.* (ed. 2) IV. I. 561 Albumoses, and especially hetero-albumoses, which are the kind commonly present in the urine, form a precipitate on the addition of nitric acid. 1958 A. ALBERT *Trends Heterocyclic Chem.* iv. 20 (*heading*) Addition to double-bonds in N-heteroaromatic six-membered rings. 1958 *Jrnl. Chem. Soc.* 3076 A wide range of new heteroaromatic systems should exist, derived from normal aromatic compounds by replacing pairs of carbon atoms, one by boron and one by nitrogen. 1959 A. ALBERT *Heterocyclic Chem.* iii. 31 (*heading*) A general discussion on heteroaromatics. 1900 E. F. SMITH tr. *von Richter's Org. Chem.* (ed. 3) II. 435 The basal element of these rings is carbon, and accordingly the members not produced by C-atoms are designated as hetero-atoms. 1949 G. B. BACHMAN *Org. Chem.* xxvii. 336 O-, S-, and NH-containing rings undergo substitution practically exclusively at the carbons holding the hetero atom. 1966 *McGraw-Hill Encycl. Sci. & Technol.* VI. 427/2 The number of heteroatoms in any one ring is commonly one or two, less commonly three or more. 1967 KATRITZKY & LAGOWSKI *Princ. Heterocyclic Chem.* v. 144 (*heading*) Four or five heteroatoms. Tetrazoles are formed by the action of nitrous acid on amidrazones, and pentazoles from the reaction of diazonium cations with azide anions. 1968 *New Scientist* 31 Oct. 268/3 Heterocyclic compounds containing nitrogen, sulphur and oxygen as the heteroatom. 1900 E. F. SMITH tr. *V. von Richter's Org. Chem.* (ed. 3) II. 435 Hetero-atomic rings. 1926 *Mineral. Abstr.* III. 186 A study of the directions of optical extinction in the lamellae, the striations on the faces, and the etch figures on the basal plane, leads to the conclusion that the twinning is of the 'heteroaxial' type. 1938 E. B. KNOPF in *Mem. Geol. Soc. Amer.* VI. vi. 84 An example of heteroaxial orienting in two phases of the deformation is the fabric of certain pencil gneisses in which the symmetry of the grain fabric in the pencils does not conform with the fabric axes of the external form of the pencils. 1894 S. H. VINES *Stud. Text-bk. Bot.* I. i. 14 In certain cases the embryo produced by the spore differs more or less widely from the adult form, and does not directly develope into it, but bears it as a lateral outgrowth; this mode of embryogeny is indirect or heteroblastic. 1932 A. HARKER *Metamorphism* xiii. 202 To rocks in which the essential constituents are of two distinct orders of magnitude Becke gave the name 'heteroblastic', in contradistinction to 'homoeoblastic'. 1954 R. L. PARKER tr. *Niggli's Rocks & Min. Deposits* vi. 237 Crystalloblastic structures: homeoblastic, heteroblastic, porphyroblastic, with porphyroblasts or possibly crystalloid phenocrysts. 1963 DAVIS & HEYWOOD *Princ. Angiosperm Taxon.* x. 342 Heteroblastic development results in the formation of adult and juvenile leaves. 1888 *Nature* 13 Dec. 151/1 Tenontogenous or desmogenous [sesamoids], like the patella, are formed heteroblastically inside of a tendon. 1935 A. GEMANT in *Philos. Mag.* XX. 933 We observe both kinds of charges on electrets. One has the opposite sign to that of the adjacent polarizing electrode, and for the sake of shortness will be denoted as heterocharge. 1965 *New Scientist* 27 May 590/2 Under a certain critical applied field the heterocharge decays to a constant value. 1965 *Jrnl. Chem. Physics* 1 Feb. 967 Both heterocharged and homocharged electrets have been made from common ice at reduced temperatures. 1895 S. H. VINES *Stud. Text-bk. Bot.* III. 512 When the calyx and corolla clearly differ from each other in colour, texture, etc., the flower is said to be heterochlamydeous. 1965 BELL & COOMBE tr. *Strasburger's Textbk. Bot.* III. 621 Perianths are of two kinds: (*a*) Homoiochlamydeous..or (*b*) heterochlamydeous, i.e. with dissimilar members, namely an outer, generally green calyx and an inner, mostly brightly coloured corolla. 1933 *Nature* 6 May 667/1 A solution of the problem of heterochrome photometry of incandescent lamps. 1889 *Ophthalmic Rev.* July 205 Liability to disease on the part of the lighter eye in heterochromia. 1964 F. C. BLODI in A. Sorsby *Mod. Ophthalmol.* III. iii. 375 There will be a gradual change in colour of the iris giving one of the varieties of heterochromia. 1911 *Ophthalmoscope* 1 July 501 Heterochromia iridium is found in two forms. In one the heterochromia is merely an anomaly, and in the other it is a symptom of a definite disease... This latter variety is best designated heterochromic cyclitis. 1940 W. S. DUKE-ELDER *Text-bk. Ophthalmol.* III. xxxviii. 3229 Heterochromic cataract..is associated with an exceedingly slow and benign inflammation of the uveal tract. 1904 T. H. MONTGOMERY in *Biol. Bull.* VI. 145 The Heterochromosomes. I offer this name to include those peculiarly modified chromosomes to which have been given the names 'accessory chromosomes'.., 'small chromosomes'..and 'chromatin nucleoli'. 1926 *Nature* 9 Jan. 50/2 Cytological investigations in the Salicaceæ, undertaken to demonstrate the presence or absence of sex- or heterochromosomes in certain species of Salix. 1968 J. A. SERRA *Mod. Genetics* III. xxiii. 552 In haploid organisms or the haploid phase the heterochromosomes are found separately in the mitoses of each sex. 1891 FOSTER *Med. Dict.* III, *Heterochthonous*, originating from without the organism. 1921 BEATTIE & DICKSON *Text-bk. Gen. Path.* (ed. 2) ix. 273 Some have defined teratomata as heterochthonous tumours derived from the inclusion of *another* individual..or the ovum or germ-cell from which such twin would have been developed. 1950 G. P. WRIGHT *Introd. Path.* xx. 374 Chorion carcinomas are not derived from the tissues of the mother, but from the tissues of a different, though at the same time fœtal, individual. For this reason such tumours are sometimes termed 'heterochthonous'. 1844 J. D. DANA *Syst. Min.* (ed. 2) vi. 443 Heteroclin..was first instituted by Breithaupt, and named..in allusion to its oblique form of crystallization. 1898 E. S. DANA *Textbk. Min.* (ed. 2) IV. 343 Marceline (heterocline) from St. Marcel, Piedmont, is impure braunite. 1884 E. COUES *N. Amer. Birds* (ed. 2) 138 Both ends of each vertebra are saddle-shaped; ..a condition which may be called heterocœlous. 1933 H. F. GADOW *Evol. Vertebral Column* XXXV. 311 The embryonic vertebrae of all Birds are at first amphicoelous, then they change through opisthocoelous into the heterocoelous or saddle-shaped type, which..represents the highest stage of interaxial joint, allowing of most excursion. 1872 *Q. Jrnl. Microsc. Sci.* XII. 367 Its minute size calls to mind *Nostoc minimum* (Currey), but in it..the heterocysts are large, whilst here..the heterocysts are but slightly wider, though longer than the ordinary cells. 1887 *Jrnl. R. Microsc. Soc.* 793 (*heading*) Heterocystous Nostocaceæ. 1951 *Proc. Linnean Soc. Lond.* CLXII. II. 195 Heterocystous blue-green algae. 1939 R. C. EVANS *Introd. Crystal Chem.* i. 8 Crystals..in which two or more different types of bond are in operation between different parts of the structure are termed heterodesmic. 1952 B. MASON *Princ. Geochem.* iv. 67 In heterodesmic structures the physical properties..are in general determined by the weakest bonds. 1963 A. H. DOERMAN in W. J. Burdette *Methodol. Basic Genet.* 34 The heterozygote is imagined to consist of a heteroduplex molecule in which every genetic site is represented twice. 1966 *Progress Nucleic Acid Res.* V. 319 The exposed polynucleotide chains can be thought to anneal during the act of rejoining to produce a heteroduplex region lying between the two recombinant segments. 1968 *Proc. Nat. Acad. Sci.* LX. 243 Each heteroduplex should thus contain a single-stranded loop in the wild-type DNA strand at the point where the deletion occurs. 1968 *Sci. Jrnl.* Nov. 5/1 The mixture [of DNA] now contains some 'heteroduplex' molecules consisting of a wild-type and a mutant strand. 1931 *Trans. Entomol. Soc. London* LXXIX. 105 Essentially different is the heterodynamic type, in which the annual cycle bears a more or less definite relation to the season. 1964 BORROR & DELONG *Introd. Study Insects* (rev. ed.) iii. 44 Most insects in temperate regions have what is called a heterodynamic life cycle; that is, the adults appear for a limited time during a particular season, and some life stage passes the winter in a state of dormancy. 1929 Heterodynamous [see *homodynamous* s.v. *HOMO-]. 1903 *Lancet* 4 Apr. 944/2 The introduction of the hetero-immune serum. 1967 C. W. H. HAVARD *Lect. Med.* vi. 147 Individuals respond immunologically to tissues of other species (hetero-immune) or to tissues of another individual of the same species (iso-immune). 1894 J. C. DACOSTA *Man. Mod. Surg.* xv. 168 Primary syphilis is not auto-inoculable, but is hetero-inoculable. 1888 E. L. KEYES *Surg. Dis. Genito-Urinary Organs* II. ii. 494 Few at the present day can be found who..consider as gonorrhœa a urethral discharge producing syphilitic chancre by hetero-inoculation. 1960 [see *homojunction* s.v. *HOMO-]. 1971 *New Scientist* 16 Sept. 628/1 The light produced in the active [region] travels into the n-type region between the two heterojunctions. 1893 Heterokinesis [see *homœokinesis* s.v. *HOMŒO-]. 1896 E. B. WILSON *Cell* ix. 305 In the second case ('heterokinesis', qualitative or differential division), the daughter-cells receive different groups of chromatin-elements, and hence become differently modified. 1892 Heterolecithal [see *homolecithal* s.v. *HOMO-]. 1896 E. B. WILSON *Cell* 336 *Heterolecithal*..having unequally distributed deutoplasm (includes telolecithal and centrolecithal). 1896 W. B. BENHAM in *Cambr. Nat. Hist.* II. x. 277 There are then three different kinds of males and of females in this one species [sc. *Nereis*], some being found at the bottom of the sea, as the large Heteronereid form, while the small Heteronereid swims on the surface. 1963 R. P. DALES *Annelids* vi. 124 In heteronereids the parapodia are greatly increased in surface area and musculature. 1875 *Encycl. Brit.* II. 67/1 Another [sexual form] which becomes transformed into a *Heteronereis* before the sexual elements are developed. 1880 F. M. BALFOUR *Compar. Embryol.* I. xii. 284 Claparède traced the passage of large asexual examples of the Nereis form into the large Heteronereis form. 1886 Heterophoria [see *exophoria* s.v. *EXO-]. 1957 *New Scientist* 9 May 38/2 Treatment aims at restoring normal functioning of the eyes, especially in what is popularly known as 'lazy' eyes and squint, but also in the more common condition of heterophoria—a tendency to squint. 1894 GOULD *Dict. Med.*, Heterophoric. 1970 *Jrnl. Gen. Psychol.* LXXXII. 109 The mean duration of lateral AM was not systemically affected by heterophoric change from zero to 28 prism diopters. 1924 *Jrnl. Genetics* XIV. 365 (*heading*) Heterophthalmic cats. 1931 *Chem. Abstr.* XXV. 2419 Benzalfluorene (IV) and I give in good yield in the fused mixt. at 130°..a white amorphous heteropolymer. 1948 C. E. H. BAWN *Chem. High Polymers* iii. 85 The individual monomers undergoing copolymerization may not polymerize alone. A copolymer formed with the latter type of monomer is often called a heteropolymer. 1952 *Jrnl. Polymer Sci.* VIII. 260 It is..recommended that the word 'heteropolymer' not be used for a copolymer in which one of the units does not polymerize by itself. 1971 *Nature* 23 July 254/2 The requirements of the RNA-DNA reaction indicate that a heteropolymer is formed, for the omission of any one of the triphosphate substrates suppresses synthesis almost completely. 1931 *Chem. Abstr.* XXV. 2418, 2 unlike units of low mol. wt., each contg. a C:C union, can ..be combined into a large mol. by polymerizing addn. The name additive heteropolymerization is suggested for such a process. 1958 *Van Nostrand's Sci. Encycl.* 132/1 Heteropolymerization is an addition polymerization. 1948 W. PIGMAN *Chem. Carbohydrates* xii. 513 The second class (heteropolysaccharides), which consists of polysaccharides giving after hydrolysis more than one monosaccharide type. 1970 Heteropolysaccharide [see *homopolysaccharide* s.v. *HOMO-]. 1891 *Jrnl. Physiol.* XII. 21 There are at least three normal proteoses formed in gastric digestion. Of these, proto and heteroproteose are first formed. 1916 A. P. MATHEWS *Physiol. Chem.* ix. 361 There are three divisions of the group: namely, the primary proteoses, including the protoproteoses and hetero-proteoses, and the secondary, or deutero-proteoses. 1905 Heteroscedastic [see *homoscedastic* s.v. *HOMO-]. 1937 YULE & KENDALL *Introd. Theory Statistics* (ed. 11) xi. 214 Arrays in which the standard deviations are equal are sometimes said to be 'homoscedastic'; in the contrary case 'heteroscedastic'. 1965 M. G. BULMER *Princ. Statistics* xii. 215 The variance of *Y* may not be constant but may depend on *x*; the regression is then said to be heteroscedastic. 1905 K. PEARSON in *Drapers' Company Res. Mem.* (Biometric Ser.) II. 23, I am thus inclined to speak of $x_1 - 1$ and x_2 as measures of heteroscedasticity and heteroclisy. 1964

Johnson & Leone *Statistics & Exper. Design* II. xvii. 321 Replication at each vertex does provide some information on possible heteroscedasticity of the residual variation. 1938 A. F. Shull *Heredity* (ed. 3) x. 92 The spermatozoa of a mammal are of two kinds, half of them containing an X chromosome, half of them a Y (or no heterosome at all in species in which Y has been lost). 1966 D. M. Kramsch tr. *Grundmann's Gen. Cytol.* ii. 102 Man has 22 autosomes and one heterosome in a haploid set. 1929 R. R. Gates *Heredity in Man* ix. 191 In 12·5 per cent. of heterospecific pregnancies an agglutinin passed from the mother's blood to that of the child. 1958 Stratton & Renton *Pract. Blood Grouping* i. 14 A mother of group O would have an incompatible or heterospecific pregnancy if the child were group A or group B. 1962 *Lancet* 5 May 965/1 Two of these components could agglutinate red cells in the presence of heterospecific serum by a mechanism previously unknown among viruses. 1969 *Nature* 6 Sept. 1021/2 The possible use of hybrids of tumour cells and heterospecific cells to stimulate an immune response. 1895 D. H. Campbell *Struct. & Devel. Mosses & Ferns* i. 6 In all of the heterosporic Pteridophytes the reduction of the vegetative part of the gametophyte is very great. 1967 M. E. Hale *Biol. Lichens* iii. 46 The identity of this heterosporic mat. 1898 *Nat. Science* June 375 Its independent appearance in distinct groups may be compared with the appearance of heterospory. 1959 *Chambers's Encycl.* XI. 613/1 Heterospory..is well developed in the seed plants, where the microspores are the pollen grains. 1901 Baldwin *Dict. Philos.* I. 96/1 Wundt uses the term Fremdsuggestion for the contrasted and usual process of suggestion from another person. The analogous Greek formation would be Heterosuggestion. 1921 *Spectator* 19 Mar. 364/2 A portion of the doctors and men of science..began to perceive that it was primarily not their suggestions, but the patient's own suggestions to his subconscious self, which produced the wonderful results... It was auto-suggestion, not hetero-suggestion. 1951 F. Hopkins in E. N. Chamberlain *Text-bk. Med.* ix. 660 The terms autosuggestion and heterosuggestion are used according as to whether the suggestion is made by oneself or others. 1913 J. M. Jones *Welsh Gram.* 72 In N. W[ales] the vowel is medium in *aw, ew, iw* before a vowel, that is the *w* is heterosyllabic. 1896 *Cambr. Nat. Hist.* II. 278 In some genera [of the family Syllidae]..there occur changes quite similar to those characterising 'Heteronereis'—that is, the posterior segments in which the genital organs exist become altered, so that the worm consists of two distinct regions, and is termed a 'Heterosyllis'. 1967 H. W. & L. R. Levi tr. *Kaestner's Invertebr. Zool.* I. xix. 496 Reproduction is complicated by alternation of sexual and asexual generations... *Syllis*, 1–5 cm... heterosyllis. 1901 Baldwin *Dict. Philos.* I. 96/1 *Heterotelic*, having or serving a foreign or external end. 1902 *Ibid.* II. 668/1 To the deist the world process is heterotelic;..to the thoroughgoing monistic idealist it is autotelic. 1904 A. F. Blakeslee in *Science* 3 June 865 According to their method of zygospore formation, the various species among the Mucorineæ may be divided into two main categories, which may be designated as homothallic and heterothallic... In the heterothallic group..zygospores are developed from branches which necessarily belong to thalli or mycelia diverse in character, and can never be obtained from the sowing of a single spore. 1959 *Chambers's Encycl.* VI. 117/1 *Sporodinia grandis* is homothallic, and a single spore from a sporangium will give rise both to sporangia and to zygospores, whereas *Mucor* is heterothallic, and a single spore gives rise only to sporangia. 1906 Heterothallism [see *homothallism* s.v. *Homo-]. 1952 *New Biol.* XIII. 107 The discovery of heterothallism proved to be of fundamental importance since it has now been shown to occur in a modified form in all the major groups of fungi. 1940 *Bot. Rev.* VI. 74 There has been progessive sexual differentiation beginning with the gametes..and extending outward from them to the gametangia and prothallia, as indicated by the successive acquirement of heterogametangy and heterothally. 1942 Heterothally [see *homothally* s.v. *Homo-]. 1919 F. Soddy in *Jrnl. Chem. Soc.* CXV. 11 Boyle's practical definition of the element..became replaced by a theoretical conception, to which..I propose to apply the term 'heterotope', meaning the occupant of a separate place in the periodic table of elements. 1959 L. W. H. Hull *Hist. & Philos. Sci.* viii. 267 It is now known that there are atoms of different weights having the same chemical properties. These are called isotopes. There are also 'isobaric heterotopes', which have the same weight but different chemical properties. 1895 *Ann. Bot.* IX. 479 The indifference manifested in the second mitosis in animals..as to whether it be heterotype or homotype, is of some theoretical interest. 1920 L. Doncaster *Introd. Study Cytol.* vi. 89 When the heterotype chromosomes split longitudinally, part of one longitudinal half of one chromosome may exchange places with a similar part of the corresponding longitudinal half of the other. 1889 *Q. Jrnl. Microsc. Sci.* XXX. 203 In another deviation, which Flemming describes as the 'homöotypic Form' (that is to say, 'a form more like the usual one than the one just described, which he names "heterotypic"'), it would appear..that longitudinal splitting may be entirely absent. 1931 W. Shumway *Gen. Biol.* vi. 149 The first maturation division is sometimes called the heterotypic division because of its novel features of synapsis and tetrad-formation. 1969 Heterotypic [see *homœotypic* s.v. *Homœo-]. 1888 Heterotypical [see *homœotypical* s.v. *Homœo-]. 1896 E. B. Wilson *Cell* ii. 60 (*caption*) Heterotypical mitosis in spermatocytes of the salamander. 1886 *Jrnl. Chem. Soc.* L. 266 These..researches have led to the isolation of another constituent of human urine, which it is proposed to call heteroxanthine. 1943 *Thorpe's Dict. Appl. Chem.* (ed. 4) VI. 206/1 Heteroxanthine appears to be a product of the metabolism of theobromine and caffeine, for when these alkaloids are administered to rabbits, dogs or men, heteroxanthine appears in the urine. 1909 G. M. R. Levinsen *Cheilostomatous Bryozoa* 74 The same heterozoœcium may appear in the same genus, even occasionally in the same species, sometimes as an avicularium, sometimes as a vibraculum... The genus *Microporella* as well as the genus *Escharina* may serve as examples of such a variable development of the two heterozoœcial forms. *Ibid.* 46 We can distinguish between four main forms of individuals (Bryozooids):.. *Heterozoœcia* (*Heterozooids*), which have no intestinal canal, and at most have a trace of a polypide in a small cell-body, furnished with a circle of fine bristles. The chamber contains a strong muscular apparatus for moving the operculum. 1959 L. H. Hyman *Invertebrates* V. xx. 325 Other types of zooids are termed collectively heterozooids and are characterized by the reduction of the polypide, which loses its nutritive and reproductive function.

heteroauxin (he:tĕro,ǭ·ksin). *Biochem.* [a. G. *hetero-auxin* (F. Kögl et al. 1934, in *Zeitschr. f. physiol. Chem.* CCXXVIII. 94), f. Hetero- + *Auxin.] A growth-promoting hormone, $(C_8H_6N)CH_2·COOH$, that occurs in some plants and micro-organisms; also called *β-indolyl acetic acid, 3-indoleacetic acid.*
1935 *Chem. Abstr.* XXIX. 195 Normal individuals excrete auxins and heteroauxin in the proportions of about 4:1. 1940 W. R. Fearon *Introd. Biochem.* (ed. 2) 446 It was found that urine was very rich in a growth-promoting factor, which on isolation proved to be chemically unrelated to auxin *a* or *b*, and was termed hetero-auxin. 1949 *New Biol.* VII. 69 Two principal plant hormones..have actually been isolated from plant tissues, viz. auxin and hetero-auxin. 1966 Nowakowski & Clarke tr. *Kretovich's Princ. Plant Biochem.* v. 175 Heteroauxin is used for accelerating root formation in cuttings of various plants e.g. citrus.

heterocaryon, -caryosis, -caryotic: see *heterokaryon, etc.

heterochromatic (he:tĕrokromæ·tik), *a.* [f. Hetero- + Chromatic *a.* 1 in sense 1; in sense 2, f. next + -ic.] **1.** Relating to or possessing more than one colour; relating to light or other radiation of more than one wavelength. Also *fig.*
1895 F. W. Oliver et al. tr. *Kerner's Nat. Hist. Plants* II. 569 Contrasting with these heterochromatic species are those with homochromatic flowers, which..invariably present the same colour and pattern. 1924 *Nature* 30 Aug. 329/2 The use of absorbing screens in heterochromatic photometry. 1932 W. S. Duke-Elder *Text-bk. Ophthalmol.* I. xxii. 861 The comparison of the luminosities of lights of different colours comprises the science of heterochromatic photometry. 1937 *Archit. Rev.* LXXXI. 198/2 Heterochromatic harmonies, or harmonies of different hue. 1969 *Nature* 11 Oct. 162/2 Along each heterochromatic channel there is..a positional as well as a wavelength variation which must be taken into account in the interpretation of line profiles.
2. *Cytol.* Of chromosome material: becoming heteropycnotic at some stage in the nuclear cycle.
1936 [see *euchromatic *a.*]. 1942 *Nature* 17 Jan. 66/2 The two properties of inertness and abnormality of nucleic acid cycle are combined in the same genes or chromomeres... These are then said to be heterochromatic as opposed to the active or euchromatic genes. 1951 G. H. Bourne *Cytol. & Cell Physiol.* (ed. 2) v. 195 On the one hand the heterochromatic regions may be thicker and more dark-staining than the euchromatic ones. ..On the other hand, a heterochromatic region or chromosome may appear thinner and stain more weakly than euchromatic regions in the same nucleus. 1956 *Nature* 10 Mar. 452/2 It usually appears as three dots arranged in a triangle, suggesting the presence of an under-stained ('heterochromatic') region close to the centromere. 1969 *Ibid.* 16 Aug. 684/2 In plants..supernumerary heterochromatic chromosomes may make pollen grains undergo additional divisions and quicken cell division.

heterochromatin (he:tĕrokrōu·mătin). *Cytol.* [a. G. *heterochromatin* (E. Heitz 1928, in *Jahrb. f. wissenschaftliche Bot.* LXIX. 764), f. Hetero- + Chromatin.] Heterochromatic chromosome material.
1932 [see *euchromatin]. 1941 *Ann. Bot.* V. 203 Large distal segments which appear as heterochromatin in mitosis, that is to say they are overstained in resting nuclei and, after freezing, understained in metaphase. 1951 G. H. Bourne *Cytol. & Cell Physiol.* (ed. 2) v. 196 Probably there are many varieties of heterochromatin and euchromatin, constituting a continuous 'spectrum', to the two ends of which the terms heterochromatin and euchromatin are somewhat arbitrarily applied. 1969 *Nature* 16 Aug. 683/1 The observations that heterochromatin formation is a variable property and represents cyclical physiological states of chromosome segments..led to a shift of emphasis from heterochromatin as a substance to heterochromatin as a state.

heterochromatization (he:tĕrokrōu:mătəizē·ʃən). *Cytol.* Also **he:terochro:matiniza·tion.** [f. prec. + -ization.] A change of state of chromosome material in which it becomes heterochromatic and the action of the genes is modified or suppressed; also, the extent to which such a change has occurred. So **he:terochro·mat(in)ized** *ppl. a.*
1941 *Cold Spring Harb. Symp. Quant. Biol.* IX. 158/2 When extra heterochromatin..is added to a cell, the heterochromatization..is lessened. *Ibid.*, The euchromatic regions which have been transferred into the vicinity of heterochromatic ones are themselves partly..'heterochromatized'. 1944 *Amer. Naturalist* LXXVIII. 207 Variable 'heterochromatinization' takes place not only in salivary gland cells but in other body cells as well. 1948 *Jrnl. Genetics* XLVIII. 80 The process of heterochromatization (heteropycnosis) is reversible. *Ibid.* 83 The active section is sometimes so highly heterochromatized that it becomes quite indistinguishable from the neighbouring inert region. 1957 C. P. Swanson *Cytol. & Cytogenetics* xiv. 469 This proposes that heterochromatin is formed from euchromatin by a process of genic degeneration... The end point of such an evolutionary trend..would be heterochromatinization of the entire Y. 1964 *Science* 10 July 130/1 Tissue..in which the heterochromatized X is of paternal origin. 1969 *Nature* 16 Aug. 683/2 The X chromosome that forms the sex-chromatin mass of normal females..is clearly heterochromatic when it forms (by facultative heterochromatization) a distinct chromocentre in the interphase nucleus.

heterochronous, *a.* Add: **b.** *Palæont.* [ad. G. *heterochron* (E. Mojsisovics *Die Cephalopoden der Hallstätter Kalke* (1893) II. 5).] Originating or formed at different periods.
1895 [see *homœomorphy]. 1913 *Q. Jrnl. Geol. Soc.* LXIX. 166 Isochronous and heterochronous homœomorphy state whether the homœomorphous species lived in the same or at different times. 1952 R. C. Moore et al. *Invert. Fossils* vi. 218 The other category includes two or more shells of different geologic age, in which one simulates the other. These are termed heterochronous homeomorphs.

he·terocy:clic, *a.* and *sb.* [f. Hetero- + Cyclic *a.*] **1.** *Bot.* = Heteromerous *a.* 2 b.
1895 S. H. Vines *Text-bk. Bot.* 500 When the whorls are heteromerous the flowers are said to be heterocyclic. 1924 Holman & Robbins *Textbk. Gen. Bot.* vii. 228 The flowers of the great majority of plants are heterocyclic.
2. *Chem.* Pertaining to or containing a ring of atoms of more than one kind. Opp. to *homocyclic, isocyclic.* Also as *sb.*, a heterocyclic compound.
1899 E. F. Smith tr. *V. von Richter's Org. Chem.* (ed. 3) I. 78 In the case of many heterocyclic compounds..the substances with open chains from which they may be theoretically deduced do not really exist. 1900 *Proc. Chem. Soc.* 6 Feb. 11 The action of bases on ethyl phenylpropiolate gives rise to heterocyclic compounds. 1936 L. J. Desha *Org. Chem.* xxiv. 479 In the most typical heterocyclic compounds, the ring systems are extremely stable. 1956 *Nature* 16 June 1116/2 The oxygenated heterocyclics were discussed with emphasis on the biochemical aspects. 1966 *McGraw-Hill Encycl. Sci. & Technol.* VI. 427/2 Heterocyclic systems are encountered in many groups of organic compounds, both synthetic and natural. 1968 *New Scientist* 31 Oct. 268/3 Classification of heterocyclics into suitable groups is somewhat of a problem.
Hence **he·terocycle,** a heterocyclic ring or compound.
1909 Webster, *Heterocycle*, a heterocyclic compound. 1931 *Jrnl. Amer. Chem. Soc.* LIII. 806 Ethylene oxide, the simplest oxygen heterocycle. 1957 E. H. Rodd *Chem. Carbon Compounds* IVa Introd. 1 Compounds containing these heterogeneous rings (heterocycles) are classed as heterocyclic.

heterodyne (he·tĕrodəin), *a.* *Electr.* [f. Hetero- + *-dyne.] Pertaining to, involving, or designating the production of a beat frequency by the combination of two oscillations of slightly different frequency, esp. as a method of radio detection in which one oscillation is the incoming signal and the other is produced in the receiver. Also (now *rare*) as *sb.*, a heterodyne receiver or its local oscillator.
1908 J. Erskine-Murray tr. *Ruhmer's Wireless Teleph.* 201 One of the most interesting of Professor Fessenden's many inventions is what he has called the 'Heterodyne' receiver. 1915 W. H. Eccles *Wireless Telegr.* 151 In receiving signals sent out by the spark sets..the heterodyne produced a telephone current variation averaging 4·65 times that due to the liquid barretter alone. 1919 *Wireless World* Mar. 663 (*heading*) Heterodyne reception. 1922 A. F. Collins *Bk. Wireless Telegraph* iii. 174 The oscillations set up by the incoming waves from the distant sending stations and those set up by the separate oscillator tube, or heterodyne..have a slightly different frequency. 1931 *B.B.C. Yearbk.* 442/2 *Heterodyne interference*, interference caused to broadcast reception by the carrier wave of an unwanted station beating with that of the wanted station. *Ibid.*, *Heterodyne reception*, a method of receiving C.W. wireless signals in which use is made of a local oscillator to 'beat' with or 'Heterodyne' the incoming C.W. 1946 *Nature* 6 July 33/1 The usual heterodyne methods of measuring a frequency..can be extended to the highest frequencies now in general use. 1957 W. Fraser *Telecommunications* xxii. 735 The heterodyne waveform when rectified produces a current which varies in amplitude at the beat frequency.
fig. 1930 *Musical Assoc. Proc.* 97 To write atonally in a harmonic form at present is to produce a thought 'heterodyne'.
Hence **he·terodyne** *v. trans.*, (*a*) to produce heterodyne interference with (a radio station) (*obs.*); (*b*) to change the frequency of (a signal) by a heterodyne process; *intr.*, to combine so as to produce beat oscillations or a difference frequency; **he·terodyning** *vbl. sb.* and *ppl. a.*; **he·terodyned** *ppl. a.*
1923 *Glasgow Herald* 15 Oct. 6 A spark station previously heterodyned by a carrier. 1925 *Ibid.* 10 Nov. 8 Glasgow can be as effectively 'heterodyned' by a German station as Radio Paris can be by a Spanish one working on an almost identical wave-length. 1926 *Encycl. Brit.* III. 1044/2 If the signal current has a frequency of 10^6 cycles (3,000 metres), to give it an audible frequency of c″–512 the heterodyning current must have a frequency of 10^6 plus or minus 512. 1933 K. Henney *Radio Engin. Handbk.* x. 274 The frequencies in this selected range are made to heterodyne with an oscillator at a frequency of 42,000 cycles. 1934 *Proc. IRE* XXII. 1269 The oscillator operated at a higher frequency than the incoming carriers and heterodyned them to 6 and 7 megacycles. 1943 C. L. Boltz *Basic Radio* xiv. 223 This sound is called the heterodyne note, and the process is called heterodyning. 1965 Wozencraft & Jacobs *Princ. Communication Engin.* vi. 493 At the receiver the signals are heterodyned back down from passband to baseband. 1967 *Electronics* 6 Mar. 288/3 In this coherent conversion system, heterodyning converts incoming reference and test signals to 45 Mhz and then to 1 kilohertz. *Ibid.* 291/1 A comparison is made between the phase of a heterodyned 1-khz test signal and the phase of a second 1-khz signal. 1972 *Science* 20 Oct. 252/2 It arises because the emission heterodynes with the collinear laser beam.
fig. 1931 T. H. Pear *Voice & Personality* iv. 44 The rudeness..heterodynes the shyness. So, while an unselective receiver, or strange listener, perceives a jumbled mixture of both, the ear which is sensitised to rudeness hears it only, while you hear only shyness.

heterogamete (he·tĕrogæ:mīt). *Biol.* [f. Hetero- + Gamete.] Either of a pair of conjugating gametes that differ in character or form.
1897 *Ann. Bot.* XI. 120 The stage of *Aphanochaete* is..the intermediate step between two motile heterogametes and the true sexuality realized in the higher types. 1949 B. B. Mundkur *Fungi & Plant Dis.* vi. 49 While the gametes are motile isogametes in *Synchytrium endobioticum*, they are non-motile heterogametes in *Olpidiopsis*.

heterogametic (he:tĕrogămī·tik), *a.* *Biol.* [f. Hetero- + *gametic *a.*] (Of a sex or its individual members) producing gametes that differ with respect to a sex chromosome. Opp. *homogametic *a.* So **he:teroga·mety,** the state or condition of being heterogametic.
1910 E. Wilson in *Science Progress* IV. ii. 572 If the two kinds of spermatozoa be designated as the 'X-class' and the 'Y-class', respectively, the eggs are all of the X-class. The male may, accordingly, be designated as the *heterogametic* sex, the female as the *homogametic*. 1932 Sinnott & Dunn *Princ. Genet.* (ed. 2) x. 215 Two kinds of sperm (male heterogamety) are formed in equal numbers, half with X, and half with Y. 1966 *Amer. Naturalist* C. 122 The fact that heterogametic males do occur suggests..that *P. nubifer* is probably in the process of changing from male to female heterogamety. 1971 [see *homogametic *a.*]. 1971 *Nature* 18 June 432/2 In animals with female heterogamety, many authors refer to the chromosome which is present only in females as W and the paired sex chromosomes of males as ZZ.

heterogamous, *a.* Add: **1.** (Earlier example.)
1839 J. Lindley *Introd. Bot.* (ed. 3) i. ii. 157 If all the flowers are hermaphrodite in the capitulum, it is homogamous; if the outer are neuter, or female, and the inner hermaphrodite, or male, it is heterogamous.
4. *Biol.* Involving, having the character of, or producing heterogametes.
1895 J. R. Green *Man. Bot.* I. i. ii. 215 The smaller are held..to be male, and the larger female. Plants with such gametes are called heterogamous. 1900 *Ann. Bot.* XIV. 662 The union of heterogamous walled aplanogametes. 1925 E. B. Wilson *Cell* (ed. 3) vii. 593 In true heterogamous forms the macrogamete becomes a large, non-motile cell. 1957 H. C. Bold *Morphol. Plants* iii. 55 These plants are the male and female gametophytes which liberate the heterogamous gametes at maturity. 1969 G. W. Prescott *Algae* i. 42 Sexual reproduction may include isogamous, anisogamous, or heterogamous methods, with or without special sex organs.

heterogamy. Add: **3.** *Biol.* **a.** The condition of having or producing heterogametes; reproduction involving heterogametes. **b.** Heterogamous reproduction.
1894 S. H. Vines *Text-bk. Bot.* iii. 275 Heterogamy:—a. Oogamy: sexual cells, oospheres and undifferentiated male cells... b. Carpogamy: no differentiated female cell. 1897 *Ann. Bot.* XI. 106 Isogamy, heterogamy, and conjugation have been observed. *Ibid.* 118 The curious heterogamy of *Aphanochaete*. 1925 E. B. Wilson *Cell* (ed. 3) vii. 584 A third type is true heterogamy... The gametes are here widely different from each other. 1966 *McGraw-Hill Encycl. Sci. & Technol.* III. 83/2 Heterogamy is also characteristic of land plants and is regarded as the most advanced type of sexual reproduction.
Hence **heteroga·mic** *a.*, characterized by heterogamy (sense *3).
1904 *Science* 3 June 866/1 In the heterogamic subdivision of the homothallic group, a distinct and constant differentiation exists between the zygophoric hyphæ and the gametes derived from them. 1904 *Proc. Amer. Acad. Arts & Sci.* Aug. 210 Two genera of the homothallic group..are heterogamic in that their gametes show a certain constant inequality in size. 1927 Gwynne-Vaughan & Barnes *Struct. & Devel. Fungi* 112 None of the heterothallic forms among the mucors is known to be heterogamic.

heterogeneous, *a.* Add: **4. a.** In various other technical usages, as *heterogeneous fusion, radiation, strain, stream.* Also *heterogeneous reactor*, a nuclear reactor in which the fuel is not uniformly mixed with the moderator and/or coolant (opp. 'homogeneous reactor'); also *heterogeneous pile*.
1883 *Encycl. Brit.* XV. 693/1 Heterogeneous strain. 1902 *Ibid.* XXVIII. 567/2 In the case of crystalline fusion it is necessary to distinguish two cases, the Homogeneous and the Heterogeneous... In the second case, the solid and liquid phases differ in composition; that of the liquid phase changes continuously, and the temperature does not remain constant during the fusion. *Ibid.* XXIX. 257/1 The nuclear divisions are what Weismann calls 'differentiating' or heterogeneous divisions. In them the microcosms of the germ-plasm are not doubled, but slowly disintegrated. 1922 W. B. Scott *Physiogr.* vi. 154 Heterogeneous streams are those which have two or more different kinds of courses, and most rivers, including almost all the longer ones, are of this class. 1947 C. Goodman *Sci. & Engin. Nucl. Power* I. ix. 320 (*caption*) Heterogeneous reactor with circulating liquid fuel. 1949 M. L. Oliphant *Atomic Age* 23 The arrangement used is a so-called 'heterogeneous pile', consisting of some tens of tons of uranium rods or slugs disposed in a calculated 'lattice' throughout a mass of some hundreds of tons of graphite. 1958 O. R. Frisch *Nucl. Handbk.* xii. 2 By far the largest number of reactors built to date have been thermal, heterogeneous reactors. 1958 *Chambers's Techn. Dict.* (ed. 3) 984/2 *Heterogeneous radiation*, that having particles of various energies and/or wavelengths.
b. Denoting the presence of more than one phase (solid, liquid, or gas) in a system or process.
1878 J. W. Gibbs in *Trans. Connecticut Acad. Arts & Sci.* III. 108 (*heading*) On the equilibrium of heterogeneous substances. 1895 *Jrnl. Chem. Soc.* LXVIII. ii. 72 (*heading*) Graphical representation of heterogeneous systems. 1928 J. K. Roberts *Heat & Thermodynamics* xviii. 356 Heterogeneous reactions in which solids and vapours take part. 1950 *Sci. News* XV. 65 Much research had been carried out on heterogeneous catalysts, that is, catalysts at whose surfaces reactions between gases are brought about.

heterogenetic, *a.* Add: **1.** (Earlier example.)
1872 *Proc. R. Soc.* XX. 264 The living units combine, they undergo molecular rearrangements, and the result of such a process of heterogenetic biocrasis is the appearance of larger and more complex organisms.
2. b. *Med.* Of a disease: produced by infection from outside the body.
1890 in Billings *Med. Dict.*
3. *Path.* [ad. G. *heterogenetisch* (Friedberger & Schiff 1913, in *Berl. klin. Wochenschr.* 25 Aug. 1558/1).] = *heterophile *a.* 1.
1918 C. H. Browning *Appl. Bacteriol.* x. 175 'Heterogenetic' antibodies (Friedberger). *Ibid.* 176 The capacity to produce a positive Wassermann reaction is probably a somewhat similar heterogenetic phenomenon. 1920 *Jrnl. Path. & Bacteriol.* XXIII. 364 The injection of organs of certain animals into rabbits leads to the development of 'heterogenetic' immune body for sheep's blood corpuscles. The tissues which act in this way are called 'heterogenetic antigens'. 1944 *Jrnl. Exper. Med.* LXXIX. 556 A heterogenetic antibody showing fixation of complement with human liver and agglutination of sheep erythrocytes was found in certain cases of acute infective hepatitis.

heterogenic (he:tĕrodʒe·nik), *a.*¹ [f. Gr. ἑτερογεν-ής (see Heterogene *a.*) + -ic; in sense 2 prob. directly f. *heterogeny 3 b.]
1. 'Occurring in the wrong sex, as a beard upon a woman' (Dorland, 1900).
2. *Biol.* Characterized by an alternation of generations; = Heterogenetic *a.* 1.
1901 *Jrnl. Exper. Med.* VI. 80 These various forms were but different phases in the cycle of a single heterogenic parasite.
3. *Med.* Derived from animals of a different species.
1911 *Jrnl. Exper. Med.* XIV. 245 Fetal tissues of the chicken grew extensively in human, rabbit, and dog plasma. In general, however,..the development of the tissues is slighter..in heterogenic than in autogenic and homogenic plasma. 1920 *Jrnl. Path. & Bacteriol.* XXIII. 366 There is established the existence of an apparently specific relationship between the lipoids of heterogenic organs and the heterogenetic antibody. 1922 *Jrnl. Exper. Med.* XXXV. 17 It would be of advantage to cultivate chicken tissue in heterogenic serum because it is much more difficult to keep alive in vitro permanently a strain of mammal fibroblasts than of chicken fibroblasts.

heterogenic (he:tĕrodʒi·nik), *a.*² *Genetics.* [f. Hetero- + *gen(e + -ic.] Having more than one allele of a particular gene.
1947 D. Lewis in *Heredity* I. 86 Pollen grains with such different alleles are described as heterogenic. *Ibid.* 88 Some types of heterogenic pollen were fully compatible on a style carrying only one of the alleles present in the pollen, but were incompatible on styles carrying the other allele only. 1967 *Hereditas* LVIII. 25 There is lack of interaction in all the twelve allelic pair constellations which are possible in the heterogenic pollen of the tetraploid.

heterogenist. (Earlier example.)
1870 *Nature* 14 July 224/1 M. Pasteur's researches..have..strongly influenced the opinions of very many scientific men on the question of the truth or falsity of the doctrines of the heterogenists.

heterogenite (hetĕrǫ·dʒinəit). *Min.* [ad. G. *heterogenit* (A. Frenzel 1872, in *Jrnl. f. prakt. Chem.* V. 404), f. Gr. ἑτερογεν-ής of different kinds, in allusion to the fact that the mineral differs in composition from some manganese oxides, which it outwardly resembles: see -ite¹.] A name applied to various hydrous oxides of cobalt.
1872 *Jrnl. Chem. Soc.* XXV. 991 Heterogenite is a product of the decomposition of smaltine or tin-white cobalt (speiss-kobalt). 1922 *Amer. Mineralogist* VII. 195 It is therefore recommended that the prior name heterogenite be applied to all *cobalto-cobaltic hydroxides*, of varying purity. 1962 *Mineral. Mag.* XXXIII. 258 Of the many names proposed for natural cobaltic hydroxide, heterogenite has a clear priority over transvaalite, stainierite, mindigite, trieuite, and boodtite. Heubachite is a nickelian and schulzenite a cuprian heterogenite.

heterogenous, *a.* Delete † *Obs.* and add later examples. (In mod. use prob. repr. the pronunc. (hetĕrǫ·dʒinəs) given by some speakers to *heterogeneous*: cf. *homogenous* (*homogeneous *a.* ¶).)
1916 [see *club necktie* (*club *sb.* 19)]. 1961 Webster, *Heterogenous* = Heterogeneous. 1971 *Nature* 20 Aug. 586/1 A very heterogenous collection of twenty articles discusses basic physiology of the small intestine. 1972 *Ibid.* 10 Mar. 78/1 A heterogenous population of schizophrenic patients.
2. *Surg.* [prob. directly f. Gr. γένος race.] Of transplanted tissue: † **a.** = *homogenous *a.* 2, *homoplastic *a.* 2. *Obs.*
1909 *Boston Med. & Surg. Jrnl.* 23 Dec. 916/1 Whether heterogenous grafts are taken from a living or from a dead subject, there is with them some likelihood of failure to form *permanent* skin.
b. = *heteroplastic *a.* 3.
1939 S. Fomon *Surg. Injury & Plastic Repair* ii. 107 Failure of heterogenous transplants. 1949 Eichelbaum & Turner in M. Thorek *Mod. Surg. Technic* (ed. 2) II. xxix. 1149/1 The heterogenous graft..has been abandoned in favor of the homogenous graft which as an autogenous graft is now in the widest use. 1965 *Jrnl. Amer. Med. Assoc.* 2 Aug. 380/1 Results with the processed heterogenous bone transplants closely parallel those with autogenous and homogenous bone transplants.

heterogeny. Add: **2.** Delete *rare*. (Later examples.)
1921 *S.P.E. Tract* v. 10 We find a heterogeny of words in use. 1927 *Sunday Express* 29 May 5/1 Every conceivable kind of article which forms the heterogeny of the shops patronised by women.
3. b. Alternation of generations, esp. of a sexual and a parthenogenetic generation.
1886 W. E. Hoyle tr. *Leuckart's Parasites of Man* 25 The theory of the heterogeny of Entozoa. *Ibid.* 96, I have for some time been accustomed to call such an alternate succession of dimorphous sexual generations by the name 'Heterogeny'. 1889 E. B. Poulton et al. tr. *Weismann's Ess. Heredity* 325 In the *Daphnidae*, heterogeny may pass into pure parthenogenesis by the non-appearance of the sexual generations. 1931 Blacklock & Southwell *Guide Human Parasitol.* xii. 105 Where one egg produces more than one adult, asexual multiplication has obviously followed the sexual multiplication, i.e. alternation of generations, or heterogeny, exists. 1946 B. Dawes *Trematoda* xiv. 501 Heterogeny, namely, the alternation of a parthenogenetic with a sexual generation.

heterogonic (he:tĕrogǫ·nik), *a.* [f. Heterogon(y + -ic; in sense 1 ad. F. *hétérogonique* (A. Pézard 1918, in *Bull. biol. de la France et de la Belg.* LII. 25).] **1.** = *allometric *a.*
1924 *Nature* 20 Dec. 895/1 Whereas in the male the ratio *abdomen-breadth : carapace-breadth* remained constant, in the female it increased continuously... Pézard (1918) has styled the growth of such an organ *heterogonic*. 1935 *Ibid.* 14 Sept. 433/2 Heterogonic growth clearly involves a regularly continuous change of proportions. 1964 H. W. Manner *Elem. Compar. Embryol.* xi. 216 Differential growth is referred to as heterogonic growth.
2. Exhibiting heterogony (sense *1 b); applied to a life-cycle in which free-living organisms alternate with parasites, and to the free-living generation of such a cycle.
1926 J. H. Sandground in *Amer. Jrnl. Hygiene* VI. 338 In the genus Strongyloides [of nematodes] heterogony may be involved in the life-cycle, but in many instances there is only one multiplicative generation and this mode of development has been called *direct* or, if we may introduce a new term to express this, homogonic in contradistinction to heterogonic or indirect. 1936 *Amer. Jrnl. Hygiene* XXIV. 83 No obvious necessity for the interpolation of a heterogonic generation in the reproductive cycle of *S. ratti* has manifested itself. 1951 L. H. Hyman *Invert.* III. xiii. 304 The eggs..may develop into free-living males and females, the offspring of which proceed to the parasitic phase so that the cycle is indirect or heterogonic.
So **he:terogo·nically** *adv.*
1932 J. S. Huxley *Probl. Rel. Growth* ii. ii. 53 The trunk grows heterogonically with reference to the head.

heterogony. Add: **1. b.** *Biol.* Alternation of generations, esp. of a diœcious and a hermaphroditic generation.
1906 P. Falcke tr. *Braun's Anim. Parasites Man* 273 In a number of Nematodes..heterogony occurs. 1908 Beattie & Dickson *Textbk. Gen. Path.* vi. 395 In other nematodes what is known as Heterogony occurs, in which there may be alteration [ed. 2, 1921, alternation] of fully developed sexual generations. 1936 *Nature* 9 May 780 Heterogony has been widely used to denote a certain type of reproductive cycle.
2. *Biol.* = *allometry.
1927 *Jrnl. Genetics* XVII. 309 (*heading*) Discontinuous variation and heterogony in Forficula. 1938 [see *allometry].
3. Phr. *heterogony of ends*: a principle enunciated by Wundt (*Ethik* 1886), according to which the development of religion and of codes of moral and social behaviour produced results that are to be distinguished from the cause of the development and were not intended at its outset; e.g. moral customs are regarded as (unforeseen) outgrowths from religious ceremonial.
1887 *Mind* XII. 286 The most general results of the author's [sc. Wundt's] investigation are a 'law of three stages' of moral development and a 'law of the heterogony of ends'. 1897 J. H. Gulliver et al. tr. *Wundt's Ethik* I. 330 The law of the *heterogony of ends*. We mean to express by this name what is a matter of universal experience: that manifestations of will, over the whole range of man's free voluntary actions, are always of such a character that the effects of the actions extend more or less widely beyond the original motives of volition, so that new motives are originated for future actions, and again, in their turn, produce new effects. 1911 *Encycl. Brit.* XVIII. 241/2 Each particular will is directed to particular ends, but..beyond these ends effects follow as unexpected consequences, and..this heterogony produces social effects which we call custom.

heterograft (he·tĕrogrɑft). *Med.* and *Biol.* [f. Hetero- + Graft *sb.*¹] A graft taken from an individual of a species different from that of the recipient; a heterotransplant.
Quot. 1909 is in the sense of *homograft.
1909 *Boston Med. & Surg. Jrnl.* 23 Dec. 918/2 Autografts succeed better than hetero-grafts. 1923 Stedman *Med. Dict.* (ed. 7) 449/2 *Heterograft*, a graft taken from an animal—not from another person, *isograft*, or from another part of the same individual, *autograft*. 1927 H. T. Karsner *Human Path.* xi. 313 The fact that chemical conditions must be nicely adjusted is indicated by the failure of heterografts and the relatively higher degree of success of autografts over homeografts. 1948 *Endeavour* VII. 165 In simpler organisms, such as *Hydra*, the graft can be of different species (heterograft). 1972 *Daily Colonist* (Victoria, B.C.) 12 Mar. 22/5 The Brooke research team looked to animals as a source of possible heterografts.
Hence **he·terografted** *ppl. a.*; **he·terografting** *vbl. sb.*
1927 H. T. Karsner *Human Path.* xi. 312 Heterografting..would throw open a large field for transplantation of various tissues and organs which cannot be removed from human donors..until death has occurred. 1961 *Biol. Abstr.* XXXVI. 2017/2 Regeneration of homografted and heterografted limbs in the stick insects. 1968 J. C. Norman et al. *Organ Perfusion & Preservation* xxviii. 382 Heterografted fox livers. *Ibid.* 383 The loss of perfusion immediately after heterografting.

heterography. 1. (Later examples.)
1908 *Daily Chron.* 6 May 4/6 On the subject of 'howlers', we all make them at times, being victims of heterophasia (or heterography). 1920 W. Perrett *Peetickay* 15 There is a pretty general consensus that unconventional spelling or heterography is bad spelling.

heterokaryotic (he:tĕrokæri,ǫ·tik), *a.* *Bot.* Also **-caryotic.** [ad. G. *heterocaryotisch* (H. Burgeff 1913, in *Ber. d. Deut. Bot. Ges.* XXX. 680), f. Hetero- + Kary(o- + -otic.] Exhibiting or resulting from heterokaryosis. Hence **he:teroka·ryon** (-kæ·ri,ǫn), a heterokaryotic cell, structure, or organism; **he:terokaryo·sis** (-kæri,ōu·sis), the condition, prevalent among fungi, in which two or more genetically different nuclei are maintained in a common cytoplasm.
1916 B. D. Jackson *Gloss. Bot. Terms* (ed. 3) 178/1 *Heterokaryotic*,..the character of spores in which both male and female nuclei exist (Burgeff); *Heterokaryosis* is the condition. 1932 *Phytopathology* XXII. 955 The term 'heterocaryosis' precisely describes the condition of a cell containing 2 or more genetically different nuclei. 1935 *Ibid.* XXV. 285 The very young spore already contains several nuclei; hence it is quite possible that each may carry different factors for cultural characters and pathogenicity. The organism may then be truly heterocaryotic. 1942 *Bull. Torrey Bot. Club* LXIX. 75 (*heading*) Heterocaryotic vigor in Neurospora. 1945 *Genetics* XXX. 13 The natural heterokaryon in *Neurospora crassa* generally contains both plus and minus nuclei. 1952 S. Emerson in J. W. Gowan *Heterosis* xii. 200 (*caption*) Heterocaryon formation resulting from hyphal fusion. 1955 *New Biol.* XIX. 20 Heterokaryosis has this advantage over sex that a single organism can contain genes from more than two parents. 1969 *Nature* 30 Aug. 961/2 The diploid nuclei were produced by rare nuclear fusion in a balanced heterokaryon formed from two different uninucleate colour mutants.

heterological (he:tĕrolǫ·dʒikăl), *a.* [ad. G. *heterologisch* (Grelling & Nelson 1907, in *Abhandl. Fries'schen Schule* II. 307), f. Hetero- + Gr. λόγος word: see Logos.] Of an adjective or other predicate: not having the property it denotes. Hence **he:terologica·lity,** the property of being heterological.
1926, 1952 [see *autological *a.*]. 1940 B. Russell *Inquiry Meaning & Truth* v. 79 'Long' is heterological because it is not a long word, but 'short' is homological. 1950 *Analysis* X. 79 When one is asked, 'Is "heterological" heterological?' no answer need be given until the notion of heterologicality is further analyzed.

heterologous. Add: **c.** (See quot. 1889.)
1889 V. H. W. Wingrave *Dunman's Gloss. Anat. Terms*, *Heterologous stimuli*, stimuli which will excite a sensory nerve when applied either to its termination or in its continuity. 1913 *Amer. Jrnl. Physiol.* XXXII. 230 Aronsohn..declared that substances in solution when poured into the nose, could be smelled... This conclusion was confirmed by Vaschide (1901), and by Veress (1903), though the latter showed that the solutions tested were rather in the nature of heterologous than homologous stimuli.
d. *Path.* and *Bacteriol.* Deriving from, associated with, or belonging to another species or type of organism or an antigen or antibody of different constitution.
1893 Parker & Rönnfeldt tr. *Weismann's Germ-Plasm* ix. 265 In the crosses between different species, the idioplasm of a cell in many stages will be composed of homologous as well as of heterologous ids. 1910 *Jrnl. Path. & Bacteriol.* XIV. 30 We have endeavoured to determine whether..the inoculation of heterologous bacteria..affects the production of specific agglutinins in animals previously immunised against the *Bacillus coli*. 1915 *Lancet* 3 Apr. 701/2 (*heading*) Heterologous transplantation: mouse tumours in rats. 1916 *Jrnl. Path. & Bacteriol.* XX. 214 Heterologous resistance as exemplified in the reactions of the sera of animals immunised with transplanted new growths from alien species. 1933 W. W. C. Topley *Outl. Immunity* vi. 91 A serum that agglutinates the same bacterium but has been produced by the inoculation of some other bacterium, differing in one or more of its antigenic components, is termed a heterologous serum. 1958 *Nature* 13 Dec. 1678/2 The multiplication of extraneous viruses in tumours grown in homologous and heterologous hosts depends on the host animal.

heterolysin (he:tĕrolǝi·sin). [a. G. *heterolysin* (Ehrlich & Morgenroth 1900, in *Berl. klin. Wochenschr.* XXXVII. 455/1), f. Hetero- + *lysin.] A hæmolysin formed in the bloodstream of an animal when blood cells from one of a different species are introduced.
1901 in Dorland *Med. Dict.* (ed. 2). 1903 *Science* 3 July 9/1 Many observations upon the effects of iso- and hetero-lysins—as for kidney and liver cells—about which there is no reasonable doubt. 1934 Zinsser & Bayne-Jones *Textbk. Bacteriol.* (ed. 7) xvi. 212 The blood cells of one animal, injected into an animal of another species, give rise to a hemolytic substance in the blood serum of the second animal, which is strictly specific for the variety of cells injected. Such hemolysins..are spoken of as *heterolysins*.

heterolysis (hetĕrǫ·lisis). [f. Hetero- + *-lysis.] **1. a.** The dissolution of blood cells by a heterolysin. **b.** The dissolution of cells of one kind by an enzyme of cells of another kind.
1902 Vaughan & Novy *Cellular Toxins* (ed. 4) vii. 129 Ehrlich suggests that the hemolytic action of the blood serum of one animal upon the corpuscles of another species be designated as 'heterolysis'. 1909 Allbutt & Rolleston *Syst. Med.* (ed. 2) VI. 783 The autolysis of bland infarcts is..due to enzymes derived from the cytoplasm..and is a much slower process than the heterolysis due to the leucocytic invasion seen in infective infarcts. 1924 R. Muir *Text-bk. Path.* i. 14 The term autolysis has been applied to the digestive softening of tissues produced by their own organisms. Such enzymes..may be produced by other cells, especially leucocytes—heterolysis. 1962 H. Hanson in A. Pirie *Lens Metabolism Rel. Cataract* 470 More significant for the specification of intracellular proteases are experiments on heterolysis, i.e. the action of cell proteases on proteins or peptides that are foreign to the cell.
2. *Chem.* The splitting of a molecule into two oppositely charged ions.
1938 C. K. Ingold in *Trans. Faraday Soc.* XXXIV. 227, I would suggest that we may as well have two [new words] and be done with it: viz., 'heterolysis' and 'homolysis', defined as follows: A:B→A+:B (*Heterolysis*) [and] A:B→A.+.B (*Homolysis*). Here the dots represent electrons. 1946 *Nature* 28 Sept. 448/2 A slow heterolysis of nitric acid cannot depend only on proton transfers. 1965 Phillips & Williams *Inorg. Chem.* I. x. 344 Due to the high energy of production of A^+A^-, heterolysis..of homonuclear species is not very probable, except in the presence of catalysts.
So **heteroly·tic** *a.*, characterized by or of the nature of heterolysis (in either sense).
1909 *Chem. Abstr.* III. 1766 Cancer tissue itself possesses no heterolytic properties. 1919 J. Ewing *Neoplastic Dis.* iii. 67 A heterolytic action of tumor ferments has been actively claimed. 1941 *Trans. Faraday Soc.* XXXVII. 604 It is necessary to distinguish between two main types of bond-fission, viz., symmetric or homolytic fission and dissymmetric or heterolytic fission, defined as follows (the dots representing electrons): R./.X (homolytic fission) [and] R/..X (heterolytic fission). *Ibid.* 608 The heterolytic substitutions under discussion are characterised by important electrical transferences in the reactant species. 1964 N. G. Clark *Mod. Org. Chem.* xxv. 514 This process is called heterolysis, and reactions involving it are termed heterolytic reactions.

heteromorphic, *a.* Add: **1. a.** In wider use in *Biol.*
1935 F. E. Fritsch *Struct. & Reprod. Algae* I. 52 The last type of life-cycle..exhibits an alternation of a large sporophyte and a generally small gametophyte. This heteromorphic (antithetic) alternation is encountered solely among Phaeophyceae. 1967 Alexopoulos & Bold *Algae & Fungi* iv. 23 In heteromorphic alternation, the two sequential organisms differ morphologically.
c. *Cytology.* Applied to homologous chromosomes that differ in size or form.
1917 E. E. Carothers in *Jrnl. Morphology* XXVIII. 449 The unusual conditions of the chromosomes in this group have made advisable the introduction of four new terms... Heteromorphic—used to designate those tetrads made up of morphologically different homologues. 1925 E. B. Wilson *Cell* (ed. 3) vi. 571 The so-called heteromorphic chromosome-pairs in which the synaptic mates are visibly distinguishable by the eye by differences of size, form, mode of spindle-attachment or structure. 1955 *Jrnl. Genetics* LIII. 593 Symmetrical separation of the pair of heteromorphic X- and Y-chromosomes in the first anaphase of meiosis.
d. *Min.* [tr. F. *hétéromorphe* (A. Lacroix 1917, in *Compt. Rend.* CLXV. 486).] (See quot. 1920.)
1920 A. Holmes *Nomencl. Petrol.* 117 *Heteromorphic*, a term applied to rocks of similar chemical composition, but of different mineral composition. 1921 —— *Petrogr. Methods* x. 410 Each of these rocks is a heteromorphic type of theralite.
e. *Zool.* (See quots.)
1931 W. C. Allee *Animal Aggregations* ii. 16 Heteromorphic colonies are formed when the divisions are unequal, as is the case with the strobila of the Scyphozoa, or during the processes of asexual reproduction of certain worms, such as *Autolytus*. *Ibid.* 23 Heteromorphic and polymorphic colonies are formed when there is a differentiation between the different members of the colony, as occurs in the hybrid Hydractinia, in which feeding, reproductive, and protective zoöids may be recognized.
3. *Biol.* Resulting from heteromorphosis.
1898 *Arch. f. Entwickelungsmech. d. Organismen* VII. 481 Neither the present nor other experiments indicate, that the influence of the organism as a whole upon the regenerating part is able to bring forth a heteromorphic structure..out of material which would normally produce something else. 1932 J. S. Huxley *Probl. Rel. Growth* vi. iii. 175 The regenerated heteromorphic appendage almost invariably is of a type which normally belongs to a more posterior segment—e.g. antenna regenerated instead of eye-stalk (Palaemon). 1966 E. D. Hay *Regeneration* ii. 75 The heteromorphic growths..may result because an insufficient mass of cells is present for correct differentiation of the skeletal pattern.
Hence **he:teromo·rphically** *adv.*, by means of heteromorphosis.
1959 *Jrnl. Biophysical & Biochem. Cytol.* V. 25 Cultures of subcutaneous areolar fibroblasts..are changed heteromorphically so as to resemble cultures of nervous tissue.

heteromorphism. Add: **2.** 'The property of replacing lost parts by new parts which are different from those that have been lost' (*Cent. Dict.* Suppl. 1909); = *heteromorphosis b.
3. Examples relating to additional senses of *heteromorphic *a.*
1895 F. W. Oliver tr. *Kerner's Nat. Hist. Plants* II. 469 One branch of the colony is concerned in the acquiring of nutrition, another in reproduction... To this differentiation among equivalent members we may apply the term Heteromorphism. 1921 A. Holmes *Petrogr. Methods* x. 407 (*heading*) Heteromorphism in igneous rocks. 1954 R. L. Parker tr. *P. Niggli's Rocks & Min. Deposits* iv. 125 Variations in primary mineralogical composition..can result from the bulk chemical composition (heteromorphism).

heteromorphosis (he:tĕromǭ·ɹfōsis, -mǭɹfōu·sis). [f. Hetero- + Morphosis; in sense b ad. G. *heteromorphose* (J. Loeb *Untersuchungen z. physiol. Morphologie d. Thiere* (1891) I. i. 10), f. Gr. ἕτερο-ς different + μόρφωσ-ις formation.] **a.** Abnormal tissue, or tissue formed at the wrong place; heteroplasia.
1891 Foster *Med. Dict.* III, *Heteromorphosis*, a malformation. The *heteromorphoses*..in Fuchs's classification of skin diseases are a genus..including nævus, ochthiasis, polytrichia, and polonychia. 1922 *Guy's Hosp. Rep.* LXXII. 200 Cells differing in structure from those characteristic of the part of the body are produced. Here we have a heteroplasia or heteromorphosis.
b. The regeneration of an organ or structure that is different from the one lost.
1894 *Jrnl. Morphology* IX. 418 To this phenomenon of the reproduction of an organ typically different from the one which had originally occupied that position, he [*sc.* Jacques Loeb] has given the name heteromorphosis. 1901 J. Loeb *Compar. Physiol. Brain* xiv. 203 The processes of heteromorphosis—that is, the transformation or substitution of one organ for a morphologically different one by means of certain external influences—force us to the same view. 1932 J. S. Huxley *Probl. Rel. Growth* vi. iii. 172 The production of axial heteromorphosis in regeneration, such as biaxial heads or tails, can also be satisfactorily interpreted in terms of the gradient hypothesis. 1940, 1966 [see *homœosis].

heteronomous, *a.* Add: **3.** (Earlier and later examples.)
a 1871 G. Grote *Eth. Fragm.* (1876) ii. 47 The will is..in a certain sense *autonomous*, not *heteronomous*. 1932 W. L. Graff *Lang.* vi. 221 These are *conditioned* sound changes, also called *dependent* or *heteronomous* because they appear to depend upon the extraneous influence of their phonetic surroundings.
Hence **hetero·nomously** *adv.*
1909 *Cent. Dict.* Suppl., *Heteronomously*, in a heteronomous manner. 1948 J. L. Adams tr. *Tillich's Protestant Era* iii. 46 Religion, if it acts heteronomously, has ceased to be the substance and life-blood of a culture and has become itself a section of it. 1966 J. A. Serra *Mod. Genetics* II. xii. 118 In a few cases, however, the colour developed heteronomously; in such cases the host influenced the colour of the implanted eye.

heteronomy. 2. (Earlier example.)
1798 Willich *Elem. Crit. Philos.* 160 Heteronomy, or a foreign legislation, is that, in which not the will itself, but something else determines us to act in a certain manner.

heteronuclear (he:tĕroniū·klĭăɹ), *a.* [f. Hetero- + Nuclear *a.*] **a.** *Chem.* Taking place on different rings in a polycyclic molecule.
1900 E. F. Smith tr. *V. von Richter's Org. Chem.* (ed. 3) II. 390 It can first be ascertained whether the substituents are in the same nucleus (isonuclear) or in different nuclei (heteronuclear). 1951 I. L. Finar *Org. Chem.* I. xxix. 586 When NO_2 or SO_3H is in the 1- or 2-position, heteronuclear substitution occurs in position 5 or 8.
b. *Physics* and *Chem.* Of a molecule: composed of atoms whose nuclei are unlike, i.e. atoms of different elements or of different isotopes of the same element.
1930 *Physical Rev.* XXXVI. 617 The words even and odd are applied..to the *complete* ψ function of any molecule, homonuclear or heteronuclear. 1940 *Ibid.* LVII. 684/2 The HD molecule is heteronuclear and there is no quantum number of total spin. 1950 J. W. T. Spinks tr. *Herzberg's Molecular Spectra & Molecular Structure* (ed. 2) I. iii. 94 This agrees with the observed infrared spectrum of HCl..and of other heteronuclear molecules.

heterophile (he·tĕrofəil), *a.* and *sb.* Also **heterophil.** [a. G. *heterophil* (U. Friedemann 1917, in *Biochem. Zeitschr.* LXXX. 334), f. Hetero- + -phil(e).] **1.** *adj.* Able to react immunologically with sera, etc., from organisms of another species.
1920 *Jrnl. Path. & Bacteriol.* XXIII. 366 The term heterogenetic is not altogether appropriate in this connection, since the important character is not that the antibody is generated by a different kind of antigen, but that it has an affinity for receptors of a species other than that in response to which it was developed, *heterophile* describes this property better. 1929 Topley & Wilson *Princ. Bacteriol. & Immunity* II. li. 748 It is now known that the so-called 'heterophile', or 'Forssman' antigen..is widely distributed among the animal species which have been examined, and in a curiously random fashion. 1935 *Arch. Path.* XIX. 841 Heterophile phenomena embrace those immunologic reactions in which the interaction of antigen and antibody is seemingly not specific in the strictly orthodox conception of the term. 1950 J. V. Dacie *Pract. Hæmatol.* xi. 150 The limits of the titre of heterophile antibodies present in health are ill-defined. 1960 I. A. Stanton *Dict. Med. Secretaries* 70/2 *Heterophile agglutination test*.., a method of testing for the presence of infectious mononucleosis. The blood of patients with this disease contains antibodies for the red blood cells of sheep.
2. *adj.* and *sb.* A polymorphonuclear leucocyte found in the blood of mammals and stained by both acidic and basic dyes; usu. as *sb.*
1938 H. Downey *Handbk. Hematol.* I. ii. 167 The special or 'heterophile' leucocytes of the animals commonly used in experimental research resemble the human neutrophile quite closely in size and morphology, but with slight variations in nuclei, and in the size and staining of the specific granules which make them distinctive. 1941 Calkins & Summers *Protozoa Biol. Research* xviii. 835 The heterophils are functional in immunity by virtue of their obvious phagocytic activities and probably because of their secretion of enzymes. 1965 P. D. Sturkie *Avian Physiol.* (ed. 2) i. 13 The heterophils in the blood of children are usually round and have a diameter of approximately 10 to 15 microns.

heterophilic (he:tĕrofi·lik), *a.* [f. prec. + -ic.] = *heterophile *a.*
1929 *Jrnl. Exper. Med.* XLIX. 497 Horse serum is known to contain a heterophilic antigen. 1966 *Jrnl. Cell Biol.* XXIX. 309 (*caption*) Thin section of heterophilic myelocyte found in rabbit bone marrow.

heterophony (hetĕrǫ·fŏni). [Gr. ἑτεροφωνία: see Plato *Leges* VII. 812D.] Simultaneous performance by two or more singers or instrumentalists of different versions of the same melody. Also **heteropho·nic, hetero·phonous** *adjs.*, different in sound.
1919 *Musical Q.* V. 599 Two forms of instrumental accompaniment were known to the Greeks, namely the homophonic and heterophonic. 1936 *Amer. Speech* XI. 299 Any phonic difference found between comparably situated sounds in heterophonous words in a given language constitutes a phonemic contrast. 1939 *Scrutiny* VII. 402 Ruth Crawford writes heterophonic music in which the various linear strands bear no relation to one another except that of proximity in time-space. 1945 *Times* 16 Mar. 6 The name for this simultaneous variation with its permitted dissonance is heterophony. 1962 *Listener* 21 June 1091/2 The pianos are brilliantly used as percussion instruments, sometimes in harp-like arpeggiated pentatonics, sometimes in wildly clashing heterophony. 1970 *Daily Tel.* 16 Nov. 9/4 Heterophony was..only just round the corner from the virelais, 'Se Ma Dame' and 'Comment qu'a moy', as treated here with instrumental doubling of the voice parts. 1971 *Guardian* 22 Feb. 8/4 The work was somewhat discursive, there being too few landmarks in the prevailing heterophonic texture to compel the listeners' attention throughout.

heteroplastic, *a.* Add: **3.** *Surg.* Said of a graft between two individuals of different species.
[1889: see *homoplastic *a.* 2.] 1898 *Arch. f. Entwickelungsmech. d. Organismen* VII. 471 There were obtained a number of larvae of normal external form, but which were composed of parts derived from two distinct species. ..The general biological interest attaching to such heteroplastic combinations is naturally great. 1908 A. Carrel in *Jrnl. Amer. Med. Assoc.* 14 Nov. 1664/1 There are several varieties of heteroplastic transplantations according to the zoologic distance which separates the host and the owner of the transplanted tissue. 1909 [see *homoplastic *a.* 2]. 1917 *Brit. Jrnl. Surg.* V. 199 Heteroplastic grafts—i.e. bones transferred from one animal to another of a different species—..show neither vitality nor proliferative capacity. 1923 [see *heteroplasty 1 b]. 1924 *Proc. Nat. Acad. Sci.* X. 69 The heteroplastic transplantation of limbs. 1959 *Arkiv f. Zool.* XII. 183 Heteroplastic transplantation would be a most valuable means of demonstrating the role of movable cells during a regeneration, because the chromosome number and morphology reveals the origin of the cells.
So **heteropla·stically** *adv.*
1928 *Proc. Soc. Exper. Biol. & Med.* XXV. 686 (*heading*) Growth of heteroplastically transplanted eyes and limbs in Amblystoma. 1932 J. S. Huxley *Probl. Rel. Growth* II. ii. 53 Eyes and other organs when heteroplastically transplanted. 1960 B. I. Balinsky *Introd. Embryol.* vi. 167 Hilde Mangold transplanted heteroplastically (from *Triturus cristatus* to *Triturus taeniatus*) a piece of the dorsal lip of the blastopore of an early gastrula.

heteroplasty (he·tĕroplæsti). [f. Hetero- + -plasty.] **1.** *Med.* and *Biol.* [ad. F. *étéroplastie* (P. F. Blandin *Autoplastie* (1836) 7)] The operation of grafting tissues between different individuals. **a.** Between two individuals of the same species.
1855 Dunglison *Dict. Med. Sci.* (ed. 12) 440/2 *Heteroplasty*, irregular plastic or formative operations, that do not admit of exact classification. 1874 *English Mechanic* 321 The name heteroplasty has been adopted for the operation of taking pieces of skin from amputated limbs, and using them to produce cicatrisation on the bodies of other subjects. 1901 A. G. Gerster in R. Park *Treat. Surgery* liv. 1154 Heteroplasty..is the operation wherein living tissues foreign to the bearer of the defect are employed. The transfer of a skin-flap from one individual to another constitutes heteroplasty. 1909 *Boston Med. & Surg. Jrnl.* 23 Dec. 915/1 (*heading*) Heteroplasty. (Grafting from another person.)
b. Between two individuals of different species.
1923 H. Neuhof *Transplantation of Tissues* i. 3 By heterotransplantation (heteroplasty, heteroplastic transplantation, zoöplasty) is meant the transference of a tissue from an individual of one species to an individual of a different species. 1929 *Ann. Surg.* XC. 929 In grafting skin derived from other animal species (heteroplasty), that of dogs, chickens or the internal pellicle of an egg may be used.
2. *Path.* = Heteroplasia.
1886 *Syd. Soc. Lex.*, Heteroplasty, same as Heteroplasia.

heteroploid (he·tĕroploid), *a.* (and *sb.*). [a. G. *heteroploid* (H. Winkler 1916, in *Zeitschr. f. Bot.* VIII. 422), f. Hetero- + *-ploid.] Having a chromosome number that is neither the haploid nor the diploid number characteristic of the species; freq., in restricted sense, = *aneuploid *a.* (see quot. 1928); also as *sb.*, a heteroploid organism. Hence **he·teroploidy,** the condition of being heteroploid.
1926 T. H. Morgan *Theory Gene* xii. 172 In so far as the addition of one or more chromosomes to, or loss from, a given group produces a new number, the word heteroploid has been used. *Ibid.* 176 In contrast to these triplo-types of Drosophila there is another heteroploid type..in which one of the small chromosomes is absent. *Ibid.* 188 The heteroploids are not so viable as the balanced types from which they arise. *Ibid.*, The occurrence of heteroploidy must be regarded as a significant genetic event. 1928 C. A. Jørgensen in *Jrnl. Genetics* XIX. 135 The term heteroploid was introduced by Winkler in 1916 to designate..not only the straight numbers 3*n*, 4*n*, 5*n*, etc., but also 2*n*+1, 2*n*+2, 3*n*−1, etc. The heteroploid numbers were divided by Täckholm (1922) into euploid..and aneuploid... I prefer to restrict the term heteroploid to the series *n*, 3*n*, 4*n*, 5*n*, etc... and to use the term *aploid* for the irregular numbers. 1934 *Nature* 3 Nov. 708/2 In the heteroploid series, all having 25 chromosomes, the anatomical differences..were due to the genic constitution of the extra chromosome. 1955 J. B. & H. D. Hill *Genetics & Human Heredity* xvii. 331 Although rare in animals, heteroploidy occurs frequently in plants. 1956 *Nature* 25 Feb. 384/2 It is of interest to try to find whether the heteroploid cells appear only in certain tissues or if they can arise anywhere in the young embryos.

heteropolar (he:tĕropōu·lăɹ), *a.* [f. Hetero- + Polar *a.*] **1.** *Bot.* [In Dict. s.v. Hetero-, where read 844 for 843.] **2.** *Electr.* Of an electric generator or its operation: using armature conductors that pass alternate north and south magnetic poles, so that the current generated is periodically reversed. Cf. *homopolar *a.* 2.
1896 S. P. Thompson *Dynamo-Electric Machinery* (ed. 5) 475 In cases where it [*sc.* a conductor] passes from being opposite a N-pole to being opposite a S-pole, the operation is said to be heteropolar. 1946 *Nature* 28 Sept. 455/1 For the majority of applications the modern heteropolar inductor alternator is the most suitable machine.
3. *Chem.* [a. G. *heteropolar* (R. Abegg 1906, in *Zeitschr. f. anorg. Chem.* L. 309).] Formed by ions of opposite sign, between which there is electrostatic attraction. Cf. *homopolar *a.* 3.
1922 [see *homœopolar* adj. s.v. *Homœo-]. 1930 *Engineering* 5 Dec. 700/3 There were, further, different types of cohesive forces... (3) Ionic or heteropolar cohesion. In the rock-salt atom, for instance, the valency electron of the sodium migrated into the chlorine structure, giving rise to electrostatic action. 1964 J. W. Linnett *Electronic Struct. Molecules* ix. 143 Values which bring the results for these heteropolar molecules to the same scale as that for homopolar molecules.

heteropycnosis (he:tĕropiknōu·sis). *Cytol.* Also **-pyknosis.** [ad. G. *heteropyknose* (S. Gutherz 1907, in *Arch. f. mikrosk. Anat.* LXIX. 495), f. Hetero- + Gr. πυκν-ός thick, dense: see -osis.] The persistence of greater than average staining in chromosomal material; the character or condition, exhibited by some chromosomes or chromosomal regions in any particular nucleus, of being more (or, for negative heteropycnosis, less) condensed and hence appearing to take up more (or, less) stain than do the majority of chromosomes or chromosomal regions. Hence **heteropycno·tic** *a.*
1925 E. B. Wilson *Cell* (ed. 3) x. 759 In the greater number of cases heteropycnosis first takes place after the final spermatogonial division. 1934 *Genetics* XIX. 467 In the latter species [*sc.* *Drosophila melanogaster*] he finds that about half of the X chromosome is heteropycnotic in the interphase. 1952 G. H. Bourne *Cytol. & Cell Physiol.* (ed. 2) v. 195 This property of heteropycnosis is especially characteristic of sex-chromosomes, such as the Y-chromosome of *Drosophila* species and the X-chromosomes of grasshoppers, but is also shown by many autosomal chromosome regions, and sometimes by whole autosomes. 1962 *Lancet* 19 May 1075/1 Once isopyknosis or heteropyknosis of a given X chromosome has been established, it is irreversible. 1963 Lewis & John *Chromosome Marker* i. i. 18 This positive heteropycnosis is later reversed and by metaphase-I the X has become negatively heteropycnotic and stains faintly. 1968 H. Harris *Nucleus & Cytoplasm* iv. 75 It is probable that other highly condensed heterochromatic or heteropyknotic regions in interphase nuclei also synthesize very much less RNA than the euchromatic regions.

heterosexual (hetĕrose·ksiuăl), *a.* and *sb.* [See Hetero- and Sexual.] **1. a.** Characterized by a sexual interest in members of the opposite sex. **b.** Pertaining to sexual relations between people of opposite sex. Also as *sb.*, a heterosexual person.
1892 C. G. Chaddock tr. *Krafft-Ebing's Psychopathia Sexualis* iii. 324 The object of post-hypnotic suggestion is to remove the impulse to masturbation and homosexual feelings and impulses, and to encourage heterosexual feelings with a sense of virility. 1920 J. Riviere et al. tr. *Freud's Coll. Papers* (1924) II. 207 To convert a fully developed homosexual into a heterosexual. 1927 *Scots Observer* 1 Oct. 15/3 A certain proportion of people..are as instinctively homosexual as the normal individual is heterosexual. 1935 *Discovery* Oct. 313/2 Homosexuality under the homosexual period of Greek life was not a sin, but it became so under heterosexual Christianity. 1966 K. Walker *Sexual Behaviour* xxv. 222 Homosexuals vary as much in physique and in temperament as do heterosexuals. 1969 *Daily Tel.* 21 Jan. 17/3 Co-educational schools probably tend to hasten heterosexual experimentation. 1972 *Sat. Rev.* (U.S.) 12 Feb. 27/1 Homosexuals, like heterosexuals, should be treated as individual human beings.
2. Pertaining to, characteristic of both sexes.
1918 *Genetics* III. 287 Studies..on ablation of the gonads in birds, combined with hetero-sexual transplantation, show that both the testes and ovary in birds produce substances that have a profound regulatory effect on growth. 1963 *Cytogenetics* II. 332 The corresponding ovaries, generally atrophied owing presumably to male testicular hormones, contained no heterosexual cells.
Hence **he:terosexua·lity,** the condition of being heterosexual; heterosexual characteristics.
1900 H. Blanchamp tr. *Féré's Sexual Instinct* viii. 183 Psycho-sexual hermaphroditism, in which there are traces of hetero-sexuality, though homo-sexuality predominates. 1965 *New Statesman* 30 Apr. 677/2 Many people are made unhappy by their heterosexuality, but I do not believe the psychotherapist exists who would try to cure that.

heterosis. Add: **2.** *Zool.* Segmentation in which the parts are different.

1902 E. R. Lankester in *Encycl. Brit.* XXV. 691/2 It becomes apparent from this enumeration that there are a good many important elements or 'meromes' in an Arthropod metamere or somite which can become the subject of heteromerism or, to use a more apt word, of 'heterosis'. *Ibid.*, The Fourth Law of metamerism (auto-heterosis of the meromes).

3. *Genetics.* The tendency of cross-breeding to produce an animal or plant with a greater hardiness and capacity for growth than either of the parents; hybrid vigour.

1914 G. H. Shull in *Zeitschr. f. induktive Abstammungs- und Vererbungslehre* XII. 127, I suggest that instead of the phrases, 'stimulus of heterozygosis', 'heterozygotic stimulation', 'the stimulating effects of hybridity', 'stimulation due to differences in uniting gametes', etc...the word 'heterosis' be adopted. The corresponding adjective 'heterotic' may also be useful in such expressions as 'heterotic effects'. **1938** *Nature* 3 Dec. 1002/2 The expression of heterosis is much influenced by various external factors. **1959** *New Biol.* XXVIII. 75 Less severe departures from normal growth in hybrids may.. involve slower development and poorer growth co-ordination, or..higher growth rates and seemingly enhanced vigour—the latter the phenomenon of heterosis, or hybrid vigour.

heterosite (he·tĕrosəit). *Min.* Formerly also **heteposite, heterozite.** [ad. F. *hétérosite, -zite* (named by F. Alluaud aîné: see L. N. Vauquelin 1825, in *Ann. de Chim. et de Phys.* XXX. 294, where the word is spelt *hétépozite*), irreg. f. Gr. ἕτερος different + -ITE[1].] A phosphate of iron and manganese, differing from purpurite in containing more iron.

1834 R. Allan *Man. Min.* 309 Heteposite. **1835** C. U. Shepard *Treat. Min.: Second Part* I. 258 Heterosite. (See Triplite). **1854** J. D. Dana *Syst. Min.* (ed. 4) II. 407 Heterosite, by Rammelsberg's analysis. **1858** J. Nicol *Elem. Min.* 241 Heterozite. **1951** C. Palache et al. in *Dana's Syst. Min.* (ed. 7) II. 676 Heterosite and purpurite are secondary minerals formed by oxidation of triphylite and lithiophilite.

heterotic (hetĕrǫ·tik), *a.* [f. Heter(osis + -otic; in sense 1 f. directly on Gr. ἑτέρωσις alteration.] **1.** Pertaining to the manipulation of differences (*nonce-use*).

1905 G. M. Fisher tr. *Höffding's Probl. Philos.* ii. 99 The advance of knowledge consists in a reduction of differences (to a 'heterotic minimum') and in an approximation to a pure description of a continuous process.

2. *Genetics.* Pertaining to or exhibiting heterosis (sense *3).

1914 *Zeitschr. f. induktive Abstammungs- und Vererbungslehre* XII. 127 A highly heterotic plant..because of its unusual vigor may develop branches from buds which in a weaker plant would remain dormant. **1944** *Bull. Torrey Bot. Club* May 267 (*heading*) Heterotic hybrids. **1967** *Amer. Naturalist* CI. 189 (*heading*) Opposite heterotic effects on male weights of reciprocal species hybrids.

heterotopy. Add: Hence **he:teroto·pically** *adv.*, with the result of becoming heterotopic.

1923 *Guy's Hosp. Rep.* LXXIII. 328 They re-acquired ..the prospective fate of being differentiated heterotopically into squamous epithelium.

heterotransplant (he:tĕrotra·ns,plɑnt), *sb.* *Med.* and *Biol.* [f. Hetero- + Transplant *sb.*] A piece of tissue or an organ taken from one individual and transplanted (or intended for transplantation) to another individual of a different species.

1918 *Jrnl. Med. Res.* XXXVII. 378 It is..improbable that the preponderance of lymphocytes in the heterotransplants..was altogether due to the bacterial infection. **1935** *Amer. Jrnl. Cancer* XXIII. 284 One of the growing heterotransplants of mouse sarcoma 37 was removed from a normal rat. **1959** *Ann. R. Coll. Surgeons* XXV. 45 So far with heterotransplants of human tissue the sex of the recipient has not appeared to influence the survival of the graft.

heterotransplant (he:tĕrotrɑns,plɑ·nt), *v.* *Med.* and *Biol.* [f. prec. sb.] *trans.* To transplant from one individual to another of a different species. So **he:terotranspla·nted** *ppl. a.*

1920 *Jrnl. Exper. Med.* XXXI. 767 The greater part of the heterotransplanted thyroid suffered..through lack of suitability of the body fluids of the host. **1959** *Cancer Res.* XIX. 633 (*heading*) Heterotransplanted choriocarcinomas. **1962** *Ibid.* XXII. 563 Two choriocarcinomas ..were successfully heterotransplanted to cortisone-treated hamsters.

heterotransplantable (he:tĕrotrɑns,plɑ·ntăb'l), *a.* *Med.* and *Biol.* [f. prec. vb., after *transplantable.*] Capable of being successfully transplanted from one individual to another of a different species. So **he:terotransplantabi·lity.**

1943 *Cancer Res.* III. 809/1 An interpretation of the significance of this demonstration necessitated a more accurate definition of the limits of heterotransplantability. **1947** Ackerman & del Regato *Cancer* iii. 70 Greene believes that failure in transplanting benign tumors suggests that heterotransplantability is a characteristic property of cancer. **1952** *Cancer Res.* V. 41/1 The heterotransplantable phase of the tumor's existence was of relatively short duration. **1971** *Nature* 16 Apr. 455/1 The newborn hamster treated with ALS seems to be more sensitive than the cortisone-treated adult for the purposes of testing heterotransplantability of cultured cells.

heterotransplantation (he:tĕrotrɑns,plɑntēi·ʃən). *Med.* and *Biol.* [f. Hetero- + Transplantation.] The operation of transplanting tissue from one individual to another of a different species.

1905 [see *homotransplantation]. **1907** *Jrnl. Exper. Med.* IX. 226 These heterotransplantations were attempted with the aim of ascertaining whether the vessels, in spite of the toxic action of the cat's blood on the dog's tissue, could..take over the function of the vessels removed. **1940** *Amer. Jrnl. Cancer* XXXIX. 170 The direct heterotransplantation of tissues is rarely, if ever, successful among higher mammals. **1971** [see *homotransplantation].

heterotropal, *a.* (Earlier example.)

1832 J. Lindley *Introd. Bot.* I. ii. 193 The radicle will point neither to the apex nor to the base of the seed, but the embry will lie, as it were, across it, or be heterotropal, as is the case in the primrose.

heterotroph (he·tĕrotrǫuf). *Biol.* Also **-trophe.** [f. Hetero- + Gr. τροφός feeder.] Any organism which requires an external supply of energy contained in complex organic compounds to maintain its existence.

1900 B. D. Jackson *Gloss. Bot. Terms* 311/1 *Heterotroph*, employed by Pfeffer to denote a pure saprophyte. **1959** *New Scientist* 7 May 1002/3 The evolutionary steps.. were first, the formation of what he calls heterotrophes, organisms that absorbed as their nutrients organic substances dissolved in the seas in which they lived. **1971** Hawker & Linton *Micro-organisms* v. 145 Fungi, most protozoa, and many bacteria are heterotrophs.

heterotrophic (he:tĕrotrǫu·fik), *a.* *Biol.* [f. Hetero- + Gr. τροφικός nursing.] Characterized by or exhibiting heterotrophy (in various senses: now only in the sense of *heterotrophy 1 c); living as a heterotroph. So **he:terotro·phically** *adv.*

1893 *Jrnl. R. Microsc. Soc.* 665 According as lichens carry on their existence with or without these lodgers, the author [*sc.* Minks] places them respectively under the head of 'autotrophic' or 'heterotrophic' lichens. **1896** *Ibid.* 326 Prof. J. Wiesner proposes the general term *Trophy* for all unilateral excesses of growth in tissues or in organs which are dependent on the position of the organ in question, the term position being understood in a wide sense, as the relation in space of the heterotrophic organ to the horizon and to its parent-shoot. **1900** A. J. Ewart tr. *Pfeffer's Physiol. Plants* I. vii. 363 Plants which are unable to assimilate carbon dioxide must obtain all their organic food-materials from without (heterotrophic or allotrophic nutrition). **1935** W. H. Brown *Plant Kingdom* xvii. 378 Some of the autotrophic bacteria can live heterotrophically. **1969** *Ecology* L. 88/1 Eleven heterotrophic aerobic bacterial types were isolated from the microcosm. **1971** I. G. Gass et al. *Understanding Earth* ix. 139/1 Fox has been able to show that proteinoid microspheres will..come into existence spontaneously on innumerable occasions. If the correct foods are available, they can proliferate heterotrophically indefinitely.

heterotrophism (he:tĕrotrǫu·fiz'm). *Biol.* [f. next: see -ism.] = *heterotrophy 1 c.

1900 A. J. Ewart tr. *Pfeffer's Physiol. Plants* I. vii. 364 All stages of transition between pure autotrophism and heterotrophism are exhibited among obligate or facultative mixotrophic plants. **1936** W. Stiles *Introd. Princ. Plant Physiol.* xiii. 285 Some autotrophic plants..when kept in the dark and provided with such organic nutrients ..exhibit a degree of heterotrophism. **1951** Werkman & Wilson *Bacterial Physiol.* 407 (*heading*) Carbon dioxide assimilation and concepts of autotrophism and heterotrophism.

heterotrophy (he·tĕrotrǫufi). [f. Hetero- + Gr. τροφή nourishment.] **1.** *Biol.* † **a.** [ad. G. *heterotrophie* (A. B. Frank 1885, in *Ber. d. Deut. Bot. Ges.* III. 143).] (In Dict. s.v. Hetero-.) (Example.) *Obs. rare.*

1891 F. P. Foster *Med. Dict.* III. 1867/2 *Heterotrophy*, the quality of obtaining nutrition adventitiously by means of a Fungus whose mycelium takes the place of root-hairs.

† **b.** [ad. G. *heterotrophie* (A. Minks 1893, in *Verhandl. d. zool.-bot. Ges. in Wien* XLII. 402).] (See quot. 1900.) *Obs. rare.*

1900 B. D. Jackson *Gloss. Bot. Terms* 123/1 *Heterotrophy*, (1) used by Minks for those Lichens living symbiotically.

c. The state or quality of living as a heterotroph; heterotrophism.

1930 W. H. Lang tr. *Strasburger's Text-bk. Bot.* (ed. 6) 259 The heterotrophy is not always..complete... Some heterotrophic Phanerogams can construct their organic carbon-compounds in the normal fashion... The converse case is frequently met with. **1949** A. Nelson *Introd. Bot.* xxv. 387 There are degrees of morphological adaptation for heterotrophy.

† **2.** *Bot.* [ad. G. *heterotrophie* (J. Wiesner 1895, in *Ber. d. Deut. Bot. Ges.* XIII. 482).] (See quot. 1900.) *Obs. rare.*

1896 *Jrnl. R. Microsc. Soc.* 326 (*heading*) Heterotrophy and anisophylly. **1900** B. D. Jackson *Gloss. Bot. Terms* 123/1 *Heterotrophy*..(2) [used] by Wiesner for the compound position of a shoot with regard to the horizon and of the mother-shoot.

heterotropic, *a.* Add: **2.** *Genetics.* Applied to chromosomes that are not paired at meiosis.

1905 E. B. Wilson in *Jrnl. Exper. Zool.* II. 508 Since there is no reason for considering the 'accessory chromosome' as in any sense accessory to the others,.. I suggest that..chromosomes of this type may provisionally be called *heterotropic* chromosomes (in allusion to the fact that they pass to one pole only of the spindle in one of the maturation-divisions). **1920** L. Doncaster *Introd. Study Cytol.* xi. 155 When..the heterotropic or X-chromosome passes undivided to one pole of the first spermatocyte division-figure, it divides equationally in the division of those secondary spermatocytes which contain it.

heterozygote (he:tĕrozəi·gǫut). *Biol.* [f. Hetero- + Zygote.] **1.** A diploid individual that has different alleles at one or more genetic loci. Also *attrib.* or as *adj.*, = **heterozygous*.

1902 Bateson & Saunders *Rep. Evol. Comm. R. Soc.* I. 126 The zygote formed by the union of a pair of opposite allelomorphic gametes, we shall call a heterozygote. **1902** W. Bateson *Mendel's Princ. Hered.* 23 This *Aa* is the hybrid or 'mule' form, or as I have elsewhere called it, the heterozygote, as distinguished from *AA* or *aa* the homozygotes. **1906** R. H. Lock *Variation, Heredity, & Evolution* vii. 186 *Plate*, Cobs borne by heterozygote plants pollinated with the recessive. **1930** R. A. Fisher *Genet. Theory Nat. Selection* 8 The heterozygote when mated to either kind of homozygote would produce both heterozygotes and homozygotes. **1949** Darlington & Mather *Elem. Genetics* v. 110 The original form of any heterozygote must be a mutation. **1971** J. Z. Young *Introd. Study Man* iii. 60 Apparently the heterozygotes have an increased resistance.

2. A bacteriophage that carries two different copies of some of its genetic information and produces two kinds of progeny when it infects a bacterial cell.

1952 Hershey & Chase in *Cold Spring Harbor Symp. Quant. Biol.* XVI. 474/1 Since there is adequate reason to call these markers allelic genes, the mottling particles are appropriately termed heterozygotes. **1968** D. S. Ray in Fraenkel-Conrat *Molec. Basis Virol.* ii. 247 The purified heterozygotes are twice the length of normal phage particles and apparently contain two DNA molecules.

Hence **he:terozygosis** (-zəigǫu·sis), the fusion of two genetically different gametes; the state or condition of being heterozygous; **he:terozygo·sity** (-zəigǫ·sĭti), the state or condition of being heterozygous; the degree or extent to which an individual is heterozygous with respect to its complement of genetic loci; **he:terozygo·tic** *a.*, resulting from or pertaining to heterozygosis; **he:terozy·gous** (-zəi·gəs) *a.*, having different alleles at one or more genetic loci.

1902 Bateson & Saunders *Rep. Evol. Comm. R. Soc.* I. 130 The determination from statistical study of zygotes must be exceedingly difficult, seeing that both resulting forms may be heterozygous. **1905** *Rep. Brit. Assoc.* 1904 585 Appearances have been seen..suggesting at first sight that a heterozygosis between two gametes, *both* extracted, may give, *e.g.*, dominance. **1911** *Amer. Naturalist* XLV. 142 The surprises of heterozygotic 'constructions' or of new combinations in F_2 may..be responsible for the case of De Vries's buttercups. **1912** East & Hayes *Heterozygosis in Evol. & Plant Breeding* 37 Decrease in vigor due to inbreeding lessens with decrease in heterozygosity. **1914** Heterozygotic [see *heterosis 3]. **1949** Darlington & Mather *Elem. Genetics* xiii. 280 (*caption*) The potential variability existing in the differences between homozygotes..must first be converted into heterozygotic potential by intercrossing. *Ibid.* xii. 261 It is only when a heterozygous species is crossed with another, heterozygous or homozygous, that its heterozygosity is shown by the mixture in the progeny. **1965** T. Dobzhansky *Heredity & Nature of Man* v. 156 Although persons who inherit this gene from both parents (homozygotes) die of fatal anemia, the heterozygous carriers may be said to show a hybrid vigor..in malaria-ridden environments.

heuchera (hoi·kərə). *Bot.* [mod.L., f. the name of J. H. *Heucher* (1677–1747), German botanist (C. Linnæus *Hortus Cliffortianus* (1737) 82).] A plant of a large genus of perennial herbs, of the family Saxifragaceæ, native to North America. Hence **heuchere·lla,** a member of a group of bigeneric hybrids between heuchera and tiarella.

1772 R. Weston *Universal Botanist* III. 387 American Heuchera or Sanicle. **1829** *Bot. Reg.* XV. 1302 Small-flowered Heuchera. **1884** J. Wood *Hardy Perennials* 143 The Heucheras bloom from May to August. **1924** R. Wright *Pract. Bk. Outdoor Flowers* II. i. 99 The newer hybrid Heucheras. **1949** L. H. Bailey *Man. Cultivated Plants* 481 Heucherella: from heuchera and tiarella. **1961** *Amateur Gardening* 23 Sept. 2/2 Hybrids, between heuchera and *Tiarella wherryi*, listed as heucherella. **1962** *Ibid.* 21 Apr. 4/2 Heucheras like open soil and an open position.

heumite (hiū·məit). *Petrol.* [a. G. *heumite, heumit* (W. C. Brögger 1898, in *Skr. udg. af Videnskabsselsk. i Christiania (Math.-naturv. Klasse) 1897* VI. 46), f. the name of *Heum* in southern Norway: see -ITE[1].] A brownish-black hypabyssal dike-rock containing alkali feldspars, biotite, and barkevikite, with nepheline, sodalite, and other rocks.

1901 *Jrnl. Chem. Soc.* LXXX. II. 169 Heumite is the name given to a dyke-rock from Heum, consisting of hornblende and felspar, with some biotite, and smaller amounts of nephelite, sodalite, diopside, &c. **1938** A. Johannsen *Descr. Petrogr.* IV. 170 The heumite from Brathagen..contains essential soda-microcline and soda-orthoclase with a little andesine.

‖ **heurige** (hoi·rigə). Also (representing G. declined forms) **heurigen, heuriger.** [South G. and Austrian G., = (adj.) new, (sb.) new wine, from the latest harvest; vintner's establishment.] **1.** The wine from the latest harvest, produced in and near Vienna.

1935 Schoonmaker & Marvel *Compl. Wine Bk.* vi. 152 Grinzing, now part of Vienna, is famous for its *vin nouveau (Heuriger)*, served in picturesque rustic cafés to the music-loving population. **1941** 'R. West' *Black Lamb* (1942) II. 490 It is delightful to drink the heuriger wine in the gardens of Grinzing. **1964** F. Bowers *Bibliogr. & Textual Crit.* I. 3 Experts..testify that a headache can be especially acute the morning after tasting *Heurige*, the new wine. **1967** A. Lichine *Encycl. Wines* 115/2 The stinging new *heurige* wine served in their wine-garden is, for the tourists who flock there, the essence of Viennese gaiety.

2. An establishment or wine-garden where this wine is served.

1934 P. Bottome *Private Worlds* iii. 27 One hot summer night at a *Heurigen*, he had met a beautiful young dancer. **1939** N. Monsarrat *This is Schoolroom* III. xvii. 400 We might..row across the lake to the tiny *Heuriger* and there drink a bottle of Niersteiner. **1965** *Daily Tel.* 23 Sept. 15/6 To be in Vienna and not visit a heuriger is like seeing Naples and not dying.

heurism (hiūə·riz'm). [f. Heur(istic + -ism.] The educational principle or practice of placing a pupil, as far as possible, in the position of a discoverer.

1919 *Times Educ. Suppl.* 17 Apr. 181/3 The whole trend of modern educational improvement is to give the pupils just the benefits that Professor Laurie demanded. Boy-scouting.., direct methods, heurism, and..'the play way'. **1920** T. P. Nunn *Education* 91 Dr. M. W. Keatinge, ..a severe critic of heurism and of the general idea of freedom in education. **1925** C. Fox *Educ. Psychol.* 214 The second general principle of method [in teaching] is that known as heurism, i.e. the method of placing the pupil..in the position of the discoverer. This method had a great vogue a generation ago. **1959** *Chambers's Encycl.* VII. 80/2 All teaching is balanced between the poles of heurism and of didactic suggestion.

heuristic, *a.* Add: **a.** (Earlier and later examples.)

1821 Coleridge *Let.* 8 Jan. (1971) V. 133, I am.. getting regularly on with my Logic—in 3 parts—..3. Organic or Heuristic (εὑρισκτικόν). **1853** *N. & Q.* I. Ser. VII. 320 *Heuristic*,..as an English scholar would write it, or *Heuristisch*, as it would be written by a German. **1955** *Sci. Amer.* July 72/3 Einstein's 1905 paper, for which (nominally) he had been awarded the Nobel prize, did not contain the word 'theory' in the title, but referred instead to considerations from a 'heuristic viewpoint'. **1967** *Listener* 28 Sept. 386/2 His [*sc.* M. McLuhan's] style is jargon-ridden—all this talk of 'heuristic probes', as if a probe could be anything but heuristic. **1973** *N.Y. Times* 2 May 36/2 The kind of criticism being written now is looser, more fluid, more ad hoc and heuristic.

b. *Educ.* (See quot. 1898.) Also **heuri·stical** *a.*

1848 W. Ross *Teacher's Man.* v. 92 The Heuristical method. **1884** in *Spec. Rep. Educ.* II. 390 in *Parl. Papers 1898* (C. 8943) XXIV. I The heuristic method is the *only* method to be applied in the pure sciences; it is the best method in the teaching of the applied sciences. **1898** H. E. Armstrong *Ibid.*, Heuristic methods of teaching are methods which involve our placing students as far as possible in the attitude of the discoverer—methods which involve their *finding out*, instead of merely being told about things. **1959** *Chambers's Encycl.* VII. 80/2 Science-teaching should always be permeated by a heuristic bias (i.e. methods of investigation must be used whenever possible).

c. *Computers.* (See quot. 1964.)

1960 *Information Processing: Proc. Internat. Conf., 1959* (Unesco) 275/1 The technique of heuristic programming is under detailed investigation as a means to the end of applying large-scale digital computers to the solution of a difficult class of problems currently considered to be beyond their capabilities. **1964** T. W. McRae *Impact of Computers on Accounting* 297 Under an 'heuristic' programming procedure the computer searches through a number of possible solutions at each stage of the programme, it evaluates a 'good' solution for this stage and then proceeds to the next stage. Essentially heuristic programming is similar to the problem solving techniques by trial and error methods which we use in everyday life. **1969** P. B. Jordain *Condensed Computer Encycl.* 235 The heuristic program should be able to judge whether the problem is closer to solution after each attempt.

Hence **heuri·stically,** in a heuristic manner; using heuristic processes.

1935 *Jrnl. Theol. Stud.* XXXVI. 314 The belief that the world's orderedness or knowability is an expression of mind, and that the category of purpose..is applicable to it otherwise than heuristically. **1963** J. Lyons *Structural Semantics* ii. 17 What is not made clear is the purpose served by the semantic criterion (except heuristically) in the first place. **1965** *Language* XLI. 507 Nonlinguistic context may be heuristically useful in establishing lexical fields. **1965** N. Chomsky *Aspects of Theory of Syntax* i. 52 It is historically accurate as well as heuristically valuable to distinguish these two very different approaches to the problem of the acquisition of knowledge. **1968** P. A. P. Moran *Introd. Probability Theory* ix. 441 On solving this problem heuristically it is found that *u*(*t*), also, is a random process.

B. *sb.* **1. a.** (Later examples.)

1945 G. Polya *How to solve It* p. vii, The subject of heuristic has manifold connections; mathematicians, logicians, psychologists, educationalists, even philosophers may claim various parts of it. *Ibid.* 102 The aim of heuristic is to study the methods and rules of discovery and invention. **1957** *Proc. Western Joint Computer Conf.* XV. 218 (*heading*) Empirical explorations of the logic theory machine. A case study of heuristic.

b. A heuristic process or method for attempting the solution of a problem; a rule or item of information used in such a process.

1957 A. Newell et al. in *Proc. Western Joint Computer Conf.* XV. 223 A process that *may* solve a given problem, but offers no guarantees of doing so, is called a heuristic for that problem. *Ibid.*, For conciseness, we will use 'heuristic' as a noun synonymous with 'heuristic process'. **1958** *IBM Jrnl. Res. & Devel.* II. 337/1 For the moment..we shall consider that a heuristic method (or a heuristic, to use the noun form) is a procedure that may lead us by a short cut to the goal we seek or it may lead us down a blind alley. **1962** Ledley & Wilson *Programming & utilizing Digital Computers* viii. 349 Such criteria are called the heuristics of the problem. The field of heuristic programming is concerned with the investigation and understanding of various aspects of heuristics, such as how they are discovered, what kinds there are. **1967** A. Battersby *Network Analysis* (ed. 2) xii. 192 It would..seem more reasonable to recalculate the float next time (6,14) was a candidate for limited resources. Some heuristics do this.

2. *pl.* The study and use of heuristic techniques in data processing.

1963 *Times* 8 June 11/2 Whether every manager needs to know about cybernetics, econometrics, and heuristics is arguable, but cost analysis, market research, production control, and other developing subjects should be within his purview. **1967** *Observer* 9 Apr. 21/3 This branch of artificial intelligence—building machines to take short cuts to a solution—is known as 'heuristics' and is being actively pursued at the moment.

hevea (hī·viă). *Bot.* [mod.L., f. native name *hevé* (F. Aublet, *Histoire des Plantes de la Guiane Françoise* (1775) II. 871).] A South American tree of the genus so called, belonging to the family Euphorbiaceæ, and having milky sap which provides rubber.

1878 W. T. Thiselton-Dyer *Let.* 17 Apr. in *Indian Forester* (1879) IV. 42 The climate of Calcutta did not prove very favorable to the Heveas, which require the conditions of growth met with in hot and moist tropical forests. **1899** *Hooker's Icones Plantarum* XXVI. 2575 The seeds of *Hevea* are..incorrectly described. **1909** J. C. Willis *Agric. in Tropics* xiii. 129 Hevea certainly forms 95% of all cultivated rubber. **1927** *Daily Tel.* 11 May 3/4 The rest would be dug up after the present season, and rubber planted in its stead, as the hevea trees remaining were widely spaced. **1935** *Nature* 16 Mar. 417/1 The exudation..is known as hevea-latex, since *Hevea brasiliensis* is only one of some four hundred plant species which elaborate juices containing rubber. **1964** J. Hutchinson *Ess. Crop Plant Evol.* viii. 167 Such an evolutionary phenomenon as *Hevea* rubber, of which some individuals of the original domestication are said to be still alive.

hew, *v.* **B. 1. b.** (Later U.S. examples.)

a **1861** T. Winthrop *John Brent* (1883) xxviii. 298, I saw a vista in new life, hewed in and took in and took up a 'claim' which I have held good. **1891** *Harper's Mag.* Aug. 451/1 How closely they hewed to the line in this respect is attested by the dying remarks of one of the men hanged.

hewed (hiūd), *ppl. a.* (under Hew *v.*). (Modern U.S. examples. Also of *hewed-log* in attrib. use.)

1793 in L. Collins *Hist. Sk. Kentucky* (1847) 517 Every purchaser or purchasers of lotts..shall build thereon a hued log house, with a brick or stone chimney. **1805** D. McClure *Diary* (1899) 14 There was a small church made of hewed logs. **1843** *Amer. Pioneer* II. 148 Two small hewed-log houses had been erected, and several cabins. **1849** *President's Mess. Congress* II. 1089 One hewed-log dwelling,..comfortably furnished cost $351. **1883** E. Eggleston *Hoosier School-boy* xvi. 106 There's the old hewed-log house..where we used to live.

hewettite (hiū·ĕtəit). *Min.* [f. the name of Donnel Foster *Hewett* (1881–1971), American geologist: see -ITE[1].] A deep red hydrated calcium vanadate, $CaV_6O_{16}\cdot 9H_2O$, occurring in nodules and as the coating of fibres or needles (see also quot. 1955).

1914 W. F. Hillebrand et al. in *Proc. Amer. Philos. Soc.* LIII. 31 The Peruvian mineral..we are pleased to name hewettite, after Mr. D. Foster Hewett, now of the U.S. Geological Survey, who has done so much to make the Minasragra occurrence known. **1955** *Amer. Mineralogist* XL. 691 There appear to be three types of material variously labelled 'hewettite' or 'metahewettite', namely, (I) hewettite, $CaO\cdot 3V_2O_5\cdot nH_2O$.., which exists in at least three hydrate forms,..(II) the unnamed sodium analogue of hewettite in the trihydrate form,..and (III) the material,..possibly a mixed sodium-calcium variety of the sodium analogue, represented by specimen 4.

hewgag. *U.S.* (Earlier and later examples.)

1850 *California Courier* (San Francisco) 6 Sept. 2/3 Beat the hong-gong; sound the hew-gag! **1855** *Vermont Free Press* 8 June (Th.), The T.I.N. Horn-et Band, with Sackbut, Psaltery, Dulcimer,..and Hugag, marched next. **1905** B. Tarkington *In Arena* 152 He had *all* the honours..; professors and students all kow-towed and sounded the hew-gag before him.

hewn, *ppl. a.* **1.** (Later example.)

1869 'Mark Twain' *Innoc. Abr.* xix. 196 A massive hewn-stone affair.

hex (heks), *v.* Chiefly *U.S.* [ad. Pennsylvanian G. *hexe*, f. G. *hexen*.] *intr.* To practise witchcraft. Also *trans.*, to bewitch, to cast a spell on.

1830 J. F. Watson *Ann. Philad.* 232 A decent storekeeper once got him to hex for his wife, who had conceited that an old Mrs. Wiggand had bewitched her, and made her to swallow a piece of linseywoolsey. **1932** B. A. De Voto *Mark Twain's Amer.* iii. 74 The cat is hexen: it is approved by Mother Hawkins, a witch. **1935** *Language* XI. 147 Belief in witchcraft is fast disappearing and with it the word [heks] *to bewitch* or *a witch* or *wizard*. A countryman may still say: *My pigs are hext.* **1956** H. Gold *Man who was not with it* (1965) xi. 97, I stood hexing the cars on Route One.

hex (heks), *sb.* Chiefly *U.S.* Also **hexe.** [Pennsylvanian G., ad. G. *hexe*. Cf. Hag *sb.*[1]] **1.** A witch. Also *transf.*, a witch-like female.

1856 G. Henderson *Pop. Rhymes* 43 'An old hexe', means an old witch, and is often applied, in a bad sense, to females of the present day. **1920** S. Lewis *Main Street* xiii. 159, I couldn't talk to you without twenty old hexes watching, whispering. **1928** *Daily Express* 10 Dec. 11/1 York County's early settlers were Germans, and their present-day descendants still remain under the spell of medieval German necromancy. One of their sacred words is 'hex', said to be corrupt German for witch. **1935** *Amer. Speech* X. 170/1 *Hex*, a witch..the *hex* trial at Lancaster.

2. A magic spell or curse.

1909 *Sat. Even. Post* 16 Jan. 7/1 'Old pal,' agreed J. Rufus, 'the hex is sure on me.' **1942** D. Powell *Time to be Born* (1943) xiii. 326 She could count on winning him... Unless Julian really could put a hex on her. **1951** M. Lowry *Let.* Apr. or May (1967) 234 Everything from how to put the hex on your more troublesome pupils [etc.]. **1952** M. McCarthy *Groves of Academe* (1953) xii. 237 She hoped..that the hex signs on the neighbouring barns would serve to ward off all evil influences from the vicinity and not, as the ignorant sometimes thought, to attract them or indicate their presence. **1958** *Times* 4 Oct. 8/7 If Ladan stayed Shehu would gather up his robes and depart, leaving a hex on us all. **1966** *Punch* 30 Mar. 458/3 'What?' says the young man, picking it up gingerly and nervously, half afraid she's put a hex on it. **1968** *Word Study* Oct. 6/1 Each of us carries around a set of shibboleths..with..totemic reverence..; meanwhile we trample blithely on spells and hexes we have never heard of.

hex (heks). Abbrev. of 'uranium hexafluoride'.

1956 S. Glasstone *Princ. Nuclear Reactor Engin.* viii. 449 For use in nuclear energy work, uranium is generally required either in the form of uranium metal, for fabrication into fuel elements, or as hexafluoride ('hex') to constitute the feed for the separation of the isotopes by the gaseous-diffusion method. **1964** M. Gowing *Britain & Atomic Energy* ii. 63 Secrecy in the project now and later was aided by the elaborate code systems used. These were different in different groups... Hexafluoride was known variously as 'V1', 'hex' and 'the working gas'. **1971** *New Scientist* 16 Sept. 617/2 Uranium hexafluoride ('hex' for short) is pumped through a cascade of.. thousands of membranes.

hexa-. Add: **he:xachlor(o)be·nzene,** C_6Cl_6, an agricultural fungicide used as a seed-dressing; **he:xachlor(o)cyclohe·xane** = **benzene hexachloride*; **he:xachlor(o)ethane** (-klōərǭu,ī·þēin, -klōəri·þēin), a toxic crystalline compound, $Cl_3C\cdot CCl_3$, having uses as an insecticide and anthelmintic and as an ingredient in smoke-producing mixtures; **hexachlo·rophane, -phene,** a diphenylmethane derivative, $CH_2(C_6HCl_3OH)_2$, a white crystalline powder used as a disinfectant, esp. for the skin; **he·xact** *a.* and *sb.* = *hexactine*; **hexactine** *a.*, also as *sb.*, a sponge-spicule having six rays; **hexactine·llidan** *a. Zool.* = Hexactinellid *a.*; **hexade·canol** = *cetyl alcohol*; **he:xadeceno·ic acid,** an unsaturated fatty acid, $C_{16}H_{30}O_2$, occurring naturally in several isomeric forms one of which is widespread in many animal and vegetable fats and oils; it has also been known as *palmitoleic, physetoleic,* and *zoomaric acid*; **hexae·thyl tetrapho·sphate** (heksă,eþīl), a viscous brownish liquid capable of irreversibly inhibiting mammalian cholinesterase; orig. used to

designate an insecticide later found to be a mixture of organic phosphates containing tetraethyl pyrophosphate as the chief active ingredient; **hexahy·drate,** a hydrate that contains six molecules of water; so **hexahydra·ted** *a.*; **hexahy·dric** *a.*, containing six hydroxyl groups in a molecule; **he·xamer** *Chem.* [*-mer], a polymeric unit or molecule made up of six monomers; **hexame·ric** *a. Chem.*, consisting of a hexamer; **hexa·merous** *a. Biol.* [Gr. μέρος part], having members in groups of six, or multiples of six; hence **hexa·merism**; **he:xametapho·sphate,** a phosphate regarded as being a salt of hexametaphosphoric acid, $(HPO_3)_6$; used loosely to designate a glassy, water-soluble sodium salt (also known as Graham's salt) having an approximate empirical formula $Na_2O\cdot P_2O_5$ and used industrially to soften water; **hexametho·nium,** a quaternary ammonium ion, $[(CH_3)_3N(CH_2)_6N(CH_3)_3]^{++}$; also, any of the salts of this ion, some of which have been used as ganglionic blocking drugs in the treatment of severe hypertension; **hexame·thylene,** cyclohexane; **hexame:thylenediamine** (-dəi,æ·mīn), a white crystalline solid, $H_2N(CH_2)_6NH_2$, used in the manufacture of nylon; **hexame:thylenete·tramine,** a colourless crystalline compound, $(CH_2)_6N_4$, having various industrial uses, esp. in the manufacture of phenol-formaldehyde resins and in many antiseptic preparations; **he·xamine** = **hexamethylenetetramine*; **hexapetalous** *a.* (earlier example); **he·xarch** *a. Bot.*, having six rays, formed from six points of origin, as in some vascular bundles; **he·xasome** *Biol.*, a set of six homologous chromosomes; so **hexaso·mic** *a.*, having six of one or more chromosomes in a nucleus that is not hexaploid.

1885 I. Remsen *Introd. Org. Chem.* 253 As the final products, hexa-chlor-benzene, C_6Cl_6, and hexa-brom-benzene C_6Br_6 are obtained. **1961** *Lancet* 22 July 176/2 Recently, several thousand cases of human porphyria have been seen in Turkey after the consumption of wheat containing 0·1–0·2% of hexachlorobenzene. **1908** *Chem. Abstr.* II. 793 Hexachlorcyclohexanes.—On continuing the action of chlorine in the sunlight, a viscous liquid is obtained. **1945** R. E. Slade *Gamma Isomer of Hexachlorocyclohexane* ('*Gammexane*') 11 The hexachlorocyclohexanes possess considerable chemical stability. **1898** *Jrnl. Chem. Soc.* LXXIV. I. 626 When a current of chlorine is passed into a mixture of pentachlorethane and aluminium chloride heated at 100°, hexachlorethane is obtained. **1936** *Discovery* Aug. 255/2 A new way of killing mosquitoes..involves the use of hexachloroethane. **1960** *Farmer & Stockbreeder* 16 Feb. 109 (Advt.), In the Nicholas Liver Fluke Drench, hexachloroethane particles are suspended in a free-flowing liquid and this ensures *accurate* dosing. Result: greater fluke kill than ever before! **1960** *Which?* Feb. 32/2 One can attempt to prevent the sweat from decomposing by using an antiseptic, such as *hexachlorophane*. **1948** *Jrnl. Amer. Med. Assoc.* 14 Feb. 471/2 The Council [on Pharmacy and Chemistry] considered the proposal of a pharmaceutical manufacturer for the use of the term hexachlorophene as a generic designation for *bis*(2-hydroxy-3,5,6-trichlorophenyl) methane... The Council voted to recognize hexachlorophene as the generic, or common, designation. **1953** *Jrnl. Parasitol.* XXXIX. 79 Hexachlorophene..is effective in removing *Raillietina cesticillus* from chickens. **1886** R. von Lendenfeld in *Proc. Zool. Soc.* 590 The Recent Families of Sponges... With hexact spicules and thimble-shaped chambers. **1887** tr. F. E. Schulze in *Challenger Rep., Zool.* XXI. 29 Regular Hexacts are all spicules in which the rays lie at right angles to one another, and are of equal length and similar form. **1940** L. H. Hyman *Invertebrates* I. vi. 327 They [*sc.* triacts] may show their origin from a hexact by the presence near the middle of knobs or branches of the axial fibre. **1900** E. A. Minchin in E. R. Lankester *Treat. Zool.* II. iii. 117 One or more rays of the hexactine..may become modified in various ways. **1940** L. H. Hyman *Invertebrates* I. vi. 327 The basic regular hexactine consists of three axes crossing at right angles, forming six rays of approximately equal length. **1887** tr. F. E. Schulze in *Challenger Rep., Zool.* XXI. 37 Certain Hexactinellidan families have typical and regular *Uncinata*, while in others they are absent. **1914** *Jrnl. Chem. Soc.* CV. II. 2251 Acetate of *l*-γ-Hexadecanol, $C_3H_6\cdot CH(O\cdot CO\cdot CH_3)\cdot C_{12}H_{25}$. **1964** *Oceanogr. & Marine Biol.* II. 177 The higher aliphatic alcohols of the wax esters [found in fish lipids] usually contain hexadecanol (cetyl alcohol) and octadec-9-enol (oleyl alcohol) as main components. **1901** *Jrnl. Chem. Soc.* LXXX. I. 252 The product was found to yield a hexadecenoic acid, $C_{16}H_{30}O_2$, melting at 36°, when fused with potassium hydroxide. **1948** A. W. Ralston *Fatty Acids* ii. 98 The fats of fresh water fish contain somewhat less 9-hexadecenoic acid than those of marine origin. **1946** *Jrnl. Econ. Ent.* XXXIX. 812/1 Hexaethyl tetraphosphate merits considerable attention as a commercial control for numerous insects causing serious damage to agricultural and garden crops. **1951** A. W. Johnson et al. in E. H. Rodd *Chem. Carbon Compounds* IA. iv. 342 A mixture of phosphates, known commercially as 'hexaethyl tetraphosphate', is formed when $POCl_3$ reacts with triethyl orthophosphate or with ethanol. **1965** A. & E. F. Grollman *Pharmacol. & Therapeutics* (ed. 6) xxxi. 823 Hexaethyltetraphosphate..is highly toxic, exerting a potent anticholinesterase action. **1908** H. C. Cooper tr. *Holleman's Text-bk. Inorg. Chem.* (ed. 3) 487 Two hexahydrates of chromic chloride, $CrCl_3\cdot 6H_2O$, are known. **1951** C. Palache et al. *Dana's Syst. Min.* (ed. 7) II. 493 [Pentahydrite is] obtained in crystals together with the hexahydrate and tetrahydrate by evaporation..of a solution of magnesium sulfate. **1880** G. Lunge *Sulphuric Acid* II. III. i. 277 (*heading*) Hexahydrated salt has been obtained by Mitscherlich from a solution of Na_2S in the air. **1951** C. Palache et al. *Dana's Syst. Min.* (ed. 7) II. 493 The minerals of this group..are isostructural with the monoclinic artificial hexahydrated sulfates and selenates of Mg, Co, Ni, and Zn. **1885** *Jrnl. Chem. Soc.* XLVIII. II. 1046 (*heading*) Reduction in hexahydric alcohols. **1964** N. G. Clark *Mod. Org. Chem.* xvi. 315 The product, sorbitol, is a hexahydric alcohol (one of the 'sugar alcohols'), which occurs naturally in many fruits. **1953** R. G. R. Bacon in E. H. Rodd *Chem. Carbon Compounds* IIA. xi. 403 A mixture of polymers..was separated, by means of a molecular still, into fractions ranging from trimers to hexamers. **1969** *Nature* 1 Nov. 493/2 The insulin hexamer..is a compact, oblate spheroid, formed by the coordination of three insulin dimers around the two zinc ions. **1940** *Jrnl. Chem. Soc.* 1169 The trimeric, tetrameric, and pentameric portions of the polymeride have been separated from one another by molecular distillation, leaving as residue a highly viscous liquid of mainly hexameric complexity. **1903** *Science* 17 July 80/2 In some species [of corals] the hexamerism becomes much obscured in later stages, while in others it is more or less distinctly preserved. **1877** T. H. Huxley *Man. Anat. Invertebr. Animals* iii. 159 The finally hexamerous Anthozoon passes through a tetramerous and an octomerous stage. **1905** I. B. Balfour tr. *Goebel's Organogr. Plants* II. 538 The first flowers of some Caryophylleae are hexamerous. **1940** L. H. Hyman *Invertebrates* I. vii. 579 In typical hexamerous anemones, the number of pairs of septa in the various cycles is then: 6 (primaries), 6, 12, 24, 48, etc. **1891** W. Ramsay *Syst. Inorg. Chem.* xxiii. 370 *Hexametaphosphates.*—These are the salts prepared by the usual methods from ordinary metaphosphoric acid. **1892** *Jrnl. Chem. Soc.* LXII. II. 1050 The uncrystallisable hexametaphosphates, $Na_6(PO_3)_6$ and $Ag_6(PO_3)_6$, are obtained from Graham's soluble sodium metaphosphate.., which is a mixture of the foregoing sodium salt with several other hexametaphosphates. **1963** A. J. Hall *Textile Sci.* vi. 292 Important sequestering agents.. include sodium pyrophosphate $Na_4P_2O_7$, and especially hexametaphosphate $(NaPO_3)_6$. **1949** *Jrnl. Pharmacy & Pharmacol.* I. 603 An antidote exists for decamethonium iodide in hexamethonium iodide..and..this substance has been suggested for use in hypertension and vascular diseases. **1964** S. Duke-Elder *Parsons' Dis. Eye* (ed. 14) xxi. 300 The operation is most conveniently done under basal anæsthesia, and the systemic administration of sympatholytic drugs such as hexamethonium may be of value in relieving congestion by lowering the general blood pressure. **1887** *Abstr. Proc. Chem. Soc.* III. 96 Hitherto all attempts to synthesise hexamethylene-derivatives have been unsuccessful. **1909** C. A. Keane *Mod. Org. Chem.* v. 64 The six carbon atom analogue of these alicyclic compounds is hexamethylene, C_6H_{12}. **1894** *Jrnl. Chem. Soc.* LXVI. I. 410 Hexamethylenediamine nitrite. **1896** *Ibid.* LXX. I. 464 Hexamethylenediamine, $C_6H_{12}(NH_2)_2$, melts at 40° and boils at 192–195°. **1962** J. K. Stille *Introd. Polymer Chem.* vi. 93 Nylon 66 is formed from the reaction of adipic acid (a six-carbon dibasic acid) and hexamethylenediamine (a six-carbon diamine). **1888** *Jrnl. Chem. Soc.* LIV. 1268 A well-cooled solution of hexamethylenetetramine. **1940** *Thorpe's Dict. Appl. Chem.* (ed. 4) IV. 27/2 The inodorous solid product, hexamethylenetetramine, $C_6H_{12}N_4$ (hexamine, urotropine), is a serviceable antiseptic in cystic affections, and is administered to typhoid 'carriers' to destroy the bacilli of the disease in the urine. **1961** *Encycl. Brit.* IX. 524/1 Nitration of hexamethylenetetramine gives the explosive, RDX. **1914** *Brit. Pharmacopœia* 171 Hexamine may be obtained by the interaction of ammonia with formic aldehyde. **1956** Ld. Amulree in A. Pryce-Jones *New Outl. Mod. Knowl.* 222 Others have used hexamine with glucose and vitamin B_1 but this treatment, again, does not seem to be generally applicable [in cases of senile dementia]. **1707** H. Sloane *Voy. Jamaica* I (Pref.), I have followed mostly the Method of Mr. Ray in his History of Plants, joining his Genera or Tribes together by the Method of Rivinus, or Number of the Petala or Leaves of the Flowers; As those which are Monopetalous first, those Dipetalous next, then the Tripetalous, Tetrapetalous, Pentapetalous, Hexapetalous, and such as have no exact Numbers of Coloured Leaves in their Flowers. **1900** B. D. Jackson *Gloss. Bot. Terms* 123/2 Hexarch. **1914** M. Drummond tr. *Haberlandt's Physiol. Plant Anat.* vii. 352 It is customary to discriminate between diarch, tetrarch, pentarch, hexarch and polyarch bundles. **1965** Bell & Coombe tr. *Strasburger's Textbk. Bot.* I. iv. 170 Pentarch..and hexarch roots also occur. **1921** A. F. Blakeslee in *Amer. Naturalist* LV. 259 The following terms are suggested to designate sets with numbers of chromosomes from 1 to 12: monosome, disome, trisome, tetrasome, pentasome, hexasome, heptasome, oktasome, enneasome, dekasome, hendekasome, dodekasome. **1922** *Genetics* VII. 527 If..the chromosomes of a tetrasomic or hexasomic set assort at random in an F_1 hybrid a 35:1 or 399:1 ratio will be found in the F_2. **1930** *Jrnl. Genetics* XXII. 146 Doubly and trebly hexasomic tetraploids are more likely to come from the derivatives of a tetraploid-hexaploid cross. **1955** J. B. & R. D. Hill *Genetics & Human Heredity* xviii. 363 Hexasomic tetraploid Globe: $4x+2I\cdot 22+2I\cdot 22$.

hexagonal, *a.* Add: **3.** *hexagonal close-packed* adj., applied to a type of crystal structure or lattice with hexagonal symmetry in which each ion or atom has twelve equidistant neighbours; so *hexagonal close-packing* (see quot. 1917); *hexagonal symmetry*, the symmetry of a figure or body that coincides with its original position after rotation about an axis through an angle of 60° or any multiple of this.

1878 Hexagonal symmetry [in Dict.]. **1917** *Physical Rev.* X. 678 The so-called hexagonal close-packing.. is one of the two alternative arrangements which the atoms would assume if they were hard spheres and were forced by pressure into the closest possible packing. **1920** *Science* 3 Sept. 228/1 The arrangement of atoms in zinc is like that in magnesium, namely: hexagonal close packed. **1966** L. G. Bassett et al. *Princ. Chem.* iii. 171 The difference between hexagonal close packing and face-centered close packing arises from the two arrangements for atoms in the third layer.

hexahydrite (hekṣăhəi·drəit). *Min.* [f. **hexahydr(ate* + -ITE[1].] The hexahydrate of magnesium sulphate, $MgSO_4\cdot 6H_2O$.

1911 R. A. A. Johnston in *Sum. Rep. Geol. Surv. Br. Dept. Mines 1910* (Canada) 257 As this is the first instance in which this salt has been recorded as occurring in a state of nature, this substance is entitled to be regarded as a new mineral, and the name hexahydrite is proposed for it, in allusion to the six molecules of water which enter into its composition. **1959** Berry & Mason *Mineralogy* xiii. 437 In dry air at ordinary temperatures epsomite loses up to $1H_2O$, changing to hexahydrite.

hexamitiasis (heksămitəi·ăsis). *Med.* [f. the generic name *Hexamita*: see *-iasis.] A disease caused by protozoa of the genus *Hexamita*; *esp.* an infectious catarrhal enteritis in turkeys often fatal to poults and caused by *H. meleagridis*.

1941 *Jrnl. Parasitol.* XXVII. 186 One essential difference in the pathology of hexamitiasis of pigeons [from that of turkeys] is the consistent catarrhal inflammation with a heavy mucous deposit in the rectum of young pigeons. **1966** T. Dalling *Internat. Encycl. Vet. Med.* III. 1432 In Great Britain, hexamitiasis appears to have been confined to a number of localized outbreaks and..the disease at present is not of great economic importance.

hexane. Add: **hexano·ic** *a.* = Caproic *a.*; **he·xanoyl** [*-oyl], the radical $C_5H_{11}CO-$ of hexanoic acid; also called *caproyl*.

1926 *Chem. Abstr.* XX. 4598/1 *Hexanoic acid*, see *Caproic acid*. **1927** *Jrnl. Amer. Chem. Soc.* XLIX. 1828 (*heading*) Some bromine derivatives of pentanoic and hexanoic acids. **1949** *Union Internat. Chim. pure et appl., Compt. Rend.* XV. 144 Rule 58.3. The Commission [of nomenclature of organic chemistry]..recommends that the carboxyl carbon be always numbered as 1 whether the name be a Geneva name or a trivial name, but not when it is a 'carbonyl' name. *Examples of numbering*: hexanoyl, caproyl [etc.]. **1952** *Ibid.* XVI. 101 Change: 'Caproyl' to 'Hexanoyl replacing caproyl'..[rule] 58.3. **1967** *Canad. Jrnl. Chem.* XLV. 2600/2 When a cyclohexane solution of hexanoyl azide and acetophenone was irradiated with light of wavelength over 3000 Å, nitrogen was rapidly evolved.

hexaploid (he·ksăploid), *a.* and *sb.* *Biol.* [f. Hexa- + *-ploid.] (Made up of somatic cells) containing six sets of chromosomes. Also as *sb.*, a hexaploid organism.

1920 G. Täckholm in *Svensk Bot. Tidskr.* XIV. 302 The diploid and tetraploid forms [of roses] of the types 1 and 2 are represented in about equal frequence in my material, the hexaploid (type 3) being rarely met with. **1921** *Ann. Bot.* XXXV. 185 The remainder of the tetraploids, the whole of the pentaploids and hexaploids, showed a partial reduction involving fourteen or twenty-eight chromosomes. **1952** *New Biol.* XIII. 32 The common bread-wheats which have 42 chromosomes, or 6 sets, are hexaploids. **1956** *Nature* 25 Feb. 384/2 Hexaploid cells in mouse blastocysts. **1970** *Watsonia* VIII. 130 Two other differences between the rhizome scales of the diploid and hexaploid were noticed.

Hence **he·xaploidy,** the state or condition of being hexaploid.

1922 *Genetics* VII. 540 There is evidence that does not support the view that tetraploidy and hexaploidy in wheat is actually due to chromosomal duplication. **1925** *Jrnl. Genetics* XV. 263 Tetraploidy and hexaploidy may play a part in the striking changes of the chromosome group from species to species.

hexatone (he·ksătǭun). *Mus.* [f. Hexa- + Tone; cf. Pentatone.] 'A gapped scale containing six notes within an octave' (Grove, 1954).

hexatonic (heksătǫ·nik), *a.* *Mus.* Characterized by hexatones.

1930 *Times Lit. Suppl.* 6 Feb. 95/1 Another good tune ..has additional sources of strength in being constituted as a ten bar sentence without any repetition of phrase, and being hexatonic in mode. **1946** R. Blesh *Shining Trumpets* (1949) ii. 44 The scalar concept scarcely enters into African music although it is constantly referred to as predominantly pentatonic and hexatonic, i.e., five and six-toned.

hexestrol, var. *hexœstrol.

hexite (he·ksəit). [See Hexa- and -ITE[1] 4.] **1.** *Chem.* [ad. G. *hexit.*] = *hexitol.

1899 E. F. Smith tr. *V. von Richter's Org. Chem.* (ed. 3) I. 540 Certain hexites have been prepared by the reduction of the corresponding glucoses..with sodium amalgam. **1938** G. H. Richter *Textbk. Org. Chem.* xx. 395 The relationship of the aldohexoses to the corresponding hexites and dicarboxylic acids.

2. A high explosive, hexanitrodiphenylamine, $[C_6H_2(NO_2)_3]_2NH$.

1931 J. F. Norris *Princ. Org. Chem.* (ed. 3) xxvi. 483 Hexite or hexil..was the high explosive used in bombs which were dropped on London. **1944** *Jrnl. Amer. Med. Assoc.* 20 May 189/2 Hexite is now being made and used but not in as large quantities as trinitrotoluene. **1951** Kirk & Othmer *Encycl. Chem. Technol.* VI. 54 Hexite is only slightly less sensitive to impact than Tetryl, but is distinctly less sensitive to the impact of a rifle bullet.

hexitol (he·ksitǫl). *Chem.* [f. Hex(ose + *-itol.] Any of a class of hexahydric alcohols that are closely related to the hexoses.

1894 *Jrnl. Chem. Soc.* LXVI. I. 395 According to theory, 10 stereoisomeric forms of hexitol should exist. **1913** T. H. Pope tr. *Molinari's Org. Chem.* II. 433 With hydrogen the hexoses form hexitols, which are not aldehydic but only alcoholic in character. **1959** A. White et al. *Princ. Biochem.* (ed. 2) xviii. 423 The cyclic hexitol *myo*inositol..occurs abundantly in nature, and is present chiefly in the form of polyphosphates in plants.

hexobarbital (heksobā·ɹbităl). *Pharm.* [f. as next: see *barbital.] The equivalent in the U.S. Pharmacopœia of *hexobarbitone.

1941 [see next]. **1943** Wood & Osol *Dispensatory of U.S.A.* (ed. 23) 511/2 Hexobarbitonum solubile... Soluble Hexobarbital; Soluble Evipal; Cyclural Sodium. **1957** T. Sollmann *Man. Pharmacol.* (ed. 8) 942/1 Hexobarbital sodium ('Evipal', 'Evipan') was introduced especially for intravenous anesthesia, as its action is brief.

hexobarbitone (heksobā·ɹbitǭun). *Pharm.* [f. Hex- (in *hexenyl*) + -o- + *barbitone.] A short-acting barbiturate, $C_{12}H_{16}N_2O_3$, with hypnotic properties; also known by proprietary names, as *Evipal, Evipan*; *loosely*, = **hexobarbitone sodium,** the monosodium derivative of hexobarbitone, a very short-acting and soluble anæsthetic usu. given intravenously or intramuscularly; also called *hexobarbitone soluble.*

1941 *Brit. Pharmacopœia 1932* (Add. 3) 8 (*heading*) Hexobarbitonum... Hexobarbitone. *Synonym.* Hexobarbital... Hexobarbitone is 5-Δ'-*cyclo*hexenyl-5-methyl-N-methyl-barbituric acid. **1943** *Brit. Med. Jrnl.* 5 June 714/2 Hexobarbitone given by mouth is an excellent hypnotic when there is no pain. *Ibid.* 20 Nov. 646/1 Cyanosis rapidly became extreme; 0·5 g. of hexobarbitone soluble (evipan sodium) was dissolved in 5 c.cm. of distilled water and administered intravenously. **1952** *Martindale's Extra Pharmacopœia* (ed. 23) I. 260 Hexobarbitone sodium has largely been superseded by thiopentone sodium which gives greater muscular relaxation. **1962** H. Burn *Drugs, Med. & Man* ix. 98 The substance hexobarbitone could be injected into a vein and the patient went to sleep with no more warning than he had when going to sleep in his own bed.

hexode, *a.* and *sb.* Add: **A.** *adj.* (Earlier example.)

1886 W. H. Preece in *Jrnl. Soc. Telegr.-Engin. & Electr.* XV. 231 A two-way mode of working, or a mode by which two messages are practically sent at the same time, will be diode working, three-way triode, four-way tetrode, five-way penthode, and-six way [*sic*] hexode.

B. *sb. Radio* [a. G. *hexode*]. A valve with six electrodes. Also *attrib.* or as *adj.*

1933 *Electronics* Mar. 76/2 The special tube which was selected..is a hexode..having a fourth grid and a redesign of all the grids. **1933** *Pract. Wireless* 27 May 358 (*caption*) Superhet circuit employing new Hexode single valve frequency changer. **1944** *Electronic Engin.* XVII. 60/1 In hexode valves the mutual conductance between the first grid potential and the anode current varies in a linear manner with the voltage applied to the third grid. **1955** J. Thomson *Electronics* vi. 90 The simplest practical form of mixer valve, but one which is not much used nowadays is the hexode. *Ibid.* 91 The hexode may be regarded as a screen-grid tetrode with two control grids.

hexœstrol (heksī·strǫl). *Pharm.* Also **hexestrol.** [f. Hex(ane + Œstr(us + -ol 2.] A synthetic œstrogen, $C_{18}H_{22}O_2$, related to stilbœstrol and used in hormone therapy.

1939 N. R. Campbell et al. in *Lancet* 5 Aug. 312/1 The œstrogenic activity of the synthetic substance 4:4'-dihydroxy-γ:δ-diphenyl-*n*-hexane (hexœstrol) has been described. **1954** *Thorpe's Dict. Appl. Chem.* (ed. 4) XI. 370/2 Hexœstrol dipropionate is of particular value for inhibiting lactation, owing to its speed of action. **1960** *Farmer & Stockbreeder* 19 Jan. 62/3 Between the two most popular synthetic oestrogens—stilboestrol and hexoestrol —the latter was to be preferred as it was less toxic to man. **1971** *New Scientist* 8 Apr. 104/1 Potent drugs—for example the hormone hexoestrol.

hexogen (he·ksŏdʒĕn). Also **hexogene.** [a. G. *hexogen.*] = *cyclonite.

1923 *Chem. Abstr.* XVII. 2364 Nitrated hexamethylenetetramine (hexogen)..and mannitol hexanitrate occupy the highest places in the order of effectiveness. **1945** *Industr. & Engin. Chem.* XXXVII. 666/2 Increasing the weight of charge in a mercury fulminate detonator ..is not nearly so effective as substituting a composition having a higher rate of detonation, such as PETN or hexogen. **1960** *Guardian* 19 Aug. 10/4 Hexogen, he added, was a very powerful explosive used by the US Army.

hexokinase (heksokəi·nēis, -z). *Biochem.* [f. *hexo(se + *kinase.] Any of various enzymes that catalyse the transfer of a phosphate group from adenosine triphosphate to glucose or other hexoses as the first step in glycolysis.

1930 O. Meyerhof in *Lancet* 27 Dec. 1418/1 By itself this activator can do nothing, but its addition to the muscle extract induces hexoses to split even more rapidly than glycogen... I propose the name 'hexokinase' for it, since it acts just as other auxiliary enzymes designated as kinases. **1954** A. White et al. *Princ. Biochem.* xvii. 404 There exists a hexokinase for fructose in yeast and brain apparently independent of the hexokinase for glucose. **1964** G. H. Haggis *Introd. Molecular Biol.* ii. 28 The breakdown of glucose to CO_2 and water involves a large number of enzymes, the first in the sequence being hexokinase.

hexone (he·ksǭun). Formerly also **hexon.** *Chem.* [See Hex- and -one; in sense 1, ad. G. *hexon, hexonbase* (A. Kossel 1898, in *Zeitschr. f. physiol. Chem.* XXV. 175).] **1.** In full, *hexone base*: any of the three basic amino-acids arginine, histidine, and lysine.

Kossel distinguished between *hexon* and *hexonbase* (see quot. 1898[1]).

1898 *Jrnl. Chem. Soc.* LXXIV. I. 612 The substances.. are called protones; these are, by the prolonged action of pancreatic juice, partially broken up into hexones (lysine, histidine, arginine). *Ibid.* 715 By the further hydrolysis of the protamines the hexon bases, histidine, arginine, and lysine, are produced, the name hexon being retained for nitrogenous substances, and obtained by the decomposition of proteids containing six atoms of carbon. **1905** J. Wade *Introd. Org. Chem.* (ed. 2) lxxiii. 555 Histidine, the principal representative of the 'hexone' bases, is an important degradation product of proteins. **1953** *Biochem. Jrnl.* LIII. 435/2 A new technique for the estimation and isolation of the hexone bases in protein hydrolysates.

2. The name for **methyl isobutyl ketone* as used industrially.

1938 T. H. Durrans *Solvents* (ed. 4) II. iii. 126 Methyl isobutyl ketone, or 2-methyl pentanone 5, known industrially as hexone, is a medium boiling solvent for nitrocellulose. **1966** *McGraw-Hill Encycl. Sci. & Technol.* X. 423/1 In the industrial process employing hexone (the Redox process), the uranium fuel is dissolved in nitric acid.

hexose (he·ksǭus, -z). *Chem.* [f. Hex- + -ose.] A monosaccharide that contains six carbon atoms; *hexose phosphate*, a phosphate derivative of a hexose, as *hexose monophosphate, diphosphate*, substances important as intermediates in many metabolic processes.

1892 E. F. Smith tr. *V. von Richter's Org. Chem.* (ed. 2) 499 The hexoses occur frequently in plants, especially in ripe fruits. **1909** *Proc. R. Soc.* B. LXXXI. 528 (*heading*) The hexosephosphate formed by yeast-juice from hexose and phosphate. **1916** A. P. Mathews *Physiol. Chem.* iv. 169 All true nucleic acids, or polynucleotides, of animal origin..have been found to contain a hexose group, or several of them. **1926** *Biochem. Jrnl.* XX. 854 Blood-plasma contains small amounts of a phosphatase similar to that found in bone. Like the latter it hydrolyses sodium glycerophosphate, hexosemonophosphate and hexosediphosphate. **1935** Tipson & Stiller in Harrow & Sherwin *Textbk. Biochem.* ii. 63 The naturally occurring hexoses (*d*-glucose, *d*-mannose, *d*-galactose and *d*-fructose) are the only sugars which undergo fermentation by yeast. **1956** *Nature* 11 Feb. 274/1 In algae it is considered that the oxidative assimilation of hexose follows the same glycolytic pathway as in higher organisms. **1960** *McGraw-Hill Encycl. Sci. & Technol.* II. 460/1 The principal alternative pathways by which sugars are dissimilated.. are known as the hexose monophosphate pathways. **1965** *New Scientist* 17 June 761/1 Food sugars absorbed from the gut are hexoses.

So **hexo·nic** *a.* [*-onic], designating an acid formed from an aldohexose by oxidation of the aldehyde group to a carboxyl group; **hexosamine** (heksǭu·sămīn) [Amine], a derivative of a hexose in which a hydroxyl group is replaced by an amino group; so **hexosami·nic** *a.*, designating the acid derived from hexosamine; **he·xosan** [-an], a polysaccharide of which the constituent monosaccharides are hexoses.

1894 G. M'Gowan tr. *Bernthsen's Org. Chem.* (ed. 2) ix. 236 By the conversion of the hexonic acids (through the hexoses) into the corresponding alcohols (mannite, etc.), the number of possible stereo-isomers is diminished. **1894** *Amer. Chem. Jrnl.* XVI. 227 If hexosans are gradually oxidized with chromic acid and afterwards distilled with acids, large amounts of furfurol are obtained. **1914** *Jrnl. Biol. Chem.* XVIII. 123 The hexosamine..was obtained directly on concentration of the product of hydrolysis of chondroitin sulphuric acid. **1915** *Ibid.* XX. 441 (*heading*) Hexosaminic acid from ribose. **1921** A. L. Smith *Lichens* v. 211 Besides these [carbohydrates], which rank as hexosans, Ulander found small quantities of pentosans and methyl pentosans. **1957** E. V. Miller *Chem. Plants* i. 7 The hexosan cellulose is the fundamental constituent of the plant cell wall. **1958** Fruton & Simmonds *Gen. Biochem.* (ed. 2) xxi. 536 The oxidation of D-galactose by *Pseudomonas saccharophila* also appears to involve the intermediate formation of hexonic acids. **1961** *New Scientist* 5 Jan. 15/1 In chronic brain diseases and manic psychoses the macromolecular hexosamine was elevated.

hex-radiate (heks,rēi·diĕt), *a.* [f. Gr. ἕξ six + L. *radiatus* rayed, Radiate.] = Hexiradiate.

1883 W. S. Kent in A. J. Adderley *Fisheries Bahamas* 37 The glass-rope, birds'-nest, and hat sponges share..the circumstance that the silicious spicules of which their skeletons are composed belong..to what is known as the hex-radiate type.

hexuronic (heksiūərǫ·nik), *a.* [f. *hex(ose + *uronic *a.*] *hexuronic acid*, any of a class of uronic acids derived from hexose; *spec.* = **ascorbic acid.*

1928 A. Szent-Györgyi in *Biochem. Jrnl.* XXII. 1409 The substance is a hitherto unknown, highly reactive isomer of glycuronic acid, so that the substance is a hexuronic acid. *Ibid.*, The isolation of the hexuronic acid from oranges and cabbages is described. **1933** *Times Lit. Suppl.* 16 Nov. 798/3 They have discovered hexuronic acid to be identical with vitamin C. **1957** *Science News* XLV. 82 Monosaccharides include hexoses such as glucose, ..hexuronic acids such as glucuronic acid and galacturonic acid with the molecular formula $C_6H_{10}O_7$, and hexosamines. **1963** D. M. Douglas *Wound Healing & Managem.* vi. 49 Szent-Györgi is credited with the chemical isolation of ascorbic acid, which he first called hexuronic acid... Szent-Györgi renamed hexuronic acid, ascorbic acid.

hexyl. Add: **hexylreso·rcinol** (also as two words), a crystalline derivative of resorcinol, $C_6H_3(OH)_2\cdot(CH_2)_5CH_3$, used as an anthelmintic and a urinary antiseptic.

1924 *Jrnl. Amer. Med. Assoc.* 20 Dec. 2007/1 Unlike a great many antiseptics and germicides, hexyl resorcinol retains its bactericidal power when dissolved in urine of either acid or alkaline reaction. **1949** Krantz & Carr *Pharmacol. Princ. Med. Pract.* xviii. 247 Hexylresorcinol capsules are composed of a solution of the drug dissolved in olive oil. **1968** J. H. Burn *Lect. Notes Pharmacol.* (ed. 9) 125 Oil of chenopodium and hexylresorcinol (by mouth) are also used for ascaris infections.

hey, *int.* (*sb.*). Add: **2. e. hey, Rube!** A rallying call or a cry for help used by circus people. As *sb.*, a fight between circus workers and the general public. *U.S. slang.* (Cf. *Reub.)

1882 *Times* (Chicago) 3 Dec. Suppl. 12/4 A canvasman watching a tent is just like a man watching his home. He'll fight in a minute if the outsider cuts the canvas, and if a crowd comes to quarrel he will yell, 'Hey Rube!' That's the circus rallying cry, and look out for war when you hear it. **1935** *Amer. Mercury* XXXV. 229/2 Heyrube: general uprising of spectators. **1939** *Sat. Even. Post* 25 Mar. 75/2 The expression disappeared forty years ago, along with the old rallying cry, 'Hey, Rube!' **1956** H. Gold *Man who was not with It* (1965) i. 6 We found ourselves with an old-fashioned hey-rube and obliged to move the show on that night. **1962** E. S. Gardner *Case of Blonde Bonanza* (1967) xv. 182 And 'Hey Rube' is a rallying cry for the circus people to unite in a fight against the outsiders? **1973** *Daily Colonist* (Victoria, B.C.) 12 Oct. 2/2 There..could be a very interesting hey Rube between incumbent Frances Elford and Ald. Brian Smith.

hi, *int.* Add: **2.** A word of greeting. *colloq.* (chiefly *N. Amer.*).

1862 M. D. Colt *Went to Kansas* ix. 143 When out on the prairie, up galloped an Indian on his pony with his saluting 'hi!' **1885** 'P. Perkins' *Familiar Lett.* (1886) 33 We would have had to walk, I believe, if a *man* hadn't come along and let out the most satisfactory 'Hi, there!' you've ever heard, and stopped a car. **1920** F. Scott Fitzgerald *This Side of Paradise* (1921) II. i. 199 *Alec*: Hi, Amory! *Amory*: Hi, Alec! Tom said he'd meet you at the theatre. **1951** J. D. Salinger *Catcher in Rye* iii. 26 He.. came in the room. 'Hi,' he said..like he was terrifically bored. **1953** H. Clevely *Public Enemy* xviii. 111 Tillie nodded to the uniformed commissionaire..and said: 'Hi, Charlie,' and they entered. **1959** I. & P. Opie *Lore & Lang. Schoolch.* vii. 116 Hi, Roy Rogers! How about a date? **1963** H. Garner in R. Weaver *Canad. Short Stories* 2nd Ser. (1968) 50 'Hi, Eric!' shouted Pete Adams..from where he was standing at the drinking fountain. **1972** Wodehouse *Pearls, Girls & Monty Bodkin* ii. 28 A musical voice in his left ear said 'Hi'. **1973** *Black Panther* 11 Aug. 2/2 (*letter to editor*), Hi, I've been following serialization of Operation Gemstone in the Panther Paper.

hi (həi), abbrev. of High *a.*, freq. used in advertising and commercial slogans. (Cf. *hi-fi.) Chiefly *U.S.*

1911 *T. Eaton & Co. Catal.* Spring & Summer 179/4 Hi-up battery. A very powerful cell for all forms of ignition work. **1930** *Engineering* 10 Jan. 63/3 This embodies the form of motor known as the Hicycle motor, that is, an alternating current induction motor, using a supply of a frequency of 180 cycles to 200 cycles. **1959** *Sears, Roebuck Catal.* Spring & Summer 87/4 Hi-Society—a glamorous new idea in lipstick! **1963** *Times* 12 Mar. p. xii/4 A new use for ball and chain. Known as the Hi-ball method, 100 acres of lightly timbered country can be cleared in a day. **1972** *Guardian* 3 Feb. 13/4 Kids prefer 'hi-riser' bicycles..with their apehanger handlebars and their drag style saddles.

hiaqua (həi·ăkwa). Also **haiqua, haigua,** etc. [Chinook Jargon, f. Nootka.] An ornament or necklace composed of tooth-shells,

formerly used as money by the Indians of the north Pacific coast of North America.
1824 G. Simpson *Jrnl.* in *Fur Trade* (1931) 96 The Ears are perforated all round and Beads or Hyaques suspended therefrom. 1862 *Nor' Wester* (Red River Settlement) 5 Mar. 4/3 Instead of the nose jewels being 'of tin' they were composed of the Hyaqua shells. a 1861 T. Winthrop *Canoe & Saddle* (1883) vii. 95 Tribes..among whom hiaqua is plenty as salmon-berries are in the woods. 1881 E. B. Tylor *Anthropol.* xi. 282 The Indians of British Columbia, whose strings of haiqua-shells..serve them..as currency. 1923 *Canad. Hist. Rev.* IV. 37 The shell *dentalium indianorum,* commonly called hiquia, is much esteemed by the coast Indians, and amongst them has fulfilled some of the functions of money. 1953 E. E. Clark *Indian Legends* 220 Hiaqua..—shell money and ornaments highly prized by the Indians of the Pacific Northwest Coast.

hiatal (həiˌēi·tăl), *a.* [f. Hiat(us + -al.] Of or pertaining to a hiatus.
1909 J. P. Iddings *Ign. Rocks* I. vi. 198 Hiatal fabrics being those in which variations in the sizes of crystals are not in continuous series, but in broken series with hiatuses, it follows that the most characteristic feature of such fabrics is the marked contrast between the sizes of some of the crystals. 1923 R. Knox *Radiogr. & Radio-Ther.* (ed. 2) I. 319 Cardiospasm (Hiatal Oesophagismus). 1957 [see *fixation 3 c]. 1961 *Lancet* 7 Oct. 810/1 In hiatus hernia, if the hiatal slit is small enough, the mucosal surfaces of the stomach may be adequately opposed.

hiatus. Add: **1. b.** Also *attrib.*, as **hiatus hernia,** a hernia in which an organ, esp. the stomach, protrudes through the œsophageal opening in the diaphragm.
1928 *Acta Radiologica* IX. 301 Diaphragmatic hiatus-hernia and oesophageal diverticulum were roentgenologically diagnosed. 1934 *Jrnl. Amer. Med. Assoc.* 24 Feb. 586/1 A hiatus hernia should always be considered as a possible diagnosis when upper abdominal symptoms of doubtful origin occur chiefly or only at night. 1958 *Sunday Times* 28 Sept. 27/3 A condition known medically as hiatus hernia..occurs when something goes wrong with the diaphragm. 1971 *Brit. Med. Bull.* XXVII. 34/1 Difficulty in swallowing is associated with..hiatus hernia.
3. Also *attrib.* and *Comb.*, as *hiatus-consonant, -filler, -glide; hiatus-filling* adj.
1945 *Mod. Lang. Notes* Dec. 550 The spelling *donmore* for *Dunmowe* may indicate the development of a hiatus-filling *r* in sandhi. 1948 D. Diringer *Alphabet* II. vi. 350 The letters..were also used as 'hiatus-consonants'. 1953 K. Jackson *Lang. & Hist. Early Brit.* 278 The hiatus-glide with native *e* was..j. *Ibid.* 367 The hiatus-filler here is y rather than j. 1968 *Language* XLIV. 454 The general outlines of this 'hiatus diphthongization' have been known for more than three-quarters of a century.

‖ **hibachi** (hi·batʃi, hibā·tʃi). Also formerly **hebachi.** [Jap. *hibachi, hi-hachi,* f. *hi* fire + *hachi* bowl, pot.] A large earthenware pan or brazier in which charcoal is burnt esp. in order to warm the hands or heat a room.
1863 R. Alcock *Capital of Tycoon* II. xvi. 379 There were also some fifty *Hebachis,* or vessels for burning charcoal and warming the rooms, corresponding with the Spanish *Brazeiro.* 1874 *Trans. Asiatic Soc. Japan* II. 132 Boys and girls assemble around the *hibachi.* 1921 *Outward Bound* Apr. 22/1 The little *kimori* was bidden to join the family circle around the *hibachi.* 1959 R. Kirkbride *Tamiko* iv. 28 The small..mat room was bare except for..the hibachi, and a scattering of china on the shelf. 1965 *Austral. Women's Weekly* 20 Jan. 27/1 The other indispensable came from a prolonged stay in Yokohama, a small serviceable iron *hibachi,* the original of the Western barbecue grill, but portable.

Hiberno-. (Further examples.)
1907 *Daily Chron.* 6 Sept. 3/1 Hiberno-Egyptian. 1908 *Westm. Gaz.* 20 Aug. 5/2 A Hiberno-Romanesque church. 1939 *Burlington Mag.* Jan. 43/1 The brilliance and precocity of the Hiberno-Saxon arts. 1953 K. Jackson *Lang. & Hist. Early Brit.* I. v. 175 This family of remote Hiberno-British princelings wished to make clear its claim to Roman status.

hibschite (hi·bʃəit). *Min.* [ad. G. *hibschit* (F. Cornu 1905, in *Tschermaks min. und petrogr. Mitt.* XXIV. 327), f. the name of J. E. *Hibsch* (1852–1940), Bohemian mineralogist: see -ite[1].] A calcium aluminosilicate hydroxide, a member of the garnet family.
1907 *Mineral. Mag.* XIV. 400 Hibschite... Minute, octahedral, optically isotropic crystals of contact-metamorphic origin, occurring in enclosures of chalk-marl in phonolite at Aussig, Bohemia. 1942 *Amer. Mineralogist* XXVII. 783 X-ray powder patterns of hibschite from the type locality..closely resemble patterns of grossularite and plazolite. 1962 W. A. Deer et al. *Rock-Forming Min.* I. 79 Hibschite, $Ca_3Al_2Si_2(OH)_4O_8$, has a cell side of 12.16Å..and eight molecules to the unit cell. 1966 [see *hydrogrossular* s.v. *hydro-].

hiccoughy, var. Hiccupy *a.*
1911 J. C. Lincoln *Cap'n Warren's Wards* i. 2 The train slowed down, in a jerky hiccoughy sort of way.

hiccup, *sb.* Add: **c.** *attrib.* **hiccup-nut** *S. Afr.,* the fruit of an ornamental shrub, *Combretum* (*Poivrea*) *bracteosum,* belonging to the family Combretaceæ; also, the plant itself; **hiccup strike** [It. *sciopero a singhiozzi*] *colloq.*, a strike normally of short duration which forms part of a series of similar and irregularly spaced strikes.
1862 Harvey & Sonder *Flora Capensis* II. 512 *P[oivrea] bracteosa*... Fruit oval or slightly obovate, indistinctly 5-angled, glabrous, 1-seeded. Called 'Hiccup-nut' in the colony. 1868 J. Chapman *Trav.* II. App. 447 The exquisite heads of scarlet flowers of the Hiccup-nut. 1899 Wood & Evans *Natal Plants* I. II. 63 The fruit is known locally as 'Hiccup Nut' and is palatable, but usually produces violent hiccough. 1951 *Dict. Gardening* (R. Hort. Soc.) II. 531/1 *C. bracteosum.* Hiccup-nut. 1950 *Times* 27 Jan., In the meantime the 'hiccup' strikes go on in the Paris region, where 12 'bus lines were out of action. 1962 *Daily Tel.* 28 June 26/3 Most are on what are known in Italy as 'hiccup strike', two-day stoppages at irregular intervals.

hiccupy *a.* (earlier and later examples).
1853 Lytton *My Novel* II. vii. x. 225 Long and loud talk recommenced, Burley's great voice predominant, Mr. Douce chiming in with hiccupy broken treble. 1968 *Listener* 11 July 55/2 Chopped-up, hiccupy sentences, often one word long. 1971 *Daily Tel.* 16 Oct. 7/6 The old Japanese anemone, mentioned in catalogues..as having the hiccupy name of Anemone hupehensis.

‖ **hic et nunc** (hīk et nuŋk), *phr.* [L., 'here and now'.] At the present time and place; in this particular situation; spatio-temporal nature of a phenomenon (see quot. 1948).
1935 *Studies in Hist. of Ideas* III. 469 One must distinguish between the type or kind of a sign, and its *hic et nunc* spatio-temporal exemplification. 1948 L. Spitzer *Linguistics & Lit. Hist.* 14 An explanation of the concrete *hic et nunc* of a historical phenomenon. 1950 I. Silone in Koestler et al. *God that Failed* 103 A political revolution, *hic et nunc.* 1966 *Listener* 20 Oct. 580/1 The presence of God in the world, *hic et nunc,* at man's level.

hick, *sb.*[1] Delete †*Obs.* and add later U.S. examples.
1916 H. L. Wilson *Somewhere in Red Gap* iii. 113 So I yelled out back to an old hick of a gardener..and he comes running. 1923 R. D. Paine *Comrades of Rolling Ocean* vi. 106, I come from a small town and that makes me a hick. My opinions aren't worth much. 1925 *Glasgow Herald* 16 Sept. 4 In the United States the 'Hicks' are still for the most part tenacious of that doctrine which assigns to speculations a 'bearish' or depressing influence on the markets. 1927 *Observer* 1 May 10/2 It is ..much easier to write a good play about hicks, boobs, hayseeds, highbrows,..and sentimentalists than about decent English people. 1928 *World's Work* Apr. 628 Broadway humorists, only a few years ago, used to make fun of Long Islanders by calling them 'hicks'. 1929 A. Conan Doyle *Maracot Deep* 18 He could not make these country hicks understand. 1970 W. Burroughs Jr. *Speed* ii. 39 The proprietor was a knobby, obliging old hick and he watched us all real close to make sure we didn't try to lift any fertilizer. 1970 J. Hansen *Fadeout* (1972) ii. 10 He was killed... They just stopped playing him. As though we was such hicks we didn't know there's such a thing as tapes these days.
b. *attrib.* or as *adj.* Like a hick, unsophisticated, provincial. *colloq.* (chiefly *N. Amer.*).
1920 S. Lewis *Main Street* xiv. 164 He graduated from a hick college in Pennsylvania. 1920 —— in *Sat. Even. Post* 11 Dec. 92/2 Why the plates turned over? That's hick-town stuff. 1921 H. C. Witwer *Leather Pushers* ix. 216 His..features wasn't bad looking in a hick way. 1936 J. Dos Passos *Big Money* 256 Tad was sitting there hanging his head, his hick hands dangling between his knees. 1952 M. McCarthy *Groves of Academe* (1953) xii. 243 He..nodded triumphantly at his friend,..who had been trying to persuade him that science fantasy was hick. 1958 *Punch* 27 Aug. 283/3 Why stick with some depressing Yank writer who takes a cool six hundred pages to chronicle the twenty-four-hour doings of a single character in a hick town? 1967 *Listener* 17 Aug. 218/3 Telly was still rather a hick affair back in 1951.

†**hickboo** (hi·kbū). *Air Force slang.* Also **hickaboo.** An air raid. *Obs.*
1919 *Athenæum* 23 May 360/2. 1925 Fraser & Gibbons *Soldier & Sailor Words* 119 *Hickaboo*.., an Air Force term for an Air Raid.

hickey (hi·ki), *sb.* Chiefly *U.S.* Also **hickie.** [Origin unknown.] **1.** Any small gadget or device; something of little consequence; = *doohickey.
Quot. 1909 is more specific than is warranted by the available evidence.
1909 Webster, *Hickey* (*a*), a device for bending a conduit, consisting of an iron pipe used as a handle fitted at one end with a tee through which the conduit is passed; (*b*) a small fitting used in wiring for electric lights, a fixture piped for gas. 1913 in Wentworth *Amer. Dial. Dict.* 291/1 S.C. Rock Hill. Common. 'Hand me that hickey.' 1928 *Papers Michigan Acad. Sci. Arts & Lett.* X. 298 *Hickie,* a word applicable to anything whose name one does not remember; 'what's its name'. 1932 *Atlantic Monthly* CXLIX. 665 We have little hickeys beside our seats to regulate the amount of air admitted through a slot in each window. 1935 *Hearst's International* Oct. 24/2 The chances are the management will be putting hickeys in their keyholes. 1940 *Sat. Even. Post* 15 June 38/2 There was a violent distaste for the fraction-of-a-cent tokens, disks of aluminum with a hole in the center. These were variously called 'Chinese money', 'hickeys', 'monkey money' and 'agony money'.
2. A pimple; a love-bite. *U.S. dial.* or *colloq.*
Said by Wentworth and Flexner to be c 1915 but printed evidence is lacking.
1934 *Amer. Ballads & Folk Songs* (1960) 19 Godamighty made a 'gator wid hickies all over his Tail. 1937 *Ten-Story Love Mag.* May 2 (Advt.), Hickies spoil everything. I know. I had 'em until I began eating Fleischmann's yeast. 1946 *Publ. Amer. Dial. Soc.* Nov. 17 *Hickey,* a small festered spot on the skin of a person. Salem. Reported 1942. 1956 H. Gold *Man who was not with It* (1965) xviii. 158 A woman is not just soul and hickie-squeezing. *Ibid.* xxv. 235 Tall..skinny, big blue hickie on the face.
3. *Printing.* (See quots.)
1940 *Correct English* (Chicago) Mar. 63/3 *Hickey,* printer's slang term for ornament. 1961 H. B. Jacobson *Mass Communications Dict.* 163 *Hickey.* 1. Slang term for slight tears or rips in wet collodion or stripfilm negatives, or for small 'runs' or blemishes in sensitized coatings. 2. A speck on the printing area of an engraving that remains after the etch. Must be routed off. 1967 E. Chambers *Photolitho-Offset* 273 *Hickeys,* faults in the printed result which show as irregular spots with white surrounding haloes, caused by dirt or hardened specks of ink.
4. See *jim-hickey.

hickey (hi·ki), *a.* *slang.* [? f. Hick *v.*] Tipsy.
Recorded in dictionaries of slang: Grose (1788), Matsell (1859), Barrère & Leland (1889), Berrey & Van den Bark (1942), etc.

hickory. Add: **2. c.** (Examples.)
1827 *Hallowell* (Maine) *Gaz.* 20 June 2/2 He was in favor of amending the Constitution, so as to let the people vote for Old Hickory. 1860 J. Parton *Life A. Jackson* I. xxxiv. 381 It was on this homeward march that the nickname of 'Old Hickory' was bestowed on the General. 1907 *Springfield* (Mass.) *Republ.* 24 Oct. 8, I should not say that Old Hickory was faultless, but Andrew Jackson was as upright a patriot as ever any nation had. 1949 B. A. Botkin *Treas. S. Folklore* p. xx, In this land..men put daring above discipline and etiquette to give us heroes like the 'Swamp Fox', 'Old Hickory', [etc.]. 1967 *Oxf. Compan. Eng. Lit.* (ed. 4) 592/1 *Old Hickory,* a nickname of Andrew Jackson.
4. a. Also applied *fig.* to members of various religious sects.
1831 *Boston* (Mass.) *Transcript* 12 Dec. 1/1 This assemblage of *Shaking* Quakers, for so many of them proved, who were only *hickory* ones till they joined the sett. 1855 *Jrnl. Discourses* II. 322 If there are any Gentiles, or hickory 'Mormons',..write it down. 1859 Bartlett *Dict. Amer.* s.v., A 'hickory Catholic'..is a flexible, yielding one. 1872 E. Eggleston *End of World* xxxix. 249 Any member of your class would do better to marry a good, faithful, honest New Light than to marry a hickory Methodist. 1940 *Sat. Even. Post* 30 Mar. 37/4 He is..referred to by the neighbors as a 'hickory Amish' because of some infraction not publicly mentioned, but most likely that of going to a movie.
b. hickory shad, the gizzard-shad (*Dorosoma cepedianum*); also, the fall-herring; **hickory shirt** (earlier examples).
1800 B. Hawkins *Sk. Creek Country* in *Georgia Hist. Soc. Colls.* (1848) III. 53 The fish taken here are, the hickory shad, [etc.]. 1871 *Amer. Naturalist* V. 398 The 'Hickory Shad'..were also filled with comminuted Crustacea. 1947 B. W. Dalrymple *Panfish* 341 Then suddenly a big buck Shad of four or five pounds, or a small Alewife or Hickory Shad. 1836 D. Harris in *Texas Hist. Assoc. Q.* (1900–1) IV. 160 Mother..made two striped hickory shirts and bags to carry provisions. 1850 L. H. Garrard *Wah-To-Yah* xii. (1927) 58 Hickory shirts. a 1861 T. Winthrop *Canoe & Saddle* (1883) iii. 33 Hickory shirts and woolen blankets are worn instead of skin raiment.

hi-coc(k)alorum, -olorum, occas. sp. of *high cockalorum*: see Cockalorum 1 and 3.
1860 *Illustr. London News* 7 Jan. 24/2 The little innocents, however, chiefly devote their energies to mud-pie manufacture and the games of Mulberry-bush, I-spy-I, Hi Cockolorum, Hopscotch, or Buttons. 1926 Fowler *Mod. Eng. Usage* 164/2 Mock Latin: bonus, bogus, hocus-pocus, hi-cocalorum (hic, hoc, horum?). 1950 C. Fry *Venus Observed* 82 The seven seas, and the milky way And morning, and evening, and hi-cockalorum are in it. 1969 I. & P. Opie *Children's Games* viii. 257 Croydon boys call it [*sc.* the game] not only 'Hi Jimmy Knacker', but 'Bung the Barrel', 'Hi Cockalorum', [etc.].

Hidatsa (hidæ·tsā). [Native name *hiratsa* willow wood lodge (Dr. Sturtevant).] A member of a group of N. American Indians; also, the language of this people. Also *attrib.*
1873 W. Matthews *Gram. & Dict. Lang. Hidatsa* p. xvi, The origin of the word Hidatsa is obscure, yet it is the name by which these Indians now designate themselves... One of their villages on Knife River was named Hidatsa; and probably when they were reduced by smallpox the majority of survivors came from that village, which then lent its name to the whole Tribe... The name Hidatsa is said by some to mean 'willows'. 1890 J. G. Frazer *Golden Bough* II. iv. 339 Some of the Hidatsa Indians explain the phenomena of gradual death..by supposing that man has four souls. 1911 —— *Ibid.* (ed. 3) I. iii. 55 Similar practices are reported among the Illinois, the Mandans, and the Hidatsas of North America. 1964 E. Bach *Introd. Transformational Gram.* v. 89 Languages..Arabic, Hidatsa. 1969 *Observer* (Colour Suppl.) 18 May 22/2 A special development in the warrior societies was found among the Mandan, Hidatsa,..and Blackfoot, which had a hierarchy of societies.

hidden, *ppl. a.* Add: **1. b.** (*the*) *hidden hand,* secret or occult influence, esp. of a malignant character.
1870 T. Taylor (*title*) The hidden hand. 1879 *Scribner's Monthly* July 326/2 Mr. Chaufrau played..the negro Wool in a dramatization of Mrs. Southworth's '*Hidden Hand*'. 1932 *Ann. Reg. 1931* II. 21 One Labour member attributed the appointment to the influence of a 'hidden hand' which was forcing the Labour Party to act against its principles. 1969 *Daily Tel.* 8 Mar. 20/3 Government action was being urgently considered against the 'hidden hand type of pressure' in public relations.

c. *Gram.* *hidden quantity* (see quot. 1898).
1898 G. M. Lane *Latin Gram.* § 2459 A vowel which stands before two consonants, or a double consonant, belonging to the same word, so that its natural quantity cannot be determined from the scansion of the word, is said to possess Hidden Quantity. 1965 W. S. Allen *Vox Latina* 65 A long vowel in such a position is sometimes said to have 'hidden quantity'.
d. *hidden reserve*: (*a*) in *Econ.* (see quot. 1965); (*b*) in general or transf. use, something kept in reserve in a concealed form.
1930 *Economist* 30 Aug. 408/1 Many of the assets in the balance sheet contain substantial hidden reserves. 1935 *Discovery* Oct. 290/2 It was not until the 'hidden reserve' of radioactivity was discovered that it was possible for the prolonged youth of the Earth to be explained. 1965 J. L. Hanson *Dict. Econ.* 213/1 If the assets of a firm have been deliberately undervalued, perhaps because the value of these assets has increased, the difference between their value as shown in the firm's balance sheet and their real value provides the firm with a 'hidden reserve', of which most shareholders will be unaware.
e. *hidden persuaders,* a term used, orig. by the Amer. writer Vance Packard (b. 1914), to describe those involved in the organization and practice of advertising; hence *hidden persuasion.*
1957 V. Packard (*title*) The hidden persuaders. 1959 *Daily Mail* 2 Apr. 1/4 This is the diet with the hidden persuader and the built-in will-power. 1960 *Guardian* 28 Dec. 8/4 The hidden persuasions. 1962 *Sunday Express* 30 Dec. 17/1 At sales time the hidden persuasion works harder—'15½ guineas slashed to £5'.

hide, *sb.*[1] Add: **2. c.** Impudence, effrontery, 'nerve'. (App. an elliptical use of 'thick hide'.)
Freq. in Australia and N.Z. but also occurs elsewhere.
1916 J. B. Cooper *Coo-oo-ee* xi. 150 Don't you think you have a hide to ask me? 1926 'J. Doone' *Timely Tips for New Australians, Hide,* a slang term denoting impudence. 1947 K. Tennant *Lost Haven* ii. 37 He *might* have told me. Just springing it on me..out of the blue... Like his hide! For two pins I'd tell him where to go. 1949 H. Wadman *Life Sentence* 9 Talk about self-confidence... What a hide! 1959 P. H. Johnson *Unspeakable Skipton* v. 34 The beast has had the hide..to dictate that to a secretary. 1961 *Coast to Coast 1959–60* 120 He wants to be a farmer. A farmer! Had the hide to try and tell *me* what farm life's like—me, born and bred on one.

hide, *sb.*[3] **II. 2.** (Later examples.) Also *attrib.*
1920 *Nature* CV. 146/2 The cock bird discovered Mr. Brook leaving the 'hide'. 1934 *Brit. Birds* XXVIII. 97, I had just seen my companion into his hide tent..when a small wader came off her nest at my feet. 1935 *Discovery* Aug. 228/1 We built our first hide..four feet away from one nest. 1940 'Gun Buster' *Return via Dunkirk* II. i. 85 The guns and vehicles went into a 'hide' in a large orchard. 1952 E. F. Davies *Illyrian Venture* xi. 220 The others had been sealed into a hide in the camp and were not discovered for four days. 1965 P. Wayre *Wind in Reeds* iv. 43 He had already started work on the construction of permanent observation huts or hides built into the sea-wall itself, from which it was possible to watch the wild geese in comfort.

hide, *v.*[1] Add: **1.** Further examples of *to hide away.* Also const. *up.* (Cf. sense 2 b.)
1854 W. Collins *Hide & Seek* III. xxiii. 227, I shall find him! I don't care where he's hid away from me. *Ibid.* xxiv. 271 She..hid it away in her bosom. 1884 'Mark Twain' *Huck. Finn* xxiv. 241 It's reckoned he left three or four thousand in cash hid up som'ers. 1891 C. Graves *Field of Tares* 109 There was a fresh canvas upon the easel, the tattered one had been carefully hidden away. 1928 E. Wallace *Flying Squad* xvi. 169 Bradley's fond of her. He hid her up once: why shouldn't he hide her up again? 1948 'N. Shute' *No Highway* vi. 168 It was impossible to hide up evidence like that.
2. c. *to hide out*: to go into hiding; to hide from the authorities. *U.S.*
1884 J. C. Harris *Mingo* 124 The revenue fellers better not git too clost ter Hog Mountain, bekaze the hidin'-out bizness is done played. 1885 'C. E. Craddock' *Prophet Gt. Smoky Mts.* ii. 44 Loneliness had made his sensibilities tender and 'hiding out' affected his spirits more than dodging the officers. 1911 R. D. Saunders *Col. Todhunter* i. 19 You got to hide out when that word is delivered, suh. 1924 F. R. Bechdolt *Tales Old-Timers* 345 A man..could hide out and hold up his herd. 1969 C. F. Burke *God is Beautiful, Man* (1970) 25 So he tries to find a pad where he can hide out.

hide-and-coop. *U.S.* = Hide-and-seek.
1850 S. Judd *R. Edney* 128 (Th.), As if religion were a game of hide and coop. 1909 *N. & Q.* 10th Ser. XII. 371/1 In 'hide and coop', each called out from his secret place a faint, long-drawn 'c-o-o-p'. 1910 S. E. White *Rules of Game* III. xii. 258 The herders grinned as the rangers came in sight. They had been 'tagged' in this 'game of hide and coop'. 1923 K. D. Wiggin *My Garden of Memory* 14 The board-piles afforded the best possible place for playing 'Hide and Coop'.

hide-and-seek. (Earlier and later (chiefly) U.S. examples of **hide-and-go-seek.**)
1724 *Essex Inst. Hist. Coll.* XXXVI. 333 At night was at Madam Brownes playing hide and goe seek with Olive Parker, Wibird, &c. 1821 J. F. Cooper *Spy* (1831) xxxvi. 405 Since when, we have been playing hide-and-go-seek with the ships. 1908 T. Roosevelt *Lett. Children* 2 Jan. (1919), Do you recollect how we all used to play hide-and-go-seek in the White House? 1925 F. Scott Fitzgerald *Great Gatsby* v. 82 At first I thought it was another party, a wild rout that had resolved itself into 'hide-and-go-seek'. 1969 I. & P. Opie *Children's Games* iv. 154 In Scotland it is often 'Hide and Go Seek'—as also in the United States. 1971 A. Bailey *In Village* (1972) xx. 202 There are games of prisoner's base and hide-and-go-seek.. with the parked cars providing a forest of hiding places.

hideaway, *sb.* Add: **2.** A small, quiet restaurant, etc., or a secluded place of entertainment.
1929 *World's Work* Nov. 40/2 The vaudeville performer on the two-a-day has played to punks in the hide-aways who turned his riot into an oil can. 1939 C. Morley *Kitty Foyle* xviii. 171 He took me to a French hideaway up on Pine Street. 1962 *Times* 9 May 7/1 Dozens of inexpensive hideaways where the entertainment ranges from progressive jazz quartets to the folk songs of the South and West. 1971 *New Yorker* 11 Sept. 99/2 (Advt.), We can take you on a tour of romantic little hideaways, golden beaches and the excitement of the bullring.
3. A place of concealment or retreat.
1930 'E. Queen' *French Powder Myst.* II. xiii. 104 This is also the conference room for directors' meetings, the Old Man's hideaway, et cetera. 1941 'R. West' *Black Lamb* II. 12 A Bulgarian leader who was lying wounded in a hideaway. 1959 *Times Lit. Suppl.* 20 Nov. 680/5 They meet in his hideaway to discuss the state of the world.

hi-de-hi (həi·ˌdiˌhəi·), *int.* An exclamation, used chiefly by army instructors to greet, or attract the attention of, their troops. Also *transf.* and *attrib.*
1941 G. Kersh *They die with their Boots Clean* II. 67 'Now, when I say *Hi-de-Hi Squad!* you shout *Ho-de-Ho!* —and shout it loud! Now: *Hi-de-Hi Squad!*' We roar: 'Ho-de-Ho!' 1943 *N. & Q.* CLXXXV. 83/2 A reviewer in *The Listener* (27 May) writes:..the hi-de-hi qualities of Dr. I. A. Richards's latest book should..make our minds spring to attention with the appropriate ho-de-ho. 1944 *N. & Q.* CLXXXVI. 57/1 It was the C.O. of a training camp who instructed his officers on seeing their men in the street to come to attention and say hi-de-hi, whereupon the men came to attention and replied ho-de-ho. 1950 J. Vedey *Band Leaders* iv. 29 He [*sc.* Harry Roy] made them sing Hi-de-hi at the May Fair.

hi-de-ho (həi·ˌdiˌhōu·), *int.* An exclamation of joy used chiefly by jazz and dance bands. Also *attrib.* and as *sb.*
1936 *Delineator* Nov. 10/6 They have not come here to dance and drink away an evening in the usual night-club hi-de-ho. 1946 Mezzrow & Wolfe *Really Blues* viii. 120 The hi-de-ho, vo-de-o-do and boop-boop-a-doo howlers that later sprouted up. 1949 L. Feather *Inside Be-Bop* iii. 23 There was plenty going on in the band besides hi-de-ho.

hideling, *a.* and *sb.* Add: Also **hidling. A.** *adj.* (Earlier examples.)
1834 R. Mudie *Feathered Tribes Brit. Isles* I. 327 The lesser white-throat is, however, a more hidling bird than the white-throat. 1839 W. MacGillivray *Hist. Brit. Birds* II. 334 So hideling are its habits that one seldom obtains a glimpse of it.

hideosity. (Earlier and later examples.)
1807 C. Wilmot *Let.* 15 May in *Russ. Jrnls.* (1934) II. 245 Nothing before our eyes but Minerals & Fossils & Animals in every State of Hideosity. 1841 C. Ridley *Lett.* (1958) 69 Never in my life have I seen such frightful hideosities of beds as they are. 1970 *New Yorker* 12 Sept. 56/1 Don't speak to me of caravans... Such hideosities.

hide-out (həi·dˌaut). orig. *N. Amer.* Also **hideout.** A hiding-place. Also *attrib.*
1885 *Century Mag.* Mar. 684/2 They guv my place the name o' Hide-out, an' they didn't conscrip' me, nuther. 1913 A. B. Emerson *R. Fielding at Snow Camp* 174 Meanwhile, the wind shrieked through the forest above their 'hideout'. 1920 B. Cronin *Timber Wolves* 76 'In rough country like this a man could bury himself for years.'..'This coast is full of hide-outs, as they call them.' 1933 *Amer. Speech* VIII. 49/2 *Hide-out country,* any trackless wilderness. 1935 G. Gorer *Africa Dances* III. iv. 225 Even if they are inspected by the local authorities..they have..neighbouring hide-outs. 1940 *Economist* 11 May 848/1 The long coastline of Norway provides innumerable hide-outs for German submarines. 1958 *Listener* 25 Dec. 1074/2 The National Gallery pictures, tucked away in their air-conditioned, thermostatically controlled, war-time hideout in the Welsh mountains. 1963 *Daily Tel.* 14 Aug. 20/6 Special squads combed the hideout house. 1973 J. Cleary *Ransom* iii. 71 She had made a mistake in choosing this place as their hide-out, but nowhere else had seemed better.

hidey (həi·di), *int.* Chiefly *Austral.* and *U.S.* Also **hidy, highdey.** [Blend of Hi and *howdey* (see How-do-ye).] A word of greeting.
1941 Baker *Dict. Austral. Slang* 35 *Hidey! Highdey!,* Hail!, How are you! 1959 I. Jefferies *13 Days* xi. 182 'Hidey,' I said. 1962 E. B. Atwood *Regional Vocab. Texas* iii. 70 *Hidy,* probably a blend of the old and new.

hiding, *vbl. sb.*[1] Add: **4.** *hiding power,* the capacity of paint or other colouring materials to obliterate certain surfaces.
1951 R. Mayer *Artist's Handbk.* 433 *Hiding power,* degree of opacity in a paint or pigment; ability to mask or conceal an underpainting. The term *covering power* is sometimes confused with it. 1966 J. S. Cox *Illustr. Dict. Hairdressing* 79/1 *Hiding power,* the power of an opaque dye or other colouring material, when applied to hair, to cover or hide its existing colour. 1967 *Gloss. Paper Ink Terms Letterpress Printing (B.S.I.)* 11 *Hiding power,* the capacity of an ink to obliterate the previously printed ink film.

hidy-hole (həi·diˌhōul). orig. *Sc.* Also **hidey-, hidie-.** [Alteration of *hiding-hole*: see Hiding *vbl. sb.*[1] 4.] A hiding-place.
1817 *Blackw. Mag.* Nov. 158/2 They're darned in some o' the queer hidy-holes about the rocks there. 1828 D. M. Moir *Mansie Wauch* ix. 79 We got James..hauled out of his hidy-hole. 1870 R. Chambers *Pop. Rhymes* 91 He had not been long in his hidy-hole, before the awful Etin came in. 1896 S. R. Crockett *Cleg Kelly* iii. 18 Tim Kelly's 'hidie holes', where he kept the weapons of his craft. 1920 *Glasgow Herald* 6 Dec. 10 The story of the Ballantrae smugglers' 'hidie-hole'. 1927 *Hutchinson's Myst. Story Mag.* IX. xlviii. 50/2 So it's obvious that there may quite easily be another snug little hidy-hole which none of us knows of. 1934 D. L. Sayers *Nine Tailors* 261 A nice little six-inch hidey-hole where nobody would ever dream of looking for it. 1955 J. Thomas *No Banners* xxxiii. 330 There's a concrete hidey-hole underneath. 1959 P. H. Johnson *Humbler Creation* xlii. 286 So this was where he came,..into this snugness, this hidey-hole. 1969 I. & P. Opie *Children's Games* iv. 157 Those who are still searching..rush to the places where they were last seen, thinking that they will be near the hidy-hole.

hielaman. b. hielaman tree (examples).
1884 A. Nilson *Timber Trees N.S.W.* 136 Heelaman tree. 1889 J. H. Maiden *Useful Native Plants Austral.* 426 *Heilaman tree.* The wood is soft, and used by the aborigines for making their 'heilamans' or shields.

hien, hsien (hyen, ʃyen). Pl. uninflected. Also **heen.** [Chinese.] An administrative division of a fu or department, or of an independent chow or district; also, the seat of government of such a division. Also *attrib.*
1837 *Penny Cycl.* VII. 77/1 The subordinate cities and districts of each province in the three ranks of Foo, Chow, and Hien, are under the charge of their respective magistrates, who take their rank and titles from the cities they govern. 1901 *Westm. Gaz.* 23 Aug. 8/2 Among the candidates was a man who has the hsien B.A. degree. During the examination it was found that he had some books with him. The examiner at once ordered his expulsion from the hall, but the men from his hsien objected. 1909 *Ibid.* 24 May 8/2 A circular has been sent to all viceroys and governors by the Pekin Cabinet ordering that a stop be put to the old custom of levying benevolences on the governors of hien cities. 1938 *New Statesman* 15 Jan. 74/1 This means in practice control of the whole administrative machine down to the smallest units, the hsien. 1959 *Chambers's Encycl.* III. 439/2 The provincial health administration..supervised the work of the *hsien* and municipal authorities. 1971 *N.Y. Times* 27 June 20 Each hsien [in China] has been given the target of becoming self-sufficient in food and light industrial products.

hierarchize, *v.* Add: (Later examples.) Hence **hierarchiza·tion.**
1931 C. Williams *Place of Lion* ii. 26 These hierarchized celsitudes are but the last traces..of the ideas which Plato taught his disciples existed in the spiritual world. 1954 B. & R. North tr. *Duverger's Pol. Parties* II. iii. 405 Party hierarchization becomes more theoretical than practical. 1961 *New Left Rev.* May–June 34/2 Management..is not so amalgamated or hierarchised. 1967 D. Cooper *Psychiatry & Anti-Psychiatry* v. 100 We have eliminated formal hierarchization to a point beyond which no similar experiment in the literature of the subject has gone. 1971 *Jrnl. Gen. Psychol.* LXXXIV. 202 By hierarchization Piaget means that there is a constant order of succession of different levels of functioning constituting a genetic sequence. 1971 G. Steiner *In Bluebeard's Castle* iv. 88 An explicit grammar is an acceptance of order: it is a hierarchization..of the forces and valuations prevailing in the body politic.

hierarchy. 4. (Later examples.)
1932 Von Wiese & Becker *Systematic Sociology* 355 Whenever persons join or otherwise enter into a plurality pattern they almost invariably take their places in an implicit or explicit hierarchy, and consciously or unconsciously expect the fact that there are ranks above and below them. 1947 *Soc. Research* XIV. 165 This hierarchy, like any other, is a social order in which human relations are determined by the degree of authority exercised by one group over another. 1961 G. G. Simpson *Princ. Animal Taxon.* i. 13 A hierarchy is a systematic framework for zoological classification with a sequence of classes (or sets) at different levels in which each class except the lowest includes one or more subordinate classes. 1963 Davis & Heywood *Princ. Angiosperm Taxon.* iii. 75 Before we go on to discuss the taxonomic hierarchy, it will be as well to list the ranks.. which are accepted in the *International Code of Botanical Nomenclature.*

hieratite (həi·ĕrătəit). *Min.* [a. It. *hieratite* (A. Cossa 1882, in *Atti d. R. Accad. dei Lincei (Trans.)* VI. 141), f. *Hiera,* ancient name of Vulcano, one of the Lipari Islands: see -ite[1].] A colourless or greyish fluoride of potassium and silicon, K_2SiF_6, occurring in stalactitic concretions in the crater of Vulcano and on Vesuvius.

The abstract cited in quot. 1882 consistently spells the word *hieralite,* although Cossa's form was *-tite.*
1882 *Jrnl. Chem. Soc.* XLII. 705 For this natural fluosilicate he proposes the name *hieralite,* from *Hiera,* the Greek name for the island Vulcano. Hieralite is associated in the concretions with selensulphur, realgar, mirabilite, [etc.]. 1951 C. Palache et al. *Dana's Syst. Min.* (ed. 7) II. 104 Hieratite is isostructural with the corresponding salts of Rb, Cs, Tl, and NH_4.

hieroglyphic, *sb.* **2. b.** (Further example.)
1853 Mrs. Gaskell *Cranford* xv. 296 About a year after Miss Matty set up shop, I received one of Martha's hieroglyphics, begging me to come to Cranford very soon.

hieromonach. Add: Also **hieromo·nk.**
1932 *Pax* Oct. 161 These are divided as follows: Maronite Antonians: 520 hieromonks, 221 monks. 1964 P. F. Anson *Bishops at Large* i. 37 Almost any alien calls himself..hieromonk..or any other Oriental ecclesiastical title.

hi-fi (həi·fəi·). [colloq. abbrev. of *high fidelity.] That part of acoustics and electronics that deals with the design, construction, and use of equipment for the recording and reproduction of sound to a fairly high standard. Also *attrib.* or as *adj.*, esp. **hi-fi equipment, set, system,** equipment for the home designed to reproduce (and sometimes to record) sound to such a standard, consisting often of several distinct units. Also *ellipt.* for *hi-fi equipment,* etc.
1950 *Audio Engin.* Aug. 24 (*title*) Hi-fi at seven-and-a-half. 1952 *Time* 22 Dec. 62/2 Until last week, most 'hi-fi' sets, which reproduce music in the home with the clarity and realism of the concert hall, were custom-made from standard parts by small radio and phonograph shops at a cost of from $150 to $2,000. 1953 *House & Garden* Dec. 132 (*heading*) What do they mean when they talk about hi-fi? 1953 *Life* 15 June 146/1 The name most generally used for this new kind of sound reproduction is high fidelity or 'hi-fi'. 1958 *Observer* 20 Apr. 10/6 The choice of a loudspeaker system is quite the most important task confronting those in search of 'Hi-Fi'. 1958 *Gramophone* June 35/1 The latest hi-fi developments. 1959 E. Fenwick *Long Way Down* x. 93 They had spent the evening working on their hi-fi system: three speakers, at different points, and miles of wiring. 1959 C. MacInnes *Absolute Beginners* 26, I put a disc on to his hi-fi. 1959 *Spectator* 20 Nov. 708/1 As recordings, they may not command the velvety tone, sumptuous presence and richly defined detail of the best of modern hi-fi. 1962 A. Nisbett *Technique Sound Studio* iii. 69 The full hi-fi treatment (in the sense, that is, of extended frequency response). *Ibid.* iv. 76 Hi-fi (a term which covers any sound equipment from medium quality upwards). *Ibid.* v. 98 Wide-range hi-fi, demanding not only high-quality transmissions but also a wide dynamic range. 1963 *Which?* Jan. 3/2 People may also want to consider hi-fi equipment. 1971 *Ibid.* June 187/1 Look for discounts in tape and hi-fi magazines. 1971 *Daily Tel.* 27 Aug. 17/6 The study thinks there is a continuing swing to quality but the lack of standards of what constitutes Hi-Fi is felt to be a disadvantage.

higgle (hi·g'l), *sb.* [f. the vb.] The adjusting of prices so that demand and supply are equal.
1908 *Daily Chron.* 5 Dec. 4/6 If they were abolished altogether the 'higgle of the market' would level freights correspondingly down.

hi·ggledipi·ggledness. Also **higgledy-piggledyness.** The quality or condition of being higgledy-piggledy.
1854 *Punch* 18 Nov. 204/1 That structural higgledipiggledness. 1918 A. Bennett *Roll-Call* I. i. 12 An agreeable and original higgledy-piggledyness!

higgledy. Abbrev. of Higgledy-piggledy.
a 1953 Dylan Thomas *Quite Early One Morning* (1954) 30 Neat and silly..they helped in the higgledy kitchen.

higgler. Add: **2. d.** (See quot.)
1930 M. Clark *Home Trade* 176 The term 'higgler' is applied in the Covent Garden market to a dealer who buys his supplies of fruit with a view to selling what he buys at a profit to any buyer, either on that market or on any other; to wholesalers or to retailers.

high, *a.* Add: **1. c.** Of clothes: high-necked.
1827 [see *canezou]. 1857 Mrs. Gaskell *Let.* 13 & 14 Sept. (1966) 471 My grey carmelite, & black moiré, high, & next to no collars. 1875 L. Troubridge *Life amongst Troubridges* (1966) x. 102, I had a chemisette to make my gown high, and no ornaments. 1937 J. Laver *Taste & Fashion* xiii. 185 In the early [eighteen-]sixties, it is interesting to note, there was less *décolletage* in good families in France than in England. The high dress was worn at dinner parties even of a formal kind. 1957 M. B. Picken *Fashion Dict.* 230/2 Boat-shaped neckline..high in front and back, wide at sides.
d. *Typogr.* (See quots.).
1683–4 J. Moxon *Mech. Exerc. Printing* (1962) 37 Head sticks..are Quadrat high, straight, and of an equal thickness all the length. 1888 C. T. Jacobi *Printers' Vocab.* 59 *High spaces,* spaces specially cast nearly type-high. 1904 Goodchild & Tweney *Technol. & Sci. Dict.* 287/2 *High,* a term applied to type or blocks which stand out in front of the rest of the type in the forme; *e.g.* new type stands higher than worn type. 1963 Kenneison & Spilman *Dict. Printing* 92 *High quads* (or *spaces*), spaces cast to the height of the shoulder of type.
4. a. Further examples of specialized meanings, esp. in *Athletics.* (Cf. sense *17 i.)
1895 High hurdle, jump [see Jump *sb.*[1] 1 b]. 1897 *Encycl. Sport* I. 50/2 High jumping may be..a gift. *Ibid.* 51/2 High kicking is very useful during the off-season. 1924 C. W. Mason *Chinese Confessions* xliv. 326 The high-jump and hurdles were my specialities. 1955 *Athletics* ('Know the Game' Series) 15 High jumping: Western roll; straddle; Eastern cut-off. 1964 M. Watman *Encycl. Athletics* 83/1 Britain possessed two world class high hurdlers in the 1930s.
b. Hence in numerous adjectival *Combs.*, as *high-back, -central, -front, -mid, -mixed, -narrow, -rising.*
1888 H. Sweet *Hist. Eng. Sounds* i. 2 So we have altogether nine positions [of the tongue]: high-back.. high-mixed..high-front [etc.]. 1924 H. E. Palmer *Gram. Spoken Eng.* I. 13 High-Rising. Nucleus-tone. 1934 J. J. Hogan *Outl. Eng. Philol.* 15, *u*: high-back rounded, as in *wolf. Ibid.* 16, *oi*: mid-back rounded + high-front. 1934 Webster I. xxviii/2 In English a high-central vowel is not usual. 1951 Z. S. Harris *Methods in Struct. Ling.* 57 High-rising [intonation] for impatient question. 1961 R. B. Long *Sentence & its Parts* xix. 422 (*heading*) High-back /u/. 1962 *Amer. Speech* XXXVII. 169 This diphthong may be described as beginning at a somewhat retracted low-front position and terminating at an open, slightly rounded, high-central position. 1964 Jakobson & Halle in D. Abercrombie *Daniel Jones* 98 The 'high-narrow' vowels are particularly short. 1964 *Language* XL. 100 The sounds..are high-mid. 1965 *Canad. Jrnl. Linguistics* Fall 65 In each case the checked high-front, high-back..and mid-back vowels are paired with phonically similar free vowels. 1966 *Publ. Amer. Dial. Soc.* XLVI. 32 Long, unrounded, high-mid front vowel.
c. *high breast wheel* (see quots.).
1880 *Encycl. Brit.* XII. 522/2 Overshot and High Breast Wheels. *Ibid.* 523/2 With greater variation of head-water level, a pitch-back or high breast wheel is better. 1888 *Lockwood's Dict. Mech. Engin.* s.v. *Breast wheel,* When the water flows in at a point above the horizontal line, the wheel is termed high breast, and when at a point below, low breast.
5. *high life* (later examples). Also *high society.*
1801 M. Edgeworth *Belinda* I. ix. 277 He had merely considered her ladyship as an object of amusement, and an introduction to high life. 1853 Mrs. Gaskell *Cranford* viii. 140 One would not have Lady Glenmire think we were quite ignorant of the etiquettes of high life in Cranford. 1892 C. M. Yonge *That Stick* I. x. 106 Utterly inexperienced as she was, even in domestic, not to say high life. 1920 D. Parker (*title*) High society. 1955 'E. C. R. Lorac' *Ask Policeman* iii. 28 Perhaps wealthy Australians were sent back to high-life schools in England in her day. 1971 'D. Halliday' *Dolly & Doctor Bird* xi. 155 The police..would spoil the leisurely high-society image. Dirty little men running over the Begum's nice holiday island.
6. a. *high art* (earlier and later examples). Also *high comedy, culture.*
1817 B. Haydon *Autobiogr.* (1926) I. xvii. 266, I had, by my public devotion to High Art, a claim on all the nobility and opulent in the kingdom. 1848 Geo. Eliot *Let.* 11 Feb. (1954) I. 247, I cannot recognize the truth of all that is said about the necessity of religious fervour to high art. 1849 —— *Let.* 20 Sept. (1954) I. 308 She is a person of high culture according to the ordinary notions of what feminine culture should be. 1856 —— in *Westm. Rev.* IX. 3 High culture demands more complete harmony with its moral sympathies in humor than in wit. 1883 'V. Lee' in P. Gunn *Vernon Lee* (1964) vii. 88 A long pseudo-medieval ballad... It felt so completely high art. 1895 G. B. Shaw *Our Theatres in Nineties* (1932) I. 106 After the exasperatingly bad acting one constantly sees at the theatres where high comedy and 'drama' prevail, it is a relief to see even simple work creditably done. 1906 E. Garnett in Defoe *Capt. Singleton* p. viii, Rembrandt's choice of beggars..for his favourite subjects seemed a low and reprehensible taste in 'high art'. 1919 G. B. Shaw *Heartbreak House* p. viii, The only part of our society in which there was leisure for high culture. 1963 *Observer* 12 May 28/3 Miss Murdoch is one of the sharpest writers of high comedy at present active in the theatre. 1964 Hall & Whannel *Popular Arts* I. ii. 55 These popular arts..were not objects of contemplation like the works of high art, but communal artifacts. 1966 D. Jenkins *Educated Society* ii. 50 High culture..tries to be creative in relation to the future and responsible in relation to the past.
d. Having a highly developed or complex organization; spec. *Biol.*, phylogenetically advanced or developed; often in the comparative degree, as *the higher apes, the higher plants.*
[1807 W. Wood *Zoography* I. p. xii, If we ascend to a higher class of beings, and contemplate the extensive range of the animal creation.] 1848 J. Lindley *Introd. Bot.* (ed. 4) xvi. 323 Plants have no circulation of their fluids analogous to that of blood in the higher animals. 1867 H. Spencer *Princ. Biol.* § 364 Every type that is best adapted to its conditions, which on the average means every higher type, has a rate of multiplication that insures a tendency to predominate. 1875 *Encycl. Brit.* II. 168/1 As man is the highest animal. 1902 *Ibid.* XXVI. 366/1 The first is an amyloin of a 'high', the second an amyloin of a 'low' type. *Ibid.* XXVIII. 343/1 The gorilla and the chimpanzee, the highest members of the apes. *Ibid.*, The embryonic stages of higher forms. 1936 E. G. Boulenger *Apes & Monkeys* 15 The apes and higher monkeys are quite as much reliant for their various needs upon the ground as among the tree-boughs. 1954 H. I. Featherly *Taxon. Terminol. Higher Plants* 157 In the evolution of the higher plants, the greatest number of changes has come about in the reproductive organs. 1964 E. Becker in I. L. Horowitz *New Sociol.* 114 Separation anxiety of the helpless higher-primate infant is the pivot for his early learning. 1966 R. & D. Morris *Men & Apes* vi. 165 This is the point at which we pass from the lower primates (the prosimians) to the higher primates (the anthropoids), and consider the evolution of monkeys, apes and man.
8. b. Of the condition of an animal or of soil: resulting from over-feeding or from too great an application of manure. Also of a crop: produced by an over-manured soil.
1834 W. Youatt *Cattle* xvi. 553 This occurs particularly in young cows after their first calving, and when they are in a somewhat too high condition. 1886 C. Scott *Sheep Farming* 116 Hill ewes are never in too high condition; the danger is all the other way. 1902 *Encycl. Brit.* XXVI. 361/1 It is better not to grow barley after roots fed off by sheep, as this rotation leaves the land..in too 'high' a condition... By taking barley as a second corn crop, the latter following roots fed off, or a 'high' crop, [etc.].
9. a. (Earlier example.)
c 1807 Jane Austen *Watsons* (1954) 344 As the partridges were pretty high, Dr Richards would have them sent away to the other end of the Table.
b. Of tobacco: moist. *U.S.*
1850 *Rep. U.S. Comm. Patents, Agric. 1849* 322 Tobacco should not be too moist, or 'high' as it is termed, when put in the stalk-bulks. 1865 *Trans. Ill. Agric. Soc.* V. 669 Care must be taken that the tobacco does not imbibe too much moisture, or get too high in case before it is bulked.
10. a. *high explosive*: see *explosive *sb.* 2.
e. Of money: lent out at a high rate of interest; dear.
1899 *Daily News* 31 Mar. 3/5 New York... Higher money.
g. *Naut.* Near the wind: designating a vessel or its head when pointing close to the wind, as in the command *no higher.*
a 1865 Smyth *Sailor's Word-bk.* (1867), No Higher! 1948 R. de Kerchove *Internat. Maritime Dict.*, No higher!
h. In card-playing: *ace high* (*king high,* etc.): having the ace (king, etc.) as highest card: said of the hand or suit, also *occas.* of the person. Cf. *ace 4.
1887 S. Cumberland *Queen's Highway* 276 Had I a 'flush' with 'king high' some one would be sure to rake in the shekels with 'ace high'. 1964 N. Squire *Bidding at Bridge* xxi. 172 You may rely upon two defensive tricks when your partner opens the bidding. These may not be in the actual suit he mentions, because it might, for example, be Knave high. 1973 D. Westheimer *Going Public* v. 79 Margo drew the low hand, a nine-high nothing. 'Poker never was my game,' he said cheerfully.
11. b. *spec.* Of a period of time: fully developed, at its peak.
1930 Baedeker *Northern Italy* (ed. 15) 622 The fine *Palazzo Larderel*.., formerly *Palazzo Giacomini,* in the High Renaissance style, by Giovanni Antonio Dosio (1558–80). 1934 *Burlington Mag.* Jan. 14/1 The usual technique of the high renaissance outside Venice. 1944 *Ibid.* Jan. 13/2 He was sufficiently adaptable to learn..some later developments of the High Renaissance. 1956 K. Clark *Nude* ix. 341 The root of high-renaissance taste. 1961 Webster (s.v. *high* adj. 1 b), The high middle ages. 1962 *Listener* 18 Oct. 601/1 This happened during the mid-sixteenth century, in the aftermath of the High Renaissance. 1964 C. S. Lewis *Discarded Image* vii. 182 Hence a modern finds those [chronicles] of the Dark Ages suspiciously epic and those of the High Middle Ages suspiciously romantic. 1965 K. Charlton *Educ. Renaissance Eng.* ii. 21 Such was the education of the High Middle Ages. 1972 *Country Life* 23 Mar. 696/3 The high Victorian Gothic style.
16. (Further examples of *high* (*old*) *time,* etc.)
1833 C. A. Davis *Lett. J. Downing* (1834) 177 Just after breakfast yesterday, I and the gineral had a high time [i.e. a heated argument] together. 1858 *Spirit of Times* 30 Jan. 345/1 Our friends..are having a real 'high old time' generally, just now, in trotting on the ice. 1869 B. Harte *Luck of Roaring Camp* (1871) 226 These are high old times, ain't they? 1870 D. J. Kirwan *Palace & Hovel* xxxi. 470 That's a werry 'igh old game is the Canteen; sort of priveet like. 1873 J. H. Beadle *Undevel. West* xxiii. 451 Santa Fe De San Francisco—so the old Spaniards named it—is a high old city. 1899 R. Whiteing *No. 5 John St.* III. xxi. 215 'Ah, they was 'igh old times!' is his final word. 1941 E. Bowen *Look at Roses* 255 Those two will be having a high old time, with the cat away. 1955 J. Potts *Death of Stray Cat* ii. 11 You probably had a high old time chasing blondes. 1962 N. Marsh *Hand in Glove* vii. 235 We picked him up ..having a high old time with the boxer bitch.
b. Phr. *high as a kite*: very drunk.
1939 *Amer. Speech* XIV. 90/2 *High as a kite,* completely drunk. a 1966 M. Allingham *Cargo of Eagles* (1968) iv. 54 He..gave them a champagne lunch in a marquee..and held a sale. By then everyone was as high as a kite.
c. Under the influence of, stimulated by, a drug or drugs. Freq. const. *on.*
1932 *Evening Sun* (Baltimore) 9 Dec. 31/4 *High,* under the influence of a narcotic. 1940 *Amer. Speech* XV. 337/1 *To be high,* to be contented from the drug. 1951 *N.Y. Times* 13 June 24/5 We would go out together and get high. I used to sleep with him whenever we got high. 1951 *San Diego Even. Tribune* 28 June A-1/4 He'd been 'getting high' on heroin week-end after week-end. 1957 C. MacInnes *City of Spades* I. x. 72 Hamilton's acquaintances.. rocking high with charge. 1961 *Spectator* 17 Nov. 712 The momentary kick when the drug is taken, when you're 'high'. 1969 *New Scientist* 29 May 455/1 It is far safer to drive a car when high on marihuana than when drunk. 1970 'D. Shannon' *Unexpected Death* (1971) i. 9 He and

his pal Roderick Drover had had some boyish fun last week sniping at an innocent driver—probably while high on something.
d. Highly interested in, keen *on.* *slang.*
1942 Berrey & Van den Bark *Amer. Thes. Slang* § 274/5 *Enthusiastic about*..high on. 1966 L. Deighton *Billion-Dollar Brain* xxii. 240 It's about Signe... She's high on you, you know. 1972 *Guardian* 30 Oct. 2/5 'I am not high on the Thieu brand of Government,' he [*sc.* McGovern] said, noting that 40,000 people had been executed..by it.
17. a. *high and dry*: further *fig.* examples. Also used in sense 'safe'. Also earlier example of the ecclesiastical sense.
1838 Dickens *Let.* 26 July (1965) I. 421, I no sooner get myself up, high and dry, to attack him [*sc.* Oliver Twist] manfully than up come the waves of each month's work. 1853 Bunsen *Let.* in *Life & Wks. Kingsley* (1901) II. 112 You know of the persecution of the Evangelicals, and High and dry against Maurice. 1881 E. W. Hamilton *Diary* 18 June (1972) I. 146 Meanwhile, Dr. Flood's successor had been appointed, and Dr. Flood was left high and dry without preferment owing to an undoubted breach of faith on the part of Duckworth. 1927 J. Galsworthy *Castles in Spain* 169 A true work of art remains beautiful and living, though an ebb tide of fashion may leave it for the moment high and dry on the beach. 1941 W. S. Maugham *Writer's Notebk.* (1949) 305 The river has flowed on and left him high and dry on the bank. The writer has his little hour..but an hour is soon past. 1960 *Times* 30 Aug. 11/6 Cella's back-heel, so deceptive, so utterly unexpected, left Rossano high and dry.
f. *high-and-mighty*: also used *absol.* *high-and-mightiness* (earlier example).
1804 M. Wilmot *Let.* 24 Apr. in *Russ. Jrnls.* (1934) I. 94 The High and Mighty then go into..another little room well heated. 1852 C. M. Yonge *Two Guardians* xiii. 246 That touch of Edmund's, which had shown her how he regarded her 'high-and-mightiness', had made her..ashamed. 1905 G. B. Shaw in *Grand Mag.* Feb. 116 Our high and mighties didn't exactly see the point. 1924 J. Galsworthy *White Monkey* II. ii, Mr. Mont's a gent.. no high-and-mighty about *him.*
i. *to be for the high jump*: see *jump *sb.*[1] 7 and *high *a.* 4 a.
j. *high, wide, and handsome* (and similar phrases), in a carefree manner, in good style (see also quot. 1971). orig. *U.S.*
1907 S. E. White *Arizona Nights* 35 Tim could talk high, wide, and handsome when he set out to. 1932 'Spindrift' *Yankee Slang* 21 *High, wide and handsome,* in good or great style. Common shout at a rodeo: 'Ride him, Cowboy, high, wide and handsome.' 1939 Wodehouse *Uncle Fred in Springtime* iii. 50 He has a nasty way of lugging Pongo out into the open and..proceeding to step high, wide and plentiful. 1958 L. van der Post *Lost World of Kalahari* (1961) vii. 155 The day was riding high, wide and handsome into the deeps of the incredible blue sky. 1971 J. Wainwright *Last Buccaneer* II. 234 The cops'll be high, wide and helpless. They won't know what in hell's hit 'em.
21. high camp, 'camp' (*sb.*[5]) of a sophisticated kind (in quot. 1963 used adjectivally); also (with hyphen) *attrib.*; so *high campery*; **high command** (see *command *sb.* 7 b); **high commission** (see Commission *sb.*[1] 7); **high commissioner** (see *commissioner 1 c); **high contrast** *Photogr.* (see quot. 1940); also (with hyphen) *attrib.*; **high country** *N.Z.*, hilly country that is difficult of access; freq. *attrib.*; *Canada* [tr. Canad. F. *haut pays*], the hinterland, the forests of N. and North-west Canada; hence **high countryman**; **high dilutionist** *Homœopathy*, an advocate of extreme dilution of medicine; **high farming,** the extensive use of fertilizers in land cultivation; **high finance,** finance concerned with large sums of money; **high forest,** a forest composed wholly or chiefly of trees raised from seed; also, in more general use, a forest of lofty trees (cf. G. *hochwald*); also *attrib.*; **high-fusing** *a.* (see quot. 1940); **high grinding** = *high milling* (Milling *vbl. sb.* 1 a); **high hook** *colloq.* or *slang*, the angler of a party who hooks the largest fish; **high-key** *Photogr.* (see quots.); **high kick** *Dancing*, a kick in the air, esp. one executed simultaneously by a row of female dancers and repeated by raising each leg in turn; also (with *high* used adverbially) **high-kick** *v.*, **high-kicking** *vbl. sb.* and *ppl. a.*; so *high-kicker*; **high lead** *Forestry* (see quot. 1957); also (with hyphen) *attrib.*, as *high-lead logging*; **high life,** a West African dance; also (with hyphen) *attrib.*; **high line,** (*a*) *Fishing* (*N. Amer. colloq.* or *slang*), the person who, or the boat which, has the best haul during a specified time; occas. a good catch; also *attrib.*; so **high liner**; (*b*) *Forestry*, an overhead cable attached to a spar tree in logging; so *high-line logging*; **high-lining** *vbl. sb.*; **high-low bed** (see quot. 1964); **high mass** (see Mass *sb.*[1] 3 a, High *a.* 19 b); **high milling** (see Milling *vbl. sb.* 1 a); **high place,** also, in *pl.*, the upper echelon of any organization; **high polymer,** a polymer with a high molecular weight; also (with hyphen) *attrib.*; **high-rise** *a.*, of a building, tall, multi-storey; also *transf.*; as *sb.*, a tall building; also occas. **hi-rise**; **high sign** *colloq.*, a surreptitious sign indicating that all is well or that the coast is clear; so **high-sign** *v.*; **high spot** (freq. in *pl.*) *slang*, the outstanding part or feature of something; *to hit the high spots,* to go to excess or extremes; to rise to a very high level; to include or touch on the most important points or places; cf. *high light 2; **high step,** (*a*) a military step in which the feet and knees are raised high; also (with hyphen) *attrib.*; (*b*) in *pl.*, a step-ladder; **high stool,** a tall stool; **high tea** (earlier and later examples); **high-tensile,** used *attrib.* of steel or other metals possessing great tensile strength; **high wine,** wine containing a high percentage of alcohol; **high wire,** a high tight-rope; **high yaller, yellow,** a half-caste of light yellow complexion; also as *adj. phr.*; **high-yield, -yielding,** designating something that furnishes or produces a large or valuable, etc., product or return.
1954 High camp [see *camp *a.* (and *sb.*[5])]. 1963 *Punch* 5 June 830/1 Gerda Charles..makes her young, with-it, high camp. 1964 *New Statesman* 6 Mar. 373/2 The..show starts depressingly, and the aura of high camp threatens to asphyxiate. 1968 *Listener* 27 June 843/2 The melodramatic trappings might have had a certain aesthetic high-camp value. 1967 *Spectator* 1 Dec. 690/3 Witness in this exhibition the high campery of the Coldstream Guards in 1821. 1940 A. L. M. Sowerby *Dict. Photogr.* (ed. 15) 157 A negative is said to have high contrast if tones but slightly different in the original subject show marked differences in the negative. 1961 G. Millerson *Technique Telev. Production* iii. 47 Subject contrast must be kept down, by preventing high-contrast surfaces appearing in the same shot. 1967 Karch & Buber *Offset Processes* iii. 64 High-contrast photography may be used. 1874 A. Bathgate *Colonial Experiences* xv. 212 Squatters whose runs include high country. 1903 S. E. White *Forest* 278 The base-line..was the only evidence of man we saw in the high country. 1930 L. G. D. Acland *Early Canterbury Runs* 1st Ser. i. 8 The severe snow storms..used up several years' profits of high country runholders. 1942 G. Campbell *Thorn-Apple Tree* 152 Some time again we winter in the high country, maybe. 1947 D. McEldowney in D. M. Davin *N.Z. Short Stories* (1953) 379 The grass was sharp and scanty on the unfertilised high country soil. 1950 *N.Z. Jrnl. Agric.* Apr. 364/1 The merino..bred..for sale to high-country runs [in Canterbury]. 1972 P. Newton *Sheep Thief* xiv. 110 He was full of questions about the country and it was obvious that his one ambition was to become a high country musterer. 1922 C. G. Turner *Happy Wanderer* 51 The high-countryman may drink his cheque. 1847 *Brit. Jrnl. Homœopathy* V. 154 We can no more reject the conclusions of the high dilutionists than we can despise those of their opponents. 1892 High dilutionist [see Dilutionist]. 1848 Mill *Pol. Econ.* I. 215 To apply the high farming of Europe to any American lands. 1894 G. B. Shaw in *Fortn. Rev.* LXI. 480 High farming cannot increase the natural rent of an acre. 1931 C. S. Orwin (*title*) High farming. 1966 *Listener* 1 Sept. 307/2 By 1939 British agriculture..had ceased to rely on the high farming we had developed in our own land. 1905 *McClure's Mag.* XXV. 48 In other words, we could eat our cake and have it, too—which is one secret of high finance. 1934 L. Mumford in W. Frank *Amer. & Alfred Stieglitz* ii. 50 Advertising, insurance, and high finance, the divine trinity that rules the world of industry. 1936 *Discovery* Sept. 280/2 It is not big guns nor even high finance that ultimately rules the world. 1879 *Encycl. Brit.* IX. 398/2 In..Germany..care is always taken that in high forest there is a good stock of self-sown trees before the old crop is entirely removed. 1927 *Forestry* I. 1. 24 The three main forestry systems, High Forest, Coppice-with-Standards, and Coppice. 1938 C. P. Ackers *Pract. Brit. Forestry* 9 High forest may be subdivided into..clear felling..and..the selective system. 1953 *Brit. Commonw. Forest Terminol.* I. 151 *High Forest Systems,* crops normally of seedling origin, either natural or artificial or a combination of both. Rotation usually long. 1955 *Jrnl. Ecol.* XLIII. 572 It will be referred to in this paper simply as 'High Forest', the adjective 'high' being used in a double sense to indicate that both the trees themselves are lofty and in the more specialized sense of the forester, to indicate that they have grown uninterrupted by coppicing or pollarding. 1959 *Times* 2 June 12/6 Growing conifers as high forest. 1911 G. H. Wilson *Man. Dental Prosthetics* viii. 313 The high fusing porcelains are practically insoluble. 1940 J. Osborne *Dental Mechanics* xxii. 412 Porcelains used in dentistry may be divided into two types according to their fusing temperatures: (1) *High fusing,* those whose fusing temperature is above the melting point of gold. (2) *Low fusing,* those whose fusing point is below 1065°C. 1956 J. N. Anderson *Appl. Dent. Mat.* xxiv. 330 For high-fusing porcelains, the furnace may be preheated. 1875 *Miller* May 55/1 (*title*) The Hungarian system of high grinding. *Ibid.*, In such mills..the whole of the high-grinding system was carried out. a 1884 Knight *Dict. Mech.* Suppl. 457/2 *High grinding,* a process of gradual reduction of the wheat by a succession of partial crushings, alternating with sifting and sorting the product. 1945 J. F. Lockwood *Flour Milling* xvii. 283 The wheat could either be rapidly and roughly reduced to flour by setting the two stones very close together, a process called low grinding, or the reduction could be made slowly and carefully with the stones further apart; this was called high grinding. 1848 in *D.A.E., High hook,* the one who catches the largest or the greatest quantity of fish. 1894 *Outing* (U.S.) XXIV. 259/2 F. was high hook with a five and a half pounder. 1918 *Photo-Miniature* XV. Mar. (Gloss.), *High-key,* a style of photographic print (portrait or landscape) consisting entirely of light tones, differing little from each other in depth. 1919 *Brit. Jrnl. Photogr. Alm.* 250 Photographs consisting almost entirely of light tones are said to be high-key. 1967 E. Chambers *Photolitho-offset* xi. 158 For high-key work, such as pencil sketches where delicate high-light detail has to be retained in offset printing, the Sears method of making a continuous tone negative..is used. 1898 G. B. Shaw *Our Theatres in Nineties* (1932) III. 336 The four beauteous ladies who, though apparently competent dancers, persist in punctuating their evolutions with graceless high kicks. 1906 E. Dyson *Fact'ry 'Ands* vii. 89 He came out on ther John's arm, 'igh kickin', 'n' singin' fit t' split. 1914 R. Brooke *Let.* 3 Sept. (1968) 613 The play was too foolish for words... Not a high-kick or a wriggle or a ragtime song in the whole thing! 1933 P. Godfrey *Back-Stage* xv. 194 Their white-washed legs..are doing energetic high-kicks and splits. *Ibid.* xviii. 224 They tap-dance, high-kick, turn cartwheels, and do the splits. 1962 A. Huxley *Island* xiii. 223 'What sort of dancing does he teach?' Mrs Narayan tried to describe it. No leaps, no high kicks, no running. 1966 'J. Hackston' *Father clears Out* 92 He had high-kicked the boss at the ..store for a fiver and had won by an ankle. 1896 W. Stevens *Let.* 31 July (1967) 9 A plain little horse though a high-kicker. 1897 Kipling *Capt. Cour.* iv. 103 There's somethin' of a sea to-night... She is a high-kicker. 1897 G. B. Shaw *Our Theatres in Nineties* (1932) III. 12 Cancanist high-kickers. 1922 Joyce *Ulysses* 359 Skirt-dancers and highkickers. 1895 High-kicking [see Split *sb.*[1] 4 b]. 1901 *Daily Chron.* 11 Dec. 6/5 Several dancers of the high-kicking and other schools. 1925 A. Philip *Crimson West* 144 He yawned sleepily. 'Got to fix a "spar-tree" for a "high-lead" to-morrow, so I better hit the hay.' 1939 Beaulieu & Barton *Appl. Lumber Sci.* (ed. 2) 35 The High-Lead is the most common method of yarding by steam. 1951 W. F. Heald *Scenic Guide to Oregon* 17 Modern 'high lead' and 'skidder' logging whisks tree sections. 1957 *N.Z. Timber Jrnl.* Aug. 59/1 *High lead,* extraction of timber by means of overhead cable attached to a high spar at hauler end. This gives the front end of logs sufficient lift to clear obstructions during haul. 1955 *Times* 23 Aug. 10/6 'High lifes', marches, songs in the vernacular of dubious propriety. 1959 A. Abbs *Ashanti Boy* vi. 212 He has a big collection of high-life records. 1959 *Guardian* 22 Dec. 5/2 Man, I've got a..West African band... They play high life. 1963 *Listener* 14 Mar. 456/1 A Ghanaian band was playing 'highlife' music for dancing. *Ibid.*, Highlife is as distinctively New African as the Kwela music played and danced to in Johannesburg at the other end of the continent. 1856 C. Nordhoff *Whaling & Fishing* xviii. 354 Several had at different times been 'high line' from Harwich. 1864 *Harper's Mag.* Feb. 367/2 Captain Aleck was determined to fish for 'high line'. 1885 J. S. Kingsley *Stand. Nat. Hist.* III. 196 The emulation to be 'high-line' for the day and for the season is extreme. *Ibid.*, In a single day a high-line fisherman has caught from ten to fifteen barrels. 1890 *Grip* (Toronto) 5 Apr. 233/2 Always 'high line', he was always 'filled up' with the split mackerel of the North Bay. 1929 F. C. Bowen *Sea Slang* 66 *High line,* on the Grand Banks a particularly good catch, also applied to the most successful fishing boat or clipper of the season. 1931 *Amer. Speech* VII. 49 The man who..attaches the 'high line' is called a 'skyrigger'. *Ibid.*, A 'high line', which is of one-inch steel, extends, often one-half mile, to either end of the 'set'. 1965 *Brit. Columbia Digest* Sept.–Oct. 19/1 Mobile spar trees, which are monstrous self-propelled cranes whose thick booms bear a multiplicity of sheaves and cables, are used in modern high-line logging to replace the spar trees used until recently as the focal point for the complicated system of cables and pulley-blocks. 1893 in *Funk's Stand. Dict.*, High-liner. 1914 W. D. Steele *Storm* 56 On the grounds he was a great 'killer', an unmerciful 'driver', and for three years running now the 'high liner' of the Old Harbor fleet. 1916 F. W. Wallace *Shack Locker* 65, I ain't a highliner this season, but we've got one thing to brag about when it comes to fishin'. 1965 *Brit. Columbia Digest* Sept.–Oct. 20/1 But many [high-riggers] just disappeared, as did the giant trees they had topped, limbed and rigged for high-lining. 1964 G. D. Cherescavich *Textbk. Nursing Assistants* vii. 52 *High-low bed,* an electrically or manually operated bed which can be raised to the height of the regular hospital bed and lowered to the height of the home bed. 1967 *Nursing Times* 18 Aug. 1088/2 There are many types of variable height beds (sometimes called high-low beds). 1918 L. Strachey *Eminent Victorians* 245 There were rumours of debaucheries in high places. 1922 G. M. Trevelyan *Brit. Hist. 19th Cent.* iii. 54 Evangelicalism brought rectitude, unselfishness and humanity into high places. 1931 F. W. Crofts *Myst. in Channel* xviii. 227 Persons in high places made comments. 1931 *Economist* 10 Oct. 656/1 Any talk at this moment in 'high places' of the abandonment of the gold standard would be quite likely to result in a stampede. 1938 *Archit. Rev.* LXXXIII. p. lvi, It is to be hoped that the contents..have percolated through to the authorities and those in 'high places' whose responsibility it is to plan ahead for the future of road transport in this country. 1971 G. Household *Doom's Caravan* ii. 54 'It was her mother's correspondence which I feared might be of interest to enemy agents.' 'That has already been taken care of in high places.' 1946 *Nature* 27 July 122/1 The most important new dielectrics are usually of the high-polymer type. *Ibid.* 19 Oct. 553/1 (*heading*) Effect of environment on the reactivity of high polymers. 1968 Greenwood & Banks *Synthetic High Polymers* i. 3 A high polymer is simply a chemical substance which is made up of giant molecules. 1954 *Archit. Rev.* CXVI. 414/2 In general form—podium and high-rise accommodation—this scheme follows the general pattern of current thought. 1958 *Listener* 20 Nov. 827/1 A point block of government offices is now going up at Wellington..and other high-rise slabs for offices and flats. 1961 *Observer* 7 Jan. 17/4 When one high-rise building is surrounded by many similar structures, height alone loses this special distinction. 1965 in *Amer. Speech* (1967) XLII. 159 Roosevelt

Grier: the high-rise football star of the L. A. Rams. **1967** *Boston Sunday Herald* 26 Mar. 1. 41/4 (Advt.), The convenience and prestige of a luxury high-rise. **1967** *Time* 20 Oct. 60 She likes the high-rise boots because 'they give my legs a sleek stocking look'. **1968** *Globe & Mail* (Toronto) 13 Jan. 45 (Advt.), Adult hi-rise building. *Ibid.* 17 Feb. 12 Block-heeled pump with high-rise vamp. **1903** R. L. McCardell *Conversat. Chorus Girl* 111 When who should peek-a-boo in but my friend!..I gave him the high sign, but he passed me up. **1928** *Sunday Dispatch* 29 July 2 He gave the pawnbroker the high-sign. **1946** Wodehouse *Joy in Morning* xxv. 229 As soon as he is in a melting mood, you give me the high sign, and I carry on from there. **1957** *Amer. Ballads & Folk Songs* 135 She was pleadin' with him, her eyes all teary and dim, As I high-signed the barkeep for mine. **1962** L. Deighton *Ipcress File* xii. 72 He gave us the high sign with thumb touching forefinger. **1962** K. Orvis *Damned & Destroyed* xxv. 185, I high-signed him to follow. **1910** W. M. Raine *B. O'Connor* 12 Here comes your train a-foggin'—also and likewise hittin' the high spots. **1926** *Spectator* 11 Sept. 373/1 Chicago was the 'high-spot' of the trip. **1927** *Daily Express* 15 Sept. 9/5 The 'high spot' of the production—cinematic equivalent to the chariot-racing scenes in 'Ben-Hur'. **1928** *Sunday Dispatch* 22 July 23/4 It looks as though the standard of racing is going to hit the high spots. **1936** *Economist* 29 Feb. 483/1 The 'high-spot' was the 20 per cent. dividend (against 10 per cent.) announced by Associated Portland Cement. **1936** 'J. Tey' *Shilling for Candles* iv. 48 Two years of hitting the high spots must have educated you to something. **1961** J. Carter *ABC for Book-Collectors* 111 'High spot' collecting is a sort of dictated eclecticism. **1964** *Publishers' Weekly* 28 Sept. 59 (*heading*), Religious books. Some Fall highspots, September through December. **1889** *Infantry Drill* 1. 1. 32 The High Step. **1894** Mrs. Alexander *Choice of Evils* II. viii. 199 He was exceedingly busy with hammer and nails, and the 'high steps', putting up fresh curtains. **1969** *Daily Tel.* (Colour Suppl.) 10 Jan. 32/1 The music [of the Celts] is the wail of the bagpipe, played to the jig and the fling and the high-step whirl. **1825** H. Wilson *Mem.* III. 76 The Duchess of Beaufort..appears never so happy nor so comfortable as when he is perched upon a high stool by her side. **1837** Dickens *Pickw.* l. 538 A high stool, four chairs, a table, and an old eight-day clock. **1892** C. M. Yonge *That Stick* I. xiii. 139 Whatever promise there may have been..must have been nipped upon the top of a high stool. **1935** G. Greene *England made Me* 1. 1 For half an hour she had sat on the same high stool, half turned from the counter. **1961** *Guardian* 14 Nov. 10/3 He sat on a high stool [at a bar]. **1831** F. A. Kemble *Rec. Girlhood* 14 June (1878) III. 49 We did not return home till near nine, and so, instead of dinner, all sat down to high tea. **1922** W. S. Maugham *Chinese Screen* xlix. 193 He thought of the high tea to which he sat down when he came home from school.., a slice of cold meat, a great deal of bread and butter and plenty of milk in his tea. **1957** *London Mag.* Nov. 53 We ate high-tea made of fresh salmon, or mushrooms we'd risen at dawn to gather. **1923** *Man. Seamanship* (*H.M.S.O.*) II. 263 All material contributing to the longitudinal strength [of a ship's hull] is of high tensile steel. **1932** *Discovery* May 146/2 High tensile irons, corrosion-resisting irons, and growth-resisting irons are now made in large quantities. **1937** *Archit. Rev.* LXXXI. 268 (*caption*) Right, a welded tubular bus frame seat, utilizing the strength of high-tensile alloy tube. *c* **1384** Heiȝe wynes [see High *a.* 8]. **1542** Boorde *Dyetary* (1870) x. 254 Hyghe wynes, as malmyse, maye be kepte longe. **1871** *Trans. Ill. Agric. Soc.* VIII. 143 The necessity would still exist for converting..corn into beef and pork and highwines. *a* **1884** Knight *Dict. Mech.* Suppl. 458/1 *High wines*, crude alcohol of higher proof than singlings. **1958** J. Carew *Black Midas* vi. 112 They drank highwine and bush rum from half-pint tumblers. **1961** Webster, High wire. **1962** *Listener* 10 May 822/3 The bulk of the poems go all out for intellectual sophistication... He is not at his ease on the intellectual high wire. **1962** *Observer* 23 Sept. 26/5 A clown picking dust off his suit on the high wire. **1923** J. Dos Passos *Streets of Night* 133 Ought to see them high yallers down there if you're stuck on girls. **1929** *Variety* 17 Apr. 51/3 She looks like a genuine high-yaller (that being her make-up in buxom mammy fashion). **1929** T. Wolfe *Look Homeward, Angel* (1930) II. xxii. 318 There's a High Yaller in here... You can have that if you want it. **1937** Wyndham Lewis *Blasting & Bombardiering* iv. viii. 241 At present I should be living in a villa just outside Paris with a Japanese cook and a Zulu butler, with three highyaller kids getting ready to go to Eton. **1947** 'N. Shute' *Chequer Board* iii. 61 [He] had played and danced with various mulatto and 'high yaller' girls back home in Nashville. **1951** M. Kennedy *Lucy Carmichael* I. iii. 25 She isn't exactly black... She's high yellow. **1969** 'J. Morris' *Fever Grass* xviii. 158 The big high yellow nodded at him with impersonal cordiality. **1957** High yield [see *low-yield* adj. s.v. *Low *a.* 23]. **1958** *Times* 12 Sept. 10/1 A high yield nuclear device was successfully exploded. **1959** *Britannica Bk. of Year* 546/1 Another new term was *high-yield explosion*, an atomic explosion powerful enough to produce various widely felt and easily registered phenomena. **1960** *Farmer & Stockbreeder* 16 Feb. 66/3 He was no 'high-yield' merchant, but he would compare his profits..with any..breeder. **1970** *Daily Tel.* 16 Feb. 17 (Advt.), The aim of High-Yield Units is to give the highest return consistent with reasonable security of capital. **1946** *Nature* 12 Oct. 522/1 High-yielding herds. **1956** *Ibid.* 3 Mar. 416/1 The cocoa trade has accepted the very high-yielding Upper Amazon selections..as conforming to its quality requirements.

b. high-liver, (*a*) (earlier example); **high-ranker,** a person of high rank.

1881 J. W. Buel *Border Outlaws* 166 All the band were known to be high livers during their periods of plenty. **1953** P. C. Berg *Dict. New Words* 93/1 *High-ranker*, a person of high rank, esp. in one of the Services. **1958** W. J. H. Sprott *Human Groups* ix. 147 When he played against a gang high-ranker, he was seldom 'on form'. **1973** J. Wainwright *Pride of Pigs* 18 High-rankers and C.I.D. wallahs bobbing in and out.

22. a. *high-accuracy, -amplitude, -carbon, -conductivity, -cost, -current, -density, -efficiency, -energy, -fat, -field, -gain, -hurdle, -impedance, -income, -intensity, -investment, -nitrogen, -octane, -order, -output, -performance, -permeability, -potency, -potential, -power, -precision, -pressure* (further examples; also as *vb.*), *-prestige, -price, -protein, -purity, -quality, -resistance, -resolution, -risk, -stability, -status, -strength, -tension, -test, -tone, -vacuum, -value, -wage*; **high-altitude,** occurring, working, or carried out at high altitudes; **high-angle** *Gunnery*, denoting the fire from guns, mortars, etc., at a high angle of elevation; hence *high-angle gun*, etc.; also *transf.*, as of a camera; **high-board,** of or relating to diving from a high diving-board; **high-duty,** (*a*) subject to heavy customs duty; (*b*) designed to perform heavy tasks; = *heavy-duty* (*Heavy *a.*[1] 24 b); **high-fashion** (see *Fashion *sb.* 9 c); **high-flash,** denoting oil whose vapour ignites only at a relatively high temperature; **high-flux,** denoting (*a*) a high density of magnetic flux; (*b*) a large number of elementary particles per second; **high-humidity** *Forestry*, of the treatment of timber by exposing it to high humidity for a specified purpose; **high-level,** situated, built, etc., in, or carried out from, a high position; denoting talks, a meeting, etc., of an exalted status or grade; in the field of *Computers*, applied to a programming language that is largely independent of any particular kind of computer and bears some resemblance to an existing language (as English) or symbolism; **high-lift,** of something that is raised high or that lifts something up high; **high-pass** *Electr.*, denoting a filter that attenuates components with a frequency less than some cut-off frequency and passes components of higher frequency; **high-sea(s),** operating or carried out on the high seas; *High Sea Fleet* = G. *Hochsee Flotte*; **high-velocity,** of high speed; *spec.* denoting a gun capable of discharging a projectile with great force and speed; also denoting the projectile so fired; **high-warp,** denoting a manner of weaving or tapestry in which the warp is vertical.

1958 *Listener* 13 Nov. 779/1 The manufacture of high-accuracy gyroscopes. **1963** B. Fozard *Instrumentation Nuclear Reactors* ix. 105 (*heading*) High-accuracy ratemeters. **1925** E. F. Norton *Fight for Everest, 1924* 86 [They] were both suffering from very bad high-altitude throat. *Ibid.* 106 This hateful duty of high-altitude cooking. **1955** E. Hillary *High Adventure* 58 We signed on ten Sherpas to act as high-altitude porters. **1966** *Electronics* 3 Oct. 181 Also in the national space lineup is project '621', aimed at developing recoverable high-altitude sounding rockets. **1956** *Nature* 21 Jan. 121/1 High-amplitude effects in reflexion amplifiers. **1879** *Man. Artill. Exerc.* I. v. 23 High-angle fire from howitzers and mortars. **1890** G. S. Clarke *Fortification* xiv. 205 Large numbers of high-angle guns which would prove most formidable to ships. **1915** *Pearson's Mag.* XXXIX. 66 High-angle trajectory. **1928** C. F. S. Gamble *N. Sea Air Station* ix. 122 The addition of some motor-cars equipped with machine-guns on 'high-angle mountings'. **1956** *Nature* 17 Mar. 502/2 (*caption*) High-angle photographs from a carbonized coking coal. **1963** *Listener* 28 Mar. 569/2 The mere novelty (for television drama) of the high-angle viewpoint gave the shot a sly edge. **1936** *Southern Counties' Amat. Swimming Assoc. Handbk.* 12 The positions in National championships are..C. D. Tomalin..Highboard diving..1st. **1959** *Times* 22 Sept. 3/3 Phelps is the European and national highboard titleholder. **1937** *Discovery* May 156/1 High carbon (hypereutectoid) steel. **1958** *Aspects of Translation* 84 The melting of the high-carbon cast iron. **1909** *Proc. R. Soc.* A. LXXXII. 232 A small, adjustable, open, high-conductivity, self-induction spiral. **1940** *Chambers's Techn. Dict.* 415/2 *High conductivity copper*, metal of high purity, having an electrical conductivity not much below that of the international standard. **1931** G. B. Ford *Building Height, Bulk & Form* Title-p., Uneconomic types of buildings on high-cost land. **1964** *Ann. Reg. 1963* 259 German farmers were high-cost producers. **1949** *Nature* 17 Sept. 485/2 (*heading*) High-current betatron conditions. **1952** *Ann. Reg. 1951* 401 High-density..central city areas. **1960** *Times* 28 Sept. 21/7 New plastics such as high-density polyethylene. **1962** *Listener* 15 Nov. 806/2 The high density life of the town. **1964** M. A. Johnson in *Oceanogr. & Marine Biol.* II. 33 The possibility of high-density turbidity currents in the ocean. **1973** *Times* 14 Aug. 2/5 Housewives doing their washing yesterday stopped the pumping of high density foam into the damaged Liberian tanker. **1920** *Ham's Yearbk.* I. *Customs* 173 In the case of high-duty goods..the issuing officer will send an advice..to the Inspector-General of Waterguard. **1923** R. T. Kent *Kent's Mech. Engin. Handbk.* (ed. 10) ix. 692 (*heading*) Table 44.—Direct-connected, electrically-driven, two-stage compressors,.. Sea level high duty. Air pressure of 100–115 lb. **1937** *Times* 13 Apr. p. xxii/2 A high-duty lubricant has been introduced to deal with high tooth pressures now usual in rear axle mechanism. **1947** *Nature* CLIX. 51 The production of still stronger high-duty irons. **1962** Simpson & Richards *Junction Transistors* ix. 202 A high-efficiency hole injector. **1938** *Nature* 29 Oct. 781/1 Excitation..is..due..to the production of numbers of fairly high-energy electrons. **1958** *Listener* 27 Nov. 871/1 Hoyle and Gold are convinced that the radio waves in these regions are emanating from high-energy particles moving in extensive magnetic fields. **1972** *Physics Bull.* Apr. 215/1 Bubble chambers are amongst the most widely used instruments in high energy physics for the detection of the tracks of ionizing particles. **1946** *Nature* 7 Sept. 350/1 A high-fat diet. **1943** *Rep. Progress Physics* IX. 184 Thermionic emission under the Schottky high-field condition. **1968** C. G. Kuper *Introd. Theory Superconductivity* vii. 130 Most of the Type II superconducting materials used for high-field magnets are rather 'dirty' metallurgically, and probably *do* contain normal inclusions. **1899** *Westm. Gaz.* 14 Feb. 4/3 A fourth high-flash oil. **1904** Goodchild & Tweney *Technol. & Sci. Dict.* 278/2 *High flash point*, oil whose vapour is only ignited at a high temperature. **1949** *Wireless World* Apr. 137/1 A high-flux version (Type R22) of the single diaphragm T2 is now available with a density of 17,500 gauss in the 1⅛ in diameter gap. **1955** *Bull. Atomic Sci.* Mar. 93/2 A high-flux heavy water reactor primarily intended to investigate radiation damage in reactor building materials. **1971** *New Scientist* 3 June 579/1 The Franco-German high-flux reactor in Grenoble. **1936** *Nature* 11 July 87/1 The apparatus was designed to detect the presence of destructive insects in timber, and consists of a sound-proof chest, a sensitive microphone and a high-gain amplifier. **1962** F. I. Ordway et al. *Basic Astronautics* v. 176 (*caption*) High-gain communications antenna. **1925** H. L. Henderson in *Bull. N.Y. State College Forestry* XXV. xvi. 65 Another method is to use the high humidity treatment... This process will balance up the moisture percentages in less than 24 hours. **1957** *N.Z. Timber Jrnl.* Aug. 59/1 *High humidity treatment*, temporarily raising the humidity of the circulating air in a kiln when drying wood requiring special treatment. **1893** *Outing* (U.S.) XXII, The following fall..Harding was third in the high-hurdle race. **1949** *Wireless World* Apr. 136/1 There seems to be a trend towards the use of high-impedance windings in high-fidelity pickups. **1937** *Amer. Speech* XII. 315/2 His store..makes little attempt to attract a high-income clientele. **1937** *Discovery* Feb. 45/1 For production of the screen image a high intensity automatic arc is being used. **1962** F. I. Ordway et al. *Basic Astronautics* iii. 68 High-intensity solar ultraviolet radiation. **1960** *Encounter* XV. iv. 10 The average rate of profit..need not be lower in a high-investment than a high-consumption economy. **1876** *Q. Jrnl. Geol. Soc.* XXXII. 185 The altitude of high-level drift on the western slopes of the Pennine chain. *a* **1890** High-level [in Dict., sense 22 a]. **1936** *Discovery* May 132/2 The high-level gravels in which Harrison discovered those eoliths that gave proof of the oldest human culture in this island. **1943** *Aeronaut. Engin. Rev.* Apr. 59/3 The removal of objections against high-level daylight raids..[is] specified. **1951** *Ann. Reg. 1950* 45 It was the will that was lacking, and high-level meetings would not alter that. **1951** *Sci. Amer.* Sept. 42/1 Some, though they have obtained degrees, can hardly be classed as capable of high-level mental work. **1959** *Daily Tel.* 8 July 10/3 The proposed policy of continual high-level talks with Russia, extending over the next few years, has its opponents. **1964** *Jrnl. Assoc. Computing Machinery* XI. 21 Emphasis has been placed on simplicity and intuitiveness while maintaining so far as possible the inherent power of a high-level programming language. **1964** K. L. Pike in D. Abercrombie *Daniel Jones* 425 High-level phonological units. **1965** N. Chomsky *Aspects of Theory of Syntax* iv. 153 Deviation from selectional rules involving high-level features is apparently more serious than deviation from selectional rules involving lower-level features. **1969** H. Perkin *Key Profession* v. 209 A long-term high-level demand for university places. **1972** *Computer Jrnl.* XV. 195/1 The main purpose of a high-level language is to make programs and programming more intelligible to human beings. **1921** *Discovery* Apr. 96/2 High-lift wings are, of course, those that give the highest lift–drift ratio. **1933** *Jrnl. R. Aeronaut. Soc.* XXXVII. 79 The Hall high-lift wing is an application of the same principle... Similar results are obtained, namely, a large increase in maximum lift and a shift backwards of the centre of pressure. **1958** *Times Rev. Industry* June 20/1 A new, high-lift boom machine. **1902** *Encycl. Brit.* XXV. 208/1 High-nitrogen foods. **1960** *Farmer & Stockbreeder* 15 Mar. 113/1, 3 cwt per acre of the high-nitrogen compound in the seedbed should not affect it. **1932** *Chem. Abstr.* XXVI. 3097 Changes in design of cracking units to enable them to produce high-octane gasoline. **1958** *Times* 6 Nov. 7/5 All high compression engines respond to high-octane, highly anti-knock fuel. **1972** M. Gilbert *Body of Girl* iv. 43 Filling the tank of an old Bentley with high-octane petrol. **1938** S. Chase *Tyranny of Words* i. 6 We..allow our language forms or symbolic machinery to fashion a demonology of absolutes and high-order abstractions. **1968** Fox & Mayers *Computing Methods for Scientists & Engineers* vii. 134 Economization was effected just by removing successive high-order terms. **1960** *Farmer & Stockbreeder* 8 Mar. 48/2 (Advt.), The best high-output combine in the world. **1963** *Listener* 14 Mar. 466/2 A living-room fire with a high-output back boiler. **1925** *Post Office Electr. Engin. Jrnl.* XVII. 311 Tests on high pass filters of three sections. **1946** *Nature* 13 July 47/1 Figures 1, 2 and 3..were taken with a high-pass filter in the circuit which attenuated the low-frequency response of the microphone. **1940** *War Illustr.* 19 Jan. 620 (*caption*) The introduction into the Fleet Air Arm of high-performance monoplanes like the Blackburn 'Skua'. **1966** T. Wisdom *High-Performance Driving* ii. 17 You need more skill for high-performance road-driving than you do for competitive motoring. **1930** *Wireless World* 21 May 538 (*heading*) High permeability alloys. **1963** B. Fozard *Instrumentation Nucl. Reactors* viii. 102 In general a tube must be placed..at least four inches away from another trochotron, a strong magnet, or a high-permeability screen. **1944** *Mod. Lang. Notes* Dec. 515 High-potency capsule. **1962** F. I. Ordway et al. *Basic Astronautics* xiii. 526 High-potency drugs such as the phenothiazines..could be incorporated. **1934** Webster, High-potential. **1954** *Essays in Criticism* IV. 313 The value of high-potential person-to-person situations. **1962** Simpson & Richards *Junction Transistors*

iv. 68 A particle such as an electron can pass through a high-potential barrier if the barrier is sufficiently thin. **1892** A. Conan Doyle *Adv. Sherlock Holmes* i. 2 A crack in one of his own high-power lenses. **1898** *Westm. Gaz.* 28 May 2/1 Modern high-power guns. **1901** *Kynoch Jrnl.* June–July 108/2 Modern high-power smokeless propellants. **1934** *Discovery* Dec. 341/2 Here the microscope is set up vertically for high-power work. **1971** *Gloss. Electrotechnical, Power Terms* (*B.S.I.*) III. vii. 21 *High-power modulation*, modulation of the carrier of a radio transmitter. **1946** *Nature* 19 Oct. 537/2 For high-precision measurements, as for observing wind-tunnel forces, the current is limited to 5 milliamp. **1960** *Times* 18 Nov. 4/6 The first essential..is a network of high-precision stations. **1872** F. W. Robinson *Wrayford's Ward* III. 207 A high-class, high-priced, high-pressure seminary. **1907** *Westm. Gaz.* 15 Oct. 1/1 The high-pressure life which he led in London. **1928** D. Brunt *Meteorol.* iv. 30 These high and low-pressure systems. **1933** *Archit. Rev.* LXXIV. 183/2 To launch a high-pressure salesmanship offensive. **1936** *Discovery* Feb. 38/2 Steam undergoes expansion in both high-pressure and low-pressure cylinders. **1940** *Topeka Jrnl.* 4 May 2 Photographers..hipressuring portrait sales. **1941** *Sat. Even. Post* 8 Feb. 54, I did not attempt to high-pressure the man too much. **1946** Mezzrow & Wolfe *Really Blues* viii. 128 He began to high-pressure us with a Chamber-of-Commerce spiel. **1958** J. Cannan *And be a Villain* vii. 165, I get so confused by high-pressure salesmen. **1962** 'A. Garve' *Prisoner's Friend* ii. 105 She didn't want to. Laurence high-pressured her into it. **1972** *Mod. Law Rev.* XXXV. 1. 24 It is surely desirable to prevent abuses in marketing, particularly fraud, high pressure techniques, and the deliberate sale of defective goods. **1956** C. W. Mills *Power Elite* xii. 282 High-prestige organizations to which the elite usually belong. **1959** V. Packard *Status Seekers* (1960) vi. 84 A few old high-prestige neighbourhoods manage..to maintain their status. **1906** *Daily Chron.* 27 Sept. 3/6 The benefits of the high-price policy which they choose to pursue. **1944** *Mod. Lang. Notes* Dec. 515 High-protein foods. **1964** 'D. Shannon' *Root of all Evil* (1966) v. 66 Sure, she gives me the high-protein diet, and I'm still learning to drink coffee without sugar. **1936** *Metals Handbk.* 925 (*heading*) Electrical properties of high-purity annealed aluminium wire. **1910** *Westm. Gaz.* 21 Apr. 12/1 Until plenty of high-quality beet is procurable. **1913** V. B. Lewes *Oil Fuel* 180 High-quality coal-gas. **1939** W. S. Churchill *Into Battle* (1941) 85 High-quality war materials. **1948** *Wireless World* Jan. 2/1 Most high-quality radio receiver units will provide an output of well over 4 volts. **1884** High resistance [see Resistance 6]. **1962** Simpson & Richards *Junction Transistors* xi. 259 When feeding a high-resistance load from a relatively low-resistance source..the potentiometer should be connected as a voltage divider. **1946** *Nature* 19 Oct. 550/2 Using the high-resolution system of the R.C.A. type E.M.U. microscope as a diffraction camera, we have attempted to find some feature of the diffraction by oxides of this type. **1970** G. K. Woodgate *Elem. Atomic Struct.* 217 High-resolution optical spectroscopy. **1951** S. A. Stouffer in Parsons & Shils *Toward General Theory Action* iv. v. 494 In a high-risk cheating situation. **1963** *Economist* 14 Dec. 1175/2 All the 'high-risk' mothers. **1969** *Times* 3 Apr. 28/6 Many companies ..had cancelled policies in high-risk areas such as the ghettoes. **1907** *Daily Chron.* 9 Oct. 4/6 The German high-sea fleet. **1931** *Times Lit. Suppl.* 10 Sept. 673/3 A lack of familiarity with the naval idiom can alone account..for calling..the fleet under Cornwallis 'the high seas fleet'. **1961** A. J. Marder *From Dreadnought to Scapa Flow* I. ix. 243 While the High Seas Fleet had concentrated.., the British Fleet was in a very different condition. **1949** *Wireless World* Mar. 16a (Advt.), High stability capacitors. **1963** B. Fozard *Instrumentation Nucl. Reactors* ix. 108 High-stability, close-tolerance components are required in the measuring circuits. **1959** V. Packard *Status Seekers* (1960) ix. 129 Some women said it made them 'feel good' just to go into a high-status store. **1965** *Language* XLI. 295 An honored or high-status person. **1940** *Chambers's Techn. Dict.* 416/1 High-strength brass. **1961** *Times* 13 Dec. 21/6 High-strength..paper bags. **1969** *Jane's Freight Containers 1968–69* 111/3 Aimed at providing a high-speed, high-strength route for 'jumbo' freight cars. **1911** *Daily Colonist* (Victoria, B.C.) 11 Apr. 10/5 Harry Gillis..was killed and ten bricklayers narrowly escaped death today when Gillis came in contact with a 30,000 volt high tension wire. **1970** *New Yorker* 10 Oct. 64/1 The denizens of the ocean, some of whom learned to produce high-tension electricity long before man. **1936** *Metals Handbk.* 439 The curves marked 'High Test' and 'Nickel' Irons in Fig. 21 were stressed to 10,000 psi. **1955** *Jrnl. Brit. Interplanetary Soc.* XIV. 159 Developments throughout the world have resulted in the production of hydrogen peroxide up to 90 per cent. strength, known as high-test peroxide or H.T.P. **1958** A. D. Merriman *Dict. Metallurgy* 126/1 *High-test cast iron*, cast iron possessing a tensile strength not less than some arbitrary value, varying with different authorities. **1959** *Chambers's 20th Cent. Dict.* Suppl. 1378/2 *High-test*, (of petrol) boiling at comparatively low temperature and so of high performance. **1923** *Bull. Sch. Orient. & Afr. Stud.* III. 1. 125 This significant tone of the future of all high tone verbs need not be specially marked in broad transcriptions. **1964** J. Carnochan in D. Abercrombie *Daniel Jones* 403 All these examples have a high-tone initial syllable. **1965** *Language* XLI. 347 A toneless base..to which I would add..three affixes (the high-tone superfix, the suffix *-a*, and the mid-tone superfix) to produce the noun. **1893** *Jrnl. Soc. Arts* 5 May 624/1 The phosphorescent glow of the novel high-frequency, high-voltage, high-vacuum lamps. **1927** *Nature* 8 Oct. 510/1 We have made a further study of the phenomena exhibited by these high-vacuum tubes, with especial reference to the spectra of the discharge. **1971** *Gloss. Electrotechnical, Power Terms* (*B.S.I.*) II. ii. 8 High-vacuum diode. **1963** *Rep. Comm. Decimal Curr.* iii. 17 in *Parl. Papers 1962–3* (Cmnd. 2145) XI. 1 Because workable cent systems with high-value major units are available..we preferred cent systems to mil ones. **1966** *Listener* 1 Sept. 307/2 High value perishable foods such as milk and meat and vegetables. **1892** F. Irwin *Fortification* (ed. 2) 43 Modern high-velocity guns. **1898** *Engineering Mag.* XVI. 112/2 These high-velocity bullets. **1934** *Discovery* June 155/2 The jet of high-velocity air is about 8 feet across, and an expenditure of energy at the rate of 400 horse-power is necessary to maintain it. **1946** *Nature* 28 Dec. 932/2 Anomalously short travel-time signifies the approach towards the surface, along the radius concerned, of a high-velocity medium. **1962** F. I. Ordway et al. *Basic Astronautics* viii. 334 The venturi principle is applied..in the aerodynamic wind tunnel to achieve high-velocity flow. **1964** Crystal & Quirk *Syst. Prosodic & Paralinguist. Features Eng.* iii. 38 Turbulent flow, with projection of high-velocity jet into pharynx. **1956** J. M. Mogey *Family & Neighbourhood* 5 Before 1920 Oxford had not been a high-wage town. **1964** S. M. Miller in I. L. Horowitz *New Sociol.* 294 The high-wage..occupations. *a* **1877** Knight *Dict. Mech.* II. 1103/1 *High-warp loom*, a tapestry loom in which the warp-frame is vertical and the weaver works standing. **1904** Goodchild & Tweney *Technol. & Sci. Dict.* 287/2 *High warp*, tapestry in which the warp takes a vertical position, *e.g.* Gobelins tapestry. **1934** *Archit. Rev.* LXXV. *caption* facing 95 For instance, Burne-Jones drew cartoons for the revival of high-warp tapestry weaving which Morris initiated.

b. *high-banked, -bosomed, -bridged, -ceilinged, -hedged, -hilled, -pooped* (examples), *-pressured, -priced* (later example), *-sided, -souled* (later examples); **high-coloured,** also *fig.*, exaggerated, forced; as, *a high-coloured description*; **high-geared,** having high gears; also *transf.*, fast-moving, active; **high-horsed** (delete † and add examples).

1899 *Daily News* 25 Dec. 5/2 A dangerous high-banked river. **1900** *Westm. Gaz.* 26 Jan. 4/3 The broad-faced, high-bosomed model is the palpable grandmother of many Rubenses. **1961** *Times* 25 Apr. 20/7 The white high-bosomed dress of Empire style. **1871** 'M. Legrand' *Cambr. Freshm.* 138 A Quixotic gentleman, of ancient lineage, in whose high-bridged and defiant nose the Indian saw a resemblance to an eagle's beak. **1907** *Daily Chron.* 30 July 4/4 It was a high-ceilinged, sombre room. **1960** *Farmer & Stockbreeder* 9 Feb. Suppl. 3/2 A high-ceilinged, rather sunless sittingroom. **1925** F. M. Ford *No More Parades* 309 There were two girls who kept a tea-shop in Poperinghe... High coloured. **1899** *Daily News* 13 Feb. 5/3 Our high-geared population. **1906** A. Bennett *Whom God hath Joined* v. 169 Gater's high-geared bicycle. **1924** H. Crane *Let.* 12 Jan. (1965) 169 Working at high speed as one does in such high geared agencies. **1904** *Westm. Gaz.* 3 Aug. 2/3 In the shady high-hedged garden. **1906** Kipling in *Tribune* 15 Jan. 2/3 A high-hedged road. **1807** J. Barlow *Columbiad* vi. 224 Gates in their van, on high-hilled Bemus rose. **1896** A. E. Housman *Shropshire Lad* 83 Westward on the high-hilled plains Where for me the world began. **1900** *Westm. Gaz.* 6 Feb. 3/2 To..ride off high-horsed on the theory that the battle had to be fought. **1928** *Observer* 22 Jan. 14/5 The high-horsed fanatics of universal Communism. **1889** O. Wilde in *19th Cent.* Jan. 54 The high-pooped galleys. **1921** W. de la Mare *Veil* 74 That midnight-stealing, high-pooped galliass, *Sleep*. **1947** *Horizon* XVI. 202 Muggleton was high-pressured and loud. **1961** *Times* 15 Nov. 19/1 High-pressured modern life. **1962** *B.S.I. News* July 19/1 It was no use building a high-priced article for 20–30 years' service if it was to be outmoded by advancing techniques. **1960** *Farmer & Stockbreeder* 26 Jan. 65/3 Force the lacerated trimmings up a delivery chute into a high-sided trailer. **1905** *Westm. Gaz.* 6 Dec. 5/2 There was no more high-souled, high-minded man than the man who was now Prime Minister. **1930** A. Huxley *Let.* 14 June (1969) 337 English reviews..have been rather snorty and high-souled about the book.

B. *sb.* **1. b.** An area of high barometric pressure. Also *transf.*

1878 *Pop. Sci. Monthly* July 310 These high and low areas, or 'highs' and 'lows' as they are technically known, travel. **1901** *Yearbk. U.S. Dept. Agric. 1900* 332 The hot wave..seemed to join forces with the permanent high over the ocean. **1958** 'N. Shute' *Rainbow & Rose* 290 A high had come along that the Met had not been able to forecast. **1966** *New Statesman* 27 May 759/1 There are.. highs and lows of political intensity: right now, we are approaching a moderate high.

c. = the 'High Street' in Oxford.

1853 'C. Bede' *Verdant Green* 1st Ser. x. 88 He at once sallied forth to 'do the High', and display his new purchases. **1912** A. Quiller-Couch in *Oxf. Bk. Victorian Verse* 860 Yet if at last not less her lover You in your hansom leave the High [etc.]. **1921** C. S. Lewis *Let.* 10 May (1966) 59 It is still pleasant to see fewer foreign visitors pacing the High with guide books. **1955** *Times* 11 Aug. 7/6 A proposal..to close Magdalen Bridge,.. preventing the High from being used as a motor thoroughfare.

d. = *high gear* (*Gear *sb.* 7 b).

1921 A. F. Hall *Handbk. Yosemite Nat. Park* 308 You may hear a driver boast that he made such and such a grade on 'high', but that is merely an admission of poor judgment. **1931** *Kansas City Star* 8 Aug., Now [all they talk about is] whether or not they were able to go up Pikes Peak on high. **1934** J. M. Cain *Postman always rings Twice* ix. 89, I went into high. **1970** *Globe Mag.* (Toronto) 26 Sept. 7/1 Lewis moved into high, knowing, but really knowing that Walter would be in the race.

e. A record, a high level exceeding that previously attained.

1926 *Chicago Tribune* 23 Jan. II. 9/1 Wheeling and Lake Erie issues resumed their advance, the common toward a new high at 49⅛. **1928** *Weekly Dispatch* 3 June 7/2 When he buys, they buy; the lot of them can create.. a new 'high' in any share in which Mr. Durant fancies. **1937** *Lit. Digest* 20 Mar. 3 (*heading*) Nazi epithets at U.S. set new high. **1939** [see *all-time* (*All *a.* 13)]. **1951** W. Stevens *Let.* 25 June (1967) 720 The whole thing has brought my morale up to an all-time high. **1953**, etc. [see *Low *sb.* 3 c]. **1959** *Encounter* Sept. 59/1 Beckett's stock has reached a steady high at Langham Place. **1964** R. D. Hopper in I. L. Horowitz *New Sociol.* 317 The gross national product has reached an all-time high. **1968** *Listener* 22 Feb. 228/2 The series manifestly represented a new high in the adaptation of fiction to film.

f. = High School (School *sb.*[1] 1 j). *N. Amer. colloq.*

1928 *Boston Even. Transcript* 30 Mar. 15/7 I'm hardly more than a schoolboy, not so very long out of Dorchester High. **1930** H. Crane *Let.* 29 Nov. (1965) 359, I left East High without even a diploma—in my junior year. **1963** H. Garner in R. Weaver *Canad. Short Stories* 2nd Ser. (1968) 49 He had graduated from technical high, and was going to university in the fall. **1968** *New Yorker* 18 May 56, I started playing drums in junior high.

g. *pl.* The higher range of audio-frequencies.

1940 in *Chambers's Techn. Dict.* 415/2. **1962** A. Nisbett *Technique Sound Studio* ii. 46 Such screens have the advantage of being dual purpose: used dead side to the microphone, they will damp down the highs: with the bright side forward, the highs will be emphasized.

h. A euphoric state induced by the taking of a drug or drugs. *slang.* (Cf. *High *a.* 16 c.)

1953 W. Burroughs *Junkie* (1972) xv. 149 Finally, the peyote came up solid like a ball of hair,..clogging my throat. As horrible a sensation as I ever stood still for. After that, the high came on slow. Peyote high is something like benzedrine high. **1956** H. Gold *Man who was not with It* (1965) i. 6 Dreamy in my high, I floated down ..to kick the smaller mark. **1967** *Listener* 3 Aug. 130/1 It is not easy..to describe the effect of a trip on LSD or a marijuana high. **1967** *Spectator* 4 Aug. 131/1, I was a drug addict..for two years. I was in an almost permanent state of high. **1969** *Times* 21 May 7/3 The two cigarettes smoked by each subject were intended to produce a 'normal social cannabis high'.

3. a. Also, the highest card in cutting for deal. *high-low-jack* (*and the game*): = All-fours 1. (See quots.)

[**1674** C. Cotton *Compl. Gamester* x. 111 This Game I conceive is called All-Fours from Highest, Lowest, Jack and Game, which is the Set as some play it. *Ibid.* 113 Sometimes you are highest, lowest, Jack, and Game.] **1814** C. Jones *Hoyle's Games Impr.* 170 All-fours..derives its name from the four chances therein, for each of which a point is scored, namely, *high*, the best trump out; *low*, the smallest trump dealt; *jack*, the knave of trumps; *game*, the majority of pips reckoned from such of the following cards as the respective players have in their tricks; viz. every ace is counted as 4; king 3; queen 2; knave 1; and ten for 10. **1818** Todd s.v. *All-fours*, The all-four are high, low, Jack, and the game. **1843** 'J. Slick' *High Life N.Y.* II. 214 Under the table..was a hull squad of playin cards..as if somebody had got beat a playing high-low-jack and the game. **1895** *Funk's Stand. Dict.*, *High-low-jack*. Same as All-fours. **1898** B. Kirkby *Lakeland Words* 72 *High-low*, a card game. High-low, Jack an' t' gam. **1911** R. F. Foster *Compl. Hoyle* 328 As High, Jack, and Game are always counted by the player holding those points at the end of the play, there can be no question about them: but serious disputes sometimes arise as to who played Low... It is even possible, if there is no other trump or counting card in play, for the Jack to be High, Low, Jack, and the Game. **1963** G. F. Hervey *Handbk. Card Games* 16 The players then turn up their tricks and score for High, Low, Jack and Game.

b. Phr. *how is that for high?*: an exclamation inviting admiration; in allusion to the card called the high in the game of high-low-jack. *U.S. colloq.*

1872 Schele de Vere *Americanisms* 326 The phrase 'How is that for high?' borrowed from a low game known as Old Sledge, where the *high* depends, not on the card itself but on the adversary's hand. Hence the phrase means, What kind of an attempt is that at a great achievement? **1887** F. Francis *Saddle & Mocassin* xviii. 315 'How's that for high, boys?' concluded the narrator, when he had told his tale. 'That's on top,' declared Black Jack; 'that takes the cake.' **1922** Joyce *Ulysses* 122 What about that, Simon?..How's that for high?

high, *adv.* Add: **2. c.** (Later examples.)

1894 *Vermont Agric. Rep.* XIV. 102 Will a colt do well..if fed high in winter? **1965** T. Capote *In Cold Blood* (1966) iii. 135 Him and Carol lived too high, kept buying stuff they couldn't nohow afford. **1967** L. J. Braun *Cat who ate Danish Modern* xv. 137 David lived high, and he gave every thing away.

9. b. *to live* (or *eat*) *high off* (or *on*) *the hog*: to live (etc.) in luxury. orig. *U.S.*

1946 *Call-Bulletin* (San Francisco) 27 May Edit. page, I have to do my shopping in the black market because we can't eat as high off the hog as Roosevelt and Ickes and Joe Davis and all those millionaire friends of the common man. **1946** *Time* 27 May 22/2 Eatin' on the Hog. In the years of political wars the Organization had grown and resistance diminished. **1956** 'E. S. Aarons' *Assignment Treason* (1967) ii. 27 He lives high off the hog. He needs money. **1966** A. Prior *Operators* vi. 69 That had been a good year, a year of living high off the hog. **1967** *Observer* 30 Apr. 11/5 Die..for what? So that the Saigonese and other civilians can live high off the hog? **1967** K. Giles *Death in Diamonds* iii. 47, I hope these Uppings eat high on the hog. **1969** *New Scientist* 6 Mar. 511/1 A cod lives quite high on the hog—until he turns up on someone else's menu.

10. a. *high-dried* (also as *sb.*), *-laced, -lying, -perched, -piled, -reared* (later examples); **high-keyed** *Mus.*, of a high pitch; also *fig.*; **high-sniffing** *colloq.*, contemptuous, disdainful; **high-strung** (further examples: now used esp. in the sense 'in a state of nervous tension').

1818 Scott *Heart of Midlothian* in *Tales of My Landlord* III. x. 265, I have always a chat with Mrs Glass when I purchase my Scots high-dried. **1858** Geo. Eliot

Scenes Clerical Life I. 48 If Mr. Barton had shaken into that little box a small portion of Scotch high-dried, he might have..[etc.]. **1893** *Funk's Stand. Dict.* I. 848/1 *High-keyed*, 1. *Mus.* high-pitched. 2. Sensitive, spirited; as, a *high-keyed* woman. **1906** *Daily Chron.* 18 June 6/3 Mr. P. Wilson Steer has several examples of his familiar high-keyed method, including a not too happy portrait of himself in a grey tweed suit. **1938** *Burlington Mag.* Feb. 75/2 Painted in thick and high-keyed colours. **1938** M. K. Rawlings *Yearling* xvi. 181 There was an excitement..that made him nervous and high-keyed. **1851** J. G. Whittier *Works* (1898) 210 Madam in her high-laced ruff. **1880** 'Mark Twain' *Tramp Abroad* II. xxxiv. 26 These wore..hob-nailed high-laced walking-shoes. *a* **1930** D. H. Lawrence *Phoenix II* (1968) 46 Even then, I lack high-laced boots and one stocking. **1934** *Discovery* June 166/2 The more high-lying burials.. contained 'reserved slip' pottery. **1958** *Times* 19 Dec. 7/1 The difficult is intonation in high-lying passages. **1969** *Daily Tel.* 12 Mar. 21/3 High-lying solos in the Mozart finale. **1863** Longfellow *Tales Wayside Inn* 45 High-perched upon the back of which there stood The image of a falcon carved in wood. **1906** *Westm. Gaz.* 16 June 12/2 Each high-perched farmhouse was changed by sunset glamour to a magic castle. **1945** W. S. Churchill *Victory* (1946) 207 Flags from high-perched windows. **1862** W. C. Bryant *Poet. Wks.* (1883) II. 312 Clouds, High-piled. **1932** V. Woolf *Common Reader* 2nd Ser. 226 Its high-piled metaphors. **1952** C. Day Lewis tr. *Virgil's Aeneid* xi. 257 Treading the high-piled embers as we perform your rites. **1896** A. E. Housman *Shropshire Lad* xxxvii. 53 The high-reared head of Clee. **1922** Joyce *Ulysses* 422 The swancomb of the gondola, high-reared, forges on. **1906** *Daily Chron.* 12 July 3/3 High-sniffing pretenders..affect to find in Mr. Meredith's poetry naught that is obscure. **1863** *Country Gentleman* 7 May 300/2 When the sire and the dam of a colt possess much spirit and are 'high strung'. **1868** W. James in *North Amer. Rev.* CVII. 322 That high-strung attitude of vigilance, suspicion, and suspended judgment. **1902** A. Bennett *Anna of Five Towns* xi. 283 You're so sensitive and high-strung. **1946** M. Lowry *Let.* 15 Sept. (1967) 125 Writers often tend to be high-strung creatures. **1956** A. H. Compton *Atomic Quest* 125 He could keep a group of high-strung specialists working smoothly together.

high altar. [OE. *heah-altar.*] The principal altar of a church.

c **950** in T. Wright *Vol. Vocab.* (1873) II. 23 *Cibborium*, þæs heahalteres ofergeweorc. **13..** *Gaw. & Gr. Knt.* 592 So harnayst as he was he herknez his masse, Offred and honoured at þe heȝe auter. *c* **1460** *Oseney Reg.* 137 Reynolde, By the grace of god Bisshop of Clone,..halowed the Chapell of Saunforde and the high auter. *c* **1553** in *Diary of H. Machyn* (1848) 399 The highe altar table. *a* **1700** Evelyn *Diary* 25 Oct. 1644, On the large high altar is a brazen vessel of admirable invention. *c* **1777** in *Cath. Rec. Soc. Publ.* VIII. 173 He lays buried in our Church at the foot of the High Altar. **1826** [see Altar 2]. **1894** C. M. Church *Chapters Early Hist. Ch. Wells* App. W. 419 High altar dedicated to St. Andrew. **1927** W. Cather *Death comes for Archbishop* ix. viii. 299 The next morning the old Archbishop lay before the high altar in the church he had built. **1954** *Oxf. Jun. Encycl.* XII. 8/2 Large churches often have subsidiary chapels with their own altars, and then the principal one is called the High Altar.

highball. Add: Also **high ball, high-ball. 1.** (Earlier example.)

1882 C. M. Chase *Editor's Run in New Mexico* 134 Mexican monte, keno, faro, high ball, etc., are the prevailing games.

2. (In full *highball signal.*) A signal to proceed given to a locomotive driver, formerly by hoisting a ball aloft. Also *gen.*, a signal to go ahead, a clear way, a straight course. *U.S.*

1897 *Chicago Record* 1 Mar. 6/1 'Milk' trains'..have 'rights' over the rails and get nothing but 'high balls'. **1909** *Sat. Even. Post* 26 June 9/1, I gave 'em the highball signal to go ahead. **1920** J. M. Hunter *Trail Drivers of Texas* 68 We had a high ball trail from there on. *Ibid.* 354 Mr. Butler and I told them [*sc.* cowboys]..to strike a high ball to town. **1931** 'D. Stiff' *Milk & Honey Route* 207 *High ball*, signal for a train to pull out of town. **1938** A. E. Parkins *South* 168 We are informed that the signal the conductor waves to the engineer to go ahead is still known as the 'high ball' on most lines. **1940** *Sat. Even. Post* 16 Nov. 18 Picking up the highball as he drifted out of the freight yards, Engineer Joe De Nobel gave her the throttle. **1959** J. Thurber *Years with Ross* iv. 63 The parlance of railroading—deadhead, highball, whistle stop, gandy dancer.

3. A drink of whisky and soda or other mineral water served with ice in a tall glass. *U.S.* Also *attrib.*

1898 *N.Y. Jrnl.* 16 Sept. 4/2 Evening dress and khaki talked much sport and a little war over 'high balls' or chicken livers. **1899** Ade *Doc. Horne* ii. 11 Lush..drank two magnificent 'high balls'. **1909** Eaton & Underhill *Runaway Place* 128 The pleasant occupation of consuming three chocolate éclairs and a high ball. **1915** Wodehouse *Something Fresh* iii. 59 Beyond Baxter, a cigar in his mouth and a weak high-ball at his side, the Earl of Emsworth took his ease. **1925** H. L. Foster *Trop. Tramp Tourists* 109 The crowd from Ohio..celebrated by drinking several highballs. **1933** N. Coward *Design for Living* III. I. 1 Who'd like a highball? **1949** F. Maclean *Eastern Approaches* I. vii. 82 The crystal chandeliers shed a cheerful light on the silver trays of highballs and old-fashioneds. **1959** *Sears, Roebuck Catal.* Spring & Summer 1394/3 Highball glasses. **1965** *Amer. Speech* XL. 77 In sophisticated drinking circles the term *high-ball* has become practically archaic... The illuminati ask for 'whiskey and water' or 'Scotch and soda'. **1973** H. Nielsen *Severed Key* viii. 92 A soiled highball glass on the counter.

hi·ghball, *v. U.S. slang.* [f. the sb.] **a.** *intr.* To give a locomotive driver a signal to proceed; also *transf.* **b.** To go or travel at speed (const. *it* or with *adv.*). **c.** *trans.* To drive (a locomotive or vehicle) at speed.

1912 *Railroad Man's Mag.* XVII. 493 The con highballed, and the manifest freight Pulled out on the stem behind the mail. *Ibid.*, She whistled twice and highballed out, They were off—down the Gila Monster Route. **1931** 'D. Stiff' *Milk & Honey Route* v. 53 Often they highball the cops and you get raided. **1935** Partridge *Slang To-Day* (ed. 2) 442 *Highball*, to travel swiftly; to depart, esp. hastily. **1941** *Amer. Speech* XVI. 233/1 *High ball*, to speed up operations. **1941** *Pop. Sci. Monthly* May 76 Its smooth power..is fully available, whether the giant is pulling away from a dead stop or highballing along at its maximum governed speed. **1943** S. K. Farrington *Railroading from Head End* 48 We highballed out at 5:05. No hurry, no lost motion, no excitement—just clocklike precision by every man jack of them. **1945** *This Week Mag.* 14 July 19/2 He highballed the big locomotive down the tracks. **1946** *Sat. Even. Post* 11 May 27/3 Everyone else had highballed..out of there. **1961** *Amer. Speech* XXXVI. 271 Imagine yourself on the seat of an enormous freighter, high-balling it down a West Coast highway. **1962** *Ibid.* XXXVII. 131 A logging train that is highballing or making a fast run. **1971** M. Tak *Truck Talk* 82 *Highball it*, to drive a tractor-trailer at or near its top speed.

high-binder. 2. (Earlier example.)

1876 *San Francisco Call* 27 Mar. 1/7 Refined ladies could no longer submit to be jostled at the church door by the Mongolian *chiffonier* or high-binder.

3. Delete def. and substitute: *transf.* Used abusively to denote a swindler, esp. a fraudulent politician.

1903 A. H. Lewis *Boss* 136 He's goin' to take copies of th' accounts that show what th' Chief an' them other high-binders at the top o' Tammany have been doin'. **1908** G. H. Lorimer *J. Spurlock* xii. 324 That's what I do mean—it [*sc.* the railroad]'s been stolen by that Bonsall bunch of high-binders. **1916** H. L. Wilson *Somewhere in Red Gap* i. 21 So I left these two lady highbinders and went on into the retail side of the Family Liquor Store. **1952** in Wentworth & Flexner *Amer. Dict. Slang* (1960) 255/2 The winter meeting of the grand inner circle of high-binders at Miami Beach.

high-boy. Add: **3.** = Tallboy 2. *U.S.*

1891 *Scribner's Mag.* X. 353 In the top drawer of a high chest of drawers, a 'high boy'. **1902** L. V. Lockwood *Col. Furnit. Amer.* 56 The common form of the flat-topped bandy-legged high-boy is made of cherry, maple or walnut. **1909** J. C. Lincoln *Keziah Coffin* i. 10 Removing towels, tablecloths, and the like from the drawers in a tall 'high-boy'. **1926** *Bulletin* 29 June 10/2 The..lady.. belonged against a background of mahogany highboys, old china..and an Adams' mantel. **1972** *Town & Country* Mar. 106/1 (*caption*) American William and Mary highboy.

highbrow, high-brow (hai·brau), *sb.* and *a. colloq.* orig. *U.S.* [Back-formation from *high-browed *a.* 2.] **A.** *sb.* A person of superior intellectual attainments or interests: occas. with derisive implication of conscious superiority to ordinary human standards.

1908 *Sat. Even. Post* 29 Aug. 27/1 It takes all sorts of men to make a party, and Mr. Hearst apparently led in a few prize-fighters with the other high-brows and reformers he accumulated. **1911** H. S. Harrison *Queed* iv. 41 Who knew but what this little highbrow was the very man they were looking for? **1914** G. Atherton *Perch of Devil* i. 41 I'll be a real high-brow in less than no time. **1921** H. S. Walpole *Young Enchanted* III. vi. 301 There was the theatre (so much better than the highbrows asserted), there were concerts. **1922** D. H. Lawrence *Phoenix II* (1968) 240 Then the highbrows come and say: 'Poor Indian, dear Indian! why, *all* America ought to belong to him!' **1925** A. P. Herbert *Laughing Ann* 86 I'll be a high-brow, but I'll look hearty, And I won't laugh at the Liberal Party. **1934** S. R. Nelson *All about Jazz* i. 18 The strangely disreputable lady 'Jazz'—disreputable because she was not sponsored by the highbrows. **1955** *Times* 23 June 11/4 The highbrows in those parts all go up in smoke or mist if you confess to liking those among their native artists who seem most typically Scottish.

B. *adj.* Of, pertaining to, or characteristic of a highbrow; intellectually superior.

1884 L. Troubridge *Life amongst Troubridges* (1966) xii. 169 Mr. Hope had suggested that we would be at some highbrow part of the Exhibition—looking at pictures I think, but Jo had said firmly, 'If I know the Troubridges they will be at the Chocolate Stall', and we were! **1914** S. Lewis *Our Mr. Wrenn* 42 All them high brow sermons. **1916** S. Leacock in 'O. Henry' *Waifs & Strays* (1919) 161 Shakespeare, except as revived at twenty-five cents a seat with proper alterations in the text, is 'highbrow'. **1917** W. J. Locke *Red Planet* xxiv. 306 She'd die of culture in the mater's highbrow establishment. **1925** *Punch* 22 Apr. 437/2 'The programmes are too highbrow,' I maintained. 'They are hopelessly beyond the intelligence of the mass, at any rate.' **1931** R. Church *High Summer* 12 Mother insists on my being highbrow and visiting all the historical places. **1943** C. Gray *Contingencies* (1942) i. 41 The choice of programme was uncompromisingly what it is customary to describe as 'highbrow', but the house was packed on both occasions. **1963** Auden *Dyer's Hand* 408 All high-brow–lowbrow frontiers of taste.

So **hi·ghbrowish** *a.*, fairly, or extremely, highbrow; **hi·ghbrowism,** the condition of being highbrow, intellectual superiority.

1921 *Glasgow Herald* 22 Jan. 4/2 This doctrine is tainted with high-browism. **1923** A. Bennett *Things that have interested Me* II. 207 The audiences were artistic and earnest, with a dash of high-browism. **1926** Whiteman & McBride *Jazz* viii. 108 Does the very word 'classical' make you nervous because it sounds so highbrowish? **1927** D. H. Lawrence *Let.* 19 May in E. & A. Brewster *Reminisc. & Corr.* (1934) 132, I can't stand high-browish..people any more. **1937** *John o' London's* 1 Jan. 585/1, I am incapable of 'highbrowism', I make no pretensions to be a literary critic. **1947** N. Cardus *Autobiogr.* I. 16 The..Bloomsbury-Chelsea highbrowism which does not understand that genius is a miracle to be revered. **1953** *Harper's Mag.* Mar. 48/2 Articles. Books. Highbrowish stories.

high-browed (hai·braud), *a.* [f. High *a.* + Brow *sb.*[1] + -ed[2].]

1. Having a lofty forehead.

1848 'Geo. Eliot' *Let.* 23 Nov. (1954) I. 273 We have brought you [*sc.* Mother Nature] many gentle maidens and high-browed, brave men. **1875** *Brit. Q. Rev.* Apr. 500 One can conjure up a vision of them: the one fair, pale, high-browed. **1891** Flügel *Eng.-German Dict.*

2. = *Highbrow *a.* orig. *U.S.*

1906 'O. Henry' in Davis & Morris *Caliph of Bagdad* (1931) xiv. 238 That, in addition to the $150 that I screwed out of the high-browed and esteemed B. Merwin during your absence. **1908** R. W. Chambers *Firing Line* ix, You were very much amused, I suppose—to see me sitting bras-dessus-bras-dessous with the high-browed and precious. **1909** H. G. Wells *Ann Veronica* vii. 121 Their very furniture had mysteriously a high-browed quality. *Ibid.* viii. 144 Goopes, she was sure, was always high-browed and slow and Socratic. **1916** —— *Mr. Britling* II. iv. § 10, I was too high-browed about this war business. **1923** A. Bennett *Things that have interested Me* II. 207 If artistic, earnest, and high-browed women only knew how to dress! **1927** *Daily Express* 3 May 3/7 Meeting a highbrowed friend.

high chair. [High *a.* 3.] A child's chair with high legs, usually fitted with a movable tray and footrest.

1848 Thackeray *Van. Fair* xliv. 400 Little Rawdon ..was perched on a high chair by the baronet's side. **1853** C. Brontë *Villette* I. ii. 23 The demure little person.. perched now on a high chair. *Ibid.* 25 She relinquished the high chair for a low stool. **1885** C. M. Yonge *Two Sides of Shield* II. i. 8 Mrs. Halfpenny [*sc.* the nanny] always put us on the high chair with our faces to the wall when we were jealous. **1913** C. Mackenzie *Sinister St.* I. i. 7 Stella was fitted into her high chair; the circular top was brought over from behind and thumped into its place with a click. **1943** C. McCullers *Heart is Lonely Hunter* II. ix. 214 Ralph banged his spoon on his high-chair tray. **1959** *Sunday Times* 25 Oct. 20/6 Spoon over-board: the tyrant in the high chair makes a point.

higher. Add: **A.** *adj.* **2.** *higher education* (examples); *Higher* (*School*) *Certificate*, an examination instituted in 1917 and replaced in 1951 by the Advanced level General Certificate of Education, taken by pupils of about 18; *Higher Thought* = *New Thought.

1866 E. Davies (*title*) The higher education of women. **1868** *Rep. Sch. Inqu. Comm.* I. 115 in *Parl. Papers 1867–8* (C. 3966) XXVIII. 1, [Endowed] Schools have been regarded as the subjects of special trusts..not as local contributions to the higher education of the country. **1877** *Encycl. Brit.* VII. 674/1 Every Jesuit college was divided into two parts, the one for higher the other for lower education. **1884** C. Bird *Higher Educ. in Germany & England* i. 5 Few [*sc.* people] realise to what an extent we are surpassed by Germany..as regards the liberal provision made for higher education. **1896** *Spectator* 30 May 767/1 The great feature of the higher education should be a very well-marked revolt of the body against the mind. **1909** Chesterton *Orthodoxy* v. 136 Of all conceivable forms of enlightenment the worst is..the Inner Light... Anyone who knows anyone from the Higher Thought Centre knows how it does work. **1909** H. G. Wells *Ann Veronica* vi. 124 Jim is up to the neck in Mahatmas and Theosophy and Higher Thought and rot. **1918** *Univ. Cambridge Local Exam. Synd. Higher School Certificate Exam. List* 6 (*heading*) Exemption from the Previous examination by means of the higher school certificate examination. **1931** (*title*) Report of the Commission on Christian Higher Education in India. **1933** *Discovery* Sept. 271/2 He is best known for his researches on the Higher Thought Processes. **1945** *Guide Educ. Syst.* Gloss. 58 *Higher School certificate* (*higher certificate*), certificate awarded on results of the examination taken at about 18 by grammar school pupils. **1949** H. McLennan *Cross-country* 169 So much for higher education in Canada. **1961** *Sunday Times* 26 Feb. 12/6 By 1970 there could be 140,000 children applying for the 70,000 places there will then be in higher education. **1963** Barnard & Lauwerys *Handbk. Brit. Educ. Terms* 106 The Higher School Certificate examination, which came into operation in 1917 and was conducted by certain university boards, was taken by pupils in grammar schools at about the age of 18. It was primarily intended to be a test of a two-years sixth form course of a somewhat specialised nature. **1968** *Listener* 6 June 723/1, I use the expression 'higher education' in the same sense as the Robbins Committee. Broadly, it covers courses for students of 18 and over which reach a standard above A-level in GCE and for which the normal entry qualifications are at least five O-level passes or the equivalent. Degree courses account for a big proportion of the field.

4. (Further examples.)

1955 *Internat. Survey Programmes Social Devel.* (U.N.) 8 Training abroad on fellowships has made an important though necessarily limited contribution to the supply of higher-level technical and professional workers in some countries. **1958** B. Abel-Smith in N. Mackenzie *Conviction* 59 The middle classes..participate with the higher

income groups in special benefits from their employment. **1960** *Amer. Speech* XXXV. 230 English open internal juncture..is defined as the boundary point between two higher level phonological units (bounded sequences). **1962** Corson & Lorrain *Introd. Electromagn. Fields* i. 24 The sum of these four terms (neglecting higher-order differentials because we are interested in the limit S→O) is equal to the right-hand side of Eq. *1–109*. **1964** Crystal & Quirk *Prosodic & Paralinguist. Features Eng.* iv. 52 One might set up the existence of higher-order patterns as a hypothesis. **1965** C. H. Springer et al. *Adv. Methods & Models* iii. 76 We might as well finish the job by deriving formulas for these *higher-order* differences. **1967** *Computers & Automation* Feb. 30/1 The single most important tool is what has become known as a 'programming language', sometimes called a 'higher level language' (to distinguish it from the normal machine codes or assembly languages..). **1968** Fox & Mayers *Computing Methods for Scientists & Engineers* iii. 32 It is..possible to start with a very accurate y_0 and use double or higher precision arithmetic, but this is lengthy.

B. *adv.* **2.** (Further examples.)

1883 'Mark Twain' *Life on Mississippi* li. 502 Explosion followed explosion..reports grew steadily sharper and higher-keyed. **1923** H. Crane *Let.* 21 July (1965) 142 Being with the largest advertising agency in the world.. will get me higher-paid positions in other places after awhile. **1969** *Punch* 15 Jan. 91/3 Allied Breweries, the International Compressed Air Corporation or other proclaimed seekers of higher-educated manpower.

higher-up. orig. *U.S.* [High *a.* 5.] One occupying a superior position or post.

1911 *McClure's Mag.* Aug. 351 Resolved to get those dynamiters and to get the 'higher-ups', if there were any behind them. **1916** C. Sandburg *Chicago Poems* 61 Higher-ups among the con men of Jerusalem. **1929** *Literary Digest* 12 Oct. 7/1 He is..insists the New York Telegram—'only the henchman of higher-ups'. **1931** L. Steffens *Autobiogr.* II. x. 254 Other 'higher-ups' confessed. **1939** 'G. Orwell' *Coming Up for Air* II. viii. 145 The mysterious higher-ups who were running the war. **1953** K. Tennant *Joyful Condemned* v. 47 She's in with all the higher-ups. **1960** H. L. Lawrence *Children of Light* iv. 63 Tell them the War Office is scared of a scandal and that the—er—higher-ups are willing to help them. **1970** K. Giles *Death in Church* ii. 38 The higher-ups were all chartered accountants from Krupps who did not like waste.

highfalutin, -ing. B. *adj.* (Earlier and later examples.)

1839 *Spirit of Times* 18 May 123/3 Them high-faluting chaps. **1941** *Coast to Coast 1941* 49 And then hear some announcer in his highfalutin voice, telling her summer was coming. **1948** *Manch. Guardian Weekly* 30 Sept. 9 When all the highfalutin and magical jargon of diplomacy is removed, you'll find the diplomats like a group of children aged about three or four. **1962** *New Statesman* 2 Mar. 308/2 This is..a pleasing unsententious compilation, not really a lecture at all. Sir Compton is never highfalutin.

high fide·lity. [f. High *a.* + *Fidelity 2 c.] In equipment used in the recording and reproduction of sound, the property of producing little distortion in the signal, so that the sound produced bears as close a resemblance as possible to the original. Also applied to the recording of electrical signals generally (quots. 1938, 1957). Also *attrib.* or as *adj.* So *higher fidelity.* Cf. *Hi-fi.

1934 *Electronics* July 223/2 If the term 'high-fidelity' is to mean anything the predicated performance should be kept at a reasonably high level. **1934** *Wireless World* 19 Oct. 318/2 With the introduction of the 'high fidelity' receiver the broadcasters, who have repeatedly criticised the manufacturers for low-quality receivers which cut off the extreme notes, now make the admission that improvement is necessary in their own transmitters. **1937** *Time* 8 Mar. 62/1 (Advt.), Victor records: higher fidelity. **1938** *Jrnl. R. Aeronaut. Soc.* XLII. 1049 Present-day high fidelity 441-line television demands modulation frequencies as high as 4 megacycles. **1940** *Chambers's Techn. Dict.* 415/2 *High-fidelity*, an inexact term generally meaning sound reproduction of a superior, but undefined, quality. **1946** *Electronic Engin.* XVIII. 54 It has not been possible up to the present to obtain full advantage of the high fidelity of the moving coil pick-up. **1955** Priestley & Hawkes *Journey down Rainbow* 27 There was talk; there were high-fidelity records. **1957** *Times* 7 Oct. 8/3 The power of the transmitters guarantees high fidelity reception of the wireless signals. **1958** *Times* 17 May 9/5 This search for the elusive 'high fidelity' is pursued with an avidity only limited by the depth of his pocket. **1972** *Sci. Amer.* Sept. 100/3 If one wishes to transmit music with high fidelity, the required bandwidth is 15,000 hertz.

high-flyer, -flier. Add: **1. c.** A variety of walnut.

1822 *Trans. Lond. Hort. Soc.* IV. 517 The Highflyer Walnut. **1824** Loudon *Encycl. Gardening* 742 Walnut.. Highflyer of Thetford, the best variety known. **1866** Lindley & Moore *Treas. Bot.* 640/2 A variety called the Highflyer Walnut, is considered the best English variety. **1880** *Encycl. Brit.* XII. 278/1.

5. d. = Flyer 5 b.

1961 in Webster. **1964** *Economist* 22 Feb. 722/1 Stocks, variously called 'glamour' issues or 'high flyers'. **1969** *Daily Tel.* 8 Feb. 6/4 The other fund may have been lucky—or clever—enough to have a number of real 'high-fliers' in its portfolio and so has had to do comparatively little switching to achieve its 100 p.c. growth. **1969** *Times* 16 July 22/4 Another of last year's high flyers came unstuck yesterday when Qualitex Yarns revealed that it had missed its forecast.

high fre·quency. [f. High *a.* + Frequency 4.] **1. a.** A frequency (see Frequency 4 b in Dict. and Suppl.) having a relatively large number of cycles in a second. Applied esp. to an electric current or voltage, an electromagnetic wave or a sound wave. Abbrev. *H.F., esp. in radio and telecommunications, where it also refers specifically to electromagnetic waves of 3–30 MHz. Often *attrib.*

1893 [see Frequency 4 b]. **1923** *Nature* 7 July 12/2 The vibrations are of frequency 200,000 per second, such as are commonly used in wireless telegraphy and telephony. Such high frequencies are neither seen nor heard, but can be detected by special methods. **1934** S. G. B. Stubbs *Electr. Encycl.* II. 612/2 *High frequency* (*H.F.*), refers to alternating currents, the frequency of which is reckoned in thousands. There is no definite line of demarcation between high and low frequency, but in general an alternating current is reckoned as of high frequency when the number of periods is greater than 10,000 per second. **1960** H. Carter *Dict. Electronics* 127 *High frequency.* (1) A general term used to distinguish signals of radio frequency from those of audio frequency. (2) A relative term used to describe frequencies at the upper end of a particular frequency band. (3) Term of specific application to radio waves in the frequency range between 3 and 30 Mc/s, i.e. of wavelengths from 100 m down to 10 m. **1962** A. Nisbett *Technique Sound Studio* iii. 79 High frequencies can be lost in a variety of ways. **1962** M. G. Say *Newnes' Conc. Encycl. Electr. Engin.* 360/1 Fig. 1 (*b*) and (*c*) show that at high frequencies above 1,000 c/s there is an increasing tendency for the current to flow in a shallow surface layer of the conductor.

b. *transf.* A high rate of occurrence, in space or in time.

1935 Huxley & Haddon *We Europeans* ix. 267 Our picture of the human species will be like a contour-map, a region of high frequency for, say, round-headedness being separated from another similar peak by a 'valley' of low frequency. **1971** *Sci. Amer.* Oct. 21/1 A high frequency of kwashiorkor is now being found among the East Pakistan refugees in India.

2. *attrib.* (usu. with hyphen). **a.** In sense 1 a, implying an applicability to an oscillation with a high frequency, as *high-frequency formula*, or an action or manner of working that involves such an oscillation.

1892 Heaviside *Electr. Papers* II. 279 The high-frequency formulae are not so generally applicable as in the case of cores. **1893** *Jrnl. Soc. Arts* 5 May 624/1 The phosphorescent glow of the novel high-frequency, high-voltage, high-vacuum lamps. **1902** *Proc. R. Soc.* LXX. 341 The special manner in which a core or rod of iron or steel placed in a varying magnetic field is affected by high-frequency oscillations transmitted from considerable distances. **1903** *Daily Chron.* 14 Nov. 5/2 High-frequency apparatus and superb laboratories are beside the point compared with air and light. **1914** R. Stanley *Text-bk. Wireless Telegr.* xii. 149 Do high and low frequency waves travel at the same speed? **1921** *Jrnl. Industrial & Engin. Chem.* July 639/2 High frequency voltage (10,000 to 20,000 cycles) is applied at the terminals of the coil. **1928** *Nature* 21 Apr. 623/1 High frequency sound waves of low intensity passed through these cells cause the protoplasm to rotate. **1955** F. E. Terman *Electronic Radio Engin.* (ed. 4) xxiv. 937 These requirements are usually more difficult to meet at the high-frequency end of the modulation range. **1960** R. W. Marks *Dymaxion World of B. Fuller* 23/1 The high-frequency hiss of the surf. **1962** A. Nisbett *Technique Sound Studio* iv. 80 Another example of high-frequency attenuation will occur in recording if the tape is in imperfect contact with the head. *Ibid.*, Where the high frequency losses are progressive, measures may be taken to equalize for them. *Ibid.* 81 For a recording head the effect of high-frequency bias is to narrow the effective gap width.

b. In sense 1 b, meaning 'occurring often or involving a rapid succession of events'.

1957 D. L. Bolinger in *Publ. Amer. Dialect Soc.* xxviii. 89 Contrast the high-frequency *suppose* with the lower-frequency *suspect*. **1965** *English Studies* Feb. 75 It is hard to judge of the size of the corpus..but it seems to be adequate at least for high-frequency items.

3. *Comb.*, as *high-frequency amplifier, choke, cinematography, condenser microphone, current, photography, response, transformer*; **high-frequency alternator,** any of several kinds of alternator designed to give an alternating current with a frequency greater than about 10 kHz; **high-frequency amplification,** amplification of signals of high frequency; in *Radio* applied *spec.* to amplification at the carrier frequency in a radio receiver; **high-frequency heating,** the heating of a substance by placing it in an alternating electric field (dielectric heating) or magnetic field (induction heating) with a frequency above that of the mains electricity supply; **high-frequency induction furnace** (see quot. 1958); **high-frequency resistance,** (*a*) the increased electrical resistance of a conductor at high frequencies, owing to the **skin effect*; (*b*) a resistance for use at high frequencies; **high-frequency treatment,** (*a*) *Med.* (see quot. 1931); (*b*) *Metallurgy*, the hardening of metals by heating them inductively and quenching.

1901 *Proc. R. Soc.* LXVIII. 514 (*heading*) The high frequency alternator. **1935** J. B. Moore in K. Henney *Radio Engin. Handbk.* (ed. 2) 564 The high-frequency alternator is one of the most used types of transmitter for long-wave transoceanic code communication. **1919** *Wireless World* Nov. 448 An eight valve amplifier, comprising five stages of high-frequency amplification, one stage of rectification, and two stages of audio-frequency amplification with transformer coupling. **1934** S. O. Pearson in S. G. B. Stubbs *Electr. Encycl.* II. 612/1 The magnification of electrical variations at the signal—or radio—frequency of the received oscillations, that is, before they are applied to the detector of a wireless receiver, is termed high-frequency amplification. **1919** *Wireless World* Feb. 628 (*heading*) The use of impedance, capacity and resistance couplings in high-frequency amplifiers. **1928** [see *Choke 6]. **1958** *Van Nostrand's Sci. Encycl.* (ed. 3) 812/1 High-frequency photography includes both still and motion-picture photography... In high-frequency (slow-motion) cinematography the exposing rate may range from 32 to several thousand frames per sec. *Ibid.* 813/1 High-frequency photography which consists of a number of high-speed photographs made in rapid succession. **1940** *Chambers's Techn. Dict.* 188/1 *High-frequency condenser microphone*, a condenser microphone in which the polarising voltage is alternating at a high radio-frequency, amplitude modulation of which is detected by a de-modulator and used for audio-frequency transmission. **1896** *McClure's Mag.* VI. 414/2 A high frequency current. **1923** E. W. Marchant *Radio Telegr.* ii. 13 When we come to deal with the high-frequency currents that are employed in Radio, their behaviour is quite different. **1925** *Jrnl. Iron & Steel Inst.* CXII. 73 High-frequency heating may be found useful in the forging and heat treatment of steel. **1967** A. K. Osborne *Encycl. Iron & Steel Ind.* (ed. 2) 218/1 The polarity of this field changes many times per second with the alternation of the current necessary for high frequency heating. **1920** *Trans. Amer. Electrochem. Soc.* XXXV. 158 Since the presentation of the above paper a development into commercial form of high-frequency induction furnaces of the oscillatory current type has been actively continued. **1958** A. D. Merriman *Dict. Metallurgy* 125/1 A high-frequency induction furnace is a melting furnace in which currents at a frequency above 500 c./sec. are used to induce eddy currents in the charge, which in turn generate heat in the material. **1958** High-frequency photography [see **high-frequency cinematography*]. **1892** Heaviside *Electr. Papers* II. 193 The high-frequency resistance is independent of the steady resistance, and must be much greater than it. **1951** A. Hund *High-Frequency Measurements* (ed. 2) ii. 96 Any amplitude differences are balanced out by properly chosen high-frequency resistances. *Ibid.* ix. 310 In *all* cases, the high-frequency resistance is that quantity which, when multiplied by the square of the effective value of the high-frequency current, gives the energy dissipated in the conductor. **1930** *Wireless World* 17 Dec. 686/2 (*heading*) High frequency response. **1923** E. W. Marchant *Radio Telegr.* iv. 56 This change in voltage may be passed on..by means of a small high-frequency transformer, to the grid of a second valve. **1931** S. R. Roget *Dict. Electr. Terms* (ed. 2) 146/2 *High frequency treatment*, curative treatment involving the use of interrupted trains of damped high frequency oscillations. **1967** A. K. Osborne *Encycl. Iron & Steel Industry* (ed. 2) 202/2 High-frequency treatment.

Highgate (hai·gĕt). The name of a hill in London, used *attrib.* in *Highgate resin*, a mineral resin similar to copal found in Highgate Hill. Also called *copalin(e)*, *copalite*, *fossil copal*.

1813 T. Thomson in *Ann. Philos.* II. 9 The colour of Highgate resin is of a dirty yellowish light brown. **1815** A. Aikin *Man. Min.* (ed. 2) 64 Fossil Copal. Highgate Resin. Colour light yellowish dirty brown. **1839** *Phil. Mag.* XIV. 91 The Highgate resin has much resemblance to copal. **1898** E. S. Dana *Textbk. Min.* (rev. ed.) iv. 543 Copalite, or Highgate resin, is from the London blue clay. **1968** *Encycl. Brit.* VI. 462/2 *Copalite*.., also termed 'fossil resin' and 'Highgate resin', a naturally occurring organic substance found as irregular pieces of pale-yellow colour in the London clay at Highgate hill.

hi·gh-grade, *a.* and *sb.* [High *a.* 22 a.] **A.** *adj.* Of a high grade or quality; *spec.* **a.** (See quot. 1909.) **b.** Denoting ores rich in metal value; *spec.* in commercial use denoting those which, owing to convenience in situation and transport facilities, can be worked at a large profit.

1878 R. J. Hinton *Handbk. Arizona* 161 The Metallic Accident is a large lode of low-grade ore, with a number of high-grade feeders. **1880** 'Mark Twain' *Tramp Abroad* I. xxiv. 230 Only the few are educated up to a point where high-grade music gives pleasure. **1890** [see High *a.* 22 a]. **1902** *Daily Chron.* 10 Apr. 7/2 Great Britain is becoming very short of high-grade ores. **1907** *Springfield Republ.* 13 May 3 (Advt.), Auction sale of high grade household furniture. **1909** *Cent. Dict.* Suppl., *High-grade*,..specifically, having more than three quarters pure blood: applied by stockbreeders to animals. **1910** *Westm. Gaz.* 4 Jan. 5/2 The high-grade nature of the material and workmanship. **1919** *Empire Rev.* 79 Its products are high-grade steels. **1929** *Daily Express* 7 Nov. 8/4 The high-grade private car. **1932** *Discovery* Nov. 357/2 The institute conducts genetical studies leading to the breeding and selection of high-grade wheats. **1939** R. R. Snapp *Beef Cattle* (ed. 3) xviii. 252 Their color indicates that they are high grade and purebred individuals of strictly beef type. **1940** *War Illustr.* 12 Apr. 360 A view of a Nazi petrol-substitute refinery where the raw material is converted into high-grade spirit suitable for use in the specially designed aeroplane engines. **1946** *Mind* LV. 114 Some human being in his right mind, that is to say, a human agent capable of relatively high-grade action. **1958** *Times Lit. Suppl.* 23 May 279/4 But, unlike other fellow-travellers, he was actually engaged in high-grade espionage.

B. *sb.* **a.** High-grade stock. **b.** (See quot. 1904 and cf. the vb.)

1882 *Rep. Maine Board Agric.* XXVI. 253 High-grades of either breed [Jersey or Guernsey]. 1904 *N.Y. Sun* 14 Aug. 11 One of the pests of gold mining in Colorado is the high-grade, which is a polite term for the ore thief. The term high grades comes from the fact that they steal only high-grade ore.

Hence **hi·gh-grade** *v. intr.* and *trans.*, to steal high-grade ore.

1907 *Westm. Gaz.* 6 June 10/1, I had been 'high grading' in the Vindicator mine. 1923 'B. M. Bower' *Parowan Bonanza* vi. 73 He..could not leave his claims and let Al Freeman..'high grade' his gold the minute his back was turned. 1927 *Blackw. Mag.* June 833/1 In Cobalt, 'high-grading' was vigorously dealt with. 1963 *Time* (Canad. ed.) 18 Jan. 10/2 Some Timmins stores have been known to accept high-graded ore in payment for grocery bills.

high-ha·ndedly, *adv.* [f. High-handed *a.* + -ly².] In a high-handed manner.

1898 N. Munro *John Splendid* xxi. 206 Seven fugitives of the clan that had come so high-handedly through their neighbourhood. 1927 *Daily Express* 26 Oct. 1/2 High-handedly putting a pistol to the heads of his opponents. 1948 A. L. Kroeber *Anthropol.* (rev. ed.) 617 Fred..treated the findings of psychology almost as high-handedly as he did those of prehistory and culture history.

high hat, hi·gh-hat. [High *a.* 1.] **1.** A tall hat; *fig.* a person of affected superiority. Also *attrib.* or as *adj.*, superior, lofty.

1889 in C. W. & P. Cunnington *Handbk. Eng. Costume 19th Cent.* (1959) 309 Is it not considered bad taste to wear anything else but a high hat with a Frock coat? 1899 A. H. Quinn *Pennsylvania Stories* 39 Houston..was under strong suspicion of having worn a high hat out to college that morning. 1923 *N.Y. Times* 9 Sept. vii. 2/1 (Stage Gloss.), High Hat—swelled head. 1924 P. Marks *Plastic Age* 139 Christmas Cove's a nice place; not so high-hat as Bar Harbor. *Ibid.* 196 We're a lot of low-brows pretending to be intellectual high-hats. 1927 *Cleveland Press* 29 Jan., We see no point in assuming a high hat attitude towards what one doesn't know. 1930 *Publishers' Weekly* 22 Mar. 1674 The shop is not 'high hat'. 1931 G. B. Stern *Shortest Night* xvi. 263 That hot-tempered young high-hat. 1932 E. Wallace *When Gangs came to London* xxi. 186 Eddy was getting high-hat..now this new and irritating situation had arisen. 1932 E. Wilson *Devil take Hindmost* xviii. 205 He gets high-hat and speaks with scorn of the Mexicans. 1953 R. Lehmann *Echoing Grove* iii. 199 Forgive me if I sound high-hat—I don't think you can possibly understand it. 1969 in Halpert & Story *Christmas Mumming in Newfoundland* 197 They looked very smart in their trimmed pants, white shirts and high hat with ribbons and tassles.

2. *Photogr.* (See quot. 1930.)

1930 *Sel. Gloss. Motion Pict. Techn.* (Acad. Motion Pict., Hollywood) 16/1 *High hat*, a very low camera stand. 1953 *Times* 2 Nov. 35/2 *High-hat*, low camera support for 'worm's eye' pictures.

3. *Jazz.* A pair of cymbals worked by the foot. Also **hi-hat.**

1934 E. Little *Mod. Rhythmic Drumming* i. xxii. 21 The 'High-Hat' pedal brings the cymbals within reach of the snare drum stroke. Rhythm can be played on these. 1948 *Metronome* Nov. 28 I'd rather use the high-hat as a back beat and break up the bass rhythms. 1949 L. Feather *Inside Be-bop* ii. 58 In the old days the drummer frequently played high hat cymbal for the second and fourth beat accents. 1956 S. Traill *Play that Music* v. 56 Keep your hi-hat cymbals flat. 1956 M. Stearns *Story of Jazz* (1957) xvii. 234 The left foot..played the high-hat, and the right foot exploded the bass drum. 1966 *New Yorker* 11 June 153/1 Barbarin..uses a high hat and ride cymbals to excellent effect. 1966 *Crescendo* Dec. 27/1 A tambourine on the hi-hat..makes a jingle sound every time the hi-hat cymbals come down. 1968 *Listener* 24 Apr. 29/1 An intriguing hi-hat backing from the excellent drummer.

hi·gh-hat, *v.* Chiefly *U.S.* [f. prec.] *intr.* To wear a high hat; to assume a superior attitude. *trans.* To treat condescendingly. So **high-hatted** *a.*

1924 H. C. Witwer in *Cosmopolitan* Apr. 68/1 'Why high hat me?' he complains. 'I'm harmless and I may be able to do you a lot of good.' 1924 H. Crane *Let.* 24 Jan. (1965) 171 It's become fashionable for the high-hatted uptowners now to buy Matisse's paintings. 1925 S. Lewis *Martin Arrowsmith* xxiii. 255 If I like 'em and old Mart high-hatted me, I'd just about come right unto letting him hear the straight truth. 1927 *Sat. Even. Post* 24 Dec. 22/3 What made me so sore..was he thinks she could high-hat me. 1929 E. C. Mersereau *Chicago* 191 Dever's dignity was mistaken by some for 'high-hatting'. 1941 N. Coward *Australia Visited* iii. 15 The true representative American..is unpretentious.. He dislikes being 'ritz'd' or 'high-hatted'. 1941 Belloc *Silence of Sea* xxxi. 193 The Americans..say of a proud man that 'he wears a high hat'... 'If you talk like that,' he was told, 'they will think you are high-hatting them.' 1965 *New Statesman* 7 May 730/2 Her ineffective efforts to make her sons 'high-hat the neighbours' and join the élite.

high-headed (stress variable), *a.* orig. *U.S.* [High *a.* 22 b.] Carrying the head high; proud, arrogant.

1837 *Southern Lit. Messenger* III. 86 One of them high-headed Roanoke planters. 1903 W. B. Yeats *In Seven Woods* 43 And that high-headed even-walking queen. 1909 R. A. Wason *Happy Hawkins* i. 10 The most obstinate, high-headed, bull-intellected thin-skin 'at ever drew down top wages fer punchin' cows. 1955 W. W. Denlinger *Compl. Boston* 112 A dog cannot display a high-headed style without well laid back shoulders.

high jinks: see Jink *sb.*¹ 3.

Highland, *a.* and *sb.* Add: **B.** *adj.* **2. a.** *Highland bonnet* = Scotch cap; *Highland Boundary Fault*, a geological fault extending across Scotland from the Firth of Clyde on the west coast to Stonehaven on the east; also called *Great Highland Fault*; *(West) Highland cattle*, a breed of small cattle from the Highlands, characterized by thick, shaggy hair and long curved horns set wide apart; *Highland dress* (examples); *Highland fling* (see Fling *sb.* 4 a); *Highland games* (see *Game *sb.* 4 d); *Highland honours* (see quot. 1858); *Highland kilt* = Kilt *sb.*; *Highland pony*, one of a breed of ponies originating in the Highlands; *Highland terrier*, a variety of terrier descended from the working terrier of the Scottish Highlands; also called *West Highland white terrier, White West Highlander.*

1760 *Aberdeen Jrnl.* 22 Jan., His Grace..appears every Day in the Highland Dress, which becomes him extremely well. 1771 [see Kilt *sb.*]. 1818 Scott *Rob Roy* II. xiv. 192 The Baillie, thus refreshed, was mounted on a small Highland poney. 1819 *Observer* 25 July 4/4 A parcel of these children dressed in the Highland kilt. 1819 *Edin. Even. Courant* 31 July 1 Lost, A Highland Terrier, that answers to the name of Brogath. 1821 *Edin. Star* 1 Feb., There is something so hearty and rapturous in the Highland honours which follow the toasts. 1822 D. Wordsworth *2nd Tour Scotl.* in *Jrnls.* (1941) II. 373 Boys dressed in their glossiest and best—little ones in Tartan with Highland bonnets. 1834 W. Youatt *Horse* iv. 59 The Highland pony is far inferior to the galloway. 1835 *Chambers's Edin. Jrnl.* 13/2 The popularity which Highland bonnets acquired from the glory of the Scottish regiments at Waterloo. 1834 W. Youatt *Cattle* iii. 66 The striking peculiarities of the Highland cattle. 1844 H. Stephens *Bk. Farm* III. 1274 The West Highland has long been famed in Scotland as a superior breed of cattle. 1846 C. St. John *Short Sk. Highland* xiv. 113 Why do Highland terriers so often run on three legs?.. I never had a Highland terrier who did not run on three legs occasionally. 1853 J. E. Millais *Let.* 7/8 Oct. in M. Lutyens *Millais & Ruskins* (1967) 93 He comes out in the Highland kilt and cap. 1858 Q. Victoria *Let.* 1 Apr. in R. Fulford *Dearest Child* (1964) 83 Alfred and Arthur in Highland dress. 1858 *Illustr. News of World* 12 June 300/3 All the loyal toasts were proposed and drunk with 'Highland honours'—a kind of demonstration which the polite reader may think somewhat ludicrous, since it consists of putting one leg on a chair and another on the table, elevating the right hand to its utmost stretch, and 'draining the wine cup to the very dregs'. 1875 *Encycl. Brit.* I. 369/1 The Kyloes or West Highland cattle. 1893 *Q. Jrnl. Geol. Soc.* XLIX. 354 The grey slate or phyllite, north of the Great Highland Fault. 1901 A. Geikie *Scenery Scotl.* (ed. 3) iii. 64 The line of the Highland boundary fault runs out to sea immediately to the north of Stonehaven. 1911 *Encycl. Brit.* VIII. 376/1 An application which was made (1906) by some of their admirers for separate classification was refused by the Kennel Club, but afterwards it was granted, the breed being classified as the West Highland white terrier. 1922 R. Leighton *Compl. Bk. Dog* xiv. 206 The Poltalloch, or White West Highlander. 1929 J. W. Gregory in Evans & Stubblefield *Handbk. Geol. Gt. Brit.* i. 10 The Middle Devonian age of the Highland Boundary Fault is shown near Loch Lomond. 1937 J. Macdonald *Highland Ponies* 19 For several years the Department of Agriculture for Scotland has been supplying Highland pony stallions for the use of the crofters. 1961 L. D. Stamp *Gloss. Geogr. Terms* 234/1 Geographically the 'Highlands of Scotland' is the name applied to the whole massif north of the Highland Boundary Faults. 1969 R. T. Wilcox *Dict. Costume* 303/2 *Scottish Highland dress*...the kilt; a kilt jacket..the sporran. 1970 *Times* 26 Feb. 44/2 Calvin was the little West Highland terrier she and Grandpa Nye had got on their wedding trip in Scotland in 1937. 1973 *Courier & Advertiser* (Dundee) 26 Feb. 5/1 (Advt.), Cattleman required for suckler herd and pedigree Highland cattle.

b. Resembling, having the characteristics of, or typical of the people of the Highlands of Scotland.

c 1780 J. R. (song title) The Highland character... No effeminate customs our sinews unbrace; No luxurious tables enervate our race. 1787 Burns in *Poems ascribed to Burns* (1801) 53 There's nothing here but Highland pride, And Highland scab and hunger. 1792 — *Poems & Songs* (1968) I. 358 In Heaven itself I'll ask no more Than just a Highland welcome. 1816 Scott *Antiquary* I. ix. 204 Rab Tull keepit a Highland heart, and bang'd out o' bed..he did follow the thing up stairs. 1818 — *Rob Roy* II. v. 94 Ay, he has a kind of Highland honesty—he's honest after a sort, as they say. 1829 *Blackw. Mag.* Sept. 400 Yet still the blood is strong, the heart is Highland, And we in dreams behold the Hebrides. 1871 L. W. M. Lockhart *Fair to See* (1872) I. ii. 58 I'm as Highland as—as—anything. 1897 R. M. Ferguson *Village Poet* 172 They'll not jew us—we're no' sae hieland.

highlander. Add: **2. b.** Highland cattle.

1771 *Caled. Mercury* 17 Aug., One Hundred Cows, mostly Highlanders, laid early on the grass in the spring to fatten. 1787 W. H. Marshall *Rural Econ. Norf.* (1795) II. 381 Highlanders, Scotch cattle of the Highland breed. 1825 Loudon *Encycl. Agric.* § 6122 The other variety of Highland cattle is the Norlands, or North Highlanders. 1834 W. Youatt *Cattle* iii. 64 There is little or no variety of breeds of cattle in the Hebrides. They are pure West Highlanders. *Ibid.* 79 The character of the Highlander must still be, that he will pay better for his quantity of food than any other breed. 1875 *Encycl. Brit.* I. 389/1 They [sc. the Pembrokes] excel the West Highlanders in this respect, that they make good dairy cattle, the cows being peculiarly adapted for cottagers' purposes.

5. *White West Highlander*: see **Highland terrier.*

hi·gh-life, -lifed, *a. U.S.* [High *a.* 16.] Full of life or spirit. (Cf. *high life* s.v. *High *a.* 21.)

1861 *Trans. Ill. Agric. Soc.* IV. 376 A long narrow-headed high-lifed brainless animal. 1902 A. D. McFaul *Ike Glidden* ix. 70 Ike told him..to always drive on the bit, because the colt was a high-life fellow. 1935 *Archit. Rev.* LXXVII. 64/2 He was busy building for 'high-life' America.

high light, hi·gh-light. [High *a.* 10, Light *sb.* 12.] **1. a.** In painting, photography, and cinematography, any of the brightest parts of a subject or a representation of it; often *pl.* Also *attrib.*

1658, 1859 [see Light *sb.* 12]. 1892 A. Brothers *Photogr.* 335 In a portrait, if well lighted, there should be a high light on the forehead and nose, for instance; they are called high lights. 1902 A. Watkins *Photogr.* (ed. 4) 64 The tone D is called the 'high light', for although it is the blackest in the negative it represents white in the original. *Ibid.* 77 It may happen that there is no white part or high light in the subject. 1921 A. Sinclair *Handbk. Photogr.* (ed. 2) 26 To clear up highlights or remove pressure marks from thick bromide papers. 1930 *Sel. Gloss. Motion Pict. Techn.* (Acad. Motion Pict., Hollywood), *High light*, object, scene, or picture having low color saturation, used in connection with tone value of white. 1931 Selden & Sellman *Stage Scenery* 158 In order to give interest to high lights and shadows. 1937 *Bell Telephone Q.* Apr. 110 The painter contributes highlight in the sense of accent. 1951 G. H. Sewell *Amateur Film-Making* (ed. 3) vi. 20 The high-light meter reads the intensity of the light falling upon the subject, instead of the light reflected from it. 1957 R. W. G. Hunt *Reprod. of Colour* vii. 73 A highlight mask is made by contact printing the transparency onto a very high contrast black and white negative material. 1966 *Gloss. Terms Offset Lithogr. Printing* (B.S.I.) 13 *Highlight stop*, a lens aperture used in half-tone photography which has a specific influence on the formation of highlight tones. *Ibid.* 17 *Highlight mask*, a mask used to retain or increase highlight contrast in the reproduction. 1971 K. H. Brown *Father* vi. 61 There were artless scarlet highlights cut from portions of the little boys' outgrown woollen underwear.

b. A bright tint in the hair; that portion of the hair that reflects the most light. Also *attrib.*

1941 *Hairdressers' Weekly Jrnl.* 1 Feb. 155/2 One can bring out high-lights over every shade of hair by the appropriate use of toning rinses. 1966 Cox *Illustr. Dict. Hairdressing* 79/2 *High-light cap*, a plastic head cap with holes in it used for drawing a strand or strands of hair through to bleach without bleaching the other hair.

2. *fig.* A 'bright', prominent, or outstanding feature or characteristic. Chiefly *pl.* Phr. *to hit the high lights*, = *to hit the high spots* (s.v. *High *a.* 21).

1855 *Knickerbocker* XLVI. 40 If we have chosen to speak of the shadows in the fair portrait, we have also respected to point out the high lights. 1905 A. Adams *Outlet* 275 It was the recognized cattle centre of Montana..but devoid of the high-lights which were to come as the trail towns. 1922 A. Waugh *Public School Life* i. 61 It would be filled with high lights, with breathless escapades, with all their high-powered modern... 1923 *Weekly Dispatch* 21 Jan. 12 There is no hitting the high lights when one is not in training. 1927 *Hutchinson's Myst. Story Mag.* Feb. 117 One by one, Dan and I met all the high lights of the town. 1931 *Times Lit. Suppl.* 3 Sept. 657/3 The highlights that are taken from some modern biographies are readable. 1932 *Morning Post* 21 Aug. 11/2 These were the highlights of to-day's practice over the course. 1940 Dalrymple & Robinson *Secret Service Chief* (1961) ii. 23 Flipping the years like pages I see many interesting and instructive highlights—and some funny ones too. 1969 *Times* 10 Feb. 6/5 The highlight of the lunch will be a speech..on the vital relationship between the industrial leader and the executive secretary. *Ibid.* 19 Feb. 25/1 The outstanding highlight of tonight's top soccer matches.

hi·ghlight, hi·gh-light, *v.* [f. prec.] *trans.* **a.** To bring into prominence, to 'feature', to draw attention to.

1934 M. Webster *Dict. Amer. Slang* x. 143 *Highlight*, to give one a prominent place on a program or a leading part in a show. *Highlighted*, marked by; featured by. 1945 Nelson & Wright *Tomorrow's House* iv. 43/2 This seemingly minor problem..highlights what should be the fundamental approach to planning. 1951 *Mind* LX. 23 The view is..made more plausible by highlighting..aspects of scientific procedure. 1952 Granville *Dict. Theatr. Terms* (ed. 2) 97 *Highlight*, to give prominence to an artiste in the billing or newspaper publicity matter, the build-up of a show, or of an individual artiste. 1955 Priestley & Hawkes *Journey down Rainbow* p. xi, Football games, night clubs, TV programmes, hotel festivities, can be ignored and certainly should not be high-lighted by the serious social critic. 1957 *Economist*

19 Oct. 192/1 The genuineness with which each holds the belief was highlit during last week's..interview. 1958 *Listener* 2 Oct. 501/1 Statistics such as these are cold and dull and do little to highlight the tremendous importance of industrial organizations in modern society. 1965 *New Statesman* 30 Apr. 670/3 Thursday's polling is unlikely to highlight any notable national trends. 1967 Singha & Massey *Indian Dances* ii. 42 A *timeras* is used to conclude a section of dance or highlight a portion in the middle.

b. *spec.* To tint or bleach portions of the hair in such a way that they catch and reflect the light. More widely in make-up: see quot. 1952. Also **high-lighting** *vbl. sb.*

1942 D. Powell *Time to be Born* (1943) vi. 132 'You've had your hair dyed'..'Not dyed. High-lighted is the new word,' said Vickey. 1952 Granville *Dict. Theatr. Terms* (ed. 2) *Highlight*, in make-up, to accentuate cheekbones and other features and effectively contrast the shade, on the face. 1959 *Punch* 3 June 752/2 For grey hair there is a treatment..which highlights the hair and dismisses yellow discoloration. 1972 *Vogue* 15 Mar. 3/2 New Pure Pearl-Light Everywhere Colour for highlighting round the eyes.

highly, *adv.* **6.** (Further examples.)

1831 J. S. Mill in *Examiner* 12 June 24 Indebted to men of highly-cultivated intellects. 1836 — in *Lond. Rev.* II. 58 The feelings of a highly-educated and sensitive girl. 1910 *Busy Man's Mag.* July 88/1 He betrays a highly-strung temperament, which is seldom his best. 1950 S. Duke-Elder *Parsons' Dis. Eye* (ed. 14) xxiv. 362 Highly myopic people. 1969 *Jane's Freight Containers 1968–69* 143/1 There is considerable evidence of highly-integrated freight services being developed.

high-muck-a-muck (hai·mʌkə·mʌk). *N. Amer. colloq.* Also **high-you-muck-a-muck.** [app. ad. Chinook Jargon *hiu* plenty + *muckamuck* food.] A self-important person, one who imagines he is more exalted than he is.

1856 *Democratic State Jrnl.* (Sacramento) 1 Nov. 3/1 The professors—the high 'Muck-a-Mucks'—tried fusion, and produced confusion. 1866 'Mark Twain' *Lett. fr. Hawaii* (1967) 32 Not if I was High-You-Muck-a-Muck and King of Wawhoo. 1869 — *Sk. New & Old* (1875) 63 High Muck-a-mucks. 1879 E. S. Wood *Jrnl.* 23 Feb. in *Oreg. Hist. Q.* (1960) [illegible] ..hi you muck a muck and here's your bill of fare. 1920 S. Lewis *Main Street* 117 He looks at me like he wants me to remember he's a highmuckamuck and worth two hundred thousand dollars. 1947 A. Philip *Painted Cliff* 14 J. B. Smith is the high muck-a-muck, the type of the mining business of British Columbia. 1947 *Chicago Tribune* 21 Dec. (Comics section) 8 There's a highmuckymuck in th' radio business vacationin' here, so we gotta be good. 1969 [illegible] 16 Apr. 14/5 Not all the Liberal high muckamucks were so warmly received.

hi·gh-powered, *a.* [High *a.* 22 b.] Having great power or drive (lit. and fig.); forceful, energetic; of good or high quality.

1903 *Daily Chron.* 1 Aug. 3/7 High-priced, high-powered, high-speed motor cars. 1927 'Contact' *Airmen's Outings* p. xv, Modern two-seaters, high-powered, fast, and reliable. 1928 *Daily Mail* 16 Aug. 5/4 This class of fraud does not require so many high-powered salesmen as the old method of selling by personal canvass. 1934 T. S. Eliot *Rock* ii. 51 Does the whole world stray in high-powered cars on a by-pass way? 1936 *Amer. Scholar* V. 85 The schools are failing, with all their high-powered modern pedagogy. 1944 *Living off Land* v. 114 High-powered microscopes. 1957 *Times Lit. Suppl.* 29 Nov. 724/2 American motor-cars that are always 'high-powered'. 1958 *Engineering* 14 Mar. 336/1 Automatic train control was developed..early this century, and, in 1920, a high-powered committee..recommended its general adoption on all British railway lines. 1958 *Spectator* 25 July 138/1 High-powered promotion men. 1960 *News Chron.* 18 Mar. 6/4 The National Executive would..set up a high-powered inquiry into those industries which have already been nationalised. 1969 *Daily Tel.* 5 Feb. 15/1 The girl graduate, however high-powered her degree, is very often unemployable.

hi·gh-ranking, *a.* orig. *U.S.* [Ranking *ppl. a.* s.v. Rank *v.*¹] Of an officer: of high rank. Also more widely: senior, important.

1922 S. Lewis *Babbitt* xvii. 231 He ought to have been a high-ranking officer, he had that natural ability to command. 1924 *Amer. Mercury* II. 173/1 Wary Wade, a high-ranking student at Ohio State, has the right angle on boxing. 1936 *Time* 13 Apr. 47/1 She was one of the high-ranking members of the Soviet ... 1940 *Times* 2 July [illegible]. 1953 *Times* 6 May 7/7 She left him..a son, Vassili, now a high-ranking officer in the Soviet Air Force. 1957 H. Roosenberg *Walls came tumbling Down* 19 We were intended as exchanges against some high-ranking Nazis. 1960 P. H. Reaney *Orig. Eng. Place-Names* vii. 163 Holderness (ERY), 'headland of the hold', a high-ranking officer in the Danelaw. 1961 *New Eng. Bible Acts* xxv. 23 Accompanied by high-ranking officers and prominent citizens. 1964 *Observer* 16 Aug. 1/3 Six sets of the master-key..in a prison..would be available..to a limited number of high-ranking officers.

hi·gh-roller. *U.S. slang.* [High *a.*, ? Roller *sb.*¹ 15.] One who spends extravagantly; one who gambles for high stakes.

1881 *Reinbeck* (Iowa) *Times* 15 Sept. 1/6 California's Speculators who invest large sums are called 'high rollers'. 1903 A. H. Lewis *Boss* xiv. 184 I'd like to learn how you moral an' social high-rollers reconcile yourselves to things. 1909 'O. Henry' *Roads of Destiny* xx. 340 There comes a party of these high-rollers that are always hunting up new places to eat in and poke fun at. 1927 K. Norris *Barberry Bush* vi. 156 'Say,' said Link, slightly drawling the last word, 'that girl is a high-roller, believe me. She's full of the Old Nick!' 1968 L. Black *Outbreak* iii. 20 He's a high roller...A gambler. 1972 *Sunday Mail* (Brisbane) 2 July 19/5 The Hughes places had included some of the chief centres for the big-money gamblers, or 'high-rollers'.

high school: see School *sb.*¹ 1 j.

hi·gh-speed, *a.* [High *a.* 22 a.] **a.** Able or fitted to work or travel at high speed. **b.** Produced by swift processes or machinery. **c.** Produced or caused by, or during, high speed.

high-speed steel, an alloy steel of such toughness and hardness that it can be used for tools cutting so rapidly as to become red-hot.

1873 [see High *a.* 22 a]. 1888 *Lockwood's Dict. Mech. Engin.* s.v., High-speed engines may be considered to embrace any engines making over 200 to 300 revolutions per minute... High-speed belting applies to belts for fans, wood-working machinery, centrifugal pumps, &c., in opposition to those for line and counter, and other slowly driving shafts. *High speed bearings*, bearings whose length exceeds their diameter by from four to six times. 1899 *Daily News* 1 Oct. 2/8 A high-speed dynamo. 1904 *Chambers's Jrnl.* 142/1 High-speed tool-steel lathes. 1908 *Daily Chron.* 6 Feb. 3/1 The crude truths that we learn from the high-speed camera. 1909 *Westm. Gaz.* 11 Mar. 4/4 [illegible]. 1912 *Chambers's Jrnl.* June 367/1 Their [sc. British makers'] introduction of high-speed steel in 1900. [illegible] 1943 *Sci. News* VII. 29 The dynamics of high-speed flow. [illegible]

high-stepping, *a.* (Earlier and later examples.)

1848 Thackeray *Van. Fair* ii. 456 Splendid high-stepping carriage horses. 1973 D. Ramsay *Deadly Discretion* 63 A millionaire..is pretty high stepping for a two-bit dancer.

highstrikes (hai·straiks). *jocular colloq. dial.* or *vulgar.* ¶. Perverted form of Hysterics.

1848 C. Selby *Jacques Strop* ii. 4 Didn't I do the high-strikes famously? 1846 D. Corcoran *Pickings* 149 She's one of the dreadfullest cases of the highstrikes I ever did see. 1899 *Pall Mall Mag.* Jan. 82 [Mrs.] Flounce fell into the highstrikes at tagging the roads after four o'clock. 1914 G. Atherton *Perch of Devil* x. 204 If you don't get us out of this quick I'll have high-strikes. 1922 Galsworthy *Windows* i. 19 Not to go and have the 'igh-strikes after. 1953 A. G. Strong *Othello's Occupation* iv. 102 Mamie put cavers along..I'll give way to the highstrikes. 1957 T. Crosbie *Forgotten Place* xvi. 155 There I was, all in among the bicycles, having hysterics—yes, I mean it, genuine old-fashioned high strikes.

hi·ghtail, hi·gh-tail, *v. colloq.* (orig. *U.S.*). [In allusion to the erect tails of animals in flight.] *intr.* To run (quickly) away, to move quickly. Freq. const. *it.*

1925 *Amer. Speech* I. 149/2 'High-tailed out of there.'.. 'High-tail' comes straight from the plains where a mustang, when startled, erects his tail in a sudden, quick gesture and runs like the wind. So to make a sudden departure is to 'high-tail'. 1928 L. R. Freeman *Nearing North* 157 A string of red-brown bodies hightailing it through the bush. 1930 *Detective Fiction Weekly* 19 Apr. 566/2 We high-tailed it for the hideout. 1953 M. Lowry *Let.* Nov. (1967) 349, I hightailed it thither anyhow, fire-extinguisher in hand. 1958 P. de Vries *Mackerel Plaza* v. 62, I was only always high-tailing it after everything in skirts, that's all. 1959 C. Williams *Man in Motion* iii. 29, I..high-tailed it in the other direction, and ducked into an alley. 1962 *Listener* 22 Mar. 524/2 The two of them high-tailed it for Oldham. 1971 *Nat. Geographic* May 721/2 Suddenly a Chinese goose, honking belligerently, high-tailed straight for me. 1973 *Caribbean Contact* Jan. 2/3 They get the point. Pronto. And high-tail it back home!

hi·gh-tone, *attrib.* or as *adj.* = next.

1897 E. W. Brodhead *Bound in Shallows* 105 They was high-tone lookin' fellers, and I'd like 'em to brag up the house. 1898 *Christian Herald* 19 Jan. 44/4 The infernal delusion that it was not high-tone for women to earn a profitable calling. 1925 F. Scott Fitzgerald *Let.* c. 27 Dec. (1964) 194 Saw Leslie also and went on some very high-tone parties. 1939 L. Hellman *Little Foxes* ii. 66 Everybody going to be high-tone rich. Big rich.

high-toned, *a.* Add: **3.** (Earlier and later examples.) Also, stylish, pretentious. Also *absol.* Cf. *high-tone* (*High *a.* 22 a).

1807 in *Western Pennsylvania Hist. Mag.* (1947) Mar.–June 56 The same editor also states that there were 'no Demo-Republicans in it, all high-toned gentlemen'. 1829 *Virginia Herald* (Fredericksburg) 28 May 2/3 He might be President if he liked; but this high-toned eulogy, he thought highly objectionable. 1854 Geo. Eliot in *Westm. Rev.* Oct. 453 That blending of the high-toned chivalry of Spain with the caustic wit and refined irony of Italy. 1866 *Washoe* (Nevada) *Eastern Slope* 11 Aug. 2/2 It is [necessary] when the high toned meet..for the purpose of cutting off debate, that they should be ready to raise the previous question at any moment. 1876 'Mark Twain' *Tom Sawyer* xxxvi. 272 A robber is more high-toned than what a pirate is..as a general thing. 1888 A. C. Gunter *Mr. Potter* iv. xxiii. 278 The Democratic Party thought Sampson Potter a more high-toned name to run for Congress than Sammy Potts. 1912 Beerbohm *Christmas Garland* p. v, Tripping off the pens of all higher-toned reviewers. 1956 Auden *Old Man's Road*, The smart crescent of a high-toned suburb. 1966 *Listener* 22 Sept. 426/1 The girls whose vicissitudes after graduating from a high-toned American college supply the film's story line.

hi·gh-up, *a. colloq.*, orig. *dial.* [f. High *a.* + Up *adv.*²] In a high or elevated position, high above the ground; also *fig.*, of high place or rank.

1868 J. C. Atkinson *Gloss. Cleveland* s.v., He's some desput high-up chap. 1899 S. MacManus *In Chimney Corners* 155 There was a lot of high-up folk being entertained. 1899 *Westm. Gaz.* 12 June 1/3 The gilt-edged ties of the nigh-up men. 1903 *Book Lover* May 3/1 In a high-up room in the Rue Lafayette. 1918 C. Wells *Vicky Van* ii. 23 A high-up Publican. 1920 J. M. Hunter *Trail Drivers of Texas* 219 Our 'high up' officers were.. somewhere in town. 1934 E. Bowen *Cat Jumps* 74 The skies were disturbed by a high-up swift rustling sigh: the summer birds flying south. 1940 'G. Orwell' *Notebooks* June in *World Review* (1950) June 27 High-up influences in England are preparing for a..sell-out. 1942 A. L. Rowse *Cornish Childhood* iv. 82 Myself sitting in the high-up baby's chair with the little tray attached in front. 1961 *Guardian* 27 May 3/7 His penetrating assessment of high-up American attitudes to Cuba.

Also *colloq.* as *sb.*, a person of high rank or status. Cf. *higher-up.

1929 W. R. Burnett *Little Caesar* vii. vi. 299 Rico got in touch with some of the high-ups. 1939 *News Review* 30 Nov. 13/1 As time went on, the Nazi high-ups took Admiral Raeder's advice. 1946 J. Richardson *Phoney Phleet* 150 Two high-ups, lordly ones, or nobs. 1949 'M. Innes' *Journeying Boy* i. 6 Only the high-ups had their heads substantially above the soapsuds. 1969 Auden *City without Walls* 84 How golden-tongued was Socrates, Who always spoke the truth, But high-ups do not like to think. 1971 *Physics Bull.* Dec. 735/2 Whitten and Poppoff, both high-ups in NASA's Ames Research Center, have filled the gap admirably despite their lack of academic background.

highveld (hai·felt, hai·velt). [Partial transl. of Afrikaans *hoëveld*, f. *hoog* high + Veld.] The inner plateau of the subcontinent of South Africa, which is from 5,000 to 6,000 feet above sea-level.

1878 A. Aylward *Transvaal To-day* 18 The High-veld... The large, bare, but healthy elevated plateau—the great watershed of the Transvaal. 1905 J. W. Gregory in *Rep. Brit. Assoc.* 399 The old rocks that form the foundation of the present high veldt of Rhodesia. 1906 Rider Haggard *Benita* v. 63 They were on the Transvaal high-veld. 1907 P. Fitzpatrick *Jock of Bushveld* 223 For perhaps a week the towering bulwarks of the Highveld were visible as we toiled along. 1954 D. D'Ewes *Mydorp* 51 Sirius and the Southern Cross.. shining with the ecstatic brightness that only the high-veld sky can offer. 1961 L. van der Post *Heart of Hunter* i. ii. 45 My first memories are of the incomparable starlight of the high-veld of Southern Africa and the far sea-sound that goes with it. *Ibid.* iii. xii. 161 The light of a high-veld sunset. 1971 *World Archaeol.* III. 178 The highveld reaches an altitude of 5,500 feet at Johannesburg. 1972 *Stand. Encycl. S. Afr.* V. 522/1 *Highveld*, Tvl. and O.F.S... It lies about 1200 to 1800 metres above sea-level, more or less between 26° and 30° S.

high-water mark. Add: **c.** *jocular colloq.* A dirty mark showing the limit to which a person has washed.

1886 in Baumann *Londinismen.* 1899 *Daily News* 31 Jan. 6/4 The high-water mark is plainly visible above a tattered scarf tied loosely round his neck.

highway. Add: **1. c.** In allusion to Matt. xxii. 9, 10, Luke xiv. 23.

1843 H. Bonar *Hymn*, '*Go labour on*' vii, Go forth into

the world's highway, Compel the wanderer to come in. 1898 *Daily News* 6 Oct. 6/7 The South London officials of the Salvation Army have..been..gathering together of late from the highways and by-paths of Lambeth those who..are entitled to be ranked as 'Hooligans'. 1908 *Westm. Gaz.* 27 Jan. 1/3 It is highways and hedges work, and we shall need van for van and lecturer for lecturer. 1948 A. J. Toynbee *Civilization on Trial* vi. 111 The United States—who had previously not only welcomed European immigrants but whose employers of labour had sought them in the highways and hedges of Europe and compelled them to come in.

2. c. *Computers.* (See quot. 1962.)

1949 D. R. Hartree *Calculating Instruments & Machines* (1950) viii. 107 A number of source-gates..on the right, and a number of destination-gates on the left, are connected by a single bus labelled 'Highway'. In this bus there is a further gate, labelled 'transfer gate', which exercises the main control over transfer of words between the various sources and destinations. *Ibid.* 108 An instruction..opens the corresponding S-gate and D-gate through selecting circuits; then at the appropriate time the main transfer gate in the highway opens. 1962 *Gloss. Automatic Data Processing Terms* (*B.S.I.*) 87 *Highway*, a major path along which signals travel from one of several sources to one of several destinations. 1964 F. L. Westwater *Electronic Computers* i. 7 The number..can either be allowed to pass on to one of the so-called 'highways' in the computer or it can be obliterated. *Ibid.* 9 Numbers are driven out on to the highways serially by admitting electronic pulses to the registers.

4. *highway robber* (later example), *robbery* (freq. in trivial use of something that is exorbitantly expensive); **Highway Code,** in Britain, an official publication containing the 'rules of the road', esp. for motor vehicles; first published in 1931, it has appeared subsequently in revised editions. Also *transf.*

1930 *Act* 20 & 21 *Geo. V* c. 45 § 1 The Minister shall..prepare a code (in this section referred to as the 'highway code') comprising such directions as appear to him to be proper for the guidance of persons using roads. 1935 *Archit. Rev.* LXXVIII. 110/3 The free issue of the Highway Code to all of his Majesty's lieges who have come (or are supposed to have come) to years of discretion is a portent. 1958 *Listener* 20 Nov. 835/1 The importance of reading the Highway Code. 1959 *Daily Tel.* 1 July 1/3 Highway Code revised for motorways. 1970 'A. Gilbert' *Death wears Mask* x. 166 And then there's the Highway Code. 1972 *Listener* 23 May 375/3 Here was a book that..had broken the Highway Code of conventional English culture. 1904 Kipling *Traffics & Discov.* 330 The other six hundred [acres] are nearly all let to folk who knew my folk..but this Turpin is a new man—and a highway robber. 1778 J. Wedgwood *Let.* 19 Mar. (1965) 218 On Monday night last there was three highway robberies between this place and Newcastle. 1853 Mrs. Gaskell *Cranford* x. 181 She indeed inclined to the idea that, in some way, the French were connected with..the burglaries and highway robberies. 1886 *Lantern* (New Orleans) 3 Nov. 2/3 Highway robbery is no name for it. 1920 A. Huxley *Limbo* 122 The organized highway robbery of Red Cross collectors. 1941 'R. Crompton' *William does his Bit* ii. 36 'I've heard people talk about highway robbery quite lately.' 'Yes, but they only mean people chargin' too much for meals,' explained Ginger, 'not the old sort.' 1967 J. B. Priestley *It's Old Country* ix. 107 Nothing on the wine list..under two-pound-ten. Highway robbery by candlelight.

highwayman. Add: **3.** Used *attrib.* or in the possessive designating a type of long loose coat or cape suggestive of the kind once worn by highwaymen.

1901 *Daily Chron.* 9 July 3/5 Sarah Bernhardt..looked very striking in a wonderful gown..half-concealed by a long paletot of white silk, made in the 'highwayman' shape, with a number of natty little capes. 1960 *Guardian* 29 July 7/1 A very dashing full length highwayman coat..with a huge cape collar. 1966 *Vogue* Nov. 88 (*caption*) Vogue's adventurers wear..highwaymen's coats.

highwood (hai·wud). [f. High *a.* + Wood *sb.*¹; cf. *high forest* (*High *a.* 21) and G. *hochwald.*] A forest of tall trees.

1900 J. Nisbet *Our Forests & Woodlands* iii. 105 Data are not yet available to indicate anything like definitely what rotation of oak in highwoods will prove most remunerative. 1904 G. A. B. Dewar *Glamour of Earth* v. 83 He turns bravely to the dripping highwood, to cut and lay in lands or rows the hazel and oak stems. 1905 J. Nisbet *Forester* I. ii. ii. 346 Highwood..woods..are not cleared..until..maturity. 1959 E. Pound *Thrones* cvii. 110 High-wood is called saltus.

higlif (iglif). [Repr. Fr. pronunc. of 'high life' (High *a.* 5 in Dict. and Suppl.).] High society.

1911 'I. Hay' *Safety Match* xiv. 222 That variegated cosmos which..Gallic students of British sociology term 'Le Higlif'. 1925 A. Huxley *Along Road* iv. 256 English *Higlif*, as seen through the eyes of an Italian touring company, was worth coming for. 1960 *Harper's Bazaar* Oct. 141/1 Token names which evoke romance or the longed-for *higlif*. 1961 *John o' London's* 24 Aug. 235/2 It sees through this neurotic *hig-lif* caper to the fear and insecurity underneath.

hi-hat, var. *High hat 3.

hijack (hai·dʒæk), *v.* orig. *U.S. slang* (now passing into general use). Also **hi-jack,** (formerly) **highjack.** [Origin unknown.] *trans.* To steal (contraband or stolen goods) in transit, to rob (a bootlegger or smuggler) of his illicit goods; to hold up and commandeer (a vehicle and its load) in transit; to seize (an aeroplane) in flight and force the pilot to fly to a new destination. Also *transf.*

1923 *Lit. Digest* 4 Aug. 51/3 'I would have had $50,000,' said Jimmy, 'if I hadn't been hijacked.' 1927 'J. Barbican' *Confess. Rum-Runner* xvii. 181 So we landed the cargo as quickly as we could, and took the chance of the cargo being seized or hijacked on shore. 1936 E. Ambler *Dark Frontier* xi. 178, I still don't see how we're going to high-jack Groom's men. 1936 J. G. Brandon *Pawnshop Murder* v. 47 Some one else has been on to this deal, given him the works and hi-jacked the stuff. 1959 A. W. Sherring *Tip Off* i. 8 A stack of old banknotes leaves Central Post Office at three this afternoon. The van will be hijacked. 1961 *John o' London's* 12 Oct. 423/3 The plot concerns an attempt to hi-jack a million dollars. 1963 *Times* 4 Sept. 5/1 A lorry driver was beaten over the head and his load of cigarettes worth nearly £50,000 hi-jacked near Isleworth. 1968 *Daily Tel.* 7 Dec. 17/1 One of our planes with 35 on board was hi-jacked and flown to Cuba. 1970 *Daily Tel. Mag.* 16 Jan. 17/2 When a virus enters a cell it hijacks it, and makes it do what it wants.

So **hi·jacking** *vbl. sb.* and *ppl. a.* Also **hi·jack** *sb.*, an instance of hijacking; also *attrib.*

1923 *Lit. Digest* 4 Aug. 55/1 So much for hijacking on the high seas. 1924 *Daily Mail* 22 Dec. (N.Y. Corresp.), The duties of American coastguards are confined to seizing rum-ships; they cannot seize a Hi-Jacking ship unless it has pirated. 1927 *Observer* 16 Oct. 15/1 Play-goers here are not interested in the humours and results of Prohibition, boot-legging, and hi-jacking. 1928 *Hearst's International* Aug. 72/1 Some hi-jacking exploit. 1929 *Daily Express* 21 Jan. 2/7 A gang of 'hijack boot-leggers' who had forced a restaurant owner to pay them £1,000. 1966 *Times* 22 Sept. 1/2 A £7,000 reward has been offered by an insurance company for information about the hi-jacking of two lorries. 1967 *Listener* 10 Aug. 179/1 The aerial hijacking of Moise Tshombe was commissioned..by the Congolese Government. 1968 *Sun* 14 Nov. 5 Hi-jack figures released by Scotland Yard. 1968 *Daily Tel.* 7 Dec. 17/1 Our airline has been hit harder by hi-jacks than any other. 1972 *Guardian* 17 Nov. 14/1 The problems of becoming a hijack haven have made their mark.

hi·jacker. Also **hi-jacker,** formerly occas. **highjacker.** [f. prec. + -er¹.] One who hijacks (illicit liquor, a vehicle, an aeroplane, etc.).

1923 *Nation* 11 July 36 There was, of course, the rush of adventurers, oil promoters, highjackers (an oil-region term for murderous robbers). 1925 *Times* 26 Oct. 13/7 A shooting affray between bootleggers and 'hijackers' (men who prey on bootleggers) took place..in a lodging-house on the west side of New York. 1936 J. G. Brandon *Pawnshop Murder* v. 43 Any lurking hi-jacker who might..have acquired an inkling of what he was carrying. 1969 *Guardian* 9 Sept. 8/1 The hijackers have put the lives of about 10,000 air passengers at risk.

‖ **hikayat** (hikai·yat). Also **Hikaiat, Hikayet.** [Malay, ad. Arab. *ḥikāya* story, narrative.] In classical Malay literature, a prose narrative combining romance and biography.

1808 *Asiatick Researches* X. iii. 177 The Cheritra or Hikaiat..is more generally written in prose, but frequently intermixed with verse. 1839 T. J. Newbold *Straits of Malacca* II. xiv. 317 From Arabia and Persia, great part of that class of compositions termed Hikayet..are borrowed. 1887 H. N. van der Tuuk in *Misc. Papers relating to Indo-China* II. viii. 3 The Malay author says in the opening that his work..gives a great many beautiful tales in the beginning, and afterwards the tale named *Hikayat Pandawa Jaya*. 1894 N. B. Dennys *Descr. Dict. Brit. Malaya* 152 There are several *Hikayats* extant. 1900 W. E. Maxwell in W. W. Skeat *Malay Magic* ii. 16 This..Malay myth occurs in the 'Sri Rama', a Malay prose *hikayat*. 1964 M. Taib bin Osman in Wang Gungwu *Malaysia* iii. xv. 211 In the written romances, we find princes and princesses bearing hindu and muslim names, hindu gods, muslim prophets and local heroes, indigenous beliefs, hindu ethics and muslim teachings: all blended together in the *hikayat* or story.

hike (haik), *v. colloq.* orig. *dial.* and *U.S.* Also †**hyke, heik.** [Of obscure origin. Cf. *Hoick *v.*¹

A possible early example of this word may be seen in the following: 1736 *Applebee's Weekly-Jrnl.* 17 July 2/1 Gowing..stood Centry 'till the Cargo amounted to as much as they could conveniently yike off with.]

1. *intr.* **a.** To walk or march vigorously or laboriously. **b.** To walk for pleasure; to go for a long walk, or walking tour, *spec.* in the country. Also, to travel by any means.

1809 S. Wesley *Lett.* (1875) 32 Adieu for the present,—we must Contrive one more Pull at Surry before I hyke over to Staffordshire. 1825 J. Jennings *Observ. Dial. W. Eng.*, *To hike off*, to go away; to go off. Used generally in a bad sense. 1868 S. Hale *Lett.* (1919) 39 This day we moved over to the Thebes side and were to have done Karnak..but..I was really sick with heiking. *a* 1872 J. M. Bailey *Folks at Danbury* (1877) 55 You've got to hike aroun', and fling some style inter the victuals. 1884 *Daily Tel.* 2 Feb. 3/1 (Farmer), We three, not having any regler homes..hike about for a living. 1886 S. Hale *Lett.* (1919) 157 You see the Churches can't do much, but Mr. Warner is wild to be heiking about. *a* 1902 *Scribner's Mag.* (Webster 1902), It's hike, hike, hike (march) till you stick in the mud, and then you hike back again a little slower than you went. 1904 *Chicago Evening Post* 23 Aug. 7 These girls had hiked up the dizzy trail along the face of Glacier to the summit. 1910 S. E. White *Rules of Game* iii. xii, 'I'm going to hike out before breakfast,' said he before turning in, 'so if you'll just show me where the lantern is, I won't bother you in the morning.' *Ibid.* v. viii, No animal in its senses would hike uphill and then down again. 1920 *Contemp. Rev.* Sept. 341 To take stick and pack, and 'hike' away from these cities for hundreds or thousands of miles. 1926 *Glasgow Herald* 25 Aug. 8 Guides in gay girlhood will hike through the hollow. 1927 A. Conan Doyle *Case-bk. S. Holmes* 149, I told him I was a busy man and could not spend my life hiking round the world in search of Garridebs. 1936 F. Clune *Roaming round Darling* xi. 101 Upon returning to Sydney I hiked out to Watson's Bay. 1937 *Amer. Speech* XII. 162/1 High school students have used *hike* to mean going to a play spot, either by automobile, by hayrack, or on foot. 1971 *Sci. Amer.* June 16/3, I occasionally take time off to hike in the mountains and ski.

2. *trans.* **a.** To force to move or go; to convey forcibly or laboriously; to pull *on, up, over*, etc.; to 'drag out'; to increase (a price, etc.). Also *fig.*

1867 S. Hale *Lett.* (1919) 32 Our side-saddles were *heiked* on to them. 1869 *Punch* 9 Jan. 8/2 If they finds any sitch thing as a jemmy about yer,..they'll hike yer off to be tried for intendin' to commit a felony. 1870 F. P. Verney *Lettice Lisle* xxiii, I'd like to hike out the whole boiling o' um. 1886 S. Hale *Lett.* (1919) 163 Tuesday he heiked us all forth early in the morning to the lake. 1899 *Strand Mag.* Apr. 454/1 We'll join hands end lay ourselves flat on the rock so thet you can hike your head over, and look all you want to. 1904 *Topeka Capital* 10 June 4 City Center kept the price of ice cream sodas at five cents until the State Sunday School convention struck town, and then the scale was hiked to ten cents. 1915 T. Burke *Nights in Town* 119 Two barearmed ladies, with skirts hiked up most indelicately behind them. 1921 *Chambers's Jrnl.* Dec. 835/2, I hiked him into a taxicab. 1927 *Blackw. Mag.* July 11/1 We flitted across the road like ghosts in the moonlight, hiking our equipage, and deposited same at the door of a wooden inn. 1929 W. P. Ridge *Affect. Regards* 117 Saw where I was paying attention..and then barged in... And apparently managed to hike me out! 1969 C. Irving *Fake!* (1970) xii. 147 To provide a testimonial which..was also so classy that it hiked the price to its limit, they hit on a scheme. 1973 *Observer* 22 July 13/1 The Bank of England hiked its minimum lending rate..to 9 per cent.

absol. *a* 1902 Kipling (Webster 1902), If you persist in heaving and hiking like this.

b. *intr.* To work upwards out of place. Const. *up.*

c 1873 Schele de Vere *MS. Notes* 488 (D.A.E.), What makes y[ou]r dress hike up so? 1890 *Amer. Dialect Notes* I. 61 The curtain hikes or hikes up. 1902 G. H. Lorimer *Lett. Merchant* ix. 119 We boys who couldn't walk across the floor without feeling that our pants had hiked up till they showed our feet to the knee,..didn't like him. 1948 *Sat. Even. Post* 4 Dec. 127/2 When I sit down, it hikes up.

Hence **hiker** (hai·kəɹ), one who hikes or goes on a hike; **hi·king** *vbl. sb.* (also *attrib.*).

1901 *Princeton Alumni Weekly* 518/1 Here I got my first chance at 'hiking' in the Philippines. 1913 F. H. Harris *Dartmouth out o' Doors* 32 While building strong physiques for themselves, the cross-country 'hikers' are providing for happier possibilities for the generations to come. 1923 *T. Eaton & Co. Catal.* Spring & Summer 51 New styles for misses, including the hiking suit and costume skirt. 1926 *Daily Colonist* (Victoria, B.C.) 20 July 18/4 Women's Tweed Hiking Breeches..of good grade materials, smart fawn and grey mixtures; well tailored and buttoned at the knee. 1927 *Daily Express* 24 May 3/5 We [of the Camping Club] have 3,000 members... Most of these are solitary 'hikers', who carry all their kit with them. 1930 *Times Lit. Suppl.* 11 Sept. 710/4 A special kind of traveller, belonging to the class of 'hikers'. 1931 *Daily Tel.* 21 Jan. 8/6 'Hikers'' Hostels at 1s. a Night. 1931 *Times Lit. Suppl.* 7 May 368/1 The sturdy young 'heroes' who accompanied him on the hiking adventures. 1931 *Daily Tel.* 21 May 16/2 The widespread hiking movement in Germany and other Continental countries. 1959 M. Shadbolt *New Zealanders* 71 He was used to hiking. 1972 *L. L. Bean, Inc. Catal.* (Freeport, Maine) Spring 8 *Sierra club cup* for hikers and campers. 1972 *Sci. Amer.* July 13/1, I enjoy cross-country skiing in the winter and hiking and bicycling the rest of the year.

hike (haik), *sb. colloq.* orig. *dial.* and *U.S.* Also †**heik.** [f. prec.] **1.** A vigorous or laborious walk; a tramp or march; a walking tour or expedition undertaken for exercise or pleasure. Also *fig.* *On hike*, on the tramp. hiking.

1865 S. Hale *Lett.* (1919) 15 I've been engaged this week in a pecunious *heik*; to wit, getting money from the ladies of the Parish to get a new gown for Dr. Hedge. 1868 *Ibid.* 45, I ascended the Grand Pyramid, Lucretia got half-way..and Susie didn't try. It is a fearful heik. *a* 1902 *Scribner's Mag.* (Webster 1902), With every hike there's a few laid out with their hands crossed. 1903 S. E. White *Forest* ii. 18 All other utensils belong to permanent camps, or open-water cruises,—not to 'hikes' in the woods. 1907 R. W. Service *Songs of Sourdough* (1908) 59 And I burrowed a hole in the glowing coal, and I stuffed in Sam McGee. Then I made a hike, for I didn't like to hear him sizzle so. 1907 R. Beach *Barrier* (1908) iv. 53 He's the feller that killed the gold-commissioner. Of course that put him on the hike again. 1916 H. L. Wilson *Somewhere in Red Gap* ix. 369 What's the matter with him and Lon taking a swift hike down to New York? 1921 *Outward Bound* June 10/1 Chinese Boy Scouts..on 'hike' on the veldt of South Africa. 1921

Blackw. Mag. Aug. 262/1 Dempsey had passed the afternoon in a 'limbering-up hike'. 1932 *News Chron.* 28 Apr. 1/2 He then began a non-stop hike up and down the corridors of the hospital that had lasted till late this afternoon. 1970 H. D. Corbin *Recreation Leadership* (ed. 3) xxiii. 333 The craving for adventure can be nurtured by a hike or an exploration perhaps more than by any other activity.

2. An increase (in prices, wages, etc.). Chiefly *N. Amer.*

1931 *Kansas City Star* 5 Aug., The hike was occasioned by the fact that cigarette butts..are now only a half inch. 1948 *Herald-Press* (St. Joseph, Mich.) 14 Aug. 3/1 There is enough unfilled demand for new cars to absorb a lot more price hikes. 1966 *Economist* 28 May 986/1 A wave of spending at the end of last year in anticipation of hikes in indirect taxes. 1968 *Observer* 28 Jan. 12/3 A 7.25 per cent price hike in two months. 1969 *Eugene* (Oreg.) *Register-Guard* 3 Dec. 1A (*heading*) Senate votes hike in tax exemptions.

Hilaria (hilē͝ə·riă). [L., neut. pl. of *hilaris* Hilarious.] A festival in honour of Cybele, celebrated at the vernal equinox. (See quots.)

1738 Chambers *Cycl.*, The *Hilaria* were solemnized with great pomp, and rejoicing. 1842 W. Smith *Dict. Gr. & Roman Antiq.* 482/2 The hilaria were..either private or public. Among the former..the day on which a person married, and on which a son was born; among the latter, those days of public rejoicings appointed by a new emperor. 1907 L. R. Farnell *Cults Gk. States* III. 301 The Attis-festival of the Hilaria. 1968 *Encycl. Brit.* II. 147/1 *April Fool's Day*..resembles the Hilaria of ancient Rome (March 25) and the Huli festival of India.

Hilary. Delete 'At Oxford now more generally called *Lent term*'.

hill, *sb.* Add: **1. b.** *hill and dale*: also, applied to any markings or groovings likened to hills and dales; *spec.* used *attrib.* to denote that manner of making gramophone records, or the records themselves, in which the undulations are cut in a vertical plane by the recording stylus. Also, applied to the alternating ridges and hollows of waste rock, etc., which are created by open-cast mining or ironstone working; also *attrib.*

1918 in Webster Add. 1929 Wilson & Webb *Mod. Gramophones* ii. 34 This form of record has several advantages over the hill-and-dale cut. 1931 *News Chron.* 20 Mar. 15/2 A graph, whose hills and dales represent maximum and minimum velocity of each of a series of strokes. 1949 *Hansard, Commons* 6 Dec. 1835 The whole countryside is disfigured by deep cuttings and large tracts of what is known as hill and dale—impassable areas of heaped limestone. *Ibid.* 1844 We do not really know enough about hills and dales to be quite satisfied in all cases. 1964 A. Nelson *Dict. Mining* 218 *Hill-and-dale formation*.., a term applied to the ridges and hollows along the surface of dumped material (usually overburden) at an opencast mine.

e. *over the hill*: having passed the prime in professional ability, physical beauty, etc. Chiefly *U.S.*

1950 *N.Y. Herald Tribune* 6 Dec. 35/2 He has lost his punch... He's a lot farther over the hill than I was when I hung up the gloves in 1927. 1952 M. R. Rinehart *Pool* xxxii. 259 The flawless skin goes, the lovely eyes fade, and she knows she is over the hill. 1957 I. Cross *God Boy* (1958) xxiii. 197 As they say about boxers who are getting on in years, she is over the hill. 1962 *N.Y. Times Bk. Rev.* 17 June 20/3 Must you feel 'over the hill' after 40? 1972 H. Kemelman *Monday Rabbi took Off* ii. 24 When a rabbi gets to be around fifty, his chances of getting another job are not so good. He's like over the hill.

3. b. (Further examples.) Also, the cluster of plants on level ground. Cf. *a hill of beans* (*Bean *sb.* 6 e).

1775 B. Romans *Nat. Hist. Florida* 120 A man ought to go through the field, and pull up those plants that look least promising leaving only three plants in each hill. 1854 *Trans. Pennsylvania State Agric. Soc.* 79 The best corn planter..marks the ground so as to keep the hills in rows in all directions. 1873 J. H. Beadle *Undevel. West* 570 Each field..contained some three hundred hills of corn. 1884 H. Butterworth *Zigzag Journ. Western States* 42 Jerry was working like a beaver, and only three hills of potatoes to the square now. 1964 A. H. Burgess *Hops* vi. 82 If rooted sets..are unobtainable, cuttings can be used for planting the [hop] garden. When this is done two or three cuttings should be planted at each hill.

d. *Her.* A charge representing a hill, usually vert.

1828 in Berry *Encycl. Her.* I. 1889 Elvin *Dict. Her.* p. lii, Three Hills, as in the arms of Brinckman. 1966 Scott-Giles & Brooke-Little *Boutell's Heraldry* (rev. ed.) 301 *Hill, or Hillock*, a green mount.

e. A nitro-glycerine factory.

1897 *Pearson's Mag.* IV. 150/2 You have now reached the bottom of the 'hill'—all nitro-glycerine factories are called 'hills'. 1921 *Dict. Occup. Terms* (1927) § 149 *Nitroglycerine hillman*, an explosive worker engaged on repetition work in nitroglycerine manufacture.

4. a. *hill-brow, -name, -slope, -wash.* **b.** *hill-station* (later examples).

1913 D. H. Lawrence *Love Poems* 40 The warm hay from The hill-brow. 1954 J. R. R. Tolkien *Fellowship of Ring* 146 The north end of the hill-brow. 1922 E. Ekwall *Place-Names Lancs.* 28 Very few hill-names, apart from those which have given names to places, are found in early sources. 1872 Tennyson *Gareth & Lynette* 181 The damp hill-slopes. 1908 *Daily Chron.* 14 May 5/4 On the north side of the valley the hill-slopes are fairly open. 1919 J. Masefield *Reynard* 97 The hill-slope [seemed] steeper. 1958 P. Kemp *No Colours or Crest* iv. 43 March-Phillipps had served in India..where he had experienced..the glitter of social life in various hill stations. 1969 *Filmfare* (Bombay) 1 Aug. 31/4 Once, while shooting at a hill station, the entire R. K. unit was staying in a quaint hillside hotel. 1936 *Nature* 29 Aug. 357/2 A hill-wash, some 11 ft. in thickness, contained large numbers of flint artefacts. 1958 F. E. Zeuner *Dating Past* (ed. 4) 158 The Middle Older Loess of the section is a complex of loessic hillwash material derived from higher up the slope.

c. Also, pertaining to the rearing and tending of sheep in hilly country.

1749 H. Purefoy in *Purefoy Lett. 1735–53* (1931) I. 162, I desire you will buy for mee ten ewes & lambs of the little short-legged horned Hill Sheep. 1841 *Penny Cycl.* XXI. 358/1 The average weight of the fleece..is now at least 3 lbs. in the hill-sheep, and nearly 4 lbs. in the lowland-sheep. 1841 Thoreau *Jrnl.* 13 Feb. (1962) I. 211/2 His hill-farm is poor stuff. 1886 C. Scott *Sheep-farming* 101 Hill farms... Hill stocks should always be fixtures on the farm. *Ibid.* 103 Hill sheep farming. *Ibid.* 116 A successful hill lambing depends very much upon.. the condition of the ewes at that period. Hill ewes are never in too high condition. *Ibid.* 123 The science of hill-herding. 1946 *Act* 9 & 10 *Geo. VI* c. 73 § 1 'Hill farming land' means mountain, hill and heath land which is suitable for use for the maintenance of sheep of a hardy kind but not of sheep of other kinds, or which by improvement could be made suitable. 1962 *Coast to Coast 1961–62* 13 Nicholas wanted to..gambol as senselessly as the new lambs in the hill paddocks. 1963 *Times* 13 Feb. 14/7 Should lamb and hill wool continue to be treated as special cases on social grounds?

d. *hill-culture, -kangaroo, -town* (further examples), *-tribe, -village.*

1936 *Discovery* June 179/2 This midden culture, which we call Sotho, differs..in nearly every respect from our Shona or Hill culture. 1950 Webster Add., *Hillculture*, a system of agriculture utilizing erosion-preventing crops that are ecologically and economically best suited for sloping or hilly (often sub-marginal) land. 1935 H. H. Finlayson *Red Centre* (1952) 40 The short-limbed, broad-chested, sturdy, hill kangaroos or euros. 1944 *Living off Land* ii. 27 A number of men are required to capture the wallaby or the euro (hill kangaroo). 1911 R. Brooke *Poems* 24 Out of the white hill-town, High up I clamber. 1972 W. Garner *Ditto, Brother Rat!* xxiii. 172 'Tell me about Vauban.'..'A dilapidated little hill town.' 1870 *Brewer's Dict. Phr. & Fable* 406/1 *Hill tribes*, the barbarous tribes dwelling in remote parts of the Deccan or plateau of Central India. 1946 *Nature* 6 July 35/1 Any hill-tribe tends to lead a more or less segregated life. 1972 *Nat. Geographic* Feb. 271/1 Although often labeled as a 'hill tribe', Thailand's Karens occupy both upland and lowland villages. 1905 *Daily Chron.* 9 Oct. 4/2 The picturesque little hill-village of Moniaive [in Dumfriesshire]. 1947 *Geogr. Jrnl.* CX. 79 By no means all hill villages..are in the pastoral zone and many are associated with a fully-developed system of common arable fields.

e. *hill-born* adj.; *hill-set a.* (after Matt. v. 14), 'set' or situated on a hill.

1911 E. Pound *Canzoni* 21 A swelling turbid sea Hill-born and tumultuous. *a* 1963 C. S. Lewis *Poems* (1964) 35 The hill-born, earthy spring,..The ripe peach from the southern wall still hot. 1906 *Westm. Gaz.* 16 June 12/2 Brown-roofed, hill-set villages. 1906 *Macm. Mag.* July 695 Ruler of his tiny hill-set principality. 1907 *Westm. Gaz.* 9 Aug. 2/4 Our hillset house of prayer.

f. hill-climb, the action of climbing hills, esp. as a test for motor vehicles; so **hill-climbing,** also *attrib.*; **hill-engraver,** in map-making, one who makes the representations of elevations on an engraved plate; so **hill-engraving; hill-map,** a map showing elevations; **hill-spur** (see Spur *sb.*¹ 11).

1905 *Westm. Gaz.* 6 June 4/2 At the hill-climb on May 27. 1907 *Ibid.* 26 Feb. 4/2 The club will organise competitions, hill-climbs, club-runs, and so on. 1971 I. Wagstaff in J. Walton *Castrol Guide Motoring Sport* x. 70 The object of a hill climb is for drivers..to reach the top of the hill in a shorter time than any other competitor. 1637 Shirley *Hide Parke* iv. sig. G2, Hill climbing white-rose, praise doth not lacke. 1861 [see Dict. 4 e]. 1900 [see *Car *sb.*¹ 1 e]. 1904 *Peel Guardian & Chron.* 23 Apr., The venue of the hill-climbing contest has not been fixed. 1908 *Westm. Gaz.* 19 Mar. 4/1 Its smooth and faultless running and wonderful hill-climbing abilities. 1931 [see *Decoke *v.*]. 1900 *Geogr. Jrnl.* June 589 The employment of hill-engravers, who are, as already stated, so much required for the completion of the hill-engraving of the 1-inch map. *Ibid.* 578 Progress of the 1-inch Hill Map of the United Kingdom. 1871 W. Morris in Mackail *Life* (1899) I. 253, I went about looking for game about the hill spurs. 1887 Meredith *Ballads & P.* 88 Down the hillspurs.

hi·ll-bi:lly. Chiefly *U.S.* Also **hillbilly, hilly-billy, -billie.** [f. Hill *sb.* 4 + Billy¹ 2.]

1. A person from a remote rural or mountainous area, *esp.* of the southeastern U.S. Also *attrib.* and *transf.*

1900 *N.Y. Jrnl.* 23 Apr. 2/5 In short, a Hill-Billie is a free and untrammelled white citizen of Alabama, who lives in the hills, has no means to speak of, dresses as he can, talks as he pleases, drinks whiskey when he gets it, and fires off his revolver as the fancy takes him. 1911 *N.Y. Sun* 10 Aug. (Funk), These two were farmers' boys and hillbillies and jayhawkers. 1932 E. Wilson *Devil take Hindmost* xxii. 236 A coarse-spoken frank humorous old hillbilly talking to neighbors. 1933 *Amer. Speech* VIII. iii. 27/2 *Hill Billy*, a rube or uncouth and stupid fellow. 1952 *History Today* July 451/1 Most of his country-men give him no grander name than 'hillbilly', a term as contemptuous as comic. 1957 *Daily Mail* 26 Sept. 8/2 At 47 the hillbilly who used to scratch a living as a dirt farmer at Greasy Creek in the Ozark Mountains has come a long way. 1966 D. Stewart in 'J. Hackston' *Father clears Out* p. x, As lively a collection of Australian hill-billies as I have ever seen. 1967 *Boston Sunday Herald* 2–8 Apr. 9/2 An Air Force man who wants a missile site, and the hill-billy fighting progress.

2. A type of American folk music. Also *attrib.*

1924 *Talking Machine World* XX. 207/1 Hill Billie Blues..Banjo. 1925 *Ibid.* XXI. 186/2 The [rural] demand is largely for Blues, Coon songs, and Hilly-Billy numbers. 1932 *Daily Mirror* 1 July 10/3, I should not be surprised to hear that the hill-billy king and his suite visit the recording studio every morning of the week. 1953 A. Moorehead *Rum Jungle* iv. 58 They wanted gramophones, and the music they liked best was hill-billy. 1955 L. Feather *Encycl. Jazz* (1956) 110 Piano with local hillbilly, Dixieland and swing bands from age 13. 1959 'F. Newton' *Jazz Scene* x. 172 The full bag of hilly-billy melodies, or the like. 1971 M. Babson *Cover-up Story* vi. 66 They may start out as Hillbilly, or Country and Western but they can be turned into Folk. 1972 *Jazz & Blues* Feb. 19/1 Fats Domino had done some real down home hillbilly-motown stuff.

hi·ll-cou:ntry. [Hill *sb.* 4.] A district composed of hills or elevated ground.

1582 [see Hill *sb.* 4]. 1789 G. White *Selborne* i. i. 2 The down..is a pleasing park-like spot..jutting out on the verge of the hill-country. 1816 U. Brown *Jrnl.* in *Maryland Hist. Mag.* (1915) X. 283 A handsome hill country in a good state of cultivation. 1865 A. D. Whitney *Gayworthys* xv. 133 The slow, ponderous vehicle went..creaking on, up into the hill-country. 1875 [see Hill *sb.* 4]. 1903 *Westm. Gaz.* 14 Jan. 2/1 The Surrey hill-country.

hilling, *vbl. sb.*² (Later U.S. examples.)

1814 J. Taylor *Arator* 233 How often he had passed over the land..in fallowing, hilling, cutting off hills, planting, replantings. 1833 B. Silliman *Man. Sugar Cane* 17 It is advisable not to plough deep, especially for the last hoeing or hilling. 1887 *Harper's Mag.* Jan. 306/1 There is a broad hilling up so as to have a slope inward toward the plants as well as away from them.

hill-man, hillman. Add: Also **hillsman.**

1854 J. D. Hooker *Himalayan Jrnls.* I. v. 136 Carriers and mountaineers... If they serve a good hills-man like themselves, they will follow him with alacrity, sleep on the cold, bleak mountain. 1920 *London Mag.* Apr. 187/2 This rain would hold. He knew it with a hillsman's knowledge. 1938 V. McNabb *Life of Our Lord* iii. 42 The hillsman from Galilee..had to pass through the country of Samaria.

hill-top. Add: **hill-top novel,** the name given by Grant Allen to those of his novels in which he expressed his views on moral and social questions, especially views not generally acceptable to his contemporaries; **hill-top surface** (see quot. 1961).

1895 G. Allen *British Barbarians* p. xi, I propose in future to add the words, 'A Hill-top Novel', to every one of my stories which I write of my own accord, simply and solely for the sake of embodying and enforcing my own opinions. *Ibid.* p. xvi, Why, once more, this particular name, 'A Hill-top Novel'? For something like this reason. I am writing in my study on a heather-clad hill-top... Everything around is fresh and pure... But away below in the valley, as night draws on, a lurid glare ..marks the spot where the great wen of London heaves and festers. 1896 *Pearson's Mag.* May 556/1 Latterly he [*sc.* Grant Allen] has doubled up his social philosophy with his fiction, producing what he calls 'Hill Top' novels. 1901 *Sketch* XXV. 256/2 The lady sportswoman is the Colonel's abomination; he regards her as a monstrous product of the divided skirt and the 'Hill-top' novel. 1930 J. Challinor in *Geography* Dec. 656 The object of the foregoing remarks is therefore to suggest that a hill-top surface, having a 'plateau quality' about it, need not necessarily be evidence of an uplifted peneplane. 1961 —— *Dict. Geol.* 100/1 *Hill-top surface*, an imaginary surface, as smooth as possible, touching the tops of the hills of a region.

hilo¹ (hī·lo). [Sp., = thread :— L. *filum.*] A thin vein of ore (see quot. 1908).

1848 *English & Foreign Mining Gloss.* 5/4 *Hilo*, a small vein or thread of ore in a lode. *hilo de la veta*, line or direction of the vein. *hilos altos*, threads or small veins of ore falling into or proceeding from the upper or hanging wall of a lode. *hilos baxos*, threads or small veins of ore proceeding from or falling into the lower wall of a lode. 1908 E. Halse *Dict. Spanish Mining Terms* 188 *Hilo* (thread), (1) *h. de la veta* or *h. del criadero*, direction of the vein or deposit..(2) small vein or thread of ore in a lode, *e.g.*, *hilos altos*, threads falling into or proceeding from the hanging wall..; in the Cauca, the name is applied to small veins of gold-bearing quartz. *Ibid.* 186 *Hebra*, (1) a thread or yarn of a hempen or other rope..; several twisted together form the *hilo*.

Hilo² (hī·lo). [Hawaiian.] *Hilo grass*, a large and coarse grass, *Paspalum conjugatum.*

1888 W. Hillebrand *Flora Hawaiian Islands* 493 The well known *Hilo grass*, which first appeared about 1840 in the district of Hilo and soon spread there,..is a native of tropical America..but is found now also in tropical Africa, the Galapagos Islands, Australia, and India. 1915 W. A. Bryan *Nat. Hist. Hawaii* 208 The well known and generally despised Hilo grass, occurs in moist, heavy soils in the lower zone. 1917 *Nature* 20 Sept. 57/2 In the moister portions of the islands large areas have been occupied by Hilo grass. **1929** W. T. Pope *Man. Wayside Plants Hawaii* 27 Vigorous cultivation of fields keeps Hilo grass in control.

hilt, *sb.* Add: **3.** Also *to the hilt*.
1950 J. D. Carr *Below Suspicion* x. 122 You may trust me to the hilt! **1965** *New Statesman* 16 Apr. 604/1 The Prime Minister..backs their decisions to the hilt. **1965** *Listener* 20 May 742/2 Every event I attended was subscribed to the hilt, and the house was full.

Himalayan. Add: **1.** **Himalayan black bear,** *Selenarctos thibetanus*.
[**1858** *Proc. Zool. Soc.* 518 Black Bear of Europeans... The species is generally distributed over the lower ranges of the Western Himalayas.] **1869** A. A. A. Kinloch *Large Game Shooting* I. 49 The Himalayan Black Bear is pretty generally distributed throughout the Himalayas. **1925** G. Burrard *Big Game Hunting* iv. 73 The Himalayan black bear is one of the most widely distributed game animals of the Himalayas. **1966** R. & D. Morris *Men & Pandas* viii. 177 The Himalayan, or Asiatic black bear (*Selenarctos thibetanus*).

Himalo- (himā·lo), used as combining form of the *Himalayas*, as in *Himalo-Chinese* adj.
1873 Elwes in *Proc. Zool. Soc.* 654 Himalayan or Himalo-Chinese Subregion. **1910** *Encycl. Brit.* III. 976/1 The Himalo-Chinese or Transgangetic province.

himself, *pron.* **IV.** (Further examples.)
1901 M. Franklin *My Brilliant Career* xxxii. 273 Pa is goin' to write a whole letter all by hisself. **1945** A. Kober *Parm Me* 146 He sits and he *shuckles* hisself and is oney one thing he's saying all the time—nutting! **1965** T. Capote *In Cold Blood* (1966) iii. 166, I talked to the deputy. Then I told the warden hisself. **1969** Widdowson & Halpert in Halpert & Story *Christmas Mumming in Newfoundland* 149 A gentleman..made hisself up with burnt cork.

hina hina (hī·na hī·na). *N.Z.* Also **hinihini.** [Maori.] = Mahoe.
1867 J. D. Hooker *Handbk. N.Z. Flora* II. 765/1 Hina-hina. *Geolog. Surv.* *Melicytus ramiflorus.* **1868** J. Buchanan in *Trans. N.Z. Instit.* I. Essays iii. 37 Hinihini or Mahoe (*Melicytus ramiflorus*). A very variable tree in size of leaves and shape of trunk, the latter angled or round. **1921** H. Guthrie-Smith *Tutira* xxii. 206 The Kowhai and fuchsia and hina hina..will have died out. **1946** J. C. Andersen in *Jrnl. Polynesian Soc.* June 151 *Hinahina*, alternative name for the tree mahoe.

hinaki (hī·nakī). *N.Z.* Pl. **hinaki.** [Maori.] A wicker eel-pot.
1845 E. J. Wakefield *Adv. N.Z.* I. viii. 252 They place eel-pots, called *hinaki*,..at the lower extremity of funnels formed by series of upright poles driven into the bed of the river. **1905** W. B. *Where White Man Treads* 256 My father and his younger brother went to lift a 'hinaki' (eel basket) into the canoe. **1921** H. Guthrie-Smith *Tutira* x. 86 The huge *hinaki* or wicker-work pot, where eels required for immediate consumption were placed. **1952** *Landfall* VI. 288 It was a well-kept looking net.. and with fine meshes. It made me feel shy about my old hinaki. **1966** *Encycl. N.Z.* II. 441/1 Traps for eels and lampreys, termed hinaki, were made from the slim stems of the mangemange.

hinau (hi·nau). Also **hino(u), inau.** [Maori.] A New Zealand evergreen tree, *Elæocarpus dentatus*, yielding a black dye; the wood of this tree. Also *attrib.*
1832 G. Bennett in *London Med. Gaz.* X. 794/2 This tree, the Elœocarpus dicera of Forster, the Inau of the natives, is found abundant on the hills of New Zealand. **1835** W. Yate *Account N.Z.* (ed. 2) ii. 49 Hinau (*Dicera dentata*)—This tree..grows to a height of sixty or seventy feet. **1845** E. J. Wakefield *Adv. N.Z.* II. 246 The *totara*, the *mai*, and the *hinau*, were found to work up into very handsome side-boards, tables, and book-shelves. **1859** A. S. Thomson *Story N.Z.* I. 156 The hinau berries are generally steeped for several days in a running stream... Hinau cakes are much esteemed. **1867** *Richmond-Atkinson Papers* (1960) II. 249 The board on the dead hinau tree announces that application for graves is to be made to E. Patten. **1883** J. Hector *Handbk. N.Z.* 130 (Morris), Hinau, a small tree about fifty feet high and eighteen inches thick in stem, with brown bark which yields a permanent blue-black dye, used for tanning. **1921** H. Guthrie-Smith *Tutira* xii. 99 Other large species in the mixed bush are hinau (*Elæocarpus dentatus*), tawa..and maire. **1968** M. Johnson *N.Z. Flowering Plants* 61 *E. dentatus*, hinau, the larger of our two species, reaches about 50 feet.

Hinayana (hīnāyā·nā). [Skr., f. *hīna* lesser, little + *yāna* vehicle.] The Lesser Vehicle, a name given to the system by the followers of the Mahāyāna, the Greater Vehicle; the Buddhism of Ceylon as distinguished from the northern or Mahāyāna Buddhism. Also *attrib.* Also **Hinaya·nism, Hinaya·nist; Hinaya·nian** *a.*
Hīnayāna was the pristine form of the faith, while the Mahāyāna represents the general one, followed by the majority of followers.
1868 J. Fergusson *Tree & Serpent Worship* 65 Mahâyâna, or as M. Julien translates it, the 'Grand Véhicule', as opposed to Hînayâna or the 'Petit Véhicule'; the distinction between the two being in almost every respect identical with that which exists between Evangelical and Mediæval Christianity. **1877** T. W. Rhys Davids *Buddhism* viii. 200 The system of the *Little Vehicle* (*Hīnayāna*). **1882** *Encycl. Brit.* XIV. 229/1 These volumes [*sc.* the Kandjur] contain about a dozen works of the oldest school of Buddhism, the Hīnayāna. **1907** D. T. Suzuki *Outl. Mahayana Buddhism* 2 Buddhism was now split into two great systems, Mahâyânism and Hînayânism. *Ibid.* 8 The distinction of Mahâyânists and Hînayânists became definite. **1945** A. Huxley *Let.* 27 May (1969) 526 In later, Mahayana Buddhism much is said about the dangers of fixed, strained, rigid concentration of the attention (such as was practised by the Hinayanists). **1951** H. Zimmer *Philos. of India* 509 'The Big Ferryboat' (*mahāyāna*), the ferry in which all may ride, in contrast to 'The Little Ferryboat' (*hīnayāna*), the way of those lonely ones, 'lights unto themselves', who steer the difficult strait of individual release. **1956** A. Toynbee *Historian's Approach Relig.* 17 The Hinayanian Buddhist gospel of self-liberation through self-extinction. **1960** 'S. Harvester' *Chinese Hammer* xi. 110 The differences between Hinayana Buddhism, the original form, and the Mahayana version. **1973** *Times* 14 Apr. (Nepal Suppl.) p. ii/3 Buddhism in the Katmandu Valley subsequently forsook the ascetic path of Hīnayāna.

hincty (hi·ŋkti), *a.* *U.S. slang.* Also **hinkty.** [Origin unknown.] Conceited, snobbish, stuck-up.
Connection with clipped forms of 'handkerchief-head' (= an Uncle Tom Negro) has been suggested but the phonetic development is incapable of demonstration.
1924 in W. C. Handy *Treasury of Blues* (1949) 144 We'll I am hinkty and I'm low down too. **1936** *Esquire* (Chicago) May 192/3 'She couldn't be mixed up in no murder trial. She's too respectable.' 'A hinkty hussy!' said Sling. **1941** *Examiner* (San Francisco) 20 July PR 2 Jack, it ain't like me to be hincty so I'll be there with my boots laced tall. **1948** *Capitol News from Hollywood* Jan. 12/1 Patrons who dropped into the hincty, ultra-ultra Circus room of Santa Monica's lavish Hotel Ambassador. **1957** J. Kerouac *On Road* (1958) 86 Wetting their eyebrows with hincty fingertip. **1969** C. Himes *Blind Man with Pistol* vi. 72 All those hincty bitches fell on those whitey-babies like they was sugar candy.

hind, *a.* Add: **A. c.** *to get on one's hind legs*: see Leg *sb.* 2 c. *to talk the hind leg(s) off a donkey*, etc.: see *talk *v.*
C. a. *hind-wing.*
1932 E. Step *Bees, Wasps, Ants Brit. Isles* p. xxiv, In flight, the fore- and hind-wings of a side usually act as one.

Hindemithian (hindəmi·tiăn), *a.* and *sb.* [f. the name of Paul *Hindemith* (1895–1963), German musician + -ian.] **A.** *adj.* Of, pertaining to, or characteristic of Hindemith, or his style of composition. **B.** *sb.* One who favours or imitates the style of Hindemith.
1941 A. Copland *Our New Mus.* 159 This last contains remarkable Hindemithian premonitions. **1954** *Encounter* June 60/1 How refreshing..to meet a Hindemithian who does not think twelve-tone music a disease. **1958** *Listener* 18 Sept. 441/1 Reizenstein..continues in a gravely Hindemithian style. **1962** *Ibid.* 21 June 1091/2 A gayer, more elastic Hindemithian counterpoint.

hinder, *a.*[1] Add: **B.** *sb.* (usu. *pl.*). Hindquarters, buttocks; hind legs.
1857 J. Scholes *Tim Gamwattle* 20 (E.D.D.), Thir is nah a barro e Smobruff uts big anouff fur iz hoindurs. **1880** J. F. S. Gordon *Bk. Chron. Keith* 55 Boasting of kissing, at their meetings, the Devil's 'hinder'. **1891** M. M. Dowie *Girl in Karpathians* xiii. 173 The painter spread his coat upon the hinders of the second horse. **1892** J. Lumsden *Sheep-head & Trotters* 268 A pull that brought the pony in a moment back upon its hinders. **1948** F. Brown *Dead Ringer* (1949) xi. 131 He stood up on his short little hinders and got himself a lawyer.

Hindki (hi·ndkī). Also **Hindeki, Hindkee.** The name of a people, and of their language, of north-west India and Afghanistan.
1815 M. Elphinstone *Acct. Kingdom of Caubul* ii. xii. 309 It has been observed that there is scarcely any part of Afghaunistaun in which the whole population is Afghaun, and that the mixture is composed of Taujiks in the West, and of Hindkees in the East. **1875** *Encycl. Brit.* I. 235/2 *Hindkis.* This is the name given to people of Hindu descent scattered over Afghanistan. They are said to be of the *Kshatri* or military caste. **1886** Yule & Burnell *Anglo-Indian Gloss.* 315/2 *Hindki* or *Hindeki*. This modification of the name is applied to people of Indian descent, but converted to Islam, on the Peshawar frontier, and scattered over other parts of Afghānistān. They do the banking business and hold a large part of the trade in their hands. **1911** *Encycl. Brit.* XVI. 80/2 The principal varieties of the northern group are Hindki (the same in meaning as Hindko) and Pōthwārī. **1919** G. A. Grierson *Ling. Surv. India* VIII. i. 333 The Lahndā of Deva Ghazi Khan goes by several names, such as Hindī, Hindkī, Jatkī, and Dērāwāl... I call it Hindkī in these pages. **1964** S. K. Chatterji in D. Abercrombie et al. *Daniel Jones* 409 Languages like..Hindki (western Panjabi or Lahndi).

hindside (həi·ndsəid). [f. Hind *a.* A.] The back part of anything. Also in quasi-combs., as *hindside-foremost*.
1862 [see Hind *a.* C. a]. **1915** E. Pound *Lett.* (1951) 91 Objectivity and again objectivity, and expression: no hindside-beforeness. **1929** W. Faulkner *Sartoris* (1932) iv. 270 Negro in a hind-side-before collar. **1931** E. Wilson *Axel's Castle* iii. 88 His hindside-foremost presentations of thought. **1934** J. Marston *Andromeda* xiv. 194 That's crude sentimentality hindside up. **1942** *New Yorker* 10 Oct. 33 What's she doing out in this rain, shoving along the beach on her hindside?

hind-sight. Add: **1. b.** *to knock* (or *kick*) *the hindsight out* or *off*: to dispose of or demolish completely. *U.S. colloq.*
1834 W. A. Carruthers *Kentuckian in N.Y.* I. 21 As sure as you saw the fire at the muzzle of his gun, so sure he knocked the creter's hind sight out. **1850** L. H. Garrard *Wah-To-Yah* (1927) xx. 248 They backed their ears preparatory to kicking the hindsights off the first man that struck them. **1872** E. Eggleston *Hoosier Schoolmaster* x. 58 Ef its rendered right, it'll knock the hind sights off of any rheumatiz you ever see. **1892** *Congress. Rec.* 1 Apr. 2843/1 The American producer.. can knock the hindsights off the producer anywhere else on the face of the earth. **1954** in J. A. Weingarten *Amer. Dict. Slang* 186/2.

hinge, *sb.* Add: **I. 1. d.** In Philately: see quot. 1883.
1883 *Stamp-Collecting & Exchange* 8/2 There is only one really satisfactory process for mounting postage labels, viz., the now almost universal hinge-system. This consists of a piece of thin paper a little smaller than the stamp itself, and affixed to the album with a drop of gum, or, if the hinge be already prepared for adhesion,.. no inconvenience or trouble will be experienced. **1892** *Stamp Collector* Apr. 33 In the earlier days of stamp collecting, before such things as gummed hinges were offered to the collecting public. **1967** *Exchange & Mart* 20 July 19/4 (Advt.), Stamps stay put without stamp hinges, in our brand new plastic stamp album. **1971** D. Potter *Brit. Eliz. Stamps* xv. 163 At the turn of the century, hinges were monstrous things, designed for their sticking power.
III. 7. hinge-ligament (see quot. 1909).
1909 *Cent. Dict.* Suppl. 589/3 *Hinge-ligament*, in bivalve mollusks, a tough, uncalcified, elastic membrane which connects the two valves. **1945** E. Step *Shell Life* (rev. ed.) 56 The hinge-ligament exerts a pulling action which tends to separate the lower edges.

hinihini, var. *hina hina.

‖ **hinin** (hī·nin). [Jap.] A member of an outcast group in Japan. Also *collect.* and *attrib.* Cf. *eta.
1884 tr. *J. J. Rein's Japan* ii. i. 329 Geshas (female dancers and singers) and Jôrôs (prostitutes)..were despised, and considered..socially below the level of the Hinin. **1891** A. M. Bacon *Jap. Girls & Women* ix. 228 The éta and hinin—outcasts who lived by begging, slaughtering animals, caring for dead bodies, tanning skins, and other employments which rendered them unclean. **1904** L. Hearn *Japan: Attempt at Interpretation* vi. 110 The banished man was most often doomed to become a *hinin*—one of that wretched class of wandering pariahs who were officially termed 'not-men'. **1970** J. W. Hall *Japan* x. 179 Tokugawa society..was conceived of..as falling into the following categories: the *kuge*, the samurai (including daimyo), priests, peasants, urban residents, and pariah (*hinin* and *eta*).

hinkty, var. *hincty *a.*

‖ **hinoki** (hinǫ·ki). Also † **finoki.** [Jap.] A large conifer, *Chamæcyparis obtusa*, native to Japan, or the timber obtained from it.
1727 J. G. Scheuchzer tr. *Kæmpfer's Hist. Japan* I. i. 118 *Finoki* and *Suggi* are two sorts of Cypress-trees, yielding a beautiful light whitish wood. **1884** tr. *J. J. Rein's Japan* ii. ii. 445 The pure and simple Shintô-temple ..is built of the white wood of the Hinoki. **1904** L. Hearn *Japan: Attempt at Interpretation* ix. 184 Such superior qualities of wood as *keyaki* or *hinoki*. **1964** *Illustr. Important Forest Trees Japan* (Japan Forest Technical Assoc.) 44/2 The hinoki cypress is abundantly found on ridges. **1965** J. Ohwi *Flora Japan* 117/1 *Chamaecyparis obtusa*...Hinoki...Much planted for timber. **1969** R. C. Bell *Board & Table Games* II. iii. 61 The best boards.. are also made of 'Gingko' (*Salisburia Adiantifolia*) and 'Hinoki' (*Thuya Obtusa*).

hino(u), varr. *hinau.

hint, *sb.* Add: **2. b.** A small piece of practical information.
1777 P. Thicknesse *Year's Journey* II. 221 (*heading*) General hints to strangers who travel to France. **1807** R. Southey *Lett. fr. England* (1951) 348 The Monthly is more miscellaneous in its contents... All sorts of heretical opinions are started here, agricultural hints thrown out, and queries propounded of all kinds. **1830** Coleridge & Southey *Devil's Walk* 24 The Devil was charm'd, for it gave him a hint For improving the prisons of Hell. **1841** S. Ellis (*title*) Family secrets; or, Hints to those who would make home happy. **1872** *Young Englishwoman* Nov. 598 Household hints for young housewives. **1926** R. Macaulay *Crewe Train* ii. x. 181 Audrey gave them household hints, about how to keep the kettle from furring, and the stove and the milk jar from smelling. **1972** *Guardian* 29 Mar. 11/5 Angela Kay's Household Hints..has a pretty good selection of useful answers to some eight hundred possible problems.

hint, *v.* Add: **1. c.** With direct speech as obj.
1843 Dickens *Christmas Carol* iii. 107 'I'm sure he is very rich, Fred,' hinted Scrooge's niece. **1891** C. Graves *Field of Tares* 41 'I have been a stranger to the neighborhood,' returned the other, 'for years.' 'Many years?' hinted Mr. Brinnilow.

2. b. With other constructions; also *absol.*
1865 *Punch* 21 Jan. 32 If I wos allowed to 'int, Ladies, I shud say, Torse Hup! **1891** Hardy *Group of Noble Dames* 77 Her husband's tutor was found to hint very strongly against such a step. **1894** G. Du Maurier *Trilby* III. vii. 178 The night above was dark, but 'star-dials hinted of morn'.

hinterland. Add: Also applied spec. to the area lying behind a port, and to the fringe areas of a town or city.
a **1910** in L. D. Stamp *Gloss. Geogr. Terms* (1961) 235 *Hinderland, Hinterland*, the region the seaborne trade of which belongs to a particular seaport or seaboard. **1922** *Geogr. Rev.* Apr. 260 The main factor which determined the selection of ports in prehistoric times was the presence of a populous hinterland of effective buyers. **1936** E. Van Cleef *Trade Centers & Trade Routes* iii. 34 The immediately contiguous territory within the continuous hinterland which in some instances contributes to the formation of the metropolitan city has been termed by the Germans, the 'Umland' or country about. **1938** A. J. Sargent *Seaports & Hinterlands* 3 A port, essentially, is a transit area, a gateway through which goods and people move from and to the sea, by way of rail, inland waterway, or sometimes by road. The region to and from which this movement is directed is commonly and somewhat vaguely described as the hinterland. **1945** E. Waugh *Brideshead Revisited* 7 Here the close, homogeneous territory of housing estates and cinemas ended and the hinterland began. **1950** *Geogr. Jrnl.* CXVI. 64 The approximate boundaries of urban spheres of influence or hinterlands. **1968** *Guardian* 23 Oct. 9 As Clydeside developed industrially so it attracted labour from its own hinterland and from famine-stricken Ireland.
2. *fig.* and *transf.*
1919 M. K. Bradby *Psycho-Analysis* (1920) 75 Unexplored territories full of mystery and danger in the hinterland of their own minds. *Ibid.* 251 The individual who is introduced to the 'hinterland' of his own conscious being. *a* **1930** D. H. Lawrence *Last Poems* (1932) 182 We are mostly unexplored hinterland. **1965** *New Statesman* 23 Apr. 646/3 The council meets in that dour ecclesiastical hinterland of Westminster Abbey, where you can buy a second-hand cassock.
3. *Geol.* (See quot. 1961.)
1937 Wooldridge & Morgan *Physical Basis Geogr.* vi. 76 The African 'hinterland' is believed to have moved northward towards the European 'foreland'. **1961** J. Challinor *Dict. Geol.* 100/2 *Hinterland*, the moving block which compresses the sediments of a geosyncline and forces them towards the foreland.

hintzeite (hi·ntsəit). *Min.* [ad. G. *hintzeït* (L. Milch 1890, in *Zeitschr. f. Kryst. und Min.* XVIII. 480), f. the name of C. A. F. *Hintze* (1851–1916), German mineralogist: see -ite[1].] = *heintzite.
1891, 1951 [see *heintzite].

hiortdahlite (hyǫ·ɹtdāləit). *Min.* [ad. G. *hiortdahlit* (W. C. Brögger 1888, in *Nyt Mag. f. Naturvidensk.* (1889) XXXI. iii. 232), f. the name of Th. H. *Hiortdahl* (1839–1925), Norwegian chemist: see -ite[1].] A rare mineral, essentially a fluoride-containing silicate of zirconium, sodium, and calcium, found in yellow triclinic crystals.
1892 E. S. Dana *Dana's Syst. Min.* (ed. 6) 377 (*heading*) Hiortdahlite. **1968** *Mineral. Abstr.* XVIII. 304/2 Hiortdahlite occurs as columnar crystals up to 3 and 4 cm long in pegmatite..in the Korgeredaba pluton in southeastern Sangilen, Tuva.

hip, *sb.*[1] Add: **2. b.** *on the hip* (further example).
1967 *Listener* 16 Nov. 628/3 You have me on the hip here a bit because I think..that all these old ideas we had are as dead as the dodo.
4. a. *hip-boot.*
1893 *Outing* (U.S.) XXII. 124/2 Gossamer hip-boots are good if of reliable stock. **1922** S. Lewis *Babbitt* x. 138 He gloated on fly-rods and gorgeous rubber hip-boots. **1969** *Sears Catal.* Spring/Summer 709 Ted Williams hip boots have nylon added to latex for long-lasting strength and comfort.
b. *hip-swaying, -swinging* adjs.; **hip-flask,** a flask for intoxicating liquor carried in a hip-pocket; **hip-hole,** a hollow dug in the ground to accommodate the hip (for greater comfort when sleeping on hard ground); **hip-huggers** *sb. pl.*, trousers that fit tightly to the hips; also **hip-hugger,** used *attrib.* of such trousers; **hip-hugging** *a.*, fitting closely to the hips; **hip-length** *a.*, denoting a garment which reaches down to the hips; **hip-line,** the outline or contour of the hips; **hip-pocket** (earlier and later examples); **hip throw,** a throw in Judo; **hip-yoke,** in dressmaking, a shaped piece extending from the waist to the hips, designed to fit the figure closely without gathers.
1923 E. Marbury *My Crystal Ball* lxxi. 352 Let these same people frequent ballrooms..and they will find the hip flasks in evidence and the consequent conditions a sorry spectacle. **1928** *Sunday Dispatch* 2 Sept. 7/1 [American *loq.*] We have become a people who think 'likker', talk it, and lead a hip-flask life! **1944** Brahms & Simon *Titania has Mother* xiii. 145 Merry moujiks drinking from hip flasks and lolling in all directions. **1973** D. Lees *Rape of Quiet Town* ii. 33 A hip-flask full of brandy. **1936** F. Clune *Roaming round Darling* xxv. 271 Then we made a fire, boiled the billy, gouged out hip-holes for ourselves, and camped till daylight. **1965** G. McInnes *Road to Gundagai* x. 167, I tried digging a hip-hole. **1967** *Boston Sunday Herald* 26 Mar. (Advt.), Zowie prints and belted hip-huggers..striped and printed. **1967** *New Yorker* 26 Aug. 82 There are separates, too, such as long jackets, stovepipe hip-hugger trousers, [etc.]. **1973** *Houston* (Texas) *Chron. Mag. People, Places, Pleasures* 14 Oct. 24/5 Today the young women wear shorts and hip-huggers, the older sit in their traditional long black dresses. **1969** *Times* 7 Nov. 14/1 Hip-hugging..trousers. **1971** B. Malamud *Tenants* 43 Willie..was dolled up in hip-hugging yellow pants and two-tone brown-and-black shoes. **1921** Hip length [see *gum-boots* s.v. *gum sb.[2] 9]. **1931** *Daily Express* 18 Mar. 5/3 Hip-length coats. **1961** *Harper's Bazaar* Feb. 75 The hip-length cardigan. **1907** *Westm. Gaz.* 30 Mar. 14/1 It [sc. a skirt] blurs over the aggressive hip-line. **1935** *Times* 2 Dec. 19/4 The coat just reaches the hipline. **1973** *Country Life* 8 Mar. 635/2 Dress in black-and-white-check wool with smooth hipline. **1880** *Cimarron News & Press* 22 July 3/2 Lee snatched Armstrong's revolver from his hip pocket and pointed it at Armstrong. **1922** S. Lewis *Babbitt* xviii. 228 Drinking together from hip-pocket flasks. **1972** *Guardian* 27 Oct. 12/1 A general election..is one option in Mr Heath's hip pocket list. **1920** *Glasgow Herald* 17 Apr. 6 The Hawaiian corps de ballet.. began a..performance of rhythmical hip-swaying dances. **1965** *Midnight* 12 July 3/1 The hip-swinging wedding party was celebrated to the strains of the watusi and frug, pounded out by Christopher's group, the Wild Ones. **1966** *Word Study* Oct. 7/1 A hip-swinging, slender-bodied blonde. **1957** Takagaki & Sharp *Techniques of Judo* ii. iv. 50 Generally, when a hip throw is employed, the opponent turns his back into you and pulls you to the right front corner by your right arm which is pulled tightly across his chest. **1960** *Oxf. Mail* 10 Mar. 8/2 He..went on to win the match with another ankle throw, with which he countered an attempted hip throw. **1931** *Daily Tel.* 21 May 6/4 Small gathers at each side of the hip yoke in front.

hip, *v.*[2] Add: **4.** To carry on the hip. *U.S.*
1818 H. C. Knight *Lett. from South* 93 Some mothers here [*sc.* in Kentucky] hip their infants, as do the Sumatrans. **1843** 'R. Carlton' *New Purchase* xx. 191 Still oftener each [log] is hipped and hipping is done by one man..who adroitly whips up the log on his hip.

hip (hip), *v.*[5] *slang* (orig. *U.S.*). [f. *hip *a.*] *trans.* To render 'hip'; to inform. (Freq. as pa. pple. in passive.) Hence **hipped** *ppl. a.* well-informed, 'with it'; (esp. with *on*) fond of, 'bitten with'.
1920 F. Scott Fitzgerald *This Side of Paradise* (1921) ii. iii. 246 I'm hipped on Freud and all that. **1925** T. Dreiser *Amer. Trag.* (1926) II. iii. xv. 191 He is still hipped over this second girl. **1927** *Daily Express* 24 Oct. 8 'New York,' as the manager of one of the largest hotels remarked lately, 'is badly "hipped" on dining in public.' **1932** *Evening Sun* (Baltimore) 9 Dec. 31/4 *Hip*, to give information. **1938** *Amer. Speech* XIII. 314/1 *Hipped to the jive*, well informed on the latest slang expressions. **1944** D. Burley *Dan Burley's Orig. Handbk. Harlem Jive* 19 Uncle is hipping a whole lot of cats as to what to do when the action gets off the track. **1945** L. Shelly *Jive Talk Dict.* 26 *Hipped spade*, smart Negro. **1947** *Esquire* Apr. 76 'Are there any squares in this outfit?' 'No, man, we're all hipped.' **1954** *Encounter* Feb. 55/1 If I admitted..to being a little bit hipped on the subject of Trotsky, I could sometimes gain an indulgent if flickering attention. **1958** J. Kerouac *Subterraneans* 90 Sand must have hipped him quietly in a whisper somewhere what was happening with the lovers. **1962** *Spectator* 20 Apr. 511 Betjeman is absolutely hipped on his subject. **1973** *Black World* Jan. 64/2, I had just about decided to find some way to hip her to contraceptives.

hip (hip), *a.* *slang* (orig. *U.S.*). [Origin unknown.] = *hep *a.* Hence **hi·p-cat** = *hep-cat; **hi·pness,** the condition or quality of being 'hip'.
1904 G. V. Hobart *Jim Hickey* i. 15 At this rate it'll take about 629 shows to get us to Jersey City, are you hip? **1926** *Detective Fiction Weekly* 16 Jan. 640/2, I sashayed for a legger an' run into a rube hip agent with a bottle and some jake which helped some. **1938** C. Calloway *Hi De Ho* 16 *Hip*, wise, sophisticated, anyone with boots on. **1944** C. Himes *Black on Black* (1973) 199 I'm a hipcat from way back. **1946** Mezzrow & Wolfe *Really Blues* xii. 226 Their hipness..bubbled up out of the brute scramble and sweat of living. **1951** *San Diego Even. Tribune* 28 June *a*–2/5 We did it because we thought it was 'hip' or smart. **1958** W. Bryant *Jive in Hi-Fi* 13 The correct word is 'hip'. It comes from a story of a fisherman warning young fishermen never to wade in deep water without hip boots on because they could run into trouble. So, when you hear the words, 'I'm hip' or 'I'm booted' it's said to let you know they have no fear of trouble or that they understand what's shaking [i.e., happening]. **1957** J. Kerouac *On Road* (1958) 10 Elmer Hassel, with that hip sneer. **1959** *Spectator* 31 July 134/2 He has a fast line of jive-patter and uses such hip endearments as 'angel-cake' and 'gorgeous'. *Ibid.* 7 Aug. 161/2 Audiences there are hip to the latest gossip. **1959** C. MacInnes *Absolute Beginners* 164 It was like getting a hip cat into a symphony concert, but I succeeded. **1959** *Observer* 4 Oct. 9/7 The only really *hip* Labour candidate. **1961** *New Left Rev.* May–June 47/1 The goofs of the second act did a lot to dispel the hipness of the first. **1961** *Listener* 9 Nov. 786/1 As Norman Mailer would say, it's 'hip' to use obscure terms and meaningless symbols. **1966** H. S. Thompson *Hell's Angels* (1967) 68 Frank was so completely hip that he went down to Hollywood and bought the blue-and-yellow striped sweatshirt that Lee Marvin wore in *The Wild One.* **1971** *Black Scholar* Jan. 22/2 As Cannonball Adderly has said 'Hipness is not a state of mind. It is a fact of life.' **1972** V. Ferdinand in A. Chapman *New Black Voices* 472 We sometimes..go in for that kind of living thinking it's hip.

hip-cat: see *hip *a.*

hipe, *sb.*[1] Also **hype.** Add quots. illustrating various kinds of hipe, as *left leg hipe, right leg hipe, swinging hipe,* etc.
1888 *Encycl. Brit.* XXIV. 690–1 The 'left leg hipe'.. consists in lifting and swinging him [*sc.* one's adversary] round to the right, then striking the inside of his right thigh with the outside of the left thigh, by which he gets off his balance and falls; the 'right leg hipe' is the same action *mutatis mutandis.* **1893** *Carlisle Patriot* 26 May 3/3 (Cumbld. Gloss.), In the third round S—— threw M—— after a tight bout with the inside hipe. In the next tussle J—— passed S—— by means of the outside hipe. **1893** Robinson & Gilpin *Wrestling* 56 Robley..has been credited with being the first introducer of the swinging hype.
So **hi·per,** one who uses the hipe in wrestling.
1823 W. Litt *Wrestliana* 168 As a hiper, he is certainly the quickest and best on the list.

hipe (həip), *sb.*[2] *Army slang.* [From the sound of the substitute commonly used for the word *arms* in such commands as 'Slope arms!'] A rifle.
1917 P. MacGill *Brown Brethren* vi. 89 He seemed to have lost all interest in his best friend, his ''ipe'. *Ibid.* xii. 173 The sun's catchin' the sniper's 'ipe. **1927** D. B. W. Lewis *On Straw* 59 Brass-Hats rich and ripe Clicked their heels together, Sloped the Army's hipe. **1942** 'N. Shute' *Pied Piper* 107 It was full of muckin' Jerries. All loosing off their hipes at Bert and me.

hipe, var. *hype *v.*[1]

hipness: see *hip *a.*

hippeastrum (hipĭæ·strə̆m). *Bot.* [mod.L. (see quot. 1821), f. Gr. ἱππεύς horseman, knight + ἄστρον star.] A member of the genus of South American bulbous plants so named, belonging to the family Amaryllidaceæ; the knight's star lily.
1821 W. Herbert in *Bot. Reg.* VII. App. 7 Other points of agreement with Crinum separate them [*sc.* the plants] still more widely from the occidental bulbs which I have heretofore called Amaryllis, but in deference to the type of Linnæus, propose now to call Hippeastrum or Knight's-star-lily. **1866** [see Amaryllis]. **1899** *Daily News* 16 Mar. 9/2 An immense stand of hippeastrums.. carried off the palm in the shape of the society's gold medal. **1938** A. G. L. Hellyer *Your Garden Week by Week* 102 Start the remaining hippeastrum bulbs into growth now. **1961** *Amateur Gardening* 30 Sept. 6/2 Another plant to be rested is the hippeastrum—or, as many still prefer to call it, the amaryllis.

hipped, *ppl. a.*: see *hip *v.*[5]

hipper[2] (hi·pəɹ). *Austral.* [f. Hip *sb.*[1] + -er[1].] (See quots.) Cf. *hip-hole* (*hip sb.[1] 4 b).
1934 *Bulletin* (Sydney) 26 Sept. 20/3 A tired man usually contents himself with a ground-sheet and blanket after digging himself a 'hipper'—a small cavity to accommodate his hips. **1945** Baker *Austral. Lang.* 106 The hipper, something soft—such as a piece of possum skin or a stuffed strip of bagging—to put under the hip when lying on hard ground.

hippety (hi·pėti). Also **-ity.** Jingling combination of Hip *v.*[1] and Hop *v.*, as in *hippety-hop, hippety-hoppety.*
1825 J. Jennings *Observ. Dial. W. Eng.* 45 *Hippety-hoppety, adv.*, in a limping and hobbling manner. **1880** *West Cornwall Gloss.* s.v., 'He goes hippety-hoppety' (walks unevenly). **1886** F. T. Elworthy *W. Somerset Word-bk.*, *Hippety-hop, hippety-hoppety*.., lame, limping in a very marked manner: applied to both man and beast. **1920** *Cornhill Mag.* Sept. 332 A row of children playing hippety-hop across a broad lawn. **1925** D. H. Lawrence *Refl. Death Porcupine* 186 'I am the captain of my soul!' Are you, old boy? Then why hippety-hop? **1950** J. Dempsey *Championship Fighting* 58 Under no circumstances take any little half-step or hippity-hop.

hippie, hippy (hi·pi), *sb.* and *a.* *slang* (orig. *U.S.*). [f. *hip *a.* + -y *suffix*[6].] **A.** *sb.* A hipster; a person, usually exotically dressed, who is, or is taken to be, given to the use of hallucinogenic drugs; a beatnik. **B.** *adj.* Of, pertaining to, or characteristic of hippies.
1953 D. Wallop *Night Light* 157 Man, I really get a bellyful of these would be hippies. **1959** *Village Voice* 18 Nov. 13 Imagine coming on so jaded,..so hippie,.. and fed up. **1965** *New Yorker* 24 July 27 All the hippies are leaving for the New World. **1967** *Daily Tel.* 21 Feb. 16/3 These people, 'writers, musicians, psychedelic popsters and hippies..' see London as a 'focal city for permissive experiments' in art and life. **1967** *Sunday Truth* (Brisbane) 2 Apr. 63/1 A hippie is the LSD Age's equivalent of a beatnik, and they turn on with marihuana, LSD, benzedrine or merely with the idea of turning-on. **1967** *Spectator* 7 July 32/3 Promises that in future they will live in accordance with the principles of love demanded by the hippies. **1967** *Guardian* 12 July 1/7 Have you ever been to the hippie district [of San Francisco]? **1967** *New Statesman* 27 Oct. 531/3 The

[demonstration] marches cannot end the war, whether they are moderate and middle-class, insurrectionist, hippy or whatever. **1968** *Times* 16 Dec. 5/3 About 1,000 Hippies from several countries have set up a 'Hippy Republic' here [S. José, Costa Rica]. **1969** *Times* 27 Mar. 1/1 Without..losing her cool, as the hippy idiom would put it. **1969** *Daily Mail* 26 Sept. 7/1 The estimated number of full-time Hippies in London is 2,000,..and more than 60,000 part-timers in Britain alone. **1969** *Times* 5 Dec. 7/1 Religion, in one form or another, is frequently a straw to which the lost generation of hippies clings. *Ibid.*, The members of the hippy commune charged with the Sharon Tate killings are no more typical of America's long-haired 'make love, not war' youth than the soldiers involved in the alleged My Lai massacre are typical of the United States Army. **1972** *New Society* 30 Nov. 496/2 The word 'hippy' is now in current usage throughout Uganda, Tanzania and Kenya, and often just refers to anyone with long hair, almost always a European. **1973** *Friend* 13 July 839/1 'Hippie' is rather a general term applied quite often to anyone young and unkempt in appearance, who is considered to have dropped out of 'straight' society, and who in general puts little premium on the values of contemporary society which he has rebelled against.
Hence **hi·ppiedom,** the condition or fact of being a hippie; the domain of hippies; **hi·ppieland,** the domain of hippies; **hi·ppi(e)ness,** the quality or characteristics of a hippie or hippies.
1967 *N.Y. Times* 5 May 42 There are two philosophical trends in hippiedom. **1967** *Economist* 15 July 217/2 Some of the music is attracting attention outside of hippieland. **1968** *Blues Unlimited* Dec. 9, I guess California, and psychedelia, and hippieness have had the influence. **1969** *Listener* 20 Feb. 250/1 *One Step Away*..is the most comprehensive documentary I have seen so far about West Coast hippiedom. **1969** *Times* 5 Dec. 7/1 The west coast with its sunshine and sparkling surf, cut off by the desert on one side and the Pacific on the other, is the hot-house of America's hippiedom. **1970** K. Platt *Pushbutton Butterfly* (1971) iv. 41 A lot of girls disappear in hippieland. **1971** *Guardian* 17 June 9/4 A group like Exodus.. with their veneer of Dutch hippiness. **1971** *Ibid.* 9 Sept. 9/3 Descriptions of hippiedom by well-meaning admirers.

hippo-. Add: **hippoma·nia** [-mania], excessive fondness for horses; so **hippoma·niac,** one affected by hippomania; **hippoma·nic** *a.*; **hippoti·grine** *a.*, pertaining to the striped members of the genus *Equus*, such as the zebra and the quagga; **hippotra·gine** *a.*, belonging to the sub-family Hippotraginæ of the family Bovidæ, a group of large African antelopes.
1961 *Spectator* 9 June 853 The combination of hippomania, secretarial-college chumship. **1963** *Daily Tel.* 8 Jan. 13/4 BBC television is to investigate present-day 'hippomania'... The word was apparently used by the Romans to describe excessive love by women for horses. **1940** V. Woolf *Roger Fry* 126 He had been thrown riding 'with that hippo-maniac Goldie'. **1963** *Listener* 7 Feb. 260/3 The splendidly hippomanic girl..who met her future husband after a toss in the hunting-field. **1875** *Encycl. Brit.* I. 259/1 Of wild horses the asinine group is characteristic of Asia, and the hippotigrine of Africa. **1947** J. Stevenson-Hamilton *Wild Life S. Afr.* vi. 51 The Cape mountain zebra (*Equus zebra*) is the smallest of the group of hippotigrine equines, standing some 4 feet at the shoulder. **1891** Flower & Lydekker *Introd. Mammals* 342 Hippotragine Section.—Includes very large African Antelopes, with long horns. **1910** H. F. Osborn *Age of Mammals* v. 337 (*caption*) A recent hippotragine type of Africa, the sable antelope (*Hippotragus niger*).

hippoboscid (hipobǫ·sid), *a.* and *sb.* [f. mod.L. *Hippoboscidæ*, f. Hippo- + Gr. βόσκειν to feed: see -id[3].] (Of or pertaining to) a member of the family Hippoboscidæ of blood-sucking flies, sometimes called louse-flies, parasitic on mammals and birds.
1891 *Insect Life* III. 357 Mr. Townsend read a paper on a remarkable new Hippoboscid..which had been taken on a bat. **1920** *Q. Rev.* July 91 Tsetse-flies..are the nearest living relations of the Hippoboscid family. **1962** W. R. Horsfall *Med. Ent.* v. 171 Nycteribiidae and Streblidae ..live in the manner of the hippoboscid flies. **1972** *Nature* 4 Feb. 249/1 Hippoboscids are at best poor fliers, with a strong tendency to wing degeneration.

Hippocratic, *a.* Add: **1.** *Hippocratic oath*, an oath comprising the obligations and professional conduct of physicians, taken by those entering upon medical practice.
1747 J. Barker *Essay Anc. Mod. Physicians* Dedication p. v, By the Hippocratic Oath, a Physician is bound to honour the Master who has instructed him in his Art. **1969** *Nature* 4 Jan. 2/1 The Hippocratic oath of the medical profession has excited the attention of several moralists.

hippodrome, *sb.* Add: **3.** A theatre used for various stage entertainments.
1899 *Daily News* 10 July 8/1 Hippodromes, as such strictly, have not hitherto been greatly in fashion in London, and they have had but a fitful existence since Mr. Batty started the memorable one known by his name, at the time of the Great Exhibition of 1851. **1902** *Encycl. Brit.* XXXI. 49/1 London in 1902 had..the 'Hippodrome' in Cranbourn Street. **1972** *Times* 8 Apr. 12/5 At the moment, 21 London theatres and halls are listed:.. they are the Duke of York's, Garrick, Her Majesty's, Coliseum, Hippodrome, Lyceum, [etc.].

hi·ppodroming, *vbl. sb.* (See Hippodrome *sb.* 2 and *v.*)
a **1867** H. Woodruff *Trotting Horse* (1868) xxxv. 288 An arrangement was entered into by means of which the former and Lancet travelled together, to trot for purses and divide the profits. It was a new sort of thing, and was ..called 'Hippodroming'. **1946** *Chicago Tribune* 1 Feb. (Grafic Mag.) 13/1 Cynics argue there is a great deal of hippodroming [in ice hockey].

hippo fly (hi·po fləi). [prob. abbrev. of Hippo(potamus, in reference to the size of the fly.] A large blood-sucking fly of the family Tabanidæ, found in central Africa.
1901 *Geogr. Jrnl.* July 75 The large biting fly called the 'hippo' fly was a great annoyance. **1929** Patton & Evans *Insects, Ticks, Mites & Venomous Animals* I. 294 The Tabanidæ have received various local names such as horseflies, seroot flies, mangrove flies, hippo flies.

hippomobile (hi·pomə̆ubīl). (*Disused.*) [f. Hippo- + Mobile *a.*] A word used in the early days of motor vehicles for a horse-drawn vehicle. So **hippomo·bilism,** the use of a hippomobile.
1904 in A. B. F. Young *Compl. Motorist* xiii. 275 A hippo-mobile was despatched. **1904** *Westm. Gaz.* 15 Nov. 4/2 A motor-van..would certainly do the journey more quickly..and presumably such a vehicle will supersede the hippomobile in time. **1905** *Ibid.* 20 June 4/2 How much longer, motorists may be tempted to ask, is such a dangerous mode of locomotion as the hippomobile to be tolerated? **1908** *Daily Chron.* 16 Jan. 4/4 Hippomobilism was out of the question. **1963** Bird & Hutton-Stott *Veteran Motor Car* 26 Even in 1899 this contrivance had an air of hippomobile antiquity.

hippus (hi·pŏs). *Ophthalm.* [mod.L., f. Gr. ἵππος horse.] Tremor of the iris; *esp.* a rhythmic contraction and dilatation of the pupil independent of the light intensity; also, a complaint of the eyes, such that they are always winking.
1684 S. Blancard *Physical Dict.* 150 *Hippus* is an Affection of the Eyes, wherein they continually shake and tremble, and now and then twinkle. **1838** H. Houston in S. Littell *Dis. Eye* 177 Oscillation of the iris is applied to the alternate contraction and expansion (hippus iridis), as well as to the trembling of the iris. **1904** L. W. Fox *Dis. Eye* xix. 428 Hippus is the term employed to designate a chronic spasm of the iris, as is seen in hysteria, neurasthenia,..and various spinal diseases. **1932** W. S. Duke-Elder *Text-bk. Ophthalmol.* I. xii. 563 In the condition of hippus a rhythmic contraction and dilatation of the pupil occurs regularly a definite number of times per minute. **1968** J. L. Smith *Neuro-Ophthalmol.* IV. iv. 60 No satisfactory explanation has been given for hippus as yet.

hippy (hi·pi), *a.*[2] [f. Hip *sb.*[1]] With large or prominent hips.
1919 F. Hurst *Humoresque* 55, I seen you widening into a thirty-eight.. You're getting hippy, girl. **1938** *Life* 5 Sept. 43 (*heading*) Hippy girls welcome the form-concealing flattery of hoopskirts. **1956** H. Gold *Man who was not with It* (1965) xix. 172 She was getting a hippy and breasty look fast. **1969** 'M. Fallon' *Fine Night for Dying* ii. 21 A small hippy woman.

hippy, var. *hippie *sb.* and *a.*

hipster[1] (hi·pstəɹ). *slang* (orig. *U.S.*). [f. *hip *a.* + -ster.] One who is 'hip'; a hip- (or hep-)cat. Also *attrib.* Hence **hi·psterism,** the condition or fact of being a hipster; the characteristics of hipsters. Cf. *hepster.
1941 J. Smiley *Hash House Lingo* 31 *Hipster*, a know-it-all. **1946** Mezzrow & Wolfe *Really Blues* 374 *Hipster*, man who's in the know, grasps everything, is alert. **1948** *Partisan Rev.* XV. 722 Carrying his language and his new philosophy like concealed weapons, the hipster set out to conquer the world. **1956** *Observer* 23 Sept. 2/5 'Hipster' is modern jazz parlance for 'hep-cat'. **1958** *Listener* 3 July 16/1 This whole vexed question of hipsterism, anger, French new realism, and so forth. **1958** *New Statesman* 6 Sept. 292/3 The anthology is valuable for a speculative essay by Norman Mailer on 'beat' or hipster culture. **1959** 'F. Newton' *Jazz Scene* 291 *Jive-talk* or hipster-talk is..an argot or cant designed to set the group apart from outsiders. **1959** N. Mailer *Advts. for Myself* (1961) 303 The exchange was called 'Reflections on Hipsterism', when it appeared in *Dissent.* I did not choose the title, and so I have altered the name of the piece. **1967** *Lancet* 15 July 150/2 The 'hipster' movement in California..seemed to be an outright rejection of accepted standards and values.

hipster[2] (hi·pstəɹ). [f. Hip *sb.*[1] + -ster.] Used esp. *attrib.* of, or pertaining to, a garment, e.g. a skirt or trousers, that extends from the hips rather than the waist. In *pl.*, such a pair of trousers.
1962 *Sunday Express* 30 Dec. 16/4 Top girls are buying camel-hair hipsters with long matching braces. **1967** *Spectator* 14 July 53/3 A shop in the Chelsea Antique Market has neatly solved the male sartorial problem at a price level to fit any hipster pocket. **1968** J. Ironside *Fashion Alphabet* 45 *Hipster*, a skirt which sits on the hips instead of the waist.

Hirado (hirā·do). Also **-ato.** The name of a small island off the west coast of the province of Hizen on the island of Kyushu in Japan, used *attrib.* to designate a rich elaborate (blue-and-)white porcelain.
The art was originally introduced from Korea about A.D. 1600.
1880 in A. W. Franks *Jap. Pottery* 96 The works were established..in accordance with the order of a prince of the Matsu-ura family residing at Hirato. Hence the articles made here are generally called Hirato ware. **1881** Audsley & Bowes *Keramic Art Japan* 154 The productions of this kiln have commonly been called Hirado ware.

hiragana (hirāgā·nā). Also **firo-, -kana, -kanna.** [Jap., f. *hira* plain + *kana* (*kanna, kari-na*) borrowed letter(s).] The cursive form of the Japanese syllabary derived from the Tsau style of Chinese ideographs: intended for use by women. Cf. *katakana.
1822 F. Shoberl tr. *Titsingh's Illustr. Japan* 122 These two kinds of poems are composed in *firokanna*, or women's writing. **1859** A. Steinmetz *Japan & her People* i. vii. 305 The other style, called *hiragana*, employs at least six characters, radically different from each other, for each sound. **1861** G. Smith *Ten Weeks in Japan* vi. 92 Sentences written in..the irregularly flowing easy current style of the Japanese Hiragana character, are to be met with in every direction. **1863** *Chambers's Encycl.* V. 687/1 The phonetic alphabet, invented about the year 810 A.D., is known as the Hiragana form of character. **1880** *Encycl. Brit.* XIII. 585/1 Each character may be written in either the *katakana* or the *hiragana* style. **1883** I. Taylor *Alphabet* I. 35 The Hirakana syllabary was derived from a cursive form of the Chinese writing called the Tsau or 'grass' character. **1928** G. B. Sansom *Hist. Gram. Jap.* 45 At some period in the development of the script, probably about the time when the *hiragana* and *katakana* syllabaries were contrived. **1973** *Physics Bull.* May 280/3 We have also extended the process by including, in addition to the 881 Chinese characters, the 50 Japanese Kata-Kana and 50 Hira-Gana characters.

hire, *sb.* Add: **5.** *hire-car, -carriage*; **hire-purchase** (further examples); used esp. *attrib.*; also as vb., and **hi·re-purchasing** *ppl. a.*; cf. *H.P.* (s.v. *H III).
1895 *Brit. Warehouseman* Feb. 30/2 The plaintiff let a piano to one Sullivan under an ordinary hire-purchase agreement. **1901** Kipling *Kim* v. 122 The woman, she kept *kabarri* shop near where the hire-carriages are. **1909** *Chambers's Jrnl.* July 421/1 Not hire-purchased, you see; I'm not to be gulled by silly advertisements. No. I found my own shop and made my own choice. **1909** *Installation News* iii. 134/1 The hire and hire-purchase of cooking, heating, and other apparatus. **1923** 'Bartimeus' *Seaways* 36 If it wasn't for my daughter Annie bein' musical an' wantin' a pianner on the 'ire purchase, I wouldn't stop aboard 'er another night. **1930** *Economist* 25 Jan. 215/2 (Advt.), Bank of England and Hire Purchase. **1947** *E. Afr. Ann. 1946–7* 28/2 (Advt.), Largest fleet of hire cars in Kenya. **1960** *Author* Summer 64/2 All up-to-date readers are hire-purchasing electronic computors to read their library books for them. **1960** *Guardian* 9 May 6/7 The true-born, hire-purchasing Briton. **1964** R. Braddon *Year Angry Rabbit* xi. 94 It was *her* hire car that rammed the car taking Dorfmann and Welch to the deportation ship. **1966** *Times* (Canada Suppl.) 28 Feb. p. ii, Planes and hire-cars whisk you to the magic. **1972** *Mod. Law Rev.* XXXV. i. 38 All three hire-purchase agreements were illegal.

hire, *v.* Add: **1.** Also in phr. *hire and fire* [Fire *v.*[1] 16]; also as *sb.* and *attrib.*
1958 *Listener* 23 Oct. 632/2 Men who can be hired and fired according to the current level of business. **1963** *Times* 6 Feb. 6/3 'Hire and fire' is no longer associated with the construction industries. *Ibid.* 24 Apr. 7/4 Sir Donald said a five-year building programme should be planned in the public sector to ensure continuity of work and stop the 'hire and fire' attitude. **1970** *Sci. Amer.* Mar. 35/3 The inability to hire and fire in order to vary the work force with fluctuations in the business cycle. **1973** *Advocate-News* (Barbados) 20 Feb. 4/2 The locals are in the 'hire and fire' positions.
3. b. (Canadian example and earlier U.S. examples.)
1776 S. Curwen *Jrnl. & Lett.* (1864) 74 The inhabitants [of Sidmouth, Devon, England] chiefly hired out to the Newfoundland traders. **1833** C. A. Davis *Lett. J. Downing* (1834) 35, I had hired out here this summer. **1969** in Halpert & Story *Christmas Mumming in Newfoundland* 26 They hired out as fishing servants.

hi-rise, var. *high-rise* (*high *a.* 21).

Hirmologion (hə̄ɹmolōu·dʒiǫn). Also **Heirmologion, Hirmologium.** Pl. **-ia.** [med. Gr. εἱρμολόγιον or f. εἱρμός (see next) + λογ-, variant of λεγ- to say.] In the Eastern Church, a book containing eucharistic prayers, hirmoi, etc.
1850 in J. M. Neale *Eastern Church, Gen. Introd.* i. 890. **1952** *Monumenta Musicae Byzantinae* IV. 1 The final publication of the whole Byzantine Hirmologium is beyond the power of one man and must be based on a complete survey of all the sources. *Ibid.* VI. i. p. xxxiv, We have a better knowledge of old Hirmologia than of old Sticheraria. **1961** E. Wellesz *Hist. Byzantine Mus.* (ed. 2) p. v, Some authors..prefer the Latin form *Hirmologium*, others the Greek spelling *Heirmologion* or *Hirmologion.*

hirmos (hə̄·ɹmǫs). Also **heirmos, hirmus.** Pl. **-moi, -mi.** [Gr. εἱρμός series, connection.] In the hymnology of the Eastern Church, a model stanza forming a pattern for the other stanzas.
1850 J. M. Neale *Eastern Church, Gen. Introd.* i. 830 There are a certain number of *Troparia*..called Hirmoi, on the model of which all other troparia, etc., are formed. *Ibid.* 835 Here we have the actual hirmos on which the third ode is arranged. **1863** R. F. Littledale *Offices East. Ch.* 214. **1880** *Encycl. Brit.* XII. 580/1 An *ode* is a song or hymn compounded of several similar 'troparia'. ..To these is always prefixed a typical or standard 'troparion', called the *hirmus.* **1959** *Chambers's Encycl.* VII. 348/2 A *contakion*, consisting (broadly speaking) of a *heirmos*, which set the pattern of what followed, succeeded by a series of *troparia.* **1961** E. Wellesz *Hist. Byzantine Mus.* (ed. 2) 14 Dom Gaisser's article on the *Heirmoi*, the model strophes of the Easter Hymn, is the first detailed study of Byzantine hymnography.

hirola, var. *herola.

Hirschsprung (hīə·ɹʃsprŭŋ). The name of Harald *Hirschsprung* (1830–1916), Danish physician, used in the possessive in *Hirschsprung's disease*, congenital enlargement of the colon, occurring esp. in boys, *spec.* such enlargement due to the absence of the ganglion cells of a segment of the lower colon or rectum.
1900 Dorland *Med. Dict* 208/2 *Hirschsprung's d[isease]*, congenital hypertrophic dilatation of the colon. **1908** *Practitioner* Sept. 456 An example of idiopathic dilatation of the colon, or—as it is called in Germany—Hirschsprung's disease, is recorded by Wagner. **1949** *Lancet* 1 Jan. 10/2 A considerable proportion of cases of megacolon (not being Hirschsprung's disease) remain in the residual 'idiopathic' group. **1966** B. C. Morson in Wright & Symmers *Syst. Path.* I. xvii. 549/1 Idiopathic megacolon..has to be distinguished from congenital megacolon (Hirschsprung's disease), in which there is a characteristic underlying histological defect. **1968** F. A. Jones et al. *Clin. Gastroenterol.* (ed. 2) iv. 121 Confusion later arose because the term Hirschsprung's disease was applied to any megacolon.

hirsel, *sb.* Add: **1. c.** The ground occupied by a flock of sheep.
1822 Scott *Fortunes of Nigel* III. ii. 50 Being in a strange country, like a poor lamb that has wandered from its own native hirsel. **1856** J. C. Morton *Cycl. Agric.* II. 723/2. **1886** C. Scott *Sheep-Farming* 122 He will be able to divide the hill into 'hirsels', and the hirsel again into 'cuts'. **1922** *Glasgow Herald* 16 Dec. 4 There they are fed for days ..till the hirsels are green again. **1944** G. Henderson *Farming Ladder* i. 28 A mountain sheep farm, or hirsel, is selected in such a manner that food and shelter can be obtained by the stock. **1965** *Punch* 3 Mar. 325/2 The Welsh fridd system, which keeps sheltered grazing fresh for lambing time, is so much better than the Scottish hirsel (where the flock grazes all the year round on one hill) that it is rapidly winning popularity.

hirsute, *a.* Add: Hence **hi·rsutism,** an abnormal degree of hairiness.
1926 in *Index-Catal. Library Surg.-General's Office, U.S. Army* VI. 753/2. **1927** *Jrnl. Amer. Med. Assoc.* 19 Mar. 969 Hirsutism and suprarenal virilism. **1959** *New Biol.* XXVIII. 132 Other serious abnormalities occur in these diseases, but hirsutism in female patients is a particularly distressing result.

hirudin (hirū·din). *Biochem.* [a. G. *her-, hirudin*, attributed to Jacobj (see F. Franz 1903, in *Arch. f. exper. Path. und Pharm.* XLIX. 362), f. L. *hirūdo* leech + -in[1].] An anticoagulant protein produced by the salivary glands of leeches which acts by combining with thrombin in the blood.
1905 *Jrnl. Chem. Soc.* LXXXVIII. ii. 339 The substance named hirudin separated from leech extract has a very variable activity. **1908** *Practitioner* Apr. 463 Coagulation is prevented by the use of 'hirudin'. **1963** R. P. Dales *Annelids* ii. 60 Leeches have one or more pairs of 'salivary' glands producing an anti-coagulant, 'hirudin', preventing the blood from clotting as it is ingested.

his, *absolute poss. pron.* Add: **b.** Used, normally in combination with or in opposition to *hers*, of an article, room (e.g. a W.C.), etc., intended for males. Occas. used before a following noun.
1949 *Good Housekeeping* June 26/2 With color contrasts in towels and washcloths, it's easy to distinguish 'his' from yours. **1950** *Charm* Apr. 82 Among your wedding gifts..may be a set of bath towels labelled with beautiful equity and foresight: 'His' and 'Hers'. **1953** *Imagination* June 115/1 Almost before Crae brought the car to a gravel-spraying stop..Ellena had the door open and was out and around the corner marked His and Hers. **1962** J. Blackburn *Gaunt Woman* x. 156 'His' was a gloomy little room, laid with dirty grey lino. **1964** *Guardian* 10 Oct. 5/2 We halve the available space..under the primitive his and hers law... There is usually a his chair and a her chair. **1965** *New Statesman* 21 May 790/3 What I call a His/Her double bedroom. **1967** *Guardian* 4 Jan. 7/4 His and Hers. J. R. L. Andersen reports on the Boat Show. Women's influence in the world of small boats is now almost dominant.

hislopite (hi·zlǫpəit). *Min.* [f. the name of the Rev. Stephen *Hislop* (1817–63), Scottish missionary and naturalist: see -ite[1].] An Indian variety of calcite coloured green by the presence of glauconite.
1859 S. Haughton in *Phil. Mag.* XVII. 18, I propose to give the name of Hislopite to the remarkable combination of Calc-spar and Glauconite found by him [*sc.* Mr. Hislop] at Nâgpur. **1893** *Rec. Geol. Surv. India* XXVI. 171 The name hislopite loses its specific value when the variation of the included so-called glauconite is..so great as the foregoing results show.

hisn, his'n. Add: (Further examples.) Also **hissn.**
1867 'Mark Twain' *Celebr. Jumping Frog* 15 It always makes me feel sorry when I think of that last fight of his'n. **1910** C. E. Mulford *Hopalong Cassidy* xx. 130 Taking the button and looking it over. 'Yep, its hissn, all right.' **1923** 'R. Crompton' *William Again* x. 179 Well it isn't his'n—it's stole stuff.

hispa (hi·spă). *Ent.* [mod.L. (C. Linnæus *Systema Naturæ* (ed. 12, 1767) II. 603), f. L. *hispidus* bristly, hairy.] A tropical leaf-beetle of the genus so named. Hence **hi·spid, hi·spine** *adjs.*, of or pertaining to the family Hispinæ, which has *Hispa* as its typical genus.
1794 P. A. Nemnich *Allgemeines Polyglotten-Lexicon der Naturgeschichte* III. 165 Hispa. [*Entomol.*]... Engl. The hispa. **1835** *Boston Jrnl. Nat. Hist.* I. 147, I found full grown larvæ of a Hispa in the leaves. **1889** E. C. Cotes *Further Notes Indian Insect Pests* 37 The Bengal Rice Hispa. **1922** D. Sharp in *Cambr. Nat. Hist.* VI. v. 282 *Hispa* is one of the most extensive of the numerous genera of Hispides. **1924** *Bull. Entomol. Res.* XIV. 245 (*title*) A new Hispid beetle. **1933** *Proc. Zool. Soc.* 669 (*title*) On the structure of larvae of Hispine beetles. **1962** R. Wyniger *Pests of Crops in Warm Climates* iii. 179 Sugar-cane hispid miner.

Hispanic, *a.* Add: Also **Hi·spanism** = *Hispanicism.*
1949 S. de Madariaga *Christopher Columbus* (ed. 2) 410 When he makes errors in Latin they are Hispanisms. **1964** Y. Malkiel in *Archivum Linguisticum* XVI. 3 Differently developed Hispanisms in Arabic.

Hispanist, hispanist (hi·spănist). [f. Hispan(ic *a.* + -ist.] A student of the literature, language, and civilization of Spain.
1934 Webster, *Hispanist*, one versed in, or devoted to, the Spanish language or the study of Spanish. **1960** *Times* 17 Oct. 20/6 His untimely death is a grievous loss to hispanists in the University of London. **1964** *Archivum Linguisticum* XVI. 12 Such an approach predominates in the work of the contemporary Hispanist.

Hispano-. Add: **Hispa:no-Ame·rican** *a.*, Spanish and American; also as *sb.*; **Hispa:no-A·rab, -Ara·bian, -A·rabic** *adjs.*, Spanish and Arabian; **Hispa:no-Go·thic** *a.*, Spanish and Gothic; **Hispa:no-Maure·sque, -More·sco, -More·sque** *adjs.*, Spanish and Moorish; **Hispa·nophil(e),** a lover of Spain and Spanish culture.
1823 T. Ross tr. *Bouterwek's Hist. Span. Lit.* 6 The nobles, who were of French or Hispano-Gothic origin. **1880** *N. & Q.* 16 Oct. 306/1 (*heading*) Hispano-Arabian poetesses. **1881** C. C. Harrison *Woman's Handiwork* ii. 104 A charm of Fortuny's studio was his Hispano-Moresco lustred pottery. **1904** W. H. Hudson *Green Mansions* 4 The nervous olive-skinned Hispano-American of the tropics. [**1906** *Daily Chron.* 19 May 3/2 A Hispanofilo to the core.] **1909** *Ibid.* 18 Nov. 1/3 The eighteen-carat gold casket presented to the King of Portugal..is oblong in shape and in the Hispano Gothic style. **1910** *Ibid.* 4 Jan. 3/1 The true Hispanophil with the cult in his veins. **1920** *Glasgow Herald* 25 Nov. 6 All good British Hispanophils. **1933** *Archit. Rev.* LXXIII. 5/1 Most people would even be eager to concede that this hispano-arabian pastry was the reverse of a great art-form. *Ibid.*, And certainly nothing could be more unlike the structureless confections of the hispano-mauresque than the type of these Kasbahs. **1936** *Burlington Mag.* Apr. 198/2 The so-called mudéjar (Hispano-Moresque) style of the kingdoms of Aragon and Castile. **1942** *Ibid.* Feb. 41/1 The two pile carpets here reproduced..provide a further link in the chain of Hispano-American art. **1959** *Listener* 12 Mar. 471/2 He belonged to a distinguished Hispano-Arab family. **1960** *John o' London's* 14 Apr. 432 Many an English Hispanophile. **1964** *Archivum Linguisticum* XVI. 3 *Plomo* may similarly have had a Hispano-Arabic predecessor. **1964** *Punch* 29 July 173/3 Recommended to all Hispanophiles. **1973** *Daily Tel.* 12 Jan. 29/1, I see happy Hispanophiles streaming from Greenock and Blackburn to the Costa del Sol.

Hisperic (hispe·rik), *a.* [ad. med.L. *Hisperica* (see def.). Cf. G. *hisperisch.*] Epithet of a variety of mediæval Latin, of which *Hisperica Famina* (probably of the 6th century) is a notable example, characterized by a highly artificial vocabulary of which the use of borrowed words of Hebrew, Greek, and other origin is a salient feature.
1904 W. P. Ker *Dark Ages* 35 The 'Hisperic' vocabulary, which is that of Apuleius, Florus, Martianus Capella, exaggerated out of all measure. **1907** *Cambr. Hist. Eng. Lit.* I. 69 The hymn..known as *Altus prosator* contains very marked specimens of Hisperic Latinity. **1931** E. J. Jones *Hist. Educ. Wales* I. 179 Hisperic words are included in the Anglo-Saxon glossaries of the tenth century. **1948** E. S. Duckett *Anglo-Saxon Saints & Scholars* i. 75 Rhyming eight-syllabled lines, again filled with uncouth Hisperic words.

hi-spy, var. Hy-spy.
1876 'Mark Twain' *Tom Sawyer* xxviii. 267 They had an exhausting good time playing 'hi-spy'.

hiss, *v.* Add: **1.** Also in *Electricity.* Cf. *hissing arc.
1961 J. Thewlis *Encycl. Dict. Physics* III. 703/1 If the current passing through a carbon arc exceeds a certain value depending on the length of the arc, the latter begins to 'hiss'.

hiss, *sb.* **1. b.** Add to def.: Now chiefly the sibilants [s] and [z].
1933 L. Bloomfield *Language* vi. 100 Our gingival spirants [s, z] are hisses or sibilants. **1953** *Archivum Linguisticum* V. 68 The distinction between hiss- (Fr. *sifflantes*) and hush-sibilants (Fr. *chuintantes*).

his-self. Add: Also as one word, representing a *colloq.* or *dial.* pronunciation of Himself.
Further examples s.v. *himself IV (above).
a **1876** E. Leigh *Gloss. Words Dial. Cheshire* (1877) 103 'He is not hissell', *i.e.* 'He is out of his mind.' **1902** *Dialect Notes* II. 316 He has hurt *hisself* mighty bad. **1938** M. Richardson in B. A. Botkin *Treas. S. Folklore* (1949) iii. i. 443 He would chop a tree by hisself. **1969** R. Pharr in A. Chapman *New Black Voices* (1972) 69 If he wants something for hisself, let him get out and work for it like we do. **1972** 'J. & E. Bonett' *No Time to Kill* x. 133 He's earned hisself a medal. **1973** *Black World* July 64/1 He can take care of hisself.

hissing, *ppl. a.* Add: **1.** (Further examples.)
1922 [see *click sb.[1] 1 b]. **1936** *Bell Syst. Techn. Jrnl.* XV. 197 If such a resistance element is in a current-carrying circuit associated with a telephone receiver or loud speaker..a steady hissing noise which sounds like that due to shot effect or thermal agitation of electric charge is heard.
d. *Comb.* **hissing adder, hissing sand-snake** (see quots.); **hissing arc,** an electric arc which emits a hissing sound.
1931 R. L. Ditmars *Snakes of World* vii. 72 *H[eterodon] contortrix* is the common eastern species [*sc.* of hognosed snake] found from Massachusetts to Florida and westward to Minnesota and Texas. It is known as the Flat-headed 'Adder', Hissing 'Adder', and other names in keeping with its antics. **1895** *Electrician* 18 Jan. 338/2 The potential difference for hissing arcs increases with length of arc, but appears to be fairly constant for a given length of arc... One talks of hissing, however, as if there were only one sort of hissing, whereas really there are at least two, with very different significations. **1941** J. D. Cobine *Gaseous Conductors* 293 The high-pressure carbon arc is seen to have two characteristics, one of hyperbolic shape for a silent arc, and the other, essentially linear, for a hissing arc. **1910** F. W. Fitzsimons *Snakes S. Afr.* iii. 54/2 Psammophis sibilans. Hissing Sand Snake... Distribution: Tropical Africa and Egypt, Rhodesia, Transvaal, Zululand, Port Elizabeth. **1954** J. A. Pringle *Common Snakes* 20 Hissing Sand-Snake. *Psammophis sibilans*..occurs in all the provinces of the Union [*sc.* of South Africa] except the Cape.

hissy (hi·si), *a.* [f. Hiss *sb.* + -y[1].] Consisting of a hiss; of a sound, resembling that of a hiss.
1905 J. Masefield *Tarpaulin Muster* (1907) 171 Snakes,..laughing in a sort of hissy chuckle. **1948** I. Brown *No Idle Words* 42 Cypress, though rather a 'hissy' word.

hist (həist), *v.* Chiefly *U.S.* Also **h'ist, hyst.** [dial. form of Hoist *v.*; cf. *heist.]
1. To provide the key for singing (a hymn, etc.). *rare.*
1857 *Harper's Mag.* Sept. 572/1 As they have no choir in the congregation, any one who considers himself qualified has authority to hist the hymns.
2. To raise aloft; = Hoist *v.* 1; also, to steal, hijack. Cf. *heist *v.*, *hoist *v.* 6.
1867 'Mark Twain' *Celebr. Jumping Frog* 18 Dan'l give a heave, and hysted up his shoulders. **1872** 'Agrikler' *Rhymes* 17 Hev a fresh cask ready histed. **1919** H. L. Mencken *Amer. Lang.* iii. 91 They still cling, in their common speech, to such forms as *h'ist* for *hoist.* **1930** *Amer. Mercury* Dec. 456/1 *Hist*, to hold up; to hyjack. 'We hist the mutt's plant for fifty cases of skee.' **1936** M. Mitchell *Gone with Wind* xxxiv. 569 You'd better hist up your skirts a little. **1938** D. Runyon *Furthermore* xiv. 290 This is one of the very first cases of histing a truckload of legal beer that comes off in this country. **1938** M. K. Rawlings *Yearling* viii. 70, I h'isted him over old Caesar's rump and away we goed.
Hence **hi·sting** *vbl. sb.*
1935 J. T. Farrell *Judgment Day* x. 213 There's been too many histing jobs pulled off lately in this neighborhood, and the sergeant has been hopping on my tail about them.

Histadrut (hi·stadrūt). Also **Histadrud, Histadruth.** [mod.Heb. *ha-histadrūṯ* the federation: in full *ha-histadrūṯ ha-kᵉlālīṯ šel hā-ʿōḇᵉdīm bᵉʾereṣ yisrāʾēl* the general federation of workers in the land of Israel.] The General

Federation of Labour in Israel, founded in 1920.

1923 P. GRAVES *Palestine* viii. 144 A young, idealistic party who work with the Ahaduth Ha-Avoda in the Histadrut Ha-Ovdim, or Workmen's Organization. **1925** L. STEIN *Zionism* v. 171 The *Histadruth* provides its members with evening classes, workmen's clubs, lectures, concerts, and travelling libraries. **1933** F. ANDREWS *Holy Land under Mandate* II. 26 The Histadruth has an active Press, its most influential paper being the daily, 'Davar', with a circulation of over four thousand. **1937** *Palestine Labour Studies* iv. 7 The Histadruth is—in aspiration and essence—an organization binding together the founders of a National Home, the builders of a country, the liberators of a people. **1952** S. SPENDER *Learning Laughter* xv. 196 The Histadrut (Trades Union Federation) is perhaps the most powerful Trades Union Organization in the world. **1972** *Times* 6 Apr. 15/7 He served in a number of senior posts in the Histadrut, the Israel Federation of Labour.

histamine (hi·stămin). *Biochem.* Formerly also **-in.** [f. *HIST(IDINE + AMINE.] An amine formed from histidine by decarboxylation, widely found in both animal and plant tissues, and having a specific action as a stimulator of gastric secretion and as a dilator of the capillaries. Hence **hi·staminase** [see *-ASE], an enzyme which inhibits the action of histamine.

1913 DORLAND *Med. Dict.* (ed. 7), *Histamin*, betaimidazolylethylamin: used like pituitrin. **1913** *Jrnl. Chem. Soc.* CIV. I. 681 Histamine..is a base which Barger and Dale separated from the intestinal mucous membrane. **1918** *Jrnl. Amer. Chem. Soc.* XL. 1723 The first 3 extracts were pale yellow in color and contained practically all the histamine. **1918** *Jrnl. Physiol.* LII. 110 In earlier papers P. P. Laidlaw and one of us [sc. H. H. Dale] referred to it by its chemical name, β-iminazolylethylamine (abbreviated for convenience to β-I.), being unwilling to coin a new name for a substance long known... Several later investigators of its action, however, have given it the name 'histamine', which is so obviously appropriate for this amine derived from histidine, that we have adopted it here. **1920** *Nature* 4 Mar. 11/2 A compound of known chemical structure, called 'histamine',.. which is able to produce a state of the circulation like that present in wound-shock. **1929** *Times* 18 Oct. 15/3 These products produce what is spoken of as 'histamine poisoning', the substance histamine being the principal toxic agent. **1930** BEST & MCHENRY in *Jrnl. Physiol.* LXX. 357 We suggest..that the substance, or system, which produces a change in structure responsible for the loss of physiological activity of histamine be designated histaminase. **1951** WHITBY & HYNES *Med. Bacteriol.* (ed. 5) 90 The hypothesis that histamine-release is the immediate cause of the symptoms of anaphylactic shock formerly rested largely on the close similarity between the symptoms of histamine poisoning and of anaphylactic shock in different animals and in vitro. **1957** *New Biol.* XXIV. 54 Histamine belongs to a group of substances known chemically as amines and physiologically as vasodilators, i.e. they dilate the blood vessels. **1965** LEE & KNOWLES *Animal Hormones* viii. 119 Histaminase, which destroys histamine, does not affect the potency of the extract and gastrin appears to be a true hormone. **1968** J. H. BURN *Lect. Notes Pharmacol.* (ed. 9) 24 Histamine is an amine which can be formed by removing the —COOH group from the amino-acid histidine.

hister (hi·stəɹ). [mod.L. (C. Linnæus *Systema Naturæ* (ed. 10, 1758) I. 358), f. L. *hister* = *histrio* actor.] A beetle of the genus so named or of the family Histeridæ of carnivorous coleoptera. Also **hi·sterid** *a.*, belonging to this family; *sb.*, a beetle of this family.

1794 P. A. NEMNICH *Allgemeines Polyglotten-Lexicon der Naturgeschichte* III. 165 Hister. [Entomol. (sic)].. *Engl.* The hister. **1839** J. O. WESTWOOD *Introd. Class. Insects* I. 182 Cadet de Vaux, in his History of the Mole, observes that, almost as soon as it is dead, it is devoured by a number of Histers. **1874** J. G. WOOD *Insects Abroad* 89 The Giant Hister. **1918** W. A. BRYAN *Nat. Hist. Hawaii* 217 The histerid beetles. **1924** A. D. IMMS *Text-Bk. Entomol.* 485 Hister and its allies frequent dung and carrion. **1965** R. E. FREEMAN tr. *Vandel's Biospeleol.* xiii. 212 Patrizi..has described another hypogeous histerid. **1966** C. SWEENEY *Scurrying Bush* iv. 32 Two adult histerid beetles, black, shiny and oval, occurred.

histidine (hi·stidīn, -in). *Biochem.* Also **-in.** [ad. G. *histidin* (A. Kossel 1896, in *Zeitschr. f. physiol. Chem.* XXII. 176), f. Gr. ἱστ-ός web, tissue + *-IDINE c.] An amino-acid; see quot. 1940.

1896 *Jrnl. Chem. Soc.* LXX. I. 582 The sulphate from.. sturgeon sperm has rather different solubilities in sodium chloride solutions, and the names *salmine* and *sturine* are suggested by the two protamines. By decomposing the latter base with sulphuric acid, a new crystalline base was prepared, which is called *histidine*. **1900** DORLAND *Med. Dict.* 301/2 *Histidin*, a substance, $C_6H_9N_3O_2$, obtainable from any protamin by the action of sulphuric acid and water. **1919** *Nature* CIV. 322/2 The diamino-acids lysin, histidin, and arginin. **1940** *Chambers's Techn. Dict.* 417/2 *Histidine*, α-amino-β-imidazole propionic acid, a protein derivative belonging to the group of hexone bases. **1963** H. BURN *Drugs, Med. & Man* (ed. 2) vi. 73 Protein is a structure rather like a house which is built from bricks, each of which is chemically an amino-acid. One of these bricks is an amino-acid called histidine.

histiocyte (hi·stiosəit). *Physiol.* [ad. G. *histiozyt* (Aschoff & Kiyono 1913, in *Folia Haematologica: Arch.* XV. 386), f. Gr. ἱστίο-ν web, dim. form of ἱστός web, tissue + -CYTE.] A large, highly phagocytic cell found in connective tissue and becoming motile when stimulated; also called *adventitious cell*, *clasmatocyte*, *macrophage*, *resting wandering cell*. So **histiocy·tic** *a.*, of or pertaining to, of the nature of, histiocytes.

1924 *Jrnl. Infectious Dis.* XXXIV. 583 These resting wandering cells or histiocytes in all their modifications throughout the connective tissues..are the most active elements in the chronic inflammatory processes and in particular in tuberculosis. *Ibid.*, It is natural that the lymphocytes should also play an active rôle in the same processes and that they join the polyblasts of histiocytic origin. **1961** *Lancet* 2 Sept. 523/2 The dark red spleen..was unremarkable microscopically except for an occasional large histiocyte in the red pulp with finely vacuolated cytoplasm. **1966** [see HISTIOCYTOSIS below].

Also **hi:stiocyto·sis** *Path.* [-OSIS], any condition characterized by a proliferation of histiocytes.

1953 V. BLOOM in *Amer. Jrnl. Path.* I. 623 For the want of a better name we would suggest 'lipoid-histiocytosis' as tending to convey the idea of a process involving the storing of lipoid material by the histiocytes throughout the body. **1953** L. LICHTENSTEIN in *A.M.A. Arch. Path.* LV. 102 The conditions previously designated eosinophilic granuloma of bone, 'Letter-Siwe disease' and 'Schüller-Christian disease' are interrelated manifestations of a single malady. The name 'histiocytosis X' is suggested as a provisional broad general description for this nosologic entity. **1966** *New Eng. Jrnl. Med.* 28 Apr. 938/1 The association of histiocytic proliferations with recurrent infections has been noted in histiocytosis X and in pigmented histiocytosis.

histo-. Add: **histoche·mically** *adv.*, by histochemical means; **hi:stocompatibi·lity**, compatibility (sense *2 b) between a grafted tissue and the recipient; so **hi:stocompa·tible** *a.*; **histo·genous** *a.*, formed by tissue; **hi·stolyse** *v.*, to subject to histolysis; so *hi·stolysing* ppl. adj.; **hi:stometa·basis** *Palæont.* [METABASIS], a state of complete fossilization in which the minute markings of grain and texture are preserved; **hi:stopatholo·gic, -ical** *adjs.*, characterized by diseased tissues; of or pertaining to histopathology; **hi:stopatho·logist**, one who specializes in histopathology; **hi:stopatho·logy**, (the study of) the tissue changes associated with a disease or disorder; **histopla·smin** [-IN¹], a sterile preparation of a culture of the fungus *Histoplasma capsulatum*, used in skin tests for histoplasmosis; **hi:stoplasmo·sis** [-OSIS], an infection due to the fungus *Histoplasma capsulatum*, endemic in parts of the United States and taking the form of either a benign transient infection of the lungs or, rarely, a disseminated, usu. fatal, disease of the reticulo-endothelial system.

1955 *Brain* LXXVIII. 327 Lafora bodies (intracellular amyloid bodies) were encountered, and were examined histochemically. **1971** *Jrnl. Insect Physiol.* XVII. 862 Approximately 10 per cent of the histochemically identifiable lysosomal phosphatase remains in tissue fragments attached to the silk. **1948** G. D. SNELL in *Jrnl. Genetics* XLIX. 87 Genes of the type postulated in the genetic theory of tumour transplantation will be here referred to as histocompatibility genes. The prefix 'histo' is used because the same genes which determine susceptibility or resistance to tumour transplants probably also determine susceptibility or resistance to tissue transplants in general. **1969** N. A. MITCHISON et al. *Organ Transplantation Today* 29 the genetic system controlling histocompatibility is now amenable to analysis. **1964** *Transplantation* II. 656/1 An inbred strain is a histocompatible donor to its F₁ offspring. **1907** *Practitioner* Sept. 455 The Histogenous Cells which originate locally as the result of local tissue proliferation. **1946** *Nature* 3 Aug. 147/2 Histogenous demarcation of infected tissue (for example, abscission in shot-hole disease of peach and demarcation by cork layer in black root rot of tobacco). **1963** R. P. DALES *Annelids* vi. 114 The muscles of the body wall are partly histolysed and those of the parapodia augmented. **1946** *Nature* 14 Sept. 367/2 The cutis acts as a physico-chemical barrier, inhibiting the penetration of the histolysing substances into the zone of regeneration from the regenerating epithelium. **1923** C. A. WHITE in *U.S. Nat. Mus. Ann. Rep. 1892* 264 The term histometabasis is applied to that condition of fossilization in which an exchange of the original substance for another has occurred in such a manner as to retain or reproduce the minute and even the microscopic texture of the original. **1917** R. S. LULL *Org. Evol.* xxv. 412 The resultant fossil retains..not only the external form but the histologic characters (histometabasis..) of the original structure as well. **1903** *Detroit Med. Jrnl.* Feb. 705/1 The histopathologic states of the inner structures of the labyrinth. **1947** *Radiology* XLIX. 292/2 Histopathological changes in organs and tissue may occur in the absence of or prior to observable changes in the blood or blood-forming tissue. **1934** WEBSTER, Histopathologist. **1969** R. VON HAAM in *Textbk. Clin. Path.* (ed. 9) xxii. 552 The histopathological change in the syphilitic lesion is a valuable addition to diagnostic methods. **1946** *Acta Med. Scand.* CXXIII. 445 Histopathological discoveries in amyotonia congenita. **1909** *Times Lit. Suppl.* 2 Apr. **1961** *Lancet* 30 Sept. 770/2 London histopathologists. **1966** *Ibid.* 31 Dec. 1450/1 The wealth of photomicrographs is likely to appeal more to the practising histopathologist than to the physicians, dentists, and students. **1896** N. WALKER tr. P. G. Unna (*title*) The histopathology of the diseases of the skin. **1908** *Practitioner* Jan. 27 The histopathology of the lesions. **1959** *Chambers's Encycl.* VI. 123/2 Fungi of this group cause systemic mycoses which may closely resemble tuberculosis in their symptomatology and histopathology. **1945** CHRISTIE & PETERSON in *Amer. Jrnl. Publ. Health* XXXV. 1135/2 Histoplasmin is the term we used to designate the antigen we were using for skin testing. It is to be hoped that this natural term will be used for any antigen which may be found satisfactory for the purpose of skin testing in relation to histoplasmosis. **1964** R. D. BAKER *Textbk. Path.* ix. 242/1 Skin tests using histoplasmin (the counterpart of tuberculin) suggest that in endemic areas histoplasmosis is a common disease, which is not usually clinically apparent. **1907** S. T. DARLING in *Maryland Med. Jrnl.* L. 125 (*heading*) Notes on histoplasmosis—a fatal disorder met with in tropical America. **1955** *Sci. Amer.* Jan. 44/3 There has been a good deal of question about the mode of spread of histoplasmosis, a lung disease widely prevalent in the Middle West. **1973** *Daily Colonist* (Victoria, B.C.) 13 May 40/1 Pigeons on the roof are suspected as the source of an illness called histoplasmosis recently suffered by two law professors.

histogram (hi·stogræm). *Statistics.* [f. Gr. ἱστό-ς mast, web + -GRAM.] A diagram consisting of a number of rectangles or lines drawn (usu. upwards) from a base line, their position along this line representing the value or range of one variable and their height the corresponding value of a second variable.

1891 in E. S. Pearson *Karl Pearson* (1938) 142 The geometrical representation of statistics.. Lecture IV. Histograms. Optical advantages of vertical over horizontal columns. **1895** K. PEARSON in *Phil. Trans. R. Soc.* A. CLXXXVI. 399 The histogram shows, however, the amount of deviation at the extremes of the curve. [*Note.* The word 'histogram' was introduced by the writer in his lectures on statistics as a term for a common form of graphical representation, i.e., by columns marking as areas the frequency corresponding to the range of their base. **1903** *Nature* 17 Dec. 149/1 We venture to protest against any such crude method of testing the goodness of fit as that of placing a frequency curve on seven or eight blocks forming a 'histogram' and judging the look of the fit. **1949** *Jrnl. R. Aeronaut. Soc.* LIII. 974/2 In Fig. 4..the best expression of the curve fitted to the histogram of the hours spent in 1,300 conversations on the London Airport R/T channel of frequency 118·1 Mcs. **1969** *Nature* 15 Nov. 655/2 The 178 *z*-values analysed are collected in bins of 0·012 and are then marked off as a histogram on a *z*-scale.

histomap (hi·sto,mæp). [f. HISTO(RY + MAP.] (See quot. 1956.)

1945 *Amer. Speech* XX. 76/2 A 'Histomap of Religion' was sent to customers and friends at Christmas time, 1943, by the Kalamazoo Vegetable Parchment Company. **1956** R. REDFIELD *Peasant Soc. & Culture* 38 A picture of their relationships would be something like the 'histomaps' we sometimes see, those diagrams of the rise and change through time of religions and civilizations.

histone (hi·stōᵘn). Formerly also **-on.** *Biochem.* [ad. G. *histon* (A. Kossel 1884, in *Zeitschr. f. physiol. Chem.* VIII. 512), perh. f. Gr. ἱστ-άναι to arrest or ἱστ-ός HISTO-: see -ONE.] Any of a small class of simple, basic proteins which are soluble in water and dilute acids but insoluble in dilute ammonia and which are most commonly found in association with nucleic acids.

1885 *Jrnl. Chem. Soc.* XLVIII. I. 572 Extraction with dilute acid isolates a substance which belongs to the class of bodies called A-peptones by Meissner... The author names this substance *histon*. **1905** C. E. SIMON *Physiol. Chem.* (ed. 2) 194 The protamins are decomposed entirely like the albumoses and peptones, while the histons are only affected in part, which coincides with the position which the histons occupy midway between the protamins and the true albumins. **1932** *Sci. News* XXIV. 37 Histones..are present in appreciable amounts in the nuclei of the tissue cells of all animal species examined. **1968** *New Scientist* 7 Nov. 303/1 What now seems much more possible is that the histones are simply the tool needed to shut down genes, and that they are instructed which particular gene..to shut down by another, specific, molecule.

historic, *a.* **1.** (Further examples.)

1907 I. ZANGWILL *Ghetto Comedies* 391 'The unconditional historic necessity will carry us on of itself towards a better social state.' 'There you go with your Marx and your Hegel!' **1921** G. B. SHAW in *Nation* 19 Feb. 705/1 Mr. Hyndman falls back..on Historic Determinism, and declares that the Bolshevists must fail because the economic conditions are not ripe. **1940** 'G. ORWELL' *Inside Whale* 22 There is no perception here of what is now called historic necessity.

historical, *a.* Add: **1.** (Further examples.)

1937 A. HUXLEY *Ends & Means* vii. 66 One word is common to all the dictatorial vocabularies and is used for purposes of justification and rationalization by Fascists, Nazis and Communists alike. That word is 'historical'. *Ibid.* 67 The dictatorship of the proletariat is an 'historical necessity'. The violence of Communism is justified because..it is being used to forward an ineluctable 'historical' process. **1965** S. BELLOW *Herzog* (1965) 99 You must sacrifice your poor, squawking, niggardly individuality..to historical necessity.

2. d. Related to or connected with history; considered from the historian's point of view; belonging to the past.

1881 E. A. FREEMAN *Hist. Geogr. Europe* I. i. 2 It is with political divisions that historical geography has to deal in the first place. **1894** G. A. SMITH (*title*) The historical geography of the Holy Land. **1923** L. D. STAMP *Introd. Stratigr.* i. 1 Stratigraphy is another name for Historical Geology. **1924** O. JESPERSEN *Philos. Gram.* ii. 31 Descriptive linguistics can never be rendered superfluous by historical linguistics. **1933** *Amer. Speech* VIII. iii. 6/2 It is the paradox of linguistic geography that a method of research strictly limited to contemporary speech has proved to be the means of revitalizing the study of historical linguistics. **1964** R. H. ROBINS *Gen. Ling.* 5 Historical linguistics is the study of the developments in languages in the course of time.

3. (Further examples.)

1816 JANE AUSTEN *Let.* 1 Apr. (1952) 452, I am fully sensible that an historical romance, founded on the House of Saxe Cobourg, might be much more to the purpose of profit or popularity. **1826** *Blackw. Edin. Mag.* Feb. 180 These historical novels may operate advantageously on the mind. **1874** GEO. ELIOT *Legend of Jubal* 193 Imagination.. Aiming at fiction called historical. **1930** O. LANCASTER *Homes Sweet Homes* 18 The dashing cloak-flinging figure of the historical fiction. **1981** *Observer* 28 Jan. 7/4 Philip Woodruff's *Colonel of Dragoons*.. a very model of what historical fiction ought to be. **1972** *Guardian* 18 Mar. 12/5 The movie..is..a great deal more than a decently-handled historical pageant.

5. (Later example.)

1936 *English Studies* Feb. 60 An interesting paper on 'Chaucer's Historical Present, its Meaning and Uses' by L. D. Benson.

7. In *Comb.*, prefixed to an adj. to denote: **a.** 'historical and..', as *historical-comparative*, *-economic*, *-sociological*; **b.** 'historically, as applied to history', as *historical-lexicographical*, *-onomatological*, *-typological*; also *historical-minded* adj., *-mindedness*.

1933 L. BLOOMFIELD *Language* i. 19 These two streams of study, the historical-comparative and the philosophical-descriptive. **1948** J. TOWSTER *Political Power in U.S.S.R.* iv. 55 The historical-economic conditions of national movements. **1965** *Language* XLI. 138 Historical-lexicographical monographs. **1914** *Fair. Vo. Chron.* 28 Jan. 14 Mar. (1927) 117 Dr. Bigg, Dr. Bryce..are historical-minded rather than thinkers. **1963** *Language* XL. 115 Egerton's linguistic work..historical-minded but not to the neglect of description. **1969** ACTON *Lect. Study History* 58 That influence for which the depressing names historicism and historical-mindedness have been devised. **1965** *Listener* 3 June 819/1 In one important aspect at least the change can be described as the growth of a new kind of historical-mindedness. **1960** *Amer. Speech* XXXV. 210 Four name maps described as a 'historical-onomatological record'. **1962** *English Studies* Feb. 73 The historical-sociological aspects of art. **1962** D. C. SWANSON in Householder & Saporta *Probl. Lexicogr.* 70 'Recent' refers to the other end of the (historical-typological) pole.

B. *sb.* (Further examples.)

1897 E. ORIENNE *Crime of One's Own* i. 7 The dreadful proliferation of Whodunnits and Historicals. **1967** H. HARRISON *Technicolor Time Machine* (1968) iv. 36 I've always been interested in doing a historical. **1971** 'A. BLAISDELL' *Practice to Deceive* iv. 53 Donaldson was fawning over a paper-back historical, when he was buzzed by the desk.

historically, *adv.* (Further examples.)

1937 A. HUXLEY *Ends & Means* vii. 67 No less 'historically' necessary and right are the brutalities of men in brown shirts. **1958** *Spectator* 27 June 849/1 One undated and long-dated Government stocks are selling at historically low prices. **1959** *Brno Studies* 1. 30 How rewarding the study of the written norm may be even for the historically-minded specialist. **1960** *Economist* 15 Oct. 285/3 The yields on copper shares are at historically high levels.

historicism (histǫ·risiz'm). [f. HISTORIC a. + -ISM; tr. G. *historismus*.] **1.** The attempt, found esp. among German historians since about 1850, to view all social and cultural phenomena, all categories, truths, and values, as relative and historically determined, and in consequence to be understood only by examining their historical context, in complete detachment from present-day attitudes.

1895 [see *historical-mindedness* s.v. *HISTORICAL a. 7]. **1920** *Contemp. Rev.* Oct. 536 If we find in him..some acute historical observation, the merit must be attributed to the historicism of the century. **1928** *Mind* XLVII. 124 Historicism..acknowledges truth only as valid in a special epoch. **1946** A. L. ROWSE *Use Hist.* v. 140 Marxism..brings us up against the question of historical relativism, or historicism. **1949** WELLEK & WARREN *Theory of Lit.* iv. 32 We must..enter into the mind and attitudes of past periods and accept their standards, deliberately excluding the intrusions of our own preconceptions. This view, called 'historicism', was elaborated consistently in Germany during the nineteenth century. **1972** *Sci. Amer.* Dec. 89/1, I was surprised, however, to find an eminent scientist embracing historicism (the theory championed by Hegel and Marx holding that history is determined by immutable forces rather than by human agency) as an explanation for the evolution of science.

2. A tendency in philosophy to see historical development as the most fundamental aspect of human existence, and historical thinking as the most important type of thought, because of its interest in the concrete, unique, and individual.

1930 J. BERLIN *Karl Marx* iii. 49 Against the scientific empiricism of the French and English, the Germans put forward the metaphysical historicism of Herder and of Hegel. **1940** *Mind* XLIX. 120 Hegel is right in teaching.. an 'absolute historicism'. **1964** C. S. LEWIS *Discarded Image* vii. 174 On this view the *differentia* of Christian historiography ought to be what I call Historicism; the belief that by studying the past we can learn not only historical but meta-historical or transcendental truth.

3. The belief that historical change occurs in accordance with laws, so that the course of history may be predicted but cannot be altered by human will; the resulting attitude to the social sciences, of regarding them as concerned mainly with historical prediction.

[**1902** C. S. PEIRCE *Coll. Papers* (1958) VIII. i. vii. 107 He may aim at hastening some result not otherwise known in advance than as that..to which some process seeming to him good must inevitably lead, such as.. whatever the historical evolution of public sentiment may decree (*historicism*).] **1940** K. R. POPPER in *Mind* XLIX. 423 Marx's emphasis on historical method in sociology, a tendency which I may call 'historicism'. **1943** F. A. HAYEK in *Economica* X. 50 (*title*) Scientism and the study of society: the 'historism' of 'historicism' and the scientistic and the historicist approaches. **1957** K. R. POPPER *Poverty of Historicism* 3, I mean by 'historicism' an approach to the social sciences which assumes that *historical prediction* is their principal aim, and..that this aim is attainable by discovering the 'rhythms'..that underlie the evolution of history. **1959** G. D. MITCHELL *Sociol.* i. 5 When we speak of historicism we refer to the attempt to discern a law governing social development.

4. Excessive regard for the institutions and values of the past; *spec.* in *Architecture*, the use of historical styles in design.

1939 *Archit. Rev.* LXXXVI. 55 If she has, in the New York Fair, done little more than to turn away from historicism to a new kind of pastiche, we can hope at least that with the new school of architects now springing up..the real reform will not be long delayed. **1942** *Ibid.* XCI. 52 In between there came a wave of European historicism, all the varieties of Victorian period imitation. **1966** *New Statesman* 25 Feb. 260/2 His [sc. J. J. Tengbom's] Högalids Church of 1916–23 and Concert Hall of 1920–26, both in Stockholm, are among the key monuments in Europe of the transition from historicism.

Hence **historicist**, an adherent or proponent of historicism (in various senses); also, one who specializes in the historical branch of a subject; also *attrib.* or as *adj.* So **historici·stic** *a.*

1937 J. ORR tr. *Iordan's Introd. Romance Ling.* iv. 298 His..field of research, namely, Indo-European philology, made him [sc. Meillet]..a historicist and comparatist. **1946** K. BURKE in W. S. Knickerbocker *XX. Cent. Engl.* 287 The modern historicist mode of thought. **1948** *Archit. Rev.* CIV. 226 Meldahl was the most important historicist in Denmark. **1949** *Mind* LVIII. 412 Guido de Ruggiero, avoided being committed to the amoralism inherent in every thorough-going historicist. **1954** *Word* X. ii. 123 A 'historicist' will be just as blind to the boundless of intimate connections which the synchronist points out between the different units of a language system. **1955** *Scott. Jrnl. Theol.* VIII. 181 The resurrection..cannot be proved by historicist methods, but it is an act. **1957** K. R. POPPER *Poverty of Historicism* ii. 41 Sociology, to the historicist, is theoretical history. **1959** G. D. MITCHELL *Sociol.* 5 The historicist tradition which we have seen in Comte. **1964** C. S. LEWIS *Discarded Image* vii. 175 The best medieval historians, like the best historians in any age, were not historicists.

historicizer (histǫ·risəizəɹ). [f. HISTORICIZE *v.* + -ER¹.] One who historicizes.

1965 *Scott. Jrnl. Theol.* IX. 406 Mark really sees Jesus and the Church as the historicizers of the Jewish eschatological hope.

historico-. Add further examples. Also used to form sbs., as *historico-novelese*, *-philology*.

1854 GEO. ELIOT tr. *Feuerbach's Essence Christ.* p. xli, This work is nothing but a faithful, rigid, historico-philosophical analysis of religion. **1900** W. A. ELLIS *Life Wagner* I. 225, I was wafted by the image of a great historico-political event. **1905** *Daily Chron.* 20 June 3/3 Pursuing his fascinating historico-biographic method, which seems to combine the movement and charm of narrative. **1906** *Daily Chron.* 9 Oct. 3/3 The King's Guardian' is an historical romance, written in historico-novelese. **1926** *Year's Work in Eng. Stud.* 1924 58 The difficulties and toils of this historico-philology. **1929** S. HOOK in *Essays in Honor J. Dewey* 164 The historico-genetic method tends to minimize origins and beginnings, since it confessedly cannot explain them in terms of spatio-temporal continuity. **1931** *Times Lit. Suppl.* 26 Mar. 242/2 To study the Austrian question from a historico-psychological standpoint. **1937** J. ORR tr. *Iordan's Introd. Romance Ling.* i. 45 Cases when the historico-comparative method may and must be dispensed with. *Ibid.* iii. 228 The whole applied from a historico-cultural angle. **1938** C. GRAY *Contingencies* (1947) 146 In..his invention of the historico-dramatic tableau he is the father of the Russian operatic composers.

historied, *a.* Add: **1.** (Example.)

1936 A. W. CLAPHAM *Romanesque Archit.* iii. 64 The historied capital, or capital carved with figure-subjects, appears first in minor features.

2. Now *literary* and chiefly *poet.*

1889 W. B. YEATS *Wanderings of Oisin* 122 You move in another dominion And hang o'er the historied stair. **1896** BELLOC *Verses & Sonnets* 63 November is the historied Emperor, Conquered in age. **1934** R. GRAVES *Claudius the God* 333 It will be a matter for laughter to you ..if your minds were ever dazzled by the historied glories of a remote past. **1943** S. SPENDER *Spiritual Exercises* 6 Within my shut skull flows a historied stream Of myths, fears, crimes.

historio-, combining form occurring in Greek (cf. Gr. ἱστοριογραφία HISTORIOGRAPHY) and now used to an increasing extent in English, as *historio-cultural*, *-patriotic*, *-pœic* adjs.

1958 W. STARK *Sociology of Knowledge* iv. 169 The historio-cultural sciences. **1967** *Listener* 16 Feb. 223/2 The story was tedious and repetitive as modern Russian historio-patriotic writing sometimes can be. **1953** *Antiquity* XXVII. 97 The abstract, age-long yearning of the Jewish nation in the Diaspora..found its historiopoeic expression in written history and poetry.

history, *sb.* Add: **2.** (Further examples.)

1928 *Daily Tel.* 14 Aug. 5/3 We have had enough of drum-and-trumpet history. **1967** *Listener* 16 Mar. 349/2 This is only one more step away from the older type of 'drum-and-trumpet' history.

3. *ancient history*: also applied *colloq.* to comparatively recent events which are regarded as nevertheless far back in a person's experience. (Further examples.)

1875 *Contemp. Rev.* XXVI. 870 The mutiny is now becoming an event of ancient history. **1908** *Busy Man's Mag.* Nov. 37/1 'Ancient history, governor,' said Woolford. 'We knew all that before.' **1910** BELLOC *Pongo* v. 92.. He reminded Dolly of the days when Consols were at 94.. All that was ancient history. **1939** C. DAY LEWIS *Child of Misfortune* III. vi. 350 Already the crisis through which they had passed was beginning to seem ancient history. **1946** J. B. PRIESTLEY *Bright Day* x. 320 I'm going to risk telling you something... It's all ancient history, but..we might as well get it straight. **1961** P. SPENCER *Full Term* i. 15 You won't get anywhere by fretting about it.. It's ancient history by now, anyway. People do odd things in drink.

4. c. *to make history*: to influence or guide the course of history; also, to do something spectacular or worthy of remembrance (see *history-maker*, *-making*, sense 9 in Dict. and Suppl.).

1862 *Chambers's Jrnl.* 1 Mar. 139/1 People engaged in public transactions are sometimes said to be making history, because they occasionally perform actions for which history condescends to impart perpetuity. **1889** *Puck* XXV. 133/2 If the hero who thinks he 'makes history' could only wake from his sleep after three hundred years and read the works of half-a-dozen..historians, he wouldn't know his own face on their pages. **1890** WILDE in *19th Century* July 137 Anybody can make history. Only a great man can write it. **1907** *Edin. Rev.* Jan. 4 The average man is of the Centre; and history in the long run is made by the average man. **1915** 'I. HAY' *First Hundred Thousand* xx. 311 We shall have a chance of making history over this, old man. **1959** N. MAILER *Advts. for Myself* (1961) 208, I had been ready to..publish..at my own expense, and try to make a kind of publishing history.

7. Delete † and add later examples.

1958 *Listener* 19 June 1024/3 It was a race that.. converted the classical 'history' into a kind of private leg show. **1963** *Ibid.* 28 Feb. 384/3 He [sc. Degas] remains ..even when he is no longer a painter of 'histories', a profoundly reactionary figure.

9. *history-master*, *-mill*, *paper*, *-play*; **history-making** *a.* (examples); also *vbl. sb.*

1898 'MARK TWAIN' in *Harper's Mag.* Mar. 538/2 On Thanksgiving Day the sitting was a history-making one. **1949** WYNDHAM LEWIS *Let.* Apr. 491 Excuse me for breaking in upon you as it were private, and partisan, history-making. **1963** AUDEN *Dyer's Hand* 278 Man is a history-making creature. **1891** W. J. GREENSTREET tr. *Guyau's Educ. & Heredity* iii. 128 The history-master might have taken us to the National Library. **1889** 'MARK TWAIN' *Connecticut Yankee* iv. 39 Sir Kay..began to fire up on his history-mill, with me for fuel. **1963** *Times Lit. Suppl.* 11 Jan. 29/1 Grist to some history-mill. **1857** J. A. SYMONDS *Let.* 8 Feb. (1967) I. 90, I was so amused yesterday with hearing the answers of some of the Sixth Form to our History Paper. **1957** N. FRYE *Anat. Criticism* 283 The Elizabethan secular *auto* eventually became the history-play.

hit, *v.* Add: **1. b.** (*b*) *to hit off* (earlier example).

1865 F. LILLYWHITE *Guide to Cricketers* 86 Messrs. Tritton and Wright hit off 25 in 20 minutes.

4. b. To strike exactly or at the proper point. Usually in phr. *to hit on* so many *cylinders*: (of an internal-combustion engine) to be running properly on so many cylinders; hence, *to be hitting on all four* or *six* (*cylinders*), to be running or working perfectly; *fig.* to be in good trim or form.

1912 [see *CYLINDER *sb.* 6]. **1928** *Sat. Even. Post* 10 Mar. 127/1 Modern science offers you a natural means to keep you 'hitting on all six'—every minute of the day.

8. *to hit home* (later example).

1938 *Times* 30 Apr. 11/4 Mr. Roosevelt says that the one lesson in events abroad that has 'hit home' is that the liberty of a democracy is not safe if the people tolerate the growth of private power to a point where it becomes stronger than their democratic State itself'.

c. To criticize, make fun of or ridicule (a person or thing): sometimes const. *at*.

1843 *Punch* 23 Sept. 131/2 Instead of an outburst of enthusiasm at the line 'Confound their politics', the waltzer is supposed to execute a *pirouette*, which is supposed to hit at our wavering propensities. **1936** *S.P.E. Tract* XLV. 190 The member of a newspaper staff who is responsible for writing the headlines prefers..*hit*..to criticize. **1945** BAKER *Austral. Lang.* vi. 121 A man who has acquired a strong dislike of another person..hits or..criticizes him. **1969** *Pen* IX. 47 He could supply her with a list of synonyms for the verb 'savage', i.e.,..hit at.

d. To occur to (a person); to affect in a particular way, to appear to; to have an impact on.

1891 KIPLING *Light that Failed* v. 78 Look at their faces. It hits 'em. **1914** G. ATHERTON *Perch of Devil* i. xxx. 175 Lucky it hit him to buy the house and send that last five thousand. **1926** 'D. N. BOWES' *Phantom Herd* v. 68, I wanted to see how it would hit you. **1928** G. B. SHAW *Doctors' Delusions* (1932) 205 Their worthlessness would not hit us in the face as the worthlessness of Dr Saleeby's figures does. **1931** GALSWORTHY *To Let* ii. xi. 212 Their manners now really quite hit you in the eye. **1937** *Econ. News* 28 Jan. 7/1 [*heading*] Finding best colours for crossing that will 'hit the eye'. **1958** *Listener* 16 Oct. 600/1 This book..was published about two years ago, and it has not yet hit the architectural profession.

e. *not to know* (or *to wonder*) *what hit one*: to be killed; to be surprised or disconcerted.

1923 J. MYNN *Jack Miner & Birds* vii. 27 He came to examine the flock near me and while his attention was riveted on them I raised up and fired, and he never knew what hit him. **1962** *Listener* 19 Oct. 689/1 Many of our less efficient firms would be hurt so hard that they could never quite know what had hit them. **1963** *Observer* 10 Feb. 24/3 They must have wondered what hit them in Fleet Street, for almost every female member of the British Press made a dead set for the hosiery counter at Galeries Lafayette.

f. fig. *to hit for six* [cf. sense 1 b (a)]: to demolish an argument, scheme, etc., to vanquish; to deal a severe blow to.

1937 *Times Lit. Suppl.* 1 May 343/4 Lawrence..was chiefly concerned to hit swots and cads and foreigners for six. **1937** I. CROSS *God Boy* (1958) xiii. 109, I had really hit her for a six and made her change her tune properly. **1961** *Oxf. Mag.* 15 June 413/1 Mr. Simon, the Secretary, hit most of the questioners for six. **1967** *Lancet* 1 July 41/2, I began to wonder if my massive and unwarranted administration of chloroform had not hit his liver—perhaps not inappropriately—for six.

g. To give or administer a narcotic drug to (a person). Also *intr.* *slang* (orig. *U.S.*).

[**1949** N. ALGREN *Man with Golden Arm* 76 It..a narcotic injection] hit all right. It hit the heart like a runaway locomotive, it hit like a falling wall.] **1953** W. BURROUGHS *Junkie* (1972) xiv. 144 'Hit me, will you, Ike?' Old Ike poked a gentle finger along the vein holding the dropper poised between thumb and finger. **1959**— *Naked Lunch* 67 The addict regards his body as an instrument to absorb the medium in which he lives, evaluates his tissue with the cold hands of a horse dealer. 'No use trying to hit there.' **1970** *N.Y. Times* 23 Feb. 26 How did you become an addict? 'You mean, who hit me first? My friend, Johnny.'

h. To kill; to rob. *slang* (orig. *U.S.*).

1955 *People* (Austral.) 19 Oct. 13/2 Dutch bellowed, 'Dewey's gotta go. He's gotta be hit.' **1968** N. GIOVANNI in A. Chapman *New Black Voices* (1972) 250, I have been robbed It looked like they knew That I was to be hit They took my tv My two rings. **1972** D. E. WESTLAKE *Cops & Robbers* (1973) xvi. 147 If they're cops, maybe it's not such a good idea to have them hit. **1973** *Publishers' Weekly* 29 Jan. 259/2 A professional killer and his 38 victims.

11. a. Freq. in modern (esp. *U.S.*) colloquial use in the sense 'to arrive at'; also, to go to (a place), go upon (a course). Phr. *to hit the trail* [less commonly *the grit*, *pike*, *road*, etc.]: to take the road, to get on the way, to go away.

1873 W. F. BUTLER *Wild North Land* xviii. 208 In the morning 'Two-poos', or the Three Tumults, sets forth to look for a moose; he hits the trail and follows it. **1888** *Amer. Speech* (1962) XXXVII. 76 *Hit the grit*, get going; get out of here. **1888** *Detroit Free Press* Oct. [Farmer], Professor Rose, who hit this town last spring, is around calling us a fugitive from justice. **1889** BARRÈRE & LELAND *Dict. Slang*, *Hit the flat*, to (cowboys), to go out on the prairies. **1893** P. H. EMERSON *On Eng. Lagoons* xii. 40, I have been hitting the road something to get here quick. **1896** *Ade Artie* xiv. 137 'A little more weather like this and we'll be hittin' the park,' he observed. **1897** *Outing* (U.S.) XXX. 374/1 Men can pass out the church door, shoulder their packs of general cussedness, and unconcernedly hit the trail to the lower [regions]. **1901** S. E. WHITE *Westerners* i. 7 Thought you wasn't on tenderfoot. Ever hit the trail? **1904** *Hartford Courant* 25 June 6 The..convention, whose delegates were so summarily ordered to hit the pike by the national committee-men. **1907** R. W. SERVICE *Songs of Sourdough* (1908) 65 It lies with thee—the choice is thine, is thine, To hit the ties or drive thy auto-car. **1918** C. E. MULFORD *Man fr. Bar-20* xiii. 131, I was a rich man until I hit town. **1925** WODEHOUSE *Carry on, Jeeves* v. 116 Jimmy Mundy.. has come to New York from itself; to force it—in his picturesque phrase—to hit the trail. **1933** Z. A. TILGHMAN *Dugout* 70, It's time we hit the road. **1931** 'D. STIFF' *Milk & Honey Route* 207 *Hitting the grit*, to be forced off a fast moving train. **1932** T. S. ELIOT *Sweeney Agonistes* 18 We hit this town last night for the first time. **1948** G. H. JOHNSTON *Death takes Small Bites* iii. 54 Go down this corridor, up the stairway at the end, straight on until you hit the second court. **1950** T. LONGSTAFF *This my Voyage* v. 97 So on May 31st Mummery and I hit the trail once more. **1973** *Guardian* 10 Apr. **1973** *Christian Sci. Monitor* 14 Apr. B6/3 Then we hit the road together, modern pilgrims making very little progress.

c. *to hit the hay*, to go to bed. Also *to hit the sack.* *slang* (orig. *U.S.*).

1912 *Dialect Notes* III. 578 *Hit the hay*, to go to bed. **1918** S. LEWIS *Babbitt* xix. 245 You probably want to hit the hay. **1930** 'SAPPER' *Finger of Fate* 184 On those two nights we all hit the hay before midnight. **1943** in J. J. Fahey *Pacific War Diary* (1963) i. 74, I hit the sack at 8 P.M. I slept under the stars on a steel deck. **1957** J. KEROUAC *On Road* (1958) 93 Terry and I..got ready to hit the sack. **1961** A. MILLER *Misfits* x. 98 Well, I don't want to hit the sack. **1965** *Times* 25 Nov. 1965/2 Ve just ignorant folks got to hit the sack. ..

d. *to hit the bricks*: (*a*) (see quot. 1950); (*b*) to go on strike. *U.S. slang.*

1931 *Amer. Speech* VI. 439 *Hit the bricks*, to be released from prison. **1947** 'Sapfever' *Lag's Feb.* 43 When you hit the bricks in those days you didn't expect to come back aboard real soon. *Ibid.* **1950** *Amer. Speech* XXV. 1940 The strike was lifted..hit the bricks. ..

19. (Later U.S. examples.)

1905 R. BEACH *Pardners* (1912) 24 We hit for camp on the run. *Ibid.* 48 So me and 'Kink' Martin..hit west. **1926** M. BROWN' *Phantom Herd* xiii. 218 When I hit for the land of orange blossoms and singing birds and sunshine.

b. *rare.* To go, pass. *U.S.*

1931 H. S. HARRISON *Queed* viii. 86 I've seen you hit by the window many's the time.

20. b. Also more widely, to become friendly, to be on good terms. (Further examples.)

1863 TROLLOPE *Rachel Ray* II. xiv. 197, I am so happy.. that you and he hit it off. **1954** T. S. ELIOT *Confid. Clerk* i. 26 Mr. Kaghan is prejudiced. He's never hit it off with Lady Elizabeth.

d. To travel at speed. *U.S.*

1930 J. P. LINCOLN *Cap'n Warren's Wards* iii. 39 They nabbed us for speeding..' Said we were hitting it at fifty an hour.

22. *hit or miss* (further examples).

1848 in *Amer. Speech* (1955) X. 40 Hit-or-miss-carpet, a carpet woven from strips of old cloth sewed together. **1864** *Harper's Mag.* June 60/1 My husband is Colonel of the Third Regiment in the Hit-or-Miss Brigade, United States Cavalry. **1927** J. ADAMS *Errors in School* 211 Hit-or-miss methods. **1936** A. L. ROWSE *Expansion Eliz. Eng.* 399 They were impulsive, chancy, amateurish, very much hit-or-miss. **1959** P. BULL *I know Face* vii. 124 It was 'there's got a Bat bit..so he's gotta hit the bricks.' **1964** *Times* 4 Oct. 11/1 The United Auto Workers hit the bricks against giant General Motors.

b. *hit and miss* = *hit or miss*. Also (with hyphens) *attrib.* and *spec.*, as **hit-and-miss governor**, a type of governor used in internal-combustion engines which causes the engine to miss one or more explosions when the speed is too high; **hit-and-miss ventilator** (see quots.).

1897 W. E. BARTON *Hero in Homespun* 377 They ripped up the new hit-an'-miss carpet for horse blankets. **1897** K. STUART *Simpkinsville* 156 Takin' 'em hit and miss, we wouldn't be two..'d'fference hardly. **1902** A. C. HARMSWORTH et al. *Motors* vii. 162 Many of these engines have not hit-and-miss governors..The tokes governor..was a hit-and-miss governor. **1908** WEBSTER, *Hit-and-miss ventilator*, a ventilator consisting of a perforated glass disk, lying flat against, and pivoted through its center to, a corresponding perforated window. **1931** *Discovery* Sept. 308/1 [Without these criteria] the procedure would be unnecessarily hit-and-miss. **1940** *Chambers's Techn. Dict.* 417/2 *Hit-and-miss ventilator*, a ventilating device consisting of a slotted plate over which may be moved another slotted plate, so that the openings for the access of air may be more or less restricted as required. **1955** W. W. DENLINGER *Compl. Boston* 156 Hit-and-miss, take-a-chance breedings are largely to blame. **1966** G. TAYLOR *Silver* iii. 53 A simple pattern, often seen on Communion Cups, consists of rows of 'hit and miss' ornament. **1972** *Morning Star* 14 July 2 The pellets of such girls are only worried about the wastage involved..and the general hit-and-miss approach to the problem.

22*. Various phrases.

a. *to hit the pipe*: to smoke opium. Also *to hit the gong*, *gow*, *stuff*: to take drugs. So *to hit cigarettes*: to smoke heavily. *U.S. slang.*

1886 'G. T. BURNES' *Profess. Criminals of Amer.* 385 Joe did not 'hit the pipe'. **1902** *Chicago Record Herald* 7 Sept. vi. 5/1 On each bunk two almond-eyed devotees of the drug may be seen 'hitting the pipe', as opium smoking is termed. **1933** *Amer. Speech* VIII. ii. 27/1 When one has contracted the [drug] habit..he is..hitting the gow. *Ibid.* 27/2 When the opium addict is smoking he is said to be hitting the gong. **1936** *Jrnl. N.Y.* 122/2 To hit the stuff, to be addicted to narcotics... The act of taking dope. **1939** *Ibid.* (1942) XVII. 206/1 Bill is hitting cigarettes some. **1955** *Ibid.* XXX. 'S. Z.' Hit the pipe. ..

b. *to hit the booze*, *bottle*, *jug*, *pot*: to drink excessively. *slang* (orig. and chiefly *U.S.*).

1889 *Oregonian* (Portland) 14 Oct. 3/1 If Dasher gets a dozen or more customers with his own appetite for hitting the booze he will have no trouble making it go. **1948** J. M. SULLIVAN *Crim. Slang* 13 *Hitting the pots*, excessive drinking. **1942** BERREY & VAN DEN BARK *Amer. Thes. Slang* § 103/22 *Drink liquor*, ..*hit the jug*. **1945** S. D. MACARTHY H. CHANDLER *High Window* (1943) 5/3 The hit the bottle for three days. **1955** M. MERRITT & WOLFE *Really Blues* 374 Hit the jug, drink heavily, often from the bottle. **1964** R. KEES, 'Is Christie *Dead Man's Folly*' iii. 40 The most incredible shirts..covered with crawling turtles and things—made me think I'd been hitting the bottle. **1957** *Landfall* XI. 38 Everyone knew he'd turn out a flop... Hit the booze and got T.B. **1965** *Times Lit. Suppl.* 25 Nov. 1068/2 We are 'wild spiders crying together'..who must crack or hit the bottle.

c. *to hit the ceiling*: see *CEILING *vbl. sb.* 5 b.

d. *to hit the headlines*: see *HEADLINE 2 c.

e. *to hit the deck*: (*a*) to go to bed; (*b*) to land an aircraft; (*c*) to fall to the ground; (*d*) to get up (from bed). *colloq.*

1918 *Sat. Even. Post* 21 Dec. 9 The sergeants and corporals emphasized the command to rise with stinging injunctions to 'Snap out of it!', 'Hit the deck!' **1935** W. DE LA MARE in *Proc. Tel. Acad.* 40 He..was 'hit the deck; he slung his hammock; he went to bed.'.. they all signify much the same thing. **1943** C. H. WARD-JACKSON *Piece of Cake* 36 Hit the deck, to land an aircraft. **1949** *Manch. Guardian Weekly* 4 Mar. 2/1 The whole House fell on its knees or went prone behind desks, as one Pacific veteran shouted out: 'Hit the deck, you damn fools!' **1956** *Amer. Speech* XXXI. 193 *Hit the deck*, wake up; get up; working; jump to the floor. **1958** F. C. AVIS *Boxing Ref. Dict.* 53 *Hit the deck*, a boxer-slang expression to fall to the ring floor. **1961** F. H. BURGESS *Dict. Sailing* 115 *Hit the deck*, take an upper-deck seats. **1966** 'J. HACKSTON' *Father clears Out* 52 I'm going to hit the deck now, ..

f. *to hit the silk*: see quot. 1942. *colloq.*

1942 *Amer. Speech* XVII. 166/2 *Hit the silk*, use a parachute. **1958** 'P. BRYANT' *Two Hours to Doom* 104 If it came to the worst, they could always drop altitude and hit the silk.

23. c. *Polo.* (See quot. 1963.)

1906 T. B. DRYBOUGH *Polo* (rev. ed.) xiv. 279 A ball over over the side-boards is out, although it rebounds in ..it is hit in by a player before it touches the ground. **1920** *Hurlingham Club Rules of Polo* (ed. 43) iii. 66 (*heading*) Explanation of terms.. 'Hit in' means 'to hit the ball into the field of play'; not to bowl it in under-hand. **1963** BLOODGOOD & SANTINI *Horseman's Dict.* 104 *Hit in*, to hit the ball into the field of play; not to bowl it in under-hand.

25*. hit up. a. To force up; to speed up. With *it*: to put on pressure; to make efforts in a certain direction.

1893 W. K. POST *Harvard Stories* 49, I could hear him objurgating Steve Hudson for hitting up the stroke. *Ibid.* 146 When you are doing better than three and a half [miles an hour], you are hitting it up pretty lively. **1902** LYNDE *Grafters* xx. 257 Two days after the Universal's triumph in the Belmont field, the Argus began to 'hit it up' boldly toward the capital. **1924** MULFORD & CLAY *Buck Peters* xiv. 40 Hit her up you all..hit it up to her. **1928** B. A. POTTE *Stretchers* (1930) 276 Back he went, while we.. When he got back with his jam we, hit it up again. **1935**

b. To make or score (runs).

1895 [see *Ut adv.* 1 g. b]. **1899** *Daily News* 9 June 6/7 They were hitting it up..until they got up to the loss of seven wickets. **1928** *Evening News* 18 Aug. 10/5 Middlesex hit up 366 in the first day's play.

c. *to hit* (a person) *up for*: to ask (someone) for. *U.S.* and *N.Z. slang*.

1917 *Chron. N.Z.E.F.* 5 Sept. 28/1 We hit him up for a loan for weeks afterwards and he always came to light too. **1937** W. THURLOW CRAIG *Paraguayan Interlude* xxv. 291, I..hit him up for a job, and got one. **1936** J. STEINBECK *In Dubious Battle* 108 Don't hit up the guys anything else but breakfast. **1957** I. CROSS *God Boy* (1958) xii. 98 'I'll have to hit my old man up for a new bike,' he said. **1972** M. J. BOSSE *Incident at Naha* iii. 135 She hit me up for bread.

hit, *sb.* Add: **1. hit-in** (in *Polo*), the hitting of the ball into the field of play. (Cf. *HIT *v.* 23 c.)

1930 *Hurlingham Club Rules of Polo* (ed. 43) iii. 63 Penalty 9..by the side *fouled*—a hit in from behind the other side. ..the defending side being free to place themselves where they choose. **1937** *Times* 16 July 5/5 Captain Morrison, after meeting a hit-in by Major Harrison put Ashdean further ahead with a stupendous shot under the pony's neck. **1959** *Times* 18 May 4/2 He came through his own line..and scored a second chukker.

b. A dose of a narcotic drug; the action of obtaining or administering such a dose. Also *attrib.*, as **hit-mark**, the scar from an injection of a drug. *slang* (orig. *U.S.*).

1951 *Amer. Speech* XXVI. 1 *Jrnl.* May 342/2 They are anxious to make a 'connection', 'score' or 'hit'. **1962** 'E. MCBAIN' *Like Love* (1964) ix. 119 A narcotics cop's definition on examining a prostitute's thighs for hit marks, even where he knows she couldn't possibly be a junkie. **1966** L. COHEN *Beautiful Losers* (1970) i. 106 They crawled among their veins for one that still carried blood, tapped the needles under the flesh, waited for the red signal of a 'hit', and then squirted the solution into circulation. **1970** *Daily Tel.* 27 Apr. 4/8 In San Francisco's Haight-Ashbury district, the hippie Mecca, the price of one 'hit' has dropped from 125 to 25 dol. **1972** *Southerly* XXXII. 103 Somebody hands me a joint and I take a hit and hand it to Marlene who takes a hit. **1972** C. WESTON *Poor, Poor Ophelia* (1973) ii. 15 You're blind! You tried to hit a for a report to see the hit marks?

c. A killing; a robbery. Also *attrib.*, as **hit-man**, a hired murderer. *slang* (orig. *U.S.*).

1970 *Sunday Truth* (Brisbane) 8 Mar. 32/5 The Mafia cringe at the way our boys carry out their hits. **1970** J. PHILIPS *Nightmare at Dawn* (1971) ix. 97 If he was..a hired gun, a hit-man. **1972** D. MACKENZIE *Sleep is for Rich* vi. 186, I got scared and called the whole thing off. Someone else must have made the hit. **1973** *Daily Tel.* 25 July 18/8 Baynard is alleged to have been a 'hit man' (assassin) for drug traffickers and to have carried out a 'contract' to kill Finley.

4. Add to def.: esp. any popular success (a person, a play, a song, etc.) in public entertainment. (Earlier and later examples.) Also *attrib.* and *Comb.*, as **hit parade**, a programme or grouping of 'hits'; **hit tune** (or **song**), a tune that proves popular.

1811 C. MATHEWS *Let.* 22 June (1838) II. 123 Mawworm was a most unusual hit, I am told. **1835** DICKENS *Let.* 9 Dec. (1965) I. 103 The insertion of another Prison Paper would decidedly detract from the 'hit' of the first. **1908** *Sears, Roebuck Catal.* 199/1 Orchestra selections.. 'Broadway Hits'. **1918** *Talking Machine News & Jrnl. Amusements* Feb. 83 (Advt.), All the song-hits of the moment. *Ibid.* 89/2 (Advt.), When a title makes a hit, we are bound to have another something like it before long. *Ibid.* Mar. p. iv (Advt.), Always a hit!! A 'Record' in a Record. **1927** *Melody Maker* Aug. 800 (Advt.), The sensational hit. Sweeping the country like a cyclone. The Doll Dance. **1932** *Amer. Speech* Apr. 252 The motto of the song writers is..'A hit is not an aesthetic triumph, it is something that sells.' **1937** *Cinema Arts* June 22 (*caption*) The Hit Parade. **1937** W. S. MAUGHAM *Theatre* xxviii. 270 I'm very pleased with her. I think she'll make quite a hit. I've half a mind to give her a contract. **1942** BERREY & VAN DEN BARK *Amer. Thes. Slang* § 580/2 A popular song which has stood the test of time; hit song or tune. **1947** R. DE TOLEDANO *Frontiers of Jazz* xvi. 172 Hit tunes of his own composition. **1948** *Manch. Guardian Weekly* 1 Jan. 13/4 The box-office of any hit-show on Broadway. **1948** F. BROWN *Murder can be Fun* (1951) i. 18 She had big blue eyes that would have been a hit on television. **1957** *Observer* 29 Sept. 13/1 It must first be said that Miss Storm has written a resounding, self-evident hit. **1957** *Times Lit. Suppl.* 4 Oct. 593/1 Hollywood now makes its smash hits out of American self-criticism. **1958** *Times* 26 May 7/6 The numbers listed in the hit parade all have a structure of professionalism. **1959** G. FREEMAN *Jack would be Gent.* vi. 129 The first dozen of you lucky kids..will be presented with my latest hit disc. **1967** *Boston Sunday Globe* 23 Apr. A28/2 Lemons and peppermints have been 'hit items' from the very first fair, and so have rides on one of the Brookline Fire Department's engines. **1967** *Melody Maker* 29 July 7/3 I'm not chasing any hit records any more. **1968** *Brit. Med. Bull.* XXIV. 245/1 Fowler (1966), using a program in Elliot 803 Autocode, has explored the effect of combining models with varying distributions of hit numbers. **1969** *Punch* 29 Jan. p. v, Don Partridge, who shot from the pavement into the hit parade, is putting on a Buskers Concert.

hit and run. 1. *Baseball.* 'A play wherein a base runner starts with the pitcher's throw as the batter attempts a hit, a sacrifice hit' (*D.A.*).

1899 *Chicago Daily News* 2 May 7/1 A rare combination for the hit-and-run game. **1904** R. H. BARBOUR *Bk. Sch. & Coll. Sports* 188 *Team batting.* The best known example of this is what is called the sacrifice hit or 'hit and run'. *Ibid.* 191 The 'hit-and-run' play may also be used when there is a man on third and a run is badly needed. **1909** *Amer. Mag.* May 35/1 Evers and Kling analyzed and discovered every hit and run signal used by the Cincinnati club. **1957** *Encycl. Brit.* III. 164/1 Since the hit-and-run play involves two players, there is usually the exchange of a secret signal between them, or a signal to both of them by the manager or one of the coaches. **1967** [see *BUNT *v.*[1] 2].

2. *Motoring.* The action of the driver of a motor vehicle who fails to stop after an accident for which he is responsible. Freq. *attrib.* Also **hit-run.**

1924 *Sci. Amer.* Sept. 181/1 With the bumper in circuit with the ignition, there would be no more 'hit-and-run' driving. **1926** *Amer. Speech* I. 460/1 The hit-and-run driver. **1933** *Bulletin* (Sydney) 11 Jan. 9 The power to inflict the death penalty on the hit-run driver who kills. **1944** E. S. GARDNER *Case of Careless Kitten* xxiv. 217 Lunk's dead, found at a road intersection a couple of blocks from his house, a hit-and-run car. **1949** 'M. INNES' *Journeying Boy* ix. 109 If Soapy had been a bit nearer the kerb, I'd have felt like a little hit and run. **1955** C. S. FORESTER *Good Shepherd* 24 He had been in trouble with the civil authorities for a hit-run automobile offence. **1965** 'L. EGAN' *Detective's Due* (1966) xii. 129 There had been a hit-run yesterday. **1972** *Times* 14 Dec. 4/8 Professor Christiaan Barnard..will suffer no after-effects from injuries sustained when knocked down by a hit-and-run motorist.

3. Used *attrib.* to denote a raid, raider, etc., using swift action followed by an immediate withdrawal.

1940 *Hutchinson's Pict. Hist. War* 2 Oct.–26 Nov. 98 Two of the three trams blasted by bombs in a Nazi 'hit and run' raid on the morning of 25th October, 1940. **1941** *Ibid.* 14 May–8 July 71 The heavily armed raiding cruiser, or hit-and-run battleship. **1955** *Times* 20 Aug. 5/2 Some of them were indulging in violence and arson, adopting 'hit and run' tactics. **1957** L. F. R. WILLIAMS *State of Israel* 30 Arab forays, which often took the form of hit-and-run tactics. **1966** [see *fire-bombing* s.v. *FIRE *sb.* B 5].

hitch, *v.* Add: **3. b.** (Later U.S. examples.)

1868 H. WOODRUFF *Trotting Horse* iv. 59 When it was found that they began to hitch and hobble, a good let-up would do more to restore the stroke than anything else. **1889** *Century Mag.* 907/2, I do not know what would happen to a man who 'hitched' in his saddle.

4. b. = *HITCH-HIKE *v.* Also *trans.*, *to hitch a lift*, etc.: to obtain a lift in a vehicle.

1931 'B. STARKE' *Touch & Go* ix. 133 She told me she had hitched her way down to New Orleans a week before. *Ibid.* xii. 192 Two lads spoke to me, and asked if I were hitching it by myself. **1948** PARTRIDGE *Dict. Forces' Slang* 94 Hitch a lift, or ride. **1959** 'G. CARR' *Swing Away, Climber* i. 17 We hitched—got lifts, you know—from Birmingham. **1960** *Sunday Express* 6 Nov. 7/5 The car in which he had hitched a lift crashed into a lorry. **1963** *Guardian* 4 Feb. 6/5 They hitch there and back.

5. e. Also *to get hitched up*, to be married, to become married. Chiefly *dial.* and *N.Z.*

1890 S. S. BUCKMAN *John Darke's Sojourn in Cotteswolds* xxii, 'Twarn't long avor we got hitched up together. **1911** 'KIWI' *On Swag* vii. 14 Elsie and I got 'hitched up'. **1944** J. H. FULLARTON *Troop Target* iii. 24 That's the fifth of the old gang to get hitched up in five months.

hitch, *sb.* Add: **1. c.** A catch in or a turn at wrestling.

1834 H. M. BRACKENRIDGE *Recoll.* ix. 94 I'll stan iny man a hitch in Butler county, if so be he'll clear me o' the la'. **1880** *Harper's Mag.* Mar. 525 How with 'ducking' heads and muffled screams you..saw them scrambling for a 'hitch'.

4. b. = **hitch-hike* sb. *colloq.*

1955 *Times* 27 Aug. 7/4 They are not asked in the middle of doing 200 miles in four hours for a half-mile 'hitch'. **1966** J. PHILIPS *Wings of Madness* (1967) II. iv. 132, I came down by bus. I thought maybe you'd give me a hitch back.

5. b. A mode of harnessing a horse or team; a vehicle with its horse or team. *U.S.*

1876 *Rep. Vermont Board Agric.* III. 143 If he can go best in one kind of a hitch, and, in that hitch, make the best time ever made by any horse. **1898** *Christian Herald* (N.Y.) 2 Mar. 167/2 Several hitches are a mule and steer together. **1905** *Springfield* (Mass.) *Weekly Republ.* 22 Sept. 12 There were also several other creditable displays, both single and double hitches. **1912** MULFORD & CLAY *Buck Peters* xxii. 201, I want a hitch of some kind,..something with speed and bottom, and the sooner the better.

8. A period of service, e.g. in the armed forces. Chiefly *U.S. slang.*

1835 *Novascotian* (Halifax) 12 Nov. 332/2 At last he said, which way are you from, Mr. Slick, this hitch. **1913** *Army & Navy Jrnl. Philippines* 4 Oct. 15/1, I had to go or else re-up For seven long years hitch. **1955** C. S. FORESTER *Good Shepherd* 24 He was new to the ship, a transfer made when they were in Reykjavik, serving his second hitch. **1957** J. KEROUAC *On Road* (1958) 257 Another hitch in prison and you'll be put away for life. **1959** *Listener* 15 Oct. 607/1 Newspapermen who did a hitch in Britain during the war. **1973** *Washington Post* 13 Jan. A3/2 In his work in intelligence, Pounder had many assignments, including a hitch as part of the White House security detail during President John F. Kennedy's Ireland trip.

9. *Comb.*: **hitch and kick, hitch-kick,** a form of high-kicking or long-jumping; **hitch-knot** = sense 6 b.

1898 *N.Y. Tribune* 6 Sept. 9/4 Michael Sweeney, the well-known jumper, broke a world's record in the hitch-and-kick by one inch. **1931** F. A. M. WEBSTER *Athletes in Action* 155 Numerous men using the 'hitch-kick', or 'mid-air-running' style, have beaten 25 ft. **1957** DUNCAN & BONE *Oxf. Pkt. Bk. Athletic Training* (ed. 2) 71 Do not spend time on learning the hitchkick at the expense of developing speed, etc. **1847** T. STODDART *Angler's Comp.* 69 In making large fly-hooks..I bring down the thread and fasten it, with a simple hitch-knot.

hitch-, in combs. = *hitching* vbl. sb. (see HITCH *v.*). *U.S.*

1899 ADE *Doc. Horne* i. 7, I jumped off my horse and threw him one end of my hitch-rein and pulled him out. **1903** A. ADAMS *Log of Cowboy* xxi. 138 Tying our horses in a group to a hitch-rack in the rear of a saloon. **1906** H. D. PITTMAN *Belle of Bluegrass C.* xiii. 187 A slim-legged yellow girl..swinging by her arms from a hitch rail. **1922** *Blackw. Mag.* June 714/1 It needs many a coat of paint, clean, and hitch-up. **1935** C. DAY LEWIS *Time to Dance* 61 A coast-to-coast hitch-up. **1954** *Encounter* Oct. 3/1 The hitch-rail enclosing the county court-house.

Hitchcock (hi·tʃkǫk). The name of L. H. Hitchcock (1795–1852) used *attrib.* to designate any one of various chairs made by him or produced in his chair factory at Barkhamsted, Conn.

c **1828** in M. R. Moore *Hitchcock Chairs* (1933) 5 Hitchcock chairs,..flag and wooden seats, warranted well manufactured. **1933** M. R. MOORE *Ibid.* 3 The first Hitchcock chairs probably had the rush seat, but very soon were added the cane and solid wood seats. **1959** L. BOGER *Compl. Guide Furnit. Styles* xxii. 373 The name Hitchcock is given to a particular type of painted American Empire open back chair. **1967** *Boston Sunday Herald* 30 Apr. 1. 23/1 (Advt.), The charm of genuine Hitchcock Colonial chairs.

hitcher, *sb.* Add: **3.** One who hitch-hikes.

1960 *20th Cent.* Nov. 476 Do you often pick up hitchers? **1972** R. QUILTY *Tenth Session* 19 When the road seemed to stretch endlessly..Bill always felt he was doing the company a service by picking up a hitcher. **1973** *Daily Colonist* (Victoria, B.C.) 1 July 4/2 The film company.. advertised for hitchers to tell their stories of experiences good and bad.

hitch-hike (hi·tʃ,həik), *v.* orig. *U.S.* Also **hitchhike.** [f. HITCH *v.* + *HIKE *v.*] *intr.* To travel by means of lifts in vehicles. Also *fig.* Hence as *sb.*, such a journey. Also **hi·tch-hiker**, one who hitch-hikes; **hi·tch-hiking** *vbl. sb.*

1923 *Nation* 19 Sept. 297/2 Hitch-hiking is always done by twos and threes. **1927** *New Masses* June 15/1 Most young janes have their heads full of a trip to Paris, or a hitch-hike thru New England. **1927** *Glasgow Herald* 6 Sept. 10 There are apparently hitchhikers in the United States who boast they can travel 500 miles free of charge without walking more than 10. **1931** 'B. STARKE' *Touch & Go* iv. 58, I..wondered how Dot would ever dare to tell the people there that she had hitch-hiked home. *Ibid.* We may charge this wicked hitch-hiker the ten cents extra that she deserves for asking for a bath towel. **1940** A. CHRISTIE *Buckle my Shoe* 182 He told amusing stories of his hitch-hikes and tramps in wild places. **1941** AUDEN *New Year Let.* 68 Kids When their imagination..and Hitch-hike a thousand miles to find The Hesperides that's on their mind. **1945** *Daily Mirror* 27 Sept., Hitch-hiking by air from London to Manila, five British Red Cross welfare sisters arrived in Canberra, Australia. **1958** *Manch. Guardian* 26 Sept. 4/4 He decided to hitch-hike around the world. **1959** *Times* 1 Oct. 9/6 Hitch-hiking appears to have replaced old-fashioned walking and has obviously graduated into a recognized pursuit, ready perhaps to be nurtured and protected by an international organization empowered to negotiate with transport ministers and police chiefs. **1973** *Black World* Apr. 91 Walton sees the need for people to realize their own cultural heritage and not hitchhike on somebody else's.

hitching (s.v. HITCH *v.* in Dict.), *vbl. sb.* (Further examples.)

1842 J. L. SCOTT *Jrnl. Missionary Tour Pennsylvania* (1843) vi. 68 When at the door they alighted, and he rode off to the '*hitching post*'. **1852** C. A. BRISTED *Upper Ten Thousand* 67 [He] pulled a hitching-strap from under the seat, and fastened his off-horse very neatly to a lamp-post. **1871** MRS. STOWE *Old Town Fireside Stories* 84 All the hitchin'-posts was full clean up to the tavern. **1884** W. D. HOWELLS *Silas Lapham* xviii. 336 He got the hitching-weight from under the buggy seat and made it fast to the mare's bit. **1920** J. GREGORY *Man to Man* ix. 103 A dozen saddle-horses were tied at the hitching-rail. **1926** J. BLACK *You can't Win* ix. 108, I..limped out to where I had an old 'swift' tied to a hitching rack. **1966** *Guardian* 4 Feb. 6/3 At weekends hitching..is widely practised. **1963** J. N. HARRIS *Weird World Wes Beattie* (1964) xviii. 216 'Oh, Gargoyle, darling,' she said, sitting down on an old hitching block at the edge of the Rosedale pavement, 'isn't it too gorgeous?'

hi·tchlessly, *adv.* [HITCH *sb.* 7.] Without a hitch.

1910 W. DE LA MARE *Private View* (1953) 26 Stealthily and hitchlessly the scenes clear, brighten, fade, and close. **1958** W. STARK *Sociology of Knowledge* viii. 325 It is thought which works in 'hitchlessly', as the Americans would say, with the established social pattern.

Hitler (hi·tləɹ). [Name of Adolf Hitler (1889–1945), chancellor of the German Reich and leader of the National Socialist (*NAZI) Party in Germany.] One who embodies the characteristics of Hitler; a dictatorial person. Also used in the possessive to designate the war of 1939–45; also *attrib.* and *Comb.*

1930 *Times* 27 Sept 10/1 The action of the Court was hailed in the Hitler camp as a great tactical success. **1931** WYNDHAM LEWIS *Hitler* I. i. 8 The Hitler Movement.. received so much advertisement in the English Press. **1932** *Times* 26 Mar. 9/5 General Groener has forbidden a rally of the Hitler Youth. **1934** E. WAUGH in *Ninety-Two Days* ii. 49 He was comic; huge feet and hands, huge mouth, and an absurd little Hitler moustache. **1934** J. SPENSER *Limey breaks In* v. 82 Sparkes wanted to be the Hitler of that kitchen. **1935** *Economist* 29 June 1478/1 Refusing to give the Hitler salute in his lecture-room. **1936** WODEHOUSE *Laughing Gas* xxiv. 257 The Hitlers and Mussolinis of the picture world. **1937** *Ann. Reg. 1936* II. 184 Membership of the Hitler Youth Movement became compulsory for all the young people of the country. **1939** *War Illustr.* 14 Oct. 138/3 Millions of leaflets warning the German people of the deadly consequences of 'Hitler's war' if it were prolonged. **1940** *Ibid.* 5 Jan. 574/3 This has been given public expression by Dr. Robert Ley addressing the Hitler Youth. **1940** M. P. PRICE (*title*) Hitler's war and eastern Europe. **1962** J. MAYO *Season of Nerves* iv. 12 A tiny chin and a sandy Hitler moustache. **1963** N. FREELING *Because of Cats* vii. 108 The building.. looks like nothing more than a Hitler-bunker. **1966** *Economist* 15 Oct. p. xii/1 The brief crop of 'Hitler babies', born in the 1930s..would soon be reaching working age—and the prospect was alarming.

So **Hitlere·sque, Hitlerian** (-lī³·r-), **Hi·tlerish** *adjs.*, of, pertaining to, characteristic of, or (somewhat) resembling Hitler; **Hitleria·na**, objects, etc., owned by or connected with Hitler; **Hi·tlerism**, the political principles or policy of the Nazi party in Germany; **Hi·tlerist**, **Hi·tlerite**, a follower of Hitler; also (both words) *attrib.* or as *adj.*; **Hitleri·stic** *a.*; **Hi·tlerize** *v.*, to make subject to Hitler, to make Hitlerite.

1930 *Times* 26 Sept. 13/2 These Hitlerite outbursts must disturb German Liberals and Socialists. *Ibid.* 27 Sept. 11/3 The most effective antidote to Hitlerism. **1930** *New Statesman* 1 Nov. 106/1 One may magnify or minimise the role of the Hitlerites, but [etc.]. **1931** WYNDHAM LEWIS *Hitler* 4 An exponent..of German National Socialism, or Hitlerism. *Ibid.* 32 The militant nationalism of the Hitlerist. **1931** *Times Lit. Suppl.* 16 Apr. 296/2 Hitlerite anti-Semitism. **1933** *Business Week* 24 May 23/1 (*heading*) Hitlerizing industry. **1934** A. HUXLEY *Beyond Mexique Bay* 113 Dislike and fear of Hitlerian Germany. **1936** N. MARSH *Enter Murderer* iv. 55 Sorry to be a bit Hitlerish, but it'll save time. **1940** *Ann. Reg. 1939* 90 The cause they had undertaken till Hitlerism had been destroyed and a liberated Europe re-established. **1941** *Common Sense* (N.Y.) Jan. 30/1 For the Hitleristic reason of grabbing everything in sight. **1943** *Times* 27 Apr. 5/3 The inherent instability of its Hitlerized Europe publicly exposed by the defection of one of the members. **1944** J. S. HUXLEY

On Living in Revol. xii. 131 In a totalitarian, Hitlerian way, or in a democratic, co-operative way. **1944** G. B. Shaw *Everybody's Political What's What?* xlii. 352 Political ignorance and idolatry will produce not only Hitleresque dictatorships but stampedes led by liars or lunatics. **1960** *News Chron.* 9 June 1/2 Mass crimes committed by Hitlerite agents. **1966** *New Statesman* 17 June 874/3 The collection of Hitleriana is small but fascinating. **1969** *Daily Tel.* (Colour Suppl.) 14 Feb. 12/3 The German ethos that produced Hitlerism. **1973** *Daily Tel.* 24 Apr. 18 Both men and women could be..detained indefinitely, without charge and without trial, during the Hitlerian war.

Hittite (hi·tait), *sb.* and *a.* [f. Heb. *Ḥitt(im*, Hittite *Ḫatti* + -ite[1]. The form *Hittite* occurs first in the Geneva version, 1560, of the Bible. The LXX has Χετταῖοι (identified as children of Heth (Χέτ), Heb. *benê Ḥêth*), the Vulgate *Hethæi*, whence *Ethei* in the Wycliffite versions, G. *Hethiter* (Luther), Eng. *Hethite* (Coverdale, etc.), F. *Héthéen*.] **A.** *sb.* **a.** In the Bible (Gen. xv. 20, etc.), one of a Canaanitic or Syrian tribe of greater or less extent, perh. an offshoot of the next. **b.** In modern archæology and philology, a member of a powerful and widespread ancient (non-Semitic) people, variously named *Khita* or *Kheta* in Egyptian, and *Khatti, Ḫatti* (see *Hattic *a.*) in Hittite and Assyrian, whose history can be traced from *c* 1900 to 700 B.C. in Asia Minor and northern Syria, or a member of a people conquered by the Hittite empire. **c.** The Indo-European language (written in cuneiform and hieroglyphs) of this people. **B.** *adj.* Of or pertaining to this people or their language.

1608 Topsell *Hist. Serpents* 93 Most fierce & cruell enemies, which should put & cast forth the Cananites, Hettites, and Cheuits. **1614** Raleigh *Hist. World* i. viii. 162 The second sonne of Canaan was Heth or Cethus: of whom came the Hethites, or Hittites. **1871** tr. *Ewald's Hist. Israel* III. 262 Among the petty Hittite (i.e. generally Canaanite), and Aramean kings. **1879** *Academy* 16 Aug. 124/1 The sculpture accompanied by inscriptions in Hittite (or Hamathite) characters..discovered at Ibreez in Lycaonia..proves that the Hittites had penetrated through the eastern barrier of Asia Minor formed by the Taurus range. **1880** *Encycl. Brit.* XI. 808/2 The discovery of a new hieroglyphic character in the Hittite inscriptions. *Ibid.* XII. 26/2 Our knowledge of the Hittite language is confined to the proper names mentioned in the Egyptian and Assyrian inscriptions. **1880** [see *Hattic *a.*]. **1884** W. Wright *Empire of Hittites* iv. 56 There can be little doubt that the Lycaonian *patois*, which continued to be the vernacular of the people till the days of Paul, was Hittite. **1884** A. H. Sayce *Ibid.* xi. 170 The bilingual inscription of Tarkondêmos in cuneiform and Hittite. **1920** A. E. Cowley *Hittites* i. 1 Until forty years ago, or less, the Hittites were still grouped with Hivites and Jebusites as an insignificant Syrian tribe unknown outside the Bible. **1926** *Year's Work in Eng. Stud.* 1924 38 We have in Hittite..the earliest Indo-European language recorded. **1952** O. R. Gurney *Hittites* vi. 130 The original home of Hieroglyphic Hittite seems on present evidence most likely to have been Kizzuwatna [1961 ed., Cilicia]. **1962** *Times* 31 Oct. 14/6 Stone stairway flanked by modern but Hittite-type lions. **1972** C. Raphael *Feast of History* i. 15 All the peoples—Amorites, Moabites, Hittites, Horites, and of course the familiar Philistines..with whom the Hebrews dealt constantly in the Old Testament.

So **Hittito·logy**, the study of Hittite philology, archæology, or history; **Hittito·logist.**

1948 D. Diringer *Alphabet* i. v. 89 Some Hittitologists accept the term *Nashili*. **1952** O. R. Gurney *Hittites* 8 There developed an entirely German science of Hittitology devoted to the study of the cuneiform tablets. **1964** *Language* XL. 149, I am not a Hittitologist.

Hittorf (hi·tǫɹf). The name of Johann Wilhelm *Hittorf* (1824–1914), German scientist, used *attrib.* and occas. in the possessive to denote various phenomena, pieces of apparatus, techniques, and concepts observed or invented by him, as *Hittorf dark space* = *Crookes dark space; *Hittorf method*, a method of finding the transport numbers of ions in an electrolyte by measuring the change in concentration of the ions near the electrodes during electrolysis; *Hittorf transport* (or *transference*) *number*, a transport number as found by the Hittorf method.

1893 T. O'C. Sloane *Stand. Electr. Dict.* 289 *Hittorf's solution*, a solution sometimes used as a resistance. It is a solution of cadmium iodide in amylic alcohol. *Ibid.* 466 *Hittorf's resistance*, a high resistance, often a megohm, composed of Hittorf's solution. **1897** M. M. P. Muir tr. *Lüpke's Elem. Electro-Chem.* i. iii. 41 (*heading*) Hittorf's transport-numbers. **1909** *Jrnl. Amer. Chem. Soc.* XXXI. 351 (*heading*) The relation between the true transference number and the Hittorf transference number. *Ibid.*, The ordinary transference number (T_H) for concentrated solutions, as obtained by the Hittorf method, is erroneous in cases where the ions are hydrated, since it is calculated on the assumption that the water remains stationary during the passage of the current. **1916** F. B. Pidduck *Treat. Electr.* xii. 487 The negative glow and the Hittorf dark space expand continually as the pressure is reduced. **1941** J. D. Cobine *Gaseous Conductors* viii. 213 Following the cathode glow is another dark space variously called the cathode dark space, the Crookes dark space, the Hittorf dark space. **1942** Glasstone *Introd. Electrochem.* vi. 114 Although the Hittorf method is simple in principle, accurate results are difficult to obtain.

hive, *sb.* Add: **7. hive-bound** *a.*, confined to a hive; **hive-moth,** an alternative name for the wax-moth or honeycomb moth.

1921 R. Graves *Pier-Glass* 30 A hive-bound bee. **1945** W. de la Mare *Burning-Glass* 67 As passive as the hive-bound bees. **1931** *Oxf. Univ. Gaz.* 17 June 703/1 Hive-moth (*Galleria*) at Nairobi.

hive, *v.* Add: **5.** (Further examples.) Now esp., to break away from, to separate from, a group. Also *trans.*, to remove from a group, a large unit, etc., to make separate.

1902 *Westm. Gaz.* 10 July 9/1 The Board is now hiving off to a mine with at least a promising name, the 'Baron Rothschild', in the Tati district. **1931** *Economist* 5 Dec. 1060/2 And even Syria Proper has been made to hive off the autonomous Governments of the Jebel Druse and Alexandretta. **1937** *Nature* 16 Oct. 659/1 Experimental psychology..has hived off from physiology. **1951** *Engineering* 28 Sept. 403/2 The..firm..was 'hived off' from the parent company. **1957** *Economist* 30 Nov. 783/2 It will be remembered that, while part of the Moroccan Liberation Army..agreed to incorporation in the Royal Moroccan army, another part preferred to hive off and disappear into remote areas. **1959** Halas & Manvell *Technique Film Animation* xix. 257 Many animators with a flair for individual work have hived off from these studios. **1959** Duke of Bedford *Silver-Plated Spoon* x. 201 The trustees were slowly hiving off part of the family estates to meet the awful burden of taxation. **1961** T. Landau *Encycl. Librarianship* (ed. 2) 146/1 Large public library systems are increasingly 'hiving off' special sections dealing with foreign literature. **1963** *Times* 20 Apr. 7/6 The territorial wings of the U.F.P. in Northern and Southern Rhodesia and Nyasaland would now 'hive off' with 'full authority to act for themselves' under new names. **1969** *New Scientist* 1 May 262/2 The large machines are beginning to sprout small sideshoots on to which specialized tasks can be hived off. **1971** *Times* 21 Dec. 14/3 Strong opposition to the British Steel Corporation's plans to hive off part of the River Don works at Sheffield..is likely to be encountered.

hives. Add: **hive-syrup** *U.S.*, compound syrup of squills.

1839 *Southern Lit. Messenger* V. 65/2 There's nothing there but a few drops of peppermint,..and some of the patent hive-syrup. **1901** T. H. Sollmann *Text-bk. Pharmacol.* 612 Syrupus Scillæ Compositus (U.S.P.) (*Hive Syrup*)... Used especially in whooping-cough. **1936** Cook & LaWall *Remington's Pract. Pharm.* (ed. 8) 304 Syrupus Scillæ Compositus... Hive Syrup, Coxe's Hive Syrup.

hiya (hai·yă), *int.* Also **hi-ya, hiyah.** [App. shortened from *How are you?* and influenced by Hi *int.* 2.] A word of greeting.

1940 'N. Blake' *Malice in Wonderland* i. ii. 31 'Hi-ya, boys and girls,' he cried. 'Hi-ya, Teddy,' the cry went back. **1940** R. Chandler *Farewell, My Lovely* xxxix. 309 Hiya, babe. Long time no see. **1946** P. Quentin *Puzzle for Fiends* (1947) 190 'Hiyah, baby,' I said. **1959** *Elizabethan* May 35/1 Robin Fawcett turned the grin on Friday and me and said 'Hi-ya'.

Hizen (hī·zen). Also †**Fisen, Fizen.** The name of a province in the north-west of Kyushu in Japan used *attrib.* and *ellipt.* to denote a class of porcelains characterized by rich decoration, delicate colouring, and fine workmanship, and including Hirado, Imari, and Nabeshima ware. Cf. *Hirado.

1727 J. G. Scheuchzer tr. *Kæmpfer's Hist. Japan* II. App. 61 In *Fisen* they have a certain white clay, of which they make all sorts of Porcellane-ware. **1859** L. Oliphant *Narr. Elgin's Miss. China & Japan* II. iii. 52 The egg-shell China is..made principally in the provinces of Fizen and Satsuma. **1875–80** Audsley & Bowes *Keramic Art Japan* I. 26 The old red, blue and gold Hizen. *Ibid.*, Old Hizen ware..includes white porcelain. **1878** J. J. Young *Ceramic Art* ii. vii. 177 The rich beauty of the coloring of Hizen porcelain is indescribable. **1965** S. Jenyns *Jap. Porcelain* iii. 89 The early Hizen blue underglaze ware..was by far the most beautiful of all Japanese underglaze porcelains.

Hjelmslevian (hielmzle·viăn), *a.* [f. the name of Louis *Hjelmslev* (1899–1965), Danish linguist + -ian.] Of, pertaining to, or characteristic of the writings or theories of Hjelmslev.

1950 *Archivum Linguisticum* II. 96 'Form' is not used here in its Hjelmslevian sense. **1961** *Amer. Speech* XXXVI. 160 An elaborate and opaque Hjelmslevian terminology. **1963** *Ibid.* XXXVIII. 139 Levin, resorting to a Hjelmslevian concept, considers two forms as 'semantically equivalent insofar as they overlap in cutting up the general "thought-mass".' **1963** J. Lyons *Structural Semantics* vi. 113 The particular (Hjelmslevian) framework within which the analysis of the Greek system is presented.

hlonipa (hlǫ·nipă), *v.* Also **hlonipha.** [a. Zulu, Xhosa *hlonipa* vb.] Among south-east African Bantu peoples of the Nguni group: to avoid in conversation the radical of the name of certain persons to whom such respect is due. In English contexts often treated as a noun. In quot. 1850 *uku-* is the Xhosa infinitive prefix.

[**1850** J. W. Appleyard *Kafir Lang.* 70 The Kafir *women* have many words peculiar to themselves. This arises from a national custom, called *ukuhlonipa*.] **1913** Pettman *Africanderisms* 212 The word *hlonipa* means that they [*sc.* women] are too bashful or polite to use such names in common everyday speech. **1936** E. J. Krige *Social Syst. Zulus* ii. 30 The behaviour towards relatives-in-law is largely bound up with the custom of *hlonipha*. **1970** *Stand. Encycl. S. Afr.* II. 96/1 Among the Nguni peoples of South Africa the relationship [between parents-in-law and children-in-law] is clearly defined in a code of conduct known as *hlonipha*. A wife must *hlonipha* her in-laws, in other words she must act humbly and respectfully toward them and shun them.

Ho (hōᵘ), *sb.*[4] [Native name, said to be a contraction of *horo* man.] **a.** One of the principal dialects of central India, belonging to the Kolarian group. **b.** One who speaks this language. Also *attrib.* or as *adj.*

1840 S. R. Tickell in *Jrnl. Asiat. Soc.* IX. ii. 997 The Ho language has no written character. **1860** F. Mason *Burmah* (ed. 2) 131 These Moondas now call themselves Hos. **1871** E. Balfour *Cycl. India* (ed. 2) II. 589/2 A Ho bridegroom buys his bride. **1905** P. Wagner tr. *Nottrott's Gram. Kol-Language* 4 The dialect of the Larka-Kols or Hos. **1906** G. A. Grierson *Linguistic Surv. India* IV. 116. **1908** H. H. Risley *People of India* 94 The Hos of Singhbhum. **1926** *Encycl. Brit.* XV. 891/2 The Hos, who are closely akin to the Mundas, also inhabit the Chota Nagpur division; in 1901 they numbered 386,000.

Ho (hōᵘ), *sb.*[5] [Native name, = a heap of dried peas.] Name of a tribe of the Ewe people living near the town of Ho in former Togoland, now part of Ghana.

1890 J. Frazer *Golden Bough* I. v. 265 The Ho tribe of German Togoland.

Hoabinhian (hōᵘăbi·niăn), *a.* Also **Hoabinian.** [f. *Hoabinh*, the name of a village in Vietnam where the first major site was found + -ian.] Of, pertaining to, or designating a Mesolithic or Neolithic culture found in parts of South-East Asia, particularly Vietnam, Laos, and Malaysia. Also *ellipt.* as *sb.*

1942 A. H. Brodrick *Little China* 63 Last of all.. comes the Hoabinhian which may stretch back into the Old Stone Age, although nothing unquestionably Palaeolithic has been found. **1971** *Nat. Geographic* Mar. 339/1 The first domestication of plants in the world was done by people of the Hoabinhian culture, somewhere in South-east Asia. **1972** *Sci. Amer.* Apr. 36/2 Since the initial discovery, many other sites containing Hoabinhian artifacts have been found. *Ibid.* 39/3 Local cultures that were distinctly different from the late Hoabinhian evolved at the start of this period. **1972** M. Sheppard *Taman Indera* 3 The culture which these pre-historic cave dwellers followed probably preceded the end of the Ice Age. It has been named Hoabinhian, after a province in North Vietnam where this culture was first recognized.

Hoadlyism (hōᵘ·dli,iz'm). Also **Hoadleyism.** [f. name of Benjamin *Hoadly* (1676–1761), Bishop of Winchester + -ism.] The opinions or policy characteristic of the latitudinarian clergy of whom Bishop Hoadly was typical. So **Hoa·dlyan** *a.*, **Hoa·dlyite.**

1800 J. Milner *Lett. to Prebendary* viii. 233 This doctrine..is absurd in the highest degree on the principles of Hoadlyism. *Ibid.* 225, I know that the Hoadlyites deny the existence of such sacramental grace. *Ibid.* 238 The Hoadlyan system..incumbers it with a great number of perplexing consequences. **1863** J. S. Brewer *Eng. Studies* (1881) 300 He may write and preach as much Hoadleyism..as he pleases. **1896** E. S. Purcell *Manning* I. 499 Its [*sc.* the Established Church's] Hoadleyism, if I may so speak, which prevailed before the Tractarian movement.

hoagie (hōᵘ·gi). *U.S. local.* [Origin unknown.] A sandwich made with a French loaf split lengthways and filled with lettuce and a variety of cold meats and cheeses. Cf. **submarine roll, sandwich*.

1967 *Amer. Speech* XLII. 280 Temporary residents of Philadelphia attending Temple University..had some knowledge of the sandwich, locally known as a *hoagie*. **1973** *Home & Store News* (Ramsey, N.J.) Mar., *Hoagies* include ham, salami, provolone cheese and shredded lettuce all stuffed into a seven-inch submarine roll. A packet of salad dressing is included with each hoagie. **1973** *Welcomat* (Philadelphia) 10 Oct. 1/1 The Council for Social Development profited from the Pine Street Sidewalk Fair, which is perhaps why five-inch 'hoagies' were sold for ninety cents.

Hoa Hao (hōᵘ:ă hau·). [Name of the village of birth of the founder, in Vietnam.] A form of nationalistic Buddhism, set up in 1939 in Indo-China by Huynh Phu So. Freq. *attrib.* Cf. *Caodaism.

1955 *Times* 2 May 12/2 The Hoa Hao commander-in-chief. **1969** *New Yorker* 20 Sept. 110/3 The scattered religious elements in the South—the Buddhists, the Catholics,..the Hoa Hao sect. **1971** I. M. Sacks in R. F. Spencer *Relig. & Change in Contemp. Asia* 54 The Hoa Hao religion was founded by the so-called 'mad monk', Huynh Phu So. The doctrines of Hoa Hao are..to honor one's parents; to love one's country; to respect Buddhism and its teachings; to love one's fellow man. ('Buddhism'..means..the teachings of Huynh Phu So.)

hoar-frost. Add: In scientific use now distinguished from rime. (Later examples.)

Hoar-frost is a crystalline deposit of ice formed by the sublimation of water vapour; rime is a more amorphous deposit formed by the rapid freezing of supercooled droplets of water when they are brought by air currents into contact with a cold surface.

1895 T. Russell *Meteorol.* iii. 53 Hoar-frost is a name given to the curious, regular figures resembling ferns that form on objects, especially on the window-panes in houses. ..Rime is a thick, heavy frost forming on objects from frozen rain or mist. **1921** A. E. M. Geddes *Meteorol.* vi. 182 An examination of these crystals shows that they have not been deposited first as water drops, for they are not frozen drops of water. A deposit of this form is called hoar frost. **1967** R. W. Fairbridge *Encycl. Atmos. Sci.* 402/1 'Frost' has several meanings, most commonly implying hoar-frost or white frost which occurs when the air has been damp, leading to direct sublimation and the building of interlocking ice crystals on exposed surfaces.

b. hoar-frost curve or **line**: a line representing the conditions for equilibrium between the solid and the vapour phases of water in the absence of the liquid phase.

1879 *Encycl. Brit.* VIII. 731/2 At this point the steam line, ice line, and hoar-frost line intersect, and it has therefore been called the triple point. **1940** Glasstone *Text-bk. Physical Chem.* vi. 459 The vapor pressure curve of ice.. is often called the hoar-frost curve.

hoary, *a.* Add: **4. c.** In names of animals having a hoary appearance (see quots.).

1781 T. Pennant *Hist. Quadrupeds* II. 398 Hoary M[armot] with the tip of the nose black..hair universally rude and long; that on the back, sides and belly cinereous at the root, black in the middle, whitish at the tip, so that the animal has a hoary appearance. **1829** [see sense 4 b in Dict.]. **1832** J. Rennie *Butterfl. & Moths* 183 The Hoary Double Crescent..frequents ash-trees. **1903** *Bull. Amer. Mus. Nat. Hist.* XIX. 539 Marmotta caligata (Eschscholtz). Hoary marmot. **1948** W. J. Stokoe *Caterp. Brit. Moths* I. 151 (*heading*) The Hoary Footman. **1954** O. J. Murie *Field Guide to Animal Tracks* 136 The hoary marmot has also sought the high mountains. **1959** Hall & Kelson *Mammals N. Amer.* I. 192/1 The hoary bat is solitary and roosts in trees and shrubs.

hob, *sb.*[1] Add: **2. b.** (Earlier and later examples.) Also *to raise hob.* Chiefly U.S.

1838 *N.Y. Mirror* 2 June 387/1 They say it's playing *hob* with the fellers in these here parts. **1905** B. Tarkington *In Arena* 23, I believe that idiot's right, he won't lose votes by playing hob with *us*. **1911** J. C. Lincoln *Cap'n Warren's Wards* vi. 88 Theoph's been raising hob because the Odd Fellows built on to their building. **1916** H. L. Wilson *Somewhere in Red Gap* iii. 120 He looked like one of them silly little critters that play hob with Rip Van Winkle..before he goes to sleep. **1927** P. Marks *Lord of Himself* xvi. 244 Carl parked the car and stowed the flask in the door-pocket before speaking, and then he murmured sympathetically, 'It [*sc.* liquor] played hob with you, didn't it, old girl?' **1935** M. de la Roche *Young Renny* xxv. 219 It's the food you eat without enjoyment that plays hob with your stomach. **1940** D. A. Lord *Our Lady in Mod. World* iii. 141 The revolutionists who are playing hob with our generation are really masters of the obsolete. **1949** *Chicago Tribune* 14 June ii. 1/1 The change in time on the new quiz programs is raising hob with getting the evening chores finished. **1967** *Electronics* 6 Mar. 352/2 The Chinese presumably could raise hob..with vlf transmission to submarines from the base at North West Cape.

hob, *sb.*[2] Add: **1. b.** One of the level supports on the top of a stove over which pots and pans, etc., are placed to be heated, etc.

1962 *Listener* 13 Sept. 411/2 The gas-cooker hobs have four self-lighting burners. *Ibid.* 411/3 Centre hob-grill. **1969–70** *Catal. Belling Electric Heating & Cooking* 59 No need to worry if anything boils over...No need to..take the pans off to lift the hob to get at the mess. Simply slide out the spillage tray from under the hob and clean it at the sink.

3. Also, a master tap.

1881 F. Campin *Mech. Engin.* 49 The taps used for making screw tools and worm wheels are called hobs. *a* **1884** Knight *Dict. Mech.* Suppl. 458/2 *Hob*, a hardened steel mandrel with a threaded portion which is fluted. **1888** *Lockwood's Dict. Mech. Engin.* 178 *Hob*, or *Hub*, a master tap.

6. *Comb.* **hob-grate,** a grate fitted with a hob or hobs.

1915 C. Mackenzie *Guy & Pauline* i. 53 Guy sat by the small hob-grate. **1959** *Times* 27 Jan. 10/6 He sat erect in a tall wooden chair beside a hobgrate.

hob, *v.*[3] Add: (Later example.) Hence **ho·bber,** one employed in driving hobnails into boots; **ho·bbing** *vbl. sb.*, the action of hob-nailing boots and shoes; **hobbing boot** = **hobbing foot*; **hobbing foot** *local*, a shoemaker's last.

1866 R. Hallam *Wadsley Jack* vii. 36 Thie fooit.. weean't skar me—noa, not if it wor a hobbin fooit. **1907** *Daily Chron.* 8 Jan. 6/2 The woman was struck on the head by a shoemaker's 'hobbing foot'. **1907** *Daily Chron.* 31 Jan. 6/7 He struck his wife on the head with an iron foot—a shoemaker's hobbing boot. **1921** *Dict. Occup. Terms* (1927) § 429 *Hobber*..drives in hobnails round rims of soles of heavy boots, by hand with a hammer. **1922** G. Blair *Haunted Dominie* 43 There's some that skimp the hammerin' upon the hobbin'-feet. **1936** T. E. Lawrence *Mint* (1955) iii. i. 166 Marching boots so hobbed that every pavement became a skating rink.

hob (hǫb), *v.*[4] [f. Hob *sb.*[2] 3.] *trans.* To cut or form by means of a hob or master tap. Hence **ho·bbing** *vbl. sb.* Also *hobbing machine*, a machine in which worm-wheels, spur and spiral gears are cut by means of a hob or master tap; *hobbing cutter* (see quot. 1940).

1892 *Lockwood's Dict. Mech. Engin.* App. 428. **1913** *Ibid.*, App. 439 *Hobbing Machines.* In these, spur and spiral gears as well as worm wheels are cut by a hob, the same hob serving for each type of gear by altering the angle of inclination of the thread. **1930** *Engineering* 17 Oct. 479/1 The defects of the present hobbing process of producing gear wheels. **1940** *Chambers's Techn. Dict.* 418/1 *Hobbing cutter*, a gear-cutting tool resembling a milling cutter or a worm gear, whose thread is interrupted by grooves so as to form cutting faces. **1943** *Jrnl. R. Aeronaut. Soc.* XLVII. 83 A hobbing machine with a properly designed solid master worm wheel will cut accurate and quiet-running gears. **1950** *Engineering* 18 Aug. 145/3 The majority of spur and helical gears are produced by hobbing.

hobbit (hǫ·bit). [See below.] In the tales of J. R. R. Tolkien (1892–1973): one of an imaginary people, a small variety of the human race, that gave themselves this name (meaning 'hole-dweller') but were called by others *halflings*, since they were half the height of normal men. Also *attrib.* and *Comb.* Hence **ho·bbitish** *a.*, resembling a hobbit, hobbit-like; **ho·bbitomane,** a devotee of hobbits; **ho·bbitry,** the cult of hobbits; hobbits collectively, or their qualities.

1937 J. R. R. Tolkien *Hobbit* i. 11 In a hole in the ground there lived a hobbit. **1947** C. S. Lewis in *Ess. presented to C. Williams* 104 *The Hobbit* escapes the danger of degenerating into mere plot and excitement by a very curious shift of tone. As the humour and homeliness of the early chapters, the sheer 'Hobbitry', dies away we pass insensibly into the world of epic. **1954** J. R. R. Tolkien *Fellowship of Ring* 7 The memoirs of the renowned Hobbits, Bilbo and Frodo. *Ibid.* 11 A few notes..are here collected from Hobbit-lore. *Ibid.* 20 The Thain was..captain of the Shire-muster and the Hobbitry-in-arms. *Ibid.* 46 It was a tendency of hobbit-holes to get cluttered up. **1955** —— *Return of King* 416 *Hobbit* is an invention. In the Westron the word used, when this people was referred to at all, was *banakil* 'halfling'. But..the folk of the Shire and of Bree used the word *kuduk*... It seems likely that *kuduk* was a worn-down form of *kûd-dûkan* [= 'hole-dweller']. The latter I have translated..by *holbytla* ['hole-builder']; and *hobbit* provides a word that might well be a worn-down form of *holbytla*, if the name had occurred in our own ancient language. **1962** *Listener* 22 Nov. 881/3 The more ambitious hobbit saga, *The Lord of the Rings*. *Ibid.*, To those who are already hobbitomanes, this book is bound to be a delight. **1966** *New Statesman* 11 Nov. 701/2 The newest and richest site of hobbitry is the American campus, where students are said to greet each other with hobbitish salutations such as 'May your beard never grow less'. **1968** *Listener* 20 June 790/3 Professor Tolkien was thinking of the average, ambling Englishman when he wrote about his hobbits. **1970** H. Perry *Human Be-In* i. 20 The consistently good people in the Tolkien books are Hobbits and they have the lowliest status of all the groups of characters in the books. The hippies thought of themselves as being or becoming Hobbits; from time to time as the winter wore on, a sign would appear in the window of one of their gathering places to this effect: Do not add to the street confusion this weekend... Be good little Hobbits and stay home.

hobble, *sb.* Add: **3.** Also (chiefly *Austral.*) *hobble chain.*

1901 'M. Franklin' *My Brilliant Career* (1966) ix. 52 The sound of camp-bells and jingle of hobble chains.. had come to these men. **1928** 'Brent of Bin Bin' *Up Country* xv. 272 He remembered the hobble-chains behind the old stable. **1959** *Listener* 15 Jan. 113/1 In a brawl, they're deadly. Anything goes—spurs, hobble chains, the lot.

4. In full *hobble-skirt.* A close-fitting skirt usually confined by a wide band below the knees and above the ankles.

1911 *Smart Set* Mar. 40 A hobble skirt is an awful habit to get into. **1912** *Punch* 3 Apr. 255/1 The continued success of the hobble..has..restricted the use of textile material. **1918** *Wireless World* Oct. 372 A feminine atrocity in a cerise muslin sheath on 'hobble-skirt' lines. **1920** M. Asquith *Autobiog.* I. 221 From the hoop to the hobble is not a more violent change than from the riding-hats of 1894 to the riding-hats of 1917. **1921** C. Torr *Small Talk at Wreyland* II. 69, I said, 'You don't go in for hobble-skirts, I see.' **1969** H. E. Bates *Vanished World* vii. 71 Ladies in flowered hats and hobble skirts.

hobby, *sb.*[1] Add: **6.** Also *hobby farm, farmer, shop, show.*

1960 Hobby farm [see *golden handshake* s.v. *golden *a.* 10 a]. **1968** *Globe & Mail* (Toronto) 15 Jan. 24/2 (Advt.), 50 acres..ideal for retirement or hobby farm. **1961** *Ann. Reg. 1960* 16 Loss relief for 'hobby farmers' was terminated. **1945** *Time* 9 Apr. 82 Prison hobby shops are not unusual. **1966** 'D. Shannon' *With a Vengeance* (1968) viii. 107 This is Mr. Seidenbaum from the hobby shop. **1967** 'W. Wright' *Shadows don't Bleed* i. 20 A girl friend of Bridget's who owns a hobby shop here in town. **1921** *Daily Colonist* (Victoria, B.C.) 9 Apr. 10/1 The first annual boys' hobby show will be held at the Y.M.C.A. this afternoon and evening...The exhibition will consist of manual training work, working models of aeroplanes, boats, etc.

hobby-horse, *v. intr.* Add: (Earlier example.) Also, to move like a hobby-horse.

1819 Keats *Let.* (1935) 315 He is not only reconcil'd to it but hobbyhorses upon it. **1958** M. Pugh *Wilderness of Monkeys* ii. 22 A sheep started at his cursing and went hobby-horsing down the hill. **1965** *Sunday Tel.* 19 Sept. 24/7 She started to hobby-horse and at the third bounce in the height of the gust of about 30 knots in went her bows. **1967** *Daily Tel.* 30 Mar. 18/6 A continual hobby-horsing which stopped it dead about every third wave it hit.

hobbyist. Add: (Further examples.) Sometimes used with a connotation of crankiness.

1935 W. de la Mare *Early One Morning* 582 Of such kind are nascent cranks and hobbyists, and some of them attain at last to the rank..of English 'characters'. **1937** Auden & MacNeice *Lett. fr. Iceland* 105 A circle where one's known Of hobbyists and rivals. **1948** F. A. Staples *Water-Color Painting* 1 A non-technical manner suited to the hobbyist or art student. **1959** N. Mailer *Advts. for Myself* (1961) 137 'That's a nice Jap machine gun.' He looked at it with the professional curiosity of a hobbyist. **1971** *Guardian* 11 Aug. 10/4 The young hobbyist was encouraged to specialise.

hobday (hǫ·bdēⁱ), *v.* [The name of F. T. *Hobday*, veterinary surgeon (1869–1939).] *trans.* To operate on (a horse) in order to improve its breathing. Chiefly as *vbl. sb.*, **ho·bdaying.**

1938 F. T. G. Hobday *50 Yrs. Vet. Surgeon* iv. 55, I contributed considerably to the successful establishment of an operation now familiarly known..as being 'Hobdayed'. **1958** J. Hislop *From Start to Finish* 171 *Hobdaying*, an operation to help the breathing of horses who are wrong in the wind, but not by putting a tube in their throat. **1963** *Times* 2 Mar. 3/4 A wind infirmity handicapped King of Saba last year and he has now been hobdayed.

hobo. For '*Western U.S.*' read 'orig. *Western U.S.*' Add further examples.

1889 *Ellensburgh* (Wash.) *Capital* 28 Nov. 2/2 The tramp has changed his name, or rather had it changed for him, and now he is a 'Hobo'. **1918** *Let.* in F. A. Pottle *Stretchers* (1929) x. 295 We had been so long separated from our organization that we had pretty thoroughly acquired a hobo frame of mind. **1925** J. Buchan *John Macnab* vii. 157 The gillies have..gathered in some wretched hobo they found looking at the river. **1928** *Punch* 15 Feb. 196/1 Few dramatic critics..could display so adequate a working knowledge of..the ways of hobos in the United States. **1959** I. & P. Opie *Lore & Lang. Schoolch.* iii. 55 Gipsies, usually known as 'gyppoes' or 'hoboes'. **1963** H. Garner in R. Weaver *Canad. Short Stories* 2nd Ser. (1968) 40 Harvest hands are like hobos, their friendships as casual as the mating of a pair of flies.

Hence **ho·bo** *v. intr.* (and quasi-*trans.* with *way*), to act the hobo; to journey or travel as a tramp; also with *it*. Also **ho·bodom, ho·boism,** the realm or world of the hobo; **hoboe·tte,** an occasional term for a female hobo.

1906 U. Sinclair *Jungle* xxv. 298 Then he explained how he had spent the last summer, 'hoboing it', as the phrase was. **1914** J. London *Let.* 28 July (1966) 426 You can scarcely find a tramp today..who has not hoboed with me. **1918** *Dialect Notes* V. 5 Hoboettes of America. **1923** H. L. Foster *Beachcomber in Orient* ix. 183 Having hoboed my way thus far, I could afford to travel as a passenger the rest of the way. **1928** *Daily Mail* 9 Aug. 11/4, I sort of hoboed my way out to San Francisco. **1930** *Publishers' Weekly* 31 May 3736/2 Any hobo temporarily sober can find a publisher to place on the market with great éclat an epic of Hobodom, provided that there is sufficient of the hobo atmosphere. **1930** *19th Cent.* June 849 These were the high days of American hoboism. **1931** 'D. Stiff' *Milk & Honey Route* xv. 172 He can never understand why the hobos want to keep out the women, whom he labels..'the hoboettes'. **1949** *Landfall* III. 136 Have I.. hoboed through fifteen of these United States only to be jailed like a common criminal? **1966** *Punch* 4 May 671/1 They give us a fleeting glimpse into their world of pop, incipient careers, hoboism and love. **1967** K. Allsop *Hard Travellin'* xvii. 206 Typical of this idiom is Benson's *Hoboes of America* pamphlet in which he loudly warns boys—and 'hoboettes' too. **1972** *Village Voice* (N.Y.) 1 June 78/3 He and his brother were hoboing it in Missouri.

hobohemia (hōᵘbohī·miă). Chiefly *U.S.* [Blend of Hobo and Bohemia.] A community of hoboes, or the district in which they live; the life of the hobo. Also **hobohe·mian** *a.* and *sb.*

1923 N. Anderson *Hobo* i. 4 This four-part concept, Hobohemia, is Chicago to the down-and-out. **1931** 'D. Stiff' *Milk & Honey Route* p. v, I have a good deal to say ..about Hobohemia, the habitat of hobos. *Ibid.* iii. 32 You will fall into your place in the Hobohemian pyramid of social status. **1936** F. Clune *Roaming round Darling* xx. 200 Nadbuck was a dinkum hobohemian. **1960** *Spectator* 2 Dec. 899 The New Lefties, the hobohemians, the *Encore* subscribers. **1973** *New Society* 3 May 257/3 An awareness of what aspects of hobohemia need consideration.

Hobson-Jobson (hǫ·bsən dʒǫ·bsən). *Anglo-Ind.* Also 7 **Hosseen Gosseen, Hossy Gossy;** 8 **Hossein Jossen, Hassan Hassan,** etc. [Corruption by British soldiers in India of Arab. *Yā Ḥasan! Yā Ḥusayn!* = O Hasan! O Husain!] **1.** Anglicized form of the repeated wailings and cries of Muslims as they beat their breasts in the *Muharram* procession; hence this festal ceremony. Also *transf.*

Hasan and Husain, grandsons of Muhammad, were killed while fighting for the faith.

1634 T. Herbert *Trav.* 167, I have seene them nine severall dayes..in the streets all together crying out *Hussan, Hussan.* **1698** J. Fryer *New Account E. India & Persia* 108 The Moors solemnize the Exequies of *Hosseen Gosseen*. *Ibid.* 357 That Liberty, which was chiefly used in their *Hossy Gossy*. **1773** E. Ives *Voy.* i. ii. 28 Their *Hassan Hassan*, in memory of the two sons of Ali by Fatima (Mahomet's daughter) being killed in one day fighting for the faith. **1817** T. S. Raffles *Hist. Java* II. 4 The ceremony of *húsen húsen*..here passes by almost without notice. **1829** *Oriental Sporting Mag.* (1873) I. 129/2 The folks makes sich a noise..shouting Hobson Jobson, Hobson Jobson. **1861** J. T. Wheeler *Madras* II. xxxii. 347 The Mussulman feast called '*Hossein Jossen*'. **1935** M. E. Houtzager *Unconscious Sound- & Sense-Assimilations* ii. 52 Hobson-Jobson, suggestive of a proper name, is the name of a native festal excitement.

2. a. Used as the title of a famous collection of Anglo-Indian words.

1886 Yule & Burnell (*title*) Hobson-Jobson, a glossary of colloquial Anglo-Indian words and phrases, and of kindred terms.

b. *the law of Hobson-Jobson*: a phrase sometimes used of the process of adapting a foreign word to the sound-system of the adopting language. So **Ho:bson-Jo·bsonism.**

1898 Morris *Austral Eng.* 287/2 The name of the shell is a corruption of this word, by the law of Hobson-Jobson. **1919** Mencken *Amer. Lang.* 41 Its variations show a familiar effort to bring a new and strange word into harmony with the language—an effort arising from what philologists call the law of Hobson-Jobson. **1934** *S.P.E. Tract* XLI. 21 There are the words ('Hobson-Jobsonisms') where the original [*sc.* Indian] form has been more or less modified in the process of Anglicization.

‖**hoc genus omne** (hǫk ge·nǫs ǫ·mni), *phr.* [L. (Horace, *Satires* I. ii. 2).] Usu. in phr. *et* (occas. *and*) *hoc genus omne*: and the whole of that class or group; and all that kind of thing (often as ornamental substitute for *et cetera*).

1748 Chesterfield *Let.* 9 Mar. (1774) I. 271 All the shops, drolls, tumblers, rope-dancers, and *hoc genus omne*. **1834** C. C. F. Greville *Mem.* (1874) III. xxiii. 95 He saw Newcastle, Winchelsea, Wetherell, and *hoc genus omne* as much the objects of idolatry as himself. **1867** Rossetti *Let.* 23 Oct. (1965) II. 641 When razzias occur on the part of organ-grinders, brass bands, *et hoc genus omne*. **1889** G. B. Shaw *London Mus. 1888–89* (1937) 94 Handfuls of earls, a dean..half-a-dozen members of Parliament, *et hoc genus omne*. **1953** *Essays in Criticism* III. 2 The Herbert Warrens and the Sidney Lees, *et hoc genus omne*.

hoch (hōχ), *sb.* [a. G. *hoch*, short for *hoch lebe* long live.] An instance of the ejaculation *Hoch!*; an exclamation of loyal approval; a cheer, hurrah. Hence **hoch** *v. intr.*, to utter a *hoch* or *hochs*; *trans.* to cheer with cries of *Hoch!*

1867 J. Paget *Let.* 20 Sept. (1901) 232 The regular German 'festive dinner'; with the speeches and songs between the courses; songs by the whole 400 of us; uproarious 'hochs'; clinking of glasses. **1870** E. G. E. Ward *Jrnl.* 23 Sept. in *Outside Paris* (1871) 18 The General was giving a dinner..to his officers, and we could hear their 'Hoch!' 'Hoch!' after the toasts. **1907** *Daily Chron.* 13 Sept. 6/6 On the whole it will be prudent for the average Englishman not to attempt a 'Hoch!' in welcoming the Kaiser to London to-day. **1909** *Ibid.* 2 June 5/6 They 'hoched' us on our way, and cheered when they got tired of 'hoching'. **1920** *Chambers's Jrnl.* June 374/2 If Britain had had a million Australian troops, they, the present gathering, would be 'hoch, hoching' in Berlin. **1921** A. S. M. Hutchinson *If Winter Comes* iii. i. 148 The 25th anniversary of the Emperor William's accession was 'Hoch'd' throughout the German Empire. *Ibid.*, Such fervent and sincere 'Hochs!' never boomed across the seas of the world.

hocheur (oʃȫr). [Fr., f. *hocher* to nod the head.] *Cercopithecus nictitans*, the white-nosed monkey or spot-nosed guenon, found in tropical Africa.

1840 tr. *Cuvier's Anim. Kingd.* 57 Next follows a group of smaller species..consisting of the Talapoin M[onkey].. the Moustache M[onkey]..the Vaulting M[onkey]..the Hocheur. **1883** *List Vert. Anim. Zool. Soc.* (ed. 8) 12 *Cercopithecus nictitans* (Linn.), Hocheur Monkey. **1905** *Westm. Gaz.* 6 Feb. 6/3 A specimen of the Hocheur monkey is among the new inmates of the Monkey House.

‖**hochgeboren** (hō·χgĕbōᵊ:rĕn), *a.* and *sb.* Also **hochwohlgeboren.** [Ger.] **A.** *adj.* = High-born *a.* **B.** *sb.* A high-born person; such people collectively.

1905 M. A. von Arnim *Princess Priscilla's Fortnight* i. 7 Was she not a *hochgeboren*, a member of an ancient house? **1930** W. S. Maugham *Cakes & Ale* iii. 38 She was a simple old lady..but she had not..forgotten that she was *hochwohlgeboren*. **1933** D. C. Peel *Life's Enchanted Cup* xviii. 238 The German ladies..were convinced that we had forced Germany to go to war, and the Hochgeboren..felt bitter towards us. **1949** O. Nash *Versus* 139 Do not overlook the fact, *hoch-geboren* Bretheren and Sisteren. **1951** J. C. Fennessy *Sonnet in Bottle* v. v. 173 He probably thinks he's as *hochgeboren* as the half-caste descendant of an English knight and a Peruvian D.P. **1972** 'J. Melville' *Ironwood* v. 74, I am *hoch geboren*, as my stuffy German grandaunts still like to put it.

hock, *sb.*[4] Add: **b. hock-cup.**

1851 *London at Table* iii. 50 (*heading*) Champagne, Hock, or Chablis cup. **1899** *Daily News* 20 May 8/7 There is a wonderful hock-cup 'made in Germany', which the knowing ones partake of. **1958** A. L. Simon *Dict. Wines* 66/2 *Hock cup.* To a bottle of Hock add 3 wine-glasses of Sherry, 1 lemon sliced, and some balm or borage.

hock, *sb.*[6] Add: (Examples.) *U.S.*

1859 G. W. Matsell *Vocabulum* 113 *Hock*, the last card in the box. **1913** C. E. Mulford *Coming of Cassidy* vii. 118 In his agitation he exposed the hock card before he realized what he was doing. **1931** G. F. Willison *Here They dug Gold* 217 The last card, the 'hock',..likewise pays nothing.

b. *from soda* (*card*) *to hock*: from the top card to the last in the dealing-box; hence, from beginning to end.

1902 H. L. Wilson *Spenders* v. 49 Young Bines played the deal from soda card to hock. **1918** C. E. Mulford *Man fr. Bar-20* ii. 21 You got me beat from soda to hock. *Ibid.* xiii. 134 'Are you in?' 'Every d—d chip; from my hat to my worn-out boots; from soda to hock.' **1925** —— *Cottonwood Gulch* xvi. 219 You've got 'em all guessin', from soda to hock. Good for you! **1938** H. Asbury *Sucker's Progress* 16 For many years a common expression was 'from soda to hock', meaning the whole thing, from soup to nuts.

hock (hǫk), *sb.*[7] *U.S. slang.* [a. Du. *hok* hutch, hovel, prison, (*slang*) credit, debt.] **a.** Phr. *in* (occas. *the*) *hock*: (*a*) in the act (of gambling); (*b*) in prison; (*c*) in pawn; (*d*) in debt. So occas. *out of hock.*

1859 G. W. Matsell *Vocabulum* 113 When one gambler is caught by another, smarter than himself, and is beat, then he is in hock. Men are only caught, or put in hock, on the race-tracks, or on the steamboats down South... Among thieves a man is in hock, when he is in prison...'If the cove should be caught in the hock he won't snickle,' if the fellow should be caught in the act, he would not tell. **1860** 'C. Martel' *Detective's Note-Bk.* 36 In about ten minutes from that time we had them 'in hock' (the cells). **1872** G. P. Burnham *Mem. U.S. Secret Service* p. vi, *In the hock*, in the act of commission; on the spot. **1883** H. C. Lukens *Jets & Flashes* 146 We deeply regret that our india-rubber armor is in hock. **1896** Ade *Artie* xviii. 169 They go back home and leave all their stuff in hock. **1898** J. London *Let.* 30 Nov. (1966) 7, I..got my watch out of hock. **1902** W. N. Harben *Abner Daniel* xxiii, I felt..ef they did git Jimmy out o' hock..without me a-chippin' in, I'd never be able to look at 'em without remorse. **1903** A. H. Lewis *Boss* 31 Well..even a crook has got to go somewhere. That is,..when he ain't in hock. **1908** G. H. Lorimer *J. Spurlock* vii. 141 He made me feel that I was doing him a favour in consenting to have my evening clothes taken out of hock. **1911** C. E. Mulford *Bar-20 Days* xiii. 141, I said pound, not pond. P-O-U-N-D; which means that it's pawned, in hock. **1913** —— *Coming of Cassidy* vii. 118 If the four lay under the Queen, Cassidy lost; if not, he either won or was in hock. **1926** J. Black *You can't Win* xxiv. 390, I was in hock to friends who saved me from a heavy sentence, provided me with work [etc.]. **1929** *Collier's* 5 Jan. 40/4 My cash was gone, and I was in hock for the next three years. **1956** B. Holiday *Lady sings Blues* (1973) i. 1 She worked her way out of hock in the hospital and took me home to her folks. **1971** M. Tak *Truck Talk* 89 *In hock constantly*, humorous expression for the financial condition of any owner-operator who has a tractor manufactured by International Harvester Company.

b. *attrib.* and *Comb.*, as **hock-game** (see quot. 1859); **hock-shop,** a pawnshop.

1859 G. W. Matsell *Vocabulum* 113 In a hock-game, if a man hits a card, he is obliged to let his money lie until it either wins or loses. **1871** *Sessions Papers* Apr. 485 That piece that I dropped in the *hock* shop. **1886** *Lantern* (New Orleans) 22 Sept. 2/3 Take the bed too, and run it into a hock shop. **1907** I. Zangwill *Ghetto Comedies* 296 The diamond necklace..stolen..and found afterwards..in a low 'hock-shop' in New Orleans. **1926** J. Black *You can't Win* xxi. 336 The average thief will walk by the hockshop and look in. The hockshop man..knows he has something 'hot', or crooked. **1969** C. Irving *Fake!* (1970) xvi. 198 He had previously pawned one of the Matisse oils..to the Mont de Piété, the French national hockshop.

hock, *v.*[3] *slang* (orig. *U.S.*). [f. *Hock *sb.*[7]] *trans.* To pawn.

1878 *San Francisco Trade Herald* Aug. 2/2 To soak—to hock—Yer upper benjamin at yer uncle's, to get the 'sugar' for a good square meal. **1902** H. L. Wilson *Spenders* xxxiii. 397 The only thing I'll do..is to hock a few blocks of the stock I bought outright. **1904** G. H. Lorimer *Old Gorgon Graham* 184 You can hock your overcoat before marriage to buy violets for a girl. **1922** H. L. Foster *Adv. Trop. Tramp* xx. 354 I've just hocked my camera, and all I've got is two dollars. **1945** G. Millar *Maquis* i. 23 You might be able to hock them if you run out of money. **1969** C. F. Burke *God is Beautiful, Man* (1970) 65 Then he went and he took everything he had—his automobile—and he hocked them.

hockelty. *U.S. slang.* Also **hocklety, hockley, hocly.** (See quots.) Cf. Hock *sb.*[6]

1843 J. H. Greene *Expos. Arts Gambling* 210 By hocklety and splitting, many men have experienced great disappointment on this same device of hocklety. *Ibid.* 166 *Hockley*, signifies the last card but one, the chance of which the banker claims, and may refuse to let any punter withdraw a card when eight or less remain to be dealt. **1867** *Bohn's Hand-bk. Games* 336 The last card but one is called hocly, and forms part of the banker's gain. **1895** Manson *Sporting Dict.* 58 (Faro), *Hock* or *Hockelty card*, the last card remaining in the box, after the deal has been made.

hockey[2]. Add: **1. b.** In N. Amer. = **ice hockey*.

The older game is referred to as *field hockey*.

1895 *Rat Portage* (Ont.) *News* 11 Jan. 1/2 Hockey is the most popular winter sport in Canada, taking the place of lacrosse. **1906** *Daily Colonist* (Victoria, B.C.) 5 Jan. 2/1 The first hockey match of the season was played here between Rossland and Nelson teams. **1953** *Canad. Geogr. Jrnl.* XLVI. 138/2 The children maintain their own open air hockey rink on the ice of Green River. **1969** Widdowson & Halpert in Halpert & Story *Christmas Mumming in Newfoundland* 162, I dressed in a hockey suit.

3. *hockey-girl, -playing, -set, -type.*

1906 *Daily Chron.* 4 Oct. 4/4 The 'hockey set' are as a rule some of the healthiest girls in college. **1909** *Ibid.* 5 May 9/2 'Dear me, no, Miss Bulliphant,' she replied in what I call the downright, hockey-girl manner. **1915** V. Woolf *Voyage Out* xiv. 211 Hockey-playing young women in Wiltshire. **1936** 'R. West' *Thinking Reed* xii. 435 You look awfully well now, well to the point of hockey-playing. **1959** *Times* 16 Feb. 11/5 The models are all looking much better fed, and without yet suggesting hockey-girls they don't any longer look like haughty hunger-strikers. **1961** *Times* 18 May 17/1 Miss Sian Reynolds as a hockey-girl St. Joan. **1963** J. T. Story *Something for Nothing* i. 17, I like the hockey type... I can't stand these sex-pots.

ho·ckeyist. *Canad.* [f. *Hockey[2] 1 b + -ist.] A person who plays ice hockey.

1895 *Athletic Life* Mar. 121 In his stead was elected..an enthusiastic hockeyist and vice-president of the Victoria club. **1902** *Canadian Mag.* Mar. 435 The Montreal and Winnipeg Clubs for some years were acknowledged to excel all other hockeyists in Canada. **1963** *National Hockey Ann.* 36/2 Shack..was actually a reluctant hockeyist and was thrust into the game by his father.

hodad (hōᵘ·dæd). *Surfing slang.* [Origin obscure.] (See quots.)

1962 *Austral. Women's Weekly* Suppl. 24 Oct. 3/2 *Ho-dad*, anyone who annoys board-riders while they surf. **1963** *Pix* 28 Sept. 62/2 *Hodad*, non-surfing beach bum. **1965** *Daily Express* 16 Aug. 6/8, I could be a 'hodad' which means a man who talks big about his surfing exploits, but jibs at riding the big waves. **1965** *N.Z. Listener* 17 Dec. 4/5 Surfers..are..antagonistic towards..ill-mannered surf bums ('hodads'). **1971** *Studies in English* (Univ. Cape Town) Feb. 26 *Ho-dad* is a much misunderstood word. Basically, it is a double contraction from the American slang word *hood*, which in turn comes from *hoodlum*.

hoddy-doddy, *sb.* **2.** Delete † *Obs.* and add later examples.

Quot. *a* 1953 is perh. influenced by sense 3, noodle, 'dodderer'.

1877 E. Lear *Laughable Lyrics*, You're such a Hoddy Doddy. *a* **1953** Dylan Thomas *Quite early one Morning* (1954) 64 Exhibitionists,..theological rhetoricians, historical hoddy-doddies, balletomanes,..windbags. **1969** R. J. White *Women of Peasen Hall* v. 56 Gardiner always refers to him as 'Dids', or a 'Hoddy-doddy'—which is Suffolk for a diminutive person.

hoddy-noddy. Delete † *Obs.* and add later examples.

1951 Dylan Thomas *Let.* 12 Apr. (1966) 356 No more of that beer-cheapened hoddy-noddy. **1970** 'M. Hebden' *Mask of Violence* (1971) xx. 183 They were..no end considerate. Shoved everybody out of the way. Damn near a punch-up with one lot of beer-cheapened hoddy-noddy.

Hodegetria (hōᵘdigi·triă). [f. Gr. ὁδηγήτρια (see below).] An iconographical variant of the Virgin and Child in which the Child is depicted on the Virgin's left arm while she indicates Him with her right hand as 'The Indicator of the Way' (the meaning of the Greek word).

According to tradition the arrangement follows that of a painting originally attributed to St. Luke, but the earliest surviving example is probably to be assigned to the 7th century. The composition was frequently copied in Italy, the Byzantine world, and Russia. Sometimes, in the process of copying, the Child was transferred from the Virgin's left arm to the right and she indicates Him with her left hand, but there is no particular significance in this variant.

1880 E. Venables in Smith & Cheetham *Dict. Chr. Antiq.* II. 1152/2 The famous *Hodegetria*..which was for so many centuries regarded with the deepest reverence by the Greeks. **1911** L. Gillet in *Cath. Encycl.* XI. 397/2 Mention must be made of the numerous icons, the various types of the Madonna (Panagia, Nicopœia, Hodegetria). **1937** *Burlington Mag.* July 18/2 The Madonna Enthroned in the central field of the Antwerp altarpiece belongs to the Hodegetria type common in Italy in the thirteenth century. **1952** D. T. Rice *Eng. Art 871–1100* v. 116 The most usual type of Virgin was that known as the *Hodegetria* or Indicator of the Way, where the Virgin holds the Child on her left arm while she points to Him with her right hand; the Child's right hand is extended in blessing. **1959** C. Cecchelli et al. *Rabbula Gospels* 49 The Madonna is a Hodegetria type, that is, inspired by a famous image venerated since the 5th century in the Church of the Guides at Constantinople.

ho-de-ho (hōᵘ·,di,hōᵘ·), *int.* An exclamation, used as the appropriate response to *hi-de-hi *int.* Also *transf.*

1941, 1943, 1944 [see *hi-de-hi *int.*].

hoden (ū·dən), *a.* *Kentish dial.* Also **hooden.** [Origin uncertain: perh. from association with *wooden* from the wooden horse's head.] Of or pertaining to the horse with wooden head and clapping jaws featured in a masquerade which formerly took place, *spec.* in Kent, on Christmas Eve. **ho·dener,** a performer in this masquerade; **ho·dening,** the name of the performance; also *attrib.*

1807 *European Mag.* LI. 358 This [mumming] is called, *provincially*, a Hodening, and the figure above described a Hoden, or Woden horse. **1887** Parish & Shaw *Dict. Kentish Dial.* 77 *Hoodening*.., the name formerly given to a mumming or masquerade. **1891** *Church Times* 2 Jan. 20/1 'Hodening' still goes on..at Deal and Walmer. **1909** P. Maylam *Hooden Horse* i. 2 Everyone springs up, saying, 'The hoodeners have come, let us go and see the fun.' *Ibid.* 7 A farm with more than one team would have a hooden horse to each team. *Ibid.* 9, I had intended to walk on to Deal and look for the hoodening parties there. **1966** G. E. Evans *Pattern under Plough* xix. 193 The hobby-horses that appear in many countryside ceremonies and ritual dances, notably the *Hodening Horse*. **1971** *Country Life* 17 June 1533/1 The Hooden Horse, a mystic man-animal found only in East Kent, will be at large in Folkestone..June 19.

Hodgkin's disease. (Earlier example.)

1865 S. Wilks in *Guy's Hosp. Rep.* XI. 56 (*heading*) Cases of enlargement of the lymphatic glands and spleen (or, Hodgkin's disease).

hodja, var. Khoja.

hodograph. Add: **1.** Hence extended to curves in which the radius represents other vector quantities.

1944 H. R. Byers *Gen. Meteorol.* viii. 212 The details of the change of the geostrophic winds with height are studied by means of a hodograph. **1961** *Aero/Space Engin.* Feb. 24/2 (*title*) The hodograph and ballistic missile trajectory problems. **1966** *McGraw-Hill Encycl. Sci. & Technol.* XIII. 508/2 These [changes] are most readily described with the aid of a hodograph of the potential gradient vector...One vector only is shown..extending from the origin of coordinates to a point on the hodograph.

hodoscope (hǫ·doskōᵘp). [f. Gr. ὁδό-ς way + -scope.] **1.** *Microscopy.* (See quot.)

1915 J. W. Evans in *Jrnl. Quekett Microsc. Club* XII. 613 It is frequently desirable to examine simultaneously the optical properties of a number of different directions in a mineral... For this purpose the microscope is.. converted into an optical instrument in which every point in the image corresponds not to a point in the object under examination, but to a direction along which light traverses that object in parallel paths. Such an instrument may be conveniently described as a *hodoscope* or path viewer, a term which is to be preferred to the word 'konoscope' employed by some authors.

2. *Physics.* An assembly of particle detectors used for observing the paths of cosmic-ray and other particles.

1950 *Physical Rev.* LXXVIII. 715/1 For the purpose of analyzing the shower phenomena in a hodoscope set it would appear desirable to use as large a number of narrow counters as possible. **1967** *Nuclear Instruments & Methods* LI. 1/1 An experiment on mu meson pair production was performed..utilizing a 192 unit hodoscope. For each hodoscope counter only the detection or non-detection of a particle (and not pulse height) was recorded.

hoe, *v.* Add: **3.** See also Row *sb.*[1] 6 b, c.

hoe (hōᵘ), *v.*[2] *U.S.* [f. Hoe-down.] To dance or play a hoe-down.

1835 *Gent's Vade-Mecum* (Phila.) 21 Mar. 3/5 'Pooh!' replied his panting rib, hoeing it off like a regular Juba, 'don't be a nigger all the days of your life.' **1909** *Cent. Dict.* Suppl.

hoe-cake. (Earlier examples.)

1745 W. Logan *Jrnl.* 12 Oct. in *Pennsylvania Mag. Hist.* (1912) XXXVI. 12 Breakfasted on Tea & Hoe Cake Bread, which we have done in common. **1745** *Ibid.* 21 Oct. 162 Got Breakfast on Tea & Hoe Cake. **1774** P. V. Fithian *Jrnl.* 15 Jan. (1900) 93 Sup'd on chocolate, & hoe-Cake, so Called because baked on a Hoe before the fire. **1780** W. Fleming in N. D. Mereness *Trav. Amer. Col.* (1916) 641, I had lived for a constancy on poor dried buffalo bull beef cured in the smook..without any addition but a piece of Indian hoe-cake.

hoed, *ppl. a.* (s.v. Hoe *v.*). (Amer. examples.)

1643 *New Plymouth Laws* 74 By ymproved lands are understood meddow land plowed land and howed lands. **1879** *Scribner's Monthly* Dec. 239/2 The owner has only to give it a year of ordinary cultivation, taking from it.. some profitable hoed crop.

hoe-down. (Earlier and later examples.)

[**1807** W. Irving *Salmagundi* 7 Mar. 98 As to dancing, no Long-Island negro could shuffle you 'double trouble', or 'hoe corn and dig potatoes' more scientifically.] **1841** *Picayune* (New Orleans) 14 Jan. 2/1 He looks and walks the character to the life, and some of his touches are of the genuine 'hoe down', 'corn-field' order. **1849** T. T. Johnson *Sights Gold Region* iv. 38 One of our party commenced a regular hoe-down, knocking his shins with heavy boots. **1855** *Knickerbocker* XLVI. 227 Rude, high-legged reels and 'hoe-downs'. **1919** T. K. Holmes *Man fr. Tall Timber* vii. 84 A medley of old-time hoe-downs and jig music. **1961** *Times* 30 Mar. 6/7 The hoe-down sequence in *Seven Brides for Seven Brothers*. **1963** *Punch* 3 July 23/3, I was invited to the last hoe-down. **1967** 'J. Munro' *Money that Money can't Buy* ix. 114 Two more cowboys appeared. ..They played hoe-down music. **1969** *Guardian* 2 Sept. 8/2 The atmosphere was that of..a hoedown in—well, perhaps in Hibbing, Minn.

hœrnesite (hō·ɹnĕzəit). *Min.* Also **hörnesite.** [ad. G. *hörnesit* (W. Haidinger, reported in *Jahrb. d. k.-k. geol. Reichsanstalt: Verhandl.* (1860) XI. 41), f. the name of M. *Hörnes* (1815–68), Austrian mineralogist: see -ite[1].] A white hydrated arsenate of magnesium $Mg_3(AsO_4)_2.8H_2O$.

1868 J. D. Dana *Syst. Min.* (ed. 5) 817 Hœrnesite... First distinguished by Kenngott in minerals from the Bannat, Hungary. **1903** *Jrnl. Chem. Soc.* LXXXIV. ii. 655 (*heading*) Crystallised magnesium phosphate and arsenate: artificial production of bobierrite and hœrnesite. **1968** I. Kostov *Mineralogy* 448 (*table*) Hörnesite. **1969** *Mineral. Abstr.* XX. 229/1 Hoernesite..is monoclinic.

Hoffmann (hǫ·fmæn). **1.** The name of Friedrich *Hoffmann* (1660–1742), German physician, used in the possessive in *Hoffmann's anodyne* (in full *Hoffmann's mineral anodyne liquor*), compound spirit of ether, a mixture of alcohol and ether with ethereal oil.

1747 R. James *New Univ. Eng. Dispensatory* 803/1 Liquor Mineralis Anodynus Hoffmanni. Frederic Hoffman's Anodyne Mineral Liquor. **1857** *Amer. Jrnl. Pharm.* May 200 For some years past the preparation sold under the name of Hoffman's anodyne has been forcibly dragged along by the materia medica list, supported alone by its unused formula and former character. **1878** W. B. Woodman tr. *von Ziemssen's Cycl. Pract. Med.* XVII. 442 The internal use of ether (Hoffmann's anodyne) may lead to poisoning. **1910** A. C. Wootton *Chron. Pharm.* I. xiii. 348 Hoffmann's 'Mineral Anodyne Liquor', the original of our Spiritus Ætheris Co., was a semi-secret preparation much prescribed by the famous inventor. **1945** D. Guthrie *Hist. Med.* xii. 217 One of his preparations, Hoffmann's anodyne (Spiritus Aetheris Co) still survives, although his extensive work..in nine volumes..has long since been forgotten.

2. The name of Friedrich *Hoffmann* (1818–1900), used *attrib.* and in the possessive to designate a form of continuous kiln consisting of a number of compartments with a common chimney, so arranged that the fire can be moved to each compartment in turn, with hot air from cooling bricks used for heating unburnt bricks.

1875 *Ure's Dict. Arts* III. 20 (*heading*) Hoffmann's continuous kiln. **1879** *Notes on Building Construction* III. 101 Hoffmann's Kiln is used chiefly in brick-manufactories on a large scale, where a great number of bricks is required annually. **1960** M. Bowley *Innovat. Building Mat.* III. viii. 163 The majority of kilns in the industry are variations of the Hoffman kiln.

3. The name of Johann *Hoffmann* (1857–1919), German neurologist, used *attrib.* and in the possessive to designate various signs, etc., discovered or described by him, as *Hoffmann('s) atrophy* = **Werdnig–Hoffmann('s) atrophy*; *Hoffmann('s) sign*: (*a*) increased sensitivity of the sensory nerves to mechanical stimulation; also called *Hoffman('s) phenomenon, symptom*; (*b*) a type of reflex action of the fingers; also called *digital reflex, Hoffmann('s) reflex*; also *ellipt.*

1900 Dorland *Med. Dict.* 600/2 *Hoffmann's s[ign]*, increased mechanic irritability of the sensory nerves in tetany. **1908** A. Gordon *Dis. Nerv. Syst.* xxi. 399 Hoffmann's Sign.—Pressure upon sensory nerve produces marked pain. **1908** A. Church *Dis. Nerv. Syst.* 916 (*heading*) Mechanical Hyperirritability (Chvostek's phenomenon, Hoffmann's phenomenon). **1910** Osler & Macrae *Syst. Med.* VII. 73 The sensory nerves may also be hypersensitive, and tapping at Valleix's points then calls forth abnormally intense sensations (Hoffmann's symptom). **1916** *Jrnl. Nervous & Mental Dis.* XLIV. 51 During a period spent at the Neurological Institute of New York, I was told of a reflex of the fingers which was designated by the term 'Hoffmann's Sign'. **1933** *Ibid.* LXXVII. 598 Of the 26 patients with the post-encephalitic Parkinson syndrome, the Hoffmann reflex was positive in 6 and the Babinski positive in 4. **1939** W. Haymaker tr. *Bing's Textbk. Nerv. Dis.* iv. 150 Increased galvanic irritability of sensory nerves, occurring at times in patients with tetany, is known as Hoffmann's phenomenon. **1950** R. N. DeJong *Neurol. Exam.* xxxvi. 593 The Hoffmann is a positive sign of definite significance, even though it may occur on occasion in the absence of organic disease.

Hofmann (hǫ·fmæn). **1.** The name of August Wilhelm von *Hofmann* (1818–92), German chemist, used *attrib.* and in the possessive to designate dyestuffs discovered and chemical apparatus and procedures devised by him, as *Hofmann('s) degradation*: (*a*) the elimination of a carbonyl group from an acid amide when it is heated in a sodium hypohalite solution, giving a primary amine; (*b*) the pyrolysis of a quaternary ammonium hydroxide to give a tertiary amine and an olefin; *Hofmann('s) exhaustive methylation*: a method for determining the structure of amino compounds by methylation followed by pyrolysis, the resulting amine being subjected successively to the same procedure until trimethylamine and olefins are obtained; *loosely*, the Hofmann degradation (sense b); *Hofmann('s) method*: (*a*) a method for finding the vapour density of a liquid (see quot. 1902²); (*b*) the preparation of amines by means of the Hofmann degradation of amides; *Hofmann('s) reaction*: either of the Hofmann degradations; *Hofmann('s) rearrangement*: the Hofmann degradation of amides; *Hofmann('s) rule* (see quot. 1954); *Hofmann('s) violet* (also *Violet*): any of several basic dyes that are salts of ethyl and methyl derivatives of rosaniline and pararosaniline and were formerly used with wool, silk, and mordanted cotton.

1869 H. E. Roscoe *Lessons Elem. Chem.* (new ed.) xxxix. 385 Triethylrosaniline, $C_{20}H_{16}(C_2H_5)_3N_3$, is manufactured for its splendid colour, and is known as Hofmann's violet. **1876** *Encycl. Brit.* V. 548/1 (*caption*) Hofmann's Vapour-Density Apparatus. *Ibid.*, Gay-Lussac and Hofmann's Methods. **1902** *Jrnl. Chem. Soc.* LXXXII. ii. 663 Sodium hypochlorite is a more valuable reagent than the hypobromite in Hofmann's reaction. **1902** J. B. Cohen *Theoret. Org. Chem.* iii. 30 Victor Meyer's method and Hofmann's method consist in ascertaining the volume occupied by a given weight of the vaporised substance. *Ibid.* xiv. 195 If a primary amine is treated by Hofmann's method with an alkyl iodide in which the alkyl group is different from that present in the amine, a mixed amine is formed. **1902** *Encycl. Brit.* XXVII. 337/2 The basis of these methods consists in causing a swelling of the cell-wall..and subsequent staining with Hoffmann's blue. **1905** Cain & Thorpe *Synthetic Dyestuffs* xiv. 90 Hofmann's violet was the first violet dyestuff of this [rosaniline] series prepared, and is now merely of historical interest. **1910** *Chem. Abstr.* IV. 583 (*heading*) Acid properties of halogen amides. Hofmann's rearrangement. **1938** L. Small in H. Gilman *Org. Chem.* II. xii. 1024 The most generally applicable method for ascertaining structure is exhaustive methylation, also known as the Hofmann degradation. **1938** *Thorpe's Dict. Appl. Chem.* (ed. 4) II. 375/1 An important reaction of carboxylic acids is their conversion to amines by the Hofmann degradation, whereby the amide is treated with bromine and alkali. **1938** G. H. Richter *Textbk. Org. Chem.* xiv. 239 (*heading*) Hofmann's exhaustive methylation. *Ibid.* 241 The decomposition of these substances on heating to a high temperature has already been indicated in the Hofmann exhaustive methylation procedure. **1949** J. R. Partington *Adv. Treat. Physical Chem.* I. vii. 760 In Hofmann's method, a uniform glass tube about 1 m. long..is supported in a glass jacket through which the vapour of a liquid, boiling in a separate vessel, is passed. *Ibid.* 761 A modified Hofmann apparatus for accurate vapour-density determinations. **1950** L. F. & M. Fieser *Org. Chem.* (ed. 2) xxxv. 865 The correct structure of the lobelia alkaloids.. was first deduced from the result of Hofmann degradation of the quaternary hydroxide. **1951** E. D. Hughes in E. H. Rodd *Chem. Carbon Compounds* IA. 183 The Hofmann rule refers to the preferential formation of that ethylene which bears the smallest number of alkyl groups. **1952** K. Venkataraman *Chem. Synthetic Dyes* I. i. 3 In 1866 Bardy oxidized a mixture of mono- and dimethylanilines to Methyl Violet, which soon eliminated the Hofmann Violets. **1954** *Van Nostrand Chem. Dict.* 356/2 *Hofmann rule*, in the decomposition of a quaternary ammonium hydroxide containing different primary alkyl radicals, the products are such that the ethylene formed will contain the least number of alkyl substituents. **1968** R. O. C. Norman *Princ. Org. Synthesis* xiv. 446 (*heading*) Hofmann rearrangement.

2. The name of Georg von *Hofmann-Wellenhof* (d. *c* 1890), Austrian bacteriologist, used *attrib.* and in the possessive in *Hofmann('s) bacillus*, a non-pathogenic bacillus similar to the diphtheria bacillus and common in the nose and throat; variously known as *Corynebacterium pseudodiphtheriticum, C. hofmanni*, etc.

[**1891** *Johns Hopkins Hosp. Bull.* II. 143/2 Kolisko and Paltauf..were able to find the pseudo-diphtheritic bacillus of von Hoffmann only very rarely.] **1897** *Jrnl. Path. & Bacteriol.* IV. 190 The cases affected with Hofmann's bacillus unassociated with diphtheria bacilli, which have come under my notice, have always recovered. **1897** *Trans. Brit. Inst. Prev. Med.* 12 Peters..does not seem to have noticed any difference of growth of the Klebs-Löffler and von Hoffmann bacilli on alkaline potato. **1959** F. S. Stewart *Bigger's Handbk. Bacteriol.* (ed. 7) xviii. 307 *C. pseudodiphtheriticum* (Hofmann's bacillus)..is often present in the nose and less frequently in the throat.

hog, *sb.*[1] Add: **1. c.** (Earlier and later examples.)

1776 W. Hooper in *Lett. James Murray* (1901) 239 That I might enjoy in my own Cabin, eat my Hogg & Hominee without anything to make me afraid. **1816** *Mass. Spy* 10 Jan. (Th.), [If a man] can be content with hog and hommany, he can live easier in Ohio. **1888** *Century Mag.* XXXVI. 261/2 Corn-bread and bacon, or, in purer vernacular, 'hog and hominy'. **1948** E. N. Dick *Dixie Frontier* 290 The monotonous diet was often referred to as 'hog and hominy'.

7. c. A person who behaves in a rude mannerless fashion without respect for the

HOG

safety or convenience of others; esp. in *road-hog* (ROAD *sb.* 12).

1906 *Daily Chron.* 2 Feb. 7/3 Showing to the astounded heathens (save the word) the latest game of 'hog-amok'. 1928 *Daily Mail* 25 July 17/4 So far we have met no 'canal hogs'. 1942 *Topeka* (Kan.) *Capital* 16 May 7/2 The Office of Price Administration made things unpleasant for 'gas hogs' tonight.

9. d. A railway locomotive used for hauling freight. *U.S. slang.*

1888 *Walla Walla* (Wash.) *Union* 24 Nov. 3/4 The 'hog' will haul nine loaded cars up the heavy Alto grade, while the ordinary road engine had a hard tussel to haul four or five. 1903 *Sci. Amer.* 23 May 392/2 In anthracite drifts steam locomotives of a small and peculiar type known as 'hogs' haul the trains. 1960 *Listener* 18 Aug. 250/2 A steam locomotive is a 'hog' or 'pig'.

e. *Forestry.* (See quots.)

1898 *Lumber Trade Jrnl.* 1 Jan. 31 (Advt.), The big slab grinding hog for grinding up slabs, edgings and mill refuse into fuel. 1904 *Dialect Notes* II. 398 *Hog*, a machine for grinding logs. 1913 WEBSTER, *Hog*, a machine with revolving knife cutters for grinding up edgings and slabs. 1957 *Brit. Commonw. Forest Terminol.* II. 93 *Hog*, a machine for reducing wood to coarse chips, usually for converting mill waste into fuel. 1969 *Timber Trades Jrnl.* 29 Nov. 57/3 Waste blocks..are often chuted..on to a conveyor which automatically takes them to a refuse hog.

f. A large, often old, car or motor-cycle. *U.S. slang.*

1967 W. MURRAY *Sweet Ride* vii. 112 The heat was on so bad we couldn't ride our bikes... Get on our hogs and them mothers'd pick us up. 1968–70 *Current Slang* (Univ. S. Dakota) III–IV. 68 *Hog*,..an old, heavy Harley Davidson motor-cycle... 1956 or 1958 Cadillac; any large car which takes up all the road. 1971 *Black Scholar* Jan. 41/1 He bought him a 'Hog' with all the accessories on it. Man, this Cadillac had air horns, white-walls, [etc.]. 1971 P. L. CAVE *Chopper* v. 45 Pulling away, he swung the hog round in a wide U-turn and went after Ethel.

10. Also, the distance-line itself, the hog-score.

1824 J. MACTAGGART *Gallovid. Encycl.* 274 Sweeping is not allowed until the stone comes over the 'hogg', unless by the person who played it. 1853 W. WATSON *Poems* 63 Stan' back at the hog wi' a besom. 1897 *Encycl. Sport* I. 258 It [*sc.* a stone] must be over the Hog, but must not touch the Stone to be guarded.

11. *like* or *as a hog on ice*, denoting independence, awkwardness, or insecurity. *U.S. colloq.*

1857 *San Francisco Call* 19 Apr. 2/3 He don't appear to care nothing for nobody—he's 'as independent as a hog on ice!' 1894 *Vermont Agric. Rep.* XIV. 124 How would a Hackney look going around the track after old Highland Gray? 'Like a hog on ice.' 1922 C. SANDBURG *Slabs of Sunburnt West* 8 Chicago fished from its depths a text: Independent as a hog on ice. 1948 *Time* 9 Aug. 18/2 They like to think of themselves as independents—independent as a hog on ice.

b. (Earlier U.S. and further examples.)

1828 in G. T. Curtis *Life D. Webster* (1870) I. 337 [Andrew Jackson] will either go with the party, as they say in New York, or go 'the whole hog', as it is phrased elsewhere. 1829 *Virginia Herald* (Fredericksburg) 28 Mar. 2/3 We all know that of late he has shown a disposition to become 'a whole hog man', but if he can swallow this, he can swallow anything. 1835 H. C. TODD *Notes Canada & U.S.A.* 46 In Virginia originated *Go the whole hog*, a political phrase marking the democrat from a federalist. 1852 *Household Words* 31 July 474/1 When a Virginian butcher kills a pig, he is said to ask his customers whether they will 'go the whole hog', as, in such case, he sells at a lower price than if they pick out the prime joints only. 1914 D. H. LAWRENCE *Prussian Officer* 207 Do you mean to say you used to go—the whole hogger? 1928 *Daily Chron.* 3 Nov. 4 The whole-hoggers argue that that statement leaves the position more ambiguous than before. 1929 S. ANDERSON in *Mercury Story Book* 234, I went the whole hog. 1964 R. H. GERHARD in D. Abercrombie et al. *Daniel Jones* 283 Bloomfield who first adopted it 'whole hog'. 1973 *Times* 28 Mar. 4/4 He does not go the whole hog with his father in his belief in the arcane and ancient mysteries of [bacon-]smoking.

13. a. hog-head *U.S. slang*, the driver of a locomotive; **hog-Latin**, bad, spurious, or mongrel Latin; **hog-line** *Curling*, the distance-line (= HOG-SCORE); **hog-tight** *a.*, said of fences which are close enough to prevent swine from forcing their way through; **hog-wild** *a.*, wild in the manner of a hog; **hog-yoke** (later U.S. example); (*b*) a quadrant.

1907 *Sunset* XVIII. 290/2 The anxious gaze of the hoghead (*Anglice*: engineer). 1931 *Illinois Central Mag.* June 30/2 To the initiated, a 'tallow-pit' is a locomotive fire-man and a 'hoghead' is the engineer. 1960 *Listener* 18 Aug. 250/2 Engineers are 'hogheads' or 'eagle-eyes'. 1810 M. DWIGHT *Journey to Ohio* (1912) 53 He pass'd us on the road, singing & screaming, advising us to go back & learn hog-latin—alias German—or dutch. 1834 C. A. DAVIS *Lett. J. Downing* 19 You shall give the address after all, only just let Seth stick a little Hog-latin into it here and there. 1930 *Daily Express* 8 Sept. 8/6 The millions now being wasted in teaching bewildered youngsters hog Latin and piano and bad Greek. 1904 *Westm. Gaz.* 13 May 3/1 There is no reason in the world why you should not mark out a 'hog' line with whitewash. 1963 *Times* 25 Feb. p. xvi/2 Briefly, the rule allows a curler to keep his grip on his rock until the first hog line is reached—a distance of 32 feet from the back in which the toe is placed. 1859 BARTLETT *Dict. Amer.* s.v., Hog-tight and horse-high, always used together, of fences that are sufficient to restrain trespassing stock. Maryland. 1879 A. W. TOURGÉE *Fool's Errand* xxx. 194 The split-board paling.. was 'horse-high, hog-tight, and bull-strong'. 1885 *Rep. Indian Affairs* 110 All of these tracts are enclosed with hog-tight fences. 1972 *Christian Sci. Monitor* 28 Sept. 16/4 The pioneers..tipped the stumps up with their roots in the air, and lined them along so they were, as the saying went, 'horse-high, hog-tight, and bull-strong'. 1904 *Dialect Notes* II. 418, I never saw such an excitement over a little thing in Arkansas as there was over that debate. They went *hog wild*. 1938 J. RICE *Somers Inheritance* III. x. 178 The fact is they're eaten up with envy because they're not getting some of the money. They're hog-wild, that's it, hog-wild. 1940 C. MCCULLERS *Heart is Lonely Hunter* (1943) I. ii. 21 This here white man had just gone hog wild. He were butting his head against the side of this brick wall. 1969 *Eugene* (Oreg.) *Register-Guard* 3 Dec. 2D/1 Arkansas has gone hog wild over its second-ranked Razorbacks this week. 1841 F. A. OLMSTED *Incidents Whaling Voy.* vi. 83 A quadrant receives the very undignified and unphilosophical name of a 'hog-yoke'. *a* 1852 F. M. WHITCHER *Widow Bedott Papers* (1883) x. 35, I ain't so fond o' pork as to eat hog yokes. 1897 KIPLING *Capt. Cour.* v. 107 The old green-crusted quadrant that they called the 'hog-yolk'. 1929 F. BOWEN *Sea Slang* 67 *Hog yoke*, the old fashioned wooden quadrant in American ships and Grand Bankers, so-called from its likeness to the wooden yoke put over hogs to prevent them breaking through fences.

b. hog-age *U.S.*, adolescence. ? *Obs.*

1848 J. MITCHELL in *Amer. Speech* (1935) X. 40 *Hog age*, between Boyhood & Manhood. 1893 FARMER & HENLEY *Slang*, *Hog-age*, the period between boyhood and manhood.

c. hog-sucker (examples.)

1883 *Bull. U.S. Nat. Mus.* XXVII. 478 *Catostomus nigricans*...Hog Sucker...United States from New York to Florida and westward to Alabama and Kansas; Great Lake region. 1888 GOODE *Amer. Fishes* 435 The..'Hog Sucker'..abounds in most waters from the great lakes southward.

hog, *v.*[1] Add: **1. a.** Also *transf.* and *absol.*

1860 R. F. BURTON *Lake Regions Central Africa* I. 85 They [*sc.* asses] hog and buck till they burst their frail girths. 1956 *Archit. Rev.* CXIX. 143/2 Owing to the eccentric placing of the prestressing wires, which cannot be avoided, there is always the tendency for the units to 'hog', i.e., to assume a permanent deflection upwards during stressing.

2. (Earlier U.S. example.)

1803 *Deb. Congr. U.S.* 19 Jan. (1851) 407/1 He did not.. believe that there would be any more danger of the ship's hogging, when lowered down..than when on the stocks.

5. For *U.S. slang* read orig. *U.S. slang* and add further examples.

1884 'MARK TWAIN' *Huck. Finn* xxvii. 275 S'pose somebody has hogged that bag on the sly? 1917 J. C. MCCORQUODALE *In Divers Moods* 16 What blinking luck!—Let's have a sup: Don't hog the lot. My Christ! it's cold. 1936 WODEHOUSE *Laughing Gas* xxii. 237 Maybe that will teach you not to go crawling to directors so that they will let you hog the camera! 1959 *Listener* 26 Mar. 566/1 He never hogs the limelight. 1960 *Woman* 9 Jan. 13/2 You've got a no-good wardrobe, piano, pram or bed, hogging precious space in the house. 1973 *Freedom* 7 July 1/4 The inquiry could go on without hogging the headlines from him.

b. *trans.* and *intr.* To behave as a road-hog; to monopolize the road. Also as *vbl. sb.*

1897 KIPLING *Capt. Cour.* vi. 129 You..go hoggin' the road on the high seas with no blame consideration fer your neighbours. 1914 'I. HAY' *Knight on Wheels* xx. 200 Now I will really hog it a bit: this is a lovely piece of road. 1925 R. J. B. SELLAR *Sporting Yarns* 135 As they were hogging it through the country-side with the speedometer hovering over the sixty mark. 1925 *Punch* 22 Apr. 432 'Frightful rate that bike we just passed was going, wasn't it?' 'Yes. They ought to have the man for "hogging".' 1926 *Chambers's Jrnl.* Dec. 875/1 Why don't you sound your hooter before hogging round corners? 1956 W. GRAHAM *Sleeping Partner* iv. 35, I hogged the road to Lewes cutting in and out among all the family 8-horse powers.

c. *trans.* To interfere with in wireless transmission, as by a more powerful instrument. So also *to hog the ether.*

1914 *Pears' Christmas Annual* 21/2 They should be hogged till doomsday..if a single ship was on fire! *Ibid.*, The operator heard. He started up as if he had been hogged himself. 1959 *News Chron.* 14 Dec. 4/6 The B.B.C., according to Mr. Collins, 'hog the ether to a shocking extent'.

d. *trans.* To eat (something) greedily.

1928 M. LOWRY *Lett.* (1967) 4 Sometime..wdst hog it over the way somewhere with me? 1932 D. H. LAWRENCE *Last Poems* (1933) 50 The only way to eat an apple is to hog it down like a pig And taste nothing. 1946 B. MARSHALL *George Brown's Schooldays* v. 24 The Bruiser did not pause to observe it, hogging down the mashed up mess in front of him.

9. *trans.* To feed swine on (a crop or crop-covered land). Also with *down* or *off*. *U.S. colloq.*

1859 H. W. BEECHER *Pleasant Talk* 93 Some of the best farmers in this region hog their corn-lands. *Ibid.* 94 Land being hogged, will be free from cut-worms. 1863 *Rep. Comm. Agric. 1862* (U.S.) 82, I was forced to hog down my crop this year. 1937 *Amer. Speech* XII. 104 *To hog down corn* means to let hogs eat unharvested corn in the field. 1948 *Clarke Co. Democrat* (Grove Hill, Ala.) 19 Aug. 7/3 A good place to plant crimson clover and rye grass is where you hogged off peanuts.

hogan (hōᵘ·gǎn). [Navajo.] The rude hut of Navajo and other American Indian tribes of the south-western United States.

1872 *Rep. Indian Affairs 1871* (U.S.) 379 When a member of a family dies, in most cases they immediately leave their hogan (or wigwam) with the dead body in it. 1904 *New York Even. Post* 2 July 2 The North American Indians in their primitive state, living in the tepees, hogans, sod-lodges and grass houses. 1927 W. CATHER *Death comes for Archbishop* VII. iii. 217 For his lodging the Bishop was given a solitary hogan. 1955 PRIESTLEY & HAWKES *Journey down Rainbow* 242 This lot have come in from their lonely hogans, the little round wooden huts with a corral or two for horses. 1973 *Times* 15 May 19/8 At night the doors to the Navajo's huts, or hogans, were often covered by a blanket.

hogback. Add: **1. b.** The sunfish, a member of the genus *Lepomis*. *U.S.*

1832 *Coll. New H. Hist. Soc.* III. 86 The hogback or sunfish, as some call it, is a very attracting thing. It is about as large as the perch.

c. Any fish with a hog-like back.

1893 in *Funk's Stand. Dict.* 1923 *Chambers's Jrnl.* Dec. 791/2 Bill, said the latter, the hog-back run is come.

2. (Earlier U.S. examples of form.)

1840 J. P. KENNEDY *Quodlibet* 26 The farm where he now lives at the foot of the Hogback. 1847 in *31st Congress 1 Sess. H.R. Ex. Doc.* No. 5. II. 731 The banks [of a river].. worn in some places into hog-backs. 1888 *Harper's Mag.* Nov. 860/1, I pushed forward across deep gulches, over high peaks and 'hog-backs'.

c. *N.Z.* (See quot. 1940.)

1933 *N.Z. Alpine Jrnl.* V. xx. 180 Dark clouds..a bevy of 'hog-backs'. *Ibid.* 235 A 'hog's back' warned that further storms were brewing. 1940 W. S. GILKISON *Peaks, Packs & Mountain Tracks* 24 He showed me a hogsback... Term applied to a particularly unwelcome cloud only too well-known to climbers, and almost invariably heralding a north-west storm.

hogen, *sb.* **2.** (Later examples.)

1905 in *Chambers's Eng. Dict.* 1963 *Punch* 20 Feb. 282/1 Hogan..is a strong Norfolk beer.

hogg, var. HOG *sb.*[1] 4 a.

1842–4 [see HOG *sb.*[1] 4 a]. 1899 *Daily News* 21 Apr. 7/4 North hoggs and Yorkshire Wold hoggs are becoming scarce. 1963 *Times* 13 May 16/7 In six lamb crops, starting as a hogg, she has produced and reared 20 lambs.

hoggery. 3. (Later example.)

1933 J. CARY *Amer. Visitor* xx. 285 A dash of hoggery now and then may even improve and refine the artistic reactions.

ho·g-ki:lling. [HOG *sb.* 12 b.] The killing of a pig. *hog-killing time* U.S., the time when pigs are killed; a time of special enjoyment; also *absol.*

1817 in A. Royall *Lett. fr. Alabama* (1830) 36 It was hog-killing day at Wills. *a* 1883 G. W. BAGBY *Old Virginia Gentleman* (1910) 96 They are the fixtures used at hog-killing time. 1879 *Harper's Mag.* Nov. 812/1, I..was as big as a dog at hog-killing. 1903 A. ADAMS *Log of Cowboy* xiii. 83 According to their report the boys had had a hog-killing-time. 1927 H. A. VACHELL *Dew of Sea* etc. 259 When I ask my friends to have a hog-killing-time with me, I foot all bills. 1933 *Amer. Speech* VIII. 49/2 *Hog-killing*, any sort of hilarious celebration or jollification. We-all shore did have a hawg-killin' time over t' th' dance t'other night. 1951 *Publ. Amer. Dial. Soc.* XIV. 37 Hog killing time is a season of severe and lasting cold weather, required for the preserving of the meat; also a time of plenty and rejoicing.

hogo. 1. b. Delete † and add later example.

1922 JOYCE *Ulysses* 368 Come near. Then get a hogo you could hang your hat on.

ho·g-pen. *U.S.* [HOG *sb.* 13.] A pen or enclosure for swine.

1640 in *Maryland Hist. Mag.* (1910) V. 374 The Neck of Land called hog penn Neck. 1663 *Springfield* (Mass.) *Rec.* I. 312 There is granted to Rowland Thomas 6 acres of the low land on hog pen dingle below ye place where hog pen was. 1695 [see HOG *sb.* 13]. 1769 in *Maryland Hist. Mag.* (1917) XII. 285 If the bounds of the Hog pen cannot be found. 1837 *Southern Lit. Messenger* III. 238 Cornwallis's cave is converted to a hog-pen. 1874 *Rep. Vermont Board Agric.* II. 512 In said basement I have my hog-pen. 1964 *Publ. Amer. Dial. Soc.* XLII. 21 All [pigs] may live in a *pigpen* or a *hogpen*.

attrib. 1850 *Rep. U.S. Comm. Patents, Agric. 1849* 122 Leached ashes, hen-house and hog-pen refuse are very valuable fertilizers.

ho·g-round. *U.S.* (See quot. 1899.)

1819 *Amer. Farmer* (Baltimore) I. 142 Bacon the hog round, 12 to 13 [dollars]. 1835 *Louisville Publ. Adv.* 14 Feb., 8000 lbs bulk pork, hog round..for sale. 1886 *Harper's Mag.* 206/2 Lard, made from hog-round. 1899 B. W. GREEN *Word Bk. Virginia* 189 Hams, shoulders and middlings have different prices, but when taken altogether at one price, it is so much hog-round.

hog's back, var. HOGBACK.

1800 in *Vermont Hist. Soc. Proc. 1920–21* (1921) 168 Whats call'd the hogs back is a ridge of mountains on the north side (of the Onion river, Vt.). 1827 J. F. COOPER *Red Rover* i, The hog's back over which the water pitches. 1834–63 [see HOGBACK 2]. 1973 *Guardian* 23 Jan. 13/1 The Prime Minister..will be there, in his retreat on the hogs-back of the Delimara peninsula.

hog-skin. 1. (Earlier Amer. example.)

1673 *Essex Inst. Hist. Coll.* L. 27 A meale trough..a hog-skin, a reele.

ho·g-tie, *v.* orig. *U.S.* [HOG *sb.*[1] 1.] *trans.* To secure by tying the four feet, or the hands and feet, together. Also *fig.*, to fetter.

1894 [see HOG *sb.*[1] 13]. 1903 A. ADAMS *Log Cowboy* xi. 75 We threw him, hog-tied him and rolled him into the water. 1905 A. H. LEWIS *Sunset Trail* i. 3 Something wherewith he might hogtie steers when in the course of duty he must rope and throw them. 1906 S. E. WHITE in *McClure's Mag.* Mar. 518/1 In time he got to be a fairly accurate and very quick shot. The same way with roping and hog-tying and all the rest. 1907 —— *Arizona Nights* III. xii. 300 With a short piece of hard rope the cow-boy always carries to 'hog-tie' cattle, he lashed her wrists together. 1910 W. M. RAINE *B. O'Connor* (1920) xx. 226 He's hogtied to the scenery long enough to do my business. 1924 C. E. MULFORD *Rustlers' Valley* xi. 136 However, just now we got to hog-tie our soarin' spirits. 1926 J. BLACK *You can't Win* xvii. 240 When I was caught in a burglary, overpowered, hog-tied, and waiting for the waggon. 1961 R. P. HOBSON *Rancher takes Wife* xv. 180 We hogtied the three..calves we were caring for and heaved them into the boat. 1968 *Word Study* Oct. 6/1 It often results in the present stupidly allowing itself to be hog-tied by the past. 1972 *Daily Tel.* 2 Nov. 19 Sir Stephen McAdden..said judges were hog-tied by stipulations banning sentences between six and 18 months on people under 21. 1973 J. ASHFORD *Double Run* xv. 132 They thought they'd got him hog-tied whereas in fact he was helping to play them for suckers.

ho·g-tie, *sb.* *U.S.* [f. the vb.] The form of securing or fettering produced by 'hog-tying'; a secure hold.

1910 W. M. RAINE *B. O'Connor* (1920) vi. 78 They sure do hate to turn loose a gringo when they have got the hog-tie on him. 1940 E. T. SETON *Trail of Artist-Naturalist* 321 The two cowmen jerked loose the hog-ties, the broncos sprang to their feet, and of course ran away.

hog-trough. (Later U.S. examples.)

1855 M. M. THOMSON *Doesticks* x. 83 After a long search [I] found him wrapped up in the colors, fast asleep with his head in a hog-trough. 1972 J. S. HALL *Sayings from Old Smoky* 87 Fifty years ago if a younger sister married first, folks would say the older sister 'had to dance in the hog trough'.

hog-wash. Add: (Further examples.) **b.** Esp. applied to inferior writings of any kind.

1773 GARRICK *Let.* 16 Nov. (1831) I. 583 The Fair Quaker, which we agreed to be *skimmed milk*, (nay, hogwash) whipped up into *syllabub*, and swallowed by a foolish audience as if substantial as *roast beef*. 1893 FARMER & HENLEY *Slang* III. 329/2 *Hogwash*..(journalists').—Worthless newspaper matter. 1912 G. B. SHAW in *Daily News* 22 May 6/5 Exactly the same 'hogwash'.. would have been lavished on the veriest dastards as upon a crew of Grace Darlings. 1930 WYNDHAM LEWIS *Apes of God* v. 161 Yes, man alive, a lousy limited edition of an intellectually-fraudulent book that no one could sell, that no one would want!..Not even as honest-to-goodness bookstall hogwash. 1939 L. DURRELL *Spirit of Place* (1969) 62 Only look at the faces of cabinet without reading their hogwash and you see that they are a pack of degenerates. 1955 A. HUXLEY *Genius & Goddess* 36 Tripe and hogwash dished out by the moulders of public opinion. 1965 *Spectator* 5 Mar. 293/1 The whole of the artistic world has been debauched by the hogwash of the do-it-yourself vogue.

Hohenzollernism (hōᵘ·entsǫːlərniz'm). [a. G. *Hohenzollern*, the name of a family originating from Hohenzollern in southern Germany which became successively electors of Brandenburg, kings of Prussia, and emperors of Germany (see PRUSSIAN B. *sb.* note) + -ISM.] The autocratic spirit or belligerent policies of the Hohenzollern dynasty. So **Ho·henzo:llernist** *a.*

1915 *Scotsman* 23 Jan. 12/2 The 'Prussianism' of to-day [is] merely 'Hohenzollernism' grafted on to a subject people. 1919 J. L. GARVIN *Econ. Found. Peace* xi. 252 Let us not think for a moment that such thoughts were associated only with Hohenzollernism, Junkerism, and militarism, and must disappear of themselves if Germany becomes an advanced democracy. 1919 G. B. SHAW *Peace Conference Hints* ii. 19 They were..far more determined to overthrow the Hohenzollernist Junkerdom than the Jingos. 1930 —— *What I Really wrote about War* (1931) 2, I did not want Hohenzollernism to win.

hohmannite (hōᵘ·mænəit). *Min.* [ad. G. *hohmannit* (A. Frenzel 1888, in *Min. und petrogr. Mittheil.* IX. 397), f. the name of Thomas *Hohmann*, mining engineer of Valparaiso, who discovered it: see -ITE[1].] A hydrous basic ferrous sulphate, $FeSO_4(OH).3H_2O$, occurring in triclinic crystals of a brownish red colour that rapidly dehydrate on exposure to air.

1888 *Jrnl. Chem. Soc.* LIV. 923 In a sample of copiapite, from Valparaiso, a new iron sulphate was found, to which the name Hohmannite has been given. 1900 E. S. DANA *Text-bk. Min.* 536 Amarantite. $Fe_2O_3.2SO_3.7H_2O$... Hohmannite is the same partially altered. 1938 *Amer. Mineralogist* XXIII. 745 On dehydration..hohmannite apparently loses the four loosely held molecules of water and forms what is here called metahohmannite. In this respect it differs from amarantite, which is much more stable under atmospheric conditions. 1968 I. KOSTOV *Mineral.* 499 Amarantite and hohmannite are triclinic-pinacoidal.

ho-ho (ho͜ho). [Chin.] *ho-ho bird*, a mythical bird of pheasant-like appearance used frequently as an emblem of courage.

1901 C. MONKHOUSE *Hist. & Descr. Chinese Porc.* II. 158 It [*sc.* the phoenix] is referred to as the *fong-hoa* or *ho-ho* bird. 1927 W. B. HONEY *Guide Later Chinese Porcelain* 84 The phoenix (*fêng-huang*, the *ho-ho* bird of old catalogues) is apparently related to the sun-bird of Near Eastern and Indian mythology, and was an emblem of the empress. 1963 *Times* 7 May 7/2 A swordguard in silver formed as a Ho-ho bird and Kirin, undercut in the round with gilt details.

Hohokam (hohokā·m), *sb.* and *a.* [ad. Pima *húhukam*] old one.] **A.** *sb.* **a.** An extinct people of North American Indians. **b.** The culture of this people, centred in Arizona and flourishing after *c* 450, characterized by irrigated agriculture and houses built in pits. **B.** *adj.* Of or pertaining to this people or culture.

1884 A. F. BANDELIER in *Archæol. Inst. Amer. Rep.* V. 80 The Casa Blanca and all the ruins of the Gila were the abode of the fore-fathers of the Pimas, designated by them as 'Vi-pi-sĕt' (great-grandparents), or 'Ho-ho-ḋom' (the extinct ones). 1912 J. W. FEWKES *Casa Grande, Ariz.* 153 The Pima name Hohókam may be adopted to designate this ancestral stock, to whom may be ascribed the erection of the casas grandes on the Gila. 1937 *Southwestern Lore* Dec. 54 As a result of investigations conducted at Winona Village certain archaeological discoveries have definitely explained the affiliation with the Hohokam Culture. 1940 E. FERGUSSON *Our Southwest* 116 What happened to the early Hohokam people is still in dispute. 1948 A. L. KROEBER *Anthropol.* (ed. 2) xviii. 809 Parallel to the Anasazi development is the Hohokam one of lowland, desert, torrid southern Arizona. *Ibid.*, Ceremonial kivas did not develop among the Hohokam, but there were dugout courts for a ritual ball game.

ho-hum (hōᵘ·hʌm), *int.* [f. HO *int.*[1] + HUM *int.*] = HUM *int.* (usu. as an expression of boredom). Also as *sb.* and *v.* As *adj.*, dull and routine.

1924 *Dialect Notes* V. 270 Oh ho hum (vex[ation] or dis[gust]). 1962 *Listener* 4 Oct. 536/3 Listing last week's dramas I find scarcely enough straw to make a useful brickbat:..the return, ho-hum, of *Maigret*. 1962 *John o' London's* 10 May 460/3 Mr. Ustinov sits there..blinking, double-talking and ho-humming. 1963 M. DUGGAN in C. K. Stead *N.Z. Short Stories* (1966) 101 His mouth served out its lying old hohums. 1969 *Daily Colonist* (Victoria, B.C.) 27 Sept. 2/3 People are pretty ho-hum on most parts of the [Vancouver] Island right now..but if anything does happen, they will be the first to cry for damages. 1973 *Jewish Chron.* 19 Jan. 14/1 So the Composers' Guild of Great Britain wants the Arts Council to twist the arms of orchestral managements to make them perform more works by British composers. Ho-hum.

hoick (hoik), *sb.* *colloq.* Also **hoik.** [See next.] **a.** *Rowing.* (See quot. 1898.) **b.** *Aeronaut.* A jerky pull (on the stick). (Cf. *HOICK *v.*[1] 2.) **c.** *Cricket.* A jerky, hoisted shot.

1898 *Encycl. Sport* II. 297/1 *Hoick*, a jerk with the arms at the beginning or end of the stroke, which prevents a steady leg drive from the stretcher. 1907 *Daily Chron.* 8 Mar. 9/1 Cambridge sacrifice everything to a terrible hoick at the finish. 1946 A. PHELPS *I couldn't care Less* vi. 43 The Magister responded to my wild hoik on the stick and came off. 1954 A. G. MOYES *Austral. Batsmen* xii. 164 He..gets a lot of runs with a stroke which Cheetham called a 'hoik'. 1956 R. ALSTON *Test Commentary* ix. 60 Lindwall's one scoring stroke was an ungainly 'hoick' for six.

hoick (hoik), *v.*[1] *slang* or *colloq.* Also **hoik.** [Perhaps orig. a local variant of *HIKE *v.*] **1.** *trans.* To lift up or hoist, often with a jerk or rapid movement. Also to haul or turn *out*. Also *transf.* and *fig.*

1898 G. NICKALLS in W. A. Morgan *'House' on Sport* 346 Until the finish, which, to be made really effective, must be honestly hoicked out. 1908 BELLOC *On Nothing* 136 Beneath him the sand sloped down until it met the sea... Every now and then Mahmoud would force a son or domestic of his to go down and hoick out a pearl. 1911 *Chambers's Jrnl.* Mar. 146/1 The patient Captain Croucher hoicked her from destruction in the nick of time. 1914 W. J. LOCKE *Fort. Youth* i. 20 He hoicked a bit of his shirt-tail from his breeches and proceeded to knot the cornelian heart secure therein. 1916 J. BUCHAN *Greenmantle* ii. 24, I had got myself adjusted to this trench business... And now you have hoicked me out. 1918 'Q' *Foe-Farrell* vi, I dashed around to the rear of the cab, collared Farrell, and hoicked him inboard. 1931 C. MACKENZIE *Buttercups & Daisies* v, Blackbirds and thrushes hoicking worms out of the moist ground. 1930 BLUNDEN *De Bello Germanico* iii. 28 His cue to 'hoik out' the unwary scrimshankers. 1934 G. B. SHAW *Too True to be Good* II. 49 The noise stops; and the bicyclist, having hoiked his machine up on to its stand..comes past the pavilion. 1952 WODEHOUSE *Pigs have Wings* v. 99 'Mr. Galahad is in the amber drawing-room.'..'Then go and hoik him out of it.' 1954 W. FAULKNER *Fable* (1955) 66 No need for them to hunt down and hoick out and execute a mere thirteen men. 1962 M. MCLUHAN *Gutenberg Galaxy* 51 This process..hoicks societies of the world of 'sacred' or cosmic space and time into the detribalized or 'profane' space and time of civilized and pragmatic man. 1972 *Country Life* 9 Mar. 548/3 Is there anything conceivably related to the art of fly fishing in hoicking out trout that have had no chance to live a natural life?

2. To force (an aeroplane) to climb steeply to a higher level. Also *intr.*, to jerk oneself *out of*, etc.

a 1918 J. T. B. MCCUDDEN *Five Yrs. R.F.C.* (1919) 287 He..hoicked out of the dive with such vim that three wing-tips at once collapsed. 1919 *Glasgow Herald* 19 Dec. 10 The pilot yanks the joystick to hoick her up. 1928 *Daily Mail* 7 May 6/4 Hoiking.—Sweeping suddenly to avoid an obstacle or a dangerous approach to earth.

hoick, *v.*[2] [Prob. a dial. variant of HAWK *v.*[3]] = HAWK *v.*[3] Hence **hoi·cking** *vbl. sb.*

1926 A. HUXLEY *Essays New & Old* 5 These frightful hoickings in the throat. 1926 —— *Jesting Pilate* I. 43 The holy man woke up and began to hoick and spit.

‖ **hoi polloi** (hoi pǫ·loi; also pǫloi·). [Gr. οἱ πολλοί, lit. 'the many'.] The majority; the masses. Also formerly in *Univ. slang*, candidates for a pass degree.

In English use normally preceded by the definite article even though *hoi* means 'the'.

[1668 DRYDEN *Dram. Poesie* 65 If by the people you understand the multitude, the οἱ πολλοί. 1791 [see POLL *sb.*[2]]. *c* 1821–2 BYRON in *Lett.* (1830) I. 633 [We] put on masques, and went on the stage with the οἱ πολλοι.] 1837 J. F. COOPER *Europe* II. 94 After which the *oi polloi* are enrolled as they can find interest. 1855 *Read & Reflect* I. 60 The *hoi polloi* [of Mauritius], as we say at Oxford, are mindless—all blank. 1895 *Brewer's Dict. Phr. & Fable* (new ed.) 613/1 *Hoi Polloi (the)*, the poll-men in our Universities, that is, those who take their degrees without 'honours'. 1905 *Daily Chron.* 29 Aug. 4/4 A couple of immense swells..staring stiffly at 'hoy-polloy'. 1932 F. L. WRIGHT *Autobiogr.* (1945) III. 256 Now it all ended in this triumph for them, hoi-polloi, rag, tag, and bobtail.

hoist, *v.* Add: **6.** *Criminals' slang.* To break into (a building) (? *Obs.*); to steal, rob. Hence **hoister**, a housebreaker (? *Obs.*); a shoplifter; a pickpocket; **hoisting** *vbl. sb.*, (esp.) shoplifting. Cf. *HEIST.

1708 J. HALL *Mem.* (ed. 4) 6 *Hoisters*, such as help one another upon their Backs in the Night-time to get into Windows. 1790 H. T. POTTER *New Dict. Cant & Flash Lang.* (1795) (ed. 2) 34 *Hoister*, a shoplifter. 1796 GROSE *Dict. Vulgar T.* (ed. 2), *Hoist*, to go upon the hoist; to get into windows accidentally left open. 1931 *Amer. Speech* VII. 109 *Heist* (or *hoist*), to hold up a person, or to rob at the point of a gun. 1936 J. CURTIS *Gilt Kid* iv. 39 What did you get done for? Hoisting? 1938 F. D. SHARPE *Sharpe of Flying Squad* xiv. 154 Gangs of women shoplifters or 'Hoisters' are to be found in Hoxton. 1960 *Observer* 25 Dec. 7/6 Various petty fiddles and con games to which Christmas trading lent itself, and of course hoisting—shoplifting. 1962 *Coast to Coast 1961–62* 21 'I know where we can hoist a car,' Mick said. 'We'll carry the stuff in it.' 1966 *New Statesman* 23 Dec. 934/2 You know Annie Ward, well she's on the hoisting racket. 1970 M. KENYON *100,000 Welcomes* ii. 10 That half-world of hustlers, hoisters, screwsmen, bogeys, bird, bent gear and tom. 1970 G. F. NEWMAN *Sir, You Bastard* ii. 45 The hoister was held under a guard a dozen strong. 1971 L. GRIBBLE *Alias the Victim* viii. 140 Cop slang. A hoister is a pickpocket or shoplifter.

hoist, *sb.* Add: **3.** Also preceded by a defining word.

a 1884 KNIGHT *Dict. Mech.* Suppl. 12/2 Pneumatic hoist. *Ibid.* 459/1 Builder's hoist. 1963 A. LUBBOCK *Austral. Roundabout* 195 Small bungalow homes with the sun-fresh washing blowing..from the rotary hoists in their back gardens. 1967 *Nursing Times* 18 Aug. 1091/2 The Winchester hoist has also an important use in home nursing.

5. Housebreaking (? *Obs.*); shoplifting. *Criminals' slang.*

1714 A. SMITH *Hist. Highwaymen* I. 143 He pursued his old Courses of going on the Top or Hoist, that is, breaking into a House in a dark Evening, by getting in at a Window one Story high, which they perform by one Thief standing on the Shoulders of another. *a* 1790 H. T. POTTER *New Dict. Cant & Flash Lang.* (1795) (ed. 2) 39 *Lift*, or *hoist*, shop-lifting, or robbing a shop. 1812 J. H. VAUX *Mem.* (1819) 180 *Hoist.* The game of shop-lifting is called the hoist; a person expert at this practice is said to be a good hoist. 1914 JACKSON & HELLYER *Vocab. Criminal Slang* 44 *Hoist*, the profession of shoplifting. 1938 F. D. SHARPE *Sharpe of Flying Squad* i. 15 Shoplifting as an art known as 'The Hoist', and its devotees are called 'Hoisters'. 1958 F. NORMAN *Bang to Rights* 72 My old woman's still out on the hoist now.

hoist-. Add: **hoist-door** (see quot.).

1881 *Harper's Mag.* Mar. 528/1 In the middle of the hall was the 'hoist-door', through which the wheat was hoisted up by a crane and stored in the loft.

Hokan (hōᵘ·kǎn). Name given to a group of languages of certain American Indian peoples inhabiting the west coast of the U.S. So **Ho·kanist**, one who is versed in these languages. Also *Comb.*, as *Hokan-Siouan*.

1913 DIXON & KROEBER in *Science* 7 Feb. 225/2 The new larger [native Indian language] families..are..Hokan, comprising Shasta, Chimariko and Pomo, probably Karok, and possibly Yana. 1920 *Internat. Jrnl. Amer. Ling.* Dec. 280/2 It is difficult for me to suggest any alternative to the hypothesis of a common origin of the Hokan and Coahuiltecan languages. 1965 *Language* XLI. 303 Two Hokanists..have examined the Sapir (1929) subgrouping labelled Northern Hokan. 1965 *Canad. Jrnl. Ling.* Spring 83 These..families had been placed in the Hokan–Siouan superstock.

hoke (hōᵘk), *v.* [Back-formation from *HOKUM.] On the stage or screen: to overplay (a part), to act (a part) in an insincere, sentimental, or melodramatic manner. Also *transf.* and with *up*.

1935 A. J. POLLOCK *Underworld Speaks* 57/1 *Hoke*, to string along; to jolly; to ridicule. 1938 M. MCCARTHY in *Partisan Rev.* Jan. 48 Actors who have been playing for a long time in the same play will..'hoke' their performances more and more. A giggle becomes a laugh; a catch in the throat, a sob. 1939 C. MORLEY *Kitty Foyle* xxv. 244 She had the guts to keep her stuff to exclusive outlets and hoke it up with all sorts of restrictions. 1940 S. LEWIS *Bethel Merriday* xxvii. 283 Mr. Nooks had ..overplayed—'hoked' is the technical word—the role of the Apothecary. 1961 *Punch* 22 Feb. 331/2 The average moviegoer..is pleased to see pictorial evidence, a little hoked up for added amusement. 1971 M. BABSON *Cover-up Story* iii. 33 Just *try* it straight..it's a mistake to hoke it up.

hokee-pokee. *rare*⁻¹. [Cf. HOKEY-POKEY.] (See quot.)

1873 J. MILLER *Life amongst Modocs* (1876) xiii. 192 One man..danced a sort of a savage hokee-pokee, and sang.

hokey, *sb.* Add: Also, *by the hokey fiddle*.

1922 JOYCE *Ulysses* 513 By the hokey fiddle, thanks be to Jesus. 1958 *Engineering* 4 Apr. 426/1 It may be five years before it pays dividends but by the hokey fiddle it'll shake us.

hokey (hōᵘ·ki), *a.* *slang* (orig. *U.S.*). Also **hokie, hoky.** [f. *HOKE *v.*, *HOKUM + -Y[1].] Characterized by hokum; sentimental, melodramatic, artificial.

1945 *N.Y. Times* 19 Aug. 3/8 Equally a part of America ..are the dull films, the tasteless, hoky confections that public taste ought to repudiate. 1968 *Surfer Mag.* Jan. 58/1 They know the films are hokie Hollywood products. 1969 *Northwest* (*Sunday Oregonian Mag.*) 14 Dec. 11/2 Not the 'hokey' set-ups I've been through a half dozen times, but a supervised transitional period. 1970 *New Yorker* 16 May 105/3 A funny and hokey reënactment of the pantomime pocket drama that Colette did in music halls all over France. 1971 *Rolling Stone* 24 June 31/4 A closing piece [on a record], 'Sometimes', is embarrassingly hokey.

hokey-cokey (hōᵘ·ki͜kōᵘ·ki). [Cf. *HOKEE-POKEE.] A kind of dance.

[1943 *Dancing Times* Sept. 570/2, I found a party-dance that was quite new to me. Locally it was (inaccurately) called the 'Hokey Pokey'...The correct name is 'Cokey Cokey'...The chorus runs:..You do the Cokey Cokey and turn around That's what it's all about.] 1966 *Crescendo* Jan. 23/1 'The next dance will be a Veleta' is about my lot, and even then I'm likely to say 'Hokey-Cokey'. 1967 *Ibid.* May 10/3 Your bandleader must present all the ritual party dances including, of course, an ever popular Hokey Cokey. 1967 *Listener* 2 Nov. 587/1 It turned out to be a strong dose of barn-dance and hokey-cokey. 1972 *Sunday Express* 2 Jan. 15/1 (*heading*) Hokey-cokey at the club forces family to sleep in tent. *Ibid.* 15/4 It was 1.30 a.m. at the county council's staff social club and they were dancing the hokey-cokey.

hokey-pokey. Add: **2.** (Later examples.)

1910 A. BENNETT *Clayhanger* II. 226 Three hokey-pokey ice-cream hand-carts, one after another, turned the corner of Trafalgar Road. 1955 *Times* 8 June 7/4 'Hokey-pokey', which, as children used to know, sold for a penny a lump and was (presumably because of its freezing touch on the tongue) the stuff to make you jump. 1970 J. BROWN *Unmelling Pot* i. 23 An Italian organ-grinder..who sold ice-cream or hokey-pokey at ½d a piece during the summer.

4. A toffee-like sweet. *N.Z.*

1939 'K. MANSFIELD' *Scrapbook* 3 We always gave him the same presents..three cakes of hoky-poky.

hokku: see *HAIKU.

Hokonui (hǫ·kōnu͜i). *N.Z.* [Maori place-name.] = MOONSHINE 4.

1947 D. M. DAVIN *Gorse blooms Pale* 29 The men outside..tasting the jar of Hokonui that Tom MacDonald had brought. 1963 *Truth* (Wellington) 17 Sept., Today's distillers of Hokonui can use electricity, smokeless fuels, to keep their spasmodic industry out of sight.

hokum (hōᵘ·kəm). orig. *U.S. Theatrical slang.* Also **hocum.** [? A blending of HOCUS-POCUS and BUNKUM.] Speech, action, properties, etc., on the stage, designed to make a sentimental or melodramatic appeal to an audience. Also *transf.* Hence *gen.*, bunkum.

1917 *Sun* (N.Y.) 5 Aug. III. 3/7 'Jasbo' is a form of the word common in the varieties, meaning the same as 'hokum', or low comedy verging on vulgarity. 1922 C. SANDBURG *Slabs of Sunburnt West* 25 Hokum—they lap it up. 1926 *N.Y. Times* 29 Aug., This may be grounding comedy, but it is not pure hokum. 1926 *Ladies' Home Jrnl.* Apr. 38 'What they tell is..bold and defiant realism.' 'Bold and defiant hokum, I should call it.' 1927 *Sunday Express* 17 Apr. 4 Channing Pollock believed that in 'The Fool' he had written a work of genius. Even when other people said it was hokum he still went on. 1928 *Publishers' Weekly* 16 June 2440 It is pure hokum to suggest that all authors are always interesting. 1928 *Sunday Dispatch* 15 July 15/1 The Adelphi..was occupied by a Mr. Sam Bernard with a musical play called 'The Belle of Bond Street'—what an outrage that 'Girl' and 'Belle' hokum must have become! 1930 *Publishers' Weekly* 15 Mar. 1559 In spite of the fact that the hokum of it all has been pointed out to them. 1937 *Daily Tel.* 26 Oct. 8/5 His story is what the film trade calls 'hokum', the recipe as before with inferior or stale ingredients. 1970 *New Yorker* 12 Dec. 125/1 Most people in Washington dismissed this statement as a piece of sentimental hokum. 1973 *Washington Post* 5 Jan. B7/4 *The Poseidon Adventure*..strictly formula hokum but reasonably diverting.

holard[2] (hǫ·laɹd). *Ecology.* [f. Gr. ὅλος whole + ἄρδειν to water.] The total water content of the soil.

1905 F. E. CLEMENTS *Res. Methods Ecol.* ii. 32 The total amount of water in the soil is divided into the available and the non-available water content. The terms suggested for these are respectively, holard.. chresard..and echard. 1926 TANSLEY & CHIPP *Study of Vegetation* vii. 127 If the echard is subtracted from the holard (total amount of water in the fresh soil) the difference is called the chresard. 1929 WEAVER & CLEMENTS *Plant Ecol.* ix. 182 Of the total water content or holard (whole amount) the larger portion can be absorbed by the plant and is consequently termed available water.

holaspidean (hǫlæspi·dĭăn), *a. Zool.* [f. mod.L. *Holaspideæ* (C. J. Sundevall *Methodi Naturalis Avium Disponendarum Tentamen* (1872) I. 53), f. Gr. ὅλος whole + ἀσπίς shield.] Pertaining to passerine birds that have a single series of large scutella on the posterior portion of the tarsus.

1885 J. S. KINGSLEY *Riverside Nat. Hist.* (1888) IV. 485 This peculiarity consists in the holaspidean tarsi, technically making them scutelliplantar, the hind surface of the tarsus being broken up into scutes similar to those covering the front part. 1907 R. RIDGWAY *Birds N. & Middle Amer.* IV. 329 In all Oscines except the Alaudidæ (in which the tarsus is holaspidean) [the tarsus] has the posterior margin contracted into a sharp or narrow ridge or edge. 1959 VAN TYNE & BERGER *Fund. Ornith.* ii. 48 Several types of passerine tarsal scutellation have been described... Holaspidean: With rear surface of tarsus covered by a single series of broad, rectangular scales.

Holbein (hōᵘ·lbəin, hǫ·l-). The name of the German painter Hans *Holbein* (1497–1543), used *attrib.* to designate embroidery, rugs, etc., embodying qualities or decoration characteristic of Holbein or his work. Hence **Holbeine·sque** *a.* [see -ESQUE], resembling the work of Holbein.

1881 C. C. HARRISON *Woman's Handiwork* I. 44 Holbein stitches appear the same on right and wrong sides of the material. 1882 CAULFEILD & SAWARD *Dict. Needlework* 252/2 *Holbein Stitch*, also known as Italian Stitch, and used in Holbein Embroidery to cover the outline patterns that form the work. *Ibid.* 253/1 *Holbein Work*..consists of an outline Embroidery executed with great care and exactitude, so that the right and wrong side of the work are alike. 1895 J. A. GRAY *At Court of Ameer* xxxi, The most skilful of the artists gave an almost Holbeinesque look to his drawings. 1904 *Westm. Gaz.* 14 Nov. 4/2 His fine Holbeinesque drawings. 1931 A. U. DILLEY *Oriental Rugs & Carpets* vi. 146 The oldest group of rugs remaining to us of the weavings of the Ottoman Turks..are the fifteenth-century products now handsomely called Holbein rugs in compliment to Hans Holbein the Younger. 1950 *Chambers's Encycl.* I. 612/2 The 'Holbein'..and 'Landsknecht' types..which are a German form of short dagger, the former so called from the sheaths being frequently decorated with Holbein's 'Dance of Death' in pierced and chased metal. 1960 *Times* 21 June 22/4 A Holbeinesque bracelet.

holcus (hǫ·lkŭs). *Bot.* [mod.L. (C. Linnæus *Hortus Cliffortianus* (1737) 468), f. Gr. ὁλκός a kind of grass.] A plant of a genus of annual or perennial grasses so called, native to Europe, Africa, and south-west Asia.

1771 P. MILLER *Abridgm. Gardeners Dict.* (ed. 6) s.v. *Holcus*..Holcus with hairy chaff, and bearded seeds. *Ibid.*, Holcus with smooth husks. 1806 J. BARROW *Voy. Cochin China* 392 All the houses were enclosed by a fence made of strong reeds, of the straw of *holcus*, or twigs of wood. 1889 [see UNHUSKED *ppl. a.*[1]]. 1915 L. H. BAILEY *Stand. Cyclop. Hort.* III. 1496/2 The name Holcus was accepted by some botanists. 1947 J. G. DICKSON *Dis. Field Crops* ix. 169 The holcus spot is characterized by tan red-bordered round to elliptical lesions on the leaves.

hold, *v.* Add: **2. b.** (Earlier and later examples.)

1744 *Laws* [of Cricket] in *New Dict. Arts & Sci.* (1755) IV. 3459/1 If the ball be held before she touches ground, though she be hugged to the body, it is out. 1868 *Baily's Monthly Mag.* July 127 Mr. Miles would have got ten wickets had there been any man in the field capable of holding a catch. 1903 G. L. JESSOP in H. G. Hutchinson *Cricket* v. 130 'Dolly' catches are much more difficult to hold than those from hard drives.

d. To keep back, detain, delay.

1891 F. H. SMITH *Col. Carter* 135 'Where did you get this?' he asked, aghast. 'From the carrier. It [*sc.* a letter] was held for postage.' 1904 *New York Times* 20 Aug. 1 The railroad has issued an order..that trains shall not be held for the..taking of baggage after the regular time scheduled for stops has expired. 1970 G. F. NEWMAN *Sir, You Bastard* viii. 203 Sneed was greeted by Sergeant Waugh, who jumped back to try and hold the lift for him.

e. To detain in custody, keep under arrest. orig. *U.S.*

1903 *N.Y. Evening Post* 19 Aug., The men were held for felonious assault, and the woman as a witness. 1906 *Springfield* (Mass.) *Weekly Republican* 14 June 1 If the New York insurance officials cannot be held for larceny, they might evidently be held for forgery or perjury. 1922 H. TITUS *Timber* iii. 32 'Why did he arrest you?' 'Oh, I dropped a cigarette out here in summer an' started a fire,..an' he held me under the fire law.' 1966 J. BINGHAM *Double Agent* xii. 183 He spoke to the Maltese police inspector. 'You would do me a personal favour if you would hold him for twenty-four hours.' 1972 J. RATHBONE *Trip Trap* viii. 89 There were no convictions, but she had been held for questioning on three occasions.

f. *Boxing.* (See quot. 1954.)

1922 N. CLARK *How to Box* xii. 191 It must be understood that it takes two to make a clinch, and unless both men are holding, the referee has not the power to call 'break-away'. 1923 T. C. WIGNALL *Story of Boxing* 318 The referee shall have power to disqualify for..holding, butting, shouldering, [etc.]. 1954 F. C. AVIS *Boxing Ref. Dict.* 53 *Hold*, to grasp an opponent with the hands—not permitted. 1960 *Times* 28 Sept. 16/7 The referee had to speak to both men for holding. 1961 *Times* 8 Mar. 17/3 Spinks appeared to be palming and holding so flagrantly.

5. b. *spec.* Of a theatre: to have capacity for (freq. with reference to the size of a theatre's seating capacity expressed in terms of the takings).

1740 C. CIBBER *Apology for Life* xii. 240 Spectators, who may remember what Form the Drury-Lane Theatre stood in, about forty Years ago, before the old Patentee, to make it hold more Mony, took it in his Head to alter it. 1812 *Dramatic Censor 1811* Apr. 218 Which..will, at the old prices, hold as much money as the modern excessively large Theatres. 1894 G. B. SHAW *Let.* 20 Mar. (1965) 421 The Avenue [theatre] holds, when full, £200. 1946 —— *Matter with Ireland* (1962) 13 It [*sc.* the theatre] held more money per square foot of ground than the classical Royal.

6. g. Phr. *to hold the stage* (or *house*): to command the attention of a theatre audience.

1889 BARRÈRE & LELAND *Dict. Slang*, *Hold the stage, to* (theatrical), is said of an experienced actor who is fully at home on the stage, and always commands the attention of the audience. 1893 FARMER & HENLEY *Slang*, *To hold the stage*, to have the chief place on the boards and the eye of the audience. 1916 *To-day* 22 July 368/1 You do not need to be a very experienced playgoer to know when an actress is holding a house. 1967 'LA MERI' *Sp. Dancing* (ed. 2) viii. 98 Many dancers assisted by only a pianist held the stage alone for the two hours of a complete evening's performance.

h. *to hold the line*: to maintain telephonic connection during a break in conversation. (Cf. *40 g.) Also *fig.*

1912 BEERBOHM *Christmas Garland* 6 It was with a certain sense of his rashness in the matter, therefore, that he now, with an air of feverishly 'holding the line', said 'Oh, as to that.' 1915 *Punch* 10 Nov. 390/1 Such are some of the miseries of holding the line. 1931 WODEHOUSE *Big Money* i. 24 'Hold the line,' he said in a low, strained voice.

7. Also *hold it!* : stay as you are; do not go on!; steady on!

1926 A. HUXLEY *Jesting Pilate* IV. 262 That's good. Hold it. 1930 AUDEN *Poems* 23 Moisten the lips and start afresh. Hold it. 1948 M. ALLINGHAM *More Work for Undertaker* xiii. 160 Oh, I say, hold it... I don't think you ought to go as far as that. 1962 A. CHRISTIE *Mirror Crack'd* xv. 167 'That'll do. Hold it. We'll have one more. ..It looks smashing,' said the photographer. 1973 E. BERCKMAN *Victorian Album* 20 'Let's go and talk to her quickly, quickly—' 'Hold it, darling,' she interrupted.

f. *to hold the road*: to continue to occupy the road; to keep to the road without skidding, etc.

1926 T. E. LAWRENCE *Let.* 27 Sept. (1938) 500 The S.S. 100 holds the road extraordinarily. 1971 P. D. JAMES *Shroud for Nightingale* ii. 43 [She] wondered whether her small car would hold the road.

8. b. (Further example.)

1934 C. LAMBERT *Music Ho!* III. 205 When a guitar hung in every negro barber's shop, and a client who was waiting would vamp about on the instrument until at a lucky *trouvaille* everyone would shout 'Hold that chord'.

11. c. Phr. *to hold one's horses*: used esp. in *imper.* (*hold your horses!*) = be patient, hold on! orig. *U.S.*

1844 *Picayune* (New Orleans) 16 Sept. 241/4 Oh, hold your hosses, Squire. There's no use gettin' riled, no how. 1847 J. S. ROBB *Streaks of Squatter Life* 24 Jest hold your hosses, boys—he'll come out directly. 1917 *Woman's Home Compan.* Nov. 58 Now Phebe..you just hold your hosses and speak a little slower. 1939 *Chatelaine* Oct. 43/4 Hold your horses, dear. 1943 HUNT & PRINGLE *Service Slang* 39 *Hold your horses*, hold the job until further orders. (Comes from the Artillery.) 1945 S. LEWIS *Cass Timberlane* (1947) xl. 272 Hey, hold your horses, Cass. Don't get sore. 1967 N. FITZGERALD *Affairs of Death* vii. 119 'I'm going in to the station *now*,' he said. 'Hold your horses,' Marr said. 'The night's young.'

d. To keep (a person) from speaking; to prevent (a person) from being troublesome. *N. Amer. colloq.*

1901 ADE *Forty Modern Fables* 244 'And I guess that'll Hold you for a While,' added the Biggest Boy in the Room. 1922 S. LEWIS *Babbitt* v. 63 'I guess that'll hold you for a while, George!' said Finkelstein. 1935 N. L. MCCLUNG *Clearing in West* xvii. 136 Maybe that would hold Miss Adams! 1965 'S. WOODS' *Though I know she Lies* xvi. 212 '*That* should hold him for a while,' said Derek with satisfaction.

e. *hold everything!*: wait! take no action!

1930 in *Amer. Speech* VI. 92. 1948 C. DAY LEWIS *Otterbury Incident* viii. 93 Hold everything now, this is the big bang coming. 1951 L. HOBSON *Celebrity* (1953) xi. 155 Hold everything; let's see.

15. e. Only in pres. pple. *holding*: 'financial', in funds. *Austral.* and *N.Z. colloq.*

c 1926 'MIXER' *Transport Workers' Song Bk.* 11 'Whatho, Jerry, how yer holding?'..'I haven't made enough this week For to pay the blooming rent.' 1930 *Bulletin* (Sydney) 29 Oct. 21/1 Whether a man was 'oldin' or whether a man was broke, Joe was a man you could bank on.

f. To be in possession of drugs for sale. *U.S. slang.*

1935 A. J. POLLOCK *Underworld Speaks* 3/2 *Are you holding?* Have you any dope to sell? 1953 W. BURROUGHS *Junkie* (1972) ii. 26 The connection was here about ten minutes ago. This character's holding, but he won't turn loose of any. 1961 R. RUSSELL *Sound* (1962) I. i. 15 Don't jump the light, baby, mother's holding, you know. *Ibid.* II. ix. 158 He was holding, just as Red had said. Santa had the sweets.

23. c. (Further examples.)

1818 SHELLEY *Let.* 30 Apr. (1964) II. 14 But this holds good, as I know, only to Milan. 1937 *Discovery* May 139/1 His words of seven years ago hold good today.

d. To continue fine, to keep from raining. (Cf. 44 i.)

1893 *Chambers's Jrnl.* 10 June 355/2 If the weather holds, we'll both take a trip.

29. d. *to hold someone's hand*, to give comfort or moral support to someone; to back someone up. *colloq.*

1935 C. ISHERWOOD *Mr. Norris changes Trains* vi. 90, I shall need your moral support. You must come and hold my hand. 1961 A. WILSON *Old Men at Zoo* iv. 220 Martha's been holding his hand in California. 1972 B. EVERITT *Cold Front* vii. 55, I 'held his hand' to the best of my ability with school-girl French and passable Italian.

30. b. *to hold up one's head* (further examples).

1808 JANE AUSTEN *Let.* (1932) 205 Her Daughter..who says as little as ever, but holds up her head & smiles. 1859 LYTTON *What will he do with It?* I. II. v. 171 But they could never again hold up their heads with the noblemen and great squires in the county. 1900 E. WHARTON *Gift from Grave* vi. 80 Why, you don't suppose if he were alive he could ever hold up his head again, with these letters being read by everybody? *a* 1953 E. O'NEILL *Touch of Poet* (1957) 33 *Nora.* You have the fine opinion av yourself! *Sara...* I've had need to have, to hold my head up, slaving as a waitress and chambermaid.

34. hold back. c. With *on*: to refrain from disclosing (something to someone).

1956 E. POUND tr. *Sophocles' Women of Trachis* 22 I'll tell the truth, I won't hold back on you.

35. hold down. c. To remain in (a position or situation); to continue to occupy (a place or post) or succeed in discharging the duties of (one's employment). orig. *U.S. colloq.*

1891 C. ROBERTS *Adrift Amer.* 92 Jumping an east bound freight.., I managed to hold it down or keep on it till I got to Alameda. 1893 *Harper's Mag.* Dec. 80/2 If a man is to 'hold down' a big ranch in northern Mexico he has got to be 'all man'. 1896 ADE *Artie* xiv. 129 I'll bet that guy up in your place don't know nothin' on earth except how to hold down his measly job. 1902 G. H. LORIMER *Lett. Merchant* v. 60 The fellow who's got the right stuff in him is holding down his own place with one hand. 1910 S. E. WHITE *Rules of Game* I. iii. 19, I didn't much think you could hold down a job here. You see there's too much doing here. 1913 F. H. BURNETT *T. Tembarom* ii, I wonder, if I ever did get his job, if I could hold it down? 1931 G. D. H. COLE in W. Rose *Outl. Mod. Knowledge* xvi. 688 There are few pleasures in life equal to that of successfully holding down a difficult and responsible job. 1936 *Punch* 14 Oct. 439/2 He never could hold down a job. 1973 A. BEHREND *Samarai Affair* i. 12 Captain Coldstream looked exactly what he was—an ex-sailor of quality now holding down an exacting shore job.

40. hold on. c. Also in jocular phrases.

1930 'SAPPER' *Finger of Fate*, etc. 35 Having to hold on by one's eyebrows whenever one moves gets a bit monotonous after a time.

e. (Earlier examples.)

1846 C. M. KIRKLAND *Western Clearings* 45 'But hold on a little till I tell ye!' interposed Master George. [For 1860 read 1848.]

g. *Telephony.* To keep the line open.

1892 KIPLING in *Times* 29 Nov. 8/1 A..millionnaire,.. clawing wildly at the telephone..'Hello!..I *told* you to hold on. What? .. No. Hold on.' 1919 V. WOOLF *Night & Day* xxiv. 327 'I'll look at my engagements...Hold on.' She dropped the machine. 1920 *Punch* 1 Sept. 176/3 'What is your number, please?'..'Just hold on a minute while I look it up.' 1920 R. MACAULAY *Potterism* III. i. 104 You mustn't ring off yet... Hold on while I tell daddy. 1949 J. B. PRIESTLEY *Home is Tomorrow* 38 (*into telephone*) Yes, I'll hold on. 1971 A. CROSS *Theban Mysteries* (1972) xi. 165 She did go to the phone, but she got the doctor's exchange, which said, 'Hold on,' and then went off the line.

41. hold out. l. To keep back; to retain or detain; (also const. *on*) to withhold (information or the like). *colloq.* (orig. *U.S.*).

1907 E. S. FIELD *Six-Cylinder Courtship* 71 If it wasn't for Bellows and Rooker, we'd hold out on him every time. 1911 H. S. HARRISON *Queed* v. 57 Surface, by clever juggling of his books had managed to 'hold out' a large sum of money in the enforced settlement of his affairs. 1916 H. L. WILSON *Somewhere in Red Gap* viii. 345, I wanted to send a postal card to the..Dye Works at Red Gap, for some stuff they had been holding out on me a month. 1923 R. D. PAINE *Comrades of Rolling Ocean* i. 13 He dumped his wages upon the sitting-room table, holding out only the price of a new pair of shoes. 1926 J. BLACK *You can't Win* ix. 112 The thief who holds out a lady's watch on his pal to give to his girl has no character. 1932 WODEHOUSE *Hot Water* i. 21 And me who had split Even Stephen with her on every deal, never chiselling, never holding out on her, no, not so much as a dime. 1944 L. A. G. STRONG *All fall Down* 99 The thought came to me that maybe the old cuss was just holding out on me. 1945 J. B. PRIESTLEY *Three Men in New Suits* v. 88 'Boss,' said Markinch, who liked to be American too, 'he's holding out on us.' 1972 'G. BLACK' *Bitter Tea* (1973) v. 81 If I find out that you've been holding out on me over this identification, I'll come down on you like a pile driver.

42. hold over. c. *U.S. colloq.* (See quot. 1889.)

1872 'MARK TWAIN' *Innoc. at Home* 18 (Farmer), You ruther hold over me, pard. I reckon I can't call that hand. 1889 FARMER *Amer.*, *To hold over one* is to have an advantage in some way or other. This particular usage probably comes from poker phraseology. 1889 K. MUNROE *Golden Days* xii. 127 Do we hold over Bowers?

44. hold up. d. (Later U.S. examples in sense 'keep back, withhold'.)

1889 *Kansas Times & Star* 4 Dec., Major Davenport is holding up the firemen's payroll for November owing to alleged irregularities. 1894 *Vermont Agric. Rep.* XIV. 70 When..a cow holds up her milk there is some disturbing element.

e. Also, to arrest the progress of, obstruct the passage of (*lit.* and *fig.*).

1904 *Philadelphia Even. Telegr.* 15 Nov. 1 Out of the 900 steerage passengers that came over on the Merion, 135 failed to pass the immigration inspectors, and were held up. 1905 *N.Y. Evening Post* 16 Mar. 1 Another landslide has occurred..and nine passenger trains are held up in the mountains. 1906 *N.Y. Herald* 5 Mar. 5 It is thought the Senate Finance Committee will seek to devise new excuses for holding up the investigation of the State Banking Department, which it has succeeded in smothering for five weeks. 1909 H. N. CASSON *C. H. McCormick* 146 One bill for $15 was held up for a week because it was not properly drawn. 1972 *Daily Hampshire Gaz.* (Northampton, Mass.) 9 May 1/3 A passing motorist asked Witkos if he was 'going to let them hold up the traffic all day?'

hold, *sb.*[1] Add: **2. b.** Also in *Judo*.

1954 E. DOMINY *Teach yourself Judo* iii. 39 Theoretically, a hold consists of nothing more than controlling your opponent's body with the weight of your own.

c. (*with*) *no holds barred*, (with) all restrictions relaxed. Also as *attrib. phr.*

1942 BERREY & VAN DEN BARK *Amer. Thes. Slang* § 217/6. 1952 *Economist* 1 Nov. 313/1 An independent [broadcasting] agency..could talk back at the Russians with no holds barred. 1958 *Times* 28 Nov. 13/6 No holds were barred, so to speak, for the Prince's unorthodox education and his own reputedly emancipated views allowed almost every hypothesis. 1961 *Economist* 6 May 524/1 Waging..a sweaty, no-holds-barred tussle all around the world without benefit of parley. 1972 *Real Estate Rev.* Winter 29/1 This allows trust managers to compete on a no-holds-barred basis.

5. b. A delay, pause, postponement. Also *attrib.*

1961 *Observer* 28 May 4/2 The long countdowns, checks and 'holds' possible at Cape Canaveral would be suicide on the moon. 1968 *Time* 27 Dec. 13 The countdown schedule had been padded with enough precautionary hold time to enable technicians to replace the oxygen without delaying the launch. 1969 *New Scientist* 27 Feb. 439/1 Unless there has been a last-minute 'hold' at Cape Kennedy, the first of a pair of *Mariner* spacecraft should now be on its way. 1971 *Daily Tel.* 4 Jan. 2/4 Concorde, counting time for 'taxi-ing' and 'holds' could arrive in New York at 7 a.m. after taking off from Heathrow at eight. 1971 *Nature* 26 Nov. 181/2 Liquid-fueled rockets ..could not be kept in the launching tower in a 'hold' status to await a transient event.

14. *Cinemat.* (See quots.)

1918 H. CROY *How Motion Pictures are Made* vii. 179 If the story demands instantaneous materialization the effect is secured by a photographic means usually known as 'the hold'. It is so called by reason of the fact that all the other characters in such a scene must hold their positions while the trick character is made to materialize. 1940 *Chambers's Techn. Dict.* 418/2 *Hold*, the retention of an image on a screen longer than is natural.

hold-all. Add: **2.** *fig.*, esp. with reference to books of the omnibus or encyclopædic kind.

1903 *Daily Chron.* 9 Nov. 3/3 There is a little of everything in the hold-all. 1904 M. DAVIES (*title*) The housewife's what's what: a hold-all of useful information for the house. 1964 [see *CETERIS PARIBUS].

hold-back. Add: **2.** (Earlier example.)

1850 *N.H. Hist. Soc. Coll.* VI. 220 The hold-backs of his harness gave way, and precipitated his gig upon the horse.

3. The act of holding back. Also *attrib.*, unprogressive.

1852 *Trans. Mich. Agric. Soc.* III. 333 A few specimens of the hold-back and stand-still class. 1888 'BUFFALO BILL' *Story of Wild West* 627 There was no brake on the wagon, and the horses were not much on the hold back.

ho·ld-down. [f. phr. *hold down* (HOLD *v.* 35).] **1.** A device to prevent material or apparatus from shifting or shaking. Also *attrib.*

1888 *Lockwood's Dict. Mech. Engin.*, *Foundation Bolts.* ..Also termed hold-down bolts. 1962 J. GLENN in *Into Orbit* 188 Then the big hold-down clamps dropped away and I could feel us start to go. 1967 KARCH & BUBER *Offset Processes* x. 465 Set the rear pile hold-down..to hold the rear edge of the sheet to the pile.

2. A judo grip.

1954 E. DOMINY *Teach yourself Judo* iii. 39 In judo contests, there are three recognised methods of obtaining a point. The first is by means of a clean throw, the second by means of a 'Hold Down'. 1956 K. MASUDA tr. *Tomiki's Judo* iii. 91 In the practice of grappling, hold-downs, strangle-holds and bone-locks are connected with one another like the three sides of a triangle, the hold-downs being the base.

holder[1]. Add: **2.** Also, a shareholder.

1848 J. J. RUSKIN *Let.* 22 Mar. in M. Lutyens *Effie in Venice* (1965) I. 17 A person of very sound Judgement on Railroad Shares...To Holders he says—if they have no other means of meeting Calls—sell.

b. *Sports.* The possessor for the time (as the winner) of a championship, cup, etc. which is open to competition.

1830 *Sporting Mag.* 2nd Ser. I. 337/1 The holder [of the sculls] shall row the best, or only, challenger, on the 10th of August. 1873 *Football Ann.* 54 Association Challenge Cup, 1872–73... Final Tie. Wanderers (holders) beat Oxford Association by two goals..to none. 1887 *Athletic Jrnl.* 9 Aug. 16 West Manchester (the winners and present holders of the Manchester Cup). 1900 *Field* 7 July 3/3 Hants County Public School Challenge Shield... This annual competition was held on the playgrounds of the holders, Churcher's College, Petersfield, on Thursday in last week. *Ibid.* 14 July 61/1 The holder of the challenge cup, Wadsley, was quite unable to do himself justice. 1928 *Daily Mail* 25 July 14/7 Middlesex, the holders, are..the only county to have won two matches in this group. 1955 *Times* 16 June 3/2 The effect on the players was clearly seen when L. Hoad, the holder, beat R. Bedard.

3. (Further examples.)

1846 R. FORD *Gatherings from Spain* xxiv. 338 A Spanish fore-finger and thumb are quite fire-browned and fire-proof, although some polished exquisites use silver holders. 1916 E. F. BENSON *David Blaize* vi. 111 A magnificent present of twenty-five cigarettes and a cherry-wood holder. 1957 J. OSBORNE *Entertainer* 16 He unwraps the cigarettes and takes out an ivory holder from his waistcoat.

holderbat (hōᵘ·ldəɹbæt). [f. HOLDER[1] + BAT *sb.*[2]] A type of bracket for fastening a pipe to a wall or other surface, consisting of two semicircular parts that are clamped round the pipe and a projection on one of the parts that is built into the wall.

1914 SAGE & FRETWELL *Text-bk. Elem. Building Constr.* xi. 225 The best method, however, is to dispense with ears and use 'holderbats' built into the wall. 1955 N. W. KAY *Mod. Building Encycl.* 331/2 The holderbat has two portions, a circular ring, with a jointed or removable portion,..and a tail which is cemented into the wall.

holdfast, *sb.* Add: **4. b.** *Bot.* An organ for superficial attachment developed by some algæ and fungi.

1841 W. H. HARVEY *Man. Brit. Algæ* p. xiv, Most Algæ are, at some period of their growth, found attached to other substances by means of a root, or at least a hold-fast. 1895 G. MURRAY *Introd. Study Seaweeds* 23 The sculpturing of outward form reaches its highest point in the differentiation..of a root-like holdfast. 1902 *Science* Jan. 59/2 Kelp hold-fasts, of which none grow in the immediate vicinity, were taken in abundance by the dredge. 1930 H. M. FITZPATRICK *Lower Fungi: Phycomycetes* iii. 45 In the typical epibiotic species (Rhizidiaceæ) the germ tube of the cytospore acts as a holdfast. 1931 L. NEWTON *Handbk. Brit. Seaweeds* p. vi, One of the main factors affecting the distribution of algæ is the securing of a suitable holdfast. 1962 C. J. ALEXOPOULOS *Introd. Mycol.* (ed. 2) x. 213 The basal part of the hypha [of members of the order Eccrinales] is in the form of a disc-like holdfast by means of which the fungus is attached to the host. 1966 F. H. BRIGHTMAN *Oxf. Bk. Flowerless Plants* 6/2 The holdfast is disc-shaped, but root-like structures also develop from the base of the stalk. 1971 *Where* July 202/2 They are the lovely coloured seaweeds or algae, with their interestingly-named parts: the 'holdfast' which forms the root, and the 'thallus' or fronds which make up the body.

holding, *vbl. sb.* Add: **1. d.** *holding up* (see quot. 1888).

1888 *Lockwood's Dict. Mech. Engin.* 178 *Holding up*, the maintaining of a firm pressure against the heads of rivets while their closing up is being effected, a holding-up hammer being used for the purpose. 1908 J. G. HORNER *Plating & Boiler-making* (ed. 2) xii. 214 Holding-up dollies were shown in Figs. 56 and 57... In girder work a rig-up like that shown in Fig. 235 is used. It consists of a heavy holding-up hammer A, on the end of a long elastic handle B.

e. *Association Football.* The obstruction of a player by taking hold of his jersey, etc.

1866 *Cassell's Illustr. Fam. Paper* 17 Mar. 509/2 'Holding' includes the obstruction of a player by the hands, arms, or body without kicking or throwing. 1967 *Association Football* ('Know the Game' Series) 33 Should the obstruction take the form of a personal foul, e.g., pushing, holding, charging unfairly or tripping, then the foul is penalised by a direct free kick.

3. c. The cards held by a player.

1929 M. C. WORK *Compl. Contract Bridge* iv. 38 With such holdings..the rebid should be made. 1959 *Listener* 10 Dec. 1054/2 The high card holding is weak.

6. b. holding ground (later example); (*b*) a site for storing floating timber; **holding operation**, an undertaking which prevents a situation from worsening, but can do little or nothing to improve it; **holding paddock** *Austral.* and *N.Z.*, a paddock where sheep or cattle are kept until required (for droving, shearing, etc.); similarly **holding pen, yard.**

1957 *Brit. Commonw. Forest Terminol.* 2 *Holding ground*, a boom site for storing timber. **1961** F. H. Burgess *Dict. Sailing* 116 *Holding ground*, the nature of the bottom and its holding quality for purposes of anchoring. **1962** *Listener* 8 Mar. 400/1 Whether it [*sc.* a White Paper] will achieve its immediate object of persuading the unions to help the economy over its next stile by holding down labour costs, we must wait and see... As a holding operation it may win some success. **1972** *Guardian* 9 Sept. 13/8 Mr Jenkins's last-minute scramble to satisfy the TUC..was only a holding operation to keep him sweet with Congress. **1933** L. G. D. Acland in *Press* (Christchurch) 28 Oct. 17/7 *Holding paddock*, a small paddock, close to yards, woolshed, or mustering hut, for holding (not feeding) sheep. **1934** *Bulletin* (Sydney) 16 May 38/4 At midday the cattle, mad with thirst, broke out of the holding paddock and vanished in a wild stampede over a high hillcrest into the vast unfenced wilderness of the hills beyond. **1941** *Coast to Coast 1941* 22 Wiggins said some steers had got out of his holding paddocks. **1950** *N.Z. Jrnl. Agric.* Oct. 35 Cattle in a holding paddock before being drafted. **1923** *Ibid.* 20 Mar. 144 The holding-pens in the shed..should never be too large. **1965** J. S. Gunn *Terminol. Shearing Industry* I. 32 *Holding pen*, one of the small pens or yards in which sheep are held, usually within the shed, under shelter, while awaiting shearing. **1950** *N.Z. Jrnl. Agric.* Apr. 377/2 For handling large herds [of cows] a crush yard is recommended, with the main holding yard either in front or at one side. **1959** *Listener* 15 Jan. 115/1 There were some good holding yards where we could put the cattle for the night.

c. *Aeronaut.* Used *attrib.*, of or pertaining to the process of 'stacking' aircraft in the air above a landing site before they come down to land.

1948 *Shell Aviation News* No. 116. 19/3 Holding procedures as an integral part of the traffic control system for jet transports should be completely eliminated. **1958** *Chambers's Techn. Dict.* 985/1 *Holding pattern*, a specified flight track..which an aircraft may be required to maintain about a holding point. *Holding point*, an identifiable point, such as a radio beacon, in the vicinity of which an aircraft under air traffic control may be instructed to remain. **1969** *Daily Tel.* 14 Nov. 1/7 Stacking over a 'holding area' while waiting a turn to land, is not uncommon, especially in poor weather conditions. **1972** *New Yorker* 16 Sept. 26/1 He turned on the plane's radio, in order to hear the traffic controller at the..airport talking to several airliners stacked in a holding pattern overhead.

holding, *ppl. a.* Add: **1.** (Later examples.)

1930 *Daily Express* 6 Nov. 16/1 The keen east wind dried up the course, which was inclined to be on the holding side. **1955** *Times* 19 May 4/4 In August last year the going was extremely holding.

b. *holding company*: a trading company which possesses the whole of, or a controlling interest in, the share capital of one or more other companies.

1906 *Daily Colonist* (Victoria, B.C.) 18 Jan. 1/5 A bill to prohibit 'holding companies' such as have lately effected the merger of various corporations..was introduced in the legislature today. **1912** *Q. Rev.* Jan. 195 The Federal Steel Company, which is technically, like the Steel Corporation of to-day, a holding company. **1928** *Britain's Industr. Future* (*Liberal Ind. Inq.*) II. viii. 93 To treat trusts, cartels, combinations, holding companies, and trade associations as inexpedient abnormalities in the economic system. **1928** *Daily Mail* 7 Aug. 18/4 This conservative finance enables the Shell, as a holding company,..to maintain its dividends in times of depression. **1930** A. Palmer *Company Secretarial Pract.* xix. 255 The Act does not specifically define a holding company, but refers to it as a company holding shares, either directly or through a nominee, in a subsidiary company. **1958** *Times* 11 Aug. 11/2 Mrs. Roebling..is director or trustee of more than 16 organizations and president of a holding company. **1972** 'E. Lathen' *Murder without Icing* (1973) xv. 131 Holland's other assets could take a long time to find. He's got everything wrapped up in holding companies.

3. *holding-down bolt, pin, ring.*

1846 *Patent Jrnl.* I. 226/2 On each side of the holding down bolts..an upright pillar is erected. **a 1877** Knight *Dict. Mech.* II. 1112/2 *Holding-down Bolt*,..one of twelve or more strong bolts, which are passed from the outside of a steam-vessel through the floor-timbers, sleepers, foundation-plate of the engine, and the bosses on the cylinders, condensers, and side-frames, and are secured by strong nuts. **1892** W. W. Greener *Breech-Loader* 19 Every gun provided with a holding-down bolt. *Ibid.* 260 Certain accessories.., such as cords, planks, and holding-down pins. **1899–1900** *Kynoch Jrnl.* Dec.–Jan. 29/2 The cones are secured to ships' decks by holding-down rings. **1930** *Engineering* 4 Apr. 440/2 Checking parts for fit and tightening holding-down and other bolts.

holdless (hǭu·ldlės), *a. Mountaineering.* [f. Hold *sb.*[1] + -less.] Affording no holds; having a smooth unbroken surface.

1922 E. R. Eddison *Worm Ouroboros* xiii. 184 Cliffs smooth and holdless as a castle wall. **1933** [see *abseil]. **1971** *Country Life* 25 Feb. 410/1 Alastair solved problem after problem..making short work finally of the hardest obstacle, an awkward holdless entry to a chimney at the start of the last tower.

hold-out. Add: **b.** The act of holding out; something that or someone who holds out; *spec.* (chiefly *U.S.*) a player, usu. in baseball, who refuses to play until he is promised higher pay.

1945 *Sun* (Baltimore) 17 Feb. 7/7 This is another year when any baseball holdouts will do their shouting in whispers. *Ibid.* (*heading*) Holdout is weapon. **1952** B. Malamud *Natural* (1963) 22 He was a holdout for $75,000 and was coming East to squeeze it out of his boss. **1964** *Amer. Speech* XXXIX. 91 Occasionally there was a hold-out, as in this lamentation: [etc.]. **1970** *Globe & Mail* (Toronto) 26 Sept. 35/1 Whatever became of another well-known holdout, Joe Kapp?..Kapp keeps waiting for the Minnesota Vikings to pay him more money.

c. Of paint or ink (see quot. 1965).

1965 *Gloss. Paint Terms* (*B.S.I.*) 32 *Hold out*, the ability of a paint film to dry to its normal finish on a somewhat absorptive surface. **1971** *Timber Trades Jrnl.* 21 Aug. 29/1 Golden Royal hardboard..was already accepted as a good painting medium, but application of a seal..meant that..certain properties, including paint hold-out, workability and strength, were improved. **1972** *Publishers Weekly* 4 Dec. 11 (Advt.), Its unique combination of high ink hold-out, high brightness, and high opacity.

hold-over. Add: **d.** Something left over; a remainder or survival. *U.S.*

1904 *Los Angeles Express* 11 Aug. 12 Doing the best it could on crackers and cheese and holdovers. **1909** 'O. Henry' *Roads of Destiny* iv. 58 She was a hold-over from the Greek classics. **1929** *Atlantic Monthly* Mar. 298 The little village of Washington in Connecticut, one of the most charming holdovers of the past that state possesses. **1951** E. Paul *Springtime in Paris* iii. 53 Like several other articles, it seemed to be a holdover from the previous tenant. **1969** *Times* 6 Mar. 23/1 Both the present assistant secretaries for international affairs in the Treasury..are Johnson hold-overs.

hold-up. (No longer restricted to *U.S. slang.*) Add: **b.** (Earlier and later examples.)

1878 F. M. A. Roe *Army Lett.* (1909) 206 The driver is their only protector, and the stage route is through miles and miles of wild forest, and in between huge boulders where a 'hold-up' could be so easily accomplished. **1904** *Daily Chron.* 23 Dec. 4/5 There are epidemics of robberies, murders and hold-ups in all the large cities. **1928** *Daily Express* 15 June 7/5 The 'hold-up' of a steamer crowded with holiday-makers on Lake Windermere. **1968** *Globe & Mail* (Toronto) 17 Feb. 2/2 Police said they had recovered $3,991 from yesterday's holdup.

c. A stoppage or check in the passage or progress of a person or thing; a temporary stoppage of traffic; a cessation, stop. orig. *U.S.*

1837 *Knickerbocker* X. 439 The wheels of the coach are shod with the preparation of iron slippers, which are essential to a hold-up. **1882** in G. H. Putnam *Mem. Publisher* (1915) 289 We don't have hold-ups [*sc.* strikes] in Leadville. **1904** *N.Y. Tribune* 15 May 2 A vote of thanks to the Tribune for its efforts to end the hold-up of the Port Chester Railroad's application for a permit to cross streets in the Bronx. **1907** *Putnam's Monthly* July 482/1 He cursed the luck of the hold-up. **1913** A. B. Emerson *R. Fielding at Snow Camp* 154 We got to sit down and wait for a hold-up [of the storm]. **1918** 'Q' *Foe-Farrell* vii. 125 There was a hold-up as we [in a taxi] neared the bridge. **1928** *Daily Express* 14 July 2/1 There had been two or three hold-ups with the points prior to my arrival at 4.50 p.m. **1964** *Ann. Reg. 1963* 224 In connexion with the various hold-ups of Allied military convoys on the autobahn..it was recalled that the D.D.R. had only temporarily handed over control of military traffic to the Soviet military. **1973** 'H. Carmichael' *Too Late for Tears* xi. 131, I nearly didn't get here. Ran into a traffic hold-up and was stuck. **1973** E. Lemarchand *Let or Hindrance* viii. 93 The hold-up over the Fortnight film while Paul King finished editing the last part.

d. An instance of extortion. Chiefly *U.S.*

1908 L. Mitchell *New York Idea* I. 15 The people insisted on electing a desperado to the presidential office—they must take the hold-up that follows. **1910** *Sat. Even. Post* 27 Aug. 6/3 Our house..cost twenty-five thousand dollars, exclusive of the plumber's little hold-up and the Oriental rugs. **1939** J. Mulgan *Man Alone* I. 14 It never was farming land. It's a hold-up, and God help the poor bastards who have to take it at that price.

e. *Bridge.* (See quot. 1959.) Cf. Hold *v.* 44 d.

1945 Phillips & Reese *How to play Bridge* III. 101 The principal device available to the declarer to prevent the establishment of an opponent's long suit is hold-up play. **1959** Reese & Dormer *Bridge Player's Dict.* 113 *Hold-up play*. A player is said to hold up when for tactical reasons he declines to play a winning card. Usually, his object is to destroy communication between the enemy hands, but there can be other and more subtle reasons for the hold-up. **1962** *Listener* 22 Mar. 534/1 This uncommon hold-up play will prevent the defence from bringing in the suit unless North has two entries.

f. (See quot.)

1945 H. D. Smyth *Gen. Acct. Devel. Atomic Energy Mil. Purposes* ix. 94 The total amount of material tied up in a separation plant is called a 'hold-up'. The hold-up may be very large in a plant consisting of many stages.

2. *attrib.* = Engaged in, involving, or characterized by forcible stopping and robbing of a person.

1881 E. W. Nye *B. Nye & Boomerang* 192, I did give him the grand bounce, and now he hath joined a hold-up outfit on the overland stage route. **1899** *Chicago Tribune* 16 Jan., The holdup gang who shot and killed policeman.. Wallner. **1930** [see *bail *v.*[2] 3]. **1959** *Times Lit. Suppl.* 20 Feb. 93/3 Unfortunately the stranger is not only a 'goy' and a drifter but he has been worse—a hold-up man, one of whose victims was once the grocer himself.

hole, *sb.* **2. b.** Delete *Obs.* and add later examples. Now usu. the cell used for solitary confinement, and hence solitary confinement itself.

1912 D. Lowrie *My Life in Prison* iv. 39 'It's a case of spending the night at the springs if you're not at your cell for the count.' In answer to my hurried inquiry about 'the springs' he informed me that he referred to 'the hole'. **1927** *Amer. Speech* II. 282/1 *Hole*, dungeon or place for solitary confinement. **1935** N. Ersine *Underworld & Prison Slang* 45 *Hole*,..the solitary confinement cells of a prison. 'Smitty just got tossed in the *hole*.' **1955** W. Gaddis *Recognitions* II. v. 488 He had..spent a fair amount of time in solitary confinement ('the hole', as it was called). **1970** G. Jackson *Let.* 25 Mar. in *Soledad Brother* (1971) 197 They're out of the hole (isolation) already.

d. A shilling. *slang.*

1934 P. Allingham *Cheapjack* iv. 38 A penny is a 'clod', and 'sprasy' means sixpence. A shilling is also a 'hole', and a two-shilling piece is a 'two-ender'. **1939** [see *bar *sb.*[1] 3 c].

3. (Later examples.)

1925 Wodehouse *Carry on, Jeeves!* iv. 81 'Mr Bickersteth is in a hole, Jeeves,..and wants you to rally round.' 'Very good, sir.' **1937** A. Christie *Murder in Mews* 218 Lawyers, even the most respectable, have been known to embezzle their client's money when they themselves are in a hole. **1970** G. F. Newman *Sir, You Bastard* v. 154 Too bad if he has, I'll be in a great big hole.

4. a. *spec.* One of the (usu. nine or eighteen) strips of land on a golf-course, consisting of a tee, fairway (and bordering rough), green and hole (sense 4 a), over which a golfer plays his ball; the play which takes place between teeing off and holing the ball; *hole in one*, the driving of the ball from the tee into the hole with only one stroke. Also *fig.*

1874 J. Blackwood *Let.* 4 Aug. in Geo. Eliot *Lett.* (1956) VI. 74 When we were a few holes out he exclaimed fervently, 'This is a great, glorious, and noble game.' **1887** J. L. Stewart *Golfiana Miscellanea* 100 The hole is won by the side holing at fewest strokes. **1891** H. G. Hutchinson *Famous Golf Links* 90 Point Garry is a long, hazardous hole. *Ibid.* 156 The third hole (135 yards) is an exact counterpart of the second. **1893** Barrie & Conan Doyle *Jane Annie* II. 41, I gives in! You have my word of honour! It's your hole. **1908** J. Braid *Advanced Golf* 252 Holes of about 360 to 380 yards. **1935** Graves & Longhurst *Candid Caddies* 28 There are all kinds of variants on the 'hole in one' story where the player has achieved this feat with a club other than the one selected by the caddie. **1971** *Daily Tel.* 12 June 1/5 (*heading*) Golfer gets two holes in one. *Ibid.*, Successive holes in one have been done only twice before in Britain. **1972** I. Stuart *Golf in Hertfordshire* 67 There are six par-three holes, all of them fair and only one over 200 yards. **1973** *Country Life* 17 May 1369/3 The final rounds of a 72-hole event.

d. *Chess.* (See quots.)

1894 J. Mason *Princ. Chess* i. 24 *Hole*, a square on the third or fourth rank, neither commanded, nor liable to be commanded, by any friendly Pawn. **1895** H. E. Bird *Chess Novelties* 115 KBP was followed by QP2, leaving at once a landing square (a nasty hole Steinitz would call it) for opponent's pieces. **1922** *Brit. Chess Mag.* Feb. 105/2 The net result of the two moves is to create a hole at White's Q Kt 4. **1955** *Chess* ('Know the Game' Series) 26/1 Another example of weakness in the pawn-formation is that of 'holes' or 'fore-posts' which may be occupied successfully by an opposing piece. **1968** O. Hindle *Further Steps in Chess* iv. 49 Holes..are squares which the defender can no longer protect with his pawns. They are thus ideal posts for attacking pieces, which can settle on them without fear of being easily driven off.

e. *Eton Fives.* A small square portion of the floor enclosed by the pepper-box and step. Phr. *to be in holes*; hence *attrib.* in *holes innings*.

1897 *Encycl. Sport.* I 399 A, who begins serving, is bound to give C—who is said to be 'in holes'—the sort of service which he prefers. *Ibid.* 400 All alike differ from Eton Courts in having no pepper-box, hole, or step. *Ibid.* 402 In the first innings of a game A (who goes in first) is said to have 'holes innings', *i.e.*, when both A and B have been put out, A will be 'in holes'.

f. *Physics.* A position from which an electron is absent: orig. a concept in the theory of the positron, now esp. a position in a semiconductor which may be regarded as a mobile carrier of a positive charge. Also *attrib.* and *Comb.*

1930 P. A. M. Dirac in *Proc. R. Soc.* A. CXXVI. 362 Only the small departures from exact uniformity, brought about by some of the negative-energy states being unoccupied, can we hope to observe. Let us examine the properties of the vacant states or 'holes'. **1933** *Ibid.* CXXXIX. 714 The few states which are unoccupied behave like ordinary particles with positive kinetic energy and with a positive charge. Dirac originally wished to identify these 'holes' with protons, but this had to be abandoned when it was found that the holes necessarily have the same mass as negative electrons. **1934** P. A. M. Dirac in *Proc. Cambr. Phil. Soc.* XXX. 150 Any unoccupied negative-energy states would be observable to us, as holes in the distribution of negative-energy electrons, but these holes would appear as particles with positive kinetic energy... It seems reasonable..to identify these holes with the recently discovered positrons. **1936** W. Heitler *Quantum Theory of Radiation* 188 Thus.. positive electrons are represented as holes in the distribution of electrons filling up the negative energy states. **1940** *Nature* 29 June 998/2 It is suggested..that in cuprous oxide the vacant lattice points and the points from which an electron is missing (positive holes) are dissociated. **1948** *Physical Rev.* LXXIV. 230/2 As a result, the current in the forward direction with respect to the block is composed in large part of holes, i.e., of carriers of sign opposite to those normally in excess in the body of the block. **1949**

[see *acceptor 3]. **1954** *Electronic Engin.* XXVI. 34 Positive charge carriers known as 'holes'... These holes are thought to have different mean life-times and mobilities in different diodes. **1957** *Ibid.* XXIX. 3 As all transistors have a finite base width all transistors must show hole storage effects due to the time taken for holes to cross the base from emitter to collector. **1962** Simpson & Richards *Junction Transistors* ii. 32 Also, because impurities or defects that trap electrons may have characteristics quite different from those of their hole-trapping counterparts, the lifetime of minority carriers may be quite different in *p*-type and *n*-type materials of similar quality. **1966** *New Scientist* 11 Aug. 317/3 Travel is limited to the distance covered before electron and hole annihilate one another.

7. b. *hole in the wall*, (an originally disparaging term for) any small, obscure place; *spec.* in the U.S., a place where alcoholic drinks are sold illegally. Applied, esp. *attrib.*, to a business that is very small, mean, dingy, or the like, or to a person running such a business.

1822 W. Hazlitt in *New Monthly Mag.* IV. 102, I had heard Mr. James Simpkins..when the character of the *Hole in the Wall* was brought in question, observe—'The house is a very good house, and the company quite genteel.' **1856** *Iroquois Republican* (Middleport, Ill.) 25 Dec. 2/3 A 'grocery'—a 'doggery'—a 'hole-in-the-wall'—is an 'odious damned spot' in any community. **1870** Dickens *E. Drood* xviii. 142 The Gate House, of which.. the Verger's hole in the wall was an appanage or subsidiary part. **1887** *Minnesota Gen. Statutes Suppl.* (1888) 248 Whoever shall attempt to evade or violate any of the laws of this state..by means of the artifice or contrivance known as the 'Blind Pig', or 'Hole in the Wall'..shall.. be punished. **1896** C. H. Shinn *Story of the Mine* 51 Many lived in 'dug-outs', which they called 'holes in the wall'. **1919** *Detective Story Mag.* 25 Nov. 129 He breakfasted at a hole-in-the-wall lunch room before starting out on his quest. **1923** D. Sells *Brit. Trade Boards System* IV. ii. 259 The emphasis which reputable employers lay upon the benefit of Trade Boards in eliminating the 'hole in the wall' employer..from the field of industry, can hardly be overstated. **1940** F. Riesenberg *Golden Gate* 212 Craft that could go into the 'holes in the wall' along the ragged Pacific Coast. **1945** E. S. Gardner *Case of Gold-Digger's Purse* (1948) xiii. 153 It's just a little place—just a little lunch counter. Sort of a hole in the wall. **1945** 'L. Lewis' *Birthday Murder* (1951) iii. 37 Sawn scorned..decadent play spots of the economically fortunate, and would insist on going to a hole in the wall infested by cockroaches, cocottes and cab drivers. **1951** C. W. Mills *White Collar* I. ii. 30 The hole-in-the-wall business, also known as a Mom-and-Pop store. **1953** W. R. Burnett *Vanity Row* viii. 60 A Bohemian section of the town..dotted with little..hole-in-the-wall cafés. **1958** *Time* 3 Feb. 23/1 To survive, most workers have to take second jobs, many of them in the innumerable hole-in-the-wall private enterprises that have sprung up. **1973** J. Goodfield *Courier to Peking* ix. 100 One of her favourite places was more a hole-in-the-wall than a shop.

c. *in holes*: perforated with holes, worn into holes.

1851 Mayhew *London Labour* II. 470/2, I can't abide this muckydam [*sc.* macadam]..it's sloppy stuff, and goes so bad in holes. **1926** A. Christie *Murder R. Ackroyd* x. 127 He wouldn't even buy new face towels, though I told him the old ones were in holes.

d. Aeronautics. *hole in the air*: an old name for an *air-pocket* (*air *sb.*[1] III. 1).

1911 G. C. Loening *Monoplanes & Biplanes* xiv. 305 The air is very variable, and even on a relatively calm day there are likely to be 'holes in the air'. **1916** H. Barber *Aeroplane Speaks* 51 Now the Aeroplane is almost over the river, and the next instant it suddenly drops into a 'hole in the air'. **1917** C. C. Turner *Aircraft of To-day* vi. 98 The terms 'air-pocket' and 'hole in the air' are frequently heard in flying circles.

e. colloq. *hole in (the) heart*: a congenital malformation of the heart in which there is an abnormal communication between the right and left sides.

1958 *Hammersmith Post* 25 July 1/4 (*heading*) Mother reassured over 'hole in heart' operation. *Ibid.*, A seven-year-old boy..is due to have a 'hole in heart' operation. **1959** *Times Lit. Suppl.* 25 Sept. 549/5 The author describes the operation for a septal defect—the condition popularly known as a 'hole in the heart'. **1961** *Listener* 2 Nov. 693/1 The oxygenator took over the duties of heart and lung in the 'hole-in-the-heart' operation. **1966** *Guardian* 17 May 3/4 Oxygen 15..is being used..for the diagnosis of the hole-in-the-heart condition.

8. *spec.* (*slang*) The mouth, the anus, or the female external genital organs. (Add further examples.)

1592 Shaks. *Rom. & Jul.* II. iv 94 This driveling Love is like a great Naturall, that runs lolling vp and downe to hid his bable in a hole. **1719** D'Urfey *Pills* IV. 72 It has a Head much like a Mole's, And yet it loves to creep in Holes: The Fairest She that e'er took Life, For love of this, became a Wife. **c 1744** in *Oxf. Dict. Nursery Rhymes* (1951) 372 Little Robin red breast, Sitting on a pole, Niddle, Noddle, Went his head, And Poop went his Hole. **1922** Joyce *Ulysses* 748 My hole is itching me. **1959** I. & P. Opie *Lore & Lang. Schoolch.* iii. 49 Habitual grumblers in London's East End receive the poetic injunction: 'Oo, shut yer moanin' 'ole'. **1966** L. Cohen *Beautiful Losers* (1970) I. 9 Don't give me this all diamond shit, shove it up your occult hole.

11. Phrases. *to put in the hole* (slang): to swindle, defraud; *to make a hole in the water*: see Water *sb.* 6 f; *to be in the hole* U.S.: to be in (usu. financial) difficulties (cf. 3); *a hole in the head*, esp. in phr. *to need (something) like a hole in the head* (cf. Yiddish *ich darf es vi a loch in kop*): applied to something not desired at all or something useless.

1812 J. H. Vaux *Vocab. Flash Lang.* in *Mem.* (1964) 243 *To put a person in the hole*, to defraud him of his due share of the booty by embezzling a part of the property, or the money, it is fenced for; this phrase also applies generally to defrauding anyone with whom you are confidentially connected of what is justly his due. **1833** *Session Papers* 3 Jan. 115/1 Miller..said they had *put him in the hole*, and he..would say where they were; by *putting him in the hole*, I understand they did not take the property away as he expected. **1890** *Centralia* (Wash.) *Chron.* 18 Sept. 3/2 His failure leaves a number of our local dealers in the hole for amounts ranging from $200 down. **1893** L. W. Moore *His Own Story* xxi. 293 What was said at that time about his being 'put in the hole', I cannot say; but I do know he held me blameless, for none of the funds, except my own share, was ever in my possession. **1897** *Boston Jrnl.* 12 Mar. 10/1 The sporting-man was $40 in the hole. **1916** *Lit. Digest* 8 Jan. 87/1 The Wards were in the hole to the extent of close to $800,000. **1926** J. Black *You can't Win* ix. 104, I thought you put me in the hole for some coin, but I found out that the people lost just what you both said. **1939** Wodehouse *Uncle Fred in Springtime* iii. 45 How in the world did you manage to get in the hole for a sum like that? **1951** in M. McLuhan *Mech. Bride* 29/2 A smart operator needs a dame like he needs a hole in the head. **1951** J. D. Salinger *Catcher in Rye* xiv. 91 The Disciples..were about as much use to Him as a hole in the head. **1955** W. Gaddis *Recognitions* I. iii. 101, I need this drink like I need a hole in the head. **1971** D. Creed *Trial of Lobo Icheka* xiii. 133 He needed Petersen about as much as he needed a hole in the head.

12. *hole-punched, -puncher*; **hole-card,** in stud poker, a card which has been dealt face down; also *fig.*; **hole-high** *a.* (see quots.); **hole-mouth(ed)** *Archæol.*, said of pottery vessels without a neck; **hole-nester,** a bird that nests in a hole; so **hole-nesting** *ppl. a.*; **hole-proof** *a.*, that will not wear into holes; **hole saw** = *crown-saw* (Crown *sb.* 35).

1908 *Sat. Even. Post* 5 Dec. 19/2 Scarcely glancing at his hole card Phelps let him take the pot, and it became Phelps' deal. **1926** C. E. Mulford *Bar 20 rides Again* xxi. 282 Beginnin' with this hand I'm bettin' five hundred blind on th' hole-card, an' seein' if I can't bring this game to a finish. **1952** J. Steinbeck *East of Eden* ix. 79 The preacher turned over his hole-card, the sure-fire card. **1971** J. Ball *First Team* (1972) xxiii. 353 We may be playing with a bust hand; we don't know if our hole card has been stolen or not. **1897** *Encycl. Sports* I. 472/2 A ball is said to be hole high when it is played on to the putting green from a distance. **1961** J. S. Salak *Dict. Amer. Sports* 228 *Hole-high* (golf), a point even with the hole but to one side or the other. **1909** *Cent. Dict.* Suppl., Hole-mouthed vase. **1960** K. M. Kenyon *Archæol. in Holy Land* v. 124 The type of jar, known as the hole-mouth jar, neckless with a simple in-curved rim, which was used for cooking and storage, may be as much as 3 feet in height. **1938** *Brit. Birds* XXXI. 242 In the present experiments three hole-nesting species..were selected. *Ibid.* XXXII. 31 In similar experiments with another hole-nester..the male attacked the male mount but.. ignored the female mount. **1953** N. Tinbergen *Herring Gull's World* x. 94 Territory in the Herring Gull most certainly has nothing to do with the reservation of a nesting site..as it has in hole-nesting birds. **1913** *Work* 17 May 102 A..cloth that will not tear—in fact, is hole-proof. **1915** *Truth* LXXVIII. 848/1 Another customer tells me her experience in regard to some 'hole-proof' hose. **1962** *Economist* 2 June 897/1 Stockings..to be ladder-proof, although not holeproof. **1956** S. Bellow *Seize the Day* (1957) ii. 42 He put the hole-punched cards in his pocket. **1961** *Lebende Sprachen* VI. 70/1 Hole puncher. **1961** Webster, Hole saw. **1967** *Catal. Black & Decker Powertools*, A drill with power to spare... Will drive holesaws up to 1¼" dia.

hole, *v.*[1] Add: **1. c.** To fire a bullet into.

1847 Trollope *Macdermots* I. iv. 59 We'll hole him till there ar'nt a bit left in him to hole. **a 1882** —— *Land-leaguers* (1883) I. ii. 34 Keep yourself from being holed as they holed Muster Bingham the other day.

6. b. *to hole (out) in one*: to achieve a 'hole in one' (see *hole *sb.* 4 a). Also *fig.*

1894 [see Hole *v.*[1] 6 c]. **1928** D. L. Sayers *Unpleasantness at Bellona Club* xii. 141 'I say we shall find a long scratch on the paint,' said Parker,..'Holed it in one, Charles.' **1939** 'N. Blake' *Smiler with Knife* iii. 49 'Oh, E.B. The E.B. printed on the flag we found in that locker?' 'Holed out in one.' **1971** *Daily Tel.* 12 June 1/5 John Hudson made golfing history in the Martini tournament..yesterday, holing in one at successive holes.

7. *to hole up*: also, to seek shelter, to seek (temporary) quarters; (*b*) to lie in wait or in ambush, to hide (chiefly *U.S. slang*).

1875 J. Burroughs *Winter Sunshine* 279 Only five days was I compelled to 'hole up' in my state-room. **1910** Mrs. H. Ward *Canadian Born* ix. 181 I'm a poor old broken-down..miner, who wants to hole-up somewhere, and get comfortable for his old age. **1912** Mulford & Clay *Buck Peters* xxvii. 235 Go slow, Tex; mebby he's holin' up on us, like he did on Buck. **1924** C. E. Mulford *Rustlers' Valley* xii. 141 Now you'll mebby have to take to th' hills an' hole up just when I need you most. **1925** —— *Cottonwood Gulch* xvi. 218 It would have been only a matter of a few minutes before they would have forced him to abandon the horse and to hole up on the defensive, to make a losing fight. **1929** Faulkner *Sartoris* IV. 282 Hole up here, you potlickin' fool. **1929** D. Hammett *Red Harvest* xviii. 179 You'll have to..take a plant on Willsson's... I hear whisper Thaler's holing-up there. **1939** R. Chandler *Big Sleep* xxvii. 240 That's the place where she's holed up. **1951** S. Lewis *World so Wide* xii. 135 We've got to begin thinking about holing up for the night. **1952** Wodehouse *Pigs have Wings* ix. 178 The poltergeist, for such he assumed it to be, appeared to have holed up behind the door that led presumably to the kitchen. **1954** 'N. Blake' *Whisper in Gloom* II. xiv. 194, I bet you Elmer's holed up in Harwich, or somewhere near it. **1961** G. Greene *Burnt-Out Case* II. iii. 37 Who would expect to find *the* Querry holed up in a leproserie? **1973** D. Jordan *Nile Green* xlv. 234 We were holed up in the flat, drinking Gold Star beer.

9. *trans.* To record by punching a hole in an allotted space in a card.

1911 *Chambers's Jrnl.* May 335/2 Not only are the old-time data, such as age,..'holed' into the card, but whether you are married or single. *Ibid.* 336/2 In this machine the data 'holed' in every tag can be all or partly recorded on another form.

holeable (hǭu·lăb'l), *a. Golf.* Also **holable.** [f. Hole *v.*[1] + -able.] Of a stroke, esp. a putt: capable of sending the ball into the hole.

1909 *Westm. Gaz.* 30 Apr. 12/2 On the green Taylor failed at a holable putt for 5. *Ibid.* 10 June 12/3 Four holeable putts which he missed. **1927** *Sunday Express* 29 May 21/7 The short eleventh was halved in three, both players missing holeable putts. **1955** *Times* 1 June 4/1 Pattinson missed only one putt that looked holeable and holed a good number that were missable.

Holi, var. Hoolee.

1910 *Encycl. Brit.* XIII. 507/2 The *Holi*, the Indian Saturnalia in the month of Phālguna (February to March). **1921** E. M. Forster *Let.* 1 Apr. in *Hill of Devi* (1953) 58 My painted escort (they had been celebrating Holi) were all that is polite. **1958** *Listener* 13 Nov. 768/1 The great spring game of Holi that they play all over northern India. *Ibid.*, Holi used to be the loveliest of festivals. **1969** *Eve's Weekly* (Bombay) 20 Dec. 67/3 When I was a young girl every year brought Holi with its gay colours.

holiday, *sb.* Add: **2. e.** Euphemistically used for: imprisonment.

1901 *Pall Mall Mag.* Feb. 197 A sentence of a month or two..a little 'holiday' with food and shelter and warmth.

4. a. *holiday centre, job, ramble, resort, tutor(ship)*; **holiday camp,** an informally run camp for a holiday; now esp. a complex of chalets, places of entertainment, etc., designed for family holidays; **holiday course** [= G. *ferienkurs*, F. *cours de vacances*, etc.], a series of lectures, classes, etc. which is held during a school or college vacation; **holiday home,** (*a*) a place where poor persons, or children whose parents cannot take charge of them, can be accommodated, sometimes at little or no cost to themselves, for a period of holiday; (*b*) a house where people spend their holidays; **holiday task,** homework to be done during the holidays.

1870 R. St. J. Corbet (*title*) The holiday camp: three days' picnic: story for boys and girls. **1927** *Cornh. Mag.* Feb. 225 Of the many thousands of children who were sent into the country..25% stayed for two weeks or more in the Holiday Camps. These Camps are an interesting feature of the Copenhagen holiday system. *Ibid.* 226 Life in a Holiday Camp is always the simplest, and is spent, so far as possible, in the open air. *Ibid.* 231 The Danes do not regard the Holiday Camp system as an ideal system. **1940** *Manch. Guardian Weekly* 1 Nov. 320/1 Then there were the Holiday Camps, cheap, social, with every modern convenience and all the modern pleasures. Their official hosts and hostesses mapped out the day with a colossal time-table of delights. **1949** M. Dickens *Flowers on Grass* vii. 181 I've got to go to a holiday camp to do some sketches of happy campers for publicity. **1958** *Times* 8 Sept. 6/1 A steel cabinet in the security block at Butlin's holiday camp at Ayr was forced at the weekend. **1944** J. S. Huxley *On Living in Rev.* I. iii. 6 The elaborate system of rest-houses and holiday centres and the equally elaborate arrangements for holiday transport. **1966** *Economist* 17 Sept. 1143/2 Very soon now a vast new 'holiday centre' will open in Aviemore itself, containing both accommodation of all kinds and prices and entertainments of a similarly wide variety. **1906** *Teacher* 30 June 616/3 The Greifswald Holiday Course..the oldest of the German holiday courses..has now been in existence for fifteen years. **1887** *Girl's Own Paper* 22 Oct. 48/3 A lady who has a large house and grounds would give a lady of small means a 'holiday home'. **1931** *Geography* Sept. 219 The hostel..is more akin to the 'dak-bungalow' or the 'cold harbour' than to the rest-camp or holiday-home. **1937** *Discovery* June p. xlvi/2 (Advt.), Schools, coaching colleges, holiday homes. **1972** *Guardian* 1 Sept. 8/6 The 'white settlers' who buy croft houses for use as holiday homes. **1973** *Ibid.* 28 May 4/1 The Welsh Language Society is planning..to try to prevent the sale of houses as second or holiday homes. **1866** 'Mark Twain' *Lett. fr. Hawaii* (1967) 91 In Honolulu it is not a holiday job to ship a crew. **1969** *Times* 14 June 18/4 (Advt.), Holiday job wanted..by 6th form girl. **1854** Sherwood & Kelly *Boys will be Boys* ii. 31 That part of the country..within the nearer reach of a holiday ramble. **1936** *Discovery* Sept. 263/1 A famous holiday-resort has been selected [as a meeting-place for the British Association]. **1827** J. Leech *Let.* in W. P. Frith *John Leech* (1891) I. i. 12, I think I shall get promoted when Dr Russell sees my Holiday Task. **1899** Kipling *Stalky & Co.* 180 They have a holiday task..which..none..will ever look at. **1912** E. W. Hornung *Fathers of Men* xi. 130 So they give you saying-lessons for holiday tasks at your school? *Ibid.* xii. 137 He'd no right to set us a holiday task of his own like that. **1930** C. Mackenzie *April Fools* vii. 138 I'm reading 'Homes without Hands' for a holiday task. **1900** *Captain* II. 375/1 When Mr. Soames asked the professor to come and be holiday tutor. *Ibid.*, Open to take a holiday tutorship.

b. *holiday-maker* (earlier example), *-making* (examples).

1792 W. B. Stevens *Jrnl.* 27 May (1965) 24 Set out in the afternoon, holiday-making to Birmingham. **1836** Dickens *Sk. 'Boz'* I. 323 The four clowns..may be all very well for the low-minded holiday makers. **1855** Geo. Eliot in *Fraser's Mag.* LII. 60/1 The good people who come to take dinner..here, by way of holiday making. **1969** *Daily Tel.* 4 Jan. 18/6 There is one valuable holiday-making aid to the Gulf and the Bay of Naples.

holier-than-thou: see *holy *a.* 5 c.

holiness. Add: **4. b.** Used *attrib.* (usu. with capital initial) to denote any of various religious sects which emphasize sanctification, spiritual purity, and perfectionism, or members, churches, etc., of any of these sects. orig. and chiefly *U.S.*

1888 *California State Gaz.* 623/2 Pasadena..contains Methodist, Baptist,..Friends, and Holiness churches. **1913** H. Kephart *Our Southern Highlanders* 271 In our day the same may be said of the Holy Rollers and Holiness People. **1928** *Amer. Mercury* Oct. 185/1 This..was first preached by the Straight Holiness sect in Kansas in the 1890's. **1940** J. B. Holt in *Amer. Sociol. Rev.* V. 740 (*title*) Holiness religion: cultural shock and social reorganisation. **1947** J. Wach *Sociol. Relig.* II. v. 189 The Methodist and the Holiness movements are especially concerned with ethical perfection. **1957** J. M. Yinger *Relig., Soc. & Individual* I. x. 282 Any description of religious trends in the United States that did not refer to the strength of the 'holiness' sects..would be incomplete. **1958** M. Argyle *Relig. Behaviour* iv. 33 Many of these sects—the Pentecostal, Holiness, Nazarene churches and others—have increased enormously in proportion to their size during this period [*sc.* 1926–1953]. **1961** B. R. Wilson *Sects & Soc.* I. v. 98 Elim is in no way as extreme as some of the Holiness cults of Tennessee and North Carolina. *Ibid.* 105 In Britain it began..in such organisations as the Faith Mission, independent Holiness missions,..and among the Plymouth Brethren. **1968** W. Phillips in P. Oliver *Screening Blues* ii. 63 Now the Holiness people, when they come in, They said, 'Boy we can make it by livin' above sin.'

holing, *vbl. sb.* Add: **1.** Also, the production of holes, e.g. in garments (cf. Hole *v.*[1] 8).

1910 *Daily Chron.* 14 Mar. 6/4 The Stockings that are actually insured against holing.

c. *Golf.* The action of holing the ball; also *attrib.*, as *holing distance, holing-out putt*.

1875 'Stonehenge' *Brit. Sports* (ed. 12) 695/1 He who succeeds in holeing in fewer strokes than his opponent wins that hole. **1901** *Scotsman* 11 Sept. 10/1 A nicely-played mashie stroke took his ball within holing distance. **1906** *Westm. Gaz.* 10 Aug. 4/2 The longer holing-out putts. **1972** *Country Life* 7 Dec. 1600/3 As often as not, the ball would finish within likely holing distance.

holism (hǫ·liz'm, hǭu·liz'm). [f. Gr. ὅλος whole + -ism.] A term coined by Gen. J. C. Smuts (1870–1950) to designate the tendency in nature to produce wholes (*i.e.* bodies or organisms) from the ordered grouping of unit structures. So **holist, holi·stic** *a.*, **holi·stically** *adv.*

1926 J. C. Smuts *Holism & Evol.* 99 The whole-making, holistic tendency, or Holism, operating in and through particular wholes, is seen at all stages of existence. *Ibid.* 127 There is a synthesis which makes the elements or parts act as one or holistically. **1927** *Brit. Weekly* 20 Jan. 418/4 The real entities of the material world must, like organisms, be creative, self-transcending, functional. They must be Holistic unities. **1931** Smuts in *Times* 2 Sept. 7/7 Instead of the animistic, or the mechanistic, or the mathematical universe, we see the genetic, organic, holistic universe. **1937** 'C. Caudwell' *Illusion & Reality* ix. 180 A large portion of reality will be conveniently removed to the sphere of religion, as among the vitalists, holists, entelechists and spiritualists generally. **1945** *Word* I. 109 Haldane suggested for this movement [in biology] the name 'holism'; others preferred to call it 'organicism'. To my mind, this new holism or organicism bears a close relationship to linguistic structuralism. **1952** C. P. Blacker *Eugenics* x. 240 The principle of monism is much the same as that of General J. C. Smuts's holism. **1959** *Times* 11 Dec. 15/1 Holism has at last penetrated departments of nutrition, and a new school of nutrition has arisen which realizes that the integration of nutrition, health and disease is a problem that must be attacked on a wide front. **1961** J. Wilson *Reason & Morals* ii. 117 There may be sane people who do not 'appreciate' (i.e. respond holistically to) works of art or to nature. **1964** *Punch* 20 May 736/1 If..we view a person as a large.. holistic, versatile, symbol processing system. **1964** *Listener* 21 May 825/1, I mentioned the unresolved conflict between the atomistic psychology of the Behaviourists and the somewhat schematic holism of the Gestalt school. **1970** *New Scientist* 9 July 96/1 Each level is equipped with its own 'laws of organization', 'intrinsic patterns' or whatever 'holistic' term you prefer to choose. **1971** *Nature* 13 Aug. 504/2 Professor J. S. Weiner has pioneered the modern holistic approach to the study of man. **1971** F. A. Stafleu *Linnaeus & Linnaeans* ii. 39 'Canon' means here 'general rule or axiom' and has overtones of 'genuine and inspired', known instrictively by a holistic, not an analytical, approach to the phenomena of life.

holk, howk, *v.* Add: **1.** (Further examples.)

1899 A. Werner *Captain of Locusts* 160 They howked a grave near the kraal, and buried their chief *pro tem.* **1926** D. H. Lawrence *Let.* 28 Dec. (1932) 676 We shall stay here if not howked out. **1950** B. Marshall *Every Man a Penny* xlix. 229 Deep in their trenches the hairy men stood, howking out the brown earth. **1955** E. Pound *Classic Anthol.* II. 90 Howk 'em up with a landing scoop.

2. (Later examples.)

1906 Kipling *Puck of Pook's Hill* 69 Dan hiked and howked with a boat-hook (the brook was too narrow for sculls). **1950** *John o' London's* 24 Nov. 621/1 The solan goose..starts howking and pecking at the rope which sustains the climber.

Holland. 1. b. Add: *Holland sauce* = *Hollandaise.

1877 E. S. Dallas *Kettner's Bk. of Table* 162 Dutch or Holland Sauce: Sauce Hollandaise. **1892** [see *dutch *a.* 3 b].

Holland[2] (hǫ·lănd). The name of J. P. *Holland* (1840–1914), the designer of a class of submarines adopted by the American navy, used as the proper name of the first submarine of this type and afterwards generically.

1899 *Westm. Gaz.* 7 Dec. 2/3 The President of the official Naval Board, and several of its members have signed a statement declaring that their 'Holland' has fulfilled all requirements in her trial trip. **1902** *Encycl. Brit.* XXXII. 576/2 The *Holland*, a smaller boat, having a length of about 59 ft., though begun after the *Plunger*, has already been completed. *Ibid.*, The latest Holland design is shown in Fig. 95. **1906** *Daily Chron.* 8 Sept. 5/3 The original Holland class of submarine.

Hollandaise (hǫlăndē̇ⁱz, ‖olandēz). [Fr., fem. of *hollandais* Dutch, f. *Hollande* Holland.] *Hollandaise sauce* (see quot. 1907); *à la Hollandaise*, served with Hollandaise sauce.

1841 Thackeray in *Fraser's Mag.* XXIII. 719/1 Turbot with lobster-sauce is too much; turbot *à la Hollandaise* vulgar. **1861** Mrs. Beeton *Bk. Househ. Managem.* 195 Green Dutch sauce, or Hollandaise verte. **1899** N. Newnham-Davis *Dinners & Diners* iv. 26 Artichokes good, though we preferred plain vinegar as a dressing to the *hollandais* one. **1907** Escoffier *Mod. Cookery* 22 Hollandaise Sauce... One and one-half lbs. of butter, the yolks of six eggs, one pinch of mignonette pepper and one-quarter oz. of salt, three tablespoonfuls of good vinegar. *Ibid.* 23 The consistence of sauces whose processes are identical with those of the Hollandaise may be varied at will. **1964** C. Willock *Enormous Zoo* ix. 169 Nile perch with Hollandaise sauce.

Hollander. Add: **b.** A South African colonist or immigrant of Dutch birth or descent. Also *attrib.*, or as *adj.*, and *Comb.*

1699 W. A. Cowley *Voy.* in W. Hacke *Collect. Voy.* (1729) v. 34 The Village inhabited by the Hodmandods, so called by the Hollanders. **1897** in H. M. Stanley *Thro' S. Afr.* (1898) v. 75, I do not blame the Boers so much as I blame the Hollanders and our Jews here. **1899** *Westm. Gaz.* 16 Oct. 7/2 The Boers who have occupied Newcastle consist of both Transvaal and Free State commandos, with 400 Hollanders. **1899** *Daily News* 2 Nov. 5/2 It has not been he, but the 'Hollander', a most unfavourable specimen of the Dutch race, who has been concerned in all the doubtful intrigues..of the last few years. **1902** *Encycl. Brit.* XXXII. 721/1 The effect of this development was the production of a body of officials in the Transvaal, partly Hollander and German, partly Boer. **1903** G. W. T. Omond *Boers in Eur.* 31 South Africa, big towns and seaports excepted, being Hollander-Boer to the core. **1934** [see *African *sb.* b]. **1971** *Rand Daily Mail* (Home Owner) 27 Mar. 16/1 Developer-builder Gard Duys, a Hollander with a soft spot for progressive architecture, is delighted with the result.

2. *Paper-making.* A beating-engine, invented in Holland, for the conversion of the bleached rags into paper-pulp. Also called *Hollander-beater*.

1878 *Design & Work* 19 Jan. 88/3 About fifty years after the invention of the 'Hollander'..alkali began to be employed for boiling the rags. **1900** Cross & Bevan *Paper-Making* (ed. 2) 172 The ordinary form of beater is fitted with a single roll, and the general arrangement of its working parts is that..described..for a 'breaking' engine. This type of beater is known as the Hollander. **1902** *Encycl. Brit.* XXXI. 458/2 One of the various forms of beating engine or 'Hollander'. **1907** Cross & Bevan *Paper-Making* (ed. 3) 179 The Hollander consists of an oblong trough, with semi-circular ends, with a partition or mid-feather running down the centre so as to form a continuous channel running round which the stuff can circulate. **1963** R. R. A. Higham *Handbk. Papermaking* xiii. 266 In the middle of the eighteenth century, the development of the Hollander beater revolutionized stock preparation methods. This beater was invented in Holland—hence the name—and it replaced the old rag stampers.

3. A Dutch clinker.

1897 Webster, *Hollander.* 2. A very hard, semi-glazed, green or dark brown brick, which will not absorb water;—called also Dutch clinker. Wagner.

4. (See quot.)

1879 *Encycl. Brit.* IX. 400/2 The largest spars [of timber] are called 'Hollander'.

hollandite (hǫ·lăndəit). *Min.* [f. the name of Sir Thomas Henry *Holland* (1868–1947), British geologist: see -ite[1] 2 b.] An oxide of barium and manganese occurring as brittle silvery grey to black crystals having a metallic lustre.

1906 L. L. Fermor in *Trans. Mining, Geol. & Metall. Inst. India* I. II. 77 For this new mineral I propose the name of *hollandite*, after Mr. T. H. Holland, F.R.S., Director of the Geological Survey of India, and President of this Institute. **1943** *Amer. Mineralogist* XXVIII. 505 Cryptomelane, hollandite, and coronadite are isostructural and form isomorphous mixtures to some extent. **1952** *Nature* 6 Dec. 974/1 The formula of hollandite, deduced from that of the psilomelane used in this investigation, is $Ba_{1\cdot 01}(Mn,R)_8O_{16}$, in good agreement with that previously proposed by Byström and Byström, $Ba_{(2-x)}Mn_8O_{16}$, where x is approximately equal to one. It has been suggested that discrete water molecules may be present in hollandite as occupants of vacant sites of otherwise perfect sub-rows.

holler (hǫ·ləɹ), *v. dial.* and *U.S.* Also **holer,** † **hollar.** [var. Hollo *v.*] *intr.* To cry out loud, to shout; to complain. In a fight: to give up, to cry 'enough'. Also, to sing a 'holler' (see next). Occas. *trans.*

1699 in *Cal. Virginia State Papers* (1875) I. 67 We gott to the River side oppisett to the ffort, & theire hollerd & Immediately they answered. **1834** S. Smith *Sel. Lett. J. Downing* 37 All hollering 'stooboy'. **1843** 'R. Carlton' *New Purchase* I. xiv. 101 Provided you knew how 'to holler', within hearing of both. **1845** W. T. Porter *Big Bear Arkansas* 41 Who hollered? Which gave up? **1852** *N. & Q.* 14 Feb. 148/2 The village boys..get some halfpence given them for their 'hollering'. **1859** *Atlantic Monthly* Aug. 239/2 Here is a boy that loves to..'holler' Fire! on slight evidence. **1883** [see Hollo *v.* 1 a]. **1898** C. M. Yonge *John Keble's Parishes* xv. 175 Curate. Have you heard the nightingale yet? Boy. Please, sir, I don't know how he hollers. Everything hollers, from a church bell to a mouse in a trap. **1901** Merwin & Webster *Calumet 'K'* viii. 155 I'll holler up to you, Max, when we're ready down below. **1904** 'No. 1500' *Life in Sing Sing* 249/2 *Hollar*, complain. **1926** J. Black *You can't Win* iv. 43 Holler before you're hurt; that's my motto. **1934** *Nat. Geographic* LXV. 624/2 Daybreak and sundown are favorite times for 'holerin''. It is an invariable accompaniment of driving the cattle home in the evening. **1936** J. A. & A. Lomax *Negro Folk-Songs* II. iii. 113 He has hollered and moaned his troubles and his observations on the ways of the world. **1940** W. Faulkner *Hamlet* IV. i. 343 And when I holler run, you run. You hear me? **1967** *Boston Globe* 30 Mar. 14/1 Everyone hollers about the damage to the children if the schools are shut one day because of a teacher–school committee disagreement. **1969** *Times* 22 July (Moon Report Suppl.) p. ii/3 When Colonel Aldrin jumped off the last step of the moon ladder..everyone in the Aldrin home was whooping and hollering. **1970** P. Oliver *Savannah Syncopators* 66 (*caption*) Arthur Crudup 'hollers' with a high-pitched voice. **1973** J. Thomson *Death Cap* xiii. 177 I'll holler you to come down.

holler (hǫ·ləɹ), *sb.*[1] *dial.* and *U.S.* Also † **hollar.** [var. Hollo *sb.*] = Hollo *sb.*; also, a complaint, a cry of protest; *spec.* in the Southern States of America, a work-song.

1825 J. Jennings *Obs. Dial. W. Eng.*, Hollar. **1886** F. T. Elworthy *W. Somerset Word-Bk.* 346 *Holler*,.. the cry given when the quarry is seen; the view-halloo. **1896** Ade *Artie* xvi. 147, I put up a holler right at the jump. **1901** 'J. Flynt' *World of Graft* 133 Some gamblers were particularly loud in making their 'hollers', and threatened to bring about an investigation. **1908** J. M. Sullivan *Criminal Slang* 13 *Holler*, plaint of a victim. **1936** J. A. & A. Lomax *Negro Folk-Songs* II. iii. 113 The holler is a way of singing—free, gliding from a sustained high note down to the lowest register. **1939** *Congress. Rec.* 5 Aug. App. 3975/1 [Will Rogers] came across the American scene with..a hoot and a 'holler', and a laugh. **1940** J. W. Work *Amer. Negro Songs* 34 Approaching his house or that of his sweetheart in the evening, or sometimes out of sheer lonesomeness, he would emit his 'holler'. *Ibid.* 35 In these 'hollers' the idiomatic material found in the blues is readily seen. **1956** M. Stearns *Story of Jazz* (1957) i. 10 The street-cry and field-holler of the American Negro are earlier examples of the same tradition. **1958** P. Gammond *Decca Bk. Jazz* i. 20 Solo work-songs of field-hands..took the form of 'hollers' or 'arwhoolies'—long meandering cries that were half-sung thoughts and half yodels. **1959** R. Condon *Manchurian Candidate* (1962) vii. 137 If you're ever around Wainwright, Alaska, you'll give me a holler. **1968** P. Oliver *Screening Blues* 4 The more primitive examples of field cries, hollers and work songs, of children's game songs and unaccompanied blues were only heard on record in the rarest of instances.

holler (hǫ·ləɹ), *sb.*[2] *U.S. colloq.* var. Hollow *sb.* 2.

1845 W. T. Porter *Big Bear Arkansas* 151 [I] putt off emediately fur watur that I node waz klose down the holler. **1947** Randolph & Wilson (*title*) Down in the holler: a gallery of Ozark folk speech. **1972** J. S. Hall *Sayings from Old Smoky* 8 Many of these stock boasts hinge on the steepness or wildness of the mountain country, the darkness of the 'hollers', even in midday, and the hardihood of the people.

Hollerith (hǫ·ləriþ). The name of Herman *Hollerith* (1860–1929), American inventor, used esp. *attrib.* in reference to the use of punched cards in accounting, statistics, etc. Hence as *sb.*, used *fig.* in reference to modern society viewed as a processing machine (see quot. 1957).

1890 *Jrnl. Franklin Inst.* XCIX. 306 The cost of compiling a census by the Hollerith system, would appear to be only one-third the cost of compiling the same census by the next best system. **1891** *Electrical Engineer* (U.S.) 11 Nov. 530/1 Not a little skill and judgment was necessary in perfecting the mechanical details of the Hollerith electric tabulating system. **1946** A. C. Clarke in *Astounding Sci. Fiction* Sept. 24/1 That wonderful battery of almost human Hollerith analyzers. **1957** *London Mag.*

June 61 The writer needs his partial disengagement from the social hollerith so as to free the one incomparable tool that he can call his own. **1958** H. T. Himmelweit et al. *Television & Child* 85 The number of Hollerith cards needed for each child. **1960** *Times* 31 Mar. 3/1 An internal Hollerith installation and external computing facilities are used extensively. **1970** *Computers & Humanities* V. 2 The first was the pre-computer phase of the 1950s, characterized by the use of Hollerith (IBM) cards and unit record equipment to sort and tally large bodies of information.

holliper, var. Oliver[2].

hollow, *sb.* Add: **1. e.** (See quot. 1940.)

1885 J. G. Horner *Pattern Making* iii. 26 Many of the best wheels are made with hollows at the roots of the teeth, for here the action of leverage on the tooth induces the greatest stress. **1924** J. McC. Wilson *Pattern-Making* iv. 28 In finishing the pattern all the angled corners are filled in either with Angled or Hollowed Fillets... Hollows are used in well-finished work. **1940** *Chambers's Techn. Dict.* 418/2 *Hollows*, fillets, or curves of small radius, uniting two surfaces intersecting at an angle.

hollow, *a.* and *adv.* Add: **7.** *hollow block, tile*; **hollow heart,** a disease of potatoes in which a cavity is formed in the centre of the tuber; **hollow-horn** *U.S.* (see quot. 1962); **hollow roll:** see Roll *sb.*[1] 11 b; **hollow wall** = *cavity wall* (*cavity 4).

1964 J. S. Scott *Dict. Building* 167 *Hollow blocks* or *hollow tiles*. Concrete or burnt clay hollow building blocks are used for making partitions or external walls, or for forming reinforced concrete hollow-tile floors. Lightweight, thermally-insulating, hollow blocks are also made of foamed slag concrete, diatomite, gypsum, etc. **1926** F. D. Heald *Man. Plant Dis.* v. 94 Hollow heart is most frequent in potatoes which have been stimulated to an excessive growth by abundant moisture. **1951** *Dict. Gardening* (R. Hort. Soc.) III. 1655/1 Hollow Heart is usually due to the tubers experiencing a dry period in which they mature and lose water so that when the rain comes the quick growth causes the inner tissues to split apart. **1805** R. Parkinson *Tour Amer.* 87 There were a few half-starved cattle; in general standing shaking with cold, and many more complaining of what they call the hollow-horn. **1825** J. Lorain *Pract. Husb.* 455 The hollow horn, a disease which seldom fails to attack half-famished cattle. **1868** *Rep. Iowa Agric. Soc. 1867* 129 Cattle have few diseases in this locality except the 'buck eye' and 'hollow horn'. **1962** J. N. Winburne *Dict. Agric.* 382/1 *Hollow horn*, an imaginary disease arising from the erroneous belief that loss of appetite and listlessness in a cow was due to hollow horns. The remedy was supposed to be (a) boring a hole in each horn ..(b) filling the cavity with salt, sugar, and pepper, and (c) plugging the hole with a wooden peg. The belief was that if the cow had hollow horn this remedy would cure her, and if she did not have hollow horn, the remedy would prevent her getting it. **1914** *Archit. Rec.* Feb. 142/2 Terra cotta hollow tile was employed in the exterior and interior bearing walls. *Ibid.* 144/2 The floor construction used was the combination system of hollow tile and reinforced concrete. **1936** *Archit. Rev.* LXXX. 144/1 Floors and roofs throughout are hollow tile and concrete, and internal walls in the ward block are of hollow partition blocks. **1823** *New Pract. Builder & Workman's Compan.* 586/2 *Hollow-wall*, a wall built in two thicknesses, leaving a cavity between, which may be either for saving materials or for preserving an uniform temperature in apartments. **1891** *Notes on Building Construction* II. 10 The hollow wall is often arranged to begin on the damp-proof course. **1942** J. A. Mulligan *Handbk. Brick Masonry Construction* 362 The building code of the City of New York uses the term 'hollow wall' instead of cavity wall.

8. *hollow-cheeked* (examples), *-chested*; **hollow-fronted, -nosed, -pointed** *adjs.*, said of a bullet with a hollow in the point to ensure expansion of the projectile on impact.

1598 Florio *Worlde of Wordes* 273/1 *Pettoruto*,..that is hollow chested. **1851** H. Melville *Moby Dick* III. xlv. 255 'Look!' replied the hollow-cheeked captain from his taffrail. **1886** W. B. Yeats *Mosada* 6 Bright-eyed, and hollow-cheeked From fasting. **1899** *Kynoch Jrnl.* Oct.–Nov. 14/2 If the ·577 pure lead hollow-fronted bullet hit a man he knew it at once. **1902** *Encycl. Brit.* XXXII. 244/1 The hollow-pointed expanding bullet with soft lead nose. **1902–3** *Kynoch Jrnl.* Dec.–Jan. 43/1 Without the mutilation so commonly caused by hollow pointed bullets. **1909** *Daily Chron.* 26 June 1/4 The other cartridges..being of nickle steel and hollow-nosed. **1920** G. Burrard *Notes on Sporting Rifles* 40 A hollow-nosed bullet. **1963** V. Nabokov *Gift* iv. 240 He listened to these hollow-chested verses.

B. *adv.* **3. hollow-ground** *a.*, ground so as to have a concave surface; so *hollow-grinding*.

1885 *Army & Navy Co-op. Soc. Price List* 1048 The guaranteed razors. Cases containing 2 Hollow Ground. **1937** R. W. Fairbrother *Text-bk. Med. Bacteriol.* ii. 10 In carrying out the examination by direct microscopy use is made of the hollow-ground slide, which is a slide with a hollow of approximately ⅛ in. ground out on one surface. **1951** R. H. Hordern *Woodworking Industry Managem.* iv. 72 (*heading*) Hollow-grinding machine. *Ibid.* 73 This..will produce a hollow-ground bevel on the cutter. There are a number of reasons why hollow grinding is preferable to straight. **1968** *Gloss. Terms Mechanized & Hand Sheet Metal Work* (*B.S.I.*) 12 *Hollow grinding*, a method of grinding a tool to produce a concave face or faces behind a cutting edge.

holloware, var. Hollow-ware.

1959 *Sears, Roebuck Catal.* Spring & Summer 575/4 Melmac Dinner Sets..Holloware in solid color. **1962** *Engineering* 22 June 819/2 Aluminium sheet clad with stainless steel..is intended basically for the domestic holloware trade. **1963** *Times* 28 May 1/7 Domestic holloware made from aluminium. **1972** *Daily Tel.* 25 Apr. 15 Pans, or 'holloware' as they are called in the trade.

Holloway (hǫ·lowēⁱ). [f. the name of Thomas *Holloway* (1800–83), their inventor and manufacturer.] *Holloway's pill*, a patent medicine used principally for laxative purposes. Also *Holloway's ointment*.

1838 *Town* 16 June 440/1 (Advt.), Holloway's universal family ointment... Holloway's external disease pill. **1849** E. Ruskin *Let.* 8 Nov. in M. Lutyens *Effie in Venice* (1965) I. 59, I get..a Holloway's Pill when I require such medicine. **1877** E. S. Dallas *Kettner's Bk. of Table* 432 This is..suggestive of the African tribes mentioned by Sir Samuel Baker, who believed in Holloway's Pills because of their rapid and irrepressible results. **1885** *Trade Marks Jrnl.* 18 Nov. 1084 Holloway's Pills... The firm trading as Thomas Holloway,..London; patent medicine vendors... Pills for human use. *Ibid.*, Holloway's Ointment... The firm trading as Thomas Holloway,.. London; patent medicine vendors... Ointment. **1939–40** *Army & Navy Stores Catal.* 396/1 Holloway's Pills—box 1/3. **1951** *Chemist & Druggist Yearbk.* 196 Holloway's Pills, Ltd... King George's Avenue, Watford. *Ibid.* 457 (*heading*) Proprietary medicines advertised to the public. ..458/1 Holloway's ointment. Holloway's pills.

holluschickie (hǫ·lŏsˌtʃiki), *collect. pl.* Also **holloschickie, holluschuckie.** [ad. Russ. *kholostyaki* pl., bachelors.] Young males of the northern, Pribilof, or Alaska fur seal, *Callorhinus ursinus*; = *bachelor 4 c.

1874, 1884 [see *bachelor 4 c]. **1893** Kipling *Seven Seas* (1896) 70 But he'll lie down on the killing-grounds where the holluschickie go. **1894** —— *Jungle Bk.* 97 They [*sc.* seals] were called the holluschickie—the bachelors. **1901** *Munsey's Mag.* XXV. 355/1 The holluschickie who have reached the age when they contemplate matrimony. **1929** *Encycl. Brit.* IX. 952/1 The young males or bachelors (*holloschickie*).

holly. Add: **3. b. holly blue,** the azure blue butterfly *Celastrina argiolus*.

1853 F. O. Morris *Hist. Brit. Butterflies* 136 Holly Blue. Azure Blue... This plain but neat species is to be found..in places where the holly abounds. **1905** *Daily Chron.* 6 Apr. 3/2 The holly-blue often flies on days when there is more hail than sun. **1927** [see *chalk *sb.* 7 b]. **1952** L. H. Newman *Transformations of Butterflies & Moths* 46 Holly Blue caterpillars vary quite considerably, but the ground colour of the body is always a shade of green. **1970** *Times* 19 Aug. 9/6, I recently chased a Holly Blue around a holly tree for at least an hour before eventually retiring exhausted, with the total day's bag of one small butterfly.

hollyhock. Add: **2. b. hollyhock disease,** = **hollyhock rust*; also, blight caused by the parasitic fungus *Colletotrichum malvarum*; **hollyhock fungus,** a fungus, *Puccinia malvacearum*, parasitic on the hollyhock; **hollyhock rust,** hollyhock fungus or the disease caused by this.

[**1865** *Gardeners' Chron.* 2 Sept. 817/2 We have received ..some information with respect to a Disease with which Hollyhocks are affected to such a degree as almost to preclude their cultivation.] **1898** W. Robinson *Eng. Flower Garden* (ed. 6) 389/1 Owing to the Hollyhock disease it is often a better plan to abandon the named kinds increased from cuttings and resort to seedlings only for stock. **1951** *Dict. Gardening* (R. Hort. Soc.) II. 1006/1 Hollyhock disease is not seen in the neighbourhood of industrial towns. **1883** W. Robinson *Eng. Flower Garden* 13/2 The Hollyhock Fungus (*Puccinia malvacearum*)..is..destructive to the Hollyhock. **1899** G. Massee *Text-bk. Plant Dis.* 252 Hollyhock rust. **1910** T. W. Sanders *Garden Foes* 227 Hollyhock Rust (Puccinia malvacearum). At one time this fungoid disease played great havoc with the hollyhock. **1951** *Dict. Gardening* (R. Hort. Soc.) II. 1005/2 The onset of Hollyhock rust..led to its almost complete disappearance as a florists' flower.

Hollywood (hǫ·liwud). [A region near Los Angeles in California, the chief production centre of the U.S. cinema business.] Generally, the American type of moving picture, its characteristics and background. Also *attrib.* (or as *adj.*) and *Comb.*

1926 A. Huxley *Jesting Pilate* II. 198 What is this famous civilisation of the white man which Hollywood reveals? **1928** H. Crane *Let.* 27 Apr. (1965) 325 She ought to be a little different than the typical Hollywood hostess. **1929** E. Wallace *Red Aces* i. 218 A high-class school at Brighton, where girls are taught to..use lipstick and adore the heroes of Hollywood. **1932** Kipling *Limits & Renewals* 143 The standardised Hollywood screech of a Producer. **1933** *Punch* 30 Aug. 225/1 An American producer says that if he had his way he could make Elstree into an English Hollywood. **1934** H. G. Wells *Exper. Autobiogr.* II. viii. 513 There was a vast editor's desk, marvellously equipped, like a desk out of Hollywood. **1935** R. Macaulay *Personal Pleasures* 137 It is not..for mechanically recorded voices, however Hollywood, to mimic this universal terrestrial passion [love]. **1937** *N. & Q.* 13 Mar. 181/2 Unlike the Japanese, they [*sc.* the Chinese] do not ape Hollywood. **1939** C. Morley *Kitty Foyle* 319 Those black and white yachting shoes..were definitely Hollywood. **1940** *Writer's Digest* June 13/2 It is at this point that the Hollywood ingenue goes Hollywood. **1942** T. Rattigan *Flare Path* I. 23 Face the music? How beautifully Hollywood! What was your idea? To get Teddy alone and say 'I love your wife'? **1959** *Listener* 10 Dec. 1048/3 A revolt against Hollywood-bourgeois values. **1970** G. Greer *Female Eunuch* 294 Her Hollywood-interior lodge.

Hence **Hollywoode·se,** the style of language supposed to be characteristic of Hollywood films; **Hollywoode·sque** *a.*, characteristic of or resembling Hollywood films; **Ho:llywoo·dian, -ean** *a.*, of, pertaining to, or characteristic of Hollywood or its films; **Ho·llywoodish** *a.*, somewhat resembling Hollywood films; **Ho·llywoodism,** characteristic style or idiom of Hollywood films; **Ho·llywoodize** *v.*, to make typically Hollywoodian; so **Ho:llywoodi·zing** *vbl. sb.*

1927 *Daily Express* 4 May 4 The cottage is so picturesque and Hollywoodesque that..it is more like a 'set' than a real house. **1928** *Ibid.* 4 May 10/2 Mr. Douglas Fairbanks..is meditating..a slap-up, original Hollywoodish sequel of his own devising. **1934** A. Huxley *Beyond Mexique Bay* 23 What frills, what flounces!..The cut was a Hollywoodian adaptation from the French. **1941** *Scrutiny* IX. 346 Puccini, that voice pervasively symbolic of the Hollywoodizing of human emotions. **1950** *John o' London's* 7 July 419/1 The modern kind of roadhouse which might adopt a Spanish wear with self-conscious Hollywoodism. **1951** F. Lawrence *Let.* 20 Jan. in *Mem. & Corr.* (1961) 298 We have a very 'Hollywoodean' place, not far from the Huxleys. **1957** *Essays in Crit.* VII. 209 One scarcely needs to pursue the poem through the Hollywoodese of 'our hearts go round'. **1959** *Encounter* July 53/2 Some deep Hollywoodean reason. **1960** *House & Garden* July 36 There is nothing Hollywoodian about Miss Caron's decorative indulgences. **1962** *Listener* 6 Sept. 342/2 Those brawny, masculine, sunburnt figures of my youth..were Hollywoodish figures. **1963** *Sunday Express* 24 Feb. 23/3 *Nine Hours to Rama* is a lamentable failure because it is permeated with Hollywoodisms. **1965** *New Statesman* 19 Mar. 462/1 So far, only the cosmopolitan and European [behaviour] has been permitted..and that, all too often, heavily Hollywoodised. **1967** *Guardian* 11 Apr. 5/1 Such incredible Hollywoodising of Malcolm X did much to create an atmosphere in which he could only be killed.

Holmesian (hǭu·mziăn), *a.* and *sb.* [f. Sherlock *Holmes*, name of the amateur detective who is the chief figure in the detective stories of A. Conan Doyle (1859–1931) + -ian.] **A.** *adj.* Of, pertaining to, or in the manner of Sherlock Holmes. **B.** *sb.* A devotee of Sherlock Holmes.

1929 'G. Daviot' *Man in Queue* vi. 62 Grant disclaimed any such Holmesian qualities. **1934** J. Carter *New Paths in Book Collecting* 34 Mr. Desmond MacCarthy is a prominent Holmesian scholar. **1940** E. Bentley *Those Days* ix. 250 If the nature of the detective-story, as of other things, is to be found in its complete development, that was the Holmesian saga. **1947** C. Waugh *Comics* iii. 40 Hawkshaw's part is to uncover such fiendish plots with the aid of magnifying glass, pipe and Holmesian costume. **1958** *Times* 19 Feb. 10/4 In the belief no Holmesian had previously done so. **1972** *Times Lit. Suppl.* 7 Apr. 391/1 The great Holmesian game trundles along with unabated vigour.

Holmgren[1] (hǭu·mgrĕn, hǭu·lmgrĕn). The name of A. F. *Holmgren* (1831–97), Swedish physiologist, used in the possessive in *Holmgren's (wool) test*, a test for colour-blindness devised by Holmgren in which the subject is asked to match differently coloured pieces of wool; also *Holmgren's wools*, † *worsteds*.

1879 B. J. Jeffries *Color-Blindness* xviii. 195 Dr. Magnus of Breslau has lately proposed a modification of Holmgren's test by letting the examined pick out from bundles of colored worsteds those which match the colors of the solar spectrum shown them at the same time. *Ibid.* xxii. 241 To further test railroad employés or others, after having decided on their color-blindness by Holmgren's worsteds,..will be..of great value to examiners. **1890** Holmgren's wools [see Green-blind *a.*]. **1932** S. Duke-Elder *Text-bk. Ophthalm.* I. xxv. 987 Holmgren's wool test, despite the strictures which have been passed upon it, is of great service. **1964** —— *Parsons' Dis. Eye* (ed. 14) xxiv. 364 Holmgren's Wools. These consist of a selection of skeins of coloured wools from which the candidate is required to make a series of colour-matches.

Holmgren[2] (hǭu·mgrĕn, hǭu·lmgrĕn). The name of E. A. *Holmgren* (1866–1922), Swedish biologist, used *attrib.* to designate a system of canals discovered by him in the cytoplasm of some cells.

1921 *Anatomical Rec.* XXII. 78 Further work upon the Holmgren canals is required to clearly demonstrate the developmental stages. **1936** *Nature* 30 May 915/2 The relation of the vacuome system to the Golgi network or the Holmgren canals. **1952** G. H. Bourne *Cytol. & Cell Physiol.* (ed. 2) vi. 250 Evidence obtained from ultracentrifugation of cells..shows that the Holmgren canals become stratified in a different position from the Golgi apparatus.

holmia (hǭu·lmiă). *Chem.* [mod.L., f. next after *erbia*, the oxide of erbium, *ceria*, the oxide of cerium, etc.; the oxide of holmium was not specifically named by either Cleve or

Soret in their papers of 1879.] The sesquioxide of holmium, Ho_2O_3, a pale yellow basic compound.

The substance obtained by Cleve was later shown (by de Boisbaudran, 1886) to be a mixture of the oxides of holmium and of a new element, dysprosium.

1880 *Jrnl. Chem. Soc.* XXXVIII. 7 The author [*sc.* J. L. Soret] considers that the new earth, holmia, discovered by Clève, is identical with an earth..to which Delafontaine gave the name philippia. **1886** [see *FRACTIONAL *a.* b]. **1924** J. W. Mellor *Inorg. & Theoret. Chem.* V. xxxviii. 697 The fractional crystallization of the ethyl sulphates enables holmia to be separated from erbia, thulia, etc., and to be partially separated from dysprosia. **1968** C. A. Hampel *Encycl. Chem. Elem.* 268/2 Holmia was not isolated into a reasonably pure compound until 1911.

holmium (hōu·lmiəm). *Chem.* [mod.L. (P.T. Cleve 1879, in *Compt. Rend.* LXXXIX. 480), f. *Holmia* Stockholm (see quot. 1879) + -IUM.] A silvery, relatively soft, metallic element of the lanthanide series which is present in monazite, gadolinite, and other rare-earth minerals and forms a series of strongly paramagnetic salts, mostly of a brown or yellow colour, in which it is trivalent. Atomic number 67; symbol Ho.

1879 P. T. Cleve in *Chem. News* 12 Sept. 126/2, I propose for this metal the name of Holmium, Ho, derived from the latinized name of Stockholm, in the neighbourhood of which so many minerals rich in yttria are to be found. **1886** [see *DYSPROSIUM]. **1893** *Jrnl. Chem. Soc.* LXIV. II. 467 The holmium oxides were obtained from strongly basic yttrium earths containing a large amount of yttrium oxide. **1924** J. W. Mellor *Inorg. & Theoret. Chem.* V. xxxviii. 696 The four rare earth elements—dysprosium, Dy; holmium, Ho; erbium, Er; and thulium, Tm—are considered as a family or sub-group of the yttrium elements. **1940** *Nature* 20 Apr. 633/1 The investigation of holmium with the mass-spectrograph has shown that the element is composed of one type of atom only, with a mass of 165. **1967** [see *DYSPROSIUM].

holmquistite (hōu·mkwistəit, -kvist-). *Min.* [ad. G. *holmquistit* (A. Osann 1913, in *Sitzungsber. d. Heidelberger Akad. d. Wiss.* IVA. XXIII. 11), f. the name of P. J. *Holmquist*, Swedish mineralogist: see -ITE[1].] A rare basic alumino-silicate of lithium, magnesium, and aluminium that is an orthorhombic member of the amphibole group and typically occurs in light blue to dark violet masses.

1914 *Chem. Abstr.* VIII. 1556 The formation of the holmquistite was no doubt a pneumatolytic process, connected with the intrusion of the well known Li-pegmatites of the region. **1930** *Amer. Mineralogist* XV. 292 Holmquistite from the North Carolina locality is composed largely of the lithia member of the amphibole group and a minor amount of a deficient silica member commonly found in basic rocks. **1966** W. A. Deer et al. *Introd. Rock-Forming Min.* II. 159 Holmquistite typically occurs at the contact of lithium-rich pegmatites with country rocks.

holo-. Add: **holoaxial** (hǫlo,æ·ksiăl) *a. Cryst.*, having or exhibiting all the axes of symmetry compatible with one another but no plane or centre of symmetry; also, more widely, having one or more axes of symmetry but no plane or centre of symmetry; **holobasi·dium** *Bot.*, = *AUTOBASIDIUM; **holobe·nthic** *a. Biol.* [Gr. βένθος depth of the sea], living in the depths of the sea at all stages of the life cycle; **holoca·rpic** *a. Bot.* [Gr. καρπός fruit, seed], designating or (of a fungus) possessing a thallus the whole of which becomes transformed into a reproductive structure at maturity; **holocephalan** (-se·fălăn), **-cepha·lian** *sbs. Ichthyol.*, a fish of the sub-class Holocephali; *adjs.*, = *holocephalous* adj. (s.v. Holo-); **ho·locrine** (-krəin), *a. Physiol.* [ad. F. *olocrine* (L. Ranvier 1887, in *Jrnl. de Micrographie* XI. 9), f. Gr. κρίν-ειν to separate], of, pertaining to, or designating a gland in which the secretion is produced by the complete disintegration of its cells; **holo-enzyme** (hǫ·lo,enzəim) *Biochem.* [a. F. *holoenzyme* (*Compt. Rend. XIIme Conf. Union Internat. de Chim.* 43)], the active form of an enzyme, consisting of the apoenzyme combined with its co-enzyme; **holomi·ctic** *a.* [ad. G. *holomiktisch* (I. Findenegg 1935, in *Internat. Rev. d. ges. Hydrobiol.* XXXII. 377), f. Gr. μικτός mixed], applied to a lake in which the full depth of water takes part in the circulation; **holomo·rphosis** *Biol.*, the perfect regeneration of a lost member or part; **holono·mic** *a. Mech.* [ad. G. *holonom* (H. Hertz *Ges. Werke* (1894) III. I. 91), f. Gr. νόμ-ος law], applied to a constrained system in which the equations defining the constraints are integrable or already free of differentials, so that each equation effectively reduces the number of coordinates by one; also applied to the constraints themselves; so **holo·nomous** *a.*, in same sense; **holopa·rasite** *Biol.* [ad. G. *holoparasit* (F. Johow 1890, in *Verhandl. Deutsch. wissensch. Ver. Santiago* II. II. 67)], an obligate parasite, unable to exist except in association with its host; so **ho:loparasi·tic** *a.*; **holopa·rasitism**, the condition of being a holoparasite; **holophytic** *a.*, substitute for def.: of, pertaining to, or designating a plant that is able to transform inorganic substances into food by photosynthesis, and so is neither parasitic nor saprophytic; (add earlier and later examples); **holopla·nkton** *Biol.* [back-formation from the adj.], a collective term for aquatic organisms that are holoplanktonic; **ho:loplankto·nic** *a.* [ad. G. *holoplanktonisch* (E. Haeckel *Plankton-Studien* (1890) iii. 25)], passing all stages of the life-cycle drifting or swimming weakly in the water; **holopneu·stic** *a. Ent.* [ad. G. *holopneustisch* (J. A. Palmén *Morphol. des Tracheensystems* (1877) vii. 78), f. Gr. πνευστικ-ός for breathing (πνεῖν to breathe)], having ten pairs of spiracles (in some cases eleven), all of which are functional; **holosa·prophyte** *Bot.* [ad. G. *holosaprophyt* (F. Johow 1889, in *Jahrb. f. wissensch. Bot.* XX. 479)], an obligate saprophyte; so **ho:losaprophy·tic** *a.*

1902 H. A. Miers *Mineral.* I. i. 45 When an axis of *n*-fold symmetry is perpendicular to *n* digonal axes and there is no other element of symmetry, the crystal may be called 'holoaxial', since it possesses all the symmetry-axes compatible with each other, and only axes. **1903** H. Hilton *Math. Crystallogr.* v. 52 In this chapter we shall investigate those finite groups of the first sort—also called holoaxial groups—which contain only 2-al, 3-al, 4-al, and 6-al rotation-axes. **1961** Terpstra & Codd *Crystallometry* iv. 129 Crystals are divided according to their true symmetry into 32 crystal classes. Of these 32 classes, 11 have symmetry elements consisting exclusively of symmetry axes: these are called the eleven holoaxial classes. [**1900** B. D. Jackson *Gloss. Bot. Terms* 124/2 *Holobasid*, an undivided basidium in Basidiomycetes (Van Tieghem).] **1928** Holobasidium [see *AUTOBASIDIUM]. **1970** J. Webster *Introd. Fungi* 279 In the toadstools and their allies the basidium is a single cylindrical cell, undivided by septa, typically bearing four basidiospores at its apex... Such basidia are termed holobasidia. **1902** *Encycl. Brit.* XXXIII. 935/2 Another hindrance to the extension of many deep-sea species is that they are holobenthic. **1916** B. D. Jackson *Gloss. Bot. Terms* (ed. 3) 181/2 Holocarpic. **1928** C. W. Dodge tr. *Gäumann's Compar. Morphol. Fungi* iii. 12 In the holocarpic forms, gametangial copulation naturally leads to the fusion of whole individuals. **1930** H. M. Fitzpatrick *Lower Fungi* ii. 24 In some of the lower families [of Phycomycetes] the entire thallus is transformed at maturity into a single reproductive organ (Olpidiaceae) or group (sorus) of them (Synchytriaceae). In such cases the organism is said to be holocarpic. **1970** J. Webster *Introd. Fungi* 62 In the Lagenidiales..the thallus is holocarpic. **1934** Webster, Holocephalan *adj.* & *n.* **1942** L. H. Hyman *Compar. Vertebr. Anat.* (ed. 2) iv. 40 The chimaeras or holocephalans are peculiar-looking fish. **1934** Webster, Holocephalian *adj.* & *n.* **1965** *Gen. & Compar. Endocrinol.* V. 434/2 The ratfish or chimera is a holocephalian. **1970** *Nature* 11 July 187/2 A holocephalian elasmobranch fish, *Hycholagus colliei.* **1905** Gould *Dict. New Med. Terms* 296/1 *Holocrine*, applied to a gland the cell of which, after having elaborated the material of secretion, falls into disuse and disappears. **1928** E. V. Cowdry *Special Cytol.* I. ii. 36 In the sebaceous glands the secretory products are elaborated by the fatty metamorphosis, destruction and discharge of the cells themselves. These are the 'holocrine' glands of Ranvier. **1939** V. B. Wigglesworth *Princ. Insect Physiol.* xi. 264 In Orthoptera, secretion is merocrine during continuous small meals, holocrine when a meal follows a period of fasting. **1949** *Gray's Anat.* (ed. 30) 1254 As the sebaceous glands produce their secretion by complete fatty degeneration of their central cells they are classed as holocrine glands. **1961** E. H. Mercer *Keratin & Keratinization* ii. 59 Some cutaneous holocrine glands of reptiles..are sac-like invaginations of the epidermis producing fatty materials. **1943** Sumner & Somers *Chem. & Methods Enzymes* i. 32 A few examples of coenzymes and holoenzymes are given in Table 11. **1950** Holoenzyme [see *CO-ENZYME]. **1971** *Nature* 15 Oct. 478/2 The rate of production of active holoenzyme from apoenzyme is enhanced by tryptophan about ten-fold. **1937** *Trans. Connecticut Acad. Arts & Sci.* XXXIII. 74 A normal (holomictic) thermally stratified lake consists of an epilimnion and a hypolimnion. **1957** G. E. Hutchinson *Treat. Limnol.* I. viii. 537 Most of the lakes discussed herein are holomictic; that is to say, when they circulate, the circulation is complete to the bottom. **1901** T. H. Morgan *Regeneration* i. 24 Under this heading [*sc.* homomorphosis] we may distinguish two cases, in one of which the entire lost part is at once, or later, replaced—holomorphosis. **1904** E. T. Whittaker *Treat. Analyt. Dynamics* ii. 33 Holonomic systems are therefore characterised by the fact that the number of degrees of freedom is equal to the number of independent coordinates required to specify the configuration of the system. **1954** R. A. Becker *Introd. Theor. Mech.* xiii. 318 Simple examples of holonomic constraints involving a single particle are those where the motion is confined to a single curve or surface. **1899** D. E. Jones & Walley tr. *Hertz's Princ. Mech.* iv. 80 A material system between whose possible positions all conceivable continuous motions are also possible motions is called a holonomous system. **1911** A. & J. G. Gray *Treat. Dynamics* x. 555 Systems are now called holonomous or not holonomous, according as the constraints are or are not defined by finite equations. **1891** *Jrnl. R. Microsc. Soc.* 70 Each of these classes, except the last, may be again divided into Holoparasites and Hemiparasites. **1903** W. R. Fisher tr. *Schimper's Plant-Geogr.* 203 Holoparasites, which live entirely at the cost of the organic substance of their host, like holosaprophytes are devoid of chlorophyll. **1965** Bell & Coombe tr. *Strasburger's Textbk. Bot.* I. iv. 197 While the semi-parasites can often at first glance hardly be distinguished from their green, wholly autotrophic relatives, the total or holo-parasites display a complete or almost complete loss of chlorophyll. **1902** *Encycl. Brit.* XXV. 439/2 Cytineæ, Balanophoreæ, Orobanchaceæ, Lennoaceæ, are families..which are characteristically holoparasitic. **1927** W. McDougall *Plant Ecol.* ix. 125 The family Scrophulariaceæ contains representatives of all gradations from complete independence to holoparasitism. **1885** Holophytic [see Holozoic *a.*]. **1900** *Ann. Bot.* XIV. 669 Thus either a saprophytic or holophytic nutrition can be maintained. **1964** Priestley & Scott *Introd. Bot.* (ed. 4) xxxvii. 593 It is usual for flowering plants to be autotrophic (holophytic). **1909** E. Warming *Oecol. Plants* xxxviii. 161 These terms 'neritic' and 'pelagic' or 'oceanic' plankton approximately correspond to Haeckel's 'meroplankton' and 'holoplankton' respectively. **1942** H. U. Sverdrup et al. *Oceans* xvii. 816 The holoplankton is composed of forms representing nearly every phylum of the animal kingdom. **1955** C. C. Davis *Marine & Fresh-water Plankton* i. 29 The life history is completed without the animals ever leaving their planktonic life. Animals of this type are classified as the holoplankton. **1893** G. W. Field tr. *Haeckel's Planktonic Stud.* in *Rep. U.S. Comm. Fisheries 1889–91* 583 Numerous organisms pass their whole life..hovering in the ocean, while with others this is not the case. The first group we call holoplanktonic. **1963** J. E. G. Raymont *Plankton & Productivity in Oceans* xiv. 371 The holoplanktonic members are also subject to seasonal breeding. **1892** J. A. Thomson *Outl. Zool.* xiii. 266 In adult aërial life, the tracheæ of the body acquire stigmata, and the insect becomes 'holopneustic'. **1947** *Trans. R. Ent. Soc.* XCVIII. 459 Aquatic holometabolous larvae which are holopneustic do not appear to exist. **1960** Richards & Davies *Imms's Textbk. Ent.* (ed. 9) I. 134 The Holopneustic Respiratory System.—This is the most primitive arrangement found in living insects, 10 pairs of functional spiracles being present. **1890** *Jrnl. R. Microsc. Soc.* 205 Herr F. Johow describes the peculiarities of structure of the 'holosaprophytes', or saprophytes destitute of chlorophyll. **1902** *Encycl. Brit.* XXV. 439/1 Angiospermous holosaprophytes are not common. **1960** W. B. Crow *Synopsis Biol.* lxxxiv. 518 The completely saprophytic genera (holosaprophytes) in Britain are the orchids *Neottia* and *Corallorhiza* and the similar *Monotropa* which is allied to the heath family. **1895** *Ann. Bot.* IX. 327 A number of holosaprophytic forms found in the tropics..constitute the Burmanniaceous genus *Thismia*.

holocaine (hǫ·lokēⁱn). *Pharm.* Also † -cain. [a. G. *holocaïn*, after *cocaïn* (now *kokain*) Cocaine.] The proprietary name of a synthetic derivative, $C_{18}H_{22}N_2O_2$, of *p*-phenetidine which resembles cocaine in its action and is used (in the form of the hydrochloride) as a surface anæsthetic for the eye; phenacaine.

1897 *Lancet* 29 May 1466/1 A new alkaloid called holocaine..described as an effective anæsthetic for ophthalmic purposes. **1899** *Amer. Jrnl. Med. Sci.* CXVII. 121 The bactericidal power of holocain gives it the advantage that solutions..will keep themselves sterile. **1910** L. W. Fox *Pract. Treat. Ophthalm.* x. 287 The pain caused by this injection is very severe, and to mitigate this the eye should be thoroughly anaesthetized with holocain, cocain being insufficient. **1955** J. Adriani *Selection of Anesthesia* x. 98 Holocaine is prepared from para-ethoxy aniline (phenetidin). **1968** J. H. Burn *Lect. Notes Pharmacol.* (ed. 9) 62 Phenacaine (holocaine) is prompt in action and is used especially for the eye.

holocaust, *sb.* **2. c.** Freq. applied to the mass murder of the Jews by the Nazis in the war of 1939–1945. Also *attrib.*

1965 A. Donat (*title*) The holocaust kingdom. **1967** N. Cohn *Warrant for Genocide* ix. 208 By the autumn of 1944 the holocaust was nearing its conclusion. **1968** *Manch. Guardian Weekly* 25 Apr. 10/4 There is now within modern history a compartment of 'holocaust studies'—dealing with the wholesale destruction by the Nazis of European Jewry. **1972** F. Forsyth *Odessa File* 306 The mausoleum of Yad Vashem,..the shrine to six million of his fellow Jews who died in the holocaust.

holocellulose (hǫlose·liulōus, -z). *Chem.* [f. Holo- + Cellulose *sb.*] A polysaccharide fraction obtained from plant material, esp. wood, by the removal of lignin and various extractives, and principally consisting of cellulose and hemicellulose.

1933 Ritter & Kurth in *Industr. & Engin. Chem.* XXV. 1250 The authors propose the word 'holocellulose' as preferable to *Skelettsubstanzen*, maintaining that the latter term does not describe the material correctly either from the physical or chemical point of view. **1946** *Nature* 14 Dec. 855/1 Holocellulose, now known to be a very important constituent of chemical pulps so far as their behaviour on beating is concerned. **1969** R. L. Janes in Macdonald & Franklin *Pulping of Wood* (ed. 2) I. ii. 52/2 In practice..all the holocellulose isolation procedures ..leave behind traces of degraded lignin and/or adsorbed solvents.

Holocene (hǫ·losīn), *a. Geol.* Also **holo-.** [ad. F. *holocène*, f. Holo- + Gr. καιν-ός new, recent, after *Eocene*, etc.] Of, pertaining to, or designating the most recent geological epoch, which began approximately 10,000 years ago and still continues and which together with the Pleistocene epoch makes up the Quaternary period; also *absol.*

[**1867** P. Gervais *Recherches sur l'Ancienneté de l'Homme et la Période quaternaire* ii. 32 Les dépôts récents que j'avais proposé, il y a plusieurs années, de nommer *holocènes*.] **1897** *Q. Jrnl. Geol. Soc.* LIII. 434 Mollusca from the Holocene deposits of the Kennet Valley at Newbury. **1927** Peake & Fleure *Apes & Men* 14 The Holocene is more often called the Recent Period. **1935** *Nature* 7 Sept. 353/1 The pauses in the contraction of the lake must be due to increased rainfall, but nothing like a 'pluvial period' is admitted in holocene times. **1963** R. Carrington *Million Years of Man* xiii. 164 In this Holocene epoch..the hunters and food-gatherers of Upper Palaeolithic and Mesolithic times adopted an entirely new mode of existence. **1971** *Nature* 24 Sept. 281/1 A considerable isostatic rise in the land must have occurred in the late Pleistocene with continuation in the Holocene.

holochoanite (hǫ:lokōu·ănəit). *Palæont.* Also **Holo-.** [f. mod.L. *Holochoanites*, altered form of *Holochoanoida* (see next): see -ITE[1] 2 a.] A nautiloid cephalopod in which the septal necks extend from the septum in which they originate as far as the next septum towards the apex; also, a member of the obsolete sub-order Holochoanites, of which such necks were characteristic.

1898 A. Hyatt in *Proc. Amer. Assoc. Advancem. Sci.* XLVII. 364 The suborders..are as follows: I. Holochoanites with the funnels of the siphuncle reaching entirely across each air chamber completely shutting off the interior of each chamber from the interior of the siphuncle. **1929** *Bull. Geol. Soc. China* VIII. 119 Stages in development of the primitive Holochoanite. **1944** E. O. Ulrich et al. *Ozarkian & Canadian Cephalopods* III. 28 Most of the Early Paleozoic holochoanites that have long straight conchs can be referred to the Endoceratidae, but a few..are being placed in the Suecoceratidae. **1952** R. C. Moore et al. *Invertebr. Fossils* ix. 342/1 Thin-section studies indicate that the elongate septal necks which are supposed to distinguish the Holochoanites occur only in a minority of them.

Hence **ho:lochoani·tic** *a.*, (having septal necks) characteristic of a holochoanite; (by some writers used to include *MACROCHOANITIC *a.*).

1905 *Bull. N.Y. State Mus.* LXXX. 339 *Piloceras newtonwinchelli* Clarke is by the structure of its ectosiphuncle not a holochoanitic form..but an orthochoanitic form. **1950** *Jrnl. Paleontol.* XXIV. 604/1 Only a small part of Hyatt's Holochoanites is actually holochoanitic. **1964** C. Teichert in R. C. Moore *Treat. Invertebr. Paleont.* K. 164/2 Endoceratidae with holochoanitic, and even macrochoanitic, septal necks and complete endocone systems also appeared in the late Early Ordovician.

holochoanoidal (hǫ:lokōu,ănoi·dăl), *a. Palæont.* [f. mod.L. *Holochoanoid-a* (A. Hyatt 1883, in *Proc. Boston Soc. Nat. Hist.* XXII. 260), f. Holo- + Gr. χοάν-η funnel + *-oida* (see -OID): see -AL.] Originally, characteristic of the group Holochoanoida (now obsolete) of nautiloid cephalopods; hence, = *HOLOCHOANITIC *a.*

1883 *Proc. Boston Soc. Nat. Hist.* XXII. 267 The first three genera appear to have holochoanoidal siphons. **1933** *Biol. Rev.* VIII. 443 There is no essential difference in the structure of the siphuncle, holochoanoidal (Foerste) or orthochoanoidal (Troedsson).

hologamy (holǫ·gămi). *Biol.* [ad. F. *hologamie* (P. A. Dangeard 1900, in *Botaniste* VII. v. 265), f. Holo- + *-GAMY.] **a.** A mode of reproduction found in certain protozoa and algæ, in which copulation consists in the fusion of whole organisms morphologically similar to the vegetative form.

1925 E. B. Wilson *Cell* (ed. 3) vii. 582 Hologamy or Macrogamy... This condition is seen in various flagellates, rhizopods, ciliates, diatoms, desmids and the Conjugatæ generally. **1940** L. H. Hyman *Invertebrates* I. iii. 99 Sexual processes are absent except in *Scytomonas* where hologamy has been observed. **1965** V. A. Dogiel *Gen. Protozool.* (ed. 2) vii. 308 Hologamy is characterized by an almost complete absence of the preparatory stage of gametocytes.

b. A mode of reproduction found in certain fungi, in which the entire thallus becomes a gametangium and fusion of two mature individuals occurs.

1928 C. W. Dodge tr. *Gäumann's Compar. Morphol. Fungi* iv. 12 This special case of gametangial copulation.. is called hologamy. **1965** J. Wilkinson tr. *Langeron's Outl. Mycol.* (ed. 2) ix. 378 Hologamy is gametangial copulation of holocarps.

Hence **ho·logamete**, a gamete that is morphologically similar to the vegetative cell and is not specially formed by fission; **hologa·mic**, **holo·gamous** *adjs.*

1925 E. B. Wilson *Cell* (ed. 3) vii. 583 Even when hologamous gametes are alike or closely similar in appearance they often display definite physiological differences that become evident at the time of conjugation. *Ibid.* 589 Gametes of the hologamic type. **1926** G. N. Calkins *Biol. Protozoa* x. 498 In the case of *Scytomonas* (*Copromonas*) *subtilis*..the evidence appears to be fairly convincing that copulation of hologametes actually does occur. **1932** Borradaile & Potts *Invertebrata* ii. 27 *Volvox*..and related forms have an anisogamy in which the female gamete is a hologamete. **1965** V. A. Dogiel *Gen. Protozool.* (ed. 2) vii. 309 Hologamous copulation takes place also in some Phytomonadina belonging to the genus *Chlamydomonas*.

hologenesis (hǫlo,dʒe·nĕsis). [ad. It. *ologenesi*: see Holo- and -GENESIS.] The name of a theory of evolution first propounded by D. Rosa (in *Ologenesi* (1918)), and later adopted by G. Montandon (in *L'Ologenèse humaine* (1928)) to account for the origin of human races.

[**1929** *Nature* 11 May 709/1 The theory of evolution which he [*i.e.* G. Montandon] applies is that formulated by Prof. Rosa of Modèna in 1918 and named by its originator 'ologenesi'.] **1931** *Ibid.* 12 Sept. 430/1 Having reached maturation, the germ plasm of the species undergoes a sudden change whereby the mother species suddenly gives rise to two daughter species... The daughter species then set out to unfold their determinants, and if the environment is favourable—for environment and selection play their part in the theory of hologenesis—then maturation stages are again reached and a further dichotomy with the production of a new species takes place. **1959** R. Altevogt tr. *Rensch's Evol. above Species Level* iv. 58 Rosa advanced a theory of 'hologenesis' according to which dichotomic branchings necessarily proceed in predetermined directions. **1959** B. Wall tr. *Teilhard de Chardin's Phenomenon of Man* III. i. 187 Man, according to these authorities, must have started simultaneously in several regions on the 'anthropoid layer' of the Pliocene era... The idea involves 'hologenesis' and therefore polycentricity.

Hence **hologene·tic** *a.*

1936 *Nature* 27 June 1055/1 An exposition of the hologenetic point of view in the origin and distribution of races. **1948** A. L. Kroeber *Anthropol.* (ed. 2) iv. 170 The hologenetic theory makes a blanket assumption in advance instead of trying genuinely to investigate each case and then seeing whether there is a common principle in them all.

hologram (hǫ·logræm). *Physics.* [f. Holo- + -GRAM.] A pattern produced when light (or other radiation) reflected, diffracted, or transmitted by an object placed in a coherent beam is allowed to interfere with an undiffracted background or reference beam related in phase to the first (or identical with it); a photographic plate or film containing such a pattern.

When suitably illuminated a photographic hologram causes a two- or three-dimensional image of the original (two- or three-dimensional) object to form in space.

1949 D. Gabor in *Proc. R. Soc.* A. CXCVII. 456 The name 'hologram' is not unjustified, as the photograph contains the total information required for reconstructing the object, which can be two-dimensional or three-dimensional. **1952** *Sci. News* XXVI. 40 In the taking of the hologram, the amplitude of the wave scattered by the object is much smaller than that of the primary wave coming from the source. **1956** *Nature* 31 Mar. 613/2 The resultant hologram..was put on to the optical bench. **1964** *Electronics Weekly* 16 Dec. 6/1 A recent demonstration..showed a toy train in accurate three-dimensional representation, using a hologram illuminated from behind by a gas laser. **1966** *Observer* 15 May 13 Every bit of a hologram contains information about the whole scene. So you can snip it into pieces, shine a laser at one of the pieces, and you will see the original scene, only somewhat fuzzier. **1967** *Applied Physics Lett.* XI. 294/2 An investigation here has shown the feasibility of using a deformable film on a solid substrate to record an ultrasonic hologram in a way similar to that in which photographic film records an optical hologram. **1968** *Ultrasonics* VI. 81/1 There may be several intermediary stages before the ultrasound hologram is in the form of a photographic transparency which can be used in the reconstruction process. **1968** *Times* 12 Dec. 15/3 Scientists at the Bell Telephone Laboratories..claim that as many as 1,000 different holograms..can be stored in a crystal of lithium niobate. **1971** R. J. Collier et al. *Optical Holography* x. 298 Holograms can be recorded in a thermoplastic film by causing its surface to deform in accordance with the light intensity variations of holographic interference patterns.

holograph (hǫ·lograf), *v.* [Back-formation from *HOLOGRAPHY, after *photograph*, *telegraph* vbs.] *trans.* To record as a hologram, to make a holographic record of.

1968 *Ultrasonics* VI. 87/1 Although Greguss has demonstrated that ultrasound holograms can be visualized by this technique, he has not published results of any attempts to reconstruct images of the original objects holographed. **1970** *Sci. Jrnl.* Aug. 17/1 The light from a pulsed laser is split into two beams, one going direct to the metal film and the other going first to the object to be holographed and then to the metal film. **1970** *Physics Bull.* Nov. 493/2 In any sort of holography..the aim is to obtain an image of the object being 'holographed'.

holographic (hǫlogræ·fik), *a.* [f. *HOLOGRAPH(Y + -IC.] **1.** (In Dict. s.v. Holograph *a.* and *sb.*)

2. *Physics.* Of or pertaining to holography; produced by, involving, or used in holography.

1964 *Physics Lett.* XIII. 306 High resolutions did not heretofore appear attainable in comparable image-forming X-ray microscopy using..holographic wavefront reconstruction methods. **1966** *Nature* 5 Mar. 1015/1 Fig. 1 shows the experimental arrangement used to produce holographic recording[s] of a piece of steel channel-section girder, and to demonstrate interference effects on distorting it. **1967** *Proc. Inst. Electr. Engin.* LV. 570/1 The extension of holographic techniques to radio frequencies offers new possibilities. **1967** DeVelis & Reynolds *Theory & Applic. Holography* i. 6 Holographic interferometry shows promise in the areas of vibration and stress analysis and turbulence studies. **1971** R. J. Collier et al. *Optical Holography* x. 294 We describe here a method of film preparation which can be easily followed in a photographic or holographic laboratory.

Hence **hologra·phically** *adv.*, by means of holography; in a holographic manner.

1966 G. W. Stroke *Introd. Coherent Optics & Holography* v. 79 (*heading*) Optical filtering with holographically matched spatial filters. **1968** *Nature* 2 Nov. 474/2 Memory might behave holographically. **1968** *Physics Bull.* Dec. 423/1 An improved interferometer for comparing the shape of a diesel fuel injector cylinder against a holographically recorded master cylinder.

holography (hǫlǫ·grăfi). [f. Holo- + -GRAPHY.] **1.** (In Dict. s.v. Holograph *a.* and *sb.*)

2. *Physics.* [f. after *photography*, *telegraphy*, etc., on the basis of *HOLOGRAM.] The process or science of producing and using holograms.

1964 *Physics Lett.* XIII. 308 A well-resolved, magnified 'image' of the scattering object has been reproduced.. without losing the general simplicity and speed which are characteristic of holography. **1965** *New Scientist* 19 Aug. 431 The technique, known as holography, relies on the fact that an optical description of an object can be stored as a diffraction pattern instead of as a photograph. **1967** *Contemporary Physics* VIII. 153 Leith and Upatnieks (1963, 1964) were the first to demonstrate the very striking three-dimensional imaging that could be obtained by holography. **1967** *Sunday Times* 12 Feb. 8/8 Holography is currently a fashionable topic in the scientific world. **1967** *Applied Physics Lett.* XI. 294/1 The advantage of using acoustic holography to visualize, in three-dimensional fashion, objects in optically opaque material has been intriguing to many investigators. **1969** H. M. Smith *Princ. Holography* i. 6 Further work on x-ray holography is still awaiting a small, monochromatic source of x-rays. **1971** *Sci. Amer.* Dec. 39 In the past 10 years holography has been used to study how objects change shape under strain, to record high-speed events in gas dynamics and to store information with high density.

holoku (holōu·ku). [Hawaiian.] A long gown with a train as worn in Hawaii.

1891 R. L. Stevenson *Island Nights' Entertainments* (1893) 177 She..stood by the track-side in her red holoku. *Ibid.* 207 Kokua concealed the bottle under her holoku. **1923** C. Cameron *Two Yrs. in Southern Seas* i. 20 Her gown was a 'holokus', the native robe of a long 'princess' style, loose and flowing with train of bright yellow. **1954** J. Sheridan in J. D. MacDonald *Lethal Sex* (1962) 155 A stately Hawaiian woman in a flowered *holoku*. **1960** *Guardian* 3 Nov. 10/4 Long-skirted dresses, called muumuus and holokus, that were adaptations of respectable female attire in Massachusetts. **1967** M. Davis *Strange Corner* (1968) xv. 111 He..dropped his eyes to the skin-tight white holoku and whistled.

holomorphic, *a.* Add: Hence **holomo·rphically** *adv.*, in such a way as to be or remain holomorphic (in sense 2); **holomorphy** (examples in *Math.*).

1957 *Pacific Jrnl. Math.* VII. 812 There exist domains ..such that all *G*-holomorphic functions can be continued *G*-holomorphically into a larger domain. *Ibid.* 820 If *D* is a domain of holomorphy, then the set *C*.. belongs to *D*. **1963** Standring & Shutrick tr. *Cartan's Elem. Theory of Analytic Functions* ii. 73 For functions of a complex variable, there is an equivalence between holomorphy and analyticity. **1966** *Mathematical Rev.* XXXI. 33/2 A closed holomorphic differential *p*-form in *x*, holomorphically varying with *y*.

Holophane (hǫ·lofēⁱn). Also **holophane.** [f. Holo- + Gr. φαίνειν to shine, appear.] A proprietary name used *attrib.* or as *adj.* to designate a type of lamp-shade that encloses the bulb but is made of glass specially fluted to refract and reflect the light in the required manner with little loss; also applied to the glass itself.

1893 Psaroudaki & Blondel *Brit. Pat.* 19,185, Our invention has for object to replace the globes, shades, reflectors..in use, by others made of transparent or crystal glass giving the..following results:— 1st. They are shining on almost the whole of their visible surface and receive for this cause the name of 'holophane' (entirely shining). *Ibid.*, For making the globe appear 'holophane'. **1911** V. Zingler in L. Weaver *House & Equipment* 116 For general dispersed lighting, a bowl or hemisphere of holophane glass, mounted on the ceiling with the lamp inside, gives a soft and pleasing light. **1927** *Brit. Weekly* 14 Nov. 193/4 The modern electric lighting with soft light and clusters of Holophane globes is a marked improvement. **1956** L. E. Jones *Edwardian Youth* x. 245 An American inventor of a prismatic glass called 'Holophane' was looking for an assistant. **1964** S. Duke-Elder *Parsons' Dis. Eye* (ed. 14) xxxvii. 559 The distribution of light from artificial sources varies greatly. It can be modified by the use of reflectors and prismatic (holophane) globes.

holophrase (hǫ·lofrēⁱz). *Philol.* [f. Holo- + Phrase *sb.*: cf. Holophrasis.] A single word used instead of a phrase, or to express a combination of ideas (e.g. *ungetatable*).

1899 E. J. Payne *Hist. New World* II. 201 This multiplication of elements denoting personality, in combination with more and more elements denoting Things, tends to the dissolution of the holophrase... The holophrase naturally follows the progression of the mind from point to point. **1914** W. R. M. Lamb *Clio Enthroned* 239 We can regard his periodic structures as a reversion..to the primitive holophrase. **1969** *Language* XLV. 325 *All gone* may be two words, or it may be a holophrase.

holophrastic, *a.* Add: Hence (as a back-formation, after *spasm*, *spastic*, etc.) **ho·lophrasm** = *HOLOPHRASE.

1862 D. Wilson *Preh. Man* II. xxv. 436 Holophrasms are common in all its [*sc.* the Algonquin] dialects. **1900** *Amer. Anthropologist* II. 615 A word-sentence may be called a 'holophrasm'.

holothuria (hǫlopiū^ə·riă). *Zool.* Pl. **-iæ, -ias.** = Holothurian *sb.*

1792 M. Riddell *Voy. Madeira* 79 These holothuriae are singularly beautiful when floating on the surface of the water in a clear day. **1816** Tuckey *Narr. Exped. R. Zaire* (1818) i. 11 The holothuria made its first appearance on the 4th instant. **1844** *Chambers's Edin. Jrnl.* 23 Nov. 323/2 It may be of small moment to you, who, mayhap, know nothing of holothurias. **1876** tr. *Beneden's Anim. Parasites* (1883) 5 Dr. Greef..found..a holothuria of a foot in length.

holotype (hǫ·lotəip). *Biol.* [f. Holo- + *TYPE *sb.*[1] 8 e.] A specimen chosen as the basis of the first description of a new species.

1897 C. Schuchert in *Science* 23 Apr. 637/2 A holotype in natural history is a particular individual deliberately selected by the author of a species, or it may be the only example of a species known at the time of original publication. **1946** *Nature* 23 Nov. 762/1 The holotype and only specimen of this was housed in the Bristol Museum. **1964** *Internat. Code Zool. Nomencl.* xvi. 77 If a nominal species is based on a single specimen, that specimen is the 'holotype'. **1966** *Internat. Code Bot. Nomencl.* ii. 18 As long as a holotype is extant, it automatically fixes the application of the name concerned.

holp, holpen, pa. t. and pple. of Help *v.* Delete 'obs. or arch.' and add: Now *U.S. dial.* Also occas. used as pres. t. and infin.

1881 'Mark Twain' *Prince & Pauper* xix. 221 Of a truth I was right—he hath holpen in a kitchen. **1890** *Dialect Notes* I. 68 *Holp*.., for *helped*. 'He holp me out of the scrape.' **1917** *Ibid.* IV. 413 *Holp* v. tr., to help. 'I axed him to *holp* me out.' **1927** *Amer. Speech* II. 357/2 He holpen me over the creek. **1931** *Ibid.* VI. 230 Such old forms of English as..'holp' for 'help', and 'effen' for 'if' may be heard in the ordinary speech of the natives [of Oregon]. **1939** Joyce *Finnegans Wake* 118 So holp me Petault, it is not a miseffectual whyacinthinous riot of blots and blurs. **1940** W. Faulkner *Hamlet* I. ii. 47 'Help him up.' So the nigger holp Ab onto the horse. **1962** E. B. Atwood *Regional Vocab. Texas* vi. 118 Items of nonstandard 'grammer' are usually considerably more frequent..in the less-educated group: for example *clum* (climbed), *throwed*, *holp(ed)*.

hols (hǫlz), *sb. pl.* Colloq. (esp. school-children's) abbrev. of *holidays* (Holiday *sb.* 2 b).

1905 H. A. Vachell *Hill* vi. 137 The governor pointed that out last hols. **1921** S. Thompson *Rough Crossing* ii. 105, I may be staying in Oxford in the spring hols. **1931** *Church Times* 25 Sept. 344/4 After next 'hols' it will be a very different little boy who will take the train at Waterloo or Victoria. **1958** *Spectator* 30 May 677/1 The House broke up for the hols.

Holstein (hǫ·lstəin, -stīn). [Name of a region in N.W. Germany.] A breed of black-and-white dairy cattle, orig. raised in Friesland. Also *attrib.*

1865 *Rep. Comm. Agric. 1864* (U.S. Dept. Agric.) 161 Holstein cattle..[have] not received that appreciation in this country to which they are entitled. **1872** *Rep. Vermont Board Agric.* I. 176 The Dutch cattle, or as I believe it is settled they are to be called, the 'Holsteins'. **1876** *Trans. Ill. Dept. Agric.* XIV. 296 S. W. Kingsley spoke in favor of Holsteins for the dairy. **1971** *Farmers Weekly* 19 Mar. 77/3 The Dutch paraded six top-grade show cows, which were not for sale, while Canada exhibited 16 Holsteins from French herds.

holster (hōu·lstəɹ), *v.* Chiefly *U.S.* [f. the sb.] *trans.* To put (a gun) into its holster. Hence **ho·lstered** *ppl. a.*

1930 *Argosy* 12 July 690/2 Both men had holstered rifles on their saddles. **1956** 'E. McBain' *Cop Hater* (1958) iv. 36 'We won't need these,' Bush said. He holstered his gun. **1972** B. F. Conners *Don't embarrass Bureau* (1973) II. 113 He holstered his weapon. **1973** R. Hayes *Hungarian Game* xvii. 108 The guard snatched at his holstered gun.

holt[2]. Add: **1.** Also *U.S. colloq.* Cf. *a-holt* (s.v. *A-HOLD *adv. phr.*).

1825 J. Neal *Bro. Jonathan* II. 60 [He cried] 'lay holt there; lay holt, every one o' you', throwing the reins behind him, into the carriage. **1848** Bartlett *Dict. Amer.*, *Holt*, for hold. Ex. 'Death has got holt of him.' **1898** E. N. Westcott *David Harum* xxii. 199 Of course you've heard the things that some folk say of him, an'.. they got some holt on your mind. **1909** R. A. Wason *Happy Hawkins* iv. 52 He'd 'a' been killed that trip if you hadn't taken holt when you did. **1930** *Amer. Speech* V. 151 *Catch holt of*, grab. 'Catch holt of my hand, quickly.' **1940** W. Faulkner *Hamlet* iv. i. 313 'Grab a holt,' the Texan said. Eck grasped the wire also.

holy, *a.* Add: **3. a.** *the holy souls*, the souls of the faithful departed, the blessed dead.

971 *Blickling Homilies* (1880) vi. 67 Drihten.. helle bereafode, & þa halgan sauwla þonon alædde, & hie generede of deofles anwalde. **1357** John de Thoresby *Lay Folks Catechism* (1901) 2 This maner of knawyng ..shuld we have had..Noght so mikell als hali saules has now in heven, Bot mikel mare than man has now in erthe. **1720** G. Stanhope tr. *St. Augustine's Pious Breathings* (ed. 5) I. xxiii. 48 (*heading*) The Happiness of Holy Souls at their departure out of this World. **1849** F. W. Faber *Jesus & Mary* 92 Pray for the Holy Souls that burn This hour amid the cleansing flames. **1898** A. G. Mortimer *Cath. Faith & Practice* II. xiv. 361 The Intermediate State, where the holy souls are waiting until their purification is accomplished. *Ibid.* 371 The joys and consolations of the holy souls in their preparation for Heaven. **1958** G. Montague *Probl. Liturgy* v. 305 It is clear..that the Holy Souls could not be properly named as the titular of a church. The Souls in Purgatory are not an object of the public veneration of the church.

b. **the Holy Name,** the name of Jesus as an object of formal devotion among Catholics, as in the Litany of the Holy Name and the festival of the Holy Name of Jesus.

See also *M.E.D.* s.v. *holi* adj.[2] 2 a.

c **1440** *Thornton MS.* f. 192 Of the vertuz of the holy name of Ihesu. **1526** [see Name *sb.* 3]. **1720** T.M. tr. *Horstius' Paradise of Soul* (1771) 409 Great are the Honours and Priviledges of the Holy Name, Jesus. **1851** J. B. Pagani *Life A. Gentili* III. xi. 196 Singing along the way the Litany of the Holy Name of Jesus. **1884** Addis & Arnold *Cath. Dict.* s.v. *Jesus*, An office of the Holy Name. **1901** G. Tyrrell *Autobiogr.* (1912) I. vii. 94 Two of the boys..would bow their heads at the Holy Name at morning prayers. **1968** R. Woolf *Eng. Relig. Lyric* v. 172 Both 'Luf es lyf' and 'My sange es in syhtyng' include verses expressing devotion to the Holy Name. *Ibid.* 173 In medieval spirituality..the devotion to the Holy Name becomes a form of devotion to Christ in His humanity. **1970** R. W. Pfaff *New Liturgical Feasts Later Med. England* iv. 63 The quite historical indulgences connected with the Name of Jesus from the thirteenth century are concerned not with the mass but with pious ejaculations mentioning the Holy Name.

4. c. Used trivially: (*a*) with *horror* or the like (orig. *U.S.*), expressing intensity; (*b*) with unfavourable implication of piety or sanctimoniousness (*colloq.*); (*c*) used with a following word as an oath or expletive, as *holy cow!*, *holy Moses!*, *holy smoke!*

Holy Joe: see quots.; *holy terror*: a person of exasperating habits or manners; *Holy Willie*: a hypocritically pious person.

[**1785** Burns *Poems & Songs* (1968) I. 74 (*title*) Holy Willie's prayer. *Ibid.* 78 Here Holy Willie's sair worn clay Taks up its last abode.] **1837** *Southern Lit. Messenger* III. 668, I have a holy horror of gossips. **1855** [see Moses 1 c]. **1860** S. Mordecai *Virginia* xxxii. 317 The Virginia Legislature had such a holy horror of banks in 1803, that they refused a charter to the petitioners. **1874** Hotten *Slang Dict.* 193 *Holy Joe*, a sea-term for a parson. **1883** G. W. Peck *Mirth for Millions* p. viii, 'Have you read "Peck's Bad Boy"!'..News agents on the Railroad cars found it almost impossible to meet the demand of those who yearned to become acquainted with this 'holy terror'. **1886** J. M. Thompson *Banker of Bankersville* 265 To get it by means of such a holy terror of exhortation. **1889** Barrère & Leland *Dict. Slang* I. 469/1 *Holy Joe* (prison and nautical), the chaplain or any religious person. **1892** Kipling & Balestier *Naulahka* i. 4 By the holy smoke, some one has got to urge girls to stand by the old machine. **1893** *Strand Mag.* VI. 105/1 Not excepting even the Dwarf, and he's, generally speaking, a holy terror. **1916** 'Taffrail' *Pincher Martin* iii. 34 Even the chaplain, the Reverend Stephen Holiman, set an example by shedding his clerical garments and trundling a barrow. The men loved seeing Holy Joe 'sweatin' himself', as they put it. **1916** G. B. Shaw *Androcles* p. xciv, The imitators of the apostles, whether they are called Holy Willies or Stigginses in derision, or, in admiration, Puritans or saints. **1917** *Dialect Notes* IV. 341 *Holy horrors*, a fright. 'It gave me the *holy horrors*.' **1920** 'Sapper' *Bull-Dog Drummond* v. 125 'Holy smoke! laddie,' he murmured. **1921** N. H. Thorp *Songs of Cowboys* (ed. 2) 73 Holy Moses and the Prophets how we split the Texas air. **1924** *Dialect Notes* V. 265 Cow: holy — (vex[ation]: New York). **1933** J. Masefield *Conway* 211 *Holy Joe*, one who is good at Scripture. **1934** J. A. Lee *Children of Poor* 130 The Holy Willies would throw a party. 'Come to our Sunday School?' **1941** Baker *Dict. Austral. Slang* 35 *Holy Joes*, prudish, narrow-minded puritans. **1942** Berrey & Van den Bark *Amer. Thes. Slang* § 194/6 Holy cow! **1941** A. L. Rowse *Tudor Cornwall* vi. 121 He must have been a holy terror to the neighbourhood. **1949** M. Lowry *Let.* 26 Mar. (1967) 177 Holy great cow, what prose is this? **1951** J. Cornish *Provincials* 40 Quit showing off. Holy cow! **1951** J. D. Salinger *Catcher in Rye* xiv. 120 They all have these Holy Joe voices when they start giving their sermons. **1958** *Listener* 18 Sept. 429/3, I cannot find justification for Mr. McCallion's term 'holy voice'. **1959** 'J. Ross' *Boy in Grey Overcoat* viii. 94 She said again, in that holy voice, [etc.]. **1960** I. Cross *Backward Sex* 40 'Holy smoke,' he gasped, 'That's a funny face.' **1967** V. Canning *Python Project* vii. 135, I said.. 'Holy Moses!' **1973** J. Wainwright *Pride of Pigs* 104 Holy cow! I forgot to switch the bloody immersion heater off.

5. a. **holy laugh** *U.S.*, a laugh by a person in a state of religious fervour; **Holy Roller** (see Roller *sb.*[1] 17 b).

1829 *Western Monthly Rev.* II. 477 Dr. Roberts is very pointed in his testimony against the abominable practice of jumping, pointing, dancing, boreing... Might he not have added the 'holy laugh'? **1833** H. Barnard *Let.* 27 Mar. in *Maryland Hist. Mag.* (1918) XIII. 328 The preacher in the midst of a fervent prayer, will all of a sudden burst out into a loud boisterous laugh... The most godly of his brethren join with him. This is called the 'Holy Laugh'. **1845** J. J. Hooper *Some Adv. S. Suggs* x. 122 Near these last, stood a delicate woman in that hysterical condition in which the nerves are incontrollable, and which is vulgarly..termed the 'holy laugh'. **1948** E. N. Dick *Dixie Frontier* 198 When it got started in an audience, everybody would be seized with hearty natural laughter. It would last for hours sometimes. This was known as the 'holy laugh'.

b. **holy basil,** the common Indian species of basil, *Ocimum sanctum*.

1880 *Encycl. Brit.* XII. 720/2 The worship of the *tulsi* plant, or holy basil, by the Hindus. **1894** A. K. Nairne *Flowering Plants W. India* 251 O[cimum] sanctum. Holy basil... Very commonly cultivated, particularly about temples and in Brahmins' gardens. **1906** T. Cooke *Flora of Presidency of Bombay* (1908) II. 440 The Holy Basil, the most sacred plant in the Hindu religion, very doubtfully indigenous.

c. Compar. *holier* in colloq. phr. *holier-than-thou*: characterized by an attitude of superior sanctity. (Cf. Isaiah lxv. 5.)

1912 T. Dreiser *Financier* lxvi. 684 The 'holier than thou' attitude, intentional or otherwise, is quite the last and most deadly offense within prison walls. **1918** *Maclean's Mag.* Jan. 45/1 His holier-than-thou attitude irritates the officials. **1922** S. Lewis *Babbitt* xix. 239 But I don't want you to think you can get away with any holier-than-thou stuff. **1928** F. Hurst *President is Born* xiii. 155 If the whole holier-than-thou house of Schuyler has got to be protected from me, dammit, I'm not going to do the protecting. **1957** R. Hoggart *Uses of Literacy* vi. 169 They counter-accuse their accusers of being 'holier than thou', of smugness, of 'hypocrisy'. **1958** *Listener* 23 Oct. 660/1 She distrusted high flights of emotion, any parade of spiritual inclinations, any holier-than-thou attitudes. **1973** *Ibid.* 4 Jan. 9/3 The Mormons were not only holier-than-thou; they were thriftier.

C. *Comb.* **b.** *holy-eyed*, *-minded* (example), *-rolling* adjs.

1922 Joyce *Ulysses* 182 An ollav, holyeyed. **1957** J. Kerouac *On Road* (1958) 221 A thin..holy-eyed..lost soul. **1902** W. James *Var. Relig. Exper.* xi. 296 The holy-minded person finds..inner smoothness and cleanness. **1965** *Punch* 20 Oct. 583/1 Sister Margaret, formidable pastor of one of those holy-rolling Harlem churchlets.

holy cross. Add: **c. holy cross toad**, a frog of New South Wales, *Notaden bennettii*, so called from a dark cross-shaped marking on the back.

1891 J. H. Rose in *Proc. Linn. Soc. N.S.W.* 2nd Ser. VI. 265 *Notaden bennettii*, the Catholic frog, or as I have heard it called the 'Holy Cross toad'. **1969** M. Burton *Animals of Australia* xiv. 120 The holy cross toad is unusual among Australian amphibians in that it comes out during the day to feed.

homalographic, *a.* Add: **1.** The form *homolographic* is the usual one. (Further examples.)

1921 *Times Lit. Suppl.* 6 Oct. 646/3 This map of the Pacific, on Mollweide's homolographic projection. **1937** *Geogr. Jrnl.* XC. 569 Equal-area projections being desirable for distribution maps, a modification of Goode's Homolographic projection—called by the compilers the 'Interrupted Mollweide's Homolographic'—is employed.

Hence **ho:malogra·phically** *adv.*

1969 N. R. Hanson *Perception & Discov.* xix. 325 We never see it sinusoidally, orthogonally, or homolographically, we just represent that world this way.

homatropine (homæ·trǫpīn). *Pharm.* Also †-in. [ad. G. *homatropin* (A. Ladenburg 1880, in *Ber. d. Deut. Chem. Ges.* XIII. 107), f. Hom(o- b + Atropine.] A synthetic alkaloid, $C_{16}H_{21}NO_3$, the tropine ester of mandelic acid, which is used chiefly as the hydrobromide in ophthalmology to dilate the pupil of the eye.

1880 *Jrnl. Chem. Soc.* XXXVIII. 410 Homatropine aurochloride..forms first as an oil, which becomes crystalline on standing. **1901** T. Sollmann *Text-bk. Pharmacol.* xi. 261 For ophthalmologic examinations the preference should be given to homatropin, since its effects set in and disappear more quickly. **1951** A. Grollman *Pharmacol. & Therapeutics* xiv. 271 Homatropine methylbromide..was introduced as a substitute for atropine in the treatment of gastro-intestinal spasm. **1970** Duke-Elder & Abrams in S. Duke-Elder *Syst. Ophthalm.* V. ix. 388 Homatropine is weaker than atropine but has the advantage of wearing off more quickly.

hombre (ǫ·mbrĕ). Chiefly *U.S.* [Sp.; cf. Ombre.] A man of Spanish descent; by extension, a man.

1846 S. S. Magoffin *Down Santa Fé Trail* (1926) 93 Not only the children, but..*hombres* (men) swarmed around me like bees. **1851** N. Kingsley *Diary* 2 Feb. (1914) 172 [I] had a fine sing in the evening with three or four other 'hombres'. **1918** C. E. Mulford *Man fr. Bar-20* viii. 79 'Friend of this hombre?' 'Yes; sort of.' **1930** [see *FIX *sb.* 5]. **1930** *London Mercury* Feb. 324 'Look here,' said Clytemnestra..'is this *hombre* worth it?' 'If you don't think so, leave him to me.' **1940** *Amer. Speech* XV. 220/2 Cowboys living a rough and hardy existence occasionally develop into 'tough hombres'. **1957** *Times Lit. Suppl.* 25 Oct. 646/4 This book describes the doings of the bad hombre. **1972** P. Cleife *Slick & Dead* xxx. 250, I had to find a tough hombre with enough guts and initiative to act on his own.

Homburg (hǫ·mbɒɹg). [Name of a town near Wiesbaden, Germany.] In full *Homburg hat.* A soft felt hat with a curled brim and a dented crown, first worn at Homburg, once a fashionable health-resort.

1894 *Country Gentlemen's Catal.* 155/1 'The Homburg Hat', as worn by H.R.H. Prince of Wales—10/6. **1901** *Sketch* 4 Sept. 254/1 The quiet gentleman in dark clothes and a Homburg hat. **1904** *To-Day* 29 June 256/1 At one time any man who wore a 'Homburg' was popularly supposed to be either an actor or an artist. **1922** E. Wallace *Valley of Ghosts* xv. 142 He..put his Homburg hat on the table. **1940, 1958** [see *Anthony Eden]. **1972** Wodehouse *Pearls, Girls & Monty Bodkin* v. 68 Characters.. who kept getting locked up in cellars under the river by sinister men in Homburg hats and raincoats.

home, *sb.*[1] and *a.* Add: **A.** *sb.* **2.** In N. America and Australasia (and increasingly elsewhere), freq. used to designate a private house or residence merely as a building.

1879 M. J. Lamb (*title*) The homes of America. **1882** *Harper's Mag.* Dec. 58/1 A lovely drive..is bordered with homes, many of which make pretensions to much more than comfort. **1889** *Kansas Times & Star* 6 July, A fine stone-front home at Twenty-seventh and Troost. *Ibid.* 5 Dec., For rent, a fine home at 1223 Broadway. **1929** *Publishers' Weekly* 7 Dec. 2661/1 Then out to see the new Ranh Brauch, a stunning private home turned over to the library. **1930** *San Antonio* (Texas) *Light* 31 Jan., Wilson wounded Elliott and his wife in a dispute Wednesday at the Elliott home in Mendota. **1955** A. Ross *Australia 55* 37 More houses (or 'homes' as a house is kindly called here) are needed. **1968** *Globe & Mail* (Toronto) 17 Feb. 1/1 Her three..sons were shot to death in their home. **1971** *Timber Trades Jrnl.* 14 Aug. 21/1 The June figures showed more private homes completed than in any month since December 1968. **1973** *Guardian* 18 May 1/6 Motorway schemes..often wipe out considerable numbers of reasonable homes in accessible areas. **1973** *Ibid.* 20 Oct. 11/6 In Beverly Hills and Bel Air, we saw the homes (never called houses) of Jane Withers, Greer Garson, and Barbra Streisand.

b. (Later example.)

1936 C. F. M. Swynnerton in *Trans. R. Ent. Soc.* LXXXIV. 520 *Home*..that portion of the tsetse-habitat used by the tsetse..for both resting and breeding.

3. *a home from home*, a place away from home which provides home-like accommodation or amenities. Also (outside Britain), *a home away from home*.

1872 in S. Walker *Whistling Commercial* (Advt.), The real comforts of a home from home. **1906** *Morning Post* 1 Feb. 7/2 To provide them with a 'home from home' while engaged in the studies which fitted them for the positions in life they were destined to fill. **1907** *Daily Chron.* 30 Nov. 3/3 The British man is a clubbable animal, and doesn't mind paying handsomely for his 'home from home'. *c* **1926** 'Mixer' *Transport Workers' Song Bk.* 21 It's like a home-away-from-home. **1961** *Times* 26 May 9/6 Durrants Hotel in George Street, for years the home-from-home of English County families. **1961** M. Beadle *These Ruins are Inhabited* (1963) xii. 165 The delicatessen that is the foreigners' home-away-from-home in Oxford. **1962** *Guardian* 6 Oct. 12/4 The idea is to provide a 'home from home' atmosphere for boys between 16 and 19.

6. (N.Z. example.)

1842 *N.Z. Govt. Gaz.* Suppl. II. 40 In accordance with instructions from home.

9. a. (Earlier example.)

1854 Dickens *Let.* 12 July (1938) II. 566 The keeping up of a 'home' at rounders.

b. *Lacrosse.* Each of the three players stationed nearest their opponents' goal.

1869 W. G. Beers *Lacrosse* (1875) xii. 191 It is essential that Goal-keeper,..and Home should be special men accustomed to those positions. *Ibid.* 195 *Home*. Should stand within eight or ten feet of the opposing goal. *Ibid.* App. 254 The players of each side shall be designated as follows:..'Home', nearest opponent's goal. **1892** *Lippincott's Mag.* XLIX. 746 Outside home, and inside home. *Ibid.* 748 To secure the ball in the 'draw-off'..and pass it to the home or attack men. **1897** *Encycl. Sport* I. 607/1 The three *Homes* must be adepts in taking short and hard catches with absolute certainty. **1964** *Lacrosse* ('Know the Game' Series) 35 First Home should make moves to as far away as Third Home level, to the goal, and round it. **1973** *Sunday Tel.* 4 Mar. 38/6 First home, Janet Roberts, with her dynamic underarm flick, deserved more than the two she obtained.

11. b. (Further Australasian examples.)

1861 T. Gilbert *N.Z. Settlers & Soldiers* 33 The..cliffs of Mokau..call to mind the chalk cliffs of dear old England—Beachy Head, and other favourite localities at 'home'. **1908** E. J. Banfield *Confessions of Beachcomber* I. ii. 77 Australians cannot with justice complain when the good old folks at home blunder..the while..so much local misapprehension prevails.

e. *Cribbage.* (See quot. 1877.)

1796 C. Jones *Hoyle's Games Improved* 294 By attending to the above Calculation any Player may judge whether he is at Home or not. **1877** *Encycl. Brit.* VI. 577/1 (*Cribbage*) Each player ought to reckon slightly over six in hand and play and five in crib, or seventeen and a half in two deals to be *at home*. A player who scores more than the average and leaves his adversary six or seven points in arrear is *safe at home*. When at home it is best to play off; when the adversary is safe at home it is best to play on.

f. Used in colloq. phrases expressing (freq. scornful) doubt or a query about the identity of a person or thing, *e.g.* 'Who is he when he 's at home?'

1887 Kipling *Plain Tales fr. Hills* (1888) 99 You.. dimonstrate to *my* frind here, where *your* frinds are whan they're at home? **1914** E. Pugh *Cockney at Home* 118 Who is Popkins when he's at home? **1930** J. B. Priestley *Angel Pavement* ii. 64 'And we can't all look like Mr. Ronald Mawlborough either.' 'Who's he when he's at home?' Mr. Smeeth inquired. **1957** M. Kennedy *Heroes of Clone* III. ii. 165 'And what's existentialist, when it's at home?' she asked. **1960** R. Collier *House called Memory* viii. 112 *Peachy*? I have no idea what you mean. What's that when it's at home? **1972** A. Ross *London Assignment* 20 'Farrow?' I said. 'Who's he when he's at home?'

g. Used of a match when the team referred to is playing on its own ground. (Cf. *AWAY *adv.* 11.)

1898 *Football Telegraph* (Kettering) 1 Jan. 3/2 Last season,..a splendid victory was achieved at home, the locals winning by 2 goals to 0. **1930** *Daily Tel.* 5 Dec. 20/3 Clapton Orient, 'at home' to Luton Town at Highbury. **1967** *Listener* 17 Aug. 223/1 They had just lost three matches at home.

13. b. (Later examples.)

1863 Queen Victoria *Let.* 5 Aug. in R. Fulford *Dearest Mama* (1968) 254 No one saw the correspondence. I think the King should look nearer home for such things. **1886** C. M. Yonge *Chantry House* II. xiv. 138 'I meant something nearer home,' said Clarence, and proceeded to ask if I did not think Lawrence Frith..smitten with Emily. **1969** *Listener* 2 Jan. 22/2 It seemed thus to be about something both larger and nearer home than Cohn-Bendit's subject.

13*. **to home.** *dial.* (also *U.S.*) = At home.

1795 B. Dearborn *Columbian Gram.* 139 Improprieties, commonly called Vulgarisms..[include] To home for At home. **1833** J. Neal *Down-Easters* I. 62 When he's to home..he's match for gab with anybody 't ever you come across. **1839** *Knickerbocker* XIV. 153, I used to be quite good at reckoning, when I was to home, in the state of New Hampshire. **1868** Lady Verney *Stone Edge* ii, I'm main sorry Master Broom ain't to home. **1873** 'S. Coolidge' *What Katy Did* xii, 'Tain't every girl would know how to take care of a fat old woman, and make her feel to home. **1910** *Dialect Notes* III. 450 (Western New York) Is your father to home? **1935** Z. N. Hurston *Mules & Men* (1970) I. ix. 192 Come on, Big Sweet, we got to go to home. **1972** J. Gores *Dead Skip* (1973) viii. 52 White meat don't turn me on. I got Maybelle and four cute kids to home.

14. *attrib.* and *Comb.* (The distinction made in the Dict. between senses A. 14 b and B. 1 is no longer valid, since present-day hyphening cannot be assumed to be a reliable guide to grammatical function.) **b.** In relation to domestic economy: *home art(s)*, *care* (later examples), *circle* (earlier examples), *daughter*, *girl*; **home-bird, homebody** orig. *U.S.*, a person, etc., who prefers staying at home to going out or travelling; **home boarder,** a day-boarder, day-boy; **homecraft**, an art or craft pursued in the home; also, the household arts; **home loan,** a loan granted to someone to assist in the purchase of a house, flat, etc., to live in; **home unit** *Austral.*, a flat or apartment, normally one owned by the occupant.

1821 J. F. Cooper *Spy* I. xi. 175 Marry him I don't think I will——unless he becomes steadier and more of a homebody. **1841** Dickens *Let.* 16 Mar. (1969) II. 238 With love to all your home circle and *from* all mine. **1850** C. Kingsley *Alton Locke* vi. 97 To..live a life of sneaking and lying under petticoat government, as all home-birds were sure to do in the long-run. **1853** Mrs Gaskell *Ruth* III. viii. 230 Leonard's remaining such a home-bird..with such a mother..will do him no harm. **1855** —— *Let. c* 20 Oct. (1966) 873 Thank you for your kind message to my home circle. **1857** A. J. Symonds *Let.* 24 May (1967) I. 110 There were two brothers homeboarders but belonging to the School. **1881** C. M. Yonge *More Bywords* (1890) 125 An excellent plan..for bringing the whole family together round our dear old mother and her home daughter. **1886** Home-bird [in Dict.]. **1902** *Encycl. Brit.* XXV. 686/1 The Home Arts and Industries Association. **1905** H. A. Vachell *Hill* xi. 228 He wished to educate his only son at Harrow as a 'Home-Boarder', or day-boy. **1914** M. Hill (*title*) Homecraft in the classroom. **1924** Wodehouse *Bill the Conqueror* ii. 32 He said this girl was one of those domestic girls, a little home-body, and might be leaving the party any moment now. **1927** Peake & Fleure *Hunters & Artists* 79 *Art mobilier*, which has been translated 'mobiliary art', 'portable art', or 'home art'. **1927** *Daily Express* 26 Feb. 5/2 Women who seek a pleasant paying homecraft. *Ibid.* 23 Mar. 5 Modern Homecraft Notions. **1934** T. Wilder *Heaven's My Destination* 53 Snappiest little home-girl in Oklahoma. **1941** *Brit. Jrnl. Psychol.* Apr. 306 Finally she had to leave her University post... 'I could no longer keep it and be a home daughter.' **1944** J. S. Huxley *On Living in Rev.* iii. 28 Ma Ferguson, pictured as a very motherly sort of home-body. **1959** S. Gibbons *Pink Front Door* iv. 52 The daughter at home, 'the home bird'. **1959** *News Chronicle* 12 Aug. 6/8 She liked to be a homebody and..lie with her head on her master's foot. **1961** R. B. Long *Sentence & its Parts* ii. 54 She isn't a home girl. **1961** *Spectator* 17 Mar. 352 The great increase of home-care cases that must follow the demolition of the mental hospitals. **1962** *Southerly* XXII. II. 92 By 'standard words' is implied those which soberly indicate what they have to indicate (like home-unit and bombora). **1966** Auden *About House* 24 A cellar never takes umbrage; It takes us as we are, explorers, homebodies. **1966** *Times* 7 Apr. 12/1 The ban on home loans by local authorities has been lifted. **1967** *Guardian* 16 May 8/4 The one school with 'home boarders'—where some boys live in the town and take 'bed and breakfast' at home. **1967** *Canad. Ann. Rev. 1966* 92 Government support to develop a complex of services, including hospital-based home care programs. **1967** E. Hunt *Danger Game* viii. 151, I know Mrs Dell lived in a home unit in Coogee. **1969** T. Parker *Twisting Lane* 48, I was a great home-bird myself. **1969** *West Australian* 5 July 48/2 (Advt.), Prestige loc[ality]..suitable flats or home units. *Ibid.*, Home unit Scarborough open for inspection... This spotless apartment..compr[ises] 2 good bedrms. [etc.]. **1972** *P.O. Telephone Directory* Sect. 102 (London Postal Area E–K) 513/3 Homecraft Supplies. **1972** *Publishers Weekly* 30 Oct. 33/2 A country of homebodies with the concerns of homebodies is quite different from an outgoing one. **1973** *Times* 27 June 23/1 (*headline*) Tempers fray in tussle between home loan chiefs and Government. **1973** *Sun-Herald* (Sydney) 26 Aug. 11/2 (*heading*) A bolt of lightning damaged a block of home units at Vaucluse.

c. (Further examples.)

1745 W. Ellis *Mod. Husb.* VI. I. 33 My Rows of broad Beans in my Home-close. **1853** C. W. Adams *Spring in Canterbury Settlement* vii. 69 The home-station was situated at the mouth of the river Motunau. **1884** W. Shepherd *Prairie Experiences* 205 Stock are always restless at first on a drive, and are striving to get back on their home-ranges. **1902** *N.Z. Illustr. Mag.* VII. 117/2 We'll keep the home-block. **1903** 'T. Collins' *Such is Life* 170 In two seconds more, Cleopatra was stretching away..towards Yoongoolee home-station, distant about sixteen miles. **1904** 'G. B. Lancaster' *Sons o' Men* 3 Lane's out-station was twelve rough miles from the home-block. **1915** D. H. Lawrence *Rainbow* i. 6 A confusion of sheds spread into the home-close. **1933** C. Day Lewis *Dick Willoughby* 9 The cowman and his herd moving into the home-meadow.

d. (Further examples.)

1797 Nelson *Let.* Apr. (1845) II. 374 Had there been no Fleet in the Channel, the French might have come up the Mediterranean and taken us all; therefore the Home Fleet certainly took care of us and covered us. **1883** *Peel City Guardian* 15 Sept., Our harbour is once more almost empty, as the Home Fleet are fishing off Douglas. **1903** *Westm. Gaz.* 30 Mar. 2/2 What military stations abroad are now reckoned as Home stations. **1904** *To-Day* 14 Dec. 162/1 The Present Home Fleet is to be called the Channel Fleet. **1906** *Daily Chron.* 24 Oct. 7/4 A distinct fleet will be constituted from the ships in commission in reserve, to be called the 'Home Fleet'. **1915** Kipling *Fringes of Fleet* 35 From the peace of the German side he had entered our hectic home-waters. **1927** W. Deeping *Kitty* i. 10 A home-service job with one of the home-service battalions. **1941** *Ann. Reg. 1940* 55 Small bodies of highly mobile and strongly armed troops, known as 'Ironsides', after the first Commander-in-Chief of the Home Forces. **1972** *Whitaker's Almanack 1973* 461/1 Sir Andrew Lewis K.C.B. (Commander-in-Chief Naval Home Command).

h. *home-breaker*, *-buyer*, *-owner* (so *-ownership*), *-seeker* (examples); *home influence*, *news*; *home-breaking* vbl. sb.

1928 *Sunday Dispatch* 2 Sept. 17/1 As a home-breaker woman is..as good as a man. **1936** N. Coward *To-night at 8.30* II. 74 You mean you'd prefer to be implicated with a professional homebreaker as opposed to an amateur one? **1907** *Daily Chron.* 18 June 3/6 Home-breaking is a more serious offence against society than house-breaking. **1969** *Times* 20 May 28/3 (*headline*) S[elective] E[mployment] T[ax] hits home buyer. **1973** *Times* 6 Oct. 1/1 A plan to help young home-buyers is likely to be announced within the next week. **1847** G. Aguilar (*title*) Home influence. **1852** C. M. Yonge *Two Guardians* xiv. 255 Marian had..weakened the only home influence..which held Caroline to the right. **1966** D. Jenkins *Educated Society* v. 208 Home influence..[is] a major factor in determining whether people will be able to take advantage of educational opportunities. **1852** Mrs. Gaskell *Let.* 2 Mar. (1966) 181, I don't think there is much home news. Last week was very quiet; and very busy with writing. **1936** *Punch* 5 Aug. 144/3 It is one of our principles that our students should *earn while they learn*, and for this reason our first practical efforts will be in the realms of the easiest department of Romantic Journalism, namely Home News. **1945** Nelson & Wright *Tomorrow's House* xvii. 203/1 There are almost 35,000,000 dwellings in the United States. Maybe you own one of them... To the homeowner who is intrigued by..tomorrow's house, several possibilities are open besides..selling the roof over his head. **1956** J. H. Greenberg in Saporta & Bastian *Psycholinguistics* (1961) 474/1 Homeowners must pay real estate taxes. **1960** *Times* 21 Nov. (Canada Suppl.) p. v/6, Canadians are the greatest home-owners in the world. **1960** *Times* 24 Oct. (Financial Rev.) p. xi/5, Thrift and home-ownership. **1972** *Guardian* 6 July 24/7 Home ownership in cities averaged 42 per cent. **1889** *Advance* (Chicago) 7 Mar. 191 The home-seekers of the 19th Century Pilgrim Fathers. **1911** *Daily Colonist* (Victoria, B.C.) 1 Apr. 12/1 (Advt.), Pandora Avenue Homeseekers. We have the finest residential buy in Victoria today. **1937** *Sunday Dispatch* 14 Feb. 18/4 The scope for the homeseeker is, in fact, extremely wide. **1963** *Times* 12 Feb. 8/1 *A Homeseeker.* Once you have found the house you are looking for, talk.. about a loan to help you buy.

i. *home-baked* (earlier examples), *-based*, *-consumed*, *-cooked*, *-cured*, *-grown* (earlier and later examples; also *transf.* and *fig.*), *-killed*, *-produced*; *home-cooking*, *-curing*, *-dressmaking*, *-nursing*, *-sewing*, *-staying* (examples); *home-duty*, *-growth*, *nurse*; *HOME HELP; *home-dressmaker*, *-grower*, *-sewer*.

1766 H. Brooke *Fool of Quality* (1870) I. 99 Nurse went upstairs with a most bountiful cut of home-baked bread and butter. **1816** Jane Austen *Emma* II. ix. 186 The finest looking home-baked apples I ever saw in my life. **1944** *Ann. Reg. 1943* 8 British and American home-based bombers made..day or night raids. **1956** *Nature* 25 Feb. 366/2 Six..schemes, applicable to the tropics, which were for home-based work..were added. **1904** *Westm.*

Gaz. 27 Dec. 4/3 A considerable proportion of export tonnage besides home-consumed manufactures..is conveyed by horse-drawn vehicles. **1923** H. CRANE *Let.* 12 Oct. (1965) 150 A very fine home-cooked chicken dinner. **1968** M. KANE *Walk of Devil* iii. 27, I know what you need... A nice home-cooked dinner in pleasant surroundings. **1934** *Amer. Speech* IX. 113/2 The tourist who enjoys strange foods..will rarely escape the home-cooking and the home-made pies that emerge from tin cans and factory-like bakeries. **1937** 'M. HILLIS' *Orchids on Budget* (1938) vi. 94 Steaks and roasts are so well prepared in the best restaurants that you might..turn up your nose at them for home cooking. **1968** O. WYND *Sumatra Seven Zero* i. 5 The other patrons..all came often to get away from home cooking. **1973** H. NIELSEN *Severed Key* i. 6 Didn't I ever tell you how much I hate home cooking? **1863** MRS. GASKELL *Cousin Phillis* in *Cornh. Mag.* Nov. 619, I handled and weighed in my fancy the home-cured ham. **1959** R. POSTGATE *Good Food Guide* 236 Breakfast was home-cured ham, thick, well grilled, with eggs and tomato. **1960** *Farmer & Stockbreeder* 16 Feb. Suppl. 8/1 When Denmark entered the trade at the end of the nineteenth century, bacon was still being produced from the whole sides in an adaptation of the traditional home-curing method. **1960** *Sunday Express* 11 Sept. 15/6 The ideal fabric for the home dress-maker. **1971** *Guardian* 10 Aug. 9/2 These costly fabrics (up to £50 a yard) are..being displayed..as temptations to home dressmakers. **1896** *Woman's Life* 4 July 138/1 Many people put up with faults in home dressmaking which they would rate in a regular dressmaker. **1966** H. YOXALL *Fashion of Life* viii. 68 When I first joined *Vogue* the early demise of home-dressmaking was predicted. **1850** MRS. GASKELL *Let.* 14 May (1966) 117 Girls, having..the home-duties of parents dependent upon them. **1825** J. S. MILL in *Westm. Rev.* III. 418 The home-grower is subject to many taxes, from which foreign corn is exempt. **1827** —— *Ibid.* VII. 173 They consume *home-grown* corn. **1895** *Cricket Handbk.* 31 It is sincerely to be hoped that the means may yet be found of recruiting the eleven from home-grown cricketers. **1935** *Discovery* June 162/1 The Great War undoubtedly had a great influence in fostering the canning of home-grown foods. **1959** *Times* 31 Mar. 11/1 Yet there was nothing unprofessional about the timing, staging, or lighting of even these most home-grown numbers. **1966** *Times* 28 Feb. (Canada Suppl.) p. xiii/3, Last summer, with a home-grown production of a musical..it played a six weeks season. **1973** J. THOMSON *Death Cap* iv. 56 A board..announcing in crude, hand-painted letters: 'For Sale. Home-grown Produce. Flowers. Fruit. Vegetables. Eggs.' **1825** J. S. MILL in *Westm. Rev.* III. 413 If the new material..be of home growth, the production of that material would open a new channel for the profitable employment of agricultural capital. **1873** E. E. ESTCOURT *Question of Anglican Ordinations* ii. 15 All the preceding seems to have been the direct home-growth of Lollardism or of Wicliffe's teaching. **1906** *Daily Chron.* 20 Sept. 6/1 Not even an expert could tell the difference between home-bred and home-killed meat unless he were on the spot. **1958** Home nurse [see *HOME HELP]. **1963** *Times* 24 May 8/5 All the old public assistance buildings replaced by pleasant old people's homes, with sufficient home nurses or home helps. **1905** *Daily Chron.* 25 Oct. 7/4 Instruction of..future mothers..in..home-nursing. **1959** *Sears, Roebuck Catal.* Spring & Summer 831/3 Illustrated family handbook of home nursing and medical care. **1905** *Westm. Gaz.* 21 Aug. 2/1 Out of deference to the agrarian interest, prohibitive Customs and 'sanitary' restrictions keep out the food which is ready on the frontiers, while the prices of home-produced flesh rise to a height hitherto unknown. **1966** *Times* 28 Mar. (Austral. Suppl.) p. v/6, Home-produced crude oil. **1964** *McCall's Sewing* ii. 19/2 Whether you are a beginner or an experienced home-sewer, every pattern should be carefully studied. **1908** *Westm. Gaz.* 13 June 13/2 Every home-sewing room should include..a skirt-board [etc.]. **1964** *McCall's Sewing* ii. 15 Without patterns, home-sewing would probably be a lost art. **1854** THOREAU *Walden* 170, I the home-staying, laborious native. **1905** *Daily Chron.* 9 June 8/5 The home-staying Englishman.

B. 1. *home comfort* (earlier and later examples).

a **1855** C. BRONTË *Professor* (1857) II. xxv. 223 To sit on a foot-stool at the fire-side—to enjoy home-comforts. **1922** JOYCE *Ulysses* 296 Their abodes were equipped with every modern home comfort. **1942** *R.A.F. Jrnl.* 13 June 15 Some have heard about the living-out system, and dream of feather beds and home comforts. **1973** 'B. MATHER' *Snowline* xix. 225 The place where we're staying—all home comforts and no questions asked.

2. a. (Further examples.)

1857 R. B. PAUL *Lett. fr. Canterbury, N.Z.* 89 If you have not decided on the site of your home station on a previous visit to the run, the first step will be to ride well over it. **1865** M. A. BARKER *Station Life in N.Z.* (1870) v. 31 By the time we reached the Home Station we were ready for luncheon. **1930** L. G. D. ACLAND *Early Canterbury Runs* v. 113 He arranged that his executors should bury him in one of the home paddocks at Cracroft. **1939** P. A. ROLLINS *Gone Haywire* 62 The ol' man's bin steadily enlargin' his home range till now it includes mos' all the head-waters o' Elk Prairie Crick. **1946** F. D. DAVISON *Dusty* ix. 95 The sheep were approaching the home paddock. **1966** *Te Reo* IX. 54 The home paddock and the night paddock which..must in the early years have been the same piece of land.

b. *Home Counties.* For def. read: the counties nearest to London, namely Surrey, Kent, Essex (and formerly Middlesex); sometimes with the addition of Hertfordshire, Buckinghamshire, Berkshire, and occasionally Sussex. Add examples.

1898 *Mddx. & Herts. N. & Q.* IV. 153 The publication, ..will..relate not only to London, Middlesex and Hertfordshire, but also to Essex, Buckinghamshire, Berkshire, Surrey and Kent; that is, to the Home Counties. **1959** I. & P. OPIE *Lore & Lang. Schoolch.* xii. 233 In London and the Home Counties the police now chase off the streets even the simple waits singing Christmas carols. **1966** *Listener* 11 Aug. 218/1 The chances are..small that a writer setting his play in outer suburbia or inner Home County will make of it more than a painful banality. **1972** J. BLACKBURN *For Fear of Little Men* ii. 29 Her accent clashed dramatically with the jargon of Home Counties suburbia.

c. (Further examples.)

1800 *Spokane Falls* (Wash.) *Globe* 26 July 1/3 Six hundred baseball cranks witnessed..one of the prettiest exhibitions of ball playing that ever took place on the home grounds. **1802** *Sporting Mag.* XIX. 219/2 They were willing to run a match with a leash of greyhounds..or six brace belonging to each county, running home and home, for a sum to be agreed on. *Ibid.* 221/2 He has refused to run you *home* and home, or in any *central county* between the two. **1887** F. GALE *Game of Cricket* 51 The principal innkeeper and a few good local players inaugurated occasionally good out and home matches, in the same season. **1889** *Seattle Post-Intelligencer* 3 July 1/8 The home team played an up hill game. **1916** E. F. BENSON *David Blaize* iv. 60 The home team took the visitors off to the dormitories to put on their flannels. *Ibid.* 62 The field was cleared for the match; the home side won the toss. **1930** H. G. WELLS *Autocr. Mr. Parham* II. iii. 109 That complete lack of information about the visitors attributed to the home team. **1955** L. A. G. STRONG *Dr. Quicksilver* 25 The creator of Zeal-of-the-Land Busy was on his home ground in the last two lines. **1962** *Listener* 4 Jan. 41/1 Three generations on my home ground were equally disgruntled. **1969** *Ibid.* 3 Apr. 473/3 The factual conclusion is that if Arsenal had fouled more in these two home games (amongst others), they might have won them. **1972** *Oxford Mail* 15 Feb. 12/6 Carlisle dropped an unexpected home point last weekend.

d. *home signal*: on railways, a stop signal marking the end of a block section, and controlling entry either to station limits or the block section ahead. (Cf. DISTANT *a.* 3 d.)

1874, 1889 Home signal [see DISTANT *a.* 3 d]. **1923** W. G. CHAPMAN *10.30 Limited* x. 79 The home signal is the second signal reached and is usually near a signal box. It is a stop signal and must not be passed at 'Danger'. **1940** A. E. TATTERSALL *Railway Signalling* i. 16 Home and starting signals only for each direction at stations on single lines which are staff or electric token posts will be necessary. **1963** KICHENSIDE & WILLIAMS *Brit. Railway Signalling* ii. 16 Stop signals are..divided into two types: 'home' signals, usually placed on the approach side of a signal box, and 'starting' signals, placed beyond or in advance of a signal box... Before a signalman can accept a train from the previous signal box, the line must usually be clear for ¼-mile beyond his home signal. **1968** L. T. C. ROLT *Railway Engin.* vi. 93 By exhibiting three, or in some cases four, 'aspects' they [*sc.* colour light signals] combine the function of home and distant signals.

e. *Home Service*, one of the programme services broadcast by the B.B.C. (see quot. 1966). Also *Home programme* and *ellipt.*, *the Home.* (On 30 Sept. 1967 the name was changed to 'Radio 4'.)

1939 *Radio Times* 8 Sept. 3/1 In addition to the Home Service..two other British programmes are going out every day. One is a short-wave service to the whole world,..the other a short-wave service meant primarily for Europe. **1947** 'G. ORWELL' *Let.* 25 Jan. (1968) IV. 276 It was done on the Eastern and African services, but in those days I wasn't well-connected enough to crash the Home. **1948** [see *BREAK *v.* 50 g]. **1960** *B.B.C. Handbk.* 238, 1939..Home Service replaced National and Regional Services. **1965** G. MELLY *Owning-Up* vi. 64 Loudspeakers in every bedroom with a control switch marked 'Light. Home. Room Service'. **1966** *B.B.C. Handbk.* 44 The Home Service serves the broad middle section of the community... It is the main vehicle for news and for the daily reporting of Parliamentary proceedings.

3. a. (Further examples.)

1765 J. WEDGWOOD *Let.* 2 Mar. (1965) 29 Our home consumption is very trifleing in comparison to what are sent abroad. **1765** in *Amer. Hist. Rev.* (1921) XXVI. 743 The planters..reside Mostely on the Borders of James and York rivers which is the best soil for tobacco Especially the Sweet sented which is so much Esteemed in England, where they keep it for their own use, or what they Call home Consumption. **1825** J. S. MILL in *Westm. Rev.* III. 418 A protecting duty, in that case, would be a premium on home production, and, therefore, injurious. **1842** *Niles' Reg.* 12 Feb. 384/2 A state convention is to be held..at Hartford, Connecticut, to adopt measures for laying before congress the claims of home industry. **1844** MILL *Ess. Pol. Econ.* i. 14 It would be for the interest..of Germany herself, to keep her linen a little below the value at which it could be produced in England, in order to keep herself from being supplanted by the home producer. **1848** —— *Pol. Econ.* II. III. xviii. 122 The value..of a foreign commodity, depends on the quantity of home produce which must be given to the foreign country in exchange for it. *Ibid.* v. iv. 400 It would seem that..taxes on exports..fall entirely on foreigners, taxes on imports wholly on the home consumer. **1876** C. M. YONGE *Womankind* xiii. 97 Factories or small home industries, such as glove or lace making. **1926** A. HUXLEY *Jesting Pilate* I. 187 We shipped the best part of a thousand lacquered kettle-drums—for home consumption, I suppose. **1937** E. SNOW *Red Star over China* vii. 253 The bulk of manufacturing..was by handicraft and home industry. **1940** *Time* 1 Jan. 29/3 Entente Cordiale.. was probably intended as French propaganda for home consumption on the present Anglo-French alliance.

b. (Further examples.)

1828 *Home Missionary Mag.* May 1 The design of the American Home Missionary Society [founded 1819] is to promote..the religious benefit of a great and growing nation. **1836** *Ibid.* Sept. 37/1 Endeavour to engage Ministers..to promote the great cause of Home Missions. **1842** *Ainsworth's Mag.* I. 232, I had occasion to accompany a home missionary into a few of the dens of London. **1844** *Ibid.* VI. 113 The manœuvres by which certain county members manage to obtain audiences of the home secretary, or the colonial secretary. **1855** MRS. GASKELL *Let.* 27 July (1966) 363 Papa..finished up his Home Mission with an address to the Students in the Chapel. **1881** E. W. HAMILTON *Diary* 25 Dec. (1972) I. 204 An office established to register the decrees of the Home Secretary and other Ministers. **1886** H. C. E. CHILDERS *Let.* 11 Nov. in S. Childers *Life* (1901) II. 245 The prerogative of mercy inherent in the Crown is dispensed..by the Home Secretary. **1938** M. C. BOATRIGHT in B. A. Botkin *Treas. S. Folklore* (1949) I. iv. 96 A Presbyterian home missionary came to a cabin and engaged a woman in conversation. **1958** S. HYLAND *Who goes Hang?* viii. 44 Home Secretary. A question to the Home Sec. **1966** *Listener* 17 Mar. 373/1 The Ministry of Defence is classified as a 'home' rather than an 'overseas' department.

4. (Further examples.) Also *transf.*

1856 *Spirit of Times* 4 Oct. 86/1 He was headed off and put out on the home base. *Ibid.*, An injudicious attempt ..to get a home run. **1867** H. CHADWICK *Base Ball Player's Bk. Reference* 138 Standing at the home base. **1920** S. LEWIS *Main St.* 300 He invariably decided that coming confinement-cases or land-deals would prevent his 'getting away from home-base for very long this year'. **1926** *Daily Colonist* (Victoria, B.C.) 22 July 12/3 Thirteen hurlers appeared. There were ten home runs, seven two-baggers and four triples. **1962** *Listener* 22 Feb. 337/2 In a big hit, what the Americans call a 'home run', the wood of which the bat is made is called upon to sustain stress several times bigger than would be required to break it in two, if maintained for a longer time. **1966** *Ibid.* 3 Nov. 644/1 The correspondent who never gets to home base comes to believe he is identifying himself with the ordinary sensible Briton.

5. b. *home key*: in *Mus.*, the basic key in which a work is written.

1959 D. COOKE *Lang. Mus.* v. 269 'They finally decide on E minor, which is, after all, the home key.' It is obvious that a modern composer need not be concerned about ending in the 'home key', after the 'progressive tonality of Mahler, Nielsen and others'. **1968** *Listener* 22 Aug. 250/1 The 'home' key of Weill's original score..is C.

home, *adv.* Add: **1. b.** (Later examples.)

1954 *Ann. Reg.* 1953 286 Americans were the target of propaganda of the 'Go Home, Yank!' type. **1973** *Guardian* 9 June 13/7 Vauxhall car workers..were..shouting, 'Yankees Go home! Bloody Americans!'

d. Also with ellipsis of *drive*, esp. in *Home, James* (*, and don't spare the horses*)!

1927 E. WALLACE *Mixer* viii. 114 'All right, Paul,' returned Mr. Sparkes...'Home, James.' 'James' grinned in the darkness, and the car moved forward. **1934** F. HILLEBRAND (*song title*) Home James, and don't spare the horses. **1964** WODEHOUSE *Frozen Assets* xi. 213 Okay, Watson, drive on. Home, James, and don't spare the horses.

2. b. *transf.* Safely or successfully at the end of (usually something arduous). Esp. in phr. *home and dry.*

Baker also records *home and dried* (*on the pig's back*) from Australia.

1930 V. PALMER *Passage* I. x. 86 You've done it this time, Lew! Home and dry on the pig's back! **1938** 'N. SHUTE' *Ruined City* ii. 26 'That's the Finnish business, then,' he said. 'We're practically home on that.' **1951** L. G. D. ACLAND *Early Canterbury Runs* 382 *Home and dry*, out of trouble. **1958** *Times* 21 Apr. 5/4 It was a most exciting finish, for apart from Miss Willcox, who, bar accidents, was 'home and dry', any one of six could easily have come second. **1962** J. BRAINE *Life at Top* iii. 46, I was in. I was home and dry. I'd got the order. **1963** *Guardian* 8 Feb. 1/1 Labour members felt after the ballot that Mr Wilson was 'home and dry'. **1965** M. WEST *Ambassadors* xi. 253 We're home and dry... Small casualties. A new Government.

3. b. (Later examples.) Also used in sport, etc., in sense shown in 4.

1903 J. A. T. BRAMSTON in Benson & Miles *Bk. Golf* 114 Be content merely to return to the fair course, and to get 'home' with the next shot. **1920** H. S. BROWNING *How to play Cards* 110 When each has dealt once, they [*sc.* the players of cribbage] should stand abreast at seventeen to eighteen, and so on throughout the game. The player who has maintained this average is said to be 'home'. **1934** W. J. LEWIS *Lang. Cricket* 130 *Home*, in various terms and phrases with reference to the batsman's ground, i.e. within the popping crease, as *to get home* in completing a run; [etc.]. **1954** F. C. AVIS *Boxing Reference Dict.* 47 *Get home*, to hit an opponent with the blow intended. **1955** *Times* 30 June 4/1 Nielsen's strength, his power of service, but not least his tactical skill in attacking Rosewall's service at all costs, just and only just got him home, when the whole issue was in doubt until the very last point. **1973** *Country Life* 17 May 1369/3 No man has won or lost until he has played home from the 14th tee.

5. b. Also with *get.*

1931 *Punch* 4 Nov. 496/1 Yet we have to admit that these songs 'get home' on us: that, singing them, we become as little children. **1958** *Church Times* 29 Aug. 3/4, I know that a lot of this will not get home among those who do not want to resolve discord.

7. d. *to write home about*: to boast of, to 'make a song about'. Usu. in negative contexts.

1914 'I. HAY' *Knight on Wheels* xxix. 291 'Anything doing at present?' 'Nothing to write home about, thanks.' **1925** A. HUXLEY *Those Barren Leaves* II. 96 There is nothing in these virtues *à la* Dickens to 'write home about'. **1930** M. KENNEDY *Fool of Family* xix, I know Bach had twenty sons, but they weren't anything to write home about. **1950** A. BARON *There's no Home* 11 Oh, this is something to write home about, all right. **1958** *Times* 3 Nov. 3/6 But for much of the day there was little to write home about as M.C.C.'s last eight wickets fell for 125 runs. **1959** *Times* 10 Aug. 6/1 Student nurses.. dismiss their share of the award as not worth writing home about. **1967** V. CANNING *Python Project* ii. 25 He has a small place in the country... Don't run away with the idea of anything worth writing home about when I say 'place'. It's a crumby little cottage.

8. a. *home-goer, -going*, also as *adj.* (later examples); *home-come* adj.; also *home-deliver* vb.

1908 *Westm. Gaz.* 12 Dec. 6/3 Who holds up to her home-come soldier's lips The babe he hath not seen. **1910** *Daily Chron.* 21 Jan. 6/6 The newsboy..handing them out to the sleepy home-goers. **1918** W. J. LOCKE *Rough Road* xix, The home-come warrior. **1937** J. JOYCE *Let.* 1 Nov. (1966) III. 408 We are sorry to hear you had such a stormy homegoing. **1958** *Time* 8 Dec. 42/2 To home-deliver fully 85% of the Sunday papers. **1967** K. GILES *Death & Mr. Prettyman* viii. 146 With a small van we could home-deliver so they don't have to leave the telly. **1973** D. CRAIG *Bolthole* i. 9 They were beginning to reach thick home-going traffic.

c. In senses 'in or at one's home', 'in one's home country', as *home-living* ppl. adj. and vbl. sb.

1881 W. D. HAY *300 Years Hence* i. 48 The disabilities under which the home-living population laboured. **1963** F. F. LAIDLER *Gloss. Home Econ. Educ.* 42 *Home living*, life within the home and family.

home, *v.* Add: **4.** *intr.* Of a homing pigeon: to fly back to its 'home' or loft after being released at a distant point; to arrive at the loft at the end of such a flight. Hence of any animal: to return to some specific territory or spot after being removed by an external agent or leaving it of its own accord. Freq. const. *to.*

1875 *Live Stock Jrnl.* 23 Apr. 57/3 Pigeons home by sight and instinct. **1895** *Cambr. Nat. Hist.* III. ii. 35 Snails and slugs possess to a considerable extent the faculty of 'homing', or returning to the same hiding-place day after day, after their night excursions in search of food. **1899** *Westm. Gaz.* 12 Apr. 9/1 The first [pigeon] homed at nine o'clock. **1904** *Daily Chron.* 25 July 9/2 Out of this vast army of birds not one homed on the day of release. **1934** E. S. RUSSELL *Behaviour of Animals* iv. 71 Arey and Crozier..relate of *Chiton tuberculatus*..that it keeps to one limited area and does not wander very far, though it does not appear to 'home' to a particular spot as does the limpet. **1956** W. H. THORPE *Learning & Instinct in Animals* xvi. 412 In one case a dog homed a 'bee-line' distance of 6 kilometres. **1958** *Observer* 26 Jan. 18/2 Migrating birds, and birds that can 'home' over great distances. **1966** R. ARDREY *Territorial Imperative* (1967) iv. 134 No random hunting or zigzag uncertainties marred the voyages. Sunfish truly home, and home to territories. **1971** *Nature* 17 Sept. 218/2 Visual recognition of their external surroundings..was used to resettle birds to 'home' to new loft-sites in place of old ones.

b. *trans.* To train (a carrier-pigeon) to fly home.

1928 *Sunday Dispatch* 29 July 22 Leatham (Downpatrick) has achieved what many thought impossible—viz., homing a bird from San Sebastian (Spain), distance over 800 miles, to the Emerald Isle, 300 miles of which, supposing the bird crossed from the northern coast of France, is over water.

5. *intr.* Of a vessel, aircraft, missile, etc.: to be set, or guided, to its target or destination, by use of a landmark or by means of a radio beam, etc. Also *fig.* Freq. const. *on* or *in on.*

1920 *Wireless World* Mar. 728/2 The pilot can detect instantly from the signals, especially if 'homing' towards a beacon. **1940** *Jrnl. R. Aeronaut. Soc.* XLIV. 569 The tanker must be equipped with D.F. gear, so that the two aircraft may 'home' on each other if visibility is poor. **1947** CROWTHER & WHIDDINGTON *Science at War* 119 Torpedoes and bombs that follow or 'home' on to their targets. **1948** [see *A.D.F.* s.v. *A III]. **1948** *Ann. Reg.* 1947 458 The equipment [for automatic take-off and control of an aircraft], which can be fitted to a standard aircraft, homes on a beam sent out by a radio beacon. **1955** C. M. KORNBLUTH *Mindworm* 53 That was near. He crossed the street and it was nearer. He homed on the thought. **1956** *Amer. Speech* XXXI. 228 A good officer could even 'home in on a bottle of whisky' placed on the landing field. **1958** 'P. BRYANT' *Two Hours to Doom* 58 Infra-red missiles which homed on the radiations given off from jet engines. **1962** F. I. ORDWAY et al. *Basic Astronautics* ix. 386 The guided vehicle then homes on the reflected signals as in the active case. **1971** *Daily Tel.* 23 Aug. 1/5 The other helicopter located the dinghy by homing in on the bleeping of the emergency distress [call]. **1971** *New Scientist* 16 Sept. 629/1 Mexico's Professor S. F. Beltran homed in on education as a critical need. **1972** *Daily Tel.* 7 June 2/8 A killer satellite is one which can home in on other objects in earth orbit and destroy them.

ho·me-along, *adv. dial.* Homewards.

1874 HARDY *Far fr. Madding Crowd* I. iv. 44 So I'll take myself home-along. **1905** E. PHILPOTTS *Secret Woman* II. vi. 154 I'll take Salome home-along presently. **1912** C. MACKENZIE *Carnival* xxxiii. 343 Well, I'm going home-along myself in November month.

home boy. Also home-boy. [f. HOME *sb.* + BOY *sb.*] **a.** A boy who is fond of staying at home. **b.** *Canad.* A boy who has been brought up in an orphanage or institution. (See HOME *sb.*[1] A. 8.) **c.** *U.S.* (See quot. 1970.)

1886 [see HOME *sb.*[1] A. 14 b]. **1913** S. A. FRANCIS *Canadian Home Boy* iii, The Canadian reader will need no explanation of the title 'Home Boy', but to the British

reader this term will convey little meaning... It denotes a boy who has been brought up in some charitable 'Home', and from whom little that is good is expected. **1932** N. M. JAMIESON *Cattle in Stall* 192 [A] lonesome little English home boy [was] playing his mouth organ softly in the dusk. **1967** *Amer. Speech* XLII. 238 *Home boy* is a slang expression particularly in vogue among students at Southern Negro colleges... *Home boy* and similar forms, such as *home girl* and *home people*, denote individuals who come from the same hometown as the speaker. **1970** C. MAJOR *Dict. Afro-Amer. Slang* 66 *Homeboy*, person from one's home town.

home-brew. Add: **2.** *Canad. Sport.* A player, *spec.* of professional football, who is native to the country, town, etc., which he or his team represents. Also *attrib.*

1957 *Star Weekly* (Toronto) 17 Aug. 9/1 The Leos still are short of homebrews and are plagued by a problem at quarterback they have never succeeded in solving. **1958** *Edmonton* (Alberta) *Jrnl.* 18 June 12/3 To make room for import talent en route from the States, Edmonton Eskimos announced the release of 16-year-old homebrew third sacker Gene Kinesewich today. **1964** *Winnipeg Free Press* 19 June 36/4 The chief economic factor is that American players, whether Canadianized or not, still draw more salary than homebrews. **1970** *Toronto Daily Star* 24 Sept. 18/2 So what has become of Mike Wadsworth..and Walt Balasuk, another homebrew tackle?

home center. *U.S.* [f. HOME *sb.*[1] 2 + *CENTRE 6 a.] A shop which sells building, hardware, and decorating materials for the home. Also *attrib.*

1966 *Building Materials Merchandiser* Dec. 59/2 Small towns everywhere can provide a good base for a modern, one-stop Home Center where hardware, paints and related items are sold together with lumber and building materials. **1971** *Hardware Retailer* June 41 Today's Home Center is a hardware store. **1972** *Hardware Retailing* May 298/2 That these products are hot sellers is borne out by the experience of thousands of home centers. **1972** *Home Center* June 14 Handyman, the Home Center chain with 26 stores in California, the northwest and Scottsdale, Ariz., will move into Texas this year. *Ibid.* 53 Today's Home Center is much more than a lumberyard, but building materials and hardware are the heart of every Home Center operation.

ho·me-co:mer. Add: [HOME *adv.* 8 a.] (Earlier and later examples.) Also in more recent use with special reference to the Isle of Man.

1540 PALSGRAVE tr. *Acolastus* sig. C1, The father reioyseth the sonne to be a safe home commer vnto hym. **1927** *Peel City Guardian* 27 May, The Homecomers will arrive at Montreal during the afternoon of June 2nd. **1930** *Ibid.* 21 June 6/1 The White Star liner, 'Doric', will anchor in Douglas Bay..with 279 'homecomers' on board. **1947** *Chicago Tribune* 1 Nov. 19/5 Illinois' embattled players..adjourned to the Champaign Country club..to escape a horde of homecomers who are flooding this university community.

home-coming, *sb.* Add: Also *attrib.*, and with special reference to the Isle of Man.

1903 *Westm. Gaz.* 11 Sept. 4/2 The only greens.. that there is any reasonable chance of getting on to at St. Andrews are the greens of the home-coming holes. **1926** *Peel City Guardian* 26 June 2/2 The Mayor of Douglas intends to do what he can..to promote a big home-coming of Manx people and their descendants for June of next year. **1927** *Ibid.* 7 May, A meeting of the Executive Committee in connection with the Manx Homecoming movement. **1935** *Chronicle-News* (Trinidad, Colo.) 16 Oct., A 'Homecoming Dance' at the Trinidad high school will be held Saturday evening following the Trinidad-Salida football game. **1947** R. ALLEN *Home made Banners* iii. 17 In the crowded beer parlors they tended to herd together, like old grads on the eve of a homecoming game.

home-coming, *a.* Add: **b.** That comes, or is coming, home.

1898 *Month* Nov. 487 The lowing of the home-coming cattle.

ho·me-croft. = CROFT *sb.*[1] 2. In accordance with a housing scheme for industrial workers, a detached cottage, with land and outbuildings for poultry and other small livestock. Also *attrib.* Hence **ho·mecrofter, ho·me-crofting** *vbl. sb.*

1859 [see HOME A. 14 c]. **1925** *Public Opinion* 7 Aug. 121/1 The industrial workers to spread out and become home-crofters as well as workers. **1925** *Spectator* 5 Dec. 1018/1 Dr. Hilda Clark, whose work in Vienna was of such immense importance, has also sent us a terse but comprehensive account of the work and its objects. We note the interesting fact that she now calls these Land Settlements 'Homecroft Holdings'. *Ibid.*, We do not think that the word 'homecrofting' had been heard of in Vienna three years ago. **1926** *Ibid.* 24 July 130/1 The land and buildings shall be used in perpetuity as 'home-crofts'. **1927** *Daily Express* 21 Nov. 3/1 Earl Beauchamp opened the first six cottages of a 'home-crofting' experiment.

home-defe·nce. [HOME *sb.*[1] 14 d.] The defence of one's native or home country; an armed force designed for this. Also *attrib.*

1642 [see HOME *sb.*[1] 14 d]. **1671** J. OGILBY *America* ii. 29 Lest if Carthage should be invaded by a foraign Enemy, it should want People for a Home-defence. **1885** *Marine Engineer* 1 July 89/2 With such a fleet.. we could dispense with the Channel Squadrons, for home defence. **1904** *Rep. R. Comm. Militia & Volunteers* in *Parl. Papers* XXX. 16 A home-defence army. **1917** 'CONTACT' *Airman's Outings* 189 Mention must also be made of the Home Defence groups, but for which wholesale Zeppelin raids on the country would be of common occurrence. **1922** *Encycl. Brit.* XXXI. 83/2 On the War Office taking over the responsibility for anti-aircraft defence from the Admiralty in Feb. 1916, a definite Home Defence organization was adopted. *Ibid.*, In June the Home Defence wing was formed to include all Home Defence units. **1934** S. BALDWIN in *Hansard Commons* 19 July 1275 Of these 41 squadrons, 33 will be allotted to Home Defence, raising the existing 42 squadrons at home to a total of 75 squadrons. **1939** *War Illustr.* 4 Nov. 237/1 Further openings for the older men will be given in two new directions..Home Defence Battalions..and an Auxiliary Pioneer Corps which will take over military pioneer work, both overseas and here. **1964** E. H. POWELL in I. L. Horowitz *New Sociol.* 342 The home-defense brigades partially filled this need.

home econo·mics. orig. *U.S.* [HOME *sb.*[1] B. 1.] The art or science of domestic economy. Hence **home eco·nomist.**

1899 (*title*) Lake Placid conference on home economics proceedings. **1926** *Chicago Drovers' Jrnl.* 5 May 3/3 Here all of the home economics work..will be housed. **1943** M. LYON *And so to Bedlam* 262 The young woman..was a graduate home economist. **1960** A. E. BENDER *Dict. Nutrition* p. v, Medical practitioners,..home economists, ..all, from time to time, step into the food field. **1963** *Weekly News* (Auckland) 27 Mar. 11/1 Miss Elsa Haglund, of Sweden, home economics officer of the [United Nations] Food and Agriculture Organisation, with headquarters in Rome. *Ibid.* 11/2 It is the job of the home economist to study..the..living patterns in any district. **1969** *Daily Nation* (Nairobi) 7 Nov. 32/1 A thorough knowledge of hygiene, nutrition and home economics is essential and applicants must be in possession of a valid driving licence. **1970** *Globe & Mail* (Toronto) 25 Sept. B 6/4 (Advt.), Applicants should have a diploma or a degree in Home Economics or Food Science and experience in the formulation of foods. **1973** *Jrnl. Genetic Psychol.* CXXII. 309 A home economics class composed of 28 eighth-grade girls.

home-fire. Used, like *hearth*, as symbolic of the home and family life, and especially popular during the war of 1914–18 in phr. *to keep the home-fires burning*: to keep the home going, to 'carry on' at home.

1892 I. ZANGWILL *Childr. Ghetto* I. I. xiii. 288 Happy fathers of happy children, men who warmed their hands at the home-fire of life. **1914** IVOR NOVELLO *Song*, Keep the home-fires burning, While your hearts are yearning. **1928** D. L. SAYERS *Unpleasantness at Bellona Club* iii. 29 Health gone—no money—heroic wife keeping the home fires burning. **1931** P. WILLIAMS *Word of To-Morrow* IV. i, When they were holding the line in France, and he and all the other old gollywogs were keeping the home fires burning, snug and comfortable. **1972** *Listener* 6 July 13/2 Famine, rapine, rape... All keep the home fires burning, spread good cheer.

ho·me-folk, -folks. *colloq.* [HOME *sb.*[1] 14 i.] The people from or near one's home, *i.e.* one's friends, relatives, or neighbours. Chiefly *U.S.*

1884 'C. E. CRADDOCK' *In Tenn. Mts.* 288 All the home-folks, an' everybody that kems hyar to sot an' talk. **1900** J. W. RILEY (*title*) Home-folks. **1907** *Westm. Gaz.* 21 Sept. 6/2 They wander from the home-folks' ken. **1909** *Ibid.* 2 Dec. 2/3 Of home-folk caught by crumbling walls. **1915** H. L. WILSON *Ruggles of Red Gap* (1917) v. 99 Then we tried his home-folks in Boston. **1955** E. POUND *Classic Anthol.* I. 19 When I come in from being out My home-folk don't want me about. **1964** MRS. L. B. JOHNSON *White House Diary* 7 Jan. (1970) 33, I greeted L. F. McCollum of Continental Oil of Houston as homefolks.

home front: see *FRONT *sb.* 5 f.

ho·me-guard, Home Guard. [HOME *sb.*[1] 14 a.] **a.** A member of a local volunteer force. *U.S.* **b.** *Hist.* In England, the Territorial Forces.

1861 *Richmond* (Va.) *Examiner* 6 Sept. 2/4 The Secessionists..attacked the..Home Guards. **1862** O. W. NORTON *Army Lett.* (1903) 123 Nothing would make me ready to fight sooner than to hear some home guard abuse McClellan. **1873** 'MARK TWAIN' & WARNER *Gilded Age* xviii. 170 He was captain of the home-guards in Hawkeye. **1891** *Century Mag.* Jan. 409 An unexpected musketry fire was opened from the Indiana side by a party of home-guards. **1896** *Congress. Rec.* 25 Apr., App. 298 Before Gen. Burnside came, the mountain men of East Tennessee organized themselves into companies called 'Home Guards'. **1909** *Westm. Gaz.* 7 Jan. 7/3 A movement has been set on foot which has for its object the presentation of colours to the 'Home Guard', as we call the Territorials. **1919** F. HURST *Humoresque* 245, I know half a dozen who have got in the home guard..and have saved themselves by volunteering from being sent to France. **1969** E. W. MORSE *Fur Trade Canoe Routes* I. ii. 15 With the Hudson's Bay Company it was the 'Home Guard' Indians, the Swampy Crees, who served as middlemen.

c. One who lives or works continuously in the same place; *spec.* (*a*) a resident logger; (*b*) a non-migrant beggar, hobo, or thief. Also *collect. N. Amer. colloq.*

1919 *Camp Worker* (Vancouver) 19 Sept. 8/3 Camp poorly organized; too many home guards. **1923** N. ANDERSON *Hobo* i. 7 The vagabond who has settled and retired, the 'home guard' as they are rather contemptuously referred to by the tribe of younger and more

adventurous men who still choose to take the road. *Ibid.* vii. 96 The home guard, like the hobo, is a casual laborer, but he works, often only by the day, now at one and again at another of the multitude of unskilled jobs in the city. **1926** J. BLACK *You can't Win* xvi. 219 The gangs [of thieves] are made up of natives and 'home guards'. **1931** 'D. STIFF' *Milk & Honey Route* iii. 37 At the other extreme from the ramblers we find a large variety of home guards who keep pretty much to one locality. **1942** R. E. SWANSON *Rhymes of Western Logger* 35 You talk of your drums! you home-guard bums should have seen the size of her 'main'! **1955** *Publ. Amer. Dial. Soc.* xxiv. 98 They will say about home guards, they'll say, 'Pay no attention to him..he's just a local character.'

d. In form *Home Guard.* The military force organized in 1940 for the defence of Great Britain and Northern Ireland against possible invasion, orig. called Local Defence Volunteers. Also a member of this force. (Disbanded 31 July 1957.) Similarly in other countries.

1940 W. S. CHURCHILL *Into Battle* (1941) 251 Behind the regular Army we have more than a million of the Local Defence Volunteers, or, as they are much better called, the 'Home Guard'. **1942** *Ann. Reg.* 1941 45 On May 14 the Home Guard celebrated its first anniversary. **1943** *N. & Q.* 10 Apr. 220/1 A company of a Home Guard battalion. **1945** *Daily Mirror* 15 Aug. 3/1 When an eighteen-year-old Home Guard decided to impersonate his brother and take his place in the Army he ran into trouble right away. **1970** *Daily Tel.* 21 Mar. 4/7 Lebanon is to establish a 'home guard' to police the border areas with Israel, where tension between Palestinian guerillas and the Lebanese Army has been high. **1970** *New Yorker* 19 Sept. 34/1 At the entrance to the hospital [in Israel], a Home Guard..sucked furiously on his unlit pipe.

home help. [HOME *sb.*[1] 14 b.] A domestic worker; *spec.* a woman made available by local authorities, etc., for help in the home.

1900 *Daily News* 13 June 8/6 The West Central Jewish Girls' Club... Its objects are educational, recreational, and religious. There are 221 members..home-helps, 17. **1939** M. SPRING RICE *Working-Class Wives* 14 While women..go into hospital..trained home helps can be provided to look after the father and children. **1958** *Times* 24 Feb. 11/1 One obvious solution to the problem is for local authorities to provide a better service of home helps and home nurses. **1958** P. TOWNSEND in N. Mackenzie *Conviction* 117 There is the nucleus of a home help service. **1973** *Times* 17 Jan. 4/2 Their response epitomizes some of the misconceptions about home helps which Mrs Clark..rebuts.

homeland. a. (Later examples.)

1941 W. S. CHURCHILL in *Second World War* (1950) III. 583 We should therefore face now the problems..of driving Japan back to her homelands and regaining undisputed mastery in the Pacific. **1963** *Ann. Reg.* 1962 314 A series of 'independent' Bantu 'homelands', where each ethnic group would have a vote. **1968** G. JONES *Hist. Vikings* III. i. 145 The political and dynastic history of the Scandinavian homelands. **1970** *Caribbean Studies* July 90 With such chronological spacing for Homeland English..it is not surprising that..English usage in colonised places also had its period of graduation before being lexicographically chronicled, and..that such chronicles should be..modest imitations of the recognised Homeland product. **1970** *Nature* 24 Oct. 311/2 One of the regions of South Africa now set aside as a 'homeland' for the native population.

b. = HOME *sb.*[1] 6.

c **1892** C. BINGHAM *Song*, The Dear Home-Land. **1899** *Daily News* 26 Oct. 7/1 Looking..at the old Home-land through the eyes of Young Australia. **1905** *Daily Chron.* 22 June 6/6 We are here to-night..a body of Canadian business men, chiefly in order that we may learn to know the people of the homeland. **1907** SCOTT & WALLAS (*title*) The call of the Homeland: a collection of English verse.

home language. [HOME *sb.*[1] B. 1 and 3.] The language spoken in one's home; one's native language; the mother-tongue.

1926 M. WEST *Bilingualism* I. i. 14 The Magh has Magh as his Home language. **1933** L. BLOOMFIELD *Language* iii. 56 Ordinarily one language is the *home language*, while the other serves a wider range. **1934** *Cape Argus* 3 May 9/6 The only thing to do, therefore, is to define the words 'home language' as the language best known and understood by the child. **1957** R. MACNAB *Emergence of Afrikaans* 4/1 Some of the descendants of those great Scottish churchmen now have Afrikaans as their home language, while their cousins have English. **1958** *Sunday Times* (Johannesburg) 21 Sept. 25/2 Applicants must be bilingual and preference will be given to those whose home language is Afrikaans.

ho·me-leave. [HOME *sb.*[1] 6.] Leave, often of fairly lengthy duration, granted to officials and others serving overseas.

1923 KIPLING *Irish Guards in Gt. War* I. p. viii, Their short home-leaves gave them sudden changes to the tense home atmosphere. **1946** *Nature* 7 Sept. 320/2 Home-leave at regular intervals and free, or at least assisted, passages for themselves and their families..should be provided for officers whose homes are not in the Colony in which they serve. **1962** *Times* 6 Aug. 9/6 The parents' frequent home-leaves. **1973** *Guardian* 30 May 15 Lieutenant Mark Phillips ..was hurried off to Germany to join his regiment... Within the fortnight he was back again on weekend leave, prompting Mr Arthur Lewis to ask a Parliamentary question..about whether officers of the Queen's Dragoon Guards got special privileges for home leave.

ho·me-life. [HOME *sb.*[1] 14 b.] Life at home or in domestic surroundings.

1846 QUEEN VICTORIA *Jrnl.* 8 June in E. Longford *Victoria R.I.* (1964) I. xiii. 184 When one is so happy and blessed in one's home life, as I am, Politics..must take only a 2nd. place. **1859** C. FOX *Jrnl.* 5 Jan. (1882) xxv. 344 Settled once more into dear, beautiful home-life. **1867** 'T. LACKLAND' *Homespun* I. 39 The sincerest pleasures of the home-life are woven closely in with those of the garden. **1871** [see HOME B. 1]. **1879** *Rep. Indian Affairs* 32 The absence of the example of the better home-life of our own people. **1898** T. N. PAGE *Red Rock* vii. 64 These men were thoroughly enjoying home life. **1906** J. LONDON *Let.* 8 Apr. (1966) 199 In addition to home-life articles..I could furnish you other descriptive articles suitable for your pages. **1933** *Punch* 31 May 608/2 The kind of thing that might quite easily wreck one's home-life altogether. **1973** E. LEMARCHAND *Let or Hindrance* xii. 148 Most of my women are married, and one doesn't want to disrupt what home life they have.

home-made, *a.* Add: (Further examples.) Also *ellipt.* as *sb.*

1898 *Westm. Gaz.* 17 Nov. 7/1 There are two fogs familiar to the Londoner—the 'home-made fog', still, cold, anticyclonic, [etc.]. **1932** BLUNDEN *Face of Eng.* 110 A box or two of popcorn and 'home-mades' in the front window of a cottage. **1934** H. G. WELLS *Exper. Autobiogr.* II. ix. 808 He [*sc.* Stalin]..preferred that it [*sc.* criticism] should be home-made by the party. **1946** A. HUXLEY *Let.* 19 June (1969) 547 Whether there are powers of evil other than our own home-made devils is an open question. **1955** [see *FARM-HOUSE b]. **1959** *Brno Studies* I. 24 The most important of such home-made digraphs is obviously *gh*, which replaced the old grapheme *h* in medial and word-final positions.

ho·me-maker. [HOME *sb.*[1] 14 h.] A housewife, esp. one in charge of the domestic arrangements (as opp. to a paid housekeeper); also, one who manages a household. So **ho·me-making** *sb.* and *a.*

1876 C. M. YONGE *Womankind* xxx. 266 Home-making is..her paramount earthly duty. **1886** [see HOME *sb.*[1] 14 h]. **1890** W. BOOTH *In Darkest Eng.* I. viii. 66 The general shiftlessness from the home-making point of view. **1895** J. R. MILLER (*title*) Home-making, or the ideal family life. **1905** *Daily Chron.* 10 May 4/5 There could not be a greater boon to the many distressed home-makers of Canada than the advent of..domestic servants from the British Isles. **1934** *G. K.'s Weekly* 11 Jan. 306/1 The breathless never-can-we-catch-up feeling ..is essential to the happiness of the true homemaker. **1954** J. STEINBECK *Sweet Thursday* 88 It was real nice. She was a home-maker. **1963** A. HERON *Towards Quaker View of Sex* ii. 19 Home-making is a satisfying outlet for many women. **1968** *Ideal Home* Nov. 42/1 Homemaking young marrieds. **1973** *N.Y. Law Jrnl.* 31 Aug. 1/2 These services might include specialized centers which can provide extensive supervision and education during the day, twenty-four-hour home-maker services, and counselling.

home market. [HOME *sb.*[1] B. 3.] The market for goods or produce in the place or country of production.

1758 HUME *Ess.* II. vi. 188 A great number of commodities are raised and perfected for the home-market. **1776** ADAM SMITH *Wealth of Nations* I. I. xi. 243 The bounty.. may..have occasioned..a greater cheapness of corn in the home-market than what would otherwise have taken place there. **1794** [see HOME B. 3]. **1813** *Niles' Reg.* IV. 274/1 Then shall the home market still be the rage for foreign export. **1843** *Amer. Pioneer* II. 214 Money became plenty, and a cash home-market was established. **1847** C. LANMAN *Summer in Wilderness* i. 14 This city is the home market for all the natural productions of a wilderness country. **1892** *Rep. Vermont Board Agric.* XII. 132 Furnishing a home market for their products. **1972** D. G. RHYS *Motor Industry* ii. 40 The immediate post-war concentration on export markets had repercussions on the home market.

home movie. [HOME *sb.*[1] B. 1.] A home-made movie; a film made of the activities of one's own circle. Also *attrib.*

1939–40 *Army & Navy Stores Catal.* 910 Home movie camera. **1941** *Time* 16 June 33/1 Once 'expense' might have been a reason for doing without home movies. **1967** *Listener* 27 July 123/2 Some home movies of a retired politician. **1968** M. RICHLER *Cocksure* x. 53 The usual home movie stuff. Mortimer mowing the lawn. **1969** *Listener* 1 May 594/1 Old newsreels, live film (as distinct from reconstructions) of men and events—home movies, for that matter—are primary sources.

home perm. [HOME *sb.*[1] B. 1.] A permanent wave in the hair produced by equipment designed for use in the home (as opp. to one prepared professionally in a hairdressing establishment). So **home-perm** *v.*, **home-permed** *ppl. a.*

1949 *Woman's Own* 29 Sept. 3/3 (Advt.), Give yourself a Toni Home Perm today! **1954** M. ALLINGHAM *No Love Lost* i. 129, I..bought..one of those cheap home perm outfits. **1955** G. FREEMAN *Liberty Man* III. vi. 173 The rows of industrious home-permed heads over the desks. **1959** [see *APPLICATOR]. **1959** *News Chron.* 11 Aug. 6/2 Home perms are now being used on many children. **1969** J. FREDMAN *Fourth Agency* i. 8 Her iron-grey hair had been home-permed.

ho·me place. *U.S.* [HOME *sb.*[1] 14 a.] The place or piece of ground where one's home is situated.

1736 in *N.H. Probate Rec.* II. 625, I give & bequeath.. my Dwelling and Devise to my Son..all my land..Known by the name of my home place. **1885** *Weekly New Mexican* 12 Feb. 4/3 They are connecting D.D.'s outlying ranches with the home place. **1931** *Amer. Speech* VII. 93 *Home-place*, the part of a farm on which the house and out-buildings are located. **1946** G. FOREMAN *Last Trek of Indians* 91 They continued in their attachment to the old home place. **1973** *Amer. Folklore Newslet.* Spring 2/2, I enjoy going back to the old home place, now deserted, and the double-log cabin falling into decay.

home plate. *Baseball.* [HOME *sb.*[1] B. 4.] The plate at the apex of the diamond at which the batter stands, and which must be touched by the base runner before a run is scored.

1875 *Chicago Tribune* 3 Aug. 7 He stole third..and reached the home-plate. **1889** 'MARK TWAIN' *Yankee* xxii. 245 No fault of hers that she couldn't fetch the home-plate. **1891** N. CRANE *Baseball* 79 *Ball*, a pitched ball, which does not pass over the home plate [etc.]. **1957** *Encycl. Brit.* III. 160/2 The outfielders are called right fielder, centre fielder and left fielder with relation to a man standing on home plate. **1970** *Globe & Mail* (Toronto) 26 Sept. 51/5 Let's appoint the umpires...Porky, you take homeplate.

homer[1]. Add: **2.** [HOME *sb.*[1] B. 4.] In *Baseball*, a home run. So **homer** *v.*, to hit a home run.

1868 *New Eng. Base Ballist* 6 Aug. 3/1 The second inning saw a change as the Champions went out for two runs, one of these a 'homer'. **1951** in WENTWORTH & FLEXNER *Dict. Amer. Slang* (1960) 265/1 Bobby Thomson, who homered..and Monte Irwin, who also homered. **1961** *Listener* 19 Oct. 594/2 A homer is a mighty hit at baseball: into the stands or out of the ground—the equivalent of a six at cricket. *Ibid.*, Babe Ruth hit his sixty homers in a season of 154 games. **1967** *Boston Herald* 8 May 16/2 Fregosi homered in the fifth and Knoop in the sixth off reliever Bob Humphreys for the only Angel extra-base hits. **1972** *N.Y. Times* 4 June v. 1/8 Johnny Bench had homered in the seventh.

3. [HOME *sb.*[1] 6.] (See quot. 1945.) *Austral.* and *N.Z. slang.*

1945 BAKER *Austral. Lang.* viii. 156 *Homer*, a wound sufficiently serious to cause a man to be sent home. **1949** E. DE MAUNY *Huntsman in Career* 180 'Don't say you've got a "homer" already, mate.' He pointed to Peter's bandaged hand. **1950** G. WILSON *Brave Company* xi. 173, I nearly did get a homer that time.

4. [*HOME *v.* 5.] A homing device.

1958 *Chambers's Techn. Dict.* Add. 985/1 *Homer*, any arrangement which provides signals or fields which can be used to guide a vehicle to a specific location, usually determined by a homer transmitter. **1959** *Daily Tel.* 1 June 9/3 Thunderbird is what is known as 'a semi-active homer'. It receives the echoes of the illuminating radar's beam in a set in its own nose and homes on to the target accordingly. **1959** 'J. WYNDHAM' & PARKES *Outward Urge* i. 29 Fix up one of the dispatch homers on it, and let it jet itself along.

ho·meroom. *U.S.* [HOME *sb.*[1] 5.] (See quot. 1961.)

1915 *Ann. Rep. Comm. Educ.* 1914 (U.S. Office of Educ.) I. iii. 45 The regular or home-room teacher feels a definite responsibility for the class wherever it may be. **1928** *Bull. Nat. Assoc. Secondary-School Principals* XIX. 16 The foundations of all guidance within the school are found in the homeroom organization. Efficiency in secondary school administration is demanding that every teacher shall be a 'homeroom' teacher. **1930** *School Rev.* Apr. 300 The home-room organization..in Detroit may be defined as a large number of one-room schools.. gathered together under one roof with a central office to facilitate their proper functioning. *Ibid.*, The home room is the unit of organization of the school and is the center of the pupil's school activities. **1961** WEBSTER, *Homeroom.* 1. A schoolroom where pupils of the same class or grade but often with different academic programs report at the opening of school and meet informally under the guidance of a teacher to conduct class business, plan and organize group activities, and discuss individual and group problems. 2. A group of pupils assigned to the same homeroom. **1966** BEREITER & ENGELMANN *Teaching Disadvantaged Children* iv. 71 The homeroom should have at least 400 square feet of floor space, and should be equipped with a piano and (like the other rooms) a chalkboard. **1968** 'R. MACDONALD' *Instant Enemy* i. 8, I got in touch with her homeroom teacher and found she'd been playing hookey all day.

home science. [HOME *sb.*[1] B. 1.] The art or science of domestic economy.

Term not used in Britain.

1912 *Calendar Univ. Otago* 79 The Home Science Department (opened in April, 1911) has been housed for the present in part of the old Mining School. **1943** J. H. MURDOCH *High Schools N.Z.* vi. 362 Home science students may substitute arithmetic for mathematics. **1963** F. F. LAIDLER *Gloss. Home Econ. Educ.* 46 The term 'Home Science' as used in New Zealand is, in general, comparable with the terms 'Domestic Science' in the U.K. and 'Home Economics' in the U.S.A. **1969** *Australian* 24 May 31/5 (Advt.), Home Science Teacher required immediately for second term, Woodstock Presbyterian Girls' School. **1969** *Hindu* 3 Aug. 2/2 (Advt.), Good academic record..preferably a B.Sc. degree in Home Science.

ho·mesite. *N. Amer.* [*HOME *sb.*[1] 2.] = HOUSE-LOT.

1911 *Daily Colonist* (Victoria, B.C.) 2 Apr. 12/2 Saanich Waterfront... Magnificent homesites of nearly two acres from $450. **1968** *Globe & Mail* (Toronto) 17 Feb. 46/6 (Advt.), Landscaped, fenced homesite.

homestead, *sb.* Add: **2. b.** Freq. in Australia and N.Z.: the residence of the owner of a sheep or cattle station; in later use also = STATION *sb.* 14 (quot. 1898).

1849 *Handbk. Suburban & Rural Districts Otago Settlement* 7 Sheep or cattle owners, who, establishing their temporary homesteads, or stations, near or in the bush, might run their flocks or herds amongst the hills. **1851** E. SHORTLAND *Southern Districts N.Z.* xiv. 263 Farmers and stock-keepers, however, who have their homesteads on the plain. **1853** C. W. ADAMS *Spring in Canterbury Settlement* vii. 70 This homestead much resembles a small English farm-house, save that the sleeping loft had seldom fewer than ten occupants. **1891** R. PRICE *Through Uriwera Country* 61 As an out-station is to the homestead of a sheep-run, so is this little fort to Te Teko. **1901** M. FRANKLIN *My Brilliant Career* (1966) viii. 39 Home to.. the dear old homestead I love so well. **1911** C. E. W. BEAN *'Dreadnought' of Darling* i. 4 Most homesteads are apt to consider themselves pretty well on the outskirts of things if they get only one mail a week. At this particular station they get two mails a year. **1930** L. G. D. ACLAND *Early Canterbury Runs* 1st Ser. i. 8 'Station' is being driven out of use in its original sense of 'a place from which to work a run' by 'homestead'... When an old fashioned squatter or station hand used the word 'homestead' he used it to signify the owner's residence as opposed to the men's quarters and other station buildings. **1933** —— in *Press* (Christchurch) 28 Oct. 15/7 *Homestead.* In the old days, the owner's residence, but only if it was some distance from the rest of the station buildings... When the Government began cutting up the runs in the 'nineties, they or their surveyors adopted the word to signify what had been formerly called *the station*, and the new settlers followed them, so that the new sense of the word is now widely used. **1941** I. L. IDRIESS *Great Boomerang* ii. 10 Fifty miles from the homestead, on the New South Wales–South Australian border. **1946** F. DAVISON *Dusty* ix. 97 People came from all over the settlement, camping in and around the old homestead. **1961** B. CRUMP *Hang on a Minute* 73 They arrived at the Paranui homestead in the late afternoon.

homestead, *v.* (Earlier and later N. Amer. examples.)

1872 *Newton Kansan* 12 Sept. 3/3 [He] had homesteaded the south-east quarter of sec. 14. **1877** H. RUEDE *Sod-House Days* (1937) 123 If he homesteads, you would have to be here inside of 6 months. **1879** *Congress. Rec.* 26 Apr. 952/1 To prove their right to pre-empt or 'homestead' their lands. **1912** J. SANDILANDS *West Canadian Dict.* 23 Any person who is the sole head of a family, or any male over 18.., may homestead a quarter-section of..land in Manitoba, [etc.]. **1952** J. STEINBECK *East of Eden* 6 There was still marginal land to be homesteaded. **1959** *Times Lit. Suppl.* 6 Nov. p. xii/4 Yoknapatawpha County is William Faulkner's just as much as if he had homesteaded there and proved his claim. **1972** *New York* 12 June 15/3 A couple who leave Lima to homestead in the jungle.

homesteader. (Further, incl. non-U.S., examples.)

1872 J. H. TICE *Over Plains* 80 As far as the eye can reach the plain is dotted with new shanties of the homesteaders and pre-emptioners. **1882** *Brandon Daily Mail* (Manitoba) 23 Dec. 4/3 Some of the neighbors found the old homesteader near his tent dead. **1906** *Daily Colonist* (Victoria, B.C.) 6 Jan. 5/5 Each route north was lined this summer and fall with homesteaders, timber cruisers and miners. **1909** *Daily Chron.* 30 June 4/4 For these homesteaders of the ultimate [*sc.* Canadian] wheat-lands..the main ordeal will be..loneliness and monotony. **1928** W. BAUCKE *Where White Man Treads* (ed. 2) 307 The nation composed of a large salting of freehold homesteaders will be the staunchest and strongest. **1950** J. JENKS *From Ground Up* xix. 204 Those who have the qualities and the desire to become homesteaders.

homesteading. Add: **2.** The granting of land according to the Homestead Act of Congress, 1862 (see HOMESTEAD *sb.* 3). *U.S.* Also, a similar settlement in Canada.

1891 *Grip* (Toronto) 13 June 377/1 Mr. Dabin moved that certain settlers in the North-West be granted the privilege of second homesteading. **1906** *Daily Chron.* 1 Oct. 4/4 There is no more homesteading—viz., free grants of land—it has to be bought. **1925** *Glasgow Herald* 10 Aug. 4 Homesteading has been a great factor in the agricultural development of Western Canada. **1936** *Scrutiny* IV. IV. 441 *Honey in the Horn* is a detailed account of life in the state of Oregon during the homesteading period (1906–8).

ho·me-stretch. *U.S.* [HOME *sb.*[1] B. 4, STRETCH *sb.* 8.] The return stretch of a course; *esp.* the stretch of a racecourse on which a race finishes. Also *fig.*

1841 *Picayune* (New Orleans) 19 Jan. 1/6 At the head of the home stretch Cowboy overtook him and..beat him out by a length. **1860** [see *BEEF *v.* 2]. **1861** *Trans. Ill. Agric. Soc.* IV. 38 On the last home stretch the steam [of the plough] became so low that it required some minutes to get up sufficient to run the furrow through. **1864** *Congress. Globe* 12 Mar. 1069/3 Already we see the slave States ..on the home-stretch to become free. **1868** H. WOODRUFF *Trotting Horse* xxiv. 207, I passed first one and then the other, and came on the home-stretch with a clear lead. **1878** *Trans. Ill. Dept. Agric.* XIV. 146 Still, a fleet horse who gathers up handsomely on the home-stretch, is not to be sneezed at by any one. **1897** [see HOME *sb.*[1] B. 4]. **1903** [see *BREAK *sb.*[1] 8 b]. **1968** *Daily Tel.* 4 Nov. 17/1 It was a furious home-stretch spurt that will last until he reaches his home in Minnesota at five a.m. on Tuesday, election day.

ho·me-town. orig. *U.S.* [HOME *sb.*[1] 14 a.] The town in which one's home is, or was originally; one's native town. Also *attrib.*

1912 *Top-Notch Mag.* (U.S.) 1 Aug. 64/2 He was killed in a pool-room row in my home town up the State. **1919** H. L. WILSON *Ma Pettengill* xi. 307 Having got the thanks of the French nation and his home-town paper. **1935** R. FROST *Let.* 21 Aug. (1964) 262 The sordidness..of home-town society all over Russia. **1938** *Times Lit. Suppl.* 6 Aug. 514/4 'Home-town' mixture, Scandinavian, Italian, ..all, whatever the race, nationality, class or occupation, end..by being good fellows. **1948** *Oakhillian* Summer 12 The other dignitaries returned to their various hometowns. **1953** [see *FAKE *sb.*[2] 3]. **1971** *Nat. Geogr. Mag.* Oct. 551/2 They [*sc.* boats] hold about a dozen people each and introduce the visitor to some of the more interesting of the hometown folk.

ho·mework. [HOME *sb.*[1] 14 b.]

1. Work done at home, esp. as distinguished from work done in a shop or factory. Also *attrib.*

a **1683** B. WHICHCOTE *Sel. Serm.* (1698) 402 Wherefore let every Man, in the first place, look after his Home-work; what he hath to do at Home. **1856** [see HOME *sb.*[1] 14 b]. **1861** G. MOORE *Lost Tribes* xviii. 364 One [apartment] more open and larger is reserved for visitors, or..is used for spinning or other home-work. **1901** *Act 1 Edw. VII.* c. 22 § 110 Prohibition of home work in places where there is infectious disease. **1907** *Rep. Sel. Comm. Home Work* in *Parl. Papers* VI. 61/1 Any instance of sweating in home work in connection with such things as ladies' jackets. **1908** *Franco-Brit. Exhib., Women's Section* 64 Frame of Quilting and Embroidery, lent by Home Work Co-operative Society. **1935** *Times Lit. Suppl.* 21 Dec. 882/1 The particularly mean swindlers who prey on the unemployed, either by 'homework' or by salaried jobs.. are duly pilloried.

b. Phr. *to do one's homework*: to do the preparatory work for a meeting, discussion, etc.

1934 R. STOUT *Fer-de-lance* iii. 40 There's three things I want to know. Or am I supposed to go up front and do my home work? **1959** *Listener* 6 Aug. 200/2 The Soviet reporter had been 'doing his homework' to some effect. **1966** M. CATTO *Bird on Wing* ii. 36, I have taken the trouble to do my homework on him. I know more about him than he thinks! **1971** *Guardian* 25 Feb. 15/2 The Government, says the report, appears not to have done its homework in considering how local radio should be financed.

2. Lessons and exercises to be done by a school-child at home.

1889 A. E. FLETCHER *Sonnenschein's Cycl. Educ.* (ed. 2) 155/1 Written home-work. **1897** C. DUKES in P. A. Barrett *Teaching & Organisation* 366 The large amount of homework which is assigned to pupils for preparation alone in the evening. **1905** *Westm. Gaz.* 28 Jan. 5/2, I think it is a wrong principle altogether to help a child with his homework. **1970** C. LACEY *Hightown Gram.* viii. 168 I've just caught two boys in the quad, copying homework.

3. *slang.* Petting; also concr., a girl-friend: used esp. in phr. *a bit* (or *piece*) *of homework.*

1942 BERREY & VAN DEN BARK *Amer. Thes. Slang* § 847/3 'Petting', 'Necking'..homework. **1945** L. LANE *How to become a Comedian* vii. 75 Bert Errol [a female impersonator], when on the stage, looks a ravishing and beautiful bit of home-work. **1948** PARTRIDGE *Dict. Forces' Slang* 95 *Homework*; a *piece of homework*,..one's sweetheart or temporary girl friend. **1953** E. AMBLER *Schirmer Inheritance* 177 'And that little bit of homework you've got with you?' 'Miss Kolin, you mean? She's an interpreter.' **1968** J. SYMONS *Man whose Dreams came True* III. v. 171 He produced a dog-eared snap of a girl in a bikini. 'How's that for a piece of homework?'

Hence **ho·me-worker, -working.**

1902 A. BALLANTYNE in T. Oliver *Dangerous Trades* vii. 98 The home-worker..has been left outside the protecting pale of the Factory Acts. **1907** *Rep. Sel. Comm. Home Work* in *Parl. Papers* VI. 19/1 The duty of seeing to the sanitary condition of home workers' premises. **1907** *Daily Chron.* 11 Feb. 4/4 The success of this home-working experiment under leasehold conditions. **1973** *Guardian* 17 Oct. 15/8 Most home workers are women..a classic case of powerless employees... They need..the flexibility of working hours that home work allows... The unions, understandably, are completely opposed to home working.

homey, homie (hōu·mi). *N.Z. slang.* [HOME *sb.*[1] 6.] An Englishman; a British immigrant, esp. one newly arrived.

1927 J. DEVANNY *Old Savage* 170 The crowd at the house were mostly 'homies'. **1939** A. E. BROWN *Farmer's Wife* III. v. 161 My fishman has a pleasant slow drawl that betrays the 'Homie'. **1939** J. MULGAN *Man Alone* 28 That Saturday night I met you and our homey here. **1953** M. SCOTT *Breakfast at Six* viii. 67 These Homies with money, they stick together. **1970** D. M. DAVIN *Not Here, Not Now* IV. i. 217 An English accent. How hard it was to remember that it was as natural to a homey as your own accent was to you.

homing, *vbl. sb.* **2.** (Further examples, *esp.* relating to animals other than pigeons.)

1901 *Cambr. Nat. Hist.* VIII. ix. 387 The same homing instinct has been observed in some females of the Green Turtle. **1907** G. B. SHAW *John Bull's Other Island* I. 16 *Broadbent*: Here you are, belonging to a nation with the.. most inveterate homing instinct in the world! and you pretend youd rather go anywhere than back to Ireland. **1922** FLATTELY & WALTON *Biol. Sea-Shore* viii. 178 There is evidence of a 'homing sense' in the common limpet and its relatives. **1939** *Copeia* III. 127 Most grown turtles (89·5 per cent) showed some homing instinct or tendency to return to territory from which they were moved. **1956** W. H. THORPE *Learning & Instinct in Animals* xvi. 412 Experiments on the homing of dogs. *Ibid.*, The homing performances of bats. **1967** GARDINER & FLEMISTER *Princ. Gen. Biol.* (ed. 2) xiv. 248 Insects respond to polarized light, and..this capacity is used in orientation and homing behavior.

3. (In sense of *HOME *v.* 5); also *attrib.*, esp. in *homing device*, an automatic device for guiding aircraft, missiles, etc.

1923 *Jrnl. Inst. Electr. Engin.* 803/2 As this method seriously affects the compass and takes a machine off its course, wing coils are only used for 'homing', i.e. flying along a radius towards a transmitting station, thus enabling the aircraft to return to its base. **1933** *Bureau of Standards Jrnl. Res.* XI. 740 In these tests the direction finder was used as a homing device. **1940** *Illustr. London News* CXCVII. 567 (*caption*) Direction-finding radio compass for work in connection with loop aerial and 'homing' beam. **1947** CROWTHER & WHIDDINGTON *Science at War* IV. 178 Another..success was gained against the acoustic homing torpedo. **1951** *Gloss. Aeronaut. Terms* (*B.S.I.*) III. 27 *Homing aids*, systems designed to guide an aircraft to an aerodrome or carrier. **1955** *Times* 28 June 8/6 There are..other means of detecting a camouflaged operations centre, such as high altitude vertical photography and 'homing' on to radio transmissions. **1957** *Oxford Mail* 20 Aug. 1/4 After launching, the weapon is guided to the target by a special homing head which picks up the ground radar beams reflected back from the enemy. **1962** *Gloss. Aeronaut. Terms* (*B.S.I.*) VI. 4 *Homing guidance*, a system wherein devices built into a missile enable it to detect and steer itself towards, or to intercept, a target.

hominid. Substitute for def.: **A.** *sb.* A member of the mammal family Hominidæ (J. E. Gray 1825, in *Ann. Philos.* XXVI. 338), of which *Homo sapiens*, man, is the only surviving species.

1889 *Cent. Dict.*, *Hominid*, one of the *Hominidæ*; a man. **1916** *Bull. Amer. Mus. Nat. Hist.* XXXV. 347 Some day..one will discover a hominid of small stature, and almost erect posture. **1925** *Glasgow Herald* 25 July 4 That the human race, with all its tentative as well as more or less realised Hominids, arose from an ancestral stock common to it and the Anthropoids. **1957** *Observer* 1 Sept. 11/1 This glittering..exhibition is..the latest in toyfairs for spoilt hominids. **1969** *Times* 17 Jan. 13/6 A fragment of the upper part of a thigh bone from a hominid which lived at least three million years ago has now been recovered from the Olduvai Gorge, in Tanzania.

B. *adj.* Of, belonging to, or characteristic of a hominid or the Hominidæ.

1916 *Bull. Amer. Mus. Nat. Hist.* XXXV. 347 The divergence of the Hominid branch occurred..from the anthropoid stem after the separation of the gibbons. **1939** C. S. COON *Races of Europe* ii. 51 During the Middle Pleistocene..a mixture took place between early white dolichocephals and one or more non-*sapiens* hominid species, including *Homo neanderthalensis.* **1971** J. Z. YOUNG *Introd. Study of Man* xxxii. 444 Hominid creatures existed a good deal earlier than had previously been supposed. **1971** *Nature* 30 July 308/1 So few identifiable parts were visible that it is remarkable it was recognized as hominid.

hominine, *a.* Add: Also pronounced (hǫ·mi-nīn). **2.** Of, belonging to, or characteristic of a hominine or Homininæ.

1959 *Cold Spring Harbor Symp. Quant. Biol.* XXIV. 244/1 The earlier forms are the more hominine... This suggests the probability that the known Australopithecines represent a somewhat specialized offshoot from the mainline of hominine evolution.

B. *sb.* [f. mod.L. *Homininæ* (G. Heberer 1949, in *Die Umschau* 1 May 258/1), the sub-family including man.] A member of the sub-family Homininæ, which is sometimes used as a division of the family Hominidæ to comprise large-brained hominids, in contrast to the small-brained ones of the sub-family Australopithecinæ.

1961 K. P. OAKLEY in *Times* 5 Sept. 13/5 Three main points of emergence in man's evolution—the first toolmakers, the first hominines, and the first men of our own species. **1963** G. G. SIMPSON in S. L. Washburn *Classification & Human Evol.* (1964) i. 29 In the Hominidae, I see no sufficient reason for having two subfamilies, especially as each has only one known genus as I and, I believe, most others now define the genera. 'Australopithecine' and 'hominine' may still be used as strictly vernacular terms for structural levels. **1971** *Nature* 6 Aug. 383/1 The bones throw some light on the structure and function of the lower limb skeleton of Middle Pleistocene hominines in East Africa.

hominist (hǫ·minist). *rare.* [f. L. *homo*, *homin-em* man + -IST.] One who advocates for men the rights and privileges conventionally accorded to women. Also *attrib.* or as *adj.*

1903 G. B. SHAW *Man & Superman* p. xviii, The wildest hominist or feminist farce is insipid after the most commonplace 'slice of life'. **1914** R. BROOKE *Let.* 10 June (1968) 592 If feminists are 'women' trying to be men, I suppose 'men' trying to be women are hominists.

hominization (hǫminəizēi·ʃən). [a. F. *hominisation* (P. Teilhard de Chardin *Le Phénomène Humain* (1948) III. i. 199), f. L. *homo*, *homin-* man + -IZATION.] The evolutionary development of characteristics, esp. mental or spiritual ones, that are held to distinguish man from other animals.

1953 J. S. HUXLEY *Evol. in Action* vi. 136 The original stock of pre-human apes differentiated into many species, all showing a trend towards what has been called hominization—the acquisition of more human characters. **1959** B. WALL tr. *Teilhard de Chardin's Phenomenon of Man* III. i. 180 Hominisation can be accepted in the first place as the individual and instantaneous leap from instinct to thought, but it is also..the progressive phyletic spiritualisation in human civilisation of all the forces contained in the animal world. **1962** W. HOWELLS *Ideas on Human Evol.* 295 (*heading*) The hominization of the masticatory apparatus, and modifications of diet.

Also **ho·minized** *ppl. a.*

1959 B. WALL tr. *Teilhard de Chardin's Phenomenon of Man* III. iii. 223 Are not the artificial, the moral and the juridical simply the hominised versions of the natural, the physical and the organic? **1973** *Times Lit. Suppl.* 27 Apr. 479/3 To supersede Rudolph Otto's notion of the holy as the numinous..and to do this for a hominized, secularized world.

hominoid (hǫ·minoid), *a.* and *sb.* [f. L. *homo, homin-* man + -OID.] **A.** *adj.* **a.** Of human form; man-like (rather than ape-like). Cf. *HUMANOID *a.* and *sb.*

1927 *Glasgow Herald* 3 Sept. 4/2 The divergence of the Hominoid and the Anthropoid branches. **1937** *Discovery* Feb. 62/1 He fixes the habitat for the first hominoid being as central Asia.

b. [ad. mod.L. *Hominoidea* (G. G. Simpson 1931, in *Bull. Amer. Mus. Nat. Hist.* LIX. 272).] Of, belonging to, or characteristic of a hominoid or the Hominoidea.

1950 *Q. Jrnl. Geol. Soc.* CV. 238 In all the hominoid species of Miocene and Pliocene age so far known..the canine is in the form of a strong pointed tooth. **1959** *Cold Spring Harbor Symp. Quant. Biol.* XXIV. 238/1 In 1949, Hürzeler in Basel re-evaluated the..*Oreopithecus* material from the 19th century, and recognized it as being hominoid. **1973** *Nature* 3 Aug. 313/1 Species of *Ramapithecus* are among the few hominoid species currently considered as possibly close to the direct line of human ancestry.

B. *sb.* **a.** An animal resembling man.

1927 *Glasgow Herald* 3 Sept. 4/2 The early hominoids.

b. *spec.* [ad. mod.L. *Hominoidea* (see above)]: a member of the superfamily Hominoidea, which includes man and the anthropoid apes.

1949 W. E. LE GROS CLARK *Hist. Primates* 74 The earliest fossil records of *true* Man, that is, of hominoids which definitely come within the family of the Hominidae, have been found in the Far East. **1950** *Q. Jrnl. Geol. Soc.* CV. 231 A great many specimens of fossil hominoids.. were collected. **1963** R. CARRINGTON *Million Years of Man* iii. 26 The smallest and most primitive of all living hominoids are the gibbons. **1967** W. E. LE GROS CLARK *Man-Apes or Ape-Men?* i. 2 The human and anthropoid ape families are now usually included in a common group, a superfamily called the Hominoidea (or, colloquially, hominoids).

hominy. **a.** (Later examples.)

1860 S. MORDECAI *Virginia* xxxi. 314 Slow as the process of 'beating hominy' is, it was a great resource, as was the eating of it for lack of hoecake. **1888** *Rep. Vermont Board Agric.* X. 30 The refuse of white corn after what is termed the 'hominy' has been removed, is more valuable as a feed for stock than yellow corn. **1922** W. G. R. FRANCILLON *Good Cookery* (ed. 2) xiv. 251 Wash and soak the hominy in water, overnight. Boil gently for half-hour or more. Stir frequently, and serve as oatmeal porridge. **1926–7** *Army & Navy Stores Catal.* 21/2 Hominy, finest pearl—bag about 5 lb. 1/4. **1959** E. TUNIS *Indians* 43/1 Wherever corn was raised in the East, hominy was an important food. **1963** R. I. MCDAVID in Mencken *Amer. Lang.* 699 In South Carolina I was taught that hominy designated what the less fortunate called grits.

b. (Further examples.)

1711 N. *Col. Rec. N. Carolina* (1886) I. 765 The planter here..dare not allow himself to partake of his own creatures except it be the corn of the country in hominy bread. **1827** J. F. COOPER *Prairie* ii, Others [were engaged] in plying the heavy pestle of a moveable hominy-mortar. *Ibid.* xvi, Giving her a morsel of venison, now and then, or a spoon around his hominy-dish. **1843** 'R. CARLTON' *New Purchase* I. xv. 111 Here were all the vulgar pots, kettles, frying-pans, hominy-block, and the like. **1876** M. N. HENDERSON *Pract. Cooking* 71 When the milk is salted and boiling, stir in the hominy grits, and boil twenty minutes. **1961** *Listener* 17 Aug. 234/2 In Dixie you are offered some strange dishes; one is called 'hominy grits'—..it tastes like a cross between porridge and tapioca.

‖**homme** (ǫm). [Fr., man.] In Fr. combinations: **homme d'affaires** (ǫm dafę̄r), a business man, an agent, a lawyer; **homme fatal**, used jocularly as the masculine equivalent of a **femme fatale*; **homme moyen** (ǫm mwayęñ), used in various phrases with defining adjective, esp. **homme sensuel moyen**, average sensual man (cf. *AVERAGE *a.* 2 b); also **homme moyen sensuel**.

1717 M. W. MONTAGU *Let.* 17 May (1837) I. 410 Every pashá has his Jew, who is his *homme d'affaires*. **1815** SCOTT *Guy M.* II. xviii. [*sic* = xvii.] 305 Dinmont..stood poking his large round face over the shoulder of the *homme d'affaires*. **1851** E. RUSKIN *Let.* 20 Dec. in M. Lutyens *Effie in Venice* (1965) II. 235 It would cost him nothing farther than a letter to his homme-d'affairs at Vienna to arrange. **1882** [see SENSUAL *a.* 4 a]. **1894** G. DU MAURIER *Trilby* III. VIII. 103 The good Taffy had constituted himself Trilby's secretary and *homme d'affaires*. **1922** C. E. MONTAGUE *Disenchantment* v. 66 Church parades, a ministration of which the average private, *l'homme moyen sensuel* of Matthew Arnold, had taken a long and glad farewell. **1928** A. HUXLEY *Point Counter Point* xiv. 266 Scientific eyes, economic eyes, *homme moyen sensuel* eyes. **1932** *N. & Q.* 8 Oct. 269/1 They represent the opinion of the *homme lettré moyen* of our day rather than give us examples of brilliant criticism. **1935** *Times Lit. Suppl.* 9 Nov. 724/4 She intrigues to interest Edward Hudson, an *homme fatal*, in her friend. **1936** C. S. LEWIS *Allegory of Love* iv. 173 Even so, long after the original reasons for the tradition have been forgotten, the *homme sensuel moyen* with his fair, large ears appears in the *Midsummer Night's Dream*. **1958** *Spectator* 7 Feb. 166/1, I find it difficult to imagine anything more nicely calculated to convince *l'homme moyen cynique* that the Parker Tribunal was an elaborate white-washing operation fixed up by the Government. **1959** *Encounter* XII. II. 32 Humbert Humbert, her *homme fatal*. **1959** *Times* 22 Sept. 11/3 A good family solicitor.. is very much the *homme d'affaires*. **1961** *Guardian* 16 Feb. 10/4 Camus finds his 'homme moyen sensuel' in Dr Rieux, who is not interested in the salvation of men but wishes to cure them. **1972** *Listener* 22 June 840/3 He works much too hard at being mysterious, at playing *l'homme fatal*.

homo. Add: (Also with pronunc. hǫ·mo.) **b.** For 'single species' read 'single living species' and add: Many other species of the genus *Homo* have been proposed, to include various fossils of extinct hominids (as *Homo neanderthalensis, H. erectus, H. habilis*, etc.). *Homo sapiens* (sē¹·pienz, sæ·pienz) [mod.L. (Linnæus *Systema Naturæ* (ed. 10, 1758) I. 20), f. L. *sapiens* wise]: the human species; the form of man represented by the surviving races and varieties. (Examples.)

1797 *Encycl. Brit.* X. 507/2 In the *Systema Naturæ*, Man (*Homo*) is ranked as a distinct genus of the order *Primates*. **1802** W. TURTON tr. *Linnæus's Gen. Syst. Nature* I. 9 Homo. *Sapiens*. Diurnal; varying by education and situation. **1864** *Rep. Brit. Assoc. 1863* II. 82 The author [*sc.* W. King] is led to regard the Neanderthal skull as belonging to a creature cranially and psychically different from man; and he proposes to distinguish the species by the name of *Homo Neanderthalensis*. **1896** KIRKALDY & POLLARD tr. *Boas' Text-bk. Zool.* 536 All Men are usually regarded as one species, *Homo sapiens*, divided into a number of races. **1924** G. ELLIOT SMITH *Evol. Man* ii. 76 The Rhodesian species was the most primitive member of the genus *Homo* at present known. **1940** *Nature* 17 Feb. 261/1 Many anthropologists, disregarding the rules of nomenclature, relegate all forms of Homo that do not differ structurally from what they loosely term 'modern man' to the sole species *Homo sapiens*. **1964** L. S. B. LEAKEY et al. *Ibid.* 4 Apr. 8/1 Genus *Homo* Linnæus. Species *habilis* sp. nov. (Note: The specific name is taken from the Latin, meaning 'able, handy, mentally skilful, vigorous'. We are indebted to Prof. Raymond Dart for the suggestion that *habilis* would be a suitable name for the new species.) **1969** LEAKEY & GOODALL *Unveiling Man's Origins* (1970) p. xiv, The word *Homo* is the scientific name for the genus of man, and includes the species *sapiens*, which is man as we know him today, and other species such as *erectus* and *habilis*. Many other species names, in this genus, have been proposed from time to time (such as *heidelbergensis* and *neanderthalensis*) but these are not now generally recognized. **1971** J. Z. YOUNG *Introd. Study Man* xxxi. 444 Most of the characteristics of *Homo* seem to have evolved well within the Pleistocene. **1971** *Sci. Amer.* Dec. 42 In a heavy brow ridge and a low forehead, the Pyrenees fossils more closely resemble *Homo erectus*, the 500,000-year-old fossil man of Java and China.

c. Used with L. or mock-L. adjs. in names imitating *Homo sapiens*, etc., and intended to personify some aspect of human life or behaviour (indicated by the adj.). *Homo faber* (fē¹·bəɹ) [H. Bergson *L'Evolution Créatrice* (1907) II. 151], a term used to designate man as a maker of tools.

1911 A. MITCHELL tr. *Bergson's Creative Evolution* ii. 146 We should say perhaps not *Homo sapiens*, but *Homo faber*. **1934** A. TOYNBEE *Study Hist.* III. 229 As *Homo Belligerans*, he focuses his convergent beams upon a single point on an aerial plane that is determined by the momentary presence of a hostile piece of aircraft. **1944** H. G. WELLS *'42 to '44* II. 100 Since poor rambling *Homo insipiens* began to put facts together and ask questions about them, he has been accumulating a vast disorder of answers. **1946** M. L. ANDISON tr. *Bergson's Creative Mind* ii. 99, I believe that it is of man's essence to create materially and morally, to fabricate things and to fabricate himself. Homo faber is the definition I propose. **1948** *Education* (Boston) Oct. 80/2 The range of possible noises that homo loquens can produce is ultimately conditioned by the structure of the human vocal apparatus and hearing apparatus. **1956** A. HUXLEY *Adonis & Alphabet* 10 There are many anthropologists who prefer to think of man as *homo faber*—the smith, the maker of tools. **1959** *Encounter* July 46/1 The Stoic philosophers.. seem to have made much of these emblems of moral dangers, turning..Odysseus into the ideal *homo viator*. **1960** E. DELAVENAY *Introd. Machine Transl.* 1 We can therefore rely on the inventiveness of *homo faber*. **1961** *Times* 25 Apr. 15/7 Symbolizing..this concept of *homo turisticus*, the new Hilton hotel..will have 500 rooms—all with a view of the Parthenon. **1962** *Daily Tel.* 7 Apr. 13 The bustling homo-sapiens was becoming homo-sedentarius, a quiet animal who sat and used his intelligence to push buttons and pull switches. **1962** M. MCLUHAN *Gutenberg Galaxy* 70 As long as *homo sedens* avoids the more potent kinds of optical conditioning..the mere shades of sacral life, as between nomadic and sedentary man, do not faze Eliade. **1963** AUDEN *Dyer's Hand* II. 88 Something managers need to be reminded of, namely, that the managed are people with faces,..that *Homo Laborans* is also *Homo Ludens*. **1964** *English Studies* XLV. (Suppl.) 244 An arraignment of Walter Pater in his quality as homo aestheticus. **1972** *N.Y. Times Bk. Rev.* 26 Nov. 22/3 *Homo lexicographicus* is a chalcenterous subspecies of mankind. **1973** HOLT & MARJORAM *Maths. in Changing World* ii. 21 On a more sophisticated level of appreciation, there seems to be evidence for an evolution of intelligence from *Homo faber*, the tool-user, to *Homo sapiens*, the wise one.

homo (hōu·mo), *sb.*² and *a.* A colloq. abbrev. of *HOMOSEXUAL *a.* and *sb.*

A. *sb.*

1929 M. LIEF *Hangover* vi. 100 Do you think Will Hays will let that play get by—with all those homos and everything? **1933** C. MACKENZIE *Water on Brain* iv. 44 There's a nasty old homo at the next table trying to catch your eye. **1967** *Listener* 21 Sept. 381/2 Sally's breathless confession to Dr Dale about hubby being a homo must have caused many a benighted bigot's heart to stop. **1973** A. S. NEILL *Neill! Neill! Orange Peel!* (rev. ed.) II. 216, I never had any symptoms of homosexuality but I wonder if some homos could date their condition to some early incident that made a girl, and subsequently all girls, taboo.

B. *adj.*

1933 E. A. ROBERTSON *Ordinary Families* xiii. 271 Round about six, fifteen and twenty are the recognized 'homo' ages in women. **1957** F. KING *Man on Rock* iv. 120 Sometimes they muttered to each other that he was 'homo'.

Hence **ho·moism** (*nonce-wd.*), homosexuality; homosexual practices.

1949 WYNDHAM LEWIS *Let.* 3 Mar. (1963) 480 Homoism died down in the 'thirties, but is so prevalent now as to be the [word] among the student or intellectual young.

homo-. Add: **ho·mocharge**, the charge on an electret polarized in the same direction as the original polarizing field; **homochlamy·deous** *a. Bot.* [Gr. χλαμύς cloak], having the outer and inner layers of the perianth alike, not differentiated into sepals and petals; **homo-chromy** *Zool.*, cryptic colouring (of an animal); **homocy·clic** *a. Chem.*, containing or designating a ring formed of atoms of a single element; **homode·smic** *a. Chem.* [Gr. δεσμ-ός bond], containing only a single kind of chemical bond; **homodyna·mic** *a. Ent.* [ad. F. *homodyname* (E. Roubaud 1922, in *Bull. Biol. de la France et de la Belg.* LVI. 470)], (of an insect, its life cycle, etc.) characterized by a continuous succession of generations throughout the year, so long as reasonably favourable conditions prevail; also **homody·namous** (*b*); **ho·modyne** [after *HETERODYNE], a name given to a radio receiver and a method of detection which employs a local oscillator tuned to the carrier frequency of the detected signal; **ho·mojunction** *Electronics* [*JUNCTION 2 b], an area of contact between different conductivity types of a single semiconducting material; **homola·teral** *a.*, on or affecting the same side of the body; **homole·cithal** *a. Embryol.* [Gr. λέκιθος yolk of an egg], (of an egg cell) having the yolk uniformly distributed throughout the cytoplasm; **homomo·rphosis** *Biol.*, the regeneration of an organ or part similar to the one lost; **homopo·lymer** *Chem.*, a polymer formed from only one kind of monomer; so **ho:mopolyme·ric** *a.*; **ho:mopolymeriza·tion** *Chem.* [a. G. *homopolymerisation* (T. Wagner-Jauregg 1930, in *Ber. d. Deut. Chem. Ges.* LXIII. 3213)], a reaction in which identical molecules become joined, forming a homopolymer; so **homopo·lymerize** *v. trans.* and *intr.*, to form a homopolymer (of); **ho:mopolysa·ccharide** *Chem.*, any polysaccharide composed of molecules of a single monosaccharide; **ho:mosceda·stic** *a. Statistics* [Gr. σκεδαστ-ός capable of being scattered (σκεδάννυμι to scatter)], of equal scatter or variation; having equal variances; so **ho:moscedasti·city**; **homosta·tic** *a. Med.* and *Biol.*, applied to transplant tissue which is inert and not actively growing in the donor's body; opp. *homovital* adj.; **homotha·llic** *a. Biol.*, (of a fungus) having no genetically controlled incompatibility system; not heterothallic; so **homotha·llism, -tha·lly**, the condition of being homothallic; **ho·motherm** *Zool.* = *homœotherm* (s.v. *HOMŒOTHERMIC *a.*); **homothe·rmic** *a. Zool.* = *HOMŒOTHERMIC *a.*; **homovi·tal** *a. Med.* and *Biol.*, applied to transplant tissue which in the donor's body contains actively multiplying cells; opp. *homostatic* adj.

1935 A. GEMANT in *Phil. Mag.* XX. 933 We observe both kinds of charges on electrets. One has the opposite sign to that of the adjacent polarizing electrode, the other has the same sign as the adjacent polarizing electrode, and will be denoted as homocharge. **1965** *New Scientist* 27 May 590/2 Under a certain critical applied field the heterocharge decays to a constant value and a homocharge does not appear. **1895** S. H. VINES *Students' Text-bk. Bot.* II. 512 When the perianth-leaves are all alike, the flower is said to be homochlamydeous. **1890** *Natural Sci.* Dec. 396 Homochromy and other protective adaptations. **1967** *Oceanogr. & Marine Biol.* V. 470 The chiton *Middendorffia caprearum* shows a conspicuous homochromy with the substratum. **1903** *Nature* 17 Sept. 475/1 The rings may be either homocyclic or heterocyclic without the character of the species being altered. **1932** H. G. RULE tr. *Schmidt's Text-bk. Org. Chem.* (ed. 2) II. i. 347 These [*sc.* carbocyclic compounds] are sometimes called homocyclic or isocyclic compounds. **1961** G. M. BADGER *Chem. Heterocyclic Compounds* i. 10 The systematic method for naming dicyclic and polycyclic compounds follows that used for homocyclic compounds. **1939** R. C. EVANS *Introd. Crystal Chem.* i. 8 Crystals..in which only one type of force occurs, are said to be homodesmic. **1966** H. D. MEGAW *Ferroelectr. in Crystals* 205 In a homodesmic structure it is incorrect to speak of 'molecule' or 'molecular weight', since the molecule is coextensive with the crystal. **1932** *Trans. Entomol. Soc.* LXXX. 195 The outstanding characteristic of this homodynamic type of development is the absence of a definite annual life-cycle, the number of generations in a year depending on the external conditions. **1964** BORROR & DELONG *Introd. Study Insects* (rev. ed.) iii. 44 Many insects, particularly those living in the tropics, have a homodynamic life cycle; that is, development is continuous and there is no regular period of dormancy. **1929** V. E. SHELFORD *Lab. & Field Ecol.* vi. 160 Homodynamic separates the higher Diptera into two categories, homodynamous and heterodynamous. **1922** STRAUSS & KRUSE *Radio Man.* iv. 149 If the local generated frequency is tuned to exactly the same frequency as the received signals..the condition of 'zero beat' is said to exist. This means of receiving is also called the 'homodyne' method. **1965** *New Scientist* 11 Feb. 344/1 The approach adopted..is to stabilise the laser at a single frequency, using an 'optical homodyne' receiver. The system uses a helium-neon laser stabilised at 6328 angstroms as both transmitter and local oscillator. **1960** R. L. ANDERSON in *IBM Jrnl. Res. & Devel.* IV. 287 Junctions between two dissimilar semiconductors will be referred to as heterojunctions, and junctions between semiconductor with different doping as homojunctions. **1966** *New Scientist* 11 Aug. 316/3 On bringing two conductivity types of semiconductor into contact, in a homojunction, the equalization of Fermi levels takes place between the borders of electrons and holes confronting each other. **1971** *Sci. Amer.* July 99/2 The structure therefore has a *p-n* junction in gallium arsenide (a homojunction). **1910** *Practitioner* July 98 Should the lesion be in or close to the red nucleus, the tremor will be on the opposite side of the body, while if any other part of the system be affected the tremor will be homolateral. **1919** *Jrnl. Exper. Zool.* XXIX. 255 At the sides of the body, those parts innervated by the pallial strands are conspicuously homolateral in their responses. **1956** *Nature* 17 Mar. 529/2 These thoracic responses also remained when the major portion of the homolateral corpora pedunculata was removed. **1892** E. L. MARK tr. *Hertwig's Text-bk. Embryol.* i. 28 The translator has been accustomed for several years to use the word homolecithal instead of alecithal, telolecithal being employed as a co-ordinate term to embrace telolecithal and centrolecithal types. **1914** W. E. KELLICOTT *Textbk. Gen. Embryol.* iii. 93 It is often difficult to distinguish the telolecithal egg from the homolecithal type. **1912** B. M. PATTEN *Found. Embryol.* iv. 78 In the egg of *Amphioxus* the yolk is relatively meagre in amount and fairly uniformly distributed throughout the cell. An ovum with such a yolk distribution is termed isolecithal (homolecithal). **1901** T. H. MORGAN *Regeneration* 23 When the new part is like that removed, or like a part of that removed, as when a leg or a tail is regenerated in a newt, the process is one of 'homomorphosis'. **1967** GARDINER & FLEMISTER *Princ. Gen. Biol.* (ed. 2) xxi. 464/1 The conditions of homo- and heteromorphosis make it apparent that in some cases an inductive effect is exerted by the tissue adjacent to the regenerating part. **1946** A. G. ROSS in Richardson & Wilson *Fund. Plastics* ix. 245 Neither the vinyl chloride–vinyl acetate copolymers nor the vinylidene chloride–vinyl chloride copolymers can be fractionated so as to yield either pure homopolymer. **1970** *New Scientist* 30 Apr. 230/3 Teflon homopolymers are good reinforcing agents for the polymer. **1971** *Nature* 26 Nov. 197/1 Combinations of synthetic primer oligomers with homopolymeric templates..allow one to distinguish the viral enzyme from other DNA polymerases. **1934** *Chem. Abstr.* XXV. 2411 The polyacetates which are..adapted to heteropolymerization show no marked tendency to homopolymerization. **1937** R. E. BURK et al. *Synthetic Resins* x. 251 Products..not easily obtained by homopolymerization. **1953** A. J. HALL *Textile Sci.* ii. 88 It [*sc.* acrylonitrile] very readily undergoes polymerisation by itself (homopolymerisation) and with other polymerisable compounds (copolymerisation). **1952** C. E. SCHILDKNECHT *Vinyl & Related Polymers* iii. 173, 2-Isopropenylthiophene would not homopolymerize on heating with peroxide catalyst, but.. copolymerized with butadiene. **1957** *Amer. N.Y. Acad. Sci.* LXIX. 334 Some amino acids..do not homopolymerize to linear peptides. **1962** *Jrnl. Polymer Sci.* LVIII. 1 The two monomers that were employed..are the most difficult to homopolymerize by a free radical mechanism to high molecular weight polymers. **1948** W. W. PIGMAN *Chem. Carbohydrates* xii. 513 Members of the first class (homopolysaccharides) give only one monosaccharide type when completely hydrolyzed. **1970** G. O. ASPINALL *Polysaccharides* i. 5 The first broad division in the classification of polysaccharides is between homopolysaccharides..and heteropolysaccharides. **1905** K. PEARSON in *Drapers' Company Res. Mem.* (Biometric Ser.) II. 22 If..all arrays are equally scattered about their means, I shall speak of the system as a homoscedastic system, otherwise it is a heteroscedastic system. **1934** *Brit. Jrnl. Psychol.* XXIV. 337 It is essential for factor studies that the correlation surfaces of the pairs of variables should be comparable. This is the case when each is homoscedastic, homoclitic, with rectilinear regression lines. **1970** *Nature* 12 Dec. 1098/1 Although the compared sample groups appear to be homoscedastic (*F* test) and results from the *t* test are significant, the size of the sample population is so small that we have relied on the nonparametric Mann–Whitney *U* test. **1905** *Drapers' Company Res. Mem.* (Biometric Ser.) II. 22 [illegible]. **1957** DIXON & MASSEY *Introd. Statistical Analysis* (ed. 2) xii. 194 The regression curve of *Y* on *X* and the regression curve of *X* on *Y* are both straight lines with homoscedasticity (constant variance) for both *X* and *Y* variables. **1953** W. P. LONGMIRE in *Jrnl. Nat. Cancer Inst.* XIV. 669 The term *homostatic graft* might be applied to inert tissues such as bone and cartilage when transferred from one individual to another of the same species or not. The term *homovital graft* might be used in reference to grafts whose cells must continue to grow and reproduce for the graft to be effective after similar transplantation. **1971** BILLINGHAM & SILVERS *Immunobiol. Transplantation* vi. 93 The long-term preservation of homostatic grafts is relatively simple. **1904** Homothallic [see *heterothallic* adj. s.v. *HETERO-]. **1967** M. E. HALE *Biol. Lichens* iii. 42 There is good reason..to suppose that lichens are homothallic. **1959** P. B. MEDAWAR in L. A. Peer *Transplantation of Tissues* II. ii. 41 Homovital grafts are those that survive transplantation..whereas homostatic grafts are progressively revitalized by the tissues of their hosts.

homocaryon, -caryosis, -caryotic: see *HOMOKARYOTIC *a.*

homocentric, *a.* and *sb.* Add: **A.** *adj.* (Later example.)

1952 G. SARTON *Hist. Sci.* I. xx. 510 The main achievement of the astronomers of this period, if not of Aristotle himself, was the completion of the theory of homocentric spheres.

2. Of rays of light or a beam of particles: diverging from or converging to a single focal point (or appearing to do so when produced).

1886 C. M. CULVER tr. *Landolt's Refraction & Accommodation of Eye* i. 13 In order that the homocentric rays may reunite in a single point..the surface must have such a form that the angles of incidence shall be everywhere the same. **1949** *Proc. R. Soc.* A. CXCVII. 456 An electron gun, combined with a suitable aperture and electron lens system, produces a coherent illumination beam, as nearly homocentric as possible. **1969** G. A. FRY *Geom. Optics* vi. 44 As long as the wave fronts are spherical and concentric, the rays all converge at the center of curvature of the wave front and constitute a homocentric bundle of rays.

Hence **homocentri·city**, the condition of being homocentric.

1959 BORN & WOLF *Princ. Optics* iv. 168 In general, the homocentricity of a pencil is destroyed on refraction or reflection.

homoclime (hǫ·moklaim). *Geogr.* [f. HOMO- + Gr. κλίμα (see CLIMATE) or Eng. CLIME.] A region or place that has a similar climate to some given region.

Orig. defined more narrowly: see quot. 1916.

1916 G. TAYLOR *Control of Settlement* (Commonwealth Bur. Meteorol. Bull. XIV) 13 Those regions in which climate, topography, and industries, etc., are similar I call 'homoclimes'. *Ibid.*, The homoclimes are British Isles, North France, much of the German Empire, etc.] **1931** A. A. MILLER *Climatology* ii. 14 Stations having similar climographs are described as 'homoclimes'. Alice Springs is the homoclime of Biskra (Algeria), Perth of Cape Town, Brisbane of Durban, etc. **1950** *New Biol.* VIII. 68 The first need is..to define the climate where the insects must be established, and to locate its homoclimes (i.e. areas of similar climate).

homocline (hǫ·moklaim). *Geol.* [f. HOMO- + *-cline*, as in *anticline, syncline*, etc.] A set of strata dipping throughout in the same general direction; *esp.* one over which the angle of dip is more or less uniform.

1915 R. A. DALY in *Geol. Survey Canada Mem.* LXVIII. 53 For convenience the word 'homocline' will here be used as a general name for any block of bedded rocks all dipping in the same direction... A 'homocline' may be a monocline, an isocline, a tilted fault-block, one limb of anticline or syncline. **1942** M. P. BILLINGS *Struct. Geol.* iii. 42 The term homocline..may be applied to strata that dip in one direction at a uniform angle. Although many homoclines are, if large areas are considered, limbs of folds, the term is useful to refer to the structure within the limits of a small area. **1966** G. J. WILLIAMS *Econ. Geol. N.Z.* xix. 344/2 The dominant structure of the Tertiary beds of the Taranaki region is that of a gently (about 4°) south or south-south-east dipping homocline.

Hence **homocli·nal** *a.*, of pertaining to, or associated with a homocline.

1916 R. A. DALY in *Bull. Geol. Soc. Amer.* XXVII. 92 The broad, useful concept denoted by the 'monoclinal' of W. B. and H. D. Rogers needs a new name... They intended to name a body of strata showing throughout dip in the same direction and for that 'homocline' [or 'homoclinal', adjective, 'homoclinal'] is the appropriate word. **1922** C. A. COTTON *Geomorphol. N.Z.* xvi. 98 A striking effect of homoclinal drifting is seen where a stream crosses the strike diagonally. **1941** —— *Landscape* x. 91 Moderately inclined strata now outcrop as homoclinal ridges. **1954** W. D. THORNBURY *Princ. Geomorphol.* ix. 225 Homoclinal ridges develop upon the dipping beds on the flanks of anticlines and synclines.

homocysteine (hǫmosi·stīn, -si·stiːin). *Chem.* [f. HOMO- b + *CYSTEINE.] An amino-acid, $HS{\cdot}CH_2{\cdot}CH_2{\cdot}CH(NH_2){\cdot}COOH$, which is important as an intermediate in the metabolism of methionine and cysteine.

1932 *Jrnl. Biol. Chem.* XCIX. 137 If the hydrolysis, however, should take place between the terminal carbon and the sulfur then methyl alcohol and homocysteine would result. **1935** *Ibid.* CXII. 149 (*heading*) The isolation of homocysteine and its conversion to a thiolactone. **1962** *Biochem. & Biophys. Res. Comm.* IX. 493 The normal pathway for the metabolism of methionine involves demethylation to homocysteine which in turn combines with serine to form cystathionine, an intermediate along the pathway to cysteine and cystine.

homocystine (hǫmosi·stīn). *Chem.* [f. HOMO- b + CYSTINE.] An amino-acid $(-S{\cdot}CH_2{\cdot}CH_2{\cdot}CH(NH_2){\cdot}COOH)_2$, which is the oxidized form of homocysteine.

1932 *Jrnl. Biol. Chem.* XCIX. 136 We have drawn the conclusion that the compound is bis-(γ-amino-γ-carboxypropyl) disulphide, the next higher symmetrical homologue of cystine. Because of its relationship we wish to suggest the name of homocystine for the compound. **1962** *Biochem. & Biophys. Res. Comm.* IX. 493 (*heading*) The identification of homocystine in the urine.

Hence **ho:mocystinu·ria** [-URIA], a rare condition, caused by a hereditary enzyme deficiency, in which homocystine is present in the urine.

1962 CARSON & NEILL in *Arch. Dis. Childhood* XXXVII. 512/2 The following abnormalities were discovered..so far undescribed abnormality in cystine metabolism, homocystinuria. **1969** *New Scientist* 3 July 10/1 Prevention of postnatal brain damage by dietary treatment has been reported in a number of other inborn errors of metabolism. Examples include galactosaemia, tyrosinosis, maple syrup urine disease and possibly homocystinuria.

Homœan (hǫmī·ăn), *a.* and *sb. Theol.* Also **Homoian** (hǫmoi·ăn). [f. mod.L. *homœ-us*, f. Gr. ὅμοιος like, similar + -AN.] **A.** *adj.* Of or pertaining to the Homœans.

1833 J. H. NEWMAN *Arians* iv. v. 364 Furthering their splitting into the Homœan and Homœousian factions. **1888** *Encycl. Brit.* XXIII. 770/1 He appears to have joined the Homoian party, which took shape and acquired influence through the patronage of Constantius in 360. *Ibid.*, The Homoian formula, 'Similis filius patri suo'. **1950** J. N. D. KELLY *Early Chr. Creeds* ix. 290 It gave expression to the new 'Homoean' formula of compromise—*like in all respects*—and strictly avoided technical terms.

B. *sb.* One of a group of Arians that developed *c* 355 and repudiated both the Homoousion and the Homoiousion, maintaining simply that in the Trinity the Son is 'like' the Father.

1896 G. P. FISHER *Hist. Chr. Doctrine* 142 The 'Homœans' would not go a step beyond the affirmation of a 'likeness',—meaning a likeness in will and active energy. **1913** *Eng. Hist. Rev.* Oct. 761 It was of course the Homoeans, and the Homoeans, who inclined towards the Nicenes. **1957** *Oxf. Dict. Chr. Ch.* 851/1 The middle party, called 'Homoeans'..aimed at avoiding dogmatic precision as far as possible. **1966** P. R. COLEMAN-NORTON *Roman State & Chr. Ch.* II. 420 Probably the Homoeans, moderate Arians, are meant.

homœo-. Add: **homœarchon** (hǫmiā·ɹkǫn), **homœoa·rchon** = *homœoarchy* (s.v. HOMŒO-); **homœobla·stic** *a. Petrol.* [ad. G. *homoeoblastisch* (F. Becke 1904, in *Compt. Rend. IX. Session Congr. géol. internat.* II. 570)], composed of grains of equal size; **ho:mœochlamy·deous** *a. Bot.* = *homochlamydeous* adj. s.v. *HOMO-; **ho·mœograft** *Med.* and *Biol.* = *HOMOGRAFT; so **ho·mœografted** *ppl.a.*, **-grafting** *vbl. sb.*; **ho:mœokine·sis** *Cytol.*, the division of a cell into cells having similar hereditary tendencies; **ho:mœo-osmo·tic, homœosmo·tic**

adjs. Physiol. [OSMOTIC *a.*], (of an animal) maintaining a more or less constant concentration of solute in its body fluids regardless of fluctuations of the concentration in the surrounding medium; usu. spelt *homoi(o)-*; so **ho:mœo-osmo·sis, homœosmo·sis**; **homœo-pola·r** *a. Chem.* [ad. G. *homöopolar*] = *HOMOPOLAR *a.* 3; **homœotype** *Taxon.* = *HOMOTYPE 2; **homœoty·pic, -ty·pical** *adjs. Cytol.* [ad. G. *homöotypisch* (W. Flemming 1887, in *Arch. f. mikrosk. Anat.* XXIX. 400)], designating the second division of meiosis.

1896 W. M. LINDSAY *Lat. Textual Emend.* 10 The homœoteleuton and homœoarchon of these lines has led to omission. **1942** *Jrnl. Theol. Stud.* XLIII. 86 In ver. 39 it seems clear that om* omitted (homoeoarchon). **1920** A. HOLMES *Nomencl. Petrol.* 118 Homoeoblastic, a term used instead of equigranular and applied to metamorphic rocks to indicate that the texture so described is due to recrystallisation. **1932, 1954** Homoeoblastic [see *heteroblastic* s.v. *HETERO-]. **1900** B. D. JACKSON *Gloss. Bot. Terms* 125/1 *Homoeochlamydeous*, used by Engler and Prantl when the perianth is uniform. **1966** Homoiochlamydeous [see *heterochlamydeous* adj. s.v. *HETERO-]. **1913** *Arch. f. Entwicklungsmech. d. Organismen* XXXVII. 263 Desquamation is approximately equally marked in both auto and homeografts. **1915** *Jrnl. Exper. Med.* XXI. 174 In..the first 2 to 5 days after transplantation, there is no noticeable difference between the auto- and homograft. **1952** *Cancer Res.* XII. 379/1 Tumor homografts between mice of unrelated inbred strains. **1930** *Physiol. Rev.* X. 562 It has been maintained that the result of homoeografting of skin largely depends upon whether donor and host do or do not belong to the same blood group. **1893** PARKER & RÖNNFELDT tr. *Weismann's Germ-Plasm* 34 These kinds of division may be spoken of as homoeokinesis and heterokinesis, that is, as a division into two daughter-idioplasms similar to or different from one another. **1906** In *Wilson Cell* ix. 305 Mitotic division is conceived (by Weismann) as an apparatus, which may distribute the elements of the chromatin to the daughter nuclei either equally or unequally. In the former case ('homoeokinesis', integral or quantitative division), the resulting nuclei remain precisely equivalent. **1939** A. KROGH *Osmotic Regulation in Aquatic Animals* 240 Homoiosmotic. **1964** *Oceanogr. & Marine Biol.* II. 307 Osmo-regulators exhibit an appreciable tendency towards homoiosmosis. **1931** *Biol. Rev.* VI. 423 In contrast with these are the 'homoiosmotic' organisms which include typically estuarine animals such as *Carcinus maenas*. **1939** A. KROGH *Osmotic Regulation in Aquatic Animals* 1 When animals maintain a total concentration of their body fluids different from that of the surrounding water they can be termed 'homoiosmotic'. **1953** E. PALMER tr. *Ekman's Zoogeogr. Sea* vi. 118 Many fishes and fresh-water animals in general are homoio-osmotic. **1967** G. E. HUTCHINSON *Treat. Limnol.* II. xviii. 153 Such animals have the power at least to some extent of controlling the osmotic pressure of their environment; they are at least partly homoiosmotic. **1921** [illegible] The latter kind of compounds, to which the greater number of simple inorganic compounds belong, is frequently called 'heteropolar' and possesses a far more typical character than the first compounds which are called 'homoeopolar'. **1923** E. N. DA C. ANDRADE *Struct. Atom* xiv. 358 We shall also make use of Abegg's terminology of homoeopolar and heteropolar compounds. **1905** SCHUCHERT & BUCKMAN in *Science* 9 June 900/2 Homoeotype—homotype..A specimen identified by a specialist after comparison with the holotype or lectotype. **1939** *Ann. Entomol. Soc. Amer.* XXXII. 694 Homoeotype: A specimen named by another than the author upon comparing it with the type. The value of a homoeotype is apparent. **1967** [see *HOMOTYPE 2]. **1889** Homoeotypic [see *heterotypic* s.v. *HETERO-]. **1925** E. B. WILSON *Cell* (ed. 3) vi. 532 (*heading*) The heterokinesis and the homoeotypic division. **1969** BROWN & BERTKE *Textbk. Cytol.* xx. 436 Previously, meiosis I was called the heterotypic mitosis, implying that it is an atypical mitosis, which it certainly is; and meiosis II was called the homeotypic mitosis, implying that it is a fairly typical mitotic division, which it is. **1935** *Jrnl. R. Microsc. Soc.* 55 He [*sc.* Flemming] finds that these cells [*sc.* spermatocytes] exhibit a remarkable dimorphism of mitosis; in the heterotypical form the chromatic formations exhibit metakinesis. The two forms, the other of which may be called homoeotypical, are sometimes found together.

homœomorph (hǫ·mi,ǫmǭɹf). Also **homeo-**. [f. HOMŒO- + Gr. μορφ-ή form.] Something that is homœomorphous or homœomorphic.

a. (In Dict. s.v. HOMŒOMORPHOUS *a.*)

b. *Palæont.* One of two or more homœomorphous fossils or fossil species.

1898 S. S. BUCKMAN in *Q. Jrnl. Geol. Soc.* LIV. 453 The compressed species of the Arietidan Epoch, which are generally classed as *Oxynoticeras*, are, as their septal details show, certainly polygenetic homoeomorphs. **1920** A. M. DAVIES *Introd. Palæont.* i. 29 The failure to discriminate between homœomorphs has frequently led to mistakes in the correlation of strata. **1952** R. C. MOORE et al. *Invertebr. Fossils* vi. 217 Some unrelated species, so similar as hardly to be distinguishable..are homeomorphs, characters are called homoeomorphies.

c. (Usu. **homeo-**.) *Math.* A figure or a topological space that is homœomorphic to some other one.

1926 *Trans. Amer. Math. Soc.* XXVIII. 3 Such a C_n is be homeomorph of an *n*-dimensional polyhedron..whose faces are all simplicial cells. **1951** M. H. A. NEWMAN *Topology of Plane Sets of Points* (ed. 2) iii. 61 Homeomorphism is an equivalence relation, and a space and its homeomorphs have all their topological properties in common. **1965** S. BARR *Exper. Topology* vi. 70 Perhaps it is a Moebius in disguise..the homeomorph of one?

homœomorphic (hǫmiǫmǭ·ɹfik), *a.* Also **homeo-**. [f. as next + -IC.] **1.** *gen.* Of the same kind or form.

1902 *Buck's Handbk. Med. Sci.* (rev. ed.) IV. 660/1 In a remarkable proportion of cases of mental and other nervous disturbances we find a history of antecedent nervous conditions, either homœomorphic, i.e. of the same order, or heteromorphic, of different type.

b. *Palæont.* = *HOMŒOMORPHOUS *a.* c.

1923 H. H. SWINNERTON *Outl. Palæont.* x. 214 These present several homoeomorphic resemblances to Ammonites.

2. (Usu. **homeo-**.) *Math.* [ad. F. *homéomorphe* (H. Poincaré 1895, in *Jrnl. de l'École polytechn.* I. 9).] Related by a homœomorphism, topologically equivalent to a complex, figure, or topological space; that is a homœomorph.

1918 O. VEBLEN *Analysis Situs* (Cambridge Colloq. Lect. V.) ii. 5 Two complexes related by a homeomorphism are said to be homeomorphic. **1926** *Trans. Amer. Math. Soc.* XXVIII. 1 The manifold condition is equivalent to demanding that the pair of points be homeomorphic to cell boundaries. **1932** *Ann. Math.* XXXIII. 350 The isomorphism..induces a homeomorphism (i.e., a one-one bi-continuous correspondence) between the points of the two group spaces. **1956** E. M. PATTERSON *Topology* i. 6 Topologically, the Möbius band is a different surface from the cylinder, which is to say that the two surfaces are not homeomorphic. **1967** F. HARARY *Seminar on Graph Theory* v. 13 A subdivision of a graph *G* is a graph obtainable from *G* by replacing some line *uv* of *G* by a new point *w* and two new lines *uw* and *wv*. Two graphs are homeomorphic if there is a third graph which can be obtained from each by a sequence of subdivisions. **1967** D. W. BLACKETT *Elem. Topology* i. 13 Any two circles are homeomorphic and..any circle is homeomorphic to any square. On the other hand, a circle and a figure eight are not homeomorphic.

Hence **homœomo·rphically** *adv. Math.*, in a way that preserves all topological properties.

1927 *Trans. Amer. Math. Soc.* XXIX. 438 Two *k*-cells are homeomorphically transformed into one another in such manner that two (*k*−1)-cells of their boundaries are similarly transformed. **1965** S. BARR *Exper. Topology* i. 10 If we draw a triangle on a lump of Plasticine, it is conceivably possible to distort it homeomorphically so as to get rid of the three angles and make it into a circle.

homœomorphism (hǫmiǫmǭ·ɹfiz'm). Also **homeo-**. [f. HOMŒO- + Gr. μορφ-ή shape + -ISM.] **1.** (In Dict. s.v. HOMŒOMORPHOUS *a.*)

2. (Usu. **homeo-**.) *Math.* [ad. F. *homéomorphisme* (H. Poincaré 1895, in *Jrnl. de l'École polytechn.* I. 9).] A one-to-one transformation of one complex or topological space on to another that is continuous and has a continuous inverse; a topological transformation; a topological equivalence between two figures.

1918 O. VEBLEN *Analysis Situs* (Cambridge Colloq. Lect. V.) ii. 5 [illegible] **1939** *Fundamenta Math.* [illegible] **1956** E. M. PATTERSON *Topology* [illegible] **1961** S. S. CAIRNS *Introd. Topology* iii. 54 [illegible] **1965** S. BARR *Exper. Topology* [illegible] **1966** LIPSCHUTZ *Theory & Probl. Gen. Topology* [illegible] Since each cell of *X* is compact and *Y* is Hausdorff, *f* is a homeomorphism onto its image.

homœomorphous, *a.* Add: **c.** *Palæont.* Exhibiting or characteristic of homœomorphy; similar in general aspect but dissimilar in detail.

1895 S. S. BUCKMAN in *Q. Jrnl. Geol. Soc.* LI. 456 Biplicate *Terebratulæ* are, at any rate in the Jurassic rocks, independent, or heterogenetic, homœomorphous derivatives of non-plicate forms. **1913** —— *Ibid.* LXIX. 167 The two species..look like enough to be confused..until they are examined in detail. **1962** *Proc. Geologists' Assoc.* LXXIII. 12 It requires only sufficient skill and discernment..to see repeatedly, in one group or another, homoeomorphous resemblances in many groups of organisms.

homœomorphy (hǫ·mi,ǫmǭɹfi). *Palæont.* Also **homeo-**. [f. HOMŒO- + Gr. μορφ-ή shape + -Y.] A superficial resemblance between two fossils or two fossil species that tends to suggest a taxonomic identity that close examination shows not to exist; esp. a resemblance that is due to convergent evolution.

1895 S. S. BUCKMAN in *Q. Jrnl. Geol. Soc.* LI. 457 Mojsisovics has called such cases, between non-contemporaneous ammonites, 'heterochronous convergence', I would rather apply the term 'heterochronous homœomorphy' to the phenomenon; and the term 'isochronous homœomorphy' would describe the resemblance between the contemporaneous *Buckmani* with its derivatives. **1952** R. C. MOORE et al. *Invertebr. Fossils* vi. 219 In spite of their resemblance, these two shells could hardly be further apart in classification. They constitute remarkable examples of homeomorphy. **1969** *Nature* 4 Jan. 15/1 The problem of homoeomorphy—the production of similar morphologies in distantly related stocks either more or less contemporaneously or at different geological times.

homœoplastic, *a.* Add: **2.** (Also *homoio-*.) *Med.* and *Biol.* = *HOMOPLASTIC *a.* 2.

1913 *Arch. f. Entwicklungsmech. d. Organismen* XXXVII. 249 Loeb did not at that time distinguish between the results of auto- and homoeoplastic transplantation. Later he found a marked difference in the results obtained after auto- and homoeotransplantation in the case of tumors. *Ibid.* 250 The homoeoplastic transplants behaved quite differently. **1938** *Amer. Jrnl. Physiol.* CXXI. 650 (*heading*) Homeoplastic transplantation of adrenal glands in rats of inbred strains. **1952** G. R. CAMERON *Path. Cell* xxvii. 527 Homoioplastic transplantation succeeds with some tissues, but it is doubtful whether success is prolonged or permanent.

Hence **homœopla·stically** *adv.*; **ho·mœoplasty** = *HOMOPLASTY.

1915 *Jrnl. Exper. Med.* XXI. 164 The difference in the growth of auto- and homeoplastically transplanted thyroid tissue. **1929** *Ann. Surg.* XC. 926 Homoplasty, homeoplasty and isoplasty mean tissue transplantation from one individual to another of the same species.

homœosis (hǫmi,ōu·sis). *Biol.* Also **homeo-**. Pl. **homœoses**. [mod.L., ad. Gr. ὁμοίωσις a becoming like, f. ὅμοιος like.] The replacement in a metamerically segmented animal, esp. in the course of regeneration, of a structure forming part of one segment by a structure characteristic of another segment; also used of an analogous process in plants (e.g. the replacement of stamens by petals). So **homœo·tic** *a.*, exhibiting or characterized by homœosis; **homœo·tically** *adv.*

1894 W. BATESON *Study of Variation* i. 85 For the word 'Metamorphy' I therefore propose to substitute the term Homœosis. *Ibid.*, The distinction between Homœotic Variation and strictly Meristic Variation is sufficiently obvious. **1913** —— *Mendel's Princ. Heredity* (ed. 2) xi. 198 A simple homœosis of the stamens and carpels. **1913** —— *Prob. Genetics* iii. 68 When a lumbar vertebra varies homoeotically into the likeness of the last dorsal and bears a rib. *Ibid.*, The consequences of such homoeoses are sometimes very extensive. **1940** R. GOLDSCHMIDT *Material Basis Evol.* 326 For a long time the phenomenon of homoeosis (called heteromorphosis by some authors) has been known as an occasional monstrosity in arthropods...The classical example is the regeneration of an antenna after removal of the eyestalk in Decapods. **1962** D. J. MERRELL *Evol. & Genetics* x. 101 The homeotic mutants cause one of a series of parts to assume the character of another member of the series. **1966** E. D. HAY *Regeneration* i. 34 A unique kind of heteromorphosis termed homeosis occurs occasionally in arthropod regeneration.

homœostasis (hǫmi,ǫ-stăsis, hǫmiostē¹·sis). Also **homeo-**. [mod.L., f. HOMŒO- + Gr. στάσις standing still, stationariness.] The maintenance of a dynamically stable state within a system by means of internal regulatory processes that tend to counteract any disturbance of the stability by external forces or influences; the state of stability so maintained; *spec.* in *Physiol.*, the maintenance of relatively constant conditions in the body (e.g. as regards blood temperature) by physiological processes that act to counter any departure from the normal.

1929 W. B. CANNON in *Physiol. Rev.* IX. 400 The coordinated physiological reactions which maintain most of the steady states in the body are so complex, and are so peculiar to the living organism, that it has been suggested (Cannon, 1926) that a specific designation for these states be employed—homeostasis. **1941** —— in *Science* 3 Jan. 8/1 The functioning of the human brain has made social homeostasis differ markedly from physiological homeostasis...An upset of constancy necessarily results. Railroads replace canals, automobiles crowd out the horse and buggy, [etc.]. **1949** KOESTLER *Insight & Outlook* xx. 279 The chemical balance of the body fluids (homeostasis). **1955** L. R. DICE *Man's Nature* ix. 113 The individual organism, to maintain its homeostasis (state of balance), must throw away not only water and salts, but even sugar if necessary. **1962** *Lancet* 6 Jan. 31/1 These clinical indications of defective renal homœostasis suggest the need for a more thorough exploration of renal function. **1962** V. C. WYNNE-EDWARDS *Animal Dispersion* xxiii. 561 The general concept of homeostasis in aquatic populations through growth-inhibitory substances..is complicated by the fact that, in some of the very same species, growth-*promoting* metabolites have also been discovered. **1964** M. MCLUHAN *Understanding Media* (1967) II. x. 109 The city, as a form of the body politic, responds to new pressures and irritations by resourceful new extensions—always in the effort to exert staying power, constancy, equilibrium, and homeostasis. **1964** N. WIENER *God & Golem* vi. 86 Science is an important contribution to the homeostasis of the community. **1971** *Nature* 20 Aug. 562/2 Drinking is essential for the homeostasis of body water.

homœostasy (hǫmi,ǫ-stăsi), anglicized form of prec.

1945 S. BRODY *Bioenergetics & Growth* x. 250 A calcium hexose monophosphate is the important intermediary in.. calcium homeostasy. **1959** *Times Rev. Industry* Jan. 42/1 The human body contains a vast hierarchy of controls which help to maintain a constant and stable state. The American physiologist W. B. Cannon termed this condition homeostasy. **1971** D. WATTS *Princ. Biogeogr.* v. 223 This notion has also been criticized by Lack on the grounds that..the stability produced from homeostacy [*sic*] could equally well result from the density-dependent factors already described.

homœostat (hǫ·miostæt). Also **homeo-**. [Back-formation from *HOMŒOSTATIC *a.*, after words like *thermostat* (see -STAT).] A homœostatic apparatus or system; something that adapts itself (within limits) to changes in its environment in such a way as to preserve a state of internal stability.

1948 W. R. ASHBY in *Electronic Engin.* XX. 382 The homeostat will adapt not only to random changes in hand settings but to *any* change in the dynamic nature of the machine. **1959** S. BEER *Cybernetics & Management* iii. 23 In a homeostat a critical variable is held at a desirable level by a self-regulatory mechanism. **1960** *20th Cent.* Mar. 269 Something goes wrong with our lives; this puts our homeostat out of order; as a result we may be unable to digest properly (and so we get stomach ulcers). **1964** J. Z. YOUNG *Model of Brain* i. 3 When we say that living things have 'needs' we emphasize that they are self-maintaining systems; 'homeostats' in modern terminology. **1971** *Nature* 22 Jan. 233/2 Beer and others have shown how a large industrial company can usefully be considered as a homeostat.

homœostatic (hǫmiostæ·tik), *a.* Also **homeo-**. [f. *HOMŒOSTASIS, after *stasis, static*.] Maintained by, involving, or effecting homœostasis; of or pertaining to homœostasis.

1929 *Physiol. Rev.* IX. 401 The adjectival form, *homeostatic*, would apply to the physiological reactions or agencies or to the circumstances which relate to steady states in the organism. *Ibid.* 417 (*heading*) The homeostatic functions of hunger and thirst. **1955** J. Z. YOUNG in B. I. Evans *Studies in Communication* 93 Since man's special homeostatic machinery is mainly social, medicine may often be called upon to assist in correcting some social incapacity. **1959** S. DUKE-ELDER *Parsons' Dis. Eye* (ed. 13) ii. 22 The pressure within the eye is maintained at its homeostatic level despite considerable variations in the other factors which tend to alter it. **1961** *Lancet* 2 Sept. 551/1 A general disturbance of the homoeostatic mechanisms regulating immune tolerance. **1963** R. P. DALES *Annelids* viii. 167 We are not yet certain about the nature of these homeostatic mechanisms which maintain the annelid body as a whole. **1964** A. RAPOPORT in I. L. Horowitz *New Sociol.* vi. 99 Man as a physico-chemical system in homeostatic equilibrium with the environment. **1964** N. WIENER *God & Golem* vi. 87 The difficulties of establishing a really homeostatic regulation of society are not to be overcome by replacing one set pattern..by an equal and opposed set pattern of the same sort. **1971** D. WATTS *Princ. Biogeogr.* v. 230 Many populations have homeostatic mechanisms within them which are designed to maintain stability through reducing the number of species which are overspecialized.

Hence **homœosta·tically** *adv.*

1959 S. BEER *Cybernetics & Management* xv. 140 These two systems are linked homeostatically as mutually vetoing systems against the two managerial criteria. **1962** J. C. WYNNE-EDWARDS *Animal Dispersion* xv. 363 It is social integration that supplies the means of controlling population-density homeostatically. **1971** *Nature* 22 Jan. 233/2 With the health sciences system and its environment homeostatically related.

homœothermic (hǫmiopǝ̄·ɹmik), *a. Zool.* Usu. **homoio-**; also **homeo-**. [ad. G. *homöotherm* (C. Bergmann 1847, in *Göttinger Studien* I. 613): see HOMŒO- and THERMIC *a.*] Maintaining an almost constant body temperature; warm-blooded; homothermous. Also **homœothe·rmal** *a.*, in the same sense.

1870 Homoeothermal [in Dict. s.v. HOMŒO-]. **1885** W. STIRLING tr. *Landois' Text-bk. Human Physiol.* I. vi. 426 Bergmann introduced the word homoiothermal animals for the warm-blooded animals. **1889** V. H. W. WINGRAVE *T. Dunman's Gloss. Anat. Terms* App. 175 *Homœothermic*,..of even temperature: applied to warm-blooded animals which maintain the same temperature, irrespective of that of the surrounding medium. **1891** *Ann. Rep. Smithsonian Inst. 1890* 407 Man, mammals, and birds are called creatures of equable temperature, homeothermic. **1903** *Phil. Trans. R. Soc.* B. CXCV. 37 Variation in production of heat is the ancestral method of homoeothermic adjustment. **1928** PEARSE & HALL *Homoiothermism* v. 31 Homoiothermal animals. *Ibid.* 33 Homoiothermic marine animals. **1965** B. E. FREEMAN tr. *Vandel's Biospeleol.* xiv. 237 Not one true cavernicole is known among the homoiothermic vertebrates, that is to say the birds and mammals. **1966** *New Statesman* 11 Nov. 697/1 Man is a homoiothermic animal... When it is cold the body attempts to reduce heat loss.

Hence **ho·mœotherm**, a homœothermic animal.

1891 *Ann. Rep. Smithsonian Inst. 1890* 411 These phenomena, which are numerous and active in animals of the higher class (homeotherms), are much less so in cold-blooded animals. **1968** D. W. WOOD *Princ. Animal Physiol.* viii. 123 In contrast to poikilotherms, homeotherms maintain their body temperature at a more or less constant level, irrespective of the environmental temperature.

homœothermism (hǫmiopǝ̄·ɹmiz'm). *Zool.* Usu. **homoio-**; also **homeo-**. [f. prec. + ISM.] The maintenance or possession of an almost constant body temperature. Also **ho·mœothermy**, in the same sense.

1903 *Phil. Trans. R. Soc.* B. CXCV. 36 During the cold weather Echidna abandons all attempts at homœothermism and hibernates for four months. **1928** PEARSE & HALL *Homoiothermism* xii. 91 Homoiothermism apparently had its origin at the time when the great dinosaurs were becoming extinct. **1961** WEBSTER, Homoiothermy, homeothermy. **1966** W. S. HOAR *Gen. & Compar. Physiol.* x. 321 Increasing complexity of organization (especially behavioral organization) makes homeothermy a necessity. **1972** *Sci. Amer.* June 71/2 Maintaining homeothermy is energetically expensive.

homœotransplant (hǫmiotrɑ·ns,plɑnt), *sb. Med.* and *Biol.* Usu. **homoio-**; also **homeo-**. [f. HOMŒO- + TRANSPLANT *sb.*] = *HOMOTRANSPLANT *sb.*

1914 *Jrnl. Med. Res.* XXX. 115 A second piece [of the thyroid] was placed into the left ear of the second or control animal as a homeo-transplant. **1930** *Physiol. Rev.* X. 551 Regenerative processes are, as a rule, more extensive in autotransplants than in homoiotransplants. **1952** *Jrnl. Nat. Cancer Inst.* XIV. 692 Tumor homoiotransplants, i.e. tumors transplanted within the species but outside the strain of origin, fail to grow, or grow temporarily and then regress.

homœotransplant (hǫ:miotrɑns,plɑ·nt), *v. Med.* and *Biol.* Usu **homoio-**; also **homeo-**. [f. prec. sb.] *trans.* = *HOMOTRANSPLANT *v.*

1926 *Amer. Jrnl. Path.* II. 117 Bone and bone marrow adjoining the xiphoid cartilage were homoiotransplanted together with the cartilage. **1930** *Physiol. Rev.* X. 561 Tumours differ from normal tissues in that they can be homoiotransplanted as well as autotransplanted.

So **homœotranspla·nted** *ppl. a.*; also **ho:mœotransplantabi·lity**.

1915 *Jrnl. Amer. Med. Assoc.* 27 Feb. 727/2 In the homoiotransplanted tissue the fibroblasts very soon form dense fibrous material. **1930** *Physiol. Rev.* X. 551 Directly around the homoiotransplanted cartilage there forms..a thick layer of fibrous tissue. **1954** *Cancer Res.* XIV. 1 (*heading*) The cytotoxicity of serum for mouse mammary cancer cells. I. The effects of admixture *in vivo* upon homoiotransplantability.

ho:mœotransplanta·tion. Usu. **homoio-**; also **homeo-**. *Med.* and *Biol.* [f. HOMŒO- + TRANSPLANTATION.] = *HOMOTRANSPLANTATION.

1913 [see *HOMŒOPLASTIC *a.* 2]. **1928** D. MARINE in E. V. Cowdry *Special Cytol.* I. xvii. 564 In man, by transplanting within the same blood group, it is probable that the average life of homeografts might be prolonged, but homeotransplantation at present has no permanent value. **1953** G. D. SNELL in Homburger & Fishman *Physiopath. Cancer* xiv. 355 It is more plausible to expect..serial transfer of normal tissue to occur in isotransplantation than in homoiotransplantation.

homo-erotic (hǫmo,ęrǫ·tik), *a.* and *sb. Psychiatry*. Also **homoerotic**. [f. HOMO- + EROTIC *a.* and *sb.*] **A.** *adj.* Pertaining to or characterized by a tendency for erotic emotions to be centred on a person of the same sex; of or pertaining to a homo-erotic person. Freq. a synonym of *homosexual*.

1916 E. JONES tr. *Ferenczi's Contrib. Psycho-Analysis* xii. 268 The development of a homo-erotic obsessional neurosis. **1917** C. R. PAYNE tr. *Pfister's Psychoanalytic Method* ix. 178 After the damming up of the homoerotic instinctive activity, a physical symptom appeared. **1936** W. S. SADLER *Theory & Pract. Psychiatry* xxxviii. 626 Many notable individuals in history have been homoerotic, among them Alexander the Great and Michelangelo. *Ibid.* 627 On getting into the case, I found that she was homoerotic—homosexual. **1959** *Jrnl. Analytical Psychol.* IV. II. 120 From his dreams in particular the repressed homo-erotic aspect of these clearly emerged. **1961** *Encounter* Mar. 75 The only enduring friendships are 'homo-erotic'. **1969** *Listener* 20 Feb. 250/3 A slick, unreal, wise-cracking comedy about New York homo-erotic life. **1971** M. ALTSCHULER in Marshall & Suggs *Human Sexual Behavior* ii. 48 One may see, in various houses, boys and young men..wrapped in each other's arms... This behavior is not specifically homosexual, but it may be called homoerotic. It is noticed only during fiestas.

B. *sb.* A homo-erotic person.

1936 W. S. SADLER *Theory & Pract. Psychiatry* xxxviii. 626 The congenital, full-fledged homoerotic is never really cured.

Hence **homo-ero·ticism, -e·rotism**, the concentration of erotic impulses on a person of the same sex.

1916 E. JONES tr. *Ferenczi's Contrib. Psycho-Analysis* xii. 253 Even superficial observation of these two kinds of homo-erotism shews that they belong..to quite different syndromes. **1936** W. S. SADLER *Theory & Pract. Psychiatry* xxxviii. 628 The well-known tendency toward a certain degree of homoeroticism in the Army and Navy. **1950** E. JONES tr. *Ferenczi's Sex in Psychoanalysis* xii. 299 Homo-erotism...The word..is in my opinion preferable to the ambiguous expression homosexuality, since it makes prominent the psychical aspect of the impulse in contradistinction to the biological term 'sexuality'.

homogametic (hǫmogămī·tik), *a. Biol.* [f. HOMO- + *GAMETIC *a.*] (Of a sex or its individual members) producing gametes that all have the same kind of sex chromosome. Opp. *HETEROGAMETIC *a.*

1910 [see *HETEROGAMETIC *a.*]. **1971** *Nature* 18 June 432/1 In the Lepidoptera, males are homogametic and females heterogametic.

Hence **homogame·ty**, the state or condition of being homogametic.

1939 C. H. WADDINGTON *Introd. Mod. Genetics* iii. 81 In a type with female homogamety, *Drosophila* for example, the non-disjunctional *XX* eggs with the extra chromosome when fertilized by *X* sperm give super-females. **1961** A. MÜNTZING *Genetic Res.* xiii. 98/2 The occurrence of homo- and hetero-gamety in dioecious plant species was first demonstrated by Correns in *Bryonia dioica*.

homogamous, *a.* Add: **c.** *Evolution.* Of or pertaining to homogamy (sense *b). Also **homogamy**, (*b*) *Evolution*, preferential breeding between individuals similar in some characteristic; inbreeding; **homoga·mic** *a.*, = *HOMOGAMOUS *a.* c.

1897 G. J. ROMANES *Darwin, & after Darwin* III. i. 5 For the sake..of securing more descriptive terms, I will coin the words Apogamy and Homogamy... Homogamy ..answers to discriminate isolation, or segregate breeding: only individuals belonging to the same variety or kind are allowed to propagate. **1903** *Biometrika* II. 481 If the male class of a given character tends to mate with a female class with generally like character, we have a tendency to homogamy. *Ibid.*, The whole range of effects from pure random matings to perfectly homogamous unions within a population is almost but not quite as important as the difference between self and cross fertilization in plants. **1907** *Fabian News* XVII. 55/2 Professor Pearson's theory of homogamic mating. **1947** *Evolution* I. 270/2 The concept of homogamy or associative mating states that within a population the most similar individuals will mate with each other. **1970** T. DOBZHANSKY *Genetics Evol. Process* iv. 100 There may, however, be assortative mating, such as some preference for mating of like (homogamy). *Ibid.* x. 328 If eggs of the two species of sea urchins..are exposed to mixtures of sperm of both species, homogamic fertilizations greatly outnumber heterogamic ones.

homogenate (hŏmǫ·dʒinēit). [f. HOMOGEN(IZE *v.* + *-ate*, after *condensate, filtrate*, etc.] The suspension of cell fragments and cell constituents that is obtained when tissue is homogenized.

1941 V. R. POTTER in *Jrnl. Biol. Chem.* CXLI. 775 Rat liver which had been freshly homogenized and diluted.. was used unless otherwise indicated. This material will be referred to as a 'homogenate'. **1948** *Biochem. Jrnl.* XLIII. 205/2 Stained smears of such homogenates consisted merely of amorphous material quite unrecognizable as belonging to the organ in question. **1962** *Lancet* 27 Jan. 191/2 Samples of lung were homogenised in 10 volumes of 0·85% saline, and the homogenate was allowed to stand for three hours. **1970** *Nature* 28 Mar. 1252/1 After homogenizing for 30 s in a chilled, loose-fitting, mechanically driven glass homogenizer, the homogenate was centrifuged at 3,000 r.p.m.

homogeneous, *a.* Add: **2. b.** *Physics.* Of light: not decomposable into light of other colours. Hence of radiation generally: monochromatic.

[**1671** NEWTON in *Phil. Trans. R. Soc.* VI. 3081 Light is not similar, or homogeneal, but consists of difform Rays, some of which are more refrangible than others.] **1783** *Phil. Trans. R. Soc.* LXXIII. 97 An oblique pencil of homogeneous rays. **1863** E. ATKINSON tr. *Ganot's Physics* VII. iv. 406 In optical researches it is frequently of great importance to procure homogeneous or monochromatic light. **1897** NICHOLS & FRANKLIN *Elem. Physics* III. vii. 73 Homogeneous light is sometimes called monochromatic light. **1913** *Proc. R. Soc.* A. LXXXIX. 246 The wave-length of a homogeneous beam of X-rays. **1942** W. B. BOAST *Illum. Engin.* i. 3 Radiant energy from a gaseous-discharge source, such as mercury vapor,..consists of..one or more homogeneous component radiations.

c. *Physical Chem.* Consisting of, occurring in, or involving a single phase.

1874 *Proc. R. Soc.* XXII. 30 The body may be either homogeneous throughout, as a continuous solid, or liquid, or gas; or it may be heterogeneous, as a mass of water and aqueous vapour (*i.e.* steam). **1878** J. W. GIBBS in *Trans. Connecticut Acad. Arts & Sci.* III. 116 By *homogeneous* is meant that the part in question is uniform throughout, not only in chemical composition, but also in physical state. **1930** W. T. HALL *Textbk. Quantitative Analysis* xi. 135 A mixture of two solid substances is not homogeneous. A solution, on the other hand, is homogeneous when it is thoroughly mixed. **1940** C. N. HINSHELWOOD *Kinetics Chem. Change* iv. 70 A reaction may be partly homogeneous and partly heterogeneous. **1947** S. GLASSTONE *Elem. Physical Chem.* xviii. 587 Homogeneous reactions..take place entirely in one phase, either gas or solution. **1966** *McGraw-Hill Encycl. Sci. & Technol.* II. 547/1 If the catalyst is in the same phase as the reactants, the process is homogeneous catalysis... A homogeneous catalyst is molecularly dispersed (dissolved) in the reactants which are, most commonly, in the liquid state.

d. *Nuclear Science.* Of a nuclear reactor: having the fuel intimately and uniformly mixed with the moderator (which if liquid may also serve as the coolant).

1947 C. GOODMAN *Sci. & Engin. Nuclear Power* I. ix. 273 The reactors are considered to be of two classes: heterogeneous and homogeneous. In the latter, the fissionable material is uniformly distributed throughout the active portion of the reactor. **1955** *Times* 10 Aug. 8/4 The so-called 'homogeneous reactor', in which the nuclear fuel is circulated in solution. **1964** M. Gowing *Britain & Atomic Energy* x. 273 The possibilities of a homogeneous heavy water pile..were it is true not promising.

3. c. *homogeneous co-ordinates*, a system in which the ratios of the co-ordinates (one more than necessary) are substituted for the co-ordinates themselves, making the equations (all except one) homogeneous.

1879 *Encycl. Brit.* X. 408/1 For the proper development of the science [of analytical geometry] homogeneous coordinates..are required. **1934** D. M. Y. Sommerville *Analyt. Geom. Three Dimensions* ii. 18 If [*X*, *Y*, *Z*] are the ordinary non-homogeneous coordinates..let $X = x/w$, $Y = y/w$, $Z = z/w$, then [*x*, *y*, *z*, *w*] are called the homogeneous cartesian coordinates. **1965** H. Eves *Surv. Geom.* II. x. 71 Homogeneous coordinates enable us to establish the important principle of duality of plane projective geometry.

¶. The spelling **homogenous** is less common than the pronunc. (hǒmǫ·dʒĭnəs), which perh. owes its currency partly to the influence of the vb. *homogenize* and its derivs.

1956 J. N. Anderson *Appl. Dent. Mat.* xx. 243 Thorough mixing of the dry ingredients..and vigorous spatulation..help to produce a homogenous mix. **1961** Webster, *Homogenous* = Homogeneous. **1964** E. Palmer tr. *Martinet's Elem. Gen. Ling.* ii. 39 No linguistic community of any great size is homogenous. **1970** *Times* 2 June (Container Suppl.) p. i/6, As general cargo is homogenized into standard boxes, it will inevitably follow in the path of the homogenous bulk trades such as oil and ore. **1971** *Nature* 17 Sept. 203/1 Fractions which were homogenous by thin-layer chromatography..were used. **1972** *Ibid.* 21 Jan. 138/2 A procedure for purification has been described that gives a product which appears homogenous.

homogenic (hǫmodʒe·nik), *a.*[1] *Med.* [f. Gr. ὁμογεν-ής of the same kind + -ic.] Obtained from an animal, or from animals, of the same species.

1911 *Jrnl. Exper. Med.* XIV. 244 Autogenic and homogenic plasma constitute the best media, but heterogenic plasma can also be used. **1922** *Ibid.* XXXV. 18 The action of different dilutions of homogenic and heterogenic sera on the rate of growth of chicken fibroblasts.

homogenic (hǫmodʒĭ·nik), *a.*[2] *Genetics.* [f. Homo- + *gen(e + -ic.] Having only one allele of a particular gene; of or pertaining to organisms that have identical alleles of some gene.

1947 D. Lewis in *Heredity* I. 101 Heterogenic (S4.6) pollen grains in a style carrying S6 and not S4 should have the same incompatibility reaction as the homogenic (S6.6) pollen grain in a style carrying both these alleles. **1966** J. R. Raper *Genetics of Sexuality in Higher Fungi* i. 3 Homogenic incompatibility, in which mating is prevented between strains having the same factor(s). *Ibid.* 4 Outbreeding is enhanced by homogenic incompatibility. **1973** *Nature* 3 Aug. 305/1 Which reacted according to the tetrapolar mechanism of the homogenic incompatibility system in the same way as known in other Basidiomycetes.

homogenization (hǒmǫ:dʒĭnəizēⁱ·ʃən). [f. Homogeniz(e *v.* + -ation.] **1.** The process of making or becoming homogeneous; the action of homogenizing. Also *fig.*

1908 *Practitioner* June 830 Methods for concentrating the bacilli..depend mainly on the homogenisation of the sputum by means of dilute alkalies or of enzymes. **1915** *Jrnl. Physical Chem.* XIX. 225 The stability of an emulsion is tremendously enhanced by homogenization. **1936** W. L. Davies *Chem. Milk* xii. 256 The homogenisation of cream, especially coffee cream, is widely practised. **1962** *Lancet* 22 Dec. 1330/2 Tissues were similarly extracted by fourfold homogenisations at 0°C in perchloric acid. **1963** *Times* 30 May 13/2 This homogenization of political discourse presents the Labour Party with some difficulties. **1964** M. McLuhan *Understanding Media* (1967) II. xxxi. 345 America long ago achieved its Common Market by mechanical and literate homogenization of social organization. **1965** P. G. Shewmon in R. W. Cahn *Physical Metall.* viii. 370 Rates of homogenization are of interest in many metallurgical problems.

2. The state produced in something that has been homogenized; uniformity of composition.

1938 D. K. Bullens *Steel & its Heat Treatment* (ed. 4) I. v. 131 When the more rapid cooling is to be used..it is desirable to attain homogenization of the austenite. **1955** P. Becher *Princ. Emulsion Technol.* v. 76 In order to achieve complete homogenization of milk..it is often necessary to employ a two-stage machine. **1962** A. G. Guy *Physical Metall.* (1963) viii. 253 At a moderate temperature the rapid diffusion of zinc in copper produces effective homogenization of cast brasses.

homogenize, *v.* Delete *rare* and add: To unite or incorporate into a single whole of uniform composition; to make uniform or similar. Also *fig.* (Further examples.)

1908 *Practitioner* June 831 Nebel homogenises the sputum with lime-water. **1957** W. H. Whyte *Organization Man* xxiv. 323 In the new middle-class rhythm of life obligations are homogenized, for the overriding aim is to have oneself precommitted to regular, unvarying monthly payments on all the major items. **1964** M. McLuhan *Understanding Media* (1967) II. x. 101 The new centralist power always takes action to homogenize as many marginal areas as possible. *Ibid.* xxiii. 244 Any community that wants to..maximize the exchange of goods..has simply got to homogenize its social life. **1965** *Times Lit. Suppl.* 25 Nov. 1063/3 To omit the commas..is in a sense to denature or homogenize the lines.

b. To subject (milk or another emulsion) to a process by which the suspended globules or droplets are broken up into smaller ones and distributed throughout the liquid, so that they have no tendency to collect into a cream.

1904 *Sci. Amer.* 16 Apr. 315/2 To the many methods of purifying, modifying, and preserving milk must now be added a process for homogenizing it so that it will keep almost indefinitely. **1913** *Pharmaceutical Jrnl.* 24 May 734/1 It has become the practice of the emulsion manufacturer to perfect his emulsions by the use of..the apparatus constructed for the dairyman and used..for homogenising milk. **1936** W. L. Davies *Chem. Milk* xii. 256 Gaulin..first conceived the idea of homogenising already existing emulsions, such as milk, so as to obtain greater stability. **1949** Kirk & Othmer *Encycl. Chem. Technol.* IV. 823 The purpose of homogenizing evaporated milk is to prevent fat separation in the manufactured product. **1971** *Nature* 29 Oct. 617/2 Machines with triple-piston pumps..are used extensively in..the dairy industry, for homogenizing milk, cream, ice-cream mix and other products.

c. To make (an alloy) more uniform in chemical composition by holding it at a high temperature for a period and then allowing it to cool slowly.

1924 [see *homogenizing* vbl. sb. and ppl. adj. below]. **1948** *Metals Handbk.* (Amer. Soc. Metals) 977/1 The [magnesium] alloys that contain aluminum may be homogenized in cast form. **1955** E. Johnson in W. C. Newell *Casting of Steel* x. 448 The practice of homogenizing alloy castings at 850°C to 950°C for one hour per inch of heaviest section is common. **1965** P. G. Shewmon in R. W. Cahn *Physical Metall.* viii. 372 This also indicates one reason why a wrought product is more easily homogenized..than a cast product.

d. To prepare a suspension of cell fragments and cell constituents from (tissue) by physical treatment in a liquid medium.

1936 *Jrnl. Biol. Chem.* XXXVI. 504 A new method for the study of tissue respiration is described in which the tissues are homogenized in a buffer medium by a high speed glass pestle. **1959** *Sci. News* LIII. 51 They killed the guinea-pig, rapidly dissected out its liver, and homogenized it in sugar solution in a Waring blendor. **1964** G. H. Haggis et al. *Introd. Molecular Biol.* ii. 31 The tissue is first homogenized, the cells being disrupted by the shearing forces of the homogenizer. **1971** *Nature* 10 Sept. 127/1 The brains were homogenized in cold distilled water (6 ml./brain) at 850 r.p.m. using a 'Teflon' pestle.

2. *intr.* To become homogeneous; to respond to homogenization.

1938 D. K. Bullens *Steel & its Heat Treatment* (ed. 4) I. v. 131 The same procedure of air cooling may be used even upon steels that would readily homogenize at ordinary annealing temperatures. **1949** Brick & Phillips *Structure & Properties of Alloys* (ed. 2) xi. 323 Small castings usually consist of very fine dendrites that homogenize quite readily with respect to C and Mn.

So **homo·genizing** *vbl. sb.* and *ppl. a.*

1913 *Pharmaceutical Jrnl.* 24 May 734/1 The actual homogenising section of the machine is a valve formed by an agate cone which is pressed into a gun-metal seating. **1924** Jeffries & Archer *Sci. of Metals* x. 353 The effect of a homogenizing treatment on the properties of an alloy. **1963** *Times* 14 Feb. 12/7 The curious homogenizing process of reaching a consensus. **1964** M. McLuhan *Understanding Media* (1967) II. xxxi. 344 The homogenizing power of the literate process had gone further in America by 1800 than anywhere in Europe.

homogenized (hǒmǫ-dʒĭnəizd), *ppl. a.* [f. Homogeniz(e *v.* + -ed[1].] **1.** Rendered uniform throughout in composition or character; *loosely*, = Homogeneous *a.* 2.

1935 G. E. Doan *Princ. Physical Metall.* iv. 140 The true solidus line of alloys of this kind is..determined..by reheating specimens of a completely homogenized alloy. **1959** *Observer* 29 Mar. 15/4 As for his [*sc.* the novelist's] 1980s these are just a little more prefabricated and homogenised than our time. **1964** M. McLuhan *Understanding Media* (1967) I. v. 60 The fragmented man creates the homogenized Western world, while oral societies are made up of people differentiated..by their unique emotional mixes. **1967** *N.Y. Times* (Internat. ed.) 11–12 Feb. 4/7 The time for studying people in different environments is running out. The world is becoming homogenized.

b. *esp.* of milk, cream, etc.: having the globules of fat reduced in size and distributed throughout the liquid; also of other emulsions (see * homogenize *v.* 1 b).

1904 *Sci. Amer.* 16 Apr. 315 Homogenized milk, a trade-name for milk which has been heated to 185°F. and forced by heavy pressure through a number of very fine openings. **1913** *Pharmaceutical Jrnl.* 24 May 734/1 Homogenised oil emulsions are permanent for an almost indefinite time. **1923** W. Clayton *Theory Emulsions* viii. 123 In America homogenised cream is extensively used in the manufacture of ice-cream, where a smooth texture is demanded. **1951** *Good Housek. Home Encycl.* 509/1 Homogenised milk is considered more quickly digestible on account of the fineness of the fat globules. **1963** *Observer* 5 May 9/5 A sales drive by dairymen to popularise homogenised milk and collect an extra halfpenny a pint is gathering momentum. **1969** *Daily Colonist* (Victoria, B.C.) 24 Sept. 9/3 (Advt.), Empress pure peanut butter. Homogenized, regular or chunk style.

2. Of various things in respect of each other: not readily differentiated; similar in nature, meaning, etc. (Used when an expected or desirable difference is not found.)

1958 *Vogue* Oct. 153 Beside this deeply particular tragedy, Mr Eliot's homogenized characters seem insufficient. **1959** N. Mailer *Advts. for Myself* (1961) 210 Each reference to yourself as individual as a carloading of homogenized words. **1970** *New York* 16 Nov. 50/2 Even these words have become murky, homogenized.

homogenizer (hǒmǫ-dʒĭnəizəɹ). [f. Homogeniz(e + -er[1].] A machine or apparatus designed to homogenize some kind of material (as milk or tissue). Also *fig.*

1886 [in Dict. s.v. Homogenize *v.*]. **1908** *N.Y. Produce Rev. & Amer. Creamery* 11 Nov. 92/2 (*heading*) A new homogenizer. **1910** *Ibid.* 9 Feb. 600/1 A machine of much interest to the dairy trade..is the Gaulin Homogenizer. **1936** *Jrnl. Biol. Chem.* XXXVI. 496 To prepare tissue for study, a measured volume of the desired buffer is placed in the homogenizer tube, which is then weighed. **1955** P. Becher *Princ. Emulsion Technol.* v. 74 A homogenizer is a device in which dispersion is effected by forcing the mixture to be emulsified through a small orifice under very high pressure. **1956** *Nature* 4 Feb. 233/2 Disintegration of spleen cells (grinding in a homogenizer at high speed) destroys their protective effect on antibody formation. **1969** J. G. Brennan et al. *Food Engin. Operations* v. 93 The many applications for ultrasonic homogenizers..include: the manufacture of salad creams, ice-cream mixes,..and baby foods. **1973** *Daily Tel.* 24 Mar. 14/4 The social and aesthetic attitudes have been passed through the homogeniser of the bureauctatic hivemind.

homogenous, *a.* Add: **2.** *Surg.* Of transplanted tissue: = *homoplastic *a.* 2.

1919 *Ann. Surg.* LXIX. 123 It is possible that further experiments with homogenous transplants from young to old and from old to young animals would answer this question. **1939** S. Fomon *Surg. Injury & Plastic Repair* ii. 107 Homogenous transplants..are usually taken from members of the same family. **1964** R. Battle *Plastic Surg.* ii. 37 Any graft, either autogenous or homogenous, that is not immediately required can be stored for use at a later date.

homogenous, var. *homogeneous *a.*

homogentisic (hǫmodʒenti·sik), *a.* [tr. G. *homogentisinsäure* homogentisic acid (Wolkow & Baumann 1891, in *Zeitschr. f. physiol. Chem.* XV. 245), f. Homo- b + Gentisic *a.*] *homogentisic acid*: a crystalline compound, $C_8H_8O_4$, which oxidizes to a black compound in air, is an intermediate in the metabolism of aromatic amino-acids, and is excreted in large amounts by persons with alkaptonuria; alkapton, 2,5-dihydroxyphenylacetic acid. So **homoge·ntisate**, a salt or ester of this acid.

1891 *Jrnl. Chem. Soc.* LX. 1129 These [crystals] consisted of a substance very similar to, but not identical with, Kirk's uroleucic acid, and the name homogentisic acid is given to it. *Ibid.*, Lead homogentisate, $(C_8H_7O_4)_2Pb + 3H_2O$, crystallises in colourless, brilliant, transparent needles and prisms. **1942** *Jrnl. Amer. Med. Assoc.* 11 July 882/1 Homogentisic acid isolated from one of the positive urines itself gave a strongly positive reaction. **1961** *Lancet* 5 Aug. 320/1 The homogentisate level in body fluids and urine. **1969** *Listener* 16 Jan. 74/3 It's now known that not only do one's genes decide whether one is to have red hair, double-jointed thumbs or a Hapsburg lip, but they also decide whether one is going to excrete homogentisic acid or not. **1970** C. N. Graymore *Biochem. Eye* vii. 478 Phenylalanine and tyrosine are converted through quinone intermediates to homogentisic acid.

homogonic (hǫmogǫ·nik), *a. Biol.* [f. Homo- + Gr. -γονος generating, γόνος offspring + -ic.] Applied to a life cycle of certain nematodes in which the offspring of a parasitic generation themselves develop into parasites, without the intervention of a free-living generation; also applied to organisms reproducing by such a cycle. Opp. * heterogonic *a.* 2.

1926 [see *heterogonic *a.* 2]. **1939** *Amer. Jrnl. Hygiene* XXX. D. 15 In those *Strongyloides* infections where both modes of larval development are commonly encountered, there is little constancy in the daily yields of heterogonic, i.e., indirect or free-living adult, in contrast to homogonic, or direct, progeny. **1951** L. H. Hyman *Invertebrates* III. xiii. 307 It is probable that any species of *Rhabdias* may employ either the homogonic or heterogonic type of life cycle. **1968** N. D. Levine *Nematode Parasites* ii. 75/1 Absence of food, too, led to homogonic development.

Hence **homogo·nically** *adv.*, by means of a homogonic life cycle.

1938 *Amer. Jrnl. Hygiene* XXVIII. 224 The production of these two types of progeny was related to the age of the single, homogonically derived parasite. **1968** N. D. Levine *Nematode Parasites* ii. 75/1 In a loam soil..40% of the larvae developed into females, 40% into males and 20% homogonically into infective larvae.

homograft (hǫ·mogrɑft). *Med.* and *Biol.* [f. Homo- (in *homogenous*, *homologous*, and *homoplastic*) + Graft *sb.*[1]] A graft taken from another individual of the same species as the recipient; a homotransplant; *homograft reaction*, the immunological reaction that causes a homograft to be rejected by the recipient's body.

1923 H. Neuhof *Transplantation of Tissues* ii. 32 Autografts of skin have a degree of permanency in both animal and clinical surgery whereas..occasional successes with homografts of skin have been more than counterbalanced by the frequent failures. **1944** *Jrnl. Anat.* LXXVIII. 182/2 The time-relations of the homograft reaction vary from one pair of animals—donor and recipient—to another. **1955** *Sci. News* XXXV. 105 Blood vessel homografts are now being used on an ever increasing scale. **1960** *News Chron.* 22 June 6/4 Within a decade, the homograft barrier will be penetrated and human 'spare-part' surgery will be a reality. **1961** *New Scientist* 15 June 632/3 Only identical twins can accept organ and tissue grafts from one another without exciting the homograft reaction. **1962** *Lancet* 27 Jan. 193/2 A woman who had received a homograft of bone-marrow from her brother. **1968** D. Longmore *Spare-Part Surg.* vi. 161 For the past three years my colleagues and I have been transplanting heart-lung homografts experimentally in animals.

Hence **ho·mografted** *ppl. a.*; **ho·mografting** *vbl. sb.* = *homotransplantation.

1923 H. Neuhof *Transplantation of Tissues* i. 3 Homotransplantation (homoplasty, isoplasty, homografting, homoiotransplantation, homoplastic transplantation, isoplastic transplantation) is the transference of a tissue from one to another individual of the same species. **1937** *Surgery* I. 558 (*heading*) Homografting of skin: with report of success in identical twins. **1952** *Jrnl. Nat. Cancer Inst.* XIV. 669 Many surgeons..have found the results of grafting with autogenous tissues..superior to the results obtained by the use of similar homografted material. **1953** *Proc. Soc. Exper. Biol. & Med.* LXXXII. 523/2 Skin homograftings have been permanently successful only between identical twins..or between members of the same family.

homographically (hǫmogræ·fikăli), *adv. Math.* [f. Homographic *a.*: see -ically.] In a homographic manner.

1860 *Phil. Trans. R. Soc.* CXLIX. 67 Four or more tetrads of points in a l ne may be homographically related to the same number of tetrads in another line. **1947** Hodge & Pedoe *Methods Algebraic Geom.* I. vi. 217 A one-to-one correspondence..is set up between the ranges,..and we say that these ranges are projectively, or homographically, related.

homography. 1. (Further examples.)

1959 E. M. Patterson *Topology* (ed. 2) ii. 21 Congruence and similarity in Euclidean geometry and homography in projective geometry are all equivalence relations. **1965** H. Eves *Surv. Geom.* II. xii. 203 A transformation of the form $x' = (ax+b)/(cx+d)$, $ad - bc \neq 0$, is called a homography (or bilinear substitution).

Homoian, var. *Homœan *a.* and *sb.*

Homoiousian, *a.* (Earlier example.)

1683 W. Cave *Ecclesiastici* 167 The Synod was divided into two principal Factions, the one of the Semiarian or Homoiousian Party,..the other of the Heterousians.

Homoiousion (hǫmoi,ɑu·ziǫn, -ɑu·siǫn, -ū·-). *Theol.* [eccl. Gr. ὁμοιούσιον, neuter form of ὁμοιούσιος: see Homoiousian *a.* and *sb.* and cf. *Homoousion *sb.* and *a.*] *The Homoiousion*, the term used to express the doctrine put forward by the Semi-Arians, that the Son is 'of like substance' with the Father; the doctrine itself: opposed to the term ὁμοούσιον (see * Homoousion *sb.* and *a.*). The masc. form homoiousios is also used.

[**1683** W. Cave *Ecclesiastici* 164 They rejected both the ὁμοούσιον and the ὁμοιούσιον [*sic*], as Expressions unknown to Scripture.] **1833** J. H. Newman *Arians* iv. 369 Acacius..proposed a creed in which the Homoousion and Homoiousion, were condemned..and his own Homoion adopted. **1873** W. Bright *Orations of St. Athanasius against Arians* p. lxvii, The Semi-arians..insisted on their own formula of the 'Homoiousion'. **1969** A. Richardson *Dict. Chr. Theol.* 347/2 In their manifesto of 358 they rejected the *Homoousios*..and proposed the *Homoiousios*.

homokaryotic (hǫ:mokæri,ǫ·tik), *a. Bot.* Also -caryotic. [ad. G. *homocaryotisch* (H. Burgeff 1913, in *Ber. d. Deut. Bot. Ges.* XXX. 680), f. Homo- + Kary(o- + -otic.] Exhibiting homokaryosis. Hence **homokaryon** (-kæ·ri,ǫn), a homokaryotic cell, structure, or organism; **homokaryosis** (-kæri,ōᵘ·sis), the condition, prevalent among fungi, in which two or more genetically identical nuclei are maintained in a common cytoplasm.

1916 B. D. Jackson *Gloss. Bot. Terms* (ed. 3) 183/2 Homokaryotic. **1921** *Phil. Trans. R. Soc.* B. CCX. 111 The contamination may be purely cytoplasmic, and the progeny of such a cell will therefore still be homocaryotic. **1928** B. D. Jackson *Gloss. Bot. Terms* (ed. 4) 443/1 Homocaryosis. **1939** *Mycologia* XXXI. 226 Although the conidia and ascospores of *B*[*otryosphaeria*] *Ribis* are multinucleate, all of the nuclei of a single conidium or ascospore have the same origin and hence are homocaryotic or genetically similar. **1949** H. W. Florey et al. *Antibiotics* II. xvi. 673 Heterokaryons are, in general, more vigorous than the homokaryons from which they originate. **1951** D. G. Catcheside *Genetics of Micro-Organisms* iv. 72 Not too much reliance must be placed in the homocaryosis of a strain maintained for some time by vegetative transfers. **1955** G. M. Smith *Cryptogamic Bot.* (ed. 2) I. xi. 414 Multinucleate mycelia of Mucorales and of other fungi may be homokaryotic..or they may be heterokaryotic. **1969** G. Sermonti *Genetics Antibiotic-Producing Microorganisms* ii. 38 In such cases a balanced combination of two (or more) genotypes enjoys a selective advantage over either homokaryon. *Ibid.* v. 165 They often continue to segregate even after homokaryosis has been established.

homological, *a.* Add: **2.** *Philos.* = *autological *a.* Hence **homologica·lity**, the property of being homological; **homolo·gically** *adv.*, (*b*) by virtue of being homological.

1940 [see *heterological *a.*]. **1952** *Mind* LXI. 85 Neither is the word *short*, simply because its meaning is predicable of its own verbal sign, homologically predicable of itself. *Ibid.* 86 This anomalous distinction between two such comparable predicates seems to be incompatible with genuine heterologicality and homologicality. *Ibid.* 88 A homological predicate, genuinely predicable of itself.

homologous, *a.* Add: **4. b.** *Cytol.* Of chromosomes: pairing at meiosis, and normally (except in the case of the sex chromosomes of some species) identical in morphology and in arrangement of genetic loci.

1903 W. S. Sutton in *Biol. Bull.* IV. 238 The double basis of hybrid characters is to be found in the pairs of homologous chromosomes of the presynaptic germ-cells. **1920** W. E. Agar *Cytol.* v. 125 It was found that there were in each diploid nucleus two chromosomes of each size... In the meiotic prophase the two chromosomes of each type, usually called homologous chromosomes, pair together to form the bivalents. **1970** Ambrose & Easty *Cell Biol.* x. 327 The nature of the forces responsible for the pairing of homologous chromosomes in the early stages of prophase I, and for their separation at the diplotene stage, is not yet understood.

c. *Med.* Derived from or involving an organism or organisms of the same species; also, involving or containing antibodies or antigens that react specifically with one another, as when an antibody has been produced by injection of an antigen.

1915 *Jrnl. Path. & Bacteriol.* XX. 76 Heterologous immunity, in which cancer cells from one species are used as antigens to immunise animals of strange species, is..different in many ways from homologous immunity. **1928** Buchanan & Fulmer *Physiol. & Biochem. Bacteria* I. iii. 361 By the use of ox serum and complement he could secure marked agglutination of bacteria in a dilution of the homologous antiserum by itself too weak to produce any trace of agglutination. **1933** W. W. C. Topley *Outl. Immunity* vi. 91 The antiserum that is produced by the inoculation into a suitable animal of a particular bacterium is frequently referred to as a homologous serum. **1946** K. Landsteiner *Specificity Serol. Reactions* (rev. ed.) i. 8 Reactions of an antibody with the corresponding antigen are said to be homologous, while heterologous reactions..are those taking place with substances other than the inciting antigen. **1958** *Immunology* I. 111 Morgan (1947) injected rhesus monkeys with homologous brain and cord in water-in-oil emulsion containing killed *Mycobacterium tuberculosis*. **1961** *Lancet* 29 July 245/1 Serum-autoantibodies against, for example, thyroid or brain can be readily elicited in rabbits or guineapigs by immunisation with the relevant homologous or, indeed autologous, tissue. **1968** J. C. Norman et al. *Organ Perfusion & Preservation* xxvii. 375 The feasibility of heterologous or homologous intermediate hosts for human resuscitative storage is untried.

homolysis (hǒmǫ·lisis). *Chem.* [f. Homo- + *-lysis.] The splitting of a molecule into two neutral atoms or radicals.

1938 [see *heterolysis 2]. **1966** W. A. Pryor *Free Radicals* ix. 119 Molecule-induced homolysis is postulated to occur when radicals are formed at an anomalously rapid rate from the interaction of nonradical species.

Hence **homoly·tic** *a.*, of the nature of or involving homolysis; **homoly·tically** *adv.*, by homolysis.

1941 Homolytic [see *heterolytic* after *heterolysis]. **1952** *Sci. News* XXVI. 57 A covalent bond may..break by homolytic fission, each of the electrons separating with one of the atoms, giving two free atoms, e.g.: H—Cl→H· + Cl·. **1964** N. G. Clark *Mod. Org. Chem.* xxv. 515 Heat and/or illumination first causes a molecule of chlorine to break homolytically into two chlorine atoms.

homomorphic, *a.* Add: **d.** *Zool.* Applied to a colony in which all the constituent individuals are alike.

1891 T. J. Parker *Lessons Elem. Biol.* xii. 137 There are no special reproductive individuals, so that the colony is *homomorphic*. **1931** W. C. Allee *Animal Aggregations* ii. 23 Homomorphic colonies have all the individuals morphologically similar and may be found among sponges and at certain times among hydroids and bryozoans.

e. *Cytol.* Applied to homologous chromosomes that do not differ in size or form; opp. *heteromorphic *a.* 1 c.

1917 E. E. Carothers in *Jrnl. Morphol.* XXVIII. 449 The unusual conditions of the chromosomes in this group have made advisable the introduction of four new terms. 1. Homomorphic—used to designate those tetrads made up of morphologically similar homologues. **1925** E. B. Wilson *Cell* (ed. 3) xii. 937 Twenty-eight male offspring have thus been examined from five matings with especial reference to three chromosome-pairs..which may be either heteromorphic or homomorphic. **1968** J. A. Serra *Mod. Genetics* III. xxiii. 533 These bodies are homomorphic sex chromosomes..not heterochromosomes.

2. *Math.* Related or produced by a homomorphism; giving rise *to* a second set under a homomorphism; that is a homomorphism.

1935 *Proc. Nat. Acad. Sci.* XXI. 482 We define a continuous homomorphic mapping π_m of $\mathfrak{H}_p{}^{m+1}(\mathfrak{G})$ into $\mathfrak{H}_p{}^m(\mathfrak{G})$. **1939** *Amer. Jrnl. Math.* LXI. 783 Two homomorphic rings. **1941** Birkhoff & MacLane *Surv. Mod. Algebra* xiii. 350 This device for getting a field as a homomorphic image of a polynomial ring is important in the discussion of algebraic numbers. **1966** *Mathematical Rev.* XXXI. 15/2 It is homomorphic to (i.e., can be contracted into, by identification of sets of connected vertices) the complete graph of order *k*. **1968** I. Adler *Groups in New Math.* xiii. 230 If there is a homomorphism that matches the members of one group with the members of another, we say that the first group is homomorphic to the second, and that the second group is a homomorphic image of the first group.

transf. **1959** S. Beer *Cybernetics & Management* vi. 49 A black Box is homomorphic with a cybernetic system, because the latter has undergone a many-one simplifying transformation (which makes it tractable) without losing its key characteristic (of indefinability).

Hence **homomo·rphically** *adv. Math.*, by a homomorphism.

1941 Birkhoff & MacLane *Surv. Mod. Algebra* xiii. 350 The direct sum *A* + *B* of two rings *A* and *B* may be mapped homomorphically on the summand *B* by the correspondence $(a, b) \to b$. **1952** Eilenberg & Steenrod *Found. Algebraic Topology* i. 7 If *G* and *H* are groups, the notation $\phi: G \to H$ means that ϕ maps *G* homomorphically into *H*. **1971** M. Herzog in Powell & Higman *Finite Simple Groups* v. 200 *G* is *p*-solvable if and only if it can be mapped homomorphically on $N_G(P)/W$.

homomorphism (hǫmomǭ·ɹfiz'm). [f. Homo- + Gr. μορφ-ή form + -ism.] **1.** (In Dict. s.v. Homomorphic *a.*)

2. *Math.* [See *-morphism.] A many-to-one (or one-to-one) transformation of one set into another that preserves in the second set the operations or relations between the elements of the first.

1935 *Duke Math. Jrnl.* I. 2 There exists an operation *F* defined topologically for all the chains and such that $F\{c_p\}$ is a homomorphism of $\{c_p\}$ into $\{c_{p-1}\}$, and of $\{c_0\}$ into the identity. **1941** Birkhoff & MacLane *Surv. Mod. Algebra* vi. 155 Under any homomorphism $G \to G'$, the identity *e* of *G* goes into the identity of *G'*, and inverses into inverses. **1959** E. M. Patterson *Topology* (ed. 2) iv. 82 In fact, there is a homomorphism between the groups (not to be confused with homeomorphism). **1965** Patterson & Rutherford *Elem. Abstract Algebra* iii. 79 A mapping *f* of a ring R_1 into a ring R_2 is called a homomorphism if $f(x+y) = f(x) + f(y)$, $f(xy) = f(x)f(y)$ for all $x, y \in R_1$. **1969** F. Harary *Graph Theory* xii. 143 If *G'* is the graph resulting from a homomorphism ϕ of *G* we can consider ϕ as a function from *V* onto *V'* such that if *u* and *v* are adjacent in *G*, then ϕu and ϕv are adjacent in *G'*.

transf. **1966** S. Beer *Decision & Control* vi. 113 A scientific model is a homomorphism on to which two different situations are mapped, and which actually defines the extent to which they are structurally identical.

homonuclear (hǫmoniū·kliăɹ), *a.* [f. Homo- + Nuclear *a.*] **a.** *Physics* and *Chem.* Of a molecule: composed of atoms whose nuclei are alike, i.e. atoms of the same element or (more strictly) the same isotope.

1930 *Physical Rev.* XXXVI. 617 In the case of homonuclear molecules (molecules composed of atoms whose nuclei are identical in charge and mass), any electron state may be either 'odd' or 'even'. **1970** *Nature* 25 Apr. 354/1 Vibrational levels of N_2^+, like those of N_2, are metastable because both molecules are homonuclear.

b. *Chem.* Taking place on the same ring in a molecule.

1938 G. H. Richter *Textbk. Org. Chem.* xxix. 596 In discussing the reactions of naphthalene it is customary to speak of homonuclear substitution if the groups are entering the same ring, and heteronuclear substitution if the radicals are entering different rings. **1951** I. L. Finar *Org. Chem.* xxix. 586 Substitution products of naphthalene... When OH..or NH·CO·CH$_3$ is in the 2-position, homonuclear substitution usually takes place in the 1-position.

homonym. Add: **1. c.** *Taxonomy.* A generic name or a binomial that duplicates a name attached to a different plant or animal.

1892 *Bull. Torrey Bot. Club* XIX. 290 Homonyms.—The publication of a generic name or a binomial invalidates the use of the same name for any subsequently published genus or species respectively. **1920** *Jrnl. Bot.* LIX. 156 Specific names should be rejected when they are homonyms. **1951** G. H. M. Lawrence *Taxon. Vascular Plants* ix. 213 A name of a taxon is illegitimate and must be rejected if it is a later homonym, that is, if it duplicates a name previously and validly published for a taxon of the same rank based on a different type. **1967** R. E. Blackwelder *Taxonomy* xxii. 463 Priority determines which homonym can be retained as an acceptable name. **1972** W. T. Stearn *A. W. Smith's Gardener's Dict. Plant Names* (rev. ed.) 13 Similar names applied to different plants are called homonyms. The rejection of later homonyms has caused a number of unavoidable but regrettable name changes.

homonymous, *a.* **1.** Delete †*Obs.* and add later examples in *Taxonomy*.

1896 Walsingham & Durrant *Rules for Nomencl.* 9 Invalid names considered merely as words are of three classes:—(1) Homonymous (*i.e.* the same name applied to different conceptions). **1964** *Internat. Code Zool. Nomencl.* 55 Homonymy does not exist between two identical species-group names originally or subsequently placed in different genera that bear homonymous names.

2. b. Add to def.: Also applied to diplopia in which images are doubled in this way. Of hemianopia: characterized by the loss of vision in the same half (left or right) of the visual field of each eye.

1882 Homonymous hemianopia [see *hemianopia]. **1884** H. E. Juler *Handbk. Ophthalmic Sci.* v. 383 This projection of the object to o' is on the same side as the deviating eye L, and the diplopia is therefore called homonymous. **1966** S. Lerman *Basic Ophthalmol.* viii. 467 If the left eye deviates inward..the patient will suffer from a homonymous (uncrossed) diplopia.

homonymy. (Later examples in *Taxonomy*.)

1896 Walsingham & Durrant *Rules for Nomencl.* 9 A name homophonous with a valid name is invalid in accordance with the rule governing homonymy. **1964** [see *homonymous *a.* 1].

homo-organic, var. *homorganic *a.*

Homoousion (hǫmo,ɑu·ziǫn, -ɑu·siǫn, -ū·-), *sb.* and *a. Theol.* [eccl. Gr. ὁμοούσιον, neut. of ὁμοούσιος: see Homoousian *a.* and *sb.*]

A. *sb. The Homoousion* (τὸ ὁμοούσιον): the term ὁμοούσιος as used, e.g. in the formula promulgated by the Council of Nicæa in A.D. 325, to express the doctrine that the Son is 'of one substance' with the Father (τῷ πατρί); the doctrine itself: opposed to the term ὁμοιούσιος (see *Homoiousion). The masc. form homoousios is also used.

[**1683**: see *Homoiousion.] **1781** Gibbon *Decl. & F.* II. xxi. 251 Their [*sc.* the Arians'] patron, Eusebius of Nicomedia,..confessed, that the admission of the *Homoousion*, or Consubstantial..was incompatible with the principles of their theological system. *Ibid.* 252 The mysterious *Homoousion*, which either party was free to interpret according to their peculiar tenets. **1833** J. H. Newman *Arians* iv. 333 The Novatians, as maintaining the Homoousion, were included in the persecution. **1875** *Encycl. Brit.* II. 538/2 At length the tenet of the *Homoiousion* was substituted for that of the *Homoousion* at the Council of Rimini (Ariminum) in 360. **1921** C. H. Turner *Catholic & Apostolic* (1931) 129 The very existence of Christianity in any full sense of the term was at stake over the Homoousion. **1969** C. D. Darlington *Evol. Man & Society* xiv. 311 Why not agree, he asked, to the *homoousion* or consubstantiality of the Father and Son? **1971** *Cath. Dict. Theol.* III. 39/2 What Nicaea had done with the homoousios to overcome Arianism, Trent was to do with transubstantiation to overcome other heresies.

B. *adj.* (Usu. in form *homoousios*.) Of the same essence or substance: = Homoousian *a.* a.

1834 *Penny Cycl.* II. 317/1 In the western part of the Roman empire, all adversaries of the doctrine of Athanasius, that the Son was *homoousios*, or of the *same* essence with the Father, were called Arians; although some of these opponents taught..that the Son was *homoiousios*, or of *similar* essence. **1936** G. L. Prestige *God in Patristic Thought* x. 197 According to the Valentinians..the abortive and degenerate fruit of the final aeon in the divine Absolute (pleroma), was homoousios with angelic ('spiritual') beings. *Ibid.* 198 A piece of marble closely resembling Mr. Gladstone..is made of different stuff from that of which Mr. Gladstone himself consisted: it is in the image of Mr. Gladstone, but not homoousios with him. **1969** A. Richardson *Dict. Chr. Theol.* 347/2 The Lord and Lifegiver could not be Homoousios with those to whom he gives life.

homophile (hǫ·mofəil). [f. Homo- + -phil, -phile.] A term for a homosexual (regarded as a person belonging to a particular social group rather than as someone who is sexually abnormal). Also *attrib.* or as *adj.*

1960 G. Westwood *Minority* v. 71 The so-called homophile organizations on the Continent provide wider facilities. In Amsterdam the 'Cultuur-en-Ontspanningscentrum' (C.O.C.) maintains clubrooms..and regular lectures are arranged on homosexuality. **1960** *News Chron.* 13 Oct. 4/6 Nor do Danes frown much on sex abnormalities. Danish 'homophiles' have their own monthly magazine. **1961** *Encounter* May 75 The relationships between homophiles seem to approximate very closely to the relationships between adult heterosexual men and women. **1963** Cory & Le Roy *Homosexual & his Society* xx. 245 The great bulk of publicity about the existence of homophile organizations—as they are usually called, somewhat euphemistically—has been, of necessity, by word of mouth. *Ibid.* 247 The participants and sympathizers in the homophile movement. **1965** *Guardian* 23 Apr. 11/5 The Dutch Society of Homophiles, set up 18 years ago as a private club for homosexuals. **1971** *Daily Colonist* (Victoria, B.C.) 15 Oct. 20/8 Homosexuals want legal marriage for tax reasons and to foil the designs of relatives, George Hyslop, founder of a Toronto homophile organization, told the Canadian Mental Health Association.

homophonic, *a.* Add: **3.** *Philol.* = Homophonous *a.* 2.

1942 [see *bimorphemic *a.*].

homoplastic, *a.* Add: **2.** *Med.* and *Biol.* Of transplantation: involving the transfer of tissue from one individual to another of the same species. Of transplanted tissue: obtained from another individual of the same species as the recipient.

[**1889** L. Ollier in *Arch. de Physiol. normale et path.* I. 168 Nous divisons les greffes en trois catégories: Les greffes autoplastiques... Les greffes homoplastiques, c'est-à-dire empruntées à un autre individu, mais à un sujet de la même espèce. Les greffes hétéroplastiques.] **1909** *Jrnl. Med. Res.* XXI. 320 He [*sc.* Carrel] also distinguishes an autoplastic transplantation, when the segment is taken from the same animal; a homoplastic, when the segment is taken from another animal of the same species; and a heteroplastic, when the segment is taken from an animal of another species. **1912** *Ann. Surg.* LVI. 381 Autoplastic bony pieces are in respect to regeneration much to be preferred to homoplastic. **1935** *Jrnl. Amer. Med. Assoc.* 8 June 2076/1 These homoplastic grafts all sloughed, while autoplastic grafts were highly successful on the same patient. **1958** *Immunology* I. 1 The homoplastic transplantation of the cortex of the adrenal gland in mice.

So **homopla·stically** *adv.*, between individuals of the same species; **ho·moplasty**, homoplastic transplantation, homotransplantation.

1912 *Ann. Surg.* LVI. 378 The best material for free bony grafts is living, periosteum covered bone, if possible from the same individual himself (autoplasy [*sic*]). Or in case this is impossible, from another individual (homoplasty). **1923** H. Neuhof *Transplantation of Tissues* ii. 31 The results have seemed to indicate that blood relationship has a bearing on the success of homoplasty. **1926** *Jrnl. Exper. Zool.* XLV. 52 Fore-limb buds of R. palustris tadpoles were transplanted homoplastically beneath the integument of the back. **1929** *Ann. Surg.* XC. 929 The lower the organism in the phylogenetic scale, the better the results obtained with homoplasty. **1942** *Univ. Calif. Publ. Zoöl.* LI. 43 Gastrular anlagen were implanted homoplastically into tadpoles.

homopolar (hǫmo,pōᵘ·lăɹ), *a.* [f. Homo- + Polar *a.*] **1.** *Bot.* (In Dict. s.v. Homo-.)

2. *Electr.* Of an electric generator or its operation: having or involving such an arrangement of magnets that the direction of the flux does not alternate with relation to the motion of the armature conductor(s), so that a direct current is generated without the use of commutators.

1896 S. P. Thompson *Dyn.-Electr. Machinery* (ed. 5) 475 (*heading*) Homopolar ('unipolar') dynamos. *Ibid.*, Where the motion is such that the conductor moves continuously past poles of one kind only, the inductive operation is said to be homopolar. **1938** Meares & Neale *Electr. Engin. Pract.* I. iv. 233 A homopolar dynamo is essentially a low voltage, heavy-current machine, but..it is generally more satisfactory to employ a motor generator driven from the ordinary supply mains. **1962** [see *Faraday('s) disc].

3. *Chem.* [ad. G. *homöopolar* (R. Abegg 1906, in *Zeitschr. f. anorg. Chem.* L. 309); cf. *homœopolar* s.v. *homœo-.] Formed by or arising out of the sharing of electrons between neutral atoms, without ionization; covalent.

1922 C. H. Desch *Metallography* (ed. 3) xv. 341 In all probability the new theory of valency based on the distribution of electrons in the molecule, will provide a satisfactory explanation of the formulæ of compounds of electrically similar ('homopolar') elements. **1930** G. Thomson *Atom* 219 In such compounds the two atoms are on an equal footing and the compounds are called homopolar. **1940** Mott & Gurney *Electronic Processes in Ionic Crystals* i. 9 The transition from polar to homopolar binding. **1972** M. H. Battey *Min. for Students* i. 9/1 The covalent (or homopolar) bond is formed when two atoms..share a pair of electrons, one electron being supplied by each atom.

homorganic (hǫmǫɹgæ·nik), *a.* Also occas. homo-organic. [f. Homo- + Organic *a.*] **1.** *Bot.* (In Dict. s.v. Homo-.) Now *rare* or *Obs.*

2. In *Phonetics* (substitute): produced by the same vocal organ or organs; sharing a specific type of articulation.

1864, 1880 [in Dict.]. **1958** Priebsch & Collinson *German Lang.* (ed. 4) II. i. 113 An affricate is a combination of a stop..with its homorganic spirant. **1959** M. Schlauch *Eng. Lang. Mod. Times* 9 A liquid or a nasal plus a homorganic voiced stop. **1960** Z. S. Harris *Struct. Ling.* 532 These occur only after voiced or voiceless consonants homo-organic with their own. **1972** *Archivum Linguisticum* III. 39 A preceding alveolar is lost, and a homorganic nasal intervenes following a bilabial or velar stop.

homosexual (hōᵘmo-, hǫmose·ksiuăl), *a.* and *sb.* [Irreg. f. Homo- + Sexual *a.*] **A.** *adj.* Involving, related to, or characterized by a sexual propensity for one's own sex; of or involving sexual activity with a member of one's own sex, or between individuals of the same sex.

1892 C. G. Chaddock tr. *Krafft-Ebing's Psychopathia Sexualis* III. 255 He had been free from homo-sexual inclinations. *Ibid.* 256 The homo-sexual woman offers the same manifestations, *mutatis mutandis.* **1897** H. Ellis *Stud. Psychol. Sex* I. i. 2 Among animals in a domesticated or confined state it is easy to find evidence of homosexual attraction, due merely to the absence of the other sex. **1914** G. B. Shaw in *New Statesman* 14 Nov. 21/2 The forty tolerated homosexual brothels of Berlin. **1921** E. J. Kempf *Psychopath.* xiii. 645 Her social interests were decidedly homosexual. **1921** *Blackw. Mag.* Jan. 134/2 What the nature of the friendship was we cannot say; it may have been homo-sexual, a love which was common among the later Greeks. **1929** R. Graves *Good-Bye to all That* iii. 40 In English preparatory and public schools romance is necessarily homosexual. **1942** *Jrnl. Compar. Psychol.* XXXIII. 160 Homosexual mountings are characteristic of the male Rhesus monkey, particularly during adolescence. **1948** A. C. Kinsey et al. *Sexual Behav. Hum. Male* xxi. 610 A considerable portion of the population..has at least some homosexual experience between adolescence and old age. **1961** E. Waugh *Unconditional Surrender* II. vii. 173 'I'm sure you aren't a pansy.' 'Pansy?' 'You're not homosexual?' Even this did not disconcert Uncle Peregrine... 'Good gracious, no. Besides the "o" is short. It comes from the Greek not the Latin.' **1963** A. Heron *Towards Quaker View of Sex* iii. 24 Not all effeminate men are homosexual, and few homosexual men can be really described as effeminate. **1965** Rosen & Gregory *Abnormal Psychol.* xviii. 381 Homosexual behavior in human females, also known as lesbianism or sapphism,..is less likely to arouse social censure than male homosexuality. **1973** *Daily Tel.* 5 Apr. 9/8 For the first time at a National Union of Students conference homosexual and lesbian students spoke out openly.

B. *sb.* A person who has a sexual propensity for his or her own sex; *esp.* one whose sexual desires are directed wholly or largely towards people of the same sex.

In non-technical contexts it is often taken to mean a male homosexual, a female one being termed a lesbian.

1912 E. Paul tr. *A. Moll's Sexual Life of Child* v. 127 An adult homosexual who as a child once did some needlework for a joke. **1932** S. Gibbons *Cold Comfort Farm* xi. 153 There were many homosexuals to be seen in Hyde Park. **1954** W. Mayer-Gross et al. *Clin. Psychiatry* iv. 179 Male homosexuals are frequently classified into the active and the passive type; female homosexuals into the masculine and feminine. **1955** *Sci. News Let.* 21 May 334/3 The idea that homosexuals are necessarily bad security risks is debunked in a report..before the American Psychiatric Association. **1969** A. Glyn *Dragon Variation* vii. 198 Uses quite a lot of cologne, doesn't he? Is he a homosexual? **1973** *Daily Tel.* 10 July 6/3 Homosexuals and lesbians make up a sizeable minority of the population.

Hence **homose·xualist**, a homosexual.

1931 A. Eiloart tr. *Heyer's Hypnosis* xiii. 204 It should be considered that a homosexualist thus suggestively transformed for weeks or even months, marries..within that time. **1933** H. Williamson *Gold Falcon* i. 19 His book..proving..that Hamlet should be played as a homosexualist in vain love with Horatio had just been published. **1961** —— *Innocent Moon* v. 100 Even Jesus..according to Nordau..was a homosexualist and should have been locked up. **1971** *Southerly* XXXI. 84 He consigned me to the lavatory-cleaning brigade together with a group of homosexualists.

homosexuality (hōᵘ:mo-, hǫ:moseksiuæ·lĭti). [f. prec. + -ity.] The quality of being homosexual, homosexual character or nature; also, homosexual behaviour or activity.

1892 C. G. Chaddock tr. *Krafft-Ebing's Psychopathia Sexualis* III. 185 (*heading*) Great diminution or complete absence of sexual feeling for the opposite sex, with substitution of sexual feeling and instinct for the same sex. (Homo-sexuality, or contrary sexual instinct.) **1892** J. A. Symonds *Let.* 21 Oct. in P. Grosskurth *J. A. Symonds* (1964) 269 There is an inborn bias toward homosexuality. **1897** H. Ellis *Stud. Psychol. Sex* I. iv. 101 Bourneville believes that 75 per cent. of the inmates of the Parisian venereal hospitals have practised homosexuality. **1948** A. C. Kinsey et al. *Sexual Behav. Hum. Male* xxi. 615 Among many clinicians this work has been taken to mean that the sex hormones control the heterosexuality or homosexuality of an individual's behavior. **1959** B. Wootton *Social Sci. & Social Path.* 15 Many citizens..prefer to look on adult homosexuality as a matter of private taste. **1965** Rosen & Gregory *Abnormal Psychol.* xviii. 381 All persons..have a greater or lesser tendency to homosexuality as well as heterosexuality. **1972** *Daily Tel.* 11 Mar. 3/2 'Gay Cambridge', the Cambridge branch of the Campaign for Homosexual Equality,..claimed that sex education at school did not adequately cover homosexuality.

homose·xually, *adv.* [f. as prec. + -ly[2].] In a homosexual manner; with respect to homosexual activities or homosexuality.

1921 J. C. Flügel *Psycho-anal. Stud. of Family* ii. 16 Homosexually disposed parents would..tend to bring up homosexual children. **1921** E. J. Kempf *Psychopath.* ii. 95 One of our patients..has been fighting strong cravings to become homosexually submissive. **1933** *Jrnl. Mental Sci.* LXXIX. 144 One was homosexually assaulted at the age of 12. **1948** A. C. Kinsey et al. *Sexual Behav. Hum. Male* xxi. 632 A homosexually experienced male could undoubtedly find a larger number of sexual partners among males than a heterosexually experienced male could find among females. **1963** A. Heron *Towards Quaker View of Sex* iii. 25 It is seldom that a public figure feels he can afford to disclose that he is in fact homosexually inclined. **1973** P. Evans *Bodyguard Man* xxiii. 145, I thought you might be homosexually attracted to each other.

homotopic (hǫmotǫ·pik), *a.* [f. Homo- + Gr. τοπικ-ός in respect to place (see Topic *a.* and *sb.*).] **1.** (In Dict. s.v. Homo-.)

2. *Math.* [ad. G. *homotop* (Dehn & Heegaard *Analysis Situs* in *Encykl. d. math. Wiss.* (1907) III. I. I. 165).] Related by a homotopy *to* another complex or path, or the mapping of which it is an image; that is a homotopy.

1918, 1930 [see *homotopy]. **1956** E. M. Patterson *Topology* i. 11 A curve which can be deformed continuously into another is said to be homotopic to it. **1961** Hocking & Young *Topology* iv. 149 We may view homotopic mappings as being members of a one-parameter family of mappings with a continuous parameter. **1967** W. S. Massey *Algebraic Topology* ii. 64 Two continuous maps $\phi_0, \phi_1: X \to Y$ are homotopic if and only if there exists a continuous map $\phi: X \times I \to Y$ such that, for $x \in X$, $\phi(x, 0) = \phi_0(x)$, $\phi(x, 1) = \phi_1(x)$.

Hence **homoto·pically** *adv.*, by a homotopy; as regards homotopy.

1930 S. Lefschetz *Topology* ii. 77 We will say that *A* and *A'* are homotopically deformable, or simply deformable, into one another, over *G*. **1952** F. Bagemihl et al. tr. *Pontryagin's Found. Combinatorial Topology* iii. 83 The mappings ϕ and θ are said to be homotopically inverse to one another. **1968** H. F. Cullen *Introd. Gen. Topology* vii. 368 The usual space E^1 of real numbers is homotopically equivalent to the trivial space consisting of 0 alone, say.

homotopy (hǫ·motǫpi, hǒmǫ·tǒpi). *Math.* [ad. G. *homotopie* (Dehn & Heegaard *Analysis Situs* in *Encykl. d. math. Wiss.* (1907) III. I. I. 164): see prec. and -y[3].] **a.** A mapping that deforms one path continuously into another in such a way that all the intermediate paths lie within the topological space of which the two given paths are subspaces (see quot. 1970). **b.** The property of being homotopic.

1918 O. Veblen *Analysis Situs* (Cambridge Colloq. Lect., Vol. 5, Pt. 2) v. 126 The term *homotopy* will be used to designate a deformation in the general sense..and two generalized complexes..will be said to be homotopic if one can be carried into the other by means of a homotopy. **1930** S. Lefschetz *Topology* ii. 77 The singular translation of *A* into *A'* just described is called a homotopic deformation, or homotopy, or simply deformation. **1951** M. H. A. Newman *Topology of Plane Sets of Points* (ed. 2) vii. 179 It is homotopies..and not identities between paths that are interesting. **1956** E. M. Patterson *Topology* i. 11 Homotopy plays an important part in modern topology. **1961** Hocking & Young *Topology* iv. 151 Theorems about homotopy are but special cases of more general theorems on the extension of mappings. **1970** C. R. F. Maunder *Algebraic Topology* ii. 25 Two continuous maps $f, g: X \to Y$ are homotopic (or '*f* is homotopic to *g*') if there exists a continuous map $F: X \times I \to Y$, such that $F(x, 0) = f(x)$ and $F(x, 1) = g(x)$, for all $x \in X$. The map *F* is said to be a homotopy, and we write $f \simeq g$ for '*f* is homotopic to *g*'.

homotransplant (hǫmotrɑ·ns,plɑnt), *sb. Med.* and *Biol.* [f. Homo- + Transplant *sb.*] **a.** A piece of tissue or an organ taken from one individual and transplanted (or intended for transplantation) to another individual of the same species.

1929 *Ann. Surg.* XC. 934 Eight days after operation, the homotransplant appeared well fixed. **1947** *Jrnl. Bone & Joint Surg.* I. 621 In order that homotransplants of bone may be available for use, methods of storing and preservation must be considered. **1963** R. Y. Calne *Renal Transplantation* iii. 33 Experimental renal homotransplants initially have the appearance and behaviour of autotransplants.

b. A homotransplantation operation.

1955 *Jrnl. Clin. Invest.* XXXIV. 329/1 Dederer..performed a homotransplant of a kidney from one puppy to another of the same litter.

homotransplant (hǫ:motrɑns,plɑ·nt), *v. Med.* and *Biol.* [f. prec. sb.] *trans.* To transplant from one individual to another of the same species.

1953 *Brit. Jrnl. Surg.* XL. 447/1 Attempts at homotransplanting kidneys in humans in France have had no more success than the animal experiments of the past fifty years. **1955** *Proc. R. Soc.* B. CXLIII. 560 When ovary and skin are homotransplanted concomitantly, the ovary may survive while the skin is destroyed. **1961** *New Scientist* 7 Sept. 593/1 The natural history of a kidney homotransplanted between genetically dissimilar people.

So **homotranspla·nted** *ppl. a.*

1939 *Proc. Soc. Exper. Biol. & Med.* XLI. 474 (*heading*) Persistence of medullary tissue in homotransplanted adrenals. **1961** *New Scientist* 7 Sept. 593/1 A homotransplanted kidney starts to function within fifteen minutes of establishing the new circulation of blood.

homotransplantable (hǫ:motrɑns,plɑ·ntăb'l), *a. Med.* and *Biol.* [f. prec. vb., after *transplantable*.] Capable of being successfully transplanted from one individual to another of the same species.

1957 *Jrnl. Nat. Cancer Inst.* XVIII. 529 It has been suggested..that chromosomal imbalance in homotransplantable tumors produces a simplified antigenic composition of the tumor cells. **1962** Feldman & Yaffe in Wolstenholme & Cameron *Transplantation* (Ciba Foundation symposia) 170 Further serial transfers through 3–4 transplant generations established homotransplantable tumour lines.

Hence **ho:motransplantabi·lity**.

1957 *Jrnl. Nat. Cancer Inst.* XVIII. 529 The only nonmalignant tissue reported as showing some degree of homotransplantability is the ovary of the rat. **1967** W. Anderson *Boyd's Path. Surgeon* (ed. 8) ix. 107 In the past it was possible to transplant a tumor only to another member of the same species (homotransplantability).

homotransplantation (hǫ:motrɑns,plɑntēⁱ·ʃən). *Med.* and *Biol.* [f. Homo- + Transplantation.] The operation of transplanting tissue from one individual to another of the same species.

1905 A. Carrel in *Jrnl. Amer. Med. Assoc.* 25 Nov. 1645/2 If the organ is replaced in the same animal from which it was removed, the operation is called an autotransplantation. If it is placed in another animal of the same species it is called a homotransplantation, while if it is placed in an animal of a different species, the operation is called a heterotransplantation. **1915** *Ibid.* 4 Dec. 1966/2 Homotransplantations between related and nonrelated rabbits. **1917** *Ann. Surg.* LXV. 702 Autotransplantation and homotransplantation of segments of veins and arteries are perfectly feasible. **1953** *Brit. Jrnl. Surgery* XL. 447/1 It is in the chronic group of cases that homotransplantation of kidneys would be necessarily first applied. **1971** *Nature* 22 Jan. 270/1 Minced skeletal muscle regenerates with autotransplantation and homotransplantation in the rat, and with heterotransplantation in the mouse.

homotype. Add: **1.** (Later example.)

1939 *Ann. Ent. Soc. Amer.* XXXII. 695 Homotype: 1. That which is constructed on the same plan or type—as metameres of the body. *Not* a nomenclatural term.

2. *Taxonomy.* A specimen identified as a type by someone other than the author of the original description, after comparison with the holotype.

1896 Walsingham & Durrant *Rules for regulating Nomencl.* 13 A specimen named by another than the author, after comparison with the type, is called a Homotype... We have added the term *Homotype* to those proposed by Mr. Oldfield Thomas. **1939** *Ann. Ent. Soc. Amer.* XXXII. 695 Homotype:... 2. Equals Homoeotype. **1967** R. E. Blackwelder *Taxonomy* xxix. 591 Specimens identified as to time or person of identification. Metatypes, homotypes (homoeotypes), and so on.

homotype (hǫ·motəip), *a. Cytol.* [f. Homo- + Gr. τύπος Type *sb.*[1]] = *homœotypic*, *-typical* adjs. (s.v. *homœo-).

1895 [see *heterotype* s.v. *hetero-]. **1907** C. E. Walker *Essent. Cytol.* v. 46 The first division following the meiotic is very commonly given a special name—the 'homotype' division. **1920** L. Doncaster *Introd. Study Cytol.* v. 67 This second division is therefore spoken of as a homotype mitosis.

homotypic, -typical (hǫmoti·pik, -ti·pikăl), *adjs.* [f. Homotyp(e + -ic, -ical.] **1.** (In Dict. s.v. Homotype.)

2. *Cytol.* = *homœotypic*, *-typical* adjs. (s.v. *homœo-).

1904 *Amer. Naturalist* XXXVIII. 741 This mitosis differs from that of the 'typical' mitoses of cells and is called 'homotypic'. **1916** Bainbridge & Menzies *Essent. Physiol.* (ed. 2) xv. 456 In the second division each chromosome splits longitudinally in the ordinary way (homotypical mitosis). **1931** J. E. Frazer *Man. Embryol.* i. 3 In the second maturation division each cell with its reduced chromosome number divides into two cells containing like numbers... Hence this division is an ordinary or 'homotypical' one. **1965** J. Wilkinson tr. *Langeron's Outl. Mycol.* ix. 378 Interkinesis is very short and the two daughter-nuclei very rapidly enter the prophase for the second meiotic (homotypic) division. **1972** *Nature* 21 Apr. 375/1 He did not believe that cancer should be thought of as a disease of cells nor of homotypic collections of cells.

3. *Ecol.* Consisting of individuals of a single species.

1930 W. M. Wheeler in E. V. Cowdry *Human Biol.* 141 Many species often assemble to form aggregations on the same tree or flower, or under the same stone, and these aggregations may be either homotypic, i.e. consisting of members of the same species, or heterotypic, when individuals of more than one species assemble. **1931** W. C. Allee *Animal Aggregations* i. 15 Homotypical associations consist of members of the same species which have arisen either sexually or asexually, which may have remained together because they are the offspring of the same parent, or which may have become accidentally associated together although of different parentage.

homozygote (hǫmozəi·gōᵘt). *Biol.* [f. Homo- + Zygote.] A diploid individual that has identical alleles at one or more genetic loci. Also *attrib.* or as *adj.*, homozygous.

1902 Bateson & Saunders *Rep. Evol. Comm. R. Soc.* I. 126 Similarly, the zygote formed by the union of gametes having similar allelomorphs, may be spoken of as a homozygote. **1902** W. Bateson *Mendel's Princ. Heredity* 23 This *Aa* is the hybrid, or 'mule' form, or as I have elsewhere called it, the heterozygote, as distinguished from *AA* or *aa* the homozygotes. **1903** —— in *Nature* 19 Mar. 463/1 The homozygotes will all have pink eyes. **1909** R. H. Lock *Rec. Progress in Study of Variation* (ed. 2) vii. 185 Let the F_1 plants, arising from the smooth yellow heterozygote grains, be crossed with the wrinkled white parent. **1927** Haldane & Huxley *Animal Biol.* ii. 67 Homozygote parents. **1930** R. A. Fisher *Genet. Theory Nat. Selection* i. 8 The heterozygote when mated to either kind of homozygote would produce both heterozygotes and homozygotes. **1949** Darlington & Mather *Elem. Genetics* xiii. 276 If we breed a homozygote, or a group of like homozygotes, the offspring will be genetically..identical both with their parents and with one another. **1965** T. Dobzhansky *Heredity & Nature of Man* v. 156 Persons who inherit this gene from both parents (homozygotes) die of fatal anemia.

Hence **ho:mozygo·sis**, the fusion of two genetically identical gametes; the state or condition of being homozygous; **ho:mozygo·sity**, the state or condition of being homozygous; the degree or extent to which an individual is homozygous with respect to its complement of genetic loci; **ho:mozygo·tic** *a.*, of or pertaining to homozygosis; homozygous; **ho:mozygo·tically** *adv.*; **homozy·gous** *a.*, having identical alleles at one or more genetic loci.

1902 Bateson & Saunders *Rep. Evol. Comm. R. Soc.* I. 152 The various homo- and hetero-zygous combinations. **1905** W. Bateson in *Rep. Brit. Assoc.* 1904 348 The other possibility is that this phenomenon is due to simultaneous homozygosis of independent allelomorphs. **1916** *Mem. N.Y. Bot. Gard.* VI. 349 Such a condition might also arise from inbreeding in which what is called 'homozygosity' might develop, giving similarity to the hereditary complex. **1925** D. F. Jones *Genetics Plant & Animal Improvement* xv. 503 Only by bringing sexually reproducing organisms to a fairly high degree of homozygosity can there be any possibility of successfully predicting the outcome in particular matings. **1927** *Hereditas* VIII. 77 Both *V. tricolor* and *V. arvensis* are homozygotic in respect of the modifying gene. *Ibid.* 147 One gene.., when homozygotically present, turns the full yellow to a yellowish white. **1942** *Nature* 10 Jan. 54/2 Such self-mating gives immediate homozygosis. **1949** Darlington & Mather *Elem. Genetics* xiii. 279 The homozygotic potential will remain as such so long as cross-breeding is absent or at least restricted to like homozygotes. **1957** C. H. Waddington *Strategy of Genes* ii. 48 Further, gene-fixation and the passage to homozygosity, will be still more delayed if the environment does not remain perfectly uniform. **1969** *Sci. Jrnl.* Dec. 86/2 Various disorders due to homozygosity for rare recessive genes..are much commoner in such inbred populations than in outbred populations. **1970** Ambrose & Easty *Cell Biol.* x. 335 White-eyed females would only be observed if they were homozygous for this recessive gene.

homrai (hōᵘ·mrəi). [Nepal.] A large black-and-white hornbill, *Buceros bicornis*, which is found from India to Sumatra.

1832 B. H. Hodgson in *Jrnl. Asiatic Soc. Bengal* I. 251 The Homrâi reaches its full size in four and [*sic*] five years; it is gregarious and sedate. **1882** D. G. Elliot *Monogr. Bucerotidæ* [s.v. *Dichoceros bicornis*] (*heading*) The Homrai. *Ibid.* 3 The Nepaulese name for this species, 'Homrai', is derived from the notes it is accustomed to utter. **1905** *Pall Mall Gaz.* 24 Apr. 6/3 This curious bird (*Dichoceros bicornis*) is also known as 'Homrai'. **1939** F. N. Chasen *Birds of Malay Peninsula* IV. 90 (*heading*) The Great or Homrai Hornbill.

hon.[1] (ǫn), abbrev. of Honourable, honorable.

1721 in D. Wilkins *Leges Anglo-Saxonicæ* List of Subscribers, The Hon. Henry Booth of the Middle Temple, Esq. **1777** *Jrnls. Continental Congress* (U.S.) X. 10 The following Gentlemen were chosen, viz. Hon. John Hancock, Samuel Adams, [etc.]. **1861** T. Hughes *Tom Brown at Oxf.* I. iii. 39 The fourth man of the breakfast-club, the Hon. Piers St. Cloud, was in his third year. **1873** 'Mark Twain' & Warner *Gilded Age* (1874) xxxiii. 304 Mr. O'Riley, still bearing the legislative 'Hon.' attached to his name. **1914** W. S. Churchill in *World Crisis* (1923) I. viii. 186 Why cannot the right hon. and learned Gentleman (Sir Edward Carson) say boldly, [etc.]. **1922** Joyce *Ulysses* 379 Over against the Rt. Hon. Mr. Justice Fitzgibbon's door. **1925** T. Dreiser *Amer. Trag.* III. xiv. 182 One Hon. Alvin Belknap, of Belknap and Jephson, of this same city. **1938** N. Marsh *Artists in Crime* ii. 13 He's an Hon., you know, and old Lord Pilgrim is doddering to the grave. **1945** N. Mitford *Pursuit of Love* ii. 14, I was a Hon, since my father, like theirs, was a lord. **1962** 'M. Innes' *Connoisseur's Case* v. 54 So old Mrs. Coulson..was an Hon?

hon.[2] (ǫn), abbrev. of Honorary *a.*

1840 *Wiltshire Topographical Soc. Laws & Regulations*, Members of the Wiltshire Topographical Society... Geo. Alexander, Esq. Architect, London, *Hon. Sec.* **1876** *Monthly Packet* Feb. notices p. [5] Your correspondent could obtain further information from the Hon. Sec. **1957** R. W. Zandvoort *Handbk. Eng. Gram.* I. ii. 34 Members wishing to resign..are requested to notify the hon. secretary before January 1st. **1966** J. Betjeman *High & Low* 44 (*title*) The Hon. Sec.

hon.[3] (hʌn), colloq. abbrev. of Honey *sb.* (sense 5 in Dict. and Suppl.).

1906 *Dialect Notes* III. 141 *Hon'*, sweetheart, darling, baby. 'Going to school, hon'?' Very common in addressing children of a tender age. **a 1911** D. G. Phillips *Susan Lenox* (1917) I. ix. 155 'Never mind, hon,' he said... 'My, but you're purty!' **1953** *Fortune* July 158/2 She saw I was reading. 'What you got there, hon?' she asked me. **1954** P. Highsmith *Blunderer* (1956) xxviii. 201 'Let's go, hon.' Bill often proposed leaving before his wife did. **1959** [see *darl]

Honan (hōᵘnā·n). The name of a province of N. China, used to designate: **a.** a variety of silk manufactured there; **b.** ceramics of the Sung dynasty, probably manufactured there.

[1878 J. J. Young *Ceramic Art* II. v. 149 The aubergine, or purple egg-plant violet, was also made under Sung, and is one of the celebrated productions of Kiun, in the province of Ho-nan.] **1923** *Daily Mail* 24 Apr. 1 Fine quality real Silk Honans, beautifully printed. **1923** Hobson & Hetherington *Art Chinese Potter* Pl. XCVIII, Bowl of conical form with rounded sides and pointed base;.. Probably Honan ware. Sung dynasty. **1960** H. Hayward *Antique Coll.* 144/2 *Honan wares*, black- and brown-glazed Chinese porcelain wares of the Sung dynasty, probably made in Honan province. **1968** J. Ironside *Fashion Alphabet* 250 *Honan*. This is a silk pongee cloth from the Honan area of China, which has a particularly uniform colour. **1972** *Times* 30 May 11/2 (Advt.), A large Honan deep bowl.

honcho (hǫ·ntʃo). *slang* (chiefly *U.S.*). Also hancho. [ad. Jap. *han'chō* group leader.] Originally, the leader of a small group or squad; hence, anyone in charge in any situation; the 'boss'. Hence as *v. trans.*, to oversee; to be in charge of.

1947 J. Bertram *Shadow of War* vii. i. 212 But here comes the *hancho*. This boat must be finished to-night. **1955** *Amer. Speech* XXX. 118 *Honcho*. 1. *n.* A man in charge. (This is a Japanese word translated roughly as 'Chief officer', brought back from Japan by fliers stationed there during the occupation and during the Korean fighting...) 2. *v.* To direct a detail or operation. **1964** *Sat. Rev.* (U.S.) 10 Oct. 82/2 Jack Bullock, who *honchoes* the Curaçao casino. **1967** *N.Y. Times* 4 June iv. 1 Mr. Komer expects to be able to name these 45 key provincial *honchos*, and he hopes to place civilians in at least a quarter of the posts. **1972** C. Weston *Poor, poor Ophelia* (1973) xiii. 77 It's out of our territory, but I'll call Pete Springer. He's honcho in that division. **1973** *New Yorker* 30 July 24/1, I was the first employee who was not one of the honchos.

honda (hǫ·ndă). *Western U.S.* Also hondo, -oo, -ou, -u. [Sp. *honda* sling.] The eye at the end of a lasso through which the rope passes to form a loop (see also quot. 1958). Also *fig.*

1887 *Scribner's Mag.* II. 508/2 The common [cowboy] terms are..*heel*, to lariat an animal by the hind leg, *hondou* (derivation unknown, though probably from the Spanish *honda*, the eye of a needle), the slip-knot of the lariat. **1894** *Dialect Notes* I. 324 *Hondoo, hondou*, the slip-knot of a reata. **1895** *Montgomery Ward Catal.* 338/1 Lariat Hondas... Hondas for lariats; firmly pressed rawhide. **1933** *Amer. Speech* Feb. 28/2 *Hondo*, a small loop made of metal or rawhide to prevent the rope from burning or cutting the hands and to hold the loop open. **1958** *Ibid.* Dec. 270 *Hondoo*, the loop plaited in the end of a rope to make a running noose. Also the metal grommet or thimble inserted in that loop. **1964** 'F. O'Rourke' *Mule for Marquesa* xi. 177 He tied the lead lines to a slip rope, dropped the loop over a rock, wedged the honda.

Honduran, Hondurean (hǫndiū°·răn, -riăn), *a.* and *sb.* [f. next: see -an.] **A.** *adj.* Of, pertaining to, or characteristic of Honduras. **B.** *sb.* A native or inhabitant of Honduras or British Honduras. So **Hondura·nean, -a·nian** *adjs.* and *sbs.*

1895 Savage & Gunter *His Cuban Sweetheart* ii. xii. 166 The Honduranean magnates. *Ibid.* iii. xiii. 184 The sunburned neck of the average Honduranean. **1902** R. H. Davis *Capt. Macklin* iii. 105 The Honduranian consul. **1911** D. Folkmar *Dict. Races or Peoples* 77 Honduran (see *Spanish American*). **1918** D. G. Munro *Five Republics Central Amer.* vi. 131 It must not be supposed, however, that the Honduraneans are necessarily inferior, intellectually or physically, to the inhabitants of the other republics. **1926** *Glasgow Herald* 17 June 10 The Honduran steamer Olancho. **1927** *Blackw. Mag.* Aug. 186/1 Living under Honduranian laws..these people, once British, are gradually being forced to give up their birthright. **1934** A. Huxley *Beyond Mexique Bay* 27 They are true-born Panamanians, Nicaraguans, Hondureans, or whatever it may be. **1941** P. E. James *Latin America* xxiii. 691 From it [*sc.* the banana world] Honduras derives financial profit through taxation but it cannot be considered to form a coherent part of the Honduran state. **1950** *Caribbean Q.* II. ii. 3 The poetry and painting of contemporary British Hondurans. **1959** W. S. Stokes *Latin American Politics* ix. 194 Even with such liberal land laws.. which has permitted practically all Hondurans..to own land, Honduras has not been able to achieve democratic government. **1972** *Guardian* 29 Jan. 1/1 Some major move up to the Honduran frontier.

Honduras (hǫndiū°·răs, hǫndiuræ·s). The name of a Central American republic and of the nearby British Honduras (now Belize), used *attrib.* to designate various plants native to the area, as **Honduras bark**, the dried bark of the tree *Picramnia antidesma*, also called *cascara amarga*, and formerly used to treat dysentery, syphilis, and other diseases; **Honduras cedar**, a local species of *Cedrela*, esp. *C. odorata*; **Honduras mahogany**, *Swietenia macrophylla*; **Honduras rosewood**, a species of *Dalbergia*, esp. *D. stevensonii*; **Honduras sarsaparilla**, the dried root of *Smilax ornata*, used in various medical preparations or as a flavouring.

1887 *Colonial & Indian Exhib., Rep. Col. Sect.* 256 'Majoe bitter'..has been lately introduced into medicine in the United States under the name of 'Honduras bark' or 'Casc[ar]a amarga'. **1931** M. Grieve *Mod. Herbal* I. 166 Cascara amarga... Synonyms. Mountain Damson Bark. Simaruba. Honduras Bark. **1875** T. Laslett *Timber & Timber Trees* xxxii. 266 The Cuba, Honduras, and Mexican Cedars are varieties of the *Cedrela odorata*. **1919** W. Winn *Timbers* i. 95 Cedar, *Cedrela odorata*, also known as..Honduras Cedar. **1950** C. W. Bond *Colonial Timbers* 73 Honduras cedar is of considerable value for decoration. **1962** R. J. Streets *Exotic Forest Trees Brit. Commonwealth* 221 'Honduras Cedar', 'West Indian Cedar'. A native of the continent of tropical America and several of the West Indies. **1803** T. Sheraton *Cabinet Dict.* 254 From this province [*sc.* Honduras] is imported the principal kind of mahogany in use amongst cabinet-makers, which generally bears the name of Honduras mahogany. **1851** Chaloner & Fleming *Mahogany Tree* v. 48 Honduras Mahogany is found to work as freely and bend as readily as English Oak. **1902** G. S. Boulger *Wood* ii. 237 Honduras mahogany, reaching 50 ft. at its first branch and 3 ft. in diam., yields logs 25–40 ft. long and 12–24 in. square, or even larger. **1969** T. H. Everett *Living Trees of World* 212/1 The Honduras mahogany is native from southern Mexico and Central America to Peru, Bolivia and Brazil. **1904** H. Stone *Timbers of Commerce* 64 Rosewood. Dalbergia sp.. Alternative Names. Honduras Rosewood. **1956** *Handbk. Hardwoods* (Forest Prod. Res. Lab.) 205 Honduras rosewood grows to a height of 50–100 ft. **1830** G. Spratt *Flora Medica* II. 212 The Honduras Sarsaparilla has a whitish or dirty brown cuticle. **1840** [see Sarsaparilla 1 b]. **1950** R. C. Wren *Potter's Cycl. Bot. Drugs* (ed. 6) 308 Honduras Sarsaparilla is generally imported in long thin bundles with few rootlets attached.

honest, *a.* Add: **1. b.** *to make an honest woman* (*of*): also used colloq., without depreciatory reference, in sense 'to marry'.

1818 Scott *Hrt. Midl.* in *Tales my Landlord* 2nd Ser. IV. ix. 201 Effie was married—made, according to the common phrase, an honest woman. **1927** W. E. Collinson *Contemp. Eng.* 108 The old phrase to make an honest woman (of a girl in trouble) is often used in jest in innocent contexts. **1968** R. Harris *Nice Girl's Story* v. 38 When are you going to be made an honest woman?

3. e. *honest broker*: a sobriquet (tr. G. *ehrlicher makler*) for the German statesman Bismarck; more generally, a representative of a country attempting to mediate diplomatically between two opposing nations, states, etc.; also *transf.*, any such mediator in industrial or other disputes.

[**1878** *Times* 21 Feb. 5/1 To my [*sc.* Bismarck's] mind, it is rather the mediation of an honourable broker who really wishes to carry on business.] **1884** W. Beatty-Kingston tr. *Busch's Our Chancellor* II. ii. 130, I [*sc.* Bismarck] don't picture to myself a peace-mediator playing the part of an arbitrator..; but a more modest one [*sc.* rôle], something like that of an honest broker, who really wants to transact business. **1884** in J. R. Ware *Passing Eng.* (1909) 154/1 *Honest broker*, matrimonial agent. 'Marriages are not all made in heaven; some of them are made by marriage brokers..though the "honest broker" does not seem to find the trade very remunerative.' **1926** Fowler *Mod. Eng. Usage* 548/1 *Honest broker*, Bismarck. **1934** *S.P.E. Tract* xlii. 88 *Honest broker*, coined by Bismarck as *ehrlicher Makler* 1878. **1967** *Guardian* 4 Feb. 9/3 (*heading*) 'Honest broker' role still possible for Britain. *Ibid.* 9/4 An 'honest broker' must have something special to offer, Britain has..immense diplomatic experience and Anglo-Saxon commonsense. **1968** G. Jones *Hist. Vikings* iv. ii. 371 A compromise was reached whereby Edmund should have Wessex and Knut the rest of the country. Among the honest brokers was Eadric Streona. **1970** *Times* 31 Mar. (Australian Suppl.) p. ii/3 On another corner of the Asian board Australia acted the comparatively successful honest broker between the United States and the kingdom of Cambodia representing Cambodia in Saigon and the United States in Phnom Penh. **1973** *Times* 9 Feb. 1/6 As Rhodesia and Zambia were not on speaking terms an intermediary or 'honest broker' had to be found.

f. *Honest Injun*: see *Injun b.

g. *Honest John*: (*a*) *colloq.* an honest man; (*b*) an American type of missile designed to carry a nuclear warhead.

1935 *Amer. Speech* X. 20/1 Any honest citizen; a hard-working fellow...*honest John*. **1954** *Birmingham* (Ala.) *News* 7 Mar. A24/5 The Army is stockpiling the plane-killing Nike units, and semi-guided field artillery rockets designated as the 'Honest John'. **1957** P. Frank *Seven Days to Never* iii. 95 The atomic cannon and Honest John rockets..began to arrive in Europe. **1973** 'H. Howard' *Highway to Murder* v. 57 One thing for sure was that she trusted me to play Honest John. **1973** *Guardian* 7 June 1/5 Britain, West Germany and Belgium are to buy the American Lance battlefield nuclear missile—to replace the obsolescent Honest John.

4. d. *honest-to-God, honest-to-goodness*: genuine(ly, real(ly. *colloq.* (orig. *U.S.*).

1913 J. London *Valley of Moon* I. x, Honest to God, Saxon, he don't like all his horses as much as I like the last hair on the last tail of the scrubbiest of the bunch. **1916** 'B. M. Bower' *Phantom Herd* iii. 45 The real honest-to-goodness-twelve-months-in-the-year West. **1916** Joyce *Portrait of Artist* v. 222 Oh, honest to God, if the crook of it caught him that time he was done for. **1918** B. Hall *Diary* in Hall & Niles *One Man's War* (1929) xxxiv. 352 The only honest-to-God aviator the Americans have ever produced—Raoul Lufbery! **1921** Galsworthy *To Let* iii. v, She was..'honest to God' indifferent to it all. **1924** W. M. Raine *Troubled Waters* xiii. 144 Few will believe it, but it's an honest-to-goodness fact. **1929** W. H. Thomson *That Terrier 'Brick'* xiii. 69 Honest-to-goodness, I didn't know that I was doing anything wrong. **1933** E. Caldwell *God's Little Acre* i. 11 'A real honest-to-God albino?' Shaw asked. 'As real as the day is long.' **1937** [see *copper-bottomed *a.*]. **1937** J. B. Priestley *Two Time Plays* p. xi, It was not until I substituted for him an honest-to-goodness exiled German professor that the play began to look right. **1945** *Tee Emm* (Air Ministry) V. 40 Plain honest-to-God engine failure. **1952** S. Kauffmann *Philanderer* (1953) iii. 50, I don't see a glimmer nowadays. Or if I do, I honest-to-goodness think it's a disappearing glimmer. *Ibid.* v. 80 The fact that we honest to God really dislike each other is thundering out loud now that the entertainment has stopped. **1959** 'J. Welcome' *Lady is Tramp* viii. 125 I'm just pure honest-to-God terrified. **1960** *Farmer & Stockbreeder* 5 Jan. (Suppl.) 4 Like all good meals, there'll be bread and butter with it—not just a substitute for butter, but the real thing, the honest-to-goodness dairy product. **1963** *Times* 4 Mar. 11/6 One recalls a very distinguished industrialist (an individualist himself) who said he preferred 'honest to God engineers and creative designers'. **1963** J. A. T. Robinson (*title*) Honest to God. **1973** J. Thomson *Death Cap* vii. 107 It's an honest-to-God mushroom and..I'd have it for tea to prove it.

5. b. Used to emphasize the truth of a statement. *colloq.* (orig. *U.S.*).

1876 'Mark Twain' *Tom Sawyer* ix. 100 Tell me, Joe,—*honest*, now, old feller—did I do it, Joe? **1901** Merwin & Webster *Calumet 'K'* viii. 160 Max..said to his sister: 'Honest, Hilda, I don't see how he does it.' **1913** A. Bennett *Regent* ii. viii. 236 'But I'm not sarcastic!' he protested. 'Honest?'..'Honest!' he solemnly insisted. **1923** R. D. Paine *Comr. Rolling Ocean* vi. 105 He is not so bad as he sounds, honest, Jud. **1928** F. B. Young *My Brother Jonathan* ii. iv, If it weren't for the life at Prince's I don't think I could stick it..honest! **1972** 'J. & E. Bonett' *No Time to Kill* viii. 103 Of course I don't know a thing, but, honest, I can't see anyone here doing these people in.

honestly, *adv.* Add: **2. b.** Used parenthetically or as an exclamation, either to emphasize the honesty of one's intentions, statements, etc., or as an expression of exasperation.

1898 G. B. Shaw *Mrs. Warren's Profession* iii. 204 Honestly, dear Praddy, do you like seeing them together? **1921** —— *Back to Methuselah* iii. 129 *Mrs Lutestring*. You were kind enough to say that I frighten you. *Burge-Lubin*. Honestly, you do. **1929** E. O'Neill *Dynamo* i. ii. 32 Honestly, I think you've got a nerve to ——. **1957** N. Marsh *Off with his Head* iii. 56 No, honestly, this is just *too* mummerset. **1970** 'D. Shannon' *Unexpected Death* (1971) ii. 24 Angel said, 'Honestly! *More* rain. I'll be going stir-crazy..cooped up in here.' **1973** *Courier & Advertiser* (Dundee) 21 Feb. 13/5 Honestly, some of these drivers should have their heads seen to!

honey, *sb.* (*a.*). Add: **4. b.** A colour resembling that of honey. Also *attrib.* and *Comb.* (see sense 6 c in Dict.).

1814 P. Syme *Werner's Nomencl. Colours* 34 Honey Yellow, is sulphur yellow mixed with chesnut brown. **1838** Honey-yellow [in Dict., sense 6 c]. **1888** C. T. Whitmell *Colour* xii. 183 Yellows may be distinguished as.. honey..gold, [etc.]. **1923** *Daily Mail* 8 May 14 In Reseda,..Champagne, Honey, Copper. **1958** *Times* 16 Aug. 4/7 The Leccesi were fortunate in having close by an inexhaustible supply of honey-gold limestone. **1959** A. K. Lang in H. Q. Masur *Murder Most Foul* (1973) 76 A confection of honey blonde hair. **1973** G. Beare *Snake on Grave* iv. 23 Her skin was deeply tanned, a smooth honey gold.

5. Now common in N. Amer., whence also in Britain and elsewhere.

1859 Mrs. Stowe *Minister's Wooing* xxiii. 215 Come to ole Candace!..Honey, darlin', ye a'n't right—dar's a dreful mistake somewhar'. **1919** C. H. Darling *Jargon Bk.* 17 *Honey*, sweetheart. **1929** M. Connelly *Green Pastures* (1930) i. v. 55 Come on, honey, an' meet de folks. **1939** [see *chin *sb.*[1] 1 d]. **1952** *Manch. Guardian Weekly* 20 Mar. 3 She doesn't have a thing that you haven't got, honey, but she has it over here. **1952** S. Kauffmann *Philanderer* (1953) i. 15 'I think you know I really understand it. But, honey, I ——.' A little pause here. **1961** J. Heller *Catch-22* (1962) xviii. 178 'Be thankful you've got me,' she insisted. 'I am, honey.' **1962** J. Ludwig in R. Weaver *First Five Years* 22 'Honey,' he said to the girl, 'let your *babbe* stay out here with the baby.' **1964** L. Nkosi *Rhythm of Violence* 27 Men are monsters!..Especially black men, honey. **1968** *New Society* 29 Aug. 305/1 'Honey' as an endearment, now rediscovered by southern Englishmen via Hollywood.

b. Anyone or anything good of its kind. *slang* (orig. *U.S.*).

1888 *Missouri Republican* 24 Feb., Dave is a honey. **1933** *Amer. Speech* VIII. iii. 35/1 [Pugilist slang] *Bearcat*, an excellent fighter, a *honey*. **1934** H. N. Rose *Thesaurus of Slang* ii. 3/1 Anything pleasing or attractive: *a honey*. **1935** [see *dilly *sb.*[2]]. **1939** *Evening News* 7 Nov. 4/5 A shot you are pleased with is a 'honey' or a 'peach' or an 'eagle'. **1946** G. Gibson *Enemy Coast Ahead* xiii. 185 On the controls she [*sc.* an aeroplane] was as light as could be. This ship was certainly a honey. **1949** *N.Y. Times* 2 Oct. 1 It is a honey of a taut melodrama. **1958** M. Allingham *Hide my Eyes* xvi. 160 It had been a honey of an evening. **1959** *Vogue* Nov. 119 The Mini-Minor is a honey for parking. **1968** *Globe & Mail* (Toronto) 3 Feb. 38/1 (Advt.), A real honey, automatic power steering, power brakes, radio.

c. (See quot. 1960.) *U.S. slang.*

1934 L. Hellman *Children's Hour* i. 27 *Martha*:.. has she always been like this? *Cardin*: She's always been a honey. Aunt Amelia's spoiling hasn't helped any, either. **1960** Wentworth & Flexner *Dict. Amer. Slang* 265/1 *Honey*, a person who is difficult to please; a difficult problem or task.

7. honey ant: substitute for def.: = *honey-pot 4; add earlier and later examples; **honey-baby** *colloq.* = Honey *sb.* 5; **honey-bucket** *N. Amer. slang*, a container for excrement; **honey-bun, honey-bunch** *colloq.* = Honey *sb.* 5; **honey-buzzard**, a buzzard-like bird, *Pernis apivorus*, native to Europe and Asia, whose chief food is the contents of bees' and wasps' nests; **honey chile** chiefly *Southern U.S.* (esp. *Black*) *colloq.*, = Honey *sb.* 5; **honey-creeper**: substitute for def.: a South American bird of the sub-family Cœrebinæ or a Hawaiian bird of the family Drepanididæ; add earlier and later examples; **honey-flow**, the secretion of honey or nectar by flowers; **honey-gilding**, (*a*) a dull gilding made from gold-leaf and honey, and used to decorate porcelain; (*b*) the process of applying such a solution; **honey-gold**, = **honey-gilding* (*a*); **honey mouse** = **honey possum*; **honey possum**, a small West Australian marsupial, *Tarsipes spenseræ*.

1868 *Amer. Naturalist* II. 382 A species of 'Honey-ant' is also found in Texas. **1874** *Ibid.* VIII. 366 The average weight of a non-producing ant is two milligrammes, that of a full honey-ant two hundred and forty milligrammes, a contrast simply immense. **1882** H. C. McCook (*title*) The honey ants of the Garden of the Gods. **1910** W. M. Wheeler *Ants* xx. 362 Honey ants have been reported from North America, South Africa and Australia. **1923** *Jrnl. Proc. R. Soc. W. Austral.* IX. 47 The geographical distribution of the various honey ants seems to point to drought as one of the most important factors in their development. **1944** *Living off Land* ii. 25 The bushman who finds honey ants can consider himself in champion class. **1948** E. Waugh *Loved One* 134 She was my honey-baby. **1959** *Times* 20 June 7/6 Most people know..what is meant by..'a honey-baby'. **1931** Brophy & Partridge *Songs & Slang 1914–18* 318 *Honey-bucket*, latrine receptacle for excreta. *Canadian.* **1962** F. G. Vallee *Kabloona & Eskimo in Central Keewatin* 48 And where do they get the money? From emptying honey buckets for the Whites and mostly sitting on their butts. **1963** *Observer* 22 Dec. 13/3 The plumbing is definitely un-American. We were warned that the 'honey-buckets' would be a draughty experience. **1969** *Beaver* (Winnipeg, Man.) Summer 6/2 A woman taxi driver tells me most houses have honey-buckets, and galvanized [bath] tubs filled by hand. **1911** *Dialect Notes* III. 544 Honey-bunch, honey-bun. **1913** *Maclean's Mag.* Mar. 58/2 'Come, honey-bun,' she enticed. **1949** *Horizon* XIX. 239 Now, honey-bun, let's talk it over. **1957** M. Summerton *Sunset Hour* iii. 46 'You might have warned me you were coming!' 'No time, honey bun.' **1969** R. Tashkent *Ambiguous Man* iii. 34 I'm sorry, honeybun—sorry. Guess I'm a little upset. **1904** G. H. Lorimer *Old Gorgon Graham* 150 Honey, Honey, Funny Honey-bunch. **1912** *Collier's* 5 Oct. 34/2 Why, can't you see, Honey-bunch, can't you see? **1937** A. Reid in *Famous Plays* (Gollancz) 721 Jenny..control yourself! *Jenny*: Very well, honey bunch! I'll try for the present. **1942** Wodehouse *Money in Bank* (1946) xiv. 122 But where does that get us, honeybunch? **1949** F. Sargeson *I saw in my Dream* 144 Honeybunch..what's it like sleeping all on your lonesome? **1678** J. Ray tr. *Willughby's Ornith.* ii. 72 The Honey-Buzzard. For bigness it equals or exceeds the common Buzzard. **1766** T. Pennant *Brit. Zool.* 67 (*heading*) The Honey Buzzard. **1825** P. J. Selby *Illustr. Brit. Ornith.* I. 23 The Honey Buzzard preys upon moles, mice, and small birds. **1971** *Country Life* 25 Mar. 705/2 Seven species breed regularly within the city limits [*sc.* West Berlin]..honey-buzzard (4–5 pairs). **1926** T. S. Stribling *Teeftallow* viii. 67 Stan up fo yo' baby... Thah you ah, honey chile, lookin' yo' baby in de eye. **1948** Mencken *Amer. Lang.* Suppl. II. vii. 125 Southern speech has suffered cruelly on the stage and in talkies, where kittenish actresses from the domain of General American think that they have imitated it sufficiently when they have thrown in a few *you-alls* and *honey-chiles*. **1957** O. Nash *You can't get there from Here* 96, I early abandoned the hopeless fight against honey-chile and you-all. **1957** Trager & Smith *Outl. Eng. Struct.* 82 There are current in popular literature allusions to 'Brooklynese' or 'honey-chile' accents. **1971** J. Yardley *Kiss a Day* viii. 153 Honey chile may well be talking through the back of her gorgeous neck. **1973** *Listener* 5 July 21/3 They [*sc.* the speakers] [were not] supposed to be rendering Deep South 'honey-chile' drawls. [**1822** J. Latham *Gen. Hist. Birds* IV. 207 Some few, indeed, of the Creepers have the tongue divided at the end and such no doubt are capable of licking honey from flowers.] **1872** Coues *N. Amer. Birds* 91 The Cærebidæ, or honey-creepers of the tropics. **1912** Brabourne & Chubb *Birds S. Amer.* I. 395 Turquoise Honey-Creeper. **1944** G. C. Munro *Birds of Hawaii* 89 (*heading*) Hawaiian Honey-creepers. Drepanididae. Hawaiian Honey-creeper Family. **1970** R. Meyer de Schauensee *Guide Birds S. Amer.* 368 Honeycreepers form a composite family of groups of birds much unlike each other... They share the habit of feeding on the pollen and nectar of flowers, as well as on berries, small seeds and insects. **1893** S. Simmins *Mod. Bee-Farm* (rev. ed.) 255 Frequent rainfalls destroy all chance of a good honey flow. **1894** *Farm, Field & Fireside* 13 July 352/3 There is no reason why a swarm hived towards the end of the honey-flow should not at once have its energies directed to storing surplus. **1955** E. Hillary *High Adventure* i. 14 All through the exciting months of the honey-flow, the dream of a bumper crop would drive us on. **1958** *Irish Bee-keepers' Manual* (Stationery Office, Dublin) xvi. 53 One or more crates of sections may be removed at any time during the honey flow, after the sections in a crate have been completely sealed. **1972** *Country Life* 3 Feb. 253/1 The worker bee at the height of the honey flow, as it is called by beekeepers, works so hard that it wears itself out and dies in a short time. **1958** M. Wykes-Joyce *7000 Yrs. Pott. & Porc.* 271/2 Honey gilding. **1960** R. G. Haggar *Conc. Encycl. Cont. Pott. & Porc.* 207/1 Honey-gilding was a more satisfactory process. **1960** H. Hayward *Antique Coll.* 129/1 Honey gilding was used at the Chelsea and Worcester porcelain factories. **1971** *Country Life* 10 June 1416/3 When heavily laid on in broad masses, honey gilding could be further enriched with light and shade patterns by chasing with a finely pointed agate. **1954** G. Savage *Porcelain* viii. 220 Gilding was often lavish, but of the brassy mercuric variety, whereas, for much of the eighteenth century, honey gold was sparingly and tastefully applied. **1965** Finer & Savage in *Lett. of J. Wedgwood* 8 The porcelain manufacturers used, first honey-gold..and later, mercuric gilding. **1923** *Austral. Zoologist* III. 148 The *Tarsipes* are known throughout the district as 'Honey Mice', which is such an excellent vernacular name, when one considers the habits detailed later on, that I venture to submit it for general use. **1965** Honey-mouse [see **honey possum*]. **1941** E. Troughton *Furred Animals of Australia* 80 The Honey Possum is readily distinguished by having three well-marked dark brown stripes along the back. **1965** *Austral. Encycl.* VII. 234/1 Honey-possum (*Tarsipes spenserae*). Restricted to the south of Western Australia, this species (which is also known as the honey-mouse), feeds on nectar, pollen, and small insects, gathered from blossoms.

b. In names of plants and fruits: **honey agaric** = **honey fungus*; **honey banana**, a West Indian name for a diploid variety of *Musa acuminata*, bearing small, sweet, thin-skinned fruit; also ellipt. *honey*; **honey fungus**, a fungus, *Armillaria mellea*, which causes a root disease in trees and shrubs, indicated by honey-coloured toadstools around affected plants and black threads like boot-laces attached to their roots; **honey-locust** (earlier and later examples); **honey mushroom** *U.S.* = **honey fungus*; **honey-ware** = Badderlocks.

1894 W. Somerville tr. *Hartig's Text-bk. Dis. Trees* I. 207 *Agaricus melleus*. The Honey Agaric. This fungus belongs to the most widely distributed and destructive of parasites. **1909** E. W. Swanton *Fungi* ii. 176 *A[rmillaria] mellea*..'Honey Agaric'. **1945** M. C. Rayner *Trees & Toadstools* ii. 35 The Honey Agaric forms its sporophores only after the tissues in which it grows have been killed. **1938** *Jrnl. Jamaica Agric. Soc.* XLII. 460 The Red banana, Honey, Apple and Fillbasket eventually found their way here. *Ibid.* 464 (*heading*) Honey [banana]. This variety is..also known by the names Sucrier and Lady's Finger... The skin is very thin and ripens to a deep yellow. The flavour of the fruit is very sweet and gives the variety its name. **1959** N. W. Simmonds *Bananas* v. 76 Notes on the principal clones. AA group. (a) 'Sucrier'. Principal synonyms..'Honey'..(West Indies). **1961** [see *fig *sb.*[1] 1 c]. **1895** W. Schlich *Man. Forestry* IV. iii. 382 The honey fungus (*Agaricus melleus*, L.)..causes a well-known disease amongst conifers. **1919** W. E. Hiley *Fungal Dis. Common Larch* viii. 153 (*heading*) The effects of the honey fungus on its hosts. **1962** *Amateur Gardening* 27 Jan. 25/1 The honey fungus, Armillaria mellea, produces thick black threads or rhizomorphs somewhat similar to black leather laces. **1971** *Homes & Gardens* Sept. 97/2 Honey fungus or armillaria.. sometimes attacks and kills the roots of trees and shrubs. **1743** J. F. Gronovius *Flora Virginica* 194 *Gleditsia*,.. Honey-locust. **1759** P. Miller *Gardeners Dict.* (ed. 7) I, s.v. Gleditsia. The first sort [of *Gleditsia*] is very common in most parts of North America, where it is known by the Title of Honey Locust. **1819** A. L. Hillhouse tr. *Michaux's N. Amer. Sylva* II. 137 In different parts of the United States, this species [sc. *Gleditsia triacanthos*] is called indifferently Sweet Locust and Honey Locust. **1838** [see *gleditschia]. **1863** [see *sweet locust* (Sweet *a.* and *adv.* C. 1 b)]. **1869** [see Locust *sb.* 5]. **1968** N. Taylor *Guide to Garden Shrubs & Trees* 135 Honey-Locust (*Gleditsia triacanthos*). A spiny-trunked native tree [*i.e.* native to U.S.A.] with small, numerous leaflets, greenish-yellow flowers, and a twisted, persistent pod. **1938** J. S. Boyce *Forest Pathology* vi. 110 *Armillaria mellea*..known as the honey mushroom, causes this disease. **1849** D. Landsborough *Pop. Hist. Brit. Seaweeds* 111 In Scotland in the Lowlands, it is by some called badder-locks, and hen-ware, which may be a contraction of honey-ware, the name given to it in the Orkney Islands. **1933** J. Gray *Lowrie* 34 He never said onything aboot tangles, hinniwirs, an' dills.

honeycomb, *sb.* Add: **5. b.** *Textiles.* Used *attrib.* of a fabric in which the warp and weft threads form ridges and indentations, producing a cell-like appearance.

1879 T. R. Ashenhurst *Weaving & designing Textile Fabrics* 250 Another cloth which may be mentioned is one known as the honeycomb cloth, which presents to the eye a series of ridges and cavities. **1913** *T. Eaton & Co. Catal.* Fall & Winter 131/3 Full Bleached English Honeycomb Quilts..fringed all round..for single beds. **1921** *Daily Colonist* (Victoria, B.C.) 5 Apr. 18/4 (Advt.), Honeycomb Towels at, each, 15c. **1929** Woodhouse & Brand *Towels & Towelling* ix. 99 The unbroken diamond in the first unit..is filled in with the 8-thread honeycomb weave. **1968** 'A. Gilbert' *Night Encounter* iii. 41 High bed with a honeycomb quilt.

c. A structure consisting of numerous intersecting surfaces designed to reduce turbulence and straighten the air flow in a wind tunnel.

1912 *Sci. Amer.* 14 Sept. 220/1 The usual method of eliminating swirls and irregularities of speed is to pass the air through a sheet metal 'honeycomb' at the front end [of the tunnel]. **1918** Cowley & Levy *Aeronautics* i. 8 At both ends of the channel proper there are placed two metal honeycombs. **1947** A. Pope *Wind-Tunnel Testing* ii. 60 If the contraction ratio of the tunnel is large, and a good honeycomb is installed, the turbulence can be low indeed. **1966** Ower & Pankhurst *Measurem. Air Flow* (ed. 4) vii. 200 The resistance coefficient of a honeycomb or a gauze is conveniently expressed in term of the loss of pressure..suffered by the air in passing through it.

d. A material consisting of a regular network of parallel, open-ended cells formed out of many bent or moulded strips (e.g. of metal or plastic) bonded together; it is usually used faced on both sides with sheeting, forming a *honeycomb sandwich*. Freq. *attrib.*

1937 *Jrnl. Franklin Inst.* CCXXIV. 282 Barkley-Grow Aircraft Corp...has fabricated an all metal honeycomb structure of great strength and lightness. **1946** *Mod. Plastics* Sept. 130/2 Standard thicknesses of honeycomb have been selected. *Ibid.*, Fire resistance may be obtained with stainless steel or laminated asbestos paper skins on asbestos honeycomb core. **1949** *Aircraft Engin.* XXI. 12/1 This flooring material is a honeycomb sandwich of aluminium alloy. **1964** Oleesky & Mohr *Handbk. Reinforced Plastics* ix. 492 Honeycomb sandwich flooring is currently being used in large computer room applications. **1966** *New Scientist* 26 May 523/3 Temperature-resistant honeycomb has been used in heat-sensitive areas of many aircraft.

6. honeycomb coil *Electronics*, an inductance coil in which the turns cross one another obliquely and adjacent ones are separated, giving a criss-cross pattern; **honeycomb moth** (earlier example); **honeycomb radiator**, a radiator for an internal-combustion engine that is pierced by numerous short tubes running from front to back through which the air passes, the ends of which give it a honeycombed appearance; **honeycomb wall**, a (brick) wall containing numerous small openings close together at regular intervals.

1922 *Wireless World* 30 Dec. p. xiv (Advt.), Gimbal type Honeycomb coil. **1959** K. Henney *Radio Engin. Handbk.* (ed. 5) iii. 9 Honeycomb coils were a type of universal winding with relatively few, widely spaced turns per layer giving a typical 'honeycomb' appearance. **1840** J. & M. Loudon tr. *Kollar's Treat. Insects* i. 75 This enemy is the caterpillar of a moth, called the..honeycomb-moth. **1904** A. B. F. Young *Compl. Motorist* iii. 55 The front of the car consists of a water-tank pierced like a honeycomb throughout its whole surface with apertures of equal dimensions; this is known as a 'honeycomb radiator'. **1919** *Jane's All the World's Aircraft* 16b A 'V' type honeycomb radiator is fitted directly behind the airscrew. **1946** A. W. Judge *Mod. Petrol Engines* vii. 261 The honeycomb radiator, which has been so widely used in automobile and aircraft work, consists of a series of thin brass tubes expanded at their ends and joined together at these ends by a soldering process. **1894** J. P. Allen *Pract. Building Constr.* (Index), Honeycomb walls. **1913** G. G. Samson *Every Man his own Builder* iii. 108 Some people build them [*sc.* sleeper walls] as 'honeycomb' walls. **1969** *New Yorker* 5 Apr. 99/1 A honeycomb wall turns into an entrance to whatever place they imagine.

honeycomb, *v.* Add: **4.** *Building.* To build as a honeycomb wall.

1908 C. F. Mitchell *Brickwork & Masonry* (ed. 2) ii. 103 To facilitate the circulation of air beneath the basement floors, these walls are usually honeycombed. **1964** E. C. Adams *Sci. in Building* I. v. 170 This [*sc.* ventilation] can be done..by 'honeycombing' sleeper walls.

honeycombing, *vbl. sb.* Add: **2.** (See quot. 1945); also **honeycombed** *a.*, of, pertaining to, or having such a defect.

1919 S. J. Record *Identification Economic Woods U.S.* (ed. 2) i. 58 When the interior finally dries, the internal strains frequently become so great that large checks open up, producing a honeycombed condition. **1938** H. E. Desch *Timber* xi. 133 When the interior dries below the fibre saturation point it..is restrained from shrinking, and interior checks may result. This condition is known as honeycombing. **1945** D. J. Schwartz et al. *Fund. Shopwork* 458 *Honeycombing*, a form of defective separation in wood which usually occurs in the interior and follows the medullary rays. It is caused by improper seasoning. **1966** A. W. Lewis *Gloss. Woodworking Terms* 14 *Check*, small seasoning cracks or shakes, invisible from the outside of the wood; also called 'honeycombing'.

honey-dew. Add: **3.** (Earlier U.S. example.)

1843 J. Lumsden *Let.* 15 May in *Amer. Memoranda* (1844) 14 The manufacturing of the nigger-head,..pigtail, honey-dew, and other varieties of the stimulating and soothing herb.

4. A colour resembling that of a honeydew melon.

1921 *Daily Colonist* (Victoria, B.C.) 19 Oct. 5/6 (Advt.), Combination dark tones and lighter hues, such as algonquin, honeydew, jade. **1949** *Brit. Colour Council Dict. Colours Int. Decoration* III. 13 *Honeydew*, a colour name.. adopted here as a more attractive name for B.C.C. standard Carrot.

5. honeydew melon, a cultivar of the musk melon, *Cucumis melo*, which has a smooth ivory or pale yellow skin and sweet greenish flesh.

1916 *Country Gentleman* 2 Sept. 1615/1 (*heading*) The honey dew melon. *Ibid.*, A new melon has recently been introduced... It has been christened Honey Dew... The Honey Dew is the result of crossing the Rocky Ford cantaloupe with a South African melon somewhat resembling a Casaba. It was propagated by John E. Gauger, of Colorado. **1923** A. Ward *Encycl. Food* 314 The Honey Dew has a smooth, somewhat warted, creamy-white to greyish skin. **1959** P. Roth *Goodbye, Columbus* (1969) iii. 46 And there were melons—cantaloupes and honey dews. **1962** Whitaker & Davis *Cucurbits* ix. 189 All 'Honey Dew' melons should have an adequate ethylene treatment. *Ibid.* x. 207 The cultivars 'Hales's Best' and 'Honey Dew' were..low in soluble solids.

honey-eater. Delete clause beginning *spec.* and substitute: **b.** An Australasian bird of the family Meliphagidæ.

1822 J. Latham *Gen. Hist. Birds* IV. 208 None of them, although the tongue be cloven into two filaments, are at all fringed at the edges, as is the case with very many of the honey-eaters. **1845** J. Gilbert in J. Gould *Mammals of Australia* I. tab. 5 It [*sc. Tarsipes rostratus*] inserted its long tongue precisely in the way in which the Honey-eaters among birds do theirs into the flower-cups for honey. **1862, 1864–5** [in Dict.]. **1901** A. J. Campbell *Nests & Eggs Austral. Birds* I. 354 The range of the splendid little Black Honeyeater extends across the southern part of Australia. **1936** [see *coach-whip 1 b]. **1964** A. L. Thomson *New Dict. Birds* 375/2 The diet of the majority of honeyeaters consists of a mixture of nectar, pollen, and insects.

ho·ney-fu:ggle, -fu:gle, *v.* *U.S. colloq.* Also -fackle, -fogle. [app. f. Honey *sb.* + Fugle *v.*[1], perh. after dial. *connyfogle* (E.D.D.).]

1. *trans.* **a.** To dupe, deceive, swindle.

1829 *Virginia Literary Museum* I. 458 *Honeyfuggle*, to quiz, to cozen. **1856** *Knickerbocker* XLVIII. 286 They go cavorting out, honey-fuggling their consciences. **1888** *Century Mag.* XXXVI. 81/2 A-tryin' to honey-fugle the varmint to git 'im to come underneath. **1902** W. N. Harben *Abner Daniel* xix, He's been tryin' to honey-fuggle the old man into a trade. **1906** *Dialect Notes* III. 141 He can't honey-fuggle him.

b. To obtain by duplicity or wheedling.

1905 D. G. Phillips *Plum Tree* xxiii, Whatever terms he could honeyfugle out of my conciliation-mad candidate. **1942** Berrey & Van den Bark *Amer. Thes. Slang* § 223/7 Coax; wheedle..honeyfogle.

2. *intr.* To act in an underhand or indirect way, in order to deceive or to obtain by duplicity.

1856 *Congress. Globe* 22 July, App. 965 Pardon me for using the word, but Sharp 'honey-fuggled' around me. **1888** *Missouri Republican* 20 Jan. (Farmer), Noonan's companion objected to this honey-fugling by knocking the demonstrative stranger down. **1906** *Nation* (N.Y.) 22 Feb. 149 'Don't honey-fugle,' he advised the committee, 'but go to the bottom in any way possible.' **1941** *Time* 26 May 16/3 Things which..Baruch had to wangle by what he called 'buttering & honey-fuggling'. **1952** T. Pyles *Words & Ways Amer. Eng.* vi. 129 Even today some of these 'tall' words, words like..*to honeyfogle*..are popularly regarded as picturesque and admirably American.

honey-guide. 1. Substitute for def.: A small tropical bird of the predominantly African family Indicatoridæ, which feeds on insects, honey, and beeswax, a habit which makes some species useful as guides to bees' nests. Add earlier and later examples.

1777 A. Sparrman in *Phil. Trans. R. Soc.* LXVII. 43 The Dutch settlers thereabouts have given this bird the name of *Honig-wyzer*, or Honey-guide, from its quality of discovering wild honey to travellers. **1900** *Ibis* 7th Ser. VI. 691 Having heard the story of the Honey-guide before, I was much interested. **1955** *Bull. U.S. Nat. Mus.* CCVIII. 1 The honey-guides are a small family of picarian birds related to the barbets, the woodpeckers, and the toucans. **1966** E. Palmer *Plains of Camdeboo* xii. 215 It is not only the honey-guides..that use other birds' nests.

honeyish, *a.* Delete † *Obs.* and add later example.

1969 P. Dickinson *Pride of Heroes* 40 There is honeyish Mrs Singleton in her dottily beautiful car.

honeymoon, *sb.* Add: **1. b.** (Later examples.)

1969 *Daily Tel.* 9 Sept. 1/2 MPs fear that the honeymoon between the troops and the civilians might be over. *Ibid.* 13 Sept. 3 It is an open secret that the honeymoon between Xerox and Rank Organisation is over. **1971** *Guardian* 15 Jan. 12/1 The Government has had its honeymoon, free from captious criticism, as Mr Wilson promised.

c. *second honeymoon*, a holiday or trip, resembling a honeymoon, taken by a couple who have been married for some time.

1872 Princess of Wales *Let.* 31 Jan. in G. Battiscombe *Queen Alexandra* (1969) viii. 118 This quiet time we two have spent here together now has been the happiest days of my life... It has been our second honeymoon. **1894** G. du Maurier *Trilby* III. viii. 152 So Taffy and his wife have come for their second honeymoon, their Indian-summer honeymoon, alone. **1910** *Nat. Police Gaz.* 16 July 3/4 They acted as if they were on their second honeymoon. **1967** 'M. Hunter' *Cambridgeshire Disaster* xxi. 137 They began what he called an extended second honeymoon, driving where she liked.

2. *attrib.* (see also in Dict. sense 1) and *Comb.*, as *honeymoon couple, period*; **honeymoon suite** = **bridal suite*.

1881 E. W. Hamilton *Diary* 26 Dec. (1972) I. 205 Lord Lyons says that Gambetta has got very well through his first short (honeymoon) session, and has displayed great vigour and tact as a parliamentary leader. **1904** *Daily Chron.* 31 Mar. 6/6 The young honeymoon couple were heard of at Newport, Mon. **1970** W. J. Burley *To kill a Cat* v. 81 A honeymoon couple stood close, arms round each other. **1953** E. Simon *Past Masters* II. i. 69 The Labour Government's honeymoon period looks like having come to an end. **1970** *Guardian* 9 Dec. 11/1 This Government's honeymoon period has been significantly shorter than previous in-coming Governments'... The package of..increased charges announced by the Chancellor on October 27 was the chief single cause of a rapid deterioration in the Government's popularity. **1956** 'E. McBain' *Cop Hater* (1958) ix. 80 The cafe still served as a sort of no-man's-land between the respectable workaday world..and the sinful..brothels... Jenny's served the same purpose as the shower stall does in a honeymoon suite. **1967** C. O. Skinner *Madame Sarah* viii. 163 Bernhardt enjoyed Chicago. Her quarters at the Palmer House roused in her continual..mirth, for she occupied the 'honeymoon suite'. **1968** B. Norman *Hounds of Sparta* xxvii. 183 'The *honeymoon* suite, for God's sake.'..'I knew nothing about that, believe me.'

honeymooner. (Earlier and later examples.)

1845 W. G. S. Cavendish *Handbk. Chatsworth & Hardwick* 196 Other honeymooners here [*sc.* at Hardwick]. **1968** *Punch* 18 Sept. 398/1 Hawker-Siddeley have dramatised for us the folly of paying a man £15,000 a year to sit around in an airport lounge with a lot of delayed honeymooners.

honey-pot. Add: **2. b.** ? orig. *Austral.* Term applied to the action of jumping into a swimming-pool, etc., with one's hands clasped round one's drawn-up legs.

1941 in Baker *Dict. Austral. Slang.* **1951** J. Frame *Lagoon* 16 She would dive backwards and do a honey-pot into the water.

3. The female pudenda. *slang.*

1719 T. D'Urfey *Pills* III. 342 For when you have possession got, Of Venus Mark, or Hony-pot. **1970** G. Greer *Female Eunuch* 265 If a woman is food, her sex organ is for consumption also, in the form of *honey-pot*.

4. In full *honey-pot ant*. An ant belonging to one of several North American, Australian, or South African genera in which some of the workers become distended with surplus food, which is regurgitated when it is needed by the rest of the colony.

1880 *Jrnl. Linn. Soc. Zool.* XV. 185 Certain individuals in each nest serve as animated honey-pots. **1927** Haldane & Huxley *Animal Biol.* xii. 291 (*caption*) A store-chamber of the honey-pot ant. **1934** A. Russell *Tramp-Royal in Wild Australia* xxxviii. 253 Worker ants and honey-pot ants. *Ibid.*, The honey-pot stores the honey in its abdomen. **1968** P. P. & M. W. Larson *Lives of Social Insects* xxi. 157 Honeypots, or repletes..serve their nest mates simply as receptacles for the colony's extra food supplies. *Ibid.* (*caption*) Honeypot ants represent an adaptation of some ant species to dry and inhospitable environments. **1970** Brown & Taylor in *Insects of Australia* (C.S.I.R.O.) xxxvii. 956/1 Some formicine ants in arid areas..store regurgitated honey-dew and nectar in the enormously distended crops of special large 'honeypot' workers.

5. Something very attractive or tempting, *spec.* an attractive girl or young woman, one who invites or attracts attention. Also *attrib.*

1929 H. A. Vachell *Virgin* iii. 58 What a honeypot she was, whether in or out of breeches and boots. **1945** A. L. Rowse *West-Country Stories* 4 His honey-pot young wife. **1972** *Daily Tel.* 7 Jan. 11/2 This mother, who is a tart, decides to set the girl up as a photographer's model, and a great honey-pot she turns out to be for the local talent who come in droves to photograph her splendid features. **1972** *Guardian* 15 Aug. 16/1 The hordes of eager students that descend on London..present too tempting a honey-pot for the smart operators to stay away from. **1973** *Times* 29 May 14/5 Concentrating facilities at centres of attraction, so-called 'honeypot' areas, and discouraging motorists from heading towards 'quiet' areas.

honeypot: see *hanepoot.

‖**hongi** (hǫ·ŋi), *sb.* *N.Z.* [Maori.] The pressing of noses together as a form of salutation. Hence as *v. intr.*

1843 E. Dieffenbach *Trav. N.Z.* I. i. ii. 61 Leaving him to indulge his natural feeling in hongi..or nose-rubbing. **1862** *Richmond-Atkinson Papers* (1960) I. xiii. 791 A great many of them hongi'd with Parris and seemed very glad to see him. **1882** T. H. Potts *Out in Open* 23 The hongi, a method of salutation by rubbing noses together, we saw frequently performed. **1905** W. B. *Where White Man Treads* 63 Our chief..as is our custom, presented his nose for the 'hongi', which the stranger, not understanding, ignored. **1941** Baker *N.Z. Slang* ii. 21 We have to thank our original inhabitants for specifying, inaccurately as might be expected, the expression *to rub noses* as the equivalent of the Maori ceremony known as *hongi*. (The nose is pressed, not rubbed.) **1949** P. Buck *Coming of Maori* (1950) iii. vi. 418 The visiting party.. pressed noses with the chief mourners. The nose pressing (*hongi*) process sometimes occupied much time. **1952** R. Finlayson *Schooner came to Atia* xvi. 83 The Maori people in New Zealand greet friends with the hongi. **1959** Tindale & Lindsay *Rangatira* xvi. 149 He greeted.. Rona-Nui by gripping both her hands and pressing his nose and forehead against hers in the hongi salutation.

Hong Kong (hǫŋ kǫŋ). [Name of a British crown colony in the South China Sea.] **1.** *Croquet.* (See quot. 1863.)

1863 M. Reid *Croquet* 10 A ball croque'd beyond the boundaries is sent to 'Hong Kong', or 'up the country'. **1897** *Encycl. Sport* I. 253/2 *Hong Kong, Off to* —. The old term for the driving of an opponent's ball from the croquet to the extreme corner of the ground. **1957** V. Nabokov *Pnin* v. 131 Madam Shpolyanski insisted it was perfectly acceptable and said that when she was a child her English governess used to call it a Hong Kong.

b. Colloq. phr. *to go to Hong Kong*, to go away, to go 'to hell'.

1880 R. Jefferies *Hodge & M.* I. ii. 40 The excitement of the day was a pleasurable sensation, and as for his master he might go to Kansas or Hong-Kong. **1921** A. Brazil *Fortunate Term* xii. 160 Opal Earnshaw may go to Hong-Kong if she likes. I don't care about her and her meannesses.

2. Used *attrib.*, *spec.* to designate a strain of the influenza virus discovered in Hong Kong in 1968, and influenza caused by it.

1911 *Encycl. Brit.* XIII. 659/1 The only legal tender is the Mexican dollar, and the British and Hong-Kong dollar, or other silver dollars of equivalent value duly authorized by the governor. **1968** *Weekly Epidemiological Rec.* 16 Aug. 411 (*table*) Influenza... A2/Hong Kong/1/68. *Ibid.* 23 Aug. 421/2 Preliminary tests of paired sera.. have shown very poor results against the new Hong Kong variant. *Ibid.* 30 Aug. 448/2 The Hong Kong strains should be classified as viruses A2. **1968** *Jrnl. Amer. Med. Assoc.* 16 Sept. 23/1 (*heading*) Experts attend Hong Kong flu talk in Atlanta. **1968** *Times* 23 Dec. 6/2 He was not feeling his best and had what scientists felt could be the symptoms of Hongkong influenza. **1970** *New Scientist* 8 Jan. 45/2 It made sad listening for those abed with Hongkong 'flu. **1973** J. Goodfield *Courier to Peking* viii. 101 When she finally got to the Hong Kong and Shanghai Bank, she..withdrew fifty Hong Kong dollars.

‖**honi soit qui mal y pense** (oni swa ki mal i pāṅs). [Fr.] 'Shame on him who thinks evil of it'; a proverb, orig. used as the motto for the Order of the Garter. (See Garter *sb.* 2.)

13.. *Gaw. & Gr. Knt.* (ad fin.), Hony soyt qui mal pence. **1589** Puttenham *Eng. Poesie* ii. 116 King *Edwarde* the thirde, her Maiesties most noble progenitour, first founder of the famous order of the Garter, gaue this posie with it. *Hony soit qui mal y pense*, commonly thus Englished, Ill be to him that thinketh ill. **1716** [see Garter *sb.* 8]. **1776** H. Walpole *Let.* 22 Aug. (1857) VI. 370, I will prevent all clamour, by adopting St. George's motto,—'*Honi soit qui mal y pense*'. **1887** *Athenæum* 5 Nov. 600/1 The maxim 'Honi soit qui mal y pense' is one which needs to be frequently invoked by the friendly narrator and critic of Samoan manners. *a* **1930** D. H. Lawrence in Lawrence & Gelder *Young Lorenzo* (1931) 262 So Tolstoi says that all nude study is bad art—Honi soit qui mal y pense. **1959** *Chambers's Encycl.* III. 510/2 The reason for the adoption of the garter and the motto *Honi soit qui mal y pense* as the emblem and motto of the order is also unknown.

Honiton (hɒ·nitən, hǫ·n-). The name of a town in Devonshire used *attrib.* to designate a type of pillow lace which is made there, consisting of floral sprigs either hand-sewn on to fine net, or joined by bars of other lace-work, as *Honiton guipure, lace, sprig*. Also *absol.* = *Honiton lace*.

1831 M. Edgeworth *Let.* 11 June (1971) 548, I.. made myself very comfortable finishing sewing the Honiton edge on my frill. **1840** *Observer* 16 Feb. 8/2 Her Majesty the Queen wore on her head a wreath of orange blossoms and a veil of Honiton lace. **1851** *Illustr. Catal. Gt. Exhib.* iii. iii. 559/1 Honiton guipure mantle. Bridal scarf and rich flounce in Honiton lace. *Ibid.* 560/1 Specimen of lace net..for the application of Brussels and Honiton sprigs. *Ibid.*, Tamboured lace scarf, imitation of Honiton, manufactured in London. **1858** Simmonds *Dict. Trade*, *Honiton lace*, a pillow, or cushion, lace made in Devonshire, remarkable for the beauty of its figures and sprigs, which are sewed on to net by the needle. **1865** F. B. Palliser *Hist. Lace* xxxii. 382 Queen Adelaide.. gave the order for a dress to be made of Honiton sprigs. **1882** *Encycl. Brit.* XIV. 188/2 Honiton pillow lace resembles Brussels lace. **1895** *Army & Navy Co-op. Soc. Price List* 15 Sept. 1126 Real Lace Handkerchiefs. Honiton. **1963** *Times* 8 Mar. 14/7 Her..crinolined profusion of Honiton lace and orange blossom under a train of silver moiré.

honk, *sb.* Add: **b.** The harsh sound of a motor-horn. Also *v. intr.*, to emit such a sound (said of the horn, the motor vehicle, or the driver); also *transf.*; *trans.*, to utter with such a sound; to cause to make the sound 'honk'; to remove or drive away by the hooting of motor vehicles. Hence **ho·nking** *vbl. sb.* and *ppl. a.* orig. *U.S.*

1895 F. Remington *Pony Tracks* 256 The irrepressible Dan begins to 'honk' on his horn. **1906** 'O. Henry' *Four Million* 51 She would honk loudly the word 'Clara'. **1910** —— *Strictly Business* (1917) v. 57 The honk of the returned motor car at the door. **1911** R. W. Chambers *Common Law* x. 312 Where now the lonely taxi honks. **1911** H. S. Harrison *Queed* i. 6 Now and then a chauffeur honked by. **1914** R. & E. Shackleton *Four on Tour* 83 The [motor] horn was honked suddenly. **1915** *Literary Digest* 4 Sept. 467/1 Nearly a dozen autos may be always seen 'honking' their way through Nome's busy thoroughfares. **1924** *Glasgow Herald* 18 Aug. 8 The thundering and purring and swishing and honk-ing of the road traffic. **1926** [see *beat *v.*[1] 3 c]. **1927** *Observer* 28 Aug. 10 In the shadiest lanes we were honked and hooted out of the way. *Ibid.* 18 Sept. 8/3 The leisured stroller..is being 'honked' off the highway. **1928** *Ibid.* 22 Jan. 10/7 'Sunstar'..is feeling fit again and proposes to honk off to Doncaster to-night. **1929** *Times* 2 Jan. 15/5 The car had been honking underneath my office window for some time. **1931** G.

Atherton *Sophisticates* iii. v. 295 The narrow thoroughfare was crowded with honking cars and taxis. **1955** *Times* 11 May 4/1 Much honking of horns from the cars round the ground. **1958** [see *bleep *sb.*]. **1969** *New Yorker* 29 Mar. 27/1 There was little honking; and our driver..was relaxed and cheerful. **1973** C. Egleton *Seven Days to Killing* ix. 99 Dodging through the honking traffic.

honker. Add: **b.** The horn of a motor vehicle. **c.** One who 'honks' (in various senses). **d.** *slang.* A nose.

1891 *Outing* (U.S.) Oct. 43/1 Though a fair 'honker', I cannot successfully imitate the constantly varying note of the snow goose. **1928** *Funk's Stand. Dict.*, *Honker*, the warning horn of a motor-car. **1948** R. Park *Harp in South* xxi. 267 It's yer own fault for having such a God-forgotten honker [*sc.* a large nose]. **1965** *New Statesman* 6 Aug. 183/3 The honker and the light-flasher drew up alongside. **1972** *Jazz & Blues* Sept. 12/1 Others in the R & B field..are just dismissed as 'honkers'.

honkers (hǫ·ŋkəɹz), *a.* *slang.* [Etym. unknown.] Drunk.

1957 R. Longrigg *Switchboard* v. 191 He stumbled out on to the lawn... 'Honkers.' 'Poor man. Pushed.' **1958** *News Chron.* 22 May 4/5 You drink wallop, sludge, or plasma (it's all ale or beer). If you get drunk you are honkers, plastered,..blotto. Or just plain paralytic. **1970** C. Wood *Terrible Hard* xii. 167 Roll on Wednesday week and we'll all get honkers on champers.

honk-honk, reduplication of *honk *sb.* b.

1908 H. G. Wells *War in Air* ii. 43 A curious, amusing, wheezing sound had got into his 'honk, honk'. *Ibid.* 58 Honk-honking and emitting weird cries. **1909** S. Ford (*title*) Honk, honk!! Shorty McCabe at the wheel. **1915** T. Burke *Nights in Town* 210 The honk-honk of motors. **1917** *Ideas* 23 Mar. 23/2 With a wild 'honk-honk!'..the motor rounded the corner. **1935** Wodehouse *Blandings Castle* ii. 44 'Good-bye.' 'Honk-honk!' said Freddie moodily.

honky (hǫ·ŋki, hɒ·ŋki). *U.S. Black slang.* Also **honkey, honkie.** [Etym. unknown; perh. a var. of *hunky* (see *hunk *sb.*[3]).] A white man; white men collectively. Also *attrib.* or as *adj.*

Disparaging in all applications.

[**1946** Mezzrow & Wolfe *Really the Blues* xii. 216 First Cat: Hey there Poppa Mezz, is you anywhere? Me: Man I'm down with it, stickin' like a honky. *Ibid.* 374/2 *Honky*, factory hand.] **1967** *Newsweek* 24 Apr. 16/1 'Go for the honkies'... The chemistry in tranquil Nashville, Tenn., spelled riot,.. Stokely Carmichael..exhorted: 'You have to go for the honkies..who are keeping you in the ghettos... Victims should never, ever apologize for their use of violence.' **1967** *Guardian* 29 Aug. 7/5 'Honky' (the white man) was using weapons like the Vietnam war. **1969** 'J. Morris' *Fever Grass* xxv. 248, I killed for the same thing you want. You're not like the rest of these honkeys. **1970** *Peace News* 17 Apr. 8/4 It is the pacifist who must declare, out of his own values, that there are neither Panthers nor Pigs, neither Niggers nor Honkies but only human beings. **1971** *Black Scholar* 35/1 You screamed on me 'bout that honky gunsel upstairs. **1971** *Black World* Mar. 69/2 Blacks should 'beware of honkies bearing gifts'. **1971** *Guardian* 1 May 9/1 Many blacks..came to see it [*sc.* the African-American Institute] as a 'honky' (white) conservative force. **1971** *It* 9–23 Sept. 8/3 The phoniness of news, TV, rich honky pop stars, etc. **1971** B. Malamud *Tenants* 46 Mary forcefully shoved him away. 'Split, honky, you smell.' *Ibid.* 133 'Now you leave this honky to me,' said Bill. 'He is my guest.'

honky-tonk (hǫ·ŋkitǫŋk). *colloq.* (orig. *U.S.*). Also **honkatonk, honkey-tonk.** [Etym. unknown.] **1.** A tawdry drinking-saloon, dance-hall, or gambling-house; a cheap night-club. Also in somewhat extended uses, and *attrib.* or as *adj.*

1894 *Daily Ardmoreite* (Ardmore, Okla.) 24 Feb. 1/4 The honk-a-tonk last night was well attended by ball-heads, bachelors and leading citizens. **1924** *Étude* Sept. 595/3 These dance resorts were known as 'Honky-Tonks'—a name, which in itself suggests some of the rhythms of Jazz. **1927** C. Sandburg *Songbag* 232 It was moaned by resonant moaners in honky tonks of the southwest. **1928** M. C. Sharpe *Chicago May* 287 *Honky-tonk*, gaudy saloon with back-room hangout. **1930** C. E. Mulford *Deputy Sheriff* xiii. 168 'This place ain't no damn' honkatonk, stranger,' reproved the bar-tender... 'Folks get throwed outa here sometimes.' **1935** A. J. Pollock *Underworld Speaks* 57/1 *Honkey tonk*, an underworld dance hall in which female entertainers are employed. **1936** *Delineator* Nov. 48/2 The inner room of a honky-tonk on a back street..New Orleans. **1940** W. Faulkner *Hamlet* iv. ii. 387 Its master whose anonymous dust lay with that of his blood and of the progenitors of saxophone players in Harlem honky-tonks. **1945** J. Steinbeck *Cannery Row* 1 Honky-tonks, restaurants and whore-houses. **1950** A. Lomax *Mr. Jelly Roll* 54 These honkey-tonks ran wide open twenty-four hours a day... Their attendance was some of the lowest caliber women in the world and their intake was the revenue from the little, pitiful gambling games they operated. **1955** A. Ross *Australia 55* 108 The town itself, a little honky-tonk in character, boasts many saloons. **1957** G. Lascelles in S. Traill *Concerning Jazz* 77 Others of possibly less talent were doing stalwart work as accompanists to the blues singers in the honky-tonks of New Orleans and St. Louis. **1962** *Daily Tel.* 31 May 19/4 A Parliamentary Bill would have to be promoted if the Norfolk Broads were to be saved from further 'honky-tonk development of the very worst type'. **1969** I. & P. Opie *Children's Games* 15 It is not only Battersea Park (the enchanted garden of our childhood) that has been turned into a honky-tonk.

2. Rag-time music or jazz of a type played in honky-tonks, esp. on the piano. Freq. *attrib.*, passing into *adj.*, as **honky-tonk piano**, an out-of-tune or tinny-sounding piano. Cf. *barrel-house 2.

1933 *Fortune* Aug. 90/2 Sometimes they spent weeks in preparation for a single recording date, yet they never sacrificed the informal, honky-tonk spirit. **1936** *Swing Music* Autumn 62/2 Superficially, 'Honky Tonk' is the musical interpretation of a train journey; fundamentally it is a twelve-bar blues. **1942** Berrey & Van den Bark *Amer. Thes. Slang* § 579/4 *Honkytonk*, primal 'swing' of the style played in the bordels of New Orleans, Memphis and St. Louis, in which a free rein is given to improvising. **1946** R. Blesh *Shining Trumpets* (1949) ix. 202 Among them were masters of the blues and barrel-house piano (or, as Morton calls it 'Honky Tonk music'). **1953** *Observer* 27 Dec., The barrelhouse piano, also known as the..honkytonk piano. **1964** *Amer. Folk Music Occasional* i. 45 They didn't play for no white folks, because the white folks didn't want that kind of music, they called it honky-tonk. **1972** *Drive* Spring 78/2 Happy, beery men thumping honky-tonk pianos.

‖**honnête homme** (onętom). [Fr.] An honest, decent man; a gentleman.

1666 W. Temple *Let.* 1 Apr. in *Wks.* (1720) II. 18 Confessor is honneste Homme. **1709** Swift *Let.* 13 Nov. in *Lett. Lit. Men* (Camden, 1843) 341, I ever thought it a mighty oversight in Courts to let the *honnete homme*, the *homme d'esprit*, and *homme de bien*, gain ground among them. **1923** Ogden & Richards *Meaning of Meaning* i. 21 The *honnête homme* may be unprepared for the lengths to which verbal ingenuity can be carried. **1931** T. S. Eliot *Charles Whibley* 10 Whibley had this discretion, that of the *honnête homme* as critic, to select subjects suited to his own temperament. **1937** A. Huxley *Ends & Means* i. 2 The *honnête homme* makes his appearance as the ideal of seventeenth-century gentlemen. **1960** J. Bayley *Characters of Love* iii. 131 Iago..is in many ways a terrible parody of the Augustan *honnête homme*. **1970** *Times* 12 May 11/1 His opening speech on tobacco, far from being a declaration on behalf of the *honnête homme*, is a means of forcing his will.

honorand (ǫ·nəræ̆nd). [ad. L. *honōrand-us*, gerundive of *honōrāre* Honour *v.*]. Someone to be honoured, spec. with an academic honorary degree.

1950 A. P. Herbert *Independent Member* lxxxiii. 478 The Honorands, the eight or nine distinguished persons who are to be voted Honorary Doctors. **1961** M. Beadle *These Ruins are Inhabited* (1963) xii. 170 The American honorand at *this* June's Encaenia. **1966** *New Statesman* 10 June 849/1 A kind of *festschrift* without an honorand. **1971** *Oxford Times* 25 June 30/1 There was brilliant sunshine for the Encaenia... For the first time for eight years there was no woman in the procession of honorands.

honorial (ǫnō°·riăl), *a.* [f. Honour, honor *sb.* + -ial.] Of or relating to titles of honour, or persons of title or rank; of or pertaining to an honour (sense 7).

1810 Evans & Britton in J. Britton *Beauties Eng. Wales* XI. Norfolk 53 (*heading*) Civil, political, and honorial history. **1828** T. Allen *Hist. County York* I. ii. vi. 327 (*heading*) Ecclesiastical and civil government, honorial history. **1943** F. M. Stenton *Anglo-Saxon Eng.* xvii. 628 On every great honour..there appear tenants holding considerable estates... The honorial barons, as tenants of this class may conveniently be called, have received less attention..than is their due. *Ibid.*, In the honorial court..the lord's baronial tenants played a leading part. **1962** H. R. Loyn *Anglo-Saxon Eng.* viii. 329 Honorial courts dealt primarily with military feudal questions.

‖**honoris causa** (ǫnō°·ris kǭ·zẽ, kau·zā). [L.: lit., for the sake of honour.] In order to honour or out of respect for a person mentioned; now used chiefly as a description of such university degrees as are conferred upon persons in recognition of certain distinctions or achievements without the customary academic examination or thesis.

1611 Coryat *Crudities* 240, I wil once more speake of our most worthy Ambassador Sr Henry Wotton, *honoris causâ*. **1626–7** in T. Birch *Crt. & Times Chas. I* (1848) I. 193 His colleagues shall be the Earl of Salisbury, *honoris causâ*, and Sir Richard Western. **1882** *Standard* 19 Dec. 2 (Stanford), Receiving the degree of D.D., *honoris causa*, from the late Dr. Sumner, Archbishop of Canterbury in 1857. **1935** *Discovery* Nov. 325/2 Uvarov was honoured by the University of Madrid with the degree of Doctor *honoris causa*. **1955** *Times* 26 May 12/5 He was instructor to the Oxford University Air Squadron and was given the degree of M.A. *honoris causa* at the end of his term of duty. **1963** [see *academician 1].

honour, *sb.* Add: **5. e.** Now, in many universities, a course of study or a series of examinations in a subject or group of subjects of a higher or more specialized character than is required for a pass or ordinary degree. (Cf. *honours degree, school* in *10.)

8. a. Also at Bridge (see quots. 1909 and 1936). Phr. *honours are even*: often used fig. to denote equality in a contest (real or imaginary).

1886 *Biritch, or Russian Whist* 4 There are five honours, viz.:—Ace, King, Queen, Knave and Ten, if trumps are declared. **1909** W. Dalton *'Saturday' Bridge* 5 Honours consist of ace, king, queen, knave, and ten of the trump suit. When there are no trumps they consist of the four aces. **1920** B. Cronin *Timber Wolves* iii. 62, 'I don't know your name?'..'But then I don't know yours, do I? That makes the honours even, don't you think?' **1927** W. E. Collinson *Contemp. Eng.* 31 The card-expressions now most prevalent in a figurative application are drawn in the main from bridge, e.g. to call one's hand (or one's bluff), to finesse, Honours are even, After you, partner, etc. **1936** E. Culbertson *Contract Bridge Complete* xxxviii. 421 An honour-card is technically a ten or higher card. **1939** N. de V. Hart *Bridge Players' Bedside Bk.* 22, I got home by playing for split honours. **1947** S. Harris *Fund. Princ. Contract Bridge* ii. i. 43 South should remember to lead a top honour from the hand which contains double honours in sequence. **1967** P. Anderton *Play Bridge* vii. 49 You drop 100 points on going down but get 100 for honours.

10. honour(s) board, a board at a school or college on which are inscribed the names of members who have gained honours; **honours degree**, an academic degree with honours or obtained in an honour school; similarly *honours examination*; **honour(s) list**, (*a*) a class list of candidates who have been successful in an examination for honours; (*b*) a list of honours conferred by the Sovereign, as at the New Year; also *fig.*; **honour-man** (earlier example); **honour-roll**, a roll of honour; **honour(s) school**, a course of study designed for those who aim at an honours (as distinguished from a pass) degree (at Oxford University, 'qui honorem ambiunt'); similarly *honour moderations*, colloq. *honour mods.*; **honour system**, a system in which examinations are completed, services paid for, etc., without, or with only a minimum of, supervision; **honour-trick** *Bridge*, a combination of cards which may be expected to win a trick.

1931 E. Waugh *Remote People* 211 They have honour boards, on which the name of one boy is inscribed every year. **1960** C. Day Lewis *Buried Day* vi. 129 My name was never inscribed on the University Honours Board in the Big School. **1851** *Oxf. Univ. Cal.* 127 A reference will be made in the lists of Honours and ordinary Degrees, to the distinctions awarded by the Moderators. **1904** *Daily Chron.* 31 Oct. 4/4, I cannot deal here with the wholly false and mischievous qualification of an 'Honours Degree' that the rich women are answerable for. **1969** *Oxf. Univ. Handbk.* 159 About one third of the students at Oxford are working for honours degrees in mathematics or natural science. [**1852** *Oxf. Univ. Commission Rep.* iii. 62 With regard to the Examination for Honours, the course of classical reading seems to have become more and more limited.] **1885** *Oxf. Univ. Cal.* p. iii, Honours Examinations for Women. **1927** C. E. Mallet *Hist. Univ. Oxf.* III. xxiii. 168 The Honours Examination at Oxford was established. *Ibid.* 170 In 1830, a Fourth Class in Honours was provided, and the Honours examination was separated from the examination for the ordinary Pass degree. **1972** *Univ. Oxf. Exam. Decrees* 33 No person whose name shall be placed in any Class List issued by the Moderators shall be admitted again as a candidate in the same Honours Examination. **1849** Thackeray *Pendennis* I. xix. 179 A man may be famous in the Honours-lists and entirely unknown to the undergraduates. **1862** *Oxf. Univ. Cal.* 273 Honours Lists issued by Moderators. **1910** H. G. Wells *New Machiavelli* (1911) iii. i. 304 The New Year and Birthday honours lists are always very sagely and exhaustively considered. **1923** J. M. Murry *Pencillings* 272 It will dawn upon me when I read my own name in capitals at the top of the Honours List, created Baron for my services to literature. **1929** *Melody Maker* Dec. 1169/1 This disc goes right into the honours list. **1964** *Granta* 2 Nov., He hoped as it was Coronation year to do rather better than usual in the New Year's honours list. **1839** J. Romilly *Diary* 12 Jan. (1967) 161 This year we have allowed our Lecture rooms to the Honor-men. **1911** *Rep. Brit. Assoc.* 219 It has been suggested that if students are not encouraged to come to the university younger the better men should be allowed to enter for Honour Mods. after six months. **1913** Honour mods. [see *Divvers]. **1922** *Rep. R. Comm. Oxf. & Camb.* 29 The marked vitality of Honour Classical Moderations bears witness to the interest still awakened by the classical languages and literatures. **1965** W. R. Ward *Victorian Oxford* x. 213 Scholarship was now examined at Honour Moderations and the final school was left unencumbered for the examination of history and philosophy. **1970** *Oxf. Univ. Cal.* 195 The prizes are awarded each year by the Moderators for Honour Moderations in English Language and Literature. **1909** *Daily Chron.* 7 Apr. 3/3 Such is the honour-roll of these sturdy spirits. **1949** Honor roll [see *bulletin 4]. **1902** *Encycl. Brit.* XXXIII. 603/2 At Oxford there are now the following 'Final Honour Schools'. **1923** *Granta* 2 Mar., At Oxford the Honours School of English claims that its 'first' is harder to win than a 'first' in any other School there. **1965** W. R. Ward *Victorian Oxford* x. 220 Rawlinson alleged that the standard in the non-classical honour schools would be less than the proposed mixed degree. **1970** *Oxf. Univ. Cal.* 194 Candidates must be members of the University reading for a Final Honour School. **1904** *Pittsburgh Gaz.* 3 Dec. 4 The most successful plan of combating the tendency of college students to cheat in examinations has been some form of an 'honor system' by which the pupil is implicitly trusted and his statement accepted that he used no dishonest aids. **1934** H. G. Wells *Exper. Autobiogr.* I. vi. 323 He had in operation an honour system of discipline that was far in advance of the times. **1936** *S.P.E. Tract* xlv. 181 *Honor system* denotes the practice, adopted at certain institutions, of conducting written examinations without supervision, the candidates being put upon their honour to use no illegitimate aids. **1958** *Sunday Times* 16 Mar. 22/5 Self-service stores must run on the honour system. **1966** *Listener* 24 Feb. 266/2 On the buses and trams an 'honour' system: you throw your money into a coin box.. and tear off your own ticket. **1969** 'E. Lathen' *Come to Dust* xiv. 140 Two young women had been discovered.. in circumstances all too clearly proscribed by the parietal rules and Brunswick's honor system. **1931** E. Culbertson *Contract Bridge at Glance* 6 A defensive honour-trick is a card, or combination of cards, which may be expected to win even if the opponents play the hand at a trump. **1959** Reese & Dormer *Bridge Player's Dict.* 116 The Culbertson system, which first popularized the honor-trick method of valuation, uses the following table for counting honor-tricks.

honourable, *a.* Add: **2. b.** (The only Lord Mayors and Provosts in the United Kingdom who are entitled to be styled 'Right Honourable' are the Lord Mayors of London, York, and Belfast, and the Lord Provosts of Edinburgh and Glasgow.)

3. *honourable mention*: see Mention *sb.* 2 e.

1866 *Lond. Gaz.* 26 June 3646/1 Grand prizes and money awards of the total value of 250,000 francs (£10,000), 100 Gold Medals,..5000 Honourable Mentions. **1869** *Bradshaw's Railway Manual* XXI. App. 99 Honourable mention.—International Exhibition, 1862.

Honved (hǫ·nvēid). [Hungarian, = *hon* home + *véd* defence.] The name given to the Hungarian second-line formation during the revolutionary war of 1848–9; later also used of the militia reserve. Also, a member of either force.

1854 E. O. S. *Hungary & its Revolutions* 429 The new army,..namely, the Honveds, National Guards, and Volunteers. **1875** *Encycl. Brit.* II. 604/2 The Hungarian militia, or 'honveds', as they are called. **1913** *Times* 7 May 7/3 It [*sc.* an ordnance factory] will supply the needs of the Honved troops and of the Hungarian quota of the joint army and navy. **1931** F. Eckhart *Short Hist. Hungarian People* vii. 192 At his [*sc.* Kossuth's] impassioned speeches, the *Honvéds* (Defenders of the Fatherland) assembled..round the standards of the King of Hungary. **1956** C. A. Macartney *October Fifteenth* I. i. 15 The Magyars did not mind taking commissions in the Honvéd, which, however, was a body not much more serious than our own militia of the day. **1959** D. Sinor *Hist. Hungary* xxix. 267 Some 152,000 Hungarian *honvéd* (as the soldiers of this war were called)..had to face the 1,200 guns of an enemy some 370,000 strong. **1962** C. A. Macartney *Hungary* vii. 163 The rank and file of the *Honvéd* were as a rule conscripted into the Imperial forces and sent to foreign stations.

honyock, honyocker, varr. *Hunyak.

hooch (hūtʃ). *colloq.* (orig. and chiefly *N. Amer.*). Also **hootch.** [Abbrev. of *Hoochinoo.] **a.** = *Hoochinoo 2. **b.** In more general use, alcoholic liquor, spirits, esp. of low quality or illegal provenance.

1897 M. H. E. Hayne *Pioneers of Klondyke* 91 The manufacture of 'hooch', which is undertaken by the saloon-keepers themselves, is weirdly horrible. **1903** *N.Y. Even. Post* 25 Sept. 3 In this bottle was some of the native spirits called 'hooch', distilled from sugar and graham flour. **1904** E. Robins *Magnetic North* ix. 161 Apart from the question of drinking raised again by the 'hootch',..they were ready to eat the more. **1907** R. W. Service *Songs of Sourdough* (1908) 18 A broken wreck with a craze for 'hooch', and never a cent to my name. **1922** G. C. F. Pringle *Tillicums of Trail* 41 I've cut out the hootch. It was getting me at White Horse. **1923** C. J. Dutton *Shadow on Glass* 200, I thought of our modern 'hooch ships' that were doing the same thing. **1927** *Punch* 20 Apr. 428/3 He knows what the poor want in the great black city of Chicago. They want cash. He knows what the thirsty want. They want hooch. **1927** *Amer. Speech* III. 167/2 *Hootch*, bad whiskey. **1953** M. Scott *Breakfast at Six* xii. 99 He's sure to have lots of hootch outside, and he'll get more expansive as the evening goes on. **1960** *Times* 5 Feb. 9/5 He apparently received consignments of empty whisky bottles from Madrid, which were filled with the 'hooch'. **1964** *Globe & Mail* (Toronto) 15 Dec. 31/8 The least the authorities could do when they make a raid would be to pack the confiscated hootch and ship to some underprivileged country. **1969** *New Yorker* 5 Apr. 97/1 The people of the city were prepared to swallow any old hootch under the rule of some wild thirst. **1970** *Times* 15 Oct. 30/3 The distilled spirits industry..wages an expensive propaganda campaign against..hooch.

hooch, hootch: see *Hoochie.

hoochie (hū·tʃi). *Mil. slang.* Also **hooch, hoochy, hootch.** [? ad. Jap. *uchi* dwelling.] A shelter or dwelling (esp. one that is insubstantial or temporary).

1952 *San Francisco Examiner* 26 Oct. 18/4 The 'hoochie' is a GI term for a bunker or a prepared defensive position. **1954** *Britannica Bk. of Year* 637/2 In its final stages, the war in Korea yielded a number of new terms, among them the British soldier's name for a dugout—a Hoochie. **1960** *Amer. Speech* XXXV. 264 Cinderella-san lived in hootchie with sisters. **1960** (*citing an Army weekly newspaper in Korea*) in *Sat. Rev.* (1968) 26 Oct. 35/3 All through our hootch, Not a creature was stirring. **1964** *N.Y. Times* 4 Sept., *Hooches*, the huts woven from banana leaves and roofed with straw or corrugated tin that are the standard housing for Vietnamese outside the cities. Some Americans have appropriated the term for their own quonset-styled barracks. **1966** *Flying* Dec. 54/2 A hootch (a house or a hut is known as a hootch in Vietnam). **1968** *N.Y. Rev. Books* 4 Jan. 4/1 Such targets as hooch lines (rows of houses along a road or canal). **1968** *TV Times* (Austral.) 6 Mar. 11/1, I was lying in a little scrap of a tent the Australians call a hoochie. **1969** I. Kemp *Brit. G.I. in Vietnam* iv. 75 Around us were scattered the makeshift 'hooches' of the A.R.V.N. soldiers, built of bamboo, wattle and mud; rectangular in shape with sloping, thatched roofs, they were..small. **1969** *Time* 5 Dec. 18/3 Calley's men in less than 20 minutes ignited 'hootches' and chased all the villagers..into groups, and shot everyone. **1971** *Fremdsprachen* XV. 207 A stereo set was blaring in an enlisted men's hootch shortly after midnight.

Hoochinoo (hū·tʃinū), and varr. [ad. Tlingit *Hutsnuwu*, lit. 'grizzly bear fort'.] **1.** A member of a small Indian tribe found in Admiralty Island, Alaska. Also *attrib.* or as *adj.* In *pl.* the tribe.

1878 Dennis in W. G. Morris *Rep. Customs Dist. Alaska* (1879) 122 On top of this there came a fight among the Hootzenoo Indians here. **1890** M. Ballou *New Eldorado* (ed. 5) 321 We pass the Indian village of Kootznahoo, occupied by a tribe of the same name, a people who have always proved to be restless and aggressive. **1915** J. Muir *Trav. Alaska* (1917) 211 They were about to set out on an expedition to the Hootsenoos to collect blankets as indemnity or blood-money for the death of a Chilcat woman from drinking whiskey furnished by one of the Hootsenoo tribe.

2. An alcoholic liquor made by Alaskan Indians, esp. the Hoochinoo tribe; also any inferior alcoholic drink (esp. whisky) in Alaska and the Canadian north-west.

1877 *Puget Sound Argus* (Pt. Townsend, Wash.) 23 Nov., I have frequently seen soldiers go to the Indian ranch for their morning drink of kootznehoo. *c* **1898** in P. Berton *Centennial Food Guide* (1966) 58/2 Whenever whisky runs short the Yukoner falls back upon a villanous decoction..known as 'hootchinoo', or 'hootch'. **1899** *Boston Jrnl.* 11 Jan. 4/5 Recently the House gave its official sanction to the word by enacting that no whisky, beer or 'hoochinoo' shall be sold in Alaska. **1937** C. L. Andrews *Wrangell & Gold of Cassiar* 49 A discharged soldier named Doyle..went to Hootznahoo, showed them how to distill a villainous compound from molasses, yeast, berries, sugar, or other compounds. It was first so called from the village, 'Hootznahoo' paraphrased as 'Hoochinoo'. **1958** P. Berton *Klondike Fever* 27 Another was to collect the excise duty on all locally made hootchinoo.

b. In full *hoochinoo still.* A still for the manufacture of hoochinoo.

1879 *Chicago Tribune* 14 May 6/3 We accidentally dropped upon a hootchenoo still in full operation. **1883** J. Wright *Among Alaskans* 150 Mr. Dennis had appointed the most reliable Indians as policemen, giving them authority, under United States revenue customs laws, to seize and destroy the hoochinoos or whisky-stills.

hoochy-coochy: see *Hootchy-kootchy.

hood, *sb.* Add: **5. j.** A waterproof folding top or cover of a perambulator, motor vehicle, charabanc, etc.; the movable cover of a typewriter or other machine.

1866 *Leisure Hour* XV. 349/1 Children are likely to be exposed for longer times to the scorching sun or the piercing wind in a perambulator (if without hood or sun-shade) than when carried in arms. **1895** *Army & Navy Co-op. Soc. Price List* 15 Sept. 1172 Perambulators..with ..reversible jointed hood. **1904** A. B. F. Young *Compl. Motorist* (ed. 2) viii. 198 It is a fine-weather vehicle, but a hood can be supplied for use in wet weather. **1912** *Motor Man.* (ed. 14) iii. 101 Complete protection can be obtained with a hood by fitting side curtains, which can be let down. **1942** *Short Guide Gt. Brit.* (U.S. War Dept.) 26 The top of the car is the hood. **1967** R. Mollon *Nursery Handbk.* (1968) i. 37 Be sure that the pram..has a good waterproof hood. **1969** C. Campbell *Sports Car* (ed. 3) ix. 226 An open cockpit and an erect windscreen completely spoil the airflow pattern over a sports car and a higher maximum speed is always given with the hood erect. **1971** *Daily Tel.* 24 Nov. 11/4 The weather during the test was too unpleasant to try the MGB as it should really be driven, with the hood down.

k. In various animals, esp. *Nautilus macromphalus* (see quots.).

1883 *Encycl. Brit.* XVI. 674/1 This part of the external annular lobe of the fore-foot [of *Nautilus*] is called the 'hood'. **1888** Rolleston & Jackson *Anim. Life* 456 In *Nautilus*..the fore-foot is divisible into an outer and inner portion. The outer portion..is thickened dorsally where it abuts against the coil of the shell and forms the hood. **1902** *Encycl. Brit.* XXV. 543/2 Movable (hinged) sclerite (so-called hood) [of *Cryptostemma karschii*] overhanging the first pair of appendages. **1932** Borradaile & Potts *Invertebrata* xvi. 527 When the animal [*sc. Nautilus*] is retracted into the living chamber the hood acts as an operculum. **1967** H. W. & L. R. Levi tr. *Kaestner's Invertebr. Zool.* I. xv. 418 The sheaths of the dorsal (anterior) tentacles are fused into a hood that is used in *Nautilus* to cover most of the aperture when the body is withdrawn.

l. A covering for the head of a horse.

a **1884** Knight *Dict. Mech.* Suppl. 461/1 *Hood*, that part of a horse blanket which covers the horse's head and neck. **1963** Bloodgood & Santini *Horseman's Dict.* 105 *Hood*, horse's head covering with eyeholes (with or without ear coverings) of serge, wool or rubber.

m. *Photogr.* (See quot. 1918.) In full, *lens hood.*

1892 [see *Finder 3d]. **1892** *Photogr. Ann.* II. 41 Have two caps for each lens, one to fit the hood and one to fit the other end of the mount. **1918** *Photo-Miniature* Mar. 25 *Lens hood*, the detachable rim of a lens-tube somewhat larger in diameter and carrying the lens-cap. Also any separate device of tubular box- or bellows-form fitted to the lens-tube, to screen the lens from strong light. **1939–40** *Army & Navy Stores Catal.* 906/2 Voightländer cameras... Fitted with a large brilliant view finder and deep hood. **1961** G. Millerson *Technique Television Production* iii. 28 (*caption*) Interchangeable camera lenses with lens hoods.

n. A protecting cover, also sometimes acting as a reflector, placed over a lamp.

1907 *Yesterday's Shopping* (1969) 260/3 Candle reading lamp. Telescopic corrugated hood. **1913** J. B. Bishop *Panama Gateway* v. vi. 382 The reflecting hood is provided with shading skirts, which prevent the glare of the lamp filament from penetrating into distance along the axis of the canal. **1939–40** *Army & Navy Stores Catal.* 277/3 The Reader. Chrome-plated with specially designed hood to shade light.

o. The bonnet of a motor vehicle. orig. and chiefly *U.S.*

1929 W. Faulkner *Sartoris* ii. vi. 145 He lifted the hood and removed the cap from the breather-pipe. **1942** *Short Guide Gt. Brit.* (U.S. War Dept.) 26 What we call the hood (of the engine) is a bonnet. **1960** *Times* 14 Sept. 12/6 You..discover that not only does the engine wear a hood instead of a bonnet but [etc.]. **1970** *Globe & Mail* (Toronto) 26 Sept. 29/5 (Advt.), A Rolls Royce whose hood was draped in white damask set with candelabra and plates.

p. A roughly shaped hat of felt, straw, or similar material for the hatter to shape by blocking or stitching.

1932 D. C. Minter *Mod. Needlecraft* 159/1 Felt or felted hoods and hats are beaten, steamed, and moulded from the flat. **1963** P. Moyes *Murder à la Mode* vii. 122 The shapeless felt 'hoods' which would eventually be steamed and seamed into smart hats.

8. hood dryer (see quot. 1966).

1962 *Guardian* 5 Dec. 6/4 Hairdryer, brush, comb and hood dryer. **1966** J. S. Cox *Illustr. Dict. Hairdressing* 80/2 *Hood dryer*, a hair dryer in the form of a hood secured to a pedestal.

hood (hud), *sb.*[2] Abbrev. of Hoodlum (in Dict. and Suppl.).

1930 *Amer. Mercury* Dec. 456/1 None of those St. Louie hoods are going to cut in here, see? **1934** J. T. Farrell *Young Manhood* xiv. 218 Jim Doyle stood by the kitchen sink, a cigar pasted in his round, jolly face, and he greeted them, calling them hoods. **1959** *Manch. Guardian* 5 Aug. 1/1 The 'News' suggests 'a special committee to greet the Kremlin's No. 1 Hood at the Washington Airport'. **1966** Wodehouse *Plum Pie* vii. 177 The hood was beating the tar out of me.

hooded, *a.* **2.** (Further examples.)

1785 Latham *Gen. Syn. Birds* III. ii. 426 Hooded M[erganser]...Size nearly that of a Wigeon. **1875** *Encycl. Brit.* III. 421/1 The Mitred Basilisk occurs in Guiana, the Hooded Basilisk in Amboyna. **1890** Ld. Lugard *Diary* 15 Mar. (1959) I. 144, I saw here [*sc.* in East Africa] the large hooded raven. **1938** *Brit. Jrnl. Psychol.* XXVIII. 334 Of the progeny of the hooded rats..I took at random six females. **1942** C. Barrett *On Wallaby* iv. 78 Hooded crows were on parade long before reveille. **1954** M. K. Wilson tr. *Lorenz's Man meets Dog* x. 107 Our hooded capuchin, Gloria,..was rather larger than a cat. **1955** *Arctic Terms* 40/1 *Hooded seal*, a large seal, *Cystophora cristata*, of the North Atlantic, dark in color with a unique inflatable muscular bag on its nose. Also called 'bladder-nose'. **1961** *Coast to Coast 1959–60* 62 I've heard that there is a place hereabouts which is a haunt of the hooded robin. **1972** *Country Life* 17 Feb. 381/1 The hooded crow, or hoodie, as it is commonly called in Scotland where the grey-mantled bird is seen as often as..the ordinary..crow.

hooden, var. *Hoden.

hoodlum. Add: (Earlier and later examples.) Now in more general use outside the U.S.

1871 *Cincinnati Commercial* 6 Sept. (Suppl.) 2/5 Surely he is far enough away here in this hideous wild of swamp, to escape the bullying of the San Francisco 'hoodlums'. **1929** F. A. Pottle *Stretchers* (1930) iv. 64 Then we were a mob of unorganized, distrustful, sick, and unshaved hoodlums, whose one burning desire was to escape from the army and go back home. **1935** A. Squire *Sing Sing Doctor* iv. 43 The hoodlums of the neighbourhood.. flattered him, and he ran along with the gang. **1952** R. Finlayson *Schooner came to Atia* 4 He's not one of that gang of hoodlums. **1959** [see *Belsen]. **1970** G. F. Newman *Sir, You Bastard* viii. 208 The man is a vicious hoodlum.

hoodlumism (earlier and later examples).

1872 *Newton Kansan* 14 Nov. 4/1 The Rev. Dr. Cunningham in a recent sermon traced the history of 'Hoodlumism'. **1920** [see *group-behaviour* (*Group *sb.* 6 b)]. **1970** G. Greer *Female Eunuch* 271 A new all-time low in political scurviness, hoodlumism.

2. hoodlum wagon *U.S.* (see quot. 1920).

1908 *Sat. Even. Post* 31 Oct. 39/2 The jolting of the hoodlum-wagon now focused the herd's attention. **1919**

H. L. Wilson *Ma Pettengill* iii. 79 The hoodlum wagon going back next morning to see what could be salvaged. **1920** J. M. Hunter *Trail Drivers of Texas* 299 A second wagon for carrying the extra beds and bringing wood and water into camps...This equipage is called the hoodlum wagon. **1968** R. F. Adams *Western Words* (ed. 2) 152/1 *Hoodlum wagon*, a cowboy's name for the bed wagon.

hoodoo, *sb.* **1.** Substitute for def.: One who practises voodoo. (Earlier, later, and *attrib.* examples.)

1875 L. Hearn *Amer. Miscellany* (1924) I. 127 Supposing you fall in love with a girl and can't get her, and that you go to one of these hoodoos, he will do something awful to her with charms. *Ibid.*, She would die..unless she could get some other hoodoo doctor to take the charm away by a counter charm. **1881** *Harper's Mag.* Apr. 738/2 Suddenly she..rushed forward with an African yell and joined in the dance as wild as any Hoodoo among them. **1946** R. Tallant *Voodoo in New Orleans* 16, I heard people say hoodoos was cannibals and used to eat babies. *Ibid.* 195, I got Adele to a good hoodoo woman and she uncrossed her.

2. (Earlier and later examples.)

1882 J. H. Beadle *Western Wilds* (ed. 2) xxxiv. 558 If you can find an Indian tradition to match it, your hoodoo is complete. **1916** H. Crane *Let.* 26 Jan. (1965) 3 Examination time is on,.. Latin and Geometry are due tomorrow. They are my hoodoos. **1945** E. Waugh *Brideshead Revisited* vii. 177 So you see things never looked like going right. There was a hoodoo on us from the start. **1958** L. van der Post *Lost World of Kalahari* viii. 195 By nightfall everyone..seemed convinced there was a permanent hoodoo on us.

3. A fantastic rock pinnacle or column of rock formed by erosion or other natural agency; an earth-pillar. Also *attrib.* orig. *U.S.*

1879 W. Whitman *Specimen Days & Collect.* (1882) 148, I had wanted to go to the Yellowstone river region—wanted specially to see..the 'hoodoo' or goblin land of that country. **1884** H. Butterworth *Zigzag Journ. Western States* 54 There is a region there called Goblin Land, full of lofty stone monuments, the remnants of erosion, called hoodoos. **1921** *Chambers's Jrnl.* June 373/2 Strange, isolated pillars, the 'hoodoos' stood like vedettes on the heights. **1940** *Canad. Geogr. Jrnl.* Feb. 84/2 The presence in Jasper Park of..earth pillars or 'hoodoos' similar to those at Banff. **1941** C. A. Cotton *Landscape* iii. 15 The picturesque minor surface-relief forms termed hoodoo columns and earth pillars are slender residual columns of unconsolidated sediment. **1968** R. W. Fairbridge *Encycl. Geomorphol.* 46/1 Occasionally, denudation of badland slopes can produce interesting features such as earth pillars and hoodoos with an overhanging 'hat'.

B. *adj.* Unlucky, bringing bad luck.

1889 *Kansas Times & Star* 17 Sept., Joe Bracken took sick Friday, September 13, but says that hoodoo date doesn't discourage him. **1904** *New York Globe* 2 Apr. 1 It is hard to find a crew for a 'hoodoo' ship. **1909** 'O. Henry' *Roads of Destiny* vii. 110 He's the hoodoo planet of the heavens. **1922** *Blackw. Mag.* Sept. 321/1 My name is Armstrong—Hoodoo Armstrong. **1926** *Scots Observer* 30 Oct. 21/2 That particular service was hoodoo.

hoodoo *v.* (earlier and later examples).

1886 *Harper's Weekly* 25 Dec., The surest way to provide against being 'hoodooed', as American residents call it, is to open one's pillow from time to time. **1888** *Judge* (U.S.) 21 July 239/2 A Washington paper..drops into the following poetry, which is sufficient to hoodoo the organization for the balance of the season. **1902** Kipling *Traffics & Discov.* (1904) 7, I wasn't going to deadhead along o' *that* crowd... 'Twould have hoodooed my gun for all time. **1914** V. Lindsay *Congo* 5 Mumbo-Jumbo will hoo-doo you.

hoo·dooism. [f. Hoodoo *sb.*] **a.** The practice of hoodoo rites. **b.** The faculty of attracting misfortune.

1881 *Harper's Mag.* Apr. 737/1 What *is* Hoodooism, anyhow? **1921** *Double Dealer* July 22/1 The white folks gaze at the negro with incredulous eye and wonder what amazing story of hoodooism will come from his lips. **1966** *Listener* 14 Apr. 542/3 Hoodoo-ism is not confined to steam locomotives. I have known a hoodoo diesel rail car.

hooer (hū·əɹ). *Austral.* and *N.Z.* [Representation (in various spellings) of a vulgar or colloq. pronunciation of *whore.*] **a.** = Whore 1. **b.** A strong term of abuse (of a man or woman).

1937 in Partridge *Dict. Slang* 401/2. **1969** D. Niland *Dead Men Running* iii. 86 The cheek of that hooer, I thought; a dirty-looking stranger. **1971** *N.Z. Listener* 22 Mar. 12/3 The dirty hua didn't even take a shower before he shot through.

hooey (hū·i). *slang* (orig. *U.S.*). [Orig. unknown.] Humbug, nonsense.

1924 P. Marks *Plastic Age* 100 My prof's full of hooey. He doesn't know a C theme from an A one. *Ibid.* 160 'Bunk!' he exclaimed, 'Hooey!' **1931** E. Linklater *Juan in Amer.* iii. v. 247 *We'll* be on Broadway if you don't talk hooey like that 'worth while' crack. **1932** Wodehouse *Hot Water* xiii. 223 Well, of all the hooey! **1934** *Discovery* Jan. 4/2 The United States of America, whose capacity for new words passes all belief, is responsible for hooey. **1935** *Punch* 10 Apr. 400/1 You have been misled, Hubert. I see it all. Somebody has been telling you the old, old story... Hooey, Hubert. Boloney. **1935** L. MacNeice *Poems* 21 Ireland is hooey, Ireland is A gallery of fake tapestries. **1948** V. Palmer *Golconda* xxv. 210 All this political hooey..doesn't affect me. **1952** Partridge *From Sanskrit to Brazil* 94 These charges of anachronism are, to put it courteously, sheer 'hooey'. **1966** Auden *About House* 21 Lip-smacking Imps of mawk and hooey Write with us what they will. **1970** G. Greer *Female Eunuch* 81 The horse between a girl's legs is supposed to be a gigantic penis. What hooey!

hoof, *sb.* **1.** *on the hoof*: delete (a butcher's phrase). Add examples. Also *transf.* and *fig.*

1818 H. B. Fearon *Sk. Amer.* 220 Cattle..are sold in this State, on the hoof, for about 3 dollars per hundred weight. **1902** *Encycl. Brit.* XXV. 186/1 The estimated dead weight of the sheep imported on the hoof for slaughter. *a* **1936** Kipling *Something of Myself* (1937) iii. 71 Why had Bret Harte, I asked, when I was prepared to supply home-grown fiction on the hoof? **1957** Wodehouse *Over Seventy* iv. 53 An august figure, weighing seventeen stone or so on the hoof. **1971** *Farmers Weekly* 19 Mar. 67/3 You can't grade hoggets on the hoof.

5. *hoof-fall, -hold*; **hoof and tongue sickness** = *foot-and-mouth disease* (*s.v.* Foot *sb.* 35); **hoof-rot** = *foot-rot* (*s.v.* Foot *sb.* 35); **hoof stick,** an instrument for manicuring the nails.

1867 *Queenstown Free Press* 22 Jan. (Pettman), We have had a great deal of hoof and tongue sickness amongst our cattle. **1910** J. Farnol *Broad Highway* ii. xlvii, Nodding sleepily with every plodding hoof-fall. **1923** H. Sutcliffe *Wrack o' Doom* ii, The broken lands that gave no hoof-hold. **1863** H. S. Randall *Pract. Shepherd* ii. 25 Scab and hoof-rot, those dire scourges of the ovine race. **1893** W. B. E. Miller et al. *Dis. Live Stock* v. 355 (*heading*) Hoof rot—foot rot. **1960** *Woman* 30 Jan. 15/2 First she lifts her cuticles gently with a hoof stick. **1970** *Observer* 8 Feb. 36/7 Keep the cuticle free from the nail with a rubber hoof stick.

hoof, *v.* Add: **1.** (Later examples.)

1877 J. Habberton *Jericho Road* i. 7 If we get stuck way up the river, so's we have to lay up all summer, and you have to hoof it in deep water. **1888** 'Buffalo Bill' *Story of Wild West* 531, I finally concluded that my prospects were good for 'hoofing' the whole distance. **1910** W. M. Raine *B. O'Connor* iv. 58 He hoofed it back to the cabin. **1923** Belloc *Sonnets & Verse* 111 A score of stout fellows who..Hoofed it amain, Rain or no rain. **1958** S. Ellin *Eighth Circle* (1959) ii. xvii. 177 They hoofed it all the way down to Barrow Street. **1972** C. Weston *Poor, Poor Ophelia* (1973) xvii. 109 Man's not very sick if he's out hoofing around that early.

b. To dance. (Also with *it.*) *slang.*

1925 *Amer. Speech* I. 36/2 A 'hoofing act' is entirely made up of step dancing. **1926** C. Van Vechten *Nigger Heaven* 13 Le's hoof, Ruby urged. Le's sit down, Anatole commanded. **1928** *Daily Express* 2 July 11/5 Mr. Tommy Nolan proposed to his partner, Miss Anna King. She accepted him, and they planned their wedding and honeymoon while 'hoofing'. **1958** 'A. Gilbert' *Death against Clock* iii. 27 A pretty nifty dancer himself in his young days and still able to hoof it quite neatly. **1972** I. Hamilton *Thrill Machine* xv. 64 She sings, she hoofs a little, she does some straight narration.

2. b. To dismiss, expel, eject. Usu. with *out.*

1893 Farmer & Henley *Slang* III. 340/2 To hoof out. **1905** *Daily Chron.* 22 Apr. 9/2 Well, at least we know for certain..that he was hoofed out of the Guards. **1915** V. Woolf *Voyage Out* xxiii. 376 They've hoofed out the prostitute. **1924** Galsworthy *White Monkey* iii. viii, A packer we had, who got hoofed for snooping books. **1928** D. L. Sayers *Unpleasantness at Bellona Club* i. 8 They'd hoof me out of the Club if I raised my voice beyond a whisper. **1959** *Punch* 6 May 600/2 They hoofed the London Museum out of Lancaster House. **1973** 'B. Mather' *Snowline* v. 60 The Bengali doctor came in at that stage and hoofed Mukkerjee out of it.

So **hoo·fer** *slang* (orig. *U.S.*), a dancer.

1923 *N.Y. Times* 9 Sept. vii. 2/1 *Hoofer*, a dancer, also a heel-beater. **1928** [see *Femme 1]. **1928** *Sunday Express* 8 Apr. 5/7 To-morrow Roy Lloyd, who was the hoofer in 'Broadway', takes up the part. **1936** 'J. Tey' *Shilling for Candles* vi. 59 The little Broadway hoofer was blossoming into the song-and-dance star. **1959** *News Chron.* 19 June 8/2 Holly..gets herself involved with a no-good hoofer in a low night club. **1969** *Daily Tel.* 17 Jan. 21/5 An orthodox tale of stage success: unknown girl hoofer becoming overnight hit by breaking all the Ziegfeld rules. **1973** *Sunday Express* 8 July 6/4 She was impressed by one of the male dancers... The one-time hoofer ended up by working for her for 40 years.

hoo-ha (hū·hā). *colloq.* Also **hoo-hah, hou-ha.** [Orig. unknown.] A commotion, a rumpus, a row.

The use in quot. 1932 seems to be without parallel.

1931 *Punch* 14 Oct. 402/1 The devil of a hoo-ha in the papers about increasing the demand for English-grown corn. **1932** T. S. Eliot *Sweeney Agonistes* 30 You've had a cream of a nightmare dream and you've got the hoo-ha's coming to you. **1937** N. Marsh *Vintage Murder* vi. 63 He came up under cover of all the hoo-hah on the stage some time after the event. **1944** 'N. Shute' *Pastoral* ix. 206 There's a bit of a hoo-hah on about your tea-party. **1954** *Times Lit. Suppl.* 24 Sept., After all the hou-ha in the *Observer* about the death of the novel, perhaps we ought to be wearing a mourning-band. **1955** E. C. R. Lorac *Ask Policeman* xvii. 187 He could cut off home after the hoo-ha died down and claim his inheritance. **1959** B. Goolden *For Richer, for Poorer* xiii. 232, I don't think Mummy will make much of a hoo-hah if she knows it's not for long. **1963** *Sunday Express* 15 Sept. 8/7 What has all the houha about the prodigious jump in productivity meant to the housewife? **1968** *Listener* 27 June 837/1 And there was a terrific hoo-ha over this because they all thought I should go and be a termination case or something. **1971** *Country Life* 27 May 1328/2 Some of these lovely irises may..be grown..successfully without much hoo-ha.

hook, *sb.*[1] Add: **1. b.** *pl. slang.* The fingers or hands. So *to get one's hooks on* or *into*: to get hold of.

1829 W. Maginn tr. *Vidocq's Mem.* IV. 261 To his clies my hooks I throw in. **1877** *Five Years' Penal Serv.* iv. 259 In a week or two a man can bring his hooks and feelers into full working trim again. **1917** E. Wallace *Just Men of Cordova* x. 169 Put your lamps over my shiners, run your hooks over me Astrakhan collar. **1926** *S.P.E. Tract* xxiv. 122 *Get one's hooks on*, get hold of. **1930** 'E. Queen' *French Powder Myst.* xxvii. 230 About these volumes... I noticed a queer hesitancy on your part when I first got my hooks into them. **1954** J. Potts *Go, Lovely Rose* viii. 41 Maybe he's eloped with that fat Lang dame. She's been trying to get her hooks into him all winter.

c. A thief, a pickpocket. *slang.* (Cf. Hook *v.* 6, Hooker[1] 1.)

1863 *Once a Week* IX. 555/1 The party who picks the pocket while the 'stiff-dropper' is attracting the victim's attention is called 'the hook'. **1885** M. Davitt *Leaves from Prison Diary* I. xi. 106 *Hooks*, these individuals, who are also known as 'gunns' and 'buzzers', in prison slang, constitute the pickpocket class in its various specialities. **1901** *Westm. Gaz.* 4 Sept. 4/1 The very same 'hook' was caught..a second time red-handed at another station. **1926** N. Lucas *London & its Criminals* xviii. 246 The 'hook' is the 'whizzer' who actually picks the pocket. **1935** A. J. Pollock *Underworld Speaks* 57/2 *Hook*, the pickpocket who does the actual stealing. **1968** G. J. Barrett *Guilty, be Damned* x. 116 We've nothing on him. But then we've nothing on half the hooks in Eastport.

2. b. on the hook: in various *fig.* uses, e.g. ensnared, in the power (of someone); in one's grasp; attached to some occupation, habit, etc. Cf. *off the hook* (sense 15 f below).

a **1635, 1730** [in Dict.]. **1927** H. Crane *Let.* 12 Aug. (1965) 304, I do hope that I can count on your assistance to the extent of the monthly amount until I can get something on my hook. **1932** L. C. Douglas *Forgive our Trespasses* xii. 233 Presently Angela was again on the hook in twenty score of composing-rooms. **1958** 'J. Brogan' *Cummings Rep.* xii. 126 She had made me wretchedly conscious of my shortcomings; that is how she had me on the hook. **1963** 'D. Rutherford' *Creeping Flesh* i. 72 'He really is on the hook.' 'The hook?' 'This drug habit.' **1970** G. F. Newman *Sir, You Bastard* ii. 60 Poor bastard might as well have been fined today as kept on the hook.

4. b. A hook upon which (in early models) the receiver rested. (The expression is still used when the reference is to the cradle upon which a telephone rests.)

1885 *List of Subscribers* (United Telephone Co.) p. iii, When your bell rings..take the Telephone *off the hook*... Unless the telephone is *on* the hook, the Subscriber *cannot call* or *be called* by the Exchange. **1921** *Conquest* Jan. 126/3 On removing the receiver from the hook, the subscriber's line is connected to a selector. **1955** W. Gaddis *Recognitions* ii. v. 508 Otto hung the receiver back on its hook. **1970** G. F. Newman *Sir, You Bastard* 262 Another phone crossed Sneed's mind, the one in his own flat with its receiver off the hook.

10. d. *Logic. colloq.* A name for the sign ⊃, used as the implication sign (cf. *Horseshoe 2 f.). Also, more commonly, a reading of the sign: thus '$p \supset q$' is read '*p*, hook, *q*'.

In colloquial use among logicians since 1955 or earlier.

1967 R. Neidorf *Deductive Forms* 65 'If...then'.. will be symbolized by a hook, ⊃. **1971** G. Hunter *Metalogic* 54 We shall call the tilde and the hook the connectives of P.

11. (Earlier examples.)

1600 Hakluyt *Voy.* III. 743 A hooke or headland. **1832** E. C. Wines *Two Years in Navy* i, We were kept off the hook, waiting either for wind or tide.

13. (Earlier and later examples.)

1896 *Badminton Mag.* Sept. 278 Leg-hitting..has found a goodly representative in the 'hook', as invented by E. M. Grace. **1904** F. C. Holland *Cricket* 26 Short-pitched balls are best disposed of by a hook. **1948** E. W. Swanton *Denis Compton* i. 14 The hook (that is the leg-side hit off the back foot) is another and more orthodox favourite. **1972** *Observer* 23 Apr. 24/8 Two regal hooks by Cowdrey were matched in the next over by two lordly cover-drives from Dexter.

b. *Boxing.* A short swinging blow with the elbow bent. Also *transf.*

1898 *Daily News* 9 Nov. 8/5 After Smith had put a left hook on the chin the issue was not in doubt. **1910** J. Driscoll *Ringcraft* 94 It only needs practice to convince anyone that the straight blow will always get there before the swing or the hook. **1929** *Evening News* 18 Nov. 16/4 With a couple of left hooks to the head. **1945** *Diamond Track* (Army Board, N.Z.) 7/1 The Division carried out the outflanking movements, the celebrated 'left hooks' which forced the enemy out of the two great strongholds. **1961** B. Fergusson *Watery Maze* x. 251 Some of the Mediterranean landing-craft had already been promised to India, to help the British mount amphibious hooks down the coast of Arakan. **1971** *Daily Express* 17 Feb. 14/7 Salah..was pinned and punished by Clark's jabs and hooks.

c. *Golf.* The act of hooking.

1890 W. Simpson in H. G. Hutchinson *Golf* vi. 204 Press ever so little when your club is turned in and yourself over-reached, and the hook is certain.

15. e. (Earlier and later examples.)

1840 H. Cockton *Life Valentine Vox* xii. 95 No man was ever able to write his own life complete. He's certain to go off the hooks before he has finished it. **1921** Galsworthy *To Let* i. i. 9 Old Timothy; he might go off the hooks at any moment. I suppose he's made his Will.

f. *off the hook*: out of a difficult situation. Cf. *on the hook* (sense 2 b above).

1864 Trollope *Small House at Allington* II. xxix. 296 'Poor Caudle!' he said to himself; 'he's hooked, and he'll never get himself off the hook again.' **1954** J. Potts

Go, Lovely Rose xii. 77 'It's an idea,' said Dr Craig...It would get Hartley off the hook, sure enough. **1962** M. Urquhart *Frail on North Circular* xxv. 140 Let Broadbent think he's off the hook and then give it another twist. **1966** *New Yorker* 25 June 49 Then he smiled, and I knew I was off the hook. **1969** A. Glyn *Dragon Variation* vii. 199 You mean she lost the kid? Well! Well, that sure lets old Walter off the hook!

16. (Earlier and later examples.)

1812 *Boston Gaz.* 23 Nov. (Th.), They forget that Rodgers himself says that he went upon his own hook. **1836** D. Crockett *Exploits & Adv. Texas* (1837) 13 But now I start anew upon my own hook. **1899** J. London *Let.* 30 Mar. (1966) 26 When I was just sixteen I broke loose and went off on my own hook. **1927** E. W. Springs *Nocturne Militaire* vi. 178 You know a man has to be crazy to go after a balloon on his own hook. **1940** M. Lowry *Let.* Spring (1967) 21 They objected to my going east on my own hook..because they would not trust me. **1952** F. Yerby *Woman called Fancy* xvi. 307 I'm not going out of this house with you on my own hook.

b. *to sling* or *take one's hook*: to go away, be off, decamp. *slang* or *dial.*

1874, 1897 [see Sling *v.*[1] 3 d]. **1885** O. Allan *Sinbad the Sailor* 22, I 'took the office' and I took my hook. **1886** M. Peacock *Tales Lindsey Folk-Speech* 106 An' soa he teks his hook back agaain to steam-hoose yard. **1890** Kipling *Barrack-Room Ballads* (1892) 34 Before you sling your 'ook, at the 'ousetops take a look. **1892** 'F. Anstey' *Mr. Punch's Model Music-Hall Songs* 130 Take your 'ook while you can. **1928** *Daily Express* 10 May 7 Magistrate: How is your husband cruel to you? Wife: He will not speak to me, and he tells me to sling my hook. **1955** L. P. Hartley *Perfect Woman* xxii. 193 Anyhow, she's gone, walked out, slung her hook. **1959** [see *Breeze *sb.*[1] 3 b].

16*. *hook, line, and sinker*: completely, without reservations.

1838 G. W. Patterson in T. W. Barnes *Mem. T. Weed* (1884) v. 60 We are gone, hook, line, and sinker. **1865** *Weekly New Mexican* 25 Aug. 1/3 Without him Chavez [*sc.* a candidate for Congress] is gone hook, line, and sinker. **1924** Wodehouse *Bill the Conqueror* ii. 59 The old man swallowed those references of yours, hook, line and sinker. **1936** N. Coward *To-night at 8.30* II. 58, I fell for it hook, line and sinker. **1945** E. Waugh *Brideshead Revisited* i. ii. 38 You, my dear Charles,.. have gone straight, hook, line and sinker, into the very worst set in the University.

17. a. *hook bolt* (examples).

1923 *Man. Seamanship* (H.M.S.O.) II. 300 It will be necessary [during salvage operations] to drill a few extra holes for hook bolts which will be used to secure the patch temporarily to the ship's side... The hook bolts are shipped in the holes ready for use. **1940** *Chambers's Techn. Dict.* 421/2 *Hook bolt*, a galvanised-iron bolt formed out of rod which is bent at one end into a hook serving as the head, and threaded at the other to take a nut; used for fixing corrugated sheeting. **1956** *Archit. Rev.* CXIX. 213/1 For fixing to metal purlins, a standard range of hook-bolts and U-bolts is available.

c. *hook-winged* adj. **d.** *hook-fishing.*

1745 Ellis *Mod. Husb.* VI. ii. 67 Those poor People.. may have the single Engines for Hook-fishing fixed within their Houses. **1841** J. Johnson tr. *Van der Donck's New-Netherlands* in *N.Y. Hist. Soc. Coll.* 2nd Ser. I. 177 Those the people call weak crabs, and they make excellent bait for hook fishing. **1851** *Illustr. Catal. Gt. Exhib.* iv. 926/1 Hook-fishing is within 3 fathoms, either in the river or in open sea. **1905** *Spectator* 7 Jan. 12/1 Every autumn the great coffee-coloured, hook-winged skua-gulls come down from the North and patrol the midway air. **1939** L. MacNeice *Autumn Jrnl.* vi. 27 A vulture hung in air..His hook-winged shadow wavered. **1968** T. Kinsella *Nightwalker* 17 Hook-winged geese or hawks.

18. hook-and-ladder *U.S.*, apparatus consisting of ladders and hooks used by firemen; often *attrib.*; **hook gauge,** an instrument for accurately determining the surface level of water and consisting of a hook and pointer attached to a fixed vernier, the hook being brought up until its tip just pierces the surface of the water; **hook-hit** = 13 b above; **hook-ladder,** a ladder with hooks at one end by which it can be suspended; **hook-pot** (see quot. *a* 1865); **hook rug** = **hooked rug*; **hook shop** *slang*, a brothel; **hook-shot** *Basketball*, a twisting shot started when the player has his back to the basket and completed as he pivots round towards the basket; **hook stroke** *Cricket*, a stroke made by hitting a short-pitched ball, after it has risen, round to leg with a horizontal swing of the bat; **hook tender** *N. Amer.* (see quot. 1905).

1821 *Minutes Boston* (Mass.) *Selectmen* XXXIX. 187 Mr. George G. Channing..declines taking command of the Fire Hook & Ladder company. **1865** *Chambers's Jrnl.* 29 July 469/2, 18 hook-and-ladder trucks. **1909** *Strand Mag.* Apr. 363 There stood the engines and the 'hook-and-ladder'. **1902** *Westm. Gaz.* 11 June 8/1 Other Americans pointed with pride to their hook-and-ladder system, which forms such an important aid to the New York fireman. **1949** *Los Angeles Times* 18 May 8/1 The Fire Department obligingly backed up a hook-and-ladder truck. **1875** *Jrnl. Franklin Inst.* XCIX. 250 The depth on the weir was observed by means of a hook-gauge. **1880** *Encycl. Brit.* XII. 477/2 The hook gauge used first by Mr. U. Boyden of Boston, in 1840. **1934** H. Addison *Text Bk. Appl. Hydraulics* xv. 304 Hook and point gauges. These are the simplest and most reliable gauges for measuring ranges of head not exceeding about 3 ft. (60 cms.) with a probable error of 1/1500 ft. (0·2 mm.). **1890** R. G. A. Allanson-Winn *Boxing* (ed. 2) ix. 43 There is another half-arm hit, called the 'hook-hit', in which the elbow is not so much bent as it is with the real 'contracted-arm'. **1919** G. B. Shaw in *Manch. Guardian* 1 Nov. 7/6 He missed that chance of a hook hit at the white chokers. **1858** Simmonds *Dict. Trade*, Hook-ladder. **1905** *Westm. Gaz.* 17 Oct. 7/1 The escape was rushed up, ladders extended, hook-ladders placed into position. **1972** *Times* 20 Sept. 3/3 Window cleaners..made their final assault on the upper storeys by jumping out of windows..by lowering a hook ladder from the summit. *a* **1865** Smyth *Sailor's Word-Bk.* (1867) 388 *Hook-pots*, tin cans fitted to hang on the bars of the galley range. **1886** R. Brown *Spunyarn & Spindrift* v. 66 A hook-pot of tea a-piece. **1896** *Idler* Mar. 173/1 There you'd see them crowding about the doors at meal-times, flourishing their hook-pots. **1913** J. Masefield *Daffodil Fields* 33 Tin dishes, sailors' hookpots. **1951** T. Capote *Grass Harp* (1952) i. 12 There was a hook rug on the floor. **1967** E. Short *Embroidery & Fabric Collage* iii. 83 The canvas [is] the one normally used for hook rugs, i.e. three holes to the inch. **1889** Barrère & Leland *Dict. Slang* I. 473/1 *Hook shop*, a brothel. **1935** A. J. Pollock *Underworld Speaks* 57/2 *Hook shop*, a house of ill fame. **1954** J. Steinbeck *Sweet Thursday* vi. 49 This kid could be pure murder in a hook-shop. **1957** *Encycl. Brit.* III. 181B/1 Farther out.. players use a one-hand shot from a stride, jump or standing position, and a hook shot which is overhead. **1969** Z. Hollander *Mod. Encycl. Basketball* 43 Washington, led by hook-shot artist Jack Nichols, defeated Oregon State. **1969** *New Yorker* 14 June 79/1 You go through Harlem and you'll see kids less than five feet tall with pretty good jump shots and hook shots. **1897** K. S. Ranjitsinhji *Jubilee Bk. Cricket* 175 Batsmen of the old school very much disliked the hook-stroke on principle. **1908** *Daily Chron.* 15 May 8/2 He began exploiting the full drive and the hook stroke. **1911** C. B. Fry in P. F. Warner *Bk. Cricket* 226 Ranjitsinhji found almost as little difficulty in making his famous 'hook strokes'. **1945** N. Cardus *Eng. Cricket* 38 Maclaren was the grand manner personified; with his hook-stroke he dismissed the fastest ball from his presence. **1893** *Atlantic Monthly* Feb. 196/1 Each man, being hired for a definite purpose, as chopper, hook-tender, barker, [etc.]..keeps closely to his own job. **1901** *Daily Colonist* (Victoria, B.C.) 3 Nov. 5/2 Seventy-five cents was taken off hook-tenders and other men not necessarily expert. **1905** *Terms Forestry* (U.S. Dept. Agric.) 40 *Hook tender*, the foreman of a yarding crew; specifically, one who directs the attaching of the cable to a turn of logs. **1966** *Sun* (Vancouver) 12 Jan. 25/5 Moore said the industry needs more managers, logging operators, hook-tenders..'and even chokermen'.

hook, *sb.*[2] *local.* Variant of Huck *sb.*[1] Also **hook-bone.** The projecting upper part of the thigh bones of cattle near the hip-joint. Cf. Huck *sb.*[1] and Huckle-bone.

1808 T. H. Horne *Compl. Grazier* (ed. 3) 9 The roof [of a bull ought to be] wide, particularly over the chine and hips, or hooks. **1844** H. Stephens *Bk. Farm* II. 161 Between the shoulders and the hook. *Ibid.* III. 1253 The broad hook-bones, with the narrow chest, are not entirely occasioned in cows by calf-breeding. **1858** C. L. Flint *Milch Cows* 17 The Ayrshire farmers prefer their dairy bulls..broad at the hook-bones and hips, and full in the flanks. **1900** *Westmorland Gaz.* 3 Feb., Advt. (E.D.D.), Strayed, two Herdwick ewes; marked red pop near shoulder and near hook. **1935** *Amer. Speech* X. 271/1 *Hooks*, hip bones of a cow. **1960** *Farmer & Stockbreeder* 8 Mar. 95/2 As is good breadth between the hookbones.

hook, *v.* Add: **3.** (N.Z. examples of *to hook off*, to make off.)

1938 F. S. Anthony in D. M. Davin *N.Z. Short Stories* (1953) 219, I hooked off on my own and rambled aimlessly about. **1940** F. Sargeson *Man & Wife* (1944) 75 If Ted saw her coming up the road he'd hook off if he could before she got near. And if he couldn't I'd hook off while they had their barney.

4. c. To make (rugs) with a hook: see *Hooked *a.* 4. *U.S.*

1882 *Harper's Mag.* Dec. 126/1 Cynthy Ann..hooked rugs from early in the morning until late into the night. **1945** B. Macdonald *Egg & I* 66 A time to repair machinery, hook rugs, patch quilts, mend harness and perform other leisurely tasks.

d. *intr.* To use a crochet needle.

1854 C. M. Yonge *Castle Builders* vi. 81 Miss Townsend ..hooked away with her crochet needle.

e. *to hook up:* to establish a link *with*, to make a connection *with*. Also const. *to*, and with direct object. Cf. *Hook-up.

1925 H. Crane *Let.* 27 Oct. (1965) 218 They want to hook the book up with an illustrious name. **1929** Wodehouse *Mr. Mulliner Speaking* v. 172 What I would propose is that we take a short cut through the fields to the station, hook up with the five-fifty express at Goresby, [etc.]. **1943** E. C. Wicks et al. *Shopwork* vi. 113 Whenever electricity is needed for any purpose, the particular job must be wired or 'hooked-up' to feed the electrical current to the necessary place. **1953** P. C. Berg *Dict. New Words* 94/1 *Hook-up*, v.t., to connect two or more broadcasting systems for the time needed to broadcast a common item on their otherwise different programmes. **1971** *Ink* 12 June 12/1 David Mercer's moving and intelligent portrait of a Marxist drama critic who can't hook up his ideology with his unresolved feelings towards his impossible working-class father. **1971** M. Tak *Truck Talk* 84 *Hook up*, to couple a tractor to a trailer. **1972** *Edmonton* (Alberta) *Jrnl.* 31 Aug. 3/4 RCMP and city police forces in Alberta have begun hooking up to a national computer system.

5. b. Usu. in pa. pple. *hooked* (*on*): addicted (to), captivated (by). *slang.*

1925 *Writer's Monthly* June 486/2 *Hooked*, to become a drug addict. **1931** D. Runyon *Guys & Dolls* (1932) vi. 115 Waldo Winchester is hooked. **1953** W. Burroughs *Junkie* (1972) 11, I drifted along taking shots when I could score. I ended up hooked. *Ibid.* vi. 61 When you are hooked, the effects are not dramatic. **1964** *Daily Tel.* 25 Nov. 22/6 The chances are that he is hooked on opium, morphine or heroin. **1965** *New Statesman* 16 Apr. 620/3 Other cities have admirers, even lovers, but Liverpool has only addicts: either you are hooked the very first time you step out of Lime Street Station to be confronted by the bulk of St George's Hall, or you never get the message. **1966** E. McGirr *Funeral was in Spain* 137 She was halfway to being hooked: one of the punks she was with was a pusher. **1967** M. M. Glatt et al. *Drug Scene* ii. 21 Once you're registered, you're hooked. It's too depressing when you're hooked, besides a girl looks terrible on heroin. **1967** *New Scientist* 25 May 478 Hopes that the millions of men and women 'hooked' on tobacco may soon be able to satisfy their craving with a 'safe' cigarette are not supported by the facts. **1970** *Daily Tel.* 8 May 3/2 Hundreds of domestic pets die each year after becoming 'hooked' on slug bait.

7. c. To solicit as a prostitute. Cf. *Hooker[1] 4. *slang.*

1959 'E. McBain' *Killer's Wedge* (1961) vi. 57 She's been in the city for almost a year, Pete. Hooking mostly. **1965** —— *Doll* (1966) v. 66 The girl was a prostitute... The girl had been hooking in the neighborhood for little more than a week. **1969** Disch & Sladek *Black Alice* v. 53 Bessie's girls didn't have to go out hooking in hotel lobbies or honkytonks, no indeedy. **1971** W. Hanley *Blue Dreams* xix. 313 A high-class hooker couldn't be entirely without redeeming social value. Especially one who..taught English and hooked on the side.

8. c. For the use in *Cricket* cf. *hook stroke* (*Hook *sb.*[1] 18). (Further examples.) Also *absol.*

1896 *Badminton Mag.* Oct. 482 Gregory, in attempting to hook Peel, put the ball straight into Richardson's hands. **1898** K. S. Ranjitsinhji *With Stoddart's Team* (ed. 3) iii. 50 He [*sc.* C. Hill] seemed able to 'drive', or 'hook', or 'glance'..with equal skill and success. **1904** [see *Cover *sb.*[1] 1 d]. **1955** [see *Boundary 2 b]. **1955** *Times* 9 May 15/2 Then, when he must have been looking ahead to a century, he was leg-before-wicket hooking at Heine. **1972** 'J. Ross' *Here lies Nancy Frail* xii. 145 I'm bloody useless with a two wood... I hook like hell with it.

d. *Boxing.* To strike (one's opponent) a swinging blow with the elbow bent (cf. *Hook *sb.*[1] 13 b). Also *absol.*

1898 *Daily News* 24 Nov. 8/3 Corbett hooked with his right hard on Sharkey's jaw. **1910** J. Driscoll *Ringcraft* 86 Wild attempts to hook him on his well protected jaw. **1973** *Times* 14 Feb. 9/3 Bugner..clubs rather than hooks.

e. *Rugby Football.* To secure (the ball) with the foot, as hooker, when it is placed in the scrummage. Also *intr.*

1906 Gallaher & Stead *Compl. Rugby Footballer* vii. 110 In Britain it is the custom to hook the ball in the scrum with the outside feet crossing over those on the inside. **1913** *Daily Graphic* 24 Mar. 15/1 D. A. Greer..may be of use to Ireland henceforward, especially as 'hooking' is his forte. **1927** Wakefield & Marshall *Rugger* 183 The front row tried trick hooking and foot-up tactics. **1955** *Times* 1 Aug. 2/3 Kroon's brilliant hooking has been a feature of the season's provincial matches.

13. *to hook Jack*: to play truant. *U.S. colloq.* (Cf. *Hookey 1.)

1877 Bartlett *Dict. Amer.* (ed. 4) 294 *Hook Jack*, to play truant. New England. **1890** *Dialect Notes* I. 22 *Hookey*, in 'to play hookey', meaning *to play truant*, used in Maine, but not usual in Boston, where the phrase was and is to 'hook Jack'. **1892** *Ibid.* 216 In all the period from 1840 to 1850 the current phrase among the boys was *to hook Jack*. **1905** J. C. Lincoln *Partners of Tide* iv. 70 The boy 'hooked Jack' for a whole day. **1967** *Publ. Amer. Dial. Soc.* xlvii. 7 *Hook jack*, 'play hookey'.

hook-billed, *a.* (Later examples.)

1785 J. Latham *Gen. Synopsis Birds* III. 495 Hook-billed Duck, *Raii Syn.* p. 150.2. **1881** *Amer. Naturalist* XV. 182 The male [of hump-back salmon] is slab-sided, hook-billed and distorted. **1911** *Encycl. Brit.* XXII. 219 Among the breeds differing in structure may be mentioned ..the hook-billed and tufted ducks.

Hooke (huk). The name of Robert *Hooke* (1635–1703), English inventor and natural philosopher, used, chiefly in the possessive, to designate his discoveries and inventions, as **Hooke('s) coupling,** a Hooke's joint; **Hooke's law,** the law, valid within the limits of elasticity, that the strain produced by a stress of any one kind is proportional to that stress; **Hooke's (universal) joint,** a kind of universal joint for transmitting rotary motion from one shaft to another.

1825 J. Nicholson *Oper. Mech.* 32 Hooke's universal joints are sometimes used to communicate motion obliquely. **1853** *Trans. R. Soc. Edin.* XX. 93 The fundamental assumption from which the following equations are deduced is an extension of Hooke's law. **1883** *Encycl. Brit.* XV. 762/2 Two Hooke's couplings. **1906** A. E. H. Love *Treat. Math. Theory Elasticity* (ed. 2) iv. 110 Many materials used in engineering structures, e.g. cast iron, building stone, cement, do not obey Hooke's Law for any strains that are large enough to be observed. **1930** *Engineering* 31 Jan. 134/1 A Hooke's joint is provided to allow for the vertical motion of the [gear] boxes. **1959** C. E. Pearson *Theor. Elasticity* v. 83 The familiar Hooke's-law experiment in which a metal wire is stretched by a tensile load. **1968** R. H. Bacon *Car* xx. 268 Most universal joints are based on the Hooke coupling.

Hence **Hookean, hookean** (hu·kiăn) *a.* [-an], obeying Hooke's law; linearly elastic.

1956 J. C. Jaeger *Elasticity, Fracture & Flow* ii. 99 (*heading*) The perfectly elastic or 'Hookean' substance.

1960 *McGraw-Hill Encycl. Sci. & Technol.* X. 476/2 On extension of polymeric materials, both rubberlike and Hookean elasticity may be present. **1968** C. G. Kuper *Introd. Theory Superconductivity* ix. 146 The elastic vibrations of a lattice of discrete point masses, coupled by Hookean springs. **1972** *Physics Bull.* Nov. 651/3 For most practical purposes alkali silicate glasses are hookean solids exhibiting a linear stress–strain curve to fracture.

hooked, *a.* Add: **4.** *hooked mat, hooked rug,* a mat or rug made on a canvas ground with woollen yarn which is pulled through with a hook. orig. *U.S.*

1880 W. D. Howells *Undiscovered Country* 415 Hooked rugs and embroidered tidies were as worthy a place in Mrs. Ford's simple house as most of the old-fashioned things. **1917** L. M. Montgomery *Anne's House of Dreams* (1926) ii. 10 Nobody seems to want anything but hooked mats now. **1932** D. C. Minter *Mod. Needlecraft* 223 (*heading*) Hooked rugs. **1960** G. Lewis *Handbk. Crafts* 91 A hooked rug is a very satisfying thing to make. **1964** Mrs. L. B. Johnson *White House Diary* 21 May (1970) 144 After the speech Mrs. Breathitt presented me with a beautiful hooked rug—an art still practiced by the old people in the area. **1969** H. A. Horwood *Newfoundland* 7 A hooked mat is a kind of tapestry made by Newfoundland women who got the art from their mothers who had it handed down to them from the first European settlers. **1970** *Globe & Mail* (Toronto) 26 Sept. 45/6 (Advt.), Superior antiques..including..hooked and braided rugs.

hooker[1]. Add: **3.** Also simply *hooker*; and in many other technical usages.

1881 *Instr. Census Clerks* (1885) 134/2 Hooker (to special trade when stated or determinable). Hooker (Undefined)..Factory Labourer. **1900** *Westm. Gaz.* 16 May 8/1 No one received injuries, the hookers having received warning by the clashing of the cage. **1919** *Camp Worker* (Vancouver) 26 Apr. 5/2 The best hooker that ever gave signals for the high-rigger, while his short-handed crew changed the haul-back without the assistance of a grass-line. **1921** *Dict. Occup. Terms* (1927) § 043 *Onsetter;..hooker, hooker-at-shaft*..; stands at bottom of shaft in coal or shale mine to push full tubs on to cage and remove empty ones. *Ibid.* § 943 *Hooker, cloth hooker, piece hooker, hooker-and-lapper.* **1966** H. Sheppard *Dict. Railway Slang* (ed. 2) 6 *Hooker*, shunter. **1967** *Amer. Speech* XLII. 291 Hookers come and attach hooks on slings from the overhead cranes, or travelers, that move over the yard transferring loads between saws and stacks.

4. A prostitute. *slang* (chiefly *U.S.*).

1845 in N. E. Eliason *Tarheel Talk* (1956) 277 If he comes by way of Norfolk he will find any number of pretty Hookers in the Brick row not far from French's hotel. **1914** Jackson & Hellyer *Vocab. Criminal Slang* 45 *Hooker*, a prostitute. **1929** T. Wolfe *Look Homeward, Angel* (1930) xxx. 435, I hope you have sense enough now to leave those old hookers alone? **1932** J. Dos Passos *1919* 43 Ain't you got the sense to tell a good girl from a hooker? **1952** J. Steinbeck *East of Eden* xlv. 504 Joe could find any hooker in any town in a few hours. **1964** *Esquire* Nov. 85/2 They were attractive but not very imaginative, they looked like hookers on horses. **1971** [see *Hook *v.* 7 c].

5. A cow or ox that 'hooks' (see Hook *v.* 11). *U.S.*

1866 *Harper's Mag.* May 816/1 He..asked 'Why that pipe [*sc.* a hookah] was like a cow?' having in mind the obvious answer that it was a *hooker*. **1885** 'C. E. Craddock' *Prophet Gt. Smoky Mts.* ii. 48 The red cow jes' hooked down the bars, bein' a turrible hooker. **1902** G. H. Lorimer *Lett. Merchant* vii. 84 You want to.. distinguish between a cow that's a kicker, but whose intentions are good..and a hooker, who is vicious on general principles.

6. *Rugby Football.* A player in the centre of the front row of the scrummage on either side who endeavours to obtain the ball by hooking it. Cf. *Hook *v.* 8 e.

1905 *Daily Chron.* 13 Sept. 7/4 The two front men—called 'hookers'—can get the ball just as well as three. **1906** [see *Back *a.* 1]. **1927** Wakefield & Marshall *Rugger* 166 Some hookers prefer to have the weight on the loose-head side. **1963** *Times* 14 Jan. 3/1 Thorne had been on the fringe of this new honour as a hooker for several years, and Davis and Owen look a sound second row. **1971** *Times* 15 Feb. 9/2 For a hooker Pullin often showed up well in the loose.

b. *Cricket.* One who uses the hook stroke.

1900 W. J. Ford *Cricketer on Cricket* 102 Like Hill he is a splendid hooker. **1911** P. F. Warner *Bk. Cricket* 208 He has a lovely shot over extra-cover's head,..and on a slow wicket is a fine puller and hooker. **1972** *Cricket World* 1. 8/2 Keith [Stackpole] has long been regarded as only a puller, hooker and cutter but, last season, developed so much that he also drove strongly.

hooker[4] (hu·kəɹ). *dial.* and *N. Amer. colloq.* [Orig. unknown.] A glass of whisky, a dram; usu. with qualifying word (e.g. *stiff*), a drink (of brandy, etc.).

1833 J. Kennedy *Geordie Chalmers* iv. 45 Ye'll be nane the waur o' a hooker after yer fricht. **1865** W. H. L. Tester *Poems* 133 Sandy liket a hooker, an' brawlie I kent, The drap creatur' wad set him a speakin'. **1887** *Grip* (Toronto) 21 May 12/2 We went in and were served out with a pretty stiff hooker each. **1906** H. Green *At Actors' Boarding House* 62 A stiff hooker of whiskey, and then another had the expected effect. **1927** *Black Mask* Feb. 37/1 It took a stiff hooker of whiskey and a lot of words to thaw her into understanding. **1930** H. Craddock *Savoy Cocktail Bk.* i. 97 The Juice of 1 Lemon. 4 Hookers Whisky. **1939** C. Morley *Kitty Foyle* xxvii. 268 She gave me such a hooker of brandy I went right to sleep. **1955** 'T. P. Kelley' *Black Donnellys* 63 Danny was quiet enough until he got about six hookers under his belt.

Hooker[5] (hu·kəɹ). The name of William *Hooker* (1779–1832), botanical artist, used in the possessive in *Hooker's green*, a bright green colour used in water-colour painting.

1853 *Dict. Archit.* (Archit. Publ. Soc.) II. 84/2 This method of imitating the mixture of ultramarine with yellow has grown into a system, but the colour is not durable: as in the cases of Hooker's green and Varley's green. **1886** *Jrnl. Bot.* XXIV. 52 His memory is associated by artists with the colour called after him, 'Hooker's Green'. **1948** F. A. Staples *Water-Color Painting* i. 2 For paints get the following water colors in tubes: Alizarin Crimson,.. Hooker's Green Dark. **1951** R. Mayer *Artist's Handbk.* ii. 51 *Hooker's green*, a mixture of Prussian blue and gamboge. Sold in two shades, yellowish and bluish.

hookey (hu·ki). Also **hooky.** [Cf. Hook *sb.* 14, *v.* 6, *v.* *13, and Hooky-crooky.] **1.** *to play hookey*: to play truant. Also *transf. colloq.* (chiefly *N. Amer.*).

1848 in Bartlett *Dict. Amer.* **1866** *Harper's Mag.* May 779/1 Kate used to..entreat him not to get feruled, nor play 'hookey'. **1867** 'Mark Twain' *Amer. Drolleries* 20 He would not play hookey, even when his sober judgment told him it was the most profitable thing he could do. **1883** E. Eggleston *Hoosier School-Boy* ix. 47 They remembered that the geography lesson was a hard one, and so they played 'hookey'. **1904** W. N. Harben *Georgians* vii. 77 'I sorter feel like playin' hookey myself,' he admitted. **1908** C. E. Mulford *Orphan* xii. 150 I'll play on them, too, when they gets home! Off playing hookey from work when we all of us aches from double shifts. **1923** Wodehouse *Inimitable Jeeves* xiv. 172 He's played hookey from the choir so often that the vicar told him..he would fire him out. **1957** W. H. Whyte *Organization Man* iii. xi. 144 Such solitary contemplation during the office day..is regarded ..as a form of hooky. **1960** I. Wallach *Absence of Cello* (1961) 98 'I like to play hooky now and then.'.. 'You can't just call it hooky.' **1965** *Globe & Mail* (Toronto) 22 Apr. 23/8 Youngsters who play hooky are..merely afraid of their classrooms.

2. blind hookey: see Blind *a.* 16. Also *fig.* or *transf.*

1852 G. C. Mundy *Our Antipodes* III. iii. 85 The process of emigration was formerly—as compared with its present gradual perfection—a very blind-hookey kind of game. **1909** J. R. Ware *Passing Eng.* 34/1 *Blind Hookey*, a leap in the dark; *e.g.*, 'Oh, it's Blind Hookey to attempt it.' **1925** *Blackw. Mag.* Aug. 286/2 It is..the common practice of politicians to play blind hookey with the great interests entrusted to them.

Hookey Walker: see Walker *int.*

hookless, *a.* Add: **b.** Of a garment: having no hooks, with its hooks missing.

1906 *Westm. Gaz.* 25 Sept. 2/1 Blouses, in various stages of hookless decrepitude.

hook-nosed, *a.* (Later examples.)

1938 [see *fawn-eyed* s.v. *Fawn *sb.*[1] 4]. **1958** L. Durrell *Mountolive* ii. 55 Cherry-starred Japan, hook-nosed Lima.

hookum (hū·kǝm). *India.* Also **hookm, hukm.** [a. Hindi, a. Arab. *ḥukm*, f. *ḥakama* (cf. Hakim).] A command, order, or instruction from a person in authority. Also *transf.* (see quot. 1925).

1838 E. Eden *Let.* 17 Dec. in *Up Country* (1866) II. ii. 13 An officer came to ask his 'hookum', or orders. **1843** Lady Sale *Jrnl. Disasters Afghanistan* 39 The troops ..instead of receiving *hookm* to enter the city, the Shah almost rudely inquired why they had come! **1858** W. H. Russell *Diary in India* 7 Oct. (1860) II. xii. 226 We had no hookum from the commissioner or deputy. **1886** [see *Dastoor, dastur(i)]. **1895** B. M. Croker *Village Tales* (1896) 224 His coat was restored to him, with a 'hookum', to say that he was free. **1925** Fraser & Gibbons *Soldier & Sailor Words* 120 *Hookum*, a regulation. The correct thing, *e.g.*, 'That's the hookum'. An old Army colloquial term.

hookum-snivey (hu·kəm sni·vi). *dial.* and *slang.* Also **hook and snivey (snivvy), hook 'em snivey, hookem snivey, hookem-snivvy, hook um snivey.** [app. orig. *hook and snivey*, prob. f. Hook *sb.* or *v.*] An imposture or deceit; also, a contrivance for undoing the bolt of a door from the outside. Also *attrib.* or *adj.*, deceitful, tricky.

1781 G. Parker *View Soc.* II. 81 He..would stand no Hook and Snivey, or Nix the Buffer. **1802** R. L. & M. Edgeworth *Irish Bulls* 129, I ranged them fair and even with my hook-em-snivey. **1823** 'J. Bee' *Dict. Turf* 98 Hook and Snivvy—practised by soldiers in quarters, when they obtain grub for *nix*, by connivance with the slavey, or her mistress. **1874** Hotten *Slang Dict.* 194 Sometimes used as an irrelevant answer, by street boys. As, 'who did that?'—'Hook um snivey'—actually no one. **1892** S. Hewett *Peasant Speech of Devon* 89, I tellee 'onesty is the best policy. Niver yü be up tü hookem-snivey ways. **1905** E. Phillpotts *Secret Woman* ii. xi, An' some lying an' doing all manner of hookem-snivey deeds. **1928** —— *Ring Fence* xvi. 132 I'd a lot rather put my thoughts into work, so as you should have a wedding worthy of you and no hookem-snivey marrying in a corner. **1950** L. A. G. Strong *Which I Never* vi. 184 'Tisn't like him to do any sort of hookem-snivvy tactics.

hook-up (hu·kʌp). orig. *U.S. colloq.* [f. phr. *to hook up*: Hook *v.* 4 b, *4 e, 5.] A connection or combination, esp. of radio or television broadcasting facilities.

1903 A. H. Lewis *Boss* 116 It'll put us in line for a hook-up with th' reform bunch in th' fight for th' town next year. **1911** H. Quick *Yellowstone Nights* vii. 191 The Golden Fountain..had no lawyer against us. It was a funny hook-up. **1922** L. D. Brigham *How to make Vacuum Wireless Receiving Sets* 39 The remainder of the hook-up is just like the other amplifier hook-ups. **1927** *Observer* 11 Dec. 16/5 National appeals are possible by wireless when the various American 'Radio Corporations' agree to a 'national hook-up'. **1929** *Lit. Digest* 18 May 79/1 These lines are the skeleton of a gigantic aerial hook-up that now seems destined to cover South America. **1930** P. W. Slosson *Gt. Crusade* x. 278 A radio hook-up brought the whole nation to the [prize] ring side. **1932** E. Wallace *When Gangs came to London* xxi. 186 There had to be either a hook-up or one side had to get out of business. **1944** E. J. Teichert *Ferrous Metall.* (ed. 2) II. xi. 227 The electrical hook-up..is of such a nature that the two motors act as a unit. **1950** Wodehouse *Nothing Serious* 16 He could not have been better informed regarding it if the facts had been broadcast on a nation-wide hook-up. **1957** *Times* 31 Oct. 11/4 This American precedent of appearing in person on a nation wide hook-up. **1959** *Punch* 28 Oct. 360/3 Are you now in favour of a Lab–Lib hook-up? **1967** *Electronics* 6 Mar. 80/1 How do you create a galvanometer..that doesn't require hours of delicate dial twiddling, trapdoor adjustments or experimental hook-ups? **1970** *New Yorker* 12 Sept. 25 Discussing layouts, viewing charts, and getting computer information via a Picturephone/Computer hookup are just a few of the areas being pursued.

hook-worm. *Zool.* [Cf. G. *hakenwurm*, mod.L. *Uncinaria*, f. L. *uncinus* hook (both coined by J. A. Fröhlich 1789, in *Naturforscher* XXIV. 136).] A parasitic nematode worm of the family Ancylostomatidæ, which infests man, other mammals, or birds, using hook-like organs to attach itself to the host's intestinal lining. Hence **hook-worm disease** = *Ankylostomiasis; **hookworm-ridden** *a.*, infested with hookworms.

1902 *Rep. Bur. Anim. Industry 1901* (U.S. Dept. Agric.) 183 (*title*) The significance of the recent American cases of hookworm disease (Uncinariasis, or Anchylostomiasis) in Man. **1902** C. W. Stiles in *Amer. Med.* III. 777 (*title*) A new species of hookworm..parasitic in man. **1909** *Times* 29 Oct. 5/6 A commission of eminent medical men to investigate the hook-worm disease. **1925** R. W. G. Hingston in E. F. Norton *Fight for Everest: 1924* iii. vi. 352 Porters should be vaccinated, and given suitable treatment for hook-worm. **1931** [see *Ankylostomiasis]. **1932** W. Faulkner *Light in August* i. 3 Department annals would not now even be remembered by the hookworm-ridden heirs-at-large. **1950** F. A. Brown *Selected Invertebr. Types* 237 It is believed that infected African Negroes brought into the United States established the hookworm in at least part of its present range. **1964** J. E. Larsh *Outl. Med. Parasitol.* xiii. 119 Long-continued anemia, characteristic in chronic hookworm disease, is most detrimental in children. **1968** N. D. Levine *Nematode Parasites* iii. 106 Prevention of hookworm infections in dogs and cats depends upon sanitation.

hooley (hū·li). Chiefly *Irish.* Also **huly, wholee.** [Origin unknown.] A noisy party, a spree (see also quots.).

1877 Bartlett *Dict. Amer.* (ed. 4) 302 *Huly*, a noise, uproar. 'To raise *huly*.' New England. **1947** *Béaloideas* XVII. 273 *Hooley*, a dance or 'spree' in a private house, often to celebrate a wedding, etc. **1950** *Ibid.* XX. 191 *Wholee, Hooley*, a party or dance in a country house. *John gave a right wholee last night.* **1960** *News-Call Bulletin* (San Francisco) 4 Jan. 16/1 She can be seen at all the best hoolies in town. **1966** 'L. Lane' *ABZ of Scouse* 50 *'ooley*, a dispute, a fight or a riot. **1966** *Listener* 12 May 687/2 Sir Laurence tries to make it one of those peculiar Irish occasions, a hooley: swinging from wild hilarity to nostalgic sadness.

hooligan (hū·ligăn). [Origin unascertained.

The word first appears in print in daily newspaper police-court reports in the summer of 1898. Several accounts of the rise of the word, purporting to be based on first-hand evidence, attribute it to a misunderstanding or perversion of *Hooley* or *Hooley's gang*, but no positive confirmation of this has been discovered. The name *Hooligan* figured in a music-hall song of the eighteen-nineties, which described the doings of a rowdy Irish family, and a comic Irish character of the name appeared in a series of adventures in *Funny Folks*.]

A young street rough, a member of a street gang. Also *attrib.* and *transf.*; *spec.* **Hooligan Navy** *U.S. Naut. slang*, the U.S. Coast Guard Service.

1898 *Daily News* 26 July 5/1 It is no wonder..that Hooligan gangs are bred in these vile, miasmatic byways. *Ibid.* 8 Aug. 9/3 The constable said the prisoner belonged to a gang of young roughs, calling themselves 'Hooligans'. **1898** *Daily Tel.* 6 Aug. (Ware), William Lineker, described as a Hooligan, sets upon an inoffensive man. **1898** *Daily Graphic* 30 Aug. 4/4 Mr. White..stated that every Saturday and Sunday nights gangs like the 'Hooligan gang' came to his house, broke the windows, glass, &c., and made disturbances. **1898** *Westm. Gaz.* 15 Sept. 1/2 The Khalifa was, after all, only a sort of Soudanese Hooligan. **1901** *Pall Mall Mag.* Feb. 198 Nobody will claim honesty as a Hooligan virtue. **1922** L. Hisey *Sea Grist* 7 Haven't even been in the Hooligan Navy? Just land lubbers. **1932** H. Walpole *Fortress* iii. 439 Crowds of roughs and hooligans, urged on by the more violent

Chartists, drove their way towards the stands with shouts and threats. 1938 AUDEN & ISHERWOOD *On Frontier* III. ii. 110, I always suspected that you and your gang of hooligans would rat. 1962 L. FARAGO *Tenth Fleet* 119 Thus was born the Coastal Picket Patrol or, as the Coast Guard called it, the Corsair Fleet. Its own personnel, mostly amateur yachtsmen, preferred to refer to it as the 'Hooligan Navy'. 1963 *Daily Tel.* 27 Dec. 1/6 Some driving he had seen amounted to 'downright hooligan behaviour'. 1971 *Guardian* 29 July 11/4 Rome's young black-shirted hooligans like to taunt the long-haired guitar-strummers by roaring around them on motor-cycles.

Hence **hoo·ligan** *v. intr.*, to act as a hooligan; also *trans.*, to treat (a person) roughly; **hoo:ligane·sque** *a.*, like a hooligan; **hooliga·nic** *a.*, resembling that of hooligans; **hoo·liganism**, the characteristic behaviour of hooligans, rough horseplay; **hoo·liganize** *v. intr.*, to act as a hooligan.

These derivatives, with the exception of *hooliganism*, are only occasional, but they are inserted here because of this additional testimony to the currency of *hooligan*.

1898 *Pall Mall Gaz.* 19 Aug. 9/3 Any unauthorized person found trespassing on the aforesaid sphere would be Hooliganised without further notice. 1899 *Ibid.* 5 Jan. 2/3 The proprietor of Lord Tennyson (in wax) says that it was a certain young man, who, with others, when called upon to desist, Hooliganed about and threw the late Laureate's head at him. *Ibid.* 1 Feb. 2/3 Larking about in the usual hooliganesque way. 1902 *Daily Chron.* 26 Sept. 6/5 Stay then your hooliganic bark. 1898 *Daily Tel.* 11 Aug. 5/7 'Hooliganism', or youthful ruffianism. 1898 *Daily Graphic* 22 Aug. 14/2 The avalanche of brutality which, under the name of 'Hooliganism'..has cast such a dire slur on the social records of South London. 1900 *19th Cent.* July 90 To strike at the very roots of truancy, juvenile crime, and Hooliganism. 1955 *Times* 14 June 6/6 In a talk with journalists he denied one by one the Government's charges of hooliganism against Catholics. 1973 *Oxford Mail* 23 Jan. 8/2 The ban had been imposed on safety grounds 'following incidents on Sunday night involving hooliganism by youngsters'. 1901 *Pall Mall Mag.* Feb. 198 The Hooligan..would Hooliganise less..if in his ruffianism he risked a cut of it [sc. the whip].

hoon (hūn), *sb.*[1] *India.* Also hun. [Hindi (Skr. *hūna*).] A gold coin, the pagoda.

1807 F. BUCHANAN *Journ. Madras* II. 310 Huns, or Pagodas. 1876 J. GRANT *Hist. India* I. xxvi. 140/1 The pagoda..was called a *hoon* by the Mohammedans, and a *varaha* by the Hindoos. 1877 J. DOWSON *H. M. Elliot's Hist. India* VII. 84 Part of the two lacs of huns (pagodas), which was the stipulated amount of his annual tribute. 1962 R. A. G. CARSON *Coins* 508 Gold was struck in two denominations, the heavier hun or pagoda and the smaller fanam.

hoon (hūn), *sb.*[2] *Austral. slang.* [Origin unknown.] A lout; a rough; a crazy person, a 'clot'; a ponce.

1938 X. HERBERT *Capricornia* xxi. 309 'You flash hoon,' he went on. 'Kiddin' you're white, eh!' 1953 BAKER *Australia Speaks* v. 124 Women whose activities are organised by hoons..are called battlers. 1959 *Telegraph* (Brisbane) 5 July 8/4 Hoon, a crackpot. 1967 *Sunday Truth* (Brisbane) 9 July 32/4 Two hoons..walked up behind him. The biggest hoon ruffled up his hair and tried to put his half-smoked cigarette in the young man's hair. 1969 *Courier-Mail* (Brisbane) 8 Jan. 2/4 Advertising these locations [sc. of radar speed traps] also advertises the locations where they are not being used, thus giving the hot-rod hoons an open go.

hoon, *v.* var. HONE *v.*[2]

1837 J. HOGG *Tales by Ettrick Shepherd* II. 235, I heard a kind o' hooning sound. 1955 D. NILAND *Shiralee* 36 Macauley felt her scrabbling over the blanket, and then sitting beside his body, hooning to herself.

hoondee, hoondi, hoondy, varr. *HUNDI.

hoonoomaun, var. *HANUMAN.

hoop, *sb.*[1] Add: **1. c.** A circular ring, often with paper stretched over it, through which acrobats or performing animals leap. Also *fig.*, esp. in phr. *to go* (or *jump*) *through* (*the*) *hoop(s)*: to undergo an ordeal or trial. Similarly *to put through the hoop(s)*.

1793 in T. FROST *Circus Life* (1875) 43 Through the Hoop on Fire, fourteen feet high, by Mr. Porter and Mr. Ducrow. 1869 B. CLARKE *Cricket Bat* 8 When a rider..has been jumping through paper hoops held up at intervals round the ring. 1875 T. FROST *Circus Life* 185 All aspirants for saw-dust honours..are required to..ride bare-backed, leap over garters, balloons, banners, &c. 1914 W. W. GIBSON *Borderlands* 55 Merry Andrew, I missed a hoop This afternoon... I've not missed A hoop since I was six. I'm forty-two. 1917 WODEHOUSE *Man with Two Left Feet* 197 It was his business to make money, and, when called upon, to jump through hoops and sham dead at the bidding of his wife and daughter. 1919 'B. CABLE' *Old Contemptible* 130 Then Tommy Dodd got hold of his sergeant and 'put him through the hoop'. 1925 FRASER & GIBBONS *Soldier & Sailor Words* 120 *Hoop, through the*: up for punishment. 1926 GALSWORTHY *Silver Spoon* II. xi. Let the girl have jumped through their hoops as much as they liked. 1930 M. KENNEDY *Fool of Family* xxxii. 275 Disagreeable thoughts were going to assail him... If he had ever let them get hold of him he would never have gone through the hoop. *Ibid.* xxx. 314 'Let's hide..upstairs...' 'No use. They'll all come up to lunch at Henry VIII's bed.' 'Oh,

well. Then it's through the hoops.' 1938 H. NICOLSON *Diary* 7 Apr. (1966) 333, I come back to find that the F.A. Committee have put Paul through the hoops also, asking whether he is 'pro-Chamberlain' or 'pro-Eden'. 1943 R. CHANDLER *Lady in Lake* (1944) xxxix. 205 She got the men that way, she could make them jump through hoops. 1958 *Economist* 1 Nov. 390/2 The most controversial new legislation may still be the bill which will oblige any British government to go through the parliamentary hoop when imposing certain economic controls. 1958 'A. GILBERT' *Death against Clock* 139 He may have jumped through the hoop right away. 1969 *Times* 12 May 15/5 Their..irritations are put through nightly hoops.

2. b. A circular wooden frame in which a cheese is moulded.

[1790 W. H. MARSHALL *Rural Econ. Midl. Counties* I. 349 The cheese vats of this district are merely 'hoops' of ash, with a boarden bottom.] 1857 *Trans. Ill. Agric. Soc.* II. 181 These vats warm, scald, and work the curd ready for the hoop without being removed. 1877 *Rep. Vermont Board Agric.* IV. 54 Most of the cheese made about here was bought and shipped almost as soon as it was out of the hoops. 1951 *Oxf. Jun. Encycl.* VII. 156/1 The curd ..is 'milled' or torn into small particles, mixed with salt, and packed into a hoop or mould.

7. (Later examples.)

1857 C. KINGSLEY *Two Years Ago* II. vi. 226 She drew off a diamond hoop, and put it quietly into his hand. 1926 J. BLACK *You can't Win* xv. 199, I go in her joint and drop a hoop to one of her frowsy little brums for nine dollars. 1970 C. MAJOR *Dict. Afro-Amer. Slang* 67 *Hoop*, ring.

8. c. *U.S. Basketball.* The (rim of the) basket. Also, a goal scored by throwing the ball through the basket.

1893 in Z. Hollander *Mod. Encycl. Basketball* (1969) 8 The baskets are strong iron hoops, with braided cord netting. 1937 F. C. ALLEN *Better Basketball* ii. 29 (*heading*) Standardized basketball nomenclature... Baskets—not Buckets,..hoops, nets, or strings. 1967 *Boston* (Mass.) *Herald* 1 Apr. 17/1 Jim Small scored the first hoop of the game. 1969 Z. HOLLANDER *Mod. Encycl. Basketball* xvii. 419 He drove for the hoop and tried to make contact while making the basket.

d. A band in contrasting colour on a jockey's blouse, sleeves, or cap. So **hooped** *a.*

1898 *Dorling's List Epsom Races* 27 May 1/3 Rose hoops, rose cap..black hooped sleeves. 1961 F. C. AVIS *Sportsman's Gloss.* 228/2 *Hoops*, narrow bands of white or coloured silks going round the jockey's blouse; also known as Rings. 1970 *Accent* June 35/1 Wives try to get colours as close to those of their husbands... The Duchess of Devonshire's silks are straw with a brown hoop on the cap. *Ibid.* 35/2 (*caption*) Maroon hooped sleeves.

e. *Austral.* A jockey.

1941 BAKER *Dict. Austral. Slang* 36 *Hoop*, a jockey. 1957 'N. CULOTTA' *They're Weird Mob* (1958) v. 72 Best hoop in the country. 1963 *Sunday Mirror* (Sydney) 20 Jan. 43/1 Glamour hoop Athol Mulley in the saddle. 1967 E. MCGIRR *Hearse with Horses* ii. 33 Old Paddy was not a great deal more successful as a trainer than I was as a hoop.

13. b. hoop-ash (earlier examples); **hoop-back,** (a chair with) a hooped back; also *attrib.* in *hoop-back chair*; **hoop-pine** (examples); **hoop-skirt** (earlier and later examples); also *fig.*; **hoop-snake,** substitute for def.: a non-venomous American snake, *Farancia abacura*, popularly believed to roll like a hoop; = *horn-snake* (b) s.v. HORN *sb.* 29; (add examples); **hoop stress,** the stress in a cylinder or in a spherical shell corresponding to the hoop tension; **hoop tension,** the circumferential tension in a transverse section of a cylinder or in a spherical shell subjected to radial pressure; **hoop-wood,** add to def.: *U.S.* the black ash, *Fraxinus nigra*, or the winterberry, *Ilex lævigata*; in Jamaica, *Zygia latifolia*.

1763 [see *BUCK-EYE 1]. 1832 D. J. BROWNE *Sylva Amer.* 133 On the Ohio it is called Hoop Ash and in Kentucky, Hack Berry. 1905 P. MACQUOID *Hist. Eng. Furnit.* II. viii. 243 The ordinary hoop-back chairs..were slow in obtaining favour. 1924 MACQUOID & EDWARDS *Dict. Eng. Furnit.* I. 244 The tall back is reminiscent of early eighteenth century walnut chairs... A deliberate attempt at Oriental effect may be seen in..the shaping of the hoop-back. 1924 *Burlington Mag.* Nov. 225/1 The chair..has a splat formed of framework..the uprights continue in the top-rail so as to form a fine 'hoop-back'. 1952 [see *comb-back s.v. *COMB *sb.*[1] 9]. 1952 J. GLOAG *Short Dict. Furnit.* 308 The two main types of Windsor chair are: the comb back and the hoop back... In the latter, the back is shaped like a hoop, into which the spindles are socketed. 1884 A. NILSON *Timber Trees N.S.W.* xxv. 63 (*heading*) Araucaria..(Natural Order Coniferæ.)..A. Cunninghamii...Moreton Bay Pine; Colonial Pine; Hoop Pine...A noble tree with a pyramidal or somewhat flattened head. 1926 D. CRONIN *Timber Wolves* i. 11 Well, what wood are they using for their matches now? I'll tell you. Their substitute is Queensland hoop pine, and I ain't heard any complaints yet. 1968 K. WEST *Gallows on Sand* xv. 153, I showed her the great hoop-pines, whose seeds had been carried by birds from the mainland (of Australia). 1969 T. H. EVERETT *Living Trees of World* 251 Another Australian species in the Richmond-river-pine or hoop-pine (*A. cunninghamii*), the latter name deriving from its bark, which has horizontal cracks in encircling bands. 1857 UNDERHILL & THOMPSON *Elephant Club* 195 Lady with hoop-skirt hails the driver. 1896 *Emporia* (Kan.) *Gaz.* 13 Aug., We have found the hoop-skirt way of failure..and found an old human hoop-skirt who has failed as a business man. 1906 *Springfield Weekly Republ.* 19 July 1 Populism was a 'hoopskirt' article of statesmanship. 1916 E. POUND *Lustra* 54 You're a very depleted fashion, A hoop-skirt, a

calash. 1784 J. F. D. SMYTH *Tour in U.S.* I. 265 From the above circumstance, peculiar to themselves, they have also derived the appellation of hoop snakes. 1840 *Southern Lit. Messenger* VI. 380/2, I never believed in the existence of hoop-snakes neither, until I went out into the western country. 1937 A. H. VERRILL *Strange Reptiles* ii. 7 Thousands of otherwise intelligent and educated persons believe in the mythical 'hoop snake' which is supposed to take its tail in its mouth and roll like a hoop with incredible speed. 1956 C. H. POPE *Reptile World* 156 Tens of thousands of country people see hoop snakes rolling about like hoops, but these snakes quickly stop rolling and crawl when a herpetologist looms in sight. 1909 WEBSTER, *Hoop stress*. 1930 *Engineering* 23 May 679/2 S is the apparent hoop stress. 1966 C. C. BARNES *Power Cables* (ed. 2) xii. 179 The maximum pressure in the cable is therefore dictated by the hoop stress developed in the sheath. 1973 *Sci. Amer.* Dec. 18/1 The tensile strength must be sufficient to withstand the 'hoop stress' resulting from centrifugal forces, otherwise the [fly]wheel would fly apart. 1896 Hoop tension [see STRESS *sb.* 5 c]. 1902 *Kynoch Jrnl.* Oct.–Nov. 17/1 Since the ratio of the inner to outer radius is now 1·595, the hoop tension at the inner surface = 62·8 tons. 1950 *Jrnl. R. Aeronaut. Soc.* LIV. 133 The earliest specimens of these [parachutes]..took account of 'hoop tensions', 'pressure across the fabric', etc. 1821 J. FOWLER *Jrnl.* (1898) 21 We get out at our ushal time; at ten miles pased a point of Rocks and a Hoop wood tree on them. 1908 N. L. BRITTON *N. Amer. Trees* 622 Winterberry—*Ilex lævigata*..is also called the Smooth winterberry and Hoopwood. 1920 FAWCETT & RENDLE *Flora of Jamaica* IV. 150 (*heading*) Z[ygia] latifolia..Horse wood, hoop wood.

hooped, *a.* Add: **1.** (See also *HOOP *sb.*[1] 8 d.)

3. Rounded like a hoop.

1852 *Trans. Mich. Agric. Soc.* III. 137 The breeder will do well to seek in his animal a proper form, viz...good, hooped ribs. 1934 M. F. MCTAGGART *Handbk. Horse Owners* II. 40 The ribs should be well hooped and deep.

4. *Comb.* **hooped-back** *a.*, said of a chair with a hooped back. (Cf. *hoop-back.)

1906 P. MACQUOID *Hist. Eng. Furnit.* III. ii. 60 In the hooped-back chair..the splat begins to show signs of subdivision. 1941 *Burlington Mag.* June 187/1 Hooped-back chairs with compass seats were produced in the provinces until after Chippendale's *Director* made its appearance. 1960 H. HAYWARD *Antique Coll.* 145/1 Hooped-back chair.

hoop-ee, *int.* (Cf. HOOP *int.* and *WHOOPEE.)

1846 J. J. HOOPER *Adv. Simon Suggs* iii. 31 Hoop-ee! won't they roll over the floor.

hoop-la (hū·plā). Also **houp-la.** [f. HOOP *sb.*[1] + LA *int.*] A game in which persons throw rings on to a surface containing a number of articles, the object being to gain any of these as a prize by throwing a ring so as to encircle it completely. Also *attrib.*

[1907 *Yesterday's Shopping* (1969) 1031/1 Hoopla!.. The players endeavour to throw balls..through circles poised on the Clown's head and hands.] 1909 *London Mag.* Sept. 26 A new game: Hoop-la! 1920 *Daily Chron.* 14 Jan. 3/5 A showman placed his 'hoop-la' near the Promenade at Whitley, Northumberland, and invited the public to throw rings (at so many a penny) for prizes. 1922 A. S. M. HUTCHINSON *Happy Warrior* v. i. 279 All Maddox's smaller-fry—coker-nut shies, hoop-las, Living Mermaid. 1924 *Other Lands* Jan. 68/2 Hoop-la stalls are a great attraction. 1927 *Blackw. Mag.* Sept. 358/1 We..finally sought refuge from the mob in a 'hoopla' stall, where we tried our luck. 1935 *Punch* 18 Sept. 319/1 One also tried one's hand at clock-golf, Hoop-la! and throwing little balls into a bucket from which they instantly sprang out again. 1959 P. H. JOHNSON *Humbler Creation* xviii. 129 His wife [had] the hoop-la [stall] at a fête. 1964 R. ROSEWALL in A. Trengove *Art of Tennis* 16 Most people do things like dealing cards, and playing table tennis and hoop-la, backhanded. 1973 P. MOYES *Curious Affair of Third Dog* x. 135 You will not turn down my heartfelt appeal to you to preside over the Hoop-La stall.

hoop-la, var. *HOUP-LA *int.*

hoopless (hū·plės), *a.* [f. HOOP *sb.*[1] + -LESS.] Having no hoop.

1885 A. MUNRO *Siren Casket* 65 Two hoopless pails.

hoop-petticoat. 2. Substitute for def.: In full, *hoop-petticoat narcissus* or *daffodil*. A plant of the species *Narcissus bulbocodium* or *N. cantabricus*, so called from the shape of the yellow or white flowers. (Earlier and later examples.)

1731 P. MILLER *Gardener's Dict.* (s.v. *Narcissus*) 35 Rush-leaf'd Daffadil, with very narrow Petals, and a large tubulous Cup, commonly call'd the Hoop-Petticoat. 1790 *Curtis's Bot. Mag.* III. 88 (*heading*) Narcissus Bulbocodium. Hoop Petticoat Narcissus. 1842 J. W. LOUDON *Ladies' Compan. Flower Garden* 190/1 The genus Narcissus is a very extensive one, embracing, as it does, the Jonquils, the Polyanthus Narcissus, the little Hoop Petticoat, the Poet's Narcissus, and the Daffodils, besides numerous others. 1926 E. A. BOWLES *Handbk. Narcissus* xix. 211 It is hard to believe that the white Hoop Petticoat was ever found so far to the north as Cantabrica. 1952 C. E. L. PHILLIPS *Small Garden* xiii. 164 The 'hoop-petticoat daffodil' (*N. bulbocodium*), 6 ins., like a little bugle with virtually no perianth. 1961 P. M. SYNGE *Collins Guide to Bulbs* 232 Bulbocodium group. These are the hoop petticoats and they are nearly all subsp., varieties or forms of *N. bulbocodium*, a very large and variable sp., the exception being the early white-flowered forms known as *monophyllus* and *foliosus* which

have now been separated under the name *N. cantabricus*. 1970 C. LLOYD *Well-Tempered Garden* 137, I dug up a congested clump of the hoop petticoat daffodil.

hooray (var. HURRAH), *int.* Used in Australia and New Zealand (in Australia also **hooroo** and other variants) in the sense 'good-bye'. (Cf. *HURROO *int.* (*sb.*).)

1898 *Bulletin* (Sydney) 4 June (*red page*), In many places the salutation 'good-day' or 'good-night' is simply 'Hooray!' 1937 J. A. LEE *Civilian into Soldier* vi. 255 His companion..threw back his 'So long,' and 'Hooray', mate.' 1941 BAKER *Dict. Austral. Slang* 36 Hooroo. 1945 — *Austral. Lang.* xiv. 251 Some authentic local equivalents [of *so long*!]..are *hooray! aroo!* and *see you!*..employed for many years, especially in rural areas, to denote 'good-bye'. 1948 D. BALLANTYNE *Cunninghams* (1963) I. vi. 55 'Bye-bye, Mr. Cunningham.' 'Hooray.' Gil said. 1966 N. HILLIARD *Maori Girl* 128 'Well, hooray.' 'Hooray.' 1963 A. LUBBOCK *Austral. Roundabout* 89 Be seein' yer soon in England. Hooray! Aroo! Good on yer, Pom. 1965 *N.Z. Listener* 25 Oct. 9/4 Best of luck on the lake. Hooray! 1969 *Coast to Coast 1967–68* 74 The others have, one by one, ..lurched off—'Hooroo!' 'Seeya, mate!'

hooroosh, var. HURROOSH *sb.* and *v.*

1836 *Knickerbocker* VIII. 208 When they were all free, they began to sky-lark and kick up a hooroosh in all quarters. 1839 J. D. HOOKER *Jrnl.* in L. Huxley *Life* (1918) I. iv. 92 He used..to set up..a hooroosh. 1894 ..set up a hooroosh and come..

hoose, hooze, sb. Delete *local*. Substitute for def.: A parasitic bronchial disease of animals; = HUSK *sb.*[2] (Earlier and later examples.)

a 1722 [see HUSK *sb.*[2]]. 1922 JOYCE *Ulysses* 309 A hoose drench for coughing calves. 1932 GAIGER & DAVIES *Vet. Path. & Bact.* xxii. 440 Parasites are a frequent cause of bronchitis in animals, particularly in cattle, sheep and swine. The condition is popularly known as 'husk' or 'hoose'. 1966 W. R. WOOLDRIDGE *Farm Animals* (ed. 2) vi. 99 Husk: Hoose: Parasitic Bronchitis. This disease is caused by a heavy infestation of lambs, up to one year of age, with the thread round worm, *Dictyocaulus filaria*. Adult sheep are fairly resistant to infection.

hoosegow (hū·sgau). *U.S. slang.* [ad. S. Amer. or Mex. Sp. *juzgao* = *juzgado* tribunal:—L. *jūdicātum*, pa. pple. of *jūdicāre* to JUDGE.] Prison.

1911 *Popular Mag.* 15 May 104/1 No thanks for th' little lady savin' th' bunch of you from th' 'hoose-gow'. 1920 *Pacific Opinion* 1 Dec. 56/1 Only the keeper and the kept in the hoosegow knew it. 1927 'J. BARBICAN' *Confess. Rum-Runner* iv. 117, I have had one dose of the hoosegow, and I can tell you I don't want any more. 1937 J. WORBY *Other Half* x. 99 We soon got to the hoose gow, and..we once again frisked for weapons. 1949 WODEHOUSE *Quick Service* xix. 257 This guy Weatherby is a right guy, and he doesn't go to any hoosegow, not while I have my strength. 1955 *Reader* Autumn 4 They would cheerfully accept ten years in the hoosegow before they would again face Ole Miss. 1962 [see *IT *pron.*]. J. JOESTEN *They call it Intelligence* vi. xxxiii. 203 Singer..has already spent..much time in the hoosegow..he was..let free. 1973 D. RAMSAY *Deadly Discretion* 163 I'm not going to answer any questions... Okay. Off we go to the hoosegow.

hoosh (hūʃ), *sb. slang.* [Origin unknown.] A kind of thick soup.

1905 E. F. SCOTT Voy. 'Discovery' I. 445 The cook.. proceeded to prepare the ingredients of the hoosh, by which term the hot, thick soup that constituted the sledging meal was generally known. 1911 *Jrnl.* 29 Nov. in *Last Exped.* (1913) I. xvi. 279 They had some of Clissold's cottage pie..there..hoosh yesterday. 1919 E. H. SHACKLETON *South* xii. 239 The hoosh-pot with our precious limpets and seaweed was kicked over in the rush. 1922 *Chambers's Jrnl.* Jan. 73/2 The thick savoury 'hoosh' of pemmican and broken plasmon biscuit.

hoosh (hūʃ), *int.* An exclamation used in driving animals, etc.

1874 HARDY *Far from Madding Crowd* I. x. 131 Saying 'Hoosh!' to the cocks and hens when they go upon your seeds. 1900 *Contemp. Rev.* Oct. 512 A loud 'hoosh' from the Kaffir roused one of the Englishmen.

hoosh (hūʃ), *v. colloq.* [f. prec. word; cf. SHOO *int.* and *v.*] *trans.* To force or turn or drive (an animal, etc.) *off* (or *out*, etc.); also *intr.*, to move (rapidly). Cf. also quot. 1943.

1908 *Athenæum* 11 Apr. 450/1, I hooshed them, hooshed them all into the shed. 1928 A. A. MILNE *House at Pooh Corner* vi. 100 'Well done, Pooh... That was a good idea of ours.'.. 'Hooshing you to the bank like that.' 'Hooshing me!' said Eeyore in surprise. 'Hooshing me? You didn't think I was hooshed, did you? I dived. Pooh dropped a large stone on me, and so as not to be struck heavily on the chest, I dived and swam to the bank.' 1933 L. A. G. STRONG *Sea Wall* xvi. 285 We could hoosh the whole lot of them off of the line, and the train could go by. 1934 A. RUSSELL *Tramp-Royal in Wild Australia* iii. 27, I untied my camel, 'hooshed' it down and mounted it. 1936 A. THIRKELL *August Folly* ix. 283 Oh, she's dressing, and Aunt Palmer hooshed me out. 1939 JOYCE *Finnegans Wake* 112 Trust her to propagate the species and hoosh her fluffballs safe through din and danger! 1943 HUNT & PRINGLE *Service Slang* 39 *Hooshing*, purely an R.A.F. word, which means landing at great speed. 1956 'A. BRIDGE' *Lighthearted Quest* ii. 37 Why do you go hooshing off to find him in this completely wild-cat way?

hooshtah (hū·ʃtā), *int.* Also **hushdar.** [Echoic.] A shout of encouragement, etc., to a camel. Hence as *sb.* Also occas. as *sb.*, a camel.

1903 R. BEDFORD *True Eyes & Whirlwind* xxxvi. 201 Their string of five camels.. were water-swollen, so that they looked like five great footballs set up on sticks. One by one Quinn and Lawler 'hooshta'ed' them to the ground. 1906 *Daily Chron.* 8 Jan. 6/7 So the camel was 'hooshtahed' down and strapped. 1911 C. E. W. BEAN *'Dreadnought' of Darling* xxiv. 211 He made a noise, probably 'Hooshta', as he went—'Hooshta' seems to be the only noise an Afghan ever does make to a camel. 1936 F. CLUNE *Roaming round Darling* xxiv. 258, I remembered ever in Cairo the cameliers used to shout 'Hooshta'.. 'hooshta' to their camels... Jack the Ripper..returned and whittered and hooshtered to Galahad, who stopped, and gently genuflected himself while the sore-headed Poet dismounted. 1942 C. BARRETT *On Wallaby* iv. 62 Many a time we've been out in the Sinai desert guarding a grazing mob of 'hooshtas'. 1964 H. M. BARKER *Camels & Outback* 74 All the time the standard oriental 'hushdar' was hissed as if it were spelt with many 's's'.

Hoosier (hū·ʒiəɹ). *U.S.* Also **hoosher.** [Origin unknown.] **a.** A nickname for a native or inhabitant of the state of Indiana.

1826 in *Chicago Tribune* (1949) 1 June 20/3 The Indiana hoosiers that came out last fall is settled from 2 ½ miles of us. 1834 *Knickerbocker* III. 441 They smiled knowingly, and said it was among the 'hoosiers' of Indiana. 1835 J. H. INGRAHAM *South-West* II. 119, I hunted several day.. for rare birds..to..wander..by. 1848 *Tribune* ..Hoosiers' and 'Suckers', 1866 *Harper's Mag.* ..These Hoosiers and Suckers are really better than the snivelling opposition. 1885 *Outing* (U.S.) Nov. 153/2 Oh, say, papa.. Did you notice that young Hoosier and his bride who sat opposite me at breakfast? 1947 *Harper's Mag.* Jan. 67/1 Other Hoosiers ridicule them as hillbillies. 1958 *Economist* 1 Nov. 417/2 Indiana, whose inhabitants go by the obscure name of Hoosiers, is a deeply conservative state.

b. An inexperienced, awkward, or unsophisticated person.

1846 J. GREGG *Diary & Lett.* 22 Aug. (1941) I. 222 Old King is one of the most perfect samples of a Hoosier Texan I have met with. Fat, chubby, ignorant, and loquacious as Sancho Panza.. we could believe nothing better as a Hoosier. 1857 E. L. GODKIN in R. Ogden *Life E. L. Godkin* I. 157 The mere 'cracker' or 'hoosier', as the poor [southern] whites are termed. 1874 J. W. LONG *Amer. Wild-Fowl Shooting* viii. 144 'Greenhorns' and 'hoosiers', as the regular hunters call such fellows. 1926 *Amer. Mercury* Jan. 64/2 The word hoosier is applied to any one who is incompetent. 1955 *Publ. Amer. Dial. Soc.* xxiv. 174 This hoosier grift is the crowd at a country fair.

c. *attrib.* Of or belonging to Indiana. **Hoosier state,** Indiana.

1839 J. PLUMBE *Sk. Iowa* 46 (Th.), The Hoosier State has reason to be proud of the magnitude and value of its waters. 1845 *Knickerbocker* XXV. 374 Three hundred miles of Hoosier mud. 1850 BARTLETT *Dict. Amer.*, *Hoosier cake*, a Western name for a sort of coarse gingerbread, which, say the Kentuckians, is the best bait to catch a hoosier with. He helped bring fond of it. 1871 E. EGGLESTON *Hoosier Schoolmaster* (1872) Pref., It has been my lot to..dwell from my earliest childhood.. 1878 J. H. BEADLE *Western Wilds* i. 18 No grammar of the 'Hoosier' language has ever been published. 1907 *Chicago Evening Post* 4 May 5 A rose festival will be given by the Indiana Society of Chicago, comprising hoosier business men of the city. 1916 *Daily Colonist* (Victoria, B.C.) 23 July 18/1 He [sc. James Whitcomb Riley] made us what nobody else has been able to do.. but tell a print like feathers tipped with toes in the Hoosier loam. 1972 *Christian Science Monitor* 28 Sept. 16/3 Some folkway survive there.

Hence **Hoo-sierism,** a peculiarity of Indiana, esp. in speech.

1843 'R. CARLTON' *New Purchase* 63 Thus the cabin lady kept on doing up her small stock of Hoosierisms. 1878 J. H. BEADLE *Western Wilds* i. 18 The native of Indiana..his speech drop some of his 'Hoosierisms'.

hoot, *v.* Add: **1. c.** To laugh. *colloq.*

1926 T. E. LAWRENCE *Seven Pillars* (1935) x. cxxii. 659 If Nasir only laughed! I cackled out like a chicken, with the wild laughter of strain... I hooted out again. 1928 S. VINES *Humours Unreconciled* xv. 201 The first time I came across it, 'Shakespeare has no bloody relation with Schiller', I just hooted. 1930 N. MAILER *Adv. for Myself* (1961) 168 The others hoot, they giggle, they are weak from the combination of their own remarks and the action of the plot. 1966 *New Yorker* 18 June 37/2 She'd mention him tragically, then hoot with laughter.

4. Also, to emit the sound of a motor-horn (said of the horn, the motor vehicle, or the driver). Also *trans.*

1912 BEERBOHM in *Seven Men* (1919) 129 Our car neither slackened nor hooted. 1927 [see *HONK *sb.* b]. 1957 A. CLARKE *Later Poems* (1961) 58 Badge and body huged guide Your cars home, hooting through our silent streets. 1966 J. BETJEMAN *High & Low* 63 Who dares to come hooting at me? I only give way to a Jag.

hoot, *sb.*[1] Add: **1. b.** A sound produced mechanically by a motor-horn, factory whistle, or the like.

1904 A. B. F. YOUNG *Compl. Motorist* (ed. 2) xii. 258 You should have a connection from the exhaust pipe led into a small reservoir and thence into the horn, so that on turning a tap a prolonged hoot will be emitted. 1927 *Scots Observer* 14 May 17/1 An imperative horn hoot

made him turn his head. 1963 H. GARNER in R. Weaver *Canad. Short Stories* (1968) 2nd Ser. 52 A long hoarse hoot of the factory whistle announced the fourth shift.

c. A laugh; a cause of laughter, a joke, a very amusing situation. *colloq.*

1942 BERREY & VAN DEN BARK *Amer. Thes. Slang* §181/4 Something humorous,..*hoot*. *Ibid.* §408/1 Humorist; amusing person,..*hoot*. 1966 *Punch* 17 Dec. 990 All the chaps shuck their clobs in a heap, and the wives have to pick a dish and go off with the owner; it's going to be an absolute hoot! 1970 'D. HALLIDAY' *Dolly & Cookie Bird* iii. 23 Jenny can imitate anybody. So can I... We'd have the whole form in hoots. 1971 *Guardian* 23 Sept. 12/1 It's a little quaint (I'm a lot of a hoot.) Dews would say) to hear him mention Peter Brook first. 1973 'D. HALLIDAY' *Dolly & Starry Bird* xiii. 188, I started talking shop, a hoot for Jacko, if he had heard me.

hoot (hūt), *sb.*[2] *colloq.* (orig. *U.S.*). [Perhaps the same as HOOT *sb.*[1] or *int.* Cf. *HOOTER[2].] The smallest amount or particle; a whit or atom. Chiefly with negative and in phrases *to give* (*care, matter*) *two hoots* (*a hoot*).

1878 J. H. BEADLE *Western Wilds* xxxviii. 615, I got onto my reaper and banged down every bolt of it before Monday night. 1923 R. D. PAINE *Comr. Rolling Ocean* xii. 214, I am glad of that even if he did tell me that as a supercargo I wasn't worth a hoot in hades. 1925 N. VENNER *Imperfect Imposter* i. v, I can't see this place gives a hoot whether I'm here or not. 1925 FRASER & GIBBONS *Soldier & Sailor Words* 120 I don't care two hoots in hell. 1926 A. P. HERBERT *She-Shanties* 56 What do I care if I don't matter two hoots how much Oxford is filed. 1926 T. E. LAWRENCE *Seven Pillars* (1935) vi. lxxx. 447 They had been there..and didn't care three tanker's hoots..for any feelings. 1943 K. TENNANT *Ride on Stranger* xix. 214, I don't care that it matters two hoots in hell if you don't function. 1947 O. SITWELL *Novels of G. Meredith* 4 The human being who is not worth a tinker's cuss,—or, in a more elegant simile, two hoots—does not exist. 1957 A. GRIMBLE *Return to Islands* vi. 78 Not that they gave a hoot for what I might say. 1961 V. NABOKOV *Gift* iv. 235 He most definitely did not give a hoot for the opinions of specialists. 1966 *Listener* 27 Oct. 613/1 Winston Churchill was idiosyncratic in that he did not care a hoot about being thought eccentric.

hoot (hūt), *sb.*[3] *N.Z. slang.* Also **9 hootoo, hout, hutu,** etc. [ad. Maori *utu* UTU.] Money paid as recompense; (as a generic term) money.

1820 J. BUTLER in Barton *Earliest N.Z.* (1927) iii. 66 He and his people were well robbed [hoyle for this] payment. 1828 W. HORTON *N.Z.* 1/3 He then seized one of our axes..saying that should be the hutu or payment for what he had done. 1859 (p. 33) L. CRAIK *New Zealanders* x. 12 What he now wanted, he said, was *hootoo*, or payment. 1845 E. MARKHAM *N.Z. or Recollections of it* (MS.) (1963) 77 The Cabin boy counted 8 Teeth marks on the Arab Revolt. 1927 *Observer* 9 Oct. 13 It doesn't matter two hoots how much Oxford is filed. 1939 JOYCE *Finnegans Wake* 316 It was the Captain's or Mr. Baxter's and the Hoot or Compensation money was a bag of Tobacco. 1842 *N.Z. Jrnl.* LXI. 117/2 Ask them (sc. Maoris) what is the *hoot* or price. 1879 J. BARR *Old Identities* xxxiii. 339 The land that's waste they'll parcel out..And sell't to ilk that's got the hoot. 1947 *Citrons*, *N.Z.E.F.* 5 Sept. 23/1 Pig Island N.C.O.'s only go for the extra couple of bob a day..the hoot is all they're chasing. 1963 'MIXER' *Transport Workers' Song Bk.* 5 He gets his 'hoot', forgets its dough. 1958 'E. HEDERSEN' *Spleticrave* 24 174 On the construction you could make a pot of hoot in no time. 1953 *Landfall* VII. 130 Put on a quid for me, for a place—I've got the hoot. 1961 B. CRUMP *Hang on a Minute* 11 14 Reckon we ought to have something to aim at, like getting a bit of hoot together to buy a little farm or a place to live or something. 1967 N. GILES *Death & Mr Prettyman* ii. 17, I got the idea of starting a chain of these places.. for blokes without much hoot and wanting a clean bed. 1970 *N.Z. Listener* 30 Jan. 12/1 'Hoot?' I said. I hadn't heard that word for money in years. I suppose in an isolated cut-off place..slang would ossify.

hootch, var. *HOOCH, *HOOCHIE.

hootchy-kootchy (hū·tʃi,kū·tʃi), *sb.* and *a.* Also **hooche(-y)-coochie, hootchie-kootchie, hootchy-kootch,** etc. [Origin unascertained.] **A.** *sb.* A kind of erotic dance. **B.** *attrib.* or *a.*, also as *adj.*, indecent, 'suggestive'.

[1890 B. HALL *Turnover Club* vii. 75, I have been told that one night 'Hoochy-Coochy' Rice, the minstrel man—they always call Billy 'Hoochy-Coochy' because he invariably says that whenever he comes on stage—entered [Charlie] Hoyt's room..and stole a new song.] 1898 F. P. DUNNE *Mr. Dooley in Peace & War* (1899) 28 He's seen th' hoochy-koochy an' th' Pammer House barber shop. *Ibid.* 29 Th' hootchy-kootchy girls dancin' before him. 1901 *Everybody's Mag.* Oct. 437/2 The Doctor was too professional to relish the hootchie coochie dance. 1904 [see *BARR *n.*[2] 2 b]. 1925 *Manch. Guardian Weekly* Aug. 103/4 That hootchy-kootchy sort of intonation. 1931 'D. STIFF' *Milk & Honey Route* xiii. 145 Enlivening them with the vitality of a hoochy-coochy dancer. 1934 F. SCOTT FITZGERALD *Tender is Night* xv. 273 There was the 'Hootchy-Kootchy'..in the dark. 1945 *Record Changer* (Fairfax, Virginia) Jan., The Chicago World Fair of 1893..gave the widest possible publicity to the new Negro dances..the cakewalk, the pasamala, the hoochie koochie, the bully dance and the bombershay. 1949 R. HARVEY *Curtain Time* v. 38 They expected a theatre man to be brassy and leering, like a sideshow barker at a hoochie-koochie tent. 1950 BLESH & JANIS *They all played Ragtime* viii. 149 A spate of exotic dances became the talk, from the hoochie-coochie to the bombashay. 1962 W. STEGNER *Wolf Willow* IV. iii. 256 A travelling group of hootchie-kootchie dancers pitched a tent in the brush and sent their impressario through town

advertising 'performances'. 1973 *Parade* 24 June 12/1 I'm trying to counteract the hootchy-kootchy aura that the dance has.

hoot(e)nanny, hootananny (hū·t(ə)næni). orig. *U.S. dial.* [Origin unknown.] **1.** A 'thingumajig'. **2.** An informal session or concert of folk music and singing. Also *transf.*

1929 *Amer. Speech* V. 151 *Hootananny*, the same as gadget. 1940 *Washington* (Seattle) *New Dealer* 25 July 4 (Advt.), The New Dealer's Midsummer Hootenanny. You Might Even Be Surprised! 1959 [see *CEILIDH]. 1962 W. SCHIRRA in *Into Orbit* 31 Don't worry about it. That's just the hootenanny valve on the watchamacallit fluttering a little. 1963 *Daily Mail* 11 Sept. 8/4 Hootenanny...is to the folk singer what a jam session is to the jazzman. 1964 *Radio Times* 5 Mar. 25/1 The Hoot'nanny Show. 1964 *Punch* 2 Sept. 348/1 A report of one of the side-shows in the psychiatrists' hootenanny held in London last week. 1964 MRS. L. B. JOHNSON *White House Diary* 13 Jan. (1970) 44, I love folk music, but the name 'Hootenanny' rather repels me. 1972 *Time* 5 June 4/2 Memorial Day: a three day national hootenanny.

hooter[1]. Add: **c.** The horn of a motor vehicle.

1908 *Lincoln Rutland & Stamford Merc.* 19 June 5 No hooter was sounded on the car when rounding this dangerous corner. 1972 M. IRVING *Mr Purpose* 10 Mike pulled up outside Joanna's house, beeped on the hooter and she appeared.

d. *slang.* The nose.

1958 F. NORMAN *Bang to Rights* 52 He held it [sc. a handkerchief] up to his face as though he was going to blow his hooter. 1972 *Times* 18 July 12/5 Derek Griffiths is a young coloured comedian with a face like crushed rubber..and a hooter to rival Cyrano de Bergerac.

hooter[2]. *U.S. colloq.* = *HOOT *sb.*[2]

1839 *Havana* (N.Y.) *Republican* 21 Aug. (Th.), Now the Grampus [sc. a vessel] stopt, and didn't buge [= budge] one hooter. 1889 *Commercial* (Cincinnati) 17 Oct., It has not harmed the Republican cause in Ohio a hooter. 1896 *Harper's Mag.* XCII. 784/1 Now I can have all I want, I don't care a hooter! 1900 E. A. DIX *Deacon Bradbury* xii, 'Do you mean that you don't know anything about the matter at all?'..'Not a hooter.'

Hoover (hū·vəɹ). Also **hoover. 1.** (With capital initial.) The proprietary name of a make of vacuum cleaner (patented in 1927). **2.** *Loosely.* (With small initial.) Any vacuum cleaner. Hence as *v. trans.*, to clean with a Hoover (or, by extension, any vacuum cleaner). Also *intr.*

1926–7 *Army & Navy Stores Catal.* 115 (Advt.), A Hoovered room..is..free from dust. 1934 *Punch* 10 Jan. 36/1 Her bodywork's smart and strikes the eye Clean-swept as though with a Hoover. 1934 S. BECKETT *More Pricks than Kicks* 67 He waddled out of the bar..into the lowly public..like a bit of dirt into a Hoover. 1939 N. STREATFEILD *Luke* 187, I was Hoovering my passage. 1940 H. G. WELLS *All Aboard for Ararat* iii. 91, I shall feel like a man trying to sell Hoover cleaners to an Arab encampment in a dust storm. 1944 M. SHARP *Cluny Brown* x. 68 Are you the one who hoovers the east corridor? 1946 M. DICKENS *Happy Prisoner* ix. 185 I've been swept out of the kitchen, and dusted out of the dining-room, and Hoovered out of the drawing-room. 1955 P. WENTWORTH *Vanishing Point* xi. 69 It was Miss Maxwell who hoovered the carpet and dusted all those innumerable ornaments. 1960 P. MORTIMER *Saturday Lunch with Brownings* 202 Louisa was vacuuming the sitting-room...'Do you mind moving, because I want to Hoover over there?' 1960 C. WILSON *Ritual in Dark* I. v. 101 The carpeting was a plain fawn colour, and looked as if it had only just been hoovered. 1971 *Engineer* 11 Nov. 66/3 How many housewives Hoover the carpet with an Electrolux?

Hooverize (hū·vəraiz), *v. U.S.* [f. the name of Herbert C. Hoover (1874–1964), food commissioner 1917–19, and President of the U.S. 1929–33 + -IZE.] *intr.* To be sparing or economical, esp. in the use of food.

1917 H. B. GROSS in *N.Y. Tribune* 13 June 10/7 It is now assured that Mr. Hoover is about to become our food regulator..and since he has..exhorted the public to exercise the utmost economy in the use of foodstuffs.. I suggest that 'to Hooverize' be universally adopted as expressing the assistance every one of us..can render in that direction. 1919 W. T. GRENFELL *Labrador Doctor* (1920) ii. 20 To make sure that there were no truants, all hands were forced to 'Hooverize'. 1932 *Blue Valley Farmer* (Okla. City) 7 Jan. 5/6 Once before he made us Hooverize When Wilson had his war.

Hence **Hoo·verizing** *vbl. sb.*

1918 *Lit. Digest* 12 Jan. 14/2 Hooverizing is commonly regarded as something new. 1918 *Harper's Bazaar* Mar. 62/2 The butter-colored straw with its horn of abundance, overflowing with many colored fruits.., is doubly significant to-day with our enforced Hooverizing.

Hooverville (hū·vəɹvil). *U.S.* [f. the name of Herbert C. *Hoover* (see prec. word) + *-ville* terminal element in many place-names.] A temporary shanty town.

The reference is to the temporary accommodation provided for unemployed workers in the economic depression of the early 1930s.

1933 *New Republic* 24 May 40/1 Hoovervilles are in a separate nation, with separate codes. 1939 J. STEINBECK *Grapes of Wrath* xix. 319 There was a Hooverville on the edge of every town. 1946 V. LINCOLN in *55 Short Stories from New Yorker* (1952) 36, I found White Creek Row. It was the town's Hooverville..a tragic, shocking, sordid shanty town. 1949 *Sat. Rev.* 6 Aug. 116 They called them 'Hoovervilles'. Evicted families lived in tin-and-cardboard shacks. 1952 *Economist* 6 Sept. 554/2 A tragic island of unemployment, a new, 1952 Hooverville. 1973 J. JONES *Touch of Danger* xviii. 108 It was all like some weird...Hooverville. They were cooking their suppers over the open fires.

hop, *sb.*[1] Add: **1. b.** Usu. in *pl.* Beer. Also (as *hop*) in *Comb.* Chiefly *Austral.* and *N.Z. slang.*

1929 W. R. BURNETT *Little Caesar* vi. iv. 218 That dame's full of hop. 1930 *Bulletin* (Sydney) 1 Jan. 11/4 The proprietor provided a beer party, and the riot that arose out of the hop-drinking led to the school's first raid. 1940 F. SARGESON *Man & Wife* 24 Before Bill came back half a dozen Maoris had shouted her, and each time she had less than half a glass of beer... She was keen enough on the hops he said, but she was like that.

3*. A narcotic drug; *spec.* opium.

1887 *Lantern* (New Orleans) 14 May 4/2 As long as a smoker can obtain his 'hop'. 1903 ADE *In Babel* 110 Me settin' around on my shoulder-blades lookin' like one o' these bamboo boys full o' hop. 1911 C. B. CHRYSLER *White Slavery* xi. 89 When a 'fiend' is full of 'hop' he is cunning as the devil. 1916 [see *COOK *v.*[1] 2 d]. 1924 G. BRONSON-HOWARD *Devil's Chaplain* vi. 97 It was he who controlled the available supply of 'hop'. 1933 C. DE LENOIR *Hundredth Man* iv. 63 'Sure,' I replied, 'but what are you going to do when you can't get a card of "hop" for love or money?' 1955 *U.S. Senate Hearings* (1956) VIII. 4161 Opium in the underworld is referred to [as]..'hop'.

4. a. (sense 3*) *hop-dream, -pipe.*

1896 H. M. BLOSSOM *Checkers* viii. 169 Half the time I think that I must be asleep, and trying to 'cash a hop-dream'. 1931 E. LINKLATER *Juan in Amer.* v. 401, I listened like I was in a hop-dream. 1934 J. M. CAIN *Postman always rings Twice* xiv. 163 'That paper..was still in the files, see?'..'You mean that hop dream she called a confession?' 1887 *Lantern* (New Orleans) 21 May 4/1 The rising smoke that curls up from the bowl of the 'hop' pipe. 1926 H. CRANE *Let.* 19 Jan. (1965) 259 He might as well be in elfin land with a hop pipe in his mouth.

b. hop-back (examples); **hop-bush** (later examples); **hop fiend** *slang* = *HOPHEAD 1; **hop joint** *slang*, an opium den; **hop-pad** *slang* = *hop joint; **hop toy** *slang*, a container used for smoking opium.

1888 F. FAULKNER *Theory & Practice Mod. Brewing* (ed. 2) ix. 145 The well-boiled wort filtered from hops and coagulated precipitate—and what this amounts to is seen under the false-bottom plates of hop-back—left on cooler floor..undergoes..evaporation. 1892 H. E. WRIGHT *Handy Bk. Brewers* i. 20 The boiling having lasted from one and a half to two hours..the copper is 'turned out' or 'struck', the boiling wort, hops and all, rushing out through an opened valve or tap into the *hop-back*, a vessel sometimes rectangular, sometimes circular in shape. 1937 'N. BLAKE' *There's Trouble Brewing* ii. 51 'Shall I send a man to look in the hop-back?' he added. 'The hop-back?' asked Nigel mystified. 'That's right. Where the wort drains into.' 1936 F. CLUNE *Roaming round Darling* xvii. 163 Lemon-bushes, minus the lemon, and a hop-bush, not hopping; but O'Malley said the old people used this bush for making bread. 1968 K. WEATHERLY *Roo Shooter* 108 A thick patch of hop bush. 1898 L. J. BECK *N.Y. Chinatown* xviii. 165 A hop fiend went on a weary stroll, In search of a friend who a pill could roll. 1911 C. B. CHRYSLER *White Slavery* xi. 89 Opium smokers, 'hop fiends' or 'hop heads' as they are called, are the fiercest of all the White Slavers. 1887 *Lantern* (New Orleans) 4 June 5/2 The police..raided them 'hop joints'. 1905 J. LONDON *Jacket* xvi. 212 Chinatown dumps and hop-joints. 1923 E. WALLACE *Clue of New Pin* xix. 167 'Running a philanthropic hop joint?' asked the other sarcastically. 1931 D. RUNYON *Guys & Dolls* (1932) i. 25 They find they are nothing but speak-easies, although one is a hop joint. 1935 A. J. POLLOCK *Underworld Speaks* 58/2 *Hop joint*, a place where opium is bought and smoked by addicts. 1946 MEZZROW & WOLFE *Really Blues* (1957) xiv. 245 A little..coal bin.. we cleaned out and converted into our hop-pad. 1881 *N.Y. Medical Record* 5 Nov. 512/1, I procured a full outfit for smoking [opium]..: A pipe..and a buffalo-horn box (hop toy) for holding the opium. 1887 Hop toy [see *COOK *v.*[1] 2 d]. 1926 J. BLACK *You can't Win* xvii. 238 At last the little horn container, the 'hop toy', is empty. 1955 *U.S. Senate Hearings* (1956) VIII. 4162 Hop toy—container for smoking opium.

hop, *sb.*[2] Add: **1. c.** *to catch* (or *take*) *on the hop*: to take unawares, to surprise, to catch in the act.

1868 *Broadside Ballad, The Chickaleary Cove* (Farmer), For to catch me on the hop..You must wake up very early in the morning. 1872 R. D. BLACKMORE *Maid of Sker* I. xxv. 301 He caught me on the hop; at a moment of rumours and serious warnings. 1887 *Brit. Med. Jrnl.* 21 May 1103/1 The attendants taking him, as it were, 'on the hop'. 1927 R. A. FREEMAN *Certain Dr. Thorndyke* I. xi, The police..caught him fairly on the hop with all the stolen property in his possession. 1947 'N. SHUTE' *Chequer Board* iii. 68 But when you catch them on the hop, then you got to be plenty tough. 1959 *New Statesman* 14 Nov. 654/2 Some months ago our second child caught us on the hop and Jean had him here in this two roomed house. 1973 A. BEHREND *Samarai Affair* viii. 83 Ships are wayward things, and she may have taken a sudden turn which caught Gosling on the hop.

d. That distance which can be or is traversed in an aircraft or motor vehicle at one stretch; one stage of a long-distance journey.

1909 *Flight* 3 July 398/1 M. Breguet has a biplane there and has made one or two short 'hops'. 1927 *Daily News* 7 June 7/1 By flying from New York to Eisleben..in one hop. 1931 D. RUNYON *Guys & Dolls* (1932) x. 214 Finally after an extra long hop in an automobile we come to the outskirts of a..little burg. 1954 A. HUXLEY *Let.* 9 May (1969) 707 We leave for Cairo the day after tomorrow.. then proceed to Beirut, Damascus and Istambul—short hops by air. 1958 *Observer* 10 Aug. 5/8 Companies.. operate their jets on a number of economically valuable short hops. 1968 K. WEATHERLY *Roo Shooter* 69 They had about three hundred miles to go, and because of the road conditions they decided to do it in two hops.

e. *Radio.* A transmission path from one point on the earth to another that involves a single reflection from some region of the atmosphere.

1939 *Proc. Inst. Radio Engin.* XXVII. 640/2 The great-circle multiple-hop mode of propagation usually observed during all-daylight-path periods with 18-megacycle signals. 1966 *Electronics* 17 Oct. 137 This 178-mile tropo hop will connect Flyingsdales [*sic*] Moor, a ballistic missile early warning radar site..to Martlesham Heath.

f. *to go on the hop, to play the hop*: to play truant. *slang.* (Cf. *HOP *v.*[1] 6 a.)

1959 I. & P. OPIE *Lore & Lang. Schoolch.* xvii. 372 A truant may also..be said to be..'hopping it' (in Rochdale 'going on the hop'). 1968 L. BERG *Risinghill* 15 Another boy said: I got the cane for playing the hop... Playing the hop, and fighting.

g. *long hop* (Cricket): see LONG *a.*[1] 18 d.

2. (Later examples.)

1892 KIPLING *Barrack-r. Ballads* 64 To dance with blowzy housemaids at the regimental hops. 1948 J. BETJEMAN *Sel. Poems* 115 You going to the Hanks's hop to-night? 1970 D. M. DAVIN *Not Here, Not Now* II. iii. 65 What about coming to the Arts Faculty bob hop on Saturday?

b. *on the hop*: on the go, with no chance to relax, busy, active; enjoying oneself.

1863 T. THOMPSON in E. Corvan et al. *Choice Collection Tyneside Songs* 129 Wiv some varry canny chiels, All on the hop and murry. 1892 E. J. MILLIKEN *'Arry Ballads* 22 A fierce-looking party, all elbows, was likeways A deal on the 'op. 1908 G. H. LORIMER *J. Spurlock* iv. 78 I'd been on the hop ever since morning, for being in love with Anita was a strenuous calling. 1923 J. MANCHON *Le Slang* 158 *To be on the hop*, être en bombe. 1952 A. HUXLEY *Let.* 29 Sept. (1969) 652 C is kept on the hop all the time, fetching, carrying, shopping etc.

3. a. Also *spec.*, as an athletic event. Also *hop, step, and long jump*. Hence *hop, step, and jumper*. (Further examples.)

1760 STERNE *Tr. Shandy* (ed. 3) I. xii. 48 Yorick.. would as often answer with a pshaw!—and if the subject was started in the fields—with a hop, skip, and a jump, at the end of it. 1906 *Westm. Gaz.* 30 Apr. 7/1 The preliminaries of the hurdles, standing high jump, hop-step, and long jumps will also be decided. 1908 *Times* 7 July 16/5 July 25..10 a.m.—Athletics—hop, step, and jump. *Ibid.* 25 July 10/1 The hop, step, and jump fell to the United Kingdom, when an Irishman, T. J. Ahearne, created a British record of 48 ft. 11¼ in. 1909 *Daily Chron.* 24 Sept. 9/3 The hop-step-and-long-jump handicap. 1928 *Observer* 17 June 28/4 The only hop, step, and jumper of quality. 1935 *Encycl. Sports* 349/2 *Hop-Step-and-Jump*, Athletic event, of very ancient practice in the North of the British Isles. Now included in the Olympic Games, it has become the object of much specialization. Known also as the hop, skip and jump. 1961 CHAPMAN & ABRAHAMS *Track & Field Athletics* x. 71 It is not often that a Hop, Step and Jump athlete is able to use the more efficient Hitch-kick long jump action. 1966 *Publ. Amer. Dial. Soc.* XLVI. 21 The paper had to be a *hop, skip, and jump* through the various sorts of annotations that Schele De Vere had made.

d. Used as adv. phr.

1906 *Smart Set* June 102/1 To go hop, skip and jump over the earth's surface.

Hop (hɒp), *sb.*[3] *Austral. slang.* [Abbrev. of *John Hop*: see *JOHN 1 c.] A policeman.

1923 D. H. LAWRENCE *Kangaroo* xvi. 356 It's our boys who've got things in hand. And handed them over to the Hops. 1933 *Bulletin* (Sydney) 8 Feb. 12 The Hops were taking the shattered body out of the water. 1959 BAKER *Drum* II. 118 *Hop*, a policeman.

hop, *v.*[1] Add: **4. c.** To jump on to (a moving vehicle); to obtain (a ride, a lift) in this way; to catch (a train, etc.). *colloq.* (orig. *U.S.*).

1909 W. STEVENS *Let.* 21 Jan. (1967) 125, I used to 'hop' coal-trains and ride up the Lebanon Valley. 1918 in F. A. Pottle *Stretchers* (1930) 214 The other day, I hopped a truck and went 'to the front'. 1929 *Lit. Digest* 30 Nov. 30/2 Boys are predominantly the ones who 'hop' rides on trucks, trains and other vehicles. 1935 M. M. ATWATER *Murder in Midsummer* xiii. 117 Before midnight he intended to hop the twelve-twenty out. 1940 J. CARY *Charley is My Darling* xlii. 167, I hopped a lorry once with Su, half-way to Twyport. 1967 C. O. SKINNER *Madame Sarah* vi. 119 She and some friends hopped a train for Liverpool and drove to the Cross Zoo.

6. a. *to hop the wag*: to play truant. *slang.*

1861 MAYHEW *London Labour* III. 113, I used to hop the wag from school. 1903 J. LONDON *People of Abyss* xxiii. 280 The boy told a certain bishop, 'At ten we 'ops the wag...' Which is to say, at ten they play truant. 1959 *Guardian* 24 Oct. 4/5 Episodic truancy during the last year at school—known as 'hopping the wag'. 1964 M. TODD *Ever Such a Nice Lady* iii. 30 The two of them had 'hopped the wag' from school one afternoon.

c. *to hop it*: to be off, go away quickly.

1914 W. OWEN *Let.* 24 Aug. (1967) 279, I should hop it, immejit. 1915 *Scotsman* 13 Jan. 7/3 The Zeppelin kept a few miles in the rear of us, and finally hopped it. 1916 'BOYD CABLE' *Action Front* 186 'Are we going to stick it here?' said one. 'Didn't the sergeant say something about 'opping it?' 1924 M. NEWMAN *Consummation* IV. xviii. 216 J. H. hopped it for all he was worth. A perfect tornado of bombs pursued him. 1934 T. S. ELIOT *Rock* i. 12 The commission bloke on the door looks at us and says: ''op it!'

d. *to hop in* (*out*): to get into (out of) a car, etc.

1914 KIPLING *Diversity of Creatures* (1917) 388 Oh, hop in and drive...We want that beer! 1933 M. DE LA ROCHE *Master of Jalna* viii. 89 Why don't you hop in,..and go with them? 1955 M. GILBERT *Sky High* xiv. 196 Hop out, Rupert. 1963 B. S. JOHNSON *Travelling People* i. 23 I'm making for Aberfyllin,..but I think I can take you about thirty miles along this road... Hop in. 1972 D. DEVINE *Three Green Bottles* 8 A car had pulled up just down the road... 'Hop in. I'll take you.'

e. With *off*. To be off, depart; *spec.*, to start on a 'hop' (*HOP *sb.*[2] 1 d) in an aircraft.

1922 C. E. MONTAGUE *Disenchantment* ii. 18 What the 'ell did you ever come to me for? 'Op off! Out of it! 1926 —— *Rough Justice* viii. 113 Molly and Auberon suddenly felt the breathless stillness of the place infringed by a low, earnest voice on the shore near them: 'It's 'im! 'Op orf!' It was the voice of an aged ferryman they knew. 1926 *Daily Colonist* (Victoria, B.C.) 7 July 5/5 The detailed story of Lieutenant Reece reveals the fact that on Friday morning..he had hopped off from Digby Island for Naden Harbor. The plane was heavily loaded, and it was not without some difficulty that he got away. 1930 *Morning Post* 5 Aug. 9/2 Given favourable weather, he will hop off for England to-morrow, his first stage to Bima..taking him across the Timor Sea. 1934 W. STEVENS *Let.* 12 Feb. (1967) 267 If all goes well, I shall hop off for Florida in a day or two.

f. Colloq. phr. *to hop into bed* (*with*): to have (casual) sexual intercourse (with). Cf. *BED *sb.* 6 c.

1951 E. COXHEAD *One Green Bottle* viii. 229 Hopping quick into bed would be all you'd think of. 1968 'J. WELCOME' *Hell is where you find It* iii. 44 His features were quite strikingly handsome... You'd think every woman he met would have only one thought and that was to hop into bed with him. 1971 C. WHITMAN *Death Suspended* i. 22 Duncan wouldn't waste too much time on her unless she was willing to hop into bed.

7. hop-off *Aeronaut.*, the take-off of an aircraft; **hop-over** *Army slang*, an assault; hence **hop over** *v.*, to attack, to go 'over the top'.

1926 *Daily Colonist* (Victoria, B.C.) 13 Jan. 1/7 The expedition planes will be..flown..to Point Barrow, where the hop-off for the Polar flight will be made. 1927 S. BENT *Ballyhoo* i. 36 To the mere hop-off by Lindbergh the *New York Times* gave three first page eight-column streamers. 1918 H. MATTHEWS in Murdoch & Drake-Brockman *Austral. Short Stories* (1951) 243 They didn't have the dash of the Australians in the hop-over. 1929 *Papers Mich. Acad. Sci., Arts & Lett.* X. 299 *Hop over*, Australian for 'go over the top'. 1933 PARTRIDGE *Words, Words, Words!* III. 194 In the hop-over, many hoped for and some got a wound sufficiently serious to cause them to be sent 'home'.

hop-dog. Add: **1.** Also a cutting tool used in hop-gardens.

1880 [in Dict.]. 1914 BLUNDEN *Waggoner* (1920) 50 Sharpened hopdog, at whose blow The stubborn cluster drops. *Ibid.* 70 Hopdog, long-handled curved knife for hop-gardens.

2. (Earlier and later examples.)

1872 C. M. YONGE *P's & Q's* iv. 30 The beautiful cocoon which the hopdog was spinning—a delicate apple-green fellow, with white tooth-brush tufts down his back, black velvet slashings..and a rose-coloured feather in his tail. 1961 R. SOUTH *Moths Brit. Is.* I. 117 As it is, or was previous to the modern 'washing', common in hop gardens at picking time, it was christened the 'hop dog'. 1968 P. JENNINGS *Living Village* 100 There are still people who use expressions such as.. *hot cat* (hairy caterpillar), *hop dog* (green caterpillar).

hope, *sb.*[1] Add: **4. a.** Freq. in negative in phr. *not a hope* (*in hell*). Also used ironically for: an expectation which has little or no chance of being fulfilled; esp. in ints., usu. expressing resignation, *some hope(s)!, what a hope!*

1899 R. WHITEING *No. 5 John St.* xxix. 297 What a hope for a night like this! 1915 F. PALMER *My Year of War* 231 'What hopes!' was the current phrase I heard among the men in these trenches. 1923 [see *HELL *sb.* 10 b]. 1929 J. B. PRIESTLEY *Good Companions* I. iv. 115 'Well, lads, wot's it yer want?' demanded their hostess. ''Cos if it's steaks and chips and feather beds, you've got a bloody hope.' *Ibid.* II. i. 249 If there was enough money behind to rent His Majesty's Theatre, it could go on better still. It amounted to *that*. 'What a hope!' she concluded bitterly. 1933 A. G. MACDONELL *England, their England* xvi. 280 Not a hope!..The dailies have gone to press ages ago. 1936 [see *EARTHLY *a.* 1 c]. 1940 'G. ORWELL' *Diary* 20 June in *Coll. Ess.* (1968) II. lvii. 352 There is a move on foot to get our police records..at Scotland Yard destroyed. Some hope! The police are the very people who would go over to Hitler. 1948 C. DAY LEWIS *Otterbury Incident* v. 53 'Well, you'd better start giving back the money..,' jeered Tuppy. 'What a hope!' 1959 S. GIBBONS *Pink Front Door* xvii. 206 Not a hope..not a single bloody ghost of a hope in hell. 1966 'K. NICHOLSON' *Hook, Line & Sinker* v. 63 He..wants to put on a good show for her. What a hope. 1966 F. HOYLE *Oct. First* vi. 60 I've given them the idea I might come up with some explanation... Some hopes. 1967 P. MOYES *Murder Fantastical* xv. 229 'His book is probably in the Lucky Dip.'.. Maud made a face. 'Some hope of finding it in that case,' she said. 1969 J. ASHFORD *Prisoner at Bar* iv. 33. I told Mrs. Green we hadn't a hope in hell, but she said it wasn't the money, it was the principle. 1971 C. EGLETON *Last Post for Partisan* xvii. 176 'Make sure you get the right mix of weapons and explosives.' 'You've got a hope.'

5. hope chest chiefly *U.S.*, a chest or box in which a young woman hopefully collects articles towards a home of her own in the event of her marriage. (Cf. *bottom drawer* (*BOTTOM *sb.* 19).)

1911 G. S. PORTER *Harvester* xx. 504 It was a big, burl-maple box, designed after the hope chests that he saw advertised in magazines. 1922 M. B. HOUSTON *Witch Man* vii. 80 The bedspread that three years before she had laid unfinished in Kaid's hope chest. 1959 'J. R. MACDONALD' *Galton Case* (1960) xi. 88 A metal box about the size of a hope chest. 1960 *New Left Rev.* Nov.–Dec. 12/2 The bride..had a good six patents dealing with biochemistry in her hope chest. 1973 *Welcomat* (Philadelphia) 10 Oct. 12 Liza Minelli has just bought her third wedding dress and packed it away along with the others in her hope chest.

hope, *v.* Add: **3. d.** Phr. *to hope against hope* [after Rom. iv. 18]: to hope where there are no reasonable grounds for doing so; to hope very much. Hence **hope-against-hope** *sb.*

1813 J. MONTGOMERY *World before Flood* 90 Hope against hope, and ask till ye receive. 1915 W. S. MAUGHAM *Of Human Bondage* lxxviii. 403 He mentioned the place and time at which they were to meet, and hoping against hope kept the appointment. 1955 G. GREENE *Quiet American* II. i. 93, I had hoped against hope that he would have gone before she returned. 1963 V. NABOKOV *Gift* i. 69 Fyodor still hoped against hope that this was a metaphysical paradox and not a traitorous *lapsus*. 1968 W. SANSOM *Grand Tour Today* ix. 181 Matisse's wonderful chapel of stained light..has nothing to do with the usual hope-against-hope that modern concrete building will 'harmonise' with older surroundings.

hoped, *ppl. a.* **1. b.** (Later examples.)

1947 *Sci. News* IV. 37 The patience in collecting and sifting evidence that goes on steadily from the moment a body is found to the hoped-for sequel when a suspect is charged. 1967 *Listener* 30 Mar. 426/1 They gave the hoped-for succession of industries from 40,000 to 9,000 years ago.

hopeful. 2. b. *sb.* (Later examples.)

1860 G. H. LEWES *Let.* 26 Sept. in *Geo. Eliot's Lett.* (1954) III. 349 My hopeful and his parent will appear at the 'Bedford' on Monday morning. 1899 F. J. CROWEST *Beethoven* 41 He committed this hopeful, only now some nineteen years of age, to the care of an old lawyer friend. 1957 W. THIELENS in R. K. Merton *Student-Physician* 133 This difference is more than partially due to *anticipation* by medical school hopefuls of the more demanding entrance requirements. 1972 *Daily Tel.* (Colour Suppl.) 27 Oct. 9/2 A dozen new applications arrive each day and are filed away with all the other hopefuls anxious to establish themselves on the circuits.

hopefully, *adv.* Add: **2.** It is hoped (that); let us hope. (Cf. G. *hoffentlich* it is to be hoped.) orig. *U.S.* (Avoided by many writers.)

1932 *N.Y. Times Book Rev.* 24 Jan. 11/4 He would create an expert commission..to consist of ex-Presidents and a selected list of ex-Governors, hopefully not including Pa and Ma Ferguson. 1965 T. L. BECKER *Political Behavioralism & Mod. Jurisprudence* p. v, Hopefully, this study, generated from the friction of highly polarized viewpoints, is the first of many steps directed towards satisfying a long-standing curiosity. 1965 *New Yorker* 27 Mar. 35/1 We asked her when she expected to move into her new apartment, and she answered, 'Hopefully on Tuesday.' 1966 in N. P. Vakar *Word Count Spoken Russian* p. viii, Professor Vakar's study will prove of enormous value... It should—and, hopefully, it will—be followed by a similar analysis of..syntactic patterns. 1967 *Lebende Sprachen* XII. 5/1 Machines will hopefully enable the scientist to find quickly the information he needs. 1969 *Language* XLV. 667 Hopefully, Tucker will publish supplements to this chapter. 1970 *Daily Tel.* 12 Feb. 21 The cost of developing a new 'Dash 50' series of engines, that hopefully will power Lockheed's 'extended range' jet, is put at around £75 million. 1970 *Sci. Jrnl.* May 27/2 By the time this issue..is on the bookstalls, *Apollo 13* should have completed its own trip to the Moon, hopefully with as little incident as its two predecessors. 1971 *Guardian* 13 Apr. 9/5 Prototype wooden rocking horses... Hopefully they will be available in the autumn at prices from £120.

hopeless, *a.* Add: **2.** Also in weakened use: ineffectual, inadequate, unable to stand up for oneself; incompetent, stupid.

1854 E. TWISLETON *Let.* 23 Oct. (1928) xiii. 245 Prussia is as *fainéant* as ever, and seems quite hopeless. 1922 W. S. MAUGHAM *Writer's Notebk.* (1949) 188 'You can't do a thing for people like that,' he said. 'They're hopeless.' 1932 R. LEHMANN *Invitation to Waltz* III. xiv. 240 But of course Mum's hopeless. She thinks virgins are sacred to all men. 1963 D. CORY *Hammerhead* ii. 26 'Who's this..?' 'That's Cary Grant, you are *hopeless*, Johnny.' 1967 O. NORTON *Now lying Dead* i. 3 'I'm *hopeless*,' she went on. 'I made a teapot once. It looked dinky. Only it wouldn't *pour*, don't you see.'

hophead (hɒ·phed). **1.** [f. *HOP *sb.*[1] 3* + *HEAD *sb.* 7 e.] An opium-smoker; a drug-addict. *slang* (orig. and chiefly *U.S.*).

1911 C. B. CHRYSLER *White Slavery* xi. 89 Opium smokers, 'hop fiends' or 'hop heads' as they are called, are the fiercest of all the White Slavers. 1915 G. BRONSON-HOWARD *God's Man* ii. 130, I told Beau to hunt up a skirt before, but you know these hop-heads—always putting things off. 1931 E. WALLACE *On the Spot* ii. 21 A hop-head will spill his friends' secrets to buy more hop. 1934 D. HAMMETT *Thin Man* xi. 72 'What's a junkie?' she asked. 'Hop-head.' 1947 E. E. CUMMINGS *Let.* 21 Sept. (1969) 180 Can someone imagine what any moujik coolie or hophead of any crevice of the Orient would sense, upon receiving such a gospel? 1959 J. CHRISTOPHER *Scent of White Poppies* ix. 144 Did you ever see a hophead when he's been kept short of what he wants? 1973 H. NIELSEN *Severed Key* vii. 75 I'll mail the letter to that hophead lawyer.

2. [*HOP *sb.*[1] 1 b.] A drunkard, a tippler. *N.Z. slang.*

[1942 *N.Z.E.F. Times* 17 Aug. 16/3 (*caption*) Private Harry Hophead seen leaving Shepheards after a brief visit (very).] 1948 D. BALLANTYNE *Cunninghams* (1963) II. ix. 166 It's Betty that can't hold the liquor... She's a real lily of a hophead. 1952 *Landfall* VI. 208 Among young people greetings like 'Hophead'..are accepted as flattery.

Hopi (hōu·pi). [Native name.] The name of a group of North American Indians living chiefly in north-eastern Arizona; also, a member of this tribe; their language. Also *attrib.*

1877 *Buffalo Soc. Nat. Sci. Bull.* 170 The title of '*Moquis*' has been applied to this confederacy by its enemies, and signifies the dying race. I understand that they usually speak of themselves as '*Ho-pees*' (our people). 1893 T. DONALDSON *Moqui Pueblo Indians* 13 The name which they call theselves by is Ho-pi, or Ho-pi-tuh-lei-nyu-muh, meaning 'peaceful people'. 1927 W. CATHER *Death comes for Archbishop* VII. i. 197 Last year, in May, he had been on his way to the Hopi Indians. 1927 D. H. LAWRENCE *Mornings in Mexico* 135 The Hopi country is in Arizona, next the Navajo country. *Ibid.*, The Hopis are Pueblo Indians, village Indians, so their reservation is not large. 1933 L. BLOOMFIELD *Language* iv. 72 The Shoshonean family..including..Comanche, and Hopi. 1937 R. H. LOWIE *Hist. Ethnological Theory* vii. 82 Swanton has shown..that in North America..many higher tribes, like the Hopi, are strictly matrilineal. 1952 A. HUXLEY *Let.* 12 Oct. (1969) 657 The ways I look at the world, are determined—determined in profoundly different ways according to whether I speak Chinese, Maori, Hopi, or English. 1956 D. ABERCROMBIE *Probl. & Princ.* i. 12 Hopi is an example of a language which lacks expression..for concepts of time. 1957 C. F. & F. M. VOEGELIN in *Int. Jrnl. Amer. Linguistics Mem. No. 14* (*title*) Hopi domains, a lexical approach to the problem of selection. 1959 E. TUNIS *Indians* 25/2 Even today an elderly Hopi will jog trot twenty miles to his cornfield, work all day, and jog home again. 1969 *Observer* (Colour Suppl.) 18 May 34/1 She is a Hopi, one of the less-known tribes. 1973 HOLT & MARJORAM *Math. in a Changing World* ii. 65 The difference between a noun and a verb to a Hopi Indian, Whorf discovered, is that a noun is more permanent than a verb.

‖ **hopo** (hōu·po). [From an Afr. language.] A trap for game consisting of two converging hedges in the form of the letter V, with a pit at the angle, into which the game is driven.

1866 J. LEYLAND *Adv. Far Interior S. Afr.* iii. 124 Two hedges are formed (called the hopo) a mile long, and the same distance at the extremity, gradually narrowing till it comes to the sides of the pit. 1932 R. CAMPBELL *Taurine Provence* 50 That natural death-trap or hopo, where the plain treacherously converges into a triangular precipice.

hopped, *a.* Add: **2.** [f. *HOP *sb.*[1] 3* + -ED[2].] Chiefly with *up*. Stimulated by, or under the influence of, a narcotic drug. Also *to hop up* v. phr., to stimulate with a narcotic drug. *U.S. slang.*

1924 G. C. HENDERSON *Keys to Crookdom* 408 *Hopped up*, intoxicated on opium. 1927 C. F. COE *Me—Gangster* iii. 43 They do their shooting when they are all hopped up with dope. 1927 E. HEMINGWAY *Men without Women* (1928) 194 I'm not drunk. I'm hopped to the eyes. 1957 *Listener* 6 June 912/2 It is possible these men were physically hopped up, or perhaps 'mushroomed-up' would be the better word, by consuming a European fungus. 1958 J. D. MACDONALD *Executioners* (1959) vii. 108 A hopped-up kid can be bought for less, but the job would be bungled. 1968 'J. WELCOME' *Hell is where you find It* vi. 80 Why didn't you go to his funeral? Too busy hopping yourself up in Paris? You do that for kicks, too, I suppose? 1973 *Guardian* 14 Apr. 10/3 Chuck Berry don't drink either but he gets hopped.

b. *hopped-up.* Excited, enthusiastic. *U.S. slang.*

1923 C. WITWER in *Collier's* 27 Jan. 28/4 He's especially hopped up about the piano down in the music room. 1935 WODEHOUSE *Let.* 4 Feb. in *Performing Flea* (1953) 79, I got all hopped up and felt that it wasn't possible to give 'em too much of this superb stuff. 1936 J. STEINBECK *In Dubious Battle* v. 69 You organize the stiffs and get 'em all hopped up with a bunch of bull. 1939 C. MORLEY *Kitty Foyle* xxii. 214 There were a lot of dames hopped up with culture and good grammar. 1973 'I. DRUMMOND' *Jaws of Watchdog* viii. 107 A hopped-up son with anarchist-pacifist connections.

3. *hopped-up.* Of a motor vehicle: having its engine altered to give improved performance. Also *transf.* *U.S. slang.*

1945 *Amer. Speech* XX. 226/2 *Hopped up*. Applied to a plane built for speed. Taken from the 'hopped up' automobiles of the high-school set. 1946 *Sat. Even. Post* 14 Sept. 14/2 Today a hot rod is a hopped-up, stripped-down flivver. 1954 *Amer. Speech* XXIX. 98 *Hopped up*,.. 'souped up'; said of a car with any added speed equipment. 1958 *House & Garden* Mar. 62 The American housewife..with her hopped-up central heating. **1971** *Islander* (Victoria, B.C.) 16 May 10/3 At the urge of the hopped-up motor in seconds they were tearing up Nanaimo Street.

hopper^{1}. Add: **1.** (Later example.)
1943 *N.Y. Times* 9 May II. 5/4 Listen, hoppers clot the action when the Duke or the Count start to beat it out.

c. *Baseball.* A ball which having been struck rebounds from the ground. *U.S. slang.*
Quot. 1914 appears to have a different sense.
1914 'B. L. Standish' in *Top-Notch Mag.* 30 Sept. 138/1 Courtney missed a hopper, though he almost fancied his bat lightly touched the whistling ball as it sped past. **1943** *Amer. Speech* XVIII. 104 *Baseball jargon*... Names for a grounder are *hopper*, [etc.]. **1968** [see *baseman].

2. (Further examples.)
a **1870** R. M. Chipman *Notes on Bartlett* 202 *Hopper*, a grasshopper, especially the ravaging locust called grasshopper at the West. **1933** *Bulletin* (Sydney) 5 July 21/2 One man says he counted 2000 hoppers [*sc.* kangaroos] in one paddock. **1946** *E. African Ann. 1946–7* 44/1 Youths have been out in the district beating and burning locusts... Besides these commendable efforts in destroying hoppers the Samburu have made a most useful war contribution. **1966** E. Palmer *Plains of Camdeboo* xv. 247 Our locust of the Karoo, the brown locust, *Locustana pardalina*, is well-known to us as a tiny dark hopper, later becoming black and orange.

10. c. *hopper window*; **e. hopper-boy** (examples); **hopper-dredge** or **-dredger**, a vessel combining the functions of a hopper and a dredger, being fitted with hoppers that receive the material dredged up and allow it to be discharged through the bottom at the place of deposit.
1787 in *Rep. U.S. Comm. Patents* (1850) 574 The other [device], denominated an hopper-boy, so constituted as to spread the meal over the floor of a mill to cool. **1813** *Niles' Reg.* V. Add. A. 6/2 Our Hopper-boy was an upright shaft revolving round with an arm. **1896** *Engin. Index* II. 120 The hopper dredge, 'Percy Sanderson', holding 1250 tons of debris and provided both with steel buckets and a suction pump. **1967** *Shipbuilding & Shipping Rec.* 6 July 21/1 (*heading*) 'McFarland'... A fully automated hopper dredge for the U.S. Army. **1876** Simons & Brown *Brit. Pat. 4382*, This Invention, which relates to improvements in hopper dredgers, has for its object..to effect the discharge of the load or contents of hopper dredgers on to the adjoining river or canal bank. **1969** R. Hammond *Mod. Dredging Pract.* iv. 101 Direct pumping ashore to disposal areas is another function performed by a hopper dredger. **1939** *Archit. Rev.* LXXXVI. 166 The windows have a side-hung casement and a long top-hung hopper window for night ventilation.

hopperdozer (hǫ·pəɹdōᵘzəɹ). *U.S.* [f. Hopper^{1}, perh. after *bull-dozer* (1876); see also quot. 1878.] A contrivance for catching and destroying insects, consisting essentially of an elongated pen or frame which is filled or smeared with some poisonous or glutinous substance and slowly drawn or pushed over the ground.
1877 *Pioneer Press* (St. Paul & Minneapolis) 19 May 5/4 (*heading of letter by A. B. Robbins*) A hopperdozer. *Ibid.* 22 May 1/5 The grasshoppers in this vicinity have hatched out very thickly... People are beginning to destroy them in various ways, principally with sheet iron hopper-dozers covered with coal tar. **1878** *Ann. Rep. U.S. Entom. Commission 1877* xiii. 390 The simple pan..known as the Robbins 'hopperdozer.' [*Note*] A word that came into very general use last year among farmers..and which doubtless takes its origin from doze, in reference to the toxic effect of the coal-tar on the locusts. **1904** W. C. Edgar *Story Grain Wheat* ii. 21 Large areas of wheat are saved by means of a machine termed in America the 'hopper-dozer'. **1932** *Chieftain* (Tecumseh, Nebraska) 5 May, Mechanical 'hopperdozers' have some value in the control of grasshoppers on fairly level fields of alfalfa. **1962** Metcalf & Flint *Destructive & Useful Insects* (ed. 4) ix. 471 On the whole, hopperdozers are much more expensive to operate and less efficient than poisoning.

hoppergrass. orig. and chiefly *U.S. dial.* = Grasshopper.
1829 *Virginia Lit. Museum* I. 458 *Hoppergrass.* This word is often used in the south for grasshopper. **1892** *Rep. Vermont Board Agric.* XII. 163 Alas! the 'hopper grasses' came and carried it away. **1934** R. Campbell *Broken Record* viii. 200 Spectacles make men look like ridiculous hoppergrasses. **1955** E. Pound *Classic Anthol.* I. 7 'Chkk! chkk!' hopper-grass, Nothing but grasshoppers hopping past. **1970** C. Major *Dict. Afro-Amer. Slang* 67 *Hopper grass*, grasshopper.

hopperings (hǫ·pəriŋz), *sb. pl.* [f. Hopper^{1} + -ing^{1}.] Gravel retained in the hopper in gold- or diamond-washing.
1893 *Westm. Gaz.* 29 Apr. 7/3 The yield of diamonds from the hopperings is 6½ carats per 100 loads. **1898** *Daily News* 27 Jan. 9/5 Washed 197 loads from the mine, 3,725 loads lumps and hopperings producing 126 carats.

hopping, *ppl. a.*^{1} Add: **1.** (Later example.)
1916 H. G. Wells *Mr. Britling* I. i. 24 The hopping inconsecutiveness of English conversation.

2. hopping-john (earlier, later, and W.Ind. examples); **hopping-mad,** for '*dial.* and *U.S.*' read 'orig. *dial.* and *U.S.*' (further examples); now usu. written without a hyphen.
1838 C. Gilman *Recoll. Southern Matron* xviii. 124 Before me..was an immense field of *hopping John*. [*Note.* Bacon and rice.] **1969** *Daily Tel.* 13 May 24/6 The dinner consisted of such things as collard greens, fried chicken, water melon, cornbread and 'hopping John', a dish of black-eyed peas and rice that is supposed to bring luck. **1970** M. Slater *Caribbean Cooking* 32 'Peas and Rice'..is cooked on every island, down from the Bahamas, where it is known as Hoppin' John, to the South American mainland. **1833** S. Smith *Life & Writings Major J. Downing* 139, I had a long talk with the General t'other day—he was hopping mad. **1915** Wodehouse *Psmith, Journalist* vii. 44 Dey was hoppin'-mad, de whole bunch of dem. **1922** [see *claim *v.* 2 c]. **1954** [see *editor 3 b]. **1960** *Guardian* 7 July 8/5 Would-be [telephone] subscribers get hopping mad. **1973** B. Graeme *Two & Two make Five* vii. 67 Old Sourpuss must be hopping mad.

3. = **hopping mad.*
1894 'Mark Twain' in *Century Mag.* Jan 338/1 Oh, my lan', ole Marse was jes a-hoppin'! **1942** I. Gleed *Arise to Conquer* vi. 57 David will be hopping. **1960** E. W. Hildick *Boy at Window* xi. 80, I was mad, real hopping! **1973** R. Hayes *Hungarian Game* xxxi. 185 'I wonder what they're fishing for.' 'Whales... And they got everyone hopping about it.'

hoppity. Add: (Later examples.)
1895 *Montgomery Ward Catal.* 235/1 The Game of 'Hopity'..is a game of skill... The particular feature of the game is the popular jumping move, pieces being allowed to jump over friend and foe alike to reach the opposite side of the board. **1969** E. H. Pinto *Treen* 222 Halma or hoppity was introduced about 1890.

hoppity^{2} (hǫ·piti). Also **-ety.** A fanciful extension of Hop *v.*^{1}, used adverbially or as *adj.*, often repeated or with the word *hop* to suggest a hopping or hobbling movement. Cf. *hippety.
1825, 1880, 1886 [see *hippety]. **1924** A. A. Milne *When we were very Young* 60 Christopher Robin goes Hoppity, hoppity, Hoppity, hoppity, hop. **1936** W. de la Mare *Wind blows Over* 46 William had been even more lively and hoppity than usual. **1943** F. Thompson *Candleford Green* viii. 125 A tame thrush which he had found in the fields with a broken wing and brought home..would follow him round the garden, *hoppity-hop*. **1963** *Times* 10 May 5/3 Two no-balls from Greenhough, whose run up, like Christopher Robin, goes hoppity hoppity.

Hoppus (hǫ·pŭs). Also **hoppus.** The name of Edward *Hoppus*, 18th-century English surveyor, used *attrib.* and in the possessive to designate a method of measuring the cubic content of round timber used in the British Commonwealth, and tabulated in his *Practical Measuring now made Easy* (1736) (known in later editions as *Hoppus's Tables* and *Hoppus's Measurer*); it involves multiplying the length in feet by the square of the quarter-girth in inches and dividing the result by 144; **Hoppus foot,** a recent name for the 'cubic foot' as arrived at by the Hoppus method, approximately equal to 1·27 true cubic feet.
[**1820** R. Monteath *Forester's Guide* xxvi. 173 It will contain 2 solid inches and one-third of an inch, according to the measurement of Hoppus.] **1894** A. D. Webster *Pract. Forestry* (ed. 2) xxv. 175 Timber-measuring is rather a vexed question, some following what is known as Hoppus's system, and others advocating that of Horton. **1924** A. C. Drummie *Pract. Forestry* xxvii. 244 The standard usually adopted is the quarter girth Hoppus measure system, with the customary allowance for bark. **1941** S. E. Wilson *Decimal Hoppus Tables* 6 When the Hoppus system is employed to measure cylindrical bodies the solid foot amounts to 4/π or 1·273 cubic feet... To call the unit..a cubic foot is accordingly inaccurate and misleading... To avoid confusion we call the 'solid foot' of this system a 'Hoppus Foot', defining it as the unit of solid contents according to the Hoppus system of measuring round timber. **1947** H. L. Edlin *Forestry & Woodland Life* xx. 144 The unit generally used in Britain for measuring round logs, trees, and plantations, is the customary cubic foot, or Hoppus foot. **1951** R. H. Hordern *Woodworking Ind. Managem.* viii. 123 The 144 divisor used in the Hoppus System will show approximately 27 per cent less cubic content than the tree actually measures. **1970** *Timber Trades Jrnl.* 14 Mar. 53/3 Average volume of the trees was 25·4 Hoppus ft, and the price reached was 3s 5½d per Hoppus ft. **1971** *Daily Tel.* 7 Jan. 12/5 The hoppus foot, a traditional measure for timber,..will be replaced next month by the cubic metre and metric tonne.

hoppy (hǫ·pi), *a.*^{1} [f. Hop *sb.*^{1} + -y^{1}.] **1.** Tasting or smelling of hops; beery.
1893 *Harper's Mag.* Feb. 458 Jest so it don't tas'e hoppy, I ain't pertic'lar; but from hoppy bread deliver me! **1964** *Listener* 26 Mar. 534/3 Over cool dinner, the sour hoppy breath.

2. Of, pertaining to, or characterized by drugs or drug-taking. *U.S. slang.* (Cf. *hop *sb.*^{1} 3*.)
1942 Berrey & Van den Bark *Amer. Thes. Slang* § 509/30 *Hoppy*, smelling of drugs. **1946** Mezzrow & Wolfe *Really Blues* (1957) vii. 98 Detroit is really a hoppy town—people must order their opium along with their groceries.

hoppy (hǫ·pi), *a.*^{2} *colloq.* [f. Hop *sb.*^{2} + -y^{1}.] Characterized by, or predisposed to, hopping; lively, full of movement; limping, lame.
1902 in *Eng. Dial. Dict.* III. 232 'To go hoppy' is to walk rather lame. Among the working classes, lame persons are often nicknamed 'Oppy', as 'Oppy Smith', which denotes a certain Smith who is somewhat lame. **1914** Joyce *Dubliners* 171 He had a game leg and for this his friends called him Hoppy Holohan. **1934** A. Woollcott *While Rome Burns* (1936) 42 Juventino Rosas..who once wrote a pleasant and rather hoppy waltz. **1942** Berrey & Van den Bark *Amer. Thes. Slang* § 53/15 *Rapid*; *quick*, hoppy. *Ibid.* § 131/11 *Crippled*; *lame*, hoppy. **1968** K. Weatherly *Roo Shooter* 119 It's a damned good thing these hoppy bastards [*sc.* kangaroos] can't understand me.

hoppy (hǫ·pi), *sb.*^{1} *colloq.* [f. Hop *sb.*^{2} + -y^{6}.] A lame man.
1904 'No. 1500' *Life in Sing Sing* xiii. 249/2 *Hoppy*, a cripple. **1909** S. Watson *Wops the Waif* iii. 5 Who-ay, Cully, here's Hoppy with the Rozin. **1962** J. Franklyn *Dict. Nicknames* 49 *Hoppy*, the nickname of anyone who walks with a limp.

hoppy (hǫ·pi), *sb.*^{2} *U.S. slang.* [f. Hop *sb.*^{1} + -y^{6}.] An opium addict. (Cf. *hop *sb.*^{1} 3*.)
1922 E. F. Murphy *Black Candle* II. i. 114 The Chinese here still furnish a large percentage of the 'hoppies'. **1924** G. C. Henderson *Keys to Crookdom* xxiv. 301 Even the 'hoppies' themselves look down on a user of cocaine. **1941** B. Hecht *1001 Afternoons in N.Y.* 129 A lush, a prosty, a hoppy, and a pain in the neck, say the police.

hop-sack. Add: **2.** Also *attrib.*, and in form *hopsac*.
1907 *Yesterday's Shopping* (1969) 831 Ascot Scarfs..made from..Hopsack shot silk. **1923** *Daily Mail* 5 July 1 Hopsac suiting,..in pretty Heather Mixture grounds. **1930** *Daily Express* 8 Sept. 11/5 Hopsack Tweed. **1968** *Surfer Mag.* Jan. 27/1 Can you tell me the approximate weight of a piece of hopsack cloth secured with resin to the front third of my board?

hop-toad. *U.S.* [Hop *sb.*^{2} or *v.*^{1}] A toad.
1827 *Massachusetts Spy* 28 Nov. (Th.), An inhabitant of the Middle States talks of 'hop-toads',—as if all toads were not hoppers. **1830** F. Trollope in *Dom. Manners* (1949) 428 Hop toads is little creatures what hop like a frog. **1844** Mrs. Stowe *Mayflower* (1849) 157 I'd always find him stopping to chase hoptoads, or off after chip-squirrels. **1847** *Knickerbocker* XXIX. 183, I have so often ..a tender leaning towards little pigs and hop toads. **1913** G. Stratton-Porter *Laddie* xiii. I like hop-toads, owls, and shitepokes. **1966** J. Dos Passos *Best Times* (1968) i. 12 We studied with passionate interest the egg-laying habits of the hoptoads that shrilled so loud.

Hopton wood (hǫ·ptǫn wu·d). Also **Hopton-Wood, Hoptonwood.** The name of a wood and neighbouring quarries close to Wirksworth, Derbyshire, used *attrib.* and *absol.* to designate a pale brownish- or greyish-white carboniferous limestone used for building and decorative purposes.
[**1811** J. Farey *Gen. View of Agric. & Minerals Derbyshire* I. 419 *List of..Building-Stone Quarries, or Delphs*... Hopton-Wood, in Middleton by Wirksworth.] **1888** G. H. Blagrove *Marble Decoration* 86 *Hopton Wood.* This stone deserves to rank as a marble, for it will take an excellent polish. **1911** J. Watson *Brit. & Foreign Building Stones* 117 Hopton Wood Stone has been employed for constructing many important buildings, including the Guildhall (1789), and the Imperial Institute (1881), in London. **1962** *Times* 3 Aug. 7/1 Any opaque stone—sandstone, bath, hoptonwood, basalt. **1967** G. Sims *Last Best Friend* vii. 62 A fine example of lapidary art..cut on a panel of Hopton-Wood stone. **1973** *Times* 14 May 14/8 Later he began to experiment with native stones, such as Hopton Wood and Ancaster.

‖ **hora** (hōᵊ·rǎ). Also **horah, horra.** [Rum. *horă*; Heb. *hōrāh*.] A Rumanian and Israeli round-dance; the music or song to which it is performed.
1878 J. W. Ozanne *Three Yrs. in Roumania* xii. 157 The *hora*, the Roumanian dance *par excellence*. **1911** *Encycl. Brit.* XXIII. 849 The popular songs of Rumania, the 'doine', the 'hora', [etc.]. **1923** *Chambers's Jrnl.* 20 Oct. 745/1 The favourite amusement of the Roumanian peasant is his dance, especially the traditional 'hora' danced in a circle. **1925** *Countries of World* V. 3431/2 Laughing girls..ready to dance the 'hora' with the men [in Rumania]. **1946** Koestler *Thieves in Night* viii. 61 [They] had formed the first ring of the horra, the stamping and swaying round-dance, a savage ring-polka. *Ibid.* 352 They were singing one of the popular horras, a folk-song with a passionate, almost hysterical tune. **1960** A. Wesker *I'm talking about Jerusalem* (1961) III. ii. 49 Let's do an Israeli dance..(*Starts doing a Zanny Hora on his own*). **1967** *Guardian* 15 July 6/2 Can you think of any other Army where officers and enlisted men would dance the hora together? **1967** *Observer* 8 Jan. 23/4 Catchy, minor-key *hora* strummings. **1969** *New Yorker* 6 Sept. 117/1 There is a scholarly argument over whether the Rumanian hora predates the Hebrew hora or vice versa. A diplomatic Rumanian-style answer is that both are of Mediterranean origin, the hora having been brought to Rumania by the Roman legions. **1971** *Observer* 17 Jan. 21/7 You were building, building, building. And on Friday nights I [*sc.* Mrs. Golda Meir] danced the Horah into the night.

horæ (hōᵊ·ri). [L., pl. of *hōra* Hour.] A book of hours (Hour 5, 6).
1875 *Quaritch's Gen. Catal.* Suppl. 51 The earlier editions of Kerver's series of Horæ. **1927** *Observer* 27 Nov. 22/4 An exquisite Flemish Horæ of about 1500. **1967** D. Diringer *Illum. Book* (ed. 2) vii. 406 An interesting

Horae of the Bodleian Library, at Oxford..was executed in France about 1430–40. A French Dominican *Horae* of the British Museum..was executed between 1425 and 1450.

horah, var. *hora.

horary, *a.* Add: **4.** *horary astrology* (examples).
1835 'Zadkiel' *Lilly's Introd. Astrol.* xxxv. 285 All these coincidences must be considered by the genuine searcher after truth as strong evidences of the truth of planetary influence, as evinced in *horary astrology*. **1911** 'Sepharial' *Man. Occultism* vi. 98, I am quite convinced from experience that there is much that is both fictitious and erroneous in Horary Astrology. **1951** M. E. Hone *Mod. Text-bk. Astrol.* xv. 272 The practice of Horary Astrology was very popular in the Middle Ages, and its best known exponent was William Lilly (1602–81). *a* **1963** L. MacNeice *Astrol.* (1964) i. 19 Casting horoscopes to answer questions is called 'horary' astrology.

horary, *sb.* Add: **2. b.** A timed programme or plan. So **hora·rium.**
1851 J. B. Pagani *Life Rev. A. Gentili* III. ii. 151 He drew up a regular horary to promote recollection and compunction. **1921** *Ampleforth Jrnl.* Jan. 139 Adapting ourselves to the School horarium we began dinner at 5 p.m. **1922** *Westm. Gaz.* 7 Oct., His scheme of breaking the Germans in 24 hours on a meticulously timed 'horary'.

Horatian, *a.* (*sb.*) Add: (Earlier and later examples.) So **Hora·tianism.**
1750 C. Smart (*title*) The Horatian canons of friendship. Being the third satire of the first book of Horace imitated. **1850** Thackeray *Pendennis* II. iii. 25 According to the Horatian maxim, a work of art ought to lie ripening (a maxim, the truth of which may, by the way, be questioned altogether). **1925** C. D. Broad *Mind & its Place* xi. 492 It is wrong to live in accordance with the Horatian ethics. **1936** F. R. Leavis *Revaluation* iv. 137 [Matthew] Green, in his Horatianism, is a good positive Augustan. **1945** Auden *Coll. Poetry* 121 The bland Horatian life of friends and wine. **1964** *English Studies* XLV (Suppl.) 217 It shows a deeply felt appreciation of ethical norms, a truly English horatianism that Cobbett, for all his lack of a classical education, inherited from the eighteenth century. **1965** B. Sweet-Escott *Baker Street Irreg.* 14 Others may wonder whether such a book *ought* to have been published, even after an interval which now exceeds the Horatian decade.

‖ **horchata** (ǫrtʃā·ta). Also **orchata.** [Sp.] A popular Spanish and Latin-American chufa-flavoured soft drink.
1859 in *Century Mag.* (1898) Dec. 312 *Orchata* is also a favorite drink; it is made from the juice of almonds, and is as white as milk. **1922** J. Hergesheimer *Bright Shawl* (1923) 63 She preferred rather than an ice, an orchata, and sipped it slowly. **1932** E. Hemingway *Death in Afternoon* xx. 271 The taste of horchata, ice-cold horchata. **1968** J. M. White *Nightclimber* ii. 9, I debate whether to order a *horchata* or a *zumo de limón*, and when the waiter arrives I ask for the latter. **1969** R. & D. De Sola *Dict. Cooking* 121/1 *Horchata*, almond-flavoured soft drink popular throughout Latin America and Spain.

horde. Add: **1. c.** *Anthropol.* A loosely-knit social group consisting of about five families.
[**1894** H. Cunow *Verwandtschafts-Organisationen der Australneger* iii. 28 In Dr. Hodgkinson's Werk 'Australia from Port Macquarie to Moreton Bay' (London 1845)..wird überall..die Bezeichnung 'tribe' gebraucht, ich habe dafür den meines Erachtens besser passenden Ausdruck 'Horde' gewählt. Unter 'tribus' ist stets die aus mehreren verwandten Horden bestehende grössere Volksgemeinschaft, der Stamm, zu verstehen.] **1896** F. H. Giddings *Princ. Sociol.* iii. 275 Practically the horde as a component of the tribe is nearly, but not quite, identical with the clan. **1918** A. A. Brill tr. *Freud's Totem & Taboo* iv. 208 Man, too, lived originally in small hordes in which the jealousy of the oldest and strongest male prevented promiscuity. **1939** *Geogr. Jrnl.* XCIV. 89 Davidson points out that the horde, a unit of about five families, in all some thirty-five persons, was the largest political unit known to the Australians. **1948** W. McDougall *Introd. Soc. Psychol.* (ed. 29) Suppl. iv. 402 That..primitive society took the form of a horde, the leader of which horde, the horde-father, actuated by his sexual jealousy, habitually treated his sons with extreme brutality.

hordeolum (hǫɹdī·ǒlǒm). *Path.* Pl. **hordeola.** [Altered form of late L. *hordeolus* a sty, dim. of L. *hordeum* barley.] A sty on the eyelid.
[**1622** R. Banister *Treat. Dis. Eyes* (ed. 2) IV. xiii. E 6 b (*heading*) Of the wart, or barly corne on the Eye-lidde, commonly called..a Stian,..in Latin, hordeum, or hordeolum.] **1806** J. Briggs tr. *Scarpa's Dis. Eye* ii. 75 The appearance of a white spot upon..the hordeolum should not induce the surgeon to be hasty in opening it. **1833** W. Lawrence *Dis. Eye* xix. 341 Some hordeola form more quickly, with greater suppuration and no slough. **1908** *Practitioner* Feb. 288 Epiphora may be due to..a hordeolum. **1966** S. Lerman *Basic Ophthalm.* viii. 459 A hordeolum (stye) is caused by a purulent inflammation in the glands of Zeis at the base of the eyelashes.

Hori (hǫ·ri), *sb. N.Z.* Also with lower-case initial. [Maori form of 'George'.] A contemptuous term for a Maori.
1933 F. E. Baume *Half-Caste* 26 'Hori', he [*sc.* the driver] said to Paul (as a negro is 'Sambo' a Maori is 'Hori' to the poor white). 'You like a drink, eh?' **1942** *National Education* (N.Z.) Mar. 59/1, I looked at the eighty husky Horis in their gum-boots, denim trousers.., tattered shirts. **1944** J. H. Fullarton *Troop Target* xi. 87 But all you Horis are related. **1960** N. Hilliard *Maori Girl* III. vii. 219 All *horis* come off the farm, I think. **1963** *N.Z. News* 23 July 6/2 The Maori representative on the Dominion Council..asked the R.S.A. to lead a move to eliminate racial and other discords fostering Communism and instanced the use of *Hori*, as a general term for a Maori, as distasteful and creating friction.

horizon, *sb.* Add: **5. b.** *Soil Sci.* Any of several layers in the soil which lie roughly parallel to the surface and are distinguishable by differences in physical properties, as colour, texture, or structure, or in chemical reaction.
A typical soil shows the following horizons (from the surface downwards): the A-horizon, generally the horizon of eluviation; the B-horizon, generally the horizon of illuviation; and the C-horizon, consisting of partly weathered parent material. Within the A-horizon and B-horizon further horizons may be distinguished (as B_1, B_2, B_3 horizons), though some writers refer to such horizons as zones or layers rather than as horizons.
1923 *Soil Sci.* XVI. 97 The soil assumes added importance as a factor when the chemical and physical differences of the separate horizons are studied in relation to root development. **1927** C. F. Marbutt tr. *Glinka's Great Soil Groups* 9 Russian investigators use the word *Horizon* rather than stratum to designate the various layers in the soil profile. **1948** White & Renner *Human Geogr.* xxiv. 406 A marked zone of lime accumulation occurs at a moderate depth in the 'B' horizon. The 'C' horizon or subsoil is brown. **1968** P. Buringh *Introd. Study Soils Trop. & Subtrop. Regions* v. 74 A much lighter reddish A_1 horizon, much lower in clay and much higher in sand, is clearly visible. **1971** E. A. Fitzpatrick *Pedology* i. 4/1 In some cases the contrast between horizons is dramatic and self-evident, while in others it is very subtle.

c. *Archæol.* A level at which a particular group of remains is found, or which is taken as representing a particular culture or cultural period.
1926 D. A. E. Garrod *Upper Palaeolithic Age* i. 34 Thanks to his journal it is possible to attempt a reconstruction of the archeological horizons which existed at the time of excavation. **1935** *Nature* 6 Apr. 550/1 Mr. Harrod was able to indicate the horizon of discovery very closely. The spear-head was found at the base of the lowest draw of peat. **1959** J. D. Clark *Prehist. S. Afr.* iv. 90 Nitrogen tests confirm that the remains are contemporary with the horizon in which they were found. **1962** R. Mason *Prehist. Transvaal* iv. 85 The Florisbad Peat 1 horizon has a radiocarbon age of more than 35,000 B.P. and may be associated with a Middle Stone Age industry. ..Four radiocarbon age estimations for Later Stone Age horizons in Matjes River Cave give ages varying from 11,250 ± 400 B.P. to 5,400 ± 250 B.P. **1973** *N.Y. Times* 15 July x. 1/1 Investigations have uncovered 15 distinct *horizons* (the archeological term for strata bearing traces of human habitation).

6*. *Embryology.* One of a numbered sequence of stages in the development of the human embryo.
1942 G. L. Streeter in *Contrib. Embryol.* No. 197. 213 (*heading*) Developmental horizons in human embryos. *Ibid.* 214 In searching for a suitable expression for the age groups under consideration..it was decided to follow the practice of other sciences and make use of the word 'horizons'. **1956** G. H. Bourne *Biochem. & Physiol. Bone* xiii. 376 Each age group or 'horizon' is characterized by a certain degree of differentiation and organization of various tissues and organs. **1968** J. B. Thomas *Introd. Human Embryol.* vi. 55 He [*sc.* Streeter] initially proposed 25 'horizons' for the human embryonal period, but this was later revised to 23. *Ibid.* viii. 80 With the appearance of these branching villi proposed horizon VII is reached.

6.** *Mining.* In horizon mining, a system of approximately horizontal tunnels lying in the same horizontal plane; the plane containing these tunnels.
1948 *Coal* Nov. 6/3 A further upper horizon was..fixed at a depth of 115 yards in the South Pit. **1960** J. Sinclair *Winning Coal* iii. 51 The coal lying between a pair of horizons is worked in such a manner that the coal flows downwards to the lower level and ventilation is ascensional. **1966** Vorobjev & Deshmukh *Advanced Coal Mining* II. xxxvi. 871 In steep and inclined coal seams (25°–9°) level division of the mining area with main workings on each horizon is preferable.

7. horizon-blue [Fr. *bleu horizon*], a light shade of blue, the colour of the uniform of the French Army during and after the war of 1914–18; such a uniform; also as *adj.*; **horizon mining,** a method of working inclined seams from approximately horizontal tunnels driven through the various strata to intersect the seams, there being several systems or 'horizons' of tunnels, one below another, connected by vertical shafts.
1919 J. Buchan *Mr. Standfast* xiv. 248 There was very little khaki or horizon-blue about. **1926** 'C. Barry' *Detective's Holiday* i. 9 'It is thus,' the man in horizon blue began...'This morning..a fisherman..discovered..the body of a man who had been murdered.' **1942** E. Paul *Narrow St.* xvii. 136 Monsieur Saint-Aulaire found himself over-stocked with horizon-blue material, the Chamber having agreed that French soldiers and officers should wear khaki in the future. **1954** W. Faulkner *Fable* (1955) 14 The whole ring of..faces was stained with a faint,..reflected horizon-blue. **1972** J. Williams *Home Fronts* viii. 137 Little girls decked out in 'horizon-blue' policemen's caps and cloaks. **1947** *Coal* Nov. 16/3 Horizon mining is planned to cut out heavy dip haulages, replacing them with locomotives. **1953** *Times* 15 Aug. 2/3 The site of the second modern horizon-mining development of the west Wales coalfield. **1963** J. Sinclair *Planning & Mechanized Drifting at Collieries* viii. 163 Probably the most important decision in planning a horizon-mining project is the position of the horizons and the vertical interval between them.

horizontal, *a.* (*sb.*) Add: **2. a.** *horizontal equivalent*, the distance between two points or two adjacent contours measured in a horizontal plane (rather than along the ground); *horizontal rainbow*, a spectrum occasionally seen on or just above the surface of a lake, appearing as an oval or as an open curve with its arms pointing away from the observer.
1889 G. W. Usill *Pract. Surveying* x. 199 The known difference of height thereof are [*sic*] called the vertical intervals, and their distance apart upon the survey are termed the horizontal equivalents. **1906** *Nature* 26 Apr. 608/2 (*heading*) A horizontal rainbow. **1916** *Monthly Weather Rev.* (Wash.) XLIV. 66/1 In general these horizontal rainbows are confined to the early hours of the day. **1952** Monkhouse & Wilkinson *Maps & Diagrams* ii. 74 If two points on a hill-side are projected on to a horizontal plane, as they are on a map, the distance between them is known as the Horizontal Equivalent (H.E.). **1957** G. E. Hutchinson *Treat. Limnol.* I. vi. 419 The horizontal rainbow, or iris, is a relatively rare phenomenon, though it has been observed on a number of lakes in many parts of the world.

b. *horizontal bar* (earlier and later examples); †*horizontal rudder* Aeronaut., an elevator on an aircraft.
1827 G. Hamilton *Elem. Gymnastics* 55 The performer, taking hold of the horizontal bar, swings backward and forward. **1843** *Lancet* 27 May 302/1 [I] made her exercise twice a day on a horizontal bar erected for the purpose. **1875** T. Frost *Circus Life* ix. 161 The rising school of young gymnasts..began to practise on..the horizontal bar, and the flying rings. **1875** *Aëronaut. Soc. Gt. Brit. 9th Ann. Rep. 1874* 56 This problem M. Renaud has solved by means of his automatic rudder... The idea occurred to him of placing a small horizontal rudder behind the sustaining planes. **1878** M. Jewry *Warne's Model Cookery* Inside front cover (Advt.), Practical Instructions on the Horizontal Bar, Parallel Bars. **1884** *Punch* 8 Mar. 117/1 Let the Ladies learn gymnastics..They'll improve too pale complexions..After practice on the ladders and the horizontal bars. **1903** A. Bennett *Truth abt. Author* ii. 16 He..taught us to fence, and to do the lesser circle on the horizontal bar. **1909** C. C. Turner *Aerial Navig.* viii. 121 The aviator must further correct this instability by control of the horizontal rudder. **1913** A. E. Berriman *Aviation* vii. 71 Originally the elevator was often called the horizontal rudder. **1964** G. C. Kunzle *Parallel Bars* i. 36 In this case it is very similar to position 2 (c)..on the Horizontal Bar.

3. a. Uniform; producing or based on uniformity. Chiefly *U.S.*
1842 *Congress. Globe* 17 Mar. 331 Was it expected that this committee would send in a horizontal tariff? **1872** *Ibid.* 28 Mar. 2016/2 The horizontal reduction of duties would do incalculable injury. **1890** *Ibid.* 9 May 4392/2 The Democratic bill made a horizontal cut of 20 per cent. **1907** *Daily Chron.* 23 Sept. 5/3 The 'Journal of Commerce' [N.Y.] says that the North German Lloyd Company announced yesterday a practically horizontal cut of 25 dols. on the eastward and westward passages. **1963** *Times* 22 May (Margarine Suppl.) p. viii, The Council of the European Community has already issued a directive in connexion with the use of colour additives. This follows what is known as the 'horizontal' system, meaning that it is of application to all food products.

b. In Industry: (see quots. 1959 and 1968). *horizontal combination, integration, merger*, an industrial merger of firms engaged in the same stages or types of manufacture; so *h. combine*, the organization resulting from such a merger.
1927 *Observer* 27 Mar. 5 Looking with a benevolent eye on horizontal combines. **1930** M. Clark *Home Trade* 204 There have been the combinations of one business with one or more of the same type. To this type the term 'horizontal combine' is usually applied. **1930** *Economist* 29 Mar. 710/2 Horizontal integration took place in the alcohol and solvents industry. **1959** *Listener* 9 July 46/2 The cotton industry is what is called a horizontal industry... As a rule the processing is done in stages in which the goods pass from one producer to another. **1960** Nanassy & Selden *Business Dict.* 96 *Horizontal combination*, formation of a business by combining two or more concerns engaged in the same kind of business. **1962** H. O. Beecheno *Introd. Business Stud.* v. 40 When business units of the same type combine..it is called horizontal integration. **1967** *Economist* 15 Apr. 253/3 He also held that all mergers must be judged by the same standards, whether they be vertical, horizontal or conglomerate. **1968** J. Ironside *Fashion Alphabet* 232 The textile and fashion industry works in two ways—horizontal or vertical... The 'horizontal' system means that in each stage of its manufacture..the textile goes through different hands. ..In the 'vertical' system, everything from weaving..to making up the garments..is done under one organisation.

c. Denoting a relationship, movement, etc., between a social group of a particular status, class, age-group, etc., and another of similar specifications, as opp. a 'vertical' relationship with a higher (or lower) authority, class, age-group, etc.
1931 H. G. Wells *Work, Wealth & Happiness of Mankind* (1932) xi. 540 The only remaining physical differences between man and woman are becoming horizontal,

i.e., differences between individuals in the same class, and not vertical differences, in which all women are put below all men, or vice versa. **1949** Koestler *Promise & Fulfilment* III. i. 289 The trend of social migrations is a 'horizontal' drift from village and agriculture to town and industry, and a 'vertical' drift from working-class to middle-class occupations. **1959** *Listener* 12 Feb. 280/2 When the children's interests turn outwards, when they develop loyalties of a horizontal sort. **1959** N. Mailer *Advts. for Myself* (1961) 374 The old exploitation was vertical—the poor supported the rich. To this vertical exploitation must now be added the horizontal exploitation of the mass by the State and by Monopoly. **1967** *Listener* 13 July 62/2 This horizontal integration of Saudi-Arabian Arab with Egyptian Arab ultimately gave way to the vertical integration of Egyptian peasant with Egyptian professional man.

d. *horizontal union* = **craft union.*
1937 H. Feldman *Probl. Labor Relations* iv. 255 Shall the mass production industries..be organized on a 'horizontal' (craft union) basis, or in 'vertical' (industrial) unions? **1950** Theimer & Campbell *Encycl. World Politics* 425/1 Unions may be craft unions, also known as horizontal unions, or industrial unions, also known as vertical unions.

4. *Mus.* (See quots. 1955 and 1970.)
1886 G. B. Shaw *How to become Mus. Critic* (1960) 122 The fact that M. Gounod has put too much sugar in it for the palate of a British Protestant might be condoned if the music were not so very horizontal. There is nearly always a pedal flowing along, and the other parts are slipping chromatically down to merge in it. **1900** C. W. Pearce *Composers' Counterpoint* iii. 28 Composers..have re-established the beautiful horizontal polyphony of the two-dimensional period, upon the perpendicular lines of the modern harmonic school. **1942** E. Blom *Mus. in Eng.* ii. 21 Vertical hearing, i.e. listening to the chordal incidence of parts rather than to their separate horizontal flow. **1955** L. Feather *Encycl. Jazz* ii. 53 Similarly 'horizontal' or 'linear' refers to the relationship of the notes or chords as they are played one after the other, read horizontally across the manuscript. **1962** *Listener* 9 Aug. 225/3 In Webern's texture the distinction between the horizontal and vertical is in a process of complete liquidation. **1970** W. Apel *Harvard Dict. Mus.* (ed. 2) 842/1 Much like woven fabric, music consists of horizontal ('woof') and vertical ('warp') elements. The former are the successive sounds forming melodies, the latter the simultaneous sounds forming harmonies.

B. *sb.* **2.** (Later examples.)
1955 *Oxf. Jun. Encycl.* VIII. 429/1 The theodolite is also used to measure vertical angles, that is, angles above or below the horizontal. **1962** *Listener* 15 Mar. 479/2 Bonnard derived his use of horizontals and verticals within the picture from Gauguin.

3. An evergreen Tasmanian tree or, in exposed positions, a shrub, *Anodopetalum biglandulosum.* Also *attrib.*, as *horizontal scrub*, the mat of vegetation formed by interlocking branches of a group of these trees.
1888 R. M. Johnston *Geol. Tasmania* p. vi (Morris), The Horizontal is a tall shrub or tree. **1891** *Australasian* 4 Apr. (Ibid.), That stuff as they calls horizontal, a mess of branches and root. **1898** Morris *Austral Eng.* 202/2 Horizontal scrub. **1927** *Blackw. Mag.* Oct. 471/2 These tentacles of horizontal were generally clothed with a thick velvety covering of damp green moss. **1936** *Discovery* Jan. 15/1 If he meets with a patch of 'horizontal scrub', he will clamber over the tree tops. **1949** D. Walker *We went to Australia* xxi. 200 Arid mountain country covered with 'horizontal scrub'. This extraordinary growth shoots upward for some twenty feet, the boughs then interlocking over very large areas; and so thickly matted is it that you can walk on the scrub some twenty feet from the ground. **1957** *Forest Trees Austral.* (Commonw. Forestry & Timber Bur.) 218/1 Associated species include myrtle beech..and shrub species such as horizontal.

4. [Fr. (*grande*) *horizontale.*] A prostitute. Also *grand horizontal*; also in French form. *slang.*
1888 E. Dowson *Let.* 13 Nov. (1967) 18, I shall let the liaison run its course—it will be very amusing & not as costly as an affair with a regular horizontale. **1909** J. R. Ware *Passing Eng.* 154/2 Horizontal. **1928** A. Philips *Boy at Bank* v. i, More than ten thousand professional 'horizontals' are light o' loves. **1963** *Times* 4 Apr. 16/2 The 'grand horizontals' were merely in the argot of the day the leading fashionable cocottes. **1967** *Observer* 8 Oct. 26/2 A *grisette* is cheaper than a regular *horizontale.* **1967** C. O. Skinner *Madame Sarah* iii. 44 La Païva, a contemporary 'grand horizontal', had attained respectability by marrying. **1970** *New Yorker* 28 Feb. 113/1 He is overshadowed throughout by Aunt Augusta, the still unretired *grande horizontale* of seventy-three.

horizontally, *adv.* Add: Later examples in senses 3 c and 4 of *horizontal *a.*
1934 *Hound & Horn* VII. 596 He [*sc.* Stravinsky] is tired of..exploiting the folk tune, horizontally, vertically, atonally, seriously or comically. **1956** M. Stearns *Story of Jazz* (1957) xii. 126 Everybody had an interesting part, for the idea was to move horizontally rather than vertically. **1958** *Listener* 9 Oct. 547/2 Differences throughout the Arab sector run both vertically and horizontally: between religious sects, social strata, settlers and nomads. **1959** *Ibid.* 13 Aug. 245/1 A society so divided vertically as well as horizontally.

Horlick (hǫ·ɹlik). The name of the British-born American industrialist W. *Horlick* (1846–1936), used in the possessive to designate the trade-name of the malted milk-powder or the drink made from this, which was first manufactured by his firm in 1883.
1891 *Trade Marks Jrnl.* 19 Aug. 784 (*heading*) Horlick's. .. A desiccated and granulated preparation of malt extract and milk as a food for infants and invalids. James Horlick,..London, E.C.; manufacturer. **1907** *Yesterday's Shopping* (1969) 510/3 Horlick's malted milk. **1932** L. Golding *Magnolia St.* II. iv. 324 The Horlick's mugs danced on the trestle tables. **1936** R. Lehmann *Weather in Streets* I. v. 152 'Hot milk?'..'No, thank you, Mum. Not even Horlick's.' **1958** J. Cannan *And be a Villain* vii. 150 A tray of 'night-caps', hot milk for Evadne, a whisky and soda for Laura, Horlick's for herself. **1966** J. Cleary *High Commissioner* iii. 53 A spoonful of Horlicks in a glass of Scotch and I'll be fine. **1973** J. Wilson *Truth or Dare* ii. 24 She..rinsed out the Horlicks mugs.

horme (hǭ·ɹmi). *Psychol.* Also **hormé.** [C. G. Jung's ad. Gr. ὁρμή impulse.] Vital or purposeful energy. Hence **ho·rmic** *a.*, of, pertaining to, or characterized by horme; **ho·rmism,** the theory of, or belief in, such purposeful energy; so **ho·rmist,** an adherent of hormism.
[*a* **1680** Cudworth *Treat. Freewill* (1838) 30 Now this love and desire of good,..is not a mere passion or *horme*, but a settled resolved principle. *Ibid.* 57 Epicurus..conceived that brutes were not merely passive to their own fancies and *hormae*, but that they could add something of their own to them.] **1915** Jung in *Jrnl. Abnormal Psychol.* IX. 396 The terminology—extraversion and introversion—depends upon my energic conception of mental phenomena. I assume a hypothetical energy which I designate as *hormé*. **1915** W. H. B. Stoddart *New Psychiatry* 4 This word 'horme' has a wide signification, applicable to all the instincts. **1920** T. P. Nunn *Education* ii. 21 To this element of drive or urge, whether it occurs in the conscious life of men and the higher animals, or in the unconscious activities of their bodies and the (presumably) unconscious behaviour of lower animals, we propose to give a single name—*horme*. **1926** W. McDougall *Outl. Abnormal Psychol.* 27 Jung says, 'I postulate a hypothetical fundamental striving which I designate *libido*'; and in a footnote..'This energy may also be designated as hormé. Hormé is a Greek word (ὁρμή)—force, attack, press, impetuosity, violence, urgency, zeal. It is related to Bergson's "élan vital". The concept hormé is an energetic expression for psychological values.' *Ibid.* 121 The vital or hormic energy of B's organism. **1927** *Contemp. Rev.* June 769 A kind of hormic determinism. **1931** R. S. Woodworth *Contemp. Schools Psychol.* 213 Purpose can properly be carried over into abnormal psychology, as has been done by the hormic psycho-pathologists, Freud especially. **1937** *Jrnl. Theol. Stud.* XXXVIII. 330 Miss Ikin is herself a thoroughly competent psychologist, with a Freudian training, but with a much wider interest in psychology of the hormic type than is usual in those whose concern has been with 'deep analysis' or psycho-analysis proper. **1944** L. Cohn in H. Treece *Herbert Read* 58 McDougall adheres to the same 'dionysian' or hormic conception of the soul which is to-day represented by Bergson, Freud, [etc.]. **1948** *Brit. Jrnl. Psychol.* June 187 The hormic-tension theory, (which explains pleasure as connected with the lowering of tension and unpleasure with its increase). **1948** W. McDougall *Introd. Soc. Psychol.* (ed. 29) Suppl. vii. 465 'Pleasure and pain are also motive forces depending upon individual experience.' ..To admit this is to combine hedonism with hormism. *Ibid.* 471 The hormist can find no clear instances that support Woodworth's thesis and can point to a multitude of instances which indicate an absence of..power. **1953** Hinsie & Schatzky *Psychiatric Dict.* (ed. 2) 653/1 Hormism is..opposed to hedonism. **1958** W. Stark *Sociology of Knowledge* 239 Though the sympherontic and hormic theories are commonly regarded as sociologies of knowledge.

hormogonium (hǭɹmogōᵘ·niǒm). *Bot.* Pl. **hormogonia.** [mod.L.: see Hormogone.] = Hormogone.
1880 W. G. Farlow *Marine Algae* in *Rep. Commissioner Fish & Fisheries 1879* (U.S. Senate Misc. Document 59, 46th Cong., 2nd Session) App. A. 12 The cells intermediate between the two heterocysts escape in the form of a small chain called a hormogonium. **1948** *New Biol.* V. 13 Most filamentous blue-green algae form hormogonia. **1965** F. E. Round *Biol. Algae* ii. 24 The short lengths of trichomes thus released are known as hormogonia.

hormonal (hǭɹmōᵘ·năl, hǫ·ɹmonăl), *a. Physiol.* [f. *hormon(e + -al.] Of, involving, or effected by a hormone or hormones; that is or acts as a hormone.
1926 *Chem. Abstr.* XX. 3029 (*heading*) The hormonal-nervous regulatory system of fat metabolism. **1949** M. Mead *Male & Female* xvi. 338 This domestic crisis..is reinforced by the hormonal instability and emotional fears that surround the menopause. **1953** *Jrnl. R. Hort. Soc.* LXXVIII. 165 The use of hormonal regulators of plant growth. **1955** R. G. Harry *Mod. Cosmeticol.* (ed. 4) iv. 95 The remark is often made that the mother-to-be has never 'looked so young' or 'so beautiful'. Undoubtedly this effect is due to hormonal influences. **1968** *Times* 11 Nov. 10/8 Apes and most monkeys have menstrual cycles, and in some species the pattern of sexual activity is known to be under hormonal control.

Hence **hormo·nally** *adv.*, by means of a hormone or hormones; as a hormone.
1939 B. Hanström *Hormones in Invert.* ix. 84 *Dixippus* shows a morphological and a physiological colour-change which are both hormonally regulated. **1955** R. I. Dorfman in Pincus & Thimann *Hormones* III. xii. 593 A group of hormonally active steroids, estrogens. **1970** *Sci. Jrnl.* June 68/1 They were both able to mate normal individuals to hormonally sex reversed animals which had exactly the same chromosomes.

hormone (hǫ·ɹmōᵘn). *Physiol.* [ad. Gr. ὁρμῶν, pres. pple. of ὁρμᾶν to set in motion (f. ὁρμή onset, impulse), with assimilation to -one.] **1. a.** Any of numerous organic compounds that are secreted into the body fluids of an animal, particularly the bloodstream, by a specific group of cells and regulate some specific physiological activity of other cells; also, any synthetic compound having such an effect.
1905 E. H. Starling in *Lancet* 5 Aug. 340/1 These chemical messengers, however, or 'hormones'..as we might call them. **1906** —— *Recent Adv. Physiol. Digest.* 75 The first products of digestion act on the pyloric mucous membrane, and produce in this membrane a substance which is absorbed into the blood stream, and carried to all the glands of the stomach, where it acts as a specific excitant of their secretory activity. This substance may be called the gastric secretin or gastric hormone. **1924** [see *growth-hormone* s.v. *growth^{1} 5]. **1930** R. A. Fisher *Genet. Theory Nat. Selection* 131 The investigation of the influence of the sex hormones has shown how genetic modifications of the whole species can be made to manifest themselves in one sex only. **1931,** etc. [see *gonadotrophic, -tropic *a.*]. **1951** A. Grollman *Pharmacol. & Therapeutics* xxvi. 581 The estrogenic hormones are responsible for certain secondary sex characteristics in the female, such as the plumage markings of some birds. **1955** *Sci. News Let.* 24 Sept. 198/3 Hormones are chemicals made by the adrenal, sex, pituitary and other body glands. **1959** A. C. Guyton *Function Human Body* i. 11 Adrenocortical hormones secreted by the two adrenal cortices..control the passage of proteins, salts, and perhaps other substances through the cell walls. **1967** *Martindale's Extra Pharmacopoeia* (ed. 25) 1277/2 Synthetic sex hormones have usually been developed from the basic steroid structure of the naturally occurring testosterone. **1968** Passmore & Robson *Compan. Med. Stud.* I. xxv. 40/1 Insulin is a powerful hormone whose actions affect the structure and function of every organ in the body. **1969** *Times* 16 June 3/8 Testosterone..is the chief of the male sex hormones known as androgens. **1970** W. B. Yapp *Introd. Animal Physiol.* (ed. 3) ii. 59 The acid of the gastric juice is secreted under the action of gastrin, a hormone secreted and liberated into the blood by the stomach wall..when it is mechanically stimulated.

b. Restricted to those compounds that have a stimulating (rather than an inhibiting) effect (cf. *chalone). Now *rare.*
1914, etc. [see *chalone]. **1955** J. T. & O. T. Lewis tr. *Houssay's Human Physiol.* (ed. 2) li. 561 The word 'hormone' is used also for substances that do not excite activity but rather inhibit it. According to Sharpey-Schafer, the term 'autocoid' would be appropriate for all chemical messengers; 'hormone' for chemical messengers that stimulate activity; and 'chalone' for those which inhibit it. This terminology has not been generally adopted.

2. Any of numerous organic compounds produced by plants which regulate growth and other physiological activities; also, any synthetic compound having such an effect.
1917 *Bot. Gaz.* LXIII. 50 In other plants the hypothetical geotropic substance might be associated with the shoot-forming hormone. **1927** *Biol. Abstr.* I. 244/2 Growth hormone of *Zea* coleoptile tips. **1951** *Jrnl. Chem. Education* XXVIII. 113 We now use plant hormones to propagate plants, prevent preharvest drop of apples,..and defoliate plants without killing the stems. **1952** Meyer & Anderson *Plant Physiol.* (ed. 2) xxviii. 555 Other terms commonly used to designate plant hormones are phytohormones, growth hormones, growth substances, and growth regulators. **1960** *Biol. Abstr.* XXXV. 4920/1 The action of synthetic plant hormones on pathogenic fungi was studied. **1966** R. M. Devlin *Plant Physiol.* xxii. 532 Bud dormancy in woody species may be regulated by some balance or ratio between a dormancy-inducing hormone and gibberellins. **1968** Y. Vardar (*title*) Transport of plant hormones. **1970** [see *cytokinin]. **1970** Wareing & Phillips *Control of Growth & Differentiation in Plants* iv. 62 Whereas the effects of most animal hormones are rather specific, a plant hormone can elicit a wide range of responses depending upon the type of organ or tissue in which it is acting.

3. *attrib.* and *Comb.*, as *hormone activity, balance, therapy, treatment, weedkiller*; *hormone-like* adj.; *hormone-controlled* ppl. adj.; **hormone cream,** a skin cream that contains one or more sex hormones.
1936 *Discovery* Nov. 362/1 Such complex subjects as inhibition, reflex action, hormone activity, etc. **1914** H. R. Harrower *Pract. Hormone Therapy* iii. 45 The intricacies of the hormone balance are fully as complicated as those of the nervous system. **1950** *Sci. News* XV. 134 Rheumatoid arthritis is related to the hormone balance of the body. **1963** A. Heron *Towards Quaker View of Sex* 54 In most mammals, the oestrous cycle of the female, hormone-controlled, is an important factor. **1938** *Encycl. Brit. Bk. of Yr. 1938* 588/1 The group of hormone and vitamin creams, etc., known collectively as 'biological' preparations. **1961** 'R. M. Dashwood' *Provincial Daughter* 85 A very good Hormone Cream which many clients find helpful after a certain age. **1962** *Punch* 12 Dec. 845/2 The Consumers' Association finds no reason for buying oestrogenic hormone creams. **1937** *Contrib. Boyce Thompson Inst.* VIII. 338 Characteristic responses of plants to hormone-like substances have been described. **1914** H. R. Harrower *Pract. Hormone Therapy* p. xii, At present the application of hormone therapy in general practice is the exception rather than the rule. **1921** *Endocrinology* V. 538 Convincing results of hormone therapy in gynecology. **1949** Koestler *Insight & Outlook* x. 138 Hormone therapy and neurosurgery aim at restoring equilibrium by action somewhere in the middle. **1955** A. Huxley *Let.* 5 Feb. (1969) 731 Hormone treatment is now being given. **1972** *Lancet* 3 June 1246/2 In 1 case hormone treatment was given

for mastitis. **1950** *N.Z. Jrnl. Agric.* Apr. 328/3 There is no danger in using hormone weedkillers on pastures.

Hence **ho·rmonize** *v. trans.*, to treat with a hormone; **ho·rmonized** *ppl. a.*, treated with a hormone; containing a hormone.
1940 *Proc. Amer. Soc. Hort. Sci.* XXXVII. 1015 Hormonized Dust gave satisfactory results with many kinds of cuttings. *Ibid.* 1016 Forty-seven days after potting, plants rooted with Hormonized Dust were 7 days ahead of untreated plants in shoot production. **1947** *Biol. Abstr.* XXI. 2043/1 Field expts. with potatoes hormonized by heteroauxin. **1950** *Ibid.* XXIV. 1611/2 The germination of seeds of endive and Brussels sprouts is not improved by hormonizing the seeds or fruits with indole-acetic acid. **1959** *Times* 16 Feb. 15/5 Its carcass did not set well on cooling, and this had led to most of the butchers' complaints about hormonized beef. **1960** *Farmer & Stockbreeder* 16 Feb. 78/1 America's top authorities confirm there is absolutely no risk with hormonized meat.

hormonic (hǭɹmǫ·nik, hǭɹmōᵘ·nik), *a. Physiol.* [f. *hormon(e + -ic.] = *hormonal *a.*
1914 [see *chalone]. **1928** *Brit. Med. Jrnl.* 18 Feb. 255/1 The functional (hormonic) over-activity of the anterior lobe of the pituitary gland. **1933** T. C. Merrill tr. *Voronoff & Alexandrescu's Testicular Grafting* 2 Animals which are too young and employed before the establishment of puberty and the production of free hormonic secretion, are unsuitable.

hormonology (hǭɹmonǫ·lǒdʒi). *rare.* [f. *hormon(e + -ology.] The study of hormones; endocrinology.
1918 *Endocrinology* II. 62 (*heading*) A bit of hormonology, with practical applications. **1944** Koestler in *Horizon* Mar. 170 The whole body of ideas had undergone a radical transformation: Relativity and Quantum mechanics, Hormonology and Psycho-analysis. **1970** *Sci. Jrnl.* June 44/1 It used to be a shibboleth of hormonology that such messenger molecules were not affected by the chemical process they influenced.

horn, *sb.* Add: **3. b.** Each of the erect and permanent bony processes, covered with hairy skin, growing on the head of a giraffe; also applied to a smaller protuberance in front of the other two.
[**1598** [implied by quot. s.v. Giraffe 1].] **1753** *Chambers's Cycl.* Suppl. s.v. *Zurnapa*, Its head is wholly of the make of the stag's, but differs in size, and has two little obtuse horns, which are not more than six fingers breadth long, and are hairy. **1840** tr. *Cuvier's Anim. Kingdom* 138 The Giraffe..is characterized by conical horns in both sexes, that are always covered with a hairy skin, and never fall... In the middle of the forehead, there is an eminence or third horn, broader and much shorter, but equally articulated by suture. **1879** *Encycl. Brit.* X. 619/2 In captivity it [*sc.* the giraffe] is said to make use of its skin-covered horns as weapons of defence. **1965** D. Morris *Mammals* 393 The Giraffe is easily the tallest of all the mammals... Both sexes have short, hair-covered horns.

5. Also jocularly, the human nose. *slang.*
1893 Farmer & Henley *Slang* III. 351/1 *Horn*, the nose. **1935** Ersine *Underworld & Prison Slang* 45 *Horn*, a man's nose, *bugle.* **1945** L. Shelly *Jive Talk Dict.* 12/2 *Horn*, the nose.

b. Also, to restrict one's expenditure, esp. of money.
1920 Galsworthy *In Chancery* I. i. 7 In the meantime, no more children! Even young Nicholas was drawing in his horns, and had made no addition to his six for quite three years. **1941** A. L. Rowse *Tudor Cornwall* xiv. 363 His will was a very cautious affair: he had to draw in his horns. **1957** I. Murdoch *Sandcastle* i. 16 If we don't get some extra money from somewhere we shall have to draw our horns in pretty sharply. No more Continental holidays, you know.

c. An erect penis; an erection. Also in phr. *to have* (*get*) *the horn*, to be sexually excited. (Not in polite use.)
1785 Grose *Dict. Vulgar T.*, *Horn Cholick*, a temporary priapism. **1879–80** *Pearl* (1970) 257 A man with light trousers, of decency shorn, Stop and talk to young ladies while having the horn. **1889** Barrère & Leland *Dict. Slang* I. 475/2 'To have the *horn*', to be in a state of sexual desire. **1922** Joyce *Ulysses* 263 Got the horn or what? he said. **1968** J. R. Ackerley *My Father & Myself* xiii. 148 He remarked to me then with a chuckle that the thing that had worried him most was that he might not be able to 'get the horn' again. **1968** L. Berg *Risinghill* 121 'Why does a boy get the "horn"?' 'The "horn" or the erection of the penis is necessary to make sure that the sperm is placed well inside the body of the woman.' **1972** *Guardian* 3 Apr. 11/3 Dirty old goat... He only bows his head to get his horn up.

6. b. *horns of consecration*: in Mycenæan art, a pictorial symbol or object, often found together with the double axe and pillar, connected with the Cretan worship of the ox.
1901 A. J. Evans in *Jrnl. Hellenic Studies* XXI. 196 The columns of the Knossian shrine apparently approach the outer edge of the openings, leaving room, however, in front of them for the 'horns of consecration'. **1939** V. G. Childe *Dawn Europ. Civilization* (ed. 3) v. 73 The cult of the Mother Goddess, associated, as in Crete, with the symbols of the dove, the double-axe, the sacred pillar and horns of consecration. **1939** J. D. S. Pendlebury *Archæol. Crete* v. 274 The double axe itself is found as a votive offering and as a cult object between the horns of consecration. **1970** Bray & Trump *Dict. Archaeol.* 149/2 Minoan religion is somewhat obscure, but includes a Mother Goddess who was worshipped in many shrines equipped with figurines..the sacred double axe and horns of consecration.

7. Delete †*Obs.* and add later example.
1942 D. Powell *Time to be Born* (1943) i. 24 Julian was almost pathologically jealous of her, fearing the final indignity of horns.

11. b. In *Golf*, the substance of which part of the face of a wooden club is made.
1743 T. Mathison *Goff* I. 5 Fenc'd with horn the head. **1801** J. Strutt *Sports & Pastimes* II. iii. 81 Goff..is performed with a bat,..the curvature is affixed to the bottom, faced with horn and backed with lead. **1839** *Chambers's Jrnl.* 22 June 173/3 The curvature, made of thorn, is affixed to the bottom, faced with horn, and backed with lead. **1890** H. G. Hutchinson *Golf* iii. 65 There is, however, something to be said in favour of dispensing altogether with the 'horn' in the case of brass-soled clubs.

c. *by the* (*great*) *horn spoon*: used as a fanciful oath or formula of asseveration. *U.S.*
1842 *Amer. Nat. Song Bk.* II. 222 He vow'd by the great horn spoon..He'd give them a licking, and that pretty soon. **1848** Lowell *Biglow P.* 1st Ser. v. 16 'I should like to shoot The holl gang, by the gret horn spoon!' sez he. **1853** *Knickerbocker* XLI. 115 'By the horn spoons!' repeated the skipper suddenly. **1897** *Outing* (U.S.) XXX. 380/2 'By the Great Horn Spoon!' the voice shouted, 'here's a chunk of civilization.' **1948** *Time* 22 Nov. 25/1 Operators had sworn by the Great Horn Spoon that they would not negotiate with Harry ('The Nose') Bridges.

13. b. *to blow* (U.S. *toot*) *one's own horn*: 'to blow one's own trumpet' (see Trumpet *sb.* 3).
1859 'Mark Twain' *Lett.* (1917) I. 43 Permit me to 'blow my horn'. **1860** G. D. Prentice *Prenticeana* 63 'Blowing your own *horn* I see,' said his comrade. **1903** A. W. Patterson *Schumann* 167 Surely these side-lights upon the straightforwardness and integrity of the man entirely free him from the calumny of ever being guilty of 'blowing his own horn'. **1940** A. E. Hertzler *Doctor & his Patients* (1941) ii. 47 He that tooteth not his own horn, the same shall not be tooted. **1949** E. S. Gardner *Case of Half-Wakened Wife* ii. 9 Gregory, on the other hand, had been reticent, inarticulate, sensitive, a man who modestly refrained from tooting his own horn and didn't like to hear others talk about themselves.

c. *French horn* (earlier and later examples).
1682 *Loyal Protestant & Domestick Intelligencer* 7 Mar. [2]/2 (Advt.), Any Gentleman may be furnished with Trumpets, French horns, Speaking Trumpets. **1771** C. Burney *Pres. St. Mus.* 149 There were two organs, and two pair of French horns. **1961** R. M. Pegge in A. Baines *Mus. Instruments* xii. 297 In England [*sc.* in the 18th cent.]..the French horn was chiefly used for the purposes of entertainment in the pleasure gardens and on the river, two performers playing duets being the usual thing. Rich men of family and fashion sometimes included in their retinues French horn players, often Negroes, to add *panache* to their equipages.

f. An instrument attached to motor vehicles, etc., which is sounded as a warning signal. Also *attrib.*
1901 *Graphic* LXIV. 268/3 The hideous toot-toot of its horn. **1914** R. & E. Shackleton *Four on Tour in Eng.* 83 The horn was honked suddenly. **1939** H. Hodge *Cab, Sir?* 19 And plenty of hornwork. The more the toots the bigger the tip. **1965** *New Statesman* 22 Oct. 594/2 The car..is taking over this enchanting city [*sc.* Rome]... The official campaign to cut down horn-maniacs appears to be a total failure. **1969** *Highway Code* 49 You must not..sound your horn at night (11.30 p.m.–7 a.m.) in a built-up area. **1973** *Sat. Rev. Soc.* (U.S.) May 42 Horn alarms: many inexpensive devices that can be hooked into the automobile horn can now be bought for less than $10.

g. A horn-shaped pastry case; an ice-cream cornet.
1908, 1960 [see *cream horn*, *cream *sb.*^{1} 7]. **1927** 'R. Crompton' *William—in Trouble* viii. 202 In one hand it held a stick of rock; in the other an ice cream horn. It licked them alternately. **1933** —— *William—the Rebel* xi. 212, I c'n eat twenty ice-cream horns. **1951** *Good Housek. Home Encycl.* 426/1 For savoury horns use a mixture such as those suggested for bouchées. **1969** *Main Cookery Bk.* (ed. 14) 172 Pastry horns can only be made using a special cone-shaped pastry case.

h. *Jazz slang.* A trumpet.
1935 *Hot News* May 5/1 He just threw his horn away and went into a pawnshop and bought another. **1938** D. Baker *Young Man with Horn* 9 And then he learned to play a horn—a trumpet, if there's anybody here who doesn't know what kind of a horn a horn is—and that was his proper medium. **1955** R. Davis in A. J. McCarthy *Jazzbook 1955* 40 Bunk was the subject of articles in the *New York Herald-Tribune* and the magazine *Time*, in which he was somewhat superlatively described as a 'genius of the horn'. **1959** G. Avakian in M. T. Williams *Art of Jazz* (1960) 68 Each of these trio cuttings ends with Bix picking up his horn to play the coda.

i. *Jazz slang.* Any kind of wind instrument.
1937 *Metronome* Jan. 25/1 Satchmo, I was only kiddin', I'll give you your horn back. **1938** [see prec.]. **1966** *Melody Maker* 30 July 8/3 Every instrument became a horn. When a guy said 'Can I bring my horn for a sit in,' you never knew whether he'd show up with a goofus or a glockenspiel. **1966** *Crescendo* Aug. 21/2 If I'm happy with the horn I've got, the mouthpiece, the set-up, the reed and everything.

j. The player of a horn (senses 13 c, *h and *i).
1945 L. Shelly *Jive Talk Dict.* 35 *The Horn*, the famous trumpeter. **1947** R. de Toledano *Frontiers of Jazz* p. ix, Did you ever try to relax while some fine horns were blowing, like for instance, Maxey, Pee Wee, and Bird? **1955** Keepnews & Grauer *Pict. Hist. Jazz* i. 14 Freddie Keppard was among the very great New Orleans horns. **1955** S. Whitmore *Solo* iv. 52 Take Buddy Bolden, if you will. A great horn. **1968** *Globe & Mail* (Toronto) 17 Feb. 26 'We've been lucky,' Harry Freedman, English horn, said. 'Ozawa has..done much to build up the orchestra.'

k. *the horn*: the telephone. *U.S. colloq.*
1945 L. Shelly *Jive Talk Dict.* 31/1 *On the horn*, telephoning. **1962** [see *flat *adv.* 3 b]. **1967** D. C. Cooke *c/o American Embassy* (1968) xi. 104 I've been on the horn half the night trying to get you. **1970** C. Armstrong *Protégé* vii. 89 I'll have to get on the horn tomorrow and poke up my contacts.

14*. a. A trumpet- or cone-shaped accessory of early gramophones and phonographs that collects sound to be recorded and amplifies the sound reproduced; a similar structure in some kinds of loud-speaker that contains the diaphragm in its throat and is designed to transmit its vibrations to the air. Also *attrib.*
1897 *Sears, Roebuck Catal.* 485/2 The Graphophone or Talking Machine is a most wonderful invention... By using the horn they can be distinctly heard in every part of a large hall. **1904** S. R. Bottone *Talking Machines* 62 The horn or trumpet which collects the sounds should be of *papier mâché*, and not of metal. **1911** *Encycl. Brit.* XXI. 468/2 The person making the record sings or plays in front of a horn or funnel. **1927** *Wireless World* 16 Nov. 664 When broadcasting first started, the only type of loudspeaker on the market was one which had..a straight conical horn. **1931** B. Brown *Talking Pictures* v. 121 Some of the first horns to be used in sound pictures were of the straight trumpet type. **1934** C. Lambert *Music Ho!* iv. 257 The old pre-electric horn recording, with its euphoniums instead of 'cellos, and its handful of Stroh violins. **1946** T. Rattigan *Winslow Boy* I. 12 He points to a gramophone—1912 model, with horn—lying on a table. **1956** C. Fowler *High Fidelity* vi. 103 Most tweeters used in high-fidelity systems employ small diaphragms which work into a horn of some sort. **1957** L. Durrell *Justine* II. 141 The same night, on the old horn gramophone..I heard some amateur's recording. **1969** *Listener* 23 Jan. 121/3 A pre-electric horn gramophone. **1970** R. D. Ford *Introd. Acoustics* v. 98 Horns are also very useful for improving the performance of loudspeakers at low frequencies.

b. *Radio.* Any hollow waveguide that increases in one or both transverse dimensions towards the open end and can consequently act as a transmitting or receiving aerial. Also *attrib.*, as *horn aerial, antenna.*
1936 W. L. Barrow in *Proc. Inst. Radio Engin.* XXIV. 1328 [The pipe can be flared into a horn-shaped radiator]. *Ibid.*, The application of horn radiators is not confined to the hollow tube system, for they may be fed by a coaxial or other lines... Thus, electromagnetic horns may be used as radiators in the wave band below ten meters. **1939** *Ibid.* XXVII. 51 The operation of the electromagnetic horn 'antenna'. **1949** H. E. Penrose *Princ. & Pract. Radar* xxii. 511 In general, the longer the opening of the horn, the more directive is the resulting field pattern. **1961** H. Jasik *Antenna Engin. Handbk.* x. 7 The pyramidal horn is frequently used as a standard horn of known gain in making measurements of other antennas. **1961** Miczaika & Sinton *Tools of Astronomer* viii. 261 Often a parabola is fed by a wave guide that terminates in a horn aimed at the disk. **1962** [see *antenna 5]. **1970** [see *despin *v.*]. **1972** *Daily Tel.* 28 June 11/8 Mounted inconspicuously in its front grille were two four-inch-square radar 'horns'—one for transmitting, the other for receiving.

19. b. Also, any cornu.
1802 C. Bell *Anat. Brain* 15 The Choroid Plexus..will be seen sinking backwards into the great inferior horn of the Ventricle. **1901** J. Berry *Dis. Thyroid Gland* i. 6 Small portions of the larynx and pharynx are embraced by the upper horns [of the thyroid]. **1957** R. T. Woodburne *Essent. Human Anat.* iv. 300/1 The coccygeal horns..articulate with the horns of the sacrum and enclose the fifth sacral intervertebral foramen. **1972** *Nature* 22 Oct. 521/1 In the spinal cord..the motoneurones of the ventral horn..are subject to a variety of inhibitory influences.

21. b. (Examples.)
1849 F. Parkman *Calif. & Oregon Trail* iv. 41 My long heavy rifle encumbered me, and the low sound it made striking the horn of my saddle startled him. *a* **1861** T. Winthrop *Canoe & Saddle* (1862) 212, I threw Klale's bridle over his neck, and grasping the horn, swung myself into the saddle. **1947** *Harper's Mag.* July 42/1 He took off his battered gray hat and rested it on the horn of his saddle.

g. *Electr.* Either of the pointed projections at the edge of a pole-piece of an electric motor or generator.
1886 S. P. Thompson *Dyn.-Electr. Machinery* (ed. 2) v. 88 The greatest amount of such eddy-currents will be generated..where the magnetic perturbations are greatest and most sudden... This should be at the leading corner or 'horn' of the pole-piece of the generating dynamo. **1923** A. S. Langsdorf *Princ. Direct-Current Machines* (ed. 3) ii. 91 Increased area is secured by means of pole shoes bolted or dove-tailed to the core in the case of solid poles, or by means of projecting tips or horns punched integrally with the sheets composing a laminated pole.

h. *Aeronaut.* (i) A short lug or lever projecting from a control surface to which the wire for moving the surface is attached.
1920 H. Woodhouse *Textbk. Appl. Aeronaut. Engin.* 319/2 *Horn-control arm*, an arm at right angles to a control surface to which a control cable is attached, for example, aileron horn, rudder horn, elevator horn, etc. More commonly called a Mast. **1928** Chatfield & Taylor *Airplane* v. 75 The cables from the horns on the ailerons are led to the stick. **1952** A. Y. Bramble *Air-Plane Flight* vii. 101 Notice the curved projecting pieces above and below on each aileron. These are called 'horns', and from these we see wires running forward into holes in the wing.

(ii) A part of an aileron or other control surface that extends across the axis of rotation over part of its length and serves to improve the balance of the surface; so *horn balance*, *-balanced* adj.

1921 *Aeronaut. Jrnl.* XXV. 539 The most common method of balancing ailerons is to have a 'horn' or projection on the aileron beyond the wing tip and forward of the aileron hinge. *Ibid.* 554 The horn method of balancing elevators. **1922** *Encycl. Brit.* XXX. 23/2 This so-called 'horn' balance proved unsatisfactory. **1939** *Jrnl. R. Aeronaut. Soc.* XLIII. 424 This effect was noticed during the war on horn-balanced rudders, and has now been used in the design of elevators. **1952** W. J. Duncan *Control & Stability Aircraft* vii. 195 A horn balance is a local protuberance of the control surface lying forward of the hinge axis... The horn may lie behind the main surface (shielded horn) or be exposed to the airstream. **1968** B. Dickinson *Aircraft Stability & Control* x. 235 If the horn extends to the leading edge of the aerofoil it is referred to as an 'unshielded' horn.

25. b. *Electr.* Each of a pair of rod conductors that diverge in a vertical plane from a narrow gap at the base, designed to extinguish any arc that forms in the gap and used to protect power lines from voltage surges; so *horn arrester*, *gap*; also, a projecting rod conductor that protects an insulator by attracting away from it any arc that forms.

1911 *Trans. Amer. Inst. Electr. Engin.* XXIX. I. 582 The relief gaps were removed before the lightning season of 1908. The grounded horn was left in place to act as a lightning rod. *Ibid.* 600 We placed the horn gaps on the towers, about 500 feet apart. **1930** *Engineering* 7 Mar. 314/2 Insulators on high-voltage lines are protected by arcing horns. **1968** P. J. Freeman *Electric Power* ix. 253 The shape of the horn gap forces the arc upwards by magnetic and thermal effects and the arc is self-extinguishing. **1969** L. Csuros in *Power Syst. Protection* (Electr. Council) III. xii. 20 The arcing horns shown on the 132 kV bushings serve the main purpose of protecting the metal fittings on..the bushing by providing a suitable anchorage for the fault arc.

27. a. *horn-call*.

1912 G. Moore *Hail & Farewell! Salve* vi. 102 If I knew Elgar, I'd write and ask him to send me a horn-call. **1954** J. R. R. Tolkien *Two Towers* i. 15 Suddenly the horn-calls ceased. **1959** D. Cooke *Lang. Mus.* ii. 57 Wagner..makes a musical demonstration of the natural 'rightness' of the harmonic series, for the horn-call is preceded by the low E flat on the basses. **1971** *Country Life* 18 Feb. 358/1 Siegfried promptly announces his arrival by horn-call and wastes no time in walking into the trap.

c. *horn-like*.

1929 D. H. Lawrence *Pansies* 26 Honking horn-like into the twilight. **1951** S. Spender *World within World* 162 The syllables which she unerringly chose to emphasize changed her speech into horn-like blasts.

28. *horn-handled* adj.

1922 Joyce *Ulysses* 619 A blunt hornhandled ordinary knife with nothing particularly Roman or antique about it. **1925** W. de la Mare *Connoisseur* (1926) 51 His horn-handled and gold-mounted umbrella.

29. horn aerial, antenna (see *horn *sb.* 14* b); **horn balance** (see *horn *sb.* 21 h (ii)); **horn-band** (examples); **horn-bug** (earlier and later examples); **horn cell** *Anat.*, any of the ganglion cells of the cornua of the spinal cord; **horn-distemper** (example); **horn-fisted** *a.*, having hands made horny by hard work; **horngarth** [Garth¹] (see quot. 1928); **horn gate** *Founding*, a horn-shaped gate (Gate *sb.*⁴ 1 a) that curves downward from a runner and then upwards into a mould cavity, discharging through its narrow end; **horn-man** (earlier and later examples); *spec.* in Jamaica among the Maroons, a man who blew the horn, giving signals; **horn-poppy**: substitute *Glaucium flavum* for *G. luteum*; (examples); **horn-pout** (earlier and later examples); **horn-rimmed** *a.*, denoting spectacles having rims made of horn; **horn-rims**, horn-rimmed spectacles; **horn-ring** (see quot. 1928); **horn speaker**, a loud-speaker that incorporates a horn; **horn spectacles** = *horn-rims*; **horn-tip** (example); **horn-worm**: substitute for def.: *U.S.*, the larva of moths of the genus *Protoparce*, which includes *P. sexta*, a pest of tobacco, and *P. quinquemaculata*, which attacks the tomato and certain other vegetables; also, the larva of other hawkmoths of the family Sphingidæ; (add later examples).

1849 J. G. Dalyell *Musical Memories Scotland* v. 170 The Russian horn-band consists of a multitude of performers whose concert comprehends the most simple music... Each instrument emits only a single note. *Ibid.* 171, I heard the Russian horn-band in this country, in the year 1833. **1938** *Oxf. Compan. Mus.* 441/1 In Russia..in the middle of the eighteenth century, proprietors of large estates established *horn bands*, much on the principle of our present day handbell ringing, each player being provided with one instrument of the appropriate size for the easy production of one note... The horns were straight ones (not circular). **1776** J. Trumbull *McFingal* (Th.), Thought horn-bugs bullets, or, through fears, Muskitoes took for musqueteers. **1869** Mrs. Stowe *Oldtown Folks* xxvii. 341 Youre saucy enough to physic a horn-bug. **1899** *Mem. Amer. Folklore Soc.* VII. 63 Horn-bugs, May-bees, May-flies, [etc.]. **1898** *Med. Chron.* IX. News 39 Collateral branches..are structures of enormous importance..representing the most direct path of nerve communication between the sensory surface..and the ventral horn cells. **1969** J. H. Green *Basic Clin. Physiol.* xx. 114/2 When the anterior horn cell sends a nerve impulse along the motor nerve, every muscle fibre in the motor unit contracts. **1843** *Knickerbocker* XXI. 254 Hence it is as important to keep the bee-moth out of hives as the horn-distemper out of cattle. **1929** F. C. Bowen *Sea Slang* 69 *Horn-Fisted*, a seaman with hands hardened with work. **1961** F. H. Burgess *Dict. Sailing* 117 *Horn-fisted*, possessing tough hands, and a character to match, through hard work. **1779** L. Charlton *Hist. Whitby* II. 96 The Horngearth..seems to have been a certain stake and yether hedge, made up in the beginning of summer by all those in Whitby-Strand who held land of the Abbot. **1890** *Hornes' Guide to Whitby* 18 About the year 1315,..the *Horngarth* was made at the town of Whitby, with wood from the abbot's forest. **1894** J. C. Atkinson *Memorials Old Whitby* 50 There is no reason whatever for questioning the conclusion that the Horngarth service..must date back to pre-conquest times. **1928** *Daily Express* 24 May 3 What is the ceremony of 'planting the Horngarth'?.. Driving in a hedge of stakes near Whitby Harbour to the sound of a horn, a custom dating to feudal times, to prevent cattle from straying into the harbour. **1909** *Hawkins' Mech. Dict.* 287/1 Horn gate. **1910** E. L. Rhead *Princ. & Pract. Ironfounding* ix. 202 Horn gates are shown in Fig. 96... The tapering form and circular sweep allow of their removal without disturbance of the sand. **1934** Laing & Rolfe *Man. Foundry Pract.* vi. 126 The semi-circular in-gates are known as 'horn-gates'. **1803** R. C. Dallas *Hist. Maroons* I. iii. 70 One of Quao's men, a hornman,..consented to accompany Captain Adair. **1957** J. Kerouac *On Road* (1958) I. iv. 201 The hornman sat absolutely motionless. **1961** F. G. Cassidy *Jamaica Talk* viii. 166 Among the Maroons an important person was the *horn-man*, who gave signals with a horn or conch. **1972** *Down Beat* 16 Mar. 26/2 Farmer and Heath are two of the best-matched hornmen at work in the idiom today. **1851** P. H. Gosse *Nat. Sojourn Jamaica* 39 The Mexican Horn-poppy (*Argemone*), the West Indian Vervain (*Stachytarpha*),..and others. **1909** *Chambers's Jrnl.* July 445/2 The wild wallflower and horn-poppy..bloom in mid-air. **1798** *Gaz. U.S.* (Philad.) 3 Aug. (Th.), The company concluded to go, for the sake of seeing a horn pout—when at last I drew one up—and behold! what was it, but a cat fish! **1832** *Coll. New H. Hist. Soc.* III. 87 On each side of their body and close to the head is a formidable weapon called a *horn*, and hence the name *Horn-pout*. **1910** *Outlook* 9 July 529 On the other side of the pond we met Sam Noyes, who was catching horn-pouts. **1943** B. Damon *Sense of Humus* 22 First he brought her a mess of horn pouts he had caught. **1894** *Idler* V. 452 Putting on a pair of horn-rimmed eye-glasses, he read it through very carefully. **1901** Kipling *Kim* i. 10 The lama mounted a pair of horn-rimmed spectacles of Chinese work. **1922** Horn-rimmed [see *beaver⁴]. **1923** *Westm. Gaz.* 12 Apr., A long-necked youth who was talking to a horn-rimmed female. **1931** R. Campbell *Georgiad* ii. 34 Women!..Who mock at horn-rimmed spectacles. **1973** *Times* 16 June 1/4 The London strip club owner..heavily disguised with..a beard and thick, horn-rimmed spectacles..was approached by three detectives. **1927** *Punch* 20 Apr. 424/3 He removed his horn-rims and began polishing them vigorously. **1959** *Encounter* July 59/1 In open-necked tennis shirt and heavy horn-rims. **1970** *New York* 16 Nov. 56/3 Junius glowers over his horn rims. **1928** *Daily Tel.* 16 Oct. 17 The horn-ring, an attachment fitted on the steering wheel so that the motorist can sound his horn without lifting his hand from the circumference of the wheel. **1962** *Which?* Apr. (Suppl.) 74/1 Horn-ring assembly came adrift from steering wheel causing horn failure. **1973** J. M. White *Garden Game* 189, I..put my hand on the horn-ring, pushed it down and held it there. The noise sounded shattering. **1928** L. S. Palmer *Wireless Princ. & Pract.* xi. 428 To reproduce such extremes without distortion is quite beyond the power of any existing horn speaker. **1957** *Encycl. Brit.* XIV. 409/1 Horn speakers may be made quite efficient. **1893** M. Beerbohm *Let.* 19 Aug. (1964) 53 An old sexton too with horn-spectacles. **1915** J. Buchan *Salute to Adventurers* xi. 163 Then he produced some papers, and putting on big horn spectacles, proceeded to instruct me in them. **1923** V. Woolf in *Dial* LXXV. 21 Mrs Dalloway, remembering Kensington Gardens and the old lady in horn spectacles. **1808** T. Ashe *Travels* III. xxxiii. 89 They sell them furs and horn tips, and receive in exchange ball powder, whiskey, tobacco, beads, ornaments, and blankets. **1784** J. Smyth *Tour U.S.* II. 132 The other [species] is the horn-worm..of a vivid green colour, with a number of pointed excrescences or feelers, from his head like horns: these devour the [tobacco] leaf. **1850** *Rep. U.S. Comm. Patents, Agric. 1849* 459 The horn-worm is deposited on the smooth or upper surface of the leaf in an egg by the tobacco fly. **1962** Metcalf & Flint *Destructive & Useful Insects* (ed. 4) xiv. 656 The winter stage of the hornworms is very often spaded up or plowed out in the spring. *Ibid.* 657 In the tomato hornworm larvae the horn is black and there are eight stripes..; while in the tobacco hornworm the horn is red and there are seven oblique stripes. **1972** *Sci. Amer.* June 73/2 The caterpillars of *Manduca sexta*..are the hornworms that feed on tobacco and tomato plants.

horn, *v.* **2.** Delete † *Obs.* and add later examples.

1952 S. Selvon *Brighter Sun* viii. 157 Look at yuh, yuh nasty dog! Yuh suspect she horning yuh! Yuh ain't have no shame? Dat poor gul don't even look at any odder man but you. **1970** 'W. Haggard' *Hardliners* i. 5 She'd given him a daughter and called it a day, horning him quite shamelessly.

3. b. *fig.* To push, as an ox with its horns. *U.S.*

1851 J. J. Hooper *Widow Rugby's Husb.* 69 You horned me off to get a chance to get gaming witnesses out of the way. **1881** *Times* (Philad.) 5 June (Th.), Mac Veagh is trying his best to horn Blaine out of the Cabinet herd, just as young buffalo bulls horn out the old ones.

c. *intr.* To push or butt *in* (*on* or *with*). *colloq.* (orig. *U.S.*).

1912 C. Mathewson *Pitching in a Pinch* 213 Many of them try hard to 'horn in' with the men who have made good as Big Leaguers. **1924** C. E. Mulford *Rustlers' Valley* xviii. 201 Why did Chet horn in on Baldy's arrest? **1927** *Bulletin* 15 Apr. 12/3 'Well, your little playmate certainly queered things,' he said. Thorn shrugged. 'I'm sorry, chief; but I couldn't help it. You saw how he horned in.' **1932** D. L. Sayers *Have his Carcase* xiv. 186 Glaisher might not like this horning in on his province. **1936** Wodehouse *Laughing Gas* xvi. 173, I suppose she felt she owed you something, after horning in on your big scene like that and trying to steal your publicity the way she did. **1939** *Airman's Gaz.* Dec., The lesson for today chicks is how to horn in on the radio racket. **1942** 'B. J. Ellan' *Spitfire!* xii. 61 Hurricanes had probably been chasing this Dornier when I had come in and attacked. Perhaps after all I was horning in on them! **1970** G. F. Newman *Sir, You Bastard* iv. 114 A proportion of detectives everywhere were at it; it was simply a question of finding them and horning in.

6. *trans.* and *intr.* To sound a horn; to signal to (someone) with a horn; to proclaim (something) loudly (as if) by sounding a horn.

1874 Hardy *Far from Madding Crowd* II. xii. 147 Jan meanwhile merging his..thoughts..in a song:—'To-morrow-to-mor—w!..To-mor-row, to-mor—'. 'Do hold thy horning, Jan!' said Oak. *Ibid.* II. xxvii. 335 'I am afeard your labour in keeping it close will be throwed away,' said Coggan... 'Labe Tall's old woman will horn it all over parish in half an hour.' **1892** G. Meredith *Poems* 77 He entreats..Compassion..For his fierce bugler horning onset. **1908** R. Broughton *Mamma* v. 45 Silence save of the nightly traffic roaring and ringing and horning past outside. **1923** W. de la Mare *Riddle* 209 The screech of its engine, horning up into the windless air. **1946** A. M. Walters *Moondrop to Gascony* xv. 199 We horned the small convoy to a stop as we approached Tanet.

hornbeam. 1. Substitute for def.: A tree of the genus *Carpinus* (family Betulaceæ), native to Asia, Europe, and North America, esp. *C. betulus*, the common hornbeam, which is native to Great Britain, or *C. caroliniana*, the American hornbeam; so called from its hard, close-grained wood. (Add earlier U.S. and further examples.)

1671 in *Early Rec. Providence, R.I.* (1893) III. 107 Boundeth on..the norwesterne Corner with a horne beame or peckled tree marked on too sides. **1791** W. Gilpin *Remarks on Forest Scenery* I. iv. 48 Very nearly allied to the beech..is the hornbeam. It grows like it, when it is suffered to grow; but it is generally seen only in clipped hedges. **1968** J. Arnold *Shell Bk. Country Crafts* xxxi. 330 When threshing was done with flails, the floors were made of hornbeam planks, and until lignum vitae came to this country hornbeam was used for bowls and still is for butchers' blocks.

hornblendite (hǭ·ɹnblendəit). *Petrogr.* Also †-yte. [f. Hornblend(e + -ite¹.] A granular rock largely or entirely composed of hornblende.

1874 J. D. Dana *Man. Geol.* (ed. 2) II. i. 70 Hornblendyte.—A very tough, granular, crystalline rock, consisting of hornblende, and hardly schistose in structure. **1901** *Jrnl. Chem. Soc.* LXXX. II. 170 Hornblendite from Brandberget, consisting essentially of hornblende. **1930** Peach & Horne *Geol. Scotl.* 26 The ultrabasic rocks (pyroxenites, hornblendites) generally form lenticles. **1967** P. J. Wyllie *Ultramafic & Related Rocks* i. 2/2 Hornblendites vary widely in mineralogy, and various combinations of hornblende, pyroxene, olivine, and biotite are known.

horned, *a.* Add: **1. c.** *fig.*

1889 W. B. Yeats *Wanderings of Oisin* 137 Between the hornèd hills. **1955** E. Pound *Classic Anthol.* III. 195 High spouts the water, from the hornèd spring.

2. b. horned adder, an African snake, *Bitis cornuta*, belonging to the viper family; **horned dace** *U.S.*, a small freshwater fish, *Semotilus atromaculatus*, of the family Cyprinidæ; **horned frog**, (*b*) a South American frog of the genus *Ceratophrys* which has horn-like projections on its eyelids; **horned helmet**, the gastropod mollusc, *Cassis cornuta*, or its shell, from which cameos are cut; **horned lizard** = *horned frog* (*a*) (*s.v.* Horned *a.* 2 b); **horned poppy**, the yellow sea-poppy, *Glaucium flavum*; **horned pout**, an American catfish, *Ictalurus* (formerly *Ameiurus*) *nebulosus*, introduced into western Europe about 1880, = *horn-pout* (Horn *sb.* 29); **horned rattlesnake**, a desert snake, *Crotalus cerastes*, found in the south-western U.S. and Mexico; = Sidewinder²; **horned screamer**, a large black and white bird, *Anhima cornuta*, distinguished by a hornlike process on its forehead and found in marshy country in the northern half of South America; **horned snake**, (*a*) = *hoop-snake* (s.v. *Hoop *sb.*¹ 13 b); (*b*) = **horned viper*; **horned toad**, (*a*) = *horned frog* (*a*)

(*s.v.* Horned *a.* 2 b); (*b*) = **horned frog* (*b*); **horned viper**, a venomous African snake, *Cerastes cornutus*, distinguished by a horny scale above each eye.

1878 A. Aylward *Transvaal of To-Day* xii. 244 The horned adder—a rather rare variety—is one of the worst of these pests. **1929** W. Rose *Veld & Vlei* 175 The Horned Adder..favours sandy localities, where, buried to the eyes, it watches for any lizards or mice that are unwary enough to approach within striking distance. **1947** J. Stevenson-Hamilton *Wild Life S. Afr.* xxxvi. 329 The Horned Adder... Also a dangerous viper, is usually found in sand regions. **1842** *Nat. Hist. N.Y., Zool.* IV. 199 The Horned Sucker is common in most of the fresh-water streams of this state... It is known under the various popular names of *Barbel*, *Dace*, and *Horned Dace*. **1896** Jordan & Evermann *Fishes Amer.* I. 222 *Semotilus Atromaculatus*... Horned-dace; Creek-chub. **1963** P. A. Parsons *Outdoor Life Compl. Bk. Fresh Water Fishing* iv. 45 They [sc. creek chubs] are often called horned dace, and are found in creeks and rivers from Montana and New Mexico to the Atlantic Coast and south to Florida. **1831** E. Griffith et al. tr. *Cuvier's Animal Kingdom* IX. 395 It is of the horned frogs, with concealed tympanum, that Gravenhorst has made his genus Stombus, but they have teeth like the others, and should not be approximated to the toads. **1902** P. Fountain *Gt. Mountains & Forests S. Amer.* xi. 290 In all outward aspects they resemble toads... They are very revolting-looking creatures, and are mostly of the species known as South American horned frogs. The common horned frog of Brazil is among them. **1962** H. R. Smyth *Amphibians & their Ways* xviii. 225 The Horned Frog possesses sharp teeth on his upper jaws and sharp bony processes on the lower jaw; what is more, he does not hesitate to use those teeth. **1863** Horned Helmet-shell [see Helmet *sb.* 5]. **1876** *Encycl. Brit.* IV. 740/1 The black helmet (*Cassis tuberosa*) of the West Indian seas, the horned helmet (*C. cornuta*) of Madagascar..are also employed [for cameos]. **1966** A. G. Melvin *Sea Shells of World* 92/2 Cassis cornuta L. Indo-Pac[ific]. Horned Helmet. Whitish... Interior is golden brown. **1806** W. Clark in R. G. Thwaites *Lewis & Clark Exped.* (1905) IV. 325 The Horned Lizzard is also common. **1844** J. Gregg *Commerce Prairies* II. 231 Horned frog..horned lizard, as those of earlier times more rationally called it. **1931** *Times Educ. Suppl.* 13 June p. iv/3 Among the new arrivals..some horned lizards known in America as 'horn toads' are of special interest. **1969** A. Bellairs *Life of Reptiles* II. xii. 498 The spiny little horned lizards (*Phrynosoma*) are able to squirt drops of blood from their eyes for several feet. **1548–1870** Horned poppy [see Poppy *sb.* I. 3]. **1909** *Daily Mail* 5 Aug. 9/3 The horned poppy is particularly plentiful in Start Bay. **1938** R. Gathorne-Hardy *Wild Flowers in Britain* vi. 41 The Yellow Horned Poppy..is named from the long curved fruit, which, like an orange tree, it bears at the same time as its blossom. **1963** W. Blunt *Of Flowers & Village* 140 Of course you know the horned poppy—*Glaucium flavum*. **1837** Horned pout [see Pout *sb.*¹]. **1839** Storer & Peabody *Rep. Fishes Mass.* 102 The Horned Pout..is highly esteemed as an article of food. **1890** K. D. Wiggin *Timothy's Quest* 126 The baby horned-pouts rustled their whiskers drowsily. **1969** A. Wheeler *Fishes Brit. Isles & N.W. Europe* 222/1 Horned Pout... A common fish in still or slow-flowing waters, it is found in many ponds, canals and lowland rivers in western Europe. Not so far reported as feral in the British Isles. **1870** *Proc. Calif. Acad. Sci.* IV. 67 The following do not occur west of this region..Horned Rattlesnake, [etc.]. **1888** *Buck's Handbk. Med. Sci.* VI. 166/1 Crotalus cerastes Hallowell, Horned Rattlesnake, 'Sidewinder'. **1965** R. & D. Morris *Men & Snakes* viii. 193 This special form of locomotion..occurs in..the horned rattlesnake or 'sidewinder' of North America. **1785, 1869–73** Horned screamer [see Screamer 3 a]. **1970** R. Meyer de Schauensee *Guide to Birds S. Amer.* 30/1 Horned screamer... General plumage glossy greenish black. **1775** J. Adams *Diary* 21 Sept. in *Wks.* (1850) II. 426 Every dip of his pen stung like a horned snake. **1784** [see **bull-snake* (*Bull *sb.*¹ 11)]. **1812** A. Plumptre tr. *Lichtenstein's Trav. S. Afr.* I. I. vii. 95 The heat of the day brought out a great many snakes; we killed two of very venomous kinds, one the horned snake, as it is called (*coluber cerastes*). **1851** D. B. Woods *Sixteen Months at Gold Diggings* 134 It answers the description of the horned snake. It is said that, taking the end of its tail in its mouth, it will form a perfect hoop with its body, rolling rapidly over till it reaches the object at which it aims. **1864** T. Baines *Explor. S.-W. Afr.* xiii. 374, I found and captured a cerastes, or horned snake, sixteen or eighteen inches long, lying in the path. **1806** *Massachusetts Spy* 16 July (Th.), A venerable Philosopher [sc. Thomas Jefferson] sitting in the middle of an immense Map, marked with vast prairies, huge rivers, and mountains of salt: surrounded by piles of Mammoth's bones, stuffed squirrel skins, and horned toads. **1883** [in Dict.]. **1914** E. G. Boulenger *Reptiles & Batrachians* iv. 70 The genus *Phrynosoma*, the Horned Toads, as these small, much flattened lizards are often called, inhabit the Southern United States and Mexico. **1956** C. H. Pope *Reptile World* 282 The grotesqueness of horned toads makes their bluffing antics ludicrous in the extreme. **1901** H. Gadow in *Cambr. Nat. Hist.* VIII. vi. 218 These 'horned toads'..make a squeaking noise when teazed. **1956** G. Durrell *Drunken Forest* vii. 134 'Horned toads..three beauties.'.. The largest..was about the circumference of a saucer, and three-quarters of his bulk seemed to consist of head. **1767** *Phil. Trans.* LVI. 287 The..horned viper of Egypt..is very rare, and scarce to be found in any of the cabinets of natural curiosities in Europe. **1863** [see Cerastes]. **1926** T. E. Lawrence *Seven Pillars* (1935) IV. xlvii. 269 This year the valley seemed creeping with horned vipers. **1957** Schmidt & Inger *Living Reptiles of World* 162/2 The sharp-tailed worm lizard..is thought erroneously by the natives to be the very poisonous young of the horned viper, *Cerastes cornutus*. **1973** 'A. Hall' *Tango Briefing* iii. 47 'Mine was a horned viper—see that?' He showed me the fang-marks.

3. b. horned cairn, a type of long barrow peculiar to Scotland.

1877 W. Greenwell *Brit. Barrows* 481 They are in all particulars, except in shape, like the 'horned' long cairns. **1907** T. R. Holmes *Anc. Brit.* I. iii. 106 Intimately related to certain chambered long barrows and the famous horned cairns, which exist only in Caithness. **1954** S. Piggott *Neolithic Cultures* viii. 238 These 'horned cairns' as they have been known since the nineteenth-century definition of the type.

Horner (hǭ·ɹnəɹ). The name of W. G. *Horner* (1786–1837), English mathematician, used *attrib.* and in the possessive (esp. in *Horner's method*) to designate a method for finding the real roots of a polynomial equated to zero by means of successive approximations.

It involves finding by trial the largest integer a_1 less than the root (x); transforming the equation to one with a root $x^1 = x - a_1$ (or $10(x - a_1)$, to avoid fractions); finding the largest one-place decimal (or integer) a_2 less than the new root; and so on, each approximation a_1, a_2, etc., providing one digit of the solution.

1842 *Penny Cycl.* XXIV. 341/1 The use of Horner's method is very much more easy than that of Newton. **1875** L. Hensley *Scholar's Algebra* xiii. 197 Horner's process..is applicable to find the roots of equations of all degrees. **1879** *Encycl. Brit.* VIII. 500/2 Horner's method (1819) gives the root as a decimal, figure by figure. **1913** W. P. Milne *Higher Algebra* viii. 352 The determination of the two consecutive integers between which the required root of $f_1(x) = 0$ lies is therefore an easy matter very early in the Horner approximating process. **1957** K. S. Kunz *Numerical Analysis* ii. 23 The Horner method was devised to reduce such pencil work. **1966** W. E. Grove *Brief Numerical Methods* i. 1 Horner's method is not used in computing, for it is designed for hand calculation.

‖ **hornero** (ǫɹnē·ro). [Sp., baker.] A South American bird of the genus *Furnarius*, esp. *F. rufus*; also called baker-bird, Oven-bird.

1880 E. Gibson in *Ibis* IV. 16 Furnarius Rufus. (Red Oven-Bird) Native name 'Hornero'... This species is..probably one of the best-known of our Buenos-Ayrean birds. **1911** J. A. Thomson *Biol. Seasons* II. 163 A masterpiece along the line illustrated by swallow and martin is the nest of some of the South American oven-birds (Furnarius)—for instance, of that species (F. rufus) which is called the 'hornero' or baker. **1956** G. Durrell *Drunken Forest* i. 21 The oven-bird appears to have more than his fair share of personality and charm...An elderly peon..solemnly told me that he would never harm an *hornero*. **1970** R. Meyer de Schauensee *Guide to Birds S. Amer.* 201 The name 'ovenbirds' has often been applied to this family [sc. Furnariidæ] because of the oven-like mud nests built by horneros.

Horner's syndrome. *Med.* [Named after J. F. *Horner* (1831–86), Swiss ophthalmologist.] A condition marked by abnormalities on one side of the face (including a contracted pupil, drooping upper eyelid, sunken eye, and a local inability to sweat) and caused by damage to the sympathetic nerves on that side of the neck.

1929 *Arch. Surg.* XVIII. 2025 The symptom-complex known variously as Claude Bernard–Horner syndrome (in France),..but usually and more rightly as Horner's syndrome. **1968** Passmore & Robson *Compan. Med. Studies* I. xxiv. 89 Interruption of the cervical part of the sympathetic chain of ganglia produces a condition known as Horner's syndrome. **1971** *Brit. Med. Bull.* XXVII. 33/2 The presence of a single underlying lesion can readily be inferred from the detection of several symptoms or signs (e.g. Horner's syndrome).

hornet¹. Add: **2.** Also, trouble, opposition. (Further examples.) Also, *to stir up a hornets' nest*.

1739–40 Richardson *Pamela* (1740) I. xxvi. 78, I rais'd a Hornet's Nest about my Ears, that..may have stung to Death my Reputation. **1921** Galsworthy *To Let* I. i. 21 An acid humour stirred in his Forsyte blood; a subtle pain divided by hair's-breadth from pleasure. If only June did not suddenly bring her hornets about his ears! **1921** H. Crane *Let.* 26 Nov. (1965) 71 But I also find that J stirred up a hornet's nest in me this summer with his words about getting away from current formulae. **1928** G. B. Shaw *Intell. Woman's Guide Socialism* lxxiv. 360 It is unlikely that a Proletarian House of Commons will suffer the nation's children to go on being taught Capitalist and Imperialist morality in the disguise of religion; and yet, the moment the subject is touched, what a hornet's nest is stirred up! **1966** *Listener* 4 Aug. 177/3 Judges have stirred up a hornets' nest in the sacred territory of 'the right to strike'. **1969** I. Kemp *Brit. G.I. in Vietnam* xiii. 200 On the third landing zone we ran into a hornet's nest. **1973** W. Tute *Resident* iv. 67 The Ambassador wants you removed. You've got a hornet's nest buzzing round your head in Whitehall.

hornfels (hǭ·ɹnfels), *sb. Petrogr.* [a. G. *hornfels*, f. *horn* horn + *fels* rock.] A fine-grained, non-schistose rock composed mainly of quartz, micas, and feldspars and formed by the contact metamorphism of an argillaceous rock. Hence **hornfe·lsic** *a.*, composed of, or having the character of, hornfels.

1854 Dana *Syst. Min.* (ed. 4) II. 246 The Hornfels of the Germans is a massive compound of quartz and feldspar breaking with a smooth flinty fracture. **1888** J. J. H. Teall *Brit. Petrogr.* xii. 374 We use the term hornfels instead of its etymological equivalent, hornstone, because in many cases the rocks termed hornfels are distinctly crystalline and not at all like what is commonly termed hornstone in this country. **1951** Turner & Verhoogen *Igneous & Metamorphic Petrol.* xv. 373 Structure is typically granoblastic (hornfelsic) or maculose. **1965** G. J. Williams *Econ. Geol. N.Z.* vii. 73/2 The conglomerates are quartzose in some places, and elsewhere contain granite, greywacke and hornfelsic components.

So **ho·rnfels** *v. trans.*, to metamorphose (a rock) to hornfels; **ho·rnfelsed** (-felst, -felzd) *ppl. a.*, **ho·rnfelsing** *vbl. sb.*

1901 *Rep. Brit. Assoc.* 634 The Old Red Sandstone is indurated and often hornfelsed to a varying distance from the margin. **1922** *Nature* 12 Jan. 62/1 Hornfelsed green rocks. **1930** Peach & Horne *Geol. Scotl.* ii. 70 The hornfelsing of the orthogneisses. **1947** *Q. Jrnl. Geol. Soc.* CII. 395 The hornfelsed graywacke is slightly silicated. **1970** G. Newall et al. *Mechanism of Igneous Intrusion* iv. 49 The granite is in contact with hornfelsed pillow lavas.

hornified, *ppl. a.* Also, adorned with horns; of a horn-like texture.

1789 J. Byng *Diary* 30 June in *Torrington Diaries* (1935) II. 113 We enter'd the old cloisters, now glazed up; and so hornified are their walls, as to give horrible presage; their being horns of all sorts of wild cattle. **1963** R. R. A. Higham *Handbk. Papermaking* ii. 36 The hornified cellulose fibres are not capable of assuming their original shape because of the permanent set that has taken place in the structural walls of the fibres.

horniness. Add: **2.** A state of sexual excitement. Cf. *horny *a.* 2 b.

1963 T. Pynchon *V.* i. 37 He was visited on a lunar basis by these great unspecific waves of horniness, whereby all women within a certain age group and figure envelope became immediately and impossibly desirable. **1969** P. Roth *Portnoy's Complaint* 102 Do I really experience this restlessness, this horniness? **1971** *Esquire* July 48/4 He establishes adolescent horniness as a condition too furtive and bewildered..to admit of real caring.

hornist. (Earlier and later examples.)

1836 *Knickerbocker* VIII. 71 Some evil-disposed citizens having no taste for music went to his honor the Mayor, and lodged grievous complaints against the distinguished *hornist*. **1906** *Westm. Gaz.* 1 Mar. 12/1 Vivier was solo hornist in the orchestra of the Imperial Opera.

hornless, *a.* Add: **2.** Without a horn (*horn *sb.* 14* a).

1909 *Talking Machine News* Oct. 337 Portable hornless machine. **1913** B. Clements-Henry *Gramophones & Phonographs* ix. 54 Some modern disc machines are hornless. **1934** N. W. McLachlan *Loud Speakers* xv. 256 A hornless speaker with baffle 4 ft. square. **1957** *Encycl. Brit.* XIV. 409/1 In the direct-radiator or hornless speakers the diaphragm is made large enough to radiate directly into unconfined air.

horn-mad, *a.* Add: **c.** Lecherous. Cf. *horn *sb.* 5 c. *slang.*

1893 Farmer & Henley *Dict. Slang* III. 356 *Horn-mad*,..sexually excited; lecherous. **1937** in Partridge *Dict. Slang*. **1951** R. Campbell *Light on Dark Horse* xvii. 251 The evil-minded and horn-mad levantine.

hornswoggle (hǭ·ɹnswǫg'l), *v. colloq.* (orig. *U.S.*). [Prob. fanciful.] *trans.* To get the better of; to cheat or swindle; to hoodwink, humbug, bamboozle.

1829 *Virginia Lit. Museum* I. 458 *Hornswoggle*, to embarrass irretrievably. **1834** W. A. Carruthers *Kentuckian in N.Y.* I. 61, I wish I may be horn swoggled, if ever I thought [etc.]. **1860** *Oregon Argus* 12 May (Th.), P. F. is going to hornswoggle the Douglas Democrats. **1904** *Boston Herald* 27 June 6 One practical working theory in advertising circles is that the ad's chief function is to hornswoggle the consumer. **1911** H. Quick *Yellowstone Nights* iii. 82 If you'll stand by..and see your old father hornswoggled out of his eye-teeth you'll never see a cent of my money. **1958** *Spectator* 11 Apr. 448/3, I am sure the manufacturers hornswoggle us and sell us a lot of useless trim and gadgets. **1968** *Guardian* 28 Feb. 18/8 In this wretched week we get a report from the Government Actuary which is quoted to hornswoggle us into increased insurance contributions. **1970** *Sunday Times* 28 June 11/3 The Americans look for value; you can't..hornswoggle them.

horny, *a.* (*sb.*) Add: **2. b.** Sexually excited; lecherous. (Chiefly used of a man.) *slang.* Cf. *horn *sb.* 5 c.

1889 Barrère & Leland *Dict. Slang* I. 476/1 *Horny*, lecherous, in a state of sexual desire, in rut. **1918** *Dialect Notes* V. 25 *Horny*, amative. **1949** H. Miller *Sexus* (1969) v. 104 Her thick, gurgling voice saying..: 'Get it in all the way...please, please do... *I'm horny*.' *Ibid.* x. 239 When I look at this thing I get horny again. **1965** J. L. Herlihy *Midnight Cowboy* (1966) II. v. 120 You *are* a gorgeous-lookin' piece, Cass. Gets a guy all horny just lookin' at you. **1968** M. Richler *Cocksure* xi. 63 When..he used to make a habit of watching the hockey games..he always felt horny. **1970** T. Lewis *Jack's Return Home* 43 The talk'd got filthier. It'd made me very horny. **1971** *Black World* Oct. 65/1 Ain't that the horny bitch that was grindin with the blind dude.

7. *horny-browed*, *fisted* (example); **horny-head** (in full **hornyhead chub**) *U.S.*, a North American cyprinid fish, *Nocomis biguttatus*; occasionally also used for *Campostoma anomalum*, a smaller cyprinid.

1916 Joyce *Portrait of Artist* iii. 158 Goatish creatures with human faces, horny browed, lightly bearded and grey as indiarubber. **1912** W. Owen *Let.* 2 July (1967) 148 A bushy-browed and horny-fisted blacksmith's assistant. **1882** Jordan & Gilbert *Synopsis Fishes N. Amer.* 212 C[eratichthys] biguttatus..Horny Head; River Chub; Jerker. **1933** *Amer. Speech* VIII. 50/1 Horny-head... A chub-like fish with short, horny protuberances on the scales, those on top of the head being largest. **1963** P. A. Parsons *Outdoor Life Compl. Bk. Freshwater Fishing* iv. 45 Hornyhead chubs, excellent bait for members of the pike family, are durable on the hook. **1965** A. J. McClane *Standard Fishing Encycl.* 891/2 Stoneroller. *Campostoma anomalum*. Also known as hornyhead and knottyhead, it is a brownish-olive minnow with a brassy luster.

B. *sb.* **2.** A policeman. *slang.*

1753 J. Poulter *Discoveries* (ed. 2) 39 *A Horney*,..a Constable. **1789** G. Parker *Life's Painter* xiv. 116 There's no horneys, traps, scouts, nor beak-runners amongst them. **1856** J. Strang *Glasgow & its Clubs* 187 Town officers..were then better known by the appellation of red-coat officers or *hornies*. **1922** Joyce *Ulysses* 160 Can't blame them after all with the job they have especially the young hornies.

3. A cow; a bullock. *Sc.* and *Austral. slang.*

1808 A. Scott *Poems* 81 Bedown the green the hornies rowt. **1879** G. Macdonald *Sir Gibbie* I. xiv. 194 Hornie—so named, indeed, because of her readiness to use the weapons with which Nature had provided her..was in fact a malicious cow. **1933** *Bulletin* (Sydney) 18 Jan. 25 Polled cattle graded better than the hornies. **1938** X. Herbert *Capricornia* (1939) xx. 273 Hornies, or beasts with ingrowing horns. **1943** Baker *Dict. Austral. Slang* 40 *Horney-steerer*, a bullock driver.

horoeka (hǫro,ī·ka). [Maori.] A small, round-headed New Zealand tree, *Pseudopanax crassifolium*, which has a juvenile form with long, toothed leaves; = *lancewood 2.

1831 G. Bennett in *London Med. Gaz.* 5 Nov. 150/1 (*heading*) The Horoeka Tree of New Zealand. This tree..has not yet been observed by botanists either in flower or fructification. **1838** J. S. Polack *New Zealand* II. 399 *Horoeka*..is generally found on elevated lands, grows to the height of thirty feet. **1868** J. Buchanan in *Trans. N.Z. Inst.* I. Essays III. 33 Horoeka (*Panax crassifolium*). A singular-looking plant in all stages of its growth. Three varieties are found at Dunedin, only distinguishable, in the young state, by the method of inflorescence. **1899** T. Kirk *Students' Flora N.Z.* 222 P[anax] crassifolium... A round-headed tree..20 ft.–60 ft. high... *Horoeka*. *a* **1939** 'R. Hyde' *Houses by Sea* (1952) 64 They cut the yellow twisted horoeka For sticks.

horomai, var. *haeremai.

horopito (hō³-, hǫropī·to). [Maori.] A small, aromatic, evergreen New Zealand tree, *Pseudowintera axillaris*; = Pepper-tree b.

1847 G. F. Angas *Savage Life* II. 17 A straggling shrub, with bright green shining leaves, resembling those of the nutmeg-tree; and a profusion of rich and delicate blossoms, looking like waxwork... The natives call this plant horopito. **1889** T. Kirk *Forest Flora N.Z.* pl. 2 (*caption*) The horopito, or pepper-tree of the settlers. **1963** Poole & Adams *Trees & Shrubs N.Z.* 40 *Pseudowintera axillaris* (J. R. & G. Forst.) Dandy. (Flowers borne in the axils of the leaves.) Horopito. Small tree reaching 8 m. **1966** *Encycl. N.Z.* II. 117/1 Horopito..is unpalatable to deer.

horra, var. *hora.

horrendous, *a.* Delete *rare* and add later examples.

1952 S. Kauffmann *Philanderer* (1953) vi. 97 The effect on the family of this drastic change in their financial status was horrendous. **1963** *Daily Tel.* 31 July 10/2 Blackmail and horrendous tales of Western avengers on their heels are probably as effective instruments as any. **1972** *Nature* 28 Apr. 434/2 Elsewhere there are horrendous accounts of how organochlorine pesticides contaminating food have killed large numbers of people.

‖ **horribile dictu** (hǫrī·bili di·ktu). [mod.L., by analogy with *mirabile dictu.] Horrible to relate.

1854 Geo. Eliot in *Westm. Rev.* Oct. 467 In some circles the effort is, who shall make the best puns, (*horribile dictu!*) or the best charades. **1883** G. Meredith *Let.* 16 Mar. (1970) II. 691 Louis Stevenson wrote to me from Marseilles in the winter... I, *horrible* [sic] *dictu*, have not yet replied to him. **1935** Huxley & Haddon *We Europeans* ii. 64 Puffed up with pride and, *horribile dictu!*, even with wealth.

horrible, *a.* (*sb.*, *adv.*) **B.** *sb.* (Further examples.)

1851 H. Melville *Moby Dick* I. xxxviii. 273 Such a waggish leering as lurks in all you horribles! **1899** [see *penny horrible* s.v. Penny 11]. **1909** *Daily Chron.* 3 Sept. 1/6 Both boys said yesterday that they had been reading 'penny horribles' about burglaries. **1917** A. Conan Doyle *His Last Bow* i. 51 There is but one step from the grotesque to the horrible.

horrid, *adv.* (Later examples.)

1867 S. Hale *Lett.* (1919) 25 I'm horrid sorry there was such a gap in my letters then. **1899** Kipling *Stalky & Co.* 135 Fags bully each other horrid. **1932** —— *Limits & Renewals* 80 The New Armies are horrid quick on the trigger.

horrifically, *adv.* (Later examples.)

1972 *Daily Tel.* 24 Feb. 2/6 The Aldershot explosion which caused the deaths of seven people went 'horrifically wrong' as an act of IRA retaliation, said Miss Bernadette Devlin. **1972** *Oxford Times* 27 Oct. 19/1 A young doctor is promised a job..if he can discover which of four horrifically insane patients is really the head of the asylum.

horrifiedly (hǫ·rifəid,li), *adv.* [f. Horrified *ppl. a.* + -ly².] In a horrified manner.

1908 R. Broughton *Mamma* v. 49 A thought which they have been horrifiedly chasing like a rat or a scorpion out of his or her own mind. **1923** J. S. Huxley *Ess. Biologist* I. 56 It is easy enough to see items on the debit side, and indeed to be so horrifiedly fascinated by it as not to have eyes for anything else. **1951** *Theatre Arts* Dec. 78 He is fondling her white throat with his fingers and she horrifiedly realizes it is these same clutching digits that have strangled his earlier victims.

horrisonous, *a.* Delete † *Obs.* and add later examples.

1901 *Daily Chron.* 31 Dec. 5/1 Sophie oft wakes on my snorting horrisonous. **1962** L. Deighton *Ipcress File* xv. 91, I listened to the ululating wail and horrisonous mewl.

horror, *sb.* Add: **3. c.** As *int.* (usu. *pl.*). An exclamation indicating shock, surprise, fear, etc.

1879 L. Troubridge *Life amongst Troubridges* (1966) xi. 152 Went to Shepherd's Bush. Oh, horror—stinking underground. *Ibid.* 153 The train went off without us! Oh, horror, no other train to Penzance. **1893** *Ladies' Home Jrnl.* Feb. 6/4 Horrors!..You don't mean that you're going to carry it any further? **1914** E. R. Burroughs *Tarzan of Apes* (1917) xvi. 137 Horrors! The lion was bounding along in easy leaps scarce five paces behind. **1928** 'Brent of Bin Bin' *Up Country* xvi. 284 After that was Miss Oswald—horrors, supposing she proposed too! **1973** *Times Lit. Suppl.* 31 Aug. 1007/3 Lord Crouch pulls strings to get..the Yard for the murder near his stately home, but horrors!—for him, anyway—when what he gets is [Inspector] Dover.

5. (Further examples.)

1805 E. Fremantle in A. Fremantle *Wynne Diaries* (1940) 4 Mar. III. 160 He [Count Barlowsky] is a little horror. **1846** E. Hall *Diary* 11 June in O. A. Sherrard *Two Victorian Girls* (1966) xvii. 162 Took the horrors for a drive, and even in the carriage Sydney and Cornelia could not behave themselves. **1849** Thackeray *Pendennis* I. xxxvii. 362 That collection of old fogies..ought to be cast in wax, and set up at Madame Tussaud's—.. In the chamber of horrors! **1859** *Macm. Mag.* Dec. 132/2 A series of magic lantern slides from some 'chambers of horrors', which he presumes to call the *Legend of the Ages*. **1889** Barrère & Leland *Dict. Slang* I. 235 *Chamber of Horrors*, the Peeresses' gallery at the House of Lords, from its being railed round as if it contained objectionable or repulsive inmates. **1891** Farmer *Slang* II. 69 *Chamber of Horrors*, sausages. **1899** *Daily Chron.* 2 Mar. 9/1 This..room..is one of terrible interest, for the 'flimsies' record the lost and overdue vessels, and the place bears the gruesome and apt title of 'Chamber of Horrors'. **1909** J. R. Ware *Passing Eng.* 69/2 *Chamber of Horrors*, the name of the corridor or repository in which Messrs Christie..locate the valueless pictures that are sent to them. **1922** Joyce *Ulysses* 613 He stowed the weapon in question away as before in his chamber of horrors, otherwise pocket. **1958** *Spectator* 6 June 746/2 Children adore reading about little horrors being taken down a peg. **1959** A. Huxley *Let.* 13 Feb. (1969) 866 Passages on infant damnation from St. Augustine... Passages on Jesus as a salesman from Bruce Barton. And so forth. A few pages of these wd constitute a stimulating Chamber of Horrors. **1959** *Times Lit. Suppl.* 10 July 409/2 A diary kept by an elderly horror whose name we never learn.

b. = *horror film.

1958 *Vogue* July 47 The American horrors pour out at an average of three a month: The Man without a Body, Back from the Dead, etc. **1958** *Economist* 6 Dec. 868/2 'Horror' is a generic term covering a wide range of films whose only common link is that they all contain a monster.

6. *horror joke*, *-loving* (example), *magazine*, *-photograph*, *story*, *-struck* (earlier and later examples); **horror comic**, a children's comic (sense *B. 2) in which the principal ingredients of the pictures and stories are violence and sensationalism; **horror film, movie, picture**, a film designed to horrify, usu. by the depiction of violence and the supernatural.

1954 *Time* 8 Nov. 60/3 Public criticism of horror comics. **1959** J. Cary *Captive & Free* lvi. 244 She had been all for the Bill putting down the horror comics, though her husband had been against it. **1964** M. Argyle *Psychol. & Social Probl.* iv. 51 There has been considerable public anxiety about the possible effect of television shows, films and horror comics on children. **1973** *Guardian* 28 Mar. 10/6 It was jokey in a horror-comic way, but I don't think horror is a reasonable reaction to a horror-comic. **1936** *Variety* 1 July 1/5 Recently showed..horror films and Sino-Japanese War cruelty shots. **1952** M. McCarthy *Groves of Academe* (1953) iii. 31 Miss Rejner and her boy-tutee..sat transfixed, as in a horror-film, watching the knob turn. **1965** *Listener* 18 Nov. 805/1 The connoisseur of the horror film knows instinctively that *Son of Dracula* will lack the blood chilling quality of *Dracula*. **1971** B. W. Aldiss *Soldier Erect* 10 My eyelids flickered like an ancient horror film, revealing acres of white-of-eye. **1963** Auden *Dyer's Hand* 372 A few years ago, there was a rage in New York for telling 'Horror Jokes'. **1909** *Westm. Gaz.* 1 May 13/2 The same horror-loving multitude flocks to its haunts of pleasure. **1939** R. Chandler *Big Sleep* x. 75 A fresh-faced kid was reading a horror magazine. **1965** Mrs. L. B. Johnson *White House Diary* 3 June (1970) 280 Tall plants called gorgonian..for all the world like those plants you see in horror movies. **1972** 'R. Crawford' *Whip Hand* I. iii. 11 A horror movie, vehicle for a Hollywood godling who had been in his grave for a decade. **1954** Koestler *Invis. Writing* xxxi. 333 He insisted on adding to the book a supplement of horror-photographs on glossy paper. **1937** *New Yorker* 9 Jan. 13/2 Mr. Arthur L. Mayer took over the..theatre, put in horror pictures (zombies and draculas), and he has made it pay every week. **1960** *Times* 14 Jan. 6/3 The world-wide success of the so-called 'horror' pictures made by Hammer Films. **1937** E. Snow *Red Star over China* I. i. 22 A torrent of horror-stories about Red atrocities. **1963** *Listener* 7 Mar. 428/1 The horror story afforded scope for the more primitive fears and desires that had gradually been squeezed out of the English novel. **1970** *Nature* 5 Dec. 900/2 Both argued that the 'horror stories' of genetic engineering are completely out of the question, at least in the foreseeable future. **1814** Jane Austen *Mansf. Park* III. vi. 134 William and Fanny were horror-struck at the idea. **1953** R. Lehmann *Echoing Grove* 23 Horror-struck, they continued to stand watching.

‖ **horror vacui** (hǫrəɹ væ·kiu,əi). *Art.* [mod.L., lit. 'the horror of a vacuum'.] The dislike of leaving empty spaces, e.g. in an artistic composition.

1845 W. Plate *Ptolemy's Knowl. Arabia* 5 Ptolemy had a tendency towards putting the inland towns too far east; but whether it was a mere *horror vacui*, or some misunderstanding, that induced him to do so, cannot be decided. **1937** *Burlington Mag.* Apr. 183/1 Pictures..showing looser compositions, less *horror vacui*, and a less trenchant treatment of light and drapery. **1954** M. Rickert *Painting in Britain* v. 108 True to the thirteenth century artist's *horror vacui*, he fills the lower corner of his miniature with more architecture..to balance that on the other side. **1960** R. Lister *Decorative Cast Ironwork* iv. 123 Castings were used in place of hand-carved or forged iron fittings,..loading them with detail with which to placate the current *horror vacui*. **1968** *Listener* 5 Dec. 774/2 A further tension builds up between the desire to create space and the desire to fill it. Sometimes the space-filling seems compulsive... *Horror vacui* struggles with claustrophobia. **1972** E. Lucie-Smith *Eroticism in Western Art* ix. 180 Nudes..fill the whole picture-space as if the artist suffered from *horror vacui*.

hors. Add: ‖ **hors concours** (or koṅkūr), *adv.*, not competing; hence, without a rival; unequalled.

1884 *Tablet* LXIII. 804/2 A work such as..would, were it shown in the *Salon*, range him Hors Concours. **1931** *Times Lit. Suppl.* 7 May 353/3 Dickens and Wilkie Collins were regarded as *hors-concours*; Trollope, however, he despised. **1941** V. Nabokov *Real Life S. Knight* (1945) xvi. 138 Most husbands are fools, but that one was *hors concours*.

horse, *sb.* Add: **1. f.** Colloq. abbreviation of Horse-power.

1904 [see Tonneau]. **1931** *Star* 8 May 13/1 Each of them with a few 'horses' in reserve. **1932** Kipling *Limits & Renewals* 341 Sign—refill, and let me away with my horses (Seventy Thundering Horses!). **1973** R. C. Dennis *Sweat of Fear* vi. 41 The Mercedes..pointed north at high speed, but there was never any chance of its outdistancing me. I had too many horses under my foot.

3. c. (Later example.)

1930 W. Faulkner *Rose for Emily* in *Coll. Stories* (1951) II. 121 So she vanquished them [*sc.* the city authorities] horse and foot, just as she had vanquished their fathers thirty years before.

4. (Further examples.)

1840 R. H. Dana *Bef. Mast* (1854) xxii. 125 Though 'a bit of a horse'..yet he was generally liked by the crew. **1844** *Southern Lit. Messenger* X. 489/2 'Huzzah!..went round the crowd, while Jeptha's..friends swore he was 'a horse'. **1847** W. T. Porter *Quarter Race Kentucky* 39 Good mornin', old hoss. **1925** J. Metcalfe *Smoking Leg* 26 There you are, old horse; don't say I never did you a good turn. **1930** D. L. Sayers *Strong Poison* iv. 48 It's your triumph at having secured a disagreement that gives you away, old horse. **1973** 'A. Hall' *Tango Briefing* i. 10 'It *is* a joke, isn't it?' 'As far as I know, old horse.'

6. a. (Later examples.)

1937 *Times* 13 Apr. (Brit. Motor Suppl.) p. vi/3 Good progress..is most noticeable with the mechanical horse ..and the trolley omnibus. **1963** *Amer. Speech* XXXVIII. 44 *Horse*..., a tractor or power unit.

c. (Examples.)

1785 J. Wesley *Let.* 17 July (1931) VIII. 281 Constant exercise. If you can have no other, you should daily ride a wooden horse, which is only a double plank nine or ten feet long, properly placed upon two tressels. **1875–98** [see Vaulting *vbl. sb.*² 2 b]. **1949** E. Williams *Wooden Horse* ii. 37 A vaulting horse, a box horse like we had at school. You know, one of those square things with a padded top and sides that go right down to the ground. **1962** *T.V. Times* 9 Mar. 22/3 Trampoline, pommelled horse and vaulting box. **1973** J. Burrows *Like an Evening Gone* iii. 40 Sporting equipment of a modest kind..a vaulting horse and a set of P.E. mats.

e. (Later example.)

1921 [see *broom *sb.* 6].

7. b. (Later U.S. examples.)

1846 D. Corcoran *Pickings* 83 One carried his saw slung on his arm, and the other had his 'horse' mounted on his shoulder. **1850** N. Kingsley *Diary* (1914) 103 Worked at hewing some sticks for horses to use the Pit Saw.

8. h. A groyne. *local.*

1852 J. Wiggins *Embanking* x. 232 Expensive works ..such as those called '*horses*' in Essex, and '*groins*' in Sussex and Hants.

i. In other uses (see quots.).

1904 Goodchild & Tweney *Technol. & Sci. Dict.* 290/2 *Horse*, (*Plast.*) the wood backing of a zinc mould, used by plasterers for running mouldings. (*Plumb.*) A wooden finial, etc., forming a core which is to be covered with lead. **1946** N. Wymer *Eng. Country Crafts* vi. 62 When the sticks are well 'cooked' the craftsman takes them, one by

one, from the sand and pulls them through a 'horse'..a wooden plank with niches cut out of the side—to straighten them. **1957** R. Lister *Decorative Wrought Ironwork* 229 *Horse*, a kind of stake..with perforations for holding other tools. **1964** W. L. Goodman *Hist. Woodworking Tools* 153 Holding his work in a vertical pedal-operated vice or 'horse'.

11. (Earlier example.)

1778 W. Pryce *Mineralogia Cornubiensis* 323 *Horse*, a portion of dead ground in a Lode, which widens like a horse's back from the spine.

b. A mud or sand bank. *dial.*

1926 H. A. Tripp *Suffolk Sea Borders* vi. 109 Below Waldringfield is a 'horse' in mid-channel—'horse' being the name given to banks that crop up with rounded backs like the back of a horse. **1929** E. A. Robertson *Three came Unarmed* ix. 149 Now the shoal-water of this coast is..full of under-water mud-banks or 'horses' which come dry or are barely covered at low tide.

14. (Earlier example.) Also *live horse*: work done and not charged for.

1770 P. Luckombe *Conc. Hist. Printing* 499 If any journeyman set down in his bill on Saturday night more work than he has done, that surplus is called Horse. **1859** Bartlett *Dict. Amer.* 247 *Live horse*, in printers' parlance, work done over and above that included in the week's bill.

14*. Heroin. *slang* (orig. *U.S.*).

1950 *Time* 28 Aug. 2/2 There are the usual thrill-seekers who take goof balls..quite often ending up as confirmed addicts of..heroin (H, horse, white stuff). **1951** *N.Y. Times* 13 June 24/3 Then one day we met another fellow and he offered us some heroin. I sniffed this too. We called it 'horse' and 'H'. **1961** *John o' London's* 16 Nov. 548 'Pot' is marijuana, and 'horse' heroin. **1962** J. Baldwin *Another Country* (1963) I. i. 14 His first taste of marijuana, his first snort of horse. **1963** L. Deighton *Horse under Water* xl. 158 Diacetyl-morphine. Which is what you would call 'heroin', or 'H', or 'horse'. **1969** *Daily Tel.* 31 Jan. 24/6 He had seen the effects of an overdose of 'horse' before. The skin becomes greenish and there was frothing at the mouth.

17. *to change* (*swap*) *horses in midstream* (*while crossing a stream*): to change one's ideas, plans, etc., in the middle of a project, process, etc.; *to talk horse* (earlier U.S. example); *to play horse with* (U.S.): to treat roughly or unceremoniously; *to hold one's horses*: see *hold *v.* II c.

1855 T. C. Haliburton *Nat. & Hum. Nat.* II. 337 Doctor, I am a borin of you, but the fact is, when I get a goin 'talkin hoss', I never know where to stop. **1864** A. Lincoln in *Compl. Wks.* (1894) II. 531, I do not allow myself to suppose that either the Convention or the League have concluded to decide that I am either the greatest or best man in America, but rather they have concluded that it is not best to swap horses while crossing the river, and have further concluded that I am not so poor a horse that they might not make a botch of it in trying to swap. **1896** Ade *Artie* xvii. 163 Do you think I'm goin' out ridin' with her and have a lot o' cheap skates stoppin' to play horse with her everywhere we go? **1904** W. H. Smith *Promoters* x. 169 You've got to have some well-matured plan ..if they try to play horse with you again. **1911** L. J. Vance *Cynthia* x. 157 Why does Madame Savaran insist on coming along to see that he doesn't play horse with her stake in the venture? **1923** —— *Baroque* viii. 49 Remember the Wop detective that used to play horse with the Black Handers. **1940** 'H. Pentecost' *24th Horse* v. 42 Don't come if you don't want to... Change horses in midstream if you want to. **1948** A. J. Toynbee *Civilization on Trial* 195 'Herodianism'..does not really offer a solution. For one thing, it is a dangerous game... It is a form of swapping horses while crossing a stream, and the rider who fails to find his seat in the new saddle is swept..to a death. **1951** H. S. Davies *Gram. without Tears* vi. 56 From the point of view of strict old-fashioned grammar, this is obviously bad; it involves a change from the singular to the plural horse in mid-stream of the sentence. **1969** *Listener* 13 Mar. 360/1 Another play which changed horses in midstream was William Ingram's *Double Take*. The long dialogue between the nervous kidnapper and his oddly calm victim was inconsequential and tense and had one thinking hopefully of Pinter.

18. (Further examples.)

1832 E. C. Wines *Two Years in Navy* I. 73 Most of us had not 'worked out our dead horses'. *Ibid.*, Dead horses are debts due to the purser on account of advances of pay. **1863** S. Butler *First Yr. Canterbury Settlement* x. 146 Some good hands are very improvident... They will come back possibly with *a dead horse to work off*—i.e. a debt at the accommodation house. **1872** *Globe* 1 Aug. 3/1 For..twenty minutes..the Premier..might be said to have rehearsed that..lively operation known as flogging a dead horse. **1886** F. T. Elworthy *W. Somerset Word-Bk.* 186 Work done in redemption of debt is called working out the dead-horse. **1907** *Westm. Gaz.* 7 Mar. 5/2 Mr. Philip S. Head, auctioneer and house agent, stated that 'Hillside' had been on his books for three years. Some people when asking for a house had stated that they did not want 'the haunted house'... His Lordship: Do you think 'Hillside' will always be 'a dead horse'? **1927** J. Sampson *Seven Seas Shanty Bk.* 45 For the first month at sea he was working for nothing—in other words he was working out the 'dead horse'. **1935** *Yachting* Dec. 82/3 *Dead horse*. The common sailor was advanced one month's pay at time of signing the articles. This usually went to his boarding-house keeper for alleged debts. During the first month out, he was said to be 'working off the dead horse'; and at the end of this period it was the custom..to make an effigy of a horse and throw it overboard with suitable ceremonies. **1970** *New Yorker* 10 Oct. 109/1 All this critical analysis would be a flogging of a dead horse. **1971** *Cabinet Maker & Retail Furnisher* 1 Oct. 14/2 If this is the case, we are flogging a dead horse in still trying to promote the scheme.

22. b. *to mount the high horse* (earlier example); *to ride the high horse* (examples); *to come, etc., off one's high horse*, to climb down, to become less arrogant.

1782 T. Pasley *Private Sea Jrnls.* (1931) 252 Whether Sir George will mount his high Horse or be over-civil to Admiral Pigot seems even to be a doubt with himself. **1809** Malkin *Gil Blas* II. vii. ¶ 5 Riding the high horse with all the arrogance of greatness. *Ibid.* IX. i. ¶ 2 Do not ride a higher horse than a thousand jockeys of quality whom I could name. **1843** Thackeray in *Fraser's Mag.* Apr. 469/2 It would be *his* turn to sneer and bully, and ride the high horse. **1887** G. R. Sims *Mary Jane's Mem.* 116 They were awfully civil, and let Mrs. Master John ride the high horse over them. **1920** A. Christie *Mysterious Affair at Styles* x. 224, I decided that I would descend from my high horse, and once more seek out Poirot at Leastways Cottage. **1928** W. S. Maugham *Ashenden* ix. 153 Come, come, my dear fellow, do not try to ride the high horse. You do not wish to show me your passport and I will not insist. **1928** *Sunday Express* 15 Jan. 6/4 The cable companies have come off the high horse at last in entering into negotiations with the wireless group. **1936** A. Christie *Murder in Mesopotamia* xix. 162 I'd like to see Sheila honest enough to come off her high horse and admit that she hated Mrs. Leidner for good old thorough-going personal reasons. **1950** W. Saroyan *Assyrian & Other Stories* 219 Only his mother felt that Mayo was not a rude boy, but his father frequently asked Mayo to get down off his high horse and act like everybody else. **1959** *Economist* 20 June 1079/1 Politicians..riding on high horses.

25. a. *to eat like a horse* (later examples); *to work like a horse* (examples); *a horse of another colour* (further examples).

1798 *Aurora* (Philad.) 27 Aug. (Th.), Whether any of them may be induced..to enter into the pay of King John I. [i.e. President Adams] is 'a horse of another colour'. **1829** G. Griffin *Collegians* II. xxii. 160 'I never tought o' dat,' said Danny... 'Dat's a horse of anoder colour.' **1853** [see *colonialism 1 b]. **1856** C. Reade *Never too Late* I. ii. 47 A gentleman is a horse of another colour than this Robinson. **1877** [see *hoss 1]. **1937** W. S. Maugham *Theatre* ii. 23 I'll give you a three years' contract, I'll give you eight pounds a week and you'll have to work like a horse. **1937** K. A. Porter *Noon Wine* 38 He never got married, for one thing, and he works like a horse. **1948** J. Carter *Taste & Technique in Book-Collecting* (1949) ii. 24 Buxton Forman's *A Shelley Library*, however, was a horse of a different colour: no mere handlist but a fully annotated and richly informative study of Shelley's original editions. **1952** 'N. Shute' *Far Country* 80 Going into the saloon for every meal, and eating like a horse. **1966** *Listener* 5 May 661/1 A horse of a somewhat different colour is that tycoon of the brush, pop-man Salvador Dali. **1971** J. Philips *Escape a Killer* (1972) I. ii. 18 She could now 'eat like a horse'.

c. *horses for courses*: a theory that each race-horse is suited to a particular race-course, and will do better on that course than on any other; also *fig.*; *horse and horse* (U.S.): equally matched, neck and neck; *the horse's mouth*: the original, authentic source of information, esp. in phr. *straight from the horse's mouth*; *horse-and-buggy* (U.S.): bygone, old-fashioned (app. used as quasi-*sb.* in quot. 1926).

a **1859** in Bartlett *Dict. Amer.* (1859) 204, I sot down to old sledge along with Jake Stebbins. It was horse and horse, and his deal. **1898** A. E. T. Watson *Turf* vii. 160 A familiar phrase on the turf is 'horses for courses'. **1908** G. H. Lorimer *J. Spurlock* i. 3 It was horse and horse between the professors. **1926** Maines & Grant *Wise-Crack Dict.* 9/2 *Horse and buggy*, young lady out of date—with long hair. **1927** K. Eubank (*title*) Horse and buggy days. **1928** Wodehouse in *Strand Mag.* Aug. 114/1 The prospect of getting the true facts—straight, as it were, from the horse's mouth—held him..fascinated. **1929** *Daily Express* 7 Nov. 18/4 Followers of the 'horses for courses' theory. **1929** A. Huxley *Let.* 1 Dec. (1969) 320 One or other of us may very likely be over..when there will be a chance of getting your news 'straight from the horse's mouth'. **1930** F. Yeats-Brown *Bengal Lancer* xii. 172 We discuss..what Sir Mark Sykes said, straight from the horse's mouth. **1934** C. Day Lewis *Hope for Poetry* vi. 29 A pandemonium of slogans,..tips from the horse's mouth, straight talks,..etc. **1944** J. Cary (*title*) The horse's mouth. **1949** *This Week Mag.* 9 Jan. 5/1 Wherever this horse-and-buggy court is held, your chances of going scot-free are slim. **1957** *Times Lit. Suppl.* 11 Oct. 609/1 She thinks that our docility, our patience, our contentedness or even complacency with charming, outmoded, horse-and-buggy ways of doing things are, as reflected in the public will, endangering our country's future. **1958** *Listener* 7 Aug. 207/1 Keats' letters remain the horse's mouth. **1958** *Ibid.* 2 Oct. 536/1 Mr. Butler [*sc.* a sculptor] spoke his own commentary: it was an odd mixture of naiveté and insight, a 'horse's mouth' statement vastly preferable to some smooth literary piece by an art critic. **1962** *Daily Tel.* 23 Oct. 15/3 (*headline*) End of 'horse and buggy' medicine. **1963** *Punch* 18 Sept. 430/1 People enjoy what they are capable of enjoying—horses for courses. **1972** *Daily Tel.* 12 July 10/5 Horses for courses is a sound adage in motoring as well as the turf, and few British motorists would look to Czechoslovakia for their car. **1972** J. L. Dillard *Black English* vi. 242 Such horse-and-buggy terms as *whiffletree* and *singletree*.

26. b. *horse-hide* (later examples), *-line*, *-market* (earlier U.S. example), *-marrow*, *-piss* (later examples; also *fig.*), *-sausage*, *-serum*, *-show*, *-team*.

1843 *Ainsworth's Mag.* IV. 116 There were..coracles or boats of horse-hides..to be seen. **1900** *Westm. Gaz.* 16 May 7/3 Horse-hide brawn is now being extensively made. **1959** *Sears, Roebuck Catal.* Spring & Summer 528/3 Work Gloves... Top grain horsehide or cowhide drivers. **1902** J. H. M. Abbott *Tommy Cornstalk* 27 If you have the last couple of hours' 'watch' on the horse-lines, you see it all. **1932** *Times Lit. Suppl.* 3 Mar. 146/3 In the night a boy on the horse-lines was killed and the flag was taken. **1934** Blunden *Mind's Eye* 79 We were transferred to some old horse-lines. **1817** V. Brown *Jrnl.* 11 Jan. in *Maryland Hist. Mag.* (1916) XI. 372 This day spent in the Horse Market trying to sell the two Horses. **1909** *Daily Chron.* 15 Apr. 4/6 The plum pudding and horse-marrow sauce. **1922** Joyce *Ulysses* 75 The sweet oaten reek of horsepiss. **1935** L. Durrell *Spirit of Place* (1969) 33 It cost 6 dracks—3d per bottle... In England I couldn't buy a bottle of horse-piss for 3d. **1973** H. Miller *Open City* xv. 170 Dominic looked straight at Michael. 'Horse piss... You're talking baloney.' **1900** *Westm. Gaz.* 16 May 7/3 Breakfast consisted of horse sausages. **1908** *Daily Chron.* 17 Jan. 5/3 The horse-sausage boat. **1909** *Practitioner* Dec. 867 The introduction of such large quantities (200 c.c.) of horse-serum. **1926** *Encycl. Brit.* II. 772/1 Particular reference may..be made to the recent therapeutic use of horse serum in the treatment of..blackwater fever. **1964** M. Hynes *Med. Bacteriol.* (ed. 8) vii. 74 The illustration is that of a rabbit immunized with horse-serum. **1856** *Porter's Spirit of Times* 181/2 The performances at the horse show..were very interesting. **1865** *Once a Week* XIII. 133/1 Within the last few years there has been a mania for shows; we have had dog shows, donkey shows, baby shows, and last, not least, horse shows. **1931** Horse-show [see *astride c *adj.*]. **1973** *Country Life* 8 Mar. 652/1 Royal Windsor horse show, Home Park, Windsor. **1817** S. R. Brown *Western Gaz.* 107 [The walls] are..wide enough on the top to admit a horse team and waggon. **1866** *Rep. Indian Affairs* 294 At Leech lake and Winnepeg, eight government oxen and two horse teams were employed ploughing during the season. **1908** *Westm. Gaz.* 27 June 6/3 Hannah's husband was drowned whilst swimming his horse-team across the flooded river.

c. *horse-barn*, *-blanket*, *-feed* (examples), *-paddock* (examples), *-pen*, *-rack*, *-road* (earlier examples), *-rod* (examples), *-shed*, *-stable*, *-trough* (earlier and later examples).

1854 M. J. Holmes *Tempest & Sunshine* xix. 266 I'd as soon be married in the horsebarn as there. **1885** *Rep. Indian Affairs* 130 The horse-barn, carpenter-shop, warehouse, and some small buildings. **1818** in *Knickerbocker* XXIX. 470 But wiser Bill Van Snort the jockey,.. Spread his horse-blanket in the manger. **1884** 'Mark Twain' *Huck. Finn* vi. 40 There was an old horse-blanket nailed against the logs at the far end of the cabin. **1967** S. Beckett *No's Knife* 23 He spread a horse blanket on the ground in a corner on the straw. **1818** J. Owen *Jrnl.* 13 Dec. in *Southern Hist. Assoc. Publ.* (1897) I. 96 No provisions to be had for horse feed. **1823** Horse-feed [see Feed *sb.* 3 b]. **1894** *Country Gentlemen's Catal.* 23/2 The Metropolitan Tram and 'Bus Companies..have not gone scientifically into the question of horse feed. **1968** R. M. Patterson *Finlay's River* 91 They had no time to linger on that good horse-feed: summer was already on its way. **1873** Trollope *Harry Heathcote* (1874) ii. 33 While they were still in the horse-paddock, Harry turned from the track. **1966** G. W. Turner *Eng. Lang. Austral. & N.Z.* vii. 144 Before the school bus was common country schools used to provide a horse-paddock for children who rode to school. **1738** in L. Chalkley *Chron. Scotch-Irish Settlement Virginia* (1912–13) II. 376 One of ye corners of Col. Carter's Horsepen. **1839** J. K. Townsend *Narr. Rocky Mts.* v. 210 Most of the men were immediately put to work making horse-pens. **1846** J. W. Webb *Altowan* I. iv. 116 Pointing out a spot for a horse-pen..he was not long in disappearing. **1633** in *Country Life* (1972) 24 Feb. 470/1 In the stable a horsracke. **1887** *Harper's Mag.* Jan. 231/2 They alighted at the horse-rack nearest the law-office. **1739–40** Richardson *Pamela* (1740) I. 185 This Pasture..is about half a Mile, and then comes to a Common, and near that a private Horse-road. **1803** D. Wordsworth *Jrnl.* 1 Sept. in *Tour Scotland* (1874) 146 There was no horse-road..but a person on foot..might make his way. **1843** W. Carleton *Traits* I. p. xiii, Beaten on the..head, with a kind of stick between a horse-rod & a cudgel. **1968** D. Braithwaite *Fairground Archit.* iii. 49 The horse rods extended below the platform, passing through radial slots. **1768** *Penn. Gaz.* 28 Jan. in *N.J. Archives* (1904) 1st Ser. XXVI. 24 There are belonging to the premises..a new large horse-shed, [etc.]. **1836** W. Dunlap *Mem. Water Drinker* (1837) I. 120 He..seated himself upon a bench under..the horse-shed in front of the house. **1849** Thoreau *Week on Concord & Merrimack Rivers* 80 Driving a poor beast to some meeting-house horse-sheds among the hills. **1959** W. R. Bird *These are Maritimes* ii. 53 We remembered seeing many little country churches with ancient horse sheds still at the rear. **1803** in W. P. & J. P. Cutler *Life & Corr.* (1888) II. 125 Keep your horse-stable free from dung. **1854** R. Glisan *Jrnl. Army Life* (1874) xii. 160 All the hotels and public buildings have carriage houses and horse stables. **1904** T. Watson *Bethany* i. 8 Among corn-cribs, cow-pens, horse-stables, pig-styes..and worm-fenced cotton fields. *c* **1826** D. W. Jerrold in M. R. Booth *Eng. Plays of 19th Cent.* (1973) IV. 106 Oh dear, and I shall go off at last without knowing the secret. I'll stay in the horse-trough. **1867** 'T. Lackland' *Homespun* I. 140 There were times.. when the horse trough was tight frozen. **1973** P. Moyes *Curious Affair of Third Dog* i. 13 The village green, with its Victorian stone horse-trough and ugly but touching war memorial.

d. *horse-broom* (examples), *-bus*, *-cab*, *-cart* (later examples), *-railway* (U.S.), *-rake* (earlier and later examples).

1840 *Picayune* (New Orleans) 22 Aug. 2/5 A new thing has appeared in the streets of New York in the shape of a horse broom for street sweeping. **1908** *Daily Chron.* 11 Nov. 5/2 Last night the Bermondsey Borough Council decided to dispense with ten horse-brooms for street-sweeping and to employ fifty men to take their place. **1905** *Westm. Gaz.* 14 Dec. 6/3 This applies equally to motor-'buses as to horse-'buses. **1963** *Times* 24 May p. vi/3 One critic described the dozen passengers sitting six a side, opposite each other, in the poorly designed, inadequately ventilated, boxlike structures of the early London horsebus as 'trussed fowls in a poulterer's window'. **1973** G. Butler *Coffin for Pandora* vi. 135, I had not walked all the way back. There was a horse bus to Folly Bridge. **1906** *Westm. Gaz.* 31 May 12/1 Ordinary horse-cab fares. **1907** *Daily Chron.* 3 July 5/2 The immediate doom of the horse cab. **1964** in S. Nowell-Smith *Edwardian England* iii. 122 Six years later, the number of motor-taxis had grown to over 6,300 and there were now fewer than 5,000 horse-cabs. **1658** *Rec. Early Hist. Boston* (1877) II. 147 Henceforth all horse-carts shall bee led by the carters with a rayne. **1821** Horse-cart [see *ASS *sb.* 4 b]. **1863** D. G. Mitchell *My Farm* 135 An active man with a sharp scythe, a light horse-cart and a Canadian pony. **1863** W. Whitman *Specimen Days* (1882–3) 31 At the end of the then horse railway route on Seventh street. **1878** *Harper's Mag.* Jan. 192 He..thought some hunting grounds might be found near the terminus of the horse-railway. **1817** in *Trans. Ill. State Hist. Soc. 1910* 147 The ground has to be cleared of the Cornstock by..cutting them down and drawing them together with a horse Rake. **1913** D. H. Lawrence *Sons & Lovers* ix. 249 He had been on the horse-rake, and, having finished, came to help her to put the hay in cocks. **1945** 'G. Orwell' *Animal Farm* iii. 24 Boxer and Clover would harness themselves to the cutter or the horse-rake (no bits or reins were needed in these days, of course).

e. *horse-barrack* (earlier examples).

1778 J. Wesley *Jrnl.* 22 June (1938) 201 A new-built horse-barrack. **1783** W. Dyott *Diary* 28 Feb. (1907) I. 10 There is a horse barracks with one troop of the 2nd horse.

f. *horse-cabby, -duffer* (Austral.), *-stealer* (later examples), *-tender, -thief, -trainer* (examples); *horse-duffing* (later example), *-eating, -hunting, -maiming, -stealing, -thieving*, sbs. and adjs.

1939 H. Hodge *Cab, Sir?* 270 The old horse-cabby. **1963** A. Lubbock *Austral. Roundabout* 161 Horse- and cattle-duffers. **1936** M. Franklin *All that Swagger* x. 92 Cattle- and horse-duffing became staple industries in the wilds of Monaro. **1910** *Daily Chron.* 14 Jan. 1/5 Saxony is not the only horse-eating part of Germany. **1960** A. Clarke *Later Poems* (1961) 87 Horse-eating helps this ill-fare state To Sunday plate. **1848** H. W. Haygarth *Recoll. Bush Life Austral.* vi. 61 Cattle-hunting in Australia is excellent sport..with less speed than in horse-hunting. **1908** *Daily Chron.* 24 Oct. 1/2 Apart from the horse-hunting we had harness to repair. **1907** *Westm. Gaz.* 6 Sept. 5/1 Another case of horse-maiming. **1908** *Daily Chron.* 14 Aug. 8/6 Three horse-maiming outrages. **1730** in *Man. Corpor. N.Y.* (1864) 677 That one Solomon Jennings hath been a notorious Horse-Stealer for many years past. **1837** W. Irving *Capt. Bonneville* (1895) I. 41 One of the most..predatory tribes of the mountains; horse-stealers of the first order. **1737** *London Mag.* Aug. 456/1 At Norwich, 2 Men receiv'd Sentence of Death for Horse-stealing. **1837** W. Irving *Capt. Bonneville* (1895) I. 43 This wild, horse-stealing tribe. **1858** T. Vielé *Following Drum* 123 Horse-stealing from the Mexicans is a..branch of their business. **1871** E. Eggleston *Hoosier Schoolmaster* (1872) vi. 70 The whole region..had the reputation of being infested with thieves, who practiced horse-stealing. **1937** *Discovery* Aug. 245/2 The horse-stealing scenes in *The Merry Wives.* **1898** W. J. Locke *Idols* xxiii. 323 Two sturdy and swarthy peasants..pausing by the horse-tender, received a voluble account of the situation. **1907** *Daily Chron.* 12 Nov. 8/1 He pushed close to the horse-tender, a Somali. **1768** *Boston Chron.* 10 Oct. 388/3 People..who have assembled..with the view of driving all horse thieves..from amongst them. **1891** M. E. Ryan *Told in Hills* II. v. 61 She intimated yesterday that he might be a horse-thief. **1893** Kipling *Land & Sea T.* (1923) 230 Murderers, horse-thieves, and cattle-lifters. **1835** R. M. Bird *Hawks of Hawk-Hollow* II. xiii. 137 Down you rogue, or I'll indict you for horse-thieving. **1874** R. Glisan *Jrnl. Army Life* xxxii. 463 A band of..horse-thieving, prairie Indians. **1945** Dylan Thomas *Let.* 28 Aug. (1966) 283 Mean, green, horse-thieving Wales. **1889** *Cent. Dict.*, Horse-trainer. **1906** *Daily Chron.* 4 Jan. 7/1 Only one British subject is reported to have been killed during the rising—a horse-trainer, who was accidentally shot. **1921** *Daily Colonist* (Victoria, B.C.) 9 Oct. 36/2 Mr Carley, of Epsom, England, licensed horse-trainer under the English Jockey Club.

g. *horse-drawn* (examples), *-driven* adj.

1890 O. Wilde in *19th Cent.* July 140 On foot, or in horse-drawn chariot, the warriors go forth to battle. **1969** *Times* 25 Apr. 11/3 The modern equivalent of the correspondents who no doubt wrote to you when the first steam train appeared and said that we should..travel as God intended, by horse-drawn carriage. **1900** *Daily News* 2 Oct. 7/1 To his efforts..the cities of Philadelphia and Chicago owe their emancipation from the..delay of horse-driven public conveyances. **1936** *Discovery* Mar. 75/1 An ancient horse-driven windlass. **1968** C. A. Doxiadis *Between Dystopia & Utopia* 8 We still cross our big cities at nine miles per hour, which was the speed of a horse-driven cart at the beginning of the century. **1973** *Guardian* 18 May 10 A petition..was taken in a horse-driven carriage to Kensington Town Hall yesterday.

h. *horse-faced* (later examples), *-headed* adj.

1916 E. Pound *Lustra* 53 The horse-faced lady of just the unmentionable age. **1937** C. S. Forester *Happy Return* 113 One of the horsefaced mannish women whom he particularly disliked. **1910** W. J. Locke *Simon* xii. 146 The horse-headed Englishman cried '*banco*'.

27. a. horse-bite, (*a*) a bite given by a horse; (*b*) *colloq.*, a rough pinch on the thigh with the hand (example is *N.Z.*); **horse-book** (*a*), a book about horses; (*b*), a betting-book; **horse brass** (see quot. 1963); **horse-bucket** (earlier U.S. example); **horse butcher**: add to def.: also, a man who sells horse-meat (*HORSE-MEAT 2); **horse-doctor** (later examples); **horse-doctoring** (example); **horse feathers** *U.S. slang*, nonsense, rubbish, balderdash; **horse-fiddle** *U.S.* (see quot. 1872); **horse-high** *a. U.S.*, (*a*) as high as a horse; (*b*) too high for a horse to jump over; also *fig.*; **horse-holder,** (*b*) *Mil.*, each of the mounted horse artillery gunners who take charge of the dismounted horses while the gun is in action; **horse-knacker** (example); **horse lot** *U.S.*, a piece of ground on which horses are pastured; **horse manure,** (*a*) = Horse-dung; (*b*) = **horse shit*; **horse opera** *colloq.* (orig. *U.S.*), a 'Western' film or television series; **horse-ride,** (*a*) a road for horse-traffic; (*b*) a ride taken mounted on a horse; **horse shit** *U.S. slang*, nonsense; **horse-sickness**: substitute for def.: an acute virus disease of horses and related animals, marked by fever, difficulty in breathing, or swelling of the head, and endemic in Africa; add earlier and later examples; **horse's neck** *slang* (orig. *U.S.*), a beverage of ginger ale flavoured with lemon-peel, with or without the addition of whisky, brandy, or gin; **horse's tail** (see *HORSE-TAIL 1 c); **horse-tailer** [Tail *v.*[1] 5] *Austral.*, one who 'tails' or follows horses; **horse-trade** *U.S.*, a deal in horses; also *fig.* or *transf.*; hence **horse-trader** (in quots. 1963 and 1972 = heroin-trader; cf. sense 14* above); **horse-trading** *vbl. sb.*, (*a*) *U.S.* dealing in horses; (*b*) *transf.*, hard or unfair bargaining; **horse-wrangler** (later examples).

1885 'Mark Twain' in *Century Mag.* Dec. 197/1 Bowers, already irritated by the pain of the horse-bite. **1949** F. Sargeson *I saw in my Dream* II. xiii. 117 He brought his hand down smartly on Len's leg, giving him a horse-bite that made him jump. **1643** in *Essex Co. Prob. Rec.* (1916) I. 30, I give to him my horse booke alsoe a pitchforke. **1909** *Times Lit. Suppl.* 21 Jan. 23/1 Colonel..Dodge..is the author of two admirable horse-books. **1962** K. Orvis *Damned & Destroyed* xiv. 91 A guy I know runs a horse-book on University. **1911** E. Lovett *Folk Lore Horse* 3 (*heading*) Horse brasses. It is almost impossible to walk through any of our towns without meeting with horses..bedecked with certain brass ornaments. *Ibid.* 7 We now turn to the other typical horse brass, viz., the crescent. **1945** 'G. Orwell' *Animal Farm* iv. 34 It consisted of a brass medal (they were really some old horse-brasses which had been found in the harness-room). **1960** 'R. East' *Kingston Black* xiv. 135 A great glittering display of silver darts trophies and horse brasses. **1963** Bloodgood & Santini *Horseman's Dict.* 107 *Horse brasses*, decorative metal ornaments in a great variety of designs hung on draft-horse harness. Probably originating in the amulets used on camels in the East and horses in Mediterranean countries to ward off the 'evil eye'. **1827** J. F. Cooper *Red Rover* II. viii. 134 There was a horse-bucket kicking about her decks. **1905** *Daily Chron.* 26 May 5/6 Whole families have been affected by the meat, and even the horse-butcher himself who sold the meat is among the victims. **1942** E. Paul *Narrow St.* vi. 48 A golden horse above the green and white awning..was the emblem of the horse butcher, M. Monge. **1810** M. L. Weems *Lett.* (1929) III. 11 A collection in which there is not a single Bible..nor Dream-book, nor Horse Doctors. **1894** *Congress. Rec.* 1 Mar. 2500/1 We found..that he was a veterinary surgeon, called in New England a 'horse doctor'. **1930** T. S. Eliot tr. *St. J. Perse's Anabasis* 65 The vast court of the horse-doctor. **1807** W. Irving et al. *Salmagundi* (1814) xv. 345 He is..resorted to as an oracle to resolve any question about..horse-doctoring. **1928** *Amer. Speech* IV. 98 Mr. William De Beck, the comic-strip comedian..assumes credit for the first actual use of the word horsefeathers. **1934** J. O'Hara *Appointment in Samarra* (1935) vi. 163 'And my orders is to see that you keep your knees together, baby.' 'Horse feathers,' she said. **1936** [see *BUTTON *v.* 3]. **1936** Wodehouse *Laughing Gas* xvi. 173 'Oh, horse-feathers!'.. The expression which she used was new to me, but one could gather its trend. Her ribald and offensive tone jarred upon me. **1966** A. A. Fair *Widows wear Weeds* xiv. 160 'We lose our licence; Sellers gets fined and maybe you get prosecuted for perjury.' 'Horsefeathers!' Bertha snorted. **1967** J. Gardner *Madrigal* ii. 21 Mostyn pointed out that..they could court-martial him *in camera*... On reflection, Boysie realised that this was all a load of horse feathers. **1807** J. Jennings *Let.* 19 Sept. in *Ind. Hist. Coll. Soc. Publ.* (1932) X. 164 The French..convened around the house of the new couple..playing on horse fiddles. **1843** *Knickerbocker* XXI. 46 The clangor of trumpets, the clattering of pans, the grinding of horse-fiddles. **1872** E. Eggleston *End of World* xlvi. 294 Bill Day had a gigantic watchman's rattle, a hickory spring on a cog-wheel. It is called in the West a horse-fiddle, because it is so unlike either a horse or a fiddle. **1911** H. Quick *Yellowstone Nights* viii. 212 In addition to the horse-fiddles and bells and horns Absalom had arranged some private theatricals. **1859, 1879** Horse-high [see *hog-tight* adj. s.v. *HOG *sb.*[1] 13 a]. **1896** W. A. White *Real Issue* 147 In the summer the field stood horse-high with corn. *a* **1930** D. H. Lawrence *Sex, Lit. & Censorship* (1955) 12 The Clean Books League, whose object was to make the law..'horse-high, pig-tight, and bull-strong'. **1972** *Christian Science Monitor* 28 Sept. 16/4 The pioneers..tipped the stumps up with their roots in the air, and lined them along so they were, as the saying went, 'horse-high, hog-tight, and bull-strong'. **1837** A. F. Oakes *Madras Horse Artillery* 18 The rear rank are horse-holders. **1875** *Man. Field Artillery Exerc.* VIII. 285 The horse-holders do not dismount. **1902** J. H. M. Abbott *Tommy Cornstalk* 35 When cavalry are dismounted for skirmishing, one man of every four—the horse-holder, or number three—is out of action. **1936** C. S. Forester *General* 7 In a long straggling line..lay the troopers of the squadron..firing away. In a gully to the rear..were the horses and horseholders. **1937** *John o' London's* 5 Feb. 762/2 [A girl] promised to wait for me. She didn't, though. Too impatient. Married a horse-knacker. **1847** W. T. Thompson in *Spirit of Times* 24 July 250/2 Way he went.., down around the house, through the horse lot, and into the old field. **1850** *Rep. Comm. Patents 1849: Agric.* 144 The man..has..no time to make manure, or to haul out and spread the little that is dropped in his horse-lot. **1889** *Harper's Mag.* June 123/2 In the horse lot she found her father putting on his coat. **1966** *Publ. Amer. Dial. Soc.* XLII. 19 *Horselot*, the enclosure around a stockbarn. **1843** S. L. Dana *Muck Man. Farmers* (ed. 2) vi. 135 The dung of pigeons is 2-7ths stronger than horse manure. **1954** A. G. L. Hellyer *Encycl. Garden Work* 127/2 Horse manure can be used safely for all plants and crops for which animal manure is desirable. **1956** *Dict. Gardening* (R. Hort. Soc.) (ed. 2) II. 1011/1 Horse manure is especially valuable in mushroom growing. **1957** J. Osborne *Entertainer* viii. 67 She's not interested in all that horse manure about Canada. **1964** *New Statesman* 1 May 674/1 Imagine his furious indignation if a similar *contretemps* had arisen (as well it might) at a Washington club because one had been reported as saying on television that federal policy on racial discrimination in the South is a load of horse-manure. **1927** *Motion Picture Classic* 2 July 26/1 Horse Opera..is an opus of the West where men are cowboys. **1948** 'J. Tey' *Franchise Affair* xix. 227 That happens only in detective-stories and the last few minutes of horse-operas. **1957** E. Hyams *Into Dream* 244 An officer who looked like a bad-man Mexican in an early horse-opera. **1958** *Times* 17 Nov. 8/6, 21 per cent. of the available time is devoted to westerns (28 of these 'horse-opera' series are being broadcast). **1903** *Daily Chron.* 6 Nov. 3/3 On one side of it there will be a horse-ride, and on the other..a gravelled walk for foot passengers. **1906** W. Owen *Let.* 15 Aug. (1967) 31 Mr Smallpage has just been for a horse-ride. **1955** M. McCarthy *Charmed Life* (1956) iii. 66 Pardon me if I say that's horse shit. **1959** W. Burroughs *Naked Lunch* 98 Gentlemen of the jury,..are we to gulp down this tissue of horse shit? **1970** *It* 12–25 Feb. 2 'This is definitely the weekend of the big bust!' 'Horseshit! You've said the same thing for the past six weekends!' **1822** J. Campbell *Trav. S. Afr. 2nd Journey* I. ii. 32 The horse sickness..was prevailing much at that time. **1899** W. J. K. Little *Sk. & Stud. S. Afr.* (ed. 2) iv. 93 Let us hope..rinderpest, horse-sickness, and the locusts will yet be conquered. **1947** J. Stevenson-Hamilton *Wild Life S. Afr.* vi. 50 The zebra..has..the immense advantage of being entirely immune from Horse Sickness and Nagana disease. **1963** Jubb & Kennedy *Path. Domestic Anim.* II. 585/1 Even in districts where horsesickness recurs annually, the distribution tends to be limited to low-lying areas such as valleys, swamps, and areas with summer rain. **1903** 'C. E. Merriman' *Lett. from Son* 177 Every man then visited the tool-house, where a tin wash-boiler filled with what they call here 'horse's neck', a savage compound of whiskey and hard cider, occupied the place of honor. **1925** J. Metcalfe *Smoking Leg* 138 A tall young man in a grey suit whose drink was horse's neck in summer and Burton in the winter. **1936** E. Ambler *Dark Frontier* xiii. 219, I ordered a Horse's Neck, remembered how bad the gin was ..and had a small beer instead. **1938** L. MacNeice *I crossed Minch* II. xv. 211 If I could sit in a garden shady With a Horse's Neck or a White Lady. **1968** *Daily Tel.* (Colour Suppl.) 13 Dec. 43/3 Brandy and soda, or brandy and some soft drink like ginger ale—the Horse's Neck—have become women's drinks. **1933** *Bulletin* (Sydney) 16 Aug. 20/2 Then the horse-tailer pulled out, and an abo. was given the job. **1954** B. Miles *Stars my Blanket* xxiii. 202 The 'horse tailers', whose job it was to look after the mob of spare horses. **1968** K. Weatherley *Roo Shooter* 100, I was only twenty at the time and working as horse-tailer for a cattle-drover. **1846** *Knickerbocker* XXVIII. 361 He was employed in..an action brought by a man against another for cheating him in a 'horse-trade'. **1902** A. D. McFaul *Ike Glidden* iii. 22 His self-confidence could not admit of a thought that he could be fooled in a horse trade. **1923** *Daily Mail* 15 June 9 [New York World] We hope that foreign Powers will not be weak enough to consent to such unprincipled horse trade. **1850** L. H. Garrard *Wah-to-Yah* vii. 109 The unfair horsetrader might have taken my scalp. **1912** I. S. Cobb *Back Home* 3 The swapping ring below the wagonyard was..clamorous with the chaffer of the horse-traders. **1932** W. Faulkner *Light in August* xiii. 283 A room where it will be quiet when her time comes, and not every durn horsetrader or courtjury that passes through the hallway. **1963** T. Tullett *Inside Interpol* iv. 44 There is another process ..that of turning the morphine into heroin, and the price then soars to £600 a pound. Even this gigantic profit is not enough for the 'horse-traders', as these criminals are known. **1972** H. C. Rae *Shooting Gallery* II. 67 And how many horse-traders are there in this part of the world? **1826** T. Flint *Recoll.* 64 Horse-trading..seems to be a favorite and universal amusement. **1853** J. G. Baldwin *Flush Times Alabama* 273 An enterprising young gentleman..engaged..in the horse-trading line. **1891** C. Roberts *Adrift Amer.* 190 We stayed in camp here for two days, during which time we did some horse-trading. **1902** A. D. McFaul *Ike Glidden* iv. 28 A horse-trading expedition. **1939** I. Baird *Waste Heritage* xix. 262 There is too damn much horse-trading going on around here. **1947** *Forum* 24 May 4/3 It was certain that some hard bargaining had taken place in that upstairs conference room, political horse-trading at which South Africa's politicians are adept. **1969** *Times* 25 Apr. 28/3 Employers in the Lancashire textile industry were accused..of 'horse trading'. **1971** *New Scientist* 20 May 460/1 A powerful ingredient in all this horsetrading is the continually deteriorating employment situation in the aerospace industry. **1902** O. Wister *Virginian* x. 109 The foreman of the 76 outfit, and the horse-wrangler from the Bar Circle-L. **1968** R. M. Patterson *Finlay's River* 102 Various packers and horse wranglers, with their pack-trains, moved with the party or made rendezvous with them as the work demanded.

b. horse ant (later examples); **horse bot:** substitute for def.: the fly *Gasterophilus intestinalis*, esp. its larva which is a parasite of horses; also **horse bot-fly**; add examples; **horse-stinger** (later examples).

1908 *Westm. Gaz.* 16 Apr. 10/3 The 'horse-ant',..

(*Formica rufa*), the big fellow which builds the great heaps, usually of pine-needles. **1945** C. P. Haskins *Of Ants & Men* ix. 167 Typical of this group [*sc.* raiding ants] is *Formica rufa*, the 'horse ant' or 'fallow ant' of England and Europe, whose great thatched mounds form a conspicuous feature of the German forests. **1840** J. & M. Loudon tr. *Köllar's Treat. Insects* I. 53 The Horse-bot, a larva proceeding from a fly resembling a humble-bee with two wings. **1925** A. D. Imms *Gen. Textbk. Ent.* III. 652 The horse bot-flies..lay their eggs on the hair. **1928** Metcalf & Flint *Destructive & Useful Insects* xxii. 779 The common horse bot may easily be told by the faint smoky spots on the wings. **1968** *Oxf. Bk. Insects* 138/2 *Horse-bot fly* (Gasterophilus intestinalis). This belongs to a separate sub-family in Muscidae—the Gasterophilinae. **1910** *Encycl. Brit.* VIII. 468/1 Dragon-fly... In many parts of England are termed 'horse-stingers'. It is almost needless to say that (excepting to other insects..) they are perfectly innocuous. **1966** 'J. Hackston' *Father clears Out* 134 Flies, mosquitos, beetles, March flies, blowflies, and horsestingers.

c. horse-bean: substitute for def.: a leguminous plant grown as food for cattle, as *Vicia faba, Canavalia ensiformis, Parkinsonia aculeata*, or their seeds; add earlier and later examples; **horse mushroom**, a species of edible mushroom, *Agaricus arvensis*, larger and coarser than the common mushroom; **horse-nicker**, a large West Indian shrub, *Cæsalpinia bonduc*, or its seeds; **horse poison**, a West Indian plant, *Isotoma longiflora*; **horse-weed** *U.S.* (examples).

1684 I. Mather *Remark. Prov.* (1890) 216 The stone weighed about seven grains, being much in the shape of our ordinary horse-beans. **1942** Castetter & Bell *Pima & Papago Agric.* 60 Of somewhat less importance were the seeds of..Jerusalem thorn or horsebean (*Parkinsonia aculeata*). **1964** J. M. Kingsbury *Poisonous Plants of U.S.* 43 *Parkinsonia aculeata*, horsebean. *Ibid.* 362 *Vicia faba* L. Fava bean, broad bean, horse bean. **1965** E. G. B. Gooding et al. *Flora of Barbados* 198 *Canavalia ensiformis* (L.) DC. (Horse bean, Overlook bean, Sword bean, etc.) is sometimes grown as a vegetable. The young pod is sliced, and eaten like French beans. **1866** Lindley & Moore *Treas. Bot.* II. 598/1 The horse-mushroom need not be excluded on account of its supposed unwholesomeness. **1890** R. D. Blackmore *Kit & Kitty* III. iii. 39 Mingling with the true Agaric some very fine 'Horse-mushrooms'. **1966** *Times* 28 Apr. 16 Horse mushrooms are bigger and stronger flavoured than proper field mushrooms. **1750** Horse-nicker [see Nicker *sb.*[4]]. **1871** C. Kingsley *At Last* I. i. 36 The grey horse-nicker-beads of our childhood. **1965** E. G. B. Gooding et al. *Flora of Barbados* 176 *Caesalpinia bonduc*... Grey nicker, Horse-nicker... Shrub, often scrambling by means of prickles. **1851** P. H. Gosse *Nat. Sojourn Jamaica* 80 One of the most venomous of plants (*Isotoma longiflora*) commonly called Horse-poison. **1955** *W. Indian Med. Jrnl.* IV. 73 *Isotoma longiflora*..Madam Fate; Star Flower; Horse Poison. **1790** L. Castiglioni *Viaggio negli Stati Uniti* II. 333 *Collinsonia canadensis*, Lin. Horse-weed. **1874** J. W. Long *Amer. Wild-Fowl Shooting* xxiii. 239 The hunter usually selects a position..amongst the high 'horse-weeds' bordering the field. **1892** B. Torrey *Foot-path Way* 72 Acres and acres of horseweed. **1963** H. A. Gleason *Illustr. Flora North-eastern U.S.* III. 475/1 *Conyza canadensis* (L.) Cron. Horseweed. Coarse annual... A weed in waste places.

horse, *v.* Add: **13. a.** To make fun of, to 'rag', to ridicule; to indulge in horseplay; to fool *about* or *around*. orig. *U.S.*

1901 *Munsey's Mag.* XXV. 407/1 Because we chose to chew his statements and remove the bones before we swallowed them, he developed the idea that we had no interest in the work and were trying to 'horse' him. **1901** F. H. Spearman *Held for Orders* 173 'Are you horsing me?' he exclaimed, raising his voice angrily. **1928** P. Buranelli et al. *Cryptogram Bk.* p. i, Always playing jokes on each other, they began to 'horse' each other cryptographically. **1928** *Amer. Speech* III. 219 *Horse around*, to indulge in ill-timed trifling or horse-play. **1939** R. Chandler *Trouble is my Business* (1950) 8 Quit horsing around. **1942** *R.A.F. Jrnl.* 2 May 15 Why must you continually horse around, Ginger? **1952** W. G. Hardy *Unfulfilled* 48 Peter horsed around and ducked Elise and she ducked him. **1954** 'W. Henry' *Death of Legend* 32 Dingus was really mad about it; he wasn't just horsing now. **1959** E. Allen *Man who chose Death* v. 49 You saw scores like him..laughing and horseing with the pretty..young Italian girls. **1959** *Punch* 10 June 776/2 The professor thought I was horsing about and came down to me. **1961** Wodehouse *Ice in Bedroom* vi. 47 When you've cleaned up pretty good, you don't want to be horsing around down in the suburbs. **1961** J. Heller *Catch-22* (1962) xxxii. 340 They were having a whale of a good time as they helped each other set up their cots. They were horsing around. **1971** *It* 2–16 June 7/1 Two black kids..were horsing around just outside the club.

b. To philander; to 'sleep *around*'.

1952 S. Kauffmann *Philanderer* (1953) ii. 32 It isn't as if I didn't love her. I'd die for her. Literally. Then why do I have to go horsing around with dames? **1956** S. Longstreet *Real Jazz* 67 'This is a respectable band,' he said, 'and there ain't goin' to be any immoral horsin' goin' on. Whoever you start sleepin' with this trip, that's how you end the tour!' **1956** C. Smith *Deadly Reaper* xxv. 201 She'd be horsing around with Nicky, giving me grounds for divorce.

horse-back. Add: **3.** (Earlier example.)

1851 J. S. Springer *Forest Life* 41 The pumpkin Pine is..found..also on abrupt ridges, called horse-backs.

5. horseback opinion *U.S.*, an opinion given (as) from horseback, without opportunity for full consideration of the question.

1879 *Congress. Rec.* 23 Apr. 728/1, I am not here as a judicial authority or oracle. I can only give a horseback opinion. **1903** A. Adams *Log of Cowboy* vi. 72 My sister gives it as a horseback opinion that she'd been engaged to this fellow nearly eight months.

horse-boat. Add: **1. b.** A type of landing-craft.

1907 *Daily Chron.* 25 Feb. 6/3 Exercises with the new form of steel horse-boat were carried out..on Saturday by an A.S.C. contingent... At the stern is a large flap which falls down to facilitate landing. **1923** *Man. Seamanship* (H.M.S.O.) II. 215 Horseboats would be excellent for landing troops except for the fact that they are unhandy.

2. For def. read: *N. Amer.* A boat whose paddle-wheels are driven by horses working a tread-mill. †*Obs.* exc. *Hist.* (Further examples.)

1823 I. Holmes *Account U.S.A.* 315 One of the horse-boats which cross the North River..is of a very peculiar construction. **1829** *Brockville* (Upper Canada) *Gaz.* 28 Aug. 2/5 In crossing the lake..the Steamboat Montreal and the horse-boat came in contact with each other. **1897** R. E. Robinson *Uncle Lisha's Outing* 257 They wended their way to the ferry just in time to see the horse boat come splashing into port, the four horses plodding their unprogressive journey on the revolving wheel. **1928** *Old-Time New Eng.* Apr. 161 Before the wreck of the *Belknap* a new type of freighter, the horse-boat, appeared on the lake. The first one of its kind was built on Long Island in 1838. It was an open scow propelled by paddle-wheels, the motive power being supplied by a pair of horses tramping a treadmill.

horse-breaker. Add: **†2.** A courtesan; a demi-mondaine; a prostitute. Freq. *pretty horsebreaker. Obs.*

1861 *Times* 28 June 12/4 They are, in the end, only fit for the company of 'horsebreakers'. *Ibid.* 12/5 If she thinks men prefer the society of 'pretty horsebreakers'.. she never was more mistaken. **1861** *Punch* 13 July 13 (*caption*) *Stout Equestrian*. 'Dou [*sic*] you know, love, I'm rather sorry I got this hat, for suppose I should be taken for a "pretty horsebreaker"!' **1865** *Public Opinion* 30 Sept. 352/2 These *demi-monde* people,..horsebreakers, hetairæ..are by degrees pushing their way into 'society'. **1865** C. Knight *Shadows of Old Booksellers* v. 112 He.. was intimate with all the high — (read 'pretty horse-breakers') in town... Some of these girls he often asked to dine with him. **1966** D. Crow *Theresa* xvi. 189 While Brewster's aim was to show her up as a pretty horse-breaker, hers was to proclaim her purity. **1970** G. Greer *Female Eunuch* 265 The vestiges of sensual innocence hung around long enough to endow us with obsolete terms like..*pretty horsebreaker*.

horse-gear. 1. (Later examples.)

1941 *Proc. Prehist. Soc.* VII. 130 The bulk of the hoard is Celtic horse-gear. **1960** G. E. Evans *Horse in Furrow* xvii. 220 An unorthodox use of a piece of horse-gear.. should be recorded.

2. (Later examples.)

1904 Goodchild & Tweney *Technol. & Sci. Dict.* 291/1 *Horse gear*, the device used for yoking horses to machinery. **1940** *Chambers's Techn. Dict.* 119/2 *Horse gear*, a device for producing mechanical power by means of a lever attached to gears, which is operated by..[a] horse walking in a circle.

horse guard. 4. (Earlier and later N. Amer. examples.)

1828 A. Wetmore *Diary* 30 May in *Mo. Hist. Rev.* (1914) VIII. 185 Formed our wagons into an oblong square, and set a horse guard. **1907** J. R. Cook *Border & Buffalo* (1938) 279 The horse-guard brought in the horses. **1971** J. McDougall *Parsons on Plains* v. 40 When you reach the horse guard..tell him to catch my horse Badger for you.

horsehair. Add: **c. horsehair snake** *U.S.*, = *horsehair-worm*.

1897 *Outing* (U.S.) XXX. 434/2 The creature referred to as a mystery is what is termed the 'horsehair snake', in reality, a hairworm. **1949** *Sci. Monthly* Jan. 56/2 Another mythical serpent, confined to the rural scene, is the horse-hair snake..the worm *Paragordius varius*. **1966** *Publ. Amer. Dial. Soc.* XLII. 19 Horsehair snake. Actually a long slender worm (*Gordius*, sp.), which spends one portion of its life in the body of a large insect and the other in shallow water, as in a watering trough. Thus arose the folk belief that these worms were originally horsehairs.

horsehead. 4. (Examples.)

1884 *Bull. U.S. Nat. Museum* No. 27. 438 *Selene argentea*..Horsehead. **1905** D. S. Jordan *Guide to Study of Fishes* II. xvii. 276 *Selene vomer*, the horse-head-fish, or look-down, is similarly but even more distorted. **1966** Leim & Scott *Fishes of Atlantic Coast of Canada* 251 Atlantic lookdown... *Selene vomer* (Linnaeus) 1758. Other common names: horsehead, moonfish.

horse-hoer. (Examples.)

1744 W. Ellis *Mod. Husb.* I. 145 It will be a difficult Point for the Horse-hougher to keep his Horse and Instrument..exact. **1935** J. Joyce *Let.* 19 Feb. (1966) III. 345 If they put on *Floradora* with George Robey in the name part the horsehoers would not notice.

horse-jockey. Add: **b.** *U.S.* One who traffics in horses. Hence **ho·rse-jo:ckeying** *vbl. sb.*

1744 A. Hamilton *Itinerarium* (1907) 31 May 3, I met one Matthew Baker, a horse-jockey. **1783** in S. E. Baldwin *Simeon Baldwin* (1919) 120 The conversation was upon News—horsejockeying—& other indifferent subjects. **1784** in *Connecticut Hist. Soc. Coll.* (1930) XXIII. 204 Ship Building is carried on with Vigor in this State, & the Horse Jockey business flourishes. **1792** J. Belknap *Hist. New-Hampshire* III. ix. 144 Few [horses] live and die on the plantations where they are bred; some are exported..; but the most are continually shifted from one owner to another, by means of a set of contemptible wretches called horse-jockies. **1866** 'Mark Twain' *Lett. fr. Hawaii* (1967) 50 The Kanaka horse jockey is fertile in imagination and elastic in conscience. *Ibid.* 288 Brown bought a horse from a native at Waiohinu for twelve dollars, but happening to think of the horse-jockeying propensities of the race, he removed the saddle and found that..recent hard riding had polished most of the hide off his back.

horseless, *a.* Add: Applied *c* 1895–1910 to motor vehicles.

1895 [in Dict.]. **1900** *T. Eaton & Co. Catal.* Christmas 5/1 Horseless carriage, large size, wind with key, best spring, 50 c. **1901** *Chambers's Jrnl.* Jan. 14/2 The three great trusts that control the making and using of these horseless machines. **1905** *Ibid.* Jan. 111/1 Farmers, who now look askance at horseless vehicles. **1973** J. Wainwright *Devil you Don't* 144 An Olde Worlde coaching town which hadn't yet got round to noticing the arrival of the horseless carriage.

horse marine[2]. Add: **2.** Phr. *tell that to the horse marines*: a colloquial expression of incredulity. Cf. Marine *sb.* 4 c.

1892 [in Dict.]. **1921** W. S. Maugham *Circle* II. 65 *Elizabeth*: He's never even kissed me. *Arnold*: I'd try telling that to the horse marines if I were you.

ho·rse-ma:stership. [See -ship.] Skill in managing horses.

1904 *Westm. Gaz.* 21 Sept. 10/1 Thereupon a trumpeter, who knew something about horse-mastership, instantly sounded the forage-call. Two minutes later every animal was standing quietly in its appointed place. **1925** E. O. Shebbeare in E. F. Norton *Fight for Everest, 1924* 365 Unless any other member of the Expedition better fitted by his knowledge of horsemastership undertakes the care of the mess mules. **1927** *Daily Tel.* 26 Apr. 16/7 The Army in India polo team..has undertaken a very difficult and interesting feat of horse-mastership. The problem to be solved is that of transporting a large stable of ponies from India to America. **1937** H. Belloc *Crusade* iii. 67 The rest of the work was foot-slogging and horsemastership. **1971** *Country Life* 30 Dec. 1829/3 The present upsurge of interest in all aspects of horse-manship and horsemastership.

horse-meat. Add: **2.** = Horse-flesh 1.

1853 S. N. Carvalho *Jrnl.* 1 Feb. in J. Bigelow *Mem. J. C. Fremont* (1856) xvii. 441 He sent a Mexican.. with cooked horsemeat. **1868** *All Year Round* 22 Feb. 252/2 Horse-meat is a common..article of food. **1870** *Food Jrnl.* 1 Dec. 620 The almost impossibility of obtaining beef and mutton naturally forced the use of horse-meat upon the people. **1904** *Westm. Gaz.* 23 Dec. 10/1 Every day the purveyors and the consumers of horsemeat increase in numbers. **1910** *Daily Chron.* 14 Jan. 1/5 The price of horse meat was 3½d. a lb. **1972** T. P. McMahon *Issue of Bishop's Blood* (1973) xi. 171 A very suave-type fellow then came to Hymie with a tale about..buying frozen horse meat to meet the chain's need for huge quantities of frozen beef.

horse-mill. (Later U.S. examples.)

1848 E. Bryant *California* xxi. 269 The flour consumed by Captain Sutter is ground by a very ordinary horse-mill. **1889** *Century Mag.* Feb. 520/2 A horse-mill—a long pole on which a man sits, and to which a horse or mule is hitched. **1971** *Country Life* 30 Sept. 848/1 Some of these groups of stone-built barns, byres and arcaded cartsheds, relieved on occasion by a circular horse-mill, [etc.].

horse plum. 2. *N. Amer.* Substitute for def.: Either of the two common wild plums, *Prunus americana* or *P. nigra*.

1790 S. Deane *New-Eng. Farmer* 222/2 The horse-plum, a very pleasant tasted juicy fruit, of a large size. **1827** *Western Monthly Rev.* I. 322 The fruit is of the size of a common horse plum. **1908** N. L. Britton *N. Amer. Trees* 494 Canada Plum—*Prunus nigra* Aiton. Also called Horse plum, this is a small, bushy tree, occurring in woods and neglected lands from Newfoundland to Alberta, and southward to Georgia.

horse-power. Now commonly written as one word. Add: **1. c.** With qualifying words, esp. **brake horsepower,** the power available at the shaft of an engine, measurable by means of a brake; **indicated horsepower,** the power produced within the cylinders, as shown by an indicator.

1859 W. J. M. Rankine *Man. Steam Engine* v. 479 Nominal Horse-power is a conventional mode of describing the dimensions of a steam engine, for the convenience of makers and purchasers of engines, and bears no fixed relation to indicated or to effective horse-power. **1881** [in Dict., sense 1]. **1887** *Encycl. Brit.* XXII. 493/2 The efficiency of the mechanism is the ratio of the 'effective' or 'brake' horse power to the indicated horse power. **1904** *Westm. Gaz.* 30 Mar. 10/2 The *Good Hope* and the *Drake*..are of 14,100 tons displacement and 30,000 indicated horse-power. **1943** A. P. Fraas *Aircraft Power Plants* vi. 110 The power required to turn the engine over, or friction horsepower,..increases rapidly with rpm. **1968** R. H. Bacon *Car* ii. 17 The power developed in the cylinder..is called the indicated horsepower or i.h.p. Some of this power is absorbed by the friction of various parts of the engine. The power remaining, that is the power that can be used for work, is called the brake horsepower or b.h.p. **1972** *Daily Tel.* 15 Mar. 11/5 The latest car is almost a hundredweight heavier while the net brake horsepower is only up by six.

4. *Comb.* **horsepower-hour,** a unit representing the work performed or energy consumed in working at the rate of one horsepower for one hour.

1899 J. Perry *Steam Engine* xvi. 250 Units of Energy used Commercially. 1 horse-power hour = ..1,980,000 foot-pounds. **1906** *Westm. Gaz.* 28 June 2/1 An output of 36 cubic feet of oxygen per horse-power hour. **1949** G. P. Sutton *Rocket Propulsion Elem.* i. 18 The specific fuel consumption is based on the horsepower output, and its units are pounds of fuel per horsepower-hour. **1963** F. D. Jones & Schubert *Engin. Encycl.* (ed. 3) 662, 1 horsepower-hour = 0·746 kilowatt-hour.

horse-sense. For '*U.S. colloq.*' read '*colloq.* (orig. *U.S.*)'. (Add earlier and later examples.)

1832 J. K. Paulding *Westward Ho!* II. 60 He's a man of good strong horse sense. **1911** H. S. Harrison *Queed* xviii. 223 A rich vein of horse-sense underlay Byrd's philanthropic enthusiasms. **1920** R. Macaulay *Potterism* II. iii. 88, I am sure both parties credited them with too much idealism and too little plain horse-sense. **1930** Chesterton *Four Felons* 252 What I say is horse-sense, for all that. **1942** *Short Guide Gr. Brit.* (U.S. War Dept.) 2 We can defeat Hitler's propaganda with..plain, common horse sense; understanding of evident truths. **1960** I. Wallach *Absence of Cello* 72 Summoning up his best horse sense (and trying to forget that the horse is an uncommonly stupid animal), Andrew said, 'I agree with Mr Clifton.' **1973** E. Lemarchand *Let or Hindrance* v. 41 All horse sense on your part, old man, but I've got a tiresome hunch that it's not going to be as straight-forward as all that.

horseshoe, *sb.* Add: **2. e.** A horseshoe bend. *U.S.*

1795 in *Amer. Speech* (1963) XXXVIII. 183 In the bend, or horse shoe..is a neck of land about 4 or 500 yards wide. *a* **1910** [see *FETCH *v.* 9].

f. *Logic.* (See quots. 1926, 1954.) Cf. *HOOK *sb.*[1] 10 d.

1926 H. M. Sheffer in *Isis* VIII. 231 The authors (1)..inform us, 'unofficially', that they are privately convinced that 'either *p* is false or *q* is true' (2) *means* '*p* implies *q*'; and they henceforth regard '*p* horseshoe *q*' as *meaning* '*p* implies *q*'. **1952** F. B. Fitch *Symbolic Logic* 15 The horseshoe symbol can be read as 'implies', but a more accurate reading is the *if-then* reading. **1954** I. M. Copi *Symbolic Logic* ii. 17 We introduce the new symbol '⊃', called a horseshoe, to represent the partial meaning common to all conditional statements. **1959** *Listener* 30 Apr. 757/2 He spent another [term] worrying about the ordinary use of the words 'if-then' whereas logicians had assured us that nothing but the 'horse-shoe' was worth talking about.

5. b. *horseshoe arch* (later example), *curve, moustache, table* (earlier examples).

1770 Duchess of Northumberland *Diary* 10 June (1926) 139 In the midst of the Room were two Horse Shoe Tables the ends of which pretty near touching form'd a Kind of Oval. **1857** C. M. Yonge *Let.* 1 Oct. (1903) viii. 212 A great horse-shoe table, holding 116 people. **1875** 'Mark Twain' in *Atlantic Monthly* Aug. 192/2 The water cuts the alluvial banks of the 'lower' river into deep horseshoe curves. **1926** F. M. Ford *Man could stand Up* I. i. 17 The gentlemen with sergeant-majors' horse-shoe moustaches. **1950** G. Brenan *Face of Spain* ii. 42 The double horseshoe arches, striped buff-white and brick-rose, arrest one by their strangeness and novelty.

d. horseshoe magnet (earlier example).

1785 G. Adams *Ess. Magnetism* (ed. 2) 419 To touch horseshoe magnets.

horse-tail. Add: **1. c.** Usu. horse's tail. A woman's hair-style in which the hair is arranged to resemble the shape of the tail of a horse; a 'pony-tail'.

1872 Trollope *Eustace Diamonds* (1873) II. xxxiv. 100 How a man can like to kiss a face with a dirty horse's tail all whizzling about it, is what I can't at all understand. **1953** R. Fuller *Second Curtain* v. 79 Her hair done in a fringe and horse's tail. **1955** G. Freeman *Liberty Man* I. iii. 48 She had blondish hair tied into a horse's tail with a piece of black ribbon. **1960** 'J. & E. Bonett' *No Grave for Lady* ii. 27 She wore jeans and a cotton sweater, her hair was in a horse-tail.

d. (See quots.)

1880 L. Higgin *Handbk. Embroidery* i. 8 'Japanese gold thread'..must..be laid on, and stitched down with a fine yellow silk, known as 'Maltese', or 'Horse-tail'. **1960** B. Snook *Eng. Hist. Embroidery* 104 A woman's court dress (1780)... Variation is obtained by the use of floss silk and horsetail, a tightly twisted silk.

horsewhip, *sb.* (Later examples.)

1843 *Knickerbocker* XXII. 56 I'd like to have some on 'em tied to a tree, and have fair play at 'em with this horse-whip. **1856** P. Cartwright *Autobiogr.* (1858) viii. 74 They came drunk, and armed with..horse-whips. **1965** S. Hult tr. *G. de Coulteray's Sadism in Movies* 111 It is horsewhips with which Christine Nordeu and Bery Baxter are armed in *Idol of Paris*.

horsfordite (hǭ·ɹsfǭɹdəit). *Min.* [f. the name of E. N. *Horsford* (1818–1893), American chemist: see -ite[1].] A brittle, silvery white antimonide of copper, perhaps Cu_5Sb.

1888 Laist & Norton in *Amer. Chem. Jrnl.* X. 62 This new mineral is named horsfordite, in honor of Professor Horsford, formerly Rumford Professor of Chemistry at Harvard University. **1923** J. W. Mellor *Inorg. & Theoret. Chem.* III. xxi. 7 The antimonide: horsfordite, Cu_5Sb. **1970** *Min. Abstr.* XXI. 340/2 The properties of cuprostibite are compared with and differ considerably from those of horsfordite, chalcostibite, tetrahedrite, and famatinite.

horst (hǭɹst). *Geol.* [a. G. *horst* heap, mass, cluster, sandbank, etc.; introduced in its geol. sense by E. Suess (*Antlitz d. Erde* (1883) I. I. iii. 167).] A block of the earth's surface which has been raised relative to the surrounding land and is bounded by faults on some or all sides.

1893 *Q. Jrnl. Geol. Soc.* XLIX. 77 We have, therefore, sunken *massifs* both west and east of the Dürrenstein; that mountain itself remains at a higher level between the two, and may be called a 'Horst' in the sense originally applied by Suess. **1904** H. Sollas tr. *Suess's Face of Earth* I. I. iii. 126 If the outer borders of two fields of subsidence approach each other so that a ridge is left between them, on both sides of which the two areas of depression descend more or less in the form of steps, then we have what we shall distinguish, making use again of a common mining word, as a *horst*. **1910** [see *GRABEN]. **1914** G. A. J. Cole *Growth of Europe* ii. 22 Far older masses have asserted themselves..as horsts, that is, as upstanding blocks from which material has been faulted down on all sides. **1942** M. P. Billings *Struct. Geol.* xi. 205 Horsts range in size from those that are only a few inches wide to those that are many miles wide. **1944** A. Holmes *Princ. Physical Geol.* xix. 416 Between the horsts of the Vosges and the Black Forest the Rhine flows through a rift valley. **1970** *Sci. Amer.* Feb. 37/3 The evidences of crustal movement are in plain view as wide-open fissures, horsts and grabens that form a classic graben structure of steps down the sides of a major depression.

Horst Wessel (hǭɹst ve·sěl). The name of the author (1907–1930) of the words of the official anthem of the German Nazi party, used *attrib.* in *Horst Wessel lied, song*. Also *ellipt.*

1937 V. Bartlett *This is my Life* x. 165 Hundreds of arms went out in the Hitler salute, hundreds of voices yelled the *Horst Wessel Lied*. **1968** R. Collin *Locust on Wind* xv. 137 This time it will be lounge suits instead of brownshirts, and no Horst Wessel while we're on television. **1968** J. Blackburn *Young Man from Lima* xvii. 173 The Horst Wessel song was one of the greatest hymns of hate that had ever been composed. **1969** *Guardian* 6 Oct. 10/6 In spite of the fears of churchmen, 'Lay, lady, lay' is likely to cause less damage than the 'Horst Wessel Lied'. **1970** *New York* 16 Nov. 35/2 Someone wondered when the cornball band..was going to..get down to the serious strains of the Horst Wessel Song.

hortal, *a.* Delete *Obs. rare*[−1] and add later example.

1926 G. C. Druce in J. J. Walker *Nat. Hist. Oxford Distr.* 119 Other hortal plants which have become established are several species of North-American Asters.

hortensia (hǭɹte·nsiă). [mod.L. (P. Commerson in A. L. de Jussieu *Genera Plantarum* (1789) 214), f. *Hortense*, adopted Christian name of the wife of J.-A. Lepaute (1720–*c* 1787), French clockmaker.] A variety of the common hydrangea, *Hydrangea macrophylla*, var. *hortensia*.

1799 *Curtis's Bot. Mag.* XIII. 438 Authors have entertained very different opinions as to what this plant really is; Jussieu following Commerson makes it an Hortensia,..Dr. Smith an Hydrangea. **1866** Lindley & Moore *Treas. Bot.* II. 598/2 Hortensia. (Fr.) *Hydrangea Hortensia*. **1906** *Daily Chron.* 9 Feb. 5/1 The bouquet was of artificial hortensias. **1961** *Amateur Gardening* 7 Oct. 25/1 A medium sized pure white 'hortensia'. **1969** *Dict. Gardening* (R. Hort. Soc.) Suppl. 42/1 Hydrangea. Cultivars of *H.* × *macrophylla*... (*a*) Hortensias. Globose corymbs of mostly sterile flowers.

horticultural, *a.* Add: **2.** Comb. *horticultural exhibition, fête, show*.

1847 F. A. Kemble *Let.* 23 June in *Rec. Later Life* (1882) III. 204 Some *unique* foreign flower, sent..to blossom at the Chiswick horticultural exhibition. **1840** J. Romilly *Diary* 16 July (1967) 197 We then went to the Horticultural Fete in the dining room..at Downing. **1843** *Ainsworth's Mag.* IV. 100 A dazzling conservatory that looked like a horticultural show. **1853** C. M. Yonge *Heir of Redclyffe* i. 3 This [camellia] is to go to the horticultural show. **1858** Queen Victoria *Let.* 21 Apr. in R. Fulford *Dearest Child* (1964) 93 We have just returned from a horticultural show in St. James's Hall. **1952** *Oxf. Jun. Encycl.* VI. 14 The most important horticultural show is the Chelsea Flower Show. **1973** P. Moyes *Curious Affair Third Dog* viii. 104 A sort of scroll proclaiming that Henry Heathfield had taken second prize for tomatoes at the County Horticultural Show.

horticulturally (hǭɹtikv·ltiŭrăli, -tʃəɹ-), *adv.* [See -ly[2].] In the way of horticulture.

1899 O. Wilde *Importance of being Earnest* II. 69, I spoke horticulturally. My metaphor was drawn from fruits. **1927** *Observer* 27 Nov. 13/3 It is not only horticulturally that the Riviera is making ready.

hortonolite (hǭɹtǫ·nŏləit, hǭ·ɹtənŏləit). *Min.* [f. the name of Silas Ryneck *Horton* (b. 1820), American amateur mineralogist + -o + -lite.] A silicate mineral, $(Fe, Mg)_2SiO_4$, yellow or greenish-yellow on fresh fracture, having a preponderance of iron over magnesium and often some substitution by manganese (see also quot. 1955).

1869 G. J. Brush in *Amer. Jrnl. Sci.* XCVIII. 19 It is proper to designate this new variety with a special name, and I propose for it the name *Hortonolite*, after Mr. Horton, who first discovered the mineral. **1939** *Amer. Mineral.* XXIV. 24 At present the names commonly accepted, in order of increasing richness [of Fe_2SiO_4] in the fayalite molecule are forsterite, chrysolite, hyalosiderite, hortonolite and fayalite. **1955** M. H. Hey *Index Min. Species* (ed. 2) 124 *Hortonolite*. 4[(Fe, Mg)$_2$SiO$_4$], with Fe > Mg (in Hortonolite proper, Fe_2SiO_4 50 to 70 mols. %)... The original Hortonolite contained 4% MnO. Olivine family. **1968** Embrey & Phemister tr. *Kostov's Mineralogy* 291 Titanolivine is rich in Ti; hortonolite is rich in iron.

‖ **hortus conclusus** (hǭ·ɹtŭs kǫnklū·sŭs). [Lat., = enclosed garden, in reference to *Song Sol.* iv. 12.] **a.** An enclosed, inviolate garden; in spiritual and exegetical tradition, the symbol of the soul, the Church, or the virginity of Mary. **b.** In Art, a painting of the Madonna and Child in an enclosed garden. Freq. *transf.*

1624 Donne *LXXX Sermons* (1640) xvii. 165 The University is a Paradise, Rivers of knowledge are there, Arts and Sciences flow from thence. Counsell Tables are *Horti conclusi* (as it is said in the Canticles). **1852** A. B. Jameson *Legends Madonna* p. xlviii, The Enclosed Garden (*Hortus conclusus*)... I have seen this enclosed garden very significantly placed in the background of the Annunciation, and in pictures of the Immaculate Conception. Sometimes the enclosure is formed of a treillage or hedge of roses, as in a beautiful Virgin by Francia. **1940** 'M. Innes' *There came both Mist & Snow* i. 7, I shall get more surely on the rails if I drop ancestry and the *hortus conclusus* of history and begin again with some account of the Priory itself. **1947** 'N. Blake' *Minute for Murder* ix. 208 People that would trample over her little *hortus conclusus*. **1956** M. Swan *Paradise Garden* xvii. 161 The protective hedges of his *hortus conclusus* were rotten and useless with disease. **1957** N. Frye *Anat. Criticism* 152 The symbol of the body of the Virgin as a *hortus conclusus*. **1963** *Listener* 21 Mar. 520/2 Nepal has long been the *hortus conclusus* of the fabulous Ranas.

Horus (hō·ɹŭs). [late L., ad. Gr. Ὧρος, ad. Egyptian *Hor*.] The name of an Egyptian deity, represented in art as having the head of a hawk, used *attrib.* to designate an image of this deity.

1851 [see *HATHOR]. **1875** *Encycl. Brit.* VII. 715/2 Hathor is associated with Horus, but her genealogical place is not clear. **1923** *Glasgow Herald* 13 Feb. 8 The horus birds. **1923** D. H. Lawrence *Birds, Beasts & Flowers* 199 The golden Great and glistening-feathered legs of the hawk of Horus. **1972** *Times* 18 May (Egypt Suppl.) p. iv/3 The falcon was sacred to Horus, the sky god who bore the sun on his head.

Hosanna. Add: **B. 2.** *attrib.* Hosanna Sunday, Palm Sunday.

1868 M. E. C. Walcott *Sacred Archæol.* 422 Hosanna (save us, we pray) Sunday, in the East and Southern Europe. **1899** J. Hastings *Dict. Bible* II. 418/2 In Christian usage, Palm Sunday, to which our Lord's entry has given name, has in certain periods and regions been called 'Hosanna Sunday' or 'Day of Hosannas', or simply 'Osanna'.

hose, *sb.* Add: **5. c.** *Golf.* The socket into which the shaft of an iron club is fitted.

1893 H. Hutchinson *Golfing* 21 A method of obviating the trouble of occasionally hitting the ball on the hose of these short-faced clubs. **1953** R. Harris *Sixty Yrs. Golf* iii. 37 The sand-track iron is a most formidable looking bludgeon... The hose or socket is 6¼ inches.

7. *hose-carriage* (earlier example), *-cart* (earlier and later examples); also **hose company** *U.S.*, a company in charge of a fire-hose; **hose-pipe** = Hose *sb.* 3; hence as *v.*, to spray (as) with a hose.

1829 *Massachusetts Laws* XI. 237 If any person shall.. damage..any Engine, Hose Carriage,..he shall be convicted. **1865** *Trans. Ill. Agric. Soc.* VI. 320 A large amount of three-inch hose always ready on hose carts. **1906** *Westm. Gaz.* 27 July 6/2 Two new..fire-stations,.. with a horsed-escape, a manual fire-escape, and a hose-cart. **1806** *Massachusetts Spy* 21 May (Th.), The efforts of several hose and fire companies at length prevailed. **1860** O. W. Holmes *Prof. Breakf.-t.* iv. 106 More widely known through the Movamensing Hose-Company, and the Wistar parties. **1948** *Times-Picayune Mag.* (New Orleans) 5 Dec. 21/2 The next and last contestant is Sound Point Protection Hose Company Number One! **1835** J. Martin *Gaz. Virginia* 139 The height of the reservoir, above these streets,..gives a jet of water by means of hose pipes, of some 60 to 80 feet elevation. **1872, 1884** Hose-pipe [in Dict.]. **1928** *Daily Express* 27 Sept. 8 A stream of the chief's choicest acid would be hosepiped his way. **1930** R. Campbell *Poems* 17 With your fountain-pen to spray the flowers, The hosepipe of your literary hours. **1940** *Flight* 5 Dec. 468*a*/1 A stripped Lewis gun as used on trawlers and such-like ships for 'hose-piping' dive-bombers and low-fliers. **1948** Partridge *Dict.*

Forces' Slang 96 *Hosepipe*, to spray liquid fire from a flame-thrower. **1973** J. Fleming *You won't let me Finish* xvii. 133 Around the cage were elephantine hose pipes to be used in case of emergency to protect the people at ground level.

hose, *v.* Add: **2.** Also with *down*. Also *fig.* and *transf.* and as *sb.*

1931 E. E. Cummings *Let.* 7 Jan. (1969) 119 Precisely as a cannon exploded, searchlights hozed the environs. **1936** [see *BALDY]. **1939** H. Hodge *Cab, Sir?* ii. 15 A shirt-sleeved washer..is hosing down a late-night cab. **1947** *Book Nine* (Caxton Press, N.Z.) 23 Treat him [*sc.* the patient] sweet, Floss. Hose him out and get him clean for us. **1961** W. Vaughan-Thomas *Anzio* v. 86 SP guns, out on the right, hosed them with fire, but they pushed on resolutely. **1964** C. Willock *Enormous Zoo* ix. 170 Roger Wheater hoses him with a blistering swathe of Swahili. **1969** *Gloss. Landscape Work* (*B.S.I.*) v. 25 *Hosing down*, the application of water by means of a hose to clean down buildings or other surfaces; or as a daily routine to control pests, dirt and humidity. **1971** *Guardian* 26 Aug. 22 Visitors to Piccadilly Circus, London, keeping out of the way of the regular hose-down.

hosier. Add to def.: Also used more generally for a men's outfitter or haberdasher.

1837 N. Whittock et al. *Compl. Bk. Trades* 289 *Haberdasher and hosier*. Each of those trades merge in the other, occasionally;..our Hosiers seldom confine their dealings to the wares which clothe our legs. **1892** P. L. Simmonds *Comm. Dict. Trade Products* 197/2 *Hosier*, one who deals in stockings, shawls, gloves, braces, laces, and under garments, etc. **1921** *Dict. Occup. Terms* (1927) 344/3 *Hosier*,..is usually also an outfitter. **1935** Wodehouse *Luck of Bodkins* xvi. 192 Monty passed a finger round his collar. A perfect fit, made to measure by the finest hosier in London, it seemed to be too tight. **1936** —— *Laughing Gas* xii. 119 In the matter of pyjamas I've always been a trifle on the choosy side. I'm not one of those fellows who just charge into a hosier's and grab anything.

hospital, *sb.* Add: **6.** *hospital nurse, tent*; **hospital bed,** (*a*) a (metal) bed as used in hospitals, higher than an ordinary bed to facilitate nursing, and freq. adjustable in several ways; also *hospital bedstead*; (*b*) an available place in hospital for a bed patient; **hospital blue**(s), the blue uniform worn by wounded soldiers in the wars of 1914–18 and 1939–45; **hospital corps,** the medical corps in the U.S. Navy; so **hospital corpsman**; cf. *CORPSMAN; **hospital letter,** a letter referring a patient for free treatment in a hospital; **hospital paper** = **hospital letter*; **hospital porter** (see *PORTER *sb.*[2]); **hospital ship,** (*b*) a ship for conveying sick and wounded soldiers to their own country or to an area remote from the battlefield; **hospital steward** (*a*) (examples); **hospital train,** a train for conveying wounded soldiers from the front to the base hospitals.

1823 C. Mathews *Let.* 7 Feb. in A. Mathews *Mem. Charles Mathews* (1838) III. 365, I slept in a bed on the road without even posts for curtains—a regular hospital-bed. **1952** *Oxf. Jun. Encycl.* X. 190/2 Small sums paid weekly during health might entitle contributors to a hospital bed in time of need. **1970** *New Yorker* 29 Aug. 55/1 He lends people hospital beds, which he happened to get at a good price. **1973** C. Mullard *Black Brit.* iv. 46 Overnight, blacks were suddenly held responsible for the unemployment figures.., lack of adequate social services, schools and hospital beds. **1860** F. Nightingale *Notes on Nursing* viii. 47 Hospital bedsteads are..very much less objectionable than private ones. **1919** 'I. Hay' *Last Million* vi. 67 Convalescent soldiers in hospital blue. **1920** J. M. Barrie *Kiss for Cinderella* (1928) III. 445 Danny, who is slightly lame and is in hospital blue. **1920** [see *AIRER]. **1957** R. Campbell *Portugal* ii. 29, I was clothed in army hospital-blues. **1973** B. Turner *Hot-Foot* iv. 20 Some day I would be spotted as a wanted man by someone who knew me either in khaki or in hospital blues. **1899** *Statutes at Large U.S.A.* XXX. 474 June 17, 1898..Be it enacted..That a hospital corps of the United States Navy is hereby established. **1945** *Amer. Handbk.* (Office War Information) xxvi. 386 Men of the Hospital Corps include pharmacists... They may be found in the amphibious units of the Marine Corps, in the dressing stations of warships, and in submarines. **1901** E. Root in Bacon & Scott *Mil. & Colonial Policy U.S.* (1916) 374 An order was made fixing the enlisted strength..exclusive of hospital corps men, at 77,287. **1943** *Sci. News Let.* 29 May 343 Soon a Hospital Corpsman with a larger kit of supplies comes along and quickly ministers to the wounded man. **1890** W. Booth *In Darkest Eng.* I. iii. 26 He had hoped to have obtained a hospital letter at the Mansion House so as to obtain a truss for a bad rupture. **1848** Thackeray *Van. Fair* lvii. 514 Women..who are hospital-nurses without wages. **1893** O. Wilde *Lady Windermere's Fan* IV. 120 You would like me to retire into a convent or become a hospital nurse, or something of that kind. **1936** A. Thirkell *August Folly* ix. 265 Jessica got up and the hospital nurse left. **1962** A. Christie *Mirror Crack'd* ii. 13 In real illness you could have a proper hospital nurse, at vast expense and procured with difficulty, or you could go to hospital. **1838** C. M. Yonge *Let.* 25 Sept. in C. Coleridge *C. M. Yonge* (1903) iv. 139 Mr. Rudd, the tall man we took the hospital paper to, is dead. **1890** —— *More Bywords* 260 Jane Cox is come for a hospital paper, ma'am. **1683** Hospital ship [in Dict.]. **1899** *Westm. Gaz.* 20 Nov. 6/2 The American hospital-ship *Maine*. **1916** 'Boyd Cable' *Action Front* 172 Swinging at top speed down the line to the base and the hospital ship and home. **1944** F. Clune *Red Heart* 12 I've seen my mates consigned to the deep on a hospital ship..in war-time. **1856** R. Glisan *Jrnl. Army Life* (1874) 350 Besides the sick, hospital steward, hospital attendants, and some three others, there will be no troops. **1895** *Outing* (U.S.) Dec. 255/2 The non-commissioned staff comprises a sergeant-major, a quarter master-sergeant, a commissary-sergeant, and a hospital steward. **1812** *Niles' Reg.* II. 131/1, 300 hospital, horsemen's and common tents. **1862** G. B. McClellan *Let.* 31 July in *Own Story* (1887) 458 They are nearly all in hospital-tents and are well provided for. **1936** C. Day Lewis *Friendly Tree* i. 14 Holding the wound shut..until he reaches the hospital-tent. **1874** B. F. Taylor *World on Wheels* I. xxiv. 162 The saddest train upon which the writer ever took passage was the Hospital Train, with its maimed and mangled burden. **1916** 'Boyd Cable' *Action Front* 172 That he might be lying warm and comfortable in the soothing ease of a bed in the hospital train. **1937** V. Bartlett *This is my Life* iv. 51, I spent my twenty first birthday in a hospital train... The Medical Officer in charge..was an old school friend.

hospitality. Add: **1. c.** Applied in conventional phr. to the admission of correspondence, etc., to a newspaper.

1913 *Punch* 16 Apr. 298, I thank you for the hospitality of your columns.

d. *to partake of* (or *enjoy*) *His* (or *Her*) *Majesty's hospitality*: to be in prison. *jocular.*

1894 *Strand Mag.* Sept. 296/2 Partaking of Her Majesty's hospitality, in a building specially erected for gentlemen who need a compulsory holiday. **1938** F. D. Sharpe *Sharpe of Flying Squad* xiv. 157 One awful thug I know, who is now enjoying His Majesty's hospitality, beat his woman up regularly.

4. *attrib.* and *Comb.* (Further examples.) Now commonly used of a room, suite of rooms, etc., in a hotel, TV studio, etc., set aside for the entertainment of guests with drinks.

1959 *Encounter* Aug. 8/1 The eminent person is ushered into..the 'hospitality' room [at the BBC]. **1960** *News Chron.* 2 Mar. 4/4 The drinks, consumed rather dismally in the hospitality room,..conspire to make a TV appearance an ordeal. **1963** J. N. Harris *Weird World Wes Beattie* (1964) ii. 20 At these conventions a lot of big companies put in hospitality suites—snake rooms, they call them. **1966** M. Hastings *Cork on Telly* i. 10 There was no time to talk to him in the hospitality room before the programme. **1969** *Observer* 8 June 38/1 The Americans led off by introducing Shatalov and Yeliseyev to the taste of bourbon from the hospitality cupboards of the private room at the American exhibition stand. **1970** *Globe & Mail* (Toronto) 25 Sept. 40/1 Opportunities ..in the ever-growing hospitality field [in motels]. **1970** *Globe Mag.* (Toronto) 26 Sept. 6/3 There aren't many bands or hospitality suites or funny hats. **1971** *Bahamas* XXIII. III. 34/1 She moved into the hospitality industry where her personality assets were plus factors. **1972** *Guardian* 20 Mar. 11/1 She said it in the Hospitality Room ..right after the show.

hospitalize (hǫ·spităləiz), *v.* [f. Hospital *sb.* + -ize.] *trans.* To place or accommodate in a hospital.

Freq. commented on as an unhappy formation.

1901 *Daily Chron.* 10 Sept. 6/2 The disease was spreading rapidly owing to the people refusing to hospitalise first cases. **1904** *Ibid.* 28 Oct. 8/3 The pauper who is hospitalised in an English casual ward. **1946** *Nature* 3 Aug. 170/1 Cases hospitalized in the Carmichael Hospital for Tropical Diseases, Calcutta. **1955** *Sci. News Let.* 15 Jan. 39/3 Surgery..was performed on 70% of the hospitalized children. **1961** *Observer* 19 Nov. 29/6 On the second day after the mutiny the ship's doctor insisted that the lives of a passenger and a wounded member of the crew depended on their being hospitalised. **1970** G. Greer *Female Eunuch* 239 The more the state undertakes to protect a man from illness..the more it has the right to sacrifice him..to hospitalize his children.

Hence **ho:spitaliza·tion,** confinement to, or accommodation in, a hospital.

1909 in Webster. **1918** A. Woollcott *Let.* 12 Jan. (1946) 40 My present brief hospitalization is traceable to eye-strain. **1932** *Nation* 25 May 604 The activities of the Legion..began with a demand for pensions and hospitalization for those disabled in war service. **1937** R. S. Morton *Woman Surgeon* vi. 70 The systematic hospitalization of the sick poor furnished a wealth of clinical material. **1967** *Spectator* 28 July 114/1 Private hospitalisation and medicine has been increased in price by this Government so that, while the rich people can cope, the moderately well off can no longer do so. **1971** Cowdry & Steinberg *Care of Geriatric Patient* (ed. 4) xxvii. 352/1 Application of new scientific knowledge to clinical medicine has profoundly affected the hospitalization requirements of elderly patients.

hoss (hǫs), dial. (also U.S.) var. of Horse *sb.*

1. = Horse *sb.* 1 and 25 a.

1815 D. Humphreys *Yankey in Eng.* v. 77 The boys.. see a ghost in the form of a white hoss; and an Indian in every black stump. **1849** N. Kingsley *Diary* (1914) 88 My supper consisted of beans, old hoss, and hard bread. **1877** J. M. Bailey *Folks in Danbury* 37 But this is a hoss of another colour. **1877** F. Ross et al. *Gloss. Holderness* 78 Hoss, a horse. **1887** T. Darlington *Folk-Speech S. Cheshire* 225 Hoss-wesh, a horse-pond. **1888** F. R. Stockton *Dusantes* 15 'These hosses won't do much at holdin back,' he said. **1889** M. Peacock *Taales fra Linkisheere* 130 In cums a great black hoss, all drippin' wi' wet. **1958** S. E. Hyman in A. Dundes *Mother Wit* (1973) 55/1 He had him a big black hoss.

2. *U.S.* = Horse *sb.* 4.

1843 in *Amer. Speech* (1965) XL. 130 Sez he old hoss, I'll eat you up jis like I would a cabbage. **1844** *Yale Lit. Mag.* X. 168 Well, hoss, I reckon I will. **1847** *Knickerbocker* XXIX. 204 We've sworn to give the first strange old hoss that comes along this road an up-and-down beating. **1847, 1857** [see Horse *sb.* 4]. **1904** W. N. Harben *Georgians* xv. 148 'Went back on us, ole hoss,' Jim said, cordially. **1940** *Amer. Speech* XV. 216/1 Varmint might be applied to a contemptible person..but in Florida hound or hoss would as likely be used.

2*. = *HORSE *sb.* 14*. *U.S. slang.*

1968 *Sun Mag.* (Baltimore) 13 Oct. 19/3, I was square to them, one step lower because I wasn't on hoss.

3. *attrib.* and *Comb.* = Horse *sb.* 26 and 27.

1837 R. M. Bird *Nick of Woods* II. iv. 63, I do [know].. all the parts injacent and outjacent, circumsurrounding the hoss-stump. **1877** J. Habberton *Jericho Road* xv. 142 Losin' your health when you started with Sam Reeves' hoss-gang. **1909** 'O. Henry' *Roads of Destiny* xvi. 263 Amos Colvin and me were thicker than two hoss thieves more than ten years. **1942** Berrey & Van den Bark *Amer. Thes. Slang* § 608/9 *Western picture*, hoss opera, opery.

host, *sb.*[2] Add: **2.** *mine host* (later examples).

1909 Wodehouse *Mike* xii. 66 That was the supreme moment in mine host's life. **1934** —— *Right Ho, Jeeves* x. 123, I entered the saloon bar and requested mine host to start pouring. **1973** J. Porter *It's Murder with Dover* vii. 67 Dover set about cross-questioning the landlord... Mine host stuck to his guns.

3. b. *Biol.* and *Med.* An animal or person that is the recipient of tissue, an organ, etc., that has been transplanted into it from another.

1906 *Brit. Med. Jrnl.* 22 Dec. 1796/1 Entire organs may be transplanted by anastomosis of their blood vessels to suitable points on the circulatory apparatus of the host. **1923** H. Neuhof *Transplantation of Tissues* i. 9 Experiments..continued from six to nine months showed the transplant completely replaced by a connective tissue derived from the host. **1958** *New Biol.* XXVII. 42 The host reacts to the antigens of the graft by producing antibodies. **1961** *New Scientist* 5 Jan. 45/3 Cells of both the donor of the graft tissue, and the recipient host, played a part in the immune reaction.

5. a. *Geol.* A mineral mass containing a different mineral; a rock containing an ore deposit or foreign rock. Freq. *attrib.*

1889 *Cent. Dict.* III, *Host*, a mineral which incloses another. **1950** E. E. Wahlstrom *Introd. Theor. Igneous Petrol.* iv. 80 Perthites are intergrowths of two feldspars. Commonly the host mineral (that is, the more abundant mineral) is potash feldspar. **1965** G. J. Williams *Econ. Geol. N.Z.* i. 1/1 These are the host-rocks of the gold mineralization west of the Alps.

b. *Physics* and *Chem.* A crystal lattice or molecular structure containing a foreign ion, atom, or molecule; *spec.* (i) a crystal or a crystalline material to which a small amount of some impurity has been added to make it luminescent; (ii) that component of a clathrate compound that encloses or surrounds the other component. Usu. *attrib.*

1939 *Trans. Faraday Soc.* XXXV. 126 It may be imagined that the function of the activating impurity is to enter the host lattice and produce therein centres of distortion. **1950** H. W. Leverenz *Introd. Luminescence of Solids* vi. 369 A host crystal should not be strongly absorbing in the spectral region where efficient luminescence emission is to be produced. **1956** *Nature* 22 Dec. 1410/1 Various instances are known of continuous crystalline 'host' lattices within which 'guest' molecules may be trapped... Among inorganic lattices the best known are the crystalline zeolites. **1961** L. F. & M. Fieser *Adv. Org. Chem.* xxii. 771 They are similar to the urea inclusion complexes..except that the host is a covalently bonded molecule with a hole in it. **1967** Vanders & Kerr *Mineral Recognition* v. 85/2 At high temperatures, foreign ions are dissolved in the solid host crystal. **1968** *Proc. Internat. Conf. Luminescence Budapest* (1966) I. 1289 Luminescence of crystalline phosphors is strongly connected with the structure of the host crystal.

6. *attrib.* and *Comb.* (see also sense *5), as (sense 1) *host country, government, population, society*; (sense 3) *host animal, cell, -parasite* (used appos.), *plant*; *host-specific* adj. (so *host-specificity*); *host-controlled, -induced* adjs.

1956 *Nature* 10 Mar. 453/1 It is reasonable to attribute these changes to the high dose of radiation delivered to the host-animal and so to identify the cells containing them as host cells. **1954** *Science* 16 July 110/1 (*heading*) Action of T_2r^+ bacteriophage on host-cell membranes. **1958** *Spectator* 22 Aug. 252/1 The life of the virus is so closely linked with that of the host cell that one cannot be attacked without injury to the other. **1965** *Ann. Rev. Microbiol.* XIX. 365 Host-controlled modification of viruses is a general term applied to those cases in which passage through certain host strains imparts one or more new, nonheritable properties to the virus without altering its genetic information content. **1959** *Manch. Guardian* 21 July 1/7 The technical staff of control posts..should consist of one-third host-country nationals. **1962** *Times* 18 Dec. 11/7 The 'host countries'..are [not] able economically to absorb the refugees permanently. **1969** H. MacInnes *Salzburg Connection* ix. 131 Switzerland is the host country. **1960** *Economist* 15 Oct. 263/1 Certain of the important 'host governments' of countries where the oil is actually produced chose not to come. **1953** *Cold Spring Harbor Symp. Quant. Biol.* XVIII. 237 (*heading*) Host-induced modifications of viruses. **1946** *Nature* 6 July 30/2 Host–parasite relations. **1964** V. J. Chapman *Coastal Veget.* iii. 77 Plants of course may exhibit host-parasite relations. **1888, 1889** Host plant [in Dict., sense 3]. **1959** Southwood & Leston *Land & Water Bugs* viii. 239 This species may be found in great abundance on its host-plant. **1971** *Guardian* 24 June 7/2 The feelings of any host population towards immigrants included suspicion, fear, and irritation. 1961 S. R. Herman in J. A. Fishman Readings Sociol. Lang. (1968) 509 Our analysis implies a cautionary note for the attitude of a host society to newcomers. The host society may legitimately wish to encourage the use of its language by the newcomers, but it may aggravate the problem of adjustment if it insists too strictly on the adherence to its language norms. 1969 Listener 13 Nov. 664/1 Just one sign of the equivocal welcome meted out by the host society. 1972 Jrnl. Social Psychol. LXXXVI. 159 The changers displayed more of behavioral and psychological affinities toward the host society than the nonchangers. 1969 New Scientist 13 Mar. 23/1 WHO workers introduced a host-specific pathogenic fungus which kills mosquito larvae. 1951 Whitby & Hynes Med. Bacteriol. (ed. 5) xxiv. 376 The viruses have the same wide range of host-specificity as the bacteria.

host, sb.⁴ Add: **2.** Also applied to the wafer before consecration (quots. 1687, 1881).

host, v.² **1.** Delete † Obs. and add later examples. Also, to receive into one's town, country, etc.; to be the host at (a party, dinner, etc.); to compère (a television show, etc.).

1939 John o' London's 7 Apr. 42/3, I am not surprised that Messrs. Auden and Isherwood came away with a sense of deep and humble respect for the people and the country who had hosted them. 1957 J. Kerouac On Road (1958) 87 He went right out and bought a pint of whisky to host me proper. 1958 I. Asimov Whiff of Death (1968) vi. 55 Those comments applying to him were read at the celebration dinner hosted (invariably) by Anson himself. 1967 Boston Sunday Herald 26 Mar. i. 12/2 Dorchester man has been hosting a series of unusual farewell parties. 1967 Boston Globe 30 Mar. 15/2 The delegation of 25 men and women..will be hosted by the Greater Boston Chamber of Commerce and Burlington town officials. 1967 New Yorker 1 Apr. 129 (Advt.), Bob Hope hosts the annual Academy of Motion Picture Arts and Sciences presentation. 1969 N. Cohn AWopBopaLooBop (1970) ix. 88 He..sold Murray the K T-shirts and hosted albums of Murray the K's Golden Gassers. 1969 New Yorker 29 Mar. 24/1 Let us host you at the Kahala Hilton, on one of Oahu's most beautiful beaches. 1969 Guardian 19 July 9/1 There will be David Frost, hosting ITV's night-long bonanza of touchdown and pop stars. 1973 Times 9 Feb. 8/8 The session was hosted by Mr William Fulbright.

b. Delete nonce-use and add later examples.

1957 Time 2 Sept. 34/2 Critic John Crosby, currently on leave from his TV syndicated column to polish up on his broadcast manners, will host. 1962 H. Kane Killer's Kiss xxviii. 216, I must do a bit more hosting—I'm expecting some special people.

hosta (hǫ·stă). [mod.L. (L. Trattinick 1812, in Archiv der Gewächskunde I. 55; formerly used by N. J. Jacquin Icones Plantarum Horti Schönbrunnensis (1797) I. 60 for a plant now included in the genus Cornutia), f. the name of N. T. Host (1761–1834), Austrian physician.] A plant of the genus so named (formerly called *funkia), native to Japan and eastern Asia and belonging to the family Liliaceæ; a plantain-lily.

[1828 Bot. Reg. XIV. 1204 Willdenow proposes to unite Hosta with Cornutia. 1930 L. H. Bailey in Gentes Herbarum II. 119 Mostly they [sc. plantain lilies] are known as species of Funkia, but Hosta is now the accepted name to those who follow the International Rules of nomenclature.] 1931 W. T. Stearn in Gardeners' Chron. 11 July 27/1 The Hostas, or Funkias, as they are commonly called, are hardy Liliaceous perennials, well-known in gardens on account of their decorative foliage. 1961 Amateur Gardening 4 Nov. 23/1 For the moist but well-drained part plants that you could grow are..hydrangeas,..hemerocallis, hostas. 1971 Country Life 25 Feb. 438/1 All hostas have good leaves, but there are some which we more particularly grow for foliage, than for flowers.

hostage, sb.¹ Add: **3.** spec. in phr. to give, etc., a hostage to fortune: to deliver one's future happiness, success, etc., into the hands of fate.

1607–12 [in Dict.]. 1732 T. Fuller Gnomologia 253 Wife and Children are Hostages given to Fortune. 1875 M. E. Braddon (title) Hostages to fortune. 1934 J. E. Neale Queen Elizabeth xiv. 235 But to enter on war was to give hostages to fortune. Her instinct was to gamble on avoiding it. 1950 W. S. Churchill Second World War IV. i. xi. 194 Once several good outfits are prepared, any one of which can attack a Japanese-held base or island and beat the life out of the garrison, all their islands will become hostages to fortune. 1965 Listener 4 Nov. 728/2 Every manufacturer who indulges in advertising is giving a hostage to fortune in that he is inviting public confidence in his goods and service, and he will rapidly go out of business if he cannot live up to his claims. 1968 Ibid. 28 Nov. 710/2 Denmark was almost a hostage to fortune by being in Nato at all.

5. attrib.

1858 F. J. Child Eng. & Scottish Ballads VI. 234 Fifteen lords in the hostage house Waiting Wallace for to see. 1905 Westm. Gaz. 26 Sept. 8/2 The hostage camps [in the Gaboon]. 1909 Daily Chron. 18 Nov. 1/7 The dragging of a man to the hostage house [in the Congo]. 1931 H. G. Wells Work, Wealth & Happiness of Mankind (1932) vii. 275 Here all over again were the murders, the mutilations..the hostage houses, and the atrocities.

hostel, sb.¹ Add: **2. b.** = *youth hostel. Hence **ho·steller**; **ho·stelling** vbl. sb.

1931 Times 21 Jan. 11/6 The Youth Hostels Association. ..It is hoped that this summer there will be 50 hostels where..a night's lodging may be obtained. 1932 Y.H.A. Handbk. Hostels 19 Much of the Y.H.A. Insurance of property in Hostels is now undertaken by the National Office. Ibid., Write to the Warden at the Hostel and quote your membership number. 1951 E. Coxhead One Green Bottle i. 26 House used to be a climbing centre, but now it's been taken for one of those youth hostels. Pity. Some of the hostellers climb, I believe. Ibid. x. 256, I treated myself to a few days' hostelling. 1961 Countryman LVIII. 434 There were always plenty of hostellers to fill the boat. 1967 Rambling & Youth Hostelling ('Know the Game' Series) 31 The YHA Map is useful for tour planning. It will enable you to see the network of hostels in the area you are visiting. 1970 P. Carlon Souvenir xii. 106 Her daughter's friends were allowed to go hostelling and hiking. 1973 Guardian 1 Jan. 6/3 In the most expensive..hostels, the overnight charge will go up from 35p to 40p for young hostellers.

hostess. Add: **2.** Also in archaic phr. mine hostess.

1832 [in Dict.]. 1962 Listener 19 July 113/3 Mine Hostess, as La Locandiera is called in Clifford Bax's otherwise excellent English translation, is one of the dozen or so greatest comedies.

b. A woman employed to entertain customers at a night-club, etc.; also in derogatory sense: a prostitute.

1931 Durante & Kofoed Night Clubs xxi. 190 Joan Sawyer—and what a girl she was!—the foremost dancer and hostess of the day, ran the shows. 1933 New Statesman 18 Mar. 331/2 If a young man dances with a 'hostess' he scurries back at the finish. 1937 C. R. Cooper Here's to Crime ix. 189 Women, reaching the dregs, become the 'hostesses' in what are known as 'jukin' joints'. 1959 Economist 11 Apr. 105/2 This question of hostesses is one of the most hotly debated in the night club business. It is also revealing since, broadly speaking, it is a management's policy towards hostesses that determines the standing of a night club. 1963 Listener 7 Feb. 260/3 American Bunny Clubs, with their Freudian fantasy-tease hostesses. 1968 Sunday Times 30 June 5/5 He..got Miss C. to admit ..'that the word hostess is a polite euphemism for prostitute'. 1970 G. Greer Female Eunuch 131 Many a prostitute, whether she calls herself a..hostess, or a common whore, imagines that she is exploiting the male sex. 1970 Guardian 22 Aug. 8/6 The French authorities have been able to organise a traditional Legion recreation centre, whose Somali hostesses are under regular medical supervision. 1972 P. A. Whitney Snowfire (1973) i. 2 Juniper Lodge in the Poconos was looking for an après-ski hostess to help with guests during the evening hours. 1973 Times 30 May 4/3 We were able to call it Europe's biggest hostess service and had an interesting offer of more than 200 girls.

c. = *air hostess; also, a woman similarly employed on a train.

1936 Jrnl. R. Aeronaut. Soc. XL. 525 On American aeroplanes for many years now we have had, as part of the staff, 'hostesses', young women of 20 to 30 years of age, generally trained nurses, whose duties are to make passengers 'feel at home' aloft. 1940 Amer. Speech XV. 213/2 Hostess..now suggests a professional person paid for her services, as the hostess at a hotel, on an airplane, or on pullman cars. 1953 R. Lehmann Echoing Grove 252 She'll be a hostess on a transglobal airline. 1958 Times Lit. Suppl. 17 Oct. 598/4 Illuminating testimonies by hostesses of various air lines. 1970 Times 17 Nov. 19/7 Five years ago..she was one of the first Gatwick hostesses. They're British Rail's answer to an air hostess.

4. hostess apron, dress, gown, pyjamas, robe, skirt, trolley.

1968 Wanganui (N.Z.) Chron. 15 Nov. 8/3 (winner of competition) Hostess apron: Mrs. S. 1, Mrs. A. J. M. 2. 1970 N. Armstrong et al. First on Moon xiv. 352 Kate Collins, wearing her mother's pink hostess apron, was passing coffee. 1951 Country Life 16 Nov., The picture or hostess dress has been re-instated for informal evenings. 1963 Times 27 Feb. 12/5 Sleeveless hostess dresses. 1968 Guardian 30 Apr. 7/1 Full-length hostess dresses, long-sleeved or sleeveless. 1938 'E. Queen' Four of Hearts (1939) x. 143 She looked ravishing in a silver lamé hostess-gown. 1945 E. Bowen Demon Lover 178 Her cretonne house-coat, the nearest thing to a hostess gown that she had. 1963 M. McCarthy Group v. 92 Instead of a dress, Kay was wearing a cherry-red velvet sleeveless hostess gown. 1945 F. & R. Lockridge Payoff for Banker (1948) vi. 58 Dorian did not have hostess pyjamas of quite this cut—quite this daring—and would hardly have worn them if she had. 1970 Daily Progress (Charlottesville, Va.) 21 Mar. C 2/3 It's time to turn to wearing the long 'hostess pyjamas' for seasonal glamour in entertaining. 1964 N.Y. Times 15 Nov. Sect. xi. 9 (Advt.), Button-front hostess robe of soft fleece. 1966 Mrs. L. B. Johnson White House Diary 6 Aug. (1970) 406, I put on my white hostess robe that Lyndon had bought me in Paris ten years ago and went in to see Luci. She was eating breakfast. 1966 Daily Tel. 19 Dec. 9/4 The kilt that grew—and grew—and grew—becomes the Christmas hostess skirt that's on view at all the cosiest fireside parties. 1963 Guardian 25 Nov. 6/3 The latest model of the L. G. Hawkins hostess trolley..in which you can carry four dishes and a joint from the kitchen.

hostess (hōu·stės), v. [f. the sb.] trans. and intr. To be the hostess at (a party, etc.); to act as hostess. So **ho·stessing** vbl. sb.

1928 Sunday Express 6 May 16 She observed herself think that possibly Athene felt political 'hostessing' to be her vocation. 1946 Lincoln State Jrnl. 24 Jan., Wednesday was the day of the party which Mrs. Herbert Folsom..hostessed..in honor of Mr. Thomas Mauck. 1951 'J. Wyndham' Day of Triffids xii. 215 She had led one of those fringe careers—modelling dresses,..hostessing for obscure clubs. 1957 J. Frame Owls do Cry xxviii. 123 It will be my first real experience of hostessing to people who really matter. 1961 Spectator 27 Oct. 605 The hell of hostessing is..that one can know ..what things make a good hostess and still be quite unable to achieve them. 1962 Aeroplane CII. 229/2 The book..details the growth of the aviation industry. ..It is vital for anyone interested in 'Hostessing', because a proper appreciation of the industry's background is invaluable in assessing the job as a whole and its importance in helping to 'sell' the airlines. 1968 C. Nicole Self Lovers ii. 22 Vanessa thought hostessing large political gatherings would grow easier with experience.

hostessy (hōu·stesi), a. [f. Hostess + -y¹.] Pertaining to, or typical of, a hostess; hospitable.

1930 J. B. Priestley Angel Pavement iii. 108 And then, all the time you were so worried and anxious about the food and the serving, you were expected to be keeping the conversation going, terribly bright and hostessy. 1940 'N. Blake' Smiler with Knife 21 She would assume a gushing, hostessy manner. 1945 Penguin New Writing XXVI. 62 Betty Mordon began by being affably hostessy towards him. 1967 D. Francis Blood Sport i. 13 Idle hostessy chat. 1973 J. Burrows Like an Evening Gone i. 18 Greta has been making hostessy interested noises.

hostile, a. (sb.) Add: **2. b.** to go hostile: to become angry. Austral. and N.Z. colloq.

1941 W. D. Haydon N.Z. Soldiers 17 Wouldn't he go hostile if he knew who pinched his bacon. 1945 Baker Austral. Lang. ii. xv. 263 To go hostile means to become angry.

3. Comb. **hostile ice** (see quot. 1966); **hostile ord.** Naval slang, an ordinary seaman who joins the Navy in wartime for the period of hostilities only.

1964 Polar Record XII. 197 Hostile ice. 1966 T. Armstrong et al. Gloss. Snow & Ice 19 Hostile ice, from the point of view of the submariner, an ice canopy containing no large ice skylights or other features which permit a submarine to surface. 1919 W. Lang Sea-Lawyer's Log iii. 37 To the active service man the 'hostile ord', as the temporary seaman is called, is a mere interloper, one who has joined the Navy to 'dodge the Kitchener', as they put it, and they do not hesitate to tell him so.

B. sb. (Earlier and later examples.)

1838 N.Y. Mirror 27 Jan. 245/3 Yesterday five Delaware chiefs, who had gone from the main army to the stronghold of the hostiles, reappeared with four Indians. 1963 Guardian 8 Sept. 6/6 There were rumours that Naga hostiles were getting help from China. 1966 A. Firth Tall, Balding, Thirty-Five iv. 48 It is most unlikely that a hostile could select such a person as himself to assist them. 1969 Hindu 3 Aug. 1/5 A strong contingent of security forces has been rushed to..Ukhrul..to intercept over 400 China-trained Naga hostiles.

hostility. Add: **1. c.** hostilities, -y only: used in the Navy to describe a seaman enlisted only for the duration of a war, or such an enlistment.

[1917 'Taffrail' Off Shore 81 Even in our small ships we carry some of the 'hostility men'.] 1942, 1961 [see H.O. s.v. *H III]. 1946 J. Irving Royal Navalese 7 Once again the 'Hostility Only' man took his silent stand beside the slender handful of us professional seamen. 1948 Hansard Commons 17 Mar. 2083 Recruiting on a hostilities only engagement was not open in the branch of their choice in the Royal Navy and Royal Marines. 1973 Times 31 Jan. 16/7 He enrolled in the Royal Navy as a 'hostilities only' rating.

hot, a. Add: **1. c.** At a high voltage, 'live'. Also in U.S. slang phrs. hot chair, seat, squat, the electric chair.

1925 Sat. Even. Post 29 Aug. 18/3 A judge sentenced a boy fifteen years old to the electric chair... A newspaper reporter said he heard the lad announce that he was not going to die 'in the hot seat'. 1927 Flynn's Weekly 1 Jan. 819/1, I never shot nobody... Lotsa times I don't carry a gat. That's one thing I try to dodge—the hot chair. 1928 J. P. McEvoy Showgirl xii. 180, I ought to get something for that don't you think? The chair maybe—better known as the hot squat. 1930 Sel. Gloss. Motion Pict. Techn. (Acad. Motion Pict., Hollywood), Hot, electrically charged, particularly when dangerous. 1937 Printers' Ink Monthly Apr. 54/2 Hot mike, a microphone in which the current is flowing. A live microphone. 1940 Chambers's Techn. Dict. 424/1 Hot, said of a conductor which is charged to a dangerously high potential. Colloquial. 1944 'P. Quentin' Puzzle for Puppets ix. 74 When they get me on to the hot seat, I won't even burn. 1958 R. Chandler Let. 11 Jan. in R. Chandler Speaking (1962) 128 That scene at the end where the girl visits him in the condemned cell a few hours before he gets the hot seat!

4. Delete †Obs. and add later examples in astrological usage.

1819 J. Wilson Dict. Astrol. 268 Aries..is a vernal, dry, fiery, cardinal,..fortunate, hoarse, bitter sign. 1889 R. Baughan Influence of Stars 12 Aries..is a hot and dry sign, and produces a lean body. 1909 Kipling Rewards & Fairies (1910) 257 Between Mars and Luna, the one red, t' other white, the one hot, t' other cold.., stands..a natural antipathy.

6. a. (Further examples.) Freq. const. for or on (= keen on, eager for), or with inf. With negative construction: unsuccessful, not good or skilful.

1862 [see *dusty a. 4 d]. 1865 G. Meredith Let. 28 Dec. (1970) I. 320, I am very hot upon Vittoria. Lewes says it must be a success. a 1877 Swinburne Lesbia Brandon (1952) iii. 75 He was always rather hot upon that song. 1924 P. Marks Plastic Age xi. 112, I didn't flunk out but my record isn't so hot. 1925 T. E. Lawrence Let. 7 Sept. (1938) 485 The Squadron Leader is hot on punishment. 1932 Blue Valley Farmer (Okla. City) 28 Jan. 2/3 For president he's not so hot. Business won't support him. 1934 J. M. Cain Postman always rings Twice vi. 57 He was all hot to show me something. 1934 N. Coward Play Parade I. p. x, It..established me both as a playwright and as an actor... Until then I had not proved myself to be so hot in either capacity. 1937 'J. Bell' Murder in Hospital vii. 133 They're quite hot on First-Aid at these race tracks and he had a tourniquet on. 1937 Tablet 23 Oct. 553/2 Reviewed long ago with hot delight. 1946 Mezzrow & Wolfe Really Blues 375 Hot for, enthusiastic about, in favor of. 1952 Manch. Guardian Weekly 8 May 5/4 American radio listener, hot for news and excitement, leans forward. 1960 S. Kauffmann If it be Love ii. i. 121 He's very hot on the Babbage Square move. He thinks it's a good idea. 1967 R. Rendell New Lease of Death ii. 22 The Victorians..were not too hot on design. 1969 D. Gray Murder on Honeymoon xix. 118 Bryce was very hot on never having a drink in the bar.

c. (Further examples.) Also of a play, book, etc.: licentious. Phr. (U.S. slang) to have (or get) hot pants, to be (or become) aroused with sexual desire. Also, hot pants, a highly sexed (young) woman. Cf. senses *10 b and *12 c.

1892 E. J. Milliken 'Arry Ballads 37 As most of our plays are now cribbed from the French, wy they're all pooty 'ot. 1898 J. D. Brayshaw Slum Silhouettes 31 She was a 'an'some gal, was Daisy..but..she wos a fair scorcher, jist abart as 'ot as they make 'em. 1908 Daily Chron. 22 June 6/5 Publishing firms..discovered that money was to be made out of what they called 'the hot novel'. 1927 K. Nicholson Barker ii. ii. 112 When you had him all hot pants you married him. 1933 D. L. Sayers Murder must Advertise viii. 148 He said to Mr. Tallboy he thought the headline was a bit hot. And Mr. Tallboy said he had a nasty mind. 1935 J. T. Farrell Judgement Day xvi. 383 A burlesque show. The hottest ones were south of Van Buren. 1936 J. Steinbeck In Dubious Battle 64 Every time the sun shines on my back all afternoon I get hot pants. 1961 S. Price Just for Record v. 36 You've got the hot-pants for some good-looking piece. 1963 M. McCarthy Group iii. 48 I've still got hot pants for her, if you want to call that love. 1963 L. Meynell Virgin Luck ii. 30, I was just a hot little bit throwing myself at the head of the nearest presentable male. 1966 C. Robertson Judas Spies iii. 31 His second wife, this blonde hot pants. 1968 K. Amis I want it Now i. 39 It would help to hold off little hot-pants, and might distract him from the thought of what he was so very soon going to be doing to her. 1968 O. Wynd Sumatra Seven Zero x. 159 'You ought to marry.' 'That can wait. I haven't got as hot pants as I look.' 1971 W. Hanley Blue Dreams xiii. 221 'I'm hot as a firecracker is what I am,' she said demurely.

d. Of an action: exceptionally good or fine. Of a person: exceptionally smart or skilled or successful (in some action or kind of work). Of dice, cards, etc.: unusually lucky or successful.

1895 S. Crane Red Badge of Courage xvii. 169 'Hot work!' cried the lieutenant deliriously. 1914 Daily Express 30 Nov. 5/7 The Deutsches have some pretty hot snipers. 1919 A. A. Milne First Plays 44 He did the ninth [hole in golf] in three... Hot? What's that for hot? 1934 J. T. Farrell Young Manhood iv. 61 I'm shakin' fair, brother. They're just hot for me this time. The dice get hot for a guy like this maybe once in his whole life. 1944 D. Runyon Runyon à la Carte (1946) 101 Here is Big Nig hotter than a stove, and here I am without a bob to follow him with... Nig can make sixes all night when he is hot. 1968 Surfer IX. 63 Cabell is really ripping it up in Hawaii now, too, but Young and McTavish are so hot it's unbelievable. 1970 Surf '70 (N.Z.) 17/3 Walsh is not the only hot surfer in New Plymouth.

7. c. Of a ball: hit or thrown hard, and difficult for the other side to deal with.

1867 Ball Players' Chron. 6 June 3/4 The hot one he sent to pitcher. 1868 H. Chadwick Game of Base Ball 41 Hot Balls. This term is applied to balls sent very swiftly to the hands from the bat, or thrown in swiftly. 1882 Australians in Eng. 37 Lockwood gave Murdoch a hot chance at point. 1886 F. H. Burnett Little Lord Fauntleroy (1887) vi. 122 He..gave a dramatic description of a wonderful 'hot ball' he had seen caught. 1917 C. Mathewson Second Base Sloan 125 The players..had not handled a ball since the summer before and the 'hot ones' made them wince and yell.

d. Associated with or affected by a trade-union dispute. orig. U.S.

1901 Denver Republican 19 Aug. 1/1 (headline) Non-union men invading 'hot' section. 1940 F. Riesenberg Golden Gate 312 Docks had become dangerously congested, for the teamsters joined with the I.L.A., refusing to haul 'hot cargo'. 1947 Seafarers Log 25 Apr. 13/2 Officers would no longer go through a picket line to move a hot ship. 1959 Economist 2 May 423/2 'Hot cargo' clauses in transportation contracts, under which a lorry-owner agrees not to handle freight coming from another employer with whom the union has a dispute.

e. Of stolen property: easily identifiable and so difficult to dispose of. In extended use: stolen. Also applied to a person wanted by the police.

1925 Collier's 8 Aug. 30/2 Stolen bonds are 'hot paper'; stolen diamonds 'hot ice'. 1926 [see hock-shop s.v. *hock sb.⁷ b]. 1931 Amer. Speech VII. 109 Hot car hustler, car thief. 1931 'D. Stiff' Milk & Honey Route 207 A person is hot when he is wanted by the law. No hobo will travel with a man who is hot. 1942 M. Schlauch Gift of Tongues (1943) 269 Jewels become 'ice', and stolen jewels..'hot ice'. 1943 P. Cheyney You can always Duck ix. 151 Look, I'm hot, see? The cops here are after me. 1943 R. Chandler Lady in Lake (1944) xi. 64 The best stunt would be to unload it on a hot car dealer. 1953 W. Burroughs Junkie (1972) iii. 36 Herman pulled out a silk dress he had under his coat—as I recall somebody unloaded a batch of hot dresses on us for three grains of morphine. 1958 'A. Gilbert' Death against Clock 70 It [sc. a stolen brooch] was too hot to touch at present. 1960 H. L. Lawrence Children of Light ii. 34 You come here, in a hot car... And the police know. 1969 Times 4 July 3/8 Many stolen works of art have been recovered recently..which suggests that thieves are finding them too hot to handle. 1973 P. Moyes Curious Affair of Third Dog xv. 207 Griselda was 'hot'. Griselda had to disappear.

f. Of a motor vehicle or aircraft: fast or powerful (esp. in relation to size). colloq. Cf. *hot rod.

[1924 T. E. Lawrence Let. 20 Feb. (1938) 456 Hot speed on a motor-bike.] 1944 Sat. Even. Post 24 June 80/2 PV's can outrun all but the fastest of Jap fighters, but they pay for their speed by landing hot. 1951 N.Y. Times Mag. 21 Oct. 59/2 Certainly, the pure jet does land a little 'hotter' than the propeller plane. 1966 T. Wisdom High-Performance Driving ii. 31 Small 'hot' machines like a Mini Cooper and a Renault Gordini. 1968 Hot Car Nov. 7 It was a 'hot little car'.

8. a. to get (or be) hot: in a game or pursuit, to come (or be) near the discovery of something concealed. Also transf.

1876 [see *cold a. 12 b]. 1882 Cassell's Bk. In-Door Amusem. (ed. 2) 29 The progress of the player is usually announced by assuring him that he is 'very cold', 'cold', 'warmer', 'warm', 'hot', 'very hot', or 'burning', according as he is far from or near to the article to be discovered. 1899 E. W. Hornung Amat. Cracksman 252 'Not there, not there,' said Raffles; 'but you're getting hot. Try the cartridges.' 1931 Times Lit. Suppl. 29 Oct. 832/4 He needs it [sc. personality] to act quickly and effectively when an appliance fails, or when an investigator gets 'hot'.

b. Read: Of a colour: intense, vivid, glowing. Add later examples.

1933 Burlington Mag. Apr. 176/2 De Vos had long admired Rubens and emulated his hot tones. 1958 Vogue Mid-Sept. (Extra Issue) 87 Hot pink velour hat. 1962 Harper's Bazaar Oct. 124/3 A new hot orange lipstick. 1967 D. Francis Blood Sport xii. 144 A..hot pink-orange tunic.

f. Metallurgy and Founding. Of metal, esp. iron: completely fluid; sufficiently above the melting point to flow readily (see also quots. 1904, 1908¹). So hot-metal, in the sense 'molten iron', is used attrib., as in hot-metal process, a steel-making process in which the charge consists wholly or chiefly of molten iron. (See also hot metal in sense 12 c below.)

1888 Lockwood's Dict. Mech. Engin. 182 Hot metal. Molten iron and brass are said to be hot when their temperature is higher than that required for the class of work for which they are intended. 1902 G. R. Bale Mod. Iron Foundry Pract. I. x. 177 In pouring..the metal should be run 'hot'.., for when the metal is poured 'dull'..it is too sluggish to expel its gases. 1904 Goodchild & Tweney Technol. & Sci. Dict. 291/1 Metal of any kind is said to be 'hot' when it is at a temperature sufficiently high for some definite operation to be carried out. Thus a smith may say iron is hot when it is just at a welding or a forging heat. 1908 B. Stoughton Metall. Iron & Steel ii. 37 The term 'hot iron' has come to be synonymous in the minds of blast-furnace foremen with iron high in silicon and low in sulphur. Ibid. xiii. 361 When I say 'hot iron' here, I mean 'fluid iron', i.e., the degree of heat above the melting-point. 1929 C. R. Hayward Outl. Metall. Pract. xxii. 463 (caption) Hot-metal mixer at South Chicago... A ladle supported by a crane is delivering metal. 1946 Firth Brown Gloss. Metall. Terms 11 In the hot-metal process, iron from the blast furnace goes through the mixer to the open hearth furnace, and constitutes 75% or more of the charge. 1951 G. R. Bashforth Manuf. Iron & Steel II. xii. 356 The cupola metal replaces the molten pig iron employed in the hot-metal open-hearth process. 1966 J. D. Sharp Elem. Steelmaking Pract. viii. 160 Ore, if used, is normally placed directly on top of the lime, or fed to the furnace immediately after the hot metal.

g. orig. U.S. Applied to jazz or highly elaborated and florid dance music with a marked beat and strong emotional appeal, freq. improvised; also to the performer or to the place where played and in other uses; opp. *cool a. 4 d; hot lick (see *lick sb. 7).

1924 Variety 9 July 9/3 The style in vaudeville jazz bands this coming season will tend toward the 'sweet' and 'hot' dance orchestras. Ibid. 24 Sept. 26-c/1 Leon Beiderbecke is a 'hot' trumpet. Ibid. 26-c/2 The boys step on it and get 'hot' at various intervals. 1926 [see *chorus sb. 6 c]. 1927 Dancing Times Apr. 117 We are in the days of the 'hot' chorus. Ibid. May 149 'Hot Music' is still so much in its infancy over here that I do not propose to deal very fully with it for the moment. 1927 Melody Maker June 531/3 A rendering from an orchestration or an extemporisation made in this way is usually termed..'hot', the word 'hot' being intended to convey lilting, dance-inspiring rhythm with the accents irregularly placed but strongly portrayed, modern, or as some call them, extreme harmonies, and phrases based on these harmonies worked round the melody. 1928 Gramophone VI. 300/2 It is not a question of 'hot' dance music or 'straight' dance music. 1933 Fortune Aug. 47/1 Continuing in the language of jazz, it may be explained that Lawrence Brown is a hot trombonist with Duke Ellington's famous Negro jazz orchestra. That is to say, he excels in spontaneous, highly syncopated solos. 1933 Punch 18 Oct. 441/3 Miss Elizabeth Welch, a coloured lady,..sings what is known as a hot jazz song about Solomon in a startling but rather fascinating way. 1934 [see *chorus sb. 6 c]. 1935 Vanity Fair (N.Y.) Nov. 38/3 Hot artists often escape from the routine of their commercial work to the night clubs of Harlem. Ibid., These are the only truly hot bands. 1938 Oxf. Compan. Mus. 777/2 Louis Armstrong (claimed by one of his admirers as 'certainly the greatest of all hot musicians..'). 1946 Mezzrow & Wolfe Really Blues xi. 195 Hugues [Panassié]..sure kept himself busy.., launching the Hot Clubs of France. 1947 [see *beat v.¹ 30 d]. 1947 Auden Age of Anxiety (1948) ii. 40 Emble did likewise but his choice was a hot number. 1949 F. Maclean Eastern Approaches i. vi. 78 In the station buffet an extremely 'hot' band with a good sense of rhythm played fairly recent jazz from New York. 1953 J. G. Moore Relig. Jamaican Negroes 126 Hot, said of drums at Cumina ceremony when the fast rim beat indicates the presence of zombies in the drum. 1956 H. Lyttelton in S. Traill Play that Music 79 Shall it be a trumpet or a cornet? I have no strong views either way—a good 'hot' style can be achieved on either instrument. Ibid. 84 'Hot' tone cannot be defined on paper... It can be heard in essence in the playing of Bunk Johnson, Tommy Ladnier ..and Louis Armstrong. Ibid. 85 Vibrato plays an important part in hot jazz, being used as a means of expression. 1970 P. Oliver Savannah Syncopators 31 The idea of 'hot rhythm' is one which Richard A. Waterman borrowed from jazz and applied to African drumming.

h. Of a Treasury bill: newly issued.

1928 Evening News 18 Aug. 11/6 The hot Treasuries were offered at 4 3-16 per cent., without finding buyers. 1929 Times 16 Nov. 18/1 The 'hot' Treasury Bills were dealt in at 5 13-32 per cent. 1929 Observer 17 Nov. 3/3 'Hot' bills yesterday morning changed hands in very small amounts at 5 5-16 per cent. 1959 Times 15 Sept. 17/5 'Hot' bills were again sold to the clearing banks.

i. Radioactive; esp. so radioactive as to be dangerous; so hot laboratory, a laboratory designed for the safe handling of highly radioactive material; also hot atom, an atom that has high kinetic or internal energy as a result of a nuclear process.

1942 Pollard & Davidson Appl. Nucl. Physics vii. 139 Almost all the 'hot' sodium was in the form of NaOH. 1945 H. D. Smyth Gen. Acct. Devel. Atomic Energy Mil. Purposes viii. 84 Later a 'hot laboratory', i.e., a laboratory for remotely-controlled work on highly radioactive material, was provided. 1946 Sci. News Let. 10 Aug. 84/1 A large part of Bikini lagoon remained..'hot' with radioactivity. 1947 Time 10 Nov. 82/2 'Hot' (radioactive) atoms have already caused plenty of trouble in laboratories. 1950 Sci. Amer. Mar. 44/3 The bizarre chemical effects sometimes produced by radioactive atoms have given rise to a fascinating new branch of investigation known as hot atom chemistry. Ibid. 47/1 The immense recoil energy of the hot carbon atoms will effect chemical reactions that would not ordinarily occur. 1955 Times 12 Aug. 6/4 The elaborate precautions needed in so-called 'hot' laboratories—those in which large quantities of radioactive materials are manipulated. 1955 Sci. News Let. 27 Aug. 134/3 At Hanford atomically 'hot' strontium is kept in large tanks until it cools down. 1958 H. Etherington Nuclear Engin. Handbk. vii. 48 The building arrangement should be such that one cannot pass from a hot to a 'cold' area without going through a clothing-change facility. 1964 M. Gowing Britain & Atomic Energy x. 286 The new laboratories at Chalk River..included a 'hot' laboratory for remote handling. 1972 Nature 25 Feb. 443/1 The Mössbauer effect has been used to elucidate..hot atom effects, nuclear lifetimes, [etc.].

9. b. Of news: sensational, striking, exciting. Phr. hot from (or off) the press: just printed.

In quot. 1945² a fanciful use.

1908 [see *headline 2 d]. 1914 Daily Express 30 Sept. 4/5 'Hot news'..must be provided for the people, and thus we learn from the Vienna 'Abendblatt' that General French is a prisoner. 1945 Koestler Twilight Bar i. 34 News? You bet it is news... Hot? You bet it is hot. 1945 Tee Emm (Air Ministry) V. 52 The gen it contains is hot from the griddle. 1955 Times 19 Aug. 2/5 But it is for novelties, hot from the press or the copyist's desk, that discontent is calling. 1958 New Statesman 3 May 557/2 Television: Curran comments that to this the response of the American press (as it is to some extent that of the British press) is to leave aside all but the hottest news because radio and television can cover it with more actuality and immediacy and to concentrate instead on..personalities. 1969 Times 5 July 8/6 There has just arrived hot off the..presses a new publication called The Open Secret.

10. b. absol. as sb. pl. Strong sexual desire. slang.

1947 in Wentworth & Flexner Dict. Amer. Slang (1960) 274/1 I'd never get the deep undying hots for that rah rah collitch [boy]. 1951 Auden Nones (1952) 18 Jack likes Jill who worships George Who has the hots for Jack. 1961 S. Price Just for Record vi. 49 One day Lamb was going to get the hots for some painted woman. 1968 M. Richler Cocksure vi. 44 Well, me, I've got the hots for Polly Morgan. 1973 Times Lit. Suppl. 10 Aug. 921/5 It is Blodgett who has the hots for Smackenfelt's mother-in-law.

11. a. hot and cold: short for 'hot and cold water' (in a hotel, etc.). Phr. to go hot and cold (all over), to go all hot and cold: to experience alternate sensations of heat and cold as from shock or embarrassment. Also used trivially.

1908 Sears, Roebuck Catal. 610/4 Brass nickel plated compression bath cocks. Combination hot and cold. 1910 Bradshaw's Railway Guide Apr. 1012/1 High-class boarding establishment... Bath (hot and cold). 1914 Barrie Admirable Crichton iii. 130 He's working out

that plan for laying on hot and cold. 1928 Daily Mail 3 Aug. 10/6 The figures given..as to my earnings..made me go all hot and cold—I wonder what I can have done with all that money! 1931 Times 16 Mar. 1/5 (Advt.), First-class guest house,..fitted hot and cold and gas fires. 1944 A. Thirkell Headmistress i. 23 'There is a basin with hot and cold,' she said. 1973 A. Price October Men v. 64 His wife had said..that she had gone 'all hot and cold' after nearly being run over.

d. Also, to give it (to a person) hot and strong (and similar phrases).

1880 E. W. Hamilton Diary 25 July (1972) I. 27 Wilfrid Blunt goes in 'hot and strong' for making the Province a sort of independence. 1881 [see *bite v. 1 b]. 1931 T. R. G. Lyell Slang 412 To give it a person hot and strong, to punish a person severely, either physically or verbally; to give a good scolding to. 1938 A. G. Macdonell Autobiogr. of Cad ii. 14 Jedediah never spent a penny on the estate unless he was legally compelled to—and then the man who legally compelled him got it hot and strong in double quick time.

g. hot under (or occas. in, around) the collar: feeling anger or resentment, agitated (cf. sense 6 b); hot and bothered: in a state of exasperated agitation; also used (with hyphens) as attrib. phr.

1895 Horse Rev. 31 Dec. 1840/3 He would storm erround' dat room an' git hot under de collar. 1906 J. London Let. 1 Nov. (1966) 217, I must confess that he got me rather hot in the collar. 1918 E. Pound Let. 4 June (1971) 138 After years of this sort of puling imbecility one gets hot under the collar and is perhaps carried to an extreme. 1919 Red Cross Mag. Dec. 3/1 He fumbles around, gets hot under the collar and falsely accuses them of being a nuisance. 1921 M. Arlen Romantic Lady iv. v. 161, I was getting very hot and bothered about the whole thing. 1923 Kipling Independence 16 It [sc. each generation] goes to its grave hot and bothered, because no new birth has been vouchsafed for its salvation. 1930 A. P. Herbert Water Gipsies 279 She was delighted, though she had been caught in that rag, and still felt hot and bothered. 1932 N.Y. Times Bk. Rev. 7 Feb. 18/4 We thought it must belong to the 'Hot-and-Bothered Virgin on a Gunboat', 'Hot-and-Bothered Virgin at the South Pole' series. 1941 Q. Jrnl. Speech Oct. 361 [speech of the Frontier] Hamilton..spoke sharply to some of the men and Ike, with the licence allowed to an old retainer, suggested that there was no need to 'get hot under the collar'. 1941 F. Reinfeld Keres' Best Games of Chess 43/2, I suddenly began to go hot under the collar as I noticed the powerful move. 1958 Listener 23 Oct. 649/2 Then she emerged, hot and bothered, glasses half down her nose. 1961 J. Barlow Term of Trial i. ii. 22 Most of the teachers..urged silence in hot-and-bothered threats. 1969 Bucks Examiner 25 July 7/1 Very few subjects can get parents as hot under the collar as education. 1970 J. Porter Dover strikes Again i. 7 Twenty-five years in the police had not tarnished Superintendent Underbarrow's basic serenity... He made it a point of honour never to get hot around the collar about anything.

h. hot dog: see as main entry in Suppl.

i. a bit hot: somewhat unreasonable.

1931 T. R. G. Lyell Slang 410 To dismiss the lad just because he forgot to post a letter is a bit hot. 1946 K. Tennant Lost Haven (1947) xiii. 204 She has to be humoured although I do think it's a bit hot her boy friend calling just as if she owned the place.

k. hot as (in hyperbolic comparisons), esp. (as) hot as hell.

1849 N. Kingsley Diary (1914) 19 Hot as blazes—glad to get under awnings. 1850 W. M'Collum California 20 It was a delightful, salubrious spot—'hot as blazes' to be sure, but fanned by gentle breezes. 1889 Farmer Americanisms 293/2 Hot as the Devil's kitchen. 1896 N.Y. Dramatic News 4 July 7/1 This afternoon was as hot as blazes. 1912 Dialect Notes III. 579 Hot as hammered hell. 1922 E. O'Neill Hairy Ape (1923) vi. 62 I'll be fire—under de heap—fire dat never goes out—hot as hell. 1935 A. J. Pollock Underworld Speaks 58/2 Hot as a 45, wanted by the police. 1935 Dylan Thomas Let. July (1966) 159 He's got a..hot-as-hell conservatory with a fountain. 1953 R. Graves Poems 29 From every-which-a-way, hot as a two-buck pistol. 1967 Word Study Dec. 4/1 One of the most common comparisons in English—is probably 'as hot as hell'.

12. a. hot-eyed, -looking, -tailed, -toned adjs.

c. **hot-ache** (later example); hence **hot-aching** a.; **hot beef**, rhyming slang for 'Stop thief!' e.g. in phr. to give (a person) hot beef; **hot bottle**, a hot-water bottle; **hot box** U.S., an overheated journal-box, esp. of a railway carriage; also fig. or transf.; **hot bricks**, chiefly in phr. like a cat on hot bricks, denoting a situation of extreme discomfort and restiveness, or expressing swiftness or nimbleness of movement; **hot bulb**, in a semi-Diesel engine, an uncooled chamber connected to the cylinder head which is maintained at a sufficiently high temperature to vaporize fuel oil injected into it prior to compression in the cylinder; any mass of metal that performs the same function in such an engine, usu. attrib.; **hot cakes** orig. U.S., (a) griddle-cakes, flannel-cakes; (b) in phr. to sell or go (off) like hot cakes, to be sold or disposed of very rapidly; to be in great demand; **hot cathode**, a cathode intended to be heated, so that electrons are emitted thermionically; also attrib.; **hot chisel**, a short thick chisel used for cutting or nicking hot metal and having a sharper edge than a cold chisel; **hot-closet** (earlier example); **hot cross bun** (see Bun sb.² 1, Cross-bun); **hot cupboard**, (a) an airing cupboard; also fig.; (b) a cupboard in which plates, dishes, etc., may be made warm; **hot diggety (dog)** U.S. slang, an exclamation of joy or surprise; cf. *hot dog 3; **hot-dip** a., involving, used for, or applied by hot dipping; **hot dipping**, the application of a coating to an article either by dipping it in a bath of hot liquid (e.g. molten metal), or by dipping it hot in a bath of liquid; so **hot-dipped** a., said of the coating; **hot flashes, flushes** pl., a menopausal symptom which manifests itself by a momentary sensation of heat, freq. accompanied by a heightening of facial colour and perspiration; **hot iron** (see sense 8 f above); **hot-iron test** Cheesemaking, a test to determine the acidity of the curd; **hot line**, a direct, exclusive communication channel between two points; spec. the direct telephone link between Washington and Moscow (and similar lines); also attrib. and fig.; **hot-making** a. colloq., embarrassing; **hot-melt** a., solid at normal temperatures but capable of being melted for application (e.g. as a coating); **hot metal**, (a) molten iron: see sense 8 f above; (b) used attrib. to designate printing machines and methods in which type made by the machine from molten metal is used; **hot money**, capital which is transferred from one country to another in order to avoid the effects of currency devaluation or to profit from high interest rates or possible revaluation; also attrib.; **hot pants**, brief shorts worn by girls and young women as a fashion in the early 1970s; cf. to have (or get) hot pants s.v. sense 6 c above; hence **hot-panted**, **hot-pantsed** adjs.; **hot-plate** (examples); also, a heating element on an electric cooker; (b) a plate with a cover for keeping food hot; (c) a low, portable heating appliance with a flat top for keeping food hot or heating vessels, food, etc., placed on it; **hot potato**, in fig. phr. (to drop something) like a hot potato; also, a ticklish subject, an embarrassing problem; rhyming slang, a waiter; **hot seat**, (a) (see sense 1 c above); (b) used as in sense 7 b in Dict.; (c) (see quots. 1933 and 1938); (d) an ejection seat in an aircraft; **hot set**, a hot chisel, esp. one with a wooden handle; **hot shift** orig. U.S., a mechanism enabling a change of gear to be made while a vehicle is moving without interrupting the drive to the wheels; **hot shoe**, a socket on a camera incorporating electrical contacts for a flash, etc.; **hot spring**, any spring whose waters issue forth at a temperature appreciably higher than that of the ground; **hot squat** (see sense 1 c above); **hot-stove** N. Amer., used attrib. of a discussion about a particular sport (e.g. baseball or ice hockey) between periods of play or in an off-season; also applied to anyone discussing a sport in this way; **hot tear** Metallurgy, a rupture produced in a casting or ingot as the metal cools and contracts; so **hot tearing** vbl. sb.; **hot top** Metallurgy, a refractory container for holding a reservoir of molten metal at the top of a mould during the solidification of an ingot; **hot tube**, in some early internal combustion engines, a metal or porcelain tube, closed at the end, which projected from one end of the cylinder and was heated externally by a flame, so that it ignited the mixture forced into it during the compression stroke; usu. attrib. in hot-tube ignition; **hot war**, an open war, involving active hostilities (opp. *cold war); hence **hot-warrior**; **hot wave**, a spell of exceptionally hot weather, a heat wave; **hot wind**, a dry, very hot wind that blows over land for large distances from the interior in Australia, the United States, and elsewhere; **hot-wire** attrib., denoting electrical instruments the indications or operation of which are based on the change in the length or resistance of a wire with changing temperature; as v. trans. (N. Amer. slang), to by-pass the ignition system of a motor vehicle (as a preliminary to stealing the vehicle); hence **hot-wired** ppl. a.; **hot with** (earlier examples).

1917 D. H. Lawrence Look! We have come Through! 155 Their scent is lacerating and repellent, it smells of burning snow, of hot-ache. Ibid. 156 What kind of ice-rotten, hot-aching heart must they need to root in! 1879 Macm. Mag. XL. 506/2 He followed, giving me hot beef (calling 'Stop thief'). 1896 A. Morrison Child of Jago x. 95 It was now that he first experienced 'hot beef'—which is the Jago idiom denoting the plight of one harried by the cry 'Stop thief'. 1973 G. Butler Coffin for Pandora vii. 157 'Hot beef, hot beef,' cried the schoolboys. 'Catch him...' 1845 Mrs. Gaskell Lett. (1966) 824 Please let Meta's feet be warm in bed (a hot bottle she has here by Dr H' direction). 1860 F. Nightingale Notes on Nursing i. 11 Hot bottles, hot bricks, or warm flannels.. should be made use of until the temperature is restored. 1967 S. Knight Window on Shanghai xxxi. 139 The Chinese don't use hot-bottles unless they're ill. 1848 Merchant's Mag. XIX. 656 Such a thing as a 'hot box' to a car has not been known,..since the sprinkler has been in use. 1873 'Mark Twain' & Warner Gilded Age xlvi. 419 A hot box had to be cooled at Wilmington. 1899 S. Hale Lett. (1919) 343 He was excellent with the horses and very careful about hot-boxes and watering. 1910 J. Hart Vigilante Girl x. 140 If that near hind box was to begin to talk different, I'd pull up and examine it right away. Mebbe I'd save a hot box by doin' it. 1971 D. J. Smith Discovering Railwayana x. 56 Hot box, axle box overheated through lack of lubricants or overwork. 1971 Flying (N.Y.) Apr. 48/3 Lovely airplane but its another hotbox. [1678 J. Ray Coll. Eng. Proverbs (ed. 2) 285 To go like a cat upon a hot bake-stone.] 1862 Whyte-Melville Inside Bar ii. 248 A well-bred, raking-looking sort of mare... Beautiful action she had, stepped away like a cat on hot bricks. 1886 'J. S. Winter' Army Society xvi, Lady Mainwaring looked..like a cat on hot bricks. 1958 'A. Gilbert' Death against Clock 165 Crook also was like a cat on hot bricks. 1961 Wodehouse Service with Smile iii. 42 Having become accustomed to this kind of thing myself..I have lost that cat-on-hot-bricks feeling which I must have had at one time. 1911 Motor Boat 26 Jan. 66/3 If you are nervous of electric ignition you should choose a hot-bulb ignition engine. 1919 W. Pollock Hot Bulb Oil Engines i. 2 All engines of the low compression type using heavy oils, and hot surface ignition of the injected spray are included under the title of hot bulb oil engines, although some makers use plates or discs, instead of bulbs. 1922 Encycl. Brit. XXXI. 519/1 Prior to starting the hot-bulb is blow-lamp-heated for 10 to 15 minutes. 1958 Van Nostrand's Sci. Encycl. (ed. 3) 823/1 It [sc. the semi-Diesel engine] has a hot bulb, which is a certain mass of metal incorporated in the cylinder head in such a way that a portion of it projects slightly in the combustion space. 1683 W. Penn Sel. Works (1782) IV. 309 Their entertainment was..twenty bucks, with hot cakes of new corn. 1839 C. F. Briggs Adventures H. Franco I. xi. 74 'You had better buy 'em, Colonel,' said Mr. Lummucks, 'they will sell like hot cakes.' 1879 Congress. Rec. 15 May 1368/1 Four per cent bonds..go off like hot cakes. 1891 Farmer Slang II. 18/2 s.v. Cake, Buckwheat and other hot cakes form a staple dish at many American tables. 1893 P. H. Emerson Signor Lippo xii. 37 It went off like hot cakes. 1908 Daily Chron. 4 Aug. 3/4 Ice creams at 3d. a time went 'like hot cakes'. 1925 J. Gregory Bab of Backwoods xi. 141 So they got the blaze going, bacon sizzling, the frying-pan balanced on the fire, hot-cakes mixed and coffee set to boil. 1930 E. Waugh Vile Bodies x. 183 Those who were fortunate enough to own cottages or public houses at the more dangerous corners..were selling tickets like very expensive hot cakes. 1949 H. Kurath Word Geogr. Eastern U.S. 34 Flannel cake is..rather uncommon in..Philadelphia, where hot-cake is in common use. Ibid. 35 Hot-cakes for griddle cakes made of flour. 1966 Publ. Amer. Dial. Soc. 1964 xlii. 16 Flapjacks, the commonest name for pancakes. Batter (or batty) cakes, flitters (or flitter cakes), fritters, griddle cakes, hot cakes are all known. 1971 Petticoat 17 July 3/1 You'll find amazing hessian bags.. selling like hot cakes for only £1.50. 1913 Physical Rev. II. 412 The idea of using a hot cathode in a Röntgen tube was not new. 1930 Engineering 31 Oct. 560/2 The hot-cathode discharge tube was free from these restrictions. 1943 Electronic Engin. XVI. 260/3 The instrument incorporates a hot-cathode, low-voltage, cathode-ray tube. 1959 Listener 12 Mar. 454/1 The hot cathode, as it was called in 1930, consisted of a small rod of alkaline earth oxide held in a tungsten coil. 1889 N. & Q. 23 Feb. 151/1 Cold and hot chisels are used for cutting cold and hot iron (or metal) respectively. 1964 S. Crawford Basic Engin. Processes xiii. 274 The process used to remove surplus metal from the forging with the aid of hot chisels or hot sets. a 1817 Jane Austen Northanger Abbey (1818) II. viii. 148 The ancient kitchen..rich in the massy walls and smoke of former days, and in the stoves and hot closets of the present. 1733 Hot cross bun [see Cross-bun]. 1825 [see Bun sb.² 1]. 1880 A. Beardsley Let. 1 Apr. (1971) 12 On Good Friday we had hot cross buns for breakfast. 1908 A. Huxley Let. 17 Apr. (1969) 26, I am almost prostrated just now by an Enourmous [sic] Hot cross Bun. 1941 C. Hole Eng. Custom & Usage iv. 43 Hot-cross buns are eaten throughout England on Good Friday... These little cakes..seemed to be descended from the wheaten cakes eaten at the Spring Festival alike by Greeks, Romans and Saxons. 1970 Simon & Howe Dict. Gastron. 90/2 Hot Cross buns are heavily spiced and eaten on Good Friday. 1930 S. Beckett Whoroscope 2 Them were the days I sat in the hot-cupboard throwing Jesuits out of the skylight. 1931 —— Proust 51 The Cartesian hotcupboard of the Guermantes library. 1950 Chambers's Encycl. VII. 256/2 One [electric cooker] for an ordinary family..would include..a hot cupboard enclosed by a drop door between the hob and the oven, and an oven. 1953 M. Dickens No More Meadows i. 20 The bathroom, which housed a boiler and a monstrous hot cupboard. 1959 Observer 12 Apr. 14/5 The bathroom..is assumed to be on first-floor level, adjoining a bedroom with access to water-heater hot-cupboard from the passage. 1967 N. Freeling Strike Out i. 11 Saskia took a plate from the hot cupboard. 1924 Dialect Notes V. 270 Hot diggity. 1927 Sun (Baltimore) 3 Apr. ii. 1. 12/2 When it comes to 'hot dog', there's no more to be said, unless it perhaps, to add a frill and make it 'hot diggety dog'. 1939 Ryerson & Clements June Mad iii. 178 I'll..get

cleaned up and into my..tux!.. Hot diggity! 1947 M. Lowry Let. May (1967) 142 All I wanted was..a single word or phrase like: O.K., Hot Diggety, or even We are not amused. 1952 M. R. Rinehart Swimming Pool xi. 104 Hot diggety dog! Ain't that something? 1923 Foundry (Cleveland) 1 June 454/3 It has been shown that the deterioration of malleable in the hot-dip galvanizing process is intimately connected with the phosphorus and silicon content of the iron. 1960 Farmer & Stockbreeder 22 Mar. 138/2 (Advt.), Rust-proofing by the Hot-dip Galvanizing process ensures complete protection against weather. 1961 New Scientist 13 Apr. 38/2 The industrial process for the hot dip tinning of steel has been practised successfully for many years. 1971 Engineering Apr. 61/2 Croda Chemicals Ltd..supply both hot-dip strippable wax coatings as well as plastics. Ibid. 63/1 These are hot-dip compounds and provide coats impervious to moisture. 1936 Jrnl. Amer. Zinc Inst. XVII. 70 We began to collect samples of hot dipped spelter coated or galvanized sheets that had been subjected to varying periods of service. 1952 Kirk & Othmer Encycl. Chem. Technol. VIII. 901 The normal hot-dipped coating consists of an alloy layer covered with a layer of pure coating metal. 1931 Amer. Machinist 8 Jan. 59/2 In hot dipping, crevices are sealed shut and filled with metal. 1954 Plastics Engin. Handbk. (Soc. Plastics Industry, Inc.) x. 300 Successive hot-dipping of objects in a tank of plastisol over the course of several hours will raise the temperature of the plastisol. 1924 A. J. Small Frozen Gold 222 Hot-eyed, livid-faced men. 1936 Times Lit. Suppl. 3 Oct. 788/4 The Herr Doktor, who was hot-eyed, not weary-eyed. 1969 Listener 5 July 21/3 For some tastes, to embark on a Doris Lessing book has always meant a tough regime of entrail-inspection and hot-eyed political pamphleteering. 1910 H. S. Crossen Diagn. & Treatm. Dis. Women (ed. 2) xiv. 851 'Hot flashes'..can hardly be classed as pathological. 1933 R. G. Hoskins Tides of Life xii. 282 This..phenomenon is accompanied by..the well-known 'hot flashes'. 1959 M. F. Willis Let. in 'I. Devi' Yoga for You (1960) 200 These benefits have continued with the extra good fortune of getting rid of seventy-five per cent of the hot flashes. 1910 W. B. Bell Princ. Gynaecol. iii. 89 The 'hot flushes'..are produced by rapid changes in the condition of the vaso-motor system. 1946 P. M. F. Bishop Gynaecol. Endocrinol. iii. 33 A premature menopause complete with hot flushes becomes established. 1948 L. Martin Clinical Endocrinol. viii. 185 The general symptoms of the menopause are well marked in over 50 per cent of women, but some entirely escape and others suffer only minor discomfort from 'hot flushes'. 1971 M. Lee Dying for Fun xxxvi. 174, I keep getting hot flushes. It's my age. 1889 Jrnl. Brit. Dairy Farmers' Assoc. V. ii. 70, I depend entirely on the hot iron test at this stage [of cheese-making]. Ibid., To determine when the curd is ready for salting the hot iron test is again resorted to. 1955 J. G. Davis Dict. Dairying (ed. 2) 209 Hot iron test. This test is held in great esteem by some cheesemakers, but is very crudely carried out. 1955 N.Y. Times Mag. 7 Aug. 10/1 To hold this breakthrough to a minimum is 'ConAd's' job. It has twelve air divisions, tied in by 'hot line' communications with one another and with the Army, Navy and Civil Defense Administration. 1962 Flight Internat. LXXXI. 401/1 'Hot line' communications are by radio. 1963 Guardian 6 Apr. 1/3 Mr Tsarapkin's acceptance of a 'hot' White House–Kremlin telephone or teleprinter line. 1963 Daily Tel. 21 June 16 Hot line. At Geneva yesterday the United States and Russia signed an agreement to set up a direct link between the Kremlin, the Pentagon and, presumably, the White House. 1963 Ibid. 6 Sept. 23/4 The British Overseas Airways Corporation yesterday opened its own 'hot line'. It is an instantaneous electronic reservations system between London and New York. 1965 New Statesman 17 Sept. 390/2 The hand [of the Pope] hesitates to reach for the hot-line to the Holy Ghost. 1966 Maclean's Mag. 17 Sept. 3 About 20 orators arose, one after another, to espouse everything from (predictably enough) free university tuition to local night shopping and radio hot lines. 1966 Melody Maker 15 Oct. 6 Is he a mystic with a direct hot-line to heaven? 1969 Times 9 Aug. 4/2 President Nixon and..the West German Chancellor, announced today that they had agreed to set up a 'hot line' for communication between Washington and Bonn. 1970 Daily Tel. 17 Mar. 13/2 'Hot lines' are becoming de rigueur these days—Washington–London–Moscow, to say nothing of London–Canberra. 1971 Ibid. 26 July 3/3 On five days during the Moon mission, ITN will have a 'hot line' over which viewers can put questions to a panel of experts. 1936 J. Curtis Gilt Kid v. 50 She wasn't quite so hot-looking as Maisie... That judy certainly had passion. 1959 Tamarack Rev. xii. 23 How could a hot-looking girl turn out so cold? 1931 Sunday Times 22 Feb. 4/1 This, to use the current phrase, hot-making play. 1958 B. Nichols Sweet & Twenties i. 34 How was it that Queen Marie delivered such hot-making effusions? 1939 Manuf. Pulp & Paper (ed. 3) V. § 4. 84 Hot-melt coatings will probably become familiar as paper coatings in the near future. 1946 Nature 30 Nov. 801/1 The production of hot-melt inks for carbon paper. 1954 Plastics Engin. Handbk. (Soc. Plastics Industry, Inc.) viii. 234 Hot-melt compounds of ethyl cellulose have become available..as another tooling material... Worn or damaged tools..can be reclaimed by remelting and recasting. 1969 New Scientist 19 June 641/3 The process is based on the use of papers precoated during production with a very thin film of 'hot-melt' plastic resin. 1960 McGraw-Hill Encycl. Sci. & Technol. X. 601/2 If type produced on 'hot metal' machines is to be used, the type is set, proved, read, and corrected. 1964 Turnbull & Baird Graphics of Communication v. 83 Hot-metal composition is done on various type-casting machines. 1971 Brit. Printer Aug. 61/1 Harris-Intertype has introduced a low-cost photo text setter, the price of which is said to be comparable to that of a modern, manual hot-metal machine. 1936 P. J. O'Brien Forward with Roosevelt xi. 191 The movements of 'refugee capital', or as President Roosevelt and other financiers..described it, 'hot money', was an important consideration in the question of currency stabilization. 1951 J. R. Winton Dict. Econ. Terms (ed. 3) 45 Hot money, a term used to describe the movement between different countries of balances and capital assets, withdrawn from one country through a loss of confidence as a result of war scares, possibilities of currency depreciation, [etc.]. 1958 Spectator 24 Jan. 117/2 Anybody who can afford to pay the 'hot' money price. 1971 Daily Tel. 3 Feb. 15 The huge volume of 'hot money' coming from abroad to chase high interest rates in Britain. 1971 New Scientist 25 Feb. 428/2 Where we once had ladies in gowns and earrings to elocute the score at us, we now have hot-panted dolly-birds who can't add up. 1972 Guardian 29 May 8/1 The miniskirted or hotpanted lass. 1970 Women's Wear Daily 23 Nov. 31/2 As for hotpants, we haven't seen anything in the market... They're going to have to be styled very imaginatively. Otherwise, they're going to look like old-fashioned short shorts. 1971 Daily Tel. 26 Jan. 4/8 'Hot pants', the fashion craze for abbreviated shorts for day or evening wear, is becoming the dominant fashion in the children's wear field, according to designers and manufacturers in New York. Ibid. 8 Nov. 12/8 Hotpants have rather quickly died a fashion death. 1971 Daily Colonist (Victoria, B.C.) 26 May 2A/1 Bibbed hot pants in easy care cotton. Styled with back zip. 1971 W. Hanley Blue Dreams xi. 178 But, she couldn't wait! Hot-pantsed little... It was all perfectly clear now! 1845 E. Acton Mod. Cook. vii. 159 The hot plates, or hearths with which the kitchens of good homes are always furnished. 1846 S. Etievant Brit. Pat. 11,096, In the Drawing,..the spit on to which the material to be roasted is placed is affixed to the hot plate J, which carries the pots for charcoal K. 1861 C. E. Francatelli Cook's Guide 3 (Advt.), Smoke Jacks, Hot Plates,.. Steam Tables, and..other description of..Cooking Apparatus. 1925 G. L. Hill Ariel Custer vii. 99 A gas hot plate with two burners. 1925 L. Stallings Plumes vii. 175 The hot plate..was set upon a bracket next the water closet. 1936 Catal. Chem. Apparatus (F. E. Becker & Co.) (ed. 25) 436 Gas Heated Hot Plate, polished smooth steel top on wrought-iron stand. 1938 Trans. Amer. Soc. Mech. Engin. LX. 70/1 Cores, crossbands and veneers..were glued on a hydraulic hot-plate press with hide or bone glue. 1966 Which? Feb. 45/2 All the other cookers had spiral hotplates with continuously variable controls. 1969 Hotplate [see *cordless a.]. 1971 Sci. Amer. Aug. 108/3, I use butyl phthalate..and heat it with small laboratory hot plates operated at a temperature of 350 degrees Fahrenheit. [1840 Picayune (New Orleans) 11 Sept. 2/2 Let's drop Nancy Daly like a hot praytee and proceed.] 1846 Spirit of Times 6 June 170/1, I dropped the divine's arm 'like a hot potatoe'. 1861 H. Rhys Theatr. Trip Canada & U.S. xi. 96 A deep growl..made me drop the article like a hot potato. 1886 B. P. Poore Perley's Reminisc. I. 448 They dropped him like a hot potato when they learned that he had accepted a place on the Republican Committee of his State. 1909 J. R. Ware Passing Eng. 155/1 Hot potato (False Rhyming—Music Hall, 1880). Waiter. 1930 W. S. Maugham Cakes & Ale xiv. 169 She dropped him, but not like a hot brick or a hot potato. 1952 M. McCarthy Groves of Academe (1953) xiii. 257 It was a very hot potato... I chose to ignore the question. 1958 Economist 20 Dec. 1062/2 It has..dropped the hot-potato question of future UN forces. 1961 C. McCullers Clock without Hands x. 199 The Judge had been distressed when Johnny agreed to take the case, was amazed at first at the way he handled it—hot potato that it was. 1963 Listener 14 Feb. 308/1 Tories continue to treat laissez-faire enterprise as a political hot potato. 1969 New Scientist 16 Oct. 116/1 The current hot potato in the sociological field is the question of poverty in Britain today. 1933 C. E. Leach On Top of Underworld xii. 185 No catalogue of the methods of con-men would be complete without an exposure of this time-honoured 'ramp' [the 'Rosary' confidence trick], and of its companion, the 'Hot Seat'. 1938 F. D. Sharpe Sharpe of Flying Squad xxviii. 286 You've got to hand it to the Hot Seat Boys. They're clever little devils. For the benefit of the great overworld public—Hot Seat Mob is the title bestowed by the Underworld on the gentlemen we so often read about under the heading of Confidence Tricksters. 1942 Time 6 Apr. 49/1 We are an entire nation of people who are trying to wage a war and everyone is trying, himself, to keep out of the hot seat. 1950 Nat. Geogr. Mag. Sept. 311/1 (heading) Jets' 'Hot Seat' saves lives. 1959 Observer 11 Oct. 21/4 The cosy climate of gracious-uxorious living shattered by hot-seat-phobia. 1961 Economist 2 Dec. 913/1 The 'hot seat' issue of Chinese representation. 1966 Listener 28 July 143/1 After fifteen months in this critical hot seat..between listeners and the BBC I am saying my farewell. 1969 Times 24 June 25/1 (headline) Return to the hot seat. 1888 Lockwood's Dict. Mech. Engin. 183 Hot set, a smith's set,..used for the nicking and cutting of hot metal. 1894 G. Hughes Construction Mod. Locomotive iii. ii. 155 Upon each side of the vertical an angle of 45° is marked, that is, nicked cold and afterwards heated and removed by a hot set. 1964 S. Crawford Basic Engin. Processes xiii. 270 The hot set is used for cutting or marking hot material, often used in conjunction with a hardie. 1971 Power Farming Mar. 15/4 The tractor of the late seventies would be similar in layout to present-day models ... The main differences would be in driver comfort,.. greater use of hot-shift transmission and quick-attach systems for implements. 1972 Good Motoring Dec. 12/1 Driver-controlled epicyclic gearboxes in which various ratios were engaged or disengaged under load ('hot shift' is the American phrase) by means of friction clutches. 1971 Amat. Photographer 13 Jan. 57/1 (Advt.), Center flash contact (hot shoe). 1972 Ibid. 12 Jan. 71/2 (Advt.), Dual flash synchronisation system accepts virtually every type of flash unit with 'hot shoe' or plug-in coupling. 1972 L. Gaunt Praktica Way 118 Their single flash contact is in the accessory shoe and is designed for use with flashguns with a contact in the foot (the so-called 'hot-shoe' contact). 1669 W. Simpson Hydrologia Chymica 154 Some other causes there are of hot springs, viz. subterraneal fires. 1780 [see Geyser 1]. 1850 [see Spring sb.¹ 2 b]. 1961 Times 18 Mar. 11/1 Hot springs are found throughout the island [sc. Iceland]. 1952 Time 18 Feb. 55 Hot stove leaguer, in the U.S., anyone who likes to talk at length..about recent developments in his favorite sport—which is usually baseball. 1959 Ottawa Citizen 25 June Suppl. 7/1 The auto industry has its own hot-stove league—the off-season gatherings of stylists, planners and engineers who talk of what might have been or what might yet be. 1962 Hockey Canada Nov. 12/2 Things we have are old and up as possible for Hot Stove sessions which are as old and as popular as the sport of hockey itself. 1965 Globe & Mail (Toronto) 13 Jan. 26/4 King Clancy.. joined Bower..on the Hot Stove panel yesterday. 1680 ? Buckhurst in Rochester Poems 78 Until her Hot-Tail'd Majesty..Had worne her Gems on Holy Days. 1933 Jrnl. Iron & Steel Inst. CXXVII. 566 The application of certain principles of casting design is important in eliminating the danger of hot-tears. 1955 Heine & Rosenthal Princ. Metal Casting xvi. 350 A rigid mold is also a potential cause of hot tears. 1945 Jrnl. Iron & Steel Inst. CLI. 380P The degree of hot tearing..can be minimised or often eliminated by using steel with a lower sulphur content. 1967 A. H. Cottrell Introd. Metall. xiii. 180 Another practical problem of shrinkage is hot tearing. 1961 New Left Rev. July–Aug. 43/1 A swinging musician with a great sense of the Blues, and 'hot-toned'. 1928 Jrnl. Iron & Steel Inst. CXVII. 831 Corrugated ingots of acid steel cast with hot tops to eliminate segregation and piping are used for the larger forgings. 1954 A. R. Bailey Text-bk. Metall. ix. 388 A 'hot top'..acts as a reservoir for feeding the ingot, the pipe eventually occurring in this portion, which is discarded. 1890 W. Robinson Gas & Petroleum Engines iii. 35 The hot tube gives a good regular ignition. Ibid., With hot-tube ignition..there is the possibility of premature and irregular explosions. 1912 Askling & Roesler Internal Combustion Engines ii. ii. 122 Ignition by hot tube is only employed for small engines. 1963 Bird & Hutton-Stott Veteran Motor Car 67, 1899 Clément-Panhard... Engine: Transverse, inclined, rear, 1-cyl., 90 × 120 mm., A.I.V. (i.o.e.), hot-tube ignition. [1768 J. Lees Jrnl. (1911) 39 The Chipawas..were at this time in a hot war with the Sous.] 1947 Newsweek 17 Nov. 25/2 The ideological war of the Communists is as ruthless and as determined..as a hot war. 1955 Times 5 July 11/6 It is a reasonable assumption that a hot war would be measured in months if not weeks. 1973 R. Hayes Hungarian Game 340 U.K. Counterintelligence..blew the cover of every Abwehrabteilung..agent in Britain during the hot war. 1966 New Statesman 22 July 116/2 It is more likely that the hot warriors..want to use rash behaviour in Hanoi to authenticate the next escalation. 1888 Encycl. Brit. XXIII. 805/1 The occasional occurrence of 'hot waves' which sweep over large areas of country, raising the temperature much above its normal height, is one of the most striking and most disagreeable features of the climate of the country [sc. the U.S.A.]. 1901 Daily Colonist (Victoria, B.C.) 23 Oct. 5/2 The 'hot wave', as the weather man calls it, was general along the coast and extended some way east. 1936 Discovery Mar. 78/2 The only periods when Buchan's spells appear to be at all true in London and the south are his cold wave of February 7th to 14th, and the hot wave from July 12th to 15th. 1968 G. R. Rumney Climatol. xi. 198 The familiar but unwelcome heat wave, or hot wave, of eastern North America. 1797 Encycl. Brit. VI. 402/1 A sign of the approaching Simoom or hot wind. 1804 Wellington Disp. (1837) III. 180 A desire to give his troops some repose, and not to expose the Europeans to the hot winds in Hindustan. 1850 R. G. Cumming Five Yrs. Hunter's Life S. Afr. I. 60 At certain seasons..northerly breezes prevail: these are termed by the colonists 'hot winds'. 1900 W. Stebbing C. H. Pearson 289 He revelled in the Victoria hot-wind days which shrivelled up everybody else. 1910 W. L. Moore Descr. Meteorol. x. 187 The 'hot winds' of our western plains [in the U.S.A.]. 1971 J. Gentilli Climates of Austral. & N.Z. v. 68 In the southern interior, especially in Victoria, these hot winds laden with dust are known as brickfielders. 1889 Telephone 15 Mar. 136/1 The actual stress in the platinum silver wire in our hot-wire voltmeter was not sufficient. 1904 Electrician Nov. 150/1 An oscillating circuit containing..a delicate hot-wire ammeter. 1914 Phil. Trans. R. Soc. A. CCXIV. 404 (heading) On the design of portable hot-wire anemometers. Ibid. 405 Standard thermometer wire is the most suitable for the purpose of hot-wire anemometry. 1922 Encycl. Brit. XXXI. 186/2 A case of great importance from its industrial application in hot wire valves is one where all the ions are negative and are emitted from the cathode. 1923 E. W. Marchant Radio Telegr. iv. 44 Other forms of detector for wireless circuits are the electrolytic, the hot wire 'barretter' and the crystal. 1933 Phil. Mag. XVI. 50 The sound recording instrument was a tuned hot-wire microphone. 1957 Haltiner & Martin Dynamical & Physical Meteorol. xiv. 218 Hot-wire anemometers are..highly sensitive instruments capable of measuring mean wind velocities over very short periods (down to 0·1 sec.). 1957 E. G. Richardson Techn. Aspects Sound II. ix. 361 The turbulence data..were measured inside the jet stream by the use of hot wire equipment. 1966 R. Thomas Spy in Vodka (1967) i. 7, I was pleasantly surprised to find my car intact. The German juvenile delinquents..can hot-wire a car in a time that makes their American counterparts look sick. 1968 Daily Colonist (Victoria, B.C.) 5 Sept. 19/2 Newton told court he yelled his intention to 'hot-wire' the Thomas car (bypass the need for an ignition key). 1962 Amer. Speech XXXVII. 269 Hot-wired, designating a motor vehicle which has had the ignition switch short-circuited by a thief. 1843 Ainsworth's Mag. IV. 437 The 'hot with' set our tongues in motion. 1856 C. Reade Never too Late III. xxx. 289 Hot with,' demanded the waiter in a sharp mechanical tone... The landlady..poured first the brandy then the hot water into a tumbler.

d. Used before numerous vbl. sbs. and pa. pples. (ppl. adjs.) that denote an industrial or manufacturing process or its result, from which arise transitive vbs. as back-formations; so **hot drawing**, the drawing of wire, tubing, or the like with the application of heat or while hot; hot-drawn adj.; hot-draw vb.; **hot gilding** (in Dict., sense 12 c); **hot moulding**, moulding in which heat is applied either to the material to be moulded (as in injection moulding) or to the mould (as in compression moulding); hot-moulded adj.; hot-mould vb.; **hot rolling**, the rolling of metal while it is at an elevated temperature; hot-rolled adj.;

hot-roll vb. Also HOT-PRESS *v.*, *HOT-WORK *v.*
1899 Hot-drawn [in Dict., 12 c]. 1910 H. P. TIEMANN *Iron & Steel* 336 With hot drawing the thickness of the walls is not generally carried below ¼ in. 1952 J. DELMONTE *Plastics Molding* xiii. 409 It is much more feasible to stock a few sizes and then, when a special tubing is required, hot-draw the tubing down to the required dimension. 1963 H. R. CLAUSER *Encycl. Engin. Materials* 685/1 Typical applications for these steels are..hot-drawing dies..and die casting dies. 1938 H. I. LEWENZ tr. *Brandenburger's Processes & Machinery Plastics Industry* i. 1 Hot moulding, where the hardening process ..takes place to some extent simultaneously with the moulding of the paste by tools, which..are heated. *Ibid.* 7 Hot-moulded articles. 1962 *Times* 13 Feb. 4/6 They have offered to build a racing eight by the hot-moulded method. 1964 N. G. CLARK *Mod. Org. Chem.* xvii. 359 The product is a colourless, transparent solid which may be machined, or hot-moulded. 1878 Hot-rolled [see *cold-rolled* (*COLD *a.* 18)]. 1888 *Proc. Inst. Civil Engineers,* XCIV. 236 (*caption*) Billet [of mild steel]. Reduced by hot rolling to wire rod. 1925 Hot-roll [see *base box* (*BASE *sb.*[1] 20)]. 1928 *Jrnl. Iron & Steel Inst.* CXVII. 840 A description of the hot-rolling, cold-rolling and drawing operations in the manufacture of nickel steel. 1955 *Times* 12 July 8/3 The new Duisburg mill will be able.. to produce 180,000 tons of hot-rolled steel products a month. 1967 A. H. COTTRELL *Introd. Metall.* xxii. 439 In steel-making..the ingots are taken from their moulds while still hot and then stored in a soaking pit or furnace ready for hot-rolling at 1000–1200° C.

hot, *adv.* **2.** (Later examples.)
1884 'MARK TWAIN' *Huck. Finn* xv. 126 As soon as I got started I took out after the raft, hot and heavy, right down the tow-head. 1922 JOYCE *Ulysses* 160 If a fellow gave them trouble being lagged they let him have it hot and heavy in the bridewell. 1924 T. E. LAWRENCE *Let.* 19 Mar. (1938) 459 *A[rabia] D[eserta]* is selling hot in U.S.A. 1927 *Melody Maker* Aug. 782/2 The verse is then taken 'hot' by the trumpet. 1928 *Ibid.* Feb. 184/2 They are..played 'hot'. 1934 T. S. ELIOT *Rock* i. 12 The shareholders ain't there, and their divvies ain't 'ardly there either, 'cos they paid too 'ot for their shares.

hot, *v.* Add: **1.** With *up*. To become hot. Also *fig.*
In quot. 1909 the passive use is somewhat unusual.
1909 W. OWEN *Let.* 23 Aug. (1967) 55, I *did* take a holiday, on account of the day being so 'hotted up'. 1936 WODEHOUSE *Laughing Gas* xiii. 146 It did not need a razorlike intelligence to show me that things were hotting up, and that flight was the only course. 1958 *Daily Mail* 3 Sept. 1/2 The cold war being fought out between Britain and Iceland in the rich fishing grounds of the northern seas hotted up yesterday. 1967 M. CHANDLER *Ceramics in Mod. World* iv. 132 The higher the frequency the more the insulator will hot up. 1969 *Times* 28 July 18/6 (*heading*) Pacific air route battle hots up.
2. (Later examples.)
1903 M. CROMMELIN *Crimson Lilies* (1913) xxiv. 176 You put something on your outside, while I hot up some tea..to put in your inside. 1928 GALSWORTHY *Swan Song* II. iv. 138 Let me hot up your stew. 1952 S. SELVON *Brighter Sun* ix. 188 Urmilla went to hot the food.
3. *fig.* or *transf.* with *up*. *spec.* (*a*) To become 'warm' or unpleasantly excited, near the point of anger. (*b*) *pass.* Of an internal-combustion engine, a car, etc.: to be 'tuned up'; to have the power increased so as to be capable of higher speeds. Also *trans.* and in *ppl. a.*, and in extended uses.
1923 WODEHOUSE *Inimitable Jeeves* x. 108 The atmosphere was consequently more or less hotted up when Cyril..breezed down centre. 1928 *Morning Post* 20 Oct. 10/7 This car..is not in any sense a 'hotted up' Morris six-cylinder. 1939 R. LEHMANN *No More Music* 56 *Jan:* Excitable girl. *Miriam:* You seem to have hotted her up nicely. 1945 *Jrnl. R. Aeronaut. Soc.* XLIX. 205/1 The Americans have had considerable success in 'hotting up' engines once intended for airline use. 1946 A. LEE *German Air Force* 13 The special hotted-up engines flown by ace pilots. 1950 *Hansard Commons* CCCCLXXII. 2036 Our present submarines are being 'hotted-up' with fast battery drive. 1958 J. CAREW *Wild Coast* xii. 161 The drummerboys hotted up the rhythm. 1958 *Spectator* 6 June 740/2 The *Observer* complains that Mr. Allsop's style has 'the hotted-up, black-and-white, metaphor- and smart-allusion-besotted quality typical of one kind of popular journalism'. 1958 *Times* 12 Aug. 7/2 The General Assembly is the place to hot things up. 1959 *Listener* 9 Apr. 628/1 Hotting up hymn tunes was the absolute end. 1960 *Guardian* 5 Jan. 4/4 A rather hotted-up radio receiver. 1962 *Listener* 5 Apr. 595/2 The hotted-up economic and ideological conflicts of the latter part of the nineteenth century. 1968 *Hot Car* Oct. 61 If you hot up your car, don't forget to step up your braking power. 1973 'M. INNES' *Appleby's Answer* xvii. 150 Far from cooling up, he might hot up, and simply jump the gun. 1973 *Guardian* 28 May 2 Sir Alec Douglas-Home.. was launching into..the need to avoid actions which would 'hot up this cod war'.

hot air. 1. *attrib.* or as *adj.* (See HOT *a.* 12 c.)
1813 M. EDGEWORTH *Let.* 26 Apr. (1971) 27 When first the hot air flues were opened..the hot air all escaped. 1841 C. CIST *Cincinnati* (Advt.), Manufacturers of Hot Air Furnaces, Stoves, &c. 1853 *Hunt's Merchant's Mag.* XXVIII. 282 Hot-air vessels are to take the place of steamships. 1854 [see HOT *a.* 12 c]. 1861 MRS. BEETON *Bk. Housek. Managem.* 1009 A better arrangement is to have a hot-air closet,..heated by hot-air pipes,..and clothes-horses on castors..for drying purposes. 1892 STEVENSON & OSBOURNE *Wrecker* i. 24, I designed..a hot-air grating for the offices. 1905 G. BACON *Balloons* i. 23 To this day large hot-air balloons inflated by the same methods employed a hundred years ago occasionally take passengers aloft. 1908 *Sears, Roebuck Catal.* 606/3 Standard American Plumbing. Hot Air and Hot Water Heating. 1940 *Chambers's Techn. Dict.* 424/1 *Hot-air heater*, one which supplies warm air through gratings in the floor or openings in the walls. *Ibid.*, Hot-air turbine. 1964 M. HYNES *Med. Bacteriol.* (ed. 8) iii. 27 The syringe, clean, assembled and lubricated, is packed in a glass or metal tube and subjected to a temperature of 160° C for 1 hour in the hot-air oven. 1970 *New Yorker* 3 Oct. 28/2 We can..move hot-air registers around simply by making holes in the floor. 1973 *Times* 8 Jan. 2/5 (*caption*) A hot-air airship, claimed to be the world's first, making its maiden flight yesterday from Newbury, Berkshire.
2. *colloq.* (orig. *U.S.*). Vaporous or boastful talk, pretentious or unsubstantial statements or claims; also *attrib.*, as in *hot-air artist* or *merchant*, one who indulges in talk of this kind.
1873 'MARK TWAIN' & WARNER *Gilded Age* xliv. 399 The most airy scheme inflated in the hot air of the Capital only reached in magnitude some of his lesser fancies, the by-play of his constructive imagination. 1900 ADE *Fables in Slang* 126 They strolled under the Maples, and he talked what is technically known as Hot Air. 1906 [see *BLUE SKY 2]. 1910 *Sat. Even. Post* 2 July 13/3 'Hot-air artists' was a phrase uncoined; the farmer called them 'jawsmiths'. 1911 H. B. WRIGHT *Winning of Barbara Worth* 327 The presence av sich..a hot air merchant..is a disgrace to any..company. 1914 [see *BULL *sb.*[4] 3]. 1920 'SAPPER' *Bull-Dog Drummond* xii. 308 Author—so-called. Hot-air factory, but useful up to a point. 1922 *Daily Mail* 20 Nov. 8 Much 'hot air' from the politicians. 1932 [see *BULL *sb.*[4] 3]. 1956 A. WILSON *Anglo-Saxon Att.* II. iii. 365 Gerald in his new mood thought only he shouldn't have poll-parroted his life away in humbug and hot air. 1963 *Times* 17 May 13/7 Anti-European slogans shouted by politicians were mere hot air. 1970 T. COE *Wax Apple* (1973) xxiv. 170, I think you're just full of hot air... I don't believe you know anything.

hot bed. Add: **3*.** *U.S. slang.* A bed, usu. in a flop-house, used continuously, day and night, by different people for limited periods. Also, the flop-house in which such beds are found.
1945 L. SHELLY *Jive Talk Dict.* 26/1 *Hot bed*, cheap flop house. 1953 POHL & KORNBLUTH *Space Merchants* (1955) vii. 77 The dorm was jammed with about sixty bunks... Since production went on only during daylight hours, the hot-bed system wasn't in use. My bunk was all mine. 1968 P. OLIVER *Screening Blues* vi. 253 In city streets where hot-bed apartments and kitchenettes were the only dwelling units available to Negroes,..the automobile parked in the street became the front parlour. 1970 C. MAJOR *Dict. Afro-Amer. Slang* 67 Hotbed,..in a flophouse a public bed for the price of 25 cents per eight hours.

hotch, *v.* Add: **3.** *intr.* To swarm.
a 1779 in *E.D.D.* 1797 *Edin. Mag.* Dec. 458 The floor i' now is just a hotchin' thrang. 1825 *Blackw. Mag.* Mar. 369 When there's sae strong a spirit of life hotchin' ower yearth and sea in this very century. 1893–4 R. O. HESLOP *Northumb. Words* II. 387 The place is fair hotchin wi' rabbits. 1909 *Daily Chron.* 17 May 4/4 'The county,' he said, in his broad Ayrshire accent, 'is fair "hotching" ..with them.' 1961 B. FERGUSSON *Watery Maze* v. 111 They found themselves delayed in Lisbon, a city then hotching with spies. 1965 J. CAIRD *Murder Reflected* ix. 108 Tripped over a stool. The place hotches with them. 1967 'H. CALVIN' *DNA Business* xiii. 148, I expect the office will be simply hotching with typists.

hotch, *sb. Sc.* [f. HOTCH *v.*] A jerk or jolt.
1721, 1824 in *E.D.D.* 1773 R. FERGUSSON *Poems* (1956) II. 124 Uncanny hotches Frae clumsy carts or hackney-coaches. 1914 N. MUNRO *New Road* ii. 24 She never mentioned it, but every time I did, I saw her give a hotch upon her chair.

hotcha (hǫ·tʃǎ). *slang* (chiefly *U.S.*). Also hotcha cha, hot-cha(-cha), etc. [Fanciful extension of HOT *a.*] **1.** Used in combination with the traditional interjection *hey nonny nonny* (cf. HEY 2).
1932 KAUFMAN & RYSKIND *Of thee I Sing* in *Famous Plays* (1933) 694 With a hey, nonny nonny, and a ha cha cha! 1937 WODEHOUSE *Lord Emsworth & Others* ii. 99 Good morning, Phipps. What ho, what ho, with a hey nonny nonny and a hot cha-cha. 1954 *Word Study* Oct. 6/1 Many of us..find Shakespeare's *Hey nonny nonny* a bit naked without the *hotcha cha*.
2. *int.* Expressing enthusiastic approval. Also as *adj.*, attractive, desirable.
1934 J. O'HARA *Appointment in Samarra* (1935) i. 21 Did you ever see her in a bathing suit? Hot-cha! 1939 R. CHANDLER *Big Sleep* ix. 71 He run Sternwood's hotcha daughter,..off to Yuma.
3. *adj.* Of jazz: hot.
1937 S. KING-HALL *King-Hall Survey 1936* ix. 93, I seemed to notice a tendency, towards the end of 1936, for what one of my daughters calls 'Hotcha' music to be replaced by 'Swing' music. 1947 R. DE TOLEDANO *Frontiers of Jazz* ii. 13 In 1934..the number of people who had anything more than a hotcha idea of jazz could be numbered on the fingers of a hand. 1957 W. C. HANDY *Father of Blues* vii. 98 Hot-cha music was the stuff we needed, and it had to be mellow. 1960 C. RAY *Merry Eng.* 60 There are hotcha gramophone records.

hotchi witchu (hǫ·tʃi wi·tʃū). Also hotchi witchi. [Romany.] The gypsy name for a hedgehog.
1843 BORROW *Zincali* (ed. 2) I. I. v. 98 They [*sc.* gypsies] have been seen..to roast hotchiwitchu or hedgehog. 1928 *Sunday Dispatch* 2 Sept. 3/3 The gypsies call the animal 'Hotchi witchu', we speak of him as the hedgehog. 1935 'X. PETULENGRO' *Romany Life* xix. 79 As we sat..eating our *hotchi-witchi*, some gentlemen.. passed near us and were curious to see what smelt so good. 1963 C. DUFF tr. *Clébert's Gipsies* v. 185 One of the dishes peculiar to Gypsies is the hedgehog..(in English Gypsy *hotchi-witchi*).

Hotchkiss (hǫ·tʃkis). The name of B. B. *Hotchkiss* (1826–85), American inventor, used *attrib.* to designate certain cannon and rifles invented by him, and a machine-gun developed by his successors.
1878 W. MORGAN-BROWN *Brit. Pat. 3017* 30 July, This Invention consists in several important improvements on the Hotchkiss revolving cannon and other machine guns. *Ibid.*, The gunner may have entire control over the laying, sighting, and firing of a Hotchkiss revolving cannon. 1880 *Encycl. Brit.* XI. 285/1 Hotchkiss Magazine Gun. 1886 *Harper's Mag.* Oct. 793/2 The types adopted by the United States navy are the Hotchkiss revolving cannon and rapid-firing single-shot guns. 1890 *Handbk. Six-Pounder Hotchkiss Quick-Firing Gun* 12 The shell is fitted with..a brass Hotchkiss band. *Ibid.*, The fuze may be either the Nordenfelt or the Hotchkiss, Mark II, pattern. 1890 G. S. CLARKE *Fortification* xiii. 191 Twelve Hotchkiss machine guns. 1899–1900 *Kynoch Jrnl.* Dec.–Jan. 31/2 The Hotchkiss Automatic Gun. 1900 [see POM-POM]. 1902 *Encycl. Brit.* XXIX. 166/1 In the Hotchkiss and also in the Skoda systems the mechanism is of the vertical breech block type. 1908 *Trade Marks Jrnl.* 3 June 906 Hotchkiss,.. Guns, gun mountings, gun carriages, and other appurtenances and accessories for guns. .. The Hotchkiss Ordnance Company, Limited,..London,..manufacturers. *Ibid.*, Hotchkiss... Ammunition (explosive). The Hotchkiss Ordnance Company, Limited, ..London,..manufacturers. 1926 T. E. LAWRENCE *Seven Pillars* (1935) III. xxviii. 168 The Frenchman began by presenting six Hotchkiss automatics complete with instructors. 1959 *Chambers's Encycl.* V. 676/2 The Hotchkiss light machine-gun..was adopted for cavalry use.

hot damn, *int. phr. U.S.* An intensified form of 'damn!'.
1936 J. STEINBECK *In Dubious Battle* 90 Hot-damn, listen. 1957 J. KEROUAC *On Road* (1958) IV. ii. 257 Hot damn, I could go with you? 1971 *Black World* Apr. 73 Pop pulled garbage in an old-law tenement building. Hot-damn..every night.

hot dog. [DOG *sb.*] **1. a.** *N. Amer. slang.* One who is skilled or proficient in some pursuit (see also quot. 1900). Also *attrib.* or as *adj. phr.*, good, superior.
Freq. with connotations of 'showing off'. In some sports, e.g. Skiing, applied to 'trick' or 'stunt' or spectacular variations of normal movements.
1896 W. C. GORE in *Inlander* Jan. 148 *Hot-dog*, good, superior. 'He has made some hot-dog drawings for ——.' 1900 *Dialect Notes* II. I. 42 *Hot-dog*. 1. One very proficient in certain things. 2. A hot sausage. 3. A hard student. 4. A conceited person. 1966 *Atlantic Monthly* Mar. 131 We had this one [basketball] player, Alston Mackintosh,.. who could hit nine out of ten from the foul line with his back to the basket. He was a real hot dog. 1968–70 *Current Slang* (Univ. S. Dakota) III–IV. 71 *Hot dog*, one who shows off by squealing his tires and gunning his motor as he drives around among his friends. A crowd pleaser who is actually obnoxious. A showoff. To show off. 1973 *Internat. Herald Tribune* 9 Feb. 13/1 'Hot dog', or acrobatic, skiing has become a fast-growing sport in the United States. 1973 *Times* 23 Feb. (Canada Suppl.) p. vii/8 The local 'hot dogs', mostly teenagers who probably learnt to ski almost as soon as they could stand up, skate over mounds of hard-packed snow 4 ft high like sprinters over hurdles. 1973 *Time* (Canad. ed.) 16 July 51/1 Cedeno pulls off so many flashy plays in the field that some National League players accuse him of being a 'hot dog'—baseballese for show-off. 1973 *Globe & Mail* (Toronto) 1 Sept. 35/1 This year..it was generally agreed that McQuay must change his hotdog antics or go. 1974 *Ibid.* 20 Feb. 34/5 Dan Genge..admits that it was those same little hills that made him what he is—one of Ontario's coolest hot-dog skiers. *Ibid.* 34/8 There has to be something of the mountain climber in the hot-dog skier. They must do something harder, tougher, higher each year. 1974 *Hockey News* (Montreal) 22 Feb. 35/1 Critics label him a 'hot dog' and a 'show-off' and several unprintable things.
b. *Surfing slang.* A particular kind of surf-board, somewhat smaller than a 'gun' (*GUN *sb.* 13**). Also *attrib.* or as *adj. phr.*
1963 *Observer* 13 Oct. 15/6 He always owns two boards at any one time, one 'hot dog' board and a 'big gun' (which is a foot or so longer) for riding really big waves. 1965 [see *GUN *sb.* 13**]. 1966 *New Yorker* 31 Dec. 28 He's a hot-dog surfer and he used to be real wigged on Zen. 1967 *Surfabout* III. VII. 38/1 Joey Hamasaki is an excellent surfer and she could possibly beat Joyce in small hot-dog surf. 1970 [see *GUN *sb.* 13**].
So **hot dogger,** a surfer who rides a 'hot dog'; **hot dogging,** the action of riding such a board (see also quot. 1963[2]). Hence (as a back-formation) **hot-dog** *v.*
1961 *Life* 1 Sept. 48 Almost every wave carries a 'hot dogger' doing tricks or sometimes even dressed in outlandish garb. 1962 T. MASTERS *Surfing made Easy* 64 *Hot dogging*, performance surfing, fast turns, quick movements, etc. 1963 *Pix* (Austral.) 28 Sept. 63 Looking good on a little wave is hard. If you can hot dog on two foot waves you are 'king'. 1963 *Observer* 13 Oct. 15/4 From all these movements..has developed the modern style of surfing

known as 'hot dogging': a spectacular, virtuoso style which concentrates on fast slides across the face of the waves, rapid turns, cut-backs, flick-offs and a repertoire of classic riding stances which have evolved in much the same way as the classic passes of the Spanish bullfight. 1966 Hot dogging [see *BOMBORA]. 1966 *Surfer* VII. IV. 39 There were plenty of hot-doggers—or trick riders as we called them then—in Hawaii when I was surfing there during the 1920's.
2. A hot sausage enclosed as a sandwich in a bread roll. orig. *U.S.*
1900 [see sense 1 a above]. 1908 'O. HENRY' *Gentle Grafter* vii. 97 Sitting on a cake of ice we ate seven hot dogs. 1920 S. LEWIS *Main St.* 304 Lining one block of Main Street were the 'attractions'—two hot-dog stands, [etc.]. 1926 *Spectator* 11 Sept. 373/1 The President of the Brooklyn team asked them to his box and gave them hot dogs. 1928 S. VINES *Humours Unreconciled* ii. 35 If one trained on hot dog, pumpkin pie, graham flour, ice-cream sundaes. 1932 E. WILSON *Devil take Hindmost* i. 1 The hot-dog stands on the motor roads sell gin. 1955 *Times* 3 Aug. 10/5 Tell the miners what Yarmouth has to offer —Tommy Trinder, Charlie Chester, Ronnie Ronalde, 'hot-dogs', bathing beauties, and all. 1957 *London Mag.* Dec. 19 There is a pier with all the usual sideshows, hot dogs, hamburger and ice cream stands. 1970 *New Yorker* 12 Sept. 50/3 She came back with a hot dog... The smell.. was..persuasive.
3. *int.* Expressing delight or strong approval. Cf. *hot diggety* (*dog*) s.v. *HOT *a.* 12 c. *U.S. slang.*
1906 *Dialect Notes* III. II. 141 Hot dog,..bravo! 1924 [see *ATTABOY *int.*]. 1934 WODEHOUSE *Right Ho, Jeeves* x. 113 Hot dog! Tell me more. 1935 M. M. ATWATER *Murder in Midsummer* xv. 138 'Hot dog!' cried Wally, his eyes shining. 'That's the old fight!' 1944 T. RATTIGAN *While Sun Shines* I. 202 Hot dog! There's some Scotch. 1965 *Amer. Speech* XL. 85 The ubiquitous *hot dog!*

hotel, *sb.* Add: **1. b.** *Hôtel de Ville* (examples).
1744 M. W. MONTAGU *Let.* 12 June (1966) II. 331 Here [*sc.* in Avignon] are 2 Consuls chose every year..and there is as much struggling for that Dignity in the Hotel de ville as in the Senate. 1789 A. YOUNG *Jrnl.* 30 July in *Trav. France* (1792) I. 149 Two bourgeois musketeers conducted me to the hotel de ville. 1797 *Encycl. Brit.* VIII. 683/1 The *hotel de ville* is what we call a *town-house* or town-hall. 1841 C'TESS BLESSINGTON *Idler in France* I. 4 It was then given to a certain Pierre Boys, in exchange for a piece of ground to erect a new hôtel-de-ville. 1846 DICKENS *Pictures from Italy* 9 Sometimes an hotel de ville, sometimes a guard-house. 1957 J. S. BROMLEY in Wallace-Hadrill & McManners *France: Govt. & Society* 147 Louis XIV made his first formal visit to the Hôtel de Ville, in 1687, to 'forgive' the City [of Paris] for its share in the Fronde. 1973 *Times* 22 Feb. 5/1 M Louis Pradel..dominates the hôtel de ville.
d. hôtel garni, a furnished apartment; an hotel or boarding-house supplying breakfast.
1774 H. WALPOLE *Let.* 7 Sept. (1857) VI. 114, I now live in dread of my biennial gout, and should die of it in an *hôtel garni*. *Ibid.* 28 Sept. 126 Take care of your papers at Paris,.. In the *hôtels garnis* they have double keys to every lock. 1858 GEO. ELIOT *Let.* 17 Apr. (1954) II. 450 He took us to two Hôtels Garnis—places where you get lodgings and attendance and coffee and nothing else. 1896 E. DOWSON *Let. c.* 12 Jan. (1967) 339, I have just risen from a bed of sickness, incomparably uncomfortable, as one in my *hotel garni*, one is reduced under such circumstances to living entirely on milk the only nourishment that one can procure. 1968 *Guardian* 28 Dec. 5/5 The cheapest alternative to *le camping*..is the *hotel garni*, providing only bed and breakfast.
e. hôtel particulier, a large privately owned town house or block of flats.
1934 JOYCE *Let.* 13 July (1966) III. 309 Right opposite is a hotel particulier of 2 storeys. 1964 *Time Off in Paris* (Observer) 16 The big Haussman-period blocks of flats (*hotels particuliers*), that are still the main form of housing in central Paris, were built each to contain different social strata, and so they still do today. 1969 N. FREELING *Tsing-Boum* xx. 144 There were still some hôtels particuliers, last bastions of privilege. 1970 *Times* 28 Nov. 12/5 Their *hotel particulier* in Paris contains countless treasures.
4. *hotel-bill, bus, clerk, garage, -keeper, -keeping, lobby, manager, omnibus, porter, prowler, register, room, tout.*
1872 GEO. ELIOT *Middlem.* IV. VII. lxx. 123 Raffles's pockets..were sure to carry..hotel-bills of the places he had stopped in. 1903 JOYCE *Let.* 26 Feb. (1966) II. 31, I was favoured with my hotel-bill on Tuesday. 1973 P. AUDEMARS *Delicate Dust of Death* xii. 175 The 10 per cent surcharge would have been added to the hotel bill. 1878 *Harper's Mag.* Jan. 194 The traveller reaches his stopping-place by hotel 'bus, carriage, or by the democratic street cars. 1923 E. F. WYATT *Invis. Gods* IV. iii. 221 He was marshalling them all toward a hotel bus. 1856 F. L. OLMSTED *Journey Slave States* (1861) V. 333 An easy and gentleman-like employment as that of hotel-clerk and bar-keeper. 1945 L. R. TRYON *Poor Man's Doctor* 185, I suppose the actor is better trained than most of us at inventing stratagems for getting past the hard and watchful eyes of hotel clerks. 1910 *Bradshaw's Railway Guide* Apr. 1101 Best Hotel Garage in London. 1829 *Virginia Lit. Museum* 336/1 The Proctor is required to make a monthly report..of any matters relating to the students or hotel-keepers which may be worthy of being noticed. 1869 W. H. H. MURRAY *Adventures in Wilderness* 35 The 'hotel guides' are paid so much per month by the hotel-keepers. 1879 *Scribner's Monthly* June 242/1 Her husband had left town suddenly on a horse belonging to the hotel-keeper. 1916 JOYCE *Portrait of Artist* (1969) i. 29 He..began to speak with the voice of the hotelkeeper. 1870 'F. FERN' *Ginger-Snaps* 247 Having then left what, in my opinion, is the perfection of hotel-keeping. 1959 *News Chron.* 19 Aug. 6/2 French small hotel-keeping at its most terrible. 1912 M. NICHOLSON *Hoosier Chron.* 182 A number of idlers in the hotel lobby regarded him with a new interest. 1964 M. MCLUHAN *Understanding Media* (1967) II. xxxi. 341 Shoddy match-wood bars and hotel lobbies. 1910 W. J. LOCKE *Simon* xv. 199 The correctly attired hotel manager in the attitude in which he habitually surveyed the lay-out of the *table d'hôte*. 1972 I. HAMILTON *Thrill Machine* viii. 34 The hotel manager led her and her group towards the lift. 1878 R. L. STEVENSON *Inland Voyage* 37 The driver of the hotel omnibus. 1910 *Bradshaw's Railway Guide* Apr. 1049 The Hotel omnibus meets the principal London trains. 1847 F. A. KEMBLE *Let.* 31 May in *Rec. Later Life* (1882) III. 190, I..despatched one of the hotel porters thither to hunt for her. 1881 LADY C. SCHREIBER *Jrnl.* (1911) II. 343 The officious Hotel porter had told a lot of other people. 1968 *Guardian* 11 Apr. 5/2 Now we are going to set off along the A 10—the hotel porter will direct you to it. 1928 M. C. SHARPE *Chicago May* 286 Hotel Prowlers, sneak thieves. 1962 K. ORVIS *Damned & Destroyed* v. 40 Hotel-prowlers and house-prowlers. 1860 in *Abraham Lincoln Q.* (1949) Mar. 262, I found the name of Mr Bates on the hotel register. 1899 E. WHARTON *Greater Inclination* iv. 99 She must take her place in the hotel register as Mrs. Garnett. 1971 P. M. HUBBARD *High Tide* i. 10 The hotel register still lay open on the desk where I had signed it. 1946 E. O'NEILL *Iceman Cometh* (1947) IV. 202 You know how it is, traveling around. The damned hotel rooms. 1972 R. LOCKRIDGE *Something up Sleeve* i. 13 The jurors had been escorted back to their hotel rooms. 1881 Hotel tout [see TOUT *sb.*[1] 2]. 1922 S. LESLIE *Oppidan* i. 7 He had picked them up at the station like a hotel-tout.

hoteldom (hǫte·ldəm). [f. HOTEL *sb.* + -DOM.] The realm of hotels, hotels collectively.
1904 *Westm. Gaz.* 1 June 9/2 Then our representative turned his attention to hoteldom. 1927 *Glasgow Herald* 28 Mar. 11/1 There is to be an addition to London hoteldom. 1964 *Sat. Rev.* (U.S.) 21 Mar. 28 The biggest thing in hoteldom lately is non-eating.

hotelier (‖otęlye, (h)ote·li,ẽi, hote·liəɹ). Also **hôtelier.** [Fr.] The keeper or proprietor of an hotel.
1905 *Westm. Gaz.* 10 Mar. 12/1 Some enterprising hotelier. 1907 *Ibid.* 7 Sept. 11/2 He grudges the lucky Swiss hôtelier his millions. 1930 *Observer* 20 Apr. 7 That prince of hoteliers, the late M. Ritz. 1956 R. BRADDON *Nancy Wake* i. 6 He was an immaculately dressed *hôtelier*, a director-manager of the Palm Beach Hotel. 1959 F. STARK *Riding to Tigris* 12 In all which countries hotelier schools exist. 1966 *Illustr. London News* 3 Sept. 11 He is on the board of 36 companies, and..can fairly claim to be the biggest hotelier in Europe. 1972 L. P. BACHMANN *Ultimate Act* iv. 33 One has to learn many languages to be a proper hôtelier,.. I attended hôtelier school in Switzerland.

hot-foot, *adv.* Add: **2.** *attrib.* or as *adj.* Acting with haste or promptitude.
1901 *Spectator* 2 Nov. 631/1 He had to make the most arduous hot-foot journeys across the country. 1904 'O. HENRY' *Cabbages & Kings* v. 90 He was private secretary of the late hot-foot president of this grocery and fruit stand that they call a country. 1940 C. DAY LEWIS tr. *Georgics of Virgil* I. 28 If you observe the hotfoot sun and the moon's phases, To-morrow will never cheat you.
B. *sb.* **1.** Prompt or rapid action or movement; a quick escape, as in the phrases *to do a hot foot* (or *foots*), *to give* (someone) *the hot foot*, *to come* (or *go*) *on the hot foot*. *U.S. slang.*
1869 *Congress. Globe* 15 Jan. 389/3 The honorable Senator..admonishes us of the importance of hot-foot in this business, if I may say so, of allowing the testimony to be taken at once. 1897 *Pop. Sci. Monthly* Apr. 833 To run from a police officer is *to do a hot foot*. 1903 'H. MCHUGH' *Back to Woods* iv. 66 Did somebody give you the hot-foot and make a quick exit? 1905 —— *You can search Me* iii. 55 If somebody ever steals his hammer he'll be doing hotfoots for the handout. 1915 H. L. WILSON *Ruggles of Red Gap* (1917) ii. 27 We'd better report to her before she does a hot-foot over here. 1926 *Flynn's* 16 Jan. 639/1, I know that th'fly was jerry because he gave me th'once over as I was comin' out and I went on th' hoot-foot... I beat it. 1929 C. F. COE *Hooch!* x. 241 You dress an' grab a cab, see? Come down here to Zuroto's on the hot foot.
2. A beating on the soles of the feet; more usually, a practical joke in which a match is put against the victim's foot and then lit. Also *fig.*
1906 A. H. LEWIS *Confessions of Detective* I. iii. 32 I'd become learned in certain mysteries, among others, the 'hot foot'... Given a man, unconscious by..rum,..you can restore him..by smartly beating the soles of his feet. 1934 D. RUNYON in *Hearst's International* Sept. 84/1 The way you give a hot foot is to sneak up behind some guy.. and stick a paper match in his shoe between the sole and the upper along about where his little toe ought to be, and then light the match. 1943 J. MITCHELL *McSorley's Wonderful Saloon* (1946) 18 Drunks reel over from the Bowery and..the kids give them hotfoots with kitchen matches. 1948 MENCKEN *Amer. Lang.* Suppl. I. v. 392 The Army also discourages the old soldiers' game of *hot-foot*, which consists in inserting matches between the soles and uppers of a sleeping comrade's shoes, and then lighting them. 1959 *Encounter* Dec. 30/2 His prose should never be quiet. It must always shock with the hot-foot.

hot-foot, *v. colloq.* (chiefly *U.S.*). [f. prec.] *intr.* To go hot-foot; to make haste. Also with *it*. Hence **hot-foo·ted** *ppl. a.*
1896 ADE *Artie* iii. 22, I hot-foots up to the dance. 1896 [see *DINKY *a.*[1]]. 1903 W. B. YEATS *In Seven Woods* 37 The hot-footed sun, And the cold sliding slippery-footed moon. 1904 *Sun* (N.Y.) 27 Aug. 10 Thousands hot footed to the corner of Broadway. 1906 *Dialect Notes* III. II. 141 *Hotfoot it*, to hasten. 1911 R. D. SAUNDERS *Col. Todhunter* ii. 34 Great Scott and Maria, you must have hot-footed it away from your vittles, young man! 1926 C. HARRIS *Flapper Anne* ii. 92 At the present nothing was further from her thoughts than marrying Sealy, but she craved the triumph of bringing him hotfooted to Milledge. 1928 *Observer* 29 Jan. 22/3 Words of such enthusiasm send one at once hot-footed to the [Crystal] palace to see for oneself. 1934 J. O'HARA *Appointment in Samarra* (1935) vii. 214 When O'Dowd did hear..he would hot-foot out to Quilty and make the sale. 1951 *Manch. Guardian Weekly* 18 Jan. 15 He short-circuits Mr. Kingley's pedestrian approach and makes a bee-line for Koestler's original like the prodigal hot-footing it home. 1970 G. GREER *Female Eunuch* 195 She hotfoots to Mexico.

hot gospeller. [GOSPELLER.] (See GOSPELLER 5 b, where defined.) Also *transf.* Hence **hot gospelling** *vbl. sb.* and *ppl. a.*; **hot-gospel** *v. trans.* and *intr.*
1562, 1874 [see GOSPELLER 5 b]. 1875 TENNYSON *Q. Mary* 271 There are Hot Gospellers even among our guards. 1899 R. WHITEING *No. 5 John St.* viii. 71 That terrible old Hot Gospeller who..screams undying hate of all institutions. 1923 W. S. CHURCHILL *World Crisis 1911–14* ii. 29 Like Henry VIII, he [*sc.* A. J. Balfour] decapitated Papists and burned hot Gospellers on the same day for their respective divergencies..from his..compromise. 1923 J. M. MURRY *Pencillings* 107, I do not think.. I can fairly be accused of advocating hot-gospelling or holy-roaring as the short road to good writing. *Ibid.* 275 He has made me bold, not to say thrasonical. I am become a hot-gospeller, a crusader. Dr. Henry Bradley himself is timid and lukewarm compared to me. 1931 R. LEHMANN *Let. to Sister* 9 Do you remember that frothing hot-gospelling temperance-man?.. He told them how he himself had found salvation when going upstairs late one night—drunk, he manfully admitted. 1933 *Times Lit. Suppl.* 19 Oct. 713/1 The hot-gospeller Spooner marries the thin-lipped Harriet. 1952 R. CAMPBELL *Lorca* i. 7 Every sub-poetaster is hot-gospelling world-messages. 1955 A. L. ROWSE *Expansion Eliz. Eng.* ix. 331 The hot-gospelling of Knox and the spread of Calvinism. 1958 *Spectator* 31 Jan. 124/3 Mr. Dulles's hot-gospelling speech. 1965 G. MCINNES *Road to Gundagai* vi. 110 In the true..hot-gospelling tradition, he ladled out the brimstone. 1968 *Times* 6 Nov. 3/4 The hot-gospeller H. H. Nininger, who preached the value and importance of meteorites. 1970 *Daily Tel.* (Colour Suppl.) 13 Mar. 47/1 He was not hot-gospelling for pop or for the American way of life. 1972 *Daily Tel.* 4 Dec. 11/3 The film..is 'Marjoe', a..deeply cynical documentary on the American weakness for hot-gospelling revivalists.

hot-house, hothouse, *sb.* Add: **3. b.** Also *attrib.*
1838 DICKENS *Nickleby* (1839) xxi. 201 Mrs. Wititterly is of a very excitable nature, very delicate, very fragile; a hothouse plant. 1840 MILL *Let.* 3 Dec. (1910) I. 119 *You* will be interested in the modern German art;..it appears to me a feeble, hot-house product. 1853 C. READE *Peg Woff.* ii. 46 She is so fresh and natural. They are all hot-house plants. 1911 G. B. SHAW *Getting Married* Pref. 156 A hothouse atmosphere of unnatural affection. 1964 *English Studies* XLV. 50 Those delicate, hot-house feelings. 1966 *Listener* 28 July 143/2 Opera will never cease to be a hothouse plant in this country until a wider public appreciates it. 1973 'E. PETERS' *City of Gold & Shadows* ii. 31 This hot-house community of time-expired settlers and pay-happy leave-men.
6. (Earlier and *attrib.* examples.)
1707 H. SLOANE *Voy. to Jamaica* I. p. ciii, One Prince, a lusty Negro, had been ill of the Yaws..and flux'd for it in one of the Chirurgeons Hot-Houses. 1788 H. MACNEILL *Observ. Treatm. Negroes Jamaica* 8 He [*sc.* a sick slave] is put into a house particularly appropriated to the purpose, (a lazaretto or *hot-house*, as it is generally called). 1790 W. BECKFORD *Descr. Acct. Jamaica* II. 17 This building has a narrow piazza in front, at the end of which is a small apartment for the nurse or hot-house woman. 1827 *Hamel, the Obeah Man* I. xxii. 244 The hothouse-keeper indulged him with a plentiful dose of grog. 1834 R. R. MADDEN *Let.* 4 Apr. in *Twelvemonth's Residence W. Indies* (1835) I. 154 The hot-house doctor is generally a negro disqualified by age or infirmity for labour in the field. He has charge of the medicines.

hot line: see *HOT *a.* 12 c.

Hotnot (hǫ·tnǫt). *S. Afr.* An abbreviated pejorative form of Hottentot.
1913 C. PETTMAN *Africanderisms* 217 *Hotnot*, a common Dutch pronunciation of Hottentot. 1939 J. S. MARAIS *Cape Coloured People* i. 31 The 'colloquial Afrikaans' for Coloured People is not, as Professor Macmillan thinks,..'Hotnots'—at any rate in the western Cape—but 'bruin mense'. 'Hotnot' is a term of contempt. 1949 *Cape Times* 8 July 8 His uncouth remarks about 'Hotnots, Coolies and Kafirs'. 1956 A. SAMPSON *Drum* xv. 201 They joked and laughed about apartheid, and about being mistaken for white men, and being called bushmen, hotnots, coolies, with compulsory wounding laughter. 1966 A. SACHS *Jail Diary* xvi. 133 That was my Grandfather's place..before the British took it away, and he even allowed a few Hotnots to sleep in the house.

hot-pot, hot pot. Add: **2.** (Earlier example.)
1851 *London at Table* I. 21 At the bottom of the table,.. let there be a hot-pot.
3. *Racing slang.* (See quots. and cf. HOT *a.* 8 e.)
1922 *N. & Q.* 23 Sept. 206/2 *Hot pot*, a horse which has been heavily backed. 1945 BAKER *Austral. Lang.* xvii. 299 'Truth' has always devoted special attention to sporting news and, as a consequence, has developed racing jargon considerably... A favourite is a hot-pot.

hot-press, *sb.* Add: Also, a similar apparatus used in making plywood (see next). Also *attrib.* (Further examples.)
1938 *Trans. Amer. Soc. Mech. Engin.* LX. 60/1 The great scarcity of hot presses in plywood plants. *Ibid.* 60/2 In the hot-press operation with resin film, the resin film is cut to dimension and laid between the sheets of veneer. 1943 SIMONDS & ELLIS *Handbk. Plastics* xi. 433 The panels are formed in these hot presses under pressures ranging from 100 to 300 lb. per sq. in. *Ibid.*, At 220° F., hot-press panels are completely bonded in from 3 to 10 minutes' cure in a hot press. 1968 KIRK & OTHMER *Encycl. Chem. Technol.* (ed. 2) XV. 903 By far the largest production [of plywood] is in a hot press in which thermosetting resins are the adhesive basis.

hot-press, *v.* Add: (Further examples.) Also, to press (veneers, etc.) between heated platens for a period in order to bond them together to make plywood.
1932 *Trans. Amer. Soc. Mech. Engin.* LIV (sect. WDI). 1/2 The necessity for hot-pressing is the chief disadvantage [of blood-albumin glues]. 1943 SIMONDS & ELLIS *Handbk. Plastics* xi. 433 Wherever the type of plywood permits it, hot pressing offers quite a few advantages to the manufacturer over cold pressing. 1957 *Times* 23 Dec. 11/2 A decision was taken to introduce hot-pressing in the veneer mills. 1968 J. ARNOLD *Shell Bk. Country Crafts* 214 After the drying, the cloth is 'hot-pressed'.
b. *trans.* To shape under pressure in a heated die or mould.
1913 J. V. WOODWORTH *Drop Forging* viii. 254 The Burdict hot-pressed nut-machine. 1938 H. I. LEWENZ tr. *Brandenburger's Processes & Machinery Plastics Industry* i. 5 Constant improvements..have taken place in the production of paste for hot pressing. 1947 KIRK & OTHMER *Encycl. Chem. Technol.* I. 561 Alloys that can be made in no other way can be produced by compressing and sintering metal powders or by hot pressing loosely compacted slugs. 1967 M. CHANDLER *Ceramics in Mod. World* vi. 173 An alternative technique for forming silicon nitride is to hot-press it.

hot rod. orig. *U.S.* A motor vehicle specially modified to give high power and speed; the driver (also **hot rodder**) of a hot rod. Hence **hot-rod** *v. intr.*, **hot-rodding** *vbl. sb.* Also *attrib.* and *fig.*
1945 *Life* 5 Nov. 87 A 'hot rod'..is an automobile stripped for speed and pepped up for power until it can travel 90 to 125 mph. Most hot rods are roadsters. 1949 *Life* 7 Nov. 123 (*caption*) Policeman holds fistful of licenses confiscated from hot-rodders. *Ibid.*, In Los Angeles and Dallas, where 'hot-rodding' is at its peak, hundreds of youngsters spend their spare time in suicidal games on wheels. 1953 J. N. LEONARD *Flight into Space* 87 All over the United States and in many other parts of the world are groups of enthusiasts who dream of hot-rodding off the tedious earth. 1953 *Time* 21 Sept. 85/1 Every day during the show he thrilled the crowds with the airborne hot-rodding that Britain encourages at Farnborough. 1955 *Sci. News Let.* 1 Oct. 213/1 Teenagers are as accident prone with shotguns as they are with hot rods. 1957 J. KEROUAC *On Road* (1958) 15 A hotrod kid came by with his scarf flying. 1958 J. D. MACDONALD *Executioners* (1959) iv. 52, I don't want her hot-rodding around in the night, going to those drive-ins. 1958 *Woman* 9 Aug. 14/1 Hot-rod road scorchers. 1958 *Observer* 30 Nov. 10/5 Bernstein and his direct knowledge of hot-rodders and what they are after. 1959 *Observer* 25 Jan. 7/4 Mike Hawthorn, champion racing driver and the idol of hot-rods all over the world. 1959 *Sunday Times* 22 Mar. 24/8 My old two-litre hot-rod after its last disastrous tune-up. 1959 *Encounter* Dec. 30/2 Almost every popular magazine nowadays is likely to be written in a hot-rod style which bursts with energy in every sentence. 1962 *Punch* 17 Oct. 559/3 The youth of these islands.. looked on the motor-cycle as their hot rod. 1967 *Listener* 23 Nov. 682/1 The Beach Boys made their reputation.. ostensibly by taking surfing and hotrodding as their subject. 1971 *Daily Tel.* (Colour Suppl.) 3 Sept. 41/1 The first hair freaks I ever saw in California were not hippies, but surfers and hotrodders.

hot-short, *a.* Add: Hence **hot-sho·rtness,** the quality or state of being hot-short.
1890 H. M. HOWE *Metall. of Steel* I. iv. 42/1 Manganese ..by bodily removing sulphur from cast-iron and probably from steel,..prevents hot-shortness, both red and yellow. 1961 S. A. HISCOCK *Lead & Lead Alloys for Cable Sheathing* vii. 221 By subtracting 20° C. from the 'cracking-temperature'..some indication will be obtained of the extrusion temperature which must not be exceeded if 'hot-shortness'..is to be avoided.

hot-shot. Restrict †*Obs.* to senses in Dict. and add: Also **hot shot, hotshot. 1. b.** An important or exceptionally capable person. Also *attrib. colloq.* (orig. *U.S.*).
1933 *Amer. Speech* Oct. 35/2 *Hot shot*, a champion, a leading contender, an excellent fighter. 1943 'B. HALLIDAY' *Murder wears a Mummer's Mask* xviii. 205 All the important critics were there—the hot shots from the East whose wire stories to their papers could make or break an actress. 1951 J. D. SALINGER *Catcher in Rye* iii. 23 He was telling us all about what a swell guy he was, what a hot-shot and all. 1952 S. KAUFFMANN *Philanderer* (1953) xiv. 234 The only relaxation I would have would be going out with the local hot-shots. 1961 J. HELLER *Catch-22* (1962) xxii. 230 How about getting us a hotel-room if you're such a hotshot? 1973 J. WAINWRIGHT *Pride of Pigs* 103 These hot-shot scientists. They love the limelight.
3. *U.S. slang.* (See quots.)
1925 *Writer's Monthly* May 486/2 *Hot shot*, a fast freight. 1931 'D. STIFF' *Milk & Honey Route* 207 *Hot shot*, a fast freight or passenger train.
4. *U.S. slang.* (See quots.)
1953 W. BURROUGHS *Junkie* (1972) 157 Hot Shot... Poison, usually strychnine, passed to an addict as junk. The peddler sometimes slips a hot shot to an addict because the addict is giving information to the law. 1971 E. E. LANDY *Underground Dict.* 106 *Hot shot*,..1. Injection of poison that user believes to be good drugs, a method of getting rid of police informers. 2. Injection of a drug that is of higher potency than the addict is accustomed to.
5. Used *attrib.* to designate a wind tunnel in which an arc discharge in a pressurized chamber is used to produce a hypersonic pulse of gas in an evacuated test chamber, the two chambers being separated initially by a diaphragm that is ruptured by the discharge.
1959 *Engineering* 13 Feb. 219/2 A wind tunnel, which will test models at the speed of satellite rockets, is expected to be in use in the United States toward the end of this year... Known as a 'hotshot' tunnel, it makes use of a powerful electric arc to pressurise the air. 1971 *Sci. Amer.* Sept. 14/2 The heated gas in the arc chamber is the working gas of a hotshot tunnel with a working section of 2·5 meters and a Mach number of about 20.

hot spot. Also **hot-spot. I. 1.** *Physiol.* One of numerous small areas on the skin that are specially sensitive to heat.
1888 W. STIRLING tr. *Landois's Text-bk. Human Physiol.* (ed. 3) xiv. 836 The chain of the 'cold-spots' usually does not coincide with those of the 'hot-spots'. 1911 *Encycl. Brit.* XXVII. 97/1 There are points [on the skin], stimulated by addition of heat, hot spots, while others are stimulated by withdrawal of heat, cold spots. 1926 S. WRIGHT *Applied Physiol.* I. 21 If all the 'hot' and 'cold' spots are marked out, the areas of skin between them are found to be insensitive to punctate thermal stimuli.
II. A spot that is hot (*lit.* and *fig.*). **2.** [SPOT *sb.*[1] 10.] A small area in a surface or body that is at a higher temperature than its surroundings.
1919 G. KAPP *Princ. Electr. Engin.* II. vi. 143 The average internal temperature of a winding can be more accurately deduced from a resistance test, whilst for the discovery of so called 'hot spots' thermo-couples must be used. 1936 D. M. ROBINSON *Dielectric Phenomena in High Voltage Cables* ii. 16 A cable subjected to excess voltage may develop local hot spots, and..frequently fails at or near the hottest of these points. 1947 *Sci. News* IV. 150 With a metal sliding on glass with a load of a few pounds, visible hot spots (temp. 520–570° C.) can be seen when the sliding speed is as low as one or two feet per second. 1958 H. ETHERINGTON *Nuclear Engin. Handbk.* XII. 17 Hot-spot effects usually require a correction in the calculation of maximum temperature. 1959 H. F. TAYLOR et al. *Foundry Engin.* x. 277 Under normal conditions of cooling, metal at the center of the *d* region is a hot spot, the last to solidify. 1970 *Times* 30 Oct. 23/2 A small.. infra-red television camera..for detecting the 'hot spots' in overhead power lines and substations which can lead to serious breakdowns. 1972 *Physics Bull.* May 284/2 This effect has been used to detect hot spots and hence structural flaws in systems as diverse as integrated circuits and the human body.
3. [HOT *a.* 7 b, SPOT *sb.*[1] 8.] **a.** A night-club or other place of lively public entertainment.
1931 D. RUNYON *Guys & Dolls* (1932) iii. 53 There are very seldom any customers in Good Time Charley's until along about five o'clock in the morning..and then it is sometimes a very hot spot indeed. 1937 J. WEIDMAN *I can get it for you Wholesale* xxv. 227 The line in Winchell's column: Martha Mills..is doing the hot spots with what prominent young manufacturer of feminine haberdashery? 1940 *Time* 15 Apr. 98/2 Eschewing Hollywood hotspots, they prefer at-homes with the quieter younger set. 1954 G. SMITH *Flaw in Crystal* iv. 39 The Wind in the Willows [a roadhouse] had been a noted hot-spot when I was a schoolboy. 1954 *Manch. Guardian Weekly* 26 Aug. 7/1 Intellectuals..went on a kind of jazz-slumming in the Harlem hot spots.
b. A place of danger; *spec.* a place where war is being actively engaged in or experienced.
1941 *Illustr. Lond. News* CXCIX. 200 (*caption*) Miss Cowles, an American journalist, whose pre-war assignments took her to most of the 'hot-spots' of Europe [*sc.* Prague, Berlin, Warsaw, etc.]. 1966 *Economist* 19 Feb. 711/2 The Administration wanted to distinguish financially between men who had been in 'hot spots' and those who had only served in non-combat zones. 1973 G. BEARE *Snake on Grave* xii. 62 You're putting yourself on the hot-spot, Sammy.
4. Specific technical applications in the literal sense.
a. In an internal combustion engine, a portion of the manifold or combustion chamber that is heated in order to facilitate vaporization of the liquid fuel; also, a region of overheating that tends to cause pre-ignition. **b.** *Astr.* A region in the sun's corona (generally associated with a solar flare) that is temporarily at a higher temperature than normal and is emitting a characteristic spectrum of radiation. **c.** *Electronics.* In a mercury-arc rectifier with a pool cathode, that part of the pool of mercury with which the arc is in contact and from which electrons are emitted.
a. 1924 *Motor Manual* (ed. 25) ii. 32 As an additional means towards obtaining efficient carburation a 'hot spot' arrangement is provided on a number of cars. 1929 *Times* 2 Nov. 4/7 The inlet manifold is jointed to the exhaust centrally, and thus a hot-spot is provided. 1934 *Jrnl. R. Aeronaut. Soc.* XXXVIII. 960 Thus making absolutely sure of high compression without hot spots acting as sparks in the combustion chamber. 1959 *Chambers's Encycl.* VII. 639/1 [In the petrol engine] pre-ignition is usually caused by an unsuitable plug or some local hot-spot, e.g. the exhaust valve head. 1968 *Practical Motorist* Dec. 459/1 Some hot spots are fitted with a thermostatically controlled flap which diverts the mixture away from the hot spot once the engine is running at normal temperature.
b. 1936 *Harvard Coll. Observatory Circ.* No. 410. 22 This region [of the chromosphere] appears to be one of abnormal excitation, possibly caused by a 'hot spot' near the limb of the sun at the time of the eclipse. 1958 *Sci. Amer.* Aug. 41/2 When the sun is observed with a radio telescope tuned to 10 centimeters, the hot spots in its atmosphere stand out brilliantly. 1966 *McGraw-Hill Encycl. Sci. & Technol.* XIII. 282/2 Coronal hot spots, indicated by emission of the yellow line of Ca XV at 5694 A above the photosphere by some 30,000 km, have been found in all spectrographic observations of the corona over a limb flare.
c. 1937 W. G. DOW *Fund. Engin. Electronics* xix. 426 The cathode 'hot spots' of mercury-pool-type mercury vapor rectifiers dodge about erratically on the mercury surface. 1971 B. SCHARF *Engin. & its Lang.* xx. 277 Where the arc terminates on the surface of the mercury a 'hot spot' occurs which forms the source of electronic emission.
5. Technical senses representing transferred senses either of *hot* or of the phr. as a whole.
a. In an ingot or casting (see quot. 1908). **b.** A particularly active part of a forest fire; so **hot-spotting** *vbl. sb.* (see quot. 1953). **c.** *Theatr., Photogr.*, etc. An area that is markedly brighter than its surroundings. **d.** *Nuclear Sci.* An area where the radiation level is much higher than in the surroundings; a local concentration of radioactivity. **e.** *Genetics.* A part of a gene especially liable to undergo mutation.
a. 1908 B. STOUGHTON *Metall. Iron & Steel* xii. 349 These localities, where the segregation is high, and which are known, when very bad, as 'hot spots', are sometimes porous or surrounded by porous parts of the casting. 1938 ROSENHOLTZ & OESTERLE *Elem. Ferrous Metal.* ii, Hot spots are irregularities occasioned by the uneven distribution of the coarse and fine parts of the charge in such a way that the coarse parts are concentrated near the furnace walls.
b. 1938 *Fire Control Notes* Oct. 26 This work includes such jobs as..burning out material between the line and the fire edge, and reduction of hot spots. 1940 *Ibid.* July 141 It is well for the scout to carry a pulaski or light ax for blazing or hot-spotting. 1953 *Brit. Commonw. Forest Terminol.* I. 77 *Hot-spotting*, checking the spread of fire on hot spots alone or at salient points, as an emergency measure employed in advance of control-line construction.
c. 1952 GRANVILLE *Dict. Theatr. Terms* 99 *Hot spot*, a bright spot in an area of uneven lighting. 1958 *New Scientist* 25 Dec. 1569/3 Some areas on the document will be reproduced as dark grey.., while on others, known colloquially as 'hot spots', there will be such a powerful reflection of light that any markings on such areas may be completely blotted out. 1971 L. B. HAPPÉ *Basic Motion Pict. Technol.* vii. 218 A brighter central area, or hot spot, is particularly obvious when a short focal length lens is used on a camera and a long focal length on the projector.
d. 1955 *Bull. Atomic Sci.* Feb. 46/2 One would expect 'hot spots' even far downwind but since the history of the bomb cloud would already be several hours 'old' by then, there would be a corresponding drop in intensity. 1955 *Sci. Amer.* Aug. 37/1 A particularly malignant feature of some of the radioisotopes is their tendency to concentrate in 'hot spots' instead of distributing themselves evenly through the bone or other tissue they invade. 1962 *Listener* 19 July 104/2 The concentration of radioactive substances into 'hot spots' may well be exaggerated in the rapidly dividing cells of the foetus.
e. 1958 BENZER & FREESE in *Proc. Nat. Acad. Sci.* XLIV. 115 A striking feature of the map is the existence of certain 'hot spots', where mutations recur with high probability. 1964 W. HAYES *Genetics of Bacteria & their Viruses* viii. 171 Out of 1,612 spontaneous mutations ..more than 500 occur at a single site or 'hot spot' in the B cistron. 1967 E. STEINER tr. *Esser & Kuenen's Genetics of Fungi* v. 300 A hot spot probably consists of a single base pair, at least one member of which is particularly sensitive to a specific mutagen.

hot stuff. a. A person or thing out of the ordinary run, something of surpassing excellence or merit; sometimes with implication of moral censure; also, specif., a woman reputed to be highly sexed. Also *attrib.*, esp. (i) sexually explicit, (ii) extremely capable or efficient. *colloq.* (orig. *U.S.*).
1889 *Kansas City* (Missouri) *Times & Star* 14 Nov., 'Miss Middleton's Lover.' Were there room for two words in that last line, 'Hot Stuff' might be appropriate. 1900 *Dialect Notes* II. 42 Hot-stuff. 1. A person of good quality; often ironical. 2. A person having merit. 1905 WODEHOUSE *Head of Kay's* 252 Kay's are hot stuff, Jimmy. 1908 *Varsity Vices* May 5/1 Gad, though, but she's tremendously hot stuff, the little devil. Stood her a bust at the Zoo. 1909 WODEHOUSE *Mike* xvii. 101 The book was obviously the last word in hot stuff. 1911 F. SWINNERTON *Casement* i. 41, I say, Trevell..you had any dealings with Jimpton? Eh? He's hot stuff..what! 1912 C. MACKENZIE *Carnival* x. 111 I'm not going to have fellows say my sister's hot stuff. 1912 *Strand Mag.* Jan. 22/2 Clarence in goal was the nearest approach to an indiarubber acrobat.. to be seen off the music-hall stage. He was, in brief, hot stuff. 1915 H. WILSON *Diary* 9 Sept. (1927) I. 249 Here is Sarrail..to urge a plan about which he knows nothing. And all because he is a Radical-Socialist. Hot stuff. 1917 A. WAUGH *Loom of Youth* II. i. 111 This side [*sc.* a Rugby football team] was certainly 'pretty hot stuff'. 1923 H. G. WELLS *Men like Gods* I. vi. 102 'Raced us from Hounslow,' said Mr. Burleigh's driver. 'Real hot stuff.' 1928 *Punch* 21 Mar. 328/1 His father had

influence.., being a frightfully hot-stuff surgeon. 1931 *Amer. Speech* Feb. 204 *Hot stuff*, new, up-to-date material or new incidents; approved as being startling, or the 'latest'. 1931 W. DEEPING *Road* viii. 86 I'm getting my new M.-B. next week. Hot stuff. She'll do eighty. 1936 E. M. FORSTER *Abinger Harvest* 43 During the interval we discussed, not whether the Scallies were good, but whether they were better or worse than the Wags. They were less hot stuff, that was admitted on all sides. 1940 GRAVES & HODGE *Long Week-End* iv. 52 Elinor Glyn was the reigning queen of popular love literature and considered 'very hot stuff'. 1944 M. PANETH *Branch Street* 64 The men say of her, 'Joan is hot stuff.' 1965 D. LODGE *Brit. Mus. is falling Down* vii. 125 A sort of novel,..the story of his affair with Mother, with just the names changed. It's hot stuff, as we used to say at school. 1965 H. PORTER *Cats of Venice* 91 Singapore boy hot stuff, hot stuff!
b. *slang.* Stolen goods. Cf. STUFF *sb.*[1] 10 b.
1924 G. C. HENDERSON *Keys to Crookdom* ix. 117 Pawnshops and second-hand stores establish a reputation for handling 'hot stuff' and there are very few such establishments that will refuse to buy from a known thief. 1962 K. ORVIS *Damned & Destroyed* vii. 51 So you're handling a bit of hot stuff as well as joy-popping?
So **hot-stuff** *v. trans.* (*Army slang*), to scrounge, steal; **hot-stuffed** *ppl. a.*; **hot-stuffer,** a scrounger.
1914 H. ROSHER *In R.N.A.S.* (1916) 36, I at once hot-stuffed one of his inlet valves and set the men to work changing it. 1929 *Papers Mich. Acad. Sci., Arts & Lett.* X. 300/1 *Hot-stuffed*, stolen. *Ibid.*, *Hot-stuffer*, a thief. 1950 E. PARTRIDGE *Here, There & Everywhere* 81 *Hotstuff*, to appropriate illicitly, to steal.

hotsy-totsy (hǫ·tsi,tǫ·tsi), *a. slang* (orig. *U.S.*). ['Coined *c* 1926 by Billie De Beck, Amer. cartoonist' (Webster).] Comfortable, satisfactory, just right. Hence **ho:tsy-to·tsiness.**
1926 B. REYNOLDS *Cocktail Continentale* ii. 29 And they sure can fix up a rip-snotin', raring, tearing, hotsy-totsy time, honey boy. 1927 DUNNING & ABBOTT *Broadway* xx. 205 They complimented Steve in the lingo of the night clubs. 'Everything is hotsy-totsy!' whooped the loud Ruby. 1927 WODEHOUSE *Small Bachelor* vi. 95 It seemed to me that I was absolutely hotsy-totsy. 1935 —— *Luck of Bodkins* xxii. 282 And a fat chance..there is of any hotsy-totsiness resulting from anything Ambrose can do. 1940 N. MARSH *Surfeit of Lampreys* (1941) xviii. 281 Daddy's all hotsy-totsy now as regards money. 1944 H. CROOME *You've gone Astray* xxi. 214 Now she thinks another man would be all hotsy totsy. 1952 [see *BABE 3 b]. 1958 L. A. G. STRONG *Treason in Egg* ix. 165 All is hotsy-totsy. 1964 W. MARKFIELD *To Early Grave* (1965) iv. 77 He became a big-time hotsy-totsy critic. 1973 J. MANN *Only Security* xii. 162 What the law allows me, is mine... So that's all hotsy totsy.

Hottentot. 1. Substitute for def.: One of the two sub-races of the Khoisanid race (the other being the Sanids or Bushmen), characterized by short stature, yellow-brown skin colour, and tightly curled hair. They are of mixed Bushman-Hamite descent with some Bantu admixture, and are now found principally in South-West Africa. Also, a member of this race. (Add further examples.)
1924 A. C. HADDON *Races of Man* (ed. 2) 41 An early migration of Hamites mixed with this population [*sc.* Bushmen] and gave cattle and elements of their language to the mixed people who in South Africa are known as Hottentots. 1930 I. SCHAPERA *Khoisan Peoples of S. Africa* III. x. 239 There is little information as to the existence of food taboos among the Hottentots in more recent times. 1965 H. B. ISHERWOOD *Racial Contours* viii. 308 Opinion seems to be gaining ground that the Bushman and the Hottentot bear so many features indicating a divergent evolutionary history that they should be classed as a separate race altogether and not grouped under the heading of Negroid. The title provisionally proposed is Khoisanoid, based on the name by which these people know themselves. 1970 P. OLIVER *Savannah Syncopators* 31 Alan P. Merriam has identified as distinct musical regions those of the Hottentot-Bushman. 1974 J. R. BAKER *Race* xvii. 318 The Hottentots, Korana, and Bushmen are not to be regarded as people adapted by natural selection to a desert life.
b. (This derogatory sense, which was based on a failure to understand an alien culture, appears now to be very rare.)
c. (In full *Hottentot fish.*) A South African marine food fish of the genus *Pachymetopon*, esp. *P. blochii.*
1798 S. H. WILCOCKE tr. *Stavorinus' Voy. E. Indies* I. 560 The Hottentot-fish, which is like a sea-bream. 1838 J. E. ALEXANDER *Exped. Disc. Int. Afr.* I. 88 The delicious Roman fish, Hottentot. 1921 *Ann. S. Afr. Mus.* XXI. 721 The Hottentot is one of the commonest Cape fishes. 1949 J. L. B. SMITH *Sea Fishes S. Afr.* 276 The Hottentot slipped away with the next wave.
d. The Khoisan language spoken by the Hottentots.
1884 [see CLICK *sb.*[1] 5]. 1910 *Encycl. Brit.* IV. 871/2 Their language..has in common with Hottentot..the peculiar sounds known as 'clicks'. 1921 E. SAPIR *Lang.* iii. 55 Certain languages, like the South African Hottentot and Bushman, have also a number of inspiratory sounds. 1953 J. B. CARROLL *Study of Language* ii. 57 He..doubts very much the validity of certain early theories that such languages..as Bantu and Hottentot are in any way related to the Semitic-Hamitic group. 1972 *Stand. Encycl. S. Afr.* V. 605/1 Afrikaans and South African English have taken over words from Hottentot, such as *buchu, dagga, kanna.*
2. *attrib.* or as *adj.* (Further examples.)
1797 *Encycl. Brit.* VIII. 684/2 The Hottentot language is..said to be a composition of the most strange and disagreeable sounds. *Ibid.* 685/1 A general opinion has prevailed that the Hottentot women have a kind of natural vail which covers the female parts. *Ibid.* 687/2 In a craal, or Hottentot village, the huts are most commonly disposed in a circle. 1811 in W. J. BURCHELL *Trav. S. Afr.* (1822) I. xv. 371 The most dangerous malady is a kind of cancerous sore or ulcer, called in the colony the *Hottentots Zeer* (Hottentot Sore). 1828 J. PHILIP *Res. S. Afr.* I. p. xviii, The missionaries..were called 'Hottentot predicants' (ministers), by way of contempt. 1924 *Internat. Jrnl. Psycho-Anal.* V. 41 It might perhaps be not without significance that three of the five patients informed me of their own accord that they possessed 'Hottentot nymphae'. 1933 I. SCHAPERA *Early Cape Hottentots* p. xii, In Hottentot mythology ǁGaunab figured as a malevolent chief. 1965 H. B. ISHERWOOD *Racial Contours* viii. 308 Associated with the Naman and talking their language are the primitive Bergdama or Hankhoin people, who are virtually under Hottentot tribal rule. 1974 J. R. BAKER *Race* vi. 97 When Van Riebeck and his little company of emigrants from the Netherlands landed at Table Bay in 1652, they were met by Hottentot herdsmen.
3. Hottentot apron (see *APRON *sb.* 5 c); **Hottentot('s) bean (tree)** *S. Afr.*, a shrub or tree of the genus *Schotia*, esp. *S. latifolia*, or its fruit; **Hottentot fish** (see sense 1 c above); **Hottentot('s) god** *S. Afr.*, a praying mantis, formerly worshipped by the Hottentots; **Hottentot('s) tea** (examples).
1801 J. BARROW *Acct. Trav. S. Afr.* I. 189 Not so with the Hottentot bean... This plant is the African *Lignum vitae*. 1833 S. KAY *Trav. Caffraria* 106 The Hottentot's bean tree. 1921 T. R. SIM *Native Timbers S. Afr.* 192 Hottentots Bean Tree. *Schotia*, sps. 1965 F. VON BREITENBACH *Indigenous Trees S. Afr.* II. 327 *Schotia afra*... Hottentot's Bean... The roasted beans are eaten by some tribes. 1785 G. FORSTER tr. *Sparrman's Voy. Cape G. Hope* I. 211 A genus of insects (the *mantis*) called by the colonists the *Hottentot's god*. 1957 W. TAPSON *Old Timer* i. 10 Flyingants and glowworms and dragonflies and hottentot-gods. 1961 L. VAN DER POST *Heart of Hunter* xii. 161 When they noticed the reverence in which the Mantis was held by some of the aborigines of the Cape, they inaccurately called him the Hottentot's God. 1962 *Cape Times Week-end Mag.* 10/1, I like Hottentot Gods, bloodthirsty little savages as they may be. 1850 L. PAPPE *Florae Capensis Medicae Prodromus* 17 *Helichrysum serpyllifolium*..goes by the name of *Hottentot's tea*,..and is much liked by the coloured people, who infuse it as tea. 1972 *Stand. Encycl. S. Afr.* V. 611/1 Hottentot tea... *Helichrysum orbiculare* = *H. serpyllifolium*... Another species, which is also called Hottentot-tea, is *H. nudifolium.*

hottie, hotty (hǫ·ti), *sb. colloq.* [f. HOT *a.* + -Y[6].] A hot-water bottle.
1947 H. WALSH *Fourth Point of Star* xx. 102, I am going to..rub my feet with meth., then get into bed with a hotty. 1956 D. M. DAVIN *Sullen Bell* vi. 40 Get a hotty for yourself. 1960 V. ANDERSON *Daughters of Divinity* vi. 51 And show her where to fill her hotty? 1967 R. HARRIS *All my Enemies* xiii. 117 You look rotten, Jenny. At least have a hottie to clutch.

hot water. Add: **1.** *attrib.* (Further examples.)
1855 MRS. GASKELL *Lizzie Leigh* 163 Pack up for each his portion of the dainty dish, and send it separately, in hot-water trays. 1872 *Young Englishwoman* Nov. 610/2 Cover for a hot-water tin. 1877 E. S. DALLAS *Kettner's Bk. of Table* 332 Hot-water paste for raised pies. 1895 KIPLING *Day's Work* (1898) 373 He turned to explore the hot-water dishes on the sideboard. 1904 GOODCHILD & TWENEY *Technol. & Sci. Dict.* 291/1 *Hot water system*, a method of warming effected by means of hot water or steam circulating in a system of closed pipes. 1921 *Daily Colonist* (Victoria, B.C.) 24 Mar. 3/1 Attractive modern home, hot water heating, lot 90 × 180, three bedrooms, [etc.] 1926 W. DE LA MARE *Connoisseur* 48 In spite of the hot-water-fountain on the counter it was..cooler in here. 1926 S. T. WARNER *Lolly Willowes* I. 46 The maid..laid the folded towel across the hot-water can. 1932 *Edinburgh Bk. Plain Cookery Recipes* 149 Hot-Water Crust. Used for Savoury Dishes. 1950 *N.Z. Jrnl. Agric.* Dec. 575/1 In almost all houses the hot-water cupboard was used for airing clothes. 1956 G. TAYLOR *Silver* ix. 191 Hot-water jugs were set on classical tripod legs with a spirit lamp between. 1970 SIMON & HOWE *Dict. Gastron.* 290/2 Hot-water crust is a stiff dough moulded to make a filling of meat or game.
b. Special comb.: **hot-water bottle,** a receptacle made of rubber, metal, or other material that may be filled with hot water and used for warming a bed, or for applying local heat to the body; **hot-water pipe** usu. *pl.*, the pipe(s) in a water-heating system.
1895 *Montgomery Ward Catal.* 107/1 Challenge Hot Water Bottles, pure rubber. 1897 [in Dict., sense 1 *attrib.*]. 1900 J. VAIZEY *About Peggy Saville* xxv. 229 I'd have a fire and an india-rubber hot-water bottle, and I'd lie and sleep. 1926–7 *Army & Navy Stores Catal.* 151/1 Aluminium hot water bottles. Round 9 in. 1932 D. C. MINTER *Mod. Needlecraft* 248/1 Hot-water bottle cover..cut to fit. 1946 G. MIKES *How to be an Alien* 25 Continental people have sex life; the English have hot-water bottles. 1972 S. HYNES *Edwardian Occasions* 169 The narrative is slacker and more trivial, begins to take note of the hot-water bottle and the nine o'clock news. 1842 *Trans. Hort. Soc.* II. 435 All the experience obtained at the Garden goes to demonstrate the great inferiority of flues to hot water pipes as a mode of heating. 1852 DICKENS *Bleak Ho.* (1853) xxviii. 274 The hot-water pipes that trail themselves all over the house..fail to supply the fires' deficiencies. 1912 BEERBOHM *Christmas Garland* 26 The faint yet heavy fragrance exhaled from the hot-water pipes. 1973 G. BUTLER *Coffin for Pandora* v. 106 The luxurious ways of Sarsen House kept all the rooms warm with hot-water pipes.

hot-work, *v.* [f. *HOT *a.* 12 d + WORK *v.* 12 e.] *trans.* To work (metal), e.g. by rolling, forging, etc., while it is hot and above the temperature at which recrystallization takes place. Usually in vbl. sb. **hot working.**
1896 H. H. CAMPBELL *Manuf. Structural Steel* xiv. 192 (*heading*) Influence of hot working on steel. 1916 D. K. BULLENS *Steel & its Heat Treatment* ii. 38 Steel which has been hot-worked down to the Ar1 point will show a finer grain. 1932 E. GREGORY *Metall.* ii. 59 The modification of structure which is brought about by hot-working is of greater importance. 1967 A. H. COTTRELL *Introd. Metall.* xviii. 313 This is one of the reasons for hot working such metals. *Ibid.* xxi. 405 This is the basis of industrial hot working processes in which metal ingots or billets are given massive deformations by comparatively small applied stresses.
Also **hot work** *sb.*, hot working.
1905 J. W. MELLOR *Crystall. Iron & Steel* 77 Hot work has no direct action upon the structure of the steel. 1928 H. M. BOYLSTON *Introd. Metall. Iron & Steel* xiii. 369 All mechanical treatment may be divided into two classes, namely, hot work and cold work. 1961 J. N. ANDERSON *Appl. Dental Materials* (ed. 2) v. 44 Hot work, on the other hand, involves simultaneous deformation and recrystallization.

houbara (hubā·rā). Also **hobara, hubara, oubara, ubara.** [mod.L. (C. L. J. L. Bonaparte *Saggio d'una Distribuzione Metodica degli Animali Vertebrati* (1832) 84), f. Arab. *hubārī* bustard.] A bustard, *Chlamydotis undulata*, found in North Africa and Asia as far east as India and Persia, and formerly included in a genus *Houbara.*
1827 J. MALCOLM *Sk. Persia* I. 56 The beautiful speckled Hubara, with his head erect and wings outspread. 1836 *Penny Cycl.* VI. 59/2 Arabs are accustomed to compare the eyes of their most beautiful women to those of the *Oubara*. 1876 *Encycl. Brit.* IV. 579/1 Two species, known indifferently by the name of Houbara (derived from the Arabic), frequent the more southern portions of the [Palæarctic] Region. 1883 *List Animals* (Zool. Soc.) (ed. 8) 517 *Houbara Bustard*. 1905 J. I. S. WHITAKER *Birds Tunisia* II. 285 North of the Mediterranean the Houbara is of merely accidental occurrence, but it inhabits North-east Africa. 1925 *Blackw. Mag.* Mar. 339/1 The hobara, or lesser bustard. 1963 *Times* 24 May 13/7 The great Houbara bustard yielded, as evidence of its continued presence..only a single stray feather in one area and a couple of sucked eggs.

Houdan (hū·dǎn). Name of a town in the department of Seine-et-Oise, France, used to designate a breed of domestic fowl characterized by black and white plumage, a heavy crest, five toes on each foot, and by its prolific laying.
1871 W. M. LEWIS *People's Pract. Poultry Bk.* 23 The Brahma, Cochin, Leghorn, Poland, and Houdan stand relatively in the position here named [as egg-layers]. 1873 C. M. YONGE *Pillars of House* I. ii. 28, I am hardly prepared to say whether it is a Hamburg or a Houdan, or a more unambitious Dorking. 1883 [see *CRÈVE-CŒUR]. 1902 *Encycl. Brit.* XXXI. 876/2 A race originated in France by crossing Houdans with Dorkings and Light Brahmas, and known as the Faverolles. 1925 *Glasgow Herald* 16 Jan. 6 During the war I had a houdan who ruled my hen run. 1960 *British Poultry Standards* (ed. 2) 72 Introduced into England in 1850, the Houdan is one of the oldest French breeds.

Houdini (hūdī·ni). The professional name (Harry *Houdini*) of an American escapologist, Erich Weiss (1874–1926), used to denote an ingenious escape, or a person who embodies the characteristics of Houdini. Also *attrib.* Hence (occas.) as *v. intr.*, to escape.
1923 *N.Y. Times* 9 Sept. VII. 2/1 *Houdini*, to get out of something, to escape. 1923 J. E. BAXTER *Locker Room Ballads* 20 With a dozen balls in play at once And each of them doing Houdini stunts. 1930 J. FAIT *Big House* 120 Don't do no Houdini, or we'll lay you out. 1946 MEZZROW & WOLFE *Really Blues* 373/1 *Do a Houdini*, disappear. 1958 [see *ESCAPOLOGIST]. 1962 L. DEIGHTON *Ipcress File* xxv. 164 There was no chance of a 'Houdini' through the boltwork. 1967 L. J. BROWN *Cat who ate Danish Modern* xiii. 117, I have to buckle the harness good and tight or he'll wiggle free... That cat's a Houdini. 1971 *Guardian* 24 Aug. 7/4 Houdini act saved injured climber in air crash.

houghmagandy (hǫχmagæ·ndi). Chiefly *Sc.* (now *rare*). Also **hochmagandy, hough-magandie.** [Fanciful formation, perh. f. HOUGH *sb.* + CANTY *a.*] Fornication.
a 1700 J. MAIDMENT *New Bk. Old Ballads* (1885) 11 And well I wot he kens the gate To play at hough-magandy. 1785 BURNS *Holy Fair* in *Poems & Songs* (1968) I. 137 An' monie jobs that day begin, May end in *Houghmagandie*. 1805 R. TANNAHILL *Poems* (1876) 44 The priest convenes his scandal court, Tae ken what houghmagandie sport Has been gaun on within the

parish. **1962** V. Nabokov *Pale Fire* 212 She would have preferred him to have gone through a bit of wholesome houghmagandy with the wench.

houhere (hau·hiəri). *N.Z.* [Maori, f. *hou* to bind together + *here* tie.] A small tree of the native genus *Hoheria*, esp. *H. populnea*; also called lacebark or ribbonwood.

1879 J. Hector *Handbk. N.Z.* 93 Houhere, ribbonwood of Dunedin. **1906** T. F. Cheeseman *Man. N.Z. Flora* 79 The Maoris apply the names *hoihere* or *houhere* to varieties *a* and *b* indifferently; the European settlers usually call all the forms 'ribbon-wood' or 'lacebark', names which are, unfortunately, also used for *Plagianthus betulinus*. **1946** *Jrnl. Polynesian Soc.* LV. 149 *Houhere*, a tree (*Hoheria populnea*), ribbonwood, lace-bark, thousand-jacket. One of the three or four deciduous Maori trees. **1968** *Landfall* XXII. 255 White edge of the sea White as the flower Of lace-bark, Te houhere, White edge of the sea Eating the land.

hound, *sb.*[1] Add: **4. e.** Used with a preceding substantive to designate a person who has a particular enthusiasm for, or interest in, the object or activity specified; esp. in **news-hound*. *colloq.* (orig. *U.S.*).

1918 *Hatchet* 7 Apr., 'Got what all figured out,' queried the news hound eagerly. **1926** *Amer. Speech* II. 45 *Comma hound*, applied to teachers of English composition. **1928** L. North *Parasites* 270 Much was made by the Zimski publicity hounds of this one hundred-per-cent Americanism of the little [film] star. **1968** *Word Study* Dec. 2/2 The enthusiast is a *bug* or a *hound*, as in *radio bug* or *hi-fi hound*. Closely related to this use of *hound* is its use as 'one who frequents', as in *tavern hound*. **1973** *Sat. Rev. Soc.* (U.S.) May 65/3 A real fun guy,..a super-duper party hound.

6. Substitute for def.: In north-eastern Canada: the old squaw or long-tailed duck, *Clangula hyemalis*. (Further examples.)

1779 G. Cartwright *Jrnl.* 19 May (1792) II. 440 There were several hounds and gulls, with some pigeons and black-divers among them. **1861** L. de Boilieu *Recoll. Labrador Life* 160 The bird called the Hound—a graceful fowl, rather larger than a teal—is very abundant. **1959** W. L. McAtee *Folk-Names Canad. Birds* (ed. 2) 14 Old Squaw [is also called] hound (the Chorus of sound from a number of these birds suggests the baying of a pack of hounds).

7. a. *hound-dog, -pup*.

1649 *Early Rec. Dedham, Mass.* (1892) III. 162 That care can be taken that the young hound doges be in time taught to hunt. **1911** R. D. Saunders *Col. Todhunter* ii. 24 I'm as hungry as a young hound-dog this very minute. **1949** *Chicago Daily News* 6 July 14/3 He's got about nine houn' dawgs. **1857** F. L. Olmsted *Journey Texas* (1861) 52 The child..five miles from a neighbor;..[with] hound-pups and negroes for playmates. **1878** J. H. Beadle *Western Wilds* xxviii. 439 What he wouldn't steal, a hound pup wouldn't pull out of a tan-yard. **1932** Kipling *Limits & Renewals* 293 Though well-meaning as a hound-pup..her face and figure were against her.

b. hound-work, the work done by the hounds in hunting.

1928 *Isis* (Oxf.) 14 Nov., Some very pretty houndwork now ensued. **1932** *Morning Post* 19 Nov. 14/4 (*heading*) Pytchley Houndwork. **1971** *Country Life* 7 Oct. 897/2 The fascination of good hound-work.

hound, *sb.*[2] **2.** *U.S.* (Earlier examples.)

1847 *Rep. U.S. Comm. Patents 1846* 264 The placing on the rear ends of the extended hounds..the adjustable sway bar. **1854** Bartlett *Pers. Narr. Explor. Texas* II. xl. 456 Mr. Flotte's large carriage got mired; and in the struggle to extricate, the tongue and hounds were broken.

hound, *v.* Add: **2.** Also with *out*, to drive away.

1922 Joyce *Ulysses* 628 Spain decayed when the Inquisition hounded the jews out. **1930** G. B. Shaw *Apple Cart* I. 41 If I attempt to fight them I shall be hounded out of public life. **1945** E. Waugh *Brideshead Revisited* ii. 50 He daren't show his great purple face anywhere. He is the last, historic, authentic case of someone being hounded out of society.

hou·ndstooth. Also **hound's tooth, hound's-tooth, hounds' tooth, houndtooth.** [f. Hound *sb.*[1] + Tooth *sb.*] A small irregular design of broken check. Also, a fabric of this design; a suit, coat, etc., of such fabric.

1936 *New Yorker* 7 Mar. 44/2 (Advt.), We carry sports jackets in checks, plaids, and hound's tooth patterns. **1951** J. D. Salinger *Catcher in Rye* ii. 32 If you're not going out any place special, how 'bout lending me your hound's tooth jacket? *Ibid.*, No kidding, you gonna use your hound's tooth to-night, or not? **1957** *Observer* 1 Dec. 11/5 Giant houndstooth designs made day dresses with jackets: so did imitation tweeds. **1959** *Woman's Own* 6 June 11/3 Clare..put on her 'good' hounds-tooth check. **1959** *Sunday Express* 14 June 14/3 Patterns are clean and singularly chic—a bold hounds' tooth check or a variety of confetti spots. **1959** *Times* 23 Sept. 5/2 The choice of cloths is famous for its variety—ranging from subdued houndstooths to full-blooded overchecks. **1959** *Guardian* 4 Dec. 6/5 In the woollens, small houndtooth designs and overchecks are liked. **1959** G. Freeman *Jack would be Gent.* i. 18 Moyra glanced at James, covertly admiring his hound's-tooth check suit. **1971** *Homes & Gardens* Aug. 57/1 The dining chairs are covered in a toffee and black houndstooth check. **1972** P. Cleife *Slick & Dead* xix. 143 Wearing my natty houndstooth instead of uniform.

houngan (hū·ngăn). Also **hougan, hungan.** [Native name in Haiti.] A priest of the Voodoo cult.

1929 W. B. Seabrook *Magic Island* 304 There are many things my father (who was one of the great *hougans* of the past generation) could do which I cannot do. He could make thunder. **1932** J. J. Williams *Voodoos & Obeahs* iii. 91 Is there a Voodoo initiation whereby a neophyte, it matters not who he is, thanks to the good will of the hougan, may be admitted to the congregation? **1937** M. J. Herskovits *Life Haitian Valley* xii. 223 Persistent failure to achieve cures will lose a *hungan* his reputation and his practice. **1941** J. G. Leyburn *Haitian People* viii. 168 While the Vodun houngan sits quietly by, a special officiant..reads or pretends to read from the Catholic prayer book. **1966** G. Greene *Comedians* I. iii. 83 A zombie who has left his grave at the command of an *houngan*.

houp-la (hū·p,lā), *int.* Also **hoop-la.** [Cf. F. *houp-là!*] An exclamation accompanying a quick or sudden movement. Also as *sb.*, commotion, ballyhoo.

1870 O. Logan *Before Footlights* xxiii. 280 But the tearful clown cries 'houp-la!' while his baby is lying dead at home. **1877** *N.Y. Tribune* 2 Mar. 7/1 The Stock Exchange to-day commenced its business of speculation with a grand 'hoop la'. **1917** H. Garland *Son of Middle Border* xxiii. 286 Hoopla! We had taken wing! **1917** *Punch* 23 May 340/2 A beam of pleasure, succeeded by a falling of the countenance, then a look of decision, ended in a 'Houp-là' as the Japanese doll descended into the basket. **1925** F. M. Ford *No More Parades* I. ii, But ..Hoopla!..He executed with his gloved right thumb in the moonlight a rapid pirouette. **1929** C. A. N. Garstin *Houp La!* xxiv. 308 'Houp la!' Bill sighed, letting his head fall back. **1948** *Carpenter* Mar. 16 Organized some three years ago amid a great hoopla of Communist propaganda and promotion, it has creaked along in a very erratic and unpredictable manner. **1973** *Guardian* 16 Mar. 10/3 There is sometimes so much surrounding hoop-la that you lose sight of the various tactics.

houp-la, var. *hoop-la.

hour. 1. b. Restrict †*Obs.* to sense in Dict. and add: In *pl.* with numerals rendered in figures (followed by those of minutes), expressing the number of hours since midnight (chiefly in the armed services and in passenger timetables). Cf. *hundred *sb.* and *a.* 1 d.

1939 *Punch* 2 Aug. 124/1 It is 21.00 hours on the last day of our month's training. **1941** *Punch* 21 May 486 A lorry is leaving R.M.P. at 0915 hours. **1967** [see *hovering *vbl. sb.* a (ii)]. **1968** A. Hammersley *Weather & Life* iv. 49, 6 hours G.M.T., i.e. 6 a.m. **1971** R. Bucknall *Trains* 46 With this time scale, 6 minutes would represent six hours, or 06.00.

c. Used to denote the distance of the sun above the horizon in the morning or afternoon. *U.S.*

1637 in *Essex Inst. Hist. Coll.* (1869) IX. 66 The keeper.. [is] to take the Cattle at the pen at Sun halfe an hour highe. **1683** in *New Hampsh. Hist. Soc. Coll.* (1866) VIII. 133 At night, about sun half an hour high. **1762** in *Narragansett Hist. Reg.* (1883–4) II. 219 We..got to the Ferry the sun about two hours high at night. **1824** *New Hampsh. Hist. Soc. Coll.* I. 244 Though the sun was an hour above the horizon, it was now as dark as midnight. **1836** *Knickerbocker* VIII. 352 The sun is two hours up yet. **1907** M. H. Norris *Veil* i. 3 The sun was an hour high when he entered a narrow road overgrown with grass.

d. (orig. *hour-mile*.) The distance which can be traversed in an hour.

[*a* **1646** J. Gregory *Posthuma* (1650) 318 One Hour-Mile of a Journie upon Hors, answereth to four English Miles.] **1785** G. Forster tr. *Sparrman's Voy. Cape G. Hope* II. 81 This place is situated at the distance of two hours (uurs) from that which we had just quitted. **1792** E. Riou tr. *J. van Reenen's Jrnl. Journey from Cape G. Hope* p. xii, Throughout the journal the word hour is to be considered as distance, and not time. Travellers at the Cape of Good Hope reckon distance by hours: one hour being supposed equal to about a league. **1798** S. H. Wilcocke tr. *Stavorinus' Voy. E. Indies* I. 58 A Dutch mile, which they in general call an hour, is about three miles and a half English. **1877** J. C. Geikie *Life & Words Christ* I. 388 Three hours from Jerusalem. **1907** in A. H. Anderson *Reading* Advts. p. xlviii a, Dulverton... 4 Hours from London, 1 Hour from Taunton and Exeter. **1920** *Country Life* 2 July 62/1 (Advt.), Radnorshire... Kington 6½ miles. Birmingham 1½ hours. An attractive period house.

e. Used as the second element, representing 'for one hour', in the names of some units of measurement, as *ampere-hour, horsepower-hour, kilowatt-hour, man-hour* (see under the first element).

f. A unit of measure of work done.

1900 *Daily News* 10 Jan. 8/3 It should be understood that the tailors' 'hour' is not one of time, but merely the word employed for an unit of calculation.

2. b. (Earlier example.)

1852 *12th Rep. Col. Land & Emigr. Comm.* in *Parl. Papers* XVIII. 151 They are not required to work 'long hours', five hours a day is what is required of them.

c. (See quot. 1955.) Cf. *children's hour* (*child *sb.* 22).

1930 *Economist* 3 May 985/1 It is argued that an advertiser who broadcasts tedious over-vulgar, or over-highbrow material in his 'hour' will eventually discover that he is losing money. **1955** M. Reifer *Dict. New Words* 102/1 *Hour*, a scheduled radio or television feature, originally one hour long; the term now refers to any length program. **1972** *Daily Tel.* 20 Apr. 12/5 The star of *The Bob Monkhouse Comedy Hour* (ITV)..is almost guaranteed to turn me off.

3. c. (Further examples.)

1816 Jane Austen *Emma* II. vii. 123, I am not fond of dinner-visiting... Late hours do not agree with us. **1821** Byron *Don Juan* III. lxvi. 36 Late hours, wine, and love are able To do not much less damage than the table. **1970** *Brewer's Dict. Phr. & Fable* (rev. ed.) 550/2 *To keep good hours*, to go home early every night; to go to bed betimes; to be punctual at one's work.

d. *to* (or *till*) *all hours*: late at night; after midnight.

1931 Belloc *Hist. Eng.* IV. I. 161 She..had had him, and one, Culpepper, in her room up to all hours. **1932** A. J. Worrall *Eng. Idioms* viii. 58, I sat up to all hours trying to finish my work. **1934** B. de Holthoir tr. *Duhamel's News from Havre* xiii. 196 She made up for lost time by sitting up till all hours of the night. **1945** E. Waugh *Brideshead Revisited* iv. 264 She sits up to all hours with her wireless. **1961** R. Chapman *Father Faber* viii. 161 He read till all hours and undertook heavy penances.

hourage (auə·rĕdʒ). *rare.* [f. Hour + -age.] The aggregate number of hours spent in working or travelling; extent or distance in hours. Also, rate of travel in hours.

1924 *Glasgow Herald* 29 Dec. 7, 52,000 individuals.. were taken for short flights, the total 'hourage' being 2400. **1968** *Times* 19 Feb. 5/2 When you've finished your optimum hourage you must stop clean and briskly.

hour-glass. Add: **d.** A marking (as on a spider) in the shape of an hour-glass.

1937 *Discovery* Dec. 368/1 Two female Black Widows, showing the 'hour-glass' markings. **1962** Metcalf & Flint *Destructive & Useful Insects* (ed. 4) xxi. 1009 (*caption*) The black widow spider, *Latrodectus mactans*. At left, the female from the underside, showing the characteristic hourglass-shaped spot.

2. Special Comb.: **hour-glass structure** *Petrol.*, a structure present in certain rocks in which the mineral crystals have the shape of an hour-glass.

1888 J. J. H. Teall *Brit. Petrogr.* vii. 159 The hour-glass structure already mentioned as occurring in the picrites is occasionally found in the normal plagioclase-augite rocks. **1932** F. F. Grout *Petrogr. & Petrol.* 100 'Hourglass' structure is occasionally noted in augite. **1959** W. W. Moorhouse *Study of Rocks in Thin Section* vi. 166 The larger grains and phenocrysts may be zoned, or they may show hourglass structure.

house, *sb.*[1] Add: **2.** *the House* (earlier and later examples).

1835 Dickens *Sk. Boz* (1836) 1st Ser. I. 3, I suppose you must have an order into the house. **1888** Mrs. H. Ward *R. Elsmere* II. III. xix. 140 If they turn us out.. there'll be nothing left but the House for us old 'uns.

b. *to bow down* (or *worship*) *in the house of Rimmon* (after 2 Kings v. 18): to pay lip-service to some principle which one does not accept; to sacrifice one's principles for the sake of conformity.

1718 Defoe in W. Lee *Life & recently discovered Writings Defoe* (1869) I. p. xiii, Thus I bow in the House of Rimmon, and must humbly recommend myself to his Lordship's Protection. **1903** Kipling *Five Nations* 104 Duly with knees that feign to quake—Bent head and shaded brow,—Yet once again, for my father's sake, In Rimmon's House I bow. **1956** N. Annan in J. Morris *From Third Programme* 150 If you bow down in the house of Rimmon you admit that its values are more important than yours.

c. *on the house*: at the expense of the tavern, saloon, etc. Also *transf.* and *fig.* orig. *U.S.*

1889 *Kansas City* (Missouri) *Times & Star* 30 Nov., The first drink Thursday was 'on the house' in the leading saloons. **1934** J. A. Lee *Children of Poor* (1949) 26 'I must have a drink.' Here, have one on the house. **1944** Auden *For Time Being* (1945) 77 A voice I'd heard before, I think, Cried: 'This is on the House.' **1958** M. Dickens *Man Overboard* xiii. 214 Laundry and cleaning were on the house. **1959** N. Mailer *Advts. for Myself* (1961) 95 One night just for the hell of it he has one on the house with the society gal, and she gets pregnant. **1967** W. Soyinka *Kongi's Harvest* 18 *Daodu*:..Naturally it's on the house. *Secretary*: No, thank you. I prefer to pay for my drinks.

4. c. (Earlier examples.) Also, in day schools, a division of the school for purposes of organization and games or other competition. Also *attrib.*

1855 J. A. Symonds *Let.* Oct. (1967) I. 64 Tom Parr who has just joined his regiment came down here today with an old house fellow. **1856** —— *Let.* June I. 73, I had them sent to the Head of the House who flogged two. **1857** —— *Let.* 15 Nov. I. 126 Yesterday I played in a house match. **1899** Kipling *Stalky & Co.* 124, I thought the house-prefects might know more about it than I did. They ought to. They're giddy palladiums of public schools. **1908** A. Huxley *Let.* Nov. (1969) 29 M'tutor, who is also my house tutor and my division beak, is a dear man. **1922** C. E. Montague *Disenchantment* vii. 93 A boy of this kidney, while looking on at a vital house match, will give his mind ease by telling a friend what 'a lot of stinkers' the other house are. **1925** *City of Oxford Sch. Mag.* Mar. 8 Kerry House hold the new cup for the winning House for the first year. *Ibid.* July 33 In the points counting for the House Shield. **1949** W. B. Gallie *Eng. School* ii. 31 He managed to infuse his ideas into the masters who coached the school's junior fifteens,

house fifteens, and so on. **1965** A. Nicol *Truly Married Woman* 76 An important football house-match was scheduled for that afternoon. **1966** P. Willmott *Adolescent Boys E. London* v. 93 After registration I took house prayers as House Captain.

f. Also, *spec.* (i) = *house of ill fame* (House *sb.*[1] 11).

c **1810** W. Hickey *Memoirs* (1960) iv. 63, I was informed with vast glee by these wild young men that..they had discovered two new houses of infinite merit. *a* **1922** T. S. Eliot *Waste Land Drafts* (1971) 5 I've kept a decent house for twenty years. **1931** R. Campbell *Georgiad* ii. 45 The old 'Matronas' of the Southern Race Can run their 'houses' with a smiling face, Business and pleasure to one end unite. **1954** P. Adler (*title*) A house is not a home. **1962** *John o' London's* 1 Feb. 115/1 The girls who had worked in 'houses' were unfitted, by temperament and training, for any other sort of life. **1968** L. Durrell *Tunc* ii. 47 'You see,' said Mrs. Henniker piteously 'what we are up against all the time? How to run a respectable house what I mean?'

(ii) A couture establishment.

1864 *2nd Rep. Children's Employment Comm.* 39 in *Parl. Papers* XXII. 1 At some houses all in the show-room are expected to wear black glacé silk. **1880** in L. de Vries *Victorian Advts.* (1968) *Messrs. Jay* import from the first houses in Paris Models of every style. **1931** S. Jameson *Richer Dust* v. 145, I have good taste. I could think of dresses. I should have to go round the various houses. **1938** D. Smith *Dear Octopus* III. i. 108, I can go to Raquelle's London house if I like. **1967** *Guardian* 24 July 4/6 Instead of showing their collections four weeks after the other houses, Givenchy and Balenciaga..show on August 3.

(iii) A printing or publishing house.

1683–4 J. Moxon *Mech. Exerc., Printing* (1962) 16 A Printing-House may admit of a twofold meaning; one..relative to the House or Place wherein Printing is used; the other..only the Printing Tools... Thus they say ..such a one has remov'd his Printing-House, when thereby they only mean he has remov'd the Tools us'd in his former House. **1871** G. Meredith *Let.* 3 July (1970) I. 448 The debt has been left unsettled owing to my having kept back my work to perfect it more. It was honourably incurred by me out of consideration for your house. **1935** *Times Lit. Suppl.* 4 Apr. 217/4 Already there obtained something approaching what is now called 'the style of the house'. **1959** N. Mailer *Advts. for Myself* (1961) 399 There was no one in the house who had guts enough to say that *Some Came Running* was a washer-woman at 1,200 pages, and could be fair at 400.

(iv) Used *attrib.*, in *house journal, magazine*, etc., a publication written for and circulated within a business firm, group, etc.

1907 *Electr. World* XLIX. 674/1 The subject of 'house organs' for manufacturing concerns was discussed. **1915** *Writer's Bulletin* Jan. 75/1 The *Hoggson Magazine*.. is an example of the artistic heights to which the 'house organ' may attain. **1925** *Writer's Monthly* June 466/1 These little magazines, or internal house organs, as they are called by the advertising fraternity, are usually of an inspirational character. **1959** *Times Rev. Industry* Mar. 5/2 Publications such as..house magazines. **1962** *Punch* 7 Mar. 375/2 Forty per cent of Sixth Form boys..read *Punch*, or so..their house-journal, *Sixth Form Opinion*, tells us. **1969** 'E. Lathen' *Murder to Go* xvii. 171 *Chicken Feed*, house organ of the Chicken Tonight organization, was not due for three weeks. **1970** T. Hilton *Pre-Raphaelites* ii. 49 *The Germ*..was the first house journal of a self-consciously avant-garde artistic group. **1970** *Vogue* May 48/2 They want..to make the cinema altogether a pleasant experience, with programmes, discussions, attractive usherettes..and a house magazine. **1971** *Nature* 5 Mar. 3/1 *Physical Review*, the house journal of the American Physical Society.

g. (Further examples.) Also, of stage or cinema performances closely following each other, *first, second house*. *House full*: the announcement posted outside a place of entertainment to indicate to the public that there is no room available; also *transf.*

1835 Dickens *Sk. Boz* (1836) 1st Ser. II. 202 There'll be a full house to-night—six parties in the boxes already. **1891** House full [in Dict.]. **1898** J. Hollingshead *Gaiety Chron.* ii. 115 No 'house full' boards were exhibited outside the theatre. **1906** J. M. Synge *Lett. to Molly* (1971) 49, I dreamed also that Tolstoi..came to our plays..and that there was a very bad house. **1906** *Daily Chron.* 23 Dec. 3/2 It would be natural to suppose that the book trade this Christmas has been a 'house full' affair. **1914** *Aeroplane* 2 Sept. 211/1 Both Services have practically put up the 'House Full' sign, and have a waiting list yards in length. **1921** G. B. Shaw *Let.* 27 Dec. in *To a Young Actress* (1960) 12 Until you can hit the boy at the back of the gallery in a three hundred pound house. **1924** in L. Warwick *Death of Theatre* (1960) vi. 62 The management have long felt that many first-class productions have been spoilt by the rush of two houses. **1930** J. B. Priestley *Angel Pavement* vii. 358 If I can get two seats for the first house tomorrow night, will you come with me? **1939** Joyce *Finnegans Wake* 62 It was after the show at Wednesbury that one tall man.. returning late..from the second house..had a barkiss revolver placed to his face. **1940** H. G. Wells *Babes in Darkling Wood* II. i. 137 He might go into some cinema... Or he might get in for the second house at the Holborn or Pavilion? **1968** *Blues Unlimited* Dec. 4 He played little guitar the first house but sang 'Two ways to skin a cat'.

h. Used *attrib.* of a permanent or resident band, jazz group, etc.

1934 S. R. Nelson *All about Jazz* iv. 76 There are many combinations which only record and play over the radio... These orchestras are often composed of prominent members of famous bands, in which case they are known as 'house' combinations. **1958** J. C. Holmes *Horn* (1959) 34 The drummer for the house band good-naturedly chased Wing's warm-up runs with precise rim-shots. **1966** *Crescendo* Nov. 23/1 He was part of the house rhythm section for Prestige Records. **1968** *Blues Unlimited* Nov. 26 The earlier Tampas are better—though hardly classifiable as 'house band'.

5. Also *son, daughter of the house*.

1802 C. Wilmot *Let.* 3 Jan. in *Irish Peer* (1920) 32 Lady Mount Cashal was handed out of the Room by Monsieur l'Abbé Siéyes, and I by Monsieur, the Son of the House. **1926** Wodehouse *Heart of Goof* ii. 54 The modern butler..looks like the son of the house. **1949** 'M. Innes' *Journeying Boy* xiii. 167 Mr Thewless..felt that the son of the house might turn out to be someone to rely upon. **1968** 'D. Torr' *Treason Line* 30 The daughter of the house, a slim, petite girl.

9. b. *Curling.* The space within the outermost circle drawn round the tee.

1883 J. Macnair *Channel-Stane* 1st Ser. 50 The stone draws past everything save the winner, which is knocked clear of the house. **1914** J. G. Grant *Complete Curler* II. vii. 91 The space within the 7-foot ring is colloquially known as the 'house' (*Scotice*, 'hoose'), or sometimes 'parish'. **1969** R. Welsh *Beginner's Guide Curling* xii. 89 A player will strike out the shot and his own stone will also roll out, leaving an empty house.

c. Lotto played (orig. in the Army) as a gambling game with special cards and checks. Also, the *house-caller* or *house-man* who organizes the game, the winning of the game or the call announcing this, or the prize given. See also *housey-housey.

1900 *Strand Mag.* Apr. 419/2 When they were not drilling they were playing 'House'. **1917** A. G. Empey *From Fire Step* 125 The two most popular games are 'Crown and Anchor' and 'House'. *Ibid.* 126 As soon as the *estaminet* is sufficiently crowded the proprietors of the 'House Game' get busy. **1918** R. D. Holmes *Yankee in Trenches* v. 60 If you get all your numbers covered, you call out 'house', winning the pot. **1919** [see *clickety-click* (*click *sb.*[1] 8)]. **1920** G. K. Rose *2/4th Oxf. & Bucks Lt. Infty.* 200 On the air floated the monotonous enumeration of 'House'. **1923** *Daily Mail* 9 June 7 The game of 'house', or 'box and numbers'. **1933** [see *clickety-click* (*click *sb.*[1] 8)]. **1936** F. Richards *Old-Soldier Sahib* iii. 69 It takes two men to work a game of House: one calls out the numbers, the other collects the money and issues the cards. **1945** *Gen* 30 June 27/1 The house-caller announced that the amount of the house was two and a half piastres short of ten pounds. **1951** *Amer. Speech* XXVI. 99/1 *House*, the man who runs the game is spoken of as the 'house' or the 'house man'. **1973** *Guardian* 12 June 20 They were in a crooked [Bingo] game... Certain cards were stacked for a quick House against others which were..distinct long-shots.

11. Also *house of accommodation* (cf. *accommodation 7 b), *house of assignation* (cf. *assignation 10).

1749 J. Cleland *Mem. Woman Pleasure* II. 12 This was the safest, politest, and, at the same time, the most thorough house of accommodation in town. **1834** *Sun* (N.Y.) 10 Apr. 2/2 Such men as Samuel Q. Wright, (a bank man) the keeper of a notorious house of assignation, and prostitution. **1861** B. Hemyng in H. Mayhew *London Labour* Extra vol. (1862) 255/1 Keepers of houses of assignation, where the last-mentioned class [sc. ladies of intrigue] may carry on their amours with secrecy. **1928** 'Brent of Bin Bin' *Up Country* x. 146 They started a house of accommodation in the most unlikely place for patrons. **1973** G. Butler *Coffin for Pandora* i. 26 Just as some of them could rise to an elegant house in Brompton, others could sink to a house of accommodation off the Haymarket or even worse.

13. *house-to-house*. Also *ellipt.* or as *sb.*, = a house-to-house collection, search, etc.

1936 'F. O'Connor' *Bones of Contention* 37 We'll make a house-to-house. **1969** B. Weil *Dossier IX* vi. 44 The house-to-house paid off. We got someone in the mews to talk. **1970** R. Rendell *Guilty Thing Surprised* xii. 142 Three of us have done a house-to-house in Myfleet. **1973** W. J. Burley *Death in Salubrious Place* v. 93 The house-to-house and the questionnaire both appear to have covered the ground pretty well.

16*. *to play* (*at*) *house*(*s*): to play at being a family and running a house (see also quot. 1968).

1871 J. H. Ewing *Flat Iron for Farthing* (1873) vii. 67 Polly and I had nothing to do.., which led us into the very reprehensible habit of 'playing at houses' in Uncle Ascott's gorgeously furnished pew. *Ibid.* 71 You know we *couldn't* play houses in the church where papa goes. **1918** 'K. Mansfield' *Lett. to J. M. Murry* (1951) 220 They always seem to think we were so very very young at the Villa Pauline—playing houses—going to bed under the table for a minute with the cloth pulled down for a blind. **1957** J. Kirkup *Only Child* ix. 120 If we were lucky, and the weather was fine, we would be allowed to have tea in our tent, and to play houses with the fireside things—the tongs, the little brush, the shovel and the ash-rake. **1959** H. Gardner *So what else is New?* xiv. 140 'Let's play house, Sally,' the boy said. **1968** *Sun* (Baltimore) 18 Sept. B7/8 He's tried several times to get me to go to his apartment. I've always refused... I'm not ready to play house yet. **1969** Gish & Pinchot *Lillian Gish* ii. 17 Whenever Dorothy and I were in the same company with Mother, our favorite game was to go to the theater early, dress in Mother's clothes, and play house on stage.

18. Also, *to pull a house over one's head*; *to throw the house out of the windows*: delete *Obs.* and add later example; *as safe as houses* (later examples); *to go* (*all*) *round the houses*: *fig.* to beat about the bush, to reach the point in a lengthy or roundabout way; *to put* (or *set*) *one's* (*own*) *house in order*: to arrange one's affairs properly.

1611 Bible *2 Kings* 20:1 And the Prophet Isaiah the sonne of Amos came to him, and saide vnto him, Thus saith the Lord, Set thine house in order: for thou shalt die, and not liue. **1861** Geo. Eliot *Silas Marner* iii. 53 You'll have less pleasure in pulling the house over my head, when your own skull's to be broken too. **1871** Hardy *Desp. Rem.* III. iv. 92, I shall be high-treasoned—as safe as houses. **1880** E. W. Hamilton *Diary* 25 Apr. (1972) I. 3 Layard has telegraphed from Constantinople that the Turk is becoming really alive to the necessity of 'putting his house in order'—(if that broken-down establishment will admit of repair). **1913** E. Phillpotts *Widecombe Fair* v. 36 They say things, and do things, and even think things, that you'd fear must throw the house out of windows, and wreck the home for evermore. **1923** J. S. Huxley *Ess. Biologist* vii. 292 The theologians..fail in the majority of cases to set their own house in order, to organize the inner reality to react with the outer. **1932** 'A. Bridge' *Peking Picnic* xix. 238 Kidnapping Frenchmen is simply too unremunerative for words, whereas we're a perfect gold mine, safe as houses. **1949** A. Wilson *Wrong Set* 49 You had better put your own house in order before you go listening to wicked lies. **1955** *Times* 31 Aug. 7/5 If the conservative parties did not put their houses in order, Japan would soon see a Socialist Government in power. **1958** F. Norman *Bang to Rights* 129 Alright but I wish you wouldn't go all round the houses. **1965** V. Canning *Whip Hand* iv. 43 You mean your little story..wasn't true?..You went a hell of a way round the houses about this. **1965** *New Statesman* 14 May 749/3 Such a policy of firmness would make it clear to all that we are determined to put our economic house in order.

19. a. *house-back, -bell, -front* (later examples), *-number, -paddock* (Austral.), *-pile, -site, -wiring*. **b.** *house-boot, -cloth* (examples), *-dress* (examples), *-frock, -gown* (earlier example), *-jacket, -linen, -plant* (earlier and later examples), *-slipper, -telephone* (also *-phone*). **c.** *house affairs* (later example). **d.** *house-servant* (earlier and later examples), *-slave, -steward* (later example).

1862 D. G. Rossetti *Let.* 22 Feb. (1965) II. 443 House-affairs get still further complicated. **1913** C. Mackenzie *Sinister St.* I. ii. 29 He..wished that he could disappear in such company round corner after corner of the world beyond the grey house-backs. **1948** C. Day Lewis *Otterbury Incident* 69 Between this path and the house-backs there are some small yards. *a* **1817** Jane Austen *Northanger Abbey* (1818) II. xiii. 249 The loud noise of the house-bell. **1834** *Chambers's Edin. Jrnl.* III. 414/1 He rung the teacher's house-bell. **1921** W. de la Mare *Crossings* 88 The far-away house-bell clangs into the room. **1652** R. Verney *Let.* in M. M. Verney *Mem.* (1894) III. ii. 46 Sir Ralph is much taken with some 'old men's house boots', called Scarfaroni. **1914** Joyce *Dubliners* 123 Then she took off her working skirt and her house-boots. **1597** T. Deloney *Jacke Newb.* (1630) sig. G3, [They] put an house-cloath about his necke in stead of a fine towell. **1934** H. G. Wells *Exper. Autobiogr.* I. iv. 149 Rolls of crash, house cloth, ticking and the like. **1897** *McClure's Mag.* X. 66 She looked charming in her long, soft house-dress. **1921** *Daily Colonist* (Victoria, B.C.) 2 Apr. 9/5 House Dresses in V-neck style with kimona sleeves, pockets and belt, open down the front and come in stripes, checks and plain colors. **1966** 'S. Ransome' *Hidden Hour* iv. 50 A woman wearing a house dress, evidently one of the tenants. **1925** *Sears, Roebuck Catal.* 28 A neat, trim gingham house or porch frock is an asset to every woman's wardrobe. **1952** C. W. Cunnington *Eng. Women's Clothing* iv. 135 House-frocks are now all-important as a result of the war economies. **1905** *Westm. Gaz.* 12 Dec. 9/1 The house-fronts of miles and miles of London streets are entirely carried on iron girders. **1963** *Times* 4 June 11/7 The drabness of house-fronts, that so deceive the casual visitor. **1832** *Chambers's Edin. Jrnl.* I. 121/3 Sometimes the fit takes the direction of a new gown for going out with on bad days,..at another time 'a house gown'. **1922** Joyce *Ulysses* 452 In housejacket of ripplecloth. **1926** *Daily Colonist* (Victoria, B.C.) 1 Jan. 11/1 (Advt.), Christmas Goods Now on Sale. Dressing Gowns, House Jackets, Ties, [etc.]. **1857** Mrs. Gaskell *Let.* 7 Dec. (1966) 492 Then we came home; and have been desperately busy ever since, looking over stores, and clothes, and house-linen, and preserves. **1905** *Westm. Gaz.* 8 Nov. 5/2 No arrangements had been made for her reception. There was no silver, no house-linen. **1895** *Montgomery Ward Catal.* 396/2 House numbers, 3 inch, made of brass and nicely nickel plated. **1972** C. Weston *Poor, Poor Ophelia* (1973) vi. 29 Casey began looking for house numbers. **1908** Mrs. A. Gunn *We of Never-Never* vi. 64 To the north-west are the stock-yards and house paddock—a paddock of five square miles. **1951** J. K. Ewers in Murdoch & Drake-Brockman *Austral. Short Stories* 335 The scrawny gums by the house-paddock. **1908** S. Ford *Side-Stepping with Shorty* ix. 137 There was no answer to the call on the house 'phone. **1935** *Archit. Rev.* LXXVIII. 73 Raymond McGrath designed standard signal lights, clocks, buzzer and house-phone mountings. **1970** 'W. Haggard' *Hardliners* i. 11 There was a house-phone on the impressive desk. **1930** M. Mead *Growing up in New Guinea* ii. 13 Around the stout house-piles the tides run. **1970** R. Lowell *Notebk.* 47 Where the Brook Trout dolphins by the housepiles, Grows common..as hamburger. **1873** *Young Englishwoman* Nov. 562/2 Those who have cultivated house plants for years. **1970** D. Bartrum *Exotic Plants for Home* ii. 29 All the house plants we buy from the florists are amenable to pot culture. **1743** W. Ellis *Mod. Husbandman* Oct. xxii. 238 There are two Sorts of Men Taskers, the Day Labourer, and the constant House Servant. **1882** W. D. Hay *Brighter Britain* I. 37 That's a new dairy-maid and house-servant my friend's just engaged. **1916** *Gilbert & Ellice Islands Protect. Rep. for 1914–15* 15 The Ellice boy, who is much inferior to the Gilbertese in all things that really matter, makes a better house servant. **1966** B. Kimenye *Kalasanda Revisited* 23 His house servant was away on leave, and the drawbacks of bachelorhood were making themselves acutely felt. **1913** J. London *Let.* 1 Mar. (1966) 373 The hedges around the house-site. **1949** M. Mead *Male & Female* ii. 40 Considerations..that one would normally consider in choosing a house-site. **1738** F. Moore *Trav. Inland Afr.* 110 Some people have a good many House-Slaves..and they live so well and easy, that it is sometimes a very hard Matter to know the Slaves from their Masters. **1962** S. Wynter *Hills of Hebron* vi. 81 His master was fond of him and made his mother a house-slave so that Cato could grow up in the big house. **1972** *Village Voice* (N.Y.) 1 June 24/1 Whitey has always pitted one black against the other. The field slaves and the house slaves. **1895** *Montgomery Ward Catal.* 512/1 Ladies' toilet and house slippers. **1965** B. Sweet-Escott *Baker St. Irreg.* vi. 161 His feet encased in a pair of black velvet house-slippers embroidered in white with the letters 'A. E.' **1922** Joyce *Ulysses* 304 The house-steward of the amalgamated cats' and dogs' home was in attendance. **1889** *Telephone* I. xxiii. 534/1 The installation of a house telephone recommends itself in large hotels. **1921** [see *camouflage *v.*]. **1950** T. S. Eliot *Cocktail Party* II. 94 The house-telephone rings. **1968** J. Fleming *Hell's Belle* i. 34 The humblest hotel in Paris has a house-telephone. **1901** *Chambers's Jrnl.* Sept. 616/1 We now propose to consider installations which require, in addition to the ordinary house-wiring, the machinery necessary for the production of the electric current. **1963** *Times* 6 May p. vii/2 The traditional type of housewiring involved separate circuits for the different kinds of load.

20. *house-cat* (earlier and later examples).

1607 Topsell *Four-f. Beasts* 107 Her tayle longer then an ordinary house cats. **1963** B. Vesey-Fitzgerald *Cat Owner's Encycl.* 54 But we do not know which species it was that first became the house-cat. **1973** R. Ludlum *Matlock Paper* vii. 61 His clumsy, long-haired house cat had knocked over a stray glass.

21. *house decoration, -decorator, -hunting* (later examples), *-letting* (examples), *-move* sb., *-moving*.

1881 C. C. Harrison *Woman's Handiwork* I. 4 Industrial arts applied to house-decoration. **1914** W. Owen *Let.* 11 Aug. (1967) 276 She has an important business.. selling House Decorations, Embroideries, and so on. **1911** G. S. Porter *Harvester* xv. 321 As a house decorator you surpass yourself. **1929** F. M. Ford *Let.* 14 Sept. (1965) 189 House-decorators find that books work into rooms with admirable effect. **1935** *Burlington Mag.* July 44/2 The colouring of cupboards and friezes and similar house-decorators' tasks. **1930** *Times Lit. Suppl.* 31 July 625/2 To engage playfully in house-hunting. **1960** *Guardian* 15 Feb. 4/2 Her recent house-hunting in London. **1839** Dickens *Let.* 19 Nov. (1965) I. 603, I am in the agonies of house-letting, house-taking, title-proving, [etc.]. **1909** *Westm. Gaz.* 1 Oct. 2/2 The Scotch House-Letting and Rating Bill. **1923** E. Bowen *Encounters* 92, I shouldn't have thought that a house-move was exactly the most leisurely time. **1961** *Times* 17 Aug. 11/2 Rearranging my books..after a house-move. **1926** R. Macaulay *Crewe Train* II. x. 184 I'm extraordinarily sleepy, with all this house-moving.

23. house-agent (earlier and later examples); **house appointment,** a position as a house-physician or house-surgeon in a hospital; **house arrest,** detention in one's house; also (with hyphen) as *vb.*; **house-author,** an author employed by a theatre; **house bill,** (*a*) a poster or programme describing a theatrical performance; (*b*) a bill of exchange drawn by a business house on itself; **house-bound** *a.* (examples); **house-boy** (examples); **house-burnt** *a. U.S.*, designating tobacco which in the course of being cured in a tobacco-house has been injured or spoilt by disease; so **house-burn** *v. intr.* and (rare) *trans.*, to become or render house-burnt; also **house-burning** *vbl. sb.*; **house call,** a visit made to a patient in his own home by a doctor, chiropodist, etc.; **house cap,** a school cap made of the colours adopted by a particular house, esp. one awarded for proficiency in games; **house-car** (examples); **house-carpenter** (later U.S. examples); **house church** (see quot. 1967); **house-cleaning** *sb.*, the cleaning of the inside of a house; hence (as a back-formation) **house-clean** *v.*; also *transf.*; **house-cleaner,** one who cleans the inside of a house; **house-coat,** a woman's informal coat-dress for wearing at home; **house colours** [*colour *sb.*[1] 6 c], colours representing a house (sense 4 c above) at a school; **house detective,** a private detective employed by a business firm, hotel, etc.; **house detention** = **house arrest*; **house dinner,** a dinner given to the staff or the occupants of a house at a school, etc.; **house-dweller,** one who lives in a house (opp. a nomad, etc.); so **house-dwelling; house finch,** a red-headed N. American finch of the genus *Carpodacus*, esp. *C. mexicanus*; **house-furnishing,** the furnishing of a house; also *pl.* in concrete sense; **house-girl,** a female domestic servant or, formerly, a slave; **house governor,** the head of administration in a hospital; **house guest,** a guest staying in a private house; **house-help** (examples); **house journal:** see sense 4 f (iv) above; **house lights,** lights on the audience side of the stage curtain in a theatre; **house longhorn, longicorn** (beetle), *Hylotrupes bajulus*, a wood-boring beetle of the family Cerambycidæ; **house magazine:** see sense 4 f (iv) above; **house-manager,** the manager of a theatre, club, concert-hall, etc.; **house moth,** either of two moths, *Hofmannophila pseudospretella* or *Endrosis sarcitrella*; **house mouse,** *Mus musculus*, which lives in buildings as well as in open fields; **house-mover** *N. Amer.*, (*a*) a person whose business is to move furniture; (*b*) a machine or apparatus for the physical removal of houses; **house Negro, house nigger** *U.S.* (*rare* exc. *Hist.*), a Negro household servant; **house officer,** a junior full-time member of the medical staff of a hospital, usually (but not always) resident; *esp.* one whose post is tenable by qualified doctors who are not yet fully registered; **house organ:** see sense 4 f (iv) above; **house-painting** (examples); **house-parent,** a house-mother or house-father acting singly or jointly as head of a community of (young) persons living together as a family; **house-parlourman,** a male servant who does work corresponding to that of a house-parlourmaid; **house-party** (earlier example); also = **house-rent party* (below); **house physician,** add to def.: now usually (in Great Britain) a house officer working in the field of general medicine; (later examples); **house-pride,** pride in one's house, desire to keep one's house clean and tidy; so **house-proud** *a.*, proud of one's house, desirous to see it always at its best (sometimes implying excessive preoccupation with it); **house-raising** (earlier and later examples); **house-rent party** *U.S.*, orig. a party aimed at raising money to pay the rent of a house; later, any 'jam' session in a house or apartment; also *house-rent stomp, strut*; **house seat,** a seat in a theatre, etc., reserved by the management for special guests; **house style,** the distinctive printing methods and regulations, including the preferred spellings and conventions of punctuation, of a publishing or printing business; also *transf.*; **house surgeon,** add to def.: now usually (in Great Britain) a house officer working in the field of surgery; (later examples); **house-tax** (earlier examples); **house-trap,** a portable bird-trap made of wire netting in which bait has been laid; **house-type,** a type of house; **housewares** *sb. pl.* (chiefly *N. Amer.*), kitchen utensils and other utilitarian household articles; **house-work,** the work required to keep a house clean and in order; **house-wrecker** = Housebreaker 2; **house wren,** the common N. American brown wren, *Troglodytes aedon*.

1843 *Ainsworth's Mag.* IV. 64 House-agents and auctioneers are their attendant sprites. **1922** T. S. Eliot *Waste Land* (1923) iii. 16 A small house agents' clerk. **1961** *Lancet* 26 Aug. 497/1 After house-appointments he graduated M.D. in 1900. **1963** *Ibid.* 19 Jan. 176/1 After holding house-appointments, he became an assistant bacteriologist in the Glasgow public-health laboratories. **1936** F. L. Schuman *Hitler & Nazi Dictatorship* Epilogue ii. 441 He was subjected to 'house arrest' for his protection. **1945** M. Allingham *Coroner's Pidgin* xv. 126 In the normal way when I put a person under house arrest and she breaks it, I pull her in. **1948** A. Koestler in *Partisan Rev.* XV. I. 33 All of them are..in fact, under a kind of curfew or house arrest. **1958** *Listener* 13 Nov. 762/2 Galileo lived the remaining years of his life under house arrest. **1963** *Economist* 31 Aug. 732/2 The Hodgsons, banned, house-arrested and persecuted in South Africa. **1970** *New Yorker* 17 Oct. 179/1 The Chinese continue to support Prince Sihanouk, who is said to be living under house arrest in Peking. **1971** *Rand Daily Mail* 4 Sept. 12/4 Father Desmond was banned and house arrested after the book was published, probably for daring to write it. **1864** P. Paterson *Glimpses Real Life* xxvii. 262 August, or even earlier, when the 'house-author' and the manager determine what it [sc. the pantomime] is to be. **1903** *Daily Chron.* 6 July 7/4 He was house-author to a theatre at Sydney. **1829** H. Foote *Compan. to Theatres* 138 The usual bills of the theatres are termed house bills. **1909** *Daily Chron.* 2 Mar. 3/2 That particular bill of exchange, the 'house bill'—a bill drawn by a firm or company on itself. **1878** *Harper's Mag.* Jan. 277/1 The rains set in furiously, and I was completely house-bound. **1960** *Sunday Express* 10 July 11/4 His ailing, house-bound wife is attended to by a home help. **1966** *New Statesman* 21 Jan. 80/3 Housebound married graduates. **1899** *Daily News* 8 June 5/7 Odd jobs as boot and knife cleaning, or where the boys are described as 'house-boys'. **1910** *Cape Times* 8 Oct. 2 House-boy; good references;.. apply 6, St. Barnabas Street. **1926** S. G. Millin *S. Africans* III. i. 74 There might be black houseboys instead of white housemaids. **1944** *Living off Land* viii. 159 He [sc. a Papuan] may be an experienced 'house-boy' in which case he will make his own arrangements for employment. **1955** B. Manvers *Shadow of Happiness* i. 19 That's my houseboy; he has an unpronounceable name, so I call him John. **1971** *E. Afr. Jrnl.* Mar. 5/1 She is procured by the African houseboys for their employer—a white expatriate. **1640** *Archives of Maryland* (1883) I. 98 Bad Tobacco shall be judged ground leafes, Second Crops leafs, notably brused or worm eaten, or leaves house burnt, sun burnt, [etc.]. **1772** *Maryland Hist. Mag.* (1919) XIV. 363 For 3 weeks past the Weather has been very unfavourable for the tob[acc]o Cured by fier very much & I suppose has House burnt all tob[acc]o not fierd. **1850** *Rep. Comm. Patents Agric. 1849* (U.S. Dept. Agric.) 321 Splitting tobacco is admired by many who contend that it cures brighter..and [is] less likely to house-burn. *Ibid.* 324 In this crop every leaf was saved, none lost by worms nor by 'house-burning' (that is suffering, or even rotting from being hung too thick). **1897** *Bradstreet's* 25 Oct. 1/4 Some of the leading growers report several crops as 'house burnt' and inclined to rot. **1966** *Publ. Amer. Dial. Soc.* XLV. 16 Tobacco will house burn in wet weather if you don't fire it. **1960** R. H. Blum *Managem. Doctor–Patient Relationship* ix. 203 The readiness of the physician to make house calls is important to patients. **1973** R. C. Dennis *Sweat of Fear* xii. 85, I opened the door, and there she was... I said, 'Doctor, I didn't know you made house calls.' **1899** Kipling in *Windsor Mag.* Dec. 33/1 'S'pose we're collared?' said Beetle, cramming his red and black house-cap into his pocket. **1907** *Daily Chron.* 15 June 4/7 The pride of a little boy who wins his house cap at school. **1856** W. Ferguson *Amer. by River & Rail* 338, I was glad to withdraw myself and my stool within the doorway of a house-car, as the covered freight-trucks are called. **1858** *Pennsylvania Rail Road Ann. Rep.* 14 The Rolling Stock..consisted..of.. Eight-wheeled House Cars. **1741** W. Stephens *Jrnl.* 1 Apr. in *Colonial Rec. Georgia* (1908) IV. Suppl. 118 The other was looked on as a Master House-Carpenter of Repute. **1758** in S. M. Hamilton *Lett. to Washington* (1899) II. 365 And all this without one farthing expence (except about nine pence per day to the best house Carpenters). **1855** *Knickerbocker* XLVI. 222 Beech timber is held in great esteem by ship-builders and house-carpenters. **1964** *New Society* 26 Mar. 5/2 Four house-churches meet regularly on week nights in people's homes in Notting Hill. They are a mixture of discussion, worship, sharing of problems and a social occasion. **1967** D. T. Kauffman *Dict. Relig. Terms* 239/1 *House Church*, church program carried out and centered in homes rather than ecclesiastical structures. **1970** *Daily Tel.* 7 Aug. 10 At the 'house church' each member is given a duplicated sheet with questions and information on the subject for discussion. **1865** Mrs. Stowe *House & Home Papers* 45 He could not come in the spring for then they were house-cleaning. **1942** Partridge *Usage & Abusage* 154/2 *House-clean* or *houseclean* 'to clean (the) house', seems to me to be a permissible—and very convenient—word. **1954** *Manch. Guardian Weekly* 1 July 2 The sub-committee should houseclean its staff. **1959** 'J. R. Macdonald' *Galton Case* (1960) xii. 93 With all the outside work I do, I don't get time to houseclean. **1959** *Time* 12 Jan. 26/3 Seldom had a government been so thoroughly housecleaned between midnight and dawn. **1905** *Daily Chron.* 16 May 5/5 A house-cleaner..who was maddened with liquor, to-day shot..his landlady. **1863** B. Taylor *Hannah Thurston* ii. 26 The first thing..was immediately to summon old Melinda..whose speciality was house-cleaning. **1928** Foy & Harlow *Clowning through Life* 294 The Chicago horror was a blessing in one respect—namely, in that it brought about a country-wide house-cleaning. **1936** *Discovery* Jan. 16/2 For one beautiful little specimen [of worked flint] the writer was indebted to the house-cleaning of a busy rabbit. **1951** H. Reichenbach *Rise Sci. Philos.* xviii. 310 A good deal of house-cleaning is necessary before a philosophy of the social sciences can be constructed. **1959** *Times* 8 June (Latin Amer. Suppl.) 1/3 The new Cuban leaders are bitter that world opinion is now outraged by their house-cleaning methods. **1965** F. Sargeson *Mem. Peon* iv. 91 A char whom she employed for house-cleaning duties. **1969** *New Yorker* 31 May 72/3 A real political house-cleaning. **1916** in *Vogue* (1966) 15 Oct. 3/1 (Advt.), Artificial Silk Sports or House Coat in pretty, quiet striped colourings. **1937** *New Yorker* 16 Jan. 48/2 The Bendel negligee department is famous for its housecoats. **1946** 'S. Russell' *To Bed with Grand Music* viii. 105 He came back..to find Deborah in a house-coat, cooking supper. **1958** 'A. Gilbert' *Death against Clock* 45 She was in a house-coat—that's what they call dressing-gowns these days. **1973** J. Wainwright *Pride of Pigs* 50 She.. slipped a housecoat over her nightdress and made her way downstairs. **1914** 'I. Hay' *Lighter Side School Life* viii. 224 Have you got your House colours? **1939** 'G. Orwell' in *Crit. Ess.* (1946) 63 That mystic world of quadrangles and house-colours. **1961** D. Bates *Fly-Switch from Sultan* xix. 106 So I started a brothel... It was run on the best public-school lines, and there were prefects and houses and in a manner of speaking there were house colours as well. **1898** *McClure's Mag.* X. 525/2 A house detective [had] observed the whole transaction. **1922** F. Scott Fitzgerald *Beautiful & Damned* II. i. 135 The group was joined by the hastily summoned house-detective. **1939** M. Allingham *Mr Campion & Others* 178 [He] was quite an ornament in the City police... When he retired he received the job of house detective here. **1969** J. Wainwright *Take-Over Man* v. 75 It's your reputation. Don't blame me if the house detective finds us and boots us out of the servants' entrance. **1958** *New Statesman* 3 May 555/2, I found an old journalist friend in his second year of house detention because he had been courageous enough to buck the army. **1870** E. L. Blanchard *Diary* 7 Apr. in C. W. Scott *Life E. L. Blanchard* (1891) II. 382 Dine at 7 at Arundel [Club], being the first 'house' dinner proposed. **1909** *Westm. Gaz.* 1 Oct. 3/3 Many of them who dwell in tents during summer and autumn become house-dwellers in Wandsworth..during winter. **1954** J. R. R. Tolkien *Fellowship of Ring* I. iv. 101 Most of the inhabitants..were house-dwellers. **1941** 'R. West' *Black Lamb* II. 103 The gypsies..would not dream of going into the church while the house-dwelling Christians were still about. **1869** *Amer. Naturalist* III. 183 About the gardens [in California] is the House Finch. **1917** T. G. Pearson *Birds Amer.* III. 7 House Finch... Other names [include] Crimson-fronted Finch; Red-headed Linnet [etc.]. **1961** O. L. Austin *Birds of World* (1962) 302/2 Very similar to the Purple Finch is the slightly smaller and brighter Mexican House Finch, a common garden bird from California southward through Mexico. **1791** Cowper *Let.* 30 Aug. in *Corr.* (1904) IV. 117 Such

[chairs] as will suit may be found probably at Maurice Smith's, of house-furnishing memory. **1858** *Leslie's Illustr. Weekly* 23 Jan. 127 Goods for Winter Use in the Housefurnishing Line. **1865** Mrs. Stowe *House & Home Papers* 61 That such is not always the case in the real home comes often from the mistakes in the house-furnishing. **1904** *Sun* (N.Y.) 9 Aug. 8 The feminine preference for garments and house furnishings over locomotives and drop forgings. **1835** M. Morrison *Let.* 5 Nov. in N. E. Eliason *Tarheel Talk* (1956) 277 We have been intending for some time to buy a house girl. **1884** J. C. Harris *Mingo* 91 Jenny, the house-girl, refused to sleep at the quarters. **1906** *Dialect Notes* III. 141 *House-girl*, maid of all work. **1945** B. A. Botkin *Lay my Burden Down* 55 Part white children sold for more than black children. They used them for house girls. **1951** R. Campbell *Light on Dark Horse* xxiii. 335 We had, at that time, two fine house-girls, the sisters Eugenia and Florentina Diaz y Medina. **1971** *E. Afr. Jrnl.* Mar. 6/1 Lawino condemns the missionaries who only wanted to make her a house-girl. **1905** M. F. Reaney *Med. Profession* iii. 39 The actual daily command is vested in the secretary or house-governor. **1934** E. MacManus *Hosp. Admin. Women* I. i. 13 The Secretary or Superintendent, who may be known either as House Governor, or by some other title, may be..a member of the Medical Staff of the Hospital—with wide experience and a gift for administration. **1961** *Times* 18 July 3/2 (Advt.), Hospital..House Governor and Secretary. **1964** G. L. Cohen *What's Wrong with Hospitals?* iii. 47 Matron will hand complaints to the Catering Officer, who refers to the House Governor, who sends a memo to the management committee. **1921** *Daily Colonist* (Victoria, B.C.) 1 Apr. 8/3 Miss Helen MacDonald..invited a number of her girl friends to meet her house guest, Miss Helen Whiteside. **1961** *Times* 27 Sept. 16/5 His house-guests hunt in the day. **1970** N. Armstrong et al. *First on Moon* iii. 61 She was conscious about getting ready for houseguests—her father and stepmother. **1835** T. C. Haliburton *Clockmaker* (1837) 1st Ser. viii. 62 Well, he roared like a bull, till black Lucretia, one of the house helps, let him go. **1958** *Listener* 14 Aug. 249/1 Foreign house-helps in London, S.W. **1970** *Canadian Antiques Collector* Feb. 24/1 Presently in came two well-dressed house-helps, one with a splendid gilt lamp..and another with a tea-tray. **1920** Wodehouse *Jill the Reckless* (1922) xviii. 260 The house-lights went up. **1957** *Oxf. Compan. Theatre* (ed. 2) 465/2 After 1765 the house lights and concealed footlights on traps cover the area in front of the proscenium. **1938** *Leaflet Forest Prod. Res.* XIV. 1 The House Longhorn beetle (*Hylotrupes bajulus* L.)..confines its attack to seasoned softwood timbers. *Ibid.* 2 This leaflet has been prepared to bring the House Longhorn beetle to the notice of architects, builders and property owners. **1962** *New Scientist* 15 Mar. 614/3 Massive timber struts..were already known to harbour death-watch beetles and the house longicorn. **1964** N. E. Hickin *Househ. Insect Pests* viii. 83 The House Longhorn Beetle is a pest of softwood. **1906** *Westm. Gaz.* 7 Nov. 10/2 Lord Selborne..was once the house-manager of the club. **1924** Wodehouse *Bill the Conqueror* xviii. 269 Give it to the house-manager at the Bijou and he'll fix you up with a couple of seats any night you want. **1968** M. Culpan *Vasiliko Affair* ii. 14 The house-manager, a courteous but wary young man in a dinner jacket. [**1897** *Proc. Dorset Nat. Hist. & Antiq. Field Club* XVIII. 147, I think that this moth is one of the most universally distributed, being found, I expect, in pretty nearly every house in the kingdom.] **1932** *Entomologist's Monthly Mag.* LXVIII. 77 (*title*) *Borkhausenia pseudospretella* and other house moths. **1966** J. R. Busvine *Insects & Hygiene* (ed. 2) xiii. 354 The house moths are probably species which originated as feeders on dry vegetable matter and have become adapted to dry animal remains. **1835** L. Jenyns *Man. Brit. Vertebr. Animals* 31 *M*[*us*] *Musculus*, Linn. (House Mouse.) — Fur dusky gray above with a tinge of yellow; beneath cinereous. **1909** *Westm. Gaz.* 17 Sept. 3/1 We do not suppose that the creation of the first house mouse awaited the building of the first house. **1916** G. E. H. Barrett-Hamilton *Hist. Brit. Mammals* II. 635 In all probability the House Mouse is of Asiatic origin. **1964** H. N. Southern *Handbk. Brit. Mammals* v. 86 Break-back traps used on a large scale for catching House Mice can produce useful figures [for population studies]. **1838** D. Stevenson *Sk. Civ. Engin. N. Amer.* 316 He and his father..had followed the business of 'house-movers' for fourteen years. **1867** *Atlantic Monthly* Jan. 106/2 Jedwort had over a house-mover from the North Village. **1959** *Times* 18 June (Suppl. Queen in Canada) p. vi/3 In all, 525 homes were lifted from their old foundations and, cradled in the steel framework of the housemovers, moved to their new locations. **1711** *Boston News-Let.* 21 May 2/2 (Advt.), A Young House-Negro Wench of 19 Years of Age that speaks English to be Sold. **1771** in *Maryland Hist. Mag.* (1919) XIV. 135 You will have it that my People are not well fed, it is true they do not live so well as our House negroes, But full as well as any Plantation negroes. **1884** J. C. Harris *Mingo* 191 The house negroes stood in mortal dread of Blue Dave. **1936** M. Mitchell *Gone with Wind* iii. 49 The house-negroes..considered themselves superior to white trash. **1880** J. C. Harris *Uncle Remus* (1884) 116 Dey er mighty biggity, dem house niggers is, but I notices dat dey don't let nuthin' pass. **1970** G. Jackson *Let.* 4 Apr. in *Soledad Brother* (1971) 207 This running dog..was transmitting the credo of the slave to our youth, the mod version of the old house nigger. **1971** K. Wheeler *Epitaph for Mr. Wynn* (1972) xii. 149 Barton called him a house nigger... By Barton's lights, I suppose he was. **1973** *Black World* June 13/1 The murder of Mrs. Mann by the Black maid in the Bullins play affirms the maid's Blackness and consummates her transition from a foot-shuffling house nigger to a proud Black woman. **1934** E. MacManus *Hospital Admin. Women* II. iii. 65 When doing 'a morning round' with a House Physician or Surgeon, the Ward Sister will put him in possession of any new facts relating to each patient before they reach that patient's bed. The Sister should stand on one side of the bed, the House Officer on the other. **1966** *Lancet* 24 Dec. 1399/1 A man holds a senior house-officer post for up to eighteen months. **1969** S. G. Hill in Milne & Chaplin *Mod. Hosp. Managem.* iii. 46 In most hospitals and in most specialties, the medical team comprises the consultant who is the senior doctor, the registrar (a doctor of some four or five years' experience), and either a junior house officer (a newly qualified doctor holding a pre-registration post) or senior house officer (a doctor with rather more experience, often from overseas). **1973** *Lancet* 17 Feb. 17 (Advt.), Burton Road Hospital, Dudley,... Applications are invited for the post of Senior House Officer in Geriatrics... Resident staff comprises of 1 Registrar and 2 Senior House Officers. *Ibid.* (Advt.), Chester City Hospital... Applications are invited for the post of House Officer (Geriatrics), pre- or post-registration... Accommodation will be available. **1875** E. Spon *Workshop Receipts* 105/1 House-Painting.—To produce the different tints, various colours are added to the white-lead base. *a* **1877** Knight *Dict. Mech.* II. 1597/2 In house-painting, the pigment most extensively used is white-lead. **1929** D. H. Lawrence *Phoenix II* (1968) 602 There were several brushes for house-painting. **1951** M. Buchanan *Children's Village* 18 There are almost 200 children in the Village now. Each house has in it 15 or 16 boys and girls, their house-parents,..and a teacher-help. **1964** *New Statesman* 10 Apr. 582/1 (Advt.), House-parents for African-Asian married students' hostel. **1923** *Daily Mail* 25 Jan. 5 A number of men are becoming house-parlourmen. **1931** *Morning Post* 10 Aug. 16/6 House-parlourman Required for maisonette. **1961** *Times* 17 Oct. 1/3 House-Parlourman wanted. **1876** Trollope *Amer. Senator* (1877) I. xxiv. 248 Partners were selected within the house party. **1956** M. Stearns *Story of Jazz* (1957) xiii. 145 Its ancestry was long obscured by labels such as 'house-party', 'rent-party', 'parlor social', or simply 'Harlem' piano style. **1968** *Blues Unlimited* Nov. 11 He was around here in town then playing houseparties. **1905** M. F. Reaney *Med. Profession* iii. 39 Finally, there is a junior staff of house physicians and surgeons and the like, working directly under the visiting staff, and responsible for the treatment of the patients in their absence. **1934** House physician [see *house officer* above]. **1962** D. Margerson *Med. as Career* vi. 42 Two pre-registration posts, those of house physician and house surgeon, must be held before full registration is granted. **1970** *New Yorker* 23 May 73/3 Alek Primrose plays..a house physician so nearsighted that he sometimes consults closets. **1908** *Daily Chron.* 3 Nov. 4/7 'A Quoi Tient la Superiorité des Anglo-Saxons?' It is in 'house pride'. **1909** *Ibid.* 7 Dec. 1/2 The 'instinct of house-pride' seems almost never entirely wanting in the home studied. **1936** *Punch* 19 Oct. 600/1 The joke of Felix's housepride continues a trifle too long. **1849** C. Brontë *Shirley* II. vii. 169 You are what you call house-proud; you like to have everything handsome about you. **1899** *Daily News* 17 Oct. 4/7 It is a bad thing for the mother of a family if she cannot be a little 'house-proud'. **1932** *Times Lit. Suppl.* 1 Sept. 603/2 His wife, house-proud as most North Country women are. **1960** *Times* 4 Mar. 13/7 Even under discouraging conditions.. you will see ample proof of house-proud families. **1704** in *Essex Inst. Hist. Coll.* (1866) VIII. 223, I was at my L[and-]Lords house raising. **1829** *Vindicator* (Montreal) 22 Dec. 3/2 A man..who, with some others, attended at a house-raising six miles from this town. **1843** *Amer. Pioneer* II. 451 The frequent necessity for united efforts at house-raisings, log-rollings, corn-huskings, &c. **1927** J. D. Freeman *When West was Young* 415 They would.. reach the West in time for a big house-raising which would be given them by the entire neighborhood. **1949** *Time* 2 May 22/2 It was just like an old-fashioned house-raising bee. **1925** *Inter-State Tattler* 27 Feb. 8/2, I am a tamer of wild women and bitterly against house-rent parties. **1926** Whiteman & McBride *Jazz* viii. 177 Big sessions of blues were held in the South among the colored people, the biggest of all at 'house rent stomps' when a negro found himself unable to pay his rent. **1938** [see *down *prep.* 2 e]. **1955** Shapiro & Hentoff *Hear me talkin to Ya* xii. 210 Joe..would bash at numerous functions and house-rent stomps along Carlisle and John Streets. **1956** M. Stearns *Story of Jazz* (1957) xv. 168 A house-rent party, an unstable social phenomenon that was stimulated by Prohibition and made necessary by the Depression. **1964** W. R. Dixon in J. H. Clarke *Harlem* 138 The legends of the house rent parties are legion. **1968** *Blues Unlimited* Nov. 22 There are the more 'functional' Texas and Chicago house-rent pianists. **1952** W. Granville *Dict. Theatr. Terms* 100 *House seats*, free seats given by courtesy of the house. **1810** *Irish Mag.* Feb. 67/2 The true riding housestile is distinguished in every page. **1940** Graves & Hodge *Long Week-end* xxv. 434 The publisher..has discovered..the 'house-style'. **1960** *Design* July 40/1 The present house style is by no means the first evidence of Watneys' interest in design. **1967** L. B. Archer in Wills & Yearsley *Handbk. Managem. Technol.* 128 The range of design activities which can be loosely termed 'house style' design. *Ibid.* 139 *House style*, the characteristic shapes and colours by which the products, paperwork, and property of a firm may be recognized. **1967** *Listener* 17 Aug. 220/2 It is the galleries which tend to have a 'house style'..which mount the most coherent exhibitions. **1905, 1962** House surgeon [see *house physician* above]. **1934** House surgeon [see *house officer* above]. **1787** in O. Browning *Despatches fr. Paris* (1909) I. 207 It is said that the Timbre, the House-Tax, and *le commerce des Grains* will be strongly oppos'd. **1825** Malthus *Diary* 17 June (1966) 232 Complaints of the weight of taxes—Capitation tax. House tax. **1833** J. S. Mill in *Monthly Repos.* VII. 580 A tradesman in Regent Street pays precisely as much house-tax (56l. 13s. 4d.) as the Duke of Devonshire pays for Chatsworth. **1939** *Brit. Birds* XXXIII. 32 Descriptions of some well-known methods such as the..house-trap, and bat-fowling have already been received. **1960** E. Ennion *House on Shore* iv. 37 House traps of one kind or another, aviaries to all appearances, are to be found in many enthusiastic ringers' gardens. **1936** *Discovery* Apr. 99/2 A distinctive Irish house-type. **1953** L. Kuper *Living in Towns* I. 7 A single house-type has been used in such a way as to create variety and interest, by the grouping of the units in different numbers and relationships. **1955** D. Chapman *Home & Social Status* vi. 84 The families living in different house-types have different social and economic characteristics. **1921** *Daily Colonist* (Victoria, B.C.) 1 Apr. 7/1 (Advt.), Extra values in reliable house wares today and Saturday. **1969** *Sears, Roebuck & Co. Spring Catal.* 565/3 Housewares. **1970** *Toronto Daily Star* 24 Sept. 29/2 (Advt.), 'Perma-Glo' 9-pc. Teflon Cookware Set..Housewares. **1971** *Sunday Express* (Johannesburg) 28 Mar. (Home Jrnl.) 9/3 (Advt.), Housewares..Grapefruit knife ..Bathroom Scales. **1841** A. Bache *Fire-Screen* 119 Mrs. Gibbs, a woman who sometimes came to assist in doing house-work, had followed Mrs. Brown into the parlour. **1871** *Rep. Indian Affairs* (U.S.) (1872) 378 While the boys are engaged in out-door work, the girls could be employed in sewing or house-work. **1928** D. H. Lawrence *Phoenix II* (1968) 531 No woman does her housework with real joy unless she is in love. **1896** A. Morrison *Child of Jago* xvii. 177 The old buildings were sold..to the house-wreckers. **1903** *Westm. Gaz.* 7 Sept. 8/2 The hand of the housewrecker. **1936** House-wrecker [see *demolition 3]. **1808** A. Wilson *Amer. Ornith.* I. 133 The House Wren inhabits the whole of the United States, in all of which it is migratory. **1848** [see Wren 1 b]. **1872** *Amer. Naturalist* VI. 275 Here the common 'house' wren is bleached and faded. **1904** E. Glasgow *Deliverance* v. i. 442 A half-finished nest which a house-wren had begun to build. **1961** O. L. Austin *Birds of World* (1962) 248/2 Most familiar of these [cavity-nesting species] is the common House Wren, a widespread species of 30-odd races which ranges from southern Canada south to Cape Horn and the Falkland Islands.

house, *v.*[1] Add: **7.** Also with *up*.

1873 J. H. Beadle *Undevel. West* i. 40 We can house up, you know, and keep warm on the prairie in winter, but we can't house up and keep cool in the timber in the summer.

house-break, *v.* Add: **2.** *trans.* To train (a domestic animal) to be clean in the house. Also *transf.*, to train (a person) to adopt a specified mode of behaviour within a house. Freq. in pa. pple. or ppl. a. *house-broken*. Chiefly *U.S.*

1900 *Daily Chron.* 27 Aug. 7/4 Malcourt, house-broken, runs to heel with the rest. **1905** T. Roosevelt *Let.* 14 May in *Works* (1926) XIX. 487 Skip is housebroken, but he is like a real little Indian. **1922** *Hotel World* 20 May 6/2, I am 'bawled out' by the dog owner saying that the management of such and such a hotel never objects to a house-broke dog being taken care of in a room. **1932** Wodehouse *Hot Water* xii. 207 House-broken husband though he was, he still had an eye for beauty. **1945** J. Steinbeck *Cannery Row* xx. 82 He didn't even house-break her [sc. a puppy]. **1948** G. Vidal *City & Pillar* II. x. 256 We weren't like all these other people here; we weren't housebroken. **1955** W. W. Denlinger *Compl. Boston* 148 Some prospective dog owners fear the ordeal of housebreaking a dog. **1961** Wodehouse *Service with Smile* (1962) ii. 36 You won't mind if I bring a friend with me?..He is house-broken and eats whatever you're having yourself.

housebreaking. Add: Formerly usu. denoting such a crime committed by day, for which *burglary* was the equivalent at night; now applied to such an act committed by day or night. (Further examples.)

1966 *8th Rep. Crim. Law Rev. Comm.* 85 in *Parl. Papers 1966–7* (Cmnd. 2977) XXXIX. 1 The present offence of housebreaking, by day at least, is not regarded as within that class, for many cases are dealt with by fine. **1966** *New Statesman* 7 Oct. 504/2 All this will be the result of abolishing a host of anomalies in the ancient law of burglary and housebreaking (a distinction which..will itself disappear). The Bill makes a bonfire of them. **1973** 'J. Patrick' *Glasgow Gang Observed* iii. 27 Under the law of Scotland 'housebreaking' covers the forced entry of 'any roofed building', i.e. shops, factories, garages, as well as private houses.

house-building. (Later examples.)

1865 Mrs. Stowe *House & Home Papers* 270 Why don't you write a paper on house-building? **1946** *Nature* 5 Oct. 462/1 The various forms of co-operative activity undertaken by the village.., such as road-making, bridge-building or house-building. **1969** *Times* 16 July 22/5 There are doubts about the growth pattern of house-building in this country.

housecraft (hau·s,krɑft). [f. House *sb.*[1] + Craft *sb.* 12.] The art of managing a house; skill in domestic duties. Also *attrib.*

1906 L. H. Yates (*title*) Modern housecraft. **1910** *Daily Chron.* 1 Mar. 9/1 Ignorance of housecraft is the favourite reproach brought against the modern woman. *Ibid.*, St. Martha's College of Housecraft. **1914** J. Collings *Colon. Rural Brit.* I. vi. 116 Among the 'optional' subjects contained in the 'Regulations for the Training of Teachers for Elementary Schools' is that of 'Rural Science' (and 'Housecraft' for girls). **1930** *Birmingham Post* 10 Dec. 4/3 Housecraft classes at the Victoria Institute. **1951** *Archit. Rev.* CIX. 285/1 As housecraft activities take up more space than science, the housecraft centre is on the first floor. **1956** W. H. G. Armytage in D. L. Linton *Sheffield* 206 This was the ancestor of the present handsome College of Housecraft opened at Totley after the Second World War.

housee: see *housey-housey.

household. Add: **8. household appliance,** a piece of equipment (e.g. a vacuum cleaner) used in the house; **household book,** a book in which household accounts are noted; **household effects,** the movable contents of a house; **household linen,** linen for the bedroom, table, etc.; **household management,** the art of running a house; **household name,** a name familiar to everyone; **household science** orig. *N. Amer.*, = **domestic science*; **household snake** = *house snake 2.

1926 *Encycl. Brit.* II. 375/2 The outstanding feature in the development of household appliances..has been the greatly extended use of those operated electrically. 1935 *Discovery* Sept. 247/2 The fame of Norwich is also spread by its mustard and starch, electrical machinery and household appliances, aircraft, ironwork and other progressive industries. 1957 *Encycl. Brit.* XI. 813/2 Modern household appliances can be divided into two main headings, electrical and nonelectrical. 1599 M. Hoby *Diary* 18 Feb. (1930) 104 Then to supper: after, I looked of the Houshould book, and then went to priuat praers. 1780 S. Pegge in *Forme of Cury* p. xxxv, The present dean of Carlisle, to whom I stand indebted for his useful notes on the Northumberland-Household Book. 1866 Trollope *Belton Est.* (ed. 3) II. vi. 145 Though she could not succeed in making retrenchments, she could and did succeed in keeping the household books. 1931 S. Jameson *Richer Dust* xviii. 503 Mrs James, pen stirring the clotted ink at the bottom of the bottle, bent close over the pages of her Household Book. 1891 [see Effect sb. 4 b]. 1933 *Burlington Mag.* May p. xv/2 The English furniture generally is of better quality than that associated with 'household effects'. 1642 Lady Sussex *Let.* in F. P. Verney *Memoirs* (1892) I. xi. 253 She gives all her linen to to of hur grandchildren not naminge hur householde linen, but in generly hur linen. 1811 Jane Austen *Sense & Sens.* I. v. 56 Household linen, plate, china, and books. 1967 E. Short *Embroidery & Fabric Collage* iii. 72 (*caption*) A basically simple design which could be adapted to almost any article of household linen. 1742 Richardson *Pamela* (ed. 3) III. ix. 50 Let your Servants, under your Direction, do all that relates to Houshold Management. 1801 Jane Austen *Let.* 14 Jan. (1932) I. 111 It will be an amusement to Mary to superintend their Household management, and abuse them for expense. 1861 Mrs. Beeton (*title*) The book of household management. 1862 Tennyson *Idylls of King, Dedication* l. 40 A Prince indeed, Beyond all titles, and a household name, Hereafter, thro' all times, Albert the Good. 1907 W. Russell *Arterial Hypertonus* v. 34 The names of Claude Bernard, Brown-Séquard, Waller, and Schiff are 'household names' in this connection. 1958 *Times* 27 Feb. 11/2 Our first encounter with so many household names in his list which were then unknown. 1969 *New Yorker* 29 Nov. 166/2 Spiro Agnew was then a household name only in his own household. 1970 *Times* 11 Mar. 21/1 (Advt.), We produce packs for a lot of household names. 1938 *Univ. Toronto Cal. 1938–39* (verso of title-p.), In the Faculty of Household Science, the University offers courses leading to the degree of Bachelor of Household Science. 1953 *Univ. London Cal. 1953–54* 464 Queen Elizabeth College... Admitted in 1928 as a School of the University for the B.Sc. Degree in Household and Social Science and for the Degree of B.Sc. (Nutrition) and B.Sc. (Household Science) in 1953. 1963 F. F. Laidler *Gloss. Home Econ. Educ.* 46 *Household science*, a comprehensive study of the scientific, technological and sociological aspects of the household. 1894 W. Crooke *Pop. Relig. & Folk-Lore N. India* vii. 276 (*heading*) The household snake. 1921 A. Evans *Palace of Minos* I. 509 Such a household snake is known, indeed, as *domachitsa* or 'house-mother'. *Ibid.*, In its homely origin, from the religious tending of the household snake, the cult itself may be supposed to be of old indigenous tradition. 1962 R. W. Hutchinson *Prehist. Crete* viii. 208 The cult of the household snake has not entirely died out even now.

housekeep, *v.* (Later examples.)

1900 Kipling in *Daily Express* 26 June 4/5 One or other of them would housekeep for him the year round. 1947 N. Marsh *Final Curtain* ii. 26 Millamant who, since the death of her husband, has house-kept for her father-in-law. 1954 *Encounter* May 20/1 The widow whom she housekeeps for.

housekeeping, *sb.* Add: **1.** Also *transf.*

1966 *Listener* 29 Sept. 448/1, I do not wish to consider ..this..in terms of national housekeeping. 1969 *Times* 2 June (Fire Protection Suppl.) p. i, One of the most important things in industrial fire protection..is what we call good housekeeping—cleanliness, tidiness, the careful removal of rubbish.

3. Used *attrib.* of a rented holiday cabin or cottage equipped for light housekeeping; similarly, *housekeeping rooms, suite*, etc., furnished accommodation with cooking facilities. *N. Amer.*

1935 M. M. Atwater *Murder in Midsummer* xx. 190 The house in which she had her tiny suite of 'housekeeping rooms'. 1968 *Globe & Mail* (Toronto) 15 Jan. 24/3 (Advt.), In sunny warm Grand Cayman. Furnished for 4, housekeeping ocean front cottages. 1968 W. Muir *Belonging* xx. 285 We had booked 'a house-keeping suite' without knowing what that was and found a sitting-room and bedroom..with a small kitchenette and a shower. 1970 *Globe & Mail* (Toronto) 26 Sept. 47/3 (Advt.), Housekeeping cabins fully equipped. 1972 *Amer. Automobile Assoc. Tour Book 1972–73* (Southwestern states) 14 (Advt.), Kinnikinnik Motor Lodge... Featuring 1 & 2 Rm. Units... Large Housekeeping Cottage. 1972 *Prince Edward Island Tourist Accommodations* 49 (Advt.), Brudenell Resort... Housekeeping chalets: single $16 to $20, double $20 to $28, extra person $4. 1973 *Kingston* (Ont.) *Whig-Standard* 5 Mar. 28 (Advt.), Single furnished light housekeeping room, refrigerator, stove and sink. *Ibid.* 8 Mar. 37/9 (Advt.), Near Kingston General Hospital, Queen's, on Union Street. All housekeeping facilities.

4. Short for **housekeeping allowance, money.*

1946 G. Kersh *Clean, Bright & Slightly Oiled* ii. 16 He never give her no housekeeping. 1966 'O. Mills' *Enemies o' Bride* ii. 20 'Do you want some money?' 'No, it doesn't matter. I haven't got the housekeeping off the Gaffer yet, that's all.' 1970 J. Porter *Rather Common Sort of Crime* ii. 25 You'll go throwing your money around but if I ask you for a bit more housekeeping, that's a different story.

5. Those operations of a computer, organization, etc., which make its work possible but do not directly constitute its performance. Freq. *attrib.*

1956 *Jrnl. Assoc. Computing Machinery* III. 269 Most of the errors..turned up in the logical or housekeeping operations, building loops, tallying, etc. 1958 Gotlieb & Hume *High-Speed Data Processing* v. 80 In programming a problem for a data-processing machine many of the instructions are not of an arithmetic nature but fall into a class called organizational or housekeeping instructions. 1962 *IBM Systems Jrnl.* I. 72 The necessary routines such as housekeeping, timekeeping, utility routines, association of equipment, etc. 1966 *Sunday Mail Mag.* (Brisbane) 2 Jan. 12/6 The airline industry calls this process 'housekeeping'. It [*sc.* inspection of aircraft fuel tanks for traces of fungus, etc.] has to be done with painstaking thoroughness. 1967 *Oxford Computer Explained* 15 Housekeeping Suites are run on demand, usually twice a week. 1970 A. Chandor et al. *Dict. Computers* 182 Housekeeping functions include the setting of entry conditions, clearing areas of store if the program expects these to be set to some initial condition,..performing standard input/output routines. 1971 *Sci. Amer.* June 66/1 The computer provides status information on the equipment and performs many routine 'housekeeping' chores, such as adjusting currents in the spectrometer magnets and logging beam currents and other quantities of interest. 1971 *Nature* 27 Aug. 662/3 Mechanized methods of library housekeeping are also considered.

6. *attrib.* and *Comb.*, as *housekeeping book, goods, keys*; **housekeeping allowance, money,** a sum of money regularly set aside for housekeeping expenses.

1914 A. Bennett *Price of Love* xvi. 309 She had received no housekeeping allowance for more than a week. 1965 in P. Jennings *Living Village* (1968) 123 Housekeeping allowance £7. 1849 Dickens *Dav. Copp.* (1850) i. 7, I kept my housekeeping-book regularly, and balanced it with Mr. Copperfield every night. 1930 A. Bennett *Imperial Palace* xxx. 200 The figures of the housekeeping-books had..startled her. 1899 *Daily News* 26 June 10/6 There is a satisfactory trade doing in damasks, and housekeeping goods. 1900 *Ibid.* 17 Sept. 2/7 Housekeeping goods and damasks. 1852 Dickens *Bleak Ho.* (1853) vi. 48 The housekeeping keys, Miss. 1885 C. M. Yonge *Two Sides of Shield* I. i. 12 She gave him money—not father's housekeeping money, but what she got for herself by writing. 1962 E. O'Brien *Lonely Girl* ix. 107 My father forgot about housekeeping money when he drank. 1966 L. Southworth *Felon in Disguise* vi. 86 Her husband had provided her with sufficient housekeeping money for her immediate needs.

housemaid. Add: **c.** *housemaid's knee* (examples).

1831 *Lond. Med. & Physical Jrnl.* LXVII. 42 The third of the cases..was of the common description: the bursa very slightly thickened, but distended with fluid, uneasy, and painful, the most ordinary state of the housemaid's knee. 1889 J. K. Jerome *Three Men in Boat* i. 3 The only malady I could conclude I had not got was housemaid's knee. 1912 Adami & McCrae *Text-bk. Path.* xii. 678 'Housemaid's knee', 'miner's elbow', and 'weaver's bottom' are all well-known forms of bursitis. 1971 *Daily Tel.* 19 Apr. 2 Occupational rheumatism known by a variety of names like weaver's bottom, housemaid's knee, and golfer's shoulder, is costing the country ..millions of pounds a year.

d. Other collocations containing the possessive form, as *housemaid's box*, a box with a container for ashes and a tray above for brushes, dusters, etc.; *housemaid's closet, cupboard*, a cupboard or small room where brooms, cleaning materials, etc., are kept, slops emptied, etc.; *housemaid's gloves*, gloves worn for protection when cleaning grates, etc.

1861 Mrs. Beeton *Bk. Househ. Managem.* 988 Her housemaid's box, containing black-lead brushes..and all utensils necessary for cleaning a grate. 1909 H. G. Wells *Tono-Bungay* III. i. 266 Housemaid's boxes it'll be a pleasure to fall over. 1950 J. Cannan *Murder Included* viii. 146 Sylvia, carrying a housemaid's box, fled at their approach. 1857 C. M. Yonge *Dynevor Terr.* I. xiv. 229 Mrs. Martha might some day let her stand in the housemaid's closet, to behold her idol issue forth in the full glory of an evening dress. 1873 —— *Pillars of House* II. xix. 162 From a housemaid's closet half-way up, Alda was bringing to light a basin. 1906 M. H. Baillie-Scott *Houses & Gardens* x. 31 The housemaid's closet with its slop sink and spaces for pails and brooms. 1911 L. Weaver *House & its Equipment* 103 In small houses where there is not a housemaid's closet. 1950 J. Cannan *Murder Included* iii. 55 There's a sort of housemaids' closet just at the head of the main staircase, and..they had an electric heater put in there so that the housemaid could make the early morning tea for everybody. 1873 L. Troubridge *Life amongst Troubridges* (1966) vi. 52 He..spent nearly all the evening sulking and sobbing in the sink (in the housemaid's cupboard). 1952 G. Raverat *Period Piece* iv. 68 They..used to cry..in the housemaids' cupboard. 1863 Dickens *Uncomm. Trav.* (1866) xxvii. 194/2 A tall straight sallow lady..who does her household work in housemaid's gloves. 1962 G. Avery *Greatest Gresham* vii. 132 Mabel..wearing housemaid's gloves, was cleaning the brass.

houseman. Add: **4.** *U.S. slang.* A burglar.

1904 'No. 1500' *Life in Sing Sing* 255/2 *Houseman*, a burglar. 1911 G. Bronson Howard *Enemy to Society* v. 147 Say a 'house man' or a 'sneak' or a 'second-story' man or a 'peteman'—anything but a 'cracksman'. 1924 G. C. Henderson *Keys to Crookdom* App. B. 408 *Houseman* burglar, prowler.

5. A person responsible for general duties in a private house, hotel, etc.

1934 in Webster. 1960 'N. Shute' *Trustee from Toolroom* 277 She went through to tell the house-man to clear away the tea. 1961 *Times* 20 July 3/1 Houseman-Handyman..Valet/Houseman. 1968 *Daily Tel.* (Colour Suppl.) 13 Dec. 24/4 His marvellous home..with a Spanish houseman and an Italian cook.

6. A house physician or house surgeon; now usually (in Great Britain) a junior doctor who, having passed all the qualifying examinations, is working for a time in a hospital under a certain amount of supervision prior to being fully registered as an independent medical practitioner.

1938 *Amer. Speech* XIII. 228/2 House men (residents and interns) and 'O.R.' (operating room) nurses take their turn being on call for emergency operations. 1942 M. Dickens *One Pair of Feet* vi. 83, I hate these Housemen—stuck-up little boys, they think they know everything. 1962 *Becoming a Doctor* (B.M.A.) 27 The law requires that newly qualified doctors shall be 'provisionally registered' for a year immediately after they have taken their Finals. They are qualified doctors, but full registration must wait upon the results of two obligatory six-month appointments in hospitals specially approved for the purpose. These hospital appointments are in medicine, surgery, or obstetrics—that is to say, in any two of these three. During this time the young doctor is what is called a 'houseman' or, more correctly, a junior house officer. 1962 D. Margerson *Med. as Career* ix. 57 What are the houseman's duties? The *complete* care of the patients. When they come into hospital he takes down the full story of their illnesses; examines them; arranges for the special tests which aid diagnosis; and prescribes the initial treatment. At least once a day.. the houseman visits all his patients. *Ibid.*, The houseman is the patient's *personal* doctor while he is in hospital. Other doctors—the consultants and his registrar assistants—will be involved in diagnosis and treatment, but none will be as close to the patient..as the houseman. 1965 N. Wymer *Behind Scenes Hospital* iii. 31 The registrars, in turn, are assisted by the house physicians and house surgeons, who, unlike their seniors, live in the hospital and therefore are known as 'housemen'. 1970 *Times* 25 Nov. 11/4 Either consultant or houseman may be called from the clinic for consultation.

Hence **hou·semanship,** the state or position of being a houseman (sense *6).

1969 *Guardian* 10 Mar. 5/4 After six years of concentrated study..the girl is hurled into her first housemanship. The pressure of her first hospital can terrify her. 1972 *Times* 31 Oct. 6/8 Many [doctors] will be completing their housemanship this year.

housemaster. Add: **2.** Also in some day schools, and *transf.*

1963 R. Pedley *Comprehensive School* iii. 124 The house in a day school simply has not got this basis... Take, for example, a housemaster who does quite a lot of administration. 1965 H. Davies *Culture & Grammar School* v. 85 Where senior house masters exist..they should be able to exercise an effective supervision over the boys in their care. 1966 *Listener* 10 Mar. 342/2 We continue to put some offenders..under the care of individual probation officers and Borstal housemasters.

Hence **hou·semasterly** *adj.*, like or characteristic of a housemaster.

1961 A. Wilson *Old Men at Zoo* ii. 112, I should have to hear myself speak in these housemasterly tones many times before I had done with the business. 1963 *Times Lit. Suppl.* 8 Feb. 87/2 The effect is at once toughly moral and academic, a house-masterly harangue.

house-mistress. Add: **2.** At a girls' school, a person responsible for certain aspects of the supervision of the pupils; in a boarding school, the mistress in charge of one of the boarding houses.

1912 A. Brazil *New Girl at St. Chad's* iii. 42 Her house-mistress would not have been ready to overlook any deficiency in punctuality. 1961 *Evening Standard* 26 July 25/6 (Advt.), Education Committee..Housemistress (non-teaching) required..Technical School for Girls. 1972 *Daily Tel.* 24 Nov. 6/6, I shall continue as housemistress and shall teach mathematics and religious education. 1973 *Radio Times* 8 Nov. 81/1 Miss Gee, housemistress of Guldeford House at Benenden School, recalls Princess Anne's time there.

house-mother. Add: (Earlier and later examples.) Also in extended uses.

1834 Carlyle *Sart. Res.* II. i, in *Fraser's Mag.* Feb. 177/2 The good Gretchen..hovered round him, as only a true housemother can: assiduously she cooked and sewed and scoured for him. 1936 *Variety* 8 July 46/1 The cynical college man who dominates his fair-haired pal, the prim house mother—they're all here and more. 1959 *Tablet* 7 Mar. 240/1 Resident Roman Catholic Housemother required to act as Deputy to Housemother-in-charge. 1970 *Observer* 25 Jan. 6/5 The house mother—a woman who calls daily to look after a family in their own home. 1971 'M. Innes' *Awkward Lie* iv. 71 Calling young women house-mothers must be one of Overbury [School]'s notably few concessions to the march of time.

house snake. For entry s.v. House sb.^{1} 23 substitute: **1.** One of several North American snakes belonging to the genera *Elaphe* and *Lampropeltis*.

1807 in *Mass. Hist. Soc. Coll.* (1815) 2nd Ser. III. 54 The milk or house snake, speckled like a rattlesnake. 1884 'Mark Twain' *Huck. Finn* xxxix. 396 We went for

the snakes, and grabbed a couple of dozen garters and house-snakes. 1958 R. Conant *Field Guide Reptiles U.S.* 171 Sometimes called 'house snake', but 'barn snake' would be more descriptive, for it would reflect the frequency with which farm buildings are entered in search of rodents.

2. A snake kept as a household god.

[1742 C. Owen *Ess. towards Nat. Hist. Serpents* 218 Ælian speaks of domestick Serpents, that were in the Houses of the Egyptians, and look'd upon as household Gods. *Ibid.*, The Arabians reputed Serpents sacred Beings... They take them into their Houses, feed and worship them as Genii, or Guardians of the Place. 1868 J. Fergusson *Tree & Serpent Worship* 25 There are house serpents which are accounted in the northern parts of Sweden as household gods.] 1894 W. Crooke *Pop. Relig. & Folk-Lore N. India* vii. 276 Should rain drive the house snake out of his hole, he is worshipped. 1935 A. Evans *Palace of Minos* IV. I. 153/2 When the house snakes died the master and mistress of the house died too. 1962 R. W. Hutchinson *Prehist. Crete* viii. 124 It was the house snake that was fed and revered as the genius, the guardian angel of the house. 1965 R. & D. Morris *Men & Snakes* ii. 47 A gallant housesnake..refused to desert the Macedonian village of Kalenova when the human inhabitants fled.

3. A southern African snake of the genus *Boædon*.

1907 R. L. Ditmars *Reptile Bk.* xxxii. 300 (*heading*) The Red Coluber; Corn Snake; Red Chicken Snake; Mouse Snake; Scarlet Racer; House Snake *Coluber guttatus*, (Linn.). 1911 *East London* (Cape Province) *Dispatch* 1 Sept. 7 These house snakes are more serviceable to man than cats, for they can follow rats and mice into their hiding places. 1932 *Discovery* XIII. 364/2 The brown house snake, *Boodon* [sic] *lineatus*, is far more efficient than a cat in keeping down rats and mice, as it penetrates into their holes and destroys the young. 1962 V. F. M. Fitz-Simons *Snakes S. Afr.* 123/1 In search of its prey it is often attracted to, and found in and about, human settlements, and has thus got the name of House Snake.

house-top. Add: **b.** *fig.* A public place; esp. (with allusion to Luke xii. 3) in phrase *to proclaim, declare*, or *cry on* or *from the house-top(s)*, to make public, to proclaim so that everyone knows. Also *attrib.*

1870 *Brewer's Dict. Phr. & Fable* 712/1 *Proclaim on the housetop*, to proclaim or make known to every one; to blab in public. 1895 *Ibid.* (rev. ed.) 632/1 To cry from the house-top. To proclaim (it) from the house-top. 1899 *Westm. Gaz.* 15 Nov. 3/2 The policy of house-top virtue. 1906 J. Jackson tr. *Thoughts of Marcus Aurelius* 35 A man who has done a good deed should be like a horse that has run its race..: in other words, he ought not to proclaim it from the house-tops. 1911 C. G. Robertson *Eng. under Hanoverians* II. iv. 345 To cry on the housetops that it was merry in England before the new industry came up. 1931 L. Birch *Pyramid* xviii. 233 Someone should come forward and cry from the housetops that all friendships that are romantic are not, for that reason, immoral. 1934 'G. Orwell' *Burmese Days* iii. 52 My friend, you do not think that. If truly you disapprove of the British Empire, you would not be talking of it privately here. You would be proclaiming from the housetops. 1955 L. P. Hartley *Perfect Woman* xxiii. 200 Don't worry; I didn't expect you to proclaim it on the house-tops, and I shan't either.

hou·se-train, *v.* [Train v.^{1}] *trans.* To train (a domestic animal or infant) to be clean in the house. Also *transf.*, esp. = *house-break v. 2 *transf.*; and *fig.*

1924 W. J. Locke *Coming of Amos* vii. 89 If you hadn't been house-trained by your excellent mother I should say that you've been making a beast of a night of it. 1928 *Punch* 25 Apr. 455/1 It [*sc.* a cheque] is house-trained and was clean about the bank. 1930 Kipling *Thy Servant a Dog* 70 It are that dash-Toby-Dog! C'm with, and house-train him! 1937 A. Huxley *Ends & Means* xii. 180 The work of the late Dr. Suttie, whose book, *The Origins of Love and Hatred*, contains an interesting chapter on the effects of early house-training upon the emotional life of human beings. 1939 S. Spender tr. *Toller's Pastor Hall* I. 41 Goethe wasn't quite house-trained either, you know. First place he was a Freemason, and second too much of a cosmopolitan. 1944 N. Streatfeild *Myra Carrol* 100 Birds aren't ever house trained. 1950 B. Russell *Unpop. Ess.* vii. 127 Some boys never learn to be what, in animals, is called 'house-trained'. 1955 M. Hastings *Cork & Serpent* xvii. 243 After thirty years of marriage, I haven't been able to house-train him yet. 1956 A. Huxley *Adonis & Alphabet* 155 It is easy to house-train a cat or a dog. 1958 [see *billetee]. 1959 *Elizabethan* Apr. 22/2 Mr. Rochester has a lot of the Wild Man in him... But he is also a good deal more house-trained than Heathcliff. 1961 *Listener* 26 Oct. 645/2 We house-train puppies on a similar basis. 1969 Auden *City without Walls* 63 Rumours ran through the city That the Tsar's bodyguard Was not house-trained. 1969 *Times* 11 July 20/5 (Advt.), Capable girl, willing help run..Highland fishing cottage and two house trained small boys. 1973 J. Wainwright *Devil You Don't* 37 You house-trained him, when he was a kid. Trained him to use the totty. Trained him in table-manners.

housey-housey (hau·si͵hau·si). Also housie-housie, housy-housy and abbrev. housey, housie, (*rare*) housee. [f. *house sb.^{1} 9 c + -y^6.] A later name for the game of 'house' (see also quot. 1964).

1936 F. Richards *Old-Soldier Sahib* iii. 69 To draw a crowd they would shout: 'Housee, housee, housee!' 1937 Partridge *Dict. Slang* 410/2 Housey-housey! 1938 in Mencken *Amer. Lang.* Suppl. (1945) I. 461/1 The game so popular in army circles in Hong Kong under the name of *tombola* is now sweeping South London as a craze called *housey-housey*. It is played for the most part by housewives who are attracted to open-door booths by a glittering display of cutlery and chromium-plated clocks. 1945 *Penguin New Writing* XXVI. 30 He was shouting numbers hoarsely from the Housey-Housey stall. 1949 S. P. Llewellyn *Troopships* 3 Men playing housie-housie (tombola). 1949 E. de Mauny *Huntsman in Career* 163 Someone started up a 'housey' school. 1957 J. Frame *Owls do Cry* 42 Like the woman down the road.. having parties every Saturday night with housey-housey and drink. 1960 *Times* (Canada Suppl.) 16 Mar. p. xv/5 Gala dances and housie-housie at night, bring people into contact. 1964 A. Wykes *Gambling* x. 249 The call used to assemble a group for a game of house was 'housey-housey'. 1967 *Stage* 2 Mar. 4/1 'Housey'-addicts never had it so good! 1971 B. W. Aldiss *Soldier Erect* 30 Desperate till now to get off the hated boat with its hated routines of exercise and housey-housey, I was suddenly reluctant to leave the shelter of a familiar place.

housing, *sb.*^{1} Add: **1.** (Later examples.)

1848 *Rep. U.S. Comm. Patents 1847* 168 The tobacco was hardly worth the housing. 1850 *Rep. Comm. Patents: Agric. 1849* (U.S. Dept. Agric.) 321 This mode is pursued easily by simply splitting..the plant..before it is cut down for housing. 1930 A. P. Herbert *Water Gipsies* viii. 96 Honest John Raven worked hard and late at the office, and sometimes brought home papers with him, or if not he talked 'housing' or threw off little anecdotes about 'National Insurance'. 1973 *Oxford Times* 9 Nov. 9/4 One problem we both have is of course that of housing.

6. b. *spec.* A massive metal frame or pillar that supports one end of a set of rolls in a rolling mill.

1869 H. S. Osborn *Metallurgy of Iron & Steel* III. iv. 771 (*heading*) View of a housing, and section of rollers and foundation. 1938 *Extension Course Ferrous Metall.* (Penn. State Coll.) (ed. 2) II. xii. 203 The housings are very heavy cast iron or steel frames, designed to support the rolls and adjusting mechanism securely in place. 1965 M. H. T. Alford tr. *Tselikov & Smirnov's Rolling Mills* v. 135 The weight of housings is taken by their feet which bear on the foundations through girders.

d. A structure that supports and encloses the bearings at the end of an axle or shaft; a journal-box. Hence more widely, a rigid case or cover that encloses and protects an axle or any other mechanism or piece of apparatus.

1889 *Cent. Dict.*, *Housing*, a housing-box; a journal-box. 1915 V. W. Pagé *Model T Ford Car* iii. 111 The bevel pinion meshes with a large bevel gear..which is attached to the differential housing. 1916 J. E. Homans *Automobile Handbk.* xvi. 177 The housings of the oil pump, water pump, [etc.]. 1922 L. Bell *Telescope* x. 242 The housing, just big enough to take in the equatorial with the tube turned low, opens on the south side and then can be rolled northward on its track..well clear of the instrument. 1935 *Times* 2 Oct. 6/5 Helical springs working in a closed oil-filled housing. 1949 Frazee & Bedell *Automotive Fundamentals* viii. 458 In conventional automobiles and trucks, the rear-axle housing supports the rear-end load on the road wheels. In addition, the housing contains the driving mechanism and carries the rear-wheel bearings. 1958 *Times* 9 May 13/7 The safety housings on the toggle switches. 1962 *Which? Car Suppl.* Oct. 139/2 The water pump to heater hose was chafing on thermostat housing. 1968 P. H. Smith *Triumph Autobk. Two* (ed. 2) vii. 101/1 The outer ends of the axle-shafts are carried in trunnion mounted bearing housings which are attached to vertical pressed steel links. 1972 *Sci. Amer.* Jan. 68 (Advt.), A large diascopic illuminator and a set of episcopic illuminators; a 4 × 5" bellows..; a mirror reflex housing; and a sensitive spot meter.

7. *housing association, problem, project, question, scheme, site, unit*; **housing development** [*development 3 d], the act or process of planning and building a group of houses (a **housing estate**) and associated services on a site; **housing list,** a waiting-list for council houses; **housing screw,** a screw by means of which the rollers in a rolling mill can be adjusted and the thickness of the metal controlled.

1957 *Act* 5 & 6 *Eliz. II* c. lvi. §189 'Housing association' means a society, body of trustees or company established for the purpose of..facilitating or encouraging the construction or improvement of houses. 1970 *Internat. & Compar. Law Q.* XIX. II. 205 Discussion inevitably ranged into such related topics as..housing associations and co-operatives. 1971 P. Gresswell *Environment* 131 The main object of Housing Associations is to provide houses to let for local people. 1951 *Amer. City* Sept. 104/1 The builders of Lakewood Park..envisioned the building of a community, including parks, play grounds, schools, churches, and a major shopping center. More than 30,000 people are already living in the 7,200 homes.. that have been completed. There is a fact more than bigness, however, that puts this housing development among the pioneers. 1959 N. Mailer *Advts. for Myself* (1961) 153 He looks down six stories into the giant quadrangle of the Queens housing development. 1920 *Times* 18 Nov. 9/4 It will afford..much-needed means of access to the L.C.C. Housing Estate at Dagenham. 1931 *Economist* 12 Dec. 1136/2 Weaker [building] societies may become mere appanages of speculative builders, agents, and other interests concerned with 'housing-estate' exploitation. 1936 W. Holtby *South Riding* I. vi. 68 More land for housing estates. 1956 J. M. Mogey *Family & Neighbourhood* 73 The housing-estate family developed a much more critical attitude. 1964 M. Argyle *Psychol. & Social Probl.* xv. 183 The Census of 1961 showed that more people than before live in suburbs and housing estates. 1952 M. Laski *Village* xiii. 181 Roy had put his name down on the housing-list at the Town Hall. 1968 L. Berg *Risinghill* 46 The bannisters broke and she fell down the stairs and broke her leg. She's been on the housing list twenty-five years. 1899 S. A. Barnett in H. Barnett *Canon Barnett* (1918) II. xlix. 311 The housing problem..is bound up with the industrial problem, with the education problem, with the social problem, and with the religious problem. 1911 G. B. Shaw *Getting Married* Pref. 117 They were content to have the whole national housing problem treated on a basis of one room for two people. 1947 *Tribune* 24 Jan. 12/1 They began talking about the housing problem. 1973 *Guardian* 18 May 16/1 One of the reasons for Britain's housing problem is that many houses are in the wrong place. 1938 *Jrnl. Social Forces* May 528 (*title*) Some eighteenth century housing projects in France. 1967 S. Sontag *Death Kit* (1968) 10 All the housing projects are unadorned boxes built of brick. 1970 *Globe & Mail* (Toronto) 25 Sept. B7/5 A $1,644,512 loan for a 139-unit apartment house in a Hamilton low rental housing project. 1899 *Daily News* 19 July 5/5 Milner had charge of the Housing Question after the boom was launched. 1903 W. Thompson *Housing Handbk.* p. v, The past twelve years, spent in..the promotion of a number of housing schemes. 1918 Lloyd George *Slings & Arrows* (1929) 199 Your housing schemes must, in the majority of cases, be schemes outside the house. 1966 J. Betjeman *High & Low* 22 We pounded through a housing scheme. 1874 *Jrnl. Iron & Steel Inst.* I. 352 The top and bottom rolls are simultaneously raised or lowered by four housing screws. 1951 *Engineering* 5 Jan. 2/1 The housing screws are of 0·5 per cent. forged carbon steel. 1926 *Daily Chron.* 13 May 2/7 Work on a Bristol housing site has stopped owing to transport difficulties. 1972 *Country Life* 30 Nov. 1487/1 Every housing site has its own unique character. 1951 M. McLuhan *Mech. Bride* 38/2 It means an extension of housing units. 1970 G. Greer *Female Eunuch* 227 No clever orientation of clean and efficient housing units..can break down the suspicion that the Oedipal unit feels.

Hova (hōu·vă, hɒ·vă). Also **Ovah.** [Malagasy.] **a.** A member of the dominant race of the Malagasy Republic (formerly Madagascar); also, in restricted use, one of the middle class, as distinct from the nobles and the slaves. Also *collect.* **b.** The language of this race. Also *attrib.*

1839 *Penny Cycl.* XIV. 259/2 The Ovahs, who inhabit the elevated plains in the interior, are in height rather above the European standard. 1858 W. Ellis *Three Visits to Madagascar* 468 In Madagascar itself different dialects exist. The spoken language of the Hovas..differs from that on the coasts. 1883 *Encycl. Brit.* XV. 171/1 The Hòva, who occupy the central province of Imérina. *Ibid.* 172/2 The majority of Hòva houses are built of layers of the hard red clay of the country. 1887 *Fortn. Rev.* Mar. 435 Only a few months ago French politicians..called the Hovas barbarians. 1911 J. G. Frazer *Golden Bough: Magic Art* (ed. 3) I. vii. 397 The Hovas and other tribes of Madagascar. 1915 J. Sibree *Naturalist in Madagascar* xxii. 299 Their [*sc.* Sàkalàva] language presents a good deal of difference from the Hova form of Malagasy, both in vocabulary and in pronunciation, yet the groundwork and the grammar is essentially the same. 1972 A. Sillery *Africa* (ed. 2) ii. 114 The most important native people, as well as the most numerous and advanced, are the Merina or Ambaniandro, commonly called Hova, whose homeland is Imerina on the plateau near Antananarivo.

hoven, *ppl. a.* Add: Also as *sb.*

1845 S. Judd *Margaret* II. v. 284 Glad you got through with the pock so well...its worse than horn-ail, hoven or core. 1877 *Rep. Vermont Dairym. Assoc.* VIII. 107 Hoven in cattle is caused by over-feeding upon succulent food like green clover. 1878 [see *bloat sb. 1 b]. 1891 [see *blowing vbl. sb.^{1} 2 d]. 1902 *Phil. Trans. R. Soc.* CXCIX. A. 400 The symptoms of 'hoven' are not unlike those of prussic acid poisoning. 1962 [see *bloat sb. 1 b].

hover, *sb.* Add: **1. a.** Also, a state of hovering. (Further examples.)

1961 *Hovering Craft & Hydrofoil* Oct. 32/2 Floatation and sea-keeping capability while floating and during transition from hover to high forward speed. 1962 *Air-Cushion Vehicles* July 16 The skirts have withstood the early trials well, most of the wear having occurred during tethered hovers on hard standing. 1967 B. W. McCormick *Aerodynamics of V/Stol Flight* v. 106 The helicopter rotor in hover or in vertical climb is relatively easy to analyze. 1969 I. Kemp *Brit. G.I. in Vietnam* iii. 61 Major Bracken eased back on the cyclic..and our speed dropped until we were almost in a hover. 1971 *Physics Bull.* Nov. 655/3 With careful design the tilting rotor propeller can be efficient in both hover and cruise.

3. Also, a shelter used in the brooding of chickens.

1907 *Elem. Sch. Teacher* Mar. 410 A hover which was made of felt was hung in the brooder. 1936 *Nature* 3 Oct. 583/1 The merits of the various systems of hovers and battery brooders are discussed.

3*. 'A floating island, or bed of reeds' (E.D.D.). Chiefly Norfolk *dial.*

1892 P. H. Emerson *Son of Fens* xiv. 120 'Have you got enough damming boards for another dam?' 'Yes... We shan't want many; there's a hover there.' 1893 H. Cozens-Hardy *Broad Norfolk* 77 *Hover*, a floating island. 1955 *Times* 3 Aug. 10/2 Now the lesser bulrush which grows on the 'hover' (excellent word to denote a table of floating vegetation) round the edge of the open water is all a-quiver.

hover, *v.*^{1} Add: **1. c.** Of a helicopter or other aircraft: to remain stationary in the air, relative to the ground.

1892 *Railroad & Engin. Jrnl.* Nov. 508/2 With 60 turns of the rubber the apparatus would just hold its own—i.e., hover in the same spot, against a wind of

9 ft. per second. 1926 J. L. Pritchard *Bk. Aeroplane* ix. 177 It had distinct possibilities of achieving what the helicopter sets out to achieve, rising, descending vertically in still air and hovering. 1935 P. H. Sumner *Aircraft* II. 63 To be of any practical use the helicopter aircraft must be capable of forward motion in addition to rising vertically and hovering. 1972 *Daily Tel.* 3 July 2 (*caption*) A Sea King helicopter..hovering over Rockall in the Atlantic.

d. Of a hovercraft: to be supported on its air-cushion, esp. while stationary; also *transf.*, to travel in a hovercraft.

1962 *Hovering Craft & Hydrofoil* Aug.–Sept. 20/1 Hovering at speeds below 10 knots..over water, the depression in the water surface beneath the craft gives a measure of 'keel' effect. 1962 *Air-Cushion Vehicles* Oct. 81/2 The rudder, however, is ineffective at low speeds or while hovering. 1967 *Jane's Surface Skimmer Systems 1967–68* 31/2 Echo sounding transducers..will remain immersed whether the craft is 'hovering' or underway. 1968 *Nature* 10 Aug. 549/1 (*heading*) Hover over from Dover. 1970 *Motoring Which?* July 111/2 A control stick adjusted the airflow from the fan, to make the craft hover or go forwards.

6. To maintain in a hovering state.

1967 B. W. McCormick *Aerodynamics of V/Stol Flight* v. 162 The gross weight that can be hovered at the power available is found by correcting the thrust for download.. and for overlap. 1969 *Aeroplane* LXXIII. 708/2 In the case of the rescue system preferred by the USA Coast Guard..the helicopter is hovered in contact with the water and a working platform is extended from the craft.

hover- (hɒ·vəɪ), the first element of *hovercraft, prefixed to other sbs. to denote things of a similar form or serving a similar purpose to the thing denoted by the sb., but having some connection with hovercraft or their principle of operation. **a.** Used to form the names of vehicles and other things that utilize an air-cushion as a means of support, as *hoverbus, -car, -ferry, -kiln, -liner, -pallet, -ship, -train, -truck, -vehicle.*

1962 *Daily Tel.* 1 Mar. 17/1 (*heading*) Hover-buses planned for south coast. 1961 *New Scientist* 31 Aug. 503 What they have in mind is a streamlined hovercar, seating 100 or more, travelling on..an elevated concrete track at speeds of around 300 m.p.h. 1970 *Nature* 17 Jan. 214/2 In the British design, the hovercar straddles a concrete beam of box-shaped cross-section. 1961 *New Scientist* 29 June 785/2 Denny's..hope by next year to produce an actual passenger-carrying hover-ferry... They ultimately envisage a hovercraft capable of carrying 50 to 70 passengers at about 20 knots. 1970 *Ibid.* 11 June 528/2 Both the hoverferry and the general purpose version have been subjected to an engineering modification programme. 1965 *Ibid.* 2 Dec. 633/1 The hoverkiln relies upon air pads to lubricate the pallets on which pottery is slowly conveyed through the baking kilns. 1962 *Listener* 16 Aug. 258/3 Plans for atomic-powered hoverliners. 1970 *New Scientist* 11 June 528/2 The original dreams of large hoverliners seem to have been shelved. 1966 *Times* 14 Oct. 11/7 The Ulster Prince, a 4,600-ton drive-on ferry.., will be equipped with 'hoverpallets'. 1967 *Jane's Surface Skimmer Systems 1967–68* 21/2 Palletised loads can be 'floated' in on hover-pallets, supplied with air from the main compressor. 1961 *Guardian* 23 June 10 The passenger-carrying 'hovership'..is a substantial advance towards bringing these curious vessels that are neither of the sea nor of the air into practical, everyday use. 1971 *Country Life* 8 Apr. 830/2 In *The Future of Ships*..D. Phillips-Birt looks at projects for 4,000 and 5,000-ton hovership freighters. 1961 *New Statesman* 14 July 67/1 The possibility of hovertrains. 1971 *Daily Tel.* 16 Apr. 6/7 A French demonstration hovertrain has been running for 18 months on an 11-mile stretch of elevated monorail outside Orléans. 1960 *Daily Tel.* 12 Aug. 11/7 A series of Hovertrucks. 1963 *Guardian* 6 June 3/3 The 'hover truck', a development from the Hovercraft which some farmers in East Anglia regard as an essential farm implement of the future. 1962 *Daily Tel.* 6 June 22/3 What we have in mind is that hover vehicles, vehicles designed to be supported on a cushion of air, shall be considered as motor vehicles whether or not adapted for use on the roads.

b. Used to form the names of things related to the operation and requirements of hovercraft, as *hoverport, -rail, -track, -way* [after *motorway*].

1967 *Guardian* 6 Feb. 3/2 A British hoverport in the south. 1967 *Jane's Surface Skimmer Systems 1967–68* 16/2 The route will be between a new hoverport in Pegwell Bay area and a French terminal..east of Calais harbour. 1962 *Engineering* 5 Jan. 6/2 A high-speed hover-rail system..could also be cheaper than motorways. 1965 *New Scientist* 5 Aug. 318/3 The hovertrack, of concrete, would be elevated, and on it would run a 'train' probably capable of carrying several passengers at 100 miles an hour. 1973 *Daily Tel.* 7 Sept. 18/2 The strong constituency interest in the future of the hovertrack centre. 1960 *Times* 12 Aug. 6/4 They will be capable of operating over land, water, marsh or broken ice, or along rough bull-dozed 'hoverways' in undeveloped areas. 1962 *Hovering Craft & Hydrofoil* May–June 15/1 It is desirable that definite 'hoverways' be established and shown on charts, especially where these traverse commercial routes where the density of shipping is high.

hovercraft (hɒ·vəɪkrɑft). Also **Hovercraft.** Pl. **-craft.** [f. Hover v.^{1} + Craft sb.^{1}] A vehicle or craft that can be supported by a cushion of air ejected downwards against a surface close below it, and can in principle travel over any relatively smooth surface (as a body of water, marshland, gently sloping land) while having no significant contact with it.

For some years from 1961 registered as a proprietary term but now in the public domain.

1959 [see *ground effect* s.v. *ground sb. 17 d]. 1959 *Daily Mail* 6 Apr. 5/2 The sea-saucer has been officially christened the Hovercraft. 1960 A. Croome *Hover Craft* iv. 60/1 When the 455 h.p. engine is started the air-cushion quickly builds up underneath and the Hovercraft rises off the ground (or water) and hovers there very steadily. 1961 C. S. Cockerell in *Hovering Craft & Hydrofoil* Oct. 8 My wife and I tried to find a name and settled for the not altogether appropriate word 'Hovercraft'. 1962 *Daily Tel.* 12 Mar. 13 British United Airways intends..to start the first Hovercraft service in the world this summer. 1965 *Guardian* 5 Jan. 3/3 Air cushion vehicles—or hovercraft, if you prefer the term. 1967 *Jane's Surface Skimmer Systems 1967–68* 24/2 Designed initially for military logistic-support duties, the BH. 8 is a twin-engined, 80 ton, open-water hovercraft with bow and stern loading doors. 1972 *Drive* Spring 75/2 From the attractive cliff gardens there are exciting views of modern hovercraft roaring in and out of Pegwell Bay.

ho·ver-height. Also **hoverheight.** [f. Hover v.^{1} + Height sb.] The distance separating the underside of a hovercraft (either stationary or in motion) from the surface below it when the vehicle is supported on its air-cushion.

1959 *Times* 2 Sept. 11/6 Even quite small Hovercraft can have a hover-height of several feet. 1961 *New Scientist* 2 Feb. 266/3 The installation of the second engine ..reduced the 'hoverheight' by 2½ inches to 8 inches. 1962 *Hovering Craft & Hydrofoil* Oct. 13/1 In rough conditions, speed can be reduced and the hover-height increased to give adequate wave clearance. 1967 *Jane's Surface Skimmer Systems 1967–68* 44/1 Its hover height can be varied to a maximum of three feet, depending on loading.

hovering, *vbl. sb.* **a.** (Further examples: see *hover v.^{1} 1 c, d.)

1916 M. A. S. Riach *Air-Screws* ix. 113 The value of (V) is zero, and the machine remains stationary. This is the condition already established for 'hovering' flight. 1960 Houghton & Brock *Aerodynamics* viii. 179 In steady hovering, the weight is balanced by the jet thrust and the force due to the 'cushion' of air below the craft. 1962 *Daily Tel.* 4 Sept. 17/1 Their hovering manœuvres, ..and the transformation in the air to 'normal' fast flying fighter planes, marked a new era in the development of aircraft.

(ii) Motion of a hovercraft while it is supported on an air-cushion: see *hover v.^{1} 1 d.

1967 *Hovercraft World* Jan.–Apr. 14/1 We waved goodbye..at 10.15 hours heading for Wexford, 75 miles away. The fog soon lifted and this leg was a straightforward piece of 'hovering'.

hove-to: see Heave v. 20 c.

hoving (hōu·viŋ), *vbl. sb.* Also **hooving.** [Hove v.^{2}] Swelling (of cheese).

1743 W. Ellis *Mod. Husbandman* May vii. 93 Others to prevent the Cheese's Hoving, will mix skim and new milk together. 1811 [see Hove v.^{2} 3]. 1858 C. L. Flint *Milch Cows* 265 Heat would make them [*sc.* cheeses] sweat..which extracts the fat, and tends to induce hooving.

Hovis (hōu·vis). The registered trade mark of a brand of flour; also, a loaf of brown bread made from this flour. Also *attrib.*

1890 *Trade Marks Jrnl.* 16 Nov. 1137/2 *Hovis.* Flour and Articles of Food, made partly or wholly of Flour, Richard Smith, Corn Flour Mills, Macclesfield, Cheshire. 1895 G. B. Shaw *Let.* 23 Dec. (1965) 584 Eat stewed fruit and hovis. 1898 *Christian World* 3 Feb. 4/5 Our representative was told that 'Hovis' was the concoction of a Cambridge undergraduate, being compounded of 'hominis' and 'vis', 'strength of man'. 1907 *Yesterday's Shopping* (1969) 510/3 Hovis Food—No. 1, for infants... No. 2, for children and invalids. 1930 S. Beckett *Whoroscope* 3 So we drink Him and eat Him And the watery Beaune and the stale cubes of Hovis. 1932 Dylan Thomas *Let.* (1966) 5 Give me a half-pint, a Hovis. 1962 L. Deighton *Ipcress File* i. 10 He had a long thin nose, a moustache like flock wallpaper, sparse, carefully combed hair, and complexion of a Hovis loaf. 1967 A. Laski *Seven Other Years* iii. 32 She liked the shapes of bread, and the different colours..some, Hovises and whole-meals, dark and interesting.

how, *adv.* (*sb.*^{3}) Add: **A.** *adv.* **2. a.** *how are you?*: (in quot. 1918) used ironically in sense 'indeed!' Also, *how goes it?* = How-do-you-do 1; *how's* (or *how are*) *things* (or, orig. Austral. and N.Z., *tricks*)?; *how do?*: = How-do-you-do.

1598 Florio *Worlde of Wordes* 41/1 How now? how goes it? go to, it is well. 1764 [see Go v. 18]. 1799 [see Do v. 19]. 1886 H. Baumann *Londinismen* 79/2 *How-do*, how d'you do? 1892 I. Zangwill *Childr. Ghetto* I. 147 'Well, how goes it, Reb Moshé?' said Reb Shemuel with his cheery smile. 1918 Joyce *Exiles* I. 2 *Beatrice.* Did he practise the piano while I was away? *Brigid.* Practise, how are you! 1926 K. S. Prichard *Working Bullocks* viii. 86 They halloed and yelled to him: 'H'lo, Red!' 'How's things?' 1928 H. Crane *Let.* 27 Mar. (1965) 321 How goes it with your translations..? *a* 1930 D. H. Lawrence *Mod. Lover* (1934) 192 Hello, you two. How's things? 1934 *Passing Show* 27 Jan. 5/4 Hey, Morrison, old socks. How's things? 1939 C. Belton *Outside Law in N.Z.* xxv. 129 'Hallo. How are things?' I greeted him. 1940 H. G. Wells *Babes in Darkling Wood* I. iv. 90 'How do, Father?' said Gemini. 1941 *Coast to Coast 1941* 195 'G'day,' he said. 'How's tricks?' 1949 D. M. Davin *Roads from Home* I. iii. 49 How's things?..How's tricks with you? 1971 B. Malamud *Tenants* 149 'I won't be coming around to say howdo this Friday, Irene,' Bill said tonelessly. 1973 L. Meynell *Thirteen Trumpeters* xiv. 211 'How's tricks?' he enquired. 'I'm fine. I slept like a top.'

b. (Earlier and later examples.)

1833 *New Sporting Mag.* V. 325 'Well thrown by Huddleston!'—'How's that?' 'Run out!' 1927 G. A. Terrill *Out in Glare* iv. 62 The ball swerved—pitched; the inner edge of his bat grazed it. 'Rap!' it had got him on the pad. ''s *that?*' cried Verlenden, unaware of the graze. 1973 A. Mann *Tiara* ii. 13 The sharp click of bat on ball, and un-Italian cries of 'Owzat!'..showed that cricket was well underway.

4. c. *how much*: used in humorous colloq. requests for the repetition of something not heard or not understood. Also, *how much?* = what? eh? (Cf. *what price* so-and-so?)

1852 F. E. Smedley *Lewis Arundel* xxxiv. 292 'Then my answer must mainly depend on the exact height of the principles.' 'On the how much?' inquired Frere, considerably mystified. 1914 C. Mackenzie *Sinister St.* II. III. i. 509 'I've got to get..a picture of Mona Lisa.' 'Mona how much?' said Alan. 'La Gioconda, you ass.' 1927 E. Bowen *Hotel* ii. 11 'She is a Hedonist.' 'A *how*-much?' 'A Hedonist.' 1928 [see *Gawd]. 1934 T. S. Eliot *Rock* I. 12 *Ethelbert*: Ain't you ever 'eard me speak o' the principles of Social Credit Reform? *Alfred*: 'Ow much? *Ethelbert*: What Major Douglas 'as to say about banks. 1938 E. Bowen *Death of Heart* II. iv. 243 'A friend of mine—could he ever come and stay here?'..'Could he how much?' said Daphne. 1949 Wodehouse *Uncle Dynamite* ii. 23 'You're like me, a gentle coffee-caddy.' 'A how much?'

d. *and how!*: excl. used to indicate that the effect of something is difficult to describe = and no mistake, very much so! orig. *U.S.*

1865 B. Taylor *Let.* 16 June in M. H. Taylor *Life & Lett. Bayard Taylor* (1884) II. xviii. 434, I finished an article for the 'Atlantic' that day. As if I were not 'a tool of the elements!' 'And how?' as the Germans say (Americanicé—'You'd better believe it!'). 1932 J. W. Drawbell *Good Time!* xvii. §3 'How's that for your orders from a typical American woman?' 'You mean it, Peggy?' 'And how!' 'Baby!' 1932 T. E. Lawrence *Lett.* (1938) 752, I want to meet Yeats and Epstein and Eliot some day and how. 1933 E. E. Cummings *eimi* 83, I have fallen, fallen And How; tumbled into exactly 180 minuteless minutes of 'materialist dialectic'. 1948 'N. Shute' *No Highway* ii. 37 'Did Mr. Honey take it seriously?' she asked. 'And how!' I said. 1963 V. Nabokov *Gift* iii. 175 One foot is buried in rich mud, the other is about to kick—and how!—the hideous, tar-black ball. 1965 *Listener* 25 Nov. 874/1 'Alas,' wrote Harrington, 'all earthly things do fail to mortals in enjoyment.' And how.

e. *how about that?*: Isn't that good, pleasing, surprising, etc.? *colloq.* (orig. *U.S.*).

1939 *Time* 25 Sept. 8/3 How about that? 1966 *Word Study* Dec. 2/2 How bout that.

f. *how's about?* = how about? (How *adv.* 4 a). *colloq.* (chiefly *U.S.*).

1952 *Holiday* Jan. 41/1 How's about a drink? 1961 Partridge *Dict. Slang* (ed. 5) Suppl. 1184/1 How's about having a drink?

7. b. *how —— can you get?*: see *get v. 33 b.

9. Also *ellipt.*, as *how to*, and often used *attrib.*, as *'how-to' discourse, 'how-to-do-it' manual*, etc. Also (in titles of books, etc.) followed by a verb.

1838 H. Martineau (*title*) How to observe: morals and manners. 1857 (*title*) How to publish: a manual for authors. 1922 Joyce *Ulysses* 634 You had to come back.. to show the understudy in the title rôle how to. 1923 A. Bennett (*title*) How to make the best of life. 1941 Beals & Brody *Lit. Adult Educ.* 453 Nearly all councils also make some provision for supplying information, such as distributing manuals of the 'how-to-do-it' variety. 1942 H. Haycraft *Murder for Pleasure* p. vii, One or two 'how-to-write-it' manuals. 1946 C. Morris *Signs, Lang. & Behavior* v. x. 143 It is 'how to' discourse: discourse informing one how to rivet, how to play the flute. 1959 *Times Lit. Suppl.* 6 Nov. p. xxxviii/4 Light fiction, popular uplift, or how-to-do-it-books. 1960 *Farmer & Stockbreeder* 19 Jan. 43/1 A how-to-cook-turkey booklet! 1968 *Sat Rev.* 27 July 26/3 Most of the how-to books at this time of year are busy with water sports. 1972 T. P. McMahon *Issue of Bishop's Blood* (1973) xii. 176 She had washed her face, and though it doesn't say so in the women's how-to-books, there are some women who look great that way.

18. *here's how!* (Later examples.)

1925 [see *bung-ho int.]. 1951 J. B. Priestley *Festival at Farbridge* II. i. 152 'Well,' said Mr. Hull, holding up his glass,..'here's how!' 1959 E. Burgess *Divided we Fall* xi. 132 Martin was clasping a tumbler half filled with whisky. 'Here's how,' said the fat man.

19. how come? *colloq.* (orig. *U.S.*) phr.: how did (or does) it come about (that)? Cf. Come v. 20.

1848 Bartlett *Dict. Amer.*, *How-come?* rapidly pronounced huc-cum, in Virginia. Doubtless an English phrase, brought over by the original settlers, and propagated even among the negro slaves. The meaning is, How did what you tell me happen? How came it? 1897 R. M. Stuart *In Simpkinsville* 230 She heard Miss Euphemia wonderin' the other day how come the right shoulder of her black silk dress to wear out. 1930 *Sat. Even. Post* 8 Mar. 12/1 These firms assert blandly to..journalists who ring them up, wishful to know how come, they have

nothing against women. 1932 T. S. Eliot *Sweeney Agonistes* 17 Hello Doris! Hello Dusty! How do you do! How come? how come? will you permit me —. 1938 M. K. Rawlings *Yearling* iv. 34 That's how come him to have appetite for a nip o' pore old Betsy. 1957 *New Yorker* 13 July 19/1 These days, the sensation of the stamp-collecting world is United Nations commemorative stamps. How come? 1958 G. Mitchell *Spotted Hemlock* xviii. 210 How come *they* didn't spot her? 1959 H. Hobson *Mission House Murder* xviii. 122 'How come you make it murder?' he asked. 1969 in Halpert & Story *Christmas Mumming in Newfoundland* 213 How come you ain't got Wallace with you tonight? 1971 *Black World* Oct. 62/1 So that's how come I asked My Man Bovanne to dance. 1973 T. Allbeury *Choice of Enemies* iv. 16 'Well, we are getting a bit like the Krauts, you know.' 'How come?'

B. *sb.*^{3} **2.** (Examples of U.S. usage.)

1845 W. G. Simms *Wigwam & Cabin* 1st Ser. 7 You joined the army again, and come in with Greene? Was that the how? 1878 J. H. Beadle *Western Wilds* ii. 27 Daddy went off at last, and that was the how of my first trip. 1916 'B. M. Bower' *Phantom Herd* xvi. 272, I calc'late that's about the how of it. 1949 *Amer. Speech* XXIV. 39 In recent years *method, explanation*, and *the how*..promise to force *modus operandi* into the archives.

how (hau), *int.*^{2} [Cf. Sioux *háo*, Omaha *hau.*] An ejaculation, orig. used by Indians of north-eastern North America in a variety of applications. Also repeated.

First noticed in the early 17th century by the French missionary Jean de Brébeuf, describing Huron oratory as he had observed it in Ontario. The Hurons showed their approval of a speech by a shouted *haau* (Jesuit Relations, documents pertaining to 1636, Thwaites, Vol. X, p. 259).

1817 J. Bradbury *Trav. Interior Amer.* 95 We were interrupted by one of the chiefs crying 'How', signifying amongst the Indians 'Come on', or 'let us begin'. 1841 G. Catlin *Lett. on N. Amer. Indians* II. 172 'How! how!' vociferated all of them, thereby approving what was said, giving assent by the word *how*, which is their word for yes. 1868 *Harper's Mag.* Feb. 301/2 The Indians..complimented 'Little Bill' with a succession of how-how-hows! 1911 *N.Y. Even. Post* 28 Jan. (Suppl.) 3 The expression 'How', used by army men in giving a toast, is equivalent to the expression, 'Here's to your health'. Some think it is merely the Indian corruption of 'How d'ye do?' abbreviated by the Indian to 'How'. Others believe the expression is derived from the Indian language direct. 1962 *Alberta Hist. Rev.* Autumn 11/1 The Blackfoot Indians usually greet a white man with 'How, How!' sometimes, 'How wa-pe'.

how (hau), colloq. abbreviation of Howitzer.

1915 *Times* 14 Apr. 7 The boom in the distance from one of our 'hows'. 1919 C. P. Thompson *Cocktails* 133 A couple of batteries of heavy Hows.

howardite (hau·ăɹdəit). [f. the name of Edward *Howard* (fl. 1802), English chemist: see -ite^1.] **1.** *Min.* A supposed silicate of iron and magnesium found in some meteorites.

1848 C. U. Shepard in *Amer. Jrnl. Sci.* VI. 253 The earthy constituent of this stone, like that of the Iowa meteorite..appears to be a tersilicate of the protoxyd of iron and magnesia, a mineral which though frequent in meteoric stones, has never yet been distinctly recognized, and which in a future paper on American Meteorites, I shall more particularly describe under the name of *Howardite*, after the Hon. Mr. Howard, that celebrated chemist, who was the first British writer whose labors contributed to elucidate the history of these extra-terrestrial bodies. 1955 M. H. Hey *Index Min. Species* (ed. 2) 127 Howardite (of Shepard). A doubtful meteoritic silicate of Fe and Mg.

2. *Astr.* [ad. G. *howardit* (G. Rose 1863, in *Physikal. Abhandl. d. K. Akad. d. Wiss. zu Berlin* 29).] Any of a group of achondritic meteorites principally composed of hypersthene and anorthite.

1881 L. Fletcher *Guide Coll. Meteorites* (Brit. Mus.) 17 Perhaps for those aerolites which contain little or no nickel-iron the division into Howardites, Eukrites, Chladnites, Chassignites, Shalkites, and Carbonaceous is the most convenient. 1915 O. C. Farrington *Meteorites* iii. 35 Of a total of nine meteorites belonging to the class of howardites, five have fallen in Russia. 1916 *Mineral. Mag.* XVIII. 36 The angrites, nakhlites, howardites, eucrites, and sherghottites..are richer in lime..than chondritic stones generally. 1971 B. Mason in I. G. Gass et al. *Understanding Earth* viii. 117 (*table*) [Achondrites are] sometimes divided into calcium-poor achondrites (aubrites, diogenites, chassignite, ureilite) and calcium-rich achondrites (angrite, nakhlite, howardites, eucrites).

how-come-ye-so, *adj. phr. archaic dial.* or *slang.* Tipsy.

1816 *Monthly Mag.* 1 July 494/2 [Degrees of intoxication.] How came you so. 1824 T. Hook *Sayings & Doings* III. (Merton) xiii. 89. She used to be—'Lord, how come ye so!' every night. 1827 J. F. Cooper *Red Rover* I. x. 258 It is quite in reason to believe your husband was..a little of what I call how-come-ye-so. 1843 *Knickerbocker* XXII. 366 We were never 'groggy', 'intoxicated',..'how-came-ye-so'..or 'tight', but once. 1904 J. C. Lincoln *Cap'n Eri* xii. 223 Drank rum by the hogshead... Well, one evenin' Labe was comin' home pretty how-come-you-so, and he fell into Jonadab Wixon's well.

how-do-ye, how-d'ye, howdy, *phr.* and *sb.* Add: Also how de. **1.** (Further examples.) Freq. in colloq. phr. *to tell* (a person) *howdy*.

c 1828 T. O. Larkin in *Calif. Hist. Soc. Q.* (1937) XVI. 25 They (Southerners) often correct Yankees in their speaking... In those words they are right, but not in.. how de for how do you do. 1837 A. Wetmore *Gaz. Missouri* 287 With a smile of welcome as she gave her hand, said, 'Howdy, Joseph.' 1850 J. Price *Let.* 17 Apr. in *Mississippi Valley Hist. Rev.* (1924) XI. 241 Tell Mother and the doctor Howdy. 1856 A. J. Hoole *Let.* 8 June in *Kansas Hist. Q.* (1934) III. 56 Give my love to.. all enquiring friends; tell..all the negroes howdie for me. 1917 J. M. Grider *War Birds* 20 Sept. (1927) 10 All the soldiers in the harbor came over to tell us howdy. 1928 L. North *Parasites* i. 30 Mr. Ashton greeted Henry a trifle more warmly... 'Howdy,' he said. 1973 J. Mann *Only Security* ii. 16, I ought really to go in and say howdy to her.

3. (Further examples.)

1931 *Times* 5 Nov. 13/4 The howdy folk of Kentucky may bring forth their best. 1932 E. Wallace *When Gangs came to London* xxiii. 224 It seems that this guy has been on howdy terms with the Big Boy ever since he came to town.

how-do-you-do. Add: Also **how d'you do. 2.** (Later examples.)

1832 J. R. Underwood *Jrnl.* in *Filson Club Hist. Q.* (1941) XV. 43 Joel Yancey..was always ready with a *how d'ye do.* 1928 D. H. Lawrence *Let.* 27 Mar. (1962) II. 1049 Goethe *began* millions of intimacies, and never got beyond the how-do-you-do stage. 1938 J. Cary *Castle Corner* iv. 213 When he came..to make love as carelessly as how d'you do, he felt betrayed and lost.

howdy: see *how-do-ye.

howe. Now the more usual spelling of How *sb.*^{2} 2.

1947 *Proc. Prehist. Soc.* XIII. 33 When we reach the 10th century there were brochs in Caithness and Orkney.. deemed eligible burial howes. 1963 *Field Archaeol.* (Ordnance Survey) (ed. 4) 45 Round burial mounds..are the commonest objects of antiquity met with in the field. They are called by different names in different parts of the country..barrow, low, howe, cairn. 1968 G. Jones *Hist. Vikings* II. iii. 117 The mightiest of all northern howes, Raknehaugen.., over a hundred metres in diameter and some eighteen metres high, is an empty cenotaph. 1971 G. M. Brown *Fishermen with Ploughs* 7 Lust builds a howe over the burning ghost.

Howeitat (hōuweitā·t). Also **Haueytat, Howeïtat, Huwaitat, Huweitat.** [Arab. (*al-*) *Ḥuwayṭāt*, Arab tribes in north-western Saudi Arabia.] **A.** *sb.* (A member of) a Bedouin tribe of northern Saudi Arabia. **B.** *adj.* Of or pertaining to this people.

a 1817 J. L. Burckhardt *Notes on Bedouins* (1831) I. 29 El Haueytat..are about three hundred horsemen, but can furnish a large body of armed camel-drivers... In seasons of drought the Haueytat approach Gaza or Hebron. 1838 J. R. Wellsted *Trav. Arabia* II. ix. 167 The Howeïtát Bedowins occupy the coast from Magnah to Jebel 'Antar. 1917 T. E. Lawrence *Let.* 10 July (1938) 225 Sherif Nasir stayed in Kaf to enrol Rualla, Sherarat and Huweitat for the Akaba expedition. 1926 —— *Seven Pillars* (1935) III. xxx. 173 With his [*sc.* Feisal's] northern neighbours, the coastal Howeitat, he had already made a beginning. 1957 H. St. J. Philby *Forty Years in Wilderness* v. 99 The tribes mentioned were Huwaitat..and some others.

however, *adv.* Add: **6.** Used by itself, or followed by points of suspension, as an interjection, or as a formula concluding, introducing, or modifying an utterance in some contextual way.

1876 Geo. Eliot *Dan. Der.* II. IV. xxxii. 319, I have not yet written a word... And I told the girls to leave it to me. However! 1901 Beerbohm *Variety of Things* (1928) 217 [I] doubt whether my creative power in caricature can be quite so strong as I had supposed. However... 1936 'N. Blake' *Thou Shell of Death* viii. 137 Comes of givin' a vote to every hayseed and short-weight grocer in the —— However, hrrumph, where was I? 1964 D. Warner *Death of Dreamer* I. i. 9 We don't want to know where the stuff comes from but where it goes *to*. However!.. 1966 I. Jefferies *House Surgeon* i. 11, I.. said, 'However. Not to interrupt what you were asking me.' 1968 M. Woodhouse *Rock Baby* vii. 66 'Our component reliability was very high indeed during trials. However.' 'Go on,' I said.

howff. (Later examples.)

1950 *John o' London's* 24 Nov. 617/1 He was just seventeen.., when he began to haunt the howffs (drinking-places) of Edinburgh's underworld. 1957 Clark & Pyatt *Mountaineering in Brit.* II. xi. 196 The crags were still relatively inaccessible—in spite of the use of boat and motor-car in conjunction with tents, howffs, or bivouacs.

howgozit (haugōu·zit). *Aeronaut.* orig. *U.S.* [Corruption of *how goes it?* (see *how *adv.* 2 a).] A graph which essentially represents the amount of fuel remaining in an aircraft in relation to that needed to complete a journey or return to the point of departure. Also *attrib.*

1941 *Jrnl. R. Aeronaut. Soc.* XLV. 308 The 'Howgozit Curve' was developed by our ocean captains... Its purpose is to present..a continuous flow of information as to the fuel reserve remaining aboard the aircraft and the fuel required for completion of the flight to destination, or back to the point of departure. 1945 H. E. Benham *Aerial Navigation* iv. 92 In flight the actual progress of the aircraft..is plotted on the 'howgozit' for comparison with the forecasted data. The slope of the actual progress line as compared with the forecasted line and its tendency toward destination or dry tanks provides a vivid picture of the flight's progress. 1960 *Aeroplane* XCVIII. 98/2 An auto-landing indicator..will provide 'howgozit' information about the progress of the final approach.

howk, *v.*: see *holk, howk v. in Suppl.

howl, *sb.* Add: **1. b.** A howling noise produced in a loud-speaker as a result of electrical or acoustic feedback; howling.

1921 *Wireless World* 10 Dec. 568/1 Ear-splitting howls associated with valve reception. 1928 L. S. Palmer *Wireless Princ. & Pract.* x. 374 The typical 'howl' to which low-frequency amplifiers are subject. 1939 H. J. Hicks *Princ. & Pract. Radio Servicing* xiii. 228 If the resistance-capacity filter in the a.v.c. line loses its effectiveness, a howl will result. 1962 E. E. Ecklund *Repairing Home Audio Systems* xiii. 246 Rumble, feedback, and howl are caused by mechanical vibrations being transmitted to the pickup.

howler. 1. b. Substitute for def.: In full, *howler monkey*. A South American monkey of the genus *Alouatta*. (Add earlier and later examples.)

1800 G. Shaw *Gen. Zool.* I. I. 72 The Allouates, or Howlers, inhabit the moist forests, in the neighbourhood of waters and marshes. 1906 E. Ingersoll *Life Anim.: Mammals* 43 The howlers, arguatos, or alluates are the largest and most powerful of South American apes and the dullest, and are peculiar in having no thumb or only a rudimentary one, and in having the hyoid bones in the throat (of the males only) widely enlarged and cavernous, so as to form a curious hollow organ, by which their voice is so increased as to be audible two miles. 1932 S. Zuckerman *Soc. Life Monkeys* xi. 192 The well-known howler monkeys (genus *Alouatta*) of the forests of the northern half of South America are usually found in small parties. 1958 J. Carew *Black Midas* ii. 26 A big white devil does sit on a treetop roarin' like a howler baboon. 1964 *Listener* 5 Nov. 710/2 Fighting is rare in wild gibbons and apparently absent in wild howler monkeys. 1967 S. A. Altmann *Social Communication among Primates* xvii. 329 Other groups of howlers avoid areas from which such calls come.

†**4.** *Telephony.* A device (used by the exchange) for producing a howling noise in a receiver in order to attract the subscriber's attention. *Obs.*

1886 *Jrnl. Soc. Telegr. Engin.* XV. 322 We supply what we call a 'howler', and whenever a subscriber leaves his tubes hanging this howler is at once put on. 1917 G. D. Shepardson *Telephone Apparatus* viii. 137 For reminding careless subscribers who neglect even to hang up the receiver..the operator..may send out a strong current of comparatively high frequency from the 'howler'.

howling, *vbl. sb.* Add: **1. b.** The emission of howls (*howl sb. 1 b) by a loud-speaker; undesirable feedback at audio frequencies in an amplifier.

1920 *Radio Rev.* Apr. 356 The resistance R_3 and condenser C_3 connected to the grid of the selected valve are increased in value until maximum amplification is obtained without 'howling'. 1939 H. J. Hicks *Princ. & Pract. Radio Servicing* xiii. 228 Open by-pass condensers across any of the tube elements..will often cause howling. 1953 J. E. Haines *Automatic Control of Heating* xiv. 340 The variation exhibits a definite rhythm—in a manner analogous to the howling of a telephone when the sound waves from the receiver are fed back to the transmitter. 1966 *McGraw-Hill Encycl. Sci. & Technol.* I. 520/2 Suppression of the low frequencies..in rooms that are excessively reverberant..increases the intelligibility of the speech and reduces the possibility of acoustic feedback (howling).

howling, *ppl. a.* Add: **1.** Spec. *howling baboon, monkey* = *howler 1 b.

1769 E. Bancroft *Ess. Nat. Hist. Guiana* ii. 133 The Howling Baboons, as they are here called, seem to be the animals which are here described by Marcgrave, and which are called by the natives of Brasil, Guereba. *Ibid.* 135 There is another Monkey, somewhat larger than the howling Monkey, which is covered with long reddish hair. 1802 [see Monkey sb. 1 b]. 1863 H. W. Bates *Naturalist on River Amazons* I. ii. 72 Morning and evening the howling monkeys make a most fearful and harrowing noise. 1887 J. G. Wood *Illustr. Nat. Hist. for Young People* 15 The Howling Monkeys are larger and not so agile as the Spider Monkeys. 1924 C. W. Domville-Fife *Among Wild Tribes of Amazons* viii. 121 On reaching camp..in the half light it was just possible to see the huddled and impaled body of a furry guaribas, or howling monkey (*simia mycetes*). 1959 *Jrnl. Mammalogy* XL. 317 (*title*) Field observations on a howling monkey society.

3. (Further examples.) Also, extreme, great. *colloq.*

1884 'Mark Twain' *Huck. Finn* xliii. 437 Le's all three slide out of here, one of these nights, and get an outfit, and go for howling adventures amongst the Injuns. 1908 *Magnet* I. 1, 'You howling ass!' shouted Bulstrode. 'I tell you he's busted my two-guinea camera.' 1933 *Times Lit. Suppl.* 27 Apr. 283/4 If his book is not a big, a very big, a 'howling' success..but I need offer no 'if's'.

4. As *adv.* In the highest degree. (Cf. *screamingly.*) *colloq.*

1895 *Century Mag.* Sept. 678/2 It's howling lonesome at the Mule Deer. 1899 Kipling *Stalky & Co.* 45 He'll be howling drunk to-night. 1928 *Sat. Even. Post* 4 Feb. 100/4 Glad! You're howling right I'm glad.

HOWLITE

howlite (hau·ləit). *Min.* [f. the name of Henry *How* (d. 1879), mineralogist of Nova Scotia, who first described it: see -LITE.] A hydrated calcium borosilicate, $Ca_2B_5SiO_9(OH)_5$, that typically occurs as white nodules forming compact structureless masses resembling unglazed porcelain.

1868 J. D. DANA *Syst. Min.* (ed. 5) 598 *Howlite*. Silicoborocalcite *H. How*...1868. Howlite *Dana*. In small rounded imbedded nodules. Texture compact, without cleavage; also chalk-like or earthy. **1917** *Amer. Mineralogist* II. 1 Not all fine grained, compact hydrous calcium borate is priceite; a number of such specimens from California localities..have been examined microscopically and all proved to be howlite. **1957** *Ibid.* XLII. 521 The mineral howlite, $H_5Ca_2SiB_5O_{14}$, has been recorded from a number of localities in California and elsewhere..and microscopic crystals have been described.., but in general the mineral is massive and very fine granular.

Howship (hau·ʃip). *Anat.* The name of John *Howship* (1781–1841), English surgeon, used in **Howship's lacuna** (earlier *lacuna of Howship*): one of the numerous microscopic depressions or pits, irregular in shape and usually containing osteoclasts, that are found on the surface of bones and bony tissue where resorption is occurring.

1876 C. S. TOMES *Man. Dental Anat.* v. 175 Microscopic examination of the bone at this point shows that the lacunae of Howship..abundantly cover its surface. **1911** T. W. WIDDOWSON *Notes on Dental Anat.* viii. 44 Upon any part of the roots of the temporary teeth, undergoing absorption, cup-shaped depressions, Howship's lacunae, occur. **1970** J. M. VAUGHAN *Physiol. Bone* ii. 43 They [*sc.* osteoclasts] may be found closely applied to bone in Howship's lacunae.

howsomever. 2. (Later examples.) *U.S. colloq.*

1896 'M. RUTHERFORD' *Clara Hopgood* xxiii. 215 He allus begins to argue with me. Howsomever, arguing isn't everything. **1929** H. W. ODUM in A. Dundes *Mother Wit* (1973) 183 Howsomever, hard times in American camps whut I'm talkin' about. **1933** E. E. CUMMINGS *Let.* 26 May (1969) 123, I fear that naught will compare with domesticity, howsomeever. **1939** *Amer. Speech* XIV. 128 The great drive for 'correctness' of the later eighteenth and early nineteenth did succeed in branding as 'vulgarisms' such hitherto acceptable forms as..*howsomever, mought, sarvent*, [etc.].

howtowdie (hautau·di). *Sc.* Also **howtoudie, how-towdy.** ['Not recorded in O.Sc. but appar. O.Fr. *hétoudeau, estaudeau*, a fat young chicken for the pot' (*S.N.D.*).] A dish whose main ingredient is a chicken (see quot. 1951).

1728 A. RAMSAY *Poems* II. 230 They all, in an united Body, Declar'd it a fine fat How-towdy. **1759** E. CLELAND *New & Easy Method Cookery* (ed. 2) iv. 91 You may do Howtoudies, or any white Fowl the same Way. **1901** *Daily Chron.* 17 Aug. 8 Howtowdie is another old Scotch dish. **1951** *Good Housek. Home Encycl.* 513/1 *Howtowdie*, a Scottish dish consisting of boiled chicken with poached eggs and spinach. **1970** SIMON & HOWE *Dict. Gastron.* 222/2 *Howtowdie*, Scots for pullet, possibly related to the old French *hutaudeau*... Eggs are poached in gravy or broth and placed around the bird on the carving dish, each on a pat of spinach.

Hoxnian (hǫ·ks,niăn), *a.* [f. *Hoxne*, name of the village in Suffolk where the type site is situated: see -IAN.] Epithet of the second (penultimate) interglacial in Britain (identified with the Mindel-Riss interglacial of continental Europe), and of a stage of the middle Pleistocene; hence, of or contemporaneous with this interglacial or stage. Also *absol.* as *sb.*, the Hoxnian interglacial or the Hoxnian stage.

1956 WEST & DONNER in *Q. Jrnl. Geol. Soc.* (1957) CXII. 86 Hoxnian Interglacial. [*Note*] A general name suggested for this interglacial period. **1963** R. G. WEST in *Proc. Geologists' Assoc.* LXXIV. 171 Evidence for sea-levels during the Hoxnian Stage is summarized in Fig. 9. **1964** K. P. OAKLEY *Frameworks for dating Fossil Man* iii. 29 In Britain the silver fir (*Abies*) is absent from Cromerian diagrams, abundant in the Late Temperate phase of the Hoxnian. **1968** R. G. WEST *Pleistocene Geol. & Biol.* xii. 276 Often they are overlain by interglacial or Flandrian beach gravels, and..they are probably mostly Hoxnian or older. **1969** *Proc. Geol. Soc.* Aug. 152 It is recommended that for the Pleistocene and Holocene of the British Isles the following ages/stages be adopted as a regional scale... Pleistocene:.. Wolstonian, Hoxnian, Anglian, Cromerian, [etc.]. **1970** *Phil. Trans. R. Soc.* B. CCLVII. 414 *Buxus* pollen has been found in the Hoxnian deposits at Birmingham. **1970** *Times* 10 Mar. 16/6 In the closing stage of the Hoxnian, the pine became the dominant forest tree.

Hoxtonian (hǫkstōu·niăn), *sb.* [f. *Hoxton*, the name of part of the borough of Shoreditch in London + -IAN.] A native or inhabitant of Hoxton; the variety of English spoken in Hoxton.

a **1935** T. E. LAWRENCE *Mint* (1950) II. xxii. 159 Adam and Eve on a raft (Hoxtonian for fried eggs on toast). **1935** G. INGRAM *Cockney Cavalcade* xii. 197 The West End—a place entirely foreign to most Hoxtonians.

hoy, *int.* Add: **B.** *sb.* **2.** *Austral.* A gambling game, resembling lotto, in which playing-cards are used. Also *attrib.*

1965 *Courier-Mail* (Brisbane) 2 Mar. 15 A hoy evening which the Royal Society of St. George planned to hold at St. George House. **1969** *Ibid.* 25 Feb. 6/10 Juliet Jones couldn't object to a few games of hoy. **1969** *Sunday Mail* (Brisbane) 24 Aug. 3/3 Police said that bingo, or hoy, which was played in the same way, was illegal in Queensland. **1971** *Tel.* (Brisbane) 3 Nov. 4/2, I have been advised that the radio competition is above board, but have had no ruling on the game hoy.

Hoyle (hoil). The name of Edmond *Hoyle* (1672–1769), author of several works on card-games (the earliest, on whist, dated 1742): often cited typically for an authority on card-playing. Phr. *according to Hoyle*, according to the highest authority, in accordance with strict rules.

1906 'O. HENRY' *Four Million* (1916) 14 The financial loss of a dollar sixty-five, all so far fulfilled according to Hoyle. **1945** A. A. OSTROW *Compl. Card Player* p. vii, It has been the custom to call books of rules on card and board games 'Hoyles', so that 'according to Hoyle' has come to mean 'according to accepted rules'. **1962** R. BARKER *Clue for Murder* v. 38 This one [*sc.* murder]'s right out of the book—strictly according to Hoyle. **1965** J. M. CAIN *Magician's Wife* (1966) xix. 147, I want our marriage to be strictly on the beam—the way it is in the books, absolutely according to Hoyle. **1971** *Melody Maker* 21 Aug. 34/7 If everything goes according to Hoyle, I'll go into semi-retirement there.

hsien, var. *HIEN.

H-test: see *H III.

huaca (wā·kă). [ad. Sp. *huacca, guaca*, from Quechua.] **a.** The name for the all-pervading spirit thought by the Peruvian Indians to be disseminated through the whole world; also, any material object thought to be the abode of such a spirit. **b.** A prehistoric Peruvian tomb or temple, usually a truncated pyramid of stone, and often of immense size.

1847 W. H. PRESCOTT *Hist. Conquest Peru* I. I. iii. 93 The subjects of the Incas enrolled among their inferior deities many objects in nature, as the elements,..great mountains and rivers... These consecrated objects were termed *huacas*,—a word of most prolific import; since it signified a temple, a tomb, any natural object remarkable for its size or shape, in short, a cloud of meanings. **1860** W. BOLLAERT *Antiquarian, Ethnol. Res. Chile* 161 A huaca was discovered, in 1830, at the entrance of the valley of Taparapacá; it was surrounded with stones. **1862** D. WILSON *Prehist. Man* I. ix. 298 The huacas or tombs of the Incas. **1875** *Encycl. Brit.* II. 452/2 The most interesting remains in Peru are those called *Huacas*; but whether they were forts, or palaces, or tombs, is not as yet clearly ascertained. **1901** A. H. KEANE *Central & S. Amer.* I. 208 Of these ruins the largest, as well as the most characteristic, are the truncated pyramids here [*sc.* Peru] called *huacas*, or burying-places. **1902** *Encycl. Brit.* XXV. 380/1 The most prolific source of Peruvian relics is the sepulchres or huacas. **1960** M. SAVILL tr. *Leicht's Pre-Inca Art & Culture* iv. 74 The ancestral worship in the land expressed itself most vividly in the veneration of the so-called Huacas, a word which signifies not only the large temples and pyramids but a host of small and insignificant objects sacred to the Indians.

huaco (wā·ko). Also **guaco.** [See quot. 1931[1].] In Peru, Bolivia, and Chile, ancient pottery and other Indian antiquities.

1931 *Connoisseur* Feb. 95 The term *huaco* is derived from the Indian word, *huaca*, meaning 'a holy place', and refers to the cemeteries and tombs from which, with few exceptions, all the examples of pre-Incaic art are obtained. *Ibid.* 97 The linear decoration of the stirrup *huacos* tend [*sic*] to portray some form of action... In the static *huacos* the legs are almost invariably crossed. **1933** *Burlington Mag.* Aug. p. xii (Advt.), The pottery includes six rare Huacos from Paracas, with green colouring. **1959** G. WOODCOCK *Incas & Other Men* IV. iv. 229 The Mochica pots—..called *huacos* by the modern Peruvians—were made specifically as grave furniture, intended to hold food and chicha for the dead.

huarache (warā·tʃi). Also **guaracha, guarache, guarachi, huaracho.** [Mex.-Sp.] A leather-thonged sandal, orig. worn by Mexican Indians.

1887 F. C. GOOCH *Face to Face with Mexicans* xii. 433 Leathern aprons and sandals of the same, called *guarachi*. **1892** *Dialect Notes* I. 190 *Huaracho*, -*s*, a kind of sandals worn by Indians and the lower classes generally. Used generally in the plural only. **1909** *Cent. Dict.* Suppl., Guaracha. **1926** D. H. LAWRENCE *Plumed Serp.* viii. 130 The dark feet in the glare of the torch looked almost black, in huaraches that had red thongs. **1928** *Funk's Stand. Dict.*, *Guaracha*, a Mexican-Indian sandal. **1943** *N. & Q.* 24 Apr. 262/1 The Mexican Indian uses a leather-thonged sandal, which he calls guarache. **1957** J. KEROUAC *On Road* (1958) I. ii. 13 My shoes..were Mexican huaraches, plantlike sieves not fit for the rainy night of America. **1963** C. BEALS *Latin Amer.* (1964) i. 11 *Guarache* sandals shuffled like a storm of autumn leaves.

Huastec (wā·stek). Also **Huasteca, Huastek, Huaxtec, Huaxteca.** [ad. Sp. *huasteco, huaxteco*.] **a.** An Indian people inhabiting parts of Mexico; a member of this tribe. **b.** The language of the Huastecs.

1845 *Trans. Amer. Ethnol. Soc.* I. 4 A comparison of near three hundred words of the Mexican, Otomi, and Huasteca, exhibits but very few, and perhaps accidental, coincidences. **1875** H. H. BANCROFT *Native Races Pacific States* I. vi. 674 The Huastecs, Huaxtecs, Guastecs, or Cuextecas inhabit portions of the states of Vera Cruz and Tamaulipas. **1914** T. A. JOYCE *Mexican Archæol.* vii. 196 The presence of a spout distinguishes Huaxtec pottery from Mexican. **1931** T. GANN in Gann & Thompson *Hist. Maya* i. 11 The Huaxtecas..were evidently a section of the people left behind in their old home. **1946** S. G. MORLEY *Anc. Maya* iii. 39 The Maya-speaking, Maya-appearing Huasteca never shared with the Maya of the Yucatan Peninsula the latter's unique culture. **1948** D. DIRINGER *Alphabet* I. vii. 123 The Huastec..already separated from the main stock in ancient times. **1955** *Amer. Speech* XXX. 126 Huastek.

hub[1]. Add: **2.** *up to the hub* (earlier examples).

1800 *Aurora* (Philadelphia) 23 May (Th.), 'This is not a half measure—I like to do things by the lump—and this bill you will allow is up to the hub.' Those who are acquainted with the slang language of the American Caucuses will be able to explain what is meant by up to the hub. **1815** D. HUMPHREYS *Yankey in Eng.* 33 I've bin up to the hub, and didn't flinch..nor won't back out now.

5. hub brake, a brake that acts on the hub of a (cycle) wheel; so **hub-braking; hub-cap**, a covering for the hub of a wheel of a vehicle.

1883 H. STURMEY *Tricyclists' Indispensable Ann.* (ed. 3) 94 There are three great classes of brakes now in use, viz.:—Tyre, ground, and hub brakes. **1936** F. J. CAMM *Every Cyclist's Handbk.* xvii. 108 The cyclist should..take great care to prevent oil entering the shell of a hub brake. **1973** *Sci. Amer.* Mar. 90/2 Two other types of brake made their appearance later. One is the coaster brake, or back-pedaling brake, which is particularly popular in the U.S. The other is the hub brake, or drum brake, of the type used in automobiles and motorcycles. **1909** *Daily Chron.* 20 Mar. 8/5 It combines hand control and hub braking. **1913** *Collier's* 11 Jan. 11. 7/1 Their wheels, perhaps, have plain hub caps. **1954** A. HUXLEY *Let.* 5 Dec. (1969) 716 One would like to find out..why..so many cubists..used forms which are identical with those obtained by photographing reflections in curved surfaces. Did the suggestion actually come from hub caps and the backs of spoons? **1957** L. DURRELL *Justine* 27 The great silver Rolls with the daffodil hub-caps. **1959** *Times* 9 Jan. 12/6, I looked forward to removing the hub-cap. **1972** J. BROWN *Chancer* xiv. 188 You name it, we found it. All this and neat little packets of H. in the hub-caps.

hubara, var. *HOUBARA.

hubba-hubba (hɒ·bă,hɒ·bă), *int. U.S. slang.* Also **haba-haba.** [Origin unknown.] Used to express approval, excitement, or enthusiasm. Also as *sb.*, nonsense; ballyhoo.

1944 in *Amer. Speech* (1947) XXII. 35 The inevitable fact is that the cry 'Haba-Haba' is spreading like a scourge through the land. **1945** *Amer. Speech* XX. 261 *Hubba-hubba*, originally gibberish, now means the spirit of double-time and eagerness; it is a verb, adjective or noun, an imprecation, warning or insult. **1946** 'S. STERLING' *Where there's Smoke* iv. 36, I suppose you think that's a lot of hubba-hubba. **1970** C. MAJOR *Dict. Afro-Amer. Slang* 68 *Hubba hubba*, expression of approval.

Hubbard (hɒ·bāɹd). Chiefly *N. Amer.* In full, *Hubbard squash*: a variety of *Cucurbita maxima* yielding large green or yellow fruits.

1868 *Mich. Agric. Rep.* VII. 349 Thos. Smith, Hamtramck [exhibited] 8 Hubbard squashes. **1898** E. N. WESTCOTT *David Harum* xxiii, Turnips, Hubbard squash, succotash. **1921** *Daily Colonist* (Victoria, B.C.) 20 Oct. 7/4 Hubbard Squash, today's selling per lb. 4c. **1925** *Brit. Weekly* 26 Feb. 520/5 It was truly an old-fashioned Thanksgiving dinner. We had..celery, onions, and hubbard squash. **1963** J. ORGAN *Gourds* xii. 121 *C. maxima* is best represented in commerce by such popular varieties as the Hubbards, Boston Marrow and Delicious. **1971** *Rand Daily Mail* 27 Mar. 22/2 Hubbard Squash (sugar pockets). Good demand. 140c to 200c.

Hubble (hɒb'l). *Astr.* The name of Edwin P. *Hubble* (1889–1953), U.S. astronomer, used in the possessive and *attrib.* to designate concepts arising out of his work, as **Hubble('s) constant** (or † **factor**), the ratio of the recessional speed of a galaxy to its distance, usu. expressed in km. per second per megaparsec; also, the reciprocal of this, usu. expressed in years and capable of being interpreted as the time that has elapsed since the galaxies first began to recede (at constant speeds) from a single point; **Hubble('s) law**, the red shifts of the spectra of distant galaxies (and hence their speeds of recession) are proportional, on average, to their distances.

1933 *Monthly Notices R. Astron. Soc.* XCIV. 159 We can consider a possible extension of Hubble's Law, and it will be seen that this may lead to a method for determining the cosmical constant and the sign of curvature in the generally accepted form of space-time. **1938** *Physical Rev.* LIII. 207/1 We can also derive a relation between *f* and Hubble's factor *u*... *u* is about 500 km per sec. per mega-parsec. or $1 \cdot 6 \times 10^{-17}$ sec.$^{-1}$. **1952** H. BONDI *Cosmol.* v. 39 The relation may be written $V = r/T$

where T is a constant..depending..on the average intrinsic luminosity of a nebula. On this interpretation the radial velocity of a nebula is proportional to its distance... The reciprocal of the constant T is known as Hubble's constant and may be evaluated to be about 540 km./sec./megaparsec. This value leads to $T = 1 \cdot 8 \times 10^9$ years. **1957** *Encycl. Brit.* VI. 502B/2 A better appreciation of the meaning of this number, H, which came to be called Hubble's constant, is given by its reciprocal $1/H$ = 5,400,000,000 years. **1971** D. W. SCIAMA *Mod. Cosmol.* iii. 45 (*caption*) The expansion of the Universe as seen from another galaxy. The recession velocity is still proportional to the distance. Thus Hubble's law does not favour the Milky Way. *Ibid.* 46 Hubble's constant is now believed to be about 10 million years. *Ibid.* 48 Any systematic evolution in the intrinsic properties of the galaxies would contribute to the observed deviation from the simple Hubble law. **1972** *Sci. Amer.* Feb. 41/2 The new value of the Hubble constant is 53 ± 5 kilometers per second per megaparsec, or almost exactly one-tenth of Hubble's original value. **1973** *Physics Bull.* Nov. 652/1 Present estimates of the Hubble constant yield a value of $R/\dot{R}$ in a range $(1\text{–}2) \times 10^{10}$ years.

hubless (hɒ·blės), *a.* [f. HUB[1] + -LESS.] Without a hub.

1970 *Official Jrnl.* (*Patents*) 31 Dec. 4577/1 Circumferentially loaded and snubbered hubless wheel surface locomotion apparatus. **1971** *New Scientist* 10 June 632/2 A load belt is wrapped round both the hubless rollers and the load hub in a tortuous path so as to apply any load from above as a circumferential load on the hubless wheels. **1971** *Daily Tel.* (Colour Suppl.) 8 Oct. 16 (Advt.), Hubless 5½" sport wheels. Radial-ply tyres.

‖**hubris** (hiū·bris). [a. Gr. ὕβρις (cf. *HYBRIS).] Presumption, orig. towards the gods; pride, excessive self-confidence.

1884 *Daily News* 28 Oct. (Ware), Boys of good family, who have always been toadied, and never been checked, who are full of health and high spirits, develop what Academic slang knows as *hubris*, a kind of high-flown insolence. **1923** J. M. MURRY *Pencillings* 272 So confident indeed did I become that I began to join in the scholarly chuckle at the vainglorious and foolhardy man—was ever a purer case of *hubris*? **1924** G. B. SHAW *Saint Joan* v. 60 You have stained yourself with the sin of pride. The old Greek tragedy is rising among us. It is the chastisement of hubris. **1950** A. HUXLEY *Themes & Variations* 259 The Greeks..knew very well that hubris against the essentially divine order of Nature would be followed by its appropriate nemesis. *a* **1963** C. S. LEWIS *Poems* (1964) 3 Walk carefully, do not wake the envy of the happy gods, Shun Hubris. **1965** *Listener* 23 Sept. 449/2 There they learned morality and conduct; the virtues of nobility and the golden mean and the menace of *hubris*. **1971** *Country Life* 11 Mar. 541/1 Not much in the way of punishment for *hubris* will be seen falling on the heads of the publicity-maniacs.

hubristic, *a.* Delete *rare* and add later examples. Also **hubri·stical.**

1904 *Westm. Gaz.* 6 May 2/1 These Chamberlainites are very hubristic. **1923** J. M. MURRY *Pencillings* 272, I was feeling at my ease, perhaps even a little hubristical myself. **1930** *Punch* 1 Oct. 381/2 This deleterious and delirious Dean, Who of the fount of Hippocrene Drank in a manner most hubristic. **1961** *Times* 18 May 17/2 It was hubristic of the band to play the National Anthem in an eighteenth century version.

hubristically (hiubri·stikăli), *adv.* [See -LY[2].] With hubris; in a presumptuous manner.

1907 *Athenæum* 19 Oct. 473/1 He was..rather inclined to treat cavalierly, not to say hubristically, the quiet people who [etc.].

Hubshee (hɒ·bʃi), *sb.* and *a.* Also 7 **Abbasie, Hobsy,** 8 **Hobshy, -ee, Habashi,** 9 **Haffshee, Hubshi.** [Pers. *ḥabšī*, Arab. *ḥabašī*, of or belonging to Habesh or Abyssinia.] **A.** *sb.* An Ethiopian, an Abyssinian. **B.** *adj.* Ethiopian, Abyssinian. In Anglo-Indian usage both *sb.* and *adj.* may be loosely applied to an African Negro (see quot. 1901).

1601 J. SANDERSON *Trav. Levant* (1931) 108 Abbasies of Ethiopia. **1698** J. FRYER *New Acct. E.-India & Persia* 147 They speak of his *Hobsies* after this manner, That with their Swords they are able to cut down Man and Horse. *Ibid.* 168 He being from an *Hobsy Cophir* made a free Denizen. **1757** J. H. GROSE *Voy. E.-Indies* 238 The Moors are also fond of having Abyssinian slaves known in India by the name of Hobshee Coffrees. **1789** *Seir Mutaqherin* III. 36 (Y.), In India Negroes, *Habissinians, Nobis* (*i.e.* Nubians), &c. &c. are promiscuously called Habashies or *Habissians*, although the two latter are no Negroes. **1834** J. FORBES *Oriental Mem.* (ed. 2) II. 473 The master of a family adopts a slave, frequently a Haffshee Abyssinian, of the darkest hue, for his heir. **1888** *Contemp. Rev.* Feb. 167 'Hubshees', who looked, though they were not, Negroes, have in India carved out thrones. **1901** KIPLING *Kim* vii. 179, I would not appear to her as a *hubshi* (nigger). **1932** —— *Limits & Renewals* 214 Whether he had been General or Sweeper—Sahib—Mussulman—Yahudi—Hubashi—or heathen.

huchen (hū·kən, hū·χən). Also **hucho.** [G.] A large salmonid fish, *Hucho hucho*, native to the Danube and its tributaries.

a **1829** H. DAVY *Consolations in Travel* (1830) iv. 204 The trout, salmon, hucho and others of the salmo genus. **1897** *Encycl. Sport* I. 14/1 Mahseer in India, the hucho in Bavaria, the monster trout of the Canadian lakes..are alike taken by the spinning rod. **1905** *Westm. Gaz.* 2/2 The huchen is excellent food. *Ibid.* 7/3 The Committee have proceeded with the experiment of attempting the introduction into the river [*sc.* the Thames] of the huchen, the non-migratory salmon. **1962** D. W. TUCKER tr. *Sterba's Freshwater Fishes of World* 56 Huchen do very well in captivity. **1971** B. J. MUUS *Freshwater Fish Brit. & Europe* 62/1 The huchen is found in the Danube and part of its tributaries, and only exceptionally in lakes.

huckleberry. Add: **b.** *huckleberry pie.*

1775 P. V. FITHIAN *Jrnl.* (1934) II. 68 We have.. boil'd potatoes & huckleberry-pie. **1869** [see **cranberry pie*]. **1947** *Mazama* Sept. 1/1 Smell that turkey, those roasting ears, and the huckleberry pies? **1972** *Punch* 1 Mar. 292/3 A sliver of freeze-dried huckleberry pie with apple pandowdy and French fries.

2. *U.S. colloq.* A small amount, degree, or extent.

1832 J. K. PAULDING *Westward Ho!* I. 182 [I once got] within a huckleberry of being smothered to death. **1920** E. BOK *Americanization of Edward Bok* 165 He always kept 'a huckleberry or two' ahead of his readers.

3. A person, *spec.* (*derog.*) a person of little consequence.

[**1835** *Gent's. Vade-Mecum* (Philadelphia) 22 Aug. 2/4 Orson, the wild man of the woods is nothing to him—not a circumstance—not a huckleberry.] **1868** *New Eng. Base Ballist* 3 Sept. 17/1 Now then, my huckleberry, look sharp! you're wrong! **1889** 'MARK TWAIN' *Connecticut Yankee* 338 The Saracen..is no huckleberry.

4. In various phrases: *to be someone's huckleberry*: to be someone's sweetheart, friend, or partner; *to be a huckleberry to* (or *over*) *someone's persimmon*: a proverbial phrase (see quots.).

1832 J. K. PAULDING *Westward Ho!* I. ix. 80 If the [broad-]horn gets broadside to the current, I wouldn't risk a huckleberry to a persimmon that we don't every soul get treed, and sink to the bottom. **1834** D. CROCKETT *Narr. Life* ix. 70 But to do this, and write the warrants too, was at least a huckleberry over my persimmon. **1856** W. G. SIMMS *Eutaw* 553 My larning ain't a huckleberry to your persimmon. **1880** A. A. HAYES *New Colorado* (1881) v. 68 The first words that we heard him speak settled his nationality, for..he sententiously remarked, 'Hi'm 'is 'uckleberry.' **1885** D. D. PORTER *Incidents Civil War* 204 'I am the fleet-surgeon of the Mississippi squadron!'..'I'm a huckleberry above that persimmon, 'cause I'm the chief cook.' **1889** [see PERSIMMON 3]. **1926** N. N. PUCKETT in A. Dundes *Mother Wit* (1973) 8/2 Sir, you is a huckleberry beyon' my persimmon. **1936** J. TULLY *Bruiser* (1946) 37 Well, I'm your huckleberry, Mr. Haney. **1951** *Publ. Amer. Dial. Soc.* xv. 56 I'll be your huckleberry.

huckster, *sb.* Add: **2. b.** An advertising agent chiefly concerned with the preparation of advertising programmes for radio broadcasting.

1946 F. WAKEMAN (*title*) The hucksters. **1947** *Britannica Bk. of Year* 840/2 *Huckster*, a radio advertising man. **1965** *English Studies* XLVI. 464 *Huckster*, broker... Also used colloquially of an advertisement copy writer.

Hence **hu·cksterism**, the theory or practice of being a huckster (usu. disparaging).

1951 *Newsweek* 27 Aug. 80 Robert Saudek, a three-time Peabody Award winner for documentaries. Saudek, a soft-spoken man without a hint of hucksterism. **1957** *N.Y. Times* 6 Jan. E 11/4 An attack on Southern schools of journalism for 'kicking the humanities around' while emphasizing 'hucksterism' and 'quick turnover' in education. **1960** *Encounter* XV. 27 One can find 'hucksterism'..among academic people in search of reputations. **1972** *Village Voice* (N.Y.) 1 June 26/2 What particularly appeals to me about the 'Jack La Lanne Show' is its serious, heartfelt hucksterism. **1973** *Observer* 2 Dec. 7/3 This high-pressure hucksterism, backed by the multi-billion investment.., has failed to impress many scientists.

huddle, *sb.* Add: **1., 1. b.** (Later examples.)

1906 *Daily Chron.* 5 Mar. 6/6 A really fine organic city and not a mere gigantic huddle of apartment dwellings. **1924** W. M. RAINE *Troubled Waters* viii. 79 Beyond the post office a great huddle of sheep was being driven forward. **1959** *Listener* 26 Feb. 384/3 Stanley Spencer's 'Temptation of St. Anthony', with its huddle of precisely drawn nudes.

4. A close or secret conference; esp. in phr. *to go into a huddle*, to hold a secret conference, to consult specially (about something). *colloq.*

1929 E. LOOKER *White House Gang* i. 13 The Gang was recruited, and..it went into a huddle, to confound all rules of deportment in high places. **1932** *Harper's Mag.* Apr. 600 'When an agent comes,' explains René, 'we have a little *conférence*. We go into what you call a huddle, yes?' **1934** E. LINKLATER *Magnus Merriman* xi. 128 Your girl-friends'll need to go into a huddle to think up something to beat that one. **1938** F. SCOTT FITZGERALD *Let.* 5 Sept. (1964) 93 I'm going into a huddle on this script and probably won't be able to write you again at length before Vassar starts. **1947** J. BERTRAM *Shadow of War* 318 He went into a huddle with one of his minions. **1959** 'A. GILBERT' *Third Crime Lucky* iii. 49 If he writes in he wants to make some of what he never earned over to us there'll be a huddle. **1972** M. YORKE *Silent Witness* v. 121 Then she got scared when he was missing and that's why she and Roy were in a worried huddle this morning.

5. *Bridge.* A period of meditation during which a player considers his next call or play.

1934 *Amer. Speech* IX. 11/1 A *huddle* is a session of silent thought indulged in by a player either during the bidding or during play. **1964** *Official Encycl. Bridge* 248/1 If the huddle is followed by a positive action, usually no harm is done to the opponents.

huddledom (hɒ·d'ldəm). [f. HUDDLE *sb.* + -DOM.] A state or condition of confusion and disorder.

1923 *Glasgow Herald* 24 May 9 The huddledoms of haunting disease, poverty, and overcrowding. *Ibid.* 31 May 6 Into this huddledom..came a youth, David Watt Torrance.

Hudson Bay (hɒ·dsǫn bēi). The name of a body of water in northern Canada (also with the possessive form *Hudson's*, from the name of the trading company, the Hudson's Bay Company), used in numerous phraseological combinations (see esp. Avis et al. *Dict. Canadianisms*), e.g. **Hudson('s) Bay blanket, Hudson's Bay tea.**

1900 J. LONDON *Son of Wolf* 181 Prince wrapped a Hudson Bay blanket about her with a mock reverence. **1912** 'R. CONNOR' *Corporal Cameron* 438 Tall, she looked..lithe and strong, her close-fitting Hudson Bay blanket coat revealing the swelling lines of her budding womanhood. **1920** A. STRINGER *Prairie Mother* 263, I sat there wrapped up in one of Dinky-Dunk's four-point Hudson-Bays. **1943** R. E. & D. SWANSON *Rhymes Haywire Hooker* (1953) 51 The stranger smiles and then he says: 'Why, them is four point Hudson Bays!' **1948** *Beaver* Mar. 14 A tea used to be made from Hudson's Bay tea or Labrador tea. **1955** W. G. HARDY *Alberta Golden Jubilee Anthol.* 85 In his last days, the old warrior was a familiar sight in his Hudson Bay blanket coat. **1961** E. E. RICH *Hudson's Bay Company* III. xxvi. 738 In 1849–50 the American government had bought some Hudson's Bay blankets to distribute as gifts to the Indians. **1968** L. J. BRAUN *Cat who turned on & Off* (1969) iii. 25 One man in a Hudson Bay blanket coat carried a small dog dressed to match. **1970** *Beaver* Winter 22 Labrador tea, also known as Hudson's Bay tea across much of Northern Canada, is a pretty evergreen shrub whose robustly aromatic leaves make it one of the most famous teas of the north country.

Hudsonian (hɒdsōu·niăn), *a.* [f. the name of an English navigator, Henry *Hudson* (died *c* 1611), discoverer of the North American bay, river, and strait which bear his name: see -IAN.] Of or pertaining to Hudson Bay and the surrounding land, *esp.* in the names of animals and the territory from Labrador to Alaska, as far south as the tree line, classified as the Hudsonian biogeographical zone.

1835 J. J. AUDUBON *Ornith. Biogr.* III. 727 The Hudsonian Godwit, *Limosa Hudsonica*..is scarcely ever found farther south along the coast than the State of Maryland. **1858** S. F. BAIRD in *Rep. Explor. Route to Pacific* (U.S. War Dept.) IX. 744 *Numenius Hudsonicus.* Short-billed or Hudsonian Curlew... Atlantic and Pacific Coasts of North America. **1871** *Bull. Mus. Compar. Zool. Harvard* II. 401 The Hudsonian Fauna doubtless embraces outlying islands of the Canadian Fauna, as the upper part of the White Mountains, and the summits of some of the higher peaks in the Adirondacks. **1939** *Beaver* June 48/1 The Hudsonian chickadees and Canada jays would come near, making life a bit more cheerful. **1947** R. T. PETERSON *Field Guide Birds* (ed. 2) 266 *Parus hudsonicus*: (1) Hudsonian Brown-capped Chickadee, *P. h. hudsonicus.* **1948** A. L. RAND *Mammals E. Rockies* 15 The Hudsonian zone in Alberta is poorly characterized by mammals. **1952** T. M. STANWELL-FLETCHER *Tundra World* 42, I was startled by a series of loud, wild 'tit-tit-tit-tit-tit' notes from a big brown bird which leapt into the air a few feet in front. A Hudsonian curlew of course. **1952** D. F. PUTNAM *Canad. Regions* 139/1 The Hudsonian Life Zone..extends from the timber limit to the south of James Bay, Lake Mistassini and Pointe des Monts on the North Shore. **1964** A. L. THOMSON *New Dict. Birds* 713/2 The smaller Whimbrel *N*[*umenius*] *phaeopus* is Holarctic (its American race being sometimes called 'Hudsonian Curlew'). *Ibid.* 715/1 The Hudsonian Godwit *L*[*imosa*] *haemastica* migrates from North America to South America.

Hudson seal. [f. the name of Henry *Hudson* (see *HUDSONIAN *a.*) + SEAL *sb.*[1]] Musk-rat fur that has been plucked and dyed to give the appearance of seal fur. Also **Hudson Bay seal.**

1914 J. W. JONES *Fur-Farming in Canada* (ed. 2) 7 When the furdressers and dyers produced a clipped and dyed muskrat skin that resembled sealskin almost perfectly, it was found that it would not sell under its own name..[the] high-priced fur is now sold as 'Hudson Bay seal'. **1920** *Eye Opener* (Calgary) 11 Sept. 1/4 The jackpot is to be the real thing in Hudson seal coats this coming winter. **1921** *Daily Colonist* (Victoria, B.C.) 2 Oct. 8/5 First Quality Hudson Seal Coat, fancy silk lined; extra large collar and cuffs of Alaska sable. **1936** D. MCCOWAN *Animals Canad. Rockies* xv. 134 When a musk-rat skin has been tanned and the coarse outer hair removed the remaining soft silky undercoat is known to furriers as Hudson seal. **1945** H. MACLENNAN *Two Solitudes* (1946) II. 219 Paul looked out the window and saw..women in black Hudson seal coats with their hands in black muffs, men with fur caps.

hue, *sb.*[1] Add: **3. c.** That attribute of a colour by which it is recognized as a red, a purple, a green, etc., and which approximately corresponds to its dominant wavelength (or to that of its complementary colour); it constitutes, along with saturation ('tint', purity, intensity) and lightness ('shade'), one of the

three attributes required for the complete specification of any colour.

In this sense hue is the quality in which different 'hues' (as distinct from 'tints' and 'shades': see SHADE *sb.* 4) differ; cf. quot. 1835 below and quot. 1859 s.v. TINT *sb.*[1] 2 a.

[**1835** G. FIELD *Chromatography* iii. 28 By mixing his colours with white, the artist obtains..tints; by mixing colours with colours, he obtains compound colours, or hues; finally, by mixing colours or tints with black, he gets..shades.] **1855** J. C. MAXWELL in *Trans. R. Scottish Soc. Arts* IV. 395 There will be two things on which the nature of each ray will depend:—(1.) its intensity or brightness; (2.) its hue, which may be estimated by its position in the spectrum, and measured by its wave length. *Ibid.* 396 Colours differ not only in intensity and hue, but also in tint; that is, they are more or less pure. **1872** —— in *Not. Proc. R. Inst. Gt. Brit.* VI. 263 Colour may vary in hue, tint, and shade... A difference in hue may be illustrated by the difference between adjoining colours in the spectrum. **1900** G. H. HURST *Colour* i. 13 The hue of a colour is that constant which is commonly denominated by the term colour, as blue, or green, or red. **1936** A. B. KLEIN *Colour Cinematogr.* i. 89 There are about 130 steps of just distinguishable difference in hue in the spectrum. **1939** M. LUCKIESH *Colour* 39 The names of colors are often taken from the hue and usually imply it. **1955** P. D. TREVOR-ROPER *Ophthalm.* x. 137 Monochromatic light may alter its apparent hue as it becomes more unsaturated, red turning to pink, orange to yellow. **1960** G. M. WYBURN *Nervous Syst.* vi. 83 Colour or hue, which is our interpretation of variations in light wavelength is comparable to the pitch of sound. **1966** R. R. COUPE *Sci. of Printing Technol.* ix. 209 To describe completely a colour, we must take into account three different properties, namely hue, saturation and lightness.

hue (hū·e), *sb.*[3] *N.Z.* [Maori.] A local name for the bottle gourd, *Lagenaria vulgaris.*

1843 E. DIEFFENBACH *Trav. N.Z.* II. iv. 49 The calabashes (hue) were..the next addition to their stock of eatables. **1868** W. COLENSO in *Trans. N.Z. Inst.* I. III. Essay. 36 The Hue, or gourd, (a species of *Cucurbitæ*), gave useful Calabashes, and vessels of several kinds and sizes, from a gill to three gallons. **1905** W. B. *Where White Man Treads* 15 Besides being a succulent delicacy when young, the matured vegetable hue, with its strong, horny rind, could be put to the uses of many utensils, as drinking cups, bowls, etc., and, most important of all, water and oil flasks. **1921** H. GUTHRIE-SMITH *Tutira* viii. 55 The land [was] usually too poor for the cultivation on a great scale of such exotics as..the *hue* (*Lagenaria vulgaris*). **1949** P. BUCK *Coming of Maori* (1950) II. i. 91 The gourd (*hue*) was grown principally to provide containers for water and for preserved birds.

hueless, *a.* **2.** (Later examples.)

c **1865** E. DICKINSON *Poems* (1955) II. 737, I sight the Aprils—Hueless to me until thou come. **1932** CHESTERTON *Chaucer* viii. 264 The sort of harsh and hueless light that can be seen in the black engravings in the old Family Bibles.

huemul, var. *GUEMAL.

huerta (hwē·ɹtă). [Sp.] A piece of irrigated land in Spain or in the Spanish-speaking areas of Latin America; also, an orchard.

1838 A. GANIHL *Mexico vs. Texas* i. 13 He was resting himself, and enjoying the cool of the evening breeze, under a spreading orange tree, in his *huerta.* **1841** BORROW *Zincali* I. II. iv. 287 The justicia will compel us to restore the ass; we have, however, already removed her to our huerta out of the town. **1859** T. R. WARREN *Dust & Foam* viii. 225 In each of these huertas is a reservoir, built of masonry, through which water is constantly flowing. **1924** *Glasgow Herald* 28 July 5 The huertas merge into a delicious confusion of flower and fruit. **1934** M. R. SHACKLETON *Europe* vii. 89 From the Ebro delta to Cape de la Nao in Valencia, the irrigated districts ('huertas', from Lat. *hortus* = a garden) are practically continuous along the coast. **1958** FISHER & BOWEN-JONES *Spain* iv. 52 Originally developed by the Moslems, the 'huertas' are small, highly cultivated plots which depend on irrigation water brought by an intricate system of channels, aqueducts and lifts.

huff, *v.* Add: **1.** Survives in phr. *to huff and puff.* (In some contexts not distinguishable from sense 4.)

1890 J. JACOBS *Eng. Fairy Tales* xiv. 69 Then I'll huff, and I'll puff, and I'll blow your house in. **1959** *Times* 11 Nov. 13/6 But it would be unrealistic to think that by huffing and puffing at him..the General can be made to change his mind. **1963** *Guardian* 1 Oct. 8/6 Mr. Liukov may huff and puff..about pernicious Western influences ..but the Bulgarians..are discovering..a sense of humour. **1967** *Listener* 13 July 57/2 Sarah [Churchill] herself never mastered the techniques of politics; she huffed and she puffed but at the end of it all few houses had been blown down. **1971** *Daily Tel.* 13 Jan. 1 Ministers were undisturbed by all the huffing and puffing against the Industrial Relations Bill at the Albert Hall last night.

7. (Earlier and later examples.)

1793 M. RISHTON *Let.* 12 Aug. in F. Burney *Jrnls. & Lett.* (1972) II. 185, I am a horrid Coward and get huffed every time we drive out. **1800** *Aurora* (Philadelphia) 18 Dec. (Th.), The Philadelphia Gazette is huffed at our stating a fact. **1906** J. M. SYNGE *Lett. to Molly* (1971) 50 Dont imagine I'm huffed or anything, little heart, I'm only weary. **1969** *Listener* 17 Apr. 535/2, I don't know whether to be huffed or pleased about it.

11. *Mil. slang.* To kill.

1919 *Athenæum* 23 May 360/2 Unmitigated slang, like..'to huff' for 'to kill'. **1925** FRASER & GIBBONS *Soldier & Sailor Words* 122 *To be huffed*, to be killed. **1933** PARTRIDGE *Words, Words, Words!* III. 197 The English synonyms for death..*huffed*; *out of mess*, dead.

huff, *sb.* Add: **1.** Survives in phr. *huff and puff.* (In some contexts influenced by sense 2.)

1961 *John o' London's* 9 Nov. 517/3 Nor do I really fancy Papermac's huff-and-puff. **1962** *Listener* 19 July 112/2 It seems a pity that this new era in telecommunications should be accompanied by an international huff-and-puff over priorities which at its worst comes perilously near to sub-lunar soap-opera. **1966** *Listener* 26 May 772/3 Even Ernesto is essentially a pallid version of those huff-and-puff baritones of Donizettian melodrama. **1967** G. SIMS *Last Best Friend* ix. 82 '—ing old woman,' Spiegl said. 'All huff and puff.' **1973** *Times* 16 Mar. 14/4 We are going through a period of a great deal of huff and puff generated by consumer organizations.

huff-duff (hɒ·f,dɒf). [Representing the pronunc. as a word of the initials *h. f. d. f.*, for *high-frequency direction-finder.*] A device for determining the direction of incoming high-frequency radio signals, enabling their source to be located when bearings are obtained by two or more such devices.

1946 *N.Y. Herald-Tribune* 16 Jan. 22/1 'Huff duff', with a range that makes it possible to detect and plot radio signals of as little as fifteen seconds' duration emitted halfway around the globe from plotting stations. **1955** C. S. FORESTER *Good Shepherd* ii. 39 Huff-Duff reports foreign transmission. **1962** L. FARAGO *Tenth Fleet* viii. 162 They used decoy subs to befuddle the Huffduff.

huffkin (hɒ·fkin). *local.* Also **uffkin.** [Origin unknown.] A kind of tea-cake made chiefly in Kent.

1790 in *Eng. Dial. Dict.* **1869** *N. & Q.* 4th Ser. IV. 76/1 Most people know what muffins and crumpets are, but in East Kent the former..are known as *uffkins.* **1887** PARISH & SHAW *Dict. Kentish Dial.*, *Huffkin, hufkin*, a kind of bun or light cake, which is cut open, buttered, and so eaten. **1928** *Daily Express* 14 Mar. 5/4 Our merry teas after net-ball, cricket, or football, without huffkins would be very 'dry'. **1952** F. WHITE *Good Eng. Food* v. iv. 199 This quantity makes 10 huffkins weighing 2¼ oz. each.

huffle, *v.* Add: **1. c.** Of the wind: to make a sound as of blowing in gusts. So **hu·ffle** *sb.*, a sudden gust of wind, or the sound made by this.

1862 W. BARNES *Poems Rural Life* 3rd Coll. 122 Where sharp-leav'd ashes' heads did twist In hufflèn wind, an' driftèn mist. **1878** HARDY *Ret. Native* I. I. iii. 70 The winds do huffle queerer to-night than ever I heard 'em afore. **1889** S. BARING-GOULD et al. *Songs of West* 9 At the huffle of the gale, Here I toss and cannot sleep. **1891** 'L. MALET' *Wages of Sin* III. VI. ii. 82 A huffle of wind, hot with the festering reek of the streets away across the river, fluttered the leaves... The wind huffled again.

huffler. Add: Still used of bargemen.

1824 *Gentl. Mag.* Aug. 111/2 We have a class of people in these parts called *Ufflers*, i.e. men in the barging line out of employ, who attend as extra help to get the craft home in our inland navigation. **1918** *Chambers's Jrnl.* Feb. 109/1 Here it is that the barge skippers and the barge mates, and the river 'hufflers' congregate. **1927** *Daily News* 27 May 6/6 If it is necessary to take on a third hand, he is generally regarded as being quite an inferior person, and is known as a 'huffler'. **1948** H. BENHAM *Last Stronghold of Sail* ii. 29 There were 'hufflers' in plenty in those days to come aboard and lend a hand.

hugaboo (hɒ·găbū:). [Fanciful, prob. infl. by BUGABOO, HULLABALOO, etc.] Portentousness, pomposity.

1930 *Eng. Rev.* Feb. 140 The silly hugaboo of minutes and protocols. **1932** S. GIBBONS *Cold Comfort Farm* i. 9 Marriages..should take place in churches, with all the usual paraphernalia and hugaboo.

hügelite (hü·gėləit). *Min.* [ad. G. *hügelit* (V. Dürrfeld 1914, in *Zeitschr. f. Kryst. u. Min.* LIII. 183), f. the Ger. family name von *Hügel*: see -ITE[1].] A yellow or brown mineral originally described as a hydrated vanadate of lead and zinc but later shown to be a hydrated arsenate of lead and uranium.

1914 *Chem. Abstr.* VIII. 310 The name hügelite is suggested for the vanadate previously described. **1916** *Mineral. Mag.* XVII. 352 Hügelite... A hydrated vanadate of lead and zinc occurring as yellow monoclinic needles on corroded galena at Reichenbach, Baden. **1962** *Amer. Mineralogist* XLVII. 418 Hügelite was described by Dürrfeld (1913) as a hydrous lead zinc vanadate... Transparent to translucent crystals are brown to orange-yellow... Microchemical tests showed Pb, U and As as main components, but no Zn, V or P... The formula is proposed as $Pb_2(UO_2)_3(AsO_4)_2(OH)_4.3H_2O$.

huggable (hɒ·găb'l), *a.* [f. HUG *v.* + -ABLE.] Such as invites hugging.

1898 D. C. MURRAY *Tales* 205 Eminently kissable little face; eminently huggable little figure. **1908** *Daily Chron.* 5 Nov. 7/5 The 'Teddy' Bear is such a huggable creature that..he almost compels a caress. **1928** L. ROSSITER *Sex Age* viii. § 5 Sorry, Jacko. But you do look so huggable tonight. **1970** *Daily Tel.* 16 Dec. 11 Huggable rag doll..with removable dress.

Hughie (hiū·i). *Austral.* and *N.Z. slang.* Also **Huey.** [Diminutive of the name *Hugh*: see -IE, -Y[6].] The 'god' of weather, especially in phr. *send her down, Hughie!*

1937 PARTRIDGE *Dict. Slang* 209/1 New Zealanders and Australians say *send her down, Hughie!* **1958** *Tararua* XII. 27 The derisive phrases, 'Send it down, Hughie' and 'Let it come down, Hughie', go back to early in this century and are variants of the British military catch-phrase 'Send it down, David'... St. Hugh has long been associated with rain. **1962** *Austral. Women's Wkly.* Suppl. 24 Oct. 3/2 Come on, Huey, send the waves up. **1967** J. CLEARY *Long Pursuit* vii. 151 'Good on you, Hughie,' he said gratefully and with true religion... Hughie was the Australian working man's vernacular for the Lord. **1971** *N.Z. Listener* 19 Apr. 57/1 Well, that night Hughie sent it down, a nor' wester followed by a southerly buster.

hug-me-tight (hɒ·g,mi,təit). [f. the phr. *hug me tight.*]

1. A woman's short close-fitting jacket, usu. made of wool. orig. *U.S.*

1860 *Godey's Lady's Bk.* Dec. 544 Hug me tight. A garment to be worn under a cloak. **1869** L. M. ALCOTT *Little Women* II. v. 68 She..made him..demand fiercely the meaning of a 'hug-me-tight'. **1910** *N.Y. Even. Post* 24 Dec., Suppl. 3 Hug-me-tights and mittens, all knit at home by grandmother. **1924** *Mod. Draper* II. 94 Articles, such as spencers, hug-me-tights, etc., which are worn above the undergarment, and under the outer garment. **1934** E. BOWEN *Cat Jumps* 130 Mrs. Archworth sat propped up in bed in a hug-me-tight trimmed with marabout. **1959** *Guardian* 25 Nov. 6/4 For that extra degree of warmth..over the shoulders..this cuffed hug-me-tight of the same tweed as the dress. **1972** F. B. MAYNARD *Raisins & Almonds* 45 A variety of cozy garments called hug-me-tights.

2. A type of buggy. Also *attrib. U.S. Hist.*

1901 W. N. HARBEN *Westerfell* i, I seed 'em takin' a ride in his new hug-me-tight buggy yesterday. **1902** —— *Abner Daniel* v, He's got a new buggy—a regular hug-me-tight. **1948** *Jrnl. Amer. Folk-Lore* Apr.–June 212 Those roads were so bad..that they kept making the buggies narrower and narrower... Some of them got so narrow they used to call them 'Hug-Me-Tights'.

Hugoesque (hiūgoe·sk), *a.* [f. the name of Victor M. *Hugo* (1802–1885), French author + -ESQUE.] Resembling the character or style of V. Hugo. Also subst. with *the.*

1893 E. SALTUS *Madam Sapphira* xiii. 164 That would be mediæval. I mean nothing so Hugoesque. **1904** *Daily Chron.* 3 Mar. 3/2 There is a touch of the Hugoesque in Rodwell. **1960** J. LODWICK *Asparagus Trench* 20 Almost Hugoesque in his unflagging pursuit of maids.

hugsome (hɒ·gsɒ̆m), *a.* [f. HUG *v.* + -SOME[1].] Such as invites hugging, huggable (see also quot. 1893).

1893 FARMER & HENLEY *Slang* III. 375/2 *Hugsome adj.* (colloquial), carnally attractive; fuckable. **1894** *Outing* (U.S.) XXIV. 417/1 A [bear's] long, straining, hugsome hug, which breaks the dog's ribs. **1942** BERREY & VAN DEN BARK *Amer. Thes. Slang* § 427.1 Hotsy-totsy, hugsome hussy, humdinger, irresistibelle.

huh, *int.* Add: (Later examples.) Also as an expression of interrogation.

1890 'O. THANET' *Expiation* ix. 166 A loud snort of contempt from the gallery betrayed that Hizzie had heard. 'Huh!' she bawled, 'you yent gwine get killed up, not long's ye kin *run*!' **1924** *Dialect Notes* V. 270 *Huh* (surp., disg., enquiry). **1937** L. B. MURPHY *Social Behavior & Child Personality* ii. 53 Agatha said, 'Want to play in the sand box, Theodore, huh? Do you?' **1940** R. CHANDLER *Farewell, My Lovely* xxi. 169 Pretty trivial in my case, huh? **1948** F. & R. LOCKRIDGE *Pinch of Poison* xvi. 151 'Listen, Mullins,' Weigand pleaded. 'Don't think, huh?' **1953** *Manch. Guardian Weekly* 20 Aug. 7/2 'I could go back there, I mean some other time.' 'But not now, huh?' **1969** K. AMIS *Green Man* iv. 182 God's purpose. Huh. I'm no more qualified than the next man to tell you what that is. **1970** *Washington Post* 30 Sept. D 4/2 'Oh boy that's just what he needs,' he said, cheerlessly. 'Tell him to take it easy, huh?' **1971** *Black World* Apr. 59 Hell, its all my fault, huh? **1972** *Canad. Jrnl. Ling.* XVII. 94 Whatever its origin and history, *huh?* is currently in widespread use in the United States. **1972** *Southerly* XXXII. 54 Because the fun part is over. They think. Huh! **1973** A. PRICE *October Men* ix. 126 'Huh!' Macready snorted derisively.

hühnerkobelite (hünəɹkōu·bėləit). *Min.* [f. *Hühnerkobel*, name of a hill near Lam in eastern Bavaria, Germany: see -ITE[1].] An olive-green phosphate of sodium, calcium, bivalent iron, and bivalent manganese, $(Na, Ca)(Fe^{++}, Mn^{++})_2(PO_4)_2$, in which iron predominates over manganese.

1950 M. L. LINDBERG in *Amer. Mineralogist* XXXV. 75 It is here proposed to give the name hühnerkobelite to the material from Hühnerkobel and from Norrö, not isostructural with true arrojadite. **1965** *Amer. Mineralogist* L. 713 (*heading*) Hühnerkobelite crystals from the Palermo No. 1 pegmatite, North Groton, New Hampshire.

huhu (hū·hū). *N.Z.* [Maori.] The beetle *Prionoplus reticularis*, or its larva (in full *huhu grub*), found in decayed timber.

1848 R. TAYLOR *Leaf from Nat. Hist. N.Z.* 5/1 Insects ..Huhu, a grub which bores into wood. **1926** R. J. TILLYARD *Insects Austral. & N.Z.* xx. 232 *Prionoplus reticu-*

laris Wh., the largest of all New Zealand beetles, measures up to two inches in length and is dark brown..; it is very common and flies to light. The larva, called 'Hu-hu' by the Maoris, is eaten as a delicacy; it bores into fallen forest timber. **1949** F. SARGESON *I saw in My Dream* I. vii. 52 The children..looked for wetas and huhus. **1956** T. SUTHERLAND *Green Kiwi* ii. 45 The timber tunnelled by the mokoroa or huhu grubs. **1960** B. CRUMP *Good Keen Man* 45 My next mate was a Maori... I once found him tearing a rotten log to pieces, scattering powdery wood all over the place as he searched for huhus. When he found one he'd pick the butter-coloured grub out with his fingers and eat it as it was. **1966** *Encycl. N.Z.* I. 187/2 Among the more conspicuous New Zealand beetles are the 'Huhu' (*Prionoplus reticularis*), a large dead-wood borer of the family Cerambycidae, [etc.].

huh-uh (ɒ·,ɒ), *int. U.S.* An expression of negation. Also **uh-huh*, **uh-uh.*

1948 F. & R. LOCKRIDGE *Pinch of Poison* xvi. 152 Mullins shook his head. 'Huh-uh,' he said. 'She ain't the type.' **1962** *Amer. Speech* XXXVII. 230 Foreign visitors to the United States, who have learned the proper affirmative and negative responses of *yes* and *no*, are often bewildered by the widespread colloquial use of *uh-huh* and *huh-uh.*

hu-hu-hu (hū,hū,hū). Also **hoo-hoo-oo.** Representing a repetition of the sound of Hoo *int.* and *sb.*

1884 [see Hoo *int.* and *sb.*]. **1911** T. E. LAWRENCE *Home Lett.* (1954) 154 With a mighty firing of guns & pistols, and the hu-hu-hu and violent tahleel of the women [*sc.* native women in Syria].

‖**hui** (hū·i). [Maori and Hawaiian.] In New Zealand, a large social or ceremonial gathering; in Hawaii, a formal club or association.

1858 J. MORGAN *Let.* 21 June in *Richmond-Atkinson Papers* (1960) I. 408 The *hui* at Rangiaohia to promote the Maori King movement broke up. **1898** M. H. KROUT *Hawaii* 18 Those present, with forty members of a royalist society called *Hui Kalaiaina*, marched to the palace. **1921** H. GUTHRIE-SMITH *Tutira* x. 81 The large *huis*—gatherings—of the Heretaunga people. **1948** KUYKENDALL & DAY *Hawaii* xxv. 274 The enterprise of oriental business men was strengthened by the local practice of forming a *hui*, the Hawaiian name for an informal syndicate which enables Chinese or Japanese members to take great advantage of a business opportunity. **1954** J. SHERIDAN in *Ellery Queen's Mystery Mag.* Oct. 15/2 Our fishing *hui*'s beginning to pay. **1959** M. SHADBOLT *New Zealanders* 20, I mean it's common knowledge about what happens at the pa when there's a big hui. There's always too much liquor there, and the boys and girls mix freely. **1960** N. HILLIARD *Maori Girl* III. vii. 220 We'll have fowls. Go to all the *huis.* **1973** *Parade* (Austral.) Sept. 11/1 The Maori custom of steeping the grain in water until it was half-rotten, and then serving it up as a special dish at a tribal hui.

Huichol (witʃōu·l). [Sp., from the native name.] **A.** *sb.* **a.** A people of Mexican Indians; a member of this people. **b.** The language of this people. **B.** *adj.* Of or pertaining to this people or their language.

1900 *Mem. Amer. Mus. Nat. Hist.* III. I. i. 22 According to the Huichol myths, corn was once deer, the deer having been the chief source of food in earliest times. *Ibid.* vi. 154 With the Huichol, the 'eye' is the symbol of the power of seeing and understanding unknown things. *Ibid.* ix. 185 Peculiar to certain rain-making feasts are a stick and a dried armadillo, which form the paraphernalia of the clown.., called in Huichol Sikwaiki. **1903** C. LUMHOLTZ *Unknown Mexico* II. v. 91 The Huichols occasionally made comments that betrayed very fair reasoning powers. **1964** E. A. NIDA *Toward Sci. Transl.* x. 228 'Glorified God' (Mark 2:12) becomes in Huichol 'said to God: You are of good heart.' **1972** *Language* XXXVIII. 847 Hockett adds Huichol which has '/i/ instead of /ü/'.

Huk (hɒk). [Abbrev. of Tagalog *Hukbalahap*, f. initial syllables of *hukbó* army + *bayan* people, country + *laban* against + *hapan* Japanese (i.e. *hukbó ng bayan laban sa hapon* people's army against the Japanese).] A guerrilla movement in the Philippines, orig. against the Japanese in World War II, later popularly identified with communism. Also *attrib.*

1947 *Britannica Bk. of Year* 840/2 *Huks*, shortening of Hukbalahaps, a Tagalog word meaning 'armed peasants', 'people's army'. **1951** H. MACINNES *Neither Five nor Three* xv. 209, I read the article..about the Philippines. It seems that the Huks have nothing to do with Communism. **1966** *Economist* 22 Oct. 359/2 Killings and ambushes by Huk guerrillas..have recently been common in central Luzon. **1967** *Ibid.* 4 Feb. 414/2 The Huks, who began as anti-Japanese guerrillas, carried on an armed rural rebellion from 1950 to 1954. **1971** W. LAQUEUR *Dict. Politics* 403 Since 1965 there has been a revival of *Huk* activity. **1972** *Times* 30 Sept. 15/4 The Huks are well-educated, ideologically-motivated men.

‖**hukilau** (hukīlau·). [Hawaiian, f. *huki* to pull + *lau* net.] An Hawaiian fishing party usually involving many people and much revelry.

1954 J. SHERIDAN in *Ellery Queen's Mystery Mag.* Oct. 20/1 Oh, we're going to a hukilau. *Ibid.* 21/2 We're about ready for the hukilau... He led us seaward, explaining to Bill that *huki* means pull and *lau* means leaf, from the *ti* leaves which are used to frighten fish into the net. **1972** *New Yorker* 8 Apr. 84/1 (Advt.), Come to a hukilau,..a sort of Polynesian fish-in.

hukm, var. *HOOKUM.

hula (hū·lă). Also † **hura; hula-hula.** [Hawaiian.] **a.** An Hawaiian dance, with six basic steps, which portrays through symbolic and imitative gestures natural phenomena, sports, and historical or mythological subjects. Also *attrib.* Hence as *vb.*, to dance the hula.

1825 W. ELLIS *Jrnl. Tour Hawaii* 59 At 4 p.m. the musicians from Kau again collected on the beach, and the dancers commenced a *hura.* *c* **1835** LOWELL SMITH in M. D. Frear *Lowell & Abigail* (1934) 102 [The public disturbance of] perpetual hulas, drumming and dancing accompanied by howling and intonations. **1851** F. A. BUCK *Yankee Trader in Gold Rush* (1930) 81 They also take advantage of this to have a grand Hoolah Hoolah, or native dance. **1853** *Putnam's Mag.* II. 21/1 And now the floor was cleared, and preparation made for the great feature of the evening, viz., the *Hula-hula.* **1866** 'MARK TWAIN' *Lett. fr. Hawaii* (1967) 70 The girls danced the lascivious hula-hula—a dance that is said to exhibit the very perfection of educated motion. *Ibid.* 170 The flower and evergreen trumpery worn by the hula girls. **1868** *Daily Territorial Enterprise* (Virginia City, Nevada) 29 Aug. 3/1 We have borne ourselves with calm fortitude at a Sandwich Island *hula-hula.* **1892** STEVENSON & OSBOURNE *Wrecker* xvii. 259, I was entertained to a sea-bathe, indiscriminate cocktails, a dinner, a *hula-hula.* **1898** [see *grass skirt* (*GRASS *sb.*[1] 13)]. **1899** W. C. MORROW *Bohemian Paris* 96 The hula-hula of the Hawaiian women lacks the grace, dash and abandon of the Turkish dance. **1919** F. O'BRIEN *White Shadows South Seas* 4 Kelly began 'Tome! Tome!' a Hawaiian hula. **1927** [see *EMPENNAGE]. **1928** L. NORTH *Parasites* 14 It appeared that you could buy anything from a home to a Hula-skirt on time-payment. **1954** 'N. BLAKE' *Whisper in Gloom* I. iii. 49 A dusky maiden in hula-hula skirt, brassiere, and little else. **1954** *Ellery Queen's Mystery Mag.* Oct. 30/1 Her *hula* was very good. *Ibid.* 31/1 David was the one who taught me to swim and to *hula.* **1956** 'N. SHUTE' *Beyond Black Stump* viii. 236 With straw hula skirts..for sale in one corner among the picture postcards. **1970** *Observer* (Colour Suppl.) 15 Feb. 26/2 An ability to hula well does not..appear to be a quality inherent in every Tahitian.

b. *hula hoop*, a tubular, plastic hoop (HOOP *sb.*[1] 3) used for spinning round the body with movements akin to those of the hula; hence *hula-hoop* v. intr., *hula-hooping* vbl. sb. and ppl. a.

1958 *Economist* 11 Oct. 144/2 In a manner reminiscent of a primitive tribal ritual, the Hula Hoop..can be made to spin round the torso, or arm or leg or neck, by a broad swaying motion resembling the hip swinging of a Hawaiian hula dancer. **1958** *Observer* 9 Nov. 10/4 Hula-hooping children. **1958** *Times Lit. Suppl.* 21 Nov. p. xxix/3 Hoops, also of ancient origin, had virtually disappeared from shops and streets, until the sudden recent craze for 'hula' hoops brought them out in a new form. **1958** *Times* 19 Dec. 12/5 If she had been a little younger he would have asked her if she had been hula-hooping the previous day. **1959** *Times* 3 Jan. 10/4 As training they are advocating two hours of 'hula-hooping' every morning. **1969** *Official Gaz.* (U.S. Patent Office) 1 July TM16/2 Wham-O Mfg. Co., San Gabriel, Calif... Junior Hula Hoop... For plastic toy hoops. **1970** *Which?* Sept. 288/2 Remember hula-hoops and dislocated hips? **1973** *Daily Tel.* (Colour Suppl.) 6 Apr. 27/2 Few 'craze' toys have as long a life as the yo-yo: the hula-hoop and the late, unlamented 'clacker-balls' are examples which had success but were soon forgotten.

hule (ū·lī). Also **ule, ulé, ulli.** [Mexican Sp. (*h*)*ule*, Nahuatl *ulli* or *olli* caoutchouc.] A Central American tree of the genus *Castilloa* (formerly *Castilla*), or the crude rubber obtained from it. So **hule-ro, ule-ro**, a collector of rubber.

1846 J. LINDLEY *Veget. Kingd.* 271 The tree Ule of Papantla, from which caoutchouc is obtained in that country. **1874** T. BELT *Naturalist in Nicaragua* 33 The Mexicans played with balls made from it, and it still bears its Aztec name of *Ulli*, from which the Spaniards call the collectors of it *Ulleros.* **1880** *Encycl. Brit.* XII. 835/2 Torquemada mentions..that an oil was extracted from the 'ulli', or rubber, by heat, possessing soft and lubricous properties... Even at that early date the Spaniards used the juice of the ulé tree to waterproof their cloaks. **1894** *Outing* XXIII. 353/1 Curious tales the huleros tell of carved rocks hidden in those fastnesses. *Ibid.*, This home of the gatherer of *hule.* **1920** *Edin. Rev.* Oct. 356 The garments he wears are splashed with ulli. **1959** J. C. T. UPHOF *Dict. Economic Plants* 79/2 *Castilla costaricana* Liebm. Hule (Moraceae).—Tree. Costa Rica. Latex from stem is source of a good rubber.

hull, *sb.*[2] Add: **1.** Also of an airship, flying boat, etc.

1918 *Aviation* 15 Mar. 231/1 A hull for flying boats having its elevational aspect determined by lines rounding off rearwardly. **1923** *Gloss. Aeronaut. Terms* (*B.S.I.*) iv. 35 *Hull*, the main flotation body of a boat sea plane. **1923** GLAZEBROOK *Dict. Appl. Physics* V. 128/2 The curves may be taken as representing the type of distribution which had been found for models of airship hulls. **1936** [see *air stewardess* (*AIR *sb.*[1] III. 4)]. **1950** *Gloss. Aeronaut. Terms* (*B.S.I.*) I. 37 *Hull*, the main structural and flotation body of a flying boat or boat amphibian. **1951** *Oxf. Jun. Encycl.* IV. 396/2 The hull of a flying boat has a planing bottom like a speedboat.

2. b. *hull down*: also *attrib.* and *fig.* Used also of a tank (see quot. 1948[2]).

1883 [in Dict.]. **1899** *Westm. Gaz.* 29 Nov. 1/3 He shook his fist at the hull-down coasting schooner. **1905** J. C. LINCOLN *Partners of Tide* xiv. 264 You've got me beat, hull down. **1933** 'L. LUARD' *All Hands* 132 We was goin' to see our Mary [Pickford]. Don't alf 'old with 'er. She's got the rest of 'em 'ull down. **1944** *Return to Attack* (Army Board, N.Z.) 18/1 Using the slight undulations of the desert to get hull down and so present the smallest target, they manœuvred for position. **1948** C. DAY LEWIS *Poems 1943–47* 29 Alas, hull-down upon hope's ashen verge Hastens the vessel that our joined hands launched. **1948** PARTRIDGE *Dict. Forces' Slang* 96 *Hull down*, a position for tanks and self-propelled guns where only the turret was visible, the rest being protected by a bank or fold in the ground. **1953** C. DAY LEWIS *Italian Visit* iv. 51 A cloud vibrating In the wash of the hull-down sun. **1960** C. S. LEWIS *Studies in Words* iv. 105 But ten years later he and Cowley are leagues apart, each 'hull down' to the other.

hullabaloo, *v.* Add: Also *trans.*; **hullabaloo·ing** *ppl. a.*

1936 M. FRANKLIN *All that Swagger* x. 93 On harvest days they were hullabalooed from bed before dawn. **1952** DYLAN THOMAS *Coll. Poems* p. ix, Ho, hullaballoing clan Agape, with woe In your beaks, on the gabbing capes!

hullo, *int.* Add: Also used in response to a telephone call and (freq. repeated) to express surprise. Cf. *HALLO *int.*, *HELLO *int.* b.

1900 C. H. CHAMBERS *Tyranny of Tears* 3 [Goes to telephone.] Hullo! hullo! [Gives them a ring up.] Are you there? **1906** *Daily Chron.* 31 May 4/7 The telephone ..we both begin with the same word: 'Hullo!' **1931** D. L. SAYERS *Five Red Herrings* xix. 193 'Hullo-ullo-ullo!' he said. 'So here we are again.' **1959** *Listener* 13 Aug. 248/2 If, when you take off the receiver, you say 'Hullo!' just think how absurd that is. Why, you might be saying 'Hullo!' to a total stranger. **1972** N. MARSH *Tied up in Tinsel* viii. 198 Hullo-ullo! I thought there was something there.

hully gee (hɒ·li dʒī·), *int.* Chiefly *U.S.* Also **holly gee.** [Corruption of *Holy Jesus.*] An exclamation of delight or surprise.

1895 E. W. TOWNSEND *Chimmie Fadden Explains* 58 And holly gee! I never knowed de Duckess could do it! *Ibid.* 69 'Holly gee!' says de mug, 'don't do dat,' he says, and slips me a fiver. **1898** 'O. THANET' *Heart of Toil* 76 Hully gee, Michael, but you are just *there*, and don't you forget it. **1907** F. H. BURNETT *Shuttle* xxiii. 238 To be treated as a gentleman by a gentleman—by 'a fine old swell like this—Hully gee!' **1936** F. CLUNE *Roaming round Darling* i. 3 We picked up a pair of wire-strainers, his leather coat, and a typewriter: then hully-gee! we were off again over the hideous Pyrmont Bridge. **1945** MENCKEN *Amer. Lang.* Suppl. I. 664 *Hully gee* (for *Holy Jesus*) was introduced by Edward W. Townsend's Chimmie Fadden and Major Max (New York, 1895), but it disappeared with the decay of the Bowery boy as an American comic type.

hully gully (hɒ·li gɒ·li). [Etym. unknown.] A dance that is a modification of the frug.

1964 *Cambr. Rev.* 10 Oct. 1/2 Two or three hours of the Hully-Gully. **1966** *Amer. Speech* XLI. 143 Mee, Charles L., Jr. Discotheque Man... [Names and briefly describes current dances: the Hully-Gully, the Frug, La Bostella, the Jerk.] **1968** D. HALLIDAY *Dolly & Singing Bird* ii. 23 We danced a Hully Gully. And then..Johnson performed a frenzied Watusi. **1969** N. COHN *AWopBopaLooBop* (1970) ix. 85 Dance-crazes bossed pop right up until the Beatles broke. There was the Hully Gully, the Madison, [etc.].

hulsite (hɒ·lsəit). *Min.* [f. the name of Alfred *Hulse* Brooks (1871–1924), American geologist: see -ITE[1].] A black borate of bivalent and trivalent iron, magnesium, and calcium in which there is some substitution of tin for trivalent iron, known only from an Alaskan locality.

1908 KNOPF & SCHALLER in *Amer. Jrnl. Sci.* CLXXV. 323 Examination of this deposit showed that an unknown mineral..was present in considerable abundance. We propose for it the name *hulsite*, in honor of Mr. Alfred Hulse Brooks, geologist in charge of the Division of Alaskan Mineral Resources. **1954** *Amer. Mineralogist* XXXIX. 524 X-ray powder photographs of ludwigite, paigeite, pinakiolite, hulsite, warwickite, sussexite, and camsellite were taken.

hulwa (hɒ·lwă). Also 7 **helwa, holway.** [a. Urdu and Arab. *ḥalwā* sweetmeat.] A kind of sweetmeat in India, Persia, etc.; = *HALVA. Cf. *HALAWI.

1662 J. DAVIES tr. *Olearius's Voy. Ambass.* 311 At Tabris, they make a certain Conserve of it [*sc.* duschab], which they call Helwa. **1698** J. FRYER *New Acct. E.-India & Persia* 358 Every Friday Night, at the Sacrament of Holway, (or Wafer made up in Sweetmeats). **1884** *Times* 30 May 8 Provisions ran short and the voyagers had to live on hulwa, a glutinous sweetmeat.

hum, *v.*[1] Add: **1.** (Later examples.)

1924 *Foundry* (Cleveland) 1 Apr. 63 (Advt.), The wheels surely are humming in the foundries. **1937** WODEHOUSE *Lord Emsworth & Others* ix. 299 We buzzed on through the pleasant countryside... The engine of the two-seater hummed smoothly. **1972** *Daily Tel.* 23 Mar. 21/2 Computers hummed and slide rules flashed as City analysts tried to come to grips with the new corporation tax yesterday.

3. Further examples of sense 'to be in a

condition of busy activity'. Phr. *to make things* (or something specified) *hum*: to bring to such a condition, to liven things up.

1884 'Mark Twain' Huck. Finn 224 He lit into that horse with his whip and made him fairly hum. 1887 M. Roberts Western Avernus iii. 34 The owner of the farm..came home, and, in American parlance, 'fairly made things hum'. 1898 [in Dict.]. 1902 E. Banks Autobiogr. Newspaper Girl xviii. 205 We took you on this paper to help us make things hum! I understand you made things hum over in England. 1906 Nation (N.Y.) 18 Jan. 44 The colleges are making athletic reform hum, as if to make up for lost time. 1911 E. M. Clowes On Wallaby x. 265 But still things are, as the American would express it, 'beginning to hum' in the irrigation line. 1914 G. Atherton Perch of Devil i. xxx, I want money to spend in Butte,..and make things hum. 1922 H. S. Walpole Cathedral i. vii. 135, I hardly need to tell you that he's not quite the man to make things hum. 1966 'J. Hackston' Father clears Out 107 By gum! Wouldn't it just make things hum? 1973 Times 16 Jan. 17/7 Our view is that if you get the logistics right and do get concentration of the best modern equipment you can really get something humming in the 1974 period.

hum, *v.*² Add: **2.** *trans.* and *intr.* To borrow (without any intention of returning); to scrounge. *Austral. slang.*

1918 Aussie Sept. 4/2 Don't shout cigarettes, hum them. 1925 Fraser & Gibbons Soldier & Sailor Words 122 To hum, to cadge. 1935 Bulletin (Sydney) 30 Jan. 21/4 Where other 'Bidgee whalers 'hummed' a town for booze, Mick 'hummed' it for tea, going from house to house with his plea: 'Missus, could y' spring a cup o' tea?' 1938 X. Herbert Capricornia (1939) xviii. 234 Gertch—you old blowbag! You're only humming for a drink. Nick off home.

hum (hɒm), *v.*³ *slang. intr.* To smell disagreeably. Hence **hum** *sb.*³, a disagreeable smell.

1902 C. H. E. Brookfield Random Reminisc. xi. 200 The burglar..finds a piece of Camembert cheese in the buffet, and exclaims: 'Lor'! it do 'um!' 1906 E. Dyson Fact'ry 'Ands xv. 197 Ther hum iv it was so fearful... Killin' cats is rotten luck. 1927 W. E. Collinson Contemp. Eng. 23 Things didn't smell, but ponged, niffed or hummed. Ibid., An awful pong or hum. 1946 I. L. Idriess In Crocodile Land xvi. 111 The abos certainly did 'hum'. 1970 Daily Tel. 31 Oct. 1/3 When the wind drops this stuff really hums.

hum, *sb.*¹ Add: **I. 1. c.** *Med.* In full *venous hum*. A continuous humming sound sometimes heard during auscultation in the upper chest and the sides of the neck, esp. in children and in cases of anæmia, and attributed to the turbulence of the flow of venous blood.

1839 J. Hope Treatise Dis. Heart (ed. 3) i. iv. 118 By the adroit management of the stethoscope..the venous murmur may often be raised, by a gradual swell, into a more or less musical hum, such as is yielded by a child's humming-top. I propose to denominate this the Venous Hum; for..this is..more intelligible than noise of the devil, by which term, derived from a plaything known to few, M. Bouillaud has designated the hum in question. 1876, 1891 [see Venous a. 2]. 1907 H. S. Anders Physical Diagnosis xiv. 346 The venous hum or bruit de diable heard over the jugulars. 1960 J. S. Butterworth et al. Cardiac Auscultation (ed. 2) ix. 81 The venous hum is usually heard only in the upright position. Ibid., The hum will disappear with finger compression of the jugular veins.

d. *Electronics.* (Usu. without *a* or *pl.*) Unwanted low-frequency variations in current or voltage (the cause of which is usually the alternating frequency of the mains) which will give rise in a loudspeaker to a steady humming sound; the sound so produced.

1929 K. Henney Princ. Radio xvi. 406 A hum output of 44 millivolts. 1934 J. H. Reyner Television viii. 90 The hum appears as a series of black bands moving slowly across the screen. 1950 A. Marcus Radio Servicing xv. 714 An antenna that is too close and parallel to a power line may pick up hum by induction. 1966 R. King Electrical Noise i. 3 Mains hum..may be generated from cathodes of thermionic valves heated by alternating current or by electromagnetic induction from near by mains frequency transformers and chokes.

III. hum-bucking coil *Electronics* [Buck *v.*⁶], a coil arranged so as to cancel the hum in another coil by providing a signal of the opposite phase.

1940 Chambers's Techn. Dict. 425/1 Hum-bucking coil. 1950 A. Marcus Radio Servicing xv. 715 Some loud-speakers, especially of the electrodynamic type, have hum-bucking coils. 1967 P. Spring Tape Recorders vii. 90 The hum-bucking coil..picks up hum in opposite phase to the hum picked up by the head and of such magnitude that the two hum signals cancel.

hum, *sb.*² Add: **2.** A persistent borrower, a scrounger. *Austral. slang.*

1919 V. Marshall World of Living Dead, The 'hum', the unskilled derelict or derelict-to-be who stands upon the 'pub' corner kerb, 'bites' all and sundry, and, at regular intervals, succeeds in getting lumbered for 'vag'. 1953 Caddie Sydney Barmaid xxxiii. 183 The speciality of the hum is the lone drunk.

hum, *sb.*³: see *hum *v.*³

hum (hiūm), *sb.*⁴ *Physical Geogr.* [Serbo-Croat, = hill.] A small, usually conical, hill characteristic of karst topography.

1921 Geogr. Rev. XI. 602 (caption) General view of a polje strewn with hums. Ibid. 604 The former limestone mass being represented by isolated fragments which Cvijic calls 'hums'. 1937 Wooldridge & Morgan Physical Basis Geogr. xix. 289 Residual limestone masses or hummocks rising above polje floors are called 'hums'. 1971 B. W. Sparks Rocks & Relief v. 204 The interfluvial areas are finally reduced to little hillocks known as hums.

huma (hū·mă). Also **Huma, Ûma.** [Hind., a. Pers. *humā* phœnix.] A fabulous bird of the east, said to be a restless wanderer but to bring luck to any person over whom it hovers.

1841 R. C. Wellesley Primitiæ et Reliquiæ 104 (heading) The Ûma, or Indian Eagle, the bird of prosperous empire. Ibid. 105 The throne of the Sultan of Mysore.. was surmounted by a representation of the Ûma, which is now deposited at Windsor Castle. Ibid. 106 A bird called by the natives the Ûma and which bore the appearance of a small eagle or vulture, built its nest..in the garden of Mr. Petrie, where Lord Wellesley resided while at Madras... The natives superstitiously ascribed much of the success of the war to the influence of the Ûma. 1858 O. W. Holmes Autocrat of Breakfast-Table i. 8, I am like the Huma, the bird that never lights, being always in the cars, as he is always on the wing. 1905 Spectator 14 Jan. 47/2 The legendary 'huma' of Eastern tradition.

human, *a.* (*sb.*). Add: **A. 3. b.** Belonging or relative to man, relating to or characteristic of activities, relationships, etc., which are observable in man, as distinguished from (*a*) the lower animals; (*b*) machinery or the mechanical element; (*c*) mere objects or events, as *human affairs, angle, chain, condition, document, element, factor, fly, interest, note, period, relations, rights, situation, story, torch, torpedo.* Also **human engineering** orig. *U.S.*, the scientific study of the interaction of man and his working environment and the exploitation of this interaction in the interests of efficiency; the application of the human sciences to the design of machines; so **human engineer; human equation**: see *equation 3 b; **human-factors engineering** = **human engineering*; so **human-factors engineer.**

1741 Hume Ess. Moral & Pol. I. 176 Such mighty Revolutions have happened in human Affairs.. as are sufficient to beget the Suspicion of still farther Changes. 1798 Human affairs [see Affair sb. 2 a.]. 1877 L. H. Morgan Anc. Society iii. i. 399 From its limited prevalence it made but little impression upon human affairs. 1949 W. L. Warner in M. Fortes Social Struct. 2 Current procedures for class stratification..require too large a field staff of experts, often making it impossible for the interested student of human affairs to use a knowledge of social class to understand his particular problem. 1941 Ann. Reg. 1940 317 The Press mirrored the age, copying from America tabloid news, the 'human angle', and..illustrations. 1908 Pop. Mechanics Jan. 15/2, 50 men formed a human chain and pulled him out. 1926 B. A. McKelvie Huldowget iii. 40 Every male in the village was a link in the human chain. 1963 Human chain [see *chain sb. 5 f]. 1814 Edin. Rev. XXII. 199 The means of bettering the human condition. 1957 P. Coveney Poor Monkey xii. 276 Lawrence's perception of the 'human condition'. 1960 C. Day Lewis Buried Day v. 87, I could not myself opt out of the human condition, as to some degree the pacifist must do. 1892 W. H. Mallock (title) A human document. 1896 Badminton Mag. 195 Regarded as a human document my guide looked dog-eared, thumbed and a trifle mildewed. 1938 Ann. Reg. 1937 320 Dr. Roberts' book is a 'human document' of great value. 1887 Scribner's Mag. I. 93/2 He managed, with masterly ingenuity, so to leave out the human element..that he gave hardly a glimpse of what it really is. 1897 Trans. Inst. Naval Architects 111 It is all very well to say the men are careless, but the human element has to be reckoned with. 1907 R. Herriot (title) The human element. A novel. 1908 Modern Business Aug. 65/1 System and organisation may be perfect on paper..but this does not necessarily lead to.. efficiency. There is another element—the human element. 1913 Pall Mall Mag. July 46/2 You must remember that in regard to the human element, we are..behind Germany. The point is that they have the human element—a large body of pilots, observers, artisans, trained during the last sixteen years. 1944 Electronic Engin. XVI. 334 The additional errors due to the human element (observation and reaction times) can be eliminated. 1957 Goode & Machol System Engin. xxx. 500 Selection and training of operators..come within the province of the human engineer. 1934 Webster, Human engineering. 1944 Amer. N. & Q. June 48 Los Angeles.. has more than its quota of 'spiritual engineers'..not to mention a 'School of Human Engineering' in a college curriculum... Such an academic unit might legitimately concern itself with anything..from anatomy and physiology to sociology. 1950 Lancet 1 Apr. 645/2 The field which has variously been described as 'fitting the machine to the man', human engineering, that part of industrial psychology not concerned with vocational guidance, etc. 1957 Goode & Machol System Engin. xxx. 481 The primary object of attention in human engineering is the man–machine link. 1970 New Scientist 23 July 199/3 Human engineering..involves a careful moulding of an educational system to nurture what is most useful and beneficial in each individual. 1921 B. S. Rowntree (title) The human factor in business. 1943 J. B. Priestley Daylight on Saturday x. 65 He was worried about his production figures... In the last resort it was the human factor that counted. 1967 L. B. Archer in Wills & Yearsley Handbk. Managem. Technol. 130 In the U.S.A. the subject [ergonomics] is called 'human engineering' or 'human factors engineering' and the man may be described as a human factors engineer. 1969 New Yorker 56/2 A group of engineers called human-factors engineers has had as much to do as anybody else with making the astronauts seem like black boxes. 1964 E. J. McCormick (title) Human factors engineering. 1919 Alameda (Calif.) Times-Star 11 Jan. 1/3 Besides being a 'Human Fly', Williams has other unusual accomplishments. 1960 Observer 25 Dec. 7/6 The climber..is likely to be agile and athletic, but there is generally no need for him to be what the Press calls a cat burglar or human fly. 1824 Byron in T. Medwin Conversations Lord Byron 237 There was another objection: all the human interest would have been destroyed, which I have even endeavoured to give my Angels. 1860 Dickens in All Year Round 28 Jan. 321/1 Figuratively speaking, I travel for the great house of Human Interest Brothers. 1912 Collier's 21 Sept. 21/3 Fu, not understanding the American newspaper idea of 'human interest', elected to think I had written a eulogy of him deliberately. 1913 E. C. Bentley Trent's Last Case ii. 23 'Prostrated by the shock,' hinted the reporter, '..human interest.' 1915 W. P. Livingstone Mary Slessor (1916) ii. vii. 46 Her simple but vivid style, the human interest of her story..made so great an impression that the ladies of Glasgow besought the Committee to retain her for a time. 1930 Wodehouse Very Good, Jeeves! x. 263 Just one of those human-interest stories, if you know what I mean. 1933 Leavis & Thompson Culture & Environment 141 What do you understand by the phrase 'human interest', and what..is vicious in the journalistic practice it derives from? 1970 Times 27 Feb. 13/4 What about that boy with blood pouring from his eye, who has now become an extra in a 'human interest' story? 1920 Fairgrieve & Young Brit. Isles p. vi, This series of elementary books is just what its name denotes—human. Everywhere the human note is predominant and the relation of man to his environment insisted upon. 1882 A. Geikie Text-bk. Geol. 901 The long succession of Pleistocene ages shaded without abrupt change of any kind into what is termed the Human or Recent Period. Ibid. 902 The Human Period is above all distinguished by the presence and influence of man. 1916 G. B. Shaw Overruled 57 Spontaneous human relations.. on the one hand and the property relation..on the other. 1926 B. Webb My Apprenticeship i. 52 A poisonous cynicism about human relations. 1954 D. Riesman Individualism Reconsidered iv. xiv. 222 Men who take courses [in] human relations in order..[to] get along with their colleagues in the office. 1967 M. Argyle Psychol. Interpersonal Behaviour x. 198 Experience with management training shows that lectures on 'human relations' are often very popular. Ibid. 199 Follow-up studies show that lectures on human relations lead to improved scores on questionnaires. 1791 T. Paine Rights of Man 110 The representatives of the people of France..considering that ignorance, neglect, or contempt of human rights, are the sole causes of public misfortunes..have resolved to set forth..these natural, imprescriptible, and unalienable rights. 1877 Independent (N.Y.) 18 Jan. 2/2 'What does that little rat know about human rights?' Pack said. 1941 'G. Orwell' in Horizon Aug. 134 An attempted definition of fundamental human rights. 1945 Charter of United Nations Art. 1 par. 3, To achieve international cooperation..in promoting and encouraging respect for human rights. 1969 New Yorker 14 June 78/2 More middle-class blacks will become involved in human rights. 1961 J. B. Wilson Reason & Morals ii. 46 The tragedy of the human situation (itself a Freudian phrase). 1872 B. Jerrold London iii. 28 The sad human stories that crowd the emigrant vessel. 1945 E. Waugh Brideshead Revisited ii. ii. 233 A woman reporter..had come..to get a 'human story' of the dangers of my journey. 1959 Times Lit. Suppl. 27 Mar. 173/3 The 'human story' and even the characters are but adjuncts of the whirling selling machine. 1969 Guardian 21 Jan. 1/1 Two more 'human torch' protests... A second Czechoslovak has tried to take his life as a political protest by setting fire to himself. 1944 Hutchinson's Pict. Hist. War 12 Apr.–26 Sept. 21/2 A new and devastating weapon called the 'human torpedo'. It can be likened to a miniature submarine. 1953 Jane's Fighting Ships 1953–54 151 Damaged in Northern waters in 1944 and repaired and reconstructed as 'Kaitan' (Human Torpedo) carrier.

c. As a subdivision of a science: that branch of the science which is concerned with the study of man, as *human ecology, geography, psychology.*

1933 Human ecology [see *ecology]. 1957 [see bio-ecology (*bio-)]. 1965 New Scientist 28 Jan. 208/3 The tag for the necessary science is 'human ecology'—the comprehensive investigation of the effects of our environment on our well-being. 1919 Fairgrieve & Young (title) Human geographies. 1936 Discovery Feb. 56/2 This map is a valuable contribution to human geography. 1959 Listener 6 Aug. 219/3 Endless information about the human geography of England at the end of the eleventh century. 1961 L. D. Stamp Gloss. Geogr. Terms 242/1 Human geography, the geographical study (the complement of physical geography) of those features, objects and phenomena of the Earth's surface which relate directly, or are due, to man and his activities. 1924 R. M. Ogden tr. Koffka's Growth of Mind 16 To bridge the gap between human and animal-psychology.

5. b. *human-tainted.*

1929 D. H. Lawrence Pansies 73 Communion with the Godhead, they used to say in the past. But even that is human-tainted now, Tainted with the ego and the personality.

B. *sb.* (Further examples.)

1832 F. Trollope Dom. Manners Amer. I. 70, I expect the sun will rise and set a hundred times before I shall see another human that does not belong to the family. 1898 H. S. Canfield Maid of Frontier i. 18 A man could ride from here to forty-mile the other side of Edwardsville and never see a human. 1902 O. Wister Virginian iv. 56 'They are just like humans,' the Virginian concluded. 1905 A. Conan Doyle Return of Sherlock

Holmes 27 You will see it often in humans. 1909 H. G. Wells Ann Veronica ii. 43 In all the species of animals the females are more important than the males; the males have to please them. Look..at the competition there is everywhere, except among humans. 1922 Joyce Ulysses 501, I always understood that the act so performed by skittish humans with glimpses of lingerie appealed to you in virtue of its exhibitionististicity. 1971 Physics Bull. Jan. 49/1 A third hailed him, as a great scientist and a superb human.

2. With *the*: (*a*) the human race, humanity; (*b*) that which is human, that which relates to man or humanity.

1841 E. B. Browning Let. 28 Aug. (1897) I. 88, I may say so now—as far as the human may say 'yes' or 'no' of their futurity. 1844 —— Poems I. 109 While the human in the minor Makes the harmony diviner. 1919 M. K. Bradby Psycho-Analysis 205 The distinctively animal gives way in order to express the distinctively human.

humane, *a.* Add: **1. d.** Applied to certain weapons or implements which inflict less pain than others of their kind, *spec.* applied to an implement for the painless slaughtering of cattle.

1904 Daily Chron. 24 May 5/3 The doctors style the bullets 'humane'. 1920 Act 10 & 11 Geo. V c. 43 §(8) (h) Any..butcher..having in his possession..any humane killer for the purpose of such business. 1927 Daily Express 6 Aug. 7/3 That the humane killer was a dangerous instrument to those who used it. 1973 Times 11 Jan. 2/6 Three veterinary surgeons..had thought he must be put down. This was done..using a humane killer.

humanics. (Later examples.)

1937 R. S. Morton Woman Surgeon xxxi. 351 The business side of practice is necessary, but its art and humanics are what carry us on. 1952 King Geo. VI & Industry: a Tribute (Industr. Welfare Soc.) p. ii, We give ungrudging honour to these pioneer students of the new science of what has been termed 'Industrial Humanics'. 1958 Progress Autumn 266/2 But, to British eyes, the American method looks superficial; it seems to rest on two doubtful assumptions... The second is that what Americans have called the 'humanics' of the subject can be reduced wholly to its 'mechanics'.

humanism. Add: **5.** *Philos.* A pragmatic system of thought introduced by F. C. S. Schiller and William James which emphasizes that man can only comprehend and investigate what is with the resources of the human mind, and discounts abstract theorizing; so, more generally, implying that technological advance must be guided by awareness of widely understood human needs.

1903 F. C. S. Schiller Humanism p. xvi, I propose.. to convert to the use of philosophic terminology a word which has long been famed in history and literature, and to denominate Humanism the attitude of thought which I know to be habitual in William James and in myself. 1904 W. James Coll. Ess. & Rev. (1920) xxxii. 450 No one can ever foresee what terms will succeed in the struggle to gain currency... 'Humanism' is perhaps too 'whole-hearted' for the use of philosophers, who are a bloodless breed; but, save for that objection, one might back it, for it expresses the essence of the new way of thought, which is, that it is impossible to strip the human element out from even our most abstract theorizing. 1907 F. C. S. Schiller Studies in Humanism 12 Humanism..is merely the perception that the philosophic problem concerns human beings striving to comprehend a world of human experience by the resources of human minds. 1945 E. A. Burtt in Humanist V. iii. 108 It may seem presumptuous, if not paradoxical, to suggest that a movement claiming the name 'humanism', and emphasizing rational comprehension as the foundation of every good achievement, might fail lamentably in its understanding of man. 1959 P. Tillich Theology of Culture ii. viii. 121 He [sc. Sartre] calls his existentialism humanism. But if he calls it humanism, that means he has an idea of what man essentially is. 1961 O. Reiser in J. S. Huxley Humanist Frame 240 A major objective of a scientific Humanism is the organization of human knowledge for the purpose of human progress. 1966 C. H. D. Clark Scientist & Supernatural v. 174 Humanism glorifies science without telling us how the laws of science arose nor how they are to save us from our innate selfishness. 1969 K. Kaunda Towards Complete Independence 43 Our philosophy of Humanism stresses above all the importance of man as an individual.

humanist. Add: **5.** *Philos.* One whose beliefs are in accordance with *humanism 5. Also *attrib.*

1903 F. C. S. Schiller Humanism p. xxi, A humanist philosopher is sure to be keenly interested in the rich variety of human thought and sentiment, and unwilling to ignore the actual facts for the sake of bolstering up the narrow abstractions of some a priori theory of what 'all men must' think and feel... The humanist, accordingly, will tend to grow humane, and tolerant of the divergences of attitude which must inevitably spring from the divergent idiosyncrasies of men. 1904 W. James in Mind XIII. 462 Bergson in France, and his disciples the physicists Wilbois and Leroy, are thorough-going humanists in the sense defined. 1949 J. Gutman in P. A. Schilpp Philos. Ernst Cassirer ii. xiii. 464 Thus as an historian and as a humanist Cassirer once again raised the standard of self-knowledge, [and] reaffirmed the doctrine that the unexamined life is no life for man. 1961 L. Elvin in J. S. Huxley Humanist Frame 272 The Humanist is content to leave it to the free play of thought, so long as thought is kept free. 1963 J. S. Huxley Human Crisis 19 Today, the new humanist vision is giving us the key idea of man as the agent for the whole future of evolution on this planet. 1966 C. H. D. Clark Scientist & Supernatural v. 175 The humanist trust in reason alone is actually unreasonable, since logic would suggest that affluence and scientific advancement ought to be accompanied by increasing mental satisfaction. 1968 A. J. Ayer Humanist Outlook 4 Present-day humanists are in fact the intellectual heirs of those nineteenth-century freethinkers.

humanistic, *a.* (*sb.*). Add: **A. 2.** Pertaining to or characteristic of humanism. (Cf. *humanism 5.)

1904 W. James Coll. Ess. & Rev. (1920) xxxii. 451 But humanistic empiricism will have many other steps forward to make before it conquers all antagonisms. 1923 B. Russell Prospects Industr. Civilization ii. xiii. 266 The distinction between the mechanistic and the humanistic conceptions of excellence is the most fundamental of all distinctions between rival sets of ideals. Ibid., The humanistic conception..regards the good as something existing in the lives of individuals, and conceives social co-operation as only valuable in so far as it ministers to the welfare of the several citizens. 1932 C. K. Ogden Jeremy Bentham 9 It was to French influences, to Fénélon, to Helvétius and to Voltaire,..that Bentham owed his first humanistic stimulus. 1961 B. Wootton in J. S. Huxley Humanist Frame 350 Plainly, what is actually happening in the world is the result of the accommodation of religions to evolving humanistic ideas and not vice versa. 1968 H. J. Eysenck in A. J. Ayer Humanist Outlook 271 The future of humanistic thought on this subject is completely bound up with the growth of psychological knowledge—without this it must remain nothing more than an alternative superstition.

B. *sb.* Delete †*Obs.* and add later example.

1952 Koestler Arrow in Blue xxvi. 240, I would shift the emphasis in popular education from stale humanistics to a lively comprehension of the mysteries of the universe and life.

humanistically, *adv.* Add: Also in relation to *humanism 5.

1904 W. James Coll. Ess. & Rev. (1920) xxxii. 451 'Radium', for example; humanistically, both the that and the what of it are creations of yesterday. But we believe that ultra-humanistically they existed ages before their gifted discoverers were born. In what shape? There's the rub! for we have no non-humanistic categories to think in. 1909 —— Meaning of Truth iii. 79 The reader would conceive the knowing humanistically. 1944 Scrutiny XII. iii. 210 The brilliant vitality of the keyboard writing.. itself tends to render this passion more directly and humanistically dramatic than the more religious (if no less intense) keyboard style of Gibbons.

‖**humanitas** (hiumæ·nitas). [L.] = Humanity.

1944 Atlantic Monthly Nov. 73 Philosophers of the ancient world would consider these to be strange limitations..upon the spiritual jurisdiction of a school which concerns itself with humanitas. 1948 E. Pound Pisan Cantos lxxxiv. 117 Yin had these three men full of humanitas (manhood). 1964 Economist 4 Jan. 37/3 The fundamental humanitas of the author. 1964 L. Hjelmslev in R. H. Robins Gen. Ling. i. 10 Linguistic theory is led by an inner necessity to recognize..man and human society behind language... At that point linguistic theory has reached its prescribed goal: humanitas et universitas. 1970 Times Lit. Suppl. 23 Apr. 442/4 The concern for a conception of society, life and humanitas that doesn't eliminate the depth in time.

humanization. Add: **c.** The treatment of cow's milk to render it suitable for consumption by infants.

1905 F. L. Dodd Municipal Milk 14 The humanization adapts the milk to infants' digestive processes. 1932 V. E. M. Bennett Welfare Infant & Child v. 70 When cows' milk is to be used for feeding the infant two modifications are always requisite... These two processes are known respectively as Humanization and Pasteurization.

humanize, *v.* Add: **5.** To treat (cow's milk) in order to make it more closely resemble human milk and suitable for consumption by infants.

1897 Amer. Jrnl. Med. Sci. CXIII. 374 Dufour.. describes a rather simple method of humanizing cow's milk. 1970 Pharmaceutical Handbk. (Pharm. Soc. Gt. Brit.) 379 In 'humanising' cow's milk, extra cane sugar or glucose is added because human milk contains more lactose than cow's milk.

humanized, *ppl. a.* Add: **3.** Of milk (see *humanize *v.* 5).

1888 Baby Dec. 15/1 This humanised cow's milk is the only natural food for artificially feeding infants, and the weakest of them thrive on it as on breast milk. 1901 Westm. Gaz. 19 Oct. 5/3 The Borough Council of Battersea has determined to undertake the supply of sterilised and humanised milk. 1947 A. B. Meering Handbk. Nursery Nurses xxii. 202 Humanised dried milk is fresh cows' milk, dried, and the content rendered as near as possible to human milk. 1955 Wohl & Goodhart Mod. Nutrition xxxiv. 919 'Humanized milks'..imitate to a greater or lesser extent the caloric proportions, the mineral content or the fat of breast milk. Ibid., The need for the humanized products is..questionable.

humanoid (hiū·mănoid), *a.* and *sb.* [f. Human *a.* (*sb.*) + -oid.] **A.** *adj.* Of human form or character; man-like; *spec.* (*a*) distinguished from anthropoid as being more human in character (cf. *hominoid *a.* and *sb.*); (*b*) as a term in Science Fiction.

1918 Mrs. D. G. Ritchie New Warden xvi. 186 Religion ..had its origin in the funk and cunning of the humanoid ape. 1922 Glasgow Herald 23 Dec. 4 Then came the separating off of the larger Anthropoid Apes, leaving the main stem humanoid. 1936 C. J. Warden Emergence Human Culture iii. 87 The humanoid stock must become specialized along human lines or forever perish from the earth. 1952 V. Nabokov Nabokov's Dozen (1959) 207 Inhabitants of foreign planets, 'intelligent' beings, humanoid or of various mythic makes. 1965 Punch 7 Apr. 525/3 Ruined machinery..leads our heroes to a planet revolving around Altair, where they meet a humanoid race. 1966 L. Cohen Beautiful Losers (1970) i. 96 It was just a shape of Edith: then it was just a humanoid shape: then it was just a shape.

B. *sb.* A humanoid animal or being.

1925 J. A. Thomson Concerning Evolution iii. 210 The humanoids and the anthropoids parting company between a million and two million years ago. 1936 C. J. Warden Emergence Human Culture iii. 87 The humanoids that managed to survive from age to age became less ape-like and more man-like as time went on. 1958 Manch. Guardian 26 Sept. 4/3 The humanoids who are accidentally brought to earth by these means inevitably turn out to be superior to us. 1965 Listener 14 Jan. 56/2 The culmination is reached when peaceful humanoids are ruled by vicious insects or lizards.

humantin (hiumæ·ntin). *Zool.* [a. Fr. *humantin* (G. Rondelet *Histoire des Poissons* (1558) xiii. viii. 301), of uncertain origin.] A spiny shark, *Oxynotus centrina*, of the family Oxynotidæ, found in the Mediterranean Sea and off the coast of Portugal.

Gamillscheg's suggestion that it is perhaps f. Fr. *humer* 'to inhale' arbitrarily modelled on Fr. *lamantin* Lamantin is chronologically inconclusive. The word is also sometimes identified with L. *humus* 'humus, soil', from the habit that the humantin has of lying in the mud on the sea-bed, but the connection remains uncertain.

[1862 T. Gill in Ann. Lyceum Nat. Hist. N.Y. VII. 376 Cuvier, in his Regne Animal, proposed the following division..Les Humantins (Centrina Cuv.). 1880–84 F. Day Fishes Gt. Brit. & Ireland II. 319 Centrina salviani...Name.—La Humantin, French. Habits.—Is supposed to inhabit great depths.] 1925 J. T. Jenkins Fishes Brit. Isles 320 The Humantin (Centrina salviani). In this shark each dorsal fin has a strong spine. 1959 A. Hardy Fish & Fisheries ix. 181 The humantin, Oxynotus centrina, an even rarer straggler to our waters from the coasts of Portugal, is noteworthy for its peculiar spines situated within its large dorsal fins.

hum-bird. (Later U.S. examples.)

1834 C. A. Davis Lett. Jack Downing 6 They both came within a humbird's eye of it. 1872 Schele De Vere Americanisms 377 The tiny Mango Hummingbird (Trochilus colubris)..is known familiarly..as Humbird or Hummer simply. 1891 R. T. Cooke Huckleberries (1896) 167, I never see a humbird fuller o' buzz than little Prudy.

humble, *a.*¹ Add: **1. c.** *your humble*: used ellipt. for 'your humble servant'.

1765 [in Dict.]. 1835 Dickens Let. 2 May (1965) I. 58 The next stage, your humble, caught them before they had changed. 1926 D. H. Lawrence Let. 4 Jan. (1962) II. 875 As for your humble, he says his say in bits. 1929 Kipling Limits & Renewals (1932) 359 'And where does he get his champagne?' 'From grateful appendices—same as your bloody 'umble,' said Scree.

3. *humble-hearted, -minded* (earlier and later examples).

1961 New Eng. Bible Matt. xi. 29 Bend your necks to my yoke, and learn from me, for I am gentle and humble-hearted. 1580 T. Pounde in H. Foley Jesuits in Conflict (1873) 109 That might soe be amonge the humble-minded. 1905 W. James Mem. & Stud. (1911) v. 77 It must be confessed that T. D. never was exactly humble-minded. 1961 New Eng. Bible 1 Pet. iii. 8 Be full of brotherly affection, kindly and humble-minded.

humble-jumble. Delete †*Obs.* and add later example. Also **hu·mble-ju·mbled** *ppl. a.*

1929 H. S. Walpole Hans Frost i. ix. 98 The ironic disappointment of fixing pretty sentences together, all, humble-jumble, on the table of his mind. 1940 Horizon Jan. 61 It is the Big Bed of Ware combined with the Procrustean bed, in which, humble-jumbled or stretched or curtailed to fit, they all lie most curiously together.

Humboldtian (humbōu·ltiăn), *a.* [f. name of K. Wilhelm von *Humboldt*, German philologist (1767–1835) + -ian.] Of, pertaining to, or characteristic of Humboldt or his work.

1901 H. Oertel Lect. Study of Lang. i. 45 Entirely Humboldtian in conception is James Byrne's General Principles of the Structure of Language (1885). 1907 E. Sapir in Mod. Philology V. i. 141 Steinthal, himself an enthusiastic follower and developer of Humboldtian views, most emphatically denies any indebtedness on Humboldt's part to Herder. 1964 R. H. Robins Gen. Ling. viii. 335 Within typological comparison, the Humboldtian tripartite division, properly understood, may be said to be a more significant system of classification than some others. 1965 N. Chomsky Aspects of Theory of Syntax 199 It seems to me that 'deep structure' and 'surface structure', in the sense in which these terms will be used here, do correspond quite closely to Humboldtian 'inner form' and 'outer form'. 1967 R. L.

Brown tr. W. von Humboldt's Conception Ling. Relativity 15 Boas also came under Humboldtian influence in another way.

humbug, *sb.* **5.** Delete *dial.* and add later examples.

1936 J. L. Hodson Our Two Englands vii. 115 A middle-aged member of the [Bradford Wool] Exchange moved about offering a paper bag of sweets; cheeks became swollen with humbugs. 1959 I. & P. Opie Lore & Lang. Schoolch. ix. 166 'Lollies' is also becoming a general term, and so is 'gob-stoppers' for 'any sweet difficult to chew', as humbugs, large aniseed balls, and fruit drops.

humbug, *v.* Add: **2.** Const. *about.* To make less progress than expected, to flounder *about*, to wallow. *local U.S.*

1840 R. H. Dana Two Yrs. before Mast xxxiv. 433 For several days we lay 'humbugging about' in the Horse latitudes, with all sorts of wind and weather. 1879 [in Dict.]. 1908 G. S. Wasson Home from Sea vi. 189 We pitch-poled and humbugged about in them latitudes till the Cap'n..was sick and tired of the whole business. 1933 P. A. Eaddy Hull Down x. 214 For several days we were kept humbugging about with light variable breezes.

humdinger (hɒ:mdi·ŋəɹ). [Origin unknown.] **1.** *slang* (orig. *U.S.*). A remarkable or outstanding person or thing, anything of notable excellence.

1905 Dialect Notes III. 62 Hum-dinger, term of admiration. 'She's a humdinger.' 1916 'B. M. Bower' Phantom Herd vi. 100 That pit'cher's a humdinger! 1926 Brit. Weekly 12 Aug. 399/1 They showed me a new game. I tell you Red, it's a humdinger. Ibid. 2 Sept. 456/3 She was a humdinger. She even puts a brand on Brangwyn, and she's no slouch. 1935 Punch 10 Apr. 399/2 Say, she knew her own mind, did that jane. A regular humdinger of a dame. 1937 N. Marsh Vintage Murder vii. 72, I reckon he's all right. Gosh, I reckon he's a humdinger. 1943 Hunt & Pringle Service Slang 40 A humdinger, any fast aircraft or vehicle; any engine which runs really well. 1958 Times 16 June 4/4 The last set was a humdinger, to use a transatlantic expression.

2. *Electronics.* A voltage divider connected across the heater circuit of a valve with the variable tap connected to a source of fixed potential, so that the hum introduced by the heater can be reduced by suitably biasing it with respect to the cathode.

1938 Admiralty Handbk. Wireless Telegr. II. Index, Humdinger. 1947 Electronic Engin. XIX. 82/3 It is.. suggested that the centre point on the 'humdinger' be returned to a suitable tapping on the cathode bias battery. 1967 P. Spring Tape Recorders vii. 90 It is customary to have a centre-tapped heater winding, or better still, a humdinger. This is adjusted until the hum level is at a minimum.

Humean, Humian, *a.* Add: Also as *sb.*

1890 W. James Princ. Psychol. I. x. 330 And if, with the Humians, one deny such a principle and say that the stream of passing thoughts is all, one runs against the entire common-sense of mankind. 1960 Guardian 2 Mar. 7/2 Most unbelievers in our society are Humeans; and one can be a Humean in this sense without ever having heard of Hume.

humect, *v.* **1.** (Later example.)

1923 E. Pound Let. 16 Jan. (1971) 184 Jock Hieland-man..ran Unto the river Liffey, Peeled off his breeches and jumped in, Humecting thus his hairy skin.

humectant, *a.* and *sb.* Restrict ?*Obs.* to senses in Dict., for **a, b** read **A, B,** and add: **A.** *adj.* **2.** Moisture-retaining.

1953 Manuf. Chemist May 202/2 The humectant action is provided by a combination of glycerin and propylene glycol. 1960 McGraw-Hill Encycl. Sci. & Technol. VI. 221/1 Because of its humectant properties, it is sprayed on tobacco before it is processed to prevent crumbling.

B. *sb.* **2.** A substance used to reduce the loss of moisture; *spec.* a food additive that does this.

Quot. 1854–67 in Dict. belongs here.

1951 M. B. Jacobs Chem. & Technol. Food & Food Products (ed. 2) III. xxxvii. 1979 Substances which retain moisture are known as humectants. They..prevent loss of moisture when incorporated into the foodstuff. 1954 Chem. & Engin. News 15 Feb. 685/2 The product can be used as..a humectant for cellulose and paper products, polishes, and tobacco. 1963 Residue Rev. II. 54 Some humectants have been used experimentally to reduce evaporation under conditions of low humidity... Rain can, of course, wash the chemical from the leaf surface. 1972 Sci. Amer. Mar. 19/3 Humectants, which are hygroscopic, offset changes in the humidity of the environment to which food is exposed.

humgruffin. Add: Also **humgruffian.**

1825 J. K. Paulding John Bull in Amer. iv. 29 Another declared she would not give a pin to save such a rude humgruffian from starvation.

humhum. *Hist.* Also 7 **hammome, hummum,** 8 **hump-hump.** [Origin obscure.] A coarse Indian cotton cloth.

1620 in W. Foster Eng. Factories India 1618–21 (1906) 193 Of sahannes and hammomes thiere are but fewe at present in towne. 1687 London Gaz. No. 2269/2, 4172 pieces of Hummums. 1695 J. F. Merch. Wareho. laid Open 22, I shall begin with a Callico called Hummums. 1745 in J. F. Watson Ann. Philad. (1830) 179 Quilted humhums, turkettees, grassetts, [etc.]. 1801 in C. Cist Cincinnati in 1841 (1841) 179 Among other goods..humhums. 1809 P. Freneau Poems I. 31 Hum-hums are here—and muslins—what you please. 1820 Massachusetts Spy 5 Jan. (Th.), The bleach rotted linen..or the sleazy humhum. 1894 A. M. Earle Costume Colonial Times 142 Hum-hum, a plain coarse-meshed Indian fabric made of cotton, much advertised in the middle of the [18th] century. We read of 'blue Hum-hums' and 'Hump-humps for Sacks' for sale in various Boston newspapers, from 1750 to 1770. 1952 Brewer's Dict. Phr. & Fable (rev. ed.) 477/1 Humhum (U.S.A.), a thin cambric material.

humic, *a.* Delete *Chem.* and add to def.: present in or of the nature of humus; rich in humus; also, formed or derived from plant remains. (Further examples.)

1891 Chem. News 22 May 248/2 The part played by humic matters in the fixation of nitrogen. 1918 Stopes & Wheeler Monogr. Constit. Coal 19/2 Potonié's separation of coals into two main groups, sapropelic and humic, is now widely adopted... But the use of the phrase 'humic coals' is much in want of elucidation, as it, too, is given different meanings by different authors. As used by Potonié and most palæontologists, it means that the coal was formed from 'humic débris', or the various largely undecayed accumulations of mixed organs of plants... When the word 'humic' is applied to coals by chemists.. it may convey the implication that 'humic acid' exists in the substance of the coal. 1936 S. A. Waksman Humus iii. 62 One may..feel justified in abandoning without reservation the whole nomenclature of 'humic acids'. 1954 W. Francis Coal v. 245 Humic substances are the major constituents of all normal coals. 1957 G. E. Hutchinson Treat. Limnol. I. ix. 609 The peaty sediments (dy) of brown humic lakes. 1960 N. Polunin Introd. Plant Geogr. xv. 497 Humic matter in solution may result in a browning of waters coming in from leached soils. 1966 B. Simpson Rocks & Minerals xx. 234 The first group is the Humic Coals in which woody tissue is of paramount importance in the composition... The second group, the Sapropelic Coals, have little woody tissue in their composition. 1971 Nature 19 Nov. 150/2 They are usually found in moist, humic soil, often near water.

humidification (hiumi:difikēi·ʃən). [f. Humidify *v.*: see -fication.] The process of making moist or humid; *esp.* the process of rendering the air humid by means of special apparatus or techniques.

1890 J. Nasmith Mod. Cotton Spinning Machinery 10 The [spinning] rooms..should be heated to a certain temperature. Closely allied to this question is that of humidification. It is not only essential to have heat, but that must be accompanied by..moisture. 1940 A. C. Noé tr. Stutzer's Geol. Coal viii. 307 A peculiar case of weathering and humidification of coal is found in karst districts, producing a sooty material... The sootiness is caused by the access of karst water. 1945 Times 31 Aug. 2/3 Air-conditioning with humidification and cleaning by electrical filtration. 1970 D. Kut Warm Air Heating viii. 124 Humidification of the air prevents the build-up of static electricity.

humidify, *v.* Add: So **humi·dified** *ppl. a.*

1921 C. V. Ekroth in A. Rogers Industr. Chem. (ed. 3) li. 1164 Humidified air keeps the exposed surfaces of the food soft and permeable. 1956 H. Williamson Methods Bk. Design xxii. 358 Specially humidified paper will have to be wrapped in waterproof material. 1971 Nature 24 Dec. 471/2 They..are maintained at 36° C in a humidified atmosphere of 5% CO_2 in air.

humidistat (hiū·mi·distæt). Also **humidostat.** [f. Humidi(ty (or Humid *a.* + -o) + -stat.] **1.** (See quot.) *rare.* ?*Obs.*

1909 Cent. Dict. Suppl., Humidostat, a small chest or cabinet lined with sheet-metal and fitted with some device for holding a damp sponge or piece of felt: used to keep cigars moist.

2. An automatic apparatus for regulating the humidity of the air in a room or building; also called a *hygrostat.*

1918 Allen & Walker Heating & Ventilation xvi. 248 In a compressed-air system of control, a diaphragm valve ..may be operated by means of a 'hygrostat' or 'humidostat', which corresponds to the thermostat of a temperature control system. 1936 Archit. Rev. LXXX. p. lxviii/1 The control consists of a thermostat, a humidistat and a change-over rotary switch which together control the supply of hot or cold humidified or dehumidified air as required. 1970 I. G. Walls Greenhouse Gardening viii. 55 Humidistats can be used..to put into operation mist nozzles in an air stream.

humidor (hiū·midǭɹ). [f. Humid *a.*, after *cuspidor.*] A box, cabinet, or room in which cigars or tobacco are kept moist; also, any apparatus, such as damp sponges, for keeping cigars, the atmosphere, etc., moist.

1903 Sun (N.Y.) 4 Mar. 6/3 (Funk), A humidor as large as a small cottage occupies most of the ground floor. 1922 Detective Mag. Nov. 100 The cigar was soft and fresh from the humidor. 1927 P. Marks Lord of Himself 198 Take a cigarette out of that humidor. 1955 Sturdevant & Brauer in J. C. Brauer Dental Asst. xx. 287 The impression should be stored in a humidor in order to minimize dimensional change. 1963 Punch 8 May 669/1 Champagne coolers and waste-paper baskets, to say nothing of door-stops and humidors. 1973 R. Hayes Hungarian Game viii. 61, I fumbled with the F.O.'s humidor for a Havanna.

humify (hiū·mifəi), *v.*² [f. Hum(us + -ify.] **a.** *trans.* To convert (plant remains) into humus. **b.** *intr.* To undergo humification. So **hu·mified** *ppl. a.*

1906 E. W. Hilgard Soils (1930) viii. 132 Excluding the unhumified while fully dissolving the humified matter. Ibid. viii. 139 Snyder..caused various substances to humify by mixing the pulverized material intimately with a soil poor in humus. 1926 Technol. Rep. Tôhoku Imp. Univ. VI. i. 10 As humification advances more and more, the cell walls are quite humified. 1936 S. A. Waksman Humus iv. 64 Plant material added to the soil has first to be 'humified' before the nutrient elements contained therein become available for plant growth. 1948 G. W. Leeper Introd. Soil Sci. xi. 136 Residues of plants are said to be humified when they lose their structure and identity in the soil.

So **humifica·tion**², the process by which plant remains are converted into humus; the state of being humified that results.

1897 Minnesota Agric. Exper. Station Bull. No. 53. 13 Analyses were made of both the soil, and the material used for humus production. After the humification process had been carried on for a year the material was weighed and analyzed. 1926 [see above]. 1958 F. E. Zeuner Dating Past (ed. 4) iii. 64 A change in the rate of peat-growth from a slow one (allowing humification to begin in the layers near the surface), [etc.]. 1968 R. G. West Pleistocene Geol. & Biol. iv. 54 The amount of humification may be measured in the field by adding a little sediment to a few ml of dilute KOH.

humilific (hiūmili·fik), *a.* (*sb.*). Now *rare* or ?*Obs.* [f. L. *humilis* Humble + -fic.] Humiliating, self-depreciating, that humiliates or tends to humble; also as *sb.*, a humble expression. (Opp. to *honorific.*)

1892 Spectator 13 Feb. 244/1 Among the Chinese and Japanese..merely honorific and humilific expressions. 1904 V. V. Banford in J. E. Hand Ideals of Sci. & Faith 120 The distinction between honorific and humilific occupations. 1905 D. Sladen Playing Game i. vii, That conversation.., so full of the 'unfortunate mistake' vein, so burdened down with Japanese 'humilifics'.

hummable (hɒ·măb'l), *a.* [f. Hum *v.*¹ + -able.] That may be hummed; apt for humming.

1941 Time 10 Nov. 54 Cole Porter's score, though never haunting, is often hummable. 1966 New Statesman 8 Apr. 512/2 Georges Delerue's music is a hummable element. 1970 Daily Tel. 19 Nov. 14/5 The soloist starts off with a most poetical and, what's more, hummable eleven note row. 1972 Guardian 11 Jan. 8/6 The hummable simplicity of Rome's melodies.

hummaul, hummaum: see Hammal, Hammam.

hummel, *a.* Add: **1. b.** Applied to a hornless stag. Also *absol.*

1907 Spectator 5 Jan. 11/1 The 'hummel' stag—that ungainly beast with no horns at all—is a better fighter than the 'switch-horn'. Ibid. 11/2 The supremacy of the 'hummels' and 'switch-horns' in battles with their own kind. 1925 J. Buchan John Macnab iv, A hummel, a great fellow of fully twenty stone. 1964 G. K. Whitehead Deer Gt. Brit. & Ireland vii. 156 Some stags never grow any antlers at all and are known as hummels. These beasts would appear to be just as capable of breeding as their antlered brethren. 1972 Country Life 17 Feb. 424/3 The hummel or antlerless stag is not welcome on the purely sporting estate.

hummer, *sb.*¹ Add: **3. b.** A person or thing of extraordinary excellence. *colloq.*

1907 C. E. Mulford Bar-20 (1914) xxiv. 229 She's a hummer—stands two hands under him, and is a whole lot prettier than that picture Cowan has got over his bar. 1919 Wodehouse Damsel in Distress (1920) xx. 235 Well, you can't get there quicker than in my car. She's a hummer. 1920 C. E. Mulford Johnny Nelson (1921) xvii. 186, I claim I was justified—an' I'll leave it to you if th' joke on Wolf wasn't a hummer? 1934 N. Scanlan Winds of Heaven xvi. 150 When the new car was swung out on to the wharf, Mike walked round it and touched it lovingly. 'She's a hummer, Dad.' 1945 Baker Austral. Lang. ii. xvi. 286 A wow or hummer [to an American] is a bonzer or big twist [to an Australian].

4. b. *slang.* False or mistaken arrest (see quot. 1963).

1932 Evening Sun (Baltimore) 9 Dec. 31/4 Hummer, false arrest. 1961 Rigney & Smith Real Bohemia p. xv, Hummer, an arrest which accidentally leads to a more serious charge. 1963 Mencken Amer. Lang. xi. 730 Hummer, any kind of charge placed against a suspect so that he can be held although there is insufficient evidence to hold him on the charge for which he is really wanted.

hummer, *sb.*² Add: **2.** A scrounger. *Austral. colloq.*

1919 W. H. Downing Digger Dial. 29 Hummer, a cadger. 1945 Baker Austral. Lang. ii. v. 108 Hummer, poler and bot-fly are additional synonyms for a cadger.

humming, *ppl. a.* **1. c.** *humming-top* (earlier examples).

1819 Keats Let. 27 Sept. (1958) II. 216, I shall..make his little Boy a present of a humming top. 1837 London Med. Gaz. 1 Apr. 8/1 He compares it to the sound of a humming-top, or some such toy, called a 'diable' in French.

hummingly (hɒ·miŋli), *adv.* [See -ly².] With a humming sound.

1908 Daily Chron. 21 Jan. 6/5 He..endeavours to follow the service and join hummingly in the hymns. 1923 Daily Mail 6 Aug. 6 However warm the sun and hummingly populous the air, they [sc. the swifts] will go.

hummus (hu·mŭs). Also **hoummos.** [ad. Turk. *humus* mashed chick-peas.] In Middle Eastern countries (and also, more recently, elsewhere) an hors d'œuvre made from ground chick-peas and sesame oil flavoured with lemon and garlic.

1955 E. David Bk. Mediterranean Food 158 Hummus... Cook the chick peas..pound them, [etc.]. 1967 Guardian 8 Dec. 6/4 Order the paste of ground chick peas, oil, and lemon which is called hummus. 1969 M. J. Philippou 101 Arabian Delights 48 Hoummos ib bandora. Chick peas in tomato sauce... Hoummos ib Taheeneh. Chick peas in Taheeneh. 1970 Simon & Howe Dict. Gastron. 223/1 Hummus bi Tahina, a widely known, traditional..Arab dish of cooked, puréed chick peas... It is served as a mezze or appetizer in Arab countries. 1973 Nation Rev. (Melbourne) III. 31 Aug. 1460/4 Passing up the usual hummus as a starter.

humoral, *a.* **1. a.** (Further examples.) Also in mod. use, contained in or involving the blood or other body fluid; involving or consisting of a chemical agent, esp. one present in the blood (such as hormones or ions).

humoral theory: (*a*) the theory that immunity is due to the presence of bactericidal substances in the blood (rather than the action of whole cells); (*b*) the theory that the transmission of nerve impulses at a synapse or a neuromuscular junction involves a chemical (rather than an electrical) agent. (See also sense 1 c in Dict.)

1898 W. S. L. Barlow Man. Gen. Path. ix. 408 It was obvious..that a purely humoral theory..is insufficient, that natural immunity cannot be explained by the presence of 'alexins' in the blood. 1924 E. D. Adrian in Brain XLVII. 400 The 'inhibiting substance' theory (and the 'humoral' theory of Loewi and others) derives all its evidence from the peripheral inhibition of cardiac or plain muscle. 1935 O. Loewi in Proc. R. Soc. B. CXVIII. 302 Some important investigations..suggest the possibility that the stimulation also of spinal nerves is transmitted by humoral means, namely, by liberation of acetylcholine. 1950 G. P. Wright Introd. Path. vii. 113 The early history of immunology contains many vigorous controversies which arose from the sharp division of opinion between the so-called 'cellular' and 'humoral' schools. 1952 W. E. Le Gros Clark Tissues of Body (ed. 3) xiii. 356 The conception of chemical (or humoral) transmitters [of nerve impulses] suggests an alternative explanation. 1968 Passmore & Robson Compan. Med. Studies I. xxvii. 2/1 The immune response can be purely humoral, mediated by circulating antibodies, or cellular, mediated by small lymphocytes, but often mixtures of both mechanisms are found. 1971 A. C. Guyton Textbk. Med. Physiol. (ed. 4) xxv. 284/2 Superimposed onto the intrinsic regulations of the circulation..are two additional types of regulation: (1) nervous and (2) humoral. 1971 Nature 26 Feb. 593/1 Human breast cancer..may well be initiated by a virus but promoted by humoral factors, in particular sex hormones.

humorism. Add: **2. b.** A humorous saying or remark.

1897 'Mark Twain' Notebook (1935) 335 In a dream I have at last encountered a humorism that actually remained one after waking. 1902 Sat. Rev. 15 Nov. 620/1 The most sparkling witticisms and the most obvious humourisms have fallen flat. 1907 Daily Chron. 26 June 7/5 His eloquent remarks were snatched up by the people he met and mistaken for brilliant humorisms. 1920 D. H. Lawrence Women in Love (1921) ii. 34 Gerald smiled grimly at this humorism.

humorize, *v.* Add: **3.** *trans.* To make humorous.

1893 W. B. Yeats Celtic Twilight 195 Our tales..hopelessly humorize the creatures [sc. water-goblins and water-monsters].

humous (hiū·məs), *a.* [f. Hum(us + -ous.] Present in or of the nature of humus; rich in humus. Cf. *humic *a.*

1866 H. Watts tr. Gmelin's Hand-bk. Chem. XVII. 473 Hermann..distinguishes eleven different humous substances. 1902 Jrnl. Chem. Soc. LXXXII. ii. 521 Oats and soy beans..were grown in humous soil. 1909 Groom & Balfour tr. Warming's Oecol. Plants xlvi. 195 The weightiest cause of the physiological dryness of the soil probably lies in the presence of free humous acids. 1968 R. G. West Pleistocene Geol. & Biol. iv. 55 Humous substances.

hump, *sb.* Add: **1. c.** Also, the flesh of the hump of other animals.

1805 M. Lewis Jrnl. 13 June in Orig. Jrnls. Lewis & Clark Expedition (1904) II. 151 My fare is really sumptuous this evening; buffaloe's humps, tongues and marrow-bones. 1823 J. Franklin Narr. Journey Shores Polar Sea 115 The meat [of the buffalo] which covers the spinal processes themselves, after the wig is removed, is next in esteem for its flavour and juiciness, and is more extensively termed the hump by the hunters. 1861 C. J. Andersson Okavango River 130 Rhinoceros hump was..a frequent and favourite dish of mine. 1863 Rhinoceros hump [see Rhinoceros 3]. 1909 Daily Chron. 5 Jan. 4/7 'Humps have arrived.' So runs the legend in an old-established shop in Green-street, Leicester-square. Ibid., A beef hump. 1913 C. Pettman Africanderisms, Hump... A favourite piece with South African housewives for salting.

d. *to live on one's hump*: to be self-sufficient, to operate from resources accumulated earlier: with reference to the camel's hump as a reserve of nourishment.

1909 Westm. Gaz. 11 Sept. 7/2 During nearly three weeks in this glorious place I have lived on my own hump.

2. b. A mound in a railway yard up which vehicles are pushed by an engine and down the other side of which they run by gravity and are switched to the proper track.

1901 Railroad Gaz. 4 Jan. 2/1 All that was necessary to take advantage of this mode of distributing cars, was to put a 'hump' in the switching track. 1911 Encycl. Brit. XXII. 842/1 Another method [of shunting], which was introduced into America from Europe about 1890, is that of the summit or 'hump'. 1921 Daily Mail Yr. Bk. 112/1 The London and South-Western Company has constructed at Feltham a new 'hump' marshalling yard. 1955 Times 12 July 3/4 They had approved new works at Perth, including the construction of a fully mechanized hump marshalling yard. 1958 Times 11 Feb. 15/3 We have carried out development on equipment..for assisting the automatic operation of hump sorting yards. 1971 D. J. Smith Discovering Railwayana x. 56 Hump yard, goods yards or sidings worked by gravity shunting methods.

c. A mountain barrier high enough to make both land and air travel difficult. Chiefly *U.S.*

1914 Sat. Even. Post 4 Apr. 10/2 There ain't a kid like him this side of the Hump [sc. mountain range in west of N. America]—nor t'other side either. 1931 'D. Stiff' Milk & Honey Route 208 'Over the hump' means to cross the mountains to the West Coast. 1936 K. Mackenzie Living Rough xv. 216 We're sure a pair of nuts riding the outside over the hump this time of the year. 1944 Time 26 June 52/1 They're flying it over 'The Hump'—the towering Himalayas between India and China. 1970 'B. Mather' Break in Line v. 64 'I think he'll be making for Rangoon, then over the bloody hills to China.'..'You really expect me to walk over the Hump?'

d. *transf.* and *fig.* The critical point (in an undertaking, ordeal, the ascent of a seaplane, etc.), esp. in the phr. *over the hump*, over the worst, well begun. (Cf. *2 c.) Also, the high point, peak (of a graph, etc.).

1914 Jackson & Hellyer Vocab. Criminal Slang 46 Hump,..the half-way point in a prison sentence. Example: 'How long have you got yet on your bit? I'm just over the hump.' 1914 Techn. Rep. Advisory Comm. Aeronaut. 1912–13 244 The floats of the flaring bow type require only about half the E.H.P. to surmount the hump. 1922 W. R. Inge Lay Thoughts (1926) ii. i. 89 If we look at a chart of the births and deaths in Germany..each war is marked by a peak in the line showing the death rate and a ravine in the line showing the birth rate. But the ravine is followed by a hump..making good the numbers lost. 1929 D. Hammett Dain Curse (1930) xxii. 253 Today won't be like yesterday. You're over the hump, and the rest of it's downhill going. 1935 P. W. F. Mills Elem. Pract. Flying v. 71 In rising from the normal semi-submerged state there is a critical point known as passing the hump, before the reaching of which point the floats are definitely water-bound. After passing the hump the floats very nearly emerge from the water and commence to hydroplane. The time taken to reach the 'hump' and the ease with which it is passed, vary greatly. 1938 Amer. Speech XIII. 188/2 Once a cocaine addict is over the hump he says he is coasting or in high. 1952 Economist 27 Sept. 771/1 The machine tool industry is probably 'over the hump' of its..task. 1959 Listener 19 Feb. 316/2 Things were very difficult with us that year, and the Americans helped us over the hump. 1960 Economist 8 Oct. 135/1 The 'hump' in imports that was desired has turned into a steady climb. 1965 Listener 16 Dec. 985/1 East German farming is getting over the hump.

3. Phr. *to give* (*a person*) *the hump* (examples).

1887 F. Gale Game of Cricket viii. 187 So let's alter the law, Without any more jaw, Or you'll give an old buffer the hump. 1910 E. M. Forster Howards End vi. 51 That tune fairly gives me the hump. 1939 T. S. Eliot Family Reunion 18 You seem to be wanting to give us all the hump. I must say, this isn't cheerful for Amy's birthday.

b. A walk or hike with a load on one's back. (Cf. Hump *v.* 2.) *Austral.* and *N.Z. slang.*

1863 J. G. Walker Jrnl. 7 Jan. (MS.) 4 It was a precious hump [over the hill for provisions]. 1890 'R. Boldrewood' Miner's Right v, We get a fair share of exercise without a twenty-mile hump on Sundays.

3*. *to get a hump on*, to hurry. (Cf. *hump *v.* 3.) *U.S. colloq.*

1892 Harper's Mag. Feb. 487/2 'We went fast enough then.' 'We do seem to be gittin' a leetle less hump on oursel's than we did then.' 1940 W. E. Wilson Wabash 231 'Let's git a hump on, Allen,' Abe said; and the two boys dipped their oars deeper into the brown water.

3.** Sexual intercourse; hence, a woman who makes herself available for sexual intercourse. *coarse slang.*

1931 G. Irwin Amer. Tramp & Underworld Slang 105 Hump,..sexual intercourse. 1969 E. R. Johnson Mongo's back in Town (1970) ii. 18 It might be a good idea to line up a Christmas hump for himself. 1969 P. Roth Portnoy's Complaint 134 Now you want to treat me like I'm nothing but some hump. 1970 'D. Craig' Young Men may Die vi. 48 It was hard to believe she could be more than an ad hoc hump for Lamartine.

4. *hump rib* (earlier U.S. examples); **hump speed** *Aeronaut.*, the speed of a seaplane or hovercraft at which the drag due to the water is a maximum (cf. quots. 1914², 1935 for sense *2 d).

1834 in Oreg. Hist. Soc. Q. (1916) XVII. 126 The tongue, the heart, the marrow bones and the hump ribs is all they use when meat is plenty. 1839 J. K. Townsend Narr. Rocky Mts. iii. 164 They..appeared to be surveying, with the keenness of morning appetite, the fine 'hump ribs' which were roasting before them. 1915 Techn. Rep. Advisory Comm. Aeronaut. 1913–14 369 It was not practicable to fit air-holes to the second step, but it is probable that a slight reduction of power at the 'hump' speed could be obtained by this means in smooth water. 1938 Jrnl. R. Aeronaut. Soc. XLII. 559 The relative distance and time required to reach hump speed depend largely on the value of the accelerating force at high speeds. 1962 New Scientist 24 May 388/1 As a hovercraft accelerates from rest there is a so-called 'hump-speed' (about 14–18 knots) above which the drag suddenly drops.

hump, *v.* Add: **1. d.** *transf.* of inanimate things.

1901 'Linesman' Words by Eyewitness (1902) 168 With the berg humping its mighty shoulders far behind them.

2. (N.Z. and U.K. examples.) Also *to hump it.* See also *bluey *sb.*

Several of the examples suggest that the object is not necessarily carried *on the back.*

1863 J. Goldie 3rd Diary 19 Feb. in J. H. Beattie Pioneers explore Otago (1947) 147 Digger custom, we humped our swag containing our house, our bed, our grub. 1863 J. G. Walker Jrnl. Mar. (MS.) 4 Humping it over from the Tiviot on our backs would not do as it was too hard work. 1864 J. C. Richmond Let. 12 May in Richmond–Atkinson Pap. II. 111 It is very hard work humping your blankets and tucker. 1865 E. R. Chudleigh Diary 16 July (1950) 193 Humping all their belonging with them. 1866 B. L. Farjeon Shadows on Snow 66 [Diggers]. The best thing we can do..is to try and hump it back again tomorrow. 1916 'Taffrail' Pincher Martin xii. 218 We'll have to hump the whole bloomin' lot out again, damn an' blast him! 1922 T. E. Lawrence Let. 7 Sept. (1938) 365, I went off to hump their swill to the camp pigs. 1924 —— Let. 20 Jan. (1938) 456 If it is the best I can do with a pen, then it's better for me to hump a rifle or spade about. 1925 Fraser & Gibbons Soldier & Sailor Words 122 To hump, to lift, to carry. Ibid., To hump it, to march with full kit, to tramp on foot. 1955 M. Gilbert Sky High viii. 112 Couldn't you hump around the heavy lectern vases. 1960 Sunday Express 6 Mar. 8/4 He..tugged out a suitcase containing his full-dress uniform, humped it across the pavement. 1971 B. W. Aldiss Soldier Erect 78, I followed behind him, humping the wireless set. 1971 N.Z. Listener 22 Mar. 13/1 He's humpin' a haversack. 1973 C. Bonington Next Horizon xii. 171 John and Dougal took the lead, while Layton and I followed, humping loads... I humped my big rucksack, taking the occasional photograph.

3. (Further U.S. and other examples.) Also, to hurry.

1884 'Mark Twain' Huck. Finn xxix. 307, I never hunted for no back streets, but humped it straight through the main one. 1906 D. Coke Bending of Twig iv. 71 'We shall have to hump ourselves for call-over,' he said..as they dashed up the hill. 1908 G. H. Lorimer Jack Spurlock i. 9 He..said to the cop on guard, 'One of them Ha'voids [= Harvard students],' and to me, 'Hump yourself.' 1928 'Sapper' Female of Species ii, Peter—your Sunbeam, and hump yourself. Ibid. xiv, That finger will connect with the trigger and the result will connect with you. So, hump yourself. 1968 M. Woodhouse Rock Baby vi. 48, I..humped myself into my coat.

b. In extended use.

1905 Smart Set Sept. 117/1 You'll have to get this machine to hump it all she knows. 1929 J. B. Priestley Good Companions i. i. 3 As you look down on Bruddersford, you feel..that it is only biding its time, that it will hump its way through somehow.

4. (Later example.)

1898 A. Beardsley Let. 16 Jan. (1971) 427 Letter writing humps me dreadfully.

5. *trans.* and *intr.* To have sexual intercourse (with). *coarse slang.*

1785 Grose Dict. Vulgar T., Hump, to hump, once a fashionable word for copulation. 1931 G. Irwin Amer. Tramp & Underworld Slang 105 Hump, to have intercourse. Ibid. 263 In 1914–1918..hump and niggle were used of both sexes, screw and shag were operatively male. 1961 J. Heller Catch-22 (1962) xiii. 131 The girls had shelter and food for as long as they wanted to stay. All they had to do in return was hump any of the men who asked them to. 1962 J. Baldwin Another Country (1963) i. i. 14 A nigger..lives his whole life, lives and dies according to a beat... He humps to that beat and the baby he throws up in there..comes out nine months later like a goddamn tambourine. 1965 M. Bradbury Stepping Westward vii. 345 Story is he humped the faculty wives in alphabetical order. 1971 W. Hanley Blue Dreams vii. 90 Gazing at her, the hem of her skirt pushed by the attitude of her dazzling legs far back along her thighs, he thought, Jesus, what would it be like to hump her?

humpback, *sb.* (*a.*). Add: **A. 3.** Also = **humpback salmon, sucker.*

1881 Amer. Naturalist XV. 182 The hump-back, taken in salt water about Seattle, shows the same peculiarities. 1913 Chambers's Jrnl. Oct. 729/1 The 'steel-head', the 'dog-salmon', and the 'humpback'. 1955 [see *Chinook b]. 1963 Vancouver Sun 13 Aug. 14/1 Humpback is the old and still popular name for the pink salmon. 1965 A. J. McClane Standard Fishing Encycl. 411/1 The humpback [sucker] prefers the slow-moving parts of larger streams.

B. humpback salmon = *gorbuscha, Haddo; **humpback sucker** *U.S.*, *Xyrauchen*

texanus, a freshwater fish of the Colorado basin.
1869 *Mainland Guardian* (New Westminster, B.C.) 25 Sept. 2/1 The Oleys or Hones appear every alternate year; they are known as the Humpback salmon. **1881** *Amer. Naturalist* XV. 177 The fact that the hump-back salmon runs only on alternate years in Puget sound..is well attested. **1891** [see *gorbuscha]. **1965** A. J. McClane *Standard Fishing Encycl.* 410/2 Humpback sucker *Xyrauchen texanus.* **1971** B. Muus *Freshwater Fish Brit. & Europe* 64/2 The humpback salmon is distinguished from other Pacific salmon by its short, 2-year life-cycle.

hump-backed, *a.* Add: *Esp.* in the names of fishes; cf. *humpback *sb.* (*a.*).
1807 tr. *Garytschev's Voy.* 28 We managed..to lay in a stock for ourselves of the hump-backed salmon, and other such fish. **1884** G. B. Goode *Fisheries U.S.: Nat. Hist. Aquatic Animals* 323 It is a frequent summer visitor all along the coast as far north as Wood's Holl, Massachusetts, where it has a peculiar name, the people there calling it the 'Hump-backed Butterfish'. **1896** Jordan & Evermann *Check-List Fishes N. & Mid. Amer.* 241 *Xyrauchen cypho.* Razor-back Sucker; Hump-backed Sucker. **1902** *Encycl. Brit.* XXVI. 391/1 The dog-salmon and the humpbacked have no commercial value. **1973** *Times* 10 Jan. 7/2 He watched two fishing boats return fully laden with hump-backed salmon.

humped, *ppl. a.* Add: Also *humped-up*.
1885 T. Roosevelt *Hunting Trips* iv. 104 The cattle..standing humped up in the bushes. **1906** W. S. Maugham *Bishop's Apron* v. 71 The coachman..sat on his box in a slovenly, humped-up fashion. **1931** G. O. Russell *Speech & Voice* xiv. 137 The high-arching and humped-up-rounding position of the tongue.

humpenscrump (hɒ·mpənskrɒmp). [*dial.* (not in *E.D.D.*), perh. f. Hump *sb.* + (*Old Father*) *Scrump*, a freq. appellation of Big Head in traditional English folk-plays; cf. Humstrum and Scrump.] A musical instrument of rude construction; a hurdy-gurdy (sense 1 a).
[**1923** R. J. E. Tiddy *Mummers' Play* 172, I old father scrump with a bell on my rump... A man..kicked a hump up on to my shoulder and there remains the hump now.] *Ibid.* 173 (*stage directions*) Instruments Tin whistle. Jew's Harp. Father Scrump carries the humpenscrump made with a tin with wires across and bridge and a stick with notches for a bow. **1933** E. K. Chambers *Eng. Folk-Play* 70 The musical instrument is generally a fiddle, or a rustic substitute called a hurdy-gurdy or humpen-scrump. **1954** L. MacNeice *Autumn Sequel* 142 The humpenscrump that charmed some forebear's ear Now makes the roast pigs caper in Cockayne.

humper (hɒ·mpəɹ). [f. Hump *v.* + -er¹.] Something or someone that humps, in the senses of the verb. (For quot. 1895 cf. *hump *v.* 3.)
1895 *Columbus* (Ohio) *Dispatch* 27 Mar. 1/3 We were coming along on time. Engine 586 is a humper, and Rankin, my fireman, was keeping her hot. **1897** Kipling *Capt. Cour.* vii. 142 Hark to her [*sc.* a ship]! She's a humper! **1961** *Spectator* 7 Apr. 471 The carrying of a second man, or 'humper', on newspaper delivery vans in London. **1967** *Times Rev. Industry* Feb. 35/2 Worked in the Smithfield meat market..earning 30s. as a humper.

humpish (hɒ·mpiʃ), *a.* [f. Hump *sb.* + -ish¹ 2.] Somewhat like a hump, somewhat squat and dumpy.
1936 W. Holtby *South Riding* ii. v. 130 Lydia stood up, a humpish stocky schoolgirl. **1939** R. Fry *Last Lectures* v. 68 They are squat and humpish renderings, often of seated forms.

humpty. Add: Also as *sb.*, a low padded cushion seat, a dumpty.
1924 *Drapery Sale Catal.*, Humptys and Cushions. **1926** *Brit. Weekly* 18 Mar. 600/2 The ladies of the village are busy making 'humpties', soft cushion seats to pull up on the rug before the peat fire. **1928** *Daily Express* 23 Jan. 5/2 Home-made Humpty.

humpy, *sb.* Add: Also **humpie**. In extended use: a hut (not necessarily one occupied by an Aboriginal). (Further examples.)
1911 *Chambers's Jrnl.* Sept. 591/2, I knew where this humpie was. **1927** M. Terry *Through Land of Promise* xiii. 161 The manager's house..had been a small humpie consisting of four stone walls capped by a tin roof. **1928** 'Brent of Bin Bin' *Up Country* viii. 130 They are big tents or humpeys of bark, with flags on top to show if they are English or Russian or some other nation. **1942** C. Barrett *On Wallaby* i. 14 The old fossicker was..seated outside a bark humpy. **1953** 'N. Shute' *In Wet* x. 348 His camp consisted of a tent for himself and his wife, and a humpy shelter made of gum tree boughs for his white ringer, Phil Fleming. **1969** *Australian* 7 June 16/1 There's a drawing in Petty's Australia Fair that shows an Aboriginal urchin in front of his family's humpy in the carcass of a Holden. **1971** *World Archaeol.* III. 168 Structures (termed 'humpies' by Europeans) which are the year-round type of dwelling used by mission-dwelling Aborigines.

humpy (hɒ·mpi), *sb.*² *Austral. slang.* Also humpie. [f. Hump *sb.* 1 + -y⁶.] A camel.
1934 A. Russell *Tramp-Royal in Wild Australia* i. 19 There's a spare riding camel in my outfit you can have... Only you'll have to rough it..rough it hard, too..same as me. But the humpie's there if you want it. *Ibid.* iii. 29 He is the despised 'humpie', the 'filthy camel'. **1945** Baker *Austral. Lang.* 214 A camel is an *ooni* or a *humpy*.

humpy, *a.* Add: **1. b.** Out of humour; melancholy, sad. Cf. Hump *sb.* 3.
1889 J. K. Jerome *Three Men in Boat* i. 10 Harris said he thought it would be humpy. He said he knew the sort of place I meant; where everybody went to bed at eight o'clock, and you..had to walk ten miles to get your baccy. **1911** 'Ian Hay' *Safety Match* xii. 194, I have got into the way of bringing you my little troubles, and turning to you generally if I felt dismal or humpy. **1920** *Glasgow Herald* 29 July 4, I mention these facts so that readers may be reassured if they are inclined to be 'humpy'... Many among us..are never so happy as when they are making themselves miserable by looking at the black side of things. **1941** 'R. West' *Black Lamb* I. 71 He had found himself tired and wounded and humpy and alone after a day's hunting.

humulene (hiū·miulīn). *Chem.* [f. mod.L. *humul-us* (in *Humulus lupulus*, taxonomic name of the hop) + -ene.] A colourless liquid sesquiterpene, $C_{15}H_{24}$, identical with α-caryophyllene and forming the principal constituent of oil of hops.
1895 A. C. Chapman in *Jrnl. Chem. Soc.* LXVII. 62 The evidence I have given as to its chemical individuality justifies me, I think, in adding to this list the sesquiterpene of the essential oil of hops, for which I venture to propose the name humulene. **1951** *Jrnl. Chem. Soc.* 22 Humulene ($C_{15}H_{24}$), the chief sesquiterpene constituent of hop oil and a minor constituent of the high-boiling terpene fraction of clove oil, has been shown to be present in the essential oil of Egyptian hashish.

humulone (hiū·miulōun). *Chem.* Also -on (-ǫn). [ad. G. *humulon* (I. W. Wöllmer 1916, in *Ber. d. Deut. Chem. Ges.* XLIX. 780), f. as prec. + -one.] A bitter, yellow, crystalline, cyclic ketone, $C_{21}H_{30}O_5$, that is an important constituent of hops and has strong antibiotic activity.
1916 *Jrnl. Chem. Soc.* CX. I. 494 A contribution to the chemistry of 'humulon' or 'a-hop bitter acid' or 'lupulic acid'. **1937** Hopkins & Krause *Biochem. appl. to Malting & Brewing* iv. 221 Both humulon and lupulon are secreted in the hop cone during growth and become more or less resinified. **1964** Kirk & Othmer *Encycl. Chem. Technol.* (ed. 2) III. 302 By far the most important constituents of the hops are the bitter substances, humulone, lupulone, and their transition products, which give beer its characteristic taste. **1967** E. Paryski tr. *Korzybski's Antibiotics* II. vi. 1447 Lupulon and humulon strongly inhibit the growth of gram-positive micro-organisms and acid-fast bacilli.

Hun, *sb.* Add: **4.** *gen.* A person of brutal conduct or character; *esp.* during and since the war of 1914–18 applied, often without animus, to the Germans (or their allies); a German. Also *attrib.*
[The immediate source of the application of *Hun* to the Germans was the speech delivered by Wilhelm II to the German troops about to sail for China on 27 July 1900. See the following examples:
1900 *Times* 30 July 5/3 According to the Bremen *Weser Zeitung* the Emperor said [27 July at Bremerhaven]:—'..No quarter will be given, no prisoners will be taken. Let all who fall into your hands be at your mercy. Just as the Huns a thousand years ago, under the leadership of Etzel (Attila) gained a reputation in virtue of which they still live in historical tradition, so may the name of German become known in such a manner in China that no Chinaman will ever again even dare to look askance at a German.' **1900** *Daily News* 20 Nov. 5/3 Herr Bebel [in the Reichstag] dwelt..at some length on the so-called Hun letters, and stigmatized the cruel and barbarous methods of European warfare in China. **1900** *Times* 21 Nov. 5/2 A great portion of the speech of the Socialist leader [Bebel] was devoted to the so-called 'Letters from the Huns' (*Hunnenbriefe*)—epistles from German soldiers in China to their relatives at home giving an account of the cruelties which have been perpetrated by the army of occupation.]
1784–5 in *Publ. Navy Rec. Soc.* (1906) XXXI. 55 Andrew Duff, Midshipman. Dead. A drunken Hun. **1862** H. Timrod *Poems* (1901) 143 Shout! let it reach the startled Huns! And roar with all thy festal guns! It is the answer of thy sons, Carolina! **1902** Kipling in *Times* 22 Dec. 9/5 In sight of Peace..With a cheated crew, to league anew With the Goth and the shameless Hun! **1914** —— in *Queen* 5 Sept. 388/2 Stand up and meet the war. The Hun is at the gate! **1915** E. Candler in *Daily Mail* 5 Apr. 4/3 She [*sc.* a Norfolk girl] told me how the eldest [brother 'at the front'] had held up three 'Huns' in a mill... She used the word 'Hun' quite naturally, with no hint of contempt or bitterness. **1916** 'Boyd Cable' *Action Front* 133 Do you suppose our friend the Flighty Hun won't have a peep at us to-morrow morning? **1916** 'Taffrail' *Pincher Martin* xiv. 269, I suppose you know, Peter,..that we were bang on the top of a Hun minefield. *a* **1918** [see *crash *v.* 6 a]. **1918** *Times* 12 Dec. 9/4 'Supposed' statements..of American 'advisers'..simply smell of Hun propaganda. **1932** [see *bit *sb.*² 4 h]. **1941** [see *crash-land *v.*]. **1942** *Tee Emm* (Air Ministry) II. 63 The squadron has, after those months of inaction, started to bag Huns. **1945** [see *abroad C *sb.*]. **1958** P. Kemp *No Colours or Crest* vi. 104 They ambushed a cartload of Huns the other day.
b. A flying cadet: see quots. *Air Force slang* (in the war of 1914–18).
1916 H. Barber *Aeroplane Speaks* 36 The Aeroplane..remonstrates..'See the Medical Officer, you young Hun.' **1918** E. M. Roberts *Flying Fighter* 233 An aeroplane..was flying over the street, but I don't know what the couple of British Huns in it were trying to do. *Ibid.* 336 Every pilot is a Hun until he has received his wings. **1925** Fraser & Gibbons *Soldier & Sailor Words* 123 The word 'Hun'..was used..for a newly-joined young officer qualifying for his 'wings', in consequence of the destructive effect on the instructional aeroplanes which young officers while learning to fly usually had.
Hence (sense *4) **Hu·ndom**, the state of being a 'Hun'; **Hu·nnish** *a.*, whence **Hu·nnishness**; **Hu·nless** *a.*, lacking Germans.
1916 F. Lawrence *Mem. & Corr.* (1961) 211, I saw you being martyred on account of my Hundom! **1918** *Punch* 27 Mar. 207/2 The Hunnish conduct of the German officer who egged on the natives. **1920** *Blackw. Mag.* Feb. 154/1 The islands were entirely Hunless. **1924** C. J. Tolley *Mod. Golfer* 7 The only piece of Hunnishness we ever encountered at Heidelberg was at the hands of an appalling doctor, who..thought fit to inoculate us against every known disease. **1928** *Manch. Guardian Weekly* 2 Nov. 350/2 Once give the better England clear evidence that Hunnishness is not the sole attribute of the German spirit, and [etc.].
5. *Comb.*, as *Hun-folk*, *-hater*, *-land*, *-talk*; *Hun-eating*, *-hunting*, *-pinching*.
1928 *Manch. Guardian Weekly* 2 Nov. 350/2 There were two Englands—(a) the impossible Hun-eating England and (b) the better England. **1923** Kipling *Irish Guards in Gt. War* I. 343 The Battalion..watched about them..the muddy-faced Hun-folk. **1920** R. Macaulay *Potterism* ii. i. 58 He would have to include among his jingoes and Hun-haters some fighting men too. **1925** Fraser & Gibbons *Soldier & Sailor Words* 123 *Hun hunting*, an Airman's phrase for going out to look for, or chase, the enemy. **1916** *Daily Mirror* 1 Nov. 4/4 Gott strafe England.., the recognised toast throughout Hunland. **1918** [see *egg *sb.* 3 d]. **1920** *Glasgow Herald* 20 Nov. 5 No such sentiments could be admitted in Hunland. **1925** Fraser & Gibbons *Soldier & Sailor Words* 124 *Hunland*, a term generally used in the War by Airmen for the country behind enemy lines, wherever it might be. **1917** A. G. Empey *Over Top* 295 '*Hun pinching*', raiding German trenches for prisoners. **1959** P. Moyes *Dead Men don't Ski* iii. 34 You ask her, Roger... You're the expert in Hun-talk.

hun, colloq. abbrev. of *honey 5 b.
1896 Ade *Artie* xiv. 126 'Look at the new hat on her.'.. 'It's a hun,' remarked Artie.

hun, var. *hoon *sb.*¹

Hunanese (hūnănī·z), *a.* and *sb.* [f. *Hunan*, name of a province of southern China + -ese.] **A.** *adj.* Of, pertaining to, or characteristic of Hunan or of the Chinese spoken there. **B.** *sb.* **a.** An inhabitant of Hunan. **b.** The dialect of Hunan.
1937 J. R. Firth *Papers in Ling. 1934–51* (1957) vii. 80 It will be found convenient to regard the Hunanese monosyllable as having one, two, or perhaps three places in which the phonetic diacritica may be said to occur. **1937** E. Snow *Red Star over China* ii. iii. 65 Li Chiang-lin was a Hunanese. **1948** J. R. Firth *Papers in Ling. 1934–51* (1957) ix. 127 With the help of Mr. K. H. Hu, of Changsha, I studied the pronunciation and phonology of his dialect of Hunanese. **1957** *Times Lit. Suppl.* 8 Nov. 677/1 The structural analysis of an element in a Hunanese dialect. **1964** L. Mitchison *Gillian Lo* v. 54 Ai-yang was clucking with excitement, her Hunanese accent..strong. **1967** *Everyman's Encycl.* VI. 643/2 Since the time of the Taiping rebellion..the Hunanese have been noted for their pride and obstinacy in admitting outside control. **1972** *N.Y. Times* 3 Nov. 24/6 It is probably correct to say that there are tricky but fruitful points of similarity between Hunanese food and the food of southern France. **1972** 'M. Hebden' *Killer for Chairman* ii. iv. 162 There was a variety of dialects among the crowd from Chekiangese and Hunanese to the difficult Szechwanese.

hunch, *v.* Add: **2. b.** To nudge (a person) so as to direct attention to someone. Also *fig.* *U.S.*
a **1852** F. M. Whitcher *Widow Bedott Papers* (1883) 76 She kept a hunchin' Miss Coon and grinnin'. **1884** 'Mark Twain' *Huck. Finn* xxv. 246 Then the king he hunched the duke, private. **1906** *Life* 1 Feb. 147 Soon some fellow hunched the Legislature, and then there had to be more or less investigating done.
c. *intr.* To push or lunge forward. *U.S.*
1911 S. E. White *Bobby Orde* (1916) xvii. 196 Bending to his task the pusher at the rear dug his toes in, while the others hunched. **1913** G. S. Porter *Laddie* viii. 232 She sat astride the foot log, and hunched along with her hands. **1925** C. E. Mulford *Cottonwood Gulch* xix. 259 They hunched closer, hugging knees under chin.
3. Also without *up*.
1906 U. Sinclair *Jungle* xviii, Hiding his hands in his pockets and hunching his shoulders together. **1906** 'K. Howard' *Old Game* ii. i. 54 She merely hunched her shoulders, swung on her heel, and marched off.

hunch, *sb.* Add: **1. b.** A hint, 'tip'. (Cf. prec. 2 b.)
1849 T. M. Garrett in *Amer. Speech* (1951) XXVI. 183/1 Another piece [of writing] gave a few hunches to the inexperienced freshman. **1901** 'H. McHugh' *John Henry* 57 The reason it's so good is because I took my hunch from Rud. Kipling's style. **1922** Z. Grey *To Last Man* ii. 36 All shootin' arms an' such are at a premium in the Tonto... An' I was givin' you a hunch to come loaded.
4. A premonition or intuitive feeling that something will happen or may be the case; a presentiment.

1904 S. E. White *Silent Places* xviii. 200 'I hope your hunch is a good one,' replied Dick. **1907** R. W. Service *Songs of Sourdough* (1908) 52 Then you've a haunch [*sic*] what the music meant. **1918** E. M. Roberts *Flying Fighter* 62 This particular night Fritz had a hunch that somebody was going to pass the place behind the screen. **1926** G. D. H. & M. Cole *Blatchington Tangle* xiii. 97, I had an awful hunch what it was. **1938** *Brit. Jrnl. Psychol.* July 7, I relied not so much on conscious thought, as on what Americans call a 'hunch'. **1955** *Times* 9 May 5/2 A churchwarden's 'hunch' could never be a wholly satisfactory substitute for professional knowledge in regard to the care of churches. **1960** M. Spark *Ballad of Peckham Rye* ii. 14 'Only a hunch,' said Dougal. 'I may be wrong.' **1973** 'H. Howard' *Highway to Murder* viii. 89 My sixth sense told me I'd got myself an extra shadow. That hunch was all I had to go on.

hunched, *a.* Add: Also with advbs.
1910 W. de la Mare *Three Mulla-Mulgars* x. 144 His little hunched-up friend. **1920** *Chambers's Jrnl.* 110/1 A long..sinuous beast that hopped in a series of hunched-up bounds. **1921** C. E. Mulford *Bar-20 Three* xxi. 267 He..clawed himself into a saddle..and rode for safety, hunched over and but half conscious.

hundi (hu·ndi). *India.* Also **hoondee, hoondi, hoondy.** [Hind. *hundī* (Skr. *huṇḍikā* bill of exchange).] A negotiable instrument, such as a bill of exchange or promissory note, used by native bankers in India and worded in the vernacular; also, money remitted by such an instrument.
1619 in W. Foster *Eng. Factories India 1618–21* (1906) 85 [They advise the dispatch of bills of exchange for rupees] hundies [17,100]. **1620** *Ibid.* 182 The exchange of rup[ees] secaus for hundies. **1810** T. Williamson *East India Vade Mecum* II. 330 Hoondies (i.e. bankers' drafts) would be of no use whatever to them. **1913** J. M. Keynes *Indian Currency & Finance* vi. 197 The *hoondees* they buy and sell to each other..are chiefly the traders' *hoondees* bearing the shroffs' own endorsements. **1930** *Economist* 12 Apr. 820/1 Bills (hundis) of the native type. **1963** *Times* 18 May 8/4 It is thought that gold smuggling gangs obtain funds by operating the 'Hundi' system among Pakistani immigrants in Britain. **1969** *Commerce* (Bombay) 26 July 150/2 Apprehensions..may push up the rate of interest in the free market from 15–18 per cent to 20–24 per cent against the hundies, promissory notes and short loans.

hundred, *sb.* and *a.* Add: **1. b.** (*b*) *the Hundred Days*: the immediate source of the phrase is the speech delivered by Louis de Chabrol de Volvic, prefect of Paris, to Louis XVIII in 1815 ('Cent jours se sont écoulés depuis le moment fatal où votre majesté quitta sa capitale'). Add examples; also *transf.*
1827 Scott *Life Napoleon Buonaparte* IX. i. 33 Here, therefore, ended that short space..that period of an Hundred Days, in which the events of a century seem to be contained. **1862** C. Knight *Popular Hist. Eng.* VIII. ii. 21 This landing in the Gulf of St. Juan on the 1st of March was the introductory scene to the great drama called 'The Hundred Days'. **1887** O. W. Holmes (*title*) Our hundred days in Europe. **1956** J. M. Burns *Roosevelt* ix. 169 The President asked for quick authorization of a civilian conservation corps... This bill interested Roosevelt himself as much as any single measure of the Hundred Days. **1965** T. C. Sorensen *Kennedy* ix. 242 'I'm sick of reading how we're planning another "hundred days" of miracles,' he [*sc.* J. F. Kennedy] said, 'and I'd like to know who on the staff is talking that up. Let's put in that this won't all be finished in a hundred days or a thousand.' **1965** Mrs. L. B. Johnson *White House Diary* 10 Apr. (1970) 257 Lyndon talked of the harvest of legislation. He said that never has there been such a hundred days. **1966** H. Wilson *Purpose in Power* i. 1 (*heading*) The first hundred days.
(*c*) *Hundred Years War*, the intermittent war between England and France from 1337 to 1453, arising out of the claim of the English kings to the French crown.
1874 J. R. Green *Short Hist. Eng. People* 275 The Hundred Years' War had ended. **1959** M. McKisack *14th Cent.* 127 Like the second world war of the twentieth century, the Hundred Years War gathered momentum slowly. Gascony was declared confiscate in May 1337 and in October Edward laid his claim to the French Crown; but there was no organized campaigning for another two years. **1961** E. F. Jacob *15th Cent.* 505 In the spring of 1453 Charles VII opened the last campaign of the Hundred Years War in overwhelming force.
(*d*) Phr., *the Hundred Flowers*: name given to a period of approximately six weeks in the summer of 1957 when certain elements of the Chinese population were invited to criticize the political system then obtaining in Communist China. (See quot. 1958².)
1958 L. F. Edwards tr. *Fauré's Serpent & Tortoise* xiv. 121 In the intellectual China of the Hundred Flowers, no one has the right to be a counter-revolutionary, but one has, to a certain extent, the right to be an idealist. **1958** *Listener* 6 Nov. 718/1 The campaign for free speech [in China] that followed the encouraging words of Mao Tse-tung—'let a hundred flowers bloom'—was evidently designed as an operation to find out what precisely were the prevailing criticisms of policy. **1959** *Ibid.* 5 Feb. 255/1 The intellectuals [in China] who blossomed with criticism during the brief Hundred Flowers Movement. **1973** *Ibid.* 2 Aug. 147/3 The humiliations which the Party had suffered during the so-called Hundred Flowers period.
d. After a numeral, used to express the two noughts in the figure representing the number of hours since midnight. Cf. *hour 1 b.
1953 P. C. Berg *Dict. New Words* 95/1 *Hundred*, the two noughts in the numerical symbols of full hours; e.g. we will meet at nine hundred hours [= 9.00 a.m.]. **1967** B. Knox *Blacklight* vi. 135 'What's the time?'..'Coming up for seven-forty-five sir.' 'Let's be formal and say near enough to twenty hundred hours... Wait till twenty-two hundred, mister.' **1973** A. Hunter *Gently French* i. 13, I got back to Elphinstone Road at about oh-one hundred hours.
2. b. Phrases: *not a hundred miles from*; *within a hundred miles of*: near, close to, in or at; also *fig.*; (*all the same in*) *a hundred years* (*hence*) (and similar expressions): gnomic formulas of consolation for present adversity; *a hundred of bricks*: see *brick *sb.*¹ 5; *a hundred to one*: a hundred chances to one; hence, an expression indicating very *slight* probability (implying 'a hundred to one against') or very *strong* probability ('a hundred to one in favour of').
1647 M. Verney in F. P. Verney *Mem.* (1892) II. xiv. 370 Tis a hundred to one pegg's husband turns them out of his house again within a fortnight. **1675** T. Jordan *Triumphs of London* 21 Though now she be pleasant and sweet to the sense, Will be damnably mouldy a hundred year hence. **1760** Sterne *Tristram Shandy* II. iv. ix. 72 What a chapter of chances, said my father... 'Twas a hundred to one—cried my uncle *Toby*. **1783** in J. Ritson *Select Coll. Eng. Songs* II. 14 We shall be nothing An hundred years hence. **1821** *Kaleidoscope* 27 Feb. 277/3 A sporting gentleman passing by a house, not a hundred miles from —— street. **1827** P. Egan *Anecdotes of Turf* 270 Within one hundred miles of the great Chancery shop of the kingdom. **1838** Dickens *Nickleby* (1839) ix. 76 As she frequently remarked when she made any such mistake, it would be all the same a hundred years hence. **1852** *Leisure Hour* I. 52/2 Scandalous transactions said to have transpired between two 'well-known' individuals 'not a hundred miles off'. **1874** L. Troubridge *Life amongst Troubridges* (1966) ix. 75 Let's look cheerful—it will be all the same a hundred years hence. **1888** 'R. Boldrewood' *Robbery under Arms* III. iv. 50 If he gets clear off..you're right. But it's a hundred to one against it. **1891** Kipling *Life's Handicap* 171 Did you ever know old Hummy behave like that before or within a hundred miles of it? **1895** A. W. Pinero *Benefit of Doubt* ii. 109 Don't fret; it'll be all the same a hundred years hence. **1903** J. M'Govan *Brought to Bay* 74 This retreat, he admitted, was not a hundred miles from the spot where they were at that moment seated. **1914** C. Mackenzie *Sinister St.* II. iv. ix. 1105 'Oh, well, it'll be all the same in a hundred years.' She picked up her white gloves, and swaggered across the crowded beerhall. **1925** W. S. Maugham *Painted Veil* i. 11, I say, you must pull yourself together. It's a hundred to one it wasn't Walter. **1955** D. Garnett *Aspects of Love* I. 27 Of course it is a hundred to one that the girl is just a tart he has picked up in Montpellier. **1968** C. Watson *Charity ends at Home* vi. 69 Certain information has reached me privately concerning the disposal of funds raised not a hundred miles from here in the name of so-called 'charity'. **1971** G. Household *Doom's Caravan* iv. 174 All the same a hundred years hence, as my Nanny used to say. **1973** M. Woodhouse *Blue Bone* xvi. 174, I don't want you, Rodway, or you, Quickie, within a hundred miles of me, ever.
c. orig. *U.S.* *a* or *one hundred per cent*: used adjectivally or adverbially with the meaning 'entire(ly), complete(ly)'. Hence *hundred-per-center*, *hundred-per-centism*.
1911 H. S. Harrison *Queed* vii. 90 You do more work in twenty-four hours than you're doing now, besides feelin' one hundred per cent. better all the time. **1918** T. Roosevelt in *N.Y. Times* 19 July 6/6 There can be no fifty-fifty Americanism in this country. There is room here for only 100 per cent. Americanism... No man who is not 100 per cent. American is entitled to the support of any party. **1923** *Westm. Gaz.* 1 Jan., An administrator is 100 per cent. successful only when he gets every individual in the factory..working as enthusiastically as if he were working for himself on his own job. *Ibid.* 9 Feb., Under a hundred per cent. disability. **1923** *Smart Set* Feb. 30 (*title*) Diary of a 100% American. **1926** W. R. Inge *Lay Thoughts* 135 Such detachment would not be possible to a 'hundred per cent. American'. *a* **1927** W. W. Woollcott (*title of poem*) I am a one hundred percent. American. **1928** *Publishers' Weekly* 26 May 2164/2, I have frequently encountered excellent accounting systems which were 100 per cent. useless. **1928** *Observer* 4 Mar. 13/2 Perhaps New York is not the place for the Hundred-per-centers. I certainly never met any. *Ibid.* 8 Apr. 8/2 He is really another victim of hundred-per-centism. **1931** G. B. Shaw *Platform & Pulpit* (1962) 232 The first thing that would occur to a real hundred per cent. American in Russia is that..it must be a splendid country to make money in. **1931** *Times Lit. Suppl.* 29 Jan. 76/3 The inevitable 'honest-to-God' hundred-per-cent. American young man..besieges and wins Valerie's heart. **1931** G. D. H. Cole in W. Rose *Outl. Mod. Knowl.* 666, I see no sign of the actual approach of this hundred per cent. American paradise. **1946** *Amer. Speech* XXI. 34/1 *Hundred per center*,..one who observes all customs and traditions without demur. **1946** *R.A.F. Jrnl.* May 169 The bomber crews had to make 100 per cent. certain of putting their H.E. loads right on the objective. **1968** M. Woodhouse *Rock Baby* xxiv. 233 You're one hundred per cent sure they'll never make any sort of bang, then?
d. *a hundred per cent*: fit, well, recovered. Freq. in negative contexts.
1960 N. Mitford *Don't tell Alfred* xiv. 155, I don't feel a hundred per cent. **1965** V. Canning *Whip Hand* xi. 131 'How's the arm?' 'It wasn't broken... It's almost a hundred per cent now.' **1965** N. Freeling *Criminal Conversation* i. viii. 52, I wasn't quite well, not ill but not quite a hundred per cent, and he did make me better. **1967** I. Hamilton *Man with Brown Paper Face* xv. 214 Actually, I'm not completely one hundred per cent. **1973** 'D. Craig' *Bolthole* ii. 30 He's been, well, off colour, yes. Not ill, but not a hundred per cent.
4. a. (Later U.S. examples.)
1838 *Knickerbocker* XI. 15 When requested..to say how much flour she should make into bread, at their first baking, she answered..'I suppose about a quarter of a hundred.' **1852** *Trans. Mich. Agric. Soc.* III. 332 To dispose of the compound of acorns, ground nuts and carrion for $2 per hundred. **1861** *Trans. Ill. Agric. Soc.* IV. 373 We want a horse sixteen hands high, that will weigh fifteen hundred.
7. (Later examples.)
1922 G. K. Chesterton in *Illustr. London News* 12 Aug. 234/1 There ought not to be anything but a plural for..the sweets called hundreds and thousands. **1932** A. Christie *Thirteen Problems* i. 22 'Cooks nearly always put hundreds and thousands on trifle, dear,' she said. 'Those little pink and white sugar things.' **1953** Dylan Thomas *Under Milk Wood* (1954) 60 Brandyballs, winegums, hundreds and thousands, liquorice sweet as sick. **1967** N. Freeling *Strike Out* 87 Little sugary pellets like hundreds and thousands.

hundredth, *a.* and *sb.* Add: **B.** *sb.* **2.** *Old Hundredth*, a hymn tune which first appeared in the Geneva psalter of 1551 and was later set to Psalm 100 in the 'old' metrical version of the Geneva Psalter (hymn 166 in 'Hymns Ancient and Modern'); the psalm itself. Also *attrib.*
[**1790** Brady & Tate *New Version Psalms* 240 The Psalms in this Version of four Lines in a single Stanza..may properly be sung as the old 100 Psalm.] **1837** T. Bacon *First Impr. Hindostan* I. vi. 153, I have often..heard the organ pealing forth the solemn notes of the old hundredth psalm. **1837** Dickens *Pickw.* xxxii. 349 Brother Mordlin had adapted the beautiful words of 'Who hasn't heard of a Jolly Young Waterman?' to the tune of the Old Hundredth. **1840** —— *Old C. Shop* xviii. 192 The dog..ground hard at the organ..and applied himself..to the Old Hundredth. **1853** Mrs. Gaskell *Cranford* xv. 298, I found out from the words, far more than from the attempt at the tune, that it was the Old Hundredth she was crooning to herself. **1934** R. Ferguson *Celebrated Sequels* 25 Our vicar's wife says that on the whole we had better stick to the Old Hundredth. **1955** M. Gilbert *Sky High* i. 7 'What's the last hymn next Sunday..?' 'Hundred and Sixty-six. Old Hundredth. You all know that.'

hung, *ppl. a.* Add: **3.** *U.S.* Of a jury: unable to agree. Cf. *hang *v.* 6 b, 17 c.
1848 E. Bryant *California* (1849) xxvi. 291 The jury..were what is called 'hung'; they could not agree, and the matters in issue, therefore, remained exactly where they were. **1962** *Listener* 10 May 799/1 No one would deny that a deadlocked jury—what the Americans call a hung jury—is a bad thing. **1965** J. Porter *Dover Two* v. 63 What with hung juries, appeals and pleas for mercy, they'd have it [*sc.* the story] on ice for a long time.
4. *slang.* Suffering from excess of liquor (or drugs). Also **hung-over** (cf. *hang-over 2), having or affected by a hang-over.
1950 R. Starnes *And when she was Bad* (1953) xvi. 76 Brafferton just came in, looking as hung over as you can get. **1952** J. van Druten *I am a Camera* (1954) 90 Enter ..wearing a dressing gown and looks hung-over. 'I say, you don't look too well this morning.' 'I've got a terrible hangover.' **1958** *Amer. Speech* XXXIII. 225 Drunks are *hung*, too. **1960** *Homes & Gardens* Aug. 70/2 It is said to have therapeutic qualities on hungover mornings. **1960** I. Shaw *Two Weeks in another Town* x. 125 He awoke late, feeling headachy and hung-over from the liquor of the night before. **1962** K. Orvis *Damned & Destroyed* vi. 48 A score or more of addicts—all hungover. **1963** H. Slesar *Bridge of Lions* i. 7, I know you're hung, Mr. Drew. **1968** J. Hudson *Case of Need* vi. 126, I was hung over from a party after the game. Really hung. Too hung. **1972** C. Drummond *Death at Bar* v. 139 Christ! You look hung over!.. What about drinks all round?

hungal, var. *hangul¹.

Hungarian. Add: **A.** *adj.* **1.** *Hungarian bonnet* or *cap*, the shell of a marine gastropod mollusc, *Capulus ungaricus*; *Hungarian grass U.S.* (*obs.*), the forage grass, foxtail millet, *Setaria italica*.
1845 *Encycl. Metrop.* VII. 288/2 In the Hungarian Bonnet, *Pileopsis Ungarica*, the tip of the shell seems to drop backwards and downwards. **1854** P. H. Gosse *Nat. Hist. Mollusca* 212 The only British species [of *Pileopsis*] is commonly known by the appellation of Torbay Bonnet... it also bears the names of Fools-cap Limpet, Cap of Liberty, and Hungarian Bonnet. **1859** *Guide Illinois Central Railroad Lands* 34 *Hungarian Grass.* This cereal, first introduced by the Hungarian exiles, is becoming a favorite with the farmers. **1883** *Rep. Indian Affairs* 23 About 12 acres were planted with Hungarian grass. **1901** E. Step *Shell Life* 220 The Hungarian Cap or Cap of Liberty (*Capulus hungaricus*) is shaped much like the cap of Liberty, the beak or apex being spirally curved back. **1971** S. P. Dance *Seashells* 103/2 A thick periostracum is a characteristic feature of some cap shells and is well seen in the Hungarian Cap.
b. Pertaining to Hungary, as *Hungarian band*, a band specializing in the performance of Hungarian music; *Hungarian point, stitch* (see quot. 1934).
1882 B. Potter *Jrnl.* 2 July (1966) 19 Mr. Edwin Lawrence had the Hungarian Band last week... They are mostly string instruments. **1894** G. du Maurier *Trilby* III. viii. 133 A large photograph..of Svengali, in

the military uniform of his own Hungarian band. **1907** *Yesterday's Shopping* (1969) 1055 Blue Hungarian Band... Handsome National Uniforms... Any number from 4 to 30 may be engaged. **1921** A. G. I. Christie *Samplers & Stitches* vii. 90 *Hungarian stitch*... The stitch illustrated in the diagram is often seen in canvas embroidery, used for filling in a background whilst, perhaps, Tent or Cross stitch works out the intricacies of the pattern. **1932** D. C. Minter *Mod. Needlecraft* 60/1 Some geometric all-over pattern, such as Hungarian stitch. **1934** M. Thomas *Dict. Embroidery Stitches* 125 *Hungarian stitch*, a Canvas Stitch which produces a most attractive pattern, and may be worked either all in one colour or in contrasting shades. **1972** *Country Life* 8 June (Suppl.) 50 Queen Anne walnut wing-chair..covered in its original Hungarian point needlework.

hunger, *sb.* Add: **4. e.** **hunger-march**, a march, undertaken usually by the unemployed, in order to call attention to their needs and claims; so **hunger-marcher**; **hunger-pain**, pain due to hunger; also *Path.* (see quot. 1905); **hunger strike**, the action of a person, esp. a prisoner, who refuses food in order to induce someone to yield to his demands; so **hunger-strike** *v. intr.*, to go on hunger-strike; **hunger-striker**, **hunger-striking** *vbl. sb.*; **hunger swarm**, the swarming of bees caused by lack of food.
1908 *Westm. Gaz.* 16 July 10/3 A statement of the purpose of the 'Hunger March'. **1939** N. Monsarrat *This is Schoolroom* i. 33 Hunger-marches, May Day processions..they were all new. **1972** M. Jones *Life on Dole* i. i. 11 Among the older people, there was grave talk of the days of mass unemployment and the Hunger Marches. **1908** *Westm. Gaz.* 16 July 10/3 He had no knowledge the 'Hunger Marchers' were coming there that day. **1922** *Ibid.* 29 Dec., Unemployed hunger marchers are persisting in their determination to see the Prime Minister. **1950** Koestler *God that Failed* i. 28 Europe trembled under the torn boots of hunger-marchers. **1820** Hunger-pain [see Hunger *sb.* 4 a]. **1905** B. G. A. Moynihan in *Lancet* 11 Feb. 341/1 If the pain does not come on for from two to four hours after a meal..the patient will often complain of what I have been accustomed to call 'hunger pain'. **1943** E. Bowen *Seven Winters* 25 The vacuum, the hunger-pain, set up in me from being unable to read. **1889** *Century Mag.* Nov. 107/2 Here I heard..the narrative of the hunger-strike of the four women in the prison of Irkutsk. **1903** H. Chisholm tr. *L. Deutsch's Sixteen Yrs. in Siberia* 78 Upon these conditions I consented not to prolong my 'hunger-strike'. **1908** *Westm. Gaz.* 18 Mar. 5/1 The Central Prisons Administration has circularised the provincial Governors regarding the so-called 'hunger strikes', which are a characteristic feature of Russian prison life. **1914** E. Pankhurst *My Own Story* iii. v. 292 She has hunger-struck in prison. She submitted herself for more than five weeks to the horrible ordeal of feeding by force. **1916** W. J. Locke *Wonderful Year* xvii. 247 'I've been to prison.' Martin..asked if she hunger-struck. **1937** Koestler *Spanish Testament* ii. 333, I had intended to stop my hunger strike as soon as my letter to the Consul had been sent off. **1970** *Times* 11 May 8/4 In February Feron went on hunger strike in protest but was taken ill and for a month existed only on drugs. **1973** *Jewish Chron.* 19 Jan. 44/5 He will stage a 48-hour hunger strike outside the Soviet Embassy. **1922** *Blackw. Mag.* Aug. 146/1 He gave his orders for the release of the hunger-strikers. **1972** *Guardian* 1 Dec. 12/1 Already the names of McSwiney and MacCaughey, hunger-strikers of an earlier era, are being conjured up. **1916** W. J. Locke *Wonderful Year* xxiii. 329 Her duties involved incendiarism, imprisonment, and hunger-striking. **1870** A. Pettigrew *Handy Bk. Bees* 150 We have known swarms starved out of their hives. Having made a few pieces of comb, and being without food, no eggs were set in them and the bees, through sheer want, cast themselves on the wide world. These are called 'hunger-swarms'. **1886** F. R. Cheshire *Bees* II. iv. 168 Bees sometimes abscond because their stores have run out, and circumstances are desperate. Such have usually been called 'hunger' or 'vagabond' swarms. **1928** C. Williams *Story of Hive* iii. 22 Bees on the point of starvation will, in a spirit of desperation, leave their hive in a body. This exodus is known as 'a hunger swarm'. **1954** C. G. Butler *World of Honeybee* xiii. 154 Another type of honeybee swarm is sometimes recognisable, the type which in Europe is usually known as a 'hunger' swarm, and which occurs when a colony is starving.

hung-over: see *hung *ppl. a.* 4.

hungry, *a.* **2. a.** Delete ? *Obs.* and add: *the hungry forties*, the decade beginning in 1840, characterized in the British Isles by much poverty and unemployment.
1905 Mrs. C. Unwin (*title*) The hungry forties. **1910** A. Baker *Poor against Rich* 34, I mention the Hungry Forties, because a lot of poor people have allowed themselves to be misled during the last election, by the fear of dear food. **1920** J. Collings in J. L. Green *Life Jesse Collings* i. iv. 29 During the 'hungry forties' eggs were sold twenty for a shilling. **1958** *Spectator* 20 June 792/1 During the famine in Ireland at the end of the Hungry Forties, it was not uncommon [etc.].

hung up (hɒ·ŋ,ɒp), *a.* and *adj. phr.* Also hung-up. [f. *hang up* (Hang *v.* 28 d).] **1.** Put into abeyance, delayed.
1878 [see Hang *v.* 28 d].
2. *slang.* Confused, bewildered, mixed-up. Also *hung-up on*, obsessed with, preoccupied with (cf. also quot. 1961).
[**1909** J. R. Ware *Passing Eng.* 156/1 *Hung up*, from the American—where personal catastrophe is referred to by this phrase.] **1945** L. Shelly *Jive Talk Dict.* 21/1 *All hung up*, completely bewildered. **1957** [see *daddy 3]. **1958** *New Statesman* 6 Sept. 294/2 Helping each other in those phases in which they are 'hung up'. **1961** Rigney & Smith *Real Bohemia* p. xv, *Hung up, to be*, one's behavior is 'stuck' in one pattern. **1966** *Sunday Times* (Colour Suppl.) 13 Feb. 35/4 *Hung up*, obsessed, cf. 'he is hung up on that girl'. **1966** *New Statesman* 1 Apr. 458/3 The U.S. is 'hung up', paralysed into inaction because it cannot reconcile the political goal of uniting Germany with the ideological necessity of maintaining Western Europe as an anti-communist fortress. **1968** *Word Study* Feb. 5/2 American students of poetry have been hung up on the lines ever since. **1969** *It* 11–24 Apr. 13/4 How we manage to generate so many good things on this hung-up, repressed little island I simply don't know. **1970** *Daily Tel.* 26 Nov. 9/2 You get so hung-up with the place you feel like going out and smashing something. **1971** *New Scientist* 4 Mar. 485/2 Roszak is very hung up on the power that science grants. **1971** B. Malamud *Tenants* 54 He was more than a little hung up, stupid from lack of sleep, worried about his work.

hunh (hɒ(n)). *U.S. dial.* (esp. in *Black English*). Used as an intensifier after a question.
1935 in Z. N. Hurston *Mules & Men* (1970) i. x. 206 You got mo' poison in yuh than dat snake dat wuz so poison tell he bit de railroad track and killed de train, hunh? **1955** W. Gaddis *Recognitions* i. iv. 155 Hunh? What do you think of that, hunh? **1970** D. L. Lee in S. Henderson *Understanding New Black Poetry* (1973) iii. 341 Can u do it, hunh? i say hunh, can u stop moving like a drunk gorilla? **1973** *Black World* Apr. 60 How about one more..for ol' times, hunh?

hunk, *sb.*¹ Add: **1. b.** A large man or woman.
[**1823** in *Dialect Notes* (1913) IV. 47 *Hunk*, bulk. A large body.] **1941** Baker *Dict. Austral. Slang* 37 *Hunk*, a large man. **1945** L. Shelly *Jive Talk Dict.* 13/1 *Hunk*, stalwart male. **1946** B. Treadwell *Big Book of Swing* 124/2 *Hunk*, very masculine male. **1957** J. Kerouac *On Road* (1958) 62, I looked at Lee Ann. She was a fetching hunk, a honey-colored creature.

hunk, *sb.*² and *a.* Add: **B.** *adj.* **b.** Colloq. phr. *to get hunk* (*with*): to get even (with). Also const. *on*.
1845 *Spirit of Times* 24 May 146 Those who lost their money on Fashion had two or three chances to 'get hunk', especially on the last day. **1903** A. H. Lewis *Boss* vii. 93 No, I don't blame Sheeny Joe... Still, while I don't blame him, it's up to us to get hunk an' even on th' play. **1949** *Boston Globe* (Fiction Mag.) 12 June 2/4 Suppose I show you how to get hunk with the cheapskates? **1950** in H. E. Goldin *Dict. Amer. Underworld Lingo* 79/1 That fink (informer) tried to get hunk on me for glomming (stealing) his broad (girl) by belching (informing) on me.

hunk (hɒŋk), *sb.*³ *N. Amer. slang.* Also **hunkey, hunkie, hunky.** [Cf. *bohunk.] A nickname applied, usually disparagingly, to immigrants to the U.S.A. from east-central Europe. Also *attrib.* Cf. *honky.
1896 *N.Y. Herald* 13 Jan. 3/4 The average Pennsylvanian contemptuously refers to these immigrants as 'Hikes' and 'Hunks'. The 'Hikes' are Italians and Sicilians. 'Hunks' is a corruption for Huns, but under this title the Pennsylvanian includes Hungarians, Lithuanians, Slavs, Poles, Magyars and Tyroleans. **1910** *Sat. Even. Post* 3 Sept. 18/1 Almost every..Hunky or Dutchman who lands in New York has in his 'kick' or wallet, the written address of some boarding house. **1914** Jackson & Hellyer *Vocab. Criminal Slang* 47 *Hunkie*, current in localities where North European laborers abound. A corruption of Hungarian, but employed to signify a Continental European who is unwashed and unnaturalized. **1928** S. Lewis *Man who knew Coolidge* i. 53 Too many foreigners—fellows with Wop names and Hunky names. **1929** *Amer. Speech* June 372 *Hunkey*, same as Bohunk. **1932** [see *goof *v.* 1 a]. **1934** J. O'Hara *Appointment in Samarra* (1935) i. v. 83 The hunkeys, the schwackies..—regional names for non-Latin foreigners—probably were inside getting drunk. **1936** *Scrutiny* V. 1/5 The twelve-hour day kept the myriads of 'hunkies' who toiled in Mellon mills out of brawls and brothels. **1939** *Archit. Rev.* LXXXV. 219/2 It has to be close to the town because most of the workmen are foreign-born hunkies and do not readily adapt themselves to living conditions in Lyndora. **1962** C. L. Barnhart in Householder & Saporta *Probl. Lexicogr.* 178 Greaser, guinea, hunky, Jap, kike. **1971** *Maclean's Mag.* Oct. 78/1, I don't know if I should get mad if someone insults the Irish, or makes cracks about Polacks or Hunkies.

hunker, *v.* For *Sc.* read 'orig. *Sc.*' Add earlier and later examples.
1720 A. Pennecuik *Streams from Helicon* i. 80 And hunk'ring down upon the cald Grass. **1902** *Dialect Notes* II. 237 (Pioneer dialect of S. Illinois.) *Hunker down.* 1. To crouch in sitting. 2. To kneel. **1907** Kipling *Actions & Reactions* (1909) 187 We heard Imam Din hunker down on the floor. One gets little out of the East at attention. **1928** Barrie *Peter Pan* ii, in *Plays* 39 Hunkering on the ground..the six are not unlike village gossips gathered round the pump. **1945** J. Steinbeck *Cannery Row* vii. 26 Mack and the boys sat on the floor, played cards hunkered down. **1946** K. Tennant *Lost Haven* (1947) ix. 129 Kelly got up, came out, and hunkered down by the sunny office wall. George hunkered down beside him. **1962** *Coast to Coast 1961–62* 85 The old woman had hunkered down, poking intently at an *hibachi*, some embers glowing darkly beneath. **1973** *N.Y. Times* 14 Aug. 34/3 One 14-year-old boy sat on a hornet when he hunkered down to get a better view of the green.

hunkerish (hɒ·ŋkəriʃ), *a.* *U.S. colloq.* [f. Hunker *sb.* + -ish¹.] Conservative, old-fashioned.
1857 *Lawrence* (Kansas) *Republ.* 2 July 3 This has in times past, been considered rather a hunkerish neighbourhood. **1888** W. Lawrence *Life Amos A. Lawrence* 139 His Kansas experience did not move Mr. Lawrence from his hunkerish sympathies in politics. **1905** *Springfield* (Mass.) *Weekly Republ.* 8 Sept. 1 The hunkerish conservatism which Mr. Cannon and the Senate stand for.

hunkerism (see under Hunker *sb.*). (Earlier and later examples.)
1845 R. Tyler *Let.* 19 Apr. in L. G. Tyler *Lett. & Times Tylers* (1896) III. 161 Every appointment made for New York or Pennsylvania or Maryland is a restoration of *Hunkerism*. **1848** C. A. Loomis *Let.* 23 Jan. in *Mich. Hist. Mag.* (1926) X. 216 Old Hunkerism is predominant, but it is hunkerism without brains. **1906** *Springfield* (Mass.) *Weekly Republ.* 6 Dec. 8 The old republican hunkerism.

hunkers, *sb. pl.* For *Sc.* read 'orig. *Sc.*' Add earlier and later examples.
1756 M. Calderwood *Journey in Eng. & Low Countries* (1842) 164 It goes down with a step, which makes the door so low, that if any body from without speaks to you, they must sit down on their hungkers. **1831** S. Lover *Legends & Stories of Ireland* 200 Up sits the fox on his hunkers. **1888** in W. B. Yeats *Fairy & Folk Tales* 109, I seen an ould woman sittin' on her hunkers. **1896** *Dialect Notes* I. 419, I had to sit on my *hunkers*. Myersville, Md., and Tannersville, Pa. **1935** A. J. Cronin *Stars look Down* i. ii. 17 Some colliers..squatted upon their hunkers against the wall. **1947** T. H. White *Elephant & Kangaroo* (1948) xx. 160 Mr. White seized Brownie by the tail—she had been sitting on her hunkers beside him.

hunkey, hunkie, hunky, varr. *hunk *sb.*³

hunky (hɒ·ŋki), *a.*¹ *U.S. slang.* [f. Hunk *sb.*² and *a.* + -y¹.] In good condition; safe and sound; all right: = Hunk *a.*
1861 'A. Ward' in *Vanity Fair* (N.Y.) 15 June 273/1 He (Moses) folded her to his hart, with the remark that he was 'a hunkey boy'. **1878** B. Harte *Man on Beach* 47 She's all hunky, and has an appetite. **1889** K. Munroe *Golden Days* xii. 125 If I'd took good care of that map.. we'd been all hunky at this minute. **1907** C. E. Mulford *Bar-20* ix. 105 That was all hunky for a while. **1926** *Bulletin* 21 Oct. 4/2 I'll be all hunky. Nurse Dainton tends me like I was made of glass.
So **hu:nky-do·ry, -do·rey** [second element of unknown origin] *a.*, satisfactory, fine.
1866 *Galaxy* 1 Oct. 275, I cannot conceive on any theory of etymology..why anything that is 'hunkee doree'..should be so admirable. **1868** in G. C. D. Odell *Ann. N.Y. Stage* (1936) VIII. 390 [Even Samuel Slater admitted that Tostee, when and if she sang, was] hunkydory. **1907** N. Munro *Daft Days* xxi, Before one marries it's hunky-dory—it's fairy all the time. **1907** M. C. Harris *Tents of Wickedness* iv. ii. 341 I've had my luncheon, and I feel better already. Oh, it's all hunky-dory. **1943** N. Balchin *Small Back Room* 172 Well, unless the electricity they taught me is all wrong, that ought to be hunky-dory. **1956** D. M. Davin *Sullen Bell* iii. iv. 233, I thought everything was hunkydory and you were well on the way to being a big executive. **1969** J. Gardner *Founder Member* ii. 17 Everythink's 'unkey dorey 'ere. No problem.

hunky (hɒ·ŋki), *a.*² [f. Hunk *sb.*¹ + -y¹.] Thick-set, solidly built.
1911 E. Ferber *Dawn O'Hara* vii. 89 Rather be hunky and healthy than skinny and sick. **1959** *Listener* 23 Apr. 733/1 A vast, hunky, surly man.

hunt, *sb.*² Add: **3*.** A hunting or oscillatory motion (see *hunt *v.* 7 b, *hunting *vbl. sb.* 1 f).
1920 *Nature* 11 Mar. 46/1 It moves backwards and forwards very slightly, and this motion we term the 'hunt'. **1934** *Brit. Jrnl. Psychol.* XXIV. 399 The 'angle of hunt', i.e. the angle of oscillation about the mean radial velocity, cannot exceed 360/*N* degrees, where *N* is the number of segments in the armature. **1937** *Jrnl. R. Aeronaut. Soc.* XLI. 410 The well behaved short period oscillation develops into an irritating hunt. **1952** A. Tustin *Automatic & Manual Control* 280 If the amplitude is not too large, and..we know how to reduce the amplitude by increasing the hunt frequency, it is possible to check a mean position of the system.
3.** *Telephony.* An operation of hunting by a selector or switch (see *hunting *vbl. sb.* 1 g).
1927 W. E. Hudson *Director Syst. Autom. Telephony* ii. 42 Relay C..is used to determine when the impulse train is finished so as to allow the automatic hunt to start. **1966** Rubin & Haller *Communication Switching Syst.* i. 31 The hunt motion is a vertical move of the wipers along the contact bank.
4. *hunt-breakfast* (earlier and later examples); **hunt ball**, a ball given by members of a hunt.
1807 *Sporting Mag.* XXXI. 40/1 The annual Hunt Ball took place at Chepstow. **1853** Mrs. Gaskell *Ruth* I. i. 13 The annual hunt-ball was to take place. **1933** A. Powell *From View to Death* iv. 113 Ungainly young men who had had a glass too much of champagne at hunt-ball suppers. **1968** A. Diment *Bang Bang Birds* x. 193 Penny told us about a hunt ball she had attended recently. **1973** K. Giles *File on Death* iv. 90 'E provides the catering for the 'unt ball at seven guineas the ticket. **1877** Trollope *Amer. Senator* II. xxiv. 254 That old farmer at the hunt breakfast. **1973** K. Giles *File on Death* v. 133 The other one..kept pawing the ladies..and falling off his horse after the Hunt Breakfast.

hunt, *v.* Add: **3. b.** (U.S. examples.)
1818 E. P. Fordham *Pers. Narr. Trav.* (1906) 221 The next day I shall cross the Little Wabash to 'hunt land'. **1834** *Visit to Texas* i. 10 An old Tennessean and his wife with their sons were going 'to hunt land'. *Ibid.* xiii. 122 He sometimes sends out three or four men to collect and mark them. This is called hunting cattle. **1891** M. E. Ryan *Told in Hills* iv. iii. 309 All were sleepy enough to hunt beds early. **1903** A. Adams *Log of Cowboy* iii. 38 Flood..suggested that all hands hunt their blankets and turn in for the night.
4. (See *huntaway *sb.*)
7. b. *intr.* Of a governor, a synchronous electric motor or generator, etc.: to run alternately faster and more slowly than the desired speed. Hence more widely of other machines, systems, etc.: to oscillate *about* a desired speed, position, or state to an undesirable extent, to jump backwards and forwards.
1877 *Proc. Inst. Mech. Engin.* 273 Siemens' interesting governor..had..a great tendency to 'hunt',..if it was first left a little behind, and then got an excess of force, it would be constantly 'hunting' or oscillating about a mean position. **1894** *Rep. Brit. Assoc. Adv. Sci.* 759 A Watt governor..does not hunt if designed for stability. **1902** *Trans. Amer. Inst. Electr. Engin.* XVIII. 374/2 The motors attempt to follow the generator exactly. If the latter pulsates, the motors pulsate also; they vibrate about a mean position, 'hunting' or pumping. **1921** M. Walker *Diagn. Troubles Electr. Machines* vi. 239 In the case of steam turbines and steam engines, it is possible for the governor to hunt in a perfectly periodic manner. **1951** S. Deutsch *Theory & Design Television Receivers* xiii. 431 If the feedback loop is underdamped, the oscillator frequency will swing below 15,750 cycles per second, whereupon the correcting voltage causes a swing above 15,750 cycles per second, etc. In other words, the oscillator will 'hunt' about the correct frequency. **1953** *Electronic Engin.* XXV. 156/1 Since the torque balance has an on-off action..it has a tendency to hunt. **1969** *Daily Tel.* 10 Jan. 26/4 The British train will be able to use existing railway tracks because of a new type of suspension... This will stop the train 'hunting' sideways. **1969** J. Argenti *Managem. Techniques* 99 If the action is too late or too weak, control will be inadequate, if too early or too strong the system will 'hunt'—i.e. swing violently above and below the standard. **1970** 'J. Earl' *Tuners & Amplifiers* iii. 74 On weak stereo signals this circuit can 'hunt' over mono and stereo in a very disconcerting manner, switching to stereo as the signal rises and back to mono as it falls.
8*. *Telephony.* Of a selector or switch: to carry out the operation of hunting (*hunting *vbl. sb.* 1 g). Const. *for*, *over*.
1924 W. Aitken *Autom. Telephone Syst.* III. lvi. 275 Dialling O..results in the starting of a free-trunk finder, which automatically hunts for the calling line. **1924** H. H. Harrison *Introd. Strowger Syst. Autom. Telephony* i. 26 The preselector or line switch..hunts to find one of ten or more idle group selectors. **1933** K. B. Miller *Telephone Theory & Pract.* III. v. 250 It is usual to adjust the speed to permit the selector to hunt over a group of 30 trunks in 1 second. **1961** *Proc. Inst. Electr. Engin.* CVII. b. Suppl. 161/2 A maximum of 1·8 millisec is required to select a channel,..and a further period of 900 microsec to hunt for a free channel.
9. b. *N.Z.* (See quot. 1933.)
1933 L. G. D. Acland in *Press* (Christchurch, N.Z.) 28 Oct. 17/7 *Hunt down*, to hunt the sheep off the higher parts of their winter country on to lower, safer spurs when snow is expected; e.g., 'We hunted down every day for a week, but no snow came.' **1961** B. Crump *Hang on a Minute* 85, I want you blokes to go round to the Snow Hut and hunt the sheep down into the valley from the open spur.

huntaway (hɒ·ntăwēi), *sb.* **a.** *Austral.* and *N.Z.* [f. vbl. phr. *to hunt away* (Hunt *v.* 4), which is further illustrated below.] (See quot. 1933.) Also (written either as *huntaway* or as *hunt away*) as *vb.*
1913 [see *heading *vbl. sb.* 4 b]. **1933** L. G. D. Acland in *Press* (Christchurch, N.Z.) 28 Oct. 17/7 *Huntaway*, a dog whose work is to drive sheep forward when mustering. As a verb the word is used in two senses, illustrated in the sentences: 'That dog hunts away well' and 'I hunt away with that dog'. **1934** *Bulletin* (Sydney) 16 May 38/3 Rock, the kelpie leading-dog, and Bruce, the nondescript little hunt-away, had never possessed any aspirations towards leadership. **1931** T. A. Harper *Windy Island* (1934) iii. 217 Vixen had turned her mob neatly over to Rough, who was hunting them away down the mountainside. **1934** J. Lilico *Sheep Dog Mem.* 27 The dogs would head, lead, huntaway, force and back. **1954** *Landfall* VIII. 221 Couple of times someone offered him big money for two good huntaways. **1961** B. Crump *Hang on a Minute* 74 They sold..two huntaway pups. **1966** Baker *Austral. Lang.* (ed. 2) iii. 73 A huntaway is sometimes known as a forcing dog. **1968** *N.Z. News* 28 Aug. 16/1 The huntaway barks loudly to shift sheep.
b. Nelson huntaway (see quots.). *N.Z.*
1941 Baker *N.Z. Slang* vi. 59 *Nelson huntaway*, a stone rolled down a hillside to move stock below instead of sending a dog out. **1949** P. Newton *High Country Days* 29 Brownie sent a boulder hurtling down the face—a 'Nelson huntaway'.

hunter. Add: **3. b.** The Jamaican cuckoo, *Hyetornis pluvialis.*
1847 Gosse & Hill *Birds Jamaica* 277 *Hunter.* Old Man.—Rainbird... The appellation of Rainbird is indiscriminately applied to both this and the preceding [*sc.* *Saurothera vetula*], as is, in a lesser degree, that of Old Man. I use a term by which I have heard it distinguished, ..perhaps derived from the perseverance with which it hunts..for its prey. **1960** J. Bond *Birds W. Indies* 116 Chestnut-bellied cuckoo. *Hyetornis pluvialis.* Local names: Old Man Bird; Hunter; Rain Bird.
5. c. hunter-spider = sense 3 a.
1867 *Amer. Naturalist* I. 409 This very large hunter-spider [*sc.* the tarantula] makes its appearance in Texas some years as early as the twenty-fifth of May.
d. hunter's green (see quot. 1957).
1872 *Young Englishwoman* Nov. 599/1 Sombre greens,..*chasseur*, or hunter's-green, myrtle, cypress. **1957** M. B. Picken *Fashion Dict.* 181/2 *Hunter's green*, dark, slightly yellowish green.
e. hunter-killer *a.*, designating a naval vessel or group of vessels equipped to locate and destroy enemy vessels, esp. submarines. Also as *sb.*
1948 *U.S. Naval Inst. Proc.* LXXIV. 505/2 The other two will be classified as 'hunter-killers'—destroyers with the prime purpose of tracking down submarines, instead of operating on convoy duty. **1950** *Jane's Fighting Ships 1950–51* 7 A new anti-submarine type of light cruiser, known as a hunter-killer ship, will be completed in 1951. **1957** *Times Lit. Suppl.* 20 Dec. 771/3 The escort carrier Guadalcanal and her 'hunter-killer' group of four escort destroyers. **1962** *Daily Tel.* 10 Dec. 18/4 Britain's first nuclear hunter-killer submarine, is expected to leave the ..yard of her builders,..on Wednesday or Thursday for sea-trials. **1972** *Sci. Amer.* July 16/3 A hunter-killer submarine is large enough to carry an array of hydrophones to produce a narrow listening beam for long-range detection.

hunterman (hɒ·ntəɹmæn). [f. Hunter + Man *sb.*¹] Used widely outside the British Isles as a local term for 'hunter, huntsman'.
1891 'Mark Twain' tr. *Hoffmann's Slovenly Peter* (1935), Behold the dreadful hunterman In all his fateful glory stand. **1907** W. Jekyll *Jamaican Song & Story* 137 Dory Dunn is a hunterman. **1913** *Chambers's Jrnl.* 22 Feb. 184/1 Nearly every village has its professional 'hunterman', whose duty it is to kill the larger game which do damage to the crops. **1922** H. B. Hermon-Hodge *Up against it in Nigeria* iv. 55 Huntermen are for the most part undependable. **1930** 'Greenhorn' *Tinker, Tailor* xi. 272 His hunterman came back to look for him and shot the python. **1933** *Amer. Speech* VIII. i. 50/1 *Hunterman*, hunter. One often sees this in country newspapers [of Ozarks]. **1954** G. Durrell *Bafut Beagles* vi. 116, I tink dis hunter man be best for all.

hunting, *vbl. sb.* Add: **1. d.** With *adv.*: also *hunting-down*.
1901 *Westm. Gaz.* 7 Sept. 5/1 Sweeping movements will give place to hunting-down tactics where the country favours the latter. **1945** *Tribune* 9 Nov. 11/1 The hunting-down of war criminals.
f. The action of a machine, instrument, system, etc., that is hunting (see *hunt *v.* 7 b); an undesirable oscillation about an equilibrium speed, position, or state.
1880 R. E. Crompton *Electr. Light for Industr. Use* 21 This causes a swinging or, as it is called by engine men, a 'hunting' action of the governor. **1894** *Rep. Brit. Assoc. Adv. Sci.* 759 If the isochronous governor works a slow-acting relay the hunting may be so serious that the steam supply alternates between complete cut-off and full supply. **1920** *Whittaker's Electr. Engineer's Pocket-bk.* (ed. 4) 223 If the damping is very small, oscillations or hunting may go on for a considerable time before the machine finally rotates steadily. **1932** *Discovery* Oct. 331/1 There was no suggestion of 'hunting' and the image remained exactly central in the [television] screen for the whole half hour. **1942** *Rev. Sci. Instruments* XIII. 218 The recording is easily accomplished at pen speeds of 12·7 cm per second across 25 cm of chart without 'hunting' taking place. **1943** *Electronic Engin.* XV. 438/2 A method for the determination of the power angle oscillations of a synchronous motor during hunting is described. **1950** *Gloss. Aeronaut. Terms* (*B.S.I.*) 8 *Hunting*, an uncontrolled oscillation about the flight path, the amplitude of which remains approximately constant. *Ibid.* 45 *Hunting*, angular oscillation of a [rotor] blade about the drag hinge. **1951** *Engineering* 6 Apr. 401/3 The train was derailed..because of a slight but periodic variation in the cant of the track which synchronised with the hunting periodicity of the engine. **1951** S. Deutsch *Theory & Design Television Receivers* xiii. 431 Hunting is revealed by a horizontal weaving or vibration of the picture. **1959** *Times* 27 Apr. (Rubber Industry Suppl.) p. vi/5 The control of the side-to-side swaying known to railway engineers as 'hunting'. **1968** *Practical Motorist* Dec. 459/1 *Hunting*, a rhythmical increase and decrease in the idling speed of an engine, caused by an over-rich mixture. **1971** *Nature* 2 Apr. 283/1 Society, regarded as a non-linear feedback system, is showing the signs of oscillation ('hunting') which one expects.
g. *Telephony.* An operation in which a selector or switch automatically goes through a group of lines until it reaches a free one and makes connection with it; now used esp. of the connection of a calling line with one of a group of outgoing lines.
1912 J. Poole *Pract. Telephone Handbk.* (ed. 5) xxxii. 528 There is no 'hunting' for disengaged lines as in the case of selectors. **1933** K. B. Miller *Telephone Theory & Pract.* III. v. 263 Automatic trunk hunting may be necessary to find an idle line. **1966** Rubin & Haller *Communication Switching Syst.* i. 31 In the Ericsson 500-line switch..the select motion is a rotation and the hunting action is radially outward to a free trunk.
3. a. *hunting boat*, *boot* (examples), *bridle*, *cap* (examples), *carpet*, *coat* (examples), *country* (later example), *excursion* (examples), *knife* (earlier and later examples), *path*, *pony*, *print*, *rifle*, *tie*.
1828 *Western Monthly Rev.* I. 577 The passengers of the hunting boats..saw him. **1894** Hunting boat [see *bidarka]. **1895** *Montgomery Ward Catal.* 521/1 Men's black English grain leather 'hunting boots', double sole, laced leg and instep. **1928** S. Sassoon *Mem. Fox-Hunting Man* 143, I was going to try on my new hunting clothes and my new hunting boots. **1957** M. B. Picken *Fashion Dict.* 181/2 *Hunting boot*, high, laced boot with waterproof sole. **1939–40** *Army & Navy Stores Catal.* p. xxxix, Hunting bridles. **1963** Bloodgood & Santini *Horseman's Dict.* 111 *Hunting bridle*, any bridle suitable for hunting. **1963** E. H. Edwards *Saddlery* iii. 32 Hunting bridles should be made of leather having plenty of substance and the width of the cheek will probably be ⅞ in. with the rein. **1814** J. Mayne *Jrnl.* (1909) 184 The postilions..do not, like our royal drivers, wear hunting-caps. **1946** M. C. Self *Horseman's Encycl.* 219 *Hunting cap*, the velvet cap worn by farmers, Masters and Servants of the hunt. **1931** A. U. Dilley *Oriental Rugs & Carpets* Pl. 9 (*caption*), Finest Extant Hunting Carpet, Middle Sixteenth Century. **1960** H. Hayward *Antique Coll.* 146/1 *Hunting carpets*, Persian carpets with elaborate hunting scenes, realistically depicted. **1789** R. F. Greville *Diary* 16 Jan. (1930) 17.. Asked Me if I had my hunting Coat with Me. **1909** *Westm. Gaz.* 9 Oct. 15/1 Ladies..may always choose a little, short coat, known to French tailors as a hunting coat. **1921** *Daily Colonist* (Victoria, B.C.) 15 Oct. 10/1 (Advt.), Pheasant Shooting Begins..Hunting Coats, with front and rear game pockets, $8.75. **1928** S. Sassoon *Mem. Fox-Hunting Man.* 136 Stephen, who was wearing a pink silk cap and a long-skirted black hunting-coat, silently received from the groom the saddle and weight-cloth. **1946** M. C. Self *Horseman's Encycl.* 219 Nothing even faintly resembling the English hunting country exists in North America. **1801** A. Mackenzie *Voy. from Montreal* 112 The Indians went on an hunting excursion. **1856** M. J. Holmes *'Lena Rivers* xxvi. 282 He had gone off on a hunting excursion. **1803** in *Minnesota Hist.* (1940) XXI. 126 When the defendant came to pierce his tent with his hunting knife..the larger part of the goods had already been moved to the spot agreed upon. **1933** B. Willoughby *Alaskans All* 3 His flannel shirt, high laced boots, the hunting-knife in his belt..made it difficult for me to realize he was a priest. **1949** *Chicago Tribune* 22 June II. 1/4 A policeman took a 6 inch hunting knife from the waist of one of the white boys. **1820** in *Minnesota Hist.* (1942) XXIII. 249 We found a hunting path which..led directly to Sandy Lake. **1821** T. Nuttall *Jrnl. Trav. Arkansa 1819* 167 We here crossed by a hunting path. **1886** *Outing* Apr. 7/1, I was riding a well-trained hunting pony. **1967** N. Freeling *Strike Out* 72 He makes these series of kind of huntin' prints. **1973** M. Mackintosh *King & Two Queens* v. 76 Eighteenth-century English hunting prints hung on the pine-panelled walls. **1856** in *Kansas State Hist. Soc. Trans.* (1890) IV. 504 The marauders were well armed with muskets,..hunting rifles,.. bowie-knives, etc. **1886** *Outing* Mar. 615 No hunting-rifles in the world possess greater accuracy. **1956** 'E. McBain' *Cop Hater* (1963) xix. 143 He said he'd shot himself while cleaning his hunting rifle. *c* **1840** Lady Wilton *Art of Needlework* xiii. 189 The Hibernian Tie: The Eastern Tie: The Hunting Tie: The Yankee Tie. **1907** *Yesterday's Shopping* (1969) p. xxxix, Hunting ties, ladies'.
b. hunting-box (earlier examples); **hunting tartan** (see quot. 1959); so **hunting Stewart tartan**, etc.
1799 *Times* 1 June 4/3 A Family Cottage, or Hunting Box, pleasantly situate in a sporting part of the country. **1814** Jane Austen *Mansf. Park* III. x. 208 Some small hunting-box in the vicinity of every thing so dear. **1855** A. Stanley *Let.* 8 Sept. (1927) iii. 74 The carpets are Royal Stewart Tartan and green Hunting Stewart. **1864** Queen Victoria *Let.* 26 Mar. in R. Fulford *Dearest Mama* (1968) 312 What is Anna's dress made of? I am going to send her a Hunting Stewart velvet. **1871** *Monthly Packet* Oct. 396 Their new winter frocks of Hunting Stewart tartan. **1959** Bain & MacDougall *Clans & Tartans* (ed. 4) 28 Hunting tartans are worn for sport and outdoor activities. Brown or some other dark hue is the predominant colour. When a Clan possessed a brightly coloured tartan it was unsuitable for hunting purposes, and hunting setts [patterns] were devised to make the wearer less conspicuous. The colours were arranged so that, when concealed in the heather, the tartan blended with the surroundings. **1969** O. Hesky *Sequin Syndicate* x. 102 A kimono in the colours of the Royal Stuart hunting tartan.

hunting, *ppl. a.* Add: **b. hunting spider** (later examples); **hunting wasp**, a wasp that preys upon other insects.
1916 A. T. de Mattos tr. J. H. Fabre (*title*) The hunting wasp. **1925** R. W. G. Hingston in E. F. Norton *Fight for Everest, 1924* 287 Beetles and hunting-spiders found a shelter on it [*sc.* a moraine]. **1928** —— *Problems of Instinct* 100 Hunting wasps..sting their victims with surgical precision. **1941** W. S. Bristowe *Comity of Spiders* II. v. 230 Some of these hunting spiders seek their prey by day and others by night. **1948** 'J. Crompton' *Hunting Wasp* i. 17 A large number of the hunting wasps sing at their work. **1964** V. B. Wigglesworth *Life of Insects* xiii. 227 The hunting wasp *Philanthus* stocks its solitary nest solely with honey-bees. **1966** E. Palmer *Plains of Camdeboo* xiv. 233 On the farm we know most of the diurnal species as jagspinnekoppe or hunting spiders.

Huntingdon (hɒ·ntiŋdən). The title of Selina, Countess of *Huntingdon* (1707–91), used in (*Lady* or *Countess of*) *Huntingdon*('*s*) *Connection* to designate a Calvinistic Methodist sect founded by her. So **Huntingdo·nian** *a.* and *sb.*
[**1773** J. Wesley *Let.* 22 Oct. (1931) VI. 51, I am afraid

Huntsman spider who adopted us and took up residence behind a painting in the living-room.

hunyak (hʌ·nyæk). *U.S.* Also honyock, -er. [f. Hungarian *a.* and *sb.* after Polack.] = *hunk *sb.*[3] (See also quot. 1941.)

Only in disparaging use.

1911 W. P. Dillingham *Dict. Races* 92 Magyar, Hungarian Hun, or Hunyak in popular language. **1919** S. Lewis *Free Air* (1924) ix. 94, I could buy out half these Honyockers! **1938** 'E. Queen' *Four of Hearts* (1939) 17 Tossing away the stockholders' dough like a hunyak on

Monthly Oct. 280/1 Hurdy-gurdy houses, with dancing-girls, music, and long bars. **1874** T. B. Aldrich *Prudence Palfrey* vii. 115 At sundown the dance-house would open,—the Hurdy-Gurdy House, as it was called. **1955** P. F. Sharp *Whoop-up Country* 192 The saloons and hurdy-gurdy houses of Benton, Macleod, and Calgary.

hu·rdy-gu·rdyish, *a.* Resembling the sound of a hurdy-gurdy; also *fig.*

1923 *Daily Mail* 22 June 6 The concertina or harmonium music is too hurdy-gurdyish. **1931** *Observer* 27 Sept. 10 A cheap, hurdy-gurdyish fellow.

hurlbutite (hɔ·ɪlbətaɪt). *Min.* [f. the name of Cornelius Searle *Hurlbut* (b. 1906), American mineralogist + -ITE[1].] A colourless or greenish-white phosphate of calcium and beryllium, $Be_2Ca(PO_4)_2$.

1952 M. E. Mrose in *Amer. Mineralogist* XXXVII. 931 The mineral here described as hurlbutite was first found as a large broken crystal on the dump at the pegmatite known as the Smith mine. **1961** *Soviet Physics—Doklady* V. 1143 The structures of danburite $CaB_2Si_2O_8$ and hurlbutite are similar to the structure of the feldspars, in particular to anorthite $CaAl_2Si_2O_8$. **1968** I. Kostov *Mineral.* 449 Fabeyite is hexagonal, occurring in acicular crystals, babefphite tetragonal,..hurlbutite orthorhombic.

hurler. Add: **5.** A pitcher at baseball. *N. Amer. slang.*

1926 *Daily Colonist* (Victoria, B.C.) 22 July 12/3 Fifty-six runs were scored in a Three-I League baseball game yesterday in which Springfield defeated Peoria... Thirteen hurlers appeared. **1965** O. Nash in *Times Lit. Suppl.* 25 Nov. 1036/5 When I am told that the hurler kicks, rocks, and either deals or delivers. **1968** *Globe & Mail* (Toronto) 10 July 26/5 But in a season dominated by pitchers, the hurlers once again reigned supreme.

Hurler[2] (hɜ·ɪlər). The name of Gertrud *Hurler*, German pædiatrician (qualified 1894), used in the possessive (and also *attrib.*) in *Hurler('s) disease, syndrome* = *gargoylism 2.

1937 W. R. Ashby et al. in *Brain* LX. 175 We share the dislike of many workers for eponymous titles, but since the other four diseases in this group have..been designated by the names of those writers who have contributed largely to their elucidation, we consider that at this juncture it would be inadvisable to depart from this practice. We suggest, therefore, in place of the title 'Gargoylism' the title 'Hurler's disease'. **1938** *Jrnl. Pediatrics* XII. 579 (*heading*) Hurler's syndrome (gargoylism). **1970** Cohen & Catcart in Keefer & Wilkins *Medicine* xxxv. 687 (*heading*) Hurler syndrome (gargoylism). *Ibid.*, The two clinical forms of Hurler's disease (one an autosomal recessive and the other a sex-linked recessive) are currently being related more specifically to these mucopolysaccharide abnormalities.

Huron (hiūə·rɒn). Also 7 Hiroon. [Fr., f. earlier *hure*, rough hair of the head.] A confederation of five Iroquoian tribes formerly inhabiting a region adjacent to Lake Huron; a member of one of these tribes; also their language. Also *attrib.* or as *adj.*

1658 F. Gorges in *Maine Hist. Soc. Coll.* (1847) II. 67 The Hiroons, who being neuters are friends both to the one [sc. the Iroquois] and the other [sc. the French]. **1756** A. Butler *Lives Saints* II. 650 The saint wrote earnestly to the general of the Society, desiring to be employed on a mission to the barbarous Hurons and Iroquois in Canada. **1782** 'J. H. St. John de Crèvecœur' *Lett. from Amer. Farmer* (1783) iv. 90 The Nattic, like the Huron, in the north-western parts of this continent, must have been the most prevailing one [sc. language] in this region. **1786** *Mem. Amer. Acad. Arts & Sci.* II. i. 125 The Huron, or Wyandot language having no affinity to the Shawanese, Delawares, and other nations. **1789** [see *Five Nations*]. **1823** J. S. Mill *Autobiogr.* (1924) 267 No one, I apprehend, would insult the understanding of this Society..by maintaining..that the Hurons and the Iroquois are the happiest and the most enlightened of mankind. **1832** 'M. Doyle' *Hints on Emigration to Upper Canada* (ed. 2) 17 The soil in the Huron territory is a rich sandy loam. **1845** [see *buckle sb.* 1 e]. **1880** A. H. Sayce *Introd. Sci. Lang.* II. ix. 291 The Hurons of North America believe that the souls of the departed turn into turtle-doves. **1933** L. Bloomfield *Lang.* iv. 72 The Iroquoian family..includes, among others, the Huron (or Wyandot) language. **1959** *Listener* 9 July 75/2 He [sc. the American writer] always feel himself to be a noble savage, a Huron, confronted by the degenerate, corrupt, or over-sophisticated stock from which he sprang. **1973** *Ibid.* 25 Oct. 549/1 The Jesuits..negotiated with the Hurons and Iroquois, as they treated with the *daimyos*.

hurrah. Add: **4.** *attrib.* or *adj.* in various *slang* or *colloq.* uses = shouting hurrah, uproarious, blindly enthusiastic; joyous, 'glad'.

1835 *Franklin Repos.* (Chambersburg, Pa.) 2 June 2/3 The New York delegation cared nothing for principles! To them..the men who could secure for the ticket and for *Martin Van Buren* the *hurra boys* was every thing! **1836** *Congress. Globe* 17 Feb., App. 115 [Some have declared] that his election had been brought about by the 'hurrah boys', and those who knew just enough to shout 'hurrah for Jackson!' **1903** N.Y. *Even. Post* 30 Oct. 16 Added to this solid element is the hurrah crowd whose enthusiasm has a venal tone. **1907** N. Munro *Daft Days* xii. 102 'And what are you doing with your hurrah clothes on?' 'I like to put on my Sunday clothes when I'm writing Charles.' **1909** *Daily Chron.* 20 July 1/2 'Hurrah-boats', as the bluejacket aptly calls excursion steamers. **1925** Fraser & Gibbons *Soldier & Sailor Words* 124 *Hurrah boats*, a Navy term for the pleasure steamers of touring trippers that go round the Fleet at Naval Reviews, usually cheering as they pass ships. **1926** *Flynn's* 16 Jan. 640/1, I had many a pal among th' touts and hurrah boys. **1928** *Daily Express* 4 Dec. 10/5 'Hurrah boys' or college students. **1964** N. Freeling *Double-Barrel* v. 176 We seem to have got quite a hurrah letter from the burgomaster. And there was a hint that I may be promoted.

Hurrian (hu·riăn), *sb.* and *a.* Also Harrian; (less freq.) Harri, Hurri; Kharri, Khurri, -ian. [f. Hittite and Assyrian *Ḫar-ri*, *Ḫur-ri* + -AN.] **A.** *sb.* **a.** Name of a widespread non-

he's a bishop now, and hush-hushes you. **1883** G. Meredith *Poems & Lyrics* 49 Not the pines with the faint airs afloat, Hush-hushing the nested dove. **1928** *Daily Express* 12 June 10/4 She might have done it long ago if she had not been firmly hush-hushed by men.

hush-money. (Later examples.)

1953 H. Miller *Plexus* (1963) x. 352 The cops will be sitting on our necks...The natural thing, under the circumstances, would be to put something aside for hush money. **1957** *Economist* 9 Nov. Suppl. 11/2 She would write her memoirs and include in them faithful records of her association with every noble lord who failed to pay hush money to the tune of two hundred pounds.

hush puppy, hush-puppy (hʌ·ʃ pʌ·pi). [f. Hush *v.*[1] + Puppy *sb.*] **1.** *U.S.* (See quots.)

1918 *Dialect Notes* V. 18 *Hushpuppy*, a sort of bread prepared very quickly and without salt. **1942** M. K. Rawlings *Cross Creek Cookery* 28 Fresh-caught fried fish without hush puppies are as men without women. **1947** *This Week Mag.* (U.S.) 4 Oct. 27/1 What's a hush puppy? You mean you don't know that Southern fried bread like a miniature corn pone—but glorified? It's made of the white cornmeal of the South, smooth and fine as face powder. **1960** *Harper's Bazaar* July 48 Crisp, brown 'hush puppies', crunchy morsels of deep-fried cornmeal batter. **1964** *Cookbk.* (Amer. Heritage) (1967) 220 Hush Puppies are usually served with fried fish. **1967** *Daily News* (N.Y.) 5 Mar. 11. 4 I'm going to eat hush-puppies, wear a snuffler and every night sing 'Silent Night'.

2. The proprietary name (*Hush Puppy*) of a lightweight soft shoe. Also *attrib.*

1961 *Trade Marks Jrnl.* 15 Nov. 1616/1 Hush Puppies. Shoes. Wolverine Shoe and Tanning Corporation. **1963** *Observer* 13 Oct. 9/1 Abbott makes 'Hush Puppies', the lightweight..shoes. **1965** *Amer. Speech* XL. 249 Hush puppies, a brand of nationally advertised shoes for men and women. **1966** 'O. Mills' *Enemies of Bride* xiii. 196 An intelligent..woman in..Hush-puppy shoes. **1969** P. Dickinson *Pride of Heroes* 38 Mr Singleton wore rubber-soled Hush-Puppies which went squeak-squeak on the shiny surface.

husi, var. *jusi.

husk, *sb.*[1] Add: **4. d.** A figure or ornament somewhat resembling a husk.

1934 *Burlington Mag.* Oct. p. xv/2 The tablet is carved with festoons, and the frieze and jambs inlaid with festoons and pendants of husks and coloured marble. **1955** R. Fastnedge *Eng. Furnit. Styles* 285 *Husk*, with 'honeysuckle' and 'wheat-ear' a favourite ornament on furniture of the Adam and Hepplewhite periods. **1971** *Country Life* 3 June 1356/3 The ground paint was decorated with motifs such as festoons of drapery and husks, interlacing hearts, urn patterns, and so on.

5. (sense *4 d) *husk design, festoon, ornament, pattern.*

1904 P. Macquoid *Hist. Eng. Furnit.* vii. 191 The sides are inlaid with the..husk design so popular at this time. **1973** *Country Life* 31 May 1567 Chestnut wood window seats..the..legs..faced by well carved husk design. **1770** J. Wedgwood *Let.* 20 Aug. (1965) 94 First, his Majesty approved of the husk festoons in particular, and I think more so than the desert pattern. **1934** *Burlington Mag.* Oct. 165/1 The back shows the honeysuckle, husk or catkin ornament. **1960** H. Hayward *Antique Coll.* 146/2 *Husk ornament*, an ornamental motif resembling the husk of a wheat ear used continually by architects and craftsmen during the Adam period. **1876** C. Schreiber *Jrnl.* 14 Nov. (1911) I. 485 A good set of Wedgwood, husk pattern.

husk, *v.*[1] Add: Also *transf.* and *fig.*

1878 B. F. Taylor *Between Gates* 182 The rough dresses of the men..out of which they husk themselves. **1892** Kipling *Barrack-Room Ballads* 199 Go husk this whimpering thief..: Winnow him out 'twixt star and star. **1910** C. E. Mulford *Hopalong Cassidy* xviii. 110 He determined to husk Meeker's body from its immortal soul.

husk, *v.*[2] Add: **2.** *intr.* Of the voice: to be or to become husky.

1922 H. Titus *Timber* xxix. 254 Her voice husked for the first time. **1958** *Sunday Times* 29 June 11/1 The birds sing louder than the crooner husking from the loudspeaker.

husker. (Earlier example.)

1780 E. Parkman *Diary* (1899) 279 Breck was very generous in treating ye huskers with Liquor.

husky, *sb.*[1] Add: **a.** (Earlier and *attrib.* examples.)

[**1743** J. Isham *Obs. Hudson's Bay* (1949) 155 Among'st the Northward Indians, and Ehuskemay's they have neither of these beasts.] **1830** in K. G. Davies *N. Quebec & Labrador Jrnl. & Corr.* (1963) 115 There was a cry that the river was full of *Hoskies* (Esquimaux). **1922** *19th Cent.* Feb. 274 As a seamstress the 'husky' woman has no equal.

c. In full *husky dog.* (Earlier and later examples.)

1852 R. Collinson *Jrnl. H.M.S. Enterprise 1850–55* (1889) 218 On his way to the ship [the dog] was kidnapped by the natives, and not being of a pure huski breed, would most likely be prized by them. **1872** *Canadian Monthly* Oct. 307/1 The 'huskie', or Esquimaux dogs..are only fed once a day. **1878** *Sask. Herald* (Battleford, N.W.T.) 18 Nov. 3/1, I had with me a 'Huskie' dog. **1886** *Colonial & Indian Exhib. Rep. Col. Sect.* 75 The original Husky has always been an animal requiring firm treatment. **1947** *New Biol.* III. 152 The most widely used type of sleigh dog is the huskie, employed mainly by Eskimos. **1970** *Islander* (Victoria, B.C.) 22 Feb. 5/1 We now had about 30 husky dogs on deck, and the noise they made when they all howled together was blood-curdling.

husky (hʌ·ski), *sb.*[2] *U.S.* [f. *husky *a.* 1 b.] A strong, stoutly-built person; one whose appearance suggests strength and force.

1864 *Old Piute* (Virginia, Nev.) 17 May, He demanded to see the Charter of the concern, which was read by the Rev. Geo. Birdsall, in his usual impressive manner, and the 'husky' accepted the apology. **1884** 'Mark Twain' *Huck. Finn* xxix. 305 It was a beautiful time to give the crowd the slip; but that big husky had me by the wrist. **1916** C. Sandburg *Chicago Poems* 60 The real huskies that are doing the work of the world. **1929** W. Heyliger *Builder of Dam* v. 46 You're going to need muscle, and he's a husky. **1945** *Jefferson Co. Republican* (Golden, Colo.) 26 Sept. 1/3 One faculty member who strayed too close to the Washington Avenue bridge was picked up bodily by ten huskies and tossed in the murky waters below.

husky, *a.* Add: **1.** (Later example.)

1905 *Daily Chron.* 14 Sept. 3/6, I have always understood that brown bread is far superior to white bread in muscle-making power and (unless very husky) in digestibility.

b. Tough and strong (like a corn-husk); big, strong, and vigorous. Also *transf. N. Amer.*

1869 Mrs. Stowe *Oldtown Folks* xvii. 191 Them wild Injuns..they're so kind o' wild, and birchy, and husky as a body may say. **1889** *Kansas City* (Missouri) *Times & Star* 1 Apr., Mike Burnett, the husky ex-fire chief. **1894** *Outing* XXIV. 447/1 He lit out of the country soon as he got husky enough to travel. **1897** *Ibid.* XXX. 364/2 A husky run down old Ontario in a gale from the West brought *Nox* into Charlotte harbor. **1906** *Eye Opener* (Calgary) Aug. 1/6 Watty himself is a husky all-round athlete. **1909** S. E. White *Rules of Game* I. ix. 51 Good food and leisure and heredity gave me a husky build. **1932** J. Dos Passos *1919* 263 Husky looking young men in khaki. **1958** *Times* 29 Nov. 9/3 If something more husky is preferred, one can camp or stay in one of the small uninhabited cabins which are to be found throughout Lapland.

huss, *sb.* Delete '*Obs.*' and add to def.: the lesser or greater spotted dogfish, *Scyliorhinus caniculus* or *S. stellaris*. (Later examples.)

1963 *Newnes Encycl. Angling* 232/2 Until recently, dogfish have been sold in the shops under the more glamorous name of rock salmon... Today, the recommended names in the retail shops are huss, flake and rig. **1965** S. Norton-Bracy in *Newnes Compl. Guide Sea Angling* 117/2 The lesser spotted dogfish (*Scyliorhinus canicula*), a smaller edition of the greater spotted, is also known by many names, probably the best known of them being robin huss. .. Many anglers seem to think it should have been 'robbing' huss, due to its habit of nipping off the bait. **1967** R. Arnold *Anglers' Handbk.* xv. 161 The most popular dogfish, from the angler's standpoint, is the bull huss, or large spotted dog, or nurse hound. **1972** A. Wrangles *Inshore Sport Fishing* v. 117 Lesser spotted dogfish... Local names. Sandy dog, dogger, rough hound, blind Jimmy, huss, etc. *Ibid.* 118 Greater spotted dogfish. .. Local names. Bull huss, nurse hound, etc.

hussar, *sb.* Add: **3.** *hussar cap.*

1846 *Knickerbocker* XXVII. 287 [He had] a smart hussar cap of green chestnut burrs. **1854** B. F. Taylor *Jan. & June* 85 Hens with hussar caps.

Husserlian (husɜ̄·rliăn), *a.* [f. name of Edmund *Husserl*, German philosopher (1859–1938) + -IAN.] Of, pertaining to, or characteristic of Husserl or his work.

1932 *Mind* XLI. 247 Coming from the same Husserlian school. **1939** E. Welch *Edmund Husserl's Phenomenology* II. ii. 78 All of these theories, on Husserlian grounds, are based on nothing but pure prejudice. **1950** *Mind* LIX. 127 As it is we have Hegelian and Husserlian phenomenology on one hand and Kierkegaardian preaching on the other. **1965** *Language* XLI. 492 It is therefore a transcendental English, an English idealized out of controllable experience, an English known only to Husserlian pure egos.

hustle, *v.* Add: **2. d.** *U.S. colloq.* To obtain, produce, or serve by hustle or pushing activity. Also with *up*.

In some contexts spec. = 'to steal' (cf. sense 2 a in Dict.).

1840 *Southern Lit. Messenger* VI. 414/2 Can't you go out to the woodpile and hustle me up a few chips to start this fire? **1908** *Grand* Dec. 614 Anita was really hungry and hustled up the luncheon in..an unromantic, business-like sort of way. **1914** R. Grau *Theatre of Science* 80 He had to write his own scenarios, direct the productions and 'hustle props'. **1926** J. Black *You can't Win* vi. 65 Don't think because you couldn't hustle a can that you ain't entitled to your coffee. **1940** S. Lewis *Bethel Merriday* iii. 35 When you grow up..you try to squirm into prison, or get a nice job hustling hash..anything to avoid going on the stage. **1950** [see *fix *sb.* 5]. **1953** W. Burroughs *Junkie* (1972) ix. 86 'Do you want to score?' he asked. 'I'm due to score in a few minutes, I've been trying to hustle the dough.' **1957** P. Mansfield *Final Exposure* ix. 134 Perhaps they can hustle up some coffee. **1967** M. M. Glatt et al. *Drug Scene* iii. 34 Hustling is a generic term which usually refers to any number of strategies addicts may use to obtain drugs. **1970** H. E. Roberts *Third Ear* 8/2 *Hustle, hustling*, to be aggressively, actively engaged in the acquisition of goods and money.

e. To sell or serve (goods, etc.), esp. in an aggressive, pushing manner. *N. Amer. slang.*

1887 *Grip* (Toronto) 5 Mar. 6/2 She hustled the hash at Gilhooley's on Blank St. **1894** T. R. Dewar *Ramble round Globe* 84 Almost every second man you meet..is now either 'hustling lumber' or farming at four or five dollars a week. **1973** *Black World* Aug. 56/2 He hustled the watch to a barber for 35 bills.

4. (Later examples.)

1903 *Westm. Gaz.* 25 Feb. 5/2 The 'Decapod' will 'hustle' in true American fashion. **1906** 'O. Henry' *Four Million* 62 Do you think I'm going to let you hustle for wages while I philander in the regions of high art? **1908** G. H. Lorimer *Jack Spurlock* vi. 87, I decided that..I'd hustle over to the nearest trust company.

5. *intr.* To engage in prostitution. *slang.*

In quot. 1954 used transitively in sense 'to "work" an area, soliciting as a prostitute'.

1930 J. Dos Passos *42nd Parallel* v. 421 She showed him a snapshot of her steady...I don't hustle when he's in town. **1954** J. Steinbeck *Sweet Thursday* i. 4 If you were hustling a state you should do honour to that state. **1957** C. MacInnes *City of Spades* II. ii. 117 'You're positive she's not hustling?' 'Muriel..is no harlot.' *Ibid.* 120 Hustling with Jumbe queers. **1959** *Listener* 10 Dec. 1048/1 She..revolted in revenge against her family, 'hustled' in Piccadilly, hated men as clients, took a ponce. **1960** *Guardian* 23 Sept. 26/7 Several clubs have been threatened with proceedings for keeping a brothel if they allow prostitutes to hustle on the premises. **1970** *Daily Colonist* (Victoria, B.C.) 8 Feb. 11/1 To put it bluntly, she was hustling, and liked the rewards in the shape of cash and jewelry.

hustle, *sb.* Add: **3.** (Further examples.) Also with *a*. (Cf. *move *sb.* 6.)

1902 W. N. Harben *Abner Daniel* ii. 19, I..told 'em to git a hustle on the'r'ses. **1908** [see *drive *sb.* 1 i]. **1968** *Globe & Mail* (Toronto) 15 Jan. 21/1 Earl Balfour is a veteran hockey player who well remembers the secret to success—hustle.

4. A swindle, racket; a means of deception or fraud; a source of income; a paid job. *slang* (orig. *U.S.*).

1963 R. I. McDavid *Mencken's Amer. Lang.* 729 *Hustle*, a racket which one pushes to get his bread. **1965** 'Malcolm X' *Autobiogr.* vi. 87 Each of the military services had their civilian-dress eyes and ears picking up anything of interest to them, such as hustles being used to avoid the draft..or hustles that were being worked on servicemen. **1969** R. Pharr in A. Chapman *New Black Voices* (1972) 63, I got me a good hustle. I write over $200 worth of numbers a day, which gives me a cool 40 bucks. **1972** *Observer* 20 Aug. 7/4, I was stark broke..from the horses..and the cards. It didn't really matter because I was on a hustle.

hustler. Add: **1.** (Further examples.) Also, a thief, a criminal; one who makes his living dishonestly or by begging; a pimp. *slang*.

1924 Jackson & Hellyer *Vocab. Criminal Slang* 47 *Hustler*,..a grafter; a pimp who steals besides. The genteel thief is designated a 'hustler'. **1926** *Clues* Nov. 161/2 *Hustler*, members of the underworld in general. **1953** W. Burroughs *Junkie* (1972) 158 *Pop corn*, someone with a legitimate job, as opposed to a 'hustler' or thief. **1957** C. MacInnes *City of Spades* II. ix. 164 They're wreckage of jazz musicians.., ponces, and other hustlers like myself... I pimp around the town, picking the pounds up where I can. **1972** *Observer* 20 Aug. 7/4 Once, the hustler was the odd man out.

2. a. For *U.S.* read 'orig. *U.S.*', and add further examples. Also, a salesman, esp. one who is energetic or aggressive.

1886 *Publishers' Weekly* 18 Dec. 965/1 Young man, a 'hustler' in every respect, wants a strictly first-class position with a 'live' book house. **1891** *Grip* (Toronto) 18 Mar. 254/1 An enterprising down East Hustler lately conceived the idea of crowding the 'Best Fifty Books' into one volume, and selling the same 'at sight' for twenty cents. **1907** F. H. Burnett *Shuttle* xxiii. 235 I'm not passing myself off as anything but an ordinary business hustler.—just under salesman to a typewriter concern. **1907** *Westm. Gaz.* 6 Sept. 4/1 His proper title is The Hustler, and his real business is to hustle from morning till night every man and woman..connected with the business. **1926** *Daily Colonist* (Victoria, B.C.) 10 Jan. 32/3 (Advt.), Hustler wanted, 25 years or older, to sell a high-grade appliance. **1954** A. M. Bezanson *Sodbusters invade Peace* 194 Chris was a hustler, and brought to our partnership a type of aggressiveness I lacked. **1973** *Black World* Apr. 38/1 He pulls in $1,500 some weeks, and he's a *small-time hustler*.

c. A prostitute. *slang.*

1924 G. C. Henderson *Keys to Crookdom* 408 *Hustler*, prostitute. **1929** D. Hammett *Red Harvest* iii. 27 'Who is this Dinah Brand?'..'A soiled dove..a de luxe hustler, a big-league gold-digger.' **1952** J. Steinbeck *East of Eden* 436 They would think she was just a busted old hustler. **1960** P. Goodman *Growing up Absurd* (1961) ix. 194 The juvenile delinquents, like the hustlers (male prostitutes), fancy themselves as movie heroes in sports cars. **1970** *Women Speaking* Apr. 4/2 Man keeps her on the defensive by a constant barrage of insulting words that describe her in sexual terms:..hustler, slut.

hustling, *vbl. sb.*[1] Add: **b.** Robbery, esp. with violence.

1823 'J. Bee' *Slang* 102 *Hustling*, forcible robbery, by two or more thieves seizing their victim round the body, or at the collar. **1826** *New Newgate Calendar* V. 337 Transported for hustling. **1972** *Observer* 20 Aug. 7/4 'Hustling' has been growing steadily for years... By 1968,..of 150 cases of bag-snatching, 87 were alleged to have been committed by black youths.

c. Prostitution, soliciting as a prostitute.

1924 G. C. Henderson *Keys to Crookdom* 408 *Hustling*, streetwalking. **1938** 'M. Benney' *Scapegoat Dances* vi. 77 Bond Street might be better. There's good 'ustling there. **1959** Anon. *Streetwalker* iv. 85, I had to use the flat for hustling. **1969** *Jeremy* I. iii. 22/2 *Hustling*, working as a male prostitute.

hut, *sb.* Add: **4.** *hut-door, -tax* (earlier and later examples); *hut circle Archæol.*, a circle of the wave front, their envelope at a later time representing a new position of it.

[1835 *Rep. Brit. Assoc. Adv. Sci. 1834* 308 The composition of the grand, or primary wave, by the union of the several secondary or partial waves, in this demonstration, has been denominated the principle of Huygens.] **1840** *Phil. Mag.* XVII. 243 (*heading*) On the application of Huyghens's principle in physical optics. **1899** W. Watson *Text-bk. Physics* III. i. 353 The construction for finding the position..of the wave-front, at a time *t* by means of the tangent to a series of circles, the radius of each of which is equal to the space passed over by the wave in a time *t*, is due to Huyghens, and is known as Huyghens's construction for the wave-front. **1900** J. C. P. Aldous *Elem. Course Physics* x. 577 The chief negative eye-piece is Huyghens' eye-piece. **1957** *Encycl. Brit.* XV. 439/1 The Huygens eyepiece..is the most common type used with the microscope. **1959** Born & Wolf *Princ. Optics* viii. 370 Fresnel was able to account for diffraction by supplementing Huygens' construction with the postulate that the secondary wavelets mutually interfere. **1970** M. V. Klein *Optics* i. 20 To follow the propagation of the optical disturbance..we have, in theory, only to make repeated applications of Huygens' principle. *Ibid.* 21 The Huygens construction.

huzoor (hə̆zuə·ɹ). Also 8 **huzzoor, huzur.** [a. Arab. *ḥuḍūr* (pronounced in India as *huzūr*) presence (employed as a title), f. *ḥaḍara* to be present.] An Indian potentate; often used as a title of respect.

1776 *Trial of Joseph Fowke* Depositions 17/2 [They] endeavour to lay their complaints before the Huzzoor. **1843** C. J. C. Davidson *Diary Trav. Upper India* I. 77 The huzoor's countenance..is as immovably tranquil as that of Boodh. **1898** *Longman's Mag.* May 80 What pleasure hath this slave in life, save to do the Huzoor's will? **1899** Kipling *Land & Sea Tales* (1923) 229 'Huzoor! (Your Highness!)' said Imam Din, stooping low. **1957** M. M. Kaye *Shadow of Moon* xxx. 438 'Be swift, my father,' said Niaz pleasantly. 'Do not keep the Huzoor waiting.'

hwyl (hū·il). Also (erron.) **hwyll.** [W.] An emotional quality which inspires and sustains impassioned eloquence; also, the fervour of emotion characteristic of gatherings of Welsh people.

1899 *Daily News* 1 Mar. 7/5 What is termed the Welsh 'hwyl', a form of eloquence which seems to exert remarkable influence on the hearers. **1928** *Observer* 15 July 11/3 The National Eisteddfod is a microcosm of Wales. There you may see preacher and ploughman, collier and clerk, all rubbing shoulders and all under the influence of the intangible and untranslatable 'hwyl' of the Eisteddfod. **1959** W. Golding *Free Fall* ii. 57 Father Anselm..was not emotional, no Welsh hwyll for him. **1970** *Daily Tel.* 16 June 7 Plaid Cymru, full of 'hwyl' and hope, are making an all out effort to repeat in Cardigan the success they achieved..in Carmarthen. **1973** *Ibid.* 13 Jan. 16 There should be a considerable degree of hwyl on board HMS Glamorgan today when 20 members of the London Glamorgan Society are entertained to lunch.

hyacinth. Add: **2. b.** (Further examples.) **Cape H.**, a plant of the genus *Galtonia*, esp. *G. candicans*; **Peruvian H.**, *Scilla peruviana*; **Roman H.**, an early-flowering variety, bred from *Hyacinthus orientalis* var. *albulus*.

[**1629** J. Parkinson *Parad.* xi. 126 The Starry Iacinth of Peru, being thought to have growne in Peru, a Province of the West Indies, but he that gave that name first unto it, eyther knew not his naturall place, or willingly imposed that name, to conceale it, or to make it the better esteemed.] **1659** T. Hanmer *Garden Bk.* (1933) 36 The Branch'd Hyacinths of Peru,..very rare here. **1673** J. Ray *Observations Journey Low-Countries* 250 In the ditches by the wayside, I observed growing wild..the lesser Grape Hyacinth, and Hepatica. **1731** P. Miller *Gardeners Dict.* s.v. *Hyacinthus*, The Hyacinth of Peru may also be rais'd from Seeds in the same manner as the common Hyacinths. *Ibid.*, Another Hyacinth..is now preserved in curious Collections of Exotick Plants; it was originally brought from the Cape of Good Hope. **1877** [see Roman *a.*[1] 14 b]. **1911** J. Weathers *Bulb Bk.* 282/1 There is a blue Roman Hyacinth that flowers somewhat later than the white form. **1917** L. H. Bailey *Stand. Cycl. Hort.* VI. 3117/2 [*Scilla*] *Peruviana*... Cuban Lily. Peruvian Jacinth. Hyacinth of Peru. *Ibid.* 3118/1 The Hyacinth of Peru is not hardy in Mass[achusetts]. **1924** J. Weathers *My Garden Bk.* xx. 348/1 Galtonia (after the S. African explorer, Francis Galton). Cape Hyacinth. **1936** T. S. Eliot *Coll. Poems* 127 Lord, the Roman hyacinths are blooming in bowls. **1956** A. M. Coats *Flowers & their Histories* I. 122 The Spire Lily or Cape Hyacinth..is on the other hand not a true hyacinth, although closely related to the family; it has therefore been renamed *Galtonia candicans*. **1961** P. M. Synge *Collins Guide to Bulbs* 162 The Roman Hyacinths tend to flower among the earliest and the spikes are rather looser and more delicate.

3. b. A variety of pigeon, characterized by its blue-black colour and white markings.

1855 *Poultry Chron.* III. 9/1 Those pretty spangled Toys..known by various names, as Porcelains, Hyacinths, Ermines, &c. **1879** L. Wright *Pract. Pigeon Keeper* 208 Victorias are simply Hyacinths of a lighter shade. **1935** R. Mannering *Lyell's Pigeon-Keeping for Amateurs* (ed. 4) 107 The Hyacinth is among those breeds which do not reveal their true colouration and markings until after the first moult. **1965** W. M. Levi *Encycl. Pigeon Breeds* 189 The Hyacinth appears to have much in common with the Suabian, and they are probably related.

hyaline, *a.* and *sb.* Add: **A.** *adj.* *hyaline cast*, a more or less transparent urinary cast composed mainly of precipitated protein; *hyaline cell*, (*a*) *Bot.*, a cell without chlorophyll, found in the leaves and stem of certain mosses; † (*b*) *Med.* (also *hyaline leucocyte*), a type of white blood-cell; = *monocyte (*Obs.*); *hyaline membrane disease* (or *syndrome*), a condition in some newborn (esp. premature) babies in which the lung spaces are lined with a hyaline membrane, causing severe dyspnœa and often early death.

1870 *Jrnl. Bot.* VIII. 229 In neither [species of *Dicranum*] do I find the beautiful chlorophyllose contents observable in the Finland specimens..nor are the longitudinal rows of hyaline cells, in the centre of the basal wing, so well defined. **1881** *Practitioner* Oct. 243 The urine contained red corpuscles, albumin, and hyaline casts. **1894** Kanthack & Hardy in *Jrnl. Physiol.* XVII. 96 The hyaline cell occurs both in blood and in the extra-vascular spaces. **1896** H. N. Dixon *Student's Handbk. Brit. Mosses* 9 The pores in the hyaline cells of the branch leaves [of *Sphagnum* species] are..occasionally found on the face of the hyaline cells. **1906** *Jrnl. Pathol. & Bacteriol.* XI. 67 The immature hyaline cell is an evenly round or oval cell. *Ibid.* 79 Others consider that the lymphocytes are the young forms of hyaline leucocytes. **1929** *Encycl. Brit.* III. 742/1 The large hyaline cell may really include two quite different cells of similar appearance. **1931** A. Piney *Recent Adv. Hæmatol.* (ed. 3) ii. 18 The large hyaline leucocyte or monocyte is not easy to place with complete certainty in either the group of granulocytes or of lymphocytes. **1938** G. M. Smith *Cryptogamic Bot.* II. iv. 87 The hyaline cells of leaves and similar cells in the cortex of a stem play an important role in the absorption and retention of water. **1953** *Adv. Pediatrics* VI. 173 If an infant does not exhibit respiratory difficulty within the first 12 hours of life I do not believe it will die of hyaline membrane disease. **1955** *Jrnl. Pediatrics* XLVII. 40/1 The triad of pathologic changes in the lungs of infants who die with the pulmonary hyaline membrane syndrome consists of atelectasis, vascular engorgement, and an eosinophilic hyaline-like membrane which lines alveolar ducts and alveoli. **1958** *New Biol.* XXVI. 102 The leaves [of *Sphagnum*] are composed of large empty 'hyaline' cells, with narrow green cells between them. **1966** *Lancet* 24 Dec. 1384/2 Mid-stream urine showed small numbers of hyaline and granular casts. **1973** *Sci. Amer.* Apr. 75/1 In the past some infants with hyaline membrane disease recovered spontaneously. Many became exhausted and died.

hyalinization (hai:ălinaizēi·ʃən). *Med.* [f. Hyalin(e *a.* and *sb.* + -ization.] A change of tissue into a homogeneous, translucent, often firm, mass somewhat resembling glass under the microscope; hyaline degeneration.

1919 *Amer. Jrnl. Med. Sci.* CLVII. 673 Undoubtedly there are other factors [affecting malignancy and benignancy] such as lymphocytic infiltration, fibrosis, hyalinization, [etc.]. **1926** *Arch. Path.* II. 338 The muscle lesion usually consists of hyaline degeneration of the contractile substance of the fibers... The hyalinization of the muscle may result in rupture with subsequent hemorrhage. **1961** *Lancet* 22 July 182/2 Histologically the aneurysmal sac was composed of dense fibrous tissue showing hyalinisation. **1963** Walter & Israel *Gen. Path.* v. 71 Hyalinisation of the walls of many small arterioles may occur as part of the aging process.

So **hy·alinized** *ppl. a.*, having the appearance that results from hyalinization.

1929 *Arch. Path.* VIII. 906 The corpus fibrosum..consists of irregularly outlined and more or less thoroughly hyalinized fibrous tissue. **1962** *Lancet* 27 Jan. 191/1 This exudate becomes hyalinised to form membranes, lining the sacs, and dense plugs in the alveolar ducts.

hyalo-. Add: **hy·alomere** *Cytol.* [*-mere], the lightly staining ground-substance of a blood platelet; **hy:alo-ophi·tic** *a. Petrol.* [ad. G. *hyaloophitisch* (B. Polenov 1899, in *Trav. de la Soc. Impér. des Naturalistes, St. Pétersbourg* XXVII. 473)] (see quot. 1920); **hy:alopilitic** (-pəili·tik) *a. Petrol.* [ad. G. *hyalopilitisch* (H. Rosenbusch *Mikrosk. Physiogr. d. Min. u. Gesteine* (ed. 2, 1887) II. 466), f. Gr. πῖλ-ος felt + *-itic* (see -ite[1])], characterized by or having needle-like microlites embedded in a glassy ground-mass; **hy·alosome** *Cytol.* [ad. G. *hyalosom* (S. M. Lukjanow 1887, in *Arch. f. mikrosk. Anat.* XXX. 551): see *-some[4]], a lightly staining cell structure resembling the nucleolus.

1936 P. E. Smith et al. *Bailey's Text-bk. Histol.* (ed. 9) vi. 150 Structurally they [*sc.* blood platelets] consist of a central granular mass (chromomere)..and a peripheral hyaline zone (hyalomere). **1969** A. W. Ham *Histol.* (ed. 6) xiv. 303/1 Most of a platelet appears to consist of a fairly clear ground substance which is colored only a very pale blue with a blood stain and is called its hyalomere. **1920** A. Holmes *Nomencl. Petrol.* 120 *Hyalo-ophitic texture*, Polenov, 1899, a texture resembling ophitic texture, in which the spaces of an open network of felspar laths are occupied by glass; a limiting case of intersertal texture. **1954** H. Williams et al. *Petrogr.* ii. 20 Where glass takes the place of pyroxene, the texture is called hyaloophitic. **1888** J. J. H. Teall *Brit. Petrogr.* 443 The normal structure of the andesites is the hyalopilitic. **1959** W. W. Moorhouse *Study of Rocks in Thin Section* v. 159 Hyalopilitic lavas are glassy with felted microlites. **1889** *Q. Jrnl. Microsc. Sci.* XXX. 168 They are therefore distinguished..as (a) 'karyosomes'..; (b) 'plasmasomes'..; (c) 'hyalosomes', which are not stained (vide Lukjanow).

hyaluronic (hai:ăliurǫ·nik), *a. Biochem.* [f. Hyal(oid *a.* (from its first being isolated from the hyaloid or vitreous humour of the eye) + *uronic *a.*] *hyaluronic acid*: a viscous mucopolysaccharide composed of acetylglucosamine and glucuronic acid units, widely found in animal tissues (e.g. in synovial fluid, in the ground-substance of connective tissue, and in the vitreous humour of the eye) as well as in bacterial capsules.

1934 Meyer & Palmer in *Jrnl. Biol. Chem.* CVII. 631 Data on the preparations of this acid for which we propose..the name 'hyaluronic acid', from hyaloid (vitreous) + uronic acid. **1957** *Sci. News* XLV. 85 *In vivo*, hyaluronic acid has the important function of offering resistance to penetration by foreign matter, including agents of infectious disease. **1969** *New Scientist* 24 Apr. 167/2 Dr W. C. McCutchen..was adamant that if the hyaluronic acid is removed from synovial fluid, it still lubricates efficiently.

Hence **hyalu·ronate**, a salt of hyaluronic acid.

1946 *Biochem. Jrnl.* XL. 583 (*heading*) The influence of hydrolysates of hyaluronate upon hyaluronidase production by micro-organisms. **1970** R. W. McGilvery *Biochem.* xxiv. 585 Solutions of hyaluronate have a high viscosity, and it is especially concentrated in the synovial fluid of the joints.

hyaluronidase (hai:ăliurǫ·nidēiz, -ēis). *Biochem.* [f. *hyaluronid-* (f. *hyaluron(ic *a.* + *-ase.] Any of the enzymes which catalyse the depolymerization of hyaluronic acid, thereby reducing its viscosity and rendering tissue containing it more permeable.

1940 Chain & Duthie in *Brit. Jrnl. Exper. Path.* XXI. 325 Hyaluronic acid is the substrate for the 'mucolytic' enzyme in spreading factor solution which accordingly is termed 'hyaluronidase'. **1951** A. Grollman *Pharmacol. & Therapeutics* xxix. 652 To accelerate the rate of absorption from the skin and subcutaneous tissues, hyaluronidase may be added to the infusion. **1967** *Martindale's Extra Pharmacopoeia* (ed. 25) 886/1 The diffusion of local anaesthetics is accelerated by the addition of 1000 units of hyaluronidase to each 20 ml. of the anaesthetic solution. **1970** Ambrose & Easty *Cell Biol.* xii. 391 The sperm acrosome secretes enzymes (including hyaluronidase) which assist penetration [of the ovum].

hyan (hai·ăn). *local.* Also **hyant, hyen, hyon.** [Origin unknown.] An acute, usually fatal, infectious disease of cattle or, occasionally, sheep, caused by the bacterium *Clostridium chauvœi*; = *black quarter* (s.v. Black *a.* 19), Blackleg 1, Speed *sb.* 10 a.

1789 *Trans. Soc. Arts* VII. 73, I..should annually have attempted to rear one hundred [calves], were it not for the disorder called here [*sc.* Scarisbrook] the *Hyon*. **1795** J. Aikin *Descr. Country round Manchester* 325 Great numbers of calves having been taken off by a disease here called the *hyon*. **1801** *Sporting Mag.* XVII. 153 Drinks to be given to young calves for striking of the *Hyen*. **1881** [see Speed *sb.* 10 a].

hyawa (hai·ăwă). Also **haiowa, hayawa, hiawa, hyawai.** [Arawak (Makuchi) *haijawa*. In Du. *hajawa* (1770).] One of several balsam-bearing trees or shrubs of Guyana, esp. *Protium heptaphyllum*. Also *attrib.*

1825 C. Waterton *Wanderings S. Amer.* 190 They paint themselves with the Roucou, sweetly perfumed with Hayawa. **1840** R. Schomburgk *Descr. Brit. Guiana* 98 The latter tree [*sc.* *Amyris ambrosiaca*], called *Haiowa* or *Sepou* by the Indians, is most abundant. **1851** *Illustr. Catal. Gt. Exhib.* IV. 980/1 Hyawai gum or incense, from the River Demerara. **1887** *Colonial & Indian Exhib. Rep. Col. Sect.* 295 Hyawa gum..is very fragrant when burnt. **1899** J. Rodway *In Guiana Wilds* 208 Then it struck him that a torch would be useful, as he saw a hyawa bush growing near. **1924** Record & Mell *Timbers Trop. Amer.* 335 The bastard cedar, incense tree, or 'hiawá' of British Guiana is *P*[*rotium*] *heptaphyllum* March.

hyawaballi (hai:ăwăbæ·li). Also **haiariballi, hiawaballi, haiowaballi.** [Arawak (Makuchi) *hyawaballi*, f. *hyawa + *-balli* resembling.] A timber tree of Guyana, *Tetragastris panamensis*.

1851 *Illustr. Catal. Gt. Exhib.* IV. 985/1 Transverse and vertical sections of the hyawaballi tree, from the River Demerara. **1887** *Colonial & Indian Exhib. Rep. Col. Sect.* 451 Hiawa-Balli. **1917** *Timehri* IV. 260 Hiawa-balli... Hard, heavy, compact, fine grain. **1943** Record & Hess *Timbers of New World* 109/2 The Haiowaballi of British Guiana is *T*[*etragastris*] *panamensis*. **1952** D. B. Fanshawe *Vegetation Brit. Guiana* 60 *Tetragastris* sp. (Haiawaballi) community. **1956** *Handbk. Hardwoods* (Forest Prod. Res. Lab.) 107 Haiariballi is a timber of British Guiana. Material received at the Laboratory showed the timber to be worthless for commercial sawing.

Hy Brasil, Hy-Brazil (hai brăzi·l). Also 9 **Brasil Rock, O'Brazil.** [Cf. Brazil[1].] Name originally applied to one of the larger islands of the Azores; subsequently and chiefly to a legendary island located off the west coast of Ireland.

[**1436** A. Bianco *Atlas* (1869) (*caption*) Y de borzil.] **1812** J. Purdy *Mem. N. Atlantic Ocean* IV. i. 128 *Brasil Rock*, in lat. 51° 10′, and long. 15° 58′..although its existence has been doubted..it was, however, seen in the year 1791. **1843** G. Griffin *Works* VIII. 210 On the ocean that hollows the rocks where ye dwell, A shadowy land has appeared as they tell; Men thought it a region of sun-

shine and rest, And they called it O'Brazil—the isle of the blest. **1899** Kipling *Stalky & Co.* p. vii, Far and sure our bands have gone—Hy-Brasil or Babylon. **1906** —— *Puck of Pook's Hill* 14 Sir Huon..setting off from Tintagel Castle for Hy-Brasil. **1948** L. MacNeice *Holes in Sky* 30 Both myth and seismic history have been long suppressed Which made and unmade Hy Brazil. **1967** V. Gielgud *Conduct of Member* xviii. 139 Want to go a long way.. Cipangu—or better, Hy-Brazil. The world was more fun when its countries had names like these.

hybrid, *sb.* and *a.* Add: **A.** *sb.* **2. b.** *Petrol.* A hybrid rock (see sense B. 2 b below).

1918 *Q. Jrnl. Geol. Soc.* LXXIV. 129 Compared with the Potter-Fell type..they would be hybrids, the former are transitional varieties. **1934** *Ibid.* XC. 599 The hybrid or its parent magma rose along the bedding-planes of the sediments. **1950** E. E. Wahlstrom *Introd. Theor. Igneous Petrol.* x. 234 The products of intermingling of these magmas originally were called hybrids, a term which..has come to have a broader meaning and now includes all rocks resulting from the assimilation or melting of solid igneous rocks by later intrusions from the same source.

c. *Physical Chem.* A hybrid orbital (see sense B. 2 d below).

1932 *Physical Rev.* XL. 62 A hybrid of 3*d*, 4*s*, and 4*p* electrons. **1962** P. J. & B. Durrant *Introd. Adv. Inorg. Chem.* v. 144 The bond angles, at a given atom, are determined by the angles between its σ hybrids. **1968** K. F. Reid *Prop. & React. Bonds in Org. Molecules* iii. 43 The second kind of hybridized orbital, termed the trigonal hybrid, arises through the hybridization of one *s* and two *p* A[tomic] O[rbital]s.

B. *adj.* **1. b.** As the first element in the names of varieties of rose, esp. **hybrid China,** a variety produced by crossing *Rosa chinensis* and *R. semperflorens*, characterized by a long flowering period; **hybrid perpetual,** a cross between *Rosa damascena* and a hybrid China rose; **hybrid polyantha** = *floribunda; **hybrid tea,** a cross between a hybrid perpetual and a tea-scented rose (*Rosa odorata*).

1837 T. Rivers *Rose Amateur's Guide* I. 20 Perhaps no plant presents such a mass of beauty as a finely grown hybrid China rose in full bloom. **1848** W. Paul *Rose Garden* II. 121 The Bourbon Perpetual..is a division embracing the varieties of Hybrid Perpetual, in which the characters of the Bourbon Rose are strikingly developed. **1859, 1890** [see Perpetual *sb.* 1 b]. **1890** *Gardeners' Chron.* 1 Feb. 132/1 Primrose Dame and Vicomtesse Folkestone are also included with hybrid Teas. **1931** M. Grieve *Mod. Herbal* II. 688/1 The most suitable are the so-called Hybrid Perpetuals, flowering from June to October. **1945, 1956** [see *floribunda]. **1951** *Dict. Gardening* (R. Hort. Soc.) IV. 1824/1 Hybrid Perpetual Roses.. originated in the crossing of the Damask Rose with the Hybrid China varieties. **1968** A. Christie *By Pricking of Thumbs* vi. 82 Got some old-fashioned roses here... Better than them new-fashioned Hybrid Teas. **1970** [see *floribunda]. **1973** *Rose Ann.* 37 The first of the new race to bear the characteristics of the hybrid tea was 'Victor Verdier' (1859), although some may claim that 'La France' (1867) was the first true hybrid tea rose.

2. b. *Petrol.* Of rock: formed by the mixing of two different magmas or by the incorporation into an intruding magma of adjacent solid rock (esp. rock of the same origin as the magma).

1904 A. Harker *Tertiary Igneous Rocks Skye* xi. 183 The processes..were of a less simple kind, mere admixture being supplemented by diffusion. The resulting hybrid rocks..are thus only in a general sense intermediate in composition between the two parent rocks, and may be abnormal in comparison with any ordinary igneous rocks formed from a single magma. **1954** H. Williams et al. *Petrogr.* vi. 110 Most diorites..are probably hybrid rocks, and many contain xenoliths that exhibit various stages of magmatic reaction.

c. *Computers.* Utilizing or involving both analogue and digital methods.

1959 E. M. Grabbe et al. *Handbk. Automation, Computation, & Control* II. xxix. 4 The purpose of the hybrid system..is to combine the advantages noted above for each of the two types of conventional computer, while at the same time obviating the disadvantages. **1964** *Ann. N.Y. Acad. Sci.* CXV. 573 More complex operations such as multiplication are not as critical to most hybrid operations and can be slower. **1968** *Brit. Med. Bull.* XXIV. 193/1 The limited accuracy of analogue computers..can be overcome by using 'hybrid' computers in which certain elements..are digitally designed to preserve accuracy.

d. *Physical Chem.* Applied to a bond or valence orbital obtained by the linear combination of two or more different atomic orbitals.

1939 L. Pauling *Nature Chem. Bond* iii. 82 The strength of the best *s-p* hybrid bond orbital. **1960** J. W. Linnett *Wave Mech. & Valency* viii. 128 The appropriate hybrid orbitals are the most successful for indicating the spatial distribution of the electrons. **1968** K. F. Reid *Prop. & React. Bonds in Org. Molecules* iii. 42 The 2*s* and the three 2*p* orbitals of the excited atom are treated in such a way that..they combine to produce four equivalent orbitals, each termed a tetrahedral or sp^3-hybrid orbital, which have their axes directed towards the four corners of a regular tetrahedron.

3. Special Combs.: **hybrid coil** *Electr.*, a type of transformer used in two-wire telephone circuits when amplification in both directions is required, having four pairs of terminals so arranged that if the impedances connected to two pairs balance, a voltage applied to a third pair divides equally between them without inducing a voltage in the fourth pair; also called a *hybrid transformer*; **hybrid swarm** *Ecol.*, a variable population caused by the hybridization of neighbouring species; **hybrid vigour** = *heterosis 3.

1925 C. A. Wright *Telephone Communication* x. 244 Conjugate alternating-current bridges or hybrid coils.. are generally used in repeater and multiplex telephone lines. **1959** K. Henney *Radio Engin. Handbk.* (ed. 5) xxviii. 33 Instead of connecting the respective input of one amplifier and output of the other amplifier directly to a line circuit, there is introduced a balancing coil or so-called hybrid coil. **1926** *Nature* 30 Oct. 623/2 (*heading*) The naming of wild hybrid swarms. *Ibid.* 624 Nor do the 'Rules' have in view the existence of the highly polymorphic hybrid swarms—in no few cases hundreds or probably thousands of distinct individuals—that are known to exist. **1947** *New Phytologist* XLVI. 229 The Oxlip is confined in Britain to a small area in East Anglia, and at the edges of this area Oxlip-Primrose hybrid populations are found; such well-defined hybrid swarms are not common amongst British plants. **1963** E. Mayr *Animal Species & Evol.* vi. 118 The barrier between two sympatric species sometimes breaks down so completely, locally or over wide areas, that the two parental species are replaced by a hybrid swarm that serves as a continuous bridge between the two parental extremes. **1969** Briggs & Walters *Plant Variation & Evol.* xi. 186 Hybrid swarms are found in which there is a remarkable range of variation. **1941** *Stand. Handbk. Electr. Engin.* (ed. 7) xxii. 2055 (*caption*) Principle of hybrid transformer. **1962** *Newnes Conc. Encycl. Electr. Engin.* 379/1 A separating device consisting of a pair of matched 'hybrid' transformers is a common method of securing two-way amplification. [**1909** E. M. East in *Amer. Naturalist* XLIII. 179 In every case an increase in vigor over the parents [*sc.* maize plants] was shown by the crosses.] **1918** Babcock & Clausen *Genetics Rel. Agric.* xii. 230 Not all species hybrids, however, display hybrid vigor. **1949** C. C. Lindegren *Yeast Cell* xxvii. 2 The degeneration or 'running out' of hybrids showing heterosis has been one of the principal problems of hybrid vigor. **1970** *Watsonia* VIII. 131 Its robust growth, vigorous vegetative spread and large fronds..suggest hybrid vigour.

hybridization. Add: **1. b.** *Petrol.* The formation of a hybrid rock.

1926 G. W. Tyrrell *Princ. Petrol.* ii. 31 There is a good deal of commingling of the magmas, with enclosure of fragments and hybridisation, along the interior contacts. **1968** B. Bayly *Introd. Petrol.* ix. 100 At the interface between a magma and a solid rock, the possibility of interpenetration exists... Where the environment is solidified magma from an earlier stage of the same magmatic event, the process is called hybridization; when, as is more common, the environment is some independent solid, the process is contamination.

c. *Physical Chem.* The mathematical combination of atomic orbitals to form a hybrid orbital.

1932 *Physical Rev.* XL. 1037, sp^3 hybridization is encountered only when the repulsions between the four electrons of the H atoms are included. **1962** P. J. & B. Durrant *Introd. Adv. Inorg. Chem.* v. 145 Table 5.1 shows some of the valence states of atoms which can be produced by the hybridisation of atomic orbitals. **1968** [see *hybrid *sb.* and *a.* A. 2 c].

d. *Biochem.* The formation of a hybrid macromolecule by artificially recombining complementary subunits (single polynucleotide strands in the case of nucleic acids and individual polypeptide chains in the case of proteins) obtained from slightly different varieties of the same molecular species or (in the case of RNA–DNA hybrids) of similar molecular species.

1959 Wolstenholme & O'Connor *Biochem. Human Genetics* (Ciba Found. Symp.) 123 Vinograd, Schroeder and Hutchinson (1959) have shown by an ingenious 'hybridization' experiment involving haemoglobin labelled with ^{14}C that the α chains of haemoglobins A and S are interchangeable and therefore similar. **1962** *Science* 21 Dec. 1329/1 It is perhaps not surprising that s-RNA is resistant to hybridization since x-ray analysis..suggests that s-RNA is a hairpin structure kept together by a highly regular system of hydrogen bonding. Until this secondary structure is disrupted there is no opportunity for pairing between s-RNA molecules and complementary sequences in the DNA. **1965** *Jrnl. Molecular Biol.* XII. 830 Hybridization in solution has one obvious disadvantage, stemming from the fact that RNA-DNA formation must compete with the re-formation of the DNA-DNA complexes. **1966** Lehmann & Huntsman *Man's Haemoglobins* xxi. 224 (*heading*) Hybridisation of abnormal human haemoglobin variants. **1968** W. A. Schroeder *Primary Struct. Proteins* 206/1 (*index*) Hybridization of ribonuclease. **1972** *Arch. Biochem. & Biophysics* CL. 407 (*title*) Hybridization of rabbit muscle and liver phosphofructokinases. **1972** W. V. Brown *Textbk. Cytogenetics* ii. 13/1 The biochemical techniques of nucleic acid hybridization, either DNA-DNA or RNA-DNA.., have recently been providing new and unexpected understanding about the DNA sequences in the nucleus.

e. *Cytology.* The fusion, by artificial means or in artificial cultures, of two somatic cells of different karyotypes to form a hybrid cell containing the nuclear material of both.

1961 *Nature* 13 May 653/2 (*heading*) Karyological demonstration of hybridization of mammalian cells *in vitro*. **1965** *Proc. Nat. Acad. Sci.* LIII. 1040 Hybridization of somatic cells *in vitro*..has since been shown to occur in mixed cultures of many different pairs of cultured mouse cells. **1970** *Nature* 18 Apr. 280/2 Somatic cell hybridization is a potentially useful technique for the introduction of genetic variability into plant species. **1970** *McGraw-Hill Yearbk. Sci. & Technol.* 214/2 Whereas these first experiments involved two closely related transformed mouse cell lines,..somatic hybridization can occur between cells derived from different species, such as mouse and man.

2. *fig.*

1960 E. R. Goodman in J. A. Fishman *Readings Sociol. of Lang.* (1968) 733 One should add that the idea of 'hybridization' or fusion of languages, which Stalin continued to use, was Marrist in origin. **1964** M. McLuhan *Understanding Media* v. 48 The crossings or hybridizations of the media release great new force and energy as by fission or fusion. **1971** *Farmer & Stockbreeder* 23 Feb. 3/1 We shall be the old Stockbreeder Redivivus: not dead, not reborn, just rejuvenated. Or, if you prefer it, reinvigorated by hybridization; for we join forces with the *British Farmer*, the journal of the NFU.

hybridize, *v.* Add: **1. c.** *Physical Chem.* To combine (atomic orbitals) mathematically so as to obtain hybrid orbitals.

1933 [implied in *hybridized *ppl. a.*]. **1939** L. Pauling *Nature Chem. Bond* iii. 96 It is found on hybridizing these orbitals that four strong bonds directed to the corners of a square can be formed. **1962** Cotton & Wilkinson *Adv. Inorg. Chem.* iii. 67 The *s* and the three *p* orbitals are hybridized to produce four sp^3 hybrids.

d. *Biochem.* To cause (a subunit of a macromolecule) to combine *with*, or become attached *to*, a complementary subunit of the same or a very similar molecular species from a different source; to cause (two such subunits) to combine together.

1959 *Jrnl. Amer. Chem. Soc.* LXXXI. 3169/1 When labelled and unlabelled hemoglobins are hybridized, the hybrids contain both labelled and unlabelled chains. **1965** *Jrnl. Molecular Biol.* XII. 829 The procedure involves immobilizing denatured DNA on nitrocellulose membrane filters, hybridizing complementary RNA to the membrane-fixed DNA, and eliminating RNA 'noise'. **1966** Lehmann & Huntsman *Man's Haemoglobins* xxi. 226 When canine haemoglobin and human haemoglobin are hybridised together, the new hybrids that result differ considerably in their mobility. **1969** *Nature* 10 May 573/1, 100 μg of RNA was hybridized with 50 μg of DNA. **1972** *Arch. Biochem. & Biophysics* CL. 407 Phosphofructokinases from rabbit muscle and rabbit liver were hybridized by dissociation at low pH followed by recombination at neutrality. **1972** S. L. Wolfe *Biol. Cell* ix. 199/1 The best evidence that transcription is asymmetric comes from experiments in which RNA is hybridized with its DNA template... The RNA will form hybrid double helices with no more than 50 percent of its template DNA... Thus the RNA is complementary to only one of the two DNA strands, and only one..serves as a template for RNA transcription.

2. c. *Biochem.* Of a protein or nucleic acid: to exchange complementary subunits in hybridization (sense *1 d). Const. *with* (or *to*).

1962 *Science* 21 Dec. 1331/2 Their sequences are unique, since they hybridize readily only to homologous DNA. **1965** *Jrnl. Molecular Biol.* XII. 830 RNA molecules possessing an extensive secondary structure will not hybridize until their own melting temperature is approached. **1969** *Times* 16 May 14/7 Each kind of RNA chemically recombines or hybridizes with the DNA segment off which it was copied. **1971** *Biochemistry* (Easton, Pa.) X. 3509/2 Hexokinases A and B can hybridize in 0·1 M sodium phosphate solution at pH values near 8. **1972** S. L. Wolfe *Biol. Cell* ix. 206/2 For example, rRNA from *Drosophila* will hybridize to some degree with chick DNA or vice versa, even though there are wide differences in base composition of the total DNA complement from the two species.

3. *fig.* (*trans.*).

1964 M. McLuhan *Understanding Media* v. 50 With literacy now about to hybridize the cultures of the Chinese, the Indians, and the Africans. **1971** *Nature* 24 Sept. 241/2 In the more remote future it may well be, when further improvements in the combustion intensity of low grade fuel mixtures are required, that they will be achieved by hybridizing the high and low temperature branches.

hybridized (hai·bridaizd), *ppl. a.* [f. prec. + -ed[1].] Obtained by hybridization; hybrid (in various senses).

1859 [see Hybridize *v.* 1]. **1926** G. W. Tyrrell *Princ. Petrol.* viii. 165 Wholesale enclosure of Dalradian quartzites and mica-schists within a large mass of norite..has led to the formation of a zone of hybridised or contaminated rocks, full of xenoliths in all stages of digestion. **1933** *Jrnl. Chem. Physics* I. 502 The valence orbital.. could be of the strongly hybridized type. **1955** [see *fine art 2]. **1959** G. E. W. Wolstenholme et al. *Significant Trends Med. Res.* (Ciba Found. Symp.) 7 The N-terminal residues of the hybridized hemoglobin S were labelled with Sanger's reagent. **1968** [see *hybrid *sb.* and *a.* A. 2 c]. **1969** *Nature* 10 May 573/2 The rat from which the hybridized RNA was obtained.

hybris (hai·bris). [ad. Gr. ὕβρις.] = *hubris.

1920 *Public Opinion* 27 Aug. 195/2 During one of these the oppressor, possessed of place and power, imagined in his hybris, that he might extend his arm across the ocean. **1929** *Encycl. Brit.* XXII. 53/1 Themis is the servant or companion of Zeus... Her opposite is Hybris (ὕβρις), insolent encroachment upon the rights of others. **1949** *Horizon* Aug. 87 Hybris means believing that you *are* a god, i.e., that you cannot suffer; pride means a defiant attempt to *become* a god. **1969** *Commonweal* 22 Aug. 524 America, like all earlier empires, is going to march to the brink of *hybris* and plunge in.

hydathode (hai·dăþōud). *Bot.* [a. G. *hydathode* (G. Haberlandt 1894, in *Sitzungsber. Akad. Wiss. Wien* CIII. I. 494), f. Gr. ὑδατ-, ὕδωρ water + ὁδός way, path.] A pore or gland which discharges water from the leaf of a plant.

1895 *Jrnl. R. Microsc. Soc.* 333 Under the name *hydathode*, Prof. G. Haberlandt designates those organs, which are frequently found in the leaves of tropical and other plants, especially designed for the storing up or excretion of water. *Ibid.*, A very simple type of hydathode occurs in the leaves of grasses. **1897** J. C. Willis *Man. Flowering Plants* I. 116 Water-pores or hydathodes are openings, resembling stomata, upon leaves or elsewhere, through which the plant excretes water. **1914** M. Drummond tr. *Haberlandt's Physiol. Plant Anat.* x. 487 Many of our native plants are provided with organs which secrete water in the liquid form. Such hydathodes are even more widely distributed among plants inhabiting the humid tropics. **1931** E. C. Miller *Plant Physiol.* vii. 390 The physiological significance of the hydathodes and of the loss of water by guttation is not definitely known. **1953** K. Esau *Plant Anat.* xvi. 433 Hydathodes are structures that discharge water from the interior of the leaf to its surface. **1967** C. D. Sculthorpe *Biol. Aquatic Vasc. Plants* iv. 90 The possibility remains that the hydathodes [of water plants] are functionless relict structures.

hydatidiform, *a.* (s.v. Hydatid *sb.* (*a.*)). Add: *hydatidiform mole*, a uterine mole (Mole *sb.*[5]) formed by the proliferation and distension of the chorionic villi; also, the condition of having such a mole in the uterus.

1859 *Lancet* 15 Oct. 397/2 A case in which a specimen of the hydatidiform mole was expelled from the uterus seven months after the birth of a first child. **1971** T. J. Deeley *Gynaecol. Cancer* xv. 228 The incidence of hydatidiform mole is about 1:700–1:3,000 known pregnancies.

hydatidosis (haidătidōu·sis). *Path.* [f. Hydatid *sb.* (*a.*) + -osis.] A pathological condition resulting from infestation with tapeworm hydatids.

1925 *Med. Jrnl. Austral.* 24 Oct. 502/1 The whole abdominal cavity may be packed with cysts of all sizes—a condition which has been called hydatidosis. **1966** *Lancet* 24 Dec. 1422/1 Subjects with pulmonary hydatidosis.. were tested.

Hyde (haid). Name of the evil personality assumed by Dr. Jekyll in R. L. Stevenson's story, 'Strange Case of Dr. Jekyll and Mr. Hyde' (1886): used allusively in reference to the evil side of a person's character. (Cf. *Jekyll.)

1887 *Puck* (U.S.) XXII. 188 Is that you, Livingston?.. No, m'dearsh, it'sh Doct' Hyde. **1915** 'I. Hay' *First Hundred Thousand* xiv. 192 But we encountered surprisingly few Hydes. Nearly all were Jekylls. **1960** *Encounter* Sept. 9/1 Normally we let Hyde loose only when for some reason he is socially acceptable. The Saturnalia existed to satisfy his demands. **1967** V. Nabokov *Speak, Memory* (ed. 2) viii. 165 One ribald and agile boy (could it be I after all—the Hyde of my Jekyll?) managed to silhouette his foot.

Hyde Park (hai·d pā·ɹk). The name of a park in central London, of which a part (known as Speakers' Corner) is traditionally the scene of 'soap-box' oratory, used allusively (freq. *attrib.*) of the type of speaker, oratory, etc., found there. So **Hyde Pa·rkian** *a.*, having the quality (of voice) of a Hyde Park orator.

1892 G. B. Shaw *Let.* 21 Apr. (1965) 337 Her voice has become much more powerful—quite Hyde Parkian in its pedal notes. **1897** *London Handbk.* 143/1 They [*sc.* 'contentious Jews'] pose as martyrs in the cause of liberty, and certainly any man well deserves such an honourable title who will listen unflinchingly to two hours of Hyde Park oratory. **1912** C. Mackenzie *Carnival* xii. 132 Love..was as incredible to her as..to a Hyde Park materialist. **1914** 'I. Hay' *Knight on Wheels* (ed. 2) xix. 186 This is a peculiarity of the Hyde Park orator. Set him on his legs, and in ten minutes he has wandered..from the point. **1934** 'G. Orwell' *Burmese Days* xvii. 245 He told Flory 'not to start talking like a damned Hyde Park agitator', and then read him a snappish little sermon. **1937** *Discovery* Aug. 254/1 Each having something to say like so many Hyde Park orators. **1965** C. Fremlin *Jealous One* xi. 85 Everyone..competing, like Hyde Park orators, for an audience for their particular problem. **1966** *Listener* 24 Feb. 287/2 A few jibes against the Christian Church more reminiscent of Hyde Park oratory than of Voltaire and Gibbon. **1972** 'E. Lathen' *The Longer the Thread* xi. 106 A small Hyde Park seemed to have sprung up on the green. A whole medley of impassioned orators was in full voice.

hydnocarpic (hidno,kā·ɹpik), *a. Chem.* [f. next + -ic.] *hydnocarpic acid*: a crystalline alicyclic acid, $C_5H_7{\cdot}(CH_2)_{10}{\cdot}COOH$, which in the form of its glycerides is one of the chief constituents of chaulmoogra oil and hydnocarpus oil.

1905 Power & Barrowcliff in *Jrnl. Chem. Soc.* LXXXVII. I. 885 The acids obtained from the respective Hydnocarpus oils consist chiefly of chaulmoogric acid and a lower homologue of the same series, the latter having been isolated..also from chaulmoogra oil. This new acid has the formula $C_{16}H_{28}O_2$, and is designated hydnocarpic acid. **1960** K. S. Markley *Fatty Acids* (ed. 2) I. ii. 215 All these oils contain one or more fatty acids having a terminal cyclopentenyl ring, especially chaulmoogric, hydnocarpic, and gorlic acid, which account for 80 to 90% or more of the total fatty acids of these oils. *Ibid.* 217 *dl*-Hydnocarpic acid has been synthesized.

Hence **hydnoca·rpate**, a salt or ester of hydnocarpic acid.

1905 *Jrnl. Chem. Soc.* LXXVII. I. 890 Ethyl Hydnocarpate, $C_{15}H_{27}{\cdot}CO_2Et$. **1927** *Proc. R. Soc. Med.* XX. 999 The veins become blocked by sodium hydnocarpate in some cases. **1953** R. A. Raphael in E. H. Rodd *Chem. Carbon Compounds* IIA. iv. 119 When ethyl hydnocarpate is reduced by the Bouveault–Blanc process, an alcohol is obtained which can be converted..to chaulmoogric acid.

hydnocarpus (hidnokā·ɹpŏs). *Bot.* [mod.L. (J. Gaertner *De Fructibus et Seminibus Plantarum* (1788) I. 288), f. Gr. ὕδνον truffle + καρπός fruit, from the resemblance of the fruit to a truffle.] **a.** A tree of the genus so called, belonging to the family Flacourtiaceæ and native to tropical Asia.

1928 *Daily Express* 20 Feb. 2/1 Hydnocarpus oil is obtained from the dried fruit of the hydnocarpus tree. **1953** N. L. Bor *Man. Indian Forest Bot.* 145 In *Hydnocarpus* the fruits are produced on the branches.

b. *attrib.*, as *hydnocarpus therapy*; **hydnocarpus oil,** a yellowish oil or soft cream-coloured fat obtained from the seeds of trees of the genus *Hydnocarpus*, esp. *H. wightiana* and *H. anthelminthica*, and used, formerly extensively, in the treatment of leprosy.

1905 Hydnocarpus oils [see *hydnocarpic *a.*]. **1927** *Proc. R. Soc. Med.* XX. 1011 Since early in 1921 we have used the preparations of chaulmoogra and hydnocarpus oils in our Hospital for Lepers, at Dichpalli, Hyderabad State. **1967** *Martindale's Extra Pharmacopoeia* (ed. 25) 494/2 Although largely replaced by the sulphones, hydnocarpus oil is still employed in the treatment of leprosy, especially in endemic areas in the East where it is cheap and readily available. **1964** R. G. Cochrane *Leprosy* (ed. 2) xxi. 377, I do not hold with many workers that the day of hydnocarpus therapy has disappeared.

hydra. 7. b. *hydra-headed* adj. (later examples).

1842 *Ainsworth's Mag.* II. 43 The Puff Literary is hydra-headed. **1899** *Daily News* 8 Feb. 7/5 The hydra-headed leadership of the Irish party. **1963** *Daily Tel.* 20 Nov. 14/2 The hydra-headed challenge of London's mounting traffic congestion.

hydralazine (haidræ·lăzīn). *Pharm.* Also **hydrallazine.** [f. hydra*zinophtha*lazine, f. *hydrazino- + *phthalazine.] A sympatholytic drug, $C_8H_5N_2{\cdot}NH{\cdot}NH_2$, used in the form of the hydrochloride, a white crystalline powder, in the treatment of hypertension.

1952 *Jrnl. Amer. Med. Assoc.* 1 Nov. 861/2 Hydrallazine Hydrochloride for 1-hydrazinophthalazine hydrochloride. **1953** *Jrnl. Pharmacol. & Exper. Therap.* CIX. 182 Hydrallazine..stimulates the heart and increases cardiac output. **1963** *Brit. Pharmacopœia* p. xxviii, British Pharmacopœia... Hydrallazine hydrochloride. International Pharmacopoeia... Hydralazine hydrochloride. **1970** A. Goth *Med. Pharmacol.* (ed. 5) xiv. 157 Hydralazine (1-hydrazinophthalazine; Apresoline) is a potent antihypertensive drug, but it can produce serious toxic reactions.

hydramnios (haidræ·mniŏs). *Path.* Also **hydramnion.** [f. Hydr- + Amnios, Amnion.] Excessive accumulation of amniotic fluid during pregnancy.

1838 F. Churchill *Outl. Princ. Dis. Females* 145 'Hydramnios', or excess of liquor amnii.., is said to result from a general serous diathesis. **1885** W. T. Lusk *Sci. & Art Midwifery* (ed. 3) xv. 288 The causes of hydramnion are hardly to be found in a single morbid condition. **1961** *Lancet* 30 Sept. 742/2 She came into hospital three times during her eighth pregnancy with..mild hydramnios.

hydrapulper (hai·drăpʌlpəɹ). *Paper-making.* Also **Hydrapulper.** [Irreg. f. Gr. ὑδρ(ο-, combining form of ὕδωρ water + Pulper.] A large vessel with a set of motor-driven rotating vanes at the bottom, designed to break up the fibres of wood pulp or other paper stock in water.

1941 *Techn. Pap. Addr. Techn. Assoc. Pulp Pap. Ind.* XXIV. 384/2 The hydrapulper is simply a cylindrical open-top tank with a dished bottom at the center of which is a heavy impeller. **1951** J. N. Stephenson *Pulp & Paper Manuf.* II. i. 67 Straw bales..are fed as rapidly as possible into a Hydrapulper..containing the required amount of chemical in hot water. **1966** *Paper Technology* VII. II. 135 These rejects are in a suspension and could possibly be extracted from the bottom of the hydrapulper tub. **1967** E. Chambers *Photolitho-Offset* xvi. 244 This acidity in uncoated paper can be traced to aluminium sulphate, which is added in the beater or hydrapulper during paper manufacture.

hydrarch (hai·drāɹk), *a. Ecol.* [f. Gr. ὕδωρ, ὑδρ- water + ἀρχή beginning.] Of a succession of plant communities: having its origin in a watery habitat.

1913 W. S. Cooper in *Bot. Gaz.* LV. 11 The terms xerarch and hydrarch are here used for the first time, for the purpose of indicating a natural and important classification of plant successions. The former is applied to those successions which, having their origin in xerophytic habitats, such as rock shores, beaches, and cliffs, become more and more mesophytic in their successive stages; the latter to those which, originating in hydrophytic habitats such as lakes and ponds, also progress towards mesophytism. **1929** Weaver & Clements *Plant Ecol.* iv. 55 Successions beginning in ponds, lakes, marshes, or elsewhere in water are termed hydrarch. **1960** N. Polunin *Introd. Plant Geogr.* xi. 324 They [*sc.* the seres] are distinguished as 'hydrarch', 'mesarch', or 'xerarch', according to whether their initiation is under damp, median, or dry conditions.

hydrase (hai·drēiz, -ēis). *Biochem.* [f. Hydr- + *-ase.] Any enzyme which catalyses an addition reaction between water and a substrate or the reverse process.

1943 Sumner & Somers *Chem. & Methods of Enzymes* IV. xvii. 296 The hydrases add water to organic compounds without causing a splitting. **1961** P. D. Boyer et al. *Enzymes* (ed. 2) V. xxviii. 455 Among enzymes catalyzing the addition of groups to double bonds (and the reverse reaction), the hydrases (or dehydrases) form the largest group.

hydrastinine (haidræ·stinīn). *Chem.* [ad. G. *hydrastinin* (Freund & Will 1887, in *Ber. d. Deut. Chem. Ges.* XX. 88), f. Hydrastin(e + -ine[5].] A synthetic alkaloid, $C_{11}H_{13}NO_3$, derived from hydrastine and sometimes employed in the form of the hydrochloride to control uterine bleeding.

1887 *Jrnl. Chem. Soc.* LII. I. 383 [Freund and Will] have examined the base hydrastinine, $C_{11}H_{11}NO_2 + H_2O$, obtained together with opianic acid where hydrastine is treated with oxidising agents. **1929** *Encycl. Brit.* XI. 961/1 The hydrolytic product, hydrastinine, is an important drug, being used as an internal styptic. **1970** S. W. Pelletier *Chem. Alkaloids* iii. 65 Oxidative hydrolysis of hydrastine with dilute nitric acid results in the formation of two fragments, opianic acid and hydrastinine.

hydrastis (haidræ·stis). [mod.L. (J. Ellis in Linnæus *Systema Naturæ* (ed. 10, 1759) II. 1088): etym. unkn.] The dried rhizome and roots of the herb golden seal or yellow root (*Hydrastis canadensis*), or an extract or tincture of them, formerly used medicinally as a bitter stomachic and to control uterine bleeding.

1865 Wood & Bache *Dispensatory U.S.A.* (ed. 12) 459 Hydrastis might probably be advantageously prepared in the form of a fluid extract. **1882** R. Bentley *Man. Bot.* (ed. 4) 425 Hydrastis is..used by the Indians of the Western States of North America to dye various shades of yellow. **1908** *Practitioner* Jan. 96 Styptol and hydrastis.. are recommended for arresting haemorrhage from the non-pregnant organ. **1951** A. Grollman *Pharmacol. & Therapeutics* 757 Color may also be imparted by the addition of..Hydrastis Tincture (N.F.) which..is soluble in alcohol and hydro-alcoholic preparations giving a yellow color. **1967** *Martindale's Extra Pharmacopoeia* (ed. 25) 592 Hydrastis..in the treatment of post-partum haemorrhage..is much inferior to ergot.

hydratable (haidrēi·tăb'l), *a.* [f. Hydrat(e *v.* + -able.] Capable of becoming hydrated.

1953 J. Davidsohn et al. *Soap Manuf.* I. xiii. 220 Animal fats..are for the most part substantially free of phosphatides or other hydratable substances. **1956** *Soil Sci.* LXXXII. 198 Hydratable products may be formed as the clay minerals decompose.

hydratase (hai·drătēiz, -ēis). *Biochem.* [f. Hydrat(e *v.* + *-ase.] = *hydrase.

1922 *Chem. Abstr.* XVI. 936 The enzymes can be divided into hydratases, hydrolases and oxido-reductases. **1953** *Adv. Enzymol.* XIV. 237 Hence the enzyme hitherto called 'fumarase' should be renamed fumaric hydratase.

hydrate, *v.* Add: **b.** *intr.* To undergo hydration, to become combined with water.

1909 in Webster. **1921** J. R. Partington *Text-bk. Inorg. Chem.* xl. 846 The residue of anhydrous $CaSO_4$ rapidly takes up water, but if the heating has been more intense the residue hydrates only very slowly. **1947** R. H. Bogue *Chem. Portland Cement* xxv. 435 Complete hydration..was obtained by mixing the compounds with water to form a plastic paste, allowing the pastes to hydrate in sealed containers for a month, [etc.]. **1962** Harris & Gruber in A. Pirie *Lens Metabolism Rel. Cataract* 382 If the block is sufficient, more sodium is gained than potassium lost and the lens hydrates in a predictable manner.

hydratuba (hai:drătiū·bă). [f. Hydra II. 6 + Tuba[1], formerly *Hydra tuba*, when the organism was thought to be a species of *Hydra*; in reference to the shape of the larva.] The polyp-like larval stage of a jellyfish of the class Scyphozoa; a scyphistoma.

1847 J. G. Dalyell *Rare & Remarkable Animals Scotl.* I. iii. 76 (*heading*) Hydra tuba, the Trumpet Polypus. *Ibid.* 77 The body of the *hydra tuba* is a hollow cone five lines in length, thick and fleshy. **1898** A. Sedgwick *Student's Text-bk. Zool.* I. iv. 159 In the *Ephyrinæ* the development is generally accompanied by an alternation

of generations; the asexual generations being represented by the *Scyphistoma* (*Hydra tuba*) and *Strobila*; but in exceptional cases it is direct (*Pelagia*). **1901** Shipley & MacBride *Zool.* iii. 64 Each [planula] fixes itself and develops into a little polyp, called a Hydratuba, not unlike a *Hydra* in appearance. **1927** *Glasgow Herald* 5 Nov. 6/2 This [*sc.* the larva] settles as a polyp-like form, the hydratuba, and this, by a kind of budding, gives origin to little saucer-like discs which grow into medusæ. **1932** Borradaile & Potts *Invertebrata* v. 155 At certain seasons the whole hydratuba is segmented by transverse horizontal furrows. **1951** *Microscope* VIII. 193 (*title*) Planula, hydratuba and ephyra stages of the common jellyfish.

hydraulic, *a.* and *sb.* Add: **A.** *adj.* **1.** Further special collocations.

hydraulic engineering (see quots.); so *hydraulic engineer*; *hydraulic gradient*: (a line representing) the variation along a pipe or channel of the head due to elevation and pressure in a liquid flowing along it; *hydraulic jump*: an abrupt change from a fast, shallow flow to a slower, deeper flow at some point in a liquid flowing in an open channel; *hydraulic radius* or *mean depth*: a length equal to the cross-sectional area of the liquid in a channel or pipe divided by the length of the wetted perimeter.

1838 *Penny Cycl.* XII. 382/1 Besides the construction of harbours for ships, the formation of the aqueducts which supplied the cities with water must have constituted an important part of the duties of the hydraulic engineer among the antients. **1916** J. Park *Textbk. Pract. Hydraulics* p. x, A valuable work of reference for all hydraulic engineers. **1835** *Rep. Brit. Assoc. Adv. Sci. 1834* 473 Practical works in hydraulic engineering of great magnitude and extent have been carried on in England. **1858** G. R. Burnell *Rudiments Hydraulic Engin.* I. i. 1 Hydraulic Engineering will be considered to include.. questions connected with building in water. *Ibid.* 2 Hydraulic engineering is principally confined to the operations in which water acts as an incompressible fluid. **1940** *Chambers's Techn. Dict.* 427/1 *Hydraulic engineering*, that branch of engineering chiefly concerned in the design and production of hydraulic machinery, pumping plants, pipelines, etc. **1956** A. H. Compton *Atomic Quest* II. 106 His.. degree in hydraulic engineering at the University of Iowa. **1881** *Encycl. Brit.* XII. 484/2 Hence [the line] CD is termed the virtual slope or hydraulic gradient of the pipe. **1935** A. J. Martin *Work of Sanitary Engin.* viii. 92 If a series of open-ended vertical pipes were carried up from the main at intervals, the water would rise in each of them up to the line of the hydraulic gradient. **1951** W. L. Russell *Princ. Petroleum Geol.* xii. 212 By using..high hydraulic gradients, oil globules and gas bubbles may be forced through these sandstones. **1922** H. E. Babbitt *Sewerage* iv. 74 The hydraulic jump will occur when a high velocity of flow is interrupted by an obstruction in the channel, by a change in grade of the invert, or the approach of the velocity to the 'critical' velocity. **1969** Chia-Shun Yih *Fluid Mech.* v. 226 Tidal bores observed in estuaries are hydraulic jumps propagating against flowing water. **1797** *Encycl. Brit.* XVIII. 891/1 Column 1..contains the hydraulic mean depths of any conduit in inches... The column is continued to 100 inches, which is fully equal to the hydraulic mean depth of any canal. **1829** *Nat. Philos.* (Libr. Useful Knowl.) I. vi. 5/2 Suppose the whole quantity of water to be spread on a horizontal surface, equal in extent to the bottom and sides of the river, when the height at which the water would so stand is called the hydraulic mean depth. **1930** *Engineering* 7 Feb. 180/1 By means of the following table, maximum scour can be calculated from calculated values of the hydraulic mean depth. **1876** L. d'A. Jackson tr. *Kutter's Mean Velocity of Discharge of Rivers* i. 2, *r* is the mean hydraulic radius, or the quotient of the water section by the wetted perimeter. **1948** D. W. Taylor *Fund. Soil Mech.* vi. 109 The ratio of volume to surface area of flow channel may be used as an alternative definition of hydraulic radius.

2. (The term is used in connection with liquids other than water.)

hydraulic buffer, a device for checking the recoil of a mounted gun, consisting of a piston in a cylinder filled with liquid; *hydraulic intensifier*, a device for obtaining an increase in pressure in a hydraulic system, usu. comprising two cylinders of different diameters containing pistons that are joined to one another.

1871 G. E. Voyle *Dict. Artillery Terms* (ed. 2) 45/1 The Hydraulic Buffer. **1876** Voyle & Stevenson *Mil. Dict.* (ed. 3) 54/2 Where guns are mounted on carriages and platforms fitted for hydraulic buffers, the buffer will invariably be kept on the platform filled with the proper quantity of oil. **1899** G. D. Hiscox *Mech. Movements* VII. 153 Hydraulic intensifier.—High pressure obtained from low pressure by differential pistons. **1959** *Chambers's Encycl.* VI. 646/1 A hydraulic buffer or brake consists of a cylinder filled with liquid, a piston and a piston rod. **1962** Walshaw & Jobson *Mech. Fluids* ii. 32 A 100 ton press with a 12 in. stroke is to be operated from a main in which the pressure is 40 lbf/in[2] gauge, via a hydraulic intensifier with a 4 ft stroke.

b. *hydraulic brake*: (*a*) a brake that utilizes the resistance to motion experienced by a piston or rotor in a chamber full of liquid; also, a hydraulic buffer; (*b*) a brake that operates by means of friction but is actuated hydraulically; (*c*) a dynamometer that works on the principle of the hydraulic brake (sense (*a*)).

(*a*) **1874** *Engineering* 11 Sept. 194/3 (*heading*) Hydraulic brakes. **1894** J. A. Ewing *Steam-Engine* ix. 269 Many governors are furnished with a dash-pot, which is a hydraulic or pneumatic brake, consisting of a piston connected to the governor, working loosely in a cylinder which is filled with oil or with water. **1902** *Encycl. Brit.* XXXI. 897/1 The buffer-stop to be seen in terminal railway stations, and the hydraulic brakes of quick-firing guns, are examples. **1959** [see sense 2 above]. **1962** D. J. Myatt *Machine Design* i. 151 Because braking torque is a function of speed of rotation of the brake rotor, electric and hydraulic brakes should not be depended upon to hold a load stationary.

(*b*) **1875** *Engineering* 8 Jan. 29/1 The hydraulic brake comprises a pump, a cistern, and an accumulator for collecting and storing the power, a regulator, and apparatus for applying that power in retarding the speed of the train. **1876** J. W. Barry *Railway Appliances* vii. 286 In the hydraulic break the pressure on the pistons is derived from the pressure of a small hydraulic accumulator. **1924** Wright & Smith *Automotive Construction & Operation* xiv. 331 Several cars have adopted the hydraulic-brake system for all four wheels. **1963** D. V. W. Francis *Morris Minor* i. 11/1 Lockheed hydraulic brakes operate on 7-in. brake drums.

(*c*) **1890** *Min. Proc. Inst. Civil Engin.* XCIX. 169 At the same meeting, Mr. William Froude gave an account of his hydraulic brake, for measuring the power of large engines. **1925** A. W. Judge *Automobile Engines* I. vi. 170 The Froude hydraulic brake..is now largely employed for automobile engine tests. In this case the power is absorbed by hydraulic resistance.

c. Of a liquid: used, or suitable for use, in hydraulic brakes or other hydraulic equipment. (Not used predicatively.)

1941 E. Molloy *Hydraulic Equipment* 86 A valve on the master-brake cylinder..allows the hydraulic fluid to escape, under pressure, to the brake cylinder in the wheel. **1951** *Adv. Chem. Ser.* V. 241 Hydraulic fluids..include those made from petroleum, synthetic oils, and aqueous solutions containing antifreeze and rust inhibitors. **1967** E. R. Braithwaite *Lubrication* iii. 161 Fire-resistant hydraulic oils represent a class of lubricants in which there is increasing interest. **1971** *Drive* New Year 33/2 It is advisable to change hydraulic fluid every eighteen months or 24,000 miles.

3. (Earlier examples.)

1829 *Arcana of Sci. & Art* 246 When the bridge of Louis XVI was constructed, much advantage was derived from mixing with the hydraulic mortar which was used, a portion of the clinkers..which had passed through the grates of the glass bottle furnace, at Meudon. **1843** *Civil Engin. & Arch. Jrnl.* VI. 157/2 Hydraulic lime will harden in a very short time. *Ibid.*, Roman cement, the most hydraulic of all mortars.

hydraulic (haidrǫ·lik), *v. U.S.* [f. *hydraulic* in *hydraulic mining*.] *trans.* To work or obtain by the methods of hydraulic mining. Hence **hydrau·lic(k)ing** *vbl. sb.*

1868 *Rep. J. Ross Browne Mineral Resources* (U.S. Treas. Dept.) 94 They can hydraulic away about 300 feet along the face of their Claim, but beyond that the hill is too deep to pay for piping. **1880** G. T. Ingham *Digging Gold* ix. 243 Two steam pumps have been..forcing water from French Creek up to these dry diggings for hydraulicing. **1892** C. F. Lummis *Tramp across Continent* viii. 122 To run a fifteen-mile pipe-line from the Sandias to Golden, and thus bring water to hydraulic the enormous areas of gold-bearing gravel. **1904** J. Lynch *Three Yrs. Klondike* (1967) ix. 127 Those grounds can only be dredged or hydraulicked. **1908** W. R. Crane *Gold & Silver* v. 379 In 1903 the Crown Mountain Mining Company..hydraulicked the material in sluices to a 60-stamp mill. **1928** W. A. Chalfant *Outposts Civilization* 42 From which it was said the owner hydraulicked $90,000 worth of gold. **1965** *Mineral Facts & Problems* (U.S. Bureau of Mines) 231 Some kaolin deposits are mined by hydraulicking.

hydraulicity. (Examples.)

1843 *Civil Engin. & Arch. Jrnl.* VI. 157/2 In constructions of this kind..the lime should possess some degree of hydraulicity, in order that it may harden before the injurious action of frost comes upon it. **1902** J. Black *Illustr. Carpenter & Builder Ser.: Plastering* iii. 46 The hydraulicity is greatest when the limestone contains thirty per cent. of clay. **1959** *Chambers's Encycl.* III. 320/2 The calcium oxide..retains all these impurities and the proportion in which they are present greatly affects the properties of the lime, notably its hydraulicity.

hydraulus (haidrǫ·lŏs). *Mus.* Also **hydraulis.** [L., f. Gr. ὕδραυλος hydraulic organ (cf. αὐλός pipe).] A type of water organ popular in classical times; = *hydraulic organ* (Hydraulic *a.* 2).

1874 W. Chappell *Hist. Mus.* I. xiii. 328 Philōn defines..a kind of 'syrinx played by the hands, which we call hydraulis'. **1903** C. F. A. Williams *Story of Organ* ii. 35 Sylvester made an organ that was played by 'warm water'; this was..the hydraulus. **1934** *Times Educ. Suppl.* 27 Jan. p. iv/2 [The Albert Hall organ] has..been heard as an accompaniment to choral singing, and, like the hydraulus in the Roman amphitheatres of old, to gladiatorial spectacles. **1970** *Oxf. Compan. Mus.* (ed. 10) 496/2 Hydraulus, hydraulis, or water organ. This is the most ancient form of organ known. *Ibid.*, In general appearance the hydraulus is like any small organ of today.

hydrazide (hai·drăzaid). *Chem.* [f. Hydraz(ine + -ide.] Any compound which may be represented as $R{\cdot}CO{\cdot}NH{\cdot}NH_2$; also, any derivative of such a compound in which univalent radicals replace one or more of the hydrogen atoms.

1888 *Jrnl. Chem. Soc.* LIV. 686 Malonyl hydrazide, $CH_2(CO{\cdot}NH{\cdot}NHPh)_2$,..crystallises..in white leaflets. **1955** Z. E. Jolles in E. H. Rodd *Chem. Carbon Compounds* IIIA. vi. 320 Hydrazo compounds and hydrazides yield the corresponding azo derivatives by oxidation.

hydrazine. Add: Now used as a rocket propellant.

1950 *Sci. News* XV. 77 One class [of propellants], called hypergols, react as soon as they come together... Such pairs are..hydrazine hydrate and hydrogen peroxide. **1955** *Sci. News Let.* 1 Oct. 212/1 Monopropellants, whose chemical structures carry both fuel and oxidizer, were listed by him as ethylene oxide, hydrazine, hydrogen peroxide and nitromethane. **1960** *Aeroplane* XCIX. 638/2 Achieving a lunar orbit would depend largely on the hydrazine engine which has yet to be tested in flight. **1969** *Times* 3 June (Suppl.) p. iii/7 Aerozine 50—an equal blend of hydrazine and unsymmetrical dimethyl hydrazine.

hydrazinium (haidrăzi·niəm). *Chem.* [f. Hydrazin(e + *-ium b.] **a.** The ion $H_2N{-}NH_3^+$ derived from hydrazine (or a substituted ion derived from this in which univalent radicals replace one or more of the hydrogen atoms). Usu. *attrib.* **b.** The less common ion $H_3N^+{-}NH_3^+$ (or a substituted ion derived from this in which univalent radicals replace one or more of the hydrogen atoms). Usu. *attrib.*

The application of the terms *hydrazinium* and *hydrazonium* has been inconsistent. Some authors have favoured using one or other of them (with or without a distinguishing numeral) for both ions, whilst some prefer *hydrazinium* for $N_2H_6^{++}$ and *hydrazonium* for $N_2H_5^+$; others would abandon the names entirely. The International Union of Pure and Applied Chemistry recommends *hydrazinium(1+)*, *hydrazinium(2+)*, respectively, for the two ions (*Nomencl. Inorg. Chem.* (ed. 2, 1971), Rule 3.17).

1927 *Chem. Abstr.* XXI. 2672 These 'acid products' really contain hydrazinium salts. **1936** *Jrnl. Amer. Chem. Soc.* LVIII. 1606/2 The ionization constants for the hydrazinium ion as an acid and hydrazinium hydroxide as a base have been determined. **1949, 1954** [see *hydrazonium a, b]. **1955** *Chem. & Engin. News* 3 Jan. 70/3 A. C. Nixon..suggests that 'hydrazonium' be retained for $H_2NONH_3^+$ [*sic*] and 'hydrazinium' be applied to $^+H_3N{\cdot}NH_3^+$. **1957, 1966** [see *hydrazonium c]. **1964** J. W. Linnett *Electronic Struct. Molecules* iv. 64 The NN bond-length in hydrazine, $H_2N{-}NH_2$, is 1·47 Å whereas that in the hydrazinium ion, $H_3\overset{+}{N}{-}\overset{+}{N}H_3$, is 1·40 Å. **1966** Cotton & Wilkinson *Adv. Inorg. Chem.* (ed. 2) xii. 335 Two series of hydrazinium salts are obtainable, those of $N_2H_5^+$ and those of $N_2H_6^{2+}$.

hydrazino(-) (haidrăzī·no). *Chem.* A combining form of Hydrazine, denoting the univalent radical $-NH{\cdot}NH_2$ (or a substituted radical derived from this in which univalent radicals replace one or more of the hydrogen atoms); also used *attrib.* as an independent word.

1907 *Jrnl. Chem. Soc.* XCII. I. 880 The present work was undertaken to determine the conditions under which the hydroxyls of phenols can be displaced similarly by the hydrazino-group $\cdot NH{\cdot}NH_2$. **1919** *Chem. Abstr.* XIII. 1304 (*heading*) Hydrazino acids. *Ibid.*, Dihydrazine hydrazinodimalonic acid. **1952** E. H. Rodd *Chem. Carbon Compounds* I. xii. 841 Hydrazinoacetic acid, $NH_2NHCH_2CO_2H$. **1953** C. C. Clark *Hydrazine* iii. 35 Hydrogenation of hydrazones and azines to hydrazino compounds is possible in many cases. *Ibid.* 36 (*heading*) Hydrazino acids and esters.

hydrazo(-) (hai·drăzo). *Chem.* A combining form of Hydrazine, denoting the bivalent radical $-NH{\cdot}NH-$ (not linked to the same atom) or a substituted radical derived from this in which univalent radicals replace one or both of the hydrogen atoms; also used *attrib.* as an independent word.

1872 *Jrnl. Chem. Soc.* XXV. 694 It also unites with hydrogen, giving rise to hydrazophenylene, $C_{12}H_{10}N_2$. **1877** *Ibid.* I. 307 On boiling the hydrazo-compound with hydrochloric acid, the isomeric dibromobenzidine..is obtained. **1907** *Chem. Abstr.* I. 498 (*heading*) Hydrazo derivatives. **1955** [see *hydrazide]. **1958** Packer & Vaughan *Mod. Approach Org. Chem.* xviii. 608 Azoxybenzene gives hydrazobenzene but not azobenzene with alkaline reducing agents.

hydrazoate (haidrăzōu·ĕit). *Chem.* [f. Hydrazo(ic *a.* + -ate[4].] = *azide.

1910 [see *azide]. **1938** *Van Nostrand's Sci. Encycl.* 583/2 Hydrazoic acid reacts..with metals..to form azides or hydrazoates (or trinitrides).

hydrazone (hai·drăzōun). *Chem.* [ad. G. *hydrazon* (E. Fischer 1888, in *Ber. d. Deut. Chem. Ges.* XXI. 984), f. Hydraz(ine + -one.] Any compound which contains the group $=N{\cdot}NH_2$ (attached to a single carbon atom) and which is thus a condensation product of an aldehyde or ketone with hydrazine; also, any substituted derivative, $=N{\cdot}NHR$ or $=N{\cdot}NRR'$, of such a compound.

1888 *Jrnl. Chem. Soc.* LIV. 590 The name hydrazone is proposed for the compounds of hydrazine with aldehydes and ketones. **1938** Allen & Blatt in H. Gilman *Org. Chem.* I. vi. 569 Oximes, hydrazones, and phenylhydrazones are utilized primarily for isolating and identifying carbonyl compounds. **1970** *Nature* 5 Sept. 1048/1 Experiments..suggested that the hydrazone side-chain of rifampicin was essential for its antiviral activity.

hydrazonium (haidrăzōu·niəm). *Chem.* [f. Hydraz(ine + *-onium.] **a.** = *hydrazinium a. **b.** = *hydrazinium b.

See the note s.v. *hydrazinium.

[**1876** *Jrnl. Chem. Soc.* II. 528 The author [*sc.* E. Fischer]

has isolated a body of the formula $C_6H_5.N_2H_2(C_2H_5)(C_2H_5Br)$, which he names phenyldiethylhydrazoniumbromide.] **1949** *Chem. & Industry* 26 Feb. 134/2 The ions derived from hydrazine are also troublesome. The one most commonly encountered, $N_2H_5^+$, is called hydrazinium in the 1940 Rules but hydrazonium in Chemical Abstracts (1945). Since the ion is essentially a substituted ammonium, the latter is strictly the more correct, but neither affords a means of distinguishing $N_2H_5^+$ from $N_2H_6^{++}$, which occurs in a few compounds. The two names could be used in fact for this purpose, but the distinction is best achieved, without confusion, by writing
$N_2H_5^+$ Hydrazonium (I) or (+1)
$N_2H_6^{++}$ Hydrazonium (II) or (+2).
1953 C. C. CLARK *Hydrazine* iii. 36 The asymmetric dialkyl hydrazines are also strong bases. They react quantitatively with active alkyl chlorides to form water soluble hydrazonium salts: $R_2NNH_2 + R'X \rightarrow R_2NNH_2.R'X$. **1954** *Chem. & Engin. News* 6 Sept. 3548/3 As a name for the ion $H_2N.NH_3^+$ there has been some division of usage between 'hydrazonium' and 'hydrazinium'; and some have applied one or other of these names to the ion $^+H_3N.NH_3^+$ as well as to the ion of single charge. The newly revised inorganic rules propose to settle the question by naming $N_2H_5^+$ 'hydrazinium (1+)' and $N_2H_6^{2+}$ 'hydrazinium (2+)'. This does not seem to us an ideal solution, to give two different ions (which are not isomers) the same name except for an affixed numeral. Might not the name 'hydrazinium' be retained for $N_2H_5^+$ and some such name as 'hydrazidiinium' coined for $N_2H_6^{2+}$? **1955** [see *HYDRAZINIUM]. **1957** [see sense c].

c. [Cf. *HYDRAZONE.] The ion $R{=}N{\cdot}NH_3^+$ that results from the addition of a proton to a hydrazone; also, any substituted ion derived from this in which univalent radicals replace one or more of the hydrogen atoms.

1957 *Chem. Rev.* LVII. 1022 The cation having the formula $R_2\overset{+}{N}NR_2$ has been given many names. Emil Fischer called the first known representative of this type a 'hydrazonium bromide' but later referred to the same compound as an 'azonium bromide'. *Chemical Abstracts* has used the names 'azinium', 'azonium', 'hydrazinium', and, most commonly of late, 'hydrazonium' for the same type. As a logical extension of the recommendations recently made for naming $[H_3NNH_2]^+$, it is suggested that the name 'hydrazinium' be adopted... It is suggested here that the name 'hydrazonium' be restricted to salts of hydrazones (including quaternary salts). **1966** P. A. S. SMITH *Chem. Open-Chain Org. Nitrogen Compounds* II. ix. 149 Quaternary hydrazinium salts can also be obtained by the hydrolysis of quaternary hydrazonium salts ($R_3\overset{+}{N}{-}N{=}CR_2X^-$).

hydric (həi·drik), *a.*[2] *Ecol.* [f. Gr. ὕδωρ, ὑδρ- water + -IC.] Of a habitat: having a plentiful supply of water.

1926 COOPER & WEESE in *Ecology* VII. 390 In order..to provide terms which shall be applicable to both plants and animals, we suggest that the adjectives 'xerophytic', 'hydrophytic' and 'mesophytic' be entirely abandoned as useless and misleading. In their place we offer the terms 'xeric', 'hydric' and 'mesic', to be defined as follows: Xeric (hydric, mesic): characterized by or pertaining to conditions of scanty (abundant, medium) moisture supply. **1947** R. F. DAUBENMIRE *Plants & Environment* iii. 148 Hydric, xeric, and mesic are commonly encountered in ecologic literature...These adjectives, if used, should be applied only to habitats. **1960** N. POLUNIN *Introd. Plant Geogr.* xi. 328 Typically this mean [between two seres] is inhabited by mesophytes and is said to be 'mesic', though relatively xeric (dry) and hydric (damp) exceptions exist. **1968** R. F. DAUBENMIRE *Plant Communities* iii. 116 Bare areas may be classified according to their characteristic water relations: wet or hydric, as a pond bottom; dry or xeric, as a rock surface exposed to the sun; or intermediate mesic, as a glacial moraine.

hydridic (həidrəi·dik), *a. Chem.* [f. HYDRID(E + -IC.] Of an atom of hydrogen: having a negative charge (like the hydrogen in ionic hydrides).

1966 PHILLIPS & WILLIAMS *Inorg. Chem.* II. xxxiii. 565 Boron-hydrogen bonds generally react with Grignard reagents as if they contained hydridic hydrogen. **1966** K. M. MACKAY *Hydrogen Compounds Metallic Elem.* v. 155 The behaviour of hydrogen in the hydride complexes reflects a change from acidic hydrogen in the carbonyl hydrides to basic, hydridic, hydrogen in the other compounds. **1968** *Inorg. Chem.* VII. 1952/2 The nature of the hydrogens in boron hydrides varies from hydridic (or negative) in diborane(6) to protonic (or positive) in the moderately strong acid decaborane(14).

hydrion (həi·drəi,ən, həi·dri,ǫn). *Chem.* [Contraction of *hydr(ogen) ion*: see *-ION[3].] The hydrogen ion or proton.

1901 J. WALKER in *Chem. News* 4 Oct. 162 The proposals which I make for naming the positive and negative radicals of salts, acids, and bases considered as ions are as follows:—...If we use with Ostwald a dot to express unit positive charge, we have for some typical kations:—Hydrion..H·. Sodion..Na·. **1921** *Phil. Mag.* XLII. 455 There are only two elements, Sir Ernest Rutherford's proton (hydrion) and electron. **1930** *Nature* 20 Sept. 434/2 The hydrion concentration (*p*H) of the rain has been recorded. **1943** *Thorpe's Dict. Appl. Chem.* (ed. 4) VI. 381/2 There is a great excess of hydroxyl ions over hydrions. **1968** PASSMORE & ROBSON *Compan. Med. Stud.* I. p. xxxii (Index), Hydrion excretion, by kidney.

hydro[2] (həi·drou). Short for HYDRO-ELECTRIC (*power, plant*). Also *attrib.* In Canada also = *hydro-electric power supply*. Cf. *HYDRO-POWER.

1916 A. BRIDLE *Sons of Canada* 185 The product of Niagara..is Hydro-Electric—familiarly abbreviated to Hydro. **1925** *Rep. Brit. Assoc. Adv. Sci. 1924* 43 Various hydro systems were formed as circumstances dictated. **1938** 'R. HYDE' *Nor Years Condemn* x. 200 Did I ever tell you about that hydro I was on, a bit South? **1939** F. P. GROVE *Two Generations* 230 The telephone and hydro wires are down. **1947** E. A. MCCOURT *Music at Close* 71 There was talk of water works, a hydro plant, a newspaper, paved streets. **1949** L. PETERSON *Chipmunk* 149 Claude wrote out the address..on the envelope of the hydro bill he'd forgotten to pay. **1952** D. F. PUTNAM *Canad. Regions* xvii. 387/2 The cities of Alberta, also, have good sources of power including coal, natural gas and hydro. **1958** *New Statesman* 11 Oct. 480/1 Geological reserves of coal are estimated to constitute about four-fifths of all Russia's coal deposits, and much of this coal lies on the surface and this favours the development of thermal-power stations to supplement the hydro-stations of the River Angara. **1962** R. F. LEGGET *Geol. & Engin.* (ed. 2) x. 354 Across the Niagara River is the..Niagara power project of the Power Authority of the State of New York, constructed a few years after the Ontario hydro project. **1963** *B.S.I. News* Feb. 14/2 The hydro developments in Scotland really got under way in the early 1950s. **1970** *Globe & Mail* (Toronto) 26 Sept. 47/2 (Advt.), T.V. outlets in each room, drapes and hydro included.

hydro-. Add: **hy:dro-alcoho·lic** *a.*, in or consisting of a mixture of an alcohol and water; **hy:dro-aroma·tic** *a. Chem.*, having one or more benzene rings which are partially or completely hydrogenated (reduced); also as *sb.*, a hydro-aromatic compound; **hy:dro-bio·logy,** the biology of aquatic plants and animals; hence **hydrobiolo·gical** *a.*, **hydrobio·logist,** one engaged in the study of hydrobiology; **hydrobi·otite** *Min.*, (*a*) a hydrated variety of biotite; (*b*) any clay composed of an intimate mixture of biotite and vermiculite; **hydroca·lumite** *Min.* [f. blend of CAL(CIUM + ALUM(INATE], a transparent, colourless to light green hydrated hydroxide of calcium and aluminium, $Ca_2Al(OH)_7.3H_2O$; **hy·drocast** *Oceanography* [contraction of *hydrographic cast* (CAST *sb.* 5)], a long cable having sampling bottles attached at intervals along it; also, a sampling operation in which this is used; **hydroce·llulose** *Chem.* [a. F. *hydrocellulose* (A. Giraud 1875, in *Compt. Rend.* LXXXI. 1106)], any of the chemically heterogeneous substances produced by the partial hydrolysis of cellulosic material; **hydroceramic** *a.*, designating porous, unglazed pottery used for cooling or filtering; **hy·drochore** [Gr. χωρεῖν to spread], a plant whose seeds are dispersed by water; hence **hydrocho·ric, -ous** *adjs.*, **hy·drochory,** the dissemination of seeds by water; **hy·droclone** [CY)CLONE] = **hydrocyclone*; **hy·drocœl(e)** (-sīl) *Zool.* [Gr. κοιλία cavity of the body], the water-vascular system of an echinoderm; also **-cele**; **hydrocy·clone,** a device in which centrifugation in a conical vessel is employed to remove or separate particles in suspension in a flow of liquid; **hy·drodrill,** a device for injecting water or fertilizers near the roots of plants; also as *v. trans.*; **hy·dro-extract** *v. trans.* [back-formation from *hydro-extractor*], to dry by means of a hydro-extractor; so *hy·dro-extracting* vbl. sb.; also **hy·dro-extraction**; **hy:droformyla·tion** *Chem.*, the catalytic addition of both carbon monoxide and hydrogen to an olefin to produce an aldehyde; **hydroga·rnet** *Min.*, any mineral whose formula is that of a garnet in which water molecules replace some or all of the silicate groups; **hy·droglider,** a form of craft designed to glide on the surface of water (see also quot. 1961); **hydrogro·ssular** *Min.*, a calcium aluminosilicate with a composition varying between that of hibschite and that of grossular (see quot. 1966); **hydrohalite** (-hæ·ləit) *Min.* [ad. G. *hydrohalit* (J. F. L. Hausmann *Handbuch d. Mineralogie* (ed. 2, 1847) II. 1458)], a hydrated chloride of sodium, $NaCl.2H_2O$; **hy:drohetærolite** (-hetīᵊ·roləit) *Min.*, a hydrous oxide of zinc and manganese similar to hetærolite; **hydroki·neter** (also -kine·ter) [Gr. κινητής, -ήρ one that sets going], a device for heating water at the bottom of large boilers by injecting surplus steam; **hydrola·ccolith** *Physical Geogr.* [from its resemblance to a LACCOLITH], an underground mass of ice in a region of permafrost which tends to increase in size and thrust up the overlying soil forming a mound; a mound so formed, esp. a pingo; so **hy:drolaccoli·thic** *a.*; **hy:drometallu·rgical** *a.*, of or pertaining to hydrometallurgy; **hydromo·rphic, -mo·rphous** *adjs. Soil Sci.*, (of a soil) developed and maintained in contact with a high water-table; (of a soil-forming process) acting in conjunction with a high water-table; **hydromu·scovite** *Min.*, a variety of muscovite containing more water and less potassium than that mineral; **hydro·philid,** a water-beetle of the family Hydrophilidæ; also as *adj.*, of or pertaining to an insect of this type; **hy·dro-plant,** plant for generating hydro-electric power; a hydro-electric generating station; **hydrosere** *Ecol.*, a plant succession having its origin in a wet habitat; **hy·dro-ski** *Aeronaut.*, a hydrofoil on a seaplane or amphibious aircraft that skims the surface of the water and provides hydrodynamic lift; **hy·drospace,** the underwater realms; **hydrospire** (earlier examples); **hydrotroilite** (-trōu·iləit, -troi·ləit) *Min.* [ad. Russ. *gidrotroilit* (M. Sidorenko 1901, in *Mém. Soc. Naturalistes Nouv.-Russie* XXIV. I. 119)], a black hydrated ferrous sulphide, $FeS.nH_2O$, occurring in the mud of lakes and inland seas; **hydrotu·ngstite** *Min.*, a hydrated tungstic acid, $H_2WO_4.H_2O$, occurring as minute green tabular crystals; **hy·drowire** *Oceanography* [contraction of *hydrographic wire*], a cable used for hydrocasts.

1887 A. M. BROWN *Contrib. Animal Alkaloids* 46 The hydro-alcoholic solution of the alkaloid was injected hypodermically into a dog of medium size. **1951** Hydro-alcoholic [see *HYDRASTIS]. **1969** *Biochim. & Biophys. Acta* CXCIV. 265 (*heading*) Optical rotatory dispersion of polyglutamic and polyuridylic acids at low temperatures in fluid hydro-alcoholic solvents. **1900** E. F. SMITH tr. *V. von Richter's Org. Chem.* (ed. 3) II. 290 Hexahydrobenzene is the parent hydrocarbon of the hydroaromatic substances. **1940** *Industr. & Engin. Chem.* Apr. 528/2 The cyclization of paraffins to hydroaromatics is preceded by dehydrogenation. **1951** I. L. FINAR *Org. Chem.* xix. 390 Many benzene derivatives may be reduced to the corresponding cyclohexane compounds, and because of this, cyclohexane and its derivatives are known as the hydroaromatic compounds. The cyclic terpenes are hydroaromatic compounds. **1933** *Geogr. Jrnl.* LXXXI. 533 It [*sc.* the Arctic Institute of the U.S.S.R.]..carries on geological, geomorphological, hydrological and hydrobiological investigations. **1932** *Ecology* XIII. 110 The fresh water hydrobiologists, especially the limnologists, have developed a third type of nomenclature based more upon the habitat than on the biotic communities. **1938** Hydrobiologist [see *HYDROLOGIST]. **1964** *Oceanogr. & Marine Biol.* II. 379 The three large brackish water lagoons on the continental coast of the Baltic..have long attracted the interest of hydrobiologists. **1972** *Nature* 28 July 194/1 It is hoped that the committee will be able to produce a register of hydrobiologists working in Britain. **1928** K. E. CARPENTER *Life Inland Waters* p. viii, The life of the ocean [has]..engrossed the energies even of followers of the new tradition in Hydrobiology. **1941** J. G. NEEDHAM in *Symposium Hydrobiol.* 3 Hydrobiology is an offshoot from the old maternal rootstock of natural history. **1965** *Math. in Biol. & Med.* (Med. Res. Council) 309 His [*sc.* Antonio Moroni's] main interests are human population genetics and hydrobiology. **1880** H. C. LEWIS in *Proc. Acad. Nat. Sci. Philadelphia* 319 Such mica exfoliates slightly when heated, is uniaxial, fusible with difficulty, and might be called Hydrobiotite for convenience. **1892** E. S. DANA *Dana's Syst. Min.* (ed. 6) 632 Hydrobiotite *H. C. Lewis*. A hydrated biotite. The name has been similarly but more definitely used by Schrauf. **1934** J. W. GRUNER in *Amer. Mineralogist* XIX. 558 Specimens 9 and 10 belong to a species for which the name *hydrobiotite* is proposed. This name was used long ago by Schrauf and others to designate biotite-like material high in water. *Ibid.* 575 X-ray diagrams are necessary to distinguish vermiculite from hydrobiotite. **1962** W. A. DEER et al. *Rock-Forming Min.* III. 251 Mixed layer clays with vermiculite as a constituent are not uncommon, the most well known being 'hydrobiotite', a random mixture of vermiculite and biotite. **1934** C. E. TILLEY in *Mineral. Mag.* XXIII. 607 In allusion to its composition as an hydrated calcium aluminate ($4CaO.Al_2O_3.12H_2O$) the name hydrocalumite is proposed. **1968** I. KOSTOV *Mineral.* 215 Apart from hydrocalumite, which is rather soft.., the other minerals are hard. **1960** *McGraw-Hill Encycl. Sci. & Technol.* IX. 267/1 Such a laboratory is located near the winches used for running out and retrieving a long string of water-sample bottles (hydrocasts). **1968** D. F. MARTIN *Marine Chem.* I. i. 8 The vessel must be stationary for the time needed to complete the hydrocast. **1971** *Nature* 7 May 37/1 In 1966, one hydrocast in this narrow and steep-sided deep revealed a temperature of 29·07° C and a salinity of 74·2‰. Attempts to place a hydrocast in the Chain deep on this cruise failed because of high winds and consequent ship drift. **1876** *Jrnl. Chem. Soc.* I. 696 Hydrocellulose is also formed when cellulose is impregnated with dilute acid and submitted to a temperature of about 100°. **1920** E. SUTERMEISTER *Chem. Pulp & Paper Making* i. 9 The formation of friable hydrocelluloses by acids is of great importance industrially for upon it is based the carbonization process for separating cotton from wool. **1956** *Nature* 18 Feb. 319/2 The hydrocelluloses produced by the action of mineral acids on cotton and wood. **1883** J. W. MOLLETT *Illustr. Dict. Art & Archæol.* 174/1 Hydroceramic (vessels), Gr., vessels made of a porous clay, in which liquids were put for the purpose of cooling them; they were a kind of *alcarazas*. **1940** *Chambers's Techn. Dict.* 428/2 *Hydroceramic*, porous unglazed pottery, used for filters and for cooling vessels. **1905** F. E. CLEMENTS *Res. Methods Ecol.* iv. 216 Hydrochores..comprise all plants distributed exclusively by water, whether the latter

acts as ocean currents, tides, streams, or surface run-off. *Ibid.* 218 Most hydrophytes are hydrochorous. **1940** *Chambers's Techn. Dict.* 428/2 *Hydrochoric*, dispersed by water. **1969** L. VAN DER PIJL *Princ. Dispersal Higher Plants* v. 61 It is difficult to describe concisely the structural modifications of hydrochory. *Ibid.*, Many hydrochores bend their fruit stalks down..whereas in *Nuphar* (not purely hydrochorous) the seeds mature above water. **1965** D. BRADLEY *Hydrocyclone* i. 1 'Hydraulic cyclone' has been abbreviated to 'hydrocyclone' and even 'hydroclone'. **1967** WHISTLER & PASCHALL *Starch* II. i. 46 The starch stream..must be further purified by passing..through hydroclones to reduce the protein content. **1888** *Phil. Trans. R. Soc.* B. CLXXIX. 266 After separating from the hydrocele the anterior body-cavity grows towards the ectoderm on the right side. **1900** E. R. LANKESTER *Treat. Zool.* III. viii. 23 Whatever may be the homologies of the hydrocoel, there is..no nephridial or other excretory system in Echinoderma. **1962** D. NICHOLS *Echinoderms* x. 121 This [sac] is called the left axohydrocoel, the anterior part being the axocoel..and the posterior the hydrocoel. **1952** *Chem. Abstr.* XLVI. 1668 Data are presented on hydrocyclones used as thickeners in starch processing, as classifiers for highly viscous and non-Newtonian liquids and as washers in ore prepn. **1962** *Engineering* 3 Aug. 146/1 The cone-shaped nozzle at the bottom of the hydrocyclones used by the National Coal Board..(for the separation and thickening of coal and shale fines from water). **1962** *New Scientist* 4 Oct. 31/3 They are using a specially-designed 'hydrodrill' to inject relatively small amounts of water into the soil, placing it directly in the plant's root zone. *Ibid.*, The vine cuttings are simply dropped into holes which have been hydrodrilled. **1928** C. E. MULLIN *Acetate Silk* xxxviii. 437 Yarns or loosely knit fabrics which are not liable to crease may be hydroextracted in the ordinary rotating cage or drum machine. **1952** E. KORNREICH *Introd. Fibres & Fabrics* viii. 143 Fabrics can also be hydroextracted by winding them on a perforated beam which is then inserted in a suitable whizzer. **1882** *Spon's Encycl. Industr. Arts* v. 1839 Centrifugal hydro-extracting machines..have been tried for separating beet-juice from the pulp. **1895** *Trans. Soc. Engin. 1894* 227 (*heading*) The principles and practice of hydro-extraction. **1912** H. H. HODGSON tr. *Masselon's Celluloid* v. 71, 100 kilogrammes of bleached pulp after hydro-extraction should weigh 60 kilogrammes. **1963** A. J. HALL *Textile Sci.* ii. 53 The water in wet viscose rayon materials is best removed by hydro-extraction. **1949** *Jrnl. Amer. Chem. Soc.* LXXI. 3051 (*heading*) Hydroformylation of unsaturated compounds with a cobalt carbonyl catalyst. **1969** S. A. MILLER *Ethylene* xiv. 1169 The OXO reaction or 'hydroformylation' is now an important industrial process. **1941** E. P. FLINT et al. in *Jrnl. Res. Nat. Bureau of Standards* (U.S.) XXVI. 14 An extension of the study revealed that silica could replace water in both $3CaO.Al_2O_3.6H_2O$ and $3CaO.Fe_2O_3.6H_2O$, and that the end products of these substitutions are grossularite garnet, $3CaO.Al_2O_3.3SiO_2$, and andradite garnet, $3CaO.Fe_2O_3.3SiO_2$, respectively... The hydrous members of the series may be termed 'hydrogarnets' to indicate their relationship to the naturally occurring garnets. **1966** W. A. DEER et al. *Introd. Rock-Forming Min.* 23 In the hydrogarnets there is replacement of SiO_4 by $2H_2O$, with vacant Si spaces in the structure. **1921** *Glasgow Herald* 23 July 7/2 The idea of using hydrogliders for passengers and for mail purposes on the lochs in the outlying districts of Scotland. **1927** *Ibid.* 26 July 9 The hydroglider which has been constructed..to accomplish the crossing of the Atlantic from Cherbourg to New York. **1961** F. H. BURGESS *Dict. Sailing* 119 *Hydroglider*, a type of boat that is designed with air-screws as its main means of propulsion. **1943** C. O. HUTTON in *Trans. & Proc. R. Soc. N.Z.* LXXIII. 174 (*heading*) Hydrogrossular, a new mineral of the garnet-hydrogarnet series. *Ibid.*, All of the isomorphous mixtures between plazolite and grossularite are termed by the writer, hydrogrossular. **1966** W. A. DEER et al. *Introd. Rock-Forming Min.* 26 Hydrogrossular has been taken as the name for members of the series $3CaO.Al_2O_3.3SiO_2$—$3CaO.Al_2O_3.6H_2O$ with a composition between grossular and hibschite (plazolite), $3CaO.Al_2O_3.2SiO_2.2H_2O$. **1861** H. W. BRISTOW *Gloss. Mineral.* 185/1 *Hydrohalite*... A hydrous chloride of sodium. **1949** *Mineral. Abstr.* X. 459 A drop of sea-water..evaporated at a low temperature..yields hexagonal flakes (½ mm.) of hydrohalite, $NaCl.2H_2O$. **1928** C. PALACHE in *Amer. Mineralogist* XIII. 308 The following data establish the characters of unaltered hetaerolite. For the partly hydrated mineral hitherto described, the name hydrohetaerolite may well be employed. **1942** [see *HETÆROLITE]. **1955** *Amer. Mineralogist* XL. 350 Hydrohetaerolite has the same structure as hausmannite, except that one sixth of the trivalent manganese occupying the octahedral sites are randomly absent, and the balance of charge is supplied by hydrogen bonds. **1883** A. E. SEATON *Man. Marine Engin.* xx. 376 (*heading*) Weir's hydrokineter. *Ibid.* 377 There are many other ways of promoting the circulation when steam is up, but none do this so efficiently during the time of raising steam as the hydrokineter. **1951** *Engineering* 20 Apr. 483/3 Surplus steam from the waste-heat boiler will be used..to keep the Scotch boilers warm and ready for service, Weir hydrokineters being fitted to the Scotch boilers to maintain a circulation. **1955** M. HOLLANDER tr. *Kuenen's Realms of Water* v. 220 (*caption*) Hydrolaccolith or hummock caused when ground-water under artesian pressure is checked by formation of layers of ice. **1961** L. D. STAMP *Gloss. Geogr. Terms* 244/2 The pingos in East Greenland but not those in the Mackenzie delta are hydrolaccoliths. **1968** R. W. FAIRBRIDGE *Encycl. Geomorphol.* 845/2 The genetic term *hydrolaccolith*, which applies to all ice-intrusions, is not just synonymous with pingo. **1970** E. WATSON tr. *Tricart's Geomorphol. Cold Environments* II. i. 78 These ice masses and the hills which they raise are called hydrolaccoliths. **1963** D. W. & E. E. HUMPHRIES tr. *Termier's Erosion & Sedimentation* 412 An arctic soil phenomenon..formed in the permafrost by a hydrolaccolithic process. **1890** WEBSTER, Hydrometallurgical. **1929** *Encycl. Brit.* VI. 406/1 Hydro-metallurgical treatment..is eminently suited for low grade ores. **1959** J. NEWTON *Extractive Metall.* vii. 436 Hydrometallurgical methods are widely employed today in the treatment of low-grade oxidized uranium ores. **1938** M. BALDWIN et al. in *U.S. Dept. Agric. Yearbk.* 991 The terms 'halomorphic', 'hydromorphic', and 'calomorphic' are not entirely satisfactory, since soil genetics rather than soil characteristics are implied. **1970** E. M. BRIDGES *World Soils* iii. 25/1 These poorly drained or hydromorphic soils frequently occur in the lower parts of the landscape. **1927** *Russian Pedol. Investigations* v. 26 Recently a single group has been formed which is..known by the name..of 'hydromorphous' soils (Neustruev). **1932** G. W. ROBINSON *Soils* xv. 301 Hydromorphous processes are those which take place under the influence of groundwater. **1889** A. JOHNSTONE in *Q. Jrnl. Geol. Soc.* XLV. 364 Margarodite, gilbertite, damourite, and sericite are mineralogists' names for varieties possessing the same composition as muscovite, differing from the latter mineral merely in containing at least about 5 per cent. of water. All of these varieties..ought to be known under one term. The common name proposed for them by the Author is hydromuscovite. **1966** W. A. DEER et al. *Introd. Rock-Forming Min.* 202 Hydromuscovites have high H_2O and low K_2O content. **1899** D. SHARP in *Cambr. Nat. Hist.* VI. v. 219 The pupae of Hydrophilides repose on the dorsal surface. **1958** F. BALFOUR-BROWNE *Brit. Water Beetles* III. 3 The mandibles of *Hydrophilus* also differ from..those of all our other Hydrophilids. **1964** R. M. & J. W. FOX *Introd. Compar. Ent.* iii. 73 The hindleg of the hydrophilid beetle is provided with a fringe of hairs. **1927** J. G. TARBOUX *Electr. Power Equipm.* ii. 31 High-voltage transmission lines must be used to connect the hydro plant to the load center. **1966** *McGraw-Hill Encycl. Sci. & Technol.* XIV. 399/2 Installed capacity of hydroplants cannot be counted upon for perpetuity because of the gradual filling of reservoirs with sediment. **1926** TANSLEY & CHIPP *Study of Vegetation* ii. 19 The earlier stages of a prisere are altogether different, according to whether the succession begins on a wet or a dry habitat... Such successions may be conveniently called *hydroseres* and *xeroseres* respectively. **1952** P. W. RICHARDS *Tropical Rain Forest* xiii. 283 During the course of the hydrosere there is a gradual change from open water to relatively dry conditions. **1967** C. D. SCULTHORPE *Biol. Aquatic Vasc. Plants* xii. 417 In the ultimate stages of the hydroseres plant debris is less and less completely decomposed. **1952** *Jrnl. R. Aeronaut. Soc.* LVI. 334/2 Somewhat allied to the hydrofoil is the hydro-ski. These obtain their lift from the water pressure on their lower surface in a similar manner to the planing lift of the [flying] boat planing bottom. **1954** *Flight* 17 Sept. 433 The Sea Dart hydro-ski fighter..which is land-based, but which uses water (or snow or ice) for take-off and landing. **1960** K. C. BARNABY *Basic Naval Archit.* (ed. 3) 448 Modifications of the submerged type consist in replacing the forward hydrofoils by partly submerged planing surfaces or 'hydroskis'. **1964** *Adv. Hydroscience* I. 2 The use of acoustic energy to perform all those functions in hydrospace for which electromagnetic energy is employed in aerospace. **1966** *New Scientist* 22 Dec. 691/1 Other categories of plot [in science fiction] include a growing preoccupation with 'inner space' or 'hydrospace'. **1869** E. BILLINGS in *Amer. Jrnl. Sci.* XCVIII. 76 In order to avoid the use of double terms, I propose to call them 'hydrospires'. *Ibid.* 77 In *Caryocrinus ornatus* there are thirty hydrospires. **1913** *Mineral. Mag.* XVI. 362 Hydrotroilite. **1957** G. E. HUTCHINSON *Treat. Limnol.* I. xi. 723 The rather poorly characterized ferrous sulfide of lake sediments, supposedly FeS, has received the name hydrotroilite. **1940** KERR & YOUNG in *Program & Abstr. 21st Ann. Meeting Min. Soc. Amer.* 9 Since this mineral appears to be an intermediate product in the alteration of ferberite to tungstite and since it resembles tungstite in many of its physical properties, the name hydrotungstite is suggested. **1963** *Amer. Mineralogist* XLVIII. 935 A similarity between the *x*-ray patterns for hydrotungstite (tungstic acid, $H_2WO_4.H_2O$) and molybdic acid ($H_2MoO_4.H_2O$). **1955** *Deep-Sea Research* III. (Suppl.) 170 A jelly bottle..will remain uncongealed long enough to permit slope determinations to be made of hydrowires. **1969** R. LANGE *Chem. Oceanogr.* v. 80 The sampler is attached to the hydrowire with a screw clamp and a snap clamp.

† **hydro-aeroplane** (həi:dro,ēᵊ·rǒplēin). *Obs.* Also (*U.S.*) **hydro-airplane.** [f. HYDRO- + *AEROPLANE *sb.*, *AIRPLANE.] An aircraft designed to land on and take off from the water; a seaplane.

1909 *Westm. Gaz.* 7 Jan. 4/2 Before the present year expires the hydro-aeroplane will be an accomplished fact. **1914** *Scotsman* 21 Dec. 8/2 A German hydro-aeroplane threw two bombs on Calais this afternoon. **1922** *Encycl. Brit.* XXX. 49/2 Its performance as a hydro-aeroplane suffered from the extra weight and resistance of the floats. **1930** *Flight* 25 July 837/1 You [*sc.* G. H. Curtiss] invented the hydroaeroplane and the flying-boat and thus opened the way for trans-Atlantic flight. **1932** CHATFIELD & TAYLOR *Airplane & its Engine* (ed. 2) xiv. 316 The term *seaplane*..is synonymous with *hydroairplane*, a word too clumsy ever to have come into common use.

hydroborane (həidro,bōᵊ·rēin). *Chem.* Also **-boran.** [ad. G. *hydroboran* (A. Stock 1926, in *Ber. d. Deut. Chem. Ges.* LIX. 2229): cf. *BORANE.] The name used by A. Stock for any of the hydrides of boron that are richer in hydrogen than others with the same number of boron atoms.

1927 *Chem. Abstr.* XXI. 541 When all of the B atoms are tervalent..the names should be 'diboran', 'triboran', 'decaboran', etc., for B_2H_4, B_3H_5, $B_{10}H_{12}$, etc. The names for the hydroborans [*printed* hydroborons] are illustrated by the following: B_2H_6 = dihydro-diboran, B_5H_9 = dihydro-pentaboran, B_5H_{11} = tetrahydro-pentaboran, etc. **1933** A. STOCK *Hydrides of Boron & Silicon* i. 18 The less stable hydrides, which are richer in hydrogen, should be called 'hydroboranes', e.g. dihydrotetraborane (B_4H_{10}).

hydroborate (həidro,bōᵊ·rēit), *sb. Chem.* [f. HYDRO- + BORATE.] = *BOROHYDRIDE.

1950 *Chem. Abstr.* XLIV. 12392/1 (Index), *Hydroborates*. See such headings as *Aluminium borohydride* and *Lithium borohydride*. **1963** *Jrnl. Amer. Chem. Soc.* LXXXV. 2725/1 Among the numerous derivatives of boranes and hydroborate ions known there are no compounds reported to date containing an NO group directly attached to one of the boron atoms. **1964** R. M. ADAMS *Boron* vi. 373 While the latter name [*sc.* borohydride] belongs to the ferrocyanide type which has long been disapproved by the IUPAC, it has become so firmly established in American usage that it will be used interchangeably with hydroborate in this chapter.

hydroborate (həidro,bōᵊ·rēit), *v. Chem.* [Back-formation from next.] *trans.* To add a borane or other boron compound to (another compound) by hydroboration.

1961 *Jrnl. Amer. Chem. Soc.* LXXXIII. 2550/1 β-Pinene..was hydroborated internally in diglyme with sodium borohydride and boron trifluoride etherate. **1964** KIRK & OTHMER *Encycl. Chem. Technol.* (ed. 2) III. 709 Dienes can be hydroborated utilizing diisoamylborane and diphenyldiborane.

Hence **hy:drobora·ting** *vbl. sb.*

1962 H. C. BROWN *Hydroboration* i. 28 The treatment of internal acetylenes, such as 3-hexyne, with the theoretical quantity of hydroborating agent results in the formation of the corresponding trivinylborane.

hydroboration (həi:dro,bōᵊrēi·ʃən). *Chem.* [f. HYDRO- + BOR(ON + -ATION.] The addition of a linked boron–hydrogen pair of atoms in a molecule across a double or triple bond between a carbon atom and another atom; for example,

$$=\!B{-}H + \overset{|}{\underset{|}{C}}{=}\overset{|}{\underset{|}{C}} \rightarrow\ =\!B{-}\overset{|}{\underset{|}{C}}{-}\overset{|}{\underset{|}{C}}H.$$

1957 BROWN & RAO in *Jrnl. Org. Chem.* XXII. 1136/2 (*heading*) Hydroboration of olefins. A remarkably fast room-temperature addition reaction of diborane to olefins. *Ibid.* 1138/2 The hydroboration reaction should provide a useful and convenient synthetic route for the transformation of olefins into organoboranes, alcohols, and other functional derivatives. **1966** STEINBERG & BROTHERTON *Organoboron Chem.* II. viii. 225 The reaction sequence can be formulated to involve hydroboration of the carbon–nitrogen triple bond followed by a hydrogen shift and final cyclization. **1971** *Nature* 20 Aug. 536/2 Many related organic syntheses depend on the hydroboration of acetylenes or dienes as the first step.

hydrocarbon. Add: **b. hydrocarbon oil,** any oil consisting chiefly of hydrocarbons.

1864 *Mechanics' Mag.* 22 Jan. 55/3 (*heading*) Petroleum and hydro-carbon oils for generating steam. **1904** G. H. HURST *Textile Soaps & Oils* iv. 115 The cheapest oils are the so-called mineral or hydrocarbon oils. **1942** *Progress Appl. Chem.* XXVII. 82 The analysis of hydrocarbon oils is a valuable aid in the control of refinery processes such as cracking.

hydrocele, var. *hydrocœl(e)* (s.v. *HYDRO-).

hydrocephalic, *a.* Add: **B.** *sb.* A person affected with hydrocephalus.

1908 A. F. TREDGOLD *Mental Deficiency* (*caption* to Plate xix), Male hydrocephalic. **1930** D. PATERSON *Sick Children* xii. 331 Fifty per cent. of all hydrocephalics suffer from convulsions. **1971** *Nature* 17 Sept. 171/1 Medicine is keeping alive idiots, hydrocephalics and cases of spina bifida.

hydrocephalus. Delete the last sentence of the definition.

hydrochlorothiazide (həi:dro,klōᵊro,þəi·āzəid). *Pharm.* [f. HYDRO- + *CHLOROTHIAZIDE.] A white crystalline powder which is a diuretic and saluretic sulphonamide drug, $C_7H_8ClN_3O_4S_2$, analogous to chlorothiazide, and which is given orally in the treatment of œdema and as an adjuvant in the treatment of hypertension.

1958 *Experientia* XIV. 458/1 Hydrochlorothiazide is the generic name of a new diuretic agent manufactured by CIBA. **1959** *Lancet* 13 June 1221/2 Hydrochlorothiazide promotes less bicarbonate excretion than chlorothiazide, but the loss of potassium after the two drugs is not significantly different. **1965** J. POLLITT *Depression & its Treatment* v. 72 In mild cases [of premenstrual depression]..Hydrochlorothiazide with potassium chloride (Hydrosaluric K) 25–75 mg. daily or on alternate days from 7–10 days premenstrually is effective. **1968** J. H. BURN *Lect. Notes Pharmacol.* (ed. 9) 105 Hydrochlorothiazide and hydroflumethiazide are ten times stronger than chlorothiazide.

hydrocolloid (həidro,kǫ·loid). Also (with hyphen) **hydro-colloid.** [f. HYDRO- + COLLOID *a.* and *sb.*] Any substance that will form a gel on the addition of water; a gel so produced. Also *attrib.* or as *adj.*

1926 *Brit. Pat.* 271,306, The subject matter of the present invention is a reversible hydro-colloid mass which may serve for..enabling castings to be taken from the oral cavity for dental purposes. **1946** C. K. TSENG in J. Alexander *Colloid Chem., Theoret. & Appl.* VI. xxxi. 717 Agar is a reversible hydrocolloid and materials made with it must be warmed to a solution and cooled to a gel. **1954** *Adv. Chem. Ser.* XI. 92 Irish moss extractive, also known as carrageenin,..is a hydrocolloid gum. **1955**

STURDEVANT & BRAUER in J. C. Brauer *Dental Asst.* xx. 287 The reversible hydrocolloids are used for tray impressions to form casts of the teeth and jaws.

Hence **hy:drocolloi·dal** *a.*

1928 *Chem. Abstr.* XXII. 1659 A mixt. of rubber, fatty substances and resins is incorporated, in fine distribution, in a hydrocolloidal mucilaginous basic substance such as sea weeds or algaceous plants. **1947** J. C. RICH *Materials & Methods Sculpture* v. 96 Commercial hydrocolloidal compositions. **1956** J. N. ANDERSON *Appl. Dental Materials* xx. 240 All hydrocolloidal gels consist largely of water enclosed rather loosely within the gel fibrils.

hy·dro-coo:ler. orig. *U.S.* [See next and -ER[1].] An apparatus for hydro-cooling, usually consisting of a water tank with cooling equipment and a conveyer.

1947 *Ice & Refrigeration* Nov. 33/2 Hydro-coolers are used commercially today to pre-cool celery, asparagus, peas,..and carrots. **1958** *Agriculture* LXV. 133 Small-scale tests were made with a small and simple hydrocooler consisting of two tanks, one containing blocks of ice and water and the other for submersion of the watercress. **1964** A. E. CANHAM *Electr. Hort.* xiii. 137 The capacity of hydro-coolers varies from the large 2 000 crate-per-day unit down to the 200 crate-per-day packaged unit.

hy·dro-coo:ling, *vbl. sb.* orig. *U.S.* [HYDRO- a.] A method of preserving the freshness of vegetables or fruit after harvesting and packing by immersing them for a time in chilled water (or, sometimes, by spraying them).

1942 *Ice & Refrigeration* Nov. 302/2 The hydro-cooling machine takes 130 boxes of asparagus at a time. *Ibid.*, After this hydro-cooling the produce is loaded into trucks. **1955** WORK & CAREW *Veg. Prod. & Marketing* (ed. 2) iv. 62 Dipping in very cold water (hydrocooling) is much more rapid. **1964** A. E. CANHAM *Electr. Hort.* xiii. 135 Hydro-cooling provides an excellent simple, effective and rapid method of cooling produce which is not seriously affected by the presence of moisture.

Hence (as a back-formation) **hy·drocool** *v. trans.*, to chill by this method; **hy·drocooled** *ppl. a.*

1945 *Refrig. Engin.* Apr. 275/3 Test boxes of..Bing cherries were handled in the following manner:..(3) hydrocooled for 7 min. in a water bath of melting ice at 32°. **1945** *Proc. Amer. Soc. Hort. Sci.* XLVI. 196 As far as cracking and decay in storage are concerned, it apparently made little difference whether the hydrocooled fruit was packaged wet or dry. **1958** *Agriculture* LXV. 133 Comparison was made between the behaviour of samples hydrocooled to 34–35°F and control samples which had received no cooling.

hydrocortisone (həidro,kǭ·ɹtizōun). *Biochem.* and *Pharm.* [f. HYDRO- + *CORTISONE.] A steroid hormone, $C_{21}H_{30}O_5$, which is produced by the adrenal cortex and involved in the regulation of carbohydrate metabolism, and which is prepared synthetically for use as an anti-inflammatory and anti-allergic agent.

1951 *Jrnl. Amer. Med. Assoc.* 22 Dec. 1631 Hydrocortisone acetate had a constant anti-inflammatory effect on the rheumatoid arthritic joint when administered locally. **1952** *Ibid.* 27 Dec. 1664/2 One eye was treated with cortisone while the more severely involved eye was treated with hydrocortisone. **1953** *Jrnl. Amer. Chem. Soc.* LXXV. 5369/1 We have isolated a fungus..which converts Compound S directly to the biologically active steroid, 17α-hydroxycorticosterone (Kendall's Compound F, hydrocortisone, Reichstein's Compound M). **1963** *Lancet* 19 Jan. 142/1 Hydrocortisone has revolutionised the treatment of itching skin diseases. **1965** *New Scientist* 1 July 32/3 The glucocorticoids, of which cortisone and hydrocortisone (cortisol) are the principal members, are..mainly concerned with carbohydrate metabolism. **1967** *Martindale's Extra Pharmacopoeia* (ed. 25) 448 Hydrocortisone is administered by mouth and by intramuscular injection and it is applied externally in ointments, creams, and lotions. **1970** R. W. MCGILVERY *Biochem.* xxiii. 565 The most active of the glucocorticoids in humans is cortisol, or hydrocortisone.

hydrocracking (həi·dro,krækiŋ), *vbl. sb.* [f. HYDRO- + *cracking* vbl. sb. (see *CRACK *v.* 23).] The catalytic cracking of crude petroleum or a heavy distillate by subjecting it to the action of gaseous hydrogen at a high temperature and pressure, so that long-chain paraffins and other hydrocarbons undergo hydrogenolysis; also called *hydrogenation cracking*.

1940 *Jrnl. Soc. Chem. Industry Japan* (Suppl. binding) XLIII. 363/1 The 'Hydro-Cracking' is a method of cracking in which mineral oil is decomposed in the presence of high pressure hydrogen. **1955** V. HAENSEL in B. T. Brooks et al. *Chem. Petroleum Hydrocarbons* II. xxix. 205 Hydrocracking produces lower boiling hydrocarbons which have higher octane numbers and lower densities than the parent hydrocarbons. **1970** *Adv. Chem. Ser.* CIII. vi. 113 When the modern version of hydrocracking was announced in 1959, capacity was 1000 barrels per day. As the decade ended capacity..was approaching 1,000,000 barrels per day.

Hence (as a back-formation) **hy·drocrack** *v. trans.*, to subject to, or convert by means of, hydrocracking; **hy·drocracked** *ppl. a.*, obtained by hydrocracking. Also **hy·drocracker,** an apparatus or plant where hydrocracking is carried out.

1940 *Jrnl. Soc. Chem. Industry Japan* (Suppl. binding) XLIII. 363/1 The hydro-cracked gasoline is reputed to be of superior quality. **1959** R. J. HENGSTEBECK *Petroleum Processing* xii. 274 Any heavy petroleum stock can be hydrocracked. **1965** *Chem. & Engin. News* 29 Mar. 32/2 (*heading*) Gulf adds hydrocracker at Port Arthur refinery. **1970** *Adv. Chem. Ser.* CIII. vi. 119 The feed is hydrocracked efficiently to jet fuel with minimum formation of lighter products. *Ibid.* 120 The hydrocracker should be operated to give maximum liquid yields.

hy:drodesu:lphuriza·tion. Also (*U.S.*) **-sulfur-.** [f. HYDRO- + DE- II. 1 + SULPHUR + -IZATION.] The removal (as hydrogen sulphide) of sulphur from crude petroleum or a petroleum product by the action of a catalyst and gaseous hydrogen at a moderately high temperature and pressure.

1950 *Industr. & Engin. Chem.* Sept. 1882/2 Several high sulfur gas oils were reduced in sulfur contents to those of gas oils from sweet crudes by hydrodesulfurization..at 750° F., 300 pounds per square inch pressure,..and 1000 cubic feet of hydrogen per barrel of charge. **1967** *Proc. 7th World Petroleum Congr.* IV. 167/2 Hydrodesulphurization has long been recognized as a means of removing sulfur [from] residual oils and asphalts.

Hence (as a back-formation) **hy:drodesu·lphurize** *v. trans.*, to subject to hydrodesulphurization; **hy:drodesu·lphurized** *ppl. a.*, **-desu·lphurizing** *vbl. sb.* Also **hy:drodesu·lphurizer,** an apparatus in which this process is carried out.

1950 *Industr. & Engin. Chem.* Sept. 1881/1 Instead of processing the gas oil, it is possible to hydrodesulfurize the entire crude. **1955** *Industr. Chemist* XXXI. 351/1 The first commercial hydrodesulphurizer applying the Shell trickle technique was recently started up. **1960** *Times Rev. Industry* Jan. 31/1 The regeneration of hydrodesulphurising catalysts for further use. **1967** *Proc. 7th World Petroleum Congr.* IV. 168/2 The hydrodesulfurized product tends to be relatively rich in nickel.

hydrodictyon (həi:dro,di·kti,ǫn). *Bot.* [mod.L. (A. W. Roth *Catalecta Botanica* (1800) II. 237), f. HYDRO- + Gr. δίκτυον net.] A green freshwater alga of the genus so called; = *water-net* (WATER *sb.* 31).

1841 W. H. HARVEY *Man. Brit. Algæ* p. viii, Among British Algæ, the only structure analogous to these exists in *Hydrodictyon*, which grows in the form of a perfect net, with regular meshes. **1872** H. C. WOOD *Contrib. Hist. Fresh-Water Algæ N. Amer.* 94 When the hydrodictyon disappear in the fall it is months before they reappear in the spring. **1927** WEST & FRITSCH *Treat. Brit. Freshwater Algae* 101 A new *Hydrodictyon*-net or *Pediastrum*-plate is formed..by the asexual method. **1967** I. MORRIS *Introd. Algae* iv. 50 *Hydrodictyon* colonies consist of large numbers of cylindrical cells joined terminally with two others to form an open net-like structure.

hy:drodyna·mically, *adv.* [f. HYDRODYNAMIC, -DYNAMICAL *adjs.*: see -ICALLY.] From the point of view of hydrodynamics.

1957 G. E. HUTCHINSON *Treat. Limnol.* I. i. 156 Schriever (1955) concludes that Croke's final theory is hydro-dynamically impossible. **1971** *Nature* 11 June 383/2 The turbulent boundary layer found near the sediment water interface may be described hydrodynamically in one of three ways.

hydrodynamicist (həi:dro,dəinæ·misist). [f. HYDRODYNAMIC(S + -IST.] An expert in or a student of hydrodynamics.

1961 *Aeroplane* CI. 553/2 Hydrodynamicists..were less interested in ships travelling in shallow water at a particular speed..than in sea-going ships or river-going ships at any speed. **1970** N. DE NEVERS *Fluid Mech.* x. 324 By 1900 the two schools had gone their separate ways, the hydrodynamicists publishing learned mathematical papers with little bearing on engineering problems and the hydraulicians solving engineering problems by trial and error.

hydro-electric, *a.* Add: Also **hydroelectric.**
1. (Earlier example.)

1827 J. CUMMING *Man. Electro Dynamics* iv. 115 Retaining the name *Hydro-electric*, for current produced by the intervention of fluids.

2. (Earlier example.)

1844 *Rep. Brit. Assoc. Adv. Sci. 1843* 39 (*heading*) On the electricity of high-pressure steam, and a description of a hydro-electric machine.

3. Generating electricity by utilizing the motive power of water; generated by, or relating to generation by, such means.

1884 *Engineering* 11 Jan. 38/3 A Hydro-Electric Syphon... A most ingenious scheme for draining the valley of the city of Mexico by means of electrical power generated by the fall of water raised by electrically driven pumps. **1899** *Min. Proc. Inst. Civil Engin.* CXXXVI. 433 (heading) Hydro-electric installation at Mont-Dore. **1904** *Electrical Mag.* I. 592/1 On the river Ain..near Lake Geneva, a hydro-electric plant has recently been completed. **1905** *Daily Chron.* 18 Sept. 4/5 Abundant hydro-electric power. **1927** CREAGER & JUSTIN *Hydro-Electric Handbk.* xxxv. 874 Hydro-electric plants are much more reliable than steam plants. **1951** T. H. CARR *Electric Power Stations* (ed. 3) II. xxiv. 579 Hydro-electric stations are usually located in wild, mountainous districts. **1955** *Bull. Atomic Sci.* Oct. 281/2 Hydroelectric power is never likely to contribute more than a small fraction of the total energy consumption of the world, since the total potential capacity is relatively limited. **1963** W. E. BALLARD *Metal Spraying* (ed. 4) xiv. 479 In hydro-electric schemes, metal spraying has often been used for the protection of the large pipes bringing the water down from a high altitude to the turbines. **1972** *Daily Tel.* 11 Apr. 13/6 Brazil wants to build a huge hydro-electric dam on the upper reaches of the River Parana.

hy:dro-ele·ctrical, *a.* [f. prec. + -AL.] = *HYDRO-ELECTRIC *a.* 3.

1925 *Trans. Soc. Engin.* 127 In New Zealand the largest hydro-electrical power scheme..is controlled by the Christchurch County Council. **1930** BAKER & CONKLING *Water Supply & Utilization* i. 17 (*heading*) Recent growth of hydroelectrical development in the United States.

hy:dro-electri·city. [f. HYDRO- + ELECTRICITY.] **1.** (In Dict. s.v. HYDRO-ELECTRIC *a.*: earlier example.)

1828 F. WATKINS *Pop. Sk. Electro-Magnetism* 61 The term hydro-electricity, applied to that which emanates from a voltaic combination.

2. Electricity generated by utilizing the motive power of water.

1904 *Electrical Mag.* I. 378/1 (*heading*) Hydro-electricity in California. *Ibid.* 592/1 (*heading*) Hydro electricity for small French factories. **1959** *Petroleum Handbk.* (Shell Internat. Petroleum Co.) (ed. 4) 19 There are..limitations to the future growth of hydro-electricity. **1972** *Sci. Amer.* Dec. 13/2 With ample supplies of coal and hydroelectricity China should be able to devote much of the oil..to a petrochemical industry.

hydrofining (həi·drofəiniŋ), *vbl. sb.* [f. HYDRO- + RE)FINING *vbl. sb.*] A catalytic process in which a petroleum product is stabilized and its sulphur content reduced by treatment with gaseous hydrogen under relatively mild conditions, so that unsaturated hydrocarbons and sulphur compounds undergo selective hydrogenation.

1931 C. ELLIS *Hydrogenation Org. Subst.* xlviii. 578 The term *hydrofining* is best applied to those treatments by hydrogen which are carried out at temperatures in the lower ranges,..where no extensive alteration of the carbon-containing molecule occurs. On the other hand, where great disturbance of carbon structure results the process is better denominated *hydroforming*. **1955** *Petroleum Refiner* May 154/1 Hydrofining was found to..improve the burning quality of West Texas No. 1 fuel oil. **1970** J. R. HUGHES *Storage & Handling Petroleum Liquids* (ed. 2) 5 Hydrofining and autofining aim to remove or reduce the content of unwanted sulphur compounds.

So **hy·drofined** *ppl. a.*, obtained by hydrofining.

1931 C. ELLIS *Hydrogenation Org. Subst.* xlviii. 577 The hydrofined lubricating oils are not as readily oxidized as the untreated oils. **1955** *Petroleum Refiner* May 153/2 The quality of the raw and hydrofined distillates was determined by inspection tests, measurement of hydrogen content, and burning-quality tests.

hydroflumethiazide (həi:droflūmeþəi·āzəid). *Pharm.* [f. HYDRO- + *flumeth-* (f. FLU(ORO- + METH(YL + *CHLOROTH)IAZIDE.] A white crystalline compound, $C_8H_8F_3N_3O_4S_2$, analogous to hydrochlorothiazide and having similar effects and uses.

1959 *Lancet* 12 Sept. 303/2 Two substances related to chlorothiazide—hydroflumethiazide and hydrochlorothiazide—have recently been produced, and are effective in much smaller quantity. **1964** A. GOTH *Med. Pharmacol.* (ed. 2) xxix. 358 Hydroflumethiazide (Saluron) has the same structural formula as hydrochlorothiazide, except that F_3C replaces Cl. **1968** [see *HYDROCHLOROTHIAZIDE].

hydrofoil (həi·drǒfoil). [f. HYDRO- + FOIL *sb.*[1], after *aerofoil*.] **1.** A plane designed to give rise to a force (other than drag) when moving through a liquid; *spec.* (*a*) a plane (usually one of two or more) attached to a vessel by means of which the hull is lifted clear of the water at speed; (*b*) one attached to a seaplane to facilitate take-off by increasing the hydrodynamic lift; (*c*) one attached at the side of a ship to act as a stabilizer.

1920 *Chambers's Jrnl.* 28 Feb. 207/2 [A boat.] Under each wing at its forward end is a series of narrow steel planes known as hydrofoils. **1933** *Aeroplane* 24 May 945/1 The possibility of using hydrofoils to cure porpoising and to revolutionise the design of flying-boats. **1939** *Jrnl. R. Aeronaut. Soc.* XLIII. 552 A special landing and take-off gear in the form of hydrofoils which will be retracted in flight. **1958** *New Scientist* 27 Feb. 13/3 The aerofoils of the sails and the hydrofoil of the keel are separated by the hull of the yacht. **1966** *McGraw-Hill Encycl. Sci. & Technol.* XIII. 212/1 Diving planes [on a submarine] are pairs of hydrofoils which extend from the sides of the ship. **1968** J. E. PLAPP *Engin. Fluid Mech.* xii. 486 The hydrofoil..is a wing designed to operate in water instead of air. **1972** *Daily Tel.* 28 Sept. 6/5 The mono hull has surface-piercing hydrofoils which lift the craft out of the water. **1973** *Nature* 12 Jan. 114/2 The use of hydrofoils for a wide variety of purposes such as propeller blades on boats, as sailboat keels, ship rudders, submarine and torpedo fins, lifting surfaces of hydrofoil boats, and shroud ring stabilizers for missiles, has prompted efforts to increase the lift-to-drag ratio.

2. A vessel fitted with hydrofoils for raising it clear of the water.

1959 *Times Rev. Industry* Nov. 41/1 This speed..will

be valuable for work on high speed motor boats or hydrofoils. **1962** *Daily Tel.* 12 Feb. 16/8 The United States Maritime Administration..believes there is a big future for hydrofoils as passenger vessels. **1968** [see *FOIL-BORNE *a.*]. **1969** *N.Z. Listener* 12 Dec. 19/1 Jaunting up Auckland Harbour on the hydrofoil. **1970** R. C. DORF *Mod. Control Syst.* x. 351 The Denison is an 80-ton hydrofoil capable of operating in seas ranging to 9 ft in amplitude at a speed of 60 knots. **1972** *Guardian* 27 May 5/3 The Thames's first modern commuter service—by Italian hydrofoil—opened yesterday.

3. *attrib.*, as *hydrofoil boat, craft, service, ship.*

1950 *Electronic Engin.* XXII. 205 A hydrofoil boat—one which has wings extending into the water to support it in motion so that the hull is out of the water—has speed and stability advantages. **1960** *Times* 19 Jan. 10/7 The United States Government has awarded a contract for the construction of a hydrofoil ship capable of carrying 100 passengers at speeds up to 90 miles an hour. **1960** *New Scientist* 15 Sept. 703/2 The hydrofoil craft is particularly suited for river work. **1964** *Weekly News* (Auckland) 8 July 30 The beginning of the hydrofoil service to Waiheke Island..means that island residents..can now make the trip between Auckland and Matiatia in 20 minutes. **1973** *Cook's Continental Timetable* June 456 Hydrofoil Services. Journey 35 minutes [from Naples] to Capri.

hydroformate (həi·dro,fǭ·ɹmēⁱt). [f. as next + *-ate*, after *filtrate, precipitate*, etc.] A product obtained by hydroforming.

1953 A. A. DRAEGER et al. in *Sci. Petroleum* V. II. 266/2 For the production of high purity aromatic chemicals, however, additional processing of the hydroformate is required. **1969** KIRK & OTHMER *Encycl. Chem. Technol.* (ed. 2) XX. 538 Then the..light hydrocarbons are removed by distillation in a 'product stabilizer'; the undistilled portion constitutes the product or 'stabilized hydroformate'.

hydroformer (həi·dro,fǭ·ɹməɹ). [f. next + -ER[1].] In an oil refinery, an apparatus or plant where hydroforming is carried out.

1941 *Refiner & Natural Gas Manuf.* May 66/2 The potential toluene production of the hydroformer amounts to about 5,000,000 gallons per year. **1959** *Times Rev. Industry* Nov. 72/3 To achieve petrol of a higher octane number a powerformer in place of the traditional hydroformer is to be employed.

hydroforming (həi·dro,fǭɹmiŋ), *vbl. sb.* [f. HYDRO- + *-forming*, after REFORMING *vbl. sb.*] In the petroleum industry, a catalytic reforming process that converts the paraffins and alicyclic compounds in low-octane petroleum naphtha to aromatic compounds by dehydrogenation at a high temperature and moderate pressure in the presence of gaseous hydrogen. Freq. *attrib.*

1931 [see *HYDROFINING *vbl. sb.*]. **1941** *Oil & Gas Jrnl.* 27 Mar. 87/1 The first commercial hydroforming plant was placed in operation recently. **1953** A. A. DRAEGER et al. in *Sci. Petroleum* V. II. 266/2 The hydroforming process is employed by the petroleum refining industry both to elevate the anti-knock quality of petroleum naphthas..for inclusion in premium quality gasolines and to produce aromatic compounds of high purity for specialized uses. **1959** *Times Rev. Industry* Aug. 112/2 Examples of..new knowledge include the technique of making high octane gasoline by hydroforming and fluid catalytic cracking. **1969** KIRK & OTHMER *Encycl. Chem. Technol.* (ed. 2) XX. 534 Toluene is produced principally from petroleum by the hydroforming of selected petroleum naphthas..which are rich in naphthenic hydrocarbons.

Hence (as a back-formation) **hy·droform** *v. trans.*, to subject to hydroforming; **hy·droformed** *ppl. a.*, produced by hydroforming.

1941 *Nat. Petroleum News* 2 Apr. R-102/1 The anti-knock quality of hydroformed gasoline has not yet been thoroughly evaluated. **1949** P. C. CARMAN *Chem. Constitution & Properties Engin. Materials* xxvii. 824 A low octane feed boiling 90°–200° C. can be hydro-formed at 480°–540° C. and 100–300 p.s.i. to give an aromatised motor fuel with an octane rating of 80. **1953** A. A. DRAEGER et al. in *Sci. Petroleum* V. II. 274/1 These factors..make it a desirable procedure to hydroform cracked naphthas in admixture with virgin naphthas.

hydrogasification (həi:drogæsifikēi·ʃən). [f. HYDRO- + GASIFICATION.] The production of methane directly from coal by treatment with hydrogen (or hydrogen and steam) at a high temperature and pressure.

1954 *Proc. Amer. Gas Assoc.* 644/2 Pressure hydrogasification produces high-Btu oil gases with satisfactory combustion characteristics. **1963** H. H. LOWRY *Chem. Coal Utilization* Suppl. xx. 996/1 The hydrogasification of coal to produce methane has an inherent thermochemical advantage. **1967** *McGraw-Hill Yearbk. Sci. & Technol.* 139/2 Up to 70% of the hydrogen needed for hydrogasification is made in place by reaction of steam with carbon. **1968** *Chem. & Engin. News* 29 July 12/3 Coal hydrogasification has reached the pilot-plant stage.

So **hydroga·sifier,** an apparatus in which hydrogasification is carried out.

1966 *Chem. & Engin. News* 18 Apr. 70/1 There is a net heat of reaction and this serves to supply the necessary heat for the hydrogasifier..to operate adiabatically.

hydrogel (həi·drǒdʒel). [f. HYDRO- + GEL(ATIN.] A gel or gelatinous precipitate in which the liquid constituent is water.

1864 [see *ALCOSOL]. **1895** THOMSON & BLOXAM *Bloxam's Chem.* (ed. 8) 123 Colloids..can generally exist in solution (the hydrosol form), but are apt to separate as a jelly (the hydrogel form) from such solutions. **1938** R. G. TORRENS *Dental Dis.* vi. 90 Gelatine is a hydrosol when warm and a hydrogel when cold. **1972** *Physics Bull.* June 336/1 Probably a wide range of factors can precipitate the clotting process including..surface texture and surface wettability as relating to hydrogel formation.

hydrogen. Add: **1. b.** An atom of hydrogen.

1920 *Jrnl. Amer. Chem. Soc.* XLII. 1431 A free pair of electrons on one water molecule might be able to exert sufficient force on a hydrogen held by a pair of electrons on another water molecule to bind the two molecules together. **1957** G. E. HUTCHINSON *Treat. Limnol.* I. iii. 196 The hydrogens, owing to their capacity to form hydrogen bonds.., will act as if they had unsatisfied single valencies.

2. a. *hydrogen* (*sc.* hydrogen bomb) *warhead*; **hydrogen bomb,** an immensely powerful bomb in which the energy released is derived from the fusion of hydrogen nuclei in an uncontrolled self-sustaining reaction initiated by a fission bomb; **hydrogen bond,** a weak bond between a strongly electronegative atom with a lone pair of electrons in one molecule and a hydrogen atom covalently bonded to another strongly electronegative atom or group in the same or a different molecule; hence *hydrogen-bonded* pa. pple. and ppl. adj., *-bonding* vbl. sb.; **hydrogen (gas) electrode,** an electrode (usu. of platinum coated with platinum black) partially immersed in a solution that contains hydrogen ions and hydrogen gas, so that an equilibrium between the ions and the molecules is established on the surface of the electrode enabling it to be used as a standard of zero potential (e.g. in measurements of other electrode potentials and of *p*H); **hydrogen ion,** the positive ion H^+ (the proton) derived from a hydrogen atom by the loss of its electron; a solvated form of this in a solution, esp. the hydrated form H_3O^+ (cf. *HYDRONIUM).

1947 *N.Y. Times* 13 Apr. IV. 9/5 Hydrogen bomb. New and improved atomic bombs were discussed at the recently held forum of the Northern California Association of Scientists. **1948** *Sci. News Let.* 17 July 35/1 This is the 'hydrogen bomb' that certain high officials in past months have vaguely..hinted may be made. **1951** C. ROBERTS *Terrace in Sun* i. 12 Would the hydrogen bomb that could reduce New York to a tangled skeleton penetrate thus far? **1954** W. S. CHURCHILL in *Hansard Commons* 30 Mar. 1840 The development of the hydrogen bomb raises strategic and political issues. **1972** *Sci. Amer.* Dec. 13/3 China exploded a fission device in 1964 and a fusion device in 1967, and it has tested several hydrogen bombs since that date. **1923** G. N. LEWIS *Valence* ix. 109 The hydrogen atom can form a loose attachment to another pair of electrons, thus forming the hydrogen bond. **1939** L. PAULING *Nature Chem. Bond* ix. 264 Although the hydrogen bond is not a strong bond..it has great significance in determining the properties of substances. **1966** *McGraw-Hill Encycl. Sci. & Technol.* XIV. 387/2 Ordinary ice consists of water molecules joined together by hydrogen bonds in a regular arrangement. **1970** AMBROSE & EASTY *Cell Biol.* iii. 110 When double-stranded DNA is heated to near 100°C, the hydrogen bonds between the two chains break and the strands separate. **1950** *Jrnl. Amer. Chem. Soc.* LXXII. 5349/2 Each residue is hydrogen-bonded to the third residue from it in each direction along the chain. *Ibid.*, The second hydrogen-bonded spiral is the five-residue spiral. **1936** *Ibid.* LVIII. 1903/2 One isomer should show complete hydrogen bonding detectable by appropriate physical methods. **1898** *Jrnl. Chem. Soc.* LXXIV. II. 89 A similar method may be used in the titration of acids and bases, if a hydrogen electrode be employed..being made of gold electrolytically coated with palladium. **1942** GLASSTONE *Introd. Electrochem.* x. 352 The hydrogen gas electrode cannot be employed in solutions containing oxidizing agents. **1964** R. G. BATES *Determination of pH* ix. 230 The hydrogen electrode is the ultimate standard for the determination of pH values, but..other electrodes reversible to hydrogen ion are commonly employed for routine pH measurements. **1896** *Jrnl. Chem. Soc.* LXX. II. 638 Etherification occurring, with or without the addition of a catalysing acid, is primarily caused by the activity of the hydrogen ions present. **1935** *Discovery* Nov. 322/1 Every process of manufacture is subject to constant scientific control, regulating temperature, humidity,..hydrogen-ion concentration. **1939** L. PAULING *Nature Chem. Bond* ix. 266 The positive hydrogen ion is a bare proton. **1942** GLASSTONE *Introd. Electrochem.* ix. 308 The hydrogen ion in solution is not to be regarded as a bare proton, but as a combination of a proton with, at least, one molecule of solvent. **1968** M. S. LIVINGSTON *Particle Physics* vi. 120 Experiments on the deflection of hydrogen-ion beams in electric and magnetic fields. **1954** in *Amer. Speech* (1957) XXXII. 137 That missile, or I.B.M. as the experts call it, will be an accurately guided rocket..capable of carrying a hydrogen warhead over a range of 4000 to 5000 miles.

b. *hydrogen cyanide*, the more usual term in modern usage for hydrocyanic acid; *hydrogen peroxide*, a colourless, viscous, somewhat unstable liquid, H_2O_2, which can act as an oxidizing and a reducing agent, is usu. prepared as an aqueous solution, and is used esp. as an oxidizing and bleaching agent, in the manufacture of peroxides and organic compounds, as a weak antiseptic, and (in concentrated form) as a rocket propellant.

1882 *Jrnl. Chem. Soc. Index of Subjects 1873–1882* 215/2 Hydrogen cyanide. **1948** *New Biol.* IV. 71 The most common fumigants in use in this country are hydrogen cyanide;..methyl bromide; [etc.]. **1872** *Jrnl. Chem. Soc.* XXV. 922 (*heading*) Determination of nitrates, nitrites, and hydrogen peroxide by solution of indigo. **1907** G. S. NEWTH *Text-bk. Inorg. Chem.* (ed. 12) II. iii. 226 When such a discoloured picture [in oils] is washed over with dilute hydrogen peroxide, the black sulphide is oxidised into the white lead sulphate. **1951** A. GROLLMAN *Pharmacol. & Therapeutics* xxv. 514 Hydrogen peroxide solution differs from most other disinfectants in the short duration of the action, which passes off as soon as all the oxygen is liberated. **1962** J. GLENN in *Into Orbit* 192 The hydrogen peroxide jets began to turn the capsule round to orbital attitude.

c. In journalistic and colloquial use: of the age, era, etc., marked by the advent of the hydrogen bomb.

1953 *Ann. Reg. 1952* 403 [Pres. Truman's] references to the bomb..firmly dated the beginning of the 'hydrogen era' as occurring in the period of the Truman Administration. **1954** *Commonweal* 10 Dec. 279/2 An unexpected wind shift and fallout of radioactive ashes made some unfortunate Japanese fishermen..the first public victims of the hydrogen age. **1959** *Times Lit. Suppl.* 29 May 315/1 His naval reforms..seem unimportant..in the hydrogen age.

3. *Comb.* **hydrogen-like** *a. Physics*, consisting (like the hydrogen atom) of a nucleus to which is bound a single negatively charged particle; characteristic of such an atom.

1927 E. N. DA C. ANDRADE *Struct. Atom* (ed. 3) ix. 190 An atom of helium from which one electron has been altogether removed, and an atom of lithium from which two electrons have been altogether removed, constitute similar systems, and may be called hydrogen-like. **1927** J. FISHER tr. *Born's Mech. Atom* iii. 155 The orbit of the radiating electron was hydrogen-like for large values of *k*, since it is situated in an approximately Coulomb field of force. **1951** S. DUSHMAN *Fund. Atomic Physics* ix. 131 A hydrogenlike atom, having a nucleus of charge *Ze* and one electron revolving about this nucleus in a circular orbit. **1969** K. ZIOCK *Basic Quantum Mech.* v. 89 Muonic atoms have hydrogen-like spectra.

hydrogenase (həidrǫ·dʒĕnēiz, -ēis). *Biochem.* [ad. F. *hydrogénase* (J. de Rey-Pailhade 1900, in *Bull. de la Soc. chim. de Paris* XXIII. 668), f. HYDROGEN + *-ASE.] Any enzyme which catalyses the addition of hydrogen to an organic substrate.

1900 *Jrnl. Chem. Soc.* LXXVIII. II. 678 The evolution of gas [from the yeast] is almost immediately arrested by the addition of sulphur, a fact which suggests that the ferments in question are closely related to philothion or hydrogenase. **1943** *Jrnl. Biol. Chem.* CLI. 384 Hydrogenase is closely related to the nitrogen-fixing system in *Azotobacter*. **1965** A. H. ROSE *Chem. Microbiol.* vii. 142 There exists a close relationship between the ability of organisms..to fix nitrogen, and the presence in the organisms of the enzyme hydrogenase. **1966** FRENKEL & COST in Florkin & Stotz *Comprehensive Biochem.* XIV. viii. 417 Hydrogenase activity can also be observed in certain algae which can carry out photoreduction of CO_2 with molecular hydrogen.

hydrogenator (həidrǫ·dʒĕnēⁱtəɹ). [f. HYDROGENAT(E *v.* + -OR.] A vessel or apparatus in which hydrogenation is carried out.

1914 C. ELLIS *Hydrogenation of Oils* xi. 192 The contents are thoroughly agitated and transferred to the hydrogenator where the actual hydrogenation takes place. **1963** *Times* 17 Jan. 7/1 The revolutionary feature of the gas recycle hydrogenator..is that it can produce high quality town gas solely from light petroleum distillate.

hydrogenic, *a.* Restrict *rare* to sense in Dict. and add: **b.** *Physics.* = **hydrogen-like* adj.

1935 CONDON & SHORTLEY *Theory Atomic Spectra* v. 137 Energy levels in hydrogenic atoms. **1960** J. C. SLATER *Quantum Theory of Atomic Struct.* I. viii. 190 The wave function of an electron in such a spherical field is very similar to the hydrogenic function given in Eq. (7–31). **1970** G. K. WOODGATE *Elem. Atomic Struct.* ii. 20 The equation therefore applies to all the hydrogenic atoms hydrogen, deuterium, tritium, muonium, positronium, etc.

hydrogenite (həidrǫ·dʒĕnəit). [ad. F. *hydrogénite* (P. Mauricheau-Beaupré 1908, in *Compt. Rend.* CXLVII. 310), f. HYDROGEN + -ITE[1] 4 a.] Either of two powders formulated to provide a convenient and portable means of generating hydrogen: (*a*) a mixture of aluminium filings, mercuric chloride, and potassium cyanide; (*b*) a mixture of ferro-silicon, sodium hydroxide, and usually also calcium hydroxide.

1908 *Jrnl. Chem. Soc.* XCIV. II. 829 Aluminium filings are mixed with a small quantity of mercuric chloride and potassium cyanide in powder. The product, to which the name 'hydrogenite' is given, has D = 1·42. **1911** *Chem. Abstr.* V. 1499 'Hydrogenite' is a compressible powder consisting of an alloy of Si and a base of soda lime, and when ignited by a suitable 'match' it reacts spontaneously..with the evolution of H. **1922** J. W. MELLOR *Comprehensive Treat. Inorg. & Theoret. Chem.* I. vii. 285 A mixture, devised by G. F. Jaubert (1910), containing 25 parts of 90–95 per cent. ferrosilicon or manganosilicon, 60 of sodium hydroxide, and 20 slaked lime, is commercially known as hydrogenite. **1963** G. S. BRADY *Materials Handbk.* (ed. 9) 678 A mixture of ferrosilicon and sodium hydroxide, called hydrogenite, which yields hydrogen gas when water is added, is used for filling balloons.

hydrogenolysis (həi:drǒdʒěnǫ·lǐsis). *Chem.* [f. HYDROGEN + -O + LYSIS, after *hydrolysis*.] The splitting of a bond accompanied by the addition of an atom of hydrogen to the atoms originally bonded (one or both of these being carbon atoms).
1931 C. ELLIS *Hydrogenation Org. Subst.* xlv. 522 For reasons of convenience we shall use 'hydrogenolysis' for all processes in which fuel material is treated with hydrogen at high pressures and high temperatures... Another general term frequently used..is 'destructive hydrogenation'. **1932** *Jrnl. Amer. Chem. Soc.* LIV. 4685 The hydrogenolysis of carbon–carbon linkages. **1971** ANDERSON & BAKER in J. R. Anderson *Chemisorption & Reactions Metallic Films* II. viii. 189 The overall hydrogenolysis and disproportionation reactions may be represented by $CH_3NH_2 + H_2 \rightarrow CH_4 + NH_3$; $2CH_3NH_2 \rightarrow (CH_3)_2NH + NH_3$.

hydrogeologist (həi:drǫ,dʒi,ǫ·lǒdʒist). [f. HYDROGEOLOG(Y + -IST.] An expert in, or student of, hydrogeology.
1935 *Geogr. Jrnl.* LXXXV. 551, I might indicate a few lines of research which..require the services of hydrogeologists. **1964** *Discovery* Oct. 7/3 An FAO hydrogeologist..has put the importance of groundwater very high.

hydrograph (həi·drǒgraf). [f. HYDRO- + -GRAPH.] †**1.** An instrument for transmitting sound under water and recording messages so received. *Obs. rare.*
1893 *Westm. Gaz.* 19 Oct. 7/3 (*heading*) Talking through the water. The wonders of the hydrograph.
2. A graph showing the variation of level, speed of flow, or another quantity at some point on a river. orig. *U.S.*
1897 *Monthly Weather Rev.* (U.S.) XXV. 129/1 Hydrographs for typical points on seven principal rivers are shown on Chart VI. **1936** *Water-Supply Paper U.S. Geol. Survey* No. 771. 71 Among the graphic devices that have been widely used in the study of stream-flow data is the hydrograph of discharge, which depicts the average flows by days,..years, or other time intervals. **1969** *Stud. & Rep. Hydrol.* I. 125 A linear distributed-system model..predicts a flood hydrograph from rainfall information and catchment characteristics.

hydroid, *sb.* Add: **2.** *Bot.* [a. G. *hydroid* (H. Potonié 1883, in *Jahrb. K. Bot. Gartens Berlin* II. 243).] An element forming part of the *HYDROME tissue of a plant.
1887 W. HILLHOUSE tr. *Strasburger's Handbk. Pract. Bot.* v. 58 The perfect wood-cells..consist only of dead cell-walls, and, as..they simulate the tracheæ, i.e. vessels, they are known as tracheïdes, more recently as hydroïdes. **1968** BELL & WOODCOCK *Diversity Green Plants* iv. 126 Surrounding a core of tracheid-like cells (sclereids), containing scattered thin-walled cells (hydroids), is a zone of cells conspicuously large in transverse section. **1971** E. V. WATSON *Struct. & Life Bryophytes* (ed. 3) ix. 126 Collectively these tissues are known as hadrom and leptom (analogues respectively of xylem and phloem) but the constituent elements are conveniently termed hydroids and leptoids.

hydrol (həi·drǫl). [f. HYDR- + -OL.] **1.** *Chem.* [f. BENZ)HYDROL.] Any substituted derivative of benzhydrol (diphenylcarbinol), $(C_6H_5)_2CHOH$; *esp.* Michler's hydrol (see *MICHLER).
1897 *Jrnl. Chem. Soc.* LXXII. I. 353 (*heading*) Condensation of hydrols with aromatic amines in presence of sulphuric acid. **1937** F. C. WHITMORE *Org. Chem.* 833 The dye intermediate, Michler's hydrol, is pp'-$(Me_2N)_2$-benzhydrol. **1956** E. H. RODD *Chem. Carbon Compounds* IIIB. xvii. 106 They are too unstable and reactive to be used as dyes but some of the hydrols are intermediates for the synthesis of triarylmethane dyes. **1971** R. L. M. ALLEN *Colour Chem.* viii. 111 The formaldehyde then condenses with two molecules of unchanged dimethylaniline, giving 4,4'bis(dimethylamino)diphenylmethane, and this is oxidised to the corresponding hydrol.
2. *Chem.* A name suggested for the simple water molecule, H_2O, as a basis for the systematic nomenclature of its polymers, $(H_2O)n$.
1900 W. SUTHERLAND in *Phil. Mag.* L. 460, I propose for international convenience to call H_2O hydrol, $(H_2O)_2$ dihydrol, and $(H_2O)_3$ trihydrol. Steam is hydrol, ice is trihydrol, and water a mixture of dihydrol and trihydrol. **1915** W. M. BAYLISS *Princ. Gen. Physiol.* viii. 234 It is to be supposed that the molecular forces, which permit the molecules of hydrol to press unusually closely together [in water], disappear when the new group constituting ice is formed, so that the latter occupies the greater volume. **1957** G. E. HUTCHINSON *Treat. Limnol.* I. iii. 196 Sutherland (1900)..supposed liquid water to consist of dihydrol H_4O_2 with trihydrol H_6O_3 in solution... Other workers believed that hydrol H_2O was also present, at least near the boiling point.
3. [perh. f. HYDROL(YSIS.] A dark viscous liquid of unpleasant taste left as a mother liquor when starch is subjected to acid hydrolysis and dextrose is allowed to crystallize out.
1926 *Jrnl. Amer. Chem. Soc.* XLVIII. 2627 The fermentable part of 'hydrol' is chiefly *d*-glucose. **1953** *Jrnl. Assoc. Official Agric. Chemists* XXXVI. 457 Hydrol, or corn sugar molasses, is obtained as a by-product in the manufacture of dextrose from starch. **1954** I. A. PREECE *Biochem. Brewing* xi. 327 In badly attenuated beers there may be a notable amount of maltotriose present.., whilst if such material as hydrol is used as a sugar adjunct it may be expected to find some gentiobiose.

hydrolase (həi·drǒlēiz, -ēⁱs). *Biochem.* [a. F. *hydrolase* (Battelli & Stern 1921, in *Arch. internat. de Physiol.* XVIII. 413), f. HYDROL(YSIS + *-ASE.] Any enzyme which catalyses the hydrolysis of a substrate.
1922 *Jrnl. Chem. Soc.* CXXII. I. 1077 Ferments are divided into three groups..namely hydratases..; hydrolases producing esterification or hydrolysis; and oxydoreductases. **1955** NEILANDS & STUMPF *Outl. Enzyme Chem.* xvi. 188 Lipases and phosphatases, being hydrolytic enzymes, logically fall into the group of hydrolases. **1970** W. H. FISHMAN *Metabolic Conjugation & Metabolic Hydrolysis* 373 The demonstration of the existence of latent hydrolases in the tadpole tail.

hydrolith (həi·drolįþ). [ad. F. *hydrolithe* (G. F. Jaubert), f. HYDRO- + Gr. λίθος stone.] A commercial name for calcium hydride as used as a convenient source of hydrogen (evolved when water is added).
1906 *Chambers's Jrnl.* 28 July 558/1 A new chemical compound somewhat akin to the calcium carbide familiar as a generator of acetylene gas has been placed on the market recently under the name of hydrolithe. **1967** G. D. PARKES *Mellor's Mod. Inorg. Chem.* xxx. 697 Calcium hydride..is a colourless, crystalline compound which has been used under the name of hydrolith for making hydrogen.

hydrological, *a.* (Further examples: cf. *HYDROLOGY 2.)
1913 R. F. FOX *Princ. & Pract. Med. Hydrol.* p. xi, Several chapters are devoted to a discussion of the *indications* for hydrological treatment in different chronic ailments and diseases. *Ibid.* x. 106 A cardinal fact in hydrological medicine. **1921** *Oxf. Index Therap.* 452 It is convenient to consider hydrological treatment from two points of view. **1957** G. E. HUTCHINSON *Treat. Limnol.* I. iv. 221 (*heading*) The hydrological cycle and the water balance of lakes.

hydrologist. (Examples.)
1938 *Times* 19 Feb. 13/2 With Papanin..on Franz Josef Land..were P. P. Shirshoff, a hydrologist and hydrobiologist, E. K. Federoff, [etc.]. **1971** *Nature* 30 July 301/1 Flood, erosion, drought and pollution..are also the chief problems facing the hydrologist.

hydrology. Add: **2.** *Med.* The branch of medicine concerned with treatment by baths and waters. *rare.*
[Cf. quots. 1670, 1716 s.v. HYDROLOGICAL *a.*] **1850** J. BELL (*title*) Dietetical and medical hydrology. A treatise on baths;..with a description of bathing in ancient and modern times. **1913** R. F. FOX (*title*) The principles and practice of medical hydrology being the science of treatment by waters and baths.

hydrolube (həi·drolᵊūb). [f. HYDRO- + *lube*, repr. first syllable of *lubricant*.] Any of various non-flammable hydraulic fluids having water and a glycol as the principal constituents.
1944 *U.S. Naval Res. Lab. Rep.* P-2273. 23 Sprays of 'Hydrolube' fluids made from a polymer thickened mixture of ethylene glycol and water required over 80% oxygen to propagate a flame. **1947** *Ibid.* P-3020. 3 The name 'Hydrolube' was selected by this laboratory early in the war..for any hydraulic fluid consisting of a polymer-thickened, corrosion-inhibited, aqueous solution having one or more glycols as major organic components. **1956** *Chem. & Engin. News* 3 Sept. 4245/3 The Navy..uses 'hydrolubes' in aircraft carrier catapult systems.

hydrolysable (həi·drǫləizăb'l). *Chem.* Also (chiefly *U.S.*) **-lyzable.** [f. *HYDROLYS(E *v.* + -ABLE.] Capable of being hydrolysed.
1908 *Jrnl. Chem. Soc.* XCIV. I. 199 (*heading*) Cacao butter, especially its non-hydrolysable constituents. **1913** *Jrnl. Amer. Chem. Soc.* XXXV. 629 Those esters which are most easily hydrolyzable with excess of alkali. **1946** L. E. WISE *Wood Chem.* vi. 150 The disaccharide..must be hydrolyzable by alkali. **1965** PHILLIPS & WILLIAMS *Inorg. Chem.* I. xiv. 522 The metals with basic oxides do not give readily hydrolysable salts.

hydrolysate (həidrǫ·lisēⁱt). *Chem.* Also (chiefly *U.S.*) **-lyzate.** [f. *HYDROLYS(E *v.* + *-ate*, after *filtrate, precipitate*, etc.] A product of, or preparation obtained by, hydrolysis.
1915 *Jrnl. Amer. Chem. Soc.* XXXVII. 1634 The hydrolysate in this instance was jet-black. **1944** L. F. & M. FIESER *Org. Chem.* xvi. 408 The determination of the proportion of the different amino acids present in a given protein hydrolyzate. **1964** *Oceanogr. & Marine Biol.* II. 149 Paper chromatography is a useful tool to establish the presence or absence of amino acids and other substances in hydrolysates and tissue extracts.

hydrolyse (həi·drolaiz), *v.* *Chem.* Also (chiefly *U.S.*) **-lyze.** [f. HYDROLYSIS, after *analyse, analysis*.] **1.** *trans.* To subject to hydrolysis; to decompose by hydrolysis.
1880 [see HYDROLYSIS]. **1902** *Westm. Gaz.* 6 Jan. 2/1 Grape sugar is formed by hydrolysing cellulose with acids. **1944** L. F. & M. FIESER *Org. Chem.* xvi. 404 Proteins are hydrolyzed by acid or alkali and by enzymes, and in the case of the simple, nonconjugated proteins, the products consist of mixtures of amino acids. **1955** *Sci. Amer.* May 37/1 When a peptide or protein is hydrolyzed—treated chemically so that the elements of water are introduced at the peptide bonds—it breaks down into amino acids. **1968** C. A. HAMPEL *Encycl. Chem. Elem.* 567/1 The trivalent praseodymium ion occurs as $[Pr(H_2O)_n]^{3+}$ and is only weakly hydrolyzed in aqueous solutions. **1971** *Nature* 17 Sept. 209/1 Crude starfish extracts were hydrolysed with hydrochloric acid.
2. *intr.* To undergo hydrolysis.
1920 *Jrnl. Biol. Chem.* XLIII. 173 The gum hydrolyzes quantitatively into levulose. **1931** LEVENE & BASS *Nucleic Acids* vii. 195 Embden and Schmidt..compared the rates of hydrolysis of the three substances and found that the muscle adenylic and inosinic acids hydrolyze at about the same rate, whereas the other adenylic acid hydrolyzes much faster. **1951** C. R. NOLLER *Textbk. Org. Chem.* v. 79 Although highly toxic to animals, it hydrolyzes rapidly to harmless phosphoric acid and ethyl alcohol when exposed to moist air. **1964** N. G. CLARK *Mod. Org. Chem.* viii. 131 Tertiary halides hydrolyse most easily and primary halides are most resistant.
Hence **hy·drolysed** *ppl. a.*, **hy·drolysing** *vbl. sb.* and *ppl. a.*
1900 PERKIN & KIPPING *Org. Chem.* (rev. ed.) x. 188 The rapidity with which hydrolysis takes place depends..on the nature of the ethereal salt and of the hydrolysing agent. **1912** *Jrnl. Biol. Chem.* XII. 297 Osborne and Guest ..obtained 59·2 per cent nitrogen in amino form in completely hydrolyzed gliadin. **1935** R. H. A. PLIMMER in Harrow & Sherwin *Textbk. Biochem.* v. 157 Arginine may be directly precipitated from the hydrolyzed solution of the protein. **1943** *Thorpe's Dict. Appl. Chem.* (ed. 4) VI. 388/2 An alcoholic solution of potassium hydroxide is sometimes used for hydrolysing purposes. **1973** *Daily Tel.* 16 Feb. 17/3 In a world of monosodium glutamate and hydrolised protein, her work is a living testimonial of a fine palate.

hydrolysis. Substitute for def.: Any reaction in which a bond is broken by the agency of water and the hydrogen and hydroxyl of the water become independently attached to the two atoms previously linked; the decomposition or splitting of a compound in this way. Also applied to the analogous decomposition of an organic compound by the action of an acid or alkali, and to any reaction between a water molecule and an ion that produces a hydrogen or hydroxyl ion. (Add further examples.)
1900 PERKIN & KIPPING *Org. Chem.* (rev. ed.) x. 188 All ethereal salts are decomposed by water, mineral acids, and alkalies, the change..being spoken of as hydrolysis... $CH_3{\cdot}COOC_2H_5 + KOH = CH_3{\cdot}COOK + C_2H_5{\cdot}OH$. **1935** R. H. A. PLIMMER in Harrow & Sherwin *Textbk. Biochem.* v. 155 Hydrolysis of proteins to the amino acids is effected by boiling with acids, or alkalis, or by the action of the enzyme trypsin. **1938** C. D. HURD in H. Gilman *Org. Chem.* I. vii. 617 In reactions in which the C—N bond is severed by hydrolysis, it is universally characteristic for the nitrogen to attract the hydrogen of water, and carbon the oxygen or hydroxyl. **1948** GLASSTONE *Textbk. Physical Chem.* (ed. 2) xii. 986 The hydrolysis must then be represented by $M(H_2O)_n^{+} + H_2O \rightleftharpoons H_3O^{+} + M(H_2O)_{n-1}OH$. **1950** P. J. DURRANT *Org. Chem.* xxix. 447 Many of the chemical reactions occurring in fermentation and in the metabolism of living organisms are hydrolyses. **1951** KIRK & OTHMER *Encycl. Chem. Technol.* VII. 741 The reaction of a nitrile with water to form an amide $CH_3CN + H_2O \rightarrow CH_3CONH_2$ is thought of as a hydrolysis, although it is also a hydration. **1964** N. G. CLARK *Mod. Inorg. Chem.* xi. 201 Hydrolysis [of nitriles] may be effected either by hot mineral acids..or by hot alkali...$CH_3{\cdot}CN + 2H_2O + HCl \rightarrow CH_3{\cdot}COOH + NH_4Cl$... $CH_3{\cdot}CN + H_2O + NaOH \rightarrow CH_3{\cdot}COONa + NH_3$.

hydroly·tically, *adv.* *Chem.* [f. HYDROLYTIC *a.*: see -ICALLY.] By means of hydrolysis; as regards hydrolysis.
1928 *Chem. Abstr.* XXII. 200 (*heading*) Ion exchange of zeolitic silicates with hydrolytically dissociated salts. **1963** F. M. DEAN *Naturally Occurring Oxygen Ring Compounds* v. 135 The furan ring can sometimes be opened hydrolytically. **1969** *Jrnl. Materials Sci.* IV. 432/2 These groups must be bonded to the silicon in a hydrolytically and thermally stable manner.

hydromagnetic (həi:dro,mægne·tik), *a.* [f. HYDRO- (in *hydrodynamic*) + MAGNETIC *a.* (in *electromagnetic*).] Of, relating to, or involving an electrically conducting fluid (as a plasma or molten metal) acted on by a magnetic field.
1943 H. ALFVÉN in *Ark. f. Matem., Astr. och Fysik* XXIXB. II. 7 As the term 'electromagnetic-hydrodynamic waves' is somewhat complicated, it may be convenient to call the phenomenon 'magneto-hydrodynamic' waves. (The term 'hydromagnetic' is still shorter but not quite adequate.) **1955** *Proc. R. Soc.* A. CCXXXIII. 310 Hydromagnetic waves in rare ionized gases are of interest in connexion with galactic magnetic fields, the heating of the corona and other astrophysical problems. **1963** ALFVÉN & FÄLTHAMMAR *Cosmical Electrodynamics* (ed. 2) iii. 73 In the sun all phenomena which are large enough to be observed visually from the earth are hydromagnetic, and the same holds for interstellar clouds. **1966** FERRARO & PLUMPTON *Introd. Magneto-Fluid Mech.* (ed. 2) i. 13 It is this coupling between the electromagnetic and mechanical

forces which characterizes hydromagnetic phenomena. **1973** *Nature* 9 Nov. 58/2 Nobody can yet give a hydromagnetic description of the process by which the neutron star captures the plasma.
Hence **hy:dromagne·tically** *adv.*, from the point of view of hydromagnetics.
1970 *Nature* 4 Apr. 48/1 Warm plasmas are considered hydromagnetically.

hydromagnetics (həi:dro,mægne·tiks), *sb. pl.* (const. as *sing.*). [f. prec.: see -IC 2.] The branch of physics concerned with hydromagnetic phenomena; = *MAGNETOHYDRODYNAMICS.
1953 T. G. COWLING in G. P. Kuiper *Sun* viii. 532 The subject of hydromagnetics is a new one. **1958** *New Scientist* 27 Feb. 12 The theory of an interaction like this, between a fluid and a magnetic field, is a central problem of the new science of hydromagnetics.

hydrome, hydrom (həi·drōᵘm). *Bot.* [ad. G. *hydrom* (H. Potonié 1883, in *Jahrb. K. Bot. Gartens Berlin* II. 243), f. HYDRO- + *-ome* as in *rhizome*, etc.] The water-conducting section of a vascular bundle.
1900 B. D. JACKSON *Gloss. Bot. Terms* 273/1 Tracheome, stated by Potonié not to be the tracheal, but the hydral system of the bundle, he therefore names it Hydrome. **1911** J. M. COULTER et al. *Textbk. Bot.* II. iii. 682 The conductive portion of the xylem is known as *hadrome* (or *hydrome*). **1929** *Encycl. Brit.* XVII. 6/1 The hydrom strand is either slightly developed or altogether absent. **1969** K. ESAU *Phloem* viii. 268 The corresponding conducting cells are hydroids (from hydrom, part of hadrom consisting of conducting cells only).

hydronium (həidrōᵘ·niǝ̆m). *Chem.* [a. G. *hydronium* (A. Hantzsch 1907, in *Zeitschr. f. phys. Chemie* LXI. 306), contraction of *HYDR(OX)ONIUM.] A name for the hydrated hydrogen ion (usu. represented as H_3O^+).
1908 *Jrnl. Chem. Soc.* XCIV. II. 15 Just as a molecule of ammonia may attach itself to a hydrogen ion, forming the ammonium ion $NH_4^{\cdot}$, so it is supposed that a molecule of water may similarly attach itself, forming the 'hydronium' ion $H_3O^{\cdot}$, so that a solution of a little water in sulphuric acid is a dissociated solution of hydronium sulphate. **1937** F. C. WHITMORE *Org. Chem.* 341 The effective catalyst is the hydronium ion, $(H_3O)^+$. **1940** *Jrnl. Chem. Soc.* 1410 When the hydrogen ion is considered to occur (in aqueous solution or in a compound) in the form $[H_3O]^+$, it is advisable to call it the hydronium ion (not hydroxonium ion). **1956** *Jrnl. Amer. Chem. Soc.* LXXVIII. 5999 (*heading*) The vibrational spectrum of the hydronium ion in hydronium perchlorate. **1959** *Nomencl. Inorg. Chem.* (*I.U.P.A.C.*) 26 The ion H_3O^+..is to be known as the oxonium ion when it is believed to have this constitution, as for example in $H_3O^+ClO_4^-$, oxonium perchlorate. The widely used term hydronium should be kept for the cases where it is wished to denote an indefinite degree of hydration of the proton, as, for example, in aqueous solution. [*The latter sentence is dropped in ed. 2 (1971).*] **1966** *Mineral. Mag.* XXXV. 1071 The existence of the individual H_3O^+ pyramidal complex, called the hydronium ion, or sometimes the hydroxonium or oxonium ion, in crystalline substances is a well-established fact.

hydrophil (həi·drǒfil), **-phile** (-fəil), *a.* [f. HYDRO- + -PHIL.] = *HYDROPHILIC *a.*
1903 *Electrician* 30 Oct. 42/1 He applied two electrodes.., contact being made by means of hydrophil cotton impregnated with a 1 per cent solution of zinc chloride. **1915** W. W. TAYLOR *Chem. Colloids* i. 7 The term lyophile has been applied to those systems in which there is marked affinity between the phases, and lyophobe to the others. When water is the dispersion medium the terms hydrophile and hydrophobe are commonly used. **1930** J. ALEXANDER *Colloid Chem., an Introd.* (ed. 3) iii. 39 Colloids of the reversible type are..said to be hydrophile or lyophile, while the irreversible colloids are hydrophobe or lyophobe. **1963** A. J. HALL *Textile Sci.* i. 10 Synthetic fibres are made from organic hydrophobic (water-repellent) polymers—the natural fibres are hydrophile (water-attractive). *Ibid.* 13 The hydrophile fibre wool..contains about 18% of moisture in its ordinary air-dry state. **1971** *Nature* 12 Feb. 489/2 The outer walls of the respiratory trumpets of mosquito pupae are hydrophil while the inner lining is hydrophuge.

hydrophilia (həidrǒfi·liă). *rare.* [f. HYDRO- + Gr. φιλία fondness.] A love of being near water.
1904 G. S. HALL *Adolescence* II. xii. 195 Others..can sit by the hour, seeing and hearing the movements of water in sea or stream. The best demonstration of the fact of this hydrophilia is the amount of cold..that it will overcome. **1959** *Times Lit. Suppl.* 31 July 450/4 A symptom of the derangement of sober, law-abiding citizens succumbing to hydrophilia is public confession... These very readable confessions can infect hardened land-lubbers with sea fever.

hydrophilic (həidrǒfi·lik), *a.* [f. HYDRO- + Gr. φίλ-ος loving + -IC.] **a.** Having an affinity for water, readily absorbing water; relating to such an affinity.
1901 *Buck's Handbk. Med. Sci.* (ed. 2) III. 694/1 The ear should be very carefully dried out, as of course the effectiveness of this solution depends upon the hydrophilic properties of the materials employed. **1954** KIRK & OTHMER *Encycl. Chem. Technol.* XIII. 919 The hydrophilic fibers can absorb and transfer water, and as such they are not naturally water-repellent. **1963** R. R. A. HIGHAM *Handbk. Papermaking* ii. 37 Because of the natural affinity of cellulose for water, paper is termed hydrophilic (water-receptive). **1971** *Nature* 19 Nov. 126/3 Attached to their hydrocarbon tails phospholipid molecules also..have hydrophilic head groups.
b. *spec.* in *Physical Chem.* [after Fr. *hydrophile* (J. Perrin 1905, in *Jrnl. de Chim. phys.* III. 85)]: applied to a hydrosol that does not readily form a precipitate, and to a gel that readily forms a sol on the addition of water or on being warmed.
1915 M. H. FISCHER tr. *Ostwald's Handbk. Colloid-Chem.* ii. 50 Different names are employed in the literature for these two sets of colloids...J. Perrin [calls them] 'hydrophilic' and 'hydrophobic' colloids. **1936** J. H. PARSONS *Dis. Eye* (ed. 8) xvi. 322 It has all the properties of a hydrophilic gel, undergoing turgescence in an alkaline, deturgescence in an acid aqueous medium. **1948** GLASSTONE *Textbk. Physical Chem.* (ed. 2) xiv. 1235 Sols of gums, starches, proteins and soaps, provide instances of hydrophilic systems. **1967** M. E. HALE *Biol. Lichens* iv. 53 The imbibition of water by a hydrophilic gel.
Hence **hydrophili·city**, hydrophilic quality.
1953 *Chem. Abstr.* XLVII. 9035 'Hydrophil[ic]ity' is defined as the ratio, by wt. of the hydrophilic to hydrophobic groups in the same mol. **1970** R. D. SWISHER *Surfactant Biodegradation* ii. 32 The nonionic hydrophilic groups have a multiplicity of elements..which have a cumulative effect; increasing their numbers in the group increases the hydrophilicity of the aggregate.

hydrophilous, *a.* Add: **c.** *Bot.* Of plants, dependent upon water as the agency of pollination or dissemination of seeds; formerly = **hydrophytic* (see quot. 1898).
1883 D. W. THOMPSON tr. *Müller's Fertilisation of Flowers* III. 567 The plants of this order [*sc.* Naiadaceæ] are anemophilous or hydrophilous. **1898** POUND & CLEMENTS *Phytogeogr. Nebraska* I. iii. 67 Hydrophilous fungi..are algae-like aquatic fungi. **1902** *Encycl. Brit.* XXV. 437/1 Dissemination is effected by the agency of water, of air, of animals—and fruits and seeds are therefore grouped in respect of this as hydrophilous, anemophilous, and zooidophilous. **1920** A. ARBER *Water Plants* xviii. 235 Certain plants, however, present transitional methods of pollination, which without being actually hydrophilous, show approaches to this state. **1967** C. D. SCULTHORPE *Biol. Aquatic Vasc. Plants* ix. 248 The ultimate adaptation to aquatic life is the formation of wholly submerged hydrophilous flowers. **1973** PROCTOR & YEO *Pollination of Flowers* viii. 283 Among the most specialised of all hydrophilous species are the Eelgrasses (*Zostera*).

hydrophily (həidrǫ·fĭli). *Bot.* [f. HYDRO- + Gr. φιλία friendship.] Pollination by the agency of water.
1920 A. ARBER *Water Plants* xviii. 236 The family Hydrocharitaceae..includes within itself all stages in the transition from entomophily to hydrophily. **1967** C. D. SCULTHORPE *Biol. Aquatic Vasc. Plants* ix. 249 The probable affinities of these few specialised plants support the belief that hydrophily and marine life are both recently acquired habits. **1973** PROCTOR & YEO *Pollination of Flowers* viii. 277 The adaptations to pollination by water (*hydrophily*) are diverse, so that it is hardly possible to speak of a syndrome of hydrophily.

hydrophobe. Add: **2.** A hydrophobic substance.
1924 H. FREUNDLICH in R. H. Bogue *Theory & Applic. Colloidal Behavior* I. xii. 320 The dyestuff sols..are so little affected by alkali salts that they cannot be classed as true hydrophobes. **1970** R. D. SWISHER *Surfactant Biodegradation* ii. 31 As an example of hydrophobes which are not derived from hydrocarbons we can cite the polyoxypropylenes.
B. *adj.* = *HYDROPHOBIC *a.* 2.
1915, 1930 [see *HYDROPHIL, -PHILE *a.*]. **1970** R. D. SWISHER *Surfactant Biodegradation* vi. 207 Increased distance between the sulfonate group and the far end of the hydrophobe group increases the speed of primary biodegradation of ABS and possibly of other surfactant types.

hydrophobia. Add: **3.** The property of a substance of being hydrophobic.
1956 *Soil Sci.* LXXXII. 163 All treated powdered clays had to overcome an initial hydrophobia which took place during the first few minutes of contact with water. **1958** J. J. BIKERMAN *Surface Chem.* (ed. 2) iii. 239 At a first approximation, hydrophoby may mean good miscibility with benzene and poor miscibility with water.

hydrophobic, *a.* (*sb.*) Add: **2. a.** Tending to repel, or not to absorb, water; pertaining to such a lack of affinity.
1938 A. D. WHITEHEAD tr. *Jordan's Technol. Solvents* i. 13 This group have a pronounced solvent power for nonpolar or weakly polar, that is hydrophobic, materials. **1947** *Jrnl. Res. Nat. Bureau of Standards* (U.S.) XXXVIII. 106/1 It is common practice..to treat fabrics intended to be water repellent with various hydrophobic compounds. **1954** KIRK & OTHMER *Encycl. Chem. Technol.* XIII. 919 With hydrophobic fibers, fabric structure to a large degree will control water repellency. **1963** [see *HYDROPHIL, -PHILE *a.*]. **1967** E. CHAMBERS *Photolitho-Offset* xiv. 205 The image is oleophilic (ink-accepting) and hydrophobic (water-rejecting), making an excellent imaging material for an offset lithographic plate.
b. *spec.* in *Physical Chem.*: applied to a hydrosol that readily forms a precipitate and on evaporation or cooling gives a solid that cannot readily be converted back into a sol.
1915 [see *HYDROPHILIC *a.* b]. **1948** GLASSTONE *Textbk. Physical Chem.* (ed. 2) xiv. 1235 Typical examples of hydrophobic sols are those of metals, sulfur, sulfides and sulfur halides.
Hence **hydrophobi·city**, hydrophobic quality.
1947 *Jrnl. Res. Nat. Bureau of Standards* (U.S.) XXXVIII. 105/1 The difference between these two energies will depend upon the relative humidity, the hydrophobicity of the surface,..etc. **1963** A. J. HALL *Textile Sci.* ii. 63 These newer fibres have several different and useful properties which are associated with their increased hydrophobicity. **1969** *Nature* 15 Feb. 637/2 A quantitative treatment for comparing the average hydrophobicities of proteins.

hydrophyte. Substitute for entry: [ad. Da. *hydrophyt*, mod.L. *hydrophyta* (J. F. Schouw *Grundtræk til en almindelig Plantegeographie* (1822) 132), f. Gr. ὑδρο- water + φυτόν plant.] An aquatic plant, or one needing a water-logged environment for its growth. (Further examples.) Hence **hydrophy·tic** *a.*
1858 A. GRAY *Introd. Struct. & Syst. Bot.* 536 *Hydrophyte*: a water-plant. **1894** F. W. OLIVER et al. tr. *Kerner's Nat. Hist. Plants* I. 75 It is usual to designate all plants that grow in water as hydrophytes or water-plants. **1905** F. E. CLEMENTS *Res. Methods Ecol.* iv. 209 The effect of these conditions is to produce a plant xerophytic as to its aerial parts, and mesophytic or even hydrophytic as to its subterranean parts. **1920** A. ARBER *Water Plants* i. 3 The ultimate term in the acceptance of aquatic conditions is reached in certain hydrophytes with submerged flowers, in which even the pollination is aquatic. **1934** *Discovery* XV. 11/1 Suitable protection [for mosquitoes] may be furnished by grass or other hydrophytic vegetation. **1955** G. M. SMITH *Cryptogamic Bot.* (ed. 2) II. v. 97 As the mat [of vegetation] becomes drier..the *Sphagnum* eventually disappears and the hydrophytic angiosperms are replaced by those of a more mesophytic type. **1967** C. D. SCULTHORPE *Biol. Aquatic Vasc. Plants* v. 147 Experimental data on the water relations of submerged hydrophytes do not lend themselves to generalisation. **1973** *Nature* 12 Jan. 88/2 A halophytic vegetation covered the basin during the past century with a hydrophytic woodland forming in the first quarter of the present century.

hydroplane (həi·drǒplēⁱn), *sb.* [f. HYDRO- + PLANE *sb.*³ (in sense 2, after *aeroplane*).] **1.** A movable horizontal plane (usually one of several) projecting from the side of a submarine and used to control movement in a vertical plane and to provide stability during motion under water.
1901 *Submarine Torpedo Boats* (Lake Torpedo Boat Company) 19 The depth of submergence beneath the surface is maintained nearly constant by hydroplanes, one or more on each side of the vessel. **1902** *Sci. Amer.* 22 Nov. 346/2 While submerged..the boat has a reserve buoyancy, and in order to totally submerge it is necessary to employ the hydroplanes. **1907** S. LAKE in *Trans. Inst. Naval Archit.* XLIX. 39 The practicability of navigating on the bottom and of opening a door for the purpose of conducting mining operations, &c, together with the hydroplanes, were the features most questioned in connection with my plans for a submarine torpedo-boat submitted to the U.S. Government in 1893. **1911** *Encycl. Brit.* XXIV. 921/1 Another technical point in the design of submarines..is the desirability or otherwise of 'bow-rudders' or 'hydroplanes'. **1919** *Jane's Fighting Ships* 118 They..have large rudder and hydroplane areas and small reserve of buoyancy for quick diving and rapid handling. **1954** K. C. BARNABY *Basic Naval Archit.* (ed. 2) vi. 93 In the submerged condition, there usually remains a small excess of buoyancy...This is overcome by means of the horizontal diving rudders or 'hydroplanes'.
2. A motor boat designed to skim the surface of the water by means of a bottom that consists in part of one or more flat surfaces sloping upwards towards the bow. Also *hydroplane boat.*
1904 *Sci. Amer.* 8 Oct. 250/3 Hydroplanes—new forms of gliding boats. This name, formed on the analogy of aeroplane, is suggested for vessels which, instead of floating in water, glide over its surface as sleighs glide over ice. **1907** *Engineering* 4 Oct. 457/2 (*heading*) The Crocco and Ricaldoni hydroplane boat. **1909** *Westm. Gaz.* 5 Jan. 4/2 We have exhibited marked enterprise in regard to the hydroplane. **1909** *Ibid.* 9 Mar. 4/1 The Alla-Va, a hydroplane boat having more than ordinary pretensions to speed, has been placed in the unrestricted racer class. **1913** W. OWEN *Let.* 28 Sept. (1967) 199 Hydroplanes are in the habit of planing over the [River] Garonne. **1936** E. G. BARRILLON in W. F. Durand *Aerodynamic Theory* VI. 137 In a hydroplane in motion, the water does not act by static pressure alone, but also by a..dynamic force analogous to that on the wings of an airplane. **1957** *Times* 8 Nov. 10/3 [Mr. Donald Campbell's] hydroplane Bluebird, which is powered by a Metropolitan-Vickers jet engine, was timed over the measured kilometre at 260·107 m.p.h. **1965** R. SHECKLEY *Game of X* (1966) xviii. 127 The hydroplane climbed out of the water, balancing on her two sponsons.
†**3.** = *HYDRO-AEROPLANE, *SEAPLANE. *Obs.* exc. *Hist.*
1911 *Daily Colonist* (Victoria, B.C.) 9 Apr. 1/1 Glenn Curtiss, the aviator, gave a successful exhibition of his new hydroplane on the surface of Salt Lake this afternoon, ascending from and descending upon the water. **1913** W. S. CHURCHILL in *Hansard Commons* 17 July 1501 We have decided to call the naval hydroplane a seaplane, and the ordinary aeroplane or school machine, which we use in the Navy, simply a plane. **1913** *Q. Rev.* Apr. 471 For reconnaissance, the work is better done by hydroplanes costing a few hundreds apiece. **1914** *Daily Express* 19 Sept. 3/4 His flights in a hydroplane attracted the attention of the King at Weymouth a couple of years ago. **1969**

K. MUNSON *Pioneer Aircraft 1903–14* 161/2 Two other intermediate designs of 1907 were another airship, the No. 16, and a wingless hydroplane, the No. 18, which underwent taxying tests on the Seine.

hydroplane (həi·drǒplēⁱn), *v.* [f. prec. sb.] *intr.* **1. a.** To travel in a hydroplane boat.
1909, 1918 [see *hydroplaning* vbl. sb. below].
b. To skim the surface of the water by the use of hydroplanes.
1914 *Techn. Rep. Advisory Comm. Aeronaut. 1912–13* 237 The machine at once hydroplaned on leaving its shed. **1928** C. F. S. GAMBLE *Story N. Sea Air Station* i. 32 Having succeeded in making his machine hydroplane on her floats. **1936** J. GRIERSON *High Failure* v. 91 Once one is hydroplaning it is much easier to go on accelerating until flying-speed is gained. **1938** C. WINCHESTER *Wonders World Aviation* I. 39 When the seaplane has gathered sufficient speed it climbs over its own wave and so hydroplanes or skims along the surface.
2. Of a motor vehicle, etc.: to aquaplane. Chiefly *U.S.*
1962 *Daily Tel.* 17 May 17/5 Flooding on the Kingston by-pass caused a car travelling at speed to 'hydroplane' and..turn completely round. **1969** C. CAMPBELL *Sports Car* (ed. 3) vii. 179 Aquaplaning (hydroplaning in America)..is in effect a high-speed skating of the tyre on a film of water when travelling on wet roads. **1973** R. HAYES *Hungarian Game* xxviii. 164 The 707 skipped once as its wheels hydroplaned on the wet runway.
So **hy·droplaning** *vbl. sb.*
1909 *Westm. Gaz.* 5 Jan. 4/3 It is due entirely to its performance at Southampton that hydroplaning has gained recognition in this country. **1918** *Chambers's Jrnl.* 20 July 541/1 The water..provides the finest possible field for motor-boating in small craft, and, I should imagine, for hydroplaning. **1922** *Encycl. Brit.* XXX. 50/2 Hydroplaning efficiency..could be sacrificed for sea-worthiness. **1938** C. WINCHESTER *Wonders World Aviation* I. 37 Running on the step is the expression used to describe the hydroplaning of a seaplane on the surface of the water. **1969** [see sense 2 above].

hydroponics (həi:drǫpǫ·niks). [f. HYDRO- + Gr. πόνος work: see -IC 2.] The process of growing plants without soil, in beds of sand, gravel, or similar supporting material flooded with nutrient solutions. Hence **hydropo·nic** *a.*, **hydropo·nically** *adv.*; **hydropo·nicist**, one who practises hydroponics; **hydropo·nicum**, the building or garden in which hydroponics is practised.
1937 W. F. GERICKE in *Science* 12 Feb. 178/1 'Hydroponics', which was suggested by Dr. W. A. Setchell, of the University of California, appears to convey the desired meaning better than any of a number of words considered. **1938** *California Monthly* Feb. 13/2 My first planting..was set in the hydroponicum on September 18, 1936. *Ibid.* 40/3 The important factors..must be understood by the hydroponicist, as they must be understood by the successful gardener. **1940** *Times* 20 Apr. 1 (Advt.), Hydroponics (crops without soil)..easy if you use Gromost hydroponic mixture. **1940** *Manch. Guardian Weekly* 17 May 394/4 Hydroponics received a great impetus in the United States shortly after the trans-Pacific air line was established. **1951** A. C. CLARKE *Sands of Mars* viii. 94 The local brew..was completely synthetic—the joint offspring of hydroponic farm and chemical laboratory. **1951** J. S. DOUGLAS *Hydroponics* iii. 30 A farm or garden devoted to soilless cultivation is usually called a hydroponicum. **1955** *Sci. News Let.* 30 Apr. 282/3 Chemical gardening or hydroponics makes a favorite exhibit. **1956** *Jrnl. Brit. Interplanetary Soc.* XV. 20 As hydroponicists have pointed out, soil is in no way a perfect medium. **1961** *Astronautica Acta* VII. 134 Men have survived over extended periods on hydroponically produced plant food. **1967** *Technology Week* 23 Jan. 48/1 Water and minerals from the Moon can be used to grow food hydroponically. **1970** *New Scientist* 5 Feb. 259/3 The potential of hydroponics..is only just beginning to be tapped. **1971** *Daily Colonist* (Victoria, B.C.) 13 May 1/4 Hydroponic grass has produced encouraging results with dairy herds, increasing their milk yield. **1973** *Listener* 13 Sept. 338/2 Growing vegetables like aubergines and sweet corn and green peppers hydroponically—that is, without soil.

hydropower (həi·dro,pauᵊɹ). Also **hydro-power** (with hyphen) and as two words. [f. HYDRO- (in *hydro-electric*) + POWER *sb.*¹] Hydro-electric power.
1933 F. F. FOWLE *Stand. Handbk. Electr. Engineers* (ed. 6) xiii. 1295 (*heading*) Steam power and its relation to hydro power. **1946** *Nature* 3 Aug. 160/1 The inexhaustible nature of hydro-power. **1969** *New Scientist* 18 Sept. 565/1 In the north western states..fossil fuel is expensive and nearly all potential hydropower sites have already been exploited. **1971** *Sci. Amer.* Sept. 157/3 Tidal energy and other forms of hydropower.

hydroskimmer (həidro,ski-məɹ). [f. HYDRO- + SKIMMER *sb.* 7.] An amphibious air-cushion vehicle, supported by peripheral jet-generated air. Orig., the name assigned by the Bureau of Ships to an experimental craft of the U.S. Navy.
1960 *Bureau of Ships Jrnl.* Feb. 22 Hydro-skimmer Model A David Taylor Model Basin. **1962** *Ibid.* Apr. 27/2 The Bureau of Ships has awarded a contract..for the design and construction of a 22-ton hydroskimmer research craft, designated SKMR-1...The craft will have four cushion fans..which will provide the lift needed. **1965** *Naval Engineers Jrnl.* June 364/1 Ground effect machines are very much in the picture of the Amphibious Force of the future. One such machine is the hydroskimmer which rides free of the surface on an air cushion... They will skim over land, sand bars, marshes, mud flats and open water with equal ease. **1970** M. W. CAGLE *Flying Ships* i. 1 One of our first tasks..is to straighten out the varied and confusing terminology which surrounds words like 'hovercraft', 'PAK-V' and 'hydroskimmer', and cryptic initials like ACV, GEM, TAC, CAB, SEV, and SES.

hydrosol (həi·drǒsǫl). [f. HYDRO- + SOL(UTION.] A sol in which the liquid constituent is water.
1864 [see *ALCOSOL]. **1895, 1938** [see *HYDROGEL]. **1937** *Industr. & Engin. Chem.* (News Ed.) 10 Mar. 104/2 Gold hydrosols nearly always contain unreduced gold compounds, which are fully reduced to metal hydrosols under suitable conditions. **1954** R. L. PARKER tr. *Niggli's Rocks & Min. Deposits* xi. 463 In mineralogy the hydrosols—i.e., the colloidal solutions with water as the dispersing medium—are almost the only sols of any importance.

hydrothermally (həidropþ·ɹmăli), *adv.* [f. HYDROTHERMAL *a.* + -LY².] By hydrothermal action.
1941 *Ecology* Oct. 448/2 The hydrothermally altered andesites and basalts of parts of the Great Basin. **1957** [see *HUTTONITE]. **1962** W. A. DEER et al. *Rock-Forming Min.* V. 329 Pure hydroxyapatite crystals up to 0·3 mm. in length have been synthesized hydrothermally by the hydrolysis of monetite..at 300°C. and 1250 lb./in.² saturated steam pressure. **1965** G. J. WILLIAMS *Econ. Geol. N.Z.* vi. 61/1 All writers presumed the gold to have been derived hydrothermally from granitic sources.

hydrotropic, *a.* Add: **2.** *Physical Chem.* [f. *HYDROTROPY.] Of, pertaining to, or produced by hydrotropy; causing a substance that is otherwise only slightly soluble in water to dissolve readily.
1916 *Jrnl. Chem. Soc.* CX. II. 555 (*heading*) Hydrotropic phenomena. **1946** *Industr. & Engin. Chem.* Apr. 382/2 Most hydrotropic solutions precipitate the solute on dilution with water. **1954** P. A. WINSOR *Solvent Properties Amphiphilic Compounds* v. 123 The reduction in solvent action on dilution..is very great with hydrotropic salts of lower molecular weight.

hydrotropically (həidro,trǫ·pikăli), *adv.* [f. prec.: see -ICALLY.] **1.** *Bot.* In a manner that results in a movement towards water.
1915 *Ann. Bot.* XXIX. 281 A disturbance of the equilibrium within the cells would thus be effected in exactly the same way as by the difference of osmotic pressure in hydrotropically stimulated roots.
2. *Physical Chem.* As regards hydrotropy.
1928 *Chem. Abstr.* XXII. 770 Hydrotropically active salts. **1951** E. HÄGGLUND *Chem. Wood* iv. 261 The extracted lignin can be precipitated from the solution by diluting the latter to a lower concentration of the hydrotropically active salt.

hydrotropy (həidrǫ·trǫpi). *Physical Chem.* [ad. G. *hydrotropie* (C. Neuberg 1916, in *Biochem. Zeitschr.* LXXVI. 107), f. HYDRO- + Gk. τροπή turn, turning.] The phenomenon whereby a substance that is only slightly soluble in water will readily dissolve in certain aqueous solutions.
1928 *Chem. Abstr.* XXII. 770 (*heading*) The applicability of Traube's rule to the phenomenon of hydrotropy. **1950** J. W. MCBAIN *Colloid Sci.* xvii. 268 Hydrotropy occurs in concentrated solutions of salts or of colloidal electrolytes. **1954** P. A. WINSOR *Solvent Properties Amphiphilic Compounds* i. 5 With the amphiphilic salts of short chain length (e.g. C_2—C_8) the 'solubilization' effect becomes marked only with their rather concentrated aqueous solutions and has, in this case, been termed 'hydrotropy'.

hydrovane (həi·drǒvēⁱn). [f. HYDRO- + VANE.] **a.** = *HYDROPLANE *sb.* 1.
1919 *Times* 22 Mar. 8/1 The submarine commander.. could do a 'crash dive', that is, go under with full weight on, hydrovanes set hard down, and taking in water ballast. **1940** 'N. SHUTE' *Landfall* ii. 49 British submarines carried identification marks upon the hydrovanes.
b. = *HYDROFOIL 1.
1920 *Glasgow Herald* 27 Apr. 7 The possibilities of the large flying boat are very great...Hydrovanes may be found..to give good results for reducing landing shocks and increasing the 'getting off' efficiency. **1936** *Jrnl. R. Aeronaut. Soc.* XL. 476 A combination of ship and hydrovane can only reduce the overall resistance to propulsion. **1939** *Ibid.* XLIII. 402 It is proposed to provide flying boats with hydrovanes. **1967** *New Scientist* 9 Mar. 460/3 The special type of hydrofoil anchor designed by Kingston..has hydro-vanes rigged in such a way as to be self-reversing in tidal currents.

hydroxamic (həidrǫksæ·mik), *a.* *Chem.* [tr. G. *hydroxamsäure* hydroxamic acid (H. Lossen 1869, in *Ann. d. Chem. u. Pharm.* CL. 315), f. *hydrox(ylamin* HYDROXYLAMINE + *am(id* AMIDE: see -IC.] *hydroxamic acid*: any of the acyl derivatives, R·CO·NHOH or $(R{\cdot}CO)_2$-NHOH, of hydroxylamine (see quot. 1966).
1875 *Jrnl. Chem. Soc.* XXVIII. 751 (*heading*) Distillation of hydroxamic acids. *Ibid.* 766 The ethers of aromatic hydroxamic acids. **1926** *Biochem. Jrnl.* XX. 1362 The hydroxy-acids present in rape oil give hydroxamic acids whose sodium salts are soluble in alcohol. **1966** MILLAR & SPRINGALL *Sidgwick's Org. Chem. Nitrogen* (ed. 3) ix. 333 The mono N-acyl derivatives of hydroxylamine are usually referred to as hydroxamic acids, a name which strictly refers to structure (I). They can clearly have the alternative structure (II) which should be called a hydroximic acid. In no case are the two isomers known as separate compounds: (I) R·C=O (II) R·COH and a com-

NHOH NOH

pound of this class..can react as if it had either structure. Commonly the term hydroxamic is used to imply either of these structures.

hydroxide. Add: **2.** *attrib.*, as *hydroxide ion.*
1955 *Chem. & Engin. News* 1 Aug. 3190/3 The name 'hydroxyl' has long denoted the group OH. In organic contexts it is undoubtedly correct; but the expression 'hydroxyl ion' is very frequently used, although possibly less logical than 'hydroxide ion'. **1959** *Nomencl. Inorg. Chem.* (I.U.P.A.C.) 30 Certain polyatomic anions have names ending in -ide. These are: OH⁻ hydroxide ion [etc.]... The OH⁻ ion should not be called the hydroxyl ion. The name hydroxyl is reserved for the OH group when neutral or positively charged.

hydroxo- (həidrǫ·kso). *Chem.* A combining form of HYDROXYL, denoting a coordinated hydroxyl group; also used *attrib.* as an independent word.
1907 *Jrnl. Chem. Soc.* XCII. II. 560 This salt and others of the same series are classified by the author [*sc.* A. Werner] as hydroxonitrosotetra-ammineruthenium salts. *Ibid.*, Precipitates of the neutral hydroxo-salts. **1940** *Jrnl. Chem. Soc.* 1411 When the hydroxyl group is bound in a complex, Werner's system of notation should be used, according to which the hydroxyl groups are designated *hydroxo-* or *ol-* groups. **1965** PHILLIPS & WILLIAMS *Inorg. Chem.* I. xiv. 532 Examples include..the formation in alkaline solution of the polynuclear cationic hydroxo-complexes of the cobalt(III) ion, e.g. $[Co(en)_2(OH)]_2^{4+}$ where en is ethylenediamine.

hydroxocobalamin (həidrǫ:ksokobǫ·lămin). *Biochem.* [f. *HYDROXO- + *COBALAMIN.] An analogue of cyanocobalamin (vitamin B_{12}) in which the cyanide ion is replaced by a hydroxide (OH^-) ion.
1950 [see *COBALAMIN]. **1961** *Lancet* 26 Aug. 483/1 For the treatment of pernicious anæmia hydroxocobalamin appears to be equal in activity to cyanocobalamin and is better retained. **1970** *Times* 24 Jan. 7/3 The condition can be successfully treated by injection of hydroxocobalamin, a special form of vitamin B_{12}.

hydroxonium (həi:drǫksōᵘ·niǝ̆m). *Chem.* [a. G. *hydroxonium* (A. Hantzsch 1907, in *Zeitschr. f. phys. Chemie* LXI. 306), f. HYDR- + *OXONIUM.] = *HYDRONIUM.
1925 *Chem. Abstr.* XIX. 2309 (*heading*) Action of ammonium chloride vapor on metals and similarity of ammonium salts and hydroxonium salts as acids. *Ibid.* 2310 This expt. provides an important confirmation of Hantzsch's hydroxonium theory. **1940, 1966** [see *HYDRONIUM]. **1956** *Jrnl. Chem. Soc.* 2913 (*heading*) Raman spectra and constitution of solid hydrates. Hydroxonium perchlorate, nitrate, hydrogen sulphate, and sulphate. **1957** G. E. HUTCHINSON *Treat. Limnol.* I. iii. 197 Three of the hydrogen atoms could be at 0·99 A. and one at 1·77 A., corresponding to the hydroxonium ion H_3O^+.

hydroxy-. Add: **3.** Special Combs.: **hydro:xyamphe·tamine** *Pharm.*, a sympathomimetic amine that is a hydroxy derivative of amphetamine but lacks its stimulant effect on the central nervous system, and is used (as a solution of the hydrobromide) as a nasal decongestant and a mydriatic; methyl tyramine, $HO{\cdot}C_6H_4{\cdot}CH_2CH(NH_2)CH_3$; **hydro:xybenzo·ic** *a.*, in *hydroxybenzoic acid*, any of three derivatives, $HO{\cdot}C_6H_4{\cdot}COOH$, of benzoic acid having a ring-substituted hydroxyl group; *spec.* the *ortho* isomer, salicylic acid; **hydro:xybuty·ric** *a.*, in *hydroxybutyric acid*, any monohydroxy derivative of a butyric acid; *spec.* the acid $CH_3CHOHCH_2$-COOH, one of the 'ketone bodies'; **hydro:xycitrone·llal**, a monohydroxy derivative, $C_{10}H_{20}O_2$, of citronellal, prepared synthetically and used extensively in perfumery to give the odour of lily of the valley; **17-hydro:xycorticoste·roid** *Biochem.*, any corticosteroid that has a hydroxyl group attached at position 17 of the steroid nucleus; **hydro:xycorticoste·rone** *Biochem.* = *HYDROCORTISONE; **hydro:xyly·sine**, an amino-acid, $H_2N{\cdot}CH_2CHOH(CH_2)_2CH(NH_2)COOH$, found principally in collagen; **hydro:xyme·thyl**, the univalent radical $—CH_2OH$, in which a hydroxyl group replaces one of the hydrogen atoms of a methyl group; **hydro:xymethyla·tion**, the introduction of a hydroxymethyl group into a compound; **hydro:xymethylcy·tosine**, a pyrimidine base, $C_4H_4N_3O$-CH_2OH, that is present in place of cytosine in the DNA of certain bacteriophages; **hydro:xypro·line**, any of the monohydroxy

derivatives, $C_5H_9NO_3$, of proline; *esp.* 4-hydroxyproline, the lævorotatory form of which is a non-essential amino-acid (strictly, an imino-acid) that is an important constituent of collagen and elastin; **hydro:xytry·ptamine**, any of the ring-substituted monohydroxy derivatives, $C_{10}H_{12}N_2O$, of tryptamine, esp. *SEROTONIN (5-hydroxytryptamine).
1948 *Arch. Otolaryngol.* XLVIII. 659 The product used ..provides..800 units of penicillin per cubic centimeter in a 1 per cent solution of hydroxyamphetamine hydrobromide (paredrine hydrobromide). **1876** *Jrnl. Chem. Soc.* II. 85 (*heading*) A new hydroxybenzoic acid. **1888** BLOXAM *Chem.* (ed. 6) 524 Salicylic or hydroxybenzoic acid, $C_6H_4.HO.CO_2H$, is prepared artificially by combining phenol with soda, and heating the product in carbonic acid gas. **1958** PACKER & VAUGHAN *Mod. Approach Org. Chem.* xxiv. 796 The most important hydroxy-benzoic acid is salicylic acid (*o*-hydroxybenzoic acid). **1879** *Jrnl. Chem. Soc.* XXXVI. 615 (*heading*) Hydroxybutyric acid, $C_4H_8O_3$. **1970** W. S. HOFFMAN *Biochem. Clin. Med.* (ed. 4) iv. 149 The so-called ketone bodies, namely, acetoacetic acid, β-hydroxybutyric acid, and acetone..are formed during the oxidation of fatty acids. **1929** *Chem. Abstr.* XXIII. 95 Pure rhodinal..was prepd. by vacuum distn. of hydroxycitronellal. **1951** P. Z. BEDOUKIAN *Perfumery Synthetics & Isolates* 231 Hydroxycitronellal has not been found in nature and is entirely a development of the laboratory. **1951** *Jrnl. Clin. Endocrinol. & Metabolism* XI. 1029 Treatment with adrenocorticotropic hormone led to an increase in concentration of the 17-hydroxycorticosteroids. **1968** *Brit. Med. Bull.* XXIV. 225/2 Simple steroid screening tests such as single determinations of 17-hydroxycorticosteroid in plasma or in collections of urine. **1943** REICHSTEIN & SHOPPEE in *Vitamins & Hormones* I. 356 17-Hydroxycorticosterone... Isolated and described as 'Substance M' by Reichstein (55) and by Kendall et al. (54), who called it 'Compound F'. **1961** *Lancet* 7 Oct. 796/1 Further reports on the urinary levels of 17-hydroxycorticosterone.. of such patients will be awaited with interest. **1925** *Chem. Abstr.* XIX. 2347 Hydroxylysine does not occur in casein or ovalbumin. **1969** *New Scientist* 3 July 11/1 One [family] excreted excessive quantities of hydroxylysine. **1937** *Chem. Abstr.* XXXI. 9888/3 (Index), Hydroxymethyl group, detn. in resol. **1964** G. H. HAGGIS et al. *Introd. Molecular Biol.* ix. 218 In 5-hydroxy-methylcytosine, a hydroxy-methyl (—CH_2OH) group replaces the hydrogen at position 5. **1948** H. WYNBERG tr. *Theilheimer's Synthetic Methods Org. Chem.* I. 147 (*heading*) Hydroxymethylation of phenols with formaldehyde. **1968** H. V. APOSHIAN in H. Fraenkel-Conrat *Molecular Basis Virol.* v. 504 Shortly after infection with a T-even phage, a new enzyme is produced which catalyzes the hydroxymethylation of dCMP at the 5 carbon atom. **1952** WYATT & COHEN in *Nature* 20 Dec. 1072/1 We have now resolved this discrepancy by isolation from *T*-even bacteriophages of a new pyrimidine base, identified as 5-hydroxymethylcytosine. **1968** H. V. APOSHIAN in H. Fraenkel-Conrat *Molecular Basis Virol.* v. 499 The unusual base, 5-hydroxymethylcytosine, is not found in uninfected *E. coli.* **1905** *Jrnl. Chem. Soc.* LXXXVIII. I. 545 (*heading*) Synthesis of hydroxypyrrolidinecarboxylic acids (hydroxyprolines). **1965** A. MEISTER *Biochem. Amino Acids* (ed. 2) I. i. 87 3-Hydroxyproline was first obtained from cattle Achilles' tendon collagen. **1967** *New Scientist* 24 Aug. 375/1 In fish from cold water the precise proportion of the imino-acid hydroxyproline in the important protein collagen..was lower than that in fish from warm water. **1949** *Jrnl. Biol. Chem.* CLXXX. 968 Tentatively, then, the constitutional formula of 5-hydroxytryptamine.. may be assigned to this vasoconstrictor principle. **1956** *Nature* 18 Feb. 332/2 The pharmacologically active substance 5-hydroxytryptamine is known to be liberated from platelets during clotting. **1956** *Chem. Abstr.* L. 5630 (*heading*) A new synthesis of bufotenine and related hydroxytryptamines. *Ibid.* 5631 6-Hydroxytryptamine.

hydroxyapatite (həidrǫ:ksi,æ·pătəit). [f. HYDROXY- + APATITE.] Calcium phosphate hydroxide, ideally $[Ca_3(PO_4)_2]_3.Ca(OH)_2$, which occurs as a rare mineral of the apatite group (in which hydroxyl replaces all or most of the fluorine of the commoner fluorapatite), and which is the principal inorganic constituent of tooth enamel and bone.
1912 *Bull. U.S. Geol. Surv.* No. 509. 100 The three formulas developed are repeated here... To the list are added those of apatite, both the fluorapatite and the hypothetical hydroxyapatite...Hydroxyapatite, $9CaO.3P_2O_5$-$CaO.H_2O$. **1917** *Jrnl. Chem. Soc.* CXI. 638 The mineral constituents of bone consist in the main of hydroxyapatite, $(Ca_3P_2O_8)_3Ca(OH)_2$, mixed with a certain amount of calcium carbonate. **1962** W. A. DEER et al. *Rock-Forming Min.* V. 333 The hydroxyapatite..in talc schist and the fluor-hydroxyapatite in chlorite schist, both from a serpentinite near Holly Springs, Georgia, are considered to be metamorphic in origin. **1964** A. WHITE et al. *Princ. Biochem.* (ed. 3) xli. 781 It seems likely that in bone, divalent cations other than Ca⁺⁺ can replace Ca⁺⁺ in the hydroxyapatite crystal lattice, whereas anions other than phosphate and hydroxyl may be adsorbed on the vast areas of surface offered by the minute crystals. **1971** *Physics Bull.* July 411/1 The enamel of a tooth is composed of hydroxyapatite crystals in prismatic form.

hydroxylapatite (həidrǫ:ksil,æ·pătəit). [f. HYDROXYL + APATITE.] = *HYDROXYAPATITE.
1927 N. H. & A. N. WINCHELL *Elem. Optical Mineral.* (ed. 2) II. viii. 128 The following formulas and names are tentatively assigned to minerals closely related to apatite. ..Hydroxylapatite (unknown alone) $3Ca_3P_2O_8.Ca(OH)_2$. **1946** *Nature* 6 Apr. 453/1 It is generally believed that bone salts are constituted either of hydroxylapatite which has absorbed calcium carbonate or of carbonato-apatite. **1951** C. PALACHE et al. *Dana's Syst. Min.* (ed. 7) II. 882 Fluorapatite, chlorapatite, and hydroxylapatite conform to the formula $Ca_5(PO_4)_3(F,Cl,OH)$. **1962** WALEY & VAN HEYNINGEN in A. Pirie *Lens Metabolism Rel. Cataract* 343 Chromatography on a column of calcium phosphate (hydroxylapatite).

hydroxylase (həidrǫ·ksilēⁱz, -ēⁱs). *Biochem.* [f. HYDROXYL + *-ASE.] Any enzyme which catalyses the hydroxylation of a compound.
1953 *Jrnl. Biol. Chem.* CCI. 187 The authors propose.. the term 11β-hydroxylase..for the enzyme whose function and properties have been described here. **1970** E. HEFTMANN *Steroid Biochem.* vii. 64 Animal tissues contain hydroxylases that mediate the introduction of oxygen into specific positions of steroids to produce α- or β-oriented hydroxyl groups.

hydroxylate (həidrǫ·ksilēⁱt), *v.* *Chem.* [f. HYDROXYL + -ATE³.] **a.** *trans.* To introduce a hydroxyl group into (a molecule or compound). **b.** *intr.* To accept a hydroxyl group.
1951 P. Z. BEDOUKIAN *Perfumery Synthetics & Isolates* 232 The tertiary carbon atom hydroxylates with greater ease than the secondary carbon atom. **1954** G. M. BADGER *Struct. & Reactions Aromatic Compounds* vii. 308 Such a reaction mixture also hydroxylates aromatic compounds. **1970** E. HEFTMANN *Steroid Biochem.* x. 97 Either pregnenolone or progesterone may be hydroxylated at C-17 by an enzyme.
So **hydro·xylated**, **hydro·xylating** *ppl. adjs.*
1900 *Rep. Brit. Assoc. Adv. Sci.* 298 The hydroxylated nucleus of the tribromonaphthol. **1930** *Chem. Abstr.* XXIV. 2988 $H_2S_2O_8$ may be used as a hydroxylating agent. **1954** G. M. BADGER *Struct. & Reactions Aromatic Compounds* vii. 308 Hydrogen peroxide becomes a very effective hydroxylating agent in the presence of mineral acid. **1961** *Engineering* 31 Mar. 462/2 Hydroxylated carboxylic acids.

hydroxylation (həidrǫksilēⁱ·ʃən). *Chem.* [f. HYDROXYL + -ATION.] The introduction of a hydroxyl group into a molecule or compound.
1879 *Jrnl. Chem. Soc.* XXXVI. 139 (*heading*) Hydroxylation by direct oxidation. **1962** H. HEATH in A. Pirie *Lens Metabolism Rel. Cataract* 364 Ascorbic acid is an essential cofactor for the hydroxylation of tryptophan to 5-hydroxytryptophan.

hydroxylic (həidrǫksi·lik), *a.* *Chem.* [f. HYDROXYL + -IC.] Of or containing a hydroxyl group.
1898 *Jrnl. Chem. Soc.* LXXIII. 997 The substance is therefore an anhydride of nitrocamphor,..but it must necessarily be derived from a hydroxylic modification. **1948** GLASSTONE *Textbk. Physical Chem.* (ed. 2) ix. 680 Hydroxylic solvents, e.g., water and alcohols. **1950** *Symposium Hydrogen Bond 1949* (R. Inst. Chem.) 4 The oxygen atom of one molecule coordinates with the hydroxylic hydrogen atom of another, thus forming the weak intermolecular bond responsible for molecular association. **1963** C. N. R. RAO *Chem. Applications Infrared Spectroscopy* iii. 175 The important bond in hydroxylic compounds arises from the O–H stretching vibration.

hydroxyprogesterone (həidrǫ:ksi,prōᵘdʒe·stěrōᵘn). *Biochem.* and *Pharm.* [f. HYDROXY- + *PROGESTERONE.] Any of several synthetic derivatives of progesterone that have a hydroxyl group in place of one of the hydrogen atoms of the steroid nucleus; *esp.* the derivative in which the hydroxyl group is at position 17, used (chiefly as the caproate ester) as a long-acting progestational compound in cases of *corpus luteum* deficiency.
1941 *Jrnl. Biol. Chem.* CXXXIX. 855 It was found to be the hitherto unknown steroid..17-hydroxyprogesterone.., a position isomer of desoxycorticosterone. **1948** W. H. PEARLMAN in Pincus & Thimann *Hormones* I. xi. 441 Erhart and co-workers..described the preparation of 12(α)-hydroxyprogesterone..from desoxycholic acid..; the substance was found to be lacking in physiological activity. **1954** *Nature* 30 Oct. 839/2 Various derivatives of progesterone by oxidation at C_{11} such as..11-α-hydroxyprogesterone and 11-β-hydroxyprogesterone..have been found to be only about 3–10 per cent as progestational as progesterone. **1962** N. APPLEZWEIG *Steroid Drugs* v. 100 When administered to pregnant women for the prevention of habitual abortion, Norlutin and 17α-hydroxyprogesterone caproate have been shown on occasion to cause masculinization of the female fetus. **1968** J. H. BURN *Lect. Notes Pharmacol.* (ed. 9) 93 The progestogen [in contraceptive pills] is either a 19-norsteroid derivative.. or a 17-hydroxy-progesterone derivative.

hydroxyquinoline (həidrǫ:ksi,kwi·nǒlīn). *Chem.* [f. HYDROXY- + QUINOLINE.] Any of the seven monohydroxy derivatives, C_9H_7NO, of quinoline, esp. *OXINE (8-hydroxyquinoline).
1881 *Jrnl. Chem. Soc.* XL. 613 An excellent yield of hydroxyquinoline is obtained by fusing the sulphonic acid with three times its weight of soda. **1925** *Chem. Abstr.* XIX. 1572 The formation of 4-hydroxyquinoline is suspected. **1937** A. W. GROVES *Silicate Analysis* vi. 131 The perfection of the separation of aluminium, iron, titanium, and zirconium from beryllium by means of 8-hydroxyquinoline now affords the improved method for the determination of beryllia which is adopted here. **1947** *Biochem. Jrnl.* XLI. 544/2 It was found that oxine alone of the seven isomeric hydroxyquinolines had chelating ability. **1970** *Watsonia* VIII. 23 Root tips..were given a pretreatment in a 0·002M solution of 8-hydroxyquinoline.

hydroxyzine (həidrǫ·ksizīn). *Pharm.* [f. HYDROXY- + PIPERA)ZINE.] A tranquillizing drug which is usu. administered as the hydrochloride, a white bitter-tasting powder, and is a complex derivative of piperazine.
1956 *Jrnl. Amer. Med. Assoc.* 16 June 604/1 Hydroxyzine (Atarax) hydrochloride is a new tranquilizing drug that is currently under clinical investigation. **1968** J. H. BURN *Lect. Notes Pharmacol.* (ed. 9) 59 Hydroxyzine appears to relieve anxiety without impairing critical faculties.

hyena. Add: **hyenoid** *a.* (example).
1945 G. G. SIMPSON *Princ. Class. & Class. Mammals* III. 224/1 The genera here set aside..are large, later Tertiary canids with heavy jaws, rather distantly convergent toward the hyenas, and so sometimes called 'hyænoid dogs'.

hygeen, hajeen (hidʒī·n, hădʒī·n). Also 6 hugiun, 8 hajan, etc., **hyghgeen,** 9 **hadjeen, hejeen, hejin.** [Arab. *hajīn* dromedary, pronounced in Egypt *hagīn* (cf. Syriac *hagīnā*, *hugānā*, in the Talmud *hōgnā*). Ult. origin uncertain.] A riding dromedary.
1600 J. PORY tr. *Leo's Geographical Hist. Afr.* IX. 338 Of camels there are three kinds; whereof the first being called Hugiun [orig. *quarum primi Hugiun nuncupati*] are grosse, and of a tall stature. **1713** *Guardian* No. 124 There has not been a Tyger, Leopard, Elephant or Hyghgeen, for some Years past, in this Nation, but I have taken their particular Dimensions. **1790** J. BRUCE *Trav.* IV. 332 If.. there was danger, [he] should return..mount a hajan or dromedary, and [etc.]. **1803** W. WITTMAN *Trav. in Turkey* 216 A smaller and more slender species of the camel, called *hedgin*, is mounted by the natives and others, and is capable of making a greater progress, on a journey, than a horse. **1830** [see *DELOUL]. **1864** J. A. GRANT *Walk across Afr.* 419 A 'Hadjeen', or riding camel, is indispensable to comfort. **1865** W. G. PALGRAVE *Narr. Journey through Arabia* I. 325 The dromedary is the race-horse of his species, thin, elegant,..light of step, easy of pace,.. though yet more often the dromedary enjoys his special title of 'hejeen' or 'dolool'. **1875** [see *DELOUL]. **1890** S. W. BAKER *Wild Beasts* II. 374 As a general rule, the hygeens are not so powerfully proportioned as those which carry baggage. **1908** *Animal Managem.* 276 The riding camel (..Hagheen, Egypt).

hygric (həi·grik), *a.* *rare.* [f. Gr. ὑγρός moist, wet + -IC.] Relating to water or moisture.
1902 *Jrnl. Nerv. & Mental Dis.* XXIX. 751 Various deviations from the normal in the hippocampal convolution contribute..support to the hypothesis that hygric illusions may be referred to that area. **1907** *Trans. Med. Soc. London* XXX. 371 The patient has complained of.. 'hygric sensibility', that is to say..whatever she touches with her hands..feels wet to her.

hygro-. Add: **hygrothe·rmograph**, an instrument that records the temperature and humidity of the air on a single chart.
1929 WEAVER & CLEMENTS *Plant Ecol.* xi. 264 It is convenient for comparison to record both humidity and air temperature upon the same record sheet... Such an instrument is called a hygrothermograph. **1969** *Ecology* L. 742/1 Hygrothermograph readings were taken..at three elevations above ground and four exposures.

hygrophilous (həigrǫ·filəs), *a.* *Ecol.* [ad. Fr. *hygrophile* (J. Thurmann *Essai de Phytostatique* (1849) I. 268), f. HYGRO- + Gr. φίλος loving.] Of plants: growing in a moist environment. Also **hy·grophile** *sb.*, a plant of this type.
1863 [in Dict. s.v. HYGRO-]. **1878** A. HENFREY *Elem. Bot.* (ed. 3) IV. i. 661 Plants are divided into *Xerophiles*, or those capable of existing in very dry climates; *Hygrophiles*, or those which can only exist in the presence of abundant moisture. **1883** [in Dict. s.v. HYGRO-]. **1903** W. R. FISHER tr. *Schimper's Plant-Geogr.* I. iii. 260 The Rain-forest is evergreen, hygrophilous in character, at least thirty meters high. **1914** M. DRUMMOND tr. *Haberlandt's Physiol. Plant Anat.* viii. 439 The degree of development of this tissue..at once shows whether any given plant is distinctly hygrophilous or xerophilous in character. **1934** H. GILBERT-CARTER tr. *Raunkiaer's Life Forms of Plants* viii. 321 More or less hygrophilous communities are succeeded by mesophilous woods of oak and beech. **1957** P. DANSEREAU *Biogeogr.* iv. 206 Hygrophilous plants are 'moisture-loving'.

hygrophyte (həi·grofəit). *Bot.* [f. HYGRO- + Gr. φυτόν plant.] A plant that grows in a moist habitat. Hence **hygrophy·tic** *a.*
1903 W. R. FISHER tr. *Schimper's Plant-Geogr.* I. 17 Typical hygrophytes have weakly developed roots, elongated axes, and large thin leaf-blades. **1932** FULLER & CONARD tr. *Braun-Blanquet's Plant Sociol.* v. 126 Hygrophytes, moisture-loving species with favourable water economy,..have morphological devices that permit the free loss of water. **1936** M. I. NEWBIGIN *Plant & Animal Geogr.* vii. 126 The actual appearance of the forests in any area depends on the competition between the hygrophytic species proper and those adapted to withstand seasonal drought. **1952** J. CLEGG *Freshwater Life* 50 The plants in the marshy area surrounding the pond are..not truly aquatic... They are usually referred to as hygrophytes.., in contrast to the true water-plants, which are called

hydrophytes. **1960** N. Polunin *Introd. Plant Geogr.* xi. 325 With further rising in level of the soil surface and relative depression of the water-table, shrubs and ultimately trees enter and in time give rise to a hygrophytic woodland. **1965** Bell & Coombe tr. *Strasburger's Textbk. Bot.* 177 In the hygrophytes we find large, thin, delicate laminae, rich in sap.

hygrostat (həi·grostæt). [f. Hygro- + -stat.] = *humidistat 2.

1915 R. C. Carpenter *Heating & Ventilating Buildings* (ed. 6) xx. 525 Some form of differential hygrostat.. controls the wet bulb temperature with respect to the dry bulb temperature so as to maintain a constant relative humidity. **1918** [see *humidistat 2]. **1937** *Archit. Rev.* LXXXI. p. lxxii/1 Heating and humidifying equipment are incorporated in each unit, the desired temperature and humidity being automatically controlled by thermostats and hygrostats. **1965** *Bull. Entomol. Res.* LVI. 265 Kitchen & Gall (1953) describe a modified Friez hair hygrostat which they used to control a humidifier for an insect breeding room.

Hyksos (hi·ksǒus), *sb. pl.* Also 7–8 **Hicsos, Hycsos.** [ad. Gr. Ὑκσώς interpreted by Manetho either as 'shepherd kings' or as 'captive shepherds', ad. Egyptian *heqa khoswe* chief of foreign lands.] A people of mixed Semitic-Asiatic stock, probably including a proportion of Habiru, who gave their name to the fifteenth Egyptian Dynasty (1650–1558 B.C.) which ruled the eastern delta. Also *attrib.* or as *adj.*

1602 T. Lodge tr. *Josephus' Works* 769 This nation was called Hicsos, which signifieth Kings shepheards,..but in other coppies I find that *Hicsos*, is not interpreted kings shepheards, but shepheards that were captiues. **1743** W. Stukeley *Abury* xiii. 78 The shepherds who quitted Egypt, under the conduct of our Hercules, call'd themselves Hycsi, as Manethon informs us in Josephus & Eusebius in chronol. The word imports royal shepherds, valiant, freemen, heroes. **1788** Gibbon *Decl. & F.* V. l. 184 Under the name of Hycsos, the shepherd kings, they had formerly subdued Egypt. **1877** *Encycl. Brit.* VII. 735/1 The invasion and conquest, at least in part, of Egypt by the Hyksos..is undoubtedly the chief cause of the obscurity of this age. **1899** A. H. Sayce *Early Israel* Introd. p. xxv, A Hyksos Pharaoh and his Hebrew vizier. **1931** J. G. Duncan *Digging up Biblical Hist.* II. 163 Hyksos pottery. **1931** M. M. Green tr. *Schneider's Hist. World Civilization* I. ii. 42 Aahmes took the capital of the Hyksos in the Delta about 1580 B.C. **1955** E. Pound *Section: Rock-Drill* xciii. 83 Where the spirit is clear in the stone as against Filth of the Hyksos, butchers of lesser cattle. **1957** [see *dark ages]. **1960** K. M. Kenyon *Archaeol. in Holy Land* vii. 184 At the period at which the Hyksos appear in Palestine and Egypt, we have on the move groups of Hurrians and Habiru, and the most probable explanation of the Hyksos is that they were recruited from such bands. **1971** *Encycl. Judaica* VIII. 1142 The Hyksos..attempted to Egyptianize and assimilate Egyptian culture. This synthesis is attested by the Hyksos religion.

hymenectomy (həimĕne·ktŏmi). *Surg.* [f. Hymen[2] + *-ectomy.] Excision of the hymen.

1931 in R. J. E. Scott *Gould's Med. Dict.* (ed. 3). **1962** *Lancet* 26 May 1117/2 After consulting a gynæcologist she underwent an operation for a hymenectomy 8 days before the hearing of the petition.

Hymettian (həime·tiăn), *a.* [f. L. *Hymettius* (f. *Hymettus*, Gr. Ὑμηττός) + -ian.] Of or belonging to Mount Hymettus in Attica, famous in antiquity for its honey and marble; hence *poet.* honeyed, sweet (cf. Hyblæan).

1601 Holland tr. *Pliny's Nat. Hist.* xvii. i. 499 Foure goodly pillars of Hymettian Marble. **1658** J. Rowland tr. *Moufet's Theater of Insects* 908 He that will make a good mixture of wine and honey, must mingle with new Hymettian Honey, old Falernian Wine. **1795** Coleridge *To R. B. Sheridan* in *Poetical Works* (1912) I. 88 Thy temples with Hymettian flow'rets wreath'd.

hymn, *sb.* Add: **2. b. hymn of hate,** the *Hassgesang* of the German poet Ernst Lissauer (1882–1937), an anti-British song; freq. *transf.*

1914 [see *hate sb.[1] 1 c]. **1915** A. Huxley *Let.* 26 Apr. (1969) 70 We're losing our heads and our senses of humour—and soon we shall be reduced to writing Hymns of Hate—then we're lost. **1916** *Anzac Book* 44 No 'Hymn of Hate' has yet been composed which would give expression to the hatred which has possessed me. **1918** G. B. Shaw *Pen Portraits* (1932) 40 Mr Chesterton, in his wildest hymns of hate, will break into a joke on his top note. **1945** H. G. Wells *Happy Turning* viii. 31 (*heading*) A hymn of hate against sycamores. **1963** E. Hyams *New Statesmanship* 8 This must not also entail blinking the facts about the Germans even if one did not join in the hymn of hate. **1966** B. Kimenye *Kalasanda Revisited* 67 She boarded the bus whilst still employed on her hymn of hate.

hymnarium (himnē·°riŏm). Pl. **-ia.** [med.L.] = Hymnary.

1924 *Glasgow Herald* 7 June 4 The extensive hymnaria attributed to the famous Bishop of Milan [St. Ambrose].

hyp, var. *hype sb.[1]

hypabyssal (həipăbi·săl), *a.* *Petrol.* [ad. G. *hypabyssisch* (attributed by H. Rosenbusch 1891, in *Tschermak's min. u. petrogr. Mitteil.* XII. 386, and by W. C. Brögger *Die Eruptivgesteine des Kristianiagebietes* (1894) iii. 123, to Brögger, 1886): see Hypo- 4.] Of igneous rock: formed from magma which has intruded into and solidified among other rocks; intermediate between plutonic and volcanic.

1895 *Mineral. Mag.* XI. 115 It [*sc.* sölvsbergite] is a true dyke-rock ('hypabyssal' rock). **1896** *Science Progress* IV. 476 Brögger has insisted on the necessity of a division intermediate between the plutonic and the volcanic, which he terms 'hypabyssal'. **1896** *Q. Jrnl. Geol. Soc.* LII. 613 The laccolitic rocks..would seem to belong to what Prof. Brögger calls hypabyssal. **1926** G. W. Tyrrell *Princ. Petrol.* vi. 106 The hypabyssal group includes the rocks of dykes, sills, and small laccoliths, etc., which occupy an intermediate position in the crust between the deepseated plutonic bodies, and the surficial lava flows. **1951** Turner & Verhoogen *Ign. & Metamorphic Petrol.* iii. 51 Some writers recognize an intermediate (hypabyssal) class to include rocks that have crystallized at moderate depth. **1969** Bennison & Wright *Geol. Hist. Brit. Isles* x. 251 The Great Whin Sill is the largest hypabyssal intrusion in Britain.

hypacusis (həipăkiū·sis). *Med.* Also **-acousis, -acusia, -akusis.** [mod.L., f. Hyp- + Gr. ἄκουσις hearing (ἀκούειν to hear).] Diminished acuteness of hearing. Cf. *hyperacusis.

1886 *Syd. Soc. Lex.*, Hypacusia. *Ibid.*, Hypacusis. *Ibid.*, Hypakusis. **1895** T. B. Hyslop *Mental Physiol.* ix. 273 Hypakusis—diminution of sense of hearing, seen in various forms of insanity. **1967** A. B. Graham *Sensorineural Hearing Processes* xii. 143 Some patients with hypacusis as against normal-hearing patients.

hypalgesia, -ic: see *hypo-.

hypanthium (həipæ·nþiŏm). *Bot.* = *hypanthodium.

1866 Lindley & Moore *Treas. Bot.* II. 611/2 Hypanthium. The fleshy enlarged hollow of the end of a flower-stalk, such as occurs in the rose, apple, or myrtle. **1887** W. Hillhouse tr. *Strasburger's Handbk. Pract. Bot.* xxx. 349 The five-celled ovary is here [*sc.* in the apple] immersed in a hollowed flower-stalk, a so-called hypanthium. **1912** H. H. Rusby *Man. Struct. Bot.* iii. 46 The enclosed portion of the calyx really is adherent, but it is not visible, since it is enclosed and concealed by the hollow torus, which is known as a Hypanthium. **1968** A. Cronquist *Evol. & Class. Flowering Plants* iii. 87 It is customary and convenient to define the hypanthium in terms of external descriptive morphology rather than on evolutionary homologies.

hypanthodium (həipænþōu·diŏm). *Bot.* [mod.L. (H. F. Link *Elementa Philosophiae Botanicae* (1824) ix. 265), f. Hypo- + Anthodium.] In certain plants, an enlargement of the receptacle, sometimes becoming fleshy and surrounding the ovary.

1832 J. Lindley *Introd. Bot.* ii. 108 If the receptacle is fleshy, and is not enclosed within an involucrum, as in Dorstenia and Ficus, it is then called by Link Hypanthodium. **1861** R. Bentley *Man. Bot.* 204 The Hypanthodium..is formed by a receptacle which is usually of a fleshy nature becoming more or less incurved.

hyparterial, *a.* Substitute for def.: Situated below the pulmonary artery. (Add examples.)

1882 *Quain's Elem. Anat.* (ed. 9) II. 511 From the continuation of the bronchus four dorsal and as many ventral hyparterial branches are given off in succession in each lung. **1921, 1962** [see *eparterial a.].

hype (həip), *sb.*[1] *slang* (orig. *U.S.*). Also **hyp.** [Abbrev. of Hypodermic.] **a.** A drug-addict. **b.** A hypodermic needle or syringe. **c.** A hypodermic injection. Cf. *hypo sb.[3]

1924 G. C. Henderson *Keys to Crookdom* xxiv. 306 Next down on the list is the 'hype' or morphine-user. Morphine is taken by hypodermic injections. **1929** [see *bang sb.[3]]. **1936** *Amer. Speech* XI. 122/2 *Hype*, the hypodermic needle used to inject narcotics. **1952** D. E. Hulburd *H is for Heroin* ii. 52 They smoked two marijuana cigarettes apiece that an old addict—'an old hype'—had sold Hortense. **1955** *U.S. Senate Hearings* (1956) VIII. 4164 Terms for morphine addicts: 'Hype', 'Hygelo', 'Head', 'Fiend', 'Needle man', 'Junky', 'Junker'. **1963** *New Society* 7 Nov. 9/1 'Once a hype, always a hype,' the American narcotics addict says. **1972** J. Wambaugh *Blue Knight* (1973) i. 11 They were dumb strung-out hypes. *Ibid.*, The tall one is wearing a long-sleeved shirt buttoned at the cuff. To hide his hype marks.

hype (həip), *sb.*[2] *slang* (orig. *U.S.*). [Origin unknown.] **a.** An instance of short-changing; a person who does not give the correct amount of change. **b.** (The usual current sense.) Deception, cheating; a confidence trick, a racket, a swindle, a publicity stunt.

1926 *Clues* Nov. 161/2 *Hype*, short change artist. **1926** [see *hype v.[1]]. **1935** A. J. Pollock *Underworld Speaks* 60/2 *Hype*, a short change artist. A person who does not give the correct amount of change to a customer. **1955** D. W. Maurer in *Publ. Amer. Dial. Soc.* xxiv. 85 Sometimes he has auxiliary *rackets*, such as the *hype* or some form of the *short con.* **1962** J. Baldwin *Another Country* (1963) ii. iv. 336 Life is a *bitch*, baby. It's the biggest hype going. **1966** C. Himes *Heat's On* xxii. 173 That was how Gus got the money...That shocked her; she had believed Gus's hype about his wife leaving him a farm. **1967** N. Mailer *Cannibals & Christians* i. 29 The hype had made fifty million musical-comedy minds; now the hype could do anything. **1968** *Sunday Times* 11 Aug. 5/1 Hype is an American word for the gentle art of getting a tune into the pop charts without actually selling any records. Its methods are various: from the crudest bribery to devious techniques for upsetting the..calculations of chart-compilers. **1969** *Listener* 25 Sept. 420/3 All the 'hype' and conning that goes into the establishing of every star. **1970** L. Sanders *Anderson Tapes* li. 135 He's been on the con or hustling his ass or pulling paper hypes. **1972** *Publishers Weekly* 6 Mar. 2/2 They carried off the biggest money-making hype in sports history.

hype (həip), *v.*[1] *slang* (orig. *U.S.*). Also **hipe.** [Origin unknown.] To short-change, to cheat; to deceive, to con, esp. by false publicity. So **hy·ping** *vbl. sb.* and *ppl. a.*

1926 Maines & Grant *Wise-Crack Dict.* 9/2 *Hype*, to overcharge. As, 'That place has a hype on this week.' **1931** G. Irwin *Amer. Tramp & Underworld Slang* 99 *Hipe*, to cheat or short change...Perhaps connected with the North-England 'hipe', to find fault with, to slander; perhaps from 'high pressure'. **1945** L. Shelly *Jive Talk Dict.* 13/1 *Hype* (v.), to try a trick. **1946** Mezzrow & Wolfe *Really Blues* 375 *Hype* v., to deliver a phony but convincing line. **1962** J. Baldwin *Another Country* (1963) iii. i. 402 He doesn't seem to be trying to hype me, not even when he talked about his wife and kids. **1968** *Sunday Times* 11 Aug. 5/1 The dominance of the charts over the pop music industry is such that many thousands of pounds may depend on these hyping operations. **1969** N. Cohn *AWopBopaLooBop* (1970) v. 51 Hype is a crucial word. In theory it is short for hyperbole. In practice, though, it means to promote by hustle, pressure, even honest effort if necessary, and the idea is that you leave nothing to chance. Simply, you do everything possible. Hype has become such an integral part of pop that one hardly notices it any more. **1969** *Sat. Rev.* (U.S.) 27 Sept. 25/1 You can't hype kids into buying things they don't want...They hate hype. **1970** *It* 27 Feb.–13 Mar. 15/1 Paying..lip-service to furiously hyping publicists. *Ibid.*, A lunchtime hyping session. **1971** *Listener* 15 Apr. 467/1 Bogus alternatives are hyped into prominence and fortune with appalling ease. **1971** *Bookseller* 23 Oct. 2053/2 In America a practice exists (known as 'hyping' in the record business) whereby a film company which has acquired the rights in a title forces it on to the bestseller list by sending young publicity men around armed with hundred-dollar bills and instructions to buy twenty or more copies from selected book shops.

hype (həip), *v.*[2] *U.S. slang.* [f. *hype sb.[1]] Usu. as **hyped** *pa. pple.* or *ppl. a.* (const. *up*): stimulated, worked up (as if from the effects of a hypodermic injection).

1938 D. Castle *Do Your Own Time* xxi. 200 Y' gotta lay off the wimmin, an' don't hype y'self up till y' goes out t' heist a joint. **1946** Mezzrow & Wolfe *Really Blues* iv. 54, I was so hyped-up I couldn't sit still. **1950** in Wentworth & Flexner *Dict. Amer. Slang* 278/2 No fireworks [in this movie], no fake suspense, no hyped-up glamour. **1970** V. Johnston *Phantom Cottage* (1971) xxi. 165 If some hyped-up character goes past at seventy miles an hour, we'll take out after him. **1973** *Publishers Weekly* 12 Feb. 64/3 Witty, intellectual fun that keeps his readers hyped up and on their mettle. **1973** *Time* 25 June 16/2 As he works, Mitchell has at times been so hyped up that Martha once asked his doctor to prescribe medication to slow him down.

hyper[2] (həi·pəɹ). *U.S. slang.* [Cf. *hype v.[1]] (See quot. 1914.)

1914 Jackson & Hellyer *Vocab. Criminal Slang* 47 *Hyper*, current amongst money-changers. A flim-flammer. **1931** G. Irwin *Amer. Tramp & Underworld Slang* 106 *Hyper*, a 'short change' artist...The logical explanation.. is that the word came from 'hyp', a contraction of hypochondria.

hyper-. Add: **I. 1. b.** *hyper-analysis.*

1942 C. S. Lewis in *Essays & Stud.* XXVII. 18 This brings us to..the psycho-analysis of psycho-analysis itself. Such a hyper-analysis..would not refer to 'really scientific people', but to the great mass of ordinary people who read psycho-analytic books with avidity and undergo their influence.

2. *hyperphrygian* (examples).

1761 Stiles in *Phil. Trans. R. Soc. 1760* LI. 713 The modes being thus augmented to fifteen..their meses will be found to stand..in the following order. Hyperlydian, Hyperæolian, Hyperphrygian or Hypermixolydian. **1922** Joyce *Ulysses* 493 It is susceptible of nodes or modes as far apart as hyperphrygian and mixolydian.

3. b. In substantives in which *hyper-* has the sense 'the analogue in a space of four or more dimensions of (what is denoted by the second element) in ordinary three-dimensional space'; as *hypercube, -cylinder, -plane, -sphere, -surface.*

1895 *Proc. R. Soc.* LVIII. p. xxxi, The manifoldness in this space..is the quadri-quadric two-dimensional amplitude common to thirteen quadric hyper-cylinders. **1903** C. M. Jessop *Treat. Line Complex* xiii. 244 Any linear equation of the form $\sum_{1}^{5} a_i X_i = 0$ singles out ∞^3 points from S_4, which will then form a space of three dimensions; the locus of these ∞^3 points will be called a hyperplane. *Ibid.* 251 In four-dimensional space, the three-dimensional quadric spaces through the intersection of S_4^2 and $X_5 = 0$..may be termed 'hyperspheres'. **1909** *Sci. Amer.* 3 July 6/2 Just as portions of our space are bounded by surfaces,..so portions of hyperspace are bounded by hypersurfaces (three-dimensional), i.e., flat or curved 3-spaces. *Ibid.* 6/3 Of these [regular hyper-solids], C_8 (or the hypercube) is the simplest, because, though with more bounding solids than C_5, it is right-angled throughout. **1955** O. Klein in W. Pauli *Niels Bohr* 100 Let..x^1, x^2, x^3, x^4 be the four space-time coordinates regarded as *c*-numbers, x^1, x^2, x^3 forming a space-like hypersurface for any given value of the general time coordinate x^4. **1966** A. Battersby *Math. in Managem.* v. 122 When the number of variables exceeds three..we could represent the process of solution by a series of three-dimensional solid bodies showing successive cross-sections of the solution space when cut by the 'hyper-plane' of P. **1968** Rosenberg & Johnson *Geom.* xiii. 520/2 If the solid cube moves in a direction 'perpendicular' to its original space, it may trace a solid hypercube. **1969** R. J. Bumcrot *Mod. Projective Geom.* ii. 30 Subspaces of dimensions 1, 2, $n-1$ are called, respectively, lines, planes, and hyperplanes. **1970** E. E. Kramer *Nature & Growth Mod. Math.* vii. 160 To say that a relation like $x^2+y^2+z^2+w^2 = 9$ is a hypersphere with radius 3 is so much easier than to state that the relation is the set of all ordered quadruples of real numbers such that the sum of the squares of these four numbers is always 9. **1972** *Computer Jrnl.* XV. 214/1 The problem of optimising a function globally over the vertices of a hypercube is encountered, for example, in hierarchical classification.

II. 4. (Further examples of the unlimited extension of this use.) **a.** *hyperarchaic, -civilized, -colloquial, -educated, -excitable, -pure* adjs. **b.** *hyperarchaism, -characterization, -determination, -dialecticism, -dialectism.* **c.** *hypercharacterize* vb.

1956 K. Clark *Nude* 380 Considering that they were spoken of as 'hyper-archaic', his restorations were remarkably self-effacing. **1956** *Archivum Linguisticum* VIII. 124 Attributable to assimilation and hyperarchaism. **1957** *Ibid.* IX. 79 If a given linguistic formation develops in such a way as to allow..one of its distinctive features to stand out more sharply than at the immediately preceding stage, one may speak of hypercharacterization (or hyperdetermination) of that feature, in the diachronic perspective. *Ibid.* 80 One may analyse Sp. dial. *Jesuso* and *Raquela* as hypercharacterized, with respect to gender, in comparison with standard *Jesús* and *Raquel*. **1915** *Times Lit. Suppl.* 13 May 160/3 Only in a hyper-civilized and introspective society such themes would be possible. **1940** O. Jespersen *Mod. Eng. Gram.* V. xxiii. 437 It sounds hyper-colloquial..when too many *don't, isn't* are substituted for *do not, is not*, etc. in reading serious prose aloud. **1960** T. B. W. Reid *Historical Philol. & Ling. Sci.* 6 Reactions such as those known as hyper-urbanism and hyperdialecticism. **1925** P. Radin tr. *Vendryès's Lang.* i. ii. 50 There are many hyper-dialectisms, for instance, in the Doric of the Pythagorean authors. **1914** J. Joyce *Dubliners* 238 The generation which is now on the wane..had certain qualities of hospitality, of humour, of humanity, which the new and very serious and hypereducated generation..seems to me to lack. **1886** *Lancet* 13 Mar. 485/2 Even normal mental impulses may cause undue motorial demonstrations if the spinal centres are hyper-excitable, as is seen in strychnine poisoning, hysteria, &c. **1972** *Nature* 10 Mar. 74/1 The animal became hyperexcitable with exaggerated startle response. **1958** *Times Rev. Industry* June 26/2 Production of hyper-pure silicon entails purifying the selected chemical to a very high degree.

IV. (In the following words *e* often replaces *æ, œ*, esp. in U.S. usage; the alternative spelling is not given for each word individually.)

hyperabdu·ction, extreme abduction (sense 3); so **hyperabdu·ct** *v. trans.*, **-abdu·cted** *ppl. a.*; **hype:raldoste·ronism** *Med.*, any condition characterized by excessive secretion of aldosterone; aldosteronism; **hyperalgic** (-æ·ldʒik) *a. Path.*, of, pertaining to, or affected with hyperalgia (hyperalgesia); **hy:perbilirubinæ·mia** *Physiol.* [Gr. αἷμα blood], an abnormally high concentration of bilirubin in the blood; **hypercalcæmia** (-kælsī·miă) *Physiol.* [Calc(ium + Gr. αἷμα blood], an abnormally high concentration of calcium in the blood; so **hypercalcæ·mic** *a.*; **hypercalcuria** (-kælsiū°·riă), **-calciuria** (-kælsi͵yū°·riă) *Physiol.* [-uria], an abnormally high concentration of calcium in the urine; **hyperca·pnia** *Physiol.* [Gr. καπνός smoke], an abnormally high concentration of carbon dioxide in the blood; so **hyperca·pnial** (*rare*), **-ca·pnic** *adjs.*; **hypercathe·xis** (pl. **-exes**) *Psychol.*, an excessive degree of cathexis; **hy:percellula·rity** *Path.*, an excess of cells at a site in the body; so **hyperce·llular** *a.*, containing more than the normal number of cells; **hy:perchloræ·mia** *Physiol.*, an abnormally high concentration of chlorides in the blood; **hy:perchlorhy·dria** *Physiol.* [Chlorhydr(ic *a.*], an abnormally high concentration of hydrochloric acid in the gastric juice; so **hy:perchlorhy·dric** *a.*; **hy:percholesteræ·mia, -cholesterolæ·mia** *Physiol.* [Gr. αἷμα blood], an abnormally high concentration of cholesterol in the blood; so **hy:percholesterolæ·mic** *a.*; **hy:percoagulabi·lity** *Med.*, an excessive tendency (of the blood) to coagulate; so **hypercoa·gulable** *a.*; **hyperda·ctyly** *Zool.* [Gr. δάκτυλ-ος finger] = Polydactyly; **hyperemo·tional** *a.*, affected by or displaying an abnormal degree of emotion; hence **hy:peremotiona·lity,** hyperemotional behaviour; **hypereute·ctic** *a.*, (of an alloy of iron) containing a higher proportion of carbon than the eutectic composition (i.e. more than about 4·3%); (in quot. 1902 = *hypereutectoid, eutectoid* itself not having been coined at that date); **hypereute·ctoid** *a.* (of steel) containing a higher proportion of carbon than the eutectoid composition (i.e. more than about 0·8%); **hy·perform** *Linguistics* [contraction of *hypercorrect form*], a hypercorrect spelling or pronunciation; **hy·perfragment** *Nuclear Physics* [*hyper(on)], a hypernucleus, esp. one produced by the breaking up of a heavier one (see quot. 1964); **hyperfu·nction** *Med.*, over-activity or over-production (in a gland or other part of the body); so **hyperfu·nctional** *a.*, **hyperfu·nctioning** *vbl. sb.* and *ppl. a.*; **hy:pergammaglo:bulinæ·mia** *Physiol.* [Gr. αἷμα blood], an abnormally high concentration of gamma globulins in the blood; hence **hy:pergammaglo:bulinæ·mic** *a.*; **hypergeusia** (-giū·ziă) *Med.* [Gr. γεῦσις taste], excessive acuteness of the sense of taste; **hy:perglobulinæ·mia** *Physiol.* [Gr. αἷμα blood], an abnormally high concentration of globulins in the blood; hence **hy:perglobulinæ·mic** *a.*; **hyperglycæ·mia** *Physiol.*, an abnormally high concentration of sugar in the blood; so **hyperglycæ·mic** *a.*; **hyperimmu·ne** *a. Med.*, subjected to, resulting from, or produced by hyperimmunization; having a high concentration of antibody; **hy:perimmuniza·tion** *Med.*, the production of a high concentration of antibody in the serum of an animal, esp. by the repeated injection of the same antigen; so **hyperi·mmunize** *v. trans.*, to produce such a condition in (an animal); **hyperi·mmunized** *ppl. a.*; **hyperinfe·ction** *Med.*, continued infection with parasitic worms owing to their larvæ developing into adult worms without leaving the body; so **hyperinfe·ctive** *a.*, causing or characterized by hyperinfection; **hyperinflation** (also hyphened), an acute form of economic inflation; **hy:perinsulinæ·mia** *Physiol.* [Gr. αἷμα blood], an abnormally high concentration of insulin in the blood; **hyperi·nsulinism** *Med.*, a condition in which the body produces excessive insulin, usu. as a result of a tumour of the islets of Langerhans in the pancreas; the presence of an excessive amount of insulin in the body; **hy:perirritabi·lity** *Med.*, increased irritability; abnormally high responsiveness to stimuli; so **hyperi·rritable** *a.*; **hyperkalæmia** (-kălī·miă), **-kaliæmia** (-kæli͵ī·miă) *Physiol.* [mod.L. *kalium* potassium (see Kali) + Gr. αἷμα blood] = *hyperpotassæmia below; hence **hyperkalæ·mic** *a.*; **hy:perkerato·sis** *Path.*, (*a*) (see quot. 1848); now *rare* or *Obs.*; (*b*) excessive development of the horny layer of the skin; **hyperkine·sia** *Path.* = *hyperkinesis* (s.v. Hyper- IV); **hyperkinetic** *a.* (examples); **hy:perleucocyto·sis, -leukocyto·sis** *Path.* = Leucocytosis; **hyperlipæ·mia** *Physiol.*, an abnormally high concentration of fats (or lipids) in the blood; any condition characterized by this; **hy:permagnesæ·mia** *Physiol.* [Gr. αἷμα blood], an abnormally high concentration of magnesium in the blood; **hyperma·nia** *Psychiatry*, (*a*) *sometimes used to mean* a severe degree of mania with partial or complete disorientation, often accompanied by violent behaviour and forming a stage in manic-depressive illness; (*b*) used *erron.* for *hypomania (quot. 1928: see also quot. 1956 for *hypermanic*); hence **hyperma·nic** *a.*; **hypermatu·re** *a. Ophthalm.*, applied to a cataract in its final stage (see quots.); so **hypermatu·rity**; **hypermeta·bolism** *Physiol.*, metabolism at a high rate; so **hy:permetabo·lic** *a.*, of hypermetabolism; **hypermobi·lity** *Med.*, abnormally great freedom of movement or flexibility in a joint; so **hypermo·bile** *a.*, characterized by or exhibiting hypermobility; **hy·permorph** *Genetics* [Gr. μορφ-ή form], any allele which is functionally more effective that the corresponding wild-type allele; so **hypermo·rphic** *a.*; **hypermoti·lity** *Med.*, excessive movement, esp. of the stomach and intestines; **hypernatræ·mia** *Physiol.* [Natr(ium + Gr. αἷμα blood], an abnormally high concentration of sodium in the blood; hence **hypernatræ·mic** *a.*; **hypernephro·ma** *Path.* [Nephr(o- + *-oma], a malignant tumour of the cortical parenchyma of the kidney; orig. such tumours were believed to derive from misplaced tissue of the suprarenal gland (whence the name) and were described in other tissues besides that of the kidney; so **hy:pernephro·matous** *a.*; **hypernu·cleus** *Nuclear Physics* [*hyper(on)], a nucleus in which a hyperon replaces one of the nucleons; a composite particle in which a hyperon is bound to one or more nucleons; hence **hypernu·clear** *a.*; **hyperosmo·tic** *a. Physiol.* = *hypertonic a. 2; const. *to*; **hyperphagia** (-fēi·dʒiă) *Med.* [Gr. -φαγία -phagy], an abnormally great desire for food; excessive eating; hence **hyperphagic** (-fæ·dʒik) *a.*, of or exhibiting hyperphagia; **hyperphalangia** (-fălæ·ndʒiă), **-phalangism** (-fălæ·ndʒiz'm), **-phalangy** (-fæ·lăndʒi) *Med.* and *Zool.* [L. *phalang-*: see Phalanx], the condition of having more digital phalanges than normal, esp. in cases where polydactyly is absent; **hyperpho·ria** *Ophthalm.*, latent strabismus in which there is a tendency for one eye to be directed above (or below) the line of sight of the other; so **hyperpho·ric** *a.*; **hy:perphosphatæ·mia** *Physiol.* [Gr. αἷμα blood], an abnormally high concentration of phosphates (or other phosphorus compounds) in the blood; so **hy:perphosphatæ·mic** *a.*; **hyperpiesia** (-pəi͵ī·ziă), **-piesis** (-pəi͵ī·sis, -pəi·ēsis) *Med.* [Gr. πίεσις pressure], high blood pressure, hypertension, esp. when without evident cause (the two words were distinguished in meaning by Allbutt: see quots.); hence **hyperpie·tic** *a.*; **hy:perpigmenta·tion** *Med.*, excessive pigmentation; so **hy:perpigme·nted** *ppl. a.*, exhibiting hyperpigmentation; **hyperpitu·itarism** *Path.* [Pituitar(y *a.* + -ism], increased hormone secretion by the pituitary body; hence **hyperpitu·itary** *a.*, of, pertaining to, or affected with hyperpituitarism; **hy·perploid** *a. Genetics* [*-ploid], having one or a few extra chromosomes (orig. also chromosome fragments) in addition to a haploid, diploid, triploid, etc., set; containing such cells; also as *sb.*, a hyperploid cell or individual; so **hy·perploidy,** the condition of being hyperploid; **hyperpnœa** (həipəɹnī·ă, -pnī·ă) *Physiol.* [Gr. πνοή breath, breathing], deep and rapid breathing; panting; so **hyperpnœ·ic** *a.*; **hy:perpotassæ·mia** *Physiol.* [Gr. αἷμα blood], an abnormally high concentration of potassium in the blood; so **hy:perpotassæ·mic** *a.*; **hyperprose·xia** *Psychol.* [Gr. προσέχ-ειν to turn (one's attention) (ἔχειν to hold, possess mentally)], the concentration of attention on one stimulus to the exclusion of all others; **hy:perproteinæ·mia** *Physiol.* [Gr. αἷμα blood], an abnormally high concentration of protein in the blood; **hyperrea·ctive** *a. Med.*, reacting unusually strongly to certain stimuli; of or pertaining to this tendency; so **hy:perreacti·vity**; **hyperrhythmical** *a.* (earlier example); **hypersa·line** *a. Biol.*, (of naturally occurring water) more salty than typical sea water; hence **hypersali·nity,** the condition of being hypersaline; **hy:persexua·lity,** a condition in which the sexual instinct is abnormally strong; **hyperso·mnia** *Med.* [L. *somnus* sleep], a condition characterized by abnormally long or frequent periods, or abnormal depth, of sleep; hence **hyperso·mnic** *a.*, of, exhibiting, or producing hypersomnia; **hyperspace,** also, any non-Euclidean space; (further examples, illustrating its currency in *Science Fiction*); **hyperspa·tial** *a.*, of or in hyperspace; **hypersplenism** (-sple·niz'm) *Path.* [Splen-], over-activity of the spleen; *spec.* a condition characterized by a general destruction of blood cells, often associated with enlargement of the spleen but in which direct involvement of the spleen is now considered doubtful; so **hypersple·nic** *a.*; **hypersta·tic** *a. Engin.*, statically indeterminate, i.e. having more members or supports than the minimum required to render it stable (and therefore requiring more than considerations of equilibrium alone for the calculation of all the internal forces and moments); of or pertaining to such a structure; **hyperste·reograph** *Photogr.*, a picture or pair of photographs taken by hyperstereoscopy; **hy:perstereo·scopy** *Photogr.*, stereoscopic photography in which the separation of the two viewpoints is greater than the distance between the eyes, resulting in a greater stereoscopic effect or exaggerated perspective; hence **hy:perstereosco·pic** *a.*; **hypersusce·ptible** *a. Med.* = *hypersensitive a. 2; so **hy:persusceptibi·lity**; **hyperte·lorism** *Med.* [Gr. τῆλ-ε at a distance + ὁρίζειν to separate from: see -ism], a condition in which the eyes are abnormally far apart, freq. found accompanying other congenital malformations of the face; **hypertro·pia** *Ophthalm.*, strabismus in which one eye is directed above the line of sight of the other; **hy:peruricæmia** (-yū°risī·miă) *Physiol.*, an abnormally high concentration of uric acid in the blood; = Lithæmia, Uricæmia; hence **hy:peruricæ·mic** *a.*; **hypervelo·city,** a speed that is (relatively) very high; usu. *attrib.*; **hy:pervitamino·sis** *Path.* [-osis], any condition caused by excessive intake of a vitamin, esp. over a prolonged period; **hypervolæ·mia** *Physiol.* [Vol(ume *sb.* + Gr. αἷμα blood], an increased volume of circulating blood in the body; hence **hypervolæ·mic** *a.*

1945 *Amer. Heart Jrnl.* XXIX. 7 The pulse in the left arm could be obliterated only by having the patient hyperabduct his arm above a 150-degree angle. *Ibid.* 6 The habit of sleeping with the arms in the hyperabducted position. **1905** Gould *Dict. New Med. Terms* 299/2 Hyperabduction. **1945** *Amer. Heart Jrnl.* XXIX. 4 The term 'hyperabduction' is used in this paper to mean that phase of circumduction which brings the arms together above the head... Actually, the term hyperabduction, although accepted in anatomic terminology, is not..an entirely logical term, for abduction is movement away from the median plane of the body, and beyond the 90° angle; the arm in so-called hyperabduction actually again approaches the median plane. **1966** J. E. Flynn *Hand Surg.* xiv. 696/1 Hyperabduction of the arm alone could stretch the subclavian artery sufficiently to produce occlusion in certain persons. **1955** Conn & Louis in *Trans. Assoc. Amer. Physicians* LXVIII. 229 What is the relationship of hyperaldosteronism to the production of renal arteriosclerosis? **1966** R. B. Scott *Price's Textbk. Pract. Med.* (ed. 10) vii. 436/2 Patients with hyperaldosteronism usually present in one of two ways, either with manifestations of hypertension or with muscular weakness and hyporeflexia sometimes sufficiently severe to cause episodic paralysis. *Ibid.* 437/1 Without treatment hyperaldosteronism is fatal, the patient usually dying of the hypertensive vascular complications. **1946** *Nature* 10 Aug. 202/1 We obtained successful results with this substance in other hyperalgic conditions, namely, cervical neuritis and trigeminal neuralgia. **1968** Cahn & Herold in A. Soulairac et al. *Pain* iv. 367 We have defined these changes as a hyperalgic state. **1923** *Q. Jrnl. Med.* XVI. 409 These latter cases are on the border line between 'physiological hyperbilirubinaemia' and the actual disease known as haemolytic (acholuric) jaundice. **1965** W. Taylor *Biliary Syst.* 647 (*heading*) Bilirubin excretion in congenital hyperbilirubinaemia. **1925** *Jrnl. Biol. Chem.* LXIII. 444 Dog 51 showed typical symptoms of hypercalcemia. **1970** C. N. Graymore *Biochem. Eye* viii. 551 Hypercalcaemia results from vitamin D poisoning, hyperthyroidism and severe renal damage. **1932** *Physiol. Rev.* XII. 605 The occurrence of..hypercalcemic symptoms. **1930** *Jrnl. Biol. Chem.* LXXXVII. p. xv (*heading*), Calcium and phosphorus metabolism in relation to certain bone diseases. I. Hypercalcuria. **1961** *Lancet* 26 Aug. 455/2, 10 of the 28 patients with hypercalciuria had no evidence of renal calcification. **1964** D. M. Dunlop *Textbk. Med. Treatm.* (ed. 9) 757 A variety of disorders which are associated with hypercalciuria tend to cause stone formation. **1908** *Amer. Jrnl. Physiol.* XXI. 140 Hypo- and hyper-capnia are abnormal conditions. **1962** *Lancet* 2 June 1183/2 The combination of hypoxia and hypercapnia is often lethal. **1908** *Amer. Jrnl. Physiol.* XXI. 141 An asphyxial (or hyper-capnial) condition of the blood supply to the spinal bulb. **1955** *Jrnl. Physiol.* CXXIX. 405 The achievement of a steady state of hypercapnic ventilation. **1962** *Lancet* 8 Dec. 1224/2 When pH was kept normal by the infusion of this organic buffer..circulation was unaltered in the hypercapnic dog. **1923** Freud in *Internat. Jrnl. Psycho-Anal.* IV. 6 Our consideration of the first case, the jealousy paranoia, led to a similar estimate of the importance of the quantitative factor, by showing that there also the abnormality essentially consisted in the hyper-cathexis (over-investment) of the interpretations of another's unconscious behaviour. **1950** J. Strachey tr. *Freud's Totem & Taboo* iii. 89 The psychological results must be the same in both cases, whether the libidinal hypercathexis of thinking is an original one or has been produced by regression. **1968** D. Rapaport et al. *Diagn. Psychol. Testing* (rev. ed.) iii. 108 The drive cathexes are kept in balance and control, harmonizing with and not encroaching upon the ego's functions, nor demanding that it employ its hypercathexes to curb them. **1955** *Bull. N.Y. Acad. Med.* XXXI. 135 Under such circumstances the marrow is hypercellular but the blood is cytopenic. **1967** J. Metcoff *Acute Glomerulonephritis* vi. 110 Some lobules may be quite hypercellular. **1908** *Lancet* 23 May 1467/2 In the older or quiescent stages [of carcinoma of the tongue] the hypercellularity disappears. **1967** J. Metcoff *Acute Glomerulonephritis* vi. 110 Mitotic figures..are easy to find in areas of hypercellularity. **1921** *Endocrinology* V. 802 One or two days before the onset of menstruation.. there is generally an absolute and relative hyperchloremia. **1969** R. L. Searcy *Diagn. Biochem.* i. 14/1 Treatment with ammonium chloride can lead to..hyperchloremia. **1891** F. P. Foster *Med. Dict.* III. 1938/2 Hyperchlorhydria. **1893** *Med. Ann.* 169 Hyperchlorhydria and hypochlorhydria are not identical with hyperacidity and hypoacidity. **1957** I. Aird *Compan. Surg. Stud.* (ed. 2) xxxiii. 710 *Hyperchlorhydria* is present in 90 per cent of duodenal ulcers. **1903** *Med. Rec.* 7 Feb. 229/2 In the last year the pain increased, and the disturbance was always of the hyperchlorhydric type. On entrance to the hospital, a small, painless tumor was clearly felt in the region of the pylorus. **1926** J. A. Ryle *Gastric Function* 119 The fractional test-meal gives hyperchlorhydric curves. **1894** Gould *Dict. Med.* 589/1 Hypercholesteremia. **1916** *Physiol. Abstr.* I. 327 (*heading*) Experimental hypercholesteræmia. **1969** R. L. Searcy *Diagn. Biochem.* xviii. 170/2 Hypercholesteremia usually..accompanies hypothyroidism. **1916** *Arch. Internal Med.* XVII. 768 In pregnancy hypercholesterolemia occurs physiologically. **1970** *Nature* 31 Oct. 465/1 Growth hormone is as efficacious as thyroid hormone in preventing hypercholesterolaemia. **1916** *Arch. Internal Med.* XVII. 784 Cells which have been bathed in and irritated by hypercholesterolemic blood. **1961** *Lancet* 7 Oct. 802/2 Cases of familial hypercholesterolæmic xanthomatosis. **1934** Webster, Hypercoagulability, -coagulable. **1962** *Lancet* 8 Dec. 1230/2 This permits one to anticipate the periods of blood hypercoagulability and thus to prevent thromboembolism successfully. **1972** *Nature* 28 Apr. 452/1 All showed adverse changes which might lead to a hypercoagulable or hyperthrombotic state compared with the non-smoker. **1902** Webster Suppl., Hyperdactyly. **1929** R. S. Lull *Org. Evol.* (ed. 2) xx. 297 As though extra toes over the normal five had been added (hyperdactyly). **1965** W. B. Yapp *Vertebrates* v. 93 The paired limbs show both more digits and more joints than usual—hyperdactyly and hyperphalangy. **1946** O. Fenichel *Psychoanal. Theory of Neurosis* xx. 478 A 'generally frigid' person has forgotten childhood emotions; the hyperemotional person is still a child. **1971** *Jrnl. Gen. Psychol.* LXXXIV. 245 Loud vocalization..is a prominent characteristic of vigorous fighting among rats and has been labelled..an index of hyperemotional behavior among normally silent species. **1958** *Science* 19 Sept. 655/2 These animals did show a gradual, but only partial, development of hyperemotionality. **1972** *Nature* 25 Aug. 454/1 According to some reports, bulbectomy also induces irritability and hyperemotionality resembling the classic septal 'rage' syndrome. **1902** *Encycl. Brit.* XXIX. 573/2 The undisturbed slow cooling from the molten state of a hyper-eutectic steel containing 1·00 per cent. of carbon. **1912** W. H. Hatfield *Cast Iron* i. 13 Hypereutectic alloys deposit primary iron-carbide along the line B'C. **1959** A. G. Guy *Elem. Physical Metall.* (ed. 2) vi. 186 As the composition changes from hypoeutectic (less than eutectic) to hypereutectic (more than eutectic) in terms of metal *B*, the primary crystals change from alpha phase to beta phase. **1911** *Encycl. Brit.* XIV. 805/2 The large massive plates of cementite which form the network or skeleton in hyper-eutectoid steels. **1966** A. Prince *Alloy Phase Equilibria* vi. 107 Hyper-eutectoid alloys on cooling from the austenite phase region deposit cementite over a range of temperature until A_1 is reached. As before, the remaining austenite then transforms to pearlite. **1933** L. Bloomfield *Lang.* xxvii. 479 This may be disclosed by isolated relic forms, or by the characteristic phenomenon of hyper-forms. **1937** *Amer. Speech* XII. iii. 168 Hyperforms are by no means always attempts to imitate city pronunciation. **1964** H. Kökeritz in D. Abercrombie et al. *Daniel Jones* 141, I have heard the hyperform [hə͵rɑs] from a colleague now deceased. **1955** W. F. Fry et al. in *Physical Rev.* XCIX. 1561 Following a suggestion of M. Goldhaber, we propose to call a nuclear fragment containing a bound hyperon or some other unstable particle, a hyperfragment. **1963** K. Nishijima *Fund. Particles* vi. 290 The study of hyperfragments offers almost the only source of getting information about the Λ-nucleon force. **1964** *Progress Nuclear Physics* IX. 172 The nucleus in which the capture occurs is usually broken up and the Λ^0-hyperon may be bound in one of the fragments that are emitted...Fragments such as these are referred to as hyperfragments. **1909** *Jrnl. Amer. Med. Assoc.* 24 July 252/2 Massalongo's supposition that the disease represents a condition of hyperfunction—hyperpituitarism—has been widely discredited. **1961** *Lancet* 16 Sept. 655/1 There was general agreement that the diagnosis of adrenocortical hyperfunction should be made preoperatively. **1962** *Circulation Res.* X. 250 (*heading*) Compensatory hyperfunction of the heart and cardiac insufficiency. **1934** Webster, Hyperfunctional. **1961** *Jrnl. Amer. Med. Ass.* 29 July 232/1 One hyperfunctional nodule proved to contain a papillary carcinoma in an adenoma. **1970** N. Simionescu *Histogenesis Thyroid Cancer* iv. 28 (*heading*) The hyperfunctional cell. **1918** *Endocrinology* II. 46 A hyperfunctioning thyroid may be poor in colloids. **1926** Hyperfunctioning [see *hypofunctioning* s.v. *hypo- II]. **1954** A. White et al. *Princ. Biochem.* xliii. 946 Hyperfunctioning of the adrenal cortex in man is seen as a result of tumors composed of cortical cells. **1961** *Lancet* 16 Sept. 655/2 There was disagreement..about whether adrenalectomy for patients with hyperplastic or hyperfunctioning glands should be total or subtotal. **1947** Dorland & Miller *Med. Dict.* (ed. 21), Hypergammaglobulinemia. **1958** *Immunology* I. iii. 245 Hypergammaglobulinaemia was a feature of the acute phase when complement levels were very low. **1971** *Nature* 31 Dec. 558/2 We have obtained evidence in support of the idea that hypergammaglobulinaemia represents an immunological host response to tumour-associated antigen(s). *Ibid.* 559/1 A hundred instances of individual immunoglobulin increases occurred in the fifty hypergammaglobulinaemic mice. **1855** R. G. Mayne *Expos. Lex. Med. Sci.* (1860) 480/1 Hypergeusia. **1888** *Encycl. Brit.* XXIII. 80/2 Increase in the sense of taste is called hypergeusia, diminution of it hypogeusia, and entire loss ageusia. **1936** *Jrnl. Clin. Invest.* XV. 475 (*heading*) Acid-base equivalence of the blood in diseases associated with hyperglobulinemia. **1966** *McGraw-Hill Encycl. Sci. & Technol.* VIII. 256/2 The diseases usually associated with hyperglobulinemia are multiple myeloma, kala-azar, Hodgkin's disease, [etc.]. **1958** Dameshek & Gunz *Leukemia* viii. 187 Hyperglobulinemic purpura. **1894** Gould *Dict. Med.* 590/1 Hyperglycemia. **1966** Wright & Symmers *Systemic Path.* I. xxiii. 693/2 It has become obvious that diabetes mellitus is a syndrome and not a disease, and that a number of diverse factors may produce prolonged hyperglycaemia. **1903** *Med. Rec.* 24 Jan. 123/1 In coma diabeticum..it is likely that the hyperglycæmic condition stands at the foundation of a diminished electrical conductivity of the serum. **1969** R. L. Searcy *Diagn. Biochem.* liii. 461/2 This theory.. could account for the hyperglycemic tendency. **1927** *Lancet* 15 Jan. 117/2 Fluids from ten different hyperimmune..rats. **1940** *Jrnl. Bacteriol.* XXXIX. 66 Mice born of hyperimmune mothers are themselves immune to intranasally administered virus. **1957** Cushing & Campbell *Princ. Immunol.* i. 24 For many laboratory procedures, or for the production of potent therapeutic serums, animals are injected for many weeks or even months. Such animals are often referred to as being hyperimmune. **1958** *Immunology* I. 82 Titres of hyperimmune sera. **1913** Dorland *Med. Dict.* (ed. 7) 445/2 Hyperimmunization. **1968** F. Haurowitz *Immunochem. & Biosynthesis Antibodies* x. 209 Hyperimmunization is the routine method used in the production of high antibody titers. **1905** *Rep. Brit. Assoc. Adv. Sci.* 553 Spreuill ..by hyper-immunising sheep with virulent blood has succeeded in producing a serum efficacious in cases of Blaauw tongue. **1968** Gell & Coombs *Clin. Aspects Immunol.* (ed. 2) xlviii. 1278 It is even possible to hyperimmunize a horse with more than one major antigen at the same time. **1927** *Lancet* 15 Jan. 117/2 A hyperimmunised rat. **1962** *Ibid.* 27 Jan. 208/2 They seem a likely source of the plasma cells which accumulate in the lung in hyperimmunised animals. **1931** E. C. Faust in *Amer. Jrnl. Hygiene* XIV. 209 In addition to the direct and indirect types of Strongyloides..there is a distinct hyperinfective type..which is responsible for the so-called 'auto-infection' (i.e. 'hyperinfection') of individuals who have once become parasitized. **1943** Craig & Faust *Clin. Parasitol.* (ed. 3) xiv. 249 In cases of hyperinfection, all or some of the rhabditoid larvæ in the lumen of the bowel metamorphose into dwarfed filariform larvæ *en transit* down the bowel, and..may produce reinfection. **1960** J. M. Watson *Med. Helminthol.* xii. 116/2 The belief formerly held that the parasitic forms had a life-span of as much as fifteen years, based on the continuance of the infection in individuals removed from all possibility of external re-infection, did not take account of the possibility of auto-infection and hyper-infection. **1931** Hyperinfective [see *hyperinfection* above]. **1936** A. C. Chandler *Introd. Human Parasitol.* (ed. 5) xvii. 359 The course of development of these larvae may follow any of three different lines..indirect, direct, and hyperinfective. **1930** F. D. Graham (*title*) Exchange, prices and production in hyperinflation. **1952** P. Einzig *Inflation* i. 23 When inflation has reached an extreme stage it may be described as 'hyper-inflation'. **1970** *Daily Tel.* 21 Dec. 3/7 The bulletin suggests a prices and incomes policy and a wealth tax, to deal with the emerging problem of hyper-inflation [in Australia]. **1924** *Jrnl. Amer. Med. Assoc.* 6 Sept. 729/2 Hypoglycemia is the result of hyperinsulinemia. **1962** *Lancet* 12 May 1003/2 Either hyperplasia or tumour of the islet-cells of the pancreas, without evidence of hyperinsulinæmia. **1924** S. Harris in *Jrnl. Amer. Med. Assoc.* 6 Sept. 729/2 It was this line of reasoning that caused me to think that there may be such a condition as hyperinsulinism. **1962** *Lancet* 13 Jan. 73/2 It seems reasonable to suppose that the characteristic hyperinsulinism immediately after these babies are born is the result of abnormal stimulation of the fœtal pancreas in utero by maternal hyperglycæmia and/or by some other factor. **1969** R. L. Searcy *Diagn. Biochem.* xxxv. 322/1 Hyperinsulinism is now a well-characterized condition known to be caused by a functioning tumor termed an insulinoma or nesidioblastoma. **1913** L. Forster tr. *Biedl's Internal Secretory Organs* ii. 61 Rudinger's contention that the condition of hyperirritability arises in the ganglion cells of the anterior cornua..did not survive the test of experiment. **1935** D. H. Shelling *Parathyroids* vi. 115 In 1876, the older Chvostek described hyperirritability of the facial nerve as a sign of tetany. **1960** *Adv. Pediatrics* XI. 107 Symptoms of acute hypernatremia are hyperirritability to stimuli despite extreme lethargy, coma, [etc.]. **1922** L. F. Barker et al. *Endocrinol. & Metabolism* I. i. 165 If the sympathetic nerve cells are hyperirritable, sympathetic action predominates in the individual. **1954** *Pediatric Clinics N. Amer.* May 347 The infant was markedly dehydrated and alternately hyperirritable and drowsy. **1949** *New Gould Med. Dict.* 483 Hyperkalemia. **1955** Elkinton & Danowski *Body Fluids* xxii. 483 Hyperkalemia is characteristic of adrenocortical insufficiency. **1961** *Lancet* 19 Aug. 399/2 Respiratory failure and hyperkalæmia are the main lethal factors. **1969** J. H. Green *Basic Clin. Physiol.* xvi. 89/1 This combination of a high blood potassium, with a high blood acid content, is termed hyperkalaemic metabolic acidosis. **1972** *Lancet* 1 July 36/2 If..the patient still tends to be hyperkalæmic, exchange resins can be given orally once or twice a day. **1841** W. Lawrence *Treat. Dis. Eye* (ed. 2) xiv. 368 Conical Cornea. Synonymes:—Sugar-loaf cornea; *staphyloma conicum*..hyperceratosis. **1848** Dunglison *Dict. Med. Sci.* (ed. 7) 442/2 *Hyperceratosis*, staphyloma of the cornea. **1907** W. A. Pusey *Princ. & Pract. Dermatol.* 88 The term hyperkeratosis is applied to those conditions of the stratum corneum in which there is an increased thickness of the horny layer with complete cornification of the cells. **1970** Jubb & Kennedy *Path. Domestic Animals* (ed. 2) II. x. 568/2 Hyperkeratosis may be..diffuse as in cattle poisoned with chlorinated naphthalenes. **1971** *Brit. Med. Bull.* XXVII. 29/2 The hyperkeratoses and pigmentation that accompanied the arsenical cancers of the hand. **1848** Dunglison *Dict. Med. Sci.* (ed. 7) 442/2 Hypercinesia. **1875** R. Fowler *Med. Vocab.* (ed. 2) 245/1 Hyperkinesia. **1935** *Jrnl. Mental Sci.* LXXXI. 835 Articulatory and respiratory hyperkinesias were the pathological basis of the coprolalia. **1961** *Lancet* 23 Sept. 683/2 He was re-admitted..with an acute choreiform illness, consisting of hyperkinesia and constant writhing movements. **1888** *Med. Chron.* VII. 391 (*heading*) The treatment of chorea and other hyperkinetic diseases with physostigmine. **1935** *Jrnl. Mental Sci.* LXXXI. 834 The onset of a hyperkinetic encephalitis was associated with tics. **1966** *Med. Ann.* 308 The hyperkinetic syndrome in children is characterized by hyperactivity, short attention span, impulsivity..and poor social adjustment. **1972** *Village Voice* (N.Y.) 1 June 36/2 Court suits can also be of help in discovering the full extent of the drugging of so-called hyperactive or hyperkinetic children. **1897** *Lippincott's Med. Dict.* 493/2 Hyperleucocytosis. **1898** *Allbutt's Syst. Med.* V. 420 In the second stage..a hyperleucocytosis occurs. **1951** *Jrnl. Clin. Endocrinol. & Metabolism* XI. 1027 Although her pneumonia was clearing..hyperleucocytosis, hypokaliemia, and the picture of metabolic alkalosis developed. **1894** Gould *Dict. Med.* 590/2 Hyperlipemia. **1936** *Physiol. Abstr.* XX. 818 In the rabbit hyperlipæmia was obtained with olive oil. **1955** H. J. Deuel *Lipids* II. v. 349 A moderate hyperlipemia (increased blood fat level) may occur. **1966** *Lancet* 24 Dec. 1379/2 The recognition that some hyperlipæmias are 'carbohydrate-induced'..further suggests that dietary carbohydrate influences serum-triglyceride. **1933** Hyper-magnesaemia [see *hypomagnesæmia* s.v. *hypo- II]. **1955** Elkington & Danowski *Body Fluids* xxii. 482 Hypermagnesemia is present. **1928** *Daily Express* 10 May 7 'He is suffering from hyper-mania, a state of unnatural excitement,' said Dr. Mould. **1945** W. S. Sadler *Mod. Psychiatry* xxxvii. 439 While three stages of mania are recognized—hypomania, acute mania, and hypermania—there is a fourth classification which has been denominated delirious mania. **1963** H. H. Kendler *Basic Psychol.* v. xiv. 510/1 A patient with hypermania, the more intense form, behaves like a raving maniac. **1956** W. H. Whyte *Organization Man* (1957) 408 A few mild neuroses conceded here and there won't give you too bad a score, and..you have the best margin for error if you err on the side of being 'hypermanic'—that is, too energetic and active. **1963** H. H. Kendler *Basic Psychol.* v. xiv. 510/1 A young soldier who exhibited at different times both hypomanic and hypermanic reactions. *Ibid.* 510/2 This hypermanic episode lasted about two weeks. **1897** *Lippincott's Med. Dict.* 494/1 *Hypermature cataract*, the final stage of progressive cataract, in which the lens substance breaks down, shrinking into a hard mass or becoming liquefied. **1962** D. G. Cogan in A. Pirie *Lens Metabolism Rel. Cataract* 294 When the entire cortex becomes liquefied the cataract is said to have become hypermature. **1904** L. W. Fox *Dis. Eye* xii. 309 The last stage is that of hypermaturity or overripeness. **1964** S. Duke-Elder *Parsons' Dis. Eye* (ed. 14) xix. 271 If the process is allowed to go on uninterruptedly the stage of hypermaturity sets in when the cortex becomes disintegrated and transformed into a pultaceous mass. **1962** *Lancet* 22 Dec. 1317/2 There is no hint of an environmental factor which could have caused this very persistent hypermetabolic state. **1971** N. R. Alpert *Cardiac Hypertrophy* 55 The particular factor that stimulates the growth of the heart acts upon the heart continuously during the hypermetabolic period. **1937** *Physiol. Abstr.* XXII. 528 It [*sc.* rectal temperature] may remain low during intense hypermetabolism. **1958** Dameshek & Gunz *Leukemia* viii. 185 Occasional cases of chronic lymphocytic leukemia are associated with extraordinary degrees of hypermetabolism (+60–+80 per cent). **1941** *Jrnl. Heredity* XXXII. 232 (*heading*) Hypermobile joints in all descendants for two generations. **1967** *Ann. Rheumatic Dis.* XXVI. 423/2 Her mother had generalized osteo-arthritis and..was probably hypermobile. **1927** *Jrnl. Amer. Med. Assoc.* 28 May 1711/2 The father's feet were normal, except for the hypermobility of the joints. **1941** *Jrnl. Heredity* XXXII. 232/2 All members of this generation show hypermobility, in varying degrees, of the joints of the fingers, thumbs, knees and elbows. **1967** *Ann. Rheumatic Dis.* XXVI. 423/2 The isolated joint hypermobility..is considered to be the result of generalized familial ligamentous laxity. **1949** Darlington & Mather *Elem. Genetics* vii. 152 The hypermorph is more efficient than the wild-type gene...The wild-type gene is hypomorphic to its hypermorphic mutant and amorphic to its neomorphic mutant. **1932** H. J. Muller in *Proc. 6th Internat. Congr. Genetics* I. 242 Since it has been found that there are reverse mutations of hypomorphic genes.., we must regard the allelomorphs thereby resulting not as hypomorphic but as hypermorphic to their immediate progenitor genes. **1966** E. A. Carlson *Gene* xiii. 112 Another type of activity exaggerated or increased the normal activity of genes; most reverse mutations would be examples of such hypermorphic activity. **1894** Gould *Dict. Med.* 590/2 Hypermotility. **1926** J. A. Ryle *Gastric Function* 83 Abnormally rapid emptying or hypermotility [of the stomach]. **1949** Koestler *Insight & Outlook* vii. 107 Pathological laughter may thus be classed among other forms of hypermotility—epileptic attacks, tantrums, tics—caused by similar release phenomena. **1932** Dorland & Miller *Med. Dict.* (ed. 16) 605/2 Hypernatremia. **1969** L. G. Wesson *Physiol. Human Kidney* xxvii. 552/1 Hypernatremia (plasma sodium concentration in excess of 150 mM/L) is observed in a variety of clinical situations. **1955** *Arch. Internal Med.* XCV. 21/1 A severe hyponatremic rather than hypernatremic acidosis. **1900** Dorland *Med. Dict.* 310/2 Hypernephroma. **1912** *Q. Jrnl. Med.* V. 157 The objects of this paper are:—(1) To classify and describe the commoner adrenal tumours... (3) To present new reasons against the hypothesis that renal hypernephromata are derived from adrenal rests. **1916** E. H. Kettle *Path. Tumours* 132 The term hypernephroma is applied to a particular group of tumours, in the belief that they are derived from suprarenal tissue. **1921** *Jrnl. Obstetr. & Gynæcol.* XXVIII. 23 (*heading*) A comparison between ovarian 'hypernephroma' and luteoma and suprarenal hypernephroma. **1923** *Guy's Hosp. Rep.* LXXIII. 193 The hypernephromata of the kidneys arise in the renal epithelium. **1967** J. S. King *Renal Neoplasia* ii. 24 The patient..had a large renal tumor..which proved to be a hypernephroma when examined microscopically. **1946** *Jrnl. Urol.* LV. 18 (*heading*) Renal adenomas in hypernephromatous kidneys: a study of their incidence, nature and relationship. **1962** *Sci. Amer.* Jan. 53/2 The discovery of hyperfragments led to a rapid development of a new field: hypernuclear physics. **1971** *Nature* 28 May 226/2 Subjects of special interest in Poland include hypernuclear physics and strong interactions. **1957** *Ann. Rev. Nuclear Sci.* VII. 473 Nuclear matter can bind Λ to form systems stable for a time comparable with the Λ mean life. Such systems are well known and are called hypernuclei or hyperfragments. **1965** R. H. Dalitz *Nuclear Interactions of Hyperons* ii. 5 The lightest Λ-hypernucleus known is $_{\Lambda}H^3$. *Ibid.* 14 Λ-Hypernuclei will generally have excited states, whose spectra will be of interest for hypernuclear physics. **1892** *Jrnl. Chem. Soc.* LXII. 1 557 This flow may be counterbalanced by subjecting the hyperosmotic solution to external pressure. **1903** *Med. Rec.* 24 Jan. 121/2 The crystalloid substances rapidly accumulate in the serum, causing it to be hyperosmotic. **1905** W. H. Howell *Text-bk. Physiol.* 885 A hypertonic or hyperosmotic solution in one whose osmotic pressure exceeds that of serum. **1964** *Oceanogr. & Marine Biol.* II. 307 Their body fluids are hyperosmotic to the surrounding water. **1941** T. C. Ruch et al. in *Amer. Jrnl. Physiol.* CXXXIII. 434 Both monkeys exhibited some type of disturbance of the chewing mechanism and a striking hyperphagia and adiposity. **1946** *Physiol. Rev.* XXVI. 549 The word hyperphagia was chosen because it does not have the subjective, psychological connotations of the terms 'hunger', 'appetite', 'satiety' and 'bulimia', and because the word 'polyphagia'..implies 'omnivorousness'... Hyperphagia is taken to mean simply, increased eating. **1969** W. Haymaker et al. *Hypothalamus* xv. 529/2 Hyperphagia and obesity have now been produced by bilateral destruction within or near the midregion of the hypothalamus in the monkey, dog, cat, rabbit, rat and mouse. **1943** *Yale Jrnl. Biol. & Med.* XV. 839 After 6 obese animals..had been completely fasted to return their weight to normal, they were again hyperphagic and became obese a second time on re-feeding. **1972** *Science* 9 June 1124/1 A hyperphagic response occurs when calcium in excess of its normal concentration is perfused..in the ventromedial region. **1899** *Jrnl. Anat. & Physiol.* XXXIII. 213 Prof. Pfitzner..read papers on brachyphalangia, hyperphalangia and on the inferior tibio-fibular joint. **1969** W. T. Mustard et al. *Pediatric Surgery* (ed. 2) II. lxxxiii. 1423 Hyperphalangia refers to an excessive number of phalanges in the longitudinal axis. **1891** Flower & Lydekker *Introd. Study Mammals* viii. 234 The Ichthyopterygia have been shown..to have gradually acquired their hyperphalangism as an adaptive character. **1959** J. J. Byrne *Hand* xv. 273 Hyperphalangism consists of an excessive number of phalanges, the thumb being most commonly involved with three phalanges. **1898** *Jrnl. Anat. & Physiol.* XXXII. p. ii (*heading*) The ossification of the terminal phalanges of mammalian fingers, in relation to hyperphalangy. **1946** R. R. Gates *Human Genetics* I. xi. 404 The fingers show considerable variation, including hyperphalangy (four joints instead of three). **1951** C. K. Weichert *Anat. Chordates* x. 485 The paddlelike limbs of plesiosaurs and ic[h]thyosaurs have a very large number of phalanges (hyperphalangy). **1886** Hyperphoria [see *exophoria* s.v. *exo-]. **1964** S. Duke-Elder *Parsons' Dis. Eye* (ed. 14) xxx. 472 It is impossible to be sure whether there is absolute hyperphoria of one eye or hypophoria of the other, the condition being relative. **1887** *Arch. Ophthalm.* XVI. 163 Only a comparatively small proportion of hyperphoric persons experience in marked degree this inability to see small objects well. **1970** *Jrnl. Gen. Psychol.* LXXXII. 111 The average period of perceived sweep..was not significantly affected by the hyperphoric condition. **1926** *Amer. Jrnl. Physiol.* LXXVI. 472 Hypercalcemia, hyperphosphatemia, cessation of kidney function and acidosis. **1969** R. L. Searcy *Diagn. Biochem.* xlvii. 418/1 Hyperphosphatemia has long been regarded as an early sign of kidney failure. **1955** H. J. Deuel *Lipids* II. iv. 324 A hyperphosphatemic reaction does not occur in dogs whose bile ducts have been ligated and transected. **1915** C. Allbutt *Dis. Arteries* I. i. 10 Lately I have preferred the etymology of hyperpiesia for the malady, and hyperpiesis for the hæmodynamic aspect of it. **1923** J. F. H. Dally *High Blood Pressure* v. 64 Hyperpiesia is the term applied by Sir Clifford Allbutt to a clinical morbid series characterised by persistently raised blood pressure (hyperpiesis) in association with hypertrophy of the heart and changes in the vessels. **1927** *Physiol. Rev.* VII. 474 (*heading*) Hyperpiesia or essential hypertension. **1951** R. Hargreaves *This Happy Breed* vii. 77 He must 'wangle' an extra half bag of coal from the Q.M. stores without provoking in the presiding demi-god an advanced condition of hyperpiesia. **1895** C. Allbutt in *Abstr. Trans. Hunterian Soc.* (1896) LXXVII. 47 The symptoms of arterial hyperpiesis are often of a functional nervous character. **1961** G. Pickering *Nature Essential Hypertension* ii. 5 His [*sc.* Allbutt's] term hyperpiesis, however, never became widely used. **1968** —— *High Blood Pressure* (ed. 2) i. 3 There remains a large residue in which no specific lesion can be found—hyperpiesis, primary hypertension, essential hypertension, high blood pressure without evident cause. **1915** C. Allbutt *Dis. Arteries* I. ix. 60 The following seemed to be a case of mixed senile and hyperpietic disease. **1920** L. M. Warfield *Arteriosclerosis* (ed. 3) viii. 187 In the hyperpietic cases the arteries undergo a transient thickening. **1890** Billings *Med. Dict.* 669/2 Hyperpigmentation. **1899** G. T. Jackson *Dis. Skin* (ed. 3) 394 Nævus Pigmentosus... A congenital, circumscribed hyper-pigmentation of the skin. **1956** D. M. Pillsbury et al. *Dermatol.* xxxviii. 868 Endocrine disturbances are.. commonly associated with hyperpigmentation such as is seen..during pregnancy, and with exophthalmic goiter. *Ibid.* 873 These areas [of the skin] are whitish and often present a well defined hyperpigmented border. **1970** Jubb & Kennedy *Path. Domestic Animals* (ed. 2) II. x. 568/2 The production of pigment in the basal cells is..a common response to injury so that acanthotic areas may also be hyperpigmented. **1909** H. Cushing in *Jrnl. Amer. Med. Assoc.* 24 July 249/1 (*heading*) The hypophysis cerebri. Clinical aspects of hyperpituitarism and of hypopituitarism. [*Note*] From an etymological point of view the terms *hyper-, hypo-, dys-*, and *a-pituitarism* are doubtless of badly mixed parentage, but there are certain obvious objections to such a combination as *hypohypophysism*. **1939** M. A. Goldzieher *Endocrine Glands* lvii. 341 The only condition to be distinguished from true gigantism, i.e. primary eosinophile hyperpituitarism, is the secondary hyperpituitarism attendant on primary insufficiency of the gonads. **1924** G. B. Shaw *St. Joan* p. xix, St Teresa's hormones had gone astray and left her incurably hyperpituitary or hyperadrenal or hysteroid or epileptoid or anything but asteroid. **1954** K. E. Paschkis et al. *Clin. Endocrinol.* iii. 31 Hyperpituitary giants may develop acromegalic features in later life. **1930** *Jrnl. Genetics* XXII. 306 In generations subsequent to the breakage it is possible for some individuals—'hyperploids'—to inherit the chromosome fragment (attached or unattached) in addition to two otherwise normal sets of chromosomes. *Ibid.* 329 Hypoploid and hyperploid individuals. **1957** C. P. Swanson *Cytol. & Cytogenetics* vi. 177 Individuals having irregular chromosome numbers are called aneuploids...The terms hyperploid and hypoploid have also been used, but less frequently. **1930** *Jrnl. Genetics* XXII. 309 Text-fig. 11 illustrates hyperploidy of parts of the *X*-chromosome. **1969** N. S. Cohn *Elem. Cytol.* (ed. 2) xvi. 373 An addition or loss of less than an entire set of chromosomes..is called aneuploidy, and it subsumes two classes, hypoploidy and hyperploidy. **1860** R. Fowler *Med. Vocab.* 157/2 *Hyperpnœa*, excessive respiration—e.g. panting. **1877** M. Foster *Textbk. Physiol.* 260 Respiratory movements become deeper.. and the rate of the rhythm is hurried... In this respect, dyspnœa, or hyperpnœa as this first stage has been called, contrasts very strongly with the peculiar respiratory condition caused by section of the vagi. **1904** *Jrnl. Physiol.* XXXI. p. xlv, The hyperpnœa of healthy men during exercise. **1962** *Lancet* 27 Jan. 172/1 Usually this significant hyperpnœa is coupled with a red suffusion of the face. **1909** *Jrnl. Physiol.* XXXVIII. 401 Where the subject had been made hyperpnœic by want of oxygen, apnœa followed after a few breaths of normal air. **1961** *Lancet* 29 July 249/2 The blood-pressure rises in the hyperpnœic phase [of breathing]. **1932** Dorland & Miller *Med. Dict.* (ed. 16) 606/2 Hyperpotassemia. **1963** J. H. Bland *Clin. Metabolism Body Water* xxi. 574/1 Muscle weakness and paralysis are commonly observed in both hypopotassemia and hyperpotassemia. **1953** *Lancet* 11 July 60/1 (*heading*) Hyperpotassæmic paralysis. **1902** A. R. Defendorf *Clin. Psychiatry* 17 Distractibility is not to be confused with 'hyperprosexia', which consists in the total absorption of the attention by a single process. **1940** Henderson & Gillespie *Text-bk. Psychiatry* (ed. 5) v. 107 Increase of attention (hyperprosexia) is less common, and is sometimes associated with a sensory hyperaesthesia. **1948** Hyperprosexia [see *aprosexia]. **1922** *Physiol. Abstr.* VII. 493 The hyperproteinæmia does not run parallel with precipitin formation. **1969** R. L. Searcy *Diagn. Biochem.* xvii. 154/1 Extreme degrees of hyperlipemia or hyperproteinemia may falsely lower serum electrolyte measurements. **1940** *Amer. Heart Jrnl.* XIX. 408 The majority of individuals with essential hypertension..manifest..marked reactions of blood pressure to various internal and external stimuli. This suggests that the mechanism for regulating blood pressure..is hyperreactive. **1955** *Sci. Amer.* Apr. 44/3 In this hyperreactive state the body responds with rapid formation of antibody to a second invasion, either by live or by killed virus. **1940** *Amer. Heart Jrnl.* XIX. 412 The vascular hyperreactivity of some patients with essential hypertension is extreme. **1970** *Clin. Sci.* XXXIX. 793 (*heading*) Vascular hyper-reactivity with sodium loading and with desoxycorticosterone induced hypertension in the rat. **1611** J. Hoskins in *Coryat's Crudities* sig. e6 Encomiological Antispasticks..rythmicall and hyperrythmicall. **1953** *Publ. Inst. Marine Sci.* III. 175 Hypersaline lagoons..occur in several parts of the world. **1964** *Oceanogr. & Marine Biol.* II. 283 Natural water containing dissolved solids in concentrations equivalent to salinities of 40 to 80‰ is referred to [in this review] as hypersaline water. **1971** D. S. McLusky *Ecol. of Estuaries* vi. 97 Hypersaline seas should not be confused with inland brines or salterns, such as the Utah Salt Lakes of America. **1957** *Publ. Inst. Marine Sci.* IV. 198 Fish have been killed by..hypersalinity. **1970** B. H. McConnaughey *Introd. Marine Biol.* i. 24 Unusually high salinities (hypersalinity) are rare in marine environments. **1915** *Amer. Jrnl. Obstetr. & Dis. Women* LXXII. 279 In many cases where dementia precox develops, a previous attack of mental disturbance has existed and the patient is to a certain extent forced by the family into the marital state on account of hypersexuality. **1964** C. W. Lloyd *Human Reprod.* xxv. 456 Temporal lobe lesions generally cause humans and monkeys to have decreased sexual responsiveness, but occasionally hypersexuality may develop. **1876** Dunglison *Dict. Med. Sci.* (rev. ed.) 523/1 Hypersomnia. **1910** *Lancet* 8 Oct. 1093/1 Dr. Albert Salmon..differentiates hypersomnia, which is an increase in normal sleep, from somnolence, apathy, and torpor,.. and from the drowsiness which occurs in old people. **1939** N. Kleitman *Sleep & Wakefulness* xxv. 361 Cerebral neoplasms have been known to produce interference with the sleep–wakefulness rhythm mainly in the direction of hypersomnia. **1966** *McGraw-Hill Encycl. Sci. & Technol.* XII. 376/1 The best known cause of hypersomnia is epidemic or lethargic encephalitis. **1929** *Jrnl. Nerv. & Mental Dis.* LXIX. 5 It is unquestionably in infundibular tumors that one encounters..the hypersomnic form of brain tumors. **1955** A. B. Baker *Clin. Neurol.* II. xxi. 1203 The hypersomnic patient closely resembles a normally sleeping individual. **1892** W. W. R. Ball *Math. Recreations & Problems* x. 191 The term hyper-space was used originally of space of more than three dimensions but now it is often employed to denote any non-Euclidean space. *Ibid.* 201 Riemann has shown that there are three kinds of hyper-space of three dimensions. **1947** I. Asimov in *Astounding Science Fiction* Mar. 117/2 Fooling around with hyper-space isn't fun... We run the risk continually of blowing a hole in normal space–time fabric. **1956** E. H. Hutten *Lang. Mod. Physics* v. 171 The propagation of the wave must be described as taking place, in most instances, in a multi-dimensional hyper-space, and not in ordinary space. **1961** *Times Lit. Suppl.* 1 Sept. 577/3 Time Travel, like hyperspace, is one of the classical Science-Fiction presumptions. **1973** *Publishers Weekly* 17 Sept. 58/3 The crew of the first interstellar voyage through hyperspace comes back as monsters. **1909** Webster, Hyperspatial. **1919** R. T. Browne *Mystery of Space* viii. 263 This is undoubtedly the weakest point in the structure of the hyperspatial geometries. **1943** C. L. Hull *Princ. Behavior* xi. 181 It seems unlikely that the Fisher-design type of experiment will yield dependable indications of the complex hyperspatial curvatures which will almost certainly be found. **1946** *Blood* I. 28 Five cases of thrombocytopenia associated with well defined splenomegaly of nonleukemic and non-neoplastic origin ('symptomatic hypersplenic thrombopenia'). **1949** Britton & Neumark tr. *Leitner's Bone Marrow Biopsy* viii. 151 Hypersplenic anæmias. **1963** Basu & Aikat *Trop. Splenomegaly* iii. 20 The clinical recognition of the hypersplenic state. **1914** *Arch. Internal Med.* XIV. 145 There may exist for the spleen conditions associated with a hyperactivity of some of its functions, let us say the function of influencing hemolysis. To such a condition the term 'hypersplenism' may be applied. **1955** W. Dameshek in *Bull. N.Y. Acad.*

Med. XXXI. 114 Who first used the term 'hypersplenism' is not accurately known, but it began to appear in Chauffard's writings from 1907 on and subsequently, and in those of Morawitz and Eppinger at a late date. **1963** Basu & Aikat *Trop. Splenomegaly* iii. 20 Hypersplenism..is a clinical term indicating non-specific overactive function of the spleen in a variety of clinical disorders. **1930** *Engineering* 3 Oct. 421/3 The method is used to solve problems arising in the design of hyperstatic systems, such as arches and portal openings, with sufficient precision. **1959** J. A. L. Matheson et al. *Hyperstatic Struct.* I. vi. 320 The behaviour of multi-storey buildings..in terms of the composite action of the floors and walls with the frame..is essentially a very complicated hyperstatic problem. **1966** J. S. C. Browne *Basic Theory of Struct.* v. 100 Extra or redundant bars will produce a truss that is hyperstatic. **1952** E. F. Linssen *Stereo-Photography* x. 147 If we take a hyperstereograph..of a mountain formation..which starts a kilometre away from us, we must beware not to include any trees or houses which are in our immediate neighbourhood. **1971** C. R. Arnold *Appl. Photogr.* xiii. 373 This tendency to produce a model effect is a well-known feature of hyperstereographs. **1939** Henney & Dudley *Handbk. Photogr.* xx. 588 The hyperstereoscopic effect..can add greatly to a stereograph's effectiveness by its strong emphasis of the depth quality. **1956** *Focal Encycl. Photogr.* 570/2 Consecutive photographs from an aerial survey series form hyperstereoscopic pairs. **1911** *Cassell's Cycl. Photogr.* 298/2 Hyperstereoscopy. **1926** A. W. Judge *Stereoscopic Photogr.* iii. 32 Hyper-stereoscopy is of much assistance in obtaining a true impression of distant hill or mountain scenery. **1958** *Newnes' Compl. Amat. Photogr.* xxvi. 231 If..we wish to take pictures of scenes such as distant mountains then, providing there are no objects in the foreground nearer than about 300 ft., we can use the long base separation method known as hyperstereoscopy. **1906** Hypersusceptibility [see '*hypersensitive a.* 2]. **1924** *Jrnl. Immunol.* IX. 86 The production of skin hypersusceptibility without infection. **1914** *Q. Jrnl. Med.* VII. 273 The so-called anaphylactic or hypersusceptible state. **1971** *Brit. Med. Bull.* XXVII. 57/1 Hypersusceptible individuals may still develop the disease despite the reduction of dust concentrations to a very low level. **1924** D. M. Greig in *Edin. Med. Jrnl.* XXXI. 560 The outstanding peculiarity of the cranial deformity for which I propose the name ocular hypertelorism, or briefly, hypertelorism, is the great breadth between the eyes. **1957** *Arch. Ophthalm.* LVII. 607/2 This is an instance of hypertelorism associated with mental retardation. **1972** *Daily Tel.* (Colour Suppl.) 22 Sept. 21/4 Jeanine..was born 28 years ago with the fish eyes, one on each side of the face, and the monstrously deformed nose characteristic of hypertelorism (Grieg's Disease). **1897** Hypertropia [see *exotropia* s.v. *exo-]. **1950** F. H. Adler *Physiol. Eye* x. 406 In a case of right hypertropia..if the right superior oblique is at fault, the head will be strongly tilted toward the left shoulder. **1894** Gould *Dict. Med.* 592/1 Hyperuricemia. **1924** *Arch. Internal Med.* XXXIV. 504 Blood uric acid values of 3·5 mg. per hundred cubic centimeters..were considered as presenting a hyperuricemia. **1970** W. S. Hoffman *Biochem. Clin. Med.* (ed. 4) xv. 756 Hyperuricemia may be due either to overproduction of uric acid or to undersecretion. **1962** *Lancet* 15 Dec. 1273/1 My own experience with three hyperuricæmic patients, two with a history of gout and one without,..lends support to Dr. Eidlitz's letter. **1955** A. E. Eggers et al. *Compar. Anal. Performance Long-Range Hypervelocity Vehicles* 24 Mar. (N.A.C.A. Rep. RM A54L10) 2 On the basis of equal ratios of mass at take-off to mass at the end of powered flight, the hypervelocity vehicle compares favourably with the supersonic airplane. **1960** *Nature* 29 Oct. 353/2 If the fused earth were buried in the manner that ejectamenta from hypervelocity impact[s] in stone are buried, then the maximum entry velocity (etc.). **1962** J. L. Potter et al. in F. R. Riddell *Hypersonic Flow Res.* 590 A small, low density, hypervelocity, continuous wind tunnel. **1964** *Bull. Amer. Physical Soc.* IX. 308/2 (*heading*) Attainability of fusion temperatures under high densities by impact shock waves of microscopic solid particles accelerated to hypervelocities. **1972** *Science* 2 June 979/2 Hypervelocity impact craters on the moon. **1928** *Biochem. Jrnl.* XXII. 1261 In the case of the fat-soluble vitamins.. several instances of supposed hypervitaminosis have been recorded. **1963** *Lancet* 5 Jan. 34/2 As in hypervitaminosis D, the increased intestinal absorption of calcium is probably responsible for the high urinary calcium. **1971** J. Z. Young *Introd. Study Man* xi. 182 A European would produce up to 800 000 I.U. [of vitamin D] per day in the tropics and might therefore suffer from hypervitaminosis, for the body has no way of detoxicating any excess. **1928** Brown & Rowntree in *Arch. Internal Med.* XXXV. 133 In view of..confusion,..terms as follows are suggested: (1) normovolemia for normal blood volume, (2) hypervolemia for increased blood volume, and (3) hypovolemia for decreased blood volumes. These terms are self-explanatory and apply only to volume states. **1964** I. N. Kugelmass *Biochem. Clinics* IV. 370 Hypervolemia in acute and subacute glomerulonephritis with pulmonary edema increases with the duration of anuria. **1948** *Amer. Jrnl. Physiol.* CLV. 338 Table 1 shows the bilateral rises in auricular pressure of 4 hypervolemic cats.

hyperacusis (-ăkiū·sis). *Med.* Also **-acousis, -acusia, -akusis.** [mod.L., f. Hyper- 5 + Gr. ἄκουσις hearing (ἀκούειν to hear), after F. *hyperacousie* (J. M. G. Itard *Traité des Maladies de l'Oreille et de l'Audition* (1821) II. 4).] Abnormally acute hearing, often resulting in pain on hearing only moderately loud sounds.

1825 J. M. Good *Study of Med.* (ed. 2) IV. 245 Hearing painfully acute and intolerant of the lowest sounds... It is the hypercousis, or, as it should rather be, the hyperacusis of M. Itard. **1894** T. M. Hovell *Treat. Dis. Ear* xlv. 656 Hyperacusia, of a pathological kind, may be associated with increased power of other special senses. **1927** Lake & Peters *Handbk. Dis. Ear* (ed. 5) xiv. 267 Hyperacousis is usually found (as otalgia) in females. **1971** D. E. Rose *Audiol. Assessment* iii. 44 Other symptoms of disorders of the ear are..increased aural sensitivity to sound (recruitment and hyperacusis).

hyperbaric (həipəɹbæ·rik), *a. Med.* [f. Hyper- 5 + Gr. βαρύς heavy + -ic.] **a.** Of a solution for spinal anæsthesia: having a greater density than the cerebro-spinal fluid.

1930 W. H. Jones in *Brit. Jrnl. Anæsthesia* VII. 110, I shall use the terms hyper-, iso- and hypo-baric, to describe solutions which have specific gravities greater than, equal to, or less than that of the cerebro-spinal fluid. **1946** J. L. Southworth et al. *Pitkin's Conduction Anesthesia* xvii. 761/2 The anesthetist should know in advance whether the anesthetic solution is hypobaric, hyperbaric, or isobaric, so that the posture of the patient can be regulated accordingly. **1962** J. Adriani *Chem. & Phys. Anaesthesia* (ed. 2) xxxi. 652/2 Hyperbaric solutions gravitate caudad if the body is inclined in the head-up, supine position after injection, while hypobaric solutions migrate cephalad.

b. At a pressure greater than normal (often, greater than the partial pressure of the gas in the atmosphere); employing or pertaining to such gas.

1963 *Lancet* 16 Nov. 1022/1 It is our firm impression that hyperbaric oxygen is the most effective method of resuscitation yet devised for the severely asphyxiated newborn infant. **1965** *Economist* 20 Feb. 780/2 Hyperbaric oxygen chambers have been used increasingly in the treatment of diseases involving oxygen deficiency. **1966** *Lancet* 24 Dec. 1394/2 A hyperbaric chamber was made from a large domestic pressure cooker. **1968** *Sci. Jrnl.* Nov. 64/2 Some kidneys survived when hyperbaric nitrogen was employed. **1972** *Lancet* 22 Sept. 639/1 Patients with advanced growths are being given hyperbaric oxygen (3–4 atmospheres absolute) for a period immediately before and during irradiating. **1973** C. Bonington *Next Horizon* xiv. 199 The following six weeks passed..in hospital, recovering from frost-bite. I spent much of the time as a guinea-pig for a new method of treatment called hyperbaric oxygen, which entailed hours spent lying in a cylinder filled with two atmospheres of oxygen. You had to wear a special anti-static tunic..apparently, in oxygen so pure and concentrated, even the slightest static electricity could have caused a fire.

hyperbolic, *a.* Add: **2. b.** *hyperbolic navigation*: navigation that utilizes the difference in the times of arrival or the phases of signals transmitted in synchronism by two radio stations to determine a hyperbola on which the receiver must lie, two intersecting hyperbolas from two pairs of stations determining its position; so *hyperbolic system*, etc.

1945 *Electronics* Nov. 94/1 Loran..is one of a family of systems known as 'hyperbolic navigation systems', which measure the relative time of arrival of two or more radio signals sent synchronously from known points. **1959** [see *Decca]. **1972** *Jrnl. Inst. Navigation* XXV. 308 The navigator has three main aids—d.f. using the world-wide chain of shore-based transmitter beacons, the short-range hyperbolic systems, mainly Decca, and his own radar.

hypercatalectic, *a.* Add: Used of Old English verse.

1813 J. J. Conybeare in *Archæologia* (1814) XVII. 265 Of the Trochaic species, with the Hypercatalectic syllable, as, Ahte ic, ealdor, stol. **1894** *Trans. Philol. Soc. 1891–4* 386 Expanded lines, called by German scholars *Schwellverse* or *Streck-verse*, are hypercatalectic.

hypercharge (həi·pəɹtʃɑːdʒ). *Nuclear Physics.* [f. *hyper(onic) charge*: see quot. 1956.] A property of hadrons that is conserved in strong interactions and is represented by a quantum number Y that is the same for all the particles of a charge multiplet (isospin multiplet), being equal to twice their average charge quantum number.

The hypercharge Y of a particle is related to its charge Q and the third component, I_3, of isospin by $Y = 2(Q - I_3)$, and to its strangeness S and baryon number B by $Y = S + B$.

1956 J. Schwinger in *Physical Rev.* CIV. 1164/2 It is very natural to suppose that the K meson, with isotopic spin ½, possesses a similar physical property [to the nucleonic charge] in the nature of a charge, which is also dynamically realized by a coupling with the field... We shall term this new property hyperonic charge Y, with Y = +1 characterizing the K^+K^0 multiplet, and Y = −1 describing the antiparticles $\bar{K}^0K^-$. As the agent for the dynamical exhibition of nucleonic and hypercharge, the vector field does not itself bear these charges. **1964** *New Scientist* 20 Feb. 460/3 According to the conservation of baryons and hypercharge, the omega-minus should be produced in collisions between K-minus mesons and protons and should decay (weakly, with a change of hypercharge) to a xi-particle and pion, or to a lambda-particle and K-minus. **1972** G. L. Wick *Elem. Particles* v. 92 All of the strongly interacting particles fall into families which are specified by three quantum numbers—hypercharge, isotopic spin and baryon number. *Ibid.*, In the final analysis, either strangeness or hypercharge will suffice as a quantum number. *Ibid.* 93 Physicists generally prefer to use strangeness—largely for historical reasons... On the other hand, some theoreticians have adopted the hypercharge as the basic quantum number. It is easier to manipulate in the equations.

hyperchromasia (həi:pəɹkromēi·ziă). Also in anglicized form **hyperchromasy** (-krōu·măsi) (*rare*). [mod.L., f. Hyper- 5 + Gr. χρῶμα, χρωματ- colour (in sense 2 repr. Chromatin): see -ia¹.] **1.** *Med.* **a.** Excessive coloration or pigmentation of the skin.

1889 *Cent. Dict.*, *Hyperchromasia*, a pathological condition marked by excess of pigment. **1908** *Practitioner* Aug. 349 A condition known as hyperchromasia, in contradistinction to achromasia, or leucodermia.

b. = *Hyperchromia.

1929 R. B. H. Gradwohl tr. *Schilling's Blood Picture* ii. 107 Special, generally designated erythrocytic blood pictures... Hyperchromasia: generally increased pigmentation. **1966** J. W. Linman *Princ. Hematol.* iii. 60 Large or excessively thick cells that stain more darkly than normal are described as being hyperchromic or as displaying hyperchromasia.

2. *Cytology.* The condition, in a cell or nucleus, of having an abnormally large amount of chromatin.

1930 *Med. Jrnl. Austral.* 22 Feb. 244/1 A detailed examination of carcinoma cells shows that hyperchromasy is the dominant cytological feature of these components. **1948** *Amer. Jrnl. Path.* XXIV. 1100 Cytologic descriptions of tumors are limited to generalities such as..hyperchromasia of nuclei. **1971** *Nature* 31 Dec. 547/2 The individual tumour cells were more rounded with nuclear hyperchromasia.

hyperchromatic (həipəɹkromæ·tik), *a.* [f. Hyper- 5 + Chromatic *a.* (in sense 2, Chromat(in) + -ic).] **1.** *Med.* **a.** Of the skin: excessively pigmented.

1894 in Gould *Dict. Med.* 589/2.

b. = *Hyperchromic *a.* 1.

1929 R. B. H. Gradwohl tr. *Schilling's Blood Picture* ii. 133 Single or many large hyperchromatic cells (megalocytes, occasional megaloblasts). **1966** [see *hyperchromasia b].

2. *Cytology.* Of a cell or nucleus: having an abnormally large amount of chromatin.

1906 E. B. Wilson *Cell* (ed. 2) ii. 68 In many cancer-cells many of the nuclei are specially rich in chromatin (hyperchromatic cells). **1930** *Med. Jrnl. Austral.* 22 Feb. 244/1 Highly differentiated, hyperchromatic squamous cells. **1966** Wright & Symmers *Systemic Path.* I. xxxii. 1128/1 Not infrequently the nucleus of the tumour cells varies considerably in size; it may be large and hyperchromatic..or it may appear as a small, irregular, pyknotic structure.

hyperchromatosis (həipəɹkrōmatō·sis). [mod.L., f. Hyper- 5 + Chromat(o-, Chromatin + -osis.] **1.** *Med.* Excessive colouring or pigmentation, esp. of the skin.

1886 *Syd. Soc. Lex.*, *Hyperchromatosis*, Auspitz's term for an excessive deposit of pigment in the epidermis. **1929** *Encycl. Medica* (ed. 2) VI. 273 Hyperchromatosis, excessive pigmentation, especially of certain skin diseases.

2. *Cytology.* [ad. G. *hyperchromatose* (R. Woltereck 1898, in *Zeitschr. f. wiss. Zool.* LXIV. 604).] An abnormally large number of chromosomes or amount of chromatin in a cell or nucleus.

1888 *Jrnl. R. Microsc. Soc.* 628 They are remarkable for their excess of chromosomes (hyperchromatosis), and for the tendency these have to group themselves in dyads and tetrads. **1913** J. Ritchie in *Textbk. Path.* 240 In malignant tumours the division of the chromosomes between the poles is often unequal... This can be recognized by the existence of differences in size and staining qualities of the nuclei, and the two phenomena are sometimes referred to as hyperchromatosis and hypochromatosis. **1966** *Nature* 31 Aug. 904/1 A hyperchromatosis and pyknosis of the nuclei which has hitherto been wrongly interpreted as a degeneration.

hyperchromia (həipəɹkrōu·miă). *Med.* [f. as next + -ia¹.] A hyperchromic condition of the blood or of an erythrocyte (see *hyperchromic *a.* 1).

1931 *Amer. Jrnl. Med. Sci.* CLXXXII. 321 There might be two types of pernicious anemia—the megaloblastic, associated with macrocytosis and hyperchromia, and the erythroblastic, associated with microcytosis and hypochromia. **1935** Whitby & Britton *Disorders of Blood* iii. 49 Hyperchromia is inevitably associated with macrocytosis. **1959** C. de Gruchy *Clin. Haemat.* iii. 43 Because it is now customary to describe the haemoglobin content of cells in terms of concentration rather than absolute amount, the term hyperchromia is best avoided.

hyperchromic (həipəɹkrōu·mik), *a.* [f. Hyper- 5 + Gr. χρῶμ-α colour + -ic.] **1.** *Med.* Characterized by or designating a colour index greater than one, or red blood cells that contain more hæmoglobin than normal and show little or no central pallor.

1924 T. R. Waugh in *Canad. Med. Assoc. Jrnl.* XIV. 114/2 The colour index is high, though total cell volume may be considerably reduced. Such a condition may be termed, therefore, hyperchromic. We find this type of anæmia especially in pernicious anæmia. **1935** Whitby & Britton *Disorders of Blood* vi. 116 The anæmia produced by a deficiency of the hæmopoietic principle is invariably macrocytic and almost always markedly hyperchromic. **1942** M. M. Wintrobe *Clin. Hematol.* vi. 125 It has been assumed that the red corpuscles in this condition [*sc.* pernicious anæmia] are 'hyperchromic' or supersaturated with hemoglobin. This is not true... The darker appearance of these corpuscles.. is due to their increased thickness. **1960** C. H. Smith *Blood Dis. Infancy & Childhood* v. 62 Hyperchromic anemias identify conditions in which macrocytes prevail and in which the color index is greater than unity.

2. Characterized by or exhibiting an increase in the extent to which light (usually, ultra-violet radiation) is absorbed; chiefly in *hyperchromic effect*.

1939 W. R. Brode *Chem. Spectroscopy* vii. 128 Ethylene..has an absorption band at 2345 Å (1950 A) in the extreme ultraviolet, but the conjugated coupling of two or more ethylene radicals together results in both hyperchromic and bathochromic effects. **1955** G. Schmidt in Chargaff & Davidson *Nucleic Acids* I. xv. 569 The quantitative degradation of DNA to mononucleotides by alkali is accompanied by an increase in absorption of approximately 20% in the ultraviolet region (hyperchromic effect). **1958** *Nature* 29 Nov. 1503/2 The hyperchromic degradation of polymers. **1968** M. W. Strickberger *Genetics* v. 73 The double-stranded form [of DNA] is called hypochromic, and the loose, single-stranded form is called hyperchromic.

Hence **hyperchromi·city,** the property of absorbing more (ultra-violet) light.

1968 *Nature* 29 Nov. 1502/1 In view of the profound importance of the anomalous ultra-violet absorption characteristics of nucleic acids in relationship to the fine structure of the macromolecule, the hyperchromicity of a number of relatively simple polynucleotide derivatives was studied. **1968** M. W. Strickberger *Genetics* v. 73 Hyperchromicity [of DNA] can be induced by heating.

hy:perconjuga·tion. *Physical Chem.* [f. Hyper- 1 b + Conjugation.] A direct interaction between the electrons of a methyl or substituted methyl group in a molecule and the electrons of an adjacent conjugated system, the former being attracted towards the latter.

1939 R. S. Mulliken in *Jrnl. Chem. Physics* VII. 340/2 In cyclic dienes..there is interaction between this [pair of electrons] and the unsaturation orbitals, resulting in what may be called 'cyclic conjugation'. This phenomenon has been..discussed by Hückel, except for the case where A is CH_2. For this case we may introduce the term 'hyperconjugation', meaning an additional conjugation beyond that ordinarily recognized. [*Note*] The term..was suggested by Professor W. G. Brown, as an alternative to 'superconjugation' at first used by the writer. **1949** *Q. Rev.* III. 229 Delocalisation of this type is known as σ-hyperconjugation, and since σ electrons are relatively tightly bound, the resulting energy of delocalisation is likely to be very small. Consequently, σ-hyperconjugation is generally ignored. **1952** *Chem. & Engin. News* 17 Nov. 4881/2 Hyperconjugation denotes resonance interaction by alkyl, and particularly methyl, substituents in systems exemplified by

$$H_3C{-}C{=}C{-} \leftrightarrow H^+H_2C{=}C{-}C^-.$$

1962 M. J. S. Dewar *Hyperconjugation* i. 13 The evidence..for the importance of hyperconjugation is much less conclusive than has been commonly supposed. **1965** Phillips & Williams *Inorg. Chem.* I. xi. 400 Hydrogen is also unable to act as a σ-donor, except possibly to a small extent in groups of hydrogen atoms as in CH_3, the so-called hyperconjugation effect.

Hence **hyperconju·gated** *ppl. a.*, exhibiting hyperconjugation.

1949 *Q. Rev.* III. 233 The carbon-carbon distance in acetaldehyde is approximately 0·04 Å less than the normal value, and..the shortening has been attributed to hyperconjugation. **1969** E. S. Gould *Mechanism & Struct. Org. Chem.* ii. 49 Hyperconjugated structures in which a shift of electron density..has left the C—H bonds with partial σ-bond character.

hypercorre·ct, *a. Linguistics.* Also **hyper-correct** (with hyphen). [f. Hyper- 4 a + Correct *a.*] Of a spelling, pronunciation, or construction: falsely modelled on an apparently analogous prestigial form. Also of a speaker using such a form.

1922 O. Jespersen *Lang.* xv. 294 Such hypercorrect forms are closely related to those 'spelling pronunciations' which become frequent when there is much reading of a language whose spelling is not accurately phonetic. **1937** *Amer. Speech* XII. 167 When James the footman says *chicking* for 'chicken' he is being hypercorrect, leaning over backwards. **1972** *Archivum Linguisticum* III. 4 Modern authors have focussed on the [French] imperfect subjunctive as a hypercorrect grammatical form which speakers use to characterize or satirize the pedantic and the pretentious.

Hence **hypercorre·ction, -corre·ctness.**

1934 Webster, *Hypercorrection*, *-ness*. **1935** *Language* XI. 145 It is only by unceasing vigilance that hypercorrection can be avoided by one whose native dialect has this phonological feature and who wishes to correct it. **1955** *Archivum Linguisticum* VII. 48 Phonemic interference may be due to..phone substitution and hypercorrectness. **1959** M. Schlauch *Eng. Lang. in Mod. Times* vi. 167 The social..pressure of superiors for hypercorrection. **1964** H. Kökeritz in D. Abercrombie et al. *Daniel Jones* 141 Salmon..tends to become ['sɑmən] by hypercorrection. **1972** *Language* XLVIII. 484 Hypercorrections like *fesk*, *mast* (for 'mask') seem to establish that nn, ss are full forms in NNE and not stylistic reductions.

hyperdrive (həi·pəɹdraiv). *Science Fiction.* Also **hyper-drive.** [f. Hyper- + Drive *sb.*; perh. suggested by *hyperspace*, *overdrive*.] A fictitious device by which a spaceship is enabled to travel from one point to another in a shorter time than light would take (usually by passing out of ordinary space into 'hyperspace' for the journey); also, the state of so travelling.

1955 B. Davenport *Inquiry into Sci. Fiction* 11 A 'hyperdrive'..may be defined simply as something that *does* enable ships to travel faster than the speed of light, no matter what Einstein says. **1959** P. Anderson *Virgin Planet* (1966) iii. 20 The ship went into hyperdrive and outpaced light. **1960** K. Amis *New Maps of Hell* (1961) i. 20 The author will fabricate a way of getting around Einstein..[by] a device known typically as the space-warp or the hyper-drive. **1965** D. Morgan in J. Carnell *New Writings in S-F* III. 144 The jolt as he came out of hyper-drive..had confirmed what he had suspected for some months. **1968** M. S. Livingston *Particle Physics* i. 7 No responsible scientist would attempt to justify support for research in this field with prediction of an 'anti-matter engine'..or a 'hyper-drive' for spaceships.

hyperemia, var. Hyperæmia.

hyperesthesia, var. Hyperæsthesia.

hyperextend (həipəɹ,ekste·nd), *v.* [f. Hyper- 4 c + Extend *v.*] *trans.* To extend, in the sense opp. to Flex *v.*, (a joint, or a part of the body moving about a joint) so as to attain an abnormally great angle. So **hyperexte·nded** *ppl. a.*

1883 *Brain* V. 348 The toes are spread out and hyper-extended. **1886** *Ibid.* IX. 169 Between this ridge and the toes there is a deep hollow, to which the hyperextended first phalanges form an almost perpendicular boundary. **1903** Tubby & Jones *Mod. Methods Surg. Paralyses* II. 218 Treatment should consist in rendering the forearm supine and hyperextending the wrist. **1927** *Jrnl. Amer. Med. Assoc.* 28 May 1711/1 The fingers could be hyperextended until they touched the dorsum of the wrist. **1963** *Lancet* 5 Jan. 6/1 An injury in which the neck has been forcibly hyperextended or flexed.

Also **hyperexte·nsion,** the state of being hyperextended.

1883 *Brain* V. 480 There is a slight hyper-extension at the torso-phalangeal, and flexion at the phalangeal joints. **1905** A. B. Jackson *Influence Growth on Deformities* iv. 85 The best result after this affection is a knee ankylosed in extension or hyperextension. **1961** *Lancet* 5 Aug. 297/1 The common pattern of deformity was flexion.. and dislocation of the hip, with extension or hyperextension of the knee. **1968** J. Winearls *Mod. Dance* (ed. 2) ii. 48 In the case of the backwards contraction, the spine is really in hyper-extension.

hyperextensible (həi:pəɹ,ekste·nsib'l), *a.* [f. prec., after *extensible*.] Capable of being hyper-extended. So **hy:perextensibi·lity.**

1946 R. R. Gates *Human Genetics* I. xi. 449 All the finger joints are hyperextensible. *Ibid.* 448 Hyperextensibility or doublejointedness of the thumbs may occur in the first joint or the second. **1961** *Lancet* 2 Sept. 526/2 There were obvious transverse palmar folds and hyperextensibility of the joints.

hyperfine (həi·pəɹfəin), *a. Physics.* [tr. G. *hyperfeinstruktur* hyperfine structure (W. Pauli 1924, in *Naturwiss.* 12 Sept. 741/1), f. Hyper- + *feinstruktur* *fine structure.] *hyperfine structure*: (the presence of) multiplets of closely spaced lines in a spectrum that are closer together than those of fine structure; *esp.* multiplets that result from the further splitting of lines by the coupling between the magnetic moment (and higher multipole moments) of the atomic nucleus and the electromagnetic field of the electrons. Also (with ellipsis of *structure*), of or pertaining to this phenomenon.

1927 *Proc. Nat. Acad. Sci.* XIII. 430 The hyper-fine structure of λ 2537 of mercury has been shown..to consist of five lines of very nearly equal intensity. *Ibid.* 431 Each of the five hyper-fine structure lines has a triplet Zeeman pattern. **1942** J. D. Stranathan *Particles of Mod. Physics* x. 400 Other details of characteristic line spectra, known as hyperfine structure, have made it necessary to assign also to the nuclei of atoms certain magnetic moments. **1950** D. Halliday *Introd. Nucl. Physics* xiii. 479 Hyperfine studies of optical spectral lines. **1953** N. F. Ramsey in E. Segrè *Exper. Nucl. Physics* I. iii. i. 358 If this fine structure is examined more closely, it is often found that each line of the fine structure can in turn be resolved into further lines or 'hyperfine structure' with a separation of the order of 1 cm⁻¹. **1954** H. Semat *Introd. Atomic & Nucl. Physics* (ed. 3) viii. 273 Two distinct types of hyperfine structure have been observed. One type..has been explained as due to the presence of two or more isotopes of the element. **1958** Condon & Odishaw *Handbk. Physics* vii. vi. 88/1 Nuclear electric quadrupole hyperfine structure in molecular spectra has yielded a number of nuclear spins and quadrupole moments. **1962** [see *fine structure 1 a]. **1965** C. M. H. Smith *Textbk. Nucl. Physics* viii. 229 If the number of hyperfine components can be counted, the nuclear spin *I* is obtained.

hyperfocal (həipəɹfōu·kăl), *a.* [f. Hyper- 1 + Focal *a.*] Applied to the distance on which a camera must be focused to give the maximum depth of field (see quot. 1957), and to the object plane at this distance.

1905 Wastell & Bayley *Hand Camera* viii. 96 The best position for the lens..is sometimes called the 'hyper-focal distance'... By dividing the hyperfocal distance by two we get the distance beyond which everything is in focus. **1957** Amos & Birkinshaw *Television Engin.* I. ix. 178 If a lens system is focused on infinity, the images are in focus..for all objects lying between infinity and the hyperfocal distance. If, however, the lens system is focused on the hyperfocal distance, images are in focus for all objects lying between infinity and half the hyperfocal distance. **1961** G. Millerson *Technique Television Production* iii. 34 We can adjust any camera lens system so that it is focused on its hyperfocal plane.

hypergamy (həipĕ·ɹgămi). *Anthrop.* [f. Hyper- + *-gamy.] A term first used by W. Coldstream, to denote the custom which forbids the marriage of a woman into a group of lower standing than her own; also *transf.*, of any marriage with a partner of higher social standing. Hence **hype·rgamous** *a.*, pertaining or relating to hypergamy.

1883 D. C. J. Ibbetson *Rep. Census Panjáb 17 Feb. 1881* I. 356 They [*sc.* the social rules]..may be referred to two laws, which I shall call the laws of isogamy and hypergamy... Mr. Coldstream writes from Hushyárpur:..For 12 years past certain classes of Khatris..have been agitating to extend the principle of isogamy, and to free themselves from the rule of contracting hypergamous alliances for their daughters. **1903** Risley & Gait *Rep. Census India 1901* § 701 *Hypergamy*, or 'marrying up' is the custom which..compels [a woman] to marry in a group equal or superior in rank. A hypergamous division, therefore, is a group forming part of a series governed by the foregoing rule. **1909** E. S. Hartland *Primitive Paternity* I. 266 The Brahmans everywhere follow a custom known as *hypergamy*, by which a man may marry or have sexual relations with a woman of lower rank, but no man of lower rank may marry into a caste above his own. **1921** *Nature* 13 Jan. 646/1 Dr. Rivers said that the term 'hypergamy' had been used loosely by both Sir Herbert Risley and Dr. W. Crooke to denote marriage between groups which differ in rank, but for the sake of clearness the term should be confined to those instances in which there was a characteristic difference between the marriage rules for the two sexes. **1929** *Encycl. Brit.* IV. 979/2 The ancient hypergamous rule which allowed a woman to be married to a man of higher class. **1951** *N. & Q. Anthropol.* (ed. 6) II. iii. 94 Sometimes there is a hypergamous system in which a socially inferior sub-caste can obtain wives from another and higher sub-caste on payment. **1957** *New Statesman* 4 May 566/2 The curse which is ruining, in fantasy if not in their own lives, these brilliant young men of working-class origin and welfare-state opportunity is what anthropologists have dubbed male hypergamy. **1966** *Ibid.* 15 Apr. 548/2 Hypergamous young men claim that the district [*sc.* Chelsea] offers more available classy girls than other art-school areas. *Ibid.* 2 Dec. 832/1 English women have long been able to go in for hypergamy.

hypergelast (həipĕ·ɹdʒilæst). [f. Hyper- 4 + Gr. γελαστής a laugher, f. γελᾶν to laugh; cf. Gelastic *a.*] (See quot. 1877.)

1877 G. Meredith *Ess. Comedy* in *New Q. Mag.* VIII. 2 We have another class of men..whom we may term hypergelasts; the excessive laughers. **1902** *Times Weekly Educ. Suppl.* 19 Dec. p. ii/2 As to the savage, he is a great laugher—a hypergelast, in fact. **1933** *Scrutiny* I. 356 Meredith's hypergelasts are enemies of the Comic spirit, but his ideal audience all laugh, in their polite drawing-room way. **1947** N. Cardus *Autobiogr.* I. 47 The word 'hypergelast'..denoting the loud vacant laughers.

hypergol (həi·pəɹgǫl). *Astronaut.* [a. G. *hypergol* (one of a series of terms ending in *-ergol*), app. f. Hyper- + Gr. ἔργ-ον work + -ol 3.] A hypergolic rocket propellant.

1947 *Jrnl. Brit. Interplanetary Soc.* VI. 104 Fuels are broadly classified in three groups, monergols, hypergols and non-hypergols. *Ibid.* 105 The distinct property of hypergols is that the reaction is self-starting. **1950** [see *hydrazine]. **1969** R. T. Holzmann *Chemical Rockets* ii. 140 The use of hypergols is both simple and safe.

So **hypergo·lic** *a.*, igniting spontaneously on contact with the oxidant or another propellant; also as *sb.*, a hypergol.

1947 *Jrnl. Brit. Interplanetary Soc.* VI. 106 A third method is to inject auxiliary fuels, hypergolic or self-igniting monergolic, into the chamber before the main fuels. **1965** *Ibid.* XX. 154/2 Ignition does not pose problems as the propellent is hypergolic. **1970** N. Armstrong et al. *First on Moon* i. 28 During the second week of June it was time to start putting hypergolic propellants into the fuel tanks. *Ibid.* iv. 75 The self-igniting..hypergolics which went into the separate control systems of the command and service module and the lunar module.

hypermarket (həi·pəɹmāɹkĕt). [f. Hyper- + Market *sb.*, translating F. *hypermarché* (f. *marché* market, after *supermarché* *supermarket).] A very large self-service store, usually situated outside a town, having an extensive car park and selling a wide range of goods.

1970 *Guardian* 1 Oct. 7/1 A proposed new 'hypermarket', a gigantic supermarket which could be the precursor of complete shops as big as whole villages. **1971** *Times* 14 July 19/4 The catalyst has been the imminent arrival in the United Kingdom of Carrefour hypermarkets. **1971** *Observer* 3 Oct. 15/2 Hypermarkets are like retail factories, vast warehouse-type buildings. **1972** *New Statesman* 28 Jan. 101/1 By 1966 four hypermarkets (i.e. units of over 25,000 sq. ft. of floor space) had been set up. These huge self-service stores have caught on in France more than in Britain—today there are 144 of them. **1972** *Daily Tel.* 23 May 22/7 Although only a quarter the size of the typical hypermarket, it has all the other ingredients—one-level parking, discount prices, a substantial non-food sector, and late opening.

hypermetric, *a.* Add: **1.** Used of Old English verse.

1892 F. J. Mather in *Mod. Lang. Notes* VII. 200 It will be well to note the occurrences of hypermetric lines in the different poems. **1906** G. P. Krapp *Andreas* p. xlvii, *Beowulf* (which contains twelve hypermetric lines). **1958** A. J. Bliss *Metre of Beowulf* 96 The distribution of hypermetric verses varies from poem to poem. **1970** M. Swanton *Dream of Rood* 61 Blocks of hypermetric verse used contrapuntally to accommodate significantly more complex thematic material.

hypermetrical, *a.* Add: Used of Old English verse.

1891 J. W. Bright *Anglo-Saxon Reader* 238 These hypermetrical half-lines occur either singly or in groups, and usually add dignity to the sense and movement of the passage. **1922** F. Klaeber *Beowulf* p. lxxi, Groups of emphatic hypermetrical types are introduced three times. **1935** A. C. Bartlett *Larger Rhet. Patterns Anglo-Saxon Poetry* 70 The hypermetrical irregularities of the other Anglo-Saxon poems. **1958** A. J. Bliss *Metre of Beowulf* 88 It is common ground that the vast majority of hypermetrical verses end with a group of syllables which is exactly equivalent to an ordinary verse.

hypermodern (həipəɹmǫ·dəɹn), *a.* [f. Hyper- 4 a. + Modern *a.*] Excessively modern; *spec.* in Chess, of or pertaining to the strategy, first used in the early 20th cent., of controlling the centre of the board with pieces at a distance.

1923 *Brit. Chess Mag.* Sept. 338 What is claimed as hyper-modern turns out to be..respectably medieval. **1945** Koestler *Yogi & Commissar* III. i. 127 His flair for hyper-modern avant-garde methods in Economy,..and Warfare. **1951** 'Assiac' *Adventure in Chess* II. v. 75 This 'hypermodern' move has been all the rage during the last few years. **1966** G. N. Leech *Eng. in Advertising* xxii. 196 These extra features..add animation to the language, imparting a racy, hyper-modern image of the product. **1970** A. Sunnucks *Encycl. Chess* 41 An exponent of the ideas of the hypermodern school of chess.

Hence **hypermo·dern, hypermo·dernist** *sbs.*; **hypermode·rnity.**

1923 J. Hart tr. *Reti's Mod. Ideas in Chess* v. 122 The Hyper-moderns are the greatest opponents of routine play. **1925** *Brit. Chess Mag.* May 219 *Humpeldinck's opening*..may be fairly described as the *dernier cri* of hypermodernity, the most up-to-date of the 'hyper-modern' openings. **1959** B. J. Horton *Dict. Mod. Chess* 89/2 The long-term plan of the hypermodernist was to allow the enemy to occupy the center and then to demolish him later.

hypernic (həi·pəɹnik, həipĕ·ɹnik). [f. Hyper- + *Nic(aragua)*.] A wood from one of several tropical American trees, esp. *Hæmatoxylon brasiletto*, or the red dye extracted from it. Also *attrib.*

1897 C. T. Davis *Manuf. Leather* (ed. 2) xlii. 567 Take 10 lbs. hypernic chips... Add the hypernic solution by degrees. **1924** Record & Mell *Timbers Tropical Amer.* 247 The wood [of *Haematoxylon brasiletto*] from Nicaragua is known..in the trade as Nicaragua wood or 'hypernic', the latter term coined to indicate a superlative quality of the wood from that country, but subsequently applied rather indiscriminately to any red dyewood or dyewood extract. **1971** R. J. Adrosko *Nat. Dyes & Home Dyeing* 26 Besides being called by the other common names for brazilwood, braziletto was also known as Nicaragua wood and hypernick.

hyperon (həi·pĕrǫn). *Nuclear Physics.* [app. f. Hyper- + *-on¹.] Any of a group of unstable sub-atomic particles that includes all the baryons apart from the proton and neutron; any strongly interacting particle with half-integral spin and a mass greater than that of the nucleons.

1953 *Compt. Rend. du 3me Congrès Internat. sur le Rayonnement Cosmique* 269 Nomenclature more frequently used during the conference... Groups of particles... H-particles (hyperons): symbol H: particles with mass intermediate between those of the neutron and the deuteron (this definition to be revised if 'fundamental' particles heavier than the deuteron are found). **1963** S. Tolansky *Introd. Atomic Physics* (ed. 5) xxiii. 391 The baryons..can best be subdivided into two groups namely (a) the familiar nucleons,..(b) the extremely unstable somewhat heavier particles, now called hyperons. **1965** *New Scientist* 18 Mar. 738/3 The SU(3) symmetry relates not just the proton and the neutron one to another, but includes also in one multiplet the six particles known as hyperons.

Hence **hypero·nic** *a.*

1956 [see *hypercharge]. **1969** *Physical Rev. Lett.* XXII. 1238/1 Σ^- hyperons are formed in the targets through the reactions $K^- + p \to \Sigma^- + \pi^+$ and $K^- + n \to \Sigma^- + \pi^0$, and upon capture by target nuclei make hyperonic atoms. **1970** *Physics Lett.* XXXIII. B. 230/2 The capture of the Σ^- hyperons takes place in atomic levels with high main quantum number *n*... The typical pattern of the Σ hyperonic X-rays is determined by the fact that the Σ^- hyperon..is a strongly interacting particle.

hyperosmolality (həi:pəɹ,ǫzmolæ·liti). *Med.* [f. Hyper- + *osmolality.] = *hyperosmolarity.

1959 *Pediatric Clinics N. Amer.* VI. 259 In severe hyperosmolality there may be a striking reduction of cerebro-spinal fluid pressure. **1963** *Amer. Jrnl. Cardiol.* XII. 654/2 Hypernatremia and hyperosmolality of serum result from loss of water without sodium. **1972** [see *hyperosmolar adj.].

hyperosmolarity (həi:pəɹ,ǫzmolæ·riti). *Med.* [f. Hyper- + *osmolarity.] The condition (in a bodily fluid, esp. serum) of having abnormally high osmotic pressure; also, the condition (in an individual) of having such serum.

1947 *Amer. Jrnl. Dis. Child.* LXXIV. 684 The less the ability of the kidney to excrete urine of high solute content, the more readily hyperosmolarity of the body fluid develops. **1958** *Jrnl. Chronic Dis.* VII. 1 The ultimate effect of hyperosmolarity of serum is dehydration of cells. **1963** *Lancet* 20 Apr. 891/1 We saw a patient with marked hyperosmolarity, due to an unusually high plasma-sodium, dehydration, and moderate ketonuria, but no loss of consciousness. **1965** *Neurology* XV. 205/1 Hyperosmolarity secondary to hypernatremia in various clinical states is known to be associated with severe neurological disturbances.

So **hyperosmo·lar** *a.*, of, exhibiting, or associated with hyperosmolarity.

1953 *Amer. Jrnl. Med.* XV. 185/1 After discontinuance of pitressin the hyperosmolar state promptly recurred. **1966** *Lancet* 26 Mar. 679/2 Eight patients with hyperosmolar non-ketoacidotic diabetic coma were new diabetics with an average age of 63 years. **1972** *Science* 19 May 815 Malone et al. stated that severe hyperosmolar dehydration could be responsible for the entire galactose toxicity syndrome. However, we have concluded..that hyperosmolality per se is not the major factor responsible.

hyperparasite. (Earlier and later examples of the various forms.)

1833 A. H. Haliday in *Entomol. Mag.* I. 482, I am not aware that any of them [*sc.* a group of ichneumons] are hyperparasitic. **1889** *Cent. Dict.*, Hyperparasite, hyperparasitism. **1897** L. O. Howard *Study in Insect Parasitism* 14 Many parasites suffer..from the attacks of hyperparasites. **1906** J. W. Folsom *Entomol.* x. 312 (*heading*) Hyperparasitism. **1914** *Entomologist* XLVII. 77 C. W. Colthrup sent me..two females which he had captured with three specimens of the hyperparasite *Hemiteles areator*. **1932** E. Step *Bees, Wasps, Ants* 185 These parasites [*sc.* Braconidæ]..are in turn attacked by still smaller species—hyperparasites. **1951** G. Lapage *Parasitic Animals* ix. 264 Many species of Hymenoptera implant their eggs inside the bodies of other parasitic insects, so that they provide good examples of hyperparasitism. **1952** A. Lysaght tr. *Caullery's Parasitism & Symbiosis* vi. 108 Sometimes, too, under new conditions, a parasite changes into a hyperparasite. **1964** T. C. Cheng *Biol. Animal Parasites* 143/1 Various species of *Nosema* and *Glugea* are capable of hyperparasitism in larval and adult helminths. *Ibid.*, One species of *Nosema*..is hyperparasitic in tapeworms.

hy:perparathy·roidism. *Med.* [f. Hyper- 5 + Parathyroid + -ism.] A condition in which there is an abnormally high level of parathyroid hormone in the blood, resulting in the loss of calcium from the bones, which become brittle.

1917 C. Voegtlin in *Surg., Gynecol. & Obstetr.* XXV. 249 A condition which might justly be termed hyperparathyroidism is unknown at the present time. **1948** Albright & Reifenstein *Parathyroid Glands* iii. 67 It is a fact, however, that one may have severe hyperparathyroidism and show no clinical, roentgenological, or histological evidence of bone disease... For all practical purposes, it usually comes down to whether or not the patient drinks milk. **1966** Wright & Symmers *Systemic Path.* II. xxxii. 1127/1 Secondary hyperparathyroidism occurs when more parathyroid hormone is required by the body than under normal circumstances. **1971** Rimoin & Schimke *Genetic Disorders Endocrine Glands* iv. 87 The symptoms of primary hyperparathyroidism are essentially those of hypercalcemia.

Hence **hy:perparathy·roid** *a.*, of or having hyperparathyroidism.

1961 *Lancet* 16 Sept. 641/1 Hyperparathyroid bone disease was the first form of the disorder to be recognised. **1966** Wright & Symmers *Systemic Path.* II. xxxii. 1130/1 'Parathyroid poisoning' is a term that has been applied to the sudden increase in the extent of metastatic calcification of organs..when a high calcium diet is forced on the hyperparathyroid patient.

hyperphoric, *a.* Add: **2.** *Ophthalm.* (See s.v. *hyper- IV.)

hy:perpolariza·tion. *Physiol.* [f. Hyper- + Polarization.] An increase in the potential difference across the membrane of a nerve fibre above the normal resting potential, so that the inside of the fibre becomes (or is) even more negative with respect to the outside.

1946 Lorente de Nó & Feng in *Jrnl. Cellular & Compar. Physiol.* XXVIII. 412 This overshooting is analogous to the temporary hyperpolarization of the membrane (positive after-potential) which occurs after the end of tetani. **1955** *Acta Physiol. Scand.* XXXV. 12 The hyperpolarization usually amounts to 25–35 mV. *Ibid.*, The rate of hyperpolarization is about 150 mV/sec. **1966** C. F. Stevens *Neurophysiol.* ii. 13 Any change which decreases the inside–outside potential difference is known as a depolarization, whereas an increase in the membrane potential is called a hyperpolarization. **1971** *Nature* 9 July 123/1 Lundberg inserted intracellular electrodes in secretory cells of this gland and observed a hyperpolarization of 5–20 mV with a latency of 200–400 ms after applying single shocks to the chorda tympani nerve.

So **hyperpo·larize** *v. trans.*, to produce such a change in (a nerve fibre); *intr.*, to undergo such a change; **hyperpo·larizing** *vbl. sb.*

1950 *Jrnl. Cellular & Compar. Physiol.* XXXV. Suppl. II. 105 It appears that presynaptic impulses may exert upon the ganglion cells a catelectrotonic (depolarizing) action as well as an anelectrotonic (hyperpolarizing) action. **1955** *Jrnl. Physiol.* CXXX. 394 On hyperpolarizing the motoneurone..the amplitude of the e.p.s.p. was only slightly affected. **1971** *Nature* 23 July 269/2 In one pair of touch cells..a depolarizing or a hyperpolarizing synaptic potential was observed when the anterior or the posterior cell was stimulated. **1973** *Ibid.* 9 Mar. 102/1 Most of these cells only hyperpolarize in response to light.

hypersensitive (həipəɹse·nsitiv), *a.* [f. Hyper- 4 a + Sensitive *a.*] **1.** Sensitive to an abnormal or excessive degree; over-sensitive.

1871, 1897 [in Dict. s.v. Hyper- IV]. **1892** *Jrnl. Mental Sci.* XXXVIII. 525 Charcot and his pupils..believe in the influence of the magnet in hysteria, where the nervous system is hyper-sensitive. **1912** D. H. Lawrence *Phoenix II* (1968) 269 This soldier poet is so straight, so free from the modern artist's hyper-sensitive self-consciousness, that we would have more of him. **1939** E. & C. Paul tr. *Stekel's Technique Analytical Psychotherapy* xx. 283 A woman brought up on religious lines, morally hypersensitive, who pursues ideal aims. **1972** *Oxford Times* 19 May 7/1 Mr Francis Barnes, defending, described Jeffries as a 'retiring and hypersensitive' man who had lived as a semi-recluse for 17 years.

2. *spec.* in *Med.*: characterized by the fact that a marked adverse bodily response may be evoked by some specific substance or agent which (in similar amounts) has no such effect on most individuals. Const. *to*.

1899 E. O. Jordan tr. *Hueppe's Princ. Bacteriol.* vi. 337 Behring has found that it is possible..to immunize animals so highly..that they..become hyper-sensitive to the toxin. **1906** *Jrnl. Amer. Med. Assoc.* 29 Sept. 1007/2 At first glance it would appear much more important for an organism to be hyposensitive than hypersensitive to infectious processes, but a closer study of the complex problems of immunity develops the curious fact that resistance to disease may be largely gained through a process of hypersusceptibility. **1922** *Jrnl. Immunol.* VII. 128 The similarity of symptoms of drug reactions with those of foreign proteins in specifically hypersensitive persons. **1935** N. P. Sherwood *Immunol.* xxiii. 465 One injection of a nontoxic dose of horse serum will render guinea pigs hypersensitive to a second injection of the antigen provided an interval of almost ten days is allowed to intervene. **1951** Whitby & Hynes *Med. Bacteriol.* (ed. 5) vii. 95 Human idiosyncrasy..differs in some important respects from anaphylaxis. The exciting agent is not necessarily an antigen; patients may become hypersensitive, for example, to drugs with a very simple chemical structure. **1964** W. G. Smith *Allergy & Tissue Metabolism* i. 7 In patients who are specifically hypersensitive to a single allergen, it is very likely that sensitisation has been brought about by a previous contact with the allergen. *Ibid.* 13 A vast literature exists on histamine..and leaves no doubt about its involvement in the hypersensitive state.

3. *Photogr.* Of a film or plate, or its emulsion: hypersensitized (see *hypersensitize *v.* b).

1937 *Discovery* May p. xliv (Advt.), Hypersensitive panchromatic roll film. **1965** M. J. Langford *Basic Photogr.* x. 179 Red-pan or hypersensitive pan materials are..sensitive to red and respond up to 6,800 Å.

hyperse·nsitiveness. [f. prec., after *sensitive, sensitiveness*.] = next. **a.** *gen.* **b.** *Med.* (See *hypersensitive *a.* 2.)

In medical use *hypersensitiveness* was at first commoner than *hypersensitivity*, but the latter is now more usual.

a. 1876 [in Dict. s.v. Hyper- IV]. **1898** W. Scheppegrell *Electr. in Dis. Nose, Throat & Ear* xxi. 218 In hyperesthetic rhinitis, Sajous recommends that each point of hypersensitiveness be destroyed by means of the electrocautery. **1917** Glueck & Lind tr. *Adler's Neurotic Constitution* (1921) 164 Every one who has become acquainted with the hypersensitiveness of neurotic subjects knows with what slight cause they feel themselves to be undervalued. **1939** E. & C. Paul tr. *Stekel's Technique Analytical Psychotherapy* xxi. 327 A persistent illusion of a bad smell, disturbances of taste,..may all be indications of an uneasy conscience; so may hypersensitiveness of certain areas of skin. **1951** I. Compton-Burnett *Darkness & Day* 21 One feels that being alive to all the troubles about one is a sign of sensitiveness and feeling, when it may be hypersensitiveness and trying for other people.

b. 1906 *Jrnl. Amer. Med. Assoc.* 29 Sept. 1009/1 As far as we know, this is the first recorded instance in which hypersensitiveness, or anaphylaxis, has been experimentally shown to be transmitted from a mother to her young. **1922** *Jrnl. Immunol.* VII. 128 The condition now known as human hypersensitiveness or allergy. **1940** Becker & Obermayer *Mod. Dermatol. & Syphilol.* vii. 75/2 The resulting inflammation is of the vesicular type, but a condition of hypersensitiveness does not have to be considered, since everyone's skin reacts to this solution. **1953** F. K. Hansel *Clin. Allergy* iii. 58 The existence of an inherited predisposition is certainly not a prerequisite to the development of hypersensitiveness.

hy:persensiti·vity. [f. as prec., after *sensitivity*.] The state or fact of being hypersensitive. **a.** *Med.* (See *hypersensitive *a.* 2.) **b.** *gen.*

a. 1914 *Q. Jrnl. Med.* VII. 273 Richet, in 1902, introduced the term 'anaphylaxis' to explain certain phenomena of hypersensitivity. *Ibid.* 275 Before hypersensitivity can manifest itself a period of time must be allowed to elapse between the first and second injection. **1929** H. G. Wells *Chem. Aspects Immunity* (ed. 2) ix. 225 One of the most spectacular phenomena discovered in immunity is that of hypersensitivity to foreign proteins. **1946** *Nature* 19 Oct. 554/2 Most human beings appear to give very similar results when subjected to the same degree of exposure [to insect bites], though special cases of hypersensitivity and severe allergy also occur. **1947** L. Schwartz et al. *Occup. Dis. Skin* (ed. 2) iv. 43 The tendency to hypersensitivity may be inherited; it may be due to..a pigment deficiency causing hypersensitivity to light; or there may be a true allergy caused by exposure to an allergen. **1964** M. Hynes *Med. Bacteriol.* (ed. 8) viii. 104 Dermal hypersensitivity [ed. 5–7 (1951–61): hypersensitiveness] to dyes, solvents, etc., is most conveniently demonstrated by a patch test. **1970** Passmore & Robson *Compan. Med. Stud.* II. xxii. 19/1 It was soon found that hypersensitivity could be evoked also by intrinsically harmless substances such as..simple chemicals.

b. 1954 J. A. Hadfield *Dreams & Nightmares* xi. 221 If that were so, hypersensitivity of hearing would be ruled out. **1956** I. Murdoch *Flight from Enchanter* x. 140 It was as if Mischa were deliberately reducing him to a state of hypersensitivity and confusion. **1970** D. W. Swanson et al. *Paranoid* iii. 61 The patient's hypersensitivity to others' feelings and failings can be disconcerting when focused on the interviewer. **1972** *Daily Colonist* (Victoria, B.C.) 27 Feb. 4/5 A hypersensitivity to criticism hangs them all up in time.

hy:persensitiza·tion. [f. next + -ation.] The action or process of hypersensitizing, or the state of being hypersensitized. **a.** *Med.* **b.** *Photogr.*

a. 1908 *Jrnl. Amer. Med. Assoc.* 15 Feb. 528/1 The remarkable phenomenon of hypersensitization or anaphylaxis. **1947** L. Schwartz et al. *Occup. Dis. Skin* (ed. 2) iv. 39 If such a worker is able to keep on working, the dermatitis clears up... In other words exposure first causes a hypersensitization, then further exposure causes a dermatitis, and still further exposure causes hyposensitization. **1959** *New England Jrnl. Med.* CCLX. 170/2 The use of human gamma globulin rather than that derived from animal plasma significantly diminishes the risk of hypersensitization to a foreign protein.

b. 1933 Carroll & Hubbard in *Bureau of Standards Jrnl. Res.* X. 212 We shall use the term 'hypersensitization' to cover any case in which increase in sensitization by a dye is produced by treatment with a material itself colorless or absorbing a spectral region different from that of the dye. *Ibid.* 213 In our experiments hypersensitization always involved an increase in relative sensitivity to the longer wave lengths. **1956** *Electronic Engin.* Feb. 78/1 The processes of hyper-sensitization (treatment before exposure) and latensification (treatment after exposure) may give an increase in relative speed of up to two or three times with normal development. **1964** J. Roubier *Odham's Pract. Photogr. & Film-Making* ii. 34 The effective speed of an emulsion can be raised by 100% or even 150% by hypersensitisation.

hypersensitize (həipəɹse·nsitəiz), *v.* [f. Hyper- 4 c + Sensitize *v.*] *trans.* To render hypersensitive. *spec.* **a.** *Med.* (See *hypersensitive *a.* 2.) **b.** *Photogr.* To increase the speed of (a photographic film or plate, or its emulsion) by immersion in a special solution, exposure to light, or other means, usually before it is exposed in the taking of a photograph (cf. *latensification). So **hyperse·nsitized** *ppl. a.*, **hyperse·nsitizing** *vbl. sb.*

1897 G. B. Shaw *Let.* 16 Apr. (1965) 746 Teddy, though hypersensitized..and petulated by more luxury than was good for him in the way of a mammy seems highly and nervously intelligent. **1914** *Q. Jrnl. Med.* VII. 277 In anaphylaxis the degree of specificity is difficult to determine, as a hypersensitized animal has an increased susceptibility to any toxin. *Ibid.* 280 A serum may have strong hypersensitizing power for passive anaphylaxis and yet have no precipitating power. **1917** E. Pound *Let.* 10 Nov. (1971) 123 Still, what the hell else are you? I mean apart from being a citizen, a good fellow.., a grouch, a slightly hypersensitized animal, etc.?? **1920** *Jrnl. Franklin Inst.* CLXXXIX. 25 (*heading*) Hypersensitizing commercial panchromatic plates. **1929** E. J. Wall *Pract. Color Photogr.* (ed. 2) xiii. 173 The exposure with these hypersensitized plates and the aesculin filter is about one-tenth of that required for the normal plate. **1954** tr. *L. P. Clerc's Photogr.* (ed. 3) viii. 178/2 The first exposure 'hypersensitizes' the material towards the second. **1969** M. J. Langford *Adv. Photogr.* viii. 171 'Hypersensitizing' or increasing the speed of an emulsion *before* camera exposure can be carried out chemically or by means of light.

hypersonic (həipəɹsǫ·nik), *a.* [f. Hyper- + *sonic *a.*, after *supersonic, ultrasonic*.] **1.** Of, pertaining to, or designating sound waves or vibrations with a frequency greater than about 1000 million Hz. (Cf. *ultrasonic *a.*)

1937 B. V. R. Rao in *Nature* 22 May 885/1 Spontaneously existing sound-waves of thermal origin of very high frequencies ('hyper-sonic waves'). **1938** —— in *Proc. Indian Acad. Sci.* A. VII. 163 It appears desirable to designate the portion of the acoustic spectrum having a frequency higher than a thousand mega-cycles per second as the 'hyper-sonic' region, while the 'ultra-sonic' region may be taken to extend from one to a thousand mega-cycles per second. **1948** *Rep. Progress Physics* XI. 205 Workers in India particularly have used *hypersonic* to denote ultrasonic phenomena (of thermal origin) at frequencies of the order of 1000 Mc/sec. or higher. **1960** *Physical Rev.* CXVII. 1248 The hypersonic absorption was studied for longitudinal and transverse waves at various frequencies up to 4000 Mc/s. **1963** J. Blitz *Fund. Ultrasonics* vi. 167 Bömmel and Dransfeld measured the attenuation of both longitudinal and shear waves in quartz at frequencies ranging from 1,000 to 4,000 Mc/s using a hypersonic technique. **1971** *Nature* 24 Sept. 238/2 Mechanical surface waves can now be generated up to 10 GHz. Mechanical waves of these frequencies (up to 100 GHz) are often termed hypersonic.

2. Involving, pertaining to, capable of, or designating speeds greater than about five times the speed of sound. (Cf. *supersonic *a.*)

1946 *Jrnl. Math. & Physics* XXV. 247 Hypersonic flows are flow fields where the fluid velocity is much larger than the velocity of propagation of small disturbances, the velocity of sound. **1958** *Engineering* 14 Mar. 347/2 Flow about bodies at subsonic, supersonic, and hypersonic speeds. **1958** *Times* 19 Dec. 11/7 Hypersonic travel is just possible in the foreseeable future. **1960** *New Scientist* 14 July 88/2 The hypersonic wind tunnel is almost more necessary than the supersonic... A tunnel capable of producing..air speeds between Mach 10 and Mach 27 (something near 18,000 m.p.h.) is now in daily use. **1960** *Nature* 6 Feb. 346/2 The transition from the supersonic to the hypersonic régime occurs at a Mach number of about 5, but all the characteristic features of the latter régime may only become well developed at a much higher Mach number of, say, 15. **1969** *Courier Mail* (Brisbane) 17 Apr. 2 The construction of a 500-seat hypersonic aircraft could cost as much as £4000 million.

hypertely (həipĕ·ɹtĭli, həi·pəɹte:li). *Zool.* [ad. G. *hypertelie* (C. Brunner 1873, in *Verh. Zool.-Bot. Ges. Wien* XXIII. 133), f. Gr. ὑπερτέλειος beyond completeness, f. τέλος end.] Extreme development of size, patterns of behaviour, mimetic coloration, etc. beyond the degree to which these characteristics are apparently useful. Also *fig.* Hence **hyperte·lic** *a.*

1895 D. Sharp in *Cambr. Nat. Hist.* V. xiii. 323 Brunner..came to the conclusion that they [*sc.* close resemblances] cannot be accounted for on the ground of mere utility, and proposed the term Hypertely to express the idea that in these cases the bounds of the useful are transcended. **1920** G. D. H. Carpenter *Naturalist on Lake Victoria* x. 196 Special procryptic colouring never fails to arouse wonder from its extreme perfection... This complimentary [*sic*] doctrine has been termed Hypertely. **1936** *Nature* 10 Oct. 603/2 We find that intraspecific selection frequently leads to results which are mainly.. useless to the species as a whole, including 'hypertelic' characters. **1937** A. Huxley *Ends & Means* xiv. 262 We are doing our best to develop a militaristic 'hypertely', to become, in other words, dangerously specialized in the art of killing our fellows. **1953** G. G. Simpson *Major Features of Evolution* ix. 282 Almost any case in which size, structure, or habit..is carried to extremes may be cited as hypertely. *Ibid.* 287 Many of the characters commonly designated as hypertelic are striking secondary sexual characters. **1965** B. E. Freeman tr. *Vandel's Biospeleol.* xi. 168 Hypertely represents the end of a long orthogenetic evolution.

hypertensin (həipəɹte·nsin). *Biochem.* [f. *hypertens(ion, -ive *a.* + -in¹.] Either of two polypeptides, of which one (*hypertensin I*) is formed in the blood by the action of renin on a protein (hypertensinogen), and the other (*hypertensin II*) is derived from it by the loss of two amino-acid residues, causes a rise in blood pressure, and stimulates the secretion of aldosterone; also, analogous polypeptides in animals. Now usually called *angiotensin*.

1939 J. M. Muñoz et al. in *Nature* 9 Dec. 980/1 This substance, which we name hypertensin, is different from adrenalin, tyramin, pitressin and urohypertensin. **1954** *Jrnl. Exper. Med.* XCIX. 282 Two types of hypertensin have been demonstrated... The first type..has been designated hypertensin I. It can be rapidly converted to a second, approximately equally pressor compound, hypertensin II. **1956** *Nature* 17 Mar. 527/2 The isolation and purification of a hypertensin peptide, resulting from the action of rabbit renin on ox serum. **1959** A. White et al. *Princ. Biochem.* (ed. 2) xxx. 750 Hypertensin I preparations of slightly different composition have been described, depending on the sources of the renin and the substrate used. *Ibid.*, Hypertensin has also been called angiotonin. A uniform nomenclature has been proposed which would designate this compound as angiotensin.

Hence **hy:pertensi·nogen** [ad. F. *hypertensinogène* (E. Braun-Menendez 1940, in *Compt. Rend. de la Soc. de Biol.* CXXXIV. 489): see *-gen], a globulin produced by the liver and present in blood, from which hypertensin I may be liberated by renin.

1941 *Amer. Jrnl. Physiol.* CXXXV. 214 Braun-Menendez has preferred the term hypertensinogen, which implies that it is the substrate on which renin acts. **1965** *New Scientist* 25 Nov. 561/3 Kidney cells, in the presence of a poor blood flow.., secreted an enzyme called renin which converted a normal constituent of plasma (hypertensinogen) into a new substance (hypertensin).

hypertension (həipəɹte·nʃən). [f. Hyper- 5 + Tension *sb.*] **1.** *Med.* Abnormally or excessively high 'tension' or pressure of a bodily fluid. **a.** Of arterial blood; *essential hypertension*, hypertension that has no apparent cause and cannot be explained as a consequence or symptom of some other lesion.

1893 *Brit. Med. Jrnl.* 4 Nov. 997/1 'Hypertension' and 'hypotension' are regarded not as indications, but as *mala in se*. **1927** *Physiol. Rev.* VII. 464 More recent writers have coined the name 'essential hypertension'. **1955** G. W. Pickering *High Blood Pressure* vi. 130 By current practice which takes not a very high figure, such as 150/100, as the lower limit of hypertension, essential hypertension becomes by far the commonest form of hypertension. **1957** *New Scientist* 9 May 23/1 Hypertension causes one in every four deaths in the United States. **1966** Wright & Symmers *Systemic Path.* I. iii. 106/1 The hypertension sometimes found in patients with hyperthyroidism is usually limited to elevation of the systolic pressure (systolic hypertension).

b. Of the intra-ocular fluid.

1918 R. H. Elliot *Glaucoma: Textbk. for Student* v. 303 The treatment of established glaucoma. In these cases the hypertension is commonly associated with evidence of ocular congestion. **1969** Duke-Elder & Jay in S. Duke-Elder *Syst. Ophthalm.* XI. ix. 630 In some cases the rise in the intra-ocular pressure has been slight and transient, to some extent possibly because the formation of aqueous may have been inhibited by the hypertension itself in a physiological compensatory mechanism.

2. A state of great (nervous or emotional) tension.

1936 *Times Educ. Suppl.* 28 Nov. 429/3 In every such case the nervous instability and hypertension should be treated by relaxation and psychological readjustment. **1947** *Year Bk. Arts in N.Z.* III. 152 Mind burst out... Man's understanding tangled, ripped. Extremes are too much with us and the hypertensions. **1953** J. Masters *Lotus & Wind* vii. 101 The opportunity for which she had been tensely waiting was upon her. In a few minutes, as this hypertension and ruthlessness faded in her, it would be gone. **1972** D. Haston *In High Places* xii. 140 Flying out of Katmandu, I felt really pleased at having a chance to come back so quickly... There was friendliness all around; urgency to do things didn't seem to exist. None of the hypertensions of the West. A relaxing atmosphere.

hypertensive (həipəɹte·nsiv), *a.* and *sb. Med.* [f. prec. + -ive.] **A.** *adj.* Of, exhibiting, or associated with hypertension, esp. of the blood; tending to raise the blood pressure.

1904 T. C. Janeway *Clin. Study Blood-Pressure* vii. 221 The bath-treatment of typhoid fever seems to have as distinct an effect on the blood-pressure curve... I cannot but feel that this hypertensive effect is evidence of an action on the vaso-motor system. **1918** *Endocrinology* II. 94 In case of the more strongly hypertensive extracts, kidney dilation and diuresis, together with augmented blood pressure, occur as the primary reaction. **1939** D. M. Dunlop *Textbk. Med. Treatm.* 654 Digitalis..does not raise blood pressure in the hypertensive patient. **1954** S. Duke-Elder *Parsons' Dis. Eye* (ed. 12) xiv. 212 This involves a rise determined..by the difficulty experienced by the sticky albuminous aqueous in escaping through the filtration channels at the angle of the anterior chamber (hypertensive iridocyclitis). **1956** *Nature* 17 Mar. 523/2 These [compounds] are highly effective in lowering blood pressure and relieving certain hypertensive symptoms. **1971** *Brit. Med. Bull.* XXVII. 39/1 Hypertensive heart disease, diagnosed on electrocardiographic and radiographic criteria, is also more common among negroes.

B. *ellipt.* as *sb.* A person with arterial hypertension.

1939 D. M. Dunlop *Textbk. Med. Treatm.* 652 Many hypertensives tend to over-eat. **1961** *Lancet* 2 Sept. 510/2 The systolic and diastolic blood-pressures were slightly higher in the hypertensives' sons. **1972** *Daily Tel.* 22 Nov. 13/8 In hypertensives, prostaglandins can lower the blood pressure.

hyperthermia (həipəɹþɜ̄·ɹmiă). *Med.* Also in anglicized form **hyperthermy** (həi·pəɹþɜ̄ɹmi) (*rare*). [f. Hyper- 5 + Gr. θέρμη heat + -ia¹.] The condition of having a body temperature substantially above the normal either as a result of natural causes or artificially induced (e.g. for therapeutic purposes).

1886 *Syd. Soc. Lex.*, Hyperthermy. **1887** A. M. Brown *Contrib. Animal Alkaloids* 143 Intoxication by the extractive matters is accompanied by hyperthermia. **1898** *Nature* 24 Nov. 95 Researches on lesions of the nervous centres produced by hyperthermy. **1921** F. A. Welby tr. *Luciani's Human Physiol.* V. ii. 82 Billroth..found that the temperature rose to 42·2° after fracture of the sixth cervical vertebra; in a similar case Simon observed a hyperthermia of 44°. **1935** *Jrnl. Amer. Med. Assoc.* 18 May 1788/2 As with all other forms of treatment for chronic infectious arthritis, the results of hyperthermia depend to a marked extent on the duration and activity of the disease. **1941** *Virginia Med. Monthly* Mar. 158/1 It was decided to give the patient another period of hyperthermy, but in view of the attack of substernal pain a medical check-up was requested. **1971** *New Scientist* 15 July 133/2 The gazelles..cannot withstand desiccation to the extent that camels do, and do not show the same degree of hyperthermia. **1971** L. B. Rowell in E. Simonson *Physiol. Work Capacity & Fatigue* vii. 149 Hyperthermia will be accompanied by high rates of sweat loss and dehydration.

So **hyperthe·rmic** *a.* [cf. Gr. ὑπέρθερμος over-warm], of or exhibiting hyperthermia.

1896 [in Dict. s.v. Hyper- IV]. **1898** W. S. L. Barlow *Man. Gen. Path.* x. 434 The symptoms presented by a hyperthermic animal when its temperature is reaching a dangerous height are those of severe distress, respiration and pulse are accelerated, and the animal lies outstretched. **1948** [see *hypothermic *a.*].

hyperthy·roidism. *Med.* [f. Hyper- 5 + Thyroid *a.* (*sb.*) + -ism.] A condition in which the thyroid gland produces more hormone than normal, resulting in an increased rate of metabolism, often with wasting of muscle and loss of weight together with restlessness and emotional instability.

1900 DORLAND *Med. Dict.* 311/1 Hyperthyroidism. **1909** G. DOCK in Osler & McCrae *Syst. Med.* VI. xvii. 431 Hyperthyroidism..seems to be the cause of the exophthalmos. **1912** *Med. Ann.* 3 Sleeplessness following..fevers, hyperthyroidism, and hysteria. **1961** L. MARTIN *Clinical Endocrinol.* (ed. 3) i. 22 A goitre and signs of mild hyperthyroidism are not uncommon [in gigantism]. **1970** S. GROLLMAN *Human Body* (ed. 2) xvi. 507 The most common form of hyperthyroidism, resulting from a diffuse increase in thyroid tissue, is known as exophthalmic goiter, Graves' disease, or Basedow's disease.

Hence **hyperthy·roid, -thyroi·dic** *adjs.*

1916 *Internat. Jrnl. Surg.* XXIX. 312 (*heading*) The etiology of the exophthalmos in hyperthyroid goitre. **1916** *Med. Times* (N.Y.) 7 July 207/1 An impression is given by the hypothyroidic eye which is just the reverse of that made by the typical hyperthyroidic, or exophthalmic organ. **1968** *Listener* 18 July 70/1, I had recently been diagnosed as hyperthyroidic. **1971** N. R. ALPERT *Cardiac Hypertrophy* 55 The pumping function of the hyperthyroid heart.

hypertonia (həipəɹtǭu·niă). *Med.* Also in anglicized form **hypertony** (həipə̄·ɹtŏni) (*rare*). [mod.L., f. HYPER- 5 + Gr. τόν-ος TONE *sb.* + -IA¹.] The condition (in muscle or muscular tissue) of being hypertonic.

1842 DUNGLISON *Dict. Med. Sci.* (ed. 3) 368/1 Hypertonia. **1881** J. ROSS *Treat. Dis. Nervous Syst.* I. I. v. 178 The condition of motor excess, or of increased tonus, may be called hypertony. **1905** *Med.-Chir. Trans.* LXXXVIII. 212 The arterial hypertonia is to be regarded as a result of the greater strain thrown on the circulatory mechanism. **1914** A. MORISON *Sensory & Motor Disorders Heart* v. 203 Exaggerated tonic cardiac action tends to be succeeded by..a minus quantity. Hypertonia necessarily yields to hypotonia. **1933** W. R. BRAIN *Dis. Nervous Syst.* i. 8 Not all muscle-groups exhibit hypertonia in equal degree in hemiplegia. **1962** *Lancet* 27 Jan. 222/1 She was re-examined..and found to have variable and bizarre hypertonia of the jaw, tongue, and neck muscles.

hypertonic (həipəɹtǫ·nik), *a.* [f. HYPER- 5 + TONIC *a.*] **1.** *Med.* Exhibiting or characterized by excessive tone or tension (in muscle or muscular tissue).

1855 R. G. MAYNE *Expos. Lex. Med. Sci.* (1860) 484/1 *Hypertonicus*, of or belonging to hypertonia: hypertonic. **1886** *Lancet* 13 Mar. 486/2 For convenience I describe the group of symptoms under the term 'hyper-tonic paresis', a symptomatic nomenclature which commits to no theory. **1907** W. RUSSELL *Arterial Hypertonus* i. 3 The degree of contraction may exceed the limits of normal variation, and when it does the term hypertonic contraction, or merely arterial contraction, will be used here. **1933** W. R. BRAIN *Dis. Nervous Syst.* i. 8 Immediately following a capsular haemorrhage the paralysed limbs are completely flaccid... After a variable interval..tone gradually returns to the affected muscles and they ultimately become hypertonic or 'spastic'. **1966** *McGraw-Hill Encycl. Sci. & Technol.* X. 528/1 When the stretch reflex is absent, the muscle is hypotonic or flaccid; when stretch reflexes are exaggerated..the muscles are hypertonic or spastic.

2. *Physiol.* Of a solution: having a higher osmotic pressure than some particular solution (usually that in a cell, or a bodily fluid). Const. *to*.

1895 *Jrnl. Physiol.* XVIII. 107 None of the water is taken up from hypertonic or isotonic solutions. **1936** A. P. MATHEWS *Princ. Biochem.* xxxiv. 364 If the solution have an osmotic pressure greater than that of the blood, it is said to be hypertonic to the blood. **1951** WHITBY & HYNES *Med. Bacteriol.* (ed. 5) vi. 73 The optimum salt concentration is usually near the range of isotonicity to body cells; markedly hypertonic saline diminishes the reaction. **1970** A. F. BRADING in E. Bülbring et al. *Smooth Muscle* vi. 172 Tissues will swell in hypotonic solutions, and shrink in hypertonic ones.

hypertonicity (həi:pəɹtǫ́ni·siti). [f. prec. + -ITY.] The condition of being hypertonic.

a. *Med.* Of muscle or muscular tissue (see *HYPERTONIC *a.* 1); = *HYPERTONIA.

1886 *Brain* IX. 231 A condition of extreme functional over-activity of the ganglion cells of the anterior cornua of the cord and medulla, causing hyper-tonicity of the entire muscular system. **1923** J. F. H. DALLY *High Blood Pressure* v. 64 Arterial hypertonicity, in larger or smaller areas, is met with in a host of disorders. **1949** *Jrnl. Neurophysiol.* XII. 371 Hypertonicity of muscles with hyperactive deep reflexes or clonus.

b. *Physiol.* Of a solution (see *HYPERTONIC *a.* 2); the extent to which a solution has higher osmotic pressure than some other.

1896 T. L. STEDMAN *20th Cent. Pract.* VII. 284 The hypertonicity of the plasma is easily calculated. **1906** *Amer. Jrnl. Physiol.* XV. 359 When the hypertonicity is slight, the degree of recovery is relatively greater than in the case of the more concentrated solutions. **1963** E. ERNST *Biophysics Striated Muscle* (ed. 2) 131 The next experiments were planned in such a way that the loss of water [from the muscle] due to the hypertonicity of the medium should be balanced by water taken up by the swelling elicited by lactic acid.

hypertonus (həipəɹtǭu·nǫs). [f. HYPER- 5 + TONUS.] **1.** *Ophthalm.* A state of increased pressure of the intra-ocular fluid; = *HYPERTENSION 1 b. Now *rare*.

1891 F. P. FOSTER *Med. Dict.* III. 1944/1 *Hypertonus*, that condition of the eye in which the intra-ocular tension is increased, sometimes without any organic disease being present. **1918** R. H. ELLIOT *Glaucoma: Textbk. for Student* v. 300 It would appear, then, that the thrombotic type of glaucoma possesses a very distinct inflammatory element; it thus comes into line with the hypertonus which we find secondary..to intra-ocular tumours.

2. *Med.* = *HYPERTONIA.

1904 *Rev. Neurol. & Psychiatry* II. 775 The arms showed the typical muscular rigidity (hypertonus) of paralysis agitans, while the legs were markedly hypotonic. **1923** J. F. H. DALLY *High Blood Pressure* v. 65 To some extent raised blood pressure may be differentiated from arterial hypertonus by instrumental methods. **1924** *Brain* XLVII. 333 General hypertonus may, however, be inferred from a slight increase of the tendon reflexes. **1950** *Physiol. Rev.* XXX. 466 After the somnolence and plastic hypertonus had passed, marked poverty of movement still persisted. **1971** RASCH & BURKE *Kinesiology* (ed. 4) iv. 83 The limb is then said to be spastic. In man such hypertonus occurs only in the antigravity muscles.

hyperu·rbanism. [f. HYPER- 5 + URBANISM.] Extreme urbanism; *spec.* in *Philol.*, the manner of speech arising from an effort to avoid provincialism; a hypercorrect form of speech or phrase resulting from this. So **hyperu·rban** *a.*, exhibiting hyperurbanism; **hyperu:rbaniza·tion** *sb.*

1925 P. RADIN tr. *Vendryès's Lang.* ii. 49 Hyper-urbanism consists in those excesses brought about by consideration for correct speech. *Ibid.*, *Plaustrum* for *plostrum*, *cauda* for *coda*.. These are hyper-urbanisms. **1928** L. BLOOMFIELD in *Language* IV. iv. 286 Hyper-urban forms occur, e.g., *foot* pronounced as [fy:t] instead of [fu:t]. **1933** —— *Lang.* xviii. 309 We may take *cōda* to be the older of the two Latin forms, and *cauda* to be a hyper-urban (over-elegant) variant. *Ibid.* xix. 330 This flavor of the [y:]-variants appears strikingly in the shape of hyper-urbanisms: in using the elegant [y:], the speaker sometimes substitutes it where it is entirely out of place. **1935** *Language* XI. ii. 106 Social climbers betray themselves by hyper-urbanisms, such as [a] in *bass* or *lass*. **1937** *Amer. Speech* XII. 175 The women who wear gloves on the most informal occasions have probably acquired the habit of never forgetting them, and so-called hyper-urbanization in speech as well as in dress, often springs from..a sense of social insecurity. **1940** C. C. FRIES *Amer. Eng. Grammar* vi. 94 There appeared in the Standard English materials some uses of *whom* which should probably be looked upon as hyperurbanisms. **1951** TRAGER & SMITH *Outl. Eng. Struct.* iii. 85 The hyperurbanism *Between you and I* for the standard *Between you and me*.

hyperventilate (həipəɹve·ntilē̆it), *v.* *Physiol.* [f. HYPER- 5 + VENTILATE *v.*, or back-formation from next.] **a.** *intr.* To breathe deeply or rapidly.

1931 *Jrnl. Neurol. & Psychopath.* XII. 14 It is..of much interest to ascertain to which changes the epileptic organism reacts with a seizure when the patient hyperventilates. **1961** *Flight* LXXX. 760/1 An experienced pilot who had..been told..that he hyperventilated and should regulate his breathing. **1968** *Everybody's* (Austral.) 12 June 31/4 Ron Taylor, perhaps Australia's greatest underwater expert, advises: Don't hyperventilate to the stage where you become dizzy. **1970** *Sci. Amer.* Feb. 56/1 They do not need to hyperventilate as much as lowlanders do when the latter go to high altitudes.

b. *trans.* To produce hyperventilation in.

1931 *Jrnl. Neurol. & Psychopath.* XII. 15 Our method has been..to hyperventilate the person in question with two or three of the methods. **1953** *Physiol. Rev.* XXXIII. 447 Brown et al...hyperventilated subjects for 24 hours in a body respirator at rates just under those producing tetany. **1968** C. OSBORNE tr. *Stenuit's Dolphin* vii. 119 When a..skin-diver hyperventilates his lungs by deep breathing.., he loads the haemoglobin of his blood with a reserve of oxygen.

hy:perventila·tion. *Physiol.* [f. HYPER- 5 + VENTILATION.] An increased or excessive exposure of the lungs to oxygen, resulting in a more rapid loss of carbon dioxide from the blood; the action of bringing this about (in oneself, by deep or rapid breathing, or in another individual).

1928 *Canad. Med. Assoc. Jrnl.* LVIII. 210/2 Such a 'hyperventilation' tetany occurring under clinical conditions has been but seldom reported. **1932** *Arch. Neurol. & Psychiatry* XXVIII. 574 We injected into each animal 0·1 mg. of strychnine sulphate per kilogram of body weight and started hyperventilation after..twenty to thirty minutes. **1951** H. DAVSON *Textbk. Gen. Physiol.* vi. 124 Special mechanisms for heat dissipation such as sweating or hyperventilation of the lungs. **1961** *Lancet* 26 Aug. 474/2 Another cause of fainting, more often related to emotional than to physical states, is hyperventilation. **1965** *Handbk. Physiol.: Circulation* (Amer. Physiol. Soc.) III. liii. 1889/2 It is not at all difficult by hyperventilation to cause a severe gaseous alkalosis with dizziness, blurred vision, mental confusion, numbness of extremities,..and muscle spasm, rarely going on to..convulsions and unconsciousness. **1970** *Sci. Jrnl.* June 84/1 The noticeable hyperventilation (heavy breathing) which occurs at orgasm shows up clearly on the tracing.

hy·phenate, *sb.* [f. *HYPHENATED *a.* 2.] A hyphenated person.

1916 *Yorks. Post* 4 Mar. 6/7 The Hyphenates throughout the country are greatly excited. **1920** *Glasgow Herald* 27 Nov. 6 This political hyphenate or composite is desirous of running a Home Rule for Scotland campaign. **1922** *Contemp. Rev.* Dec. 693 The 'hyphenates'—Irish and Germans, Poles and Russians and Italians—..joined in the condemnation of Wilsonism.

hyphenated, *a.* Add: **2.** Applied to persons (or, by extension, their activities) born in one country but naturalized citizens of another, their nationality being designated by a hyphenated form, e.g. *Anglo-American, Irish-American*; hence, to a person whose patriotic allegiance is assumed to be divided. Also in extended use. orig. *U.S.*

1893 FARMER & HENLEY *Slang* III. 386/2 *Hyphenated American*, a naturalised citizen, as German-Americans, Irish-Americans, and the like. **1900** *Daily News* 15 Aug. 3/1 My opponents were of the hyphenated variety—Dutch-Americans and Irish-Americans predominating. **1904** *Westm. Gaz.* 3 Jan. 3/2 American politics, where men who call themselves Irish-Americans, German-Americans, Dutch-Americans, and so on, are contemptuously referred to as 'hyphenated Americans'. **1907** *Nation* (N.Y.) 7 Nov. 410 Some of these hyphenated American journals. **1915** *Lit. Digest* 4 Sept. 462/1 Hyphenated residents will continue to insist that American newspapers should be strictly neutral. **1948** *Manch. Guardian Weekly* 3 June 7/1 'Hyphenated Americans'—undigested immigrant stock. **1965** B. SWEET-ESCOTT *Baker St. Irreg.* i. 37 It was thought that, with the whole of western Europe under Nazi domination, hyphenated Americans might provide recruits for work in occupied territory. **1973** *Times* 17 Oct. 6/8 This did not go down well with the Greek community here [*sc.* in the U.S.], or with other groups of 'hyphenated Americans'.

hyphening (həi·fəniŋ), *vbl. sb.* [f. HYPHEN *v.* + -ING¹.] The action of the vb. HYPHEN.

1929 *Conc. Oxf. Dict.* p. ix, A consequence of this reformed hyphening is that the presence of a hyphen in such a compound [as *tipsy cake*] assures the reader that the word-stress falls on the first part.

hyphenism (həi·fəniz'm). *U.S.* [f. HYPHEN + -ISM; cf. *HYPHENATED *a.* 2.] The state of being a hyphenated American; the attitude or conduct involved or implied by this.

1930 P. W. SLOSSON *Great Crusade* (1931) xi. 288 The years..so marked by hyphenism and the echoes of Old World feuds.

hyphomycetes (həi:foməisī·tīz, -ts), *sb. pl.* Also in *sing.* form **hyphomycete.** *Bot.* [mod.L. (E. M. Fries *Systema Mycologicum* (1821) I. p. xxx), f. Gr. ὑφή web + μύκητες fungi.] Imperfect fungi of the group Hyphomycetes; filamentous moulds bearing naked, asexual spores.

1836 M. J. BERKELEY in J. E. Smith *Eng. Flora* V. II. 328 Hyphomycetes... Sporidiferous flocci naked (not included in a uterus or seated on a proper receptacle). **1857** —— *Introd. Cryptogamic Bot.* 297 The species contained in the division *Hyphomycetes*, consist of Fungi which, like *Mucorini*, are known under the common name of moulds. **1887** H. E. F. GARNSEY tr. *A. de Bary's Compar. Morphol. & Biol. Fungi* i. 1 In the more simple Fungi the branched hypha alone constitutes the thallus; such forms are termed Hyphomycetes, Filamentous Fungi (Fadenpilze), or Haplomycetes. **1930** C. THOM *Penicillia* iv. 24 A hyphomycete genus such as Penicillium..is an aggregation of species with a common type of asexual fruiting. *Ibid.* xi. 147 Penicillium is characterized and discussed here as a 'form-genus' in the great aggregate of form genera commonly known as the Hyphomycetes. **1971** M. B. ELLIS *Dematiaceous Hyphomycetes* 7 About 50 per cent of the specimens sent to the Commonwealth Mycological Institute are hyphomycetes.

hypnago·gically, *adv.* [f. HYPNAGOGIC *a.*: see -ICALLY.] In a hypnagogic manner.

1957 P. MCKELLAR *Imagination & Thinking* iii. 40 Confusion of hypnagogically imaged music with the noises of reality also occurred, and more than one of our subjects descended the stairs believing that the wireless had been left on.

hypnoanalysis (hipno,ănæ·lisis). Also **hypno-analysis.** [f. HYPNO- (taken as combining form of *hypnosis*) + ANALYSIS.] Psychoanalysis performed while the subject is under hypnosis; psychotherapy that combines psychoanalysis with hypnosis.

1920 J. A. HADFIELD in H. C. Miller *Functional Nerve Dis.* v. 71 The method which I venture to name 'Hypnoanalysis' consists simply in hypnotizing the patient and inducing him to speak of his troubles. **1949** KOESTLER *Insight & Outlook* x. 146 The less drastic forms of psychotherapy like psycho-, narco-, and hypnoanalysis..are all primarily means of releasing functional components of the psychoneural apparatus which had been repressed by faulty integration. **1963** W. S. KROGER *Clin. & Exper. Hypnosis* xlvii. 323/1 Hypnoanalysis is particularly indicated for psychoneurotics who do not respond to brief hypnotherapeutic procedures.

Hence **hypnoa·nalyst**, one who uses hypnoanalysis; **hy:pnoanaly·tic** *a.*, of or involving hypnoanalysis.

1922 *Jrnl. Abnormal Psychol.* XVI. 344 (*heading*) A hypnoanalytic study of two cases of war neurosis. **1945** *Diseases Nervous Syst.* VI. 374/1 The hypnoanalyst..is thrust into a form of relationship with his patient that transcends even the transference situation obtaining in routine psychoanalysis. **1947** LECRON & BORDEAUX *Hypnotism Today* xiii. 229 Basically, hypnoanalytic treatment is modified psychoanalysis with inclusion of hypnotism for brevity. **1960** I. BENNETT *Delinquent & Neurotic Children* iv. 124 Lindner (1944) undertook the hypno-analytic treatment of a criminal psychopath. **1963** W. S. KROGER *Clin. & Exper. Hypnosis* xlvii. 323/2 The process utilizes free associations, dreams, analyses and recollections, all of which are interpreted by the hypnoanalyst.

hypnogogic, var. HYPNAGOGIC *a.*

1906 E. SALTUS *Vanity Square* II. i. 154 The hypnogogic hallucinations continue. **1933** *Scrutiny* I. 380 In this hypnogogic state the mind makes no selection. **1962** *New Scientist* 2 Aug. 267/3 The next section contains a quantity of fascinating material on such topics as..'night jerks', which occur in the hypnogogic state (his spelling 'hypnagogic' is not a useful innovation). **1972** [see *HYPNOPOMPIC *a.*].

hypnoid (hi·pnoid), *a.*² *Psychol.* [a. G. *hypnoid* (Breuer & Freud 1893, in *Neurol. Centralblatt* XII. II. 43), f. Gr. ὕπν-ος sleep + -OID.] Applied to a state of consciousness characterized by heightened suggestibility or dissociation, such as occurs in hysterical conditions.

1898 B. SIDIS *Psychol. of Suggestion* xxiii. 234 By the term 'hypnoid' I indicate the coexistence of two or more fully independent functioning constellations of moments-consciousness, such as is presented in the phenomena of automatic writing and of hysteria. **1902** W. JAMES *Var. Relig. Exper.* xvi. 413 To the medical mind these ecstasies signify nothing but suggested and imitated hypnoid states. **1924** J. RIVIERE et al. tr. *Freud's Coll. Papers* I. 34 Splitting of consciousness..exists in a rudimentary fashion in every hysteria and..the tendency to this dissociation—and therewith to the production of abnormal states of consciousness, which may be included under the term 'hypnoid'—is a fundamental manifestation of this neurosis. **1951** R. BRUN *Gen. Theory Neuroses* III. 333 Fantasies in hypnoid conditions, such as constantly occur in hysteria, will even more readily produce regressive excitations in the most disparate organs.

Also **hypnoi·dal** *a.*, in the same sense.

1898 B. SIDIS *Psychol. of Suggestion* xxiii. 239 In hypnoidal states past, outlived experiences heave up into the upper consciousness. **1921** *Discovery* Nov. 294/1 A similar [half-waking] state can be produced artificially and is called light hypnosis or the hypnoidal state. **1970** R. R. MONROE *Episodic Behavioral Disorders* ii. 45 Abrupt alteration in awareness, such as [is] seen in petit mal or hypnoidal states.

hypnopædia (hipnopī·diă). Also **-pedia.** [f. HYPNO- + Gr. παιδεία education.] The exposure of a sleeping subject to lessons played on a radio, tape recorder, etc.; teaching or learning by this method, 'sleep-learning'.

1932 A. HUXLEY *Brave New World* ii. 27 The principle of sleep-teaching, or hypnopædia, had been discovered. **1959** *Listener* 26 Feb. 385/1 Subliminal persuasion, hypnopaedia, brain-washing. **1969** *New Scientist* 30 Jan. 216/1 Sleep learning or hypnopedia, as its practitioners prefer to call it, is now acquiring a new status among Soviet teaching circles. **1971** *Nature* 24 Sept. 290/1 The opening chapters dispel misconceptions many people have about hypnopaedia.

Hence **hypnopæ·dic** *a.*, of or involving hypnopædia; **hypnopæ·dically** *adv.*

1932 A. HUXLEY *Brave New World* iv. 82 The sort of words that suddenly make you jump..they seem so new and exciting even though they're about something hypnopædically obvious. *Ibid.* x. 173 Listening unconsciously to hypnopædic lessons in hygiene and sociability. **1970** *Globe & Mail* (Toronto) 26 Sept. 7/1 The Russians are also learning..how to avoid the fatigue brought on by repeated hypnopedic sessions.

hypnopompic (hipnopǫ·mpik), *a.* [f. HYPNO- + Gr. πομπ-ή sending away (f. πέμπειν to send) + -IC.] That accompanies the process of awakening from sleep.

a **1901** F. W. H. MYERS *Human Personality* (1903) I. p. xvii, To similar illusions accompanying the *departure* of sleep, as when a dream-figure persists for a few moments into waking life, I have given the name *hypnopompic*. *Ibid.* iv. 125 Equally remarkable are the hypnopompic pictures. **1925** *Proc. Soc. Psychical Res.* XXXV. 331 Of these four examples only the first is hypnagogic; Herschel's was a day-vision (at the breakfast-table), and the other two are hypnopompic. **1972** *Science* 16 June 1203/3 The ASC's [*sc.* altered states of consciousness] experienced by almost all ordinary people are dreaming states and the hypnogogic and hypnopompic states, the transitional states between sleeping and waking.

Hypnos (hi·pnǫs). [Gr. Ὕπνος (ὕπνος sleep) (see below).] Name of the god of sleep in Greek mythology.

1906 T. E. LAWRENCE *Home Lett.* (1954) 40 It is eleven o'clock now and I ought to have thoughts on Hypnos. **1938** W. DE LA MARE *Memory* 38 What of the strange world that teems—Where brooding Hypnos reigns—with dreams? **1970** *Oxf. Classical Dict.* (ed. 2) 535/2 Throughout antiquity Hypnos was usually thought of as a winged youth who touches the foreheads of the tired with a branch..or pours sleep-inducing liquid from a horn.

hypnotherapy (hipnoþe·rāpi). [f. HYPNO- (taken as combining form of *hypnosis*) + THERAPY.] Psychotherapy that involves the use of hypnotism.

1897 *Lippincott's Med. Dict.* 496/1 *Hypnotherapy*, the therapeutic use of hypnotism. **1907** *Alienist & Neurologist* XXVIII. 447 Constant current..cephalic galvanization..is a valuable addition to hypnotherapy, medical or otherwise, in psychiatry. **1931** A. EILOART tr. *Heyer's Hypnosis & Hypnotherapy* xiii. 204 Occasionally impotence or spasm of the vagina..are suitable subjects for hypnotherapy. **1947** LECRON & BORDEAUX *Hypnotism Today* xii. 196 Direct persuasive suggestion under hypnosis was the type of hypnotherapy used by the old medical practitioners of hypnotism. **1958** *Sunday Times* 17 Aug. 15/5 Hypnotherapy has a rightful place in medical treatment. **1960** *Spectator* 25 Nov. 838 Scientific training is attempted in the..two-year Hypnotherapy Centre course.

Hence **hypnothe·rapist**, one who employs hypnotherapy. Also **hy:pnotherapeu·tic** *a.*, of or involving hypnotherapy; **hy:pnotherapeu·tically** *adv.*

1892 *Jrnl. Mental Sci.* XXXVIII. 522 (*heading*) Hypno-therapeutic treatment. **1944** BRENMAN & GILL *Hypnotherapy* (1947) iv. 79 Erickson..who frequently collaborates with psychoanalysts in his hypnotherapeutic work, often utilizes the insights of psychoanalysis. **1958** *Spectator* 15 Aug. 236/1 Hypnotherapist and hypnoanalyst. **1963** Hypnotherapeutic [see *HYPNOANALYSIS]. **1963** W. S. KROGER *Clin. & Exper. Hypnosis* xlvii. 323/2 Age-regression was hypnotherapeutically induced in several patients as an emergency measure to prevent suicide. **1970** *Daily Tel.* 8 June 11/4 Hypnotherapists have found that migraine has psychological causes.

hypo, *sb.*² (Earlier example.)

1855 [see PRINT *sb.* 13].

hypo (həi·po), *sb.*³ *slang.* [Abbrev. of HYPODERMIC.] A hypodermic needle or injection; a drug-addict.

1904 *San Francisco Chron. Suppl.* 30 Oct. 4 (caption to picture showing a morphine addict) The 'Hypo'. **1925** *Writer's Monthly* June 486/2 Hypo, a hypodermic syringe or a hypodermic needle. **1926** J. BLACK *You can't Win* xii. 159 'Vag these two hypos', said the cop to the desk man. **1936** G. K. ZIPF *Psycho-Biol. of Lang.* ii. 31 *Hypo* may be a truncation of a *hypodermic injection*..or it may be an abbreviation of '*hyposulfite of soda*'. **1942** *R.A.F. Jrnl.* 3 Oct. 30 Horrible rumours about the ease with which a hypo. needle will break. **1953** W. BURROUGHS *Junkie* (1972) ii. 21 An eyedropper is easier to use than a regular hypo, especially for giving yourself vein shots. **1956** H. GOLD *Man who was not with It* (1965) ii. 18 He.. dragged his fingers along the little scabs from the hypo. **1973** J. WAINWRIGHT *Devil you Don't* 89 The night medic ..held the loaded hypo.

hypo (həi·po), *v.* *slang.* [f. *HYPO *sb.*³] To administer a hypodermic injection (to). Also *fig.* Hence **hy·poing** *vbl. sb.*

1925 *Flynn's* 7 Feb. 489/2 *Hypo*, to use a hypodermic syringe. **1945** *Variety* 24 Oct. 4/4 (*heading*) Many new pix to hypo bond preems [i.e. new pictures offered as inducements for bond premieres]. **1956** C. D. SIMAK *Strangers in Universe* (1958) 14 Apparently not too dangerous. Not with every single soul hypoed and immunized and hormoned to his eyebrows. **1960** *Time* (Dom. ed.) 25 Jan. 90/3 Because of continuing hypo-ing, his arms and legs become abscessed. **1968** *Listener* 31 Oct. 567/3 This impulse is very much hypo'd up during an election year.

hypo-. Add: **II.** (In the following words *e* often replaces *æ*, *œ*, esp. in U.S. usage; the alternative spelling is not given for each word individually.) **hypalgesia** (-dʒī·siă) *Med.* [Gr. ἄλγησις sense of pain], diminished sensitivity to pain, hypalgia; so **hypalge·sic** *a.*, exhibiting or tending to produce hypalgesia; **hypo-aci·dity** *Physiol.*, a deficiency of acid constituents, esp. in the gastric juice; **hypo-acti·vity** *Physiol.*, diminished activity, *esp.* diminished secretory activity of a gland; **hypoæsthesia** (həi:po,ĕsþī·ziă) *Path.* = HYPÆSTHESIA; hence **hypoæsthe·tic** *a.*; **hy:po-albuminæ·mia** *Physiol.* [Gr. αἷμα blood], an abnormally low concentration of albumins in the blood; **hypoalge·sia** *Med.*, = *hypalgesia*; **hypocalcæmia** (-kælsī·miă) *Physiol.* [CALC(IUM + Gr. αἷμα blood], an abnormally low concentration of calcium in the blood; hence **hypocalcæ·mic** *a.*; **hypoca·pnia** *Physiol.* [Gr. καπνός smoke], an abnormally low concentration of carbon dioxide in the blood; **hy:pochloræ·mia** *Physiol.*, an abnormally low concentration of chlorides in the blood; **hy:pochlorhy·dria** *Physiol.* [CHLORHYDR(IC *a.*], an abnormally low concentration of hydrochloric acid in the gastric juice; so **hy:pochlorhy·dric** *a.*; **hypocho·rdal** *a.* *Zool.* [CHORD *sb.*¹], ventral to the notochord or spinal cord; **hypoesthesia,** var. *hypoæsthesia* above; **hypoeute·ctic** *a.*, (of an alloy of iron) containing a lower proportion of carbon than the eutectic composition (i.e. less than about 4·3%); (in quot. 1902 = *hypo-eutectoid, eutectoid* itself not having been coined at that date); **hypoeute·ctoid** *a.*, (of steel) containing a lower proportion of carbon than the eutectoid composition (i.e. less than about 0·8%); **hypofu·nction** *Med.*, diminished or insufficient activity or production (in a gland or other part of the body); so **hypofu·nctional** *a.*, **hypofu·nctioning** *vbl. sb.*; **hy:pogammaglo:bulinæ·mia** *Path.* [Gr. αἷμα blood], an abnormally low concentration of gamma globulins in the blood; also, a disorder of which this condition is characteristic; **hypoge·nitalism** *Path.*, hypogonadism; also, underdevelopment of the genitalia; **hypogeusia** (-giū·ziă) *Med.* [Gr. γεῦσις taste], diminished acuteness of the sense of taste; **hypoglycæ·mia** *Physiol.*, an abnormally low concentration of sugar in the blood; so **hypoglycæ·mic** *a.*, of or exhibiting hypoglycæmia; tending to reduce the blood-sugar level; **hypogo·nadism** *Path.*, the reduction or absence of gonadal activity, esp. of hormone secretion; so **hypogo·nadal** *a.*; **hypoischium** (-i·skiǒm) *Zool.* [ISCHIUM], a small cartilaginous or bony process that projects backwards from the ischial symphysis in the pelvic arch of many reptiles and some other vertebrates, supporting the ventral wall of the cloaca; hence **hypoi·schiac, -i·schial, -ischia·tic** *adjs.*; **hypokalæmia** (-kălī·miă), **-kaliæmia** (-kæli,ī·miă) *Physiol.* [mod.L. *kalium* potassium (see KALI) + Gr. αἷμα blood], = *hypopotassæmia* below; hence **hypokalæ·mic** *a.*; **hypokine·sia, -kine·sis** *Path.*, abnormally decreased muscular movement; **hy:poleucocyto·sis, -leukocyto·sis** *Path.* = *leucopenia* (s.v. LEUCO-); **hy:pomagnesæ·mia** *Physiol.* and *Vet. Sci.* [Gr. αἷμα blood], an abnormally low concentration of magnesium in the blood, important in cattle as the cause of grass tetany; hence **hy:pomagnesæ·mic** *a.*; **hypometa·bolism** *Physiol.*, metabolism at a low rate; **hy·pomorph** *Genetics* [Gr. μορφ-ή form], any allele which is functionally less effective than the corresponding wild-type allele; so **hypomo·rphic** *a.*; **hypomoti·lity** *Med.*, diminished movement, esp. of the stomach and intestines; **hyponatræ·mia** *Physiol.* [NATR(IUM + Gr. αἷμα blood], a lower than normal concentration of sodium in the blood; hence **hyponatræ·mic** *a.*; **hy:po-osmo·tic, hyposmo·tic** *adjs.* *Physiol.* = *HYPOTONIC *a.* 1; const. *to*; **hypophalangia** (-fălæ·ndʒiă), **-phalangism** (-fălæ·ndʒiz'm), **-phalangy** (-fæ·lăndʒi) *Med.* [L. *phalang-*: see PHALANX], the congenital absence of one or more digital phalanges; **hypopho·neme** [*PHONEME] *Linguistics*, in the terminology of stratificational grammar, a phonological unit (see quots.); so **hypophone·mic** *a.*, **hypophone·mically** *adv.*; **hypopho·ria** *Ophthalm.*, latent strabismus in which there is a tendency for one eye to be directed below the line of sight of the other; **hy:pophosphatæ·mia** *Physiol.* [Gr. αἷμα blood], an abnormally low concentration of phosphates in the blood; so **hy:pophosphatæ·mic** *a.*; **hy:pophosphata·sia** *Path.*, a familial congenital disease associated with an abnormally low level of alkaline phosphatase in the body and defective bone development; **hypopitu·itarism** *Path.* [PITUITAR(Y *a.* + -ISM], diminished hormone secretion by the pituitary body; hence **hypopitu·itary** *a.*, of, pertaining to, or affected with hypopituitarism; **hypopla·nkton,** plankton found in the layer of water directly above the bottom of the ocean; **hypopleu·ra, -pleu·ron** *Ent.* (pl. **-pleura**), the region on the thorax of Diptera underneath the metapleuron and above the middle and posterior coxæ; so **hypopleu·ral** *a.*; **hy·poploid** *a.* *Genetics* [*-PLOID], having one or a few chromosomes (orig. also chromosome fragments) missing from a haploid, diploid, triploid, etc., set; containing such cells; also as *sb.*, a hypoploid cell or individual; so **hy·poploidy,** the condition of being hypoploid; **hy:popotassæ·mia** *Physiol.* [Gr. αἷμα blood], an abnormally low concentration of potassium in the blood; so **hy:popotassæ·mic** *a.*; **hy:poproteinæ·mia** *Physiol.* [Gr. αἷμα blood], an abnormally low concentration of protein in the blood; so **hy:poproteinæ·mic** *a.*; **hy:poprothrombinæ·mia** *Med.* [Gr. αἷμα blood], an abnormally low concentration of prothrombin in the blood; a disorder so characterized; so **hy:poprothrombinæ·mic** *a.*; **hy·poscope** *Mil.* [-SCOPE, after *periscope*], a form of periscope for attachment to a rifle or for use as a hand instrument; **hyposecre·tion,** diminished secretion; **hyposmotic**: see *hypo-osmotic* above; **hy:posthenu·ria** *Med.* [Gr. σθέν-ος strength + -URIA], the secretion of urine of abnormally low specific gravity; **hypotaurine** (-tǭ·rīn) *Chem.* [a. F. *hypotaurine* (Chatagner & Bergeret 1951, in *Compt. Rend.* CCXXXII. 450)], a pale yellow crystalline amino-acid, $NH_2(CH_2)_2SO_2H$, found in some higher organisms; 2-aminoethanesulphinic acid; **hypothermal,** (*c*) *Petrol.*, of, pertaining to, or designating mineral and ore deposits formed by hydrothermal action at relatively high temperature and pressure; **hypotrichosis** (-trikǭu·sis) *Path.* [ad. G. *hypotrichose* (R. Bonnet 1892, in *Anat. Hefte* I. I. viii. 235), f. Gr. τρίχωσις growth of hair, f. τριχοῦν to cover with hair (θρίξ, τριχ- hair)], partial or complete absence of hair; hence **hypotricho·tic** *a.*; **hy:povitamino·sis** *Path.* [-OSIS], any condition caused by vitamin deficiency; **hypovolæ·mia** *Physiol.* [VOL(UME *sb.* + Gr. αἷμα blood], a decreased volume of circulating blood in the body; hence **hypovolæ·mic** *a.*; **hypoxæmia** (həipǫksī·miă) *Med.* [ad. F. *hypoxémie* (P. A. Piorry *Traité de Méd. pratique* (1847) III. 123), f. *ox-ygène* + *-émie* (Gr. αἷμα blood)] = *ANOXÆMIA; **hypoxia** (həipǫ·ksiă) *Med.* [OX(YGEN + -IA¹] = *ANOXIA; hence **hypo·xic** *a.*, of or pertaining to hypoxia; deficient in oxygen.

1881 J. ROSS *Treat. Dis. Nervous Syst.* I. iii. 84 Eulenberg has proposed the term hypalgesia or hypalgia, to indicate diminution of painful reaction, while limiting analgesia to its abolition. **1906** *Jrnl. Nerv. & Mental Dis.* XXXIII. 324 (*heading*) Hypesthesia and hypalgesia and their significance in functional nervous disturbances. **1971** P. C. LUND *Spinal Anesthesia* vii. 318 Sharp needles are..utilized to determine the level of hypalgesia which precedes the development of analgesia. **1911** STEDMAN *Med. Dict.* 405/2 Hypalgesic. **1916** L. F. BARKER *Monographic Med.* IV. 137 The effect of summation of stimuli should..be noticed, by drawing a sharp needle lengthwise over an analgesic or hypalgesic area. **1935** *Discovery* Aug. 226/2 One very fortunate property which such a generator appears to possess is its pain-relieving virtue, or hypalgesic action, a very useful condition when treating post-operative cases. **1900** DORLAND *Med. Dict.* 311/2 Hypoacidity. **1902** *Encycl. Brit.* XXXI. 551/2 Hyperacidity from lactic may obscure hypoacidity of hydrochloric acid. **1943** E. URBACH *Allergy* (1944) ii. 67 Gastric hypo- or anacidity is often observed. **1910** *Bull. Johns Hopkins Hosp.* XXI. 127/2 Conditions therefore simulating grades of hypoactivity. **1914** *Arch. Internal Med.* XIV. 145 Hypo-activity of the thyroid and pituitary. **1965** B. E. FREEMAN tr. *Vandel's Biospeleol.* xxi. 351 The majority of the follicles of the thyroid..show signs of hypoactivity. **1906** *Jrnl. Nerv. & Mental Dis.* XXXIII. 324 Hypoesthesia is the term heretofore employed to express this condition, but its awkward form at least excuses the employment of hypesthesia as a more euphonious and therefore more satisfactory expression. **1967** D. SINCLAIR *Cutaneous Sensation* viii. 148 'Hyperaesthesia' and 'hypoaesthesia' are similarly misused. **1940** *Lancet* 17 Feb. 303/2 Complete recovery [from frostbite] may apparently take place, but after a variable interval neuralgic pains may begin. The skin is usually hypoæsthetic. **1937** *Acta Med. Scand.* XCI. 336 A simple method for the determination of hypoalbuminemia and hypoproteinemia..is afforded by the determination of the specific gravity of serum. **1962** *Lancet* 6 Jan. 52/1 This loss of protein may be significant in the pathogenesis of the hypoalbuminæmia of kwashiorkor. **1929** DORLAND & MILLER *Med. Dict.* (ed. 15) 584/1 Hypo-algesia. **1945** *Jrnl. Clin. Invest.* XXIV. 505 A patient had 'hypoalgesia' to pin prick on parts of his left hand. **1968** A. SOULAIRAC et al. *Pain* 36 The marked hypoalgesia recorded in this animal was associated with a double right lesion. **1925** *Jrnl. Biol. Chem.* LXVI. 345 Hypocalcemia was produced..by thyroparathyroidectomy. **1960** *Farmer & Stockbreeder* 22 Mar. 135/2 What is the glucose dosage for young pigs with hypocalcæmia? **1962** A. SORSBY in A. Pirie *Lens Metabolism Rel. Cataract* 298 Congenital cataract..can be caused by such frankly environmental disturbances as..maternal hypocalcæmia. **1935** D. H. SHELLING *Parathyroids* vi. 148 Other means of demonstrating hypocalcemic tetany are now available. **1908** Hypocapnia [see *hypercapnia* s.v. *HYPER- IV]. **1961** *Lancet* 26 Aug. 475/1 The combination of extreme hypoxia with hypocapnia may well be fatal. **1927** *Amer. Jrnl. Med. Sci.* CLXXIII. 649 (*heading*) Acute intestinal obstruction: mechanism and significance of hypochloremia and other blood chemical changes. **1963** H. L. BOCKUS et al. *Gastroenterology* (ed. 2) I. xxviii. 646/2 If hypochloremia and alkalosis are present, gastric retention and vomiting have probably preceded the bout of bleeding. **1893** Hypochlorhydria [see *hyperchlorhydria* s.v. *HYPER- IV]. **1971** J. SONG *Path. Sickle Cell Dis.* xviii. 355 The usual hypochlorhydria present in this disease may account for some of the gastric manifestations. **1921** *Chem. Abstr.* XV. 894 When the concn. varies between 0.010 and 0.012 sp. gr. the indications are that it [*sc.* the stomach] contains dissolved alimentary residues and tends to be hypochlorhydric. **1971** J. SONG *Path. Sickle Cell Dis.* xviii. 356 Many individuals present hypochlorhydric states of a like degree. **1901** *Gray's Anat.* (ed. 15) II. 96 The future vertibræ..are soon joined across the middle line on the ventral aspect of the notochord by a hypochordal cartilaginous bar. **1962** M. JOLLIE *Chordate Morphol.* vi. 153 This splint is the ventral, perichondral ossification of the hypochordal cartilage. **1902** *Encycl. Brit.* XXIX. 572/2 They are called hyper-eutectic or hypo-eutectic according as this excess is cementite or ferrite, i.e., according as their carbon-content is above or below the 0·90 per cent. which the eutectic itself contains. **1926** W. E. WOODWARD *Metallogr. Steel & Cast Iron* i. 27 In a 2·0% C steel 0·3% C (= 4·5% Fe_3C) will have been required to form the eutectic portion of the hypo-eutectic alloy. **1959** Hypoeutectic [see *hypereutectic* adj. s.v. *HYPER- IV]. **1911** *Encycl. Brit.* XIV. 805/2 This ferrite flows around and immediately heals over any cracks which form in the small quantity of cementite interstratified with it in the pearlite of hypo-eutectoid steels. **1966** A. PRINCE *Alloy Phase Equilibria* vi. 107 The structure of a hypo-eutectoid Fe—Fe_3C alloy is one of ferrite with pearlite, the latter appearing in characteristic form. **1905** GOULD *Dict. New Med. Terms* 303/2 Hypofunction. **1913** L. FORSTER tr. *A. Biedl's Internal Secretory Organs* 53 Vassale thinks that the new formation of tissue points to a hyper-function of the gland, the wasting of the colloid to a hypo-function. **1920** *Endocrinology* IV. 344 Hypofunction of the thyroid. **1972** *Lancet* 12 Aug. 299/2 There was a high frequency of sexual hypofunction and testicular atrophy among male patients. **1933** A. W. ROWE *Differential Diagn. Endocrine Disorders* viii. 116 'Hyperfunction' indicates a condition..in direct antithesis to..the known hypofunctional state. **1961** *Jrnl. Amer. Med. Assoc.* 29 July 232/2 Of 43 hypofunctional nodules, only 2 proved to be due to carcinoma. **1926** J. S. HUXLEY *Ess. Pop. Sci.* 291 Whenever we can trace the effect of a hypo- or hyper-functioning of one of these [ductless] glands, we find that it affects..a complex of characters..related to the performance of a single function. **1954** A. WHITE et al. *Princ. Biochem.* xliii. 936 In circumstances of adrenal cortical hypofunctioning..there is a failure of normal renal tubular reabsorption of sodium. **1955** *Jrnl. Amer. Med. Assoc.* 13 Aug. 1344 (*heading*) Hypogammaglobulinemia associated with a severe wound infection. **1970** PASSMORE & ROBSON *Compan. Med. Stud.* II. xviii. 103/1 Individuals with hypogammaglobulinaemia..produce little or no detectable circulating antibody and are vulnerable to bacterial invasion, but are not so susceptible to viral infection. **1972** *Lancet* 27 May 1151/2 Patients with the common variable type of severe hypogammaglobulinæmia exhibit lymphocytes with surface immunoglobulins. **1917** STEDMAN *Med. Dict.* (ed. 4) 458/2 Hypogenitalism. **1922** L. F. BARKER *Endocrinol. & Metabolism* I. 157 Obesity is a frequent manifestation of hypogenitalism, either the physiological hypogenitalism of the menopause or the acquired form due to disease or to the surgical removal of the ovaries. **1964** L. MARTIN *Clinical Endocrinol.* (ed. 4) vii. 222 Hypogenitalism [in males] means abnormally small size or underdevelopment of the male external genitalia which need not necessarily include testicular failure. **1888** Hypogeusia [see *hypergeusia* s.v. *HYPER- IV]. **1969** C. PFAFFMANN *Olfaction & Taste* 578 Treatment with D-penicillamine had produced hypogeusia (a decrease in taste acuity). **1894** GOULD *Dict. Med.* 594/2 Hypoglycemia. **1911** *Jrnl. Biol. Chem.* X. 160 Recent investigations on the production of hypoglycaemia. **1960** *Farmer & Stockbreeder* 9 Feb. 102/1 Hypoglycæmia is a symptom rather than a disease on its own. **1923** *Jrnl. Physiol.* LVII. 318 The blood became hypoglycæmic. **1965** J. POLLITT *Depression & its Treatment* vi. 78, 20–80 units of soluble insulin before breakfast may be necessary to produce a mild hypoglycæmic reaction. **1970** PASSMORE & ROBSON *Compan. Med. Stud.* II. vi. 18/1 Today, approximately one-third of the total diabetic population is being treated with an oral hypoglycaemic agent. **1933** *Med. Jrnl. & Rec.* CXXXVII. 457/2 The vast majority of hypogonadal patients were within normal weights. **1944** R. S. HOTCHKISS *Fertility in Men* iv. 90 Disproportional height span of legs and torso suggest the hypogonadal state. **1961** W. C. YOUNG *Sex & Internal Secretions* (ed. 3) I. v. 348 Hypogonadal disorders of man. **1918** STEDMAN *Med. Dict.* (ed. 5) 469/1 Hypogonadism. **1933** *Jrnl. Amer. Med. Assoc.* 7 Jan. 70/2 A method for the assay of blood and urine for testicular hormone..has been used as a laboratory test for hypogonadism. **1966** R. B. SCOTT *Price's Textbk. Pract. Med.* (ed. 10) vii. 450/1 The term female hypogonadism implies a deficiency of both the ovulatory and hormone secretory functions of the ovary. **1970** PASSMORE & ROBSON *Compan. Med. Stud.* II. xii. 11/2 In children hypogonadism leads to delayed puberty. **1951** C. K. WEICHERT *Anat. Chordates* x. 485 A posterior prolongation from the ischial symphysis in *Sphenodon* and in many lizards and turtles is called the hypoischiac process, or cloacal bone. **1910** PARKER & HASWELL *Text-bk. Zool.* (ed. 2) II. 354 In the Chelonia..both pubes and ischia meet in ventral symphyses, and epipubic and hypoischial cartilages may be present. **1959** W. MONTAGNA *Compar. Anat.* v. 116 In lizards an epipubic cartilage projects anteriorly and an hypoischial cartilage projects posteriorly from the symphysis. **1897** W. N. PARKER tr. *Wiedersheim's Compar. Anat. Vertebr.* (ed. 2) 117 In Hatteria there is a marked epipubis and a hypoischiatic process continuous with the epipubic cartilage. *Ibid.* 118 A longitudinal fibro-cartilaginous ligament, continuous anteriorly with the plug-like epipubic cartilage and posteriorly with the hypoischium. **1925** J. S. KINGSLEY *Vertebr. Skeleton* 265 Squamata... The hypoischium, usually movable, is well developed and may be cartilage or bone in the adult. **1949** *Jrnl. Clin. Invest.* XXVIII. 409 (*heading*) Some observations on the development of hypokaliemia during therapy of diabetic acidosis. **1951** *Dorland's Med. Dict.* (ed. 22) 713/2 Hypokalemia, hypokaliemia. **1972** *Lancet* 1 July 36/2 Hypokalæmia may be seen in any stage of renal failure. **1953** *Jrnl. Clin. Invest.* XXXII. 538 (*heading*) The effect of potassium in nephrectomized rats with hypokalemic alkalosis. **1962** *Lancet* 1 Dec. 1145/1 During this period the patient became hypokalæmic. **1886** *Syd. Soc. Lex.*, Hypokinesia. *Ibid.*, Hypokinesis. **1927** I. S. WECHSLER *Clin. Neurol.* IV. 391 Instead of hypokinesis there may be hyperkinetic phenomena. **1970** *Nature* 4 Apr. 21/1 Parkinsonism is.. characterized by tremor, rigidity of the limbs and poverty of movement (hypokinesia). **1897** *Lippincott's Med. Dict.* 498/1 Hypoleucocytosis. **1898** [see *leucopenia* s.v. LEUCO-]. **1930** H. DOWNEY in E. T. Bell *Text-bk. Path.* xxviii. 599 In pernicious anemia we see an example of neutrophile hypoleukocytosis. **1933** *Jrnl. Clin. Invest.* XII. 982 (*heading*) Clinical manifestations of hypo- and hyper-magnesaemia. **1971** *Arable Farmer* Feb. 70/2 A high level of potash in the soil..can lead to hypomagnesaemia (grass staggers) in dairy cows. **1960** *Times* 28 Nov. 16/5 Hypomagnesaemic tetany was common on sheep that were rapidly transferred back from good pasture to poorer hill grazing. **1932** DORLAND & MILLER *Med. Dict.* (ed. 16) 6111/2 Hypometabolism. **1962** T. L. SOURKES *Biochem. Mental Dis.* xxiv. 302 This syndrome has been variously termed metabolic insufficiency, non-myxedematous hypometabolism, and euthyroid hypometabolism. **1932** H. J. MULLER in *Proc. 6th Internat. Congr. Genetics* I. 235 Scute-1 is therefore a hypomorph. **1946** *Nature* 12 Oct. 520/1 This mutant allele is therefore a hypomorph to the normal allele. **1962** I. H. HERSKOWITZ *Genetics* xxiv. 210/1 Mutants having a similar but lesser effect than the normal gene are called hypomorphs. **1932** H. J. MULLER in *Proc. 6th Internat. Congr. Genetics* I. 235 Apricot, like eosin, is a mutant gene which produces an effect similar to that of the normal allelomorph, but a lesser effect... It is..like a lesser-normal. I therefore call it a 'hypomorphic' mutant. **1962** I. H. HERSKOWITZ *Genetics* xxiv. 210/1 We can represent the relationship between the normal gene and its hypomorphic mutants diagrammatically. **1900** DORLAND *Med. Dict.* 312/2 Hypomotility. **1914** C. G. STOCKTON *Dis. Stomach* ix. 183 (*heading*) Diminished gastric motion, hypomotility, gastric atony. **1970** *Radiology* XCIV. 303/2 An upper gastrointestinal examination..failed to show any abnormality, except for generalized hypomotility of the stomach and small intestine. **1935** DORLAND & MILLER *Med. Dict.* (ed. 17) 649/1 Hyponatremia. **1969** L. G. WESSON *Physiol. Human Kidney* xxvii. 554/1 Hyponatremia may be defined somewhat arbitrarily as a plasma sodium concentration less than 130 mM/L in man. **1955** *Arch. Internal Med.* XCV. 21/1 Infants who present hyponatremic acidosis. **1905** W. H. HOWELL *Text-bk. Physiol.* 885 A hypotonic or hyposmotic solution is one whose osmotic pressure is less than that of serum. **1957** B. T. SCHEER et al. *Rec. Adv. Invertebr. Physiol.* 237 The antennal secretion of *P. crassipes* is slightly hypo-osmotic to the blood in 50% sea water. **1963** R. P. DALES *Annelids* v. 104 The ability to form a hyposmotic urine. **1971** W. J. MCCAULEY *Vertebr. Physiol.* i. 9 If it has a lower osmotic pressure, it is said to be a hypoosmotic solution. **1916** *Genetics* I. 90 Various types of developmental malformation of the hands and feet have been described under such terms as..hypophalangia. **1911** STEDMAN *Med. Dict.* 411/2 Hypophalangism. **1965** *Arch. Internal Med.* CXV. 581/2 The present family is unique in that hypophalangism is limited to the fourth digits and associated with symphalangism. **1929** R. R. GATES *Heredity in Man* viii. 154 Brachyphalangy combined with hypophalangy (less than five fingers) was transmitted for six generations. **1966** S. M. LAMB *Outl. Stratif. Gram.* 18 Such cases have particularly attracted the attention of linguists in hypophonemic systems... The hypophonemic and hypersememic strata might be called the phonetic and semantic, respectively. *Ibid.* 19 The hypophonemic system appears not to have a sign pattern. *Ibid.* 28 The tactics of the hypophonemic stratum of a language specifies how hypophonemes (i.e. phonological components) are arranged in segments and clusters. **1968** J. ALGEO in *South Atlantic Bull.* XXXIII. ii. 2 The distinctive features of sound, the hypophonemes in Lamb's terminology, are relatively easy to study because there are so few of them—only about twelve to fifteen in most languages. Sample hypophonemes are *plosion, spirancy, nasality, labiality,* and *unvoicing*. **1969** *Language* XLV. 303 Such alternations would be treated as alternate realizations of these phonons in terms of the units of the lower phonological stratum, the hypophonemes. *Ibid.* 307 In Figure 6 the first vowels of /glosó/ and /dəbé/ would be treated as the same, as /Vo/ hypophonemically. ..They would be the same only when viewed as hypophonemic signs, which include non-distinctive as well as phonemic elements. **1932** L. C. MARTIN *Introd. Appl. Optics* II. iv. 143 In Hypophoria one eye turns downwards. **1964** [see *hyperphoria* s.v. *HYPER- IV]. **1935** D. H. SHELLING *Parathyroids* vi. 138 Fish has found hypophosphatemia as well as hypocalcemia. **1962** *Lancet* 2 June 1169/1 The other forms of rickets and osteomalacia are of the vitamin-D-resistant type, and are characterised by persistent hypophosphatæmia. **1946** M. R. EVERETT *Med. Biochem.* (ed. 2) viii. 628 The Fanconi syndrome (intractable hypophosphatemic rachitis accompanied by acidosis and renal glycosuria). **1968** R. F. PITTS *Physiol. Kidney & Body Fluids* (ed. 2) xiii. 237/2 One or the other parent is hypophosphatemic. **1948** J. C. RATHBUN in *Amer. Jrnl. Dis. Children* LXXV. 831 It was therefore decided to call this disease 'hypophosphatasia' to single out the remarkably low alkaline phosphatase levels. **1957** *Amer. Jrnl. Med.* XXII. 730/1 There is now good evidence that hypophosphatasia is a specific genetically determined metabolic disease characterized by three salient features: (1) abnormal mineralization of bone, (2) diminished alkaline phosphatase activity, and (3) increased urinary excretion of phosphorylethanolamine. **1909** Hypopituitarism [see *hyperpituitarism* s.v. *HYPER- IV]. **1921** *Glasgow Herald* 10 Sept. 4/5 There were several causes of dwarfism; sometimes disorders of the thyroid gland were the cause, but other varieties were produced by hypo-pituitarism. **1961** *Lancet* 30 Sept. 760/2 Prof. H. L. Sheehan showed that, in severe postpartum hypopituitarism, patients who had occasional uterine bleeding had just as great a destruction of the anterior pituitary as those who had permanent amenorrhœa. **1921** *Endocrinology* V. 800 A presentation of five cases of preadolescent hypopituitary infantilism. **1955** R. H. WILLIAMS *Textbk. Endocrinol.* (ed. 2) ix. 604 The hypopituitary dwarfs usually show marked retardation of their epiphyseal development. **1902** *Encycl. Brit.* XXXIII. 933/2 It is possible that the plankton immediately over the bottom [of the ocean] may prove to be sufficiently distinct to be separately classed as hypoplankton. **1903** *Nature* 5 Nov. 23/2 There is evidence that certain forms [of Copepoda] are confined to the bottom, and form part of a true hypoplankton. **1942** H. U. SVERDRUP et al. *Oceans* xvii. 814 The swimming powers of many animals put them midway between the plankton and the nekton, and many forms..live both on or near the bottom and are sometimes called hypoplankton. **1955** C. C. DAVIS *Marine & Fresh-Water Plankton* i. 28 The hypoplankton consists of plankters living near the bottom. **1884** *Trans. Entomol. Soc. Lond.* 503 *Hypopleura*, a distinct piece above the two last pairs of coxæ, and behind the sternopleura, from which it is separated by a suture. **1951** COLYER & HAMMOND *Flies Brit. Isles* 24 Spiracles or apertures of the tracheae (breathing-tubes) are located before the mesopleuron and behind the hypopleuron respectively. **1951** L. S. WEST *Housefly* ii. 28 The hypopleura lies behind the middle coxa, the sternopleura just in front. **1884** *Trans. Entomol. Soc. Lond.* 511 Hypopleural bristles. **1930** *Jrnl. Genetics* XXII. 306 Other individuals—'hypoploids'—may fail to inherit the fragment. *Ibid.* 313 Viable heteroploid or hypoploid zygotes. **1957** Hypoploid [see *hyperploid* s.v. *HYPER- IV]. **1930** *Jrnl. Genetics* XXII. 329 The phaenotypic effects of hypoploidy and hyperploidy of every portion of the chromatin. **1969** Hypoploidy [see *hyperploidy* s.v. *HYPER- IV]. **1932** DORLAND & MILLER *Med. Dict.* (ed. 16) 612/2 Hypopotassemia. **1949** *Jrnl. Clin. Invest.* XXVIII. 409/1 Hypopotassemia may appear during therapy of diabetic acidosis or coma. **1963** Hypopotassemia [see *hyperpotassæmia* s.v. *HYPER- IV]. **1950** *Jrnl. Amer. Med. Assoc.* 16 Dec. 1328 A resultant hypopotassemic, hypochloremic alkalosis. **1953** *Lancet* 11 July 60/1 The more familiar hypopotassæmic paralysis. **1934** *Jrnl. Biol. Chem.* CV. 327 (*heading*) The effect of nutritional hypoproteinemia on the electrolyte pattern and calcium concentration of serum. **1961** *Lancet* 5 Aug. 299/1 Hypoproteinæmia may be due to impaired synthesis of protein (especially albumin) in malnutrition or liver disease, or..to abnormal loss of protein in starvation, after injury, and from discharges. **1935** *Clin. Sci.* II. 60 Hypoproteinæmic. **1942** *Jrnl. Amer. Med. Assoc.* 3 Jan. 22 The disturbed osmotic relations in the hypoproteinemic dog. **1966** J. W. LINMAN *Princ. Hematol.* v. 156 Hypoproteinemic dogs or rats. **1936** *Jrnl. Exper. Med.* LXIII. 798 Titration of prothrombin, however, revealed a very marked hypoprothrombinemia. **1961** *Lancet* 19 Aug. 390/1 Subacute intestinal obstruction associated with excessive hypoprothrombinæmia due to oral anticoagulant therapy. **1962** *Ibid.* 27 Jan. 177/1 Two types of hereditary hypoprothrombinæmia are known to exist. **1942** *Chem. Abstr.* 20 Nov. 7087 The danger of the hypoprothrombinemic hemorrhage was very slight. **1955** *Arch. Internal Med.* XCV. 2/2 The opportunity to study various types of congenital hypothrombinemic states repeatedly over a period ..has furnished data on their clinical course. **1902** *Daily Chron.* 16 July 9/1 The 'Hyposcope' competition..the peculiarity of which is that, by an optical contrivance, the marksman, completely under cover, may fire round a corner, so to speak, at an enemy. **1915** *Illustr. London News* 20 Feb. 236/1 A trench-periscope (or, to give it its correct name, a hyposcope)... The Hyposcope is on the principle of the camera-obscura. **1909** *Jrnl. Amer. Med. Assoc.* 24 July 251/2 A condition of hyposecretion of this part of the gland. **1939** B. J. E. IHRE *Human Gastric Secretion* vii. 95 A reduced rate of secretion (hyposecretion). **1900** DORLAND *Med. Dict.* 313/2 Hyposthenuria. **1909** J. B. HERRICK in Osler & McCrae *Syst. Med.* VI. vi. 126 Unless hyposthenuria be counteracted by polyuria, renal insufficiency must result. **1971** J. SONG *Path. Sickle Cell Dis.* xv. 284 Hyposthenuria in sickle cell anemia was considered a reversible renal defect by Keitel et al. **1951** *Chem. Abstr.* XLV. 6232, $H_2N(CH_2)_2SO_2H$, tentatively named hypotaurine. **1965** A. MEISTER *Biochem. Amino Acids* (ed. 2) I. i. 75 This amino acid [*sc.* L-cysteinesulfinic acid], and the product of its decarboxylation, hypotaurine,..have been found in the free state in rat brain. **1966** *Biochim. et Biophys. Acta* CXVII. 495 (*heading*) The occurrence of hypotaurine and other sulfur-containing amino acids in seminal plasma and spermatozoa of boar, bull and dog. **1922** W. LINDGREN in *Econ. Geol.* XVII. 293 The terminology proposed..would be as follows:..A. Hydrothermal deposits. *a.* Epithermal. Formed by ascending hot waters near the surface in or near effusive rocks at relatively low temperature and pressure. *b.* Mesothermal. Formed by ascending hot waters in or near intrusive rocks at intermediate temperature and pressure. *c.* Hypothermal. Formed by ascending hot water in or near intrusive rocks at high temperature and pressure. ..The prefix 'hypo' has been substituted for 'kata' to correspond with Ransome's now generally accepted terms of 'hypogene waters', the suggestion implied being that the hypogene waters are principally derived from the region of the hypothermal deposits. **1969** BENNISON & WRIGHT *Geol. Hist. Brit. Isles* x. 247 The latter [mineral veins] are of two phases, hypothermal and mesothermal, usually of different and characteristic trend. **1970** PARK & MACDIARMID *Ore Deposits* (ed. 2) xii. 294 Many minerals of the igneous metamorphic zone continue without interruption into the hypothermal zone. **1896** T. L. STEDMAN *20th Cent. Practice* V. 575 Bonnet..suggests designating any lack of hair through error of development, hypotrichosis. **1968** A. J. ROOK et al. *Textbk. Dermatol.* II. xlvi. 1377 Congenital hypotrichosis of sufficient degree to cause social embarrassment..is not uncommon. **1937** *Jrnl. Biol. Chem.* CXVIII. 627 The administration of cystine stimulated hair growth in the hypotrichotic rat. **1923** STEDMAN *Med. Dict.* (ed. 7), Hypovitaminosis. **1946** *Nature* 7 Sept. 342/2 The excretion of aneurin was decreased on account of deficient renal function, a fact not signifying hypovitaminosis in this case. **1925** Hypovolemia [see *hypervolæmia* s.v. *HYPER- IV]. **1935** HARROW & SHERWIN *Textbk. Biochem.* xv. 413 Simple hypovolemia occurs in obesity and in certain types of renal edema. Polycythemic hypovolemia occurs in conditions of.. water deprivation. **1965** R. P. MOREHEAD *Human Path.* xxi. 501/2 Deficient water absorption leads to hypovolemia, dehydration, or both. **1952** *Jrnl. Amer. Med. Assoc.* 6 Sept. 11 The question of whether the normovolemic or the hypovolemic patient or animal should be used in evaluation of plasma expanders. **1961** A. C. GUYTON *Textbk. Med. Physiol.* (ed. 2) xxxvii. 482/2 One of the most common types of shock..is that caused by hemorrhage; this is called hemorrhagic shock and is a type of hypovolemic shock. **1886** *Syd. Soc. Lex.*, Hypoxæmia. **1936** *Brain* LIX. 115 The hypoxaemia of arterial and of arm vein blood encountered in a large proportion of epileptics is an expression of the stagnant physical and mental state which so often accompanies epilepsy. **1971** PORTER & KNIGHT *High Altitude Physiol.* 36, 25 per cent of the reported cases of chronic mountain sickness have some type of pathology which *per se* produces hyperventilation and hypoxaemia. **1941** *Ann. Internal Med.* XIV. 1245 During hypoxia..blood flow is increased. **1967** *New Scientist* 26 Jan. 195/1 Today, the single most important cause of perinatal deaths is intrauterine hypoxia, in which the foetus becomes starved of oxygen because of impaired

metabolism in the placenta. **1970** *Sci. Amer.* Feb. 53/2 Life on the mountains is made rigorous not only by hypoxia but also by cold. **1958** C. C. ADAMS et al. *Space Flight* 243 The hypoxic zone, less than three miles up, where the decreased oxygen pressure brings human psychological and physiological discomfort. **1966** *Lancet* 24 Dec. 1381/2 Polycythæmia secondary to hypoxic lung disease. **1970** *Sci. Amer.* Feb. 56/1 The mountain dwellers' metabolism also appears to be affected by the hypoxic conditions.

hypoacusis (həi:po,ăkiū·sis). Also **-acousia, -acousis.** [f. HYPO- 4: see *HYPACUSIS.] = *HYPACUSIS.

1947 ACKERMAN & DEL REGATO *Cancer* vii. 336 A unilateral diminution in the sense of hearing, hypoacousia, is very commonly found accompanying tumors of the nasopharynx. **1961** STEDMAN *Med. Dict.* (ed. 20) 740/2 Hypoacusis. **1969** *New Scientist* 10 July 53/2 Their ears have been done in by discotheques and it's not expressions of dumb insolence they're displaying but the symptoms of hypoacousis.

hypo-allergenic (həi:po,ælədʒe·nik), *a.* orig. *U.S.* Also **hypoallergenic.** [f. HYPO- 4 + *ALLERGENIC *a.*] Having little tendency to cause an allergic reaction; specially prepared or treated so as to cause no reaction in persons allergic to the normal product.

1953 *Jrnl. Soc. Cosmetic Chemists* Aug. p. x (Advt.), Investigations indicate that Modulan is hypo-allergenic. **1953** *Arch. Otolaryngol.* Nov. 541 The cosmetic-sensitive patient must mingle in society with persons who are not wearing hypoallergenic powder. **1957** S. KRAMER in E. Sagarin *Cosmetics* xxxviii. 879 Hypo-allergenic cosmetics are an important contribution to the allergic woman in that they enable her to continue the use of cosmetics. **1962** L. H. CRISP *Clin. Immunol. & Allergy* 187 Hypoallergenic milk..is milk denatured by heating. **1972** *Times* 22 Aug. 6/5 Ashley clothes fit in with hypoallergenic cosmetics,..ethnic dress, conservation and home grown food.

hypobaric (həipobæ·rik), *a.* *Med.* [f. HYPO- 4 + Gr. βαρ-ύς heavy + -IC.] Of a solution for spinal anæsthesia: having a lower density than the cerebro-spinal fluid.

1930, etc. [see *HYPERBARIC *a.* a]. **1971** P. C. LUND *Princ. & Pract. Spinal Anesthesia* vii. 300 When administering hypobaric local anesthetic solutions in large volumes..the rate of injection is relatively unimportant but the positioning very important.

hypocentre (həi·posentəɹ). Also (*U.S.*) **-center.** [f. HYPO- 2 + CENTRE *sb.*] **1.** The focus of an earthquake, the point within the earth where it originates.

1905 C. DAVISON *Study Rec. Earthquakes* i. 3 The region within which the displacement occurs is sometimes called the hypocentre, but more frequently the seismic focus, or simply the focus. **1938** [see *CENTRUM 2]. **1972** *Sci. Amer.* May 58 The first seismic waves to leave the region of the break (the hypocenter) are waves of alternate compression and rarefaction.

2. = *ground zero* (*GROUND *sb.* 18).

1960 *Observer* 29 May 26/1 The Hiroshima survivors are unusual in having experienced a single whole-body exposure, the dose varying according to distance from the hypocentre (point directly beneath centre) of the explosion. **1962** J. F. LOUTIT *Irradiation* ii. 75 The incidence of leukemia..is high for those who were exposed near to the hypocenter.

hypochondria. **2.** Add to def.: Now identical in meaning with HYPOCHONDRIASIS (q.v. in Dict. and Suppl.); it remains the commoner term among laymen. (Further examples.)

1839 F. WINSLOW *Physic & Physicians* II. 155 Cowper's madness, most undoubtedly, originated from some bodily ailment... All through his disorder, the digestive organs were impaired... Such, indeed, was the true source of his hypochondria. **1899** J. MACPHERSON *Mental Affections* viii. 147 The two affections [*sc.* melancholia and hypochondria] are different in so far as the emotional depression, combined with the intellectual disturbance of melancholia, is not present in hypochondria. **1928** R. D. GILLESPIE in *Guy's Hosp. Rep.* LXXVIII. 409 The term hypochondria appears always to have referred to preoccupation with complaints of illness, and usually of bodily illness. **1955** *Times* 4 Nov. 5/4 The fantastic array of purges, nerve sedatives, tonics,..processed this, irradiated that, and impregnated the other shows that hypochondria is widespread.

hypochondriac, *a.* and *sb.* **A.** *adj.* **1.** Delete ? *Obs.* and add earlier and later examples.

1599 R. SURFLET tr. *A. Du Laurens' Discourse Preservation of Sight* 125 The Hypochondriake disease..[is] a drie and hote distemperature of Mesenterium, the liver and spleene. **1965** W. G. KLOPFER in B. B. Wolman *Handbk. Clin. Psychol.* 830/2 Hypochondriac symptoms commonly occur and may, if no discernible cause for the symptom is found, be due to exaggerated needs for attention and other psychological desires.

b. (Earlier example.)

1599 R. SURFLET tr. *A. Du Laurens' Discourse Preservation of Sight* 131, I have seene two Hypochondriake persons..raging mad.

B. *sb.* **1.** (Further examples.)

1888 R. ROOSE *Nerve Prostration* I. xv. 348 The hypochondriac is always dwelling upon his symptoms, and constantly talks about his health. **1916** M. H. FUSSELL *Monographic Med.* V. XII. 781 Hypochondria is more likely to be confounded with neurasthenia [than with hysteria], but as the hypochondriac believes he is ill and constantly talks about his symptoms, and hysteriacs also..complain of various symptoms there may be confusion. **1932** H. S. WALPOLE *Fortress* IV. v. 693 She was no hypochondriac, but from a kind of outside consideration she summoned her forces. Had she a headache? Did her eyes smart? How was her throat? **1955** *Sci. Amer.* Apr. 104/3 He was a hypochondriac and a crank, chronically dyspeptic and unamiable. **1973** *Times* 27 Sept. 15/1 George S. Kaufman..was..an obsessive card player, compulsive womanizer, necrophobe, hypochondriac.

2. (Earlier example.)

1599 R. SURFLET tr. *A. Du Laurens' Discourse Preservation of Sight* 126 The other part where the Hypochondriake breedeth, is the liver.

hypochondriasis. Add to def.: Now regarded as a condition characterized by a morbid preoccupation with one's bodily health together with unfounded beliefs and exaggerated anxieties about real or imagined ailments, usually the symptom of a neurotic disorder. (Further examples.)

1798 A. CRICHTON *Inquiry Mental Derangement* II. 339 Hypochondriasis, therefore, is chiefly characterized by erroneous notions relating to the patient's own frame, and by painful corporeal feeling. **1855** *Asylum Jrnl.* I. 214/1 Hypochondriasis and melancholia monomania were not clearly distinguished by physicians until recent years. **1905** M. CRAIG *Psychol. Med.* vii. 114 Though hypochondriasis is usually found in patients in whom no known bodily disease can be diagnosed, it may be associated with organic disease, the hypochondriacal symptoms being.. the patient's misinterpretation of true physical signs. **1956** H. P. LAUGHLIN *Neuroses Clin. Pract.* x. 451 Sigmund Freud very early included Neurasthenia, along with Anxiety Neurosis and Hypochondriasis, as the so-called 'actual neuroses'. *Ibid.* xi. 500 In some psychiatric quarters today Hypochondriasis is regarded simply as a symptom complex, or merely as a manifestation which is present as a part of many other emotional reactions, rather than as a separate diagnostic entity.

hypochoristic (həi:pokǒri·stik), *a.* Erroneous (but increasingly used) form of HYPOCORISTIC *a.*

1931 G. STERN *Meaning & Change of Meaning* 262 Bob has been considered a hypochoristic shortening of *Robert*. **1933** L. BLOOMFIELD *Lang.* xxiii. 424 It seems..that forms like *Bob, Dick* existed as common nouns, perhaps with symbolic connotation, before they were specialized as hypochoristic forms of *Robert, Richard*. **1958** A. S. C. ROSS *Etymology* 167 The hypochoristic form..of a name.. is the 'familiar' form, as MnE *Liz* for *Elizabeth*.

hypochromasia (həi:pokrǭumē̆i·ziă). *Med.* [mod.L., f. HYPO- 4 + Gr. χρῶμα, χρώματ- colour: see -IA¹.] = *HYPOCHROMIA b.

1929 R. B. H. GRADWOHL tr. *Schilling's Blood Picture* III. 258 Regarding the color index, marked decrease is called hypochromasia. **1930** H. DOWNEY in E. T. Bell *Text-bk. Path.* xxviii. 586 In addition to hypochromasia, poikilocytosis may develop. **1942** M. M. WINTROBE *Clin. Hematol.* ii. 68 Hypochromia, hypochromasia or 'achromia' are terms used to describe cells in which this normal pallor is increased.

hypochromatic (həipokrǭumæ·tik), *a.* *Med.* [f. HYPO- 4 + CHROMATIC *a.*] **a.** Of the skin: deficient in pigment.

1894 in GOULD *Dict. Med.* 593/2.

b. = *HYPOCHROMIC *a.* 1.

1929 R. B. H. GRADWOHL tr. *Schilling's Blood Picture* II. 112 Remissions, often very extensive, are striking... The blood picture may appear normal again and may become temporarily hypochromatic. **1930** H. DOWNEY in E. T. Bell *Text-bk. Path.* xxviii. 585 In this type of anemia one may find all gradations from cases showing abundant signs of regeneration to the simple hypochromatic types. **1971** V. F. FAIRBANKS et al. *Clin. Dis. Iron Metabolism* (ed. 2) iii. 129 Hypochromatic changes in the erythrocyte ..may occur in a variety of unrelated disorders including iron-deficiency anemia.

hypochromatosis (həi:pokrǭumătǭu·sis). *Cytology.* [mod.L., f. HYPO- 4 + CHROMAT(O-, CHROMAT(IN + -OSIS.] An abnormally small amount of chromatin or number of chromosomes in a cell or nucleus.

1913 [see *HYPERCHROMATOSIS 2]. **1919** *Encycl. Medica* (ed. 2) VI. 311 Hypochromatosis is a deficiency in chromatin—thus nuclear hypochromatosis is the condition of the nucleus of a cell in which there is a reduction in the number of chromosomes.

hypochromia (həipǒkrǭu·miă). *Med.* [f. as next + -IA¹.] **a.** (See quot. 1890.)

1890 GOULD *New Med. Dict.* 201/2 *Hypochromia*, abnormal pallor or transparency of the skin, occurring in certain skin diseases. **1968** A. J. ROOK et al. *Textbk. Dermatol.* I. xxii. 724 The lesions always retain their tendency to merge together producing progressive atrophy of the skin with tendency to hypochromia and achromia.

b. A hypochromic condition of the blood or of a red blood cell (see *HYPOCHROMIC *a.* 1).

1931 [see *HYPERCHROMIA]. **1966** J. W. LINMAN *Princ. Hematol.* v. 177 Because of the clear-cut relationship between iron deficiency and hypochromia, it has often been assumed erroneously that this morphologic abnormality must be the result of iron lack or a block in iron incorporation. **1972** *Nature* 10 Mar. 71/2 Lead seems to affect red cells chiefly by..interfering with haemoglobin synthesis, as judged by..hypochromia of the red cells.

hypochromic (həipokrōu·mik), *a.* [f. Hypo- 4 + Gr. χρῶμ-α colour + -ic.] **1.** *Med.* Characterized by or designating a colour index less than one, or red blood cells that contain less hæmoglobin than normal and show an increased central pallor; esp. in *hypochromic anæmia.*
1924 T. R. Waugh in *Can. Med. Assoc. Jrnl.* XLVII. 114/1 Such anaemias consequently show considerable variation in their blood pictures. If the response is slight ..the red cells are small, and stain poorly,..and the color index is low. This type is therefore hypochromic. **1935** Whitby & Britton *Disorders of Blood* vi. 126 Hypochromic anæmia, especially, is more often a symptom than a disease. **1958** G. C. de Gruchy *Clin. Haematol.* ii. 42 In the tail of the film the cells are often distorted and flattened, and hypochromic cells may actually appear normochromic. **1966** J. W. Linman *Princ. Hematol.* v. 157 Most hypochromic anemias are caused by iron lack.
2. Characterized by or exhibiting a decrease in the extent to which light (usually, ultra-violet radiation) is absorbed; chiefly in *hypochromic effect.*
1946 *Ann. Rep. Progr. Chem.* XLII. 118 For substituents attached to the carbonyl group, the effects are quite different, both $\lambda_{max.}$ and $\epsilon_{max.}$ being usually decreased (hypso- and hypochromic effects). **1959** *Jrnl. Amer. Chem. Soc.* LXXXI. 6003/1 In general the oxidation of a methylthio group to a methyl sulfone involves a hypsochromic shift of from 10 to 40 mμ... This hypsochromic shift usually is accompanied by a definite hypochromic effect. **1968** [see *hyperchromic *a.* 2].
Hence **hy:pochromi·city**, the property of absorbing less (ultra-violet) light.
1958 *Nature* 29 Nov. 1502/1 In view of the zero hyperchromicity of polyguanylic acids at alkaline *p*H's, the variation of hypochromicity with *p*H was examined for a number of derivatives. **1960** D. Shugar in Chargaff & Davidson *Nucleic Acids* III. xxx. 59 When the extinction of a given oligonucleotide is lower than that of its constituent mononucleotides, it is 'hypochromic' or exhibits 'hypochromicity'.

hypocone (həi·pokōun). *Zool.* [f. Hypo- + Cone *sb.*¹] An external cusp on the inner back corner of a mammalian upper molar tooth.
1888 H. F. Osborn in *Amer. Naturalist* XXII. 1072 The first 'secondary' cusps (hypocone—hypoconid), added to the upper and lower molars of the primitive triangle, modify the crown from a triangular to a quadrangular shape. **1891** [in Dict. s.v. Hypo-]. **1933** A. S. Romer *Vertebr. Paleontol.* xii. 248 In the upper molars the tooth tends to square itself up usually by the addition of a fourth cusp, the hypocone, at the inner back corner. **1968** R. Zangerl tr. *Peyer's Compar. Odontol.* 187 In the upper jaw a talon formed in that a second lingual cusp developed next to the protocone, a so-called hypocone.

hypoconid (həipokōu·nid). *Zool.* [f. *hypocon(e + *-id⁵.] A cusp on a mammalian lower molar tooth corresponding to the hypocone on an upper molar.
1888 H. F. Osborn in *Amer. Naturalist* XXII. 1075 There is no evidence as to the origin of the hypoconid, which as a rule preceded the hypocone. **1919** J. H. Mummery *Microsc. Anat. Teeth* i. 36 In man the trigonid is represented by the protoconid and metaconid only,.. and the five cusps are made up of these and three cusps of the talonid—the hypoconid, entoconid, and hypoconulid. **1968** R. Zangerl tr. *Peyer's Compar. Odontol.* 187 In the lower jaw three cusps developed on the talonid: counting labio-lingually, hypoconid, hypoconulid, and entoconid. **1970** *Nature* 25 July 356/1 The crest connecting the entoconid and the hypoconid was continuous.

hypoconulid (həipokōu·niulid). *Zool.* [f. *hypocon(e + *-ul-* + *-id⁵.] An intermediate cusp between the principal ones on the heel of a mammalian lower molar tooth.
1897 H. F. Osborn in *Amer. Naturalist* XXXI. 1002 The talonid widened into a basin-like shelf supporting an outer cusp, the 'hypoconid'; an intermediate cusp, the 'hypoconulid', and an inner cusp, the 'entoconid'. *Ibid.* 1003 Why notice such a detail as the posterior intermediate cusp or hypoconulid? **1933** A. S. Romer *Vertebr. Paleontol.* xii. 248 A hypoconulid may also appear in the heel. **1972** *Nature* 24 Mar. 180/1 *Oligopithecus* as well as the other Fayum catarrhines share the distinct lingually placed and somewhat prominent hypoconulid.

hypocoristic, *a.* Add: Also as *sb.*
1889 in *Cent. Dict.* **1930** *Times Lit. Suppl.* 1 May 361/2 Mere riddles..where there is no question of..stop-voicing of hypocoristics. **1953** K. Jackson *Lang. & Hist. Early Brit.* ii. 555 The AS. personal names *Cata, Ceatta*..are from Pr[imitive] W[elsh] hypocoristics. **1957** R. W. Zandvoort *Handbk. Eng. Gram.* ix. ii. 303 The technical term or attributive 'pet' is 'hypocoristic' (adj. and noun).

hypodermic, *a.* Add: **1.** (Earlier example.)
1863 *Lancet* 17 Oct. 444/1 Many..speedily furnished the journals with their experience of the 'hypodermic treatment'.
b. Also, a hypodermic injection or syringe.
1893 *Funk's Stand. Dict.*, *Hypodermic*, a hypodermic syringe or injection. **1907** I. McIsaac *Primary Nursing Technique* vii. 104 Hypodermics are given in the chest or fleshy part of the arm or thigh. **1969** *Daily Tel.* 11 Apr. 28/5 He..preferred a hypodermic of nicotine to a cigarette inhaled. **1970** *Ibid.* (Colour Suppl.) 18 Sept. 18 Divers.. began to use large hypodermics designed to inject a 10 c.c. dose of formalin, enough to kill a starfish within hours.
c. *fig.* (adj. and sb.).
1901 *Harper's Mag.* CII. 786/1 Novelty is at a ruinous premium, and amusement a hypodermic to be taken in large doses, ever increased. **1903** *Monthly Rev.* Jan. 44 The admission of clergymen to the schools at certain hours for the purpose of administering a sort of hypodermic injection of religion is futile. **1936** W. Plomer *Visiting Caves* 46 The hypodermic steeple Ever ready to inject The opium of the people. **1959** *Listener* 12 Feb. 300/3 Admirers may find his appreciation of Waugh's more hypodermic humour respectful rather than hilarious. *Ibid.* 5 Nov. 796/1 The professional intimate, the confidential heart-worm with the hypodermic technique, is one of the horrors of television.

hypodermically, *adv.* (Earlier example.)
1863 C. Hunter in *Lancet* 17 Oct. 444/1 The alkaloids of belladonna, aconite, and other medicines were first employed hypodermically by myself.

hypodigm (həi·podəim, -dim). *Taxonomy.* [ad. Gr. ὑπόδειγμα example.] The material on which the description of a species is based.
1940 G. G. Simpson in *Amer. Jrnl. Sci.* CCXXXVIII. 418, I therefore propose the term 'hypodigm' (pronounced hy'-podim, from the Greek ὑπόδειγμα, 'token, example'). All the specimens used by the author of a species as his basis for inference, and this should mean all the specimens that he referred to the species, constitute his hypodigm of that species. *Ibid.*, The hypodigm, whether it include one specimen or a thousand, is a sample from which the characters of a population are to be inferred. **1953** E. Mayr et al. *Methods & Princ. Syst. Zool.* xii. 237 A hypodigm is all the available material of a species. This term is mentioned here because it is occasionally used in the paleontological literature. **1963** Davis & Heywood *Princ. Angiosperm Taxon.* i. 11 The hypodigm changes with our knowledge of the species. **1972** *Nature* 24 Mar. 180/1 The sixteen teeth which make up the hypodigm of this taxon [sc. *Purgatorius unio*] have, however, been correctly allocated.

hypogamy (həipǫ·gămi). *Anthrop.* [f. Hypo- + *-gamy.] The marriage of a woman into a lower caste or into a tribe of lower standing than her own. Hence **hypo·gamous** *a.*, pertaining or relating to hypogamy. Cf. *hypergamy.
1946 J. H. Hutton *Caste in India* v. 48 Hypogamy, on the other hand, is associated with a bride-price. *Ibid.*, Hypogamous marriages..are *pratiloma*, against the grain, that is, against what is natural or proper, since the status of the bride is in this case higher than that of the bridegroom. **1949** R. K. Merton *Social Theory* (1951) i. 60 This pattern, which we may call caste hypogamy, is not institutionalized, but it is persistent. **1956** R. Pieris *Sinhalese Social Organization* v. i. 177 A very rare case of *hypogamy*, that is, women marrying *below* their caste. **1957** *New Statesman* 4 May 566/3 The fictions of D. H. Lawrence have several examples of the lady marrying downwards (hypogamy, in anthropological vocabulary).

hypoid (həi·poid), *a.* and *sb. Mech.* [Said to be a contraction of *hyperboloid* or of *hyperbolic paraboloid* (in reference to the shape of the teeth on the wheel).] **A.** *adj.* **a.** Applied to a kind of gear similar to a spiral bevel gear but having the pinion offset from the centre-line of the wheel, so that it can be used to connect shafts whose axes do not intersect; it is commonly employed in motor vehicles to transmit the power from the propeller shaft to the axle.
1926 *Jrnl. Soc. Automotive Engin.* June 575/1 Hypoid-gears are tapered gears having offset axes. **1935** *Times* 2 Oct. 6/5 The rear axle is fully-floating, and final drive is by hypoid spiral bevel gears. **1937** *Jrnl. Applied Mech.* IV. A-31 The behavior of lubricants in the region of boundary lubrication has become of added importance due to the recent adoption of hypoid gears in automobiles. **1941** F. D. Jones *Engin. Encycl.* I. 676 The chief advantages of hypoid gears are noiseless operation, increased load-carrying capacity, the possibility of high reduction and low numbers of teeth, long life, and high efficiency. **1969** K. Ball *Rover 2000 1963–1969 Autobook* vii. 64/1 The unit contains a hypoid crownwheel and pinion for the final drive from the power input shaft to the half shafts.
b. Suitable for or employing a hypoid gear.
1937 *S.A.E. Jrnl.* (*Transactions*) XLI. 557/2 Six or seven months ago the lubrication of hypoid axles in the field was a much more serious problem than it is today. *Ibid.* 563/1 One lubricant submitted by a manufacturer as an extreme-pressure hypoid lubricant was found to be a 'straight' mineral oil when analyzed. **1963** *Times* 3 May 10/6 A four-speed, all synchromesh gearbox..incorporates the hypoid final drive. **1969** G. M. Mitchell *Jowett Javelin, Jupiter 1947–1953 Autobook* vi. 65/1 For lubrication of this axle, hypoid lubricant only should be used.
B. *sb.* A hypoid gear.
1935 R. Trautschold *Stand. Gear Bk.* ix. 144 Hypoids are machined in practically the same manner as spiral gears. **1962** D. W. Dudley *Gear Handbk.* ii. 12 Sufficient offset of the hypoid permits straddle mounting of the pinion and the gear.

hypolimnion (həipoli·mniǫn). Pl. **hypolimnia.** [f. Hypo- 2 + Gr. λιμνίον, dim. of λίμνη lake.] The lower, cooler layer of water below the thermocline in a stratified lake.
1910, 1936 [see *epilimnion]. **1957** G. E. Hutchinson *Treat. Limnol.* I. v. 341 The hypolimnia of all lakes. **1960** *New Scientist* 31 Mar. 773/3 The hypolimnion.. has become deoxygenated as the vegetation decomposes and provides ideal living conditions for sulphate-reducing bacteria. **1971** *Nature* 26 Feb. 596/1 In summer months.. the upper layers warm up more quickly than the lower regions and a sharp division in temperature—a thermocline—is formed...The reservoir or lake becomes divided into a lower, anaerobic cool layer or hypolimnion and an upper, warm aerobic epilimnion.
Hence **hypolimne·tic** [cf. Gr. λιμνήτης living in marshes], **hypoli·mnial** *adjs.*, of or within the hypolimnion.
1928 *Proc. Linn. Soc.* CXL. 101 On account of the depth the hypolimnetic body of water is great compared with the epilimnetic. **1940** Hypolimnial [see *epilimnion]. **1964** *Oceanogr. & Marine Biol.* II. 126 Alsterberg (1927) suggests that hypolimnetic (upper) water movements are confined to thin horizontal laminae, with currents in alternate directions.

hypomania (həipomē·niă). *Psychiatry.* [mod.L., ad. G. *hypomanie* (E. Mendel *Die Manie* (1881) ii. 38): see Hypo- 4.] A minor form of mania, often part of the manic-depressive cycle, characterized by elation and a feeling of well-being together with quickness of thought.
1882 *Jrnl. Nerv. & Mental Dis.* IX. 432 This description ..belongs not to acute mania properly so-called, but to the hypomania of Mendel,..or the so-called subacute mania of asylum reports. **1892** D. H. Tuke *Dict. Psychol. Med.* I. 618/2 *Hypomania*, a name given to subacute attacks of mania, which are marked by an initial melancholia, retardation of the flow of ideas, and consequently as incoherence, restlessness, increased self-consciousness with delusions of a grandiose character and perversion of sexual instincts. **1904** T. Johnstone tr. *Kraepelin's Lect. Clinical Psychiatry* vii. 60 This combination of symptoms..we designate by the name of Mania, or, if the individual disturbances are only slightly developed,.. by that of Hypomania. **1912** B. Hollander *First Signs of Insanity* xviii. 225 The chief mental characteristic of this disease, known as hypomania, is a loss of mental inhibition and consequently a rapid, ill-regulated, and easily disconnected train of thought. **1927** Henderson & Gillespie *Text-bk. Psychiatry* vii. 121 While such conditions as hypomania, acute mania, and delirious mania can readily be recognised,..the differentiation of these states is not by any means clean cut. **1963** N. H. Pronko *Textbk. Abnormal Psychol.* x. 367 The mildest degree of manic excitement is termed hypomania, and concerns essentially speed and direction of thought rather than thought content. **1971** *Brit. Med. Bull.* XXVII. 77/2 Is a mild hypomania, with some euphoria and flight of ideas, but no delusions or gross behavioural disturbance, to be called psychotic or not?

hypomaniac (həipomē·niæk). [f. prec., after Maniac *a.* and *sb.*] = *hypomanic *sb.*
1910 B. Hollander *Mental Symptoms Brain Dis.* iii. 25 The good spirits of the hypomaniac seem, excepting for occasional slight abatement, to be inexhaustible. **1965** Rosen & Gregory *Abnormal Psychol.* xiv. 291/2 Like simple depressives, many hypomaniacs receive no treatment.

hypomanic (həipomæ·nik), *a.* and *sb.* [f. as prec., after *manic *a.* and *sb.*] **A.** *adj.* Of or affected with hypomania.
1927 Henderson & Gillespie *Text-bk. Psychiatry* vii. 126 The hypomanic elation showed the usual characteristics—overactivity (with erotic tendencies coming to the fore), an infectious and excessive gaiety, overtalkativeness..and lack of sustained attention. **1941** S. H. Kraines *Therapy Neuroses & Psychoses* xv. 359 Early in the attack she appeared to the casual observer to be merely an active, vivacious girl;..she was hypomanic only in relation to her accustomed and usual behavior. **1965** [see *cyclothymic* s.v. *cyclo-].
B. *sb.* A person affected with hypomania.
1932 *Brit. Jrnl. Psychol.* XXIII. 155 The expansive hypomanic greets us with the sunny smile of happiness. **1938** S. Beckett *Murphy* ix. 168 A hypomanic teaching slosh to a Korsakow's syndrome. **1954** W. Mayer-Gross et al. *Clin. Psychiatry* v. 196 Hypomanics are realistic, quick to grasp opportunities, versatile and often rather superficial.
Hence **hypoma·nically** *adv.*
1958 M. Argyle *Relig. Behaviour* ix. 107 Hysterics tend to become hypomanically excited at revival meetings.

hyponym (həi·ponim). [f. Hypo- + Gr. ὄνομα name.] **1.** *Taxonomy.* A name made invalid by the lack of adequate contemporary description of the taxon it was intended to designate.
1904 *Bull. Torrey Bot. Club* XXXI. 258 A specific or subspecific name is a hyponym when it has not been connected with a description identifiable by diagnostic characters or by reference to a type specimen, figure or locality. **1904** *Science* 25 Mar. 509/2 Some are hyponyms, never having been associated with a recognizable binomial species. **1946** D. B. Swingle *Textbk. Syst. Bot.* (ed. 3) ix. 225 A name not so described or identified is a hyponym and nonvalid for the group for which it was intended. *Ibid.* 226 Homonyms and hyponyms result in the formation of synonyms, for these names must be replaced by usable terms, leaving them as synonyms.
2. *Linguistics.* One of two or more words related by hyponymy.
1963 J. Lyons *Structural Semantics* iv. 69, I say that *scarlet* is a hyponym of, or is included in, *red.* **1965**

Language XLI. 509 Hyponyms are placed under the head word.

hyponymy (həipǫ·nimi). *Linguistics.* [f. Hypo- + Gr. ὄνομα after Synonymy, etc.] (See quot. 1963.)
1955 C. E. Bazell in *Litera* II. 34 There is a relation of hyponymy when one word may invariably be replaced by a second word, but not vice-versa, without change of meaning. **1962** F. W. Householder in Householder & Saporta *Probl. in Lexicogr.* 280 Thorough investigation of semantic structure is desirable, and where we have already sufficient knowledge to indicate relationships of incompatibility, hyponymy..etc., this should be done in the most economical way possible. **1963** J. Lyons *Structural Semantics* iv. 69 Hyponymy is the relation that holds, for instance, between *scarlet* and *red*, or between *tulip* and *flower*, in English...It may be defined in terms of unilateral implication. Thus, *X is scarlet* will be understood (generally) to imply *X is red*; but not conversely. **1970** A. Cameron et al. *Computers & O.E. Concordances* 91, I would draw your attention to the sketchy attempt to state the sense-relations of Old English words in terms of ..hyponymy (or inclusion of sense: ⊂ sign).

hy:poparathy·roidism. *Med.* [f. Hypo- + Parathyroid + -ism.] A condition in which there is an abnormally low level of parathyroid hormone in the blood, resulting in hypocalcæmia and hyperphosphatæmia with consequent tetany and other signs of neuro-muscular excitability.
1910 Ochsner & Thompson *Surg. & Path. Thyroid & Parathyroid Glands* xx. 345 In a patient suffering greatly from subtetanic hypoparathyroidism..tetany has for two years been averted. **1938** Smith & Gault *Ess. Path.* ix. 746/1 The relationship of tetany to hypoparathyroidism is dependent primarily upon a drop in blood calcium. **1966** Wright & Symmers *Systemic Path.* II. xxxii. 1126/1 The commonest cause of hypoparathyroidism is, of course, accidental removal of the glands during thyroidectomy. **1970** *Med. Ann.* 199 The tragedy of hypoparathyroidism lies in the lifelong burden of its control.
Hence **hy:poparathy·roid** *a.*, of, resulting from, or having hypoparathyroidism.
1910 Ochsner & Thompson *Surg. & Path. Thyroid & Parathyroid Glands* xii. 205 To the internist the question of a hypoparathyroid etiology in the various tetanies became of interest. **1950** Cope & Hamlin in S. Soskin *Progr. Clin. Endocrinol.* ii. 110 Following operative correction of hyperparathyroidism, patients may have chronic tetany of continued hypoparathyroid origin. **1970** C. N. Graymore *Biochem. Eye* iv. 309 Lens lesions indistinguishable from hypoparathyroid cataract.

hypopharynx. Add: **2.** *Anat.* The lower, laryngeal part of the pharynx (into which the larynx opens), extending from the epiglottis to the top of the œsophagus.
1907 *Lancet* 25 May 1421/1, I propose to give a description of a new and excellent method of inspecting the laryngeal part of the pharynx, the hypopharynx as it is also called. **1954** W. H. Hollinshead *Anat. for Surgeons* I. viii. 405/2 The laryngeal pharynx or hypopharynx extends from just above the level of the hyoid bone superiorly to the cricoid cartilage inferiorly, narrowing rapidly to become continuous with the esophagus. **1962** *Lancet* 28 Apr. 901/2 The patients were placed in the lateral head-down position owing to bleeding from the hypopharynx.

†**hypo·physal,** *a. Obs.* [f. Hypophys(is + -al.] = *hypophysial *a.*
1877 *Gray's Anat.* (ed. 8) p. cxxx, Others refer the hypophysal part of the pituitary body to epiblastic elements derived from the buccal part of the epiblast only. **1892** C. S. Minot *Human Embryol.* (1897) xxvi. 574 The hypophysal diverticulum now elongates and its upper end expands to a considerable vesicle.

hypophyseal, var. *hypophysial *a.*

hypophysectomy (həi:pofise·ktŏmi). *Surg.* [f. Hypophys(is + *-ectomy.] Excision of the hypophysis.
1909 Reford & Cushing in *Bull. Johns Hopkins Hosp.* XX. 106/1 The effects of total canine hypophysectomy. **1939** M. A. Goldzieher *Endocrine Glands* I. 270 Adult amphibia show atrophy of both ovaries and testes after hypophysectomy. **1962** *Lancet* 15 Dec. 1235/1 Women with advanced breast cancer who subsequently fail to respond to adrenalectomy or hypophysectomy excrete significantly smaller amounts of urinary 11-deoxy 17-oxosteroids than do women who respond. **1967** S. Taylor et al. *Short Textbk. Surg.* xv. 189 Adrenalectomy and hypophysectomy are being performed much less often now than a few years ago, because remissions are unpredictable and often short-lived.
Hence **hy:pophyse·ctomize** *v. trans.*, to deprive of the hypophysis, perform hypophysectomy on; usu. as **hy:pophyse·ctomized** *ppl. a.*
1910 *Bull. Johns Hopkins Hosp.* XXI. 131/1 Two of the partially hypophysectomized animals survived for 5 months. **1912** H. Cushing *Pituitary Body & its Disorders* i. 11 Of a litter of three puppies..one, the largest, was partially hypophysectomized, one was fed daily..with powdered extract of the whole gland..: the third and smallest was kept as a control. **1934** C. J. Wiggers *Physiol. Health & Dis.* (1935) lxvii. 1055 Hypophysectomized tadpoles neither grow nor metamorphose. **1936** *Jrnl. Path. & Bacteriol.* XLII. 403 A tendency to maintenance of body weight after grafting of pituitary cultures into hypophysectomised rats has been taken as evidence of replacement of pituitary function. **1966** W. S. Hoar in Harris & Donovan *Pituitary Gland* I. vi. 246 The Amphibians are relatively easy to hypophysectomize. **1970** *Nature* 31 Oct. 464/2 Administration of growth hormone to hypophysectomized rats essentially prevented the appearance of any hypercholesterolaemia.

hypophysial (həipofi·ziăl), *a.* Also **hypophyseal** (həipofi·ziăl, həi:pofisī·ăl). [f. Hypophysi(s + -al; *hypophyseal* by alteration of *hypophysial.*] Of or pertaining to the hypophysis.
1882 *Jrnl. Linn. Soc.* (*Zool.*) XVI. 133 In the Mammalian series I have to observe that, in the lower and smaller members, as the brain loses in relative size and complexity, the pineal or conarial and pituitary or hypophysial bodies and connections show a relatively larger size. **1893** *Jrnl. Path. & Bacteriol.* I. 360 In the dog there is a more or less symmetrical lateral arrangement of the hypophyseal fissures. **1909** H. Cushing in *Jrnl. Amer. Med. Assoc.* 24 July 254/1 Symptoms of hypophyseal origin doubtless occur in association with many diseases in which they are overlooked. **1912** —— *Pituitary Body & its Disorders* iii. 233 In clinical conditions of hypophyseal deficiency somnolence is a conspicuous feature. **1943** G. W. Corner in *Science* 15 Jan. 68 In the 1864 revision of Webster..epiphyseal appears before epiphysial; in the revision of 1909 apophyseal enters the lists; in the current (1934) revision hypophyseal appears, and the spelling with -eal is preferred in all three cases...Possibly 'epiphyseal' goes back to the days when the noun was often written 'epiphyse', or possibly some writers thought it was advisable to make the adjectives from the genitives of the Greek nouns (e.g., *epiphyseos, hypophyseos*). Much more likely, however, the spelling has been influenced by an American trend in the pronunciation... At the present time (and as far as my observation goes, for decades past) American speakers almost universally place the primary accent on the fourth syllable, e.g., hypophyséal... In spite of the fact that '-eal' is philologically irregular, I make bold to suggest that it be adopted as standard in American scientific writing,..in conformity with our well-nigh general pronunciation of the three words in question. **1955** G. W. Harris *Neural Control Pituitary Gland* v. 124 The pathway by which a stimulus passes from the hypothalamus to the anterior pituitary appears to be by means of the hypophysial portal vessels. **1962** *Gray's Anat.* (ed. 33) 1038 The hypophysis receives its blood supply from the internal carotid artery through a superior and an inferior hypophyseal artery on each side.

hypophysiotrophic (həi:pofiziotrōu·fik), **-tropic** (-trōu·pik, -trǫ·pik), *a. Physiol.* [f. Hypophysi(s + -o + *-trophic, *-tropic.] Regulating the activity of the hypophysis.
1962 B. Halász et al. in *Jrnl. Endocrinol.* XXV. 147 Histological and functional evidence has been obtained which defines what may be called the 'hypophysiotrophic' region of the hypothalamus. **1968** C. B. Jørgensen in Barrington & Jørgensen *Perspectives in Endocrinol.* viii. 489 The existence of more diffuse 'hypophysiotropic' regions within the hypothalamus has been supported by the results of recent studies. **1970** *Nature* 25 Apr. 322/2 These results allow us to claim the preparation, for the first time, of a hypothalamic hypophysiotropic releasing factor.

hypophysis. Add: Pl. **hypophyses** (həipǫ·fisīz). **3.** [First used by S. T. Soemmerring (*De Corporis Humani Fabrica* (rev. ed., 1798) IV. 70).] Add to def.: A small endocrine organ that is attached or adjacent to the hypothalamus in the brain of vertebrates, has two lobes distinct in origin, nature, and function (cf. *adeno- and *neurohypophysis), and produces a number of hormones, some of which regulate the activity of other endocrine organs; in man it lies within a cavity of the sphenoid bone under a covering of *dura mater* and is attached by the infundibulum to the undersurface of the brain. (Examples.)
The application of the term has varied. Some writers have restricted it to the anterior or the posterior lobe only (differentiating the hypophysis from the pituitary body); others have taken it to include not only the two lobes but also the infundibulum.
1825 A. Monro *Elem. Anat. Human Body* II. vi. ii. 371 The Pituitary Gland, or Hypophysis, is a small oblong-shaped body, which is inclosed by the dura mater, and situated in the tela sphenoidalis. **1877** W. Turner *Introd. Human Anat.* I. v. 213 This ventricle is prolonged downwards into a funnel-shaped process, the infundibulum, which is connected with the pituitary body, or hypophysis cerebri. **1898** W. H. Howell in *Jrnl. Exper. Med.* III. 246 The hypophysis cerebri is usually described as consisting of two lobes. One, the large anterior lobe, is distinctly a glandular structure... Properly speaking the term hypophysis cerebri should be restricted to this lobe, and this significance is now given to it by morphological writers, although in human anatomy it is still commonly employed to include the so-called posterior lobe as well. *Ibid.* 247 Extracts were made of both the hypophysis cerebri, or anterior lobe, and the infundibular body. **1899** F. H. Gerrish *Text-bk. Anat.* 534 The hypophysis has been found greatly enlarged in cases of giantism, and hence has been supposed to sustain a relation to the stature of the individual. **1906** J. P. McMurrich *Devel. Human Body* (ed. 2) xiv. 428 At its extremity the hypophysis comes in contact during the fifth week with the enlarged extremity of Rathke's pouch..and applies itself closely to the posterior surface of this..to form with it the pituitary body. **1915** A. M. Paterson *Man. Embryol.* II. i. 110 The pituitary body has a double origin... The tuber cinereum, infundibulum, and posterior lobe (hypophysis cerebri) are outgrowths from the ventral aspect of the diencephalon. The anterior lobe of the pituitary body is derived from Rathke's pouch. **1919** W. B. Bell *Pituitary* 2 The word 'hypophysis', strictly speaking, refers to the epithelial portions of the Pituitary—the pars anterior and the pars intermedia—and should only be used in this connexion. **1926** G. R. de Beer *Compar. Anat. Pituitary Body* ii. 26 The term *pituitary body* should be used to denote the well-defined anatomical unit consisting of four parts—anterior, intermedia, nervosa, and tuberalis. Other structures, such as the infundibulum and the tuber cinereum, may be grouped with the pituitary body to form a functional unit, the pituitary complex. The term *hypophysis* is often loosely used as synonymous with pituitary body, but wrongly, since it definitely refers only to the epithelial constituent of the body. **1936** *Jrnl. Morphol.* LX. 127 A cytological study of the hypophyses of more than 100 female bats. **1940** *Res. Publ. Assoc. Res. Nerv. & Mental Dis.* XX. 22 The mammalian hypophysis consists of three major divisions: lobus glandularis, lobus nervosus and the infundibulum or neural stalk. This classification has been recommended by the International Commission on Anatomical Nomenclature (1935). **1944** E. T. Bell *Text-bk. Path.* (ed. 5) xxviii. 755 The hypophysis exercises a certain amount of control over the other glands of internal secretion and the growth and activity of various organs by means of specific hormones which it elaborates. **1954** A. White et al. *Princ. Biochem.* xlv. 957 The hypophysis is one of the most important endocrine glands in the body, exerting a profound influence over other endocrine structures and thereby regulating a large portion of the endocrine activity of the organism. **1960** B. I. Balinsky *Introd. Embryol.* xviii. 452 The agent necessary for activating the thyroid gland is produced in the anterior (epidermal) lobe of the hypophysis.

hypoplastic (həipoplæ·stik), *a. Med.* [f. *hypoplasia* (s.v. Hypo- II) + *-plastic.] Of an organ or tissue: undersized at maturity owing to insufficient growth; *hypoplastic anæmia*, anæmia that is due to an insufficient production of red blood cells by the bone-marrow.
1877 tr. *H. von Ziemssen's Cycl. Pract. Med.* XVI. 543 The sexual organs will remain relatively hypoplastic, or will be late in arriving at functional maturity. **1906** C. P. Emerson *Clin. Diagn.* v. 515 By hypoplastic anæmia is meant one due to insufficient blood formation. **1915** *Brit. Dental Jrnl.* XXXVI. 174 In studying the etiology of hypoplastic teeth, development and calcification may be divided into three distinct periods. **1918** *Amer. Jrnl. Med. Sci.* CLVI. 49 The aorta was described as hypoplastic in 101 cases...In 71 of these cases hypoplasia was determined by actual measurement of the relative width of the pulmonary artery and the aorta. **1938** Diamond & Blackfan in *Amer. Jrnl. Dis. Children* LVI. 464 In the past seven years we have had the opportunity to see an intermediate type of anemia—hypoplastic rather than completely aplastic. This condition has been characterized by slowly progressive anemia.. with the production of a small and inadequate number of reticulocytes from bone marrow which shows moderate hypoplasia. **1961** L. Martin *Clinical Endocrinol.* (ed. 3) vii. 199 The breasts were large but the nipples hypoplastic. **1963** M. C. G. Israëls *Diagn. & Treatm. Blood Dis.* viii. 59 Senile hypoplastic anæmia is really a premature diminution of bone-marrow activity.

hyposensitization (həi:posensitəizē·ʃən). *Med.* [f. Hypo- 4 + Sensitization, after *hypersensitization.*] The process of diminishing the sensitivity of a hypersensitive individual (as by the introduction of the allergen in a series of gradually increasing doses); a state of diminished sensitivity so produced.
1922 R. A. Cooke in *Jrnl. Immunol.* VII. 241 On account of the confusion that must result from the use of the well defined term 'desensitization' to designate clinically lessened sensitiveness in allergy it is suggested that the latter be referred to as a 'hyposensitization'. **1941** J. W. Thomas *Allergy in Clin. Pract.* i. 11 When complete avoidance is impossible or impractical, hyposensitization with inhalants..should be considered. **1953** F. K. Hansel *Clin. Allergy* xxxix. 750 The terms desensitization and hyposensitization are used interchangeably to designate a lessening of sensitization or an increase of tolerance. **1971** O. Swineford *Asthma* ix. 148 Treatment of allergic asthma by injecting extracts of allergens is called hyposensitization.
Hence **hypose·nsitize** *v. trans.*, to subject (an individual) to such a process; also *absol.*; **hypose·nsitized** *ppl. a.*
1931 W. T. Vaughan *Allergy & Appl. Immunol.* xx. 249 If contact with the allergen then produces no symptoms, he is..desensitized. If his symptoms are improved but not entirely relieved, he is hyposensitized. **1939** —— *Primer of Allergy* vii. 67 We can hyposensitize with practically any of the allergens which are inhaled (except chemicals). **1944** E. Urbach *Allergy* x. 249 Coca has succeeded in hyposensitizing a patient with turpentine allergy..by injecting turpentine dissolved in sterile almond oil every seven days. **1971** O. Swineford *Asthma* ix. 148 How does one hyposensitize a patient?

Hypospray (həi·pospreī). Also **hypospray.** [f. Hypo(dermic *a.* + Spray *sb.*²] The proprietary name of a kind of jet injector (see *jet *sb.*³ 9).
1947 *Life* 24 Nov. 65/1 Known as the Hypospray, this new device 'blasts' a microscopically small jet of medicinal fluid into the body tissues. **1948** *U.S. Pat. Off. Gaz.* 7 Sept. 69/1 *Hypospray.* R. P. Scherer Corporation, Detroit, Mich... For Hypodermic Injection Devices. **1948** *Amer. Practitioner* III. 206/1 The Hypospray may be used to

give either intramuscular or subcutaneous injections. **1956** *Ann. Rheumatic Dis.* XV. 231/2 A practical feature of the hypospray injection is the simplicity of the technique. The physician can inject several joints in a short time without the preliminary preparation required for needle injection. **1969** *Ibid.* XXVIII. 61/2 The Hypospray was considerably less painful and more acceptable to patients.

hypostasis. Add: **7.** *Genetics.* [Back-formation from *hypostatic *a.* 3] The inhibition of the expression of one gene by the action of another non-allelic (epistatic) gene.
1917 *Genetics* II. 615/1 (Index), Hypostasis. **1962** I. H. Herskowitz *Genetics* vii. 53/1 Genes whose detection is hampered by nonallelic genes are said to be hypostatic, i.e., to exhibit hypostasis. As dominance implies recessiveness, so epistasis implies hypostasis.
8. *Linguistics.* The citing of a word, word-element, etc., as an example, a model, etc. Also, a linguistic element thus referred to.
1933 L. Bloomfield *Lang.* ix. 148 Hypostasis, the mention of a phonetically normal speech-form, as when we say, 'That is only an *if*', or 'There is always a *but*', or when we talk about 'the word *normalcy*' or 'the name *Smith*'. One may even speak of parts of words, as..'the suffix *-ish* in *boyish*'. **1940** *Language* XVI. 238 When the sign is combined with a morpheme or is used in another grammatical category (hypostasis)..it is said to be characterized or positivized. **1961** *Lingua* X. 175 All I want to say is that the subject of this paper is..how to analyse linguistic signs occurring in *suppositio materialis* or (as I shall henceforth say) in hypostasis. **1963** *Ibid.* XII. 211 Sometimes hypostasis forms are used in other syntactical functions than subject, object or part of an adverbial adjunct, but at least those three functions are the most frequent in English. **1967** K. L. Pike *Lang. in Rel. Human Behavior* (ed. 2) 108 Spelling words aloud is a form of hypostasis. *Ibid.* 484 This is treating sentences in hypostasis.

hypostatic, *a.* Add: **3.** *Genetics.* Of, causing, or affected by hypostasis (sense *7). Const. *to.*
1907 [see *epistatic *a.*]. **1961** A. Müntzing *Genetic Res.* vi. 58/2 We may also say that B is hypostatic to A. **1965** J. A. Serra *Mod. Genetics* I. iii. 62 The effect of one gene, the epistatic gene, is superimposed on the effect of another, the hypostatic gene, either by obscuring the phenotypic effect of the hypostatic gene, or by inhibiting its effect.

hypotension (həipote·nʃən). *Med.* [f. Hypo- 4 + Tension *sb.*] Abnormally low 'tension' or pressure of a bodily fluid. **a.** Of arterial blood.
1893 [see *hypertension 1 a]. **1938** *Lancet* 31 Dec. 1510/2 Hypotension is sometimes accompanied by.. postural giddiness. **1966** Wright & Symmers *Systemic Path.* I. iii. 110/1 Moderate degrees of hypotension may.. occur in patients with chronic wasting diseases, especially when they are confined to bed for long periods.
b. Of the intra-ocular fluid.
1909 A. M. Ramsay *Diathesis & Ocular Dis.* v. 46 The easiest..method..is to puncture the sclerotic with a broad needle... This causes immediate hypotension. **1969** Duke-Elder & Jay in S. Duke-Elder *Syst. Ophthalm.* XI. x. 724 The subject of ocular hypotension has excited much less interest and speculation than that of hypertension.

hypotensive (həipote·nsiv), *a. Med.* [f. prec. + -ive.] Of, exhibiting, or associated with hypotension, esp. of the blood; tending to lower the blood pressure.
1904 T. C. Janeway *Clin. Study Blood-Pressure* vi. 153 Typhoid fever is more frequently hypotensive in the average case than the other acute diseases, pneumonia least. **1927** *Medicine* VI. 147 It is also true that many hypotensive subjects possess great bodily vigor. **1951** A. Grollman *Pharmacol. & Therapeutics* xi. 213 It is also used..to combat a drop in blood-pressure occurring during spinal anesthesia and in other acute hypotensive states due to vasomotor failure. **1961** *Lancet* 29 July 222/2 Patients with severe arterial hypertension were given hypotensive drugs only if they had been so treated previously. **1966** *Ibid.* 26 Mar. 677/1 A few hours after admission the patient became hypotensive and vasopressor drugs had to be used. **1973** *Times* 2 Oct. 15/2 One of the most common fatal side-effects is that it becomes hypotensive (low blood pressure).

hypothalamo-hypophysial (həipoþæ:lămo,-həipofi·ziăl), *a.* Also **-eal.** [f. *hypothalam(us + -o + *hypophysial *a.*] Of, pertaining to, or connecting the hypothalamus and the hypophysis; applied *spec.* to a tract of nerve fibres that runs from the hypothalamus to the neurohypophysis.
1934 *Arch. Neurol. & Psychiatry* (Chicago) XXXII. 217 The entity of the fibers traversing the infundibulum and entering the posterior lobe of the hypophysis is called the hypothalamo-hypophyseal fasciculus ('faisceau hypothalamo-hypophysaire'). **1950** G. W. Harris in *Jrnl. Physiol.* CXI. 347 The hypophysial stalk has two main component parts, the hypothalamo-hypophysial tract of nerve fibres associated with the neurohypophysis, and the hypophysial portal vessels associated with the adenohypophysis. **1971** N. G. Sutton *Anat. Brain & Spinal Medulla* v. 78 A series of hypothalamohypophyseal portal vessels descend in the form of numerous blood sinusoids to end among the cells of the adenohypophysis. **1972** *Nature* 5 May 15/1 These connexions..may prove to be an important component of the pathways which subserve the marked effect of light on the hypothalamo-hypophysial activity.
Also **hypothala:mico-hypophy·sial, -eal** *a.*, in the same sense.
1934 *Trans. Coll. Phys. Philadelphia* II. 223 Diabetes insipidus and Fröhlich's syndrome are due to disturbances of this hypothalamico-hypophyseal mechanism. **1944** J. Hoffman *Female Endocrinol.* xv. 226 A neuro-endocrine unit which has recently attracted much attention is the so-called hypothalamico-hypophyseal system. **1961** *Lancet* 2 Sept. 522/2 Lesions have been found in the hypothalamicohypophyseal system in only a few cases of primary diabetes insipidus.

hypothalamus (həipoþæ·lămʊs). *Anat.* [mod.L. (W. His (at Waldeyer's suggestion) 1893, in *Arch. f. Anat. u. Physiol.* (*Anat. Abth.*) 159), f. Hypo- + Thalamus.] The lower part of the diencephalon of the brain in vertebrates, lying below and in front of the thalamus and forming the floor and part of the wall of the third ventricle; in mammals it acts as the chief co-ordinating region of the autonomic nervous system and helps to regulate the hormonal activity of the adenohypophysis.
Writers vary in counting as part of the hypothalamus (*a*) the infundibulum and neurohypophysis, and (*b*) the subthalamic tegmental region.
1896 *Jrnl. Compar. Neurol.* VI. 309 (*in list of anatomical names*) Hypothalamus. **1899** L. F. Barker *Nervous Syst.* xlvi. 666 These various fibres..pass through the pedunculus cerebri..to reach the diencephalon, where most of them in all probability terminate in the hypothalamus, in the thalamencephalon, or in the nucleus lentiformis. **1909** *Gray's Anat.* (ed. 17) 849 The hypothalamus..includes the subthalamic tegmental region and the structures which form the greater part of the floor of the third ventricle, viz. the corpora mamillaria, tuber cinereum, infundibulum, pituitary body, and optic commissure. **1942** L. H. Hyman *Compar. Vertebr. Anat.* (ed. 2) xiv. 434 The hypothalamus reaches its greatest development in fishes, where it is an important correlation center for olfactory, gustatory, and other sensory impulses. **1944** J. Hoffman *Female Endocrinol.* xv. 231 Proof that the hypothalamus regulates the gonadotropic function of the anterior hypophysis is still lacking. **1956** A. C. Guyton *Textbk. Med. Physiol.* l. 633/1 When a surgeon operates in the region of the hypothalamus, simply tugging on the tissues is likely to cause such intense changes in heart rate, blood pressure, blood glucose level, body temperature, etc., that the operative mortality is approximately 40 per cent from these factors alone. **1962** T. W. Torrey *Morphogenesis Vertebr.* xviii. 510 In all vertebrates it is the hypothalamus which exercises control over such truly involuntary actions as temperature regulation, sexual reactions, breathing rate, emotional responses, and the rhythm of sleep. **1964** Parker & Haswell *Text-bk. Zool.* (ed. 7) II. 245 [In the dogfish, *Scyliorhinus*] the hypothalamus in the floor of the diencephalon is well-developed and may, as in 'higher' vertebrates, be concerned with the regulation of various unconditioned reflexes and visceral functions. **1968** A. van Tienhoven *Reprod. Physiol. Vertebr.* viii. 207/1 The hypothalamus thus performs the task of being an intermediary between the nervous system, of which it is part, and the endocrine system, to which it can send 'hormonal' information. **1968** *Times* 10 Oct. 8/5 One part of the brain which contains noradrenaline is the hypothalamus, which has the function of controlling body temperature by monitoring the temperature of the blood and directing the body systems to lose or conserve heat. **1972** *Sci. Amer.* Nov. 24/1 The pituitary gland is attached by a stalk to the region in the base of the brain known as the hypothalamus.
Hence **hypotha·lamic** *a.*, of or pertaining to the hypothalamus.
1899 L. F. Barker *Nervous Syst.* xlvi. 683 In man the fibres of the main mass of white matter in the hypothalamic region..do not pass through the hilus into the thalamus. **1938** J. Beattie in W. E. Le Gros Clark et al. *Hypothalamus* 100 Sleep..is due to a damping-down of hypothalamic activity. **1942** O. Larsell *Anat. Nerv. Syst.* iii. 28 The thalamus and hypothalamus are bounded from each other by the hypothalamic sulcus. **1954** T. L. Peele *Neuroanat. Basis Clin. Neurol.* xiv. 310/2 Hypothalamic lesions..have resulted in a loss of cyclical sexual activities and genital atrophy. **1968** A. van Tienhoven *Reprod. Physiol. Vertebr.* viii. 249/1 Some of the hypothalamic hormones, e.g. oxytocin and vasopressin, are stored in the neurohypophysis.

hypothecate, *v.* Add: **2.** *trans.* = Hypothesize *v.* 2.
1906 *Nature* 7 June 136/1 Mr. Cowell hypothecated a resisting medium through which the earth travels. **1912** R. Fry in *Gt. State* ix. 271 Mr. Wells's Modern Utopia ..hypothecates a vast superstructure of private trading. **1915** E. B. Holt *Freudian Wish* i. 4 One will best..not hypothecate to this end any such thing as 'psychic energy'. **1920** E. Pound *Let.* 12 Sept. (1971) 161 You are talking through your hat when you suggest that I..was ever ass enough to have picked 'La Figlia' for the fantastic occasion you hypothecate. **1952** *Pediatrics* IX. 724 One had to hypothecate the existence of a mutation of organisms.

hypothermia (həipoþɔ̄·ɪmiă). *Med.* Also in anglicized form **hypothermy** (həi·poþɔ̄ɪmi) (*rare*). [f. Hypo- 4 + Gr. θέρμ-η heat + -ia¹.] The condition of having a body temperature substantially below the normal, either as a result of natural causes or artificially induced (e.g. for cardiac surgery).
1886 in *Syd. Soc. Lex.* **1887** A. M. Brown *Contrib. Animal Alkaloids* 143 Intoxication by animal alkaloids is accompanied by hypothermia. **1898** W. S. Barlow *Man. Gen. Path.* x. 441 Emphysema and some other forms of pulmonary disease..are associated with a slight degree of hypothermia. **1903** *Jrnl. Nerv. & Mental Dis.* XXX. 574 The same toxic agent which acts on the nervous system producing the condition of epilepsy may be the cause of the hypothermy in these cases. **1937** *Brit. Encycl. Med. Pract.* III. 499 The reduction of the metabolic processes in cretinism is shown, as in myxoedema, by hypothermia. **1955** *Sci. News Let.* 18 June 389/2 When patients are given 'frozen sleep', or hypothermia, for operations inside the heart, they can be quickly warmed to normal by diathermy. **1964** *Courier-Mail* (Brisbane) 18 July 2 His problem probably had been 'hypothermia'—low body temperature—quite common in cold weather in elderly people who do not keep themselves sufficiently warm. **1966** *New Statesman* 11 Nov. 697/1 The experts concluded that 'hypothermia is a serious though unspectacular condition with a very high mortality rate'.
So **hypothe·rmic** *a.* [cf. Gr. ὑπόθερμος somewhat hot], of or exhibiting hypothermia.
1898 W. S. L. Barlow *Man. Gen. Path.* x. 440 A general sluggishness of nerve and of muscle in hypothermic persons and animals is always noticeable. **1948** A. R. Moritz in W. A. D. Anderson *Path.* vi. 143 The severity of injury caused at any given temperature tends to be proportional to the duration of the hypo- or hyperthermic episode. **1961** *Lancet* 2 Dec. 1216/2 The complete absence of residual signs of cerebral dysfunction was particularly noteworthy in one patient who remained hypothermic and unconscious for seven days.

hypothetic. **B.** *sb.* **2.** Restrict *obs.* to senses in Dict. and add: *sb. pl.* The making of hypotheses; hypothesizing.
1890 A. Lang *Life Sir Stafford Northcote* (ed. 2) II. xiii. 89 That belongs to the science of hypothetics, and anyone may sincerely believe that matters might have been kept quiet by a sagacious and well-informed policy. **1958** *New Biol.* XXV. 15 Perhaps we shall be helped to estimate its importance by an exercise in hypothetics.

hypothe:tico-deduc·tive, *a. Philos.* [f. Hypothetic(al *a.* + -o + Deductive *a.*, prob. as ad. It. *ipotetico-deduttivo* hypothetical-deductive (M. Pieri 1900, in *Mem. d. R. Accad. d. Sci. di Torino* XLIX. 173).] Making use of or consisting in the testing of the consequences of hypotheses (i.e. seeing whether the consequences are consistent with observation) as a means of determining whether the hypotheses themselves are false or can be accepted.
1912 *Philos. Rev.* XXI. 642 The type of reasoning that takes place in the hypothetico-deductive fields of thought. **1929** H. A. Wolfson *Crescas' Critique of Aristotle* i. 25 The Talmudic hypothetico-deductive method of text interpretation. **1949** tr. *H. Weyl's Philos. Math. & Nat. Sci.* I. i. 27 Pure mathematics, in the modern view, amounts to a general hypothetico-deductive theory of relations; it develops the theory of logical 'molds' without binding itself to one or the other among the possible concrete interpretations. **1952** J. O. Wisdom *Found. Inference Nat. Sci.* xxiv. 223 Many difficulties in the nature of the causal relation and inductive inference are obviated by using the scheme of explanation provided by the hypothetico-deductive system, according to which a hypothesis is accepted when it has been 'tempered' by severe testing and has not been falsified. **1953** R. B. Braithwaite *Scientific Explanation* i. 9 It is this hypothetico-deductive method applied to empirical material which is the essential feature of a science. **1957** G. H. von Wright *Logical Probl. Induction* (ed. 2) 208 The theory of induction cannot, in the name of the hypothetico-deductive method, be banished from holding a prominent place within the methodology of science. **1963** *Listener* 12 Sept. 378/1 This alternative interpretation of the nature of the scientific process..is sometimes called the 'hypothetico-deductive' interpretation, and this is the view which Professor Karl Popper..has persuaded us is the correct one. **1971** J. Z. Young *Introd. Study Man* xxi. 283 It is by adopting the 'hypothetico-deductive' system that men have been able to make forecasts much more far-reaching and reliable than those of any animal.
Hence **hypothe:tico-dedu·ctively** *adv.*, by the hypothetico-deductive method.
1953 R. B. Braithwaite *Scientific Explanation* ix. 299 The latter of these propositions is establishable hypothetico-deductively without reference to the establishment of the former.

hypothy·roidism. *Med.* [f. Hypo- 4 + Thyroid *a.* (*sb.*) + -ism.] A condition in which the level of thyroxine in the blood is abnormally low resulting in a decreased metabolic rate and which when severe causes cretinism (if the condition was congenital) and myxœdema (if acquired).
1905 Gould *Dict. New Med. Terms* 304/1 Hypothyroidism. **1909** G. Dock in Osler & McCrae *Syst. Med.* VI. xviii. 447 The known results of hypothyroidism or athyroidism are: 1. Congenital myxœdema... 2. Infantile myxœdema... 3. Spontaneous myxœdema of adults. **1955** *Sci. News Let.* 24 Sept. 207/3 The tragic effects of hypothyroidism in babies, such as dwarfing and mental retardation, may be prevented by early diagnosis and adequate treatment. **1961** L. Martin *Clin. Endocrinol.* (ed. 3) iii. 111 Transient neonatal hypothyroidism may also result from maternal overdosage with antithyroid drugs during pregnancy. **1966** Wright & Symmers

Systemic Path. II. xxxi. 1103/2 It is customary to refer to the milder forms as hypothyroidism and to reserve the term myxœdema for the severer clinical varieties.
Hence **hypothy·roid, -thyroi·dic** *adjs.*
1909 G. Dock in Osler & McCrae *Syst. Med.* VI. xviii. 448 Many other conditions, especially certain forms of infantilism and obesity, are classed by some writers as hypothyroid states. *Ibid.* 455 Hypothyroid infants. **1916** *Med. Times* (N.Y.) 7 July 207/1 The hypothyroidic eye is dull, seemingly small, apparently sunken, expressionless, in short, featurally insignificant. **1968** Passmore & Robson *Compan. Med. Stud.* I. xxv. 18/2 The hypothyroid person is characteristically cold, sluggish and constipated and often has mild anaemia.

hypotonia (həipotōu·niă). Freq. in anglicized form **hypotony** (həipǫ·tŏni). [mod.L., f. Hypo- 4 + Gr. τόν-ος Tone *sb.* + -ia¹.]
1. *Ophthalm.* A state of reduced pressure of the intra-ocular fluid.
1886 *Syd. Soc. Lex.*, *Hypotony*... Applied by Nagel to the globe of the eye when less resistant than normal. **1892** A. Duane tr. *Fuchs's Text-bk. Ophthalm.* II. vii. 360 Diminution of the intra-ocular pressure (hypotonia) is found in very diverse affections of the eyeball. **1951** H. S. Sugar *Glaucomas* xxiv. 371 Hypotony, particularly after the trephining operation, may result from too rapid drainage into the conjunctiva. **1966** S. Lerman *Basic Ophthalm.* iv. 246 Another major cause of postoperative hypotony..is a serous detachment of the ciliary body.
2. *Med.* The condition (in muscle or muscular tissue) of being hypotonic.
1886 *Syd. Soc. Lex.*, *Hypotony*, defective tone of a part, or an organ, or a structure. **1907** *Practitioner* Oct. 547 Undue muscular strain on the ankle, knee, hip, and spine, which, together with hypotonia, tends to break down the long arch. **1914** [see *hypertonia]. **1933** W. R. Brain *Dis. Nervous Syst.* i. 12 A lesion of this path causes hypotonia, which is manifested in flaccidity and a diminished resistance to stretching of the affected muscles. **1962** *Lancet* 6 Jan. 22/1 Her growth was stunted and the muscles showed a marked hypotonia with genu recurvatum.

hypotonic (həipǫtǫ·nik), *a.* [f. Hypo- 4 + Tonic *a.*] **1. a.** *Physiol.* Of a solution: having a lower osmotic pressure than some particular solution (usually that in a cell, or a bodily fluid). Const. *to.*
1895 *Jrnl. Physiol.* XVIII. 114 The passage of a salt from a hypotonic fluid into the blood-plasma. **1946** *Nature* 9 Nov. 665/2 The fluid obtained in this way was hypotonic to the cœlomic fluid. **1951** H. Davson *Textbk. Gen. Physiol.* vii. 163 If the plasma surrounding the cells is steadily diluted (i.e., made hypotonic), the latter increase in volume until they finally burst. **1970** [see *hypertonic *a.* 2].
b. *Ophthalm.* Of the eye: having a reduced intra-ocular pressure.
1918 R. H. Elliot *Glaucoma: Textbk. for Student* ii. 19 The case of a hypotonic eye (with a Schiötz reading corresponding to a tension of 8 to 15 mm. of Hg.) such as may be met with after a trephining or other sclerectomy operation. **1966** S. Lerman *Basic Ophthalm.* iv. 246 A hypotonic eye indicates that a leaking wound may be present.
2. *Med.* Exhibiting or characterized by diminished tone or tension (in muscle or muscular tissue).
1904 [see *hypertonus 2]. **1908** *Practitioner* Oct. 560 The lower extremities have been thin, powerless, and extraordinarily hypotonic. **1966** [see *hypertonic *a.* 1].

hypotonicity (həi:potǫni·siti). [f. prec. + -ity.] The condition of being hypotonic.
a. *Physiol.* Of a solution (see *hypotonic *a.* 1 a): the extent to which a solution has a lower osmotic pressure than some other.
1906 *Amer. Jrnl. Physiol.* XV. 367 Osmotic changes in the direction of hypotonicity. **1939** A. Krogh *Osmotic Regulation Aquatic Animals* 66 Enid Edmonds..found a slight but definite hypotonicity, viz. ocean water 622 mM., blood of the crab 577 mM. **1972** *Lancet* 2 Dec. 1160/2 Hypotonicity of the plasma was observed only on the 3rd and 4th postoperative days, when the daily intake of isotonic dextrose exceeded 1500 ml.
b. *Med.* Of muscle or muscular tissue (see *hypotonic *a.* 2); = *hypotonia 2.
1910 A. Abrams *Diagn. Therapeutics* v. 711 This hypotonicity of the muscles..may also be noted in executing other unaccustomed movements of the muscles of the hand. **1934** C. J. Wiggers *Physiol. Health & Dis.* (1935) xlix. 765 Gastric hypotonicity does not necessarily interfere with emptying the stomach, because the tone of the pyloric sphincter is also reduced, and consequently less intragastric pressure is required to expel the chyme. **1959** E. B. Smith et al. *Princ. Human Path.* lv. 912/1 The hypercalcemia causes interesting clinical findings incident to..hypotonicity of muscle.

hypotonus (həipǫtōu·nǝs). [f. Hypo- 4 + Tonus.] **1.** *Ophthalm.* = *hypotonia 1. Now *rare.*
1891 F. P. Foster *Med. Dict.* III. 1955/2 *Hypotonus*, that condition of the eye in which the intra-ocular tension is below normal, without being of necessity accompanied by any organic disease of the eyeball. **1918** R. H. Elliot *Glaucoma* ii. 22 If too large and too free a channel is cut in the ocular tunic, the escape of fluid is so rapid that a condition of hypotonus results.
2. *Med.* = *hypotonia 2.
1904 *Rev. Neurol. & Psychiatry* II. 776 The legs showed very marked hypotonus. **1928** J. F. H. Dally *Low Blood Pressure* ii. 22 Hypotonia (vascular hypotonus) represents a dynamic and physical state of diminished tonus of the smooth muscle in the walls of arteries and veins. **1939** W. Haymaker tr. *Bing's Textbk. Nervous Dis.* xi. 318 Hypotonus of the quadriceps is sometimes so marked in tabes that the heel can be brought up to the buttock.

hy:poventila·tion. *Physiol.* [f. Hypo- 4 + Ventilation.] A diminished or insufficient exposure of the lungs to oxygen, resulting in a reduced oxygen content of the blood or an increased carbon dioxide content (or both).
1932 *Arch. Neurol. & Psychiatry* XXVIII. 580 To combat this physiologic hypoventilation, the animals were subjected to artificial respiration of normal frequency and depth. **1954** A. White et al. *Princ. Biochem.* xxvii. 715 Hypoventilation of whatever origin (morphine poisoning, pneumonia, pulmonary edema, etc.) has the opposite effect and lowers extracellular pH. **1961** L. Martin *Clinical Endocrinol.* (ed. 3) ii. 55 A cardio-respiratory syndrome has been described by Berlyne (1958) in cases of extreme obesity, of which alveolar hypoventilation is the basic defect.

hypsi-. Add: **hy·psiconch** (-kǫŋk), **hypsiconchic** (-kǫ·ŋkik), **-conchous** (-kǫ·ŋkəs) *adjs. Anthropol.* [Gr. κόγχ-ος eye-socket], having orbits that are high in relation to their width, with an orbital index of 89 (formerly, 85) or more; so **hy·psiconchy**, the condition of being hypsiconchic; **hypsithe·rmal** *a. Geol.* [ad. It. *ipsotermico* (A. Chiarugi 1936, in *Nuovo giorn. bot. ital.* XLIII. 55)], designating that period of the geologically recent past (*c* 7000 to *c* 600 B.C.) when relatively warm conditions prevailed in the northern hemisphere; also *absol.*
1920 H. H. Wilder *Lab. Man. Anthropometry* 67 Hypsiconch..[Orbital index of] 85+. **1960** M. F. A. Montagu *Introd. Physical Anthropol.* (ed. 3) 606 Hypsiconch..[Orbital index of] 85·0 – ×. **1902** Hypsiconchic [see *chamæcephalic* adj. s.v. *chamæ-]. **1960** J. Comas *Man. Physical Anthropol.* vii. 409 Hypsiconchic..[an orbital index of] 89·0 and over. **1885** *Jrnl. R. Anthropol. Inst.* XIV. 71 Hypsikonchous..[Orbital index of] 85·1 and over. **1965** *Dorland's Med. Dict.* (ed. 24) 717/2 *Hypsiconchous*, having an orbital index over 85. **1902** *Biometrika* I. 460 In both sexes there is sensible correlation between the palate and orbital indices, hypsiconchy being associated with brachystaphyline characters. **1957** Deevey & Flint in *Science* 1 Feb. 182/2 The long, warm interval spanned by Danish pollen zones V through VIII, which has been dated from approximately 7000 B.C. to approximately 600 B.C., we propose to call the *hypsithermal* interval. We have changed the spelling of Chiarugi's *ipsotermico* to conform with the English style of Greek adjectives and to express the customary distinction between *hypsi-*, high, and *hypso-*, a height. **1957** *Bull. Geol. Soc. Amer.* LXVIII. 1895 The period from 7000 B.C. to 600 B.C. is now known as the hypsithermal interval, rather than by its former but less suitable designations (postglacial climatic optimum, thermal maximum). **1967** Martin & Wright *Pleistocene Extinctions* 135 The onset of widespread aridity [in Australia], which Gill (1955) equates with the Climatic Optimum or Hypsithermal of the Northern Hemisphere. **1968** R. W. Fairbridge *Encycl. Geomorphol.* 1051/2 Modern deposits may have obtained their maximum growth during the hypsithermal phase of postglacial time.

hypsochrome (hi·psokrōum), *a.* and *sb.* [f. as next.] **A.** *adj.* = next.
1892 *Jrnl. Soc. Chem. Industry* 31 Oct. 807/2 Groups causing deepening are distinguished as 'bathochrome', whilst those to which heightening of the colour is due are termed 'hypsochrome'. **1908** *Jrnl. Chem. Soc.* XCIV. I. 477 (*heading*) Measurement of the effect of certain hypsochrome and bathochrome groups on the colour of azobenzene. **1917** Fort & Lloyd *Chem. Dyestuffs* xiv. 130 Acylation..always gives hypsochrome properties, whether acting upon amines or hydroxy groups.
B. *sb.* A hypsochromic atom or group.
1917 Fort & Lloyd *Chem. Dyestuffs* xiv. 130 Groups or atoms that cause an intensification of colour when introduced into compounds have been called Bathochromes, and those that decrease the colour of a compound Hypsochromes.

hypsochromic (hipsokrōu·mik), *a.* [ad. G. *hypsochrom* (M. Schütze 1892, in *Zeitschr. f. physik. Chem.* IX. 136), f. Hypso- + Gr. χρῶμ-α colour: see -ic.] Causing or characterized by a lightening of colour, or a shift of the absorption spectrum towards shorter wavelengths.
1892 *Jrnl. Chem. Soc.* LXII. I. 562 Definite atoms or groups of atoms on entering a molecule cause..a characteristic rise [*read* fall] ('bathochromic groups'), or fall [*read* rise] ('hypsochromic groups'), of the tint. **1932** S. J. Gregg tr. *Eggert's Physical Chem.* 559 The hydrogen atom itself is hypsochromic in nature, for on removing the unsaturated character of a chromophoric group by adding hydrogen to it the absorption band migrates toward the ultra-violet. **1946** [see *hypochromic *a.* 2]. **1957** *Jrnl. Biol. Chem.* CCXXIX. 716 In addition to the reduction in extinction there is a hypsochromic shift of 2 to 3 mμ accompanying the formation of the polymers. **1972** Rys & Zollinger *Fund. Chem. & Applic. Dyes* x. 153 As a rule, the vatting of indigo derivatives..produces a hypsochromic shift.

hypsograph (hi·psograf). [f. Hypso- + -graph.] = *hypsographic curve.*
1937 *Geogr. Jrnl.* XC. 60 The verticality in all the hypsographs at low elevations is explained by the high coastal cliffs and deep V-shaped valleys which prevail.

hypsography. Add: Hence **hypsogra·phic** *a.*, hypsographical; *hypsographic curve*, a curve showing the area or proportion of the earth's (solid) surface, or of a part of it, above any given elevation or depth.
1895 *Geogr. Jrnl.* V. 577 The author points out how the generalized hypsographic curve of the Earth's surface defines the continental plateau. **1937** [see *clinographic *a.*]. **1971** *Nature* 16 July 181/2 Kuenen was able to show from the world hypsographic curve that under present conditions a eustatic rise of 100 m would flood about ⅕ to ¼ of the continents.

hypsometric, *a.* Add: *hypsometric curve* = *hypsographic curve.*
1924 J. G. A. Skerl tr. *Wegener's Orig. Continents & Oceans* ii. 28 If the whole earth be divided into square kilometres and these are arranged in a series according to their height above sea-level, the well-known..hypsometric curve of the earth's surface is obtained. **1954** W. D. Thornbury *Princ. Geomorphol.* xxi. 531 A hypsometric curve obtained in this way permits comparison of forms of drainage basins of different sizes and altitudes.

hyraceum (həiərē·siǝm). Also **hyracium.** [f. mod.L. Hyrax, once used as the name of a genus including these animals.] A secretion produced by the African rock hyrax, *Procavia capensis*, formerly used as a fixative for perfume.
1866 Brande & Cox *Dict. Sci., Lit. & Art* II. 182/1 Hyracium. An article imported, as a substitute for castor, from the Cape of Good Hope, and derived from one of the species of *Hyrax*. **1892** P. L. Simmonds *Commercial Dict. Trade Products* (rev. ed.) Suppl. 462/2 Hyraceum, a secretion of the Cape badger, at one time considered to have medicinal properties. **1923** W. A. Poucher *Perfumes & Cosmetics* I. 3 Hyraceum is a secretion having a most disagreeable odour of excreta and urine, and is obtained from a species of monkey [*sic*]. It is occasionally used as a substitute for Castor... This material is not recommended. **1966** C. Sweeney *Scurrying Bush* ii. 34 The faeces of the rock rabbit..contains a substance called hyraceum, which is..incorporated in various perfumes. **1971** D. J. Potgieter et al. *Animal Life S. Afr.* 394/1 These deposits [of the dried urine of the dassie] contain hyraceum, a valuable material used in perfumery.

hyracotherium (həiə:răkoþīə·riǝm). Also **H-.** [mod.L., f. Hyraco- + Gr. θηρίον wild beast.] An extinct mammal of the genus so called; a primitive type of horse.
1840 R. Owen in *Proc. Geol. Soc.* III. 163 The resemblance of the molar division..in the new genus, for which the name of Hyracotherium is proposed, and the Chæropotamus, is sufficiently close. *Ibid.*, The incisor teeth with the ossa intermaxillaria are wanting in the specimen of the Hyracotherium. **1851** [in Dict. s.v. Hyraci-, hyraco-]. **1904** *Daily Chron.* 4 Jan. 9/1 Illustrations were given of a four and a five-toed horse, the extinct hyracotherium, no bigger than a Newfoundland dog. **1931** *Discovery* XII. 32/1 The evolution of the horse from the little hyracotherium of the Eocene period—a creature not much larger than a cat—is admirably shown. **1955** *Sci. News Let.* 12 Feb. 104/1 The ancient horse, *Hyracotherium*, was not very much of a horse by modern standards. It was about the size of a shepherd dog and, unlike modern horses, it had four functional toes.

Hyrcan (hǭ·ɪkăn), *sb.* and *a.* Also 6–7 Hircan, 6 Hyrcane. [ad. L. *Hyrcānus*, a. Gr. Ὑρκανός.] = next.
Hyrcan tiger, after L. *Hyrcanæ tigres* (Virgil *Æn.* IV. 367).
1567 W. Barker tr. *Xenophon's Discipline of Cyrus* IV. sig. O2ᵛ, Cyrus beholding the feates of the Medes and Hyrcanes, did as it were, rebuke him selfe. **1584** B. Rich *Second Tome Simonides* sig. C1, These Souldiers, like to Hircan Tigers, reuenge them selues on their owne bowelles. **1592** Daniel *Delia* xviii, But yet restore thy fearce and cruell minde, To Hyrcan tygers, and to ruthles Beares. **1602** T. Lodge tr. *Josephus' Workes* 755 Conspiring with the king of the Hyrcans to passe into Media. **1605** Shakes. *Macb.* III. iv. 101 The arm'd Rhinoceros, or th'Hircan Tiger. **1911** in W. James *Mem. & Stud.* xv. 395, I took the Hyrcan tiger by the scruff And tore him piecemeal.

Hyrcanian (həɪkē·niăn), *sb.* and *a.* [f. L. *Hyrcānia* (Gr. Ὑρκανία) + -an. Cf. prec.] **A.** *sb.* A native or inhabitant of Hyrcania, an ancient region on the Caspian Sea. **B.** *adj.* Of or pertaining to this region.
Hyrcania was the wild region *par excellence* to the ancients.
1567 W. Barker tr. *Xenophon's Discipline of Cyrus* VIII. sig. C8, Of the Hyrcanians, the Colonells son. **1596** Shakes. *Merch. V.* II. vii. 41 The Hircanion deserts. **1602** —— *Ham.* II. ii. 472 The rugged Pyrrhus like th'Hyrcanian Beast. **1607** [see *Cimbrian *a.* and *sb.*]. **1671** Milton *P.R.* III. 317 The Hyrcanian cliffs Of Caucasus. **1777** J. Richardson *Dict. Persian, Arabic & Eng.* I. 1172/2 The red Hyrcanian or Tabristan willow. **1820** Shelley *Ode to Liberty* viii, in *Prometh. Unb.* 213 From what Hyrcanian glen or frozen hill,..Didst thou lament

the ruin of thy reign? 1824 Carlyle Let. 4 Dec. (1909) II. 44 Frightful as the Hyrcanian Tiger. 1838 Penny Cycl. XII. 419/2 Josephus..mentions a king of the Hyrcanians in the time of Vespasian. 1885 Encycl. Brit. XVIII. 603/1 In [A.D.] 59 the Hyrcanian ambassadors were able to return home. Ibid., The Hyrcanians were still independent c. 155. 1973 R. L. Fox Alexander the Great ix. 141 In the Caicus valley..the colonists from distant Hyrcania..lived on in the land called the Hyrcanian Plain, where Cyrus had settled them two centuries earlier. Ibid. xi. 160 Medes, Armenians, Hyrcanians, North Africans and Persians themselves..fled through the stockade.

Hy-spy. (Later examples of the form 'I spy'.)
1890 W. James Princ. Psychol. II. xxiv. 421 It is the same instinct which leads a boy playing 'I spy' to hold his breath when the seeker is near. 1906 Folk-Lore XVII. 97 Key Hoy. Possibly a modification of 'I Spy'. 1963 Times 15 May 15/7, I lament the passing of our daily games of catch-as-catch-can in the cupboards, hide-and-seek behind the wardrobe and I-spy under the piano!

hyssop. Add: **4.** *hyssop-heavy, -laden* adjs.
1899 W. B. Yeats Wind among Reeds 52 The hyssop-heavy sponge, the flowers by Kidron stream. 1881 O. Wilde Poems 229 No need have we of hyssop-laden rod.

hysteresis. Delete '*Electr.*' and substitute for def.: A phenomenon observed in some physical systems, by which changes in a property (e.g. magnetization, or length) lag behind changes in an agent on which they depend (e.g. magnetizing force, or stress), so that the value of the former at any moment depends on the manner of the previous variation of the latter (e.g. whether it was increasing or decreasing in value); any dependence of the value of a property on the past history of the system to which it pertains. (Further examples.)
1881 J. A. Ewing in Proc. R. Soc. XXXIV. 40 All changes of magnetisation produced by slow or fast, continuous or discontinuous, changes of the magnetising force exhibit hysteresis. 1903 Nature 17 Dec. 160/2 In the physical theory of stress to strain, or twisting couple to twist produced, rocks exhibit a marked hysteresis. 1906 Biochem. Jrnl. II. 72 The slow change in osmotic pressure observed in colloidal solutions, such as that leads to normal conditions in gelatine after a short period at a higher temperature,..indicates a kind of hysteresis in such solutions, or a very slow return to equilibrium after the state of aggregation has been disturbed. 1931 E. S. Hedges Colloids xv. 197 The hysteresis in the case of agar is..far more striking, gelation occurring at about 40° and melting at about 85°. 1939 L. F. Bates Mod. Magnetism ix. 279 Temperature hysteresis, i.e. the ferromagnetic loses its ferromagnetism at one temperature on heating and regains it at a temperature below it on cooling. 1950 J. W. McBain Colloid Sci. xi. 166 There is a good deal of hysteresis, that is, a time lag between the cooling and the setting to be expected of the jelly. 1956 J. F. D. Smith in McPherson & Klemin Engin. Uses of Rubber vi. 130 An inspection of load-deflection diagrams reveals that the loading line may not be the same as the unloading line, for hysteresis plays an important part in flexometers. 1965 A. P. Boresi Elasticity Engin. Mech. iv. 103 Whenever a body exhibits the phenomenon of hysteresis—that is, of returning to its original size and shape only slowly or not at all—its behavior is not perfectly elastic.
b. = *hysteresis loss.
1896 Min. Proc. Inst. Civil Engin. CXXVI. 216 If B = 2,500 were taken as the limit of the cycle the hysteresis of this 'record' specimen would be only 0·16 watt per lb.
2. *Comb.*, as **hysteresis curve** = *hysteresis loop; **hysteresis loop**, a graph showing how the value of some property of a hysteretic system varies as the agent causing it is varied from one value to another and back again, having the form of a closed curve whose area is a measure of the loss of energy in the cycle; **hysteresis loss**, the energy dissipated as heat in a system as a result of hysteresis.
1894 Rep. Brit. Assoc. Adv. Sci. 577 The three stages of magnetic displacement each have a sharply defined position on the hysteresis curve. 1954 C. Zwikker Physical Prop. Solid Materials xii. 208 Permanent magnet materials are chosen for having a large area of loop on the B–H hysteresis curve. 1896 F. Bedell Princ. Transformer iii. 32 Curves of magnetization for a complete cycle, or 'hysteresis loops', as they are called, are shown in Fig. 33. 1897 A. G. Webster Theory Electr. & Magn. ix. 394 The hysteresis-loop..has an important physical significance. 1946 Rubber in Engin. (H.M.S.O.) iv. 69 The stress-strain curve for rubber on retraction does not follow the same course as during extension, but forms a hysteresis loop. 1966 McClintock & Argon Mech. Behavior Materials i. 6 Under cyclic straining, any kind of inelastic strain leads to a hysteresis loop on a stress-strain plot. 1893 Proc. R. Soc. LIV. 76 Great permeability does not necessarily imply small hysteresis losses. 1927 T. F. Wall Applied Magnetism xv. 233 A simple and rapid means for measuring the hysteresis loss in a transformer. 1962 A. Edwards in D. Hadfield Permanent Magnets vi. 294 If the flux density in a magnet continually increases and decreases in use, there is hysteresis loss at every cycle and some eddy-current loss.

hysteresial *a.* (Earlier and later examples.)
1887 Rep. Brit. Assoc. Adv. Sci. 1886 551 The hysteresial dissipation of energy per unit volume of iron is the same whether the magnetic circuit be open or closed. 1971 J. A. C. Harwood in C. M. Blow Rubber Technol. & Manuf. iii. 69 Stress softening, a hysteresial phenomenon observed at moderate and high extensions, is probably also viscoelastic in origin.

hysteretic (histəre·tik), *a.* [Prob. f. Hysteresis after such pairs as *synthesis, synthetic* and *prosthesis, prosthetic*; but cf. Gr. ὑστερητικός 'which comes on later'.] Of, pertaining to, or exhibiting hysteresis.
1892 Trans. Amer. Inst. Electr. Engin. XI. 25 Two other sets of determinations of the hysteretic loss of energy, for the frequency of 170 complete periods per second, were made on two laminated horse shoe magnets. Ibid. 43 This figure shows well the three characteristic forms of hysteretic curves. 1931 S. R. Williams Magnetic Phenomena i. 60, η is called the hysteretic constant or coefficient of hysteresis loss, which varies from one ferromagnetic body to another. 1958 C. L. Mantell Engin. Materials Handbk. xxxii. 10 Hysteretic properties of elastomers also affect their utility in applications where resilience is important. 1971 Nature 15 Jan. 155/3 A slow change in the concentration of a metabolite may still be rapid compared with the hysteretic adjustment of the activity of an enzyme that it controls.
Hence **hystere·tically** *adv.*, by means of or as a result of hysteresis.
1904 Electr. World & Engin. 30 July 163/2 The actual condenser dissipates energy hysteretically in the dielectric. 1956 Aeronaut Q. VII. 60 A general theory of small hysteretically damped vibration.

hysterical, *a.* and *sb.* Add: **B.** *sb.* **3.** = Hysteric *a.* and *sb.* B. 2.
1892 A. W. Pinero Magistrate iii. i. 136 (The sound of a shriek from Agatha and Charlotte.) Lugg...Don't notice them. They're hystericals. They're mild now to what they have been. 1922 M. Sadleir Excursions in Victorian Bibliogr. 2 This is a book about first editions, and will be read only by the initiate. If we be hystericals, we have at least our weakness in common. 1950 E. Hemingway Across River xxxix. 229 'I've stopped [crying],' she said. 'I'm not an hysterical.'

‖ **hysterica passio** (histe·rikă pæ·sio). [L., = 'hysteric passion'. See Hysteric.] = Hysteria 1.
1603 S. Harsnet Declaration Popish Impostures vi. 25 Maynie had a spice of the Hysterica passio, as seems from his youth. 1605 [see Mother sb. 12]. 1934 W. B. Yeats King of Gt. Clock Tower 22 But popular rage Hysterica passio dragged this quarry down. 1963 Listener 17 Oct. 626/2 Before watching 'The Mersey Sound'..I took tranquillizers to forestall a fit of hysterica passio.

hystericky, *a.* (Earlier U.S. example.)
1823 J. F. Cooper Pilot II. xiv. 239 In order that the women need not be 'stericky in squalls.

hysteriform, *a.*² (Examples.)
1887 W. Phillips Man. Brit. Discomycetes 384 Stictis hysterioides. Desm. Immersed, closed, hysteriform, then erumpent. 1957 Snell & Dick Gloss. Mycol. 79/1 Hysteriform..Elongated, boat-shaped and cleft, resembling the sporocarps of the genus Hysterium.

hystero-¹. Add: **hystero-rrhaphy** *Surg.* [*-rrhaphy] = hysteropexy (s.v. Hystero-¹).
1887 H. A. Kelly in Amer. Jrnl. Obstetr. XX. 34 It is my purpose here formally to propose, and to formulate rules for the adoption of a new operative procedure in the treatment of..intractable cases of retroflexion, and of prolapsus uteri... In accordance with the principles here laid down, the term 'Hysterorrhaphy' is used to define the suspension by suture of a viciously posed uterus. 1953 R. W. Te Linde Oper. Gynecol. (ed. 2) vi. 107 Ventrofixation, hysterorrhaphy and hysteropexy were terms applied to the earliest operation in which an attempt was made to suspend the retroplaced uterus. The chief interest of this operation is now historical.

hystrichosphere (hi·striko,sfiəɹ). *Palæont.* [ad. mod.L. *hystrichosphæra* (O. Wetzel 1933, in *Palaeontographica* A. LXXVIII. 32), f. Gr. ὕστριξ, ὑστριχ- porcupine + σφαῖρα ball.] Any of numerous kinds of microscopic planktonic fossil organisms characterized by a spherical or oval shape with numerous short thin projections.
1957 Q. Jrnl. Geol. Soc. CXII. 416 In the clays, dinoflagellates and hystrichospheres were abundant and formed a considerable proportion of the combustible organic matter. 1963 Palaeontology VI. 83 The hystrichospheres having spines closed distally exhibit an overall range in shell diameters from 5 μ to 240 μ.

hyther (həi·þəɹ). [See quot. 1907.] A quantity determined from temperature and humidity and intended to represent the discomfort attributable to their combined effect; also, a unit on a scale of 0 to 10 expressing this.
1904 W. F. Tyler in Jrnl. Balneology VIII. 25 A number of persons..were requested to estimate daily at noon the degree of 'hyther' on a scale of 0 to 10. This word hyther was introduced to indicate the sensation caused by a warm climate, and supposed to be due to the combined effect of heat and humidity. 1907 — in Monthly Weather Rev. June 268/1, I consider the term 'sensible temperature' to be rather misleading, temperature being only one factor in the subjective effect. It was for this reason that..I coined the word hyther from hydro and thermos. 1931 A. Castellani Climate & Acclimatization ii. 39 It was observations like these that caused Tyler to attempt to correlate personal sensations with meteorological data, and to formulate his 'hyther' degrees, by which he meant the degree of discomfort caused by high air temperatures associated with high relative humidity. 1937 Nature 9 Jan. 79/1 Their sensations [of temperature] were recorded on an arbitrary scale extending from 1 to 10; the numbers on the scale have been called hythers, hyther 10 being taken to represent an 'unbearable condition'.

hythergraph (həi·þəɹgrɑf). *Climatology.* [f. Gr. ὑ-ετός rain + θέρ-μη heat + -graph.] A climograph having temperature and either humidity or precipitation as coordinates; usu., one in which the mean monthly values of these coordinates are plotted for each month of the year, the plotted points for each successive month then being joined by straight lines.
1918 T. G. Taylor Austral. Environment viii. 30 The two chief controls..are temperature and rainfall... To the graph representing these controls I have given the name *hythergraph*—from the Greek words for rain and heat. [Note] Tyler has used the word 'hyther' to express the joint effect of humidity and temperature. 1940 Ecology XXI. 189/1 Though shown..as being an area of tall grass, the hythergraphs of stations in the area depart somewhat from the typical grassland figure and approach diagrams of the deciduous forest. 1950 Conrad & Pollak Methods in Climatol. (ed. 2) vii. 221 The annual variation of the combined rainfall–temperature element is represented by the graphs shown in Fig. 58. These curves are called hythergraphs.

I

I. Add: **I. 2. b. i-mutation, i-umlaut** (also *i/j-mutation*, etc.) *Philology*, the fronting influence of an *i or *j on the vowel of a preceding syllable in one and the same word; also, the result of this. So **i-mutated, i-umlauted** ppl. adjs.
1870 F. A. March Compar. Gram. Anglo-Saxon Lang. 13 i-umlaut of ō: fōt, fēt(e). 1891 A. L. Mayhew Synopsis Old Eng. Phonol. 41 In North. and Mercian oe = e, the i-umlaut of o. 1906 H. C. Wyld Hist. Study Mother Tongue i. 10 This particular kind of change, known as i-mutation, occurs in hundreds of words in O.E., though, as a rule, the i or j which caused the fronting, disappeared. 1908 J. & E. M. Wright Old Eng. Gram. iii. 28, a was the only vowel which underwent i-umlaut in OS. and OHG. 1927 Englische Studien 10 Nov. 81 There was, by the side of OE. scēat..an i-mutated variant sciete or scyte with the same sense. 1927 E. V. Gordon Introd. Old Norse 246, ø, was the w-mutation of ę or (rarely) a late i-mutation of ǫ. 1945 S. Einarsson Icelandic i. v. 30 The I-Shift (..i-umlaut, i-mutation) is so called because it was caused by an i or a j—now often lost—in the ending of a word. 1953 L. F. Brosnahan Some Old Eng. Sound Changes 63 The phenomenon of i- or j- mutation.

II. 7. a. In *Physics I* (rarely i) is the symbol of the quantum number of nuclear spin. [Adopted by Back and Goudsmit 1928, in *Zeitschr. f. Physik* XLVII. 175.]
1930 Pauling & Goudsmit Struct. Line Spectra xi. 203, i is a new quantum number, the nuclear spin quantum number. 1932 Bacher & Goudsmit Atomic Energy States 20 The spectrum of bismuth, for which the nuclear moment I is 4½, is an interesting example of this type of hyperfine structure. 1966 D. H. Whiffen Spectroscopy iii. 22 Intrinsic nuclear angular momenta are quantised and may be expressed as Iħ where I..is called the spin quantum number. 1967 [see *F III. 1 k].
b. Occas. used as the symbol of the quantum number of isospin (more commonly *T).
1953 Progress Theoret. Physics IX. 420 In general, selection rules are intimately connected with the conservative quantities which we shall inquire for a system involving Fermions. Those are the total angular momentum J and the total isotopic spin I of the system. 1962 A. Ramakrishnan Elem. Particles & Cosmic Rays i. 31 We use the symbol t for the isotopic spin operator of a system of particles and τ for a single particle, their eigenvalues being denoted by T and I respectively.

III. I., Intelligence (see also I.Q. below); **I.A.A.,** indoleacetic acid (indolylacetic acid); **I.A.E.A.,** International Atomic Energy Agency; **I.A.T.A.,** International Air Traffic (or Transport) Association; **I.B.A.,** Independent Broadcasting Authority; **I.B.M.,** intercontinental ballistic missile; International Business Machines (used to denote the computers made by this firm); **I.C.,** in charge, in command; integrated circuit; (*Linguistics*) immediate constituent; **I.C.A.,** Institute of Contemporary Arts; **I.C.A.O.,** International Civil Aviation Organization; **I.C.B.M.,** intercontinental ballistic missile; **I.C.E.,** internal combustion engine; **I.C.F.T.U.,** International Confederation of Free Trade Unions; **I.C.I.,** Imperial Chemical Industries; **I.C.S.,** Indian Civil Service; **I.D.,** identification, identity (card); **I.D.A.,** International Development Association; **I.D.B.,** illicit diamond buyer, buying; **I.E.,** Indo-European; **I.F., i.f.,** intermediate frequency (see *intermediate a. and sb. A. 3); **I.F.F.,** Identification, Friend or Foe; **I.F.R.,** Instrument Flight Rules; **I.G.Y.,** International Geophysical Year; **I.L.O.,** International Labour Organization; **I.L.P.,** Independent Labour Party; **I.L.S.,** Instrument Landing System; **I.M.C.O.,** Intergovernmental Maritime Consultative Organization; **I.M.F.,** International Monetary Fund; **I.P.A.,** International Phonetic Alphabet (or Association); **i.p.s.,** inches per second; **I.Q.,** intelligence quotient (see *intelligence sb. 8); **I.R.,** infra-red; **I.R.A.,** Irish Republican Army; **I.R.B.M.,** intermediate range ballistic missile; **I.R.O.,** International Refugee Organization; **I.R.S.,** Internal Revenue Service (*U.S.*); **I.S.O.,** Imperial Service Order; International Organization for Standardization; **I.T.A.,** Independent Television Authority; **I.T.A., i.t.a.,** initial teaching alphabet; **I.T.U.,** International Telecommunication Union; **I.T.V.,** Independent Television; **I.U., i.u.,** international unit; **I.U.(C.)D.,** intra-uterine (contraceptive) device; **I.V., i.v.,** intravenous(ly); also as *sb.*, an intravenous drip, injection, etc.; **I.W.W.,** Industrial Worker(s) of the World.
1917 'Contact' Airman's Outings iv. 87 An air reconnaissance is essentially the observer's show; its main object being to supply the 'I' people at headquarters with private bulletins from the back of the German front. 1925 Fraser & Gibbons Soldier & Sailor Words 125, I, the Service abbreviation for 'Intelligence', i.e. information of military value. 1972 G. Lyall Blame the Dead xiii. 86 'What were you in?' '"I" Corps.' 1947 Jrnl. Biol. Chem. CLXIX. 465 Indoleacetic acid (IAA) tends to stimulate growth in the light. 1969 New Scientist 7 Aug. 272/1 Isolated bacteria were cultured and shown to form IAA from the amino acid tryptophan. 1958 Times 20 Jan. 6/4 Mr. Robert McKinney, United States member of the I.A.E.A. board of governors. Ibid., The United States last year already declared its intention to match all contributions of materials to the I.A.E.A. made by other countries up to June, 1960. 1958 P. Noel-Baker Arms Race p. xvii, IAEA, International Atomic Energy Agency (established in Vienna, October 1957). 1963 Times 28 Sept. 6/7 Sir Roger Makins, chairman of the British Atomic Energy Authority and chief delegate to I.A.E.A. 1931 Flight 20 Mar. 255/1, I..submitted therefore to the International Air Traffic Association, IATA, a suggestion for organising, at the earliest possible moment, a general European air mail net. 1962 Daily Tel. 11 Sept. 12/3 The continuance of rate-cutting could not be tolerated if IATA was to continue its work. 1970 Internat. & Compar. Law Q. 4th Ser. XIX. i. 125 It is something of a surprise that the United States Government and the International Air Transport Association (IATA)..should be proposing a system of absolute liability. 1971 Guardian 12 Nov. 1/8 The Sound Broadcasting Bill..authorises the new stations under the control of the Independent Television Authority—renamed the Independent Broadcasting Authority... The IBA could be advanced up to £2 millions to set up the services. 1971 Times 12 Nov. 8/5 The IBA would have the same obligation in radio to devote sufficient time to accurate and impartial news. 1954 I.B.M. [see *ballistic a. d]. 1955 Ann. Reg. 1954 402 Reports from the United States referred to a rocket called I.B.M. (intercontinental ballistic missile), a wingless rocket-shaped device already perfected to travel 2,500 miles and expected to have a range of 5,000 miles in due course. 1955 R. J. Schwartz Compl. Dict. Abbrev. 89 I.B.M., International Business Machines. 1956 S. Bellow Seize the Day (1957) ii. 42 When he saw the two sums punched out so neatly on the cards he cursed the company and its IBM equipment. 1956 A. Huxley Adonis & Alphabet 109 Thanks to finger-printing, punched cards and IBM machines, they know practically everything about practically everyone. 1963 I. Fleming On H.M. Secret Service i. 16 Bond's mind ticked and whirred, selecting cards like an IBM machine. 1970 Amer. Jrnl. Physics XXXVIII. 1294/2 Computations on the jth particle velocity..have been carried out on the Oberlin College IBM 360/44. 1928 T. E. Lawrence Let. 19 July (1938) iv. 615 No, I am not adjutant, to this camp. Just typist, and i/c files, and duty rolls. 1947 R. S. Wells in Language XXIII. ii. 81 We aim in this paper to replace by a unified, systematic theory the heterogeneous and incomplete methods hitherto offered for determining immediate constituents (hereafter abbreviated IC, plural ICs). 1953 Language XXIX. 88 Shannon has conducted experiments in ordinary English orthography, and the reviewer has conducted similar ones, with the proper audiences, in terms of phonemic notation, the results of which bear on the stated correlation between IC-analysis and information theory. 1958 Spectator 15 Aug. 219/1 If you were i/c security, it was obviously necessary to flush the lavatories of spies. 1962 L. Deighton Ipcress File viii. 51 The above named article of War Department property..should be returned to officer i.c. special issue room. 1962 B. M. H. Strang Mod. Eng. Struct. vi. 79 They are immediate constituents (ICs), i.e. the forms that directly go to make up that which is under analysis. They themselves have ICs. 1965 Canad. Jrnl. Ling. Fall 45 Chomsky develops..IC analysis by his grammatical model of 'phrase structure + transforms'. 1965 Listener 11 Nov. 763/2 The commanding officer; the 2 I.C.; the adjutant. 1966 Electronics 17 Oct. 87 The major problems in using IC's. 1969 New Scientist 18 Dec. 601/3 The IC memory is three times faster than the conventional memory. 1970 'J. Earl' Tuners & Amplifiers ii. 28 In a few years' time the majority of radio tuners will carry ICs as well as a few transistors and junction diodes. 1971 D. Crystal Linguistics iv. 212 In IC analysis, however, such disambiguation was impossible. 1958 Listener 20 Nov. 842/1 At the I.C.A. there is an exhibition of three collagists. 1969 Ibid. 27 Mar. 436/3 The ICA has taken us aback by giving some public performances of a radio work, reproducing it stereophonically in a darkened theatre. 1947 Times 17 May 3/5 (headline) Italy nominated for I.C.A.O. 1955 Sci. Amer. Jan. 94/3 The specialized agencies of the United Nations..ICAO (International Civil Aviation Organization). 1963 Thorn Electr. Industr. Group Profile 25 The system has been approved..by the ICAO. 1955 Newsweek 30 May 13 The Air Force is now calling the Intercontinental Ballistic Missile the ICBM. 1956 Spaceflight Oct. 24/1 The relatively small margin in performance between the I.C.B.M. and a satellite vehicle suggests that great use will be made of it in the future. 1965 I. Fleming Man with Golden Gun ii. 28 Their U-boat fleet and their ICBMs. 1950 Chambers's Encycl. XV. 586/2 I.C.E.: see Internal Combustion Engine. 1958 Listener 20 Nov. 835/2 The present advanced state of the internal combustion engine, or 'I.C.E.' as my log-book calls it. 1968 Economist 25 May 45/3 But now a modern version of the steam engine has appeared as the major threat to the ICE. 1955 Times 30 May 4/4 First, the I.C.F.T.U. will continue its fight to ensure that all workers' rights are respected. 1968 Telegraph (Brisbane) 18 Apr. 10/2 Mr Mick Jordan..represents the International Confederation of Free Trade Unions (ICFTU)—a world-wide anti-Communist union body. 1934 H. G. Wells Exper. Autobiogr. II. viii. 638 Brunner Mond & Co. was only the embryo of I.C.I. 1964 M. Gowing Britain & Atomic Energy 1939–1945 ii. 75 I.C.I. offered to take over..the Halban and Kowarski research. 1931 Times Lit. Suppl. 14 May 390/3 The late Ross Scott, I.C.S. 1957 J. Masters Far, Far the Mountain Peak iii. 27 If Peter has really made up his mind to go to the I.C.S.—it will be wonderful. 1971 Shankar's Weekly (Delhi) 18 Apr. 22/4 The Prime Minister called the ICS Secretaries of the Central Departments some weeks ago and admonished them. 1955 R. J. Schwartz Compl. Dict. Abbrev. 90 Id, identification. 1963 T. Pynchon V. xiii. 373 Pig was understandably nervous, trying simultaneously to salute, produce ID and liberty cards. 1965 New Statesman 3 Dec. 880/3 'ID's'..are pretty obscure to English readers as translations..of..papiers (identity documents). 1968 A. Diment Bang Bang Birds v. 75, I had the usual range of forged driver's licences, ID cards, credit chits. 1970 Globe & Mail (Toronto) 28 Sept. 7/1 Once inside I was forced to produce my driver's licence, draft card, student I.D. 1971 Leader (Durban) 7 May 1/1 The loss of the money was not important. I am more concerned about my ID card, as I am presently applying for a house in Unit 10. 1972 J. Ball Five Pieces Jade ii. 21 Tibbs was politely asked for his ID. He produced his police credentials. 1961 Ann. Reg. 1960 472 The major development among international agencies was the establishment of the International Development Association (I.D.A.). 1965 Economist 26 June 1512/2 The World Bank's 'soft-loan' subsidiary, the International Development Association (IDA), already lends for up to fifty years. 1884 M. A. Carey-Hobson At Home in Transvaal II. xlii. 520 'The fellow had money there, with which he turned I.D.B.' 'What's that?'..'Illicit diamond buyer.' 1886 W. M. Kerr Far Interior I. i. 15 In spite of the vigilance of the detective department a great deal of illicit diamond buying is successfully carried on; hence the well-known 'IDB', which refers to the illegal trade. 1909 H. G. Wells Tono-Bungay ii. i. 122 Barmentrude..used to be an I.D.B.—an illicit diamond buyer. 1917 New Statesman 17 Nov. 150/1 To represent the typical Bolshevik as a German agent..is just as clever and fair as to try to make out that an I.D.B. from the Rand..is the type of a British Imperialist. 1972 P. Driscoll Wilby Conspiracy (1973) xiii. 163 He does a bit of IDB on the side. So what? He's a jeweller. 1894 V. Henry Compar. Gram. Eng. & Ger. v. 113 The greater part of such roots as began with the group in question exhibited already in the I.-E. period a peculiar alternation. 1964 R. H. Robins Gen. Ling. viii. 307 The I-E language family is represented all over the world today. 1927 H. J. Round Shielded Four-Electrode Valve viii. 77 It should be possible to do with only one stage of intermediate frequency on account of the gain per stage in H.F. and I.F. 1956 Tibbs & Johnstone Frequency Modulation Engin. (ed. 2) ix. 387 Second channel interference can be avoided by choosing the i.f. to be greater than half the band of frequencies to be covered. 1963 J. A. Walston Transistor Circuit Design xxiii. 321 The difference frequency (IF frequency)..must be such that the transistor will function as an amplifier. 1945 Electronic Engin. XVII. 686 An I.F.F. unit can be briefly described as being a transmitter-receiver device installed in friendly aircraft whose purpose is to reply to the interrogation of the friendly Radar station. 1961 Listener 30 Nov. 909/1 The I.F.F. radar identification sets in our bombers. 1948 Jrnl. R. Aeronaut. Soc. LII. 90/1 The biggest factor affecting reliable running is the time difference between operations under clear and low visibility conditions, C.F.R. (Contact Flight Rules) and I.F.R. (Instrument Flight Rules) as they are called. 1964 Times Rev. Industry Apr. 40/3 Under IFR, electronic navigation equipment defines the position, and facilitates landing in bad weather or at night. 1955 Sci. News Let. 15 Jan. 42/1 Scientists from at least 39 countries, including Russia, are now making plans for coordinated research efforts during 1957–58 in a world-wide investigation of the earth, its seas and air. The many-pronged attack, aimed at a better understanding of the planet we live on, is known as the International Geophysical Year, or IGY. 1964 Economist 11 Jan. 128/3 The IGY lasted 30 months. 1924 B. Webb Diary 8 Jan. (1956) i. 2 The P.M. to meet Thomas, the French head of the I.L.O. at Geneva. 1969 Listener 1 May 614/3 We're going to give legislative backing to the ILO conventions on the right to join trade unions. 1893 G. B. Shaw Let. 24 Apr. (1965) 390 My remarks..were not levelled at the I.L.P. 1917 A. Huxley Let. 12 Nov. (1969) 136, I suppose it would pain the poor Duke too much if he sat in Parliament as a member of the I.L.P. 1932 Auden Orators iii. 104 The Simonites, the Mosleyites and the I.L.P. 1946 Koestler Thieves in Night ii. 91 Max, who has an enormous, sniffing tapir-nose and an unkempt I.L.P.-mane. 1959 Listener 22 Jan. 179/3 A number of intellectuals and I.L.P-ers. 1947 Shell Aviation News No. 108. 3/1 I.L.S. (Instrument Landing System), G.C.A. (Ground Control Approach), ground radar and flight radar are all proven, available, and should be installed. 1966 New Scientist 13 Jan. 65/1 The special Trident has been trying the ILS of other airports,..and has actually made 'hands-off' landings at them. 1954 Chambers's Encycl. World Survey 42/2 The Intergovernmental Maritime Consultative Organization (IMCO). 1970 Globe & Mail (Toronto) 28 Sept. 6/2 After the Torrey Canyon disaster the Intergovernmental Maritime Consultative Organization (IMCO) turned to the machinery of the international convention on safety of life at sea to try to establish rules that would ease the threat of pollution. 1948 G. Crowther Outl. Money (rev. ed.) ix. 330 The main purpose of the I.M.F...is to provide countries that have deficits with the foreign currencies they

require to cover those deficits. 1965 New Statesman 23 Apr. 632/2 A determination to invoke the scarce currency clause in the IMF agreement to legalise discrimination against chronically surplus countries. 1933 L. Bloomfield Lang. vi. 103 There has arisen a convention of transcribing British English, not by the symbols..in accord with the principles of the IPA alphabet. 1954 Pei & Gaynor Dict. Ling. 105 I.P.A., The International Phonetic Alphabet. 1961 Amer. Speech XXXVI. 201 The modified IPA symbols used in the [linguistic] Atlas. 1970 Publ. Amer. Dial. Soc. 1968 l. 5 The phonetic notation used in transcribing the responses of the informants is..a finely graded phonetic alphabet based on that of the IPA. 1959 W. S. Sharps Dict. Cinematogr. 104/2 I.P.S., abbreviation for inches per second. 1968 Times 29 Nov. p. ii/1 The rise of the tape recorder was attested by the publication of commercial tapes at 7½ i.p.s. 1922 R. S. Woodworth Psychol. xii. 274 Brightness or dullness can also be measured by the intelligence quotient, which is employed so frequently that it is customarily abbreviated to 'I.Q.' 1948 A. Huxley Let. 3 June (1969) 582 Cecil Burt sees a drop in the average intelligence of the British population..of 5 IQ points before the end of the present century. 1959 N. Mailer Advts. for Myself (1961) 150 Any man in the infantry or cavalry who has a good I.Q. is sure to have his name turned up..whenever a new typist is needed. 1968 Scottish Daily Mail 16 July 2/1 The questionnaire is a tongue-in-cheek parody of the IQ tests which the U.S. Government gives would-be employees. 1972 Science 20 Oct. 232/2 The IQ tests ignore much in us that is artistic, contemplative, and nonverbal. 1957 Which? Autumn 7/2 An investigation into the effects, on bottle-makers, of the infra-red and ultra-violet (often referred to as I.R. and U.V.) radiations. 1967 Electronics XL. 127/1 A scope tracing that shows i-r energy as a curve derived from the video signal, with the amount of energy determining vertical deflection. 1921 G. B. Shaw Matter with Ireland (1962) 245 The I.R.A. is flushed with success. 1932 Morning Post 23 Aug. 10/3 A force of 200 men of the I.R.A. have seized Donamon Castle. 1939 J. B. Priestley Let People Sing ii. 24 So they thought he was the I.R.A., eh? That explained the bomb, of course. 1959 New Statesman 7 Nov. 615/2 The IRA is now really discredited; young men in the dreary pubs which offer the only way out of the drearier provincial towns of Ireland must find other amusements than plotting. 1971 Guardian 11 Aug. 10/2 The IRA and the Provisionals use the South as a sanctuary. 1957 Economist 30 Nov. 774/2 IRBMs are to begin flowing to Europe by late 1958. 1960 Ibid. 30 July 460/3 In April the official defence policy was laid in ruins with the abandonment of Britain's IRBM, Blue Streak. 1947 Times 15 May 5/7 Resettlement will still remain one of the main features of the I.R.O. 1948 Hansard Commons 11 Mar. 1531/1 I.R.O. consented to act as our agents. 1955 Sci. Amer. Jan. 95/1 IRO (International Refugee Organization). 1963 Listener 7 Mar. 412/1 The I.R.S. takes good care that the United States citizen abroad knows just where he stands, tax-wise. 1964 Financial Times 12 Mar. 24/3 The attitude of the I.R.S. in 1958, after they had caught up with Mr. Wilson, was to say: 'We're not concerned with your troubles.' 1972 New York Law Jrnl. 22 Aug. 4/4 (heading) IRS issues rulings on political dinners. 1902 Encycl. Brit. XXXI. 340/1 The members of the order have the distinction of adding the letters I.S.O. after their names. 1909 Whitaker's Almanack 118 Thos. H. Sanderson Sanderson, G.C.B., K.C.M.G., I.S.O. 1946 N.Y. Times 2 Nov. 22/6 Howard Coonley..has been elected president of the new International Organization for Standardization, formation of which has been completed by delegates from twenty-five nations meeting in London... Gustave L. Gerard..has been chosen vice president of the new organization which will be known informally as ISO. 1969 Jane's Freight Containers 1968–69 400/1 The equipment covered by this plan is standard ISO 20 ft steel containers with fork-lift pockets, and standard 40 ft ISO aluminium containers. 1955 Ann. Reg. 1954 385 While viewing through I.T.A. stations would not be possible for some time, competitive bidding for 'stars' went on actively. 1962 Rep. Comm. Broadcasting 1960 1 in Parl. Papers 1961–2 (Cmnd. 1753) IX. 259 There is a distinction to be drawn..between the ITA and independent television (ITV). The ITA is the Authority, the public corporation set up by the Television Act, 1954: independent television comprises not only the ITA but also the programme companies. 1965 Guardian 2 Feb. 3/8 Mr Gordon Walker.., adviser to the ITA foundation,..would have to learn more about the initial teaching alphabet. 1967 New Statesman 6 Oct., All the mothers know everything about O- and A-levels, have taken the measure of the 11-plus.., and some have heard of i.t.a. 1950 Chambers's Encycl. XI. 470/1 The International Telecommunications Union (I.T.U.). 1962 B.B.C. Handbk. 113 The BBC also participates in the work of the International Telecommunication Union (ITU), a specialized agency of the United Nations with its headquarters in Geneva. 1958 'A. Gilbert' Death against Clock 93 All they talk about is what they saw on ITV last night. 1958 Spectator 27 June 827/3 As the General Election approaches, both BBC and ITV must know where they stand. 1969 Listener 24 Apr. 559/3 While shepherds washed their socks by night And turned on ITV, The Angel of the Lord came down And switched on BBC. 1950 Chambers's Encycl. XIV. 347/2 The League of Nations standard of requirement for vitamin B_1 is 300 I.U. per day. 1951 Good Housek. Home Encycl. 339/1 The chief food value of apricots lies in their roughage and in their vitamin A content—approximately 1,000 i.u. per serving. 1962 Lancet 6 Jan. 12/1 A small bottle..of some 60 ml. capacity, filled to the top with normal saline and containing 1000 i.u. of heparin. 1963 New Scientist 19 Dec. 716/3 A simple, cheap, safe and effective method of birth control..known as IUCDs (intrauterine contraceptive devices). 1966 New Statesman 18 Mar. 370/1 The IUCD consists of a small loop or coil of plastic material which is inserted into the uterus. 1965 New Scientist 27 May 606/3 When as occasionally happens, conception occurs and gestation proceeds with the IUD in situ, [etc.]. 1967 Time 7 Apr. 73 The IUD's underlying principle traces back to an old practice of Arab cameleers: putting a round, smooth stone in the womb of a female camel at the start of a long trade journey, to avoid the economic loss of having the animal get pregnant. 1973 Guardian 29 June 13/3 With the nationalisation of birth control virtually every GP in the country will be inserting IUDs. 1951 Dorland's Med. Dict. (ed. 22) 766/2 I.V., abbreviation for intravenously (by intravenous injection). 1961 Amer. Speech XXXVI. 145 I.V.,..an intravenous infusion. 1970 New Yorker 21 Nov. 64/2 One of the doctors from Surgery will be coming down soon to put in your I.V. and a stomach tube. 1971 Guardian Weekly 24 Apr. 5/1 The bedside IV feeding bottle. 1972 Nature 8 Dec. 351/1, 3 African green monkeys were inoculated i.c. (0·2 ml.) and i.v. (0·3 ml.). 1917 B. Hall Diary 25 July in Hall & Niles One Man's War (1929) 278 The Government had some trouble in Los Angeles with the I.W.W. 1919 H. L. Wilson Ma Pettengill vii. 212 Even the youngest [girl]..had tenaciously held out for a grown man's pay, which made her something even worse than a Bolshevik; it made her an I.W.W. 1920 M. Beer Hist. Brit. Socialism II. iv. xviii. 356 Similar views..led in 1905 to the formation of the Industrial Workers of the World (I.W.W.). 1957 Encounter Apr. 65/1 That strange and unique contribution of America to anarcho-syndicalism, the 'Wobblies' (officially the Industrial Workers of the World, or IWW..) organised great masses of unskilled workers and led strikes..that were as much social rebellions as economic conflicts. 1969 Taft & Ross in Graham & Gurr Violence in Amer. viii. 285 Unlike the other national federations.., the IWW advocated direct action and sabotage.

I, *pers. pron.* Add: **I. 2.** (Further examples.)
1744 J. Steuart Letter-Bk. (1915) 449 The postscript to your letter..gave my wife and I unexpressable joy. 1866 Harper's Mag. Jan. 162/2, I have heard him..make a bet that 'between you and I' is correct, and refuse to be convinced of his error. 1959 N.Z. Listener 25 Sept. 11/2, I have heard 'between you and I, old man', and 'people like you and I', from graduates in some arts other than the art of speech. 1971 Guardian 20 Aug. 24/4 'There were two grilles between Eugene and I, and we must have been about six feet apart,' she said. 1972 J. Rossiter Rope for General Dietz ii. 24, I was sure they were looking for Michael and I. 1973 Oxford Mail 27 Aug. 4/4 After showing photographer Bill Radford and I her stitching skill she went back to the tea table.
II. 6. The narrator of a work of fiction, appearing on his own account. Also *attrib.* or as *adj.*
1946 'G. Orwell' in Observer 10 Feb. 3/3 The 'I' of the story describes himself as a Democrat. 1962 John o' London's 19 Apr. 372/1 The tendency for novelists to move away from the 'I' kind of storytelling. 1965 English Studies XLVI. 390 The point of view of the 'I' narrator is perfectly maintained throughout the tale. 1969 Listener 30 Jan. 151/3 The 'I' of David Martin's tense and elusive story.
III. I AM, I am (later examples). Also in weakened colloq. use: a (self-)important person.
1915 D. H. Lawrence in Signature 18 Oct. 8 David dancing naked before the Ark, asserting the oneness, his own oneness, the one infinity, himself, the one God, I AM. 1926 S. T. Warner Lolly Willowes iii. 184 Jim thought himself quite a Great I AM. 1928 D. H. Lawrence Lady Chatterley x. 131, I am a cypher. You are the great I-am! as far as life goes. a 1940 F. Scott Fitzgerald Last Tycoon (1949) vi. 138 'Get one that can talk—tell him to bring one of his books along.' He spoke as if he wanted to meet a member of the 'I am' cult. 1954 W. Faulkner Fable 57 Lifting its voice against the Absolute, the ultimate I-Am. 1965 N. Gulbenkian Pantaraxia xi. 227 Cyril Radcliffe..did not take the short-cut favoured by so many of his colleagues who say..: 'I am the great I am, Queen's Counsel.' 1970 D. Francis Rat Race i. 13 He had none of the 'I am' aura which often clings around the notably successful.
Also **I and Thou, I-and-Thou.** In the theology of M. Buber: describing a relationship between two people, as opposed to that between a person and an object. Also *attrib.*
1937 R. Smith tr. Buber's Ich-Du (title) I and Thou. 1958 D. M. Baillie Out of Nazareth xxii. 157 Divine realities can only be known in a personal 'I-and-Thou' relationship. 1968 L. Berg Risinghill 64 Such teachers had never seen their pupils as fellow human beings before, as Martin Buber's 'I' and 'Thou'.

-i, *suffix*, a termination used in the names of certain Near-Eastern and Eastern peoples, as *Iraqi, Israeli, Pakistani*.

-ian, *suffix.* Add: **2.** *Min.* [Abstracted from the adjs. *magnesian, manganesian*.] Used to form, from the (Eng. or L.) names of the elements, adjectives having the sense 'having a (small) proportion of a constituent element replaced by (the element concerned)' (see quot.).
1930 W. T. Schaller in Amer. Mineralogist XV. 568 Can a uniform, clearly understandable scheme of nomenclature be adopted to express a minor and variable isomorphous replacement of an essential chemical element of a mineral by another analogous element?.. The writer has concluded that the ending ian, or oan if it is desired to indicate a lower valency, is the most satisfactory, and its consistent use is here advocated... If the chemical element has only one valency or the author does not wish to bring up the question of valency, ian should be used.

-iana, *suffix.* See Ana *suff.* and add examples. Cf. also *Australiana, *Canadiana, etc.
1679 [see *bibliographical a.]. 1718 [see Shakesperiana]. 1728 J. Smedley (title) Gulliveriana: or, a fourth volume of Miscellanies, being a sequel to the three volumes published by Pope and Swift. To which is added, Alexanderiana, etc. 1776 [see *Johnsoniana]. 1800 (title) Walpoliana. 1838 (title) Railroadiana. A New History of England, or Picturesque..Sketches..Descriptive of the Vicinity of the Railroads. 1879 G. J. Finch-Hatton Voices through Many Years III. 85 Gladstoniana. Ibid. 118 Grevilliana. 1890 Century Mag. Aug. 515/2 A number of these 'whaleiana' hang in the 'Captains' Room'. 1898 W. Graham Last Links with Byron 120 Several writers of Keatsiana follow that most inaccurate of writers,..Leigh Hunt. 1902, etc. [see *Edwardiana sb. pl.]. 1929 Daily Tel. 22 Jan. 6/5 The personal papers of James Boswell..are being published... Sixteen or eighteen volumes will eventually be needed to hold all the new Boswelliana. 1952 J. Carter A B C for Bk.-Collectors 19 Boswelliana, Railroadiana, Etoniana... Harveiana and Dickensiana..Wiseiana. 1972 Country Life 27 Jan. 223/2 In his..biography William Butterfield ..Dr Paul Thompson has hauled in Butterfieldiana in shoals..churches, colleges, schools, hospitals, cottages.

ianthinite (i,æ·-, əi,æ·nþinəit). *Min.* [ad. Du. *janthiniet* (A. Schoep 1926, in *Natuurwetenschappelijk Tijdschrift* VII. 97), f. Gr. ἰάνθιν-ος violet-coloured: see -ite¹.] A hydrated oxide of uranium found as orthorhombic crystals of a dark violet colour.
1927 Mineral. Abstr. III. 232 Small (to 2 mm.) black crystals with a violet tinge and semi-metallic lustre were found in veinlets in pitchblende from the Kasolo mine in Katanga, and are named ianthinite. 1954 Amer. Mineralogist XXXIX. 1018 Ianthinite reportedly contains only quadrivalent uranium and has a formula of $2UO_2.7H_2O$.

Iapygian (əiăpi·dʒiăn), *a.* and *sb.* [f. L. *Iāpygius*, f. *Iāpyx, -ygem*, a son of Dædalus said to have ruled over southern Italy: see -ian.] **A.** *adj.* Of or pertaining to the ancient natives and district of Iapygia, the name given by the Greeks to the peninsula of Apulia in southern Italy. **B.** *sb.* **a.** A native of Iapygia. **b.** The language of the Iapygians; = *Messapian sb. b.
1773 J. Langhorne tr. Denina's Diss. Anc. Republics Italy 26 The Iapygyans and Messapians [lost] fifty thousand infantry. 1864 P. Smith Hist. World II. 138 It is here..that we find traces of the Iapygian race, in the peninsula called by the Greeks Messapia. 1880 Encycl. Brit. XIII. 443/2 The peninsula which stretches eastward towards Greece was inhabited by a people termed by the Greeks Messapians or Iapygians. 1882 Ibid. XIV. 327/1 Inscriptions have been found in considerable numbers, written in a language known as Iapygian or Messapian. 1888 King & Cookson Princ. Sound & Inflexion Gr. & Latin 30 Of the Iapygian in the extreme south and the Ligurian in the north, very little is known. 1959 Chambers's Encycl. VII. 778/1 In Apulia..the Iapygian tribes of Messapii, Daunii and Peucetii established themselves.

iarfine (ī·aɹfinĭ). *Irish Hist.* [Ir., f. *iar* after + Fine *sb.*²] One of the four branches of the Irish clan structure comprising the men in the third grade of relationship to the chief. Cf. *geilfine, *indfine.
1875 H. S. Maine Lect. Early Hist. Inst. 209 The eldest member of the Iarfine moved into the Indfine. 1879 Anc. Laws Ireland IV. p. xlix, In all the Brehon Law Tracts there are references to an existing organization, generally known as the Geilfine system, and to the four classes designated as the Geilfine, Deirbfhine, Iarfine, and Indfine. 1921 E. MacNeill Celtic Ireland x. 162 If the *Iarfine* and the *Indfhine* existed as communal family groups,..the explanation is to be found in the reluctance to disturb the family holdings when the *Derbfhine* had run its course. 1967 F. J. Byrne in Moody & Martin Course Irish Hist. iii. 49 The brehon lawyers drew up a very elaborate scheme of the different degrees of relationship. The *geilfhine*..was the normal family group... And the *iarfine* and *indfhine* the second and third cousins respectively.

-iasis, *suffix*, the form in which -asis *suff.* always occurs as a living mod.L. suffix.

Iatmul (yæ·tmul). [Native name.] A people of New Guinea, living near the Great Sepik River; their language. Also *attrib.* or as *adj.*
1932 G. Bateson in Oceania II. 245 (title) Social structure of the Iatmül people of the Sepik River. Ibid. 248 In the Iatmül language as in English, adjectival words precede the nouns which they qualify. In Iatmül there is no equivalent of the English conjunction 'and'. Ibid. 249, I have thought it most convenient to use the simple word Iatmül to denote that part of the linguistic group which lives on the banks and close to the Sepik river. 1943 S. W. Reed Making of Mod. New Guinea i. 22 The Iatmul of the Middle Sepik River, who number approximately 10,000 persons, live in twenty-odd villages on either side of the river. Ibid. 23 In an isolated mountain enclave west of the Iatmul area and north of the Sepik River dwell the Kwoma. 1949 G. Bateson in M. Fortes Social Struct. 35 The sayings, actions, and organization of the Iatmul had certain characteristics. 1949 M. Mead Male & Female iii. 52 A Iatmul head-hunter, calling in his age-grade to rape his recalcitrant wife into submission. Ibid. App. 403 We had to analyze the language..and this was true also..of Mr. Bateson's original work in Iatmul. 1951 R. Firth Elem. Social Organiz. i. 22 The chronic state of the Iatmül culture.

iatrochemist. (Earlier example.)
1668 E. Maynwaring Medicus Absolutus ix. 78 The Compleat Chymical Physitian. This Iatro-chymist and Hermetick Philosopher, is educated from his Youth in all necessary learning..that he may be introduced into the Medical Art legitimately.

iatrochemistry (əi,ætro,ke·mistri). [f. Iatro- + Chemistry, after the family of mod.L. words beginning *iatrochem-, iatrochym-* (cf. Chemic *a.* and *sb.*).
These words appear to have originated in the (translated) work of Paracelsus (d. 1541): 1573 I. Dalhemius tr. Paracelsus' Chirurgia Magna I. ii. xiii. 21 Verum quia Iatrochymista [*so in* Opera Omnia (1658), *but here printed* latrochymista] sum: vtrumq3 enim scio & Medicinam & chemiam. They afterwards occur in medical works in L. by other writers, e.g. G. Phædro Opuscula iatro-chemica quatuor (1611); D. Burnet Iatrochymicus, siue de præparatione et compositione medicamentorum chymicorum artificiosa (1616).]
The theory or school of thought that existed in the 16th and 17th centuries and regarded medicine and physiology as subjects to be understood in terms of the chemistry of the time (see Chemiatric *a.*).
1830 T. Thomson Hist. Chem. I. v. 201 The most eminent of all the English supporters of iatro-chemistry was Thomas Willis, who was a contemporary of Sylvius. 1881 Roscoe & Schorlemmer Treat. Chem. III. i. 4 Towards the sixteenth century, the cultivators of this science [sc. alchemy]..exhibited activity mainly in two directions, in the first place in the prosecution of the branch science of metallurgy, and secondly, in the development of iatrochemistry. 1909 P. J. Hartog in Dict. Nat. Biogr. XIII. 176/2 Mayow stands immeasurably above such men as Willis and Sylvius, with their medley of half-digested Cartesianism and iatrochemistry. 1958 L. Thorndike Hist. Magic & Exper. Sci. VIII. xxvii. 117 The alchemy and iatrochemistry and medicine of the late seventeenth century differed little from that of the early seventeenth century, especially in Germany.

iatrogenic (əi,æ:trodʒe·nik), *a. Med.* [f. Iatro- + *-genic.] Induced unintentionally by a physician through his diagnosis, manner, or treatment; of or pertaining to the induction of (mental or bodily) disorders, symptoms, etc., in this way.
1924 E. Bleuler tr. Brill's Textbk. Psychiatry xiii. 502 Not entirely unimportant, unfortunately, is the iatrogenic origin of neurotic manifestations. 1948 L. Kanner Child Psychiatry (ed. 2) ix. 143 Difficulties, arising from medical clumsiness in the handling of patients, are common enough to have originated the diagnosis of iatrogenic, or physician-determined, conditions of health. 1952 A. Huxley Devils of Loudun vii. 219 Like the sulpha poisoning and serum-fevers of the present, the Loudun epidemic was an 'iatrogenic disease', produced and fostered by the very physicians who were supposed to be restoring the patients to health. 1970 Brit. Med. Bull. XXVII. 13/2 The epidemic of iatrogenic deaths in asthmatic children shows the need for continuous monitoring of vital statistics. 1971 Sci. Amer. June 99/1 Although it is not common in this country, iatrogenic goiter—goiter caused by medical treatment—is becoming a more significant factor. Sulfonamides prescribed for urinary-tract infections..and many iodine-containing compounds administered as expectorants in the treatment of asthma are potentially goitrogenic. 1973 Guardian 18 Jan. 12 Drug induced (iatrogenic) conditions are on the increase.
So **iatro·geny**, the iatrogenic induction of a disorder.
1927 Henderson & Gillespie Text-bk. Psychiatry xiv. 416 Too often we find that in the causation of a psychoneurotic illness there has entered a very large element of 'iatrogeny'. 1940 Hinsie & Shatzky Psychiatric Dict. 275/1 When the physician..gives any diagnosis that serves as the nucleus around which the patient builds a neurosis or psychosis, the condition is known as iatrogeny. 1973 Interfaces May 45 The biographers of Alfred Nobel have not dealt satisfactorily with the question of his full grasp of the powers of iatrogeny (prize in physiology and medicine).

iatromathematically, *adv.* (Later example.)
1889 H. E. Handerson tr. Baas's Outl. Hist. Med. 503 Edward Barry thought that the age of a man could be calculated Iatro-mathematically from the frequency of the pulse.

iatromathematics. Restrict † to sense in Dict. (s.v. Iatromathematical *a.*) and add: **b.** The iatromathematical theory or school of thought.
1830 T. Thomson Hist. Chem. I. v. 209 He was a zealous supporter of iatro-mathematics, and as such a professed antagonist of the iatro-chemists. 1889 H. E. Handerson tr. Baas's Outl. Hist. Med. 504 This was accomplished..at a time when within its borders Iatro-chemistry and Iatro-mathematics were still in the perfection of their bloom. 1961 J. R. Partington Hist. Chem. II. xii. 442 Some steps in the direction of Iatromathematics (probably influenced by Galileo) were taken by Santorio Santorio (Sanctorius Sanctorius, 1561–1636),..who discovered insensible perspiration..by living on the platform of a large balance, on which he worked and took his meals.

iatromechanical, *a.* (Earlier and later examples.)
[1801 K. Sprengel Versuch einer pragm. Geschichte d. Arzneikunde (ed. 2) IV. xiv. 500 Die Schule..heisst die iatromathematische oder iatromechanische.] 1856 C. G. Comegys tr. Renouard's Hist. Med. viii. x. 533 After the death of the celebrated professor of Leyden [sc. Boerhaave], the iatro-mechanical doctrine fell apart. 1971 Nature 30 July 351/1 Long established Aristotelian doctrines began to crumble before the assault of Baconian empiricism and atomism modified by Descartes, Gassendi and Boyle: the emergence of the iatrochemical and iatromechanical 'schools' of medicine completed the rout.

iatromechanics (əi,ætro,mĭkæ·niks), *sb. pl.* (const. as *sing.*). [f. Iatro- + Mechanics.] = *iatromathematics b.
1886 Syd. Soc. Lex., Iatromechanics, the same as Iatromathematics. 1889 H. E. Handerson tr. Baas's Outl. Hist. Med. 496 Subsequently an effort was made to bring Iatro-chemistry into accord with Iatro-mechanics. 1926 C. G. Cumston Introd. Hist. Med. xvii. 274 Specificity..was to end in a sort of Kabbalism or in union with chemistry or physics.., and, from the latter, were soon to develop the new systems of Iatro-chemistry and Iatro-mechanics. 1953 M. H. Fisch in E. A. Underwood Sci., Med. & Hist. I. iv. 546 It was perhaps on the strength of the Neapolitan edition of this his major work, a contribution to iatromechanics along Cartesian lines, that Santanelli became professor of medicine at Naples in 1708.
So **iatromecha·nic, -mechani·cian, -me·chanist**, one belonging to the iatromathematical school.
1856 C. G. Comegys tr. Renouard's Hist. Med. viii. x. 548 The Iatro-chemists and the Iatro-mechanics had attempted in vain to explain the functions of organized bodies by the general laws of matter. Ibid. xi. 576 The essence of inflammation consists..with the Iatro-mechanician, in the obstruction of the vessels. 1899 F. S. Lee tr. Verworn's Gen. Physiol. i. 18 The hopes of the iatromechanics and iatrochemists of being able completely to resolve vital phenomena into physics and chemistry were not fulfilled. 1926 C. G. Cumston Introd. Hist. Med. xix. 320 Lancisi, Hecquet and Baglivi.—These three names are those of the leaders of organicism at the beginning of the XVIIIth century. They were supposed to be Iatro-mechanicians with a smattering of Hippocraticism and Galenism. 1943 Garrison & Morton Med. Bibliogr. 6 Hoffmann of Halle was the most important of the Iatromechanists. He believed an ether-like 'vital fluid' to be present in the nervous system and to act upon the muscular system, giving them 'tonus'. 1955 E. H. Ackerknecht Short Hist. Med. x. 111 The iatrochemists were never accepted to the extent that the iatromechanists were.

iatromechanism (əi,ætro,me·kăniz'm). [f. Iatro- + Mechanism.] = *iatromathematics b.
1885 Index-Catal. Library Surg.-General's Office, U.S. Army VI. 769 (subject heading) Iatro-mechanism. 1886 Syd. Soc. Lex., Iatromechanism, the doctrine of the Iatromechanics. 1926 C. G. Cumston Introd. Hist. Med. xxi. 351 Towards the end of the XVIIth century Iatro-mechanism so completely counterbalanced Iatro-chemistry that the mechanics of the living being rather than the composition of the humours became the object of study.

iatrophysical, *a.* Substitute for def.: = Iatromathematical *a.* b. (Earlier and later examples.)
1883 Encycl. Brit. XV. 810/2 The iatro-physical school of medicine grew out of physiological theories. 1954 R. H. Major Hist. Med. I. 506 Borelli was unquestionably the perhaps unwitting founder of a new school of medicine—the Iatro-physical School, whose most extreme advocates sought the explanation of all medicine in physics or in mechanics, just as the Iatro-chemists found the ultimate explanation in fermentation.

iatrophysicist (əi,ætro,fi·zisist). [f. next, after *physics, physicist*.] = Iatromathematician.
1889 H. E. Handerson tr. Baas's Outl. Hist. Med. 497 In therapeutics the Iatro-physicists..managed in accordance with the principles of genuine (Hippocratic) experience. 1917 A. H. Buck Growth of Med. xxviii. 366 It became customary to employ the terms, 'iatrochemists' and 'iatrophysicists' in speaking of the partisans of the two schools of medicine (the iatrochemical and the iatrophysical or iatromechanical). 1957 Encycl. Brit. XV. 201/1 The language and the theories of the iatrophysicists, the iatrochemists and the vitalists..have long been discarded.

iatrophysics (əi,ætro,fi·ziks), *sb. pl.* (const. as *sing.*). [f. Iatro- + Physics.] = *iatromathematics b.
1886 in Syd. Soc. Lex. 1889 H. E. Handerson tr. Baas's Outl. Hist. Med. 501 William Cockburn (about 1696) too embraced Iatro-physics eclectically. 1928 C. Singer Short Hist. Med. iv. 127 Numerous fresh theories arose, of which the more important can be classed under the three headings Iatrophysics, Iatrochemistry, and Vitalism. 1945 D. Guthrie Hist. Med. xi. 204 Sydenham..rejected entirely the ideas of iatrophysics or iatro-chemistry. 1961 J. R. Partington Hist. Chem. II. viii. 297 Domenico Sanguinetti and Joseph del Papa..were practically the only Italian physicians of the time to oppose the chemical theory and prefer Iatrophysics.

Iban (ī·bæn, ībæ·n), *sb.* and *a.* [Native name.] **A.** *sb.* A people of Sarawak, also known as the Sea Dyaks; a member of this people, and the name of their language. **B.** *adj.* Of or pertaining to this people.
1911 F. W. Page-Turner in Sarawak Mus. Jrnl. I. 133 A child which they adopted and named Diang Idah who is the origin of the tribe called Iban. Ibid., Tiang Laju is thus the origin of the Iban race as the grand-parents of Diang Idah came from there. 1960 Guardian 9 Nov. 10/3 The population includes Ibans (also known as Sea Dayaks). 1962 B. Harrisson Orang-Utan ii. 53 An Iban Dayak from the Sebuyan area in the Second Division of Sarawak. 1967 B. Sandin Sea Dayaks of Borneo p. xix, 'Iban' is a term of Kayan origin which did not come into general use until quite late in the nineteenth century, while the term 'Sea Dayak' was imposed by the Brookes... The Ibans originally had no term which recognised their indisputable ethnic unity. 1969 Franciscan Missionary Herald XXXIV. 35 Mother St. Robert and Sister Otteran speak Iban. Ibid. 36 The interior of an Iban Long-house.

Ibanag (ī·bănăg), *sb.* and *a.* [Native name.] **A.** *sb.* The name of one of the peoples inhabiting northern Luzon in the Republic of the Philippines, of a member of this people, and of their language. **B.** *adj.* Of or pertaining to this people.
1885 Encycl. Brit. XVIII. 753/1 The other tribes of the Philippines—the Ilocanes, Pampangos, Pangasinanes, Ibanags or Cagayans,..&c. 1900 F. H. Sawyer Inhabitants Philippines 252 The Ibanags inhabit the Babuyanes and Batanes Islands. 1901 Rep. Philippine Comm. III. 405 In order to state very briefly how the remaining Philippine languages or dialects are related we select from among them some of the principal ones... These are the Ibanag and Ilocano, of North Luzon. 1924 D. P. Barrows Hist. Philippines (rev. ed.) 11 The valley of the Cagayan [is inhabited] by a people commonly called Cagayanes, but whose dialect is Ibanag. 1937 Publ. Inst. Nat. Lang. (Manila) I. 3 The Institute..has taken unto itself the task of studying as many..languages as it could possibly manage, and has included..Ibanag and Ivatan. 1942 J. R. Hayden Philippines 864 Ranging in number from between 700,000 and 800,000 to about 60,000 are the Bikol,..Ibanag, and Sambal groups. 1958 G. F. Zaide Hist. Filipino People ii. 11 The Malayan Filipinos..comprise the..Ilokanos, Bikols, Kagayans (Ibanags), Pampangans.

Iberian, *a.* and *sb.* Add: **A.** *adj.* **3.** Pertaining to the Iberians of Britain (cf. *B. *sb.* 3).
1880 Encycl. Brit. XII. 605/2 Extreme exponents of the theory do not hesitate to speak of the Iberian ancestors of the people of England. 1907 T. R. Holmes Anc. Brit. 65 The race to which they [sc. neolithic inhabitants of Britain] belonged is often called the Iberian, though there is no reason to believe that its British representatives belonged to the Iberian rather than to some other branch of the Mediterranean stock.
B. *sb.* **3.** A neolithic inhabitant of Britain, considered as one of a branch of the continental Iberians.
1880 W. B. Dawkins Early Man in Brit. 322 The Silures, identified by Tacitus with the Iberians, were left only in those fastnesses which were subsequently a refuge for the Welsh against the English invaders. 1900 W. A. Dutt Norfolk 7 The Iceni..were probably mentally as well as physically superior to the Iberians. 1920 H. F. Henderson Relig. in Scotl. i. 11 The Iberians absorbed the Celts without serious dilution of their original characteristics. 1957 G. Ashe King Arthur's Avalon i. 15 Throughout a long stretch of years the inhabitants of Britain were dark little Iberians.

iberis (əibiᵊ·ris). [mod.L. (J. J. Dillenius in Linnæus *Systema Naturæ* (1735)), prob. f. Gr. Ἴβηρες Iberians, as several species come from Spain, but cf. Gr. ἰβηρίς, L. *iberis* a kind of cress.] A low-growing herb or sub-shrub of the genus so called, native to southern Europe and western Asia, and bearing flattened heads of small white, pink, or purple flowers; = Candytuft.
1768 P. Miller Gardeners Dict. (ed. 8) s.v. Iberis 7, Iberis with roundish crenated leaves. 1788 Mawe & Abercrombie Every Man his Own Gardener (ed. 12) 582/2 Round leaved alpine Iberis, Evergreen linear-leaved Cretan Iberis. 1871 W. Robinson Hardy Flowers viii. 25 How pretty and useful even as tiny evergreen shrubs, are the Iberisis! 1931 M. E. Stebbing Hardy Flower Gardening x. 170 Catmint, *Iberis*, and similar half-shrubby plants should have each shoot pulled off with a downward jerk. 1971 A. Scott-James Down to Earth xi. 128 Iberis (both annual and perennial), valerian,..polyanthus and sedum have all grown contentedly in my garden.

Ibero- (əibiᵊ·ro), combining form of Iberian *a.* and *sb.*, with the meaning 'Iberian and'.
1891 Rhys in Academy 26 Sept. 268/2, I believe Picts and Iberians to have belonged to one and the same family, which I have ventured to call Ibero-Pictish. 1896 A. H. Keane Ethnology 378 margin, The Ibero-Berber problem. 1900 tr. J. Deniker's Races of Man 285 Tawny white skin, black hair. Short stature, dolichocephalic Ibero-insular. 1920 Glasgow Herald 24 Sept. 6 The Ibero-American Republics. 1927 [see *Getulian A. adj. b]. 1955 Proc. Prehist. Soc. XXI. 50 As for the micro-burins of the Capsian and Iberomauretanian, it seems possible to relate them to ours. 1955 Archivum Linguisticum VII. 68 The immense resources of Ibero-Romance have rarely been tapped. 1963 Economist 7 Dec. 985/3 The new Ibero-French relationship. 1964 C. F. & F. M. Voegelin in Anthropol. Ling. Nov. VI. viii. 1 (title) Languages of the world: Ibero-Caucasian and Pidgin-Creole. 1971 Language XLVII. 232 Klimov rejects the migrational theory for the Ibero-Caucasian languages, regarding them as autochthonous.

Ibibio (ibī·bī·o), *sb.* and *a.* [Native name.] **A.** *sb.* A people of Southern Nigeria; a member of this people; their language. **B.** *adj.* Of or pertaining to this people.

1822 J. Adams *Sk. Voy. Afr.* v. 77 Three-fourths of all the negroes sold at Bonny were Heebos, the remaining fourth was composed of..the Ibbiby. **1862** H. Goldie *Dict. Efik Lang.* v. p. xlix, A few Ibibio words are inserted. **1890** —— *Calabar & its Mission* i. 13 By far the greater part of the oil exported is produced by the tribes behind, especially by Ibibio. **1905** C. Partridge *Cross River Natives* ii. 33 The district of Calabar is peopled by..the Efiks, the Ibibios, and the Ekois. *Ibid.* ii. 43 Most of the native labour is drawn from the Ibibio country. **1915** D. A. Talbot *Woman's Mysteries of Primitive People* xv. 213 Among the Ibibios different funeral rites are prescribed according to age, position, and manner of death. **1919** H. H. Johnston *Compar. Study Bantu & Semi-Bantu Lang.* I. 814 My information regarding..*Ibibiω*, and the allied *Kwō* dialect..has been chiefly obtained from the Rev. Hugh Goldie's *Dictionary of the Efik Language.* **1925** J. A. MacCulloch *Mythol. All Races* VII. 111 The head-pad..figures in some curious magical ceremonies of the Ibibio. **1932** *Africa* V. 503 One West African language, viz. Ibibio, spoken by about three-quarters of a million people in the Calabar province of Southern Nigeria. **1936** J. G. Frazer *Aftermath: Suppl. Golden Bough* i. 18 Among the Ibibios of Southern Nigeria, 'old women may not touch soup made in deep pots, lest they receive too much nourishment therefrom, which will cause them to live beyond the allotted span'. **1958** J. S. Coleman *Nigeria* i. 31 The Ibo and Ibibio languages belong to different branches of the large Niger–Congo linguistic family. **1960** *Times* 5 Oct. 18/7 Astonishing..naturalism in the very large Ibibio mask in wood.

Ibicencan (ibiþe·ŋkăn), *sb.* and *a.* Also **Ibicenco, Ibizenco.** [f. Sp. *ibicenca* native or inhabitant of Ibiza, *ibicenco* pertaining to Ibiza + -an.] **A.** *sb.* A native or inhabitant of Ibiza, an island off the Mediterranean coast of Spain. **B.** *adj.* Of or pertaining to Ibiza, esp. in *Ibicencan hound* (= *Ivicene *sb.*).

1911 J. E. C. Flitch *Mediterranean Moods* ix. 188 The Ibicencos have never naturalised the guitar. *Ibid.* 199 They sang chiefly in Catalan, and the song had a different character from that of the Ibicenco peasants. **1952** E. Whelpton *Balearics* xv. 189 The Ibicencans..were for centuries considered to be savage barbarians. **1959** *Encounter* Oct. 39 The Ibicencan hounds—pale rib-thin beasts. **1969** C. Irving *Fake!* (1970) xii. 139 He liked the Ibizencos, too, a friendly and dignified island people. **1970** *Globe Mag.* (Toronto) 26 Sept. 12/1 There is a native strain of dogs on the island, the Ibicencan hounds, said to be descended from the hunting hounds of the ancient Egyptians. **1972** *Times* 3 Feb. 25/6 One of the most attractive apartment developments..in typical Ibicenco style.

-ibility [F. *-ibilité*, L. *-ibilitātem, -tās*], termination of abstract sbs. from adjs. in -ible.

ibis. Add: **2.** *Angling.* The name of a type of artificial fly; now more usu. applied to a sort of red-dyed feather used in making this type of fly.

1863 *Harper's Mag.* Oct. 691/2 He trailed his 'ibis' lightly across the dark eddy at the edge of the foam. **1931** *Hardy's Anglers' Guide* (ed. 53) 75 Wet flies for brook trout... Ibis. **1961** A. C. Williams *Dict. Trout Flies* (ed. 3) 294 Tail: Red ibis feather. *Ibid.* 299 Tag: Bright red wool, or scarlet ibis. **1973** *Country Life* 7 June 1585/3 The butcher is a well-tried trout fly... It is made from what is known as dyed ibis (a hen's feather, dyed bright red), silver tinsel, [etc.].

3. A fashion colour (see quot.).

1927 *Daily News* 9 May 2/3 A skirt of satin..in the new pale apricot known as 'ibis'.

4. *attrib.* and *Comb.*, as *ibis-headed, -red* adjs.

1910 *Daily Chron.* 19 Feb. 6/2 Thoth, god of wisdom, was ibis-headed. **1907** *Westm. Gaz.* 29 Oct. 4/3 A pretty blouse of white lace, so arranged with ibis-red velvet as to have the effect of a smart little bolero. **1909** *Ibid.* 21 June 5/2 An ibis-red coat and skirt.

Ibiza (ibī·þă, ivī·þă). The name of one of the Balearic Islands, used esp. *attrib.* to denote a local breed of dog, the *Ivicene.

1935 *Hutchinson's Dog Encycl.* II. 924/1 (*caption*) The Ibiza hound..is reputed to refuse to interbreed with any other type of dog. **1948** B. Vesey-Fitzgerald *Bk. Dog* I. 73 An Ibiza hound, one of the rarest breeds I know and found..only in the Balearics. *Ibid.*, The Ibiza is not, at first sight, prepossessing.

Ibizan (ibī·þăn, ivī·þăn), *a.* [f. *Ibiz(a + -an.] = *Ibicencan *a.* **Ibizan hound** = *Ivicene *sb.* Also as *sb.*, the language of Ibiza; an Ibizan hound.

1911 J. E. C. Flitch *Mediterranean Moods* viii. 169 An Ibizan oasis. **1936** M. K. Sheppard *Cottage in Majorca* iii. 51 Their gold ornaments were similar to those of the Ibizan women. **1952** E. Whelpton *Balearics* xv. 199 See the local dances and hear the strange Ibizan music. *Ibid.* xvi. 201 Ibizan, like Mallorquin, is based on Catalan. **1971** F. Hamilton *World Encycl. Dogs* 371 As a hunter the Ibizan Hound is unequalled. **1972** *Country Life* 10 Feb. 329/2 When the sporting varieties were exhibited the classes were almost entirely filled with pharaoh hounds and Ibizan hounds. *Ibid.* Suppl. 17 Ibizans..Irish Terriers..Jack Russell Terriers. **1972** *Listener* 13 July 55/2 Irving, with help from another Ibizan writer, put together the book.

Ibo (ī·bo), *a.* and *sb.* Also **Ebo, Igbo.** [Native name.] **A.** *adj.* Of or pertaining to the Ibos (see below). Cf. Eboe in Dict. and Suppl.

Some of the examples refer to Ibos in the U.S.A. and the West Indies.

1732 *South Carolina Gaz.* 20/1 Stolen..an old Ebo Negro Man;..had on a blue Negro Cloth Frock. **1774** E. Long *Hist. Jamaica* II. iii. ii. 403 The Ebo men are lazy, and averse to every laborious employment; the women performing almost all the work in their own country. **1799** [see Negro 3 a]. **1822** J. Adams *Sk. Voy. Afr.* iii. 41 Breeché, in the Heebo language, signifies gentleman. **1834** [see Eboe]. **1884** *Encycl. Brit.* XVII. 319/1 Soudan and Guinea...Ibo group. **1899** E. A. Wise in *Niger & Yoruba Notes* Nov. 37/1 We are morally pledged to do this by having a Mission in the Ibo country for over 40 years. **1950** D. Jones *Phoneme* 21 The Igbo language of Nigeria. **1951** R. Firth *Elem. Social Organiz.* v. 165 Some of the Ibo people of South-Eastern Nigeria construct elaborate series of clay figures. **1960** *Spectator* 21 Oct. 616 The squalor and nobility of life in an Ibo tribe. **1968** *Listener* 19 Sept. 353/1 The Ibo officer who had just murdered the Premier of the Northern Region. *Ibid.*, The Ibo leader, Ojukwu, and his five or six million Ibo are now concentrated within a narrowing portion of their former region.

B. *sb.* **1. a.** A Negro people of the lower Niger in Africa; also, a member of this people.

1757 *St Jago Intelligencer* 14 May, 1 Ebo, 1 Angola, 1 Mundingo. [**1789** O. Equiano *Life* I. i. 18 Mahogany-coloured men from the south west of us: we call them *Oye-Eboe*, which term signifies red men living at a distance.] **1822** J. Adams *Sk. Voy. Afr.* iii. 40 To this nation the Heebos express a strong aversion. *Ibid.* 41 The Heebos, in their persons, are tall and well-formed. **1822** *Amer. Beacon* (Norfolk, Va.) 3 Sept. 2/1 (Th. Suppl.), Monday Gell is an Ebo, and now in the prime of life. **1836** F. H. Rankin *White Man's Grave* I. v. 106 Shortly after arriving, when Settlers and Maroons were to me as equally black and undistinguishable as Soosoos and Ibbos, I innocently inflicted deep injury on the sensitive mind of the laundress by inquiring why she had omitted to bring home some particular article of dress. **1911** *Encycl. Brit.* XIV. 223/2 The Ibo are a strong well-built Negro race. **1954** M. Gluckman in *Institutions Primitive Soc.* vi. 70 In the past, an Ibo in Nigeria could only travel safely in distant parts to trade by following chains of relationship from place to place. **1960** *Guardian* 15 July 15/3 The Ibos (or better, the Igbos) live mainly in the Eastern Region [of Nigeria]. **1961** *Listener* 30 Nov. 901/2 The intensely individualistic and vital Ibo in the south-east [of Nigeria]. **1973** *Black World* Jan. 9/1 Another example is the figure of Ikenga—god of fortune among the Igbos—in whose left hand is a skull.

b. The language of this people, which constitutes one of the major language groups of Nigeria.

1880 Mrs. G. Sturge tr. *Burdo's Niger & Benueh* viii. 141 'The King, our master,' one of them said to me in the language of Ebo. **1883** R. N. Cust *Sk. Mod. Lang. Afr.* I. xi. 223 Ibo or eboe: commences at the apex of the Delta of the Niger...There appear to be four dialects. **1950** Forde & Jones *Ibo & Ibibio Speaking Peoples* 11 *Igbo* is one of the Kwa languages. **1955** [see *Fanti *sb.* and *a.* a]. **1958** J. S. Coleman *Nigeria* i. 18 Before the British occupation..the present Eastern region consisted of small semiautonomous communities of Ibo- and Ibibio-speaking peoples. **1962** *Amer. Speech* XXXVII. 227 A case of disagreement taken from Ibo, a tone language. **1968** Chomsky & Halle *Sound Pattern Eng.* 378 In a language such as Turkish there are four classes of harmonizing words, rather than two as in Nez Perce or Igbo.

ibogaine (ibō^u^·gă,īn). *Chem.* [a. F. *ibogaïne* (Dybowski & Landrin 1901, in *Compt. Rend.* CXXXIII. 749), f. *iboga*, Congolese name and specific epithet (E. H. Baillon 1889, in *Bull. de la Soc. linn. de Paris* I. 782) of the shrub (see def.): see -ine⁵.] The principal alkaloid, $C_{20}H_{26}N_2O$, of the shrub *Tabernanthe iboga* of equatorial Africa, a colourless crystalline compound that is a pentacyclic indole derivative and acts as a stimulant of the central nervous system when ingested, producing intoxication.

1902 *Jrnl. Chem. Soc.* LXXXII. i. 114 The active principle, ibogaine,..is present in the bark and wood and particularly in the roots of the plant. **1955** *Amer. Jrnl. Psychiatry* CXII. 467 Concerning ibogaine, the natives of French West Africa do not ascribe to it any hallucinogenic property. 'When questioned they insist that it has an action identical with that of alcohol without impairing the reason.' **1960** *Acta Crystallogr.* XIII. 553 Ibogaine, $C_{20}H_{26}N_2O$, is an alkaloid in which an indole ring system is attached to a seven-membered nitrogen containing ring, two sides of which form part of an adjoining tricyclic *iso*-quinuclidine ring system.

Ibsenism (i·bsəniz'm). [f. the name of Henrik *Ibsen* (1828–1906), Norwegian dramatist and poet + -ism.] The dramatic principles and aims characteristic of the writings of Ibsen and the Ibsenites, which examined and criticized social conventions. So **Ibs(c)e·ne, Ibs(c)e·nity** *nonce-wds.* (with play on *obscene, obscenity*); **Ibsene·sque, Ibse·nian, I·bsenish** *adjs.*, of, pertaining to, or resembling the style or views of Ibsen; **I·bsenist, I·bsenite,** an admirer or imitator of Ibsen; also as *adjs.* = **Ibsenian* adj.

1889 E. Dowson in *Lett.* (1967) 432 The brave little band of Ibsenites. **1890** G. B. Shaw *London Music 1888–89* (1937) 283 A reprobate who greatly prefers Ibsenism to Walter Besantism. **1891** —— (*title*) The quintessence of Ibsenism. *Ibid.* 141 When one of the more specifically Ibsenian parts has to be filled, it is actually safer to entrust it to a novice than to a competent and experienced actor. *Ibid.* App. 158 Without being necessarily an Ibsenist, a critic may see at a glance that abuse of the sort quoted..is worthless. *Ibid.* 159 Mr William Archer expressly guards himself against being taken as an Ibsenist doctrinaire. **1892** *National Observer* 17 Dec. 107/1 When the din of political factions is silent, and Ibscenity has faded into a literary curiosity. **1893** *Ibid.* 7 Jan. 190/2 Her story is amateurish, sentimental, Ibsene. But Ibsenity is in the air. **1893** *Athenæum* 16 Dec. 857/3 The suicide of the woman..is nothing if not Ibsenish. It is unheroic, unromantic, ineffective, insignificant. **1895** G. B. Shaw in *Sat. Rev.* 9 Nov. 618/1 The material is what we now call Ibsenite: the technique is that of Scribe. **1902** Chesterton *Lunacy & Lett.* (1958) 39 The resistance of the conventional mind to Ibsenism..is fundamentally right. **1902** M. Beerbohm in *Sat. Rev.* 24 May 644/2 Ellida..is the usual Ibsenist heroine, propounder of the regular Ibsenist ideas. **1902** W. B. Yeats *Let.* 4 Dec. (1954) 386 He [*sc.* Joyce]..did not knock at the gate with his old Ibsenite fury. **1905** *Daily Chron.* 7 July 8/4 She takes what might be called an Ibsenian view of humanity. **1906** M. Beerbohm in *Sat. Rev.* 5 May 552/2 A strong-minded, Ibsenesque heroine, with a contempt for social conventions. **1906** W. Stevens *Let.* 27 Apr. (1967) 91 You always were an Ibsenist, without knowing it. **1912** R. Macaulay *Views & Vagabonds* vi. 112 She thinks they're Ibsenesque, but really they're like Miss Yonge in a fit of religious doubt. **1916** *Everyman* 5 May 54/2 Their passion for the erotic and Ibsene. **1916** T. MacDonagh *Lit. in Ireland* 18 The drama of modern Ireland, in English, is..not free from the faults of impressionism, of quasi-scientific Ibsenism, of unreal gloom and of shallow cynicism. **1928** *Radio Times* 16 Mar. 564/2 Two distinguished Ibsenites..will be heard from London tonight. **1957** A. Miller *Coll. Plays* (1958) Introd. 12 When *All My Sons* opened on Broadway it was called an Ibsenesque play. **1966** *Punch* 9 Feb. 206/1 The first love—for all her professed Ibsenist rationalism—won't let him go. **1970** *Daily Tel.* 30 Oct. 12/2 The play..remained Scandinavian in..its Ibsenite thesis that the truth is dangerous to man's precarious happiness. **1972** *Listener* 7 Sept. 310/1 The formation of the 'new woman' of the Ibsenite and Shavian period. **1973** *Times* 8 June 11/6 This meeting between Dr Miller and Charles Darwin, for example, had something unmistakably Ibsenish about it.

icaco (ikā·ko). [a. Sp. *icaco, hicaco*, f. Taino *hikako.*] A small tree, *Chrysobalanus icaco*, of the family Rosaceæ, native to tropical America and the West Indies; the fruit of this tree. Also called Coco-plum.

1752 P. Miller *Gardeners Dict.* (ed. 6) s.v. Chrysobalanus, This Genus of Plants is titled by Father Plumier *Icaco* which is the Indian name of this Fruit. **1756** P. Browne *Civil & Nat. Hist. Jamaica* 250 (*heading*) Chrysobalanus... Icaco... The Cocco Plumb Tree. **1852** T. Ross tr. *A. von Humboldt's Personal Narr. Trav. Amer.* II. xvii. 136 Hedges of bead-trees encircled groups of icacoes laden with fruit. **1887** C. A. Moloney *Sk. Forestry W. Africa* 347 Icaco or Cocoa Plum of the West Indies. **1943** Record & Hess *Timbers of New World* 454/2 The best known species [of *Chrysobalanus*] is the Coco Plum or *Icaco, C. icaco* L., which has a large edible fruit so wrinkled as to suggest the face of a monkey.

Icarian, *a.*[1] (Later examples.)

1936 E. Sitwell *Sel. Poems* 249 Eagle-winged Icarian flights. **1972** *Daily Tel.* (Colour Suppl.) 1 Dec. 16/1 In the view of some social philosophers and historians, space flight is an Icarian venture at its best—and an extravagance at its worst.

Icarus (i·kărŏs). *Gr. Myth.* The name of the son of Dædalus, who attempted to fly by means of artificial wings fastened with wax (see Icarian *a.*[1]): used allusively.

[**1589** Greene *Menaphon* (Arb.) 53, I feare..in the height of my thoughts soaring too high, to fall with wofull repenting Icarus.] **1591** Shakes. *1 Hen. VI*, iv. vi. 56 Then follow thou thy desp'rate Syre of Creet, Thou Icarus, thy Life to me is sweet. *Ibid.* vii. 16 There di'de My Icarus, my Blossome, in his pride. **1594** Nashe *Unfort. Trav.* I3ᵛ, These insolent fancies are but Icarus feathers, whose wanton waxe melted against the sun. **1694** D'Urfey *Don Quix.* ii. Ep. Ded. sig. A1ᵛ, The roving Icarus in Poetry, By you is levell'd, when he soars too high. **1924** B. Russell (*title*) Icarus, or, The future of science. **1931** *Times Lit. Suppl.* 24 Sept. 714/2 Much that he [*sc.* D'Annunzio] has given is not pure gold, but decorative lumber comparable to the ill-assorted trophies of the Vittoriale from which, an Icarus who has ceased to fly, he makes well-calculated sallies.

ice, *sb.* Add: **2. c.** Phrases. *on ice*: (*a*) kept out of the way until wanted, in reserve; in custody, in prison; (*b*) of a venture, game, etc.: sure of being achieved or won, a certainty; *to cut no ice*: to carry no weight, to fail to impress; hence *to cut ice*: to impress, to make an effect.

1890 A. C. Gunter *Miss Nobody* xx. 231 For Election. Gussie de P. Van Beekman... On ice! **1894** P. L. Ford *Hon. Peter Stirling* 328 They say she's never been able to find a man good enough for her, and so she's keeping herself on ice. **1895** J. S. Wood *Yale Yarns* 12 Such speeches! Eloquence cut no ice at *that* dinner. **1897** *Scribner's Mag.* Sept. 305/1 And it don't cut no ice with you whether folks call you inconsistent or not. **1916** J. Buchan *Greenmantle* ix. 117 Because the German mercantile marine was laid on ice till the end of the war, they had turned him on to this show. **1917** A. Conan Doyle *His Last Bow* 291 It cuts no ice with a British copper to tell him you're an American citizen. **1924** A. Huxley *Let.* 28 Oct. (1969) 235, I was very glad..to hear that you liked *Those Barren Leaves.* It cuts more ice, I think, than the others and is more explicit and to the point. **1930** W. S. Maugham *Cakes & Ale* iv. 48 There was a softness in Roy's voice such as I imagined he would use if he were telling a prospective father that his wife was about to gratify his wishes. It cut no ice with me. **1930** G. B. Shaw *Apple Cart* i. 7 Oh, sit down, man, sit down. Youre in your own house: ceremony cuts no ice with me. **1931** *Sat. Rev. Lit.* (U.S.) 18 July 978/2 Among the words and phrases common among racketeers, not yet in general use..there are the following:..*on ice*, in the penitentiary. **1932** E. Bowen *To North* xxiii. 247 Sheer man-to-man envy of Markie for cutting so much ice. **1933** D. L. Sayers *Murder must Advertise* xix. 322 Their idea is to put you on ice quietly till they've had time to settle up their affairs. **1936** E. S. Gardner *Case of Sleepwalker's Niece* xiv. 131, I figured that and the record of the telephone call would be enough to put the case on ice. **1936** Wodehouse *Laughing Gas* xviii. 195 Take him back to his room and keep him there on ice till it's time to go to the studio. **1944** W. S. Maugham *Razor's Edge* vii. 276, I haven't signed on the dotted line yet, but it's on ice. The fella I'm going in with was a roommate of mine at college..and I'm dead sure he wouldn't hand me a lemon. **1945** *Chicago Daily News* 4 Oct. 12/1 They ..accumulated enough runs in the first inning to put the game on ice. **1953** W. Burroughs *Junkie* (1972) x. 103 'I sent for the wagon,' said the guy with the pipe. 'We'll take them over to the third precinct and put them on ice.' **1954** Koestler *Invis. Writing* xxxiv. 369 He will for a considerable time be 'put on ice'—isolated from any contact with other members. **1955** *Times* 12 May 14/5 Burns demanded that the purse be handed over before he entered the ring. He ended a bitter argument by declaring: 'That cuts no ice with me. I want the referee to hold the money.' **1957** A. Grimble *Return to Islands* 75 The problem that his resignation had left on ice, for whomever it might concern. **1965** *New Statesman* 14 May 771/2 Presumably the book, finished in 1957, was put on ice, for Mr Wood can hardly have hoped to get away in Beaverbrook's lifetime with much of the discussion of his 'brash brutality'. **1973** 'I. Drummond' *Jaws of Watchdog* xii. 154 Scotland Yard could not keep him on ice that long. He would have to be brought to a court to be charged. **1973** J. Porter *It's Murder with Dover* vii. 63 MacGregor remembered..that logical argument didn't cut much ice with Dover and he abandoned it.

d. A piece or pieces of ice placed in a drink, or into which a bottle, etc., is placed to cool the contents.

1833 C. Redding *Hist. Mod. Wines* xiv. 316 Before drinking, the wine should be kept an hour in ice. **1846** A. Soyer *Gastronomic Regenerator* 701 When the syrup is a little cool, taste if palatable, place a little upon some ice, and if strong enough fill your mould, which place in ice. *a* **1922** T. S. Eliot *Waste Land Drafts* (1971) 61 Where's a cocktail shaker, Ben, here's plenty of cracked ice. **1927** E. Glyn *'It'* xii. 110 A perfect dinner had been ordered—the champagne was on ice.

3. (Further examples.)

1884 'Mark Twain' *Huck. Finn* xxxii. 333 'How'd you get your breakfast so early on the boat?' It was kinder thin ice, but I says: 'The captain see me standing around.' **1904** [see *coin *sb.* 6]. **1938** E. Bowen *Death of Heart* iii. ii. 343 The *idea* of her never leaves me quiet, and by coming into this room she drives me on to the ice. **1962** J. G. Bennett *Witness* xxii. 287 Thus, without knowing it, I was treading on very thin ice.

4. b. = **ice pigeon.*

1881 J. C. Lyell *Fancy Pigeons* 81 The smooth-legged chequered or spangled ones are known in this country as Ural ice, while the rough-legged spangled birds are called Siberian ice.

c. Diamonds; jewellery. *slang* (orig. *U.S.*).

1906 H. Green *At Actors' Boarding House* 26 Her in evenin' clothes and a bunch of ice on her hands. **1915** G. Bronson-Howard *God's Man* iv. iv. 281 Along comes a guy..a piece of ice in his tie that made Tiffany's front window look like a hardware exhibit. **1924** Wodehouse *Leave it to Psmith* ix. 184 Diamonds, Eddie. A necklace. ..Some of the best ice I've saw in years. **1925, 1942** [see *hot a. 7 e]. **1936** J. G. Brandon *Pawnshop Murder* i. 2 The glitter of stolen 'ice' or other jewels spread out upon the table. **1942** M. Schlauch *Gift of Tongues* (1943) 269 Jewels become 'ice'. **1959** *Listener* 23 Apr. 706/2 'Shiners' and 'ice' to the light-fingered boys, the diamond is known to the gemmologist as the hardest..of all minerals. **1961** Wodehouse *Ice in Bedroom* i. 11 Yes, someone got away with her bit of ice all right. **1972** 'H. Howard' *Nice Day for Funeral* i. 30 Prager caught sight of five hundred grand in cracked ice.

d. Profit from the illegal sale of theatre, cinema, etc., tickets. *U.S. slang.*

1927 *Theatre Mag.* Sept. 30/2 Thousands of tickets for special attractions in the large movie houses are sold over the box-office counter to speculators by the treasurers of the houses, their charge, or 'ice', running to as much as $1 a ticket on the 'sell-outs'. **1960** *Observer* 30 Oct. 1/17 'Ice'..is the money from the sale of hot tickets..and enables..box office clerks to buy themselves Cadillacs. **1964** *Economist* 25 Jan. 313/2 Kick-backs—'ice' as it is called on Broadway—on theatre tickets whose prices are marked up illegally.

e. Protection money. *slang.*

1948 E. L. Johnson in E. L. Irey *Tax Dodgers* (1949) xiii. 229 Willie..said, 'The extra hundred sixty is ice.' 'Ice' is argot for graft or protection fees. **1951** *Economist* 29 Sept. 747 Gross..who had confessed to paying this sum in 'ice' for the protection that made it possible for him to earn $100,000 a year. **1951** E. Kefauver *Crime in Amer.* (1952) xvii. 186 When the combine's books

finally were seized, examination disclosed recorded payments totalling $108,000 for the service known as 'juice', which is the California gambling profession's euphemism (in Florida the term is 'ice') for 'protection' money.

5. b. (Earlier and later examples.)

1723 J. Nott *Cook's & Confectioner's Dict.* sig. B4 Make Ice with the White of an Egg, powder'd Sugar, Orange or Lemon Flowers. **1885** C. M. Yonge *Nuttie's Father* II. v. 61 How dreadfully hard the ice on the wedding cake was, so that when Annaple tried to cut it the knife slipped.

7. a. *ice-dancer* (so *ice-dancing* vbl. sb.), *-merchant, -show, -wagon; ice-cutter* (examples). Also *ice-clear* adj.

1946 S. Spender *European Witness* ii. iii. 146 The ice-clear light of that part of Germany. **1948** C. Day Lewis *Poems 1943–1947* 87 Toward my expectation's bed They move in a hushed, ice-clear trance. **1970** R. Lowell *Notebk.* 204 Once or twice, blurt your ice-clear sentence. **1791** J. Long *Voy. & Trav. Indian Trader* 120 The fishing party consisted of..natives of Canada, who, being provided with axes, ice-cutters..set off. **1969** *New Scientist* 13 Mar. 574/2 A new type of ice-breaker is needed for breaking up solid ice. We have devised one and christened it an ice-cutter. **1925** E. Law *Dancing on Ice* iii. 24 To the question what should ice-dancers do to acquit themselves properly, the obvious answer is, that they should first learn exactly what the valse-figure is. **1969** *Times* 15 Nov. 10/8 Only one skater now remains of that team of six brilliant ice dancers. **1925** E. Law *Dancing on Ice* v. 46 Invite the independent judgment of..some one who knows and understands the whole theory and practice of ice-dancing. **1969** *Times* 15 Nov. 10/8 The world ice dancing champions..will not be competing. **1864** *Chambers's Jrnl.* 99/1 The men take the ice to ice-merchants, who are ready to buy it in any quantity. **1973** *Post Office Telephone Directory Section 471, London Yellow Pages Classified (North)* 177/2 (*heading*) Ice and cold storage companies and ice merchants. **1950** *Oxf. Jun. Encycl.* IX. 432/2 Of recent years 'ice shows' have been made popular in the U.S.A. and Canada. **1966** *Listener* 29 Dec. 959/1 Original plays..and an ice-show. **1865** J. D. Burn *Three Years among Working-Classes U.S.* 304 The ice-waggons may be seen with their crystal loads flying about the towns in all directions from May to the end of September. **1898** J. London *Let.* 30 Nov. (1966) 6 Saturday I worked on an ice wagon. **1905** *Sketch* LI. 38/1 The earliest on his rounds was the man with the ice-waggon, who put down on the door-step of each house a block of ice. **1971** M. Tak *Truck Talk* 88 *Ice wagon*, a refrigerated trailer. **1972** *News & Observer* (Raleigh, N. Carolina) 30 Dec. 4/2 We don't hear much [nowadays] about..ice wagons, branch-lines.

b. *ice-barrier* (examples), *-blockade, -bridge, -cake* (examples), *-cover, -crystal, -edge, -face, -flake, -fringe, -hump, -lake* (examples), *-margin, -ridge* (examples), *-spicule, -waste* (examples).

1874 G. Campbell *Let.* 14 Feb. in *Log Lett. from 'Challenger'* (1876) ii. 99 We had hoped to see the great ice-barrier, that endless wall of ice two hundred feet in height which fringes the southern continent. **1934** I. W. Hutchison *North to Rime-Ringed Sun* x. 106 The ice-barrier was thickening fast, and as it seemed impenetrable under the westerly gale..we cast the *Trader's* anchor. **1939** L. MacNeice *Autumn Jrnl.* xxiv. 95 The waters of life are free of the ice-blockade of hunger. **1969** in Halpert & Story *Christmas Mumming in Newfoundland* 32 The winter ice-blockade which made the inshore fishery a limited seasonal operation. **1792** E. P. Simcoe *Diary* 15 Feb. (1911) vii. 77 Coll. Simcoe and I were going to walk on the ice bridge. **1880** 'Mark Twain' *Tramp Abroad* II. xl. 115 A young porter..started across an ice-bridge which spanned a crevasse. **1909** *Westm. Gaz.* 2 Sept. 9/2 After twenty days we found an ice-bridge over the water between the continental ice and the Polar ice. **1870** *Canad. Illustr. News* 26 Mar. 334/1 One ice-cake after another struck her boat. **1923** R. Frost *New Hampshire* (1924) 65 The seal yelp On an ice cake. **1953** *Canad. Geogr. Jrnl.* July 150/1 A strip of open water stretched between shore and ice-cake which filled most of the bay. **1960** S. Plath *Colossus* (1967) 44 Farther out, the waves will be mouthing icecakes. **1966** T. Armstrong et al. *Illustr. Gloss. Snow & Ice* 21 *Ice cake*, a floe smaller than 10 m across. **1882** A. Geikie *Text-bk. Geol.* iii. ii. 416 On the ice-worn surface of Norway singular cavities..known as 'giants' kettles'..have had an origin under the massive ice-cover which once spread over that peninsula. **1958** Priebsch & Collinson *German Lang.* (ed. 4) ii. 22 The ice-cover made the whole of Scandinavia and Northern Germany uninhabitable. **1849** Thoreau *Week Concord Riv.* 394 It matters not through what ice-crystals it is seen. **1919** D. H. Lawrence in *Eng. Rev.* June 485 Some, blonde, blue-eyed, northern, are evidently water-born, born along with the ice-crystals and blue, cold deeps. **1956** *Nature* 18 Feb. 321/2 Collisions occurring between ice crystals and small hailstones. **1947** G. Rawson *Arctic Adventures* ix. 187 The cutter reached the ice edge, and a man sprang out of it. **1966** T. Armstrong et al. *Illustr. Gloss. Snow & Ice* 22 *Ice edge*, the boundary at any given time between open water and sea, river or lake ice of any kind, whether drifting or fast. **1856** E. K. Kane *Arctic Explorations* I. xvi. 187, I had to walk through the broken ice, which rose in toppling spires over my head, for nearly fifty yards, before I found an opening to the ice-face, by which I was able to climb down to them. **1898** J. O. Maund in W. A. Morgan *'House' on Sport* 276 Above the ice face snow slopes lying at a much less rapid angle led to the final rock peak. **1915** E. Pound *Cathay* 23 Hung with hard ice-flakes, where hail-scur flew. **1937** *Discovery* July 220/1 Two of them..were the first to land on the ice-flake. **1902** *Spectator* 25 Oct. 604/1 Persecuted until they were practically driven off the seas, they took refuge furthest north along the ice-fringe. **1966** T. Armstrong et al. *Illustr. Gloss. Snow & Ice* 23 *Ice fringe*, a very narrow ice piedmont, extending less than about 1 km inland from the sea. **1910** W. de la Mare *Three Mulla-Mulgars* 67 Floating like a cork among the ice-humps. **1961** *Times* 24 Apr. 14/7, I cannot tell whether an icehump is two inches high or a dangerous two feet. **1934** Ice-lake [see *geochronological *a.*]. **1957** G. E. Hutchinson *Treat. Limnol.* I. i. 8 The earliest postglacial Baltic was an ice lake. **1889** G. F. Wright *Ice Age N. Amer.* xx. 482 The buried vegetable deposits under consideration do not mark a warm climate, but a climate much colder than the present—such a vegetation, in fact, as would naturally flourish near the ice-margin. **1958** F. E. Zeuner *Dating Past* (ed. 4) 22 As the ice-margin retreated the varves followed it. **1892** C. T. Dent et al. *Mountaineering* vi. 223 Nothing tends more to weary and render him [*sc.* a mountaineer] careless than some hours of step-cutting on an ice-ridge. **1929** F. Smythe *Climbs & Ski Runs* xv. 288 We gained the foot of the rock pitch separating the third ice-ridge from the fourth. **1957** J. Blish *Fallen Star* vii. 104 The ice-ridge on our left screamed, broke free and reared skyward. **1881** *Nature* 10 Feb. 338/2 The ice-spicules are built up 'in the teeth' of this current. **1962** F. I. Ordway et al. *Basic Astronautics* iii. 90 Three in number, they [*sc.* the rings of Saturn] are very thin in comparison to their width, and are made up of millions upon millions of..particles, pebbles, grains of dust, and perhaps ice spicules. **1905** *Westm. Gaz.* 9 Dec. 16/1 There is neither the sport nor the game that cheers the Northern ice-wastes. **1964** F. Warner *Early Poems* 80 Lonely ice-wastes.

c. *ice-coated, -covered* (later example), *-locked* (examples) adjs.

1880 'Mark Twain' *Tramp Abroad* II. xxxiv. 32 They came to an ice-coated ridge. **1928** *Observer* 15 July 22 Ice-coated ships. **1956** *Nature* 17 Mar. 508/1 The lofty ice-covered interior of Antarctica. **1866** J. G. Whittier *Snow-Bound* 50 Wide swung again our ice-locked door. **1907** *Westm. Gaz.* 21 Oct. 2/3 Ice-locked Polar snows. **1972** S. Burnford *One Woman's Arctic* i. 13 Now the strait was ice-locked.

d. *ice-dammed* (see sense 8 below), *-hearted* (later example) adjs.

1950 G. Barker *News of World* 49 Just as the ice-hearted stars Stand around like avatars. **1960** S. Plath *Colossus* (1967) 23 At the source Of your ice-hearted calling—Drunkenness of the great depths.

e. *ice-breaking* (earlier and later examples), *-cutting* (examples), *-making* (earlier example).

1824 *Canad. Mag.* III. Dec. 541 He accosted us all gaily, without any of that long ice-breaking conversation about the weather, which generally occupies the first half-hour of our stage-coach journeys. **1956** *Nature* 31 Mar. 600/2 The two expeditions will share a large ice-breaking sealer. **1970** *Daily Tel.* 21 Mar. 15 Canadian seal-hunting ships with ice-breaking capacity. **1854** Thoreau *Walden* 229 The only obvious employment, except wood-chopping, ice-cutting, or the like business. **1908** *Westm. Gaz.* 29 May 2/1 The ice-cutting looks so like harvesting or hay-making. **1960** J. J. Rowlands *Spindrift* 75 Few remember the days of ice-cutting. **1864** *Chambers's Jrnl.* 13 Feb. 101/1 The attention of some millions of persons was attracted, in the International Exhibition of 1862, to two ice-making machines of a very remarkable character.

8. ice-age, also *fig.* and *Comb.*; **ice-apron** (earlier example); (*b*) (see quot. 1958); **ice-arm,** an arm or projecting portion of ice; **ice-ax** (usu. **ice-axe**), (*b*) an implement for cutting ice for domestic purposes; = **ice-pick* (*a*); **ice-block,** (*a*) a block of ice; also *fig.*; (*b*) Austral. = **ice-lolly*; **ice-blue,** a very pale blue; **ice-box** (further examples); also (*U.S.*), a refrigerator (see also quot. 1971); also *transf.*; **ice-bucket** = *ice-pail*; **ice-capade** [jocular blend of Ice *sb.* + Escapade], an event, show, etc., that takes place on ice; **ice-car,** a refrigerating van adapted for the transport of perishable goods; **ice-cart,** a cart in which ice is conveyed for delivery; **ice-cave** (examples); (*b*) a hollow in the ice at the lower end of a glacier; (*c*) a kind of small domestic refrigerator (*disused*); (*d*) a cave hollowed out of ice as a shelter; **ice-cellar** *N. Amer.*, a cellar kept cool by blocks of ice and used to preserve food; **ice-chest** (earlier and later examples); **ice-chimney,** a chimney (Chimney *sb.* 8) formed in ice; **ice-claw** = Crampon *sb.* 3; **ice-cloud,** a cloud consisting of ice crystals; **ice colour,** any of a class of azo dyes which are insoluble in water, being formed directly on the fibre by impregnating it with one component of the dye (naphthol or a naphthol derivative) and then immersing it in a solution of the other (a diazo compound); so called because the solutions originally used needed to be kept ice-cold; **ice contact** *Physical Geogr.*, a surface or deposit that was originally formed in contact with a body of ice; usu. *attrib.*, esp. in *ice-contact slope*, a (usually steep) slope so formed; **ice-craft** (later examples); **ice-crop,** the yield of ice in a single winter or from a certain place; **ice cube,** a small block of ice formed in a mould in a refrigerator and used to chill drinks; **ice-dam,** a dam across a river formed by a glacier; so **ice-dammed** *a.*; **ice-dyke,** a narrow crevasse filled with ice columns; **ice-farm,** (*a*) a place where ice is formed by allowing water to freeze in specially hollowed-out beds, or in shallow earthenware pans; (*b*) a place where

naturally frozen ice is stored; **ice-fishing** (examples; see also quot. 1907); hence [as a back-formation] *ice-fish* v. intr.; *ice-fisherman*; **ice-flowers,** delete †*Obs.* and add later examples; (*b*) (see quot. 1955), also in *sing.*; **ice-front,** the margin of a glacier, ice-shelf, or ice-sheet; **ice-gorge** (earlier example); **ice-green,** a very pale green; **ice guard,** (*a*) (see quot. 1905) *U.S.*; (*b*) *Aeronaut.*, a wire grid that may be fitted in the intake of an aero-engine, so that any ice forms on it rather than in the engine; **ice-hammer,** (*a*) a hammer for breaking ice to be used in drinks; (*b*) a hammer used in mountaineering (see quot. 1932); **ice-harvest,** the ice-crop; the period during which the ice-crop is gathered; **ice hockey,** a game developed from field hockey but played on ice; also *attrib.*; **ice-jam,** the blocking of a channel with broken ice; the jam so formed; also *fig.*; **ice-lane** (see quot.); **ice line,** in a phase diagram of water, a line representing the conditions of temperature and pressure at which ice and water will be in equilibrium in the absence of water vapour; **ice-lobe,** a portion of a continental ice-sheet that projects from the main area; **ice-lolly,** a water ice on a stick; **ice-machine** (examples); **ice-maiden** *colloq.*, a 'cold' or unresponsive woman; **ice-maker** (later example); (*b*) = *ice-machine*; **ice-mould,** a hollow utensil used in shaping ice; **ice-needle,** (*a*) a strong needle used to break up a lump of ice; (*b*) any elongated, needle-like ice crystal; **ice-pack** (later examples); (*b*) a pack (Pack *sb.*[1] 11) prepared with ice; **ice-pan,** a small slab of floating ice; **ice-pick,** (*a*) a small domestic tool with a sharp spike designed for breaking up ice (e.g. for drinks); (*b*) in Mountaineering, a pick (Pick *sb.*[1]); **ice pigeon,** a breed of domestic pigeon whose prevailing colour is a pale bluish lavender; **ice-pitcher** (examples); **ice-piton,** a piton used to assist climbing on ice; **ice-plate,** a small, usu. glass, plate on which ice-cream is served; **ice-point,** a temperature at which ice and water are in equilibrium; *spec.* the temperature (0°C.) at which ice is in equilibrium with water saturated with air and under standard atmospheric pressure, formerly taken as a primary fixed point but now replaced for this purpose by the triple point; **ice-pole** *Canad.*, a long pole used by seamen for levering against ice-floes, etc.; **ice-push,** lateral pressure exerted on a shore as a sheet of floating ice expands following changes in temperature; also, an ice-rampart formed as a result; **ice-rampart,** a ridge of beach material along a shore-line which has been forced up by the lateral movement of floating ice; **ice-rink** (see Rink *sb.*[2] 3); **ice-run,** a stretch of ice prepared for tobogganing; also *fig.*; **Ice Saints** (see quot. 1922); **ice-scape** [after Landscape], (a picture of) ice scenery; **ice-scour, -scouring,** the action of an ice-sheet or glacier in eroding the land and modifying and producing landforms; so **ice-scoured** *a.*; **ice-screw,** also, an ice-piton (q.v.) which is screwed, rather than hammered, into the ice; **ice-shed,** a divide between two expanses of moving ice; **ice-shelf,** a floating sheet of ice permanently attached to a land mass; **ice-skate** = Skate *sb.*[2] 1; also as *vb.*; so *ice-skater, ice-skating* vbl. sb.; **ice-spirit,** frost as a nature-spirit; **ice sport,** a sport taking place on ice; **ice step,** a step cut into ice; **ice-storm,** a storm of freezing rain that leaves a deposit of ice on trees, etc.; **ice-tongue,** any body of ice that projects from a glacier, iceberg, or ice-sheet, esp. one that is relatively long and narrow (see also quot. 1893); **ice-tray,** a tray used in a refrigerator for making ice cubes; **ice-wedge,** a vertical wedge-shaped mass of ice in the soil of a permafrost region; **ice-white** *a.*, having a whiteness like that of ice; **ice wool** = *eis wool.

1957 C. Day Lewis *Pegasus* 24 Cold chisels of wind, ice-age-edged. **1966** *Listener* 10 Mar. 338/2 With..the certainty by 1950 that America and Russia both possessed atomic weapons, the world entered the new ice-age of the cold war between the big two. **1973** A. Price *October Men* ix. 122 The temperature was perhaps slightly less arctic now he had said his piece, but that was no sure sign that a second..ice age was not about to set in. **1871** *Scribner's Monthly* II. 170 It [has been] necessary to construct enormous breakwaters, having ice-aprons of strong oak timber. **1958** Armstrong & Roberts in *Polar Record* IX. lix. 93 *Ice apron*, a thin mass of snow and ice adhering to a mountain side. **1928** *Daily Tel.* 4 Sept. 11/5 He..had mistaken the Frederikshaab ice arm for the Sukkertoppen ice arm. **1960** J. J. Rowlands *Spindrift* 69 It was the magic of his [*sc.* an ice-vendor's] skill in using an ice-axe that enthralled me. **1963** I. Deutscher *Prophet Outcast: Trotsky* v. 504 He [*sc.* Trotsky] grappled with the murderer, bit his hand, and wrenched the ice-axe from him. **1853** Ice-block [in Dict., sense 7 b]. **1864** G. M. Hopkins *Notebks. & Papers* (1937) 27 Those wastes where the ice-blocks tilt and fret. **1957** J. Blish *Falling Star* ix. 120 We built Wentz's igloo with..the one Keystone ice-block at the summit. **1958** *Church Times* 3 Jan. 3/1 Hopes of some thaw in the international ice-block rose with the publication of warm greetings of peace and goodwill sent by the Russian leaders. **1962** J. R. Bernard in *Southerly* XXII. 97 Without loss we add to..the meaning of..ice-block that of frozen confection. **1966** Baker *Austral. Lang.* (ed. 2) xiii. 290 *Ice-block*, a small block of coloured and sweet-tasting ice on a stick. **1935** *N. & Q.* 5 Jan. 7/2 From a draper's catalogue for the coming winter sales, I cull a few names of colours to me at least new: Ice-blue, [etc.]. **1952** 'J. Ross' *Yellow Drawing-Room* ix. 138 An ice-blue satin skirt. **1958** A. Wilson *Middle Age of Mrs Eliot* ii. 219 The walls [were] distempered ice blue. **1970** W. Smith *Gold Mine* xxv. 57 His office was in white and ice-blue. **1846** *St. Louis Reveille* 9 Sept. 4/5 Everything requisite for funerals, such as Hearse, Carriages,..Ice, Ice-boxes. **1877** [see *bavaroise]. **1908** Wodehouse & Westbrook *Globe by the Way Bk.* 13/2 His brain worked like a buzz-saw in an ice-box. **1927** *Rev. Eng. Stud.* Oct. 435 This healthy linguistic instinct is seen in *e.g.* the substitution of *raincoat* for *mackintosh*, *ice-box* for *refrigerator.* **1938** D. Castle *Do Your Own Time* v. 45 Scavengers..cut down the hanged men, place them in cheap coffins, and cart them to the 'ice box', as the morgue is known in prison. **1943** Wyndham Lewis *Let.* 31 Mar. (1963) 352 We are freezing out here [*sc.* in Canada] slowly, in this icebox of a country. **1963** *Amer. Speech* XXXVIII. 173 *Icebox*, a co-ed engaged to a young man in a distant college who refuses to date at all while at college. **1971** M. Tak *Truck Talk* 87 *Ice box.* (1) A refrigerated trailer used for hauling produce and perishables. (2) The bunker for ice in a bunk-and-blower type cooling system in an insulated trailer. **1919** Barrie *Alice Sit-by-the-Fire* ii. 61 Supper for two, champagne in an ice-bucket. **1929** E. Hemingway *Farewell to Arms* xxv. 276 The champagne in the ice-bucket and our glasses on the table. **1939** N. Monsarrat *This is Schoolroom* ii. vii. 176 A magnum of GH Mumm ready to hand in an ice-bucket. **1959** N. Marsh *False Scent* (1960) iv. 94 Gantry tipped some [water] out of the ice bucket. **1941** *Time* 10 Feb. 67/1 Another touring frostbite fiesta called *Ice-Capades of 1941*. **1963** *Times* 7 Mar. 3/6 Football.. finally broke out of the strangling cocoon of snow, ice and mud to be rediscovered as a thing of excitement and calculated skill almost forgotten during the recent ice-capades. **1909** *Chambers's Jrnl.* Aug. 560/2 In Canada there is also a special ice-car service for the carriage of butter to Montreal. **1842** *Knickerbocker* XX. 205 Before an omnibus or hotel or restaurant or ice-cart had assumed its popular cognomen. **1864** T. L. Nichols *40 Yrs. Amer. Life* I. 247 Every morning the ice-cart comes round. **1873** *Young Englishwoman* July 334/1 Ice-carts call as regularly as does the baker. **1889** A. B. Marshall *Cookery Bk.* p. xx (Advt.), Marshall's patent ice cave.. will freeze a quantity of water placed in the inner cave into a solid mass. **1897** *Geogr. Jrnl.* June 670 The term 'ice-cave'..should especially apply to the hollows in the ice at the lower end of glaciers, whence the glacier waters make their exit. **1911** *Madame* 20 May 318/1 Various forms of iced pudding, which, even if you do not happen to possess one of A. B. Marshall's ice caves, are still quite possible to prepare with very little trouble. **1926–7** *Army & Navy Stores Catal.* 167/3 Ice caves with loose shelves. Japanned iron. **1930** F. Smythe *Kangchenjunga Adventure* ii. 45 In lieu of tents, ice caves were carved in the solid ice at Camps Eight and Nine, large enough to hold six to eight persons. **1933** J. Buchan *Prince of Captivity* i. iii. 95 He would draw terrible pictures of an ice-cave at Gundbjorns Fjord, and two dead men. **1950** *Chambers's Encycl.* VII. 359/1 The so-called ice caves (not to be confused with caves in glacier ice) are caves in solid rock which, although situated below the line of perpetual snow, nevertheless contain large deposits of ice. **1961** R. M. Patterson *Buffalo Head* iv. 136 We explored..finding..mountain sheep, ice caves, the fairest of alpine rock gardens, [etc.]. **1771** J. R. Forster tr. *Kalm's Trav. N. Amer.* III. 232 Some of the people of quality make use of ice-cellars, to keep beer cool. *Ibid.*, These ice-cellars are commonly built of stone, under the house. **1865** Milton & Cheadle *N.W. Passage by Land* vi. 82 They even went down into the ice-cellar, where the meat is kept. **1883** *Harper's Mag.* July 261, I visited one of the..largest beer factories, and took copious notes about..the ice cellars colder than Siberia ever dared to be. **1921** *Chambers's Jrnl.* 21 May 395/2 Seeds of maple and wheat have been observed growing into blocks of ice in an ice-cellar. **1841** C. Cist *Cincinnati in 1841* (Advt.), Manufacturer of packing-boxes, ice-chests, trunk and segar boxes, &c. **1935** 'J. Guthrie' *Little Country* xxiv. 359 Young Merryweather..thought the place like an ice chest. **1972** *Even. Telegram* (St. John's, Newfoundland) (Advt. Suppl.) 27 June 1 Foam plastic ice chest 99¢. **1929** F. Smythe *Climbs & Ski Runs* xv. 293 The ice-chimney continued for some distance with great difficulty. **1934** *Discovery* Mar. 60/2 We were obliged to negotiate some ugly ice-chimneys. **1955** J. E. B. Wright *Technique Mountaineering* v. 97 Ice chimneys may be pitches in couloirs and icefalls. **1920** G. W. Young *Mountain Craft* vii. 286 Any man who wishes to make big ascents is well advised if he begins early to learn how to use ice-claws (or crampons). **1954** W. Noyce *South Col* v. 79 Crampons or ice-claws are sets of metal spikes on frames tied to the feet to assist walking up ice. **1955** G. Band *Road to Rakaposhi* xii. 141 Climbers have managed for so long without these useful 'ice-claws' in the Himalayas. **1883** R. H. Scott *Elem. Meteorol.* 403 (Index), Ice clouds. **1900** W. Allingham *Man. Marine Meteorol.* xvi. 154 Halos and other refraction phenomena afford proof that cirro-stratus is an ice cloud. **1963** D. Irving *Destruction of Dresden* iii. i. 106 Ice clouds were blanketing Europe. **1903** C. Salter tr. *G von Georgievics's Chem. Dye-Stuffs* 82 This method is employed to produce dyeings of considerable beauty and fastness, which now play a great part in dyeing cotton piece goods and in calico printing (Ice colours). **1968** E. N. Abrahart *Dyes* vi. 161 A small range of Ice Colours, so called because the preparation of the diazo solutions by the dyer needed ice, was built up. **1896** *Amer. Geologist* XVIII. 152 The tracing on the ground of the ice-contacts shows that other morainal belts come into the region about Wickford from the southeast. *Ibid.*, The accompanying map..exhibits by special designations the position of the ice-contact slopes or moraine terraces. **1968** R. W. Fairbridge *Encycl. Geomorphol.* 438/2 Ice contact deposits outline the holes and tunnels in the last wasting of the basal ice. **1970** C. A. Lewis *Glaciations Wales* ii. 29 The fresh ice-contact slopes (15°–20°) and re-entrant features noted on the inner flanks of the Cors Geirch terraces are present also on their eastern side. **1923** G. D. Abraham *First Steps to Climbing* iv. 45 We find the Mount Everest Expedition largely composed of rock-climbing specialists whose knowledge of snow and ice-craft is almost negligible. **1955** E. Hillary *High Adventure* 16 Harry was New Zealand's outstanding climber, with a tremendous reputation for brilliant ice-craft. **1853** A. Bunn *Old Eng. & New Eng.* I. ii. 31 Content ourselves by observing that the ice-crop (as it is drolly called).. proved to be a fair average one. **1864** *Chambers's Jrnl.* 100/1 Producing an ice-crop which will pay all expenses and leave a profit. **1929** M. Lief *Hangover* vii. 128 She dashed into the kitchen and came back with a bowl of ice-cubes and some more bottles. **1939** *Vogue's Cookery Book* 122 Ice cubes for summer drinks can be made decorative by freezing cherries..inside them. **1949** *Consumer Reports* June 250/1 Four medium-size ice-cube trays. **1962** J. Braine *Life at Top* iv. 71 She took out the ice cube tray. **1883** *Proc. Amer. Assoc. Adv. Sci.* XXXII. 207 Among the most interesting results of the author's survey in Ohio, was the demonstration of the existence of an ice-dam across the river at Cincinnati. **1935** W. J. Miller *Introd. Physical Geol.* (ed. 3) xiv. 396 Existing ice-dam lakes are not common, and few, if any of them, are large. During the Ice Age, however, thousands of them formed and lasted only as long as the ice dams existed. **1914** W. B. Wright *Quaternary Ice Age* iii. 71 Two main chains of ice-dammed lakes in the Cleveland valleys to the north of Pickering are described by Professor Kendall. **1965** G. de Boer in A. Small *Fourth Viking Congress* 207 The whole flat floored, hill girt hollow, is readily seen for what it is, the bed of a former ice-dammed lake. **1905** W. H. Sherzer in *Smithsonian Misc. Coll.* XLVII. 468 Ice dykes. These consist of narrow crevasses, two to fifteen inches across, completely filled with columnar ice. **1889** *Pall Mall Gaz.* 6 Feb. 3/1 When the winter fairly sets in the scene on an ice-farm is a busy one. **1908** *Sci. Amer.* 25 Jan. 58/2 Natural ice making in the tropics—the peculiar 'ice farms' of hot Bengal. **1963** *Brit. Columbia Digest* Nov.–Dec. 31 It'll be nothing less than wonderful if I manage to make one good steelhead trip and icefish in three lakes all winter. **1960** M. Sharcott *Place of Many Winds* vii. 123 The ice-fisherman..earns a few extra cents a pound. **1963** *Times* 12 Mar. 9/6 In most cases, the ice fisherman will avail himself (at a modest fee) of the services of a resort owner who ploughs roads across the ice and clears sites for the ice houses. **1890** T. H. Bean in *Forest & Stream* XXXV. 417 (*title*) Ice fishing in arctic Alaska. **1907** J. G. Millais *Newfoundland* i. 9 Twenty years at the 'ice fishing' (seal hunting)..will try the strongest man. **1970** *Globe & Mail* (Toronto) 25 Sept. 35/6 (Advt.), Sand beach, floating dock, very private, ice fishing. **1909** E. Shackleton *Heart of Antarctic* II. 341 Ice-flowers occurred on freshwater ice at Clear Lake. **1911** J. Masefield *Jim Davis* iii. 26 The frost had covered the window with ice-flowers, so that we could not see through the glass. **1955** *Arctic Terms* 42/1 *Ice flower*, a delicate tuft of frost or rime, resembling a fern or flower, that occasionally forms on surface sea ice around a salt crystal nucleus. **1965** P. Wayre *Wind in Reeds* vi. 79 All along the tide mark the frost had formed delicate ice-flowers. **1890** *Bull. Geol. Soc. Amer.* I. 201 Further inland, where plains are found up to altitudes of a thousand or more feet above sea level, I think the water in which they accumulated was fresh water, temporarily ponded by the ice front. **1957** G. E. Hutchinson *Treat. Limnol.* I. xv. 833 The receding ice front passed Hertford about 19,500 years ago. **1966** B. B. Baker et al. *Gloss. Oceanogr. Terms* (ed. 2) 83/1 *Ice front* (also called *front, ice cliff, ice face, ice wall*), (1) the seaward facing, cliff-like edge of an ice shelf (so called by the British Antarctic Place-names Committee), (2) any vertical wall of ice. **1862** *Congress. Globe* 2896/1 The island is..below the bend in the Delaware, and hence mainly out of danger from ice gorges. **1925** E. Sitwell *Troy Park* 40 Leaves like a starry crown Are clear as the splintered star ice-green That is a crown for a negro queen. **1934** L. B. Lyon *White Hare* 16 The washed sky opened like an arctic rose, ice-green. **1938** W. de la Mare *Memory* 62 Skies ice-green. **1905** *Forestry Bureau Bull.* (U.S.) No. 61, 40 *Ice guards*, heavy timbers fastened fan shaped about a cluster of boom piles at an angle of approximately 30 degrees to the surface of the water. They prevent the destruction of the boom by ice, through forcing it to mount the guards and be broken up. **1947** *Jrnl. R. Aeronaut. Soc.* LI. 298/2 In the early stages of the war some Mosquitos had flown home from Sweden through bad icing weather. The machines had reached home, but there was a complaint that they had lost 2 lb. of boost because the ice guards had iced up completely. **1907** *Yesterday's Shopping* (1969) 599/4 Electro-plated ice tongs and ice hammer. **1932** *Mountaineering Jrnl.* Dec. 100 The ice-hammer is used for chiselling steps and handholds as well as for driving in pitons. **1933** G. D. Abraham *Mod. Mountaineering* ix. 175 Young Continental experts..have evolved the ice hammer which is used for driving in the pitons. **1953** J. Hunt *Ascent Everest* i. iv. 38 The more familiar gear..rope and line, pitons, snaplinks, icehammers and axes. **1864** *Chambers's Jrnl.* 100/2 The season of the ice-harvest being short and uncertain. **1884** [in Dict., sense 6]. **1904** *Westm. Gaz.* 17 Mar. 2/1 Men with horses were ploughing the ice-harvest of the river. **1883** *Boy's Own Paper* 13 Oct. 30/1 For ice hockey the ball is from six to seven inches in circumference. **1898** *Daily News* 28 Nov. 8/7 The first ice hockey match at Niagara took place on Saturday. **1907** *Westm. Gaz.* 4 Dec. 7/2 An ice-hockey match at Prince's last night. **1909** *Ibid.* 18 Jan. 12/4 The great ice carnivals, skating championships, and ice-hockey matches. **1940** Auden *Another Time* 26 Superb at ice-hockey, a prince at the dance, He's fierce as the tigers, secretive as plants. **1970** *Guardian* 5 Jan. 14/5 Canada yesterday withdrew from this year's world ice hockey championships after losing their battle to allow professionals to take part. **1846** R. H. Bonnycastle *Canada & Canadians in 1846* II. 3, I have mentioned that, in the spring of 1845, an ice-jam, as it is called here, occurred, which suddenly raised the level of the Niagara thirty and forty feet above its ordinary floods. **1863** [in Dict., sense 7 a, s.v. *ice-jam*]. **1909** *Westm. Gaz.* 23 Apr. 8/2 The great ice-jam at Niagara. **1924** M. H. Mason *Arctic Forests* 246 In the depth of winter they travelled by dog-sled over the rough ice jams of Bear River. **1959** *Washington Post* 3 Feb. A.16/1 A willingness to explore new ideas could help break the East–West ice-jam. **1962** R. B. Fuller *Epic Poem on Industrialization* 158 The inevitable Social-economic ice jam. **1893** Kipling *Seven Seas* (1896) 29 Down a cruel ice-lane, That opened as he sped, We saw dead Henry Hudson Steer, North by West, his dead. **1879** *Encycl. Brit.* VIII. 731/2 At this point the steam line, ice line, and hoar-frost line intersect, and it has therefore been called the triple point. **1937** M. W. Zemansky *Heat & Thermodynamics* xi. 177 In investigating the ice line of water at very high temperatures, Bridgman and Tammann discovered four new modifications of ice. **1893** *Jrnl. Geol.* I. 131 Moraines formed by the Great Miami ice lobe. **1954** W. D. Thornbury *Princ. Geomorphol.* xvi. 384 The edges of the ice caps were probably never straight for any great distance, but in addition to many minor reentrants and projections along their margins there were numerous larger protrusions or ice lobes down lowlands. **1949** *Ice Cream Topics* June 12 Ice lollies or iced lollies..sell at 1d. or 2d. and capture the kiddy trade, being cheaper than cones and wafers filled with ice cream. **1957** *Times* 22 Aug. 8/6 They.. drip ice-lollies on the desk. **1970** *Daily Tel.* (Colour Suppl.) 2 Jan. 24/4 Ishmahil licked at a spoon of caviar as if it were an ice-lolly. **1850** T. Masters *Short Treat. Production Ice* iii. 19 The patent ice machine and its various modifications. Before giving a detailed description of the ice machine,..it will be as well to mention a few of the prominent advantages it possesses. **1873** C. M. Yonge *Pillars of House* II. xvii. 129 Is he awake? I have brought some more ice... I have a little ice-machine for Indian use. **1897** 'Mark Twain' *Following Equator* iii. 62 The ice-machine has traveled all over the world. **1973** *Country Life* 8 Feb. 345/1 There are enough hotels..to suit any pocket. And swimming pools, air-conditioning and ice-machines are usually included. **1953** Dylan Thomas *Under Milk Wood* (1954) 61 The butcher's unmelting icemaiden daughter veiled for ever from the hungry hug of his eyes. **1968** V. Canning *Melting Man* v. 134 Now stop doing an ice-maiden act on me. Write it off to experience. **1970** B. Turner *Another Little Death* xx. 118 Her ice-maiden act was a cover for frustrated lust. **1927** *Daily Tel.* 11 May 18/3 New patent non-chemical refrigerator and ice-maker. **1969** K. Amis *Green Man* i. 12 The ice-maker had broken down. **1970** *Cape Times* 28 Oct. (S.A. Fishing Rev.) 6/6 (Advt.), South African Agents for Skokie international seawater ice makers. **1846** R. Ford *Gatherings from Spain* vii. 74 The leading animal is furnished with a copper bell with a wooden clapper..which is shaped like an ice-mould. **1864** *Chambers's Jrnl.* 101/1 A continuous current through the cistern containing the ice-moulds. **1873** *Young Englishwoman* July 334/1 We put in a block as large as the tin will hold, and then with an ice-needle, price one shilling, break up the rest of the ice. **1928** *Funk's Stand. Dict.*, *Ice-needles*, n.pl., a deposit of ice, especially in gravelly soil, in the form of vertical needles. **1937** *Jrnl. R. Aeronaut. Soc.* XLI. 598 A further very rare case occurs when at a high altitude in the 'ice needle clouds'..a light accretion of hoar frost..forms on the aeroplane. **1965** H. Riehl *Introd. Atmosphere* v. 100 Ice needles are long thin crystals forming on very cold winter days through sublimation... Floating leisurely in the air, they provide a magnificent spectacle when the sun is shining on them. **1970** R. J. Small *Study of Landforms* x. 323 Another process which leads to the upheaval of material in the active layer [of the soil] is the development of small localised ice-masses and ice-needles ('pipkraker'). **1900** *Daily Chron.* 12 Nov. 5/6 The condition of Lord Roberts's daughter is somewhat serious. Ice-packs have been applied. **1926** *Daily Colonist* (Victoria, B.C.) 3 Jan. 3/3 The letter informed the President [Coolidge] that the [Detroit Aviation] society was about to join in an attempt to explore the ice pack between Point Barrow and the Ice Pole. **1930** *Times Educ. Suppl.* 25 Jan. p. iv/1 Rocky outcrops and ice-packs. **1955** *Sci. Amer.* Apr. 52/3 The Atka saw very little of the drifting ice pack that surrounds the continent. **1973** R. Ludlum *Matlock Paper* xiv. 129 There's a nurse in there with ice packs and stuff if localized pain bothers her. **1901** *Geogr. Jrnl.* July 40 The ice-pans appear to drift capriciously backward and forward, and, without any apparent cause, they will select some unexpected course. **1926** *Blackw. Mag.* July 67/1 An awful journey through a country devoid of human beings, across treacherous moving ice-pans. **1934** I. W. Hutchison *North to Rime-Ringed Sun* xii. 120 Suddenly out of the mist, upon an ice-pan, stood the little shrivelled figure of an old Eskimo hunter of seals. **1963** *Calgary Herald* 20 Sept., Turquoise ice pans (last year's ice) cluttered the water just off shore. *a* **1877** Knight *Dict. Mech.* II. 1169/1 *Ice-pick* (Domestic), an awl-shaped tool to break ice into fragments. **1879** F. R. Stockton *Rudder Grange* i, It is not probable that I can sell that ice-pick after you have used it for ten years. **1883** [in Dict. sense 7 a, s.v. *ice-crusher*]. **1937** E. A. M. Wedderburn *Alpine Climbing* ii. 29 A Hammer for driving in ring-spikes may be combined with an ice-pick; this weapon is..useful for cutting steps..in very steep ice. **1953** E. S. Gardner *Case of Green-Eyed Sister* (1959) viii. 97 J. J. Fritch was killed by repeated stabs with an ice-pick. **1960** *News Chron.* 11 June 2/8 Jacques Mornard, ice-pick assassin of Leon Trotsky. **1881** J. C. Lyell *Fancy Pigeons* 81 The

Ice Pigeon. This variety derives its name from its beautiful lavender blue colour, considered by the German fanciers to resemble blue ice. **1969** C. R. Hill *Pet Library's Pigeon Guide* viii. 151 (*caption*) White barred blue Ice Pigeon, young cock. A German exhibition breed. *Ibid.* 152 Ice pigeon, blue spangled, old hen. **1865** *Nation* (N.Y.) 3 Aug. 159/3 (Advt.), At this season of the year nothing adds more to one's comfort than to drink freely of the contents of our new pattern richly double-plated ice pitchers. **1883** 'Mark Twain' *Life on Mississippi* xviii. 221 Take that ice-pitcher down to the texas-tender. **1932** *Mountaineering Jrnl.* Dec. 100 Ice-pitons are made of wrought iron not too soft and not too brittle, 7 to 10 inches long and saw-toothed along both edges. **1954** W. Noyce *South Col* vi. 110 We then fixed the bridge and tied it down with ice pitons. **1956** C. Evans *On Climbing* vii. 114, I should be reluctant to trust a piton for this, since ice-pitons are less certain in their hold than rock-pitons. **1876** Lady C. Schreiber *Jrnl.* (1911) I. 443 Small ice-plates have now become the object of our pursuit. **1902** H. James *Wings of Dove* v. x. 178 The very servant who came to receive Milly's empty ice-plate. **1907** *Yesterday's Shopping* (1969) 944 (*caption*) Ice plates. **1903** *Phil. Trans. R. Soc.* A. CC. 108 The Comité International adopted as the normal scale of temperature the scale of a constant-volume hydrogen thermometer, in which the pressure at the ice-point was 1000 millims. of mercury. *Ibid.*, The coefficient for hydrogen, at a pressure of 100 millims. of mercury at the ice-point, is given..as 0·00366254. **1941** *Temperature* (Amer. Inst. Physics) i. 10 A limitation of this scale [*sc.* Kelvin's thermodynamic scale] is that if we make the fundamental interval from the ice point to the steam point 100 degrees, the actual temperature of the ice point is subject to experimental determination. **1966** *Units & Standards of Measurement: Temperature* (Nat. Physical Lab.) (ed. 2) 18 Changes made in 1960 to the text of the International Temperature Scale of 1948... The triple-point of water was given formal status as one of the defining fixed points of the scale... Its value was given as 0·01°C (Int. 1948) and the ice point appeared among the secondary fixed points with the value 0·000°C. **1851** W. P. Snow *Voy. 'Prince Albert'* 154 The slackest and thinnest part of the floe, or fragment, was cut into with the axes and chisels until some fortunate blow or prise of the ice-pole rent and loosened it. **1906** J. Lumsden *Skipper Parson* 107 This useful instrument also serves as an ice pole, enabling the daring sealer..to leap from 'pan to pan'. **1970** *Globe & Mail* (Toronto) 28 Sept. 4/4 Men working on the ice..should hold a boat-hook or small ice-pole in their hands. **1911** *Jrnl. Geol.* XIX. 157 He has never detected any evidence of ice push against shores as a result of expansion. **1939** P. G. Worcester *Textbk. Geomorphol.* xii. 383 Although unimportant on most sea shores, the shores of many lakes that freeze over in winter are profoundly affected by ice push. **1957** G. E. Hutchinson *Treat. Limnol.* I. vii. 532 This sheet may then exert pressure on the shore, forcing gravel and stones landward and building an ice push or ice rampart. **1969** J. L. Davies *Landforms Cold Climates* iv. 63 Seasonally frozen lakes in the tundra and elsewhere may form ice-push ramparts around their edges. **1901** E. R. Buckley in *Trans. Wisc. Acad. Sci., Arts & Lett.* XIII. 142 The diurnal and weekly changes of temperature during the winter months cause a sufficient expansion and contraction of the ice covering the inland lakes of Wisconsin to shove up the sand, gravel, boulders, and sod along the shores into peculiar ridges, known as ice ramparts. **1968** R. W. Fairbridge *Encycl. Geomorphol.* 546/1 Along the margins of ice-covered lakes and seas and the shores of the Arctic Ocean the effects of floating ice pressure (under wind stress) may be observed, leading to the building of large pressure-ridges in the beach gravels and other littoral deposits... The terms ice-shore ridges and ice ramparts are also sometimes used. **1886** *Field* 13 Mar. 310/1 Nor is it less strange that so few ice rinks are found in England. **1930** *Daily Express* 16 Aug. 3/1 By the middle of October there will be at least nine ice-rinks in London alone. **1953** X. Fielding *Stronghold* 277 The surface of this winding watercourse was like that of a shattered ice-rink. **1900** *Daily News* 30 Nov. 5/2 The new ice-run for tobogganers..is almost ready for use. **1910** H. G. Wells *Hist. Mr. Polly* vi, Mr. Polly swerved a little from the conversational ice-run upon which he had embarked. **1895** *Brewer's Dict. Phrase & Fable* (new ed.) 643/2 *Ice Saints*, those saints whose days fall in..'the black-thorn winter'. **1922** *Meteorol. Mag.* LVII. 177 The quasi-periodic occurrence of a cold spell lasting for a few days early in May is a well-known popular belief... On the Continent three 'Saint Days', those of St. Mamertius, St. Pancras and St. Gervais, falling on May 11th, 12th and 13th..are popularly known as the *Eisheiligen*, or 'Ice Saints'. **1936** *Times* 13 Feb. 14/1 May will be remembered mainly on account of the severe frost during the period 12th to 19th, an impressive vindication of the firmly rooted tradition of the 'Ice Saints'. **1969** *Guardian* 7 June 7/5 There was the normal cold iceburst in mid-May, but it came a week later than the proverbial 'Ice Saints' days on the Continent. **1904** J. D. Hooker *Let.* 3 Dec. in L. Huxley *Life J. D. Hooker* (1918) II. 457 His landscapes, seascapes and ice-scapes are most interesting. **1936** J. Grierson *High Failure* ix. 208, I had never before experienced the hypnotic splendour of the ice-scape. It seemed as though I were flying in a dream. **1969** *Sunday Times* 9 Feb. 6 The whole icescape was awash with light. **1936** Finch & Trewartha *Elem. Geogr.* xvii. 364 The surface configuration of plains where ice scour was predominant is characterized by rounded rock hills and broad open valleys with comparatively low local relief. **1968** R. W. Fairbridge *Encycl. Geomorphol.* 502/2 The basins of the Great Lakes..were formed by a combination of stream erosion during Mesozoic to Pleistocene times and glacial ice scour in the Pleistocene. **1936** Finch & Trewartha *Elem. Geogr.* xvii. 364 The drift of ice-scoured plains commonly is neither deep enough nor continuous enough to be tillable except in patches. **1954** W. D. Thornbury *Princ. Geomorphol.* xvi. 385 In areas where the surface over which the ice caps moved was mountainous..the result was not an ice-scoured plain, but a general smoothing off..of the topography. **1901** *Science* 5 Apr. 552/1 Ice-scouring during maximum glaciation reached far up the mountain slopes above the trough walls. **1957** G. E. Hutchinson *Treat. Limnol.* I. i. 57 The Saint Gotthard lakes..are noted..as having been formed by ice-scouring. **1965** D. Bathgate in *Scottish Mountaineering Club Jrnl.* XXVIII. 109 At the shortest point I surmounted the steep part, fixed an ice screw, and then traversed back across the centre of the ice to about twenty feet above the belay. **1968** P. Crew *Encycl. Dict. Mountaineering* 71/1 Ice-screws are very effective as they can be used on most types of ice..with a reasonable degree of security and they are very easy to remove after use. **1971** C. Bonington *Annapurna South Face* x. 116 It's difficult to get an ice-screw started, rather like an ordinary screw in hard wood—you first have to tap out a little hole to allow the thread to get a purchase. **1894** J. Geikie *Gt. Ice Age* (ed. 3) 830 The ice-shed in Sandinavia did not coincide with the water-parting.. **1932** E. G. Woods *Baltic Region* 159 When the ice-shed lay east of the watershed..a considerable amount of ice and also water flowed along the originally eastward-sloping valleys towards the west. **1957** J. K. Charlesworth *Quaternary Era* I. iv. 77 The iceshed, more than 1000 miles (1600 km) long, seems to lie behind South Victoria Land and Dronning Maud Mountains. **1914** T. W. E. David in *Geogr. Jrnl.* XLIII. 606 If the meaning of the term 'shelf' can be extended to include old pack ice, old bay ice, 'schollen-eis', piedmonts aground or afloat, glacier tongues, etc., it may be termed the ice shelf coast, or, as it is hardly a true coast at all, simply ice shelf. **1940** *Beaver* June 22/1 The men worked with feverish energy to repair the damage caused by the treacherous ice shelf. **1958** *Times* 13 Jan. 9/6 We..made our very laborious way..on to the flat going of the Ross ice shelf itself. **1958** J. H. Zumberge *Elem. Geol.* x. 180 Masses of ice that break off from the edge of the various ice shelves form the huge tabular icebergs unique to the Antarctic region. **1897** *Sears, Roebuck Catal.* 97/2 Ice skates. **1912** T. Dreiser *Financier* 36 He was an adept at turning all sorts of practical tricks, such as..taking the agency for the sale of a new kind of ice-skate from an ice-skate company. **1948** *Evening News* 2 Jan. 2/6 Two nephews are down on the creek ice-skating. **1950** *Oxf. Jun. Encycl.* IX. 432 There are three principal kinds of ice skates: those for figure-skating, for hockey, and for racing. **1937** Ice-skater [see *audition *v.* 2]. **1957** *Encycl. Brit.* XX. 730/1 An additional valuable impetus in bringing figure skating before the public was the ice-skating carnival. **1973** E.-J. Bahr *Nice Neighbourhood* vi. 61 We told him about the museums, free exhibits, outdoor concerts, ice-skating, bicycling. **1897** E. L. Voynich *Gadfly* iii. viii. 353 He might have recalled some splendid and fearful ice-spirit of the mountains. **1900** *Month* Jan. 85, I took you for that evil thing, the ice-spirit, who freezes the limbs of our people. **1901** (*title*) Ice sports. **1908** *Daily Chron.* 27 July 4/4 Boating in summer and ice sports in winter. **1898** J. O. Maund in W. A. Morgan *'House' on Sport* 276 It took but a moment before our ice steps were filled with these hailstones. **1908** *Westm. Gaz.* 25 Jan. 14/1 We stood in the ice-steps. **1931** *Discovery* Feb. 41/1 To cut ice-steps all along so great a ridge will be an affair not of hours but of a day or two. **1876** 'Mark Twain' *Punch, Brothers, Punch!* (1878) 17 We have to credit the weather with..the ice-storm. **1886** J. Geikie *Outl. Geol.* 50 By repeated thawings and regelations the branches and boughs are gradually loaded with ice and snow, and becoming top-heavy, the trees are liable to fall, even when no wind is blowing. Should one be overthrown, it collides against its neighbour, and this in turn falls upon another, until shortly the trees are seen crashing to the ground in all directions. This is what is known in North America as an ice-storm. **1899** *Daily News* 20 An 'ice storm' in Somerset..reminded me of a sudden hailstorm. **1921** R. Frost *Mountain Interval* 29 But swinging doesn't bend them down to stay. Ice-storms do that. **1965** *Kingston* (Ontario) *Whig-Standard* 13 Dec. 19/1 Nearly every cloud has a silver lining and Sunday's ice-storm was no exception. **1968** Ice storm [see *freezing *ppl. a.* 1]. **1893** *Funk's Stand. Dict.*, *Ice-tongue*, a steep, narrow cliff of ice, rising high above glacial névé, and extending upward toward the higher mountain-peaks. **1896** *Amer. Geologist* XVIII. 155 Between the ice-tongue which filled the cove and the hills on the west, coarse gravels were deposited. **1919** E. Shackleton *South* viii. 129 At the head of an ice-tongue that nearly closed the gap through which we might enter the open space was a wave-worn berg. **1968** R. W. Fairbridge *Encycl. Geomorphol.* 673/1 Where the borders of these ice sheets roughly coincide with the coastal mountain ranges, the ice spills through them in the form of ice rivers called outlet glaciers or ice tongues. **1936** *New Yorker* 29 Feb 37/2 (Advt.), Then think of ice-trays—stacks of them. **1962** *Which?* June 176/2 We filled the ice trays with cold water..and set the thermostats to coldest. **1965** M. Spark *Mandelbaum Gate* iv. 109 A scarred, lop-sided oil refrigerator..stood in the passage outside, from which anyone who wanted beer took it, depositing the money in the ice-tray. **1915** *Jrnl. Geol.* XXIII. 642 The constant association of ice wedges with definite loci of frost cracks. **1970** C. A. Lewis *Glaciations Wales* ii. 29 Near Traian (328365) a large fossil ice-wedge..cuts the deposits in a small quarry. **1928** *Daily Express* 10 Oct. 3/3 Where an ice-white salmon stream flows through a gorge. **1931** *Daily Tel.* 21 May 6/3 A gown of very heavy ice white satin. **1882** Ice wool [see *eis wool]. **1926** *Daily Colonist* (Victoria, B.C.) 10 Jan. 18/1 Ice Wool Scarves in a good assortment of colors, stripes or plain shades, with contrasting stripe borders.

ice, *v.* Add: **1. a.** *to ice up.* Also, to hold fast with ice.

1899 C. J. C. Hyne *Further Adventures Capt. Kettle* xii. The boats are frozen on to the chocks... Did you never see a boat iced up before?

4. b. To kill. *U.S. slang.*

1969 *New Yorker* 15 Feb. 51 A friend of his had come to his apartment..in clothes that were spattered with blood, and announced, 'I just iced two girls.' **1973** *Guardian* 6 Mar. 14/3 A would-be assassin who considers it his mission to 'ice the fascist pig police'.

5. b. Esp. of aircraft: to become covered with ice (and thus rendered ineffective). Const. *up.* Chiefly in *pass.*

1928 *Aviation* 16 Apr. 1032/2 Once a plane has become iced-up, two alternatives for clearing away the accumulation of glazing may be available. 1940 *Times* (Weekly ed.) 10 Jan. p. ii/4 During the operations a snowstorm was encountered and the aircraft became badly iced-up, in addition to being subjected to anti-aircraft fire. 1943 *Aeronautics* Mar. 60/1 The wings and controls may be iced up. 1947 *Sci. News* IV. 72 A ship has often been thoroughly de-greased by wind-swept rain and spray before it becomes iced-up. 1950 T. Longstaff *This my Voyage* ii. 24 By now we were all looking like Arctic travellers, well iced-up.

-ice, -icè (isi), *suffix*[2], in med.L. forming adverbs from adjs., as *Anglice, *Gallice, *ironice*, Scotticè, Scoticè, and hence used occas. to form jocular nonce adverbs on English stems, as *golficè*.
1743 Pope *Dunciad* I (*footnote to l. 23*), *Ironicè*, alluding to Gulliver's representations of both. 1886 *Golficè* [see *divot *sb.* 2].

iceberg. Add: **2.** Delete 'Arctic'. (Further examples.)
1830 *Edin. Encycl.* XVII. 12/1 The floating iceberg remains to be considered... In many parts of the Antarctic regions, they are met with in vast numbers, and of a prodigious size. 1961 A. Defant *Physical Oceanogr.* viii. 274 The flat-topped Antarctic icebergs immerse to greater depths.

b. *transf.* Used allusively with reference to the larger portion of an iceberg being unseen (and hence a largely unknown quantity, problem, etc.).
1961 in Webster. 1964 *Observer* 26 July 10/4 This.. situation is illustrated by what is..called the iceberg of disease. Above the surface is the illness we know about. 1965 *Listener* 21 Oct. 614/2 These were only the visible part of the iceberg. Most of the organs of the left..were greatly influenced. 1969 M. Gilbert *Etruscan Net* II. ii. 109, I believe that Broke's been made the victim of an elaborate frame-up. I think, to employ a well-known metaphor, that all we can see at the moment is the tip of the iceberg, and that there is depth beyond depth below it. 1973 *Daily Tel.* 19 Feb. 2/1 They believe the number of prosecutions for such offences represent only 'the tip of the iceberg', and that many undischarged bankrupts..are violating the law undetected.

4. iceberg lettuce *U.S.*, a crisp light-green lettuce.
1893 *Burpee's Farm Annual* 28 As long as our supply lasts we will send a sample packet of the iceberg lettuce free for trial. 1904 W. W. Tracy *Amer. Varieties Lettuce* 56 *Iceberg* [*lettuce*], a decidedly crisp variety, strictly cabbage-heading, large, late, slow to shoot to seed. 1933 F. M. Farmer *Boston Cooking-School Cook Bk.* (rev. ed.) xxviii. 420 Iceberg or California Lettuce. Cut in halves or quarters. Remove hard center. 1960 'I. Devi' *Yoga for You* (1965) 189 One head romaine or any other green salad, except iceberg lettuce, which contains almost no chlorophyll. 1966 *McGraw-Hill Encycl. Sci. & Technol.* VII. 478/2 Crisphead or iceberg lettuce is the most widely grown type. 1970 'J. Morris' *Candywine Devel.* xx. 223 He took a loaf of French bread, an iceberg lettuce and three onions from the refrigerator.

ice-boat. Add: **3.** A fishing-vessel equipped with facilities for the refrigeration of fish. *N. Amer.*
1878 *Saskatchewan Herald* (Battleford) 29 July 4/1 The crew of the Lady Ellen are building an ice-boat for the fishing trade this winter. 1941 E. J. Kammer *Socio-Econ. Survey Marshdwellers Louisiana* viii. 118 Ice boats are larger than the ordinary trawl boat and are used only for transporting shrimp. 1970 *National Fisherman* Sept. 18-A/4 They were ice boats, designed for a 10-ton capacity.

ice-breaker. Add: **1.** (Earlier example.)
1819 D. Thomas *Trav. Western Country* 247 Notwithstanding these precautions, and that of placing ice-breakers to the south, [the bridge] was only saved from destruction the ensuing winter by the intrepidity of.. one of the proprietors.

2. b. *transf.* Cf. *to break the ice* (Ice *sb.* 2 b).
1883 'Mark Twain' *Life on Mississippi* xxxix. 365 They closed up the inspection with a few words—having used it, evidently, as a mere ice-breaker and acquaintance-ship-breeder—then they dropped into business. 1904 *Daily Chron.* 27 Feb. 4/6 If you must use an ice-breaker, the pianola is decidedly effective..as a cure for shyness. 1961 Bowman & Ball *Theatre Lang.* 178 *Icebreaker*, a fast song for chorus girls—Musical comedy. 1963 Barnard & Lauwerys *Handbk. Brit. Educ. Terms* 110 *Ice-breaker*, a term used to describe a preliminary to a series of tests. It is designed to accustom the candidate to the experience which he is about to undergo, but its results are not counted in the ultimate assessment of the tests. 1963 M. McCarthy *Group* i. 19 The recipe was an ice-breaker..everyone tasted it and agreed that it was the maple syrup that made all the difference. 1968 *Daily Tel.* 15 Nov. 16/7 Swearing, in addition to its cathartic effect and as a means of non-violent assault.., is also an ice-breaker. 1973 *Sun-Herald* (Sydney) 26 Aug. 80/3 Then they went on to 'icebreakers'—short talks about themselves, reading a message they'd selected from a book.

ice-cream. Add: (Earlier and later examples.)
1744 in *Pennsylvania Mag. Hist. & Biogr.* (1877) I. 126 Among the rarities..was some fine ice cream, which, with the strawberries and milk, eat most deliciously. 1751 H. Glasse *Art of Cookery* (ed. 4) 333 (*heading*) To make Ice Cream..set it [*sc.* the cream] into the larger Bason. Fill it with Ice, and a Handful of Salt. 1957 *Min. Agric. Food Standards Comm. Rep. Ice Cream* (H.M.S.O.) 4 The bulk of ice cream now produced in this country is made from margarine and other non-milk fats... We recommend that special provision should be made to reserve the description 'dairy ice cream'..for ice cream in which the fat content is wholly milk fat.

attrib. (Further examples.)
1821 *National Advocate* (New York) 3 Aug. 2/2 Among the number of ice cream gardens in this city, there was none in which the sable race could find admission and refreshment. 1844 J. Cowell *30 Yrs. among Players* II. iii. 64/2 With the ice-cream profits, he purchased bricks and mortar, and built the *Chatham Theatre*. 1851 A. O. Hall *Manhattaner* 46 How the ice cream saloons resound with clattering spoons. 1854 *Rep. Trans. Pennsylvania State Agric. Soc.* 363 Three ice cream freezers. 1873 J. H. Beadle *Undevel. West* xxviii. 623 Two months vigorous courting will cost more than that—particularly in the ice-cream season. 1878 N. A. Donnelley *Lakeside Cook Bk.* 30/2 Ice Cream Cake. 1880 E. W. Wilcox *Buckeye Cookery & Pract. Housek.* 83 Ice-cream cake... Make good sponge-cake... Fill with ice-cream just before serving. 1881 C. C. Harrison *Woman's Handiwork* III. 223 Finger-bowls and ice-cream plates were ruby Bohemian glass. 1884 *Milnor* (Dakota) *Teller* 27 June, An ice cream parlor where the dudes and dudines sip.. congealed milk and sugar. 1886 *Mobile* (Alabama) *Daily Reg.* 23 Apr. 2/3 (Advt.), Drink Ice Cream Soda. 1904 *Proc. R. Soc.* LXXIII. 504 Using the modified ice-cream mixer described by Moody. 1905 Ice-cream cake [see *angel-cake]. 1909 *Sat. Even. Post* 15 May 11/2 The remainder is about equally divided among popcorn, ice cream cones, and candy. 1934 *Archit. Rev.* LXXVI. 159/1 (*caption*) 40-gallon ice-cream fountain. 1947 *Ibid.* CI. 212/1 Across the road, though, competition has set in, not only from the tricycle ice-vendors but from the Ice Cream Parlours. 1950 G. Greene *Third Man* ii. 16 The American zone, which you couldn't mistake because of the ice-cream parlours in every street. 1961 E. Wilson in R. Weaver *Canad. Short Stories* (1968) 2nd Ser. 14 How about an ice-cream cone? 1961 C. Willock *Death in Covert* iii. 73 The liveried servants held flambeaux like ice-cream cornets. 1966 J. Cleary *High Commissioner* i. 19 An ice-cream van went slowly by, its bell tinkling. *a* 1966 M. Allingham *Cargo of Eagles* (1968) xv. 162 An ice cream tricycle was already doing good business. 1968 P. Jennings *Living Village* 204 The dreadful hoarse tintinnabulations, as though a giant were banging a dulcimer, which herald the urban ice cream van. 1972 *Guardian* 2 Sept. 9/2 He fetched his mother an ice-cream cornet.

Hence **ice-crea·mer**, an (Italian) ice-cream vendor; so (*derogatory*) an Italian.
1901 E. W. Hornung *Black Mask* 110 He had every low-down Neapolitan ice-creamer on my tracks. 1940 N. Mitford *Pigeon Pie* iii. 41, I remembered that there are Chinks and Japs and Fuzzy Wuzzies and Ice Creamers and Dagos, and so on. 1949 L. P. Hartley *Boat* i. 3 Italians are not people according to our ideas... Ice-creamers, some call them, but it's too polite for them.

iced, *ppl. a.* Add: (Further examples.)
1673 J. Ray *Observations Journey Low-Countries* 267 Many..with a barrel at their backs and glasses in their hands, crying *Acqua ghiacciata*, or *Acqua nevata*, *i.e.* Snowed water or iced water. 1777 P. Thicknesse *Year's Journey* I. xxviii. 240 Their chocolate, lemonade, iced water, fruits, &c. are their chief luxuries. 1831 B. H. Smart *Outl. Sematology* iii. 237 The thirsty wight who, in a state of profuse perspiration, calls for a glass of iced-water, may know there is danger in the draught. 1848 Iced champagne [see *ball supper* s.v. *ball *sb.*[1] 4]. 1877 *Forest & Stream* VIII. 411/1 Courtney..became very sick immediately after taking a glass of iced tea after his dinner, and was unable to row. 1879 M. E. Braddon *Vixen* III. vii. 187 Afternoon tea at Ashbourne included iced coffee. 1880 *Amer. Punch* Jan. 4/1 Some were talking of.. the cooling and invigorating influences of 'iced tea'. 1902 H. James *Wings of Dove* v. x. 172 The small cup of iced coffee she had vaguely accepted from somebody. 1913 W. Stevens *Let.* 7 July (1967) 179 A spread of chicken salad, ..watermelon, iced-tea etc. 1930 A. Bennett *Imperial Palace* xvii. 109 The cocktail jugs, the iced-water jugs. 1952 W. M. Miller *View from Stars* (1965) 42 She..drank a glass of iced tea. 1954 *Good Housek. Cookery Bk.* (rev. ed.) 453 (*heading*) Iced coffee. 1958 *Listener* 14 Aug. 229/1 Fifteen of the family hang over the back and laugh at you through their iced lollies. 1960 *News Chron.* 6 Aug. 1/1 Prince Charles bought an iced-lolly. 1972 H. Osborne *Pay-Day* iv. i. 163 Every year..another hotel with an iced-water tap rises out of the mud.

2. Of a cake, bun, etc.: covered with icing. Of preserved fruit: = Glacé *a.* 2.
1858 P. L. Simmonds *Dict. Trade Products*, *Iced*, cakes frosted with sugar. 1866 Mrs. Beeton *Preserves & Confectionery* 5 Iced Apples, or Apple Hedgehog... Cover the apples very smoothly all over with the icing. *Ibid.* 89 Iced currants... Lay them to dry on paper, when the sugar will crystallize round each currant. 1892 *Encycl. Pract. Cookery* I. 240/1 Iced cake. 1973 *Fortnum & Mason Christmas Catal.* 4/2 Iced and decorated Christmas cake.

b. (See quots.) Cf. Frosted *ppl. a.* 3 b.
1829 J. C. Loudon *Encycl. Plants* 1100/1 Iced..covered with particles like icicles. 1900 B. D. Jackson *Gloss. Bot. Terms*, *Iced*, having a glittering papillose surface, as *Mesembryanthemum crystallinum*.

ice-house. Add: **2.** A hut made of ice or snow.
1857 G. F. McDougall *Eventful Voy. 'Resolute'* 426 The remains of two ice houses yet existed, but were rapidly thawing away, under the influence of the heat of the sun. 1958 *Listener* 25 Sept. 482/2 They sat marooned for four days in an icehouse 14,000 feet up.

Iceland. Add: **Iceland falcon,** *Falco rusticolus islandicus*, a variety of gyr-falcon native to Iceland; **Iceland gull,** *Larus glaucoides*, a grey and white Arctic gull.
1771 *Gentl. Mag.* XLVI. 297/1 The Iceland Falcon..is a noble and stately bird. 1822 *Mem. Wernerian Nat. Hist. Soc.* IV. 181 They [*sc.* the Shetland fishermen] have distinguished this bird by the name of Iceland *Scorie*, (or the Young Iceland Gull); *Scorie* being the general Shetlandic appellation for the young of several species of the gull family. 1843 W. Yarrell *Hist. Brit. Birds* I. 27 Those specimens obtained from Iceland were called exclusively Iceland Falcons. *Ibid.* III. 461 The Iceland Gull sometimes makes its appearance in winter at the mouth of the Elbe. 1901 H. H. Slater *Man. Birds Iceland* 31 The Iceland Falcon is a remarkably handsome bird. 1927 M. U. Hachisuka *Handbk. Birds Iceland* 40 The Iceland Falcon is a national emblem. 1930 *Ibis* 415 The Iceland Gull apparently fills the same 'niche' in Godthaab Fiord as the Herring-Gull..in the British Isles. 1956 D. A. Bannerman *Birds Brit. Isles* V. 21 In Ireland the Iceland gyr-falcon has been recorded five times. *Ibid.* 22 The Iceland falcon was not nearly as uncommon as had been supposed. 1962 *Ibid.* XI. 334 In flight the Iceland gull looks more graceful and has more rapid wing-beats.

iceman. Add: **1.** (Earlier example.)
1851 W. P. Snow *Voy. 'Prince Albert'* 302 Ten men formed the number of the working seamen; there were no 'icemasters', nor regular 'ice-men': but most of the sailors were long accustomed to the ice.

2. (Earlier example.)
1845 *Times* 10 Feb. 5/4 The ice in the Serpentine yesterday was not above an inch thick, and through the exertions of the icemen of the Royal Humane Society, no persons ventured on except a few boys.

3. (Earlier example.) Also, one who delivers ice for domestic use.
1844 *Maysville* (Kentucky) *Eagle* 7 Sept. 1/3, I do wish an ice man would come this morning. 1870 'F. Fern' *Ginger-Snaps* 179 Let no grocer boy or ice-man fondly hope to retain the celestial spark, while he briefly deposits his wares in my kitchen. 1959 N. Mailer *Advts. for Myself* (1961) 230 We played our games. I was the iceman and she was the housewife.

ice-master. 1. (Earlier example.)
1851 [see *iceman 1].

Icenian (əisī·niăn), *sb.* and *a.* Also **Icenæan** (əisinī·ăn). [See -ian.] **A.** *sb.* A member of the Iceni, an ancient British tribe inhabiting the district roughly corresponding to modern Norfolk and Suffolk.
1598 R. Greneweу tr. *Tacitus' Annales* xiv. x. 209 The chiefest of the Icenians..were dispossessed of al their ancient inheritance. 1670 Milton *Hist. Britain* II. 55 The Icenians, a stout people untouch'd yet by these Warrs. 1864 Tennyson *'Boädicea'* in *Enoch Arden* 169 Hear Icenian, Catieuchlanian, hear Coritanian, Trinobant!

B. *adj.* **1.** Of or pertaining to the Iceni or the district they inhabited. Also **Ice·nic** *a.*
1757 J. Dyer *Fleece* III. 72 This method still Norvicum favours, and the Icenian towns. 1830 *Forby's Vocab. E. Anglia*, *Mem.* p. xxxix, With only one more extract I will close what remains to be said respecting the Icenian Glossary. *c* 1873 A. D. Bayne *Royal Illustr. Hist. E. Eng.* I. 393 Some Icenic names are supposed to remain in several towns of Norfolk and Suffolk. *Ibid.* 395 There are thousands of pits in many places, and these are supposed to have been the foundations of Icenian huts. 1900 W. A. Dutt *Norfolk* 39 Some authorities have suggested that that important Icenic settlement was at Caistor. 1921 R. A. S. Macalister *Text-bk. Europ. Archæol.* I. 158 To this type of flint, or to the supposed industry which it represents, has been given the name *Icenian*. 1962 T. C. Lethbridge *Witches* vii. 95 Hiccafrith becomes the Sun husband of the Icenaean moon and horse goddess, Ma Gog.

2. *Geol.* Applied to the Norwich Crag, Chillesford Beds, and Weybourne Crag of Norfolk and Suffolk (sometimes, to the Norwich Crag alone) and to the period when they were deposited, formerly regarded as late Pliocene but now held to be early Pleistocene; occas. used as the epithet of a stratigraphical stage in Britain. Also *absol.*
1896 *Q. Jrnl. Geol. Soc.* LII. 782 He [*sc.* H. B. Woodward] was..glad that Mr. Harmer now agreed that the beds belonged to one formation; and if it were desirable to use a term that should correspond with the other group-names used by the Author, he would suggest that the old term 'Icenian' be used for this Norwich Crag Series. 1900 F. W. Harmer in *Ibid.* LVI. 721 For the deposits hitherto known as Norwich Crag.., which extend..from Aldeburgh in Suffolk to Horstead and Burgh in Norfolk, a distance of more than 40 miles in one direction, and 20 miles, from Hoxne to Southwold, in another, I adopt the name Icenian, originally proposed for the Crag-formation generally by S. P. [*sic*] Woodward. *Ibid.*, The Icenian Period. 1931 Gregory & Barrett *Gen. Stratigr.* xvi. 228 The Sicilian Series is represented in England by the Norwich Crag, Chillesford Beds, and Weybourne Crags, which were grouped by Harmer as the Icenian. 1957 J. K. Charlesworth *Quaternary Era* II. xxxii. 697 The impoverished state of many Icenian shells may (doubtfully) have been due to a freshened North Sea which resulted when the Scandinavian ice..blocked the northern outlet of that sea. 1968 R. G. West *Pleistocene Geol. & Biol.* xiii. 337 A major time of extinction [of species of mollusc] was after Icenian Crag times, and before the Hoxnian temperate stage.

ice-plant. Add: Also used in Tasmania to refer to two species of *Tetragonia*.
1889 J. H. Maiden *Useful Native Plants Austral.* 63 *Tetragonia implexicoma*... Called 'Ice Plant' in Tasmania. Baron Mueller suggests that this plant be cultivated for spinach. 1898 E. E. Morris *Austral Eng.* 429/1 Spinach, New Zealand, *n. Tetragonia expansa* Murr...called also *Iceplant*, in Tasmania. 1944 *Mod. Jun. Dict.* (Whitcombe & Tombs) 204 Ice plant... The Tasmanian name for a plant allied to 'New Zealand spinach', and to the mesembryanthemum or 'pig-face'.

icer (əi·səɹ). [f. Ice *v.* + -er[1].] One who ices; *spec.* a worker who prepares icing and applies it to the surface of cakes, pastry, etc.
1909 Webster, *Icer*, one who ices. 1921 *Dict. Occup. Terms* (1927) §433.

ice-water. (Later examples.)
1803 C. Wilmot *Let.* 6 Mar. in *Irish Peer* (1920) 160 Ice water is their [*sc.* the Neapolitans'] greatest luxury. 1846 *Knickerbocker* XXVIII. 187 They may pour in a large spoonful of that [*sc.* brandy], and then fill it up with ice-water. 1889 *Harper's Mag.* Sept. 560/1 Here were found..the huge brown hogsheads for ice-water. 1906 *N.Y. Even. Post* 25 July 4 A glass of ice-water placed before us the moment we sit down to breakfast;..a pitcher of ice-water sent to our bed-rooms. 1910 W. de la Mare *Three Mulla-Mulgars* xxi. 280 Colder than ice-water. 1969 *New Yorker* 6 Sept. 35/1 She will..arrange them, filling the Chinese bowl with ice water, carefully clipping the stems. 1972 *Straits Times* 25 Nov. 10/6 They were sitting at an ice water stall when they saw four men approaching them from the direction of Sembawang Circus.

ice-work. Add: **3.** *Mountaineering.* Climbing on icy surfaces; the techniques of such climbing.
1856 A. Wills *Wanderings High Alps* xiv. 288 Our ice hatchet..was..better adapted to the mere ice-work we had then to perform. 1892 C. T. Dent et al. *Mountaineering* iv. 125 For a snow expedition—that is, one in which snow and ice work will probably form the chief difficulties—the numbers of a party may be largely increased even to eight or ten. 1940 F. S. Chapman *Helvellyn to Himalaya* iv. 89, I was not much good on really difficult rock, my experience of step-cutting and ice-work was small. *Ibid.* v. 98 At last, in early September, by developing an entirely new technique of ice-work, the ridge was reached.

ice-worm (əi·s'wɜ̄ɹm). [f. Ice *sb.* + Worm *sb.*] **a.** A small oligochaete worm, *Mesenchytræus solifugus*, found in North American glaciers and ice fields; also called glacier-worm and snow-worm. **b.** *Canada.* An imaginary creature that first 'appeared' during the Klondike gold rush.
a. [1885 J. Leidy in *Proc. Acad. Nat. Sci. Philadelphia* 408 The little worms of the ice appear to be an undescribed species. 1886 *Bull. Washburn Coll. Lab. Nat. Hist.* I. 186 What was popularly supposed to be an Ice Worm was found in the ice used in Salina, Kansas, toward the latter part of the summer of 1885.] 1904 G. Eisen in C. H. Merriam *Harriman Alaska Exped.* XII. 61 It is not impossible that the various glaciers of Alaska contain several species of black ice worms. 1949 *Nature* 24 Dec. 1098/1, I found a considerable number of the oligochaetous annelids known as ice-worms. 1970 *Nature* 14 Feb. 587/1 During the day, ice worms, which are black and about three quarters of an inch long, were found clumped together 6–12 inches deep under the drainage furrows of the névé.
b. 1901 *Klondike Nugget* 10 Apr. 2/2 In a country where ice worms abound there is no telling but that deadly serpents may also be found. 1964 *Edmonton Jrnl.* 11 July 27/1 The Klondike ice worm, immortalized in Robert Service's The Ice Worm Cocktail, has been imported to Edmonton.

Ichabod (i·kăbǫd). Name given by Eli's daughter-in-law to her son, used as an exclamation of regret, in allusion to 1 Sam. iv. 21 (she named the child Ichabod, saying, 'The glory is departed from Israel').
1702 I. Mather (*title*). Ichabod, or A discourse, shewing what cause there is to fear that the glory of the Lord, is departing from New-England. 1812 Bogue & Bennett *Hist. Dissenters* IV. vi. 383 The orthodox dissenter would inscribe, 'Ichabod, the glory is departed'. 1901 'A. Hope' *Tristram of Blent* xxi, 'Bring me some cold beef,' he commanded, and the waiter brought it with an air that said 'Ichabod' for the Imperium. 1904 'H. S. Merriman' *Last Hope* xxi, 'Ichabod,' he said, with a short laugh. 1915 *N. & Q.* 6 Feb. 110/1 At one time the Scriptural name Ichabod was used, presumably with a knowledge of its derivation, with the sense of alas! regretting the good old times. 1941 J. Gore *King George V* xi. 125 An income-tax, still adjusted from time to time in pennies below a shilling, left to the newly rich and the owners of ancient wealth, newly recruited, plenty over for luxurious living and only a scintilla of justification for crying *Ichabod*. 1957 A. MacNab *Bulls of Iberia* xiii. 139 Belmonte—the first torero ever to wear a bowler hat, ichabod!
Hence **Ichabo·dian** *a.*, regretful, lamenting.
1887 *Daily News* 1 Dec. 2/1 Dirges were sung with an Ichabodian refrain.

Ich dien (iχ[y] dī·n). [a. G. *ich dien* = I serve.] Used as the motto of the Prince of Wales, adopted with the crest of ostrich feathers after the battle of Crécy (1346), from John of Luxembourg, King of Bohemia, who was killed in the battle. So *concr.* for the Prince of Wales himself, and allusively.
Spelt *ich diene* on the tomb of the Black Prince, Edward Prince of Wales, at the time of his burial at Canterbury in 1376.
a 1529 Skelton *Sp. Parrot* in *Poetical Works* (1843) II. 5 *Ic dien* serueth for erstrych fether. 1545 in *Catal. Seals Dept. MSS. Brit. Mus.* (1892) II. 232 Henry VIII. Ich Dien. 1677 F. Sandford *Geneal. Hist. Kings Eng.* III. iv. 182 Prince Edward..deplumed his Casque of those Ostrich Feathers, which..became his Cognizance,.. with Scroles containing this Motto, ich dien, that is, *I serve*. 1780 H. Walpole *Let.* 23 Sept. (1858) VII. 441 If *Ich Dien* does not wear one, he at last,..boudes those who voted against the Admiral. 1923 D. H. Lawrence *Birds, Beasts & Flowers* 170 That pale fragment of a Prince up there, whose motto is *Ich dien*. *a* 1930 —— *Phoenix* (1936) 588 The whole world screams *Ich dien*.

ichebo, -u, ichibo, -u, varr. Itzebu.

I Ching (ī·,tʃiŋ). [Chin., lit. = Book of Changes.] The name of an ancient Chinese divination manual, based on symbols known as the eight trigrams and sixty-four hexagrams.
1876 T. McClatchie tr. *Classic of Change* p. v, The mere translating of the Yih King is not without difficulties; but to decipher the *system* taught therein, is impossible without some knowledge of other pagan systems. 1952 *Musical Q.* XXXVIII. 128 Cage has often used the I-Ching, an old Chinese method of throwing coins or marked sticks for chance numbers, like our use of dice. 1957 *Encycl. Brit.* V. 519/2 Wên Wang..is said to have produced the *I Ching*, or *Canon of Changes*, a volume based upon the trigrams and later viewed with great veneration and incorporated into the Confucian canon. 1965 L. T. Culling *Incredible I Ching* 8 There was no language attached to the I Ching of Fu Hsi—only eight figures, or Trigrams. 1968 *Listener* 19 Sept. 377/1 By means of chance operations—the use of the I-Ching, filling in imperfections in the manuscript paper, tossing coins—the parameters of pitch (and therefore form) were removed from the domains of the composer's will and taste. 1970 Feng & Kirk *Tai-Chi* 7/2 Although it was an ancient document when Confucius first encountered it, I Ching is traditionally awarded the status of a 'Confucian' classic. 1971 R. van Over *I Ching* I. ii, The text of the *I Ching* was an early valuable reference for ministers and leaders in ancient China as well as the average citizen. 1972 *Last Whole Earth Catalog* (Portola Inst.) 433/3 And then, way back under the bed, face down on the floor, opened at the hexagram called Youthful Folly, D.R. found their *I Ching*. 1973 *Listener* 15 Feb. 209/1 Margaret is 'into' astrology, and consults the *I-Ching* each morning.

ichneumoned (iknĭū·mənd), *a.* [f. Ichneumon 2 + -ed[2].] Infested with parasitic ichneumon flies.
1897 W. F. Kirby in R. Lydekker et al. *Conc. Knowledge Nat. Hist.* 576 Ichneumoned larvæ generally attain their full growth, and then die. 1944 W. J. Stokoe *Caterpillars Brit. Butterflies* 99 A very large proportion of any [Large Tortoiseshell caterpillars] collected are almost certain to be found 'ichneumoned'.

ichor. Add: **2.** (Further examples.) Now chiefly *poet.*
1930 Blunden *Poems* 128 Meanwhile the woods with ichor in their limbs Wake in a dance of slow religious love. 1960 S. Plath *Colossus* (1967) 77 The ichor of the spring Proceeds clear as it ever did From the broken throat, the marshy lip. 1970 R. P. Warren *Incarnations* 15 The great-gashed navel's cup Pours forth the ichor that had filled it up.

4. *Geol.* A fluid or 'emanation' from a magma which is held to cause granitization of rock.
1926 J. J. Sederholm in *Bull. de la Comm. Géol. de Finlande* XII. lxxvii. 89 The writer proposes to introduce, instead of the word granitic juices, the term granitic ichor, preliminarily with no more strictly defined signification than that possessed by the word juice. It will soon be possible..to give to the term a stricter definition. 1934 *N.Z. Jrnl. Sci. & Technol.* XV. 354 The minerals described..constitute an assemblage typical of an area mineralized by granitic ichor. 1965 A. Holmes *Princ. Physical Geol.* (ed. 2) viii. 183 When he [*sc.* Sederholm] was urged to define his 'ichor' in more material terms, as he often was, he suggested 'a magma containing much water in a gaseous state'.

ichthammol (i·kþæmǫl). *Pharm.* Also † **ichthamol.** [f. Ichth(y)ol with insertion of Amm(onia).] Ichthyol (as defined in Suppl.).
1907 *Brit. Pharm. Codex* 89 Ammonii Ichthosulphonas. Ammonium Ichthosulphonate. Synonyms.—Ichthamol; Ammonium Sulpho-ichthyolate. Ammonium ichthosulphonate consists of the ammonium salts of the sulphonic acids prepared from ichthyol—the oily product of the destructive distillation of a greyish bituminous schist.. found in the Karwendel Mountains. 1950 [see *ichthyol]. 1956 D. M. Pillsbury et al. *Dermatol.* xxvi. 388 In a patient with atopic dermatitis..our own single preference for an initial ointment is one containing 2 to 3 per cent ichthammol in zinc oxide ointment USP. 1967 *Martindale's Extra Pharmacopoeia* (ed. 25) 1406/2 Ichthammol has occasionally been administered by mouth..as an expectorant and intestinal antiseptic but it is irritant to the gastric mucosa.

ichthyo-. Add: **ichthyoacanthoto·xism** (i:kþio,ăkæ:nþo-) [Gr. ἀκανθο- thorn (f. ἀκή point) + Tox-[1]], poisoning resulting from a venomous sting or bite by a fish; **i:chthyosa:rcoto·xism** [Sarco- + Tox-[1]], poisoning resulting from the ingestion of a fish whose flesh is poisonous; **ichthyoto·xism** [Tox-[1]], poisoning resulting from the natural poison or venom of a fish.
1953 B. W. Halstead in *Copeia* I. 32/2 The following nomenclature and classification is proposed: Ichthyotoxism is the general term which would be used to include the forms of intoxication resulting from contact with both 'poisonous' and 'venomous' fishes... Ichthyotoxism may be of two types: (a) Ichthyosarcotoxism... (b) Ichthyoacanthotoxism. 1962 K. F. Lagler et al. *Ichthyol.* iv. 131 The field of ichthyotoxism includes the various forms of intoxication resulting from eating poisonous fishes (ichthyosarcotoxism) or being stung by venomous fishes (ichthyoacanthotoxism). 1953 Ichthyosarcotoxism [see *ichthyoacanthotoxism* above]. 1960 Ichthyosarcotoxism [see *ciguatera]. 1962 K. F. Lagler et al. *Ichthyol.* v. 160 The strongest evidence as to the origin of ichthyosarcotoxism now points to the feeding habits of fishes. [1898 V. C. Vaughan in T. L. Stedman *20th Cent. Pract.* XIII. 33 (*heading*) Fish poisoning (ichthyotoxismus).] 1900 Dorland *Med. Dict.* 315/2 Ichthyotoxism. 1922 *U.S. Naval Med. Bull.* XVII. 201 Poisoning by inherent fish poisons is called 'ichthyotoxism'. 1953 Ichthyotoxism [see above].

ichthyol. Add: Also, the dark, viscous liquid obtained by sulphonating the distillate and neutralizing the product with ammonia, in which form the substance has been most commonly employed (see quot. 1950). (Earlier and later examples.)
Ichthyol has been registered as a proprietary name.
1884 *Trade Marks Jrnl.* 1 Oct. 908 Ichthyol... Ichthyol Gesellschaft Cordes, Hermanni & Co., Hamburg; merchants... Chemical substances prepared for use in medicine and pharmacy. 1900 A. R. Cushny *Text-bk. Pharmacol. & Therapeutics* II. 379 Ichthyol is the ammonia salt of a sulphonic acid derived from the tar of a bituminous shale which is found in the Tyrol. 1907 [see *ichthammol]. 1932 C. J. Miller *Clin. Gynecol.* v. 105 Eroded and inflamed cervices are often temporarily improved by local applications of tincture of iodine..or some similar agent, after which tampons saturated with ichthyol and glycerine are inserted into the vagina. 1950 *Jrnl. Soc. Chem. Industry* LXIX. 107/1 There is confusion in the nomenclature: the British Pharmacopoeia defines ichthammol as 'the ammonium salts of sulphonic acids of an oily substance, prepared from a bituminous schist, together with ammonium sulphate and water'. The oily substance..is commonly known as ichthyol. The B.P.C. however, states that ichthyol is a proprietary name for a brand of ichthammol. Unna..urged the use of the unsulphonated oil and the restriction of the name ichthyol to this material... When [sulphonated and] neutralized with ammonia, the product was variously known as ammonium ichthyosulphonate, the original name of ichthyol, and finally as ichthammol: this is the ichthammol of the B.P. 1962 *Brit. Jrnl. Plastic Surg.* XV. 278 All burns have had the same basic treatment—exposure and regular painting of the burned areas and surrounding skin with glycerin and ichthyol.

ichu (ī·tʃū). Also **icho, ychu.** [Quechua.] An alpine grass, *Stipa ichu*, growing on the uplands of the Andes, where it is used for fodder and thatching.
[1781 H. Ruiz *Relación del Viaje* (1931) 137 Se mantiene porción de ganado bacuno y carneros de Castilla con el Icho y con el corto, pero abundante pasto.] 1891 E. B. Clark *Twelve Months Peru* 136 The coarse *ychu* grass, growing in tufts upon the mountain slopes. 1921 *Glasgow Herald* 23 Apr. 10/1 The great ichu-covered steppes of the plateau. 1950 T. H. Goodspeed *Plant Hunters in Andes* iv. 119 Herds of llamas and alpacas graze the ichu. 1964 A. R. Steele *Flowers for King* ii. 42 Other desired specimens rating special mention [in 1776] were cinchona, the source of quinine; *icho* grass from the high mountain plateaus, useful for matting and cordage. *Ibid.* vii. 99 The scientists..swung perilously upon a raw-hide rope bridge one cold midnight by the light of *icho* grass flares.

icing, *vbl. sb.* Add: **1.** Also *fig.* in phr. *the icing on the cake*, the 'trimmings'.
1969 *Listener* 3 Apr. 468/3 All this theology is icing on the cake... '*Pas sérieuse*'..is an epithet I would apply to the whole book. 1970 *Globe & Mail* (Toronto) 26 Sept. 12/1 The missionaries have a dubious legacy, says Father Gallagher. 'We had been giving them the icing and not the cake, the Christian tools with which they need to meet the circumstances in life they face.' 1973 R. Perry *Nowhere Man* v. 100 It was quite a neat ploy, the icing on the cake.

2. b. Preservation by means of ice; refrigeration with ice.
1883 R. F. Walsh *Irish Fisheries* (Fish. Exhib. Publ.) 17 A rate for fish carrying is then struck; this includes icing.

3. Also *Aeronaut.* (esp. with *up*), the formation of ice on an aircraft. (Further examples.)
1929 R. Duncan *Air Navigation & Meteorol.* (ed. 3) xii. 124 What goes on during the action known as icing-up, or the accumulating of ice-coats on the plane's surfaces. 1937 *Evening News* 29 Jan. 8/3 The latest theory connects the very fine rain with one of the most dangerous phenomena in flying: icing-up of the wings. 1945 *Tee Emm* (Air Ministry) V. 48 Icing-up caused him to lose

control of the aircraft. *Ibid.* 53 Icing may occur, which requires hot air and..reduces available engine power. 1955 *Times* 12 Aug. 5/1 The Hull Trawler Officers' Guild are meeting soon to discuss the icing up of trawlers, and it is expected that they will make recommendations to the trawler owners. 1957 *Economist* 21 Sept. 965/1 This warns pilots to fly no higher than 16,000 feet whenever they meet conditions likely to bring on the Britannia's particular form of icing trouble. 1958 'N. Shute' *Rainbow & Rose* viii. 295 There was no icing on the aircraft. 1966 T. Armstrong et al. *Illustr. Gloss. Snow & Ice* 27 *Icing*, the accumulation of a deposit of ice on exposed objects, e.g. aircraft, ships, aerials, instruments. Icing may be produced by the deposition of water vapour or by the freezing on impact of droplets in the air.

4. *icing sugar*: finely powdered sugar.
1889 A. B. Marshall *Cookery Bk.* iii. 41 *Royal Icing.*—To two and a half pounds of icing sugar put seven or eight whites of fresh eggs and half a tablespoonful of lemon juice strained. 1896 J. T. Law *Grocer's Manual* 232/1 Icing sugar is that [sugar] which is ground to a very fine, impalpable powder, resembling flour, and should consist of pure cane sugar; for beet sugar..is not capable of high crystallisation. 1907 *Yesterday's Shopping* (1969) 42 Sugars... Loaf..Granulated..Icing. *c* 1938 *Fortnum & Mason Price List* 58/2 Icing sugar..per tin 1/2. 1970 Simon & Howe *Dict. Gastron.* 365/1 Icing sugar..is used to make icing..and meringues. 1972 J. Wilson *Hide & Seek* i. 19 Where's the icing sugar?..There's only the gran and castor there.

ickle (i·k'l), *a.* A hypocoristic form of Little *a.*: in childish use. Also **i·ckly.**
1846 Dickens *Dombey* (1848) i. 5, I came down from seeing dear Fanny, and that tiddy ickle sing. 1905 E. M. Forster *Where Angels fear to Tread* viii. 278 Good ickle quiet boysey, then. 1906 E. Dyson *Fact'ry 'Ands* xiv. 184 Oo's mummy's ickle sly-boots, oo is—*oo is!* Baby's a baddy baddy 'icky bubb-bubb. 1936 'G. Orwell' *Keep Aspidistra Flying* i. 26 A Peke, the ickle angel pet, wiv his gweat big Soulful eyes and his ickle black nosie—oh so ducky-duck! 1937 R. Macaulay *I would be Private* I. i. 19 Was it a nice ickly boy, then? 1968 'P. Hobson' *Titty's Dead* xi. 121 She changed her role. Now she was Daddy's ickle girl.

icky, ikky (i·ki), *a.* and *sb. colloq.* [Origin uncertain.] **A.** *adj.* **a.** *Jazz.* Ignorant (of true swinging jazz and liking the 'sweet' kind). *U.S.* **b.** Sweet, sickly, sentimental; hence a general term of disapproval: nasty, repulsive, sticky, etc.; also, ill, sick.
1935 *Vanity Fair* (N.Y.) Nov. 71/2 If the straight music is also oversweet, the term *icky* (a pseudo-baby-patter word, meaning 'little') is frequently employed to denote this. 1938 D. Baker *Young Man with Horn* IV. vii. 277 Smoke Jordan tried hard to get him to..maybe take a vacation, Florida's nice. 'Get yourself wheeled up and down like an icky banker?' 1939 Joyce *Finnegans Wake* 555 His wrinkly waste of methylated spirits.. and pulverised rhubarbarorum, icky. 1942 D. Powell *Time to be Born* (1943) x. 229 'A pet-shop on the first floor? But Miss Haven! Really!' gasped Miss Finkelstein. 'How *icky*!' 1945 L. Shelly *Jive Talk Dict.* 26/2 *Icky vicki*, stupid gal. 1952 S. Kauffmann *Philanderer* (1953) 13 'It's just that—oh, I don't know—now everything's so icky.'..Another of her dubious charms. The high-school words. 1959 S. Bellow *Henderson Rain King* xiv. 200 Under the thickened rain clouds, a heated, darkened breeze sprang up..choky, sultry, icky. 1964 C. Chaplin *Autobiogr.* xxi. 352 He must hide his blindness... His stumblings and bumpings into things make the little girl laugh joyously. But that was too 'icky'. 1964 *Harper's Bazaar* Nov. 110/2 Roast chestnuts or icky home-made fudge. 1967 H. Hunter *Case for Punishment* vi. 108 She wears the most *fright*-ful cardigans. Always some sort of *ikky* colour—to go with everything, I suppose. 1968 J. Hudson *Case of Need* IV. iii. 239 I'm not hungry... The food tastes icky. 1970 *New Yorker* 28 Nov. 122/2 We pick out the icky things in men and call them male, and then we say that the woman who has them..we say she's a male-identifier. 1972 R. Quilty *Tenth Session* 21 A group of ton-up boys dipped their icky fingers in the sugar bowls. 1972 M. Woodhouse *Mama Doll* ix. 117, *I* showed *him*..lots of gore, you know, and he went all green and icky and dashed off.

B. *sb.* **1.** A person who is ignorant of true swinging jazz and likes the 'sweet' kind. *U.S.*
1937 *Amer. Speech* XII. 180 Musicians unable to swing, and who therefore can only play corn are called 'long hairs' and those who enjoy listening to their music are 'Ickies'. This term implies definite bovine qualities. 1938 *Better English* Nov. 51/2 *Icky*, one who is not hip but thinks he is. 1939 *Words* Oct. 108/1 *Ickies*..are hicks who'll never be able to tell a sweet from a swing arrangement. 1955 L. Feather *Encycl. Jazz* x. 346 *Icky* (*Obsolete*), person with poor musical taste.

2. Something which is sickly, disagreeable; sickness.
1969 D. Francis *Enquiry* vii. 88 Poppy's got the morning ickies again. I'll be glad when this lousy pregnancy's over.

Also in extended (and possibly unconnected) expressions, as **icky-boo** *a.*, ill (in quot. 1970[2] written *icky-poo*); **ickylickysticky** *a.* (nonce-wd.), unpleasantly sticky.
1920 'Sapper' *Bull-Dog Drummond* ix. 234 Can it be that my little pet is feeling icky-boo? Face going green—slight perspiration—collar tight. 1922 Joyce *Ulysses* 570 They blow ickylickysticky yumyum kisses. 1930 'Sapper' *Finger of Fate* 188 The jolly old tum-tum is not icky-boo or anything like that. 1970 'B. Mather' *Break in Line* v. 57 Call the airline office..and tell 'em you're feeling an icksy bit icky-boo and want a stopover. 1970 *New Yorker* 14 Nov. 55 (*caption*) If any..engine conkouts, or fires make you feel icky-poo, you just come and tell.

icon. Add: **3. b.** *Philos.* (See quot. 1934.) Also *transf.*
a 1914 C. S. Peirce *Coll. Papers* (1931) I. III. iii. 195 It has been found that there are three kinds of signs which are all indispensable in all reasoning; the first is the diagrammatic sign or *icon*, which exhibits a similarity or analogy to the subject of discourse. *Ibid.* 196 There may be a mere relation of reason between the sign and the thing signified; in that case, the sign is an *icon*. 1934 *Mind* XLIII. 497 An *icon* is a sign which represents its object by virtue of having some character in common with the object: the colour of a colour-card as representing the colour of the object which it resembles is an icon, and a map as representing spatial relations is an icon. 1949 *Poetry* (Chicago) Jan. 234 *Icons*, images, which are the aesthetic signs of the poem, analogous to the symbolic signs of scientific discourses; they have, as signs, semantic objects, or refer to objects, and, in addition, as iconic signs, resemble those objects. 1954 W. K. Wimsatt *Verbal Icon* (1967) p. x, The term *icon* is used to-day by semeiotic writers to refer to a verbal sign which *somehow* shares the properties of, or resembles, the objects which it denotes.

Iconian (əikōᵘ·niăn), *a.* and *sb.* [-an.] **A.** *adj.* Of or pertaining to Iconium (mod. Konya), a town in southern Asia Minor where St. Paul preached, or to the church established there (Acts xiii. 5, xiv. 1–7). **B.** *sb.* A native or inhabitant of Iconium.
1899 W. M. Ramsay in *Expositor* Aug. 112 There were strife and wrangling and jealousy between the Antiochean Church and the Iconian Church about precedence and comparative dignity. 1911 —— *First Christ. Cent.* xxv. 168 St. Paul addressed the Iconian audiences in Greek. *Ibid.* 171 The Iconians clung to their Phrygian character as opposed to the Lycaonian.

iconic, *a.* Add: **c.** *Semiotics.* Pertaining to or resembling an icon (sense *3 b). Also *transf.*
1939 C. W. Morris in *Kenyon Rev.* I. IV. 415 The aesthetic sign..is an iconic sign (an 'image') in that it embodies these values in some medium where they may be directly inspected (in short, the aesthetic sign is an iconic sign whose designatum is a value). 1949 [see *icon 3 b]. 1956 E. H. Hutten *Lang. Mod. Physics* ii. 15 Sometimes, the sign is similar to the thing it stands for, in the manner in which a picture represents, and we have iconic signs. 1964 T. W. McRae *Impact of Computers on Accounting* v. 132 There are many kinds of model. The one described above is an *iconic* model, that is a physical representation of the original item. 1965 C. H. Springer et al. *Adv. Methods and Models* i. 6 He might use..an *iconic* model, which doesn't *act* like the real thing (as the analog model does) but only *looks* like it. 1966 M. Pei *Gloss. Ling. Terminol.* 118 *Iconic*, characterized by a symbolism which purports to present an image of the object described (Chinese pictographs). 1970 *English Studies* LI. 279 Non-roman notations are generally 'iconic', i.e. 'the symbols are not arbitrary signs, but in some way resemble what they stand for'. 1971 *Language* XLVII. 416 There is..growing evidence that language contains many elements which are iconic—that is, imitative of non-linguistic reality.

ico·nically, *adv.* [f. Iconic *a.*: see -ically.] In an iconic manner.
1946 C. Morris *Signs, Lang. & Behavior* 193 Such conflict is presented iconically in the music itself. 1973 *Times Lit. Suppl.* 5 Oct. 1188/4 The rhythm..depicts the distance iconically, since the farther away the goal, the fewer cycles of the dance occur in a given period.

iconicity (əikǫni·siti). [f. Iconic *a.* + -ity.] The quality or fact of being iconic (in various senses).
1946 C. Morris *Signs, Lang. & Behavior* 191 Spoken language contains some sounds which are clearly iconic ('onomatopoetic'); the extent of its iconicity is a difficult matter to determine. 1971 *Language* XLVII. 425 In some cases, there is iconicity between language and a non-human communication system. 1972 *Sci. Amer.* Sept. 91/2 The only element of genuine representation (also called iconicity) in such a case is the actual shape of the geographical features, although even these are normalized according to given rules of transformation to allow a part of the globe to be shown on a flat map.

iconism. Add: Revived in *Semiotics* in sense 'the quality or fact of being an icon or intentional sign'.
1971 *Language* XLVII. 417 Among homotherms—that is, warm-blooded vertebrates, such as birds and mammals—iconism is less salient. Yet it seems to be present, at least latently, in the 'intention movements' exhibited..by herring gulls who are about to take flight and who seem, by spreading their wings, not only to be preparing for flight themselves but also to be inviting neighboring gulls to fly with them.

icono-. Add: **i:conodu·le** = Iconodulist; **iconopho·bia**, hatred of images; also **i·conophobe, iconopho·bic** *adjs.*, of or pertaining to one who hates images.
1893 *Funk's Stand. Dict.*, *Iconodule*, *iconodulist*, one who serves images; an image-worshipper; iconolater. 1900 'Odysseus' *Turkey in Europe* 230 The division of Asia representing Iconoclasts and Europe Iconodules is almost without exception. 1901 E. Gosse in *Daily Chron.* 22 Nov. 3/3 The sentiment of the author is vehemently on the side of the Iconodules. 1939 A. Toynbee *Study of Hist.* IV. 595 The indomitable Iconodule Patriarch Germanus found a worthy successor in the reigning Patriarch Nicephorus. 1967 H. Chadwick *Early Church* xviii. 283 The iconodules replied: (a) we venerate not the icons but those whom they depict [etc.]. 1958 *Times Lit. Suppl.* 23 May 278/3 Saxl himself believed that such an approach to history was natural, and indeed inevitable, in an age of illustrated papers and the film and television; but the English are notoriously iconophobe, and it may still be necessary to insist on this point. 1926 H. Read *Eng. Stained Glass* ii. 16/1 Free from the iconophobia which infested Southern Christianity, the Romanesque builders freely developed the art of sculpture. 1963 Auden *Dyer's Hand* 359 The *Mayflower* carried iconophobic dissenters.

iconographical, *a.* Add: (Further examples.)
1958 *Times Lit. Suppl.* 12 Dec. 716/3 The book is a *tour de force* in iconographical method. 1962 *Daily Tel.* 20 Nov. 17/4 The space between is filled entirely by Nicolas Untersteller's superb, sparkling glass, dark red and blue in patterns that are both good abstracts and also tell an iconographical story. 1962 *Listener* 27 Dec. 1087/1 Parallel with the abundance and high quality of abstract art is a great deal of what one may call iconographical art. Let me remind you of a few names: Bacon, Balthus, Giacometti, Dubuffet, Asger Jorn, de Kooning.
So **iconogra·phically** *adv.*
1959 *Times* 17 Jan. 9/1 Iconographically they [paintings] are of great interest in relation to the usages of the Eastern Church.

iconography. Add: **2.** (Further examples.) Also *transf.*
1939 E. Panofsky *Stud. Iconology* i. 3 Iconography is that branch of the history of art which concerns itself with the subject matter or meaning of works of art, as opposed to their form. 1957 *Times Lit. Suppl.* 8 Nov. 680/1 The iconography of the watermark is a new aspect of the study of the migration of symbols at present so popular among art historians. 1960 *Listener* 8 Dec. 1041/2 The fashion of so-called 'iconography' at this moment has produced many cumbersome interpretations. 1962 *Ibid.* 27 Sept. 473/2 The iconography of a work of art is its dramatic structure seen in terms of characters, situations, and images that are related to a specific religious, social, or historical context. 1969 *Ibid.* 17 Apr. 533/2 Now people are interested in iconography, in saying what the subject represents, and offering infinitely ingenious interpretations of every picture. 1970 *Oxf. Compan. Art* 555/1 *Iconography*, a term in art history..extended in the 20th c. to cover the whole descriptive investigation of the subject matter of the figurative arts... Iconography..studies the development of the themes which artists use, for instance the transformation of the images of planets in astrological manuscripts, the rise of genre painting, the origins of still life, and the use of political satire.

iconological, *a.* (Further examples.)
1938 E. Panofsky in *Jrnl. Warburg Inst.* I. 23 From the purely iconological point of view the Ottawa picture may be compared to the two Cassone panels. 1958 *Times Lit. Suppl.* 23 May 278/3 On the contrary, the uniquely valuable and fruitful idea embodied in the Warburgian programme is the iconological idea; it is of universal application, and thus of service to all the special historical disciplines mentioned.

iconology. 1. (Further examples.)
1939 M. Praz *Stud. 17th-Cent. Imagery* I. iv. 184 Iconology during the period of enlightenment in philosophy and in literature, takes the place held by emblematics during the age of the Jesuits and the Baroque. 1949 Wellek & Warren *Theory of Lit.* xi. 125 The conceptual and symbolic meanings of works of art ('Iconology'). 1956 E. Panofsky *Meaning in Visual Arts* 31 The discovery and interpretation of these 'symbolical' values (which are often unknown to the artist himself and may even emphatically differ from what he consciously intended to express) is the object of what we may call 'iconology' as opposed to 'iconography'. 1958 *Times Lit. Suppl.* 23 May 278/3 If the Institute were not called after its founder, *tout court*, and had to find a compendious title to describe its activities, it might surely best be called an Institute of Iconology, as being a body dedicated to the study and interpretation of historical processes through visual images.

iconometer (əikǫnǫ·mɪtəɹ). *Photogr.* [f. Icono- + -meter.] (See quots.)
1894 E. L. Wilson *Cycl. Photogr.* 194/1 *Iconometer*, view-meter; a pocket instrument which quickly indicates what kind of objective to use..from a given standpoint, or..the standpoint suitable for a view with a given objective. 1918 *Photo-Miniature* Mar. 25 *Iconometer*, a view-meter of 'direct-vision' pattern, i.e., consisting of an open frame with an eyehole or lens fixed behind it. 1919 *Brit. Jrnl. Photogr. Alm.* 244 A view-meter, or iconometer, is a separate accessory for ascertaining the picture produced by any given lens and size of plate.

iconometry (əikǫnǫ·mɪtri). [ad. F. *iconométrie* (A. Laussedat 1892, in *Ann. du Conservatoire d. Arts et Métiers* IV. 374), f. Gr. εἰκονο- Icono-: see -metry.] The process of taking measurements from photographs of an area and using them to make a map or survey of it.
1898 *Ann. Rep. U.S. Coast & Geodetic Survey 1897* II. 628 Iconometry means the measuring of dimensions of objects from their perspectives ('Bildmesskunst'), and this term could well be applied to those graphic constructions which serve to convert perspectives into horizontal projections; iconometry is the reverse of perspective drawing. 1923 D. Clark *Plane & Geodetic Surveying* II. vi. 244 In plotting the map, the distances and elevations

required must be obtained from the perspective dimensions on the photographs. The process—termed iconometry—is therefore the reverse of perspective drawing. 1934 R. E. Davis et al. *Surveying* (ed. 2) xxvii. 787 The iconometry of aerial photographic surveying is much less accurate than that of terrestrial work.

Hence **iconome·tric, -me·trical** *adjs.*, employing or forming part of this process; **iconome·trically** *adv.*, in or by means of iconometry.
1898 *Ann. Rep. U.S. Coast & Geodetic Surv. 1897* II. 631 The correct orientation of the picture traces forms the most important part of iconometric plotting. *Ibid.* 630 If..two different perspectives..of the same object.. are given, the dimensions and the position of the object with reference to the two stations may be determined iconometrically. 1906 J. A. Flemer *Elem. Treat. Phototopogr. Methods* i. 6 It is not easy to make free-hand sketches of landscapes geometrically accurate enough to be used iconometrically in place of the landscapes. *Ibid.* vi. 121 (*heading*) Graphical iconometrical plotting methods. 1934 R. E. Davis et al. *Surveying* (ed. 2) xxvii. 775 (*heading*) Iconometric interpretation of the stereoscopic view. 1944 A. L. Higgins *Higher Surveying* iii. 219 [Photogrammetry.] Office Work. Apart from the preparation of negatives and prints, this consists of (1) plotting the triangulation, (2) iconometrical plotting, and (3) topographical mapping.

iconoscope (əikǫ·nǫskōᵘp). [f. Icono- + -scope.] †**1.** [a. F. *iconoscope* (E. Javal 1866, in *Compt. Rend.* LXIII. 927).] (See quot. 1890.) *Obs.*
1866 *Chem. News* 7 Dec. 273/2 (*heading*) On a new instrument, the iconoscope, intended to give relief to plain images examined with the two eyes. 1890 Billings *Med. Dict.* 678/2 *Iconoscope*, an instrument for suppressing binocular parallax. It makes real objects appear flat like pictures, but..gives to flat pictures a relief like that obtained by a monocular view, by removing those binocular sensations that keep the observer reminded of the flatness of the picture. It may be described briefly as a small telestereoscope reversed. 1900 C. Weiland tr. *Tscherning's Physiol. Optics* xxiii. 321 The iconoscope of Javal resembles somewhat an inverted telestereoscope. *Ibid.*, Looking through the iconoscope the relief is more marked than when simply closing one eye.

2. A kind of television camera tube (now little used) in which the target plate that receives the image consists of a mosaic of photo-emissive material on an insulating sheet that is backed with a conducting sheet, the video signal being obtained from the variation in the current flowing to or from this latter sheet as the mosaic is scanned with an electron beam.
The term was registered in the U.S.A. as a trade name in 1935 but it is now a generic term in the public domain.
1933 V. K. Zworykin in *Jrnl. Inst. Electr. Engin.* LXXIII. 437/1 The device has been named the 'iconoscope', and it consists of a vacuum tube containing an electron-emitting gun and a photo-sensitive surface of a unique type. This photo-sensitive surface is scanned by an electron beam from the gun, which serves as a type of inertialess commutator. 1935 [see *definition 5 c]. 1953 Amos & Birkinshaw *Television Engin.* I. iv. 68 Iconoscope camera tubes have given satisfactory results in television services over a number of years. 1961 G. Millerson *Technique Television Production* 50 A familiar effect with the iconoscope camera-tube, shading appears as gradual darkening or lightening over parts of the picture. 1966 *McGraw-Hill Encycl. Sci. & Technol.* XIII. 464/1 The iconoscope was used in early live television broadcasting but is now used only in motion picture reproduction... The vidicon is replacing the iconoscope as a film reproducer.

b. Applied to a modified form of the instrument intended as an infra-red telescope or detector.
1946 *Electronic Engin.* XVIII. 317 Another instrument used by the Germans was an infra-red iconoscope. It does not differ in general principle from the iconoscope used in television, the only difference being in the photo-sensitive layer. Whereas in television, a mosaic layer..is used, the infra-red instrument uses a semiconducting layer, the resistance of which changes on irradiation. 1949 A. R. Weyl *Guided Missiles* 97 The final installation would have been the 'Electric Eye' iconoscope target-finding device of Rambauske which was, however, not operationally developed when the War came to an end.

iconotropy (əikǫnǫ·trǫpi). [f. Gr. εἰκών, εἰκον- Icon + τροπή a turn, turning: cf. Trope.] The misinterpretation by one cult of the icons, etc., of another (earlier) cult, so as to bring the beliefs and myths depicted into accord with those of the later cult. Hence **iconotro·pic** *a.*, of a myth, tradition, etc., suggestive of an origin in such a misinterpretation.
1946 R. Graves *King Jesus* 355 A similar technique of misinterpretation—let us call it iconotropy—was adopted in ancient Greece as a means of confirming Olympian religious myths at the expense of the Minoan ones which they superseded... In iconotropy the icons are not defaced or altered, but merely interpreted in a sense hostile to the original cult. *Ibid.* 356 The story of Lot and the Sodomites suggests the same ancient icon from which Herodotus derived his iconotropic account of the sacking of the Temple of the Love-goddess Astarte at Ascalon by the Scythians. 1955 —— *Greek Myths* I. 21 If some myths are baffling at first sight, this is often because the mythographer has accidentally or deliberately misinterpreted a sacred picture or dramatic rite. I have called such a process 'iconotropy'... Greek myth teems with iconotropic instances. 1958 *Observer* 7 Sept. 18/8 Professor Webster appears to subscribe to Mr. Robert Graves's theory of iconotropy, or the misinterpretation of myths from visual sources.

icos-, icosa-, icosi-. Add: **b.** In *Chem.* (usu. spelt **eicos-**) denoting the presence in a molecule of twenty atoms of some element (usu. carbon). [App. first so used in G. *eikosylen* (Lippmann and Hawliczek 1879, in *Ber. d. Deut. Chem. Ges.* XII. 72).] So **ei·cosane** [-ane 2 b], a hydrocarbon, $C_{20}H_{42}$, of the paraffin (alkane) series, esp. the normal isomer; **eicosa·nic** *a.*, in *eicosanic acid*, = **eicosanoic* adj.; **eicosano·ic** *a.*, in *eicosanoic acid*, a saturated fatty acid, $C_{19}H_{39}COOH$, of which the normal isomer (also called arachidic acid) is a waxy solid present in small amounts in many natural oils and fats; **eicose·nic** (-ī·nik) *a.*, in *eicosenic acid* [tr. G. *eikosensäure* (M. Bodenstein 1894, in *Ber. d. Deut. Chem. Ges.* XXVIII. 3403)], = **eicosenoic* adj.; **eicoseno·ic** *a.*, in *eicosenoic acid*, an unsaturated fatty acid, $C_{19}H_{37}COOH$, of which one isomer, 9-eicosenic acid (gadoleic acid), is a minor constituent of many fish oils and another, 11-eicosenic acid, occurs in the wax of certain plant seeds; **eicoso·ic** *a.*, in *eicosoic acid*, = **eicosanoic* adj.
1889 G. McGowan tr. *Bernthsen's Text-bk. Org. Chem.* 34 (*table*) $C_{20}H_{42}$. Eicosane. 1948 A. W. Ralston *Fatty Acids* xi. 882 The following are the melting and boiling points which have been reported for the higher normal saturated hydrocarbons: nonadecane, $C_{19}H_{40}$, 32°, —; eicosane, $C_{20}H_{42}$, 38°, $b_{0\cdot5}$ 148°; [etc.]. 1966 *Nomencl. Org. Chem.* (I.U.P.A.C.) (ed. 2) A. 6 The first four saturated unbranched acyclic hydrocarbons are called methane, ethane, propane and butane. Names of the higher members of this series consist of a numerical prefix and the termination '-ane'... Examples:..6 Hexane. 7 Heptane... 20 Eicosane. 1923 *Chem. Abstr.* XVII. 2560 Normal eicosanic acid is found in nature in the oil of *Nephelium lappaceum* J. as a glyceride. 1953 Heilbron & Bunbury *Dict. Org. Compounds* (rev. ed.) II. 464, *n*-Eicosanic Acid (Arachidic acid, *n*-nonadecane-1-carboxylic acid, eicosoic acid, eicosanoic acid). 1924 *Chem. Abstr.* XVIII. 4318/2 (Index), Eicosanoic acid. See Arachidic acid. 1948 A. W. Ralston *Fatty Acids* i. 44 The vegetable oils generally contain less than 1% of eicosanoic acid. *Ibid.*, Coffee bean oil contains 2·11% of eicosanoic acid..and peanut oils from 3 to 4%. 1895 *Jrnl. Chem. Soc.* LXVIII. I. 127 Icosenic acid, $C_{20}H_{38}O_2$, is formed when behenolic acid is heated with fused caustic potash at 250–270°. 1953 Heilbron & Bunbury *Dict. Org. Compounds* (rev. ed.) II. 464 Δ^{10}-Eicosenic Acid (Eicosenoic acid). 1936 *Jrnl. Chem. Soc.* 1755 The fatty matter present in the seeds of *Simmondsia californica* Nutt. ..is a mixture of wax-esters, and not glycerides... The chief acid is $\Delta^{11:12}$-eicosenoic. 1951 A. W. Johnson et al. in E. H. Rodd *Chem. Carbon Compounds* IA. ix. 640, 9-Eicosenoic acid, gadoleic acid, $CH_3(CH_2)_9CH{:}CH(CH_2)_7COOH$ is a common component of fish and marine animal oils such as herring and shark liver oil. 1954 E. W. Eckey *Vegetable Fats & Oils* xiv. 435 Investigations made after eicosenoic acid had been isolated from rapeseed oil and shown to exist in substantial proportion in hare's ear mustard oil confirmed the presence of this C_{20} monoethenoic acid in rapeseed oil and indicated that it occurs generally in oils of the Cruciferae. 1923 *Chem. Abstr.* XVII. 4447/2 (Index), Eicosoic acid (eicosanic acid), *n*-. 1951 I. L. Finar *Org. Chem.* ix. 144 Some still higher acids are found in waxes: arachidic (eicosoic), $C_{20}H_{40}O_2$ (m.p. 77°), behenic (docosoic), $C_{22}H_{44}O_2$ (m.p. 82°), [etc.].

icosidodecahedron. (Examples.)
1911 *Encycl. Brit.* XXII. 28/2 The icosidodecahedron.. is a 32-faced solid, formed by truncating the vertices of an icosahedron so that the original faces become triangles. 1939 H. S. M. Coxeter *Ball's Math. Recreations & Ess.* (ed. 11) v. 136 The compound of five cubes has the 30 facial planes of a triacontahedron. Reciprocally, the compound of five octahedra has the 30 vertices of an icosidodecahedron. 1972 *Science* 12 May 654 (*caption*) Icosidodecahedron, its first two stellations, and its final stellation.

ictal (i·ktăl), *a. Med.* [f. Ict(us + -al.] Of, pertaining to, or caused by an ictus (sense *2 c).
1950 Penfield & Rasmussen *Cerebral Cortex of Man* ix. 161 The ictal paralysis of normal function passes over into postictal paralysis. 1968 Schmidt & Wilder *Epilepsy* ii. 35 Ictal emotional experience is frequent among patients with temporal lobe epilepsy.

ictas, ictus, varr. *iktas *sb. pl.*

ictero- (i·ktĕro), combining form of Gr. ἴκτερος jaundice, as in **i:cterogene·tic** (*rare*), **icteroge·nic** *adjs.*, causing jaundice.
1897 *Allbutt's Syst. Med.* IV. 38 This drug is the most notable of all icterogenetic poisons. 1903 Dorland *Med. Dict.* (ed. 3) 331/1 Icterogenic. 1944 *Lancet* 16 Sept. 365/1 Icterogenic yellow fever vaccine. 1962 *Lancet* 19 May 1057/2 'The pill' should not be overlooked when the medicine-chest is scrutinised for potentially icterogenic drugs.

ictus. Add: **1.** Used of Old English verse.
1823 J. Bosworth *Elem. Anglo-Saxon Gram.* 246 [quoting J. J. Conybeare] The ear is satisfied, not by the number of syllables, but by the recurrence of the accent, or ictus, if one may call it so. 1888 A. H. Tolman in *Publ. Mod. Lang. Assoc.* III. 21 March..declares that 'the time from each ictus to the next is the same in any section'. 1953 F. P. Magoun in *Speculum* XXVIII. 458 The first down-beat or ictus in the off-verse does not here alliterate with the preceding on-verse.

2. c. A stroke, seizure, or fit. Also in some mod.L. phrases.
1890 Gould *New Med. Dict.* 204/1 *Ictus*, a stroke or attack of disease coming without premonition. 1890 Billings *Med. Dict.* I. 679/1 *Ictus sanguinis*, apoplexy. 1908 A. Church *Dis. Nervous Syst.* 487 Especially severe attacks of tickling in the throat, arrest of respiration, unconsciousness, and epileptic attacks have been described as *ictus laryngeus*. 1931 I. S. Wechsler *Textbk. Clin. Neurol.* (ed. 2) IV. 345 The clinical course [of apoplexy] may be conveniently divided into (1) the acute apoplectic stroke or ictus, and (2) the stage of paralysis. 1939 W. Haymaker tr. *Bing's Textbk. Nervous Dis.* xvii. 444 The most striking symptom of apoplectic stroke (*ictus apoplecticus*) is sudden loss of consciousness. 1961 *Lancet* 29 July 223/1 The conscious level chosen for stratification of patients in the trial was that at twenty-four hours after the ictus.

icy, *a.* Add: **4.** *icy-clear.*
1922 W. de la Mare *Down-adown-Derry* 93 Fleet-foot deer Lap of its waters icy-clear. 1925 Blunden *Masks of Time* 41 Icy-clear The air of a mortal day shocks sense.

id[2] (id). *Psycho-analysis.* [A use of L. *id* it, as a rendering of G. *es* it, which was adopted by Freud (*Das Ich und das Es* (1923)) following its use in a similar sense by G. Groddeck (*Das Buch vom Es* (1923)).] The inherited instinctive impulses of the individual, forming part of the unconscious and, in Freudian theory, interacting in the psyche with the ego and the super-ego. Also *attrib.*
[1917 Freud *Briefe* 5 June (1960) 316 [*To Georg Groddeck*] Ich muß Anspruch auf Sie erheben, muß behaupten, daß Sie ein prächtiger Analytiker sind, der das Wesen der Sache unverlierbar erfaßt hat. Wer erkennt, daß Übertragung und Widerstand die Drehpunkte der Behandlung sind, der gehört nun einmal rettungslos zum wilden Heer. Ob er das 'Ubw' [*sc.* Unbewußte, 'unconscious'] auch 'Es' nennt, das macht keinen Unterschied.] 1924 J. Riviere et al. tr. *Freud's Coll. Papers* II. xxi. 250 The essay..describes the various allegiances the ego owes, its mediate position between the outer world and the id, and its struggles to serve all its masters at one and the same time. [*Translators' note.*] To translate the German '*es*', which means 'it' and thus implies the impersonality of the mind apart from its ego, the Latin '*id*' has been selected. *Ibid.* 254 Keep in mind this dissection of the mental apparatus that I have proposed, namely, into ego, super-ego and id. 1927 J. Riviere tr. *Freud's Ego & Id* ii. 27 We need feel no hesitation in finding a place for Groddeck's discovery in the fabric of science. I propose to take it into account.. by following Groddeck in giving to the other part of the mind, into which this entity [*sc.* the ego] extends and which behaves as though it were Ucs [*sc.* unconscious], the name of Id (Es). [*Note*] Groddeck himself no doubt followed the example of Nietzsche, who habitually used this grammatical term for whatever in our nature is impersonal and, so to speak, subject to natural law. 1942 *Essays & Stud.* XXVII. 12 Dreams..that fulfil the much darker wishes of the Id. 1943 H. Read *Educ. through Art* vi. 176 The super-ego is the direct representative of the unconscious, of the id, and hence the possibility, indeed, the inevitability, of a conflict with the ego. 1952 Shaffer & Lazarus *Fund. Concepts Clin. Psychol.* vi. 188 The forces which keep the id impulses in check as an adaptation to the pressures of the outside world comprise the ego. 1957 J. Braine *Room at Top* xxvi. 208 Roy, a quiet type normally, seemed to become, as Charles said, all Id when he'd had one over the eight. 1961 R. W. Lundin *Personality* i. 21 The id is entirely unconscious, having no contact with reality except through the ego. One may liken the id to the primitive or animal nature of man. 1962 R. Fine *Freud* xi. 156 The id is the source of all drives, the reservoir of instincts. 1965 C. M. & S. Grossman *Wild Analyst* xii. 109 It became clear, when the two books [*sc.* Groddeck's *Das Buch vom Es* and Freud's *Das Ich und das Es*] were translated, that it had been wise to distinguish between the 'Id' and the 'It', because Freud's concept of the Id, broadened as it was, was still not as broad as Groddeck's concept of the It. *Ibid.* 110 He [*sc.* Groddeck] gave Freud the Id as a gift—he, too, had borrowed it, from Nietzsche. 1967 R. R. Greenson *Technique & Pract. Psychoanal.* I. i. 20 The combing of her hair stirred up repressed id impulses which brought her into conflict with her ego and super-ego... There were indications that her ego already was relatively depleted and her id lacked adequate discharge possibilities... As a consequence the fantasies mobilized by the hair combing increased the id tensions to a point where they flooded the infantile defenses of the ego and involuntary discharges took place, eventuating in acute symptom formation.

-id[2]. Add: **b.** *Astr.* Added to the name of a constellation to form the name of a meteor in a shower having its radiant point in that constellation, as *Andromedid, Leonid, Lyraid, Perseid*; also more widely used (cf. *Bielid).
c. Used as a terminal element in the names of epic poems, as Æneid, *Heracleid, Thebaïd.

-ID

-id, *suffix*⁴, in the nomenclature of mammalian teeth, used to indicate a structure forming part of a tooth in the lower jaw. Cf. *HYPOCONULID.

1897 H. F. Osborn in *Amer. Naturalist* XXXI. 1006 The suffix -*id* is employed arbitrarily to distinguish the elements of the lower molars from those of the upper. 1949 A. S. Romer *Vertebr. Body* x. 304 The names of specific cones are formed by adding..prefixes..and, where necessary, by suffixes: -ul(e) indicates a minor cusp, and -id a lower jaw element.

Idæan (əidī·ăn), *a.* Also Idaian. [f. L. *Idæus*, Gr. Ἰδαῖος (f. Ἴδα, Ἴδη) + -AN.] Of, belonging to, or dwelling on Mount Ida, either (*a*) a mountain in Asia Minor near the ancient Troy; or (*b*) the chief mountain in Crete, the birthplace of Zeus.

1590 Spenser *F.Q.* II. vii. 55 Here eke that famous golden Apple grew..For which th'Idæan Ladies disagreed. *Ibid.* III. vi. 6 Like as Cupido on Idæan hill. a 1649 Drummond of Hawthornden *Works* (1711) 7/1 Trembling Woods of Trees..Which make Idæan Woods in every Crook. 1810 Scott *Lady of Lake* I. xxvi. 32 Where Ellen's hand had taught to twine The ivy and Idæan vine. 1820 Shelley *Prometh. Unb.* III. i. 9? The Dædal Fountains of..Idæan Jove. 1876 Gladstone *Homeric Synchr.* 123 Teucer, son of Scamander and of an Idaian Nymph. 1891 *Public Opinion* 27 June 568/1 The sad dwellers on Idæan plains. 1970 *Oxf. Classical Dict.* (ed. 2) 540/1 *Idaean dactyls*, literally the Fingers of Ida, but whether the Phrygian or the Cretan Ida and whether their name refers to craftsmanship, dwarfish size, or something else, the ancients were in doubt.

id al-fitr: see *ID-UL-FITR.

Idalian (əidē·liăn), *a.* [f. L. *Idalius*: see -AN.] Of or belonging to the ancient town of Idalium in Cyprus, where Aphrodite was worshipped.

1599 Nashe *Lenten Stuffe* 34 Those debonaire Idalian nimphs and their spangled trappings. 1697 Dryden tr. *Virgil's Æneis* I. 955, I mean to plunge the Boy in pleasing Sleep, and, ravish'd, in Idalian Bow'rs to keep. 1799 T. Campbell *Pleasures of Hope* II. 55 Some cottage-home..With peace embosom'd in Idalian bow'rs! 1832 Tennyson *Œnone* in *Poems* 60 Idalian Aphrodite oceanborn. 1928 J. H. Mozley tr. *Statius* I. 187 Golden Venus..on her way from the height of Eryx to the Idalian groves.

iddingsite (i·diŋzəit). *Min.* [f. the name of Joseph P. *Iddings* (1857–1920), American geologist + -ITE¹.] A red-brown to orange-brown silicate of calcium, magnesium, and trivalent iron having an indefinite composition and formed as an alteration product of olivine.

1893 A. C. Lawson in *Bull. Dept. Geol. Univ. Calif.* I. 31 The common characteristic of all facies of these eruptive rocks is the presence, as a phenocryst of a mineral which..has received not little attention... The most extended and satisfactory note that has yet appeared regarding it is by Prof. J. P. Iddings... For this reason and also in recognition of Professor Iddings' eminent services to the science of petrography, it is proposed to name the mineral iddingsite. 1900 H. E. Gregory in *Bull. U.S. Geol. Survey* No. 165. 181 This alteration product again appears somewhat like the iddingsite found in the California trachyte. 1962 *Amer. Mineralogist* XLVI. 96 The optical homogeneity of many 'iddingsites' suggests that, even if the material is not to be regarded as a definite compound with relatively fixed chemical constitution, there is some structural control over the alteration process. 1966 W. A. Deer et al. *Introd. Rock-Forming Min.* 5 The compositions of both iddingsite and bowlingite show considerable variation.

iddy-umpty (idi,ʌ·mpti). Also **iddy-iddy-umpty.** Conventional verbal representation of the dots and dashes of the Morse code.

1904 *Punch* 24 Jan. 60/1 An 'Iddy Umpty' Idyll. 1914 *Daily Express* 15 Dec. 4/5 To see men practising the 'iddy-umpty', as they call it, with the back of a sheath-knife on the top of an empty tobacco-tin in lieu of a regulation 'dummy-key'. 1924 *Glasgow Herald* 25 June 10 For my sins of commission and of omission—as far as the worship of that fetish 'Iddy-Umpty' was concerned—I became for a time an inmate of the great signalling camp at Swanage. 1928 Fraser & Gibbons *Soldier & Sailor Words* 126 *Iddy (or Itty) Umpty*, an expression first used in India in teaching the dot-and-dash Morse system to native troops. An 'Iddy Umpty' in that way came to be used as a term for a signaller.

-ide. Add: The use of this suffix has been greatly extended in organic chemistry, notably in the generic names of various kinds of naturally occurring compounds, as *GLYCOSIDE, *PEPTIDE, SACCHARIDE (qq.v.); it is used *spec.* to form the names of glycosides from those of the corresponding sugars (as *galactoside* from *galactose*, *furanoside* from *furanose*).

2. *Chem.* Used to form *LANTHANIDE and later (by analogy) *ACTINIDE, signifying a similarity in properties to lanthanum and actinium, respectively.

idea, *sb.* Add: **4.** *big idea*: the purpose, intent. Freq. in ironic phr. *What's* (or *what is*) *the big idea?* orig. *U.S.*

1908 G. H. Lorimer *Jack Spurlock* vii. 151 That's not the Big Idea, I know; it's the idiotic one, but the market for idiocy is unlimited. 1917 R. W. Lardner *Gullible's Travels* (1926) iii. 83 Then we done a little spoonin' and then I ast her what was the big idear. 1927 A. P. Herbert *Plain Jane* 52 But now I'm not wanted no more Unless it's for scrubbing a floor, And if that's what a person is for—Well, what's the Big Idea? 1933 M. Lowry *Ultramarine* v. 220 What's the big idea not telling me before? 1937 A. Christie *Death on Nile* I. i. 35 You're crazy! What's the big idea? 1951 M. McLuhan *Mech. Bride* 43/1 Latch onto our big idea index for deep consolation? 1962 P. Gregory *Like Tigress at Bay* vii. 76 Jill entered, her face pale. 'What was the big idea?'

9. b. *to have no idea*: (*a*) not to anticipate or expect (a situation or occurrence); (*b*) to be unable to comprehend; usu. in phr. *you have no idea.*

1852 E. Ruskin *Let.* 17 Apr. in M. Lutyens *Effie in Venice* (1965) II. 298 In two days he got it done and they are grateful you have no idea. 1866 [in Dict.]. 1916 'Taffrail' *Pincher Martin* vii. 114 He's that conceited, you've no idea.

c. Colloq. phr. *to get* (or *have*) *ideas* (*into one's head*): to conceive notions of a particular kind, usu. undesirable or harmful; *spec.* to entertain a notion or intention of being rebellious, violent, etc.

c 1848 F. A. Kemble *Let.* in *Rec. Later Life* (1882) III. 322 A young boy..brought up in a girl's convent, and taken out for a week, during which he..sups and gets tipsy at the mess, and, in short, 'gets ideas' of all sorts. 1932 H. C. Wyld *Universal Dict. Eng. Lang.*, *To get ideas into one's head*, to cherish illusions. 1935 J. C. Squire *Reflections & Memories* 10 Babus would get ideas into their heads, but the Mutiny had taught its lesson and the redcoat had the situation well in hand. 1941 I. Baird *He rides Sky* 146 That's the second happy couple I've seen busted up in a month and it's cured me if I ever had ideas. I'd no more marry with a war on than jump over the moon. 1955 W. C. Gault *Ring around Rosa* vii. 82 Don't get any ideas, Brock Callahan. There are times when I simply—I mean, there's a definite therapeutical need for some form of release in a society as hectic as ——. *Ibid.* xiii. 156 Don't get any ideas, Callahan. This is an easy trigger.

d. Used after a possessive to denote a person's conception of an ideal, typical, or adequate example of the person or thing specified.

1903 G. B. Shaw *Man & Superman* III. 111 Is that your idea of a woman's mind? I call it cynical and disgusting materialism. 1907 —— *John Bull's Other Island* I. 7 Now thats my poor English idea of a whisky and soda. 1919 E. O'Neill *In Zone* in *Moon of Caribbees* (1923) 22 If this is your idea of a joke I'll have to confess it's a bit too thick for me to enjoy. 1933 —— *Ah, Wilderness!* (1934) I. 21 Gosh, he's always reading now. It's not my idea of having a good time in vacation. 1969 *Listener* 10 July 39/3 He would not be everyone's idea of a military dictator.

e. An idea worthy of consideration or capable of realization; a possibility; usu. in phr. *it's* (or *that's*) *an idea.* *colloq.*

1914 G. B. Shaw *Misalliance* 27 Thats an idea. Thats a new idea. I believe I ought to have made Johnny an author. 1919 —— *Inca of Perusalem* in *Heartbreak House* 197 Thats an idea. I will. 1942 A. Christie *Body in Library* i. 19 It *might* be. It's an idea, Jane. 1973 K. Giles *File on Death* i. 16 'I suppose I can take my Sergeant.'..'It might be an idea...Your Sergeant might wheedle his way where Chief Inspectors fear to tread.'

12. *idea-monger*; **idea(s) man,** a creative, inventive, or ingenious man.

1840 H. Reeve tr. *A. de Tocqueville's Democracy in Amer.* III. I. xiv. 123 For some few great authors..you may reckon thousands of idea-mongers. 1909 *Englishwoman* Apr. 305 Ibsen..was not merely an ideamonger, but a dramatist. 1923 *Glasgow Herald* 25 Jan. 4/2 Mr. Wells is a prolific idea-monger. 1938 'E. Queen' *Four of Hearts* (1939) i. 10 You're an idea man, and that's what they pay off on in Hollywood. 1940 *Ann. Reg.* 1939 363 Bryan Wallace was appointed Ideas Man to the Government. 1954 Koestler *Invis. Writing* xxxi. 333 He looked like the nonchalant impresario and idea-man of the great Comintern variety show. 1958 [see *copy-writer*]. 1960 *Guardian* 16 Nov. 7/3 David Bean, the ideas man, has specialized in exposing traps laid for consumers. 1967 *Ibid.* 17 Feb. 8/5 Dilettante ideas-men like Teilhard de Chardin.

ideal, *a.* and *sb.* Add: **A.** *adj.* **1. c.** *Sociol.* **ideal type** [ad. G. *idealtypus* (M. Weber 1922, in *Grundriss d. Sozialökonomik* I. i. 3), f. IDEAL *a.* + *typus* TYPE *sb.*¹], a hypothetical construct made up of the salient features or elements of a social phenomenon, or generalized concept, in order to facilitate comparison and classification of what is found in operation. Also (with hyphen) *attrib.* Hence **ideal-typical** *a.*, of or pertaining to an ideal type; **ideal typology,** the concept of ideal types.

1928 P. A. Sorokin *Contemp. Sociol. Theories* xii. 677 The outlined 'spirit of modern capitalism' is an example of one of the 'ideal types' of Max Weber. 1936 Wirth & Shils tr. *Mannheim's Ideology & Utopia* iv. 189 The pure types..of the utopian mind are constructions only in so far as they are conceived of as *ideal-types*. *Ibid.* 204 Max Weber always insisted that his general typology was created in order to characterize ideal-typical tendencies, and not immediately perceivable unique constellations. 1947 Henderson & Parsons tr. *Weber's Theory Social & Econ. Organiz.* I. i. 84 As a type ('ideal type') which has the merit of clear understandability and lack of ambiguity. 1949 R. K. Merton *Social Theory* xiv. 329 The Puritan ethic, as an ideal-typical expression of the value-attitudes basic to ascetic Protestantism generally, so canalized the interests of seventeenth-century Englishmen. 1962 T. B. Bottomore *Sociol.* ii. 33 Weber's exposition of his 'ideal type' method. 1964 Gould & Kolb *Dict. Social Sci.* 311/2 *Ideal-type analysis* denotes a method of sociological analysis associated with the name of M. Weber. 1964 I. L. Horowitz *New Sociol.* 42 Accounts of history that..were at best 'ideal-typologies' with strong subjective biases. *Ibid.* 456 The idealized or ideal-typical individual entrepreneur as against the political collectivity.

2. b. *ideal language* (Philos.): a supposed language which would mirror the world perfectly (cf. *logically perfect language).

1922 B. Russell in tr. *Wittgenstein's Tractatus* Introd. 8 The whole function of language is to have meaning, and it only fulfils this function in proportion as it approaches to the ideal language which we postulate. 1944 M. Black in P. A. Schilpp *Philos. B. Russell* 251 The 'ideal language' is, by definition, the symbolism which would be entirely free from the philosophical defects which Russell claims to find in ordinary language. 1953 G. E. M. Anscombe tr. *Wittgenstein's Philos. Investigations* I. § 81 It may look as if what we were talking about were an *ideal* language. 1963 R. Carnap in P. A. Schilpp *Philos. R. Carnap* 29 When we found in Wittgenstein's book statements about 'the language', we interpreted them as referring to an ideal language; and this meant for us a formalized symbolic language. 1964 M. Black *Compan. to Wittgenstein's Tractatus* xx. 133 In this section W. seems to be subscribing to the ideal of an 'ideal language'. But cf. 5.5563 (ordinary language is perfectly in order). 1967 *Encycl. Philos.* VII. 361 The frequently recurring project of an ideal language is to be found for the first time in the very first extant treatise on language. 1973 A. Kenny *Wittgenstein* iv. 70 In an ideal language,..to each element of the propositional sign would correspond a single object in the world.

4. b. *ideal construction* (Philos.): a mental conception formed by abstracting properties found in experience and recombining or developing them; the process of forming such a conception.

1874 G. H. Lewes *Foundation of Creed* I. 288 Hume did not clearly understand that Science is essentially an ideal construction very far removed from a real transcript of facts. 1877 —— *Physical Basis of Mind* I. i. 8 This unity is only recognised in an *ideal construction* which lets drop all concrete differences. *Ibid.* III. i. 314 Science..is the systematisation of Experience under the forms of ideal constructions. 1883 F. H. Bradley *Princ. Logic* I. ii. 75 Ideal constructions connected, by an inference through identity of quality, with the real that appears in present perception. 1890 W. James *Princ. Psychol.* I. xiii. 533 We have a *conception* of absolute sameness,..but this..is an ideal construction got by following a certain direction of serial increase to its maximum supposable extreme. 1901 G. F. Stout *Man. Psychol.* (ed. 2) IV. vi. § 7. 531 The external world as an ideal construction is a social product. 1917 J. Gibson *Locke's Theory of Knowledge* iv. 78 The nature of ideal construction as the discovery of possible alternatives admitted by the nature of some universal. 1946 *Mind* LV. 153 Suppose, however, that Euclidean points are ideal constructions.

6. b. *Geom.* [ad. F. *idéal*, introduced in this sense by J. V. Poncelet (*Traité des Propriétés proj. des Figures* (1822) I. ii. §§ 50 ff.).] Having no proper existence in real Euclidean geometry as the thing so designated, but introduced into projective or complex geometry in order to do away with what would otherwise be exceptions to generalizations; chiefly in *ideal point*, the single point (at infinity) at which two parallel lines are regarded as intersecting; similarly *ideal line, plane*, the single line (or plane) at infinity that is regarded as containing all the ideal points (or lines) of a plane (or of space).

1879 *Encycl. Brit.* X. 389/1 We may say that all points at infinity in a line *appear* to us as one, and may be replaced by a single 'ideal' point. 1885 C. Leudesdorf tr. *Cremona's Elem. Projective Geom.* xxi. 226 The segment *HH'* has been called an ideal chord of the conic... Accepting this definition we may say that a diameter contains the middle points of all chords, real and ideal, which are parallel to the conjugate diameter. 1937 B. C. Patterson *Projective Geom.* i. 4 Such considerations lead us to assume the existence, in each plane, of one and only one ideal line. It is the locus of all the ideal points of the plane, and it is also the line of intersection of the plane with all parallel planes. *Ibid.* 5 We assume..the existence of an ideal plane of space, the locus of all ideal points and ideal lines of space. 1962 W. T. Fishback *Projective & Euclidean Geom.* iv. 32 We created the projective plane by adding ideal points and an ideal line to the Euclidean plane.

7. ideal case, one perfect or supremely excellent of its kind; **ideal copy** *Bibliogr.*, the most complete and perfect copy possible of an issue of a printed book, as properly described in a descriptive bibliography from the examination and analysis of multiple particular copies; **ideal fluid,** a hypothetical fluid that has no viscosity (no internal friction) and is incompressible; **ideal gas,** a hypothetical gas

(which actual gases approach more or less closely in their behaviour) for which the product of the pressure and the volume (of a given mass) is proportional to its absolute temperature; **ideal home,** used, esp. in titles as *Ideal Home Exhibition*, *Ideal Home Magazine*, in the sense 'a well-designed functional house (and its contents)'; **ideal observer** *Philos.* (see quot. 1957); **ideal state,** an imaginary perfectly constituted political community, harmonious and stable; **ideal utilitarianism,** in ethics, any form of utilitarianism which takes other intrinsic goods besides pleasure as ultimate ends, together constituting an ideal end; so **ideal utilitarian,** an adherent of ideal utilitarianism; also as *attrib. phr.*, of or pertaining to such a theory.

1847 W. Whewell *Philos. Inductive Sci.* (ed. 2) XI. v. 45 A body left to itself will move on with unaltered velocity;..(taking this as our Ideal Case) we find that all actual cases are intelligible. 1961 E. Nagel *Struct. of Sci.* xiii. 463 A..device commonly employed in the physical sciences is to formulate a law for a so-called 'ideal case'... For example, Galileo's law for freely falling bodies is formulated for bodies moving in a vacuum. 1949 F. Bowers *Princ. Bibliogr. Descr.* ii. 113 An *ideal copy* is a book which is complete in all its leaves as it ultimately left the printer's shop in perfect condition and in the complete state that he considered to represent the final and most perfect state of the book... [Footnote] Nothing is invested in the description of an ideal copy. Instead, all the evidence to be gained from the examination of numbers of copies is analyzed..in order to discover what was the actual most perfect form of the book achieved by the printer within an issue. 1952 J. Carter *ABC for Bk.-Collectors* 101 Though it is possible for an individual example of a book to conform to it, the 'ideal copy' is a sort of Platonic archetype, exhibiting the final intention of the printer and publisher in the completion of printing. 1961 *Bull. Inst. Hist. Res.* [illegible] 1906 F. Soddy [illegible] 1958 L. Stevenson *Fluid Mech.* [illegible] An ideal fluid..is a fluid without viscosity. 1962 [illegible] 1964 *Listener* [illegible] Ideal Home Exhibition. 1967 K. Giles *Death in Diamonds* [illegible] A Savoy..Ideal Home..something about a slightly Mayfairish touch of sophistication to the otherwise Ideal Home format. 1966 *New Yorker* 30 Nov. 58/3 What do you think about the ideal home? 1972 R. Perry *Fall Guy* v. 86 The furniture wasn't exactly a Big Wave of the Ideal Home..but..a labour-saving device..a necessity in this Ideal Home. 1935 *Burlington Mag.* Jan. 3/2 The plans rather suggest an Ideal Home. Has it been a success..? 1925 A. Huxley *Those Barren Leaves* I. iv. 41 Agreeing..was a heaven of fake..and 'Ideal State' in the Ideal Home Exhibition. 1957 R. Firth in *Philos. & Phenomenol. Research* XII. 317 (title) Ethical absolutism and the ideal observer. *Ibid.* 1952 *Moral Philos.* 390 The..'ideal observer' theory of the meaning of ethical terms..maintains that 'X is good' can be translated into some such statement as 'If there were an omniscient, disinterested and dispassionate observer he would approve of X'. 1959 R. B. Brandt *Ethical Theory* xi. 173 We must explain further the properties of the 'ideal observer'. 1971 T. D. Campbell *Adam Smith's Sci. of Morals* vi. 128, I shall argue..that to present Smith's theory as a form of Ideal Observer theory is a mistake. 1972 J. Rawls *Theory of Justice* § 30. 185 Suppose that the ideal observer is thought of as a perfectly sympathetic being. 1874 W. Wallace tr. *Hegel's Logic* Prolegomena xix. p. cxlvi, The measure dominates the conception of Plato's ideal state. 1892 B. Jowett tr. *Plato's Dialogues* (ed. 3) V. Index 442 Ideal state, the, difficulty of. 1901 R. L. Nettleship *Lect. Republic of Plato* (ed. 2) vi. 131 An outline is given of the institutions of the ideal state. 1931 L. R. Palmer tr. *Zeller's Outl. Hist. Greek Philos.* II. iii. 126 The Syracusan Hermocrates was to describe the degeneration from the original ideal state to the present. 1946 A. Gray *Socialist Tradition* iii. 63 The actual description of life in the ideal state—the social gadgets—may appear trivial and puerile. 1952 K. R. Popper *Open Soc.* (ed. 2) I. iii. 21 In believing in such an ideal state which does not change, Plato deviates radically from the tenets of historicism. 1967 *Encycl. Philos.* VI. 330/2 In the *Republic*,..Plato delineates his famous Ideal State, or 'Callipolis'. 1970 J. Passmore *Perfectibility of Man* xii. 258 Kant looked forward..to an ideal State, or, in his later writings, to an 'ethical Commonwealth'. 1907 H. Rashdall *Theory of Good & Evil* I. vii. 184 This view of ethics, which combines the utilitarian principle that Ethics must be teleological with a non-hedonistic view of the ethical end, I propose to call Ideal Utilitarianism. According to this view actions are right or wrong according as they tend to produce for all mankind an ideal end or good, which includes, but is not limited to, pleasure. 1930 W. D. Ross *Right & Good* ii. 19 The theory of 'ideal utilitarianism', if I may for brevity refer to the theory of Professor Moore.. *Ibid.* 23 The 'ideal utilitarian' theory can only fall back on an opinion..that one of the goods is the greater. 1959 R. B. Brandt *Ethical Theory* xiv. 355 Universal personal pluralism (often called 'ideal utilitarianism'). *Ibid.* xv. 385 Some ideal utilitarians (for example, Hastings Rashdall) think that qualities of character like veracity, sexual purity, and temperance have great intrinsic value. 1970 J. N. Findlay *Axiological Ethics* iii. 46 There is no reason why an ideal utilitarian may not sometimes place so high a value on certain actions..as to let them outweigh all consequences.

B. *sb.* **1.** (Earlier example.)

1796 J. A. Nitsch *Gen. View Kant's Princ. concerning Man* 55 Materialism, Idealism, Spiritualism, and Scepticism..are merely ideals, which can only be approached, but never reached.

3. *Math.* [a. G. *ideal* sb., introduced in this sense by R. Dedekind (in P. G. L. Dirichlet *Vorles. über Zahlentheorie* (ed. 2, 1871) Suppl. x. 452) after the adjectival use in *ideale zahl* ideal number (E. E. Kummer 1846, in *Ber. über die zur Bekanntmachung geeigneten Verh. d. K. Preuss. Akad. d. Wiss. zu Berlin* 87).] A subring that contains all products of the form *rx* and *xr*, where *r* and *x* are elements of the ring and of the subring, respectively; also called a *two-sided ideal*; *left* (or *right*) *ideal*, a subring that contains all products of the form *rx* (or *xr*).

1898 *Bull. Amer. Math. Soc.* IV. 228 The relation between Dedekind's ideals and Kronecker's forms is discussed. 1927 *Encycl. Brit.* XXI. 856/1 It is a fundamental theorem that every ideal can be resolved into the product of a finite number of prime ideals, and that this resolution is unique. 1937 A. A. Albert *Mod. Higher Algebra* (1938) xi. 253 Right ideals (right invariant subrings) are defined analogously, and we call right and left ideals [illegible] 1941 Birkhoff & MacLane *Surv. Mod. Algebra* [illegible] 1953 E. T. Bell *Math.* [illegible] 1962 L. Fox? [illegible] The study of rings, particularly with regard to the structure or morphology of rings. Ideals entered modern algebra through the theory of algebraic numbers..in the 1870s, but it was only in the 1920s and 1930s that their deep relevance for much of algebra and algebraic geometry was recognized. 1966 F. M. Hall *Introd. Abstr. Algebra* II. vii. 183 An important example of an ideal is the subring of multiples of *n* in the ring of integers. *Ibid.*, Even when *R* is commutative not all subrings are ideals.

idealistic, *a.* Add: **2.** *Sociol.* In the theory of P. A. Sorokin, a type of culture which is a synthesis of spiritual and material values. Cf. *IDEATIONAL *a.* 2 and *SENSATE *a.*

1937 [see *IDEATIONAL *a.* 2]. 1937 R. Linton? [illegible] 1965 C. P. & Z. K. Loomis *Mod. Social Theories* (ed. 2) 456 European culture is classified as idealistic during the 12th to 14th centuries. 1967 T. Parsons *Sociol. Theory & Mod. Soc.* xiv. 381 The idealistic pattern is conceived as intermediate between the two, not in the sense of a simple 'compromise', but rather of a synthesis which can achieve a harmonious balance between the two principal components.

ideate, *a.* **2.** Concerned with ideas as opp. to reality (cf. IDEA *sb.* 8).

1966 *New Statesman* 23 Sept. 434/3 A best seller, its characters mere contrivances and its talk vacuously ideate. 1968 *Listener* 11 Jan. 49/1 It is not the absence of ideas but the absence of things..which most diminishes Williams's poems. He preached upon 'things' till it argued him ideate.

ideational, *a.* Add: **1.** (Later examples.)

1894 Creighton & Titchener tr. *Wundt's Lect. Hum. & Animal Psychol.* xix. 294 In both cases we are only dealing with a particular consequence of the principle of *ideational unity*. 1916 *Proc. Nat. Acad. Sci.* II. 631 Despite widespread interest in the evolution of reasoning, the comparative study of ideational behaviour has been neglected. 1958 *Times Lit. Suppl.* 12 Sept. 515/1 These regulators..are apprehended by, for instance, consciousness as 'archaic images' or 'ideational instincts'. 1970 *Jrnl. Gen. Psychol.* Oct. 144 The technique might differentiate more clearly for response style than for ideational content. 1971 J. Z. Young *Introd. Study Man* xxxv. 489 It has also been called the 'area of ideational speech' or indeed 'word store'.

2. *Sociol.* A term used orig. by P. A. Sorokin (see quot. 1937) to describe a type of culture based on spiritual values and ideals, whose material needs are the minimum necessary to forward those ideals. (See also *IDEALISTIC *a.* 2 and *SENSATE *a.*)

1937 P. A. Sorokin *Social & Cultural Dynamics* I. i. 67 Of these two systems one may be termed *Ideational* culture, the other *Sensate*. *Ibid.* 68 Some [cultures] have contained a balanced synthesis of both pure types. This last I term the *Idealistic* type of culture. (It should not be confused with the Ideational.) 1944 H. P. Fairchild *Dict. Sociol.* 148 *Ideational*,..a type of culture which exalts the spiritual above the material values. 1964 *Economist* 8 Aug. 563/1 Conservatism is not, like socialism, liberalism or democracy, an 'ideational' ideology but 'situational'. 1966 P. A. Sorokin *Sociol. Theories Today* xi. 258 The phase of growth or 'spring' of Danilevsky-Spengler-Toynbee's civilizations is similar in several traits with Sorokin's ideational. 1970 G. A. & A. G. Theodorson *Mod. Dict. Sociol.* 194 Rather than stressing the manipulation of the empirical world to improve the quality of life, the ideational culture emphasizes adjustment to the existing world.

ideationally (əidi,ēɪ·ʃənăli), *adv.* [f. IDEATIONAL *a.* + -LY².] By means of ideation.

1890 W. James *Princ. Psychol.* II. xix. 127 Under ordinary circumstances, the entire brain probably plays a part in draining any centre which may be ideationally active. 1910 R. R. Rusk in *Brit. Jrnl. Psychol.* III. 379 The dissociation of the perceptually excited elements of consciousness from the ideationally excited factors. 1934 *Mind* XLI. 350 The interpretation..relates it [sc. the sensation] ideationally to other items.

ideatum (əidi,ē·təm). *Philos.* [mod.L., neut. of *ideātus*: see IDEATE *a.* and *sb.*] = IDEATE *sb.*

1708 Berkeley *Works* (1948) I. 100 The distinction between Idea and Ideatum I cannot otherwise conceive than by making one the effect or consequence of Dream, reverie, Imagination, the other of sense & the Constant laws of Nature. 1889 Scotus Novanticus *Metaphysica Nova et Vetusta* (ed. 2) 86 Yet there is between mind and object perfect equivalence between the idea and the ideatum. 1920 S. Alexander *Space, Time & Deity* II. 84 The object of which we are conscious as an idea or ideatum. 1933 *Mind* XLII. 303 Here the relation between a mode-factor in one attribute and its correlate in the other attribute is the absolutely unique relation of an idea to its ideatum.

‖ **idée** (ide). The French word for 'idea' used in certain French phrases, as:

idée fixe (ide fiks): a fixed idea (see FIXED *ppl. a.* 2) (an obsession).

1836 H. Greville *Diary* 20 Feb. (1883) 88 The King..has some *idée fixe* about marrying the Duke of Orleans. 1877 L. W. M. Lockhart *Mine is Thine* (1878) I. vii. 48 It was evident..the attraction of the heart would require to be something out of the common run if it were to dislodge this *idée fixe*. 1922 Joyce *Ulysses* 443 I am sure he has an *idée fixe*, Haines said. 1941 Auden *New Year Let.* II. 29 Prefer our *idées fixes* to be True of a Greed Reality. 1953 C. Day Lewis *Italian Visit* II. 28 Then fast, faster Drawn by the magnet of his *idée fixe*, Head down, tail up, he's charging the horizon. 1965 D. Lodge *Brit. Mus. is falling Down* viii. 133 [illegible] 1972 'M. Innes' *Appleby's Answer* vi. 60 Not an *idée fixe* or [illegible] 1974 *Times Lit. Suppl.* 15 Feb. 162/1 Impatience in the search makes it fatally easy to freeze a hypothesis into a rigid system of *idées fixes*.

idée maîtresse (ide mɛtrɛs): a leading idea (cf. *master-idea* s.v. MASTER *sb.*¹ 25 b).

1939 *Times Lit. Suppl.* 16 Dec. 729/1 Mystified by M. Romains's apparent reluctance to disclose his *idée maîtresse*, one could not always avoid an injured and carping note of appreciation. 1958 *Listener* 7 Aug. 193/2 The *idée maîtresse* of his outlook. [illegible] the *idée maîtresse*, of the Enlightenment.

idée mère (ide mɛːr): = *mother idea* (s.v. MOTHER *sb.*¹ 16 a).

1841 [illegible] 1963 J. W. [illegible] the *idée mère*, the primitive element, in the formation of the notion of justice, was conformity to law. 1908 [illegible] *idées mères* of this couplet. 1932 R. Fry *Let.* (1972) II. 653 It's a remarkable book even if it's hypothetical and he may have got an *idée mère*.

idée reçue (ide rəsy): a generally accepted notion or opinion (cf. RECEIVED *ppl. a.* 3 b).

1937 E. Bowen in *New Statesman* 13 Mar. 418/1 He, too, lines up *idées reçues*. 1957 E. Wilson *Piece of my Mind* vii. 105 The foolish old *idée reçue* that Greek literature is the real thing and Latin a second-rate imitation. 1964 *Listener* 9 Jan. 92/1, I sympathized with his dismay that he should be considered right wing, because he did not accept left wing *idées reçues*. 1970 *Times* 23 Feb. 12/1 Your sort of pianist is always..unwilling to accept *idées reçues*, even when they come with most authoritative credentials. 1974 *Times Lit. Suppl.* 22 Mar. 303/2 It sweeps the *idées reçues* of sociologists, town planners and other rationalists into the dustbin.

idempotent, *a.* Also pronounced (əiːdĕmpōᵘ-tĕnt). Add def.: Of a quantity or element *a*: having the property that $a \times a = a$, where × represents multiplication or some other (specified) binary operation. Also applied to

an operator or set for which this is true for any element *a* and to statements expressing this fact. (Further examples.)

1937 A. A. Albert *Mod. Higher Algebra* (1938) iii. 88 A matrix *E* is called idempotent if $E^2 = E$. 1937 *Duke Math. Jrnl.* III. 629 We recall that $A \supset B$ if and only if $A = (A, B)$ and $B = [A, B]$, and that union and cross-cut are associative, commutative, and idempotent operations. 1940 W. V. Quine *Math. Logic* 56 A binary mode of statement composition..is said to be..idempotent if ⌜φ ≡ .φ*φ⌝ is true for all statements φ. 1941 Birkhoff & MacLane *Surv. Mod. Algebra* xi. 313 All of these except for the idempotent laws and the second distributive law correspond to familiar laws of arithmetic. 1941 *Mind* L. 274 The element is only idempotent with respect to the combining relation defined as the combining relation of the group. 1950 W. V. Quine *Methods of Logic* (1952) § 1. 3 '*pp*' reduces to '*p*'. Conjunction is idempotent, to persist in the jargon. 1959 E. M. McCormick *Digital Computer Primer* 181 It is further apparent..that $A + A = A$ and..that $A \times A = A$. These are sometimes referred to as the idempotent laws. 1967 A. Geddes tr. *Dubreil & Dubreil-Jacotin's Lect. Mod. Algebra* i. 22 If every element of *E* is idempotent, the composition law is called idempotent and *E* is called an idempotent set.

B. *sb.* An idempotent element; also in more restricted use (see quot. 1958).

1941 Birkhoff & MacLane *Surv. Mod. Algebra* i. 6 Prove that the following rules hold in any integral domain:..(h) the only 'idempotents' (that is, elements *x* satisfying $xx = x$) are 0 and 1. 1958 S. Kravetz tr. *Zassenhaus's Theory of Groups* (ed. 2) 182 The element *e* is called an idempotent if $ee = e$ and if *e* is not a zero element. 1960 C. E. Rickart *Gen. Theory Banach Algebras* i. 35 Let 𝔄 be a Banach algebra and let *e* be a proper idempotent in 𝔄 (that is, $e \neq 0$, 1 and $e^2 = e$).

Hence **idempotence** (stress variable), **idempo·tency,** the property of being idempotent.

1940 *Mind* XLIX. 461 The truth is that Eddington, in spite of all that he says about getting all the mathematics he wants out of the idempotency of the *J* symbols, employs them in accordance with the laws of ordinary algebra whenever he thinks fit. 1940 W. V. Quine *Math. Logic* 60 In the case of conjunction and alternation, repetition of components has..been seen to be immaterial (idempotence). 1957 P. Suppes *Introd. Logic* ix. 205 Equations (9) and (10) express what is usually called the idempotency of union and intersection. 1959 K. R. Popper *Logic Sci. Discovery* 351 *p* (*aa*, *b*) = *p* (*a*, *b*)...This is the law of idempotence, sometimes also called the 'law of tautology'. 1960 P. Suppes *Axiomatic Set Theory* ii. 27 The next three theorems assert the commutativity, associativity, and idempotence of union. 1968 *New Scientist* 16 May 339/1 Idempotency..occurs if an operation produces no change in the number or set on which it operates.

idem sonans (əi·dem sōᵘ·nænz). *Law.* [L., lit. = sounding the same.] Identity of sound in pronunciation; the occurrence in a document of a material word or name misspelt but having the sound of the word or name intended. Also *adj.*, homophonous *with*.

1848 Wharton *Law Lexicon* 304/2 The courts will not interfere in setting aside proceedings on account of the misspelling of names, provided..there is an *idem sonans* between the pronunciation of the right name and that which is inserted in the proceedings; as Lawrance, instead of Lawrence, Reynell for Reynolds, Beniditto for Benedetto. 1856 *Newsp. & Gen. Reader's Compan.* II. § 1749 The verb was unluckily *idem sonans* with another word. 1919 H. L. Mencken *Amer. Lang.* viii. 273 In America, with a language of peculiar vowel sounds and even consonant-sounds struggling against a foreign invasion unmatched for strength or variety..the legal rule of *idem sonans* is of much wider utility than anywhere else in the world.

-idene, *suffix. Chem.* [Prob. taken from *ETHYLIDENE (ad. F. *éthylydène* (A. Lieben 1858, in *Compt. Rend.* XLVI. 663), f. *éthylène* with insertion of the *-yd-* of *aldehyde*).] Forming the names of bivalent organic radicals in which both valencies derive from the same atom. Cf. *-YLIDENE.

1927 *Chem. Abstr.* XXI. 4576/1 *-idene* added to any radical usually means a double bond at point of attachment. 1966 *Nomencl. Org. Chem.* (I.U.P.A.C.) (ed. 2) A. 16 Bivalent and trivalent radicals derived from univalent acyclic hydrocarbon radicals whose authorized names end in '-yl' by removal of one or two hydrogen atoms from the carbon atom with the free valences are named by adding '-idene' or '-idyne', respectively, to the name of the corresponding univalent radical.

ident. Colloq. abbrev. of IDENTIFICATION (or *identification bracelet*), IDENTIFY *v.* orig. *U.S.*

1952 *Jewelers' Circular-Keystone* Sept. 65 (Advt.), Popular 'Idents' that build profitable sales. 1955 R. J. Schwartz *Compl. Dict. Abbrev.* 90 *Ident*, identification (US Army). 1965 'D. Shannon' *Death-Bringers* (1966) i. 12 The group being showed thousands of mug shots at Records, hoping for a possible ident. 1966 M. & G. Gordon *Undercover Cat prowls Again* (1967) xiv. 95 It was so dark he couldn't make a positive ident until the girl was halfway along. 1967 R. J. Serling *President's Plane is Missing* iii. 55 'Washington Center, Air Force One. Squawk ident.' 'Roger, identing.' 1970 P. Laurie *Scotland Yard* ix. 196 A searcher..ploughs through the four thousand or so forms..in the slim hope that one will match, giving him an 'ident'.

identical, *a.* Add: **2. c.** *identical points* = *corresponding points* (*CORRESPONDING *ppl. a.* 1 b).

1841 W. Mackenzie *Physiol. Vision* xvi. 253 (*heading*) Corresponding or identical points of the retina. 1880 L. Owen tr. *Giraud-Teulon's Elem. Treat. Function of Vision* I. ii. 16 The same object being depicted upon the two retinae, at homologous points, must give rise to a single sensation... This..has been called the doctrine of identical points. 1932 S. Duke-Elder *Text-bk. Ophthalm.* I. xxvii. 1028 Points on the two retinæ from which images are projected to the same place in the common visual field are called corresponding (or identical) points.

d. *identical twin*, one of a pair of twins who, as a result of being monozygotic, are of the same sex and very similar to one another in appearance; usu. *pl.* Opp. **fraternal twin*. Similarly *identical triplet*.

1889 S. Schönland tr. *Weismann's Ess. Heredity* vi. 381 Under conditions of nutriment which are as identical as possible, *two* egg-cells develope into unlike twins, *one* into identical twins; although we cannot yet affirm that the latter result invariably follows. 1938, 1941 [see *FRATERNAL *a.* c]. 1964 M. Argyle *Psychol. & Social Probl.* vi. 77 The best method of studying the extent of genetic factors is by means of identical and fraternal twins. 1972 A. Christie *Elephants can Remember* xiv. 190 There was a project on hand..to follow up the general lives of selected pairs of identical twins. 1973 *Oxford Times* 6 Apr. 8 On Monday the first LP by the only identical triplet sisters act in British show business was released.

B. *sb.* **1.** *pl.* Identical things.

1696 J. Sergeant *Method to Sci.* 264 We can as easily define their Abstract Notions as we can the other, (or rather much more easily) and consequently Reduce them to their Identicals. 1903 J. Gott *Lett.* (1918) 195 Most of the books..worry me with endless and subtle refinements and hair-splitting distinctions between identicals. 1943 W. V. Quine in *Jrnl. Philos.* 4 Mar. 113 One of the fundamental principles governing identity is that of *substitutivity*—or, as it might well be called, that of *indiscernibility of identicals*.

2. An identical twin.

[1932 A. Huxley *Brave New World* i. 8 'Can you tell us the record for a single ovary..?'..'Sixteen thousand and twelve; in one hundred and eighty-nine batches of identicals.'] 1938 [see *FRATERNAL *a.* c]. 1964 M. Argyle *Psychol. & Social Probl.* vi. 77 If it is found that identicals are more alike in some respect than fraternals, this suggests that the condition [*sc.* a mental disorder] is to some extent inherited.

identifiability (əide·ntifəi,ăbi·liti). [f. IDENTIFIABLE *a.* + -ITY.] The property of being identifiable.

1898 W. James in R. B. Perry *Tht. & Char. W. James* (1935) II. 369 The identifiability of these objects in different fields. 1959 P. F. Strawson *Individuals* i. 17 The possibility that the identifiability of particulars of some sorts may be in some *general* way dependent on the identifiability of particulars of other sorts.

identification. Add: **1. b.** Esp. in *Psychol.*, the (freq. unconscious) adaptation of one's ideas and behaviour to fit in with those of a person or group seen as a model. (Further examples.)

1913 A. A. Brill tr. *Freud's Interpr. of Dreams* iv. 126 Identification is a highly important factor in the mechanism of hysterical symptoms; by this means patients are enabled in their symptoms to represent not merely their own experiences, but the experiences of a great number of other persons. 1930 W. Healy et al. *Struct. & Meaning Psychoanal.* v. 240 Identification is the unconscious molding of a person's own Ego after the fashion of one that has been taken as a model. 1950 *Brit. Jrnl. Psychol.* XLI. 176 It was..clear from her stories and her behaviour at school that J. W. succeeded in making a 'good identification'. 1964 *Listener* 21 May 825/2 The participatory emotions..are mediated by processes variously known as empathy, *rapport*, projection, and identification. 1972 *Sci. Amer.* Jan. 36/3 Girls often are socialized in early childhood to satisfy their achievement needs passively by identification with the accomplishments of their father or their brothers.

4. A document such as a passport, driving licence, health card, or a disc or mark that serves to identify a person, or indicates his nationality, military unit, etc. Also *collect.*

1947 *Amer. College Dict.* 599/1 *Identification*, something that identifies one: *have you any identification?* 1958 P. Kemp *No Colours or Crest* vi. 100 We got good identifications—all from the 1st Alpine Division. 1964 Mrs. L. B. Johnson *White House Diary* 10 June (1970) 164 What we should have done was to ask them all to wear some identification. 1965 M. Bradbury *Stepping Westward* v. 272 Got any identification? 1970 *Globe & Mail* (Toronto) 28 Sept. 7/1 Once inside I was forced to produce my driver's licence, draft card, student I.D., and all other identification I had on me.

5. *attrib.* and *Comb.* [cf. * IDENTITY 10], as *identification badge, bracelet, card, mark, papers*; **identification beacon** *Aeronaut.* (see quot.); **identification disc,** a disc carried or worn by a person (usu. one in the armed forces), giving his name, etc., as a means of identification; **identification lamp, light,** a light (e.g. on an aircraft) that provides a means of identification; **identification panel,** a sign used to indicate the position of ground troops to friendly aircraft; **identification parade,** a parade of persons from among whom a suspect is to be identified; **identification patch** (see quot.); **identification plate,** the registered number plate of a motor vehicle; **identification signal** (see quot. 1918); **identification tag** *N. Amer.* = **identification disc*; also *transf.*

1945 Identification badge [see *film badge* (*FILM *sb.* 7 b)]. 1951 *Gloss. Aeronaut. Terms* (*B.S.I.*) III. 24 *Identification beacon*, a beacon displaying a coded light and identifying a geographical point. 1969 *New Yorker* 30 Aug. 57/2 (Advt.), The 17-jewel movement attaches to a rich gold..identification bracelet. 1970 *Ibid.* 24 Oct. 48/2 He gave her his identification bracelet, from which he had had eight links removed. 1908 *Westm. Gaz.* 13 Feb. 5/2 An identification-card with his photograph on it. 1969 *New Yorker* 10 May 29/1 Residents will show identification cards to gain admittance. 1915 R. Brooke *Let.* Mar. (1968) 665 Round my neck with my identification-disk. 1915 *Sphere* 11 Dec. p. iv/3 Copies of the soldiers' identification discs in gold. 1930 T. B. Bruce *Missing* 13, I burnt all letters and papers..keeping only my identification disc. 1935 H. G. Wells *Things to Come* 15 An identification disk—his introduction card so to speak—will be carried. 1932 *Gloss. Aeronaut. Terms* (*B.S.I.*) (proofs) IX. 7 *Identification lamp*, a lamp mounted on an aircraft for purposes of recognition. 1933 *Gloss. Aeronaut. Terms* (*B.S.I.*) XIII. 78 *Identification light*, a light on or near a beacon having a character differing from, but serving to identify, it. 1946 R. A. McFarland *Human Factors Air Transport Design* xii. 608 Accidents may arise from misinterpretation of identification lights on other aircraft. 1897 E. L. Voynich *Gadfly* III. i. 253 For you to go there just now, with all your identification marks, would be to walk into a trap with your eyes open. 1901 *Westm. Gaz.* 24 Apr. 3/2 He found fifteen other bodies, searched them for their identification marks. 1904 *Ibid.* 30 Aug. 4/3 The identification marks of the cartridges of all the known armies of the world. 1942 B. A. Shields *Princ. Flight* vii. 231 When an aircraft is licensed, it is assigned an identification mark. *Ibid.*, The identification mark signifies the type of flying in which the airplane is permitted to engage. 1944 *Ann. Reg.* 1943 251 The Germans also resented the wearing..of..colours of the R.A.F. aircraft identification marks. 1918 E. S. Farrow *Dict. Mil. Terms* 305 *Identification panels*, in aëroplane balloon signalling, panels which are displayed at the sound signal of the aircraft or upon the initiative of the command post. 1957 P. Kemp *Mine were of Trouble* viii. 139 We had white identification panels spread on the crest in front of us, to indicate our forward positions. 1903 *Westm. Gaz.* 28 Dec. 7/2, 300 men who had no workmen's identification papers. 1918 Identification papers [see *GOODS *sb. pl.* 1 b]. 1927 W. E. Collinson *Contemp. Eng.* 78 The shortcomings of the identification parades at the police-station. 1965 W. Soyinka *Road* 54 Perhaps..if you promised not to look in his face..so that you could not recognise him at an identification parade. 1966 A. Sachs *Jail Diary* xi. 109 Perhaps they are taking me to another cell, or they are going to put me on an identification parade. 1972 *Police Rev.* 10 Nov. 1472/3 You think it is necessary to hold identification parades in this case. 1918 E. S. Farrow *Dict. Mil. Terms* 305 *Identification patch*, tags placed upon the backs of the men's coats when advancing behind a barrage. 1901 *Motor-Car World* II. 74/1 Identification-plate. 1906 *Westm. Gaz.* 26 Aug. 8/2 The Commission recommends that identification plates should 'be rigidly fixed in an upright position'. 1909 *Ibid.* 24 June 4/1 Five marks for identification plates. 1918 E. S. Farrow *Dict. Mil. Terms* 305 *Identification signals*..which will identify the authority sending the communication..are assigned from division headquarters and each consists of one letter and one numeral. 1946 *Happy Landings* July 9/1 The identification signal could still be heard. 1918 Identification tag [see *dog tag* (*DOG *sb.* 18 a)]. 1918 E. S. Farrow *Dict. Mil. Terms* 305 An identification tag by which he can be identified if killed or wounded. 1960 J. J. Rowlands *Spindrift* 146 A small boy with an identification tag tied to his lapel walked silently beside the stewardess. 1964 Gould & Kolb *Dict. Social Sci.* 244/2 Some students suggest that there is a great concern among those desirous of upward mobility to discard their ethnic identification tags.

identificational (əidentifikēɪ·ʃənăl), *a. Linguistics.* [f. IDENTIFICATION + -AL.] Relating to being identified or not; involving identification.

1933 L. Bloomfield *Lang.* xii. 203 The class-meaning is, roughly, 'identificational character of specimens'. 1964 E. A. Nida *Toward Sci. Transl.* iii. 44 A communication..can be called identificational, for the source..exhibits a high degree of identification with the receptor.

identificatory (əidentifikēɪ·təri), *a.* [f. IDENTIFICAT(ION + -ORY².] Serving to bring about identification (in various senses).

1943 *Internat. Jrnl. Psycho-Anal.* XXIV. 97/1 Thus identificatory thinking is employed for the purpose of avoiding what is unpleasurable and obtaining what is pleasurable. 1949 Koestler *Insight & Outlook* xii. 175 This identificatory rite survives with great vigour to our day in the ceremony of the eucharistic meal. 1959 P. F. Strawson *Individuals* ii. 82 Kant was very careful to empty this 'I' of referential, identificatory force. 1965 *Philos.* XL. 336 The relevant rules of difference are of course those of space-time and its trustworthiness as an identificatory medium. 1972 *Computers & Humanities* VII. 121 Each line includes identificatory information such as page number, abbreviated poem title, line number, and whether the line is a variant or not.

identifier. Add: **1.** (Later example.)

1959 E. Fenwick *Long Way Down* ii. 15 Where the hell's your identifier?

b. One who identifies himself with a cause, group, etc.

1966 *New Statesman* 8 July 56/3 He [*sc.* Mark Twain] lacked the moral or imaginative resources of other identifiers, like Dickens or Balzac.

2. a. That which identifies.

1894 'Mark Twain' in *Century Mag.* June 237/2 There was never a twin..that did not carry from birth to death a sure identifier. 1907 *Yesterday's Shopping* (1969) 633/2 Gamekeepers' identifiers... will burn for a period of 3 to 4 minutes. 1973 A. Quinton *Nature of Things* iii. 61 The view that the identity of things through time is due to the presence in them of an identifying component or substance, an identifier, as I shall call it.

b. *Linguistics.* A linguistic element that has the function of identifying.

1946 [see *DESCRIPTOR]. 1965 *Language* XLI. 73 The noun phrase..may be regarded as a string consisting of identifier tagmeme, manifested here by *a.*; [etc.].

c. *Computers.* A sequence of characters arbitrarily devised to identify or refer to a set of data, a location in a store, or a point in a program.

1958 *Communications Assoc. Computing Machinery* Dec. 11 Strings of letters and figures enclosed by delimiters represent new entities. However, only two types of such strings are admissible: 1. Strings consisting of figures ζ only represent the (positive) integers G (including 0) with the conventional meaning. 2. Strings beginning with a letter λ followed by arbitrary letters λ and/or figures ζ are called identifiers. They have no inherent meaning, but serve for identifying purposes only. 1960 *Computer Jrnl.* III. 67/2 An identifier may be used in an ALGOL 60 program as a simple variable. This means that the program, when ultimately translated and run on a computer, will associate a particular storage location with that identifier. The number held in this store at any stage in the calculation is called the current value of the variable. 1962 R. V. Oakford *Introd. Electronic Data Processing Equipment* vii. 245 An INPUT declaration defines one or more input-data sets, each of which is given a name in the form of an identifier. 1967 Cox & Grose *Organiz. Bibliogr. Rec. by Computer* II. 19 Such a file sequence carried an 8-character identifier and a sequence-number. 1968 Corlett & Tinsley *Pract. Programming* ii. 14 Reference is made to the memory of a computer by giving names or identifiers to variables... Identifiers are also used to denote labels which mark particular points in a program. 1973 *Computers & Humanities* VII. 144 The first ten columns contain identifiers for manuscript, book, question, and line number.

identify, *v.* Add: **1. b.** *to identify oneself with*: also, to model oneself on, esp. unconsciously; to feel oneself to be associated with or part of; freq. *absol.* with ellipsis of the refl. pron. Also occas. *intr.*, to perform or undergo such a process with regard to something unspecified. (Further examples.)

1913 A. A. Brill tr. *Freud's Interpr. of Dreams* iv. 126 If she has put herself in the place of her friend, or, as we may say, has identified herself with her friend. *Ibid.* 127 An hysterical woman identifies herself most readily..with persons with whom she has had sexual relations. 1940 'G. Orwell' *Inside Whale* 51 Sam Weller, Mark Tapley, Clara Peggotty..identify with their master's family. 1955 *Publ. Amer. Dial. Soc.* XXIV. 5 The teen-agers who almost compulsively identify with this semicriminal subculture. 1958 *Observer* 2 Feb. 14/3 Readers can immediately identify with her nice puzzled hero. 1958 M. Argyle *Relig. Behaviour* v. 40 Other investigators have studied the beliefs of children who..'identify' themselves with their parents—i.e. wish to be liked by them, wish to resemble them. 1959 *Listener* 31 Dec. 1174/2 An engaging series of attempts and failures to 'identify', as cricket-master at a prep school, or as a journalist on a go-getting daily. 1967 G. Steiner *Lang. & Silence* 81 Because we are trained to give psychological and moral credence to the imaginary..we may find it more difficult to identify with the real world. 1968 *Blues Unlimited* Sept. 8 Finally Tina came on and tore the joint up. She signified, the women identified and the men just drooled. 1969 *Times* 17 Oct. 18/5 Everyone identified madly and Biba's knew no failure. 1972 *Where* Jan. 18/2 Thus the parents, in conversation at home, are able to identify themselves with the place and people under discussion.

3. To discover, perceive; to localize. *colloq.*

1922 D. H. Lawrence *England, my England* 45 After a lapse and a new effort, he identified a pain in his head.

identikit (əide·ntikit). Also **identi-kit.** [Blend of IDENTI(TY + KIT *sb.*¹.] A composite picture of a person whom the police wish to interview assembled from features described by witnesses. Also *transf.*

1961 *Observer* 12 Mar. 5/7 About forty police forces in this country are now testing an American device called an 'Identi-Kit', which is used to translate witnesses' descriptions of a person into visual terms. 1961 *Spectator* 1 Sept. 277 The identi-kit must depend..on the memory of the witness being questioned. 1962 *Times* 19 Apr. 13/2 The identi-kit method of political detection is not really very plausible here. 1963 *Listener* 19 Sept. 416/2 At their worst the genres of contemporary fiction provide no more than a kind of 'identikit' novel, prefabricated from ready-made elements. 1967 *Spectator* 29 Sept. 359/1 One at least managed to build up an identikit description of the soul: 'A most wonderful, delicate, small thing'. 1969 G. Greene *Trav. with my Aunt* II. iii. 245, I don't resemble whatever identikit portrait you have of me. 1969 *Times* 13 Mar. 13/1 If one were looking at that time for an Identikit of an England side to defend the World Cup in Mexico in 1970 one would scarcely have recognized the person we were looking for. 1971 'J. Ashford' *Bent Copper* vi. 43 He'd called in the D.C. from H.Q. who specialised in Identikit and..the D.C. had built up several faces. 1973 *Times Lit. Suppl.* 9 Mar. 255/1 The identikit of this regime [in Greece] is predictably unpleasant.

‖ **Identitätsphilosophie** (ide·ntitēɪts,filǫsōfi). Also **i-.** [G., identity-philosophy.] The term used for a system, propounded by, among others, F. W. Schelling (1775–1854), that assumes the fundamental identity of spirit and nature.

1866 H. Sidgwick in A. & E. M. Sidgwick *Henry Sidgwick* iii. 151, I am coming more and more to the opinion that the whole 'Identitäts-philosophie' (Fichte, Schelling, and Hegel) is a monstrous mistake. 1905 W. James *Meaning of Truth* (1909) v. 128 Humanism, here, is only a more comminuted *identitätsphilosophie*. 1905 —— *Ess. Radical Empiricism* (1912) iv. 134 This the post-Kantian idealists..acknowledged by calling their doctrine an *Identitätsphilosophie*. 1938 *Mind* XLVII. 281 Spinoza's doctrine is not an Identitätsphilosophie; that is to say, the one and only substance has no underlying identical nature, which is only manifested in different ways in the attributes.

identity. Add: **6.** Also *attrib.*, as *identity formula, relation, sentence.*

1940 W. V. Quine *Math. Logic* 232 *I* is..the identity relation $\hat{x}\hat{y}\,(x = y)$. 1965 B. Mates *Elem. Logic* ix. 146 It will be useful to introduce a couple of obvious conventions for writing identity-formulas. *Ibid.* 149 The identity relation among the elements of one domain will be different from that among the elements of another. *Ibid.*, Thus every identity-sentence would be either trivial or absurd. 1967 *Encycl. Philos.* IV. 123/1 When we utter an identity sentence such as 'Venus is the morning star', what we wish to express..is that the terms 'Venus' and 'the morning star' both mean the same thing. 1970 J. D. Carney *Introd. Symbolic Logic* vii. 160 The identity relation has some rather special properties.

7. (*old*) *identity*: a person long resident or well known in a place. *N.Z.* and *Austral.*

1862 *Otago, Its Goldfields & Resources* 9 The exclusive spirit of the 'old identity'. 1874 A. Bathgate *Colonial Experiences* iii. 26 The term 'old identities' took its origin from an expression in a speech made by one of the members of the Provincial Council, Mr E. B. Cargill, who, in speaking of the new arrivals, said that the early settlers should endeavour to preserve their old identity... A comic singer [R. Thatcher] helped to perpetuate the name by writing a song. 1879 W. J. Barry *Up & Down* xx. 197 The 'old identities' were beginning to be alive to the situation. 1889 *Bulletin* (Sydney) 28 Sept. 8/1 Many of the old identities of '52 and '53 will remember the license-hunting and shanty-raiding days. 1893 *Auckland Weekly News* 9 Dec. 7 Both these old identities are in possession of all their faculties to a wonderful degree. *Ibid.* 28 Another old identity passed away on Dec. 1 in the person of Mr. Thomas Hunt. 1929 'M. Barnard Eldershaw' *House Is Built* (1945) v. 111 He was the sort of man who becomes an old identity almost at once, so that the residents of the Parramatta Road..soon thought they had been seeing him drive past in his indescribably sailorly fashion all their lives. 1942 'M. Innes' *Daffodil Affair* II. ii. 46 Ron's dad was a well-known identity Cobdogla-way. 1944 *Mod. Jun. Dict.* (Whitcombe & Tombs) 205 In Australia and New Zealand a very old resident in a place is called an 'old identity'. 1962 J. R. Bernard in *Southerly* XXII. II. 97 We [Australians] add to..identity that of outstanding local citizen. 1970 *N.Z. Woman's Weekly* 9 Nov. 19/1 Havelock North identity Mrs C. E. Turner-Williams..at 98 stitches happily on.

8. *Math.* **a.** An element of a set which, if combined with any element by a (specified) binary operation, leaves the latter element unchanged.

1894 *Bull. Amer. Math. Soc.* I. 61 Given an (abstract) group G_n..with elements s_1 = identity, s_2, s_n. 1937 R. D. Carmichael *Introd. Theory Groups of Finite Order* xiii. 395 For every *a* we have $ai = a = ia$. Then we call *i* an identity with respect to the rule of combination of the group. *Ibid.* i. 17 Since the identity plays the role of unity in multiplication, it is often denoted by the symbol 1. 1941 Birkhoff & MacLane *Surv. Mod. Algebra* i. 2 The number zero has the characteristic property that it leaves unaltered any number to which it is added; hence we say that zero is an 'identity element' for addition. By formal analogy, the 'unity' 1 is an identity for multiplication. 1966 Meyer & Hanlon *Fun with New Math* i. 12 The number one is the multiplicative identity, for the product of it and any other number leaves the second number unchanged. 1971 E. C. Dade in Powell & Higman *Finite Simple Groups* viii. 254 The associativity in *G* easily implies that of multiplication in *FG*, and the identity 1_G of *G* is also the identity for *FG*.

b. A transformation that gives rise to the same elements as those to which it is applied.

1910 Veblen & Young *Projective Geom.* I. iii. 65 The correspondence which makes every element of the system correspond to itself is called the identical correspondence or simply the identity, and is denoted by the symbol 1. 1959 E. M. Patterson *Topology* (ed. 2) ii. 20 If $A \subset B$, the transformation $i: A \to B$ defined by $i(a) = a$ is a one-one transformation called an inclusion; in particular, if $A = B$, the inclusion $i: A \to A$ is called the identity. 1961 H. S. M. Coxeter *Introd. Geom.* ii. 29 If the product of two transformations is the identity, each is called the inverse of the other, and their product in the reverse order is again the identity.

9. *S. Afr.* (See quot. 1924.)

1924 E. H. Brookes *Hist. Native Policy S. Afr.* iii. 62 Most modern thinkers on the Native question argue as if there were no *via media* between the principle which refuses to acknowledge any real difference between Europeans and Natives, the policy of identity as we may call it,..and the principle which insists on the subordinate position of the Native in the body politic, the policy of subordination. 1961 *Listener* 30 Nov. 898/2 These influences..led in South Africa to the policy sometimes known as 'identity', of regarding all men as much the same. *Ibid.*, The earlier British policy of identity broke down.

10. *attrib.* and *Comb.* with the meaning 'that serves to identify the holder or wearer', as *identity bracelet, card, certificate, disc, papers, patch*; also **identity element** *Math.* = *IDENTITY 8 a; **identity matrix** *Math.*, a matrix in which all the elements of the principal diagonal are one and the remainder zero, so that its product with another matrix gives that matrix; **identity parade** = **identification parade*.

1968 J. Ironside *Fashion Alphabet* 167 *Identity bracelet*, a gold or silver chain with a flat space for the owner's name. 1973 G. Sims *Hunters Point* xiii. 124 On his wrists a gold watch and a gold identity bracelet. 1900 *Westm. Gaz.* 2 Jan. 3/1 When troops are going on service each man has issued to him what is known as a field dressing and an identity card. 1931 *Times Lit. Suppl.* 1 Jan. 2/2 He..forged an identity card, and procured a pistol. 1940 *Ann. Reg.* 1939 101 Some 65,000 enumerators, who..issued identity cards for all the persons mentioned in the forms. 1953 C. Day Lewis *Italian Visit* i. 14 The identity cards that inform us Not who we are or might be, but how we are interchangeable. 1961 *Daily Mail* 20 July 9/3 Millions of people may soon have to carry special medical identity cards... A card..would contain information about their illnesses..any special drugs they were taking. 1972 *Daily Tel.* 23 Nov. 6 The BBC is to tighten up security at its London studios and offices by issuing identity cards to staff. 1918 *Act* 8 *Geo. V* c. 6 § 11 Every person who receives, detains or has in his possession any identity certificate, life certificate, or other certificate. 1909 *Daily Chron.* 15 June 5/5 Rations for three days, ammunition, field bandages, and identity discs were issued to the men. 1911 *Punch* 15 Mar. 181/1 By the March Army Orders the identity discs issued to officers and men in war time are in future to be issued to the former in peace time. 1915 'I. Hay' *First Hundred Thousand* vi, Its called an Identity Disc. Every soldier on active service wears one. 1919 J. B. Morton *Barber of Putney* i, In due course came vaccination and inoculation, and identity discs. 1956 R. St. B. Baker *Dance of Trees* vi. 79 When the top soil of Tel Fara was excavated we found modern spurs, identity discs, even a copy of the *Tatler*..reminders of the days of Allenby. 1902 *Trans. Amer. Math. Soc.* III. 486 There exists a left-hand identity element, that is, an element i_l such that, for every element *a*, $i_l a = a$. 1966 May & Moss *New Math for Adults Only* vi. 33/2 Zero is the identity element in addition and one is the identity element in multiplication. 1941 Birkhoff & MacLane *Surv. Mod. Algebra* viii. 197 Corresponding to the identity transformation $y_i = x_i$ is the $n \times n$ identity matrix *I*, which has entries 1 along the principal diagonal (upper left to lower right) and zeros elsewhere. 1908 *Daily Chron.* 21 Feb. 4/6 The 'identity papers', which every man and woman in Prussia must carry about with them. 1955 'A. Gilbert' *Is she Dead Too?* viii. 141 Put a gorilla in a set of ginger reach-me-downs and you could put up the pair of 'em in an identity parade and no one could tell the difference. 1973 E. Lemarchand *Let or Hindrance* vii. 69 We may ask you to come along to an identity parade. 1959 M. Levin *Eva* 37 The Ukrainians didn't have to wear identity patches, since the Germans considered them allies.

b. *Philos. attrib.* and *Comb.* as *identity doctrine, sign, thesis*; **identity theory,** the materialist theory that physiological and mental perceptions are identical; hence *identity theorist*, a person professing belief in the identity theory.

1920 S. Alexander *Space, Time & Deity* II. 9 The statement..is a species of the identity doctrine of mind and body, maintaining that there are not two processes, one neural, the other mental, but one. 1950 W. V. Quine *Methods of Logic* (1952) 211 It is the use of the identity sign between variables, rather than between singular terms, that is fundamental. 1965 Hughes & Londey *Elem. Formal Logic* xxxvii. 258 We need one further symbol, which is written '=' and is known as the identity sign. 1951 G. Humphrey *Thinking* viii. 245 Inspired by the behaviourists one group of advocates of what may be called the 'identity theory' has stressed the importance of the so-called implicit speech movements which occur during thinking. 1966 *Amer. Philos. Q.* III. 227/2 Some philosophers..infer that the Identity Theory is an empirical theory. *Ibid.* 233/2 Some Identity Theorists are anxious to eliminate mental properties. 1967 *Encycl. Philos.* V. 339/1 The identity theorist uses the familiar philosophical distinction between significance and reference..to make the claim that mentalistic and physicalistic expressions..will turn out as a matter of empirical fact to refer to or denote one and the same thing, namely physical phenomena. 1954 H. Feigl in P. A. Schilpp *Philos. R. Carnap* (1963) 259 The prima facie implausibility of the identity thesis arises..mainly from the psychological incompatibility of images such as of nervous tissue..with the qualities of some data of consciousness. 1967 *Philos. Rev.* LXXVI. 201 In recent years, a number of philosophers have argued in favor of materialism in the form of an identity thesis—that is, a thesis to the effect that mental events are identical with certain physiological events.

c. Belonging or relating to identity (sense 2), as in **identity crisis,** a phase of varying severity undergone by an individual in his need to establish his identity in relation to his associates and society as part of the process of maturing. Also *transf.*

1954 *Jrnl. Amer. Psychoanal. Assoc.* II. Apr. 327 George Bernard Shaw arranged for himself a psychosocial moratorium at the age of twenty when his identity crisis led him to leave..his family, friends and familiar work. 1959 *Listener* 29 Oct. 746/2 The prolonged identity crisis of this great young man. 1965 *Times Lit. Suppl.* 25 Nov. 1078/4 A sympathetic study of 'identity crisis' might appear peculiarly relevant to many Americans. **1968** *Internat. Encycl. Social Sci.* VII. 63/2 An era's identity crisis is least severe in that segment of youth which is able to invest its fidelity in an expanding technology. **1971** R. A. Carter *Manhattan Primitive* (1972) xv. 137 Girl is on the verge of a breakdown, in deep identity crisis. **1971** M. McCarthy *Birds of America* 110 His college tutor, a stupid Freudian, had advised his mother that Peter had an 'identity problem'. **1974** *Times Lit. Suppl.* 19 Apr. 409/1 A middle-aged cuckold with piles and an identity crisis.

ideo-. Add: **i:deogene·tic** *a.*, producing ideas or images; **i:deokine·tic** *a. Path.*, denoting that form of apraxia in which the sufferer retains the motor ability to perform an action or movement and understands a request to perform it, but cannot perform it on request; **ideophone** (later example); (b) a term used principally in Bantu linguistics to refer to particular classes of onomatopœic and sound-symbolic words found in these languages; hence **ideopho·nic** *a.*

1904 *Jrnl. Philos.* 21 July 412 In the ideogenetic thinking of artists, the word-symbols are not used. **1908** *Jrnl. Nervous & Mental Dis.* XXXV. 636 There is a great variety of abnormal manipulations..of objects, which are described under the head of apraxia. These may be divided into (1) manifestations of ideomotor apraxia (ideo-kinetic, of Liepmann, formerly motor apraxia). **1914** H. Liepmann in *17th Internat. Congr. Med.* XI. II. 100 Both limb-kinetic, and particularly ideokinetic apraxia, generally have ideational defective reactions admixed with them. **1933** W. R. Brain *Dis. Nervous Syst.* i. 95 Apraxia has been analysed by Liepmann into limb-kinetic apraxia.., ideo-kinetic apraxia, due to a dissociation between ideational and kinaesthetic processes, and ideational apraxia. **1947** F. B. Walsh *Clin. Neuro-Ophthalm.* i. 75/1 In ideokinetic apraxia there is a transferring of movement to other parts of the body, omission of movements, and the production of amorphous movements. **1909** *Cent. Dict.* Suppl., *Ideophone.* In *phonetics*, the auditory symbol of a word or phrase that is perceived as a whole and thus constitutes a single idea. Ideophones are distinguished as *sensory* or *motor*, according as the sound or group of sounds corresponding to the word or phrase is heard or spoken. **1935** C. M. Doke *Bantu Ling. Terminol.* 118 *Ideophone*, a vivid representation of an idea in sound. A word, often onomatopoeic, which describes a predicate, qualificative or adverb in respect to manner, colour, sound, smell, action, state or intensity. The ideophone is in Bantu a special part of speech, resembling to a certain extent in function the adverb. **1953** W. J. Entwistle *Aspects of Lang.* xi. 360 The use of ideophones by Zulus and other Bantus shows that the quality of an action is of more interest to them than its specific nature. **1954** G. V. Smithers in *Archivum Linguisticum* VI. 73 Some English Ideophones. *Ibid.* 82 The term *ideophone* may as fittingly be applied to the English words of both groups [onomatopoeic and imitative] as to those in other languages. **1955** L. W. Lanham *Study of Gitonga of Inhambane* ix. 220 Certain irregular usages of the tenses given above are observable when ideophones are incorporated as predicative stems. **1964** *Afr. Lang. Stud.* V. 87 Examination..brought to light, both in Swahili and in kindred languages of the Coastal area, a number of what have been..called ideophones. **1954** *Archivum Linguisticum* VI. 83 It seems that certain types, at least, of ideophonic root can be struck out in various languages at various periods. **1962** G. Fortune *Ideophones in Shona* 37 Ideophonic forms, ideophonic constructions as well as ideophonic phonemes..are peculiar to 'free expression' as contrasted with 'formal speech'. **1964** *N. & Q.* Oct. 372/2 An infallible sign that the word is an ideophonic formation.

ideogram. (Further examples.) Also used in *transf.* senses, esp. of figurative diction.

1940 E. Pound *Let.* 16 Jan. (1971) 333 Early characters were pictures, squared for aesthetic reasons. But I think in a well-brushed ideogram the sun is seen to be rising. **1951** H. Kenner *Poetry E. Pound* 89 The Anglo-Saxon scholar's term for just such a vivid figure is 'kenning': the particulars by which the person or object in question is known. 'Whale-road', 'soul-bearer', are both ideogram and metaphor. **1959** N. G. L. Hammond *Hist. Greece* i. 34 The ideograms (signs which portray objects) and the symbols for numbers narrow the field of its interpretation somewhat, but much is left open to doubt. **1962** W. Nowottny *Lang. Poets Use* iv. 78 The particulars which inhabit these schemes, though extraordinarily difficult to summate, permit themselves to be assimilated to a common ideogram of decline (of the year, of the day, of a fire). **1964** E. Palmer tr. *Martinet's Elem. Gen. Ling.* v. 152 The members of an American board of examiners showed that they agreed neither on the position of the accent in *ideogram* nor on the quality of the first two vowels of the word. **1972** *Times* 29 June 16/3 Traditional graphic symbols and ideograms..were used extensively by the Manding.

ideogrammic (idio,græ·mik), *a.* [f. Ideogram + -ic.] Of the nature of an ideogram; expressed by means of symbols. Also **ideogramma·tic** *a.*

1929 W. J. Locke *Ancestor Jorico* i. 15 The swift ideogrammic air-speed which Toby and Jones have invented between themselves is a mystery. **1951** H. Kenner *Poetry E. Pound* 84 Joyce's catalogue of Bloom's books in *Ulysses* is the simplest possible application of the ideogrammic method. **1962** Y. Malkiel in Householder & Saporta *Probl. Lexicogr.* 22 Special features worthy of mention include: the use of abbreviations or peculiar ideogrammatic classifiers to mark for a given word its grammatical or semantic category. **1962** M. McLuhan *Gutenberg Galaxy* 22 No pictographic or ideogrammic or hieroglyphic mode of writing has the detribalizing power of the phonetic alphabet. **1967** *Punch* 18 Jan. 80 To others, the 'No Entry' sign (a highly ideogrammatic white bar on a red disc) variously indicated a pedestrian crossing, a level crossing, a Belisha beacon, the end of a speed limit, a first aid post, and a major road ahead.

ideograph. (Further examples.)

1911 *Encycl. Brit.* XV. 172 The usual charge for advertisement is from 7d. to one shilling per line of 22 ideographs (about nine words). **1951** H. Kenner *Poetry E. Pound* 89 The Chinese ideograph, like the metaphor, deals in exceedingly condensed juxtapositions. **1972** *Computers & Humanities* VI. 259 There is a basic corpus of 2,444 morphemes, each corresponding to a single Chinese logograph (ideograph).

ideographic, *a.* (Further examples.) Of ideographs: representing ideas pictorially or figuratively.

1948 D. Diringer *Alphabet* 174 An outside initiative has suggested the replacing of the Japanese ideographic-syllabic script by the Latin alphabet. **1955** P. Heron *Changing Forms of Art* 109 This long picture..was new; it represented, as far as Picasso is concerned, the defeat of the plastic by the ideographic. **1964** M. A. K. Halliday et al. *Linguistic Sciences* 49 The Chinese script is not ideographic: the symbols do not represent ideas, they represent formal items of the language. **1972** *Computers & Humanities* VI. 260 Because of an ideographic writing system extending back for 3,500+ years and still shared by all the Chinese languages, cognates can be visually identified.

ideological, *a.* Add: **2. b.** Of or relating to an ideology (sense *4).

1914 *Atlantic Monthly* June 775/1 If Mr. Mencken's earnest seekers after truth wish to evolve ideological schemes of municipal taxation..then, indeed, the newspaper discussions of these questions would be bewildering to these visionary workers in the realms of pure reason. **1925** M. Eastman *Since Lenin Died* iv. 32 Without realising this, you cannot penetrate beneath the ideological surface of the dispute which followed. **1937** *Times* 2 Nov. 17/2 Japan will be an absentee at Brussels... So will Germany.... The inconvenience, to say no more, of ideological attitudes has been very swiftly illustrated. **1939** *Times* 2 Nov. 8/2 The attacks on Great Britain and France for waging ideological warfare (is not Communist warfare ideological?) and at the same time being actuated by imperialistic motives are an obvious contradiction in terms. **1940** E. Pound *Cantos* lv. 56 But his brat was run by his missus And they had an ideological war. **1952** *Ann. Reg.* 1951 4 The rulers of Russia inherited imperialism and added to it an ideological imperialism. **1963** *Daily Tel.* 14 Oct. 12/2 Though it has been left this year to Albania to sponsor Communist China's membership, it will still be supported by her newly declared ideological foe, the Soviet Union, and by the victim of her aggression, India. **1969** *Guardian* 11 Sept. 7/2 Nor was he ever 'ideological', in the sense required by New Left dogmatists.

Hence **ideologically** *adv.* (later examples).

1957 P. Worsley *Trumpet shall Sound* iii. 67 It.. represents a conscious step towards the establishment of forms of organization, which..were independent both organizationally and ideologically. **1970** *Daily Tel.* 10 Jan. 12 As China's strength and numbers grow she becomes less prepared..to play second fiddle to Russia ideologically. **1971** *Ibid.* 3 Apr. 10/5 The Soviet system remains ideologically and politically committed to the destruction of our way of life.

ideologist. Add: **3.** A proponent or adherent of an ideology (sense *4).

1888 Marx & Engels *Manifesto of Communist Party* 14 Just as..the nobility went over to the bourgeoisie, so now a portion of the bourgeoisie goes over to the proletariat, and in particular a portion of the bourgeois ideologists, who have raised themselves to the level of comprehending..the historical movement. **1937** *Daily Herald* 20 Jan. 1/7 We are not content to see Europe arming feverishly under the contending standards of rival ideologists.

ideologue. (Later examples.)

1955 *Times* 6 June 7/7 From outside it is so easy to think of the Russians as a nation of ideologues. **1961** *Spectator* 9 June 826 When the plotters and ideologues come up for trial. **1972** *Sat. Rev.* (U.S.) 20 May 34/2 Women's Liberation ideologues and strategists have advanced some fine, socially refined arguments. **1972** *Observer* 10 Dec. 36/2 One of the stated principles of this outfit's chief ideologues is to steer clear of the politicos.

ideology. Add: **4.** A systematic scheme of ideas, usu. relating to politics or society, or to the conduct of a class or group, and regarded as justifying actions, esp. one that is held implicitly or adopted as a whole and maintained regardless of the course of events. Also *Comb.*

1909 *Westm. Gaz.* 4 May 10/2 It may be worth while giving some account of the ideology behind the German proposal, and of the details as worked out in the Conservative programme, bearing in mind that it is the scheme of a reactionary Agrarian party. **1936** Wirth & Shils tr. K. Mannheim (*title*) Ideology and Utopia. **1939** Auden in *I Believe* (1940) 22 It is despair at finding a solution to this problem which is responsible for much of the success of Fascist blood-and-soil ideology. **1955** E. Shils in *Encounter* V. 52 (*title*) The end of ideology? **1966** D. Jenkins *Educated Society* iv. 177 The processes of ideology-formation can go on even in the most high-minded of circles. **1970** D. D. Raphael *Probl. Pol. Philos.* i. 17 Ideology..is usually taken to mean, a prescriptive doctrine that is not supported by rational argument.

ideoplasm (i·dĭ,oplæz'm). *Spiritualism.* [f. Ideo- + -*plasm* after *ectoplasm.] = *ectoplasm. So **i:deopla·smic** *a.*; **i·deoplasmy** (see quot. 1961).

1926 A. Conan Doyle *Hist. Spiritualism* I. i. 7 Ectoplasm..has also been called 'ideoplasm', because it takes on in an instant any shape with which it is impressed by the spirit. **1961** W. H. Salter *Zoar* vi. 63 The hypothesis of 'ideoplasmy', that is to say, the view that materialisations are produced from the medium's energy and a substance ('ectoplasm') supplied by him with the assistance perhaps of the sitters, and that they take form in accordance with the thoughts of those present. *Ibid.* vi. 69 A clumsy attempt, whether ideoplasmic or fraudulent, to imitate the established conception of a spirit?

ideoplastic (idĭ,oplæ·stik), *a.* [ad. F. *idéoplastie* (J. P. Philips *Cours théorique et pratique de Braidisme* (1860) ii. 44), adj. *idéoplastique*, f. Ideo- + Plastic *a.* 5.] Denoting those physiological or artistic processes which are supposed to be moulded or modified by mental impressions or suggestions; also, pertaining to the suggestive function of the imagination; so **ideopla·stically** *adv.*, in a manner influenced by mental or imaginative impressions; **ideopla·sty, ideopla·sy,** imagination in its suggestive capacity, esp. as modifying certain physiological functions or processes.

Somewhat specialized uses in spiritualistic writings are not clearly distinguishable from the above. Durand de Gros (see quot. 1901) and J. P. Philips (see etym.) are names of the same person.

1901 Baldwin *Dict. Philos. & Psychol.* I. 507/2 *Ideoplastic*, applied to the physiological functions considered as liable to modification from suggested ideas (used originally by Durand de Gros). *Ibid.*, *Ideoplasy*, suggestion operative in the production of physiological changes. **1919** A. Conan Doyle *Vital Message* App. 209 We accept Dr. Geley's statement that they are 'ideoplastic'. **1929** *Encycl. Brit.* I. 51/1 For half a century no perceptible progress was made on the idea thrown out by Durand (1855, 1860) that a clear distinction should be effected between the 'ideoplastic' and hypotaxic phenomena. **1935** *Burlington Mag.* Mar. 110/1 As regards the ideoplastic character of pre-Greek art..it is called ideoplastic because the artist renders what he *knows* about the object rather than what he sees. *Ibid.* 121/2 He built up his works ideoplastically. **1943** H. Read *Educ. through Art* vi. 136 Later writers..relate this two-fold distinction to Verworn's classification of primitive art as physioplastic and ideoplastic. **1960** *Times Lit. Suppl.* 12 Aug. 576/2 The 'ideoplasty' presumed to cause the body of a medium to extrude and model ectoplasm into the image, telepathically received, of old Aunt Kate. **1961** R. Crookall *Supreme Adventure* ii. i. 60 The substance composing their environment resembles ectoplasm in being ideo-plastic and responding automatically to their thoughts, feelings, expectations, hopes and fears.

-idin, *suffix. Chem.* [f. -id(e + -in¹.] Used to form the names of the anthocyanidins, as in *cyanidin, delphinidin, pelargonidin, peonidin.*

-idine, *suffix. Chem.* [f. -id(e + -ine⁵.] Used to form the names of many organic compounds containing nitrogen which, with few exceptions (as *guanidine*), contain one or more rings; esp.: **a.** Certain amino derivatives (*a*) of simple monocyclic aromatic hydrocarbons, as *cumidine, cymidine, mesidine, toluidine, xylidine*, or of derivatives of such hydrocarbons, as *cresidine*; (*b*) of symmetrical bicyclic aromatic hydrocarbons, as *benzidine, naphthidine, tolidine.* **b.** Certain aminophenol ethers, as *anisidine, phenetidine.* **c.** Certain heterocyclic compounds with nitrogen in the ring (the use of the suffix in some cases implying that the ring is saturated), as *piperidine, pteridine, pyridine, pyrrolidine.* (Hence, in mod. systematic nomenclature, forming the suffixes -*iridine*, -*etidine* and -*olidine*, as in *aziridine.*) **d.** Pyrimidine nucleosides, as *cytidine, thymidine, uridine.* **e.** Certain alkaloids, as *anhalidine, pilocarpidine, quinidine.*

idio-. Add: **idiochro·mosome** *Cytology* = **sex chromosome*; **idioglo·ssia** [Gr. ἰδιόγλωσσος of distinct tongue], a form of dyslalia in which the person affected consistently makes substitutions in his speech sounds to such an extent that he seems to speak a language of his own; **idiographic** *a.*, (*b*) concerned with the individual, pertaining to or descriptive of single and unique facts and processes (opp. Nomothetic *a.*); **idiolalia** (-lēi·liă) [*-lalia] = *idioglossia* above; **idiophone,** a percussion instrument that consists simply of elastic material (as metal, wood, etc.) capable of producing sound (as opp. to *membranophone in which stretched skin is used as the agent of sound); **idiopho·neme** *Linguistics*, a phoneme in individual speech; hence **idiophone·mic** *a.*; **idiore·tinal** *a.*, applied to what is seen when the eyes are shut and there is no external stimulation of the retina; **idio(r)rhy·thmic** *a.* (later example); also as *sb.*; **idioventri·cular** *a. Med.*, proper to the ventricle alone; used of the rhythm of contraction set up within the ventricle when the normal auricular stimulus to ventricular contraction is blocked.

[illegible] clarify the diagnosis of..idioventricular rhythm with inadequate cardiac output.

idioblast (i·dioblɑst). [f. Idio- + -blast.] **1.** *Bot.* (In Dict. s.v. Idio-.)

2. *Cytology.* [a. G. *idioblast* (O. Hertwig *Zelle und Gewebe* (1893) I. ix. 272).] A hypothetical structural unit of living protoplasm. *Obs.* exc. *Hist.*

1893 *Nature* 2 Feb. 315 He [*sc.* O. Hertwig]..suggests the employment of the term 'Idioblasts' for the minute elementary particles, which Darwin called 'gemmules' in his hypothesis of pangenesis. **1925** E. B. Wilson *Cell* (ed. 3) 1134 *Idiosome*, the same as *idioblast, plasome, pangen* etc.

3. *Petrol.* [a. G. *idioblast* (F. Becke 1904, in *Compt. Rend. IX Sess. Congr. Géol. Internat.* II. 564)], a mineral crystal within a metamorphic rock which has developed its own characteristic crystal faces.

1920 A. Holmes *Nomencl. Petrol.* 122 *Idioblast*, Becke, 1903, a term applied to pseudoidiomorphic crystals, such as garnet, occurring in metamorphic rocks. **1962** T. F. W. Barth *Theoret. Petrol.* (ed. 2) 288 The majority of the minerals in metamorphic rocks are irregular in outline, xenoblasts; but some minerals are frequently bounded by their own crystal faces, idioblasts.

So **idiobla·stic** *a. Petrol.*, (of a mineral crystal within a metamorphic rock) having its own characteristic crystal faces; (of a crystal face) having its own characteristic form; *idioblastic order, series*, a ranking of minerals expressing their relative ability to develop idioblastic crystals when competing with each other.

1908 *Q. Jrnl. Geol. Soc.* LXIV. 482 Most of the hornblende existing in the amphibolite..is clearly secondary, and from its idiomorphic forms would be called 'recrystallized'... For such cases Prof. F. Becke has proposed the term idioblastic. **1954** H. Williams et al. *Petrogr.* ix. 166 It is possible to list metamorphic minerals in a generalized sequence—the crystalloblastic series (idioblastic order)—such that each tends to develop idioblastic surfaces against any other mineral placed lower in the series.

idiogram (i·diogræm). *Cytology* and *Med.* [ad. Russ. *idiogramma* (S. Navashin: in *Zhurn. Russk. Bot. Obshch.* (1921) VI. 171 he is reported as having used the term in his lectures for many years): see Idio- and -gram.] = *karyotype *sb.* 1b: usually, a diagrammatic or systematized representation of a chromosome complement (of one cell or of many) indicating the number of chromosomes, their relative lengths, the position of the centromeres, etc.

1927 *Genetics* XII. 64 The relative size of the chromosomes, peculiar shape, and especially the presence of satellites (S. Nawaschin, 1912) and constrictions (Sakamura 1915, 1920) furnish criteria for distinguishing the members of a given complement, or using the terminology of S. Nawaschin, they characterize the idiogram or the specific arrangement of the diploid nuclear plates. **1934** L. W. Sharp *Introd. Cytol.* (ed. 3) ix. 128 The diagrammatic representation of a karyotype, as in Fig. 70, is called an idiogram (S. Nawaschin, 1921). **1957** C. P. Swanson *Cytol. & Cytogenetics* (1958) v. 118 The shape and size of chromosomes are his guideposts, and their constancy has enabled him to determine for purposes of comparison the karyotypes or idiograms (haploid complements) of many plants and animals. *Ibid.* xiii. 448 When represented in diagrammatic fashion..the karyotype is usually referred to as an idiogram. **1966** D. M. Kramsch tr. *Grundmann's Gen. Cytol.* ii. 108 All the chromosomes of one set form the karyotype specific for each species and with such a chromosomal idiogram it is possible in certain cases to demonstrate relationships between the species. **1969** R. R. Eggen in Davidsohn & Henry *Todd-Sanford Clin. Diagn.* (ed. 14) xxxii. 1224 (*caption*) Idiogram of a normal human cell... The male karyotype differs from the female in that male cells normally have a single X chromosome and therefore have only 15 group C chromosomes. **1971** *Nature* 11 June 887/1 The average forms of all chromosomes of a complement are defined by an idiogram, based on a large number of karyotypes. **1973** *Lancet* 24 Feb. 420/1 Strictly speaking the actual pictures [of chromosomes] are karyotypes, and an idiogram is a diagram of the chromosome state of an individual.

idiolect (i·diolekt). *Linguistics.* [f. Idio- after Dialect.] The linguistic system of one person, differing in some details from that of all other speakers of the same dialect or language.

1948 B. Bloch in *Language* XXIV. 7 The totality of the possible utterances of one speaker at one time in using a language to interact with one other speaker is an *idiolect*. **1948** R. A. Hall Jr. in *Studies in Linguistics* VI. ii. 31 Language exists in individuals, as a set of habits which each individual possesses (an *idiolect*). **1953** C. E. Bazell *Ling. Form* 96 It must not be supposed that such [linguistic] systems are necessarily less determinate than for instance that of a single idiolect as recorded over a short space of time. **1953** J. B. Carroll *Study of Lang.* ii. 10 Indeed each member of a speech community may be said to possess his own *idiolect*, his own personal variety of the language system. **1953** *Internat. Jrnl. Amer. Ling.* XIX. ii. Suppl. 40 Hockett defined 'idiolect' as the individual's total repertory of speech habits over a short period of time. **1958** A. A. Hill *Introd. Ling. Struct.* ii. 13 The English which is described in the personal dialect of a single speaker or, to use the technical term, a single idiolect. **1964** M. A. K. Halliday et al. in J. A. Fishman *Readings Sociol. of Lang.* (1968) 158 A person's idiolect may be identified, through the lens of the various registers, by its grammatical and lexical characteristics. **1964** R. H. Robins *Gen. Ling.* ii. 51 The lower limit of dialect division comes down to the individual speaker, and for this limiting case of dialect the term *idiolect* (the speech habits of a single person) has been coined. **1970** W. Labov in *Rep. 20th Round Table Meeting on Ling. & Lang. Stud.* 89 So we are dealing not with the idiolect of the investigator, but the idiolect of one isolated boy whose position in the community is quite uncertain.

Hence **idiole·ctal** *a.*, of or pertaining to an idiolect; **idiole·ctally** *adv.*

1953 *Internat. Jrnl. Amer. Ling.* XIX. ii. Suppl. 37 Utterances are swiftly and with assurance identified despite idiolectal differences. **1958** *Archivum Linguisticum* X. ii. 146 Who..pronounces *cow* and *bough* with different diphthongs, unless idiolectally and idiosyncratically? **1965** *Language* XLI. 502 Idiolectal diversity is an inevitable result of the..productivity inherent in every single individual's linguistic habits. **1972** *Ibid.* XLVIII. 314 We believe these variations to be idiolectal rather than dialectal.

idiom. Add: **3. b.** A characteristic mode of expression in music, art, or writing; an instance of this.

1921 J. B. McEwen *First Steps Mus. Comp.* 5 To put it in somewhat colloquial terms, the composer of ancient music wrote melody, the composer of modern music writes tunes. It is no part of my purpose to make comparisons between these two idioms. *Ibid.*, An interesting reversion to the contrapuntal idiom. **1923** H. Crane *Let.* 9 Feb. (1965) 121 Tate has a whole lot to learn when he finds his way out of the Eliot idiom. **1927** *Grove's Dict. Mus.* (ed. 3) II. 537/2 The folk-songs of all nations have been cultivated..for the sake, mainly, of their undoubted freshness and spontaneity of idiom as compared with pseudo-classical models. *Ibid.* 538/1 The study of this melodic music has suggested many harmonic idioms of notable freshness and beauty. **1939** *Burlington Mag.* Aug. 90/1 Buildings and industrial products which are now the accepted 'idiom' of design throughout the modern world. **1955** *Times* 9 May 3/1 We in this country have had experience of Anglo-American cooperation in film-making, and, whatever may be said in its favour from the practical, economic point of view, it certainly tends to blur and weaken the natural idiom and character of the countries involved. **1957** S. Dance in *Jazzmen Concerning Jazz* 43 The three great names in the preservation of jazz in the pure New Orleans idiom..were King Oliver, Jelly Roll Morton and Louis Armstrong. **1958** B. James in P. Gammond *Duke Ellington* ii. 145 Ellington's music had its origins in the New Orleans style, as has so much else worthwhile in the jazz idiom.

4. (Further examples.)

1866 Geo. Eliot *Let.* 15 Aug. (1955) IV. 301, I took unspeakable pains in preparing to write Romola—neglecting nothing I could find that would help me to what I may call the 'Idiom' of Florence, in the largest sense one could stretch the word to. **1870** G. M. Hopkins *Jrnls. & Papers* (1959) 195, I noticed it [*sc.* snow]..sketched in intersecting edges bearing 'idiom'..I have no other word yet for that which takes the eye or mind in a bold hand, not being beauty nor true inscape yet gives interest. **1936** R. Campbell *Mithraic Emblems* 46 To form the idiom of her flesh I faceted in clearest thought An arctic crystal whose mesh Of frosty rays the sun is caught.

5. *Comb.* **Idiom Neutral,** an international language based on Volapük, devised chiefly by W. Rosenberger, and first published in English in 1903.

1903 W. Rosenberger (*title*) Idiom neutral. **1907** W. J. Clark *Internat. Lang.* ii. v. 99 The famous linguistic club of Nuremberg is remarkable for having gone through the evolution from Volapük to Idiom Neutral *vià* Esperanto. **1922** A. L. Guérard *Short Hist. Internat. Lang. Movement* ii. vi. 137 Idiom Neutral is frankly based on the principle of greatest internationality, at least so far as root-words are concerned. **1949** M. Pei *Story of Lang.* (1952) vi. iii. 443 In 1902 an academy of Volapük experts devised a radical simplification of their tongue, which they rechristened Idiom Neutral.

idiomaticity (idiomăti·siti). [f. Idiomatic *a.*: see -icity.] The quality or state of being idiomatic.

1965 *Language* XLI. 504 Mrs Palmer's translation is.. accurate, without sacrificing freedom or idiomaticity. **1971** T. F. Mitchell in *Archivum Linguisticum* II. 57 Although such correspondences [as *make up to = flatter*, *make it up to = compensate*, etc.] are usually substantial, they are apparently not a necessary condition of idiomaticity.

idiomorphism (idiomǭ·ɹfiz'm). *Min.* [f. Idiomorph(ic *a.* + -ism.] The condition of being idiomorphic.

1908 *Q. Jrnl. Geol. Soc.* LXIV. 482 Professor F. Becke.. points out that, in the formation of the crystalline schists, the relative idiomorphism is not to be considered as denoting the order in which they have formed. **1951** Turner & Verhoogen *Ign. & Metamorphic Petrol.* xii. 56 Idiomorphism seems..a doubtful criterion of early crystallization in the case of certain consistently idiomorphic accessories, such as apatite.

idiosome (i·diosōᵘm). [f. Idio- + *-some⁴.] **†1.** *Biol.* A supposed ultimate unit of living matter. *Obs.*

1894 C. O. Whitman in *Biol. Lectures 1893* (Wood's Hole, Mass., Marine Biol. Lab.) 123 It will find the unit of organization, growth, development, not in cell-formation, but in those ultimate elements of living matter, for which *idiosomes* seems to me an appropriate name. **1899** [see Pangene].

2. *Cytology.* [Proposed (as F. *idiosome*) by C. Regaud (*Arch. d'Anat. microsc.* (1910) XI. 343) as a better word than *idiozome.*] = *idiozome.

1918 *Amer. Jrnl. Anat.* XXIV. 37 (*heading*) The development of the idiosome in the germ-cells of the male guinea-pig. **1934** L. W. Sharp *Introd. Cytol.* (ed. 3) xiv. 216 In another series of forms (mollusks, amphibia, other vertebrates) the Golgi bodies are closely aggregated about the centrioles where their lightly staining substance flows together to form the idiosome. **1953** O. E. Nelsen *Compar. Embryol. Vertebr.* iii. 126 The idiosome (idiozome) is a rounded body of cytoplasm which, in many animal species, takes the cytoplasmic stain more intensely than the surrounding cytoplasm.

idiosyncrasy. Add: **1.** *spec.* An individual's hypersensitivity *to* a drug or other substance which is ingested or inhaled or which otherwise comes into contact with the body. (Further examples.)

1887 *Brit. Med. Jrnl.* 20 Aug. 431/1 Dr Daniel Bernouilli, of Basle, reports an instance of idiosyncrasy with respect to antipyrin. **1893** *Edin. Med. Jrnl.* XXXVIII. 627 No very tenable theory has been put forward to explain the reason of idiosyncrasy to drugs. **1912** *Jrnl. Amer. Med. Assoc.* 13 Apr. 1088 The view that the peculiar idiosyncrasies, so-called, with respect to certain articles of diet—strawberries, cheese..cow's milk, etc.—are forms of anaphylactic reactions is steadily growing. **1922** *Jrnl. Immunol.* VII. 128 Peculiar drug reactions, known as idiosyncrasies, were well recognized, but they were not identified as allergies until 1916. **1932** *Jrnl. Soc. Chem. Industry* 13 May 440/2 Mention was made..of the frequent lack of adequate proof of the supposed irritant action of certain dyes and of the difficulties due to idiosyncrasy. **1971** *Lancet* 25 Sept. 698/2 Idiosyncrasy to CS has not been reported, and the old and young do not seem to be at exceptional risk.

idiot, *sb.* Add: **4. idiot asylum,** a term formerly used for a hospital for the mentally ill; **idiot board** *colloq.*, a prompting board held before a television speaker but not projected on the film; **idiot box** *colloq.*, a television set; also *transf.*; **idiot card** *colloq.*, = **idiot board*; **idiot fringe,** (*a*) a fringe of hair in a style once worn (see quots.); (*b*) (*occas.*) = **lunatic fringe*; **idiot light** *colloq.*, a warning light, usu. red, that goes on when a fault occurs in a mechanical or electrical device; **idiot sheet** *colloq.*, = **idiot board*; **idiot stick** *U.S. slang*, a shovel; **idiot stitch,** tricot-stitch, the easiest stitch in crochet work.

1866 J. MacGregor *Thousand Miles in Rob Roy Canoe* (ed. 2) vi. 104 Close to the inn was the idiot asylum, an old castle with poor demented women in it. **1880** 'Mark Twain' *Tramp Abroad* I. xxv. 250 They meant an asylum—an *idiot* asylum. **1952** *Newsweek* 4 Aug. 51/2 The Republicans and the Democrats got their 'idiot boards' free. **1961** G. Millerson *Technique Television Production* x. 194 (*caption*) Held-up cue card, idiot board, goof sheet. **1971** *Radio Times* 4 Nov. 6/4, I can never work with an auto-cue or idiot board. I learn my lines before the show. **1959** P. Bull *I know the Face* xi. 193 The rationing period, when my very existence depended on the magic idiot-box. **1965** *Lancet* 2 Jan. 46/2 Often they may be found, in semi-hypnotic state, watching the 'idiot box' with its endless stream of images and fullness of sound, all signifying nothing. **1972** J. Flower *Cobweb* ii. 72, I thought you spent all your time with the idiot box. **1973** *Ann. N.Y. Acad. Sci.* CCXI. 282 We assume that the modern general purpose computer is an idiot-box capable of performing only the simplest of routines. **1959** *Globe & Mail* (Toronto) 18 Sept. 1/7 No one held up an 'idiot card'—a prompter's card for actors who have forgotten their lines. **1960** B. Keaton *Wonderful World of Slapstick* (1967) 238 They had worked out an interesting technique for these foreign-language remakes. They used 'idiot cards'. On these the foreign words are spelled out in phonetic English and held up beyond camera range. **1969** *New Yorker* 27 Sept. 86/3 We had all the written questions put on idiot cards, and then the people read them before the camera. **1886** H. Baumann *Londinismen* 81/1 *Idiot-fringe*, fransenartig auf die Stirn herabgekämmtes Haar der Fabrikmädchen, Hökerweiber u.s.w. **1923** J. Manchon *Le Slang* 130 *Fringe*, le devant des cheveux dans la coiffure à la chien (habituelle chez les femmes de la basse classe). On dit aussi idiot-fringe. **1965** *Spectator* 5 Mar. 286/1 An article on a small and dying idiot-fringe sect of nationalists in North Germany. **1968** E. McGirr *Lead-Lined Coffin* iii. 173 He watched the idiot lights in the dashboard..for the warning red which would betoken he had lost the oil. **1971** M. Tak *Truck Talk* 88 *Idiot light*, a small light sometimes found on the front end of a tractor, the light turns on when the ignition is started. **1961** A. Berkman *Singers' Gloss. Show Business* 29 *Idiot sheet*, a large placard or paper roll, on which is [*sic*] printed the actual lines to be spoken by an announcer or performer. The printed characters are large enough to be seen from quite a distance from the camera. **1962** R. Bretz *Techniques Television Production* 487 When cue cards first came into use for full script purposes, they were known derisively as 'idiot sheets'. **1942** Berrey & Van den Bark *Amer. Thes. Slang* § 75/32 *Shovel*, idiot stick. **1968** R. F. Adams *Western Words* (rev. ed.) 158/1 *Idiot stick*, what the miner calls his shovel; sometimes shortened to *idiot*. **1882** Caulfeild & Saward *Dict. Needlework*, *Idiot stitch*, one of the names given to Tricot Stitch.

idiozome (i·diozōᵘm). *Cytology.* Also **idiosome** (q.v. above). [ad. G. *idiozom* (F. Meves 1896, in *Anat. Hefte* (Abt. 2) VI. 315), f. Idio- + Gr. ζῶμ-α loin-cloth, band, girdle.] A rounded structure present in the cytoplasm of developing germ cells in members of most animal groups, in certain of which it forms part of (or is associated with) the Golgi apparatus in spermatids and spermatocytes and develops into the acrosome of the spermatozoon.

1899 *Zool. Jahrbücher* (*Abt. für Anat.*) XII. 14 At the pointed end of the cell lies a spherical or oval body, homogeneous in appearance, which is stained more deeply by the action of Hermann's fixative than the enveloping cytoplasm... Meves..proposes for it the name 'Idiozom', which I shall adopt in the present paper... In the idiozome of the resting spermatogonium I have been unable to detect any granules which could be regarded as centrosomes. **1920** L. Doncaster *Introd. Study Cytol.* vii. 102 The acrosome is described in some forms as arising from a mass of denser protoplasm known as the idiozome. **1928** C. W. Metz in E. V. Cowdry *Special Cytol.* II. xxxvi. 1282 The Golgi bodies..usually appear first in the form of small granules or rods, which during the growth stages typically collect about the idiozome. **1952** G. H. Bourne *Cytol. & Cell Physiol.* (ed. 2) vi. 254 The Golgi apparatus in developing male germ and other cells is frequently associated with a differentiated area of cytoplasm near the nucleus, called the idiozome.

idle, *a.* (*sb.*). Add: **3. b.** (See quot. 1956.)

1956 J. Masters *Bugles & Tiger* ii. 46 The word 'idle' meant anything the staff considered unsoldierly. We *were* idle; we had idle haircuts, idle rifles, idle bicycles; we *did* idle salutes, idle jumps. **1959** *News Chron.* 4 Aug. 1/4 'Idle' is a Brigade [of Guards] adjective that describes everything that is not perfection in execution. A bootlace undone is an idle bootlace. **1963** D. Walder *Bags of Swank* vi. 65 Ransome looked at the lecturer's hip pocket now revealed as undone. 'Idle and naked,' he said loudly to Lilburne.

4. (Further examples.) Freq. in phr. *the idle rich.*

1865 Mill *Auguste Comte* 160 He allows of no idle rich. **1900** B. Matthews *Confident To-Morrow* 178 Mr. Dircks wishes to shift the burdens of the worthy poor upon the shoulders of the idle rich. **1926** *Encycl. Brit.* I. 527/1 We must remember as a rule the 'idle rich' do not represent idle riches. **1938** *Times Lit. Suppl.* 11 June 403/3 The crew, Reds of various shades, mutiny and drive ashore the idle-rich passengers. **1960** C. Day Lewis *Buried Day* viii. 171 The idle rich, the boss class, fleeing..from the wrath to come. **1964** M. Argyle *Psychol. & Social Probl.* xv. 185 There has been a decline of the old 'idle rich' upper middle class, living partly on unearned income and passing on wealth and advantages to its children. **1971** *Daily Tel.* 20 Oct. 10/4 More than 25,000 Coventry workers were idle yesterday as a result of the dispute.

5. c. *Electr.* Of a wire on an armature: having no electromotive force induced in it. Of a component of an alternating current: 90° out of phase with respect to the voltage; wattless, reactive.

1884 S. P. Thompson *Dynamo-Electr. Machinery* iii. 33 Where the coils are working in series, it has been considered advantageous to arrange the commutator to cut out the coil that is in the position of least action, as the circuit is thereby relieved of the resistance of an idle coil. *Ibid.* vii. 126 The advantage originally claimed for this construction, namely, that it allows less of the total length of wire to remain 'idle' on the inner side of the ring, is rather imaginary than real. **1904** R. M. Walmsley *Electr. in Service of Man* II. vi. 1069 This current.. contributes nothing to the power, and is therefore often referred to as the idle current. **1908** Slingo & Brooker *Electr. Engin.* (new ed.) ix. 354 The portions connecting the horizontal limbs are always idle, inasmuch as they do not cut, but only slide through, the lines of force.

d. Of money: out of circulation.

1931 *Times Lit. Suppl.* 19 Feb. 124/2 It may be thought that saving cannot exceed investment because idle money automatically becomes the basis of bank credit. **1965** Seldon & Pennance *Everyman's Dict. Econ.* s.v. Dishoarding, A distinction is made between 'active' money in circulation and financing current transactions and 'inactive' money held in idle balances.

8. a. *idle-handed* (later example), *-minded* (so *-mindedness*).

1899 *Westm. Gaz.* 21 Aug. 2/3 The empty-headed and idle-minded exist in both sexes. **1899** *Leisure Hour* Dec. 153 Hence the dull lives of many children of the poor, their occasional trend towards mischief from sheer idle-mindedness. **1917** Kipling *Years Between* (1919) 47 But the idle-minded overlings who quibbled while they died. **1927** T. Wilder *Bridge San Luis Rey* 100 Even the busiest mother stands for a moment idle-handed. **1928** *Oxford Poetry* 39 Because this place is full of moneyed young men And indolent phallophil idleminded girls.

b. idle Dick, Jack *S. Afr.*, formerly used as a local name for the grass-bird, *Sphenœacus afer*; **idle-fellow,** formerly, a fellow of a college who had no formal duties; so **idle fellowship.**

1901 A. C. Stark *Birds S. Afr.* II. 168 *Sphenœacus natalensis.* Natal Grass-Bird... 'Idle Dick' and 'Lazy Dick' of English Colonists. **1919** R. Frost *Let.* 8 Aug. (1964) 132, I am going..to Ann Arbor to become an idle-fellow of the University of Michigan for one year. **1884** in J. R. Ware *Passing Eng.* (1909) 157/1 Much has been said against what are called idle Fellowships. **1909** J. R. Ware *Ibid.* 156/2 *Idle fellowships* (*Oxford and Cambridge*), the old as distinct from the new fellowships. Parliamentary action swept away towards the end of the 19th century most of these fatal sinecures. **1884** R. B. Sharpe *Layard's Birds S. Afr.* (rev. ed.) 281 It..will suffer itself to be taken with the hand rather than rise again; for this reason it has acquired the name of *Idle Jack* or *Lazy Dick.* **1893** A. Newton *Dict. Birds* 458 Idle Jack, a local name in the Cape Colony for *Sphenœacus africanus* (Grass-bird).

B. *sb.* **4.** [f. the vb.] Idling (of an engine); idling speed.

1939 C. H. Fisher *Carburation & Carburettors* iv. 96 If the throttle is closed completely the adjustment of the idling mixture is rendered too sensitive, hence it is usual to give a very small opening of the throttle when adjusting the idle. **1943** A. P. Fraas *Aircraft Power Plants* vii. 134 As the throttle is opened farther..the idle needle is withdrawn very rapidly and so has no effect on fuel flow at powers above an idle. **1966** *McGraw-Hill Encycl. Sci. & Technol.* II. 478/2 Because of increased piston and other friction with a cold engine, greater throttle opening as well as more fuel is required for idle at that time. **1972** *Practical Motorist* Oct. 160/3 Start the engine and set the tick-over to 650 rpm, using the air screw. Now turn the jet adjusting screw one way or the other until the smoothest idle is achieved.

5. *attrib.* (in sense *B. 4: cf. *idling *vbl. sb.* 2), as *idle jet, needle, nozzle, power, range, stroke.*

1943 A. P. Fraas *Aircraft Power Plants* vii. 119 A high metering head..is used to induce a flow of fuel through an idle jet. **1943** Idle needle [see sense *B. 4]. **1968** C. F. Taylor *Internal-Combustion Engine* II. vi. 200 Further opening of the throttle gradually exposes the idle nozzle to the full manifold depression, which may be as much as 10 psi..below atmospheric pressure in a normal idling engine. **1946** R. H. Thorner *Aircraft Carburetion* ii. 65 The richest mixture is required at the carburetor during the lowest idle power. **1939** C. H. Fisher *Carburation & Carburettors* iv. 95 Since most engines idle with a manifold depression of about 16" of mercury,..any good carburettor can be made to deliver a wide band of mixture strengths covering the idle range. **1896** W. Norris *'Otto' Cycle Gas Engine* ii. 6 The idle strokes of the 'Otto' cycle are far from theoretically correct.

idle, *v.* Add: **1.** (Later example.)

1938 M. K. Rawlings *Yearling* ix. 75 The woodbox was low and Jody idled outside to fill it.

3. (Earlier and later examples.)

1788 E. Sheridan *Jrnl.* (1960) 138 And to compleat all, they beg'd him to see another person who idled him two hours more. **1968** *Globe & Mail* (Toronto) 5 Feb. 21/7 Gilchrist was..idled with a leg injury. **1972** *Nat. Geographic* Sept. 322 (*caption*) *Idled by war*, workers await the call to return to the Karnaphuli Rayon and Chemicals plant in Chandraghona.

4. a. *intr.* Of an engine: to run while disconnected from a load or out of gear, so that it performs no external or useful work; also, to run very slowly.

1916 [implied in * idling *vbl. sb.* 1]. **1920** V. W. Pagé *Useful Hints Motorists* iii. 78 Turn petrol adjustment to the right..until motor idles smoothly. **1925** A. W. Judge *Carburettors & Carburation* iii. 37 The ideal carburettor should:..(4) Enable the engine to run very slowly when 'idling', without undue waste of fuel. **1932** Chatfield & Taylor *Airplane & its Engine* (ed. 2) viii. 169 Airplane engines must be able to idle, that is run very slowly, in order to keep the landing speed as low as possible. **1934** *Boys' Mag.* XLVII. 23/2 One after another the four engines were started, 'revved' with a deafening roar singly and all together, and then left quietly 'idling'. **1953** G. E. M. Anscombe tr. *Wittgenstein's Philos. Investigations* I. § 132 The confusions which occupy us arise when language is like an engine idling, not when it is doing work. **1965** P. H. Smith *High-Speed Two-Stroke Petrol Engine* x. 212 The engine is..idling at tick-over speed. **1970** *Commercial Motor* 25 Sept. 64/2 Neither vehicle had much mileage on the clock which was probably the reason why the engines idled badly.

b. *trans.* To cause (an engine) to idle; *to idle down*, to slow down and idle (an engine).

1925 E. W. Knott *Carburettor Handbk.* i. 41 It is surprising what a small degree of throttle opening is necessary to pass the requisite amount of mixture to 'idle' an engine. **1938** J. Steinbeck *Long Valley* 17 She heard him drive to the gate and idle down his motor. **1938** *Amer. Speech* XIII. 131/2 In case of some delay, the engineer may idle down or slow down the tractor. **1972** 'H. Buckmaster' *Walking Trip* 163 David slowed the car and idled it by the side of the road.

idler. Add: **3. b.** A wheel or roller that when in contact with a moving belt, tape, or the like transmits no power but serves to support it, guide it, or make it taut. Freq. *attrib.*

1899 G. D. Hiscox *Mech. Movements* ii. 37 A single belt, with two idlers, for tightening and directing the half twist of the belt. **1908** [see *belt conveyor* (*belt *sb.*¹ 6)]. **1951** *Wire Ropes in Mines* (Inst. Mining & Metall.) 290 Supporting idlers are necessary to prevent severe whipping and vibration, if the distance is great between the drum and head sheave. However, if it is at all possible, the installation of idler sheaves should be avoided. **1962** A. Nisbett *Technique Sound Studio* 256 On a tape deck, the idler presses the tape against the capstan when the drive is switched on. **1969** *Times* 2 May (Suppl.) p. viii/7 This means that idler rolls made to this design will operate on conveyors at speeds up to and in excess of 1,000 ft. a minute. **1970** *Jrnl. Soc. Dyers & Colourists* LXXXVI. 87/1 The web follows an up-and-down zigzag path over idler rollers. **1972** *Reader's Digest Repair Manual* 551/4 Incorrect speed is generally due to a worn idler wheel.

4. *Railways.* (See quots.)

1909 *Cent. Dict.* Suppl., *Idler*, in *railroading*, an empty car; an empty. **1962** *Amer. Speech* XXXVII. 133 *Idler*, an empty car which is coupled to another car having a load of logs longer than the car, thus permitting the load to be coupled into a train.

‖ **idli** (i·dli). [Tamil.] A steamed cake of rice and black gram, popular in southern India.

1958 R. K. Narayan *Guide* i. 16 Give me coffee and *idli*, please, first thing in the day. **1961** K. Nagarajan *Chron. Kedaram* iii. 64 There was no lack of mirth or old-fashioned cheer; plenty of *pongal* and *idlis* were consumed. **1971** *Hindustan Times Weekly* (New Delhi) (Suppl.) 4 Apr. p. iv/3, I learnt Tamil..even got initiated into the mysteries of *idlis*. **1971** *Illustr. Weekly India* 25 Apr. 49/1 Dev closes the generation gap between Hema and him by the handsome device of welcoming this *idli*-white 'It Girl' with open arms. **1972** *New Yorker* 26 Aug. 28/1 Sell this poor child two *idlies.* Give him freshly made ones, not yesterday's. *Ibid.*, The *idlies*—so soft and pungent, with green chutney spread on top.

idling, *vbl. sb.* (s.v. Idle *v.*). Add: (Further examples, corresponding to *idle *v.* 4.)

1916 R. T. Nicholson *Bk. of Ford* 151 You will run very economically, but you will find starting and 'idling' very difficult. **1925** A. W. Judge *Carburettors & Carburation* iv. 57 For slow running, or 'idling', two adjustments are provided. **1949** I. Katz *Princ. Aircraft Propulsion Machinery* xx. 216 Idling is desirable from the standpoint of warm-up prior to flight. **1966** *McGraw-Hill Encycl. Sci. & Technol.* II. 478/1 Slight intake or exhaust valve leaks can make smooth idling impossible.

2. *attrib.* (esp. designating parts of an engine concerned with idling), as *idling adjustment, condition, jet, mixture, passage, screw, speed.*

1924 B. G. Elliott *Automobile Repairing* x. 133 Carburettors which have idling adjustments may be adjusted to overcome loading. **1940** *Chambers's Techn. Dict.* 437/1 *Idling adjustment*, a setting of the slow-running jet and throttle position of a carburettor, so as to give regular idling. **1943** A. P. Fraas *Aircraft Power Plants* vii. 131 The model *H* idle system depends on a small hole in the tip of the metering pin to restrict the fuel flow under idling conditions. **1929** *Motor World* 11 Jan. 23/2 In the new Zenith a special idling jet, fed by its own independent compensating jet, is employed. **1942** B. A. Shields *Air Pilot Training* xiii. 216 The engine is supplied with fuel for starting and idling by a special arrangement called an idling jet. **1959** *Motor Manual* (ed. 36) iii. 55 Here the idling mixture is adjusted manually. **1925** A. W. Judge *Carburettors & Carburation* iv. 100 There is a slow running, or idling passage, of the usual pilot jet type. **1972** *Drive* New Year 99/2 The throttle-stop, or idling screw, will be found where the accelerator cable meets the carburettor; the mixture screw is at the base of the carburettor. **1932** Chatfield & Taylor *Airplane & its Engine* (ed. 2) viii. 139 The idling speed is generally about 200 r.p.m. when the airplane is at rest on the ground. **1973** M. Woodhouse *Blue Bone* xii. 128, I..opened the hood, screwing up the throttle control rod a couple of turns to raise the idling speed.

idling, *ppl. a.* (s.v. Idle *v.*). (Further examples, corresponding to *idle *v.* 4.)

1936 E. S. Gardner *Case of Stuttering Bishop* (1937) iii. 41 The ever-present throbbing undertone of sound.. from idling motors. **1968** [see *idle nozzle* s.v. *idle *sb.* 5].

Ido (ī·do). [In this language, = offspring.] An artificial language, based on * Esperanto, selected by the 'Delegation for the Adoption of an Auxiliary International Language' (founded at Paris in 1901), and made public in 1907. Hence **I·dist, I·doist,** a student or speaker of Ido; also *attrib.*

The official name is 'Linguo Internaciona di la Delegitaro (Sistema Ido)'.

1908 *Daily Chron.* 7 Sept. 7/2 The new language has been named 'Ido', and its inventor [M. de Beaufront, of Geneva] claims that it is easier to learn and is more methodical than Esperanto, its parent. **1916** H. G. Wells *Mr. Britling* I. v. § 10 'There would be no English, no Germans, no Russians. Just Esperantists.'..'Or Idoists,' said Herr Heinrich. **1922** O. Jespersen *Lang.* xviii. 347 Anyone who has written much in Ido [etc.]. **1926** *Encycl. Brit.* III. 906/1 In 1907 two Frenchmen, Messrs. Couturat and de Beaufront, produced a modification of Esperanto which they named simplified Esperanto. Owing to Esperantist protests, the 'linguo internaciona' was renamed Ido... The Idists claim that Ido is Esperanto rendered more scientific and natural. *Ibid.*, The chief Idist grammarian, Dr. Max Talmey, advocates what he calls 'Improved Ido' or 'Ilo' or 'Arulo'. **1928** O. Jespersen *Internat. Lang.* i. 24 The practical experiences of Volapükists, Esperantists and Idists..have shown that the fears of sceptics are groundless with regard to pronunciation. **1946** H. Jacob *On Choice of Common Lang.* 79 The British section of the Ido-union. **1947** —— *Planned Auxiliary Lang.* 57 Idists practising the language for many years.

idolize, *v.* Add: **1.** Also *absol.*

1919 M. K. Bradby *Psycho-Anal.* 65 For instance, one person may be prone to jealousy, another to idolise.

Idomenian, -enean (əidomī·niăn, əidomenī·ăn), *a.* and *sb.* [f. *Idomeneus*, f. Gr. Ἰδομενεύς, a king of Crete + -an.] **A.** *adj.* Of or belonging to a race imagined by Thomas Reid, an 18th-c. metaphysician, to have no sense but sight, and to believe that space has only two dimensions. **B.** *sb.* A member of this race.

1764 T. Reid *Inquiry Human Mind* vi. 252 'The Idomenians,' saith he, 'are many of them very ingenious, and much given to contemplation.' *Ibid.* 255 The geometry of the Idomenians agrees in every thing with the geometry of visibles. *Ibid.* 257 A person of great genius, who is looked upon as having had something in him above Idomenian nature. *Ibid.*, The Idomenian faculties were certainly intended for contemplation. *Ibid.* 258 Every Idomenian firmly believes, that two or more bodies may exist in the same place. **1871** A. C. Fraser *Life Berkeley* x. 400 The invisibility of that sort of distance can thus be proved even to the Idomenian. **1890** W. James *Princ. Psychol.* II. xx. 214 One of Reid's Idomenians would frame precisely the same conception of the external world that we do, if he had our intellectual powers.

idryl (i·dril, əil). *Chem. Obs.* exc. *Hist.* [a. G. *idryl* (C. Bödeker 1844, in *Ann. d. Chem. u. Pharm.* LII. 102), f. Idr(ialin 2 + -yl.] **a.** A mixture of fluoranthene and other hydrocarbons (see quot. 1952) orig. obtained from the mercury ores of Idrija, in north-west Yugoslavia, and thought to be a single compound. **b.** = *fluoranthene.

1845 C. Bödeker in *Chem. Gaz.* 15 Feb. 73 One of the products obtained in the working of the bituminous mercurial ores in Idria is the so-called *Stupp*, a soft black mass mixed with globules of mercury. It was supposed to contain idrialine, but in its stead another hydrocarbon was found, idryle C^3H, and a more accurate examination has proved that idrialine..is an oxide of idryle. **1866** H. Watts tr. *Gmelin's Hand-bk. Chem.* XVII. 477 Idryl. $C^{42}H^{14}$... The crystals are nearly colourless, with a tinge of yellowish-green... Idryl assumes a golden-yellow colour in cold oil of vitriol. **1877** *Chem. News* 16 Nov. 222/2 G. Goldschmidt, 'Idryl'. This body, found by Bödeker in Idria, is ascertained to consist of several hydrocarbons... Besides chrysen, pyren, anthracen, phenanthren, a new hydrocarbon, $C_{15}H_{10}$, was isolated, and receives the name originally applied to the mixture. **1892** Roscoe & Schorlemmer *Treat. Chem.* (new ed.) III. vi. 523 Fluoranthene, $C_{15}H_{10}$, is found in coal-tar, and is also identical with idryl, one of the constituents of 'stuppfett'. **1926** H. G. Rule tr. *Schmidt's Text-bk. Org. Chem.* ii. xvii. 513 Fluoranthene or Idryl, m.p. 110°, b.p. 250° (at 60 mm.). **1952** *Chem. Rev.* L. 483 In 1844 Boedeker..distilled these ores, with production of a similar hydrocarbon which he named idryl. Much later Goldschmiedt..showed..that idryl consisted of anthracene, phenanthrene, chrysene, pyrene, and a hitherto unknown hydrocarbon, $C_{16}H_{10}$, to which he gave the name idryl. Simultaneously with Goldschmiedt's discovery, Fittig and Gebhard..isolated..a hydrocarbon which was named fluoranthene and was claimed to be identical with Goldschmiedt's idryl.

‖ **id-ul-fitr** (i·d,ul,fit'r). Also (a better form) **id al-fitr.** [Arab. *'īd al-fiṭr.*] The Feast of breaking the Ramadan Fast, or Lesser Bairam, celebrated on the 1st of the month of Shawwāl: one of the two major festivals in Islam.

1734 G. Sale tr. *Koran* Prelim. Disc. vii. 150 The first of them is called, in Arabic, *Id al fetr*, i.e. The feast of breaking the fast, and begins the first of Shawâl, immediately succeeding the fast of Ramadân. **1832** G. A. Herklots tr. *Shurreef's Qanoon-e-Islam* xxiv. 261 The *Eed-ool-fitr* (or feast of alms), called also the feast of *Rumzan*, is observed on the first day of the month *Shuwal.* **1836** E. W. Lane *Acct. Manners & Customs Mod. Egyptians* II. 238 It is also called *'Eed el-Fitr* (or the Festival of the Breaking of the fast). **1896** E. Sell *Faith of Islām* (ed. 2) vi. 319 A very usual form of the Khuṭbah of the 'Īdu'l-Fiṭr which is preached in Arabic is as follows... We praise and thank him for the 'Īdu'l-Fiṭr, that great blessing, and we testify that beside him there is no God. **1909** *Daily Chron.* 18 Oct. 6/4 Undisturbed by the hum of the traffic in Holborn, the Imam lifted up his voice, and..droned out the Idul-Fitr prayer. **1970** *Cambr. Hist. Islam* II. 907 *'Īd al-Fiṭr*, 'the Feast of the Breaking of the Fast' or *al-'Īd al-Ṣaghīr* (the Small Feast), held after the end of Ramaḍān, the month of fasting. **1972** *Mainichi* (Japan) *Daily News* 6 Nov. 11/4 (Advt.), Kobe Muslim Mosque. Idul-Fitr prayers will be held at 9.00 a.m. on Tuesday, November 7th.

Idumæan (əidiumī·ăn, id-), *sb.* and *a.* Also -ean. [f. L. *Idūmæa*, a. Gr. Ἰδουμαία, f. Heb. *Edōm* Edom + -an.] **A.** *sb.* A member of the race inhabiting Idumæa. **B.** *adj.* Of or belonging to Idumæa or Edom, an ancient kingdom situated between Egypt and Palestine.

c **897** Ælfred tr. *Gregory's Pastoral Care* (1871) 387 Ezechiel...cwæð ðæt hie wolden weorðan forlorene & oferwunnene mid orsorgnesse, swa swa Idumeas wæron. **1382** Wyclif 3 *Kings* xi. 1 Forsothe kyng Salomon to brennyngly louede many hethen wymmen, the douȝter forsothe of Pharao, and Moabitis, and Amonytis, and Ydumees, and Cidonees, and Ethees. **1537** tr. *Original & Sprynge of Sectes* 55 Idumeans. **1602** Lodge tr. *Josephus' Workes* 679 Simon the sonne of Cathla,..spake vnto the Idumæans from a tower, and commanded them to cast down their armes. **1737** W. Whiston tr. *Josephus' Works* 381 He fell upon the Idumeans, the posterity of Esau, at Acrabattene. **1838** *Penny Cycl.* XII. 437/1 The Idumæans marched to the assistance of Jerusalem when it was besieged by Titus. **1863** *Chambers's Encycl.* V. 343/1 The family was of Idumean descent. *Ibid.* 499/2 The Idumæan, Antipater. **1880** *Encycl. Brit.* XI. 754/1 Herod was the name of a family of Idumæan origin. **1936** C. Roth *Short Hist. Jewish People* viii. 78 The age-long enmity between the Jews and Idumaeans (Edomites), which had found its expression in the raids of Judah the Maccabee, culminated in the final subjection of the country. **1968** P. Needham tr. *Gelzer's Caesar* v. 251 The Idumaean Antipater, the minister of the Jewish High Priest Hyrcanus.

idyll, idyl (and derivs.). Now also commonly with pronunc. (id-). (See Fowler *Mod. Eng. Usage* (1926) 253, R. Bridges in *S.P.E. Tract* (1929) xxxii. 403, and A. C. Gimson *Everyman's Eng. Pronouncing Dict.* (1969) 241.)

idyllic, *a.* Add: Hence **idy·llicism.**

1941 L. MacNeice *Poetry of Yeats* v. 86 This snob idyllicism. **1964** *Economist* 25 Jan. 294/1 There is little idyllicism left.

Idzo, var. *Ijo *sb.* and *a.*

ieie (ī·e,ī·e). Also ieie. [Hawaiian.] A climbing screw-pine of the genus *Freycinetia*, esp. *F. arborea*, whose prop-roots yield a fibre. Cf. Kie-kie.

1883 C. F. G. Cumming *Fire Fountains* I. viii. 128 Vines without number, the most notable being one called the *ie.* **1903** R. C. L. Perkins in *Fauna Hawaiiensis* I. 400 All those [birds] that were utilised could readily be caught by a bait of flowers, excepting perhaps the Ou, and this bird probably not less easily on account of its fondness for the ripe ieie. **1915** W. A. Bryan *Nat. Hist. Hawaii* xvi. 211 Another plant peculiar to the lower woods..is the ieie, a climbing shrub. **1970** S. Carlquist *Hawaii* xvii. 337 (*caption*) The flowers of the ieie are surrounded by short leaves bright orange in color.

if, *conj.* (*sb.*). Add: **3. a.** *if I were you*: (examples).

1814 Jane Austen *Mansf. Park* I. vi. 109 If I were you, I should not think of the expense. *Ibid.* 112 'Mr Rushworth,' said Lady Bertram, 'if I were you, I would have a very pretty shrubbery. One likes to get out into a shrubbery in fine weather.' **1869** Trollope *Phineas Finn* II. xli. 30 'Of course you'll go,' said Phineas. 'I should, if I were you.' **1974** D. Gray *Dead Give Away* vi. 65 I'd lay off stirring up trouble for a bit if I were you.

4. a. (Later examples.)

1848 [see Twitch *v.*¹ 8 a). **1965** *New Statesman* 16 Apr. 598/3 If Mr Stewart is top of the Tory pops, a few ministers are also high up in the charts. **1967** *Listener* 17 Aug. 205/1 If my father's people were mill-workers, my mother's people were agricultural workers. **1969** *Ibid.* 24 Apr. 585/1 If Mozart was a life-long admirer of J. C. Bach, his views on Clementi were disparaging, to put it mildly.

6. Also *if any* (see Any *a.* 2); *if anything*, if in any degree, perhaps even; *if not, why not* (see Why *adv.* 4 b).

1711 [see Any *a.* 2]. **1836** G. C. Lewis *Let.* 15 July (1870) 54 The writer says that the wages are not.., if anything, the King's wages are rather the lowest. **1861** H. Spencer *Social Statics* xxviii. 359 If anything, we were comparatively deficient in some respects. **1873** Lytton *Kenelm Chillingly* II. iv. vi. 123 Shall I ever be in love? and if not, why not? **1909** 'Vaile *Mod. Golf*' v. 92 If anything, touch the grass first. **1921** *Wireless World* IX. 187/1 L. M. T...asks..(3) If a diagram he sends is correct and, if not, why not. **1931** Belloc *Hist. Eng.* IV. ii. i. 260 If anything the idea of King-worship was even more exaggerated in her reign than in the case of Henry or of Edward. **1944** R. A. Esdaile *St. Martin in the Fields* ii. 30 If anything the destruction was greater than in 1547. **1967** *Listener* 2 Feb. 261/1 At the start of every month I have to send him an account..of my earnings, if any.

7. (Later examples.)

1846 *Swell's Night Guide* 49 And, so help me never! if his nibs didn't go and dossed with her the same night. **1914** *Sat. Even. Post* 4 Apr. 10/1 'If it ain't Frisco Red!' exclaimed one prone figure. **1925** T. Dreiser *Amer. Trag.* I. xvii. 145 'Oh, Gee, well, ain't that the limit?'..'If this aren't the grouch.' *Ibid.* II. iii. 184 Well, by jing, if it ain't Tom.

8. c. as if: also followed by a clause containing the present tense.

Further examples in Fowler *King's Eng.* (ed. 3) pp. 165–6.

1693 Dryden tr. *Persius' Satires* I. 61 As if 'tis nothing worth that lies conceal'd. **1751** H. Fielding *Amelia* II. vii. 110 [illegible] **1862** Thackeray *Philip* II. viii. 173 As if a coarse woman..has a right to lead a guileless nature into wrong! **1963** D. Storey *Radcliffe* xxxvi. 367 As if everything that goes on is ..simply imitating some distant and incoherent idea.

g. if and when, in reference to a future time but with a strong element of doubt.

1926 Fowler *Mod. Eng. Usage* 254/1 *If & when.* Any writer who uses this formula lays himself open to entirely reasonable suspicions on the part of his readers... There is the suspicion that he is a timid wordsman who thinks he will be safer with a second word in his left hand. **1946** G. B. Shaw *Matter with Ireland* (1962) 283 If and when the situation becomes grave enough to convince Americans that I have no alternative, I will recognize you? **1963** *Amer. Speech* XXXVIII. 255 If and when a study of local words in Missouri appears, we will be able to trace the same patterns in that state too.

10. *Comb.* **if-clause,** a clause of condition or supposition introduced by the word *if*; **if-shot, if-stroke** *Cricket*, 'a stroke considered to be unsound and likely to cause the batsman's dismissal if the ball is hit' (Lewis).

1893 Turner & Hallidie *Primary Eng. Gram.* 113 The if-clause tells us when, or under what circumstances, desert would be a paradise. **1904** C. T. Onions *Adv. Eng. Syntax* 37 A Complex Sentence consisting of an Adverb Clause of Condition (the If-Clause, sometimes called the Protasis) and a Principal Clause (sometimes called the Apodosis) is called a Conditional Sentence. **1926** Fowler *Mod. Eng. Usage* 576/2 It [*sc.* the word *were*] is entirely out of place in an *if*-clause concerned with past actualities & not answered by a *were* or *would be* in the apodosis. **1964** *English Studies* XLV. 85 The meaningful weight of the complete sentence may move from the *if*-clause to the headclause. **1966** G. N. Leech *Eng. in Advertising* vi. 61 Favourite openings are..*if* clauses. **1897** K. S. Ranjitsinhji *Jubilee Bk. Cricket* iv. 165 In its worst form this [slipping the ball] is commonly known nowadays as the 'if-stroke'. Originally it was called the 'but-stroke', after its great exponent, the Sussex wicket-keeper [Butt]; but some wag suggested that it should be called in preference the 'if-stroke', because if you hit the ball you are nearly sure to be out. **1920** D. J. Knight in P. F. Warner *Cricket* 36 It is essentially an 'if' shot, and must, to a certain extent, be unsound. **1922** *Cricketer Ann. 1922–23* 62 We will not indulge in 'cowshots' or 'ifshots'.

ifé (i·fe). Also **ife.** [Native name.] A tropical African plant, *Sansevieria cylindrica*, of the family Liliaceæ, which yields a fibre formerly used as a substitute for hemp.
1859 *Curtis's Bot. Mag.* LXXXV. 5093 About three years since there were received at the Foreign Office, and transferred to the Admiralty, samples of a peculiar fibre and cordage under the name of Ifé, said to be derived from a new plant at the Portuguese settlement, Angola. **1866** Lindley & Moore *Treas. Bot.* II. 617/2 Ife. An Indian name for *Sanseviera* [sic] *cylindrica*. **1892** P. L. Simmonds *Commercial Dict. Trade Products* (rev. ed.) 462/1 Ife, a name in India for the *Sansievera* [sic] *cylindrica*, a plant useful for its fibre.

Ife (ī·fe). The name of a town in Western Nigeria, the religious centre of the Yoruba people, used *attrib.* to designate the art of the Yoruba people, *spec.* the bronzes and terracottas of which the first examples were found there in 1912.
[**1913** R. Blind tr. *Frobenius' Voice Afr.* I. xiv. 293 Various comparatively coarse stone monuments are raised above the earth in Ifé. Below it we found old sculptured stonework..bronze work far exceeding that of Benin in its perfection and representative skill, and at last some terra-cottas amazing to those who inspect them.] **1939** *Burlington Mag.* LXXV. 152/2 The Ife terra-cottas and bronzes show the most diverse racial characteristics, from the Semitic..running through all the ranges to the Negroid. **1949** Fagg & Underwood in *Man* XLIX. 2 (*caption*) The exhibition of Ife bronzes at the British Museum, July–September 1948. **1959** *Chambers's Encycl.* II. 609/2 All the Ife heads are technically and artistically on the highest level. The portrait-like treatment, with absolutely correct proportions, is extremely realistic and the modelling is of a beautiful sensitiveness and vitality. As works of art they rank even higher than the more impersonal and conventional Benin heads. **1960** *Times* 5 Oct. 18/7 The features of the Ife heads suddenly appear again.

iff. A written form of abbreviation of the phrase 'if and only if', always read as 'if and only if', used in *Math.* and *Logic* to introduce a condition that is necessary as well as sufficient, or a statement that is implied by and implies the preceding one.
1955 J. L. Kelley *Gen. Topology* vii. 232 *F* is equicontinuous at *x* iff there is a neighborhood of *x* whose image under every member of *F* is small. **1961** R. R. Stoll *Sets, Logic, & Axiomatic Theories* i. 5 Two sets are equal iff (if and only if) they have the same members. **1964** T. O. Moore *Elem. Gen. Topol.* i. 2 We use 'iff' as an abbreviation for 'if and only if'. **1965** B. Mates *Elem. Logic* viii. 137 $\phi \in \Delta$ iff ϕ is derivable from Δ. **1971** G. Hunter *Metalogic* 16 Hereafter we abbreviate 'if and only if' to 'iff'. **1972** *R. Inst. Philos. Lect.* V. 34 An integer *n* is prime iff the only integers which divide it without remainder are itself and one.

iffy (i·fi), *a.* orig. *U.S.* [f. If *conj.* + *-f-* + -y[1].] Of a question, proposal, prospect, etc.: full of 'ifs', contingent, doubtful.
1937 *World this Week* 9 May, Very 'iffy', Mr. Roosevelt might characterize such talk. **1941** *Chicago Daily News* 17 Jan., The President had been asked the status of some proposal, or of some event..whether this event was likely to happen... The president replied that the whole thing was 'iffy'. **1941** Baker *Dict. Austral. Slang* 37 *Iffey*, uncertain, unsound, 'chancey'. **1941** *Time* 14 July 15/3 His chances of pulling out a plum were rather iffy. **1963** *Weekly News* (Auckland) 5 June 39/1 This gamble..is far too 'iffy' to be classed as a prudent business procedure. **1963** *Times* 12 June 12/7, I have no comment to make on it. As Mr. Roosevelt said, this would be giving hypothetical answers to 'if-y' questions. **1971** E. Fenwick *Impeccable People* xiii. 72 We knew this was rather an iffy tenant, morally speaking, before we rented. *Ibid.* 73 Now..she looks a little iffier. **1973** *Listener* 20 Sept. 364/2 If the Police do their job, if the new law is effectively enforced, these are very 'iffy' points indeed.

Ifugao (i·fugau), *sb.* and *a.* [Native name.] **A.** *sb.* The name of a people of northern Luzon in the Republic of the Philippines, of a member of this people, and of their Malayo-Polynesian dialect. **B.** *a.* Of or pertaining to the Ifugao people. Also **I·fugaon.**
1859 J. Bowring *Visit to Philippine Islands* viii. 171 This race [*sc.* Itaneg] has a mixture of Chinese blood, the Ifugaos of that of the Japanese. **1875** F. Jagor *Trav. Philippines* v. 55 [Dialects of] Nueva Vizcáya: Gaddan, Ifugao, Ibilao, Ilongote. **1900** F. H. Sawyer *Inhabitants Philippines* I. xxviii. 271 The Ifugaos, who bear a strong resemblance to the Japanese, inhabit a territory in central Nueva Vizcaya. *Ibid.* 272 The Spaniards built and garrisoned a chain of forts in the Ifugao territory to keep them in order. **1901** A. G. Robinson *Philippines* x. 166 Following these tongues [*sc.* Visayan, Ilocano]..are the Pampangan..Ibanag, Ifugao, Ibilao [etc.]... Most of them are dialects, though sufficiently different to be practically unintelligible to those of other groups. **1914** D. C. Worcester *Philippines Past & Present* xx. 535, I assured him that we were friends who had come to get acquainted with the Ifugaos. *Ibid.* 536 An Ifugao climbed down from above. *Ibid.* 537 My mission to the Ifugao country was to establish kindly relations with the people. **1944** W. E. Harney *Taboo* (ed. 3) 21 The Melville Island native shield..may have come from the shields of the 'Ifugao' of the Philippine Islands. **1964** E. A. Nida *Toward Sci. Transl.* iii. 50 In Ifugao in the Philippines and in Moré, a language of the Haute Volta, 'hair' and 'feathers' are included under the same term. **1968** R. Nelson *Philippines* 131 (*caption*) The Ifugaos are talented and artistic woodcarvers. *Ibid.* 133 (*caption*) An Ifugaon peasant..leaving his tobacco field.

igarape (igarapē·). Also **-apé, -ipe.** [Pg.] In South America, a tributary or other water-channel wide enough for a canoe.
1853 A. R. Wallace *Trav. Amazon* ii. 41 Sometimes we would start down the igaripé in the montaria, not returning till late in the afternoon. *Ibid.* vii. 168 The word *igaripé*, applied to all small streams, means 'path of the canoe'. **1860** Mayne Reid *Odd People* 38 Travelling can only be done by water, either upon the great rivers, or by the narrow creeks (igaripes) or lagoons. **1866** L. Agassiz in *Atlantic Monthly* July 58/1 We turned..into a narrow stream, which has the character of an Igarapé in its lower course. **1933** *Discovery* Jan. 24/2 It is intersected by innumerable water-ways, the igarapés, along which the tide ebbs and flows. **1944** S. Putnam tr. *E. da Cunha's Rebellion in Backlands* ii. §1. 59 The filling of the river [*sc.* Amazon] brings a stoppage of life. Caught in the meshes of the *igarapés*, man displays a rare stoicism.

Igbirra (i·gbirǎ). Also **Igbira.** [Name of an area in Kabba province, Northern Nigeria.] The name of a tribe in Northern Nigeria; a member of this tribe; also, their language.
1863 J. Thomas *Jrnl.* 24 July in *Jrnls. & Notices Native Missionaries River Niger* (1864) 88 He went to our school-room and offered a prayer in Igbira. **1880** Mrs. G. Sturge tr. *Burdo's Niger & Benueh* xv. 265 (*heading*) Attack by the Igbiras. **1883** R. N. Cust *Sk. Mod. Lang. Afr.* I. xi. 226 Williams, a Negro Catechist..has lately made Translations into Igbira, some of which are used in Divine Service. **1925** C. K. Meek *Northern Tribes Nigeria* II. vii. 114 The Igbira mummify the body by pouring in gin or beer. **1958** J. S. Coleman *Nigeria* i. 24 Some of these [pagan tribes] were organized into unconsolidated village groups..others (for example, Igbirra, Bede, Bachama, and Igala) had achieved a certain measure of tribal unity. **1967** *Encycl. Brit.* XIII. 233/1 The native authority for Igbirra division, inhabited mainly by a tribe of the same name, consists of an elected council representative of the five Igbirra clans.

Igbo: see ***Ibo** *a.* and *sb.*

iggerant: see ***ignorant** *a.* 5.

‖ **iggri, iggry** (i·gri), *int.* Also **iggoree** *adv.* [Representing Egyptian colloq. Arab. pronunc. of *ijri*, imper. of *jarā* to run.] Hurry up! Also as *sb.* in phr. *to get an iggri on.*
1919 W. H. Downing *Digger Dial.* 29 *Iggoree*, quickly. **1925** Fraser & Gibbons *Soldier & Sailor Words* 127 *Iggry* (*iggri*),..a phrase in use in the Egyptian Expeditionary Force. 'Iggry Corner' at Bullecourt was so named by Australian troops who had been stationed in Egypt, as being an exceptionally dangerous locality from shell fire, where it was necessary to move rapidly. **1946** *Penguin New Writing* XXVIII. 173 'Come on, Chalky,' he pleaded, 'get an iggri on!'

igloo. Add: **3.** *transf.* A small dome-shaped building or construction (see quots.).
1956 W. A. Heflin *U.S. Air Force Dict.* 262/2 *Igloo*.., a dome-shaped or rounded structure, usually made of reinforced concrete and earth, normally used for storage of explosives. **1969** *Jane's Freight Containers 1968–69* 438 (*caption*) A moulded fibreglass igloo unit as used by Delta Airlines. **1970** *Observer* 19 Apr. 6/8 The nerve gas is to be stored in 100 concrete igloos. **1970** *New Scientist* 21 May 382/2 The main advantage is that the igloos, unlike most other emergency shelters, can be made on the site; hence there are no major transport problems. **1972** *Evening Telegram* (St. John's, Newfoundland) 28 June 12/7 Igloos are spun-glass shells shaped to fit the bellies of jet-liners or the fuselages of all-cargo DC-8s. The use of the system is one of the main reasons for the rapid growth in air cargo operations.

ignimbrite (i·gnimbrəit). *Geol.* [f. L. *ign-is* fire + *imbr-is*, *imber* shower of rain, storm-cloud + -ite[1].] Any pyroclastic rock, typically a welded tuff, deposited from or formed by the settling of a *nuée ardente.
1932 P. Marshall in *N.Z. Jrnl. Sci. & Technol.* XIII. 200 The type of rocks formed in this way varies greatly, but it is suggested that they should all be included in a separate group, for which the name 'Ignimbrite' seems satisfactory. **1959** A. McLintock *Descr. Atlas N.Z.* 11 Early in the Pleistocene Period, huge eruptions there formed a plateau of rhyolitic rock known as 'ignimbrite', some 10,000 square miles in extent. **1962** E. A. Vincent tr. *Rittmann's Volcanoes* ii. 80 Ignimbrites, or welded tuffs, are formed in enormous quantities from overflowing glowing clouds in fissure eruptions of very acid magmas. The incandescent ash particles are intimately fused to one another and attain a largely pseudo-liquid state, especially in the deeper portions of the deposit, so that the massive rock which results often shows columnar jointing and is easily confused with a rhyolitic lava. **1969** C. Ollier *Volcanoes* vii. 73 Ignimbrite will be used here in the sense of Cook (1966). This is a rock unit term, and should not be used as a petrological term: several petrological types can give rise to ignimbrites though rhyolite and andesite predominate. Neither does the term imply any post-depositional alteration such as welding, though this may be present. **1970** *Nature* 12 Sept. 1125/1 This..is matched in volume only by the large welded tuff sheets of the central North Island of New Zealand, the type locality for ignimbrite.

ignitability, -ibility. (Later examples.)
1925 Coward & Wheeler *Ignition of Firedamp* 20 The ignitibility of any mixture by an electric spark is inversely proportional to the rate of movement of flame in that mixture. *Ibid.* 21 The relative ignitibilities of a wide range of inflammable methane-air mixtures. **1953** Kirk & Othmer *Encycl. Chem. Technol.* XI. 326 Ignition temperature, ignitibility, stability, and hygroscopicity are important in determining the certainty of functioning [of pyrotechnic compositions]. **1966** *Lancet* 24 Dec. 1404/1 Comparing circumstances in pure oxygen at 1 atmosphere with those in air at the same pressure one finds that there is a thousandfold increase in the ignitability of most substances.

ignite, *v.* Add: **4.** *trans.* and *intr.* To strike (an arc).
1917 *Proc. IRE* V. 298 It is assumed that the arc ignites every third time on the edges. **1919** E. W. Stone *Elem. Radiotelegr*]. (1920) viii. 152 The arc is now 'struck' or ignited by this potential. **1933** [see *ignitron].

igniter. Add: **b.** (Further examples.)
1886 D. Clerk *Gas Engine* viii. 205 The first few explosions cause a condensation of water upon the points and the spark then fails... The igniters then require to be uncoupled. **1902** *Min. Proc. Inst. Civil Eng.* CXLIX. 453 The hydrocarbon vapour sucked in from the carburettor..causes the igniter to glow brightly. **1905** H. J. Spooner *Motors & Motoring* 46 The principal features of the sparking plug or igniter. **1907** *Westm. Gaz.* 18 Nov. 10/2 The low-tension plugs and igniters are neatly placed in the sides of the cylinder heads. **1950** J. V. Casamassa *Jet Aircraft Power Syst.* xiv. 206 When the engine reaches a predetermined speed, the starter, ignition coils, and the flame igniters are automatically cut out. **1950** *Sci. News* XV. 78 One class [of propellants], called hypergols, react as soon as they come together, a property which eliminates the need for an igniter of some sort. **1954** J. W. T. Walsh *Textbk. Illuminating Engin.* (ed. 2) xii. 172 A special form of distance controller..switches on the igniter just before the gas is admitted to the burner. **1962** F. I. Ordway et al. *Basic Astronautics* x. 417 The solid rocket motor is composed of a cylindrical combustion chamber, a nozzle, a solid propellant grain, and an igniter.
2. *Electronics.* Also **ignitor.** A small rod of some refractory material that dips into the pool of mercury in an ignitron and serves as an auxiliary anode for restriking the arc in each cycle.
1933 *Electronics* VI. 166/1 At the present stage of development [of the ignitron] the instantaneous current required by the igniter to start the arc reliably is from 5 to 30 amperes. **1937** A. V. Eastman *Fund. Vacuum Tubes* iv. 107 One of the major problems in the design of Ignitrons is to make certain that the anode will pick up the discharge as soon as the igniter strikes. **1952** Young & Bueche *Fund. Electronics & Control* vi. 253 Because the ignitor can be controlled to cause initiation of the arc at any point in the positive half cycle, the average anode current flow can be controlled as in the case of thyratrons. **1962** C. Susskind *Encycl. Electronics* 361/1 In pulse-type circuits where capacitors are discharged, the ignitor can be made to start the discharge in less than 1 μsec.

ignition. Add: **2. b.** *spec.* A means of producing the spark in an internal-combustion engine; an ignition system, or the device that activates it. (Further examples.)
1906 *Daily Chron.* 12 May 3/5 All good cars nowadays have the two independent ignitions, the accumulator and coil, and the magneto driven by a cog-wheel on the engine shaft. **1961** W. Hartley *Motorist's Home Repair Bk.* v. 70 There is no need to do more than insert the starting handle and have the engine turned over slowly (with the ignition switched off, of course!) while a petrol-moistened piece of silk is held against the commutator. **1972** J. Gores *Dead Skip* (1973) xiv. 97 A..Mustang with the driver's window open and the key in the ignition.
3. *Electronics.* The striking or initiation of an arc.
1919 E. W. Stone *Elem. Radiotelegr.* (1920) viii. 152 The potential existing across the arc at this instant is termed the extinction voltage as distinguished from that at ignition. It is less than the ignition voltage. **1927** J. G. Tarboux *Electr. Power Equipm.* iv. 128 The ignition anode is now drawn up by the spring..and at the point of rupture with the mercury an arc is started. **1945** *'Electr. Engineer' Ref. Bk.* x. 3 The action of the mercury-arc rectifier requires the cathode spot to be produced first by some auxiliary device before the arc can be established. This is generally effected by drawing an arc between the cathode and a small auxiliary anode, a process termed ignition. **1962** *Newnes Conc. Encycl. Electr. Engin.* 380/1 The ignitron differs from the other mercury-arc rectifiers..in the method used for ignition of the arc.
4. *attrib.* and *Comb.* (esp. in terms relating to internal-combustion engines and motor

vehicles), as †*ignition box*, †*chamber*, *circuit*, *system*; **ignition advance**, the extent to which the moment when the spark occurs precedes the moment of greatest compression of the mixture in a cylinder of an internal-combustion engine, usu. expressed in terms of the angle between the two corresponding positions of the crank-shaft; **ignition coil**, an induction coil in an internal-combustion engine for converting a low-voltage current into one whose voltage is sufficient to produce the spark; **ignition key**, a key for operating an ignition switch in the form of a lock; **ignition lag**, the delay following the occurrence of the spark (in a spark-ignition engine) or the injection of the fuel (in a compression-ignition engine) before the pressure first begins to rise as a result of combustion; **ignition plug** = **sparking plug*; **ignition point** = **ignition temperature*; **ignition rating** (see quot. 1940); **ignition switch**, the switch by means of which the ignition circuit of a motor vehicle may be closed or opened (thereby allowing the engine to be started, or stopping it); **ignition temperature**, the lowest temperature at which a combustible substance in air will ignite and continue to burn; **ignition tube**, (*a*) *Chem.*, a small cylindrical vessel of heat-resistant glass in which a substance may be heated for purposes of analysis; †(*b*) a *hot tube* (*hot *a.* 12 c) in some early internal-combustion engines.
1908 H. E. Wimperis *Internal Combustion Engine* viii. 282 It is permissible to set the ignition to take place..slightly before the dead centre... This reduces the arc of ignition advance throughout which the magneto is called upon to generate an effective spark. **1946** A. W. Judge *Mod. Petrol Engines* ii. 32 For weaker mixtures the flame rates are lower so that a greater ignition advance is necessary than for correct or slightly rich mixtures. **1900** G. D. Hiscox *Horseless Vehicles* xi. 184 The ignition of the charge is effected by heating the nickel tubes projecting about 2½ inches from the rear ends of the cylinders into the ignition box. **1894** B. Donkin *Text-bk. Gas, Oil, & Air Engines* I. xi. 140 At a given moment, a zig-zag passage in the slide valve is brought opposite the ignition chamber, and opens communication between it and the admission port into the cylinder. **1907** R. B. Whitman *Motor-Car Princ.* vi. 74 While wire is sometimes used on ignition circuits for the return as well as the lead, the most usual method is to utilise the metal of the engine to return the current to its source. **1900** G. D. Hiscox *Horseless Vehicles* vii. 127 (*heading*) Electric ignition coils. **1953** I. Frazee et al. *Automotive Fuel & Ignition Syst.* vii. 247 The ignition coil must produce sufficient voltage and current to ignite the fuel mixture at the various engine speeds encountered. **1933** D. L. Sayers *Murder must Advertise* ix. 155, I have the ignition keys of both cars. **1934** *Punch* 3 Oct. 388/3 Stepping into Humbottle's car am surprised to hear burst of gruff laughter... This probably explained by absence of ignition-key. **1962** J. Braine *Life at Top* xix. 190 She turned the ignition key; the car jerked forward convulsively, then stopped. 'Put it in neutral first,' I said. **1972** *Police Rev.* 1 Dec. 1577/3 They removed the ignition keys and sent a message by radio for an officer to attend with the breath test equipment. **1932** *Fuel* XI. 187/1 At a definite temperature the ignition lag becomes infinitely great and ignition no longer takes place. **1900** G. D. Hiscox *Horseless Vehicles* vii. 130 (*heading*) Electric ignition plug. **1902** A. C. Harmsworth et al. *Motors* viii. 161 The ignition plugs may be examined to see that they are not coated with oil. **1933** I-B. O. Sneeden *Introd. Internal Combustion Engineering* ix. 164 The number of ignition plugs per cylinder is not limited to one. Two are often used. **1887** *Proc. Amer. Acad. Arts & Sci.* XXII. 483 In..trials, which I made with sound pine wood, I soon found that the ignition point was greatly affected by the way in which the sample was heated. *Ibid.* 486 The ignition point of the pine wood varied from 330°..to 218°. **1922** Glazebrook *Dict. Appl. Physics* I. 338/2 The ignition points of oils are frequently required in connection with internal combustion engine problems. **1922** *Wireless World* IX. 727/1 Extensive use was made of portable accumulators..on motor cycles. The intermittent discharge through a trembler contact gave an apparent duration of double the hours at..a certain current. This led to the ignition rating of portable cells which is..misleading..for wireless work. **1940** *Chambers's Techn. Dict.* 437/2 *Ignition rating*, a special rating (in ampere-hours, q.v.) employed for accumulators used for supplying ignition systems; it is generally twice the continuous rating at a low discharge rate. **1952** *Ellery Queen's Mystery Mag.* XIX. 31/2 There was a car parked in front of the entrance, a Ford coupe, 1937... The key was in the ignition switch. **1902** A. C. Harmsworth et al. *Motors* viii. 159 The magneto ignition system..consists of a magneto-electric rotary machine, combined with a series of mechanical contact-breakers. **1943** A. P. Fraas *Aircraft Power Plants* viii. 140 The so-called battery ignition system has been used almost universally in automobiles... All the larger and many of the smaller aircraft engines make use of magneto ignition systems. **1956** H. E. Milburn *Motor-Cars To-Day* vi. 115 A normal ignition system consists of an induction coil, a condenser, a contact breaker, a distributor, a sparking plug for each cylinder of the engine, a switch, and the necessary connecting wires. **1881** *Jrnl. Iron & Steel Inst.* II. 679 (*heading*) Ignition temperature of mixed gases. **1897** F. Grover *Pract. Treat. Mod. Gas & Oil Engines* xix. 196 (*heading*) Ignition temperatures of explosive gaseous mixtures. **1968** *Fuel* XLVII. 119 According to D. W. van Krevelin the ignition temperature of coal depends on experimental conditions such as furnace design, way of heating, particle size, oxygen concentration and coal rank. **1874** F. Clowes *Elem. Treat. Pract. Chem.* II. 30 (*heading*) Small ignition tubes. **1890** W. Robinson *Gas & Petroleum Engines* vii. 229 The average life of ignition tubes in the Differential was 180 hours, and in the Cycle engine 120 hours. **1962** A. Atkinson *Compl. Pract. Chem.* xii. 272 Heat a little of the substance in an ignition tube or in a small dry test-tube until no further change occurs.

ignitor, var. *igniter 2.

ignitron (ignəi·trǫn). *Electronics.* [f. Ignit(e *v.*, Ignit(ion + *-t)ron.] A kind of mercury-arc rectifier capable of handling large currents and having a pool cathode, a single anode, and an igniter to initiate the arc afresh in each cycle (the timing of this being used to control the output).
1933 D. D. Knowles in *Electronics* VI. 164/1 The Ignitron* is the latest arrival in the rapidly growing family of industrial electronic tubes. It is a tube in which conductivity is established by igniting the arc. [**Note*] Pronounced like the word 'ignite'. **1951** *Engineering* 14 Sept. 323/3 The ignitrons, which are water-cooled,..act as an electronic contactor which controls the current in the primary of the welding transformer. **1962** *Newnes Conc. Encycl. Electr. Engin.* 380/2 The larger ratings of ignitron are constructed with a double-walled stainless steel envelope, water being circulated in the jacket so formed for cooling. **1967** H. Cotton *Adv. Electr. Technol.* xv. 726 Typical applications of the ignitron are (*a*) small-power spot welding, (*b*) large-power control of variable-speed d.c. motors, such as certain types of colliery winder.

ignorance. Add: **3.** (In full *the time* or *days of ignorance*; tr. Arab. *jāhilīyah* state of ignorance, f. *jāhil* ignorant.) The period of Arabian history previous to the teaching of Muhammad.
1788 Gibbon *Decl. & F.* V. l. 185 Of the time of ignorance which preceded Mahomet, seventeen hundred battles are recorded by tradition. **1895** A. Menzies *Hist. Relig.* 213 The Arabs called the period before Islam the 'time of ignorance'; in that period they considered their race had no history. **1904** W. P. Ker *Dark Ages* 14 The student of heroic poetry may admire the temper of the Arabian Dark Ages—the 'Ignorance'. **1937** P. H. Hitti *Hist. Arabs* vii. 87 The term *jāhilīyah*, usually rendered 'time of ignorance' or 'barbarism', in reality means the period in which Arabia had no dispensation, no inspired prophet, no revealed book.

ignorant, *a.* Add: **1. b.** (Later examples.)
1875 'Mark Twain' in *Atlantic Monthly* Jan. 71/1 This fellow had..an ignorant silver watch and a showy brass watch chain. **1892** —— *Amer. Claimant* Pref., It ought to be the ablest weather that can be had, not ignorant, poor-quality, amateur weather.
5. *dial.* and *colloq.* Ill-mannered, uncouth.
¶ Sometimes written as *iggerant* in imitation of vulgar speech.
1886 R. E. G. Cole *Gloss. Words S.W. Lincs.* 71 *Ignorant*, ill-mannered. **1886** F. T. Elworthy *West Somerset Word-Bk.* 363 *Ignorant*, wanting in manners. The usual description of a rough, uncouth lout. **1946** K. Tennant *Lost Haven* (1947) xvii. 273 He used the word 'ignorant' in the country sense of knowing nothing of good manners. **1965** *Listener* 22 July 137/3 He writes what he *thinks* 'a Puerto Rican' is thinking (very elemental, Latin, iggerant, dumb, baffled, passionate). **1966** 'L. Lane' *ABZ of Scouse* 49, I jes' can't stan' that feller, 'e's plain bloody 'iggerant. **1968** *New Society* 22 Aug. 266/1 Ignorant, meaning 'bad-mannered', is non-U.

‖ **ignotum per ignotius** (ignǭ·tǝm pǝɪ ignǭ·ʃiǝs or ignǭ·tiǝs). [late L., lit. the unknown by means of the more unknown.] An attempt to explain what is obscure by something which is more obscure, leading to 'confusion worse confounded'.
c **1386** Chaucer *Canon's Yeoman's Tale* 1457 And Plato answerde vnto hym anoon, 'Take the stoon that Titanos men name.' 'Which is that?' quod he. 'Magnasia is the same,' Seyde Plato. 'Ye, sire, and is it thus? This [is] *ignotum per ignocius* [v.r. *ignotius*]. What is Magnasia, good sire, I yow preye?' *c* **1450** Lydgate *Secrees* (1894) 588 With goldeyn Resouns in taast moost lykerous, Thyng per ignotum prevyd per ignocius. **1584** R. Scot *Discov. Witchcr.* III. xvii. 67 Confuteth that opinion by a notable reason, called *Petitio principij*, or rather, *Ignotum per ignotius*. *a* **1734** R. North *Examen* (1740) II. i. 28 When he drops his own Authority, and brings Fact to confirm all, the vouching that Fact by his own pure Parole, is a Cheat termed *Ignotum per ignotius*. **1888** *Athenæum* 22 Dec. 843/2 When Arabic names are twisted and mis-copied..the identification of sites resolves itself into a case of 'ignotum per ignotius'. **1931** *Times Lit. Suppl.* 30 July 590/3 [He] even goes so far on one occasion as to explain an Egyptian *chiaoush* as being analogous to an Indian *chobdav*, which to some readers may be a case of *ignotum per ignotius*. **1935** *Ibid.* 2 May 289/1 A distinct failure to escape the imputation of *ignotum per ignotius*.

Igorot (ī·gǒrǭt). Also **Igolot(e), Igorrot(e), Ygorrote.** [ad. Sp. *Ygolote* (A. de Morga, 1609), f. the native name.] Name of a people inhabiting northern Luzon in the Republic of the Philippines. Also as *collect. sing.* and *attrib.*
1821 W. Walton tr. *T. de Comyn's State of Philippine Islands* p. xli, An expedition was also sent against the Igorrots, inhabiting the mountains of the interior, where gold is obtained. *Ibid.* ii. 32 The Igorrot Indians, who occasionally come down from the mountains to barter with the Christians, use certain coarse jars or vessels of copper. **1840** *Penny Cycl.* XVIII. 88/2 The Ygorrotes, who inhabit the mountains east of the Gulf of Lingayen, are distinguished by a peculiar physiognomy and a lighter colour. **1875** F. Jagor *Trav. Philippines* 131 The Ygorrotes are not Christians. **1885** *Encycl. Brit.* XVIII. 753/1 The Igorrotes or Igolotes proper (for the name is by many writers very loosely applied to all the pagan mountain tribes of Luzon) inhabit the districts of Bangued, Lepanto, Tiagan, Bontoc. **1898** D. C. Worcester *Philippine Islands* 438 One of the Igorrote peoples is believed to be descended from the followers of the Chinese invader Limahong. **1903** Barrows in *Philippine Jrnl. Sci.* Oct. 796 The powerful and numerous Igorot of northern Luzon. **1914** D. C. Worcester *Philippines Past & Present* I. 343 This once prosperous little Igorot hamlet had been burned by the Spaniards. **1925** K. Mayo *Isles of Fear* 256 The Igorots proper, however, number only about 70,000. **1957** *Encycl. Brit.* XII. 75/1 There are..two broad groupings of the Igorot as a whole. **1967** Wernstedt & Spencer *Philippine Island World* III. ix. 349 The peoples and cultures of the North Luzon Highlands... Popular reference adopted the term 'Igorot' for the whole population,..although the term properly is only the Tagalog word for 'mountain-dwelling' or 'mountaineer'.

iguana. 2. A name used in Africa for a large monitor lizard of the genus *Varanus*, esp. *V. niloticus*, the aquatic Nile monitor.
1753 N. Owen *Jrnl. Slave-Dealer* (1930) 32 They [*sc.* the Bulums] eat alegators, guanas and long worms. **1801** J. Barrow *Acct. Trav. S. Afr.* I. v. 346 None of the people with me could testify to have seen any other species of that genus [*sc.* crocodile] frequenting the water, except Iguanas, from six to ten feet in length. **1803** T. Winterbottom *Acct. Native Africans Sierra Leone* I. iv. 69 Although they have several species of lizards, they use only one as an article of diet, the *guana*, lacerta iguana, which they esteem delicate food. **1834** T. Pringle *Afr. Sk.* II. vi. 210 One of the deep lagoons formed by the river, and which the [Moravian] brethren have named the Leguan's Tank, from its being frequented by numbers of the large amphibious lizard called the leguan or guana. **1875** J. J. Bisset *Sport & War* xx. 179 Hence [*sc.* from under water] the 'Iguana', a small kind of crocodile, proceed on shore at night and take chickens from the hen-roosts. **1900** H. A. Bryden *Animals Afr.* xv. 174 In South Africa this reptile [*sc.* the Nile Monitor] is often miscalled an 'iguana'. **1947** J. Stevenson-Hamilton *Wild Life S. Afr.* xxxv. 315 There are two large monitor lizards or leguaans of the genus *Varanus* found in South Africa. These are sometimes called 'iguanas', though the true iguanas are almost all confined to the New World, and the popular term may be merely a corruption of this name. **1964** J. P. Clark *Three Plays* 30 They struggled like Two iguanas till outspent, they stopped.

iiwi (ī,ī·wi). [Hawaiian.] A Hawaiian bird, the honeycreeper *Vestiaria coccinea*, whose red feathers were formerly used to make the cloaks of native chiefs.
1779 J. King *Jrnl.* Mar. in Cook *Voy. Pacific Ocean* (1784) III. vi. 119 The birds of these islands are as beautiful as any we have seen... Another is of an exceeding bright scarlet colour;..its native name is *eeeeve*. **1890** Wilson & Evans *Aves Hawaiienses* I. 1 Vestiaria Coccinea. Olokele or Iiwi. *Ibid.* 3 The call-note of the 'Iiwi' is peculiar. **1899** A. H. Evans in *Cambr. Nat. Hist.* IX. vii. 564 The splendid feather-cloaks of the Hawaiian kings..were of old chiefly composed of the plumage of the 'Mamo' (*Drepanis pacifica*) and the 'Iiwi' (*Vestiaria coccinea*). **1915** W. A. Bryan *Nat. Hist. Hawaii* xxxi. 430 The beautiful red iiwi..is infested with three genera [of bird-lice]. **1944** G. C. Munro *Birds Hawaii* 93 With bright scarlet body, black wings and tail, and rose colored, inch long, curved bill, the iiwi is one of the most beautiful of the Hawaiian native birds. **1970** S. Carlquist *Hawaii* xi. 198 Iiwis are still relatively common in a number of forest areas of the Islands.

Ijo (ī·dʒo), *sb.* and *a.* Also **Ejo, Idzo, Ijaw.** **A.** *sb.* The name of a tribe inhabiting the Niger delta, on the coast of Nigeria; a member of this tribe; the language of this tribe. **B.** *adj.* Of or pertaining to the Ijo tribe.
1856 W. B. Baikie *Narr. Voy. Niger* iii. 40 The people are of the same tribe as those who inhabit the tract of country up to the Rio Formoso, where however they are called Ejo or Ojo, by which name also they are known at Abó, at Brass, and even at Bonny. By English palm-oil traders they are often termed Jo-men. **1883** R. N. Cust *Sk. Mod. Lang. Afr.* I. 220 Within the Idzo Language-Field many of the Ibo-speaking Races reside. **1906** A. G. Leonard *Lower Niger* II. i. 52 So extremely touchy and sensitive are these people—the Ijo particularly so,—so easily piqued and disturbed, even by the mildest of chaff, that they have no hesitation in taking their own lives or the lives of others, on the spur of the moment. *Ibid.* I. iv. 42 An examination of the six or more dialects of Ijo and the five of Bini..shows that they are dissimilar not only to each other, but to the other neighbouring tongues. **1926** P. A. Talbot *Peoples S. Nigeria* IV. iii. 82 Ijaw is a very primitive negro tongue, perhaps the most ancient in West Africa, and..has not the smallest affinity with any other. **1958** J. S. Coleman *Nigeria* xviii. 390 Demands for a separate state were made by minority groups in the Middle Belt in the Northern Region, by the Ijaw peoples of the Delta area, and elsewhere. **1966** M. Crowder *Story Nigeria* v. 80 The Ijo exported dried fish and salt, which they panned in the salt water creeks, to the peoples of the hinterland in exchange for vegetables and tools. **1971** P. Young in J. Spencer *Eng. Lang. W. Afr.* 180 The Nigerian poet and novelist,

Gabriel Okara. His novel *The Voice* exhibits a highly idiosyncratic syntax obtained, according to Okara, by a direct substitution of the syntax of his native Ijaw for normal English syntax.

ijolite (ī·yoləit). *Petrogr.* [ad. G. *ijolith* (Ramsay & Berghell 1891, in *Geol. Fören. i Stockholm Förh.* XIII. 304), f. *Ijo*, Sw. rendering of Finnish *Ii*, the name of a village and district on the Finnish coast near Oulu and also the initial element in the names of local geographic features, as *Iijärvi, Iijoki, Iivaara*: see -lite.] A plutonic igneous rock consisting essentially of nepheline and pyroxene and containing no felspar.
1897 A. Harker *Petrol.* (ed. 2) 52 The 'ijolite' of Ramsay and Berghell from Finland, a nepheline-pyroxene-rock free from felspar but sometimes rich in garnet. **1962** W. T. Huang *Petrol.* iv. 139 Much of the ijolite on Alnö Island, Sweden, and Magnet Cove, Arkansas, contains abundant melanite garnet.
Hence **ijoli·tic** *a.*, resembling or of the nature of ijolite.
1938 *Mineral. Abstr.* VII. 36 Masses of limestone and ijolitic intrusives. **1954** H. Williams et al. *Petrogr.* iv. 72 Most of these ijolitic rocks..owe their characteristics largely to contamination of magma by assimilation of limestone. **1966** R. L. Johnson in Tuttle & Gittins *Carbonatites* I. 221 Here there are a number of carbonatite complexes, which include granites, syenites, nepheline syenites and small amounts of ijolitic rocks within their limits.

‖ **ikat** (i·kat). [Mal., lit. 'to tie, fasten'.] A technique of fabric decoration common in Indonesia and Malaysia, in which warp or weft threads, or both, are tied at intervals and dyed before weaving; also, a fabric of this kind.
1931 C. F. Iklé (*title*) Ikat technique and Dutch East Indian ikats. **1942** *Ciba Rev.* Aug. 1586 The ikat process can be applied to the warp, the weft, or to both systems of the threads of a fabric. **1953** C. A. Gibson-Hill *Malay Arts & Crafts*, § Hand-woven cloths. A variant form (of cloth) is sometimes introduced by a tie-and-dye technique (*ikat chuai*), in which short lengths of the woof are tied round with dye-proof strips of banana leaf and the skein then immersed in the dye. **1959** *Chambers's Encycl.* IV. 703/1 A..technique known as *ikat* is practically limited to Indonesia. In this method the warp threads of a cloth are set out before weaving and tied at intervals before dyeing. **1968** *Encycl. Brit.* XII. 180/1 In the textiles in *batik* (wax-resist) and *ikat* (pre-dyed thread)..Indonesian art has shown a continuing vigour.

‖ **ikbal** (i·kbal). [Turkish.] A member of the harem of an Ottoman Sultan.
1910 *Encycl. Brit.* XII. 950/2 The so-called Odalisks..are subdivided according to the degree of favour..into *Ikbals* ('Favourites') and *Geuzdés* (literally the 'Eyed' ones), those whom the sultan has favourably noticed. **1937** *Times Lit. Suppl.* 16 Jan. 36/3 The Sultan selected his favourites, or *ikbals*. **1956** A. D. Alderson *Struct. Ottoman Dynasty* ix. 80 If the relationship showed any signs of permanence, she was promoted to the rank of 'Ikbal' (Fortunate). **1962** J. Fleming *When I grow Rich* iii. 40 She had ceased to be an *ikbal* (the name given to those members of the harem who actually slept with the Shadow of God upon Earth).

ike (əik), colloq. abbrev. of *iconoscope 2.
1937 *Amer. Speech* XII. 101 As the microphone is a *mike* so the television iconoscope is an *ike*. **1947** L. A. Sposa *Television Primer* 224 *Iconoscope* (*Ike*), a cathode-ray camera pickup tube developed by RCA.

ike: see ***ikey** *sb.* and *a.*

‖ **ikebana** (ikībā·nā). [Jap., f. *ikeru* to keep alive, arrange + *hana* flower.] The art of Japanese flower arrangement in which flowers are formally displayed according to strict rules, sometimes with other natural objects.
1901 F. Brinkley *Oriental Series: Japan* III. i. 9 Another remarkable outcome of the Military epoch was the art of flower arrangement. The name applied to it, *ike-bana*, or 'living flower', explains at once the fundamental principle..that the flowers must be so arranged as to suggest the idea of actual life. **1934** A. Koehn *Art Jap. Flower Arrangement* 2 The word 'ikebana' is used for all forms of Japanese Flower Arrangement. **1960** Koestler *Lotus & Robot* II. vii. 191 It has its charm; and so has ikebana, the Art of Flower Arrangement. **1965** W. Swaan *Jap. Lantern* xix. 227 Chrysanthemums are particularly popular for *ikebana*. **1966** *New Statesman* 3 June 809/1 On a pole in the street..I saw a signboard that advertised a local *ikebana* school. **1967** 'G. Black' *Wind of Death* x. 197 There was no *ikebana* flower arrangement.

ikey (əi·ki), *sb.* and *a.* *slang* and *dial.* Also **ike, iky, I-.** Familiar abbreviated form of the Jewish name *Isaac* (also **ikeymo**, f. *Isaac* and *Moses*), used typically for: a Jew or someone taken to be or resembling a Jew; also, a (Jewish) receiver, moneylender, etc.; *transf.*, a loafer; a tip, information; (*Austral.*) a bookmaker. As *adj.*, (*a*) artful, crafty, knowing, 'fly'; (*b*) having a good opinion of oneself, 'stuck-up'.
Derogatory and offensive in all uses as applied to persons.
1835 Dickens *Sk. Boz* (1836) 1st Ser. II. ii. 44 'Let me alone,' replied Ikey, 'and I'll ha' vound up..in five seconds.' **1864** Hotten *Slang Dict.*, *Ikey*, a Jew 'fence'. **1870** Leybourne in Farmer & Henley *Slang* (1896) IV. 2 My name it is ikey Bill, A Whitechapel Covey am I. **1881** *Punch* 10 Sept. 110/1 'Arf ikey of course, put-up bizness. **1887** Parish & Shaw *Dict. Kentish Dial.* 83 *Ikey*, proud. **1889** Barrère & Leland *Dict. Slang*, *Ikey* (popular), a Jew;..Also said of any one who thinks himself knowing, smart, and has a great opinion of himself. **1892** Chevalier in Farmer & Henley *Slang* (1896) IV. 2 Artful little ikey little ways. **1897** I. Scott *How I stole 10,000 Sheep in Austral. & N.Z.* 33 Jim now hit on a rather 'Iky' way to do the ride to Wellington. **1906** Russell & Rigby *Making of Criminal* i. 7 His clothes are so very poor that he does not like..'to show himself up' by appearing in them, especially if they are of a kind only affected by the professional loafer or 'ike'. **1913** D. H. Lawrence *Sons & Lovers* ix. 241, I want people to think we're awful swells. So look ikey. **1922** Joyce *Ulysses* 458 Three cheers for Ikey Mo! **1927** T. Prentis *Music-Hall Memories* 16 Sez as I'm as ikey as the Dook of Boocle-oo. **1934** *Bulletin* (Sydney) 18 Apr. 11/1 She laid the odds, as smartly and acutely, I'll bet, as any trousered Ikey Mo. **1936** J. G. Brandon *Dragnet* xiii. 132 'E passed the ike, that there was somethink on there. **1941** *Penguin New Writing* III. 69 Go indoors an' 'elp yer sister with the washing-up, you lazy young ike. **1942** P. Abrahams *Dark Testament* I. xiii. 72 Any guy his pal. Chink, ike. **1954** J. Symons *Narrowing Circle* xxx. 135 I'm a Hackney Jew, Dave. At school they called us Ikeymoes and Jewboys. **1966** F. Shaw et al. *Lern Yerself Scouse* 36 *Yer very ikey*, you are very dandified. **1969** *Private Eye* 6 June 14 (*caption*) Pull your head in, ikey mo! **1972** R. Player *Oh! Where are Bloody Mary's Earrings* vi. 168 The Ikeys will win—you can stake your life on that.
Hence **ike, i·ky** *v.* (see quot. 1932); **i·keyness**, the quality of, or an act of, being ikey.
1911 D. H. Lawrence *White Peacock* vii. 484, I haven't been to see them lately—can't stand Meg's ikeyness. **1932** *Amer. Speech* VII. 401 *Iky*, to jew down the price; to cheat. 'He ikied me out of my turn.' **1960** Wentworth & Flexner *Dict. Amer. Slang*, *Ike v.*, to cheat; to lower the price by haggling; to 'Jew down'.

iktas (i·ktās), *sb. pl.* *N. Amer. colloq.* Also **ectas, ictas, ictus, iktics, iktus.** [Chinook.] Goods; belongings; things.
1856 *Democratic State Jrnl.* (Sacramento, Calif.) 4 Oct. 3/1 They are getting short of blankets and other *ictas*. *a* **1861** T. Winthrop *Canoe & Saddle* (1863) iv. 53 My motley retinue followed me humbly, bearing 'ikta', my traps, and their own plunder. **1892** in *Brit. Columbia Hist. Q.* (1941) Oct. 302 They chiefly took their pay in blankets and provisions and other ectas—the balance in coin. **1951** R. P. Hobson *Grass beyond Mountains* 101 We got eighteen horses, a summer's grub and all the ictus we need for the time being. **1965** S. G. Lawrence *Forty Yrs. on Yukon Telegraph* vi. 34, I must get some iktics together and hide.

ikunolite (ikū·noləit). *Min.* [Named after the *Ikuno* mine in Hyōgo prefecture, Japan, where it was first found: see -lite.] A bismuth sulphide, Bi_4S_3, that contains some selenium in place of sulphur and occurs as grey foliated masses.
1959 A. Kato in *Mineral. Jrnl.* II. 398 The chemical properties and X-ray studies clarified this mineral to be a new species of the composition $Bi_4(S, Se)_3$ with rhombohedral symmetry. The writer has called this mineral ikunolite after the locality name. **1961** *Mineral. Abstr.* XV. 43/2 Ikunolite resembles joséite-A (Bi_4TeS_2) in structure, crystallography, and optical properties. **1962** *Amer. Mineralogist* XLVII. 1431 The bismuth sulfide mineral ikunolite (Bi_4S_3) was first described by Kato (1959)...A second occurrence of this mineral has recently been found in specimens from the bismuth-molybdenite pipes at Kingsgate in the New England district of New South Wales.

-il, -ile, *suffixes.* Add: **2.** Following cl.L. and med.L. ordinal numerals of the type *quartīlis, quintīlis, sextīlis* (Eng. Quartile, etc.), *-ile* (-əil) is used in *Statistics* to form substantives denoting (*a*) those values of a variate that divide a population into the indicated number of groups, equal in size, and (*b*) the groups themselves; so *octile* and *quartile* (1879), *decile* (1882), *percentile* (1885).

Ila (ī·lǎ), *sb.* and *a.* Also **Ba-ila.** [Native name.] **A.** *sb.* **1.** An African of a Bantu people in Zambia (formerly Northern Rhodesia); also used as collect. sing. = this people. **2.** The Bantu language of this people. **B.** *adj.* Of or pertaining to this people or their language.
1907 E. W. Smith *Handbk. Ila Lang.* i. 1 The Ila language is spoken by the Baila,..a people living in North-West Rhodesia. *Ibid.* i. 3 In Ila, as in all Bantu languages, alliteration..is not an accident of style, it is the very essence of the language. **1911** *Encycl. Brit.* III. 359/2 Next is a group which might be styled the Subiya-Tonga-Ila, though some authorities think that Tonga and Ila deserve to be ranked as an independent group. **1920** Smith & Dale (*title*) Ila-speaking peoples of Northern Rhodesia. *Ibid.* p. xxvii, They are very mixed, but now the language of them all is Ila. **1956** W. V. Brelsford *Tribes N. Rhodesia* viii. 55 The Ila have a great and deserved reputation as fighting men, but their comparative paucity in numbers and the possession of great herds of cattle made them a constant prey to the bigger tribes of Barotseland. **1959** *Chambers's Encycl.* I. (*caption*, facing p. 140) A chief of the Ila peoples, photographed wearing ivory bracelets and *impande*, the insignia of office. *Ibid.* II. 109/2 The Ba-ila live in independent communities. **1960** R. C. Bell *Board & Table Games* I. 121 *Chisolo.* This is played by the Ba-ila speaking peoples of Northern Rhodesia. **1970** P. Oliver *Savannah Syncopators* 26 The Ila and Tonga people are neither of them West African,..being of the Middle Zambesi 2,300 miles away.

‖ **Ilag** (ī·lag). [Ger. abbrev. form of *Internierungslager*.] A prison-camp for civilian internees in Nazi Germany.
1941 Wodehouse *Berlin Broadcasts* in *Performing Flea* (1961) i. 261 An Oflag is where captured officers go. Stalags are reserved for N.C.O.s and privates. The civil internee gets the Ilag.

ilala (ilā·la). *S. Afr.* Also **lala.** [Zulu.] A fan palm, *Hyphæne natalensis*, native to the coastal region of Natal.
1868 J. Chapman *Trav. S. Afr.* II. 464 *Ilala*, Hyphæne. **1884** E. P. Mathers *Trip to Moodie's* 29 There is a natural drink..which..goes by the name of kaffir beer..it is the exudation of a native palm tree (*ilala*). **1911** *Encycl. Brit.* XIX. 253/2 Of palms there are two varieties [in Natal], the ilala (*Hyphaene crinita*), found only by the sea shore and a mile or two inland, and the isundu (*Phoenix reclinata*). **1954** T. V. Bulpin *Ivory Trail* iv. 44 He..passed the hot hours in plaiting a hat from lala palm leaves. *Ibid.* xi. 114 The elephants..also had a liking for this potent lala palm wine. **1966** D. Varaday *Gara-Yaka's Domain* viii. 89 He ruled the troop like a despot from his rocky keep above a growth of ilala palms.

Ilamba, var. *Lamba *sb.* and *a.*

ilb (ilb). Also **ailb, elb.** [Arab.] A spiny tree of the genus *Zizyphus*, esp. *Z. spina-Christi*, found in North Africa and the Middle East.
1894 *Kew Bull.* 330 *Zizyphus Lotus* Linn. A tree 20–30 feet high. Very common in Hadramaut. Native name, 'Ailb'. Extends through North Africa and South Europe. **1914–16** E. Blatter *Flora of Aden* 158 *Zizyphus spina Christi*... Arabic name: Elb. **1940** F. Stark *Winter in Arabia* 118 It is only because of the English peace that he can count on the safe and quiet possession of his own waterless strip of 'ilb trees and plough. **1963** *Times* 12 Mar. 12/6 One looks out over fields of millet and barley, waving date palms and sturdy *ilb* trees.

Ilchester (i·ltʃestəɪ). The name of a town in Somerset, used *attrib.* or *ellipt.* as *sb.* to designate a cheese mixed with beer, chives, and spices.
1963 *Guardian* 11 Jan. 7/7 Mr Ken Seaton, landlord of the Ilchester hotel, Ilchester, has developed a new cheese, made with beer and chives and spices. He calls it Ilchester. **1965** T. Fitzgibbon *Art Brit. Cooking* 132 Ilchester cheese is a recent commercially blended cheese made in Ilchester, Somerset, from Cheddar cheese, Worthington beer, spices and chives. **1965** *Harrods Food News* May, Ilchester is Cheddar with Beer and Spices. **1967** T. A. Layton *Wine & Food Soc. Guide Cheese & Cheese Cookery* ii. 68 Ilchester cheese is the creation of a Yorkshireman, Ken Seaton, who hit upon the recipe almost by accident only six years ago. **1971** *Selfridge Christmas Food Catal.* 10/2 Ilchester cheese with Beer and Chives. 4 oz. pot, each 17½p.

ileal (i·liǎl), *a.* [f. Ile(um + -al.] Of, within, or supplying the ileum.
1893 in *Funk's Stand. Dict.* **1895** R. T. Morris *Lect. Appendicitis* v. 100 (*heading*) The experimental production of ileal intussusception with carbonate of sodium. **1902** D. J. Cunningham *Text-bk. Anat.* 802 The ileal branch [of the superior mesenteric artery]..turns upwards and to the left in the lowest part of the mesentery. **1934** Lake & Marshall *Surg. Anat. & Physiol.* xxxiii. 521 An appreciable degree of obstruction of the small gut ('ileal stasis'). **1971** *Brit. Med. Bull.* XXVII. 26/2 Epidemics of ileal obstruction were first recognized in 1964.

ileo-. Add: **i:leocolo·stomy** *Surg.* [*-stomy], the operation of joining, and creating a passage between, a part of the ileum and a part of the colon so that the intervening part of the intestines is bypassed; the connection so formed; **i:leo-ileo·stomy** *Surg.* [*-stomy], the operation of joining, and creating a passage between, two parts of the ileum so that the intervening part is bypassed; the connection so formed; **i:leo-si·gmoid** *a.*, between or connecting the ileum and the sigmoid flexure of the colon; **i:leosigmoido·stomy** *Surg.* [*-stomy], the operation of joining, and creating a passage between, the sigmoid flexure of the colon and a part of the ileum so that the intervening part of the intestines is bypassed; the connection so formed; **ileostomy**, substitute for def.: the operation of attaching the ileum to the abdominal wall and constructing an artificial anus so that the

intestinal contents are evacuated directly from the ileum; also, the opening or artificial anus thus made; (examples).
1887 *Trans. Internat. Med. Congr. 9th Session* I. 485 An ileo-colostomy or ileo-rectostomy..should be done in all cases of irreducible ileo-cæcal invagination. *Ibid.* 475 Intestinal anastomosis, by making an ileo-colostomy by lateral apposition of the ileum to colon below the invagination. **1967** J. H. Garlock *Surg. Alimentary Tract* ix. 427 In the same category, one may include the bypassing operation of ileocolostomy. **1887** *Trans. Internat. Med. Congr. 9th Session* I. 479 Two adjacent coils of the ileum were united by making an ileo-ileostomy. **1889** [in Dict.]. **1901** N. Senn *Pract. Surg.* xxiii. 833 Jejuno-ileostomy and ileo-ileostomy by apposition with decalcified bone-plates..is an operation almost devoid of danger. **1903** Ileo-sigmoid [see **colo-colic* adj.]. **1968** J. C. Goligher et al. *Ulcerative Colitis* xiii. 237 The first such operation involving an ileo-sigmoid anastomosis was recorded by Lilienthal as early as 1901. **1892** *Lancet* 16 Apr. 864/2 (*heading*) Ileo-sigmoidostomy (Senn's method) for intestinal obstruction due to malignant disease of the hepatic flexure of the colon. **1964** H. E. Bacon *Cancer Colon, Rectum & Anal Canal* xv. 806/1 Ileosigmoidostomy..shortcuts the entire colon and renders it more or less functionless. **1890** Billings *Med. Dict.* I. 682/1 Ileostomy. **1891** T. Bryant in *Lancet* 3 Jan. 1/1 The operation I am about to describe may well be called 'ileostomy'. **1926** *Brit. Jrnl. Surg.* XIII. 711 Against ileostomy, however, we urge the loss of an important sphincter and portion of the digestive tract. **1968** J. C. Goligher et al. *Ulcerative Colitis* xiii. 236 The creation of a terminal ileostomy. **1968** Colcock & Braasch *Surg. Small Intestine in Adult* xiii. 179 The great majority of patients with ulcerative colitis who need surgery will require a complete removal of the colon and rectum and an ileostomy. *Ibid.* 181 A temporary ileostomy appliance is cemented to the skin over the ileostomy. **1971** M. Sparsberg (*title*) Ileostomy care.

Iliac (i·liæk), *a.*[2] [ad. L. *Iliacus*, a. Gr. Ἰλιακός: see Iliad.] Pertaining or relating to ancient Ilium; Ilian.
1878 Gladstone *Homer* 37 The other epics of the Iliac Cycle differ in their narrative from the Iliad.

iliacus (iləi·ǎkǝs). *Anat.* [late L.: see Iliac *a.* (*sb.*).] = *iliac muscle* (s.v. Iliac *a.* 2). Also † *iliacus internus, iliacus muscle.*
1615 H. Crooke *Descr. Body Man* x. vi. 744 The thighes are each of them bent by two [muscles] called Psoas and Iliacus. *Ibid.* xxxviii. 811 The second Bender is called Iliacus internus the inward haunch-Muscle. **1726** W. Cheselden *Anat. Humane Body* (ed. 3) II. ii. 123 Iliacus internus, arises from the concave part of the ilium, and from its lower edge, and passing over the ilium near the os pubis, joins the former muscle [*sc.* psoas magnus], and is inserted with it, to be employed in the same action. **1733** G. Douglas tr. *Winslow's Anat. Expos. Struct. Human Body* I. iii. x. 53 The Iliacus and Psoas thus united pass under the Ligamentum Falloppii. **1875** *Encycl. Brit.* I. 840/2 The thigh can be bent on the abdomen by the action of the psoas, iliacus, and pectineus, which lie in front of the joint. **1967** G. M. Wyburn et al. *Conc. Anat.* vi. 158 The iliacus and psoas muscles pass behind the inguinal ligament and the iliacus joins the tendon of the psoas.

iliahi (īli,ā·hi). [Hawaiian.] One of several trees of the genus *Santalum*, esp. *S. freycinetianum*, which grow in Hawaii and yield an aromatic wood.
1825 W. Ellis *Jrnl. Tour Hawaii* 247 *Iliahi*, sandalwood. **1888** W. Hillebrand *Flora Hawaiian Islands* 389 *Santalum*...The following species are weakly defined, and all furnish an equally fragrant wood..while the name of the tree is 'Iliahi'. **1915** W. A. Bryan *Nat. Hist. Hawaii* xv. 219 The iliahi furnished the first article of export which attracted commerce to the islands. **1970** S. Carlquist *Hawaii* xiv. 269 Typical of this low stature is the coastal iliahi, or sandalwood (*Santalum ellipticum* var. *littorale*), which is a low, rounded shrub with succulent leaves, gray-green in color. Inland sandalwoods are all trees.

Ilian (i·liǎn, əi·liǎn), *a.* (*sb.*). [f. *Ili(um* + -an.] Of or pertaining to any of the successive towns of Ilium in the Trojan Plain; also as *sb.*, an inhabitant of Ilium.
1582 R. Stanyhurst tr. *Virgil's Æneis* I. (Arb.) 26 Whilst stood the great Ilian empyre. **1626** G. Sandys tr. *Ovid's Met.* XIII. 222 To th'Ilian Court I went. **1847** J. Leitch tr. *Müller's Anc. Art* 483 Coins of the Ilians. **1869** Tozer *Highl. Turkey* I. 44 The site of the ancient Village of the Ilians. **1876** Gladstone *Homeric Synchronism* 34 He appears as the representation of the Dardanian Branch, with a sidelong jealous eye towards the predominating Ilian House of Priam. **1881** *Jrnl. Hellenic Stud.* II. 18 After the victory, he [*sc.* Alexander] gave substantial proof of gratitude to the Ilian gods and heroes. **1888** *Encycl. Brit.* XXIII. 579/2 The temple of the Ilian Athene. *Ibid.*, It was politic to recall the legend of Roman descent from Æneas. Lucius Scipio and the Ilians were alike eager to do so. **1927** W. W. Tarn *Hellenistic Civilisation* ii. 70 Antigonus I..created three sectional Leagues: the Ionian..the Ilian..and the Islanders. **1941** M. Rostovtzeff *Social & Econ. Hist. Hellenic World* I. iii. 154 Some of the leagues had existed before, such as the Ionian, Aeolian, Ilian (?), Lycian, and Carian leagues in Asia Minor. **1957** *Encycl. Brit.* XXII. 505/1 In their temple of Athena the Ilians showed him arms which had served in the Trojan war.

Iliat (i·liǎt). Also **Eylat, Ilat, Iliaut, Iliyat, Illyat.** [Turkish *īlāt*, pl. of *īl* country, wandering pastoral tribe (cf. *rūm-eyli* 'country of the Romans', Rumelia).] One of several tribes of nomads scattered throughout Persia.
1840 *Penny Cycl.* XVII. 472/1 The wandering tribes of Persia are comprehended under the general term of *Iliyats* or *Ilat*, and are found in every part of Persia. **1865** *Chambers's Encycl.* VII. 420/2 The nomad or pastoral tribes, or eylats (*eyl*, a clan), often spelt *illyats*, are of four distinct races—Turkomans, Kurds, Lûurs, and Arabs. **1888** *Encycl. Brit.* XXIII. 661/1 They are known by the name of Ilāt or Iliyāt. **1902** D. G. Hogarth *Nearer East* 160 The nomad 'Iliats' who wander here and there.

iligant (i·ligǎnt), *a.* ¶ Used, chiefly as an Irishism, for Elegant *a.* (sense ¶ 8). See also *illigant.
1822 M. Edgeworth *Let.* 4 Mar. (1971) 361 We have an excellent job landau..and quite *iligant* coachman horses and *all*—for one pound per day. **1833** [see *elegant *a.* 8]. **1846** *Punch* XI. 79 A large assortment of most iligant blunderbuss's. **1888** [see Elegant *a.* 8].

ilima (ilī·mǎ). [Hawaiian.] A shrub of the genus *Sida*, esp. *S. fallax*, bearing yellow or orange flowers.
1888 W. Hillebrand *Flora Hawaiian Islands* 43 *Sida*... A large genus, distributed over the warmer regions of the entire globe.—Nat[ive] name of all species: 'Ilima'. **1915** W. A. Bryan *Nat. Hist. Hawaii* xv. 209 Two species of ilima occur in the lower zone throughout the group. Their bright yellow flowers, so much used in leis, are well known. **1920** *Glasgow Herald* 16 Apr. 8 Three magnificent wreaths of ilima, the Hawaiian royal flowers. **1970** S. Carlquist *Hawaii* xiv. 269 *Sida fallax*, the ilima, is a mat with yellow-orange flowers and finely hairy white leaves, common near the shores of all the islands.

ilk, *a.*[1] Add: **3.** ¶ Also, by further extension, = kind, sort.
1790 J. Fisher *Poems* 155 Ilk ane a cap an' cloak o' silk Has got, as if she was a lady, An' that indeed, o' nae sma' ilk. **1881** J. A. Morgan *Shakes. Myth.* i. 36 Milton was the enemy of all the ilk. **1897** *Evesham Jrnl.* 23 June (E.D.D.), The horses most famous for marvellous exploits must have blood as well as bone, but only certain districts of the Green Isle can produce this ilk. **1899** *Westm. Gaz.* 29 June 3/2 Two very new hats of quite another 'ilk'. *Ibid.* 6 July 3/2 A coat of this 'ilk' is quite another matter from the coat of the tailor costume. **1969** *Times* 8 May 8/6 This habit is confined to Tory backbenchers like..Rear-Admiral Morgan Giles and others of that ilk. **1973** E. McGirr *Bardel's Murder* ii. 42 One doesn't like or dislike a fellow of that ilk... He was a kind of barrow boy in a shop.

ill, *a.* and *sb.* Add: **A.** *adj.* **8. b.** = Sick *a.* 2.
1928 C. F. S. Gamble *Story N. Sea Air Station* 244 Before they could be attended they were given too much food and were violently ill as a result. **1929** A. Conan Doyle *Maracot Deep* 153 The sight was so horrifying that we were all ill.

ill-. Add: **A. III. 4. ill-favour** *v. trans.*, to treat badly, to be inimical to or hostile towards.
1899 F. J. Crowest *Beethoven* 128 His environment generally throughout his early life ill-favoured the contemplative mood. **1908** *Westm. Gaz.* 21 Aug. 2/1 Fortune ill-favoured them in many skirmishes.
7. a. *ill-acquired* (example); **b.** *ill-constructed, -pleased*; **c.** *ill-adapted* (example), *-adjusted, -armed* (examples), *-assorted* (earlier and later examples), *-balanced* (examples), *-concealed* (examples), *-conducted* (example), *-considered* (examples), *-directed* (examples), *-equipped, -governed* (examples), *-informed* (examples), *-lit* (example), *-performed, -qualified, -supported* (example).
1848 Mill *Pol. Econ.* II. v. §1. 293 The same reckless prodigality with which they squandered any other part of their ill-acquired possessions. **1878** W. James *Coll. Ess. & Rev.* (1920) 53 If his powers correspond to the wants of this social environment, he may survive, even though he be ill-adapted to the natural or 'outer' environment. *Ibid.*, Individuals who, by their special powers, satisfy these desires are protected by their fellows and enabled to survive, though their mental constitution should in other respects be lamentably ill-'adjusted' to the outward world. **1903** *Daily Chron.* 25 June 4/1 In this ill-adjusted world men have to take what they can get. **1807** J. Barlow *Columb.* VI. 225 Ridgefield and Compo saw his valorous might, With ill armed swains put veteran troops to flight. **1942** W. S. Churchill *End of Beginning* (1943) 32 China, ill-armed or half-armed, has..withstood the main fury of Japan. **1814** Ill-assorted [see Assorted *ppl. a.*]. **1938** W. S. Churchill *Into Battle* (1941) 29 Three..Ministers of the Crown have been..prejudiced by being given..ill-assorted tasks. **1864** G. M. Hopkins *Poems* (1948) 119 The clouds come like ill-balanced crags. **1956** K. Clark *Nude* vii. 284 Could fill an ill-balanced nature with destructive envy. **1895** B. M. Croker *Village Tales* 38 As time wore on, there actually arose an ill-concealed jealousy of their old corps. **1960** *Farmer & Stockbreeder* 8 Mar. 63/3 The report of the Cook Committee was awaited with 'ill-concealed impatience'. **1839** Mill in *Westm. Rev.* Apr. 497 There is a notion abroad that they are the ill-conditioned and ill-conducted portion. **1835** —— in *London Rev.* II. III. 116 In the English aristocracy there has surely been..crude and ill-considered legislation enough. **1956** E. E. Evans-Pritchard *Nuer Relig.* iii. 96 It was because this was not appreciated by British administrators that the role..of the prophets was very largely misunderstood and their treatment of them ill-considered. **1847** Mill *Let.* 19 June in *Works* (1963) XIII. 721 The long paper..is full of unfinished & ill constructed..sentences. **1944** R. Chandler *Let.* 26 Jan. in *R. Chandler Speaking* (1966) 42 A column and a half of respectful attention will be given to any fourth-rate, ill-constructed, mock-serious account of the life of a bunch of cotton pickers in the deep south. *a* **1849** Poe *Works* (1865) I. 418 Some feeble and ill-directed efforts resulted in complete failure on their part, and, of course, in total triumph on mine. **1947** *Mind* LVI. 132 They have thus been led to offer facile, but ill-directed, rebutments of the refutation. **1956** *Nature* 10 Mar. 446/2 At present we are ill-equipped to do so. **1962** W. Nowottny *Lang. Poets Use* i. 5 The exhausted and ill-equipped army. **1685** tr. *Arnauld & Nicole's Logic* I. iii. 68 Such a Presumption and Rashness is a sign of an ill-govern'd and ill-qualified mind. **1860** Ruskin *Unto this Last* in *Cornh. Mag.* Sept. 281 In the inactive or ill-governed nation, the gradations of decay and the victories of treason work out also their own rugged system of subjection and success. **1824** Ill informed [see Informed *ppl. a.* 2 b]. **1961** *New Eng. Bible Rom.* x. 2 To their zeal for God I can testify, but it is an ill-informed zeal. **1905** W. Holman Hunt *Pre-Raphaelitism* I. iii. 46 With small and ill-lit studios, and without means to pay models, he [*sc.* Haydon] could never do justice to his intellectual conceptions. **1890** W. James *Princ. Psychol.* I. x. 369 The only service that transcendental egoism has done to psychology has been by protests against Hume's 'bundle'-theory of mind. But this service has been ill-performed. **1946** A. L. Bacharach *Brit. Mus.* i. 12 The Churches, who never seemed to realise the paralysing effect their badly written, ill-performed music was having on their congregations. **1861** Trollope *Orley F.* (1862) I. xxi. 161 An angry ill-pleased wife is no pleasant companion for a gentleman on a long evening. **1685** Ill-qualified [see **ill-governed*]. **1733** A. Baxter *Inquiry Human Soul* II. xxvi. 81 The Atheist's ill-supported, tottering world.
IV. 8. *ill-visaged* adj.
1865 G. M. Hopkins *Poems* (1948) 142 And John shall lie, where winds are dead, And hate the ill-visaged cursing tars. **1894** Ill visaged [see Visaged *a.*].
B. ill assurance, want of assurance; **ill-formed** *a.* (earlier example and context in Locke); so **ill-formedness.**
1905 *Macm. Mag.* Dec. 151 As they grew out of the restive sensitiveness of youthful ill-assurance,..they turned with gratitude towards their parent. *c* **1909** D. H. Lawrence *Collier's Friday Night* (1934) ii. 51 Ernest (flushing up at the sound of her ill-assurance). *a* **1672** A. Bradstreet *Several Poems* (1678) 236 Thou ill-form'd offspring of my feeble brain. **1690** Locke *Hum. Und.* IV. iv. §16 They..make bold..to destroy ill-formed and mis-shaped Productions. **1972** *Language* XLVIII. i. 82 These have to do with (1) dialect variations and (2) degree of ill-formedness. **1972** A. Makkai *Idiom Struct. Eng.* 84 Let us now examine a few cases of lexemic illformedness.

Illano (ilyā·no). Also **I(l)lanon, I(l)lanum.** [Native name.] A member of a Moro people of Mindanao in the Republic of the Philippines; also, the language of this people. (Cf. Moro[2].) Hence **Illano·an** *a.*
1779 T. Forrest *Voy. New Guinea* xii. 302 They possess an island in the very heart of the Philippines, called Burias, where has been a colony of Illanos, for many years. **1821** W. Walton tr. *T. de Comyn's State of Philippine Islands* viii. 231 Long have the inhabitants of the Philippines deplored..the ravages committed on their coasts and settlements by..the Malanos, Ilanos and Tirones Moors and others. **1840** *Penny Cycl.* XVIII. 87/1 He was informed that the Illanos from that island [*sc.* Magindanao] had formed a settlement on Burias. **1848** F. S. Marryat *Borneo* 21 Pirate's Bay, so called from its being a favourite resort of the Illanoan pirates. **1898** F. T. Bullen in *National Rev.* Aug. 857 There is yet another small section of the inhabitants of this Archipelago, who were once the terror of the surrounding seas... These have long been known as 'Illanons'. **1900** F. H. Sawyer *Inhabitants Philippines* 364 The Moros Illanos..inhabit the country between the Bay of Iligan and Illana Bay, also round Lake Lanao, the Rio Grande and Lake Liguan. **1957** *Encycl. Americana* XIX. 472a/1 Some of the best-known Moro groups are the..Lanao Filipinos (divided into the Maranaos and Ilanums or Iranums). *Ibid.* XXI. 751/1 The principal languages and dialects with the number of persons speaking each were divided into the following groups:..Ilanon-Lanao-Lanao..109. Ilanon-Lanao-Maranoy..170,195.

illatinate (ilæ·tinět), *a.* (*sb.*). *rare.* [f. Il-[2] + Latin *a.* and *sb.* + -ate[2], after Illiterate *a.*] Having no knowledge of Latin; ignorant of Latin. Also as *sb.*
1922 S. Leslie *Oppidan* xxv. 345 They appeal even to the illatinate. **1941** E. Marsh tr. *Horace's Odes* Preface p. vii, Unless the version can give the illatinate reader some notion of Horace's quality as a poet, it is a superfluity.

illative, *a.* and *sb.* Add: **A.** *adj.* **4.** *Gram.* Denoting the case expressing motion into.
1890 A. S. Gatschet *Klamath Indians* 483 The suffix marks as well (1) a motion or direction toward an object or into a place or country, as (2) a stay or rest at or in a place, region, country. It is more frequently used in the former sense, and hence I have called the case the illative case. **1958** A. S. C. Ross *Etym.* 167 *Illative case*, the case of 'motion towards', as in Finnish *Helsinkiin* 'to Helsinki'. **1959** [see *elative *a.*]. **1968** C. J. Fillmore in Bach & Harms *Universals in Linguistic Theory* 9 Redden..finds five case indices in Walapai..and identifies each of these with terms taken from the tradition of case studies:..-*l* is illative/inessive, [etc.].
B. *sb.* **1.** Delete †*Obs.* and add examples.

1972 *Jrnl. Illinois State Hist. Soc.* LXV. 245 Nor should Illinoisans be surprised that politics is ever present.

illipe(e, var. ILLUPI (in Dict. and Suppl.).

illiquid, *a.* Add: **b.** Of an asset, investment, etc.: not easily or readily realizable. Hence **illiqui·dity,** the character of being illiquid.

1913 *Globe* 24 Oct. 6/4 Is cottage property a safe investment for the Reserve Fund? It is an illiquid asset, and a sudden drain on a fund thus locked up might be very inconvenient. 1927 *Glasgow Herald* 18 Apr. 9 The illiquidity of credit in Japan. 1930 *Times* 26 Mar. 26/7 Slow and illiquid loans and investments. 1948 G. CROWTHER *Outl. Money* (rev. ed.) ii. 80 There is some difference between having a variety of small loans in different degrees illiquid, and being closely associated..with one or two large concerns that are known to be doing badly. 1971 *Nature* 5 Feb. 363/2 Some of the officers of the British Association seem to be hoping that they can solve their chronic problem of what is called illiquidity by merging either with the Royal Society of Arts or with the Royal Institution. 1972 *Real Estate Rev.* Winter 27/2 Illiquidity is generally valued at about 1·5 percent to 2·0 percent in most analyses of the capitalization rate, so the equity trust shareholder holds his own quite nicely in the return contest upon closer inspection. 1973 M. WOODHOUSE *Blue Bone* xv. 167 One is liquid or illiquid at awkward times, just as with any other business.

illite (i·ləit). *Min.* [Named after the state of *ILL(INOIS: see -ITE¹.] Any of a group of clay minerals that belong to the mica group and are characterized by a lattice that does not expand through the absorption of water; *esp.* a mineral structurally and chemically similar to muscovite but differing in having more water and silicon and less potassium. Also used as a generic term for these minerals.

1937 R. E. GRIM et al. in *Amer. Mineralogist* XXII. 816 There remains only the alternative of giving a new name to the mica occurring in argillaceous sediments, and the term *illite*, taken from the State of Illinois, is here proposed. It is not proposed as a specific mineral name, but as a general term for the clay mineral constituent of argillaceous sediments belonging to the mica group. 1944 *Mineral. Mag.* XXVII. 60 Illite is not a stable mineral under humid temperate conditions. *Ibid.* 61 According to Fleischer, bravaisite may be the specific mineral species in illites. 1953 R. E. GRIM *Clay Mineral.* iii. 36 Grim, Bray, and Bradley gave the general formula for illites as $(OH)_4K_y(Si_{8-y}\cdot Al_y)(Al_4\cdot Fe_4\cdot Mg_4\cdot Mg_6)O_{20}$. In muscovite *y* is equal to 2, whereas in illite *y* is less than 2 and frequently equal to 1 to 1·5. According to the formula, illites would include both trioctahedral and dioctahedral types, and no attempt was made to differentiate between biotite and muscovite types of crystallization. At the present time the name illite is generally used, and will be used herein, for clay-mineral micas of both dioctahedral and trioctahedral types and of muscovite and biotite crystallizations. 1957 R. C. MACKENZIE *Differential Thermal Investigation of Clays* x. 260 Studies by Jackson et al. (1952) and Barshad (1954) suggest that illite itself should be considered as an interstratified mineral with mica and montmorillonite layers. 1959 C. S. HURLBUT *Dana's Man. Min.* (ed. 17) v. 463 Illite is the chief constituent in many shales. 1962 W. A. DEER et al. *Rock-Forming Min.* III. 213 The structure of illite is essentially that of a mica in that it contains layers with a plane of octahedrally coordinated cations sandwiched between two inward pointing sheets of linked $(Si, Al)O_4$ tetrahedra. *Ibid.* 214 A less common illite in which sodium is the inter-layer cation is called brammallite (Bannister, 1943). 1972 *Nature* 28 Jan. 220/1 The essential point about the quickclays is that they contain non-swelling clay minerals such as illite, chlorite and vermiculite, but not montmorillonite.

Hence **illi·tic** *a.*, containing, composed of, or characteristic of illite.

1949 KIRK & OTHMER *Encycl. Chem. Technol.* IV. 31 The illite group of minerals occurs abundantly in marine shales and in many soils, particularly in soils derived from illitic shales. 1963 D. W. & E. E. HUMPHRIES tr. *Termier's Erosion & Sedimentation* vi. 153 On the recent volcanic rocks of the northwest coast of Sumatra, H. Erhart (1954) has observed illitic soils between sea level and a height of 4,000 feet. 1972 *Nature* 28 Jan. 220/1 There are also, however, quickclays containing more than 80% particles finer than 2 μm, mainly of illitic nature.

illiteracy. Add: Also used more generally in sense: ignorance, lack of understanding (of any pursuit, activity, etc.). Cf. next.

1965 W. LAMB *Posture & Gesture* x. 128 Our 'illiteracy' in physical behaviour communications is related to our puppet-like proclivities. 1970 *Nature* 11 Apr. 96/1 The responsibility for carrying out the act was left to local authorities, many of which are apparently more alarmed at being thought to provide 'sex on the rates' than by the widespread sexual illiteracy in the British population. 1973 *Times Lit. Suppl.* 15 June 660/5 Combating what was not yet called physical illiteracy and teaching girls and women ease and freedom of carriage and movement.

illiterate, *a.* Add: Also used more generally in sense: characterized by ignorance or lack of learning or subtlety (in any sphere of activity). Cf. prec.

1953 *Ann. Reg. 1952* 397 The pre-war type of speculative builder's house, which by its illiterate design.. had been largely responsible for the poor reputation of English suburban architecture. 1956 C. S. LEWIS *Let.* (1966) 268 One must first distinguish the effect which music has on..people like me who are musically illiterate and get only the emotional effect. 1962 *Sunday Times* (Colour Suppl.) 10 June 4 The 'traditionalists', who are in the main musically illiterate.

‖ **illiterati** (ilitĕrā·ti, -ē·təi), *sb. pl.* [ad. L. *illitterātī*, pl. of *illitterātus*.] Illiterate, unlearned, or uneducated people. Cf. ILLITERATE *sb.*

1788 H. WALPOLE *Let.* 4 July (1905) XIV. 51 A lower species, indeed, is that of the scribes..who every night compose a journal for the satisfaction of such *illiterati*. 1822 *Blackw. Mag.* XI. 608 Ye're the most tinklertongued pack of illiterati. 1962 *Listener* 22 Nov. 877/1 Weekly periodicals for the younger *illiterati*.

illocution (ilokiū·ʃən). *Philos.* [f. IL-¹ + LOCUTION.] An act such as ordering, warning, undertaking, performed in saying something. Hence **illocu·tionary** *a.*

1955 J. L. AUSTIN *How to do Things with Words* (1962) viii. 99, I shall refer to the doctrine of the different types of function of language..as the doctrine of 'illocutionary forces'. *Ibid.* 101 Act (B) or Illocution: He urged (or advised, ordered, &c.) me to shoot her. *Ibid.* ix. 109 It is the distinction between illocutions and perlocutions which seems likeliest to give trouble. *Ibid.* 113 We must avoid the idea..that the illocutionary act is a *consequence* of the locutionary act. 1955 [see *EO IPSO *advb. phr.*]. 1963 M. FURBERG (*title*) Locutionary and illocutionary acts. 1964 *Philos. Rev.* LXXIII. 58 Describing and evaluating ..are only two among hundreds of kinds of illocutionary force. 1972 J. RAWLS *Theory of Justice* §62. 406 Nor do I oppose the view that a certain illocutionary force is central to 'good'. 1973 *Times Lit. Suppl.* 5 Oct. 1161/5 The illocutionary act was the act performed by a speaker *in* saying something, such as the act of asking or answering a question.

illogic. (Later examples.)

1955 *Bull. Atomic Sci.* Apr. 131/1 So there was no illogic or softheadedness in the Commission's concern about the dangers inherent in the investigative process or in clearance determinations. 1965 *Sunday Times* (Colour Suppl.) 23 May 21/1 There's no room for illogic in an Ellery Queen mystery. 1972 *Nature* 1 Sept. 54/2 With the same illogic with which the Englishman will eat cow but not dog, will hang game but not chicken.

ill treatment, ill-treatment. [ILL *a.* 2, TREATMENT I. Cf. ILL-TREAT *v.*] Bad or unfavourable treatment; rough handling; harsh or unsympathetic dealings.

1667 T. SPRAT *Hist. R. Soc.* 401 A just occasion of lamenting the ill Treatment which has bin most commonly given to Inventors. 1676 LOCKE *Jrnl.* 23 Aug. in *Ess. Law Nature* (1954) 275 Making yourself an enemy to all by ill-treatment. 1677 —— tr. *Nicole's Ess.* (1828) 112 Ill treatment and persecutions would certainly fall to their lot. 1713 [see ILL *a.* 2 a]. 1811 A. GRAYDON *Mem.* 292 In revenge for some real or supposed ill-treatment. 1818 *Public Ledger* 8 Apr. 3/2 Alledged ill-treatment during a short confinement in that prison. 1864 E. B. PUSEY *Daniel* i. 21 Who revenge on man their illtreatment at the hand of man. 1879 G. MEREDITH *Egoist* II. v. 94 He contemplated her with an air of stiff-backed ill-treatment. 1905 CHESTERTON *Heretics* 79 A permanent possibility of selfishness arises from the mere fact of having a self, and not from any accidents of education or ill-treatment.

illuk (i·luk). [Sinhala.] The name used in Sri Lanka (Ceylon) for a coarse grass, *Imperata cylindrica*; = *LALANG.

1864 G. H. K. THWAITES *Enumeratio Plantarum Zeylaniæ* v. 369 *I[mperata] arundinacea*... Common in the hotter parts of the island. Nom. vulg. 'Illook'. The leaves make an excellent thatch. 1931 E. K. COOKE *Geogr. Ceylon* II. x. 164 The grass of Talawa is usually long and coarse, very different from the average patana grass; it usually reaches about three or four feet in height. The best known species are mahna and illuk. 1950 [see *CHENA]. 1956 R. W. SZECHOWYCZ in S. Haden-Guest et al. *World Geogr. Forest Resources* xxii. 485 There is a total absence of undergrowth [in the savanna forest zone of Ceylon], and the herb layer consists of grasses, of which *Imperata cylindrica* (illuk) is the most prominent.

illume, *sb.* (Later *poet.* example.)

c 1882 E. DICKINSON *Poems* (1955) III. 1069 Better an ignis fatuus Than no illume at all.

illuminance (iliū·minăns). *Optics.* [f. L. *illūmin-āntem* (see ILLUMINANT *a.* and *sb.*) + -ANCE.] The amount of luminous flux per unit area; = ILLUMINATION 1 b.

1943 D. H. JACOBS *Fund. Optical Engin.* iv. 66 We have derived equations that give the illumination (or illuminance) only at points on the axis of the system. 1949 H. MARGENAU et al. *Physics Princ. & Applications* xliv. 621 Illuminance is expressed in lumens/ft² (foot candles) or lumens/m² (luxes). 1955 R. C. BROWN *Textbk. Physics* IV. I. 902 The illumination (sometimes called 'intensity of illumination' and also illuminance) of a surface is defined as the luminous flux incident per unit area on the surface. 1961 F. W. WEYMOUTH in Hirsch & Wick *Vision Aging Patient* iv. 41 Since the extent of dark adaptation is..curtailed with age, the performance at low levels of illuminance begins to decline even before the age of 40 years. 1970 *Nature* 24 Jan. 347/2 Each pattern..was projected..at an average screen illuminance of 3 millilamberts.

illuminate, *ppl. a.* and *sb.* **A.** *adj.* **2.** Delete †*Obs.* and add later example.

1909 *Westm. Gaz.* 19 May 4/2, I once knew a follower of the Rosy Cross, an illuminate member.

B. *sb.* (Later example.) Also *spec. pl.* = ILLUMINATI b.

1906 *Edin. Rev.* July 49 The influence of the German 'illuminates' on the French lodges. 1946 G. G. SCHOLEM *Major Trends Jewish Mysticism* (rev. ed.) iv. 121 The outspoken illuminates and ecstatics.

illuminate, *v.* Add: **1. c.** To direct a beam of any kind of radiation at (an object or region): used esp. of radio waves and microwaves in connection with radar and telecommunication.

1942 [implied in *ILLUMINATED *ppl. a.* 1 b]. 1947 CROWTHER & WHIDDINGTON *Science at War* I. 6 The power radiated proved ample to detect aircraft, flying at a height..sufficient to bring them within the space 'illuminated', at the range of 75 miles. 1948 POLLARD & STURTEVANT *Microwaves & Radar Electronics* iv. 126 The..antenna pattern from a 30-inch-diameter dish illuminated at 3·2 centimeters..is shown. 1957 R. WATSON-WATT *Three Steps to Victory* 470 Of the secondary radiations, excited by 'illuminating' the craft by ground installations emitting light, heat, sound or radio-waves, the first two are excluded by atmospheric absorption. 1973 *Physics Bull.* Mar. 149/1 A cold cathode discharge source..provided a beam of electrons which could illuminate an object such as an aperture or a wire grid.

7. b. (For Dict. entry read:) *intr.* To take fire, to light up; of a town, etc.: to be decorated with lights as a sign of festivity or celebration. Also, to become excited (see also quot. 1926).

1706 DEFOE *Review* 2 Mar. 108 The other Gentleman and his Man coming, and holding a real Candle up to Decide it, Affirm'd the Doctors Candle was Lighted; the Grave Gentleman went on, with an *Ay, I think it does begin to Illuminate.* 1801 [in Dict.]. 1843 *Punch* IV. 256 If Stockport and Paisley do not..illuminate when they shall hear the glad tidings, then is the spirit of manufacture dead to gratitude. 1871 'MARK TWAIN' *Sk. New & Old* (1875) 176 For eight-and-forty hours no soul in all the barony but did dance and sing, carouse and illuminate, to celebrate the great event. 1926 WOOD & GODDARD *Dict. Amer. Slang* 24 *Illuminated*, lit, drunk. 1927 E. H. C. MOBERLY BELL *Life & Lett. C. F. Moberly Bell* 96 The people illuminate, and go mad with frenzy for Gordon, and curse the Mahdi. 1972 *Jrnl. Social Psychol.* LXXXVII. 90 Subjects were asked to estimate the probability that an alleged peer..would shock them when a light illuminated.

illuminated, *ppl. a.* Add: **1. b.** Made, or being, the target of (non-visible) radiation of some kind.

1942 J. C. SLATER *Microwave Transmission* vi. 275 The illuminated region can be considerably greater than the distance between poles. 1950 J. D. KRAUS *Antennas* xii. 345 An antenna with a uniformly illuminated circular aperture 10 wavelengths in diameter has a gain of 600 or nearly 28 db with respect to a ½-wavelength dipole antenna. 1966 TOLSTOY & CLAY *Ocean Acoustics* vi. 198 The dimensions of the illuminated area are much larger than the acoustic wavelength.

5. a. Also as *sb. rare.*

1923 E. POUND *Let.* 12 May (1971) 187 Price 25 dollars per copy, and 50 and 100 bones for Vellum and illuminateds.

illuminating, *vbl. sb.* (s.v. ILLUMINATE *v.*). Add: Also *attrib.*, as **illuminating engineering,** the branch of engineering and applied science concerned with the design, installation, and modification of artificial lighting; so **illuminating engineer.**

1906 (*title*) Transactions of the Illuminating Engineering Society. 1907 *Installation News* Nov. 3 That latest addition to the ever-growing family of specialists—the Illuminating Engineer—is providing the electrical industry and press with a good deal to think about. 1930 *Engineering* 31 Jan. 143/2 The development of the science of illuminating engineering has been continuous during the past few decades.

illuminatingly, *adv.* (Later examples.)

1965 *Language* XLI. 216 The adjective structures illuminatingly discussed by Lees. 1972 *Nature* 28 Jan. 204/2 Another examination scheme has also proved illuminatingly effective.

illumination. Add: **1.** (Further examples, corresponding to *ILLUMINATE *v.* 1 c.)

1942 J. C. SLATER *Microwave Transmission* vi. 275 Vertically the illumination falls off gradually in intensity as the poles are approached. 1948 POLLARD & STURTEVANT *Microwaves & Radar Electronics* iv. 127 A horn feed illumination is the most widely favored. 1966 TOLSTOY & CLAY *Ocean Acoustics* vi. 199 It is convenient to measure the scattered signal relative to the signal reflected by a mirror-like surface when the illumination factor, source position, receiver position, etc., are the same.

b. For 'reflected from' read 'incident on'. (Further examples.)

1943, 1955 [see *ILLUMINANCE].

illu·minatively, *adv. rare.* [f. ILLUMINATIVE *a.* + -LY².] In an illuminative way.

1925 T. DREISER *Amer. Trag.* (1926) II. III. xxviii. 349 These hard, white-painted walls brightly lighted..by incandescent lamps in the hall without at night—yet all so different from Bridgeburg,—so much more bright or harsh illuminatively.

‖ **Illuminé** (ilümine). Also with lower-case initial. [Fr.: see ILLUMINEE.] One of the Illuminati.

1794 A. PAGET *Let.* in *Paget Papers* (1896) I. 20 It would be highly interesting to Your Lordship to get some information relative to the people called *Illuminés*. 1799 [see ILLUMINEE]. 1816 SCOTT *Antiq.* I. xiii. 289 A simple youth whispered me that he was an *Illuminé*, and carried on an intercourse with the invisible world. 1927 R. FRY *Let.* 19 Aug. (1972) II. 605 Simon Bussy..said.. that I was an *illuminé* who imagined such things and then got excited about them.

illuminized (iliū·minəizd), *ppl. a. rare.* [f. ILLUMINIZE *v.* + -ED¹.] Initiated (see ILLUMINIZE *v.* 2).

1920 *19th Cent.* July 104 A great Masonic meeting which was held in 1786 at Frankfurt—whither the headquarters of illuminised Freemasonry were removed after the famous Congress.

illuminometer (iliūminǫ·mītəɹ). [f. ILLUMIN(ATION + -O + -METER.] A photometer, *esp.* one for measuring the illumination of surfaces (rather than the intensity of light sources).

1895 HOUSTON & KENNELLY in *Operator & Electr. World* 9 Mar. 309/2 The question at issue is..the degree of illumination actually produced, and this question cannot be decided by the photometer, but requires the use of an illumination measurer or illuminometer. *Ibid.* 310/2 We call our instrument an illuminometer and have filed applications for a patent on the same. 1908 *Illuminating Engineer* (London) I. 714/2 It was proposed to use it [*sc.* Houston and Kennelly's photometer] for candlepower photometry as well as for the measurement of illumination. The unfortunate name 'illuminometer' was given to the instrument, and..this mongrel word is sometimes used at the present day. 1953 KIRK & OTHMER *Encycl. Chem. Technol.* XI. 326 The luminous intensity of the flame..can be measured by an illuminometer consisting of a barrier-layer photocell connected to a micro-ammeter. 1971 *Jrnl. Gen. Psychol.* LXXXIV. 125 Exposure times of 40, 150, and 3500 msec and luminances of three and 30 fc, as read with a Macbeth Illuminometer, were employed.

illupi. Add: So *illipe butter*, any of various vegetable fats.

1904 J. LEWKOWITSCH *Oils, Fats, & Waxes* (ed. 3) II. xiv. 702 The commercial fat is a mixture of Mowrah seed oil with Mahua butter or illipe butter, the fat from *Bassia latifolia*. 1911 *Encycl. Brit.* XX. 47 Mahua butter, Illipé butter. 1951 A. E. BAILEY *Industr. Oil & Fat Products* (ed. 2) vi. 141 Mowrah fat, or illipé butter, which is obtained from the Indian plant *Bassia longifolia*, resembles shea butter in being relatively unsaturated, and correspondingly soft.

illusion. Add: **4. b.** *the argument from illusion* (Philos.): the argument that the objects of sense-experience, usually called ideas, appearances, or sense-data, cannot be objects in a physical world independent of the perceiver, since they vary according to his condition and environment.

1932 H. H. PRICE *Perception* ii. 27 It is commonly held that the *Argument from Illusion* (as it is called) is sufficient to refute Naïve Realism. 1936 A. J. AYER *Lang., Truth & Logic* viii. 228 The so-called argument from illusion. 1940 —— *Found. Empirical Knowl.* i. 3 The answer is provided by what is known as the argument from illusion. 1959 J. L. AUSTIN *Sense & Sensibilia* (1962) iii. 20 The primary purpose of the argument from illusion is to induce people to accept 'sense-data'. 1971 A. FLEW *Introd. Western Philos.* 353 We can characterize the Argument from Illusion as the attempt to show that..what we actually see is never things but only the appearances of things.

5. (Earlier and later examples.)

1857 *Lawrence* (Kansas) *Republican* 28 May 3 Lace, Gimp, Straw, Silk, Blend, Hair and Illusion Bonnets. 1865 F. B. PALLISER *Hist. Lace* xxxv. 423 M. Doguin, who afterwards used the fine silks, and invented that popular material first called 'zephyr', since 'illusion'. 1869 L. M. ALCOTT *Little Women* II. xiv. 207 She put on Flo's old white silk ball dress, and covered it with a cloud of fresh illusion. 1873 *Young Englishwoman* Jan. 50/1 White muslin, trimmed with white lace, illusion veil. 1902 *Daily Chron.* 13 Dec. 8/4 It was trimmed with ruches of black tulle illusion. 1967 *Boston Sunday Herald* 30 Apr. v. 2/3 The headpiece was a lace mantilla with an illusion veil.

6. *attrib.* and *Comb.*, as *illusion effect*; *illusion-disturbing, -proof* adjs.

1909 W. M. URBAN *Valuation* ix. 275 The elimination of illusion-disturbing moments is a conscious process. 1971 *Jrnl. Gen. Psychol.* LXXXV. 77 Differences in the magnitude of the illusion-effect were introduced. 1924 G. B. SHAW *Saint Joan* p. xvi, Napoleon or any other illusionproof genius.

Hence **illu·sionless** *a.*, not characterized by illusions.

1897 G. B. SHAW in *Academy* 16 Jan. 67/2 Illusionless conversation..when the old people in Ibsen..tell each other the frozen truth. 1964 S. BELLOW *Herzog* 229 His face was illusionless, without need of hypocrisy.

illusional (iliū·ʒənăl), *a.* [f. ILLUSION + -AL.] Pertaining to, characterized by, or subject to illusions.

1900 DORLAND *Med. Dict.* 317/1. 1923 *Daily Mail* 1 Mar. 4 On reception into prison he was suspicious, illusional, and confused. 1942 *Art Digest* 15 Jan. 30 She also goes in for illusional effects, such as an interwoven lost-and-found sensation in *Malaga Cathedral* and the vaporous beach scene *Le Bar*.

illusioned (iliū·ʒənd), *ppl. a.* [f. as prec. + -ED².] Full of illusions.

1920 *Glasgow Herald* 6 July 6 A fervent but illusioned admirer. 1921 GALSWORTHY *To Let* III. vi. 267 Even his love felt tainted, less illusioned, more of the earth. 1971 *Guardian* 11 Oct. 10 We remain totally illusioned..about the general good intentions of those who practise the art of government.

illusionism. Add: **2.** The use of illusionary effects in art or sculpture.

1951 M. L. WOLF *Dict. Arts* 345/2 *Illusionism*, the effort in painting or sculpture to create as nearly as possible an illusion of visual reality. 1960 [see *FANTASIST]. 1961 M. LEVY *Studio Dict. Art Terms* 62 *Illusionism*, the practice of Trompe L'Oeil painting. 1962 *Times* 6 Apr. 17/3 It [*sc.* the attitude and style of pop-art] is a gaudy glamorization of its subject which deliberately crashes its visual gears from abstract pattern right through to *trompe-l'oeil* illusionism.

illusionist. Add: **2.** (Earlier example.)

1850 *Punch* XIX. 215/2 The novel trick of shifting the lectern, will be introduced by that celebrated illusionist, the Rev. Mr. Cozens.

Hence **illusioni·stic** *a.*, pertaining to illusionism or the illusionists.

1911 *Encycl. Brit.* XXIII. 482/1 The lighting is carefully calculated with a view to illusionistic effect under the local conditions. 1938 *Mod. Lang. Rev.* Oct. 549 The tendency..to employ an illusionistic realism for the purpose of sensationalism..is what lies behind Wölfflin's conception of 'baroque'. 1944 *Burlington Mag.* Aug. 199/1 Compared with this the Byzantine solution may still be called illusionistic. 1958 *Times* 24 Sept. 3/1 They [*sc.* Dali's pictures] are irritatingly artificial concoctions with passages..quite amazingly inept for a specialist in illusionistic devices. 1968 *Times* 21 Dec. 19/1 There was an upsurge of decorative architecture in Bavaria, Franconia and Swabia. It was illusionistic and what the writer calls 'a-tactonic'.

illustrate, *v.* **3.** Delete † *Obs.* and add examples.

1843 DICKENS *Mart. Chuz.* (1844) ii. 10 Charity..did so well set off and illustrate her sister! 1917 *Mod. Lang. Rev.* XII. 205, I do not desire to illustrate my version of 'the Fause Knight', but merely to claim that it throws new light on the subject matter of the original ballad.

4. (Later examples.) Still *rare.*

1858 J. H. NEWMAN *Mission of Benedictine Order* in *Sel. Ess.* (1902) 191 The famous Congregation of Cluni, illustrated by St. Majolus, St. Odilo, Peter the Venerable. 1931 [see *DITHER *sb.* b]. 1952 G. SARTON *Hist. Sci.* I. xi. 277 We now come to the greatest mathematician of the century, the first man to illustrate the name of Hippocrates.

6. Also *absol.*

1863 O. M. MITCHELL *Astron. Bible* 168 The subject is difficult. Permit me to illustrate.

illustrating, *ppl. a.* (Later example.)

1840 M. EDGEWORTH *Let.* 30 Dec. (1971) 574 The *illustrating* speech of the Irishman to the waiter of 'I'd wish to have the ham and the butter separate'.

illustrational, *a.* Delete *rare* and add examples. Also as *sb.*

1926 R. FRY *Transformations* 147 Rodin's concern is with the expression of character and situation, it is essentially dramatic and illustrational. 1955 P. HERON *Changing Forms of Art* III. xi. 168, I believe Sutherland's phantasy is essentially illustrational, poetic, non-plastic. 1959 *Listener* 6 Aug. 206/1 Everything in fiction as in life had to satisfy his [H. James's] 'appetite for the illustrational'.

illustrious, *a.* Add: **3. c.** *most illustrious*: the special epithet of the Order of St. Patrick.

[1783 *London Gaz.* 4–8 Feb. 1/1 The King has been pleased to order Letters Patent to be passed under the Great Seal of the Kingdom of Ireland, for creating a Society, or Brotherhood, to be called Knights of the Illustrious Order of St. Patrick.] 1858 B. BURKE *Bk. Orders Knighth.* 102 (*heading*) The Most Illustrious Order of Saint Patrick. 1971 *Whitaker's Almanack* 259 The Most Illustrious Order of Saint Patrick... *Ribbon*, Sky Blue. *Motto*, Quis separabit?.. (No conferments since 1934).

illustriousness. Add: **2.** [tr. G. *durchlaucht.*] With possessive adjective as a title of dignity or honour given to German princes.

1929 'S. N. D.' *Sir W. Howard Visct. Stafford* iv. 48 His Electoral Illustriousness, the Prince Palatine of the Rhine.

illuvial (iliū·viăl), *a. Soil Sci.* [f. IL-¹ + -*luvial*, as in ALLUVIAL, ELUVIAL *adjs.*] = *ILLUVIATED *ppl. a.*; also, resulting from illuviation.

1924 *Geol. Mag.* LXI. 451 Three main horizons are generally recognized,..the A or eluvial horizon, the B or illuvial horizon, and the C horizon, which consists of the parent material. 1932 G. W. ROBINSON *Soils* iv. 92 We may also distinguish a type of clay which has been enriched by illuvial accumulation. 1952, 1963 [see *ELUVIAL *a.* 2].

illuviation (iliūviē̆·ʃən). *Soil Sci.* [f. prec. + -ATION.] The deposition of salts or colloids in a soil horizon from percolating water which has removed them from another, generally superior, horizon. So **illu·viated** *ppl. a.*, having received material by illuviation.

1928 *Bull. Amer. Soil Survey Assoc.* IX. 31 Illuviation. *Ibid.* 37 The illuviated horizons of the solum. 1932 *Forestry* VI. 28 The different horizons are designated by capital letters...'B' the horizon of illuviation or deposition. 1949 W. W. WEIR *Soil Sci.* (ed. 2) vi. 116 Horizons, commonly topsoil layers, that have lost materials through eluviation are described as eluvial or eluviated; and horizons, commonly subsoil layers, that have received the materials, illuvial or illuviated. 1955 F. E. BEAR *Chem. of Soil* i. 37 In pedology, the term eluviation has been applied to the loss of material from the surface horizon, and the term illuviation, to the gain of material by the subsoil horizon.

ill-wish, *v.* Add: Also *absol.* Hence **ill-wish** *sb.*, the evil or misfortune wished.

1853 T. Q. COUCH in J. Brand *Observations Pop. Antiq.* (1870) III. 101 The witch's malice, or, as it is termed, 'the ill-wish'. 1899 S. BARING-GOULD *Bk. of West* II. 16 A bard, and after him a saint, might not ill-wish unless he had been refused a just request. *Ibid.*, If he ill-wished unjustly, then it was held that the ill-wish returned on the head of him who had launched it.

illy, *adv.* For '*dial.*' read 'chiefly *U.S.*', and add later examples.

1903 *Atlantic Monthly* July 116 Which is far better than if it had rushed into illy considered legislation. 1904 CLAIBORNE *Old Virginia* 268 (Th.), I dropped one of my gauntlets,..which I could illy afford to lose. 1925 T. DREISER *Amer. Trag.* (1926) II. xlvi. 57 By contrast with Sondra, how illy-dressed in the blue traveling suit..she had equipped herself for this occasion. 1927 C. A. & M. R. BEARD *Rise Amer. Civilization* II. 213 To meet a crisis of this kind labor was illy prepared. 1965 *Sat. Rev.* (U.S.) 5 June 28/1 *The Blue Hen's Chick* is no graceful memoir of a life well or illy spent.

Illyrian (ili·riăn), *a.* and *sb.* [f. L. *Illyrius*, a. Gr. Ἰλλυριός.]

A. *adj.* **1.** Of or pertaining to the Illyrians or to ancient Illyria (or Illyricum), a country lying east of the Adriatic and at an early period extending northwards to the Danube.

1553 N. GRIMALDE tr. *Cicero's Bookes of Dueties* (1558) II. f. 88ᵛ Bargulus, the Illirian robber, of whom mention is made in Theopompus. 1593 SHAKES. *2 Hen. VI*, IV. i. 108 Bargulus the strong Illyrian Pyrate. 1678 J. DAVIES tr. Appian (*title*) The History of Appian, of Alexandria. In Two Parts. The First consisting of the Punick, Syrian, Parthian, Mithridatick, Illyrian, Spanish, and Hannibalick, Wars. 1776 GIBBON *Decl. & F.* I. i. 27 The provinces of the Danube soon acquired the general appellation of Illyricum, or the Illyrian frontier. 1797 *Encycl. Brit.* II. 714/2 The Heneti, an Illyrian people. 1880 *Ibid.* XII. 709/1 The Danube..was the limit of the Illyrian tribes towards the north. 1935 HUXLEY & HADDON *We Europeans* vi. 179 The Dinaric (Illyrian) type [of people] is found on both sides of the northern Adriatic, the Illyrian mountain system with extensions to the north and south. 1953 [see *BANDKERAMIK].

b. Of or belonging to Illyria, a former division of Austria-Hungary, since 30 October 1918 forming part of Yugoslavia.

Illyrian Provinces, a division, made by Napoleon Bonaparte in 1809, of various Austrian territories lying north and east of the Adriatic, and abolished in 1814–15, after which it was made a nominal kingdom of the Austrian Empire.

1820 C. KELLY *Hist. French Revolution* I. xiii. 614 The treaty of Vienna has given to France a great extent of territory on the Adriatic sea: this territory is formed into the Illyrian republic. 1838 *Penny Cycl.* XII. 445/2 Napoleon,..in 1809, gave to several tracts of territory ceded by Austria..the name of the Illyrian Provinces. 1845 *Encycl. Metrop.* XIII. 1084/2 Meanwhile the war in the Illyrian provinces..received a fresh complication from the secession of Bavaria from the French alliance. *Ibid.* XX. 559/1 The Emperor of Austria rules the Illyrian Kingdom with uncontrolled authority. 1921 *19th Cent.* May 856 The Illyrian provinces..provided a strong bulwark against Austria. 1965 *New Cambr. Mod. Hist.* IX. xi. 331 The Illyrian provinces taken from Austria in 1809 remained directly under the control of Napoleon through a governor-general.

c. In literary use, pertaining to the regions lying along the east coast of the Adriatic.

This corresponds to the literary use of *Illyria*, which was retained even when the name had no political significance.

1852 M. ARNOLD *Empedocles on Etna* I. ii. 41 The Adriatic breaks in a warm bay Among the green Illyrian hills... There those two live, far in the Illyrian brakes. 1853 TENNYSON *To E. L.* in *Poems* 352 Illyrian woodlands, echoing falls Of water. 1950 J. C. FENNESSY *Way to Sea* xvii. 131 Lapis lazuli blue as the Illyrian sea.

d. Used in the names of breeds of dog that originated in this part of Yugoslavia, as *Illyrian hound, Illyrian sheepdog.*

1935 *Discovery* Oct. 310/2 The fierce dogs of the Yugoslav South..are now officially called Illyrian Sheepdogs. 1948 A. LOKAR in B. Vesey-Fitzgerald *Bk. Dog* II. 519 The Illyrian Sheepdog is..the best-known Yugoslav sheep-herding dog and is in big demand in the hills in summer, and in the valleys when the sheep and goats come down for winter feeding. 1964 E. F. DAGLISH tr. *Schneider-Leyer's Dogs of World* 198 The Illyrian Hound is not directly descended from the oft-mentioned Celtic Hound. 1971 F. HAMILTON *World Encycl. Dogs* 376 Illyrian Hound. This medium-sized hound is named for chief functions of the tuned radio-frequency input amplifier is to prevent simultaneous reception of two stations in this way. By tuning this amplifier to the desired signal, the undesired or 'image' frequency is discriminated against. **1940** *Amat. Radio Handbk.* (ed. 2) 24/2 Interference is still likely to result from an incoming signal of such a frequency as to produce the correct intermediate frequency. This is known as 'image' or 'second channel interference'. **1950** K. HENNEY *Radio Engin. Handbk.* (ed. 4) xvii. 802 Since the function of the converter is to produce the difference between applied frequencies, it cannot distinguish between the signal and the image and produces i-f output from each. **1962** *B.B.C. Handbk.* 130 The selectivity of the receiver is improved and this reduces 'second channel', alternatively called 'image', interference. This is generally characterized by an irritating whistle of constantly changing pitch, sometimes accompanied by unwanted morse signals and programme modulation. Much of the interference experienced in the short-wave broadcast bands is due to such image effects. **1967** R. L. SHRADER *Electronic Communication* (ed. 2) xviii. 392/1 A second RF amplifier will reject the image very well. However, at frequencies in the 30-MHz range, for example, even the two RF amplifiers may not reject images satisfactorily.

8. *image-apprehension, -association, -brilliance, -complex, -field, -formation, -motif, -pattern, -sound, -substitute, -type, -world; image-crowded, -laden, -ridden, -seeing* adjs.; **image cluster** (see CLUSTER *sb.* 3 b); **image converter,** an image tube, *esp.* one for converting an invisible image formed by infrared or other invisible radiation into a visible one; **image dissector,** a kind of television camera tube in which a photo-emissive surface receives the image and the corresponding pattern of emitted electrons is deflected in a scanning pattern to and fro across a point anode, producing the video signal; **image frequency**: see sense 7** above; **image iconoscope,** a kind of television camera tube combining the iconoscope and the image dissector, the target plate receiving not the optical image (as in the former) but a pattern of emitted electrons produced by the image at a photo-emissive surface (as in the latter); **image intensifier,** an image tube or other device in which an image is formed by light or other radiation on a photo-emissive or photoconductive surface and the resulting flow of electrons utilized to produce a corresponding visible image of increased brightness; **image interference**: see sense 7** above; **image orthicon,** a kind of television camera tube in which a flow of electrons, produced as in an image tube, strikes a thin glass sheet and forms on it a pattern of positive charges corresponding to the picture, the video signal being derived from the variation this produces in a scanning electron beam that strikes the other side of the sheet (neutralizing the charge at that point) and returns to the electron gun and associated electron multipliers with an intensity reduced in accordance with the magnitude of the neutralized charge; **image toy,** a small decorative figure in earthenware, esp. one made in the 18th century by John Astbury (see *ASTBURY) or Thomas Whieldon; **image tube,** an electron tube in which an image, formed by light or other electromagnetic radiation on a photo-emissive surface, causes it to emit a corresponding flow of electrons which may be used to reproduce the image in a different form (as in an image converter or an image intensifier).

1962 I. M. CROMBIE *Exam. Plato's Doctrines* I. iii. 120 The criteria employed in calling things ducks do not constitute more than an image-apprehension of duckdom. *a* **1930** D. H. LAWRENCE *Apocalypse* (1932) 97 The curious image-association. The oriental loved that. **1950** *Essays & Stud.* III. 39 The striking image-associations of this passage were noted by W. Clemen in *Shakespeares Bilder*. **1946** *Nature* 19 Oct. 533/2 Under the title 'geometrical properties of visual instruments' are discussed such questions as field-size..and image-brilliance in different parts of the field. **1946** E. A. ARMSTRONG *Shakespeare's Imagination* 184 As no two poets employ the same image clusters, therefore works of doubtful provenance can be assigned to a poet with certainty if it contains clusters..characteristic of writings known to be authentic. **1961** *N. & Q.* Apr. 156/2 Particular stress is put on imagery and image-clusters in the cases of *Edward III* and *The Two Noble Kinsmen*. **1963** *Ibid.* Sept. 332/1 One's faith in image-clusters as evidence for authorship tends..to be diminished by such a coincidence. **1966** *English Studies* XLVII. 302 A Bible-inspired image-complex in Vaughan's poems. **1946** *Electronic Engin.* XVIII. 157/1 The principle of the infra-red image convertor is fairly well known. **1950** P. PARKER *Electronics* xvii. 843 An image converter is a device which converts an image formed by light-rays on a photo-cathode into one formed by electron beams... The name image converter is generally kept, however, for tubes in which the electron image is formed on a fluorescent screen. **1952** *Electronic Engin.* XXIV. 307/1 The case of an image convertor of the ME1200 type will enable infra-red photographs to be taken using normal high-speed emulsions. **1959** *Proc. IRE* XLVII. 905/1 A group of American astronomers have undertaken the development of an image-converter tube which permits the electron image to emerge through a thin membrane or foil to expose an external photographic plate. **1968** L. LEVI *Appl. Optics* vi. 268 Image converters greatly enhance night vision..and have in this capacity served in both military and zoological applications. **1911** W. B. YEATS *Plays for an Irish Theatre* p. ix. We feel our minds expand convulsively or spread out slowly like some moon-brightened image-crowded sea. **1934** P. T. FARNSWORTH in *Jrnl. Franklin Inst.* CCXVIII. 411 Means for producing these saw-tooth currents and means for synchronizing them between the 'Image Dissector', or transmitting tube, and the receiving tube, are discussed. **1968** *Brit. Med. Bull.* XXIV. 261/1 A non-storage camera tube of the type known as an 'image dissector' also possesses some very favourable properties, particularly in so far as resolution is concerned. *Ibid.* 261/1 The only instruments of real interest at the moment seem to be those capable of automatic measurement of optical density at many different points of an image field. **1923** J. S. HUXLEY *Ess. Biologist* ii. 81 Even in the most 'intelligent' of birds or mammals, the power of image-formation is very probably absent, and the power of concept-formation..certainly so. **1972** *Jrnl. Social Psychol.* LXXXVII. 37 Mental imagery and image formation have recently become a topic of concern again. **1939** *Proc. IRE* XXVII. 547/1 These image iconoscopes are practical working tools, advanced well beyond the laboratory stage. **1951** AMOS & BIRKINSHAW *Television Engin.* (rev.) I. iv. 75 Image iconoscopes, in common with all high-velocity tubes, have tended to be superseded by image orthicons in nearly all television services. **1939** L. M. MYERS *Electron Optics* viii. 449 We might term the device an image intensifier. *Ibid.* 450 An image intensifier was indeed the unrealised and unrealisable dream of the mechanical optical television engineer. **1954** *Radiology* LXIII. 870 An ideal image intensifier would receive only information-bearing X-ray signals from the subject and construct therefrom an image of arbitrary size and brightness. **1959** *Proc. IRE* XLVII. 909/1 An entirely different principle of image amplification is used in the solid-state image intensifier... In the simplest form of this device, a phosphor layer is placed between two conducting plates to which an electric field of about 100 v is applied. If an ultraviolet image is focused on the screen from the front, a marked increase in light emission is produced. **1969** *New Scientist* 22 May 485/1 At Herstmonceux experiments are in progress on a number of different types of image intensifier to find the most suitable system to be used in conjunction with the 100-in. Newton telescope. **1904** *Westm. Gaz.* 27 Aug. 6/2 This heavily-scented, image-laden atmosphere. **1943** D. GASCOYNE *Poems 1937–1942* 33 Blows back With long-held burning breath through eyeholes bored By image-laden rays. **1937** E. ELLIS-FERMOR *Some Recent Res. Shakes. Imagery* 23 This [sc. Kolbe's *Shakespeare's Way*]..has some illuminating suggestions about the underlying image-motifs in the plays. **1945** *Birmingham* (Alabama) *News* 29 Oct. 9/2 The product of RCA engineers..the device is known as an 'image orthicon'. **1946** *Proc. IRE* XXXIV. 428/2 The image orthicon derives its increased sensitivity over the iconoscope and orthicon from (1) the higher photosensitivity of a continuous photocathode relative to that of an insulating mosaic; (2) the multiplication by secondary emission of the electron current to the target and (3) the use of an electron multiplier for the signal current. **1953** AMOS & BIRKINSHAW *Television Engin.* I. v. 101 The image orthicon tube may be regarded as a combination of an orthicon tube with an image stage similar to that used in the image-iconoscope. **1971** H. E. ENNES *Television Broadcasting* i. 20 There are three basic types of pickup tubes used in television cameras: the image orthicon, the vidicon, and the lead oxide. The image orthicon is used primarily in monochrome studio and field cameras for live pickups. **1949** C. DAY LEWIS *Poetic Image* 24 Its image-pattern is so skilfully composed from certain theme-images. **1949** A. M. FARRER *Rebirth of Images* i. 21 The evidence for the unity of the Johannine writings..lies in the identity of image-pattern in the Gospel and the Apocalypse. **1925** AUDEN & ISHERWOOD *Dog beneath Skin* i. 33 Our impulses are unseasonal and image-ridden. **1939** A. HUXLEY *Holy Face* 21 Image-seeing and poetical. **1920** D. H. LAWRENCE *Purnogr. & So On* (1936) 73 It is, for one thing, the imagination, which is dangerous imagination. **1925** I. A. RICHARDS *Princ. Lit. Crit.* 123 But the degree of correspondence between the image-sounds, and the actual sounds that the reader would produce, varies enormously. *Ibid.* 120 Something takes the place of vivid images in these people and..provided the image-substitute is efficacious, their lack of mimetic imagery is of no consequence. **1957** MARKOWITZ & HAGAR *Conc. Encycl. Eng. Pottery & Porcelain* 115/1 *Image toys*, earthenware, stoneware or porcelain figures. **1960** H. HAYWARD *Antique Coll.* 147/1 *Image toys*, contemporary designation of mid-18th cent. Staffordshire pottery figures. **1971** *Country Life* 2 Dec. 1555/1 Pioneer maker of image toys in Staffordshire was John Astbury (1688–1743), his colours restricted to the browns and whites of his burnt clay. **1936** *Jrnl. Optical Soc. Amer.* XXVI. 187/2 The construction of the photosensitive cathode to be used in any given image tube will, of course, depend upon the spectral region in which maximum sensitivity is desired. **1940** ZWORYKIN & MORTON *Television* iv. 91 The image tube is of importance because it can be combined with the Iconoscope to make a television pickup tube which is many times more sensitive than the normal Iconoscope. **1969** *New Scientist* 10 July (Optics Suppl.) 21/1 Objects illuminated with non-visible radiation..can be seen with the aid of these devices. The image tube can also intensify very faint images..so that they become visible. **1971** *Nature* 3 Sept. 37/1 The difficulties associated with studying very faint [celestial] objects are very great, because of the increasing difficulty in detecting photons against the natural and man-made noise, even with the use of image-tube techniques. **1925** I. A. RICHARDS *Princ. Lit. Crit.* 123 If this were not the case the absence of glaring differences between people of different image-types would be astonishing. **1904** *Daily Chron.* 8/2 Immersed in this innocent, harmless, image-world. **1953** R. MANHEIM tr. *Cassirer's Philos. Symbolic Forms* I. 78 All live in particular image-worlds, which do not reflect the empirically given, but which rather produce it in accordance with an independent principle.

image, *v.* Add: **1. c.** (Later examples.)

1957 A. C. CLARKE *Deep Range* xv. 129 The familiar rocky terrain was imaged on TV and sonar screens. **1970** *Physics Bull.* Nov. 490/2 Figure 1 shows the simplest possible optical system which includes both a parallel beam in which the working space can be placed (s) and a lens which can image it on to a receptor.

4. a. (Later examples.)

1880 J. MCCOSH *Emotions* I. i. 11 The mind of man has the power of imaging or representing in old forms by the memory, and in new forms by the imagination, whatever it has at any time known or experienced. **1924** T. H. TROTTER *Music & Mind* v. 56 Whether or not it is necessary for the listener to image in his mind the scene to be represented is a moot point. **1972** *Science* 16 June 1208/1 Some recall past events by imaging the scene.

imaged, *a.* **1.** (Later examples.)

1889 W. B. YEATS *Wanderings of Oisin* II. 23 Between the lids of one The imaged meteors had shone and run. **1975** W. EMPSON *Poems* 13 Drink deep the imaged solid of the bone.

imageless, *a.* Add: (Later example.)

a **1930** D. H. LAWRENCE *Last Poems* (1932) 214 The gods are nameless and imageless.

b. Special Comb.: *imageless thought* (see quot. 1934); so *imagelessly* adv.

1909 G. F. STOUT *Analytic Psychol.* I. i. iv. 85 An imageless thought is no absurdity, however opposed such a conception may be to..those who..regard consciousness as a kind of picture gallery. **1909** E. B. TITCHENER *Lect. Exper. Psychol. Thought-Processes* iii. 113 Binet..gives illustrations of imageless thought that must undoubtedly be classed with the conscious attitudes. **1920** S. ALEXANDER *Space, Time & Deity* I. 213 Observation..has been concerned with..the existence of 'imageless thinking', which seemed so inconceivable to some earlier psychologists. **1921** B. RUSSELL *Analysis of Mind* xi. 226 Similar remarks apply to the general idea of 'imageless thinking', concerning which there has been much controversy. **1934** R. H. WHEELER *Sci. Psychol.* 151/2 *Imageless thought*, an idea, thought, or train of thinking which is wholly lacking in sensory content. **1963** R. S. WOODWORTH *Contemp. Schools Psychol.* (ed. 3) iv. 74 Watson poured an increasing scorn on the 'imageless thought' controversy. **1953** J. B. CARROLL *Study of Lang.* iii. 74 One outcome of Wundt's work..was the discovery, around 1900, of 'imageless thought'—a kind of subjective behaviour, noted in the process of thinking, which could not be described as..sensation or perception. **1973** *Science* 11 May 630/2 This causal attitude toward definition means that he need never ask whether..there are 'imageless' concepts or thoughts.

image-maker. Add: **b.** *fig.* (in sense *5 b of image). So **i·mage-making** *sb.* and *adj.* (various senses).

a **1930** D. H. LAWRENCE *Last Poems* (1932) 83 The gibe of image-making love. **1953** S. SPENDER *Creative Element* 50 For Yeats, spiritualism put him in touch, as he thought, with the image-making collective unconscious of all civilizations. **1960** *Guardian* 7 Oct. 17/4 [Owing] to the failure of the image-makers..neither candidate seems to have put across a strong or provocative personality. **1967** *Listener* 3 Aug. 140/3 Darwin and Einstein..must have been endowed with an extraordinary capacity of image-making to be able to see the world in a startlingly new shape. **1969** *Times* 1 Aug. 6/7 The image-makers said that he was the family's best politician. **1971** *Daily Tel.* 23 Mar. 13/1 The population at large are as inwardly proud of their image-making ex-servicemen as..the image-makers as Ned Kelly, the waterless outback and Aussie-rules football. **1972** E. LUCIE-SMITH *Eroticism in Western Art* xx. 262 Since erotic art obeys the necessity to be specific, photography..offered powerful competition to more traditional kinds of image-making.

imager. Add: **2.** (Later example.) Also, one who forms a mental image or images; something that produces an image (e.g. in radar).

1960 A. HUXLEY *Let.* 17 July (1969) 893 Your work with eggers sounds very interesting. Have you any idea why some people visualize and others don't? **1963** *Listener* 28 Mar. 547/2 One regular hypnagogic imager has amusing images that would form excellent subject matter for a Walt Disney cartoon. **1967** C. WEBSTER *Word & Symbol* 10 Poets have always regarded themselves as the bearers of truth and imagery of reality. **1971** *Daily Tel.* (Colour Suppl.) 10 Dec. 34/1 The Army regards radar as an integral part of the night-vision family; just as important as image intensifiers or thermal imagers.

imagic (i·mǽdʒik), *a.* [f. IMAGE *sb.* + -IC.] Imagic-like.

1937 'C. CAUDWELL' *Illusion & Reality* 237 The manifest content is imagic phantasy. **1953** *Essays in Crit.* III. 431 'Soundlessly' shares with 'Spawning' the imagic honours. **1957** N. FRYE *Sound & Poetry* 136 The 'New Criticism', chiefly concerned with the imagic and referential meaning (semantic) levels.

imaginability. Delete *rare*[−1] and add later example.

1953 G. E. M. ANSCOMBE tr. *Wittgenstein's Philos. Investigations* I. 120 There is a lack of clarity about the role of *imaginability* in our investigation.

imaginal, *a.*[1] Restrict † *Obs. rare* to sense b and add later examples. Also, of or pertaining to a mental image.

Quot. a 1901 is perh. a transferred use of IMAGINAL *a.*[2] *a* **1901** F. W. H. MYERS *Human Personality* (1903) I. Gloss. p. xviii, *Imaginal*,..metaphorically applied to transcendental faculties shown in rudiment in ordinary life. **1925** [see *ATTITUDE 5]. **1925** J. E. TURNER *Theory Direct Realism* iv. 42 The imaginal and ideal elements, although of course they are undoubtedly present in the percipient's mind, never reveal themselves to him in their proper nature. **1935** *Brit. Jrnl. Psychol.* Apr. 445 The imaginal retention of a previously perceived common element. **1951** G. HUMPHREY *Thinking* ii. 32 Pure intellection as contrasted with thinking on imaginal terms. *Ibid.* iv. 129 Sensory aspects of experience which we may call imaginal, affective and cognitive.

Hence **ima·gin-ally** *adv.*

1925 I. A. RICHARDS *Princ. Lit. Crit.* xviii. 157 As the eye wanders imaginally from point to point the relations between the parts of the picture-space change.

imaginary, *a.* Add: **1. a.** *imaginary museum* = *musée imaginaire* (see *MUSÉE 2).

1963 *Times Lit. Suppl.* 23 Aug. 641/5 Mr. Skelton's readers are given hardly any notion of the imaginary museums in which poets roam and of which they too must be habitués if they are to appreciate poems more than superficially. **1967** *Listener* 25 May 679/1 Marino's *Galeria*, the 'imaginary museum' where one assembles one's favourite works of art from a number of different locations.

imagination. Add: **6.** *imagination-consciousness, -game, -image, -mill, -process, -world; imagination-liberating, -manufactured, -stunning* adjs.

1901 E. B. TITCHENER *Exper. Psychol.* I. 1. 2 An imagination-consciousness, our mind as it is when we are imagining something. **1926** E. BOWEN *Ann Lee's, & Other Stories* 53 But the imagination-game palled upon her. **1890** W. JAMES *Princ. Psychol.* II. xviii. 69 Imagination-images..feel subject to our spontaneity [etc.]. **1933** R. TUVE *Seasons & Months* i. 38 It was not the imagination-liberating theory of the Neoplatonic nature. **1902** *North Amer. Rev.* Dec. 768 The [Christian] Science..secures to him life-long immunity from imagination-manufactured disease. **1890** *Harper's Mag.* Dec. 411 His imagination-mill was hard at work in a minute. **1890** W. JAMES *Princ. Psychol.* II. xviii. 72 The imagination-process can then work over into the sensation-process..In other words, the imagination can be centrally originated. **1892** 'MARK TWAIN' *Amer. Claimant* x. 88 The imagination-stunning material development of this century. **1904** *Daily Chron.* 19 Oct. 8/1 This glimpse into the imagination-world of London.

imagine, *v.* Add: **5. c.** *colloq.* To believe or suppose. Also used with apoiopesis in phr. *can you imagine?*

1934 N. MARSH *Final Curtain* x. 150 We all opened our letters yesterday morning, at breakfast. Can you imagine? I got down first and really—such a shock! **1952** E. GRIERSON *Reputation for Song* iii. 20 'Is your mother coming down?' 'I imagine so.' **1962** *Guardian* 25 May 10/4 Naively imagining that MI 5 was only bloodhounding those with suspected Cliveden or Mosley ideas. **1968** D. DEVINE *Sleeping Tiger* i. 12 Peter borrowed the Jag to bring her here and he scraped it on the gate! Can you imagine? **1971** C. BONINGTON *Annapurna South Face* iii. 28 At times, though, in these hectic weeks of organisation, as I imagine happens with any expedition, I have wondered whether we would ever make our deadline for packing all our gear ready to go to India. **1973** *Listener* 20 Dec. 842/2 This prospect must bring a lot of cheer to the Speaker... I imagine that after Mr Ford's swearing-in he slept the sleep of the just.

imaging, *vbl. sb.* (see under IMAGE *v.*). (Later examples.)

1920 S. ALEXANDER *Space, Time & Deity* I. 25 In imaging the act of mind is provoked from within. **1920** J. LAIRD *Study in Realism* ix. 67 Imaging is a different bodily margin from perceiving. **1945** *Mind* LIV. 331 'Imagination' sometimes means the forming and contemplating of mental images, visual, auditory or other; this is more appropriately called 'imaging'. **1953** H. H. PRICE *Thinking & Experience* viii. 236 All these people, when asked to think of something, report the spatial locations of things..would be completely at a loss if the power of visual imaging suddenly deserted them. **1971** *Sci. Amer.* Aug. 83/1 'Imaging' is a central process in which verbal information is transformed through visual imagery, for example, Cicero suggested learning long lists (or speeches) by placing each member of the list in a visual representation of successive rooms of a well-known building.

Imagism (i·mǽdʒiz'm). Also imagism. [f. IMAGE *sb.* + -ISM.] **1.** Name given to a movement in poetry, originating in 1912 and represented by Ezra Pound, Amy Lowell, and others, aiming at clarity of expression through the use of precise visual images.

In the early period often written in the Fr. form *Imagisme*.

1912 E. POUND *Let.* 1 Aug. (1971) 38 I should like the name 'Imagisme' to retain some sort of a meaning. It stands, or I should like it to stand for hard light, clear edges. *Ibid.* 12 Aug. (1971) 39 If you want to drag in the word Imagisme you can use a subtitle 'an anthology devoted to Imagisme, vers libre and modern movements in verse' or something of that sort. **1913** *Poetry* (Chicago) Mar. 198 (*title*) Imagisme. Some curiosity has been aroused concerning *Imagisme*... Editor's Note—In response to many requests for information regarding *Imagism* and the *Imagistes*, we publish this note by Mr. Flint... *Imagism* is not necessarily associated with Hellenic subjects, or with *vers libre* as a prescribed form. **1915** E. POUND *Let.* Jan. (1971) 48 If I had acceded to A. L.'s [sc. Amy Lowell's] proposal to turn 'Imagism' into a democratic beer-garden, I should have undone what little good I had managed to do by setting up a critical standard. **1917** *North Amer. Rev.* CCV. 106 The third characteristic of The New Poetry, and particularly of Imagism, is what might be metaphorically described as faithfulness to the architectural line. **1924** T. MAYNARD *Our Best Poets* 198 Imagism brings together, with an indulgent catholicism, those who are really as brilliant exactness, and those who use only cadence. **1929** A. NOYES *Return of Scarecrow* iv. 25 The sharp-edged imagism..with which our younger writers are experimenting today. **1931** G. HOUGHES *Imagism & the Imagists*. **1967** *Listener* 2 Mar. 297/2 Pound may have believed that his imagism ..was an alternative to symbolism, but there is no reason now for us to call it 'Imagism'.

2. *Philos.* (See quot. 1953.)

1952 R. I. AARON *Theory of Universals* ii. 20 It might be argued that what Hobbes was attacking..was imagism rather than conceptualism, the notion [*i.e.* imagism] that there must be an image before us whenever we universalize. **1953** H. H. PRICE *Thinking & Experience* viii. 234 The theory that thinking consists in operating with mental images..has no generally accepted name, I am going to call it 'Imagism'.

Imagist (i·mǽdʒist). Also imagist and in Fr. form **Imagiste.** [f. as prec. + -IST.] **1.** An adherent of Imagism (sense 1). Also *attrib.* or as *adj.*

1912 E. POUND *Let.* Aug. (1971) 10 I send you all that I have on my desk—an over-elaborate post-Browning 'Imagiste' affair and a note on the Whistler exhibit. *Ibid.* Oct. (1971) 11 I've had luck again, and am sending you some *modern* stuff by an American, I say modern, for it is in the laconic speech of the Imagistes, even if the subject is classic. **1913** [see *IMAGISM 1]. **1914** R. ALDINGTON in *Egoist* 1 June 202/1 (*title*) Modern poetry and the Imagists. *Ibid.* 202/1 Why do we call ourselves 'Imagists'? ..'I think it is a very good and descriptive title... Let me say from memory what I, as an Imagist, consider the fundamental doctrines of the group... We do not say 'O how I admire that exquisite, that beautiful..woman'.. but we present that woman, we make an 'Image' of her, we make the scene convey the emotion. **1915** *Egoist* 1 May 70/2 One of the first 'Imagist' poems by T. E. Hulme, 1909 [etc.]. **1959** *Spectator* 1 ..*Isle of la Belle Rivière*, was written in what is now called imagist verse, in 1908. **1934** *Edin. Rev.* July 101 In much of the work of the imagists..we find a more or less conscious..and more or less effective yielding to that influence. **1931** [see *GROUND sb. 11 b]. **1931** [see *IMAGISM 1]. **1960** AUDEN *Homage to Clio* 42 No 'imagist' poem can be more than a few words long. **1970** *English Studies* LI. 269 This period also saw the birth and death of the more obviously revolutionary groups such as the Vorticists, Imagists, [etc.].

transf. **1962** *Times* 3 May 18/4 Two of the most ripproaring imagists of European action-painting. **1962** *Listener* 27 Sept. 484/1 The Prerafaelites, the Symbolists, the Kafkarians, the Imagists.

2. *Philos.* An adherent of imagism (sense 2). Also *attrib.* or as *adj.*

1948 *Mind* LVII. 483 He [sc. Ewing] backs it up with his criticisms of Behaviourist accounts of Belief and Verbalist and Imagist accounts of Thinking. **1953** H. H. PRICE *Thinking & Experience* viii. 239 The Imagist does not deny that words have meaning, but he holds that they have it only indirectly, as substitutes for images. *Ibid.* 241 The starting point of the Imagist theory..is private thinking, and private thinking of the 'free' symbol-using kind. **1972** *Science* 12 May 630/2 Thus he can adopt an imagist theory of meaning after carefully listing several objections to it which are never answered.

Hence **imagi·stic** *a.*, of or pertaining to Imagism (both senses); **imagi·stically** *adv.*

1916 E. POUND *Let.* 17 Apr. (1971) 76 Some of the things [sc. poems] seem to me 'just imagistic', neither better nor worse than a lot of other imagistic stuff that gets into print. **1960** *Chamb. Jrnl. Lit. July* (1965) 63 In an imagistic way [this] singularly seems to agree with the substance of your opinion. **1960** *Kenyon Rev.* 277 The words 'fog' and 'bloody'..must be taken not only.. imagistically but symbolically. **1964** *Mind* LIII. 216 Imagistic, literary, associative, or other kinds of meaning. **1965** *Listener* 14 Feb. 500/3 This imagistic language is carried to its logical conclusion in the controversial ending to the film. **1969** *Jrnl. Eng. & Gmc. Philol.* LXVIII. 174 The reason is imagistically indicated in the immediately preceding lines. **1973** *Times Lit. Suppl.* 2 Nov. 1346/3 He works for the most part imagistically, spacing small, autonomous chunks of perception around a page, resolutely subduing 'theme' to the eye-stopping images which compose it.

imago. Add: **2.** *Psycho-analysis.* A subjective image of someone (esp. a parent) which a person has subconsciously formed and which continues to influence his attitudes and behaviour. So *father-imago, mother-imago.*

1921 *19th Cent.* Feb. 214 Since 1914 every constituent element that has been supposed to go to the making of great men—spacious times, tension, supreme effort, turmoil, battle, revolution—has abounded, but the imago has not emerged.

1916 B. M. HINKLE tr. *Jung's Psychol. of Unconscious* (1918) 492 Here I purposely give preference to the term 'Imago' rather than to the expression 'Complex', in order ..to invest this psychological condition, which I include under 'Imago', with living independence in the psychical hierarchy... 'Imago' has a significance similar on the one hand to the psychologically conceived creation in Spitteler's novel..and on the other hand to the ancient religious conception of 'imagines et lares'. **1919** M. K. BRADBY *Psycho-Anal.* 59 That web of ideas and emotions which is woven in the course of the child's life round the image of the parent or 'parent imago'. **1924** J. RIVIERE et al tr. *Freud's Coll. Papers* II. xxviii. 313 If the physician should be specially connected in this way with the father-imago (as Jung has happily named it) it is quite in accordance with his actual relationship to the patient. **1927** W. E. COLLINSON *Contemp. Eng.* 107 Most educated people will by now have heard of the Oedipus complex and will have a nodding acquaintance with libido and imago and may have, with distressing results, tried on themselves the method of free-association. **1948** M. KLEIN in S. Lorand *Psycho-Analysis Today* 65 The super-ego of the child does not coincide with the picture presented by its real parents, but is created out of imaginary pictures or *imagos* of them which it has taken up into itself. **1956** R. F. C. HULL tr. *Jung's Coll. Wks.* V. iv. 57 In most of the existing religions it seems that the formative factor..is the father-imago, while in the older religions it was the mother-imago. **1967** BRUSSEL & CANTZLAAR *Chambers's Dict. Psychiatry* 121 *Imago*, in Jung's *analytical psychology*, a conception of another person that one acquires in infancy or childhood and carries through to adulthood in the unconscious.

imagy (i·mėdʒi), *a.* [f. IMAGE *sb.*: see -Y[1].] Of the nature of an image.

1937 *Aristotelian Soc. Suppl. Vol.* XVI. 213 We could suggest that an event has two sorts of constituents, *sensible* ones and *imagy* ones. **1953** H. H. PRICE *Thinking & Experience* vii. 201 The type-word 'cat' has *imagy* tokens as well as overtly perceptible ones.

imambara (imā·mbārā, imāmbā·rā). Also imambarah, -barra, -bra, imaum-. [Hind., f. Arab. IMAM + Hind. *bārā* enclosure.] In India, a building in which Shiite Muslims assemble at the time of Muharram; the gardens, courtyards, etc., surrounding such a building; also, any large tomb.

1837 E. EDEN *Jrnl.* 28 Dec. in *Up Country* (1866) I. ix. 87 In the afternoon we went to see the Emaumberra and Rooma Durwanee, two of the most magnificent native buildings I have seen yet. **1867** J. FERGUSSON *Hist. Archit.* II. v. viii. 702 In Lucknow there..is..one building especially, the Imambara, which..is not unfit to be spoken of in the same chapter as the earlier buildings. **1883** *Encycl. Brit.* XV. 49/2 The immense Imámbára, or mausoleum of Asaf-ud-daulá [at Lucknow]. **1886** YULE & BURNELL *Hobson-Jobson*, Imaumbarra. **1907** *Westm. Gaz.* 4 Apr. 10/2 The imambra connected with the Mohammedan morgue at Lucknow. **1955** J. TERRY *Charm Indo-Islamic Archit.* 40 (*heading*) Imambara, Lucknow, 18th century. A large hall built by a king of Oudh for famine-relief work. **1964** A. A. A. FYZEE *Outl. Muhammadan Law* (ed. 3) 319 An imāmbāra is a private apartment set apart by a member of the Ithnā 'Asharī Shiite faith for the performance of certain ceremonies at Muḥarram and other times; it is not a public place of worship like a mosque.

Imam Bayildi (imā·m ba·yildi). Also Imam Baildi. [Turk., lit. = the priest fainted (supposedly from pleasure at, or because of the cost of, the dish).] A dish consisting of aubergines stuffed with an onion-and-tomato mixture and cooked in oil.

1935 M. MORPHY *Recipes of all Nations* 767 This vegetable [sc. the aubergine] is extremely popular in Turkey, and one of the commonest ways of preparing it is called *Imam Baildi*..'the Swooning Imam'—the Imam having fainted with delight when he first partook of this wondrous dish. **1952** HOWE & ESPIR *Sultan's Pleasure & other Turkish Recipes* 75 *The fainting imam* (Patlıcan Imam Bayıldı). There is an amusing story connected with this traditional dish. **1958** I. ORGA *Turkish Cooking* 114 *Aubergine Imam Bayildi*... Take the onion mixture and fill the aubergines with this then lay them side by side in a wide-bottomed saucepan. **1969** R. STOUT *Death of Dude* (1970) xii. 176 Eggplant stuffed with a purée which the Turks call *Imam Baïldi*—'Swooning Imam'. Onions browned in oil, tomatoes, garlic, salt and pepper. **1972** J. RATHBONE *Trip Trap* x. 111 'Some more Imam Bayildi?' He motioned to the aubergine stew. **1973** *Guardian* 26 Jan. 11/1 A dish of aubergine..named Imam Bayildi or The Imam Fainted.

Imari (imā·ri). The name of a town in the north-west of the Japanese island of Kyushu, used *attrib.* and *ellipt.* to denote a type of Hizen porcelain.

1875–80 AUDSLEY & BOWES *Keramic Art Japan* I. 5 The productions of these numerous factories are usually exported from the seaport of Imari, and are therefore commonly known as Imari ware. **1878** J. J. YOUNG *Ceramic Art* vi. 175 The chief kinds of [Japanese] porcelain are the Hizen (also called Imari and Arita), [etc.]. **1890** B. H. CHAMBERLAIN *Things Japanese* 284 The second variety of Kutani ware may often be mistaken for 'old Japan' (i.e., Imari porcelain). **1902** *Encycl. Brit.* XXIX. 725/1 In the term 'Hizen porcelains' are included not merely the richly decorated Imari ware—the 'Old Japan' of Western collectors [etc.]. **1954** H. GARNER *Oriental Blue & White* 63 The so-called 'Old Imari' wares. **1954** G. SAVAGE *Porcelain* 43 The *Imari* style is also Japanese, and was based on native textiles and brocades. **1969** *Guardian* 16 July 18/2 Japanese Kakiemon and Imari patterns were copied at Canton for shipment to Europe. **1970** *Ashmolean Mus. Rep. Visitors 1969* 47 Oil jug, Japanese export Imari porcelain, overglazed in red and gold. **1971** *Times* 5 Apr. 14/5 Delicate [chamber] pots in Imari porcelain and Wedgwood.

imbalance (imbæ·lāns). [f. IM-[2] + BALANCE *sb.*] An unbalanced condition; a lack of

proper proportion or relation between corresponding things.

Orig. a technical term in Ophthalmology but now used generally in many subjects and contexts.

1898 *Ophthalmic Rec.* VII. 87 Some advocates of operative interference for ocular imbalance in the functional neuroses fail to recognize the fact that heterophoria, or even squint, may be only a symptom. **1930** G. HAMILTON *Med. Social Terminol.* 25 *Intellectual imbalance*, the state of an individual with special abilities or disabilities, markedly competent in some respects and deficient in others, but not well integrated or compensated. **1934** *Scrutiny* III. III. 230 It is a matter of common observation that a high degree of artistic discrimination can go along with an extreme paucity or imbalance of general emotional life. **1937** *Nature* 16 Jan. 90/2 The operating mechanism [of the human constitution] is the autonomic nervous system, which, especially in sensitive subjects is liable to ..pass into imbalance or dysfunction, and eventually permanent disease. **1949** S. DUKE-ELDER *Text-bk. Ophthalm.* IV. xlvi. 3964 Operative treatment is the alternative measure to correct a heterophoria, and is particularly applicable in essential imbalance when the cause is anatomical. **1949** JANIS & FADNER in H. D. Lasswell et al. *Lang. Pol.* II. viii. This Coefficient of Imbalance is intended to be applicable to all types of communications..except those in which the communication is arbitrarily restricted to specified symbols. **1952** *N.Y. Times* 6 May 28/2 The trade imbalance and the drain of gold and hard currency reserves have lessened. **1953** *Manch. Guardian Weekly* 7 May 3/3 Which will prevent their own budgets from remaining in disastrous imbalance. **1954** *New Biol.* XVI. 80 While the maintenance of an adult organism is achieved by the exact balance of synthetic and degradative processes, growth is realized by a degree of imbalance in which synthetic processes predominate. **1957** *Economist* 7 Sept. 780/1 Increasing the existing imbalance between male and female employment. **1957** *English Studies* XXXVIII. 97 A remarkable corrective to this imbalance appeared in the writings of Henry Sweet. **1957** *Listener* 26 Dec. 1080/3 Shows clearly the imbalance created by an exodus of young persons (predominantly females) from the rural areas. **1961** *Lancet* 22 July 211/1 Imbalance in the mental diet results in greed. **1962** *Ibid.* 2 June 1167/2 Though the mammalian body has striking powers of recovery from the effects of long-maintained hormonal imbalance, there must be limits to the stresses which can safely be imposed. **1967** *Boston Sunday Globe* 21 May H 3/1 The imbalance problem [proportion of black and of white people] could be solved in such a setup... The 20 or so schools in a park would be able to share many facilities and personnel. **1969** *Times* 5 Aug. 9/5 The imbalance in the world's financial system has become grotesque. **1970** *New Society* 5 Feb. 222/1 Imbalances between homes and people emerge. **1973** *Sci. Amer.* Sept. 133/3 The marked increase in the average salary reflects both the imbalance of supply and demand for health workers and, more important, the highly desirable increased value placed on such workers.

imbauba (imbǭ·bā, imbā,ū·bā). [a. Pg. *imbaúba, umbaúba*, f. Tupi *ambauba, umbauba*.] A Brazilian tree of the genus *Cecropia*, esp. *C. peltata*, which has a hollow stem and yields a softwood timber; also called *trumpet-tree, trumpet-wood*. Cf. *CECROPIA 1.

1849 R. SPRUCE *Notes of Botanist on Amazon & Andes* (1908) I. i. 39 Many leaves are grey or hoary beneath, as in the Cecropias (or Imba-úbas, as they are called by the Indians). **1924** RECORD & MELL *Timbers Trop. Amer.* 145 There are various species of *Cecropia* in Brazil, at least twelve in the Amazon region, and they are generally known by the name of 'imbaúba' or 'embaúba'. **1927** R. R. GATES *Botanist in Amazon Valley* iv. 83 Another characteristic tree fringing the bank..in front of the main forest, is imbauba. **1931** B. MIALL tr. *Guenther's Naturalist in Brazil* xvii. 319 Another kind of Aztec ant inhabits the interior of the trunk of the Imbauba-tree.

imbecilic (imbĭsi·lik), *a.* [f. IMBECILE *sb.* + -IC.] Characteristic of an imbecile; idiotic.

1918 D. FLATAU *Yellow English* xxxv, Looking round with an imbecilic grin. **1927** *Daily Express* 3 Dec. 4 Courteline..collected..the most hideous and imbecilic portraits he could find. **1960** 'A. BURGESS' *Right to Answer* ii. 18 There was a kind of imbecilic helper in the public bar. **1968** *Punch* 7 Feb. 194/1 Extrovert delegates prance about singing, blowing horns and shouting imbecilic slogans.

Hence **imbeci·lically** *adv.*, in an imbecile manner.

1946 B. MARSHALL *George Brown's Schooldays* 28 Pretended not to pray or looked imbecilically heathen.

imbed, *v.*, **imbedded,** *ppl. a.*, varr. EMBED *v.*, EMBEDDED *ppl. a.* (Later examples.)

1950 C. R. HINE *Machine Tools for Engineers* xii. 240 The fine, sharp abrasive particles become imbedded in the lap and it is ready for use. **1969** *N.Y. Rev. Bks.* 16 Jan. 4/4 Some of the divergences between Freud and Jung are, however, better attributed to the fact that they chose opposed solutions to the problem of how to imbed the idea of the unconscious into already existing traditions of Western thought. **1971** *Nature* 18 June 437/2 It will be noted that the regions of the three brightest sources are further imbedded in very dense areas; in fact, they are close to the centroids of the Coma cluster. **1971** POWELL & HIGMAN *Finite Simple Groups* iii. 174 The process simply imbeds one Chevalley group in another. **1972** *Physics Bull.* May 284/3 Indeed, the liquid crystalline properties inherent in this bilayer structure are almost certainly important for the organization of molecules imbedded in the membrane as well as for controlling transport through it.

Also **imbe·ddedness.**

1952 T. PARSONS *Social System* 361 Philosophical investigation, as distinguished from the general imbeddedness of philosophical problems..in any system of action.

imberb (imbə̄·ɹb), *a. rare.* [ad. F. *imberbe*, f. L. *imberbis* (see IMBERBIC *a.*).] Beardless.

1923 A. HUXLEY *Antic Hay* iii. 42 A face of such childish contour and so imberb that he looked like a little boy playing at grown-ups.

imbibitional (imbibi·ʃənăl), *a.* [f. IMBIBITION + -AL.] Of, pertaining to, or resulting from imbibition.

1916 *Science* 6 Oct. 502/2 (*heading*) Imbibitional swelling of plants and colloidal mixtures. **1924** *Jrnl. Agric. Sci.* XIV. 204 (*heading*) Remarks and observations on imbibitional soil moisture. **1931** E. C. MILLER *Plant Physiol.* iv. 166 Very strong imbibitional forces may be developed within the plant even when the cell walls and protoplasm contain considerable water. **1959** *Chambers's Encycl.* VI. 331/2 In the early stages of water absorption by the barley grain,..the imbibitional pressure may reach 900 atm.

imbonga, imbongo, varr. *MBONGO.

imbruement. (Earlier example.)

1859 J. S. BLACKIE *Let.* 20 Sept. (1909) 143 To the gratification of the old Adam, but with no imbrument [*sic*] to the new.

imbuya (imbwī·yā). Also **imbuia, embuia.** [ad. Pg. *imbuia*, f. the local name for the tree.] A Brazilian timber tree, *Phoebe porosa*, or the wood obtained from it.

1919 *Jrnl. Forestry* XVII. 156 Embuia is the timber de luxe of the four southern Brazil States. **1929** *Tropical Woods* XVIII. 17 Imbuia, or Embuia, one of the best known hardwood timbers of southern Brazil, is used in the United States to a limited extent as a substitute for Walnut, and is sometimes sold as Brazilian Walnut. *Ibid.* 19 The wood of Imbuia varies from yellowish or olive to chocolate-brown, either plain or beautifully variegated and figured. **1947** J. C. RICH *Materials & Methods Sculpture* x. 289 *Imbuya* is a brown wood imported from Central and South America. It is occasionally called Brazilian walnut because of the similarity of its color to walnut. It is softer than black walnut and is usually available in logs and thin planks. **1956** *Handbk. Hardwoods* (Forest Prod. Res. Lab.) 115 Imbuya is also known as 'Brazilian walnut' or embuia (Brazil). **1969** B. J. RENDLE *World Timbers* II. 100 Imbuya is variable in colour, from yellowish-brown to chocolate-brown, and may be plain or variegated. **1971** *Cape Times* 13 Feb. 14/6 (Advt.), Solid imbuia kist with brass trims: excellent condition.

imerinite (imərī·nəit). *Min.* [a. F. *imerinite* (A. Lacroix *Minéral. de la France* (1910) IV. 787), f. *Imerina* (F. *Imérina*), name of a region in central Madagascar: see -ITE[1].] A colourless or pale blue hydrated silicate of sodium, magnesium, and iron, $Na_2(Mg,Fe^{II},Fe^{III})_6Si_8O_{22}(O,OH)_2$, related to richterite.

1913 *Mineral. Mag.* XVI. 363 Imerinite... A soda-amphibole containing only a small amount of sesquioxides and so allied to soda-richterite. **1963** W. A. DEER et al. *Rock-Forming Min.* II. 353 The names imerinite (Lacroix, 1921) and szechenyite (Krenner, 1900), used for members of the richterite series from Madagascar and Burma respectively, serve no useful purpose and should be abandoned.

imhofite (i·mhǫfəit). *Min.* [ad. G. *imhofit* (G. Burri et al.: see quot. 1965), f. the name of Josef *Imhof*, 20th-c. Swiss mineral collector: see -ITE[1].] A soft, white sulphide of thallium and arsenic occurring as tiny monoclinic plates.

1965 G. BURRI et al. in *Chimia* XIX. 409 [*English summary of an article in German.*] The description of a new thallium arsenosulfosalt, imhofite, found in the Lengenbach quarry, is given. **1969** *Mineral. Abstr.* XX. 15/2 The structures of the following complex sulphides are summarized:..imhofite, proustite, [etc.]... The greater part of these minerals are from Lengenbach, Binn valley, Switzerland.

imidazole (imi·dǎzōul). *Chem.* [ad. G. *imidazol* (Hantzsch & Weber 1887, in *Ber. d. Deut. Chem. Ges.* XX. 3119), f. *imid* IMIDE + Az(o- + *-ol* *-OLE.] **a.** A colourless, crystalline, heterocyclic compound consisting of the five-membered ring NH—CH=N—CH=CH.

Also called *GLYOXALINE. **b.** Any of the derivatives of this compound.

1892 *Jrnl. Chem. Soc.* LXII. 313 Derivatives obtained by displacing a CH group in these by N are designated imidazole = $C_3H_3(NH)N$, oxazole = $C_3H_3(O)N$, and thiazole = $C_3H_3(S)N$ respectively. **1900** E. F. SMITH tr. *V. von Richter's Org. Chem.* (ed. 3) II. 480 The glyoxalines or imidazoles. **1936** L. J. DESHA *Org. Chem.* xxiv. 493 Pyrazole and imidazole, $C_3H_4N_2$, are isomers differing in the relative positions of the nitrogen atoms. **1946** [see *GLYOXALINE]. **1951** A. GROLLMAN *Pharmacol. & Therapeutics* xi. 216 Certain imidazole derivatives also exert actions similar in some respects to the sympathetic amines. **1953** R. H. WILEY in H. Gilman *Org. Chem.* IV. viii. 787 The imidazole nucleus is found naturally in the amino acid histidine..; in the purines..; in biotin..; and in histamine. **1968** [see *GLYOXALINE]. **1968** L. A. PAQUETTE *Princ. Mod. Heterocyclic Chem.* vi. 185 Imidazole and pyrazole..readily form salts with metal ions.

imide. Add: [First formed as F. *imide* (A. Laurent 1835, in *Ann. de Chim. et de Phys.* LIX. 400).] Substitute for def.: Any compound containing the group —NH— (or the substituted form —NR—) attached either to two atoms of a metal or to one or two carbon atoms (which strictly should form part of an acidic group or groups). (Further examples.)

1857 W. A. MILLER *Elem. Chem.* III. 231 The imides are bodies formed from the amidated acids by depriving these compounds of two equivalents of water. **1892** E. F. SMITH tr. *V. von Richter's Chem. Carbon Compounds* (ed. 2) 365 The imides result by substituting the divalent acid radicals for two of the hydrogen atoms of ammonia. **1950** N. V. SIDGWICK *Chem. Elements & their Compounds* I. 666 Imides proper R·C(OH)(=NH) (cyclic imides like phthalimide are of course of a different type) are scarcely known except as possible tautomeric forms of amides; their alkyl derivatives, the imino-ethers R·C(O·R)(=N·R'), change easily and irreversibly into the disubstituted amides R·C(=O)NR$_2$. **1951** C. R. NOLLER *Textbk. Org. Chem.* xiii. 195 The monoacyl derivatives of ammonia, primary amines and secondary amines, having the general formula $RCONH_2$, RCONHR, or $RCONR_2$, are known as amides. Diacyl derivatives of ammonia, $(RCO)_2NH$, and of primary amines, $(RCO)_2NR$, also are known and are called imides. Except for the cyclic imides..their preparation usually is more difficult. **1962** P. J. & B. DURRANT *Introd. Adv. Inorg. Chem.* xix. 695 The imides, for example lithium imide Li_2NH, contain the group >NH. Lithium, calcium, germanium, tin, and lead form imides.

imidic (imi·dik), *a. Chem.* [In first quot. ad. F. *imidique* (A. Haller 1895, in *Compt. Rend.* CXX. 1194), but in later use prob. independently formed: see IMIDE and -IC.] Of the nature of an imide; in mod. use applied to organic acids of the type R·C(NH)OH and their derivatives.

1895 *Jrnl. Chem. Soc.* LXVIII. I. 648 Compounds containing the group NHRR' might be called imidic acids. **1919** *Decennial Index Chem. Abstr. 1907–1916* 3325/1 Imidic acids. **1951** I. L. FINAR *Org. Chem.* ix. 161 Imidic esters, which are also known as imino-ethers, are best prepared by passing dry hydrogen chloride into a solution of an alkyl cyanide in anhydrous alcohol. **1965** C. R. NOLLER *Chem. Org. Compounds* (ed. 3) xiv. 272 The tautomeric form of an amide, RC(=NH)—OH, is a nitrogen analog of a carboxylic acid and is known as an imidic acid. Although the imidic acids cannot be isolated because of the greater stability of the amide form, their derivatives are easily prepared.

iminazole (imi·nǎzōul). *Chem.* [f. *IMIN(E + AZ(O- + *-OLE.] = *IMIDAZOLE.

1901 *Jrnl. Chem. Soc.* LXXX. I (*heading*) Preparation of substituted iminazoles. **1926** H. G. RULE tr. *Schmidt's Text-bk. Org. Chem.* III. iii. 578 The ring system of the iminazoles, like that of the pyrazoles, consists of three carbon and two nitrogen atoms... Iminazole, the parent compound of the series, is formed..by the action of ammonia on glyoxal, and hence is also known as glyoxaline. **1968** [see *GLYOXALINE].

imine (i·mīn). *Chem.* [ad. G. *imin* (A. Ladenburg 1883, in *Ber. d. Deut. Chem. Ges.* XVI. 1150), formed by altering *amin* AMINE (cf. IMIDE, AMIDE).] Any compound containing the group =NH (or the substituted form =NR) attached to one carbon atom that forms part of a non-acidic organic group, or containing it in symmetrical compounds of the type R·NH·R; also applied (unsystematically) to other compounds in which it is attached to two carbon atoms (strictly, atoms forming part of non-acidic groups), esp. when they are part of a ring (as in ethylene imine, $(CH_2)_2NH$).

1883 *Jrnl. Chem. Soc.* XLIV. 910 (*heading*) Imines. **1889** G. M'GOWAN tr. *A. Bernthsen's Text-bk. Org. Chem.* 194 If two hydrogen atoms in a molecule of ammonia are replaced by a divalent alcohol radicle, 'Imines', e.g. ethylene imine, $(C_2H_4)NH$", result. **1909** *Proc. Chem. Soc.* XXV. 309 Some confusion exists owing to the present system of nomenclature adopted for the imino-compounds. These substances, which possess the group C:NH, are obviously just as different from the secondary amines having the group C>NH (C) as the ketones with the group C:O are different from the ethers with the group C>O (C), yet the general name of imine is applied to both these classes. **1935** *Jrnl. Amer. Chem. Soc.* LVII. 2328/1 (*heading*) The preparation of ethylene imine from monoethanolamine.

1950 N. V. SIDGWICK *Chem. Elements* I. 666 Imines of the type (C)(C)>C=N·H rarely occur and are always unstable. **1967** I. L. FINAR *Org. Chem.* (ed. 5) I. viii. 181 Aldehydes react with primary amines to form imines (Schiff bases)...R·CHO + R'NH$_2$ → R·CH=NR' + H_2O. **1970** AMBROSE & EASTY *Cell Biol.* viii. 267 The sugar molecules form imine linkages (=NH) with the partially dissociated protein complexes.

imino(-) (imī·no). *Chem.* Comb. form of *IMINE; also used *attrib.* as quasi-*adj.* So *imino-chloride, -compound; iminosulphonic acid*; **imino-acid,** any organic acid that contains an imino-group; also (now *rare*), an imidic acid; **imino-ester** or **-ether,** any compound that contains —C(NH)OR and is consequently an ester of an imidic acid; more correctly called an *imido-ester*; **imino-group,** the group =NH as it occurs in imines (in quot. 1906 it denotes what is more correctly called an *imido-group*).

1903 *Jrnl. Chem. Soc.* LXXXIV. I. 692 The formation and properties of some iminoacid anhydrides of the type of the hypothetical iminoformic anhydride, NH:CH·O·CH:O. **1937** Imino-acid [see *imino-ether* below]. **1953** FRUTON & SIMMONDS *Gen. Biochem.* iii. 49 The hydrolysis of a protein leads to the formation of a variety of amino acids...The compounds having the general formula shown are termed α-amino acids, whereas proline and hydroxyproline are more correctly designated α-imino acids; for convenience, however, these two cyclic compounds are also called amino acids. **1961** *Jrnl. Clin. Invest.* XL. I. 843/1 Since nearly all of the hydroxyproline of the body is found in collagen, it has been suggested..that the urinary excretion of the imino acid may be an important index of collagen metabolism. **1967** *New Scientist* 24 Aug. 375/1 The precise proportion of the imino-acid hydroxyproline in the important protein collagen. **1900** *Jrnl. Chem. Soc.* LXXVIII. I. 295 The authors think it probable that aliphatic substituted imino-chlorides of aromatic acids are readily decomposed into alkyl chlorides and aromatic nitriles or their polymerides. **1904** *Ibid.* LXXXV. 1726 (*heading*) The formation and reactions of imino-compounds. **1924** C. HOLLINS *Synthesis Nitrogen Ring Compounds* vii. 203 The yield of imino-compound.. was very small. **1908** *Jrnl. Chem. Soc.* XCIV. I. 419 The catalysis of imino-esters. **1935** H. J. LUCAS *Org. Chem.* xxi. 314 In the presence of anhydrous hydrogen chloride, nitriles add alcohols to form imino esters: $CH_3C{\equiv}N + HOC_2H_5 \rightarrow CH_3C(=NH)OC_2H_5$. **1951** C. R. NOLLER *Chem. Org. Compounds* xiii. 242 The corresponding *O*-alkyl derivatives, RC(=NH)—OR, are known and are called imido esters (less correctly imino esters or imino ethers). **1897** *Jrnl. Chem. Soc.* LXXII. II. 804/2 (Index), Imino-ethers. **1937** TAYLOR & BAKER *Sidgwick's Org. Chem. Nitrogen* (rev. ed.) v. 154 The imino-ethers, which can also be regarded as esters of imino-acids and are sometimes called imino-esters have the general formula R·C(OR'): NH. **1906** *Jrnl. Chem. Soc.* LXXXIX. II. 1837 The members of the former class contain the grouping CO<(NH·CO·)(NH·CO·), whereas in no member of the latter class is this grouping present... Each of the imino-groups contained in the above-mentioned grouping, being connected with two carbonyl groups, will be possessed of acidic properties. **1966** NOWAKOWSKI & CLARKE tr. *Kretovich's Princ. Plant Biochem.* i. 16 Proline, strictly speaking, is not an amino acid, as it contains an imino group (=NH). **1896** *Jrnl. Chem. Soc.* LXX. II. 911/1 (Index), Iminosulphonic acid.

Examples of the general use (without hyphen) as quasi-*adj.*

1901 *Proc. Chem. Soc.* XVII. 61 A liquid base of 'imino' odour. **1937** F. C. WHITMORE *Org. Chem.* I. 226 This product loses water to form the imino analog of formaldehyde, $H_2C{=}NH$.

imipramine (imi·prămīn). *Pharm.* [Rearrangement of some elements of *dimethylaminopropyl-iminodibenzyl*, a systematic name for this compound: see -INE[5].] A tricyclic tertiary amine, $C_{19}H_{24}N_2$, given orally as the hydrochloride, a white crystalline compound, in the treatment of endogenous depression.

1958 *Amer. Jrnl. Psychiatry* CXV. 459 (*heading*) The treatment of depressive states with G 22355 (imipramine hydrochloride). **1960** *Brit. Med. Jrnl.* 30 Jan. 348/2 We have..noted the development of jaundice and also hypomania in patients receiving therapeutic doses of imipramine. **1963** *Lancet* 23 Mar. 638/2 The administration to pregnant rabbits of imipramine ('Tofranil') led to the development of fœtal abnormalities. **1964** HAFLIGER & BURCKHARDT in M. Gordon *Psychopharmacol. Agents* I. iii. 77 Imipramine has no influence upon monoamine oxidase. *Ibid.* 89 The most frequent side effects of imipramine..include dryness of the mouth, sweating, disturbance of accommodation, constipation, insomnia, and giddiness. **1965** *Nursing Times* 5 Feb. 187/1 Either electro-convulsive therapy or one of the antidepressant drugs (such as imipramine—Tofranil) will probably be used. **1965** J. POLLITT *Depression & its Treatment* iv. 58 Imipramine is effective mainly in Type S classical (retarded) depressions, and in depressed patients of middle age and over. It is less valuable in atypical depression, and in sensitive or hysterical personalities the ordinary mild side effects may cause bitter complaint. **1972** *Brit. Med. Jrnl.* 1 Apr. 45/2 These fairly substantial figures do not support the view that imipramine is liable to cause fetal abnormalities if given to pregnant women.

imitation. Add: **1. c.** *Psychol.* The adoption, whether conscious or not, during a learning process, of the behaviour or attitudes of some specific person or model.

[**1807** WORDSWORTH *Poems* II. 153 The little Actor cons another part..As if his whole vocation Were endless imitation.] **1895** J. M. BALDWIN *Mental Development* xii. 351 First..biological or organic imitation... Second: we pass to psychological, conscious, or cortical imitations. *Ibid.* 352 *Plastic Imitation.* This phrase is used to cover all cases of reaction or attitude, toward the doings, customs, opinions of others, which once represented more or less conscious adaptations..but which have become what is ordinarily called 'secondary automatic' and subconscious. **1899** H. C. WARREN tr. *Tarde's Social Laws* 42 Giving the word imitation the very wide meaning accorded to it..by Mr. Baldwin..one might regard imitation as the fundamental fact, not only of social and psychological life, but of organic life as well, where it would appear as the necessary condition of habit and heredity. **1903** E. C. PARSONS tr. *Tarde's Laws of Imitation* p. xiv, By imitation, I mean every impression of an inter-psychical photography..willed or not willed, passive or active. **1924** F. H. ALLPORT *Social Psychol.* x. 239 Before the rise of a really critical science of behavior the term 'imitation' enjoyed wide repute in social theory... Our treatment of imitation must therefore be mainly negative. **1946** D. MCCARTHY in L. Carmichael *Manual of Child Psychol.* 497/1 An interesting controversy..on the problem of the relationship between imitation and comprehension of language. **1968** *Internat. Encycl. Social Sci.* VII. 96/1 Learning by vicarious experience has historically been referred to as 'imitation'.

5. (Further examples.)

1840 H. REEVE tr. *A. de Tocqueville's Democracy in Amer.* III. I. xi. 100 Imitation-diamonds are now made which may be easily mistaken for real ones. **1871** *Post Office Directory Leather Trades* 66 (*heading*) Imitation leather makers... Kid leather dressers... Kid reviver makers. **1902** *Encycl. Brit.* XXVIII. 611/1 This is a very different thing from the imitation diamond so common in shop windows. Here the chemist has only succeeded in making a paste or glass..wanting the hardness and 'fire' of the real stone. **1904** GOODCHILD & TWENEY *Technol. & Sci. Dict.* 303/1 *Imitation parchment*, ordinary paper passed through a bath of sulphuric acid, which has the peculiar effect of 'toughening' the fibres. **1904** *Westm. Gaz.* 6 Oct. 4/2 Contempt of imitation jewels, imitation furs, imitation lace. *Ibid.*, A lovely coat of..imitation sealskin. **1916** E. POUND *Lustra* 53 The small child in the soiled-white imitation fur coat. **1929** D. H. LAWRENCE *Pansies* 124 Will the Proustian lot go next? And then our English imitation intelligentsia? **1937** E. J. LABARRE *Dict. Paper* 159/1 Imitation art paper is a highly finished printing [paper] prepared by the addition of a heavy percentage of china clay to the pulp and a water-finish. **1940** *Chambers's Techn. Dict.* 552/1 *Mock leno*, a fabric in which openwork effect is produced by a grouping of threads, which, however, do not cross, as they do in leno and gauze fabrics. Also called imitation gauze. **1954** *Paper Terminol.* (Spalding & Hodge) 32 *Imitation kraft*, a quality of wrapping paper made from unbleached sulphite, mechanical pulp and waste papers and coloured brown to give the appearance of Kraft. **1957** *Encycl. Brit.* IV. 775/2 Imitation gauze weaves.. are..largely utilized in the production of..embroidery cloths. *Ibid.* XIII. 850 A/2 The first feasible imitation leathers were based on patents issued in the United Kingdom in 1851. *Ibid.* XIX. 635/1 Imitation rum is produced by flavouring a neutral spirit..with high-ester Jamaican rum or with artificial essences. **1963** R. R. A. HIGHAM *Handbk. Papermaking* 282 *Imitation art*, paper which is highly finished by the action of super-calendering and water finishing and which contains a high percentage of china clay in the furnish.

imitativeness. (Later example.)

1973 *Daily Tel.* (Colour Suppl.) 9 Feb. 7/2 The imitativeness in these films is even more striking than their lack of artistry.

imma, var. *IMMY.

immanence. Add: (Further examples.) Also *attrib.*, as **immanence philosophy,** a theory evolved in Germany at the end of the nineteenth century that reality exists only through being immanent in conscious minds.

1901 BALDWIN *Dict. Philos. & Psychol.* I. 520/2 The immanence-philosophy (philosophy of the immediately given or science of pure experience) is the doctrine of a group of recent German thinkers. **1931** W. R. B. GIBSON tr. *Husserl's Ideas* II. ii. 133 Apart from perception, we find a variety of intentional experiences which essentially exclude the real immanence of their intentional objects. **1953** D. H. FREEMAN tr. *Dooyeweerd's New Critique Theoret. Thought* I. I. i. 112 It appears, that also modern phenomenology and Humanistic existentialism move in the paths of immanence-philosophy. **1970** D. M. LEVIN *Reason & Evidence in Husserl's Phenomenology* i. 16 If we suppose that the absence of spatial profiles could be a sufficient condition for immanence, then it would seem that we should have to consider mathematical entities and axioms as immanent objects.

immanental (imăne·ntăl), *a.* Delete *rare* and add later examples. Also, of or pertaining to philosophical immanence.

1897 W. M. URBAN *Hist. Princ. Suff. Reason* i. 8 The postulate of an immanental logic..becomes a permanent element. **1909** —— *Valuation* i. 9 The immanental reference..is a present state, referring, not beyond the present state, but to something more deeply implicit. **1920** A. S. PRINGLE-PATTISON *Idea of God* (ed. 2) 219 This lower pantheism..is common in the popular cults of the East, where the immanental unity of the divine is little more than the idea of a teeming nature. **1930** C. J. WRIGHT *Miracle in Hist.* 214 Theism can only abide with the recognition of the immanental activity of God in His universe. **1955** *Scottish Jrnl. Theol.* VIII. 88 This process is illustrated in religions which tend towards cosmic pantheism, impersonal fatalism, Platonic idealism, immanental piety, rigorous legalism, and ego-centric eudaemonism.

immanentism (i·măněntiz'm). [f. IMMANENT *a.* + -ISM.] Belief in immanence, esp. the immanence of the Deity. So **i·mmanentist** *a.*, holding or characterized by this belief; also as *sb.*, one who believes in the immanence of the Deity.

1907 *Hibbert Jrnl.* July 919 Immanentism..explains away rather than explains that irrational fact of experience which we call evil. **1916** C. C. MARTINDALE *Life Mgr. R. H. Benson* II. 392 To this Immanentist school would thus belong St. Teresa, Dame Juliana of Norwich, ..and Francis de Sales. **1918** M. D. PETRE *Modernism* x. 207 He has been charged with immanentism. **1930** *Times Lit. Suppl.* 14 Aug. 648/2 Mr. Wright stands for a modern, liberalizing and immanentist theology. **1931** *Ibid.* 22 Oct. 812/2 The pure phenomenology which resolves Being into Becoming by a sheer immanentism. **1945** *Mind* LIV. 275 Immanence and transcendence are logical complementaries, and..few thinkers can long afford to remain mere immanentists or mere transcendentists. **1952** *Mind* LXI. 102 Aristotle criticised Platonic doctrine, attacking the theory of Ideas from the 'immanentist' standpoint. **1965** *Rev. Eng. Stud.* XVI. 94 Mr. Miller speaks of the feeling that God is near at hand as God's 'immanence'... The Victorian age was notoriously a period in which immanentist theologies were fashionable.

immature, *a.* Add: **2. d.** *Ophthalm.* Of a progressive cataract: characterized by a marked but incomplete opacity, with the lens usually swollen and its superficial layers still largely transparent.

1850 B. E. BRODHURST *Of Crystalline Lens & Cataract* 57 The terms mature, and immature,..are well adapted to express the stage of progress at which any particular cataract may have arrived. **1904** L. W. FOX *Dis. Eye* xii. 311 The special difficulties..in removing a cataract before maturity are that parts of the cortex, clear at the time of operation, will remain adherent to the capsule of the lens, and later undergo the process of opacification... Some operators, however, operate on immature cataracts, washing out the tenacious material with a syringe. **1970** A. H. KEENEY *Ocular Exam.* ix. 143/2 Early (incipient or immature) cataracts cause changes in refraction..marked by increasing myopia at an age when refraction should either be stable or show increasing hyperopia.

e. Of a soil: not having a fully developed profile. Of a soil profile or its parts: not fully developed.

1926 C. F. MARBUT in Tansley & Chipp *Study of Vegetation* vii. 139 In every region there are hillside soils as well as alluvial and colluvial deposits in which the texture profile will not be well developed because of the short time during which the material has been subjected to the forces of soil development. Such soils are immature or have imperfectly developed profiles. **1927** N. M. COMBER *Introd. Sci. Study Soil* xiii. 139 Glinka divides all soils in the first instance into two groups:—(1) Immature or endodynamomorphic soils, in which the processes of formation have not had full play..(2) Mature or ektodynamomorphic soils, in which the processes of formation have had full play. **1963** D. W. & E. E. HUMPHRIES tr. *Termier's Erosion & Sedimentation* vi. 138 When erosion is greater than the rate of formation, the soil is immature. **1968** R. W. FAIRBRIDGE *Encycl. Geomorphol.* 273 These soils (lithosols) are immature and without soil profiles.

immediacy. Add: **4.** *pl.* Immediate needs.

1923 H. G. WELLS *Men Like Gods* III. iv. 288 The old things and the foul things, customs, delusions, habits,.. base immediacies, triumph over us!

immediate, *a.* Add: **3. b.** *immediate constituent* (Linguistics): a grammatical subdivision of a sentence, phrase, or word, which can sometimes be analysed into further such constituents; in the case of a word, so as to reveal its morphological structure. (Opp. **ultimate constituent*.)

1933 L. BLOOMFIELD *Lang.* xiii. 210 The principle of immediate constituents will lead us..to class a form like *gentlemanly* not as a compound word, but as a derived secondary word, since the immediate constituents are the bound form *-ly* and the underlying word *gentleman*. **1943** *Language* XIX. 79 In separating immediate constituents, one attempts to disturb as little as possible the relationship between the meaning of the parts of the combination and the meaning of the combination as a whole. **1958** A. A. HILL *Introd. Ling. Struct.* viii. 127 Immediate constituent analysis is the process of segmenting a complex construction by successive single cuts. **1961** R. B. LONG *Sentence & its Parts* 491 Immediate Constituents. This term is often used of what are here called simply components. **1963** F. G. LOUNSBURY in J. A. Fishman *Readings Sociol. of Lang.* (1968) 45 Linguistic analysis proceeds by the method of 'immediate constituents', i.e., by division of a larger unit into two immediate constituents... In the end, a systematic pursuit of a different set of policies in immediate-constituent division would, in fact, produce a different grammar of the same language. **1963** J. LYONS *Structural Semantics* ii. 14 It has sometimes been assumed that all the sentences of a particular language can be analysed syntactically in terms of the immediate-constituent, or phrase-structure model. **1964** R. H. ROBINS *Gen. Ling.* vi. 240 The processes of immediate constituent analysis.. by which the longest and most complex sentences can be reduced by analysis to successive expansions of one of a few simple basic sentence structures. **1967** F. P. DINNEEN *Introd. Gen. Ling.* ix. 263 In analyzing the sentence

Poor John ran away, Bloomfield found that it contains five morphemes: *poor, John, ran, a-*, and *way*. They are also the ultimate constituents of the sentence, but the immediate constituents are *Poor John* and *ran away*. **1971** P. Gaeng *Introd. Princ. Lang.* v. 91 The sentence *The rebellious students walked to the dean's office* consists of two main parts—two immediate constituents—namely, *the rebellious students* and *walked to the dean's office*. Each part, in turn, consists of two parts, and each of these consists of two parts, until by cutting the sentence into smaller and smaller groupings, we reach the level of single words or morphemes, the ultimate constituents.

4. c. *immediate access store*: in a computer, a store whose access time is negligible compared with the time required for other operations.

1960 G. N. Lance *Numerical Methods for High Speed Computers* i. 5 The memory..can usually be separated into distinct parts. Firstly, there is the high-speed or immediate access store. **1964** F. L. Westwater *Electronic Computers* iv. 79 Magnetic core stores are often referred to as immediate access stores (I.A.S.).

Immelmann (i·məlmæn). Also erron. **Immelman.** The name of F. *Immelmann* (1890–1916), a German fighter pilot, used alone or *attrib.* in *Immelmann turn*, to designate an evasive manœuvre in the air. Also as *v. intr.*, to execute this manœuvre.

1917 B. K. Adams *Amer. Spirit* (1918) 27 Next I tried the so-called Immelman turn. **1918** J. M. Grider *War Birds* (1927) 206 As I half rolled on top of him, he half rolled too and when I did an Immelman, he turned to the right and forced me on the outside arc and gave his observer a good shot at me. **1919** *Conquest* Dec. 68/1 One of the most useful stunts employed during the war was the 'Immelman turn', its name being that of the aviator who introduced it. In this manœuvre the machine rears up suddenly, turns sideways over the vertical, and emerges in the opposite direction. **1923** W. T. Blake *Flying* 43 *Immelmann turn.* This manœuvre is more commonly termed a 'half-roll' in England. **1934** V. M. Yeates *Winged Victory* III. viii. 254 He could turn better than the Pfalz, and felt he was winning the duel. After its tail, round and round. It straightened and he fired but it immelmanned away. He went after it, but it dived away all out. **1942** *R.A.F. Jrnl.* 27 June 18 Three outside snap rolls, a flick at 110 m.p.h., a stall off an Immelman and a power inverted spin. **1952** J. Steinbeck *East of Eden* xiv. 131 It..made Immelmann turns..and flew over the field upside down. **1967** *Boston Sunday Herald* 14 May (Comic Section), Wow! A loop th' loop! How about a Immelmann turn?

immensikoff (ime·nsikɒf). ? *Obs.* [See quot. 1896.] Jocular name for a heavy overcoat.

1870 D. J. Kirwan *Palace & Hovel* xxxiv. 504 The chorus..of a popular street and music-hall song, which every one is now humming in London..as follows: '..I fancy I'm a Toff; From top to toe I really think I looks—Immensekoff.' **1889** *Pall Mall Gaz.* 25 Sept. 6/1 Heavy swells clad in Immensikoffs, which is the slang term, I believe, for those very fine and large fur robes affected by men about town. **1896** Farmer & Henley *Slang* IV. 3/1 *Immensikoff*, a fur-lined overcoat. From the burden of a song, 'The Shoreditch Toff', sung (*c* 1868) by the late Arthur Lloyd, who described himself as Immensikoff, and wore an upper garment heavily trimmed with fur. **1911** A. Bennett *Hilda Lessways* I. vii. 70 His white muffler and large overcoat (which Dayson called an 'immensikoff').

immersal (imɜ̄·ɹsăl). *rare.* [f. Immerse *v.* + -al.] = Immersion 2.

1901 Greenough & Kittredge *Words* 98 Theological and philosophical studies are also pursued with vigor, and this means an immersal in Latin.

immersement. (Later example.)

1903 W. James *Let.* 29 Dec. in R. B. Perry *Tht. & Char. W. James* (1935) II. 331 Your letter finds me in my nineteenth day of immersement, with grippe, still weak as a 'cat'.

immersion. Add: **3.** (Later example.)

1971 *Nature* 17 Dec. 406/1 None of the light curves showed any signs of an atmosphere on Io: in all cases the curves were flat just before and after occultation with abrupt changes in intensity at immersion and emersion.

5. (sense 1) **immersion foot**, a condition similar to trench foot caused by prolonged exposure of the feet to wet and usually cold conditions; **immersion heater**, a heater (usually electric) whose element may be immersed in the liquid to be heated; *esp.* one having a thermostatic control and designed to be fixed inside a domestic hot-water cylinder; **immersion suit**, a garment designed to give the wearer buoyancy when in the water and to provide insulation from the cold.

1941 *Lancet* 6 Dec. 690/1, I have never seen a case of immersion-foot, and for its adequate description we must await the reports of those whose war experience has brought them greater opportunities of observing it. **1967** *New Scientist* 25 May 449/3 In the Pacific during the second World War a warm water variety of immersion foot was common. **1969** J. McM. Mennell *Foot Pain* v. 104 Immersion foot is similar to trench foot, but the wet environment seems to be more important than the cold; it may occur with relatively warm immersion. **1914** M. Lancaster *Electr. Cooking, Heating, Cleaning* 208 The water in cylinder A..is heated by the immersion heater B. *Ibid.* 209 An additional immersion heater could be fitted, current for which would pass through the meter,..to be switched on if at any time the demand for hot water were much beyond the ordinary requirements. This auxiliary immersion heater could be controlled automatically by a thermostat..switch. **1935** *Jrnl. R. Aeronaut. Soc.* XXXIX. 455 The lubricating oil being kept at a temperature by immersion heaters in the oil tanks. **1936** *Archit. Rev.* LXXX. p. lxii/3 In many parts of the country..automatically controlled immersion heaters are being fitted as auxiliary heaters to fuel fired boilers. **1944** T. A. Longmore *Med. Photogr.* 140 The immersion heater is an electrically heated poker or element which is placed in the developer to raise its temperature and is withdrawn before the solution is put to its normal use. **1951** *Good Housek. Home Encycl.* (1956) 76/1 Portable immersion heaters and boiling rings. **1958** *House & Garden* Mar. 70/2 The Agamatic also provides hot water—in the summer an immersion heater takes over. **1963** *Times* 8 Jan. 5/5 Mrs. Henderson told him that she decided not to have a bath, left the immersion heater on, and went to bed. **1951** R. H. Davis *Deep Diving & Submarine Operations* (ed. 5) I. xiv. 275 Immersion suits can be worn..to protect the escaper from the cold, and..to keep him afloat. **1968** *New Scientist* 15 Feb. 348/1 The immersion suit consists essentially of a double-layer rubber suit which can be inflated... Not only does this provide flotation but also excellent insulation.

immigrant, *sb.* (Later *attrib.* examples.)

1969 *Times* 18 July 4/8 Wolverhampton's Grove School..was described as the '90 per cent immigrant school'. *Ibid.*, There was some criticism..at this high proportion of immigrant children. **1971** *Economist* 12 June 31/2 Those [*sc.* children] born in England to immigrant parents cease to be classified as immigrant schoolchildren after their parents have been here 10 years, while those born overseas remain within the category no matter how long they have been in England. **1973** *Times* 9 Nov. 2/4 Allowance must be made for immigrant children to adjust to a new social and educational environment.

immigration. Add: Also *attrib.*

1872 *Atlantic Monthly* Apr. 456/1 Natives of Europe.. not included in the immigration reports [etc.]. **1879** *Bradstreet's* 10 Dec. 2/3 It is our idea that immigration societies are doing us no good. **1880** D. M. Gordon *Mountain & Prairie* 298 Such companies, spurred into activity by the prospect of profitable land sales, will probably be more zealous than Government immigration agents. **1890** *Stock Grower & Farmer* 25 Jan. 7/2 Col. Edward Haren, of the immigration department of the Santa Fe, is in the city on his return from Albuquerque. *Ibid.* 22 Feb. 3/2 This territory has never had an immigration 'boom'. **1904** F. Bradshaw *Alien Immigration* 121 When the alien has passed the Immigration Department his troubles are not yet over. **1905** *Act 5 Edw. VII* c. 13 §1 An immigrant shall not be landed in the United Kingdom from an immigrant ship except at a port at which there is an immigration officer appointed under this Act. **1906** *Daily Chron.* 21 May 1/7 Mr. Seddon, Premier of New Zealand, has challenged the right of the British Government to interfere with Australasian immigration legislation. **1907** *Westm. Gaz.* 30 Jan. 8/2 The Premier [of Australia] has..liberalised the immigration regulations. **1922** *Encycl. Brit.* XXXII. 854/1 By the Immigration Act passed in 1921 the number of immigrants admitted from any one country in the year July 1 1921 to June 20 1922, was restricted to 3% of the persons of that nationality resident in the United States in 1910. **1926** *Ibid.* III. 21/1 Since the opening of immigration [into Palestine] with the promulgation of the Immigration Ordinance (1920) [etc.]. **1949** Koestler *Promise & Fulfilment* iv. 40 It is conceivable that they could have achieved sufficient pressure at least to mitigate the immigration bar of 1939. *Ibid.* vi. 56 Except the small number of those who already held pre-war immigration certificates. *Ibid.* 60 The majority..were Zionists..who at the eve of the war had been waiting for their turn on the immigration quota. **1969** *Times* 19 July 8/3 The controversy over Britain's immigration policy. **1971** 'D. Halliday' *Dolly & Doctor Bird* ii. 25 He carried a Turkish passport through Immigration Controls. **1973** P. Geddes *Ottawa Allegation* v. 63 The immigration officer..took his time over Fender's passport.

absol., the immigration checks or authorities. *colloq.*

1966 F. Hoyle *Oct. First* i. 5 We got into London airport more or less on time. Quickly we were into the reception hall and through immigration. **1972** J. Potter *Going West* 17 He produced his passport and transit card for immigration.

2. *collect.* The body of immigrants. *U.S.*

1852 H. Stansbury *Exped. Valley Gt. Salt Lake* 126 In the autumn, another large immigration arrived under the president, Brigham Young, which materially added to the strength of the colony. **1857** *Trans. Illinois Agric. Soc.* II. 365 The immigration was generally a moral, correct people. **1948** *Sat. Rev.* (U.S.) 17 July 20/1 A far vaster immigration..began pouring through the city portals.

immiscible, *a.* Add: Usu. *spec.* of a liquid: incapable of forming a true solution *with* or *in* another liquid. (Later examples.)

1934 A. J. Mee *Physical Chem.* x. 436 (*heading*) Vapour pressure of a mixture of immiscible liquids. **1964** D. F. Eggers et al. *Physical Chem.* viii. 276 Steam distillation is frequently used to carry over organic substances immiscible in water.

immiserization (imi:zəraizēⁱ·ʃən). Also (slightly earlier) **immiseration.** [f. Im-¹ + Miser(able *a.* + -ization; tr. G. *verelendung*.] The act of making or becoming progressively more miserable; pauperization, impoverishment. So **immi:serifica·tion**, in the same sense; **immi·serize** *v. trans.*, to impoverish.

1942 J. A. Schumpeter *Capitalism, Socialism & Democracy* (1943) iii. 22 The glowing indictment of 'exploitation' and 'immiserization'. **1948** R. Strausz-Hupé in *Philos. of Sci.* X. 270 Fifty years ago the Revisionists pointed out the fallacies of Marx's theory of the immiseration of the proletariat. **1971** J. Vaizey *Social Democracy* 37 The general trend of real wages, after 1850, was upwards, and the general immiserisation of the proletariat was not brought about. *Ibid.* 48 The international industrial system had been created; the proletariat was not immiserised; and the modern nation-state had been created. **1975** D. McLellan *Marx* iii. 44 Marx was thinking in terms of trends and projected into the future tendencies that he saw in contemporary society. One of the most important of these trends was the immiserization of the proletariat. Marx was usually chary of claiming that the proletariat would become immiserized in any absolute sense.

immittance (imi·tăns). *Electr.* [f. Im(pedance + (ad)mittance.] Admittance or its reciprocal, impedance: used when it is desired to refer to both quantities without making a distinction between them.

1957 D. Woods in *Proc. Inst. Electr. Engin.* CIV. c. 507/2 Immittance is the basic parameter of any a.c. measuring system for electrical quantities. [*Note*] Immittance is used to convey the general idea of impedance or admittance. **1960** Clement & Johnson *Electr. Engin. Sci.* xi. 341 Each element *R, L,* and *C* of a network has an immittance to sinusoidal current. **1969** P. M. Chirlian *Basic Network Theory* viii. 496 We can eventually calculate the input current $\mathbf{I}_1(s)$. This will be expressed as $\mathbf{V}_2(s)$ times some function of the immittances of the ladder network.

immobilism (imōᵘ·biliz'm). [ad. F. *immobilisme* (also used).] A policy or attitude of extreme conservatism or opposition to progress.

1949 *Time* 17 Oct. 34 Successful 'immobilism' (patient compromise, appeasement, moderation). **1955** *Times* 5 May 10/7 The Radicals..wished to remain a party of the left; social and political *immobilisme* (opposition to progress) was the best ally of Communism. *Ibid.* 24 Aug. 6/1 There could be no question of allowing themselves to be swayed by a question of superstition or of fetishism, which would be contrary to national dignity. But in any case, M. Faure added, it had been clearly agreed that they would not adopt a policy of immobilism. **1959** *Economist* 21 Feb. 662/2 This outright *immobilisme* [among farmers] condemns itself. **1961** *Encounter* XVII. ii. 21 The notorious *immobilisme* of French society. **1967** *Economist* 8 Apr. 110/2 Mr Callaghan will have made a sad present to industrial immobilism if he throws any such opportunity for reducing surtax away.

immobilize, *v.* Add: **2.** *Soil Sci.* and *Bot.* To convert (a plant nutrient) from a form in which it can be utilized by a plant to one in which it cannot; to assimilate and thereby render unavailable to (other) plants. Also *absol.*

1951 *Proc. Soil Sci. Soc. Amer.* XV. 168/2 Fertilizer and nonfertilizer nitrogen must have been absorbed and immobilized in proportion to their relative available concentrations in the soil. **1952** L. M. Thompson *Soils & Soil Fertility* ix. 135 If there is a deficit of nitrogen in the added organic matter, the microbes will immobilize ammonia and nitrates from the soil. **1958** *U.S. Dept. Agric. Yearbk.* 1957 760/1 The addition of fresh straw or sawdust to the soil may greatly increase the number of bacteria. These remove available nitrogen and phosphorus from the soil and immobilize them within their cells. **1967** Fried & Broeshart *Soil–Plant Syst.* vii. 240 Soluble phosphates that are applied to the soil are generally immobilized. **1971** *Nature* 9 Apr. 403/1 *Becium homblei*..grows on soils rich in copper, and it has been shown to accumulate heavy metals. Other plants also have this characteristic, and their ability to immobilize in insoluble form..has been noted.

immobilization (further examples, corresponding to the foregoing sense of the vb.).

1951 *Proc. Soil Sci. Soc. Amer.* XV. 166/2 Immobilization is here used to denote the process of conversion of inorganic nitrogen to organic combinations either through plant uptake or through microbial absorption. **1952** L. M. Thompson *Soils & Soil Fertility* ix. 132 Either plants or microorganisms may convert NH_4^+ or NO_3^- to organic form (chiefly protein). The conversion of inorganic compounds to organic form is called immobilization. **1956** *Proc. Soil Sci. Soc. Amer.* XX. 217/2 The tie-up (immobilization) or release (mineralization) of soil nitrogen depends very much on the chemical composition of the material undergoing decomposition. **1970** *Analytical Biochem.* XXXIII. 341 The physical entrapment of enzymes in insoluble matrices seems to provide a general method of immobilization.

immobilized (imōᵘ·biləizd), *ppl. a.* [f. Immobilize *v.* + -ed¹.] Rendered immobile or stationary.

1923 *Edin. Rev.* Jan. 179 The left and centre of the enemy was held... It was now possible to attempt a decisive attack on an immobilised enemy.

immoralism (imǫ·răliz'm). [ad. G. *immoralismus* (Nietzsche), f. Immoral *a.* + -ism.] The reverse or negation of moralism; a system of thought or practice which rejects moral law.

1907 *Athenæum* 23 Mar. 348/1 The system..of Nietzsche, with all its blasphemy and immoralism. **1918** *Hibbert Jrnl.* Apr. 378 The sickly social idealism.. which treated the most healthy immorality so much more harshly than it treated the most unhealthy immoralism. **1929** B. H. Streeter *Primitive Church* v. 179 The theoretic basis of Gnostic immoralism was a distinction between the ultimate Good God and the more or less evil Creator of the material universe. **1973** E. R. Dodds *Anc. Concept of Progress* 105 Nietzsche himself recognized the affiliation of his immoralism to the sophistic movement.

immortable (imǭ·ɹtăb'l), *a.* [f. Immort(al *a.* + -able.] Having the capacity to live after death. So **immortabi·lity.**

1922 J. Y. Simpson *Man & Attainment of Immortality* xiii. 275 The contention that eternal life..is morally conditioned, that man, in short, is immortable rather than immortal. **1930** S. D. McConnell (*title*) Immortability. An Old Man's Conclusions. **1950** S. Eddy *You will survive after Death* 3 We may have at least..'immortability'—a fitness in the quality of human personality for survival.

immortal, *a.* **2. b.** (Later examples.)

1928 E. C. Webster *Pot Holes* 3, I am as fond of Burns as any, and have read a good deal of his poetry,..but I am not one of those who believe that the Immortal Memory can only be preserved by a yearly pickling in alcohol. **1959** *Times* 17 Apr. 15/3 His record of devotion to the 'Immortal Memory'—a toast which he had proposed all over Scotland and England—was typical of this special cult which the wandering Scot has carried all over the globe. **1973** *Listener* 15 Mar. 344/2 The Johnson celebration..the toast to 'the immortal memory'.

immortelle. Add: (Further examples.) Also *attrib.* and *Comb.*, *transf.*, and *fig.*

1883 'Mark Twain' *Life on Mississippi* xlii. 431 A milder form of sorrow finds its inexpensive and lasting remembrancer in the coarse and ugly but indestructible 'immortelle'—which is a wreath or cross or some such emblem, made of rosettes of black linen, with sometimes a yellow rosette at the conjunction of the cross's bars,—kind of sorrowful breastpin, so to say. **1890** A. Martin *Home Life Ostrich Farm* i. 21 Pink and white *immortelles*, gladioli, ixias, and irises of all kinds abound. **1929** D. H. Lawrence *Pansies* 6 Anyhow I offer a bunch of pansies, not a wreath of *immortelles*. I don't want everlasting flowers. **1936** *Times Lit. Suppl.* 21 Nov. 935/1 Poor little immortelles surviving from that old gorgeous efflorescence of linguistic misuse. **1960** R. G. Haggar *Conc. Encycl. Cont. Pott. & Porc.* 67/2 *Blaublümchenmuster*, (German) blue flower pattern: 'aster' or 'immortelle' pattern. *Ibid.* 239/2 Early [Klösterle] wares were decorated in blue with popular Meissen 'Immortelle' and 'bird on rock' designs. **1963** V. Nabokov *Gift* iii. 153 Even if he had put on the light the immortelle-like yellowness of daytime electricity would have been no help at all. **1968** E. Lovelace *Schoolmaster* i. 7 The immortelle holds its scarlet blossoms still. **1970** *New Yorker* 22 Aug. 38/3 The old man was himself like a trophy of immortelles. **1971** E. M. Roach in J. Figueroa *Caribbean Voices* I. 22 The giant immortelles Splash fire on the hills.

immram (i·mram). Also **imram.** Pl. **im(m)rama.** [ad. O.Ir. *imram* (mod.Ir. *iomramh*), f. *imm-rá* to row around.] Any of the stories of fabulous sea voyages written in Ireland in the seventh and eighth centuries. Also *attrib.*

1895 A. Nutt in K. Meyer *Voy. of Bran* iv. 161 The *imrama* literature has been investigated by Professor Zimmer with all his wonted acuteness. **1917** *Mod. Philology* XV. 450 The *imram* is a sea-voyage tale in which a hero, accompanied by a few companions, wanders about from island to island, meets Otherworld wonders everywhere, and finally returns to his native land. **1948** M. Dillon *Early Irish Lit.* vi. 124 Of the seven immrama mentioned in the two lists of sagas, only three have come down to us. **1951** G. Turville-Petre *Heroic Age of Scandinavia* ix. 94 It is by no means improbable that the *Immrama* contain distorted descriptions of scenes, which the hermits had witnessed on their travels. **1962** *Guardian* 14 Dec. 7/4 The suggested relation between 'Gulliver's Travels' and the Early Irish *immram*. **1964** G. Jones *Norse Atlantic Saga* i. 7 The best known of the Irish *Imrama*..records the travels of St. Brendan. **1967** Dillon & Chadwick *Celtic Realms* viii. 200 The briefer account of Brendan's voyage is included in the *Vita*,.. and we thus possess two apparently independent accounts of the saint's *immram*.

immune, *a.* (*sb.*). Add: **2. a.** Also *transf.* and *fig.*, wholly protected *from* something injurious or distasteful; not susceptible or responsive *to* something.

1898 Mercier in *Brit. Med. Jrnl.* 3 Sept. 586/1 There is for every insane person a certain sphere of conduct for which he ought to be entirely immune from punishment. **1900** *Daily News* 5 July 3/2 A man whose achievements should render him immune from all mud throwing. **1922** D. H. Lawrence *England, my England* 235 Among the graves, she felt immune from the world. **1944** A. Holmes *Princ. Physical Geol.* xvii. 367 No place can be regarded as permanently immune from shocks. **1947** *Jrnl. R. Aeronaut. Soc.* LI. 293 Ice guards..proved of considerable value during the war when used on aircraft which themselves were not completely immune from icing. **1955** *Sci. Amer.* June 96/3 The magnetic-core memory..is relatively immune to unwanted electrical disturbances. **1973** *Human World* Feb. 8 The vision of the future that is to carry them through is the high-rise block of flats, the motorway... And if the 'underprivileged' prove immune to sense and prosperity? Well, thinks Mr Maddox, they can't. **1973** *Sci. Amer.* Feb. 83/1 The system is extremely complicated and therefore would be rather expensive and not as reliable or immune to functional failure as one would like. **1973** *Daily Tel.* 7 Mar. 18 Orwell was a bad poet and immune to the arts (though a most likeable man). **1974** *Ibid.* 16 Feb. 9/1 The white pawns are immune from Black's bishop.

b. as *sb.* Substitute for def.: An immune individual. (Further examples.)

1909 *Rep. Brit. Assoc. Adv. Sci.* 764 All extracted immunes [*sc.* wheat plants immune to yellow rust] should breed true to this feature. **1951** Whitby & Hynes *Med. Bacteriol.* (ed. 5) viii. 105 After an epidemic the community remains free from that disease until the proportion of immunes declines and the density of susceptibles is once more raised to pre-epidemic level.

3. *Med.* (Only in *attrib.* use.) Relating to immunity or its development; serving to bring immunity about.

1907 Muir & Ritchie *Man. Bacteriol.* (ed. 4) xix. 484 Various substances..remove the opsonic property from a normal serum, while they have no effect on an immune-opsonin. **1928** L. E. H. Whitby *Med. Bacteriol.* ii. 19 The immune substances produced by the animal are termed antibodies. **1946** K. Landsteiner *Specificity Serological Reactions* (rev. ed.) i. 4 The immune antibodies ..react as a rule only with the antigens that were used for immunizing and with closely similar ones. **1953** S. Raffel *Immunity* iv. 40 (*heading*) Mechanisms of acquired immunity. Antibody as a specific immune mechanism. **1969** *Times* 24 Mar. 4/7 Antilymphocytic serum..combats the particular immune defence mechanism responsible for rejecting tissue grafts.

b. Specific collocations: *immune body* = *antibody; *immune globulin*, (*a*) a preparation containing antibodies obtained from normal individuals or from ones immunized against a specific disease, and suitable for use as an antiserum; (*b*) = *immunoglobulin; *immune response*, the reaction of the body to the introduction into it of an antigen; *immune serum*, serum which contains antibodies, esp. one which can confer immunity to the corresponding antigen on a recipient; = *anti-serum b.

1899 Muir & Ritchie *Man. Bacteriol.* (ed. 2) xix. 485 Ehrlich has recently applied his theory of antitoxines to the lysogenic action of sera towards bacteria and red corpuscles... His observations show that the body specially developed in the blood of the animal treated—the 'immune-body', enters into firm combination with the red corpuscles. **1900** P. Ehrlich in *Proc. R. Soc.* LXVI. 443, I have sought..to make clear the mechanism concerned in the action of these two components—the stable, which may be designated 'immune body', and the unstable, which may be designated 'complement'—which, acting together, effect the solution of the red blood corpuscles. **1937** Immune body [see *amboceptor]. **1971** Herbert & Wilkinson *Dict. Immunol.* 91 *Immune body*, obsolete synonym for antibody. **1935** *Jrnl. Amer. Med. Assoc.* 17 Aug. 493/1 The use of Immune Globulin (Human) in the modification of measles has been known for a considerable period. **1948** H. J. Parish *Bacterial & Virus Dis.* ix. 67 Placental extract or 'human immune globulin' is an extract of placental globulins and has approximately the same potency as adult serum. **1958** —— *Antisera, Toxoids, Vaccines & Tuberculins* (ed. 4) ix. 78 Large pools of adult plasma normally contain a variety of protective antibodies..located mainly in the gamma fraction of the globulin or, as it is sometimes called, the immune globulin. **1966** *Lancet* 24 Dec. 1403/2 A technique of immunodiffusion of serum through agar gave a sensitive and fairly accurate measure of the concentrations of the three main classes of immune globulins—γ G, γ M, and γ A. **1953** *Jrnl. Nat. Cancer Inst.* XIV. 755 The over-all pattern presented is one of interference with an immune response of the host. **1963** Gell & Coombs *Clin. Aspects Immunol.* i. 5 Before we can understand much about 'immune' responses and their results, protective or damaging, on the host, we need to know more about the separation of antibody responses from the 'cellular' responses which result in delayed (non-antibody dependent) sensitivity. **1964** *New Scientist* 1 Oct. 10/2 The 'immune response' is an important feature of a vertebrate animal's defence mechanism. **1902** R. T. Hewlett *Man. Bacteriol.* (ed. 2) v. 140 For the lysis of a given quantity of bacteria a certain amount of immune serum is necessary. **1946** K. Landsteiner *Specificity Serological Reactions* (rev. ed.) i. 7 Sera that contain antibodies as the result of the injection of antigens are called 'immune sera' (antisera). **1955** *Sci. Amer.* Mar. 65/2 In experimental work, antibody, or what is known as immune serum, is produced by injecting virus into an animal. **1970** Gold & Peacock *Basic Immunol.* vii. 247 Chiefly noted in man in connexion with immune serum therapy, serum sickness has also been widely studied as an experimental disease in laboratory animals.

immu·ne, *v. rare.* [f. the adj.] *trans.* To render immune.

1849 G. S. Faber *Let.* 16 May in R. Chapman *Father Faber* (1961) xi. 220, I think if a little experience does not immune me to the row..I *must* go to the back. **1928** Hardy *Coll. Poems* 431 The vision That immuned me from the chillings of misprision.

immunize, *v.* Add: **2.** *intr.* Of an organism or substance, regarded as an antigen: to produce immunity in an individual into which it is introduced.

1942 *Jrnl. Bacteriol.* XLIII. 405 Strains we have classified as weakly antigenic (in so far as they fail to immunize significantly against homologous virus injected intracerebrally). **1951** Whitby & Hynes *Med. Bacteriol.* (ed. 5) viii. 108 It is possible to kill bacteria without so altering their antigenic structure that they no longer immunize against living bacteria. **1973** *Nature* 30 Mar. 330/1 The ability of lactating mammary gland to immunize against D1 and D2 mammary tumour growth.

immunizer (i·miunəizəɹ). [f. Immunize *v.* + -er¹.] That which renders immune; *occas.*, one who uses or advocates immunization.

1927 *Daily Express* 18 June 9/2 'Immuniser' for Cancer. ..It may well be that chemical and medical research will discover this natural immuniser, which will strengthen resistance to cancer in all individuals. **1931** *Amer. Jrnl. Cancer* XV. 627 Formalin..is a less potent immuniser than antiserum. *Ibid.*, No single mode of treatment is entirely satisfactory both as a 'destroyer' and as an 'immuniser'. **1950** G. B. Shaw *Farfetched Fables* 90 Lying advertisements of panaceas, prophylactics, elixirs, immunizers, vaccines, antitoxins, vitamins, and professedly hygienic foods. *Ibid.*, They are in fact exalting every laboratory vivisector and quack immunizer above Jesus and St James.

immuno- (imiū̆·no, i·miuno), used as comb. form of Immune *a.*, Immunity, *Immunology, and related words. (In the following words secondary stresses are in general left unmarked, since they vary in the manner indicated above.)

immuno-assay (imiū̆no,æ·se), a bio-assay performed by means of immunological methods; **immunobio·logy** = *immunology (see quot. 1970); so **immunobiolo·gic, -biolo·gical** *adjs.*; **immunoche·mistry,** chemistry as applied to immunology; the chemistry of immunological phenomena; so **immunoche·mical** *a.*, of or pertaining to immunochemistry; using the methods of immunochemistry; **immunoche·mically** *adv.*; **immunoche·mist,** a student of or expert in immunochemistry; **immunodiffu·sion,** diffusion of immunologically active substances; a technique for investigating antigens and antibodies by observing any precipitates that may form when initially separate portions of them are allowed to intermingle by diffusion through a gelatinous or other medium; **immuno-ele:ctrophore·sis,** a technique for characterizing the proteins in a mixture (such as serum) by first separating them by electrophoresis and then subjecting them to immunodiffusion (in the same or a different medium); so **immuno-ele:ctrophore·tic** *a.*, **-phore·tically** *adv.*; **immunofluore·scence,** a method of demonstrating antibodies (or antigens) in microscopic preparations by introducing corresponding antigens (or antibodies) labelled with a fluorescent dye; fluorescence emitted by such preparations; so **immunofluore·scent** *a.*, of, pertaining to, or involving this method; **immunogene·tic** *a.*, of or pertaining to immunogenetics; so **immunogene·tically** *adv.*; **immunogene·tics,** the related study of immunology and genetics, either as a branch of genetics in which immunological methods and knowledge are employed, or as the study of the genetic aspects of immunological phenomena and substances; **immunohæmato·logy** (*U.S.* **-hematology**), the immunology of the blood; so **immunohæmatolo·gic, -lo·gical** *adjs.*; **immunopatho·logist,** a student of or expert in immunopathology; **immunopatho·logy,** the pathology of the immune response; the study of immunological phenomena and substances in relation to pathology; hence **immunopatholo·gic, -lo·gical** *adjs.*; **immunoprophyla·xis,** the prevention of disease by immunization; so **immunoprophyla·ctic** *a.*, of or pertaining to immunoprophylaxis; *sb.*, an agent that prevents (a) disease by producing immunity; **immunosuppre·ssant,** an agent which has an immunosuppressive effect; also *attrib.*; **immunosuppre·ssed** *a.*, (of an individual) rendered unable to react immunologically to an antigen; **immunosuppre·ssion,** the suppression of the immune response of an organism; **immunosuppre·ssive** *a.*, suppressing the immune response of an organism; **immunosympathe·ctomy** [*-ectomy, used *loosely*], the destruction of many of the sympathetic ganglia of a new-born animal by injection of an antiserum for the appropriate nerve growth-factor; so **immunosympathe·ctomized** *ppl. a.*, treated in this way; **immunothe·rapy,** treatment of disease by the production of immunity (whether by the introduction into the individual of appropriate antibodies, etc., or by the stimulation in it of an immune response); **immunotransfu·sion,** the transfusion of blood which has been previously immunized against the recipient's infection.

1959 *Nature* 21 Nov. 1648/2 We have previously reported on the immuno-assay of beef insulin. **1969** R. Hall et al. *Fund. Clin. Endocrinol.* xiv. 258/1 The hormone that reacts with anti-insulin serum in the immunoassay technique for insulin accounts for only a small part of the total ILA in plasma. **1930** *Jrnl. Amer. Med. Assoc.* 12 Apr. 1188/2 [*tr. a Finnish title*] Immunobiologic conditions in tuberculosis. **1959** *Biol. Abstr.* XXXIII. 1822/1 A few statements on the reflex mechanism of immunobiological processes. **1966** E. D. Day *Found. Immunochem.* vii. 87 One definition [of *hapten*] is immunobiological. **1957** (*title*) Journal of microbiology, epidemiology and immunobiology. **1970** Alexander & Good *Immunobiol. for Surgeons* i. 1 In its older and classical meaning, it [*sc.* immunology] was the study of immunity, the processes by which organisms defend themselves against infection... More recently, cellular immunity has been recognized as being important in processes which have to do with recognition phenomena, self-characterization, growth and development, heredity, aging, cancer, and transplantation. With this expansion, immunology has exceeded the limits of its original meaning, and immunobiology has become a preferable term for this expanding field. **1925** C. H. Browning *Immunochem. Stud.* 15 The immunochemical properties of serum. **1948** Kabat & Mayer *Exper. Immunochem.* i. 5 The application of immunochemical methods has extended far beyond the study of immunity to disease and has become a valuable tool in the characterization of proteins and polysaccharides. **1960** *Jrnl. Immunol.* LXXXV. 37 (*heading*) Immunochemical studies of human serum Rh agglutinins. **1961** Webster, Immunochemically. **1966** *Lancet* 31 Dec. 1435/1 In certain human antiserums, dog insulin is immunochemically distinguishable from pork insulin although both have the same aminoacid sequence. **1948** M. Heidelberger in Kabat & Mayer *Exper. Immunochem.* p. vi, The immunochemist is in possession of a store of marked molecules, antigens and antibodies, each as distinctively marked with respect to the other as if it contained a radioactive tracer element. **1970** *Nature* 18 July 229/2 Immunochemists from ten laboratories cross matched almost seventy antibodies to the proteins of the vertebrate eye lens when an international working party on crystallin immunochemistry met in..Edinburgh University. **1971** Nossal & Ada *Antigens, Lymphoid Cells, & Immune Response* i. 2 The key discoveries about lymphocytes and antibody-producing cells..have been less fully digested by immunochemists. **1907** S. Arrhenius *Immunochem.* p. vii, I have given to these lectures the title 'Immuno-chemistry', and wish with this word to indicate that the chemical reactions of the substances that are produced by the injection of foreign substances into the blood of animals, *i.e.* by immunisation, are under discussion in these pages. **1956** *Nature* 3 Mar. 426/2 Virulent and protective avirulent strains have been studied comparatively by the methods of immunology, immunochemistry and biochemistry. **1970** Immunochemistry [see *immunochemist* above]. **1959** *Nature* 30 May 1512 (*heading*) Cellulose acetate as a medium for immunodiffusion. **1966** *Lancet* 24 Dec. 1403/2 A technique of immunodiffusion of serum through agar gave a sensitive and fairly accurate measure of the concentrations of the three main classes of immune globulins. **1971** *Nature* 8 Jan. 119/2 Antibody studies were made using immunodiffusion techniques set up with cerumen suspensions and with IgA and IgG antibody. The IgA or IgG antibody was then placed in the centre well; diffusion was allowed to take place..and precipitin lines were recorded. **1958** *Federation Proc.* XVII. 530/2 (*heading*) Starch gel immunoelectrophoresis. **1964** G. H. Haggis et al. *Introd. Molecular Biol.* ii. 25 (*caption*) The identification of protein fractions present in human plasma using immunoelectrophoresis. **1968** H. Harris *Nucleus & Cytoplasm* v. 104 The antigens..were isolated as antigen-antibody complexes by immuno-electrophoresis. **1955** Williams & Grabar in *Jrnl. Immunol.* LXXIV. 158 (*heading*) Immunoelectrophoretic studies on serum proteins. **1970** J. T. Barrett *Textbk. Immunol.* v. 114 A valuable modification of gel precipitation tests is the immunoelectrophoretic procedure of Grabar and Williams. **1961** A. J. Crowle *Immunodiffusion* iv. 101 According to Ryback (1959), plasmin exists immunoelectrophoretically as two zones in the beta region. **1960** *Jrnl. Biophysical & Biochem. Cytol.* VII. 43 (*heading*) Observations of measles virus infection of cultured human cells. I. A study of development and spread of virus antigen by means of immunofluorescence. **1961** *Lancet* 16 Sept. 663/2 The nuclear immunofluorescence obtained with heated and unheated sera was compared. **1971** tr. *K. Federlin's Immunopath. Insulin* 32 In direct immunofluorescence, the tissue is first covered with a layer of labelled antibody or antigen, then freed from the uncombined agent by several washings, and finally mounted with a coverslip. **1959** *Proc. Soc. Exper. Biol. & Med.* CI. 289 (*heading*) Quantitative determination of infectious units of measles virus by counts of immunofluorescent foci. *Ibid.* 290/2 The cells were..prepared for immunofluorescent microscopy. **1970** Harris & Sinkovics *Immunol. Malignant Dis.* i. 37 Other immunofluorescent studies by Morton have demonstrated antibodies to osteosarcoma in the serum of patients with this tumor. **1936** Irwin & Cole in *Jrnl. Exper. Zool.* LXXIII. 85 (*heading*) Immunogenetic studies of species and of species hybrids in doves. **1970** W. H. Hildemann *Immunogenetics* iii. 86 The newest area in which immunogenetic characterization of microorganisms has provided substantial insights involves the many viruses capable of inducing cancer. *Ibid.* vii. 224 (*heading*) Immunogenetic concepts of cancer and aging. **1971** *Nature* 12 Nov. 103/1 The paternal strains, A/J and A2G respectively, are immunogenetically identical at the major H-2 histocompatibility locus. **1947** M. R. Irwin in *Adv. Genetics* I. 133 The term 'immunogenetics' was proposed by the author some years ago to designate studies in which the technics of both genetics and immunology were employed jointly... The term indicates the study of genetic characters as yet only detectable by immunological reactions. **1965** P. L. Carpenter *Immunol. & Serol.* (ed. 2) ix. 265 The use of serologic techniques in genetics is expanding. Genes control the formation of antigenic substances, so detection of antigens provides an objective and useful tool for study of genes. The term immunogenetics is applied to this field of research. **1971** J. A. Bellanti *Immunol.* iii. 60 Immunogenetics includes all those processes concerned in the immune response which may have a genetic basis. In the past, the term has been largely restricted to mean genetic markers on immunoglobulin polypeptide chains. **1954** *Amer. Jrnl. Clin. Path.* XXIV. 1333 It is just 50 years ago that the first immunohematologic test was introduced by Donath and Landsteiner. **1959** H. S. Lawrence *Cellular & Humoral Aspects Hypersensitive States* v. 133 (*heading*) Immunohematologic disease. **1967** *Biol. Abstr.* XLVIII. 172/2 (*heading*) Significance of immunohematological methods of investigation. **1950** *Brit. Med. Jrnl.* 16 Sept 673/1 (*heading*) Immuno-haematology. **1954** *Amer. Jrnl. Clin. Path.* XXIV. 1334 Serologic technics applied to the study of diseases of blood added up to what is called immunohematology, a separate and distinct subdivision of hematology... Immunohematology encompasses diseases of blood of which the causes, the pathogenesis, or the clinical manifestations have been shown to be determined by an antigen–antibody reaction. **1972** Immunohematology [see *isoagglutinogen* s.v. *iso-]. **1960** *Federation Proc.* XIX. 208/2 (*heading*) An immunopathologic study of avian nephrotoxic nephritis in the rabbit. **1959** Grabar & Miescher *Immunopath.* 17 Attention has been focussed on the immunopathological consequences of leucocyte isoantigens. *Ibid.* 41 Much data of interest to immunopathologists. **1970** *Nature* 19 Sept. 1196/2 Studies on the amino-terminal sequences of a number of myeloma and pathological immunoglobulin chains also evoked much interest among the immunopathologists. **1959** Grabar & Miescher *Immunopath.* 13 Immunopathology presumably covers all immune phenomena associated with general pathology—the majority of the reactions of course being physiogenic and beneficial to the host—others again being inconsequential or even harmful. **1971** K. Federlin (*title*) Immunopathology of insulin. **1960** *Biol. Abstr.* XXXV. 168/2 (*heading*) Poliomyelitis: the present status of some epidemiological and immunoprophylactic problems. **1964** D. F. Gray *Immunol.* x. 95 Immuno-prophylactic procedures include not only the long-term protection afforded by active immunization against a number of the epidemic and endemic diseases to which urban man is prone, but also short-term passive protection against immediately anticipated infection. **1972** *Lancet* 21 Oct. 876/1 Administration of B.C.G. to mice who no longer have palpable disease does not prolong their life... This is true whether or not B.C.G. had been given as an immunoprophylactic. **1964** *Canad. Jrnl. Public Health* LIII. 346 (*heading*) Advances in the immunoprophylaxis of smallpox. *Ibid.*, In the last twenty years progress in this field [*sc.* smallpox] has been less dramatic than in other fields of immunoprophylaxis of virus diseases, such as poliomyelitis or measles. **1972** *Lancet* 21 Oct. 875/2 (*heading*) Immunoprophylaxis and immunotherapy of leukæmia with B.C.G. **1965** *Jrnl. Immunol.* XCV. 1019/1 Immunosuppressant compounds such as chloramphenicol.., actinomycin D.., 6-mercaptopurine..and corticosteroids. **1970** M. C. Vale tr. *Nezlin's Biochem. Antibodies* v. 282 One of the most active immunosuppressants is cyclophosphamide. **1967** *Jrnl. Nat. Cancer Inst.* XXXVIII. 754/1 Pretreatment..with this alkylating agent resulted in..increased growth and earlier tumor deaths in the immunosuppressed animals. **1965** *Jrnl. Immunol.* XCV. 1019/1 The mechanism of acriflavine-induced immunosuppression remains unknown. **1968** *New Scientist* 13 June 557/1 Reports of increasing success in combating graft rejection, following improvements in tissue typing and immunosuppression. **1963** *New England Jrnl. Med.* CCLXVIII. 1315 (*heading*) Prolonged survival of human-kidney homografts by immunosuppressive drug therapy. **1968** *Observer* 5 May 3/3 To stem any rejection of the new heart, Mr West is now receiving immuno suppressive treatment with a drug. **1970** Balner & Beveridge *Infections & Immunosuppression Subhuman Primates* 194/2 Most of the immunosuppressive agents currently in use are entirely nonspecific, that is to say they depress all aspects of immune reactivity and thereby the immunological defences against infectious microorganisms as well. **1961** Levi-Montalcini & Angeletti in Kety & Elkes *Regional Neurochem.* VII. 369 The injected and untreated animals did not differ from each other. Immunosympathectomized mice became pregnant, nursed and took care of the litter as controls. **1964** *Nature* 26 Dec. 1315/1 'Immunosympathectomized' rats, in which extensive irreversible atrophy of the peripheral sympathetic system is produced by injecting, during the first few days of life, an antiserum to nerve growth factor.., appear healthy, grow, reproduce and have normal gastro-intestinal function. **1962** *Internat. Jrnl. Neuropharmacol.* I. 163 Immunosympathectomy performed in newborn mice, results in a striking depletion of NA in the heart of the same adult animals. **1972** J. B. Martin in Steiner & Schönbaum *Immunosympathectomy* xii. 196 The use of immunosympathectomy in analyzing the importance of adrenergic innervation of individual glands is limited to some degree by the incomplete effect produced by the antisera to nerve growth factor. Thus, immunosympathectomy should be valuable in investigations of the role of sympathetic innervation of those glands (pineal, pituitary, thyroid) in which denervation is nearly complete. It is of limited use in studies of the abdominal glands or genital tract. **1913** R. L. Crockett in *N.Y. State Jrnl. Med.* XIII. 213/2 For some time I have been trying to discover what place (if any) vaccine therapy or, as I prefer to call it—immuno-therapy—had in the branches to which I limit my practice. **1937** *Jrnl. Amer. Med. Assoc.* 19 June 2171 Further investigations on active immunotherapy of whooping cough. **1969** *Daily Colonist* (Victoria, B.C.) 9 Oct. 40/4 He said immunotherapy—injection of materials into the body to stimulate development of tumor-resistant anti-bodies—had been successful with eight of his patients. **1971** J. A. Bellanti *Immunol.* xix. 510 Immunotherapy is passive immunization through the use of serum or gamma globulin which confers temporary protection to one host by the introduction of antibodies actively produced in another. **1919** A. E. Wright in *Lancet* 29 Mar. 500/2 The therapeutic method here employed is..a combined method of serum therapy and transfusion. We may perhaps call it 'immuno-transfusion'. **1941** Kolmer & Tuft *Clin. Immunol.* x. 256 To increase the amount of protective antibody in the donor's blood, the donor may be injected with vaccine prior to the withdrawal of his blood. This type of transfusion is referred to as immunotransfusion.

immunogen (imiū̆·nodʒĕn). *Biol.* and *Med.* [f. *immuno- + -gen.] **1.** An antigenic substance, or a preparation of it, believed to reside in the ectoplasm of the bacterial cell and to be removed by washing.

Immunogen has been registered as a proprietary name in Great Britain and the U.S.A.

1923 *Trade Marks Jrnl.* 31 Jan. 181 Immunogen... Parke Davis & Company. **1923** *Official Gaz.* (U.S. Patent Office) 18 Dec. 550/1 Immunogen... Immunizing Agents Used for the Prophylaxis and Treatment of Diseases of Bacterial Origin. **1926** Horder & Ferry in *Brit. Med. Jrnl.* 31 July 179/2 As it would appear from these results ..that the antigenic or immunizing portion of the bacterial cell is more ectoplasmic than endoplasmic in origin, it is proposed to call this type of antigen an 'ectoantigen', and for products prepared in such a manner as to contain only this ectoantigen the designation 'immunogens' is suggested, to distinguish them from other antigenic products already in use. **1928** *Jrnl. Amer. Med. Assoc.* 15 Dec. 1914/2 There is no difference between vaccines and immunogens. Their sphere of action is the same. Generally speaking it would be best to prepare vaccines or immunogens from 'autogenous material', but when that is not practicable commercial preparations may be used. **1929** J. W. Bigger *Handbk. Bacteriol.* (ed. 2) vi. 88 'Immunogens' (Horder & Ferry) are the washings of bacteria. They are stated to contain the immunizing substances of bacteria without the toxins. **1934** *Jrnl. Amer. Med. Assoc.* 22 Sept. 939/2 A little girl..was given..doses of catarrhal immunogen (Parke-Davis) on biweekly visits. *Ibid.*, Immunogen (Parke, Davis & Co.) may be a mixture of antigenic substances. It is not on the accepted list of New and Nonofficial Remedies.

2. Any substance that elicits an immune response or produces immunity in the recipient (see quots.).

1959 A. D. Bussard in *Ann. Rev. Microbiol.* XIII. 280 The following terms will be used..in the course of this review: immunogenicity: ability of a substance to direct the formation of a specific antibody; antigenicity: ability of a substance to react with an antibody under a given set of conditions. An immunogen is a substance exhibiting immunogenicity and an antigen, a substance exhibiting antigenicity. **1971** S. O. Freedman et al. *Clin. Immunol.* i. 3 Although the terms immunogen and antigen are frequently used interchangeably, they are not necessarily synonymous. Immunogenicity may be defined as the capacity of a substance to initiate a humoral or cell-mediated immune response, whereas antigenicity may be defined as the capacity of a substance to bind specifically with the antibody molecules whose formation it has elicited... Employed correctly, the term immunogen specifies that a substance acts at the afferent limb of the immune response... Most conventional antigens possess both immunogenic and antigenic capacities. **1972** *Science* 2 June 1028/3 Pneumococcal polysaccharides are generally considered weak immunogens since they elicit poor primary antibody responses.

immunogenic (imiū̆nodʒe·nik), *a. Biol.* and *Med.* [f. *immuno- + *-genic.] Of, pertaining to, or possessing the ability to elicit an immune response.

1933 *Jrnl. Amer. Med. Assoc.* 24 June 2013/2 (*heading*) An immunogenic paradox. **1934** *Index Medicus* XV. 637/2 Appearance of impedin in production of specific opsonin by local application of immunogenic salves to skin. **1942** *Jrnl. Bacteriol.* XLIII. 397 There is ample evidence in the literature to suggest that fixed virus strains differ in their immunogenic properties. **1962** *Lancet* 5 May 965/1 This incidence had been reduced by early vaccines, but more recent batches had not been sufficiently immunogenic. **1971** *Nature* 4 June 286/3 Why cells become more immunogenic after neuraminidase treatment is not yet clear.

Hence **immunogeni·city,** immunogenic property. Also **immunoge·nesis,** the formation or production of antibodies; bodily processes, collectively, that constitute an immune response.

1944 *Science* 16 June 496 Minute amounts..exhibited marked antityphoid immunogenicity. **1948** Biester & Schwarte *Dis. Poultry* (ed. 2) 567 The criteria of immunogenesis and pathogenesis. **1950** *Jrnl. Bacteriol.* LIX. 263 Textbooks..commonly state that a high degree of encapsulation is responsible for the reportedly poor immunogenicity of this fungus. **1956** *Jrnl. Immunol.* LXXVI. 217/1 It is demonstrated that the immunogenicity of immune globulin is suppressed by the simultaneous injection of other serum components. **1960** A. Hofman et al. tr. A. N. Gordienko (*title*) Control of immunogenesis by the nervous system. **1968** D. Osoba in B. Cinader *Regulation Antibody Response* xii. 232 (*heading*) The regulatory role of the thymus in immunogenesis. **1970** *New Scientist* 19 Mar. 543/1 Irradiation with gamma rays, which render it non-infective without.. loss of mobility or immunogenicity.

immunoglobulin (imiū̆:no-, i:miuno,glǫ·biū̆lin). *Biochem.* and *Med.* [f. *immuno- + Globulin.] Any of the group of proteins present in the serum of vertebrates which are characterized by their structure, in being a mixture of larger ('heavy') and smaller ('light') polypeptide chains, linked usu. by disulphide bonds, and by their function in that all known antibodies are immunoglobulins.

1959 J. F. Heremans in *Clin. Chim. Acta* IV. 643 The data..seem to point to the existence of a system of closely related, though not identical, proteins which are capable of acting as antibodies. These are: (a) γ-globulin (7S, low carbohydrate content, heterogeneous mobility)..; (b) β_{2A}-globulin (also 7S, high carbohydrate content, high mobility)..; and (c) β_{2M}-globulin (19S, high carbohydrate content, high mobility). The outlined similarities in nature and function clearly call for the adoption of a common name for all these substances. A word such as 'immunoglobulins' would seem to be suitable. **1965** *New Scientist* 11 Mar. 626/1 The body's 'immune' defence mechanism against bacteria, viruses and other foreign materials depends largely on the presence in the blood serum of a family of proteins known as immunoglobulins. **1970** Gold & Peacock *Basic Immunol.* iii. 93 A new nomenclature was proposed in which the prefix γ or Ig stood for immunoglobulin and a suffix denoted the particular class of immunoglobulin. Thus the 7S, 160,000 M.W., gamma-globulins became IgG or γG, the macroglobulins became IgM or γM, and the third class of immunoglobulins, IgA or γA... Two other classes of human immunoglobulin, IgD..and IgE.., are now known. **1970** Rothschild & Waldmann *Plasma Protein Metabolism* xvi. 259 The immunoglobulins are a group of structurally related proteins produced by plasma cells and lymphocytes. **1970** *Nature* 12 Dec. 1040/1 These studies have shown that immunoglobulins are composed of light chains (molecular weight 23,000) and heavy chains (55,000) which are linked through disulphide bonds. **1971** *Ann. N.Y. Acad. Sci.* CXC. 8 The lamprey has 6·6S and 14S immunoglobulins which have similar light and heavy chains.

immunology (imiunǫ·lŏdʒi). [f. Immun(ity + -ology.] The science which treats of the phenomena and causes of immunity (sense 5).

1910 *Jrnl. Amer. Med. Assoc.* 5 Mar. 828/1 Relations between pharmacology, immunology and experimental therapy. **1911** *Ibid.* 25 Feb. 578/2 The science of immunity, or immunology, would explain the mechanism by which the animal body is enabled to resist disease. **1947** *Nature* 4 Jan. 15/2 The hope that they [*sc.* incipient cancers] may elaborate specific antigenic substances, and thus provide serological tests, goes back to the first years of immunology. **1963** Gell & Coombs *Clin. Aspects Immunol.* p. xvii, Immunology has two aspects, its use as a tool and its investigation as a biological phenomenon.

Hence **immunolo·gic** (chiefly *U.S.*), **-lo·gical** *adjs.*, of or pertaining to immunity or immunology; **immunolo·gically** *adv.*, from the point of view of immunology; as regards the phenomena, properties, etc., of immunity; **immuno·logist,** an expert in or student of immunology.

1912 *Jrnl. Exper. Med.* XVI. 635 Hemolysins, precipitins, or other bodies included in the province of the immunologist. **1914** Billings & Irons *Forchheimer's Therapeusis Internal Dis.* V. iii. 121 (*heading*) Immunological reactions in diagnosis. **1919** Immunologic [see *allergy a.]. **1929** R. T. Hewlett in *Syst. Bacteriol.* (Med. Res. Council) III. 375 Dickson had previously observed that the Nevin cheese strain differed immunologically from other strains with which he was working. **1955** *Sci. News Let.* 12 Mar. 168/1 Because of his lack of gamma globulin, the Minnesota boy has what his doctors term almost complete 'immunologic paralysis', meaning the mechanism in his body that should help him develop immunity, or resistance, to disease has been paralyzed. **1960** *New Biol.* XXXI. 103 It seems..that any potentially antigenic material which comes into contact with the cells during the stage of immunological immaturity fails subsequently to cause antibody formation. **1963** *Times* 25 Jan. 7/7 In such a case the twins are immunologically tolerant of each other's tissues. **1967** W. O. Weigle *Natural & Acquired Immunologic Unresponsiveness* i. 1 The evidence available suggests that the acquisition of an immunologic unresponsive state to self is not genetically determined, but is acquired early in life before maturation of the immune mechanisms. **1968** *Sunday Tel.* 28 Jan. 4/4 Dr. M. C. Botha, the immunologist who conducted the tests. **1973** J. Goodfield *Courier to Peking* x. 120 You hear the voice of the true immunologist. *Ibid.* 126 The question was a highly technical one and understanding it demanded a thorough background in immunological theory.

immuration (imiurēⁱ·ʃən). [f. Immur(e *v.* + -ation.] = Immurement.

1895 Pollock & Maitland *Hist. Eng. Law* I. II. ii. 427 Stephen Langton seems to have condemned two of the laity to that close imprisonment which was known as immuration. **1959** J. L. M. Trim in Quirk & Smith *Teaching of Eng.* iii. 77 [Speech training] may easily lead to..an increasing immuration of the individual instead of the liberation which education should bring. **1963** *Yale Rev.* Winter 291 The first, the cloistered family, guarded the purity of the flesh and preserved the ideal of chastity through the Dark Ages by immuration.

immy (i·mi). Also **imma, immie.** [perh. f. Imitation.] **a.** (See quot. 1928.) **b.** A type of marble used by children.

1928 *Funk's Stand. Dict.*, *Immy*, a choice marble made in imitation, as of a cornelian or an agate. **1936** *Fortune* June 36/2 Like glass for stained-glass windows and water tumblers, marble glass has a soft lime base, which makes an immie so resilient. **1941** Baker *Dict. Austral. Slang* 37 *Imma*, a type of marble. **1952** [see *agate *sb.* 1 b]. **1972** *Sat. Rev.* (U.S.) 29 Jan. 8 Shooting marbles was played by four or five boys kneeling around a circle... In the center there were a lot of marbles—'immies', as they were known.

imp. Abbreviation of *imperative, imperator, imperatrix, imperfect, imperial, impersonal, important, imprimatur, imprint, improvement.*

impact, *sb.* Add: **b.** Now commonly the effective action of one thing or person upon another; the effect of such action; influence; impression. Esp. in phr. *to make an impact* (*on*).

1946 *Sat. Rev. Lit.* (U.S.) 28 Dec. 15/1 The impact of the images, their skilful juxtaposition, and the bold page layouts make words superfluous. **1952** B. Russell (*title*) The impact of science upon society. **1958** *Church Times* 8 Aug. 7/1 The story..is presented by means of narrative and dramatic episodes in a manner familiar to all radio-listeners, but it is the lighting which makes the great impact. **1965** *Listener* 26 Aug. 297/1 However much you give them, you are not going to make a significant impact on growth, though you may make an impact in the charitable sense. **1966** *Economist* 10 Dec. 1144/3 What has had an impact on food distributors, apparently, is the opening of an investigation by the Federal Trade Commission into supermarket games and stamps. **1967** E. Short *Embroidery & Fabric Collage* i. 18 The most dynamic colour combination if used too often loses its impact. **1969** Ld. Mountbatten in *Times* (India Suppl.) 13 Oct. p. i/1 He [*sc.* Gandhi] made such an impact on me that his memory will forever remain fresh in my mind. **1973** *Daily Tel.* 5 Mar. 6/2 The main impact of the campaign will be made by full-page newspaper advertisements.

2. Special Comb.: **impact crater,** a crater or a hollow in the ground believed to have been produced by the impact of a meteorite; **impact extrusion,** a process for producing tubular objects in which metal in a die is struck by a punch that fits into it and forces the metal between their two surfaces and out of the die; **impact head** = **impact pressure*; **impact load,** a load imposed suddenly and for a short time, as when one body strikes another; **impact loading,** (the application of) an impact load; **impact pressure,** the total pressure in a moving fluid in the direction of flow, being equal (in the case of a fluid of negligible viscosity) to the sum of the dynamic pressure and the static pressure; **impact resistance** = **impact strength*; **impact strength,** the ability of a solid to withstand an impact or shock; strength as measured by an impact test; **impact test,** any of various tests for measuring the resistance of a body to suddenly applied stress in which it is broken, usually by a blow, under standard conditions; **impact tube,** a thin tube (usu. rigid with a right-angled bend) which may be placed in a flow of fluid with an open end facing upstream, so that the impact pressure in the fluid may be found by measuring the pressure in the tube; cf. **Pitot tube*.

1895 G. K. Gilbert in *Bull. Philos. Soc. Washington* XII. 265 The inquiry has followed three lines. First, an investigation of the ellipticity of lunar craters; second, an experimental investigation of the relation between incidence angle and ellipticity of impact craters; third, a more refined investigation of the orbital relations affecting the incidence angles of moonlets. *Ibid.* 291 Does the earth exhibit impact craters? If not, then erosion and sedimentation have destroyed them. **1965** R. B. Baldwin *Fund. Survey Moon* vii. 66 If it can be shown that the impact craters on earth are..similar to lunar craters.. we will be on firm ground in considering that the lunar craters..were formed by the impacts..of meteorites. **1967** *Listener* 20 Apr. 521/2 Another impact crater is that at Wolf Creek in Australia, where the diameter is half a mile. **1935** *Metal Industry* 11 Oct. 373/1 A new..heavy duty press..for the cold impact extrusion of aluminium tubes and shells in one operation. **1963** H. R. Clauser *Encycl. Engin. Materials* 340/1 Parts produced by impact extrusion are essentially longitudinally oriented, e.g., collapsible tubes, cans, etc. **1928** G. Martin *Treat. Chem. Engin.* xvi. 3 A connection between the density of a fluid W and the power required to move it through a pipe of area A when the impact head is l_1 in. of water and the dynamic head is l in. of water. **1924** E. E. Mann *Introd. Pract. Civil Engin.* x. 166 Wind pressure on roofs is of the nature of an impact load. **1928** C. F. S. Gamble *Story N. Sea Air Station* 10 Her two stream-lined gondolas..were designed to be capable of sustaining severe impact loads when alighting on water. **1963** D. A. Firmage *Fund. Theory of Struct.* iii. 56 Impact load is only a minor portion of the total load on any bridge. **1948** Courant & Friedrichs *Supersonic Flow & Shock Waves* iii. 240 The basic problem of wave propagation in a bar of elastic-plastic material is concerned with the motion resulting from impact-loading, i.e., from a velocity being suddenly imparted to one end of the bar and then maintained there. **1973** *Sci. Amer.* Feb. 85/1 It has been equipped with a grille structure that distributes the impact loading more uniformly. **1919** A. B. Eason *Flow & Measurem. Air & Gases* xiv. 237 The increase in the value of the impact pressure in the case of wind blowing directly on an opening will be due to the fact that more of the momentum of the air is destroyed than when air blows on the small area of the Pitot tube mouthpiece. **1966** Daily & Harleman *Fluid Dynamics* vi. 128 For liquids of small viscosity, Eq. (6–71) can be used to compute the velocity from the stagnation or impact pressure measured on the blunt nose of a probe in a steady flow. **1934** B. Stoughton *Metall. Iron & Steel* (ed. 4) xiii. 404 Impact Resistance of Steel.—The resistance of steel to shock decreases very much with lowered temperature. **1958** C. L. Mantell *Engin. Materials Handbk.* III. 12 The impact resistance of standard malleable iron, measured by the Charpy test using a notch, 0·394-in. square bar, and 0·079-in. depth of notch is about 16·5 ft-lb. **1904** *Proc. Inst. Mech. Engin.* IV. 1227 The more or less consistent relation that appeared to exist in Messrs. Sankey and Kent-Smith's tests between impact strength and reduction of area. **1939** *Proc. Amer. Soc. Testing Materials* XXXVIII. II. 39 There are..8,000,000 molded phenol plastic telephones in use... The extent to which..breakage occurs..is determined largely by the impact strength of the molding material. **1952** Wood & Von Ludwig *Investment Castings for Engineers* x. 209 Both beryllium copper and aluminum bronze will develop higher tensile strength and hardness when heat treated, but they will not have good impact strength, whereas manganese bronze is one of the toughest cast metals available. **1967** M. Chandler *Ceramics in Mod. World* iv. 118 The impact strength of all ceramic materials..is rather low. **1899** W. C. Unwin *Testing Materials of Construction* (ed. 2) 239 (*heading*) Example of an impact test. **1915** [see *Charpy]. **1918** *Machinery* 31 Jan. 477/1 Tensile impact tests, in which a sudden tensile stress is applied to a specimen by means of a falling weight, have also been practised..in recent years. **1943** F. D. Jones *Engin. Encycl.* (ed. 2) II. 680 The torsion impact test breaks the specimen by twisting. **1971** B. Scharf *Engin. & its Lang.* iv. 24 The most common impact test in this country [*sc.* Great Britain] is the Izod test, in which a notched test piece fixed at one end is broken by a blow from a pendulum hammer, the energy absorbed in fracturing the specimen being recorded... Another method widely used in other countries and increasingly also in this country is the Charpy impact test, in which a notched test piece supported at both ends is broken by a blow from a striker, on the face opposite to and immediately behind the notch, the energy absorbed in fracturing the specimen being recorded. **1916** *Trans. Amer. Soc. Mech. Engin.* XXXVII. 1410 For high pressures, the manometer attached to the impact tube can be replaced by a mercury column or steam gage attached to a receiver in the main where velocity is largely reduced. **1934** J. H. Perry *Chem. Engineers' Handbk.* 689 The length and shape of the tip, so long as the opening faces upstream, usually have little effect upon the head indicated by an impact tube. **1966** Daily & Harleman *Fluid Dynamics* ix. 178 The actual readings from impact tubes will depend both on the viscous effect and on the size of the pressure-sensing hole in the probe tip.

impact, *v.* Add: **3.** *intr.* **a.** To come forcibly into contact with a (larger) body or surface. Const. various preps.

1916 [see impacting *ppl. a.* below]. **1929** 'Seamark' *Down River* vi. 172 Something impacted with a soft thud against Lingard's temple. **1945** *Jrnl. Sci. Instrum.* XXII. 191 A jet of air issuing from a slot and impacting on a plane surface. **1962** F. I. Ordway et al. *Basic Astronautics* v. 201 The Soviet Lunnaya Raketa was launched early in the afternoon of September 12, 1959 and impacted onto the Moon's surface just after midnight on September 14, Moscow time.

b. *fig.* To have a (pronounced) effect *on*.

1935 W. G. Hardy *Father Abraham* 370 For there was about them an air of eagerness and of shuddering expectation which impacted on his consciousness and fascinated even while it repelled him. **1956** *Oxf. Mag.* 8 Nov. 81/1 The Magazine..is not the place for consideration of national and international events except in so far as they impact on Oxford.

4. *trans.* To cause to impinge or impact *on*, *against*, etc.

1945 *Jrnl. Sci. Instrum.* XXII. 187 Experimental results for the efficiency of jets in impacting particles are correlated. **1964** K. Stewart in White & Smith *High-Efficiency Air Filtration* ii. 57 All impactors make use of the inertia effect which particles exhibit when the gas stream in which they are suspended is constrained to turn abruptly. The particle under suitable conditions cannot follow the stream lines and is impacted against a collecting plate. **1972** J. O. Ledbetter *Air Pollution* A. v. 187 An aerosol moving toward an obstacle may impact particles on the obstacle.

Hence **impacting** *ppl. a.*, impinging, colliding.

1916 'Boyd Cable' *Action Front* 95 No ping and smack of impacting lead hailed about them. **1961** *Sci. Amer.* Nov. 58/2 The impacting bodies may have been asteroids or comets. **1972** *Daily Tel.* 17 Apr. 6/8 These particles.. cannot be measured or analysed from the Earth's surface. On the Moon, however, the impacting particles leave trails in the detector. **1973** *Nature* 13 July 68/2 Craters of the size of St Magnus Bay and The Firth would be formed by impacting meteorites of masses about 1 million tons.

impacted, *ppl. a.* Add: **b.** Applied *spec.* to fæces lodged in the intestine (cf. *impaction 2); also *transf.*, applied to (a part of) the intestine when so blocked.

1844 *Boston Med. & Surg. Jrnl.* XXX. 309 (*heading*) History of a case of impacted colon. **1850** *Lancet* 19 Jan. 80/2 The bowel is found in a state of great distention from an impacted mass closely adhering to its walls. **1875** *Cincinnati Med. News* VIII. 353 (*heading*) Remarkable case of impacted colon, with suppression of urine. **1902** J. P. Tuttle *Treat. Dis. Anus* (1903) xiv. 545 Distressing symptoms are relieved either by the loosening up of an impacted fæcal mass, or possibly by the undoing of a volvulus or intussusception. **1972** F. A. Jones in Jones & Godding *Managem. Constipation* iv. 128 In some patients, impaction may have already led to severe symptoms, with acute distress, and it is then necessary to organize the immediate removal of the impacted fæces.

c. Applied to a bone fracture in which the broken parts are driven together so as to become locked.

1850 J. A. Orr *Princ. Surg.* II. xi. 153 Impacted fracture is when one broken extremity of the bone is driven into and lodged in the other. **1921** Baetjer & Waters *Injuries & Dis. Bones & Joints* v. 102 When the fracture is just behind the head [of the femur] or in the middle of the neck, impaction is relatively rare. Fracture at the base is generally impacted. **1967** E. L. Ralston et al.

Handbk. Fractures vii. 116 Impacted fractures [of the humeral neck] even with angulation of 25 to 30 degrees are best treated with the use of a sling and swathe and early active exercises.

d. Applied to a tooth which, owing to obstruction by another tooth or by bone, fails to erupt properly and remains partly or wholly within the jaw-bone.

1876 H. Moon in T. Bryant *Pract. Surg.* (ed. 2) I. xiii. 546 In all cases where the impaction of a lower wisdom tooth is a source of irritation, the impaction should be at once got rid of... The serious results which may attend purulent inflammation about an impacted wisdom tooth, will receive notice later. **1928** H. Prinz *Dis. Soft Struct. Teeth* i. 29 Impacted teeth usually do not cause painful symptoms unless they meet on their path of retarded eruption an obstruction or they exert pressure upon nerve fibers. **1971** Costich & White *Fund. Oral Surg.* viii. 93/2 The maxillary third molars frequently fail to erupt but may not necessarily be regarded as impacted teeth.

2. That has impinged upon or struck something.

1952 *A.M.A. Arch. Industr. Hygiene & Occup. Med.* V. 464 Although the size of impacted particles can be measured under a microscope, the impaction principle.. is used more as a method of sampling than as a method of determining the particle-size distribution.

b. That has been struck by an impacting body; also *fig.* (*U.S.*) of an area: affected by a larger demand than usual on public services, esp. schools.

1924 in *Sci. Amer.* (1974) July 12/1 One need only study a large raindrop falling into a still pool of water. There is first a surging outward of the impacted water. **1963** *Economist* 25 May 777/2 The..scheme for aid to 'impacted' areas (where schools are over-loaded). **1967** *Compton Yearbk.* 232/1 Funds were also earmarked for.. federally 'impacted' areas, that is, areas where the families of federal workers had swollen school enrollments. **1970** *Time* 6 Apr. 12 Nixon..proposed that $1.5 billion in federal funds be made available to 'racially impacted areas'..to help desegregating school districts meet their special needs. **1971** *Nature* 16 July 162/2 The shape of a newly formed impact crater is caused by the sudden release of the kinetic energy of the impacting mass within a small volume somewhat below the original impacted surface.

impacter, var. *Impactor.

impaction. Add: **I. 1.** (Further examples, corresponding to *Impacted *ppl. a.* 1 c, d.)

1876 [see *Impacted *ppl. a.* 1 d]. **1921** [see *Impacted *ppl. a.* 1 c]. **1957** J. G. Bonnin *Textbk. Fractures* i. 8 Impaction is important in aiding fixation and indicates, as a rule, that little displacement has occurred. **1972** D. E. Waite *Textbk. Pract. Oral Surg.* xi. 141 They also vary widely in degree of impaction; some are partially erupted, while others are completely encased in bone.

2. *spec.* in *Med.* **a.** The lodging of a mass of (usu. hardened) fæces in the intestine so that defecation is prevented or impeded; hence, the obstruction of (a part of) the intestine in this way.

1853 *Assoc. Med. Jrnl.* I. 606 (*heading*) Impaction of the rectum from unground wheat. **1866** *Clin. Lect. & Rep.* (London Hospital) III. 193 Three cases of obstruction of the bowels..produced by the impaction of hardened fæces in the rectum, and colon. **1902** J. P. Tuttle *Treat. Dis. Anus* xiv. 543 In simple constipation and in impaction there is always a channel for the escape of gases from the bowels. **1943** Niles & Martin in E. J. Stieglitz *Geriatric Med.* xxxv. 599 Rectal impaction is much more common in the aged. *Ibid.*, Colonic impaction (scybala) is seen occasionally in old people, especially among those who are bedridden. **1972** [see *Impacted *ppl. a.* 1 b].

b. *concr.* A mass of (usu. hardened) fæces lodged in the intestine so as to impede defecation.

1902 J. P. Tuttle *Treat. Dis. Anus* xiv. 542 The author has known a patient to suffer from a continuous diarrhœa for six weeks..apparently from no other cause than an impaction of fæces in the sigmoid flexure. **1931** M. C. Pruitt *Mod. Proctology* xviii. 350 It is important to determine..whether the impaction is hard or soft. **1958** A. F. R. Andresen *Office Gastroenterol.* 442 Small impactions can usually be induced to pass by means of a cleansing enema.

II. 3. The process of causing something to impinge or impact on something else (cf. *Impact *v.* 4); also, the action of so impinging (cf. *Impact *v.* 3).

1945 [see *Impactor 2]. **1952** *A.M.A. Arch. Industr. Hygiene & Occup. Med.* V. 476 Larger and heavier particles are thrown onto a collecting surface in front of the jet, while smaller and lighter particles escape impaction. **1956** P. L. Magill et al. *Air Pollution Handbk.* xiii. 32 The impaction of aerosol particles on cylinders has been given considerable attention since it provides an insight into the functioning of fibrous filters. **1972** J. O. Ledbetter *Air Pollution* A. vi. 230 Impaction devices.. that depend upon the wind to carry out the impaction.

impactite (impæ·ktəit). *Geol.* [f. Impact *sb.* + -ite[1], after *tektite.] Any piece of glassy material formed in or around a meteorite crater by the heat of impact.

1940 V. E. Barnes *N. Amer. Tektites* in *Univ. Texas Publ.* no. 3945, p. 558 Spencer's meteorite splash origin.. is valid for the formation of certain glasses. Glasses of this type will be distinguished in general from most of those now included under tektites. These meteorite splashes should be given a distinctive name such as 'impactites'. This name was suggested by Dr. H. B. Stenzel. **1960** I. Vidziunas tr. *Krinov's Princ. Meteoritics* vii. 435 This presence of numerous fragments of impactites in the area of meteoritic craters served as a basis for considering these objects to be fragments of fused terrestrial quartz sand. **1964** *New Scientist* 16 Jan. 160/1 The expert can easily distinguish tektites from..impactites (silica glass found around some meteorite craters).

impactive (i·mpæktiv, impæ·ktiv), *a.* [f. Impact *sb.* + -ive.] Of, pertaining to, or characterized by impact; having an impact.

1934 F. Scott Fitzgerald *Tender is Night* 5 Feeling the impactive scrutiny of strange faces, she took off her bath-robe and followed. **1942** W. Faulkner *Go down, Moses* 197 They faced one another, not close yet at slightly less than foils' distance, erect, their voices not raised, not impactive, just succinct. **1955** *Financial Times* 17 Jan. (Packaging Suppl.) 6/4 Mechanical tests have been devised for studying vibration, impactive shocks and compression loads in stacking. **1969** *Esquire* Feb. 20 Even more impactive, maybe, is The Spot, where the waitresses wear what they please.

impactor (impæ·ktəɹ). Also impacter. [f. Impact *v.*: see -or, -er[1].] **1.** A device or machine that delivers impacts or blows.

1916 *Chambers's Jrnl.* Dec. 830/2 The impactor golf-machine is a new invention. **1950** J. H. Perry *Chem. Engineers' Handbk.* (ed. 3) 1129/2 The Impactor..is a reversible hammer crusher without a discharge grating or cage. **1966** *McGraw-Hill Encycl. Sci. & Technol.* V. 471/1 To increase the effectiveness of the forging blows, the impacter forging machine was developed.

2. An impinger, esp. one in which the particles are deposited on a dry surface rather than in a liquid.

The distinction between impactors and impingers explained in quot. 1945 is rarely made.

1945 K. R. May in *Jrnl. Sci. Instrum.* XXII. 187 (*heading*) The cascade impactor: an instrument for sampling coarse aerosols. *Ibid.* 188 The method which has been adopted for depositing the sample is direct impaction of the particles on to glass slides which may be coated with a suitable medium...The terms 'impactor' and 'impaction' were suggested by Prof. J. H. Gaddum to avoid confusion with 'impinger' instruments... In 'impingers' a fine jet of air is directed at very high speed on to a flat surface to obtain the maximum efficiency of deposition of small particles. In the case of 'impactors' jet speeds are lower and larger particles are dealt with. **1956** P. L. Magill et al. *Air Pollution Handbk.* x. 30 Most cascade impactors are now operated using high air velocities. *Ibid.* 29 A single-stage impactor..is often useful because of its simplicity of operation. **1964** [see *Impact *v.* 4]. **1971** *Nature* 12 Feb. 501/1 Partition of single uredospores between the first and second stage of a cascade impactor indicated a terminal velocity of 0·6 cm s⁻¹.

impair, *sb.*[2] Add: In roulette (with pronunc. aṅpɛ̄r), an odd number, or a number marked 'impair'.

1850 *Bohn's Hand-bk. Games* 348 (Roulette) The impair wins, when the ball enters a hole numbered impair. **1891** 'L. Hoffmann' *Cycl. Card & Table Games* 626 If he places his money on Impair, he bets that the ball will drop into an odd number. **1902** *Encycl. Brit.* XXXII. 304/1 Pair indicates even numbers, impair odd numbers. **1966** 'W. Haggard' *Power House* xii. 125 The croupier was paying out. Mortimer was on the *Impair* side. **1973** L. Meynell *Thirteen Trumpeters* iv. 66 His right hand was.. stretching out to place his stake on the next throw (a green on *pair*)... '*Impair*' was called.

impaired, *ppl. a.* Add: **2.** Of a driver or his driving: adversely affected by the influence of alcohol or narcotics. *Canad.*

1951 *Act* (Canada) 15 *Geo. VI* c. 47 §14 Driving while ability to drive is impaired. **1957** (*title*) Report on impaired driving tests (Crime Detection Laboratories of the Royal Canadian Mounted Police). **1967** W. S. Avis et al. *Dict. Canad. Eng., Senior Dict.* 573/2 *Impaired driver*, one whose driving ability has been impaired by alcohol or narcotics. **1970** *Toronto Daily Star* 24 Sept. 37/1 Ange Gardien..was charged with impaired driving. **1972** *Evening Telegram* (St. John's, Newfoundland) 24 June 1/1 A police spokesman said the car received only slight damage. The driver was arrested and charged with impaired driving. **1973** *Kingston* (Ontario) *Whig-Standard* 18 Apr. 15/2 Another motorist..was fined $175 and prohibited from driving for four months on a charge of impaired driving. **1973** *Daily Colonist* (Victoria, B.C.) 26 Apr. 41/3 Georg Edward Haines..was fined $350 following his plea of guilty to a charge of being impaired early Wednesday in Victoria while in care or control of a vehicle. *Ibid.*, Edward Weiland..pleaded guilty to a two-count Victoria charge of impaired driving and refusing to take a breath-analysis test. **1974** *Kingston* (Ontario) *Whig-Standard* 16 Jan. 5/4 A snowmobile operator was one of five persons assessed penalties ranging from $175 to $200 each in county court Tuesday on impaired driving charges.

impala (impā·lă, -pæ·lă). Also impalla, mpala. = Pallah.

1875 W. H. Drummond *Large Game & Nat. Hist. S. & S.-E. Afr.* vii. 330 The roibok or impalla..is about the size of a small reed-buck doe, though more slenderly made. *Ibid.*, These impalla..could easily distance any dog I possessed. **1886** W. M. Kerr *Far Interior* I. ii. 29 It was here [*sc.* Boatlanama] that I shot my first antelope —a fine impala (*Æpyceros Melampus*), with a good head. **1888** P. Gillmore *Days & Nights by Desert* xvii. 136 (*caption*) The Mpala Antelope (*Æpyceros Melampus*). **1896** [see Pallah]. **1907** P. Fitzpatrick *Jock of Bushveld* 302 We sat like statues as the impala walked out from its stall between Teddy's knees. **1931** *Times Lit. Suppl.* 3 Dec. 983/4 Graceful photographs of impalla, eland, and other antelope. **1947** J. Stevenson-Hamilton *Wild Life S. Afr.* xiii. 87 The impala (*Æpyceros melampus*) —Discovered by..Lichtenstein more than a hundred years ago, and variously termed impala, pallah, and rooibok, this antelope may claim to be one of the most beautiful and graceful members of the existing African fauna. **1965** A. Nicol *Truly Married Woman* 44 It was, after all, only a red-buck, an impalla, that they were afraid of. **1971** L. H. Matthews *Life of Mammals* II. xiv. 394 Impala in East Africa do not move far from their restricted home range.

impale, *v.* Add: **4. c.** *fig.* To transfix (a person) with one's gaze. Hence impa·ling *ppl. a.*

1877 *My Mother-in-Law* vi. 60 Mrs. Pinkerton devoted herself to impaling me with her eyes once in a while. **1903** *Critic* XLIII. 349/2 There was an impaling fierceness in his eyes.

impalement. Add: **4. b.** The act or fact of being impaled upon rocks, the spikes of a gate, or the like.

1874 *Belgravia* Aug. 175 There was..one tall church-steeple which by the celerity of its approach appeared.. anxious that I should be impaled on its apex... I declare that the grotesqueness of the position of impalement— all legs and wings, like a cockchafer—..visibly occurred to me. **1885** *Austral. Med. Jrnl.* New Ser. vii. 436 A case of laceration of the rectum and jejunum by accidental impalement. **1887** *Graphic* 19 Mar. 307/2 His ship was rescued after impalement on a rock. **1921** *Contemp. Rev.* Aug. 272 Do you remember climbing the gate and just avoiding impalement? **1971** *Brit. Med. Jrnl.* 26 June 748/1 Perforation of the bladder following rectal impalement is extremely rare.

impaler. (Later example.)

1969 *New Yorker* 6 Sept. 106/2 Ceauçescu..probably the most popular Rumanian national leader since Vlad IV, called the Impaler, who successfully fought the Turks in the mid-fifteenth century.

impalla, var. *Impala.

imparting, *vbl. sb.* (in Dict. s.v. Impart *v.*). (Later example.)

1952 *Mind* LXI. 309 Lying..is the deliberate imparting of false information in order to deceive.

impasto. Add: **2.** *Ceramics.* (See quots.)

1903 M. L. Solon *Hist. Old French Faience* 188 *Impasto*, clay or enamel colours laid so thickly on to the ware as to stand up in relief from its surface. **1960** R. G. Haggar *Conc. Encycl. Cont. Pott. & Porc.* 233/1 '*Impasto blue*', inky blue pigment which is applied thickly and stands up in slight but palpable relief on early 'oak-leaf' pots made at Florence.

impastoed (impɑ·stoᵘd), *ppl. a.* Encrusted with paste.

1923 *Blackw. Mag.* May 641/2 [He] thrust the point of his palette-knife under an impasto'd mass of paper.

impaternate (impătə̄·ɹnět), *a. Biol.* [f. Im-[2] + Patern(al *a.* + -ate[2].] Produced parthenogenetically by a female without fecundation by a male.

1934 in Webster. **1936** *Nature* 11 July 78/1 If the queen bee is diploid and heterozygous for a recessive factor, she..would be expected to produce impaternate haploid drones, equal numbers of which would show the dominant or the recessive character. **1965** [see *deuterotoky* s.v. *deutero-].

impayable, *a.* **3.** (Later examples.)

1906 G. Meredith *Let.* 23 Nov. (1970) III. 1579 His [*sc.* Whistler's] tales of his student life in Paris..were *impayable*. **1954** P. Bottome *Against Whom?* xxiii. 177 As a patient, she is unsatisfactory..as a girl she is *impayable*!

impeccableness (impe·kăb'lněs). [f. Impeccable *a.* + -ness.] The character or condition of being impeccable.

1696 J. Sergeant *Method to Sci.* i. vi. 64 Original Sin, Impeccableness in the Saints in Heaven, Obdurateness in Sin in the Divels. **1901** F. H. Burnett *Making of Marchioness* ii. xiv. 233 With her ruby and her coronets and her lodging-house street, she is of an impeccableness —she does not even know she could be doubted. **1952** *New Yorker* 15 Nov. 166/2 They are inclined to be seduced by pieces whose chief charm rests in a thin, careful impeccableness.

impedance. Substitute for def.: The overall opposition to an electric current, arising from the combined effect of resistance R and reactance X and measured by the ratio of the e.m.f. to the resulting current (peak or r.m.s. values); it may be represented as a scalar quantity whose value is $\sqrt{(R^2+X^2)}$ or as a complex number $R+jX$. (Further examples.)

1923 E. W. Marchant *Radio Telegr.* ii. 15 If the conductor is a long straight wire it offers comparatively little obstruction or 'impedance' to the passage of a current; while, if it is wound up into a coil it will offer a very high 'impedance' to the passage of the high-frequency current. **1926** A. T. Dover *Theory & Pract. Alternating Currents*

iii. 49 The sides OC, CE, OE of triangle OCE are proportional to the resistance, reactance, and impedance respectively. On account of this feature the triangle OCE, when drawn to an ohm scale, is called the impedance triangle of the circuit. Impedance is therefore a complex quantity, i.e. it is only completely specified when its magnitude and inclination, or alternatively its two perpendicular components with respect to the current, are given. **1930** M. G. Malti *Electr. Circuit Analysis* vii. 87 A sine voltage $E_{11}=70-j50$ is impressed on a series circuit of impedance $Z_L=15$ cis 30° ohms. What is the complex expression for the current? **1931** Moyer & Wostrel *Radio Handbk.* ii. 76 The term $\sqrt{(R^2+X^2)}$, known as the impedance of the circuit, takes the place in alternating-current calculations of the resistance in direct-current work. **1948** A. L. Albert *Radio Fund.* v. 128 If the impressed voltage is divided by the input current of *any* line, a value of impedance is obtained, and for *any* line this is called the input impedance. *Ibid.* 129 The characteristic impedance, usually designated by Z_0, is the input impedance of a line infinite in length. **1961** H. Jasik *Antenna Engin. Handbk.* xxxi. 2 A variable load impedance connected to a source will receive the maximum possible power from the source when it is adjusted to equal the complex conjugate of the impedance of the source. **1962** C. Susskind *Encycl. Electronics* 422/2 The impedance of dynamic, ribbon, carbon, and magnetic microphones is nearly always less than 40,000 ohms and is essentially resistive.

b. Something that has impedance and may be made part of a circuit.

1935 Campbell & Childs *Measurement of Inductance, Capacitance, & Frequency* xiii. 260 The unknown impedance Z is put in series with the known resistance R, and across them is connected a potential divider with slider Q. **1966** *McGraw-Hill Encycl. Sci. & Technol.* VII. 39/1 Standard resistors, capacitors, and inductors are often used as comparisons for unknown impedances.

2. *Mech.* and *Acoustics.* Any of several analogous properties of oscillatory mechanical systems that represent the force, pressure, etc., necessary to produce a given speed, rate of flow, etc., esp. *mechanical impedance*, the ratio of the force on an oscillating body or particle to the resulting velocity; *specific acoustic impedance*, in a wave, the ratio of the (excess) pressure at any point to the resulting particle velocity (i.e. the mechanical impedance per unit area of the wave-front); *acoustic* (or *acoustical*) *impedance*, the ratio of the average (excess) pressure over an imaginary surface in a wave to the resulting rate of volume flow across it (i.e. the specific acoustic impedance, averaged over the surface, divided by the area of the surface).

The qualifying adjs. are often omitted in contexts where there is no danger of ambiguity.

1919 A. G. Webster in *Proc. Nat. Acad. Sci.* V. 275 (*heading*) Acoustical impedance, and the theory of horns and of the phonograph. *Ibid.*, The term 'impedance'..has been productive of very great convenience in the theory of alternating currents of electricity. Unfortunately, engineers have not seemed to notice that the idea may be made as useful in mechanics and acoustics as in electricity. In fact, in such apparatus as the telephone one may combine the notions of electrical and mechanical impedance with great advantage. *Ibid.*, If we have any oscillating system into which a volume of air X periodically enters under an excess pressure p, I propose to define the impedance by the *complex* ratio $Z=p/X$. **1927** I. B. Crandall *Theory Vibrating Syst.* ii. 71 In most parts of the text it is convenient to take the mechanical impedance as the ratio of maximum force to maximum velocity. Strictly, the velocity is analogous to current density rather than to total current... The *acoustic* impedance per unit area, *divided by the area*, is what corresponds most closely to the electrical impedance. **1934** N. W. McLachlan *Loud Speakers* i. 4 The difference between acoustical and mechanical impedance must not be confused. In the one case waves travel in some form of conduit or channel which impedes their progress. The greater the area the smaller the impedance. In the mechanical form something is being driven, so the greater the area driving the medium the greater the impedance opposing motion. **1936** P. M. Morse *Vibration & Sound* vi. 192 A consideration of the behavior of the specific acoustic impedance of a plane wave will enable us to work out the details of its interaction with various mechanical systems. **1953** *Sci. News* XXIX. 11 Whenever in the layers lying below the bed of the sea there is a sudden change in the acoustic impedance (that is in the product of the density of the material and the velocity of sound waves in it), there is a reflecting boundary. **1955** Hueter & Bolt *Sonics* ii. 33 The concept of acoustic impedance is useful in the analysis of lumped systems, such as cavity resonators, sirens, jets, etc. **1960** R. B. Lindsay *Mech. Radiation* ix. 218 The excess pressure in a sound wave is taken to correspond with the electromotive force, and the volume current (the product of the particle velocity ξ and the area of the wavefront S) is assumed to be analogous to the electric current. Hence it is natural to define the acoustic impedance of a wave as $Z=P\epsilon/\dot{X}$, where $\dot{X}$ = volume current $=S\xi$. With the use of the complex notation for wave quantities the impedance will usually be complex. We first examine..the case of a plane harmonic wave of angular frequency w...The impedance of such a wave is then $Z=\rho_0V/S$, a real quantity. The numerator is called the specific acoustic impedance of a plane wave and denoted by Z_s. It is the impedance for a unit area of wavefront. **1961** Bickley & Talbot *Introd. Theory Vibrating Syst.* xi. 139 The example is that of torsional waves on a shaft... The [angular] velocity is proportional to the transmitted torque... Now in the electrical analogy the ratio of force to velocity is that of e.m.f. to current, a ratio which the electrical engineer calls impedance. It is therefore convenient to call the ratio of transmitted torque to angular velocity (and the analogous quantity in other cases) the transmission impedance. **1963** C. T. Morrow *Shock & Vibration Engin.* I. vi. 127 A high impedance indicates that even a large force results in little motion; it suggests massiveness and rigidity. A low impedance indicates that motion is easy to produce; it suggests lightness and flexibility. A resistive impedance indicates that the energy supplied at the point is absorbed, as in a dashpot... A reactive impedance indicates that energy supplied is stored, as in..a compressed spring.

3. Special Comb.: **impedance bond,** a kind of rail bond used to connect electrified rails in adjoining signalling sections, having a low resistance (so that the direct traction current can pass unhindered) and a high inductance (so that the alternating signalling currents are confined to their respective sections); **impedance-matching,** the adjustment of impedances in such a way as to minimize the power reflected or the reduction in the power transferred that occurs when an oscillatory current or other wave meets a change in impedance; also *attrib.*

1926 Harding & Ewing *Electr. Railway Engineering* (ed. 3) xix. 258 The development of the two-rail signal system making use of impedance bonds..marked the beginning of a new era in the development of the new railway signal systems. Practically all of the automatic block signals which are being installed on both alternating- and direct-current railways at present are of this general type. **1967** G. F. Fiennes *I tried to run Railway* v. 49 Installing impedance bonds in the track. **1929** E. Mallett *Telegr. & Telephony* vii. 160 Impedance matching is necessary in order that the maximum available power may be absorbed where it is required. **1934** *Jrnl. Inst. Electr. Engin.* LXXV. 803/2 Impedance-matching transformers have to be used at both ends of the screened lead-in feeder. **1968** *Listener* 22 Feb. 235/3 The bony levers [of the middle ear] reduce the amplitude of vibration and so serve as what an engineer would call an 'impedance-matching transformer'. This gives efficient transfer of energy from the air to the denser fluid in the cochlea of the inner ear.

impeller. Add: **b.** Also occas. **impellor.** A part of a machine or apparatus designed to impart motion to a fluid by rotation, esp. in a restricted space (as in a centrifugal pump or compressor).

1890 P. R. Björling *Pumps* 190 The pump being charged with water, the impeller or fan is set in motion at a great speed, imparting centrifugal motion to the water contained in the impeller, and so driven into the casing or body of the pump. **1923** *Daily Mail* 13 July 12 Cooling is by the thermo-syphon system assisted by a water impellor. **1934** *Archit. Rev.* LXXV. 203 Fresh air..is drawn into the eyes of the fans—the largest of which has a capacity of 641,000 cu. ft. of air a minute and an impellor diameter of 28 ft. **1942** R. A. Beaumont *Aeronaut. Engin.* 100/1 The mixture, thrown outwards at high velocity by the rapidly revolving impellor, passes through diffuser vanes into a chamber, from which it is delivered to the induction pipes. **1947** A. W. Judge *Mod. Gas Turbines* viii. 206 The compressor has a double-sided impeller with twenty-nine vanes on each side which enables the maximum air intake area to be obtained for a given tip diameter. **1956** McCabe & Smith *Unit Operations Chem. Engin.* vi. 282 In these machines mixing is done by a mechanically driven impeller, which creates a flow pattern in the liquid. *Ibid.* 283 The three main types of mixing impellers are paddles, turbines, and propellers. **1966** *McGraw-Hill Encycl. Sci. & Technol.* V. 334/2 A fluid coupling consists of an impeller on the input or driving shaft and a runner on the output or driven shaft,.. Impeller and runner are bladed rotors, the impeller acting as a pump and the runner reacting as a turbine.

impellingness (impe·liŋněs). *rare.* [f. Impelling *ppl. a.* + -ness.] The quality of being impelling.

1922 F. H. Burnett *Robin* i. 1 A certain impellingness of mood suggested that exercise would be a good thing.

impellor, var. *Impeller.

imperative, *a.* and *sb.* Add: **A.** *adj.* **1. b.** *imperative logic* (Philos.): the theory of logical reasoning based on the commands and obligations contained in the imperative mood.

[**1839** Mill *Let.* 4 Nov. in *Works* (1963) XIII. 412 Above all mine is a logic of the indicative mood alone—the logic of the imperative, in which the major premiss says not *is* but *ought*—I do not meddle with.] **1939** *Philos. of Sci.* VI. 453 (*heading*) A logic of the doubtful. On optative and imperative logic. **1952** R. M. Hare *Lang. Morals* i. ii. 27 It is important to realize that *modal* imperative logic is as distinct from the logic of simple imperatives as in the case of the indicative mood. **1958** *Analysis* XVIII. 50 A thoroughgoing reduction of deontic logic to imperative logic would require Procrustean amputations.

B. *sb.* **2. b.** (Earlier example.)

1796 F. A. Nitsch *Gen. View Kant's Princ. concerning Man* 195 An Imperative..which is founded upon reason itself..is a Categorical Imperative which represents an action as necessary in itself.

imperativism (impe·rătiviz'm). *Philos.* [f. Imperativ(e *a.* and *sb.* + -ism 2.] Reasoning based on the concept of obligation contained in the imperative mood (see Imperative *a.* 1 and *1 b). Hence **impe·rativist,** one who bases his reasoning on a concept of obligation; as *adj.*, of or pertaining to reasoning based on such a concept.

1926 R. B. Perry *Gen. Theory of Value* iii. 79 They have been driven to adopt a metaphysical 'voluntarism' or 'imperativism'. **1950** S. E. Toulmin *Exam. Place of Reason in Ethics* iv. xiii. 189 The 'imperativist' is prepared to consider the present account because it represents for him a sophisticated imperative theory. **1952** *Mind* LXI. 96 The imperativist analysis of moral indicatives cannot be correct. **1960** *Philos. of Sci.* XXVII. 374 This exposure tells against the 'imperativist' interpretation of law-sentences. **1963** W. Sellars in Castañeda & Nakhnikian *Morality & Lang. of Conduct* 163 A rough approximation to more sophisticated analyses of the imperativist type. **1965** *Philos. Rev.* LXXIV. 108 It is Zink, and not the regiment of imperativists, emotivists,.. and noncognitivists, who is faithful to the insights of Moore and Wittgenstein.

imperception. (Later examples.)

1961 W. R. Brain *Speech Disorders* xiii. 143 Hughlings Jackson..first recognized both agnosia and apraxia. He called the former 'imperception'. **1968** P. McKellar *Experience & Behaviour* x. 260 The selective imperception of the virtuous actions or motivation of people we dislike.

impercipience. (Later examples.)

1905 *Westm. Gaz.* 30 Sept. 4/1 It is only our physical or mental impercipience that leaves the sluggish..mind an easy prey to the promptings of vulgarity. **1925** A. Quiller-Couch *Charles Dickens* 71 A lost child, mooning incuriously along the hedgerows with an impercipience rivalling that of a famous Master of Trinity. **1971** *Country Life* 1 Apr. 785/2 H. G. Wells making one of his terrifying comments (terrifying for impercipience).

imperfect, *a.* (*sb.*). Add: **A.** *adj.* **8. b.** Of a stage in the life cycle of a fungus: not producing or not known to produce sexual organs. Of a fungus, having (apparently) no sexual stage: belonging to the group designated *imperfect fungi* (or formally, in mod.L., *Fungi Imperfecti*), in which are included all those fungi which, because a sexual stage is missing or unknown, cannot be assigned to other taxa.

1895 M. C. Cooke *Introd. Study Fungi* xxii. 259 The group now under consideration is analogous, in external features, to the *Pyrenomycetes*, but wholly deficient of asci. The perithecia, or pseudoperithecia, include only stylospores, and have been assumed to be imperfect representatives, or imperfect stages or conditions, of the *Pyrenomyceteae*, and hence called 'imperfect capsular fungi'. **1898** *Jrnl. R. Microsc. Soc.* 660 (*heading*) Rabenhorst's cryptogamic flora of Germany (Fungi Imperfecti). **1908** *Ibid.* 626 Similar cultures were successfully carried through with *Gnomoniella tubiformis* on alder leaves, of which the 'imperfect' form was proved to be *Leptothyrium alneum*. **1952** C. J. Alexopoulos *Introd. Mycol.* xiii. 312 A great many fungi are known which have septate mycelium and which, so far as anyone has been able to discover, reproduce only by means of conidia. Since these fungi apparently lack a sexual stage (perfect stage), we call them, commonly, 'imperfect fungi', and technically, Fungi Imperfecti. **1971** P. H. B. Talbot *Princ. Fungal Taxon.* v. 77 The phase associated with asexual spores or sterile mycelia is known as the imperfect state of the fungus, while that associated with production of zygotes or of spores resulting from any type of sexual process is the perfect state.

10. *imperfect induction* (Philos.): a term signifying induction (see Induction 7) from an incomplete set of instances, usu. used in contrast to the notion of perfect induction.

[**1843** Mill *Logic* I. iii. ii. § 1. 352 The induction is asserted not to be perfect, unless every single individual of the class A is included in the antecedent.] *a* **1856** W. Hamilton *Lect. Metaphysics & Logic* (1860) III. 325 This Imperfect Induction they [*sc.* logicians] held in contingent matter to be contingent. **1870** W. S. Jevons *Elem. Lessons Logic* xxv. 213 The assertion that all the planets move in one direction round the sun..is derived from Imperfect Induction; for it is possible that there exist planets more distant than the most distant-known planet Neptune. **1914** C. Read *Logic* (ed. 4) xv. 197 Imperfect Induction..is..the method of showing the credibility of an universal real proposition by an examination of *some* of the instances it includes. **1957** J. Passmore *100 Yrs. Philos.* i. 22 Conventionally, two sorts of induction had been distinguished: perfect and imperfect... We are driven back upon 'imperfect' induction, as..the only sort of inductive inference.

11. *imperfect competition* (Econ.): competition diluted by elements of monopoly so that individual producers or consumers are able to exercise some control over the market price.

1881 F. Y. Edgeworth *Math. Psychics* ii. 48 This condition, though not spontaneously generated by imperfect as by perfect competition, should be introduced *ab extra*. **1933** J. Robinson (*title*) The economics of imperfect competition. **1937** *Q. Jrnl. Econ.* LII. 529 'Imperfect competition' is a more familiar expression in England, while the term 'monopolistic competition' is more familiar in the United States. **1948** E. H. Chamberlin *Monopolistic Competition* (ed. 6) p. ix, Monopolistic Competition is a fusion of the..theories of monopoly and competition, whereas Imperfect Competition contains no monopoly. **1961** *Rev. Econ. Stud.* XXVIII. 182 Writers on imperfect competition in the product and factor markets have paid little attention to this question. **1969** D. C. Hague *Managerial Econ.* iv. 87 Imperfect

competition, that is to say..markets where there is not either pure competition or monopoly.

B. *sb.* **3.** *pl.* Goods of which the quality is not high enough for them to be sold to the public, except at a reduced price.

1952 *Amer. Speech* XXVII. 264 Textile products which ..do not come up to standard quality are referred to as *imperfects, seconds,* and *run-of-the-mill*. **1962** E. Godfrey *Retail Selling & Organization* ii. x. 95 The retail buyers.. buy up manufacturers' and wholesalers'..factory imperfects. **1962** S. Strand *Marketing Dict.* 358 *Imperfect*, merchandise below standard... In many cases imperfects are useful products, but because of a manufacturer's flaw..they are removed from prime merchandise. **1969** *Observer* 9 Nov. 1/8 (Advt.), 'Imperfects' offered at a much reduced price.

imperfectibility. (Later example.)

1971 *Nature* 31 Dec. 523/1 The same principle of imperfectibility applies to all sex chromosome mechanisms.

imperfection. Add: **4. a.** *Printing. pl.* Letters that are wanting in a fount; types cast to make up a deficiency in a fount.

1681–5 Fell *Let. to Marshall* 24 Oct. (MS.), The compositor upon Mr. Junius his lexicon wants several imperfections, that we cannot supply without his Matrices. **1683–4** J. Moxon *Mech. Exerc. Printing* (1962) 344 When the Founder has not Cast a proportionable number of each sort of Letter, the wanting Letters are called Imperfections, as making the rest of the Fount unperfect. **1771** P. Luckombe *Hist. & Art of Printing* 243 Less occasion to cast imperfections, which often prove very hurtful to a new fount of letter; as they are seldom exact to the prior sorts..: so that, was it not for the eagerness of the Compositor,..many a sort, cast for perfecting, would be returned. **1808** C. Stower *Printer's Gram.* 56 It should be an invariable rule with master printers to examine imperfections before they go into the hands of the compositor. **1888** C. T. Jacobi *Printers' Vocab.* 61 *Imperfections*, short sorts required to perfect a typefounder's bill for a fount of a certain weight. **1924** *Southward's Mod. Printing* (ed. 5) I. xx. 124 The fount should be carefully examined with a view of ascertaining whether there are any imperfections—the founder's storekeepers sometimes making mistakes in the apportionment of particular letters. **1962** Davis & Carter in J. Moxon *Mech. Exerc. Printing* 344 (*footnote*) Typefounders charge for sorts at a higher rate than for founts; but sorts to supplement a fount, if ordered within three months of delivery of the fount, are called 'imperfections' and charged at fount-price.

b. *Bookbinding.* A surplus or missing sheet of a work.

1683–4 J. Moxon *Mech. Exerc. Printing* (1962) 315 He Doubles or Quires up all the other Heaps and..writes upon them Imperfections of (the Title of the Book), and Writes on it the Signature of the Sheet that is Wanting. **1790** A. Smith *Let.* 25 May in *Sotheby Catal.* (19 July 1937) lot 74, The bookbinder informed me..that one of the copies is imperfect, wanting the sheet E. I will beg the favour of you to send down the imperfection by the first parcel you send to Scotland. **1835** J. Hannett *Bibliopegia* i. 13 If any sheet is wanting or belongs to another volume, or is a duplicate, the further progress of the work must be suspended, till the imperfection is procured or exchanged. **1888** C. T. Jacobi *Printers' Vocab.* 61 *Imperfections*, sheets required by a binder to make good books imperfect through bad gathering, collating, or spoiled sheets. **1963** Kenneison & Spilman *Dict. Printing* 95 *Imperfections*, sheets rejected by the binder and returned to the printer to be replaced.

imperfective, *a.* (*sb.*). Add: **A.** *adj.* **2.** (Further examples.) Also, by extension, of a similar form or aspect in some non-Slavonic languages.

1912 [see *durative *a.*]. **1924** [see *aspect *sb.* 9 b]. **1955** *Word* XI. 546 In general it shows the action as completed (perfective) or incomplete (imperfective) relative to the time of the action of the main verb. **1957** R. W. Zandvoort *Handbk. Eng. Gram.* i. ii. 33 The aspect expressed by the present participle..is called *imperfective* or *durative*. **1958** H. G. Lunt *Fund. Russian* 59 The imperfective aspect does not say anything about the end of the action. **1972** *Language* XLVIII. 169 By accident, some presents with imperfective meaning contained an *e*, e.g. *bher-e-* 'carry'.

B. *sb.* **2.** An imperfective verb, case, or aspect.

1939 *Language* XV. 230 In a space of ten lines..there are four present imperfectives. *Ibid.*, The imperfectives, whether present or past, definitely indicate a repeated or a continuing action. **1949** *Ibid.* XXV. 403 There is only one present tense, namely an imperfective. **1949** *Archivum Linguisticum* I. ii. 176 Imperfectives, if not iterative, become perfective by prefixing a preposition. **1962** K. Katzner *Russian Review Text* 183 The Imperfectives are called the *Determinati* and the *Indeterminati*. **1962** *Word* XVIII. 17 These 'secondary' imperfectives enter, in turn, into a derivational relation with the 'primary' imperfectives. **1965** *Canad. Jrnl. Linguistics* Spring 117 Agentives are apparently built upon future duratives, not upon imperfectives. **1966** *Jrnl. Linguistics* II. ii. 249 The distribution 52% perfectives, 48% imperfectives, confirms that the relative frequency..of the Russian marked and unmarked members differs sharply. **1972** Hartmann & Stork *Dict. Lang. & Ling.* 20 Ancient Greek, for example, had perfective, imperfective, and aorist.

Hence **imperfectiviza·tion** *Gram.*, the making of a verbal form or tense 'imperfective'; the condition of being 'imperfective'.

1943 *Language* XIX. 273 The statement of 'means of imperfectivization' is a list of suffixes. **1962** *Word* XVIII. 15 Its function is basically grammatical, i.e. that of imperfectivization. **1966** *Jrnl. Linguistics* II. ii. 249 A formal distinction, as regards morphological potentiality, secondary imperfectivization, is made.

imperial, *a.* and *sb.* Add: **A.** *adj.* **2. c.** Designating certain decorations or orders.

1878 *London Gaz.* (Suppl.) 4 Jan. 113/1 The Queen has been graciously pleased..to institute and create an Order of Distinction, to be styled and designated 'The Imperial Order of the Crown of India'. **1902** *Encycl. Brit.* XXXI. 340/1 The Imperial Service Order was..instituted on 26th June 1902, to commemorate King Edward's coronation. *Ibid.*, The Imperial Order of the Crown of India is conferred for like purposes as the order of the Indian Empire. **1971** *Whitaker's Almanack* 261 The Imperial Service Order..consists of the Sovereign and Companions (not exclusively male) to a number not exceeding 1325 of whom 750 may belong to the Home Civil Services and 575 to Overseas Civil Services.

d. *Imperial Defence*: defence of Great Britain and of its dependent territories.

1897 G. S. Clarke (*title*) Imperial defence. **1902** *Encycl. Brit.* XXVI. 401/1 The appointment in 1879 of a royal commission to consider the question of Imperial defence, which presented its report in 1882, led to a considerable development and reorganization of the system of Imperial fortifications. **1910** *Ibid.* III. 254/1 His [*sc.* A. J. Balfour's] institution of the permanent Committee of Imperial Defence, and of the new Army Council (1904), were reforms of the highest importance. **1938** *Ann. Reg. 1937* 67 The national defence was being quite adequately organised by the Committee of Imperial Defence.

e. Designating a policy or an institution concerned with the development of commerce between the constituent parts of the British Commonwealth (formerly the British Empire); esp. in *imperial preference*, a system of tariff concessions granted by members of the British Empire or Commonwealth to one another.

1902 *Encycl. Brit.* XXVI. 397/2 The foundation of the Imperial Federation League—in 1884. *Ibid.* XXXIII. 393/1 The British Empire League, and the Imperial Trade Defence League endeavour to promote inter-Imperial trade. *Ibid.* 681/2 On 4th July she laid the foundation stone of the Imperial Institute. **1912** J. S. Huxley *Individual in Animal Kingdom* ii. 54 All the wheat in Canada, with Imperial Preference to help, would not keep her [*sc.* the English nation] from starvation. **1922** *Encycl. Brit.* XXX. 1016/2 At the end of April [1917] Mr. Bonar Law announced..that the Imperial War Cabinet had accepted the principle of Imperial Preference. *Ibid.* 1025/2 The main feature of the budget [1919] was the establishment at last of imperial preference. **1927** *Daily Tel.* 5 Mar. 8/7 Appreciation of the work of the Imperial Economic Committee and the Empire Marketing Board for the development of the market for Dominion produce in Great Britain. **1931** G. C. Tryon (*title*) Short history of imperial preference. **1958** *Listener* 18 Sept. 407/2 Both [*sc.* Australia and New Zealand] have sought to give foreign suppliers a better competitive position in their own markets by reducing imperial preferences to oil the wheels of reciprocity. **1971** A. Shonfield in A. Bullock *20th Cent.* xiv. 331/2 The Imperial Preference system was successful in helping the British industrial recovery in the 1930s.

8. b. or **B. 5. b.** (The old measurements have been replaced: see quots.)

1952 E. J. Labarre *Dict. Paper* (ed. 2) 130/1 *Imperial*. Now standardized for writings and printings at 30″ × 22″ and for wrappings at 29″ × 22½″, for boards at 30″ × 22½″ and their multiples. **1968** *Specification Sizes Papers & Boards (B.S.I.)* 14 (*heading*) Table 2. Writing and printing papers and offset cartridges... Imperial..22 × 30. *Ibid.* 15 Table 3. Ledger papers... Imperial..22 × 30. *Ibid.* 17 Table 7. Paste boards; and duplex, triplex and ivory boards... Paste imperial..22½ × 30. *Ibid.*, Table 8. Drawing cartridges... Imperial..22 × 30.

10. *imperial elephant, mammoth* [tr. of *Elephas imperator* (J. Leidy 1858, in *Proc. Acad. Nat. Sci. Philadelphia* 10), which was later included in the genus *Mammuthus*]: a fossil mammal, *Mammuthus imperator*, found in Pleistocene remains in south-western North America; *imperial pigeon*, a large fruit-eating pigeon, esp. one of the genus *Ducula*, found in south-eastern Asia, including India, and the Pacific region.

1913 W. B. Scott *Hist. Land Mammals W. Hemisphere* xii. 485 The largest of American proboscideans was the Imperial Elephant. **1910** H. F. Osborn *Age of Mammals* vi. 442 The Columbian and imperial mammoths were for a time at least contemporaneous with the mastodon. **1945** A. S. Romer *Vertebr. Paleont.* (ed. 2) xxi. 416 *M[ammuthus] imperator*, the imperial mammoth of southern North America, was more advanced in size and dental development. **1966** Imperial mammoth [see *baluchitherium]. **1864** T. C. Jerdon *Birds India* III. 455 *Carpophaga sylvatica*... Imperial Pigeon of Europeans in the South of India. **1895** *Jrnl. Bombay Nat. Hist. Soc.* X. 360 The green Imperial Pigeon..may be frequently met with nearly all over North Cachar. **1913** E. C. S. Baker *Indian Pigeons & Doves* 103 Hodgson's Imperial Pigeon is less quarrelsome than most of the family. **1934** 'G. Orwell' *Burmese Days* xiv. 211 Flo [*sc.* a dog]..came running excitedly up..with the big imperial pigeon in her mouth. **1967** D. Goodwin *Pigeons & Doves of World* 384 The large species of fruit pigeons are often termed the imperial pigeons because of their impressive and majestic-looking if somewhat ponderous appearance.

B. *sb.* **8.** (Earlier examples.)

The earlier examples confute the attribution of the style to the Emperor Napoleon III (who became Emperor in 1852).

[**1829** Balzac *La Maison du Chat-qui-Pelote* in *Œuvres* (1938) I. 64 Sa figure..était encore animée par de petites moustaches relevées en pointe et noires comme du jais, par une impériale bien fournie,..et par une forêt de cheveux noirs assez en désordre.] **1835** S. Horsley *Let.* in R. B. Gotch *Mendelssohn & his Friends in Kensington* (1934) 192 What with his black hair longer than ever, a *beard* which he is now cultivating to a great length, mustachios and an imperial,..made you suspect the fact of his having escaped across the country from some wandering menagerie. **1838** H. Mozley *Let.* 2 Nov. in D. Mozley *Newman Family Lett.* (1962) 77 Mr. Sidney Herbert..is..a silly looking coxcomb, with a most disfiguring imperial. **1839** *Blackw. Mag.* Oct. 507/2 An imperial—*i.e.* a dirt-coloured tuft of hair.

imperialism. Add: **2.** (Earlier and U.S. examples.)

1878 J. Chamberlain *Let.* 15 Oct. in J. L. Garvin *Life J. Chamberlain* (1932) I. 267 This infernal Afghan business is the natural consequence of Jingoism, Imperialism, 'British interests', [etc.]. **1899** H. H. Bancroft *New Pacific* viii. 145 The word imperialism is used in this connection in a modern, American sense, as applicable to the empire of industry as well as to domain... It [implies].. the extension of political and commercial influence, particularly in the Pacific. **1899** Carnegie in *North Amer. Rev.* Jan. 5 Imperialism implies naval and military force behind; moral force, education, civilization are not the backbone of Imperialism. **1914** *Cycl. Amer. Govt.* II. 152/1 As used in American politics, imperialism is employed to designate the policy on which the United States has embarked of acquiring territory not a part of the United States proper, nor contiguous to it.

3. Used disparagingly. In Communist writings: the imperial system or policy of the Western powers. Used conversely in some Western writings: the imperial system or policy of the Communist powers.

1918 *Manch. Guardian* 13 Dec. 7/4 The Menshevik and the small bourgeois parties have published a declaration calling on workers all over the world to rally to the support of the Russian Revolution against the Imperialism attacking it. **1939** Varga & Mendelsohn (*title*) New data for V. I. Lenin's 'Imperialism, the highest stage of capitalism'. **1957** C. Hunt *Guide to Communist Jargon* xxiv. 82 The term imperialism has largely replaced Capitalism in the Communist vocabulary. *Ibid.* 83 The essential features of imperialism are the concentration of capital, the merging of industrial and banking capital into 'finance capital' and the division of the world between national and international monopolies. **1957** *Encycl. Brit.* VI. 135/2 They [*sc.* the Communists] regarded the cause of World War II to be not German aggression but British and French imperialism. **1964** Gould & Kolb *Dict. Social Sci.* 319/2 It is sometimes said that Russian control of East European countries is 'Russian imperialism'.

imperialist. Add: **1.** (Later example.)

1900 *Westm. Gaz.* 14 Sept. 2/2 It is interesting to note a new use of the word Imperialist in to-day's *Daily News*: 'A troop of American cavalry surprised 300 Imperialists at Shaho...' This is from Pekin—clearly Imperialist is to be the way of distinguishing the Chinese Imperial troops from rebel Boxers.

3. (Later U.S. example.)

1900 *Congress. Rec.* 11 Jan. 766/2 The trouble with these imperialists is that they confound the Government of the United States with their puny President.

3*. An advocate of 'imperialism' (sense *3).

1963 [see *capitulationism]. **1969** *Times* 8 Aug. 1/2 'We all know,' Mr. Katushev said, 'how violently the imperialists are fighting on all fronts..to give support to anti-socialist forces.'

4. (Later examples.) Also *Comb.*

1937 [see *capitalist]. **1957** C. Hunt *Guide to Communist Jargon* xxiv. 83 Lenin does not seem to have regarded the actual possession of an empire as essential to it [*sc.* imperialism], so that, according to his theory, the United States is an imperialist Power, as communist propaganda daily represents it. **1967** H. Arendt *Origins Totalitarianism* (new ed.) v. 130 The imperialist-minded businessmen were followed by civil servants who wanted 'the African to be left an African'. *Ibid.* viii. 257 The imperialist-inspired 'parties above parties' had never known how to profit from popular hatred of the party system as such. **1969** *Times* 8 Aug. 1/2 President Ceausescu [of Rumania] argued that the socialist world was strong enough to safeguard itself against imperialist tactics. *Ibid.* 21 Aug. 7/3 The Soviet Union stood revealed for what it now is—an imperialist power. **1973** D. Miller *Chinese Jade Affair* xviii. 174 The K.M.T. group.. would probably suggest that K. Lawson was engaged in a typical Western imperialist plot.

imperialistic, *a.* (Later examples.)

1933 N. Waln *House of Exile* iii. iii. 223 His remarks were a stirring appeal to the citizens of China to unite against the 'Imperialistic Foreign Devil'. **1958** [see *colonial *a.* 1 a]. **1969** *Listener* 24 Apr. 568/3 They show the unwavering will of the people to make up for all the damage done by these imperialistic traitors in the shortest possible time.

imperishable. Add: Also as *sb.*

a **1910** 'Mark Twain' *Mysterious Stranger* (1916) iii. 27, I am of the aristocracy of the Imperishables. **1964** *New Statesman* 3 Apr. 514/2, I have seen *Oliver Twist*,..and a number of Jules Verne's imperishables.

impersistent, *a.* (Earlier and later examples.)

a **1866** J. Grote *Exam. Utilitarian Philos.* (1870) xix. 324 We ought..to know whether some races are strong and persistent,..others weak and impersistent, so as to yield to others and die out. **1965** G. J. Williams *Econ. Geol. N.Z.* iii. 27/2 Various impersistent shoots were worked in the Pandora and South Keep-it-Dark mines.

impersonalism (impə̄·ɹsənăliz'm). [f. Impersonal *a.* + -ism.] The character of being impersonal; the absence of personal contacts. So **impe·rsonalist,** one who is, or aims at being, impersonal; **impersonali·stic** *a.*

1899 *Speaker* 9 Dec. 263/1 The weak point in the armour of the impersonalist is the dedication. **1908** *Daily Chron.* 26 May 3/4 The workmen are getting the impersonalism of Socialism without its humanity. **1920** F. M. Ford *Let.* 30 June (1965) 109 There is the whole open question of Impersonalism to discuss. **1932** J. C. Powys *Glastonbury Romance* xxvi. 907 It was the appearance of Dave Spear's figure now—for the impersonalist had decided to disobey Zookey—that drove the tipsy giant to his next move. **1968** *Austral. Jrnl. Philos.* XLVI. 257 The existence of such principles, principles which are perfectly consistent as 'impersonalistic' principles are actually used but which would be inconsistent if 'impersonalistic' principles were used in the way being supposed, does not, however, give impersonalism any sort of advantage over nonimpersonalism.

impersonally, *adv.* **2.** (Earlier example.)

1854 A. G. Henderson tr. *Cousin's Philos. of Kant* vii. 178 There is a state..where the reason manifests itself almost entirely impersonally.

impersonalness (impə̄·ɹsənălnes). [f. Impersonal *a.* + -ness.] Impersonal quality; absence of personality.

1871 P. Brooks *Let.* 11 Jan. in A. V. G. Allen *Phillips Brooks* (1908) 247 When I see a small audience I lose the impersonalness of the thing. I think of individuals and that always puts me out. **1955** P. Heron *Changing Forms of Art* 233 In his painting it never matters one jot when his beautiful girl has a distinctly anonymous air. Her impersonalness is part of her charm. **1956** A. Toynbee *Historian's Approach to Relig.* iv. 44 The impersonalness of an oecumenical empire as an institution makes itself felt in the remoteness of its metropolis from the daily life of the great majority of its subjects.

imphee. (Earlier examples.)

1857 *Country Gentleman* 11 June 379/2 A plant bearing the name of *Imphee*, or *Imphey*, or *Imphye*..which it is alleged is identical with the Chinese Sugar Cane, has been introduced by Mr. Leonard Wray, from Southern Africa. **1862** T. Baines *Jrnl.* 5 Apr. in *Explor. S.-W. Afr.* (1864) xiv. 438, I..spent the intervening time with a circle of old fellows, who gave me imphi (holcus saccharatus) stalks to chew.

impi. (Earlier and later examples.)

1862 G. H. Mason *Zululand* xv. 200 There is always an 'Impi', (or army,) preparing for an attack on some neighbouring district. **1970** *Cape Times* 28 Oct. 2/9 Lands rich in grass and game and savage impis. **1971** *Rand Daily Mail* 4 Dec. 3/4 Dressed in full tribal regalia of leopard skin and feathers he led the dancing and singing impis who paid homage to their king.

impinge, *v.* **3.** Delete *Obs.* and add example.

1910 *Practitioner* July 109 The striker's thumb..impinges the skull of his opponent.

impinger (impi·ndʒəɹ). [f. Imping(e *v.* + -er[1].] Any of various instruments for collecting samples of the particles suspended in air (or another gas), this being either drawn into a liquid or directed in a jet against a flat surface so that some particles are deposited.

1922 *Rep. Investigations U.S. Bureau of Mines* No. 2392. 3 The analytical procedure necessary when using this new impinger-bubbler apparatus is almost identical with that employed with the Palmer apparatus. **1932** *Jrnl. Industr. Hygiene & Toxicol.* XIV. 301/1 The Greenburg-Smith impinger was introduced in 1922. **1936** Drinker & Hatch *Industr. Dust* vii. 116 The impinger flasks, stoppers, etc., should be washed..in..the sampling fluid. **1945** [see *impactor 2]. **1972** Blakeslee & Reckner in R. D. Ross *Air Pollution & Industry* v. 270 Dry impingers are capable of depositing particles in distinct size ranges on surfaces such as microscope slides. Several units are connected in series, each containing an orifice which is aimed at the surface.

impinging, *ppl. a.* (Later example.)

1955 *Sci. News Let.* 14 May 308/1 Mercury vapor inside the vacuum tube gives the glow as its atoms are excited by impinging electrons.

implant, *v.* Add: **1. b.** *Med.* Surgically to place or insert (tissue, or something inorganic) in the body: used esp. when what is inserted does not correspond with what is naturally found at the site that receives it.

1886 W. D. Younger *Implantation of Teeth* 8, I have since tried implanting teeth which have been extracted for weeks and months. **1887** *Lancet* 12 Feb. 334/2 In his early attempts he used fresh teeth, which he obtained from other dentists..and endeavoured to keep alive by implanting them in cocks' combs. **1919** *Jrnl. Amer. Med. Assoc.* 26 July 301/1 On the other hand, the pathologic opaque [corneal] graft, implanted in sound tissue, grows normal and in time becomes transparent. **1927** *Lancet* 15 Jan. 120/2 Animals not infrequently recover from tumours implanted into them. **1941** *Jrnl. Amer. Med. Assoc.* 27 Sept. 1070/1 In January 1940 a 200 mg. tablet of testosterone was implanted. After another month, hair appeared on the upper lip..and the penis further increased in size. **1952** *Brit. Jrnl. Radiol.* XXV. 423/2 A method of implanting radioactive Ta^{182} wire for the treatment of patients with carcinoma of the bladder is described. **1963** *Lancet* 12 Jan. 78/1 The amnion was prepared and implanted by the method of Troensegaard-Hansen (1956). In 8 cases where symptoms were unilateral the implant was placed in the affected leg. **1964** *Ann. N.Y. Acad. Sci.* CXI. 1063 During the past 3 years the pacemaker has been implanted in 43 patients ranging in age from 39 to 85 years.

3. b. Delete *rare* and add to def.: to furnish *with* by insertion or implantation. (Further examples.)

1919 *Lancet* 29 Mar. 490/2 We are dealing with serum implanted with gas-gangrene bacilli. *Ibid.* 493/2 An agar surface implanted with a serophytic organism. **1956** *Brit. Jrnl. Radiol.* XXIX. 509/1 When the whole area.. has been 'implanted' with the introducing needles, their stilettes are removed and tantalum wires..are passed down their lumina. **1968** *Canad. Jrnl. Physics* XLVI. 671/1 A silicon sample implanted at room temperature with As ions. **1971** *Nature* 18 June 454/2 Sprague-Dawley male albino rats were stereotaxically implanted..with stainless steel cannula systems for the injection of.. various amines.

4. *intr. Embryol.* To be or become implanted; to undergo implantation (sense *6).

1954 *Contrib. Embryol.* XXXV. 219/1 It appears.. that the human blastocyst in the same stage of development implants earlier than that of the macaque. **1963** C. G. Hartman *Mechanisms Conception* vii. 349 After transfer to pseudopregnant hosts, the blastocysts implanted..with about the same frequency of success as blastocysts transferred without cultivation. **1967** Strong & Corney *Placenta in Twin Pregnancy* ii. 16 They may implant at adjacent sites so that the placentae in growing would fuse to form a dichorionic, but single placenta. **1970** *Sci. Jrnl.* June 48/1 Fertilization takes place in the Fallopian tube and some days later the fertilized egg implants into the lining of the uterus.

implant (i·mplɑnt), *sb.* [f. Implant *v.*] **a.** Anything implanted, esp. within the body.

1890 *Sat. Rev.* 15 Nov. 551/2 It seemed to tell of an ineradicable implant of commercial hypocrisy. **1911** *Chem. Abstr.* V. 2499 However, since after extirpation of these implants the hypersensitiveness persisted, it would seem that sessile receptors are not essential. **1919** *Jrnl. Amer. Med. Assoc.* 26 July 301/1 A disk is cut from the opaque cornea and another from the transparent periphery, and the disks are transposed. The transparent implant being impermeable, like all autografts, becomes invaded by the abnormal elements surrounding it. **1952** *Brit. Jrnl. Radiol.* XXV. 421/1 A radon-seed implant has proved to be a useful method of treatment for early carcinoma of the bladder. **1961** J. N. Anderson *Appl. Dental Materials* (ed. 2) viii. 79 They [*sc.* chrome-cobalt alloys] can..be used as an 'implant' beneath the gum to which a denture may be fastened. **1963** [see *implant *v.* 1 b]. **1968** *New Scientist* 11 Jan. 80/1 In recent years surgeons have been fitting an increasing number of implants to fix and repair damaged bones. **1968** *Brit. Med. Bull.* XXIV. 242/1 Radium implants for radiation therapy. **1968** *Canad. Jrnl. Physics* XLVI. 667/1 Since the implanted atoms are confined to a surface region less than 0·1 μ thick, a well-defined 'impurity' peak is observed in the scattering spectra. [*Note*] In the case of phosphorus implants..the mass difference between P and Si is too small for the 'impurity' peak to be resolved.

b. An act or operation of implanting something; an implantation.

1941 *Jrnl. Amer. Med. Assoc.* 27 Sept. 1069/1 Our more recent multiple 50 mg. pellet implants were performed with a trocar. **1970** *Times* 28 Apr. 5/4 French doctors today made the world's first implant of an atomic powered heart simulator into a human being.

implantable (implɑ·ntăb'l), *a.* [f. Implant *v.* + -able.] Capable of being implanted (in the body).

1960 C. N. Smyth *Med. Electronics* 253 (*heading*) An implantable pacemaker for the heart. **1965** *Adv. Biol. & Med. Physics* X. 367 It is conservatively estimated that over 4000 implantable cardiac pacemakers have been used to date with gratifying results. **1972** *Physics Bull.* June 336/2 Since there is no sign of complex implantable functioning organs at the present stage of medical engineering technology, devices which replace heart, lung, kidney or liver remain outside the body.

implantation. **5. b.** Add def.: The (or an) operation of implanting something in the body (see *Implant *v.* 1 b). (Earlier and later examples.)

1885 M. Hay tr. *H. von Ziemssen's Handbk. Gen. Therapeutics* II. 399 The method described by Bruns, and characterised by him as dry injection or implantation... In this, the drugs are likewise applied in thin cylinders or plugs, for the introduction of which..a special form of implantation needle is used. **1886** W. D. Younger (*title*) Implantation of teeth and pericemental life. **1887** *Lancet* 12 Feb. 334/1 Implantation is the ingrafting of a natural tooth into an artificial socket. **1929** *Jrnl. Amer. Med. Assoc.* 1 June 1900/2 In cases..in which it is not possible to remove the oviducts without removing the ovaries, Michel cuts off a piece of one of the resected ovaries..and sutures it to the uterus... He obtained good results in six out of seven cases in which he used his method of autoplastic implantation of a portion of an ovary. **1938** *Lancet* 10 Sept. 606/2 A very prolonged effect of certain androgens and œstrogens could be obtained by a single implantation, under the skin, of pure dry hormone in the form of crystals or compressed tablets. **1963** *Ibid.* 12 Jan. 77/2 Troensegaard-Hansen (1956) described the treatment of intermittent claudication by implantation of human amnion into the thigh. **1969** H. A. Salhanick et al. *Metabolic Effects Gonadal Hormones* 723 The most direct experimental approach to this problem is by the intracranial implantation of small amounts of crystalline steroids.

6. *Embryol.* The attachment of the fertilized ovum (blastocyst) to the wall of the uterus.

1902 A. Keith *Human Embryol.* viii. 96 The implantation of the ovum in the decidua is in the posterior wall of the uterus in over 60% of cases. **1936** F. J. Taussig *Abortion* iv. 70 In about ten days from fertilization, the ovum is ready for implantation. **1969** Klopper & Diczfalusy *Foetus & Placenta* iii. 62 Implantation begins on the 20th day of the cycle, and probably takes several days to complete... The great majority of implantations in the case of the human take place in the fundus of the uterus and in the posterior wall.

7. *Physics.* The introduction of ions into a crystalline structure by bombardment with an ion beam.

1965 *Nuclear Instruments & Methods* XXXVIII. 169/2 Implantations reported in this paper were made in a 24 inch radius calutron. **1967** *Canad. Jrnl. Physics* XLV. 4053 Because of the nonequilibrium nature of the implantation process, the relative number of impurities on substitutional and interstitial sites may differ from that observed following conventional thermal diffusion. **1973** *Sci. Amer.* Apr. 65/3 The accelerated-ion technique offers fairly precise control of both the number of ions implanted and the depth of implantation.

implanted, *ppl. a.* (In Dict. after Implant *v.*) Add: **2.** In sense corresponding to Implant *v.* 3 b (in Dict. and Suppl.): having something implanted or inserted into it; *spec.* containing implanted ions.

1965 *Nuclear Instruments & Methods* XXXVIII. 172/1 Fig. 7 shows variation in the sheet resistance of an implanted area with temperature. **1970** *New Scientist* 15 Oct. Suppl. 17/2 The Japanese..have already announced an implanted microwave transistor. **1973** Wilson & Brewer *Ion Beams* iv. 300 The oxide serves..to protect against compensation of the implanted region during a subsequent diffusion.

implausibility (implǭzĭbi·lĭti). Delete † *Obs. rare* and add later examples.

1926 B. Freeman *Towards Answer* 7 Perhaps the thing is an impossibility or an implausibility. **1966** G. N. Leech *Eng. in Advertising* ii. 17 The other two examples are also freakish..because of their contextual implausibility.

implausible, *a.* **2.** (Later examples.)

1971 *Nature* 5 Feb. 408/1 It is implausible that such animals should have preceded in time the small and structurally different species *Hipparion nagriensis* Hussain. **1974** *Daily Tel.* 20 Feb. 13/2 The play was..a highly stagey and implausible piece.

implausibly, *adv.* (Later examples.)

1928 *Music & Lett.* July 234 Men..able to..write categorically though implausibly. **1965** *Chicago Tribune* 10 Aug., The principal river..was variously named Konomick, Killamick, Calamick and sundry other versions which some claim, implausibly, meant 'white beaver'.

implead, *v.* **1.** Delete 'Now only *arch.* or *Hist.*', and add later examples.

1957 *Listener* 19 Dec. 1020/1 Can one implead the State concerned? *Ibid.*, At present you cannot implead a State in United Kingdom Courts except under the Warsaw Convention. **1964** *Welsh Hist. Rev.* II. 47 It was further stipulated that a similar arrangement would be adopted if Owain or Llywelyn or their heirs wished to implead any of the King's subjects. **1973** *N.Y. Law Jrnl.* 1 Aug. 1/7 The plaintiff sued the appellant for damages for fraudulent acts under the contract. The appellant impleaded the plaintiff's president on his guarantee.

implement, *v.* Delete 'Chiefly *Sc.*' and add examples of sense 1 a.

1909 *Westm. Gaz.* 30 Aug. 4/3 [The] council has been prepared to implement that agreement. **1950** C. Morris *Social Case-Work Gt. Brit.* 7 With the post-war flood of social legislation, social workers are required..to help to implement the laws. **1964** *Ann. Reg. 1963* 102 Henceforward the bargaining..continued until late in the year, with some decisions being announced and implemented piecemeal. **1969** *Times* 3 Sept. 11/6 Three years later Armageddon found him [*sc.* Churchill] ready at the Admiralty with plans and authority to implement that policy. **1972** *Daily Tel.* 27 June 2/7 The provision of the Act of which they are most apprehensive..has yet to be implemented by the Government.

implementation (i:mplĭměntēⁱ·ʃən). [f. Implement *v.* + -ation.] The action of implementing; fulfilment.

1926 *Spectator* 16 Oct. 627/1 The Irish delegation will seek the implementation of co-equality with the States of the British Commonwealth. **1944** *Mind* LIII. 184 Professor Collingwood's *Principles of Art* attempted an implementation and restatement on the epistemological side. **1951** R. Firth *Elem. Social Organiz.* ii. 45 In a democratic society consent of the individual is ultimately necessary for the implementation of programmes. **1955** *Bull. Atomic Sci.* June 223/3 As a possible mode of implementation, I suggest that a civilian-directed Scientists Corps (SC) be set up within the Defense Department for an emergency. **1965** *Listener* 27 May 775/2 Such plans..were fundamentally sound in the practical potential of implementation. **1973** *Daily Tel.* 16 Mar. 6/4 The successful implementation of this policy is vital. The cost of failure is incalculable.

implicans (impli̇kæ·nz). *Logic.* The pl. form used is **implicants.** [L., pres. pple. of *implicāre* (see IMPLICATE *v.*).] In implication (see *IMPLICATION 2 c), the active proposition; the proposition that implies. Cf. also IMPLICATE *sb.* 2. in Dict. and Suppl.

1921 W. E. JOHNSON *Logic* I. ii. 30 In the implicative function 'If *p* then *q*', *p* is the implicans and *q* the implicate. **1922** *Ibid.* II. x. 211 We shall take..the implicants and disjuncts to stand for particular propositions. **1930** L. S. STEBBING *Mod. Introd. Logic* v. 70 It is clear that the order of the implicans and the implicate is not indifferent. **1937** D. J. B. HAWKINS *Causality & Implication* 61 This factor consists in the implicans being given in reality, as opposed to merely given in thought. **1953** I. M. COPI *Introd. Logic* viii. 229 The constituent statement between the 'if' and the 'then' is called the *antecedent* (or the implicans). **1963** J. LYONS *Structural Semantics* vii. 189 Sentences of the form Np [etc.]..with their transforms and implicants..are well integrated in the field.

implicate. B. *sb.* **2.** (Later examples.)

1900 [see *extraconscious* (*EXTRA- 1)]. **1921** W. E. JOHNSON *Logic* I. iii. 35 From an implicative, combined with the affirmation of its implications, we may infer the affirmation of its implicate. **1937** D. J. B. HAWKINS *Causality & Implication* 61 What factor..must be present in order that the implicate should be dependent in being on the implicans. **1946** C. MORRIS *Signs, Lang. & Behavior* i. 22 A sign which is more general than another sign,..is an analytic implicate of the other sign. **1971** *Jrnl. Gen. Psychol.* Apr. 222 A theorem deducible only from the conjunction of axioms as their only implicate is, therefore, most probative of the theory that contains them.

implication. Add: **2. c.** *Logic.* A relationship between propositions such that the one implies the other; also, a proposition asserting such a relationship. Also *attrib.*

1906 B. RUSSELL in *Amer. Jrnl. Math.* XXVIII. 202 The subject which comes next in logical order is the theory of *formal* implication. **1922** W. E. JOHNSON *Logic* II. vi. 152 When a formula of implication is used as a premiss in the process of deduction, its implicans must first be formally certified in order that its implicate may be formally certified. **1932** LEWIS & LANGFORD *Symbolic Logic* v. 93 The dot preceding the implication-sign. **1947** H. REICHENBACH *Elem. Symbolic Logic* § 6.24 The expression to the left of the implication sign is called implicans. **1952** P. F. STRAWSON *Introd. Logical Theory* iii. 85 The futility of identifying conditional statements with material implications is obvious. **1954** I. M. COPI *Symbolic Logic* ix. 286 The proposition..may have its implication sign deleted. **1957** A. N. PRIOR *Time & Modality* i. 1 Moh Shaw-Kwei has attempted to lay down the conditions which entitle an operator to be considered as an implication-operator. **1963** W. SELLARS in Castañeda & Nakhnikian *Morality & Lang. of Conduct* 178 In other words, shall-statements, unlike implication statements, are in the object-language. **1968** J. LYONS *Introd. Theoretical Ling.* x. 446 Implication, in the sense in which it has been defined here, is in principle objectively testable.

implicational (impli̇kēi·ʃənăl), *a.* *Logic.* [f. IMPLICATION + -AL.] Of, concerned with, or using implication.

1881 H. MACCOLL in *Phil. Mag.* XI. 40 (*title*) Implicational and equational logic. **1881** J. VENN *Symbolic Logic* xviii. 377 In this case the implicational mode of expression certainly tells its tale more simply and obviously. **1906** B. RUSSELL in *Amer. Jrnl. Math.* XXVIII. 198 Thus both *p* and not-*p* may be replaced, in implicational formulae, by equivalences. *a* **1943** R. G. COLLINGWOOD *Idea of Hist.* (1946) 262 This implicational relation is a compulsive one. **1951** J. ŁUKASIEWICZ *Aristotle's Syllogistic* 22 It is always easy to deduce from an implicational thesis the corresponding rule of inference. **1955** A. N. PRIOR *Formal Logic* 49 Łukasiewicz has shown that a single axiom for the 'implicational calculus' must contain at least 13 letters. **1964** *Language* XL. 264 Most of his universals are 'implicational', of the type 'if a language has a category of gender, it always has a category of number'. **1970** *Ibid.* XLVI. 551 These studies reveal that socially significant linguistic features occur in an implicational series such that the presence of some feature A in the speech of a certain individual means that the speaker will also be found to use features B, C, and D. *Ibid.* 552 Since implicational analysis requires binary decisions..how can such..decisions be made? **1971** *Newslet. Amer. Dial. Soc.* Feb. 16 Implicational analysis..'attempts not to describe a set of speech acts but to model the idealized competence of the persons involved in those speech acts'. **1972** *Computer Jrnl.* XV. 292/2 One can define the set of all strings which are proofs in implicational calculus.

Hence **implica·tionally** *adv.*, in an implicational manner.

1922 W. E. JOHNSON *Logic* II. v. 108 Such a trio of equations are taken to be implicationally independent of one another. **1964** *Current Res. & Devel. Sci. Documentation* XIII. 220 Extensive investigations are also being carried out on procedures of proof in implicationally ordered formal systems.

implicative. Add: **A.** *adj.* (Later examples.) Also, containing at least implications or implications only.

1910 WHITEHEAD & RUSSELL *Principia* I. i. 7 The Implicative Function is a propositional function with two arguments *p* and *q*. **1930** L. S. STEBBING *Mod. Introd. Logic* v. 71 The implicative proposition cannot be simply converted. **1955** A. N. PRIOR *Formal Logic* 61 A purely implicative formula. **1967** *Listener* 5 Oct. 422/3 Substitution of implicative features for defining features—that is, the use of emotive language—is the trade-mark of propaganda. **1971** *Language* XLVII. 341 For reasons that will soon become apparent, I suggest the terms 'implicative' and 'non-implicative' verbs.

Hence **impli·cativeness,** the quality of being implicative.

1932 *Times Lit. Suppl.* 2 June 411/3 Professor Trout's chapters on the systolic and diastolic implicativeness of 'religious rapport' and 'religious dominance'. **1951** J. MILES *Primary Lang. Poetry* 493 Richards allies emotion..to order, and so supports its implicativeness in shape and form and connotation.

implicatively, *adv.* Delete ? *obs.* and add later example.

1953 PARTRIDGE *Shaggy Dog Story* iii. 84 Many years ago *Collier's Weekly* published, implicatively as reminiscence, a 'shaggy dog' that must..rank very high.

implicit, *a.* Add: **2.** *implicit definition* = *contextual definition.*

1959 K. R. POPPER *Logic Sci. Discovery* iii. 72 Sometimes the axioms are described as '*implicit definitions*' of the ideas which they introduce. **1961** E. NAGEL *Struct. of Sci.* v. 95 The fundamental assumptions of the theory provide only implicit definitions for the theoretical notions employed in them. **1973** A. QUINTON *Nature of Things* ix. 279 We can thus define a logical term as one whose meaning is wholly specified by implicit definitions.

impliedly, *adv.* (Later examples.)

1964 *Mod. Law Rev.* XXVII. III. 266 Furthermore, it must be noted that union officials could equally be drawn into a 'conspiracy to intimidate impliedly'. **1970** [see *CORPS 2 b].

implode, *v.* Add: **1.** Also *trans.* and *fig.* Hence **implo·ded, implo·ding** *ppl. adjs.*

1913 J. MURRAY *Ocean* v. 97 Only those parts of the structure would be burst inwards ('imploded') into which water could not enter rapidly enough to equalise the pressure on the two sides, say, of an iron plate. **1963** *Observer* 17 Mar. 3/3 The 'jet' could be in the outer part of the galaxy, and might represent part of the imploding star which has broken away. **1964** M. MCLUHAN *Understanding Media* I. iii. 35 In our present electric age the imploding or contracting energies of our world now clash with the old expansionist and traditional patterns of organization. *Ibid.* v. 51 Individualism is not possible in an electrically patterned and imploded society. **1973** *Sci. Amer.* Mar. 46/2 Applied evenly around the surface, these forces would suddenly implode the pellet to a density 100 times higher than that of lead.

implore, *v.* **1. c.** Delete *rare* and add further examples.

1850 F. E. SMEDLEY *Frank Fairlegh* xv. 136 'Gentlemen, don't ring the bells, pray,' implored the old man. **1887** M. CORELLI *Thelma* III. III. ii. 251 'Let me go with thee!' he implored, in broken accents. **1891** HARDY *Group Noble Dames* 101 'Oh, take it away—please take it away!' she implored.

2. Later examples; const. *for* (a thing), *of* (a person).

1870 H. SMART *Race for Wife* x, She flopped down on her knees, and implored for mercy. **1904** L. T. MEADE *Love Triumphant* ii. 15, I implored of Granny to let us leave the cottage.

implosion. Add: **3.** *fig.* (as the opposite of *explosion*).

1960 J. G. BALLARD in D. Knight *100 Yrs. Sci. Fiction* (1969) 350 The population of Sumatra, for example, has declined by over fifteen per cent in the last twenty years... Do you realize that only two or three decades ago the Neo-Malthusians were talking about a 'world population explosion'? In fact, it's an implosion. **1964** M. MCLUHAN *Understanding Media* I. vii. 71 The rush of students into our universities is not explosion but implosion. *Ibid.* II. x. 92 Our speed-up today is not a slow explosion outward from center to margins but an instant implosion and an interfusion of space and functions.

implosive, *a.* Add: **2.** *fig.* (Cf. prec. word.)

1964 M. MCLUHAN *Understanding Media* II. xi. 111 The implosive (compressional) character of the electric technology. *Ibid.* xix. 185 The implosive speed of the airplane. **1967** *Listener* 8 June 744/1 Television has an implosive effect on a culture.

Hence **implo·siveness.**

1953 C. E. BAZELL *Ling. Form* 42 If all initial occlusives are explosive and all final occlusives are implosive, it is obvious that two distinct conventions (explosiveness of initials and implosiveness of finals) need not be postulated.

impolder (impǫ̈u·ldər), *v.* [ad. Du. *inpolderen*: see IM-¹ and POLDER¹.] *trans.* = *EMPOLDER *v.* Hence **impo·ldering** *vbl. sb.*

1898 D. S. MELDRUM *Holland & Hollanders* 209 Into the sea that it is proposed to impolder, there falls, at Kampen, the river Ysel. **1899** *Pall Mall Gaz.* 13 Apr. 4/2 This impoldering and pumping, the raising and keeping of dykes and dams. **1901** *Speaker* 9 Mar. 633/2 It was about this time that Haarlem Mere was impoldered. **1929** *Encycl. Brit.* XI. 648/1 A great part of the Netherlands has now been impoldered. *Ibid.* 648/2 The largest impoldering scheme on record has now been commenced.

‖ **imponderabilia** (impǫ:ndĕrăbi·liă), *sb. pl.* [neut. pl. of mod.L. *imponderābilis,* things that cannot be weighed.] Imponderables, imponderable factors.

1925 W. J. LOCKE *Great Pandolfo* xxiv. 300 A man beaten not by Fortune, not by the hostility of material influences, but by spiritual imponderabilia almost on the borderland of sanity and unreason. **1933** *Times Lit. Suppl.* 27 Apr. 283/2 The *imponderabilia* that constitute 'Toryism'. **1938** E. POUND *Let.* 12 May (1971) 316, I think Eliot would prefer your emendation. At any rate we are on ground of imponderabilia. **1942** E. WAUGH *Put out More Flags* ii. 111 Every week we have chamber music. There are certain *imponderabilia* at the Old Mill which, to be crude, have their market value. **1957** *Times* 17 Dec. 9/7 Tactically, the chiefs are in a weak position: but nobody can say what psychological *imponderabilia* may not be raked up if the attack is carried further.

imponderable, *a.* and *sb.* Add: **A.** *adj.* **a.** *fig.* (Later example.)

1959 *Manch. Guardian* 29 Jan. 5/5 It is not so much the calculable cost but the possible, imponderable one if things go wrong.

b. (Later example.)

1963 D. W. & E. E. HUMPHRIES tr. *Termier's Erosion & Sedimentation* x. 194 They [*sc.* stratification joints] seem to result from those 'imponderable' particles..which remain in suspension.

B. *sb.* (Later examples.) Now chiefly in sense 'something that cannot be estimated'.

1927 *Sunday Express* 22 May 12/4 It is not always possible to show gratitude by a gift... If we wish to keep ourselves free from the dreadful disease of ingratitude we should strive to remember these imponderables of life. **1938** S. BECKETT *Murphy* 20 Murphy's respect for the imponderables of personality was profound. **1952** M. MCCARTHY *Groves of Academe* (1953) vi. 110 'Fitness to teach' was an imponderable which he had no intention of pretending to weigh. **1960** *Times* 24 Oct. (Financial Rev.) p. xvii/5 Here the largest imponderable is first whether sufficient wool can be produced to meet the demand and secondly the challenge of synthetics. **1963** D. W. & E. E. HUMPHRIES tr. *Termier's Erosion & Sedimentation* xviii. 355 He [*sc.* A. Lombard] distinguishes between the 'imponderables' which are the salts in process of precipitation, ..and the 'ponderables' which sink rather quickly. **1964** M. GOWING *Britain & Atomic Energy* ix. 267 The British contribution was largely made up of imponderables. **1969** *Times* 17 Oct. 10/4 (*heading*) Farm imponderables in calculating cost of entry to Europe.

impoof (impū·f). Also **empofo, impofo, impoofo, impophoo.** [Zulu *i-mpofu,* f. *mpofu* tawny.] The common African eland, *Taurotragus oryx.*

1785 G. FORSTER tr. *Sparrman's Voy. Cape Good Hope* II. xiv. 205 In one of the places above referred to, I have mentioned that it [*sc.* the eland] is called by the Caffres *empofos*; I have since found..that it is likewise called by the same nation *poffo.* **1834** *Penny Cycl.* II. 89/1 The Canna,..improperly called *eland* or elk by the Dutch colonists of South Africa, and *impoof* by the Caffres. **1839** W. C. HARRIS *Wild Sports S. Afr.* x. 83 During the day I killed another impoofo, which actually measured nineteen hands two inches at the shoulder. **1875** *Encycl. Brit.* II. 101/2 The eland or impophoo (*Boselaphus Oreas*) is one of the largest of the antelopes. **1884** *Cassell's Nat. Hist.* III. 21 Writing on the hunting of these creatures, known in South Africa as the *Impoofo*, the same author [*sc.* W. Cornwallis Harris] remarks that, 'notwithstanding the unwieldy shape of these animals, they had at first greatly exceeded the speed of our jaded horses'. **1900** SCLATER & THOMAS *Bk. Antelopes* IV. 198 *Eland* of the Dutch at the Cape... *Impofo* of the Amandabele, Zulu, and Kafirs... *Mpofu* (Swaheli). **1964** E. P. WALKER et al. *Mammals of World* II. 1419 (*heading*) Elands; Elande, Eland Antilope, Impofo, Pofu, Siruwa (native names).

impoon (impū·n). [Zulu *i-mpunzi.*] The grey duiker, *Sylvicapra grimmia,* a common small antelope of southern Africa.

1839 W. C. HARRIS *Wild Sports S. Afr.* 386 The Duiker. Duikerbok of the Cape Colonists. Impoon of the Matabili. **1868** *Chambers's Encycl.* X. 570/1 *Impoon* (*Antilope* or *Cephalopus mergens*),..a small species of antelope, very plentiful in South Africa, in wooded districts. **1895** SCLATER & THOMAS *Bk. Antelopes* I. 205 *Duiker* and *Duiker-bok* of Dutch and English colonists... *Impunzi* or *Impuzi* of Matabili (also of Zulus and Swazis).

import, *sb.* Add: **3. b.** (Later examples.)

1963 A. MAIZELS *Industr. Growth & World Trade* vi. 150 An analysis has been made of the magnitude of the import-substitution that has in fact taken place in the industrial and semi-industrial countries since 1913. **1969** *Times* 13 Jan. 11/2 What many people wanted to know was how much £100 m. of import substitution was worth to the balance of payments. **1971** D. E. WESTLAKE *I gave at the Office* (1972) 121 The import duty you pay now,... The export duty you pay on the way out.

4. b. import–export: *a.*, of a business: engaged in both importing and exporting goods.

1955 G. GREENE *Quiet American* I. iii. 49 'Really? What kind of business?'..'Import, export.' **1965** 'W. HAGGARD' *Hard Sell* ii. 11 He had an import-export business, mostly import.

importance. Add: **2. b.** One who is important; an important person.

1896 A. MORRISON *Child of Jago* ii. 26 The Importances from the platform came to find the tea. **1907** 'MARK TWAIN' *Christian Sci.* II. v. 147 To place the Virgin first, the Saviour second, and Mrs. Eddy third, seems to.. make it an ascending scale of Importances, with Mrs. Eddy ranking the other two and holding first place.

c. An important thing.

1938 R. GRAVES *Coll. Poems* 129 And old importances came swimming back—Wine, meat, log-fires, a roof over the head.

important, *a.* Add: **1. b.** *spec.* Of antiques or the like: very valuable.

1904 H. JAMES *Golden Bowl* II. xiii. 368 She had passed her arm into his, and the other objects in the room, the other pictures, the sofas, the chairs, the tables, the cabinets, the 'important' pieces, supreme in their way, stood out, round them. **1969** *Times* 18 Mar. 18/1 (Advt.), A highly important jewelled bridge. **1973** *Country Life* 15 Nov. 76 An important tortoise shell and ormolu English Bracket Clock..by Robert Hodgkin, London, c. 1720.

4. Preceded by *more* or *most*: used as a kind of sentence adjective. Cf. *IMPORTANTLY *adv.* 1.

This construction is discussed in R. Quirk et al. *Gram. Contemp. Eng.* (1972) 5.26 (p. 255).

1964 N. SPINRAD in D. Knight *100 Yrs. Sci. Fiction* (1969) 270 What were these quasi-stellar objects and perhaps even more important, how were they giving off so much energy? **1965** J. C. DAVIS *Adv. Physical Chem.* x. 449 The carbon atom in its ground state is not completely described by hydrogen-like orbitals. More important, a carbon atom in a molecular configuration hardly resembles a free carbon atom. **1968** R. H. W. BROWN *Gardening Complete* vii. 192 It is a mistake to follow this advice too rigidly. One must wait until the soil is damp enough and, more important, warm enough. **1972** *Physics Bull.* Oct. 577/1 The participants must be fed with a stream of information at the crucial times... Most important of all, the foreign guests must be assured that the hosts will ease all problems of entry into their country. **1972** *Sunday Times* 22 Oct. 15/6 But, most important of all, it is now clear that impounding of income which has been leaked in an unauthorised manner means [etc.]. **1973** *Sci. Amer.* Jan. 7 (Advt.), But most important of all, we begin by giving you the training you need to [etc.]. *Ibid.* 13/1 It can be readily synthesized from coal, oil or natural gas. More important, it can be produced simply by splitting molecules of water. **1973** *Daily Tel.* 13 Jan. 25/8 But, more important, a linked policy can be encashed—surrendered—before maturity date and the savings kept a high proportion of his savings returned to him.

5. *Comb.*, as *important-looking* adj.

1925 F. SCOTT FITZGERALD *Great Gatsby* iii. 45 On a chance we tried an important-looking door. **1926** 'C. BARRY' *Detective's Holiday* 42 Another important-looking person.

importantly, *adv.* **1.** (Later examples.)

Now esp. common as a kind of sentence adverb preceded by *more* or *most*; in some contexts it is interchangeable with *important* and so has the function of a quasi-adj. Cf. *IMPORTANT *a.* 4.

1938 C. WILLIAMS *He came down from Heaven* ii. 22 The main point is..The first outrage against *pietas*, and (more importantly) the first imagined proclamation of *pietas* from the heavens. **1941** *Jrnl. R. Aeronaut. Soc.* XLV. 309 Just as importantly, the chart is of interest to the aircraft designer. **1961** H. WILLIAMSON *Day Shakespeare Died* viii. 88 More importantly, Shakespeare, though using Holinshed as his main source, occasionally used Hall as the direct source of various passages. **1966** M. SPARK *Mandelbaum Gate* xiv. 287 It appeared that Ruth assured Barbara to be someone importantly on her side. **1966** *Listener* 26 May 791/1 Edward Dahlberg's mother is a sizeable part of his literary capital. He dealt with her importantly in two novels. **1969** *Nature* 1 Nov. 477/1 Most importantly, when the particles of the pair are brought together, they annihilate. **1972** *Ibid.* 31 Mar. 200/1 And, most importantly with an internal lipid bilayer, a membrane. **1972** *Times* 12 Apr. 16/5 Perhaps more importantly, income not applied to exclusively charitable purposes is not exempt from taxation. **1972** *Daily Tel.* 31 Oct. 14/6 But, importantly in this case, there is a well-built girl attendant who is chased about the stage by someone bearing a striking resemblance to the will-eyed non-speaking member of the Marx Brothers team. **1972** *Times Higher Educ. Suppl.* 17 Nov., It will of course be recognized as a great modern dictionary, as we shall see presently; but more importantly,..for all of the indications it gives of having registered the full impact of our so-called permissive age, is the way it preserves certain antique myths.

importation. Add: **3.** *Logic.* The inference that if a proposition implies that a second proposition implies a third, then the first and second together imply the third; the converse of *EXPORTATION 4.

1903 B. RUSSELL *Princ. Math.* ii. 16 If *p* implies that *q* implies *r*, then *pq* implies *r*. This is the principle of importation. **1918** C. I. LEWIS *Survey Symbolic Logic* iv. 231 This theorem contains Peano's..Principle of Importation. **1957** P. STRAWSON *Introd. Logic* ii. 23 and it is sometimes known as the law of importation, 'p⊃(q⊃R)]⊃(p & Q⊃R)'. **1966** HUGHES & LONDEY *Elem. Formal Logic* xv. 197 It is known as the Law of Importation.., since its effect is to 'import' the antecedent of the consequent into the antecedent of the whole wff.

Hence **importa·tional** *a.*, of or relating to importation.

1935 *Mind* XLIV. 154 Importational logograms (analogous to definitions) are dispensed with.

importee (impǭɹtī·). [f. IMPORT *v.* + -EE¹.] A person imported from abroad.

1885 CARLYLE *Fredk. Gt.* I. iv. vi. 445 Painter Pesne, a French Immigrant, or Importee,..was sent for. **1888** *Scottish Leader* 19 Sept. 5 It was strange, we thought, that the most recent census, taken in 1946, counted..3,500 East Indians, descendants of indentured importees.

importune, *v.* Add: **5. b.** To solicit for purposes of prostitution.

1847 *Act* 10 & 11 *Vict.* c. 89 § 28 Every Person who.. commits any of the following Offences..may be committed to Prison... Every common Prostitute or Nightwalker loitering and importuning Passengers for the Purpose of Prostitution. **1943** C. E. VULLIAMY *Polderoy Papers* 79 Even the prostitutes no longer 'importune', but hand you politely their cards. **1958** *Times* 17 Dec. 11/4 A severe national law already punishes any man who persistently solicits or importunes for immoral purposes.

importuning, *vbl. sb.* and *ppl. a.* (Later examples.)

1958 *Times* 17 Dec. 11/4 The bill ignores the graver offence of importuning. This is mainly an activity of men. **1969** *Daily Tel.* (Colour Suppl.) 10 Jan. 7 Advertisements from importuning homosexuals.

impos., colloq. abbreviation of IMPOSSIBLE *a.*

1924 GALSWORTHY *White Monkey* I. ix, 'If you're tired we could cut that.' 'My dear! Impos.!'

imposer. Add: **b.** *Printing.* One who imposes (see IMPOSE *v.* I d).

1921 *Dict. Occup. Terms* (1927) § 522 *Stone hand*, imposer; imposes type, which has been set up in page form, together with blocks, etc., in a definite order, and arranges type in correct sequence in chase or iron frame, for printing in sheets.

impossibilism (impǫsi·bĭliz'm). [f. IMPOSSIBLE *a.* + -ISM.] Belief in ideas, especially on social reform, which cannot reasonably be put into practice. So **impossibilist** (impǫsi·bĭlist, -pǫsi·b-) *a.*, of or pertaining to such views; also as *sb.*, one who holds such impracticable views.

1901 G. B. SHAW *Let.* 14 Dec. (1965) 146 We detect.. the..anarchical impossibilism to which our poor lonely secluded geniuses are driven. **1902** — *Let.* 22 Aug. (1965) 362 They had better circulate it among the impossibilists. **1906** *Speaker* 3 Mar. 592/1 Even amongst those who were but recently impossibilist politicians saner views are prevailing. **1906** *Westm. Gaz.* 13 Nov. 2/1 Impossibilist policy is impossibilist politics. **1909** *Daily Chron.* 17 Apr. 16/3 Impossibilists in Labour Politics. **1910** *Daily News* 5 Feb. 4/2 There is no fixed Socialist policy, and the Labour party has been pushed into the impossibilist attitude. **1917** CHESTERTON *Short Hist. Eng.* 76 The most practical politics a crusade, a case for the impossibilist, impossibility. **1921** *Public Opinion* 29 July 102/1 In a world largely controlled by fanatics, dreamers, and impossibilists, the one thoroughly practical policy is that of the League of Nations Union. **1930** *New Statesman* 18 Nov. 712/2 An exposure of Nazi impossibilism.

impossibility. Add: **4.** *Comb.* **impossibility theorem** (see quots.). (Earlier known as the *possibility theorem.*)

[**1950** K. J. ARROW in *Jrnl. Pol. Econ.* LVIII. 342 The above formulation of the problem... If we are to make no prior assumptions are made about the nature of individual orderings, there is no method of voting which will remove the paradox of voting discussed in Part I, neither plurality voting nor any scheme of proportional representation, no matter how complicated.] **1957** LUCE & RAIFFA *Games & Decisions* xiv. 333 (*heading*) Conditions on the social welfare function and Arrow's impossibility theorem. **1960** *Q. Jrnl. Econ.* LXXIV. 509 (*heading*) Proof of the Arrow impossibility theorem. **1967** R. J. ARROW in Laslett & Runciman *Philos., Politics & Society* 3rd Ser. 228 The following general theorem may be stated: There can be no constitution simultaneously satisfying the conditions of Collective Rationality, the Pareto Principle, the Independence of Irrelevant Alternatives, and Non-Dictatorship. This theorem is the Impossibility Theorem. The proof falls into two parts. It is first shown that if an individual is decisive for some pair of alternatives, then he is a dictator, contrary to the condition of Non-Dictatorship. Hence, no individual is decisive for any pair of alternatives, and the Impossibility Theorem itself then follows easily with the aid of the Pareto Principle. **1969** D. BLACK in *Jrnl. Law & Econ.* XII. ii. 227 The Impossibility theorem is based on the general case and apart from restrictions on the members' preferences, no committee procedure may be able to satisfy certain conditions which, Arrow suggests, a procedure might reasonably be required to meet, and that whichever committee procedure we may choose will, for certain sets of schedules, infringe one or more of the apparently reasonable conditions he specifies. **1971** W. LEE *Decision Theory & Human Behavior* iv. 103 Unhappily, reasonable conditions for grouping individual preferences ranging from individual rankings are inconsistent with one another, i.e., in general there may be no social ranking conforming to the desired conditions, a conclusion known as Arrow's impossibility theorem.

impossible, *a.* and *sb.* Add: **B.** *sb.* **2.** With def. article: that which is or seems impossible.

1895 BAILEY *Festus* 266 He only holds Perfections, which are but the impossible To other beings. **1895** CYRUS MARTINEAU-CRABBEN *Liberation of Italy* xix, 290 Garibaldi..had always demanded the impossible of his men. **1904** *Daily Chron.* 6 May 7/5 The history of Germany.. had been a triumph of the impossible. **1916** HUNEKER *Ivory, Apes & Peacocks* 34 All three were consumptives...; all three suffered from the nostalgia of the impossible. **1930** F. NANSEN in *Penguin Dict. Mod. Quots.* (1971) 166/2 The difficult is what takes a little time; the impossible is that which takes a little longer. **1972** *Pacific Nov.* 31 Remember the Festival motto:—be realistic—demand the impossible.

impost, *sb.*² **1.** Add *attrib.* examples.

1850 J. H. PARKER *Gloss. Terms Grecian, Roman, Italian, & Gothic Archit.* (ed. 5) I. 258 It is better..to designate the mouldings as impost mouldings. *c* **1863** *Dict. Archit.* (Archit. Publ. Soc.) III. 15/1 The decorative impost, or point at which the ornamental impost moldings are placed, is frequently below the springing... In some archways the impost point is ornamental with horizontal moldings. **1901** R. STURGIS *Dict. Archit.* II. 464 *Impost block*, a member which gives direct support to one side of an arch, or to the adjoining parts of two arches.

impotency. 2. b. (Later examples.)

1972 *Oxford Times* 28 July 7/1 An Oxford doctor is holding special surgeries for male undergraduates suffering from impotency. *Ibid.* 7/2 Male impotency is very common.

impo·tentizing, *ppl. a. rare*⁻¹. [f. IMPOTENT *a.*: see -IZE, -ING².] That renders one impotent.

1920 JOYCE *Let.* 24 Oct. (1957) I. 149 Moly could also be..absinthe the cerebral impotentising (?!) drink of chastity.

impound, *v.* Add: **b.** *spec.* To confine and store (water) in a reservoir; to confine water so as to form (a reservoir).

1861 [in Dict., sense 2]. **1893** TURNER & BRIGHTMORE *Princ. Waterworks Engin.* ix. 185 The entire site should be closely surveyed at every foot of elevation, in order to determine the extent of the works required to impound the desired quantity of water. **1937** (*Discovery* June 186/2 It [*sc.* the dam] impounds 1,400,000,000 gallons of water. **1959** *Chambers's Encycl.* IV. 355/2 The lake impounded by the Grand Coulee is 150 mls long. **1966** G. M. FAIR et al. *Water & Wastewater Engin.* I. ii. 6 Numerous reservoirs are impounded by throwing dams across the stream valley.

Also **impounding reservoir,** a reservoir whose function is to store sufficient water to ensure an uninterrupted supply in times of relative drought; **impoundage, impoundment**: delete † and add later examples.

1875 *Encycl. Brit.* II. 225/1 With but little artificial addition, Loch Katrine, Loch Venachar, and Loch Drunkie were converted into impounding reservoirs, the first for the supply of the city, and the two latter for compensation. **1889** [in Dict.]. **1893** TURNER & BRIGHTMORE *Princ. Waterworks Engin.* ii. 15 The first-mentioned requirement is satisfied by the formation of 'impounding-reservoirs', the office of which is to gather the irregular natural yield of surface-water, in order that it shall be supplied at a uniform rate; the second is met by the construction of 'service-reservoirs', tanks and cisterns, from which water is distributed as required for the hourly consumption of the public. **1954** FAIR & GEYER *Water Supply & Waste-Water Disposal* viii. 188 In the absence of adequate natural storage, engineers resort to the construction of impounding reservoirs or, more rarely, to the excavation of storage basins. *Ibid.* 194 Allowances for evaporation from the water surface that is created by the impoundage. *Ibid.* 195 When more than one reservoir is developed on a stream, the overflow from each impoundage becomes available to the reservoir next below. **1957** *Encycl. Brit.* XXIII. 432/2 Virtually all public regulatory authorities require that the impoundment be sufficiently great to provide for the release of compensation water for downstream users. **1973** *New Yorker* 28 Apr. 29/3 There is some room for impoundment, but not to the tune of twelve billion dollars.

impracticably, *adv.* (Later examples.)

1966 *McGraw-Hill Encycl. Sci. & Technol.* X. 234/2 The tube would have to be made impracticably long to avoid cooling of the plasma by the ends. **1974** *Daily Tel.* 11 Mar. 9/2 These [proposals] are now generally recognised as impracticably complicated.

impractical, *a.* Delete *rare* and add further examples. Also = IMPRACTICABLE *a.*

1925 T. DREISER *Amer. Trag.* (1926) I. I. ii. 11 For Clyde's parents had proved impractical in the matter of the future of their children. **1930** *Amer. Speech* IV. 331 'Impractical'..is more commonly used than either 'unpractical', or 'impracticable', as it seems to squint and have both meanings. **1933** J. T. ADAMS *Epic of Amer.* iv. 112 So impractical was American cotton culture considered that..England seized eight bales. **1947** F. W. FELLER *Instrument & Control Manual* vi. 171 The number of units to be controlled in a single plant all tend to render hand control impractical if not impossible. **1962** E. GODFREY *Retail Selling & Organiz.* ii. 21 On a busy ground floor, this would be impractical. **1964** K. H. ROBINS *Gen. Ling.* i. 2 As an impractical ideal he [*sc.* the general linguist] would know something about every language. **1970** *Daily Tel.* 4 May 9/7 The scheme was reckoned to be very best to transferable pensions, which the Government believes to be impractical at the moment. **1973** *Sci. Amer.* Mar. 113/2 The second calculating method..is too complicated and impractical to explain here.

impracticality (impræktikæ·lĭti). [f. IMPRACTICAL *a.* + -ITY.] The character of being impractical; impracticableness.

1916 H. S. WALPOLE *Dark Forest* I. iv, The Russian character..with its lack of restraint, its idealism, its impracticality. **1926** J. BUCHAN *Dancing Floor* II. x. 187 The impracticality of an entrance..at that point.

So **impra·cticalness.**

1905 *Macm. Mag.* Nov. 55 He was accused of vagueness, impracticalness, generality.

impractically (impræ·ktikăli), *adv.* [f. IMPRACTICAL *a.* + -LY².] In an impractical manner; to an impractical or impracticable degree; not practicably.

1947 *Math. Tables & Other Aids to Computation* II. 359 Long sequences of accurate computations which would be impractically lengthy by any other slower means. **1959** E. M. MCCORMICK *Digital Computer Primer* viii. 115 It is generally impractically expensive to obtain the very large storage. **1974** *Sci. Amer.* Jan. 93/1 The forces toward the left between these lines are impractically high; normal playing is confined to the area toward the right.

imprecise, *a.* Add: So **impreci·seness.**

1907 *Athenæum* 9 Mar. 282/3 He [*sc.* Henry James] must..deck it with the most elaborated precisions of impreciseness. **1943** A. L. ROWSE *Spirit Eng. Hist.* i. 11 An impreciseness which characterises the English mind.

imprecision. Delete *rare* and add later examples.

1954 W. STEVENS *Coll. Poems* 353 The romance of the precise is not the elision Of the tired romance of imprecision. **1971** *Nature* 4 June 274/1 For one thing, it pointed to strictly legal imprecisions in the legislation which at present regulates the use of transplants in Britain.

impredicable, *a.* Add: (Later examples.) Hence **impre:dicabi·lity,** the condition of being impredicable.

1906 P. LOWELL *Mars & its Canals* viii. 95 Even on Mars nothing in the way of weather is absolutely predicable but impredicability. **1937** A. SMEATON tr. *Carnap's Logical Syntax of Lang.* iv. § 60 The so-called 'impredicable' concept is that of the property of not being applicable to oneself. It has been thought that if one is to be able to apply one's properties to themselves, we must be prepared for contradictions to arise concerning the property of 'impredicable'. *Ibid.* IV. § 60a. 212 Hence a definition of the form given for 'impredicable' is obviously impossible. **1965** F. SOMMERS in M. Black *Philos. in Amer.* 272 For example, the term *clean* is impredicable of the equator. *Ibid.* 273 It becomes unnecessary to introduce special 'type' restrictions to account for impredicability.

impredicative (impre·dikǎtiv), *a.* [f. IM-² + PREDICATIVE *a.*] Of a proposition, thing, etc.: not definable except in terms of a totality of which it is itself a part. Hence **impre:dicatively** *adv.*; **impredicati·vity,** the state or quality of being impredicative.

1937 A. SMEATON tr. *Carnap's Logical Syntax of Lang.* iv. § 44. 162 A thing is usually called impredicative (in the material mode of speech) when it is defined (or can only be defined) with the help of a totality to which it itself belongs. **1944** K. GÖDEL in P. A. Schilpp *Philos. B. Russell* ii. 136 What an impredicative definition really requires is to construct a notion by a combination of a set of notions to which the notion to be formed itself belongs. **1963** W. V. QUINE *Set Theory* § 34. 242 He [*sc.* Poincaré] called the suspect procedure impredicative. **1965** C. D. PARSONS in M. Black *Philos. in Amer.* 196 Such extensions..allow impredicatively defined classes. *Ibid.* 197 This can only be because the mathematics itself involves impredicativity.

impreg (i·mpreg). Also **Impreg.** [Abbrev. of *IMPREG(NATED *ppl. a.* 2.] A type of wood impregnated with a synthetic resin to improve its dimensional stability and resistance to distortion or decay.

1942 *Fortune* Oct. 160/2 F[orest] P[roducts] Laboratory makes a special 'impreg' by treating wood with raw resin so that the resin actually penetrates the wood cells. **1953** HUNT & GARRATT *Wood Preservation* (ed. 2) xi. 365 The shrinkage and swelling of wood..can be reduced..by first impregnating the wood (in thin sheets) with a solution of unpolymerized phenol-formaldehyde resin, then drying the wood at moderate temperatures, and finally heating the dry wood at about 300°C to polymerize the resin in place; the product formed is called Impreg. **1953** STAMM & HARRIS *Chem. Processing of Wood* (1954) viii. 226 Impreg was manufactured during World War II only for military uses. **1967** R. H. FARMER *Chem. in Utilization of Wood* viii. 119 Uncompressed resin-impregnated wood was first made in the United States, where it is called Impreg.

impregnability. (Earlier example.)

1847 W. SMITH tr. *Fichte's Characteristics Present Age* 30 The supposed impregnability of the mode of thought which we have now described arises precisely in this way.

impregnant, *a.*¹ (*sb.*). Restrict 'Now *rare*' to sense in Dict. and add: **B.** *sb.* A substance used for the impregnation of something else. (Cf. sense 2 in Dict.)

1933 H. BENNETT *Chem. Formulary* I. 441/1 Leather Soles, Impregnant for. **1948** KIRK & OTHMER *Encycl. Chem. Technol.* II. 546 The bitumens employed as felt impregnants are usually soft, low-melting materials of high fluidity in the molten condition. **1955** SEYMOUR & STEINER *Plastics for Corrosion-Resistant Applications* xii. 191 The sealing effect of the impregnant..is dependent upon an increase in molecular size of the resin used. **1969** *New Scientist* 30 July 223/3 Inside present day automotive storage battery cells..a waffle-shaped network of lead grids..is impregnated with a material..The purpose of the impregnant is that it keeps the battery alive.

impregnated, *ppl. a.* Add: **2.** *spec.* **impregnated wood,** (*a*) wood saturated with a preservative; (*b*) = *IMPREG.

1942 *Amer. Jrnl. Bot.* XXIX. 552/1 A brown mold has been observed..on small pieces of impregnated wood. **1944** *Fortune* Oct. 180/2 It [*sc.* 'compreg'] consists of layers of resin-impregnated wood that have been compressed together. **1944** *Amer. Speech* XIX. 92 Plywood manufacturers..developed a technique by which the glue..is forced through the entire texture of all the plies in a laminated sheet, and 'impregnated wood' is the special name of the resultant product. **1951** MACTAGGART & CHAMBERS *Plastics & Building* x. 114 This phenomenon [*sc.* impregnation of wood with synthetic resin] has resulted in the development of two types of impregnated wood called generally Impreg and Compreg. **1963** A. D. WOOD *Plywoods of World* iv. 141 Improved and impregnated wood is more difficult to work than solid timber.

impregnation. Add: **2.** *spec.* The saturation of wood with a preservative.

1872 *Jrnl. Chem. Soc.* XXV. 186 It may be stated that the impregnation of wood with sulphate of copper, or with creosote oils, or their vapours, is of service in rendering the wood three or four times as lasting as unprepared wood. **1927** *Chem. Abstr.* XXI. 3887 The process of impregnation was to be universally adopted, the danger of the exhaustion of the world's supply would be deferred. **1946** CARTWRIGHT & FINDLAY *Decay of Timber* xiii. 298 Impregnation treatments are of two kinds. **1953** J. H. BLAKE *Seasoning & Preservation of Timber* 18 Impregnation, strictly the saturation of wood with a preservative... Generally used to describe treatments giving a high loading of preservative in the wood, e.g. pressure treatments.

impresa. Restrict † *Obs.* to sense 2 in Dict. and add later examples of sense 1.

1865 F. B. PALLISER *Hist. Lace* 435 Then follow three pages in *intaglio* ... the impresa of the printer, a lion rampant, holding a sword in his fore paws. **1971** *English Studies* LII. 122 The last *impresa* in the supplementary chapter is almost certainly a personal device of Daniel's dedicatee, Sir Edward Dymoke.

impress, *v.*¹ Add: **2. c.** *Electr.* To apply or establish (an e.m.f. or a potential difference) by some external means.

1881 [implied in next]. **1918** *Wireless World* VI. 145 A certain steady voltage is impressed on the grids. **1930** [see *INSULATE *v.* 2]. **1948** A. L. ALBERT *Radio Fund.* v. 128 If a voltage is impressed across the input terminals.. the voltage will force a current into the line.

impressed, *ppl. a.*¹ Add: **2.** *Electr.* Of an e.m.f. or potential difference: applied by some external means.

1881 *Jrnl. Soc. Teleg. Engin.* X. 271 Let *M* be the induction through the coil when its plane is at right angles to the lines of force of the external field, *a* the angle turned through from this plane at time *t*...; then *M*ω sin *a*t is the impressed E.M.F. in the coil. **1967** *Jrnl. Soc. Teleg. Engin.* XXII. 408/1 A small current flows into the excited winding sufficient to produce a counter voltage equal to the impressed voltage. **1948** [see *IMPEDANCE 1 a]. **1973** *Sci. Amer.* Oct. 125/1 Leakage through the capacitor should not exceed 10^{-12} ampere at an impressed potential of 10 volts.

impression, *sb.* Add: **2. e.** *Dentistry.* A negative copy of the teeth or oral cavity (from which a positive cast or model may be made) formed by bringing them into intimate contact with some substance that will take their shape.

1839 C. A. HARRIS *Dental Art* xxi. 350 Models of this kind are obtained by taking a wax impression of both jaws at the same time. **1878** C. HUNTER *Mech. Dentistry* i. 7 When the composition has become sufficiently hard, the impression is withdrawn from the mouth, and cold water should be allowed to flow over it. **1940** J. OSBORNE *Dental Mech.* i. 1 The introduction..to the subject is the technique necessary for the accurate construction of a model, or positive likeness of the patient's mouth, from an impression or negative likeness.

3. d. *Bibliogr.* In bibliographical classification and description, a subdivision of an edition, denoting all the copies printed at one time; chiefly applicable to books of the nineteenth and twentieth centuries.

1927 R. B. MCKERROW *Introd. Bibliogr.* II. iii. 175 When dealing with modern books, 'edition' and 'impression' as a rule are the same thing, for the early printer normally distributed his type immediately it had been printed from, though there were..exceptions to this. **1949** F. BOWERS *Princ. Bibliogr. Descr.* xi. 379 In its purest sense an *edition* of a book consists of all copies printed at any time or times from one setting of type, or its equivalent in the form of plates or stereotype rolls; i.e., it is the sum of all impressions from one setting... All the copies of any single edition are not necessarily printed at any one time but may accumulate from a series of separate impressions removed from each other in date... Copies of each impression compose a part of an edition. **1972** P. GASKELL *New Introd. Bibliogr.* 313 It was not unusual in the nineteenth century for stereos to be used for ten successive impressions, and for electros to be used for as many as thirty; while, if a set of plates was kept as a 'mother' from which further sets could be made, the number of successive impressions of an edition that could be printed from plates was virtually unlimited.

6. b. Esp. in phr. *first impression(s).*

1700 CONGREVE *Way of World* iv. i. 52 How shall I receive him? In what figure shall I give his Heart the first Impression? There is a great deal in the first Impression. **1753** H. JACKSON *Hist. Rev. Pennsylvania* (1759) 270 It must have been while he was under the first Impressions given him by the Governor to our Disadvantage. **1843** DICKENS *Mart. Chuz.* (1944) v. 57 First impressions, you know, often go a long way, and last a long time. **1847** [in Dict.]. **1924** E. O'NEILL *Welded* 1. 90 The first impression of her whole personality is one of charm, partly innate, partly imposed by years of self-discipline.

d. An imitation or impersonation of a person or thing, done by a comedian as a form of entertainment.

1953 J. LAURIE *Vaudeville* 99 Some [beginners] just stuck to the regular 'impressions' and went through show biz getting by. **1969** *Times* 7 Nov. 13/3 An American entertainer..joked, sang, went on singing, and did impressions. **1971** D. NATHAN *Laughtermakers* ii. 46 Peter would come in and do a few impressions of Kenneth Horne and others. *Ibid.* xiii. 227 Later on I'm going to do one or two impressions—I've got some good bird impressions, I eat ...

7. In modern use, often implying that the belief or idea is mistaken, esp. in *under the impression.*

1860 RUSKIN *Unto this Last* (1862) iv. 131, I believe that many of our merchants are seriously under the impression that it is possible for everybody, somehow, to make a profit in this manner. **1865** — *Sesame* i. § 86 Generally, we are under an impression that a man's duties are public, and a woman's private. **1949** W. F. HOOK *Lives Abps.* V. xxi. 356 Under the impression that they had there seen the imperial eagles.

9. *Comb.*: **impression compound,** any impression material manufactured from a number of different ingredients, esp. one that is a non-elastic thermoplastic solid; **impression material,** any substance used for taking dental impressions.

1903 *Dental Rec.* XXIII. 415 Do not think a good impression of a full denture cannot be taken in impression compound. **1904** J. H. PROTHERO *Prosthetic Dentistry* iii. 22 Other impression materials..are furnished by the dental supply houses and are called impression compounds. The usual claim made is that they are composed of materials that can be dried after the impression is taken. **1934** F. W. FRAHM *Princ. & Technics Full Denture Construction* iv. 42 In recent years..many substances have been added to our list of materials and is listed under the trade name of 'Dentocoll'. It is a hydro-colloidal, possessing unusual plasticity, some elasticity and a slight compressibility. **1965** PHILLIPS & SKINNER *Elem. Dental Materials* v. 58 The formulas of the modern impression compounds are 'trade secrets', any discussion of composition must be very general. In general, compounds are a mixture of waxes, thermoplastic resins, a filler, and a coloring agent. **1976** C. HUNTER *Mech. Dentistry* I. 2 Wax as an impression material is now seldom used, composition (Godiva, or Stent) or plaster of Paris being now almost invariably employed. **1965** PHILLIPS & SKINNER *Elem. Dental Materials* iv. 33 If a rigid impression material has been used (i.e., plaster, compound, etc.), the mix of dental stone is poured into the impression carefully, preferably under vibration.

impressional, *a.* Delete *rare* and add later examples.

1920 S. ALEXANDER *Space, Time & Deity* II. 158 Impressional intensity, as Mr. Stout calls what Hume described as vivacity. **1964** K. & E. COOMARASWAMY *Refs.* & Engin. Council *Graphic Arts Terminol.* Ibid. The principal advantages of the Mallander factories are stated to be..the excellent control of plate inking and dampening and impressional squeeze in order to obtain the highest quality image with little..stock waste.

impressionism. Add: **4.** *Mus.* A style of composition, originating in the late 1880s with Debussy, characterized by its harmonic system, esp. in the use by Debussy of the whole-tone scale, and departing from the strong and direct structure and themes of the Romantic composers. Also used for a type of jazz with similar 'atmospheric' characteristics.

1889 G. B. SHAW *London Music 1888–89* (1937) 128 There is a great deal in Mefistofele that is mere impressionism; and the impressionism is fluttering as it is enchanting when it is successful. **1908** W. H. DALY *Debussy* 31 So far as one can classify Debussy's use of a 'programme' at all, it is necessary to interpret the term in conjunction with painting, and style it impressionism. **1922** *Musical Opinion* May 691/1 Let it be thoroughly understood that impressionism is no more a matter of technique, of using dots, blobs and squares..any more than musical impressionism means the use of the whole-tone scale with its implied harmonies. **1934** C. LAMBERT *Music Ho!* I. 35 Impressionism, as I have said, is a term easily misused, and one may doubt the logic of its use as a musical term at all; but in association with the work of Debussy and his followers is so widespread that one may conveniently use it as a generic label for that period of disruption in music which Debussy was the dominating figure. **1947** A. EINSTEIN *Music Romantic Era* xvi. 69 The first symphonic poem *Les Éolides* (1876), inspired by Leconte de Lisle's poem, is a perfect example of impressionism, long before Debussy had invented the term. **1958** B. ULANOV *Hist. Jazz in Amer.* (1958) xii. 133 It was another sixteen years before the impact of Impressionism was again to be felt in jazz. **1958** *N. Stearns Story of Jazz* (1957) xii. 146 The Debussyesque impressionism of Smith's 'Morning Air'. **1969** *Times* 27 Feb. 13/1 Mr. Fox..was keenly aware that impressionism for Debussy was not just a matter of vagueness.

impressionist. Add: (Earlier and later examples.) Also *transf.*, an impressionistic painting. Also *transf.*

1876 H. JAMES *Parisian Sk.* (1958) 131 An exhibition for which I may at least claim that it can give rise..to two dangerous perversities of taste is that of the little group of the Irreconcilables—otherwise known as the 'Impressionists' in painting. **1933** *Burlington Mag.* Dec. 276/2 The Impressionist movement in England will never count for much. **1958** L. DURRELL *Mountolive* xi. 208 They would take a slow turn up and down the picture-gallery, with its splendid collection of Impressionists.

attrib. and as *adj.* **1876** H. JAMES *Parisian Sk.* (1958) 131 The 'impressionist' doctrines strike me as incompatible, in an artist's mind, with the existence of first-rate talent. **1893** R. FRY *Let.* 20 Sept. (1972) I. 154, I don't quite

make out whether Elsie Howard is Impressionist; I suppose so, but I thought she was Ruskinian. **1894** G. MEREDITH *Let.* 5 July (1970) III. 1163 Beware of a hurried habit of mind that comes of addiction to Impressionist effects. **1969** *Listener* 28 Aug. 274/3 A letter from the aged Earl Attlee in his own impressionist typewriting, full of splendidly sinister words like 'Ghsmberlian' and 'recoll3ftian'.

2. *Mus.* A composer of impressionistic music. Also *attrib.*

1908 W. H. DALY *Debussy* 10 It is convenient, even if it is an incomplete definition of his altogether novel attitude towards music, to describe Debussy as an impressionist. He is something more than an impressionist as the term is commonly understood, although in his work there is not a little which recalls the methods and the points of view of the masters of impressionist painting. **1927** [see *BITONALITY]. **1947** C. GRAY *Contingencies & Other Ess.* iii. 91 The so-called impressionists were anticipated by him [*sc.* Liszt] in many of their most characteristic effects and procedures, sometimes by as much as half a century. **1948** *Penguin Music Mag.* Oct. 46 The half-Tristanesque, half-impressionist Nocturne. **1952** B. ULANOV *Hist. Jazz in Amer.* (1958) xii. 132 The French Impressionist composers and their American disciples and imitators made a great impression upon Bix's generation of jazz musicians. **1955** R. BLESH *Shining Trumpets* (ed. 3) xii. 281 His borrowed effects from jazz, the Impressionists, and the French Romantics. **1959** D. COOKE *Lang. Mus.* i. 2 Medieval music was largely architectural in conception: the romantics were much concerned with the literary; the impressionists with the pictorial; modern music has swung back again to the architectural.

impressionistic, *a.* Add: (Further examples.) Also, in a general sense: subjective, unsystematic (formed directly from IMPRESSION *sb.* 7 and only indirectly influenced by senses of *impressionism, -ist*).

1891 G. MEREDITH *Let.* 15 Apr. (1970) II. 1025 You have at times..insisted on your impressions. That is, you have put on your cap, sharpened your pencil, and gone afield as the Impressionistic poet. **1900** *Atlantic Monthly* LXXXVI. 78 As for the impressionistic writer about literature—he is apt to concern himself very little with this historical origin of a work of art. **1908** W. H. DALY *Debussy* 32 Debussy's music..deals..in suggestions... There is something swift, vague, and elusive, but strangely vivid and satisfying... But the broad, impressionistic methods are not taken to avoid difficulties of definition. **1909** W. JAMES *Pluralistic Universe* ii. 52 Impressionistic philosophizing, like impressionistic watch-making or land-surveying, is intolerable to experts. **1915** J. HUNEKER *Ivory, Apes & Peacocks* 38 Laforgue..was an ardent advocate of the Impressionistic painters. **1921** *Times Lit. Suppl.* 24 Feb. 114/1 The impressionistic reporter who asked Mr. Edison what he considered the chief mark of a truly valuable invention. **1933** *PMLA* XLVIII. 598 Such rather impressionistic terminology has been found suggestive and helpful in teaching. **1934** S. R. NELSON *All about Jazz* v. 101 All these are impressionistic music of the programme type. **1947** A. EINSTEIN *Music Romantic Era* vii. 69 Berlioz..made the two middle movements nothing but picturesque scenes—most finely impressionistic, most genuinely akin to French *plein air* painting. **1955** R. BLESH *Shining Trumpets* (ed. 3) xii. 268 An impressionistic montage of solo moods. **1958** R. A. BONE *Negro Novel in Amer.* ii. 68 The style is appropriately impressionistic, full of hyphenated adjectives aimed at vivid impressions of Harlem life. **1962** *Listener* 25 Jan. 195/3 Delius's impressionistic elusiveness. *Ibid.* 196/1 Its impressionistic string, harp and horn texture. **1973** *College English* XXXIV. 1103 Prosodic studies have tended toward tentative, impressionistic assertions and *ad hoc* methodology.

2. *Phonetics.* Non-systematic, subjective, non-structured; determined by the recorder's impressions of speech sounds, not by the sound-system of the language, dialect, etc., being recorded.

1939 H. KURATH *Handbk. Ling. Geogr. New England* iv. 122 The field workers' phonetic notations..are not phonemic, but on the contrary intentionally phonic, that is *impressionistic.* **1940** *Amer. Speech* XV. 145 This paper makes no claim to presenting a detailed impressionistic analysis of PM [i.e. Piedmont, U.S.] phonetics; it attempts merely to indicate a relative phonemic distribution. **1948** R. A. HALL *Leave Your Lang. Alone!* II. vi. 41 Trying to describe sounds in auditory, impressionistic terms is likely to give about as accurate results as would, say, describing chemical elements in terms of their smells. The impressions we get through our senses of hearing and smell just can't be stated in clear and analyzable enough terms to be of any use in scientific work. **1960** D. JONES *Outl. Eng. Phonetics* (ed. 9) 349 Systematic transcriptions have to be distinguished from transcriptions made on a general phonetic basis without reference to the needs of any particular language. The latter may be described as 'non-systematic' or 'impressionistic'. **1962** H. ORTON *Survey Eng. Dial., Introd.* i. 18 The merits and demerits of the impressionistic method need not be re-stated here. **1963** *Amer. Speech* XXXVIII. 127 The recording of the speech sounds is phonic (impressionistic) in intent. All premature phonemicization is strictly avoided.

impressionistically (impreʃəni·stikăli), *adv.* [f. IMPRESSIONISTIC *a.*: see -ICALLY.] In an impressionistic manner (in various senses); from the point of view of an impressionist.

1909 W. JAMES *Pluralistic Universe* iii. 92, I make no claim to understanding it, I treat it merely impressionistically. **1924** GALSWORTHY *White Monkey* II. ii, When you smile, Miss Collins, I see you impressionistically. **1924** J. A. HAMMERTON *Countries of World* III. 1507/1 Their flowers..painted (impressionistically but with unerring truth of impression) in their frescoes. **1926** W. J. LOCKE *Old Bridge* I. i, I try to express myself..impressionistically. **1960** H. ORTON in *Orbis* IX. 337 The fieldworkers transcribed the informants' responses phonetically, and impressionistically, in the phonetic alphabet of the International Phonetic Association. **1964** F. BOWERS *Bibliogr. & Textual Crit.* II. iii. 51 The hypothesis is factually, not impressionistically, based.

impressionize (impre·ʃənaiz), *v. rare.* [f. IMPRESSION *sb.* + -IZE.] **a.** *trans.* To make an impression on; to introduce impressions into, to portray as an impression or set of impressions. **b.** *intr.* To gather impressions.

1894 F. M. ELLIOT *Roman Gossip* i. 15 He had.. imagined this *mise en scène* to impressionise his fellow-citizens. **1894** H. JAMES *Notebks.* (1947) 160, I must picture it [*sc.* a story to be written], summarize it, impressionize it, in a word—compress and confine it by making it the picture of what I see. **1905** D. SLADEN *Playing Game* I. i. 16 Instead of thinking Japan a God-forsaken country..he was impressionizing in an indolent æsthetic way.

impressive, *a.* Add: **4.** *Comb.*, as *impressive-looking* adj.

1904 *Daily Chron.* 21 July 4/5 Even in these impressive-looking statistical tables little bits of cheerfulness obstinately obtrude themselves. **1925** T. DREISER *Amer. Trag.* (1926) II. xl. 10 Arabella Stark..in a large and impressive-looking car, was waiting.

imprest, *sb.*¹ **1. b.** Delete † *Obs.* and add later examples.

1957 F. KING *Man on Rock* iii. 81 He drew up a log-book for the car, checked the postage imprest. **1958** E. A. ROBERTSON *Justice of Heart* iv. 48 Off you go, see the foreign editor, get an imprest—don't imagine you can squander money on this trip, though!

‖ **imprévu** (æ̃nprevü). Also **imprevu.** [Fr., f. IM-² + *prévu*, pa. pple. of *prévoir* to foresee.] The unexpected, the unforeseen.

1854 *Punch* 9 Sept. 106/2 Had he well read the people.., their love of *l'imprévu*? **1858** GEO. ELIOT *Let.* 2 Mar. in J. W. Cross *George Eliot's Life* (1885) II. viii. 13 Perhaps we may go to Dresden, perhaps not: we leave room for the *imprévu.* **1925** G. B. SHAW *Let.* 15 Dec. in *To a Young Actress* (1960) 88, I am prudent and foresee everything. My life lacks the imprevu. **1969** R. HARPER *World of Thriller* II. 65 Stendhal..understood that without the '*imprévu*' there is no charm in human existence.

imprimatur. Add: **2.** (Later examples.)

1955 *Bull. Atomic Sci.* June 209/2 His calculations and conclusions bear no imprimatur of the Atomic Energy Commission. **1973** *N.Y. Law Jrnl.* 31 Aug. 3/1 It is well-settled contract law that courts do not give their imprimatur to such arrangements.

¶ Used confusedly = IMPRINT *sb.* 3. (Quot. 1971 is *fig.*)

1970 *Daily Tel.* 7 May 13/2 The agent, not the candidate, is the one liable to fines..if he..issues one word of election literature without his own and the printer's imprimatur on it. **1971** *Nature* 7 May 40/2 The site near Kültepe has been identified as the Assyrian karum Kanis and most of the tablets can thus be considered to bear the implicit imprimatur 'found at Kanis'.

‖ **imprimatura** (imprīmatū·ra). Also **imprimitura.** [ad. It. *imprimitura.*] A thin priming or ground, frequently coloured, applied to an artist's canvas or panel.

1951 R. MAYER *Artist's Handbk.* 434 *Imprimatura,* a veil or thin glaze of colour applied to a ground as a preliminary coating. Term not in very wide use. **1958** M. L. WOLF *Dict. Painting* 137 *Imprimatura,* in painting, a type of glaze applied as a toner to the canvas or other ground..an underpainting, intended to relieve the monotony of a white or gray background. **1967** J. N. BARRON *Lang. of Painting* 99 *Imprimatura,* a thin glaze or veil of color brushed or rubbed over the white of a ground... It may also be used at the same time to reduce, if necessary, the absorbency of the ground.

imprint, *sb.* **3. b.** (Further examples of extended uses.)

1972 *Times Lit. Suppl.* 27 Oct. 1276/5 (Advt.), Kahn & Averill (imprint of Stanmore Press Ltd.). **1973** *Ibid.* 2 Feb. 121/4 (Advt.), Diana Burfield who, before her resignation last summer after seventeen years with the imprint, was editorial director of Tavistock [Publications].

4. (With capital initial.) An old-style type face, named after the periodical for which it was designed.

1913 *Imprint* Jan. p. vi, The newly designed type in which our pages are presented to the reader was cut by the Lanston Monotype Company at our instance... The type has been christened Imprint Old Face. **1934** A. F. JOHNSON *Type Designs* iv. 120 The first acceptable book type to be cut after the Old Style of 1860 was the 'Monotype' Imprint of 1913. **1966** P. M. HANDOVER in K. Day *Bk. Typogr. 1815–1965* 160 Imprint was a reformed Caslon, regularised on the principles that had inspired old style which now resulted in another more distinguished face.

imprint, *v.* Add: **5.** *Animal Behaviour.* To bring about in (a social animal, usu. a young one) a state of habitual recognition of or trust in another animal or an object, which may thus come to be regarded as a parent; const. *to* or *on* the object of recognition. Also, of an animal or thing: to become established as an object of recognition or trust in the behaviour pattern of (a young animal) (quot. 1967): see *IMPRINTING *vbl. sb.* 2. Usu. as *pa. pple.* **imprinted.**

1951 *Ibis* XCIII. 259 Young Partridges which had been caught after the fields had been mowed but were still only a few hours old had nevertheless already become definitely imprinted to their normal parents. **1956** W. H. THORPE *Learning & Instinct in Animals* vi. 116 The parent Cichlid fish may become imprinted to the young as well as the young to the parent. **1963** *Ibid.* (ed. 2) xv. 414 If young birds are kept together in groups they are harder to imprint than if they are kept singly. **1966** R. & D. MORRIS *Men & Apes* iii. 57 Monkeys imprinted on human beings as babies will readily accept them as sexual partners on reaching maturity. **1967** M. ARGYLE *Psychol. Interpersonal Behaviour* i. 23 Dogs raised by humans, may be imprinted by them, and human babies who are reared by wolves may fail to be imprinted by humans. **1972** *Nature* 2 June 287/2 Many hand-reared birds become imprinted on their human handlers if isolated from their parents at an early age. **1972** *Sci. Amer.* Aug. 25/3 We also took wild ducklings from their natural mother 16 hours after hatching and tried to imprint them to humans.

i·mprinted, *ppl. a.*² [f. IMPRINT *sb.* + -ED².] Bearing a bookseller's own imprint (IMPRINT *sb.* 3).

1926 *Publishers' Weekly* 17 July 181/2 A month before publication 300,000 imprinted postcards had been asked for. **1927** *Ibid.* 25 June 2371 Imprinted brochures are now ready for distribution.

imprinting, *vbl. sb.* Add: **2.** *Animal Behaviour.* The establishment of a behaviour pattern of recognition and trust, usu. directed at its own species, during a critical period of susceptibility in a (young) social animal, esp. in birds.

1937 K. LORENZ in *Auk* LIV. 262 This process of acquiring the biologically 'right' object of social reactions by conditioning them, not to one individual fellow-member of the species, but to the species as such, is so very peculiar that I have thought it necessary to use a particular word to describe it. I have called it 'Prägung' in German, which I propose to translate into English by the term 'imprinting'. **1953** J. S. HUXLEY *Evolution in Action* iv. 102 Young geese which have been hatched in an incubator will attach themselves to birds of other species or even to human beings and follow them about as if they were their real parents. This so-called 'imprinting' has to take place during a critical period soon after hatching, only takes a minute or so, and is then irreversible. **1967** M. ARGYLE *Psychol. Interpersonal Behaviour* i. 17 There may be 'imprinting' during the first year of life: the infant becomes attached to the dominant moving object in its environment, and does its best to follow that object. **1970** *Primate Behavior* I. 130 Imprinting data suggest that in birds early experiences before or during a critical period can influence social attachment in a relatively permanent fashion.

imprisonable, *a.* Add: Of an offence, etc.: for which a person can be imprisoned.

1971 *Daily Tel.* 31 July 11/2 Before a court recommends deportation of a Commonwealth citizen who has been convicted of an imprisonable offence, the offender must be given at least seven days clear notice.

improbable, *a.* Add: **2.** Also, that does not 'look the part'.

1958 *Times* 18 Apr. 11/7 An immense arched building of blue painted woods decorated with an improbable metal dove. *Ibid.* 14 May 15/1 As if it were miraculous that this gentle and improbable individual should exist at all.

improperium (imprǫ̆pīᵊ·riŏm). Pl. **-ia.** *R.C. Liturg.* [late L., = reproach.] *pl.* A series of antiphons with responses forming part of the liturgical service of Good Friday, expressing the reproach or sorrowful remonstrance of Christ with the Jewish people; *sing.* one of these antiphons with its response.

1880 GROVE *Dict. Mus.* II. 1/1 The *Improperia* are sung, very softly, and without any accompaniment whatever, by two Antiphonal Choirs. **1884** ADDIS & ARNOLD *Cath. Dict.* 405/1 During the adoration the 'Improperia' are sung, each improperium being followed by the Trisagion in Greek and Latin. **1959** *Collins Mus. Encycl.* 331/2 *Improperia,* part of the Roman liturgy for Good Friday.

improve, *v.*² **2. b.** (Later example.)

1906 L. L. BELL *Carolina Lee* 293, I could refuse an offer to improve my land, denuded and mortgaged as it is.

improved, *ppl. a.* Add: **5.** *improved wood,* thin sheets of wood attached to each other by films of synthetic resin, thus improving its resistance to distortion, shrinkage, etc.

1940 *Jrnl. R. Aeronaut. Soc.* XLIV. 673 It..could be expected that..composite wood plastics, like..'improved' wood, such as commonly used in the modern aeroplane manufacture, would be suitable for the construction of petrol containers. **1957** *N.Z. Timber Jrnl.* Aug. 59/2 Certain commercial treatments to prevent movement owing to moisture variation in wood, have resulted in what is termed 'improved wood'. **1963** A. D. WOOD *Plywoods of World* iv. 138 Improved wood is the product of a combination of veneers..interleaved with a synthetic resin glue film.

improvement. 2. a. and **b.** Restrict † *Obs.* (exc. in *U.S. dial.*) to senses in Dict. Add further N. Amer. and N.Z. examples in sense: the turning of farmland to better account by the erection of buildings, fences, etc.; also *concr.*, the buildings, fences, etc., themselves.

1769 *Quebec Gaz.* 16 Feb. 3/1 The Possessors of such Concessions shall be entitled to such Part of them as shall be proportioned to the Improvements they have made thereon. 1841 W. Deans *Let.* 25 Mar. in J. Deans *Pioneers of Canterbury* (1937) 33 Mr Molesworth let a town acre of his for £240..for 14 years, buildings and improvements to remain at the end of the lease. 1856 'J. Phoenix' *Phoenixiana* (1859) xxxiii. 202 Three other small buildings, unoccupied, a fence, and a grave-yard, constitute all the 'improvements' that have been made at the 'Playa'. 1891 R. Wallace *Rural Econ. Australia & N.Z.* xv. 225 Many [N.Z.] settlers naturally prefer the lease, as a perpetual leasehold is practically as good as a freehold, while it leaves a settler free to invest his capital in improvements. 1908 *Indian Laws & Treaties* (U.S.) III. 382 Any person who..shall be an actual resident upon any one such lot and the owner of substantial and actual improvements thereon. 1927 *Amer. Speech* May 358 *Improvements and land* (noun phrase), a farm with its buildings and cleared land. 1930 L. G. D. Acland *Early Canterbury Runs* viii. 180 The Studholmes had the Terrace Station... When they sold it about 1862..the only improvements on it were a shepherd's hut and a set of sheep yards. 1949 *Lubbock* (Texas) *Morning Avalanche* 23 Feb. II. 3 Good 8 inch irrigation well, large loan, good improvements, possession. 1958 *New Yorker* 6 Sept. 37, I remember it as it originally was, for my brother and I, aged eight and six, accompanied my father when he went out to make the first 'improvements'.

7. *attrib.* and *Comb.*; **improvement lease**, in the United States and Australia, a lease granted with conditions of improvement to be made by the lessee.

1840 W. Sewall *Diary* 15 Aug. (1930) 218/1 Henry and myself repairing fence west side of improvement ditch. 1849 E. Chamberlain *Indiana Gazetteer* (ed. 3) 34 They rented land on improvement leases, by which they were to have the use of from ten to twenty acres from seven to ten years, and often at the end of that time they were able to buy land for themselves. 1895 *Act* (New South Wales) 58 *Vict.* no. 18. § 26 (*heading*) Improvement leases. 1900 *Daily News* 16 May 5/3 The new improvement undertakings of the Council. 1909 *Commonwealth Law Rep., High Court Australia* VII. 1, Sec. 26 of the *Crown Lands Act* 1895..authorizes the Governor to grant 'improvement' leases of Crown lands which by reason of inferior quality, heavy timber, or other cause are not suitable for settlement until improved. 1909 *Daily Chron.* 30 Apr. 6/4 The improvement values added by enterprise. *Ibid.* 7/7 Housing and improvement schemes. 1909 *Westm. Gaz.* 19 May 2/2 The sale of sites in improvement areas. *Ibid.* 4/1 The trifling amount of improvement-sites which their predecessors have acquired. 1964 *Ann. Reg.* 1963 128 Then..the Court decided that the Council of Ministers could not transfer municipal powers to improvement boards. 1972 *Guardian* 15 June 14 Improvement grants..are overwhelmingly taken up by better-off owner-occupiers and speculative property developers.

improver[1]. Add: **2. c.** (Earlier example = *dress-improver* s.v. Dress *sb.* 4 a.)

1872 *Young Englishwoman* Oct. 554/1 The improver consists of a thick calico foundation stiffened with whalebone..furnished with strings to tie round the waist.

(ii) Any substance or preparation added to a foodstuff by a manufacturer or processor in order to improve it in some respect (e.g. in texture or keeping quality).

1902 Leffmann & Beam *Food Analysis* 364 Improvers and Preservatives.—Mixtures of potassium nitrate, sodium chlorid, and other mineral preservatives with a little coloring-matter..are sold for improving the appearance of meat. 1925 Mojonnier & Troy *Technical Control Dairy Products* (ed. 2) xiii. 300 Several commercial products commonly known by the general term 'ice cream improvers' are in common use. These consist of rennet or pepsin mixed with certain powders such as milk sugar. These products react upon the casein in the mix, causing an increase in the viscosity. 1927 *Manch. Guardian Weekly* 11 Mar. 183/2 Indiscreet and provocative references in the press to 'improvers' and bleaching agents used in the preparation of flour for bread-making. 1960 W. J. France *Breadmaking & Flour Confectionary* xi. 109 Barbadoes and Demarara.—These sugars because of their colour, are useful brown bread improvers, adding a characteristic flavour to brown breads. 1970 Fox & Cameron *Food Sci.* viii. 165 This ageing period can be dispensed with if the flour is treated with a minute quantity of one of a number of oxidizing agents which are called flour improvers. *Ibid.*, The first four of the improvers listed increase the whiteness of the flour by bleaching the carotene and xanthophyll, which..give the flour a slight yellow tinge.

improvisation. Add: **1.** *spec.* of Old English verse.

1928 W. W. Lawrence *Beowulf & Epic Trad.* 3 We must agree to judge *Beowulf*,..not as the improvisation of an untutored minstrel, but as a well-considered work of art. 1960 *English Studies* XLI. 5 The use of traditional diction is one thing; improvisation is something else again. The two need not go together and in *Beowulf* they most emphatically do not.

improvisa·tional, *a.* [f. Improvisation + -al.] Of or relating to improvisation, impromptu.

1923 *Theatre Arts Mag.* Oct. 328 He is always true to the great school of acting of the Italian improvisational comedy. 1958 *Times* 15 Dec. 3/4 His playing of Schubert's great B flat major sonata..was far too improvisational and dreamy. 1970 P. Oliver *Savannah Syncopators* 19 Jazz, at least in its early phases, was primarily a group music, using brass and wind instruments with rhythm background, employing improvisational techniques both collectively and in solo. 1972 *Jazz & Blues* Nov. 6 Orchestral textures, improvisational complexity, harmonic innovation. 1973 *Black World* Mar. 26 A stock feature of chanted sermons: the highly rhythmic, imaginative and improvisational rendering of the Word of God.

Hence **improvisa·tionally** *adv.*

1946 R. Blesh *Shining Trumpets* (1949) ii. 33 The leader chants, often improvisationally, in strong declamatory phrases. 1963 *Times* 30 Apr. 15/4 Prokofiev's Seventh Sonata was also somewhat improvisationally showy instead of stable and strong.

improvisator. Add: *spec.* of Old English verse.

1915 W. W. Lawrence in *PMLA* XXX. 400 A direct quotation of the lay sung by the improvisator. 1928 —— *Beowulf & Epic Trad.* 261 Each individual poet..adapting, with the adroitness of the improvisator, bygone legends to the demands of special occasions.

improvise, *v.* Add: **1, 3.** *spec.* of Old English verse.

1892 J. Earle *Deeds of Beowulf* 136 The minstrel did not merely narrate, but improvised in alliterative verse. 1915 W. W. Lawrence in *PMLA* XXX. 400 A thane of the king, skilled in story..entertains the company, improvising a song in honor of Beowulf. 1948 P. F. Baum in *Mod. Philology* XLVI. 76 The minstrel..begins by improvising in verse an account of Beowulf's adventures which he had just learned about and had had little opportunity to work up beforehand. This..presupposes something different from the *Beowulf*, which certainly does not give the impression of improvising. 1961 W. Whallon in *PMLA* LXXVI. 310 The oral poet cannot pause; he must improvise continuously with no apparent effort.

improviser. Add: *spec.* of Old English verse.

1948 P. F. Baum in *Mod. Philology* XLVI. 76 Its meter and the variations exhibit by turns both the ease and freedom of the improviser and the careful workmanship of the artist.

impsonite (i·mpsŏnait). *Min.* [f. *Impson*, the name of a valley in Pushmataha Co., Oklahoma + -ite[1].] An asphaltic mineral similar to albertite.

1901 G. H. Eldridge in *Ann. Rep. U.S. Geol. Survey* XXII. I. 265 This differs from albertite, however, sufficiently, it is thought, to warrant a distinctive name, and for this, impsonite is suggested, after the valley in which it is chiefly found. 1951 E. N. Tiratsoo *Petroleum Geol.* xi. 292 Impsonite..seems to be a kind of end-product in the process of bitumen formation, having a very high proportion of fixed carbon and being infusible. 1965 E. T. Degens *Geochem. Sediments* v. 263 One may distinguish between the asphaltic, e.g. wurtzilite, elaterite, albertite, and impsonite, and the nonasphaltic pyrobitumens.

impulse, *sb.* Add: **2. c.** *Aeronaut.* *specific impulse*: the ratio of the thrust produced in a rocket engine to the rate of consumption of propellant (expressed as mass, or weight, per second).

Equivalent to the impulse (sense 2 b) obtained per unit mass, or weight, of propellant.

1947 *Jrnl. Brit. Interplanetary Soc.* Mar. 101 The most important requirement is a low consumption, or to use a term more commonly employed in rocketry, a high 'specific impulse'; the specific impulse being the thrust obtained from the consumption of one unit of propellant mixture per second. 1950 *Sci. News* XV. 76 The most useful measure of the efficiency of a rocket is called 'specific impulse'. 1962 F. I. Ordway et al. *Basic Astronautics* x. 422 In the liquid fuel reactor hydrogen is bubbled through liquid uranium compounds. With this method specific impulses on the order of 1500 lb-sec/lb may be attained. 1971 P. J. McMahon *Aircraft Propulsion* iii. 116 In the foot-pound-second system, specific impulse has the units lbf-sec/lb. For many years it was the practice to define the specific impulse as the thrust divided by the *weight* flow rate of propellants. Using this definition the units of specific impulse became 'seconds'. *Ibid.* x. 298 The specific impulse of a cordite type propellant will be of the order of 2 000 N-s/kg when operating with a chamber pressure of 6 000 kN/m² and exhausting to 100 kN/m².

4. c. *Dancing.* (See quot. 1949.)

1949 Shurr & Yocom *Mod. Dance* 190 *Impulse*, the impetus or impelling force used to initiate a movement sequence, such as a hip contraction or a hip release. 1968 J. Winearls *Mod. Dance* (ed. 2) ii. 64 Thus with continued forward impulses the pelvis moves in a backwards–downwards–forwards–upwards–circle with an accent at the bottom of the circle.

4*. *Electr.* A sudden, momentary change in voltage or current from an otherwise steady (or slowly varying) value. (More commonly *pulse*.)

1883 E. Atkinson tr. *Ganot's Elem. Treat. Physics* (ed. 11) x. vi. 850 As they are all connected together we get, not so much a series of separate impulses, as a continuous series of currents. 1904 *Daily Chron.* 10 Dec. 7/2 It [*sc.* a tape] is inserted in the aperture of the transmitting instrument, and by the perforations the electrical impulses are created and recorded at the receiving station. 1943 A. L. Albert *Fund. Telephony* ix. 206 The vertical movement of the selector switch is controlled by the electric impulses received from the subscriber's dial. 1971 H. E. Ennes *Television Broadcasting* vii. 347 The circuit works by virtue of the fact that spurious noise impulses normally are much narrower than the desired sync pulses.

5. *attrib.* and *Comb.* **a.** Simple *attrib.*

1901 *Daily Chron.* 3 Sept. 3/7 A genuine chivalrous impulse-desire—that natural desire for companionship. 1929 D. H. Lawrence *Let.* 1 Oct. (1962) II. 1204 You are working all the time from wrong impulse-sources. 1949 M. Mead *Male & Female* xvii. 355 Modern psychology and modern literature emphasize the importance of impulse gratification.

b. Special Comb.: **impulse clock, dial**, a secondary clock operated by electrical impulses transmitted at regular intervals by a master clock; **impulse coupling** = *impulse starter*; **impulse-reaction turbine**, a turbine comprising two (or more) stages, one working on the principle of the impulse turbine and the other on that of the reaction turbine; **impulse starter**, a mechanical device which may be fitted to the magneto of an ignition system to cause its rotor to turn in a series of jerks instead of continuously, resulting in an increased voltage that facilitates the production of a spark at low speeds or when starting; **impulse tube**, a tube serving to expel a torpedo; **impulse turbine**, a turbine in which the working fluid undergoes no drop in pressure in the rotor, this being driven solely by the change it causes in the direction of flow.

1923 Langman & Ball *Electr. Horology* v. 82 Clocks coming under this section are generally..designated as dials, impulse clocks, secondary clocks, journeyman clocks, sympathetic clocks, or step by step movements. 1951 S. J. Wise *Electr. Clocks* (ed. 2) iv. 67 An impulse or repeater clock is a device which receives the timed electrical impulses transmitted by a master clock, and translates them, through its wheelwork, into seconds, minutes and hours. 1916 Impulse coupling [see *impulse starter* below]. 1943 A. P. Fraas *Aircraft Power Plants* viii. 147 To ensure a good spark at cranking speeds the magnetos for many of the smaller engines are fitted with an impulse coupling. 1931 F. Hope-Jones *Electr. Clocks* ii. 6 Circuits of electrical impulse dials, in which a master clock transmits impulses every minute or half-minute to propel the hands. 1940 —— *Electr. Timekeeping* i. 6 After many years of futile attempts to apply electricity to horology, inventors turned their attention to systems of electrical impulse dials, an obviously sane and effective method of indicating uniform time throughout a large building. 1929 T. M. Naylor *Steam Turbines* i. 4 Combination turbines or disc and drum turbines, as they are often called, are a combination of impulse and reaction types of turbine. The first part of the turbine is impulse, and the remainder of the turbine is reaction, so that this type of turbine might be called impulse-reaction. 1951 *Engineering* 5 Oct. 438/3 The high-pressure turbine is of the impulse-reaction type. 1916 V. W. Pagé *Automobile Starting* iii. 224 The device..is known as the Eisemann impulse starter coupling. This may be attached to any model of Eisemann magneto and is said to have no effect upon its regular operation except at slow speeds, when it causes the armature to rotate in a series of jumps instead of at a uniform speed. 1940 W. E. Crook *Electr. in Aircraft* vii. 93 The impulse starter..enables the engine to start on its own magnetos. It is a purely mechanical piece of apparatus, consisting essentially of a spring-loaded pawl and ratchet gear. 1877 *Illustr. London News* 14 Apr. 339/3 It is fired by what is called an 'impulse-tube', which..discharges the torpedo into the water. 1878 *Cassell's Family Mag.* 312/2 Direction is given to the torpedo by means of an iron impulse-tube built into the vessel. 1885 *Marine Engineer* 1 Sept. 144/2 The fish torpedoes lie side by side. Immediately behind them..are a couple of 'impulse tubes'. 1881 *Encycl. Brit.* XII. 524/1 In some turbines the whole available energy of the water is converted into kinetic energy before the water acts on the moving part of the turbine. Such turbines are termed Impulse Turbines, and they are distinguished by this that the wheel passages are never entirely filled by the water. 1906 W. H. S. Garnett *Turbines* iv. 42 Impulse turbines..are unsuited for the development of high speed motion from low falls. ..For running, on the other hand, at low speeds under a high fall, the impulse turbine cannot be surpassed. 1971 P. J. McMahon *Aircraft Propulsion* v. 163 The extreme case of a zero reaction stage wherein all the pressure drop occurs in the nozzle blades is known as an 'impulse turbine' in accordance with steam turbine practice. A pure impulse turbine would rarely be used for an aero-turbine engine.

c. Of or pertaining to a purchase or purchases made on impulse, usu. at the point of sale of displayed goods, as *impulse buyer, buying*, etc.

1959 *Times Lit. Suppl.* 29 May (Children's Books) 5 [Children's books] snarl 'impulse buyers' in the supermarket. 1959 *Sunday Express* 26 July 10/4 The Opposition Leader's Lady and Judy O'Grady are '*impulse buyers*' under the skin. 1959 *News Chron.* 18 Nov. 4/2 What the traders call 'impulse buying' is increasingly popular. 1962 *Sunday Express* 8 July 15/2 Top dressers..plunge with the odd '*impulse buy*'. 1963 *Punch* 27 Nov. 772 New Prime Minister..impulse-buys sixty Phantom IIs. 1964 *Punch* 6 May 655/3 Trolleys full of impulse-bought bargains. 1965 *Mod. Law Rev.* XXVIII. v. 557 None of the reasons given for minimising the importance of 'impulse sales' apply with special force to the book trade. 1967 L. J. Braun *Cat who ate Danish Modern* x. 88 These are little boutique items for the impulse buyer. 1968 'S. Jay' *Sleepers can Kill* vi. 66 He goes into a shop..and buys a boat... An impulse buy, if ever there was one. 1972 *New Statesman* 26 May 709/3, I impulse-bought some crumpets the other day, because they were on the counter at the dairy.

impulse, *v.* Delete 'now *rare*' and add later examples. Also *intr.*

1931 *Times* 27 June 11/5 The interference is due..to..sparks in stays of masts, and to loose metallic contacts, which impulse the receiver in the same way as heavy atmospherics would do. 1936 *Nature* 12 Sept. 445/2 The law and governmental forms..clearly lag behind even economic developments as impulsed by scientific discovery. 1943 *Gloss. Terms Telecomm.* (*B.S.I.*) 62 *Impulsing signal*, a signal carrying the selective information to steer the call in the desired direction. 1949 E. C. Berkeley *Giant Brains* iii. 41 This type of relay has the property of staying..in either position until the opposite coil is impulsed. 1960 *Lang. & Speech* III. 140 (*title*) Recurrently impulsed resonators in speech and psychophysical studies. 1972 J. Quartermain *Rock of Diamond* xiii. 72 Her small voice..impulsed through the network of cables.

impulsive, *a.* Add: **4.** *Electr.* Consisting of, or of the nature of, an impulse or impulses.

1920 *Whittaker's Electr. Engineer's Pocket-Bk.* (ed. 4) 136 The passage of the spark in such a case appears to involve an impulsive rush of electricity, and the setting up of electric pulses in the neighbourhood of the spark gap can be demonstrated. 1940 *Chambers's Techn. Dict.* 440/2 *Impulsive current*, a current which comprises one or more impulses in one direction round a circuit, as in dialling. 1947 D. G. Fink *Radar Engin.* xi. 127 The impulsive type of noise (ignition interference and static) so commonly encountered on lower frequencies is generally absent in radar. 1962 *Newnes Conc. Encycl. Electr. Engin.* 466/2 Although the secondary winding [of a magneto] may have 10,000 turns, its voltage is not sufficiently impulsive, however. The contact-breaker is therefore arranged to short-circuit the primary.

impulsivity (impʌlsi·vĭti). [f. Impulsive *a.* + -ity.] The character of being impulsive or acting on impulse, without reflection or forethought; impulsiveness. Hence **impu·lsivist**, one who acts on impulse.

1891 C. Lombroso *Man of Genius* 348 The psychology peculiar to the epileptic—impulsivity, double personality, childishness. 1895 tr. *M. Nordau's Degeneration* 120 Moral insanity, however, is not present in Verlaine. He sins through irresistible impulse. He is an Impulsivist. 1925 *Public Opinion* 24 Apr. 391/2 A man of exceptional ability and yet of a febrile impulsivity. 1969 D. A. Winn in Caplan & Lebovici *Adolescence* 256 Impulsivity was prominently a part of the suicidal behavior of most of the 60 adolescents in this study. 1971 *Jrnl. Gen. Psychol.* Oct. 273 This relationship would seem to indicate an underlying impulsivity as being a modest contributor.

impunitive (impiū·nitiv), *a.* *Psychol.* [f. Im-[2] + Punitive *a.*] Adopting an attitude of resignation towards frustration; characterized by blaming neither oneself nor others unreasonably. Contrasted with *intropunitive *a.* and *extrapunitive *a.*

1938 S. Rosenzweig in H. A. Murray *Explorations in Personality* vi. 587 He may experience emotions of embarrassment and shame, making little of blame and emphasizing instead the conciliation of others and himself to the disagreeable situation. In this case he will be more interested in condoning than in condemning and will pass off the frustration as lightly as possible by making references, even at the price of self-deception, to unavoidable circumstances. This type of reaction may be termed 'impunitive'. 1954 G. W. Allport *Nature of Prejudice* xxi. 349 Some frustrated people..are so detached and philosophical about life's frustrations that they blame no one; they are impunitive. 1958 M. Argyle *Relig. Behaviour* viii. 90 A related personality variable is that of punitiveness: people are said to be extrapunitive if they react to frustration by aggression directed outwards, intropunitive if the aggression is directed inwards, and impunitive if they do not react aggressively at all.

Hence **impu·nitively** *adv.*, in a way characteristic of an impunitive individual.

1958 M. Argyle *Relig. Behaviour* viii. 91 The humanitarians on the other hand responded impunitively. 1969 M. D. Vernon *Human Motivation* ix. 148 Unacceptable motivational tendencies..may be treated impunitively and denied entry to consciousness.

impure, *a.* (*sb.*) Add: Hence **impu·rist**, one who is not a purist.

1937 V. Woolf *Writer's Diary* 3 Apr. (1953) 280 The purists and the impurists. 1959 'F. Newton' *Jazz Scene* vi. 113 The jazz public has always been divided, but before the modernist revolution normally only into 'purists' and 'impurists'.

impurify, *v.* Delete †*Obs.* and add later examples.

1904 *Daily Chron.* 19 July 3/7 Impure aeration that..takes place in dirty, fœtid cowsheds impurifies it [*sc.* milk]. 1928 D. H. Lawrence *Let.* 18 Feb. (1962) II. 1040, I am having to impurify and dilute it [*sc. Lady Chatterley's Lover*] for the market.

impurity. Add: **3. b.** An impurity atom; *esp.* an atom of dopant present at a normal lattice site in an impurity semiconductor.

1931 A. H. Wilson in *Proc. R. Soc.* A. CXXXIV. 279 Electrons on a foreign atom [in a semiconductor] do not take part directly in conduction. They must first be transferred by the effect of the lattice vibrations to an atom of the pure substance. In this case the main function of the impurities is to provide electrons for the upper unoccupied energy bands of the crystal, while acting as scatterers is only a secondary function. 1950 W. Shockley *Electrons & Holes in Semiconductors* i. 12 The conductivity [of the silicon] arises from the presence of arsenic atoms which are termed 'impurities', even though added deliberately in the otherwise pure silicon. *Ibid.* 14 Impurities with a valence of five are called 'donor impurities' because they donate an excess electron to the crystal; those with a valence of three are called 'acceptor impurities', since they accept an electron from somewhere else in the crystal.., thus leaving a hole to conduct. 1952 J. S. Koehler in W. Shockley et al. *Imperfections in nearly Perfect Crystals* vii. 206 The maximum value..of this force occurs when the impurity is two or three atomic distances from the dislocation. 1959 R. A. Smith *Semiconductors* iii. 45 We must distinguish two types of impurity, substitutional impurities, which replace atoms of the host crystal on their lattice sites, and interstitial impurities which occupy positions in between the lattice sites. 1972 F. J. Bailey *Introd. Semiconductor Devices* i. 21 By the addition of donor impurities to the silicon, large numbers of free electrons become available as current carriers.

4. Special Comb.: **impurity atom**, an atom that differs from the bulk of those present in a substance in being of a different element; **impurity level**, an energy level in a semiconductor that is due to an impurity atom and generally lies either just above the highest filled (valence) energy band (in the case of an acceptor) or just below the lowest empty (conduction) band (in the case of a donor); **impurity scattering**, scattering of current carriers by impurity atoms in a crystalline solid; **impurity semiconductor**, a semiconductor in which most of the carriers of electric current are electrons and holes from impurity atoms.

1939 A. H. Wilson *Semi-Conductors & Metals* i. 3 The [electrical] resistance in a metal is caused by the scattering of the electrons by irregularities in the crystal; these irregularities may be due to the presence of impurity atoms and strains or to the temperature motion of the atoms. 1949 *Physical Rev.* LXXV. 866/1 A substitutional impurity atom from the fifth group has one more valence electron than is required to fill the four valence bonds with neighboring silicon atoms. 1970 W. Bollmann *Crystal Defects* iv. 37 The elementary point defects are vacancies and interstitial atoms, and also impurity atoms. 1972 F. J. Bailey *Introd. Semiconductor Devices* i. 21 The doping level refers to the ratio of impurity atoms to silicon atoms. For a transistor this may typically be $1:10^8$, i.e. one impurity atom for every hundred million silicon atoms. 1973 *Sci. Amer.* Feb. 93/2 If other atoms, which can be referred to as impurity atoms if their concentration is relatively small, are introduced into a helium discharge, the random collisions in the discharge will mix the impurity atoms with the helium atoms. 1933 R. H. Fowler in *Proc. R. Soc.* A. CXL. 507 Semi-conductors with impurity levels full of electrons at low temperatures owe their conductivity to the excitation of electrons to band 1 both from the impurity levels and from band 2, the former predominating at ordinary temperatures. 1964 J. M. Ziman *Princ. Theory Solids* vi. 169 In a semiconductor there may also be scattering from a neutral impurity—where, for example, an electron has settled in a donor impurity level, or a hole is resident on an acceptor level. 1946 *Physical Rev.* LXIX. 258/2 (*heading*) Theory of impurity scattering in semiconductors. 1965 Lindmayer & Wrigley *Fund. Semiconductor Devices* vii. 259 In crystals used for devices, thermal scattering and impurity scattering dominate the collision processes. 1946 *Trans. Amer. Inst. Electr. Engin.* Nov. 713/3 At low temperatures the conductivity of different samples varies by large factors. In this region silicon is said to be an impurity semiconductor. 1950 W. Shockley *Electrons & Holes in Semiconductors* i. 12 (*heading*) Impurity semiconductors; donors and acceptors.

imputation. Add: **4.** An economic theory of value (see quot. 1965); also (freq. *attrib.*) a form of taxation levied on company profits, usu. in phr. *imputation system, tax*. Cf. *impute *v.* 6.

1893 C. A. Malloch tr. *F. von Wieser's Natural Value* III. ii. 77 So far as this method succeeds in founding, upon the imputation of the return, a valuation of goods and a plan of production..it is the height of practical wisdom. To show that imputation in this sense is both allowable and practicable take one single case. 1931 A. Gray *Devel. Econ. Doctrine* xii. 354 The most characteristic part of Wieser's contribution to the school lies in his doctrine of *Zurechnung*, which has been acclimatized in this country as 'imputation' or 'attribution'... In a sense the higher goods have no value until a value is 'imputed'..to them. 1934 F. A. von Hayek in C. Menger *Coll. Works* I. p. xv, He answers the problem of imputation..by saying that such quantities of the different factors as can be substituted for each other..must have equal value. 1938 E. Roll *Hist. Econ. Thought* viii. 388 In the theory of distribution Menger is responsible for posing what is known as the problem of imputation; that is the problem of the value of goods of a higher order. 1965 Seldon & Pennance *Everyman's Dict. Econ.* 210 *Imputation*, the process of attributing value to productive resources in accordance with their contribution to the value of their products... The theory of imputation argued that the value of factors ('higher-order' goods) was *in all cases* determined by (imputed from) the value of the final ('lower-order') goods to whose production they contributed. 1971 *Daily Tel.* 17 Nov. 19/2 The Government..preferred to replace it not with the French *imputation* or tax credit method but with a two-rate system... The *imputation* system..would mean that all company profits..would be taxed at the rate of 50 p.c. 1972 *Accountant* 23 Mar. 366/1 The Chancellor..would introduce a new tax based upon the imputation system. *Ibid.*, Under the imputation system, the company will pay tax on its total profits. 1972 S. Marcus *Finance Act 1972* i. 2 Under the new 1973 method, known as the 'imputation system', all profits (except capital gains) will be taxed at the same rate, whether they are distributed or not. 1973 *Daily Tel.* 31 Mar. 19/5 Dividend covers..have been adjusted to take into account the new corporation 'imputation' tax system. *Ibid.* 2 Apr. 20 The 'imputation' method of taxing company profits..becomes the technical core of investment.

impute, *v.* Add: **6.** *Econ.* To attribute or assign (value) to a product or process by inference from the value of the products or processes to which it contributes.

1893 C. A. Malloch tr. *F. von Wieser's Natural Value* III. viii. 96 To each single item or quantity is imputed the smallest contribution which..can be economically aimed at by the employment of this particular item or quantity. 1893 W. Smart in C. A. Malloch tr. *F. von Wieser's Natural Value* p. xv, This determination of imputation by equations of return tells us nothing more than that certain shares *are* imputed to certain elements. 1945 E. Heimann *Hist. Econ. Doctrines* VIII. ii. 197 A problem..had greatly concerned the Austrians from Menger on, namely how separate shares in the product can be imputed to the cost factors when they invariably co-operate in production. 1950 Dingwall & Hoselitz tr. *Menger's Princ. Econ.* III. ii. 139 We logically impute this importance to the goods on whose availability we are conscious of being dependent. 1965 Seldon & Pennance *Everyman's Dict. Econ.* 210 The value of factors..was *in all cases* determined by (imputed from) the value of the final..goods to whose production they contributed. 1973 *N.Y. Law Jrnl.* 2 Aug. 4/4 If interest is not provided for, the Internal Revenue Service may impute interest to any deferred payments.

imputed, *ppl. a.* Add: **3.** *Econ.* Estimated, valued in relation to something else. Spec. *imputed price, value* (see quots.).

1909 Webster, *Imputed value*, Econ., the value that a thing has merely for its utility in the production of something else;—called also *derived value*. 1929 *Encycl. Brit.* XXII. 960/2 Imputed price..is an estimate of the amount of money for which a given article or a given quantum of goods could be sold or bought. 1973 *Daily Tel.* 27 Mar. 23 (Advt.), With imputed tax credits this is equivalent to a franked payment of 6·5625 p per share.

imputrescibility. (Later example.)

1972 C. Lévi-Strauss in P. Maranda *Mythology* xiii. 275 They are operators, which make it possible..to express..a set of equivalences connecting life and death, vegetable foods and cannibalism, putrefaction and imputrescibility, [etc.].

imram, var. *immram.

imshi (i·mʃi). *Services' slang.* Also **imshee**, **imshy**. [Local Arabic (Berggren).] Be off, go away. Also as *vb.*

1916 *Anzac Book* 135/2 And the King-of-all-the-Huns said, 'It is enough, Imshee!' *Note*, Imshee is the Arabic for 'go away'. The Australasian Corps, which had so far employed it only to street hawkers in Cairo, used this war-cry on April 25. *Ibid.* 136/1 So they imsheed. 1919 *Athenæum* 28 July 664/2 'Imshy', go away (generally corrupted into something like 'hampshire'). 1942 *Word Study* Dec. 6/2 '*Imshi*,' said the drongo; 'you're *shikkered*!' 1965 G. McInnes *Road to Gundagai* iv. 70 Now, imshee-allah!

‖ **imu** (ī·mū). [Hawaiian.] 'In the Hawaiian islands, a pit used for baking meat or vegetables by means of heated stones' (*Cent. Dict.* Suppl., 1909).

1928 J. C. Andersen *Myths & Legends Polynesians* 296 When he reached Makila, on the confines of Lahaina, he saw a number of people heating an *imu*, or ground oven. 1954 *Ellery Queen's Mystery Mag.* Oct. 43/2 Next came the ritual of putting the pigs into the *imu*. *Ibid.* 47/1 *Imu* pork, white and succulent inside crackling brown skin.

i-mutation: see *I I. 2 b.

in, *prep.* Add: **1.** Examples of *in a ship, vessel*. Also (U.S.) *in school*, attending a school, receiving education at a school = (U.K.) *at school* (cf. School *sb.*[1] 1 b).

In former times *in school* was also used in Britain in the sense 'attending a school': see quot. *c* 1205 s.v. School *sb.*[1] 1 b. Also *c* 900 *Bæda's Eccl. Hist.* (1890) 190/12 Sum leornungmon in scole Scotta cynnes; *a* 1350 *Harley Lyrics* (1948) 63 Whil y wer a clerc in scole.

1848 J. F. Cooper *Jack Tier* I. iii. 80 *In* a vessel is as correct as *in* a coach, and *on* a vessel, as wrong as can be; but you can say *on board* a vessel, though not 'on the boards of a vessel'. 1916 'Taffrail' *Pincher Martin* xiv. 248 'When I was in the old *Somerset*, in nineteen-nine,' somebody would start the ball rolling, 'we had a fellow who'——. 1942 *Short Guide Gt. Brit.* (U.S. War Dept.) 8 The tales of Scott and Robert Louis Stevenson which many of you read in school. 1972 R. Quirk et al. *Gram. Contemp. Eng.* 310 He's {at school (BrE) / in school (AmE)} (= 'He attends/is attending school'). He's in school (= (in BrE) 'He's actually inside the building—not, *eg* on the playing fields').

4. Esp. of a gradient.

1726 [in Dict.]. 1830, etc. [see *one 5 b]. 1840 [see Grade *sb.* 10]. 1861, 1868 [see Gradient *sb.* 1]. 1869 *Bradshaw's Railway Manual* XXI. 318 The gradients and curves are generally favourable, the steepest gradient being 1 in 82½. 1923 *Michelin Guide Gt. Brit.* (ed. 7) facing p. 277, Gradients on roads are shown thus:..1 in 20 to 1 in 14.

1 in 14 to 1 in 10. 1 in 10 and over. 1973 E. Course *Railways S. Eng.: Main Lines* i. 29 Over the nineteen and a half miles from Redhill to Tonbridge the maximum gradient was 1 in 250.

5. b. In phrases implying incidental distribution, e.g. *in parts, in places*.

1905 [see *curate 2 b]. 1922 D. H. Lawrence *England, my England* 132 And I sensed I was a prisoner, for the snow was everywhere deep, and drifted in places. 1924 A. D. Sedgwick *Little French Girl* II. v, The long iron staircase down the face of the cliff was almost as steep as a fire escape in places. 1973 *Listener* 8 Feb. 167/2 The Appeal Court..found the [Warhol] film dull, dreary, and offensive in parts.

10. *in-work* (nonce-wd.), one who has work.

1924 Galsworthy *White Monkey* I. xii, The out-of-works and the in-works.

b. (Further examples.) Cf. also Bud *sb.*[1] 4, Flower *sb.* 10, Foal *sb.* 1 b, In-calf *a.*, *in-foal *a.*, *in-pig *a.*, Leaf *sb.* 3.

1813 M. Edgeworth *Let.* 6 Apr. (1971) 10 The coffee tree in *red berry*... The palm tree in *fruit and flower*... The banana in *fruit*. 1882 [see Pod *sb.*[1] 1]. 1972 *Hilliers' Man. Trees & Shrubs* 83 Corylus avellana 'Contorta'... A winter feature when in catkin.

12. d. Often dependent upon a superlative or a commendatory epithet: within the sphere of (a particular class or order of things). *colloq.*

1866 Ruskin *Crown Wild Olive* ii. § 53 The newest and sweetest thing in pinnacles. 1879 [see Thing *sb.*[1] 7]. 1911 W. J. Locke *Glory of Clementina Wing* ii, I may not be the latest thing in dandyism. 1966 G. N. Leech *Eng. in Advertising* ix. 92 ABC: the first name in entertainment. 1974 *Radio Times* 3 Jan. 58/1 The most dazzling cruises in holiday history.

17*. With a following sb. forming attrib. phrases: *in-car*, within a car; *in-career*, of training, etc., received while in employment; *in-churn*, of a method of machine-milking direct into a churn; *in-company*, of training, etc., received while in the employment of a company; *in-depth* (see *depth I. 3 c); *in-person* (cf. Person *sb.* 11); *in-pile*, within a nuclear reactor; *in-plant*, within a 'plant' or factory; *in-process* (cf. Process *sb.* 1), of any activity, etc., that is in process; *in-process gauging* (see quot. 1968); *in-sack*, within a sack; *in-service* (cf. Service[1] 1), of training, etc.: received by a person while engaged on some activity; of an object: relating to its reliability, maintenance, etc., while in use. Cf. *in-college *a.* (Cf. analogous uses mentioned near end of In- *pref.*[1])

1968 *N.Y. Times* 7 Apr. 1/4 The sound problem was eventually solved with in-car speakers. 1971 *Daily Tel.* (Colour Suppl.) 4 June 39/1 These damp cloths are part of something the manufacturers..don't seem to have heard of; in-car luggage. They seem to think you can put everything in the boot. 1973 *Times* 13 Feb. 24/1 A Lucas spokesman said yesterday: 'I can confirm that we shall be entering the in-car entertainment market this year.' 1968 *New Scientist* 3 Oct. 31/2 Whether in-career re-education will be best inside or outside universities is a matter for debate. 1970 *Physics Bull.* June 242/2 The engineers' survey..includes unemployment and in-career training as well as remuneration figures. 1970 *Nature* 28 Nov. 814/2 In-career retraining may become very important in the future. 1955 J. G. Davis *Dict. Dairying* (ed. 2) 745 Probably the most important development has been the introduction of the 'In-Churn System'. In-churn milking passes the milk direct from the cow to the churn. *Ibid.*, In-churn recording..is carried out by means of weighing scales. *Ibid.*, In-churn cooling. 1960 *Farmer & Stockbreeder* 16 Feb. 39/3 This new..unit..provides you with modern in-churn milking. 1966 *Ann. Rep. Travelers Insurance Co.* (*Hartford, Conn.*) 1966 26/2 Our in-company training programs. 1969 *Timber Trades Jrnl.* 13 Dec. 35/2 There were twin pillars to training—in-company training and, for young people in particular, further education. 1970 *Times* 28 Apr. 26/7 Having already completed most of the in-company training for the introduction of decimalization. 1955 L. Feather *Encycl. Jazz* i. 21 Although the white jazzmen rarely found opportunities for expressing themselves freely on 'in-person' jobs, the work..on..recording sessions compensated. 1957 S. Dance in S. Traill *Concerning Jazz* 37 No experience of jazz can be so exciting or so illuminating as the in-person performance. 1959 *Spectator* 9 Oct. 469/1 On any one TV appearance Macmillan and Gaitskell must have been seen by more people than the sum total audience of their in-person tours. 1972 *Jazz & Blues* Nov. 18/1 Her recordings and in-person work illustrate that she is an artist who is always willing to experiment. 1960 *Times Rev. Industry* Dec. 16/2 Zirconium..is useful for 'in-pile' equipment, such as fuel element supports, tubes for control equipment, flexible hose and packing pieces. 1961 *Times* 10 May 2 The work includes:—out-of-pile and in-pile testing. 1963 B. Fozard *Instrumentation Nucl. Reactors* iii. 33 The second type of measurement is made with in-pile detectors. 1943 *Atlantic Monthly* Sept. 55 Few of them participate in in-plant training and upgrading programs. 1958 *Technology* Feb. 414/2 The proportion of in-plant to total training is..low. 1959 *Times* 5 Feb. 2/5 They involve an application of this industry in in-plant technical scale studies. 1967 *Jane's Surface Skimmer Systems 1967–68* 62/2 The power source can be an in-plant air supply system or [etc.]. 1971 *Timber Trades Jrnl.* 21 Aug. 26/3 In-plant treatment will now be extended to other '1800' components. 1925 *Nat. Assoc. Cost Accountants Yearbk.* 24 Divide the average 'in process' inventories into the amount of transfers to finished stores. 1967 *New Scientist* 20 Apr. 140/1 In-process gauging, on the machine tool itself, could halt..appalling waste. 1968 *Gloss. Terms Air Gauging* (*B.S.I.*) 13 *In-process gauging*, gauging carried out during processing, e.g. measurement of a workpiece whilst it is being machined. 1971 *Computers & Humanities* VI. 41 In-process corrections, however, are very difficult to make since holes cannot be erased. 1971 *Gloss. Terms Quality Assurance* (*B.S.I.*) 6 *In-process inspection*, product inspection carried out at various discrete stages in manufacture. 1958 *Times* 24 Nov. 15/4 For drying grass seed.., the in-sack drier had many advantages. 1960 *Farmer & Stockbreeder* 23 Feb. 100/1 (Advt.), Heat for..in-sack grain drying. 1928 *Rep. Comm. Educ., U.S. Dept. Interior* 30 June 6 The movement for improving preservice and inservice training of teachers for rural schools. 1960 *Guardian* 13 July 5/4 Development of in-service training..for staff nurses. 1963 F. F. Laidler *Gloss. Home Econ. Educ.* 48 *In-service training*, the continuing education and training given to a person after he/she has begun to work in a particular occupation. 1964 M. A. K. Halliday et al. *Ling. Sci.* 264 He [*sc.* the primary school teacher] needs an appropriate training in his new task, either during his initial period of training as a teacher or by means of in-service training. 1967 *Technology Week* 23 Jan. 43/2 (Advt.), Such data may well reveal over-design or design deficiencies, thus providing opportunities for improvement of safety characteristics, in-service reliability. 1972 *Lebende Sprachen* XVII. 72/2 The in-service performance of the trio is likely to dictate the specification for production models which are due to appear before the end of 1971.

21. d. (Further examples.) Not restricted to explicitly negative sentences.

1924 C. Mackenzie *Heavenly Ladder* xvi. 223 Mark had never been near his house in a year. 1957 R. A. Heinlein *Door into Summer* (1960) ix. 143 The place smelled like a vault that has not been opened in years. 1971 *Daily Tel.* 1 June 4/8 The first bridge across the Bosphorus in 2,300 years..is now being built. 1972 'E. McBain' *Sadie when she Died* xiii. 42 Arlene said that she had not played tennis in three years. 1973 *Sci. Amer.* Jan. 53/1 When Mariner 9 reached Mars on November 13, 1971, the greatest dust storm in more than a century was raging.

25. (Further examples.)

1810 *Q. Rev.* Feb. 193 If a man has it in him, he can do anything any where. 1841 Lytton *Night & Morning* (ed. 2) I. I. v. 103, I will work for you day and night. I have it in me. 1846 G. E. Jewsbury *Sel. Lett. to Mrs. Carlyle* (1892) 224, I did care for him once, long and well—better than I have it in me to care for any man now. 1892 I. Zangwill *Childr. Ghetto* III. 52 That girl's got it in her, I can tell you. She'll take the shine out of some of our West-Enders. 1895 H. James *Notebks.* (1947) 408, I didn't know I had it in me. 1919 Beerbohm *Seven Men* 119 He looked to me to 'do something big, one of these days', and that he was sure I had it 'in' me. 1924 *Isis* (Oxf.) 30 Jan. 16/2 He may become a fine actor—he has it in him. 1928 Foy & Harlow *Clowning through Life* 297, I didn't believe he had it in him. 1938 R. Finlayson *Brown Man's Burden* 79 They didn't think Kay had it in him to do it. 1958 *Listener* 13 Nov. 786/2 As between draughts and chess this is outweighed by the fact that there is more 'in' chess. 1960 M. Spark *Bachelors* i. 2 'You must have it in you,' said Ronald, 'going all the way to Piccadilly for herbs.' 1973 'E. McBain' *Hail to Chief* ii. 30 If you could find it in yourself to go over to the hospital and identify your brother.

b. *nothing, not much, little*, etc., *in it*: little or no difference between competitors or any persons or things that are compared. orig. *Racing slang.*

1914 in *Concise Oxf. Dict.* 1927 *Observer* 18 Dec. 19/3 The first round there was nothing much in it. In the second round Angus..punched Mansfield round the ring. 1929 S. E. Thomas *Elem. Econ.* (ed. 4) xxix. 523 While in the course of a year Britain imports considerable quantities of gold, she also exports almost equally large quantities, and on balance there is usually very little in it.

c. *in it*: an advantage (to be received from something). Usu. in phr. *what was* (or *is*, etc.) *in it for* (someone).

1963 T. Parker *Unknown Citizen* v. 140 He seemed to have an inbred suspicion of any kind of offered help, he wanted to know why people were giving it, what was in it for them. 1968 *Guardian* 2 Apr. 11/1 The 'Washingtonologists' in Moscow must be getting their files out to see what is in it for the Soviet Union—and for the world. 1971 'A. Gilbert' *Tenant for Tomb* ii. 39, I can't see what there was in it for Mrs Plum.

27. (Later examples.)

1902 H. James *Wings of Dove* VI. xx. 298 'You scarcely call him, I suppose, one of the dukes.' 'Mercy, no—far from it. He's not, compared with other possibilities, "in" it.' 1907 F. H. Burnett *Shuttle* xxxviii. 381 'Hope you had a fine time, Mr. Selden?' 'Fine! I should smile. Fine wasn't in it.' 1912 A. Bennett *Matador* 272 We were completely outshone. I tell you, we were not *in* it, not anywhere near being in it! 1913 F. L. Barclay *Broken Halo* vi. 69 In fact, the Egyptian dynasties weren't in it! She was positively antediluvian! 1915 A. Huxley *Let.* Oct. (1969) 82 At present I share Balliol with one..man..who rather repels me at meals by his..habit of shewing satisfaction with the food: Sir Toby Belch was not in it. 1960 L. Cooper *Accomplices* IV. ii. 224, I thought the Party knew all the technique there is about handling people, but they're not in it with the Church. 1964 H. E. F. Donohue *Conversations with Nelson Algren* ii. 74 All people are killers, potentially. Tigers aren't in it with people. 1966 'J. Hackston' *Father clears Out* 140 A fight in the snow is a tame affair and not in it with a hot summer contest. 1968 *Globe & Mail* (Toronto) 15 Jan. 9/4 We just weren't in this one... Nobody was going to beat them today.

38. in so far: still conventionally written thus (*Hart's Rules for Compositors*, ed. 37, 1967, p. 75) but also freq. as a single word or with hyphens. Cf. In so far (as main entry).

1940 *Economist* 6 July 13/2 Insofar as it ensures that trade between the two participating countries shall be reciprocal, it is an extension of compensation trade. 1948 J. Steinbeck *Russ. Jrnl.* (1949) 10 The C-47's are a little run down insofar as upholstery and carpeting go. 1959 B. Wootton *Social Sci. & Social Path.* viii. 267 Differentiation between the one and the other will be called for only insofar as it affects the kind of treatment that is likely to be helpful. 1969 *Times* 31 Oct. 29/5 Enforcement, insofar as salaries are concerned, is costing nothing. 1971 *Watsonia* VIII. 205 The results of it will be used here in-so-far as they affect the classification of the group.

in, *adv.* **6. b.** Delete † *Obs.* and add later examples. Cf. *count *v.* 2 b.

1884 'Mark Twain' *Huck. Finn* xxxi. 321 As long as I was in, and in for good, I might as well go the whole hog. 1893 —— in *Century Mag.* Jan. 342/1, I could n't venture it now; I was in too deep.

d. (Earlier example.) Also of a batsman given 'not out' by the umpire.

1744 *Laws* [of Cricket] in *New Dict. Arts & Sci.* (1755) IV. 3459/1 Laws for the strikers, or those that are in. 1844 *Blackburn Standard* 17 July, The bowler asked 'in or out?' 1871 'Thomsonby' *Cricketers in Council* 32 Men who run with their bats in the air are constantly run out in cases where they would have been safely 'in' if they had adopted the contrary practice. 1898 K. S. Ranjitsinhji *With Stoddart's Team* (ed. 3) iv. 70 He was given 'in' by the umpire when appealed to.

i. (Later examples of sense 'in fashion'.)

1923 *Ladies' Home Jrnl.* Sept. 50 Her hostess, in black silk crêpe, rejoices that trains are once more 'in', knowing the value of long lines and loose draperies. 1954 L. MacNeice *Autumn Sequel* 131 Accomplishments were in, enthusiasm out. 1959 *Encounter* Sept. 60/2 Beckett is a fashionable reputation which is still 'in'. 1965 M. Morse *Unattached* i. 24 Perhaps 'being at a party' is a qualification for being 'in'?

j. Of a school: in session, in progress. *Sc.* and *N.Z.*

1812 P. Forbes *Poems, chiefly in Sc. Dial.* 95 On Saturday, nae school being in. 1895 W. C. Fraser *Whaups of Durley* iii. 27 We would be stopped by a shout, 'The schule's in'. 1949 F. Sargeson *I saw in my Dream* I. vi. 40 But I don't remember nothing about when school was in.

k. well in. (*a*) *Racing.* Applied to a horse which has been treated leniently by the handicapper. (*b*) In comfortable or easy circumstances. *colloq.* orig. *Austral.* Also, profitably engaged in speculation.

(*a*) 1854 J. Mills *Life Race-Horse* xvii. 111 The handicapper..considerately classed me among the middle ones, and awarded 6 st. 12 lb. as my burthen. 'He's vell in,' said my owner,..'very vell in.' 1894 G. Moore *Esther Waters* xxx. 247 Are the 'orses he backs what you'd call well in? 1898 A. E. T. Watson *Turf* i. 16 A horse which is well in in a little handicap.

(*b*) 1891 [see Well-in *adj. phr.*]. 1902 Webster Suppl., *Well in*, engaged in a profitable speculation in stocks; said of a speculator whose purchases have risen considerably in value on his hands. Hence, in a general sense, prosperous; well off; well to do. (*Colloq., Australia.*) 1913 M. Roberts *Salt of Sea* 158 If you ain't lucky you're bound to be dishonest,..if you means to be well in all the time.

l. Of fortune or luck: favourable. (Cf. Out *adv.* 23.)

1901 A. E. W. Mason *Clementina* i, His luck for the moment was altogether in. 1912 'Saki' *Chron. Clovis* 187 Her fellow-gamblers were always ready to entertain her..when their luck was in.

9. in with. a. (Later examples.)

1925 'R. Crompton' *Still—William* vi. 112 So far County had persistently resisted the attempts of Mrs. Bott to 'get in' with it. 1942 E. Paul *Narrow St.* xxiv. 213 Naturally, the Prime Minister was in with Stavisky, too. 1964 P. M. Hubbard *Picture of Millie* ii. 15 We..go along to the Carrack for a drink..occasionally, but we're not really in with the people staying there.

12. a. in-maintenance, maintenance for a person living in a workhouse or the like; **in-patient**: now freq. used *attrib.*

1860 in C. S. Davies *Hist. Macclesfield* (1961) v. 267 In-Maintenance of paupers £1027 5.5 d.9. 1885 *Encycl. Brit.* XIX. 475/2 Relief given in a workhouse is termed 'in (or indoor) maintenance' relief. 1905 *Daily Chron.* 1 Sept. 2/5 For every £1 spent on out-relief in 1902–3 no less than £7 12s. 4d. was expended on in-maintenance. 1959 *Times* 13 Jan. 3/2 Inpatient therapy. 1965 *Mod. Law Rev.* XXVIII. v. 580 Persons who..are receiving in-patient treatment for mental disorder.

b. in-side *Cricket*, the side which is batting.

1837 D. Walker *Games & Sports* 224 Batters, belonging to the In-side. 1882 *Australians in Eng.* 16 Matters are going wrong with the in-side.

15. *in on* (cf. *on *adv.* 13): participating in; being (one of a group) in possession of knowledge concerning (something).

1923 A. Christie *Murder on Links* viii. 101 You don't mean—that you're in on *that*? 1928 E. Wallace *Gunner* xxiii. 189 As you're in on this, Gunner, you'd better see what I've said. 1935 M. M. Atwater *Murder in Midsummer* xxv. 235 He'll want to be in on this. 1946 [see *ground floor b]. 1953 W. Burroughs *Junkie* (1972) ix. 83 They are in on narcotics, and they are connected with Communism. *Ibid.* xii. 121 If any one makes a good score, she puts out a grapevine to find out who was in on the job. 1958 B. Nichols *Sweet & Twenties* 197, I was very much 'in' on the birth of this song. 1959 *Listener* 12 Feb. 283/1 It does not work for the American reader, who is not in on the secret. 1970 R. Lowell *Notebk.* 221 Anyway you should be in on it. Only In imagination can we lose the battle. 1973 'M. Innes' *Appleby's Answer* xv. 128 Don't imagine I have the slightest wish to be in on your muckraking.

in, *a.* Add: **2. a.** Fashionable, sophisticated;

esoteric. Cf. *in *adv.* 6 i, *in-group, *in-joke, *in-reference.

1960 *Spectator* 14 Oct. 555 A personable young *strippeuse* at Vegas (as we 'in' people call Las Vegas). 1961 *Harper's Bazaar* Apr. 138/2 The *dahlings* of the profession—beloved by 'in' audiences who adore a coterie joke. 1965 *Melody Maker* 13 Feb. 10 Record companies release more discs in the belief that folk is the new 'in thing'. 1969 C. F. Burke *God is Beautiful, Man* (1970) 107 He's got all the in stuff on and a big stick pin. 1969 *Daily Tel.* 2 July 19/6 The audience—totally committed to the 'pop' ideal and fully conversant with its idiomatic 'in' vocabulary—reacted suitably. 1970 O. Norton *Dead on Prediction* ii. 41 It *is* the in place. You'd be surprised who you meet there.

b. With hyphen and so passing into In- *pref.*[1]

1961 *Guardian* 29 Apr. 12/7 You will quickly find in N.Y. that the in-thing to do is to pronounce..Broadway. 1965 *Punch* 17 Nov. 721/1 Gnomes, it judged, are now in-people—possibly even top people. 1968 *Daily Tel.* 15 Nov. 16/7 The pace-setters here are upper-class girls who treat swearing as a snobbish in-thing to do. 1969 J. Braidwood *Ulster Dialect Lexicon* 17, I am informed that it has become the in-word in all Northern Ireland Ministry of Agriculture pamphlets. 1969 *New Scientist* 9 Oct. 74/1 The words 'computer' and 'education' must be two of the most overworked in-words of the decade. 1969 *Word* Feb. 38/1 Change is a contemporary theme; change is one of those in-words at the moment. 1969 *Time* 11 Apr. 55 The magazine's critics still point to its smug, In-crowd perspective. 1970 *Times* Suppl. 9 May 26/4 The in-crowd calls it [*sc.* Casablanca] 'Casa', and I offer the information here for anyone who can use it to advantage.

in, *sb.* Add: **2. ins and outs. b.** Those who are constantly entering and leaving the workhouse. Cf. *in-and-out class*, etc. (s.v. In and out, in-and-out *adv.* 4 in Dict. and Suppl.).

1884 *Daily News* 10 Dec. (Ware), There are considerable numbers of paupers..who find the workhouse a convenient retreat on emergency... They are known familiarly as 'the ins-and-outs'. 1896 *Rep. Poor Law Schools Comm.* xi. 71 The fluctuating class of children whose parents frequently discharge themselves from the workhouse and in a few days seek readmission. These cases are known among Poor Law officials as 'ins and outs'. 1905 *Rep. Brit. Assoc. Adv. Sci.* 467 The 'ins and outs' of Great Britain have characteristics which may be described as nomadic.

3. a. An introduction to someone of power, fame, or authority; influence with such a person. *colloq.* (orig. *U.S.*).

1929 E. D. Sullivan *Look at Chicago* (1930) ii. 21 His strong 'in' with police, built largely at the outset, with their organization's money. 1930 *Amer. Mercury* Dec. 454/1 *An in*, an introduction; to place in a position to bribe. 'Get me an in with the skipper of that precinct.' 1940 R. Chandler *Farewell, my Lovely* xxxvi. 276 It stands to reason that he had an in with the city government, but that don't mean they knew everything he did or that every cop on the force knew he had an in. 1947 J. Steinbeck *Wayward Bus* iv. 44 If her cousin was Clark Gable, why, that was an 'in' you couldn't beat. 1950 'J. Tey' *To love & be Wise* xix. 248 A girl has a more difficult time getting an 'in' in a racket. 1958 N. F. Leopold *Life plus 99 Yrs.* iii. 58 He had some sort of special 'in' with Warden Westbrook of the jail and was allowed more privileges than the other reporters. 1961 J. Heller *Catch-22* (1962) xxi. 210 The only colonel he trusted was Colonel Moodus, and even he had an in with his father-in-law. 1962 J. Wain *Strike Father Dead* v. 210 He had always been meaning to base himself in Paris for a while; all he had been waiting for was what he called 'an in'. Well, now he had got the ideal in. 1966 J. B. Priestley *Salt is Leaving* xiii. 177, I have an in with a couple of the directors. 1973 E. McGirr *Bardel's Murder* i. 18 He wondered if she had an 'in' with some manufacturer.

b. *to be on the in*: to have inside information. *U.S. colloq.*

1936 J. Steinbeck *In Dubious Battle* vi. 78 'You a big guy?' 'I'm on the in,' said the boy. 1942 Berrey & Van den Bark *Amer. Thes. Slang* § 480/4 *On the in*, smarted up, wise (to), wised (up).

in. Latin preposition. Add: **in abse·ntia**, in (his, her, or their) absence.

1886 *Edin. Univ. Cal.* 1885/6 141 Conferred *in absentia*. 1938 *Times Lit. Suppl.* 5 Feb. 88/3 The clergy in general are likewise condemned, though *in absentia*. 1955 *Times* 11 July 9/3 A Goa military court has sentenced *in absentia* Peter Alvarez, former president of the Goa National Congress. 1961 L. Mumford *City in History* xiii. 386 Residences of royal power *in absentia*, like Londonderry, Philippeville, and Christiansand. 1965 *Listener* 7 Oct. 544/2 *The Occult Diary*..goes on principally to describe his relationship, both actual and telepathically *in absentia*, with his third wife. 1972 D. Bloodworth *Any Number can Play* xi. 92 The renegade..has been condemned to death *in absentia*.

in a·ctu, in practice (as opp. to theory or potentiality).

[*a* 1680 S. Charnock *Works* (1684) II. 171 Some say..we are active *in actu exercito*, but not *in actu signato*. 1749 Chesterfield *Let.* 5 Dec. (1932) IV. 1453, I can only allow him *in actu primo* (to talk logic) and seldom *in actu secundo*.] 1902 W. James *Varieties Relig. Experience* xviii. 451 Whatever we may be *in posse*, the very best of us *in actu* falls very short of being absolutely divine. 1905 —— *Ess. Radical Empiricism* ix. 239 Radical empiricism, unable to close its eyes to the transitions caught *in actu*, accounts for the self-transcendency or the pointing..as a process that occurs within experience. 1907 —— *Pragmatism* vi. 222 Health *in actu* means, among other things, good sleeping and digesting.

in a·ntis *Class. Arch.* [see Anta], denoting a building in which the side walls are prolonged beyond the front and the pilasters terminating them are in line with the columns of the façade.

1848 W. Smith *Dict. Greek & Roman Antiquities* (ed. 2) 1105/2 There were never more than ten columns in the end portico of a temple; and when there were only two, they were always arranged in that peculiar form called *in antis*. 1875 *Encycl. Brit.* II. 388/2 Temples in Antis, with a portico of two or four columns in front. 1955 L. Woolley *Alalakh* ii. 71 It is tempting to restore this as a temple *in antis*, with two columns between the projecting walls (or between reveals from those walls) on the analogy of Niqme-pa's palace front. 1973 *Country Life* 20 Sept. 763/3 Its original first-floor portico, with Ionic columns *in antis*, survives.

in contuma·ciam, applied to sentences given against persons in contempt.

1892 in *Stanford Dict. Anglicised Words* 460/2. 1918 *Wireless World* VI. 156 It was only because they were for the moment beyond reach of the Italian Courts that they were condemned *in contumaciam*. 1923 *Westm. Gaz.* 28 Dec., Several of these verdicts were passed in contumaciam.

in co·rpore = *in vivo.

1906 *Rep. Brit. Assoc. Adv. Sci.* 1905 552 The nature of the substance contained in the serum was discussed. *In vitro* it has little power. *In corpore*,..the amount..necessary..is of no consequence. 1969 *Nature* 12 July 189/1 This is the first reported case of *in corpore* fertilization and development in the sea urchin, *Arbacia punctulata*.

in di·stans, at a distance (see *actio in distans).

1890 W. James *Princ. Psychol.* I. ii. 47 This blindness was probably due to inhibitions exerted *in distans*. 1909 —— *Pluralistic Universe* viii. 311 Remote professorial minds operating *in distans* upon conceptual substitutes for him [*sc.* God] alone.

in extenso (later examples).

1906 *Rep. Brit. Assoc. Adv. Sci.* 1905 257 Star Streaming. By Professor J. C. Kapteyn. (Ordered by the General Committee to be printed *in extenso*.) 1965 *Mod. Law Rev.* XXVIII. v. 618 Nearly every section is quoted *in extenso*.

in loco, (*b*) in a (*or* the) place; locally.

1671 Locke *Essay Draft B* (1931) § 94 p. 200 If it be said that it [*sc.* the soul] cannot change place, because it has none, for spirits are not *in loco* but *ubi*, I desire that distinction may be put into English or any other language and made intelligible. 1908 *Practitioner* Jan. 22 Some toxin either generated in loco..or reaching the skin from some distant focus of disease.

in no·mine, in the name (of): applied to (*a*) a motet or antiphon in fugal style, probably so called because originally used of a composition set to a text in which these words occurred, e.g., the Introit 'In nomine Jesu', the Psalm 'Deus, in nomine tuo'; (*b*) a free fugue in which the answer does not exactly correspond with the subject.

1636 C. Butler *Princ. Musick* 91 The *In-nomine's* of Parsons, Taverner, D. Ty, etc. 1876 Stainer & Barrett *Dict. Mus. Terms* s.v., The *in nomines* which exist are chiefly the production of composers of the 16th century. 1970 W. Apel *Harvard Dict. Mus.* (ed. 2) 412/2 Purcell's two 'In nomine' compositions represent a late attempt at revival of the form.

in nu·ce, in a nutshell, in a condensed form.

1854 Geo. Eliot tr. *Feuerbach's Essence Christianity* II. xxii. 214 The religious man is happy in his imagination; he has all things *in nuce*; his possessions are always portable. 1948 L. Spitzer *Linguistics & Lit. Hist.* iii. 100 A reduction *in nuce* of the general aesthetics of Racine. 1972 *Times Lit. Suppl.* 7 Apr. 382/1 A grand summation and cosmology *in nuce*. 1973 *Times Lit. Suppl.* 26 Oct. 1307/1 The idea of the modern family as social democracy *in nuce*.

in pe·ctore = *in petto* (see Petto).

1858 N. Wiseman *Recoll. Last Four Popes* vii. 333 The Pope made this speech..in this form: 'Moreover, *we create* a cardinal of the Holy Roman Church,..whom, however, we reserve *in pectore*.' 1876 *Encycl. Brit.* V. 98/1 The change which Paul III. introduced consisted in confining the secret of the unpublished nominations [of cardinals] to his own breast, keeping it '*in pectore*'. 1963 *Times* 29 May 10/2 One suggestion about the current affairs discussed with Cardinal Cicognani is that the Pope is preparing to reveal the identities of the three Cardinals whom he created *in pectore* in 1960—a device by which the Pope may keep to himself the names of certain new appointments if he feels that the moment is inopportune to publish them.

in propria persona (later examples).

1851 Geo. Eliot *Let.* 28 Jan. (1954) I. 344 When am I to have Mr. Bray's promised letter? or to see him in propria persona. 1958 *Times* 19 Aug. 11/2 *Così fan tutte* was performed in the great hall of the Residenz, where Leopold and Wolfgang Mozart themselves had often been *in propriis personis*. 1967 *Listener* 23 Mar. 408/3 In Figaro he [*sc.* Rossini] has created the ideal character from which he can speak *in propria persona*.

in re, (*a*) in reality (see sense 21 in Dict.); (*b*) *Logic*, = *extra dictionem (opp. *in voce*); (*c*) *Metaph.*, (of universals) dependent for their existence on the existence of the particulars that instantiate them, as Aristotle held; having real or objective, not merely mental, existence, but not separately from particulars (cf. *ante rem); also **in re·bus**; (probably taken by modern writers from Duns Scotus (*c.* 1264–1308)); (*d*) in the matter of, referring to, = Re *sb.*[2]

(*a*) 1602 W. Watson *Decacordon* 145 Wherein the Iesuits..had any speciall commoditie or gaine *in re* or *in spe* thereby. *a* 1680 S. Charnock *Works* (1684) II. 853 Their Sacraments and ours were the same *in re*, though diverse in signs.

(*b*) 1847 [see *in dictione]. 1906 [see *extra dictionem].

(*c*) 1879 W. James *Coll. Ess. & Rev.* (1920) 112 If they begin with a clear nominalistic note, they are sure to end with a grating rattle which sounds very like *universalia in re*, if not *ante rem*. 1904 [see *ante rem]. 1907 W. James *Pragmatism* vi. 221 Like wealth, health also lives *in rebus*. 1927 [see *ante rem]. 1952 R. I. Aaron *Theory of Universals* ii. 26 Locke here denies the *In Re* theory... The white itself is not *in re* but 'in the mind'. 1953 H. H. Price *Thinking & Experience* i. 10 The traditional Aristotelian doctrine of *universalia in rebus*. *Ibid.* ii. 56 There *are* universals, existing and subsisting *in rebus*.

(*d*) 1877 *Times* 18 Jan. 11/4 Court of Bankruptcy... In re B. and L. Harris. This was an adjourned sitting for public examination. The bankrupts, Messrs. Benjamin and Lawrence Harris, were merchants. 1886 *Athenæum* 20 Nov. 671/2 The alleged 'misrepresentation' *in re* Squeers *v.* Bentley. 1896 E. Terry *Let.* 7 Dec. in *Ellen Terry & Bernard Shaw* (1931) 136 What do you mean by saying (in re The Philanderer) it is dull and bestial? 1955 *Times* 29 June 3/7 *In re* Hillier ([1954] 1 W.L.R. 700), but it did not occur to counsel in that case. 1972 *Times* 22 Feb. 14/5 In *In re Scarisbrick* ([1951] Ch 622) the Court of Appeal held that the distinction between a public or charitable trust and a private trust depended on [etc.].

in se *Philos.*, in itself. So **inseity** (in,si·ĭti), in-itselfness.

1868 W. James *Let.* 5 Apr. in R. B. Perry *Tht. & Char. W. James* (1935) I. xv. 269 To the Greeks a thing was evil only transiently and accidentally... Bystanders could remain careless and untouched—no after-brooding, no disinterested hatred of it *in se*, and questioning of its right to darken the world. 1879 —— *Coll. Ess. & Rev.* (1920) 94 Substance *in se* cannot be directly imaged by feeling. 1909 —— *Meaning of Truth* vii. 167 Useful to test truth by, the matrix of circumstance, he thinks, cannot found the truth-relation *in se*. 1899 A. E. Garvie *Ritschlian Theol.* ii. 48 The thing which we represent for ourselves is an existence in itself (inseity). 1940 *Mind* XLIX. 177 The *otherness* of 'self' and 'other' need in no wise conceal the inseity of the 'other' from the 'self'.

in situ (earlier example). Also *attrib.*

1740 W. Stukeley *Stonehenge* iv. 21 Eleven of them are standing *in situ*. 1912 *Proc. Amer. Philos. Soc.* LI. 490 *In situ* forests occur frequently in shale beds. 1940 *Archit. Rev.* LXXXVII. 102/2 The foundations to receive the superstructure are formed of concrete in-situ posts. 1968 *Daily Tel.* 4 Nov. 9/6 Constructed entirely in in-situ concrete..it resembles a fortified gateway to the main hospital buildings. 1971 *Nature* 24 Dec. 432/1 The Department of the Environment, which has responsibility for *in situ* historical monuments in England.

in sta·tu nasce·ndi, in the process of creation, formation, or construction.

1890 W. James *Princ. Psychol.* II. xvii. 11 Black can only be felt in contrast to white..and in like manner a smell, a taste, a touch, only, so to speak, *in statu nascendi*, whilst, when the stimulus continues, all sensation disappears. 1927 B. Malinowski *Sex & Repression in Savage Society* IV. i. 180 And let us clearly and explicitly recognize that we can never observe it [*sc.* culture] in *statu nascendi*. 1948 *Sci. News* VII. 112 The ENIAC has been in use for some time, the other machines are still *in statu nascendi*.

in sta·tu pupilla·ri, as a pupil or ward; under scholastic discipline; at the universities, designating all who have not the degree of Master.

1855 *Newsp. & Gen. Reader's Compan.* § 571 A young Englishman..while still in statu pupillari. 1860 *Once a Week* 21 July 95/2, I fully admit that in later years we are all of us apt to grow sentimental about the traditions of our respective schools—I merely deny that we do so whilst we remain *in statu pupillari*. 1862 Thackeray *Philip* III. x. 214 Other young women who are kept by over-watchful mothers too much *in statu pupillari*. 1882 *Standard* 25 Dec. 5 (Stanford), Academic and urban magnates, fellows, and tutors have predominated over guests who are *in statu pupillari*. 1903 'Sigma' *Personalia* 172 One of those dusky potentates *in statu pupillari*, who were nearly always represented at the Master's dinners. 1930 *Sunday Times* 12 Oct. 26/2 The possession of a motor-car makes it easy for a person *in statu pupillari* to spend a large part of his existence elsewhere than in the University. 1965 *Rep. Comm. Disciplinary Powers Vice-Chancellor & Proctors* (Univ. Oxf.) 8 Difficulties might arise if graduates *in statu pupillari* who, very often, do the same sorts of things as undergraduates, such as running university clubs, are not bound by the same regulations as their undergraduate colleagues.

in terrorem (later examples).

1960 *Times* 12 Apr. 13/7 They operate mainly *in terrorem*; the potential tax avoider will not incur the cost and complication of a transaction unless he can be reasonably sure of its result. 1970 *Internat. & Compar. Law Q.* XIX. II. 214 Perhaps no more would be necessary than a criminal sanction against rent usury operating *in terrorem*, as exists in German law. 1971 *Daily Tel.* 1 July 4/6 This idea that the Treaty of Rome, like diamonds, is forever, is held *in terrorem* over people.

in toto (further examples).

1811 G. Constable *Let.* 31 Dec. in J. Constable *Corr.* (1962) I. 73 If my opinion was requested it would not be to give up your female acquaintance in toto. 1954 M. Beresford *Lost Villages* vi. 186 In fact we cannot be all that certain that it was all that much less *in toto*. 1955 *Times* 4 May 22/6 These amounts, *in toto*, fall short of our requirements if the business is to continue to expand as it is doing. 1965 *Mod. Law Rev.* XXVIII. 595 Lord Cameron's reasoning is applicable *in toto* to the situation in *The Acrux*.

in u·tero, in the uterus or womb, unborn.
1713 W. CHESELDEN *Anat. Humane Body* IV. iii. 170 It seems highly necessary, that the Ducts thro' which the Body receives Nourishment after the Birth, shou'd be kept open by a Fluid passing that way whilst it is in Utero. 1728 CHAMBERS *Cycl.* s.v. *Generation*, For that Dr. Harvey could never discover any thing of it *in utero*. 1795 W. TURNBULL in *Mem. Med. Soc. London* IV. 364 (*title*) A case where small-pox was communicated from the mother to the child in utero. 1871 A. MEADOWS *Manual of Midwifery* (ed. 2) III. ii. 156 The placenta, which..still remains for awhile in utero. 1901 [see STOMATITIC *a.*]. 1964 M. CRITCHLEY *Developmental Dyslexia* xiii. 75 Any theory of minimal brain damage—whether or not sustained *in utero*—is also unconvincing. 1966 *Lancet* 24 Dec. 1403/2 Stiehm et al. conclude that raised levels of γM or γA globulins at birth are presumptive of in-utero infection.

in vacuo (later examples).
1937 [see *ELECTRONIC *a.* 1]. 1942 *Tee Emm* (Air Ministry) II. 140 It is more correct to regard a running fix as a means of approximating the position 'in vacuo' so to speak, when more precise methods are unobtainable. 1955 *Times* 25 May 15/3 Whatever meaning people might ascribe to the word 'tramp' *in vacuo*.

in vi·no ve·ritas, truth comes out under the influence of alcohol; a drunken person tells the truth.
1594 LYLY *Mother Bombie* in *Wks.* (1902) III. III. iii. 199, I perceiue sober men tel most lies, for *in vino veritas*. 1616 T. ADAMS *Divine Herball* 27 And though the Prouerbe be, *In vino veritas*; yet as drunke as he is, you shall neuer haue truth break out of his lips. 1831 DISRAELI *Young Duke* III. IV. vi. 50 There was Cogit, who, when he was drunk, swore that he had had a father; but this was deemed the only exception to *in vino veritas*. 1927 D. H. LAWRENCE *Mornings in Mexico* 174 They say: *in vino veritas*. Bah! They say so much! 1936 N. MARSH *Death in Ecstasy* xii. 145 We had a clear case of *in vino veritas*. *a* 1953 E. O'NEILL *Long Day's Journey* (1956) IV. 145 Got to tell you now. Something I ought to have told you long ago... Not drunken bull, but 'in vino veritas' stuff.

in vitro (vī·tro). *Biol.* [lit. 'in glass'.] In a test tube, culture dish, etc.; hence, outside a living body, under artificial conditions; also *attrib.*, performed, obtained, or occurring *in vitro*.
1894 GOULD *Dict. Med.* 623/2 *In vitro*, in the glass; applied to phenomena that are observed in experiments carried out in the laboratory with microörganisms, digestive ferments, and other agents, but that may not necessarily occur within the living body. 1901 *Jrnl. Exper. Med.* V. 355 Serum obtained by immunising with one race did not necessarily give more than a trace of reaction in vitro and none whatever in vivo when tested with another race. 1912 BROQUET & SCOTT tr. *Burnet's Microbes & Toxins* x. 193 The neutralisation in an ordinary test-tube of a toxin by an antitoxin was one of the first and most brilliant '*in vitro*' experiments in immunity. 1925 C. H. BROWNING *Immunochem. Stud.* 14 These alcoholic extracts possess the property of reacting with heterophile antibody *in vitro*. 1955 *Sci. Amer.* June 88/3 Streptomycin has little or no activity against fungi *in vitro*, but it..controls blue mold of tobacco..and rot of sugar beets—all fungus diseases. 1962 *Lancet* 5 May 936/2 The in-vitro results show that fucidin is very active ..in sterilising cultures of staphylococcus at higher concentrations. 1974 *Nature* 22 Nov. 302/1 Onset of rapid haemoglobin formation in the blood islands of the developing chick blastodisc, both *in ovo* and *in vitro*, commences at the stage of the 6- to 7-somite embryo.

in vivo (vī·vo). *Biol.* Within the living organism; also *attrib.*, performed, obtained, or occurring *in vivo*.
1901 [see **in vitro*]. 1947 *Sci. News* V. 78 Some chemical interchange between the plant and the bacteria must take place *in vivo* and be largely responsible for nitrogen fixation. 1962 VAN HEYNINGEN & WALEY in A. Pirie *Lens Metabolism Rel. Cataract* 336 Proteolytic enzymes are often characterized by their action on substrates which are not known to be the natural substrates of the enzyme *in vivo*. 1973 *Nature* 16 Feb. 457/2 The *in vivo* experiments were also carried out using mice bearing a 14 day old tumour.

21. Other phrases: *in abstracto* (examples); *in articulo mortis* (later examples); *in camera* (examples); *in concreto* (examples); *in excelsis* (examples); *in flagrante delicto* (examples); also (colloq.) *in flagrante*; *in pa·ri mate·ria*, in a like case or position; *in puris naturalibus* (earlier and later examples); *in sæcula sæculorum* (later examples).
1602 W. WATSON *Decacordon* ix. 310 Which if he can bring to passe..then shall the French be so fleeced in *abstracto*..to be distracted out of their wits. 1630 [see *in concreto* below]. 1884 W. JAMES *Ess. Radical Empiricism* (1912) xii. 269 Let us fall back from all concrete attempts and see what we can do with his notion of through-and-throughness, avowedly taken *in abstracto*. 1920 D. H. LAWRENCE *Lost Girl* vii. 127 She existed *in abstracto* as far as he was concerned. 1928 O. JESPERSEN *Internat. Lang.* I. 22 Those who have spoken in favour of the idea *in abstracto*. 1933 *Times Lit. Suppl.* 13 Apr. 256/2 Towards the religious life of India Dr. Söderblom is kinder than many critics, but he, of course, is dealing with it *in abstracto* and not in its cruder manifestations. 1951 PARSONS & SHILS *Toward Gen. Theory Action* i. 99 The actor *in abstracto* is simply a set of properties by which he can be classified; in action he is involved in a system of relationships. 1967 *Word* XXIII. 385 We remember that the verb expresses nothing but the verbal idea in abstracto without containing any actualizing elements such as tense, mode, or voice. 1617 J. CHAMBERLAIN in T. Birch *Cour & Times James I* (1848) II. 1 The late lord chancelor left this world, being visited *in articulo mortis*, or not full half an hour before. 1825 SCOTT *Talisman* in *Tales Crusaders* IV. xv. 344 Nor did I mention it, save *in articulo mortis*,..to yonder reverend hermit. 1929 *Encycl. Brit.* VI. 231/1 Those under discipline were allowed to receive the eucharist when *in articulo mortis*. 1872 E. BROWNING *Exposition Laws Marriage & Divorce* 5 The Judge may, and usually does sit *in camera* to hear suits for nullity of marriage, where the matters to be disclosed are unfit for the public ear. 1882 *Standard* 26 Dec. 5/7 The case is one that in England would be heard *in camera*. 1955 *Times* 25 June 11/7 All documents filed in the Court be confidential and kept secret, and that every application for an adoption order should be heard and determined *in camera*. *Ibid.* 14 July 5/1 A naval court martial held *in camera* to-day in H.M.S. Victory at Portsmouth. 1966 P. MOLONEY *Plea for Mersey* 34 Fearing that the evidence might be a bit lurid the judge ordered 'Clear the court!' The evidence will be heard in camera.' The accused immediately asked 'What the Hell's "in camera"..?' 1970 *Guardian* 8 Aug. 9/2 The expurgated version of the in-camera sessions. 1602 W. WATSON *Decacordon* ix. 310 Though by the law Salique the Lady Infanta may be..put from her..lawfull claime to the whole kingdome of France, *in concreto*. 1630 T. ADAMS *Workes* 633 And the Popes haue so wrought it and brought it about now, that they will not onely *in abstracto* bee had in reuerence; but *in concreto* be feared with obseruation. 1798 A. F. M. WILLICH *Elem. Critical Philos.* 39 A bare idea of a possible science, which is no where given *in concreto*. 1885 W. JAMES in *Mind* X. 41, I may satisfy him that the words mean for me just what they mean for him, by showing him *in concreto* the very animals and their arrangements, of which the pages treat. 1902 —— *Varieties Relig. Experience* ii. 31 Not a deity *in concreto*, not a superhuman person, but the immanent divinity in things..is the object of the transcendentalist cult. 1936 C. S. LEWIS *Allegory of Love* vi. 289 The symbol ..fits all, and gives *in concreto* a characteristic of our life. 1602 W. WATSON *Decacordon* i. 21 Though to vs vnknowne to be of the same church triumphant *in excelsis*. 1882 *Athenæum* 23 Dec. 854 It is an uncritical guide-book *in excelsis*. 1927 *Melody Maker* Sept. 923/3 By all means strive to become a Frankie Trumbauer, or a Jimmy Dorsey, for that is musicianship and style *in excelsis*. 1965 *New Statesman* 30 Apr. 690/1 Tchaikovsky is the Victorian Artist *in excelsis*, grand, noble, compassionate. 1612 T. SHELTON tr. *Cervantes' Hist. Don Quixote* III. viii. 190 All was done in *Flagrante*, there was no leisure to giue me torment, the cause was concluded. 1772 'JUNIUS' *Lett.* II. lxviii. 314 A person positively charged with feloniously stealing, and taken *in flagrante delicto*, with the stolen goods upon him, is not bailable. 1876 tr. *P. J. van Beneden's Animal Parasites* 2 The sharper passes for an honest man as long as he has not been taken *in flagrante delicto*. *a* 1930 D. H. LAWRENCE *Phoenix* (1936) 17 But at last she caught him *in flagrante*. 1942 E. PAUL *Narrow St.* ii. 16 He saw, not his cringing wife and the imaginary lover he had always sworn to catch *in flagrante delicto*, but his swarthy waiter and a strange girl. 1972 D. BLOODWORTH *Any Number can Play* i. 2 A shocked Font-le-Baume caught them *in flagrante*. 1867 WHARTON *Law Lexicon* (ed. 4) 485/2 *In pari materiâ*, dealing with the same subject-matter. 1932 *Times Lit. Suppl.* 16 June 436/3 The comparison..on page 106 is not quite *in pari materia*: the three and a half million Churchfolk are adults, the thirty-five the whole population. 1955 J. L. AUSTIN *How to do Things with Words* (1962) ix. 112 Even the minimum physical action..is, being a bodily movement, *in pari materia* with at least many of its immediate and natural consequences. 1602 W. WATSON *Decacordon* vii. 204 As inclined to seeke for good to eschewe euill, and wishing after *summum bonum*, if *in puris naturalibus* they could haue obtained it. 1859 *Harper's Mag.* June 46/2 The natives..may be found not only in their primitive state, but even *in puris naturalibus*. 1930 D. H. LAWRENCE *Nettles* 18, I thought it was a commonplace That a man or a woman in a state of grace In puris naturalibus, don't you see, Had normal pudenda, like you and me. 1716 POPE *Let.* 9 July (1956) I. 347 We begin to wish you had the singing of our Poets..to yourselves, in *Sæcula Sæculorum*. 1841 THACKERAY *Misc. Ess.* (1885) 219 So Pride and Hatred continue *in sæcula sæculorum*. 1890 [see **atom-like*]. 1940 *Penguin New Writing* I. 10, I thought of the British Raj as an unbreakable tyranny, as something clamped down, in saecula saeculorum, upon the will of prostrate peoples.

in-, *pref.*[1] Add: **b.** *Geom.* Representing INSCRIBED *ppl. a.* 3, as in *in-centre*, *in-circle*, *in-sphere*.

-in, *suffix*[1]. Add: Also used systematically to form the names of certain unsaturated six-membered heterocyclic monocyclic compounds having no nitrogen atom in the ring, as *dioxin*. Cf. *-INE[5].
1928 [see *-INE[5]]. 1940 in PATTERSON & CAPELL *Ring Index* 21. 1957 E. H. RODD *Chem. Carbon Compounds* IVA. 4 Six membered rings in their least hydrogenated forms have names ending in '-in' when non-nitrogenous and '-ine' when nitrogenous.

-in, *suffix*[3]. The adverb IN used as a suffix originally designating a communal act of protest by Negroes in the United States against racial segregation (cf. *SIT-IN); subsequently indicating any group protest or large gathering for some common purpose. Examples are very numerous: e.g. *apply-in*, *BE-IN, *bury-in*, *chain-in*, *cook-in*, *drive-in*, *eat-in*, *FISH-IN, *hate-in*, *join-in*, *kiss-in*, *kneel-in*, *lie-in*, *LOVE-IN, *marry-in*, *mill-in*, *pedal-in*, *play-in*, *pray-in*, *read-in*, *scrub-in*, *sew-in*, *shout-in*, *sit-in*, *sleep-in*, *solve-in*, *stall-in*, *stand-in*, *study-in*, *sweep-in*, *swim-in*, *teach-in*, *wade-in*, *walk-in*. Chiefly attached to verb stem, but also to adjectives, e.g. *fat-in*, *nude-in*, and to substantives designating a participant in the protest or gathering, e.g. *kneeler-in*, *sitter-in*, *wader-in*.
1960 *Newsweek* 16 May 34/1 Into the already-roiled waters of the South, Negroes will wade this summer in a campaign to break down segregation at public beaches—a wade-in counterpart to the widespread lunch-counter sit-ins of recent weeks. 1960 in *Amer. Speech* (1961) XXXVI. 282 Negro college students have initiated a new 'kneel-in' campaign..by attending services at white protestant Atlanta churches. 1961 in *Ibid.*, He called for *walk-ins* in art galleries and museums, *drive-ins* at segregated motels and roadside ice cream stands, *sit-ins* in court rooms, *study-ins* at segregated schools, and *bury-ins* to integrate cemeteries. *Ibid.*, Negro teen-aged boys in an impromptu swim-in at an undesignated beach drew a crowd of 300 shoving, shouting Memorial Day bathers and boaters yesterday. *Ibid.* 284 A Chattanooga, Tenn., Negro stand-in demonstrator says his son's life has been threatened. 1961 *Guardian* 26 May 11/1 The United Presbyterian Church of America...recommended a 'kneel-in' campaign as a manifestation of the belief in the right of all people to worship regardless of race. 1961 *N.Y. Times* 9 Nov. 37 Last night, twenty-four students gathered in the campus library for an all-night 'read-in' demonstration. 1963 *Time* 30 Aug. 12 Demonstrators.. prostrate themselves before bull-dozers at construction-site 'lie-ins'. *Ibid.*, The 'pray-in' at churches. 1964 *Economist* 25 Apr. 376/1 The threatened 'stall-in' of thousands of motor cars [by Negroes]. 1965 *N.Y. Times* 28 Mar. 2E/8 There have been sit-ins, lie-ins, stand-ins, eat-ins, shop-ins, sleep-ins, swim-ins, and sing-ins. 1965 *Economist* 19 June 1401/2 This week Mr Johnson countered the university 'teach-ins', protesting against his foreign policy, with a cultural 'play-in' demonstrating his respect for the arts. 1966 *Daily Tel.* 12 Aug. 11/3 William Bryden-Smith, aged 10, who wrote to us, wants to take part in the cook-in. 1967 *Ibid.* 3 Mar. 23/7 A 'kiss-in' to protest against Michigan University's stern regulations on 'public displays of affection' was described by students last night as the most enjoyable form of demonstration yet devised. 1967 *New Statesman* 17 Mar. 356/3 Last week police arrested scores of teenagers at a rave-in, and left-wing Catholics staged a pray-in. 1967 *Times* 28 Mar. 4/7 It took the police three hours to clear a 'mill-in' at the intersection of Haight and Ashbury streets, celebrated as the chief resort of the bearded and sandalled 'hippies' who travel to San Francisco from all over the United States to signify their dissent from modern society. 1967 *Observer* 11 June 10 If everyone was fat there'd be no war. No one would pass the physical.—A speaker at the New York Central Park 'Fat-in'. 1967 *Listener* 10 Aug. 188/3 This is a very exciting inversion of psychedelic soulfulness, a hate-in. 1967 *World Study* Dec. 7/1 *Chain-in*, demonstrators locking themselves to a city hall pillar,..until the mayor listened to their grievances. *Ibid.* 7/2 *Stand-in*, demonstrators lining up at a theater ticket booth until given admission to the theater, not simply to a segregated section in the rear of the house but to any area where a white patron—or a member of any race—may sit. 1968 *Lebende Sprachen* XIII. 68/1 Their action fits into a wave of unofficial, unconnected *nude-ins* so far this year in Golden Gate Park, starting with freebeachers dancing nude at the great be-in. 1968 *Listener* 26 Dec. 849/1 Charge of the Light Brigade, 20th-century style: a lie-in at Porton Microbiological Research Establishment. 1969 *New Yorker* 3 May 31/1 Another rally was held at the Campus Center, followed by what was termed a 'mill-in' at the Army, Air Force, and Marine recruiting site. 1969 *Daily Tel.* 29 Jan. 1/6 About 20 of the militants..ended their vigil yesterday with a 3 a.m. 'swim-in' in the basement pool. *Ibid.* 2 July 18/3 Practical tests in revolutionary rhetoric ('shout-ins'), wall-defacement and anti-Establishment violence. 1970 *Ibid.* 2 Mar. 16 To teach-in ..and sit-in..have now been added walk-in and work-in. The first means the occupation of premises outside undergraduates' recognised territory, the second a teach-in during vacation. 1971 *Guardian* 28 Sept. 15/3 A student sleep-in began last night. 1973 *Daily Tel.* 3 Dec. 13/8 College catering would be disrupted by students alternately boycotting canteens and then holding mass eat-ins.

in absentia: see *IN *Lat. prep.*

inacceptable, *a.* Delete *rare* and add later examples.
1908 *Westm. Gaz.* 22 Oct. 7/1 The Turkish Foreign Minister replied that the proposal was inacceptable. 1934 A. S. C. Ross in *Neuphilol. Mitt.* XXXV. 129 The perpetuation of a Victorian prudishness (inacceptable in philology beyond all other subjects). 1938 E. BEVAN *Symbolism & Belief* v. 108, I am in the company of others who find Professor Alexander's philosophy in this respect inacceptable. 1970 H. BRAUN *Parish Churches* xii. 160 Such dreariness being inacceptable to the parish churchman..Gothic forms began..to represent the appropriate Anglican architecture.
Hence **i:nacceptabi·lity**, the quality or condition of being inacceptable.
1922 *Glasgow Herald* 13 May 8 Whatever..France's attitude..might be, the inacceptability of the memorandum as a whole appears to render equally null in her eyes this detail. 1957 J. S. HUXLEY *Relig. without Revelation* (rev. ed.) iv. 75 The inacceptability, to my growing intellectual interest, of any Christian theology proffered to me.

inactivate (inæ·ktivēɪt), *v.* [f. INACTIV(E *a.* + -ATE[3].] *trans.* To render inactive. So **ina·cti-vated** *ppl. a.*
1906 *Jrnl. Physiol.* XXXIV. p. xxxvi, If only the decomposition of the substrate is considered, heat inactivates enzymes completely. *Ibid.*, The presence of inactivated enzyme has a marked influence on the reaction

between active enzyme and substrate. 1927 HALDANE & HUXLEY *Animal Biol.* ix. 189 When unaccustomed proteins enter their system, they [*sc.* higher animals] can.. destroy or inactivate them. 1949 *Jane's Fighting Ships* 1949–50 358 She has been 'inactivated', and is laid up in a state of preservation as a potential fighting ship. 1955 *Times* 11 June 6/4 It said that the vaccine is composed of three types of viruses, mixed together after they had been inactivated or killed, but in some cases live viruses were found in the final mixture after tests. 1962 *Lancet* 5 May 941/1 The pattern of serological response to inactivated poliovirus vaccines has been..well documented. 1971 J. PHILIPS *Escape a Killer* (1972) II. ii. 136 The moment we inactivate the fence, we sound an alarm.
Also **inactiva·tion**, the process of inactivating.
1906 DORLAND *Med. Dict.* (ed. 4) 353/2 *Inactivation*, the destruction of the activity of a serum by the action of heat or other means. 1936 *Brit. Jrnl. Psychol.* XXVI. 394 We shall employ the term 'impulse' in describing our experiments as implying the conscious activation and inactivation of the motor apparatus. 1946 *Nature* 27 July 121/1 The primary factor in heat damage seems to be an inactivation of the cell. 1968 *Economist* 6 July 16/3 Does the splendid nonsense word coined by the Americans, the 'inactivation' of Khe Sanh, signal another step in de-escalation?

inactivator (inæ·ktivēɪtǝɹ). [f. prec. + -OR.] That which inactivates; also, an individual considered in respect of his or her speed of metabolizing and so inactivating a drug in the body.
1944 *Biol. Abstr.* XVIII. 2662/3 (Index), Inactivator of vit. B_1 in diets of foxes. 1946 *Nature* 10 Aug. 201/1 The inactivator or inactivators involved are susceptible to heat and do not pass through a Seitz filter. 1960 *Times* 11 Nov. 17/2 Individuals could be divided into two categories: rapid and slow inactivators of the drug. 1970 PASSMORE & ROBSON *Compan. Med. Stud.* II. ii. 14/1 Patients who are 'slow inactivators' of isoniazid are more likely to develop peripheral neuritis if given the usual dose.

inactive, *a.* Add: **2.** *Chem.* [tr. F. (*moléculairement*) *inactif* (J. B. Biot 1840, in *Ann. de Chim. et de Physique* LXXIV. 403).] Not rotating the plane of polarization of polarized light. Often qualified by *optically*.
Pasteur adopted the term from Biot (*Jrnl. de Pharm. et de Chim.* (1848) XIII. 449).
1853 L. PASTEUR in *Chem. Gaz.* 1 Sept. 323 The latter [body]..resists isomeric transformation, and remaining without alteration in the quinicine, gives this its feeble deviation to the right. The other group, which..is very active, becomes inactive when the quinine is heated so as to become converted into quinicine; so that quinicine is nothing but quinine in which one of the active constituent groups has become inactive. 1857 W. A. MILLER *Elem. Chem.* III. v. 334 It [*sc.* another modification of tartaric acid] has been termed by Pasteur, inactive tartaric acid, in allusion to its want of action upon polarized light. 1905 *Jrnl. Physiol.* XXXII. p. xxxix (*heading*) The formation of inactive arginine by enzymes from proteids which yield optically active arginine on hydrolysis with acids. 1961 L. F. & M. FIESER *Adv. Org. Chem.* iii. 83 Properties of the two active and two inactive forms of tartaric acid are given in Table 3.2. A noteworthy point is that the *dl*-form, racemic acid, melts at a higher temperature than the optically active components.

in actu: see *IN *Lat. prep.*

inadequate, *a.* Add: Also as *sb.*, an inadequate person; one whose personality is in some way insufficient to meet the expectations of society.
1962 *Guardian* 7 Nov. 8/4 How can prison help a social inadequate through his troubles? 1963 T. PARKER *Unknown Citizen* 166 The majority of them are in fact the inadequates of whom Charlie Smith is typical. 1966 *Listener* 18 Aug. 226/1 The people that we are accustomed to call inadequates or weak characters. 1971 *Guardian* 26 Mar. 14/1 So-called 'inadequates' are the victims of society which is inadequate.

in-a-door, *adv.* = INDOORS *adv.* (App. only in Blunden.)
1932 BLUNDEN *Face of England* 14 We keep in-a-door unless compelled forth. 1937 —— *Elegy* 28 Hasten here in-a-door. 1949 —— *After Bombing* 36 And dog and cat run in-a-door.

inadvisable, *a.* Delete *rare* and add later examples.
1953 *Times Lit. Suppl.* 25 Sept. 613/2 The idea..is that it would be inadvisable for Germany to annex parts of Austria. 1953 *Times* 3 Oct. 3/4 Coventry isolation hospital is now receiving only cases which it would be inadvisable to send..to the isolation hospital at..Birmingham.

inagglu·tinable, *a.* [f. IN-[3] + *AGGLUTINABLE *a.*] Incapable of being agglutinated (*by*).
1919 *Lancet* 4 Oct. 607/2 This inagglutinable strain was isolated from a case a few weeks before..experiments commenced. 1934 *Jrnl. R. Anthrop. Inst.* LXIV. 34 Every individual whose corpuscles were inagglutinable by the A and B sera. 1951 WHITBY & HYNES *Med. Bacteriol.* (ed. 5) xii. 203 It [*sc.* the Vi antigen] is not affected when the organisms are killed with 75 per cent alcohol, although they may then become inagglutinable by a Vi antiserum.
Also **i:nagglutinabi·lity**, the property of being inagglutinable.
1925 C. H. BROWNING *Bacteriol.* viii. 169 Occasionally typhoid bacilli when they have been recently obtained from the living body are not susceptible to agglutination by antiserum, but such inagglutinability usually soon disappears. 1948 *Biol. Abstr.* XXII. 897 Inagglutinability of red cells occurs.

inajá (inădʒā·). [Tupi.] In full **inajá palm**. A palm tree, *Maximiliana martiana* (*regia*), native to the Amazon region.
1849 R. SPRUCE *Notes of Botanist on Amazon & Andes* (1908) I. i. 45 The pinnate fronds of the..Inajá (*Maximiliana regia*) reach sometimes 40 feet in length. 1853 A. R. WALLACE *Trav. Amazon* ii. 33 Here also grew the Inajá, a fine thick-stemmed species, with a very large dense head of foliage. 1860 MAYNE REID *Odd People* 82 These nuts [*sc.* palm-nuts] are the fruit of several kinds of palms, but the best are those afforded by two magnificent species,—the 'Inaja' (*Maximilliana* [sic] *regia*), and the 'Urucuri' (*Attalea excelsa*). 1866 LINDLEY & MOORE *Treas. Bot.* II. 726/1 M[*aximiliana*] *regia*, the Inajá Palm of the Amazon, has a trunk a hundred or more feet high. 1927 R. R. GATES *Botanist in Amazon Valley* viii. 174 The Inajá..is one of the common palms of the Amazon. It produces enormous masses of ovoid fruits the size of a lemon. 1966 E. J. H. CORNER *Nat. Hist. Palms* xii. 283 As a major feature of vegetation, they [*sc.* South American palms] call for botanical investigation in every respect. For example, the babaçu.., cohune.., piassava.., and inajá.

inaka, var. *INANGA.

inamorato. (Later examples.)
1941 E. R. EDDISON *Fish Dinner* (1968) xiii. 221 How would you like Shelley for your *inamorato*? 1973 'M. INNES' *Appleby's Answer* iii. 29 You must invite him to luncheon here. And invite me as well. I'd love to meet the *inamorato*.

in and in, in-and-in, *adv.* and *sb.* Add: **A.** *adv.* **b.** Entirely in, sharing fully.
1926 J. BLACK *You can't Win* x. 131 We know you are 'right'...That's why you are declared 'in and in' with the works.
2. (See quot.)
1926 *Paper Terminol.* (Spalding & Hodge) 14 *In and in*, a method of packing reams too large conveniently to travel flat. The ream is divided in half, and the two portions clasped in and in to each other.
B. *sb.* **3.** *slang.* (See quot.)
1935 *Evening News* 29 June 3/2 The '*in-and-in*' is simply the point at which the swindler apparently risks his own money with that of the dupe.

in and out, in-and-out, *adv.* Add: **4.** *attrib.* (quasi-*adj.*). (Further examples.) Also, *in-and-out bolt* (earlier example); *in-and-out boy*, *man*, someone in and out of prison; a burglar; *in and out family*, formerly, a family constantly entering and leaving a workhouse; *in-and-out running* (earlier example); *in and out work*, work which is not continuous; also, irregular or unlawful practice.
1841 R. H. DANA *Seaman's Manual* 111 *In-and-out*, a term sometimes used for the scantline of the timbers, the moulding way, and particularly for those bolts that are driven into the hanging and lodging knees, through the sides, which are called *in-and-out bolts*. 1855 GEO. ELIOT in *Fraser's Mag.* June 699/2 Heavy-looking in-and-out corridors, such as one found only in German inns. 1885 *Referee* 26 Apr. 1/2 Now and again in-and-out running on the part of a horse subjects his owner to considerable annoyance. 1888 'R. BOLDREWOOD' *Robbery under Arms* III. xviii. 277, I began to hear that there was a deal of in-and-out sort of work about my getting my freedom. 1903 *Westm. Gaz.* 29 Sept. 10/1 His engagements are only for particular plays—'in and out' work. 1904 *Daily Chron.* 14 June 9/1 One notable 'In and Out' family entered and discharged itself sixty-two times from a London workhouse in one year. 1906 E. DYSON *Fact'ry 'Ands* viii. 97 It was in 'n' out sorter work. 1910 E. M. FORSTER *Howards End* v. 38 Oh, heavens! I've knocked the In and Out card down. 1936 H. G. WELLS *Anat. Frustration* xv. 178 That does not close the Jewish problem for you. It merely brings you back to the fundamental age-long problem of this nation among the nations, this in-and-out mentality, the essential parasitism of the Jewish mycelium upon the social and cultural organisms in which it lives. 1937 C. PRIOR *So I wrote It* xvi. 191 Among the boys I knew, very few had either the courage or skill to tackle 'live gaffs' by night. Most of them were in-and-out boys. They did their eighteen months in Wandsworth or Pentonville, had a run of a month or so and went back to do a twenty-one or even a lagging. 1939 H. HODGE *Cab, Sir?* 221 An 'in and out job' is a passenger who comes back to his starting point. 1959 *Times* 31 Dec. 11/3 The discovery was announced in 1925 and met with a very in-and-out reception. 1960 *Encounter* Mar. 77/1 The *In-and-Out* Game is played...'Everyone' knows that Hemingway is not so *In* as Faulkner. 1961 PARTRIDGE *Dict. Slang* (ed. 5) II. 1143/2 *In-and-out man*, an opportunist thief. 1961 *Times* 14 Apr. 5/4 Two clubs who have had an in-and-out season meet at Old Deer Park. 1970 *Guardian* 12 Sept. 11/2 Israeli casualties were..high...The alternative ..would be a quick 'in and out' operation. 1972 *Times* 15 Dec. 14/3 'In and out' records through lapsing..are not uncommon.
B. *sb.* *The 'In' and 'Out'*, the name of the Naval and Military Club in London.
1925 FRASER & GIBBONS *Soldier & Sailor Words* 127 *The In and Out*, the Naval and Military Club, 94, Piccadilly. So called familiarly from the words 'In' and 'Out', painted on the pillars of the approach to the courtyard in front. 1967 *Guardian* 21 Feb. 3/6 Two London clubs, the 105-year-old Naval and Military, better known as the 'In and Out', and the 92-year-old Devonshire, may merge.
So **in-and-outer**, one who is only moderately skilled or is erratic in performance; one who holds office intermittently.
1905 *Outing* (U.S.) Feb. 572/2 Whippets run in more consistent form than horses, and there are few 'in-and-outers'. 1934 E. B. MARKS *They all Sang* 210 He was an in-and-outer..and this was one of his 'out' periods. 1952 *Assoc. Press* 8 Oct., Reynolds has 30 knockouts among his 52 victories but he has been an in-and-outer. 1960 *Farmer & Stockbreeder* 29 Mar. 53/3 There are a great many 'in-and-outers' who know mighty little of the art of breeding. 1967 *Economist* 5 Aug. 496/1 The British unfamiliarity with the American concept of 'ins and outers' —men who alternate between private careers and public service.

inanga (ī·naŋă). Also **inaka** (the South Island form). [Maori.] **1.** The New Zealand name for a small fish, *Galaxias attenuatus*, the young form of which is called whitebait.
1845 E. J. WAKEFIELD *Adventure N.Z.* II. 100 This fish is called *hinanga*, and resembles Blackwall whitebait in size and flavour. 1896 *Australasian* 28 Aug. 407/3 About the same size as this fish is the 'inaka' much used for bait. *a* 1939 'R. HYDE' *Houses by Sea* (1952) 64 Brown women drying out inanga. 1944 *Mod. Jun. Dict.* (Whitcombe & Tombs) 209 *Inanga, Inaka*, the Maori name of the small fish called 'whitebait' in New Zealand; inanga in the north, inaka in the south. 1959 TINDALE & LINDSAY *Rangatira* x. 93 Every year shoals of the tiny, delicious inanga fish came up the creeks from the sea. 1962 *Post-Primary School Bull.* (N.Z.) XV. i. 26 Whitebait belongs to the species *Galaxias attenuatus*... The adult fish, called Inanga, reach a length of about four inches.
2. An evergreen shrub or small tree, *Dracophyllum longifolium*, belonging to the family Epacridaceæ and native to New Zealand.
1889 T. KIRK *Forest Flora N.Z.* 215 Mr. Charles Traill informs me that it [*sc. Dracophyllum longifolium*] is termed 'inaka' by the Maoris on Stewart Island. 1910 L. COCKAYNE *N.Z. Plants* 76 In similar situations [*sc.* along the coastline] the inuka (*Dracophyllum longifolium*) and the smaller New Zealand flax (*Phormium Cookianum*) are common. 1929 W. MARTIN *N.Z. Nature Bk.* viii. 126 The Inanga ranges in size from a small shrub..to a tree 15 to 30 ft. high. 1961 H. H. ALLAN *Flora N.Z.* I. 532 *D. longifolium*... Inanga... The Forsterian specimens I have seen are scrappy. 1968 *N.Z. Listener* 15 Mar. 6/5 Lying in its sheltered cove..complete with its own rata tree, inaka bushes, and great cushy heaps of golden-green moss.

ina·ngulated, *ppl. a.* *Ent.* [f. IN-[3] + ANGULATED *ppl. a.*] Angled inwardly.
1898 *Proc. Zool. Soc.* 441 The inner stripe more or less strongly inangulated below median vein, the outer stripe zigzag.

in antis: see *IN *Lat. prep.*

inapparent, *a.* Delete †*Obs.* and add later examples.
1898 J. CAIRD *Univ. Sermons* 371 These are the invisible realities, inapparent, impenetrable to sense. 1960 *New Scientist* 16 June 1521/3 This vaccine is then fed to the individual and produces an inapparent or mild infection which leads to immunity. 1970 *Ibid.* 15 Jan. 101/1 One of the most difficult variables in biological research—namely, inapparent infections, which can cause differing and misleading responses to experimental treatments.

inappositeness (inæ·pǒzitnės). [f. INAPPOSITE *a.* + -NESS.] The character or quality of being inapposite.
1893 E. SALTUS *Madam Sapphira* iii. 50 With an inappositeness which afterward was to occur to Nevius as curious. 1894 W. J. LOCKE *At Gate of Samaria* (1895) xii. 134 The words estranged them still further. They were pathetic in their ludicrous inappositeness.

inarticulacy (inaɹti·kiŭlăsi). [f. INARTICULA(TE *a.* + -CY.] Inarticulateness.
1921 *Observer* 11 Sept. 10/2 That it has not done so..we must ascribe..to the inarticulacy of the scientist himself. 1927 M. SADLEIR *Trollope: a Comm.* 41 He has virtually entreated her forgiveness for his inarticulacy. 1951 A. BARON *Rosie Hogarth* 81 Perhaps it was this, more than his inarticulacy,..that had held him back. 1970 *Listener* 30 July 157/2 Prone to miscarriages, inarticulacy, and general indecision. 1971 *Daily Tel.* 9 Oct. 9/6 'Y'know' and 'I mean' and other apologies for inarticulacy.

inaugural. **B.** *sb.* Add: (Earlier U.S. example.) Also, an inaugural lecture at a university.
1832 *Reg. Deb. Congress U.S.* 5 May 2778, I turn now to the other points in the inaugural. 1958 *Listener* 18 Sept. 429/3 That well-known story about Seeley's inaugural. 1965 J. A. W. BENNETT in J. Gibb *Light on C. S. Lewis* 44 The mere quotability of his Cambridge Inaugural may lead us to forget his positive achievement. 1967 *Listener* 26 Jan. 139/2 Enright's famous inaugural at the University of Singapore.

inauguration. Add: **4.** (Later U.S. examples of *Inauguration Day*.)
1893 K. D. WIGGIN *Polly Oliver's Problem* (1894) xvii. 185 As it chances to be a presidential year, we will celebrate Inauguration Day. 1948 *Denison* (Texas) *Herald* 1 July 4/3 That was when Inauguration Day was changed from March 4 to January 20, for Roosevelt's second term.

inauthentic, *a.* Delete *rare* and add later example. **inauthenticity** (later examples).
1964 S. BELLOW *Herzog* (1965) 81 The cant and rant of pipsqueaks about Inauthenticity and Forlornness. 1970 G. GREER *Female Eunuch* 152 The altruism of women is merely the inauthenticity of the feminine person carried over into behaviour. 1972 *Times Lit. Suppl.* 24 Mar. 342/2 This does not prevent the two schools of Zen from attacking each other as being inauthentic.
Hence **inauthe·ntically** *adv.*
1965 *Philos. Rev.* LXXIV. 215 The philosophers he is attacking..have written inauthentically.

i·n-basket. [Cf. IN *adv.* 12 a.] In an office, etc.: a basket or tray for incoming correspondence and other documents. Cf. *IN-TRAY.
Sometimes written as two separate words with *in* regarded adjectivally.
1940 *Amer. Speech* XV. 247 His *incoming* mail is put in an *in-basket*. 1948 'N. SHUTE' *No Highway* v. 130, I sighed and pulled my IN basket towards me, full of the arrears of work. 1957 J. BRAINE *Room at Top* xxviii. 217 My in-basket was full. 1970 *New Yorker* 26 Sept. 40/3 She sat down beside him with pencil and paper and let him churn up the contents of his 'in' and 'out' baskets.

inbe (inbī·), *v.* *poet. nonce-wd.* [f. IN-*pref.*[1] + BE *v.* A. 1 ¶ (centre column p. 716). Cf. L. *inesse*.] To be within.
1921 HARDY *Late Lyrics* (1922) 70 Where such inbe, A dwelling's character Takes theirs.

in-between. Add: Hence **in-betwee·ner**, **inbetwee·ner**, a person who takes up an intermediate position (chiefly *fig.*).
1924 *Contemp. Rev.* Apr. 459 The undeserving and the in-betweeners far outnumber the deserving. 1927 C. SHAW *Let.* 17 May in Knightley & Simpson *Secret Lives Lawrence of Arabia* (1969) xvii. 256 In Ireland...We had The Gentry and The People: nothing else. You will say 'but the in-betweeners?' They belonged to the people. 1939 C. E. SMITH in Ramsey & Smith *Jazzmen* 247 The inbetweeners are still there but the musician has become an individual. 1942 O. NASH *Good Intentions* (1943) 162 He was a born in-betweener,... He always thought that there was much to be said on both sides. 1970 *S.A. Panorama* Feb. 25/1 He joined the London Polytechnic Studios. He started as an 'in-betweener'—the name given to the most junior apprentice.

inbind (inbəi·nd), *v.* [f. IN-[1] + BIND *v.*] To bind within (usu. in *pass.* and *fig.*); to bind within a book or manuscript (cf. *INBOUND *a.*[2]).
1888 RUSKIN *Praeterita* (1889) III. i. 32 The most stern practical precept of that doctrine still holding me,—it is curiously inbound with all the rest,—was the Sabbath keeping. 1913 D. H. LAWRENCE *Sons & Lovers* x. 316 He had never been very closely inbound into the family. 1932 *Ampleforth Jrnl.* Spring 133 A transcription of the fragment inbound in the Sarum Missal in the monastery library.

inboard, *adv.*, *prep.*, and *adj.* Add: **A.** *adv.* (Earlier and later examples.)
1830 J. F. COOPER *Water Witch* III. vii. 216 Assured of the position of his enemy, he returned in-board. 1851 H. MELVILLE *Moby Dick* III. xlix. 306 Two of them clutched the gunwale..and..hurled themselves bodily in-board. 1912 BELLOC *This & That* xxxviii. 284 We picked up the little buoy..and we got it in-board. 1958 *Engineering* 28 Mar. 393/1 Movable projectors are housed in the two cylindrical frameworks of the dumb-bell assembly inboard of the star globes. 1962 S. CARPENTER in *Into Orbit* 161 You make sure the abort handle is *inboard and locked*. 1963 *Times* 12 Feb. 15/6 Inboard-mounted disc brakes. 1967 *Autocar* 5 Oct. 7/2 The Triumph 2000 port face is 'inboard' of the holding-down studs. 1971 *Times* 16 Feb. 8/6 The unsprung weight has been reduced by mounting the front and rear suspension units inboard.
C. *adj.* (Earlier and later examples.) Also, applied to parts of vehicles, aircraft, etc.
1847 WEBSTER, Inboard cargo. 1886 *Forest & Stream* 13 May 316/3 The Hampton flattie is best described as a beamy sharpie with an 'inboard jib'. 1893 *Funk's Stand. Dict.*, Inboard stroke of the piston. 1909 *Cent. Dict. Suppl.*, In *mech.*, toward the inside; toward the main center or center-line: as, an inboard stroke of the piston; an inboard bearing. *Ibid.* s.v. *Profile*, *Inboard profile*, in ship-building, a plan which shows the internal arrangements of a vessel by a longitudinal vertical section at the centerline. 1921 *Rudder* July 5 Even in large sailing boats the use of an inboard engine is sometimes not advisable. 1936 *Discovery* Dec. 380/1 We were five boats in all, some inboard launches, others with outboard motors, which passed the inspection of the river police. 1945 *Times* 30 June 4/5 A new type of drive couples the engines to contra-rotating airscrews, 16 ft. in diameter. Those driving the inboard engines can be reversed on landing to act as brakes. 1948 'N. SHUTE' *No Highway* iii. 66 Stop the inboard engines [of the aircraft] and turn back. 1951 *Yachting Monthly* Oct. 202 Her modest inboard-cutter sail plan is easily handled by two. 1956 LOOMIS *Hotspur Story* 183 With her new inboard rig *Hotspur* is now a single-headsail cutter. 1957 M. SWAN *Brit. Guiana* 202 In the old days when the mining men, the pork-knockers, went up the river they would paddle against currents which an inboard engine can now only just fight. 1959 E. K. WENLOCK *Kitchin's Road Transport Law* (ed. 12) 19/2 Inboard brakes, that is to say brakes acting on the axle shafts. 1971 *Power Farming* Mar. 57/3 A unique feature is the individual shaft bottom drive to each cutting disc, relieving the strain on the inboard drive components compared with other designs where the cutters are powered in train. 1972 *National Observer* (U.S.) 27 May 9/3 The inboard portions of the belts are short to prevent the lap belt from slipping over the hip bone onto the stomach, where the belt might cause serious injury in an accident.

inborn, *ppl. a.* Add: **2.** *inborn error of metabolism*: any disorder or abnormality that is due to a hereditary fault in the metabolic processes of the body, generally attributable to the lack or alteration of some enzyme.
1908 A. E. GARROD in *Lancet* 4 July 3/1 Quite unlike that of the above metabolic diseases is the course of the anomalies of which I propose to treat in these lectures and which may be classed together as inborn errors of metabolism. 1909 —— (*title*) Inborn errors of metabolism. 1935 HARROW & SHERWIN *Textbk. Biochem.* xxviii. 716 The disease [*sc.* cystinuria], frequently referred to as an 'inborn error of metabolism' is rare. 1968 PASSMORE & ROBSON *Compan. Med. Stud.* I. xi. 24/2 Hereditary lack of the necessary enzyme, phenylalanine hydroxylase, leads to an inborn error of metabolism, phenylketonuria, in which phenylalanine..is transaminated extensively to phenylpyruvic acid... Presence of phenylpyruvic acid in the blood and central nervous system leads to mental retardation.

inbound, *a.*[1] (Further examples.)
1967 *Jane's Surface Skimmer Systems* 1967–68 63/1 Inbound freight is sorted on arrival at terminal. 1971 D. POTTER *Brit. Eliz. Stamps* xiii. 144 Clever collectors with new issues of stamps purchased at the all-night post office in London, would rush to stations within 50 miles or so of London to post their covers on inbound services. 1973 D. WESTHEIMER *Going Public* ix. 127 Drivers on the inbound lanes slowed to a crawl.

i·nbound, *a.*[2] [f. *INBIND *v.*; cf. BOUND *ppl. a.*[2]] Of a leaf, gathering, etc., in a manuscript or book: bound within.
1953 D. C. C. YOUNG in *Scriptorium* VIII. 15 There is an inbound quaternion of blanks.

inbreath (i·nbreþ). [f. IN *adv.* + BREATH.] A drawing in of the breath.
1921 R. GRAVES *Pier-Glass* 26 The deep in-breath, The breath roaring out. 1936 J. G. BRANDON *Dragnet* iv. 37 It was accompanied by a hissing inbreath from Ferradi. 1956 J. LOTZ in L. T. White *Frontiers of Knowledge in Study of Man* xiv. 212 In Swedish..the word *ja*, 'yes', is very often spoken on the in-breath.

in-build (inbi·ld), *v.* Also **inbuild**. [f. IN-[1] + BUILD *v.*] *trans.* To build in (see BUILD *v.* 2 b in Dict. and Suppl.). Chiefly as **i·n-bui:lt** *ppl. a.* = built-in (*BUILT *ppl. a.* 1 b).
1920 C. C. MARTINDALE in C. Hess *God & Supernatural* x. 342 Christ speaks of His faithful as pillars in-builded into that Temple which is Himself. 1923 *Daily Mail* 12 June 5 (Advt.), The difference between good and bad waterproofs being inbuilt, therefore invisible. 1961 *Times* 27 Mar. 5/3 The triphenyl tin compounds possess an inbuilt safety factor. 1961 *John o' London's* 7 Dec. 636/3 An in-built sense of humour. 1961 B. R. WILSON *Sects & Society* 7 The sect often in-builds a hard core of suspicion. 1965 *Catholic Herald* 5 Mar. 5/4 Any nation has an inbuilt resistance to immigrants. 1969 *Jane's Freight Containers* 1968–69 582/1 Basically it is a bi-directional roller beam with an inbuilt hydraulic jacking arrangement. 1971 *Daily Tel.* 15 Mar. 17/5 They are ten times brighter than the present traffic lights but contain an in-built device to ensure that no dazzle occurs.

in-by(e, *adv.* Add: **b.** *attrib.* (Examples.) Also used *absol.* or as *sb.*
1824 J. HOGG *Private Mem. Justified Sinner* 227 Gie up your crooning, or I'll pit you to an in-by place. 1894 J. CUNNINGHAM *Broomieburn* vi. 88 The inbye hand, Jock, would emerge from his bed in the stable-loft. 1918 *Border Standard* 18 May 2 Louping-ill or trembling is proving very destructive amongst in-by or park lambs. 1940 *Geogr. Jrnl.* XCVI. 108 Parallel galleries also served as inbye and outbye roads. 1958 *Rep. R. Comm. Common Land 1955–58* (Cmnd. 462) 274 *In-by land*. The term is widely used in the North of England and derives from the Scandinavian word (*by*) for a farm. Hence it means the fenced-in land nearest the homestead. 1960 *Farmer & Stockbreeder* 26 Jan. 56/1 Glanilyn Farm,..which has some 140 acres of in-bye... It was wasteful to use in-bye land. 1961 *Guardian* 18 Dec. 6/4 A large part of the 84 acres will be able to support pasture and contribute to the in-bye in the valley. 1971 *Country Life* 20 May 1259/1 Ewes and lambs on in-bye land at Lartington in Yorkshire.

Inc. U.S. abbreviation of INCORPORATED *ppl. a.*
1906 *Country Life in Amer.* May 16/1 (Advt.), The Engleside Company, Inc., Owners. 1928 *Publishers' Weekly* 16 June 2451/1 The retail business will be known as 'Dutton's, Inc.' 1936 MENCKEN *Amer. Lang.* (ed. 4) 244 An Englishman writes *Ltd.* after the name of a limited liability (what we would call *incorporated*) bank or trading company, as we write *Inc.* 1973 R. T. ELSON (*title*) The world of Time Inc.: the intimate history of a publishing enterprise.

Incaic (iŋkēɪ·ik), *a.* = INCAN *a.* Also **Incaean** (iŋkă,ī·ăn), **Incarian** (iŋkēǝ·riăn) *adjs.*
1880 G. W. CABLE *Grandissimes* iv. 23 Possibly between the two sides of the occipital profile there may have been an Incaean tendency to inequality. 1909 *Cent. Dict. Suppl.*, Incaic, Incarian. 1926 *Glasgow Herald* 30 Jan. 8 The ruins of an ancient Incaic village. 1937 *Jrnl. R. Anthrop. Inst.* 321 The resemblance to possibly pre-Incaic pottery from the vicinity. 1963 *Times* 9 Feb. 9/7 An intact group of thousands of Incaic and pre-Incaic tombs, some dating from about 1200 B.C., was found only a few weeks ago.

incandescent, *a.* Add: Also *sb.* An incandescent lamp or burner.
1908 S. FORD *Side-Stepping with Shorty* 38 It was dark, and about half a million incandescents had been turned on. 1925 C. R. COOPER *Lions 'n' Tigers* v. 143 The great, empty building, where only a few incandescents gleamed dully. 1971 *General Electric Investor* II. II. 5 Lucalox is giving Washington double the light output of its former mercury lighting and six times the levels of its original incandescents.

incant, *v.* Restrict †*Obs.* to senses a and b in Dict. and add: **c.** *trans.* To raise by incantation. **d.** To chant, intone.
1926 *Chambers's Jrnl.* June 359/1 A little sorceress, talking to some spirit which she had incanted. 1959 *Times* 23 Sept. 3/7 The umpire incanting the score. 1961 M. SPARK *Prime of Miss Jean Brodie* ii. 22 They sat in the twilight eating toffees and incanting witches' spells.

incapacitant (inkăpæ·sitănt). [f. INCAPACIT(ATE *v.* + -ANT[1].] A substance that can be used to incapacitate a person for a time without wounding or killing him.
1961 *Today's Health* Mar. 76 If the Chemical Corps succeeds in standardizing effective incapacitants, it will represent a significant advance. 1963 *Listener* 7 Feb. 238/2 The nature of the military 'incapacitants'..includes such 'psychic poisons' as lysergic acid diethylamide, LSD-25, which produces extreme mental confusion. 1968 *New Scientist* 29 Feb. 465/1 BZ, one of the standard US incapacitants produces dizziness, heart palpitation, urinary retention and constipation. 1970 *Daily Tel.* (Colour Suppl.) 20 Feb. 19 Incapacitants are also intended for 'humane' use; their idea is to render an enemy temporarily helpless and to put him out of action for short periods.

Incaparina (inkăpărī·nă). [f. the initials of *I*nstitute of *N*utrition of *C*entral *A*merica and *P*anama + Amer. Sp. *f*)*ariña* powdered manioc f. L. *farina* flour, meal.] A preparation of vegetable protein, used as a dietary supplement.
1960 N. S. SCRIMSHAW in *Proc. Conf. Cottonseed Protein for Animals & Man* 18/1 The name Incaparina has now been adopted as a generic name to refer to any vegetable mixture developed by INCAP suitable for feeding to young children and containing at least 25% protein of a quality comparable to that of milk and other products of animal origin. 1964 *Listener* 28 May 866/2 'Incaparina', a mixture of cottonseed meal, maize, and several other ingredients, has been used widely in Guatemala. 1965 S. M. CANTOR in J. M. Leitch *Food Sci. & Technol.* III. 460 One of the most successful protein mixtures, Incaparina, a food product of plant origin, was developed by the Institute of Nutrition of Central America and Panama... Incaparina is an uncooked, dry powder which may be heated, flavoured, and consumed alone as a liquid, or it may be added to other foods. 1970 M. PYKE *Food Sci. & Technol.* (ed. 3) xi. 212 The nutritional value of this preparation, which is marketed under the name of 'Incaparina', is sufficiently high to allow its being fed to young children as a substitute for milk.

incapsulate, *v.* Add: Also *fig.* (see *ENCAPSULATE *v.*).
1939 R. G. COLLINGWOOD *Autobiogr.* x. 113 But this secondary life is prevented from overflowing into my primary life by being what I call incapsulated, that is, existing in a context of primary or surface knowledge which keeps it in its place and prevents it from thus overflowing. *Ibid.* 114 Historical knowledge is the re-enactment of a past thought incapsulated in a context of present thoughts. 1954 D. RIESMAN in *Amer. Jrnl. Sociol.* Jan. 382/2 Institutions incapsulate them or awaken them or destroy them. 1955 M. BELOFF *Foreign Policy & Democratic Process* 50 It may simply be that the older nations of Europe have retained, incapsulated within them, sufficient relics of their predemocratic leadership to provide the necessary ballast for the democratic sails. 1962 W. NOWOTTNY *Lang. Poets Use* vii. 146 To offer such a definition is to attempt to incapsulate a number of critical trends. 1972 *Country Life* 6 Jan. 56/3 For his ministers the problem was often to incapsulate him [*sc.* William IV] or walk round him.

incapsulation. (Later *fig.* examples.)
1934 PRIEBSCH & COLLINSON *German Lang.* viii. 344 Though German has recently turned away from such *Einschachtelung* or 'incapsulation', it still retains as a regular feature the relegation of the finite verb to the end-position of the subordinate clause. 1959 *Times Lit. Suppl.* 27 Feb. 110/3 This comes perilously near to incapsulation.

in-car *attrib.*: see *IN *prep.* 17*.

in-career *attrib.*: see *IN *prep.* 17*.

incarn, *v.* **2.** (Later examples.)
1904 HARDY *Dynasts* I. I. iii. 17 Did I incarn in moulds of all mankind. 1907 *Westm. Gaz.* 7 Feb. 8/2 Incarned as 'Superman' by G. B. S.

incarnational (inkaɹnēɪ·ʃǝnăl), *a.* [f. INCARNATION + -AL.] Of or relating to the theological doctrine of incarnation.
1912 F. VON HÜGEL *Eternal Life* viii. 166 Even the most general incarnational doctrine..must, then, be Superstition or Fanaticism for him. 1942 D. JENKINS *Nature of Catholicity* i. 12 This..widely influential tendency, with its affinities with historical Romanticism and the so-called 'Incarnational' theology, is a development of the Anglican doctrine of catholicity in the direction of Liberalism. 1954 D. L. SAYERS *Introd. Papers on Dante* 122 This Way, though it is perhaps more typically Western and might appear to be more typically Catholic and Incarnational than the other, has, I believe, never been fully mapped by any mystical theologian—unless we count Dante. 1971 *Daily Tel.* 29 July 6/3 His Christian faith is essentially this-worldly, 'incarnational', to use one of his favourite words.

incarnationist. Add: Also *attrib.* or as *adj.* So also **incarna·tionalist**.
1903 *Q. Rev.* Apr. 519 Wherever the Incarnationist idea originated, it did not originate in Hellenism. 1939 A. TOYNBEE *Study of Hist.* IV. 625 The Christology which Dr. Conybeare calls 'Incarnationist' ought properly to be called 'Conceptionist'. 1962 *Listener* 11 Jan. 68/2 This 'incarnationalist' type of doctrine—to be culled from almost every page of so very orthodox a teacher as our own Henry Scott Holland, for example—urges that the divine for us must mean the vision of a new humanity.

incendiary, *a.* and *sb.* Add: **A.** *adj.* **1. b.** (Further examples.) Used esp. of a type of aerial bomb that ignites on impact.
1885 E. S. FARROW *Mil. Encycl.* I. 666/1 (*heading*) Incendiary fire-works.—The incendiary preparations are ..incendiary-match, and hot-shot. *Ibid.* 666/2 Incendiary-match is made by boiling slow-match in a saturated solution of niter, drying it, cutting it into pieces, and plunging it into melted fire-stone. It is principally used in loaded shells. 1911 *Aero* May 37/1 The following are reckoned..to be the principal offensive uses of the war-aeroplane: (1.) Attacking supply stores and setting them on fire with incendiary bombs. 1915 *Lancet* 12 June 1249/2 The incendiary bomb may cause a serious outbreak of fire. 1917 H. WOODHOUSE *Textbk. Naval Aeronaut.* (1918) xix. 120/2 The Zeppelins also dropped incendiary bombs intended to set places on fire. 1918 E. S. FARROW *Dict. Mil. Terms* 307 *Incendiary grenade*, a form of grenade designed to scatter molten metal upon bursting. 1935 *Jrnl. R. Aeronaut. Soc.* XXXIX. 164 Reference is made to incendiary bombs with magnesium alloy case and thermit filling. 1940 *Illustr. London News* 5 Oct. 435/2 The oil bomb, which may be of various sizes, is filled in some cases with petrol, thus becoming a tremendously powerful incendiary bomb. 1941 *Ann. Reg.* 1940 69 Though large numbers of incendiary bombs were dropped the damage done by fire was kept within fairly narrow limits.
B. *sb.* **1. c.** Short for *incendiary bomb*.
1940 *Flight* 19 Dec. 522/2 The pilot found his objective at once and his incendiaries started four large fires. 1942 *R.A.F. Jrnl.* 13 June 24 You may have tried to put out an incendiary bomb by heaping..sand on it... No amount of sand will smother an incendiary. 1958 *Times Lit. Suppl.* 28 Mar. 176/1 In 1941 Lambeth Palace Library was heavily bombed. Incendiaries fell in the middle of the seventeenth-century Great Hall which was its centre.

incendiate, *v.* Add: So **ince·ndiated** *ppl. a.*, set on fire. *rare*[−1].
1922 JOYCE *Ulysses* 702 The carbonised remains of an incendiated edifice.

incendijel (inse·ndidʒel). orig. *U.S.* Also (*corruptly*) **incenderjel**, **incinderjell**. [f. INCENDI(ARY *a.* + JEL(LY *sb.*] An inflammable jelly used in incendiary weapons, composed of polystyrene, petrol, and benzene.
1966 *Air Force/Space Digest* (U.S.) Mar. 45/2 Nearly 700 day and night sorties were flown by B-57, A-1E, F-100, and F-8 aircraft, which dropped more than 450,000 pounds of incendijel. 1966 *New Statesman* 8 Apr. 496/1 The new jelly—or what US pilots refer to as 'incendijel'—is replacing the old soap-jelled gasoline napalm as used in Korea. 1967 (*title*) Modification kit for the MXY 377/E 32 incendijel mixing and transfer unit (U.S. Air Force Armament Lab.) (AD 830 504). 1967 *N.Y. Rev. Bks.* 20 Apr. 5/1 The Americans do not dissemble what they are up to. They do not seem to feel the need, except through verbiage: e.g., napalm has become 'Incinder-jell'. 1968 *Listener* 22 Feb. 244/2 The issues and living horror of the war disappear in a deadened, bureaucratic language—'incinderjell', 'the other side', 'body counts'. 1968 V. W. SIDEL in S. Rose *Chem. & Biol. Warfare* iii. 44 The name napalm has been retained as a generic one for weapons of this type. Recently the term has also been applied to a gel consisting of petrol, benzene, and polystyrene which is also called 'incenderjell' or Napalm-B.

incendivity (insendi·viti). [f. the stem of L. *incend-ere* (see INCEND *v.*) + -IVITY.] The ability to effect ignition or set on fire.
1919 *Jrnl. Chem. Soc.* CXV. 103 These results clearly establish the fact that the incendivity of a spark does not depend on the total energy of the spark. 1972 *Physics Bull.* Aug. 454/3 Later laboratory measurements..of the incendivity of spark discharges showed that if the uniform field strength in a tank roof space away from any protrusions was 500 kV m^{-1}, then brush discharges having sufficient energy to ignite an inflammable hydrocarbon-air mixture could occur at metal protrusions.
Hence (as a back-formation) **ince·ndive** *a.*, of or pertaining to incendivity; capable of effecting ignition.
1959 *Rep. Investigations U.S. Bureau of Mines* No. 5463. 19 Chemical composition alone is not sufficient to define the incendive character of an explosive. 1972 *Physics Bull.* Aug. 456/1 Differences between the incendive properties of metal to metal sparks and liquid to metal sparks. 1973 *Ibid.* Mar. 145/3 It draws attention to the risks of incendive sparks at or near deck openings resulting from static electricity.

incentive, *a.* and *sb.* Add: **A.** *adj.* **3.** Of or pertaining to a system of payments, concessions, etc., to encourage harder work or a particular choice of work.
1943 *Reader's Digest* Aug. 11/1 Mr. Charles E. Wilson.. is urging war industries to adopt 'incentive pay'—that is, to pay workers more if they *produce* more. 1948 *Ann. Reg. 1947* 283 In an effort to increase foreign trading various incentives were offered to exporters..a proportion of the foreign exchange..for the purchase of raw materials,..and 'incentive' goods for their workers. 1951 *Engineering* 2 Mar. 245/2 The body of the book is concerned with a description of the ['Armstrong Merit Sharing'] scheme...A series of 'requirements' is laid down to which it is considered that any incentive scheme should conform. 1952 'VIGILANS' *Chamber of Horrors* 72 *Incentive bonus*, a bonus *in advance* as an encouragement, for workers, to work. 1957 *Encycl. Brit.* XXIII. 272/1 For a substantial number of wage earners compensation is defined in terms not of time units but of output, under various types of incentive systems. *Ibid.* 272/2 The essential characteristic of an incentive wage rate structure is that payment depends on output rather than work time. 1967 G. F. FIENNES *I tried to run Railway* iv. 31 Work Study became synonymous with incentive payments. 1970 T. LUPTON *Managem. & Social Sci.* (ed. 2) ii. 53 These controls were turned mainly to the manipulation of the incentive payment system.
B. *sb.* **2.** An incentive payment, scheme, etc.
1948 [see adj. sense 3 above]. 1956 HICKMAN & KUHN *Individuals, Groups & Econ. Behavior* ii. 50 Debate about this central question has hinged in large part on the issue of incentives. *Ibid.* 60 Gordon enumerates several nonfinancial incentives of some apparent importance, including power, prestige, and security. 1960 H. C. WALLICH *Cost of Freedom* iii. 81 This leaves us with incentives in the narrower sense as the third element in the triad of forces upon which we rely to call forth initiative and effort. *Ibid.* 89 Labor and management performance is not the only place where we must look for the effects of incentives.

incentre, in-centre (i·nsentǝɹ). *Geom.* Also (*U.S.*) -center. [f. *IN-[1] b + CENTRE, CENTER *sb.*] The centre of an inscribed circle.
1903 E. H. ASKWITH *Course Pure Geom.* i. 12 The pedal triangle has for its incentre the orthocentre of the original triangle. 1904 HALL & STEVENS *School Geom.* III. 204 Given the base and vertical angle of a triangle, find the locus of the in-centre. 1963 R. A. ROSENBAUM *Introd. Projective Geom. & Mod. Algebra* i. 3 Prove that the bisectors of the angles of a triangle are concurrent. (The point of concurrency is the center of the inscribed circle, abbreviated 'incenter'.)

incest. **3.** *attrib.* and *Comb.* (Further examples.)
1819 SHELLEY *Let.* 16 Nov. (1964) II. 531 The incest scene of Amon & Tamar is tremendous. 1921 D. H. LAWRENCE *Psychoanal. & Unconscious* i. 20 At the root of almost every neurosis lies some incest-craving. *Ibid.* ii. 31 And this brings us finally to incest, even incest-worship. 1933 LD. RAGLAN *Jocasta's Crime* i. 6 Having eliminated reason, instinct, and religion as bases of the incest taboo, what have we left? The answer is magic. *Ibid.* ii. 15 There are no stronger upholders of the incest taboo than those very Australian tribes who believe that the child has no physical connexion with the father. 1946 KOESTLER *Thieves in Night* 162 The intimacy of life in the smaller Communes acts as a gradually materialising incest-barrier. 1949 —— *Insight & Outlook* xxiii. 323 Examples of such archetypes or primordial experiences are..the incest-motif. 1949 M. MEAD *Male & Female* ix. 199 Usually the primary incest-taboos are extended in various ways. 1963 A. HERON *Towards Quaker View of Sex* 56 Incest thoughts appear in human dreams. *Ibid.* 59 The sexual nature of this relationship is reinforced and with it feelings of incest-guilt, especially in the child. This situation may cause the boy or girl later to shirk contact with the opposite sex for fear of violating the incest barrier.

incestuous, *a.* Add: **2. c.** *Fig.* use of sense 2 a.
1869 D. G. ROSSETTI *Let.* 27 Aug. (1965) II. 727 'Solemn poetry' belongs to the class of phrases absolutely forbidden I think *in* poetry. It is intellectually incestuous, —poetry seeking to beget its emotional offspring on its own identity. 1971 *Listener* 2 Sept. 307/3 Systematising.. the old-boy network..would almost undoubtedly exacerbate the incestuous intolerance of the present scene.

inch, *sb.*[1] Add: **2. a.** Esp. in phr. *within an inch of one's life* (or † *skin*): almost to the point of death; so as to be nearly killed; freq. hyperbolically and *fig.*
1726, 1839 [in Dict.]. 1854 B. P. SHILLABER *Life & Sayings Mrs. Partington* 81 I'll be tempered to whip you within an inch of your skin. 1896 W. D. HOWELLS *Impressions & Experiences* 74 The defendant..had invited her to come down the street to a certain point, and be beaten within an inch of her life. 1909 L. M. MONTGOMERY *Anne of Avonlea* xii. 126 The grammar class were parsed and analysed within an inch of their lives. 1932 'E. M. DELAFIELD' *Thank Heaven Fasting* III. i. 250 She's always bullied Cecily within an inch of her life. 1939 L. M. MONTGOMERY *Anne of Ingleside* i. 12 He said that..everybody else would be dressed within an inch of her life. *Ibid.* iv. 30 If I had talked to my parents like that..I would have been whipped within an inch of my life.
3. e. *to give him* (or *them*, etc.) *an inch..*: example of extended use.
1973 *Times* 21 Feb. 3/1 If you turn your back for an instant or give them an inch they will park their cars on it.
4. a. *inch-wide* (earlier and later examples).
1873 *Young Englishwoman* Apr. 194/2 Inch-wide lace. 1950 W. DE LA MARE *Inward Compan.* 89 From inch-wide eyes I scan their..flames. 1964 C. DENT *Quantity Surveying by Computer* iii. 31 The new 1-inch-wide eight-channel paper tape now in use on some computers.
d. **inch-rule** (earlier example); **inch-taped** *a.* (nonce-wd.), covered with inch-tape; **inch-worm** (examples); also *fig.*
1850 DICKENS *Dav. Copp.* lx. 595 Neither will you find him measuring all human interests..with his one poor little inch-rule now. 1939 DYLAN THOMAS *Map of Love* 6 'His mother's womb had a tongue that lapped up mud,' Cried the topless, inchtaped lips. *a* 1861 T. WINTHROP *Life in Open Air* (1863) 123 All the green inch-worms vanish on the tenth of every June. 1881 *Harper's Mag.* Oct. 656/1 A wriggling inch-worm,..awaiting..an opportunity to measure the length of your nose. 1949 *Sat. Even. Post* 12 Mar. 33/1 One evening the Main Line local hunched its cars together like an inchworm and skidded to a halt. 1954 BORROR & DELONG *Introd. Study Insects* xxvi. 522 The larvae of geometers are the familiar caterpillars commonly called inchworms or measuring-worms. 1959 G. MATTINGLY *Defeat of Spanish Armada* xx. 218 There was only one offset to the exasperation of this inchworm progress. 1970 R. LOWELL *Notebk.* 235 Have you ever seen an inchworm crawl on a leaf,..Feeling for something to reach something?

inchastity. (Isolated later example.)
1972 *Times Lit. Suppl.* 25 Aug. 1000/4 It is difficult..to imagine how sanctions against inchastity could be re-imposed.

Inche (i·ntʃǝ). Also **Enche**, **Enc(h)ik**, **Inchi**. [Mal. *enche'*, *enchek*, *enchik* master, mistress.] In Malaysia, a prefixed title signifying respect, used for persons with no other special distinction; equivalent to 'Mr.'
The current official spelling (since 1972) is *Encik*.
1834 P. J. BEGBIE *Malayan Peninsula* ii. 83 It now only remains to mention the fate of Inchi Oowan Saban. 1897 *Jrnl. Straits Branch R. Asiatic Soc.* July 85 The following account of the method of rice cultivation..was written..by Inche Muhammad Ja'far. 1913 RANEE OF SARAWAK *My Life in Sarawak* xviii. 160 Inchi Sawal was a great stickler for grammar. 1937 *Jrnl. Malayan Branch R. Asiatic Soc.* XV. i. 63 Inche Abdul Raffar, Collector and Deputy Registrar, Ulu Selangor. 1951 *Proc. Legis. Council Fedn. Malaya* B159 Enche Abdullah: Sir, I beg to second the motion. 1972 *Straits Times* 4 May 1/5 Inche Ghafar clarified today that he had not made any offer for the formation of a coalition government.

in-churn *attrib.*: see *IN *prep.* 17*.

incidence. Add: **4.** *angle of incidence*, (*b*) the angle which the chord of an aircraft wing makes with the direction of the undisturbed air current.
1908, etc. [see *ANGLE *sb.*[2] 1 ¶]. 1948 *Sci. News* VII. 28 The pilot is able to increase or decrease the lift by altering the angle of incidence. *Ibid.*, As the angle of incidence increases, the lift also increases and the aircraft is able to climb, but if the tilt is made too large the flow on the upper surface separates and eddies are formed.
8. Special Comb.: **incidence wire** *Aeronaut.*, on a biplane (see quot. 1916).
1916 H. BARBER *Aeroplane Speaks* 143 *Incidence wire*, a wire running from the top of an interplane strut to the bottom of the interplane strut in front of or behind it. It maintains the 'stagger' and assists in maintaining the angle of incidence. 1928 V. W. PAGÉ *Mod. Aircraft* v. 175 These wires are called 'incidence wires', as they keep the planes in the proper angular relation to each other.

incident, *sb.*[1] Add: **1. b.** An occurrence or event, sometimes comparatively trivial in itself, which precipitates or could precipitate political unrest, open warfare, etc. Also, a particular episode (air-raid, skirmish, etc.) in war; an unpleasant or violent argument, a fracas.
1913 *Ann. Reg. 1912* I. 441 He had invariably done everything France wanted him to do and, especially at the time of the Agadir incident, had rejected German.. advances. 1920 W. S. BLUNT *My Diaries* II. v. 138 Bramley..had reported the incident in a serious light, and Cromer had taken it up seriously, seeing in it..a danger to the British occupation. 1920 T. E. LAWRENCE *Let.* 8 Aug. (1938) 313 Our communications are very bad, our defence positions all have both flanks in the air, and there seem to have been two incidents lately. 1930 *Economist* 30 Aug. 392/1 When some special 'incident' has occurred, a Mandatory Power frequently furnishes the Commission with additional information. 1937 V. BARTLETT *This is my Life* xiv. 245 There were very few 'incidents'. I visited almost every district of Berlin,..and I saw no man beaten..but I came back..overwhelmed with shame that people could be proud of so much bullying. 1937 L. BROMFIELD *Rains Came* I. iv. 32 One impotent little man

from Clapham, who was insolent to the Maharani, received for his pains a dismissal from the Civil Service for having made an 'incident'. *Ibid.* lv. 242 It was threatening to become an 'incident' which might unsettle the peace of India. **1938** *Encycl. Brit. Bk. of Yr. 1938* 296/1 Such incidents are bound to form an integral part of large-scale warfare under modern conditions. **1943** Hunt & Pringle *Service Slang* 40 There are no occasions, occurrences, or events in an airman's life. Anything that happens to him is an 'incident'..why, nobody knows. **1945** *Ann. Reg. 1944* 113 They [*sc.* the Home Guard] had given valuable aid to the Civil Defence Services in dealing with air-raid incidents. **1955** *Ann. Reg. 1954* 15 The number of 'incidents' in the Canal Zone increased. **1959** R. Collier *City that wouldn't Die* xi. 211 A warden doesn't write off an incident until he has personally made certain there is no one else on the premises. **1960** Partridge *Charm of Words* i. 23 If a business man speaks of incidents when he means quarrels, he has been influenced by journalism. **1973** G. Greene *Honorary Consul* i. i. 23 The Governor didn't want any incidents.

5. *attrib.* and *Comb.* (also *incidents*). **incident book,** a book in which all relevant information is kept at an incident room; **incident office, post, room,** names for a centre set up by the police close to the scene of a crime, accident, or disaster, from which all operations are controlled and monitored.

1967 Baker & Wilkie *Police Promotion Handbk.* iv. xxiii. 302 It will be clear from the list of duties that the manning of an Incident Post at the scene of a major disaster will call for a sufficiency of staff. **1971** 'J. Fraser' *Death in Pheasant's Eye* xxiii. 132 Three night men on duty in the Incidents Room. *Ibid.*, The Incidents Book was on Inspector Coates' table... The Incidents Book was the Bible of any murder investigation; every action was telephoned to the man who looked after the book. **1972** L. Lamb *Picture Frame* xviii. 154 Mr Glover's incident office is in a red corrugated-iron hut. **1973** *Times* 12 Mar. 1/5 The murder hunt is being led by Detective Superintendent Brian Weight, who set up an incident room at Aylesbury police station.

incident, *a.*[1] Add: **7. a.** Of light: further examples in photographic contexts.

1951 G. H. Sewell *Amateur Film-Making* (ed. 2) ii. 21 The makers of other meters have adopted the Incident-light principle and have provided modifying attachments for their instruments. **1952** J. F. Dunn *Exposure Meters* i. 22 The term 'incident light' is used instead of 'illumination' when we wish to consider only the strength of the light being received *from* the source irrespective of the angle of inclination of the surface receiving it. **1956** *Nature* 4 Feb. 231/1 Blocks of soil were..directly examined with the aid of the incident-light equipment of a Leitz 'Ortholux' microscope. **1962** M. L. Haselgrove *Photographer's Dict.* 130 Light falling on a surface is said to be incident on the surface. In most photographic writing this term used without qualification refers to the light incident on the subject to be photographed.

incidental, *a.* Add: **5.** Special collocations: *incidental advertisement* (see quot.); *incidental music*, music played as an accompaniment or 'background' to a play or film, or to a radio or other performance or entertainment; *incidental number*, a piece of incidental music; also *transf.*

1931 *Times Lit. Suppl.* 20 Aug. 636/3 'Incidental' advertisements, advertisements..which are printed in a separate gathering from the body of the book and sewn in at either the end or the beginning. **1864** in H. J. Byron *Orpheus & Eurydice* 2 (*heading*) The incidental music selected and arranged by Mr. Frank Musgrave. **1928** *Melody Maker* Feb. 214 (Advt.), Liber's incidental music. **1938** *Oxf. Compan. Music* 464/1 Incidental music to plays has always been an important side-line of the art and business of the composer. **1904** W. D. Adams *Dict. Drama* I. p. vi, Musical Composers, the latter ranging from the writers of operas and operettas to the providers of 'incidental numbers' for plays. **1912** E. Wylie (*title*) Incidental numbers.

incidentalist (inside·ntălist). [f. Incidental *a.* + -ist.] One who describes or insists on what is merely incidental and not essential.

1904 *Daily Chron.* 19 Nov. 3/1 Mr. Palmer may be described as an 'incidentalist'. **1924** H. E. Fosdick *Mod. Use Bible* 163 Folk who insist on that kind of literal inerrancy in ancient documents are not Fundamentalists at all; they are incidentalists.

incidentality (i:nsidentæ·lĭti). *rare.* [f. Incidental *a.* + -ity.] The quality of being incidental.

1791 *Deb. Congress U.S.* (1834) II. 1942 Hence the incidentality of this authority to the mere existence of Government is inferred.

incidentally, *adv.* Add: **2.** In point of fact: used to accompany a not immediately pertinent statement.

1925 T. Dreiser *Amer. Trag.* (1926) I. i. ii. 15 Incidentally by that time the sex lure or appeal had begun to manifest itself. **1926** Fowler *Mod. Eng. Usage* 264/2 *Incidentally* is now very common as a writer's apology for an irrelevance. **1961** A. Huxley *Let.* 8 Jan. (1969) 902 Pure perceptual receptivity is the basis, incidentally, of many Tantrik exercises. **1972** R. Quirk et al. *Gram. Contemp. Eng.* 667 *Incidentally* adds explicitly that what is being said is a digression, if only slight, and an afterthought: The airlines charge half-price for students. Incidentally, I have already bought my ticket to New York.

incinderjell, var. *incendijel.

incipient, *a.* Add: **2.** *Comb.* **incipient species,** a group of plants or animals in the process of becoming sufficiently distinct to be described as a full species.

1859 Darwin *Origin of Species* ii. 52 A well-marked variety may be called an incipient species... It need not be supposed that all varieties or incipient species necessarily attain the rank of species. **1942** E. Mayr *Systematics & Origin of Species* vii. 156 Some subspecies are incipient species, or subspecies are potentially incipient species. **1963** Davis & Heywood *Princ. Angiosperm Taxon.* iii. 100 There is a tendency to regard subspecies as incipient species. **1971** *Nature* 2 Apr. 275/2 In *D[rosophila] paulistorum* there are a number of incipient species (usually called 'semispecies').

incipit. (Further examples.)

1963 [see *explicit b]. **1973** *Times* 2 Nov. 6/1 The *incipits*, or titular opening phrases, of more than 200 literary works current in Sumer in the early second millennium B.C.

incisal (insəi·zăl), *a.* *Dentistry.* [f. *incis-* stem of Incisor, etc. + -al.] Of, pertaining to, or designating the cutting edge or surface of an incisor or a canine tooth.

1916 M. Dewey *Dental Anat.* ii. 44 The incisal edge of the central incisor is formed by the junction of the lingual and labial surfaces of the tooth. **1924** J. F. Hovestad *Pract. Dental Porcelains* xiv. 121 Frequent accidents to anterior teeth..causing loss of incisal corners or edges, have given the dentist great trouble. **1956** J. N. Anderson *Appl. Dental Materials* xxii. 300 The incisal edge of an anterior tooth is mainly enamel. **1969** *Gloss. Terms Dentistry* (*B.S.I.*) 80 *Incisal guidance*, the influence on mandibular movements exerted by the contacting surfaces of the mandibular and maxillary anterior teeth.

incise, *v.* Add: **1. c.** *trans. Geol.* Of a river: to cut (a channel or valley) in an underlying landform. Also *absol.* Usu. as **incised** *pa. pple.* (cf. next).

1893 *Science* 17 Nov. 278/1 The process by which the present ravines are forming is not a direct continuation of the process by which the gentler slopes of the upland were formed. The former are incised in the latter. **1896** *Nat. Geogr. Mag.* VII. 190 With the uplift of the region the meandering river would proceed to incise its channel beneath the uplifted surface. **1926** *Jrnl. Geol.* XXXIV. 31 Green River has a highly meandering course which is deeply incised in the rocks of the plateau. **1954** W. D. Thornbury *Princ. Geomorphol.* vi. 147 Heavily loaded streams are more likely to cut laterally as they incise their valleys than those with lesser loads. **1972** *Science* 27 Oct. 409/2 Meandering rivers that have incised in bedrock and yet have maintained a sinuous pattern may be of two basic types.

incised, *ppl. a.* Add: **3.** *Geol.* Of the channel of a stream, esp. a meander: cut abnormally deeply into underlying deposits or bedrock. Also, of a landform: cut by channels.

1899 W. M. Davis *Physical Geogr.* ix. 254 Incised meanders and cut-off spurs occur on the Allegheny river above Pittsburg. **1906** *Bull. Geogr. Soc. Philadelphia* IV. iv. 1 (*heading*) Incised meandering valleys. *Ibid.* 9 The Potomac river..exhibits some well-defined incised meanders in a valley of monoclinal strata. **1944** A. Holmes *Princ. Physical Geol.* xi. 198 In Utah, where recent uplift has made possible the development of many deeply incised meanders.., there are several examples of such arches. **1968** R. W. Fairbridge *Encycl. Geomorphol.* 332/2 (*caption*) Incised etchplain. *Ibid.* 548/2 A meandering river valley that has cut down its bed into the bedrock, because of uplift or lowered base level, is called incised, intrenched, entrenched, inclosed or ingrown.

incision. Add: **5*.** *Geol.* The cutting down and deepening of its channel by a river; a channel so made.

1906 *Bull. N.Y. State Mus.* No. 92. 333 So relatively inconspicuous are the incisions in this upland, that..the sky line will appear a nearly level one. **1914** *Jrnl. Geol.* XXII. 473 Lack of flats along the stream indicates that incision is still in progress. **1970** R. J. Small *Study of Landforms* ii. 65 A change of climate, leading to..a condition of stream underloading, will be accompanied by incision of the rivers into these deposits.

incitory (insəi·tŏri), *a.* *rare.* [f. *incit-* stem of Incite *v.*, etc.: see -ory[2].] Having the quality of inciting; stimulative, provoking.

1941 A. Warren in N. Foerster et al. *Lit. Scholarship* iv. 147 The second [group] holds art to be incitory, an invitation to action. **1949** Wellek & Warren *Theory of Lit.* iii. 27 Is some literature incitory and some cathartic? **1950** *N.Y. Times* 20 Apr. 1/8 He had used his employes for espionage and to gather material for 'fictitious reports of incitory character' for the voice of America.

inclinometer. **2.** (Earlier and later examples.)

1852 *Mechanics' Mag.* LVII. 416/2 An improved apparatus, instrument, or means for ascertaining or setting off the slope or level of drains, banks, inclines, or works of any description..which the patentee calls an 'inclinometer'. **1913** *Captain* Sept. 1069/2 An inclinometer, to show the angle of ascent or descent. **1919** H. Shaw *Text-bk. Aeronaut.* xiii. 169 Inclinometers of the spirit-level type, or constructed on the principle of the pendulum,.. are inaccurate when the machine is subjected to an acceleration as in turning. **1960** [see *gradiometer a].

include, *v.* Add: **3. b.** Const. *out*: to exclude (oneself or someone). Hence, pleonastically, *to include* (someone) *in*.

This colloquial expression, which freq. indicates strong feelings of cynicism or disillusion, is attributed to the American film producer Sam Goldwyn.

1937 A. Johnston in *Sat. Even. Post* 8 May 6/1 An ordinary man, on deciding to quit the Hays organization, might have turned to his fellow producers and said, 'Gentlemen, I prefer to stand aloof,' or, 'Gentlemen, I have decided to go my own way.' Sam [Goldwyn] said, 'Gentlemen, include me out.' **1938** *Hansard Commons* 8 Nov. 18/1 It may be that the First Commissioner of Works ..will now label the 'Aye' Lobby the 'Sez you' Lobby, and the other the 'Include me out' Lobby. **1946** Wodehouse *Joy in Morning* xvi. 140 Include me out... Nothing doing. **1958** G. Mitchell *Spotted Hemlock* x. 101, I shall suggest giving you lunch and include him in. **1958** *Times Lit. Suppl.* 1 Aug. 435/1 He sees that our young men are angry; they do not see visions or dream dreams; they want neither to go back nor move forward: they only ask to be included-out of the social and historical process. **1959** M. Steen *Woman in Back Seat* i. vii. 128 Oh, darling, please include me out! I can't bear affairs of that kind. **1967** Wodehouse *Company for Henry* ii. 39 You surprise me. A free meal, and he made no attempt to include himself in? **1969** V. Gielgud *Necessary End* iii. xv. 120 I'd like nothing better—in the immortal words—than to include myself out. **1971** *Daily Tel.* 13 Oct. 11/5 Half the in-jokes included me out, but I revelled in the way the actors caricatured famous folk. **1972** G. Bromley *In Absence of Body* iii. 33 Looking for clues? If so, include me in.

included, *ppl. a.* Add: **d.** *Linguistics.* (See quot. 1933.)

1933 L. Bloomfield *Lang.* xi. 170 When a linguistic form occurs as part of a larger form, it is said to be in included position; otherwise it is said to be in absolute position and to constitute a sentence. **1949** E. A. Nida *Morphol.* (ed. 2) 76 Some morphemes occur in included position, either partial or complete. **1952** C. C. Fries *Struct. of Eng.* xi. 253 These are the signals of 'included' sentences... In some sentences there are devices that signal the inclusion of two or more separate sentences within the structural pattern of a single free sentence unit. **1954** Pei & Gaynor *Dict. Ling.* 98 *Included position*, the position had by a word, phrase or other linguistic form when it is part of a larger form and does not constitute a sentence in itself. **1962** B. M. H. Strang *Mod. Eng. Struct.* v. 82 Linguistic structures may always be either ..absolute or included.

e. Bot. *included phloem, sapwood* (see quots.).

1933 *Tropical Woods* XXXVI. 2 Included Phloem.—Phloem strands or layers included in the secondary xylem of certain dicotyledonous woods. (To replace *Interxylary Phloem.*) **1937** *Ibid.* L. 11 Strands of included phloem usually isolated, but occasionally linked by parenchyma. **1969** K. Esau *Phloem* 3 In some genera a third category of phloem is found, *included phloem*, which consists of strands or layers embedded in the secondary xylem. **1933** *Tropical Woods* XXXVI. 3 Included Sapwood.—Masses or concentric zones included in the heartwood, which retain appearance and technical properties of sapwood. **1956** F. W. Jane *Struct. Wood* x. 212 Sometimes, heartwood is formed irregularly and patches of pale coloured wood are enclosed within the heartwood, a feature referred to as included sapwood.

inclusion. Add: **2.** More generally in technical use (e.g. *Cytology*, *Metallurgy*): any discrete body or particle which is recognizably different or distinct from the groundmass or relatively solid and homogeneous substance in which it is embedded. (Further examples.)

1896 E. B. Wilson *Cell* i. 15 The lifeless inclusions in the protoplasm have been collectively designated as *metaplasm* (Hanstein) in contradistinction to the living *protoplasm*. **1897** *Jrnl. Morphol.* XII. Suppl. 14 (*heading*) Discontinuous elements or inclusions. **1904** *20th Rep. Bureau Animal Industry, U.S. Dept. Agric.* 149 Borrel considers that his researches show that the microbe of sheep pox is ultramicroscopic and that the cellular inclusions described as parasites of vaccinia..cannot be the true cause of the disease. **1913** *Jrnl. Iron & Steel Inst.* LXXXVII. 655 The various kinds of slag inclusions occurring in steel..may be classified as follows: 1. Those.. dispersed throughout the metal, but mostly near the surface. 2. Those..dispersed throughout the whole mass of the metal. 3. Small inclusions..occurring between the crystals of the metal. **1939** A. Johannsen *Descr. Petrogr. Igneous Rocks* (ed. 2) I. iii. 39 Von Leonhard used the term *Porphyr-Struktur* for that texture in which crystals, crystal fragments, grains, or flakes lie in a dense, unbroken groundmass. The inclusions, he said, are usually different from the groundmass and do not touch each other. **1960** F. C. Steward *Plant Physiol.* IA. i. 11 The metabolically active inclusions are the mitochondria, the microsomes, and the chloroplasts. **1966** *Nature* 28 May 879/2 Ice specimens..prepared with inclusions of fine air bubbles. **1967** A. H. Cottrell *Introd. Metall.* xxi. 390 There may additionally be local groupings of dislocations created round large foreign inclusions by thermal shrinkage strains.

3. *Math.* Usu. *inclusion map(ping), function.* A mapping of a set *A* into a set *B* con-

taining *A* which maps each element of *A* on to itself.

1949 *Ann. Math.* L. 956 The symbolism $f:(X', A') \subset (X, A)$ is read: *f* is the inclusion map of (X', A') into (X, A). **1956** E. M. Patterson *Topology* ii. 20 If $A \subset B$, the transformation $i: A \to B$ defined by $i(a) = a$ is a one-one transformation called an inclusion. **1962** B. Mendelson *Introd. Topology* (1963) i. 30 Let $A \subset X$. The function $i: A \to X$..is called an inclusion mapping or function. *Ibid.* iii. 111 Let the topological space *Y* be a subspace of the topological space *X*. Then the inclusion mapping $i: Y \to X$ is continuous. **1964** Sze-Tsen Hu *Elem. Gen. Topology* i. 8 Consider the case $X \subset Y$. Then, the function $i: X \to Y$ defined by $i(x) = x \in Y$ for every $x \in X$ is called the inclusion function of *X* into *Y*... We write $i: X \subset Y$.

inclusive, *a.* Add: **1. b.** (Further examples.)

1910 *Bradshaw's Railway Guide* Apr. 1061 Bath Hotel ..Inclusive Terms from 7/6 per day. *Ibid.* 1067 Specimen Menus and very moderate Inclusive Rates per return. **1970** *Times* 21 Nov. 22/2 The first British inclusive holidaymakers are going out to the new summer resorts. **1972** *Times* 30 Sept. 11/4 Inclusive holidays offered by various tour operators.

inclusivity (inklusi·vĭti). *rare.* [Inclusive *a.* + -ity.] The quality of being inclusive.

1939 E. Pound *Let.* 2 Sept. (1971) 326 Re yr. extension of contents: The *real* work of a time is never done by more than four or five people with a fringe of occasional compositions. I suspect inclusivity. **1955** H. Read *Icon & Idea* ii. 36 But 'actuality' does not imply inclusivity.

incoagulable, *a.* Add: (Later example.) Hence i:ncoagulabi·lity, the property or state of being incoagulable.

1915 *Jrnl. Med. Res.* XXXII. 452 The leech extract incoagulable blood in vivo differs markedly from blood made incoagulable by receiving it into leech extract in vitro. *Ibid.* 454 In anaphylactic blood and in leech extract blood the probable explanation of a part at least of the incoagulability of the blood consists in a definite effect of these conditions on the blood platelets. **1964** W. G. Smith *Allergy & Tissue Metabolism* ii. 17 Other workers demonstrated a deficiency in fibrinogen, platelets, and prothrombin as a cause of the incoagulability.

incoherence. Add: **4.** *Physics.* The property (of waves, or of phenomena involving them) of being incoherent (sense *5); lack of a definite or stable phase relationship between waves at different points (in space or in time).

1938 *Physica* V. 785 In the usual treatment of interference and diffraction phenomena, there is nothing intermediate between coherence and incoherence... The first term is understood to mean complete dependence of phases, the second complete independence. **1958** *Physical Rev.* CI. 7/2 This is a classical type of incoherence, such as would result if we had an assembly of classical oscillators with random initial phase. **1971** K. R. Barnes *Optical Transfer Function* ii. 7 Spatial incoherence implies that the phases of the monochromatic components of the light from all points of the object are changing at random.

incoherent, *a.* Add: **5.** *Physics.* Producing, involving, or consisting of waves that have no definite or stable phase relationship with one another.

1929 A. Sommerfeld *Lect. Wave Mech.* vi. 91 In addition to secondary radiations of the same frequency as the incident light, we have also radiations of altered frequency... These modified radiations bear obviously no phase relationship with the exciting radiation, and might therefore be described as 'incoherent radiations'. **1953** C. E. Hall *Introd. Electron Microsc.* vii. 155 If two incoherent sources are observed with a microscope, the intensity at the image plane is the sum of the intensities from each one taken separately. **1959** Born & Wolf *Princ. Optics* vii. 255 In beams from different sources, the fluctuations are completely uncorrelated, and the beams are said to be mutually incoherent. *Ibid.* x. 506 The light vibrations arising from different elements of the source may be assumed to be statistically independent (mutually incoherent), and of zero mean value. **1966** *McGraw-Hill Encycl. Sci. & Technol.* XII. 57/2 It is also useful to distinguish between coherent and incoherent scattering; the distinction is made on the basis of the ability of the scattered wave to interfere with the incident one. Inelastic scattering is always incoherent.

incoherently, *adv.* (Further examples, corresponding to *incoherent *a.* 5.)

1938 *Physica* V. 795 For equal illuminating and observing apertures the result will be that two points just separable by the objectives are incoherently illuminated. **1961** *Encycl. Dict. Physics* I. 723/1 Incoherently radiating oscillators may very quickly become coherent through their interaction with their common radiation field.

incohesion (inkohī·ʒən). [f. In-[3] + Cohesion.] Want of cohesion.

1882 H. Spencer *Princ. Sociol.* II. v. iii. 278 Our own Indian Empire..held together by force in a state of artificial equilibrium, threatens some day to illustrate by its fall the incohesion arising from lack of congruity in components. **1922** *Public Opinion* 17 Mar. 244/2 The interminable vacillation, procrastination and incohesion of the Government at home. **1924** *These Eventful Years* I. 157 All these regimes are dependent on British support as against the incohesion of tribes and sects.

i·n-co:llege, *a.* [*in *prep.* 17*.] Residing within the buildings of a college; of or pertaining to teaching or administration within the precincts of a college. Cf. Out-college *a.*

1845 J. Pycroft *Collegian's Guide* vi. 131 There is a list of the eleven in-collegemen. **1899** in Dict. s.v. In- *pref.*[1] ad fin., In-college residents. **1945** G. B. Grundy *55 Yrs. at Oxf.* 108 An in-college Bursar. **1966** *Rep. Comm. Inquiry Univ. Oxf.* II. 122 In-college teaching. *Ibid.* 143 In-college teacher.

income, *sb.*[1] Add: **6. a.** *national income*: the income of a nation as a whole, *spec.* the aggregate amount available for distribution among the agents of production.

1878 *Encycl. Brit.* VIII. 258/1 The income tax returns given in the preceding tables furnish important materials for ascertaining, if only approximately, the national income of England. **1925** S. E. Thomas *Elem. Econ.* xvi. 214 The total of the national income represents not only the reward which flows to land, capital, labour and enterprise: it is also the total available in the hands of all members of the community for purchasing goods and services. **1931** *Times Lit. Suppl.* 19 Feb. 124/2 The national income may be divided into the income (wages, salaries and interest) of the producers of capital and consumable goods respectively. **1964** Gould & Kolb *Dict. Social Sci.* 452/2 The term national income is used in a generic sense to refer to the net value of all economic goods and services produced by a nation during a particular time, usually a calendar year... In a more specific sense...it denotes the aggregate of all income payments accruing to the factors of production. **1971** A. Shonfield in A. Bullock *20th Cent.* 327/1 Many of the European nations were used to earning up to one-quarter of their national income through sales abroad.

b. (Later examples.)

a **1902** S. Butler *Notebks.* (1912) i. 12 All progress is based upon a universal innate desire on the part of every organism to live beyond its income. **1939** T. S. Eliot *Family Reunion* ii. ii. 105 It is as if I had been living all these years upon my capital, Instead of earning my spiritual income daily. **1953** A. Huxley *Let.* 16 Nov. (1969) 688 He was a retired business man, living beyond his intellectual income.

7. *income account, bracket, level*; *income-earning* adj. Also, **income funds, investment, share, stock,** investments regarded primarily as a source of income; **income group,** a section of the population graded according to income; **incomes policy,** a policy introduced in the U.K. by the Labour Government of 1964–70 for the control of inflation by attempting to restrict increases in wages, salaries, dividends, etc.; any similar programme.

1869 *Bradshaw's Railway Manual* XXI. 419 Add balance to credit of income account. **1940** Income bracket [see *bracket *sb.* 5 c]. **1947** *Partisan Rev.* XIV. 482 The hard core of the stratum that lives off ideas..are the graduates of the fashionable Eastern colleges whose social origin is in the upper income-bracket groups. **1969** *Times* 30 Sept. 11/8 Good citizenship is not decided by income brackets. **1972** *Listener* 6 Apr. 467/1 A poor cabbie loses his girl because her family don't like his income bracket. **1909** *Daily Chron.* 22 Feb. 1/4 No doubt many an 'old alibi' has won the pension for some young wage-earning or income-earning person. **1946** Koestler *Thieves in Night* 88 Of these, 6,624 Work Days were spent on income-earning labour. **1969** *Times* 15 Nov. 16/1 The neglected income funds are..coming back into favour. **1934** B. J. Newman in *Encycl. Social Sci.* XIV. 93 s.v. *Slum*, The worst structural and sanitary conditions and the most degraded occupancy, usually by the lowest income groups, of any given period. **1936** *Discovery* Apr. 98/2 For the income-groups below the adequacy level an increased consumption of milk, butter, eggs, fruit, vegetables, and meat is desirable. **1957** J. Braine *Room at Top* vii. 62 She possessed the necessary face and figure and the right income group. **1970** *Listener* 21 May 687/1 The boom in Swiss industry after the war led to..a flood of foreign workers, mostly Italians and Spaniards of the lowest income-groups. **1900** *Westm. Gaz.* 22 Oct. 9/3 This would give a good prospect of dividend for the ordinary shares, and so make those shares not a bad 'income' investment. **1928** M. Dobb *Wages* vi. 113 The potential supply of lawyers or doctors..will be almost entirely confined..to children of parents above the income-level which makes possible a somewhat costly public school and university career. **1951** M. McLuhan *Mech. Bride* (1967) 117/2 What would you say was the income level of this family group? **1955** T. H. Pear *Eng. Social Differences* 136 According to occupation or income-level. **1900** *Westm. Gaz.* 22 Oct. 9/3 Looking upon their shares as 'income shares' only. **1958** *Spectator* 27 June 849/3 A number of so-called 'income' shares in the consumer goods trades. *Ibid.*, The yields on 'income' stocks. **1965** *New Statesman* 19 Mar. 434/1 When he first goes into battle over a wage claim or a price increase—when an incomes policy is first translated into action. **1966** *Listener* 17 Mar. 391/1 The kind of incomes policy they advocate requires unemployment and short-time working to make it effective. **1969** H. Perkin *Key Profession* v. 181 In terms of an incomes policy a long-term change in relativities was desirable in the national interest in order to restore the universities to a position from which they could compete. **1972** *Listener* 24 Aug. 239/2 The publicly and privately expressed views of the Bank of England that a formal incomes policy of some kind was needed.

income-tax. Add: Also *attrib.*

1844 A. W. Kinglake *Eothen* viii. 111 The fear that my party might be a company of Income-tax commissioners. **1844** *Punch* VII. 206 Internal things. The Income-Tax papers. **1862** Trollope *Small House at Allington* (1864) I. iv. 36 Mrs. Cradell..who had also succeeded in getting her son into the Income-tax Office, had placed him in charge of Mrs. Roper. **1878** [see *income *sb.*[1] 6 a]. **1879** Geo. Eliot *Let.* 23 July (1956) VII. 185, I have filled up one income-taxpaper for this district. **1902** *Encycl. Brit.* XXXIII. 194/2 An Income Tax Code. **1909** *Westm. Gaz.* 24 Sept. 5/3 Returns may be demanded in every income-tax district in which a person may reside. **1916** *Act* 6 & 7 *Geo. V* c. 24 § 37 Income tax relief on war insurance premiums. **1920** A. E. Housman *Let.* 3 Sept. (1971) 178 The additional trouble in filling up my Income Tax return. **1927** Bowley & Stamp *Nat. Income 1924* 17 'Income-Tax' income exceeds ordinary income under certain heads, since it includes *inter alia* various losses which do not come out in the assessment. **1941** *Time* 1 Dec. 84/2 Now appealing a three-year sentence for income-tax evasion. **1950** T. S. Eliot *Cocktail Party* ii. 105 I'd like to see *you* filling up an income-tax form. **1959** *Listener* 2 July 23/2 These customers are known in agriculture as 'income-tax farmers'. **1966** *Ibid.* 5 May 663/1 A composer worth his salt should be able to set..an income-tax demand to viable music.

incoming, *ppl. a.* Add: **e.** Of game: approaching the sportsman.

1892 W. W. Greener *Breech-Loader* 211 It often happens that incoming and motionless ground game is shot over.

incommunicado (i:nkǒmiūnikā·do), *a.* (or *adv.*). Also ‖ **incomunicado.** [Sp. *incomunicado*, pa. pple. of *incomunicar* to deprive of communication.] Having no means of communication with other persons; isolated; in solitary confinement.

1844 G. W. Kendall *Narr. Santa Fé Exped.* II. 255 Now that I was incomunicado—now that all intercourse with my friends was cut off,..my situation became irksome in the extreme. **1911** R. H. Davis *Once upon a Time* 57, I asked the official concerning Judge Rojas. 'Oh, yes,' he said readily. 'He is still *incomunicado*.' **1934** J. M. Cain *Postman always rings Twice* x. 109 They can hold you forty-eight hours incommunicado. **1941** 'R. West' *Black Lamb* II. 310 This was an island: parts of it were even now *incommunicado*. **1952** M. McCarthy *Groves of Academe* (1953) v. 96 The defendant or victim in such cases as mine ought to be held incommunicado till his well-wishers have concluded their efforts. **1955** *Times* 4 May 10/6 The Dutch prisoners were arrested at the beginning of 1954 and held *incomunicado* for a considerable time. **1956** 'A. Bridge' *Lighthearted Quest* 42 Liners' captains, in her experience, kept themselves *incommunicado* except when on the high seas. **1962** I. Murdoch *Unofficial Rose* vi. 57 Randall had been practically incommunicado for ten days. **1970** *Observer* 13 Sept. 4/7 The prisoners will be..detained again under the Terrorism Act by the Special Branch, permitted..to hold any person *incommunicado* for any length of time.

in-company *attrib.*: see *in *prep.* 17*.

incompatibility. Add: **3.** *Pharm.* The condition (of drugs) of being incompatible (sense *5); an instance of this.

1825 R. Best *Tables Chem. Equivalents* 29 This table refers only to chemical incompatibility. The physician must decide for himself, on the propriety of combining those substances which it designates as incompatible with each other. **1877** R. Farquharson *Guide Therapeutics* 14 Incompatibility may be of different sorts, and is generally divided into chemical and physiological. **1948** Gross & Greenberg *Salicylates* ii. 16 Incompatibility of salicylic acid has been reported with compounds of bismuth. **1969** J. A. Bevan *Essent. Pharmacol.* lxv. 628 The whole problem of drug incompatibility is changing as our knowledge of drug metabolism enlarges. *Ibid.* 629 In Table 65-2 some common therapeutic drug incompatibilities are listed. **1970** A. & E. F. Grollman *Pharmacol. & Therapeutics* (ed. 7) i. 20 Incompatibilities between drugs may also occur when they are mixed, as a result of physical or chemical changes.

4. *Biol.* **a.** The incapacity of cells or tissue from one individual to tolerate those of some other individual when an organic union of some kind is formed between them, esp. in grafting and transplantation, in the transfusion of blood, and in parasitism.

1904 *Mass. Agric. Exper. Station Techn. Bull.* No. 2. 15 After a close study of a large number of these defective unions, the writer has reached the opinion that they are almost always due to the incompatibility of stock and cion. **1916** J. Loeb *Organism as a Whole* iii. 46 A lesser though still marked degree of incompatibility exists also in lower animals for grafts from a different species. **1916** *Mem. N.Y. Bot. Garden* VI. 429 Between bloods of members of any one class there is no incompatibility in the form of agglutination. **1927** *Jrnl. Agric. Res.* XXXIV. 675 Reciprocal grafts between the pigmented and the nonpigmented varieties showed no incompatibility whatever, as they produced well-established unions. **1935** N. P. Sherwood *Immunol.* xii. 270 In discussing the incompatibility of species not closely related, Loeb mentions the rigid specificity requirements for successful skin grafting or organ transplantation. **1957** Mahlstede & Haber *Plant Propagation* xvi. 277 Many theories have been proposed relative to the causes of incompatibility in grafted plants. **1970** *Woman's Own* 3 Jan. 26/3 Expectant mothers will be immunized against the development of the Rhesus incompatibility which can affect their babies. **1971** *Canad. Jrnl. Bot.* XLIX. 304/2 Parasite/host combinations are referred to by the corresponding genotypes under study which specify compatibility or incompatibility of the relationship.

b. Inability to succeed in sexual reproduction under circumstances where fertile gametes are produced and brought together; *orig.* used mainly of failure of crossing between different species, but now usu. *spec.* such inability (occurring in many species of

fungi and angiosperms) between individuals belonging to the same species, which is genetically controlled and which usually acts to promote outbreeding.

1905 *Biol. Bull.* VIII. 320 The incompatibility was less marked... In general..the eggs crossed much more readily than did the California form. **1913** W. Bateson *Probl. Genetics* xi. 240 Whether the incompatibility between species is to be associated with that of the self-steriles also cannot be positively asserted. **1916** A. B. Stout in *Mem. N.Y. Bot. Garden* VI. 335 The term incompatibility has been, of course, used to characterize a wide range of causes of failure in reproduction, but it can well be limited in its application to those causes existing in the plants themselves which prevent fertilization in and between plants with normal reproductive organs and gametes. *Ibid.* 336 We may distinguish two quite distinct types of incompatibility: 1. Anatomical incompatibility... 2. Physiological incompatibility. **1954** *Adv. Genetics* VI. 236 The widespread distribution and high frequency of incompatibility in the higher plants and fungi indicate that it is one of the most important outbreeding mechanisms in plants. **1966** J. R. Raper *Genetics of Sexuality in Higher Fungi* iv. 52 The terms 'incompatibility' and 'incompatibility factors' as applied to this and other systems of self-sterility in the fungi are of more recent origin. **1970** *Bot. Gaz.* CXXXI. 139 (*heading*) Self- and interspecific incompatibility in the Convolvulaceae.

incompatible, *a.* Add: **5.** *Pharm.* Of a drug: reacting or interfering *with* another (specified) substance in such a way that the two should not be mixed or prescribed together; unsuited to simultaneous administration to a patient.

1812 J. A. Paris *Pharmacologia* p. vii, The incompatible substances, i.e. all those which are capable of destroying its properties, or rendering its flavour, or aspect, unpleasant, or disgusting. **1855** R. G. Mayne *Expos. Lex. Med. Sci.* (1860) 519/1 *Incompatible*, applied to substances which act chemically on each other, and which therefore cannot with propriety be prescribed together. **1881** R. Farquharson *Guide to Therapeutics* (ed. 2) 24 Infusions containing tannic acid are incompatible with metallic salts generally. **1898** E. W. Lucas *Pract. Pharmacy* xliv. 299 Incompatible substances cannot exist together in solution without mutual decomposition. *Ibid.* 301 Sodium bicarbonate is incompatible with solution of strychnia. **1917** E. A. Ruddiman *Incompatibilities in Prescriptions* (ed. 4) p. iii, The second object of the writer is to furnish the student of pharmacy with a list of incompatible prescriptions in such form that he may find out for himself what the trouble is. **1970** Goodman & Gilman *Pharmacol. Basis Therapeutics* (ed. 4) 1716/1 Cationic substances and anionic substances..are often incompatible with each other.

6. *Biol.* **a.** Exhibiting or causing incompatibility (sense *4 a). Const. *with*.

1904 *Mass. Agric. Exper. Station Techn. Bull.* No. 2. 14 When the two members are unlike in nature and in some way physiologically incompatible (whatever that may mean), the wound does not heal readily, owing to some sort of irritation which continues to be felt at this point. **1918** *Jrnl. Immunol.* III. 99 A patient of group I, for example, requires a donor of group I, the blood of all other groups being incompatible. **1936** *Jrnl. Pomol.* XIV. 360 Later, the sour orange also proved to be incompatible as a stock with imported varieties of this species. **1962** J. D. Smyth *Introd. Animal Parasitol.* xxxii. 371 Physiological resistance. This type of resistance is due to some aspect of the host physiology being incompatible with that of the invading parasite at some stage in its life history. **1966** Wright & Symmers *Systemic Path.* I. iv. 151/2 For transfusions..under no circumstances should the donor's red cells be incompatible with the recipient's plasma. **1971** *Canad. Jrnl. Bot.* XLIX. 303 (*heading*) Transfer of ³⁵S from wheat to the powdery mildew fungus with compatible and incompatible parasite/host genotypes.

b. Having or exhibiting incompatibility (sense *4 b); unable to cross.

1905 *Biol. Bull.* VIII. 323 No eggs segmented, but neither did the eggs in the check experiment in sea water, showing that the eggs or the sperm were poor, or else incompatible. **1913** W. Bateson *Probl. Genetics* xi. 239, I first tried Cinerarias, which are usually self-sterile, but I found no incompatible pairs of plants. **1916** *Mem. N.Y. Bot. Garden* VI. 419 The parent species were cross-incompatible. **1937** Gwynne-Vaughan & Barnes *Struct. & Devel. Fungi* (ed. 2) 5 Often, in these self-incompatible fungi, the sexual apparatus has partially or wholly disappeared. **1967** Briggs & Knowles *Introd. Plant Breeding* xv. 187 In *Gasteria*, pollen germination and tube development were not affected in incompatible pollinations.

incompetence. Add: **2. b.** *Med.* Inability to function correctly; *esp.* inadequacy of a valve or sphincter properly to regulate the passage of liquid or solid matter.

1876 [in Dict., sense 2]. **1890** F. Taylor *Man. Pract. Med.* 458 Incompetence of the aortic valves gives rise to a murmur during the dilatation of the ventricle. **1900** Dorland *Med. Dict.* s.v., *Pyloric i[ncompetence]*, passage of food undigested into the intestine. **1939** Dible & Davie *Path.* xxxii. 535 If there has been much necrosis, contraction, or deformity of the valve curtain, a definite degree of incompetence will be left. **1950** *Amer. Jrnl. Obstetr. & Gynecol.* LIX. 69 The symptoms of the incompetence of the internal os are chiefly those of habitual abortion. **1970** *Med. Jrnl. Austral.* 25 July 179/1 The incidence of round ligament vein incompetence in women greatly exceeds that of short saphenous vein incompetence. **1971** *Gut* XII. 102 Of 19 patients... 17 (89%) exhibited duodenogastric reflux of barium indicating pyloric incompetence. **1971** *Biol. Abstr.* LII. 8355/1 Gonadotropic hormones, the prostate and an experimental model of its incompetence.

incompetency. Add: **1. b.** *Med.* = *incompetence 2 b.

1865 J. H. Bennett *Clin. Lect. Princ. & Pract. Med.* (ed. 4) vi. 577 (*heading*) Incompetency of aortic valves. **1892** W. Osler *Princ. & Pract. Med.* v. 611 An acute dilatation of the left ventricle with relative incompetency of the mitral segments. **1950** *Amer. Jrnl. Obstetr. & Gynecol.* LIX. 69 The degree of incompetency [of the internal os of the cervix] may vary.

incompetent, *a.* Add: **2. b.** *Med.* Unable to function correctly: used esp. of a valve or sphincter. (Cf. *incompetence 2 b.)

1863 W. Braithwaite in *Retrospect Med.* XLVII. 69 If..the aortic valvular segments are rendered incompetent, we have the aortic regurgitant current. **1879** [in Dict., sense 2]. **1939** Dible & Davie *Path.* xxxii. 537 The aortic valves are incompetent and allow a certain quantity of blood to flow back into the left ventricle during diastole. **1950** *Amer. Jrnl. Obstetr. & Gynecol.* LIX. 69 An abortion due to an incompetent internal os. **1971** *Amer. Jrnl. Digestive Dis.* XVI. 307 Eight patients had free gastroesophageal reflux through a weak, incompetent sphincter. **1971** *Amer. Jrnl. Dis. Children* CXXI. 481/2 The immunologic system may be incompetent or contribute to the etiology of the disease.

5. *Geol.* Of rock or a stratum: apt to flow or to be crushed when laterally compressed; incapable of forming a simple fold and supporting any overlying strata without being distorted by plastic flow. Also applied to structures and processes dominated by such strata.

1891–92 B. Willis in *Ann. Rep. U.S. Geol. Survey* (1893) XIII. ii. 250 If the thrust be not powerful enough to raise the load there will be no uplift; or if the layers be so plastic that they yield to the thrust by swelling, then the principal result of deformation is change of form other than by simple flexure, and it assumes some phase of flowing. This is incompetent structure. **1923** *Jrnl. Geol.* XXXI. 506 Several attempts were made to reproduce the overfolds..of the Alps by applying rotational stresses to models in which competent layers were placed between incompetent layers. **1949** C. M. Nevin *Princ. Struct. Geol.* (ed. 4) iii. 54 Folding has been divided into two groups: competent folding, where the dominantly horizontal pressures are transmitted by the competent beds acting as a strut; and incompetent folding where the forces are mostly vertical and the incompetent beds react passively. *Ibid.* 55 Immediately beneath the arch formed by a folded relatively competent bed a formation as incompetent as shale may rise as a broad simple arch, even at a considerable depth. **1967** A. I. Levorsen *Geol. Petroleum* (ed. 2) viii. 358 In these cases the salt mass acts as an incompetent formation, rising as a result of the deformation of the enclosing rocks. *Ibid.* 378 Folds in the relatively incompetent salt.

incompletability. Add: **b.** *Logic* and *Math.* The property, inherent in certain kinds of logical or mathematical system, of being incapable of providing a proof of every true proposition that can be formulated within the language of the system, no matter how many of these propositions be taken as extra axioms of the system. Cf. *incomplete *a.* 2 b.

1940 W. V. Quine *Math. Logic* Pref. p. vii, Gödel's theorem regarding the incompletability of logic and arithmetic is derived along novel lines. **1950** —— *Methods of Logic* (1952) § 42. 248 Church's argument makes essential use of Gödel's theorem of the incompletability of number theory. **1962** [see *Gödel].

incomplete, *a.* Add: **2.** Philos. *incomplete symbol* (see quot. 1910).

1910 Whitehead & Russell *Principia Math.* I. iii. 69 By an 'incomplete' symbol we mean a symbol which is not supposed to have any meaning in isolation, but is only defined in certain contexts. **1919** B. Russell *Introd. Math. Philos.* xvii. 182 Classes are in fact, like descriptions, logical fictions, or (as we say) 'incomplete symbols'. **1930** L. S. Stebbing *Mod. Introd. Logic* ix. 156 The notion of an incomplete symbol is required in order to define what is meant by a 'logical construction'. **1956** J. O. Urmson *Philos. Analysis* iii. 30 Russell is now writing as though to show that 'X' is an incomplete symbol is tantamount to showing that there are no X's. **1967** R. A. George tr. *Carnap's Logical Struct. World* ii. § 27. 48 These other signs we call, after Frege, incomplete symbols. *Ibid.* 49 An incomplete symbol designates..a quasi object.

b. *Logic* and *Math.* Of a formal logical or mathematical system: containing true propositions for which no proof of validity is possible using only the formal rules of the system. Cf. *incompletability b.

1932 Lewis & Langford *Symbolic Logic* xi. 365 It is to be observed that there is an important respect in which each of these sets is incomplete and therefore open to further determination. **1937** A. Smeaton tr. *Carnap's Logical Syntax of Lang.* iii. 100 Now, however, Gödel has shown that not only all former systems, but all systems of this kind in general, are incomplete. **1958** Nagel & Newman *Gödel's Proof* vi. 58 Gödel showed that *Principia* [*Mathematica*], or any other system within which arithmetic can be developed, is essentially incomplete. **1970** J. van Heijenoort tr. *Gödel's Completeness & Consistency* in *Frege & Gödel* 107 If *S* contains *Z*, *S* is incomplete, that is, there are in *S* propositions..that are undecidable on the basis of the axioms of *S*, provided that *S* is ω-consistent.

incompleteness. Add: **b.** (Further examples, in *Logic* and *Math.*, corresponding to *incomplete *a.* 2 b.) Also *attrib.* in **incompleteness theorem,** the mathematical proof of incompleteness (cf. *Gödel).

1932 Lewis & Langford *Symbolic Logic* xi. 365 This point of incompleteness can be brought out by contrasting examples. **1937** A. Smeaton tr. *Carnap's Logical Syntax of Lang.* iii. § 34a. 100 In spite of this necessary incompleteness of the method of derivation.., the method retains its fundamental significance. **1955** K. R. Popper in P. A. Schilpp *Philos. of R. Carnap* (1963) 200 Gödel, by his two famous incompleteness theorems, had proved that one unified language would not be sufficiently universal for even the purposes of elementary number theory. **1957** P. Suppes *Introd. Logic* iv. 70 Gödel's theorem on the incompleteness of elementary number theory is probably the most important theorem in the literature of modern logic. **1962** B. Meltzer tr. *Gödel's On Formally Undecidable Propositions* 62 The true source of the incompleteness attaching to all formal systems of mathematics, is to be found..in the fact that the formation of ever higher types can be continued into the transfinite. **1971** G. Hunter *Metalogic* 257 This is Gödel's second incompleteness theorem.

incompletive (inkǒmplī·tiv), *a.* (*sb.*) *Gram.* [f. Incomplet(e *a.* + -ive.] An aspect of the verb indicating incompletion of an action or process; = Imperfective *a.* (*sb.*) 2.

1944 E. A. Nida *Morphol.* II. ix. 130 This imperfective aspect, which may also be called 'atelic' or 'incompletive', is often associated with the future time, with the negative or with the potential. **1964** —— *Toward Sci. Transl.* ix. 200 In the Old Testament, however, the differences between the Hebrew completive and incompletive forms are essentially contrasts between kinds of action. **1964** *Language* XL. 77 Two aspect classes of the active voice are also marked by simulfixation: the incompletive, marked by the structural absence of a simulfix, and the completive, marked by the simulfix for active voice. **1968** *Ibid.* XLIV. 293 There are four aspects in Huixtec: timeless, incompletive, completive, and perfective.

incomunicado, var. *incommunicado.

inconclusible. Delete †*Obs.* and add later example.

1930 T. S. Eliot *Ash-Wednesday* ii. 18 Conclusion of all that Is inconclusible.

Inconel (i·nkǒnel). A proprietary name of various alloys which contain nickel (70–80%), chromium (12–19%), and iron (usually between 5% and 8%), and are useful for their strength and their resistance to corrosion and oxidation at high temperatures.

1933 *Official Gaz.* (U.S. Patent Office) 12 Sept. 263/2 International Nickel Company, Inc... *Inconel.* For Nickel Alloys and Alloys of Nickel, Chromium, and Iron. Claims use since Nov. 25, 1932. **1935** *Chem. Abstr.* XXIX. 7257 (*heading*) Corrosion resistance of nickel, of copper-nickel alloys, of nickel silver and of Inconel. **1938** *Jrnl. R. Aeronaut. Soc.* XLII. 962 A further development is the alloy 'Inconel' containing about 80 per cent. Ni., 13 per cent. Cr., 7 per cent. Fe., which can be hardened by work, but not by heat-treatment... Its resistance to heat-oxidation at very high temperatures finds application in silencers and exhaust manifolds. **1966** *McGraw-Hill Encycl. Sci. & Technol.* VII. 265/2 The Inconels (Nimonics in England) are advantageous in cyclic heating and cooling service where large temperature gradients exist. **1967** D. Fishlock *New Materials* ii. 23 Inconel X-750 (75 per cent nickel) is today used extensively for the structure of the 4104 m.p.h. X-15 rocket aircraft..whose skin..is designed to withstand 650°C.

incongruent, *a.* (Later examples.)

1940 *Mind* XLIX. 349 Evident empirical facts (*i.e.*, rotational motion and incongruent counterparts) which Newton and Kant designated as rendering the relational theory untenable. **1951** A. Gardiner *Theory of Speech & Lang.* (ed. 2) 232 Grammar is, in the main, concerned with linguistic form in congruent function, and treats of incongruent function only in so far as this is building up new forms in which such function will be congruent. **1966** *Amer. Philos. Q.* III. 179/1 Kant made use of incongruent counterparts to establish a theory of space. **1972** *Jrnl. Social Psychol.* LXXXVI. 121 When a man's life style is incongruent with the demands of a task he must perform he will experience stress.

incongruently, *adv.* (Later examples.)

1933 *Amer. Jrnl. Sci.* XXV. 277 Fayalite melts incongruently with separation of iron. **1970** *Nature* 19 Dec. 1144/1 The ruby laser permits fairly fast deposition rates even of 'difficult' materials and, in particular, allows preparation of thin films of compounds that normally evaporate incongruently.

inconnu (i·ŋkǒnïu, æňkonü). [Fr., unknown.] **1.** A game fish, *Stenodus leucichthys*, belonging to the family Salmonidæ and found in Alaska and north-west Canada.

[**1806** S. Fraser *Lett. & Jrnls.* (1960) 231 The Indians also state that there are plenty white fish uncou some trout carp Jub, &c. in the fall of the year.] **1833** G. Back *Jrnl.* 14 Aug. in *Narr. Arctic Land Expedition* (1836) iii. 101 La Prise set a net, which..produced a few white fish, a trout, and, what surprised the Indians, an inconnu. **1844** J. H. Lefroy *In Search Magnetic North* (1955) 99, I tried the Inconnu there, but it is far inferior to the Trout. **1905** D. S. Jordan *Guide to Study of Fishes* II. iv. 67 The Inconnu, or Mackenzie River salmon, known on the Yukon as 'charr'..belongs to this genus [sc. *Stenodus*]. **1943** *Beaver* Mar. 28 Inconnu (*Stenodus Mackenzii*), generally called 'conny', is a fish peculiar to the Mackenzie River. **1965** A. J. McClane *Standard Fishing Encycl.* 424/1 Inconnu have fair food value and are utilized to a large extent by the Eskimos.

2. (æňkonü.) An unknown person; a stranger. Also fem. **-e.**

1807 *Salmagundi* 4 Apr. 142 When we toast a frenchman, we merely mean one of those *inconnus*, who swarmed to this country, from the kitchens and barber's shops of Nantz, Bourdeaux, and Marseilles. **1829** G. Griffin *Collegians* (ed. 2) I. xiii. 266 The lovely inconnue..felt her heart beat somewhat quickly. **1865** 'Ouida' *Strathmore* I. v. 76 The jewels that sparkled on the hands of the fair inconnue. **1920** F. S. Fitzgerald in *Smart Set* July 15/2 Edith murmured a conventional 'Thanks loads—cut in later,' to the *inconnu*.

inconquerable, *a.* Delete †*Obs.* and add later examples.

1905 *Daily Chron.* 4 Oct. 5/1 An inconquerable penchant to conspiracy. **1921** *Edin. Rev.* July 134 That rather ticklish matter, the government and command of the British Empire Navy,..does not present inconquerable obstacles to be overcome. **1946** W. S. Churchill *Secret Session Speeches* 29 The inconquerable, the inexhaustible adaptiveness and ingenuity of the British mind.

inconsciently (inkǫ·nʃiəntli), *adv.* [f. Inconscient *a.* + -ly[2].] Unknowingly.

1913 *Glasgow Herald* 18 Oct. 11/1 They stood, inconsciently perhaps, for some of the priceless intangible things without which the millennium would be a sterile and bitter anti-climax. **1929** R. Bridges *Testament of Beauty* iv. 126 Held by the inborn love of Beauty inconsciently Of preference to imitate the more beautiful things.

inconsequential, *a.* Add: Also as *sb.*

1936 L. C. Douglas *White Banners* xiii. 277 Exchanging inconsequentials and blandly taking each other's measure. **1936** *Mind* XLV. 334 The three protagonists may actually hold a few inconsequentials in common. **1957** *Economist* 2 Nov. 436 (Advt.), News never buried under frothy inconsequentials.

inconseque·ntialness. [f. Inconsequential *a.* + -ness.] Inconsequentiality.

1931 *Daily Express* 22 Sept. 10/6 The fact is, the profligacy of war expenditure, the high pay, the inconsequentialness of it all, led every one into bad, extravagant habits. **1956** *Essays in Crit.* VI. 117 Her English..having a quality of inconsequentialness.

incontinent, *a.* **3.** (Further examples.)

1901 G. B. Shaw *Admirable Bashville* ii. 309 To begin my life a speechless babe, hairless, incontinent, Hobbling upon all fours, a nurse's nuisance. **1973** *Times* 11 June 14/2 Mrs Jones is doubly incontinent and vomits several times a day.

in contumaciam: see *in *Lat. prep.*

inconvertible, *a.* Add: **2. b.** (Earlier examples.)

1847 A. De Morgan *Formal Logic* iv. 58 The universal affirmative..and the particular negative..are not necessarily convertible, and are generally called *inconvertible*. **1849** W. Thomson *Outl. Laws of Thought* (ed. 2) liii. 216 The judgment O is usually considered inconvertible by the ordinary method. **1857** W. Spalding in *Encycl. Brit.* XIII. 606/1 They hold O to be inconvertible.

Hence as *sb.*, a proposition which cannot be converted.

1847 A. De Morgan *Formal Logic* iv. 62 As to inconvertibles, contranominal and converse are terms of the same meaning.

incoordinated, *a.* (Examples.)

1885 F. Warner *Physical Expression* 71 Co-ordinated movements and inco-ordinated movements. **1911** W. James *Mem. & Stud.* ix. 218 There was no appearance of general dismay and little of chatter or of inco-ordinated excitement.

incorporate, *v.* Add: **3. b.** *spec.* To admit a graduate of another university *ad eundem*.

incorporation. Add: **2. c.** The action of incorporating (cf. *incorporate *v.* 3 b) into another university.

1966 *Rep. Comm. Inquiry Univ. Oxf.* I. 404 Degrees in absence and by incorporation.

incorporatorship (inkǭ·ɹpǒrēitəɹʃip). [Incorporator 2.] The position of an incorporator.

1873 'Mark Twain' & Warner *Gilded Age* xlii, It would be more money in my pocket, in the end, than my brother-in-law will get out of that incorporatorship, fat as it is.

in corpore: see *in *Lat. prep.*

incorrectitude (inkǒre·ktitiūd). *rare.* [f. In-[3] + *correctitude.] The state of being in the wrong in one's conduct or opinions; incorrectness.

1898 *Westm. Gaz.* 4 Oct. 2/2 We are certain that it is not wise to adopt this position of positive incorrectitude. *Ibid.* 10 Nov. 2/3 He is brought round to own the incorrectitude of calling Mr. Gladstone a spider. **1947** *New Biol.* III. 163 It furnishes an everyday illustration of the incorrectitude of regarding different species as being necessarily intersterile.

incorrigibility. **b.** Delete † (*obs.*) and add: of not being open to proof or disproof. So **inco·rrigibilist,** one who adheres to an incorrigible theory.

1956 E. H. Hutten *Lang. Mod. Physics* vi. 226 To ascribe absolute incorrigibility to them [*sc.* sentences] is misleading since it suggests that they are never questioned within any context. **1966** *Amer. Philos. Q.* III. 101/2 The supposed incorrigibility of first person pain reports. *Ibid.*, Part of the guile of the incorrigibilist is not to disclose which reasons..he is prepared to take into account as relevant.

incorrigible, *a.* (*sb.*) **A.** *adj.* **3.** Delete †*Obs.*, and add: not verifiable; that cannot be proved false.

1946 A. J. Ayer *Lang., Truth & Logic* (ed. 2) 10 What may be said to verify them [*sc.* basic propositions] conclusively is the occurrence of the experience to which they uniquely refer... Propositions of this kind are 'incorrigible'..it is impossible to be mistaken about them except in a verbal sense. **1956** —— *Probl. Knowl.* 54 These experiential statements..are taken as basic because they are held to be 'incorrigible'. *Ibid.* 55 Experiential statements are not incorrigible in the sense that once they have been discovered to be true they cannot subsequently be denied.

B. *sb.* **2.** Something not open to verification.

1936 H. H. Price *Truth & Corrigibility* 28 Innumerable judgements..will have to be admitted as incorrigibles.

incorrigibly, *adv.* Add: Also in sense of *incorrigible *a.* 3.

1956 A. J. Ayer *Probl. Knowl.* 56 Such conditions as make it reasonable for me to claim that the statement is incorrigibly known.

in-country. Restrict *Sc.* to sense in Dict. and add: **2.** Used *attrib.*: in the country; in a contextually specified country. Cf. *in *prep.* 17*.

a **1953** Dylan Thomas *Prospect of Sea* (1955) 9 Between the incountry fields and the incoming sea. **1963** *Times* 6 Feb. 6/7 Cornwall..has agreed to give all technical assistance that is needed on payment by the Scillies of travelling and subsistence expenses, and to charge for the use of its educational and welfare institutions at the 'in-country' instead of the 'out-country' rate. **1966** *N.Y. Times* 1 May iv. 3 In South Vietnam, in what is called in Saigon the 'in-country' war, development efforts have been concentrated upon types of weapons best utilized in jungles and rice paddies. **1969** *Daily Tel.* 20 Nov. 5/3, 2,500 American Marines are to leave South Vietnam.. over a five-day period starting from today. This will reduce America's 'in-country' military strength to 484,000 troops. **1973** *Ibid.* 13 Mar. 5/2 American in-country troop strength [in Vietnam] stands at 7,170 men.

increase, *v.* Add: **6. b.** (See quot. 1957.)

1840 J. Gaugain *Lady's Assistant* 96 Increase a stitch on each wire, by knitting the last stitch in the common way; knit it again from the back part of the loop (this is the way to increase without making a hole). **1872** *Young Englishwoman* Nov. 607/1 The increasing and decreasing may..take place at the ends or in the middle of the work. In increasing in the middle, the increase is effected by taking up stitches. **1944** A. Thirkell *Headmistress* iii. 60 They..had to take off all the stitches and unravel back to where they ought to have begun increasing and pick up all the stitches again. **1957** M. B. Picken *Fashion Dict.* 182/2 *Increase*, in knitting, crocheting, tatting, etc., to add to number of stitches in row, pattern, or round so as to enlarge the piece. **1971** *Vogue's Guide to Crochet* 14 Increasing a stitch means adding a stitch, and decreasing a stitch means losing it. *Ibid.*, Care must be taken not to decrease or increase in such a way as to leave an uneven edge.

increase, *sb.* Add: **5. b.** (See *increase *v.* 6 b.)

1872 [see *increase *v.* 6 b]. **1971** M. Hamilton-Hunt tr. *Mon Tricot Knitting Dict.* 20 An increase is the method of making an extra or supplementary stitch in the course of the work... The single increases are known as simple increases... Double increases are used for darts, etc.

increasing, *vbl. sb.* (Later examples.)

1872 [see *increase *v.* 6 b]. **1932** D. C. Minter *Mod. Needlecraft* 70/1 *Increasing*... After knitting a stitch.. knit again into back of same loop. **1972** *Where* May/June 133/1 There has been a 78 per cent increasing in the principal venereal diseases, that is to say gonorrhoea, syphilis and chancroid.

increasing, *ppl. a.* Add: Esp. in *law of increasing return(s)*: the observed fact that in certain manufactures and industries the expenditure of labour or capital up to a certain point produces a more than proportionate corresponding return. Cf. *diminishing *ppl. a.* 1 b.

1890 A. Marshall *Princ. Econ.* I. iv. xiii. 379 While the part which Nature plays in production conforms to the Law of Diminishing Return, the part which man plays conforms to the Law of Increasing Return. **1925** S. E. Thomas *Elem. Econ.* x. 114 The Law of Increasing Returns..states that the expansion of an industry, in which there is no dearth of the necessary agents of production, tends to be accompanied by increasing returns.

increment. Add: **4. c.** *Forestry.* The increase in the quantity of wood produced by a tree or group of trees during a limited period; the value of this increase. Also *attrib.*, as **increment borer** [tr. G. *zuwachsbohrer* (M. R. Pressler *Zur Forstzuwachskunde* (1868) 19], a kind of auger with a hollow bit, used to measure the magnitude of this increase in individual trees, by removing a piece of wood in which annual rings can be examined; **increment boring,** the cylinder of wood removed from a tree by an increment borer.

1889 W. Schlich *Man. Forestry* I. ii. 167 The increment laid on by an individual tree does not by itself govern the increment produced per acre, because the latter is represented by the increment per tree, multiplied by the number of trees per acre. **1895** *Ibid.* III. i. 13 In the case of standing trees, the measurements are made with Pressler's Increment Borer. **1905** *Terms Forestry & Logging* (U.S. Dept. Agric. Bureau Forestry) 14 *Increment*, the volume or value of wood produced during a given period by the growth of a tree or of a stand. **1938** Weaver & Clements *Plant Ecol.* (ed. 2) ii. 34 The increment borer..removes a small core of wood from circumference to center, without injuring the tree. **1942** *Amer. Jrnl. Bot.* XXIX. 553/2 (*caption*) An increment boring from a butt treated western red cedar pole. **1967** T. E. Avery *Forest Measurements* xii. 210 In climates where tree growth is characterized by annual rings, ages of standing trees are usually determined by extracting a radial core of wood with an increment borer. **1972** N. D. G. James *Arboriculturalist's Compan.* xiv. 159 In forestry the measurement and calculation of increment can be of considerable economic importance.

6. *attrib.* and *Comb.*

1909 [see *budget *v.* c]. **1909–10** *Act* 10 *Edw. VII* c. 8 § 2 (1) The increment value of any land shall be deemed to be the amount (if any) by which the site value of the land, on the occasion on which increment value duty is to be collected.., exceeds the original site value of the land as ascertained in accordance with the general provisions of this Part of this Act as to valuation. **1971** *Jrnl. Gen. Psychol.* Jan. 68 At a 30-msec interval the difference threshold for a *decrease* in pulse interval would be 30 msec, while the difference threshold for an *increase* in pulse interval would be five msec. These two types of difference thresholds will be referred to as decrement and increment thresholds, respectively.

incrispated, *ppl. a.* For *Obs. rare*[−1] read *rare* and add later example.

1922 Joyce *Ulysses* 695 Incrispated black hairs.

in-crowd: see *in *a.* 2 b.

incubate, *v.* **3. b.** Add to def.: To maintain at a constant degree of warmth that will favour growth or continued survival (e.g. of micro-organisms); more widely, to maintain under given conditions in a controlled or artificial environment. (Further examples.)

1912 *Jrnl. Exper. Med.* XVI. 171 One volume of a 5 per cent. suspension of erythrocytes in isotonic salt solution was mixed with two volumes of serum in capillary pipettes. The pipettes were incubated at 38°C for two hours and then put in an ice box for twenty-four hours. **1938** [see *incubation 4*]. **1947** *Growth* XI. 232 The ability to ferment galactose occurs exclusively in cells that have been grown on galactose, or have been incubated in a galactose solution for a few hours. **1948** *Biochem. Jrnl.* XLIII. 538/2 The requisite amount of homogenate..is then added and incubated at 37° for 1 hr. **1962** Harris & Gruber in A. Pirie *Lens Metabolism Rel. Cataract* 373 The second method of study has been that of incubating the lens under various conditions at 37°C for a period of time prior to analysis. **1969** *Clin. Sci.* XXXVII. 99 The diffusion of glucose and potassium between erythrocytes and their medium was impaired when the red cells were packed by centrifugation and incubated for 48 hr without agitation. *Ibid.* 409 Rat liver slices were incubated in serum obtained from normal volunteers and from uraemic patients. **1970** *Sci. Jrnl.* May 19/3 The inoculated samples were incubated for up to 27 days at various temperatures—0°, 5°, 10° and 20°C. **1970** *Biol. Abstr.* LI. 11404/2 By incubating the seeds for 8 days at 15 C and then for 3 days at 25 C in a non-sterilized soil coming from a rice field..seed resistance to rotting may be estimated. **1971** *Nature* 19 Nov. 154/1 Because uterine secretions are slightly alkaline, a copper strip was incubated for 18 h in *p*H 8 carbonate buffer.

incubate (i·nkiubēit), *sb.* [f. the vb. + -ate[3], after *filtrate*, *precipitate*, etc.] A preparation, or material, that has been incubated.

1959 *Jrnl. Amer. Chem. Soc.* LXXXI. 4109/2 Aldosterone was the most abundant steroid found in incubates of adrenals from the American bullfrog, *Rana catesbeiana*. **1972** *Nature* 22 Dec. 470/2 Biosynthesized ¹⁴C-labelled 11 KT was detectable in the control but not in the Cd-damaged tissue incubate. **1973** *Biochem. Jrnl.* CXXXI. 406/2 The incubates were extracted with equal volumes of ethyl acetate.

incubation. Add: **1.** More widely, the protection of its eggs by an animal, or the provision of conditions that favour their development. Also, the embryonic development of an animal within an egg.

1835–6 R. B. Todd *Cycl. Anat. & Physiol.* I. 785/2 When the process of incubation begins [in the crayfish] the surface of the yolk is first seen to become covered with star-like or serrated spots. **1905** J. Wyman in D. S. Jordan *Guide to Study of Fishes* I. xi. 170 Among the Siluroid fishes of Guiana there are several species which, at certain seasons of the year, have their mouths and bronchial cavities filled either with eggs or young, and, as is believed, for the purpose of incubation. **1931** J. R. Norman *Hist. Fishes* xvi. 320 The period of incubation [of ray eggs] lasts from four and a half to nearly fifteen months. **1960** T. H. Waterman *Physiol. Crustacea* I. xiii. 440 As in female vertebrates, the ovary of the crustaceans..secretes a hormone which prepares the mother for incubation. **1962** K. F. Lagler et al. *Ichthyol.* x. 299 To carry egg protection to its highest degree, some fishes have evolved various types of internal incubation or gestation. *Ibid.*, In a Brazilian catfish..the male parent develops an enlarged lower lip to form a pouch in which labial incubation of the eggs takes place.

4*. The process, or an instance, of incubating anything in a controlled or artificial environment (see *incubate *v.* 3 b).

1928 *Jrnl. Amer. Med. Assoc.* 3 Nov. 1338/1 The maturity-provoking hormone is definitely somewhat more stable to incubation in the presence of acid. **1938** *Biochem. Jrnl.* XXXII. 450 Batches of 100 flasks of Czapek-Dox solution..were sown with..*Helminthosporium leersii*..and incubated in the dark at 24°. The rate of growth was slow and incubation was continued for 90 days... The mycelium was separated by filtration. **1969** *Clin. Sci.* XXXVII. 409 Incubation of slices in uraemic serum had no effect upon glycogenolysis or on glycogen synthesis, but utilisation of glucose was inhibited. *Ibid.* 410 Slices from the same animal served for incubations both with uraemic and with control sera. **1971** *Nature* 19 Nov. 154/1 An 18 h incubation of a copper strip in 2·5 ml. of saline solution alone yielded a Cu^{2+} concentration of 4×10^{-6} M.

5. *attrib.* and *Comb.* Also *incubation drive*, *fever*, *-period* (further examples); **incubation-patch**, an area of the ventral surface of a bird's body that swells and loses its feathers during the incubation of eggs, as a way of providing them with the necessary warmth; usu. called a *brood-patch*.

1953 N. Tinbergen *Herring Gull's World* xvii. 155 It is likely that his birds, by being robbed so often, had on the average a lower incubation-drive than our birds. **1954** D. A. Bannerman *Birds Brit. Isles* III. 133 Colonel Meiklejohn suggests that this cock may have been suffering from incubation fever and built the nest to satisfy its craving. **1952** J. Fisher *Fulmar* iv. 95 Fulmars have only a single incubation-patch, into which two eggs could probably not fit. **1926** H. M. Kyle *Biol. Fishes* 392/1 (*index*) Incubation period. **1940** *Jrnl. Biol. Chem.* CXXXIV. 250 In a prolonged incubation period there was observed nearly complete hydrolysis. **1943** *Biol. Abstr.* XVII. 2253/2 Av. incubation period was 380 hrs. at about 15·5°C [for the tide-pool cottid]. **1961** E. Cameron *Cockroach* ii. 24 The incubation period [of *Periplaneta americana*] is a fairly long one, and varies a good deal according to the temperature, and the relative humidity of the environment. **1969** A. Bellairs *Life of Reptiles* II. x. 444 The incubation period of eggs, from the time of fertilisation to that of hatching, is exceedingly variable. **1973** *Nature* 30 Mar. 329/2 Following a 24 h incubation period, the medium and cells which had not stuck to the well surface were decanted.

incubator. Add: **1. a.** Also, any of certain other animals having particular patterns of behaviour to keep their eggs at a higher temperature than the surrounding environment.

1969 A. Bellairs *Life of Reptiles* II. ix. 430 Of the six or more species which have acquired the brooding habit, only the Indian python is now known on good evidence to be a true incubator.

5. *Comb.* **incubator-bird** *Austral.* = Megapode, megapod.

[**1896** F. G. Aflalo *Sk. Nat. Hist. Austral.* 152 These most interesting birds do not incubate their eggs...for instead..we find them deposited in a perfectly planned, thoroughly heated incubator.] **1943** C. Barrett *Austral. Animal Bk.* xvii. 148 There are numerous species of Megapodes... All are 'incubator-birds', burying their large eggs in huge nest-mounds, to be incubated by the heat generated by decaying vegetation, or by solar heat. **1963** *Times* 12 Mar. (Austral. Suppl.) p. x/7 There are several kinds of megapode or incubator-birds in Australia, of which the best-known is the lowan or mallee-hen.

incudal (i·nkiūdăl, inkiū·dăl), *a.* [f. L. *incus, incudem* anvil + -al.] Of or pertaining to the incus.

1890 in Billings *Med. Dict.*

incudate (i·nkiūdēit, inkiū·dĕt), *a.* [f. as prec. + -ate².] In rotifers, designating a type of mastax in which the mallei are reduced or absent and the rami enlarged and curved.

1886 Hudson & Gosse *Rotifera* I. iii. 29 The typical trophi may, then, be named as follows:...Incudate. Mallei evanescent; rami highly developed into a curved forceps. **1896** M. Hartog in *Cambr. Nat. Hist.* II. viii. 211 In Asplanchnidæ the rami are large and hooked, constituting the 'incudate' mastax. **1967** P. A. Meglitsch *Invertebr. Zool.* viii. 263 (*caption*) Lateral and dorsal views of incudate trophi.

incudo- (inkiū·do), before a vowel **incud-**, combining form of Incus 1, in terms denoting the association of the incus with another part, as *incu:do-ma·lleal*, *incu:do-stape·dial*, *incu:do-tympa·nic* adjs.

1884 P. McBride *Guide to Study of Ear Dis.* Plate I (*heading*) Long process of the Incus, incudo-stapedial joint and posterior ramus of the stapes. **1894** Gould *Dict. Med.* 608/1 Incudo-... In composition, signifying relationship to the incus. I.-malleal... I.-orbicular... I.-stapedial... I.-tympanic. **1908** *Practitioner* Jan. 123 The lax membrane resting upon the inner tympanic wall and incudo-stapedial articulation. **1943** Fischer & Wolfson *Inner Ear* v. 191 The capsule of the incudo-malleolar joint is divided [in the operation described]. **1967** *Coll. Papers Surg. Mayo Clinic* LVIII. 373 The most common ossicular injury sustained during stapedectomy is dislocation of the incudomalleal joint.

incumbent, *sb.* **2.** Delete 'Now *rare*', and add later examples.

1904 W. Osler *Aequanimitas* v. 82 His son..held the chair for nearly the same length of time, and the remainder of the period has been covered by the occupancy of John Goodsir, and his successor..the present incumbent. **1940** W. Faulkner *Hamlet* i. iii. 61 They had waited about the store to see what would happen when he arrived who..must have still believed himself the incumbent. *Ibid.* ii. i. 95 The incumbent, the Professor at that time, was an old man. **1966** P. Green tr. *Escarpit's Novel Computer* vii. 92 The training division of A.I.M.R. was still at Brive, and boasted a general manager's office; but the incumbent was in fact a low-ranking deputy-manager. **1971** *Nature* 17 Sept. p. x (Advt.), Applications are invited for the Harry Bolus Chair of Botany... It is required that the incumbent should promote work in both experimental and field botany. **1972** *Daily Tel.* 29 Nov. 8 The victory by Mr James Conway..in a ballot for general secretary of the Engineering Union is being challenged... Mr Conway, the incumbent, received 169,806 votes.

incumbent, *a.* Add: **7. b.** Occupying or having the tenure of any post or position.

1972 *Nature* 28 Apr. 417/2 The incumbent head of the Institute of Theoretical Astronomy would..expect to have more than an equal say in the appointment of his partner. **1972** *Science* 22 Sept. 1087/3 The MSU trustees..appointed the Pontiac school's incumbent dean..to be dean of the College of Osteopathic Medicine at MSU. **1973** *Sci. Amer.* Mar. 43 (Advt.), Well beyond buffhood is the incumbent partisan of astronomy's cause at the Kodak Research Laboratories.

incunabulist (inkiunæ·biŭlist). [f. Incunabula *sb. pl.* 2 + -ist.] One who collects or is interested in incunabula.

1921 S. Gaselee in *Library* Sept. 135 To all incunabulists it [*sc.* the letter *H*] is almost indissociably the abbreviation for Hain's *Repertorium*. **1928** *Ibid.* Sept. 162 The incunabulist has to stop at the threshold of the year 1501. **1949** F. Bowers *Princ. Bibliogr. Descr.* ix. 339 Incunabulists view the information in colophons with suspicion until it is checked by other evidence. **1970** *Times Lit. Suppl.* 14 Aug. 906/2 It is well known from the work of the incunabulists that the typographical equipment of a fifteenth-century printer was identifiably unique.

in-curl (i·nkɔ̄ɪl). *Curling.* [f. In *adv.* + Curl *sb.*] = *inturn *sb.* 4.

1903 *Westm. Gaz.* 31 Jan. 3/1 The secret of the game is to be able to play the 'out-curl or in-curl' as the skip may direct.

incurve (i·nkɔ̄ɪv), *sb.* [f. Incurve *v.*] In baseball and softball, the bending or curving of a ball inwards (i.e. across the front of the batter); the course of such a ball; a ball pitched so as to curve to the right.

1886 H. Chadwick *Art of Pitching & Fielding* 14 It is essential to change the direction of the curve from an 'out-curve' to an 'in-curve'. **1906** *Spalding's Base Ball Guide* 112 An In-curve..is a ball which curves in towards the batsman as he stands in his position. **1949** M. Wadlow in Smalley & Dennis *Official Softball—Track & Field Guide* 17 To throw an incurve, the ball is held in the same manner as for the fast ball.

incut, *ppl. a.* (Later example.)

1973 C. Bonington *Next Horizon* viii. 114 All the climbing [on the Avon Gorge] is similar: balancing up on sloping holds, using small, incut ledges for hand-holds, [etc.].

indamine (i·ndămīn). *Chem.* [f. Ind(o-² + Amine.] A blue dye, $NH_2{\cdot}C_6H_4{\cdot}N{:}C_6H_4{:}NH$ (also called *phenylene blue*); also, any of the derivatives of this compound, which form a group of blue and green dyes now important only as intermediates for safranine dyes.

1888 *Jrnl. Chem. Soc.* LIV. 949 (*heading*) Indamines and indophenols. **1903** C. Salter tr. *G. von Georgievics's Chem. Dye-Stuffs* 265 When di-p.-amidodiphenylamine is oxidised with potassium bichromate, there is formed a blue dye-stuff, known as Indamine or Phenylene Blue... From this are derived a whole group of dye-stuffs, which are, therefore, named 'Indamines'. **1961** L. F. & M. Fieser *Adv. Org. Chem.* xxvi. 852 The highly pigmented N-phenyl derivatives of quinonimine and quinonediimine, indophenol and indamine, are relatively stable, crystallizable substances.

indanthrene (i·ndănþrīn). Also Indanthren(e. [f. Indo-² + Anthr(a- + -ene.] **a.** (With lower-case initial.) Indanthrone, $C_{28}H_{14}N_2O_4$ (in quot. 1921 used for the oxygen-free parent compound, $C_{28}H_{18}N_2$: cf. quot. 1920 s.v. *indanthrone). **b.** (Usu. with capital initial.) Any of a large and important class of vat dyes derived from or containing indanthrone or other compounds based on the anthraquinone nucleus.

Indanthren(e) has been registered as a proprietary term in Great Britain.

1901 *Trade Marks Jrnl.* 23 Oct. 1063 Indanthrene. **1901** *Jrnl. Soc. Dyers & Colourists* XVII. *facing* p. 302 Indanthrene X is a blue paste insoluble in water. *Ibid.*, Indanthrene S is a coppery coloured paste which is soluble in hot water. **1903** *Jrnl. Chem. Soc.* LXXXIV. i. 446 The substance 'A', of the German Patent 135407,..can be purified by dissolving it in concentrated sulphuric acid and allowing the solution to gradually absorb water; well-formed, blue needles separate which, in the case of 'Indanthrene C', have the composition $C_{28}H_{10}O_4N_2Br_4$, and, in the case of indanthrene itself, the composition $C_{28}H_{12}O_4N_2$. **1920** [see *flavanthrone]. **1920** [see *indanthrone]. **1921** E. de B. Barnett *Anthracene & Anthraquinone* xvi. 342 The first cyclic azine of the anthraquinone series to be prepared was *trans. bisang.*-anthraquinonedihydro azine. This was placed on the market under the name Indanthrene Blue, and the name 'indanthrene' has come into general use in the literature. The word 'indanthrene', however, is a registered trade name (B.A.S.F.) and is applied to many vat dyes which are not azines. Indanthrene Blue is an anthraquinone derivative and ketonic in structure, and in order to denote its ketonic nature the name should terminate in -one. In the following pages, therefore, the word 'indanthrone' is used to denote the ketonic hydroazine, indanthrene (without a capital) being used for the parent, oxygen free hydroazine (*trans. bisang.*-dihydroanthrazine). Where 'Indanthrene' is used as a registered trade name it is spelt with a capital. **1922** *Trade Marks Jrnl.* 8 Nov. 2019 Indanthren. **1952** K. Venkataraman *Chem. Synthetic Dyes* II. xxx. 861 With the amalgamation of the German dyemakers into the IG in 1924, Indanthrene was adopted as the group name for vat dyes with the maximum all-round fastness. **1961** [see *flavanthrone]. **1971** R. L. M. Allen *Colour Chem.* x. 163 If hydroxyl groups are introduced into the molecule of Indanthren Yellow GK in positions 4 and 8 the shade becomes violet; this product is manufactured as Indanthren Brilliant Violet BBK.

indanthrone (i·ndănþrōᵘn). *Chem.* [f. *indanthr(ene + -one.] **a.** A blue compound, $C_{28}H_{14}N_2O_4$, which is composed of two anthraquinone nuclei linked by two imino-groups and is the parent compound of an important group of vat dyes. **b.** Any of the derivatives of this compound.

1920 F. A. Mason tr. *G. von Georgievics's Text-bk. Dye Chem.* (ed. 2) 449 Indanthrene (or more correctly *Indanthrone*, which expresses the quinoid character of the compound, whereas the name generally used would correspond to an oxygen-free compound) is a dianthraquinone-dihydroazine. **1921** [see *indanthrene]. **1921** E. de B. Barnett *Anthracene & Anthraquinone* xvi. 343 By far the most important method of obtaining the indanthrones is by fusing the β-aminoanthraquinones with caustic alkali. **1954** I. L. Finar *Org. Chem.* (ed. 2) I. xxxi. 689 Indanthrone (Indanthrene Blue R; Bohn, 1901) is prepared by fusing 2-amino-anthraquinone with potassium hydroxide in the presence of potassium chlorate or nitrate at 250°. **1959** *Jrnl. Amer. Chem. Soc.* LXXXI. 3762/2 Strong amide resonance contributes greatly to the resonance hybrid of the indanthrones.

indazole (i·ndăzōᵘl). *Chem.* [ad. G. *indazol* (Fischer & Kuzel 1883, in *Ann. d. Chem.* CCXXI. 264), f. *indol* Indole *sb.*: see Azo- and *-ole.] **a.** A colourless crystalline compound, $C_7H_6N_2$, in which a benzene ring is fused to a pyrazole ring. **b.** Any of the derivatives of this compound.

1884 *Jrnl. Chem. Soc.* XLVI. 441 Indazole crystallises in colourless needles. **1967** Katritzky & Lagowski *Princ. Heterocyclic Chem.* 138 Indazole is the trivial name for the benzopyrazole nucleus. *Ibid.* 140 Certain *o*-toluenediazonium salts cyclize spontaneously to indazoles.

indecent, *a.* Add: **3.** (Later example.)

1965 *Times* 17 Feb. 17/2 After verdict and before sentence, while the jury were still in court, they had come to the conclusion that neither the films nor the brochure were obscene, but had concluded that they were indecent.

4. Special collocations: *indecent assault*: an assault (sense 3) of a sexual nature, but not involving rape or attempted rape. Used *colloq.* as a euphemism for rape (Webster, 1934); *indecent exposure* (see *exposure 1 f).

1861 *Act* 24 & 25 *Vict.* c. 100 § 52 Whosoever shall be convicted of any indecent Assault upon any Female..shall be liable..to be imprisoned for any Term not exceeding Two Years, with or without Hard Labour. *Ibid.* § 62 Whosoever..shall be guilty..of any indecent Assault upon any Male Person, shall be guilty of a Misdemeanor. **1938** J. Curtis *They drive by Night* xvii. 190 You want to get pinched for indecent assault, do you? **1956** *Act* 4 & 5 *Eliz. II* c. 69 § 14 It is an offence, subject to the exception mentioned in subsection (3) of this section, for a person to make an indecent assault on a woman. *Ibid.* § 15 It is an offence for a person to make an indecent assault on a man. **1969** C. Allen *Textbk. Psychosexual Disorders* (ed. 2) xxii. 458 The Act does not provide a definition of 'indecent assault'... In the majority of cases a man..has, without invitation, put his hand

up the clothing of an unsuspecting female. **1974** *Guardian* 24 Jan. 7/3 Mr Breredon, at present serving four years for two cases of indecent assault.

indecently, *adv.* (Later examples.)

1961 New Eng. Bible *Rom.* i. 27 Males behave indecently with males. **1970** *Times* 8 Jan. 2/8 As she walked along a lonely path..a boy threw her to the ground, indecently assaulted her and stabbed her.

indecomposable, *a.* Add: (Earlier and later examples.)

1807 R. Kirwan *Logick* I. iii. v. 212 It has been laid down..that water was an element, and that elements were indecomposable. **1968** P. A. P. Moran *Introd. Probability Theory* ix. 406 A distribution containing all its probability concentrated at two points is indecomposable. **1971** Powell & Higman *Finite Simple Groups* iii. 140 A root system Δ with base Π is *indecomposable* if it is impossible to split up Π into two non-empty subsets which are orthogonal to each other.

Hence i:ndecomposabi·lity.

1950 *Mathematical Rev.* May 377/1 (*title*) Proof of the indecomposability of a certain graph. **1968** P. A. P. Moran *Introd. Probability Theory* ix. 409 We now consider an example of the indecomposability of absolutely continuous distributions.

indeedy (indī·di, indi·di), *adv. colloq.* (orig. *U.S.*). [f. Indeed *adv. phr.* + -y *suffix*⁶.] Used as an emphatic affirmative (or negative), esp. after *yes* (or *no*): indeed, certainly.

1856 *Knickerbocker* XLVIII. 620 'Is thy eye not opened?' 'Yes, indeedy,' says I. **1872** 'Mark Twain' *Roughing It* 336 He never shook his mother... No indeedy..he looked after her and took care of her. **1932** W. Cather *Obscure Destinies* 165 I'd like to, powerful well, Mrs. Harris. I would, indeedy. **1961** *Sunday Times* 26 Mar. 13/1 He's also got two episcopal rings—'One's a spare I can use for the washing-up.' Yes, indeedy. **1965** N. Freeling *Criminal Conversation* i. ix. 62 'People always lie to policemen.' 'Yes indeedy.' **1966** 'D. Shannon' *With a Vengeance* xi. 147 'A little lower than the angels,' he said. 'Yes indeedy.' **1967** J. Porter *Chinks in Curtain* xii. 124 Yes, indeedy! That was a good question. **1970** L. Sanders *Anderson Tapes* xxviii. 73 No, indeedy.

indefinability (indĭfəinăbi·liti). *Logic.* [f. Indefinable *a.* (*sb.*) + -ity.] The quality of being indefinable; incapability of definition in simpler or more fundamental terms.

1903 B. Russell *Princ. Math.* § 80 The point which is chiefly important..is the indefinability of propositional functions. **1908** W. R. B. Gibson *Probl. Logic* 82 The indefinability of the proper name. **1936** *Mind* XLV. 101 He proceeds to consider the nature of such good, criticising Dr. Moore's theory of indefinability.

indefinable. B. *sb.* Delete *rare* and add later examples.

1904 *Mind* XIII. 132 On the indefinables of philosophy, especially, much new light is thrown by M. Couturat's labours. **1927** F. A. Lindemann in R. J. S. McDowall *The Mind* 137 Though the so-called physical laws will be in general consistent with reality, there is no certainty that the indefinables employed in their statement..have any fundamental significance beyond their appeal to the mental preferences..of the physicist. *Ibid.* 139 The commonest indefinables used in physics are space and time. **1932** —— *Physical Significance of Quantum Theory* 12 The indefinables upon which all our thought processes depend. *Ibid.* 14 The three indefinables commonly used in physics are length, time, and mass. **1948** *Mind* LVII. 409 What we ignored was the less exact and less elegant work of insuring that the defining but undefined general terms upon which our explanations finally depended were themselves understood. The application of these indefinables we knew could be explained only by examples but the work of providing and arranging those examples and surrounding them with comment bringing out their interconnexions we could not bring ourselves to do. **1963** *Times* 26 Jan. 4/7 Halewyn declares his allegiance to evil and gallops away to his destruction; his talk is a series of verbal conjuring tricks with indefinables.

indefinition (indefini·ʃən). [f. In-³ + Definition.] A condition of being indefinite, of lacking definition.

1888 *Pall Mall Gaz.* 14 Nov. 5/1 This negative quality of indefinition leaves a sense of uncertainty. **1940** Blunden *Poems 1930–40* 57 About the stern defining phrase A gay indefinition plays. **1958** C. Rabin in *Aspects of Translation* 128 Turkish or Arabic with marking 'individuation + indefinition' rather than definition.

indelible, *a.* Add: **a.** Also *indelible pencil*.

1885 *Encycl. Brit.* XVIII. 490/2 In the indelible and copying pencils which have come into use in recent years, the colouring matter is an aniline preparation mixed with clay and gum. **1916** *Daily Colonist* (Victoria, B.C.) 23 July 9/1 All the newspaper offices in San Francisco yesterday received a communication written in Roman scrip [*sic*] with an indelible pencil, many of the words being heavily underscored. **1961** *Lebende Sprachen* VI. 69/2 Indelible pencil, der Tintenstift.

indemn (inde·m), *v. rare*⁻¹. [See Indemn *a.*] *absol.* To indemnify.

1906 Hardy *Dynasts* II. i. viii. 177 To your ally, the Tsar, I must refer you... He can indemn.

indene (i·ndīn). *Chem.* [f. Ind(o-² + -ene, or as a contraction of Indonaphthene.] **1. a.** = Indonaphthene (which name *indene* has completely superseded). **b.** Any of the derivatives of this compound.

1888 *Jrnl. Chem. Soc.* LIV. 1303 (*heading*) Indene-derivatives. **1892** E. F. Smith tr. *V. von Richter's Chem. Carbon Compounds* (ed. 2) 903 The formation of the carboxyl derivatives of indene..proceeds in a manner analogous to the formation of alkyl indenes. **1956** Rodd & Van Alphen in E. H. Rodd *Chem. Carbon Compounds* IIIB. xx. 1260 (*heading*) Substituted indenes. **1968** L. K. Arnold *Introd. Plastics* ix. 111 Coumarone and indene, found in certain coal tar fractions, are polymerized to produce resins which range from those liquid at room temperature to others melting at 300° F.

2. Special Comb.: **indene resin**, any thermoplastic resin made from indene; usually = **coumarone-indene resin*.

1900 *Jrnl. Chem. Soc.* LXXVIII. i. 657 The indene resin decomposes at 290–340°. **1947** [see *coumarone b].

indentation. Add: **5.** Special Comb.: **indentation hardness**, hardness as determined by one of the indentation tests; **indentation test**, any of various tests for determining the hardness of a solid by making an indentation in a sample under standard conditions and measuring either its size or the distance travelled by the indenter.

1918 *Proc. Inst. Mech. Engin.* Oct.–Dec. 487 Researches on indentation hardness. **1956** B. W. Mott *Micro-Indentation Hardness Testing* i. 9 The general definition of indentation hardness..is the ratio of the load applied to the surface area of the indentation. **1968** D. A. Smith *Addition Polymers* x. 447 The testing of fabricated polymer components..may include evaluation of tensile stress-strain and flexural properties.., indentation hardness, and the examination of the effects of temperature on rigidity. **1897** *Min. Proc. Inst. Civil Engin.* CXXIX. 334 (*heading*) A new indentation test for determining the hardness of metals. **1956** B. W. Mott *Micro-Indentation Hardness Testing* i. 2, (1) Static indentation test: A steady load is applied to an indenter..and the hardness is calculated from the area or depth of indentation produced. (2) Dynamic indentation test: A ball, cone or a number of small spheres is allowed to fall from a definite height and the hardness number is obtained from the dimensions of the indentation and the energy of the impact.

indenter² (inde·ntəɪ). Also **-or.** [f. Indent *v.*² + -er¹, -or.] Something that produces indentations; *spec.* a small hard sphere, pyramid, or similar object used for producing an indentation in a solid (as in an indentation test).

1929 *Proc. Inst. Mech. Engin.* i. 384 If only a fraction of the attention..had been devoted to the economical production of high quality cones or pyramids as indenters, they would now be in a more satisfactory position with regard to hardness testing. **1948** Spielmann & Elford *Road Making & Administration* (ed. 2) vii. 265 Indenters, Crimpers, and Key Cutters are no longer manufactured. **1950** *Engineering* 4 Aug. 102/3 Mechanical testing..includes..measurement of the shape of impressions made by Knoop hardness indenters. **1966** *Nature* 28 May 879/1 A hard steel indenter (sphere, cone, or pyramid) was pressed with a known load for a specified time on to the surface of a smooth block of ice. **1971** B. Scharf *Engin. & its Lang.* iv. 23 All three are indentation tests in which a weighted steel ball or other 'indentor' is applied to the test piece. The hardness is assessed according to the size of the impression (Brinell and Vickers) or the travel of the indentor (Rockwell).

indentor², var. *indenter².

indentured, *ppl. a.* **1.** (Further examples.)

1952 S. Selvon *Brighter Sun* v. 72 He had come from India to work as an indentured labourer on the white man's plantations. **1961** [see *East Indian *sb.* 2]. **1969** S. M. Sadeek *Windswept & Other Stories* 29 My grand parents were indentured immigrants, never-the-less pioneers.

independable (indĕpe·ndăb'l), *a.* [f. In-³ + Dependable *a.*] Not dependable; untrustworthy; not to be depended *upon*.

1802 Bentham *Let.* 27–30 June in *Works* (1843) X. 388/1 All Frenchmen are independable upon. **1897** *Geogr. Jrnl.* Feb. 122 This lady is..capricious, independable, and exacting. **1921** S. Graham *Europe* xvii. 210 English action is so sluggish and so independable.

independence. Add: **1. c.** Corresp. to *independent *a.* 3 d.

1902 *Bull. Amer. Math. Soc.* VIII. 296 No attempt seems to have been made to prove the independence of the postulates employed to define a group. **1941** O. Helmer tr. *Tarski's Introd. Logic* § 56. 194 (*heading*) Independence of the axioms of the simplified system. **1954** I. M. Copi *Symbolic Logic* vi. 179 The axioms of a deductive system are said to be independent (or to exhibit independence) if no one of them can be derived as a theorem from the others. **1955** A. N. Prior *Formal Logic* 233 This is the method regularly used for establishing the 'independence' of the axioms in a given set, i.e. for showing that no axiom in the set is superfluous, in the sense of being derivable from the others as a theorem. **1956** J. H. Woodger tr. *Tarski's Logic, Semantics, Metamath.* 390 On account of the non-ramifiability the independence proof for such sentences cannot be carried out..by an interpretation in logic. **1962** W. & M. Kneale *Devel. of Logic* xii. 692 Bernays..turned to investigate the independence of the axioms of the calculus of propositions in *Principia Mathematica*.

3. *Independence Day* (earlier and later examples); also *transf.*

1791 J. Hiltzheimer *Diary* 4 July (1893) 170 This being Independence Day, the Governor invited several of the neighbors to dine with him. **1967** R. Lowell *Near Ocean* 17 Another summer! Our Independence Day Parade, all innocence Of children's costumes, helps resist The communist and socialist. **1970** M. Slater *Caribbean Cooking* 45 Strangers to the islands may well be taken by surprise on..Independence Day. **1972** *Times* 9 Oct. (Nigeria Suppl.) p. i/6 In his Independence Day speech last week General Gowon disclosed that the census would take place..next year.

independent, *a.* and *sb.* Add: **A.** *adj.* **3. b.** *independent suspension*.

1930 *Engineering* 7 Feb. 162/3 A special chapter on independent suspension systems. **1963** R. F. Webb *Motorists' Dict.* 135 *Independent suspension*, a form of suspension where each wheel is completely independent of the others and no connecting axle beams are used. **1973** *Country Life* 15 Dec. 1581/2 For absolute comfort and stability at very high speeds there is all round independent suspension.

d. Of one of a set of equations, axioms, or quantities in respect of the others: incapable of being expressed in terms of, or of being derived or deduced from, the others; hence applied to a set of axioms, etc., all of which have this property; *linearly independent*, (of each of a set of equations or quantities) incapable of being expressed as a linear combination of the others, i.e. satisfying no relation of the form $a_1x_1 + a_2x_2 + \ldots + a_nx_n = 0$ (where x_i are the quantities and a_i arbitrary constants) unless $a_1 = a_2 = \ldots = a_n = 0$.

1740 N. Sanderson *Elem. Algebra* I. ii. 105 If a problem be justly proposed, it ought to have as many independent conditions..as there are unknown quantities to be discovered by them. **1798** J. Wood *Elem. Algebra* (ed. 2) 73 These equations must also be independent, that is, not deducible one from another. **1875** *Encycl. Brit.* I. 541/1 A problem is limited when the conditions furnish just as many independent equations as there are unknown quantities to be determined: if there be fewer, the problem is indeterminate; but if there be more, the problem in general admits of no solution. **1885** [see *Legendre]. **1902** *Trans. Amer. Math. Soc.* III. 142 Hilbert states..that his body of axioms consists of independent axioms, that is, that no one of the axioms is logically deducible from the remaining axioms. **1931** L. J. Rouse *College Algebra* v. 69 The equations $2x + y = 5$ and $x - y = 4$ cannot be reduced to the same form and are therefore independent. **1941** O. Helmer tr. *Tarski's Introd. Logic* § 39. 131 We strive to arrive at an axiom system which does not contain a single superfluous statement..which can be derived from the remaining axioms... An axiom system of this kind is called independent (or a system of mutually independent axioms). **1944** A. Church *Introd. Math. Logic* I. i. 25 An axiom of a logistic system is said to be independent if, in the system whose axioms and rules consist of all axioms and rules of the original system except that one, the suppressed axiom is not a theorem. **1959** G. & R. C. James *Math. Dict.* 107/1 The numbers 3 and π are linearly independent with respect to rational numbers, since $a_1{\cdot}3 + a_2{\cdot}\pi$ can not be zero if a_1 and a_2 are rational numbers, not both zero. Since $-1{\cdot}3 + (3/\pi)\pi = 0$, 3 and π are linearly dependent with respect to real numbers. **1961** Powell & Craseman *Quantum Mech.* v. 117 Two solutions of Eq. (5–47) (or, more generally, any two functions of *x*) are linearly independent if the equation $C_1\psi_1 + C_2\psi_2 = 0$ cannot be satisfied identically in *x* for any choice of the constants C_1 and C_2 except $C_1 = C_2 = 0$. *Ibid.* 118 Two linearly independent solutions, ψ_1 and ψ_2, are a complete set in the sense that every solution of Eq. (5–47) can be expressed as a linear combination of ψ_1 and ψ_2. **1965** Hughes & Londey *Elem. Formal Logic* xviii. 132 Since A4 is non-independent, the axiom set for PM could be reduced by one. But no further reduction of this sort is possible; neither A1 nor A2 nor A3 nor A5 is a consequence of the other three under Substitution and Detachment, and these four are therefore said to be independent axioms.

5. b. *Independent Labour Party* (abbrev. I.L.P.: see *I. III): the title of the political organization founded at Bradford in January 1893 by James Keir Hardie as an offshoot of the Social Democratic Federation, for the support of parliamentary candidates of approved socialistic views; orig. as opp. the Conservative and Liberal parties, later distinct from the Labour Party.

c **1888** *Scottish Labour Party Manifesto* in M. Beer *Hist. Brit. Socialism* (1920) II. xv. 300 The formation of a distinct, separate, and *Independent Labour Party*. **1902** *Encycl. Brit.* XXXII. 668/2 Attempts had been made to influence politics directly by means of an Independent Labour Party..which bound itself to support only candidates of sound socialist views. **1922** *Ibid.* XXXII. 507/1 The Labour party..included the Independent Labour party and the Fabian Society and one or two smaller Socialist bodies. **1953** D. E. Butler *Electoral Syst. in Brit.* ii. v. 154 The total of 5 [members of parliament from outside the three major parties] for 1935 is made up of 4 Independent Labour Party members from Glasgow—where a substantial part of the Labour party had split away—and 1 Communist. **1955** *Times* 24 May 14/3 An election manifesto issued last night by the Independent Labour Party reaffirms the party's belief

that workers' control 'is an essential part of Socialism'. **1971** Butler & Pinto-Duschinsky *Brit. Gen. Election 1970* v. 112 The failure of those other traditional spokesmen of the left, the Independent Labour party and the Socialist Party of Great Britain, was still more complete.

d. Of schools: receiving no grant from the government and not subject to the control of a local authority.

1944 *Act* 7 & 8 *Geo. VI* c. 31 § 70 The Minister shall appoint one of his officers to be Registrar of Independent (i.e. private) Schools. **1957** *Encycl. Brit.* VII. 989/1 The Independent Schools association..; its official publication is the *Independent School*. **1966** *Rep. Comm. Inquiry Univ. Oxf.* II. 45 Independent schools are subdivided into independent boarding and independent day schools according to whether the majority of pupils were boarders or day pupils.

e. *Independent Television* (*Authority*) (abbrev. I.T.A., I.T.V.: see *I. III): a corporation, independent of direct government control, engaged in commercial television broadcasting in Great Britain; also, the channel carrying their programmes; renamed in 1972 the *Independent Broadcasting Authority* (abbrev. I.B.A.: see *I. III), and widened to include commercial radio broadcasting.

1954 *Act* 2 & 3 *Eliz. II* c. 55 § 1 There shall be an authority, to be called the Independent Television Authority..whose function shall be to provide, in accordance with the provisions of this Act, and for the period of ten years from the passing of this Act, television broadcasting services, additional to those of the British Broadcasting Corporation. **1958** *Times Lit. Suppl.* 15 Aug. p. xl/1, This brings sharply into the picture the B.B.C.'s rival, Independent Television. **1959** *Chambers's Encycl.* XI. 349/2 (*heading*) Independent Television Authority. *Ibid.*, This corporation is controlled by directors appointed by the government and leases facilities to privately financed companies, which draw their revenues from advertisements. **1960** *B.B.C. Handbk.* 165 The Postmaster General issued a broadcasting licence, for television only, at a later stage to the Independent Television Authority, which was set up under the Television Act of 1954. **1971** [see *I.B.A.* s.v. *I. III]. **1973** *Times* 13 Dec. 4/5 The Independent Broadcasting Authority will hold meetings today..about advertising which will be lost as a result of the shorter hours of transmission announced yesterday. *Ibid.* 15 Dec. 2 BBC and independent television agreed last night to spread their closing hours.

6. b. *independent float*: in Critical Path Analysis, the amount of 'float' or leeway in any one activity which can occur without affecting the timing of the whole operation.

1963 R. E. McGarrah *Production & Logistics Managem.* viii. 211 'Independent float' pertains to those non-critical activities whose leeway is not affected by the starting or completion time of its preceding or succeeding activities. **1964** K. G. Lockyer *Introd. Critical Path Analysis* v. 48 *Independent float*, the time by which an activity can expand without affecting any other activity either previous or subsequent. **1967** A. Battersby *Network Analysis* (ed. 2) App. 4. 335 *Independent float* is so called because it is what remains if all preceding jobs finish as late as possible and all succeeding jobs begin as early as possible. **1967** S. Woodgate in Wills & Yearsley *Handbk. Managem. Technol.* 80 Independent float is the minimum spare time available under any condition, i.e. either early or late. **1968** *Gloss. Terms Project Network Analysis* (*B.S.I.*) 8 *Independent float*, earliest date of succeeding event minus latest date of preceding event minus activity duration. (If negative, the independent float is taken as zero.)

7. *independent seconds* (*watch*): see quots.

a **1877** Knight *Dict. Mech.* II. 1179/1 *Independent seconds-watch*, a watch in which the action of the center seconds-hand is independent of the regular going works of the watch... For great nicety in timing, quarter and fifth second watches are now made. **1962** E. Bruton *Dict. Clocks & Watches* 93 *Independent seconds*, clock or watch with seconds hand..which jumps from one second to the next, i.e. it is dead beat.

in-depth *attrib.*: see *depth I. 3 c.

inderborite (indəɪbōᵊ·rəit). *Min.* [f. *Inder*, the name of a lake in Kazakhstan + Bor(on + -ite¹.] A colourless to white hydrated borate of calcium and magnesium, $CaMgB_6O_{11}{\cdot}11H_2O$.

1941 G. S. Gorshkov in *Compt. Rend.* (*Doklady*) *de l'Acad. des Sci. de l'URSS* XXXIII. 255 The new mineral is an analogon of hydroboracite and has been named inderborite after the place of finding. **1967** E. L. Muetterties *Chem. Boron & its Compounds* iii. 188 The binary salt minerals, in the system $CaO.MgO.B_2O_3.H_2O$ (hydroboracite and inderborite) are probable derivatives of the colemanite-meyerhofferite trimer.

inderite (i·ndərəit). *Min.* [f. as prec. + -ite¹.] A colourless hydrated magnesium borate, $Mg_2B_6O_{11}{\cdot}15H_2O$.

1937 Boldyreva & Yegorova in *Mat. Central Sci. Investig. Geol. Prospecting Inst.* (U.S.S.R.) Gen. Ser. No. 2. 52 Inderite, a new hydrated magnesium borate—$Mg_2B_6O_{11}{\cdot}15H_2O$ has been found in the Kzyl-Tau deposit (Western Kasakhstan, Inder Mountains) in the form of fine white, sometimes slightly pinkish kidney-shaped nodules. **1946** *Amer. Mineralogist* XXXI. 71 A second occurrence of inderite ($Mg_2B_6O_{11}{\cdot}15H_2O$)..has been discovered. **1967** E. L. Muetterties *Chem. Boron & its Compounds* iii. 161 NMR analyses have been performed on the minerals inderite, $Mg_2B_6O_{11}{\cdot}15H_2O$..borax..and tincalconite.

indestructible, *a.* Add: Used *subst.* An indestructible thing.

a **1861** T. Winthrop *Life in Open Air* (1863) 17 My pair of these indestructibles [*sc.* socks] will outlast my last legs. **1880** J. Ross *Hist. Corea* x. 332 The things used in the evening sacrifice are to be removed,..but spirits and indestructibles may remain. **1898** G. Meredith *Odes French Hist.* 76 Refreshful chatter, laughter, gaillard songs. So like Earth's indestructible they were.

indetectable, *a.* Delete *rare* and add later examples.

1930 W. de la Mare *On the Edge* 207 In hope of detecting that shadow's indetectible motion! **1961** D. G. James *Matthew Arnold* i. 27 There always remained in him something evasive and indetectable.

indeterminacy. Delete *rare* and add later examples.

1919 J. M. Keynes *Econ. Consequences Peace* v. 208 The Allies recognised the inconvenience of the indeterminacy of the burden laid upon Germany. **1947** P. A. M. Dirac *Princ. Quantum Mech.* (ed. 3) i. 4 There is an unavoidable indeterminacy in the calculation of observational results, the theory enabling us to calculate in general only the probability of our obtaining a particular result when we make an observation. **1953** C. F. Hockett in Saporta & Bastian *Psycholinguistics* (1961) 46/1 Thus, if there is no indeterminacy, no element of choice, there can be no transmission of information. **1962** *Listener* 15 Feb. 290/1 Most murderers are now subject to a higher degree of indeterminacy in their sentence than would be the case with those who were given the maximum sentence which I am canvassing. *Ibid.* 27 Sept. 472/2 Contemporary composers such as Stockhausen are often attacked for their deliberate indeterminacy—a word which has become a technical term. **1963** D. A. Firmage *Fund. Theory of Struct.* v. 118 To determine the degree of indeterminacy..remove reaction components one at a time and see if the structure is stable against all possible forces. **1971** *Archivum Linguisticum* II. 50 It will be seen that the inclusion of a single additional item serves to reduce in some measure the indeterminacy we have noted as attending the parts and the whole of words.

b. Physics. *principle of indeterminacy* (or *indeterminacy principle*) = **uncertainty principle*.

1928 A. S. Eddington *Nature Physical World* x. 220 It was Heisenberg again who set in motion the new development in the summer of 1927, and the consequences were further elucidated by Bohr. The outcome of it is a fundamental general principle which seems to rank in importance with the principle of relativity. I shall here call it the 'principle of indeterminacy'. The gist of it can be stated as follows: a particle may have position or it may have velocity but it cannot in any exact sense have both. **1938** W. S. Maugham *Summing Up* 289 Of late the Principle of Indeterminacy, by bringing to view certain events to which apparently no causes can be assigned, has cast a doubt on the universal efficacy of those laws upon which science has hitherto been based. **1964** D. Rosenthal *Introd. Properties Metals* iv. 68 There is the unavoidable lack of resolution caused by the interaction between waves and particles and deriving from the fundamental behavior of matter and energy (the indeterminacy principle). **1959** *Listener* 1 Oct. 519/1 Heisenberg's principle of indeterminacy showed that the process of observing could affect the observation. **1971** *Physics Bull.* June 329/2 It led Heisenberg to formulate his principle of indeterminacy, postulating that a measurement of one property of an entity, such as its momentum, involved an inevitable uncertainty in our knowledge of the conjugate property, such as its position.

indeterminate, *a.* (*sb.*) Add: **2. c.** Also (in *Engin.*), of a structure or its stresses, = *hyperstatic* adj. (s.v. *hyper- IV): usu. qualified by *statically*.

1905 I. Hiroi (*title*) The statically-indeterminate stresses in frames commonly used for bridges. **1919** Pippard & Pritchard *Aeroplane Struct.* xxii. 280 The principle of Least Work..enables the engineer to find the stresses in structures which are statically indeterminate, i.e. in structures with too many members. **1926** Parcel & Maney *Elem. Treat. Statically Indeterminate Stresses* p. v, This book has grown out of the authors' needs in teaching the subject of Indeterminate Structures during the past fifteen years. **1963** D. A. Firmage *Fund. Theory of Struct.* v. 118 A truss can..be determinate, indeterminate, or unstable with respect to the system of bars. **1964** J. H. Faupel *Engin. Design* ii. 130 In a statically indeterminate system the basic equations of statics (ΣF and $\Sigma M = 0$) are insufficient for obtaining a solution.

e. *indeterminate sentence*, a sentence in a criminal case which leaves the prisoner's release dependent on his conduct and on the probability of his amendment; *indeterminate vowel*, the 'obscure' vowel (ə), mid-mixed-wide; = Sheva 2.

1873 E. C. Wines *Rep. Internat. Penitentiary Congr. 1872* xxxviii. 273 It is extremely doubtful whether society will, or could, ever become reconciled to so great a change on the sudden as that from determinate to wholly indeterminate sentences. **1888** King & Cookson *Princ. Sound & Inflexion Gr. & Latin* 70 The same indeterminate vowel appears in Sk., where it is called a *svarabhakti* vowel. **1894** W. M. Lindsay *Latin Lang.* 257 The obscure or indeterminate vowel (like the Hebrew shᵊva). **1962** *Listener* 15 Feb. 290/1 Indeterminate sentences have been common in western Europe and the United States for some time. **1972** *N.Y. Law Jrnl.* 14 Nov. 6/5 People, &c., v. Manuel Santos—Indeterminate sentence.

indeterminism. Add: **2.** = Indeterminacy.

1928 *Chem. Rev.* V. 472 Because of the ambiguity resulting from the Heisenberg indeterminism principle, the future of a dynamical system can never be predicted with certainty. **1957** *Encycl. Brit.* XV. 680/1 Because *h* is so small, the indeterminism caused by the uncertainty principle is of no consequence in ordinary experience. **1969** R. B. Fuller *Operating Man. Spaceship Earth* v. 65 Heisenberg's principle of 'indeterminism' which recognized the experimental discovery that the act of measuring always alters that which was being measured turns experience into a continuous and never-repeatable evolutionary scenario.

indeterminist. Add: Also *attrib.* = indetermini·stic *a.*, of or pertaining to the doctrine of indeterminism.

1902 W. R. B. Gibson in H. Sturt *Personal Idealism* iii. 160 The Indeterminist, like the Britisher, is king of his own castle, and woe to the combatant who fights the battles of Freedom within that..enclosure. Of such a kind is the indeterministic challenge of Professor James. **1903** A. E. Taylor *Elem. Metaphysics* iv. iv. 376 The essence of the indeterminist position is the denial of the principle affirmed alike by the doctrine of self-determination and, in an unintelligent travesty, by the determinist theory that conduct results from the reaction of 'character' upon circumstances. **1907** W. James *Pragmatism* 117 The chaplet of my days tumbles into a cast of disconnected beads as soon as the thread of inner necessity is drawn out by the preposterous indeterminist doctrine. **1925** A. G. Hogg *Redemption from this World* 243 Bushnell, accepting a very indeterministic type of libertarianism, postulates an unresolved plurality of 'powers' or uncaused causes. **1936** *Times Lit. Suppl.* 11 Jan. 38/4 A very lucid exposition of the indeterminist philosophy. **1940** *Mind* XLIX. 465 An erroneous belief that acceptance of the new 'indeterministic' view of physical law would make it less likely that the sun would rise tomorrow. **1969** *Adv. Hydroscience* V. 102 Indeterministic models would exclude those possessing deterministic and probabilistic properties.

index, *sb.* Add: **2.** (Further examples: cf. *index *v.* 5.)

1863 J. Watson *Theory & Pract. Art of Weaving* vi. 209 After the wheel is turned, the next process is to divide its circumference into as many divisions as will make up the number of teeth required; this is done by an index which is fixed on the spindle of the lathe. **1879** J. J. Holtzapffel *Turning & Mech. Manipulation* IV. v. 118 The index, is a steel spring or rod terminating in a point, which is inserted in any required series of holes, in any of the circles of the division plate, to retain the mandrel for the time, at rest, in certain relative positions. For example, to divide the work into 12 parts; the point of the index is placed successively in the holes 8. 16. 24. 32. etc. of the 96 circle..and while the mandrel is arrested at these points, the work is marked.

5. d. *Computers.* A set of items each of which specifies one of the records of a file and contains information about its address.

1962 *Gloss. Terms Automatic Data Processing* (*B.S.I.*) 27 *Index*, a sequence or array of items with keys, used to identify or locate records. **1970** O. Dopping *Computers & Data Processing* xvi. 261 The index can contain the addresses of all individual records, but if the file is ordered, it is usually more economic to make the index cruder. **1971** R. L. Boyes et al. *Introd. Electronic Computing* viii. 201 An index is simply a shorthand substitute for the original information and is used to assist in the location of a given record... The general form of an index..will contain these two items: 1. The index term. This is the shorthand description of a stored record... 2. Record identification or location. This may be a document number or the physical address of the record described by the index term.

8. d. *Computers.* A quantity which is fixed in relation to the set of operations laid down by a program but which assumes a prescribed sequence of values as the program is run; *spec.* (*a*) one held in an index register and used to modify the addresses of instructions; (*b*) one in a DO statement (in Fortran) or a FOR statement (in Algol) that is used to control the number of repetitions of a sequence of instructions. Freq. *attrib.*, esp. denoting the portion of an instruction specifying the appropriate index register (see also *index register* in 10 below).

1957 D. D. McCracken *Digital Computer Programming* viii. 99 Instructions which call for an index to be added are written with the one or two following the address. **1959** J. W. Carr in E. M. Grabbe et al. *Handbk. Automation, Computation & Control* II. ii. 51 When the values of the bound variables (usually indices) that assume a sequence of different values during the course of a problem solution change..such variables are changed in actual, although not notational value. By a change in such an index, therefore, no change is made in the flow diagram notation, although the actual value of free or floating variables will change. *Ibid.* 55 Many artificial instruction codes..use such index registers to speed up hand programming. A certain portion of every instruction word is used to designate just how that particular instruction is to be modified with respect to one or more such special locations, which have been filled with specified values of an index. **1961** Leeds & Weinberg *Computer Programming Fund.* vi. 177 Flow diagrams will be generally easier to follow if we can represent our control logic in enumerative terminology. To do this we make use of an index rather than an actual computed quantity on the flow diagram to show the count. **1962** Y. Chu *Digital Computer Design Fund.* xii. 454 The amount of change of an address, called the index value, is stored in an index register. *Ibid.*, The number in the index field [of an instruction] designates the index register selected. **1962** Huskey & Korn *Computer Handbk.* xx. 29 The index is not always added, so there is an address modifier..which determines whether the index is to be added to the address or not. **1966** B. A. M. Moon *Computer Programming* vii. 117 Within the range of the DO no statement is permitted which alters the value of the index. **1969** V. J. Calderbank *Course on Programming in FORTRAN IV* iv. 36 The DO statement automatically causes execution of all the statements following it up to and including the statement labelled *n* for values of *i* from m_1 in steps of m_3... The counter *i* is sometimes referred to as the index of the loop. **1969** Index bit [see *index *v.* 6]. **1969** C. W. Gear *Computer Organization & Programming* ii. 53 There are four items of information to be specified—the start of the loop in memory (X), the initial value of the index (0), the increment (1), and the end condition on the index (999). **1970** O. Dopping *Computers & Data Processing* vi. 101 The character..in the last position means that the content of the corresponding index register is to be added to the address... The last position of the instruction can be called the 'index tag'.

e. *Computers.* One of a continuous sequence of numbers each of which specifies one of an ordered set of items.

1962 R. V. Oakford *Introd. Electronic Data Processing Equipment* iv. 101 Assume that 15 independent quantities are stored in memory registers 0016 through 0030 and that another 15 independent quantities are stored in registers 0031 through 0045. The location of the *i*th register in the first set may be designated as A_i, while that of the second set may be designated as B_i; then *i* can be considered as an index that assumes the values 1, 2, 3,..., 14, 15. Thus A_1 is equal to 0016. **1972** H. S. Stone *Introd. Computer Organization* vi. 120 The FORTRAN statement dimension x(100) creates an array named x with 100 elements such that the first has index 1 and the last has index 100.

9. e. *Econ.* A number showing the variation (increase or decrease) in the prices or value of some specified set of goods, shares, etc., since a chosen 'base' period (often represented by the number 100), as a retail price index, a cost-of-living index, etc. Cf. *Dow-Jones.

1886 *Rep. Brit. Assoc. Adv. Sci. 1885* 872 The index for quantity is the same as that for value in the standard year (1883); that for 1884 is arrived at by dividing the value index by the price index, and is shown in the last column. **1922** *Encycl. Brit.* XXX. 759/1 Suppose that the modification of diet (margarine instead of butter, decrease of sugar and eggs and increase of other foods) reduces the food index to 260,..and the index is 200 instead of 305. **1927** [see *cost *sb.*² 1 e]. **1942** J. R. Hicks *Social Framework* xv. 160 The most famous of all British index-numbers is the cost-of-living index published by the Ministry of Labour. The basket of goods on which this index is based is supposed to be that consumed in a week by a representative working-class family. It is thus an index of very fundamental importance... It covers a large part of the field which would be covered by the ideal index which we should desire to have for measuring the national income in real terms. **1955** *Times* 31 Aug. 9/3 Their members have expressed lack of confidence in the index as a measurement of their living costs. **1958** *Spectator* 22 Aug. 261/2 Industrial equity shares touched bottom—161·5 for the index. **1969** *Daily Tel.* 13 June 3 The index would have dropped say 12 points, War Loan would have continued its downhill march. **1972** *Accountant* 28 Sept. 398/1 When the heaviest Index fall in a day is accompanied by a reduction in the number of Stock Exchange recorded bargains to under 8,000, it can only be assumed that the vast bulk of the 8,000 deals was the same selling way. **1973** *Daily Tel.* 12 Apr. 21/3 The provisional price index of goods manufactured for the home market rose by just over 0·25 p.c., compared with 0·5 p.c. in both January and February.

9*. [f. *index *v.* 5.] A movement from one predetermined position to another during the indexing of a work-piece.

1962 G. H. DeGroat *Metalworking Automation* v. 120 (*caption*) Another 'homemade' automated machine is this eight-spindle Boremetic with two banks of four spindles each. This one bores, chamfers, and grooves servo valve bodies, finishing two parts per index at 100 pieces per hour. **1964** *Automobile Engineer* LIV. 200/2 After each index, the table is positively locked.

10. *index arm*, *crank*, *pin*, *spindle* (all parts of an index head or used in indexing (sense *2)); **index board**, a type of heavy paper as used for index cards; **index card**, a card for a card-index file; **index centre** *Engin.*, each of the centres (sense 5) that support work for indexing; **index circle** *Engin.*, one of the circles of holes on an index plate; **index figure** *Econ.* = sense 9 e above; **index fossil** = *guide fossil* (*guide *sb.* 13); **index head** *Engin.*, an attachment used with a milling machine or gear-cutting machine that holds the work and enables it to be readily and accurately indexed between successive operations; **index horizon** *Geol.*, a horizon distinguished by certain groups of fossils found within it, or other characteristics which make it an indicator of a particular stratigraphic position; **index map**, a relatively small-scale map which is so marked as to act as an index to a series of more detailed maps; **index number**, (*a*) = sense 9 e above; (*b*) a number in an index; *spec.* the registration number of a motor vehicle; **index plate**, (*a*) in Dict.; (*b*) *Engin.*, a disc that contains regularly spaced holes arranged in concentric circles, which represent different divisions of a circle and determine the possible angular positions of work being indexed; (*c*) a plate bearing the registration number of a motor vehicle; **index register** *Computers*, a register whose contents may be added to or subtracted from the address portion of an instruction before the instruction is executed and then (by means of a second instruction) increased or decreased by a prescribed amount, so enabling the first instruction to be used for a series of identical operations on a series of different operands.

1919 H. Thompson *Mod. Engin. Workshop Pract.* xi. 173 By turning this index arm and spindle, motion is given to the worm and worm-wheel. **1937** E. J. Labarre *Dict. Paper* 160/1 *Index board* or *Bristol* also termed Fourdrinier Bristol is (1) (card-) board resembling a heavy ledger specially adapted for this purpose...(2) also a size of board 30½" × 25½". **1962** F. T. Day *Introd. to Paper* 116/2 Index boards are made in white and tints with an even and well-finished surface... Stock size is 20½ in. × 35½ in., also cut sizes. **1928** *Funk's Stand. Dict.*, Index card. **1947** *Partisan Rev.* XIV. 469 He placed the books down on the main desk, stuck the envelope of index cards and cross-references under his arm and walked out. **1971** *Jrnl. Gen. Psychol.* LXXXV. 52 Ss were supplied with 3 × 5 inch white unlined index cards. *a* **1884** Knight *Dict. Mech.* Suppl. 434/1 The tool-post can be removed from the sliding table, and index centers, milling vise, or any milling fixture put on, required for milling. **1913** *Lockwood's Dict. Mech. Engin.* (ed. 4) 439 *Index centres*, the head, and the tail stock between which work is carried to be pitched or indexed. **1953** L. E. Doyle *Metal Machining* ix. 214 Index centers provide means for spacing cuts accurately around a workpiece. **1902** *Internat. Library of Technol.* II. § 15. 23 For convenience of measuring fractional parts of a turn of different values, as ⅔ of a turn, 1⁄16 of a turn,..etc., the index plate is provided with several concentric index circles, each circle having a different number of holes. **1950** J. Martin in A. W. Judge *Machine Tools & Operations* III. iii. 142 In ordinary plain indexing, the use of 30 holes in a 42-hole index circle would give 42/30 × 40 = 56 divisions in work. **1905** T. R. Shaw *Machine Tools* vi. 415 Forty revolutions of the index crank are required to make one complete revolution of the spindle. **1964** S. Crawford *Basic Engin. Processes* vi. 172 Indirect Indexing... When using this method of indexing the worm is permanently engaged with the worm wheel, and the workpiece is rotated by means of the index-crank. **1927** Index figure [see *costing *vbl. sb.*]. **1930** *Engineering* 3 Jan. 23/3 The index figure of 100 being given to Great Britain in both cases. **1900** C. R. Eastman tr. *Zittel's Text-bk. Palaeontol.* I. 4 Having determined the chronological succession of the clastic rocks by means of their superimposition and their characteristic or index-fossils, they may be divided up into still smaller series. **1933** R. C. Moore *Historical Geol.* xiv. 186 In precise correlation of fossil-bearing strata it is important to recognize and differentiate species that appear only in a given bed, or a short succession of beds, for the occurrence of the same species elsewhere points to equivalence in age of the containing strata. Such fossils may be termed *index fossils*. **1968** J. R. Beerbower *Search for Past* (ed. 2) viii. 207 Relatively few groups of organisms provide most of the index fossils. **1902** Index head [see *indexing *vbl. sb.* 2]. **1923** R. C. H. Heck *Mechanics of Machinery: Mechanism* v. 234 Indexing is done mostly on the milling machine and on gear cutting machines. For general service a distinct appliance called the index head is used. **1961** L. E. Doyle et al. *Manuf. Processes* xxiv. 586 A dividing or index head is a mechanical device for dividing a circle accurately into equal parts. **1956** W. Edwards in D. L. Linton *Sheffield* 13 No marine horizons are known, but a widespread index-horizon with *Euestheria*, the Low '*Estheria*' Band, overlies a split-off lower leaf of the Silkstone Coal. **1969** Bennison & Wright *Geol. Hist. Brit. Isles* ix. 224 The marine horizons (called marine bands) are sometimes of great lateral extent and act as vitally important index horizons. **1869** Index-map [in Dict.]. **1932** *Discovery* May 153/1 As large a selection of the recorded information as the smaller scales will carry is issued on the standard scale of one inch, and with further selection and reduction, on the ¼ inch scale of the 'index map'. **1875** W. S. Jevons *Money & Mechanism of Exchange* xxv. 332 A table containing the Total Index Number of prices, or the arithmetical sum of the numbers expressing the ratios of the prices of many commodities to the average prices of the same commodities in the years 1845–50. **1886** *Rep. Brit. Assoc. Adv. Sci. 1885* 871 The index number for the price of each article in 1883 is 1 or 100, according to the use or otherwise of the decimal point. **1887** *Encycl. Brit.* XXII. 466/1 The only matter connected with price which it is necessary to refer to here is the theory of the index number. **1893** *Jrnl. Soc. Arts* 3 Feb. 211/2 This total index number..merged all prices high and low in a single figure. **1900** A. L. Bowley *Wages in U.K. in 19th Cent.* xii. 95 (*caption*) Index numbers, showing rate of Change of Wages in the London Building Trades. **1928** J. W. F. Rowe *Wages in Pract. & Theory* 14 Index numbers based on changes in the nominal weekly rates in these industries afford a general guide to the character of wage fluctuations. **1942** [see sense 9 e above]. **1966** *McGraw-Hill Encycl. Sci. & Technol.* III. 233/1 The distribution of coal by rank, geologic age, and district is indicated in Fig. 4. The index numbers, which refer to coal districts, are grouped by continent and country in the accompanying list. **1973** *Daily Tel.* 11 July 6/6 He admitted owning a car which had an index number identical to one Miss Mallalieu had noted. **1905** T. R. Shaw *Machine Tools* vi. 413 If the index pin does not come exactly opposite a hole, there is an adjustment by means of two screws. **1879** *Jrnl. Franklin Inst.* CVIII. 106 The making of practically perfect index plates for gear cutting machines is a different matter from graduating circles for astronomical instruments. **1902** Index plate [see **index circle*]. **1923** R. C. H. Heck *Mechanics of Machinery: Mechanism* v. 235 At the other end of the worm shaft [of the index head] is the index crank C, with handle H and plunger pin Q which can be let into any hole in index plate P. **1950** J. Martin in A. W. Judge *Machine Tools & Operations* III. iii. 131 By the use of worm-gearing, the indirect dividing head can space work up to 360 divisions, using standard index plates. **1973** *Daily Tel.* 15 Sept. 2/5 He noticed that the index plates..had been hurriedly removed from another vehicle. **1955** R. K. Richards *Arithmetic Operations in Digital Computers* xi. 348 In these machines..each instruction specifies an index register as well as an operation and an address. For each operation, the number stored in the indicated index register is automatically added to the address, and the sum is then the actual address which is used. **1957, 1970** Index register [see *indexing *vbl. sb.* 3]. **1970** O. Dopping *Computers & Data Processing* vi. 101 In most modern computers address modification and counting is facilitated by index registers. *a* **1884** Knight *Dict. Mech.* Suppl. 494/2 The centers are shown..attached to the index-spindle.

index, *v.* Add: **1.** Also *transf.* (cf. *index *sb.* 5 d). (Further examples.)

1969 C. W. Gear *Computer Organization & Programming* vi. 242 Each time that a file was completed, the system would index it; that is, its name would be placed in a table of file names..with an indication of where it was physically located. **1969** P. B. Jordain *Condensed Computer Encycl.* 242 If the cylinder index becomes too long, a master index may be created to index it.

5. *Engin.* **a.** *trans.* (Cf. quot. 1879 s.v. *index *sb.* 2.) To rotate (work to be machined, or a machine part) through a given aliquot part of a complete turn; to position in accordance with intermittent motion of this kind; hence, to transfer or move from one predetermined position to another in order that different locations may be machined or different operations performed. (Cf. *indexing *vbl. sb.* 2.)

1902 [implied in *indexing *vbl. sb.* 2]. **1913** [see *index centre* s.v. *index *sb.* 10]. **1936** Colvin & Stanley *Drilling & Surfacing Pract.* xxii. 242 Turning the indexing crank without this geared connection indexes the spindle in any desired number of graduations, or parts of a circle. **1951** H. C. Town in *Gen. Engin. Workshop Pract.* (ed. 2) iii. 112/2 A spring-loaded plunger..locates the turret in any one of four positions... To index the turret, the ball handle is revolved and the screw lifts the turret clear of the locating plunger, so that it can be rotated to the next..position, and then locked in position. **1953** L. E. Doyle *Metal Machining* ix. 214 The workpiece is turned by means of the worm and wheel and is indexed by the pin that registers in holes in the face of the worm wheel. *Ibid.* xiii. 298 On a dividing head with a 40 to 1 ratio, one full turn of the crank is needed to index a 40 tooth gear from one tooth space to the next. **1959** *Machinery* XCIV. 511/2 Spacing of the holes axially along the length of the rod is accomplished by indexing the rod vertically in 0·026-in. steps by means of an accurate lead-screw. **1966** *McGraw-Hill Encycl. Sci. & Technol.* VII. 407/2 When the turret [of the lathe] is indexed for successive operations, the saddle acts as a guide for the ram in its strokes to and from the work.

b. *trans.* To produce or obtain (a desired number of divisions or operations) by indexing.

1900 *Machinery* (N.Y.) Nov. 88/2 For indexing prime numbers we must use other than the one-hole basis. **1923** R. C. H. Heck *Mechanics of Machinery: Mechanism* v. 237 The test or criterion of ability to index any number *n* is expressed by putting Eq. (89) into the form $c = 40 \times h/n$. **1936** Colvin & Stanley *Drilling & Surfacing Pract.* xxii. 242 The index sector is a great convenience in counting holes to index the required number of divisions. **1961** L. E. Doyle et al. *Manuf. Processes* xxiv. 587 Several means are available for indexing numbers not obtainable with standard plain indexing, especially large numbers.

c. *intr.* To move or travel during indexing.

1901 *Machinery* (N.Y.) Jan. 147/1 To divide into 91 parts, index forward, on the front side of the plate, six spaces on the 39 circle; then index forward on the back of the plate, 14 spaces on the 49 circle. **1953** L. E. Doyle *Metal Machining* xiii. 298 The crank must be turned $40 \div 36 = 1\frac{4}{36} = 1\frac{1}{9}$ turns to index from one space to another on the gear. **1966** *McGraw-Hill Encycl. Sci. & Technol.* VII. 409/1 A horizontal, circular table holding the rotating chucks indexes under the vertical spindles with a different operation being performed at each station. **1968** Boothroyd & Redford *Mechanized Assembly* ii. 8 With continuous transfer the work carriers are moving at constant speed whilst the workheads index backwards and forwards. *Ibid.* 13 Reciprocation of the transfer bar over a distance equal to the spacing of the workheads will cause the work carriers to index between the workheads.

6. *Computers.* To modify (an instruction or its address) by causing the contents of a specified index register to be added to the address before the instruction is executed; to provide with a number that brings about such modification; also, to carry out (a repetitive sequence of operations) by this means.

1962 Y. Chu *Digital Computer Design Fund.* xii. 454 Its contents are used to modify the address of the instruction to be indexed. **1962** R. V. Oakford *Introd. Electronic Data Processing Equipment* iii. 89 The counter may be added to the operand address of an instruction to index the repeated performance of an operation on a sequence of registers. **1969** C. W. Gear *Computer Organization & Programming* ii. 52 We can indicate this in our program writing by using Y to mean the unindexed address Y (that is, the index bits are 0) and by using Y,I to mean the address Y indexed by index register I. **1972** BERGMAN & BRUCKNER *Introd. Computers & Computer Programming* vii. 204 Since this instruction is indexed, its effective operand is 000 (operand) + 251 (contents of index register) = 251 (effective operand and address of X1).

indexation (indeksēi·ʃən). *Econ.* [f. INDEX *v.* + -ATION.] An adjustment in rates of payment in money (e.g. wage-rates, bond prices, etc.) to reflect changes in the value of money by means of an index of such changes. Cf. *INDEX *sb.* 9 e.

1960 *Spectator* 29 Apr. 602 The system of 'indexation' by which wages and prices rose in an officially sponsored spiral. **1972** *Times* 22 July 16/3 If the current rate of inflation continues it will be essential to introduce some form of indexation into the capital gains tax system. **1973** *Daily Tel.* 10 Feb. 31/1 The most likely form of control is by 'indexation' of business leases, which would mean that business rents would be able to rise no faster than the wholesale price index, the cost of living index or such index as the Government cares to regard as a fair indicator of the trend of inflation.

indexed, *ppl. a.* Add: **3.** *Computers.* Modified by or executed by means of an index (sense *8 d).

1957 D. D. MCCRACKEN *Digital Computer Programming* viii. 98 Examples are presented which compare an indexed loop with the nonindexed form. **1972** M. D. FREEDMAN *Princ. Digital Computer Operation* ii. 23 An indirect address can be indexed. We will illustrate the concept of an indexed indirect address by an example.

indexible (inde·ksĭb'l, i·ndeks-), *a.* Also **-able.** [f. INDEX *v.* + -IBLE, -ABLE.] Capable of being indexed.

1951 *Industrial Equipment News* Dec. 100 (*caption*) Indexable Kennametal tip has 4 cutting edges. **1959** C. V. L. SMITH *Electronic Digital Computers* ii. 45 Most type B instructions are 'indexable'. This means that, when any such instruction is executed, the address actually referred to is the address given decreased by the contents of the index register specified by the tag bits. **1966** C. J. SIPPL *Computer Dict. & Handbk.* 145/2 *Index register*... Used with indexible instruction addresses during execution. **1971** *Computers & Humanities* VI. 68 By 'indexible' we mean [to denote] a character string that..is likely to be used as a criterion for selective retrieval and/or sorting.

indexical, *a.* **c.** (Later examples.)

a **1914** C. S. PEIRCE *Coll. Papers* (1932) II. II. iii. 164 Some indices are more or less detailed directions for what the hearer is to do in order to place himself in direct..connection with the thing meant....Along with such indexical directions of what to do to find the object meant, ought to be classed..*selective* pronouns. *Ibid.*, Other indexical words are prepositions, and prepositional phrases. **1957** *Lingua* VII. I. 27 Within the same vowel system, diaphonic differences are considerable, and appear to be highly indexical, both socially and regionally, in differentiating accents within England. **1970** *English Studies* LI. 276 Beside these indexical features a medium also has aesthetic properties.

indexing, *vbl. sb.* (in Dict. s.v. INDEX *v.*). Add: **2.** *Engin.* The intermittent rotation of work through aliquot parts of a complete turn in order that some operation may be performed on it at equal angular intervals; also, the movement of work or of a machine part or tool from one predetermined position to another during machining operations.

1902 *Internat. Library of Technol.* II. § 15. 21 Direct indexing is done by the aid of an index plate fastened direct to the index-head spindle; that is, the index plate is moved to obtain the divisions. In indirect indexing, the index plate is normally stationary, and the index-head spindle is rotated by the use of suitable gearing. Indirect indexing is divided into..simple and compound indexing. **1930** *Engineering* 30 May 693/3 During the return stroke, indexing of the worm takes place. **1936** COLVIN & STANLEY *Drilling & Surfacing Pract.* xxii. 243 The essential thing to remember in all indexing is that the worm wheel on the spindle has 40 teeth and that the worm is a single thread. This means that for every turn of the crank the spindle is rotated $\frac{1}{40}$ of a revolution. **1957** S. E. RUSINOFF *Manuf. Processes* (ed. 2) xii. 489 In straight in-line indexing, the work piece moves intermittently from one machining station to the next in a straight line; a group of machine tools is assembled into one machine. **1961** L. E. DOYLE et al. *Manuf. Processes* xxiv. 587 Linear indexing may be done on a general-purpose milling machine for such jobs as cutting rack teeth. **1964** S. CRAWFORD *Basic Engin. Processes* vi. 172 When using this method of indexing the number of possible divisions is limited to the three-hole circles available on the front index plate. **1966** *McGraw-Hill Encycl. Sci. & Technol.* VII. 408/1 Turret indexing, actuation of the collet, feeding of stock, and spindle clutch operations are automatic.

3. *Computers.* The automatic increasing or decreasing of the address portion of an instruction, following each execution, by means of an index register, so that a succession of operands is operated on.

1957 D. D. MCCRACKEN *Digital Computer Programming* viii. 106 All we really want from the indexing is the sum of the index register contents and the indicated address. *Ibid.* 109 Locations 1200, 1208, 1216, etc., up to 1272 contain ten numbers...Write a program, using indexing, to compute the sum of the squares of these numbers and store it in 1307. **1961** LEEDS & WEINBERG *Computer Programming Fund.* iv. 101 Many machines have special circuitry designed to facilitate common operations associated with repetition...In mathematics, indexes, or subscripts, are used to keep track of such processes. As an outgrowth of these terms, the words *indexing* and *index arithmetic* have been given to the type of computer instruction which does this job, although indexing may be thought of as a special type of control which also takes over certain tasks of counting from the regular arithmetic instructions. **1970** HULL & DAY *Computers & Probl. Solving* 197/2 Index registers have a second important use in what is called automatic address modification or indexing. This occurs when the address of an instruction is modified, just before the instruction is executed, by having the content of an index register added to it.

4. *attrib.* and *Comb.*, as *indexing pin*; **indexing head** = *index head* (*INDEX *sb.* 10).

1901 *Machinery* (N.Y.) Jan. 145/1 A dividing or indexing head...With the tailstock shown..comprises what is commonly known as a pair of index centers. **1944** *Model Engineer* 17 Aug. 157/1 The method..is a modification of the indexing head of a milling machine. **1960** H. W. PORTER et al. *Machine Shop* (ed. 2) viii. 294 The indexing head..comes equipped with three index plates. **1964** S. CRAWFORD *Basic Engin. Processes* v. 117 A spring-loaded indexing-pin in the face of the compound-slide accurately locates the post in any one of the four positions.

indfine (i·ndfinĭ). *Irish Hist.* Also **indfhine, innfine.** [O.Ir. (mod.Ir. *innfhine*) f. *ind* end + *fine* family, FINE *sb.*[2]] One of the four branches of the Irish clan structure comprising the men most distantly related to the chief. Cf. *GEILFINE, *IARFINE.

The fourth branch is the *derbfine* (not in Dict.).

1875 H. S. MAINE *Lect. Early Hist. Inst.* 209 The eldest member of the Iarfine moved into the Indfine; and the eldest member of the Indfine passed out of the organisation altogether. **1893** P. W. JOYCE *Short Hist. Ireland* I. ix. 70 A *Gelfine* organisation when complete consisted of seventeen men all related to each other:..the *gelfine* group..consisting of five; the *derfine*, the *iarfine*, and the *innfine*, of four each. **1921, 1967** [see *IARFINE].

India. Add: **6.** *India calico, carpet, cotton* (example), *muslin* (examples), *shawl, silk* (examples); **India Office** (earlier examples); **India tag,** a type of tag which is used to fasten papers together and consists of a cord with a small metal bar at either end.

1805 *Times* 7 Nov. 1/1 Stout India calico, full ell-wide. *c* **1702** C. FIENNES *Journeys* (1947) 346 A pladd bed lined with Indian callicoe and an India carpet on the bed. **1931** A. U. DILLEY *Oriental Rugs & Carpets* Pl. 32 (*caption*) India Carpets of Seventeenth Century from the Palace of the Maharaja at Jaipur. **1881** C. C. HARRISON *Woman's Handiwork* III. 176 Scarves of India cotton worked in tarnished gold. **1796** M. EDGEWORTH *Parent's Assistant* (ed. 2) II. 167 Oh, Miss Eden, your beautiful India muslin!—take care of the chimney sweeper. **1852** E. TWISLETON *Let.* 10 July (1928) ii. 27, I intend to appear in my India muslin. **1929** D. H. LAWRENCE *Pansies* 39 A yard of India muslin is alive with Hindu life. **1869** *Bradshaw's Railway Manual* XXI. 394 Government Director of the Indian Railway Companies, India Office, Whitehall, S.W. **1880** E. W. HAMILTON *Diary* 24 Apr. (1972) I. 2 Cabinet-making has been going on all day... Lord Hartington [is to go] to the India Office. **1822** D. WORDSWORTH *Jrnl.* 23 Sept. (1941) II. 374 The ladies.. wore gorgeously embroidered India shawls. **1837** DICKENS *Mudfog Papers* (1880) 159 His view of the ladies within being obstructed by the India shawls. **1756** B. FRANKLIN *Writings* (1905) III. 294 When you incline to buy.. India silks. **1816** SCOTT *Antiq.* III. vii. 142 His India silk handkerchief. **1881** C. C. HARRISON *Woman's Handiwork* I. 48 The India silks manufactured for Mr. Louis Tiffany, by a well-known firm in Connecticut, from cocoons imported by themselves. **1912** *List Articles Authorised to be Supplied by H.M.S.O.*, Tags, Treasury (Insertion), Tags, India (cross-bar). **1963** R. L. COLLISON *Mod. Business Filing & Archives* ii. 47 The documents are then secured to the file by what is known as a Treasury or India tag—i.e. a cord with a metal tag at each end.

indialite (i·ndiălait). *Min.* [f. INDIA + -LITE.] The hexagonal dimorph of cordierite.

1954 MIYASHIRO & IIYAMA in *Proc. Japan Acad.* XXX. 746 The present paper is concerned with the finding of a new hexagonal mineral from Bokaro coalfield, India, $(Mg, Fe^{+2})_2Al_4Si_5O_{18}$ in composition, polymorphic with cordierite. This mineral has hitherto been identified as cordierite; but..it is a distinct species, to which the name 'indialite' is proposed. **1972** M. H. BATTEY *Mineral. for Students* 254/1 Indialite, a dimorph of cordierite, is isostructural with beryl.

Indian, *a.* and *sb.* Add: **A.** *adj.* **1. b.** (Further examples.)

c **1702** Indian calico [see *India carpet*]. **1718** J. STEUART *Letter-Bk.* (1915) 64, I doe not mean you should goe to the expense of bying Indian chints. *c* **1793** JANE AUSTEN *Volume First* (1933) 9 Your sentiments so nobly expressed on the different excellencies of Indian & English Muslins. **1794** A. YOUNG *Trav. France* (ed. 2) I. xix. 548 They print a great quantity of Indian callicoes. **1798** Indian shawl [see SHAWL *sb.* 2]. **1830** M. EDGEWORTH *Let.* 13 Dec. (1971) 449, I have..laid out fifteen guineas on—an Indian shawl. **1851** J. F. ROYLE *On Culture & Commerce of Cotton in India* I. 22 Indian cotton is well known to have certain good qualities of its own. **1863** A. J. MUNBY *Diary* 12 May in D. Hudson *Munby* (1972) 160 Curtains of old Indian chintz. **1873** C. M. YONGE *Pillars of House* IV. xlii. 218 Cherry cleared her large sofa, and covered him up with her Indian silk quilt. **1879** *Queen* 1 Mar. (Advt.), An entirely new under garment successfully introduced in Silk, Merino, Gauze Merino, and Indian Gauze. **1898** A. BENNETT *Man From North* xxviii. 241 Its square of Indian carpet over Indian matting. **1910** *Encycl. Brit.* VII. 258/1 The Indian cottons are usually of short staple. **1937** J. LAVER *Taste & Fashion* xii. 178 A combination chemise and knickers of Indian gauze. **1938** *Decorative Art* 83 A modern Indian rug in green. **1971** *Habitat Catal.* 76/2 Indian cotton durries..simple tough cotton rugs. **1974** 'G. BLACK' *Golden Cockatrice* xi. 190 Blood had reached an Indian rug.

2. (Further examples.) *spec.* In reference to the endurance of tortures and hardship by North American Indians.

1737 J. WESLEY *Let.* 22 July (1931) I. 225 When..He shall have chosen one or more to magnify Him,..not with a stoical or Indian indifference, but blessing and praying for their murderers. **1817** COLERIDGE *Biogr. Lit.* I. x. 185 Three week's truly Indian perseverance in tracking us. **1822** WORDSWORTH *Eccl. Sk.* 67 The shrouded Body, to the Soul's command, Answering with more than Indian fortitude. **1916** C. A. EASTMAN *From Deep Woods to Civilization* iii. 32 He took out his Bible and hymn-book printed in the Indian tongue. **1931** F. J. STIMSON *My United States* i. 4 Indian camps were not unusual in the vacant lots outlying Dubuque. **1934** *Beaver* June 9 When word came that they had landed, I left the living room for the Indian shop. **1940** W. FAULKNER *Hamlet* II. 98 A man who was not thin so much as actually gaunt, with straight black hair..and high Indian cheekbones. **1942** *Chicago Tribune* 24 Nov. 12/3 This stone marks the site of an ancient Indian village and chipping station. **1966** *Oxf. Compan. Amer. Hist.* 405/2 Since Indian tribes acted as buffer states between French and British colonies and were invaluable allies in time of war, colonial governors always sought to make treaties of friendship with the Indians. **1969** *Islander* (Victoria, B.C.) 5 Oct. 10/3 An Indian express was sent ahead to Fort Colville. **1970** D. BROWN *Bury my Heart at Wounded Knee* ii. 15 Many of the Mexicans had Indian blood. **1971** *Times* 30 Sept. 12/3 There has been a resurgence of interest in the Indians.. partly (among the young) to embrace aspects of Indian life on behalf of the counter-culture. **1971** *Times* 15 Oct. 14/4 Dr. Robert Euler..has directed salvage archeology at prehistoric Indian sites in the area to be mined. **1973** *Freedomways* XIII. 81 The book is rich with examples of Indian culture and social life as practiced by the Oglala.

4. a. Indian antelope = *black-buck* (*BLACK *a.* 19); **Indian elephant,** the smaller of the two existing species of elephant, *Elephas maximus*; **Indian English,** the form of English used by inhabitants of India for whom English is not a native language; cf. *BABU; **Indian hay** *U.S. slang*, marijuana; **Indian head** (see quot. 1957[2]); **Indian lotus,** an aquatic plant, *Nelumbo nucifera*, native to Asia, and bearing fragrant white or pink flowers; also called Egyptian or sacred lotus; **Indian tea,** tea grown in India or Sri Lanka, especially in Assam and the Darjeeling district; cf. sense 4 b below; **Indian work,** Indian handicraft, *spec.* drawn-thread work on muslin.

1888 Indian Antilope [see *black-buck* (*BLACK *a.* 19)]. **1964** E. P. WALKER et al. *Mammals of World* II. 1457/1 Blackbucks, Indian Antelopes... This animal occasionally lives in herds of several hundred members. **1607** TOPSELL *Four-f. Beasts* 192 The Indian Elephants are greatest, strongest, and tallest. **1965** D. MORRIS *Mammals* 338 Smaller than the African Elephant, the bull Indian Elephant rarely reaches 10 feet high at the shoulder. **1907** G. C. WHITWORTH (*title*) Indian English: an examination of the errors of idiom made by Indians in writing English. **1934** R. C. GOFFIN in *S.P.E. Tract* XLI. 31 We have touched on the preference of Indian English for the archaic in vocabulary and phrase. **1971** *Shankar's Weekly* (Delhi) 4 Apr. 22/3 The best of Indian English et al you gave. **1939** *Jrnl. Amer. Med. Assoc.* 1 July 4/1 In America the name marihuana..is used. It has numerous picturesque names, such as muggles, reefers, Mary Warner, Indian hay, the weed and tea. **1969** *Sci. Amer.* Dec. 17 In the U.S. it [*sc.* marijuana] is variously called the weed, stuff, Indian hay,..and other names. **1911** *Daily Colonist* (Victoria, B.C.) 9 Apr. 24/2 New Wash Dress Materials in Muslins..Indian head and Linen Suitings. **1957** M. MCCARTHY *Memories Catholic Girlhood* vi. 117 Blue Indianhead 20 yds. **1957** M. B. PICKEN *Fashion Dict.* 183/1 *Indian head*, trade name for sturdy, firm, cotton material of linen-weave, made in many lovely colors. Used for work, play, and sports clothing and for home furnishings. **1901** L. H. BAILEY *Cycl. Amer. Hort.* III. 1065/1 (*caption*) The Indian Lotus, Nelumbium speciosum of the trade, but properly Nelumbo nucifera. **1963** W. BLUNT *Of Flowers & Village* 104 The seeds of the Indian Lotus 'rattle in their sockets like teeth in the jawbone of a skull'. **1884** E. MONEY *Tea Controversy* (ed. 2) 10 A pound of Indian Tea of any grade will give considerably more infusions of a like strength than a similar pound of Chinese. **1893** *Illustr. London News* 11 Nov. 618/2 Indian tea shipping warehouses. *c* **1938** *Fortnum & Mason Price List* 6 State whether China or Indian Tea is preferred. **1960** *Good Housek. Cookery Bk.* 451/2 Household teas,.. are usually full-flavoured blended Indian teas. **1969** *Times* 13 Oct. (India Suppl.) p. x (Advt.), Darjeeling, Assam, Nilgiris. Three of the greatest names in Indian Teas. **1865** F. B. PALLISER *Hist. Lace* xxi. 250 There is also a good specimen of that description of drawn muslin lace, commonly known under the name of 'Indian work', but which appears to have been made in various manners. **1882** CAULFEILD & SAWARD *Dict. Needlework* 157/2 Fig. 299 is of a later description of Drawn Work, and would be known as Indian Work, as its foundation is muslin. **1900** E. JACKSON *Hist. Hand-Made Lace* 150 Drawn-work.. was known all over Europe as Hamburg Point, Indian work (when executed in muslin). **1920** A. K. ARTHUR *Embroidery Bk.* ix. 90 Some of the elaborately worked pieces of Indian and Persian work, where tinsel braids are freely employed, are things to marvel at.

b. Indian agent (see *AGENT *sb.* 4 b); In-

dian apple (examples); **Indian awl** (see quot. 1941); **Indian bean** (examples); **Indian blanket,** orig. a blanket made by or for N. Amer. Indians, often used as a cloak; now also a blanket made in imitation of this; **Indian bureau** (or **Bureau**), in N. and S. America, a bureau concerned with the affairs of American Indians; *spec.* in the U.S., the Bureau of Indian Affairs; **Indian currant** = *coral-berry* (CORAL *sb.*[1] 9); **Indian devil,** a N. American name for either the wolverine or the cougar; **Indian file** (earlier and later examples); **Indian gift, giver** (later examples); so **Indian giving; Indian mound** [MOUND *sb.*[3] 4 e], in N. America, a mound or earthwork erected in former times by Indians as a burial place, fortification, etc.; **Indian paint,** a N. American perennial herb with reddish sap and thick red-tinted roots; **Indian paint-brush,** a herbaceous plant of the genus *Castilleja*; **Indian path** (earlier and later examples); **Indian pear** *N. Amer.*, a tree or shrub of the genus *Amelanchier*, or its edible fruit, a fleshy red or purple berry; **Indian pipe** (examples); also **Indian pipe-stem; Indian pony,** a type of pony descended from horses originally brought to America by Spanish colonists; **Indian reservation, reserve** (see RESERVATION 3 b, RESERVE *sb.* 5 b); **Indian rice**: substitute for def.: a North American aquatic grass, *Zizania aquatica*, or one of several similar plants resembling rice; add examples; **Indian sign(s),** a (usually faint) track or trail, etc., that reveals the presence of Indians; also a smoke-signal or other signal used by Indians; phr. *to put* (or *have*) *the Indian sign on* (someone) (see quot. 1944); **Indian tea,** any one of several N. American plants whose leaves are used to make a drink resembling tea; = *Labrador tea*; **Indian tobacco** (earlier and later examples); also applied to other plants; **Indian turnip** (earlier and later examples).

1766 W. JOHNSON *Let.* 23 Jan. in R. Rogers *Jrnls.* (1883) 216 As Commandant and Indian Agent, it will be extremely difficult to check him. **1816, 1901** [see *AGENT *sb.* 4 b]. **1974** A. MACLEAN *Breakheart Pass* iii. 44, I asked the Governor here to appoint me Indian agent for the territory; I settle differences..allocate reservations, try and stop the traffic in guns and whisky. **1847** F. PARKMAN in *Knickerbocker* XXIX. 310 The rich flowers of the Indian-apple were there in profusion. **1931** W. N. CLUTE *Common Names of Plants* 35 The May-apple (*Podophyllum peltatum*) was known to the settlers as Indian apple, but it is really a berry. **1821** G. SIMPSON *Jrnl. Occurrences in Athabasca Dept.* (1938) 142 Awls, Indian, doz. **1922** *Beaver* July 10/2 It requires but three tools to build a canoe: an axe, a 'crooked knife', and a square or Indian awl. **1941** *Beaver* Sept. 38 *Indian awl*, a square two-ended awl for bark, leather or wood. **1843** J. TORREY *Flora N.Y.* II. 25 Catalpa. Indian Bean...About habitations...The Catalpa is more esteemed for ornament than for use. **1933** [see *cigar-tree*]. **1969** T. H. EVERETT *Living Trees of World* 298/2 The Indian-bean has many more flowers in each cluster than the Western catalpa. **1764** in *New Jersey Archives* (1902) 1st Ser. XXIV. 350 There are a blue Great Coat, and an Indian Blanket missing. **1782** *Quebec Gaz.* 19 Dec. 4/1 For Sale..A Large assortment..Indian blanket Rugs. **1807** *Salmagundi* 13 Feb. 49 The shawl..thrown over one shoulder, like an Indian blanket. **1927** W. CATHER *Death comes for Archbishop* I. iii. 31 The earth floor was covered with thick Indian blankets; two blankets, very old, and beautiful in design and colour, were hung on the walls like tapestries. **1962** J. BRAINE *Life at Top* xxvi. 278 There's a cigarette burn in my Indian blanket. **1966** *Islander* (Victoria, B.C.) 27 Feb. 7/1 We smelled perfume and talcum from many Indian blankets. **1831** C. ATWATER *Remarks Tour to Prairie du Chien* 142 Taking advantage of frauds committed on the Indians, by persons connected with the Indian bureau, the Factory system gave place to the present system. **1922** D. H. LAWRENCE *Phoenix II* (1968) 239 The Indian Bureau is supposed to do the cherishing. **1972** *Buenos Aires Herald* 2 Feb. 7/1 Officials of Brazil's Indian Bureau here believe that numerous small tribes of Indians living along the projected new highway have never seen an outsider. **1785** H. MARSHALL *Arbustrum Amer.* 82 *Lonicera Symphoricarpos.* Indian Currants, or St. Peter's Wort. This hath a shrubby stalk, which rises from four to five feet high. **1806** LEWIS & CLARK *Orig. Jrnls. Lewis & Clark Expedition* (1905) V. 327 Deep purple berry or the large Cherry of the Current Species which is common... The engagees call it the Indian Current. **1866** [see CURRANT 3]. **1948** *South Sierran* Feb. 2/2 'Indian currant' was also in bloom. **1851** J. S. SPRINGER *Forest Life & Forest Trees* 66 A dangerous specimen of the feline species, known by woodsmen as the 'Indian devil', had prowled from time immemorial. **1901** W. M. THOMPSON *In Maine Woods* 60 The cougar, or 'Indian devil', is sometimes seen. **1937** H. H. LANGTON in P. Campbell *Trav. Interior Parts N. Amer.* 71 The Wolverine..has long been extinct in the province [*sc.* Nova Scotia], where it is remembered as the 'Injun devil'. **1965** *Wildlife Rev.* (Victoria, B.C.) Mar. 19 Cougars, also known as panthers, pumas, catamounts, mountain lions, and Indian devils, are large unspotted cats. **1758** in *Essex Inst. Hist. Coll.* (1881) XVIII. 179 They march'd in Indian file. **1873** G. M. GRANT *Ocean to Ocean* 189 As the line of march had to be in Indian file, we soon exchanged the undemonstrative 'good-bye' with him. **1922** *Beaver* Sept. 6/2 The long camp fire gave sufficient light to see the dancers, who followed one another in Indian file. **1971** *Ceylon Times Weekender* (Colombo) 3 Oct. 5/5 As the family came trotting along towards the crocodile in Indian file, the lean, rangy patriach was in the lead. **1879** B. F. TAYLOR *Summer-Savory* xxvi. 207 She is glad it [*sc.* the check] is an 'Indian gift', that the conductor did not present it to her outright. **1892** H. C. BOLTON in *Jrnl. Amer. Folk-Lore* V. 68 If an American child, who has made a small gift to a playmate is indiscreet enough to ask that the gift be returned, he (or she) is immediately accused of being an Indian-giver, or, as it is commonly pronounced Injun-giver. **1904** *N.Y. Herald* 10 Sept. 5 Later he took the position of the 'Indian giver' and wanted the money back. **1939** *Time* 23 Oct. 4/1 Call us Indian-giver. **1965** R. MANHEIM tr. *Grass's Dog Yrs.* I. 12 'You threw my knife.' 'It was my knife. Don't be an Indian giver.' **1971** M. MCCARTHY *Birds of America* 29 He wanted the little violin which Hans, an Indian giver, took back to the store. **1837** W. IRVING *Capt. Bonneville* II. vi. 71 His experience in what is proverbially called 'Indian giving' made him aware that a parting pledge was necessary on his own part. **1962** B. SPOCK *Problems of Parents* (1968) II. 50 The one-year-old who has yelled bloody murder during his physical examination may, ten minutes later..sweetly hand him a toy and then take it back. This latter trick may look like Indian giving. **1791** in *Mass. Hist. Soc. Coll.* (1794) 1st Ser. III. 24 There is an Indian mound, the base of which is about three hundred paces round, and rises in a conic form about one hundred feet. **1869** B. HARTE *Luck of Roaring Camp* 186 Cattle and sheep are gathered on Indian Mounds waiting the fate of their companions whose carcases drift by us. **1949** *Illinois State Archaeol. Soc. Jrnl.* Jan. 7/2 When the term, 'Indian Mound', is mentioned, one naturally thinks of a large ceremonial Mound which covers one to three acres of ground and is perhaps five or ten or even 20 feet in height. **1950** *Caribbean Q.* II. II. 24 Scientific excavation of the so-called Indian Mounds has brought to light masses of..pottery ware, celts. **1803** A. F. M. WILLICH *Domestic Encycl.* (Amer. ed.) IV. 442/1 Sanguinaria Canadensis, called commonly Puccoon, blood-wort, red-root, Indian paint, turmeric. **1931** W. N. CLUTE *Common Names of Plants* 26 The bloodroot..is also called puccoon-root, Indian paint. **1892** *Jrnl. Amer. Folk-Lore* V. 101 *Castilleia coccinea*..Indian paint-brush. **1923** 'B. M. BOWER' *Parowan Bonanza* ii. 21 He plucked a bright red 'Indian paint brush' from beside a rock. **1959** *Calgary Herald* 31 July 10/1 There is no finer feature of the autumn landscape in Alberta..than the roadside aster,..Indian paint brush and wild geranium. **1969** *Islander* (Victoria, B.C.) 5 Oct. 7/1 Sometimes the red-orange of the Indian paintbrush was like a blanket spread beneath the pines. **1634** W. WOOD *New Englands Prospect* II. vii. 71 An Indian path (which seldome is broader than a Cart's rutte). **1793** W. BENTLEY *Diary* 26 Aug. (1907) II. 51 An Indian path was visible through the town from the N.W. corner. **1794** A. THOMAS *Newfoundland Jrnl.* (1968) iv. 63, I shall resume my journey..endeavouring as well as I can to give you an idea of an Indian Path in Newfoundland. **1939** G. H. EVANS *Pigwacket* 3 To the consideration of this important gateway, through which the old Indian path entered Pigwacket, further attention will be directed. **1796** *Descr. Prince Edward Island* 5 A fruit in this Island, called the Indian Pear, is very delicious. **1856** W. E. CORMACK *Narr. Journey across Newfoundland* (1874) 19 On the skirts of the forest, and of the marshes, are found.. Indian pear [etc.]. **1873** G. M. GRANT *Ocean to Ocean* 156 The sasketoon are simply what are known in Nova Scotia as 'Indian pears'. **1956** T. H. RADDALL *Wings of Night* 12 Then came the long reach of scrub woods, wire birch and poplar mostly, with blossoming clumps of Indian pear, like patches of snow on the slopes. **1822** A. EATON *Man. Bot.* (ed. 3) 357 *Monotropa uniflora*, birds nest, indian-pipe. **1884** 'C. E. CRADDOCK' *In Tennessee Mts.* I. 60 The ashy Indian pipes silvered the roots of the trees. **1962** *Maclean's Mag.* 10 Mar. 17/1 It was a palely sprouting parasite called Indian pipe, or corpse plant, which extrudes almost overnight from rotting compost and lives only briefly. **1870** *Amer. Naturalist* III. 6 The Indian Pipestem will be found rarely in low woods. **1758** P. STEVENS *Jrnl.* 13 July in N. D. Mereness *Trav. Amer. Colonies* (1916) 314 He is given us for an Indian pony in his place. **1865** *Nor'Wester* (Red River Settlement) 21 Sept. 2/3 Our hardy Indian ponies found many purchasers at good prices. **1869** [see *BRONCO a]. **1965** D. M. GOODALL *Horses of World* 211 The descendants of these [Spanish] horses, known as Mustangs, migrated through Mexico to the United States, scattering far west and north-west. Some were captured by the Indians and subsequently became Indian ponies. **1819** D. THOMAS *Trav. Western Country* 12 This tract is an Indian reservation. **1949** *Kansas Hist. Q.* Feb. 2 They learned that Council Grove was situated on an Indian reservation and was not available for settlement. **1954** H. EVANS *Mist on River* 43 She said that this fall or next..she would try to get him into an Indian reservation school. **1792** in *Rep. Bureau Arch. Ontario* (1905) III. 215 Every farm of good land, in that part of the Country, is comprehended within the Indian reserve. **1818** F. HALL *Trav. Canada & U.S.* 131 The river Credit is an Indian reserve, well stocked with salmon. *a* **1843** [see RESERVE *sb.* 5 b *attrib.*]. **1958** *Edmonton* (Alberta) *Jrnl.* 18 June 31/7 Conditions on the two Indian reserves have improved greatly. **1822** *Minutes of Council, N. Dept. of Rupert's Land*, in *Publ. Hudson's Bay Rec. Soc.* (1940) III. 22 All the Indian rice and corn he may collect. **1872** SCHELE DE VERE *Americanisms* 409 The Indians of the Northern regions ..depend largely upon..this perennial plant, from whence it is also known as Indian rice. **1933** E. C. JAEGER *California Deserts* xiii. 165 Indian rice (*Oryzopsis hymenoides*) is frequent in porous soils, particularly on blown sand. **1805** P. GASS *Jrnl.* 30 May (1807) 93 We see a great many fresh Indian tracks or signs as we pass along. **1854** J. R. BARTLETT *Pers. Narr. Explor. Texas* I. iv. 95 They had seen 'Indian sign', and pointed out..the well-known Indian signal of a puff of smoke suddenly rising from the earth. **1866** *Rep. Indian Affairs* (U.S.) 188 On the 19th, proceeding again to the Malheur river, but found no fresh Indian signs. **1873** J. H. BEADLE *Undevel. West* xxxiv. 738, I rode around our camp, a circle of about six miles, looking for Indian signs. **1910** R. GRAU *Business Man in Amusement World* 97 He has also been careful to protect himself against 'the Indian Sign'. **1912** *McClure's Mag.* XXXIX. 235/1 He sure put the Indian sign on Tommy Ryan that time. **1916** *Boy Scouts' Year Bk.* (N.Y.) 93/2 He had proved that the Indian sign wasn't infallible... After all, then, the Indian sign was a kind of superstition. *a* **1918** G. STUART *40 Yrs. on Frontier* (1925) I. 115 Plenty of Indian signs, saw signal fires on the mountains to westward. **1929** *Bookman* Sept. 62/2 Owen Johnson always has had what is known as the 'Indian Sign' on me. I never won a bet from him and never will. **1944** R. F. ADAMS *Western Words* (1945) 83/1 To put the Indian sign on someone meant to hex or curse him with some kind of witchcraft, also to get him where you want him. **1948** V. PALMER *Golconda* x. 79 What silly notions about ourselves we can coddle! I believe the old boy's put the Indian sign on you, Neda. **1973** *Listener* 5 June 6/3 We were riding very high in terms of our legislative programme. We had the Indian sign on the Congress, so to speak. **1709** J. LAWSON *New Voy. Carolina* 91 This plant [*sc.* Yaupon] is the Indian Tea, us'd and approv'd by all the Savages on the Coast of Carolina. **1771** G. CARTWRIGHT *Jrnl.* 6 Mar. (1792) I. 100 As I judged, that Indian tea was of the same nature with the herbs which are recommended by that author [*sc.* Dr. Brookes], I had some gathered from under the snow in the woods, and gave her a pint of the strong infusion of that plant sweetened with sugar. **1794** A. THOMAS *Newfoundland Jrnl.* (1968) iv. 71, I breakfasted with Mrs Harty on Indian Tea, the growth of Newfoundland. **1925** A. HEMING *Living Forest* 131 The old woodsman infused some of the Indian tea he had gathered. **1938** *Beaver* June 23/1 Light springing feet of youths and maidens would make little impression on the tough bushes of Indian Tea. *a* **1618** Indian tobacco [in Dict., sense 2]. **1872** SCHELE DE VERE *Americanisms* 415 Indian Tobacco (*Lobelia inflata*), occasionally used instead of tobacco by virtue of its acrid leaves. **1950** C. P. LYONS *Milestones on Mighty Fraser* 89 Indian tobacco..was used by the Thompson Indians from time immemorial. **1806** *Deb. Congress U.S.* 19 Feb. 1142 Indian turnip, wild carrot, wild onion. **1856** [see TURNIP *sb.* 2 b]. **1873** J. MILLER *Life amongst Modocs* x. 128 He [*sc.* winter] cut down the banners of the spring that night, lamb-tongue, Indian turnip and catella. **1894** [see TURNIP *sb.* 2 b]. **1911** G. S. PORTER *Harvester* xix. 459 He..brought her samples of ginger leaves, Indian hemp, queen-of-the-meadow, cone-flower, burdock, baneberry, Indian turnip, [etc.]. **1949** *Nature Mag.* Apr. 178 A few of these, like Indian turnip or jack-in-the-pulpit, cowslip and milkweed, may be considered mildly inedible.

c. Chess. (i) *Indian problem* (see quots. 1878). Also *ellipt.*

1846 *Chess Player's Chron.* 96 We now publish the names of those amateurs who have sent us the correct solution of our Indian problem. **1878** S. LOYD *Chess Strategy* 96 Its entire difficulty consists in the skill with which the fact of its being an *Indian* is concealed. *Ibid.* 97 The theme of the Indian problem culminates in a stale-mating position, which has been anticipated by preparing an ambush for allowing the defence a move that may expose him to a discovered mate. *Ibid.* 100 The old Indian problem... The leading feature of the problem does not consist in the discovered check, but in the unexpected and apparently useless withdrawal of the two pieces to a remote quarter of the board, the one intersecting the protection of the other so as to allow the adverse King a move. **1913** A. C. WHITE *Sam Loyd* 287 Two of Loyd's best Indians are shown in Nos. 402 and 403.

(ii) Used to denote openings in which a player seeks to control the centre of the board with knights, fianchettoed bishops, etc., rather than by advancing his centre pawns. Now esp. *Indian defence*, where Black plays thus; also *ellipt.*

King's (or *Queen's*) *Indian defence*: an Indian defence in which Black's king's (or queen's) bishop is fianchettoed. See also *NIMZO-INDIAN *a.*

1896 H. F. CHESHIRE *Hastings Chess Tournament 1895* 369 Openings,..Hungarian,..Indian,..King's Gambits. **1899** E. E. CUNNINGTON *Mod. Chess Primer* ix. 181 Or Black may play the 'Indian Defence'. **1925** *Chess Budget* I. 117/2 A third case was in a small club tournament with the new form of Indian defence. **1929** *Chess Amateur* XXIII. 202/1 Indian Defence. Played recently in the semi-final of the Three Counties' Championship Tournament. **1942** H. GOLOMBEK *50 Great Games Mod. Chess* 67/1 (*heading*) Queen's Pawn, Queen's Indian Defence. **1950** *Hoyle's Games Modernized* (ed. 20) IV. 410 The Indian Defences (those in which Black does not play P–Q4 at an early stage). *Ibid.* 411 The *Catalan system*—a modern combination of the King's Indian and Queen's gambit. **1958** *Listener* 13 Nov. 803/3 If you favour complications, you are most likely to be suited by the queen's pawn opening and by the Sicilian and King's Indian Defences. **1964** I. A. HOROWITZ *Chess Openings* 638 King's Indian Defense... Introduced in the Leipzig tournament of 1879.

B. *sb.* **1.** Delete 'Now *rare*' and add examples in the sense of (before 1947) a native or inhabitant of the Indian sub-continent and (after 1947) of the Republic of India.

1885 H. J. S. COTTON *New India* ii. 11 The attitude of Englishmen to Indians is not of a character to inspire confidence. **1896** in *Macm. Mag.* Sept. (1906) 820 Purely Indian-raised troops. **1912** E. M. FORSTER *Let.* 25 Dec. in *Hill of Devi* (1953) 17 In the evening we went to the Tennis Club—all Indians—and drove back through the tidy little town. **1923** *Glasgow Herald* 20 Feb. 6 The possible future absorption by Indians of the bulk of Indian Army commissions and military appointments. **1946** J. H. HUTTON *Caste in India* iii. 27 It is the cultivated and educated Bengali who has been probably more than any other Indian the interpreter of the Englishman to the Indian. **1954** *Chambers's Encycl. World Survey* 210/1 Indians have long shown a genuine flair for the parliamentary system of government. **1965** N. C. CHAUDHURI

Continent of Circe iv. 89 No sensible Indian will have a moment's hesitation in saying which is preferable. **1971** M. EDWARDES *Nehru* xx. 328 The machinery that was to move Lal Bahadur Shastri, a very different sort of Indian, into the office of Prime Minister had already been put in gear.

c. One of the indigenous inhabitants of the Philippine Islands; applied esp. to one who has been converted to Christianity.

1697 W. DAMPIER *New Voy. round World* xi. 307 He fell in with these Philippine Islands, and anchored at Luconia; where he warr'd with the native Indians, to bring them in obedience to his master the king of Spain. **1817** J. MCLEOD *Narr. Voy. H.M.S. Alceste Yellow Sea* 171 The religion of the Indians under the immediate control of the Spaniards is Christianity; but at Mindanao and the other islands..it is said to be a mixture of Mahomedanism with ..Pagan rites. **1885** *Encycl. Brit.* XVIII. 752/1 Cigars they [*sc.* the Negritos of the Philippines] often smoke with the burning end between the teeth—a practice occasionally observed among the civilized Indians. **1905** F. W. ATKINSON *Philippine Islands* x. 264 The Spaniards used the term Indian in speaking collectively of these seven Christian tribes and the word Filipino in speaking of any one born in the Islands without distinction as to religious belief. **1967** D. MANNIX *Sporting Chance* ii. 35 This outfit was made by the Zubanoan Indians who live in the mountains of Mindanao. **1968** R. NELSON *Philippines* ii. 31 The majority of the people, called Indians by the Spanish, had an olive complexion.

† **d.** A member of one of the indigenous peoples of Australia and New Zealand. *Obs.*

1769 J. BANKS *Jrnl.* 21 Oct. (1896) 191 We applied to our friends the Indians [*sc.* Maoris] for a passage in one of their canoes. **1770** *Ibid.* 28 Apr. 264 During this time, a few of the Indians [*sc.* Australian Aboriginals]..remained on the rocks opposite the ship, threatening and menacing with their pikes and swords. **1825** B. FIELD *Geogr. Mem. New South Wales* 437 Some of the Indians have also seriously applied to be allowed convict-labourers..till the maize and cabbage that have been planted to their hands are fit to gather. **1830** A. F. GARDINER *Friend of Australia* xiii. 244 It is the observation of some writers, that the system pursued in Australia for educating the children of the Indians, is not attended with success.

2. (Later examples.) Also, examples of *American Indian*. Cf. *AMERIND, AMERINDIAN *sbs.*

1846 J. HALL *Wilderness & War Path* 1 The life of the American Indian is not so destitute of the interest created by variety of incident, as might be supposed by a casual observation of the habits of this singular race. **1926** S. G. INMAN *Probl. in Pan Americanism* ii. 68 The visitor to South America will see nothing more pitiful than the Indians of Peru and Bolivia, descendants of great civilizations now forgotten. **1931** F. J. STIMSON *My United States* xviii. 187 Harvard College was founded in 1636 to teach the Puritans to be preachers and to teach the Indians to be Christians. **1933** *Recent Social Trends in U.S.* (President's Res. Comm. Social Trends) I. xi. 590 Doubts as to the educability of the Indian have been dispelled by the increasing number of those creditably completing college courses and by the measures of mental tests which indicate intelligence of a high rank. **1962** *Canada Month* May 29 North American Indians number more than 500,000: 180,000 in Canada, more than half of whom live in B.C., Saskatchewan, and Ontario, and 350,000 in the U.S. **1963** *Amer. Speech* XXXVIII. 271 The military-like terms might have been introduced by American Indians who went to Haskell after completing their compulsory military service. **1965** J. BALDWIN in *Penguin Dict. Mod. Quots.* (1971) 10/2 It is a great shock at the age of five or six to find that in a world of Gary Coopers you are the Indian. **1965** *Globe & Mail* (Toronto) 26 May 3/4 Those whose names are in a register kept by the federal Indian Affairs Branch are officially Indians; others are not... The Indians have special rights. **1971** D. HEFFRON *Nice Fire & Some Moonpennies* i. 10 Hey ya know, you look like a real Indian with that head-band, you know that? **1971** *Times* 23 Aug. 10/3 In every measurable way Indians are at the bottom of the great American pile. *Ibid.* 10/6 The modern plight of the American Indian is almost entirely traceable to treachery of the white man.

b. *Red Indian* (later examples).

Still used commonly outside N. Amer. but rarely in the U.S.A. and Canada. The *Dict. of Canadianisms* (1967) labels the use 'erroneous' exc. in respect of the extinct Beothuk people in Newfoundland. The usual term in N. Amer. is 'Indian'.

1891 *Trans. R. Soc. Canada* IX. II. 124 The name Red Indians..is the translation of the Micmac name for them, Maquajik, which means redmen or red people. **1897** E. R. YOUNG *On Indian Trail* 11 Romantic missionary work among the red Indians will soon be a thing of the past. **1903** G. B. SHAW *Revolutionist's Handbk.* viii, in *Man & Superman* 212 Museums are set up throughout the country to encourage little children and elderly gentlemen to make collections of corpses preserved in alcohol, and to steal birds' eggs and keep them as the red Indian used to keep scalps. **1922** JOYCE *Ulysses* 760, I hope he hasnt long greasy hair hanging into his eyes or standing up like a red Indian. **1957** *Chronicle-Herald* (Halifax, Nova Scotia) 16 Aug. 29/8 Since then I have always liked the Red Indians. **1964** *Newfoundland Q.* Summer 12/3 The Journal takes us back to pioneer days in Labrador and Newfoundland to a day when the wigwams of the Red Indians could be seen on the shores of Exploits Bay. **1973** *Times* 5 June 8/7 There he is—a voluminously white-clad figure with headdress..and a face..which might be that of a Red Indian Chief.

c. One of the 'Indians' in a child's game. Cf. *COW-BOY 3 d.

1883 W. W. NEWELL *Games & Songs Amer. Children* i. 26 The players were divided into Indians and hunters, the former uttering their war-cry... The game ended with the extermination of one party or the other. **1941** in R. D. Abrahams *Jump-Rope Rhymes* (1969) 148 One little, two little, three little Indians. Four little, five little, six little Indians, etc. **1969** I. & P. OPIE *Children's Games* v. 181 The attraction of 'Stalking' ('Gang Stalking', 'Shadowing', 'Indians') is that the hunters can become the hunted.

4. For def. read: Any one of the languages spoken by American Indians. (Add later examples.)

1894 M. EELLS *Father Eells* 91 A school was also kept in Indian, the lessons being prepared on paper, hung up on the side of the house and read and recited. **1946** T. M. STANWELL-FLETCHER *Driftwood Valley* 114 Be that as it may, I have pronounced and spelled it, just as we do all the various other localities, as it sounds to us in Indian. **1958** *Camsell Arrow* (Edmonton, Alberta) Jan.–Feb. 1 Although [the sisters] are French and the children speak Indian, English is the language of the school.

7. Short for *Indian tea* (see sense 4 a above).

1748 M. W. MONTAGU *Let.* 17 July (1966) II. 407, I have planted a great deal [of tea] in my Garden, which.. has succeeded very well. I cannot say it is as strong as the Indian, but has the advantage of being Fresher and at least unmix'd. **1907** *Yesterday's Shopping* (1969) 1 Tea ..[Blend] Congou and Indian. **1933** E. A. ROBERTSON *Ordinary Families* xiii. 278, I ordered fresh tea..and pressed the visitor to make up her mind..whether she preferred Indian or China. **1938** [see *CHINA[1] 6]. **1958** J. CANNAN *And be a Villain* i. 39 'Couldn't you find the Earl Grey?'.. Primrose said, 'Indian 's better for pulling you together.' **1968** G. BUTLER *Coffin Following* ix. 198 You will take tea? Will you have Indian or China?

Indiana (indi,æ·nă). The name of a state in the U.S.A., used esp. *attrib.* to designate objects, etc., from or connected with that state; esp. *Indiana limestone*, an oölitic limestone sometimes known as Bedford limestone.

1858 O. H. SMITH (*title*) Early Indiana trials; and sketches. **1894** *Country Gentlemen's Catal.* 113 Shirts.. in our special Indiana Gauze Oxford. **1934** WEBSTER, Indiana limestone. **1947** J. C. RICH *Materials & Methods Sculpture* viii. 219 Indiana limestone can be secured in buff to gray colors and is an attractive, fairly soft, and easily worked stone. **1948** MENCKEN *Amer. Lang.* Suppl. II. 616 An aged Indiana poetaster named John Finley. **1965** *Harper's Bazaar* Feb. 41 Mean black blazer in rubber-proofed Indiana, 7 gns. **1969** R. MAYER *Dict. Art Terms* 193/2 *Indiana limestone*, the limestone most widely used by American sculptors... Bedford Stone is another Indiana limestone.

Indianan: see *INDIANIAN.

Indian corn. Add: Also *attrib.*

1642 in *Essex Inst. Hist. Coll.* (1863) V. 219/2 Other poore people shalbe repayed..at the next Indian corne harvest. *a* **1738** W. BYRD *Hist. Dividing Line* (1929) 209 We encampt near one of these Indian Corn Fields, where was excellent Food for our Horses. **1751** J. BARTRAM *Observations Pensilvania to Ontario* 60 This repast consisted of 3 great kettles of Indian corn soop, or thin homony, with dry'd eels and other fish boiled in it. **1778** in *Pennsylvania Mag. Hist. & Biogr.* (1902) XXVI. 32 Went into winter quarters in Newport, in old empty houses.. and the food worse,—little bread and that made of rice and Indian corn meal. **1818** F. HALL *Trav. Canada & U.S.* xxviii. 324 The traveller..suddenly mounting a little rise, close to a poor cottage with its Indian corn patch..finds himself opposite to the Capitol of the Federal city. **1888** W. WHITMAN *November Boughs* 76 On the floor of the big kitchen, toward sundown, would be squatting a circle of twelve or fourteen 'pickaninnies', eating their supper of pudding (Indian corn mush) and milk. **1964** *Cookbk.* (Amer. Heritage) (1967) 163 Indian corn stew.

Indianesque (indiăne·sk), *a.* [f. INDIAN *a.* 2.] Of an Indian type.

a **1861** T. WINTHROP *John Brent* (1883) iv. 36 This was the Indianesque Saxon who greeted me. *a* **1861** —— *Canoe & Saddle* (1883) ii. 12 Indianesque, not fully Indian, was her countenance. **1882** J. W. STEELE *Frontier Army Sk.* (1883) 84 In all that is peculiarly Indianesque, she excels her master. **1896** *Month* Jan. 113 Most elaborate and beautiful examples of Indianesque architecture. **1942** *Chicago Tribune* 15 Nov. (Pict. Sect.) 6/2 (*caption*) Indianesque.

Indianian (indi,æ·niăn). Also **Indianan.** [f. *INDIANA + -IAN.] A native or inhabitant of Indiana.

1784 T. JEFFERSON *Writings* (1894) III. 401 Should.. the Indianians and Kentuckians take themselves off. **1835** C. F. HOFFMAN *Winter in West* I. 226 The term 'Hooshier', ..has now become a *soubriquet*, that bears nothing invidious with it to the ear even of an Indianian. **1863** W. WHITMAN *Specimen Days* (1882–3) 48 Some unconscious Indianian, or from Ohio or Tennessee. **1877** J. HABBERTON *Jericho Road* x. 97 Small parties of discharged Illinoisans and Indianians had frequently passed through. **1900** *Congress. Rec.* 5 Feb. 1508/2 The nation..mourns the death of this distinguished soldier, and Indianians feel it as a personal loss. **1944** *Amer. N. & Q.* Mar. 188 Its advertisements..were answered by a number of Indianans. **1947** *Amer. Speech* XXII. 250 Indianan is now used officially in the State. **1948** MENCKEN *Amer. Lang.* (Suppl.) II. 620 *Hoosier*, at the start, did not signify an Indianan particularly, but any rough fellow of what was then the wild West.

Indianism (i·ndiăniz'm). [-ISM.] Action or policy devoted to the interests of Indians; advocacy of (North American) Indians.

1651 W. FRENCH in *Strength out of Weakness* (1652) 37 All the while I went on in Indianisme I was going from God. **1871** *Rep. Indian Affairs* (1872) 181 We were in our original Indianism. **1887** W. TIREBUCK in Longfellow *Hyperion, Kavanagh and Trouvères* Introd. p. xiii, Those who stood between the culture of *Hyperion* and *Kavanagh*, and the wild Indianism of Aimard and Cooper. **1923** *Spectator* 29 Sept. 422/1 In India it works out in unrest.. an exaggerated sense of Indianism—the nationalistic idea. **1950** *Caribbean Q.* II. III. 14 It is in the Antilles, in Cuba and Santo Domingo that one finds the most abundant, and some of the best manifestations of literary Indianism. **1969** *Indo-Asian Culture* Oct. 25 It was after the coming of the Buddha that Indianism travelled fast.

Indianist. Add: **2.** *attrib.* or as *adj.* Of or pertaining to American Indians.

1950 *Caribbean Q.* II. III. 14 Perhaps the finest *Indianist* writing was produced in Brazil by Goncalves Dias and José de Alencar. **1960** *Guardian* 22 Nov. 9/6 The new generation of [Mexican] artists and architects were obsessed and inspired by..their own Indianist past.

Indianization (i:ndiănaizēi·ʃən). [f. INDIANIZE *v.* 2 + -ATION.] The process of making Indian in character or composition; *spec.* the replacement of Europeans or other foreigners by native-born Indians in positions of authority.

1918 *Pall Mall Gaz.* 29 June 5/2, I do not expect as large an Indianisation of the central Cabinet as of the provincial Cabinets. **1922** *Q. Rev.* July 142 The more rapid Indianisation of the public services. **1924** J. F. BRYANT (*title*) Gandhi and the Indianisation of the Empire. **1928** *Manch. Guardian Weekly* 29 June 513/1 The Indianisation of the Indian army. **1930** *Times Educ. Suppl.* 1 Mar. 92/4 Quicken the process of 'Indianization' by recruiting the superior services..in India. **1930** *Economist* 14 June 1322/1 The rapid progress of Indianisation in the 'transferred' fields. **1969** *Eve's Weekly* (Bombay) 20 Dec. 51/1 The Indianisation of the Roman Catholic Church has come to stay in India. **1973** *Black Panther* 5 May 6/1 (*heading*) 'Indianization' at Wounded Knee.

Indianize, *v.* Add: **2.** (Earlier and later examples.) So **I·ndianized** *ppl. a.*

1702 C. MATHER *Magnalia Christi Americana* VII. 68/2 On March 18. the French with Indians, being half one, half t'other, half Indianized French, and half Frenchified Indians..fell suddenly upon Salmon Falls. **1782** 'J. H. ST. JOHN DE CRÈVECŒUR' *Lett. from Amer. Farmer* xii. 293 Many an anxious parent..went to the Indian villages where..their children had been carried in captivity... They found them so perfectly Indianised, that many knew them no longer. **1924** E. M. FORSTER *Passage to India* xxiv. 225 The invocation of Mrs. Moore continued... The syllables..became Indianized into Esmiss Esmoor. **1935** *Times* 19 June 7/4 If they really meant to Indianize the Army, did they mean to do it because they thought the Indian Army would be more efficient when it had got rid of British officers, or because they thought it would please the Indian sepoy? **1936** *Discovery* Nov. 337/1 The capital of a great Indianised empire extending over the Malay Peninsula and the islands of Java and Sumatra from the 8th to the 12th century A.D. **1963** *Times* 2 May 15/2 The house party, so to speak, includes an English doctor, an Austrian musician who has become Indianized, and an American sociologist and his wife. **1969** *Eve's Weekly* (Bombay) 20 Dec. 51/4 Their Indianised service is generally well-attended, and very little opposition has been met with.

Indianness (i·ndiănnĕs). [f. INDIAN *a.* + -NESS.] The quality or state of being Indian, or of displaying Indian characteristics.

1934 A. HUXLEY *Beyond Mexique Bay* 192 He had had ..enough..education to make him aware of his own Indianness. **1967** *Evening Standard* 20 Sept. 10/1 The Indianness of our summer was largely, I think, the creation of the flower children. **1969** *Guardian* 5 Mar. 7/1 She exploits her Indian-ness where she has met most disapproval. **1970** *New Yorker* 18 Apr. 103 After mastering the meaning of Negritude and *machismo* they would have to grapple with the meaning of Indianness. **1971** *Illustr. Weekly India* 11 Apr. 6/2 That non-Hindu Indians abroad tend to shake off their Indianness.

Indianologist (indiănǫ·lŏdʒist). [f. INDIAN *sb.* 2.] A student of, or authority on, the American Indian.

1894 *Nation* (N.Y.) 31 May 417/3 His ears, in the view of Indianologists, were a sure mark of aboriginal origin. **1947** *Time* 3 Mar. 37/1 The R.D.C. preferred the route that followed the old telegraph line strung diagonally across the great Brazilian plateau by General Candido Nariano Rondon, a famed Indianologist.

Indian rubber: see INDIA-RUBBER.

Indian summer. Add: **a.** (Earlier and later examples.) Also *transf.* in other countries.

1778 'J. H. ST. JOHN DE CRÈVECŒUR' *Sk. 18th-Cent. Amer.* (1925) 41 It [*sc.* snow] is often preceded by a short interval of smoke and mildness, called the Indian Summer. **1878** G. MEREDITH *Let.* 9 Oct. (1970) II. 563 Perhaps you and Mrs. Carr will do us the honour to come and see the Indian summer here. **1906** W. MARRIOTT *Hints to Meteorol. Observers* (ed. 6) 66/2 *Indian Summer*, fine weather for a few days about September 30, in North America. **1939** [see *GUPTA *a.* and *sb.*]. **1958** HAYWARD & HARARI tr. *Pasternak's Dr. Zhivago* I. iv. 110 The end of a hot golden autumn had turned into an Indian summer. **1960** J. RAE *Custard Boys* I. vi. 69 Meanwhile the Indian summer continued warm and dusty on the trodden earth of the farmyard. **1962** *Sunday Times* 16 Sept. 1/3 An Indian summer in the West Country brought peak holiday traffic jams in Devon yesterday.

fig. (Later examples.)

1918 GALSWORTHY in *Five Tales* (*title*) The Indian summer of a Forsyte. **1930** V. SACKVILLE-WEST *Edwardians* iii. 100 Meanwhile she was quite content that Sebastian should become tanned in the rays of Sylvia's Indian summer. **1962** N. DEL MAR *Richard Strauss* I. ix. 418 The works of his Indian Summer when, in the last five years of his life, inspiration came to him once more.

b. (Further examples.)

1856 MRS. STOWE *Dred* II. 221 It is a calm, still, Indian summer afternoon. **1927** M. DE LA ROCHE *Jalna* xvii. 190 The drowsy Indian-summer heat still continued, but the air had become heavier. **1961** M. BEADLE *These Ruins are Inhabited* (1963) vii. 97 Homeward bound in a lackadaisical Indian-summer mood.

India-rubber, India rubber. Add: **2*.** = RUBBER *sb.*[1] 11.

1794 EARL HOWE *Let. to Capt. Sir R. Curtis* 20 Feb. (MS. in Henry E. Huntington Library) f. 1[v], If India Rubbers means pieces of the elastic Gum for rubbing out pencil lines &c., I have enough of that. **1857** M. J. HOLMES *Meadow-Brook* v, Taking my India-rubber, I erased it [*sc.* the writing]. **1910** 'I. HAY' in *Granta* 11 June, He concluded by imploring me as a personal favour to purchase an india-rubber and erase the lot. **1939–40** *Army & Navy Stores Catal.* 376/2 Artists' india-rubber... Kneaded rubber for Pastel, Chalk and Pencil.

3. *india-rubber ring.*

[**1852** MRS. GASKELL *Cranford* (1853) v. 82 Indian-rubber rings, which are a sort of deification of string.] **1939–40** *Army & Navy Stores Catal.* 812/3 Indiarubber rings. For use on hockey sticks.

b. (Further examples.)

1900 *Daily News* 23 July 3/3 (*heading*) An indiarubber enemy. **1929** E. WALLACE (*title*) The india-rubber men. **1929** F. BOWEN *Sea Slang* 72 *India rubber man*, the physical training instructor. **1949** 'M. INNES' *Journeying Boy* vii. 78 Mr. Wambus..is known as the Great Elasto, the India-rubber Man.

Indic, *a.*[1] Delete *rare* and add further examples. Also, designating the Indian branch of the Indo-Iranian languages, including the dead languages Sanskrit, Prakrit, and Pali and the living languages Hindi, Bengali, Marathi, etc.

1877 A. H. KEANE tr. *Hovelacque's Sci. of Lang.* 320 Hindui..Aryan-Indic. **1893** in *Funk's Stand. Dict.* **1909** *Indogerman. Forsch.* XXV. 177 Although the rule is not without exception even in the older Indic and Hellenic documents, it nevertheless holds true in general for the languages claimed. **1911** L. BLOOMFIELD in C. F. Hockett *Leonard Bloomfield Anthol.* (1970) 7 The Indic and the Iranian languages differ greatly in their treatment of the IE. stopped consonants. **1935** A. TOYNBEE *Study of Hist.* (ed. 2) I. 87 And so we have identified the society 'apparented' to the Hindu Society. Let us call it 'Indic'. **1937** *Discovery* Sept. 286 The American School of Indic and Iranian Studies. **1973** *Amer. N. & Q.* 83/1 Eliot began Ph.D. studies at Harvard, majoring jointly in the Indic and Philosophy Departments.

indicate, *v.* Add: **1.** *pass.* Of a course of action, treatment, etc.: to be pointed out or suggested as desirable or necessary. Also *transf.*

1880 [in Dict.]. **1886** *Brit. Med. Jrnl.* 17 Apr. 750/1 Recourse to this method is more particularly indicated when the teeth to be replaced are front teeth. **1907** *Lancet* 25 May 1462/1 Considering the great analogy which is presented between syphilis and leprosy..it seemed indicated to try on this patient the treatment by large doses of atoxyl. **1919** WODEHOUSE *Damsel in Distress* xv. 184 Strategy, rather than force, seemed to the curate to be indicated. **1922** W. R. INGE *Lay Thoughts* (1926) 235 Whenever a patient consults a doctor, the latter should always order some drug, even if drugs are not really indicated. **1946** M. PEAKE *Titus Groan* 353 My dear ladies,..I feel that some hot coffee is indicated, but what do *you* feel? **1971** *Nature* 31 Dec. 557/1 Further investigations of the directions of propagation and the presence of pacemaker function are certainly indicated. **1972** *Sci. Amer.* May 80/2 Periodic measurement of arterial and central venous pressure, blood acidity and dissolved gases may be indicated in the management of patients with heart, kidney, lung or metabolic disorders.

indication. Add: **2. b.** *Mining.* Something which indicates the presence of valuable ore, oil, etc. *U.S.*

1855 *Southern Californian* (Los Angeles) 28 Mar. 1/7 Mineral indications in Southern California..differ in many respects with the mineral indications of the earth's surface in Central, or Northern California. **1862** 'MARK TWAIN' *Lett.* (1917) I. 80 We went and looked at the ledges, and both of them acknowledged that there was nothing in them but good 'indications'. **1873** J. H. BEADLE *Undevel. West* xviii. 326 My comrades are off down the mountain side picking at 'indications', and mapping out the 'run of the country rock'. **1877** B. HARTE *Story of a Mine* 394 Luckily the fertile alluvium of these valleys..offered no 'indications' to attract the gold-seekers. **1948** *Duncan* (Oklahoma) *Daily Banner* 2 July 1/3 The location was one of three staked by the Ohio Oil Co. on indications given by the Palmer No. 1 Leard-Amerada.

indicator. Add: **1. a.** (Earlier and later examples.)

1819 (*title*) The indicator. **1907** *New Pictorial & Descr. Guide to Malvern* 56 Many of the more prominent objects and principal sites may be identified by..the help afforded by the Indicator, or Toposcope, erected on the summit of the hill. **1927** *Glasgow Herald* 4 June 9 (*headline*) Mountain Indicator. *Ibid.*, The indicator..points out over ninety places famous in Border song and story. **1967** *Listener* 10 Aug. 167/2 There's a recent book which displays, at a glance, what's distinctive about the scientific analysis of international affairs. It's the *World Handbook of Political and Social Indicators.*

d. *Philos.* = *token-reflexive word*; see quots. Also *indicator-word.*

1951 N. GOODMAN *Struct. Appearance* xi. 290 Roughly speaking, a word is an indicator if..it names something not named by some replica of the word... Among the commonest indicators are the personal indicators, the spatial indicators, and the temporal indicators. Of the personal indicators, an 'I' or 'me' normally refers to its own utterer. **1952** A. J. AYER in *Mind* LXI. 444, I shall refer to predicates instead of properties and to individual signs, to which I shall give the name of indicators, instead of individuals. **1954** *Mind* LXIII. 380 Like Professor Ayer I shall..speak of indicators and predicates rather than of individuals and properties. **1960** W. V. QUINE *Word & Object* § 21. 101 The indicator words: 'this', 'that', 'I', 'you', 'he', 'now', 'here', 'then', 'there', 'today', 'tomorrow'.

2. b. Delete '*esp.* a chemical re-agent' and add: *spec.* a substance which may be added to a solution to indicate whether the concentration of hydrogen ions or of some other ion in the solution is above or below a particular value, esp. by giving different colours for the two conditions. (Further examples.)

1874 *Jrnl. Chem. Soc.* XXVII. 191 (*heading*) Alizarin as an indicator in volumetric analysis. **1902** *Encycl. Brit.* XXVIII. 12/2 Para-nitro-phenol has colourless molecules, but an intensely yellow negative ion. In neutral, and still more in acid solutions, the dissociation of the indicator is practically nothing, and the liquid is colourless. **1930** *Jrnl. Amer. Chem. Soc.* LII. 2347 The use of various organic substances..as irreversible oxidation-reduction indicators in the titration of trivalent antimony with ceric sulfate was proposed by Rathsberg. **1938** R. E. OESPER tr. *Böttger's Newer Methods Volumetric Chem. Analysis* i. 19 The fluorescent indicators are related to the ordinary acidimetric color indicators. However, a change of the pH value of the solution produces in them no color change, but a fluorescence appears or disappears. **1939** P. J. DURRANT *Gen. & Inorg. Chem.* ix. 214 An indicator may be used for determining the *p*H value of a solution, or for deciding at what stage during an acid-alkali titration the *p*H value of a solution has reached a certain value. **1966** *Ward Lock's Compl. Gardening* v. 78 Certain chemical indicators are available for testing a soil in the field as to its approximate pH. **1970** R. U. BRUMBLAY *First Course Quant. Analysis* vi. 83 Adsorption indicators are organic compounds with rather complex molecules which undergo a change in color due to a slight structural change which occurs when they are adsorbed on the surfaces of colloidal particles.

(ii) An isotope (usually a radioactive one) used as a tracer (see *TRACER[1]).

1919 [see *ISOTOPIC *a.* a]. **1926** R. W. LAWSON tr. *Hevesy & Paneth's Man. Radioactivity* xv. 122 In problems of this kind, where the radio-element is not the object but the agent of the investigation, we say that the radio-elements serve as 'indicators'. **1938** *Ibid.* (ed. 2) xviii. 168 In those rare cases in which inactive isotopes are readily accessible they can be used as indicators in fundamentally the same manner, though other methods of measurement must be used... Such experiments have been carried out with deuterium ^2D and the oxygen and nitrogen isotopes ^{18}O, and ^{15}N. **1943** *Thorpe's Dict. Appl. Chem.* (ed. 4) VI. 432/1 The hydride of bismuth was discovered by using thorium-C as indicator and causing hydrogen to be generated in an apparatus in which this radioactive indicator had been placed. The gas which escaped was found ..to have radioactive properties. **1956** E. DE B. BARNETT *Mechanism Org. Chem. Reactions* i. 13 The use of an isotopic indicator or tracer often gives useful information as regards mechanism.

c. A geological clue to the presence of gold.

1894 R. A. F. MURRAY in A. W. Howitt *Miners' Handbk.* 5 Where the gold ceases is usually near and above the line of reef or vein whence it was derived...'Indicators' or small veins of pyrites, ironstone, and often thin bands of peculiar slate, intersected by small quartz veins, should..be carefully looked for. **1943** VON BERNEWITZ & CHELLSON *Handbk. Prospectors* (ed. 4) xvii. 200 A lead-bismuth mineral is an indicator in the Quartzburg district of Idaho. The gold is intimately associated with it, yet free.

d. *Ecology.* A group of plants or animals whose presence acts as a sign of particular environmental conditions. Also *attrib.*

1906 E. W. HILGARD *Soils* 545 Its [*sc.* alkali-heath's] perennial, deep-rooting habit of growth, and flexible, somewhat wiry rootstock, which enables it to persist even in cultivated ground, render it a valuable plant as an alkali indicator. **1920** F. E. CLEMENTS *Plant Indicators* iv. 112 These [areas of disturbance] furnish an enormous amount of indicator material. **1949** W. C. ALLEE et al. *Princ. Animal Ecol.* IV. xxix. 567/2 Typical or characteristic organisms..are biotic indicators. **1964** J. M. MACLENNAN tr. *Viktorov's Short Guide Geo-Bot. Surveying* vi. 122 The composition, structure, and other features of the plant cover may often serve as indicators of various environmental conditions. **1970** *Nature* 25 July 381/1 They [*sc.* chaetognaths] have also frequently been used as 'indicator species' for water masses.

e. A board or device in a railway station used to indicate the times and platform numbers of arriving and departing trains.

1913 LD. MONKSWELL *Railways Gt. Brit.* iii. 163 A large indicator showing the platform at which each train will depart, is displayed high up on the side of one of the buildings flanking the concourse. **1914** H. M. HALLSWORTH *Elem. Railway Operating* iii. 46 Since there are trains for different destinations frequently departing from the same platform at or about the same time passengers are liable to board the wrong train. This is a danger which must be provided for by suitable indicators. **1961** *Trains Illustr.* Jan. 22/2 (*caption*) The new train indicator board for the Cambridge main line departure platforms. **1972** *Travelling* Winter 43/3 New automatic departure and arrival indicators will be installed. **1973** *Railway Mag.* Mar. 151/2 A new form of train departure indicator is replacing the printed sheets at many stations on BR.

f. In *Cryptography* (see quot.).

1961 SHULMAN & WEINTRAUB *Gloss. Cryptogr.*, *Indicator*, a means of showing a change of key or encipherment, usually with a letter of the alphabet agreed upon in advance.

3. g. A device fitted to a motor vehicle to indicate an intended change in direction. Cf. **direction indicator*.

Now usu. consisting of two pairs of lights, one pair placed at the front and the other at the rear of a vehicle, that flash on one side or the other when operated by the driver.

1932 [see *FLASHING *ppl. a.* 1 c]. **1958** *Observer* 17 Aug. 15/6 Triggers under the steering-wheel work the self-cancelling indicators and the headlamp flasher. **1962** *Which? Car Suppl.* Oct. 133/1 All the indicators' flasher times fell within the legal limits. *Ibid.* 133/2 The rear indicators could be seen from the side. **1973** R. LEWIS *Blood Money* vi. 69 A police car..turned into the gateway, indicator flashing.

4*. *Math.* [tr. F. *indicateur* (E. Prouhet 1846, in *Nouv. Ann. de Math.* V. 176).] = TOTIENT.

1919 *Amer. Math. Monthly* XXVI. 290 E. Prouhet.. defined the term indicator of *n* as the number $\phi(n)$ of the positive integers less than *n* and prime to *n*. **1939** H. N. WRIGHT *First Course in Theory of Numbers* i. 11 Other names for $\phi(m)$ are the totient of *m* and the indicator of *m*. **1948** O. ORE *Number Theory* v. 110 We shall consider the problem of finding how many of the numbers 1, 2, 3, ..., $m-1$, *m* are relatively prime to *m*. This number is usually denoted by $\phi(m)$, and it is known as Euler's ϕ-function of *m* because Euler around 1760 for the first time proposed the question and gave its solution. Other names, for instance, indicator or totient have occasionally been used.

5. indicator lamp, light, a luminous signal indicating operating conditions (see quots. and sense 3 g above); **indicator switch,** a switch for an indicator light.

1961 *Which?* Dec. 313/2 For most cookers, we found that the indicator light going out gave a satisfactory indication that the oven was nearly at its steady temperature... The *Falco Royalty* had no indicator lamp. **1962** *Ibid. Car Suppl.* Oct. 133/2 Both front and rear indicator lights were visible from the side in the VW 1500. **1959** *Motor Manual* (ed. 36) vi. 183 When the indicator switch is moved, either for a right- or left-hand turn, current is fed to the lamps [etc.].

in dictione (in diktĭǭu·ni), *phr. Logic.* [L. tr. of Gr. παρὰ τὴν λέξιν (Aristotle, *Sophistical Refutations*, ch. 4).] Of fallacies: resulting from the linguistic expression used; due to ambiguity, division, etc. (Opp. *EXTRA DICTIONEM *phr.*)

1826 R. WHATELY *Elem. Logic* iii. § 1. 135 The division of Fallacies into those in the words *in dictione*, and those in the matter *extra dictionem*, has not been, by any writers hitherto, grounded on any distinct principle. **1847** A. DE MORGAN *Formal Logic* xiii. 241 The Aristotelian system of fallacies contains two subdivisions. In the first, which are *in dictione*, or *in voce*, the mistake is said to consist in the use of words: in the second, which are *extra dictionem*, or *in re*, it is said to be in the matter. **1852** [see *EXTRA DICTIONEM *phr.*]. **1870** J. MCCOSH *Laws Discursive Thought* III. § 82. 172 Fallacies from the days of Aristotle have been logically divided into those *In Dictione* and those *Extra Dictionem*, or, to use a better mode of expression, into those in Form and those in Matter. **1916** H. W. B. JOSEPH *Introd. Logic* (ed. 2) xxvii. 578 The fallacies *in dictione* are so many different forms of error that may arise through the double meanings of language. **1970** [see *EXTRA DICTIONEM *phr.*].

indie (i·ndi), colloq. abbrev. of INDEPENDENT *a.* and *sb.* (used esp. in the film industry in the United States).

1942 BERREY & VAN DEN BARK *Amer. Thes. Slang* § 217/5 *Indie*, independent. *Ibid.* § 587/4 *Indie*, an independently owned theater. *Ibid.* § 605/23 *Indie*, an independent exhibitor. **1961** A. BERKMAN *Singers' Gloss. Show Business* 29 *Indie*, independent movie producer. **1962** *Amer. Speech* XXXVII. 149 On names for 'indies'—independent motion picture companies. **1970** *New Yorker* 15 Aug. 65/1 An indie producer met a dope in a bar. An indie producer is an independent producer.

indifference. Add: **3. b.** *Psychol.* **indifference point** [tr. G. *indifferenzpunkt*], a position or value between two continua of experience, such as a temperature that is experienced as neither warm nor cold, or a feeling-value that is neither pleasant nor unpleasant.

1887 A. SETH *Hegelianism* ii. 57 Schelling proceeds to define the Absolute as the indifference-point of subject and object. **1890** W. JAMES *Princ. Psychol.* I. xv. 616 An 'Indifference-point'; that is to say..a time which we tend to estimate as neither longer or shorter than it really is, and away from which, in both directions, errors increase their size. **1901** BALDWIN *Dict. Philos. & Psychol.* I. 533 *Indifference point*,..the theoretical point at which neither of two contrasted sense or other qualities, which are supposed to depend on the same sort of stimulation, is experienced. **1938** R. S. WOODWORTH *Exper. Psychol.* xviii. 445 The 'indifference point', a phenomenon observed for many years in judgments of magnitude.

8. indifference curve (occas. **line of indifference**) *Econ.*, a graph, the co-ordinates of which are the quantities of alternative goods and services that would leave the consumer indifferent in choosing between them because he judges them of equal value. Also **indifference map** (see quot. 1972).

1881 F. Y. EDGEWORTH *Math. Psychics* I. 21 It is evident that X will step only on one side of a certain line, *the line of indifference*, as it might be called. [illegible]

indigency. Restrict †*Obs.* to senses 1 and 3, and add later examples of sense 2.

1906 *Daily Chron.* 25 Sept. 6/7 The Government has set up an Indigency Inquiry Commission. 1924 *Glasgow Herald* 7 Feb. 7 Unemployment and indigency..existed on a scale that was entirely disproportionate to the size of the white population.

indigenization (indi·dʒĭnaizē·ʃən). [f. INDIGENOUS *a.* + -IZATION.] The act or process of rendering indigenous or making predominantly native; adaptation or subjection to the influence or dominance of the indigenous inhabitants of a country; *spec.* the increased use of indigenous people in government, employment, etc. Also **indi·genist**, an advocate or supporter of indigenization; also *attrib.* or as *adj.*; **indi·genize** *v. trans.*; **indi·genized** *ppl. a.*

[illegible]

indigent, *a.* (*sb.*) **B.** *sb.* Delete †*Obs.* and add later examples.

1903 *Westm. Gaz.* 15 Jan. 6/2 Mr. Chamberlain..gratefully accepted the offer to provide accommodation for the indigents. 1905 *Daily Chron.* 19 Sept. 5/1 The farmers submitted a lengthy list of subjects for redress, including ..settlements for indigents, &c. 1922 W. S. MAUGHAM *On Chinese Screen* viii. 37 He was a man whose purse was always open to the indigent. 1972 *N.Y. Law Jrnl.* 31 Oct. 1/5 The Legal Defense Panel administers the program of assistance to indigents who are criminal defendants.

indigest, *v. trans.* Add: (Later example.)

1938 S. BECKETT *Murphy* 81 On this part of himself that I am about to indigest may the Lord have mercy.

b. *intr.* Also, to fail to digest.

1857 MRS. GASKELL *Let.* Dec. (1966) 489, I don't see exactly what you do in America. You indigest, all of you, and some of you make money at a great rate. 1954 W. FAULKNER *Fable* 337 'Then we will starve,' the first said. 'Or indigest,' the third said.

indigo. Add: **C. 1.** and **2.** *indigo-planter* (earlier example); **indigo weed** (earlier examples).

1772 J. HABERSHAM *Let.* 12 Aug. in *Coll. Georgia Hist. Soc.* (1904) VI. 202 We have had a great Quantity of Rain fall, which must hurt the Indigoe Planters. 1785 *Mem. Amer. Acad. Arts & Sci.* I. 473 *Indigofera*... Indigoweed. .. A durable pale blue may be obtained from the leaves and small branches. 1852 *Trans. Mich. Agric. Soc.* III. 197 My timber is generally oak, with some hickory, indigo weed, tea weed.

indigoid (i·ndigoid), *a.* (*sb.*) [a. G. *indigoid* (P. Friedländer 1908, in *Ber. d. Deut. Chem. Ges.* XLI. 773): see INDIGO *sb.* (*a.*) and -OID.] Of a dye: similar to indigotin in chemical structure, *spec.* in having another atom or atoms, esp. of sulphur, in place of one or both of the imino-groups. Hence as *sb.*, an indigoid dye.

1908 *Jrnl. Chem. Soc.* XCIV. I. 371 (*heading*) Indigoid dyes. *Ibid.*, The author applies the term 'indigoid' to dyes which are related to indigotin in that the imino-groups of the latter are substituted by a sulphur or other bivalent atom or group. 1939 J. B. CONANT *Chem. Org. Compounds* (rev. ed.) xxx. 560 Indigo, its halogenated derivatives, and certain related compounds of analogous structure are sometimes referred to as the indigoids. 1946 *Biochem. Jrnl.* XL. 669/1 This communication describes the isolation and identification of the indigoid pigments indirubin and indigotin from the acid-treated urine of a case of sprue. 1952 K. VENKATARAMAN *Chem. Synthetic Dyes* II. xxxiii. 1004 Apart from indigo itself, the thioindigoids constitute a much more important series than the indigoids. 1963 A. J. HALL *Textile Sci.* iv. 178 All these new synthetic dyes related in chemical composition to indigo are classed as indigoid dyes. 1966 KIRK & OTHMER *Encycl. Chem. Technol.* (ed. 2) XI. 562 The nomenclature of the indigoids is based on the indole.., or thionaphthene, part of the molecule.

Indio (i·ndio). [Sp. and Pg.] A member of one of various indigenous peoples of America and E. Asia in those areas formerly subject to Spain or Portugal; *spec.* (*a*) in Brazil and Mexico, an Indian, distinguished as an *Indio bravo*, if he had retained his independence, and *Indio manso* or *Indio fidele*, if he had come under European domination; (*b*) = *INDIAN *sb.* I c. Also **Indiano.**

1836 *Penny Cycl.* V. 365/1 All the aborigines, who lead an independent and roving life, are called in Brazil Indianos bravos, or Gentios, in contradistinction to the Indianos mansos (domesticated Indians), who have settled among, or in the neighbourhood of the Europeans. 1839 *Ibid.* XV. 158/1 The Indios Bravos generally live on the [illegible]

indirect, *a.* Add: **1. d.** *indirect lighting* (see quot. 1925).

[illegible]

2. a. *spec.* *indirect aggression*, aggression by one nation by other than military means; so *indirect aggressor*; *indirect evidence* = *circumstantial evidence* (CIRCUMSTANTIAL *a.* 1); *indirect rule*, a system of government in which the governed people retain certain administrative and legal, etc., powers.

1824 T. STARKIE *Pract. Treat. Law of Evidence* I. III. 475 These positions lead immediately to an inquiry into the nature and force of indirect or circumstantial evidence. 1865 Indirect evidence [in Dict.]. 1922 F. D. LUGARD *Dual Mandate Brit. Trop. Afr.* x. 199 The Governor of the Gold Coast..observed: 'The chiefs are keenly appreciative of our policy of indirect rule, and of the full powers they retain under their native institutions.' 1928 —— (*title*) Representative forms of government and 'indirect rule' in British Africa. *Ibid.* 19, I propose in this chapter to discuss..'Indirect Rule'—though 'Dependent Rule' would seem a more suitable term,—more especially in..the conditions of tropical Africa. 1931 *Economist* 28 Mar. 667/2 He [*sc.* Sir Donald Cameron]..submitted that this dream would be shattered if the policy, inaugurated in Tanganyika in 1925, of developing that territory on the lines of indirect rule by the mandatory and direct economic production by the natives were allowed to develop. 1939 H. NICOLSON *Diary* 20 July (1966) 406 The Ambassador is.. so interested in convincing them how right is the Soviet definition of 'indirect aggression' that he forgets to offer them any tea. 1940 B. WARD *Russian Foreign Policy* 28 Mutual guarantees, the definition of indirect aggression—all these were trivial points compared with the principal obstacle, the Polish guarantee. 1957 P. WORSLEY *Trumpet shall Sound* 261 This became particularly urgent, when..the growing inadequacy of direct methods of rule ..brought about the introduction of Indirect Rule in many territories. 1958 *Hansard Commons* 16 July 1245 The question is one of perverting nationalist feelings and perverting those who wish to overthrow the established order of society so that they serve to further indirect aggression. 1958 *Listener* 7 Aug. 185/1 Much has been said about President Nasser's propaganda machine, which has given rise to a new term, 'indirect aggression', which in Western eyes, it seems, may now be held as justification for military intervention. 1958 *Spectator* 8 Aug. 184/3 Nor was this member of the Baghdad Pact the only indirect aggressor against France. 1959 JOWITT *Dict. Eng. Law* II. 960/1 *Indirect evidence*, proof of collateral circumstances from which a fact in controversy, not directly attested by witnesses or documents, may be inferred. 1959 *Spectator* 21 Aug. 236/2 The invention of 'indirect rule', the system of governing a territory by allowing the existing tribal authorities to continue to administer tribal law under the restraint of a British Resident. 1962 *Listener* 18 Oct. 593/2 The legacy of Britain's policy of indirect rule in this Region of Nigeria is clearly visible in the comparatively static nature of these societies.

d. *Metallurgy.* Designating a process by which wrought iron or steel is obtained from the ore through the intermediate stage of pig iron (the usual method).

1869 H. S. OSBORN *Metall. Iron & Steel* II. ii. 274 In ancient times iron was extracted from the ore as malleable iron. This is called the direct, in contra-distinction to the present method of producing cast iron and afterward malleable, which latter is called the indirect method. 1967 W. H. DENNIS *Found. Iron & Steel Metall.* i. 13 Gradually a technique was developed of removing the derived impurities... This process was designated fining..and resulted in wrought iron... The method involved a two-stage process: (a) Reduction of iron ore to make pig iron, and (b) remelting and purifying the pig iron to make wrought iron and hence was an indirect process in contrast to the former direct process of producing iron blooms from the ore in one stage.

4. b. *Biol.* Of nuclear or cell division: mitotic.

[1879 W. FLEMMING in *Arch. f. path. Anat. u. Physiol.* LXXVII. 3 Bei anderen..Fällen,..lassen sich Bilder, die anscheinend einer directen Kerntheilung entsprachen, einer indirecten (s.u.) zudeuten.] 1880 *Jrnl. R. Microsc. Soc.* III. 51 In discussing the changes undergone by the nucleus in cell-division, Professor W. Flemming distinguishes two methods of division which have been described by various observers, the direct and the indirect. 1888 [see *AMITOTIC *a.*]. 1909 J. R. GREEN *Hist. Bot.* II. i. 181 The terms indirect, and direct, nuclear division were introduced by Flemming in 1879, and were long in favour. 1925 E. B. WILSON *Cell* (ed. 3) ii. 116 Mitosis (indirect division).

5. Of or pertaining to the work and expenses which cannot be apportioned to any particular job or undertaking, pertaining to overhead charges and subsidiary work. (Cf. *DIRECT *a.* 6 f.)

1903, 1922 [see *DIRECT *a.* 6 f]. 1925 R. J. H. RYALL *Primer of Costing* 49 Labour may be employed in..repairing machinery..or in supervising the direct workers. ..Such labour is classified as Indirect Labour. 1966 *New Statesman* 19 Aug. 256/1 The Americans..are more inclined to lay down precise standards for the number of indirect workers who ought to be employed in a particular location. 1974 *Times* 4 Feb. 15/4 Its first offer..covers.. some 4,300 'indirect' workers—men who service the production areas—in the body plant.

6. *indirect fire*, gunfire aimed at a target which cannot be seen (see quot. 1918).

1879 *Man. Siege & Garrison Artill. Exerc.* I. 23 Indirect or curved fire from guns or howitzers. *Ibid.* 24 Breaching by indirect fire would, as a rule, be by demolition and not by the formation of regular cuts. 1918 E. S. FARROW *Dict. Mil. Terms* 309 *Indirect Fire*, when the target cannot be seen, and guns are aimed by means of calculations, from map, or by bearings. *Indirect Laying Fire*, when the gun is laid for direction on an aiming point or on aiming points and elevation adjusted by sight clinometer. 1962 *Ordnance Technical Terminol.* (U.S. Army Ordnance School) (AD 660 112) 162/2 *Indirect fire*, gunfire delivered at a target that cannot be seen from the gun position or firing ship.

indiscernibi·lity. [f. INDISCERNIBLE *a.*: see -ITY.] The quality or condition of being indiscernible.

1878 S. H. HODGSON *Philos. of Reflection* II. 140 Indiscernibility in point of content is therefore the final test of truth in concrete reasoning. 1892 W. WALLACE tr. *Hegel's Logic* (ed. 2) 417 The principle of individuation or indiscernibility is: 'If two individuals were perfectly alike

[etc.]'. 1936 *Mind* XLV. 245 This latter kind of immediacy is unmediatedness, or indiscernibility of any mediation. 1953 W. V. QUINE *From Logical Point of View* viii. 139 One of the fundamental principles governing identity is that of *substitutivity*—or, as it might well be called, that of *indiscernibility of identicals*. 1955 A. N. PRIOR *Formal Logic* 264 Even if we do not admit..the 'identity of indiscernibles', we must admit the indiscernibility of identicals. 1962 W. & M. KNEALE *Devel. of Logic* x. 604, $x=y \supset (fx \supset fy)$ may perhaps be called the principle of the indiscernibility of identicals. 1973 A. QUINTON *Nature of Things* vi. 153 The indiscernibility of dreams..does not cast doubt on necessary truths.

indiscretion. Add: **2.** (Further examples.) Also, an accidental or ('calculated indiscretion') a supposedly accidental revelation of an official secret, etc.

1929 T. S. ELIOT *Dante* 63 The *Vita Nuova* is neither a 'confession' nor an 'indiscretion' in the modern sense. 1930 [see *CALCULATED *ppl. a.* 1]. 1931 *Economist* 5 Dec. 1066/1 Socialist 'indiscretions'..took the form of the publication of certain alleged confidential information as to advances to French banks and foreign Governments from the French Treasury. 1955 *Bull. Atomic Sci.* Mar. 84/2 We have useful men denied the opportunity to contribute to our scientific efforts because of their youthful indiscretions. 1961 *Spectator* 26 May 742/1 He is psychologically indiscretion-prone.

indiscutable (indiskiū·tăb'l), *a.* [f. IN-³: see DISCUTABLE *a.* Cf. F. *indiscutable* indisputable.] = INDISCUSSIBLE *a.*

1933 *Proc. Brit. Acad.* XVIII. 171 In every field, a writer may without injustice give plain warning to the world that he regards certain questions as ultimate and indiscutable. 1959 *Encounter* XII. IV. 53 The book is *indiscutable*.

indispensable, *a.* (*sb.*) Add: **B.** *sb.* (Later examples.)

1901 *Westm. Gaz.* 12 Dec. 3/1 An indispensable to the complete success of the lace blouse is a chiffon lining. 1965 *Austral. Women's Weekly* 20 Jan. 25 Inevitably in a life of constant moving one picks up indispensables.

c. (Earlier example.)

a 1828 J. BERNARD *Retrospections Stage* (1830) II. iv. 116 Black silk indispensables, and stockings.

indisseverable, *a.* Delete †*Obs.* and add later example.

1950 A. L. ROWSE *England of Elizabeth* p. viii, We next have to tackle the Church, not as a system of belief, but as a social institution—indeed as the whole of society regarded from one aspect, inextricably entwined with secular life at every level, indisseverable from it.

indisseverably, *adv.* Delete † and add later examples.

1935 W. DE LA MARE *Early One Morning* 5 The fresh and virgin waters are so rapidly and indisseverably involved with the rest. 1952 *Scottish Jrnl. Theol.* V. 307 For [Bouyer] as for Preiss eschatology and Christology are woven indisseverably together as the background of ethics.

in distans: see *IN *Lat. prep.*

indistinguishable, *a.* Add: **1.** Also as *sb.*

1903 *Daily Chron.* 24 Nov. 4/5 All this contention and uncertainty might be avoided if we abolished..the arti-

very little to do with analysis, but..lives a sort of parasitic existence at its expense..; we cannot assent to any interference with its correct application as meaning the opposite of Group Psychology. 1951 E. E. EVANS-PRITCHARD *Social Anthropol.* iii. 45 There are various and particular objections to each of these successive attempts to explain social facts by individual psychology. 1959 L. RADL in Adler & Deutsch *Ess. Individual Psychol.* 162 At this point the close similarity between the Existentialist doctrine and Individual Psychology once again becomes strikingly apparent.

c. individual variable *Logic*, a variable that ranges over individuals. Cf. sense B. *sb.* 2 b in Dict.

1937 A. SMEATON tr. *Carnap's Logical Syntax of Lang.* 195 A *v is called an individual variable. 1952 P. F. STRAWSON *Introd. Logical Theory* v. 130 To the variable '*x*' and other variables of the same type..we give the name 'individual variables'. 1954 I. M. COPI *Symbolic Logic* iv. 67 The small letter '*x*'—called an 'individual variable'—is a mere *place marker* which serves to indicate where an individual constant may be written for a singular proposition to result. 1965 HUGHES & LONDEY *Elem. Formal Logic* xxiii. 169 We shall call such variables individual-variables (meaning thereby, of course, not that the variables are themselves individuals—whatever that might mean—but that they stand indifferently for the names of individual things).

individualism. Add: **1.** (Earlier example.)

1827 L. T. REDE *Road to Stage* 59, I beg to disclaim, in these observations, any individualism; several talented persons may be found connected with such establishments.

2. (Earlier example.)

1851 MILL in *Westm. Rev.* LVI. 87 Socialism as long as it attacks the existing individualism, is easily triumphant.

6. *Bot.* [ad. G. *individualismus* (K. von Tubeuf *Pflanzenkrankheiten* (1895) I. ii. 102).] A type of symbiosis in which the product of the relationship differs from either of the component organisms. Now *rare*.

1897 A. SCHNEIDER in *Minnesota Bot. Stud.* I. 944 The best known and perhaps the most typical form of complete individualism is represented by the higher lichens. 1913 *Mycologia* V. 102 It is supposed that the relationship is becoming closer and closer, and that finally it will be so intimate that neither symbiont will be able to live independently. Then will the individualism be perfect. 1967 P. GRAY *Dict. Biol. Sci.* 268/1 *Individualism*, a type of symbiosis, in which the aggregate differs from any of its components. A lichen is a case in point.

individualistically (i:ndividiu,ăli·stikăli), *adv.* [f. INDIVIDUALISTIC *a.*: see -ICALLY.] In an individualistic manner; from the individualistic standpoint.

1894 *Internat. Jrnl. Ethics* Oct. 42 The trumpery decorations of the present-day individualistically arrayed establishment. 1922 A. G. HOGG *Redemption from this World* 245 In India the problem has been conceived individualistically, while by the Hebrews it was..conceived socially. 1925 *Contemp. Rev.* Aug. 234 They find themselves at variance of purpose with other [*sic*] less individualistically inclined. 1938 R. G. COLLINGWOOD *Princ. Art* xiv. 324 This activity..is performed not only by the man whom we individualistically call the artist, but partly by all the other artists..'influencing' him.

individuated, *ppl. a.* Add: **1. b.** Denoting a person who has been through the process of

i. 8, I sat before the examiner and the two experts for my oral examination on The Individuation Process, which is the essence of analysis.

5. c. *Bot.* [tr. G. *individualismus* (K. von Tubeuf *Pflanzenkrankheiten* (1895) I. ii. 102).] = *INDIVIDUALISM 6.

1897 W. G. SMITH tr. *Tubeuf's Dis. Plants* viii. 87 This unification of two living beings into an individual whole, I have designated 'Individuation'.

Indo-¹. Add: *Indo-Arabian, -Arabic, -Austral*; **Indo-Abyssinian** *a.*, of or pertaining to both the Dravidians of India and the Hamites of north-east Africa; also as *sb.*, one of these peoples; **Indo-African** *a.*, of or pertaining to India and Africa, *spec.* applied to a supposed former continent now covered by the Indian Ocean; relating to Indians and Africans in South Africa; **Indo-Anglian** *a.*, of or pertaining to literature in English written by Indian authors; also as *sb.*, a writer of such literature; **Indo-British** *a.* (example); **Indo-Chinese** *a.* (later example); **Indo-Hittite** (see quot. 1930); **Indo-Malayan** *a.*, pertaining to India and Malaya; *spec.*, denoting an ethnological region comprising Sri Lanka, the Malay peninsula, and Malayan islands; **Indo-Oceanic** *a.*, pertaining to the East Indian islands and the islands of the Pacific Ocean; **Indo-Pacific** *a.*, relating to the Indian Ocean and the adjacent parts of the Pacific Ocean; also, relating to the group of languages, usually called Malayo-Polynesian, spoken in the islands of these waters; **Indo-Pakistan, -Pakistani** *adjs.*, pertaining to India and Pakistan or to their inhabitants; also **Indo-Pak** colloq. abbrev.; **Indo-Saracenic** *a.*, pertaining to the products of mixed Indian and Saracenic origin; **Indo-Scythian, -Scythic** *adjs.*, pertaining to India and Scythia; also as *sb.*; also **Indo-Scyth**, a person from these regions; **Indo-Teutonic** (earlier and later examples: still *rare*).

1896 A. H. KEANE *Ethnol.* viii. 170 Again, what is to be made of the expression 'Indo-Abyssinian', or even 'Abyssinian' at all as an ethnical term. *Ibid.* x. 229 Considerable sections of the Indo-African Continent..must have persisted far into the tertiary epoch. *Ibid.* xii. 295 It is admitted by all ethnologists that Asia is the original home of the Mongolic division, a fact which harmonises with the view that the vanished Indo-African Continent was the cradle of mankind. 1971 *Illustr. Weekly India* 18 Apr. 24/3 Dr Naicker and Dr Dadoo forged an Indo-African political alliance. 1883 in K. R. Srinivasa Iyengar *Indian Writing in English* (1962) i. 3 Indo-Anglian. 1935 A. R. CHIDA (*title*) Anthology of Indo-Anglian Verse. 1943 K. R. SRINIVASA IYENGAR (*title*) Indo-Anglian literature. 1962 *Times Lit. Suppl.* 10 Aug. 596/3 Authors such as R. K. Narayan, Dom Moraes, Balachandra Rajan (now called 'Indo-Anglians') find their public in the West, rather than inside India itself. 1969 *Sunday Standard* (Bombay) 3 Aug. (Mag. Sect.) p. vii/7 Anita Desai is one of the most competent amongst the small band of Indo-

INDUCT

nature of the inducing substance. **1966** E. R. M. Kay *Biochem.* xxvii. 354 One form [of repressor] is active as a repressor, but combines with an inducer if present in the form of an inducing substrate, thereby becoming an inactive form so that the structural gene may form 'messenger' RNA normally. **1968** M. W. Strickberger *Genetics* xix. 401 The 'inducing' trigger that produces lysis in lysogenic strains..involves a change in the activity of the phage from a quiescent prophage state to a proliferative vegetative state.

induct, *v.* Add: **3. b.** *U.S.* To bring into military service.

1934 Webster, *Induct*, to enroll for military service in compliance with a draft law, as the selective service act of 1917. **1940** *Congress. Rec.* 6 Sept. 11675/2 Men..who are voluntarily inducted pursuant to this act. *Ibid.* 11676/1 The word 'inducted' I maintain means any of them [service men] because they are taken in... They are inducted either after they volunteer or after they are conscripted. **1967** *Boston Sunday Herald* 26 Mar. II. 7/7 Muhammad (Clay) was to be inducted—supposed to be inducted—into the Army April 11.

inductance. Substitute for def.: That property of a circuit or device by virtue of which any variation in the current flowing through it induces an e.m.f. in the circuit itself (self-inductance) or in another conductor (mutual inductance): without qualification usu. the former. Also, the magnitude of this, as measured by the ratio of an induced e.m.f. to the rate of change of the inducing current. (Earlier and later examples.)

1886 O. Heaviside in *Electrician* 12 Feb. 271 Conductivity and conductance are mathematically related in the same manner (except as regards 4π) as inductivity and what it is naturally suggested to call Inductance. The Inductance of a circuit is what is now called its coefficient of self-induction... When the mutual coefficient of induction of two circuits is to be referred to, it will of course be the mutual inductance. **1889** J. A. Fleming *Alternate Current Transformer* I. ii. 42 The..inductance of a circuit is, speaking generally, a quality of it in virtue of which a finite and steady electromotive force applied to it cannot at once generate in it the full current.., and when the electromotive force is withdrawn time is required for the current strength to fall to zero. **1928** Sterling & Kruse *Radio Manual* i. 20 The inductance of a circuit conductor, coil or of any apparatus is a property of that thing just as resistance is one of its properties. **1943** C. L. Boltz *Basic Radio* v. 88 In most radio work the only inductance anywhere worth considering is certainly that in coils. Nevertheless we must not forget that any conductor whatever has some inductance. *Ibid.* 90 Two coils have inductances of 2H and 4H. If they are coupled together so that only 50 per cent. of the flux is linked with all the turns, what is the value of the mutual inductance? **1957** *Encycl. Brit.* VIII. 293/2 When direct current is used, the inductance has no effect while the current is steady, but it delays the establishment of current when the circuit is first completed. **1962** Corson & Lorrain *Introd. Electromagn. Fields* vi. 232 In rationalized m.k.s. units, inductance is measured in webers/ampere, or in henrys. *Ibid.* 233 We shall now..calculate the self-inductance of a long solenoid ..and the mutual inductance between two coaxial solenoids. **1964** Goodier & Meynell *Electr.* iv. 49 A circuit will have an inductance of 1 henry if a current in it changing at the rate of 1 ampere per second induces an e.m.f. of 1 volt.

b. = *inductor 3 d.

1908 J. A. Fleming *Elem. Man. Radiotelegr.* ii. 66 One form which the inductance may take is that of a loop of one or a few turns of insulated wire. **1928** Sterling & Kruse *Radio Manual* i. 20 The most commonly employed inductance at radio frequencies consists of a single layer coil wound as an air core solenoid. **1962** D. F. Shaw *Introd. Electronics* i. 9 The coupling between inductances may reach a value close to unity if they share a ferromagnetic core.

2. Special Comb.: **inductance coil**, an inductor (sense *3 d) in the form of a loop or coil.

1902 *Encycl. Brit.* XXXIII. 232/1 The Slaby-Arco arrangement consists at the transmitting end of an inductance coil elevated above the ground; one end of this coil is connected to the earth..and the other end to a condenser, the opposite terminal of which is connected to one secondary terminal of an induction coil. **1923** E. W. Marchant *Radio Telegr.* iv. 38 The coherer may, in a simple circuit, be conveniently placed across the inductance coil.

inductee (indʌktiː). *U.S.* [-ee¹.] A person inducted into military service. Also *transf.*

1941 *Ann. Reg. 1940* 278 The Act provides that no 'inductee' may be required to serve outside the Western Hemisphere. **1956** W. H. Whyte *Organization Man* (1957) v. 58 Universal organization training..so effectively emphasises the group spirit that there is little danger that inductees will be subverted into rebelliousness. **1958** S. Ellin *Eighth Circle* (1959) II. xii. 137 You sound like an army doctor asking an inductee about his sex life.

induction. **2.** Delete *rare* and add *attrib.* examples.

1962 *B.B.C. Handbk.* 162 Induction courses were also continued during the year for all senior members of staff joining the Corporation to acquaint them with its purpose, organization, and basic procedures. **1962** E. Godfrey *Retail Selling & Organization* xi. 121 It may be very useful to provide a special week or fortnight of induction training for them [sc. juniors]. **1965** *New Statesman* 30 Apr. 678/3 The organisation of refresher or induction courses. **1966** *Ibid.* 14 Jan. 49/1 Many firms run so-called 'induction' classes for new entrants, teaching them something about the company and its welfare provisions, perhaps giving them a brief run-down on the reasons for deductions from their wage packets, and even occasionally dealing with safety and hygiene.

4. d. *U.S.* Introduction into military service (cf. *induct *v.* 3 b). Also *attrib.*

1934 in Webster. **1940** *Congress. Rec.* 6 Sept. 11676/1 Any person..shall be afforded an opportunity to volunteer for induction. **1951** *N.Y. Herald-Tribune* 26 Dec. 11/3 You label this procedure of impressing R.O.K. Army members into your army as voluntary induction... It is nothing more than forced induction, impressment. **1967** *Boston Sunday Herald* 26 Mar. II. 7/7 Clay has been ordered to appear for induction in the Army on April 11. **1973** C. Himes *Black on Black* 209 Here is your induction papers... I hope the army likes you bettern I does. **1973** *Times Lit. Suppl.* 19 Oct. 1269/1 One summer the dreaded Induction Notice comes and he goes to war.

9. *spec.* Induction of labour.

1840 *Lancet* 7 Nov. 225/1 (*heading*) Induction of premature labour. **1916** G. P. Shears *Obstetrics* xxvi. 567 By the induction of abortion is meant the artificial interruption of pregnancy during the first twenty-eight weeks; i.e., before the fœtus becomes viable. The artificial interruption of pregnancy at any subsequent period is known as the induction of labor. **1962** *Lancet* 6 Jan. 6/2 The failure-rate was decidedly high when induction of labour was started after the calculated date for confinement had passed. **1968** D. C. Bethea *Introd. Maternity Nursing* ix. 120 Inductions may also be done for the convenience of the mother and/or the doctor.

b. *Embryol.* The determination of the development or differentiation of an embryonic region into a particular morphogenetic pattern by the influence or activity of another embryonic region; an instance of this.

1928 *Biol. Abstr.* II. 686/2 In the small yolk-plug stage both the median and paramedian parts of the posterior ⅗ of the gut roof are capable of induction. **1935** *Discovery* May 136/2 If..an organisation centre is grafted out of its usual place in an egg into new surroundings it will cause those new surroundings to develop into a complete embryo or complete organ. This 'induction' of a new embryo involves both sorts of embryological change; the production of new sorts of tissues..and the arrangement of those tissues. **1950** L. G. Barth tr. *Brachet's Chem. Embryol.* x. 397 The middle layer..gave good inductions in 16 per cent of the cases. **1958** B. M. Patten *Found. Embryol.* vi. 134 Experimental studies..have yielded extraordinarily interesting information as to the way one part of a developing embryo may influence the differentiation of other parts. When this occurs it is spoken of as induction. **1960** B. I. Balinsky *Introd. Embryol.* vi. 169 The result may be expressed as a percentage of successful inductions. **1962** T. W. Torrey *Morphogenesis Vertebr.* xviii. 481 Neural induction has been shown to occur in vertebrates other than amphibians.

c. *Biochem.* An increase in the rate at which an enzyme is synthesized by a cell (esp. in a micro-organism), or the initiation of its synthesis, as a result of the exposure of the cell to some specific substance (the inducer).

1947 *Growth* XI. 242 Where the enzymes have not been obtained in pure crystalline state.., the evidence must come mainly from a study of the specificity of the phenomenon of induction. **1951** *Biochim. & Biophys. Acta* VII. 599 These observations are incompatible with all hypotheses which imply that the induction is connected..with the activity of the enzyme. **1953** Cohn, Monod et al. in *Nature* 12 Dec. 1096/1 It might prove unpractical to abandon the use of the term 'enzyme adaptation' altogether at this stage; but we should like to suggest that.. a more accurate and significant terminology be employed. We therefore propose the following terms and designations; previously used terms are placed in parenthesis. A relative increase in the rate of synthesis of a specific apoenzyme resulting from exposure to a chemical substance is an 'enzyme induction' (enzyme adaptation). Any substance thus inducing enzyme synthesis is an enzyme 'inducer'. An enzyme-forming system which can be so activated by an exogenous inducer is 'inducible', and the enzyme so formed is 'induced' (adaptive). Although many compounds can act both as inducer and substrate, the terms are not equivalent. Certain substrates for induced enzymes are not inducers, while some inducers cannot function as substrates of the enzymes the formation of which they elicit. **1966** E. R. M. Kay *Biochem.* xxvii. 357 By many mechanisms of feedback control, repression, and induction, enzyme levels can be regulated in accord with the metabolic demands of the cell. **1971** *Nature* 26 Nov. 177/2 Substrate induction of enzymes (that is, their synthesis in response to the presence of their substrates) is now commonplace in microorganisms and not infrequent in higher animals.

d. *Biol.* The initiation of the lytic cycle in a bacterium carrying a prophage; the process of inducing a bacterium that contains a prophage.

[**1950** A. Lwoff et al. in *Ann. de l'Inst. Pasteur* LXXIX. 817 Entre l'induction et la libération du bactériophage, il s'écoule de quarante-cinq à quatre-vingts minutes. **1951** *Jrnl. Bacteriol.* LXII. 302 High titer stocks of this phage were obtained from K12 by induction of phage production with UV (the Lwoff effect).] *Ibid.* 304 The plaque count rises..to a number equal to the colony count before induction. **1953** *Cold Spring Harbor Symp. Quant. Biol.* XVIII. 101/2 In 1950, Lwoff, Siminovitch and Kjeldgaard showed that..irradiation of cultures of lysogenic *Bacillus megatherium* with ultraviolet light greatly increased the proportion of bacteria producing phage... This phenomenon is called induction. **1959** Jacob & Wollman in Burnet & Stanley *Viruses* II. ix. 326 It makes it possible to compare phage development in the same phage-bacterium system, whether after infection of sensitive bacteria, or after induction of lysogenic bacteria. **1959** *Jrnl. Molecular Biol.* I. 177 When the reverse mating (♂ non-lysogenic × ♀ λ-lysogenic) is performed, zygotic induction does not occur. **1968** Echols & Joyner in H. Fraenkel-Conrat *Molecular Basis Virol.* vi. 557 The treatments which produce induction of wild-type prophages are rather diverse (e.g., UV irradiation, growth in mitomycin C, thymine deprivation of a thymine requiring bacterium).

10. (Earlier example.)

1801 *Encycl. Brit.* Suppl. I. 572/2 (*marginal note*) A neutral body attracted, because rendered electrical by induction.

b. (Also *magnetic induction.*) Magnetic flux or (more commonly) flux density, by virtue of which an electric current experiences a magnetic force; as flux density, it is a vector quantity whose magnitude at any point is the magnetic force exerted per unit length on a conductor carrying unit current in the direction that gives rise to the maximum force, and whose direction is normal to those of the current and the force. Also (*electric induction*), electric flux or flux density, = Displacement 2 d (now somewhat *rare*). *Line of induction*, one of the imaginary lines conceived as representing, by their direction and number, the induction at each point throughout a region.

The C.G.S. unit of magnetic induction (flux density) is the gauss; in the International System of Units the tesla (= 10,000 gauss) is used.

1855 J. C. Maxwell in *Trans. Cambr. Philos. Soc.* (1864) X. 49 The unit cells in this case are portions of space in which unit of magnetic quantity is produced by unity [*sic*] of magnetizing force. The length of a cell is therefore inversely as the intensity of the magnetizing force, and its section is inversely as the quantity of magnetic induction at that point. *Ibid.* 50 If a closed conductor move transversely to the lines of magnetic induction.. there will be no current. **1861** —— in *Phil. Mag.* XXI. 168 The total amount of magnetic induction through a closed surface surrounding the pole of a magnet, depends entirely on the strength of that pole. **1873** —— *Treat. Electr. & Magn.* I. I. ii. 77 If *dS* is the element of the surface, the electric displacement through *dS* will be..$KR \cos \epsilon \, dS/4\pi$. Since we do not at present consider any dielectric except air, $K = 1$. We may..avoid introducing at this stage the theory of electric displacement, by calling $R \cos \epsilon \, dS$ the Induction through the element *dS*. *Ibid.* 85 We have used the phrase Lines of Force because it has been used by Faraday and others. In strictness, however, these lines should be called Lines of Electric Induction. *Ibid.* II. III. ii. 24 The three vectors, the magnetization $\mathfrak{I}$, the magnetic force $\mathfrak{H}$, and the magnetic induction $\mathfrak{B}$ are connected by the vector equation $\mathfrak{B} = \mathfrak{H} + 4\pi\mathfrak{I}$. **1879** *Encycl. Brit.* VIII. 75/2 For 'number of lines of force' may of course be substituted the equivalent expressions, 'induction through the circuit', or 'surface integral of magnetic induction'. **1885** O. Heaviside in *Electrician* 4 Sept. 311/1 There is a definite magnitude called by Maxwell 'the magnetic induction', which may well be called simply 'the induction'. *Ibid.* 311/2 We pass to electric displacement, the analogue of magnetic induction (noting by the way that it had better not be called the electric induction..but be called the displacement). **1922** Glazebrook *Dict. Appl. Physics* II. 449/2 *Magnetic flux*, the total amount of magnetic induction through a circuit, measured by the number of lines of induction which are linked with the circuit. **1925** F. B. Pidduck *Treat. Electr.* (ed. 2) iii. 96 It is of some importance..to inquire what happens when the polarisation is not proportional to the electric force. We now require two vectors, P, E, not in general parallel to each other, to express the state of the medium at any point... If we define a vector $D = E + 4\pi P$, $\operatorname{div} D = 0$. The vector D is called the electric induction. **1938** G. P. Harnwell *Princ. Electr. & Electromagn.* ix. 278 By analogy with the introduction of the electric field E in electrostatics it is convenient to introduce a vector B, known as the magnetic induction, which determines the force on a current element. The element of induction is defined by the equation $dF_1 = i_1 \, dl_1 \times B_2$. **1957** B. I. & B. Bleaney *Electr. & Magn.* v. 116 Both a magnet and a current-carrying coil are said to produce a magnetic induction B, which exerts forces on other coils or magnets. *Ibid.* 128 With a magnetic pole, H is the force vector, while the introduction of a uniform magnetic medium throughout the whole of space leaves the magnetic induction B due to a pole unchanged. In the case of a current, B is the force vector and introduction of a magnetic medium leaves H unchanged. If the magnetizable matter does not fill the whole of space, then it is the surface integral of B, the total normal induction, which remains unchanged in magnetostatics. **1962** Corson & Lorrain *Introd. Electromagn. Fields* v. 179 If the current *I* is distributed in space with a current density J amperes/meter², then *I* becomes *J da* and must be put under the integral sign... Thus, in the general case, the magnetic induction B at a point in space is given by $B = (\mu_0/4\pi)\int_\tau (J \times r_1)/r^2 \, d\tau$, where the integration is carried out over any volume τ which includes all the currents.

12. (Earlier example.)

1857 Bunsen & Roscoe in *Phil. Trans. R. Soc.* CXLVII. 381 The act by which the resistance to combination is diminished, and the combining power thus brought into greater activity, we call Chemical Induction.

13. (sense 10) **induction accelerator** = *betatron; **induction coil**, add to def.: a coil in which an electric current is induced; an inductance coil; *Teleph.*, a transformer in a telephone comprising two coils with a common core; (earlier and additional examples); **induction compass** = **inductor compass*; **induction furnace**, a furnace for melting metal by means of induction heating; **induction generator**, an induction motor driven at

a greater speed than its synchronous speed, so that it acts as a generator; **induction hardening**, hardening of ferrous metal by means of induction heating followed by quenching; **induction heater**, an apparatus for the induction heating of objects; **induction heating**, in which an alternating current is made to induce heating currents in the substance or object to be heated or (less commonly) in its container; **induction motor**, an a.c. electric motor in which the torque or force is due to the interaction between a moving magnetic field produced by stationary primary windings and currents induced by this field in moving secondary conductors; (sense 12) **induction period** *Chem.*, the time elapsing between the initiation of a chemical reaction and the production of detectable amounts of the product or products; **induction valve** (earlier U.S. example).

1940 D. W. Kerst in *Physical Rev.* LVIII. 841/2 Of several suggestions which have been made for naming the apparatus, induction accelerator seems to be the shortest descriptive one. **1958** Condon & Odishaw *Handbk. Physics* IX. ix. 154/2 To make this device practical, it is necessary only to restrain the beam to a closed path around the flux and to maintain it in a stable orbit over some thousands of revolutions. Credit for the solution of the latter problem goes to D. W. Kerst, who built the first successful induction accelerator or 'betatron' in 1940. **1837** *Mag. Pop. Sci.* III. 110 A lever has also been applied to move..the induction-coil up and down along two magnetic bars. **1885** R. S. Culley *Handbk. Pract. Telegr.* (ed. 8) IX. 328 The current from the battery does not itself pass out to line, but through a local circuit formed by the primary wire of an induction coil. This coil has a core made of soft iron wires,..and is wound with two wires one over the other... Every variation of the battery current in the primary, produces a corresponding current in the secondary, but of a much higher potential; this last goes out to line, and acts on the distant receiver. *Ibid.* 329 The microphone, induction coil and battery, form the sending portion of the telephone. **1891** J. W. Urquhart *Dynamo Construction* 9 (*caption*) The induction coils combined with the field magnet. **1943** A. L. Albert *Fund. Telephony* vii. 142 When the receiver is removed..the battery causes direct current to flow through the transmitter and the primary of the transformer, often called in telephony an induction coil. **1966** *McGraw-Hill Encycl. Sci. & Technol.* VII. 72/2 Still another type of induction coil, called a reactor, is really a one-winding transformer designed to produce a definite voltage drop for a given current. **1925** *Mech. Engin.* XLVII. 796/2 Since to keep on the correct course it is only necessary to keep the pointer of the induction compass indicator on zero, it is easier to read than the magnetic compass. **1931** B. Jones *Avigation* ii. 18 The induction compass is a distant-reading magnetic compass, the part indicating the heading to the pilot being at a considerable distance from the part affected by the earth's magnetism. **1906** A. Hiorth *Brit. Pat.* 28,960 (*heading*) Improved electrical induction furnace with electrodes. **1951** G. R. Bashforth *Manuf. Iron & Steel* II. ix. 251 When..it was decided to dispense with the iron core of the early induction furnaces, it was necessary to increase the frequency of the current in the primary coil. **1904** G. T. Hanchett *Alternating Currents* xiv. 173 The induction generator cannot generate its own magnetizing current, but must receive a reaction from the line which will permit the magnetizing current of displaced phase to flow in its fields. **1952** G. V. Mueller *Alternating Current Machines* viii. 284 In an induction generator driven above synchronous speed with a negative slip..the rotor conductors are moving faster than the magnetic field of the stator. **1941** *Metals & Alloys* Nov. 687/1 Induction hardening has been applied to the surface of certain steel parts for modern tractor and road-building machinery. **1968** E. N. Simons *Outl. Metall.* III. ii. 142 The three main forms of surface hardening steel parts are carburizing, cyaniding, and nitriding... There are also the processes of flame hardening and induction hardening. **1919** *Electr. World* 29 Mar. 634/1 (*heading*) High-power-factor induction heaters. **1930** *Engineering* 9 May 611/2 The electric resistance [heaters]..heated the chilled surface of the roll and were not so good as the induction heaters, which generated heat within the body of the roll. **1919** H. P. Tiemann *Iron & Steel* (ed. 2) 153 Induction furnace (induction heating). **1937** *Metals & Alloys* May 149/1 Applications of induction heating..now being installed.. include..hardening of metal surfaces by quenching after induction heating. **1954** J. W. Cable *Induction & Dielectric Heating* xii. 319 Many types of furnaces use induction heating as an indirect source of heat... Installations using graphite crucibles, which extract energy from the magnetic field and transfer it to the charge, fall in this general classification. **1959** *Engineering* 13 Feb. 210/1 Plastics to plastics welding is most easily performed by high frequency induction heating. **1962** G. R. Bashforth *Manuf. Iron & Steel* (ed. 2) IV. ii. 58 Induction heating differs from the conventional methods of heating metallic components by virtue of the fact that the heat is generated within the material itself without the surface of the material coming into contact with the heating media. **1897** *Electrician* 17 Sept. 688/1 (*heading*) A 400 H.P. induction motor. **1918** P. Kemp *Alternating Current Electr. Engin.* xxvi. 394 On account of the fact that it must run at a speed rather less than that of synchronism, the induction motor is sometimes termed an asynchronous motor. **1962** *Newnes Conc. Encycl. Electr. Engin.* 492/1 In its basic form the induction motor is essentially a constant-speed motor: the variation in speed from no load to full load when running normally near synchronous speed is only a few per cent. **1971** E. R. Laithwaite *Linear Electr. Motors* i. 7 (*caption*) Imaginary process of unrolling a conventional motor to obtain a linear induction motor. **1973** *Nature* 9 Feb. 359/2 It is now feasible both to levitate and to propel a hovertrain using only a linear induction motor. **1902** *Proc. R. Soc.* LXX. 74 The induction and deduction periods follow as a necessity from the same general thermodynamic conceptions. **1924** H. S. Taylor *Treat. Physical Chem.* II. xviii. 1219 The induction period was not a function of the oxygen content of the gas—this only retarded the reaction velocity. **1953** Frost & Pearson *Kinetics & Mechanism* viii. 156 The duration of the induction period, arbitrarily taken as the time to reach the point of inflection on the C [*sc.* concentration of product] versus *t* [*sc.* time] curve..is easily seen to be equal to the time for B [*sc.* concentration of intermediate] to reach its maximum value. **1970** [see *inhibition 3 b]. **1847** *Rep. Comm. Patents 1846* (U.S.) 87 The induction valve is then closed, and an expansion valve simultaneously opened.

inductionist (indʌ·kʃənist). [f. Induction + -ist.] An adherent of inductive methods in philosophy or science. Cf. *inductivism.

1893 in *Funk's Stand. Dict.* *a* **1915** J. C. Wilson *Statement & Inference* (1926) II. IV. i. 589 Now clearly the argument is demonstrative, and the inductionists in their opposition of induction to deduction would be obliged to call it deduction.

inductionless (indʌ·kʃənlės), *a. Electr.* [f. Induction + -less.] Possessing no inductance.

1902 *Encycl. Brit.* XXX. 600/1 The wattmeter can best be standardized by employing it to measure the known power taken up in an inductionless circuit, such as a bank of incandescent lamps. **1908** J. Skelton tr. *Kolbe's Introd. Electr.* II. vi. 347 The self-induction is thus almost entirely stopped (inductionless winding).

inductive, *sb.* Restrict † *Obs.* to sense in Dict. and add: **2.** = *inductionist. *rare.*

1877 F. H. Laing *Ld. Bacon's 'Philosophy' Examined* xii. 110 The inductives themselves are forced..to employ ..all these words, which perpetually occur in their writings.

inductive, *a.* Add: **4.** (Further examples.)

1828 Mill in *Westm. Rev.* IX. 140 They talk in high-flown language, not always conveying very precise ideas, of a supposed system of *inductive* logic, which is to supersede the syllogistic, and really to accomplish still more than the other even attempts. *Ibid.* 150 An inductive logic would be highly useful as a supplement to the syllogistic logic, not to supersede it. **1865** —— *Auguste Comte* 58 Comte's determined abstinence from the word and the idea of Cause, had much to do with his inability to conceive an Inductive Logic. **1934** A. C. Ewing *Idealism* iv. 170 An inductive logic which dispenses with such arguments. **1964** F. Bowers *Bibliogr. & Textual Crit.* II. i. 36, I am aware that *inductive* and *deductive* as applied to reasoning have acquired philosophical half-lights that may in the end cause a subtle mind to deny that there is any difference between them. **1964** E. Mendelson *Introd. Math. Logic* 9 In the course of this deduction, $P(n, y_1, \ldots, y_k)$ is called the inductive hypothesis.

5. (Earlier example.) Also *inductive coupling*, coupling between two electric circuits or devices that is due to their mutual inductance; so *inductive-coupled* adj., coupled in this way; *inductive reactance*, reactance due to inductance.

1832 Faraday in *Phil. Trans. R. Soc.* CXXII. 137 The inductive force was of course greater. **1907** Inductive coupling [see *direct coupling* s.v. *direct *a.* 6 i]. **1911** *Whittaker's Electr. Engineer's Pocket-Bk.* (ed. 3) 124 The two may be added algebraically, the capacity reactance being considered as opposite in sign to the inductive reactance. **1913** A. F. Collins *Man. Wireless Telegr.* (ed. 3) vii. 126 (*heading*) Diagram of an inductive-coupled resonator. **1954** E. Molloy *Radio & Television Engineer's Ref. Bk.* vii. 7 Figs. 7 and 8, with closely-coupled inductive or direct coupling, give no harmonic voltage reduction in the aerial circuit other than that afforded by the aerial loading coil. **1960** E. Hughes *Electr. Technol.* xi. 339 The inductive reactance is proportional to the frequency. **1966** *McGraw-Hill Encycl. Sci. & Technol.* III. 521/1 Inductive-coupled circuits have a common magnetic flux linking the two circuits.

b. Possessing inductance.

1902 *Encycl. Brit.* XXX. 599/2 In series with the inductive circuit another non-inductive circuit is joined. **1920** *Whittaker's Electr. Engineer's Pocket-Bk.* (ed. 4) 129 The current through an inductive circuit of negligible resistance lags 90° behind the impressed e.m.f. **1962** [see *capacitive *a.*].

7. *Embryol.* Of, pertaining to, or producing induction (sense *9 b).

1931 J. Needham *Chem. Embryol.* I. 578 The inductive power of the cells of the dorsal lip is not abolished by drying them. **1962** Saxén & Toivonen *Primary Embryonic Induction* i. 7 The blastoporal lip of an early gastrula, and that part of an older one, have qualitatively different inductive properties. *Ibid.*, Until 1932, the real nature of the inductive action presented an enigma. **1967** T. W. Torrey *Morphogenesis Vertebr.* (ed. 2) xviii. 359/1 When extracts are treated with proteolytic enzymes, inductive ability is destroyed.

8. Comb. *inductive-minded* adj.

1905 W. James in *Mind* XIV. 191 The one condition of understanding humanism is to become inductive-minded oneself, to drop vigorous definitions, and follow lines of least resistance 'on the whole'.

inductively, *adv.* **2.** (Earlier example.)

1848 W. S. Harris *Rudimentary Electr.* iii. 52 The cover, being insulated, does not take up the electricity of the plate, but is acted upon inductively.

inductivism (indʌ·ktiviz'm). [f. Inductive *a.* + -ism.] The preference for, use of, or belief in the superiority of, inductive as opposed to deductive methods; the belief that scientific laws can be inferred from observational evidence. Opp. *deductivism.

a **1866** J. Grote *Treat. Moral Ideals* (1876) xviii. 425 The matter is complex..on account of the exceeding inappropriateness of the assumption of inductivism for anything in its way so idealist as Utilitarianism is. **1951** *Mind* LX. 43 Professor Popper..attacks frequently what he calls 'observationalism' or 'inductivism'. **1956** E. H. Hutten *Lang. Mod. Physics* vi. 229 If we take induction as a psychological process of discovery and, at the same time, as a logical method of proof, we end up with an insoluble problem... Neither Hume's scepticism nor Mill's belief in the uniformity of nature are solutions of the inductive problem, but rather they are attempts to banish a riddle. There is, to-day, an ever-increasing number of philosophers as well as of scientists who reject this inductivism, and they advise us to drop the term 'induction' altogether when describing scientific method. **1959** K. R. Popper *Logic Sci. Discovery* I. i. 30 The view that a hypothesis can only be empirically *tested*..might be called 'deductivism', in contrast to 'inductivism'. **1962** [see *deductivism]. **1965** J. W. N. Watkins *Hobbes's Syst. Ideas* ii. 34 (*heading*) Repudiation of inductivism. **1972** A. J. Ayer *Probability & Evidence* iii. 74 Popper.. explicitly rejects what he calls inductivism. **1972** *Nature* 10 Nov. 110/1 The author then delineates in turn three theories of the logical structure of science—inductivism, Popperian falsificationism and positivism.

inductivist (indʌ·ktivist), *sb.* and *a.* [f. Inductive *a.* + -ist.]

A. *sb.* One who follows or upholds inductivism or inductive methods; one who holds that the method of science is inductive. Cf. prec. word.

1940 K. R. Popper in *Mind* XLIX. 421 Jeans was.. originally an inductivist, that is, he thought that theories are obtained from experience by some more or less simple procedure. **1962** *Listener* 22 Mar. 513/2 The kind of sociologist whom Sartre despises, the cautious inductivist, collects facts but has no theoretical equipment to understand them. **1968** A. J. Ayer *Origins Pragmatism* 99 The inductivist was having a run of failures. **1968** J. J. C. Smart *Between Sci. & Philos.* vi. 196 The inductivists, who wish to justify inductive procedures.

B. *adj.* Of, pertaining to, or employing inductivism or inductive methods; implying that the method of science is inductive.

1945 *Mind* LIV. 3 Certain 'inductivist' accounts of scientific procedure seem to assume that relevant evidence, or relevant data, can be collected in the context of an inquiry prior to the formulation of any hypothesis. **1956** E. H. Hutten *Lang. Mod. Physics* vi. 268 To believe that we learn by induction is part of the inductivist myth which identifies a psychological process with a logical method. **1960** E. H. Gombrich *Art & Illusion* ix. 321 This inductivist ideal of pure observation has proved a mirage in science no less than in art. **1968** J. J. C. Smart *Between Sci. & Philos.* vii. 247 It is doubtful.. whether such an inductivist account of geometry will do.

inductomeric (indʌ:ktŏme·rik), *a. Chem.* [f. Inducto- + *-meric* as in *electromeric *a.*] Of, pertaining to, or designating the ability of an atom or group to become polarized along a saturated bond by an external electric field (e.g. that of another molecule).

1933 C. K. Ingold in *Jrnl. Chem. Soc.* 1124 It cannot be doubted that a counterpart of two coexisting polarisations obtains also for polarisability effects, and that an inductive polarisability or inductomeric effect completes the scheme annexed. **1943** A. E. Remick *Electronic Interpretations Org. Chem.* v. 61 The extent to which the inductomeric effect is called into play..will depend on the strength of the polarizing field and on the polarizability of the bond in question. **1953** C. K. Ingold *Struct. & Mech. Org. Chem.* ii. 72 By inductomeric polarisability is understood the polarisability which atoms and groups in saturated combination exhibit along the lines of their bonds... For isoelectronic atoms with completed valency shells, inductomeric polarisability thus depends on electronegativity. **1956** E. de B. Barnett *Mechanism Org. Chem. Reactions* i. 4 If a powerful electron-repelling group such as a hydroxyl ion approaches a hydrogen atom of a nitromethane molecule..the resulting inductomeric effect is so great that a hydrogen atom is completely protonized. **1959** E. S. Gould *Mechanism & Struct. Org. Chem.* vii. 208 A number of workers..prefer to differentiate inductive influences in reactants and products from those electronic shifts that occur in the activated complex as a result of the electrical demands of one reagent on another. The former are termed inductive effects, the latter inductomeric effects.

inductor. Add: **3.** (Earlier and later examples.) *Esp.* one which induces an e.m.f. or current in another part (as in an inductor alternator).

Delete quot. 1871, which appears (with corrected date) under sense c below.

1867 W. Thomson in *Proc. R. Soc.* XVI. 67 To stems connected with the inside coatings of two Leyden phials are connected [cylindrical] metal pieces, which..I shall call inductors and receivers. *Ibid.* 68 Suppose now a small positive charge of electricity be given to the first jar. Its inductor electrifies negatively each drop of water breaking away in its centre from the continuous uninsulated water above. **1892** *Electrician* 13 May 37/2 Electric currents are produced by revolving the magnetic inductors in proximity to the magnet. **1909** R. B. Whitman *Motor-Car Princ.* (rev. ed.) 320 The Remy magneto is of this type, and Fig. 17 is a diagram of the revolving core, or inductor, with the coil surrounding it. **1915** W. H. Eccles *Wireless Telegr.* 204 This alternator is of the inductor type... The inductor or rotor is a chrome-nickel steel disc about a foot diameter with 300 slots cut ⅛ in. apart near its edge. **1941** A. W. Judge *Aircraft Engines* II. ix. 338 The polar inductor magneto has stationary magnets as well as coils, and the changes of magnetic flux are obtained by rotating soft-iron inductors between the poles of the magnets. **1956** D. Warburton-Brown *Induction Heating Pract.* i. 8 In any induction heating arrangement there are three main components, namely: (a) a high-frequency generator; (b) a work-coil or inductor; (c) a work-piece.

b. A conductor or device in which an e.m.f. or current is induced; **earth inductor**, a device for investigating the earth's magnetic field, consisting essentially of a coil of wire that can be rapidly turned about an axis in its own plane so that a current is induced in it proportional to the component of the field normal to the axis of rotation.

1837 tr. Gauss in *Mag. Pop. Sci.* III. 109 A few weeks ago I had my inductor increased again, (from 3527 to about 6800 convolutions,) and now its effects are much stronger. The sensations it produces by the current being transmitted through the body..are not only very perceptible, but, when the inductor is rapidly moved, painful almost beyond endurance. **1883** E. Atkinson tr. *Ganot's Elem. Treat. Physics* (ed. 11) x. vi. 852 The inductor itself..consists of a drum-shaped frame of soft iron wire covered with a layer of insulating material, and fixed to an axle which..is rotated... Machines of this class give continuous currents, but alternators..are also constructed. **1883** *Encycl. Brit.* XV. 240/2 This is the principle of Weber's 'earth inductor', by means of which the horizontal and vertical components of the earth's force can be measured, and in consequence the declination and inclination determined. **1901** Sheldon & Mason *Dynamo Electr. Machinery* iii. 45 By inductor is meant that part of the winding conductor which lies on the face of the armature that sweeps past the pole pieces, and.. in which E.M.F. is induced. **1940** R. R. Lawrence *Princ. Alternating-Current Machinery* (ed. 3) i. 1 Any direct-current generator, with the exception of the unipolar generator, is in fact an alternator in which the alternating voltage set up in the armature inductors is rectified by means of a commutator. **1966** *McGraw-Hill Encycl. Sci. & Technol.* IV. 338/2 The earth inductor has almost completely supplanted the dip circle throughout the world for precise measurement of magnetic inclination. **1973** *Sci. Amer.* Feb. 101/2 A unipolar inductor is a magnetized metal sphere. One terminal of an external circuit is attached to one of the sphere's poles of rotation and the other terminal is a stationary brush in contact with the sphere's equator. When the sphere is spun, a galvanometer in the circuit registers the passage of an electric current.

c. An induction coil (Ruhmkorff coil).

1872 J. & C. Lassell tr. *Schellen's Spectrum Analysis* xxx. 157 By connecting the binding screws 1, 2 on one side with the inductor, and on the other side..with the platinum wire *b* of the first vessel, and a_1 of the last vessel, ..the electric current may be made to pass through all the liquids. **1904** *Electr. World & Engin.* XLIV. 513/2 A very imposing view..is given by the seven induction coils arranged upon steps in the centre of the cabinet, beginning at the bottom, with an inductor giving a 40-inch spark and ending at the top with one of 10-inch spark.

d. A device (commonly a coil) possessing inductance or used on account of its inductance.

1928 Sterling & Kruse *Radio Manual* i. 20 Iron Core Inductance.—This form of inductor is made by winding many turns of wire on an iron core. **1950** K. Henney *Radio Engin. Handbk.* (ed. 4) iii. 124 Straight wires are used as inductors in h-f applications where the inductance must be very low. **1963** Williams & Prigmore *Electr. Engin.* v. 129 A standard mutual inductor.. can be made by winding coil No. 1 on a long straight core and winding No. 2 round the mid-portion of No. 1.

4. *Chem.* Any substance which while reacting with one substance (the 'actor') increases the rate at which this reacts with a second substance (the 'acceptor'); a substance that has an accelerating effect on a reaction but differs from a catalyst in being consumed.

1903 *Jrnl. Chem. Soc.* LXXXIV. II. 277 The substance taking part in both these reactions..is termed the 'actor'; the substance taking part only in the primary reaction.. is the 'inductor'; the substance taking part only in the secondary reaction is the 'acceptor'. **1918** *Chem. Abstr.* XII. 111 A number of reduction reactions which take place at ordinary temps. in sunlight, fail to proceed even at more elevated temps. in the dark. The reactions may be induced to take place in the dark by the addition of small amts. of certain oxidizing agents ('inductors'). **1937** *Thorpe's Dict. Appl. Chem.* (ed. 4) I. 12/2 *Acceptor*, a substance..which normally is not oxidised by oxygen (or reduced by hydrogen) but is oxidised (or reduced) when in presence of another substance termed the inductor.. which itself is undergoing oxidation (or reduction) by a third substance the actor. **1966** A. G. Sykes *Kinetics Inorg. Reactions* ix. 205 Arsenite ions are..effective in inducing the reaction between Cr^{VI} and iodide, and, at high iodide concentrations, a ratio of two equivalents of iodide to one of the inductor are likewise involved.

5. *Embryol.* A region of an embryo, or a substance produced by such a region, capable of causing induction (sense *9 b).

1929 *Biol. Abstr.* III. 1495/2 The action of the inductor, whatever it be, manifests itself in the activation of a supernumerary embryonic field. **1946** L. B. Arey *Developmental Anat.* (ed. 5) ix. 163 The specific, morphogenetic effect brought about by a chemical stimulus transmitted from one embryonic part to another is known as an induction or evocation. The part exerting this influence is an inductor or organizer, and the chemical substance emitted is an evocator. **1963** E. J. W. Barrington *Introd. Gen. & Compar. Endocrinol.* vi. 152 Germ cells that enter the cortex become female, those that enter the medulla become male, and in embryological terminology these two regions are said to act respectively as female and male inductors. **1967** T. W. Torrey *Morphogenesis Vertebr.* (ed. 2) xviii. 359/1 An inherent difficulty in the identification of the 'natural' inductor lies in the minute amount available for analysis.

6. Special Comb.: **inductor alternator**, an alternator in which both armature and field windings are stationary, the current being produced by the periodic variation in the magnetic flux through the armature windings as successive teeth of a rotating inductor pass by; **(earth-)inductor compass**, any of various kinds of compass in which the earth's magnetic field is made to induce in a coil an electric current whose strength depends on the relative orientation of the coil to the field; **inductor generator** = **inductor alternator*.

1893 W. P. Maycock *Electr. Lighting* II. viii. 239 Kingdon's Inductor Alternator. **1940** R. R. Lawrence *Princ. Alternating-Current Machinery* (ed. 3) 3 An inductor alternator is usually characterized by large armature reaction, relatively high magnetic density, small air gap and greater weight than alternators of the other types. The difficulties in the design of a satisfactory inductor alternator have caused this type of alternator to go out of use. **1965** J. Hindmarsh *Electr. Machines* viii. 482 Another special case of single-phase generator is the inductor alternator used to provide high-frequency supplies in the range 1000 to 10,000 cycles/sec for use in induction furnaces. Here, all the windings are on the stator. **1922** *Rep. U.S. Nat. Advisory Comm. Aeronaut.* No. 128. 44 The earth inductor compass..was developed by Dr. Paul R. Heyl and Dr. Lyman J. Briggs of the Bureau of Standards. **1926** *Encycl. Brit.* Suppl. I. 19/2 The earth inductor compass, developed in America, depends upon the measurement of the electromotive force induced by the earth's magnetic field in a coil rotating about a vertical axis. **1927** Lindbergh in *Sci. Monthly* XXV. 91/1, I also had a magnetic compass; but it was the inductor compass which guided me so faithfully that I hit the Irish coast... The inductor compass was so accurate that I really needed no other guide. **1931** M. F. Schoeffel in P. V. H. Weems *Air Navigation* vii. 93 Although the armature of an inductor compass is gyroscopic, yet, since it is also pendulous, it tends to..bank with the plane. **1943** Redpath & Coburn *Air Transport Navigation* iv. 80 The principle of the inductor compass is to generate electric currents utilizing the earth's magnetic field to operate suitable indicating instruments in the cockpit. **1958** *Van Nostrand's Sci. Encycl.* (ed. 3) 386/2 The earth-inductor compass was designed..for use on aircraft, but has been rendered obsolete by the aperiodic and gyro-flux-gate instruments. **1940** *Chambers's Techn. Dict.* 446/1 Inductor generator. **1957** *Encycl. Brit.* VIII. 148/1 For a given number of poles and a given speed of rotation, an inductor generator delivers twice the frequency of the present common types of A.C. generators. In the early years of radio telegraphy the inductor generator was the best source of the frequencies required.

indulge, *v.* Add: **1. d.** (Later example.)

1951 Auden *Nones* (1952) 39 How jocular the bells as They Indulge the peccant shore.

8. *intr.* (Without preposition.) To gratify a desire, appetite, etc.; to take one's pleasure; *spec.* to 'partake', i.e. (too) freely of intoxicants (*colloq.*).

a **1718** T. Parnell *Poems on Several Occasions* (1722) 125 Wretch that I was! I might have warn'd the Dame, Yet sat indulging as the Danger came. **1953** P. C. Berg *Dict. New Words* 96/2 *Indulge*, to take alcoholic liquors without restraint. **1973** P. O'Donnell *Silver Mistress* xv. 250 Tarrant..took out his cigar case. He had not indulged all night.

indumentum (indiume·ntəm). *Bot.* Pl. -ta. [a. L. *indumentum*; see Indument.] The covering of hairs on part of a plant, esp. when dense, e.g. the covering of the lower surface of the leaves of many species of rhododendron. Cf. Indument 2.

1847 J. Lindley *Elem. Bot.* (ed. 5) p. xlix, Indumentum: The hairy covering of plants, of whatever kind. **1858** A. Gray *Introd. Struct. & Syst. Bot.* (ed. 5) 537/1 Indumentum: any hairiness or downy covering. **1930** J. B. Stevenson *Species of Rhododendron* 364 Under surface [of leaves of *R. eriogynum*] at first clad with a thin white flaking stellate indumentum easily rubbed off and quickly falling away. **1950** J. M. Cowan *Rhododendron Leaf* i. 1 On looking at the leaves of Rhododendrons, it will be seen that the under surface is often clothed with a hairy or scale-like covering, now loose and of a fine texture like a spider's web, now dense and compact like a thick felt or wool, now a mere sprinkling of brownish dots on a green background. This leaf covering, which is very variable, is known as the indumentum. **1963** Davis & Heywood *Princ. Angiosperm Taxon.* v. 155 To be of use for critical discrimination, indumentum must, of course, be described in detail, each type of hair present (and its disposition) being separately recorded. **1970** *Watsonia* VIII. 104 *H[ieracium] termifrons* occurs in two well-marked forms differing in shape of head and indumentum of the phyllaries. **1972** *Biol. Abstr.* LIV. 68/1 *X[imenia] caffra* differs from *X. americana* in that a dense or sparse indumentum is often present.

induna. Add: Also 9 tuna, 9– indoona. (Earlier and later examples.)

1835 A. Smith *Diary* 19 June (1940) II. 79 Masalacatzie has two grades among his chiefs, viz.: *numzan* and *tuna*, the former the highest. **1837** F. Owen *Diary* (1926) 28 A regiment is stationed at each town under several Indoonas or Captains. *Ibid.* 60 The King..was seated in his hut on a chair: his Indoonas were also present. **1955** E. A. Ritter *Shaka Zulu* xiii, These [soldiers] were now harangued for a considerable time by their fighting *indunas* or officers and given a discourse on new tactics. **1971** *Daily Dispatch* (S. Afr.) 24 May 1 In the Libode district unoccupied huts were set on fire by tribesmen. Recently a headman and his induna in the district were murdered.

b. *transf.* A person, especially a black person, in authority; a pundit.

1953 P. Lanham *Blanket Boy's Moon* II. iv. 116 A big factory, where he had obtained the job of *Induna* or head-boy. **1953** F. Robb *Sea Hunters* iv. 28 On deck.. Ndwe, the Induna or bossboy, heaved on the vang and centred the boom over the gaping hatchway. **1970** *News/Check* (S. Afr.) 4 Sept. 9 This followed the attack on the Press by rugby induna Dr. Danie Craven for blowing up the incidents of rough play in the second Test. **1970** W. Smith *Gold Mine* xxix. 72 The Old One, the Shangaan Induna, lived in a Company house. **1971** in *Towards Dict. S. Afr. Eng.* 42 They have an induna there to hold you down while they X-ray you.

Hence **indu·naship**, the office or dignity of an induna.

1955 in M. Gluckman *Judicial Process among Barotse* iii. 87 This is indunaship—this is ruling.

Indus (i·ndəs). *Astron.* [L.] = Indian *sb.* 5.

1838 *Penny Cycl.* XII. 467/1. **1910** *Encycl. Brit.* VII. 13/1 Johann Bayer, a German astronomer..published a *Uranometria* in 1603, in which twelve constellations, all in the southern hemisphere, were added to Ptolemy's forty-eight, viz. Apis (or Musca) (Bee),..Indus (Indian), [etc.]. **1964** D. H. Menzel *Field Guide Stars & Planets* iv. 113 Indus (the Indian). Although the name is masculine, Flamsteed drew a female figure.

indusium. Add: **1. b.** Also *erron.* induseum. The thin layer of grey matter covering the upper surface of the corpus callosum. In full *indusium* (or *induseum*) *griseum*.

1890 Billings *Med. Dict.* I. 693/2 Indusium griseum. **1908** *Quain's Elem. Anat.* (ed. 11) III. I. 401 The lateral striæ are similar, but their corresponding grey matter is even more rudimentary than the indusium. **1948** A. Brodal *Neurol. Anat.* x. 330 Smaller contingents [of fibres] appear to reach it from the induseum griseum. **1963** Zeman & Innes *Craigie's Neuroanat. Rat* (ed. 2) vii. 147 The rest [of the hippocampal formation] curves around it and becomes continuous with the indusium. **1966** Hassler & Stephan *Evolution of Forebrain* 104 The supracallosal gyrus (or induseum griseum).

industrial, *a.* and *sb.* Add: **A.** *adj.* **c.** Of a quality suitable for industrial use.

1904 Goodchild & Tweney *Technol. & Sci. Dict.*, *Industrial soaps*, a term used to describe that class of soap used for special purposes, such as ox gall soap, which is useful for scouring woollen goods and cleaning carpets, soap for silk dyers, fulling soap, etc. **1904** *Chemist & Druggist* LXV. 852/2 Industrial Alcohol Committee.. appointed to inquire into the use of duty-free alcohol for industrial purposes. **1905** *Ibid.* LXVI. 630/2 There is only one way in which the influence of the spirit-duties can be satisfactorily counteracted in favour of industrial alcohol. **1906** *Act 6 Edw. VII* c. 20 § 4 The expression 'industrial methylated spirits' means any methylated spirits (other than mineralized methylated spirits) which are intended for use in any art or manufacture within the United Kingdom. **1968** J. Ironside *Fashion Alphabet* 89 *Industrial*, applied to special fabrics or garments specially designed for use in particular industries.

d. Characterized by highly developed industries.

1911 C. G. Robertson *Eng. under Hanoverians* II. iv. 346 Napoleon..failed to see that he fought not with a nation of shopkeepers—a commercial State—but with a nation of capitalists and artisans—an industrial State. **1948** S. Lilley *Men, Machines & Hist.* vi. 72 England was transformed from one of the most backward to one of the most rapidly advancing commercial and industrial countries of Europe. **1953** J. D. Bernal *Sci. & Industry in 19th Cent.* vi. 171 Only the industrial countries of Europe and the newly industrialized parts of America contributed to modern science.

e. In specific uses.

industrial accident, an accident occurring in the course of one's employment, esp. in a factory; *industrial action*, action such as a strike, a go-slow, working to rule, etc., taken by industrial or other workers; *industrial archæology*, the study of the equipment and workings of industry of former times; so *industrial archæologist*; *industrial art*, art applied to the design of industrial products; so *industrial artist*; *industrial assurance*, a form of life assurance for industrial workers, mainly to cover funeral costs, with premiums payable in small regular instalments; *industrial court*, a court for the settlement of industrial disputes; *industrial design*, design as applied to industrial products; so *industrial designer*; *industrial disease*, a disease contracted in the course of one's employment, esp. in a factory; *industrial dispute*, a dispute between employers and employees; *industrial espionage*, spying directed towards discovering the secrets of a rival industrial company, manufacturer, etc.; *industrial estate*, an area of land devoted to factories and other industrial enterprises; *industrial fatigue*, fatigue in industrial workers; *industrial frequency* (see quot. 1940); *industrial injury*, an injury occurring in the course of one's employment, esp. in a factory; *industrial insurance*, (*a*) = **industrial assurance*; (*b*) insurance for industrial workers against injury or absence from work; *industrial park* (chiefly N. Amer.) = **industrial estate*; *industrial proletariat*, the section of the proletariat that is employed in industrial work; *industrial property*, the collective name applied to commercial rights derived from patents, designs, trade marks, etc.; *industrial psychology*, psychology as applied to all aspects of human involvement in industry; so *industrial psychologist*; *industrial relations*, relationships between employers and employees; *industrial revolution*, a rapid development in industry; *spec.* (freq. with capital initials) the development which took place in England in the late eighteenth and early nineteenth centuries, chiefly owing to the introduction of new or improved machinery and large-scale production methods; *industrial spy*, a person engaged in industrial espionage; *industrial union*, a union of all workers in an industry irrespective of their craft or occupation; so *industrial unionism, unionist*; *Industrial Workers of the World*, a labour organization advocating syndicalism which enjoyed its greatest support in the western United States during the early twentieth century.

1910 *Encycl. Brit.* IX. 361/1 By a law of..1910, *Sweden* adopted the principle of the personal liability of the employer for industrial accidents. **1922** *Ibid.* XXXI. 698/1 The former has legalized deductions for hospital benefits on approval of the Industrial Accident Commission. **1968** *Brit. Med. Bull.* XXIV. 256/1 A 40-year-old man with a history of several industrial accidents..presented after a fall at work. **1971** *Times* 17 Mar. 1/8 The Times regrets that, in common with other national newspapers, it will probably be unable to publish tomorrow because of industrial action. **1972** *Guardian* 15 June 26/1, I fear..that if you imprison individual dockers we are going to get full scale industrial action. **1954** M. Rix in *Country Life* 28 Oct. 1501/1 This canal..has a special interest for industrial archæologists. **1951** *History Today* July 59/2 The most fascinating subject of all is what might be called the industrial archaeology of the area. **1971** K. Hudson (*title*) A guide to the industrial archaeology of Europe. **1850** *Punch* 29 June 10 Mind where you fix your show... Where Fashion rides and drives House not industrial Art, But 'mid the busy hives Right in the City's heart. **1851** *Illustr. Lond. News* 21 June 605/3 Premiums for works of industrial art were offered. **1863** J. B. Waring (*title*) Masterpieces of industrial art and sculpture at the International Exhibition, 1862. **1902** *Encycl. Brit.* XXV. 687/1 The awakening of interest in industrial art—sharply separated by pedantic classification from fine art—which began about the middle of the 19th century. **1930** *Times* 7 May 11/4 Industrial Artists. An Association is to be formed of artists engaged in industry. **1896** *Act 59 & 60 Vict.* c. 26 (*title*) Collecting Societies and Industrial Assurance Companies Act. **1920** *Rep. Industr. Assurance Comp.* (Cmd. 614) 2 The business of Industrial Assurance consists in the assurance of small sums, payable for the most part on the death of the life assured, in consideration of the payment of weekly premiums. **1935** *Economist* 2 Mar. 497/2 Following the terrible disaster at the Gresford Colliery, a sum of £10,000 was paid to some 200 families under industrial assurance policies. **1919** *Act 9 & 10 Geo. V* c. 69 § 14 This Act may be cited as the Industrial Courts Act, 1919. **1973** *Listener* 15 Nov. 660/1 The leaders of the AUEW see their own refusal to recognise the Industrial Court as part of the historic struggle of trade-unionists for the rights of the working man. **1934** H. Read (*title*) Art and industry, the principles of industrial design. **1967** L. B. Archer in Wills & Yearsley *Handbk. Managem. Technol.* 122 There is a range of products in which aesthetic appearance and convenience in use are very important, such as in furniture, domestic appliances, and office machinery. Design of this kind is called 'industrial design'. **1940** H. Van Doren *Industr. Design* 27 Industrial designers who take their work seriously cannot afford to play the prima donna. **1972** F. MacCarthy *All Things Bright & Beautiful* 147 The hero of the piece was the industrial designer. **1906** *Act 6 Edw. VII* c. 58 § 8 (*heading*) Application of act to industrial diseases. **1974** *Guardian* 20 Mar. 1/8 A working class Yorkshire family, whose father contracts cancer through an industrial disease. **1907** *Times* 1 Feb. 4/5 The Minister of Labour [in Canada] has brought in an Industrial Disputes Investigation Act, which provides for the constitution of conciliation boards. **1973** *Times* 13 Dec. 18/8 Men for whom wage claims and industrial disputes are tools to be used in..the destruction of the existing political order. **1962** L. Deighton *Ipcress File* viii. 52 His reports concern industrial espionage. **1972** K. Benton *Spy in Chancery* ii. 22 He runs an industrial espionage service. **1953** P. C. Berg *Dict. New Words* 96/2 *Industrial estate*, a trading estate. **1972** M. Jones *Life on Dole* ix. 68 The Council declared this property to be an industrial estate. **1974** *Times* 14 Jan. 2/5 A secondary modern school on an industrial estate. **1914** *Rep. Brit. Assoc. Adv. Sci.* 176 What increase..has occurred in general morbidity in recent years, and to what extent this can be ascribed to industrial fatigue. **1950** *Chambers's Encycl.* VII. 543/1 In 1918 the Industrial Fatigue Research Board was formed. **1940** *Chambers's Techn. Dict.* 446/1 *Industrial frequency*, a term used to denote the frequency of the alternating current used for ordinary industrial purposes, usually 50 or 60 cycles. **1958** *Engineering* 14 Mar. 341/1 Two [railway] systems..which are able to utilise widely spaced substations and light overhead conductors, are the Swedish system at 16⅔ cycles, and the French (and now British) system at industrial frequency (50 cycles) and 25 kV. **1933** M. Correll in *U.S. Women's Bureau Bull.* No. 102 (*title*) Industrial injuries to women in 1928 and 1929. **1940** *Bull. U.S. Bureau of Labor Statistics* No. 667. p. ix, Efficient accident prevention can be promoted by administrators of workmen's compensation laws by prescribing types of reports to be submitted in cases of industrial injuries which can be used in analyzing accident causes. **1946** *Industr. Welfare & Personnel Managem.* XXVIII. 214 Important changes in Industrial Law have taken place recently as a result of such measures as the Industrial Injuries Act (which supersedes the Workmen's Compensation Acts). **1971** *Morning Star* 8 Apr. 3 Industrial injuries, on average, had been halved. **1911** *Encycl. Brit.* XIV. 671/2 The system of industrial insurance was introduced into the United States in 1876. **1920** S. & B. Webb *Hist. Trade Unionism* (rev. ed.) ix. 507 The life assurance agents—principally those employed in 'industrial' insurance—number 100,000. **1920** *Rep. Industr. Assurance Comp.* (Cmd. 614) 3 There is little thought for the development of industrial insurance upon the sound economic lines by which it might become a valuable instrument. **1955** *Barron's Nat. Business & Financial Weekly* 10 Oct. 13 (*title*) Industrial parks; planned factory districts are attracting more customers. *Ibid.* 15/3 The combined shopping center and industrial park. **1957** *Urban Land* Apr. 5/2 Several questions indicated a strong local interest in the development of industrial parks as a means of stimulating development. The Panel stated that such parks are becoming increasingly popular and effective in attracting the smaller manufacturing plants and distribution warehouses. **1963** *Amer. City* July 95/2 The new industrial park..offers the prospective industry a tract of graded land located on a wide, paved and landscaped boulevard. The attractive 'price of admission' also includes curbs, gutters and storm drains, water and sewer mains, gas, electric and telephone lines and access to railroad sidings. **1972** *Evening Telegram* (St. John's, Newfoundland) 5 Aug. 10/1 Murray and Davis Properties Ltd. has announced the opening of their new..building.. in the industrial park area. **1887** F. K. Wischnewetzky tr. *Engels's Condition of Working-Class in Eng.* i. 15 (*heading*) The Industrial Proletariat. **1930** G. B. Shaw *John Bull's Other Island* Pref., in *Wks.* XI. 71 The growing political power of the industrial proletariat organized in trade unions. **1972** G. Wigg *George Wigg* i. 23 The tiny red brick houses, typical of the growth of an industrial proletariat. **1884** in Hertslet *Treaties* (1890) XVII. 408 The International Convention for the Protection of Industrial Property, concluded at Paris on the 20th March, 1883. **1952** Lauterpacht & Oppenheim *Internat. Law* (ed. 7) II. II. ii. 330 Enemy assets..are to be returned—with the important exception of industrial property (*i.e.*, patents, designs, trade marks and trade names, etc.). **1921** J. Drever *Psychol. of Industry* iv. 46 Even when the factors affecting industrial efficiency and success are physical, the work of the industrial psychologist may still be valuable in tracing out the physiological and psychological results of physical conditions. **1936** *Discovery* Sept. 279 There are too many salients in the front line of social progress and it is the duty of the industrial psychologist to smooth these away. **1964** M. Argyle *Psychol. & Social Probl.* viii. 114 It is one of the major tasks confronting industrial psychologists today to discover ways of organizing such work that will make it an acceptable human activity. **1971** A. K. Korman *Industr. & Organizational Psychol.* i. 12 The industrial psychologist..continues to be actively concerned with the..techniques and methods which will increase the effectiveness of manpower utilization in the modern organization both from the company's and from the individual's point of view. [**1913** H. Münsterberg (*title*) Psychology and industrial efficiency.] **1917** B. Muscio (*title*) Lectures on industrial psychology. **1932** *Discovery* Nov. 374/1 Industrial psychology is an awkward term for the improvement of the human factor in industry, but its value is unquestionable. **1970** D. P. Schultz *Psychol. & Industry* i. 1 The field of industrial psychology includes a complex of activities covering all facets of the relationship between man and his work. **1904** S. A. Barnett et al. in H. Barnett *Canon Barnett* (1918) II. xli. 258 Luxury..leads to cruelty in our industrial relations. **1972** *Times* 4 May 4/1 Members of Parliament are likely to be given greater latitude by the Speaker in discussing..matters that have been brought before the Industrial Relations Court. **1973** *Listener* 25 Oct. 553/3 There were no shop-stewards, solicitors or industrial relations managers present. *Ibid.* 15 Nov. 660/2 The Labour Party is committed to the repeal of the Industrial Relations Act. **1848** Mill *Pol. Econ.* II. III. xvii. 119 The opening of a foreign trade..sometimes works a complete industrial revolution in a country whose resources were previously undeveloped. **1884** A. Toynbee (*title*) Lectures on the Industrial Revolution in England. **1911** C. G. Robertson *Eng. under Hanoverians* II. iv. 341 The manufacturer of the Industrial Revolution is the modern master who provides capital, owns his mill or factory,..and creates and maintains a market. **1938** H. Granville-Barker *Quality* 3 We sometimes refer to the Industrial Revolution as if it were a thing of the past. **1957** G. E. Wright *Biblical Archaeol.* viii. 120/2 The Philistine defeat..meant *an industrial revolution.* Philistine power was broken and the secret of the iron smelting process became common property. **1973** *Guardian* 4 June 9/6 It is no good being nostalgic for society before the Industrial Revolution. **1959** K. Vonnegut *Sirens of Titan* (1962) iii. 73 He..had a superb system of industrial spies. **1972** G. Lyall *Blame the Dead* xii. 77 This letter..would be a useful guide to any industrial spy trying to penetrate your organisation. **1923** J. D. Hackett in *Managem. Engin.* May 344/1 *Industrial union*, a union of all workers within a plant or within an industry, irrespective of occupation or craft, and outside the control of the employer. **1928** *Britain's Industr. Future* (Liberal Industr. Inquiry) III. xiv. § 3. 155 The Industrial Unions are a modern development, inspired by the idea of enabling all the workers in an industry..to present a united front against their employers. **1937, 1950** Industrial union [see *horizontal *a.* 3 d]. **1905** *Socialist* Oct. 4/3 Industrial Unionism is the name applied to that form of trades unionism which has sprung into existence as a direct outgrowth of modern industrial conditions under which whole industries are practically owned and controlled by capitalists through the medium of a trust or combine. **1912** [see Syndicalism]. **1920** S. & B. Webb *Hist. Trade Unionism* (rev. ed.) ix. 659 The revolutionary Industrial Unionism and Syndicalism preached by James Connolly and Tom Mann..between 1905 and 1912 did not commend itself to the officials of the Trade Unions. **1905** *Socialist* Oct. 4/3 The Industrial Unionist calls upon the workers to organise in a manner consistent with the economic conditions with which they are surrounded... The plan of the industrial unionist..calls for the joining of all in the one national body of the industry. **1917** G. Harvey *Industr. Unionism & Mining Industry* v. 148 Analysis of the Arrangement of Industries (based on Industrial Unionist literature). **1905** *Industrial Workers of the World: Constitution & By-Laws* 4 This Organization shall be known as 'The Industrial Workers of the World'... And shall be composed of thirteen International Departments, subdivided in industrial unions of closely kindred industries..for representation in the departmental administration. **1912** *Century Mag.* July 473/1 Counsels of violence were emphatically rejected, despite the opposition of the ideas of the Industrial Workers of the World. **1920, 1957** [see *I.W.W.* (*I. III)]. **1962** G. Woodcock *Anarchism* xiv. 466 After 1905 the anarchists who were interested in labor organization tended to join the Industrial Workers of the World.

B. *sb.* **3.** A joint-stock industrial enterprise.

1908 *Westm. Gaz.* 20 Jan. 11/1 A Colonial Industrial. The prospectus of the Vryheid (Natal) Railway, Coal, and Iron Company, Limited, has now made its appearance. **1909** *Ibid.* 3 June 13/4 A large falling off is shown in the earnings of that well-known industrial, Wm. Cory and Son.

industrialization (indʌ:striălaizēi·ʃən). [f. Industrialize *v.* + -ation.] The process of industrializing or fact of being industrialized; also, the conversion of an organization into an industry.

1906 *Westm. Gaz.* 26 Oct. 2/1 A creed..that commends itself to all whom the 'industrialisation' of American politics has revolted. **1911** G. H. Mair *Eng. Lit.: Mod.* ix. 225 If we want a picture of the great fact of modern Scotland, its industrialisation, it is to Galt we must go. **1923** *19th Cent.* Jan. 47 It is easy to obtain an exaggerated idea of the pace at which the industrialisation of the country by means of indigenous agency is taking place. **1928** *Daily Tel.* 20 Mar. 12/6 The present rulers of Russia have staked everything upon the policy of industrialisation under State auspices. **1953** *Encounter* Oct. 78/2 The process of industrialisation produces certain social types. **1963** P. Trench in *Industrialized Building Syst. & Components* Oct. 10/1 It must be some ten years ago that Roger Watters..and I first talked industrialisation. I can't remember who coined the word—it may well have been neither of us... It is now the U word. *Ibid.* 10/2 Industrialisation as a process must cover everything from the use of ready mixed concrete through modular co-ordination to the closed proprietary building system. It is in fact too wide a term to be of real use but since we have no other to describe the change in the nature of building which is going on we must accept it. **1969** *Observer* 12 Jan. 9/2 Art education is a safeguard against over-organisation, industrialisation, bureaucratisation and dehumanisation. **1969** H. A. Frey tr. *Schmidt & Testa's Syst. Building* 34/2 *Industrialization of building* in the proper sense, and in contrast to building rationalization, has to do not only with selected operations, but comprises the entire construction process from the planning stage to execution. **1973** *Listener* 20 Dec. 857 Industrialisation played a big part in the drive to paint the map pink.

industrialize, *v.* Add: **b.** *intr.* To become industrial.

1965 D. E. C. Eversley in Glass & Eversley *Population in Hist.* ii. 60 Countries which did not industrialize shared in this process because they began to act as granaries for areas not self-sufficient in food. **1971** *Daily Tel.* 16 Oct. 10/4 The pace at which it [*sc.* Australia] is being forced to industrialize is turning large sections of the wide brown land into a polluted wasteland.

Hence **indu·strializing** *vbl. sb.* and *ppl. a.*

1925 *Glasgow Herald* 16 Nov. 9 It appears to be the general conclusion that any industrialising schemes apart from the water-power projects cannot succeed. **1959** *Daily Tel.* 10 Mar. 13/1 Fairly good industrial production increases occurred in new industrialising countries like Pakistan, the Philippines, Formosa and South Korea. **1967** *Economist* 10 June 1105/3 Rapidly mounting food deficits are one sign of an industrialising society. **1971** A. Shonfield in A. Bullock *20th Cent.* 326/1 The end of colonial empires would start a race with the newly industrializing nations in which the Old World would soon be overtaken.

industrialized, *ppl. a.* (in Dict. s.v. Industrialize *v.*). Add: *industrialized building*, building in which industrial methods are employed, esp. prefabrication, mechanization, and standardization. Hence applied to buildings built in accordance with such methods.

1963 *Daily Tel.* 4 Feb. 18 Prefabricated units developed under 'industrialised building' methods will be made available to small as well as big builders. 1963 *Ibid.* 2 May 14 The suggestion is that spare shipbuilding capacity should be turned over to the manufacture of component parts for industrialised housing. 1963 *Industrialised Building Syst. & Components* Oct. 5 Industrialised building is not the mere substitution of prefabrication for traditional methods or the adoption of 'systems'. We see it as the application of modern industrial methods to the planning and execution of construction projects. 1965 R. B. White *Prefabrication* III. vi. 302 'Industrialized' building, or 'system' building,..is usually taken to mean a considerable degree of prefabrication. 1967 *Times Rev. Industry* Feb. 23/2 The National Building Agency..suggests that the number of industrialized houses in the public sector should be increased to at least 165,000. 1968 *New Scientist* 23 May 388 Industrialized buildings by their nature are less continuous than traditional structures. The degree of continuity depends on the techniques adopted to join together the prefabricated units.

industry. Add: **5. b.** *Archæol.* A collection of prehistoric implements of the same age found at an archæological site and used as evidence of the original technique of working; also, the technique so revealed.
1911 *Jrnl. R. Anthrop. Inst.* 458 (*title*) On the classification of British Stone-Age industries. 1952 [see *Gravettian *a.* and *sb.*]. 1959 J. D. Clark *Prehist. S. Afr.* ii. 39 A single living or workshop site where a number of specimens all of the same age are associated is called an 'industry', while the term 'culture' is usually employed to describe a number of 'industries' all of the same type, and of which the distribution is more than purely local.
c. Preceded by a personal name or the like: scholarly or diligent work devoted to the study of a particular author or subject; also, the practice of a profitable occupation.
1965 *New Statesman* 9 Apr. 575/1 The Pindar industry began fairly early in antiquity, as we can still see in the surviving scholia. 1966 *Listener* 10 Nov. 685/1 The Shakespeare industry..is a very old one; I have an eighteenth-century snuffbox made from Shakespeare's mulberry tree. 1969 *Daily Tel.* 24 Apr. 22/3 The way to tackle 'Ulysses' is to plunge into it headfirst,..which is what we all had to do before the Joyce industry began. *Ibid.* 5 Aug. 14/2 The brisk pick-up of business in the abortion industry has greatly alarmed many people.
7. Comb. *industry-wide* adj.
1946 *Nature* 30 Nov. 797/1 Research should mainly be on an industry-wide basis, and for the benefit of the industry as a whole. 1958 *Listener* 24 July 116/2 It is possible to preserve industry-wide bargaining. 1971 *Publishers' Weekly* 2 Aug. 43/3 The American National Standards, Inc., the official agency for certifying industry-wide standards in the United States.

indwelling, *ppl. a.* Add: **b.** *Med.* Of a catheter, electrode, or other device: more or less permanently fixed in position either within the body or leading from the interior to the exterior of the body.
1932 Dorland & Miller *Med. Dict.* (ed. 16) 252/1 Indwelling catheter. 1962 *Lancet* 5 May 950/1 Infection nearly always followed on drainage by indwelling catheter. 1964 *Ibid.* 26 Dec. 1349/1 All the very immature babies..were fed from birth by indwelling tubes. 1968 *Jrnl. Thoracic & Cardiovasc. Surg.* LV. 555 (*heading*) Thoracic duct cannulation in the dog: a study of thrombotic occlusion of indwelling cannulas. 1972 *Amer. Jrnl. Physical Med.* LI. 113 (*title*) Electromyographic study of the anterolateral abdominal musculature utilizing indwelling electrodes.

-ine, *suffix*[5]. Add: also used systematically to form the names of certain six-membered monocyclic compounds having a nitrogen atom in the ring, as *azine*. Cf. *-in[1].
1928 *Jrnl. Amer. Chem. Soc.* L. 3078 In the field of six-membered [heterocyclic] rings are found names corresponding to the above systematic names for five-membered [heterocyclic] rings, but with the suffix -*ine* or -*in* replacing -ole (or -ol), as: triazine, oxazine, thiodiazine, dioxin (the latter being non-nitrogenous). Thus the ending -ine (or -in), although regarded as properly the ending for bases, has a specific sense in which it indicates a six-membered ring. 1940 in Patterson & Capell *Ring Index* 21. 1957 [see *-in[1]].

in-earnestness (inə̄·ɹnéstnės). *rare*[−1]. [f. In *prep.* + Earnestness; cf. Earnest *sb.*[1] 2 a.] Seriousness, serious intention.
1879 G. M. Hopkins *Lett. to R. Bridges* (1955) 89, I do avoid them [*sc.* inversions], because they weaken..the earnestness or in-earnestness of the utterance.

‖ **inédit** (inedi). [Fr.; cf. *inedita.] An unpublished work. Also *fig.*, something secret or unrevealed.
1910 *Encycl. Brit.* XII. 231/1 The inner, undiscovered, minute truths of contemporary existence, the *inédit* of life. 1922 A. E. Housman *Lett.* 25 Oct. (1971) 206, I know you bibliophiles and your passion for *l'inédit* regardless of merit. 1958 *Times Lit. Suppl.* 18 Apr. 212/4 Harmon is not an unknown figure, nor his journal strictly an 'inédit'. *Ibid.* 17 Oct. 592/2 There are magazines anxious for *inédits*.

inedita (ine·ditā). [mod.L., neuter pl. of L. *ineditus*, f. *in-* In-[3] + *editus*, pa. pple. of *edĕre* to give out, Edit *v.*] Unpublished writings.
1886 *Encycl. Brit.* XXI. 141/1 The luminous exposition of the grammar and the happy choice of the pieces in the chrestomathy..all inedita. 1948 *Mind* LVII. 522 Scholars ..brought to light a number of *inedita* and checked..his writings.

ineducability (ine:diūkăbi·lĭti). [f. Ineducable *a.* + -ity.] The condition of being ineducable.
1918 *Dial* (Chicago) 23 May 492 Carlyle in despair over the ineducability of nineteenth-century minds. 1927 Carr-Saunders & Jones *Survey Social Struct. Eng. & Wales* 214 The criterion of mental deficiency for adults..is social inefficiency, while for children..it is ineducability. 1964 M. Critchley *Developmental Dyslexia* xiii. 77 Perhaps the patient and his parents have resigned themselves to a state of hopeless ineducability, and no longer importune doctors and teachers.

inée (īnēi·). Also **onage, onaye.** [Fr., ad. Fang *ene*, Mpongwe *onai*.] An arrow-poison made from the seed of *Strophanthus hispidus*.
1874 Lindley & Moore *Treas. Bot.* II. Suppl. 1323/1 *Onaye* or *onage*..also called Inée and Kombé. 1887 *Encycl. Brit.* XXII. 608/1 The inée or onaye poison of the Gaboon, the kombé poison of equatorial North Africa, ..are..derived from members of this genus.

ineffability. (Later examples.)
1902 W. James *Var. Relig. Exper.* xvi. 381 *Ineffability*. .. *Noetic quality*... These two characters will entitle any state to be called mystical, in the sense in which I use the word. 1922 Joyce *Ulysses* 709 The supernatural character of Judaic scripture: the ineffability of the tetragrammaton: the sanctity of the sabbath. 1961 M. Laski *Ecstasy* xxii. 243 When *ineffable*, etc. is attached to a noun, 'supreme of its kind'..seems always to be implied. The *ineffability* device is particularly common among religious writers.

ineffable, *a.* Add: **1. b.** Applied to a person.
1832 [in Dict., sense 1]. 1961 Webster, Ineffable bungler. 1969 *Spectator* 14 Mar. 325/3 The ineffable Mr George Thomson.

ineffectual, *a.* Add: **c.** Also as *sb.*
1925 G. Greene *Babbling April* 4 You snobbish intellectual, Suburban ineffectual, Can't you feel that shimmy in the air?

inegalitarian (iniɡælitē̈ə·riăn), *a.* and *sb.* [f. In-[3] + *egalitarian *a.* (and *sb.*).] **A.** *adj.* Favouring, pertaining to, or marked by inequality. **B.** *sb.* One who denies or opposes equality between persons, = Inequalitarian.
1955 T. H. Pear *Eng. Social Differences* x. 211 The views of fervent English egalitarians and inegalitarians. 1961 *Guardian* 1 May 6/7 Madame Lefauchaux is dealing with their inegalitarian deficiencies. 1971 P. Worsthorne *Socialist Myth* ix. 219 The forces of democracy, *in practice*, compel society towards a shape that is highly inegalitarian. 1972 *Guardian* 15 June 14/5 The present distribution of financial aid [for housing]..is highly inegalitarian in its effects. 1972 *Daily Tel.* (Colour Suppl.) 24 Nov. 20/1 Paradoxically, Australia has a most remarkably inegalitarian educational system.
Hence **inegalita·rianism.**
1966 *Economist* 15 Oct. p. xxviii/1 This increased the inegalitarianism of what rapidly became a very inegalitarian German tax system. 1971 P. Worsthorne *Socialist Myth* v. 70 The Labour party has come to terms with the economic purpose of inegalitarianism.

inelastic, *a.* Add: **1. b.** Of a collision (esp. between sub-atomic particles), or the scattering of one particle by another: involving a reduction in the total kinetic energy of the particles or bodies that come together, or a change in their internal energies.
1847 L. D. B. Gordon tr. *Weisbach's Princ. Mech. Machinery & Engin.* I. iv. iv. 302 The vis viva lost by inelastic impacts is equivalent to the sum of the products of the masses and the squares of their loss or gain in velocity. *Ibid.* 310 These two general formula [*sic*] also embrace the laws of perfectly elastic and perfectly inelastic impact. 1907 J. H. Jeans *Theoret. Mech.* ix. 238 When the contact between the surfaces of two bodies is of such a nature that they do not rebound at all after impact, it is said to be perfectly inelastic. 1938 *Physical Rev.* LIII. 795/1 The results of experiments on the inelastic scattering of such fast neutrons are especially suitable for theoretical investigation because the number of excited states in which the nucleus may be left when the neutron is reemitted will be large enough to make statistical considerations valid. 1942 *Ibid.* LXI. 129/1 The scattering of fast neutrons by nuclei is at least partly inelastic in the case of medium and heavy masses like Fe, Ag, and Pb. 1958 W. K. Mansfield *Elem. Nucl. Physics* iv. 30 The compound nucleus formed in inelastic collisions disposes of its surplus energy, the kinetic energy and binding energy of the incident neutron, by several different processes involving the ejection of particles or radiation from the nucleus. *Ibid.* 33 For inelastic scattering to occur the neutron must have sufficient energy to leave the target nucleus in an excited state. 1962 *Gloss. Terms Nucl. Sci.* (*B.S.I.*) 103 In inelastic scattering the scattered particle or photon loses energy by exciting the struck nucleus. 1966 J. Harwood *Introd. Mech.* xi. 138 A collision between two balls of lead or putty would be inelastic.
2. b. *Econ.* Varying less than in proportion to changes in price; (more loosely) unresponsive to changes in price: applied either to demand (for commodities, money, labour, etc.) or to supply. So **i:nelasti·city.**
1890 A. Marshall *Princ. Econ.* III. iii. 167 There may be..violent changes..in the price of a thing which is not necessary, if it is perishable and the demand for it is inelastic. 1913 *Q. Rev.* Oct. 520 The demand for gas is comparatively inelastic. 1925 S. E. Thomas *Elem. Econ.* iv. 37 Demand is said to be elastic when a rise or fall in the price causes a more than proportionate rise in the amount demanded. On the other hand, demand is said to be inelastic when a fall or rise in price causes relatively little rise or fall in the amount demanded. *Ibid.*, The Elasticity and Inelasticity of Demand. 1929 *Ibid.* (ed. 4) 51 Even in the case of necessaries, the degree of inelasticity may vary considerably. 1969 D. C. Hague *Managerial Econ.* II. iii. 57 Demand at all of these numerical elasticities less than one is often described as being inelastic. 1973 *Lancet* 14 Apr. 815/1 Russell found that the demand for cigarettes is inelastic... This means that if prices rise by 1% demand falls, but by an amount less than 1%; total revenue from taxation would then increase just as long as demand remained inelastic.

inelastically (inflæ·stikăli), *adv.* [f. prec.: see -ically.] In a manner characteristic of inelastic bodies; with a reduction in the total kinetic energy.
1938 *Physical Rev.* LIII. 796/1 Inelastically scattered neutrons. 1942 J. D. Stranathan *'Particles' of Mod. Physics* vi. 229 It is possible that a few atoms..may occasionally collide inelastically, that one atom may spend a part of its kinetic energy in raising the electron of another atom to a higher energy level. 1966 D. G. Brandon *Mod. Techniques Metallogr.* 112 It introduces a proportion of inelastically scattered electrons into the transmitted beam. 1969 *Physics Bull.* Mar. 86/2 The primary electron transfers an amount of energy V_e to the atom and then leaves again, being observed externally as an inelastically reflected electron of energy $V_p - V_e$.

ineluctably, *adv.* (Later examples.)
1922 Joyce *Ulysses* 214 That lies in space which I in time must come to, ineluctably. 1939 —— *Finnegans Wake* I. 120 Those throne open doubleyous..reminding uus ineluctably of nature at her naturalest.
Also **ineluctabi·lity**, the condition of being ineluctable.
1943 *Mind* LII. 11 The limitation of the ability of a man to achieve salvation..may be, and often is, hypostasized to seeming ineluctability.

inemotivity (i:nėmə̄uti·viti). [f. In-[3] + Emotivity.] Lack of emotional sensibility.
1894 W. James in *Psychol. Rev.* I. 529 M. Sollier thinks ..that in complete *inemotivity* the visceral reactions themselves do not take place. 1902 —— in *Encycl. Brit.* XXXII. 66/1 We must remember that the patient's inemotivity may have been a co-ordinate result with the anæsthesia of his neural lesions, and not the anæsthesia's mere effect.

inenarrable, *a.* Delete '† *Obs.*' and add later examples.
1914 R. Brooke *Coll. Poems* (1918) 18 The inenarrable godhead of delight. 1923 A. Huxley *Antic Hay* xv. 213 Those Mohammedan ecstasies that last..six hundred inenarrable years apiece. 1936 —— *Eyeless in Gaza* xlvii. 525 The scent of the flowers was like the brief and inenarrable revelation of something more than earthly. 1967 *Listener* 5 Jan. 37/3 The music has an inenarrable greatness which quite transcends the occasion of its composition.

inenu·bilable, *a.* *rare.* [f. In-[3] + L. *ēnūbil-āre* to make clear (see Enubilate *v.*) + -able.] That cannot be cleared of clouds or mist, or (*fig.*) of obscurity; indistinct; inexplicable.
1903 *Sat. Rev.* 7 Feb. 169/1 This business of the avalanche is treated by the critics as something quite inenubilable. 1911 Beerbohm *Zuleika D.* xii. 191 There is nothing in England to be matched with what lurks in the vapours of these meadows, and in the shadows of these spires—that mysterious, inenubilable spirit, spirit of Oxford.

inepticality (ineptikæ·lĭti). *rare.* [f. Inept *a.* + -ical + -ity.] = Ineptitude.
1923 E. E. Cummings *Let.* 15 Sept. (1969) 102 A friend of mine called 'Slater' 'Brown', looks not too much like J. Christ and has attacks of total inepticality and is a very pleasant person.

inequable, *a.* Delete *rare* and add later examples.
1924 *Glasgow Herald* 21 Nov. 11 The inequable incidence of the regulations. 1926 A. E. Ellis *Brit. Snails* 31 An important obstacle in the way of fresh-water colonization is the inequable temperature of rivers and lakes.

inequalitarian. Add: Also as *adj.*
1949 *Mind* LVIII. 207 Consider an inequalitarian society in which there are rich people who buy 'luxuries' and poor people who buy only 'necessities'. 1966 *Punch* 13 Apr. 521/2 The ritualistically observed Sunday is inequalitarian... While the rich can enjoy their private swimming-pools..and so on, the poorer elements are denied access even to the inadequate public amenities.

inequi·valence. [f. In-[3] + Equivalence *sb.*] Lack of equivalence.
1879 A. Macfarlane *Princ. Algebra of Logic* 53 (*heading*) The signs of inequivalence > and <.

inequivalent, *a.* Delete † *Obs.* and add later example.

1954 I. M. Copi *Symbolic Logic* v. 127 A similar pair of inequivalent propositions may be written as [etc.].

inerasable, *a.* (Earlier example.)
1811 Shelley *Let.* 17 May (1964) I. 90 Nor do I think her Xtianity of the most inerasible nature.

inert, *a.* Add: **1. c.** *inert gas*: (*a*) As an ordinary use of the adj. with *gas*: any gas that is (relatively) inert. (*b*) Usu. as (*the*) *inert gases* (now apprehended as a special collocation, analogous to the terms *alkaline earths* and *rare earths*): any of the elements of group 0 of the periodic table, viz. helium, neon, argon, krypton, xenon, and radon, all of which are colourless, odourless, and tasteless gases which were formerly thought to be completely unreactive chemically, forming no compounds (though compounds of some of the gases are now known). Cf. *noble gas.
(*a*) 1885 W. Macgregor *Gas Engines* v. 127 Nitrogen retards the combustion of hydrogen and that of carbonic oxide... The inert gas at the same time lowers the temperature of combustion. 1911 *Encycl. Brit.* XIX. 715/1 Nitrogen is a very inert gas: it will neither burn nor support the combustion of ordinary combustibles. 1966 *McGraw-Hill Encycl. Sci. & Technol.* V. 553/2 The gas contains about 27% carbon monoxide and over 70% of inert gases (CO_2 and N_2), giving it the lowest heating value..of any of the commercially used fuel gases. 1973 *Sci. Amer.* Dec. 22/3 The flywheel and the generator-motor would operate in an atmosphere of inert gas (hydrogen or helium).
(*b*) [1898 W. Crookes in *Proc. R. Soc.* LXIII. 411 Professor Ramsay and Mr. Travers have discovered two other inert gases accompanying argon in the atmosphere. These are called Neon and Metargon.] 1902 G. S. Newth *Text-bk. Inorg. Chem.* (ed. 9) II. iv. 232 This property of nitrogen of uniting directly with magnesium was utilised in effecting the separation of the nitrogen of the air from the small quantities of argon and other 'inert gases' contained in the atmosphere. 1927 J. W. Mellor *Comprehensive Treat. Inorg. & Theoret. Chem.* VII. xlviii. 906 The five gases—helium, neon, argon, krypton, and xenon—are colourless, and without odour. They are chemically indifferent and are hence called the inert gases or the rare gases, or the noble gases of the atm. 1939 H. J. Reich *Theory & Applications Electron Tubes* xi. 396 An external electrode may also be used to initiate breakdown of a mercury pool tube containing a small amount of inert gas. 1950 N. V. Sidgwick *Chem. Elements* I. 10 Apart from the molecular ions occurring in the gas, there is in no case satisfactory evidence of the existence of chemical compounds of any of the inert gases. 1961 G. A. Cook *Argon, Helium & Rare Gases* I. i. 1 Unlike the atoms of oxygen, nitrogen, and some of the common gaseous elements, the atoms of the inert gases do not combine to form stable diatomic molecules. 1962 *Proc. Chem. Soc.* 218/2 Although inert-gas clathrates have been described, this compound [*sc.* xenon hexafluoroplatinate (v), $Xe^+[PtF_6]^-$] is believed to be the first xenon charge-transfer compound which is stable at room temperatures.

inertia. Add: **1. c.** *Photogr.* The exposure corresponding to the inertia point, from which the Hurter and Driffield speed of an emulsion may be calculated.
c 1886 F. Hurter in W. B. Ferguson *Photogr. Res. Hurter & Driffield* (1920) 12 Supposing that a source of diffuse white light of intensity one acting directly upon a plate needed a time *t* to so far alter a bromide of silver gelatine film on that plate that an impenetrable black deposit of silver was caused upon it on development... That time *t*..measures the inertia of the plate. 1888 Hurter & Driffield *Brit. Pat.* 5545, Upon one of the fixed pair of scales..we mark what we call and hereinafter define as the 'inertia' or slowness of the plate. 1899 C. F. Townsend *Chem. for Photographers* (ed. 2) iv. 73 Several of the dots are in a straight line; this is the period of correct exposure. The straight line is prolonged to meet the base line, and the number read off, which gives the 'inertia' of the plate. 1927 C. B. Neblette *Photogr.* ix. 238 The inertia is an inverse measure of the speed of the plate: that is to say, a slow plate has a high inertia while a rapid plate has a low inertia. *Ibid.*, The precise significance of the inertia as a measure of speed is somewhat difficult to define. The exposure which it represents is not the 'threshold exposure' (the minimum exposure necessary to produce a measurable density) nor does it indicate the maximum [? *read* minimum] exposure which will give proper rendering of the gradations of the subject, but an exposure somewhere between these extremes. 1955 E. F. Teal tr. *Lobel & Dubois's Sensitometry* 96 Characteristic curves are drawn for a range of development times using a developer containing no restraining bromide. The straight line portions of the curves intersect on the log exposure axis at *E*, called the inertia point. Distance *OE* is the inertia *i* and the H. & D. speed is defined as 34/*i*, *i* being measured in log candle-metre-seconds.
3. Special Comb.: **inertia governor** *Engin.*, a governor which operates by virtue of both centrifugal force and inertia (1934 in Webster); **inertia point** *Photogr.*, the point in which the straight-line portion of the characteristic curve cuts the horizontal (log exposure) axis when produced; **inertia reel** *Motoring*, a reel which enables a safety belt looped around it to be self-adjusting, esp. in *inertia reel (safety) belt*; also the belt served by the reel; **inertia selling**, the supply of goods to persons who have not requested them, in the hope that the recipients will not take the necessary action to refuse them; **inertia starter** *Aeronaut.*, a starter (Starter 6) which utilizes the energy stored in a flywheel.
1907 Sheppard & Mees *Investigations Theory Photogr. Process* II. vi. 221 The reciprocity failure may be considered to start at much the same point relatively to the inertia points in the two plates. 1955 [see sense 1 c above]. 1957 K. M. Hornsby *Sensitometry in Pract.* ii. 9 Hurter and Driffield regarded the inertia point as a fixed characteristic of the emulsion, and they used it as a criterion by which to specify the sensitivity of emulsions. This criterion is not now recognised as an unchangeable characteristic of the emulsion since..fog will shift it to the left..and a developer containing bromide does not give a constant inertia point for all development times. 1962 *Daily Tel.* 14 Aug. 16/6 The webbing runs off a small inertia reel mounted at the foot of the door pillar. 1962 *B.S.I. News* Dec. 10/1 With the exception of inertia-reel models..all the belts on display seemed to bear a Kite-mark label. 1968 *Economist* 6 July 49/2 Not so usual are inertia reel belts, the sort that are always held at the correct tension (many injuries to wearers of seat belts may arise because they were too loose). 1970 *Motoring Which?* Apr. 43/2 Daimler V-8 250..radio (with power operated aerial), automatic (inertia reel) safety belts; two years old. 1971 *Guardian* 25 Jan. 7/7 Spaghetti belts are replaced by inertia reels. 1968 *Times* 25 Apr. 14/3 Mrs. Butler..asked if the President of the Board of Trade would take powers to control inertia selling campaigns... She deplores this exploitation of inertia sales promotion... It will be difficult for people who are subjected to inertia selling. 1970 *Times* 13 Jan. 4/6 The Daily Mail reported on the banning..of inertia selling advertisements... Inertia selling was the practice by which companies sent unsolicited goods, then pestered people for payment if they were not returned. 1970 *New Statesman* 13 Feb. 217/1 Now the Consumer Council is urging another dairy monopolist to abandon the practice of slipping homogenised milk, at 1d extra on to doorsteps .., and billing for it if it is accepted. The company denies that this is inertia selling, but does it matter what you call it? 1972 *Guardian* 14 Oct. 1/4 Disreputable trading practices which victimize the housewife. Among these are 'inertia selling' and 'pyramid selling'. 1929 *Flight* 7 Mar. 181 (*title*) Inertia starters for aero engines. 1931 D. Garnett *Grasshoppers Come* 89 He..got the engine to fire with the inertia starter. 1958 *Times* 1 July p. iv/5 Simms Motor Units Ltd. have entered a Tensec inertia starter which enables Diesel engines to be started by hand with the minimum effort.

inertial, *a.* Add: **b.** Applied to a frame of reference in which Newton's first law of motion holds, i.e. a frame in which a body continues in a state of rest or of uniform motion in a straight line unless that state is altered by an external force.
1887 [in Dict.]. 1914 L. Silberstein *Theory of Relativity* i. 5 The 'fixed-stars' system of reference... We will call [this]..following the modern habit, the inertial system, or sometimes, also, the Newtonian system of reference. 1918 A. S. Eddington *Rep. Relativity Theory Gravitation* viii. 83 The measurement of the rotation of the earth detects something of the nature of a fundamental frame of reference—at least in the part of space accessible to observation. We shall call this the 'inertial frame'. 1924 *Physical Rev.* XXIII. 543 Lange and Mach have done much to spread a relative conception of motion. Lange introduced the name 'inertial system'. 1952 C. Møller *Theory of Relativity* ii. 36 The concept of simultaneity between two events in different space points consequently has an exact meaning only in relation to a given inertial system. 1953 E. T. Whittaker *Hist. Theories Aether & Electr.* II. v. 159 In Einstein's general theory, the velocity of light at any place has always the value *c* with respect to any inertial frame of reference for this neighbourhood. 1959 J. Aharoni *Special Theory Relativity* i. 3 Let *K* be an inertial frame of reference, then, according to the principle of relativity, in every set of axes *x′*, *y′*, *z′* which moves along a straight line and with constant velocity relative to *K* it is equally possible to adjust the clocks as in *K*, and all the frames of reference so obtained are inertial and are completely equivalent to each other not only with respect to the law of inertia, but also with respect to any other law or physical relation. 1970 *Nature* 17 Oct. 273/1 The experimental observation, by Michelson and Morley and others, of the isotropic constant velocity of light in inertial frames.
2. Special collocations: *inertial guidance*, (automatic) control of the course of a vehicle or vessel by a system employing the principle of inertial navigation; *inertial mass*, mass as measured by the ratio of the force on a body to the resulting rate of change of its momentum; cf. *gravitational mass; *inertial navigation*, navigation in which the course of a vehicle or vessel is calculated automatically by a computer, without the need for external observations or equipment, from its acceleration at each successive moment, this being measured by accelerometers whose orientation is gyroscopically controlled; *inertial system*, (*a*) (see sense b above); (*b*) a system for carrying out inertial guidance.
1955 *Aviation Age* Jan. 28 (*heading*) Dependable inertial guidance systems can be found. 1956 *Time* 30 Jan. 40/1 One [guiding system]..is 'inertial guidance'. Its heart is a subtle instrument that senses every force that acts on the flying missile... This information goes to a computer. 1958 *Economist* 16 Aug. 521/1 Navigation under the ice requires a further modern development—inertial guidance to replace conventional types of navigational aid which become unreliable near the pole or under water. 1962 F. I. Ordway et al. *Basic Astronautics* ix. 383 The heart of the inertial guidance system is the stabilized platform with its gyroscopes and accelerometers. 1920 R. W. Lawson tr. *Einstein's Relativity* xix. 65 If now, as we find from experience, the acceleration is to be independent of the nature and condition of the body and always the same for a given gravitational field, then the ratio of the gravitational to the inertial mass must likewise be the same for all bodies. 1955 Inertial mass [see *gravitational mass]. 1954 *Aviation Age* Sept. 34/2 The idea of 'absolute' guidance—which evolves rather naturally from advanced work in inertial navigation—is startlingly promising. 1957 *Sci. News Let.* 27 Apr. 259/1 Inertial navigation is of particular importance to the military because it is jam-proof. 1969 *New Scientist* 28 Aug. 418/2 Inertial navigation systems ..depend..on the fact that an acceleration integrated twice, gives distance run and therefore position. 1952 K. W. Gatland *Development of Guided Missile* iii. 56 Attempts are being made to develop the inertial system, e.g., the technique used in the A-4 rocket. 1962 F. I. Ordway et al. *Basic Astronautics* ix. 385 A possible solution to the problem of overcoming accumulated error over long flight times is an inertial system supervised by a continuous series of fixes on celestial bodies. 1970 *New Scientist* 1 Jan. 22/2 The inertial system, already tested by BOAC during some 500 hours of flying, has been working *en route* with an average error of only one nautical mile an hour.

inertialess (inə̄·ɹʃiălės), *a.* [f. Inertia + -less.] Having no inertia; responding instantaneously to any change in the forces acting on it.
1927 *Jrnl. Franklin Inst.* CCIV. 592 This correction.. causes the servo-mechanism to deliver the position which an ideal inertialess watthour meter would have. 1933 [see *iconoscope 2]. 1937 *Discovery* Nov. 329/1 These beams of inertialess particles are regulated in large vacuum tubes. 1961 G. Millerson *Technique Television Production* ii. 19 A small gun in the camera-tube generates a continuous beam of electrical particles (electrons). This fine inertialess 'pointer' explores the charge pattern on the camera-tube screen. 1971 *Physics Bull.* July 402/1 Amongst potential advantages of laser machining are the following: (1) Light is inertialess; hence high 'tool' velocities with very rapid stopping and starting becomes [*sic*] possible.

inertially (inə̄·ɹʃiăli), *adv.* [f. Inertial *a.* + -ly[2].] By means of or as a result of inertia or inertial forces.
1957 *Astronautics* Dec. 27/1 (*title*) Servo considerations in an inertially stabilized reference system utilizing air-bearing gyros. 1971 *Sci. Amer.* June 21/1 Energy-releasing fusion reactions..can be initiated and to some extent controlled within an 'inertially confined' plasma. 1971 *Ibid.* Dec. 57/1 The catalyst is inertially removed from the product vapors in a cyclone separator.

inescapably (inėskē̈i·păbli), *adv.* [f. Inescapable *a.* + -ly[2].] Inevitably; undeniably. Also **inescapabi·lity**, inevitability, undeniableness.
1881 *Academy* 24 Dec. 468/2 The single word 'inescapably' (i.e., inevitably)..has crept in we know not how. 1945 *Mind* LIV. 210 Thus the inescapability of some sort of realism is proved once again! 1963 *Times* 24 May 13/7 The very clarity and inescapability of the story which the landscape tells has helped to create the beginnings of a will to remedy these evils.

inessential, *a.* (*sb.*) **B.** *sb.* (Later examples.)
1902 *Daily Chron.* 22 Apr. 3/3 If we ignore the inessentials of place, costume, speech and employment, friendship is the same to-day as it has been for a myriad of yesterdays. 1927 *Observer* 22 May 8 Stripped of its inessentials (which are too roughly handled) his story comes down to this [etc.]. 1960 J. Betjeman *Summoned by Bells* 96 All the inessentials of the Faith.

ineuphonious (inyufō̈u·niəs), *a.* [f. In-[3] + Euphonious *a.*] Not euphonious.
1887 *Lancaster* (Pennsylvania) *Daily Examiner* 7 Apr., The name is too remote, too foreign and too ineuphonious. 1921 *Public Opinion* 15 July 62/3 Their own ineuphonious patois..communicates its harshness to the voice itself. 1927 *Brit. Weekly* 14 July 337/3 In spite of the..drawback of a somewhat formless, ineuphonious style.

inevictable (inīvi·ktăb'l), *a.* *rare.* [In-[3].] That cannot be evicted. Hence **inevi·ctably** *adv.*
1895 W. Stevens *Let.* 23 July (1967) 6 Some unnameable smathering of greasy fritters,..and of course the inevictable applesauce. 1954 W. Faulkner *Fable* (1955) 155 The scarlet-spurting stump inevictably aloft. *Ibid.* 166 The inevictable establishment in coeval space of the sum of his past.

inevitable, *a.* Add: In extended use: that cannot fail or is bound to occur, appear, be used, etc.; that is inherent (in) or naturally belongs *to* (see also quot. 1893). Hence *sb.* (with *an* or *pl.*), an inevitable fact, event, truth, etc.; a person who, or thing which, is necessarily chosen or employed.
1879 F. W. Farrar *Life St. Paul* Pref., The English version..only requires the removal of errors which were inevitable to the age in which it was executed. 1888 W. Whitman in *Century Mag.* (1911) Dec. 255/1 Grant was one of the inevitables: he always arrived; he was as

invincible as a law. 1893 *Funk's Stand. Dict.*, *Inevitable*..jocularly, customary; usual; as, the *inevitable* row with the cabman; the *inevitable* hash for breakfast. 1901 *Westm. Gaz.* 24 Apr. 4/2 Here at length are some true inevitables. 1903 *Ibid.* 28 Aug. 3/2 After the Canadian Arch it was only an 'inevitable' that there should be a rush to the Dominion. 1927 *Observer* 17 July 9/4 A further twenty volumes of Everyman's Library... The 'Areopagitica' was another of the inevitables. 1932 *N. & Q.* 6 Feb. 107/1 Illustrations of French wit;..of the 'inevitable' phrase, that gift to the world past all praise. 1932 H. Read in H. J. & H. Massingham *Great Victorians* 400 Patmore, at this stage of his inspiration, was no inevitable poet. 1936 L. C. Douglas *White Banners* viii. 167 Paul greeted them amiably, exchanged the inevitables with the comely Mrs. Edmunds [etc.]. 1965 *Listener* 22 July 112/2 It seemed that it was only a matter of time before..the United States would bow gracefully to the inevitable. 1974 *Daily Tel.* (Colour Suppl.) 4 Jan. 19/3 Walter Harris was awarded a light diet of prunes, apricots, grapefruit, broth, and the inevitable yoghurt.

inexpectation. (Later example.)
1946 M. Peake *Titus Groan* 390 He had made so rapid and nimble a detour of the stone table that he surprised Steerpike, appearing with such inexpectation beneath the boy's nose.

inexpellable (inekspe·lăb'l), *a.* *rare.* [f. In-[3] + Expellable *a.*] Incapable of being expelled.
1911 Beerbohm *Zuleika D.* iii. 33 He loved her, and he could not help seeing her... Inexpellable was her image.

inexpertise (inekspəɹtī·z). [f. In-[3] + *expertise.] Lack of expertise.
1926 *Chicago Even. Post* 12 Nov., It results in part from ignorance of foreign politics and international relations, and from inexpertise in discussing them. 1961 J. Sherwood *Half Hunter* vi. 74 The élite were performing their loops and spins in an atmosphere of solemnity and calm, oblivious of the revolving mass of inexpertise around them. 1963 *Times* 9 Mar. 11/3 If one grows impatient in performance, it is more often with vocal than compositorial inexpertise.

inexplicably, *adv.* (Later examples.)
1865 G. M. Hopkins *Poems* (1967) 162 Meadows to them inexplicably dear. 1903 G. B. Shaw *Man & Superman* p. x, Inexplicably forcing you to range the hero with his enemy the statue on a transcendent plane. 1922 Joyce *Ulysses* 680 Was the proposal of asylum accepted? Promptly, inexplicably, with amicability, gratefully it was declined.

inexquisite (inekskwi·zit), *a.* *rare.* [f. In-[3] + Exquisite *a.*] Not exquisite.
1922 Joyce *Ulysses* 264 The bar where bald stood by sister gold, inexquisite contrast.

in extenso: see also *in *Lat. prep.*

inface (i·nfē̈is). *Geomorphol.* Also **in-face.** [See quot. 1896.] The steep scarp-face of a cuesta.
1896 W. M. Davis in *Science* 15 May 732/2 The surrounding rims of harder stratified rocks offer interesting examples of outer slope and inface, with inner subsequent valleys, all in concentric circular arrangement. [*Note*] The invention of this excellent term, the abbreviation of 'inward facing escarpment', should be credited to Mr. L. C. Glenn, of Darlington, S.C. 1939 A. K. Lobeck *Geomorphol.* xiii. 451 Each cuesta has a steep inface and a gentle back slope, down the dip of the beds. 1954 W. D. Thornbury *Princ. Geomorphol.* v. 133 One of the common regional expressions of gently or moderately dipping rock is the cuesta. This has an abrupt escarpment or inface on the up-dip side and a more gentle backslope or dip slope extending in the direction of the regional dip.

infall. Add: **4.** (A) falling upon or into (esp. a planet) from an outside source.
1899 *Edin. Rev.* Oct. 328 The waning atmospheric stock of carbon is reinforced by meteoric infalls. 1961 *New Scientist* 23 Feb. 465/3 Cosmic infall dust..would also provide a source of very finely divided ferrous material [on the moon]. 1969 *Nature* 20 Dec. 1160/1 Relative to the galactic gas the average velocity of infall would be about 500 km s⁻¹. 1971 I. G. Gass et al. *Understanding Earth* iii. 67/2 At present the energy of infall of a meteorite or a space rocket is very large.
b. Material that falls or has fallen.
1960 *New Scientist* 18 Feb. 387/2 There is..some evidence of it in the spectra of comets, and fragments from these continue to form part of the meteoroidal infall. 1962 F. I. Ordway et al. *Basic Astronautics* v. 188 On airless worlds surface materials may be..mixed with cosmic infall, including meteoritic debris.

infalling (i·n,fǭ:liŋ), *a.* Also **in-falling.** [In *adv.* 11 a.] Falling into or towards something (specified or understood).
1954 J. R. R. Tolkien *Two Towers* iv. iv. 259 They washed themselves and drank their fill at the in-falling freshet. 1969 *Nature* 20 Dec. 1160/1 The gravitational field of the galaxy, as well as the infalling gas from the universe, will sweep the halo clouds back into the galactic layer. 1971 I. G. Gass et al. *Understanding Earth* iii. 67/2 As the Earth grew and its gravitational attraction became greater so the energy released by an infalling body increased.

infant, *sb.*[1] Add: **5. a.** *infant prodigy.*
1831 [see Prodigy 3 c]. 1924 R. M. Ogden tr. *Koffka's Growth of Mind* ii. § 4. 49 Infant-prodigies who fail to live up to their early promise. 1939 F. Pratt *Secret & Urgent* i. 25 He was Jean François Champollion, an infant prodigy.
b. *infant industry* (see quot. 1914).
1870 *Congress. Globe* App. 29 Mar. 240/3 But, argue our defenders of monopoly, let us protect our infant industries, and when they have grown to manhood..they will need no further protection. 1906 *Daily Chron.* 17 Oct. 6/5 The argument for Protection there is not at all the infant-industry argument. 1914 *Cycl. Amer. Govt.* II. 176 *Infant Industry*. This term is applied to the need of protecting new industries in order to give them opportunity to compete with older foreign establishments.
6. *infant mistress*, a woman teacher of infants at an elementary school; *infant mortality*, the death of infants, *spec.* of those less than a year old; *infant welfare*.
1877 W. R. Alger *Life Edwin Forrest* I. 147 The rate of infant mortality may be reduced to one per cent of its present murderous average. 1918 *79th Ann. Rep. Registrar-General 1916* (Cd. 8869) 35 Infant welfare organisations might well devote special attention to the first days of the life of illegitimate children. 1921 *N.Z. Educ. Gaz.* 1 Nov. 12/2 (Advt.), *Infant-mistress*—South Wellington (Grade VII G): £310–£320. 1922 *Encycl. Brit.* XXX. 650/2 *Infant Welfare Centres*.—The first task has been to coördinate the work at the Infant Centre and the visitation of the mothers in their own homes. *a* 1930 D. H. Lawrence *Phoenix II* (1968) 18 Last week the infant mistress did not come up, so I was alone. 1939 M. Spring Rice *Working-Class Wives* p. vii, The high rates of infant mortality..in the early years of the present century. *Ibid.*, Better methods of mothercraft through the influence of the Infant Welfare Centre. 1962 *Guardian* 9 Mar. 8/2 The orders might contain..infant welfare foods (distributed as a voluntary service). 1963 B. Pearson *Coal Flat* iii. 48 'The first infant mistress's job that's going,' she thought.

infanteer (infăntīə·ɹ). *slang.* [f. Infant(ry + -eer.] An infantryman.
1944 J. H. Fullarton *Troop Target* xxiii. 174 A lone infanteer..watched it all from a slit trench. 1961 *Guardian* 14 Nov. 8/3 Did he and his fellow-Gunners never talk of Guardees or Infanteers?

infanticipate (infănti·sipē̈it), *v.* Chiefly *U.S.* [f. Infant *sb.*[1] 1 + Ant)icipate *v.*] *intr.* To be in the state of expecting a child. Hence **infanti·cipating** *ppl. a.* and *vbl. sb.*
1934 W. Winchell in *News & Post* (Baltimore) 28 May 10/1 The J. Clark Baldwin, 3ds,..are infanticipating. 1941 in *Amer. Speech* (1942) XVII. 271/2 'Storkettes' for 'infanticipating' friends of yours. 1961 *N.Z. Listener* 8 Sept. 34/4 Although I agree that 'infanticipating' will be very hard to rival. 1963 *New Musical Express* 10 May 12/1 Frank Sinatra jnr. dating actress Jana Taylor... Infanticipating—singer Gene McDaniels' wife.
Hence **infanticipa·tion**, the state of expecting a child; the child that is *in utero*.
1934 W. Winchell in *News & Post* (Baltimore) 2 July 8/2 The Alan Mowbray infanticipation is due in late October. 1939 in *Amer. Speech* (1940) XV. 218/1 Cameraman Leonard Smith shot Miss O'Sullivan behind fern fronds, through leafy screens, at respectful distances, permitted his camera to drop no hint of her..infanticipation.

infantile, *a.* Add: **2.** *Geol.* Of a landscape: in the earliest stage of the cycle of erosion. Of a land form or feature: characteristic of such a landscape.
1885 *Proc. Amer. Assoc. Adv. Sci.* XXXIII. 429 Just as the surface of the deposit rises above its base-level of erosion..a smooth, unbroken plain is revealed... The smoothness of the surface and the shallow lakes are indeed truly infantile features, retained only during the earliest life of the plain, and soon lost in its further development. 1941 C. A. Cotton *Landscape* xvii. 191 The theoretical distinction between 'infantile' forms developing on a peneplain as it is slowly uplifted and the 'senile' forms it exhibited before uplift was first made by Walther Penck. 1968 R. W. Fairbridge *Encycl. Geomorphol.* 1110/2 Initial or infantile (i.e., uneroded) forms of mountains composed of such materials have indeed had no real existence, because they have been destroyed while relatively slow or intermittent upheaval has been in progress.
3. Special collocations: *infantile mortality* = *infant mortality; *infantile paralysis*, poliomyelitis (which affects chiefly the young).
1859 W. Moore (*title*) On infantile mortality, and the establishment of hospitals for sick children. 1901 *Daily Chron.* 14 Nov. 5/5 In England the term 'infantile' mortality applies only to the deaths of children under one year of age. 1911 G. B. Shaw *Getting Married* 135 The high birth-rate of the very poor is counterbalanced by a huge infantile-mortality in the slums. 1843 *Lancet* 27 May 301/1 There is a disease of very considerable frequency—I mean infantile paralysis—to which, I think, so much attention has not been given as its importance merits. 1916 *Daily Colonist* (Victoria, B.C.) 5 July 1/2 Twenty-five children died from the epidemic of infantile paralysis..during the 24 hours ending at 6 o'clock tonight. 1955 *Sci. News Let.* 16 Apr. 242/2 No matter how it is called, poliomyelitis, infantile paralysis, or polio for short, it is a scourge that has long been a crippler and killer. 1957 *Economist* 7 Sept. 847/1 Experience with the Salk vaccine during the past two years has convinced the National Foundation for Infantile Paralysis that the enemy is in full retreat.

infantilism (infæ·ntiliz'm). *Path.* [prob. ad. F. *infantilisme*: see Infantile *a.* and -ism.] Infantile or childish condition; *spec.* **a.** The state or condition of being physically undeveloped. **b.** *Psychol.* A condition in which infantile behaviour patterns persist, owing to some emotional repression in early life, and are dominant over more appropriate reactions.
[1871 F. V. Faneau de la Cour (*title*) Du féminisme et de l'infantilisme chez les tuberculeux.] 1895 W. D. Morrison in Lombroso & Ferrero *Female Offender* Introd. p. xvi, Sexual peculiarities, such as feminism in men, mascul[in]ism in women, and infantilism in both. 1897 tr. *T. Ribot's Psychol. of Emotions* 422 The formula which..sums up and explains the unstable is this: psychological infantilism. *Ibid.*, The term infantilism is equally applicable to the congenital and the acquired forms. The former have never left their childhood behind, the latter return to it. 1903 *Lancet* 30 May 1526/1 A case of Infantilism in a child, aged ten years, who had not grown since four years old. Her weight was 26 pounds and her height was three feet. 1909 Robertson & Mackenzie tr. *Tanzi's Text-bk. Mental Dis.* xxii. 670 There is then an arrest of the genetic instinct at the infantile stage—an erotic infantilism. 1924 J. Rivière et al. tr. *Freud's Coll. Papers* I. xiv. 277 The 'infantile sexual traumas' were in a sense supplanted by the 'infantilism' of the sexuality in these cases. 1924 W. B. Selbie *Psychol. Relig.* xv. 297 The worship of God is a form of infantilism, and survives because it meets a certain elementary need and satisfies the sense of dependence which man never altogether loses. 1943 S. R. Slavson *Introd. Group Therapy* vii. 224 Children whose parents are in conflict with each other are not only fixed in their infantilism as a result, but use it to exploit the two parents. 1950 R. H. Williams *Textbk. Endocrinol.* ii. 49 Gonadal hypoplasia with dwarfism..is sometimes referred to as infantilism or the Lorain–Lévi syndrome. 1953 A. W. Spence *Clin. Endocrinol.* xli. 495 The word 'infantilism' was first used in medical literature by Lorain (1871) when he wrote the preface of the thesis of one of his pupils, Faneau de la Cour, on feminism and infantilism in tuberculous patients. 1962 R. H. Williams *Textbk. Endocrinol.* (ed. 3) xv. 933 The differentiation of persistent sexual infantilism from constitutionally delayed adolescence.

infantilistic (infæ:ntĭli·stik), *a.* [f. Infantile *a.* + -istic.] Pertaining to, exhibiting or characterized by infantilism; abnormally immature.
1930 C. G. Seligman *Races Afr.* ii. 51 Rather should they be considered to represent an early human type 'infantilistic' both physically and mentally. 1936 *Discovery* June 167/2 Negritos, a pygmy, infantilistic, dark-skinned type.

infantility. Delete † *Obs. rare*[−1] and add examples. Also, an instance of infantile behaviour.
1919 M. K. Bradby *Psycho-Anal.* iii. 31 In these [*sc.* unreasoned convictions] we shall find his kinship with the primitive and with his own infantility. 1921 *Discovery* May 133/2 Its chief characteristics [*sc.* of the 'Personal Unconscious'] are its infantility and its compensatory character. 1928 *Daily Tel.* 28 Aug. 5/2 With amazing veracity and power, Tolstoy portrays all that infantility of the Russian race. 1930 A. Huxley *Brief Candles* 66 He laughed at every naïveté or impertinence she uttered.. and..led her on into fresh infantilities.

infantilization (infæ:ntĭləizē̈i·ʃən). [f. Infantil(e *a.* + -ization.] The action of prolonging or perpetuating a state of infancy.
1939 *Nature* 25 Feb. 325/2 Following Bolk's thesis of infantilization he would regard man as the permanent baby amongst mammals. 1943 D. M. Levy *Maternal Overprotection* iv. 53 Infantilization consists in the performance of activities in the care of a child beyond the time when such activities usually occur. 1970 G. Greer *Female Eunuch* 98 In pushing the masochistic role as the proper role for woman, psychology reinforces the infantilization which has gone on ever since she was born. 1971 Jonas & Klein (*title*) Man-child: a study of the infantilization of man.
Hence **infa·ntilized** *ppl. a.*, made the object of infantilization; **infa·ntilizing** *vbl. sb.* and *ppl. adj.*
1943 S. R. Slavson *Introd. Group Therapy* i. 8 The period of forbearance on the part of the group therapists is much shorter with pampered and infantilized children. 1943 D. Levy *Maternal Overprotection* v. 73 Maternal overprotection becomes largely an infantilizing process. 1965 *Listener* 29 Apr. 626/2 In this context, the infantilizing mother, who holds herself ever indispensable, is a kind of murderer.

infantinely, *adv.* Change example reference to: 1833 Mill *Lett.* (1910) I. 77.

infant-school. Add: (Earlier and further examples.) Also *infants* (or *infants'*) *school*.
The usual form is now *infant school*.
1824 S. Wilderspin *Importance Educating Infant Poor* 33 Rules to be observed by the Parents of Children admitted into the Spitalfields Infant School. *Ibid.* 50 An infant school may be regarded..as a combination of the school and nursery. 1839 [see *Coleridgian *a.* and *sb.*]. 1921 Mencken *Amer. Lang.* (rev. ed.) 120 An English boy whose father is unable to pay for his education goes first into a *babies' class*..in a *primary* or *infants'* school. 1945 *Guide Educ. Syst. Eng. & Wales* (Min. Educ.) 59 *Infants School* (*Infants Department*), primary school for children of about 5 to 7 years, including in some cases classes for children of 3 and 4. 1969 I. & P. Opie *Children's Games* xii. 331 Even beyond Infant School the girls sometimes play 'Mothers and Fathers'.

infare, *sb.* **2.** (Delete *western*, and insert earlier U.S. examples.) Also *transf.*
1744 J. MacSparran *Diary* (1899) 18 Dr. Hazard and Betty Gardiner went to..Billy Hazard's weding. They

are both gone again..to the Infair. 1794 in *Amer. Pioneer* II. 223 An Infair was given to-day by Mason, to a fellow named Kuykendall, who had..run off with Mason's daughter..a few weeks ago. 1872 Schele de Vere *Americanisms* 236 The minister is said to settle,..a ceremony which..in many churches is made the occasion of much ceremony, called an installation or infare, because resembling an old-fashioned wedding festival.

infatuate, *ppl. a.* Add: Hence as *sb.*, an infatuated person.
1934 in Webster. 1949 *Scrutiny* XVI. 210 The earlier criticism of *Antony and Cleopatra* tended to stress the downfall of the soldier in the middle-aged infatuate.

infauna (i·nfǭnă). [ad. Da. *ifauna* (C. G. J. Petersen 1913, in *Beretn. f. d. Danske biol. Station* XXI (in *Fiskeri-Beretn.* 1912). 15), f. In-[3] + Fauna.] A collective term for animals that live just beneath the surface of the sea bed.
1914, 1964 [see *epifauna]. 1969 G. Vevers tr. *Friedrich's Marine Biol.* v. 254 The infauna, that is the animals living in the substrate, is either liberosessile, semisessile, or more or less motile.

infeasibility. (Later examples.)
1955 *Bull. Atomic Sci.* Mar. 78/2 That such a development is not now in sight does not offer conclusive proof of technical infeasibility. 1965 H. I. Ansoff *Corporate Strategy* (1968) viii. 130 This example is based on an actual diversification history in which the infeasibility was not recognized.

infectious, *a.* Add: **2. c.** In the names of various diseases, as *infectious hepatitis*, an acute infectious virus disease characterized by hepatitis and jaundice; *infectious mononucleosis*, an acute disease (also called glandular fever) chiefly affecting young adults, characterized by fever, swelling of the lymph nodes, and leucocytosis.
1920 Sprunt & Evans in *Bull. Johns Hopkins Hosp.* XXXI. 410 (*heading*) Mononuclear leucocytosis in reaction to acute infections ('infectious mononucleosis'). 1945 *Amer. Jrnl. Med. Sci.* CCX. 561 (*heading*) Homologous serum hepatitis and infectious (epidemic) hepatitis. 1970 Passmore & Robson *Compan. Med. Stud.* II. xviii. 11/1 Infectious hepatitis is usually spread by faecal contamination from a patient or convalescent carrier of the disease. 1970 A. J. Zuckerman *Virus Dis. Liver* xiii. 149 Infectious mononucleosis is an endemic disease affecting principally adolescents and young adults.

infective, *a.* Add: **2. b.** *infective hepatitis* = *infectious hepatitis.
1939 G. M. Findlay et al. in *Trans. R. Soc. Trop. Med. & Hygiene* XXXII. 578 To those cases where the jaundice is associated with sensitivity phenomena the term 'allergic jaundice' might be applied, while for those in which hepatitis is the primary lesion 'epidemic' or possibly 'infective hepatitis' is probably preferable to the name 'common infective hepatic jaundice' used by Rolleston and McNee (1929). 1959 [see *hepatitis]. 1971 *New Scientist* 25 Mar. 676/1 Infective hepatitis..results from liver cell damage caused by a virus (or viruses) usually taken in with food or drink, and picked up from the faeces of an already infected person.

‖ **infectum** (infe·ktə̆m). [L. (Varro *De Lingua Latina* ix. xcvii).] (See quot. 1954[2].)
1954 L. R. Palmer *Latin Lang.* II. ix. 266 In Latin the three aspects of the IE. verb were reduced to two, for the verbal system shows a contrast only between the *infectum* and the *perfectum*. 1954 Pei & Gaynor *Dict. Ling.* 100 *Infectum*, the aspectual category introduced by the Roman grammarian Varro (1st century B.C.), including the present, preterit and simple future tenses. 1965 *Amer. Speech* XL. 113 The infectum verbal form.

in-feed (i·nfīd). Also **infeed.** [In *adv.* 11 d.] **a.** The action or process of supplying a machine; *spec.* in centreless grinding, movement of the work-piece part-way into the space between the two wheels followed by its withdrawal, in contrast to its passage right through. **b.** A mechanism that carries out this process. Freq. *attrib.*
1901 *Manning, Maxwell & Moore Catal.* 636 The upper in-feed tool is fluted. 1926 *Automotive Industries* 2 Sept. 385 The back carriage has only a straight in-feed and no longitudinal motion except for setting tools. 1937 Colvin & Stanley *Grinding Pract.* v. 80 Work having shoulders or heads..is ground by the infeed method. 1960 *Times Rev. Industry* Sept. 39/3 Two independently variable rates of in-feed are provided. 1964 S. Crawford *Basic Engin. Processes* vii. 190 At the end of each traverse the wheel is fed in towards the work..; this movement is referred to as in-feed, and it controls reduction of work diameter. 1968 *Gloss. Terms Offset Lithogr. Printing* (*B.S.I.*) 35 *In-feed roller*, a roller aiding or controlling the forward movement of a sheet to the printing or other processing units. 1971 *Timber Trades Jrnl.* 14 Aug. 57/2 On model KDK 36, with an electric infeed, the speeds are 36 and 72 ft/min.

infer, *v.* Add: **4.** (Further examples.)
This use is widely considered to be incorrect, esp. with a person as the subject.
1813 J. S. Stanhope in A. M. W. Pickering *Mem.* (1902) II. 377 He said nothing that could in any way infer the necessity of a retreat of the allies. 1814 Scott *Waverley* II. xviii. 284 They are..more benign in demeanour than their physiognomy or aspect might infer. 1946 M. Peake *Titus Groan* 373 That he had fulfilled his intention of inveigling his enemy to the place of his own choosing must surely infer that the initiative once again lay with him. 1969 Bennison & Wright *Geol. Hist. Brit. Isles* i. 5 A. Holmes deprecated the use of the term 'Absolute' age as inferring an accuracy which is unwarranted. 1970 *Private Eye* 2 Jan. 12, I can't stand fellers who infer things about good clean-living Australian sheilahs. 1973 *Daily Tel.* 30 June 14, I have seen references..to the watering of Ascot racecourse, inferring that the water has been taken from public mains at a time when economy is being urged on all consumers.

inferably (i·nfərăbli), *adv.* [f. Inferable *a.* + -ly[2].] By inference; = *inferribly *adv.*
1903 *Harvard Psychol. Stud.* I. 340 The extent of this differentiation—and inferably the definition of rhythmical synthesis—corresponds to the reported musical aptitudes of the subjects.

inference. Add: **1. b.** *inference rule*, in a system of logic: any rule permitting inferences of a specified form.
1962 Clark & Welsh *Introd. Logic* ii. 93 We lay down the following cluster of inference rules. 1964 Kalish & Montague *Logic* 14 From the second and third lines of the derivation we may infer 'Q' by means of an inference rule. 1965 B. Mates *Elem. Logic* vi. 93 We need a reasonably small group of simple inference-rules.
2. (Further examples.) Also, an implication; the conclusion that one is intended to draw. Cf. *infer *v.* 4.
1933 D. L. Sayers *Hangman's Holiday* 147 'I don't know if you realise, Mr. Egg,' observed the inspector, 'the bearing, or, as I might say, the inference of what you said just now.' 1972 P. H. Kocher *Master of Middle-Earth* (1973) v. 82 These four are named 'first', with the inference that they deserve priority. 1973 *Daily Tel.* (Colour Suppl.) 5 Oct. 7/2 The main inference of the propaganda is that unless we adopt the metric system Britain will lose export orders.

inferiority. Add: **2.** *attrib.* and *Comb.* **inferiority complex**, generalized and unrealistic feelings of inadequacy caused by a person's reactions to actual or supposed inferiority in one sphere, sometimes compensated for by aggressive self-assertion; *colloq.*, exaggerated feelings of personal inadequacy; also *inferiority feeling*.
1922 A. G. Tansley *New Psychol.* (rev. ed.) xix. 214 Thus the 'inferiority complex' may account for a whole series of well-known human traits. 1924 C. Mackenzie *Heavenly Ladder* xxiii. 288, I could psycho-analyse all Bloomsbury now. They all suffer from an inferiority complex. Either they feel themselves intellectually inferior to Newton or physically inferior to Sandow or morally inferior to Christ. 1925 *N.Y. Times* 20 Sept. ix. 12/2 Those psychic disturbances which Freudians attribute to repressed sex impulse Adler attributes to a deficiency in the mechanism of self-assertion—to the 'inferiority complex', which today is on the tongue of thousands who have no idea of what they are talking about. *Ibid.* 12/3 Dr. Adler replied: 'Individual psychology holds that the most important key to the understanding of both personal and mass problems is the so-called sense of inferiority, or inferiority complex, and its consequences.' 1926 W. McDougall *Outl. Abnormal Psychol.* xxvii. 433 The inferiority complex is an important factor in some neurotics. 1933 W. J. H. Sprott tr. *Freud's New Introd. Lect. Psycho-Anal.* 88 A writer who brings in the expression 'inferiority-complex' thinks he has satisfied all the demands of psycho-analysis.... As a matter of fact the phrase 'inferiority-complex' is hardly ever used in psycho-analysis. 1947 N. L. Munn *Psychol.* xi. 224/2 Alfred Adler..popularized the terms 'inferiority and superiority complex' which are regarded as outcomes of, respectively, unsuccessful and successful exploitation of the 'will to power'. 1954 L. Carmichael *Manual of Child. Psychol.* (ed. 2) xviii. 1118/2 Such a child..will develop an inferiority complex and will soon slow down in his attempts to cope with his environment. 1974 *Daily Tel.* 9 Mar. 15/3 Mrs Thompson seemed to have an inferiority complex. Strangers were made to feel unwelcome. 1934 H. C. Warren *Dict. Psychol.* 137/2 Inferiority feeling. 1937 C. Day Lewis *Starting Point* i. ii. 32 I'd not be sure that I wasn't doing it from false motives—from envy..or inferiority-feeling. 1961 J. A. C. Brown *Freud & Post-Freudians* iii. 39 In order to compensate for inferiority feelings, each child develops..his own particular strategy.

inferrability (infə̄ɹăbi·lity). Also **inferability.** [f. Inferrable *a.* + -ity: see Inferrible *a.*] = Inferribility.
1914 C. D. Broad *Perception* ii. 128 This inferrability of one attribute from another. 1924 W. E. Johnson *Logic* III. viii. 104 The term 'one-one correspondence' is understood as equivalent to reciprocal inferability.

inferribly (infə̄·rĭbli), *adv. rare.* [f. Inferrible *a.* + -ly[2].] By inference. (Cf. *inferably *adv.*)
1908 J. M. Robertson *Did Shaks. write Titus Andron.?* 198 It would inferribly be his latest play, as it has 69 double-endings to 461 lines of blank verse. 1913 —— *Baconian Heresy* 18 When they were written we know not ..but it was inferribly before 1623.

infestation. Add to def.: Also, the state or condition of being infested.

infield, in-field, *sb.* Add: **3.** (Examples.)
1867 H. Chadwick *Base Ball Player's Bk. Reference* 138 The In-Field.—That portion of the field within the base lines. 1897 *Encycl. Sport* I. 76/1 Short Stop..is also called upon to back up all the positions of the in-field. 1906 *Spalding's Base Ball Guide* 15 The 'infield' team comprising the three base players and short stop. 1912 C. Mathewson *Pitching in a Pinch* ii. 23 Devoe beat out an infield hit. 1970 *New Yorker* 3 Oct. 32/1 Soon there will be nobody around who remembers muddy infields.
4. *Cricket.* The part of the playing area near the wicket; *collect.*, the fieldsmen stationed there (as opposed to the Outfield). **b.** = *in-fieldsman.
1898 G. Giffen *With Bat & Ball* vii. 99 In the in-field the soil had been well watered and was fairly hard... But what of the out-field? *Ibid.* xi. 188 He is a brilliant in-field. 1954 A. G. Moyes *Austral. Batsmen* iii. 46 Few bowlers..relish seeing the ball flying back over their heads, forcing them to weaken the in-field to stop the fours. 1960 E. W. Swanton *W. Indies Revisited* vii. 167 May's in-field was spread so deep for Sobers as to invite him to take a single almost anywhere.
5. *U.S.* The area enclosed by a race-track. Also *attrib.*
1923 E. Hemingway *Three Stories & Ten Poems* 36 You could see them [*sc.* horses] way off across the infield all in a bunch starting on the first swing like a lot of little toy horses. 1929 —— *Farewell to Arms* xx. 138 We..walked across the infield and then across the smooth thick turf of the course to the paddock. 1934 in B. A. Botkin *Treas. S. Folklore* (1949) II. iv. 404 His cry is echoed by a mighty 'They're off' from the packed stands and the infield crowd. *Ibid.* 407 They buried Black Gold there in the infield of the track. 1966 *Publ. Amer. Dial. Soc.* 1964 xlii. 6 *Infield*, area inside the racing course.
B. *adv.* In or towards the centre of a playing-field.
1959 *Times* 18 Feb. 5/1 Evans..came infield and burst through the centre. 1960 V. Jenkins *Lions Down Under* viii. 116 A brilliant try by Malcolm Thomas, who ran down the touch-line..before cutting in-field to touch down under the posts.

infielder, in-fielder (i·n,fī:ldəɹ). [f. Infield *sb.* + -er[1].] **1.** *Baseball.* One of the players on the in-field.
1867 H. Chadwick *Base Ball Player's Bk. of Ref.* 138 The in-fielders include the first six players of a nine. 1897 *Encycl. Sport* I. 76/2 Third Baseman..must be more on the alert..than the other in-fielders. 1905 *McClure's Mag.* June 123/2 Hutchinson, an in-fielder, had played in the same league with Johnson. 1967 *Boston Sunday Herald* (TV Mag.) 7–13 May 12/1 He hit a line drive right at the pile of baseballs, causing them to scatter in all directions. Every infielder came up with a different ball.
2. *Cricket.* = next.
1927 M. A. Noble *Those 'Ashes'* 193 Macartney..was going very quickly, picking out the gaps in the field or lofting the ball over the heads of the in-fielders. 1930 C. V. Grimmett *Getting Wickets* ii. 49 When a new batsman comes in, have your in-fielders close enough to save the singles.

in-fieldsman (i·n,fī:ldzmæn). *Cricket.* [f. In *adv.* + Fieldsman.] A fieldsman placed close to the wicket.
1910 *Westm. Gaz.* 19 Mar. 18/1 If the ball got past the in-fieldsmen there were men on the boundary to save the four. 1928 *Daily Tel.* 17 July 17/5 Nor was Freeman the only bowler who had his infieldsman too far away.

infight, *v.* Restrict † *Obs. rare* to sense in Dict. and add: **2.** To fight or box at close quarters; also *fig.* and *transf.* (cf. next).
1916 J. B. Cooper *Coo-oo-ee* xi. 156 Jack glued his chin to his chest and 'smothered', watching his opportunity to in-fight. 1966 T. Pynchon *Crying of Lot 49* vi. 164 We face this anarchy of jealous German princes, hundreds of them scheming, counter-scheming, infighting.

in-fighting, *vbl. sb.*, **in-fighter.** Add: Also *fig.* and *transf.*
1907 J. London *Iron Heel* ii. 38 He was unused to this fierce 'infighting', as Ernest called it. 1928 L. Woolley *Sumerians* ii. 52 Two [spears] have plain butts and are intended for in-fighting. 1930 H. G. Wells *Autocracy of Mr. Parham* iv. i. 274 The temperament and tradition of both navies disposed them for attack and in-fighting. 1936 G. B. Shaw *Millionairess* iv. 193 *Epifania*. You are two stone heavier than I; and I cannot keep my head at infighting as you can. 1960 Auden *Homage to Clio* 79 Untrained in a ruthless Verbal in-fighting. 1970 W. Smith *Gold Mine* xxxi. 78 She would use even the dirtiest in-fighting to see that Rod was not overlooked. 1972 *Listener* 24 Aug. 241/1 The cheerful in-fighting that characterizes intellectual exchange between clinician and medical scientist. 1973 *Time* 25 June 38/2 Inside the Politburo, he is known as a tough infighter.

infill, *v.* Delete *rare* and add later examples. Also *intr.*
1958 *Antiquity* XXXII. 110 This had been infilled with weathered flints. 1971 *Farmer & Stockbreeder* 23 Feb. 39/1 He is renovating the drainage system, in-filling and levelling, so 230 acres can be ploughed. 1971 P. Gresswell *Environment* 133 It is possible to 'infill' between two distant houses. 1972 *Daily Tel.* 23 May 15/4 Get the outline first, with basic shrubs...Then 'infill' with smaller shrubs and roses. 1973 *Nature* 23 Mar. 227/1 Channels, cut by streams rejuvenated during regression of the sea, have been infilled by alluvial and terrestrial clastics.

infill (i·nfil), *sb.* = Infilling *vbl. sb.* (various senses).
1929 *Geogr. Jrnl.* Jan. 56 While it is not theoretically impossible for great stretches of the present flood-plain to have been formerly aggraded, followed by removal of the infill [*sic*]..such an explanation involves difficulties. 1958 *Antiquity* XXXII. 110 The basis of the infill of the enclosure. 1958 *Archit. Rev.* CXXIII. 102/2 Infill panels of brick, precast concrete or coloured glass. 1966 J. S. Cox *Illustr. Dict. Hairdressing* 83/1 *Infill*, (1) The dressed hair enclosed within a specified area. (2) Knotting hair within a specified area. 1969 *New Scientist* 15 May 352/3 Sound absorbing infill is there primarily for heat and fire resistance. 1970 *Observer* 15 Mar. 7/6 Local landowners..stand to make small fortunes out of selling their acres to the infill excavators. *Ibid.*, Welsh infill could not be used on the M4. 1971 *New Scientist* 6 May 317/1 The full four-runway airport will undoubtedly need more infill than could be provided from..dredging.

infilling, *vbl. sb.* Add: (Later examples.) Also *fig.*
1895 J. MacNeil *Spirit-Filled Life* xiv. 100 The Lord Jesus is prepared to grant us a fresh infilling, a 'refilling' of the Holy Ghost. 1958 *Times Rev. Industry* May 46/1 Aluminium curtain walling in conjunction with a steel framed building..reduces labour charges..incurred by the use of the normal masonry infilling. 1961 *Guardian* 27 May 5/5 In its earliest form, watercolour was little more than infilling for sketches in ink and bistre. 1962 *Listener* 1 Feb. 228/3 Those coloured earths which in French gardens so often provide the infilling for *broderies*. 1970 *Cabinet Maker & Retail Furnisher* 30 Oct. 201/2 Infillings of horse hair..and hand-springing are used exclusively. 1972 L. Alcock *By South Cadbury* viii. 178 A plausible reconstruction of the building would suggest an infilling of wattle and daub for the walls.
b. *spec.* in *Town Planning*.
1943 Forshaw & Abercrombie *County of London Plan* 35 Five types of units of dispersed industrial population: 1. Infilling of gaps in incomplete schemes within the County. 1944 P. Abercrombie *Greater London Plan 1944* 35 A rounding off of the communities by infilling on the backlands and vacant frontages. 1964 *Daily Tel.* 17 Mar. 23/2 No part of his plan could possibly be regarded as infilling. By which they meant one or at most two houses pushed into existing gaps. 1971 P. Gresswell *Environment* 134 Infilling in villages is in many cases both inevitable and desirable. 1974 *Times* 5 Feb. 10 New building, other than infilling, was more extensive on the mainland.

infiltrate, *v.* Add: (Now usu. pron. with stress on initial syllable.) **4.** *Mil. trans.* and *intr.* To penetrate (enemy lines) by the gradual or surreptitious movement of small numbers of troops; to move (one's own troops) surreptitiously into the enemy's lines. Also *fig.*, esp. for the purpose of political subversion.
1934 in Webster. 1944 *Times* 1 Apr. 8/1 Skilfully infiltrating through the chain of Japanese outposts and garrisons, the force penetrated hostile territory as far as the Shan States. 1956 A. H. Compton *Atomic Quest* 125 Some of its branches had been infiltrated by Communists. 1958 [see *fifth column]. 1972 *Daily Tel.* (Colour Suppl.) 21 Jan. 27/1 There CIA agents succeeded in infiltrating several of its groups. 1972 *Daily Tel.* 30 Mar. 4/6 Terrorist groups started infiltrating from neighbouring African States. 1972 *Sunday Express* 21 May 17/5 Security men fear that Moscow would take the chance to infiltrate agents into Britain. 1972 *Daily Tel.* 19 June 2/1, I was paid..about £500 for infiltrating the IRA network in London. 1974 *Ibid.* (Colour Suppl.) 1 Mar. 7/2 Equally insidious is the way in which Communists..are infiltrated into the sphere of education.

infiltration. Add: **1. d.** The gradual penetration of one people into another.
1904 *Westm. Gaz.* 14 Nov. 5/1 In the interior of the Empire the French work of gradual 'infiltration' will proceed by not less efficacious means. 1927 Peake & Fleure *Priests & Kings* 54 As time went on there seems to have been an ever-increasing infiltration of Southern Steppe-folk from the desert. 1930 J. L. Myres *Who were the Greeks?* ii. 55 The southward infiltration of Albanian and Slav into districts formerly Romanized.
e. *Mil.* The gradual or surreptitious penetration of enemy lines by small numbers of troops.
1930 *Economist* 16 Aug. 313/1 They thus succeeded..in reaching the outlying quarters of Peshawar, albeit in small numbers, by a process of nocturnal 'infiltration'. 1933 B. H. L. Hart *Future of Infantry* 27 We profited from the lesson taught us by the remarkable success, at our expense, of the new infiltration or soft spot tactics in..1918. 1967 *N.Y. Times* (Internat. Ed.) 11–12 Feb. 1/6 At a background briefing early in November, the American command made available infiltration figures covering the year through Sept. 30.
fig. (Cf. *infiltrate *v.* 4.)
1940 *Economist* 15 June 1036/1 The Nazis have developed the technique of infiltration to such a pitch that [etc.]. 1941 *Ann. Reg. 1940* 209 Great uneasiness was caused in the country by the infiltration..of thousands of able-bodied young Germans in the guise of tourists. 1949 Koestler *Promise & Fulfilment* II. v. 281, I wonder whether an American don is the right match for the propaganda and infiltration experts of the Soviet Union. 1958 *Times* 20 Jan. 5/7 Alleged Communist infiltration into the Oxford branch of the National Union of Railwaymen. 1973 P. Evans *Bodyguard Man* v. 44 You're an ex-Special Branch man, supposedly a professional at infiltration techniques.
4. **infiltration anæsthesia, anæsthetization** of an area by the injection into it of a local anæsthetic; **infiltration capacity**, the maximum rate at which soil in a given condition can absorb water; **infiltration rate**, the rate at which soil absorbs water; infiltration capacity.
1897 *Med. Rec.* LI. 44/1 (*heading*) The method of infiltration anæsthesia: its technique and..advantages. *Ibid.* 45/2 Cocaine anæsthesia lasting from two to five minutes, infiltration anæsthesia from fifteen to twenty minutes. 1958 J. H. Burn *Lect. Notes Pharmacol.* (ed. 5) 36 Lignocaine..is a useful anaesthetic for producing nerve block and for infiltration anaesthesia. 1933 R. E. Horton in *Trans. Amer. Geophysical Union* XIV. 446 'Infiltration-capacity' will be used to denote the maximum rate at which rain can be absorbed by a given soil when in a given condition. 1952 *Proc. Soil Sci. Soc. Amer.* XVI. 85/2 The term 'infiltration capacity'..has been an object of some controversy because it implies that an extensity is involved, whereas an intensity such as *infiltration rate* would be more apt. 1940 *Ibid.* V. 400/2 It is unfortunate that the terms 'infiltration-capacity' and 'infiltration rates' have sometimes been confused... There may be an infinite variety of rates but there is only one capacity at a particular time for a particular soil. 1952 *Ibid.* XVI. 88/1 Infiltration rate [soil]: The maximum rate at which soil, in a given condition at a given time, can absorb rain. 1957 *Soil Sci.* LXXXVII. 338 After infiltration begins, increasing the initial moisture content reduces the infiltration rate.

infiltrator (i·nfiltrēitər). [f. Infiltrate *v.* + -or.] One who infiltrates (esp. in sense *4 of the vb.).
1944 *Infantry Jrnl.* (U.S.) May 28 Our men had orders not to leave the lines in case of any infiltration, because our rear elements would take care of the infiltrators. 1949 Koestler *Promise & Fulfilment* I. xv. 165 Arab infiltrators from neighbouring countries were executing 'considerable administrative control' in Samaria. 1953 *Wall St. Jrnl.* 1 Sept. 1/3 Police already have arrested nearly 4,000 infiltrators from beyond the Suez zone. 1957 L. F. R. Williams *State of Israel* x. 100 The 'economic infiltrators' who slip across the frontier to harvest crops on fields once their own. 1959 *Economist* 27 June 1154/1 The much greater confidence existing at Salisbury between union officials, shop stewards and the rank and file..is always the first target of communist infiltrators on the spot. 1970 *Guardian* 2 May 11/5 The Mekong..seldom poses problems to Communist infiltrators. They cross by raft or skiff at night.

infiltrometer (infiltrɒ·mitər). [f. Infiltr(ation + -ometer.] An apparatus for measuring the rate at which soil can absorb water.
1940 P. B. Rowe *Construction, Operation & Use of North Fork Infiltrometer* (U.S. Dept. Agric.) 1 Laboratory tests of the infiltrometer and the infiltration studies in which it has been employed since it was first developed in 1934 indicate that it affords a practical and reliable means of measuring the infiltration capacity of soils under natural conditions. 1958 R. K. Frevert et al. *Soil & Water Conservation Engin.* iii. 47 Infiltrometers using simulated rainfall cover areas varying from less than 2 to nearly 100 square feet. 1971 *Agronomy Jrnl.* LXIII. 306/2 High infiltrometer tests were made in the spring and fall.

‖**infima species** (i·nfimă spī·ʃīz). Pl. **infimæ species.** [f. L. *infima* nom. fem. of *infimus* lowest (see Infimous *a.*) + Species *sb.*] The lowest species of a classification or division; *concr.*, an 'infimous' person (?*obs.*).
1645 J. Howell *Lett.* I. xii. 23 Being contented to be the *infima species*, the lowest in the predicament of your frends. 1843 Mill *Logic* II. III. xxii. 134 Such generalizations..ought to be grounded upon an examination of all the *infimæ species* comprehended in them. 1961 T. Landau *Encycl. Librarianship* (ed. 2) 283/1 Any one species may, in its turn, be capable of further division into species, when for that purpose it becomes a genus—and so on down to the most minute species required, the *infima species*.

infimum (infəi·mɒm). *Math.* [L., = lowest part, neut. of *infimus* lowest (see prec.).] The largest number that is less than or equal to each of a given set of real numbers; an analogous quantity for a subset of any other ordered set.
1940 G. Birkhoff *Lattice Theory* ii. 16 We shall use the words 'supremum' and 'sup' synonymously with l.u.b.; similarly, we shall use 'infimum' and 'inf' and 'counterpart' synonymously with gr.l.b. 1949 S. Lefschetz *Introd. Topology* i. 27 We shall also use on occasion the supremum and infimum of a nonvoid set *A* of real numbers, written sup *A*, inf *A*. 1964 W. J. Pervin *Found. Gen. Topology* i. 15 In the natural ordering for the set of natural numbers, 2 is the infimum of the set of even numbers. In the case of the ordering <1, 3, 5, ..; ..., 6, 4, 2>, there is no infimum to the set of even numbers even though they are bounded below. *Ibid.*, The rational numbers ordered by size are not order-complete, since the subset consisting of all rationals which are positive and have squares greater than 2 does not have an infimum, even though it is bounded below by 0. 1968 E. T. Copson *Metric Spaces* i. 14 The infimum of a subset *A* is its greatest lower bound, and is denoted by inf *A*.

infinite, *a.* Add: **1. d.** *infinite regress* (see quots.).
1836–7 [see Regress *sb.* 1 *fig.*]. 1934 A. C. Ewing *Idealism* iv. 149 If we once view relations as terms we are involved in Bradley's infinite regress. 1946 P. Harrison *Oxf. Marmalade* i. iii. 27 George's criterion of niceness, however, might be described as an infinite regress. Every girl he met was nice, until he met another, and she was nicer. 1968 E. H. Gombrich *Art & Illusion* (ed. 3) x. 268 Are we not led into what philosophers call an infinite regress, the explanation of one thing in terms of an earlier which again needs the same type of explanation? 1973 A. Quinton *Nature of Things* 79 On account of this 'axiom'..solves the problem of the infinite regress of justification.
5. Read: *Mus.* Applied to a form of musical structure which can be repeated infinitely.
1869 [in Dict.]. 1876 Stainer & Barrett *Dict. Mus. Terms* s.v. *Canon*, The above is also an *infinite canon*, because, anyone having such a remarkable desire as to play it for ever, could do so. 1880 Grove *Dict. Mus.* s.v. *Canon*, Many canons lead back to the beginning and thus become 'circular' or 'infinite'. 1959 *Collins Mus. Encycl.* s.v. *Canon*, If each part, on coming to the end of the melody, goes back to the beginning again and repeats, the result is a 'perpetual' or 'infinite' canon.

infinitesimal, *sb.* and *a.* Add: Hence **infinitesimalist**, one who supports the method of infinitesimals (sense A. 2).
1863 [see Momentarian]. 1937 *Mind* XLVI. 227 Berkeley's penetrating criticism of the postulates of the fluxionists and infinitesimalists of his day.

infinitism (infi·nitiz'm). [f. Infinite *a.* + -ism.] The belief that God, or the world, is infinite; (*a*) that there is an actual infinite. So **infi·nitist**, an exponent or adherent of infinitism; also *attrib.* or as *adj.*
1897, 1900 Infinitist [see *finitist *sb.*]. 1902 W. James *Var. Relig. Exper.* 523 He [*sc.* God] is assumed..to be 'infinite'... Nevertheless..religious experience..cannot be cited as unequivocally supporting the infinitist belief. 1922 Infinitism [see *finitism].

infinitive, *a.* and *sb.* Add: **B.** *sb.* **3.** *attrib.* and *Comb.*, as *infinitive-adjunct*, *-splitter*, *-splitting*.
1957 R. W. Zandvoort *Handbk. Eng. Gram.* I. i. 9 In the fourth example the infinitive stands in apposition to the noun; in the fifth the infinitive adjunct is semi-adverbial. 1927 *Glasgow Herald* 1 Nov. 8/7 A competition..to discover the most distinguished infinitive-splitters. 1926 Fowler *Mod. Eng. Usage* 447/1 They were obsessed by fear of infinitive-splitting.

infinitize (i·nfinitəiz), *v.* [f. Infinite *a.* + -ize.] *trans.* To render infinite. Also as *vbl. sb.*
1913 E. Underhill *Mystic Way* 62 They aspire to infinitise life and to define infinity. 1930 [see *finitize *v.*].

infinity. Add: **4. b.** In *Photogr.* also used of any distance, or the range of distances, at which an object is effectively in focus when the lens is set for the greatest possible distance. (Earlier and further examples.)
1867 Sutton & Dawson *Dict. Photogr.* 122 In every lens there is..a certain distance of a near object from it, between which and infinity all objects are in equally good focus. 1910 *Photogr. for Beginners* (Country Life Ltd.) ii. 13 These are called 'fixed focus' cameras, and the lens is 'set at Infinity', which means that, provided the object to be photographed is not nearer the camera than, say, twenty feet, everything in the picture will appear sharp. 1929 R. H. Goodsall *Beginner's Guide Photogr.* v. 22 The next thing is to adjust the lens to correct focus... All subjects over 100 feet away will be at 'infinity', smaller distances are marked on the scale. 1939 J. M. Blair *Pract. & Theoret. Photogr.* ix. 89 The hyperfocal distance is the least distance at which a lens may be focussed when objects at infinity are still in focus. 1950 G. L. Wakefield *Your Camera Lens & Shutter* ii. 27 In photography, an object at 100 feet away from the lens would be normally considered as being at infinity. Infinity cannot be rigidly defined in the photographic sense, but it becomes a greater distance as the focal length of the lens increases. 1974 *Trafford Catal. Spring & Summer* 890/1 Kodak 100 pocket camera outfit. Fine 3-element lens gives sharp colourful prints or slides from 4 ft. to infinity.

infirm, *a.* **1. b.** (Later example.)
1952 *Mind* LXI. 83 Clearly the argument is infirm.

infirmation. (Later example.)
1953 W. V. Quine *From Logical Point of View* ii. 37 Method of empirical confirmation or infirmation.

infixation (infiksēi·ʃən). *Gram.* [f. Infix *v.* + -ation.] The action of infixing; the state of being infixed.
1921 E. Sapir *Lang.* iv. 76 A peculiarly interesting type of infixation is found in the Siouan languages, in which certain verbs insert the pronominal elements into the very body of the radical element. 1964 E. Bach *Introd. Transformational Gram.* iii. 48 And the infixation of the object marker..should be brought about by a low-level rule in the grammar.

inflammability. (Further examples: see next and cf. *flammability.)
1966 Wallace & Wilkinson *Res. Burns* 634 Standards of non-inflammability to which textile fabrics must conform. 1968 E. Miller *Textiles* vi. 133 The necessity for some form of treatment which would reduce inflammability risk in fabrics has been on record since the seventeenth century at least.

inflammable, *a.* (*sb.*) (Further examples, illustrating the continued currency of the word alongside *flammable *a.*, q.v.)
1962 Muir & Barclay *Burns & their Treatm.* Foreword, It is unfortunate that children and the elderly should bear the brunt of our apparent unwillingness to..elaborate methods of rendering clothing less inflammable. 1968 *Which?* Mar. 92 Recently..the Toilet Preparations Federation recommended to their members that inflammable hair preparations should carry a warning. 1968 E. Gale *From Fibres to Fabrics* xiv. 151 To render fabrics non-inflammable, they may be treated with mineral salts. 1972 *Sci. Amer.* Mar. 54/3 If a resulting spark has enough energy inflammables are ignited and explosions are set off.

inflammably, *adv.* (Later example.)
1922 Joyce *Ulysses* 716 The same concupiscence, inflammably transmitted first with alarm, then with understanding.

inflammatorily, *adv.* (Earlier example.)
1840 Dickens *Let.* 2 Oct. (1969) II. 131 Yours inflammatorily and despondingly Charles Dickens.

inflatable, *a.* Add: **B.** *sb.* An object, e.g. a dinghy, a toy, etc., which is capable of being inflated.
1954 *Official Directory—American Toy Fair* 46 Phonographs, inflatables, swimming accessories, [etc.]. 1962 *Engineering* 3 Aug. 151 No less than 17,000 Allied aircrew lives were saved by inflatables. 1970 *Guardian* 30 July 9/6 Mark Fisher..and Simon Connolly..are now preparing a DIY inflatable dome kit to order... It makes up into an inflatable 12 ft in diameter by 8 ft high. 1971 A. Bullock *20th Cent.* 245/1 The sculptor has a variety of..plastics with which to make inflatables or multiples. 1971 *New Scientist* 27 Apr. 284/3 Inflatables help children move without fear. 1971 *Ink* 12 June 11/4 Children's Rights Day: Action Space with inflatables and events. 1973 *New Society* 11 Jan. 70/2 A huge plastic inflatable for people to jump around on.

inflate, *v.* Add: **4.** Also *intr.*, to resort to, exhibit, or produce (monetary) inflation.
1940 *Economist* 27 Jan. 136/1 The most direct..method of inflating, if it cannot be avoided, would..be the deliberate creation of additional credit. 1965 *New Statesman* 31 Dec. 1021/1 Even if all countries inflated at the same rate, some problems would remain. 1971 *Sunday Times* 24 Oct. 44/6 More recently..all types of Southern property have been inflating faster than anywhere in the country. 1973 *Time* 25 June 23/2 In the supermarket..prices have been inflating at an annual rate of 25% or more.

inflation. Add: **6.** *spec.* An undue increase in the quantity of money in relation to the goods available for purchase; (in lay use) an inordinate rise in prices. (Earlier, later, and *attrib.* examples.)
1838 D. D. Barnard *Speeches & Rep.* 195 The property pledge can have no tendency whatever to prevent an inflation of the currency. 1841 in X. D. MacLeod *Biogr. F. Wood* (1856) 75 We have been periodically visited by panics, revulsions, and distresses, inflations and reactions. 1863 W. S. Jevons *Serious Fall in Value of Gold* i. 13 The inflation of credit must be checked by the well defined boundary of available capital. *Ibid.* 14 A revulsion occasioned by a failure of the national capital must cause..a collapse of credit, and of any inflation of prices due to credit. *Ibid.* 25 It is impossible to account for this permanent change [in prices] by any excessive speculation, inflation of currency, or credit. 1870 W. W. Fowler *Ten Yrs. Wall St.* 315 Used ten thousand shares of the new stock to load up the bulls at these inflation prices. 1874 *N.Y. Tribune* 26 Nov. 5 An Inflation Party [has been organized] reaffirming, in effect, the financial plank of the Indiana Independents. 1922 *Encycl. Brit.* XXX. 984/1 Inflation had the effect of reducing the pre-war unit of value. 1949 *Times* 10 Sept. 5/6 Inflation is used to describe the situation in any country where there is an excess of currency and credit in relation to the work to be done, an excess of purchasing power and effective demand in relation to the goods available, with prices and wages, and prices again, rising in consequence. 1967 *Economist* 16 Dec. 1143/3 That fearsome thing called 'inflation psychology' is taking hold. 1972 *Accountant* 5 Oct. 434/5 He made a plea for a Dutch recommendation on inflation accounting in external reporting. 1973 *Sun* 18 Jan. 16 The Premier named inflation as public enemy No. 1.
8. **inflation-proof,** *v. trans.*, to protect from the effects of monetary inflation; so *inflation-proofing*; **inflation-rubber**, a removable rubber sleeve inside each teat cup of a milking machine which, as it is rhythmically inflated and deflated, squeezes the cow's teats; also *ellipt.* **inflation**.
1973 *Times* 26 Nov. 14/6 The Chancellor could profitably devote the budget..to inflation-proofing those of our laws and institutions which have been created on the false assumption that monetary values are constant. *Ibid.* 27 Nov. 18/6 Neither of them enjoys any automatic inflation-proofing beyond the moment of their retirement. 1974 *Guardian* 23 Jan. 10/1 The cuts affect universities in three ways: delayed building, a loss of inflation-proofing in current spending, and doubt about student grants. 1950 *N.Z. Jrnl. Agric.* Feb. 114/2 Old rubberware such as used inflations must never be allowed to accumulate..close to the dairy. *Ibid.* Oct. 301/3 Inflation rubbers give quicker milking if they retain their tension. 1960 B. Crump *Good Keen Man* 102 His bellowing discourse on the good and bad brands of inflation-rubbers for milking machines.

inflationary (inflēi·ʃənări), *a.* [f. Inflation + -ary¹.] Of, pertaining to, characterized by, or involving (monetary) inflation.
1920 *Glasgow Herald* 21 Aug. 7 The transition from an inflationary to a deflationary period in prices. 1921 *Spectator* 28 May 677/1 The enormous Government loans, with their inflationary influence. 1930 *Time & Tide* 1 Nov. 1356 France is uneasy about the inflationary effects of the gold she has collected. 1931 *Economist* 12 Dec. 1118/1 Progressive depreciation of sterling..would..make the beginning of an all-round inflationary spiral a certainty. 1943 *Ann. Reg. 1942* 31 Increases in basic wages had not been of a kind calculated to produce the inflationary spiral. 1945 Webster Add., *Inflationary gap*, the gap between an increasing purchasing power, that is, money available for spending, and a shrinking or static supply of civilian goods available for purchase, which as it widens sets in motion an inflationary spiral. 1970 T. Lupton *Managem. & Social Sci.* (ed. 2) ii. 46 The inflationary pressure generated by collective bargaining at the workplace in conditions of full employment.

inflationism (inflēi·ʃəniz'm). [f. as prec. + -ism.] The condition or fact of being inflated; the policy of inflating the currency. Also *transf.* and *fig.*
1919 J. M. Keynes *Econ. Consequences Peace* 223 The inflationism of the currency systems of Europe has proceeded to extraordinary lengths. 1930 *New Statesman* 9 Aug. 572/2 The book is a tract against inflationism. 1938 C. Connolly *Enemies of Promise* ix. 93 The deflationary activities of the Cambridge critics..have replaced the inflationism of Bloomsbury. 1973 *Spectator* 3 Mar. 280/3 His real enemy is not the dialectical materialism of the Marxist militants but the dialectical inflationism of the Tories' own monetary policy.

inflectable (infle·ktăb'l), *a. Gram.* [f. Inflect *v.* 3 + -able.] Capable of being inflected.
1958 W. N. Francis *Struct. Amer. Eng.* 594 Noun,..a lexical word which..is inflectable with the plural and possessive inflections (-es) and ('s). 1963 F. G. Lounsbury in S. Koch *Psychol.* VI. 563 The determination of the particular form of an inflectable word. 1965 *English Studies* XLVI. 30 An inflectable possessive pronoun.

inflexional, inflectional, *a.* Add: **2.** *Geom.* Of or pertaining to a point of inflexion.
1862 G. Salmon *Treat. Analytic Geom. Three Dimensions* x. 182 We shall call the two lines which meet the surface in three coincident points, the inflexional tangents at the point. 1926 S. Ganguli *Theory Plane Curves* (ed. 2) II. i. 32 Through each point there pass four lines, each of which passes through two..points of inflexion. These lines are called inflexional lines. 1966 J. H. Cadwell *Topics in Recreational Math.* x. 100 Still more special is an inflectional tangent which both touches and crosses a curve.

inflight (i·nfləit), *a.* Also **in-flight.** [f. *in *prep.* 17* + Flight *sb.*¹] Within or during a flight.
1945 *This Week Mag.* 25 Mar. 20/3 The Air Transport Command furnished quarters..and provided another 2,000,000 'inflight' meals. 1952 *N.Y. Times* 21 Dec. x27/3 Thousands of in-flight refueling operations have been completed without a single accident. 1964 *Punch* 11 Nov. 712/2 World airlines..are offering..in-flight movies and TV. 1969 *Sunday Times* (Colour Suppl.) 17 Aug. 18/3 There are six big screens for the inflight movies which no American airline can now do without. 1972 *Lebende Sprachen* XVII. 72/1 By the spring of 1970 in-flight results should be available for comparison with wind-tunnel predictions.

influence, *sb.* Add: **4. c.** *under the influence*: affected by alcoholic liquor; intoxicated, drunk.
[1866 Mayne Reid *Headless Horseman* xix. 110 If not absolutely intoxicated, it could be seen that the ex-officer of volunteers was under the influence of drink.] 1879 'Mark Twain' *Lett.* (1917) I. 367 Nobody got in the least degree 'under the influence', and we had a pleasant time. 1922 Joyce *Ulysses* 297 Lowest blackguard in Dublin when he's under the influence. 1925 Wodehouse *Carry on, Jeeves!* vii. 159 Boat-Race Night. Then, if ever, you will see Bertram under the influence. 1940 L. A. G. Strong *Sun on Water* 77 If a man under the influence tries to give you too much, sure, you slip it back into his pocket when he isn't looking. 1960 H. & M. Williams *Double Yolk* in *Plays of Year* XXI. 47 The police sergeant..asked if I'd been drinking... I'd had a drink—and then he said I must go to the station with him, as he'd have to charge me with driving under the influence.
8. **influence line** *Engin.*, a graph showing how the resultant moment, stress, or other quantity at a given point of a structure varies with the position of the applied (constant) load producing it; **influence pedlar** (or **peddler**) *U.S.* (see quot. 1968); hence *influence peddling*.
1902 *Encycl. Brit.* XXVI. 377/1 In dealing with the action of travelling loads much assistance may be obtained by using a line termed an influence line. Such a line has for abscissa the distance of a load from one end of a girder, and for ordinate the bending moment or shear at any given section, or on any member, due to that load. 1936 Pippard & Baker *Analysis Engin. Struct.* xiv. 319 We shall consider the truss shown in Fig. 14.18 and find the influence line for the force in a diagonal bracing member as a load rolls along the bottom chord. 1972 R. C. Coates et al. *Structural Analysis* ii. 51 A bending moment diagram (shear force diagram) shows graphically the value of the bending moment (shear force) at *all* sections of a beam under a force whose position is *fixed*; an influence line for bending moment (shear force) shows graphically the value of the bending moment (shear force) at a *single* section for all possible positions of a movable force. 1949 *N.Y. Times* 14 Aug. E7/5 The Investigations subcommittee of the Senate Committee on Expenditures in the Executive Departments which is inquiring into the activities of the 'five percenters' and the 'influence peddlers' has discovered that selling influence may be unethical, but it is not always illegal. 1968 W. Safire *New Lang. Politics* 204/2 *Influence pedlar*, one who has, or claims to have, the contacts and 'pull' supposedly necessary to get government contracts and favors from public officials, for a fee. 1971 *Wall St. Jrnl.* 20 July W1/3 The acknowledged dealer in stolen securities said part of the $100,000 he paid went to Nathan Voloshen, a convicted Washington influence peddler. 1972 *N.Y. Times* 22 Sept. 43 In 1956 a Senate investigation showed that Mr. Chotiner was engaged in influence peddling.

influenceable, *a.* (Later example.)
1970 G. Greer *Female Eunuch* 94 Deutsch..drew an extraordinary picture of woman as the ideal life-companion... 'They seem to be easily influenceable.'

influent, *a.* (*sb.*) **2.** Delete †*Obs.*, insert *arch.*, and add later example.
1922 T. Hardy *Late Lyrics* 93 No influent star endeared me, Unknown, unrecked, unproved!
B. *sb.* **2.** *Ecology.* An organism which affects the ecological balance of a plant or animal community.
1926 V. E. Shelford in *Ecology* VII. 389 Professor [F. E.] Clements kindly suggested the term *influent* to cover those organisms which have important relations in the biotic balance and interaction. 1935 [see *biome]. 1938 Weaver & Clements *Plant Ecol.* (ed. 2) xviii. 478 It is more convenient to employ a distinct term and call them [*sc.* animals] influents in reference to their abundance and corresponding importance.

influential, *a.* (*sb.*) **B.** *sb.* Delete *rare* and add later examples.
1837 W. Dyott *Diary* 16 July (1907) II. 257 And was there strongly entreated by the influentials in the South Division of the county to allow himself to be put in nomination with Lord Ingestre for the south. 1899 'Mark Twain' in *Forum* (N.Y.) Mar. 29 It was their official duty to entertain the influentials after some sort of fashion. 1957 R. K. Merton *Social Theory* (rev. ed.) 128 Two types of influentials—the local and the cosmopolitan. 1965 D. V. Glass in Glass & Eversley *Population in Hist.* i. 19 Manuals of domestic medicine, intended not only..for the practitioner but also for the clergy and for other local 'influentials'.

influenzal, *a.* (Later examples.)
1955 *Sci. News Let.* 19 Mar. 190/2 Meningitis may also be caused by a germ called Hemophilus influenzae. This form is called influenzal meningitis, but has nothing to do with influenza. 1969 *Daily Tel.* 18 Dec. 15/3 There had been a sharp increase this month in influenzal-type illness in South-East England. 1972 *Ibid.* 20 Jan. 7 A total of 180 people died during the week ending Jan. 7 from influenza and influenzal pneumonia.

info (i·nfōu), *colloq.* abbrev. of Information.
1913 *Sat. Even. Post* 15 Feb. 8/2, I can slip you the info. 1925 Wodehouse *Sam the Sudden* xiii. 93 So you've only to pool your info' to bring home the bacon? 1954 'N. Blake' *Whisper in Gloom* ii. 29 Then we pass the info. to the police, and they raid the joint. 1968 J. C. Holmes *Nothing More to Declare* 28 Legman..the walking dossier of scandalous info about the sex habits of politicians. 1971 *New Scientist* 9 Sept. 582 (*heading*) Generating info for schools.

in-foal (i·n,fōul), *a.* [*attrib.* use of phrase *in foal*: see In *prep.* 10 b.] That is in foal.
1929 *Chilean Rev.* No. 31. 59/1 To prevent joint-ill in foals, these stations feed a teaspoonful of potassium iodide crystals twice per month to each in-foal mare. 1948 *Vet. Rec.* 25 Dec. 679/1 The in-foal mares were run out during the day until near to foaling.

infolio. Add: Also *attrib.*
1897 W. J. Locke *Derelicts* xx. 346 She..took up a great in-folio black-letter.

inform, *v.* Add: **5.** Const. *about*, *on*.
1775 Sheridan *Rivals* Pref. p. viii, For on subjects on which the mind has been much informed, invention is slow of exerting itself. 1863 G. Macdonald *D. Elginbrod* III. iii. x. 170 Do not hesitate to inform us on all possible subjects. 1880 Disraeli *Endymion* II. lxi. 255 It must be a mind..fairly informed on the questions involved in the wealth of nations. 1888 Mrs. H. Ward *R. Elsmere* III. xliii. 254 The Frenchman..had been informed about him. 1944 P. Cheyney *They never say When* xi. 172 It seems an amazing thing that you didn't inform someone about this. 1963 S. Broström *Increasing Frequency Preposition 'About'* viii. 323 'Inform about' is an accepted construction in sentences like 'Did he inform you about it?', 'I wasn't informed about it', i.e. in sentences where the preposition is followed by *it*.
c. With quoted words as object.
1877 *My Mother-in-Law* xii. 112 'Baby is not very well, Charlie,' Bessie informed me. 1881 Mrs. J. H. Riddell *Senior Partner* I. xiv. 290 'Ye're out of your mind, Janet,' Mr. McCullagh informed her.

informal, *a.* Add: **1. c.** *N.Z.* and *Austral.* Of a vote or voting-paper: not in due form, spoilt, invalid.
1957 *Wanganui Herald* 2 Dec. 5/2 (*table*) *Cotterill* 6121 *Mrs MacLean* 4626 *Marks* 3167. There were 39 informal votes on the preliminary count. 1965 *Parliamentary Handbk. Austral.* 359 Victoria, 1951. No. on Rolls, 1,388,116. No. who Voted, 1,332,339. No. of Informal Ballot-papers, 90,887. 1966 *Official Year Bk. Western Austral.* 105 Number of Informal Ballot Papers, 922.

d. Pertaining to or in respect of the non-compulsory admission of a patient to a mental hospital, esp. by his or her own volition.
1959 *Act 7 & 8 Eliz. II* c. 72 § 5 (*heading*) Informal admission of patients. *Ibid.*, Whenever possible admission and treatment should be on an informal basis and..compulsion should only be resorted to where it is absolutely essential in the interests of the patient and for his safety or that of the public. 1964 B. Ackner *Handbk. Psychiatric Nurses* (ed. 9) 3 The Mental Health Act of 1959..permitted mentally ill patients to be admitted on an 'informal' basis without any documentation. 1968 K. O'Hara *Bird-Cage* vii. 65 'I don't understand this about informal patients.' 'It should have been explained to you. We don't call it certification now, but we do still need powers to protect patients from themselves.' 1970 *Times* 9 Sept. 16/6 (Advt.), Is there a Christian community who would accept a maladjusted 29-year-old girl? She is an informal patient in a mental hospital. 1972 *Guardian* 11 Aug. 11/3 An informal patient can sign for himself; the next of kin of a detained patient must give his permission.

informant. Add: **B.** *sb.* **2. c.** A person from whom a linguist, anthropologist, etc., elicits information about language, dialect, culture, etc. Used esp. in *Dialect Geography*. Also *attrib.*
1889 A. J. Ellis *On Early Eng. Pronunc.* V. 2 Where I was unable to obtain vivâ voce or palaeotypic information, I had the same difficulty as before in interpreting the informants' orthography. 1902 *Amer. Anthropologist* IV. 732 To quote the words of my Indian informant, 'the ceremonies of the other shrines were like branches of this shrine'. 1917 *Internat. Jrnl. Amer. Ling.* I. i. 1 They were obtained by dictation from a few informants. 1933 L. Bloomfield *Lang.* xix. 324 The forms were collected in each case from a single informant by means of a questionnaire of some two-thousand words and phrases. 1936 *West Virginia Univ. Stud.* I. 51 The first requirement is that both potential informants be natives. 1943 *Language* XIX. i. 42 The language must be learned from the lips of a native informant, whose sole function is to talk in his own language. 1944 *Amer. Speech* Apr. 135 The danger inherent in partial reporting, especially when based on..inadequate informant work. 1953 J. B. Carroll *Study of Lang.* vi. 173 A linguistic scientist..directed the classroom teaching process and used native informants as models for drill purposes. 1963 J. Lyons *Structural Semantics* iv. 76 The linguist can satisfy himself..by going around and exasperating several tobacconists with his 'informant-technique'. 1964 R. H. Robins *Gen. Ling.* ix. 355 The informant is not a teacher, nor a linguist; he is simply a native speaker of the language willing to help the linguist in his work. 1971 D. Crystal *Ling. Interlude* 137 These characteristics of the informant sample would have to be made..explicit.

informatics (infǫɹmæ·tiks). [tr. Russ. *informátika* (A. I. Mikhailov et al. 1966, in *Nauchno-tekhnicheskaya Informatsiya* XII. 35), f. Information: see -ics.] (See quot. 1967.) Cf. *information science* (*information 8). Hence **informa·tical** *a.*, **informati·cian.**
1967 *FID News Bull.* XVII. 73/2 Informatics is the discipline of science which investigates the structure and properties (not specific content) of scientific information, as well as the regularities of scientific information activity, its theory, history, methodology and organization. 1970 *Times* 2 Sept. 9 It was agreed..that an introduction to Informatics should form an integral part of general education. 1972 *Jrnl. Librarianship* IV. 177 The name Informatics satisfies several criteria for the designation of a new discipline. *Ibid.*, Other terms can be derived from it, such as Informatician for a person who is engaged in activities in this field..and the adjective informatical, to describe the attributes of the field. 1973 *Times Lit. Suppl.* 28 Sept. 1133/1 The problem falls into two parts: the preparation of decisions, which is a matter of informatics, and the making of the decisions themselves, which is a matter of 'politics'.

information. Add: **3.** (Further examples, illustrating a contrast with *data*.)
1970 [see *datum 1 d]. 1970 O. Dopping *Computers & Data Processing* i. 14 In administrative data processing, a distinction is sometimes made between data and information by calling raw facts in great quantity 'data', and using the word 'information' for highly concentrated and improved data derived from the raw facts.
c. Separated from, or without the implication of, reference to a person informed: that which inheres in one of two or more alternative sequences, arrangements, etc., that produce different responses in something, and which is capable of being stored in, transferred by, and communicated to inanimate things.
Information in this sense may at the same time be, or be regarded as, information in the following sense.
1937 *Discovery* Nov. 329/1 The whole difficulty resides in the amount of definition in the [television] picture, or, as the engineers put it, the amount of information to be transmitted in a given time. 1944 *Jrnl. Sci. Instrum.* XXI. 133/2 Information is conveyed to the machine by means of punched cards. 1953 J. C. Eccles *Neurophysiol. Basis Mind* i. 1 We may say that all 'information' is conveyed in the nervous system in the form of coded arrangements of nerve impulses. 1953 Watson & Crick in *Nature* 30 May 965/2 In a long molecule many different permutations are possible, and it therefore seems likely that the precise sequence of the bases is the code which carries the genetical information. 1958 *Spectator* 4 July 22/3 The complex molecules carrying genetic information from one generation to the next. 1961 *New Scientist* 26 Jan. 201/2 The colour information is added to a conventional black-and-white signal on an amplitude and phase modulated sub-carrier located in the vision band. 1962 *Listener* 10 May 817/2 The fertilized ovum of a particular animal is not, in any obvious way, like that animal; yet its development will proceed along certain lines only. It contains the information characteristic of that particular kind of animal. 1962 *Times* 5 July 15/7 A disc is apt to give slightly inferior quality towards the centre, where the information is more crowded. 1971 R. M. Dowben *Cell Biol.* v. 97 Genetically transmitted information precisely determines the amino acid composition of all proteins synthesized by each cell.
d. As a mathematically defined quantity (see quots.); now *esp.* one which represents the degree of choice exercised in the selection or formation of one particular symbol, sequence, message, etc., out of a number of possible ones, and which is defined logarithmically in terms of the statistical probabilities of occurrence of the symbol or the elements of the message.
The latter sense (introduced by Shannon, quot. 1948²), though foreshadowed earlier) is that used in information theory, where information is usually regarded as synonymous with entropy.
1925 R. A. Fisher in *Proc. Cambr. Philos. Soc.* XXII. 709 What we have spoken of as the intrinsic accuracy of an error curve may equally be conceived as the amount of information in a single observation belonging to such a distribution. *Ibid.* 710 If we regard the sum $S\{\frac{1}{m}(\frac{\partial m}{\partial \theta})^2\}$ where m is the expectation in any one class [and θ is the parameter], as the amount of information supplied by the data. 1928 R. V. L. Hartley in *Bell Syst. Techn. Jrnl.* VII. 540 What we have done then is to take as our practical measure of information the logarithm of the number of possible symbol sequences. *Ibid.* 541 The information associated with two characters will be $\log s^2$. 1935 R. A. Fisher in *Jrnl. R. Statistical Soc.* XCVIII. 47 One could, therefore, develop a mathematical theory of quantity of information from these properties as postulates, and this would be the normal mathematical procedure. *Ibid.*, As a mathematical quantity information is strikingly similar to entropy in the mathematical theory of thermo-dynamics. 1948 N. Wiener *Cybernetics* iii. 76 Thus a reasonable measure of the amount of information associated with the curve $f_1(x)$ is: $\int_{-\infty}^{\infty} (\log_2 f_1(x)) f_1(x)\,dx$. The quantity we here define as amount of information is the negative of the quantity usually defined as entropy in similar situations. [The definition..is not the one given by R. A. Fisher for statistical problems, although it is a statistical definition.] 1948 C. E. Shannon in *Bell Syst. Techn. Jrnl.* XXVII. 379 We have represented a discrete information source as a Markoff process. Can we define a quantity which will measure, in some sense, how much information is 'produced' by such a process, or better, at what rate information is produced? Suppose we have a set of possible events whose probabilities of occurrence are $p_1, p_2, \ldots, p_n$. These probabilities are known but that is all we know concerning which event will occur. Can we find a measure of how much 'choice' is involved in the selection of the event or of how uncertain we are of the outcome? If there is such a measure, say $H(p_1, p_2, \ldots, p_n)$, it is reasonable to require of it the following properties: [etc.]. *Ibid.* 394 We shall call $H = -\sum p_i \log p_i$ the entropy of the set of probabilities $p_1, \ldots, p_n$... The quantity H has a number of interesting properties which further substantiate it as a reasonable measure of choice or information. 1949 W. Weaver in Shannon & Weaver *Math. Theory Communication* 99 The word *information*, in this theory, is used in a special sense that must not be confused with its ordinary usage. In particular, *information* must not be confused with meaning. In fact, two messages, one of which is heavily loaded with meaning and the other of which is pure nonsense, can be exactly equivalent, from the present viewpoint, as regards information. *Ibid.* 100 Information in communication theory relates not so much to what you *do* say, as to what you *could* say. That is, information is a measure of one's freedom of choice when one selects a message. 1953 D. Gabor in W. Jackson *Communication Theory* i. 2 'Information' in the exact sense of communication theory is far more restricted than the vague concept which goes by this name in everyday life. It may also be mentioned that this definition has nothing to do with the value of information. It is a measure of the minimum effort or cost by which the message can be transmitted, not of its importance or consequences. 1953 J. B. Carroll *Study of Lang.* vii. 200 Information (in the special sense required in communication theory) may be measured in bits. 1953 C. F. Hockett in Saporta & Bastian *Psycholinguistics* (1961) 45/2 The keynote of the quantification of information is the matter of choice of any message, for actual transmission at a given time, from a fixed repertory of possible messages. 1956 L. Brillouin *Sci. & Information Theory* p. x, Information is a function of the ratio of the number of possible answers before and after, and we choose a logarithmic law in order to insure additivity of the information contained in independent situations... This definition cannot distinguish between information of great importance and a piece of news of no great value for the person who receives it. 1957 Kendall & Buckland *Dict. Statistical Terms* 138 In a specialised sense in the theory of estimation, the amount of information about a parameter θ from a sample of n independent observations drawn at random from a population with a frequency function $f(x, \theta)$ is defined as $nE(\partial \log f/\partial\theta)^2 \equiv n\int_{-\infty}^{\infty} (\partial \log f(x,\theta)/\partial\theta)^2 f(x,\theta)\,dx$. 1968 J. Lyons *Introd. Theoret. Ling.* ii. 84 Another important statistical notion has to do with the amount of *information* carried by a linguistic unit in a given context; and this also is determined by (or is generally held to be determined by) its frequency of occurrence in that context. 1968 P. A. P. Moran *Introd. Probability Theory* i. 53 In statistical theory 'information' is usually 'information about a particular parameter' of a probability distribution, and is measured by the reciprocal of the square of the standard deviation of some estimator of that parameter. 1970 O. Dopping *Computers & Data Processing* i. 19 Data carrying different amounts of information, containing different frequency of occurrence of different symbols has less information per symbol than another (hypothetical) language with the same number of symbol values but with equal probability of occurrence of them all.
8. *attrib.* and *Comb.*, as *information content*, *desk*, *explosion* [*explosion 4 b], *flow*, *gap* [*gap *sb.*¹ 6 a], *office*, *service*, *storage*, *system*, *transfer*, *work*; *information-carrying*, *-gathering* (so *-gatherer*), *-giving*, *-seeking* vbl. sbs. and ppl. adjs.; **information bureau**, an office where information is given and questions are answered; also *fig.*; **information officer**, a person engaged in the provision of specialized information; **information processing**, the processing of information so as to yield new or more useful information; data processing; **information retrieval**, the tracing of information stored in books, computers, or other collections of reference material; **information room** (see quot. 1958); **information science**, (that branch of knowledge which is concerned with) the procedures by which information, esp. that relating to technical or scientific subjects, is stored, retrieved, and disseminated; hence **information scientist**, a person employed in providing an information service, or one who studies the methods used to do so. Also *information theory.
1882 R. Wallace *Flying Fifty-five* vii. 44 Well, Jebson... You're a pretty fine information bureau! You told me that Patience hadn't a ghost of a chance. 1906 *Aslib Prospectus*, The objects of the Association are..to develop the usefulness and efficiency of special libraries and information bureaux under whatever title they may function. 1968 *Listener* 4 July 31/3 The information bureau of the Disabled Living Activities Group. 1962 *Science Survey* IV. 58 The information content of a message is defined in terms of the frequency... 1972 J. Z. Young *Introd. Study Man* p. v, The 'spectacular' recent information that biochemistry has provided about the large molecules in the body, and especially about the information-carrying properties of the nucleic acids. 1928 *Bell Syst. Techn. Jrnl.* VII. 541 For example, in the Baudot System..the number of primary symbols is..; and the information content of one selection is log 32. 1963 C. J. C. Wilson *Television Engin.* xii. 426 The information-content of a television image has been evaluated solely from the point of view of what is transmitted. 1965 *Language* XLI. 585 This concept of information-content..is of special interest because of various correlations with vocabulary, information-content, etc. 1967 *Economist* 11 Nov. 627/3 A national Referral Centre for Science and Technology is trying to build up a world-wide 'information desk' for advice on where and how to obtain information. 1973 D. Mackenzie *Postscript to Dead Let.* 23, I..put the key on the hook under the information desk and left it at the Information Desk. 1964 *New Statesman* 13 Mar. 396/2 The 'population explosion' has collided with the 'information explosion'. Vastly more people and more kinds of people are chasing vastly more information about more kinds of things. 1972 *Jrnl. Librarianship* IV. 161 The advent of ISR roughly coincided with the first commercial applications of computers and it was then thought that very rapid handling of coded data was all that was needed to cope with the 'information explosion'. 1953 C. F. Hockett in Saporta & Bastian *Psycholinguistics* (1961) 64/1 Energy flow is power; information-flow is entropy; money-flow (at least in one direction) is income. 1965 H. I. Ansoff *Corporate Strategy* (1968) i. 19 Product-market characteristics create operating needs, and these, in turn, determine the structure of authority, responsibility, work flows, and information flows within the firm. 1969 *Daily Tel.* 11 Jan. 12/8 Bold human causes..will not be served by ignoring the new technologies which space research is encouraging; and Britain would be well advised to close the information gap which seems to be developing. 1971 K. Hopkins *Hong Kong* iii. 95 Mr. Woo's speech was an example of the many and repeated expressions of concern by members of the public and of Government about a so-called 'information gap' between Government and the people. 1964 M. McLuhan *Understanding Media* (1967) ii. xxviii. 302 Man the food-gatherer reappears incongruously as information-gatherer. 1971 J. Z. Young *Introd. Study Man* xxiv. 317 One of man's many paradoxes is that although with him each individual organism is more important than in other animals as an information-gatherer for the species, yet his manner of life is largely controlled by his fellows. 1964 M. McLuhan *Understanding Media* (1967) ii. xiv. 149 In the age of instant information man..assumes the role of information-gathering. 1967 Cox & Grose *Organiz. Bibliogr. Rec. by Computer* 70 A subject-specialist studies the information needs and information-gathering habits of a group of teachers. 1908 *Westm. Gaz.* 1 July 6/3 The fish products of Canada, states one of the numerous information-giving tablets, are worth thirty million dollars a year. 1927 J. Adams *Errors in School* iv. 122 Instruction must be distinguished from mere information-giving. 1890 W. Booth *In Darkest Eng.* App. p. xiv, We shall also be glad, through the information office of Labour Department, to give you..further information. 1918 E. S. Farrow *Dict. Mil. Terms* 310 Information officers..send to their own commanders all information of military importance to them. 1935 *Aslib Rep. Proc. 12th Conf.* 38 (*heading*)

B. Fullman... (Information Officer, British Non-Ferrous Metals Research Association). 1947 *Jrnl. Documentation* II. 240, I am not a librarian at all; I am not even a trained information officer. 1970 *Aslib Proc.* XXII. ix. p. ii (Advt.), Vacancy for Scientific Information Officer at the Commonwealth Forestry Bureau. 1958 *Automation* Mar. 65 (*heading*) Information processing. 1959 *Unesco Bull. Libr.* XIII. 226 Nearly 2,000 electronic computer experts took part in the International Conference on Information Processing organized by Unesco in Paris from 15 to 20 June. 1964 T. W. McRae *Impact of Computers on Accounting* vii. 190 Even today few companies segregate 'information processing' or even 'data processing' under a separate cost head. 1970 O. Dopping *Computers & Data Processing* i. 11 Many speak of the advent of mechanized information processing as the second industrial revolution. *Ibid.* 15 When both input and output are data, that is, digital information consisting of a great number of records in standardized layout, the information processing is usually called data processing. 1950 C. N. Mooers *Theory Digital Handling Non-Numerical Information* (*Zator Techn. Bull.* No. 48) 5 The requirements of information retrieval, of finding information whose location or very existence is a-priori unknown, now requires that it be possible by some efficient technique to specify a selection of complexes Cj by means of *any* set or combination of descriptors chosen in *any* way from the vocabulary ((aj)). 1958 *Listener* 11 Dec. 983/1 Only a week or two ago there was a conference on information retrieval in Washington. 1963 *Publishers' Weekly* 23 Sept. 34/2 At the Oxford store, the feature which so far has attracted the most attention is the free bibliographical information retrieval service. 1963 *Cambr. Rev.* 12 Oct. 24/1 A book miscatalogued..is a book lost: and they [*sc.* librarians] thereby justify greater and greater expenditure on more and more elaborate systems of 'information retrieval'. 1972 *Computers & Humanities* VII. 61 Prof. D. Raj Reddy offers a set of exercises in statistics, natural language processing, language translation, poetry concordance, and information retrieval to interested readers. 1934 J. Moylan *Scotland Yard* (ed. 2) v. 132 At Scotland Yard there are Information and Operation Rooms from which the wireless cars are directed. 1940 R. Morrish *Police & Crime-Detection* ii. 28 Every Force has its 'Information Room', to which members of the public should report by telephone anything suspicious. 1958 A. Garfitt *Bk. for Police* I. iii. 77 An Information Room is established at some [police] headquarters and is the centre through which information, particularly as to crime and suspected crime, can be disseminated by wireless, teleprinter or telephone. 1970 P. Laurie *Scotland Yard* i. 16 The first floor carries the electronic complexities of the Information Room. 1960 *Computers & Automation* IX. 39/2 Moore School of Electrical Engineering, University of Pennsylvania... Prof. Saul Gorn, Chairman, Computer and Information Sciences Curriculum. 1962 *Conf. on Training Science Information Specialists 1961–62* (Georgia Inst. Technol.) 115 Information science..investigates the properties and behavior of information, the forces governing the flow of information, and the means of processing information for optimum accessibility and usability. 1963 *Library Jrnl.* LXXXVIII. 4161/1 The information sciences are conceived as: 1) the study of the properties, structure, and transmission of specialized knowledge; and 2) the development of methods for its useful organization and dissemination. 1971 C. W. Hanson *Introd. Science-Information Work* 2 'Information science' can be used to imply..the exploitation of scientific and technical information of all kinds... On the other hand, it is often used to imply the application of science and technology..to handling information generally. 1958 *Nature* 4 Jan. 20/1 A meeting will be held on January 23..to discuss terms of inauguration of an Institute of Information Scientists. The aims of the proposed Institute would include the promotion of high standards in scientific and technical information work, the promotion of educational courses, and the establishment of qualifications for those engaged in such information work. 1963 *Aslib Proc.* XV. 100 These are post-graduate courses for those about to become information scientists. 1972 *Jrnl. Librarianship* IV. 169 The American protagonists of IS generally say that people involved in this science should be called information scientists. 1956 J. Klein *Study of Groups* x. 140 The whole elaborate process of information-seeking, evaluation and decision. 1935 E. S. Hedges in *Aslib Rep. Proc. 12th Conf.* 35 An information service which distributes in-coming information to interested quarters can be more effective than one which merely renders the information available on request. 1950 *N.Y. Times* 20 Apr. 1/8 Mr. Kolarek..has been in Czechoslovakia since September, 1945, serving first as assistant and later as chief press attaché and information service director. 1968 B. E. Holm *How to manage your Information* iii. 55 The Dow Chemical Company is one of the many organizations which provides information services to its engineers. 1950 Information storage [see *information transfer* below]. 1972 *Jrnl. Librarianship* IV. 161 Somewhat later, it was realized that, to 'retrieve' information from a place, it obviously had to be stored prior to the retrieval, so the term was augmented to 'Information storage and retrieval' (ISR). 1953 C. F. Hockett in Saporta & Bastian *Psycholinguistics* (1961) 64/2 If it is necessary to maintain some analogy between an information-system and a power-system, then entropy can better be compared to voltage. 1964 T. W. McRae *Impact of Computers on Accounting* iii. 82 The objective of an information system..is to note all of the events happening within the organization being controlled, to extract those events which require to be reported and to report them to the controlling authority fast enough for compensating action to be possible. 1969 D. C. Hague *Managerial Econ.* i. 18 The information system [of a firm]..will be partly a rather mechanical system for providing routine reports about things like production, costs, sales or profits. It will also be partly a much less formal arrangement whereby those within the firm pass on information..to those who need it. 1950 *Amer. Scientist* XXXVIII. 278/2 A consideration of the effects of information storage and information transfer on physical, chemical, biological, psychological, and sociological systems..may help in understanding and predicting many of the aspects of our universe. 1964 G. H. Haggis et al. *Introd. Molecular Biol.* x. 279 Each operator with its associated structural genes forms a coordinated unit of information-transfer to which Jacob and Monod have given the name *operon*. 1935 B. Fullman in *Aslib Rep. Proc. 12th Conf.* 38 Organised information work is at present only in its infancy. 1959 *Aslib Proc.* XI. 290 The role of the text-book in technical information work is usually a fundamental one. 1972 *Jrnl. Librarianship* IV. iii (inside front cover), The *Journal of Librarianship* is an independent quarterly journal dealing with all aspects of library and information work.

informational, *a.* (Later examples.)
1967 M. McLuhan *Medium is Massage* 138 Subtle electric informational media. 1971 *Nature* 2 July 66/2 The 'monogenomic' viroplast consists of viral genome, informational RNA, a complex of specific viral enzymes and structural proteins. 1971 *Jrnl. Gen. Psychol.* LXXXV. 207 Innovative substitutions of informational types of thinking in psychology. 1972 *Jrnl. Social Psychol.* LXXXVI. 111 The influence of interpersonal evaluations may result from their informational value for the individual.

informationally (infǫɹmē¹·ʃənăli), *adv.* [f. Informational *a.* + -ly².] As regards information.
1964 Y. R. Chao in D. Abercrombie et al. *Daniel Jones* 41 Those aspects of speech which are informationally important but often acoustically weak. 1965 *Math. in Biol. & Med.* (*Med. Res. Council*) iv. 133 Here is a system [*sc.* the nervous system of higher animals] that reacts more reliably and predictably to informationally rich stimuli than to 'simple' ones. 1973 *Physics Bull.* May 281/1 Informationally, the incident speech signal requires about 50 000 bit/s to specify it.

informationless (infǫɹmē¹·ʃənlės), *a.* [f. Information + -less.] Without information; carrying or conveying no information.
1965 [see *droppable *a.*].

information theory. [See *information 3 d.] The quantitative theory, based on a precise definition of information and on the theory of probability, of the coding and transmission of signals and information.
1950 D. M. MacKay in *Phil. Mag.* Mar. 290 (*heading*) The formalisation of information theory. 1950 W. G. Tuller in *Trans. Amer. Inst. Electr. Engin.* LXIX. 1612/1 The statistical theory of communications, developed over the past few years and often called information theory, can be of real assistance in the design of communication systems. 1950 L. Brillouin in *Amer. Scientist* XXXVIII. 594 (*heading*) Thermodynamics and information theory. 1955 H. Quastler *Information Theory in Psychol.* i. 8 It is basic to information theory that any event is evaluated against the background of the whole class of events which could have happened. Information theory proposes to measure the effect of operations by which a particular selection is made out of a range of possibilities. Choice, specification, discrimination, recognition, are examples of such operations. 1960 E. Delavenay *Introd. Machine Transl.* 131 The statistical study of language in so-called information theory bears mainly on the frequency of reference of graphemes and phonemes. 1961 *New Scientist* 26 Jan. 200/2 Normal television amplitude modulation of a carrier takes no advantage of the fact that the rate of change of picture content is relatively slow, and that because of this much redundant information is continuously transmitted at the cost of considerable bandwidth. Unfortunately no system of television picture transmission has yet been devised which is more logical from an information theory point of view and does not also require a complex computing apparatus in the receiver. 1962 J. R. Pierce *Symbols, Signals & Noise* vi. 107 The two great triumphs of information theory are establishing the channel capacity and, in particular, the number of binary digits required to transmit information from a particular source and showing that a noisy communication channel has an information rate in bits per character or bits per second up to which errorless transmission is possible despite the noise. 1964 M. A. K. Halliday et al. *Ling. Sci.* i. iv. 104 Information theory, which has a place in the quantitative description of a language, implies nothing about the relative efficiency of languages or the effectiveness of language activity. 1968 *Unesco Bull. Libr.* XXII. 62 When Shannon and Weaver evolved their information theory, they..were guilty of an unfortunate use of terminology. They were concerned, of course, not with a theory of information but a theory of signals, the message-carrying capacity of a symbol, a telephone wire, or any other medium or channel of communication. 1970 O. Dopping *Computers & Data Processing* i. 21 Information theory deals largely with what happens when a random interference ('noise') is superimposed on the desired signal. 1972 *Jrnl. Librarianship* IV. 164 It is worth noting that Shannon never referred to his theory as an 'Information theory'.
Hence **information theorist.**
1953 C. F. Hockett in Saporta & Bastian *Psycholinguistics* (1961) 57/1 Neither the information theorist nor the linguist would claim that codes (*d*) and (*d'*) are identifiable with code (*a*) in quite the same way as codes (*b*) and (*c*). 1972 J. L. Dillard *Black English* i. 25 Research on the degree of intelligibility between the white man's English and Black English would be an expensive proposition... Ideally,..not only linguists but also anthropologists, psychologists, and an information theorist should be involved.

informative, *a.* Add: **2. b.** *Bridge.* = *informatory *a.* b. So **informatively** *adv.*
1921 A. E. M. Foster *Auction Bridge made Clear* 105 There is an 'informative' double which some English players adopt... It is the double of a bid of one of a suit. 1925 —— *Auction Bridge Play & Probl.* 53 Thus Z bids 'one No Trumps' and A doubles informatively... The Informative Re-double is never left in. 1931 —— *Auction Bridge made Clear* (rev.) 103 It is no longer an 'informative' double, but a 'business' double.

informativeness (infǫ·ɹmătivnės). [f. Informative *a.* + -ness.] The quality or condition of being informative.
1924 C. Mackenzie *Old Men of Sea* v. 68 With the words Mr. Harper managed to assume an air of such bland informativeness that..his attitude reminded me of a conjurer who entreats a member of the audience to mount the stage. 1942 *Mind* LI. 60 It has been shown that the informativeness of memory plays a double part in the process of proof. 1955 *Essays in Crit.* V. 369 One of their number would write with such moral informativeness that he would catch up with Shakespeare.

informatorily (infǫ·ɹmătərili), *adv. Bridge.* [f. Informatory *a.* + -ly².] Informatively; in order to give information. Cf. next.
1928 *Observer* 29 Apr. 25 We should be placed in the ridiculous position that a player could not double informatorily, just because he happened to hold the suit that has been called against him. 1928 *Daily Express* 23 July 4 Unless you are prepared for any answer partner may make, do not double informatorily. 1929 W. Buller *Reflections Bridge Player* 178 Do you double two of a suit informatorily?

informatory, *a.* Add: **b.** Bridge: *informatory double*, a double which is intended to give information to one's partner, as distinct from a 'business double' which is for the purpose of scoring penalty points. So *informatory pass*.
1926 Foster & Hervey *Auction Bridge Informatory Doubles* 82 This Informatory Pass would be made against (say) a Spade bid on such a hand as [etc.]. 1927 [see *business 21 d]. 1929 W. Buller *Reflections Bridge Player* 176 'The Informatory Double' is not strictly *set* convention. 1950 *Hoyle's Games Modernized* (ed. 20) 55 If a player holds a good hand against a call of 1 or a suit, but is without sufficient strength or length in any other suit to justify him in overbidding, he doubles the original call. This is known as the *informatory double* (or *informatory raise*). 1964 G. F. Hervey *Handbk. Card Games* 161 When a player has bid one of a suit, it is usually better for the opponent to make an informatory or take-out double rather than a weak overbid. An informatory double is a request to partner to bid his best suit.

informedly, *adv.*² (Later examples.)
1922 *Times Lit. Suppl.* 10 Aug. 516/3 He deals briefly, but informedly, with the innumerable West African tribes. 1938 *Ibid.* 3 Sept. 573/2 There is in existence no other book dealing so thoroughly and so informedly with the problems of deafness.

info·rmedness. [f. Informed *ppl. a.* + -ness.] The fact or quality of being informed; knowledgeableness.
1946 *Scrutiny* XIII. iv. 268 The sheer informedness about society..impresses us with its range. 1960 C. Day Lewis *Buried Day* x. 221, I came to rely more and more ..upon the good sense, the informedness and seriousness of Charles Fenby. 1971 D. Crystal *Ling.* i. 11 Topics such as these are..discussed..with varying degrees of informedness.

informosome (infǫ·ɹmosōᵘm). *Biol.* [f. Inform(ation + -o + *-some⁴.] A cellular particle composed of messenger RNA and associated protein, the latter being thought to protect the messenger RNA from ribonucleases.
1964 A. S. Spirin et al. in *Zhurnal Obshchei Biologii* (U.S.S.R.) XXV. 338 mRNA is transferred into the cytoplasm..in the form of specific complexes with protein; these ribonucleoprotein particles, named 'informosomes', were isolated and proved to exist as real complexes. 1971 *Nature* 15 Oct. 448/2 Over the past several years evidence has been accumulated which suggests that messenger RNAs occur in cell nuclei in the form of ribonucleoprotein particles—so-called informosomes.

infra (i·nfră), *adv.* [L.] Below, underneath, further on.
1740 J. Grassineau *Mus. Dict.* 107 Hypo, *infra*, *below*; this word when joined to the name of any interval or mode, &c. shews that it is lower than it was without, as *Hypo diapason* an octave lower. 1888 *Encycl. Brit.* XXVIII. 702/1 See *infra* in regard to rotary printing. 1888 Rolleston & Jackson *Forms Animal Life* (ed. 2) 459 The cell has been..supposed by various authorities to be muscular, nervous, or composed of connective tissue. See the original authorities, *infra*. 1955 A. G. Dickens *Robert Holgate* i. 3 Cf. *infra*, p. 24.

infra-. Add: **i:nfra-ato·mic** *a.*, subatomic; **infraba·sal** *Zool.*, any of a series of plates forming a ring beneath the basals in crinoids; also as *adj.*; **infra-ba·ss** *Mus.*, = *sub-bass* (1909 in *Cent. Dict.* Suppl.); also *transf.*; **infra-Chri·stian** *a.*, somewhat less than Christian; **infracortical** *a.* (examples); **infracostal** *a.* (examples); **infragla·cial** *a.*, subglacial; **infrahuman** *a.* (earlier and later examples); **i:nfra-mole·cular** *a.*, at a level of organization below that of a molecule; **infra-ra·tional** *a.*, below what is rational;

infra-speci·fic *a.*, (applied to a category) at a lower taxonomic level than a species; **i:nfra-umbi·lical** *a. Anat.*, situated below the umbilicus.

1923 J. S. Huxley *Ess. Biologist* i. 55 The infra-atomic world of electrons. **1966** I. Asimov *Fantastic Voyage* viii. 90 It was not merely radioactivity that had to be sensed, but radioactive particles that had themselves been miniaturized; and that, because of their incredibly tiny, infra-atomic size could pass through any ordinary sensor without affecting it. **1890** *Ann. & Mag. Nat. Hist.* V. 318 The atrophy of infrabasals is we see a very gradual process. *Ibid.*, The distinction between an infrabasal ring of 5 plates and one of 3, is of far inferior importance. **1962** D. Nichols *Echinoderms* ii. 22 In some forms an additional whorl, the infrabasals, is intercalated between the basals and the centro-dorsal. **1958** J. Blish *Case of Conscience* (1959) ii. x. 104 The infrabass of the buried city's thunder shook the glass in front of him. **1906** W. R. Inge *Truth & Falsehood in Relig.* ii. 63 It is not justifiable to take examples of infra-Christian survivals in Christianity, and use them to discredit the religion of Christ. **1917** J. Denney *Christian Doctrine of Reconciliation* ii. 51 As an infra-Christian mode of thinking, it sometimes curiously flawed what was otherwise pure Christian truth. **1890** *Cent. Dict.*, Infracortical. **1895** *Psychol. Rev.* Mar. 117 In man the consciousness attached to infra-cortical centres is altogether subliminal, if it exist. **1925** *Lancet* 8 Aug. 274/2 Tremor is an involuntary movement belonging to the old motor system (infra-cortical, subpallial). **1858** Gray *Anat.* 732/2 (Index), Infra-costal muscles. **1867** *Quain's Elem. Anat.* (ed. 7) I. 243 The subcostal or infracostal muscles are small bundles lying on the inner aspect of the thoracic wall. **1922** Joyce *Ulysses* 695 A cicatrice in the left infracostal region below the diaphragm. **1894** J. Geikie *Gt. Ice Age* (ed. 3) vii. 91 All such infra- or intra-glacial deposits..occur somewhat partially. **1957** J. K. Charlesworth *Quaternary Era* II. xliv. 1253 The British infraglacial beach, though much narrower than the Norwegian strandflat, marks a steady level of the sea over a considerable time. **1847** J. Wilson *Lands of Bible* I. iv. 105 The gods of the Egyptian pantheon, human, superhuman, and infrahuman. **1970** *Jrnl. Gen. Psychol.* July 42 Motor responses at infra-human level. **1972** *Science* 5 May 541/2 Microelectrode recording has shown that tilt detectors in the infrahuman visual system are each turned 20 deg or so on either side of a preferred orientation. **1899** *Phil. Mag.* XLVIII. 462 All her [*sc.* Nature's] activities at infra-molecular degrees of proximity. **1919** A. N. Whitehead *Enquiry Princ. Nat. Knowledge* ii. 18 We may penetrate below the molecule to the electrons and the core which compose it, and thus obtain infra-molecular equations. **1933** *Mind* XLII. 265 In the concept of life, one may stress either its unconscious, infra-rational, chaotic fecundity or the conscious order..of its historic manifestations. **1935** *Downside Rev.* LIII. 451 Thus the indefinable element would be irrational in the sense of infra-rational. **1939** *Entomol. News* L. 198 In practice they propose infra-specific names—polynomials as well as trinomials—in proper Latin form. **1970** *Watsonia* VIII. 42 The geographical significance of this variation should not be lost sight of and may eventually receive taxonomic recognition at some infra-specific level. **1900** Dorland *Med. Dict.* 322/1 Infra-umbilical. **1906** *Practitioner* Dec. 781 A supra-umbilical and infra-umbilical zone. **1967** *Gray's Anat.* (ed. 34) 641 In its infra-umbilical portion the linea alba is narrow.

infract, *a.*[1] Substitute *arch.* for † *Obs.* and add later example.

1901 *North Amer. Rev.* Feb. 314 The illusion renews itself in the great moments, but I wish it could be kept infract in the small.

infra-red, *a.* Substitute for entry (in Dict. s.v. Infra-):

infra-red (infră,re·d), *a.* and *sb.* Also as one word (without hyphen). [f. Infra- + Red *a.* and *sb.*] **A.** *adj.* **1.** Lying beyond the red end of the visible spectrum: the epithet of electromagnetic radiation (and of the part of the spectrum containing it) which has a wavelength greater than that of red light (about 0·7–0·8 microns) and (in modern use) less than that of the shortest microwaves (of the order of 1000 microns, i.e. 1 mm.); it is invisible, and most of the radiation from bodies below red heat is emitted in this form.

1881, 1896 [in Dict.]. **1926** *Encycl. Brit.* I. 946/2 Extending beyond the red end of the visible spectrum there are 9 octaves of infra-red radiation overlapping with very short Hertzian waves... A. Glacolewa-Arkadiewa.. has created Hertzian waves of only 0·082 mm. in wave length. **1929** *Punch's Almanack for 1930* 4 Nov. p. xxxvii (Advt.), These are like the health-giving rays of heat emitted by the sun—those short infra-red rays that.. enfold you with their beneficial warmth. **1932** Hardy & Perrin *Princ. Optics* xi. 234 Photographs of landscapes taken by infrared light have the general appearance of night scenes because the sky appears dark and the high reflectance of chlorophyl gives foliage the appearance of intense local lighting... Photographs of extremely distant objects are made possible by infrared light because of its greater penetration through atmospheric haze. **1935** *Practitioners Libr. Med. & Surg.* VIII. i. viii. 168 Infra-red radiation..produces the sensation of heat when it comes in contact with the body. This form of radiant energy is produced..by any heated body. **1939** *Jrnl. R. Aeronaut. Soc.* XLIII. 1012 Heating elements such as the steam radiator, which does not glow, emit infra-red energy of long wave-lengths (far infra-red). The filament lamp, like the sun, is a source of near infra-red energy. This is the portion of the spectrum which lies chiefly in the wavelengths slightly longer than the red visible radiation. **1957** *House & Garden* Dec. 99 An electric Rotary Spit that automatically turns joints, steaks, chops, or poultry under penetrating infra-red heat.

2. a. Involving, producing, or pertaining to infra-red radiation or its use.

1910 *Photogr. Jrnl.* Oct. 320 The idea of photographing landscapes through the infra-red screen. *Ibid.* 330 In the infra-red photograph the shadows are practically black and the sky is very dark. **1929** *Brit. Jrnl. Photogr.* 29 Mar. 183/2 By comparing the results of ordinary photography..with those of infra-red photography. **1935** *Practitioners Libr. Med. & Surg.* VIII. i. viii. 169 The wavelength used in infra-red therapy is not of particular importance. It is mainly a determining factor in the penetration or the depth which the heat reaches. **1951** Mrak & Mackinney in M. B. Jacobs *Chem. & Technol. Food & Food Products* (ed. 2) III. xxxiii. 1787 Infrared lamps have been installed in a plant for natural dried cod to reduce the moisture from about 60 to 43%. **1954** 'J. Christopher' *22nd Century* 197 He went through to the kitchen, switched the infra-red heater off and collected his breakfast. **1963** F. C. Webb *Biochem. Engin.* xvii. 482 Small catering-size infra-red heaters are in limited use for grilling. **1968** *Times* 9 Dec. 7/2 Infra-red galaxies.. are just one of several types of cosmic objects being studied in the booming new science of infra-red astronomy.

b. Sensitive to infra-red radiation.

1932 *Discovery* Sept. 292/1 These infra-red plates are not as a rule made sensitive to the green and orange, although they retain their sensitivity to blue. **1938** *Encycl. Brit. Bk. Yr. 1938* 500/1 New films for aerial photography included a fast infra-red film. **1961** *Daily Mail* 20 July 6/3 Midas picks up a rocket by detecting the intense heat of its exhaust flame with an infra-red eye. **1973** *Daily Tel.* (Colour Suppl.) 11 May 44/3 Another student..wants the astronauts to photograph volcanoes with infra-red film which records temperature differences on the Earth rather than grades of light.

B. *ellipt.* as *sb.* The infra-red part of the spectrum; *near, far infra-red,* the part close to, or far from, the visible spectrum.

1881 *Phil. Mag.* XI. 167 Experiments supposed to prove the existence of lines in the infra-red. **1887** *Encycl. Brit.* XXII. 376/1 Becquerel finds lines in the infra-red at 11,420. **1923** Glazebrook *Dict. Appl. Physics* IV. 533/2 Such solids as rock-salt have a very strong absorption band for light in the far infra-red. **1937** Jenkins & White *Fund. Optics* i. 13 On the long-wavelength side of the visible lies the infrared, which may be said to merge into the radio waves at about 4 × 10⁻² cm. **1960** Conn & Avery *Infrared Methods* p. v, By restricting attention to the range from 1 to 25μ we have limited the interest to the 'near infrared' but even this embraces a range of some four to five octaves. **1966** Houghton & Smith *Infra-Red Physics* i. 1 The infra-red is that region of the electromagnetic spectrum which lies between the visible and microwave regions, i.e. between wavelengths of about 8000 Å (= 0·8 μ) and 1 mm. **1973** *Sci. Amer.* Feb. 89/2 The laser's output was in the infrared.

infrasonic (infrăsǫ·nik), *a.* Also **infra-sonic.** [f. Infra- + *sonic *a.*, after *supersonic, ultrasonic,* as tr. F. *infra-sonore* (E. Esclangon 1925, in *Mém. de l'Artillerie Française* IV. 647).] **A.** *adj.* **1.** Of, pertaining to, or designating sound waves or vibrations having a frequency below the audible range (i.e. less than 15–30 Hz).

1927 E. G. Richardson *Sound* x. 236 The lower pitch limit is about 16 vibrations per second. Slow vibrations.. remain unperceived as tones if their rate of pulsation falls below this limit... Such 'infra-sonic' waves have been extensively studied by Esclangon. **1945** *Electronic Engin.* XVII. 328/3 These will not run well..at infra-sonic frequencies. **1960** *McGraw-Hill Encycl. Sci. & Technol.* XII. 506/1 Many problems in vibration, oceanography, seismology, and the dynamic behavior of elastic materials are analyzed by treating the phenomenon being studied as sound waves of infrasonic frequency. **1965** *Aerospace Med.* XXXVI. 817/2 It has been estimated that the very large super boosters of the future..will produce their maximum noise energy in the infrasonic range (below 20 cps). **1965** T. N. Davis in W. N. Hess *Space Sci.* v. 227 Infrasonic waves with periods ranging from 10–110 seconds are observed at auroral and middle latitudes. These appear only on nights of moderate or strong auroral activity.

2. Designating speeds below that of sound; subsonic. *rare*⁻¹.

1942 *Jrnl. R. Aeronaut. Soc.* XLVI. 85 Equations of general application are derived, proving that both for infrasonic and for ultrasonic (supersonic) velocities, an extremum of the cross-section is possible.

infrasound (i·nfrăsaund). [f. Infra- + Sound *sb.*[3], tr. F. *infra-son* (1906 in Robert).] Sound, or a sound, of infrasonic frequency.

1930 A. B. Wood *Textbk. Sound* ii. 213 The 'ondes de bouche' (from the mouth of the gun) consists principally of waves of very low frequency (of the order 1 p.p.s.), described by Esclangon as Infra-Sounds. **1965** *Aerospace Med.* XXXVI. 824/1 Very low sonic frequency noise and moderate levels of infrasound are commonly encountered in conventional aerospace operations. **1970** B. W. Aldiss *Moment of Eclipse* 66 This was infrasound. The plant was emitting slow air vibrations at less than ten hertz. **1971** *Observer* 28 Nov. 3/1 Many cars and lorries travelling at sustained high speeds produce such intense levels of low-frequency noise or 'infrasound' inside them that drivers can experience symptoms very similar to those produced by heavy drinking. *Ibid.*, In a wide range of passenger cars travelling at motorway speeds the sound energy is largely concentrated as intense infrasounds—most of which can be felt rather than heard.

infrastructure (i·nfrăstrʌktiŭɹ). [Fr. (1875 in Robert), f. Infra- + Structure *sb.*] A collective term for the subordinate parts of an undertaking; substructure, foundation; *spec.* the permanent installations forming a basis for military operations, as airfields, naval bases, training establishments, etc.

1927 *Chambers's Jrnl.* 14 May 374/2 The tunnels, bridges, culverts, and 'infrastructure' work generally of the Ax to Bourg-Madame line have been completed. **1950** W. S. Churchill in *Hansard Commons* CDLXXVI. 2145 In this Debate we have had the usual jargon about 'the infrastructure of a supra-national authority'. **1951** *European Rev.* Oct. 2/1 This new term 'infrastructure'.. denotes fixed military facilities such as airfields, base installations and transport systems. **1956** D. Noakes tr. *Hodeir's Jazz* 197 What I call the infrastructure is the regularly produced two- or four-beat meter (2/2 or 4/4 measure) that characterizes any jazz performance. **1957** T. Kilmartin tr. *Aron's Opium of Intellectuals* iv. 133 Thirty years ago, the dominant school of thought in the Soviet Union undertook..the task of analysing the infrastructure of society. **1960** *Times* 9 Dec. 14/2 Part of the Nato infrastructure programme. **1971** *Inside Kenya Today* Mar. 15/1 A.I.D. assistance will be focused on Vihiga Division and will..upgrade the infrastructure of roads and other social services. **1971** J. Spencer *Eng. Lang. W. Afr.* 31 A very complex infrastructure of scores of vernacular languages.

So **i:nfrastru·ctural** *a.*

1963 *Economist* 13 Apr. 130/2 Very low interest rates.. for various forms of infrastructural development. **1967** *Ibid.* 23 Sept. 1104/2 Big, infrastructural programmes—the building of dams, power stations, roads, railways, harbours and airports.

infu·riant, *sb.* [f. pr. ppl. stem of med.L. *infuriāre*: see Infuriate *v.*] Something that infuriates; an object, fact, condition, etc., which excites to anger or passion.

1953 K. Amis *Lucky Jim* viii. 87 The sight of Welch's 'bag' and fishing-hat on a nearby chair, normally a certain infuriant, only made him hum his Welch tune. **1960** *Guardian* 7 Dec. 8/3 There would be a reasonable chance of the races working happily together if only the infuriant of Federation were removed.

infuse, *v.* Add: **1.** (Later examples.)

1908 *Amer. Jrnl. Physiol.* XXI. 144 Saline was infused into a vein. **1972** *Sci. Amer.* May 75/1 Intravenous feeding had its beginning in 1843, when the French physiologist Claude Bernard infused sugar solutions into the veins of animals.

5. Delete † *Obs.* and add: In wider use, to impregnate, pervade, imbue (*with* some quality, opinion, etc.).

1900 *Westm. Gaz.* 29 Jan. 1/3 The cant with which the political history of the war is infused and suffused. **1928** *Observer* 5 Feb. 11/4 The splendid camaraderie and corporate spirit which infuse a newspaper staff.

infusoriform, *a.* Add: [repr. G. *infusorienartig* infusorian-like (A. Kölliker 1849, in *Ber. von der k. zootomischen Anstalt zu Würzburg* II. 61).] Usu. *spec.* designating or pertaining to a stage in the life-cycle of species of the order Dicyemida (phylum Mesozoa), which comprises parasites of certain cephalopods. (Further examples.)

1877 *Q. Jrnl. Microsc. Sci.* XVII. 143 The infusoriform embryo probably distributes the species by transmitting the parasite from one cephalopod to another. **1883** *Mitt. Zool. Station zu Neapel* IV. 41 The infusoriform embryos desert the parent before the development of vermiform embryos begins. **1940** L. H. Hyman *Invertebrates* I. iv. 236 The infusoriform stage is an asexual larva. **1964** T. C. Cheng *Biol. Animal Parasites* vii. 179/1 The infusoriform larvae escape from the parent rhombogen and leave the host. **1967** H. W. & L. R. Levi tr. *Kaestner's Invertebr. Zool.* I. ii. 21 Only after fertilization do the ova detach, each developing into a short, egg-shaped dispersing infusoriform dicyemid.

infusorigen (infiusǭ·ridʒĕn). *Zool.* [f. *infusori-* (repr. Infusoriform *a.*) + -gen.] A hermaphroditic group of cells formed within a rhombogen during the life-cycle of species of the order Dicyemida (phylum Mesozoa) and giving rise to infusoriform larvæ.

1883 C. O. Whitman in *Mitt. Zool. Station zu Neapel* IV. 38 The entire cell-group may be called the Infusorigen, a term used by Van Beneden as a synonym for Rhomb gen. *Ibid.* 39 There arise in the course of the history of every Infusorigen two nuclei. **1940** L. H. Hyman *Invertebrates* I. iv. 239 After having given off a succession of pseudo-eggs, the infusorigens degenerate. **1964** T. C. Cheng *Biol. Animal Parasites* vii. 179/1 Most authorities on the Mesozoa now agree that infusorigens are hermaphrodites. **1972** *Sci. Amer.* Dec. 95/3 Some of the axoblasts, instead of developing into vermiform embryos, develop into a structure that remains within the axial cell of the adult vermiform and may be thought of as a hermaphroditic gonad. The term infusorigen has been applied to this structure, which in a sense is the only organ the Mesozoa possess.

infusorioid (infiusǭ·ri,oid), *a.* [f. Infusori(a *sb. pl.* + -oid.] Resembling an infusorian.

1853 A. Henfrey tr. *Braun's Phenomenon of Rejuvenescence in Nature* 281 Active, Infusorioid structures.. occur not unfrequently in the interior of decaying cells of green fresh-water Algæ.

ingathering, *vbl. sb.* Add: Applied *spec.* to the congregating of the Jews in Israel.

1952 S. Spender *Learning Laughter* iii. 36 But to-day there is another kind of paradox... This might be called the paradox of the Ingathering. **1964** *Economist* 16 May 703/1 The other chief fact of Israel, politically as well as economically, is the determination to make the in-gathering of Jews..viable and safe. **1971** *Listener* 14 Jan. 36/3 When Israel became a state on 15 May 1948 the first principle written into its proclamation of independence affirmed that 'the state of Israel will be open to Jewish immigration and the in-gathering of the exiles'. **1973** *Guardian* 12 Mar. 10/3 Young Israelis..feel no mission to the Jews of the Diaspora. The ingathering of the exiles is a remote slogan.

ingenium (indʒī·niəm). [L., = mind, intellect.] Turn of mind; genius; talent.

1879 W. James *Let.* 10 Oct. in R. B. Perry *Tht. & Char. W. J.* (1935) II. 17 Wundt..certainly is not a first-class ingenium. **1886** G. Macdonald *What's Mine's Mine* I. xii. 225 It [*sc.* a poem] will serve to show something of Ian's youthful ingenium. **1920** T. P. Nunn *Education* 205 To school a boy in the tradition of one of these ancient occupations is to ensure (if it suits his *ingenium*) that he will throw himself into his work with spirit. **1921** *Glasgow Herald* 26 Nov. 5 His scientific ingenium was as keen as ever.

ingénue (æ̃ʒenü). [Fr., fem. of *ingénu* Ingenuous *a.*] An artless, innocent girl or young woman; also, the representation of such a character on the stage, or the actress who plays the part. Also as *adj.* = Ingenuous *a.* Hence ingénueism.

1848 Thackeray *Van. Fair* ii. 454 When attacked sometimes, Becky had a knack of adopting a demure ingénue air, under which she was most dangerous. **1857** G. A. Lawrence *Guy Liv.* xxv. 239 Mary herself could hardly play the *ingénues*, when in mature age. **1883** J. Hawthorne *Dust* I. viii. 134 Was this lady more or less of a woman of the world than he had imagined? Was there not, after all, something of the *ingénue* about her? **1923** C. Mackenzie *Seven Ages of Woman* iii. 112 'My dear, ingenue is a charming and attractive quality; but do not be too ingénue. **1930** *Daily Express* 6 Oct. 1/3 French actresses grow in popularity and ingénueism the older they become. **1931** *Times Lit. Suppl.* 5 Mar. 161/4 Her French audiences, still demanding the ingénue, find her enigmatical and disgusting. **1958** *Times* 12 Nov. 3/5 The Nanetta..brings a vibrant line to music more ingénue than the singer. **1967** C. O. Skinner *Madame Sarah* viii. 171 Every one of them, with the exception of an eighteen-year-old ingénue..declared to stay. **1973** *Daily Tel.* 16 May 13/1 And we were willing to play out with the quaint tale of the philandering husband and the sweet little ingénue Nanette.

in-goal (in,gǭ·l). *Rugby Football.* [f. In *prep.* + Goal *sb.* 3.] (See quot. 1897.)

1897 *Encycl. Sport* I. 430/2 Those portions of the ground immediately at the ends of the field of play and between the touch-lines, produced to the dead-ball lines, are called in-goal. The goal-lines are in-goal. **1935** [see *fair *a.* 10 d]. **1960** V. Jenkins *Lions Down Under* xix. 197 The penalty-try awarded when E. J. Faire obstructed A. J. F. O'Reilly in the in-goal area.

in-God-d, *v.* [cf. *engod* (En-*pref.*[1] B. 2).] *trans.* To deify, make divine; to take into God or into the godhead. So **in-Godding** *vbl. sb.*

1891 *Church Times* 20 Nov. 1128/1 The Humanity already is in-Godded in the Person of the Second Person of the Holy Trinity. **1957** D. L. Sayers *Further Papers on Dante* 187 Every creature..possesses it as to the will which, however much perfected or (in Dante's words) 'in-godded', is never absorbed or lost in God. **1959** *N. & Q.* July–Aug. 301/2 Dante expounded in the *Comedy* a mystic Way of Affirmation for the 'in-Godding' of man based on the vision of the beloved.

ingoing, *vbl. sb.* Add: **2.** The sum paid by a tenant or purchaser for fixtures, etc., on taking over business or other premises.

1930 *Daily Chron.* 4 May, Furniture Business..for Sale;..ingoing about £200. **1959** *Daily Tel.* 13 May 19/3 The principal Fully-Licensed Family and Commercial Hotel.. Ingoing £1,300.

ingoing, *ppl. a.* Add: **2.** Penetrating, thorough.

1928 *Blackw. Mag.* May 645/1 Whether he was quite so ingoing as this would have been is far from sure. **1928** E. C. Butler *tr. Grou's Meditations Love of God* p. v, He is so ingoing, so searching, while his every spiritual writer with his salt, very exacting. **1930** *Vatican Council* II. xxviii. 244 To this, the most ingoing response to the appeal of the Council, a very..

ingot. Add: **3.** **ingot iron,** iron which contains too little carbon to temper and is nearly pure by industrial standards, differing from wrought iron in containing no slag; **ingot stripper,** a machine for separating an ingot from the mould containing it; **ingot structure,** the arrangement of crystals in an ingot.

1877 *Trans. Amer. Inst. Mining Engin.* V. 20 Mr. A. L. Holley, Chairman of the International Committee, appointed by the Institute to consider the nomenclature of iron and steel, offered the following report... That all compounds of iron with its ordinary ingredients, which have been cast from a fluid state into malleable masses, and which are not sensibly hardened by being quenched in water, while at a red heat, shall be called ingot iron. **1938** J. Newton *Introd. Metall.* vii. 178 Ingot iron and wrought iron both have very low carbon, and their physical properties approach those of pure iron. *Ibid.* xvi. 499 Ingot iron is commercially pure iron, with total impurities of 0·25 per cent., which is made by a special basic open-hearth process. **1962** A. G. Guy *Physical Metall. for Engineers* v. 138 Ingot iron in the form of galvanized or enameled sheets is used for such purposes as roofing and siding. **1904** J. W. Hall in Harbord & Hall *Steel* 250 The ingot stripper is a most efficient machine.. saving..all damage to the moulds from sledging to remove the ingots. **1957** Camp & Francis *Making, Shaping & Treating of Steel* (ed. 7) xv. 295 (*caption*) Schematic representation of the action of an ingot stripper in removing the moulds from (left) big-end-down ingots and (right) big-end-up ingots. **1938** E. Gregory *Metall.* ii. 49 The original ingot structure exerts a profound influence on the behaviour of the material during forging, rolling, etc. **1952** J. Wulff et al. *Metall. for Engineers* xvi. 316 Some knowledge of how steels are melted is useful for interpreting ingot structure.

†Ingersollian (iŋgǝsǭ·liăn), *a. Obs.* [f. the name of the American agnostic, Robert Green *Ingersoll* (1833–99) + -ian.] Imbued with the tenets of R. G. Ingersoll. So **I·ngersollism,** the doctrines or tenets of Ingersoll.

1883 G. R. Wendling (*title*) Ingersollism; from a secular point of view. **1892** Stevenson & Osbourne *Wrecker* ix. 179, I don't know if you quite believe in prayer, I'm a bit Ingersollian myself.

Ingin. A U.S. colloq. spelling of Indian. (Cf. *Injun.)

1863 *Early Rec. Groton, Mass.* (1880) 82 If any Ingins can prove a lagiall [= legal] title. **1869** B. Harte *Luck of Roaring Camp* 15 They're mighty rough on strangers, and they worship an Ingin baby. **1870** J. C. Duval *Adventures Big-Foot Wallace* xlvi. 303 Whenever he can get to where the Ingins are..and may be..

ingle, *sb.*[2] Delete † *Obs.* and add later examples.

1926 T. E. Lawrence *Seven Pillars* (1935) lxxxi. 448 Abd el Kader called them whoresons, ingle's accidents, sons of a bitch,..jetting his insults broadcast to the room-full. **1962** H. Nicolson *Monarchy* v. 54 The Romans were startled by the arrival of this Asian ingle as their Emperor.

ingliding (i·ŋglaidiŋ), *ppl. a. Phonology.* [f. In *adv.* + Gliding *ppl. a.*] Gliding towards the central vowel sound /ǝ/, as in words like *air, here,* and *poor,* and in U.S. regional pronunciations like *wood* (wuǝd), *bell* (beǝl), *stem* (steǝm), *pal* (pæǝl); = *centring *ppl. a.* Also **i·nglide** *sb.*

1948 R. I. McDavid in *Amer. Speech* XXIII. 203 An apparent tendency to replace the low-country ingliding diphthongs in *date, boat* [deᵊt, boᵊt] with the up-country upgliding type [de·ɪt, bo·ʊt] also suggests a reversal of the trend in prestige values. **1956** D. W. Reed in A. A. Hill *First Texas Conference on Problems of Linguistic Analysis in English* (1962) 3 According to the Trager and Smith analysis there are nine pure vowels—/V/—and nine combinations of vowel and length or inglide. *Ibid.* 93, I think in-glide is a better name than central glide, since in-glide names a direction, central names only an area. **1959** H. Whitehall..*Dial. Soc.* xxxii. 76 The other..longer diphthongs in English. *Ibid.* 112 Off with an ingliding diphthong that is heard with equal frequency. **1961** Kurath & McDavid *Pronunc. Eng. in Atlantic States* iii. 101/2 Before voiceless stops, inglides are less common and briefer, and in words of more than one syllable they are infrequent. **1964** *Amer. Speech* XXXVII. 70 Stressed free vowels are upgliding, and all stressed checked vowels are ingliding. **1967** *Ibid.* XLII. 129 An ingliding diphthong [iǝ]. **1969** *Canad. Jrnl. Linguistics* Fall 64 The free segments are predominantly monophthongal or ingliding. **1972** H. Kurath *Studies in Area Linguistics* vi. 73 Elsewhere ingliding [oǝ ~ ɔǝ] or even monophthongal [o ~ ɔ] are current.

ingot. …

ingotism (i·ŋgǫtiz'm). [f. Ingot + -ism.] The presence of many large dendritic crystals in an ingot or casting.

1881 H. M. Howe et al. in *Proc. Amer. Soc. Testing Materials* VIII. 185 By ingotism we mean that extremely coarse structure which exists in unannealed ingots and steel castings. **1925** *Jrnl. Iron & Steel Inst.* CXI. 525 In killed steels the ill effect, if any, of a too high casting heat is pronounced ingotism. **1957** Camp & Francis *Making, Shaping & Treating of Steel* (ed. 7) xx. 395/2 Ingots exhibiting ingotism tend to crack excessively during rolling unless light drafts are employed for the first few passes in the rolls.

ingrain, *a.* (*sb.*[2]) Add: **B.** *sb.* **2.** That which is ingrain or inherent.

1899 J. Milne *Romance of Pro-Consul* vi. 49 The natives of the Australian North-West were a fine race physically, and, he judged, had an ingrain of Malay blood. **1918** P. T. Forsyth *This Life & the Next* v. 53 It is the body as what might be called the ingrain, the tissue, the physiognomy of eternal love.

ingrateful, *a.* **3.** (Later example.)

1913 F. Thompson *Works* II. 163 We take ingrateful, for a blinded while, Thine ignorant, sweet smile.

ingredience, *sb.* Restrict † *Obs.* to sense 1 and add later examples of sense 2.

1926 A. N. Whitehead *Sci. & Mod. World* (1926) x. 237 This complete ingredience in an occasion, so as to yield the most complete fusion of individual essence with other eternal objects in the formation of the individual emergent occasion, is evidently of its own kind and cannot be defined in terms of anything else. **1955** *Scottish Jrnl. Theol.* VIII. 426 Nor can it [*sc.* history] be seen as a total process given meaning by the ingredience of non-historical truth.

ingredient, *a.* and *sb.* **A.** *adj.* **b.** (Later examples.)

1933 *Theology* XXVI. 331 The distinction between the realm of possibility and that of actuality, between 'eternal objects' and the 'actual occasions' into which the eternal objects are ingredient. **1957** G. Ryle in C. A. Mace *Brit. Philos. in Mid-Cent.* 241 He has to declare that his subject-matter consists..of the sentences and their ingredient words in which arguments are expressed [etc.].

Ingres (æ̃ŋgr). The name of J. A. D. *Ingres* (1780–1867), French painter, used *attrib.* in **Ingres paper** [tr. F. *papier Ingres*], a French mould-made drawing-paper; also used to describe thick mottled paper.

1910 *Winsor & Newton Catal.* in H. Macbeth-Raeburn *Sketchers' Oil Colour Manual* 37 Drawing papers..Ingres paper (imitation Michallet). **1925** V. Blake *Way to Sketch* ix. 79 Ingres or Michelet, either white or tinted—say the blue-grey note—is useful for more finished sketches, in crayon or carbon pencil. **1941** [see *Conté]. **1968** P. Nuttall *Picture Framing* ii. 14 There is also a wide variety of papers suitable for this purpose (*sc.* mounting), such as ..Ingres paper, textured papers, silk wallpaper, marbled papers, and so on.

ingress, *v.* **1.** Revived in the U.S.

1963 V. Nabokov *Gift* iii. 178 Boris Ivanovich, horribly smiling, squeezed sideways into the room..then, ingressing with his shoulder, he would shut the door tightly behind him and sit by Fyodor's feet. **1970** N. Armstrong et al. *First on Moon* Gloss. p. xii, Ingress, to enter the spacecraft.

ingressant (ingre·sănt), *a. nonce-wd.* [f. ppl. stem of L. *ingredī* to enter + -ant[1].] Entering, in-going.

1947 Auden *Age of Anxiety* (1948) vi. 126 His [*sc.* God's] Good ingressant on our gross occasions Envisage our advance.

ingressive, *a.* Add: (Later examples.) **c.** *Phonetics.* Of or pertaining to utterances made while breathing in. Also as *sb.*, an ingressive verb or sound.

1931 G. O. Curme *Syntax* xix. 377 The ingressive aspect is often expressed by *begin, commence,* or start in connection with an infinitive or gerund. *Ibid.* 378 The ingressive idea is often expressed by the ingressives *get, grow, fall, turn, wax,* [etc.]. **1932** *Jrnl. Eng. & Gmc. Philol.* XXXI. 251 The former is called the ingressive aspect: 'He woke up (i.e., entered upon the waking state) early.' **1943** [see *effective *sb.* 3]. **1943** K. L. Pike *Phonetics* vi. 88 Sounds thus made are *ingressives* (or *rarefactives*). *Ibid.*, Ingressive lung-air sounds, and clicks, are produced in this way. **1952** [see *conceptualistically *adv.*]. **1963** *Amer. Speech* XXXVIII. 52 Phonemes..can be produced with an ingressive as well as an egressive air stream. **1969** M. M. Firestone in Halpert & Story *Christmas Mumming in Newfoundland* 66 Some are able to 'talk like a janney'—ingressive utterances at a high pitch. **1972** R. K. Waugh *Introd. Ling.* iii. 32 Sometimes ingressive air, that is, air going to the lungs, may also be used. *Ibid.* 36 Ingressives are rare indeed. In English confined perhaps to certain kinds of exclamatory sounds.

ingre·ssively, *adv.* [f. Ingressive *a.* + -ly[2].] In an ingressive manner.

1921 H. Poutsma *Characters of Eng. Verb* i. 2 The actions expressed by verbs..may be..(1) indefinitely durative, ..(2) ingressively durative, i.e. with the initial stage of the action more distinctly thought of than the rest. **1928** — *Gram. Late Mod. Eng.* (ed. 2) I. 2 The combinations into which they [*sc.* copulas] enter are a) indefinitely durative, b) continuatively durative, or c) ingressively durative. **1969** M. M. Firestone in Halpert & Story *Christmas Mumming in Newfoundland* 66 'Yes' and other words..are indeed normally uttered ingressively in this area.

i·n-group. [Cf. *In *a.* 2.] A small group of people, within a wider context, whose common interest tends to exclude others; also *attrib.* Hence *ingroupiness, ingroupness*; **i·n-grouper,** a member of an in-group. Cf. *out-group.

1907 W. G. Sumner *Folkways* i. 12 Thus a differentiation arises between ourselves, the we-group, or in-group, and everybody else, or the others-groups, out-groups.

1932 H. Becker *Syst. Sociol.* ii. xi. 182 Members of such groups may indeed manifest a certain loyalty and consideration for all fellow-members, i.e., for the 'in-group'. **1939** *American Imago* Nov. 24 In-groupers must learn to tolerate..anxiety. **1942** *New Statesman* 11 July 26/3 The very progress of civilisation has laid the foundation for a vast extension of in-group mutual dependency and mutual support. **1947** *Partisan Rev.* XIV. 478 The sense of in-groupness of the bohemian intellectual provides a source of psychological security. **1950** B. Wootton *Testament Social Sci.* vi. 127 For many primitive peoples..what the sociologist calls the 'in-group' consists of a relatively small tribal community, whereas the Christian is taught..to treat all men as brothers. **1964** M. Argyle *Psychol. & Social Probl.* x. 134 It is well known that residential courses are the scene of violent in-group feelings and wild enthusiasm. **1964** *Rev. Eng. Stud.* Aug. 338 Two masterly impersonations: of the shy poet hauled before dons, and of the puzzled philistine, ready with cautious, deflationary, in-group jokes. **1965** *Listener* 4 Mar. 345/2 Mr. Gelber's mixture of sentimentality, in-groupiness, and Village dourness, can be more than a little grating. **1967** J. Gardner *Madrigal* iii. 65 Cheery as an in-grouper at party time. **1970** *Jrnl. Gen. Psychol.* Oct. 259 An error was considered..ingroup intrusion if the response was a word paired with a similar stimulus. **1971** *Listener* 18 Nov. 674/3 Their closed or in-group speech made me feel out of it. **1972** J. L. Dillard *Black English* vi. 230 American Blacks are quick to perceive pronunciation differences on the part of West Indians who migrate to cities like New York and are somewhat slow to accept them into the in-group.

ingrown, *ppl. a.* Add: **c.** *Geol.* Applied to an incised meander having a characteristic asymmetrical cross-section (see quot. 1954) as a result of lateral erosion and movement of the bed as it was being cut.

1914 J. L. Rich in *Jrnl. Geol.* XXII. 470 The In-grown Meander Valley is one whose stream, which may or may not have inherited a meandering course from a previous cycle, has developed such a course or expended its inherited one. Thus, as the stream sunk its channel lower and lower into the bed-rock, the meanders were continually growing or expanding. The term 'in-grown' has been chosen to express this idea. **1954** W. D. Thornbury *Princ. Geomorphol.* vi. 145 Two types of incised or inclosed meanders are generally recognized: (1) entrenched or intrenched meanders.., which show little or no contrast between the slopes of the two valley sides of a meander curve, and (2) ingrown meanders.., which exhibit pronounced asymmetry of cross profile with undercut slopes on the outside of the meander curves and slipoff slopes on the inside. **1960** B. W. Sparks *Geomorphol.* ix. 225 Ingrown meanders are more slowly incised, due to less rapid downcutting or to more resistant rocks.

ingubu (iŋgū·bo). *S. Afr.* Also 9 **ingoobu, ingouboo, ingubo.** [Nguni; cf. Fanagalo *ngubo*, Bantu-Botatwe *ingubo*, blanket, clothes.] Applied to articles of dress offered for sale to the native inhabitants of Natal.

1833 S. Kay *Trav. Caffraria* i. 37 He maintained that every thing around him, mountains, rivers, grass, cattle, and even his *ingubu*, 'beast-skin garment', proved the truth of what had been said respecting the being of a God. **1837** F. Owen *Diary* (1926) 25 They [*sc.* the natives] asked in exchange for their fowls, Indian corn and pumpkins, either handkerchiefs, blankets or 'ingubo' i.e. a mantle or carosse. *Ibid.* 77 He..abruptly asked me what was the use of giving all that ingoobu to the children, alluding to the Kilts of Dingareen with which I have clothed the boys. **1860** W. Shaw *Story of Mission in S.-E. Afr.* 406 A Kaffir wears this ingubu, or 'kaross', with the hairy side next to his skin, throwing it over his shoulders, from whence it hangs down to his ancles. **1899** G. Russell *Hist. Old Durban* ix. 187 Cast-off articles of European attire, known to the Natives as 'Ingouboos'.

inguinally (i·ŋgwinăli), *adv.* [f. Inguinal *a.* + -ly[2].] By or in the groin.

1908 *Practitioner* Aug. 255 There are also certain other disadvantages to which the inguinally retained testis is liable. **1966** C. A. W. Guggisberg *S.O.S. Rhino* iii. 68 The mammae, two in number, are situated inguinally.

Ingush (i·ŋguʃ, ingu·ʃ). Also **Ingoush.** Pl. **Ingoushee, Ingush, Ingushes.** [a. Russ. *Ingúsh*, the name of the former autonomous area of Ingush.] **a.** One of a North Caucasian people, forming the minor part of the population of Checheno-Ingushetia. Also *attrib.* or as *adj.* **b.** The North Caucasic language of this people.

1902 *Encycl. Brit.* XXX. 1/2 The Kabardian aristocracy, who were possessed of feudal rights over the Ossets, the Ingushes, the Abhazes, and the mountain Tatars. **1908** J. F. Baddeley *Russian Conquest Caucasus* v. 86 The Ingoush elders..[were] summoned to Mozdók. *Ibid.* xxviii. 468 The Russians..decided to gather the Ingoushee..into a few large settlements. **1910** *Encycl. Brit.* V. 548/1 Although the Ingushes speak a Chechen dialect, they have recently been proved to be, anthropologically, quite a distinct race. **1954** Pei & Gaynor *Dict. Ling.* 101 *Ingush*, a Chechen dialect (Eastern Caucasian group of the North Caucasian family of languages). **1957** [see *Chechen]. **1958** *Everyman's Encycl.* III. 189/2 A number of peoples (Chechens, Ingushes, Balkars, Karachays, and Kalmyks) were deported from N[orth] C[aucasus] to Central Asia in 1943 for alleged collaboration with the Germans, and only rehabilitated in 1957.

Ingvaeonic (iŋvi,ǫ·nik), *sb.* Also **Inguaeonic** (-gw-), **Ingweonic.** [f. L. *Ingaeuones* (Tacitus), a Germanic tribe.] From Tacitus's division of the Germanic people into Ingaeuones, Istaeuones and Herminones, the name applied to the hypothetical language from which the earliest recorded dialects of West Germanic except Old High German descended. Sometimes used synonymously with *Anglo-Frisian. Also *attrib.*

[**1907** H. M. Chadwick *Orig. Eng. Nation* ix. 222 The identification of the Inguaeones with the Anglo-Frisian group rests on the assumption that languages of this type were once spoken in the western Baltic, a hypothesis for which no solid evidence has been produced.] **1933** L. Bloomfield *Lang.* iv. 58 We conclude that English is an offshoot of an *Anglo-Frisian* (or *Ingweonic*) dialect area, which must have been fairly extensive before the migration to Britain. **1939** *Trans. Philol. Soc.* 87 Sporadic cases of ō for 'Ingvaeonic' ā occur in OSax. **1948** *Neophilologus* XXXII. 176 The pronounced Inguaeonic characteristics of early Low German sources. *Ibid.* 181 The oldest Germanic language of the Dutch area is thus understood to have been primitive Inguaeonic. **1948** *Trans. Philol. Soc.* 1947 14 The original Germanic language of the Low German area was not in any essential matter distinguished from Frisian... In that original state it formed with English a loose unity..having in a common articulation potentialities for common developments... This loose unity we may call Ingvaeonic. **1959** A. Campbell *Old Eng. Gram.* § 4 This West Germanic without Old High German is often called 'Ingvaeonic', because in Tacitus' threefold division of the Germans the Ingvaeones lie near the sea.

inhabited, *ppl. a.* Add: **b.** Historiated, e.g. *inhabited scroll,* an arabesque pattern of foliage in which figures, birds, etc., appear.

1952 D. T. Rice *Eng. Art 871–1100* v. 149 On [a font] at Alphington in Devon there is an inhabited scroll border which suggests the influence of a manuscript of late-eleventh-century type. **1954** M. Rickert *Painting in Brit.: Middle Ages* ii. 36 This motif..is worked into running scroll patterns enclosing at intervals lively figures of birds and animals—the so-called inhabited scroll. **1959** *Listener* 1 Oct. 538/3 The exuberance of a St. Alban's inhabited scroll. **1970** M. Swanton *Dream of Rood* 12 The narrower sides of the shaft are more purely decorative, carved with the so-called 'inhabited vine-scroll'. This is a Middle Eastern motif deriving from models like the Ravenna throne.

inhalant, *a.* Add: Also, concerned with inhalation.

1875 Huxley & Martin *Course Elem. Biol.* 105 These 'inhalent' and 'exhalent' currents go on, so long as the animal is alive and the valves are open. **1935** Twenhofel & Shrock *Invertebr. Paleontol.* ix. 314 In the earliest and most primitive pelecypods..the edges of the mantle are entirely free, but posteriorly they are folded in such a way as to produce an upper exhalant channel separated from a lower inhalant one. **1968** R. D. Purchon *Biol. Mollusca* iv. 164 *S[olen] delesserti* can also perform a swimming escape reaction by expulsion of a series of jets of water from the inhalant siphon.

inhalator (i·nhălēitǝɹ). orig. *U.S.* [f. Inhale *v.* + *-ator.] = Inhaler 2.

1929 *Lit. Digest* 30 Mar. 79/2 The most effective arrangement of all is to see that the city fire or police department and the hospital ambulances have inhalators. **1947** *Chicago Tribune* 17 July 32/2 Inside the ambulance is an inhalator. **1949** *Chicago Daily News* 4 May 1/4 (*caption*) [He] is administered oxygen by members of Inhalator Squad 2. **1956** *New Gould Med. Dict.* (ed. 2) 600/2 *Inhalator*.., a device for facilitating the inhalation of a gas or spray. Used for providing oxygen or oxygen–carbon dioxide mixtures for respiration in resuscitation. **1974** J. Wainwright *Evidence I shall Give* xvi. 61 It was a nasal inhalator..made of white plastic and about the size of a lipstick tube..with a tiny hole at its pointed end.

inhalatorium (inhē̆ilătǭ^ǝ·riǝm). *Med.* Pl. **inhalatoria.** [f. Inhale *v.* after Sanatorium.] A building or room used for the treatment of respiratory complaints by vaporized medicaments.

1906 *Chambers's Jrnl.* 347/2 In the medical institute called the Inhalatorium special rooms are set apart for the use of patients, who sit for half-an-hour at a time breathing an atmosphere charged with the vapour suited to their special complaints. **1912** *World* 7 May 697/1 The inhalatoria and gurgling-rooms. **1966** *Punch* 2 Feb. 161/2 So heigh-ho for the tap-room, the ambulatory, the graduated walks, the grandiose inhalatorium.

inhale, *v.* Add: **1.** Also *spec.* of tobacco smoke, and *absol.*, as in *do you inhale?*

1933 E. O'Neill *Ah, Wilderness!* (1934) iii. i. 85 Say, you oughtn't to inhale like that! Smoking's awful bad for girls, anyway. **1970** *New Yorker* 17 Oct. 36/3 He reached for a package of Celtique cigarettes, took one, lit it, and inhaled deeply.

i·nhale, *sb. U.S.* [f. the vb.] The action of inhaling (tobacco smoke).

1934 J. O'Hara *Appointment in Samarra* (1935) v. 133 Not holding the cigarette very expertly, but taking appalling inhales. **1954** W. Faulkner *Fable* 179 Drawing the cigar to life in one slow inhale-exhale. **1959** N. Mailer *Advts. for Myself* (1961) 45 He lit a cigarette, and then after the first couple of inhales, he felt the anger coming back in him.

i·nheld, *ppl. a. rare.* [f. In *adv.* + Held *ppl. a.*] Held within.

1903 T. Hardy *Dynasts* I. i. vi. 53 His lips with inheld laughter grow deformed.

inheritance. Add: **5.** *inheritance tax* (or *taxation*) (later examples); *spec.*, orig. *U.S.*, a tax on inherited property levied on individual beneficiaries, varying according to their degrees of relationship to the testator, rather than on the estate before its distribution.

1895 E. R. A. Seligman *Ess. Taxation* v. 133 The inheritance tax to-day scarcely needs defence. It is found in almost every country... In the United States..there is now a decided movement toward the progressive inheritance tax. **1903** S. F. Weston *Princ. Justice in Taxation* viii. 289 We shall not attempt..to discuss the various theories of the inheritance tax. **1911** *Encycl. Brit.* IX. 465/2 The really vital change was the extension in 1894 of the old Probate Duty into a comprehensive impost... This 'Inheritance Tax'—to give it its scientific title—operates as a complementary property tax. **1929** A. Comstock *Taxation in Mod. State* xi. 154 Examples of inheritance taxation may be found in antiquity and in the Middle Ages... Since 1914 inheritance taxes have had a less spectacular rôle than income taxes and sales taxes. **1937** M. Newcomer in C. Shoup *Stud. Current Tax Probl.* 16 The annual burden of the estate and inheritance taxes has been taken to be the amount of the premium for life insurance sufficient to cover these taxes at death. **1965** *Listener* 2 Dec. 881/2 The positive case for inheritance taxation has become stronger. **1972** *Daily Tel.* 22 Mar. 32/6 This looks a pretty good start towards the possible inheritance tax, in substitution for estate duty.

inhibit, *v.* Add: **4.** *Psychol.* (See *inhibition 4.) Extended from sense 3.

1876 W. James *Coll. Ess. & Rev.* (1920) 30 A representation arises in a mind, but..it is inhibited by another which confronts it. **1943** C. R. Griffith *Princ. Syst. Psychol.* xvi. 596 A strong connection between two elements *a* and *b* inhibits the formation of connections between *a* and some other element *c* or *d*. **1957** Partridge *English gone Wrong* i. 22 *Inhibition* and the adjective *inhibited* and the verb *inhibit* properly denote the restraint that one psychical activity (for instance, thought) imposes upon another (as it might be fear) and also, derivatively, any psychical impediment to the free workings of body or mind or of both; it is this latter sense which has become debased to mean—as if that were a deplorable thing—the dictates of a decent self-restraint and the promptings of a natural modesty.

Hence in *Psychol.*, **inhibited, inhibiting** *ppl. adjs.*, **inhi·bitedness.**

1942 A. L. Rowse *Cornish Childhood* ii. 44 Because of his reserve, his essential inhibitedness. **1961** J. A. Brussel *Layman's Guide Psychiatry* viii. 80 This is best accomplished by removing or modifying the inhibiting factors which have blocked the individual's personality development. **1963** A. Heron *Towards Quaker View of Sex* i. 7 This still repressive and inhibited outlook towards sex. **1967** Hilgard & Atkinson *Introd. Psychol.* (ed. 4) xxii. 556/1 With neurotics who are too inhibited to discuss their feelings spontaneously, more directive methods are usually necessary. **1969** E. Mirel in P. Solomon *Handbk. Psychiatry* xxxii. 389 The 'pathologically shy' children who manifest passivity, inhibited initiative and motor action.

inhi·bitingly, *adv.* [f. Inhibiting *ppl. a.* + -ly[2].] In an inhibiting manner.

1941 *Scrutiny* X. 178 The nature and circumstance of the rescue leave each exquisitely and inhibitingly scrupulous about taking advantage of the other's helplessness or chivalry. **1965** *Sat. Rev.* (U.S.) 30 Oct. 91 He has also fought—rigidly, inhibitingly—with his sense of humor.

inhibition. Add: **3.** (Later examples.)

1906 C. S. Sherrington *Integrative Action Nervous Syst.* iii. 84 Classical examples of inhibition are those of the vagus nerve on the heart, and of the *corda tympani* on the blood-vessels of the submaxillary region. **1927** G. V. Anrep tr. *Pavlov's Conditioned Reflexes* iii. 43, I consider it advisable to give a brief description of inhibition of centres as observed in the field of unconditioned reflexes. **1967** R. F. Thompson *Found. Physiol. Psychol.* vii. 168 These more limited hypotheses still imply that inhibition tends to act near the region of the cell body where the spike discharge is initiated.

b. *Chem.* (See quot. 1902[2].)

1902 S. W. Young in *Jrnl. Amer. Chem. Soc.* XXIV. 299, I will..use the word 'inhibition' to cover the phenomena in point. *Ibid.* 302 Inhibition, *i.e.* a marked reduction of the reaction rate under the influence of minute quantities of foreign substances. **1923** [see *inhibitory *a.* 2]. **1956** *Nature* 3 Mar. 432/2 Anti-competitive (uncompetitive) inhibition, in which the inhibitor combines with the enzyme-substrate complex but not with the enzyme, has been considered theoretically. **1970** G. Odian *Princ. Polymerization* iii. 221 Polymerization is completely stopped by benzoquinone, a typical inhibitor, during an induction or inhibition period.

4. *Psychol.* A voluntary or involuntary restraint or check that prevents the direct expression of an instinctive impulse; also *colloq.*, in looser use, an inner hindrance to conduct or activity.

1876 W. James *Coll. Ess. & Rev.* (1920) 32 Doubt itself is an active state, one of voluntary inhibition or suspense. **1897** J. Adams *Herbartian Psychol.* 257 It is this work of inhibition that causes the peculiar feeling of effort that marks all voluntary attention as opposed to involuntary. **1916** A. A. Brill tr. *Freud's Wit & its Relation to Unconscious* iv. 206 One cannot possibly consider the amount of the pleasure so great as to believe that it has the power to annul deep-rooted inhibitions and repressions. **1932** E. Bowen *To North* xiv. 141 Blurred by the inhibitions of Pauline, upon which his sister dwelt with such gusto. **1936** *Discovery* Aug. 254/1 To guard against auto-suggestions and personal inhibitions of various kinds. **1965** A. L.

Fisher tr. *Merleau-Ponty's Struct. Behaviour* i. 18 That the brain possesses a general power of inhibition would be accepted. **1973** W. J. Burley *Death in Salubrious Place* iii. 56 It was light enough to see the boy's embarrassed shrug. Georgie had no such inhibitions. 'She had a thing about Vince.'

inhibi·tionism. *Psychol.* [f. Inhibition + -ism.] (See quot. 1934.) Also *gen.* A tendency towards inhibition.

1934 H. C. Warren *Dict. Psychol.* 138/2 *Inhibitionism*, the view that character is a function of the inhibition of instinctive tendencies. **1952** *Archit. Rev.* CXII. 195/2 For men of his type *fin de siècle* sensitivity and twentieth century inhibitionism are equally abhorrent.

inhibitive, *a.* (*sb.*). Delete *rare* and add later examples of the adj.

1899 W. James *Talks to Teachers* xv. 181 A familiar example of the paralyzing power of scruples is the inhibitive effect of conscientiousness upon conversation. **1902** *Jrnl. Amer. Chem. Soc.* XXIV. 299 An 'inhibitive agent' is then a substance producing an 'inhibition', or having an 'inhibitive effect'. **1944** G. B. Shaw *Everybody's Pol. What's What?* xxiii. 207 He [*sc.* Pavlov] thought he had discovered that reflexes have negative phases as well as positive ones, and can be classed as Excitatory or Inhibitive. **1952** Kirk & Othmer *Encycl. Chem. Technol.* IX. 4 Colloids apparently owe their inhibitive action to their colloidal nature. **1963** A. Heron *Towards Quaker View of Sex* i. 10 The emphasis on morality has so often gone with a cold and inhibitive attitude. **1970** G. F. Newman *Sir, You Bastard* i. 22 Permission was given to live at home provided the distance wasn't inhibitive. **1971** *Mod. Law Rev.* XXXIV. vi. 655 The principal purpose of the Act is inhibitive.

inhibitor. Restrict *rare* to sense in Dict. and add: **2.** That which inhibits.

1902 W. James *Varieties Relig. Experience* xi. 265 Danger is for most men the great inhibitor of action. **1908** W. McDougall *Introd. Social Psychol.* iii. 55 [Fear] is thus the great inhibitor of action. **1955** *Sci. Amer.* Apr. 72/2 Many seeds have water-soluble germination inhibitors in their covering. **1973** *Nature* 12 Jan. 140/1 The hydroid may contain a substance which inhibits its nematocysts from discharging and stinging other members of the colony. An occasional hydroid may provide *Pagurus* with such an inhibitor.

b. *Genetics.* A gene whose presence prevents the expression of some other non-allelic gene.

1911 *Jrnl. Genetics* I. 190 The Brown Leghorn..never produces pigmented birds and we..regard it as entirely without the factor *P*. But it possesses the inhibitor factor *I*. **1911** R. C. Punnett *Mendelism* (ed. 3) vii. 70 Probably we ought to regard the beardless as a bearded wheat in which there is an inhibitor that stops the beard from growing. **1925** D. F. Jones *Genetics in Plant & Animal Improvement* iv. 75 A fifth factor, *I*, when present, prevents any color from appearing, no matter what other factors are there. It is called an inhibitor. **1949** F. B. Hutt *Genetics Fowl* vi. 154 The first evidence that the inhibitor of dermal melanin is sex linked was found by Davenport (1906). **1965** J. A. Serra *Mod. Genetics* I. iii. 60 Inhibitors..completely hinder the manifestation of the gene with which they interact.

c. Any substance which (often in small quantities) slows down or effectively prevents a particular chemical or biochemical process or diminishes the activity of some reactant or catalyst (e.g. in corrosion, the formation of gum in petrol, or enzymic reactants).

1914 S. E. Sheppard *Photo-Chem.* vii. 289 The actual inhibitor in the case of ammonia is probably NCl_3. **1924** F. O. Rice in H. S. Taylor *Treat. Physical Chem.* II. xiv. 923 A type of negative catalysis in which the inhibitor combines with one of the reactants to form a molecular compound... The rôle of the inhibitor in all these reactions would therefore be that of a competitor for one of the molecular species undergoing change. **1925** *Jrnl. Soc. Chem. Industry* XLIV. 163 T/1 (*heading*) Water-line corrosion of iron and steel, with special reference to the action of the so-called 'inhibitors' of corrosion. **1935** *Jrnl. R. Aeronaut. Soc.* XXXIX. 791 An attractive method of preventing ice formation..is the use of some form of inhibitor which may be added to the fuel. **1938** W. V. Thorpe *Biochem. for Med. Students* x. 130 In many instances the inhibition of an enzyme can be regarded as the result of chemical combination between the enzyme and the inhibitor. **1951** I. L. Finar *Org. Chem.* I. iii. 44 Gum formation in cracked gasolines is prevented by the addition of inhibitors, which are mainly phenols or aromatic amines. **1962** H. L. Kern et al. in A. Pirie *Lens Metabolism Rel. Cataract* 385 Ouabain is a relatively specific inhibitor of cationic transport. **1962** *Lancet* 27 Jan. 192/1 It seems likely that the tissue activator may be 'unavailable' because an inhibitor is present, but this conclusion must be tentative until more is known about the nature and precise measurement of the inhibitors of fibrinolysis. **1968** R. O. C. Norman *Princ. Org. Synthesis* xv. 480 It is advisable to add a small quantity of an inhibitor to the more readily polymerized olefins..to prevent polymerization during storage. **1971** P. J. McMahon *Aircraft Propulsion* x. 298 Small quantities of various inhibitors may be used to restrict the rates of reaction in certain places. An inhibitor could for example prevent case temperatures from becoming excessive by cutting down the burning rate of propellants immediately adjacent to the case surface. **1972** *Nature* 18 Feb. 398/2 A generalized metabolic inhibitor..elicited 0% drug response by itself.

inhibitory, *a.* **2.** Delete *Physiol.* and add: More widely, that inhibits or checks anything; producing inhibition. (Further examples.)

1879 W. James *Coll. Ess. & Rev.* (1920) 129 Positivism takes a middle ground, and with a certain consciousness of the beyond, abruptly refuses by an inhibitory action of the will to think any further. **1901** B. Hollander *Revival of Phrenol.* i. 36 The frontal lobe, as the seat of the reasoning faculty, is an inhibitory apparatus against the lower and more instinctive natural impulses. **1902** *Jrnl. Amer. Chem. Soc.* XXIV. 306 The inhibitory actions in question are quite closely confined to reactions in which free oxygen is involved. **1923** *Jrnl. Physical Chem.* XXVII. 325 The inhibitory power of water in the esterification of acids in alcoholic solutions..represents a complex case of the Titoff type of inhibition. **1944** G. B. Shaw *Everybody's Pol. What's What?* xxiii. 205 Some [*sc.* conditioned reflexes] are too cruel for civilized people to tolerate, and from being what Pavlov calls excitatory have become inhibitory. **1959** *Metabolism* VIII. 101 Calcium gluconate exerted a significant inhibitory effect on insulin degradation.

inhistoricity (i:nhistǫ̆ri·sĭti). *rare.* [In-[3].] Lack of historicity.

1930 C. J. Wright *Miracle in Hist.* 8 The main fact that emerges for our study is that all of these hypotheses postulate a large amount of inhistoricity in the narratives.

inhomogeneity (inhǫ:modʒĭnī·ĭti). [In-[3].] **1.** Something that is not homogeneous with its surroundings; a local irregularity or departure from uniformity.

1899 J. Ward *Naturalism & Agnosticism* I. iv. 117 The former consists of smallest inhomogeneities,—a finely grained structure, as we say in English. **1936** *Jrnl. R. Aeronaut. Soc.* XL. 595 A local slip occurred (owing to a stress-concentration effect at a local inhomogeneity or flaw) in the zone separating the part of the crystal which had slipped from that which had not. **1955** *Jrnl. Brit. Interplanetary Soc.* XIV. 20 Just as the visual twinkling of stars gives information about atmospheric irregularities.. so the 'twinkling' of radio stars in radio wave-lengths can show up 'inhomogeneities' in the ionosphere, especially at the top. **1956** *Nature* 10 Mar. 487/1 Crystallographic cracking can be found in such an alloy, although its analysis is complicated by the presence of gross inhomogeneities and intermetallics. **1959** *New Scientist* 22 Jan. 167/1 The solidified ribbon of glass has a certain amount of distortion which cannot be avoided arising from small differences in viscosity due to chemical and thermal inhomogeneities. **1959** *Wiltshire Archaeol. & Nat. Hist. Mag.* LVII. 176 The sensitivity of the [electrical resistivity] method depends on..the size and depth below the surface of the inhomogeneity. **1971** I. G. Gass et al. *Understanding Earth* i. 36/1 Gneissose banding is also developed from original inhomogeneities in the rock such as bedding.

2. The property of being inhomogeneous; lack of homogeneity.

1916 *Sci. Abstr.* A. XIX. 154 The inhomogeneity of the field in the canal-ray tube. **1921** [see *graininess]. **1930** *Proc. R. Soc.* A. CXXIX. 221 All the evidence..tends to show that this tail is due, not to initial inhomogeneity of velocity, but to scattering or absorption. **1938** R. W. Lawson tr. *Hevesy & Paneth's Man. Radioactivity* (ed. 2) xx. 189 The condition of initially complete uniformity in distribution is not satisfied, and it is just this inhomogeneity that is utilized..for the concentration of the isotopes. **1942** *Jrnl. Biol. Chem.* CXLVI. 459 Further evidence for the inhomogeneity of ferritin is the variability of the iron, phosphorus, and nitrogen content of different crystallized ferritin samples. **1962** *Listener* 12 July 62/3 Information on fundamental crystal size and crystal shape, on lattice strain, on inhomogeneity.

inhomogeneous (i:nhǫmodʒī·niǝs), *a.* [In-[3].] Not homogeneous. **a.** Not of uniform nature throughout; composed of diverse constituents; heterogeneous.

1904 *Jrnl. Physical Chem.* VIII. 425 The 58·5 percent alloy is homogeneous when annealed at 720° and quenched. It becomes inhomogeneous..if annealed at 685° and quenched. **1938** R. W. Lawson tr. *Hevesy & Paneth's Man. Radioactivity* (ed. 2) ii. 34 At the moment of their emission the α-rays from a single radioactive substance all possess the same velocity... As soon as they have traversed a sheet of an absorbing substance, however,..the pencil of rays begins to be inhomogeneous. **1956** *Nature* 25 Feb. 380/2 The relatively great variations are due partly to the inhomogeneous microstructure of the muscular tissue. **1962** Corson & Lorrain *Introd. Electromagn. Fields* xi. 400 We shall not attempt here a rigorous discussion of wave propagation in such inhomogeneous media.

b. *Math.* Consisting of terms that are not all of the same degree or dimensions.

1943 Margenau & Murphy *Math. Physics & Chem.* vii. 237 One remarkable feature of an inhomogeneous equation..is that it may not possess solutions for every value of *k* even though the homogeneous equation, with the same boundary condition, has solutions. **1946** L. Brillouin *Wave Propagation* vi. 109 This is an inhomogeneous differential equation as it stands. **1957** L. Fox *Numerical Solution Two-Point Boundary Probl.* viii. 266 In § 5 we solved the approximate finite-difference form of this problem by considering two trial solutions, $y^{(1)}$ and $y^{(2)}$, of which the former satisfied the inhomogeneous finite-difference equations and the correct initial condition, and $y^{(2)}$ satisfied the homogeneous forms..and had initial value zero. **1962** W. B. Thompson *Introd. Plasma Physics* iv. 57 Of the two possible choices for pressure, the first leads to slightly simpler results, although it does make eq. (4.5.8) inhomogeneous.

Hence **i:nhomoge·neously** *adv.*, in an inhomogeneous manner; unevenly, irregularly.

1909 in Webster. **1937** *Q. Jrnl. Geol. Soc.* XCIII. 582 A monomict, inhomogeneously, orientated, heteroaxial quartz-B-tectonite with partial recrystallization. **1966** D. G. Brandon *Mod. Techniques Metallogr.* i. 4 The difficulty comes when the features of interest are too small to be resolved at a low magnification and too widely or inhomogeneously dispersed to include a representative area..at a high magnification. **1973** *Sci. Amer.* Dec. 70/3 Owing to the random distribution of impurity ions..the energy levels are inhomogeneously broadened over the volume of the medium.

inhour (i·n,auǝɹ). *Nuclear Science.* [f. *in(verse) hour*: so named because if the reactivity is small it is inversely proportional to the corresponding reactor period (to a first approximation).] A unit for expressing the reactivity of a nuclear reactor, being the reactivity of one having a reactor period of one hour (i.e. in which the neutron flux increases by a factor *e* in one hour).

1947 H. L. Anderson et al. in *Physical Rev.* LXXII. 17/1 The unit of [control] rod position was given the name inhour (from 'inverse hour', symbol: ih), with the significance that when the control rod is displaced from the critical position by 1 inhour, the pile will have a period of (very nearly) 1 hour. *Ibid.* 21/2 The inhour is useful as a measure of rod displacement because it is a measure of pile reactivity which is independent of the position of the control rod. **1954** R. Stephenson *Introd. Nucl. Engin.* vii. 269 There is no simple relationship between inhours of reactivity and pile period observed. Thus 2 inhours of reactivity do not make the pile period equal to 2 hr, nor do they make the pile period exactly one-half an hour, although for small reactivities the number of inhours is about directly proportional to the reactivity. **1966** D. Jakeman *Physics Nucl. Reactors* ix. 340 Equation 9.69 is referred to as the inhour equation and is used to define a unit of reactivity called the inhour. This is the amount of reactivity to give a reactor period of 1 hour and for a U^{235} system is equal to a reactivity of $2{\cdot}62 \times 10^{-5}$.

in-house (i·nhaus), *a.* and *adv.* [In *pref.*[1]] **A.** *adj.* Of or pertaining to the internal affairs of a business or institution, etc., as distinguished from its relations with groups or persons external to itself. **B.** *adv.* Internally; without outside assistance.

1956 W. A. Heflin *U.S.A.F. Dict.* 268/1 *In-house research*, research done within the Air Force, not by contract. **1966** *Electronics* 14 Nov. 25 Under the new arrangement it's expected that more of the work will be done in-house at the Marshall Space Flight Center. **1967** *Ibid.* 6 Mar. 8/3 Although some electronic equipment makers do produce their own integrated circuits—or at least maintain an in-house capability—most still buy on the open market. **1967** Karch & Buber *Offset Processes* iii. 47 The type may be set 'in-house' or obtained from a composition house. **1968** *Lebende Sprachen* XIII. 4/1 A relatively small number of stock microcircuits..made by outside suppliers or by his own in-house facilities. **1971** *Meta* XVI. 141 The translation assignment came from a large pharmaceutical company with an in-house staff of translators. **1971** *New Scientist* 27 Apr. 251/2 More R and D should be put out to firms, thus further reducing the highly expensive 'in house' staff. **1971** E. F. Schoeters in B. de Ferranti *Living with Computer* viii. 67 This does not mean that the day of the in-house computer is coming to an end. **1972** *Science* 5 May 500/3 Postdoctoral fellows, who will be recruited to come to the academy as in-house resident scholars. **1973** R. W. Burchfield in McDavid & Duckert *Lexicogr. in English* 100 Making full use..of in-house photocopying apparatus.

inhu·manism. [In-[3].] Lack of humanism; inhumanity.

1907 W. James *Pragmatism* i. 20 You find empiricism with inhumanism and irreligion. **1933** *Archit. Rev.* LXXIII. 207/2 The dogged enthusiasm of Eric Gill.. undermined the hard-headed business-man's equally dogged belief in the sacrosanct inhumanism of graphic *laissez-faire*. **1960** H. Read *Forms of Things Unknown* xi. 178 The problem is mass-suffering, mute and absurd: in one word—inhumanism.

inhumanita·rian, *sb.* and *a.* [In-[3].] **A.** *sb.* One who does not accept the views and practices of humanitarianism. **B.** *adj.* Not accepting, or disregarding, the views and practices of humanitarianism.

1936 R. Frost *Let.* 25 July (1964) 282, I hate to be done out of it by a hard-boiled inhumanitarian. **1947** *Mind* LVI. 170 The ideals..which had inspired the French Revolution, have been dexterously transformed into justification of absolute monarchy and inhumanitarian nationalism.

inhumorous (inhiū·mǫ̆rǝs), *a.* [f. In-[3] + Humorous *a.*] Not humorous; lacking in humour. So **inhu·morously** *adv.*

1898 *Contemp. Rev.* Aug. 194 Burne Jones was sincere in his art, not fanatically or inhumorously, but quietly and subtly. **1920** *Blackw. Mag.* Aug. 138/1 Many Englishmen ..allow themselves to be convinced by hearsay that Scotsmen are inhumorously inclined. **1926** W. J. Locke *Old Bridge* ii. vi. 78 'Life is real and life is earnest'—but so is the drivelling existence of the inhumorous ant.

inio- (i·nio), also before a vowel **ini-**, combining form of Gr. ἰνίον occipital bone, occiput, and of Inion[1], used in a few medical terms, as **i:nience·phalus** [Gr. ἐγκέφαλος brain] = next; also, a monster exhibiting iniencephaly; **i:nience·phaly,** an abnormality in which part of the brain protrudes through an opening in the occiput and which is generally accompanied by spina bifida and retroflexion of the spine; so i:niencepha·lic *a.*; i:nio-glabe·llar *a.*, extending from the inion to the glabella.

1893 *Trans. Edin. Obstetr. Soc.* XVIII. 227 A sagittal section of an iniencephalic female fœtus. 1958 R. A. Willis *Borderland Embryol. & Path.* iv. 158 (*caption*) Paramedian section of the 2-cm. iniencephalic embryo described in the text. [1836 I. G. Saint-Hilaire *Hist. Gén. et Particulière des Anomalies* II. 308 (*heading*) Iniencéphale, *Iniencephalus*.] 1857 Dunglison *Dict. Med. Sci.* (rev. ed.) 499/2 *Iniencephalus*, a monster whose encephalon is in great part in the cranium, and in part out of it, behind, and a little beneath the cranium, which is open in its occipital portion. 1905 *Jrnl. Obstetr. & Gynaecol.* VIII. 236 (*heading*) Iniencephalus. 1925 *Surg., Gynecol. & Obstetr.* XLI. 182/2 My own specimen is a large full term female fetus, apparently perfectly developed in every way with the exception of the craniovertebral axis, which shows the characteristic features of iniencephalus. 1951 *Jrnl. Obstetr. & Gynaecol.* LVIII. 463/2 Iniencephalus is a rare condition. [1836 I. G. Saint-Hilaire *Hist. Gén. et Particulière des Anomalies* II. 311 Les trois observations d'iniencéphalie que possède la science suffisent en effet pour fournir les élémens d'une caractéristique exacte.] 1902 *Encycl. Medica* XII. 139 This retroflexion of the foetus is often combined with defective development of the lower part of the occipital bone, when the name iniencephaly is sometimes given to it. 1968 H. Kalter *Teratology Cent. Nervous Syst.* vi. 165 Animals with craniorachischisis sometimes also had iniencephaly. 1803 J. Barclay *New Anat. Nomencl.* 146 If lines be drawn between every two of the different aspects, they will constitute the four following diameters: The Dextro-sinistral, the Corono-basilar, the Inio-glabellar, and the Inantinial. 1903 *Science* 30 Oct. 554/2 An inio-glabellar line can be drawn which will correspond very closely to the lower boundary of the cerebrum.

initial, *a.* and *sb.* Add: **A.** *adj.* **1. d.** Math. *initial condition*, each of a set of conditions giving the values (*initial values*) of dependent variables or their derivatives for a single set of values of the independent variables.

1834 W. Hamilton in *Phil. Trans. R. Soc.* CXXIV. [250 No general solution has been obtained assigning (as a complete solution ought to do) 3n relations between the n masses.., the 3n varying coordinates.., the varying time t, and the 6n initial data of the problem, namely, the initial coordinates.., and their initial rates of increase..; the quantities called here initial being those which correspond to the arbitrary origin of time.] 273 The problem of integrating these [differential] equations consists in proposing to assign, by their means, six relations between the time t, the masses m_1 m_2, the six varying coordinates..and their initial values and initial rates of increase. 1890 A. R. Forsyth *Theory Differential Equations* I. iii. 82 He [*sc.* Jacobi] shewed that the introduction of 'initial values' of the variables..renders it possible to take the integrals of the first subsidiary system in a form, which leads immediately to the transformation of the equation. 1902 *Ibid.* IV. i. 4 The conditions, as to the arbitrarily assigned values to be acquired at ζ by w and its derivatives, are called the initial conditions; the values are called the initial values. 1920 H. T. H. Piaggio *Elem. Treat. Differential Equations* iv. 53 As t usually denotes time and x and y rectangular coordinates, a condition such as z = 0 when t = 0 is called an initial condition, while one such as z = 0 if x = 0, or if x = l, or if y = x, is called a boundary condition. 1957 L. Fox *Numerical Solution Two-Point Boundary Probl.* i. 5 Boundary-value problems can always be solved in theory, and often in practice, by a combination of initial-value problems, the extra initial conditions being chosen more or less arbitrarily but finally adjusted to satisfy the prescribed boundary conditions. 1968 Fox & Mayers *Computing Methods for Scientists & Engineers* iii. 30 We need two conditions, and here there are two main possibilities. In the first, giving the so-called initial-value problem, we are provided with two adjacent values y_s and y_{s+1} for some s, or possibly y_s and some linear combination of y_{s-1}, y_s, and y_{s+1}...The second possibility, giving the so-called boundary-value problem, is the specification say of y_0 and y_n, the values at the two ends of some range.

2. b. *initial teaching alphabet*, a 44-letter phonetic alphabet, originally known as the 'Augmented Roman' alphabet, devised by Sir James Pitman (b. 1901) to assist the teaching of reading and writing.

1962 J. A. Downing *I.T.A. Reading Exper.* (1964) 14 Sir James Pitman's new Initial Teaching Alphabet..has been evolved from his grandfather's 'Phonotypy' and from the 'Nue Spelling' of the Simplified Spelling Society. 1964 *Daily Tel.* 20 Mar. 19/1 The initial teaching alphabet scheme, in use experimentally for three years to make children read more easily, is to get a Government grant.

B. *sb.* **2. c.** *initial-word*, an acronym.

1939 *Jrnl. Inst. Journalists* Jan. 19/3 For one man who says 'London County Council' a thousand say 'Ellceecee', and euphony demands..that this curious initial-word, and L.M.S. and F.J.I., should all be preceded by 'an'.

4. *Bot.* An initial cell (cf. A. *adj.* 1 c).

1914 M. Drummond tr. *Haberlandt's Physiol. Plant Anat.* ii. 84 A stratification of the meristem due to the vertical seriation of the initials is sometimes evident. 1938 *Nature* 10 Dec. 1042/2 Douin..has now shown that *P. asplenioides* develops axillary branch initials, the upper developing into normal branches. 1955 *Jrnl. Ecol.* XLIII. 51 Two small..trees, bearing catkin initials, were transplanted. 1970 Raven & Curtis *Biol. Plants* ii. 129/2 These initials, or growth-initiating cells, appear to surround a group of cells in which no cell division takes place.

initialese (iniʃălī·z). [f. Initial *sb.* 2 + -ese.] Abbreviation by using the initial letters of the words to be shortened.

1955 *Amer. Speech* XXX. 110 In a dinner speech to the National Institute of Social Sciences given in New York in November, 1952..Mr Lovett frankly advocated the convenience of what he called *initialese*, a system of using initials and contractions, in order to save time. 1961 *Engineering* 17 Nov. 654 ASLE is initialese for American Society of Lubrication Engineers. 1962 *Guardian* 31 Jan. 8/2, 2,000-odd abbreviations listed in 'International Initialese'.

initialization (iniʃăləizē·ʃən). *Computers.* [f. next + -ation.] The action or process of initializing; the computer operations involved in this.

1957 D. D. McCracken *Digital Computer Programming* xiii. 166 The programmer sitting at his desk with no more information than the trace may wish to know what happened in an early section of the program which set up the initialization. 1961 Leeds & Weinberg *Computer Programming Fund.* iii. 75 By initialization we mean the program steps which prepare the program to carry out its function. In this instance the initialization would probably correspond to rewinding a magnetic tape to its starting point. 1969 G. B. Davis *Computer Data Processing* xii. 271 The general steps for programming a loop are the following: 1. Initialization. 2. Execution. 3. Modification. 4. Test for termination.

initialize, *v.* Restrict *rare* to senses in Dict. and add: **2.** *trans.* (*Computers.*) To set to the value, or put in the condition, appropriate to the start of an operation. Const. *to*.

1957 D. D. McCracken *Digital Computer Programming* xi. 146 [Instructions] 18 and 19 initialize the address of the instruction with which numbers are brought in from temporary storage. 1961 N. Chapin *Programming Computers for Business Applic.* vii. 188 The failure..to initialize the switch would produce errors. 1963 P. M. Sherman *Programming & Coding Digital Computers* vii. 124 Another data instruction..is used to 'initialize' the ADD instruction to its initial value. 1972 *Computer Jrnl.* XV. 205/2 To trace from a base point *P*, the trace routine initialises *A* to the address of and *M* to the mode of the object pointed to by *P* and executes the following code.

Hence ini·tialized *ppl. a.*, ini·tializing *vbl. sb.* and *ppl. a.*

1957 D. D. McCracken *Digital Computer Programming* vi. 75 The preliminary steps which set up the loop and are not repeated constitute the initializing section of the loop. 1968 P. Wegner *Programming Lang.* iv. 234 Algol contains a small subset of..initialized identifiers. 1969 Maisel & Wright *Introd. Electronic Digital Computers* v. 74 Initializing procedures may also provide for such things as changes in the withholding tax rates or the Social Security laws. 1971 K. R. Britting *Inertial Navigation Syst. Analysis* i. 1 An appropriately initialized inertial navigation system is capable of continuous determination of vehicle position and velocity without use of external radiation or optical information.

initiand (ini·ʃi,ænd). [ad. L. *initiandus*, gerundive of *initiāre* Initiate *v.*: see *-and².] One who is about to be initiated (in quot. 1969, one who initiates).

1915 *Edin. Rev.* Jan. 127 The initiands are taken away from human society, often to a mountain, sometimes to a forest. 1931 K. E. Kirk *Vision of God* 473 A prayer of the initiand follows, in which he addresses the gods. 1969 R. Manheim tr. *Corbin's Creative Imagination in Ṣūfism* Introd. 60 To have him as a master and initiand is to be obliged to *be* what he himself is.

initiate, *v.* **1. b.** Delete ? *Obs.* and add later examples.

1963 S. Tolansky *Introd. Atomic Physics* (ed. 5) xxv. 423 If pure deuterium gas can be raised to a temperature of the order of 500 million degrees C., then a thermonuclear reaction should initiate. 1971 *Nature* 8 Jan. 111/2 Neutrons with incident wave vectors initiating within the cross-hatched area of the example shown satisfy the conditions for diffraction.

initiation. Add: **2. b.** *initiation ceremony, rite.*

1899 Spencer & Gillen *Native Tribes Cent. Austral.* vii. 212 All Australian natives, with rare exceptions, have to pass through some initiation ceremony before being admitted to the secrets of the tribe. 1935 B. Malinowski *Coral Gardens* II. vi. 234 The instruction may take place in the course of initiation ceremonies. 1951 R. Firth *Elem. Social Organiz.* ii. 47 Economic, ritual, and recreational affairs..are often difficult to disentangle within a complex institutional sequence of events, such as an initiation ceremony. 1916 H. B. Alexander *N. Amer. Mythol.* xi. 243 A myth which seems clearly reminiscent of initiation rites. 1937 R. H. Lowie *Hist. Ethnol. Theory* (1938) xi. 180 Boys' initiation rites probably involve circumcision. 1974 'S. Woods' *Done to Death* 186 Tribal initiation rites in one of the Indian reservations in the United States.

initiator. Add: (Later examples.)

1943 [see *articulator 4]. 1971 I. F. Hancock in J. Spencer *Eng. Lang. W. Afr.* 117 The process of creolising in the direction of a language other than the initiator language of the pidgin form has been rather misleadingly called *relexification*.

b. An explosive or device used to detonate the main charge.

1915 A. Marshall *Explosives* xxix. 417 Of all these explosives silver azide, mercury fulminate, and the aldehyde are the only ones that have a sufficiently high acceleration to be of any use as initiators of detonation. 1944 *Compt. Rend. (Doklady) de l'Acad. des Sci. de l'URSS* XLIV. 18 One might naturally expect that the flash-point ..would be lower for the initiators than for secondary explosives. 1962 *Ordnance Technical Terminol.* (U.S. Army Ordnance School) (AD 660112) 164/1 *Initiator*, a device used as the first element of an explosive train, such as a detonator or squib... It generally contains a small quantity of a sensitive explosive. 1964 M. Gowing *Britain & Atomic Energy* ix. 264 His criticism enlivened discussions on bomb assembly, and he participated very actively in the design of the initiator.

c. *Chem.* Any substance which starts a polymerization reaction.

1940 in *Chambers's Techn. Dict.* 1951 Frith & Tuckett *Linear Polymers* ii. 50 In any large-scale production of a polymer, catalytic initiation is nearly always used... The word 'catalyst' in this connection is rather a misnomer, as it is almost always destroyed in starting off polymerisation; 'initiator' is less open to objection, but has never really established itself. 1959 Cram & Hammond *Org. Chem.* xxv. 573 There are three principal classes of free-radical initiators: 1. Compounds..which undergo thermal decomposition... 2. Photosensitizers... 3. Redox systems. 1972 Billingham & Jenkins in A. D. Jenkins *Polymer Sci.* i. i. 19 Azo-bis-isobutyronitrile is very frequently used as an initiator for research studies on radical polymerization.

in-itse·lfness. [f. phr. *in itself* (In *prep.* 22) + -ness.] The quality or state of being independent of any relation to other entities.

1906 S. S. Laurie *Synthetica* I. ix. 136 Mere sentience admits of the in-itselfness (which is the for-itselfness) of the object being revealed. 1917 A. S. Pringle-Pattison *Idea of God* xx. 399 It was the aloofness—the in-itselfness, as we might call it—of his [*sc.* Mr. Bradley's] Absolute, which made the stronger impression on contemporary thought.

inject, *v.* Add: **1. c.** *transf.* in scientific contexts: *spec.* (*a*) to introduce or feed (an alternating current or voltage) *into* a circuit or device; (*b*) to introduce (charged atomic or subatomic particles) *into* an accelerator; (*c*) to introduce (charge carriers) *into* a region of a semiconductor device.

1939 *Amat. Radio Handbk.* iii. 43/1 The triode is arranged as an oscillator injecting its oscillations into the common cathode lead of the R.F. pentode. 1945 *Jrnl. Appl. Physics* XVI. 583/1 The electrons are injected with a voltage ranging from 30 to 70 kv and, if allowed to remain in the 66-inch diameter circular orbit for the entire quarter cycle, they circle the magnetic flux about 250,000 times. 1949 Ryder & Shockley in *Physical Rev.* LXXV. 310/2 When terminal 1 is negative, no holes enter from it and holes injected from terminal 2 are spread over a wide area where their concentration is so small as to produce an inappreciable lowering of resistance. 1950 D. Halliday *Introd. Nucl. Physics* ix. 342 Protons are injected into the cavity at 4-mev energy, from a pressure-type electrostatic generator which serves as an ion source. 1956 L. P. Hunter *Handbk. Semiconductor Electronics* iii. 10 When such a junction is biased in the forward direction..electrons (*A*) are injected into the P-type region and holes (*B*) are injected into the N-type region. 1962 *Proc. IRE* L. 1784/2 At any voltage..there will be excess charge injected into the insulator. 1966 *McGraw-Hill Encycl. Sci. & Technol.* IX. 583/1 Ions are injected into the accelerator by an electrostatic machine. 1968 Marton & El-Kareh *Electron Beam & Laser Beam Technol.* 64 The beam formed by the gun is injected into a cylindrical drift tube immersed in a uniform axial magnetic field. 1970 J. Earl *Tuners & Amplifiers* vi. 139 Hum currents can be easily injected into the system with a resulting very loud 'roar' from the loudspeaker. 1972 *Physics Bull.* Mar. 175/2 A high energy electron beam is injected into a CO_2/N_2 mixture to produce uniform volume ionization of the gas [in the laser]. 1973 *Nature* 16 Feb. 444/1 Plutonium isotopes..have been injected into the stratosphere as a result of atmospheric nuclear weapons tests.

d. *Astronautics.* To put *into* (an) orbit.

1961 C. T. Morrow *Symposium Ballistic Missile & Aerospace Technol.* I. 207 Methods by which a satellite can be injected into the 24-hour equatorial or stationary orbit. 1964 Mueller & Spangler *Communication Satellites* xii. 242 Launch operations..will probably entail..a 99 per cent probability for each satellite that it will be injected into orbit, following successful launch of a single booster. 1970 *Sci. Jrnl.* Aug. 10/4 The extra load prevented it from injecting the 260 kilogramme Satellite Test Vehicle (STV) into orbit. 1970 *Nature* 10 Oct. 154/2 Energy considerations make it difficult to consider seriously proposals to inject unwanted material into orbit.

2. Delete 'Now *rare*'. Also (esp. from sense 1 b of Injection), to insert, introduce.

1950 *N.Y. Times* 20 Apr. 1/6 Senator Harry P. Cain..sought to inject a note of caution into the debate. 1956 *Britannica Bk. of Year* 493/1 *Inject*, used in the sense of to insert, to interpolate. 1958 *Ann. Reg.* 1957 443 Films of events in the news could be 'injected' directly into the national (London) news bulletin. 1965 *New Statesman* 30 Apr. 696/1 It would inject some urgency into the lives of those..teams who plod through the season without any real hope of promotion or fear of relegation. 1969 *Listener* 2 Jan. 4/2 The raid injects new factors. 1970 *Nature* 12 Dec. 1019/1 The British government's decision to tighten the purse strings by refusing to inject a large sum of money into either the European Airbus or its rival, the BAC 3-11. 1972 *N.Y. Law Jrnl.* 31 Oct. 15/5 Issues would be injected in the consolidated proceedings which would prejudice the rights of one of the claimants.

injectable, *a.* Add: **2.** Suitable for injection into the body. Hence as *sb.*, a substance suitable for injection; *spec.* a drug or medicine that may be injected directly into the bloodstream.

1960 *Antibiotics Ann.* 1959–60 462 Thirty-three gonococcal infections in 31 unselected men have been treated with injectable tetracycline at St. Luke's Clinic. 1967 *N.Y. Times* 1 Aug. 27 The injectables are most often used by persons who formerly took amphetamines orally. 1969 *Daily Tel.* 3 July 23/3 An injectable form of the pill might be more acceptable in developing countries and help to damp down the population explosion. 1970 *Sci. Jrnl.* Aug. 8/3 Pharmacists have voluntarily restricted the supply of injectable amphetamines to hospitals. 1973 *Austral. Humanist* xxvi. 2/1 'Bad news' (he did not explain further) was 'coming up' about Copper 7 and injectables.

injected, *ppl. a.* Add: **1. b.** In sense corresponding to *Inject *v.* 1 c.

1949 W. Shockley et al. in *Bell Syst. Techn. Jrnl.* XXVIII. 346 The electronic structure of the germanium is modified in the neighborhood of the emitter point by the presence of the injected holes. 1971 *Solid-State Electronics* XIV. 268/1 The current changes from ohmic to space charge limited when the density of the injected majority carriers exceeds that of the thermally generated ones.

injection. Add: **1. b.** (Further examples, relating to internal-combustion engines: see *fuel injection* s.v. *Fuel *sb.* 3 b.)

1894 B. Donkin tr. *Diesel's Rational Heat Motor* ii. 60 The injection of the combustible powder–dust coal takes place gradually and continuously, during part of the stroke of the piston. 1921 W. H. Berry *Mod. Motor Car Pract.* i. 32 Mechanical injection would permit of greater power development. 1933 A. W. Judge *High Speed Diesel Engines* iii. 25 Advancing..the moment of injection is equivalent to..advancing the ignition in petrol engines. 1966 *McGraw-Hill Encycl. Sci. & Technol.* V. 556/1 In a diesel engine, a fuel pump starts injection at the proper engine crank angle and meters the required quantity of fuel through the nozzle.

c. *transf.* in scientific contexts (see *Inject *v.* 1 c).

1945 *Jrnl. Appl. Physics* XVI. 593/2 The electron injection and orbit shift circuits were instantaneous in their action. 1949 Ryder & Shockley in *Physical Rev.* LXXV. 310/1 Pronounced lowering of resistance can result from transistor action, i.e. the injection of holes into the N-type material from a metal contact. 1949 W. Shockley et al. in *Bell Syst. Techn. Jrnl.* XXVIII. 345 We shall discuss..evidence that holes are actually introduced into n-type germanium by the forward current of an emitter point... We shall refer to this important process as 'hole injection'. 1950 D. Halliday *Introd. Nucl. Physics* ix. 360 Injection may be done by taking advantage of the necessary radial stability of the betatron orbit. 1952 *Adv. Electronics* IV. 225 A method for the phase control of microwave tubes..utilizes the injection of power via the output circuit of the oscillator to be controlled. 1958 W. Ehrenberg *Electr. Conduction Semiconductors & Metals* xii. 322 The injection of minority carriers is the basic effect in junction rectifiers and transistors. 1962 *Proc. IRE* L. 1781/1 Under double injection, that is, the simultaneous injection into the insulator of electrons from a cathode and holes from an anode, space-charge limitations are..partially overcome. 1971 *Nature* 9 July 77/1 Cyclotrons have an inherent advantage over tandem Van de Graaffs—there is no need to produce a negative ion for injection. 1971 *Physics Bull.* Aug. 461/1 Such [acoustic] waves are obtained either by deliberate injection from a transducer or by amplifying some of the thermally generated waves already present.

d. *Astronautics.* The placing of a spacecraft in, or its entry into, a particular orbit or trajectory; the time when this occurs; freq. *attrib.*, as **injection point,** the point at which a spacecraft enters a new orbit or trajectory.

1959 *IRE Trans. Military Electronics* III. 150/1 We will assume the vehicle's orbit consists of three phases..: first, escape hyperbola..; second, the sun-centered ellipse controlled only by sun's gravity and using injection conditions taken from the escape hyperbola leaving earth; and third,..the approach hyperbola. *Ibid.* 151/1 The properties of the trajectory after injection. 1960 *Ibid.* IV. 152/1 In order to have zero relative inclination, the injection point must be exactly in the plane of the moon's orbit. 1961 C. T. Morrow *Symposium Ballistic Missile & Aerospace Technol.* I. 208 An initial powered flight phase leads to injection of the vehicle into a coast trajectory. 1963 S. Lees *Air, Space, & Instruments* 100 For each..location there is a readily computable impulsive velocity change (termed the 'injection velocity')..which will place the vehicle in a hyperbolic path with the desired terminal velocity. *Ibid.*, Geographical restrictions tend to limit the choice of injection points. 1966 E. Burgess *Assault on Moon* iii. 73 A 1 mile error in position at injection can cause a miss of 650 miles.

e. *fig.*

1968 *Listener* 7 Nov. 601/1 Colonel Ojukwu has had sizable injections of capital from European sympathisers. 1970 *Daily Tel.* 7 Oct. 7 An immediate financial injection into the hard-pressed agricultural industry, worth..£54 million, is to be made by the Government. 1972 *Guardian* 22 July 1/8 The report will ask for a massive injection of Government money into the docks industry.

4*. *Math.* Also *injection map(ping)*. A one-to-one transformation, esp. (formerly) an inclusion.

1950 S. MacLane in *Bull. Amer. Math. Soc.* LVI. 488 If S is a subgroup of G (notation $S \subset G$), then the injection κ..of S into G is that homomorphism of S into G with $\kappa(s) = s$ for every $s \in S$. 1963 D. Bushaw *Elem. Gen. Topology* 147 If X is a subset of Y, the function $j: X \to Y$ defined by $j(x) = x$ is called the injection map from X to Y. The injection map from X to X is called the identity map on X. 1968 E. T. Copson *Metric Spaces* vii. 86 If the inverse image of each point of E_2 is either empty or consists of a single point of E_1, the mapping $f: E_1 \to E_2$ is said to be an injection or a one-to-one mapping. If $f: E_1 \to E_2$ is an injection, $f(x_1) = f(x_2)$ implies $x_1 = x_2$; and $x_1 \neq x_2$ implies $f(x_1) \neq f(x_2)$.

5. Also freq. in terms relating to the injection of fuel into the combustion chamber of an internal-combustion engine, as *injection nozzle, period, pressure, system, time*; *injection valve* (further example). Also **injection laser,** a laser in which radiation is produced in a suitably shaped semiconductor crystal by the recombination of electrons and holes at a *p–n* junction as a result of the injection of a large enough number of electrons to produce a population inversion; **injection moulding,** a process for making moulded articles from plastics or other materials by forcing the heat-softened substance through an orifice into a cold, closed mould; hence **injection mould, injection-moulded** *ppl. a.*; **injection point** *Astronautics*: see sense 1 d above.

1963 *Proc. IEEE* LI. 602/1 (*heading*) Doping of semiconductors for injection lasers. 1966 Smith & Sorokin *Laser* i. 14 In contrast with the other types of lasers, the injection laser is almost two-dimensional, all the light originating from a region within a few microns of the junction plane. 1970 *New Scientist* 16 July (Telecommunications Suppl.) 16/1 The light emitted from the injection laser depends on the current passed through it. 1945 H. Barron *Mod. Plastics* xvi. 343 Flash is not formed in a well-made injection mould. 1947 Johnson & Daniels in P. I. Smith *Pract. Plastics* xiii. 187/1 The second moulding is in thermoplastic resin, injection moulded. 1969 T. C. Thorstensen *Pract. Leather Technol.* xv. 250 The expansion of the direct-molded sole and injection molded sole shoe is already placing requirements on the leather with regard to oil content and adhesion. 1932 *Chem. Abstr.* XXVI. 4419 (*heading*) Working plastic substances by the injection molding process. 1945 H. Barron *Mod. Plastics* xvi. 344 The essential feature in injection moulding is to force the plastic into the mould at a sufficient speed that the mould is completely filled before the material sets by contact with the cold metal. *Ibid.*, Materials which are nowadays universally fabricated by injection moulding include cellulose acetate, ..polystyrene; methyl methacrylate resins, etc. 1967 M. Chandler *Ceramics in Mod. World* 10 He [*sc.* the ceramist] now employs shaping methods such as dry-pressing and injection molding, which are not traditional to his craft. 1973 A. Parrish *Mech. Engineer's Ref. Bk.* xvi. 5 Injection moulding is a particularly important process for producing complex mouldings, primarily from thermoplastic powders or granules but also from thermosetting powders. 1900 B. Donkin *Text-bk. Gas, Oil, & Air Engines* (ed. 3) xxiii. 416 Connected to the injection nozzle are two valves, the admission and an overflow. 1946 A. W. Judge *Mod. Petrol Engines* viii. 326 For larger cylinders two injection nozzles located on opposite sides of the cylinder would be used to give a better admixture of fuel and air. 1971 B. Scharf *Engin. & its Lang.* xv. 221 The injection nozzle is normally a spring-loaded needle valve. 1916 A. Garrard *Gas, Oil, & Petrol Engines* vii. 140 A strong spring tends to keep the valve closed, and a cam on the cam shaft operates..to hold it open during the injection period. 1934 *Engineering* 3 Aug. 111/2 The fuel pumps..differ from those fitted on most direct-injection engines in that the termination of the injection period is kept constant, the effective stroke of the pump being varied so as to alter the commencement of the injection period. 1930 *Engineering* 11 Apr. 472/1 It was therefore decided to determine the injection pressure which would give a short injection time, in conjunction with good atomisation. 1941 *Nature* 26 July 105/2 The injection system, as compared with many present carburettors, has some slight but definite advantages. 1968 C. F. Taylor *Internal-Combustion Engine* II. xii. 566 A few Diesel-engine builders design and manufacture their own injection systems. 1930 Injection time [see *injection pressure* above]. 1916 A. Garrard *Gas, Oil, & Petrol Engines* vii. 139 Fig. 65 shows the form of injection valve and pulverizer adopted by the majority of makers for all kinds of fuel except the very heaviest crude oils.

injective (indʒe·ktiv), *a. Math.* [f. Inject *v.* (or L. *inject-* ppl. stem) + -ive.] Of the nature of or pertaining to an injection (sense 4*).

1952 Eilenberg & Steenrod *Found. Algebraic Topology* i. 8 A set of homomorphisms $i_\alpha: G_\alpha \to G$, $\alpha = 1, \ldots, n$, determine a homomorphism $i: \Sigma^n_{\alpha=1} G_\alpha \to G$. ..If i is an isomorphism of ΣG_α onto G, then the set $\{i_\alpha\}$ is called an injective representation of G as a direct sum. 1965 J. J. Rotman *Theory of Groups* ix. 184 (*heading*) The injective property. 1966 Sze-Tsen Hu *Introd. Gen. Topology* i. 8 A function $f: X \to Y$ is said to be one-to-one or injective iff, for every point $y \in Y$, the inverse image $f^{-1}(y)$ is either empty or a singleton.

injector. Add: **1.** Also, a device for injecting fuel into the combustion chamber or its intakes in an internal-combustion engine (or into the furnace of a steam engine, quot. 1890). (Further examples.)

1890 W. Robinson *Gas & Petroleum Engines* iv. 96 This oil is used to produce steam in stationary boilers. The oil is forced into the furnace in the form of fine spray by steam or compressed air, through injectors, and mixed with the air required for the combustion. 1912 R. B. Whitman *Gas-Engine Princ.* v. 72 This pressure vaporizer, or injector, is used on 2-cycle engines, the pressure being obtained from the crank case... An injector of this kind permits the use of kerosene. 1914, 1962 [see *fuel injector* (*Fuel *sb.* 3 b)]. 1932 *Compression Ignition Engines* ix. 115 In some buses run by a big city in the North only three cases of choked injectors have occurred in a year's running. 1947 *Jrnl. Brit. Interplanetary Soc.* VI. 107 The rocket motor can be divided into..the propellant injectors, combustion chamber and expansion nozzle. 1963 *Adv. Space Sci. & Technol.* Suppl. I. 173 In a liquid bipropellant rocket engine the injector serves the same function as the carburetor of a gasoline engine. In addition to atomizing and mixing the liquids, it meters the flow to the combustion chamber.

b. Something that injects (in sense *1 c), e.g. particles into an accelerator, or charge carriers into a semiconductor device.

1945 *Jrnl. Appl. Physics* XVI. 593/2 The polarity of the charge on the capacitors was made so as to make the second..half-cycle the period of electron acceleration thereby allowing charging of the capacitors in the injector and orbit shift circuits during the first half-cycle. 1953 *Proc. IRE* XLI. 1715/1 The efficiency of the diodes as hole injectors is lower than that usually considered desirable. 1962 [see *emitter 2]. 1972 *Physics Bull.* Mar. 144/2 The Linac injector is designed to accelerate protons ..along its drift tube structure to energies of 19·2 MeV.

in-joke (i·n,dʒōᵘk). [cf. *In *a.* 2.] A joke enjoyed or appreciated by only a limited group of people. Cf. *in-reference.

1964 *Economist* 31 Oct. 518/2 Professor Yamey..has published the first independent assessment of the Resale Prices Act..for those who like in-jokes. 1966 *Punch* 4 May 657/1 The dialogue peppered with British upper-class in-joke slang of the most blatant appeal to Lancashire slum-dwellers and Turkish primary schools. 1968 M. Allingham *Cargo of Eagles* vi. 82 Those who can read between the lines will find some amusement at what are called 'in-jokes' today. 1971 M. Babson *Cover-up Story* ii. 24 They were laughing at me... It was in-joke laughter, and I was on the outside looking in.

Injun (i·ndʒən). Also **Injin.** **a.** Colloq. and U.S. dial. form of Indian *sb.* 2; also *attrib.* (Cf. *Ingin.)

1812 Col. J. Cocke in *Salem Gaz.* 28 Aug. 1/2 The people of Tenessee is antious to have orders commanded out for us to march against the injuns on the Wabash. 1850 Mayne Reid *Rifle Rangers* I. i. 5 Four till one! Injuns!—murder!—help, hyeer! 1853 *Ibid.* (ed. 2) I. iii. 24 Thur's a mighty grist o' venturin', I heern; beats Injun fightin' all holler. 1868 M. I. Carrington *Ab-Sa-Ra-Ka* 83 Better not go *fur*. There is *Injuns* enough lying under wolf skins, or skulking on them cliffs. 1872 [see *heap *sb.* 4 d]. 1889 K. Munroe *Golden Days* 118 No more attention was paid to the shooting of an 'Injun' than if he were a coyote. 1911 R. D. Saunders *Col. Todhunter* vii. 104 But you're sure about it, too, ain't you? She ain't doin' no Injun-givin' in your case? 1937 [see *Indian devil* s.v. *Indian *a.* 4 b]. 1959 I. & P. Opie *Lore & Lang. Schoolch.* viii. 134 In the United States such a child, who succumbs to the temptation of wanting back, is termed an 'Injun-giver'. 1973 *Nature* 13 Apr. 485/2 The alternative to receipt of a reprint is a slog through hundreds of miles of Injun territory. 1973 R. Thomas *If you can't be Good* (1974) iii. 30 She had..long hair..glossy blue-black in color. Injun hair, she called it.

b. In various allusive uses and phrases: *honest Injun*, honour bright: perh. orig. an assurance of good faith extracted from Indians; *to play Injun*: to act like an Indian; to avoid being seen or captured; of children playing, to pretend to be Indians.

[1676 J. Talcott *Let.* 8 June in S. Judd *Hist. Hadley* (1905) xv. 169 We sent 27 women and children to Norwich under conduct of some of those we call honest Indians.] 1876 'Mark Twain' *Tom Sawyer* ii, Ben, I'd like to, honest Injun; but [etc.]. 1887 F. Anstey *Paleface & Redskin* in *Graphic* 31 Dec. 728/3 'Are you sure..on your honour?' he asked eagerly. 'Honest Injun!' said Lambert. 1887 H. Frederic *Seth's Brother's Wife* II. 160 'Is what you've be'n tellin' me here honest? Don't lie to me.' 'Honest Injun,..every word.' 1891 H. C. Bunner *Short Sixes* 90 'Hope to die—Honest Injun—cross my breast!' said the boy. 1896 G. B. Shaw *Let.* 8 Sept. in *Ellen Terry & Bernard Shaw* (1931) 54 A thing she would never have done if she had not forgiven him quite thoroughly—honest Injun. 1902 S. E. White *Blazed Trail* xix. 140 'Our compact holds now, honest Injin, doesn't it?' asked the boy anxiously. 1904 'A. Dale' *Wanted: a Cook* 98 But, Archie, this is all true. It is, honest Injun. 1918 C. E. Mulford *Man from Bar-20* xiv. 140 So they're combin' th' country an' patrollin'. Hereafter an' henceforth I've got to play Injun for all I'm worth. 1922 F. Hamilton *P.J.: Secret Service Boy* ii. 85 Bar rot, Mr. Ambrose. Real honest Injun? 1922 Joyce *Ulysses* 295 'Are you codding?' says I. 'Honest injun,' says Alf. 1950 L. A. G. Strong *Which I Never* i. 12 'You've invented him.' 'Which I never, sir,..' 'Honest Injun?'

injunct, *v.* Add: Now in somewhat more general use. (Earlier U.S. example and later U.K. examples.)

1872 Schele de Vere *Americanisms* 653 Violent contractions, derived from well-known and well-formed words, like burgled, injuncted, and excurted. 1900 *Westm. Gaz.* 12 Sept. 2/2 Sir Edward Clarke is very likely right in thinking that the Court would hesitate to injunct a man for pirating his own speech. 1957 *Times* 23 Nov. 3/2 Counsel said that Mr. Fielding wanted the Court to say that as he had first claim on Steele's services he should be injuncted from entering a contract with anyone else. But Steele could not be injuncted from going to South Africa.

injunctive, *a.* Add: **2.** *Gram.* Applied to the form of a verb (in Vedic, Hittite, etc.) having secondary personal endings and expressing injunction. Also as *sb.*

1910 A. A. Macdonell *Vedic Gram.* vii. 316 The unaugmented forms of past tenses used modally, are sometimes called improper subjunctives, but they are more suitably termed injunctives, as they appear to have originally expressed an injunction. 1927 E. A. Sonnenschein *Soul of Grammar* § 73 What a pity that no one thought of calling the [subjunctive] mood ἐπιτακτική—a term which would have been translatable by the Latin *iniunctivus*, 'injunctive', i.e. 'enjoining'. 1965 *Language* XLI. 1, *-si* imperatives derive from root injunctives, to which could be suffixed *-i* or *-u* to form imperatives. 1971 F. R. Adrados in *Archivum Linguisticum* II. 97 Ambrosini believes that *s* [in Indo-European] was the marker of intransitivity; Pariente calls it an injunctive characteristic. *Ibid.* 99, I think we can ascribe to older Indo-European, verbal forms of a type similar to the injunctive, well attested in Vedic and Hittite.

injured, *ppl. a.* Add: **1.** Esp. in phr. *injured innocence*, the offended attitude of one who is undeservedly accused of something; freq. with the implication that the accusation is in fact just; also occas. used to designate a person adopting such an attitude.

1713 Addison *Cato* iii. 36 Lucia, thou injur'd Innocence! 1827 M. Wilmot *Jrnl.* 3 July in *More Lett.* (1935) 268 She next told us of an act of her benevolence towards a poacher caught in the fact, and perhaps too severely punished... Our Chezy dreadnought..speaks of him as injured Innocence. 1869 Trollope *Phineas Finn* I. xvi. 132 Phineas assumed a look of injured innocence, as though his father was driving him too hard. c 1900 H. A. Jones in M. R. Booth *Eng. Plays of 19th Cent.* (1969) II. 364, I decline to give a certificate of injured innocence to any young person who misses her last train. 1971 J. D. MacDonald *Seven* (1974) v. 101, I have watched the same game a lot of other times... Just don't give me injured innocence.

injuria (indʒūᵊ·riă). *Law.* [L.] An invasion of another's rights; an actionable wrong. Cf. *damnum.

1876 *Wharton's Law Lexicon* (ed. 6) 477/1 *Injuria*, injury; a wrongful act done. 1898 *Encycl. Laws Eng.* VI. 485 *Injuria* can only be defined as an infringement of a legal right. 1972 *Times* 23 Feb. 27/8 The jury should be neither encouraged nor allowed to look beyond as generous a solatium as was required for the injuria simply to give effect to feelings of indignation.

injurious, *a.* Add: **4.** *injurious affection* (Law): a term used of a situation in which part of a person's land is acquired compulsorily under statutory powers and the remaining part is reduced in value, either because it is a smaller piece or because of what has been done on the land compulsorily acquired; also, of other situations in which an owner seeks compensation for the deleterious effect on his property of the exercise of statutory powers; *injurious falsehood* (Law): an actionable falsehood, a false statement claimed to have caused damage to the plaintiff in respect of his office, profession, trade or business, etc.

[1845 *Act* 8 & 9 *Vict.* c. 18 The damage, if any, to be sustained by the owner of the lands by reason of the severing of the lands from the other lands of such owner, or otherwise injuriously affecting such lands.] 1867 *Law Rep.* (Queen's Bench) II. 239 The injurious affection of the house by the vibration, smoke, and noise. 1889 *Law Rep.* (Appeal Cases) XIV. 159 The acts complained of as an injurious affection were not done on the land taken. 1909 Halsbury *Laws Eng.* III. 41 In assessing compensation for..injurious affection all damage that can be reasonably foreseen should be taken into account. 1932 *Act* 22 & 23 *Geo. V* c. 48 Account shall be taken of any additional injurious affection of the property. 1947 *Act* 10 & 11 *Geo. VI* c. 48 The compensation (if any) to which that person would be entitled for such injurious affection if the..land were compulsorily acquired. 1965 *Act Eliz. II* c. 56 § 10 This section shall be construed as affording in all cases a right to compensation for injurious affection to land. 1971 *Country Life* 6 May 1109/2 A claim for compensation on account of injurious affection is now made under the provisions of sections 7 and 10 of the Compulsory Purchase Act. 1972 *Daily Colonist* (Victoria, B.C.) 10 Feb. 38/3 The report also suggests that the basic formula for compensation [in case of expropriation] be based on the market value of the property expropriated plus damages for 'injurious affection'. 1907 J. W. Salmond *Law of Torts* xv. 426 The second form of actionable misrepresentation, namely that which we have termed Injurious Falsehood. 1928 *Ibid.* (ed. 7) xv. 582 The most important example of the wrong of injurious falsehood is the use of fraudulent or misleading trade names. 1933 *Law Jrnl. Rep.* CII. 191 A false statement detrimental to the plaintiff's business, but not defamatory, carelessly made in the belief it was true, will not support an action for injurious falsehood. 1955 *Rep. Patent, Design & Trade Mark Cases* (Patent Office) LXXII. 160 The amendment of the writ has been such as to raise the cause of action known as 'injurious falsehood'. 1967 J. G. Fleming *Introd. Law of Torts* xi. 218 The tort of injurious falsehood, partaking of elements familiar to defamation and deceit. 1973 J. D. Heydon *Economic Torts* IV. 66 Injurious falsehood. This tort is committed where the defendant maliciously publishes to a third party written or oral falsehoods about the plaintiff in his trade which are calculated to produce and do produce actual damage. 1974 *Trans. Philol. Soc. 1973* 19 The only possibility of action would appear to be in the tort of injurious falsehood, but for a trade-mark proprietor to succeed he would have to show malice on the part of the publishers or editor of the dictionary.

injury, *sb.* Add: **4.** *attrib.* and *Comb.* **injury-feigning** *vbl. sb.* and *ppl. a.*; **injury time,** the extra time allowed in a game of football or the like to make up for time spent in attending to injuries.

1925 J. S. Huxley in *Brit. Birds* XIX. 93 The Purple Sandpiper which nests on the..tundra has an 'injury-feigning' performance which must be hard to beat for elaboration. *Ibid.* 94 In regard to 'injury-feigning' the Avocet is one of the most spectacular of birds. 1932 D. Lack in *Ibis* 282 Injury-feigning is, therefore, assumed to be primarily a partial paralysis due to anxiety. 1948 *Brit. Birds* XLI. 237 The cock several times gave 'lure displays' of the 'injury-feigning' type, creeping away from the dummy with wings dragging and depressed and tail fanned. 1960 *Sunday Times* 18 Dec. 20/1 The winning score..did not come until the 43rd minute of the second half in what, in the absence of any official designation, is termed 'injury time'. 1971 *New Society* 1 July 24/3 One of the better ideas in the competition is their tie-breaker. If there is a tie the teams play off to an instant death by goals scored from penalties kicked immediately after extra injury-time.

ink, *sb.*¹ Add: **2. a.** *ink fever* (nonce-word); **d.** *ink-blue, -like* (later example), *-purple*; *ink-shine* v. (nonce-word). Also **e.** (in Chinese calligraphy, etc.) *ink-brush, -painting, -sketch, -squeeze, -stick, -study.*

1963 *Times* 16 Feb. 11/6 Baby ink-blue mussel shells arranged in flower-like clusters with sprays of dried seaweeds in between. 1971 *Guardian* 14 Dec. 9/2 In black, tan, bottle green, honey, and ink blue. 1951 R. Firth *Elem. Social Organiz.* v. 164 A traditional Chinese painter works with a definite theory about the use of the ink-brush. 1922 T. E. Lawrence *Home Lett.* (1954) 355 This long-drawn-out battle over my narrative of the campaigns of Feisal has put an ink fever into me. I find myself always going about trying to fit words to the sights & sounds in the world outside me. 1933 W. de la Mare *Fleeting* 49 Whose waters..ink-like, ebon,..flow. 1925 R. Fry *Chinese Art* 9 Many ink-paintings ascribed to him [*sc.* Li Lung-mien] are extant, and a few may be originals. *Ibid.* Plate 9 (*caption*) Ink painting on silk. 1954 *Oxf. Jun. Encycl.* XII. 225/1 The greatest master of ink-painting [in Japan] was Sesshū (1420–1506). 1935 E. Bowen *House in Paris* ii. vii. 159 Toppling ink-purple clouds. 1922 Joyce *Ulysses* 216 The jet beads of her mantilla inkshining in the sun. 1906 S. W. Bushell *Chinese Art* II. xiii. 136 She [*sc.* the Lady Kuan] was a clever painter of flowers and her rapid ink sketches of peonies, prunus-flowers and orchids were admirable. 1910 *Brit. Mus. Guide Exhib. Chinese & Japanese Paintings..in Print & Drawing Gallery* 32 But the typical painting of the [Ashikaga] period was the ink-sketch of landscape, bird or flower. 1935 *Burlington Mag.* Oct. 185/2 The illustrations are all line-blocks made from 'rubbings' (or, more correctly, 'ink squeezes'). 1926 F. B. Wiborg *Printing Ink* ii. 50 The Chinese never keep liquid ink... Many ink-sticks are provided with a rounded notch at the lower end to secure a firmer hold for the finger, while the upper part to be rubbed is rounded. 1935 Chiang Yee *Chinese Eye* viii. 199 A certain Li T'ing-Kuei of the 'Five dynasties' period supplied all the leading calligraphists with ink-sticks, and he compounded them from ten parts of pine-smoke to three of powdered jade and one of gum. 1939 *Burlington Mag.* Jan. 47/1 Inksticks and experience in making inks. 1936 *Ibid.* Nov. 236/1 Chinese ink-studies.

4. ink-blot, a blot of ink; also *fig.*, and *attrib.*, esp. in **ink-blot test** *Psychol.*, a projective test in which the subject's imaginative reactions to a random ink-blot shape are analysed and used as a guide to his personality; also *ellipt.* as *ink-blot*; **ink-cap**: substitute for def.: a fungus of the genus *Coprinus*; add further examples; **ink disease,** a fungal disease caused by species of *Phytophthora*, esp. *P. cambivora*, affecting chestnut and occasionally other trees, making the surface of their roots and sometimes trunks a darker colour; **ink-feed,** the duct which carries the ink to the nib of a fountain pen; also, the feeding of ink through this duct; **ink-jerker** *U.S.*, **-spiller** = *ink-slinger*; **ink-mirror,** a surface of ink used in clairvoyance in place of a crystal; **ink-slab** (examples); (*b*) a container for ink; *spec.* in the Far East, a slab on which ink is mixed ready for writing; **ink-wash** *Japanese Painting* (see Wash *sb.* 4 b); **ink weed** *Austral.* and *N.Z.*, a perennial herb of the genus *Phytolacca*, which bears black berries containing a reddish juice (= Poke *sb.*⁴ 2 a).

[a 1500 ? Lydgate *Lavenders* 18 Wasshe withe wyne the fervente ynkes blote.] 1928 J. J. B. Morgan *Psychol. Abnormal People* iv. 145 The Rorschach test is a special development of the ink-blot test. The ordinary ink-blot test has not yielded very significant results heretofore. Rorschach, some years ago, developed a series of ink-blots which have been more successful than those used previously. 1931 *Psychol. Abstr.* V. 268 An explanation of the use of the Rorschach ink-blot test as a measure of intelligence. 1940 R. S. Woodworth *Psychol.* (ed. 12) v. 151 A class of tests sometimes called the 'fantasy tests'... Most used is the inkblot... A variety of things can be seen in such a blot. 1955 *Publ. Amer. Dial. Soc.* xxiv. 13 The interpretation given ink-blots by criminals reveals their unconscious motivations. 1965 *Sun* 20 May 2/2 American military planners are weighing up an 'inkblot' strategy to clear South Vietnam of Communist guerillas. 1967 E. Short *Embroidery & Fabric Collage* i. 28 *Ink Blots and Dribbled Paint.* These methods are not as haphazard as they would at first seem as the hand is guided to a great extent by one's intuitive sense of design. 1968 D. Rapaport et al. *Diagn. Psychol. Testing* (rev. ed.) ix. 272 From the perceptual point of view it *appears* that the Rorschach inkblots are 'unstructured' perceptual raw material. 1972 *Jrnl. Social Psychol.* Dec. 303 Human movement responses..may be interpreted as an index of social approach during the administration of an inkblot test. 1927 Gwynne-Vaughan & Barnes *Struct. & Devel. Fungi* 301 The genus *Coprinus*, the ink cap, is one of the commonest forms with black spores. 1963 Lange & Hora *Collins Guide Mushrooms & Toadstools* 136 *Coprinus*—'Ink caps'. Characterised in almost all species by the gradual 'auto-digestion' of the gills, and sometimes the cap, into a black ink-like fluid. 1923 *Rev. Appl. Mycol.* II. 188 The so-called 'ink' disease of chestnuts constitutes a serious damper to French sylviculture. 1932 *Forestry* VI. 182 The fungus causing the ink disease of chestnut, has a very wide distribution. 1968 F. G. Browne *Pests & Dis. Forest Plantation Trees* ii. 915 It [*sc.* *Phytophthora cambivora*] is one of the pathogens..associated with ink disease, a severe malady of *Castanea*. 1907 *Westm. Gaz.* 23 Oct. 11/1 The Patent Spoon-Feed has successfully overcome the ink-feed difficulty. 1935 *Discovery* Jan. 15/2 He must decide whether to alter the colour of the ink, its consistency, the rate of ink-feed. 1865 *Harper's Mag.* 683/2 This rattle-brained scribbler, this miserable ink-jerker. 1905 E. F. Benson *Image in Sand* ii, Abdul had..tried him with the simple experiment of the ink-mirror, and found him extraordinarily sensitive. 1890 C. T. Jacobi *Printing* 288/2 *Ink slab*, the table on which ink is distributed, either at press or machine. 1895 *Montgomery Ward Catal.* 210/2 Slate ink slabs... 5 inch square with heavy glass cover. 1911 *Encycl. Brit.* XXII. 354/2 A second..roller conveys the ink from this drum to the distributing table or ink slab. 1938 *Burlington Mag.* Aug. 90/2 The inkslabs and inkstones, so sought after by collectors. 1963 Kenneison & Spilman *Dict. Printing* 100 *Ink slab*, that part of certain printing presses, consisting of a large, flat, steel bed, from which the distributing rollers..take ink. 1969 *Korean Folklore & Classics* I. 23 He..picked up his teacher's ink-slab. 1881 *Punch* 10 Sept. 110/2 To think people ain't got more savvy than what these inkspillers enjoy. 1936 *Burlington Mag.* Oct. 162/2 The spirited ink-wash technique. [1906 T. F. Cheeseman *Man. N.Z. Flora* 1085 *Phytolacca octandra*... Ink-plant; Poke-weed.] 1913 *N.Z. Dept. Agric. Jrnl. Agric.* VII. 369 Inkweed, or pokeweed (*Phytolacca octandra*), is a poisonous plant. 1933 *Bulletin* (Sydney) 15 Nov. 28/4 Inkweed is hard to burn or eradicate if it gets a hold. 1962 N. C. W. Beadle et al. *Handbk. Vasc. Plants Sydney Distr.* 160 Weed of waste ground. Introd. from Trop. Amer. Ink Weed. P[*hytolacca*] *octandra*.

ink (iŋk), *sb.*⁴ *Sc.* [var. of Ing.] *pl.* Low-lying grassland subject to flooding by spring tides. Also (in *sing.*) *attrib.*

a 1692 A. Symson *Large Descr. Galloway* (1823) 138 Down the river, about a mile from the Church..a large plott of fine fir-planting, over-looking a rich ink ground. 1802 *Farmer's Mag.* Aug. 331 Eighty acres..consisted of a rich sea marsh, or *inks*, as we call them here, almost a true level,..about 4 or 5 acres,..16 inches lower, being a younger marsh, and nothing but what we call ink grass growing upon it. 1824 J. Mactaggart *Gallovid. Encycl.* 280 *Inks.* On muddy, level shores,..pieces of land overflowed with spring tides, and not touched by common ones..; on these grow a coarse kind of grass. 1848 *Scottish Jrnl. Topogr.* II. 234/1 *Spurlings* to net in the inks of the Cree. 1899 *Galloway Advertiser* 27 July 1/3 Extent, 240 acres or thereby of Carse Land of excellent quality, together with a very large extent of 'Inks' or Shore Pasture. 1974 *Scottish Field* Apr. 15/4 Some of the sunsets especially, behind the Inks and Wigtown Sands, are as fine as you can see anywhere.

ink, *v.* Add: **2.** Also *fig.*; *ink over* (earlier example).

1803 *Lett. Miss Riversdale* I. 319 The Prince took down the notes in pencil..and promised to ink them over for Lady Belfont. 1929 A. C. & C. Edington *Studio Murder Myst.* iii. 23 It was inked into his mind's eye, so that even when he shut his eyes..he could not shut out that awful picture. 1952 C. Day Lewis tr. *Virgil's Aeneid* iii. 59 A streaming night inked out The sky. 1959 *Times* 26 Jan. 6/1 The Yeames is a picture surely so inked into the national memory that it would need more than condemnation of its taste to eradicate it.

inkily (i·ŋkili), *adv.* [f. Inky *a.* + -ly².] In an inky manner; like ink.

1894 Stevenson & Osbourne *Ebb-Tide* i. vi. 111 The sea.., inkily blue.

inkle, *v.* **1.** Delete † *Obs.* and add examples. (In these uses a back-formation from Inkling 2.)

1901 S. Butler *Erewhon Revisited* 42 People like being deceived, but they also like to have an inkling of their own deception, and you never inkle them. 1904 Hardy *Dynasts* I. i. vi. 57 Thou art young, and dost not heed the Cause of things Which some of us have inkled to thee here.

inkosi (iŋkōᵘ·si). *S. Afr.* Also **enkosi, inkhasi, inkos, inkose(e),** and with capital initial. [Zulu. Cognate forms are found in other Bantu langs. (see quot. 1937); also Bondei, Zegua, Nguru *m-gosi*, Karanga *a-hosi* man, Gogo *mu-gosi* chief.] **a.** The royal title of a Zulu ruler. **b.** A chief, lord; in which sense also used as a title of respect.

[1824 W. J. Burchell *Trav. S. Afr.* II. xiv. 364 The different members of his family, and the *kōsies* or subordinate chieftains, formed round us a circle two or three deep. 1827 G. Thompson *Trav. S. Afr.* x. 118 Calling the king, Kousie, which is not his name, but his title, *kousi* signifying king or principal chief in their language.] 1835 A. Steedman *Wanderings S. Afr.* I. x. 256 Among the Zoolahs the title of *Inkose* is solely confined to the principal Chief. 1836 N. Isaacs *Trav. E. Afr.* (1937) II. 245 When the monarch is firmly seated on his throne..he becomes an absolute king, or 'Inquose'. 1846 J. C. Brown tr. *Arbousset & Daumas's Narr. Tour N.-E. of Cape Good Hope* (1852) xxvii. 423 As if a Zulu Inkhosi could show clemency! 1899 B. Mitford *John Ames* x. 96 Policeman he want to see Inkose. 1905 *Westm. Gaz.* 8 June 2/1 All the members of the kraals concerned will..form,..with the 'inkosi', his several wives and their brothers and sisters and children and dependent relatives, a formidable audience. 1910 J. Buchan *Prester John* xii. 215 Courage, Inkoos; in an hour's time you will be free. 1937 I. Schapera *Bantu-Speaking Tribes S. Afr.* viii. 174 At the head of the whole tribe is the Chief (Nguni, *inkosi*; Shangana-Tonga, *hosi*; Venda, *khosi*; Sotho, *morêna*, *kxosi*). 1948 A. Paton *Cry Beloved Country* III. ii. 220, I have thought, inkosi, that we should try to keep some of them in this valley.

Hence **inkosikazi** (9 **inquosegose**, etc.), (*a*) the wife of a chief; (*b*) native name for a white married woman.

1835 A. Steedman *Wanderings S. Afr.* I. x. 256 The Chief having many wives..the sovereignty devolves on the offspring of the *Inkose kosi*, female chieftain, or queen. 1836 N. Isaacs *Trav. E. Afr.* (1937) II. 63 All the inquosegoses were present. 1866 H. Robertson *Mission Life among Zulu-Kafirs* 103 Look, Inkosikazi, here is Mary putting this in my best trowsers. 1878 H. A. Roche *On Trek in Transvaal* 246 He [*sc.* the washing Kaffir] acquits himself at his task better than the *Inkosigas* who bungles hers so sadly. 1948 A. Paton *Cry Beloved Country* II. viii. 175 My heart holds a deep sorrow for you, and for the inkosikazi, and for the young inkosikazi, and for the children. 1969 I. Vaughan *Last of Sunlit Yrs.* iii. 25 'Nkosikaas,' Moses had asked, 'what sort of a table is this?'

inky, *a.* Add: **6. inky cap** = *ink-cap* (*Ink *sb.*¹ 4).

[1891 M. C. Cooke *Brit. Edible Fungi* vi. 47 The 'Inky Mushroom' (*Coprinus atramentarius*) received that name because, when it becomes old, the gills melt away into a thick, black, inky fluid, which may be used as ink.] 1923 J. Ramsbottom *Handbk. Larger Brit. Fungi* 81 One of the chief characteristics of *Coprinus* is the so-called 'deliquescence', which gives the popular name 'inky cap' to these fungi. 1967 W. P. K. Findlay *Wayside & Woodland Fungi* 149 (*heading*) *Coprinus:* Inky Caps.

inky (i·ŋki), *sb.* Colloq. abbrev. of *incandescent lamp*. Also **i·nkie.**

1929 *Photoplay* (Chicago) Apr. 31/2 *Inkys*, incandescent lights, the silent lights used for talking pictures in contrast to the old noisy arc lights. 1936 C. B. DeMille in *Words* Oct. 6/2 An incandescent lamp is an 'inkie'. 1959 W. S. Sharps *Dict. Cinematogr.* 103/2 *Inkie* or *Inkie-Dinkie*, a small incandescent lighting unit used for local lighting. 1970 *T.V. Times* (Austral.) 1 Apr. 8/3 Put a Ted Lewis on the Inky Dinky—insert a little, dented, pipe-like piece of tin in front of the small incandescent spot light to diffuse the beam.

‖ **inkyo** (i·ŋkyo). Also **inkiyo.** [Jap., f. *in* (in the) shade, retired + *kyo* to dwell.] In Japan, the act of resigning or renouncing one's office or position; one who has thus abdicated or resigned. Also as *adj.*

1871 A. B. Mitford *Tales Old Japan* II. 122 *Inkiyô*; abdication. The custom of abdication is common among all classes, from the Emperor down. 1896 L. Hearn *Kokoro* iii. 40 Old men and women likewise—the *inkyō* of the vicinity. *Ibid.* xii. 224 The aged *inkyō*, whose sight and bearing begin to fail, talks cheerily of the impending change that is to provide him with a fresh young body. 1911 B. H. Chamberlain *Jap. Poetry* 178 Little wonder that heads of families became *inkyo*..that is, retired from active life as early as possible. 1958 G. B. Sansom *Hist. Japan to 1334* x. 199 It was common for the head of a great institution or a great house to retire at an early age... This custom, known as Inkyo (which means a sheltered or passive life) has not entirely disappeared.

inky-pinky (i·ŋki,pi·ŋki). *Sc.* Also **inkie-pinkie, inker-pinker,** etc. [Etym. obscure; see *Sc. Nat. Dict.*] Small beer.

The word occurs in versions of the Hallowe'en play *Galatians*; see E. K. Chambers, *English Folk-Play* (1933), 55. There are other senses and forms, on which see the *Sc. Nat. Dict.*

1835 J. Maidment *Galatians* 4 Inky Pinky about seventy or eighty years since was used by the brewers in Stirlingshire to designate the smallest kind of beer. 1842 R. Chambers *Pop. Rhymes Scotl.* 69/2, I have a little bottle of *inker-pinker* in my pocket.

inlaid, *ppl. a.* Add: **2. b.** Of linoleum or the like: decorated with a design that is set into the surface.

1908 *Westm. Gaz.* 30 May 7/2 The floors will be covered with Greenwich inlaid linoleum. 1959 *Sears, Roebuck Catal.* Spring & Summer 780 Easy-care inlaid vinyl... Toughest plastic used in floorcoverings. *Ibid.* 731 Inlaid linoleum.

in-lamb, *a.* (Later examples.)

1968 J. Arnold *Shell Bk. Country Crafts* 90 In-lamb ewes can be accommodated and cared for. 1972 *Country Life* 15 June 1580/1 When grazing is finished the in-lamb ewes are brought indoors.

inland, *sb.*, *a.*, and *adv.* Add: **A.** *sb.* **2.** (Later examples.)

1913 L. V. Kelly *Range Men* 71 Canny men and good traders, built posts in the great inland. 1934 A. Russell *Tramp-Royal in Wild Austral.* iii. 29 The camel..will long continue to be in many parts, the great utility animal of the Inland. 1941 I. L. Idriess *Great Boomerang* xxxi. 235 New cities, new industries, and a great, far-spread population in our inland, where population is needed so urgently. 1969 'A. Garve' *Boomerang* ii. 39 Our inland is still very empty country, and a lot of it isn't easily accessible. 1973 *Nation Rev.* (Melbourne) 31 Aug. (Suppl.) 1/1 He enjoys studying the unique wildlife of the inland, and..the people who live and work there.

B. *adj.* **1. a.** *inland ice* (*sheet*), the ice which forms a permanent cover or ice-cap over most of Greenland; the region over which this ice extends. Hence, more widely, any mass of ice of comparable extent and thickness underlain by rock; *Inland Sea*, (*b*) (see quot. 1891).

1853 H. Rink in *Jrnl. R. Geogr. Soc.* XXIII. [149 Let us call the group of peninsulas and islands..the *outskirts* of the land [*sc.* Greenland], and the compact continent to the E. the *inland*.] 151 The exclusive origin of the icebergs from the inland ice, through the icy friths, has been mentioned. 1871 A. B. Mitford *Tales Old Japan* I. 173 *Shikoku*, one of the southern islands separated from the chief island of Japan by the beautiful 'Inland Sea'. 1876 W. E. Griffis *Mikado's Empire* I. v. 55 The 'Inland Sea' (Séto Uchi) is a name which has been given by foreigners, and adopted by the Japanese, who until modern times had no special name for it as a whole. 1880 *Encycl. Brit.* XI. 166/2 The Danes divide Greenland into two physical divisions—the 'outskirts' and the 'inland ice'. The first comprises the coast-lying land, the latter the interior. 1891 Chamberlain & Mason *Handbk. Travellers Japan* (ed. 3) 357/2 The Inland Sea is the name given to the water space lying between the Main Island on the North and the islands of Shikoku and Kyūshū on the South. 1895 *Jrnl. Geol.* III. 244 During the climax of the glacial period, when the Scandinavian 'inland-ice' invaded the low grounds of middle Europe, those low grounds supported an Arctic-alpine flora. 1898 *Q. Jrnl. Geol. Soc.* LIV. 200 In addition to these 'inland ice-sheets' there are in Spitsbergen glaciers of the ordinary Alpine type. 1953 G. Williamson *Changing Greenland* xiv. 167 Six-sevenths of Greenland's 840,000 square miles is locked in the implacable embrace of the Inland Ice. *Ibid.* xx. 254 Before leaving Disko Bay, tourists may..make an excursion to the fringe of the Inland Ice. 1958 G. B. Sansom *Hist. Japan to 1334* xiv. 300 Before him lay the Inland Sea route to the Straits of Shimonoseki. 1966 T. Armstrong et al. *Illustr. Gloss. Snow & Ice* 27 *Inland ice sheet*, an ice sheet of considerable thickness and more than about 50,000 square km in area, resting on rock. 1969 M. Smeeton *Misty Islands* ii. 19 We entered the Inland Sea through the Hayasui Seto. 1973 *Nature* 5 Oct. 251/2 Although the strong echo from the upper surface of the ice [in Antarctica] shows little variation in strength, the normal bottom echo from inland ice..shows strong fading along the flight line.

inlander. Add: (Canadian and Australian examples.)

1843 *Standing Rules* (Hudson's Bay Co.) Index, Rations and regales to Inlanders, Scale of. 1933 [see *binghi, Binghi]. 1934 A. Russell *Tramp-Royal in Wild Austral.* iii. 33 The amazing feat accomplished by a young inlander. 1935 A. R. Evans *Reindeer Trek* 179 He had rushed terrified from the spot to tell a tribe of passing Inlanders. 1944 F. Clune *Red Heart* 61 Andrew Hume begged his parents to allow him to accompany the famous German inlander. *Ibid.* 72 Several inlanders and explorers had a go. 1968 *Globe & Mail* (Toronto) 13 Feb. 3/3 Hannah Bearskin's parents are Inlanders.

inlaut (i·nlaut). *Philol.* [G.] A medial or internal sound; a sound which occurs in the middle of a word.

1892 G. Dunn in *Classical Rev.* Feb. 1/2 Latin d as inlaut and auslaut frequently represents Indo-Germanic dh. 1950 E. M. Uhlenbeck in E. P. Hamp et al. *Readings in Ling. II* (1966) 251 For the different positions (anlaut, inlaut, auslaut) it must be ascertained separately what phonemes are permissible.

inlaw (i·nlǭ), *sb.*² *nonce-wd.* (See quot.)

1880 G. M. Hopkins *Note-bks. & Papers* (1937) 314 His [*sc.* the universal being's] *inlaw*, the law of his being is unlike mine.

-in-law. Add: (Further examples of *in-law* in more general use; also *attrib.*) Hence **i·n-lawship,** the state of being an in-law.

1912 D. Canfield *Squirrel Cage* I. i. 10 Her mother felt the usual in-law conclusion about her daughter's life. 1926 C. Sidgwick *Sack & Sugar* i. 9 Eva had invited her future in-laws, male and female, to five o'clock a day or two before her wedding. *Ibid.* vii. 77 He is lost to every-one but his wife and his in-laws. 1939 N. S. Colby *Remembering* i. 19 Robert's sister Catherine has to mix in-law amenities with adoring love. 1952 A. Grimble *Pattern of Islands* 205 A retired..policeman, married to a local woman and on a visit to his in-laws. 1954 M. Gluckman in E. E. Evans-Pritchard *Inst. Primitive Society* vi. 69 This immediately established links of in-lawship with people who were, by standards of blood-ties, his enemies. 1957 V. W. Turner *Schism & Continuity in Afr. Soc.* ix. 258 Links of in-lawship between villages. 1964 Gould & Kolb *Dict. Social Sci.* 368/1 An indefinite number of affinal relatives or 'in-laws'. 1965 G. Melly *Owning-Up* xv. 190 He was living with his wife and baby daughter in his in-laws' semi out at Mill Hill. 1970 *Daily Tel.* 8 May 17 How difficult and unnatural are in-law relationships! 1970 G. F. Newman *Sir, You Bastard* iii. 91 His in-laws bought the furniture for the new house. 1972 R. Milner in W. King *Black Short Story Anthol.* 374 The small salary added to his wife's keeping his in-law landlords quiet as he sweated through his first year of accounting.

inlay, *sb.* Add: **2. b.** *Dentistry.* A filling of gold, porcelain, or other suitable material which is pre-formed to the required shape and then cemented into a cavity.

1888 *Dental Cosmos* XXX. 542 One of the chief obstacles to success..has been the difficulty of exactly fitting the inlay to the tooth. 1921 J. B. Parfitt *Operative Dental Surg.* xix. 175 An 'inlay' is a filling which is constructed outside the patient's mouth and then cemented into place in the tooth cavity. 1963 C. R. Cowell et al. *Inlays, Crowns, & Bridges* ii. 3 Originally gold inlays were prepared by a technique similar to that used for porcelain inlays. 1973 *Which?* Mar. 78/2 Your dentist would not be allowed to fill your mouth with gold inlays when ordinary amalgam fillings would do.

inlaying, *vbl. sb.* Add: **3.** *attrib.* **inlaying machine,** a machine used in the manufacture of inlaid linoleum; **inlaying-saw** (see quot.).

1908 *Westm. Gaz.* 30 May 7/2 A scheme for increasing the speed at which our inlaying machines are worked. a 1877 Knight *Dict. Mech.* II. 1189/2 *Inlaying-saw*, a saw used in piercing stuff for buhl-work.

inleak (i·nlīk). [f. In *adv.* 11 d + Leak *sb.*] Leakage into the inside of something.

1909 *Cent. Dict. Suppl.*, *Inleak*, the leaking of a gas or liquid into an enclosed space or pipe. 1970 *Sci. Jrnl.* Mar. 40/3 Boiloff rate of about 0·2 per cent per day is several times as high as for most above ground insulated tanks, and some frozen earth tanks have failed as a result of excessive heat inleak. 1971 *Physics Bull.* Dec. 722/3 Techniques of constructing large vacuum insulated vessels are well established, but even so the heat inleak (typically a few tens of watts) is such that large closed cycle refrigerators are needed.

i·n-leakage. [f. In *adv.* 11 d + Leakage.] = prec.

1905 *Trans. Inst. Naval Archit.* XLVII. II. 410 It was merely a race between the hand-pump and the in-leakage of water whether she was ever raised to the surface or not. 1963 B. Fozard *Instrumentation Nucl. Reactors* iv. 40 Thereafter a steady flow of gas sufficient to prevent in-leakage of air is maintained.

inlet, *sb.* Add: **4*.** *Anat.* The upper opening into a cavity of the body; used orig. of the pelvis and later of the thorax (both as cavities of the skeleton) and of the larynx.

1828 J. Quain *Elem. Anat.* ii. 69 The central line or axis of the inlet, differs very decidedly from that of the outlet;..both therefore decussate towards the centre of the pelvic cavity. 1906 A. M. Buchanan *Manual of Anat.* I. 210 The true pelvis..presents a brim or inlet, a cavity, and an outlet. 1960 E. Gardner et al. *Anat.* xxix. 339/1 The thoracic cavity communicates with the front of the neck by the superior thoracic aperture, or thoracic inlet. *Ibid.* lxxxii. 938/2 The inlet or auditus of the larynx..leads from the laryngopharynx into the cavity of the larynx.

5. *inlet-cam, -chamber, -nipple, -pipe.*

1901 L. M. Waterhouse *Conduit Wiring* 56 Metal inlet and outlet nipples. 1903 *Architect* (Suppl.) 24 Apr. 23/2 A 5-inch pipe is carried from the inlet-chamber to the reservoir. 1907 *Westm. Gaz.* 1 Oct. 7/1 Connecting the nozzle to the inlet-pipe of the balloon, the further supply of gas was added to that already within the envelope. 1908 *Ibid.* 9 Jan. 4/1 The inlet-cam being drawn out of position..the inlet-valves become completely closed.

inlet, *ppl. a.* Add: **2.** *Needlework.* Ornamented *with* lace, etc., let in or inserted. So **i·nletting** *vbl. sb.*

1901 *Daily Chron.* 31 Aug. 8/3 Flowing skirts inlet with lace. 1904 *Ibid.* 3 May 8/5 The inletting and trimming of a serge or cloth gown with Irish lace is quite usual.

i·n-letter, *sb.*² [f. In *adv.* 12 a + Letter *sb.*¹ 4.] An incoming letter. Cf. *in-basket, *in-tray.

1955 C. E. Carrington *Rudyard Kipling* p. vi, Every in-letter was destroyed as soon as it was answered. 1973 *Times Lit. Suppl.* 9 Nov. 1359/1 In-letters, in particular, have been sacrificed: there is only one from Margaret Gillett..as against more than 200 from Smuts to her.

i·n-line, *sb.* and *a.* Also **inline.** [f. phr. *in line* (cf. *In *prep.* 17*).] **A.** *sb. Printing.* (See quot. 1958.) Also *attrib.* or as *adj.* orig. *U.S.*

1923 *Amer. Type Founders Co. Specimen Bk.*, Cheltenham Inline. 1931 Bastien & Freshwater *Printing Types* 133 Erbar Inline...The inline..is a masterpiece of balance and design. 1953 Berry & Johnson *Encycl. Type Faces* 263 Inline fat face capitals and figures. The white line occupies only a small portion of the main strokes. 1958 *Ibid.* (ed. 2) 269 Outline or open letters should be those in which the whole interior of the stroke has been removed, shaded letters those which have a white line running down one side or the other, and Inline letters those with a white line running through the centre of the strokes. 1973 *Publishers Weekly* 1 Jan. 47/1 The third release is Neuland [type-face], which comes in a standard, two outlines, an inline and a black.

B. *adj.* **1.** (Composed of parts) arranged or situated in a line. **a.** Applied to internal-combustion engines in which the cylinders are arranged in one or more rows (in contrast to radial engines); usu. restricted to those in which the cylinders are vertical (so excluding V engines). Also *ellipt.* or as *sb.*

1929 V. W. Page *Mod. Aviation Engines* II. xlvi. 1886 Engines of the in-line type and both static and rotary radial two cycle forms continue to receive attention. **1934** *Discovery* Dec. 353/1 The tendency..is to develop..the large in-line engine..composed of four banks of cylinders forming an H, and the corresponding radial engines with two circles of cylinders one immediately behind the other. **1949** L. Katz *Princ. Aircraft Propulsion Machinery* i. 13 The principal cylinder arrangements are: 1. Inline—Single crankshaft, one cylinder bank, one piston per crankpin. 2. Inline-inverted.—Inverted version of inline to ease problems of installation and facilitate larger propeller swing in small aircraft. 3. Opposed-cylinder.. 4. V... 5. V-inverted [etc.]. **1958** R. D. Blacker *Basic Aeronaut. Sci.* ix. 143/2 In-line engines consist of one or more rows of cylinders placed one behind the other. The rows of cylinders may be arranged in an 'X' or 'V', as well as in a single line. **1961** J. Mackerle *Air-Cooled Motor Engines* x. 200 Twin cylinder engines are arranged in in-line parallel twins, V-engines or horizontally opposed. **1969** K. Munson *Pioneer Aircraft 1903–14* 22 Wright Flyer III, ca. summer/autumn 1905. *Engine*: one 20 h.p. (approx.) Wright 4-cylinder water-cooled in-line. **1970** *Commercial Motor* 25 Sept. 56/2 A 370 bhp version of the Cummins 335/350 bhp six-cylinder in-line was in production. **1971** P. J. McMahon *Aircraft Propulsion* xi. 312 By the early 1930s..the inline vee..was beginning to offer strong opposition. *Ibid.*, Even though the radial made a comeback..the inline always had this fundamental advantage of a lower frontal area.

b. *gen.*

1965 A. D. & K. H. V. Booth *Automatic Digital Calculators* (ed. 3) x. 134 (*caption*) (*a*) Crossed film cryotron; (*b*) in-line cryotron. **1968** *Sci. Jrnl.* Oct. 29/3 Aerial elevation and azimuth are..shown together with range on in-line digital indicators on the control unit.

2. Taking place or situated as an integral part of a continuous, usu. linear, sequence of operations or machines (as in an assembly line); involving or employing such a sequence.

1958 E. E. Eckersley *Automation in Pract.* xi. 172 In straight in-line indexing, the work piece moves intermittently from one machining station to the next in a straight line. **1967** *Electronics* 6 Mar. 47/1 Production volume of monolithic integrated circuits has reached a point where automatic in-line testing and sorting will pay off in reliability. **1967** *Times Rev. Industry* May 60/2 The accommodation is designed for all the latest production techniques, including automatic inspection, bulk palletisation, in-line decoration and mechanical packing. **1968** Boothroyd & Redford *Mechanized Assembly* ii. 8 An in-line assembly machine is one where the work carriers are transferred in line along a straight slideway. **1971** *Engineering* Apr. 73/1 From the point where the operator selects the proper conductor wires, a portable electro-hydraulic in-line joining machine..completes the cycle in under 18 seconds. **1971** *Physics Bull.* July 401/2 A typical problem in a steel mill is the in-line measurement of the roundness and diameter of steel rods, which are both hot and vibrating as a consequence of the production process.

3. *Computers.* **a.** Applied to a subroutine that is written, in full, directly into a program wherever it occurs. Now *rare*.

1958 Gotlieb & Hume *High-Speed Data Processing* vi. 107 A subroutine may be incorporated into a routine in either of two ways. If the instruction sequence is of reasonable length it may be inserted directly into place in the routine of which it forms part... A subroutine used in this manner is called an open or in-line subroutine. If a subroutine consists of a long sequence of instructions, or if it must be used in several different places in the routine, it is desirable to store it separately..and enter it by means of a jump.

b. Applied to data processing in which input data are processed in the order in which they are produced or obtained, without being first sorted into batches.

1959 J. Jeenel *Programming for Digital Computers* ix. 419 Random-access storage would permit input data to be processed efficiently in the chronological order in which they arise. This type of processing, which lends itself particularly well to certain commercial applications, is frequently referred to as 'in-line processing', as opposed to 'batch processing'. **1964** T. W. McRae *Impact of Computers on Accounting* i. 17 An in-line processing system updates all of the records on the same run, and the input data do not require sorting.

c. = *on-line *a.* b.

1969 E. M. McCormick *Digital Computer Primer* ix. 135 The input-output equipment of a computer is sometimes referred to as peripheral. If operated and controlled by the computer itself, it is in-line or on-line; if operated independently of the computer, it is off-line. **1971** N. Chapin *Computers* viii. 152 On-line peripheral equipment [or in-line, as it is sometimes called] operates under the direction of the control unit of the automatic computer.

in loco: see *in *Lat. prep.*

in-lot. **2.** (Examples.)

1779 in J. R. Robertson *Petitions Early Inhabitants Kentucky* (1914) 51 [We] pray that every Actual settler..may be entituled to Draw a free lott;..the lotts to consist of half acre in lott and five acre out lott. **1790** in *Amer. Pioneer* (1842) I. 72 Nathaniel Massie doth bind and oblige himself his heirs, &c., to make over and convey..one in-lot in said town. **1819** E. Dana *Geogr. Sk. Western Country* 74 The in-lots 62½ by 87½ feet each, were sold at public auction. **1837** W. Jenkins *Ohio Gazetteer* 109 The regular in-lots are ninety nine feet in front, extending back one hundred and ninety eight feet. **1948** E. N. Dick *Dixie Frontier* 148 The area in and around one of these stations was plotted and each settler could hold one or more 'in lots' or building plots on the townsite and one or more 'out lots' or farming areas.

in-maintenance: see *in *adv.* 12 a.

in-migrant (i·nmai͡grănt), *sb.* and *a.* orig. *U.S.* [In *adv.* 11 e.] **A.** *sb.* One who migrates from one place to another in the same country. **B.** *adj.* Migrating from one place to another in the same country.

1942 *Fortune* Oct. 194 About two-thirds of the 120,000 in-migrants will be without new housing. **1943** *New England Electr. News* Aug. 32 If a demand for the houses exists among the eligible in-migrant workers in a community. **1962** *Amer. Speech* XXXVII. 16 After 1900, the largest number of in-migrants came [to New York] from the Mid-Atlantic states. *Ibid.* 23 The in-migrant Negroes from the South concentrated in their neighborhoods too.., the largest concentration being the Harlem section in uptown Manhattan. **1963** *New Society* 10 Oct. 26/1 In-migrants to Aberdeen contained a much higher proportion of..university trained women than the native population. **1966** *Publ. Amer. Dial. Soc.* 1964 XLII. 29 Poor in-migrant Southerners.

in-migration (i·nmaigrē^i^·ʃən). orig. *U.S.* [In *adv.* 12 a.] The action of moving from one place to another within the same country, e.g. from one state to another in the United States.

1942 *N.Y. Herald Tribune* 17 May N1 An immigration of approximately 3,000 new workers. **1957** *Economist* 28 Sept. 1031/1 Nowadays, with the tide of immigration from Europe a fading memory, American cities are growing by grace of what sociologists call 'in-migration'—movements of people from other parts of the United States. **1971** *Sci. Amer.* July 183/1 Of the 12 states in this region only three showed an excess of in-migration over out-migration. **1972** *Real Estate Rev.* Winter 21/1 The heavy in-migration of people from the mainland following the attainment of statehood made the demand for apartment-type housing acute. **1973** *Daily Colonist* (Victoria, B.C.) 11 Oct. 5/1 Land use—controls on highrise development and limits to 'in-migration', the influx of outsiders to the Victoria area.

i-n-milk, *a.* [attrib. use of phr. *in milk*: cf. Milk *sb.* I c.] Of a cow: in a condition to yield milk.

1958 *Times* 11 Dec. 12/7 If an in-milk cow laid down on the grass she was likely to be milked by a hedgehog. **1966** *Farmer & Stockbreeder* 11 Jan. 53/1 Best bid of 108 gs. was made..for The Pynes Herds' in-milk Jersey heifer, Lindfield Lady Jane.

inmix (inmi·ks), *v.* [f. In *adv.* + Mix *v.*] *trans.* and *intr.* = Immix *v.*

1920 G. Meredith *Vane Enamoured* in *Mod. Love* 99 Then shall those nobles of the earth and sun Inmix unlike to waves on savage sea. **1909** —— *Celt & Saxon* xvi. 237 Celt and Saxon are much inmixed with us. **1931** Belloc *Ess. Catholic* xvi. 218 It was badly inmixed with motives in no way Catholic.

innards (i·nădz), *sb. pl.* Dial. and vulgar alteration of *inwards* (see Inward *a.* and *sb.* B. 1 b) 'entrails'. Now in common *colloq.* use. (Marshall, 1787 (see *E.D.D.*) has only *in-wards*.) Also *transf.* and *fig.*, the inside (of anything).

1825 J. Britton *Beauties Wiltshire* III. 375 *Innards*, the entrails of a hog. **1874** 'S. Beauchamp' *Grantley Grange* I. ii. 29 It's summut I' his innards, or his yud. **1878** Trollope *Is He Popenjoy?* III. i. 7 The Marquis was still in bed. His 'in'ards' had not ceased to be matter of anxiety to Mr. Walker. **1896** S. Baring-Gould *Dartmoor Idylls* viii. 193 I'm terrible bilter in my in'ards. **1905** Kipling *Traffics & Discov.* (1904) 38 There was the cutter's innards spread out like a Fratton pawnbroker's shop. **1921** *Wireless World* 15 Oct. 439/1 The instrument is assembled from a Mk. III ebonite top.., the parts of an aeroplane 'remote control', etc... Its 'innards' were collected from many different firms at all sorts of prices. **1929** G. Mitchell *Mystery of Butcher's Shop* iv. 47 Damned nuisance about the head... He's left us everything else, including the innards. **1932** J. T. Farrell *Young Lonigan* i. 29 His innards made slight noises, as they digested fourteen and the digesting a juicy beefsteak. **1934** *Mind* XLIII. 234 The music is so bound up with the feelings and impressions of their minds, so spun out of the composers' 'innards', that it..is inseparable from the feelings and impressions themselves. **1936** *Discovery* May 132/1 The 'innards' of the atom. **1937** *Evening News* 5 Feb. 8/2 The best larder for a good kill was his innards, and the savage's sound logic in overeating passed down the Middle Ages as Feast Days. **1941** Wyndham Lewis *Let.* 12 Nov. (1963) 310 Next, the true innards of Fascism are uncovered. **1962** *Listener* 16 Nov. 825/1 Here we are in Hogarth, at thirty-five, exploring the dark innards of the town. **1962** *Ibid.* 8 Mar. 405/2 The whole thing [*sc.* the jury system] can only live by the inner sense in which we regard it all. **1971** *Physics Bull.* Jan. 27/3 Theoreticians who could not care less about planetary atmospheres but deal purely with the innards of the CO_2 molecule. **1974** *Observer* (Colour Suppl.) 13 Jan. 27/1 The next time I slid the clamp up it wouldn't grip, the sheath of the nylon rope came down with it, and the white innards stretched thin over the lip of rock.

innatism (i·nē^i^tiz'm). [f. Innate *a.* + -ism.] Innate ideas, or belief in them.

1909 in Webster. **1953** *Scottish Jrnl. Theol.* VI. 441 There are discussions on innatism and ontologism.

innelite (i·nəlait). *Min.* [ad. Russ. *innelit* (S. M. Kravchenko), f. *Inneli*, Yakut name for the Inagli river: see -ite¹.] A yellow-brown complex silicate of barium, near $Ba_4Ti_3Si_4O_{14}(OH)_{14}Na_2SO_4$, found as tabular crystals in pegmatites in the Inagli massif, South Yakutsk, U.S.S.R.

1960 *Geochem.* 741 Innelite—new barium silicate (after S. M. Kravchenko's data). *Ibid.* 745 The RE [*sc.* rare-earth] ratios in the strontium mineral lamprophyllite and the barium mineral innelite are characterized by a high relative content of La (37–44% of the total RE). **1963** *Doklady Earth Sci.* CXLI. 1297/1 Innelite..was discovered in 1957 in aegirite-akermanite-microcline pegmatites of the Inagli massif (south Yakutia). *Ibid.* 1298/1 In comparison with all other known barium silicates, innelite contains the greatest quantity of barium.

inner, *a.* (*sb.*²) Add: **1.** **b.** (Further examples.)

1875 Geo. Eliot *Let.* 12 Feb. (1956) VI. 121 Because we have suffered from acquaintance that makes us only the more glad to have friends, and you are one of the inner circle. **1926** H. Crane *Let.* 20 June (1965) 262 An 'inner circle' of literary initiates. **a 1930** D. H. Lawrence *Etruscan Places* (1932) iii. 78 Here in the tombs everything is in its sacred or inner-significant aspect. **1973** *Times* 26 Mar. 9/1 It smacks too much of the confidential procedures of an inner circle for many churchmen to feel at ease with it.

c. *Printing.* In sheet work, designating the forme containing the type pages from which the inner side of the sheet is printed and including the type page for the second page of the printed sheet.

1755 J. Smith *Printer's Gram.* 229 (*caption*) The Inner Form of a Sheet in Quarto. **1841** W. Savage *Dict. Art of Printing* 422 *Inner form*, the forme that has the second page in it; it is always worked before the outer forme, except there be some particular reason for the contrary. **1888** C. T. Jacobi *Printers' Vocab.* 63 *Inner forme*, the pages of type which fall on the inside of a printed sheet in 'sheet' work—the reverse of 'outer' forme. **1892** A. Powell *Southward's Pract. Printing* (ed. 4) ix. 159 The forme containing the first page is always called by printers the outer forme; the one which contains the second page the inside or inner forme. **1946** A. Monkman in H. Whetton *Pract. Printing & Binding* x. 61/2 So far as the four-page [impression] schemes are concerned, therefore, it is only necessary to remember that if the job is to be worked as sheet work, pages 1 and 4 will be the outer forme and pages 2 and 3 the inner forme. **1965** *Library* XX. 14 In Table I, the data on choice of forme are abstracted from the list of books, showing for each the number of inner and outer formes printed first... Here is a grand total of 3,338 sheets [printed 1600–1800], of which 3,002, or 73 per cent were printed inner forme first.

f. *inner light*: in Quaker use (see Light *sb.* 7 b and quot. 1957).

1856 [see Inner *a.* 2]. **1909** Chesterton *Orthodoxy* v. 135 The Quaker doctrine of the Inner Light. **1957** *Oxf. Dict. Chr. Ch.* 692/1 *Inner light*, the principle of Christian certitude, consisting of inward knowledge or experience of salvation, which is upheld by the Society of Friends.

g. *Phonetics.* Denoting a sound articulated in a part of the mouth nearer the throat than that designated by the unqualified term.

1867 A. M. Bell *Visible Speech* 61 If the breath within the mouth be compressed behind the articulating organs while an *inner closure* is held, a distinct, and in some cases, a powerfully percussive effect will be produced on the abrupt separation of the organs. **1888** H. Sweet *Hist. Eng. Sounds* 5 Most of these [point and blade consonants] admit also of 'inner' and 'outer' varieties.

h. *Inner Circle*: name of one of the lines of the London (underground) railway system.

1869 *Bradshaw's Railway Manual* XXI. 217 Metropolitan District. Incorporated..(29th July, 1864), to construct a series of lines to complete an inner circle of railway north of the Thames. **1882** *Times* 24 July 10/4 The Inner Circle would connect..with the railways south of the river. **1884** *Times* 12 Feb. 11/3 As to the Inner Circle line, by the 1st of June, or certainly by the beginning of the second half of the current year, that great work would be finished. **1911** *Encycl. Brit.* XVI. 944/1 This company combines with the Metropolitan District to form the Inner Circle line, which has stations close to all the great railway terminii north of the Thames. **1938** G. Greene *Brighton Rock* iv. iii. 177 He could feel his blood pumped from the heart and moving indifferently back along the arteries like trains on the inner circle. **1966** J. Chamier *Cannonball* i. 11 Planes whizzing around a damn sight quicker than the Inner Circle.

i. *inner tube*: in a pneumatic tyre, a separate tube, inside the cover, which is inflated with air.

1895 *Montgomery Ward Catal.* 556/3 Pneumatic Tires.. Inner Tubes complete with valve stem and valve. **1902** *Encycl. Brit.* XXXIII. 535/1 In most tyres for cycles and motor-cars, an inner tube of indiarubber is made separate from the outer cover. **1902** A. C. Harmsworth et al. *Motors* x. 213 Half the number of spare covers and inner tubes are required as compared with the requirements when the wheels are of different sizes. **1904** A. B. F. Young *Compl. Motorist* (ed. 2) ix. 250 The piercing of the outer cover and inner tube by a nail or other puncturing agent. **1912** *Motor Manual* (ed. 14) iii. 106 The inner tube has become nipped between one of the security bolts and the cover. **1923** *Michelin Guide Gt. Brit.* (ed. 7) 883 Covers, inner tubes or pneumatic tyres.

1967 P. Freeling *Strike Out* 42 He had done two hundred kilometres a day, on rough country roads with spare inner tubes slung round the neck.

j. *inner Cabinet* (or *cabinet*): an informal term for a group of decision-making people within a ministerial Cabinet or similar group.

1900 *Westm. Gaz.* 13 Nov. 2/2 No one imagines that this Committee of twenty really decides critical matters of high policy. Those are deputed to the 'inner Cabinet'. **1936** H. Nicolson *Diary* 24 Feb. (1966) 245 J. H. Thomas says that our group is not consulted; that there is an inner Cabinet which discuss things between themselves. **1970** *Times* 3 Mar. 2 The meeting was also told that Hebdomadal council, the university's 'inner cabinet', had appointed a committee to listen to the views of the students' elected representatives. **1972** *Guardian* 11 Jan. 20/1 The TUC 'inner cabinet'—the finance and general purposes committee.

k. *inner product* (Math.) [tr. G. *inneres produkt* (H. Grassmann *Die lineale Ausdehnungslehre* (1844) p. xi)]: so named because an inner product of two vectors is zero unless one has a component 'within' the other, i.e. in its direction]: the sum of the products of corresponding components of two real vectors $(a_1, a_2, \ldots, a_n)$ and $(b_1, b_2, \ldots, b_n)$, i.e. the number $a_1b_1 + a_2b_2 + \ldots + a_nb_n$; in a complex vector space, the number $a_1\bar{b}_1 + a_2\bar{b}_2 + \ldots + a_n\bar{b}_n$, where $\bar{b}_i$ is the complex conjugate number of b_i; (see also quot. 1966).

1920 T. Muir *Theory of Determinants* III. i. 7 The theorem on the inner product of two magnitudes each of the mth 'Stufe' and consisting of m simple factors. **1922** E. H. Neville *Prolegomena Analytical Geom.* iv. i. 192 Let an ordered set of three numbers be called a triplet, and let the number $fp+gq+hr$ be called the inner product of the triplets (f, g, h), (p, q, r). **1941** Birkhoff & MacLane *Survey Mod. Algebra* vii. 181 Physicists often speak of the inner product as a 'scalar product' of two vectors. **1966** A. L. Rabenstein *Introd. Ordinary Differential Equations* vi. 156 The inner product of $f(x)$ and $g(x)$ with respect to the weight function $w(x)$ on the interval (a, b) is defined to be $(f, g) = \int_a^b w(x)f(x)g(x)dx$. *Ibid.* 157 If the inner product of $f(x)$ and $g(x)$ is zero,.. then $f(x)$ and $g(x)$ are said to be orthogonal with respect to the weight function $w(x)$ on the interval $a<x<b$. **1968** E. T. Copson *Metric Spaces* ix. 139 In order to avoid confusion between multiplication of a vector by a scalar and the scalar product of two vectors, the scalar product of two vectors is often called their inner product. *Ibid.* 140 A vector space on which an inner product is defined is called an inner product space.

l. *inner quantum number* (Physics) [tr. G. *innere quantenzahl* (A. Sommerfeld 1920, in *Ann. d. Physik* LXIII. 231)]: a quantum number now identified with that of the total angular momentum of an electron, *j* (*J II. 6 c).

1923 H. L. Brose tr. *Sommerfeld's Atomic Struct. & Spectral Lines* vi. 364 If we wish to exclude the forbidden lines by a principle of selection, we must..introduce a new quantum number; we call it the inner quantum number and designate it by n_i. **1926** *Proc. R. Soc.* A. CXI. 84 Each term..in general will be a multiple term consisting of several members with different values of the 'inner quantum number' *j*. **1967** W. R. Hindmarsh *Atomic Spectra* ii. 18 The regularities of the multiplet structure of spectra were considered in some detail by Sommerfeld... He introduced an 'inner' quantum number to distinguish the various states of a multiplet, and suggested that it may be connected with a property of the electrons in inner shells (the core electrons). The true explanation of the doublet structure of the terms of alkali metal atoms is provided by the concept of electron spin.

m. *inner reserve* (Finance): a secret reserve not disclosed in a balance-sheet and due to an understatement of certain capital assets.

1930 *Daily Express* 16 Aug. 10/1 Former Inner Reserves are now brought from the Assets in which they were hidden and are grouped in an exposed Reserve on the Liability side of the Sheet. **1955** *Times* 10 May 18/5 Your directors have now decided to transfer a part of these inner reserves in order to increase the contingencies reserve.

n. *inner-directed* adj. (Sociol.): a term coined by D. Riesman to designate persons whose behaviour and goals are directed by the standards and ideals which they formed early in life; also postulated as a cultural stage in a society. (See quot. 1950.) Cf. **other-directed* and **tradition-directed* adjs. Hence *inner direction*.

1950 D. Riesman et al. *Lonely Crowd* i. 9 The society of transitional population growth develops in its typical members a social character whose conformity is insured by their tendency to acquire early in life an internalized set of goals. These I shall term inner-directed people and the society in which they live a society dependent on inner-direction. *Ibid.* 16 The inner-directed person becomes capable of maintaining a delicate balance between the demands upon him of his life goal and the buffetings of his external environment. **1959** *Spectator* 4 Sept. 307/2 A criticism renewed by sociology—he [*sc.* C. Wilson in *The Age of Defeat*] seems to imply—can help to renew literature by restoring 'the hero', and 'the hero' will re-accredit in real life the image of the 'inner-directed' man. **1959** *Listener* 3 Sept. 363/2 Mr. Wilson discerns a similar awareness of the difference between 'inner-direction' and 'other-direction' in the existentialist writings of Camus and Sartre. **1961** M. Singer in B. Kaplan *Studying Personality* 51 The influence of parents and teachers, so vital in the formation of 'inner-direction', is being superseded by the influence of 'peer-groups' and the mass media. **1964** M. Argyle *Psychol. & Social Probl.* xv. 186 The life of managers is also changing: as in America the inner-directed individualist is being replaced by the other-directed organization man, who fits in easily with the ideas of others, and subordinates his interests to those of the concern. **1968** P. McKellar *Experience & Behaviour* xi. 288 The capacity to 'go it alone' is characteristic of the inner-directed personality. **1972** *Jrnl. Social Psychol.* LXXXVI. 224 Low authoritarian subjects were more inner-directed.

o. *inner space* [after **outer space*]: (*a*) the regions between the surface of the earth and outer space; (*b*) the regions below the surface of the sea; (*c*) [cf. sense 2] the part of one's mind or personality that is not normally experienced or within one's consciousness.

1958 *Times* 29 Mar. 7/4 We seem to need..names for the parts where the atmosphere is still a drag, where the Earth's gravitation is dominant... Tentatively it might be suggested that these be called..inner space. **1966** I. Asimov *Fantastic Voyage* i. 10 We would pin him into an X-ray rocket and fire him through inner space. **1967** *Times* 7 Oct. 10/3 The Sea Dragon's [*sc.* a submarine's] captain..had radioed in advance that he considered 'this voyage has proved the feasibility of protracted flights in "inner" space'. **1969** *Sci. Amer.* Apr. 64/2 There is a remarkable similarity between many of the problems faced by the astronaut in 'outer space' and those of the aquanaut in 'inner space'. (*c*) **1968** *Sat. Rev.* (U.S.) 25 Sept. 28/2 Must this inner space continue to be peopled with imaginative dragons of strange color and dropping off places that confine the moral venture to the shallow water of one's own mainland or adjacent islands of narrow self-interest? **1961** 'J. Duncan' (*title*) Exploring inner space: personal experiences under LSD-25. **1968** A. Diment *Bang Bang Birds* viii. 143 The Indian and Chinese prophets..knew a thing or two about inner space and the turned-on mind. They did it on contemplation through and mushroom juice. **1969** *Daily Tel.* 20 Feb. 16/7 It is they..who are the investigators of what J. G. Ballard terms 'inner space'—the remoter recesses of man's mind where images of crisis are spawned.

p. *inner city*: the central area of a city, esp. regarded as having particular problems of overcrowding, poverty, etc. Also *attrib.* (see sense 6 below). orig. *U.S.*

1968 *Sat. Rev.* (U.S.) 16 Nov. 95 The twin concepts of decentralization and community control of the schools developed in response to the failure of schools in the inner city. **1973** *Black Panther* 17 Mar. 11/1 I'm..interested in getting a little more practical and down to present social policies in the cities, in the inner-cities; the continuing and ever occurring crisis in the inner-cities, where large numbers of people are trapped in a cycle of poverty. **1974** *Times* 19 Jan. 10/2 The problems of the inner city—a work area where almost everyone has gone home.

2. (Further examples.)

1854 Geo. Eliot tr. *Feuerbach's Essence Christianity* i. 2 The inner life of man is the life which has relation to his species. **1860** J. W. Palmer tr. *Michelet's Love* 118 A feeling that the woman's inner self will not be reached, her soul not attained. **1880** W. James *Coll. Ess. & Rev.* (1920) 217 The point of application of the volitional effort always lies within the inner world, being an idea or representation. **1886** Inner self [see *climate *sb.* 3 b]. **1899** W. James *Talks to Teachers* ii. 15 There is a stream, a succession of states, or waves, or fields (or of whatever you please to call them), of knowledge, of feeling, of desire, of deliberation, etc., that constantly pass and repass, and that constitute our inner life. **1902** —— *Varieties Relig. Experience* i. 7 Often they [*sc.* religious leaders] have led a discordant inner life, and had melancholy during a part of their career. **1905** A. Lang *Adv. among Books* 122 She [*sc.* Mrs. Radcliffe] delighted in descriptions of scenery, the more romantic the better, and usually drawn entirely from her inner consciousness. **1915** V. W. Brooks *World of H. G. Wells* v. 106 The force of a work of art does not reside in its 'inner meanings'. **1927** B. Russell *Outl. Philos.* ii. 20 We all have an inner life, open to our own inspection but to no one else's. **1930** *Amer. Jrnl. Psychiatry* 1019 The former is derived from *persona* meaning the essential or inner self. **1944** Auden *For Time Being* (1945) 35 The manifestations of the inner life should always remain so easy and habitual. **1952** Gerth & Martindale tr. *Weber's Anc. Judaism* p. xi, He displayed an inner-worldly, stoic attitude in the face of death. **1953** R. G. Davis *Ten Mod. Masters* p. xiv, Even a fairy-tale or fantasy must have its inner logic. **1971** *Daily Tel.* 6 Aug. 9/7 This is a girl with an intense inner life. **1974** *Listener* 17 Jan. 76/1 Each [Buddhist monk].. inhabits his private inner world, and yet they're in harmony with each other.

b. *inner speech* (*form*): see quots. Also *inner linguistic* (or *language*) *form*, *inner form*.

[**1885** D. G. Brinton in *Proc. Amer. Philos. Soc.* XXII. iv. 319 Besides the grammatical form of a language, Humboldt recognized another which he called its *internal form*.] **1888** H. A. Strong tr. *Paul's Princ. Hist. Lang.* xx. 460 The influencing force extends merely to what Humboldt and Steinthal have described as the *inner language form* ('innere Sprachform'). *Ibid.* 471 A language suffers influence in its *inner linguistic form* principally in the mouths of those who speak it as a foreign tongue. **1901** H. Oertel *Lect. Study Lang.* i. 64 This is the 'outer speech form', the external, phonetic aspect of the speech symbols. The 'inner speech form' is the definite arrangement of the prelinguistic psychical material into definite groups, the coherence of each group being secured by labelling each with one definite sound-tag. **1930** J. R. Firth *Speech* v. 44 The early discussions on inner speech did not touch the bigger question of the extensive motor accompaniment of thought. **1934** Webster, *Inner speech*, *Psychol.*, use of words or word images in thinking, without audible or visible speaking. *Ibid.*, *Inner speech form* (trans. of G. *innere Sprachform*, used by Humboldt and Steinthal), the mental concept or image associated with a word prior to its use or upon hearing or reading it, as the concept of a quadruped associated with the word 'animal'; abstractly, the quality by which a word evokes such a mental picture. **1970** H. C. Shands *Semiotic Approaches to Psychiatry* 10 Inner speech patterns are constructed throughout the developmental period in human beings.

3. Also, *the inner woman*.

1857 Trollope *Barchester T.* III. x. 184 She ate and drank, and as the inner woman was recruited she felt a little more charitable. **1858** Hawthorne *Passages from Fr. & It. Note-Bks.* (1871) I. 190 To behave as her inner woman prompts. **1892** *Gentlewoman's Bk. Sports* I. 44 After refreshing the inner woman, I was all for trying the Sandhills again.

6. Various phrases used *attrib.*

1908 *Daily Chron.* 22 Jan. 3/3 You may browse at will among the epistles or the notes, feeling that you are always with informed, inner-circle folk. **1909** *Westm. Gaz.* 14 Apr. 10/2 What colour of glass must be used for the front door and inner-court doors? **1927** J. Adams *Errors in School* 32 An idea does not merely mean the inner-world of his own mind. **1933** *Amer. Speech* VIII. 27 Inner-circle Guardian Angel. **1936** C. E. Bazell *Ling. Form* 57 The rare cases of inner-word sequence-relevance may be dealt with analogously. **1937** C. Hunt *Guide to Communist Jargon* xxv. 88 According to the *Political Dictionary*, inner-party democracy is the consistent application of the principles of 'democratic centralism', though, to be more accurate, it stands for its democratic as opposed to its dominant centralized aspect. **1960** H. Edwards *Spirit Healing* vii. 63 Spiritual healers have long known that the origin of organic diseases most often lay in inner-self disharmonies. **1961** *Observer* 8 Oct. 10/3 It is doubtful if Mr. Gaitskell himself had thought through the problem of inner-party democracy. **1964** F. Bowers *Bibliogr. & Textual Crit.* iv. ii. 112 Any separate small pile of inner-forme sheets. **1968** *Sun* (Baltimore) 4 July 4/1 A highly qualified explanation of the inner-city language superiority (which had disappeared by the third grade) was, Mrs. Entwisle thought, the unrestricted time which small inner-city children spend in front of television sets. **1970** *New York* 16 Nov. 42/2 From here the city spreads in a wheel-spoke design through seven inner-city black neighborhoods. **1972** *Guardian* 26 Feb. 6/8 Camden's housing problems have often been in the spotlight for revealing inner-city trends. **1972** *Ibid.* 5 Jan. 5/2 The solution to the problem of the inner city child eludes us.

innerly, *adv.* Add: **1.** (Examples from D. H. Lawrence.)

1917 D. H. Lawrence *Look! We have come Through!* 50, I have been so innerly proud, and so long alone, Do not leave me, or I shall break. **1923** —— *Ladybird* 81 If you are true to me, innerly, innerly true, he will not hurt us.

innervation. Add: Also, the supply of nerve fibres to, or disposition of nerve fibres within, an organ or part. (Further examples.)

1879 *Jrnl. Physiol.* II. 342 More recently Severini, in his able monograph on the innervation of the blood-vessels, has laid great weight on the contractility of the capillaries. **1908** *Westm. Gaz.* 8 July 2/1 It has been found that the density of the cutaneous innervation—i.e., the number of sensitive nerve terminations in the unit of surface—is greater in small animals than in large. **1910** *Jrnl. R. Microsc. Soc.* 154 Innervation of tympanum.—Agostino Gemelli describes..the tympanal ramifications (1) of the auriculo-temporal branch of the trigeminal, and (2) of the nerve of Jacobson. **1945** *Amer. Jrnl. Physiol.* CXLIV. 477 It is tacitly assumed that if part of the innervation of a muscle is permanently destroyed, the remaining motor units..continue their normal function. **1967** Gardner & Osburn *Struct. Human Body* iv. 121/2 The nerve supply to a muscle is referred to as its innervation.

2. *Psychol.* = Kinæsthesis.

1880 W. James in *Anniversary Mem. Boston Soc. Nat. Hist.* 4 Wundt..adopts the term *Innervationsgefühl* to designate the former [*sc.* the feeling of force exerted] in relation to its supposed cause, the efferent discharge. Feelings of innervation have since then become household words in psychological literature. **1898** G. F. Stout *Man. Psychol.* I. ii. vi. 192 According to Bain, there is a direct sense of energy put forth which is independent of any results the putting forth of energy may produce. This peculiar modification of sensory consciousness has been called the sense of effort, or the innervation-sense. **1904** E. B. Titchener tr. *Wundt's Princ. Physiol. Psychol.* iii. 57 (*heading*) General principles and problems of a mechanics of innervation. **1924** J. Riviere et al. tr. *Freud's Coll. Papers* I. 63 The conversion may be either total or partial, and it proceeds along the line of the motor or sensory innervation that is more or less intimately related to the traumatic experience. **1953** Hinsie & Shatzky *Psychiatric Dict.* (ed. 2) 667/1 The expressive innervations are involuntary, even though they can be influenced, up to a point, by volition.

inning, *vbl. sb.* Add: **4.** (Earlier and later examples in Cricket and Baseball.)

1735 in H. T. Waghorn *Cricket Scores* (1899) 10 London ..got 67 notches the first innings. **1856** *Spirit of Times* 6 Dec. 229/1 After the first inning is played, the turn commences at the player who stands next to the one on the list who lost the third hand. **1955** *Times* 9 May 15/2 Afterwards he played the sort of innings that not surprisingly drives some spectators to distraction. **1968** *Washington Post* 4 July C1/7 The righthanded sinkerballer faced his greatest challenge in the sixth inning.

b. (Earlier example.)

1836 Dickens *Pickw.* (1837) xxiii. 238 It's my innings now, gov'rnor, and as soon as I catches hold o' this here Trotter, I'll have a good 'un.

Inniskilling (iniski·liŋ). Also **Enniskillen, Enniskilling, Inniskellen, Inniskillen.** The

name of the county town of Fermanagh in Northern Ireland, used *attrib.* or *absol.* to designate a regiment originally raised for the defence of that town in 1689. So **Inniski·lliner,** a member of this regiment.

[**1690** A. Hamilton (*title*) A true relation of the actions of the Inniskillingmen, from their first taking up of arms in December, 1688 for the defence of the Protestant religion, and their lives and liberties.] **1715** in E. S. Jackson *Inniskilling Dragoons* (1909) ii. 35 A detachment of Greys and Inniskillings arrived in Edinburgh from Stirling. **1797** *Encycl. Brit.* IX. 243/2 Its [*sc.* Inniskilling's] inhabitants distinguished themselves..in the wars of Ireland at the revolution, out of which a regiment of dragoons, bearing the title of the *Inniskilleners*, was mostly formed. They form the 6th regiment of dragoons in the British Army. **1817** G. Jones *Battle of Waterloo* II. 55 The second heavy brigade of cavalry..consisted of.. the 6th, or Inniskillings. **1822** M. Edgeworth *Let.* 11 Feb. (1971) 347 Will you inquire for Wilkie what is put on the caps of the Inniskellen 6th Dragoons... If you know of any Dragoon that would be better than the Inniskillen tell me. **1853** J. H. Stocqueler *Mil. Encycl.* 141/1 *Inniskilliners*, the officers and soldiers of the 6th dragoons and the 27th foot are so called, from the two regiments having been originally raised at Inniskilling, a town of Ulster. **1893** J. C. Ropes *Campaign of Waterloo* xvi. 300 In rear of the left wing..stood the Union brigade..composed of the Royal Dragoons, the Scots Greys, and the Inniskilling Dragoons. **1968** *Encycl. Brit.* IX. 185/1 At Newtownbutler, in 1689, the Protestant Enniskillen men severely defeated a superior Roman Catholic army of James II and began the victorious tradition of the 'Inniskillings', now represented in the British Army by the Royal Inniskilling Fusiliers and the 5th Royal Inniskilling Dragoon Guards. **1971** G. Blaxland *Regiments Depart* vi. 204 The previous December..the Inniskilling Dragoon Guards were relieved by the 1st Royal Tank Regiment.

innit, vulg. form of *isn't it*.

1959 M. Gilbert *Blood & Judgment* i. 17 That's right, innit? **1962** N. Marsh *Hand in Glove* iv. 105 Dead right, innit? **1965** *Guardian* 3 Apr. 6/3 It's some place to go innit? **1973** J. Wainwright *Touch of Malice* 56 That's a bloody good reason, innit?

innocence. **6.** (Examples.)

1821 W. P. C. Barton *Flora N. Amer.* I. 119 Fairy-flax-Bluett. Innocence. Venus' Pride. **1863** *Rep. Comm. Agric. 1862* (U.S. Dept. Agric.) 159 The 'Bluets', 'Innocence', 'Dwarf Pink', with, perhaps, some other common name, is one of the prettiest ornaments of our spring meadows. **1892** *Jrnl. Amer. Folk-Lore* V. 97 *Houstonia cærulea*, innocence. Boston, Mass. **1954** C. J. Hylander *Macmillan Wild Flower Bk.* 389 This familiar wild flower, also known as Innocence and Quaker-ladies, is a tufted or matted plant.

innocent, *a.* Add: **2.** *innocent party* [Party *sb.* 11], in matrimonial proceedings, the person adjudged to be innocent.

Since the Divorce Reform Act 1969 the usage has been legally obsolete in England, since that Act abolished the concept of a matrimonial offence as a ground for divorce and substituted for it the concept of irretrievable breakdown of the marriage.

1729 G. Jacob *New Law-Dict.*, *Divorce*: In Divorces for Adultery, several Acts of Parliament have allowed the Innocent Party to marry again. **1835** *Tomlins's Law Dict.* (ed. 4) I. (s.v. Divorce), The commissioners appointed by Henry VIII. and Edward VI...recommend divorces *à mensa et thoro* to be abolished, and complete divorces to be allowed for adultery, desertion, bad treatment, &c., the innocent party to be allowed to marry again. **1948** J. H. S. Bossard *Sociol. of Child Devel.* xvi. 369 One principle usually observed is that custody goes to the so-called innocent party. **1958** *Daily Mail* 3 July 4/8 When are we going to hear the last of that time-worn phrase so beloved of newspaper columnists and the legal profession—'innocent party'? **1959** Jowitt *Dict. Eng. Law* I. 67/2 The Matrimonial Causes Act, 1857..created a Court for Divorce and Matrimonial Causes..which would grant to the innocent party a divorce *a mensa et thoro* on the ground of the other's adultery.

5. **b.** *innocent conveyance*, a conveyance which does not have any tortious operation, one which does not create a discontinuance or result in forfeiture.

All conveyances are now innocent by statute in England and in the United States.

1811 E. B. Sugden *Gilbert's Law of Uses & Trusts* (ed. 3) 232 A conveyance by lease and release is like a bargain and sale, and covenant to stand seised what is termed an innocent conveyance. **1841** H. J. Stephen *New Comm. Laws Eng.* I. 508 The other conveyances can, in their nature, pass no more than the grantor might lawfully transfer. For this reason, they have received, by way of distinction from a feoffment, (and others now abolished of the like nature,) the appellation of *innocent* conveyances. **1848** Wharton *Law Lexicon* 322/2 *Innocent conveyances*, a covenant to stand seised, a bargain and sale, and release, so called, because since they convey the actual possession by construction of law only, they do not confer a larger estate in property than the person conveying possesses. **1937** W. S. Holdsworth *Hist. Eng. Law* VII. 357 Both a bargain and sale and a covenant to stand seised were, unlike a feoffment, 'innocent' conveyances.

6. (Further examples.)

1799 Malthus *Diary* 30 June (1966) 108 He was a remarkably meek & innocent looking man. **1860** F. & J. Greenwood *Under Cloud* III. xiv. 307 The jovial, innocent-hearted actor. **1895** A. W. Pinero *Second Mrs. Tanqueray* i. 42 Paula..is..beautiful, fresh, innocent-looking. **1925** Blunden *Eng. Poems* 19 So innocent-gay was her look. **1955** E. Bowen *World of Love* xi. 212 Banks were innocent-blue with scabious. **1959** J. L. Austin *Sense & Sensibilia* (1962) i. 4 In philosophy it is often good policy, where one member of a putative pair falls under suspicion, to view the more innocent-seeming party suspiciously as well. **1962** Y. Olsson in F. Behre *Contrib. Eng. Syntax* 87 Like innocent-looking people, even the most innocent-looking personal pronouns are deeper than one would be led to believe.

in nomine: see *in *Lat. prep.*

innovate, *v.* **2.** Delete † *Obs.* and add later *Comm.* examples.

1967 *Times Rev. Industry* Oct. 86/2 Nylon..was first invented in 1928, but not innovated until 1939. **1972** *Physics Bull.* Feb. 67/1 (Advt.), We've been innovating electrometer values like these for over 20 years.

3. *spec.* in *Comm.* (Later examples.)

1965 *New Statesman* 9 Apr. 561/1 There is a willing market to offset a more reluctant British industry, whose failure to innovate fast enough has certainly made the British computer industry less competitive. **1971** *Physics Bull.* Dec. 707/2 The view has been put forward that the very large firms, the dinosaurs of the 20th century, do not truly innovate and, indeed, may hinder innovation because they are so inflexible.

innovating *ppl. a.* (later examples).

1972 C. Layton *Ten Innovations* I. i. 7 Study is needed of other possible tax incentives for small innovating enterprises. *Ibid.* 12 The small innovating firm crosses a whole series of thresholds as it grows in size.

innovation. Add: **5.** *Comm.* The action of introducing a new product into the market; a product newly brought on to the market.

1939 J. A. Schumpeter *Business Cycles* I. iii. 84 Innovation is possible without anything we should identify as invention, and invention does not necessarily induce innovation. **1958** J. Jewkes et al. *Sources Invention* ix. 249 It seems impossible to establish scientifically any final conclusion concerning the relation between monopoly and innovation. **1962** E. M. Rogers *Diffusion of Innovations* v. 124 It matters little whether or not an innovation has a great degree of advantage over the idea it is replacing. What does matter is whether the individual perceives the relative advantage of the innovation. **1967** J. A. Allen *Sci. Innovation & Industr. Prosperity* ii. 8 Innovation is the bringing of an invention into widespread, practical use...Invention may thus be construed as the first stage of the much more extensive and complex total process of innovation.

6. **Innovation trunk,** a kind of wardrobe trunk.

1912 *Bag, Portmanteau & Umbrella Trader* 20 Nov. 18/1 The Innovation Trunk Company..makes a striking show of wardrobe trunks. They also specialise in the refitting of ordinary wardrobes according to their principle. **1913** A. Bennett *Regent* iii. 83 Many parcels and boxes, comprising diverse items in the equipment of a man-about-town, such as tie-clips and Innovation trunks. **1915** E. Phillpotts *Angel in House* i. 17 Robert and Manservant enter through the main entrance carrying Innovation trunk.

innovational *a.*, also in *Comm.*

1959 J. P. Lewis *Business Conditions Analysis* v. xxiv. 534 The insights of economics do not illuminate the process of innovation very much... On the optimistic side of the innovational outlook, it can be argued, [etc.]. **1960** L. S. Silk *Research Revolution* iii. 50 In the past, the United States has had three great innovational pushes.

innovative, *a.* Add: (Later examples.) *spec.* in *Comm.* in sense of *innovation 5.

1970 N. Armstrong et al. *First on Moon* i. 20 The Air Force..sounded more exciting and more innovative. **1971** *Sci. Amer.* Mar. 14 His division is now involved with ..innovative building processes and systems. **1971** *Times* 6 Sept. 12 (Advt.), Well-known American company is introducing an innovative line of electro-optical measurement systems into the European Machine Tool and Metal Working industries. **1972** *Lebende Sprachen* XVII. 72/2 In order to release scarce manpower for the later phases of the innovative process, including marketing.

Hence **i·nnovatively** *adv.*, in an innovative manner; involving innovation; **i·nnovativeness.**

1962 E. M. Rogers *Diffusion of Innovations* vii. 195 The innovativeness continuum may be divided into adopter categories. *Ibid.* x. 285 The present chapter is an attempt to demonstrate two means of predicting innovativeness. **1971** *Bull. Amer. Assoc. Univ. Professors* Sept. 337 My other two innovatively run courses..are seemingly as successful as courses could be. **1973** *Black World* Mar. 23 The conventions of folk music might be employed in the traditional manner..or innovatively, as is Miss Brooks' 'The Sermon on the Warpland'.

innovatory, *a.* (Later examples.)

1965 *Listener* 23 Sept. 464/1 In these with-it days of innovatory cinematic trends. **1967** J. A. Allen *Sci. Innovation & Industr. Prosperity* ii. 25 There has been little deliberate innovatory effort over a long period. **1971** *Nature* 2 Apr. 301/3 Historians have been so impressed by the innovatory nature of modern science. **1972** *Physics Bull.* Mar. 151/2 Remembering that the major innovatory phase took place before 1939, most of this gain has resulted from sustained research and development by physicists.

in nuce: see *in *Lat. prep.*

Innuit (i·nuit). Also **Inuit.** [ad. Inupik Eskimo *inuit* people, pl. of *inuk* man.] An Eskimo; the Eskimos collectively. Also *attrib.* or as *adj.*

1765 C. Drachard in *Ethnohistory* (1972) XIX. 136 They [*sc.* the Labrador Eskimos] also by way of eminence in contra-distinction to the Europeans call themselves Innuit (the Men). **1774** B. La Trobe *Brief Acct. Mission Esquimaux Indians Labrador* 10 Formerly, they..looked upon the Europeans as upon dogs, giving them the appellation, Kablunets, that is, Barbarians, but called themselves Innuit, which signifies men. **1850** [see *Eskimo *sb.* 1]. **1860** Mayne Reid *Odd People* 88 They [*sc.* the Esquimaux] generally call themselves 'Inuit' (pronounced enn-oo-eet), a word which signifies 'men'. **1864** *Spectator* 31 Dec. 1506 The Innuits believe in a supreme Being called Anguta, whose daughter Sidne is the creator and the tutelary deity of the Innuit people. **1864** C. F. Hall *Life with Esquimaux* I. 122 A highly-intelligent Innuit.. was boat-steerer. **1895** [see *Eskimo *sb.* 1]. **1903** J. London *People of Abyss* xxvii. 313 In Alaska, along the banks of the Yukon River, near its mouth, live the Innuit folk. **1919** W. T. Grenfell *Labrador Doctor* (1920) vii. 140 With the influx of white settlers from Devon and Dorset, Scotland and France the 'Innuits' were driven farther and farther north. **1963** *North* (Ottawa) May–June 34 Without us the Innuit go hungry. **1973** *Sci. Amer.* Sept. 196/3 Nearly 400 sculptors of quality are represented here from the *inuit*.

innumeracy (iniū·mĕrăsi). [f. In-³ + *numeracy.] The quality or state of being innumerate.

1959 *15 to 18: Rep. Cent. Advisory Council for Educ.* (*Eng.*) (Ministry of Educ.) I. xxv. 271 If his numeracy has stopped short at the usual Fifth Form level, he is in danger of relapsing into innumeracy. **1960** *Times Rev. Industry* June 59/3 Handicap on British business was the 'innumeracy' of the population. **1965** *New Scientist* 5 Aug. 348/1 The word 'innumeracy', to describe an intellectual disability in regard to mathematics (and, by reference, to science) has become one of the vogue words of our time. **1970** *Daily Tel.* 22 Dec. 8 Lord Snow was complaining a decade ago about the vice of innumeracy.

innumerate (iniū·mĕrĕt), *a.* and *sb.* [f. In-³ + *numerate *a.*] **A.** *adj.* Unacquainted with the basic principles and ideas of mathematics and science.

1959 *15 to 18: Rep. Cent. Advisory Council for Educ.* (*Eng.*) (Ministry of Educ.) I. xxv. 270 When we say that a historian or a linguist is 'innumerate' we mean that he cannot even begin to understand what scientists and mathematicians are talking about. **1967** *Times Rev. Industry* June 108/3 'Why are so many of us innumerate?' asked Lord Annan... The answer..was quite simple. For years, girls and boys have given up maths at 15 or 16, thus maiming themselves intellectually. **1969** *Physics Bull.* Sept. 383/2 In this short book, intended for the general reader (though not for the innumerate one), a distinguished engineer describes how computers work. **1971** *Daily Tel.* 13 Oct. 2/7 It alleges that one student in every four entering colleges has had no sixth form experience, and that up to one-fifth are 'practically innumerate'.

B. *sb.* One who is innumerate; freq. (with *the*) in *pl.* sense.

1971 *Daily Tel.* 1 Feb. 15 The old gibe that 'you can prove anything with figures' is perhaps not heard so frequently now. It was the classic defence of the innumerate. **1971** *Nature* 2 Apr. 306/1 The conflict is between the technological and the humane, and derives from various related sources: first, the fable that educational research can be a science as pure as classical physics..and finally the innumerate's fascination with statistics. **1972** *Ibid.* 10 Mar. 55/1 At the other extreme stand the recalcitrant innumerate, proclaiming themselves the last defenders of humanism.

innutritious, *a.* Add: Also *transf.*

1905 *Sydney Morning Herald* 15 Feb. 4/6 The..teacher is offered for his mental food these innutritious products as the bread of life. **1916** E. V. Lucas *Vermilion Box* cxxv. 141 As a rule I have found that the soldier who sits opposite one on railway journeys is an innutritious person, whether he has been to the front or not.

inoculant (inǫ·kiŭlănt). [f. Inocul(ate *v.* + -ant¹.] A substance suitable for use in inoculation; *spec.* in *Metallurgy*, a substance with which molten metal is inoculated.

1911 *Experiment Station Rec.* XLVII. iii. 234 State laws concerning the sale of seeds and legume inoculants. **1944** *Jrnl. Iron & Steel Inst.* CL. 144 A The second purpose was to measure the effect of increasing-percentages of steel and varying amounts of inoculants. **1960** *Ibid.* CXCV. 222/1 The use of graphitizing inoculants, usually based on silicides, is described. **1962** A. G. Guy *Physical Metall. for Engineers* v. 149 Typical inoculants are ferrosilicon and calcium-silicon, very small amounts of which are effective in reducing flake size.

inoculate, *v.* Add: **3.** For 'a person' read 'an individual' and add to def.: To introduce (cells or organisms to be cultured) *into* a culture medium or its container. (Further examples.)

1928 L. E. H. Whitby *Med. Bacteriol.* iii. 46 The loop or needle must be sterilized in the flame before being charged with the material to be inoculated. **1939** K. L. Burdon *Med. Microbiol.* xx. 273 At this temperature it [*sc.* agar] is still liquid and yet cool enough so that the organisms to be inoculated will not be killed by its heat. **1964** Wheeler & Volk *Basic Microbiol.* viii. 89/1 The pour-plate method consists of inoculating the mixed culture into a test tube containing melted agar. **1970** Passmore & Robson *Compan. Med. Stud.* II. xviii. 50/1 When bacteria are

inoculated into a fresh fluid medium there is little or no increase in their number for a period.

b. Add to def.: To introduce infective material into (a plant) or cells or organisms for culture into (a culture medium or a vessel containing one). (Further examples.)

1886 H. M. Biggs tr. *Hueppe's Methods Bacteriol. Investigation* iv. 171 It [*sc.* the nutrient solution] is then inoculated with a few drops of the mixture of bacteria to be tested. **1920** E. F. Smith *Introd. Bacterial Dis. Plants* ii. 112 In studying a particular disease, the student will.. seek to inoculate those parts of the plant which naturally develop the disease. **1925** [see *inoculum]. **1933** K. M. Smith *Rec. Adv. Study Plant Viruses* viii. 196 The plant tissue is macerated..and then rubbed over the foliage to be inoculated by means of the swab. **1956** *Nature* 18 Feb. 302/2 When White Burley tobacco seedlings in a glasshouse are inoculated with the virus from cowpea..they become systemically infected. **1962** F. J. Baker *Handbk. Bacteriol. Technique* v. 56 When inoculating broth tubes, care must be taken not to spill the contents.

4. *Metallurgy.* To add a small quantity of some substance to (metal, esp. iron, about to be poured) in order to produce a smaller grain size or otherwise to modify the microstructure of the cast metal.

1931 *Proc. Inst. Brit. Foundrymen* XXIII. 96 Small grain-size and high density can be..achieved by 'dosing' or inoculating the alloy so that prior to the main solidification taking place the alloy contains numerous evenly and finely-dispersed nuclei to form centres of crystallisation. **1933** *Jrnl. Iron & Steel Inst.* CXXXVIII. 640 A bath of iron which would normally cast white is prepared and is 'inoculated', by the addition of suitable proportions of nickel and silicon, to cause graphitisation. **1963** C. H. Samans *Metallic Materials Engin.* vi. 316 When it is desired to improve the structure of the cast iron and, consequently, its mechanical properties, the metal often is inoculated just before pouring. **1971** *Daily Tel.* 4 Nov. 5 (Advt.), Semi-continuous casting is standard practice in the [aluminium] industry and, unless the melt is inoculated or grain-refined to produce a fine-grained equiaxed structure, the process has an inherent tendency to grow massive columnar crystals.

inoculated, *ppl. a.* Add: **3.** *Metallurgy.* Applied to cast iron whose properties have been improved by inoculation.

1932 *Proc. Inst. Brit. Foundrymen* XXIV. 137 What the author says about the germ theory and inoculated irons is exceedingly interesting. **1956** *Jrnl. Iron & Steel Inst.* CLXXXIV. 89/2 The inoculated cast irons are more elastic, they have smaller damping capacities, and their electrical and magnetic properties are close to those of steels.

inoculation. Add: **2.** **b.** Also, the (usually intended) introduction of infective material into a plant or of cells or organisms to be cultured into a culture medium. (Further examples.)

1886 H. M. Biggs tr. *Hueppe's Methods Bacteriol. Investigation* iv. 160 Inoculations are made by picking out, with a platinum needle..a particle from a pure culture.. and introducing it quickly into the solution. **1910** Hiss & Zinsser *Text-bk. Bacteriol.* viii. 141 For the inoculation of solid media and the making of stab cultures, a straight 'needle' or wire should be used. **1920** E. F. Smith *Introd. Bacterial Dis. Plants* iii. iii. 165 The inoculations may be made by spraying or by touching the leaf-tip with an infected platinum needle. **1933** K. M. Smith *Rec. Adv. Study Plant Viruses* viii. 194 The second method of artificial virus transmission is that of inoculation, using this term in its restricted sense of actual application of the virus-containing sap to the plant tissue. **1958** Pelczar & Reid *Microbiol.* viii. 81/1 After inoculation of the media.. and following a suitable period of incubation, it is possible to determine the cultural characteristics of the organism being studied.

2*. *Metallurgy.* The addition of an inoculant to molten metal, esp. iron (see *inoculate *v.* 4).

1932 *Proc. Inst. Brit. Foundrymen* XXIV. 122 After remelting the graphite particles were restored artificially, when the metal solidified pearlite-flake graphite. The process of putting graphite back into the melt is conveniently referred to as 'inoculation'. **1950** *Jrnl. Iron & Steel Inst.* CLXVI. 260/1 D. J. Reese gave some details of the production and properties of nodular cast iron produced in the U.S.A., using the technique of magnesium inoculation. **1963** B. Harocopos *Princ. Struct. Metall.* viii. 104 The graphite may also be made to appear as spheroids in a ferritic matrix by inoculation with an alloy of silicon, magnesium and zirconium..and the resulting iron has a U.T.S. of 32 t.s.i., elongation 15%, in the annealed state. **1968** E. N. Simons *Outl. Metall.* iv. 88 In the inoculation of cast iron in Britain calcium silicide is added to the ladle or the cupola spout at the rate of about 120 oz./ton.

inoculist. (Earlier U.S. example.)

1776 in *Narragansett Hist. Reg.* (1886–7) V. 352 Practitioners and Inoculists may be served with them [*sc.* medicines], if applied for soon.

inoculum (inǫ·kiŭlŏm). Pl. **inocula.** [mod.L., f. L. *inoculāre* to Inoculate *v.*] (A quantity of) infective material used for or capable of inoculating an organism or culture medium.

1902 J. W. H. Eyre *Elem. Bacteriol. Technique* xv. 267 (*heading*) The preparation of the inoculum. **1925** C. H. Browning *Bacteriol.* iii. 47 The needle, charged with the inoculum, is introduced into the tube and..the broth..is now said to have been 'inoculated'. **1949** N. G. Heatley in H. W. Florey et al. *Antibiotics* I. iv. 212 The reason why small inocula (down to single cells) do not normally grow out was thought to be due to the inhibitory power of a number of substances. **1950** J. C. Walker *Plant Path.* viii. 347 Since the fungi overwinter on debris, sanitary measures are helpful in reducing spring inoculum. **1970** W. H. Smith *Tree Path.* xxiii. 243 Early in an epidemic.. the absolute rate of increase of disease is small,..as only a limited quantity of inoculum is available. **1973** *Nature* 16 Feb. 456/1 It is highly tumorigenic (inocula of 10^2 cells produce fatal tumours in the animals).

inodorous, *a.* Add: **b.** [In-³.] Malodorous; having an unpleasant smell. Also *fig.* or *transf.*

1823 *Reading Mercury* in *Spirit of Public Jrnls. M.DCCC.XXIII* (1825) 206 The smell at this time was certainly somewhat inodorous, and was like almost any thing but a nosegay. **1858** *Sat. Rev.* 27 Mar. 308/1 Cerebrated in inodorous fallacies, he has gone to corrupt amid old arithmetical dross and the rotting refuse of deceased paradoxes. **1861** C. M. Yonge *Young Step-Mother* viii. 98 Brown, inodorous materials for petticoats, blouses, and trowsers.

inoperancy (inǫ·pərănsi). *rare*⁻¹. [f. In-³ + Operancy.] Failure to operate or function.

1936 T. S. Eliot *Coll. Poems 1909–1935* 189 Evacuation of the world of fancy, Inoperancy of the world of spirit.

inoperculate, *a.* **2.** *Bot.* Of an ascus or sporangium: lacking an operculum and therefore opening by splitting. Also as *sb.*, a fungus having this characteristic.

1879 W. Phillips tr. E. Boudier in *Trans. Woolhope Naturalists' Field Club* (1887) 202 The second [section of the family] I would call *Inoperculate Discomycetes*, or simply *Inoperculæ*, because the exit of the sporidia takes place by a small hole, formed at the extreme summit of the asci, with its margin more or less elevated, but without any appearance of an operculum. **1913** *Trans. Brit. Mycol. Soc.* IV. 402 In the inoperculate species, the spores have a tendency to septation. **1929** *Trans. Brit. Mycol. Soc.* XIV. 267 The epithecium, overlying the hymenium ..occurs in several inoperculate genera. **1943** *Mycologia* XXXV. 585 The outstanding characters of this Panamanian discomycete are unlike those of any of the stromatic inoperculates. **1950** E. A. Bessey *Morphol. & Taxon. Fungi* iii. 45 In this order [*sc.* Chytridiales] the zoospores or motile gametes escape through an exit papilla or tube whose apex softens and permits the motile cells to push out (the inoperculate series), or they escape through a sort of cap that opens like a trap door, the so-called operculum (the operculate series). **1970** J. Webster *Introd. Fungi* i. 25 In the inoperculate chytrids..the sporangium forms a discharge tube which penetrates to the exterior of the host cell.

inordinacy. (Later example.)

1943 C. Brogan *Who are 'The People'?* xi. 119 It is merely another instance of inordinacy.

inordinancy. For † *Obs.* read *rare.* Add earlier and later examples.

a **1617** P. Baynes *Christian Lett.* (1620) sig. N7ᵛ, No such inordinancy of griefe. **1955** R. Niebuhr *Self & Dramas of Hist.* (1956) 29 Hobbes..could regard the..reason ..as the cause of the inordinancy of human ambitions.

inordination. Delete † *Obs.* and add later example.

1883 G. M. Hopkins *Sermons & Devotional Writings* (1959) 133 The inordination, ugliness of sin in the frame and world of Creator and creature.

inorganic, *a.* Add: **B.** *sb.* An inorganic chemical.

1945 *Chem. & Engin. News* 10 Jan. 103 We offer: inorganics—gallium, germanium, indium metals & salts in quantity. **1968** *New Scientist* 23 May 391/1 The production of inorganics was complemented by the equally distinct manufacture of organic chemicals. **1971** *Nature* 31 Dec. 515/1 Plastics increased in volume by 16·8 per cent a year; basic organics by 13·7 per cent and inorganics by 6·1 per cent.

inositol (inōᵘ·sitǫl). *Biochem.* [f. Inosit(e + -ol.] Modern name of Inosite: any of the nine stereoisomers of hexahydroxycyclohexane, $(CHOH)_6$; *spec.* that isomer (also called *meso-inositol* or *myoinositol*) which is a member of the vitamin B complex, occurs in many animal organs esp. muscle and (often as its hexaphosphate) in plant leaves and seeds, and promotes the growth of bacteria and yeasts.

1891 Roscoe & Schorlemmer *Treat. Chem.* (new ed.) III. 214 Inositol forms large transparent monosymmetric crystals, which possess a sweet taste and are soluble..in about 6 parts of water, yielding an optically inactive solution. **1936** A. P. Mathews *Princ. Biochem.* v. 69 Inositol is found in wheat bran and in many vegetables, where it occurs often in the form of 'phytin', which is the hexa-phosphoric acid-ester of inositol. **1941** *Jrnl. Biol. Chem.* CXL. 465 The ability of substances related to mesoinositol to replace this compound in the nutrition of the mouse and of yeast has been examined. It was found that *d*-inositol, *l*-inositol..[etc.] were inactive for both species. **1948** *Adv. Carbohydrate Chem.* iii. 45 The comparative rarity of the other inositols has justified the general retention of the name inositol to denote *meso*-inositol. **1954** Cantarow & Schepartz *Biochem.* vi. 214 No manifestations attributable to inositol deficiency have been recognized in man. This may be due, in part at least, to its synthesis by intestinal bacteria. **1970** R. W. McGilvery *Biochem.* xxiv. 599 Phosphatidylinositides are compounds containing residues of inositol, a structural isomer of glucose that can be made from glucose-6-phosphate as inositol-1-phosphate.

inotropic (əin-, inotrōᵘ·pik, -trǫ·pik), *a.* *Physiol.* [ad. G. *inotrop* (T. W. Engelmann 1896, in *Arch. f. ges. Physiol.* LXII. 555): see Ino- and *-tropic.] Modifying the contractility of muscle.

1903 *Sci. Amer. Suppl.* 4 July 22992/3 The author describes..as inotropic such [influences] as lessen or destroy contractibility. **1971** *Nature* 25 June 531/1 An automated system for monitoring both the chronotropic and inotropic effects of chemical and physical agents.

So **ino·tropism**, modification of the contractility of muscle.

1902 [see *dromotropic *a.*]. **1971** *Nature* 25 June 531/1 Although inotropism is defined as a change in force, by measuring changes in optical density, we can measure indirectly the relative elongation and contraction of the cells during each contraction cycle.

in pari materia: see *in *Lat. prep.* 21.

in pectore: see *in *Lat. prep.*

in-people: see *in *a.* 2.

in-person *attrib.*: see *in *prep.* 17*.

in petto: see Petto.

in-phase (stress variable), *attrib. phr. Electr.* [f. phr. *in phase* (phase *sb.* 3).] That is in phase; of or pertaining to signals that are in phase.

1914 H. Pender *Amer. Handbk. Electr. Engineers* 1297 The active or in-phase component of the current in a circuit is that component which is in phase with the voltage across the circuit. **1940** *Amat. Radio Handbk.* (ed. 2) xii. 186/1 Besides the desired equal and opposite currents in the two wires, there are two more in-phase currents flowing in the parallel circuit. **1962** A. Nisbett *Technique Sound Studio* 264 If a monophonic signal is fed to an in-phase pair [of loudspeakers] the sound will appear to come from between and behind them. **1968** *Brit. Med. Bull.* XXIV. 253/2 Special techniques, such as in-phase rejection,..were adopted in its design to overcome interference from extraneous voltages.

in-pig (in,pi·g), *a.* [attrib. use of phrase *in pig*: see In *prep.* 10 b.] Of a sow: that is in pig; pregnant. Cf. In-calf *a.*, *in-foal *a.*

1950 *Farming* June 179 In-pig sows can of course receive a proportion of their ration as bulky food. **1959** *Times* 31 Aug. 5/3 For the first time since the..national in-pig and litter tests began..the three major awards.. have been won by the same sow. **1969** *Times* 6 Jan. 7/7 The December sample figures for England and Wales indicate a slight cutback south of the Border since September in in-pig gilts, but the number of sows in pig is still rising. **1974** *Listener* 21 Mar. 334/2 Philip had just lost all his pigs from swine vesicular disease, and was meditating buying in-pig gilts and starting again.

in-pile, in-plant, in-process *attrib.*: see *in *prep.* 17*.

in potentia: see Potentia (in Dict. and Suppl.).

in propria persona: see *in *Lat. prep.*

input, *sb.* Restrict *Sc.* to sense in Dict. and add: **2.** **a.** That which is put in or taken in, or which is operated on or utilized by any process or system (either material or abstract).

1893 *Phil. Trans. R. Soc.* B. CLXXXIII. 228 The pressure in the large systemic veins becomes raised during vagus action, because the quantity of blood which passes from them into the right ventricle (which we may refer to as the 'input' of the heart), in a given time, is diminished. **1929** *Prosp. Mona Copper Co.* 6 With Copper at £70 per ton on the basis of an input of 600 tons daily..a gross annual profit of about £100,000 could be expected. **1966** T. Lupton *Managem. & Social Sci.* iv. 88 To Rice..an organization is an 'open system'. It takes in inputs from the environment, converts them, and sends outputs back into the environment. **1971** *Sci. Amer.* Sept. 111/1 These carcasses are retrieved in the spring, and the meat is considered one of the more flavorsome food inputs. **1971** J. B. Carroll et al. *Word Frequency Bk.* p. vi, The Corpus is drawn from written, and edited, published materials. There is no direct oral input.

b. Energy supplied to a device or system; *spec.* an electrical signal that enters an electronic device.

1902 *Encycl. Brit.* XXVII. 574/2 The useful return or 'output' at the terminals of a large machine may amount to as much as 95 per cent. of the mechanical energy which forms the 'input'. **1931** *Daily Express* 21 Sept. 7/4 An advanced form of band-pass tuning, providing a special selector circuit between the input from the aerial and the first valve of the set. **1933** E. T. A. Rapson *Electr. Transmission & Distrib.* xiv. 163 Accurate metering of the power input to and output from the Grid is essential. **1943** C. L. Boltz *Basic Radio* xiv. 216 The input is applied between control grid and cathode. **1970** J. Earl *Tuners & Amplifiers* vi. 142 This same switch also receives an input from the a.m. section.

c. *Econ.* The total of resources necessary to production, including raw materials, use of

machinery, and manpower, which are deducted from output in calculating assets and profits. (Cf. *output.) Also *attrib.*
1926 J. D. Black *Production Economics* III. xi. 277 The term input..will be used..to refer to the amounts of the production elements that are used in turning out any product... In the present illustration, as the inputs of seed increase,..the outputs of grain increase. **1947** *Bull. U.S. Bureau of Labor Statistics* No. 913. 11 Most persons, in using the term 'productivity', have meant the physical output obtained for a given physical input. **1953** Steiner & Goldner *Productivity* ii. 5 What do we mean by input? A typical product is a combination of raw materials, machinery, workers' time, power, and many other factors. Each of these is called an input. Input items are combined in the manufacturing process into products or output. Should the unit of input be one worker, or one hour of labor time, or one machine, or a ton of raw materials or a kilowatt hour of electricity? Any of these could be an input although each is different. **1958** *Economist* 15 Nov. 592/3 When the effect of other variables has been allowed for, the farmer is found to be using, even at low inputs, 4s. worth of concentrates to produce a gallon of milk which he sells for 3s. 1½d... Concentrate inputs beyond £45 per cow per year..have no additional effect..upon milk yield. **1959** *Oxf. Univ. Gaz.* 16 Mar. 796/1 The farmer also uses up large quantities of 'industrial inputs' (equipment, motor fuel, fertilizers, &c.), representing goods and services which could, directly or indirectly, have been exported if the British farmer had not used them, or which, in some cases, have to be imported. **1971** [see *input price* (5 a below)]. **1972** *VAT: Gen. Guide* (H.M. Customs) 16 Those goods and services are called his inputs, and the tax on them is his input tax. **1972** *Accountant* 13 Apr. 471/1 Historically, Britain's indirect taxes had been collectible at a single point and from a restricted clientele; VAT, on the other hand, would be all pervasive. For any person receiving taxable 'inputs', zero-rating would be found preferable, 'if at all possible' to exemption. *Ibid.* 28 Sept. 402/3 Companies with only internal transactions should ensure that they did not lose relief for VAT suffered (input tax).

d. *Computers.* Data or program instructions that are fed into or processed by a computer; also, the physical medium on which these are represented.
1948 *Math. Tables & Other Aids to Computation* III. 7 The 'input' for a computational problem (i.e., the information available before the start of the computation) consists of two kinds of elements: numbers, and 'orders'. *Ibid.* 9 The tapes which contain the input for any problem are classified into three groups. **1949** D. R. Hartree *Calculating Instruments & Machines* (1950) vii. 80 Input and output for this machine are expressed in standard teletype code, with a coded symbol for the operation required. **1964** A. Lytel *Fund. Data Processing* viii. 165 Punched paper tape can be read and used as a computer input. **1967** D. Wilson in Wills & Yearsley *Handbk. Management Technol.* iii. 45 It is sometimes necessary to obtain a detailed listing of all the input to determine where the error has occurred. **1973** *Time* 13 Aug. 20/2 Business gave its own donation at the office, with the computer talk of 'inputs'..and 'print-outs'.

e. *Psychol.* The resources of mental and sensory stimuli available to an individual.
1954 *Canad. Jrnl. Psychol.* VIII. 70 The maintenance of normal, intelligent, adaptive behaviour probably requires a continually varied sensory input. **1959** *Amer. Jrnl. Psychiatry* CXV. 1110/1 These studies suggest that maintaining adequate sensory input during space missions will be less of a problem than providing adequate information input. **1972** *Jrnl. Social Psychol.* LXXXVI. 220 Individuals who can tolerate diverse inputs from the environment may not be markedly affected by success or failure.

f. *Linguistics.* (See quot. 1966.) Freq. as *input string.*
1961 H. A. Gleason *Introd. Descr. Ling.* (rev. ed.) xii. 173 It is normally stated in the form of rules which may be applied to one of the pair—an *input*—altering it to produce the other—an *output*. **1966** M. Pei *Gloss. Ling. Terminol.* 126 Input, in transformational grammar, the term applied to a construction that is transformed into another..which is called the *output* (input: 'he goes'; output: 'he does go', 'he does not go', etc.). **1969** W. A. Cook *Introd. Tagmemic Analysis* ii. 42 The transformational rule is simply a rule of change. This rule has an input string, a rule of change, and an output string. With kernel sentences as input, it is possible to set up a series of optional rules that will produce the output, the derived sentences. **1971** R. Fowler in *Archivum Linguisticum* II. 136 These rules..are typical of the local transformations which follow base constituent-structure rules on the present grammatical model. Their inputs and outputs are concatenated sets of syntactic..features, and their effect is to replace or add one feature in one set.

3. A place where, or device through which, an input enters a system, esp. an electronic device.
1929 J. H. Morecroft *Elem. Radio Communication* vii. 228 Either of these..would give a beat frequency of 50 kc., which is then 'detected' and supplied to the 'input' of the I.F. amplifier. **1933** *Boys' Mag.* XLVII. 108/2 Connect pick-up to 'input' and loud speaker to 'output'. **1946** *Math. Tables & Other Aids to Computation* II. 100 A flip-flop has two inputs and two outputs. **1963** Gould & Ellis *Digital Computer Technol.* iv. 33 Data passing from the input to the computer proper, or from this to the output, can be marshalled, sorted and coded..to a large extent independently of the rest of the equipment. **1971** *Hi-Fi Sound* Feb. 105/1 This recorder has inputs for microphone, radio and magnetic and/or ceramic pickup cartridges.

4. The action or process of putting in or feeding in.
1947 *Math. Tables & Other Aids to Computation* II. 356 No means of numerical input or output other than the keyboard and the display panel are provided. **1948** *Ibid.* III. 7 The speed of input is well in balance with the computing speed. **1955** *Sci. Amer.* Jan. 69/3 If a block of iron were magnetized as a single large domain..it..would require the input of a considerable amount of energy. **1964** T. W. McRae *Impact of Computers on Accounting* i. 15 The basic idea behind this method of input is to print the characters on the original document in a special type of magnetic ink. **1973** *Nature* 13 Apr. 440/1 What is lacking ..is a steady input of information on research and development on other fuels.

5. *attrib.* and *Comb.* **a.** simple attributive, as *input circuit, device, impedance, price, routine, tape, terminal, transformer, unit.* **b.** constituting input, as *input current, data, information, signal, voltage.*
1921 *Wireless World* 25 June 214/2 The input circuit varies according to the receiver circuit to which the connection is made. **1940** *Amat. Radio Handbk.* (ed. 2) iv. 69/1 The first general axiom is to arrange matters so that the output stages are as far as possible from the aerial or input circuits. **1930** Field & Weill *Electro-Plating* 38 Fig. 11, showing the form of the input and output currents. **1948** *Math. Tables & Other Aids to Computation* III. 7 The Harvard machines use punched cards for most input data. **1948** *Ann. Computation Lab. Harvard Univ.* XVI. 248 The design of input and output devices for electronic digital computers poses a specialized problem in electrical communications. **1968** *Brit. Med. Bull.* XXIV. 191/1 The basic configuration of any computer consists of a store, a suitable input and output device, and a control mechanism. **1928** *Times* 23 Mar. 20/1 A certain input impedance which had the effect of increasing or decreasing the resistance of the tuned circuit. **1949** E. C. Berkeley *Giant Brains* x. 175 Since the input information must be carefully verified, we shall need a second magnetic-tape device. **1956** Input information [see *buffer² 1 d]. **1971** *Daily Tel.* 19 May 17/2 Input prices—the price of basic materials and fuel—have risen by more than 3 p.c. since the start of the year. **1954** *Math. Tables & Other Aids to Computation* VIII. 32 The function of the Ferut Input Routine is to read information from tape, perform certain alterations on routines or numerical data and store the routines or data in assigned locations in the machine. **1962** *Gloss. Terms Automatic Data Processing* (B.S.I.) 42 *Input routine*, a routine, sometimes stored permanently in the computer, to control the readings of programs and data. **1950** *Mind* LIX. 440 It will seem that given the initial state of the machine and the input signals it is always possible to predict all future states. **1949** E. C. Berkeley *Giant Brains* iii. 27 We shall need one register to read the input tape and to store the number or operation recorded on it. **1919** *Wireless World* Dec. 505/1 It will ..be convenient to provide two 'input' terminals, two terminals to which the..accumulator may be connected, and two for..an external plate battery. **1946** Input terminal [see *control *sb.* 3 f]. **1919** *Wireless World* Dec. 506/1 The leaky grid condenser..is brought into use, the input transformer..being isolated. **1962** *Gloss. Terms Automatic Data Processing* (B.S.I.) 81 *Input unit*, that portion of an a.d.p. system used only for input. **1966** B. A. M. Moon *Computer Programming* i. 6 Typical input units are punched card, paper tape and magnetic tape units. **1940** *Amat. Radio Handbk.* (ed. 2) ii. 28/2 When an A.C. input voltage is applied to the grid of a valve an amplified A.C. voltage appears across the output load in the anode circuit.

c. Also **input–output** (*input-output, input/output*), usu. *attrib.*
1914 H. Pender *Amer. Handbk. Electr. Engineers* 961 With large motors it is desirable to use a generator as a load in making an input-output test. **1947** *Math. Tables & Other Aids to Computation* II. 363 Another important element of the automatic computer which affects the compromise is the input-output mechanism. **1953** *Economist* 26 Sept. 870/1 An input-output analysis of the British economic structure..will be ready by 1956. *Ibid.* 870/2 Mr. Roger Keyes..recently ordered the suspension of work on American input-output data; the colleagues he left behind..used to share his mistrust of input-output. **1964** Gould & Kolb *Dict. Social Sci.* 452/2 Input–output tables show the interrelations among the major industry groups of the economy... Tables or matrices are constructed which show the goods-and-services inputs and outputs of each on a 'from-whom to whom' basis. **1964** T. W. McRae *Impact of Computers on Accounting* ii. 38 The introduction of magnetic tapes..facilitates the compact storage and fast input–output of large files. **1967** *Technology Week* 23 Jan. 11/1 (Advt.), Sigma 5..does foreground real-time control..and high-speed input/output. **1970** *Sci. Amer.* Oct. 94/2 Ordinarily it is difficult to measure the input-output relations of an ecosystem, particularly those involving nutrients. **1970** *Sunday Times* 29 Nov. 68/6 Input-output tables are brought into play to help estimate that trading profits per vehicle exceed manual pay for the men who make it.

input, *v.* Restrict †*Obs.* to senses in Dict. and add: **3.** (Stressed **i·nput**.) *Computers.* To supply or feed in (data, a program, etc.) *to*; to feed into. Pa. pple. **i·nput,** (less commonly) **i·nputted.**
1946 *Nature* 12 Oct. 503/2 These switches are connected up so that for any two-figure argument *x* from 00 to 99, input to the function table, the value of the function for that argument is output in the form of pulse groups on the appropriate digit lines. **1953** A. D. & K. H. V. Booth *Automatic Digital Calculators* viii. 62 It is possible to input up to 300 decimal digits per second. *Ibid.* 74 Working instructions do not have to be input after each shut-down. **1964** F. L. Westwater *Electronic Computers* vi. 104 Magnetic tape provides a fast means of inputting information. **1965** K. Nicol *Elem. Programming* iii. 14 When all your program has been input and compiled control is transferred from the compiler program to the machine code instructions of your program. **1967** W. F. Bauer in W. J. Karplus *On-Line Computing* iv. 81 He inputs the data directly into the console by electronically filling out a form which the computer provides on the cathode-ray tube. *Ibid.* 80 Data verification is done by the computer's reflecting back to the user on the cathode-ray-tube scope exactly what has been inputted. **1968** *Brit. Med. Bull.* XXIV. 222/1 At convenient intervals, the day lists are input to the computer. **1970** O. Dopping *Computers & Data Processing* xv. 230 A series of records which are to be input to a computer or have been output from it is called a file.

inquest, *sb.* Add: **2. b.** *grand inquest*: also *transf.*
1903 *Daily Chron.* 3 July 7/1 The First Lord of the Admiralty..described it as a grand inquest of the nation.
3. c. Delete 'Now *rare*'. Now used *colloq.*, a discussion or investigation of a game, event, etc., after it has taken place.
1932 *News Chron.* 29 Feb. 8/5 She never in any case holds inquests. You can't make the next shot [in Golf] good by worrying over the last. **1934** *Punch* 3 Jan. 22/3 My intention was to wait for the inevitable inquest and then say..'I don't play much bridge you know.' **1967** J. Symons *Man who killed Himself* I. i. 12 'What made you double that heart call?' Clare asked... Mr. Payne wagged a finger. 'Now now. No inquests.' **1970** *Times* 20 Apr. 1/5 The Apollo 13 astronauts..tonight left Hawaii to return to Houston where they will soon begin the long inquest into the spacecraft failure.

inquiline, *sb.* (*a.*) **1. a.** Delete †*Obs.* and add later example.
1914 C. Mackenzie *Sinister St.* II. IV. iv. 926 Half the inquilines of a night and even some of the less transient lodgers ultimately escaped owing her money.
3. *attrib.* or as *adj.* (Later example.)
1958 *Times Lit. Suppl.* 30 May 299/2 The inquiline figures painted so vigorously by Sir Osbert come to life with extraordinary clarity.

inquinate, *v.* Delete †*Obs.* and add example. Hence **i·nquinating** *ppl. a.*
1914 C. Mackenzie *Sinister St.* II. IV. iv. 914 Street followed street, each one..being a little less able to resist the corrosion of a persistently inquinating migration. **1918** H. W. Steed in *19th Cent.* Dec. 987 It is surely enough that some of these stipulations should..have needlessly protracted the War, without their being allowed to inquinate the peace.

inquiration (inkwəiərēi·ʃən). *dial.* [f. Inquire *v.* + -ation.] Inquiry.
1789 C. Smith *Ethelinde* III. 136 If so be you would acquaint me where I may make enquiration after your frinds or hern, it would come to the same thing. **1839** C. Clark *John Noakes* v, If they their inquirations make In winter time, some will Condemn that place as no great shakes. **1850** Dickens *Dav. Copp.* li. 516 A decent woman as spoke to her about..making secret inquiration concerning of me and all at home. **1886** Hardy *Mayor Casterbr.* II. xiv. 196 'Suppose we make inquiration into it, Christopher,' continued Longways; 'and if we find there's really anything in it, drop a letter to them most concerned.'

inquirendo. Add: **2.** An investigation.
a **1846** G. Darley in B. W. Procter *Procter* (1877) 286 Confound your prose lunatics who leave you no time for inquirendos upon poetic ones! **1897** L. I. Guiney (*title*) Patrins, to which is added an Inquirendo into the Wit and Other Good Parts of His Late Majesty King Charles the Second.

inquiry, enquiry. Add: **4.** (Earlier and later examples of *inquiry room.*) Also *inquiry agency, agent.*
1850 J. Gallaher *Western Sk.-Bk.* 243 He came into the inquiry room, and told me..that he had been living in sin. **1892** Kipling & Balestier *Naulakha* xvii. 204 See here, young woman, do you run a private inquiry agency? **1901** *Daily News* 5 Feb. 4/4 The Rev. F. B. Meyer..hoped that each church would continue its..band of inquiry-room workers. **1922** *Kelly's Directory Liverpool* 1181/3 Ramage & Kelly private inquiry agents. **1971** K. Gottschalk in B. de Ferranti *Living with Computer* iv. 31 Society is being confronted with..ultraviolet photography practised by inquiry agents.

in re: see *in *Lat. prep.*

i·n-re:ference. [cf. *in *a.* 2.] A reference understood by only a limited group of people. Cf. *in-joke.
1967 *Punch* 15 Feb. 243/2 The pieces..have clearly journeyed from a lost civilisation, although..in-references to..period West End lions, have been omitted. **1968** *Listener* 26 Sept. 421/2 Peel's linking comments are liberally sprinkled with in-references to musicians and to long-playing records destined, one suspects, for infinitesimal sales.

in-rigger (i·nri:gəɹ). [f. In *adv.* + Rigger¹ 4.] A boat having the rowlocks formed in the gunwale.
1893 J. H. Clasper in *Westm. Gaz.* 9 Oct. 7/3 The Düsseldorf boat was an in-rigger—for which orders are very seldom given nowadays.

‖ **inro** (i·nro). [Jap., f. Chin. *yin* seal + *lung* basket.] An ornamental nest of boxes, connected by a thin cord, made of lacquer, ivory, or the like, in which seals, medicines, and other necessaries can be carried, formerly worn by the Japanese at the girdle.

1617 W. Adams *Let.* 10 Nov. in *Trans. Asiatic Soc. Japan* (1898) XXVI. 207 Your Inro or metsin boxe Skinro told me he would sent it me from Meaco. **1882** *Century Mag.* Dec. 228/2 Gilded pictures of wave, sky, cloud, field, and house, seen on box and tray, *inro* and scroll. **1911** *Connoisseur* Mar. 209/2 Among the *objets d'art* most associated with old Japan are the *inro*, or little medicine cases which the Japanese used invariably to carry about with them. **1960** *Times* 2 Jan. 9/4 Since these garments [sc. kimonos] were without pockets, the Japanese carried such belongings as ink, seals, and medicines in lacquer boxes called inros. **1971** *Times Lit. Suppl.* 20 Aug. 998/5 The variety and wit of the subjects used in inrō decoration defy description. **1972** *Country Life* 30 Nov. 1500/1 Carved netsuke—the Japanese toggles by which the inro was prevented from falling from the belt.

inrun, *sb.* Add: **2.** [tr. G. *anlauf.*] In ski-jumping, the distance from the start to the point of taking off; an approach trestle.
1949 F. Elkins in Elkins & Harper *World Ski Bk.* 103 A group of skiers were preparing their jumping hill for a meet when a small figure was seen leaving the top of the wooden inrun. **1963** *Amer. Speech* XXXVIII. 203 Some of the English terms are literal equivalents of terms used by German-speaking skiers and might be called loan translations...*inrun Anlauf.*

in-running, *ppl. a.* Add to def.: That runs into a river or the sea.
1931 *Times Lit. Suppl.* 1 Oct. 738/2 'Australian Literature'..must contribute to the great stream of English literature a new in-running river.

in-sack *attrib.*: see *in *prep.* 17*.

inscape (i·nskēip), *sb.* [Origin unknown; perh. f. In *adv.* 12 + Scape *sb.*³, or ad. Inshape.] The individual or essential quality of a thing; the uniqueness of an observed object, scene, event, etc. (see quots.). Hence **i·nscape** *v. trans.*; **i·nscaped** *ppl. a.*
1868 G. M. Hopkins *Jrnls. & Papers* (1959) 127 His [sc. Parmenides'] feeling for instress, for the flush and foredrawn, and for inscape is most striking. *Ibid.* 129 The way men judge in particular is determined for each by his own inscape. *Ibid.* 174 Two plants especially with strongly inscaped leaves cover the mountain pastures. *Ibid.* 177 The whole cascade is inscaped in fretted falling vandykes. **1879** —— *Lett. to R. Bridges* (1955) 66 Design, pattern, or what I am in the habit of calling 'inscape' is what I above all aim at in poetry. Now it is the virtue of design, pattern, or inscape to be distinctive and it is the vice of distinctiveness to become queer. **1886** —— *Let.* 7 Nov. (1956) 373 The essential and only lasting thing left out—what I call *inscape*, that is species or individually-distinctive beauty of style. **1919** R. Fry *Let.* 29 Apr. (1972) II. 450 His [sc. G. M. Hopkins's] aesthetic—his 'inscape'; that's what we are after, however much we miss it. **1938** D. Gascoyne *Hölderlin's Madness* 35 All is an inscape And yet separates Thus shelters the Poet. **1944** *Downside Rev.* LXII. 185 The prefix 'in-' of 'inscape' is the operative part. 'Inscape' is the perception that comes only with contraction to a point. The inscape of a scene is not its correspondence with an externally conceived pattern; it is that scene experienced as absolutely unique, knit together in that oneness which is nameable only by relation. **1944** [see *instress *sb.*]. **1945** C. Williams *All Hallows' Eve* vii. 113 He forgot Simon..he forgot Lester... The inscape of the painting became central. **1948** W. A. M. Peters *G. M. Hopkins* i. 1 'Inscape' is the unified complex of those sensible qualities of the object of perception that strikes us as inseparably belonging to and most typical of it, so that through the knowledge of this unified complex of sense-data we may gain an insight into the individual essence of the object. **1953** W. H. Gardner in G. M. Hopkins *Poems & Prose* 229 *Twindles*..a portmanteau word inscaping 'twists' and 'dwindles'. **1970** *Country Life* 26 Feb. 484/2 In Manchester there is the fabric of buildings and structures which contribute by their reality to the inscape of the place.

inscenation (insīnēi·ʃən). [prob. after G. *inszenierung.*] Theatrical representation, *mise en scène.*
1897 G. B. Shaw in *Sat. Rev.* 13 Nov. 514/2 Maeterlinck's plays, requiring a mystical inscenation in the style of Fernand Knopf, would be nearly as much spoiled by Elizabethan treatment as by Drury Lane treatment. **1900** W. A. Ellis *Life Wagner* I. I. iii. 69 Geyer took an active part in the inscenation of his piece, in which he himself played Painter Klaus. **1963** *Times* 1 Mar. 13/1 A new inscenation of *Trovatore.* **1971** *Times* 8 June 8/2 Britten's setting of the Chester miracle play, and Mr Graham's inscenation of it, strive to recover the rough-and-tumble primitive gusto.

i·nscientist. *nonce-wd.* [cf. Inscience.] A non-scientist.
1909 W. Tuckwell *Pre-Tractarian Oxford* vi. 150 He knew nothing of Science or of Microscopes... So he came to all our Meetings, the one avowed Inscientist amongst us.

inscribable, *a.* (Example.)
1879 A. Macfarlane *Princ. Algebra of Logic* 14 The characters 'regular' and 'inscribable in a circle'.

inscript (inskri·pt), *v. rare.* [f. L. *inscript-*, ppl. stem of *inscrībere* to Inscribe *v.*; or back-formation from Inscription.] *trans.* To inscribe. Hence **inscri·pted** *ppl. a.*
1923 *Public Opinion* 16 Feb. 155/3 The statement at the head of this article might usefully be inscripted in all Theological Halls. **1923** *Blackw. Mag.* Feb. 151/1 It does not require inscripted monuments to perpetuate the memory of British deeds.

inscription. Add: **9.** *inscription maritime* [Fr.], the French naval system of recruiting; a list of men who may be called to serve in the French navy.
1902 *Encycl. Brit.* XXVII. 499/1 This arrangement is purely for the embodiment of the men of the Inscription Maritime. *Ibid.* XXXI. 103/1 For the purpose of the Inscription Maritime the Newfoundland fisheries were kept up at considerable expense to the nations. **1905** *Westm. Gaz.* 3 Aug. 10/1 A system called 'maritime inscription', which..furnishes a contingent of about 4,700 naval recruits every year.

in se: see *in *Lat. prep.*

insect, *sb.* Add: **4. a.** *insect-drone, -repellent.* **b.** *insect control, -eater* (later examples); *insect-borne, -feeding, -pollinated, -proof* adjs.; *insect-like* adj. or adv. (examples).
1909 R. W. Boyce *Mosquito or Man?* iv. 23 It is Dr. Beauperthuy whom we must regard as the father of the doctrine of insect-borne disease. **1946** *Nature* 21 Dec. 913/1 Analogy with filariasis elsewhere would suggest that the infection is insect-borne. **1972** *Ibid.* 21 Jan. 135/2 To prevent the spread of insect-borne diseases. **1936** *Discovery* Feb. 44/1 The legal insistence on insect control is lax or non-existent until there is an actual outbreak of some pest causing serious financial loss. **1951** A. W. A. Brown (*title*) Insect control by chemicals. **1902** W. de la Mare *Songs of Childhood* 30 Is it for fear the birds are flown, And shrills the insect-drone? **1939** 'N. Blake' *Smiler with Knife* xi. 168 The insect-drone of a lawn-mower. **1908** *Westm. Gaz.* 22 Feb. 16/1 There is a class of small mammals, mostly of nocturnal habits, that come under the order of Insectivora, or insect-eaters. **1936** *Discovery* July 212/2 Bee-eaters, swallows, swifts, and other insect-eaters. **1909** *Westm. Gaz.* 23 Apr. 4/2 The migratory, insect-feeding birds from the South..begin their nesting work. **1772** G. White *Let.* 9 Mar. in *Selborne* (1789) II. xii. 147, I..believe that many of the swallow kind..do, insect-like and bat-like, come forth at mild times. **1929** D. H. Lawrence *Pansies* 118 Working men Pale and mean and insect-like, scuttling along And living like lice. **1930** R. Campbell *Adamastor* 55 Faint, insect-like and thin it came, The wistful sound those heroes made. **1911** F. O. Bower *Plant-Life* 96 In a family (Ranunculaceae) as a rule insect-pollinated. **1953** J. S. Huxley *Evolution in Action* i. 34 Insect-pollinated flowers. **1908** *Japan Chron.* 1 July 4/6 It [sc. a kind of paper] is said to be capable of being worked into all sorts of patterns, to be insect-proof and damp-proof. **1946** *Nature* 21 Sept. 417/2 Two insect-proof cubicles in the glasshouse were filled with healthy young turnip and Chinese cabbage plants. **1953** Scott & Fisher *Thousand Geese* v. 50 We had brought effective insect-repellents, so we were not much troubled by the biting elements of the insect population. **1971** L. Payne *Even my Foot's Asleep* xvi. 214 A musky, incense-type perfume..probably an insect repellent.

insectual (inse·ktiuăl), *a.* [f. Insect *sb.* + -ual as in *conceptual.*] Like an insect, small.
1912 Beerbohm *Christmas Garland* 61 That swarm of things insectual. **1965** *New Statesman* 26 Nov. 838/3 To him the attacks..are mere 'insectual backbiting'.

insecure, *a.* **1.** Delete †*Obs.* and add: *esp.* in *Psychol.*
1935 F. B. Holmes *Exper. Study Fears of Young Children* xiii. 278 Karl is very insecure and clings to adults. *Ibid.* 284 The fearful children were more frequently described as being dependent upon adults for help..and as appearing generally insecure. **1941** Pritchard & Ojemann in *Jrnl. Exper. Educ.* X. 114/1 The term 'insecurity' and its correlative 'desire for security' appear extensively in child development literature... We need methods by which we can discriminate between the relatively secure and the relatively insecure children. **1947** A. T. Jersild *Child Psychol.* (ed. 3) vii. 271 In a study..of children who were rated by their teachers as being 'insecure', it was found that such children..exhibited a greater tendency to be apprehensive. **1954** A. H. Maslow *Motivation & Personality* iii. 38 He would not have taken this attitude unless he felt rejected and disliked (insecure). **1960** R. D. Laing *Divided Self* iii. 44 The ontologically insecure person is preoccupied with preserving rather than gratifying himself. *Ibid.* 45 Three forms of anxiety encountered by the ontologically insecure person. **1967** M. Argyle *Psychol. Interpersonal Behaviour* i. 29 Individuals who are 'insecure', i.e. uncertain about how to evaluate themselves, are particularly anxious to receive approval from others. *Ibid.* vii. 126 Adolescents, who have only just formed a tentative self-image, are particularly sensitive to the reactions of others, and are 'insecure' in this sense. **1969** W. Mayer-Gross et al. *Clin. Psychiatry* (ed. 3) xi. 640 Any sudden change..may produce an emotional crisis, especially in the insecure or over-sensitive child.

insecurity. **1.** Delete † and add: *esp.* in *Psychol.*
1917 Glueck & Lind tr. *Adler's Neurotic Constitution* (1921) p. x, A sickly girl..in her consciousness of an unusual insecurity leans upon her father and in so doing strives to become superior to her mother. **1932** W. H. Burnham *Wholesome Personality* ix. 329 The one outstanding condition menacing the mental health of..every youth, is some form of the emotion of fear, if not acute fear, at least a sense of insecurity. **1937** K. Horney *Neurotic Personality* ii. 36 The inner insecurity expressed in this dependence on others is the second feature that strikes us in neurotics on surface observation. **1942** E. Fromm *Fear of Freedom* v. 178 The automatization of the individual in modern society has increased the helplessness and insecurity of the average individual. **1942** B. Klopfer *Rorschach Technique* III. x. 240 Rorschach signs of insecurity and anxiety, as opposed to signs of a balanced personality structure. *Ibid.* 241 Often the language used by the subject..reveals to some extent open insecurity or anxiety. **1969** R. C. Carson *Interaction Concepts of Personality* ii. 32 Beyond infancy the experience of anxiety..has the character of a drop in self-esteem or an increase in felt insecurity. **1969** I. Stevenson *Psychiatric Exam.* iv. 62 Insecurity accompanied by a need to impress the examiner..may also lead the patient to bring forth unnecessary details. **1971** G. E. Gardner *Emerging Personality* III. vi. 132 First of all, a higher level of anxiety, a greater feeling of insecurity, often appears to beset the nonlearning boy.

inseity: see in se s.v. *in *Lat. prep.*

inselberg (i·ns-, i·nzəlbəɹg). *Geomorphol.* Also Inselberg. Pl. -bergs, -berge. [G., lit. 'island mountain'.] An isolated hill or mountain which rises abruptly from its surroundings, typically a plain in a hot, dry region.
[**1898** W. Bornhardt in *Zeitschr. der deutschen geol. Ges.* (*Verhandl.*) L. 71 Näher und ferner in ganz unregelmässiger Vertheilung erheben sich aus ihr die merkwürdigen Inselberge.] **1907** *Q. Jrnl. Geol. Soc.* LXIII. 166 Except around a few clustered 'island-hills' (Insel-berge) the drainage-gradients throughout this great basin are peculiarly low. **1913** *Rep. Brit. Assoc. Adv. Sci. 1912* 476 A striking feature of these kopjes and inselberge is that they rise at intervals from an apparently level or undulating plain. **1918** *Q. Jrnl. Geol. Soc.* LXXIV. 34 The bare rounded inselberg of gneiss stands alone, like an island in an undulating sea of vegetation. **1937** Wooldridge & Morgan *Physical Basis Geogr.* xx. 310 The association of Inselberge and flat rock-plains, with or without a thin veneer of sand or gravel, is..an established fact of observation. **1954** W. D. Thornbury *Princ. Geomorphol.* xi. 295 Others have applied inselberge rather indiscriminately to any island-like hill which stands conspicuously above its surrounding, such as the so-called sugarloafs of tropical rainy climates. **1960** B. W. Sparks *Geomorphol.* xi. 257 Many African landscapes..consist of a series of isolated, steep-sided inselbergs rising from an almost flat plain. **1969** Bennison & Wright *Geol. Hist. Brit. Isles* xii. 272 Two important features of Triassic geography were the hills of Charnwood and the Mendips. These were inselbergs rising above the general level of the Triassic landscape but gradually buried as the deposits accumulated.

inseminate, *v.* Add: **2.** To impregnate with semen, by natural or artificial means.
1923 *Vet. Jrnl.* LXXIX. 172 The application of artificial insemination on fox-farms could give a new impetus to the development of this industry... One male can easily inseminate twenty and more females. **1943** *Lancet* 7 Aug. 176/2 To avoid the dangers of in-breeding..the Jockey Club allowed no thorough-bred to be inseminated except by the covering stallion. **1958** *Times* 17 Nov. 15/6 They will now go to A.I. centres and will be used to inseminate a number of Friesian..cows.

insemination. Add: **c.** = **artificial insemination.*
1923 *Vet. Jrnl.* LXXIX. 171 Apart from this kind of artificial insemination of females with 'natural sperm'..it is also necessary to indicate the possibility of insemination with the so-called 'artificial sperm'. **1944** *Jrnl. Obstetr. & Gynaecol.* LI. 527/1 One donor was usually used for 4 to 8 inseminations carried out within short intervals from each other. *Ibid.* 528/1 One single insemination with spermatozoa from a donor was followed by pregnancy. **1959** *Chambers's Encycl.* I. 652/1 When the husband is sterile and the wife fertile, insemination with semen obtained from another donor has been used with success. **1974** *Times* 21 Jan. 14/5 Increased beef inseminations may be connected with the subsidized switch out of dairying.

insensitive, *a.* Add: **2. c.** Of a mathematical or physical quantity: (relatively) unaffected in value by variations in some related quantity. Const. *to.*
1968 Fox & Mayers *Computing Methods for Scientists & Engineers* iii. 31 As *r* decreases, the results become increasingly insensitive to small changes in the given condition. **1970** *Nature* 25 July 334/2 The energy of interaction turns out to be rather insensitive to ε(ω) at these frequencies if the layer materials are of similar density.

insentience. (Later examples.)
1924 J. M. Murry *Voyage* xii. 224 All was as it must be: happiness for one, pain for another, and for yet another insentience like his own. **1936** D. H. Lawrence *Phoenix* 344 The insentience of armed, bullying men, in face of living, sentient things. **1951** M. McLuhan *Mech. Bride* 13/1 It would be hard to know where to begin to peel back the layers of insentience..in such an ad. **1972** *Sci. Amer.* May 33/1 Whereas the insentience of the viscera has long been received as physiological dogma with a status comparable to the circulation of the blood, the insentience of muscle has often been called in question and probably cannot be regarded as universally accepted even today.

insequent (i·nsīkwĕnt, insī·kwĕnt), *a.*² *Geomorphol.* [f. In-³ + *-sequent* in Consequent, Subsequent *adjs.*] Of a stream, stream valley, or drainage pattern: having a course or form that appears haphazard and exhibits no apparent relation to the form or structure of the land.

1897 W. M. Davis in *Science* 2 July 24/1 Then the side streams, growing headwards, are accidentally located; and streams of this class have been called *autogenetic* by McGee. *Insequent* may prove to be a more satisfactory name for such streams, as it is of the same etymological family as *consequent, subsequent* and *obsequent*... As *insequent* has proved servicable [*sic*] in my lectures during the past winter, it is now submitted for trial by others. **1939** P. G. Worcester *Textbk. Geomorphol.* viii. 155 Streams that develop their valleys on flat-lying sediments or on massive rocks, such as granites, without strong structural control are called insequent streams. **1954** W. D. Thornbury *Princ. Geomorphol.* v. 114 Insequent valleys are those whose courses are controlled by factors which are not determinable. **1968** R. W. Fairbridge *Encycl. Geomorphol.* 1075/2 Such a condition, one of literally no structural control, is manifested by an insequent drainage pattern.

insert, *sb.* Add: **a.** and **b.** (Examples.)
1893 in *Funk's Stand. Dict.* **1907** *Installation News* Dec. 1/2 There are three of these loose inserts. One is a pamphlet..; the second is an advance price sheet [etc.]. **1928** R. B. H. Bell *Life Abundant* 142 This little book would not be complete without an insert on the Art of Prayer.
c. An object of one material around which another material (as concrete, plastic, or metal) sets or solidifies, or which is forced into it after it has set.
1913 G. A. Hool *Reinforced Concrete Construction* II. iv. 152 These castings are made in convenient lengths and the slot in the bottom makes it possible to place hangers or bolts at any desired location along the length of the insert. The casting can be anchored as securely in the concrete as may be necessary. **1933** L. F. Rahm *Plastic Molding* i. 10 Where the production of work is large enough to justify it, special machines may be developed for the simultaneous staking-in of multiple inserts, instead of molding these in the article. **1934** H. Chase *Die Castings* iv. 145 Inserts are usually knurled on the surface in contact with metal cast around them. **1967** B. Harocopos tr. *Technol. Gravity Die-Casting* ix. 121 In die-casting, it is possible to produce articles comprising cast-in-place inserts of ferrous metals, bronze, brass or, less frequently, of aluminium. **1968** *Gloss. Formwork Terms* (*B.S.I.*) 16 *Insert*, a piece of timber or other material cast into the concrete surface usually to provide a fixing.
d. A shot inserted into a cinema film, taken after the filming of a particular sequence.
1916 'B. M. Bower' *Phantom Herd* xvi. 269 He made all of his 'close-ups', his inserts and sub-titles. **1949** A. Huxley *Let.* 6 Mar. (1969) 593 Hitchcock..now shoots continuously a whole reel at a time, doing everything without cutting, getting the necessary close-ups and inserts..by camera movements and movements of the actors. **1957** *B.B.C. Handbk.* 119 Items presented from a central studio may be combined with 'live' or filmed inserts originating from anywhere in Britain. **1965** *Movie* Spring 26 The insert shots representing Hamp's mental images (these recall the joke insert of the mother dropping dead in *Shoot the Pianist*). **1970** *New Yorker* 26 Sept. 123/1 Keaton doesn't care much for inserts. 'I like long takes, in long-shot,' he says. 'Close ups hurt comedy.'
e. Misc. uses.
1922 M. B. Houston *Witch Man* xiv. 180 She glanced quickly through the sheets of paper lying there, even at the insert in the typewriter. **1950** *Jrnl. Acoustical Soc. Amer.* XXII. 655/1 (*heading*) Magnetic insert earphone insertable in the ear of the user... This small telephone receiver is of the earphone type such as is used with hearing aids. **1955** *Gloss. Acoustical Terms* (*B.S.I.*) 24 *Insert earphone*, an earphone of small dimensions associated with a fitting for insertion into the auditory meatus. **1961** *Times* 29 Aug. 13/7 Special inserts telling the story behind this new cigarette are contained in each packet. **1962** *B.B.C. Handbk.* 51 The BBC contributed 116 programmes and received a total of 199 complete programmes or inserts from the network. **1968** *Bodl. Libr. Rec.* VIII. 62 The printing of book-form catalogues is the principal aim of the project; these catalogues are to be maintained up to date by insert sheets. **1970** *Globe & Mail* (Toronto) 26 Sept. 52/1 (Advt.), Men's Leather Palm Wool Gloves. Expertly fashioned of a bulky knit wool with slip-resistant leather palm inserts. **1971** D. Potter *Brit. Eliz. Stamps* 11 The latest area to have taken up Queen Elizabeth British issues is Western Europe, and the Post Office has responded with overseas agencies and translation inserts in their packaged sets. **1972** *N.Y. Law Jrnl.* 31 Oct. 4/8 Such warranties can be found in many places: in the advertisements of the product, in the circulars or package inserts accompanying it.

insertable, *a.* (Later example.)
1971 *Timber Trades Jrnl.* 14 Aug. 58/1 DCE..will exhibit their Dalmatic insertable filters in a new 'knocked down' kit form on stand 21.

inserting, *vbl. sb.* (in Dict. s.v. Insert *v.*). Add: *spec.* = Insertion 2 b. ? *Obs.*
1847 Webster *Inserting*, something set in, as lace, etc., into garments. **1879** *N.Y. Fashion Bazar* 22 Nov. 10/2 The latest is a ruff of fine plaitings of Breton lace, sometimes four rows upon a narrow inserting, again put on a shell shape. **1886** *Harper's Mag.* Nov. 836/1 An elaborate trousseau made chiefly of tucks and insertings and edgings.

insertion. Add: **1. b.** *Astronautics.* = *injection 1 d; also **insertion point** = **injection point.*
1962 J. Glenn in *Into Orbit* 192 The computers..had indicated that the insertion of the capsule was good for a minimum of seven orbits. A. Shepard *Ibid.* 174 During the first four and a half minutes of launch, before we reach the insertion point and the 'Go' or 'No Go' decision as to orbit. **1963** C. McLaughlin *Space Age Dict.* (ed. 2) 86 *Insertion point.* That point where a spacecraft acquires a centrifugal force equal to the gravitational field force and goes into orbit.
2. b. (Earlier example.)
c **1840** Lady Wilton *Art of Needlework* xvi. 267 Patterns, without any edging, were seemingly designed for what we should now call 'insertion' work or lace.
4. *insertion stitch*; **insertion loss** *Electr.*, the decrease in the power delivered to a load (or in the voltage across it or the current through it) as a result of the insertion of a four-terminal device or network between it and the source, expressed (usu. logarithmically, in decibels or nepers) in terms of the ratio of the power, etc., without the network in place to that with it; similarly **insertion gain**, the negative of the insertion loss when expressed in logarithmic units.
1930 T. E. Shea *Transmission Networks & Wave Filters* ii. 49 A negative insertion loss is an insertion gain, and corresponds to an increase in load current amplitude as the result of inserting a network in a circuit. **1964** V. Uzunoglu *Semiconductor Network Analysis & Design* v. 69 The insertion gain of an amplifier connected between a source and load impedance (both being specified) is defined as $P_{ins} = 10 \log P_{OUT}/P_R$, where P_R is the power which would be delivered to the load if the amplifier were removed. **1930** T. E. Shea *Transmission Networks & Wave Filters* ii. 49 Insertion loss measures the actual change in load current caused by the insertion of a network. **1971** Kim & Meadows *Mod. Network Analysis* vi. 241 The behavior of a two-port coupling network such as a filter or equalizer..for use in a communication, signal-processing, or control system is frequently studied or specified in terms of an insertion loss defined in terms of voltage or power ratios. **1932** D. C. Minter *Mod. Needlecraft* 51/2 Various..insertion stitches may be formed by working an edging stitch, as braid edging or Antwerp edging. **1934** M. Thomas *Dict. Embroidery Stitches* 128 This simple insertion stitch consists of a row of braid edging stitch worked along both edges of the material to be joined. **1967** *100 Embroidery Stitches* (J. & P. Coats Ltd.) 35 Buttonhole insertion stitch..consists of groups of four buttonhole stitches worked alternately on each piece of fabric to be joined.

in-service *attrib.*: see *in *prep.* 17*.

insetter² (i·nse:təɹ). [f. Inset *v.* + -er¹.] A person who, or device which, insets sheets.
1891 *Pall Mall Gaz.* 27 Oct. 7/2 Compositors, printers, ..stereotypers, insetters. **1960** *Economist* 16 Apr. 275/1 An electronic bulwark called the insetter, which will enable national newspapers to carry full colour pictures—and advertisements—this autumn. *Ibid.*, It is the insetter device that has the job of correcting paper tension and high-speed 'wobble'.

‖ **inshallah** (inʃā·lā), *int.* Representing Arab. *in šā' Allah* if Allah wills (it), a very frequent pious ejaculation among Muslims.
1857 J. Bowring *Kingdom & People of Siam* II. xvi. 304 Inshallah! Such promptitude was, I believe, never before exhibited in an Asiatic Court. **1867** 'Ouida' *Under Two Flags* II. iii. 74 But,—Inshallah! we endure only for a while... Allah is great! we can wait. **1909** M. Diver *Candles in Wind* xviii. 183 Guns—Inshallah! The guns of the Maharajah. **1911** T. E. Lawrence *Let.* 21 May (1938) 104, I have been photographing this last week—and will more next. Developing too inshallah. **1922** *Ibid.* 20 Nov. 384 I'm hoping to find a regular means of dodging up to London... In a few days, insh'allah. **1959** W. Burroughs *Naked Lunch* 40 We aren't a matriarchy here, Insh'allah. **1971** *Shankar's Weekly* (Delhi) 11 Apr. 4/2 Nadir Shah Yahya Khan may fancy himself as the Scourge of Dacca. But people's voices wild with pain, Inshallah! surely they will rise again.

inshoot (i·nʃūt). *Baseball.* [f. In *adv.* + Shoot *sb.*¹] The act of causing the ball to move rapidly inward, as a ball that is pitched with a curve; a ball which moves thus.
1892 *Outing* (U.S.) Jan. 302/1 An old ball player..taught Harry to pitch and to try some curves and 'in shoots' of his own device. **1897** *Encycl. Sport* I. 74/2 A movement of the hand, an elevation of the head by the latter [sc. the catcher], lets the pitcher know that this ball is to be an in-shoot, the other an out-curve. **1904** *Sci. Amer.* 16 July 42/3 The right-handed pitcher delivers his 'outshoot' with much greater effect of incurve at the plate than he can accomplish with his 'inshoot'... The speed of the ball for 'outshoot' and 'inshoot' is the same. **1940** H. L. Mencken *Happy Days* 230 When I ventured on an inshoot it was apt to be recovered, not by the catcher, but by the third baseman.

inside, *sb.*, *a.*, etc. Add: **A.** *sb.* **2. a.** (Further examples.)
1808 Jane Austen *Let.* 20 Nov. (1952) 233 We mean..to go one night to the play. Martha ought to see the inside of the Theatre once while she lives in Southampton. **1819** R. Woodhouse *Let.* 20 Sept. in *Keats Lett.* (1958) II. 165 He parted with me at the Coach door—I had the inside all to myself. **1969** *Times* 12 Nov. 10/7 It took one war to get a foothold in the treaty ports. That still left the 'inside'.
e. The inner history, the real facts; also, a person in possession of such information. Phr. *to be on the inside*, to have knowledge that is not generally available.
1904 W. H. Smith *Promoters* v. 101 I'll give it to you straight, for I happen to know the inside. **1926** *Flynn's* 16 Jan. 637/1 Also he spills th' info as to how many insides they is; that is, how many bargain counters has one of our gang behind it. **1932** *Daily Express* 28 Jan. 15/5, I have chatted with men who are believed to be on the inside, and they have informed me that there will certainly be changes at forward and in the three-quarter line. **1959** *Economist* 18 Apr. 252/3 This desire to be 'on the inside' is no doubt personal as well as nationalist.
4. Also, *to know* (something) *inside out*: to know (it) extremely well.
1921 A. Huxley *Let.* 24 Aug. (1969) 201 Maria and I have just come back from..Rome... What a place! It inspires one at once with a kind of passion to know it utterly and inside out. **1967** N. Freeling *Strike Out* 31 A restaurant—that's a simpler affair, and Marguerite knows it inside out.
5. In various games, a position on the field; a player in that position.
1886 W. Arnott in B. James *England v Scotland* (1969) ii. 39 The Corinthians..have perfected the tactics of the three insides. **1901** W. H. Pickering *Hockey for Ladies* ix. 33 Left inside should be able to shoot for she gets as many chances as any other forward. **1905** E. E. White *How to play Hockey* vii. 47 Touch line tactics make it imperative that the two insides should get out of position. **1935** *Encycl. Sports* 289/1 (*Association Football*) The game is begun by one of the centre-forwards, who usually taps the ball gently to either of his insides. *Ibid.* 517/2 (*Rugby*) One common fault with three-quarters..is running on a slant; it is a pathetic sight to see a fine wing man so bottled by his own 'insides' that he is forced into touch almost as soon as he has the ball. **1967** J. Potter *Foul Play* vi. 79 George and Boozy moved up on the German insides like a pair of avenging demons.
6. (See quot. 1927.) *U.S.*
1899 B. Tarkington *Gentleman from Indiana* i. 11 Presently the 'Herald' announced a news connection with Rouen, and with that, and the aid of 'patent insides', began an era of three issues a week. *Ibid.* xiv. 249 We must buy 'plate matter' instead of 'patent insides'. **1927** *Amer. Speech* II. 242/1 Only in small country papers does one find 'patent insides'. The country editor frequently buys four pages of his paper already printed, filled with 'features', fiction, and advertising. He has only to fill the four remaining pages with local news and advertising.
B. *adj.* (Further examples.)
1824 E. Weeton *Let.* 31 May–2 June in *Jrnl. of Governess* (1969) II. 276 No consideration could have induced me to travel inside the coach; the guard offered me an inside seat in the night, but I declined it. **1849** Thackeray *Pendennis* I. xxii. 206 He would not take an inside place in the coach, but sate up behind with his friend the Guard. **1855** D. K. Clark *Railway Machinery* I. 93/1, ⅜ inch lap in Sharp's inside-cylinder engines exhausts as well as the 1¼ and 1⅛ inch of lap, which Mr. Sinclair has found necessary in his outside-cylinder engines on the Caledonian Railway. **1857** Mrs. Gaskell *Let.* 13 Sept. (1966) 471 We..found to our dismay that there was no inside places. However we got tilted up to the top of the coach behind. **1907** F. H. Burnett *Shuttle* xxiii. 226 We didn't come over on one of the big liners... Took a cheap one, inside cabin, second class. **1959** B. J. Farmer *Murder Next Year* ii. 7 A modern bathroom and an inside w.c. **1963** *Guardian* 7 Mar. 7/2 There are 1,045 berths on this ship, and 16 of them will be available at £35—in four-berth 'inside' cabins on C deck. 'Outside' four-berth cabins cost £55 a berth. **1971** *Ibid.* 1 Dec. 12/5 Some 200,000 families with severely disabled people need improved accommodation because they lack an inside lavatory.
b. (Earlier U.S. and later *fig.* examples.) Also of a person travelling inside a coach (cf. *sb.* 3 in Dict.).
1807 Southey *Lett. from Eng.* II. xxxvii. 125 These coaches..carry four inside passengers, two outside. **1841** E. Hall *Diary* 17 Dec. in O. A. Sherrard *Two Victorian Girls* (1966) iv. 48, I did feel very lonely and miserable as the coach drove off with the solitary inside passenger. **1886** W. D. Howells *Minister's Charge* xxv. 368 Look at the Blue-book,..it's the apotheosis of farm-boys, mechanics, insidemen, and I don't know what. **1946** K. Tennant *Lost Haven* (1947) viii. 119 The 'inside men', who fished the lakes..also appropriated any convenient grassy slope. **1948** 'N. Shute' *No Highway* vi. 165 He's an inside man... He's deeply interested in research, and he doesn't concern himself very much with user problems.
c. (Examples of *inside track* in fig. use.)
1857 *Richmond* (Virginia) *Whig* 5 Sept. 2/1 In a word, 'Gizzard-Foot' has the inside track for the Senatorship. **1882** W. James *Let.* 23 Apr. in R. B. Perry *Tht. & Char. W. James* (1935) I. 794 Whoever does it gets the inside track for promotion here on Bowen's withdrawal. **1931** L. Steffens *Autobiogr.* IV. xvii. 782 We cannot any more govern them or exploit or have the inside track in them. **1967** *Times Rev. Industry* Mar. 90/3 The real strength of the Labour Party does not lie in Transport House but on the inside track possessed by trade union officials, shop stewards and others,..all particularly important in influencing the male vote. **1972** *Accountant* 21 Sept. 359/1 At this stage in the contest, President Richard M. Nixon, as the incumbent, clearly enjoys the advantages of the inside track.
d. *fig.* Coming from 'the inside'; inner; not generally available.
1888 *Daily Inter-Ocean* 20 Feb. (Farmer), A secret service officer..claims to have inside information as to the facts in the case. **1896** S. Leavitt *Our Money Wars* ii. 11 Sometimes a few lines of inside history are worth whole books of that usually printed. **1912** C. Mathewson *Pitching in a Pinch* ix. 184 Behind this game is some 'inside' history that has never been written. *c* **1919** H. C. Witwer *Smile a Minute* i. 32 He wanted some inside dope for his paper. **1923** D. L. Sayers *Whose Body?* xiii, 'Peruvian Oil..hasn't paid a dividend for umpteen years.' 'No..but it's going to. I've got inside information.' **1924** H. Croy *R.F.D. No. 3* 157, I knew a fellow on the inside and we used to pal around together and I got a lot of inside dope. **1932** H. Walpole *Fortress* III. 447 John..had been most entertaining. If not of Parliament he was near it enough to have plenty of inside information. **1936** *Discovery* Apr. 129/2 His biography..gives the inside story of Coué in his years of fame. **1938** Wodehouse *Code of Woosters* vi. 139, I would be able to get together with Gussie..and learn the inside dope. **1950** D. Riesman et al. *Lonely Crowd* ix. 199 Some inside-dopesters actually crave to *be* on the inside, to join an inner circle or invent one; others aim no higher than to *know* the inside. **1959** *Encounter* Dec. 16/1 The influence of the..gossip-columnists and 'inside dopesters' has steadily risen. **1970** *Daily Tel.* 20 Apr. 17 The newest 'inside' book on the monarchy. **1972** 'H. Carmichael' *Naked to Grave* v. 59, I can get all the credit for an inside story... The way you described it to me is just tailor-made for my column. **1972** P. D. James *Unsuitable Job* iii. 90 If you want the inside dope on Garforth House, you should ask him. **1973** D. Lees *Rape of Quiet Town* i. 10 As soon as they hear I'm from the *Pictorial* they'll fall over themselves to give me the inside gen.

e. Special collocations: **inside centre** *Rugby Football*, the centre playing immediately outside the half-backs; **inside clinch** (see quot.) and cf. Clinch *sb.*¹); **inside country** *Austral.* (see quot. 1959); **inside forward**, in association football and hockey, either of the two players, called the inside left or right (see below), of the forward line; the position of such a player; **inside half** = **scrum-half*; **inside job** *slang*, a crime committed in a house, etc., by, or with the help of, a resident or servant, etc., in the building; **inside left, right**, in association football and hockey, a player playing between the outside left or right and the centre forward; the position of such a player; **inside lining** *slang* (see quots.); **inside man** *U.S. slang*, one involved in any of various special roles in a confidence trick or robbery; **inside right** (see *inside left*, above); **inside squatter** *Austral.*, one who lives within the margin of the settlements; **inside stand** *slang* (see quot. 1935).
1936 H. B. T. Wakelam *Rugby Football* ix. 171 Returning again to outside lines-up, Now and again..we come across a side which, really imitating the five-eighth game, plays two outside-halves, or even an 'inside' and an 'outside' centre. **1960** E. S. & W. J. Higham *High Speed Rugby* ix. 111 When the inside centre receives the ball, he proceeds..to straighten his run. **1969** *Sun-Herald* (Sydney) 13 July 45/5 Barry Honan's play at inside centre was rich encouragement for the side. **1886** *Encycl. Brit.* XXI. 591/2 Inside Clinch. The end [of a rope] is bent close round the standing part till it forms a circle, when it is securely seized..thus making a running eye. **1911** C. E. W. Bean *'Dreadnought' of Darling* xxxv. 317 But, be the 'inside' country never so tame..there will always be a huge stretch of country 'outside' which cannot by any known means be closely settled. **1959** Baker *Drum* 119 *Inside country*, well-populated country near or in coastal areas, specifically in contrast to inland or outback. **1897** *Windsor Mag.* Dec. 25/2 Each of the two inside-forward positions. **1897** *Encycl. Sport* I. 418/1 Try to feed the centre and inside forwards when in your opponent's territory. **1965** *Men's Hockey* (Know the Game) (rev. ed.) 30/1 If the inside forwards are close to the centre, the opposing centre half can cover all three. **1921** E. H. D. Sewell *Rugby Football* iii. 69 In our view *all* backs should look at the inside half when the ball is being put in the scrum. **1949** *Rugby League Football* ('Know the Game' Series) 8 Scrum half back or inside half. **1969** *Programme* (Llanelli v. Swansea 1 Apr.) 6 Jim Lamb, Mewnwr. Inside-half. **1973** *Sunday Tel.* 18 Mar. 38/2 Smith, for England, was forceful at inside half. **1908** 'O. Henry' *Gentle Grafter* xi. 142 The police are calling it an inside job. **1924** G. S. Dougherty *Criminal as Human Being* 187 One of my assistants..made an investigation that convinced him an 'inside job' had been committed by the servants. **1925** A. Christie *Secret of Chimneys* xii. 121 How long have you had the idea that it might be an inside job? **1926** J. Black *You can't Win* xxi. 334 It was an inside job..done by the storekeeper to beat his creditors. **1933** D. L. Sayers *Murder must Advertise* xv. 249 You seem convinced that the murder of Victor Dean was an inside job. **1972** 'M. Innes' *Open House* II. x. 94 Wasn't there something factitious about the whole affair? Didn't it match the hoary old formula of the inside job disguised as an outside job? **1897** *Encycl. Sport* I. 517/1 This is a comparatively easy task for inside right, but by no means so for inside left, who cannot reach out to hit the ball as it is rolled in. **1969** B. James *England v Scotland* ii. 51 Scottish inside-left J. Macpherson..collapsed in a heap. **1851** H. Mayhew *London Labour* I. 20/1 He was 'going to get an inside-lining' (dinner). **1935** A. J. Pollock *Underworld Speaks* 61/2 *Inside lining*, an exceptionally good meal. **1935** *Amer. Speech* X. 21/1 (s.v. *Steerer*), *Steerer*, an *inside man* or tipster who locates prospects for robbers or safeblowers. **1937** *N.Y. Times* 22 Dec. 22 *Inside man*, a spy placed in a plant as an employé. **1938** F. D. Sharpe *Sharpe of Flying Squad* xxviii. 288 The 'mug finder' books a room and spots a likely name in the register... When the 'mug's' name is announced in the restaurant by the page, he is followed to the telephone by the 'inside man' and identified. **1940** *Amer. Speech* XV. 119/1 Inside-men are highly specialized workers; they must have a superb knowledge of psychology to keep the mark under perfect control during the days or weeks while he is being fleeced. **1906** *Westm. Gaz.* 16 Mar. 5/1 Bloomer developed into a forward of exceptional skill in the inside-right position. **1969** B. James *England v Scotland* iii. 64 Chadwick provided an opening for inside-right Goodall to score England's second goal. **1881** A. C. Grant *Bush-Life in Queensland* II. xxxii. 171 Stations were formed for nearly a hundred and fifty miles outside John's run, and he began to regard himself as quite an inside squatter. **1959** Baker *Drum* 119 *Inside squatter*, a farmer or large land-owner in a fairly well-populated district. **1932** Wodehouse *Hot Water* xiii. 227 Oily's got himself into the house, and he's planning to let Soup in when he's good and ready... It's what's known as the inside stand. **1935** *Punch* 4 Dec. 637/3 The 'inside stand', as the business of insinuating a member of a gang into the doomed house is called. **1960** Wodehouse *Jeeves in Offing* v. 54 The butler turned out to be one of a gang of crooks, planted in the house to make it easy for them to break in. The inside stand, it's called.

C. *adv.* **1. b.** *fig.* In a position to have private information. *rare.*
1870 *Congress. Globe* 3 Feb. 1022/1, I ask the gentleman from Ohio to name the ships which he says have been sold for a song. The gentleman is inside on all these matters.
c. *slang.* In prison.
1888 *Referee* 14 Oct. 1/4 There dashes past a once member of the dangerous classes, who has been 'inside' many a time and oft, but who, having run into a bit of ready, will now go straight. **1925** E. Wallace *King by Night* xxvi. 116 You've been 'inside', and you're going in again unless you can explain..what you're doing here. **1958** *Listener* 6 Nov. 743/2 Only a very insensitive reader could reach the end of the book without feeling that he had shared the author's daily, even hourly, existence 'inside'—..the effect comprising a powerful, and frequently nauseating, picture of what life in prison, and in a Borstal, is like. **1959** 'L. Bruce' *Our Jubilee is Death* xvi. 149 She was afraid of me going inside again. **1972** C. Drummond *Death at Bar* ii. 54 Over the years she had been convicted three times, spending in all four years 'inside'.
2. (Earlier and later examples.)
1824 E. Weeton *Jrnl.* 21 July (1969) II. 309 The Liverpool fares..were all 4£ inside and 2£ out. **1847** *Punch* XII. 14 Has he [sc. the bus conductor] directions to say..'Full inside'? **1905** *19th Cent.* 817 To adopt an elusive bush idiom, the railways bring the country 'inside'. **1930** L. Munday *Mounty's Wife* xvii. 209 After twenty years, almost all of which we had spent 'inside', as the North is always called. **1945** Baker *Austral. Lang.* 59 When a man from the far interior comes to the city he says he is coming inside. **1957** *Arctic Spotter* (Edmonton, Alberta) Oct. 9 Not once did I meet anyone who was sorry to be 'inside'. **1970** 'D. Halliday' *Dolly & Cookie Bird* v. 66 You didn't know Daddy like I did. He was an awful old softie inside.
3. *inside of.* (Earlier example.)
1839 *Spirit of Times* 27 July 246/1 There are dozens of horses..that can trot their mile in harness inside of three minutes.
D. *prep.* (Later examples.) Now esp. in titles and headlines to indicate special or intimate knowledge.
1924 A. J. Small *Frozen Gold* i. 39, I hear all about it inside twenty-four hours. **1936** J. Gunther (*title*) Inside Europe. **1942** R. Benchley (*title*) Inside Benchley. **1972** Everson & Fitzgerald (*title*) Inside the city.

in-side: see *in *adv.* 12 b.

i:nside-ou·tness. [f. *inside out* (Inside A. *sb.* 4) + -ness.] The state of being inside out.
1919 R. Bridges in *S.P.E. Tract* II. 40 The insideoutness, topsy-turviness, and preposterousness of Mr Jones' method is incredible. **1960** *Encounter* XIV. v. 70 The inside-outness Criticus.

insider. Add: (Earlier and later examples.) Also *attrib.* and *Comb.*
In quot. 1957¹ with play on title of Colin Wilson's work *The Outsider.*
1848 W. Armstrong *Stocks* 7 Insiders are those by whom and through whom all transactions are made in and about the Exchange. **1902** H. L. Wilson *Spenders* xxx. 355 Shepler's back of all three [stocks]. The insiders are buying up now, slowly and cautiously, so as not to start any boom prematurely. **1913** *Q. Rev.* July 256 At any rate, as regards the original 10,000 shares bought by Sir Rufus Isaacs, they took part in it as 'insiders' exploiting the ignorance of the public. **1923** F. Waldo *Down Mackenzie through Gt. Lone Land* 248 The Outsider cannot know: the Insider never can make clear to him the grip that holds, the urge that stirs and never sleeps. **1942** *R.A.F. Jrnl.* 2 May 2 Estimates concerning the output of Japanese airplane industry vary. Insiders consider it to be from 1,500 to 2,500 planes a year. **1957** *Observer* 8 Sept. 10/2 They are fools who compare him [*sc.* Anouilh] with Pinero, a born insider who upheld conformity and bade us compromise with imperfection. **1957** C. Pepler *Riches Despised* ii. 27 To the outsider in fact all these religions would have seemed more or less alike... Indeed the insiders, the Israelites, found themselves all too easily drawn to the high places of the mountains. **1958** *Times Lit. Suppl.* 30 May 293/3 At any moment earnest pedantry may break through to reveal a very young man's novel with nothing much to say and not quite enough 'insider' knowledge to avoid a '*gaffe*'. **1966** *Economist* 12 Feb. 642/2 The main thrust of the Act was to apply to smaller US companies the same financial reporting, proxy-solicitation and insider-trading rules that have long been applied to larger US corporations. **1972** *Observer* 10 Sept. 11/3 The softness of our line compared with that of the US shows up clearly in the area of insider trading. In Britain it is not illegal to use confidential information to make a profit in the stock market. **1973** *Times* 8 June 1/2 The Stock Exchange has reiterated its view that 'insider dealing' in a company's shares should be made a criminal offence. In a memorandum on company law reform presented to the Department of Trade and Industry, it says the police should be empowered to call for the disclosure of true beneficial ownership when they suspect that insider trading has taken place under the cloak of nominee shareholdings.
2. *U.S. slang.* A pocket or pocket-book.
1846 *National Police Gaz.* (U.S.) 12 Sept. 5/1 The mode adopted to get an 'insider' or what may be better understood, the book from an inside coat pocket is as follows. **1896** I. K. Friedman *Lucky Number* 154 Britch is used to designate the front-pocket; gerve, vest-pocket; insider, inside coat-pocket; [etc.]. **1925** H. Leverage in *Flynn's* 7 Feb. 489/2 *Insider* (a double insider), an inside vest pocket. **1955** D. W. Maurer in *Publ. Amer. Dial. Soc.* XXIV. 115 An *insider* is the long flat wallet carried inside the breast-pocket. **1970** C. Major *Dict. Afro-Amer. Slang* 69 *Insiders*, one's pockets.
†**3.** An inside passenger. *Obs.*
1854 B. F. Taylor *Jan. & June* 170 'No Room For Two!' was the exclamation of some insider, the other morning. **1892** *Harper's Mag.* Jan. 257/1 The exhilarating pace, the smooth roads, and the juxtaposition of the insiders tended, in a high degree, to the promotion of enjoyment.

insight, *sb.*¹ Add: **2. d.** *Psychol.* In studies of behaviour and learning, the sudden perception of the solution to a problem or difficulty; applied to animals, giving an indication of their capacity for ideas and reasoning. In *Psycho-analysis*, perception of one's mental condition. Also *attrib.* and in *Comb.*, as *insight-giving, -learning, therapy.*
1909 W. James *Meaning of Truth* xiii. 260 His insistent desire to have a world of that sort is felt by him to be..an altogether peculiar *insight-giving passion* to which..he would be *stupid* not to yield. **1916** R. M. Yerkes *Mental Life Monkeys & Apes* iii. 68 The curve of learning plotted ..would..be described as an ideational, and possibly even as a rational curve; for its sudden drop..to the base line strongly suggests, if it does not actually prove, insight. *Ibid.* 87 This young orang utan..strove persistently, and often vainly, to gain insight. **1925** E. Winter tr. *Köhler's Mentality of Apes* vii. 194 Since, however, we have to decide whether chimpanzees ever behave with insight, [etc.]. *Ibid.* 198 Hence follows this characteristic: to set up as the criterion of insight, the appearance of a complete solution with reference to the whole lay-out of the field. **1949** *Brit. Birds* XLII. 355 These are not necessarily instances of insight learning, but they do represent persistent attempts to reach the milk when the top of the bottle was no longer visible to the bird. **1949** M. Mead *Male & Female* 449 Society is the patient. Those who have been in some way hurt or distorted give us many valuable insights into what is wrong with it. **1956** W. H. Thorpe *Learning & Instinct in Animals* v. 100 Thus insight-learning seems to be a kind of action by hypothesis, and has often been held to be evidence of ideational processes. **1964** M. Argyle *Psychol. & Social Probl.* x. 129 It has been found possible to reduce racial prejudice by means of brief insight-therapy. **1965** A. D. Weisman *Existential Core of Psychoanal.* vi. 139 Insight in psychoanalysis often seems to be the *result* of a resolved conflict, not the *cause* of its resolution. **1967** R. R. Greenson *Technique & Pract. Psychoanal.* I. i. 44 Only rarely does insight lead very quickly to a change in behavior.

insighted, *a.* (Later examples.)
1879 G. M. Hopkins *Let.* 22 Feb. (1955) 71 The thought of the last tercet is truly insighted. **1881** —— *Let.* 16 Sept. (1935) 55 'The Old Bishop' is a fine and insighted picture.

i·nsightful, *a.* [f. Insight *sb.*¹ + -ful.] Characterized by insight. Also **i·nsightfully** *adv.*
1907 Galsworthy *Country House* II. i. 104 As if she had been guilty of thoughts too insightful, Mrs Pendyce blushed. **1932** *Brit. Jrnl. Psychol.* Jan. 196 The letters so distorted cannot have been insightfully apprehended. **1934** *Ibid.* July 5 When the problem passes the threshold of insightful understanding these answers are chosen more frequently. **1945** *Jrnl. Compar. Psychol.* XXXVIII. 367 (*title*) The relation of previous experience to insightful problem-solving. **1951** *Mind* LX. 518 Behaviour which we might call 'knowing' or 'insightful'. **1955** *Sci. Amer.* Apr. 109/1 Myers was a goose, but he was also brilliant, insightful and high-minded. **1957** R. K. Merton *Student-Physician* 177 Physicians and educators have often commented insightfully on their own experiences. **1967** *Language* XLIII. 744 A sensible and insightful account of the history of linguistic thought. **1970** D. L. Emblen *Peter Mark Roget* (1971) xv. 276 A remarkably insightful—if somewhat cavalier—critic. **1973** *N. & Q.* Jan. 24/1 Mr. Shippey has written an insightful and in many ways exciting study for the non-specialist university student and teacher.

insignia, *sb. pl.* Add: Further examples of *insigne* used (correctly) and of *insignia* used (erroneously) as sing.
¶ The erroneous use is discussed by L. Pound in *Amer. Speech* (1956) XXXI. 156 f.
1912 R. A. Knox in *The Blue Bk.* July 124 The seal, and symbol, and secret of Watson is, of course, his bowler. It is not like other bowlers——it is a priestly vestment, an *insigne* of office. **1948** W. R. Benét *Reader's Encycl.* 850/1 The men of Lord Louis Mountbatten's Southeastern Asia command wore it [*sc.* the figure of a phoenix] as an insigne in World War II. **1971** *Times* 24 Dec. 13/2, I saw not a single racer at Sestrière bearing an insignia that seemed out of place.

insignis (insi·gnis). [a. L. *insignis* remarkable, used as the specific epithet of *Pinus insignis* (D. Douglas in J. C. Loudon *Arboretum et Fruticetum Britannicum* (1838) IV. 2265), the former name of the tree.] The Monterey pine, *Pinus radiata*, which is native to southern California and widely cultivated elsewhere.
[**1866** 'Senilis' *Pinaceæ* iii. 128 Although not so beautiful as Insignis, yet it [*sc.* the radiated-coned pine] is a useful ornamental Pine.] **1920** C. Coltman-Rogers *Conifers* i. 41 The Insignis is a tree that undoubtedly favours sea air. **1931** E. Maxwell *Afforestation in Southern Lands* xxxix. 175 The Insignis thrives well under widely different degrees of rainfall. *Ibid.* 179 The quality of Insignis

timber varies greatly according to the way it has been grown and its age. **1957** *Handbk. Softwoods* (Forest Prod. Res. Lab.) 41 Pine, radiata... Other names. Insignis pine (general), insignis (S. Africa), Monterey pine (U.S.A.).

insinuate, *v.* **8.** Delete †*Obs.*
(Still used in the Commissions issued by the Bishop of Winchester to the Deans of Jersey and Guernsey as his Commissaries.)

insinuendo (insiniu,e·ndo). [A 'portmanteau' blending of INSINUATION and INNUENDO.]
A tasteless word.—Ed.
1885 B. MATTHEWS in *Longman's Mag.* Dec. 151 Could I not damn with faint praise and stab with sharp insinuendo?—to use the labor-saving and much-needed word thoughtlessly invented by the sable legislator of South Carolina. **1906** *N. & Q.* 3 Mar. 171/1 An old Yorkshire friend of mine..used the following words frequently. He thought they were good English:—'Disastrophe' = disaster + catastrophe, 'Insinuendo' = insinuation + innuendo. **1909** *Daily Chron.* 9 June 6/7 It was a sable legislator who howled back with scorn the 'insinuendos' of a political opponent. **1921** C. MACKENZIE *Rich Relatives* ix. 216 Anyone more cunning I've never seen. Nasty insinuendos, enough to make anyone sick! **1966** P. MOLONEY (*title*) A plea for Mersey or the gentle art of insinuendo.

insist, *v.* Add: **3. c.** With quoted words.
1888 MRS. H. WARD *R. Elsmere* II. xxxix. 174 'And rather than try,' he insisted, 'you will go on believing [etc.].' **1906** W. S. MAUGHAM *Bishop's Apron* iv. 47 'Do you care for me at all?' he insisted.

in situ: see *IN *Lat. prep.*

insolubilize (insǫ·liŭbiləiz), *v.* [f. L. *insolūbil-is* INSOLUBLE *a.* + -IZE.] *trans.* To render incapable of dissolving. So **inso·lubilized, inso·lubilizing** *ppl. adjs.*
1897 *Daily News* 4 Oct. 6/4 The colouring matter remaining attached to the paper, and held there by the insolubilised gum. **1904** *Buck's Handbk. Med. Sci.* (rev. ed.) VIII. 351/2 Both the iron and magnesium compounds must be removed from the stomach as soon as possible since the insolubilized arsenic may be again resorbed through the solvent action of the fluids of the body. **1947** *Jrnl. Soc. Chem. Industry* LXVI. 417/2 A small proportion of solids..became insolubilized during boiling. **1962** J. T. MARSH *Self-Smoothing Fabrics* xx. 333 Hence the main effect of heat is to insolubilise the resin. *Ibid.* vii. 100 This compound had some previous application in Germany as an insolubilising agent for the protein binder in the pigment printing process. **1967** E. CHAMBERS *Photolitho-Offset* xiii. 191 The image consists of an organic resin insolubilised by light. **1971** *Nature* 6 Aug. p. iii (Advt.), The world's most comprehensive selection of insolubilized biochemicals. **1972** *Enzymologia* XLII. 275 Treatment with glutaraldehyde could have insolubilized crystalline catalase in two fashions.
Hence **inso:lubiliza·tion,** the process of insolubilizing.
1926 *Chem. Abstr.* XX. 838 According to the French pat. 413,007, addition of an organic acid to a bath containing gelatin and CH_2O prevents the insolubilization, which is restored by NH_3. **1946** *Nature* 28 Dec. 925/2 Reactions of rubber and olefinic systems generally with sulphur and other reagents which are known to produce insolubilization. **1971** *Ibid.* 6 Aug. p. iii (Advt.), Supports for the insolubilization of proteins.

insomniac (insǫ·mniæk). [f. INSOMNI(A + -AC.] One who suffers from insomnia. Also *attrib.* or as *adj.*
1908 *Lancet* 8 Feb. 407/2 The urinary secretion is increased, not diminished. The latter phenomenon is, like all the phenomena of sleeplessness, most evident in the neurotic insomniac. **1930** H. CRICHTON-MILLER *Insomnia* ii. 21 Most insomniacs wish it to be believed that the onus of their insomnia lies with their circumstances and environment. **1939** G. GREENE *Lawless Roads* iv. 126 A Pennsylvanian with pouchy insomniac eyes. **1953** E. HYAMS *Gentian Violet* ix. 176 Jim's absence of mind might cause thousands of back-street chemists to have seven instead of four forms to complete when purchasing barbiturates; and compel insomniac neurotics to pass sleepless nights. **1957** P. I. ROSENTEUR *Morpheus & Me* xvi. 270 Our system of individual enterprise has made it possible for anyone, whatever his original social or economic status, to work and worry himself right into the front ranks of the insomniac army. **1959** *Listener* 23 July 121/2 Baghdad has been an insomniac's paradise. **1973** L. MEYNELL *Thirteen Trumpeters* xiv. 214 A Mickey Finn, guaranteed to put the worst insomniac in the world fast asleep in five minutes.

inspan, *v.* Add: (Earlier examples.)
1834 T. BOWKER *Jrnl.* 25 Dec. in *Towards Dict. S. Afr. Eng.* (Dict. S. Afr. Eng. Dict. Committee) (1971) 43 Arrive at [M]erais after sunset find the waggons inspanned ready for going away. **1838** F. OWEN *Diary* (1926) 118 At length, our oxen being inspanned, the waggon being loaded with the chief necessaries belonging to both families..we left the station.
b. *fig.* or *transf.*
1914 KIPLING in *Geogr. Jrnl.* Apr. 373 One man, apparently without effort, inspans the human equivalent of 'three blind 'uns and a bolter' and makes them do miracles. **1928** *Sunday Express* 8 July 10/5 There are hundreds of keen young players with the player's eyesight available for this lining business. Is it beyond the wit and the capacity of the Wimbledon authorities to inspan them? **1939** R. CAMPBELL *Flowering Rifle* I. 15 Our great Victory,..in her car, Which all the way from Portugal to France She inspans in her thundering advance. **1949** *Cape Times* 13 Sept. 8/9 To rescue the Coloured man, all forces will have to be inspanned to raise him economically. **1971** *Rand Daily Mail* 29 June 15 Why, Mrs Barton often gets on the telephone and inspans private householders to help out.

in-spawn (i·n,spǭn), *a.* [attrib. use of the phrase *in spawn* (cf. IN *prep.* 10 b).] That is about to spawn.
1908 *Westm. Gaz.* 14 Mar. 11/3 Netting the in-spawn dace. **1923** *Daily Mail* 13 Feb. 12 After the fish was landed it disgorged an in-spawn trout of 2 lb.

inspection. Add: **1.** (See quots. and cf. *INSPECTORSHIP.)
1861 *Act* 24 & 25 *Vict.* c. 134 § 192 Every Deed or Instrument made or entered into between a Debtor and his Creditors..relating to the Debts or Liabilities of the Debtor, and his Release therefrom, or the Distribution, Inspection, Management, and winding-up of his Estate. **1869** *Act* 32 & 33 *Vict.* c. 71 § 125 (4) The special resolution, together with the statement of the assets and debts of the debtor, and the name of the trustee appointed, and of the members, if any, of the committee of inspection, shall be presented to the registrar. **1883** *Wharton's Law Lex.* (ed. 7) 419/2 Inspection, Deed of.
6. *attrib.* and *Comb.* (Further examples.)
1753 in *Maryland Hist. Mag.* (1908) III. 366 Which made me apprehend they intended some Opposition to the Inspection Law. **1773** *Ibid.* (1907) II. 358 To compel all the Owners or Makers of Tobacco to send it to certain Inspection Houses whence it cannot be again removed till it is put on Board a Ship. *a* **1884** KNIGHT *Dict. Mech.* Suppl. 501/2 *Inspection car*, one used by the officers of a railway while inspecting the track. **1890** *Railways of Amer.* 146 It would require a separate article to give even a brief description of the different kinds of cars which are now used... Inspection-car, [etc.]. **1903** *Westm. Gaz.* 12 Jan. 7/2 The covering of the electrical inspection box at the Thames-street corner of Bennett's-hill was blown off this morning by the fusing of the wires. **1904** GOODCHILD & TWENEY *Technol. & Sci. Dict.* 311/2 *Inspection chamber*, a chamber at the junction of drains to allow of inspection. **1906** *Daily Chron.* 27 Aug. 3/4 To see that these things are observed..the firm employs inspection engineers. **1907** *Westm. Gaz.* 13 Nov. 8/2 Free access to the valves and inspection-doors of the crank chamber is given. **1907** *Ibid.* 11 Dec. 3/2, I have found when towing..that the garage people have a knack of putting difficulties in the way of drivers..using the inspection pits. **1908** *Ibid.* 2 Apr. 4/2 There are two large inspection doors in the crank-case. **1908** *Installation News* II. 87/2 The straight through type of inspection box should be inserted in the conduit run at short intervals. *Ibid.* 148/2 The other remarkable reductions [in price] are..Inspection Fittings [etc.]. **1908** *Westm. Gaz.* 20 Nov. 4/2 At the rear it [*sc.* an automobile] has a large inspection-cover. **1909** *Daily Chron.* 18 Sept. 1/6 The houses were flooded by water and filth from the sewers through the stone slabs covering the inspection chambers being displaced by the flood. **1930** *Engineering* 21 Nov. 658/2 The front inspection lamp consists of an electric bulb with the switch above it. **1946** A. PHELPS *I couldn't care Less* x. 76 Improperly secured inspection panels blew off my wings. **1951** *Good Housek. Home Encycl.* 187/1 Cast-iron manhole covers flush with the ground, technically termed 'inspection covers'. **1967** *Gloss. Sanitation Terms* (*B.S.I.*) 26 *Inspection chamber*, a shallow manhole. **1972** M. GILBERT *Body of Girl* iv. 44 An open-fronted workshop..with two inspection pits.

inspectorship. Add: **b.** *attrib.*, as **inspectorship deed** (also *deed of inspectorship*), see quots. and cf. *INSPECTION 1.
1861 *Act* 24 & 25 *Vict.* c. 134 § 191 As to Trust Deeds for Benefit of Creditors, Composition and Inspectorship Deeds executed by a Debtor. **1883** *Wharton's Law Lex.* (ed. 7) 419/2 *Inspectorship, Deed of*, an instrument entered into between an insolvent debtor and his creditors, appointing one or more person or persons to inspect and oversee the winding up of such insolvent's affairs on behalf of the creditors. **1902** *Encycl. Brit.* XXX. 151/2 Estates, however, continued to be wound up under deeds of arrangement, whether in the form of an assignment, or composition, or inspectorship deed. **1959** JOWITT *Dict. Eng. Law* I. 980/2 *Inspectorship, Deed of*,..the repealed Bankruptcy Act, 1869.

in-sphere (i·n,sfīəɹ). *Math.* [f. *IN-[1] b + SPHERE *sb.*] A sphere that touches all the faces of a given polyhedron.
1886 G. S. CARR *Synopsis Pure & Appl. Math.* I. II. 890 (Index), In-sphere of a tetrahedron. **1939** H. S. M. COXETER *Ball's Math. Recreations & Ess.* (ed. 11) v. 132 With each of these polyhedra we may associate three concentric spheres: one (the 'circum-sphere') through all the vertices, one touching all the edges, and one (the 'in-sphere') touching all the faces.

inspirate, *v.* Delete †*Obs.* and add later examples.
1939 L. H. GRAY *Foundations of Lang.* 406 They [*sc.* Hottentot and Bushman] are characterised by inspirated consonants. **1964** P. DELATTRE in D. Abercrombie et al. *Daniel Jones* 48 It [*sc.* *oui*] can be either expirated or inspirated.

inspira·tionalism. [f. INSPIRATIONAL *a.* + -ISM.] = INSPIRATIONISM.
1911 W. DE MORGAN *Likely Story* 343 Authors who may be said to belong to the school of Inspirationalism. **1961** B. R. WILSON *Sects & Society* I. i. 19 Checks on the possibly disruptive tendencies on [*sic*] radical inspirationalism.

inspirator. 1. Delete † *Obs.* and add later examples.
1903 K. M. ABBOTT *Old Paths New Eng.* 99 The first power-loom was set up by Francis Cabot Lowell, the inspirator of cotton manufacture. **1904** *Westm. Gaz.* 19 Nov. 10/3 He might characterise him as the great inspirator of the Opposition.

inspired, *ppl. a.* Add: **5. b.** Phr. *inspired guess*, a guess not based on fact or known information (cf. *educated guess*).
1914 G. B. SHAW *Misalliance* p. xlix, In a single brain..we get the inspired guess of the man of genius. **1969** J. ARGENTI *Managem. Techniques* 120 Improving the accuracy of a forecast usually calls for the careful and detailed analysis of past records plus an estimate or an inspired guess as to how future trends will differ from those of the past. **1972** N. TINBERGEN in *Proc. R. Soc.* CLXXXII. 389 The effects of other behaviours are so difficult to trace that even now no more than inspired guesses are possible.

inspissator (i·nspisēitəɹ). [f. INSPISSAT(E *v.* + -OR.] An apparatus for thickening or coagulating serum or other body fluids by heat.
1897 MUIR & RITCHIE *Man. Bacteriol.* ii. 48 (*caption*) Blood serum inspissator. **1928** F. W. TANNER *Pract. Bacteriol.* i. 24 Some media..must be sterilized at low temperatures. For this purpose a serum coagulator, or inspissator, is used. The media are heated at from 57° to 60° C. for different lengths of time. **1951** WHITBY & HYNES *Med. Bacteriol.* (ed. 5) iii. 22 A very rich solid medium may be made by coagulating serum or blood into a solid mass. The coagulation is effected by heating to 75° C. in an inspissator.

inst., abbrev. of INSTANT *a.* 2 b.
1771 J. WEDGWOOD *Let.* 7 Sept. (1965) 113 Letters of the 3rd 4th and 5th inst. **1954** CROOKS & DAWSON *Dict. Typewriting* (ed. 6) 146 The principal commercial abbreviations beginning with this letter [*sc.* the letter I] are—..Inst. [etc.].

install (i·nstǭl), *sb.* [f. INSTALL *v.*[1]] Something installed or placed in. (Only G. M. Hopkins.)
1871 G. M. HOPKINS *Jrnl.* (1959) 21 Apr. 207 These are not ribs; they are a 'wracking' install made of these two realities—the frets,..and the whiter field of sky shewing between. **1874** *Ibid.* 23 May 244 True bold realism but quite a casual install of woodland. *Ibid.* 245 Happy use of openings, accidental installs, people's feet, hands etc seen through. *c* **1883** —— *Sermons & Devotional Writings* (1959) II. iii. 146 For accidental being, such as that of the broken fragments of things or things purely artificial or chance 'installs', has no true and intrinsic oneness or true self.

installation. Add: **2.** Also *attrib.*
1898 *Engineering Mag.* XVI. 48 The installation cost is a matter of estimate. **1908** *Installation News* II. 56/2 Competent to act as installation inspector. **1921** *Dict. Occup. Terms* (1927) § 690 *Installation engineer*.., a technical engineer who interviews prospective consumers and points out advantages of electric light or power and advises them as to installation, etc. *Ibid.* § 692 Installation inspector. **1962** *Gloss. Terms Automatic Data Processing* (*B.S.I.*) 52 *Installation time*, time spent in installing, commissioning, testing, and approving equipment.

installer (in Dict. s.v. INSTALL *v.*[1]). (Later examples.)
1926 [see *check-back]. **1968** *Listener* 23 May 683/3 It's worth boring the installer with endless questions when you get any new piece of equipment. **1971** *Sci. Amer.* Sept. 25/1 In Chicago I was successively a telephone installer, a postal clerk, [etc.].

instalment[2]. Add: The spelling *instalment* is now usual in the U.K. and *installment* in the U.S. **3.** *attrib.* (freq. in recent use), as *instalment credit, plan, system*, etc.
1876 *Los Angeles Daily Herald* 4 Oct. 3/6 Lots for Sale on the Installment Plan. **1887** *Courier-Jrnl.* (Louisville, Kentucky) 18 Jan. 3/7 Installment men and agents generally will find just what they need by addressing Installment Dealers' Supply Co. **1894** *Vermont Agric. Rep.* XIV. 94 They..sell the horse on the installment plan, getting enough down to pay all the horse is worth, and holding notes for the rest. **1895** KIPLING *Day's Work* (1898) 156 'It's the Governor,' said the skipper. 'He's been selling her on the instalment plan.' **1904** 'SAKI' *Reginald* 52 They're getting there on the instalment system—so much down, and the rest when you feel like it. **1909** 'O. HENRY' *Roads of Destiny* iii. 47 When they get money they exhibit a strong tendency to spend it..instead of giving it to the instalment man. **1923** R. D. PAINE *Comrades of Rolling Ocean* xi. 187 'This is still a valuable ship,' explained Torrance, 'even if she is floating around the high seas on the installment plan.' **1927** W. E. COLLINSON *Contemp. Eng.* 112 To pay by the instalment system. **1927** *Sat. Even. Post* 24 Dec. 28/2 Upon that premise..is based the vast present-day development of installment selling. **1935** *Economist* 3 Aug. 234/2 The United States itself is the leader of a new, hard, materialistic civilisation..whose priests are the instalment-seller and the advertising expert. **1942** *Short Guide Gt. Brit.* (U.S. War Dept.) 31 *Instalment plan*, hire-purchase system, or hire system. **1948** *Manch. Guardian Weekly* 8 Jan. 4/1 The President says inflation can be beaten..by putting an end to instalment buying. **1952** *Ibid.* 15 May 3/1 The Federal Reserve Board was a little tight-lipped in explaining its move..to suspend controls on hire purchase, or what is known here as instalment credit. **1955** *Times* 15 Aug. 11/5 The Federal Reserve..has expressed mild concern over the rapid rise of consumer instalment (hire purchase) debt. **1965** M. SPARK *Mandelbaum Gate* iv. 104 On his discharge he bought a car on the instalment plan. **1967** *N.Y. Times* (International Edition) 11–12 Feb. 3/3

The measures include..more severe installment purchase restraints and reduction in travel allowances. **1972** *Accountant* 17 Aug. 206/2 Instalment credit..group services and group funding, are the subject of special reviews. **1973** *N.Y. Law Jrnl.* 20 Feb., The assumption of installment payments by appellant did not serve to release decedent.

instant, *a.* Add: **4. c.** Of a processed food: that can be prepared for use immediately. Also *transf.* and *fig.*, hurriedly prepared or carried out, etc.
1912 *Ladies' Home Jrnl.* Oct. 71/4 (Advt.), Instant Postum..is regular Postum in concentrated form—made in the cup—no boiling required. **1915** E. B. HOLT *Freudian Wish* ii. 87, I wish I had..drunk less of that hot-wash that my wife calls instant coffee. **1924** *Ladies' Home Jrnl.* Oct. 198/1 G. Washington's Delicious Instant Coffee... The coffee ready to drink when dissolved in hot water. **1957** D. KARP *Leave me Alone* xi. 151 Your grand new world of jet airplanes, nylon stockings, frozen food, instant coffee and brain-washing. **1957** *Amer. Speech* XXXII. 313 Instant lather..instant shaving lather. **1958** *Woman* 18 Oct. 4/3 In these days of 'ready-mix' cakes, 'instant puddings' and other time-saving boons to the busy housewife. **1959** *News Chron.* 26 Sept. 3/5, I tried the new instant tea... I measured out a half teaspoonful of the instant powder, poured on hot water and stirred in milk. *Ibid.* 30 Oct. 6/4 British food firms are being chary about introducing 'instant meals'... Instant bread comes as small frozen pebble shapes which fluff up to fresh crisp rolls after a few minutes in the oven. **1961** H. TRACY *Season of Mists* vii. 84 Sit you down, and I'll make the coffee, nice real instant coffee, none of your messy old grounds. **1962** *Listener* 25 Jan. 185/1 Their habit of producing, on all sorts of occasions, bits of what may be called instant poetry. **1962** J. TERRAINE in *Jrnl. R. United Service Inst.* May 140 There are all too many [historical works] which fall into the category of what I call 'instant history', rapidly composed on the strength of a very fresh approach to the subject, generally devoid of any true period sense, and loaded with comment from false premises. **1963** *Sunday Times* 17 Feb. 29/6 September 3 [*sc.* 1939] is the roundest of capsules: instant peace, instant war, instant history... Instant history may be surface-glossing, but it has a jerky, filmic excitement about it. **1963** *Listener* 7 Mar. 427/3 'This post-war propaganda, piling corpse on corpse, heaping horror on futility, seems bound to fail...' Yet it is back with us now, in the hands of the 'instant historians'. **1965** *Hair Do* July 57/1 Instant curls: for the woman who longs for short, casual curls..the brief luxurious wig is the perfect hair do. **1965** *Punch* 13 Oct. 552/3 A bumper pack of Instant Art for the..young executive—with money. **1966** *Economist* 9 Apr. 161/1 This phrase 'hundred days' is itself out of fashion, because 'instant government' is now condemned as a derogatory term. **1967** *New Scientist* 9 Feb. 352/1 The preparation of instant food is very simple in principle. It consists essentially of precooking the food ..then proceeding to a complete or partial dehydration. **1967** *Observer* 26 Mar. 9 Music and 'instant sex' take subsidiary place to drugs as the real focus of the hippy subculture. **1967** U. SEDGWICK *My Learn-to-cook Book* 56 Fruit is an instant food. **1969** *New Yorker* 29 Nov. 167/1 How could Agnew be so confident of the *Register*'s objectivity and so certain that it would find the speech worthy of an editorial—tantamount to 'instant analysis'. **1973** *Times* 22 Feb. 1/3 Rising demand for mueslis and instant porridges.

instant, *sb.* Add: **7.** An 'instant' beverage (see prec.); *spec.* instant coffee.
1954 *N.Y. Times Mag.* 19 Sept. vi. 52 The 'instants'—soluble coffee, soluble cocoa. **1963** 'A. GILBERT' *Ring for Noose* xi. 132 Julie..asked if she could have coffee. 'I'll make you a cup of Instant, dear,' said kind-hearted Sally. **1968** J. PORTER *Dover goes to Pott* iii. 40 She offered the two detectives a cup of coffee..with..the assurance that there was plenty more instant in the tin. **1973** 'D. HALLIDAY' *Dolly & Starry Bird* i. 4 He had the kettle on and the Instant on the table.

instantaneous, *a.* Add: **1. b.** *Photogr.* Applied to an exposure whose duration is brief and predetermined by means of a shutter mechanism; orig. applied to those sufficiently brief for a moving object to be photographed (in contrast to the time exposures that were usual), and later to those for which the camera may be held in the hand (see quots.). Also *transf.*, applied to a shutter designed for taking such exposures, a photograph taken with one, etc. Now chiefly *Hist.*
1851 *Athenæum* 6 Dec. 1286/2 (*heading*) On the production of instantaneous photographic images. **1858** SUTTON & WORDEN *Dict. Photogr.* 229 Photographs obtained in a fractional part of a second of time are said to be instantaneous. **1867** SUTTON & DAWSON *Dict. Photogr.* 156 Mr. England and others use a guillotine sort of shutter, with a slot across it, which falls immediately in front of the sensitive plate... In some respects this is the best instantaneous shutter that has yet been devised. **1884** *Gaiety Theatre Programme* Oct. in L. de Vries *Victorian Advts.* (1968) 99/3 Superior Sets, comprising..Rectilinear Lens, with Instantaneous Shutter, suitable for taking instantaneous pictures of yachts sailing. **1891** W. E. WOODBURY *Encycl. Photogr.* 356 Instantaneous dry plates are those coated with a very highly-sensitive emulsion that will become sufficiently impressed by the rapid exposure of the image upon it. An instantaneous lens is one that admits a large amount of actinic light. *Ibid.* 357 Instantaneous Photography.—In the very earliest days of photography this term was applied to what would now be considered very slow work indeed. We now usually apply this term when the exposure does not exceed one second. **1902** *Encycl. Brit.* XXXI. 698/2 It [*sc.* an efficient shutter] should be adjustable for variable instantaneous and for prolonged or 'time' exposures. **1939** EMANUEL & DASH *All-in-One Camera-Bk.* 33 It is usual to divide the long series of commonly used exposures into two parts. Exposures of 1/50 or 1/100 second or less belong to the instantaneous group, whereas the slower shutter speeds, such as 1/10 or 1/5 second, and so on, count as time exposures. *Ibid.*, If your hand is reasonably steady you will be able to give an exposure of 1/25 without moving the camera. A picture of this kind, made without the help of a tripod or other fixed support, can be counted as a snap—an instantaneous exposure. **1973** B. JAY *Victorian Cameraman* 57 Instantaneous exposures were not possible when Frith first began his photographic business.
fig. **1867** [in Dict., sense 1].
c. Applied to a gramophone record in which the grooves are made directly by the cutting stylus during recording, and to the making of such records.
1937 *Electronics* Sept. 65/3 (Advt.), Disclube protects the delicate sound impressions in the record groove... A 2 oz. bottle of Disclube will be sent..to any person or firm equipped to make instantaneous recordings. **1949** FRAYNE & WOLFE *Elem. Sound Recording* xiv. 263 Besides its commercial applications, instantaneous recording finds considerable use in the home and as an adjunct in some of the educational fields, such as voice and music training, although..magnetic recording is becoming a serious competitor in certain fields. *Ibid.* 270 There is little practical difference between the best instantaneous records and the best pressings..when both are played with good light-weight reproducers.
d. *instantaneous* (*water-*)*heater* = GEYSER 2.
1935 *Plumber & Decorator* Mar. 57/2 Instantaneous water heaters are of two types—the multi-point, supplying hot water to several taps in different parts of the house; the single-point 'geyser' supplying hot water at the place where it is fixed. **1940** *Chambers's Techn. Dict.* 450/2 Instantaneous water-heater. **1959** *Chambers's Encycl.* VII. 258/1 Instantaneous heaters may be multipoint, supplying hot water to a number of taps, or single point for one tap only.

instantial, *a.* Delete *rare* and add later examples. Spec. *instantial premiss* (see quot. 1933). Hence **instantia·lity,** the state of being instantial.
1921 W. E. JOHNSON *Logic* I. x. 160 This entirely distinct and peculiar use of the term 'existential' has given rise to endless confusion; and..the term should be entirely discarded and replaced by some such term as *instantial*, or, more accurately, indeterminately instantial. **1922** —— *Logic* II. x. 210 The conclusion is a generalisation of a certain premiss or set of premisses which..may be spoken of as 'the instantial premiss'. **1933** C. A. MACE *Princ. Logic* xii. 243 Induction may be defined as a form of inference in which, given instantial premisses (i.e. premisses concerning 'instances' or particular cases), we draw a conclusion involving some measure of generalization of these premisses. **1961** E. NAGEL *Struct. of Sci.* iii. 31 The above premisses..contain a number of singular or instantial statements. **1966** J. ELLIS in C. E. Bazell *In Memory of J. R. Firth* 81 Contextual meaning is either potential or instantial (or, less unambiguously, 'potential' or 'actual'). *Ibid.*, Between these extremes we might recognize a cline of potentiality/instantiality.

instantiate (instæ·nʃi̯ēit), *v.* [f. INSTANCE *sb.* (L. *instānti(a)* + -ATE[3].] *trans.* To represent by an instance. Also **insta·ntiative** *a.*, of or pertaining to such instances; **insta·ntiating** *ppl. a.*
1946 H. H. PRICE in *Proc. Brit. Acad.* XXXII. 117 This act of recognizing..is at the same time the verifying of a proposition, the discovering of a fact or truth. The proposition in question is an existential proposition (it might also be called an instantiative proposition). **1949** J. R. JONES in *Philos. Rev.* LVIII. 162 He expressly implies that they are instantiated in the concrete things to which they belong. **1951** J. HOLLOWAY *Lang. & Intelligence* ii. 18 Two apples..both instantiate the single universal redness. **1954** I. M. COPI *Symbolic Logic* iv. 100 The instantiating constant '*b*' occurs in the premiss. **1959** P. F. STRAWSON *Individuals* iv. 130 C is a member of a set of complete concepts K, such that all and only the members of K are, in fact, each uniquely instantiated. **1965** A. PLANTINGA in M. Black *Philos. in Amer.* 214 God could instantiate *P* by instantiating P_1. **1972** *Language* XLVIII. 350 Thus the ambiguity of the distinctive-feature relations is mirrored by the ambiguity..of the implementation rule, thereby instantiating one of the chief consequences of the iconic function.

instantiation (instænʃi̯ēi·ʃən). [f. as prec. + -ATION.] The action or fact of instantiating; representation by an instance. Also *attrib.*
1949 J. R. JONES in *Philos. Rev.* LVIII. 162 It is the view that instantiation is necessarily a dyadic relation. **1953** H. H. PRICE *Thinking & Experience* i. 18 It is not easy to see how the doctrine of *universalia in rebus* can make room for this important and familiar notion of instantiation. **1954** A. J. AYER *Philos. Ess.* i. 18 The ∃x.. has no descriptive force; it serves only to make an instantiation claim. **1954** I. M. COPI *Symbolic Logic* iv. 98 The instantiation rules *UI* and *EI* must be reformulated. **1956** E. H. HUTTEN *Lang. Mod. Physics* vi. 234 The second alternative, i.e. by instantiation, similarly suggests that we use the law as an empty schema. **1964** M. BLACK *Compan. to Wittgenstein's Tractatus* lxiii. 293 W. [*sc.* Wittgenstein] modifies the convention by demanding that visibly different variables shall be treated *differently* for the purpose of instantiation. **1972** *Language* XLVIII. 343 But a more difficult task also confronts the linguist: discovering the universal laws underlying language as a whole, of which specific rules are instantiations.

instantize (i·nstăntəiz), *v.* [f. INSTANT *a.* + -IZE.] *trans.* To make (foodstuffs) available in instant form (see *INSTANT *a.* 4 c). So **i·nstantized** *ppl. a.*
1962 R. J. CLARKE in *Proc. 1st Internat. Congr. Food Science & Technol.* (1965) IV. 5 'Instantized' milks, with lowered bulk density are making their appearance. **1970** *Americana Ann.* 58 The department contracted..for the purchase of 'instantized' nonfat dry milk for distribution in welfare programs. **1970** *New Scientist* 24 Dec. 560/1 The formulated, instantised, convenience foods will no longer look like meat, milk, cereal or vegetable.

instar (i·nstāɹ), *sb.* *Ent.* [mod.L. (L. H. Fischer *Orthoptera Europæa* (1853) I. 37), a. L. *instar* form, figure, likeness.] Any of the stages in the life of an insect or other arthropod between successive ecdyses, including the stage between hatching from the egg and the first ecdysis; also (and orig.), an individual arthropod at a particular stage.
1895 D. SHARP in *Cambr. Nat. Hist.* V. v. 158 It may be well to adopt a term suggested by Fischer, and call the Insect as it appears at hatching the first instar, what it is as it emerges from the first ecdysis the second instar, and so on. **1925** A. D. IMMS *Gen. Textbk. Entomol.* II. 176 The intervals between the ecdyses are known as stages or stadia, and the form assumed by an insect during a particular stadium is termed an instar. **1932** METCALF & FLINT *Fund. Insect Life* vi. 156 The molts occurring during the growing period divide this life stage..into a number of sharply separated sizes or steps that are called instars. **1964** R. M. & J. W. FOX *Introd. Compar. Ent.* viii. 268 Most spiders provide a cocoon in which the first two instars are spent in passive existence. **1970** *Nature* 24 Oct. 382/1 In Jamaica the fifth instar larvae of the New World hawkmoth..exhibit four basic colours.

in statu nascendi: see *IN *Lat. prep.*

in statu pupillari: see *IN *Lat. prep.*

instep[1]. Add: **3. b.** The arched part of a boot or shoe between the heel and the sole.
1826 *Kaleidoscope* VI. 223/2 A boot-jack of this description..has an equal purchase on the instep and heel. **1913** E. C. BENTLEY *Trent's Last Case* v. 106 On each [shoe], in the angle between the heel and the instep, he detected a faint trace of red gravel. **1927** 'E. BRAMAH' *Max Carrados Mysteries* 160 He slyly inserted a nail in the angle of the instep.

instigatrix. For '† *Obs. rare*⁻⁰' read '*rare*' and add later example.
1902 *Dublin Rev.* July 105 Lady Macbeth, the instigatrix of the murder.

i·nstinctless, *a.* [INSTINCT *sb.* + -LESS.] Without or lacking instinct.
1890 W. JAMES *Princ. Psychol.* I. ii. 77 Schrader gives a striking account of the instinctless condition of his brainless pigeons. **1947** A. EINSTEIN *Mus. Romantic Era* xiii. 181 This return [to antiquity]..was no longer inspired by the reverential Classicism of the 18th century, but rather by the indiscriminate and instinctless 'learnedness' of the 19th.

instinctual (insti·ŋktiuăl), *a.* [f. INSTINCT *sb.* (L. *instinctu(s)* + -AL.] Of or pertaining to, involving or depending upon, instinct.
1924 G. KNOX *Land of Afternoon* 294 She possessed the female's instinctual power to project this force. **1925** J. RIVIERE et al. tr. *Freud's Coll. Papers* IV. 62 A stimulus of instinctual origin does not arise in the outside world but from within the organism itself. **1934** H. C. WARREN *Dict. Psychol.* 140/1 *Instinctual fusion*, the theory that every mental process is the result of a fusion of the life instinct and the death instinct. **1937** H. READ *Art & Society* vii. 263 This passage clearly implies an opposition between instinctual (imaginative, creative) activity and practical, mechanical activity. **1945** KOESTLER *Yogi & Commissar* III. iv. 228 Its instinctual root is probably the feeling of insecurity. **1955** J. C. FLÜGEL *Stud. Feeling & Desire* iv. 100 In civilised societies suicide is hardly ever considered as occurring as a result of a natural or instinctual impulse. **1957** R. F. C. HULL tr. *Jung's Coll. Works* I. 132 Here it is not so much a lack of ethical feelings..as an excess of instinctual drives. **1966** R. ARDREY *Territorial Imperative* (1967) ii. 46 The pattern..is common to the species and is instinctual.

institute, *sb.*[1] **4.** (Further examples without defining word, esp. as shortened form of *Women's Institute*.)
1924 KIPLING *Debits & Credits* (1926) 116 She told me there was a whist-drive that afternoon at the Institute. **1939** M. SPRING RICE *Working-Class Wives* v. 111 A woman in another Essex village speaks of 'work for the Institute'. **1959** M. M. KAYE *House of Shade* vi. 74 She misses the Institute and the Girl Guides.

institution. 8. (Further examples.)
1902 *Daily Chron.* 7 May 3/7 Institution life year after year is like pushing a stone uphill. **1905** *Ibid.* 27 Sept. 4/5 To any high-spirited woman, the tyranny of institution life must be almost unbearable. **1930** J. B. PRIESTLEY *Angel Pavement* v. 210 That institution atmosphere..was rather depressing. **1956** [see *APPROVED *ppl. a.* 5]. **1960** I. BENNETT *Delinquent & Neurotic Children* iv. 129 Failures in the socialization process..will occur more frequently among institution children. **1963** F. F. LAIDLER *Gloss. Home Econ. Educ.* 48 *Institution administration*, carrying out the management of, or the executive duties concerned with institutions, e.g. Colleges, Hospitals, etc.

institutional, *a.* Add: **1. c.** Of religion: organized into or finding expression through institutions (a church, ordained ministers, ritual). Cf. *INSTITUTIONALISM (*a*).
1908 F. VON HÜGEL *Mystical Element of Relig.* p. vi, The Infinite can still find room for the Historical and Institutional elements in Religion. **1924** A. E. J. RAWLINSON *Authority & Freedom* vi. 136 The Christianity of history is a sacramental and institutional religion. **1941** A. C. BOUQUET *Compar. Relig.* vii. 99 But if the world is unreal, and Deity unknowable, ordinary institutional religion, with its sacrifices and celestial bargaining, becomes a mere fraud.
d. *Linguistics.* (See quots.)
1958 T. HILL in *Orbis* VII. 454 A new branch of systemic linguistics dealing with the types of relation that arise in use between tongues, and between them and their users... As its aim is to confront tongues and communities as social institutions, it might be called *Institutional Linguistics*. **1964** M. A. K. HALLIDAY et al. *Ling. Sci.* i. 16 There is one aspect [of the study of how language works] ..which does not strictly fall within 'descriptive linguistics': the study of language in relation to those who use it. This, since it is really the study of language as an institution, is recognized as a separate branch of the linguistic sciences and has come to be known as 'institutional linguistics'. *Ibid.*, Whereas in a descriptive statement we may note..that a speaker has used a transitive and not an intransitive clause.., in an institutional statement we may note that he has spoken English and not French.., that he spoke conversationally and not formally.
3. (Later examples.)
1942 *Times Lit. Suppl.* 14 Mar. 127/1 Born in a new country at a time when the desire for freedom from old institutional ways of living..was burning like a flame in many lands. **1960** I. BENNETT *Delinquent & Neurotic Children* vii. 290 Personality was in some ways similar to that of a deprived institutional child. **1963** F. F. LAIDLER *Gloss. Home Econ. Educ.* 50 *Institutional management*, the study of all branches of the administration of large residential or non-residential establishments, including all branches of food service, menu planning, food purchase, preparation and service, equipment and personnel administration. **1966** R. BARTON (*title*) Institutional neurosis. **1972** *Jrnl. Social Psychol.* LXXXVI. 64 The Christian Conservative has a parallel in the 'Institutional Restraint' type in Gordon's analysis. **1972** *Times* 16 Oct. 12/5 In most cases they no longer require medical or nursing care and suffer mainly from 'institutional neurosis' due to their long stay in the protective environment of the hospital ward.
b. Of advertising, etc.: that lays stress on the business firm or institution rather than on the product itself.
1919 T. RUSSELL *Commercial Advertising* 258 Some manufacturers and commercial firms have used what is called 'institutional advertising': instead of advertising their wares they advertised their firm. **1930** *Harvard Advertising Awards, 1924–28* (Harvard Univ. Grad. School of Business Admin.) 75 In 1923, General Motors.. accepted institutional advertising as a business tool. **1957** CLARK & GOTTFRIED *University Dict. Business & Finance* (1967) 10/1 Institutional..Advertising..is aimed at keeping the name of a company or group before the public. **1967** *Economist* 4 Feb. 448/2 A..dignified 'institutional' sales approach.

institutionalism. Add: *spec.*, (*a*) the principles of institutional religion; (*b*) the system of housing people in institutions; the characteristics of life in an institution.
1907 *Daily Chron.* 17 Oct. 8/4 There is nothing of 'institutionalism' about life at the Browning Bethany Homes. **1909** G. TYRRELL *Christianity at Cross Roads* Pref. p. xx, The time has come..for a criticism of categories—of the very ideas of religion, of revelation, of institutionalism, of sacramentalism, of theology, of authority, etc. **1909** *Daily Chron.* 28 Oct. 5/3 The great impediment to success in the work of saving fallen women was the machinery of institutionalism. **1927** *Observer* 21 Aug. 7 The Society has set its face against any tendencies towards 'institutionalism', against any attempt..to take children away from their own homes. **1930** W. R. INGE *Christian Ethics & Mod. Probl.* i. 16 The unquestionable advantages which this fanatical institutionalism confers upon the Church as an organisation. **1958** *Times Lit. Suppl.* 19 Sept. 531/1 Institutionalism is usually followed by repeated further stays in institutions.

institutionalist. Add: Also, one who favours the retention of an institution or institutions.
1920 G. B. SHAW in *New Commonwealth* Suppl. 2 Jan. 3/1 That distinction between the churchman, between the person the Dean of St. Paul's calls the institutionalist, and the genuine out and out Protestant mystic, will always cause a certain division. **1957** O. R. MCGREGOR *Divorce in England* v. 134 'Institutionalists' who insisted on its [*sc.* the matrimonial offence's] retention.

institu:tionaliza·tion. [f. INSTITUTIONALIZE *v.* + -ATION.] The condition or state of being or becoming institutionalized; the action of institutionalizing.
1951 E. E. EVANS-PRITCHARD *Social Anthropol.* iii. 49 This order is brought about by the systematization, or institutionalization, of social activities so that certain persons have certain roles in them and so that the activities have certain functions in the general social life. **1964** M. ARGYLE *Psychol. & Social Probl.* vi. 85 The traditional [mental] hospital was run rather like a prison: inmates were treated as if they were incapable of taking an active or responsible place in the community, and it has been found that the longer they have been inside the more they suffer from 'institutionalization'—i.e. they become apathetic..and become quite dependent on the hospital. **1964** I. L. HOROWITZ *New Sociol.* 21 If we confront problems of social development, we cannot rest content with their institutionalization in one minuscule town. **1969** H. PERKIN *Key Profession* i. 2 The universities are..the institutionalization of innovation in the arts. **1970** G. GREER *Female Eunuch* 234 The alternative is not the institutionalization of parental functions. **1972** *Physics Bull.* Aug. 445/2 The institutionalization of science has brought it, and particularly its R and D aspects, decisively into the realm of government and politics.

institutionalize, *v.* Delete *rare* and add: (Later examples.) Freq. as *ppl. a.*
1935 B. MALINOWSKI in M. Black *Importance of Lang.* (1962) 80, I have defined magic as the institutionalized expression of verbal optimism. **1952** W. J. H. SPROTT *Social Psychol.* ix. 187 In some simple societies it [*sc.* homosexuality]..is institutionalized;..those showing a persistent preference for their own sex are given a special position. **1959** *Daily Tel.* 3 Apr. 10/2 Only a firm and institutionalised American commitment to Western Europe could achieve this end. **1962** W. NOWOTTNY *Lang. Poets Use* viii. 200 It means that the tears of the saviour have been metamorphosed into an institutionalized church. **1966** *Listener* 17 Nov. 749/3 That it should issue from a highly institutionalized company should not surprise us. **1971** A. QUINTON in A. Bullock *20th Cent.* 261/2 Marxism, institutionalized in the Communist world, has developed in a direction precisely opposite to the one he hoped for.
b. To house, train, or bring up in an institution; to subject (a person) to institutional life. In *pass.*, to show signs of the influence of institutional life.
1905 J. BURNS in *Daily Chron.* 1 Dec. 4/4 He has been 'institutionalised', and I never yet knew the average man survive that pauperising ordeal. **1924** *Glasgow Herald* 24 May 9 The day of the institutionalised boy or girl is past. **1949** M. MEAD *Male & Female* x. 204 Our techniques for diagnosing or institutionalizing the criminally insane are inadequate. **1959** B. WOOTTON *Social Sci. & Social Path.* iii. 119 In which either a parent had been permanently institutionalized or the child concerned had been sent to live with relatives. **1969** *Daily Tel.* 3 May 21/8 Because he was hopelessly institutionalised he was unable to look after himself when free. **1970** G. F. NEWMAN *Sir, You Bastard* v. 150 Sneed's doctor was sympathetic to his views, and thought the best plan was to put the child in an institution and have another. He offered to go and see Angie and help Sneed to institutionalize the infant. **1971** *Oxford Times* 15 Oct. 1/9 [He] had been in approved schools, prison and mental hospitals for much of his life and had become 'institutionalised'.
c. *Linguistics.* Usu. in *pa. pple.* or as *ppl. a.*: recognized or accepted by the speech community.
1949 J. R. FIRTH in *Trans. Philol. Soc. 1948* 128 It is especially helpful that there *are* things called English words and Arabic words..; indeed, English words and Classical Arabic words are firmly institutionalized. **1961** Y. OLSSON *Syntax Eng. Verb* ii. 24 Punctuation..and spacing..mark off units which are institutionalized, that is to say, recognized by the speech community. **1962** *Listener* 27 Sept. 467/2 Both in grammar and vocabulary Pidgin has deeply engrained distinctive features which are quite institutionalized, as we say.

institutionize (institiū·ʃənəiz), *v.* [f. INSTITUTION + -IZE.] *trans.* To render institutional; to institutionalize.
1903 G. TYRRELL *Church & Future* 61 In regarding the 'institutionising' of Christianity as a corruption..Protestantism seems to me to ignore universal and natural laws. **1903** W. JAMES *Mem. & Stud.* (1911) xiv. 335 The institutionizing on a large scale of any natural combination of need and motive always tends to run into technicality and to develop a tyrannical Machine.

i·n-store, *a.* [f. phr. *in store*: cf. *IN *prep.* 17*.] Of or relating to goods, etc., held in store; that is situated or takes place in a store.
1961 *Progressive Grocer* Oct. 38 The in-store advancement training program is given much credit for the success of Safeway Stores in his division. **1968** *Observer* 22 Dec. 11 What are euphemistically described as 'in-store wastages', can mean, he says, 'losing a whole truckload of stuff before it even gets through the loading bay.' The in-store thief could be someone extracting a little money from each till when it is being cashed up, a 16-year-old on the stocking-counter, or a loading bay hand working in collusion with a truck driver. *Ibid.*, There are no niceties about in-store security. **1972** *Police Rev.* 10 Nov. 1467/2 (Advt.), Investigation of all in-store security.

instress (i·nstres), *sb.* [f. IN *adv.* 12 + STRESS *sb.*] In the theories of Gerard Manley Hopkins: the force or energy which sustains an inscape (see quots.). Hence **i·nstress** *v. trans.* and *intr.*; **i·nstressed** *ppl. a.*; **i·nstressing** *vbl. sb.*
1868 [see *INSCAPE *sb.*]. **1873–4** G. M. HOPKINS *Note-bks. & Papers* (1937) 226 You can without clumsiness instress, throw a stress on/a syllable so supported. **1875** —— *Jrnls. & Papers* Feb. (1959) 263 Standing before the gateway I had an instress which only the true old work gives from the strong and noble inscape of the pointed-arch. **1876** —— *Wr. Deutschland* in *Poems* (1967) 53 His mystery must be instressed, stressed. **1881** —— *Note-bks. & Papers* (1937) 349 This song of Lucifer's was a dwelling on his own beauty, an instressing of his own inscape. **1944** W. H. GARDNER *G. M. Hopkins* i. 11 In the vagaries of shape and colour presented by hills, clouds, glaciers and trees he discerns a recondite pattern—'species or individually-distinctive beauty'—for which he coins the word 'inscape'; and the *sensation* of inscape (or, indeed, of any vivid mental image) is called 'stress' or 'instress'. **1948** W. A. M. PETERS *G. M. Hopkins* i. 14 The original meaning of instress..is that stress or energy of being by which 'all things are upheld'..and strive after continued existence.

in-stroke. Add: Also **instroke. 2.** The stroke which carries the piston away from the crankshaft and further into the cylinder of an engine.
1902 *Encycl. Brit.* XXVIII. 183/2 Suction during an entire outstroke of the piston;..compression during the following instroke. **1922** GLAZEBROOK *Dict. Applied Physics* I. 320/1 In the Otto cycle the inlet valve usually opens slightly before the end of the in-stroke of the working piston.

instruction. Add: **4. c.** *Computers.* An expression in a program or routine, or a sequence of characters in a machine language, which specifies an operation (esp. a basic operation) and freq. also one or more operands, and results in its performance by the computer.
1947 GOLDSTINE & VON NEUMANN in J. von Neumann *Coll. Wks.* (1963) V. 82 The control scans the coded instruction in the selectron memory as a rule linearly. **1948** *Math. Tables & Other Aids to Computation* III. 121 The terms defined in the following short glossary are used in their technical sense in this discussion... 2. Word: A group of digits..stored in coded form in a single memory position. .. 5. Instruction: A word directing the machine to perform a particular operation. **1949** D. R. HARTREE *Calculating Instruments & Machines* (1950) 67 A single instruction of this type specifies two locations, one in the store and one in the arithmetical unit, and an operation. **1951** [see *COMMAND *sb.* 1 d]. **1958** M. PHISTER *Logical Design Digital Computers* xi. 342 An instruction (also called a command or an order) is a computer word which identifies which of a limited number of operations the computer is to carry out, and the computer functions by executing a sequence of these instructions, one at a time. **1959** E. M. MCCORMICK *Digital Computer Primer* x. 139 The number of instructions which can be executed by a computer represents a compromise between the designer's and user's requirements. The fewer the instructions, the easier it is to design and build the machine, but the more difficult (at least time-consuming) it is to code for it. *Ibid.* 141 Some computers have two separate instructions in one word... However, most computers, binary or decimal, use one instruction per word. **1962** B. A. GALLER *Lang. Computers* xii. 193 In the IBM 650 computer, for example, a typical [machine] instruction might be 1512150123 in which the first two digits (15) indicate that the contents of the location specified by the next four digits (1215) are to be added to the lower half of the accumulator register. The next four digits (0123) give the address of the location in storage in which the next instruction to be executed is stored. **1967** [see *COMMAND *sb.* 1 d]. **1969** P. B. JORDAIN *Condensed Computer Encycl.* 253 The word instruction is preferable to the words command and order, sometimes used synonymously. Command should be reserved for electrical signals. **1970** O. DOPPING *Computers & Data Processing* vi. 98 When local control units have their own instructions, these instructions can be called commands to distinguish them from the instructions to the central control unit.
5. *attrib.*, as *instruction book*.
1895 *Montgomery Ward Catal.* 123/2 Tissue Paper Flower Outfits..Containing..12 half sheets assorted tissues,..and instruction book. **1926–7** *T. Eaton & Co. Catal.* Fall & Winter 300/3 Easy Method Instruction Books for Various Instruments. **1970** *Which?* Aug. 237/1 The instruction books generally gave good information for setting up and using the machines.

instructional, *a.* Add: **3.** spec. *instructional film* (also ellipt.), *set, television* (U.S.).
1933 (*title*) Guide to instructional and educational films available for use by educational and social organisations in Great Britain (Central Information Bureau for Educational Films). **1944** R. MANVELL *Film* I. 36 The case of the close-up in documentary..is rather different... So also is the obvious importance of close-up in the instructional film, where processes are being explained and emphasised. **1959** HALAS & MANVELL *Technique Film Animation* 13 We are seeing a considerable expansion of the animated film into every kind of use from the television commercial to the highly specialized instructional film. **1962** *Listener* 25 Jan. 193/1 The earnest reportage, the non-theatrical instructionals, and the forward-looking films. **1971** *Jrnl. Gen. Psychol.* Jan. 59 Kazsuk and Bartley supported these findings in a subsequent investigation of the effect of instructional sets on brightness matching. **1972** *Jrnl. Social Psychol.* LXXXVI. 155 For individual cooperative choices, instructional set..and trial blocks.. were significant. **1966** *Sat. Rev.* (U.S.) 19 Nov. 88 After more than a decade of intensive effort and the expenditure of hundreds of millions of dollars, instructional television seems to have arrived. **1968** *Economist* 21 Dec. 31/3 The receiving stations will still fill the rest of the day with instructional television (which is what ITV stands for in America) of a quality which Mr McGeorge Bundy of the Ford Foundation has described as staggeringly mediocre.

instructive, *a.* Add: **b.** Denoting the case used in some languages, e.g. the Ugro-Finnish group, to express means.
1857 *Trans. Philol. Soc.* 34 Nominative..genitive.. dative..instructive..affective. **1890** C. N. E. ELIOT *Finnish Gram.* 22 The instructive is formed by simply adding the letter *n* to a root, and expresses the means by which anything is done. **1896** [see INESSIVE *a.*]. **1954** PEI & GAYNOR *Dict. Ling.* 102 *Instructive*, in certain languages (e.g., Finno-Ugric languages), a case having the same denotation as the English *by means of*.

instructor. Add: **b.** (Earlier and later examples.)
1722 in B. Peirce *Hist. Harvard Univ.* (1833) 232 Voted ..that Mr. Judal Monis be *improved* as an instructor of the Hebrew language in the College. **1873** in *Sci. Amer.* (1973) May 11 There are at present 500 students, and the faculty consists of 40 professors and instructors. **1891** *Univ. Chicago Official Bull.* I. 11 Lecturers and teachers.. shall be classified as follows... (6) The Instructor. **1947** [see *full professor* (*FULL *a.* 12 d)].
Hence **instructo·rial** *a.*
1952 [see *assistant professor].

instrument, *sb.* Add: **2.** (Further examples.) Also applied to devices whose primary function is to respond to a physical quantity or phenomenon, esp. by registering or measuring it, rather than to accomplish an effect, and which may function with little direct human intervention and be of complicated design and construction.
1672 [see BAROMETER]. **1839** [see INDICATOR 3]. **1845** *Encycl. Metrop.* IV. 68/1 By the term *Electroscope* we understand an instrument which enables us to observe the presence of free electricity. **1864** *Rep. Brit. Assoc. Adv. Sci. 1863* 147 Electric currents are most simply compared by 'electro-dynamometers' (20)—instruments which, unlike galvanometers, are practically independent of the intensity of the earth's magnetism. **1876** [see INDICATOR 3]. **1884** *List of Subscribers* (London & Globe Telephone Co.), The newest and most improved Telephonic Instruments and Apparatus. *Ibid.*, Subscribers..must not allow non-subscribers to use their instruments for the transmission of messages. **1888** S. R. BOTTONE *Electr. Instrument Making* (ed. 2) 116 Ammeters.—These instruments are intended to measure the amount of current in ampères passing through any given circuit. **1889** PREECE & MAIER *Telephone* xxix. 464 An instrument of such marvellous sensitiveness as the telephone. **1910** R. W. A. BREWER *Art of Aviation* xvi. 213 Other instruments will be carried to indicate the speed of the machine relatively to the air and to the land. **1924** P. J. RISDON *Wireless* xxxviii. 304 These bells are kept ringing until the operator gets to his instruments, adjusts his headpiece and is ready to receive the direction signals giving the position of the vessel in distress. **1947** E. W. F. FELLER *Instrument & Control Manual* xii. 280 The primary purpose of an emergency instrument is to sound an alarm or shut down equipment upon some abnormal operation. **1950** A. MARCUS *Radio Servicing* xiii. 585 The vacuum-tube voltmeter..is rapidly becoming one of the most useful instruments employed in radio servicing. **1957** *Encycl. Brit.* XII. 447/1 Instruments which make a record on paper of the magnitude of an electrical quantity with time are available in considerable variety. **1959** K. ULLYETT *Jaguar Compan.* x. 170 Polished figured walnut instrument panel features revolution counter and speedometer..and separate instruments for oil pressure, water temperature, fuel gauge and ammeter. **1963** GODDARD & BROWN *Pract. Chem.* iii. 57 In modern analytical techniques, heavy reliance is placed upon the use of instruments, such as pH-meters, potentiometric titrators,.. spectrographs, polarimeters, refractometers, etc. **1967** D. P. DAVIES *Handling Big Jets* viii. 231 The attitude indicator is the primary flight instrument for turbulence flying.
6. *instrument-carrying* adj.; **instrument board** = *instrument panel*; **instrument panel,** a surface on which gauges, dials, etc., of measuring or indicating instruments are grouped together (as in a motor vehicle or aircraft).
1917 'CONTACT' *Airman's Outings* 18 Not till a pilot can fly his bus unconsciously does he keep place without repeated reference to the throttle and instrument-board. **1926** *Amer. Speech* I. 686/1 Automobile nomenclature... Instrument board [American], facia board [English]. **1930** P. WHITE *How to fly Airplane* iv. 65 The first endeavor of the pilot-instructor is to familiarize the beginner with..the functions of the various controls, and with the instrument board. **1935** *Discovery* Feb. 44/2 The Leningrad University Observatory is stated to be building a high-altitude instrument-carrying rocket for stratospheric work. **1959** *Daily Tel.* 1 May 1/1 The programme will certainly enable Britain to put three or four instrument-carrying satellites into space. **1933** *Discovery* Feb. 59/2 An instrument panel..is incorporated in the camera. **1935** *Economist* 7 Dec. 1140/1 The use of plastics in the motor accessory field will undoubtedly increase considerably in the future... Another noteworthy development with which plastics are connected is the combination instrument panel... By grouping all the instruments in a single panel the space saved can be utilised to give larger storage pockets. **1958** W. R. BARRETT et al. in H. W. Cremer *Chem. Engin. Pract.* V. 126 Large [air-compressor] installations usually have an instrument panel on which are fitted all pressure gauges and a temperature indicator. **1972** E. H. J. PALLETT *Aircraft Instruments* iii. 43 A more effective and standardized grouping has now been adopted; this is known as the 'basic T'... It constitutes a system by which various items of related flight information can be placed in certain standard locations in all instrument panels.
b. With reference to the use of, esp. dependence on, instruments in the flying of aircraft, as *instrument conditions, runway, weather*; **instrument flying,** flying in which the pilot makes no observation of the ground but depends entirely on the instruments in the aircraft; so *instrument flight*; similarly **instrument approach** or **landing,** an approach (*APPROACH *sb.* 13) or landing in which the pilot depends entirely on instruments and a ground-based radio guidance system (an *instrument landing system*).
1947 *Engineering News-Record* 16 Oct. 532/3 Runway E, for instrument approach is 10,000 ft. long and will be paved 8,000 × 200 ft. **1957** *Encycl. Brit.* I. 230/1 During World War II another system was developed for instrument approaches. This..used a talk-down technique. **1957** Instrument conditions [see *instrument landing* below]. **1943** Instrument flight [see *instrument flying* below]. **1956** W. A. HEFLIN *U.S. Air Force Dict.* 270/2 *Instrument flight*, a flight made by using instruments, without visual reference to the ground. **1928** STERLING & KRUSE *Radio Manual* xiii. 506 While instrument flying may enable a pilot to keep his craft at a safe altitude and in a generally correct direction, the hazard of getting far away from the course..is ever present. **1943** *Instrument Flight* (U.S. Bureau of Aeronautics (Navy Dept.)) p. iii, Instrument flying is as logical and easy as contact flying. **1938** *Proc. IRE* XXVI. 681 At present, the major airlines are planning to install a number of instrument landing systems. **1942** J. B. HOAG *Basic Radio* xxxiv. 288 At certain airports, a radio wave is transmitted ..to provide a glider path for 'blind' or instrument landing. **1957** *Encycl. Brit.* I. 230/1 In the United States the Civil Aeronautics administration developed a system for landing approaches under instrument conditions, commonly known as ILS (instrument landing system). **1947** *Shell Aviation News* CXIII. 3/1 It will have three runways: an instrument runway (8,500 ft. × 250 ft.) and two non-instrument runways. **1951** *Gloss. Aeronaut. Terms* (*B.S.I.*) III. 23 *Instrument runway*.., a runway..equipped with non-visual aids for take-off and landing. **1949** *Britannica Bk. of Year 1948* 37/1 Seventy-nine per cent. of all aircraft..landed during instrument weather with no traffic delays.

instrument, *v.* Add: **3.** *trans.* To equip or provide with instruments (for measuring, recording, controlling, etc.).
1949 *Trans. Soc. Instrument Technol.* I. v. 21/1 The extent of instrumentation..is greater than it is in comparable British plants, although the most modern of these are just as comprehensively instrumented as their opposite numbers in the U.S.A. **1959** *Instrument Pract.* XIII. 194/1 The methods [of analysis] are used specifically for production control and merely to instrument them for laboratory use may be to dally unnecessarily halfway. **1961** *New Scientist* 16 Mar. 661/1 The system will be equipped throughout for automatic control and instrumented for safety. **1962** *B.S.I. News* Feb. 27/2 The whole equipment will be scientifically instrumented. **1962** F. I. ORDWAY et al. *Basic Astronautics* iv. 125 Explorer II ..was instrumented primarily to detect and measure high-energy gamma rays. **1970** *Physics Bull.* Mar. 107/1 It would be too costly to instrument and monitor the whole process.
So **i·nstrumented** *ppl. a.*, equipped with or using instruments.
1947 *Shell Aviation News* CIX. 13/1 In a large hangar there [is]..the completely instrumented fuselage of a modern bomber. **1954** *Trans. Soc. Instrument Technol.* VI. 47/1 An analysis of the total number of instruments employed on a well-instrumented ironmaking plant. **1957** *Times* 10 Oct. 10/1 The first fully instrumented satellite was planned for launching in March. **1967** *Guardian* 19 Oct. 1/7 The Russian success in carrying out the first instrumented landing on another planet. **1972** *Lebende Sprachen* XVII. 72/2 This is the launch day scheduled for UK-3, the third British-instrumented satellite to be lifted into orbit by an American scout rocket.

instrumental, *a.* and *sb.* Add: **A.** *adj.* **7.** *Psychol.* A term used to describe the type of learning where a particular response is the instrument by which the organism is taught to alter its environment.
1940 HILGARD & MARQUIS *Conditioning & Learning* iii. 51 When the occurrence of the reinforcement is contingent upon the organism's behavior the procedure may be termed instrumental conditioning. *Ibid.* 52 As a reference experiment for instrumental reward training we may select a study by Grindley. **1956** B. R. BUGELSKI *Psychol. of Learning* iv. 58 'Instrumental' learning covers all other types. It is called 'instrumental' because the organism is learning how to affect its environment to bring about some change. **1964** M. ARGYLE *Psychol. & Social Probl.* x. 128 Instrumental learning consists of the blind stamping-in of rewarded responses. **1969** R. H. SCHUSTER in D. P. Hendry *Conditioned Reinforcement* viii. 194 The complication of three factors is necessary—simple association, instrumental reinforcement, and a decay process. *Ibid.* 195 Whether the stimuli precede or follow an instrumental response. **1971** J. L. GEWIRTZ in R. Glaser *Nature of Reinforcement* viii. 288 He assumes that instrumental conditioning might be more effective for the acquisition of responses connoting complex skills. **1972** *New Yorker* 26 Aug. 32/1 Edward L. Thorndike..is credited with the first rigorous investigation of trial-and-error, or instrumental, learning.
B. *sb.* **4.** (See quot. 1945.)
1940 *Swing* July 17/1 Bob Mersey's *Blue Ink* is another slightly *Wham*-like instrumental. **1945** *Music Library Assoc. Notes* 2nd Ser. VII. I. 45/1 *Instrumental*, composition written for instrumental performance, solo or group. Also, any performance without benefit of a vocal. **1949** L. FEATHER *Inside Be-Bop* i. 9 Dizzy..changed it from a slow ballad to a jump-tempo instrumental. **1972** *Jazz & Blues* Sept. 11/3 'Jump' instrumentals, normally featuring the tenor of Herb Hardesty.

instrumentalism (instrʊme·ntăliz'm). *Philos.* [f. INSTRUMENTAL *a.* + -ISM.] The pragmatic theory of John Dewey (1859–1952) that thought exists as an instrument of adjustment to the environment; *spec.* that terms of thought and meaning are relative to the function they perform and that their validity or truth is determined by their efficacy.
1909 *Philos. Rev.* XVIII. 396 By instrumentalism is meant that element of pragmatism which has grown out of the application of the evolutionary method to logical problems. **1929** J. DEWEY *Experience & Nature* iv. 151 'Instrumentalism' is a theory not about personal disposition and satisfaction in knowing, but about the proper objects of science. **1931** A. WOLF in W. Rose *Outl. Mod. Knowl.* 549 Pragmatism, instrumentalism, and fictionism..treat beliefs as instruments of life, and to be valued accordingly. **1948** B. RUSSELL *Human Knowl.* 75 There is another kind of 'meaning', which gives occasion for pragmatism and instrumentalism. **1968** J. J. C. SMART *Between Sci. & Philos.* v. 142 Instrumentalism does allow theoretical concepts to be quite free constructions of the theorist.

instrumentalist. Add: **3.** *Philos.* One who advocates the theory of instrumentalism. Also *attrib.*
1909 *Philos. Rev.* XVIII. 397 From the instrumentalist standpoint, the inquiry, *What is reality?* appears..futile. **1913** W. CALDWELL *Pragmatism & Idealism* i. 17 Professor Dewey has also written many..short studies upon the application of an instrumentalist conception of philosophy to education and to social questions. **1940** B. RUSSELL *Inquiry into Meaning & Truth* viii. 154 There are some schools of philosophy—notably the Hegelians and the instrumentalists—which deny the distinction between data and inferences altogether. **1965** J. D. NORTH *Measure of Universe* App. 421 It would be misleading to describe the views represented in this book as 'pragmatist' or 'instrumentalist'.

instrumentally, *adv.* Add: **3.** *Gram.* In or by the instrumental case.
1846 M. MONIER-WILLIAMS *Elem. Gram. Sanscrit Lang.* 160 Instrumentally Dependent..those [compounds] in which the relation of the first word (being in the crude) to the last is equivalent to that of an instrumental case.

instrumentation. Add: **4.** The design, construction, and provision of instruments for measurement, control, etc.; the state of being equipped with or controlled by instruments; also, such instruments collectively. orig. *U.S.*
1931 *Instruments* Jan. 11 Mr. Schroeder's book..fills an even greater need than that filled by our industrial instrumentation manual. **1932** M. F. BÉHAR *Man. Instrumentation* I. p. xi, I confidently predict the early recognition of Instrumentation as a distinct branch of engineering and as a distinct field of scientific management. **1944** *Jrnl. R. Aeronaut. Soc.* XLVIII. 153 The paper..emphasised quite clearly that instrumentation was coming more and more to the fore in the testing of aircraft. **1949** [see *INSTRUMENT *v.* 3]. **1952** [see *AUTOMATIZE *v.* 2]. **1958** *Engineering* 28 Mar. 387/2 As the age of automation develops it is bound to bring with it increased instrumentation and increased efforts to measure both more rapidly and more accurately by systems of remote control. **1959** *Daily Tel.* 13 Mar. 15/6 The instrumentation on the ground and in the missile recorded the flight satisfactorily. **1959** *Instrument Pract.* XIII. 195/1 In America we find a marked trend towards the instrumentation of a variety of physico-chemical estimations, such as phosphate and sulphur assays. **1963** B. FOZARD *Instrumentation Nucl. Reactors* xiii. 157 Instrumentation of a nuclear power reactor is commonly undertaken on a very large scale. **1973** *Physics Bull.* May 273/3 Josephson's work..has sparked off very fruitful developments in the instrumentation field, leading for example to highly sensitive voltage detectors.

insulant, *a.* Restrict † *Obs. rare* to sense in Dict. and add: **B.** *sb.* Any substance or medium which insulates (electrically, thermally, etc.).
1934 in WEBSTER. **1946** *Electronic Engin.* XVIII. 280/1 The growth of the electronics and high frequencies industries has been phenomenal. With this growth there has been a demand for more efficient insulants. **1959** *Times* 27 Apr. (Rubber Industry Suppl.) p. viii/5 The best thermal insulant is a vacuum. *Ibid.*, Most thermal insulants..work on the principle of trapped air. **1971** *Engineering* Apr. 112/1 (Advt.), A dense yet feather-light carpet of pure mineral fibre... A natural thermal and acoustic insulant.

insular, *a.* Add: **4. b.** *Palæogr.* (See quots.)
1908 W. M. LINDSAY *Contractions in Early Latin Minuscule MSS.* 1 The most fertile source of error..is the unfamiliarity of the writers with the contractions used in the Irish or pre-Carolingian script... The correct term is Insular, for English MSS. are included and Welsh too. **1913** F. W. HALL *Compan. Classical Texts* 167 Insular hands..i.e. Irish and Anglo-Saxon; a peculiar type of the half-uncial developed in the sixth century. **1960** G. A. GLAISTER *Gloss. Bk.* 195/1 *Insular hand*, the name given to the Hiberno-Saxon script widely used in England until the Norman Conquest for non-Latin texts. Its origins may be traced to 6th-century Ireland. An example is the first London Charter, 1066, which may be seen in the Guildhall Library. **1960** E. A. LOWE *Eng. Uncial* 14 By Insular symptoms we mean features and practices peculiar to Anglo-Saxon (and Irish) scribes. **1971** T. A. M. BISHOP *Eng. Caroline Minuscule* p. xiii, The most extensive repertories of Insular abbreviations in Caroline minuscule are MSS. of probably Continental origin.

insulate, *v.* Add: **3.** (Earlier and later examples.) Also used with reference to sound.
1742 J. T. DESAGULIERS *Diss. Electr.* 2 They must be insulated, that is, they must not be suspended from..

any Bodies but what are Electricks *per se.* **1927** Davis & Kaye *Acoustics of Buildings* ix. 178 The practice rooms have been so satisfactorily insulated, that it is impossible to hear any sound either through the floors or the partitions. **1955** *Oxf. Jun. Encycl.* XI. 175/2 The need for insulating floors against noise has long been understood.

insulated, *ppl. a.* Add: **3.** (Earlier and later examples.) Also used with reference to heat and sound (cf. Insulate *v.* 3 in Dict. and Suppl.).

1772 H. Cavendish in *Phil. Trans. R. Soc. 1771* LXI. 650 Any number of bodies, insulated and communicating with each other by conducting substances. **1777** T. Cavallo *Compl. Treat. Electr.* 3 A body resting intirely upon non-conductors is said to be insulated. **1964** W. Markfield *To Early Grave* (1965) x. 176 An appetizing store, where..you can take home their potato salad in an insulated bag. **1970** *Guardian* 24 Aug. 14/2 Insulated containers for the fruit and food trade. **1970** C. Duerden *Noise Abatement* vii. 115 Construct a special sound insulated chamber. **1974** A. Ross *Bradford Business* 75 A length of insulated cable..snaked across the floor to a three-pin socket.

insulating, *ppl. a.* Add: (Earlier and later examples.). Also used with reference to sound.

1767 J. Priestley *Hist. & Present State Electr.* 200 Upon the subject of insulating bodies, he observes, that when cakes of sulphur..are made use of for this purpose, they ought to be well cooled before they are used. **1893** *Funk's Stand. Dict.*, Insulating tape. **1910** *Hawkins' Electr. Dict.*, 219/1 *Insulating tape*, tape, usually adhesive, rendered non-conducting by being saturated with an insulating compound, for the purpose of covering..exposed parts of insulated electrical conductors. **1927** Davis & Kaye *Acoustics of Buildings* ix. 171 Suitable insulating material can be..interposed between columns, girders, cross-beams etc., to counteract the transmission of sound. **1940** *Chambers's Techn. Dict.* 451/1 *Insulating oils*, special types of oil..used for oil-immersed transformers, circuit-breakers, etc. **1945** *Archit. Rev.* CVI. 308 Roof of steel decking covered with half-inch insulating board and mineral-faced felt. **1955** *Oxf. Jun. Encycl.* VIII. 399/1 The defence against noise lies in planning wisely..and also in using sound insulating and sound absorbing materials. **1963** H. R. Clauser *Encycl. Engin. Materials* 352/1 The selection of the proper insulating varnish for an electrical unit, such as a motor, generator, solenoid coil, or a transformer, is becoming of increasing importance as the efficiency and ratings of electrical units are improved.

insulation. Add: **1.** (Earlier and later examples.) Also, an island.

1848 E. Bryant *California* xi. 157 The waters surrounding these insulations could be traced between them as far as the eye could reach. **1871** *Scribner's Monthly* II. 7 Their smooth sides, uniform width and height..considered in connection with the courses which had wrought their insulation, excited our wonder and admiration.

3. (Earlier and later examples.) Also used with reference to sound.

1767 J. Priestley *Hist. & Present State Electr.* 515 It is advisable that there should be no sharp edges or angles about the rubber [of the electrical machine], for that would make the insulation of it..ineffectual. **1913** *Chem. Abstr.* VII. 689 (*heading*) Incombustible and refractory materials for insulation from sound, heat and cold. **1955** *Oxf. Jun. Encycl.* VIII. 399/2 Double partitions, ..'floating floors',..and independent or suspended ceilings are all forms of construction used for sound insulation. **1972** L. L. Doelle *Environmental Acoustics* xiv. 173 Bare concrete slab..gives satisfactory insulation against airborne noises.

b. (Further examples.)

1927 Davis & Kaye *Acoustics of Buildings* ix. 171 Felt-like insulation may..be introduced into the..structure of the building, to assist in isolating noise. **1969** *Sears Catal.* Sprihg/Summer 963 Fiber glass insulation..keeps water hotter and jacket cooler.

4. *insulation meter, tape, test, tester.*

1909 *Cent. Dict.* Suppl., Insulation-meter. **1920** *Talking Machine News & Jrnl. Amusements* Feb. 77/2 Electrical insulation tape in a metal horn. **1903** *Whittaker's Electr. Engineer's Pocket-Bk.* 276 (*heading*) Insulation tests. **1923** *Nature* 13 Jan. 63/2 The 'Meg' insulation tester.., a remarkably light and cheap megger.

insulative (i·nsiŭlēitiv, -ātiv), *a.* [f. Insulat(e *v.*: see -ive.] Of, pertaining to, or as insulation.

1945 *Sci. News Let.* 23 June 397/3 The insulative concretes vary in weight from one-third to one-half that of ordinary gravel concrete. **1971** *Nature* 4 June 331/1 The fur was very spiny with virtually no soft underfur and thus probably offers little insulative protection. This may be an adaptation to the tropical environment. **1973** *Ibid.* 26 Jan. 240/1 Their fur is less than half as long..and has a correspondingly lower insulative quality.

insulator. For 'telegraph wires' read 'telegraph or telephone wires, or power lines'. Also used with reference to sound. (Further examples.)

1814 G. J. Singer *Elem. Electr.* iii. iii. 278 Insulation may..be partially preserved by coating all the glass insulators with sealing wax. **1847** Brett & Little *Compendium Improvements Electr. Telegraphs* 22 The insulators are made of earthenware, and secured direct to the poles. **1927** Davis & Kaye *Acoustics of Buildings* ix. 173 The usual principle is to have double walls..and to fill the space..with sound absorbents or insulators. **1950** Knudsen & Harris *Acoustical Designing in Archit.* xi. 246 If the porous blocks are not plastered, they may be very poor insulators: sound 'leaks' through the interstices. **1957** *Encycl. Brit.* XIV. 115I/1 On steel tower transmission lines the lightning voltage which can exist on the conductors..depends on the lightning flashover value of the supporting insulators.

insulin (i·nsiŭlin). *Biochem.* Also † **insuline.** [f. L. *insul-a* island (because it is produced by the islets of Langerhans) + -in[1].] **1.** A polypeptide hormone (the amino-acid composition differing slightly from species to species) which is concerned with carbohydrate metabolism in man and some other vertebrates, being produced by the islets of Langerhans and having effects that include the removal of sugar from the blood (so that a deficiency of insulin causes diabetes mellitus) and the promotion of protein synthesis and fat storage.

The name *insulin(e)* was proposed for the hormone on three separate occasions, each time independently: see quot. 1926.

[**1909** J. De Meyer in *Archivio di Fisiol.* VII. 96 Le produit de la sécrétion interne du pancréas (non denommé encore).., s'il dérive, comme nous le pensons, des ilots de Langerhans pourrait être appelé *Insuline.*] **1914** E. A. Schäfer *Introd. Study Endocrine Glands* 84 The results of pancreas-extirpation and pancreas-grafting can..be best explained by supposing that the islet-tissue produces an autacoid substance which passes into the blood and affects carbohydrate metabolism and carbohydrate storage in such a manner that there is no undue accumulation of glucose in the blood. Provisionally it will be convenient for description purposes to refer to this hypothetical autacoid as insuline. **1916** —— *Endocrine Organs* xvii. 128 The islet tissue produces an autacoid which passes into the blood and affects carbohydrate metabolism...Provisionally it will be convenient to refer to this hypothetical autacoid as insuline. **1922** F. G. Banting et al. in *Proc. & Trans. R. Soc. Canada* XVI. v. 27 (*heading*) The preparation of pancreatic extracts containing insulin. **1922** F. G. Banting et al. in *Amer. Jrnl. Physiol.* LXII. 175 Purified alcoholic extracts of pancreas, for which we suggest the name *insulin*, when injected subcutaneously into normal rabbits cause the percentage of sugar in the blood to fall within a few hours. **1923** A. E. Housman *Let.* 16 May (1971) 213 When I saw the invention of Insulin I..expected to hear you were cured already. **1926** E. A. Schäfer *Endocrine Organs* (ed. 2) II. xlix. 343 To this autacoid the name insulin is applied. [*Note*] The term was introduced by de Meyer (Arch. di fisiol., vii., 1909). In ignorance of this it was employed as a convenient term to denote the autacoid of the islet tissue in the first edition of this work, published in 1916. It was independently adopted by the Toronto workers [*viz.* F. G. Banting et al.] in 1922. **1956** *Arch. Biochem. & Biophysics* LXV. 427 The amino acid sequences in pig and sheep insulins were compared with that of cattle insulin...The only differences found were in the three residues occupying positions 8, 9, and 10 in the glycyl chain. **1959** *Ann. N.Y. Acad. Sci.* LXXXII. 340 Most commercial preparations are mixtures of beef and pork insulin. **1968** Passmore & Robson *Compan. Med. Stud.* I. xxv. 43/1 Insulin promotes all the known pathways of glucose disposal and transformation including glycogen storage, fat formation, total oxidation and the use of the hexose-monophosphate pathway.

2. *attrib.* and *Comb.*, as **insulin shock,** hypoglycæmia resulting from excessive insulin in the body, producing nervousness, weakness, sweating and in extreme cases coma, **insulin coma; insulin (shock) treatment,** a treatment for mental illness consisting of a course of artificially produced insulin comas (see quot. 1951).

1942 M. Dickens *One Pair of Feet* viii. 167 A policeman, who had mistaken an Insulin coma for a drunken stupor. **1959** *Times* 26 Mar. 15/5 Not for them the hashish-dream and the insulin-coma. **1972** *Chem. Abstr.* LXXVII. 135684 After more than 45 min in an insulin coma, administration of glucose..to the affected animals did not interrupt the coma. **1925** *Ann. Clin. Med.* III. 381/1 In diabetics whose blood sugars were high, that is above 300 mgm., symptoms of insulin shock frequently have been observed when their blood sugars were rapidly reduced to within usual normal range. **1936** *Jrnl. Nerv. & Mental Dis.* LXXXV. 504 Insulin shock stimulates metabolism in general and liver function in particular. **1938** *Arch. Neurol. & Psychiatry* (Chicago) XXXIX. 1 (*heading*) Insulin shock treatment of schizophrenic patients. **1951** F. Hopkins in E. N. Chamberlain *Text-bk. Med.* ix. 692 Insulin shock treatment..consists of the induction of coma by the injection of insulin. When the optimum dose for the individual case has been ascertained, a coma is induced each day for five or six days a week until a course of 50 or thereabouts has been given. **1958** H. Beckman *Drugs* xxxiv. 351/2 Epinephrine is used adjuvantly in combating insulin shock. **1940** *Ann. Internal Med.* XIV. 393 The insulin treatment of schizophrenia owes its origin to an accidental observation made by Sakel in his treatment of morphine addicts. **1958** I. Murdoch *Bell* xxvi. 302 Catherine had been having insulin treatment and was continually under the influence of drugs.

Hence **i·nsulinase** [*-ase], an enzyme or enzyme system that breaks down insulin; **i·nsulinized** *ppl. a.*, treated with insulin.

1928 *Amer. Jrnl. Physiol.* LXXXIV. 571 In both of these fishes the normal sugars are often even lower than the reduced sugars of the insulinized trout, scup and menhaden. **1949** Mirsky & Broh-Kahn in *Arch. Biochem.* XX. 8 Pending further investigation, it is proposed..to characterize the active principle in these extracts as 'insulinase'. **1959** *Metabolism* VIII. 99 The liver is rich in an enzyme system designated 'insulinase', which causes cleavage of insulin; this system may be comprised of more than one enzyme. **1966** *Chem. Abstr.* LXV. 1157 The increased sensitivity of aged humans and animals towards hypoglycemics is probably due to decreased insulinase activity. **1969** *Physiol. Chem. & Physics* I. 355 (*heading*) The subsequent accumulation of labeled glucose by insulinized frog muscle at 0°C.

insult, *sb.* Add: **1. d.** *Med.* Anything which tends to cause disease in or injury to the body or to disturb normal bodily processes; also, the resulting reaction, lesion, or injury.

[Cf. quots. 1603, 1610 for sense 1.] **1904** Stedman *Dunglison's Dict. Med. Sci.* (ed. 23) 581/1 *Insult*, injury; trauma exciting a morbid process. **1959** S. Duke-Elder *Parsons' Dis. Eye* (ed. 13) xvii. 228 The avascularity of the sclera and the lack of reaction of its dense fibrous tissues to insult whether traumatic or infective, make diseases of this tissue relatively rare. **1961** *Acta Psychiatrica et Neurol. Scand. Suppl.* CL. 110 Patients with cerebrovascular insults. *Ibid.* 112 It was possible that the insult had been provoked by the anti-hypertensive therapy. **1962** Harris & Gruber in A. Pirie *Lens Metabolism Rel. Cataract* 373 Changes in the level of inorganic and organic phosphate within the lens may follow certain metabolic insults. *Ibid.* 375 The older the lens, the less is it able to withstand the insult which cold induces. **1970** G. R. Taylor *Doomsday Bk.* vii. 158 The US population was exposed to 'severe chronic lead insult' (insult being a technical term in medicine). **1971** *Nature* 23 July 276/2 Left handedness may be caused by neurological insults associated with prenatal or birth trauma. **1971** *Sci. Amer.* Oct. 118/2 Congenital heart abnormalities are the most serious of the commoner defects. Often caused by rubella or similar insults in the early months of pregnancy, they affect about one birth in 60.

2. Freq. in phr. *to add insult to injury.*

1748 E. Moore *Foundling* v. v. 60 This is adding Insult to Injuries. **1805** *Deb. Congress U.S.* 31 Jan. 1072 It was adding insult to injury, and expenses to both, as it regarded the claimants. **1807** *Ibid.* 18 Nov. 933 Were the laws of a free and respectable State to be evaded by such shameful expedients as these? This was adding insult to injury. **1853** H. T. Riley tr. *Phædrus' Fables* 429 You wanted to revenge the sting of a tiny insect with death; what will you do to yourself who have added insult to injury? **1928** A. Huxley *Point Counter Point* ix. 155 It shocked him that one should lie in bed while other people were up and working. To get up late was somehow to add insult to injury. **1970** B. Spock *Decent & Indecent* 19 The behavioural sciences then added insult to injury. Psychologists have given man the impression that he responds like a laboratory rat. **1972** *Times* 12 Dec. 22/7 They submitted defective contracts, made arithmetical errors adding up to several thousand pounds and, to add insult to injury, charged their fees on the wrong and higher scale.

4. Used *attrib.* in expressions denoting contests in verbal insult, characteristic of U.S. Blacks.

1964 *Amer. Folk Music Occasional* I. 81 One of the standard routines of both the Negro and blackface minstrel show was the insult-dialogue. Though this theatrical expression is no longer widely current, this kind of dialogue has retained an interest... Here..is an insult-routine used in a street corner situation. **1968** P. Oliver *Screening Blues* vi. 246 Dr Abrahams makes acknowledgment of the dozens as played by young Negroes in the army but had not himself noted much use of the insult game among females or adults generally. **1969** *Language* XLV. 602 The children I studied engaged in constant verbal play..(for instance, their ritualized insult game).

insurance. Add: **4. e.** The act or system of insuring employed persons against sickness or unemployment, esp. in accordance with the National Insurance Acts of 1911, 1920, 1946, and 1965, which require certain wage-earners to make weekly payments supplemented by their employers, in return for which they are entitled to State assistance in sickness, unemployment, etc.

1878, etc. [see *National Insurance* s.v. *national *a.* 5]. **1911** *Times* 28 Mar. 10/3 The preparation of the Sickness and Invalidity Insurance Bill. *Ibid.* 5 May 14/3 If he had divided his bill into two—one dealing with unemployment and one with invalidity insurance. *Ibid.* 14/5 The burden imposed by State insurance must necessarily fall on manufacturers. **1911** *Act 1 & 2 Geo. V* c. 55. 337 National Health Insurance. **1912** [see sense *5]. **1920** *Act 10 & 11 Geo. V.* c. 30 § 48 (1) This act may be cited as the Unemployment Insurance Act, 1920.

5. (sense 4) *insurance adjuster, agency, agent, commissioner, company* (examples), *man, officer, policy* (examples), *premium*; (sense *4 e) *insurance act, benefit, card, committee, stamp.*

1755 Insurance premium [see Assurance 5]. **1784** in H. M. Brooks *Days of Spinning-Wheel in New Eng.* (1886) 62 The Gentlemen forming this Insurance Company, whose names are inserted in each Policy. **1866** C. N. Emerson *Internal Revenue Guide* 73 Insurance agents shall pay ten dollars. **1869** 'Mark Twain' *Innoc. Abr.* xxxviii. 409 If her [*sc.* Smyrna's] 'crown of life' had been an insurance policy, she would have had an opportunity to collect on it. **1874** B. F. Taylor *World on Wheels*, etc. ii. ii. 199 He was an insurance agent—a retired doctor, who growing weary of saving lives with pills, had taken to insuring lives with policies. **1879** *Harper's Mag.* July 215 The insurance men..would insure the lives of the hands who were at work there. **1881** *Instructions to Census Clerks* (1885) 83 Insurance Company's officer, manager, actuary, secretary,..clerk. **1883** 'Mark Twain' *Life on Mississippi* xliii. 436 Insurance-agency business, you know; mighty

irregular. **1889** *Cent. Dict.*, *Insurance commissioner*, in some of the United States, a State officer who in behalf of the public maintains a supervision over the affairs of insurance companies. **1911** *Act 1 & 2 Geo. V.* c. 55 § 15 The regulations made by the Insurance Commissioners. *Ibid.*, The Insurance Committee for each county or county borough. *Ibid.* § 115 This Act may be cited as the National Insurance Act, 1911. **1912** *Chemist & Druggist* LXXX. 950/2 Cards and stamps for health insurance under the National Insurance Act are now procurable at post offices. **1912** *Punch* 31 July 99/3 Mr. Masterman has laid it down that it is the wife's duty, and not that of the husband, to lick the servants' insurance stamps. **1913** *Ibid.* 15 Jan. 49/1 As the 15th of January approaches, bringing fulfilment of 9d. for 4d. through operation of Insurance Act. *Ibid.* 13 Aug. 148/3 Somebody come to see about an insurance card or something. **1915** W. Owen *Let.* 22 June (1967) 189 Am feeling quite independent of Insurance Policies just now. **1923** D. H. Lawrence *Birds, Beasts, & Flowers* 86 Ah Phoenix, Phoenix, John's Eagle! You are only known to us now as the badge of an insurance Company. **1926** Fowler *Mod. Eng. Usage* 741/1 The injustice of throwing on the landlord in whose house they happen to be resident the cost of a large additional insurance benefit for those who are sick. **1929** J. B. Priestley *Good Companions* I. i. 32 He..threw an insurance card and some money on the table. **1930** *Morning Post* 7 Aug. 11 The employers at four factories agreed to take upon themselves the charge of the insurance stamp which the men refuse to pay. **1933** Insurance agent [see *confidence *sb.* 10]. **1933** *Radio Times* 14 Apr. 98/1 Meltonian Cream..is..an insurance policy for shoes. **1934** Insurance adjuster [see *adjuster]. **1945** N. L. McClung *Stream runs Fast* xii. 99 But one day, an insurance man, hearing that Wes had sold his drug store came out to offer him an agency, and Wes became an agent for the Manufacturers' Life Insurance Company. **1958** *Listener* 23 Oct. 634/2 The insurance officer denied that this was an industrial accident. **1961** *Ibid.* 10 Aug. 219/2 An insurance adjuster who also acts as a private detective. **1972** J. Gores *Dead Skip* (1973) xiv. 101 Harvey E. Wyman was red-faced and jovial... He was also, unlike so many small insurance agents Ballard had met, very smart. **1973** W. McCarthy *Detail* i. 56 You should check the people you choose more carefully... They must also have special diets. Just an insurance policy.

insure, *v.* Add: **4. d.** (Cf. *insurance 4 e.)

1911 *Act 1 & 2 Geo. V* c. 55 § 1 All persons so insured (in this Act called 'insured persons').

‖ **insurrecto** (insɒre·kto). [Sp.] An insurgent or rebel. Also *attrib.* or as *adj.*

1907 *Cablenews* (Manila, Philippines) 21 Aug. 5/7 Villa, then Colonel of the insurrecto army. **1910** *Sat. Even. Post* 15 Oct. 17/2 I'll declare an amnesty for him and all his insurrectos. **1930** J. Dos Passos *42nd Parallel* I. 1 Up to where them insurrectos was afightin' fit to kill. **1947** *Sat. Even. Post* 8 Mar. 18/3 Later he saw a bit of fighting in the Philippines, north of Manila, against Aguinaldo's insurrectos.

inswept (i·nswept), *a.* [f. In *adv.* + Swept *ppl. a.*] Of the frame of a motor vehicle: narrowed at the forward end or at the side.

1907 *Westm. Gaz.* 12 Nov. 12/3 The frames will be inswept from the dash. **1908** *Ibid.* 30 Jan. 4/1 The frame..is inswept in front and gracefully upturned in the rear.

inswinger (i·nswiːŋəɹ). *Cricket.* [In *adv.*] A ball bowled with a swerve or swing from the off to leg in its flight; also, the bowler of such a ball. So **i·nswing,** the swerve or swing imparted to such a ball; a ball bowled in this manner; as vb., to bowl an inswinger; hence **i·nswiːnging** *ppl. a.*

1920 *Times* 29 May 7/2 He [*sc.* Mr. Robertson-Glasgow] again seemed to rely too much on his 'inswinger' on the leg side, and he would surely do very much better if he could be certain of pitching his inswinging ball on the off stump. **1920** E. R. Wilson in P. F. Warner *Cricket* 67 The first right-handed 'in-swinger'..who was in-swinging as early as 1895. **1924** N. Cardus *Days in Sun* 49 Jacques, of Hampshire..was a bowler commanding an in-swinging flight and an off-break. *Ibid.* 254 Nor would it be accurate to describe this ball as an in-swinger, for the true in-swinger swings from almost the first few yards of its flight through the air. **1927** *Observer* 17 Apr. 17/3 Last year he [*sc.* Mr. Allom] specialised in the 'inswing'. **1933** *Times* 28 Jan. 11/5 The leg-theory has been bowled in this country for a number of years, starting with 'in-swing' bowling. **1953** R. Warner *Escapade* 74 He bowls in-swingers himself, though as a matter of fact he takes most of his wickets with his leg-breaks. **1955** *Times* 5 July 4/1 D. J. Smith, in spite of an in-swinger's action, was making the ball move from leg, and C. S. Smith was finding umpire Pothecary hard to satisfy. **1958** *Times* 22 Oct. 14/1 As it is, it may be many a long day before he has better figures as a medium-pace inswinger. **1962** *Times* 2 Aug. 3/3 Buxton, who bowled his inswing deceptively.

‖ **intacta** (intæ·ktă). [fem. of L. *intactus* (see Intact *a.*).] A shortening of L. *virgo intacta* a woman of inviolate chastity, used as adj. to denote: unaffected, not spoiled or sullied, esp. in *fig.* senses.

1941 H. G. Wells *You can't be too Careful* iii. xvi. 198 Edward Albert attempted an ironical whistle, but Mrs Butter held her position, intacta. **1960** A. West *Trend is Up* (1961) vi. 244 There's no sex angle to it. The kid is *intacta*, so's the woman for that matter. **1966** *New Statesman* 25 Feb. 255/3 The *People*'s 5½ million readers found their favourite Sunday paper *intacta* this week: the carefully balanced mixture of sex, exposure and sport ..had not been tampered with.

intaglio. Add: **3.** *attrib.* as *intaglio cylinder, engraving, impression, method, principle, printer, process, type, work*; **intaglio print,** an impression of a plate cut in intaglio; **intaglio printing,** the group of processes used to print intaglio plates.

1859 *Abridgments of Specifications relating to Printing* (Patent Office) 354 The invention consists..in producing intaglio-graphic printing and other plates from forms of intaglio types by taking a casting in plaster of Paris, or other suitable material. *a* **1877** Knight *Dict. Mech.* II. 1192/1 *Intaglio-type*, a process..depending upon the production of a friable surface of oxide of zinc on a metallic plate under hydraulic pressure... At drying, the surface is brushed over,..leaving an intaglio impression. **1886** R. A. M. Stevenson tr. *Delaborde's Engraving* iii. 55 It is his [*sc.* Finiguerra's] invention..of the art of printing intaglio engravings, or rather of the art of engraving itself, that has made him immortal. **1888** C. T. Jacobi *Printers' Vocab.* 66 *Intaglio*, printing, such as from copper-plate—the reverse of 'relief' printing. **1914** E. H. Richter *Prints* i. 8 The different intaglio processes produce their blacks in different ways. **1917** E. Pound *Lustra* 181 Give up the intaglio method? **1921** *Dict. Occup. Terms* (1927) § 529 *Intaglio printer*..places metal intaglio cylinder in position in machine, sets roll of paper and adjusts machine for even printing. **1930** D. Strang *Printing of Etchings & Engravings* 1 The term 'etchings' will often..be intended to include all the other processes which are commonly employed by the artist in the making of intaglio prints. **1933** T. S. Barber in W. Atkins *Art & Pract. Printing* IV. i. 8 Printing was done by the intaglio method of inking and wiping. **1959** *Chambers's Encycl.* V. 341/2 Intaglio work, when the design is cut into the block or plate. **1965** Zigrosser & Gaehde *Guide to Collecting Orig. Prints* iv. 46 Intaglio Process... A general term descriptive of all techniques employing the intaglio principle of duplication, where the design is incised below the surface of the plate. **1967** V. Strauss *Printing Industry* i. 28/1 Intaglio printing is the name of a process family comprising a variety of printing methods which all use printing-image carriers with the printing areas sunken, or embedded, in the depth of the carrier material. **1972** W. Chamberlain *Thames & Hudson Man. Etching & Engraving* i. 11 In the context of intaglio printmaking, the term 'etching' normally refers to both the action of corroding lines, etc. into a metal plate with acid, and the inked, paper impression taken from the surface of the plate.

intake, *sb.* For 'Chiefly *Sc.* and *north. dial.*' read 'orig. *Sc.* and *north. dial.*' Add: **1.** (Further examples.)

1940 *Economist* 9 Nov. 590/1 Thanks to the heavy intake of raw wool this year..it has been possible to meet military and almost all export requirements without stinting the home consumer. **1955** *Times* 10 May 18/3 The intake of orders for the first four months is higher than ever before. **1971** *Nature* 2 July 63/3 Thus there is the intake and evaluation in one computerized centre of duplicate magnetic tapes from many sources.

b. (One of) a group of entrants to the army, a school, a trade, etc.

a **1943** B. Webb *Our Partnership* (1948) ii. 79 It was..among educational ladders..the most elaborate in its organisation of 'intakes' and promotions. **1943** *Times* 10 Dec. 2/1 That is evident from the moment when new intakes arrive and at once are interviewed by the..commanding officer. **1946** *News Chron.* 8 Aug. 1/4 It is understood that they will be part of the new intake of the U.S. Army. **1958** *Technology* May 66/2 The intake pattern can be worked out for transfer at appropriate stages. **1970** *Nature* 28 Nov. 798/1 The school should be functioning by 1975, with an intake of 100 students.

3. *attrib.*, as *intake crib, tunnel, well.*

1909 *Daily Chron.* 21 Jan. 1/7 A fire which occurred at the construction works of a waterworks intake crib on Lake Michigan. **1909** *Westm. Gaz.* 21 Jan., This new intake tunnel was thrust further out into the lake. **1964** R. Perry *World of Tiger* xi. 160 A tiger had attacked a tapir at a reservoir near Kuala Lumpur, and..both had fallen into the dry 'intake' well.

4. b. Short for *air-intake* (*air *sb.*[1] B. II).

1946 *Flight* 1 Aug. 115/2 (*caption*) A Heinkel single-jet proposal with the intake between two nacelles. **1959** *Listener* 30 July 164/1 He dealt with the ice that started forming on the engine intakes of the Vickers Vimy machine.

7. *attrib.* and *Comb.* (see also 3 above).

1921 *Dict. Occup. Terms* (1927) § 449 *Intake man*, grain intake man (grain milling); at a signal from men on ship, barge, etc., alongside wharf that suction pipe is inserted in grain, starts air pump, which draws up grain through pipe by suction. **1940** *Chambers's Techn. Dict.* 451/2 *Intake belt course*, a projecting course of stone or bricks, serving as an intake at a place where the thickness of a wall is diminished. **1941** *B.B.C. Gloss. Broadc. Terms* 15 *Intake report*, analysis summarizing the content of broadcasts directed to listeners in a particular country, in its own language, from a selected number of stations. **1957** *Times* 2 July (Agric. Suppl.) p. vi/2 A typical old barn now houses the grain intake pit, cleaner, pre-dry bin, continuous dryer, and sectional storage bins. **1958** *U.S. Dept. Agric. Yearbk. 1957* 760/2 *Intake rate*, the rate, usually expressed in inches per hour, at which rain or irrigation water enters the soil. **1961** B. Fergusson *Watery Maze* v. 127 Those ports on the West Coast which had become intake valves for our life's blood from across the Atlantic. **1966** Intake chimney [see *Cordtex]. **1972** *Classification of Occupations & Directory Occupational Titles* (Department of Employment) III. 478/1 *Intakeman* (grain, sugar and similar materials). Checks supplies of grain, sugar and similar materials into storage silos, transfers materials to processing departments and maintains stock records.

intaker. Restrict †*Obs.* to sense in Dict. and add: **2.** (See quot. 1921.)

1921 *Dict. Occup. Terms* (1927) § 368 *Drawer*, drawer-in,..healder, in-taker,..attaches weaving beam to drawing-in frame, and draws each warp yarn, separately, with a hook, through eye (or loop) of heald, and through dent of reed in loom. **1960** *Classification of Occupations* (General Register Office) 49/3 In-taker.

intaking, *vbl. sb.* Restrict † *Sc. Obs.* to senses in Dict. and add later examples in senses 1 and *1 b of Intake *sb.*

1905 C. Kernahan *Visions* 283 The soft intaking of a baby's breath. **1959** *Listener* 5 Mar. 402/2 A period of intensive learning—of intaking rather than outputting. **1966** I. Jefferies *House-Surgeon* iv. 88, I just went to the intaking chief.

intangibility. Add: **b.** Inviolability.

1783 C. J. Fox *Memorials & Corr.* (1853) II. 102, I beg of gentlemen to be aware of the lengths to which their arguments upon the intangibility of this charter may be carried. **1929** *Times* 13 Aug. 10/2 There has been too much talk..of the intangibility of the Young Plan.

intangible, *a.* Add: **B.** *sb.* Anything intangible; spec. (in *pl.*) = *intangible assets*, i.e. assets (e.g. goodwill, rights, etc.) which cannot easily or precisely be measured.

1914 *Cycl. Amer. Govt.* III. 496/1 The term 'personal property'..includes..visible property and intangibles. **1930** *Economist* 29 Mar. 710/1 Net tangible assets may be defined as total assets less 'intangibles' (goodwill, patents, etc.), current liabilities, and funded debt. **1933** *Discovery* Oct. 313/2 Scientific changes were coming in so thick and fast that other factors in social life—the intangibles of credit, the improvements in political and international ideas—were unequal to the task of accommodating them. **1949** *Here & Now* (N.Z.) Oct. 30/3 The intangibles—the many local developments—being not reducible to statistics, the food of all bureaucracy, count for nothing. **1957** *Economist* 19 Oct. (Suppl.) 1/2 The success of individual motor producers will depend mainly upon intangibles such as the success of their design policy.

‖ **intarsia** (intā·ɹsiă). Also **-io.** [It. *intarsio.*] = Tarsia. Also *attrib.*, *transf.* and *fig.* So **intarsiatore** (intāːɹsiătōə·re), a worker in intarsia; **intarsiatura** (intāːɹsiătūə·rā), pl. **-e,** = *intarsia.

1863 A. Jameson *Legends of Monastic Orders* (ed. 3) 275 The fine intarsiatura in the Choir of San Francesco di Assisi. **1867** *Ecclesiologist* XXVIII. 216 Hidden under the intarsio pavement. **1868** C. C. Perkins *Italian Sculptors* 262 A celebrated wood-carver and 'intarsiatore' named Luchino Bianchini..helped them to carve the presses for the sacristy. *Ibid.*, Luchino Bianchini..made the woodwork about its great portal, as well as the intaglios and intarsiature of the choir at San Lodovico. **1892** A. M. Clerke *Familiar Stud. Homer* x. 266 Some rusty dagger-blades..skilfully ornamented in coloured metallic intarsiatura. **1894** *Daily News* 6 Dec. 5/2 Humorous intarsia showing Polyphemus..feeling the backs of the sheep. **1896** *Q. Rev.* Oct. 471 The intarsias of the choir-stalls of S. Maria Maggiore at Bergamo. **1906** *Westm. Gaz.* 4 July 2/1 The Brunellese looked critically at the intarsia chests of drawers. **1913** Mrs. H. Ward *Mating of Lydia* iv. xix. 389 The gleaming reflections on lacquer and intarsia, on ebony or Sèvres. **1919** H. F. Jones *Samuel Butler* II. 67 The seats of the stalls in the church of Santa Maria Maggiore at Bergamo are ornamented with intarsia work. **1945** *Burlington Mag.* Aug. 191/2 The treatment of the *intarsia* is closely paralleled in the panelling of coeval rooms in Swiss museums. **1957** *Textile Terms & Defs.* (Textile Inst.) (ed. 3) 54 *Intarsia.* (1) Weft-knitted plain, rib or purl fabrics containing designs in two or more colours... (2) A motif design in stitch and/or colour. **1958** *Listener* 11 Sept. 388/3 Of the poems, with their tessellated intarsia of natural scenery, natural passion and liturgical imagery, perhaps the most revealing on the subject of Zhivago's destiny is the first. **1970** *Times* 28 Feb. (Sat. Suppl.) p. vii/4 The most startling form of marquetry was perspective picture making in wood, known as intarsia. **1973** *Guardian* 10 Apr. 13/3 Sweater with intarsia thistle motif.

integral, *a.* and *sb.* Add: **A.** *adj.* **1.** (Further examples of uses in technical contexts.)

1923 Glazebrook *Dict. Appl. Physics* V. 165/1 This cylinder has an open-ended steel barrel with integral fins. **1958** *Chambers's Techn. Dict.* 987/1 *Integral stiffeners*, the stiffening ridges left when an aircraft skin panel is machined from a solid billet. **1968** *Gloss. Formwork Terms* (*B.S.I.*) 16 *Integral facing*, a special facing concrete or mortar cast simultaneously with the backing concrete so as to be monolithic with it. **1972** [see *integrally *adv.* b].

2. (Further examples of uses in technical contexts.)

1945 H. D. Smyth *Gen. Acct. Devel. Atomic Energy Mil. Purposes* xii. 132 Two 'integral experiments' (experiments on assembled or integrated systems comprising fissionable material, reflector, and perhaps moderator also) may be described. **1953** C. Wallace *Photographer's Pocket-Bk.* 112 In modern colour materials the colours are achieved by building up on a suitable base..an 'integral tri-pack' of three separate emulsions.

4. d. *integral domain*: see *domain *sb.* 4 d.

integralism (inte·grăliz'm). [f. Integral *a.* 1 + -ism.] A name sometimes adopted for a philosophical or political, etc., doctrine or theory which involves the concept of an integral whole.

1871 S. P. Andrews *Primary Synopsis Universology* xii. 178 Integralism is the new and final philosophy; the

all-sided and complete reconciliation of all possible sectarian divisions in all spheres; not as extinguishing individual differences, but as softening, co-ordinating, and utilizing them. **1939** *Times* 18 Feb. 17/4 The counterpart of the Nazi movement—Integralism—has its supporters. **1964** D. G. MacRae in J. H. Plumb *Crisis in Humanities* 127 By 'integralism' is meant a set of beliefs that involve one in claiming that social structures form a 'seamless web' in which every institution and social position is linked to every other and is part of a unique, interconnected configuration. **1969** D. M. Smith *Italy* (ed. 2) vii. xxi. 255 The Rome party congress of 1906 had recorded a victory for 'integralism'..yet revolutionary socialism continued to spread.

integralist (inte·grălist), *sb.* [f. Integral *a.* 1 + -ist.] One who favours a policy or doctrine of integralism. Also *attrib.* or as *adj.* Cf. *integrist.

1907 I. Zangwill *Ghetto Comedies* 412 Russia is to be saved..by the Integralists, who alone maintain the purity of the Social Revolutionary programme. **1922** *Glasgow Herald* 23 Jan. 11 The so-called 'integralists' who held that every good Catholic should see eye to eye with the Holy Father in everything. **1930** *Times Educ. Suppl.* 3 May 197/2 The schools or coteries of the last few decades—the symbolists,..integralists,..and so forth. **1938** *Sun* (Baltimore) 12 May 1/1 To smash completely the outlawed integralist Greenshirt organization. **1967** C. Seton-Watson *Italy from Liberalism to Fascism* vii. 267 His [*sc.* Ferri's] contribution to the restoration of party unity was ..inspired by the 'integralist' formula of 'Neither to right nor to left, but straight ahead'. *Ibid.* xi. 437 So far from being an integralist, he had during his years at Bologna come under suspicion of modernist sympathies. **1968** R. K. Merton *Social Theory* (rev. ed.) iii. xv. 529 He [*sc.* Sorokin] adopts an 'integralist' conception of truth.

integrally, *adv.* **b.** Delete † and add later examples.

1936 Colvin & Stanley *Turning & Boring Pract.* ix. 131 Gisholt lathes have been materially changed in design and construction. Headstocks are now cast integrally with the bed. **1952** S. E. Rusinoff *Forging & Forming Metals* iv. 55 Small steam hammers have the anvil and frame cast integrally, but large hammers have a separate anvil. **1972** *Sci. Amer.* Jan. 49/2 (Advt.), An integral peak-reading meter lets you optimize record level without using a scope. Options include a 5 to 30 foot loop adaptor, an interrupting voice channel, and an inverter for 12 or 28 VDC..all integrally mounted.

integrand (i·ntĭgrænd). *Math.* [ad. L. *integrand-us*, gerundive of *integrāre* (see Integrate *v.*): see *-and[2].] An expression that is to be integrated.

1897 H. F. Baker *Abel's Theorem* xviii. 561 The integrand of the Abelian integral u_r is single-valued on the Riemann surface. **1937** *Proc. Cambr. Philos. Soc.* XXXIII. 374 It is natural to approximate by expanding the integrand in powers of z. **1968** Fox & Mayers *Computing Methods for Scientists & Engineers* ix. 178 We illustrate this process..by considering the computation of the integral $I = \int_0^1 (0{\cdot}92 \cosh x - \cos x)\,dx$,... Table 9.3 gives the tabulated values of the integrand and its differences.

integraph (i·ntĭgrɑf). [ad. F. *intégraphe* (B. Abdank-Abakanowicz 1885, in *La Lumière électr.* 17 Oct. 111/1), f. *intégral* Integral *a.* and *sb.*, *intégrer* to Integrate *v.*: see -graph.] Any of various kinds of apparatus which mechanically draw a curve representing the variation in the integral of some given curve or function as a limit or parameter varies.

1885 *Min. Proc. Inst. Civil Engin.* LXXXII. 162 The machines that he [*sc.* B. Abdank-Abakanowicz] had called briefly 'Integraphs' traced these curves mechanically. **1902** *Encycl. Brit.* XXX. 582/1 While an integrator determines the value of a definite integral, hence a mere constant, an integraph gives the value of an indefinite integral, which is a function of x. **1927** *Jrnl. Franklin Inst.* CCIII. 64 Mechanical integrators usually evaluate the definite integral between given fixed limits... The present machine, which we have called an integraph, since it records the result of an integration in the form of a plot or graph, has therefore been developed to evaluate $F(x)$ against x from the expression $F(x) = \int_a^x f_1(x)f_2(x)dx$ where f_1 and f_2 are known functions, formal or empirical. **1931** *Ibid.* CCXII. 77 The Photo-Electric Integraph..extends the range of practical solution of mathematical problems through its usefulness in the evaluation of integrals having a variable parameter within the integrand. **1961** S. Fifer *Analogue Computation* IV. xxv. 968 The development of the harmonic analyzer is closely associated with the development of two devices, (1) the planimeter,..and (2) the integraph, which draws a graph of the indefinite integral of a function.

integrate, *a.* Add: **2.** *Psychol.* Of, pertaining to, or designating people with strong eidetic imagery (particularly in the theories of Jaensch).

1930 O. Oeser tr. *Jaensch's Eidetic Imagery* iii. 93 In these individuals functions that later are separate still interpenetrate one another to a high degree and influence each other. That is why we call them 'integrate'. The integrate type is an earlier one from the evolutionary point of view. *Ibid.* 105 The integrate and disintegrate types are true fundamental forms of human existence corresponding, in a sense, to the fundamental forms discovered by biology. **1931** *Brit. Jrnl. Psychol.* July 94 When the child is in an integrate state, it should not be forced to behave in a disintegrate manner. **1943** H. Read *Educ. through Art* iv. iv. 81 Younger children and primitive peoples belong to an earlier evolutionary type which Jaensch calls 'integrate'.

integrate, *v.* Add: **2. b.** To bring (racially or culturally differentiated peoples) into equal membership of a society or system; to cease to segregate (racially). Also *intr.*, to become integrated. (See *integration 1 c.)

1948 *Richmond* (Virginia) *Times-Dispatch* 5 Aug. 1/8 (*headline*) Democrats 'integrate' Negroes for campaign. **1949** W. E. Barker in *Jrnl. Racial Affairs* I. i. 25 In the same way it can be seen that were South Africa to try to integrate her widely differing races, she would only create far greater problems than such a policy could ever solve. **1962** *Daily Tel.* 2 Aug. 10/2 It might be supposed that in doing these things for its people [i.e. of Tristan da Cunha], something had also been done to them: that they had..been 'integrated'. Not so. They want to go home, with an intensity and unanimity of desire. **1964** Mrs. L. B. Johnson *White House Diary* 10 June (1970) 163 A girl who was one of the first students to be integrated at Little Rock was praised by her counselor. **1964** L. Nkosi *Rhythm of Violence* 46 Why don't Indians in this country ever 'integrate'? **1966** *New Statesman* 22 Apr. 575/1 Those children who came knowing some English integrated well, but, when we threw those who knew none into the maelstrom, they sank... No one was going to integrate without first being able to communicate. **1969** *Times* 30 Apr. 8/3 The Americans intend to make the scheme permanent ..but we advised the experiment to see how well they integrate. **1972** *Nature* 24 Mar. 133/1 Old people, sick people and isolated people need access to a telephone if they are to be fully integrated with the rest of society.

integrated, *ppl. a.* Add: Also of a personality in which the component elements combine harmoniously.

1941 *Brit. Jrnl. Psychol.* Apr. 298 Among the individuals studied..it was possible to distinguish..*integrated* personalities, in which conflicting drives had been reconciled and were now functioning in harmony with each other. **1945** *Psychol. Rev.* LII. 65/2 Later we shall see that an integrated personality does indeed have a kind of 'autonomy' which gives it a certain limited immunity to the Law of Effect. **1950** O. H. Mowrer *Learning Theory & Personality Dynamics* i. ii. 59 When, however, opposing impulses are more nearly matched, the submerged one is likely to exert its influence..by diminishing the smoothness and efficiency of the main stream of integrated behavior. **1954** G. W. Allport *Nature of Prejudice* v. xx. 339 What vanishes in an integrated personality are the racial bogies and traditional scapegoats who have nothing, really, to do with life's woes. **1963** A. Heron *Towards Quaker View of Sex* v. 48 Idealism can be a sign of spontaneous and selfless devotion in an integrated personality. **1973** *Lancet* 24 Feb. 441/2 The plaintiff was well integrated and had learned to live with the problem.

b. Uniting in one system several constituents previously regarded as separate; *integrated circuit*, a small unit or package, often no larger than a button, which is made as a single indivisible structure (such as a chip) and is electrically equivalent to a conventional circuit of many separate components.

1947 F. C. Simmons in *Jrnl. Forestry* XLV. 347/2 At the roadside, or at the using plant, the electric chain-saw, or one of the highly efficient circular-saw cutting-up plants,..can divide these tree lengths into raw material for the products to which they are best suited. This type of logging, which we call 'integrated logging' should result in higher returns from the logging job. **1954** Webster Add., *Integrated logging*, a system of logging planned to remove in one cutting all usable timber and to separate the primary products and distribute them to industries where they will bring the highest returns. **1958** J. T. Wellmark in *Aviation Age Res. & Devel. Technical Handbk. 1958–1959* F-6/1 One promising technique uses the 'integrated electronic device' concept. *Ibid.*, The true 'integrated device' can do the jobs of both active and passive components. **1959** *IRE Trans. Electronic Computers* VIII. 103/1 Some specific examples of integrated logic circuits..are shown in the figures below. **1962** *Electronics, Reliability & Microminiaturization* I. 184/1 In integrated circuits the [circuit] element may be a region in a block of the material rather than a separate device. **1962** A. Battersby *Guide to Stock Control* x. 96 Looking ahead in the field of Integrated Data Processing, we can foresee the situation in which the machine will prepare the advice note and invoice, classify and analyse the sales, calculate ROL and ROQ for each item and go on to send out orders automatically. **1964** *Times Rev. Industry* Apr. 49/2 It will be an integrated mill—trees will go in one end and fine paper will emerge from the other. **1967** *Times Rev. Industry* Mar. 46/1 Where once the circuit designer had to employ comparatively large components mounted on some sort of backing sheet..now he can often put in an integrated circuit, in which the separate components are commonly created in miniscule form on the surface of thin wafers of silicon by complex processing. **1969** A. C. Tickle *Thin-Film Transistors* i. 3 The first advanced all-thin-film integrated circuit, in which *both* active and passive devices were produced by thin-film deposition techniques, was produced by Weimer in 1965. It was designed for scanning an image sensor and contained 360 thin-film transistors, 180 diodes, 360 resistors, and 180 capacitors. **1970** F. C. Fitchen et al. *Electronic Integrated Circuits* i. 4 The monolithic integrated circuit is an IC whose [circuit] elements are formed in situ upon or within a semiconductor substrate with at least one of the elements formed within the substrate... In a multichip integrated circuit, the elements are formed on or within two or more semiconductor chips that are separately attached to a substrate. **1970** 'J. Earl' *Tuners & Amplifiers* i. 16 The majority of Class B and Class AB amplifiers..are of the so-called 'integrated' kind. This means that the preamplifiers and control unit..are built into the same housing—usually on the same chassis—as the power amplifiers and that a common power supply feeds both sections. **1973** *Physics Bull.* May 297/3 The continual recurrence of the same names in different fields of study is..in no way to be confused with, or used to justify, the contemporary rush to some illusory 'integrated science' concept.

c. Of institutions, groups, etc., which are not divided by considerations based on race or culture (see *integrate *v.* 2 b).

1948 *Richmond* (Virginia) *Times-Dispatch* 5 Aug. 1/8 This will be an integrated rather than a segregated operation. **1956** *N.Y. Times* 1 Oct., Approximately 2,400,000 Negro and 6,500,000 white pupils remained in segregated classes. Integrated school districts numbered 780; segregated numbered 3,000. **1958** *Newsweek* 29 Sept. 22 (*title*) Integrated schools or none. **1959** *Spectator* 11 Sept. 319/1 The newly integrated junior high school at Little Rock reopened, but not before three bombs had gone off. **1965** B. Sweet-Escott *Baker St. Irreg.* vi. 167 My assignment at A.F.H.Q. was with the planners. It was my first introduction to a really integrated staff... It was difficult to have a word in private with any of the British members of it because it was a rule that Americans and British should share rooms. **1974** *Times* 14 Jan. 12/4 Southern schools are now more integrated than their northern counterparts.

integrating, *vbl. sb.* and *ppl. a.* (in Dict. s.v. Integrate *v.*). Add: (Further examples.) *integrating circuit* Electr., a circuit whose output is the integral, with respect to time, of the input; *integrating factor* Math., an expression by which a differential equation may be multiplied to turn it into an exact equation (and therefore integrable as it stands); *integrating meter*, a meter which indicates the total amount of one quantity (e.g. electric charge passed) by effectively integrating, with respect to time, another (e.g. electric current).

1948 L. Jánossy *Cosmic Rays* ii. 43 Some authors make use of an integrating circuit which allows [one] to read the average counting rate at any instant. **1961** H. J. Reich *Functional Circuits & Oscillators* iv. 18 Integrating circuits find applications in..electronic instruments and controls,..in analog computers and in circuits for the generation of linearly rising voltages. **1973** M. Woodhouse *Blue Bone* iii. 24 The integrating Dekatron counter I was designing. **1859** G. Boole *Treat. Differential Equations* iv. 55 To every differential equation of the form $Mdx + Ndy = 0$, pertain an infinite number of integrating factors, all of which are included under a single functional expression. **1962** T. M. Apostol *Calculus* II. v. 241 A differential equation may have more than one integrating factor. **1902** *Encycl. Brit.* XXX. 597/1 All the above forms of house meters are called continuously integrating meters, in that the operation of recording or obtaining the time-integral of the current or power is continuous. **1943** *Gloss. Terms Electr. Engin.* (*B.S.I.*) 75 *Integrating frequency meter* (master frequency meter), an instrument for integrating the number of cycles through which the supply voltage has passed, and enabling this to be compared with the number through which it would have passed had the frequency been maintained at the prescribed value. **1952** E. Molloy *Electr. Instruments* 37 The induction disc principle is now adopted as the basic pattern for all types of alternating current integrating meters throughout the world.

integration. Add: **1. b.** *Psychol.* The combining of diverse parts into a complex whole; a complex state the parts of which are distinguishable; the harmonious combination of the different elements in a personality. Also *attrib.*

1855 H. Spencer *Princ. Psychol.* iii. xiv. 481 Progress in integration has been a necessary accompaniment of progress in speciality and complexity. **1893** J. M. Baldwin *Elem. Psychol.* 36 Integration, therefore, represents a structural change in the direction both of simplicity and of complexity. **1931** *Brit. Jrnl. Psychol.* July 25 A more adequate psychological theory..speaks of the 'integration' of the constituent sensations into a perception. **1937** L. T. Hopkins *Integration* i. 2 Integration must be the shorthand word to describe the process involved in this intelligent ongoing, interacting, adjusting behavior. **1938** L. P. Thorpe *Psychol. Found. Personality* ix. 434 Perhaps we are warranted then in using the word integration with the understanding that it stands for a wholeness in personality which gives direction to the coordination of parts. **1943** H. Read *Educ. through Art* iv. iv. 81 Jaensch's next step is to relate his classification to the degree of integration which the individual establishes between his mental imagery and the external world. **1963** Langner & Michael *Life Stress & Mental Health* xvi. 460 Integration involves the incorporation of the 'thou shalt nots' as well as the 'thou shalts', the acceptance of the middle-class rules of the game as well as the goal of winning.

c. The bringing into equal membership of a common society those groups or persons previously discriminated against on racial or cultural grounds.

1940 T. J. Haarhoff *S. Afr. & Crisis Mod. Civilization* 19 For the great task that awaits us in South Africa is a task of integration, of making the Union into a unity. **1949** *Jrnl. Racial Affairs* I. i. 25 Although..assimilation would destroy the racial differences in South Africa, it would necessitate the prior integration of the cultures—a very difficult and uncertain process. Apart from that one argument, there is practically no other logical reason in support of either racial or cultural integration. **1951** J. D. L. Kruger *Bantustan: Study in Pract. Apartheid* iii. 15 In fact it is difficult if not impossible to think of a

single aspect of integration which could be regarded as beneficial to the white population. **1955** *Ann. Amer. Acad. Pol. & Social Sci.* CCCII. 25/1 The frontier for race relations has been shifting more and more to the housing field... The degree of integration in the schools now depends very largely on the residential pattern. **1968** *Listener* 26 Dec. 855/3, I define integration not as a flattening process of assimilation but as equal opportunity accompanied by cultural diversity in an atmosphere of mutual tolerance. **1970** *Times* 23 Mar. 13/3 It seemed that all we had to do to achieve integration was to sit down at enough lunch counters together.

integrational (intĭgrēi·ʃənăl), *a.* [f. Integration + -al.] Of or pertaining to integration.

1937 G. W. Allport *Personality* (1938) iii. xiii. 354 From each condensed portrait (called by the experimenter 'integrational hypothesis'), the judges predicted what responses the eight subjects would make. **1957** V. W. Turner *Schism & Continuity in Afr. Soc.* p. xxi, Ndembu ritual..compensates for the integrational deficiencies of a politically unstable society. **1960** J. B. Carroll in Saporta & Bastian *Psycholinguistics* (1961) 339/2 She tested decoding..and encoding functions with respect to semantic, grammatical, and integrational responses.

integrationist (intĭgrēi·ʃənist), *sb.* and *a.* [f. Integration + -ist.] **A.** *sb.* An adherent or advocate of integration, esp. political or racial.

1955 *N.Y. Times* 1 June 29/3 The initial reaction, including segregationists and integrationists, overridingly stressed the court's ruling. **1956** *Atlantic Monthly* Nov. 49/1 It is because there the adolescent and 'unprejudiced' mind can be reached that the integrationists have chosen the Southern schools as their primary target. **1959** *Listener* 27 Aug. 305/1 It is becoming less and less possible for the 'integrationists' to paper over the cracks by references to General de Gaulle's 'silences'. **1963** *Economist* 23 Feb. 695/2 This racial appeal is disliked by integrationists. **1971** *Black Scholar* Jan. 52/1 In the work of black authors who are integrationists a tacitly separatist or ethnically independent element appears frequently.

B. *adj.* Of, or pertaining to, persons or policies favouring integration, esp. political or racial.

1956 *Newsweek* 21 May 17/1 A political unknown named Sumter Lowry, who ran on the single plank of preserving segregation, got 130,000 votes. An integrationist candidate got 5,000. **1958** *Times* 11 Nov. 8/1 The solitary list is strongly integrationist, and is led by Azem Ouali, a prominent member of the Algeria-Sahara public safety committee. **1968** *Ann. Amer. Acad. Pol. & Social Sci.* CCCLXXVI. 199/2 These Negroes fervently embraced an integrationist ideology. **1973** *Black World* Mar. 34 The integrationist tendencies of the Negro intellectuals. **1973** E. Bullins *Theme is Blackness* 4 The militant integrationist syndrome.

integrative, *a.* (Later examples.)

1906 C. S. Sherrington (*title*) The integrative action of the nervous system. **1937** [see *dominative *a.* 1 c]. **1937** G. W. Allport *Personality* (1938) ii. v. 138 Whatever condition makes for mental health is called 'integrative', whatever condition makes for mental difficulty is called 'disintegrative'. **1953** J. S. Huxley *Evolution in Action* iv. 91 What we may call integrative emotions, like love. **1957** M. Banton *W. Afr. City* ix. 179 Up to this point the integrative effects of the companies have been stressed. **1967** *Amer. Pol. Sci. Rev.* LXI. 91 Far from finding a stagnation of integrative processes since 1958, I would argue that..European integration may have moved into full gear only *since* 1958. **1972** *World Archaeol.* III. 231 He [*sc.* V. G. Childe] saw culture as an essentially integrative device.

integrator. (Earlier and later examples.)

1876 W. Thomson in *Proc. R. Soc.* XXIV. 269, I have made many attempts to plan a mechanical integrator which should give solutions by successive approximations... We have the instrument founded on my brother's disk-, globe-, and cylinder-integrator. **1884** S. P. Thompson *Dynamo-Electr. Machinery* iv. 59 Fig. 48 gives the curve as integrated..for me from Fig. 47 by the aid of the very ingenious curve integrator of Mr. C. Vernon Boys. **1931** *Jrnl. Franklin Inst.* CCXII. 450 There has been an enormous change in technique since the time when Sir William Thomson first suggested..that the integrators developed by his brother could be connected together and thus forced to produce solutions of differential equations. **1938** *Math. Gaz.* XXII. 343 Integrators... Any continuously variable gear can act as an integrating mechanism. **1955** T. L. Martin *Electronic Circuits* 618 Waveform relationships in integrators and differentiators. **1964** *Ann. N.Y. Acad. Sci.* CXV. 571 Until recently, analog computers have depended on the use of relays for mode control, switching, and resetting of integrators.

integrist (i·ntegrist). Also **intégriste.** [F. *intégriste.*] = *integralist. Hence **i·ntegrism** = *integralism.

1907 *Dublin Rev.* Jan. 38 The Carlists and Integrists, who muster half a dozen deputies between them. **1938** *Downside Rev.* LVI. 154 Benedict XV..told Cardinal Billot that he wanted to hear no more talk of 'integrism'. He demanded that..where Rome had not decided, all should be free to state their views without being attacked as heretical. **1969** R. Manheim tr. *Corbin's Creative Imagination in Ṣūfism* Introd. 41 We shall let Ibn 'Arabī describe the encounter between the integrist Aristotelian master and the young man. **1970** J. Ardagh *New France* xi. 577 The tradition of integrism will not die easily in France. *Ibid.*, The unreconciled integrists represent only a small proportion of practising Catholics. **1971** *Month* May 146/2 [The] silent assembly of *intégristes* at Versailles in November 1970. *Ibid.* July 17/1 The French *intégriste* who attended one of the sessions and refused the kiss of peace.

integro-differential (iːntĭgroˌdifĕre·nʃăl), *a.* *Math.* [ad. It. *integro-differenziale* (V. Volterra 1909, in *Atti d. r. Accad. dei Lincei: Rendiconti* (*Classe di sci. fisiche*) XVIII. i. 167).] Involving both integral and differential quantities.

1914 *Trans. Amer. Math. Soc.* XV. 215 A large part of the theory of integral and integro-differential equations may be reduced to the corresponding theory of algebraic and differential equations by the introduction of convenient symbolism. **1923** *Bull. Amer. Math. Soc.* XXIX. 210 Integro-differential invariants of one-parameter groups of Fredholm transformations. **1930** M. Long tr. *Volterra's Theory of Functionals* 31 As it has the characters of both integral and differential equations, it will be called an integro-differential equation. **1958** E. M. Grabbe et al. *Handbk. Automation, Computation, & Control* I. ix. 18 The method..is applicable to 'integro-differential equations' such as the following: $a_0dx/dt + a_1x + a_2\int_0^t x\,dt = f(t)$. **1964** N. N. Hancock *Matrix Analysis of Electr. Machinery* v. 70 The fundamental equations of an electro-dynamic system are inherently integro-differential equations.

intellectual, *a.* and *sb.* Add: **A.** *adj.* **1. b.** (Further examples.)

1786 A. Seward *Let.* 25 Mar. (1811) I. 130 Those who are not interested in its anecdotes, can have little intellectual curiosity and no imagination. **1881** *Atlantic Monthly* May 597/1 He talked in a way..lively enough to sap his own intellectual integrity. **1891** A. James *Diary* 24 June (1965) 216 Owing to my curious, given my inheritance and surroundings, complete absence of intellectual curiosity. **1891** W. James *Let.* 21 Sept. in R. B. Perry *Tht. & Char. of W. James* (1935) II. 174 What a strange thing an intellectual *atmosphere* is! **1896** —— *Will to Believe* (1897) 9 Mr. Balfour gives the name of 'authority' to all those influences, born of the intellectual climate, that make hypotheses possible or impossible for us. **1903** G. B. Shaw *Man & Superman* ii. 50 That tone of intellectual snobbery. **1933** *Week-end Rev.* 11 Feb. 142/1 Intellectual integrity is necessary precisely in proportion as it is difficult. **1939** L. MacNeice *Autumn Jrnl.* xii. 49 Spiritually bankrupt Intellectual snobs. **1965** H. A. Gleason *Ling. & Eng. Gram.* 35 In the intellectual climate of the late nineteenth century this was most attractive. **1971** D. Crystal *Ling.* iv. 187 We must..take account of the intellectual climate of the time.

B. *sb.* **4.** (Further examples.)

1847 J. J. Ruskin *Let.* 2 Sept. in M. Lutyens *Ruskins & Grays* (1972) vi. 50, I want you to stand well with Lockhart and the Intellectuals. **1903** *Sat. Rev.* 19 Dec. 760/2 We are compelled to rank higher the mind of the average young man of fashion than the mind of the average 'intellectual' at those literary tea-parties. **1931** [see *culture *sb.* 5 d]. **1937** [see *brain trust, brains trust]. **1960** *Times Lit. Suppl.* 12 Aug. 513/2 The English have a great respect for brute facts; and the intellectual in politics often looks to them like a man busily engaged in brushing unpleasant facts under the carpet. **1974** *Times* 15 Feb. 15/8 Russian history has set a pattern of alienated intellectuals.

intellectualist. Now freq. attrib. or as adj.

1890 W. James *Princ. Psychol.* II. xxii. 325 The traditional intellectualist philosophy has always made a great point of treating the brutes as wholly irrational creatures. **1941** J. S. Huxley *Uniqueness of Man* ii. 76 Men..of an intellectualist and academic type.

intellectualistic, *a.* Add: So **intelleːctuali·stically** *adv.*

1907 W. James *Pragmatism* iii. 121 Yet dark tho they be in themselves [*sc.* the words *God*, *free-will*, etc.], or intellectualistically taken, when we bear them into life's thicket with us the darkness *there* grows light about us. **1909** —— *Pluralistic Universe* ii. 72 The substituted conceptions are treated intellectualistically, that is as mutually exclusive and discontinuous. **1927** L. Stein *A.B.C. of Æsthetics* i. 10 No one can get much good out of my book who reads it intellectualistically.

intellectually, *adv.* Add: Used with an adj. as a quasi-Comb.

1859 J. A. Symonds *Let.* Oct. (1967) I. 211, I want to get Puller to organize a club with me, introducing 5 other intellectually-pursuited men. **1923** J. S. Huxley *Ess. Biologist* i. 8 Intellectually-minded men. **1941** *Mind* L. 236 Mr. Clive Bell in his excellent and intellectually snobbish book *Civilization*..makes this blunder. **1951** M. McLuhan *Mech. Bride* 58/2 The intellectually creative man with whom the future of mankind always rests.

intellectualness. In quot., for '1884' read '1854'.

intelligence, *sb.* Add: **7. a.** *spec.* Information of military value.

1799 G. Harris *Diary* 4 Apr. in Wellington *Disp.* (1837) I. 24 If our intelligence is true, his [*sc.* Tippoo Sultaun's] whole army are in a complete state of terror. **1818** [in Dict.]. **1899** *McClure's Mag.* Mar. 473/2 The swift single cruisers, the purveyors of intelligence. **1925** Fraser & Gibbons *Soldier & Sailor Words* 125 'Intelligence', *i.e.* information of military value. The use of the word as a military technical term dates from the 16th Century, but in the War of 1914–18 it was used to denote specially the department of the General Staff dealing with information. **1974** *Times* 15 Jan. 14/3 The first question that everybody asked was why the intelligence of the Arab armies massing on the borders..was misread?

c. Delete † and add later examples.

1915 Kipling *France at War* 21 The Intelligence with its stupefying photo-plans of the enemy's trenches. **1949** A. Christie *Crooked House* xii. 95 A person who has something to hide can't really afford to talk *at all*. The blokes knew that in Intelligence during the war. **1957** [see *courier *sb.* 1 b]. **1963** A. Orlov *Handbk. Intelligence* i. 10 Stalin in 1932 ordered intelligence to discontinue sending him quarterly surveys of foreign countries. **1974** *Listener* 31 Jan. 142/1 Tizard has managed to get it through that someone should be seconded to British Intelligence for a while.

d. (sense 7 c in Dict. and Suppl.) *intelligence agency, corps, officer, operator, service*; **intelligence department,** a department of a state organization or of a military or naval service whose object is to obtain information (esp. by means of secret service officers or a system of spies).

1951 Intelligence Agency [see *C.I.A.* s.v. *C III. 3]. **1960** J. Blish *Galactic Cluster* 116 This intelligence agency is my sole source of income... I have every right to operate a private investigation bureau. **1963** A. Orlov *Handbk. Intelligence* i. 7 The American intelligence agencies monitor as many as five million words daily from foreign radio broadcasts alone. **1961** Intelligence Corps [see *intelligence operator* below]. **1974** P. McCutchan *Call for Simon Shard* xiv. 132 Hedge had been in the war, a captain in the Intelligence Corps. **1875** *Encycl. Brit.* II. 573/1 The Intelligence Department [of the Army]..under a Deputy-Adjutant-General. **1895** *Whitaker's Almanack* 230 Naval Intelligence Department. **1928** W. S. Maugham *Ashenden* i. 2 The Colonel..was known in the Intelligence Department..by the letter R. **1885** Intelligence officer [in Dict.]. **1901** *Westm. Gaz.* 31 Dec. 10/1 You are intelligence officer to the new cavalry brigade. **1914** R. Brooke in E. Marsh *Rupert Brooke* (1918) 132 Intelligence Officer in H.M.S. *Vengeance.* **1946** E. M. Zacharias (*title*) Secret mission: the story of an intelligence officer. **1961** *Times* 12 Sept. 2/6 The Intelligence Corps... Vacancies exist for training in the trades of:—Staff Intelligence Operator and Counter Intelligence Operator. **1930** B. Miall tr. *Berndorff's Espionage* i. 16 At Nice..he continued to work for the French Intelligence Service. **1932** R. H. B. Lockhart *Mem. Brit. Agent* iv. iii. 234 A Colonel in our Intelligence Service. **1956** A. L. Rowse *Early Churchills* 262 Of this Marlborough, whose intelligence-service was always of the first order, was well aware. **1974** *Listener* 31 Jan. 142/1 The Führer..said he had a secret weapon. Immediately Neville Chamberlain..asked the Intelligence Services what the secret weapon was.

8. *attrib.* and *Comb.*, as **intelligence quotient** [ad. G. *intelligenz-quotient* (W. L. Stern, 1912)], a number arrived at by means of intelligence tests and intended to express the degree of intelligence of an individual in relation to the average for the age-group, which is fixed at 100; abbrev. *I.Q.* (*I. III); so **intelligence test, tester, testing.**

1921 C. Burt *Mental & Scholastic Tests* 151 If a child's mental age be divided by his choronological age, the quotient will state what fraction of ability the child actually possesses... This fraction may be termed..the child's 'intelligence quotient'. **1922** R. S. Woodworth *Psychol.* xii. 274 Brightness or dullness can also be measured by the intelligence quotient. **1944** H. Read *Educ. Free Men* iii. 13 Truth, we say, is not found exclusively in the possession of those with a high 'intelligence quotient'. **1953** *Sci. News* XXIX. 45 Isolated concepts, from sex-motivation to intelligence quotients, are taken up, puffed up, and what may be limited but useful notions become ridiculed, because of the disproportionate importance attached to them. **1971** *Nature* 2 Apr. 306/1 The worship of the Intelligence Quotient, mercifully a-dying, is still not entirely dead. **1972** Kagan & Havemann *Psychol.* (ed. 2) xiv. 473 The intelligence quotient, or I.Q., is simply the relationship of mental age to chronological age; it is obtained by the formula I.Q. = MA/Chronological age × 100... This is the general principle for computing the I.Q. on the Stanford–Binet. In actual practice..the I.Q. is usually determined from tables that make it possible to compare the child's raw score with the scores made by other children of the same age. This latter method is also the one used with all other intelligence tests... This statistical method of computing the I.Q. is valuable because the concept of mental age..is not meaningful for adults. **1914** *Eugenics Rev.* Apr. 42 General ability, estimated by intelligence tests, is largely hereditary. **1957** *Technology* Mar. 10/1 The trade school, however, is well equipped to sort wheat from chaff—each candidate is given the latest types of intelligence and aptitude tests. **1927** A. Huxley *Proper Stud.* 65 The intelligence-testers would isolate..the sum of the activities of the whole mind. **1962** H. J. Eysenck *Know your own I.Q.* 31 From the point of view of the intelligence tester it is very undesirable to have mixed groups to deal with. **1972** J. L. Dillard *Black English* i. 28 Some intelligence testers..have suffered from lack of valid information about Black English in standard sources on American dialects. **1919** *Elem. School Jrnl.* Sept. 26 (*title*) Intelligence testing as an aid to supervision. **1958** [see *child welfare* (*child *sb.* 22)]. **1972** J. L. Dillard *Black English* i. 28 An intelligence-testing procedure which is completely invalid because of its cultural and linguistic bias.

intelligentsia (intelidʒe·ntsiă; formerly also intelige·ntsiă). Also (formerly) **intelligenzia.** [f. Russ. *intelligéntsiya*, ad. L. *intelligentia* Intelligence *sb.*] The part of a nation, orig. in pre-revolutionary Russia, that aspires to intellectual activity; the class of society regarded as possessing culture and political initiative.

1907 M. Baring *Year in Russia* vii. 77 They [*sc.* the revolutionaries] fear that if the question of a Republic is brought forward there will be a general massacre of the educated bourgeoisie, the so-called 'Intelligenzia'. **1910** —— *Landmarks Russian Lit.* iii. 68 Chekov has depicted the pessimism and the ineffectiveness of the 'intelligenzia'. **1914** *Round Table* Dec. 115 The importance and meaning of the so-called *intelligentsia*. **1916** H. G. Wells *Mr. Britling* I. ii. 62 They are the sort of equivalent of the Russian Intelligentsia, an irresponsible middle class with ideas. **1921** A. Huxley *Let.* 31 May (1969) 197 The English colony [at Florence] is a queer collection; a sort of decayed provincial intelligentsia. **1922** C. E. M. Joad *Highbrows* vi. 224 Those waifs and strays of the intelligentsia who had resolutely refused to participate in the war. **1922** C. Sidgwick *Victorian* xxviii. 211 He told me ..that he belonged to the Intelligentsia and that he was out to shoot capitalists. **1924** Galsworthy *White Monkey* I. ix, It was not the intelligentsia, but just intellectual society, which was gathered there. **1940** Wodehouse *Eggs, Beans & Crumpets* 75 It was a painful shock to the intelligentsia..when they discovered that their old friend was not going to prove the geyser of easy money they had anticipated. **1949** I. T. Sanders *Balkan Village* i. 7 The intelligentsia, as they were called by the [Bulgarian] peasants, were the most influential group in the community... I was sure to find several of them playing cards... The mayor..could watch..the municipal building... The priest could look..to the church just beyond. The village doctor's husband was near in case his wife needed him. **1956** R. Redfield *Peasant Society & Culture* ii. 61 To the administrative and cultural intermediaries between local life and wider life the word 'intelligentsia' has long been applied. **1971** H. Seton-Watson in A. Bullock *20th Cent.* 139/1 The revolutionary propensity of the intelligentsia has been definitely correlated with the extent of the cultural gap between the educated élite and the mass of the people.

Intelsat (i·ntelsæt). Also **INTELSAT.** [An acronym f. the name of the organization, *In*ternational *Tel*ecommunications *Sat*ellite Consortium.] **a.** An international organization of member countries formed in 1964 to establish and operate a worldwide system of commercial communication satellites. **b.** A communication satellite owned by this organization.

1966 *Aviation Week* 7 Mar. 125/3 International Telecommunications Satellite Consortium (Intelsat) is already on the road to becoming a profit-making venture by 1970. It now includes 48 nations... U.S.'s Communications Satellite Corp. serves as manager for the consortium. **1967** *Wall Street Jrnl.* 6 Jan. 4/3 The successful weekend orbiting of Intelsat II. **1969** *Listener* 5 June 779/1 In communications we have accepted INTELSAT, for example, a global organisation which runs the satellite communications system. **1971** *New Scientist* 3 June 552/1 The settlement now reached mostly meets the criticism.. that Intelsat was American dominated. **1972** *Sci. Amer.* Sept. 111/1 The latest Intelsat has a capacity of 5,000 voice channels or 12 television channels.

intemporal, *a.* Delete †*Obs. rare*⁻¹ and add later examples.

1911 J. Ward *Realm of Ends* xiv. 306 The intemporal world of ideas. **1962** tr. *J. L. Borges's Labyrinths* (1964) 262 There was no other sound than the intemporal one of the crickets.

intendant, *sb.* Add: **3.** The administrator of an opera house or theatre (cf. G. *intendant*); see also quot. 1903.

1903 R. Hughes *Mus. Guide* I. 174/2 *Intendant*,..director, conductor. **1958** *Spectator* 1 Aug. 166/2 The vital demarcation between intendant and administrator is overgrown by the cult of the amateur run riot. **1961** *Times* 30 Jan. 14/2 In each city, the fortunes of opera, ballet, and drama depend largely on an intendant who occupies his position for a comparatively few years before moving elsewhere. **1966** *New Statesman* 18 Nov. 758/2 Very often the Intendants hardly understand the meaning of the word 'choreography',..and automatically scrap each ballet at the end of the season. **1973** *Times* 11 Apr. 12/5 In most of the big towns the Intendant can attract the audience he needs only with a fairly rapid turnover of plays.

intendedly, *adv.* (Later example.)

1972 *Daily Tel.* 3 Feb. 14 Your implication that priests present at the illegal, though intendedly peaceful, protest march have no right to be 'mixed up' with their people is absurd.

intense, *a.* Add: **4. b.** Also, manifesting intense emotion or excitability, esp. in æsthetic or intellectual contexts.

1817 Keats *Let.* 28 Dec. (1931) I. 76, I...went..to see 'Death on the Pale Horse'. It is a wonderful picture, when West's age is considered; but there is nothing to be intense upon, no women one feels mad to kiss, no face swelling into reality. **1879** W. D. Howells in *Atlantic Monthly* Jan. 38/2 'Why Miss Blood you are intense!' 'I don't know what you mean by that,' said Lydia. 'You like to take things seriously. You can't bear to think that people are not the least in earnest, even when they least seem so.' **1880** G. du Maurier *Eng. Soc. at Home* pl. 49 Fair Æsthetic (suddenly, and in deepest tones, to Smith, who has just been introduced to take her in to Dinner). 'Are you *Intense*?' **1897** H. Ellis *Stud. Psychol. Sex* I. 88 Miss M., aged 29, the daughter of English parents (both musicians) who were both of what is described as 'intense' temperaments. **1925** *Punch* 6 May 487 *Intense Lady.* 'Tell me—have you ever been psychoed?'

intensifier. Add: (Further examples.) *spec.* (*a*) *Gram.* = Intensive *sb.*; (*b*) = *hydraulic intensifier* (s.v. *hydraulic *a.* and *sb.* A. 2); (*c*) = *image intensifier* (s.v. *image *sb.* 8).

1892 *Lockwood's Dict. Mech. Engin.* (ed. 2) 219 *Intensifier*, a device frequently employed in place of a hydraulic accumulator, for converting a low water pressure into a higher. **1915** E. H. Barton *Introd. Mech. Fluids* xii. 191 In connection with an hydraulic pressure system an intensifying accumulator or intensifier may be used. In this device a piston works in the low-pressure cylinder and a connected piston rod, or ram, in the high-pressure cylinder. ..The pressure is magnified in the ratio of the areas of ram and piston. *Ibid.*, The intensifier presents an analogy to the hydraulic press. **1931** G. Stern *Meaning & Change of Meaning* 338 Intensifiers used ironically, that is, as down-toners. 'A lot you know about that!' **1939** L. M. Myers *Electron Optics* viii. 449 The term amplifier, in contradistinction to converter, indicates an increase in intensity. With the same reasoning we might term the device an image intensifier. Perhaps the term intensifier may be the best of all. However, we shall employ the usually accepted term converter, because there is always a conversion effect. **1940** C. C. Fries *Amer. Eng. Gram.* 205 The pressure to add *-ly* to intensifiers modifying adjectives is especially strong in Standard English. **1951** W. Empson *Struct. Complex Words* i. 26 The modern feeling about *quite* is entirely different from that about a discredited intensifier (*awfully, frightfully*). **1959** *Proc. IRE* XLVII. 907/2 Although the resolution obtainable is obviously much too low for military use, it is still of interest to the nuclear physicist as a simple and inexpensive intensifier for scintillation photography. **1964** *Language* XL. 39 Intensifier is the name given here to a small class of modifiers, e.g. *mere, utter*. **1970** R. H. Warring *Fluids for Power Syst.* i. 17 Intensifiers are almost invariably used for the production of pressures over 50 000 psi. **1971** G. Ansre in J. Spencer *Eng. Lang. W. Afr.* 162 The word classes: adjective, qualifier, quantifier, pluraliser and intensifier respectively.

intensify, *v.* Add: **inte·nsifying** *ppl. a.*, esp. in **intensifying screen**, a fluorescent screen placed in contact with the film or plate when a radiograph is taken in order to increase the effect on it of the X-rays.

1866 R. W. Thomas *Mod. Pract. Photogr.* 14 (*heading*) Intensifying solution. **1878** W. Abney *Treat. Photogr.* x. 71 These intensifying solutions may be applied to the image either before or after fixing. **1879** [see *Intensifier b]. **1903** Pusey & Caldwell *Pract. Application Röntgen Rays* vi. 144 The ratio of the exposure necessary with a single intensifying screen and photographic plate to that which is necessary with the same plate without the screen is about 1 to 4 or 5. **1940** K. S. Low *Metall. & Industr. Radiol.* iv. 30 Intensifying screens when placed in contact with the film will by fluorescence under the action of the radiation supplement the action of x-rays on the film, and thus shorten exposure periods. **1968** *Kodak Med. X-Ray Catal.* 8 High speed, fine grain X-ray film for use with intensifying screens.

intensional (inte·nʃənăl), *a. Philos.* [f. Intension 5 + -al.] Related or pertaining to the intension, or the attributes contained in a concept. Cf. *extensional *a.* 2.

1883 F. H. Bradley *Princ. Logic* 162 Dismissing for the present the intensional reading, let us consider interpretation in Extension. **1903** B. Russell *Princ. Math.* vi. § 66. 67 Thus every predicate (provided it can be sometimes truly predicated) gives rise to a class. This is the genesis of classes from the intensional standpoint. **1949** S. I. Hayakawa *Lang. in Thought & Action* (1952) xv. 253 By intensional orientation, 'capitalist', 'Bolshevists', 'farmers', and 'working men' 'are' what we say they are. **1956** A. Church *Introd. Math. Logic* (rev. ed.) I. 28 We shall not have occasion to use variables whose values are propositions, but we would suggest the term propositional variable for these. **1969** *Philos. Q.* XX. 34 The star notation can be made the groundwork of a system of intensional propositional logic in which a wide class of formal principles can be proved.

intensionalist (inte·nʃənălist), *a.* and *sb. Philos.* [f. prec. + -ist 3 b.] **A.** *adj.* Of or pertaining to the intensional attributes of a concept. **B.** *sb.* One who considers a concept from the standpoint of its inner attributes.

1948 *Mind* LVII. 298 No one, I think, ever intended to deny the interdependence of meaning and intention... Although some 'intensionalists' have spoken as if it were denied. *Ibid.*, The confusion just described is a mild and limited form of what we might describe as 'the intensionalist error'. **1962** *Times Lit. Suppl.* 16 Mar. 187/3 In a level-headed and sometimes ingenious discussion...the author shows themselves uncompromisingly 'intensionalist'. **1963** R. Carnap in P. A. Schilpp *Philos. R. Carnap* 919 The basic ideas underlying my intensionalist thesis are simple.

intensionality (intenʃənæ·liti). *Philos.* [f. *intensional *a.* + -ity.] The state or fact of being intensional. Cf. *extensionality.

1937 A. Smeaton tr. *Carnap's Logical Syntax of Lang.* iv. § 71. 259 The difference between the extensionality and intensionality of a language has nothing to do with the difference between the formal and the material treatment. **1944** K. Gödel in P. A. Schilpp *Philos. B. Russell* 138 Chwistek..has shown that the system of simple types becomes contradictory if one adds the 'axiom of intensionality' which says (roughly speaking) that to different definitions belong different notions.

intensionally (inte·nʃənăli), *adv. Philos.* [f. *intensional *a.* + -ly².] By way of intension, in an intensional manner.

1883 F. H. Bradley *Princ. Logic* 167 If again you emphasize the connection of the differences, you take the judgment intensionally. **1903**, etc. [see *extensionally *adv.*]. **1953** K. Britton *J. S. Mill* vi. 189 This judgement also must be understood intensionally: i.e. as a connexion of universals which holds without reference to the mere spatio-temporal positions of the instances. **1970** *Philos. Q.* XX. 43 We may take it that the expression so obtained is formally valid intensionally.

intensity. Add: **1.** (Further examples.)

1834 *Phil. Trans. R. Soc.* CXXIV. 222 The term intensity.. is immediately referable to..the operation of electric power, or the whole of the total force in a given direction up to the point of discharge. **1879** [see *saturation 4]. **1933** L. Bloomfield *Lang.* ix. 156 The second more specialized type of connotation..is intensity. The most characteristic intense forms are exclamations. **1939** Brooks & Warren *Understanding Poetry* 21. 167 The effect of this condensation in poetry is a sense of greater intensity than is usually found in prose fiction. **1961** J. H. Goodier *Dict. Painting & Decorating* 143 *Intensity*, the purity of a colour, sometimes called the 'saturation' or, in American terminology, the 'chroma'.

b. (Later example.)

1890 G. F. Stout *Analytic Psychol.* I. i. v. 110 The word 'intensity' in..psychology..is usually regarded as synonymous with the liveliness or vivacity of which Hume has so much to say.

3. *attrib.* and *Comb.*

1844 in H. M. Noad *Lect. Electr.* (ed. 2) ix. 401 The trifling decomposing effect from the intensity-inductor. **1877** Knight *Dict. Mech.* II. 1192/1 An intensity battery is one in which the elements are coupled up together, to give a current known as high-tension. **1927** E. G. Richardson *Sound* x. 238 Beside the frequency limits there are intensity limits to the sounds which the ear can perceive. **1932** D. Jones *Outl. Eng. Phonetics* (ed. 3) xxi. 277 Contrast-emphasis may be applied to almost any word. **1953** [see *affective *a.* 2 c]. **1972** *Science* 16 June 1236/1 The audio frequency Doppler signals from the detector were amplified sufficiently to provide a 30- to 40-volt signal for intensity-modulating the cathode-ray tube.

intensive, *a.* (*sb.*) Add: **5. b.** Suffixed to sbs. to form adjs. with the sense 'intensively using the thing specified', as *capital-intensive, labour-intensive*.

1957 K. A. Wittfogel *Oriental Despotism* vi. 218 The replacement of labor-intensive irrigation farming by labor-extensive cattle breeding. **1959** *Listener* 22 Oct. 666/1 We have the highly capital-intensive process of textile manufacture. **1970** *Times* 2 June (Container Suppl.) p. iii/4 The latest sophisticated container systems add up to a capital-intensive system of some magnitude. **1972** *Guardian* 29 June 15/5 The developing world..is beginning to see the case for labour-intensive farming.

8. **intensive care**: a form of medical treatment in which a patient is kept under concentrated and special observation; so *intensive-care unit*, etc.

1963 *Lancet* 19 Jan. 169/2 Our medical staff found the medical intensive-care unit so valuable that they requested..a separate surgical intensive-care unit. **1965** *Listener* 16 Sept. 401/1 We would like to see intensive-care units in all large modern hospitals. **1967** *Spectator* 11 Aug. 159/2 This method of dealing with a serious cardiac emergency is known as intensive care. **1973** J. Goodfield *Courier to Peking* x. 123 First, one of the general wards..and then to our new intensive care unit.

intention, *sb.* Add: **6. b.** In literary criticism: the aim or design which a critic detects in a writer's work.

1946 Wimsatt & Beardsley in *Sewanee Rev.* LIV. 469 Intention has obvious affinities for the author's attitude toward his work, the way he felt, what made him write. **1959** *Times Lit. Suppl.* 20 Feb. 97/1 Intention, in Mr. Wimsatt's use of the word, does not mean what it means in Dr. Richards's distinction between sense, tone, feeling, and intention... It means what we might have reason to think that the author thought he was up to.

12. b. (Further examples.)

1886 *Echo* 30 Nov., In the Communion Service a '*special intention*' was made known by the introduction of words implying that the 'sacrifice was received in memory' of the dead. **1966** *New Statesman* 22 Apr. 608/1 The text of the Mass itself, with its special intention in capitals.

15. Special Comb.: **intention movement** [tr. G. *intentionsbewegung* (O. Heinroth)], a movement or action on the part of an animal which itself performs no function except to reveal or signal that a further movement or action may follow or is contemplated; **intention tremor**, a tremor which is manifested when a voluntary movement is performed.

[**1910** O. Heinroth in *Jrnl. für Ornith.* LVIII. 122 Eine Modifikation des eigentlichen Locktones, wie sie kurz vor dem Auffliegen hervorgebracht und dann mit den schon beschriebenen Intentionsbewegungen verbunden wird, ist ein längeres und fein ausklingendes 'Hu'.] **1950** K. Z. Lorenz in *Symposia Soc. Exper. Biol.* IV. 242 We know of two physically distinct ways, by which non-social, mechanically effective endogenous activities may develop into social releasers: in one case the so-called 'intention movement' [*Intentionsbewegung*, Heinroth], in the other the so-called 'displacement activities'. *Ibid.* 243 Intention movements..are very reliable indicators for the present 'mood' of an animal. **1953** N. Tinbergen *Herring Gull's World* xvi. 163 A bird may rapidly change from pecking to the intention-movement of brooding. **1961** A. J. Berger *Bird Study* v. 158 The first intention movement preceding walking or hopping often is a 'bow'. **1887** Vickery & Knapp tr. *A. von Strümpell's Textbk. Med.* 593 The tremor in multiple sclerosis comes on only with intended movements, 'intention tremor'. **1969** *Times* 6 Feb. 4/1 All but one suffered from defects in their nervous systems. The commonest of these seems to have been intention tremors, which are tremors occurring when a voluntary movement is made.

intentional, *a.* Add: **1.** **intentional fallacy**: in literary criticism, the fallacy that the meaning or value of a work may be judged or defined in terms of the writer's intention.

1946 Wimsatt & Beardsley in *Sewanee Rev.* LIV. 468 (*title*) The intentional fallacy. ..The question of 'allusiveness', for example,..is certainly one where a false judgement is likely to involve the intentional fallacy. **1954** W. K. Wimsatt *Verbal Icon* i. 6 It is not so much a historical statement as a definition to say that the intentional fallacy is a romantic one. **1958** *Listener* 9 Oct. 578/2 A studied defence of what has been labelled—and dismissed—as 'the intentional fallacy'.

intentionalism (inte·nʃənăliz'm). [f. Intentional *a.* + -ism.] The doctrine that a literary work or some other work, etc., is the result of conscious intention or design. So **inte·ntionalist**, one who propounds such a doctrine; **intentionali·stic** *a.*, of, pertaining to, or characterized by intentionalism.

1878 W. Affleck tr. *Janet's Final Causes* i. vi. 215 God has come forth from Himself as well in pantheism as in creationism or intentionalism. **1946** Wimsatt & Beardsley in *Sewanee Rev.* LIV. 485 Allusiveness in poetry is one of several critical issues by which we have illustrated the more abstract issue of intentionalism. *Ibid.* 488 Allusiveness would appear to be in some recent poems an extreme corollary of the romantic intentionalist assumption. **1954** W. K. Wimsatt *Verbal Icon* i. 21 The use of biographical evidence [in criticism] need not involve intentionalism. **1958** *Listener* 16 Oct. 595/1 There is a suspicious silence on the whole issue of intentionalism among the classical English critics. **1958** I. P. Hungerland *Poetic Discourse* vi. 162 The Intentionalists have neglected such situations. **1958** M. C. Beardsley *Aesthetics* 27 There are other critics who tend to shift back and forth between the work and its creator, never quite clear in their own minds when they are talking about the one and when about the other. They mingle the evidences of intention with the evidences of accomplishment, and sometimes decide what the work is or means primarily on external evidence. This is to practice intentionalistic criticism.

inter, L. prep. Add: **inter partes** (Law), of an action: relevant only to the two parties in a particular case (see quots. 1966); of a deed or the like: made between two parties; **inter se** (later examples); **inter vivos**, between living persons (esp. of a gift as opposed to a legacy).

1816 Maule & Selwyn *Rep. Cases King's Bench* III. 308 A deed *inter partes* cannot operate as a release to strangers. **1906** *Daily Colonist* (Victoria, B.C.) 26 Jan. 10/2 Mr. Bodwell..contended that this must not be considered an action *inter partes*; in form it may have been, but in substance it was a propounding of the will. **1960** *Times* 4 Aug. 11/3 On July 29 the injunction was granted in the *inter partes* form. **1966** Black & Brown *Outl. Eng. Law* v. xvii. 160 In modern terms such actions are Actions in Rem or Actions in Personam or Inter Partes. *Ibid.* 161 An action..*inter partes* produces a judgment which is binding only as between the parties to the case and not upon third parties. **1968** *Law Rep.* 106 The old common law rule that in an indenture inter partes the covenantee must be named as a party to the indenture to take the benefit of an immediate grant or the benefit of a covenant. **1971** *Mod. Law Rev.* XXXIV. vi. 605 The Court gives decisions on preliminary points of law..when such questions are referred to it by municipal courts or tribunals in the course of proceedings *inter partes*. **1888** G. F. Stout *Stud. Philos. & Psychol.* (1930) i. 42 Forms of combination among presentations other than those by which sense-perceptions are connected *inter se*. **1971** *Mod. Law Rev.* XXXIV. vi. 598 Suppose that..the *inter se* doctrine of Commonwealth relations had emerged as a credible body of legal principle. **1972** *Times* 14 Mar. 12/2 The essential link between members of a volunteer unit was not their contractual relationship *inter se* but their common military service to the Crown. **1837** T. Lewin *Pract. Treat. Law Trusts & Trustees* vi. 86 The Bank of England cannot be made a trustee, for the Company will not enter notice of instruments *inter vivos* upon their books. **1949** W. B. Leach *Cases Law of Wills* 50 In 1931 decedent had created a large inter vivos trust. **1955** *Times* 8 June 4/3 Transfer of property through inheritance or *inter vivos* gifts or settlements should reckon as realization. **1963** *Economist* 16 Mar. 976/1 Death duties are..largely avoided by transfers *inter vivos*. **1969** *Daily Tel.* 1 Oct. 20/6 Half the Scandal 'empire' is owned by Lord Reid-haven following his mother's foresight in transferring the property outside the *inter vivos* period, which exonerates the assets from death duty. **1972** *Accountant* 28 Sept. 387/3 Benefits..should be mentioned in reply to the question in the Inland Revenue Affidavit about gifts *inter vivos*.

inter., **inter** (i·ntəː), abbrev. of Intermediate *a.* = *intermediate examination* (in arts, etc.), often used in ordinary colloquial speech.

1893–4 *London Univ. Guide* 39 Inter. Arts Honours. *Ibid.* 40 Inter. Science and Prel. Sci. Exams. **1913** W. Owen *Let.* 28 Sept. (1967) 199 He was a student of Nottingham College, but failed to pass Inter-Arts. **1948** L. Walmsley *Master Mariner* ii. i. 134 He had entered for a degree in science, had passed his 'inter' with the same ease as he'd passed his matric, and was sitting for his final in the coming term.

inter-, *prefix.* Add: **1. a.** *interfile* (*interfiling* vbl. sb.), *-lot* vbs.

1959 J. Lawler *H. W. Wilson Co.* viii. 111 Specially trained women..handle the task of interfiling the metal linotype slugs. *Ibid.* 112 The Production Department staff interfiles the new and old slips of copy. **1962** V. Malkin in Householder & Saporta *Probl. Lexicogr.* 15 Interfiling..of authentic word cards. **1967** Cox & Dews in Cox & Grose *Organiz. Bibliogr. Rec. by Computer* ii. 16 Anonymous works are frequently interfiled by title in author catalogues. **1967** C. J. Duncan *Ibid.* 41 The insertion of special sorts or 'pi-characters' has been the traditional inter-filing solution adopted. **1969** *Bulletin* (Sydney) 22 Nov. 28/1 'Inter-lotting', a practice gratuitously rendered by woolbrokers, consists in selecting a number of what would be star lots, and bracketing them into one big lot to sell as such.

b. *inter-influence* (vbl. sb. *-influencing*), *-lend* vbs.

1921 E. Sapir *Lang.* viii. 184 In many such cases it is perfectly clear that there could have been no dialectic interinfluencing. **1961** L. F. Brosnahan *Sounds of Lang.* v. 215 Genetic and linguistic interinfluencing of such groups favours the reverse trend. **1968** *Listener* 5 Sept. 319/3 Two of the very few important libraries which do not 'inter-lend' are those of Oxford and Cambridge.

2. a. *inter-availability, -behaviour, -celebration, -fertility, -influence, -racialism, -substitutability; inter-available, -behavioural, -fertile, -responsive, -sterile, -substitutable* adjs.

1920 *Glasgow Herald* 3 July 6 The issue of a circular cancelling the inter-availability of practically all railway tickets between Scotland and England. **1933** *Ibid.* 27 Jan. 11 The deputation intends..to press for the inter-availability of these tickets on all systems. **1935** *Punch* 21 Aug. 204/3 'Interavailability of Tickets between the G.W., L.M.S. and L. & N.E. Companies'—*G.W.R. Pamphlet.* **1973** *Daily Tel.* 16 Feb. 2/6 Tickets are inter-available and British Rail fares will now reflect the London Transport increase. **1938** J. R. Kantor in *Jrnl. Philos.* XXXV. 449 The interbehavioural hypothesis signifies that the event known..consists of the concrete interbehaviour of specific individuals with things. **1939** *Mind* XLVIII. 111 The inter-behavioural hypothesis encourages us properly to evaluate every form of scientific and logical work. **1921** *Spectator* 26 Mar. 391/2, I should be glad to see inter-celebration, under present conditions... **1960** Partridge *Charm of Words* i. 23 This interinfluence is unavoidable. [...] use at about the same period..and the possibility of their mutual relationship and inter-influence has continued to tease the brains of scholars. **1931** *Amer. Speech* VII. 78 A writer in the *Congregationalist* says 'Interracialism, like love, service, and brotherhood, is a splendid word which has been cheapened by overuse.' How many acquaintances have you whose diction suffers from over-use of the word interracialism? **1906** *Macm. Mag.* Apr. 434 The mystic net-work, inter-responsive as a delicate system of nerves. **1955** R. Blesh *Shining Trumpets* (ed. 3) ii. 30 It is a thing..as sensitively interconnected and inter-responsive, as the branches of a tree. **1916** *Mem. N.Y. Bot. Garden* VI. 352 It was found that at least three varieties are strongly inter-sterile. **1956** *Nature* 21 Jan. 142/1 The mating type of the hybrid shows no change, that is, the hybrid males remain intersterile with *O* females, as in the original strain *H*. **1954** I. M. Copi *Symbolic Logic* vii. 224 Of the list of elementary valid argument forms..the last ten were equivalences whose intersubstitutability was assumed. **1957** P. Geach *Mental Acts* 90 We cannot define synonymy as a supreme degree of equivalence, intersubstitutability *salva veritate* in all contexts. **1964** I. M. Copi *Symbolic Logic* iii. 44 The list contains all those logical equivalences which are certified as intersubstitutable. **1957** P. Geach *Mental Acts* 102 In an 'extensional' logic..predicates or relative terms would be freely intersubstitutable provided that they held good of the same objects.

3. a. *interglyph, -mutule.*

1875 *Encycl. Brit.* II. 404/1 The interglyphs are each one-seventh of the whole tablet or triglyph. **1865** *Dict. Archit.* (Archit. Publ. Soc.) IV. 149/1 The temple to Ceres at Pæstum has no mutules or triglyphs. The example at Albano has no intermutules. **1901** R. Sturgis *Dict. Archit.*, *Intermutule*, the space between two mutules, as in an architrave.

4. a. *inter-cavernous, -consonantal, -electronic, -follicular, -marginal, -micellar, -morainic, -nebular, -villous* adjs.

1890 Webster, Inter-cavernous. **1968** Passmore & Robson *Compan. Med. Stud.* I. xi. 9/1 The two cavernous sinuses intercommunicate across the midline by means of the anterior and posterior intercavernous sinuses. **1931** *Amer. Speech* VII. 19 His inter-consonantal *a* is perhaps Southern. **1964** J. W. Linnett *Electronic Struct. Molecules* i. 11 In chromium..the six electrons are spread among the five 3d and one 4s orbitals to reduce the effect of inter-electronic repulsion. **1888** *Syd. Soc. Lex.*, *Interfollicular*, situated between follicles. **1968** J. H. Burn *Lect. Notes Pharmacol.* (ed. 9) 88 Thyrocalcitonin is a hormone formed in the interfollicular cells of the thyroid gland. **1858** *Phil. Trans. R. Soc.* CXLVIII. 280 Intermarginal cavities [of sponges]. **1900** E. R. Lankester *Treat. Zool.* III. xiv. 246 In some genera intermarginal plates occur between these two series. **1905** J. H. Parsons *Dis. Eye* x. 171 The palpebral conjunctiva is said to commence at the anterior edge of the intermarginal part of the lid, but from this point to the posterior margin of the edge (the intermarginal strip) and for about 2 mm beyond (to the sulcus subtarsalis) there is a transitional zone. **1937** *Chem. Abstr.* XXXI. 5155 The dimensions of the intermicellar spaces decreases considerably on shrinkage. **1962** J. T. Marsh *Self-Smoothing Fabrics* xv. 27 They..concluded that the resin must be in the inter-micellar spaces of the cellulose. **1894** J. Geikie *Gt. Ice Age* (ed. 3) xxxv. 595 The inter-morainic lakes which occupy similar positions at the base of the Alps in North Italy. **1957** G. E. Hutchinson *Treat. Limnol.* I. i. 90 A few intermorainic bogs still exist but the lakes in general have disappeared. **1929** J. H. Jeans *Universe around Us* ii. 142 The vast stretches of internebular space. **1957** *Jrnl. Brit. Interplanetary Soc.* XVI. 22 Calculations on the feasibility of long-range interstellar travel—including internebular journeys. **1890** Billings *Med. Dict.* I. 710/2 *Intervillous lacunæ*, irregular vascular spaces connected with the maternal blood-vessels, surrounding the foetal villi in the placenta. **1964** *Gray's Anat.* (ed. 33) 115 In the early stages of placental development the blood in the foetal vessels is separated from the maternal blood in the intervillous space.

b. *inter-testamental* adj.

1956 *Jrnl. Theol. Stud.* VII. 292 Professor Stauffer..is apt to find in the N.T. references to lost intertestamental literature. **1973** *Times Lit. Suppl.* 7 Dec. 1516/2 De Vermes begins by ransacking the inter-testamental and rabbinic literatures (especially the Dead Sea Scrolls, the Targumim and the Palestinian Talmud).

c. *inter-African, -American, -cameral, -Caribbean, -coastal, -cultural, -dialectal, -ecclesiastical, -ethnic, -generational, -governmental, -perceptual, -professional, -racial* (later examples; so *-racially* adv.), *-regional, -territorial* (later examples).

1955 *Nature* 25 Feb. 366/2 An Inter-African Committee for Social Sciences. **1960** *Guardian* 6 Oct. 10/6 An expression of inter-African solidarity. **1973** *Black World* May 61/2 Those whom the truth would incriminate hid behind the walls..of amnesty, reconciliation and inter-African dialogue. **1938** *Sun* (Baltimore) 5 Dec. 2/5 In this and in other matters of inter-American interest, I anticipate working with the representatives of Ecuador. **1961** *Ann. Reg. 1960* 532 The Inter-American system is incompatible with any form of totalitarianism. **1964** *Illustr. London News* 18 Jan. 81 President Johnson..sent his Assistant Secretary of State for Inter-American Affairs..to the Canal Zone. **1972** *Buenos Aires Herald* 4 Feb. 8/4 Dr. Bruno Quijano met yesterday with Charles Meyer, assistant secretary of state for inter-American affairs. **1973** *Sunday Advocate-News* (Barbados) 25 Feb. 15/7 Barbados strongly opposes any case of racial discrimination..and for that reason opposes the admission of Portugal in the Inter-American system. **1929** *Times* 29 Oct. 16/1 Nobody believes that the question will..go to the intercameral conference before the Special Session ends. **1935** *Economist* 13 July 60/2 The Public Utility Holding Company Bill..the House refused to the Senate, which has sent it to an inter-cameral conference. **1971** *Jamaican Weekly Gleaner* 10 Nov. 5/5 Mr. Arnold Foote..lauded the function as a 'gesture of inter-Caribbean friendship'. **1973** *Caribbean Contact* Jan. 6/3, I believe that this magazine would help bridge the gap between Caribbean peoples and so foster inter-Caribbean harmony. **1927** *Contemp. Rev.* Oct. 493 Already in inter-coastal traffic New York is surpassed. **1969** *Jane's Freight Containers 1968–69* 57/2 Intercoastal Services via the Port of New York. **1937** *Theology* XXXV. 347 Our present consideration of intercultural contacts. **1955** *Sci. Amer.* Apr. 84/2 In the interest of intercultural understanding various U.S. Government agencies have hired anthropologists. **1972** *Ibid.* Nov. 82/1 If pictorial recognition is universal, do pictures offer us a lingua franca for intercultural communication? **1959** M. Schlauch *Eng. Lang. in Mod. Times* i. 28 An inter-dialectal survey. **1920** *Q. Register* Nov. 339 A great moment of international and inter-ecclesiastical opportunity. **1959** H. Wolff in *Anthropol. Ling.* Mar. 34 (*title*) Intelligibility and inter-ethnic attitudes. **1971** *Sunday Nation* (Nairobi) 11 Apr. 13/3 There was one section of the population in Uganda,

inter-ethnic in composition, which was not using English as the primary qualification for professional ascent. This section was the armed forces of Uganda. **1964** S. Lieberson in J. A. Fishman *Readings Sociol. of Lang.* (1968) 551 We find intergenerational linguistic unity of nationality groups is less than the unity within the immigrant groups themselves. **1968** *Listener* 29 Feb. 259/1 The intergenerational differences amount to a mutation. **1971** *Black Scholar* June 29/2 The black revolution will then become an intergenerational revolution. **1927** *Glasgow Herald* 20 Jan. 8 When it came to negotiation, agreement could not be obtained by inter-Governmental intervention. **1946** J. S. Huxley *Unesco* ii. 48 This important group of human activities..has never previously been adequately dealt with by any inter-governmental organisation. **1969** *Jane's Freight Containers 1968–69* 142/3 Some of the problems requiring action at the inter-governmental level ..have been indicated above. **1973** *Times* 21 May 4/7 The inter-governmental convention establishing the European University Institute was signed by the six founding members of the Community in April, 1972. **1932** W. T. Stace *Theory of Knowledge & Existence* vii. 160 What it is or is not during inter-perceptual periods makes no difference to me as a practical person. **1934** *Mind* XLIII. 151 Why must we believe that causation continues to operate during inter-perceptual intervals? **1960** *Times* 15 Feb. 11/7 The third Interprofessional Conference in March will examine some cardinal moral issues. **1971** *Optometry Today* (Amer. Optometric Assoc.) 13 Naturally, interprofessional referrals are a two-way street. **1905** *Athenæum* 30 Sept. 430/1 Inter-racial cordiality. **1953** E. H. Brookes *S. Afr. in Changing World* v. 105 Thus the services of Americans to South Africa in the interracial field are spanned across a century. **1960** *Spectator* 22 July 128 There is a large, brand-new 'inter-racial' hotel. **1968** *Blues Unlimited* Dec. 12 The local interracial Dirty Blues Band. **1972** *Publishers Weekly* 7 Feb. 37 (Advt.), Grace Halsell..describes what happens to interracial couples when they are joined in a love affair or marriage. **1964** *Punch* 26 Aug. 290/1 Anger..fomented internally and directed inter-racially. **1972** *Publishers Weekly* 14 Aug. 40/2 He hustled a basketball scholarship, lost it for dating interracially. **1945** Interregional [see *expressway]. **1962** H. R. Loyn *Anglo-Saxon Eng.* ii. 77 Gold was the essential commodity for interregional exchange in the Mediterranean context. **1964** *Ann. Reg. 1963* 102 Agreement..that joint authorities..would continue to provide services on an inter-territorial basis was reached. **1967** *Economist* 17 June 1234/2 The main points of the treaty are: the establishment of an East African development bank; interterritorial tariffs on goods originating in one country and exported to another, [etc.].

5. *inter-arrival, -bank(s), -borough, -caste, -centre, -church, -city* (examples; also *absol.*), *-class* (examples), *-dealer, -electron, -faith, -family, -fibre, -library, -nucleon, -office, -particle, -party, -species, -stream, -trial, -union, -university* (later examples), *-valve, -village, -zone.*

1962 J. Riordan *Stochastic Service Syst.* i. 3 For this, as for the Poisson, the intervals between demands (inter-arrival intervals) are independent random variables. **1968** P. A. P. Moran *Introd. Probability Theory* iii. 179 Much more complicated queueing systems with general inter-arrival and service time distributions are considered. **1907** *Daily Chron.* 29 Sept. 9/2 The Inter-Banks Team Race, in which sixteen banks in London were represented. **1966** *Economist* 12 Nov. 718/2 The big broker in unsecured 'interbank' sterling deposits. **1973** *Daily Tel.* 9 Feb. 21 Overnight money in the interbank market rose to a peak of 13 p.c. earlier this week. **1905** *Westm. Gaz.* 8 Nov. 5/1 The officials of the inter-borough railways. **1928** *Daily Express* 19 May 9/5 The first inter-borough competition of its kind ever organised in London. **1908** *New Reformer* II. 62 It is a case of inter-caste and inter-provincial marriage. **1925** E. S. Jones *Christ of Indian Road* xiii. 243 We sat down to an intercaste dinner—a hundred high-caste Hindus, a hundred outcastes, a hundred Indian Christians, a few Mohammedans, and several of us of the West. **1964** *Ann. Reg. 1963* 358 Among social measures which came into force were the banning of polygamy and permission for inter-caste marriages. **1946** *Nature* 20 July 97/2 Each centimetre on the plotting-board of the apparatus corresponds with 'the cone intercentre distance', that is, the distance between the centre of one foveal cone and that of its next-door neighbours. **1970** *New Scientist* 31 Dec. 596 Saturation signalling involves a large volume of inter-centre communication. **1905** *Daily Chron.* 22 Nov. 1/7 At an inter-Church Conference in New York 17,000,000 Church members of all denominations were represented. **1926** *Brit. Weekly* 22 July 327/1 Dr. Clark has toured the world to build this inter-church, international organisation. **1909** *Westm. Gaz.* 3 June 14/1 In the first inter-city match played between Edinburgh and Glasgow. **1940** A. L. Albert *Electr. Communication* (ed. 2) vii. 174 Connections are made between the two exchanges by means of toll lines... In some locations this has been designated intercity service. **1955** *Railway Mag.* May 334/1 The 9 a.m. 'Inter-City' from Paddington averaged 18½ min. late at Snow Hill. **1968** *Daily Tel.* 12 Nov. 26/7 A three-months' trial scheme..was introduced by British Railways on inter-city trains in September. **1970** A. Dekker *Divers Diamonds* xxii. 145 The GPO intercity microwave link towers. **1972** *Guardian* 13 Mar. 13/1 The expression 'Inter-City' was a British invention and its success is demonstrated by the adoption of the title in such countries as Germany, Japan, Holland, and now the United States, to describe express business trains. **1973** *Times* 29 Oct. 14/8 On the inter-city to Sunderland I was one of the first people in the dining car. **1909** *Post-Intelligencer* (Seattle) 18 Apr. 1/2 The interclass crews and girls' crews work at other hours during the day. **1950** T. H. Marshall *Citizenship & Social Class* I. iv. 57 A divided educational system, by promoting both intra-class similarity and inter-class difference, gave emphasis..to a criterion of social distance. **1971** P. J. Keating *Working Classes in Victorian Fiction* iii. 71 A romance of corrupted character and inter-class love. **1972** *Listener* 27 July 107/2 Venice was untouched by the inter-class antagonisms..that troubled the domestic peace of other states. **1968** *N.Y. Times* 19 Feb. 60 It appears to be technically feasible to use a central computer to record and report interdealer quotations. **1970** *Washington Post* 30 Sept. D 11/4 The following..are representative interdealer prices. **1964** J. W. Linnett *Electronic Struct. Molecules* i. 11 Clearly, for the lowest energy state, it is advantageous to reduce inter-electron repulsion as much as possible. **1967** *Economist* 6 May 574/2 It is joining with 39 other firms and an interfaith group to provide 1,500 new jobs. **1970** *Toronto Daily Star* 24 Sept. 37/1 An interfaith group which supports tough anti-obscenity laws. **1973** *Jewish Chron.* 2 Feb. 13/5 The Israel Interfaith Committee. **1946** *Nature* 10 Aug. 204/2 The successful production of hybrid plants from interspecific, intergeneric, and even interfamily crosses has naturally led to investigation of the events which occur in some of the crosses which habitually fail. **1965** J. E. Cross in *English Studies* Apr. 96 The provincial and interfamily strife..between..Deira and Bernicia. **1962** J. T. Marsh *Self-Smoothing Fabrics* iv. 39 In a fabric with a high cover factor, *i.e.* a tightly woven material, the friction between fibres is apt to be high enough to restrict inter-fibre movement. **1928** J. A. McMillen (*title*) Selected articles on interlibrary loans. **1938** *Times* 1 Jan. 10/1 All England and Wales is now covered by schemes for the inter-library lending of books. **1968** *Language* XLIV. 211 The publications..are available to any American library on inter-library loan. **1953** *Physical Rev.* XCI. 1529/2 The meson theory, which predicts strong internucleon interactions. **1934** Webster, Inter-office. **1938** *Times* 29 Sept. 19/1 War Loan..was being quoted 95 middle in inter-office dealings. **1969** *New Yorker* 20 Sept. 181/1 An interoffice memo cast in cablese. **1972** *Accountant* 6 Apr. 456/1 On the sports side, it was hoped to revive the inter-office cricket tournament and play matches with other students' societies and the law students. **1946** *Nature* 6 July 13/1 The resistance increased, indicating that the width of the inter-particle contacts had diminished. **1953** *Physical Rev.* XCI. 1527 The leading term..depends only on the interparticle distances. **1962** W. B. Thompson *Introd. Plasma Physics* vii. 148 The simple theory presented here may be extended by including in a phenomenological way the effect of interparticle collisions. **1909** *Westm. Gaz.* 8 Mar. 2/1 A break in the party seems inevitable. The issue of this inter-party struggle will be one of great moment for South Africa. **1957** C. Hunt *Guide to Communist Jargon* xxx. 103 He [*sc.* Marx] held that, under the capitalist system, parliaments were simply the instruments by means of which the bourgeoisie maintained its domination, and that inter-party conflicts within this framework had no real significance. **1920** *Proc. Amer. Soc. Hort. Sci. 1919* 50 (*heading*) Inter species pollination of plums. **1962** *Punch* 14 Mar. 443/1 The possible consequences of inter-species communication between man and dolphin. **1902** Webster Suppl., *Interstream* a. (Phys. Geog.), between streams. **1968** D. S. Simonett in R. W. Fairbridge *Encycl. Geomorphol.* 559 Many geomorphologists today use the term [*sc.* interfluve] for the interstream area to imply a discrete landscape or geomorphic unit. **1971** *Jrnl. Gen. Psychol.* Jan. 125 Within each order, 21 plates consisting of three different spatial arrangements of each level of number were also shown randomly with an intertrial interval of 10 seconds. **1972** *Jrnl. Social Psychol.* Dec. 225 The timer was preset so that the signals for all shock options were of three seconds duration, with a 15-second intertrial interval. **1969** *Daily Tel.* 10 Feb. 19/5 Because of bitter inter-union rivalry, the original purpose of the strike..has become almost totally obscured. **1973** *Times* 17 Jan. 16/8 Eliminating the possibility of inter-union disputes in one factory. **1956** *Nature* 3 Mar. 405/1 In 1944 the Secretary of State for the Colonies invited British universities to set up an Inter-University Organization to assist the development of higher education in the Colonies. **1970** *Internat. & Compar. Law Q.* XIX. i. 181 Dr. Verheul is now Head of the private international law department of the T. M. C. Asser Institute in The Hague, an inter-university institute which is engaged in comprehensive documentation in the field of international law. **1909** *Cent. Dict.* Suppl., *Inter-valve*, in steam-engines, noting the space which is between the throttle- and slide-valves. **1921** *Wireless World* 2 Apr. 17/2 The 5-step Amplifier, with intervalve resistance couplings, is of the usual type. **1921** *Ibid.* 14 May 120/1 Substitute an intervalve transformer for your telephone transformer. **1949** M. Mead *Male & Female* 413 The biggest inter-village economic exchange. **1960** Inter-zone [see *dust *v.*¹ 12 c]. **1961** *Times* 8 June 5/2 They gained a 5–1 winning lead over Denmark in the inter-zone final.

6. **interacinar** (-æ·sinăɪ) = *interacinous* (s.v. Inter- 6); **intera·trial**, situated between the atria of the heart; **inter-ca·rdinal**, of points of the compass: lying midway between the cardinal points; also as *sb.*; **interco·rtical**, situated within the (or a) cortex (properly **intracortical*); **interge·nic** [*genic *a.*], taking place or existing between neighbouring genes; **interpa·lpebral**, situated between the eyelids; **interpla·nar** *Cryst.*, existing between the planes of a crystal lattice; **interplical** (-pləi·kăl), situated between folds (see Plica 2); **interpopula·tional**, occurring or existing between populations or groups; **interpro·ximal** *Dentistry*, situated between adjacent teeth; on or affecting the surfaces bounding such a region; **interpu·pillary**, existing between the pupils of the eyes; **interto·nic**, occurring between two tones or stresses; **intertube·rcular**, placed or situated between tubercles; **intervari·etal**, formed or obtained from, or occurring between, (members of) different varieties (Variety 6 b); **intervei·nal** *Bot.*, situated or occurring between the veins of a leaf; **interxy·lary** *Bot.*, situated within the secondary xylem.

1900 *Bull. Johns Hopkins Hosp.* XI. 205/2 Interacinar cell islets have been studied by Harris and Gow. **1961** *Lancet* 29 July 258/2 A high prevalence of pancreatic arteriosclerosis..could have caused the associated insular and interacinar fibrosis and hyalinisation. **1911** Stedman *Med. Dict.* 431/1 *Interatrial*, between the atria of the heart. **1967** G. M. Wyburn et al. *Conc. Anat.* iii. 89/1 On the interatrial septum there is a shallow depression, the fossa ovalis. **1909** Webster, Intercardinal, adj.—n., an intercardinal point. **1961** F. H. Burgess *Dict. Sailing* 121 *Inter-cardinal points*, the half-cardinals. **1969** *Gloss. Terms Magnetic Compasses & Binnacles* (*B.S.I.*) 5 *Inter-cardinal points.* These are north-east, south-east, south-west, north-west and may be marked with the capital letters NE, SE, SW, and NW respectively. **1902** *Proc. Zool. Soc.* II. 215 Cortex [of the sponge] fibrous throughout,..without intercortical cavities. **1924** R. M. Ogden tr. *Koffka's Growth of Mind* ii. 55 According to Edinger, morphological changes in the brain are indicated by..the growth of intercortical pathways. **1933** J. Rosett *Intercortical Syst. Human Cerebrum* iv. 26 In the course of cerebral evolution the intercortical systems increase in number and in bulk. **1941** *Cold Spring Harbor Symp. Quant. Biol.* IX. 161/1 The breaks of minute rearrangements would in all probability differ from them in being only intergenic. **1951** G. H. Bourne *Cytol. & Cell Physiol.* (ed. 2) v. 225 It..assumes that there are no non-genic portions of the chromosome (matrix, intergenic connexions). **1968** *Canad. Jrnl. Genetics & Cytol.* X. 50 A UV-sensitive mutant of *Aspergillus rugulosus* unrelated to intergenic crossing over. **1892** A. Duane tr. *Fuchs' Textbk. Ophthalm.* xii. 467 That part of the cornea and of the scleral conjunctiva which ordinarily is not covered by the lids is called the interpalpebral zone... When the eyes are a little screwed together..the interpalpebral zone diminishes in size. **1934** E. Wolff *Path. Eye* ii. 29 Pinguecula, or interpalpebral spot, is a slightly raised yellowish area placed next to the inner or outer margin of the cornea. **1932** *Amer. Mineralogist* XVII. 549 (*caption*) The values given are the calculated interplanar spacings. **1948** K. Lonsdale *Crystals & X-Rays* iii. 77 There is now, however, no way of measuring interplanar angles. **1970** A. J. C. Wilson *Elem. X-Ray Crystallogr.* ii. 19 All the interplanar spacings capable of producing lines in an X-ray powder photograph can be obtained by consideration of the distance between successive planes of the lattice points. **1900** *Proc. U.S. Nat. Museum* XXIII. 398 The term interplical ridge has been applied to the summit of the radial convexities of the wing and interplical groove the bottom of the alternating concavities. **1903** *Phil. Trans. R. Soc.* B. CXCV. 161 The thickened chitin bands, which..are in contact by those edges directed towards the interplical space, are here divaricated. **1971** *Listener* 30 Dec. 907 There are no known IQ tests which are capable of overcoming the *inter*populational cross-cultural barrier. **1971** *McGraw-Hill Yearbk. Sci. & Technol.* 253 Interpopulational developmental comparisons..helped to clarify what in the past seemed to be an aberrant course of leaf development. **1897** E. C. Kirk *Amer. Textbk. Operative Dentistry* iii. 97 The counteraction of the tongue and cheek..drives the finer particles of the food into the interproximal spaces. **1908** G. V. Black *Operative Surg.* 211 (*heading*) Interproximal wear and its treatment. *Ibid.* 213 If there has been much loss of interproximal gum tissue..watch the case for a few weeks. **1960** Kerr & Ash *Oral Path.* vii. 115 Interproximal caries involve all of the teeth, but the molars and bicuspids are involved more frequently than the incisors. **1972** S. Garfield *Teeth* iv. 88 The small spaces between adjacent teeth around the contact points are called Contact Areas or Interproximal Spaces. *Ibid.* ix. 133 Inside the ridges between the roots of each adjacent tooth the bone is called Interproximal Bone. **1907** J. H. Parsons *Dis. Eye* xxvi. 525 With an interpupillary distance of 60 mm. this angle is about 2°. **1962** L. S. Sasieni *Princ. & Pract. Optical Dispensing* v. 106 (*heading*) Inter-pupillary distance. **1909** *Cent. Dict.* Suppl., Intertonic. **1953** K. Jackson *Lang. & Hist. in Early Brit.* ii. 268 Syncope of intertonic syllables in Vulgar Latin. **1893** *Jrnl. Anat. & Physiol.* XXVII. 260 The level which is expressed by our inter-tubercular line. **1967** G. M. Wyburn et al. *Conc. Anat.* ii. 53/1 The muscle is inserted..into the lateral lip of the intertubercular groove on the front of the humerus. **1916** *Mem. N.Y. Bot. Garden* VI. 352 No essential differences were noted in intervarietal fertility and fruitfulness. *Ibid.*, Intervarietal crosses. **1951** *New Biol.* XI. 115 In *Paramecium aurelia* there are seven varieties, and each variety contains types I and II... Intervarietal mating takes place very rarely and then only between type I of one variety and type II of the other. **1934** Webster, Interveinal. **1946** *Nature* 9 Nov. 663/2 The virus strain used produced a bright yellow interveinal mottle in the tomato. **1970** *Bot. Gaz.* CXXXI. 152/1 This strip [of leaf] was five interveinal areas wide at the upper margin. **1889** *Ann. Bot.* III. 293 The phloëm-islands, or interxylary phloëm-strands, are formed centripetally by certain portions of the normal cambium. **1969** K. Esau *Phloem* vi. 213 In..the interxylary or included phloem, the cambium..forms periodically some phloem increments towards the interior of the stem so that the phloem becomes embedded in the xylem.

interacinar: see *inter- *pref.* 6.

interact, *sb.* Add: *attrib.*

1908 *Daily Chron.* 3 July 3/5 At the 'private' or roofed theatres, the performance of interact music was the rule. **1909** *Morning Leader* 29 May 4/6 At the 'private theatres', such as the Blackfriars, they not only had interacts but interact music.

interact, *v.* (Further examples.)

1967 M. Argyle *Psychol. Interpersonal Behaviour* viii. 144 Children and adolescents are very limited in their social techniques, and may be able to interact with other children, and parents, but not with other adults... Most mental patients apart from schizophrenics are able to interact, but are much less successful in forming permanent relationships. **1972** *Jrnl. Social Psychol.* LXXXVII. 7 Very few Ss [*sc.* subjects] left their chair in order to interact with *E* [*sc.* the examiner]; those who did were ignored.

interaction. Add: (Further examples.) *spec.* in *Physics*, referring to the action between atomic and subatomic particles. Also *attrib.*

1930 *Proc. R. Soc.* A. CXXIX. 4 The simplified interaction which couples the nuclear system and the electron. **1951** T. Parsons et al. in Parsons & Shils *Toward Gen. Theory Action* ii. iv. 190 The specific interaction systems of ego. **1955** H. B. G. Casimir in W. Pauli *Niels Bohr* 130 The electrostatic interaction energy is e^2/a where *a* is a distance between electrons. **1959** *Listener* 9 July 52/1 There are four distinct types of force through which the manifold transformations of matter and energy arise. These forces are known as the strong, the electromagnetic, the weak and the gravitational interactions. **1964** *Language* XL. 242 An interaction cycle between the mother and child that amounts to a pair of reciprocal transformations. **1967** M. Argyle *Psychol. Interpersonal Behaviour* vii. 128 If A presents himself as an upper-class person but B reacts to him as a working-class person, interaction will not proceed smoothly.

interactionism (intəræ·kʃəniz'm). *Philos.* [f. Interaction + -ism.] The theory that in the causal relations between mind and body the causal influence runs in both directions, in sensation from body to mind and in volition from mind to body. So **intera·ctionist** *sb.*, an adherent of interactionism; *adj.*, of or pertaining to this doctrine.

1902 *Pop. Sci. Monthly* 459 Interactionism has to maintain, in this concrete form of the 'survival theory', that the mental process as such is an aid to evolution. **1903** C. A. Strong *Why the Mind has a Body* 5 Whether this interactionist tendency in biology be significant or ephemeral, we need not inquire. *Ibid.* 23 What the interactionist..takes for an action of mind on body is..an action of the brain upon the rest of the body. *Ibid.* 33 Interactionism seems almost to necessitate two juxtaposed realities exchanging influences, and thus to imply a metaphysical dualism. **1934** *Brit. Jrnl. Psychol.* Jan. 273 The annoyance may not be a response to being struck with a pin... We will not try here to settle the quarrel between such a parallelistic interpretation and the interpretation of an interactionist. **1941** *Mind* L. 86 He argues for interactionism, as against psychophysical parallelism. **1956** J. O. Urmson *Philos. Analysis* ii. vii. 110 One [metaphysician] advocates psycho-physical parallelism, the other interactionism. **1972** *Jrnl. Social Psychol.* Aug. 214 Pursuing the explanations of the symbolic interactionist somewhat further, Mead's theory suggests..maximum category overlap.

interactive, *a.* (Later examples.)

1972 *Computers & Humanities* VI. 195 A method which will provide the student with a rapid interactive feedback. **1972** *Science* 2 June 1019/1 A complicated and interactive set of physical, chemical, biological and human processes act in concert to yield a spectrum of measured values.

inter-African: see *inter- *pref.* 4 c.

inter-agency. (Later examples.)

1969 *R. & E. Coordinator* (Res. & Engin. Council Graphic Arts Industry) Apr. 4/1 The Government uses millions of microfiche internally for storage and retrieval systems and for interagency communications. **1973** *Times* 24 May 8/4 On July 5, 1970, Mr Nixon said, he appointed Mr Hoover chairman of an inter-agency committee to discuss a need for better intelligence operations.

inter alia: see Inter Lat. prep.

inter-allied (intərălai·d, -æ·ləid), *a.* [f. Inter- 1 b + Allied.] Existing or constituted between allies or allied forces. So **inter-ally** (-æ·ləi), *a.*

1919 J. M. Keynes *Econ. Consequences Peace* 96 The supreme authority is to be in the hands of an Inter-Allied Rhineland Commission. *Ibid.* 240 The settlement of inter-Ally indebtedness. **1925** A. Toynbee *Survey Internat. Affairs 1920–23* 1 The four Treaties..had provided for the establishment of a number of inter-Allied bodies to perform temporary or permanent executive duties. **1942** W. S. Churchill in *Hutchinson's Pict. Hist. War* 18 Mar.–9 June 30/2 The task of national self-preservation and of inter-allied duty. **1973** *Times* 29 Oct. 15/2 This public airing of an inter-allied dispute on a day of acute international tension was tactless to say the least.

inter-American: see *inter- *pref.* 4 c.

interanimate, *v.* Add: Hence **interanima·tion**, mutual animation.

1925 I. A. Richards *Princ. Lit. Crit.* xxxi. 237 Conflicts, resolutions and interanimations. **1966** *English Studies* XLVII. 200 The interanimation of the complex parts.

inter-arrival, -atrial, -availability, -available, -bank(s): see *inter- *pref.* 5, 6, 2 a, 5.

interbed, *v.* (Earlier example.)

c **1806** D. Wordsworth *Jrnl.* (1941) I. 252 These two islands..were intermingled with the water, I might say interbedded and interveined with it.

inter-behaviour, -behavioural: see *inter- *pref.* 2 a.

interbellum (intəɪbe·lə̆m), *a.* Also **interbella.** [f. Inter- 4 b + L. *bellum* war.] Of or with reference to a period between two wars, e.g. between the wars of 1914–18 and of 1939–45. So **interbe·lline** *a.*

1940 Beerbohm *Mainly on the Air* (1946) 82, I feel sure that even in the inter-bella period a river would have done Bloomsbury no end of good. **1958** *Oxf. Mag.* 22 May 454/1 The interbelline English Don..did his best not to take 'Eng. Lit.'..too seriously. **1961** *Ann. Reg. 1960* 462 Nathalie Sarraute..was noticeably closer than he to the English experimental novel of the interbelline years. **1969** P. Anderson in Cockburn & Blackburn *Student Power* 231 The German emigration, coming from a philosophical culture that was quite distinct from the parish-pump positivism of interbellum Vienna, avoided this island.

inter-borough: see *inter- *pref.* 5.

intercalarium (i:ntəɪkălē̆ᵊ·riə̆m). *Zool.* Pl. **-ia.** Also **interca·lare, interca·lary.** [mod.L., neut. sing. of L. *intercalārius* Intercalary *a.*] **1.** An element found between adjacent neural arches in the vertebral column of elasmobranchs and certain other fishes.

1887 *Proc. Zool. Soc.* 31 Intercalaria are absent in the hæmal tube [of the shark, *Carcharodon rondeletii*]. **1922** J. F. Daniel *Elasmobranch Fishes* iii. 49 Each dorsal intercalary [is perforated] by the dorsal root of the same [*sc.* spinal] nerve. **1925** J. S. Kingsley *Vertebr. Skeleton* 28 (*caption*) Intercalaria. *Ibid.*, The ring with arches is called the centrum, the one without is an intercentrum or intercalare. **1967** *Jrnl. Linn. Soc.* (*Zool.*) XLVII. 186 In *Polyodon*, dorsal intercalaries may be absent in the abdominal region, but they are usually present in the caudal.

2. In cypriniform fishes, an ossicle forming part of the Weberian apparatus linking the inner ear with the swimbladder.

1893 *Phil. Trans. R. Soc.* B. CLXXXIV. 83 Like the scaphium [in Macrones] the intercalarium consists of ascending and horizontal processes united at nearly a right angle. **1962** K. F. Lagler et al. *Ichthyol.* viii. 248 These ossicles are derived from the apophyses of anterior vertebrae; the hindmost of them, the tripus, touches the anterior wall of the gas bladder and is connected with a ligament to the next bone, the intercalare. **1970** *Amer. Mus. Novitates* No. 2428. 16 The intercalarium is attached to the tripus by a very dense, elongate ligament.

intercalate, *v.* Add: **3.** *intr.* To become part of a sequence or array as an extraneous interpolation; to become intercalated *in* or inserted *into*.

1960 Ubbelohde & Lewis *Graphite* vi. 141 The easy exchange that is observed between halogen atoms when iodine monochloride or monobromide intercalate in graphite lends some support to this suggestion. **1970** *Nature* 24 Oct. 322/2 The dye, ethidium bromide, which intercalates into DNA, binds to tRNA chiefly at one site. **1973** *Sci. Amer.* Apr. 22/1 Large numbers of ethidium bromide molecules can intercalate in a nicked duplex loop or a linear duplex.

intercalate (intəɪkæ·lĕit), *sb.* [f. the vb.] **a.** An atom or molecule, or a substance, that enters between the layers of the crystal lattice of another substance, esp. graphite. Freq. *attrib.*

1964 *Proc. R. Soc.* A. CCLXXIX. 291 Fuller understanding of interaction forces between the intercalates and the macro-aromatic molecules is..needed to interpret some highly interesting properties of crystal compounds in the direction of the *c* axis. **1966** *Ibid.* CCXCI. 332 Short range order within any single layer of intercalate clearly resembles that in a liquid or glass. **1968** *Ibid.* CCCIV. 26 Entry of intercalate molecules from the edge of basal planes of crystallites must be kept sufficiently slow to mitigate mechanical strains..before the intercalate has spread uniformly through any layer. **1971** *Nature* 2 July 43/2 Hindrances to charge wandering across the layers in the *c*-axis direction of such crystal compounds must depend extensively on the electron affinity and the local repulsion potentials of the intercalate molecules.

b. A compound formed by foreign atoms or molecules entering between the layers of a crystal lattice, esp. of graphite.

1968 *Proc. R. Soc.* A. CCCIV. 26 First sequence intercalates may eventually be formed on saturation. **1969** *Ibid.* CCCIX. 300 For preparing intercalates with quite low anion uptakes..the mounted specimen was actually immersed in concentrated nitric acid.

inter-cameral, -cardinal, -Caribbean, -caste, -cavernous, -celebration: see *inter- *pref.* 4 c, 6, 4 c, 5, 4 a, 2 a.

intercellular, *a.* Add: Hence **interce·llularly** *adv.*, between cells.

1935 E. A. Bessey *Text-bk. Mycol.* (1939) x. 251 The mycelium is long, slender and branching, growing intercellularly within the host. **1950** J. C. Walker *Plant Path.* viii. 346 The fungus develops in the epidermal wall..and it also proceeds intercellularly.

intercept, *sb.* Add: **1.** *spec.*, of a ball passed or thrown to an opponent.

1954 in Webster Add. **1960** T. McLean *Kings of Rugby* xi. 143 Meredith made an intercept.

3. *Navigation.* The angular difference between the calculated and the observed zenith distances of a heavenly body.

1901 J. R. Walker *Explanation Method Obtaining Position at Sea* 15 Let the difference between the calculated and observed zenith distance be *k* miles nearer (say) at the first observation and *l* miles further at the second. Let the extremities of these intercepts be A and B. **1939** *Geogr. Jrnl.* XCIV. 254 Nowadays the best way of obtaining latitude and longitude with a theodolite is the quadrantal and intercept method. **1969** G. Richards *Sextant Observations* iii. 92 The difference in value between the true altitude and the tabulated altitude is the intercept, and it can be 'Towards' (the body) or 'Away' (from the body).

4. A conversation, message, code, etc., that is picked up or discovered by the use of a concealed microphone, by listening to a radio communication, etc.; a device for achieving this.

1942 *Sun* (Baltimore) 10 Dec. 1/3 Distributing these copied excerpts to Government departments and officials who might have an interest in such 'intercepts', as they are called. **1945** *Ibid.* 30 Nov. 3 In the War Department.. the intercepts were shown only to Secretary Henry L. Stimson, [etc.]. **1958** *Manch. Guardian* 13 Jan. 4 The 'intercepts' (transcripts of recordings obtained by wire-tapping)..were used as evidence in the hearing. **1967** 'W. Haggard' *Conspirators* xii. 123 We can't monitor every wavelength round the clock..but we've picked up one or two intercepts between Kaunas and this diplomat. **1972** R. Ludlum *Osterman Weekend* i. 32 Intercepts have been placed on all telephones. **1973** *N.Y. Times* 22 June 35 Given torrents of intercepts..cryptanalysts could not reach a solution for thousands of years.

interception. Add: **1. d.** The action of closing in on and trying to destroy an enemy aircraft or missile. Also *attrib.*

1941 in R. W. Zandvoort et al. *Wartime English* (1957) 102 The toll of enemy raiders increases as the R.A.F. improves its methods of interception. **1955** *Bull. Atomic Sci.* Mar. 79/2 There is a great deal more that can be done to set up an effective warning and interception system. **1955** *Sci. News Let.* 15 Oct. 243/1 Once in the air, the missile's miniature radar and its own electronic computer—a baby version of the one that launched it—would keep it on an interception course. **1958** C. C. Adams et al. *Space Flight* 64 Surface-to-surface and intercontinental missiles are, generally, a means for delivering atomic and hydrogen bombs as a sort of 'interception-proof' air weapon.

interceptor. Add: **b.** *Aeronaut.* A fast aircraft which is designed specifically for the interception of hostile aircraft. Also *attrib.*, as *interceptor fighter, 'plane.*

1930 *Flight* 27 June 691/2 For a normal fighter, tanks to carry fuel for a flight of..2½ hours are sufficient. The new class of 'interceptor fighters' require even less endurance than that. **1934** *Times* 26 June *Air Suppl.* p. xvi/1 The modern interceptor was evoked by the fast day bomber. **1935** *Economist* 12 Oct. 708/1 There is no direct defence against air attack—'the bomber will always get through'—there never can be an adequate force of interceptor 'planes. **1948** 'N. Shute' *No Highway* xii. 294 One of the jet interceptor fighters coming into squadron use. **1966** *Electronics* 17 Oct. 104 The considerations will be presented in terms of a typical modern interceptor system. **1973** *Times* 14 May 12/6 The Warsaw Pact is also superior in light bombers, interceptors, and ground attack aircraft.

interchange, *sb.* Add: **1. b.** *Cytology.* Reciprocal exchange of chromosome segments, esp. between non-homologous chromosomes.

1927 *Jrnl. Genetics* XVIII. 198 In other words, chromosomes I and IX seem to have undergone interchange of a terminal segment in the ancestry of the isomorphic *B* strain. **1963** Lewis & John *Chromosome Marker* i. iii. 66 Individuals which are heterozygous for an interchange are known as interchange heterozygotes. **1968** R. Rieger *Gloss. Genetics & Cytogenetics* 440 An interchange of segments between homologous chromosomes is called 'fraternal', while one between nonhomologous chromosomes is called 'external'.

5. A junction of two or more highways on separate levels, allowing vehicles to change from one to another without lines of traffic crossing.

1944 *Sun* (Baltimore) 6 Dec. 7-0 An accident on the Pennsylvania Turnpike near the New Stanton interchange. **1954** Hewes & Oglesby *Highway Engin.* viii. 207 An interchange not only offers grade separation between the two traffic arteries, but in addition provides easy routes for vehicles transferring from one through facility to the other. **1958** H. M. Sherrard *Austral. Road Pract.* xix. 365 Much ingenuity has been displayed in the design of such intersections or 'traffic interchanges'. **1962** *Amer. Speech* XXXVII. 267 A multilevel freeway interchange. **1968** *Listener* 29 Aug. 267/3 It was an important junction and interchange. **1970** *Daily Tel.* 14 Nov. 1/7 The Berrygrove interchange of the A41 and the M1. **1973** D. Westheimer *Going Public* ix. 127 A four-mile long.. traffic jam..backing up to the San Diego Freeway interchange.

interchanger. Add: **2.** A heat exchanger.

1896 A. J. Wallis-Tayler *Refrigerating & Ice-Making Machinery* viii. 136 An interchanger was also sometimes provided, wherein the air that had done duty in the storage or cold chambers was utilised for further reducing the temperature of the compressed air. **1949** M. Davies *Physical Princ. Gas Liquefaction* vi. 86 The tubular type of interchanger carrying the compressed gas through a pipe mounted within a wider one which carries the expanded gas, is..more common..than the Hampson Spiral.

inter-church, -city, -class, -coastal: see *INTER- *pref.* 5, 4 c.

intercolonial, *a.* (Earlier and later examples.)
1843 J. Osborne *Guide Madeiras* 199 (*heading*) Intercolonial voyages. 1905 *Daily Chron.* 14 July 5/5 The Intercolonial Railway to Montreal.

intercolumn. (Later example.)
1934 H. P. Clunn *Face of London* (ed. 5) 28 The City of London Club, built in 1832–3 with..a Doric order of seven inter-columns.

intercom (i·ntəɹkǫm). Also **inter-com, intercomm** (with hyphen), **intercomm.** [Colloq. abbrev. of INTERCOMMUNICATION.] A system of intercommunication by radio or telephone between or within aircraft, offices, vehicles, etc. Also *attrib.*
1940 C. Olsson in Michie & Graebner *Their Finest Hour* iv. 61 The others behind me were gossiping as usual on their 'intercoms'. 1941 [see *DECK *sb.*[1] 3 e]. 1941 *War Illustr.* 20 Oct. 215/1 The rear gunner, I remember, called up on the inter-com., and said, 'I hope you chaps see the next one before I do.' *Ibid.* 30 Dec. 383/1 Unable to talk to the others over the 'intercom.' because my mouthpiece was not working, I stuck to the controls. 1942 T. Rattigan *Flare Path* I. 101 He even moaned to me over the intercomm. because he'd shot down a Messerschmitt. 1943 *Electronic Engin.* XVI. 140 A is valve set for three-fold communication: tank–commander; tank–tank; intercomm. in tanks. 1949 *Ibid.* XXI. 109 Electricians will welcome new chapters on Intercoms (loud-speaking telephones). 1951 J. Steinbeck *Log from 'Sea of Cortez'* (1958) p. xxv, An intercom phone between the basement and the upstairs office. 1964 M. McLuhan *Understanding Media* II. xxiv. 236 The close teamwork and tribal loyalty now demanded by electrical intercom again puts the Japanese in positive relation to their ancient traditions. 1967 [see *INTERCOMMUNICATION 4]. 1972 G. Durrell *Catch me a Colobus* iv. 73 The intercom system that we have all over the zoo. 1972 J. Potter *Going West* 8 The intercom announced the departure of Flight BA 531.

intercombination (i:ntəɹkǫmbinēi·ʃən). [f. INTER- + COMBINATION.] (See INTER- 2 a.) *spec.* in *Physics*, an electronic transition between atomic states of different multiplicities (i.e. having different spin quantum numbers); also *ellipt.* for **intercombination line**, a spectral line so produced.
1930 Ruark & Urey *Atoms, Molecules & Quanta* xx. 705 When intercombination lines occur, the intensity rules considered in this section must be modified. 1934 O. W. Richardson *Molecular Hydrogen* iii. 46 There may be intercombination lines between the singlet and triplet states but if so they must be very faint. 1937 J. W. T. Spinks tr. *Herzberg's Atomic Spect. & Atomic Struct.* ii. 79 Terms of the triplet system of He practically do not combine with the terms of the singlet system, and conversely. That is, a prohibition of intercombinations is observed. 1941 *Rev. Mod. Physics* XIII. 75 Intercombinations may occur with appreciable intensity only if the molecule contains some heavier atoms. 1950 *Discussions Faraday Soc.* IX. 16 To identify spectroscopically the long-lived luminescence and converse absorption bands as intercombinations, use is made of the characteristics of the spin-orbit coupling process. 1970 G. K. Woodgate *Elem. Atomic Struct.* vii. 132 When S changes one speaks of intercombination lines.

intercommunal (intəɹkǫ·miunăl), *a.* [f. INTER- 4 c + COMMUNAL *a.*] Existing or occurring between communities or races. Hence **interco·mmunalism.**
1909 in Webster. 1960 S. Foot *Emergency Exit* xii. 98 The inter-communal strife was at its worst. Turk and Greek were going for each other. 1971 *Black Scholar* June 51/1 With the establishment of society through intercommunalism, the entire social contract must be altered. 1973 *Black World* Dec. 19/2 We believe in intercommunalism—the relatedness of all people. 1974 *Black Panther* 9 Feb. 8/3 The Maryland Pen Intercommunal Survival Collective is calling for a community-based united front to halt the injustices and tortures before it is too late.

intercommunication. Add: **4.** *attrib.* Cf. *INTERCOM.
1911 M. Hird in L. Weaver *House & its Equipment* 124 With an 'intercommunication' system of telephones in the house, room after room can be..easily 'rung up'. 1967 *Lebende Sprachen* XII. 137/1 *Intercommunication system* (*intercom*), a system of wiring which enables two-way communication between teacher and student(s).

intercommunion. Add: **3.** Participation in the sacrament of Holy Communion by members of different religious denominations.
1921 [see *inter-celebration* s.v. *INTER- 2 a]. 1931 W. Temple *Thoughts on Probl. of Day* iii. 99 It is perfectly clear that the authors of the Memorandum never contemplated such action as formal Intercommunion. 1936 A. M. Ramsey *Gospel & Catholic Ch.* i. 8 To the one 'intercommunion' is meaningless without unity of outward order; to the other 'intercommunion' seems the one sensible and Christian way towards unity. 1966 *Church of Eng. Newspaper* 3 June, The laity expressing disquiet at what may generically be described as 'acts of intercommunion'. 1971 *World Council of Churches: Faith & Order, Louvain* 63 The whole area of question has generally been referred to in the past as the question of 'intercommunion', but that one word cannot cover the whole range and has become seriously ambiguous. It will be better to find terms which can exactly describe the different practices and their ecclesiological significance, among which the term 'intercommunion' may find its precise and particular place. 1973 *Times* 17 May 21/3 (*heading*) Bishops a stumbling block in intercommunion talks.

i:nterconne·ctedness. [f. *interconnected*, pa. pple. of INTERCONNECT *v.* + -NESS.] The property or state of being interconnected.
1922 A. G. Hogg *Redemption from this World* vi. 191 We labour hardest to perceive the interconnectedness of events. 1952 *Mind* LXI. 285 Logical interconnectedness ..is the mark of genuine knowledge. 1952 S. Spender *Learning Laughter* xi. 149 The inter-connectedness of Western and Eastern influences. 1959 *Africa* Apr. 142 The inter-connectedness of political and ritual status.

interconne·ctor. [f. INTERCONNECT *v.* + -OR.] Something that interconnects, *spec.* (see quot. 1940).
1930 *Engineering* 13 June 771/3 The substitution of a supply from Ardnacrusha..has necessitated the erection of a sub-station..into which three of the inter-connectors from Fleet-street to Pigeon House Fort are looped. 1940 *Chambers's Techn. Dict.* 452/1 *Interconnector*, a feeder which serves to interconnect two substations or generating stations, and along which energy may flow in either direction. 1962 *Newnes Conc. Encycl. Electr. Engin.* 598/2 An attempt to adjust..power flow only on the tie between two generating stations will result in an undesirable change of power on the remainder of the interconnectors. 1971 *Nature* 13 Aug. 470/1 (*caption*) Photograph of four diodes in a monolithic array... Contact is made to the devices by aluminium interconnectors.

inter-consonantal: see *INTER- *pref.* 4 a.

intercontinental, *a.* Add: **b.** Capable of travelling or of being sent from one continent to another; esp. in the designation *intercontinental ballistic missile* (abbrev. **I.C.B.M.*).
1956 *Spaceflight* I. 22/2 The terrible threat implicit in the alliance of inter-continental ballistic missiles and thermonuclear warheads, to which the whole world will stand utterly defenceless, is evident for all to see. 1957 *Jane's Fighting Ships 1957–58* 478 The combination of Regulus 1 guided missiles and submarines has given the United States Navy an intercontinental missile capability today instead of in the years to come. 1968 *Times* 16 Dec. 7/2 The second stage of the Saturn V rocket, with a thrust of one million lb., was comparable with the thrust of the Atlas rocket first developed as an inter-continental ballistic missile. 1969 *Guardian* 23 June 10/2 The Russians did not have a true intercontinental bomber until 1954.

interconversion (intəɹkǫnvə·ɹʃən). [INTER- 2 a + CONVERSION.] The process of converting each of two or more things into the other(s).
1865 *Fortn. Rev.* I. 441 It shall be shown..in what these molecular movements themselves consist..and how their mutual interconversion is effected. 1911 *Ann. Rep. Progr. Chem.* VIII. 65 The interconversion of the hydroxy-acids was realised in accordance with the following scheme. 1946 *Nature* 2 Nov. 610/2 These reaction systems..which must include the interconversions of dietary, bacterial and protozoan protein, through the agency of ciliates. 1970 P. J. Wheatley *Chem. Consequences Nucl. Spin* iv. 80 The half-life of interconversion [of ortho- and para-hydrogen] is not known accurately.

interconvert (intəɹkǫnvə·ɹt), *v.* [f. INTER- 1 b + CONVERT *v.*, or as back-formation from prec.] *trans.* To convert into one another.
1953 S. F. Mason *Hist. Sci.* xviii. 167 The ordinary and extraordinary rays were interconverted when the crystals were placed at right angles. 1955 *Sci. News Let.* 7 May 297/1 Other enzymes in the muscle interconvert these two forms [of phosphorylase] and keep them in equilibrium. 1971 *Sci. Amer.* Aug. 47/1 Lester Friedman and John G. Miller..'interconverted' *R*-carvone and *S*-carvone into their enantiomers and then back again.

intercooler (i·ntəɹkūləɹ). Also **inter-cooler.** [INTER- 2 b.] An apparatus for cooling gas heated by compression, esp. before it is compressed a second time (as in a multi-stage compressor or the cylinders of a supercharged engine).
1900 *Engin. Mag.* XIX. 679 A complete system of jacket water and a very large intercooler are used, the total capacity of the machine being 3,000 cubic feet of free air per minute compressed to 125 pounds' pressure. 1903 [see *AFTERCOOLER]. 1932 Chatfield & Taylor *Airplane & its Engine* (ed. 2) viii. 234 Compressing air by a supercharger..increases its temperature considerably and it is sometimes desirable to cool this air before it reaches the engine. This is accomplished by means of an intercooler which usually takes the form of a small radiator inserted in the air passage between the supercharger and the engine. 1944 [see *AFTERCOOLER]. 1963 A. W. Farrall *Engin. for Dairy & Food Products* viii. 202 The booster compressor..discharges its gas into an intercooler, which in turn is piped to the suction of a standard high-pressure refrigeration machine. 1970 *Motor Boat & Yachting* 16 Oct. 39/2 By installing a type of heat exchanger, called an intercooler, the mixture leaving the turbo-charger is cooled before it enters the cylinder and this increases the efficiency.

intercooling (i·ntəɹkūliŋ), *vbl. sb.* Also **inter-cooling** (with hyphen). [f. the vb. *intercool* contained in prec. + -ING[1].] The cooling of gas between successive compressions; the use of an intercooler.
1902 G. D. Hiscox *Compressed Air* xii. 187 The value of proper intercooling. 1923 J. M. Ford *Compressor Theory & Pract.* viii. 153 The intercooling of the air or gas..is imperfect in that the temperature is not reduced to the initial value between the several stages. 1951 Cohen & Rogers *Gas Turbine Theory* i. 7 If..the compression process is carried out in two or more stages with intercooling, the work of compression may be reduced appreciably. 1970 *Commercial Motor* 25 Sept. 56/3 Output of about 350 bhp, naturally aspirated, would be possible with 600 bhp when turbocharged with inter-cooling.
Hence (as a back-formation) **i·ntercool** *v. trans.*, to equip or provide with an intercooler; **i·ntercooled** *ppl. a.*
1944 E. W. F. Feller *Air Compressors* x. 348 (*caption*) Power saved by intercooling a multistage centrifugal compressor. 1947 *Shell Aviation News* No. 109. 22/3 A wide range of performance at both moderate and high altitudes is provided by the..inter-cooled and after-cooled supercharger. 1970 *Motor Boat & Yachting* 16 Oct. 39/2 A number of diesel engines these days are offered in turbo-charged form, which pushes up their horsepower, and then intercooled as well as turbocharged for a further increase in power. 1971 *Engineering* Apr. 103 (Advt.), The five stage intercooled turbo compressors absorb 1555 bhp delivering 22,700 Kg/hr.

intercorrelate (intəɹkǫ·rĭlēit), *v.* [f. INTER- 1 b + CORRELATE *v.*] *trans.* and *intr.* To correlate with one another.
1909 *Amer. Jrnl. Psychol.* XX. 368 We find..that efficiency in marking A's on a sheet of printed capitals, efficiency in finding circles or hexagons or isosceles triangles on a sheet of printed geometrical forms and efficiency in finding misspelled words are in adults all very closely intercorrelated (to ·8 or more), but are by no means so closely correlated to general intellect. 1955 T. H. Pear *English Social Differences* 11 These [types] are so intercorrelated that they can be separated only abstractly. 1970 *Jrnl. Gen. Psychol.* LXXXII. 171 If the DAS correlates as highly with each of these scales as they intercorrelate with each other, then it could be argued that the DAS measures anxiety in general rather than death anxiety in particular. 1971 *Nature* 25 June 538/1 Faculty responses to each of these items were highly intercorrelated. 1972 *Jrnl. Social Psychol.* LXXXVII. 69 Tests were intercorrelated, and factor analyses carried out.

intercorrelation (i:ntəɹkǫrĭlēi·ʃən). *Statistics.* [f. INTER- 2 a + CORRELATION.] Correlation (sense *1 c) that relates each of a number of variates with one another.
1901 *Psychol. Rev.* VIII. 540 The laboratory mental tests show little inter-correlation in the case of college students. 1904 *Amer. Jrnl. Psychol.* XV. 92 The three observations of the same objective series presented the extraordinarily small inter-correlation of 0·22. 1922 *Jrnl. Exper. Psychol.* V. 68 Intercorrelations of a number of variables may be efficiently solved on the adding machine by means of transmutation of gross scores into class numbers with the aid of standard grouping tables. 1935 [see *FACTORIALLY *adv.*]. 1961 *Lancet* 12 Aug. 359/2 Intercorrelations were calculated between the four types of test set. 1970 *Jrnl. Gen. Psychol.* LXXXIII. 125 The usual statistical information regarding the tests was obtained, including means, standard deviations,..and intercorrelations.
Hence **i:ntercorrela·tional** *a.*
1970 *Jrnl. Gen. Psychol.* LXXXIII. 157 These additional data..plus the original measures were subjected to an intercorrelational analysis.

intercortical: see *INTER- *pref.* 6.

intercourse, *sb.* **2. d.** (Later examples.)
1919 M. K. Bradby *Psycho-Anal.* III. ix. 118 Witches were examined during their trials for evidence of their fleshly intercourse with the devil. 1922 Joyce *Ulysses* 402 He..did not scruple..to attempt illicit intercourse with a female domestic. 1963 A. Heron *Towards Quaker View of Sex* i. 6 The incidence of extra-marital intercourse is great. 1973 S. Fisher *Female Orgasm* i. 26 Some of the decline in intercourse frequency and responsiveness..may be caused by the..physiological decline of their husbands.

intercrop (intəɹkrǫ·p), *v.* [INTER- 1.] To raise a crop among plants of a different kind, usually using the space between rows. Hence **intercro·pping** *vbl. sb.*; **i·ntercrop** *sb.*, a crop so raised.
1898 W. Robinson *Eng. Flower Garden* (ed. 6) v. 94 Some kind of inter-cropping would give an excellent result in the flower garden also. 1935 H. F. Macmillan *Tropical Planting & Gardening* (ed. 4) iv. 30 The intercrop or catch-crop may retard the growth of the principal crop. 1945 in R. W. Zandvoort et al. *Wartime English* (1957) 133 In a West-country orchard..rows of intercropped potatoes flourish beneath the fruit trees. 1951 J. S. Douglas *Hydroponics* vi. 89 Any well-planned hydroponicum will allow plenty of scope for careful intercropping. 1966 Webster & Wilson *Agric. in Tropics* x. 230 An intercrop is clearly undesirable on account of competition for soil moisture. 1969 G. Wrigley *Tropical Agric.* (ed. 2) ii. 119 The production of an acre intercropped was equivalent to an acre and a half cultivated in a pure stand.

intercrystalline (intəɹkri·stăləin), *a.* [INTER- 4 a.] Situated or occurring between crystals, esp. those which form a metal.

1901 *Phil. Trans. R. Soc.* A. CXCV. 295 Where the quantity of impurity present is sufficiently great, this eutectic can be seen under the microscope forming an inter-crystalline cement. 1923 Glazebrook *Dict. Appl. Physics* V. 361/1 Intercrystalline cracking near the melting-point sometimes takes place under very low stresses. 1963 *Science Survey* III. 319 The final failure of the metal is essentially mechanical and it usually occurs by the propagation, through the body of the metal, of intercrystalline (or intergranular) cracks.

inter-cultural: see *INTER- *pref.* 4 c.

intercut, *v.* Restrict †*Obs. rare* to sense in Dict. and add: **2.** *Cinematogr. trans.* To insert (a scene or shot) into an existing one by cutting. Const. *with.* Also *intr.* and *transf.* So **i·ntercutting** *vbl. sb.*
1953 K. Reisz *Technique Film Editing* i. 33 Inter-cut with a number of slow-moving shots of Kerensky proudly ascending the stairway, are separate titles describing Kerensky's rank. *Ibid.* ii. 133 The explosion itself is conveyed by the rapid intercutting of frames of the submarine and of a cone of water thrown up by a depth-charge. 1954 *Encounter* Aug. 52/1 Frank Norris.. employed the method of ironic contrast, intercutting the death of a destitute widow from starvation with description of the sumptuous dinner given by a rail-road king obliquely responsible for her condition. 1957 Manvell & Huntley *Technique Film Music* iii. 38 The military advance to drum taps and trumpet, beautifully intercut in track and picture to a rising string crescendo for the masses. 1958 *Listener* 6 Nov. 730/2 There was an intrusive guitar, an impromptu ballad, a ticking clock, and much elaborate montage, inter-cutting street voices with voices on the radio. 1962 *Ibid.* 27 Sept. 474/2 *La Notte* abounds with high-angled shots intercut with the reverse low angles which create a complex of vertiginous effects. 1966 *Punch* 9 Nov. 691/1 I intended to do all the characters' complete dialogue before I attempted inter-cutting (the TV phrase). 1970 I. C. Jarvie *Towards Sociol. of Cinema* ix. 128 Resnais..intercuts scenes from the heroine's memories..with her present *affaire*.

interdefinable (intəɹdĭfai·năb'l), *a. Logic.* [f. INTER- 2 a + DEFINABLE *a.*] Of constants, etc.: that can be defined interchangeably with each other. Hence **interdefinabi·lity**, the state or quality of being interdefinable; **inter-defini·tion**, one of two or more definitions that are interchangeable.
1948 Ambrose & Lazerowitz *Fund. Symbolic Logic* iii. 41 With the exception of '~', which is absolutely primitive, all the symbols for the logical constants are interdefinable. *Ibid.* 56 The possibilities of interdefinition of the relatively primitive constants. 1951 *Mind* LX. 265 The interdefinability of the constants of the system. 1955 A. N. Prior *Formal Logic* I. 9 Evaluation of truth-functional forms, and the inter-definability of truth functions. 1958 H. B. Curry et al. *Combinatory Logic* I. v. 155 (*heading*) Interdefinability of simple combinators. 1965 *Philos. Rev.* LXXIV. 522 'Ought' and 'must' are interdefinable. 1972 H. B. Curry et al. *Combinatory Logic* II. xiii. 221 In § 2 we considered interdefinitions among the basic arithmetical combinators.

interdental, *a.* Add: **2.** (Later examples.)
1933 L. Bloomfield *Lang.* vi. 98 Contact..can be made against the edges of the upper teeth (interdental position). 1943 K. L. Pike *Phonetics* ii. 123 An interdental sound is one in which the tip of the tongue is placed between the upper and lower teeth.
B. *sb. Phonology.* A sound formed by placing the tip of the tongue between the teeth.
1953 C. E. Bazell *Ling. Form* iv. 65 The distribution of voiced and voiceless inter-dentals in English is quite different from that of other voiced/voiceless pairs. 1961 R. B. Long *Sentence & its Parts* xix. 430 This obstruction can occur..at the front teeth, as for the interdentals.

interdentally (intəɹde·ntăli), *adv.* [f. INTERDENTAL *a.* + -LY[2].] In an interdental position; between the teeth.
1920 *Practitioner* Jan. 115 The neck of the tooth..is smeared by a thin shallow flap of gum, continuous interstitially (interdentally) with the gum pad. 1939 I. Birkenfeld *Toothbook* i. 1 Irregular occlusal openings to the interproximal spaces (between the teeth) through which food was forced interdentally.

interdepartmentally (intəɹdīpāɹtme·ntăli), *adv.* [f. INTERDEPARTMENTAL *a.* + -LY[2].] Between or among departments.
1901 *Westm. Gaz.* 7 June 2/2 This is the bane of Government offices, both departmentally and interdepartmentally. 1906 *Ibid.* 30 Oct. 2/1 But there are strong grounds for believing that the question of the attitude of the Government towards the proposals now before the Berlin Conference has not been made a Cabinet question; in other words, that it has been dealt with departmentally. 1963 *Times* 18 Apr. 9/4 One result is that interdepartmentally External Affairs staff frequently display a lack of understanding of other departments and programmes. 1966 *Economist* 10 Dec. 1155/1 In actual fact the case is still being argued interdepartmentally.

inter-dialectal: see *INTER- *pref.* 4 c.

interdiction. Add: **4.** The interruption of supply operations by aerial bombing. Freq. *attrib.*
1944 *B.B.C. War Rep.* 1 Nov. (1946) xiv. 283 The enemy railways were harassed day and night by what was known as 'interdiction'—or, in other words, rail-cutting by air attack. 1955 *Bull. Atomic Sci.* Feb. 56/3 The inability of the Air Force interdiction campaign—Operation Strangle—to bring about the collapse of Communist armies in Korea has obscured the true potential of air interdiction. 1963 *Listener* 21 Feb. 331/1 Overwhelming air support, invaluable as it was in an interdiction role, sometimes proved a clumsy weapon when used in close co-operation with ground forces. 1966 *Guardian* 26 Sept. 9/1 Using a bombing technique known as 'interdiction in depth', Navy planes had destroyed two locomotives, 225 goods wagons. 1973 *Times* 19 Apr. 18/8 Sensitive Washington spokesmen do not like the term bombing. They find it jarring. In Vietnam it was interdiction, armed reconnaissance and protective reaction.

interdigital, *a.* Add: **2.** *Electronics.* Having the form of or consisting of two similar series of parallel strips, those of each series forming part of a single structure and interdigitating with those of the other series: used of a kind of transducer.
1967 *IEEE Trans. Electron. Devices* XIV. 185/2 The operation of an interdigital pattern whose fundamental frequency was 15 MHz has been observed at odd integral multiples up to the seventh on AT-cut quartz. 1968 *Ibid.* XV. 586 (*heading*) Frequency response of an interdigital transducer for excitation of surface elastic waves. 1972 *Sci. Amer.* Oct. 52/3 In recent years the technology of acoustic waves has expanded rapidly with the development of the interdigital transducer, an efficient type of transducer for converting the electrical signal into an acoustic surface wave and reconverting the acoustic wave back into an electrical signal.

interdigitate, *v.* **2.** Delete *rare*⁻⁰ and add to def.: In *Geol.* = INTERSTRATIFY *v.* 1. (Examples.)
1969 Bennison & Wright *Geol. Hist. Brit. Isles* viii. 160 In north Devon marine and non-marine beds occur in the same sequence. They are interdigitated due to alternating expansions and contractions of the area of marine deposition. 1974 *Sci. Amer.* Feb. 64/1 The last step..involves interdigitating a set of thick filaments in the spaces between the thin filaments.

interdine (intəɹdəi·n), *v.* [INTER- 1 a.] *intr.* Of members of different castes and tribes: to eat a meal together.
1932 G. S. Ghurye *Caste & Race in India* iv. 73 Evidently interdining, like intermarrying, was a mark of equality. 1933 Rahmat Ali *Now or Never* (1934) 4 We [*sc.* Muslims] do not inter-dine [with Hindus]; we do not inter-marry. 1960 Koestler *Lotus & Robot* II. x. 228 In India, inter-marriage, and even inter-dining, between different castes was unthinkable.

interdisciplinary (intəɹdi·siplinări), *a.* [INTER- 4 c.] Of or pertaining to two or more disciplines or branches of learning; contributing to or benefiting from two or more disciplines.
1937 *Jrnl. Educ. Sociol.* Dec. 251 Programs of study submitted should provide..for training of an interdisciplinary nature. 1956 J. Lotz in L. White *Frontiers of Knowledge in Study of Man* 217 There is also considerable interest in interdisciplinary studies involving structural linguistics. 1957 E. Bott *Family & Social Network* ii. 35 Ten years ago interdisciplinary research was very much in vogue. 1965 *Listener* 18 Nov. 788/2 The work..is again interdisciplinary and does not fall into any of the established academic categories. 1970 *Guardian* 12 Nov. 5/6 The academic policies put forward—essentially a two-year general interdisciplinary degree. 1972 *Language* XLVIII. 487 Child language acquisition has proved to be one of the more important interdisciplinary areas of the past decade.

inter-dominion (intəɹdomi·nyən), *a.* [f. INTER- 5 + *DOMINION 2 b (*c*).] Occurring, or carried on, between the self-governing dominions of the British Commonwealth.
Used esp. in Australia and N.Z.
1949 *Australia 1949* 776/1 Totalisator investments for the three days of the Auckland Inter-Dominion Championships totalled £506,498/10/–. 1950 *Sydney Morning Herald* 12 Jan. 10/7 Sir Nigel will now go to Melbourne for the inter-Dominion championships in February. 1966 *Weekly News* (N.Z.) 27 Apr. 39 (*caption*) Australian and New Zealand crews competed for an interdominion championship under conditions similar to those at the Cowes Regatta. 1968 *Wanganui* (N.Z.) *Chron.* 15 Nov. 6/9 The 1971 inter-Dominion championship series have been allocated to the New Zealand Metropolitan Trotting Club.

inter-ecclesiastical: see *INTER- *pref.* 4 c.

inter-electrode (i:ntərĭle·ktrōud), *a. Electr.* Also **interelectrode.** [INTER- 5.] Existing between two or more electrodes; said esp. of electrical quantities pertaining to the space between electrodes.
1922 Glazebrook *Dict. Appl. Physics* II. 902/1 Certain types of valve are constructed to reduce the inter-electrode capacity to a minimum. 1925 E. B. Wedmore *Electr. Engineers' Data Bks.* III. 162 For high frequency amplification it is desirable to use valves having a low interelectrode capacity and also a low amplification factor. 1930 *Daily Express* 9 Sept. 11/1 The new..Valve has a greater effective amplification because its inter-electrode capacity is lower. 1962 W. B. Thompson *Introd. Plasma Physics* v. 76 In order to relate the resistance to the resistivity of the plasma, the tube was filled with a solution of known conductivity and the inter-electrode resistance measured. 1964 Crookson & Sonsiad *Fund. Microwave Electronics* ii. 19 When the cathode is heated, the negative charge of the emitted electrons depresses the potential in the interelectrode region.

inter-electron: see *INTER- *pref.* 5.

interest, *sb.* Add: **11.** (sense 4) *interest-bound* adj.; (sense 7) *interest-awaking, -compelling* adjs.; (sense 10) *interest-equalization, -rate*; *interest-free* adj.; **interest group**, a group of individuals possessing a common identifying interest.
1901 *Daily Chron.* 18 Nov. 6/3 The 'great retrograde, tyrannical, interest-bound party'. 1902 *Ibid.* 17 July 6/4 Its interest-awaking value. 1902 *Ibid.* 6 Aug. 3/2 As interest-compelling..as the story of a medieval queen. 1908 A. F. Bentley *Process Govt.* xii. 300 The deeper-lying interest groups of society. 1936 Wirth & Shils tr. *Mannheim's Ideology & Utopia* III. i. 136 The hitherto constantly emphasized interest-bound nature of political thought. 1945 E. Blunden *Return to Husbandry* iii. 18 This demands ample credit of an interest-free nature. 1957 M. Swan *Brit. Guiana* iv. 78 Houses which had been built with interest-free loans or with other forms of state assistance. 1959 E. Pound *Thrones* xcvii. 42 Byzance lasted longer than Manchu because of an (%) interest-rate. 1962 *Economist* 25 Aug. 683/2 The interest-wise signs of change..in the growth of interest-group organisations. 1963 *Daily Tel.* 4 Oct. 2/8 The interest-equalization tax to raise the cost of foreign borrowing in New York. 1964 R. Wilkinson *Gentlemanly Power* ix. 48 The parliamentary Conservative party has come to represent different interest-groups, farmers..manufacturers, small professional men..stockbrokers, shopkeepers and elderly widows. 1966 *Times* 28 Feb. (Canada Suppl.) p. viii/5 The United States's interest-equalization tax of 15 per cent, to be paid by any United States resident buying foreign stocks. 1972 *Sat. Rev.* (U.S.) 6 May 38/3 The company store charged exorbitant prices, but extended interest-free credit. 1974 *Times* 18 Feb. 14/5 It may..be difficult with such a budget deficit to prevent interest rates from rising.

interesterification (i:ntəɹ,esterifĭkēi·ʃən). [f. INTER- 2 a + *ESTERIFICATION.] The exchange of alkoxy or acyl groups between an ester and another compound, sometimes used to modify the properties of margarine and other fats.
1941 T. P. Hilditch *Industr. Chem. Fats & Waxes* (ed. 2) iv. ii. 305 Suggestions have..been made..with the objects either of interchanging the acyl radicals between the triglyceride molecules of the mixture of fats used, or of introducing a certain amount of butyroglycerides into the margarine fats. Brief reference may be made to these 'interesterification' processes. 1961 *Times Rev. Industry* July 8/2 Interesterification..brings about a molecular rearrangement in the fat. This manifests itself in changes in the melting characteristics and crystal type.
Hence (as a back-formation) **i:ntereste·rify** *v. trans.* and *intr.*, to subject to or undergo interesterification; **i:ntereste·rified** *ppl. a.*
1950 Kirk & Othmer *Encycl. Chem. Technol.* V. 817 Some esters will interesterify with each other at 100° C. or lower with no catalyst added. 1958 C. Placek in *Mod. Chem. Processes* V. 84/2 In baking performance, directly interesterified lard has all the desirable properties of vegetable shortening. *Ibid.* 88/1 Until operating conditions (equilibrium) are reached, material appearing at the end of the run is usually not interesterified to the desired point and is recycled after catalyst separation.

interesting, *ppl. a.* Add: **3.** *(to be) in an interesting condition, situation, state*: (to be) pregnant; also *to be interesting*; *interesting event*: a birth.
1748 Smollett *R. Random* II. lxix. 335 So that I can not leave her in such an interesting situation, which I hope will produce something to crown my felicity. 1838 Dickens *Nickleby* (1839) xxix. 286 Mrs. Lenville (who, as has been before hinted, was in an interesting state). [see Situation 7 b]. 1899 *Westm. Gaz.* 27 June 6/3 'Interesting event'. Another daughter! 1928 W. B. Maxwell *We Forget because we Must* II. iii. 44 I'm afraid I seem to make heavy weather of this approaching confinement. 1930 Galsworthy *On Forsyte 'Change* 171 Winifred, beginning to be 'interesting', owing to the approach of a little Darlie, kept her eyes somewhat watchfully on 'Monty'. 1970 K. Giles *Death in Church* ii. 49 Her little maid got into An Interesting Condition and the young fellow was willing to solemnise it.

interestingly, *adv.* Add: Used as a sentence adverb.
1972 *Times* 13 May 10/4 Interestingly Balanchine is no longer called a director of New York City Ballet. 1973 *Nature* 18 May 118/2 Interestingly, the results presented were not in accord with the rather precise quantum calculations.

inter-ethnic: see *INTER- *pref.* 4 c.

interface. Add: **2.** *transf.* and *fig.* **a.** A means or place of interaction between two systems, organizations, etc.; a meeting-point or common ground between two parties, systems, or disciplines; also, interaction, liaison, dialogue.

1962 M. McLuhan *Gutenberg Galaxy* 141 (*heading*) The interface of the Renaissance was the meeting of medieval pluralism and modern homogeneity and mechanism. 1962 *Evening Star* (Washington, D.C.) 18 Aug. 1/6 Interface..seems to mean the liaison between two different agencies that may be working on the same project. *Ibid.* 1/7 The Defense Communications Agency.. was made responsible for the resolution of interface problems. 1964 A. Battersby *Network Analysis* viii. 116 Interfaces: events should be established at stages where the work passes from one department to another—these stages are known as interfaces. 1965 H. I. Ansoff *Corporate Strategy* (1968) vi. 107 Functional organizations, such as research, development, finance, and marketing, have a strong interface with the outside environment. 1965 *Internat. Sci. & Technol.* Oct. 30/1 The advantages of high-speed transport were piddled away at the nodes or interfaces: from bus to train, train to train, city terminal to airport terminal, check-in counter to loading gate, and so on. 1967 *Technology Week* 23 Jan. 75/1 The interface across which the engineer-scientist and the biologist can interact is a broad one. 1967 *Times Rev. Industry* Feb. 27/1 The third interface between government and the marketing system is with the intermediate firm supplying either other intermediate firms or the consumer. 1967 *Economist* 16 Sept. p. ix/1 The North Sea and Channel ports form the biggest frontiers in world trade—or the biggest interface, in the language of the modern transport man, meaning the place where the greatest quantity of international cargo changes its mode of transport. 1970 *Nature* 23 May 684/1 The interface between physics and music is of direct relevance to..the psychological effects of hearing. 1970 *Interior Design* Dec. 767/4 Educationalists are convinced that the need for the interface of lecturer and student will not diminish. 1972 *Sci. Amer.* Nov. 51/3 The issue of insanity as a defense in criminal cases..is at the interface of medicine, law and ethics.
b. (An) apparatus designed to connect two scientific instruments, devices, etc., so that they can be operated jointly.
1964 *Ann. N.Y. Acad. Sci.* CXV. 574 The collection of components which connects the analog and digital computers to each other, and which controls and converts the data, is generally termed the 'interface'. 1966 *Electronics* 3 Oct. 130 If a flight carries special equipment, then modular interfaces can easily be designed to adapt the general-purpose computer to the equipment. 1973 T. Allbeury *Choice of Enemies* xvi. 79 Programs are written in a computer language... If you wanted to use one of the IBM languages on an ICL machine, you'd have to have what's called an interface to make the two different things compatible. 1973 *Physics Bull.* Apr. 242/3 Scobie and Wellum..have built interfaces for two pulse height analysers.

interface (i·ntəɹfēis), *v.* [f. the sb.] **1. a.** *trans.* To connect (scientific equipment) *with* or *to* so as to make possible joint operation.
1969 *Computers & Humanities* IV. 76 Professor Louis Delatte..publishes..various computer-prepared indices to classical texts, using a Selectric typewriter interfaced with his own local computer. 1969 *Physics Bull.* Sept. 367/2 The prospect of interfacing each device specifically with each computer on each application becomes formidable in these circumstances in terms of effort and cost. This is avoided using the CAMAC technique of interfacing the device to the dataway, via a module, and the dataway to the computer, via the controller. 1970 *Sci. Jrnl.* Mar. 17/4 Their movements were monitored by a series of illuminated photoconductor cells, which were interfaced to the PDP-8/S computer. 1973 *Nature* 6 Apr. 402/2 A 'Perkin-Elmer 900' and a 'Hewlett Packard 7610A' chromatograph,..interfaced with a 'Perkin-Elmer PEP-1' gas chromatography data system, were used. 1973 *Physics Bull.* Apr. 240/2 The memory uses 'static' circuitry and no clocking is required which makes it easy to use and interface to any system.
b. *intr.* for *pass.*
1969 *New Yorker* 11 Jan. 42 Inflated space units, which have to 'interface'—a space-age verb meaning, roughly, to coordinate—with equipment in the cabin. 1971 *Physics Bull.* Jan. 42/3 The minimum system can be attached to 16 devices..; the largest can 'interface' with about 2000 remote sensing/control devices.
2. *intr.* To come into interaction *with.*
1967 M. McLuhan *Medium is Massage* 88 A strange bond often exists among antisocial types in their power to see environments as they really are. This need to interface, to confront environments with a certain anti-social power, is manifest in the famous story 'The Emperor's New Clothes'. 1968 *Lebende Sprachen* XIII. 4/1 Before turning to a discussion of how this management system..interfaces with functional organization let us try to define what we mean by project management. 1973 *LSA Bull.* Mar. 14 Mr. Hamp, the LSA delegate to UNESCO, reported on ways which he felt the Linguistic Society could interface with the United States National Commission.

interfacing (i·ntəɹfēi:siŋ), *vbl. sb.* [f. INTER- 2 a + FACING *vbl. sb.*; cf. prec.] The action of the verb *INTERFACE; also *concr.* (see quot. 1964). Also *attrib.* or as *ppl. a.*
1964 Margolis *Compl. Bk. Tailoring* 180 An interfacing is a reinforcing or shaping fabric used between the outer fabric and the facing or the lining. 1968 J. Ironside *Fashion Alphabet* 89 Very similar to interlining but interfacing is usually only placed along edges or in collars, cuffs, etc. 1968 *Lebende Sprachen* XIII. 4/1 There will be a large or small group, depending on the complexity of the interfacing problems. *Ibid.* 104/1 (*heading*) Terms in PERT planning interfacing network. 1969 *Physics Bull.* Sept. 367/1 The use of a common highway permits many sources and acceptors of data (devices) to be connected to the computer..via the CAMAC compatible modules. This type of interfacing would be unnecessary..if the number of devices were small and the computer had sufficient input-output channels which were already matched to each device.

inter-faith, -family: see *INTER- *pref.* 5.

interfere, *v.* Add: **2.** Also in *Broadcasting*: to transmit a signal which is received simultaneously *with* the signal sought; to cause or emit interference (sense 4* a). (Further examples.)
1904 in J. Erskine-Murray *Handbk. Wireless Telegr.* (1907) x. 179 From 11.27 till noon the receiving ship 'Hancock'..and the experimental station in building §75..interfered continually. 1928 L. S. Palmer *Wireless Princ. & Pract.* iii. 49 Stations transmitting on the same wavelength but lying in different directions from the receiving station can be prevented from interfering. 1940 L. R. Lohr *Television Broadcasting* iii. 37 If television images were capable of being received beyond the horizon, the received images..would interfere with the transmissions of other stations. 1960 *Which?* Apr. 72/2 Does it [*sc.* the vacuum cleaner] interfere with radio or TV?
4. c. *Chess.* Of a piece: to obstruct the line of action of another piece. (Cf. *INTERFERENCE 1 b.)
1913 A. C. White *Sam Loyd* 303 The White pieces can interfere in all kinds of ways with the Black pieces, and the Black pieces can interfere with each other with varied and beautiful results. 1926 H. Weenink *Chess Probl.* 38 It will be noticed that..Re6 in turn interferes with the line of force of the Bg4, shutting off its command of d7 and c8. 1930 White & Hume *Valves & Bi-Valves* 139 The moves of the checking Valve interfere, not..on an actual line of the Black Queen, but on her possible (potential) line of check next move. 1937 T. R. Dawson *Caïssa's Wild Roses in Clusters* vi. 24/2 One Black piece..interferes on two squares on the line of another piece... Each of two Black pieces..interferes with the other.
d. Const. *with*: to molest or assault sexually.
1948 D. Ballantyne *Cunninghams* (1963) iii. 22 The former Mayor..who was kicked out of his church for interfering with a youngster in a Sunday School class. 1956 L. McIntosh *Oxford Folly* viii. 119 'All the girls I used to fall for,' said Julian, 'were incredibly beautiful, and as hard as nails. But they tolerated me, because I was much too timid ever to "interfere" with them.' 1968 A. MacLeod *Dam* v. 51 She had reason to suspect that Sandra had been 'interfered' with. 1972 *Observer* 3 Sept. 33/1 In a few cases, like Byron's, a nanny sexually 'interfered' with her charge.
6. *U.S. Football.* To interpose between the player with the ball and a would-be tackler so as to help the former. *Baseball.* To obstruct a runner between two bases; also, to obstruct a catcher or fielder who is trying to take or throw the ball. (Cf. *INTERFERENCE 1 c.)
1920 W. Camp *Football without a Coach* 51 The full-back and the right half must interfere for their companion. 1969 *Official Baseball Rules* 39 The batter..is entitled to first base without liability to be put out..when..the catcher or any fielder interferes with him.

interference. Add: **1. b.** *Chess.* Obstruction of the line of action of one piece by another. Also *attrib.*
1913 A. C. White *Sam Loyd* 303 There are many forms of interference play which have nothing to do with avoiding stalemate... But interference has a far wider scope than the cutting off of one White man by another. 1926 H. Weenink *Chess Probl.* 39 In both problems there is mutual interference of the black Rook and Bishop. 1931 G. Hume in A. C. White *Probl. by my Friends* 210 By forgoing the interference of the Black Bishop with the Black Pawn, a second flight-square has been obtained. 1947 T. R. Dawson *Caïssa's Fairy Tales* 7/2 Rc4, which is Black interference permitting 2 Sb6 mate... The interference and pin ideas create nice new task record objectives. 1963 J. Bochkor tr. *Bán's Tactics of End-Games* II. 115/2 We can also make use of line interference by sparing a hostile piece that is closing a line and is thus harmful to its own camp.
c. *U.S. Football.* (*a*) The act of interposing between a runner and a tackler to obstruct the latter; (*b*) see quot. 1895; (*c*) a player or players who obstruct the tackler or tacklers. *Baseball.* The act of obstructing a runner between two bases.
1894 *Outing* (U.S.) XXIV. 112/2 The special feature of American Rugby arises from the principle of interference to aid the man running with the ball. 1895 G. J. Manson *Sporting Dict.* 61 *Interference*, using the hands or arms in any way to obstruct or hold a player who has got the ball. 1920 W. Camp *Football without a Coach* 59 To amount to anything at all interference must be perfectly timed... On a play between tackle and guard..the interference must reach that point prepared to take care of the tackle, the guard and even the backfield men, too. 1922 D. Canfield *Rough-Hewn* xxvi. 241 Where *was* the ball? Sometimes it came straight through and the next minute on the same formation swung outside—and Neale uselessly buried under the interference. 1927 H. G. Salsinger in *Secrets of Baseball* 147 Interference plays, too, are scored as they probably have been made. 1969 *Official Playing Rules Nat. & Amer. Football Leagues* 68 Interlocked interference means the grasping of one and another by, or encircling body to any degree with, hands or arms by offensive players.
d. Sexual assault or molestation.
1968 M. Culpan *Vasiliko Affair* v. 63 You'll get the pathologist to examine the body. But I'd say no interference. 1972 J. Symons *Bloody Murder* xii. 160 Before the War..the rape would have been mentioned delicately. 'Any sign of—interference?' 1973 'D. Shannon' *No Holiday for Crime* (1974) ii. 27 Not raped, for ninety-nine per cent sure—no interference.
4*. Various scientific and technical senses.
a. *Broadcasting* and *Telecommunications.* Disturbance of the transmission or reception of signals by the intrusion of extraneous signals; hence, signals collectively or radiation which causes such disturbance, or the effects by which it is perceived (e.g. unwanted sounds in radio reception).
1887 *Electrician* 7 Oct. 462/1 Strong signals were received on the copper [telephone] wires, although they were completely isolated from any possible interference. 1888 *Operator & Electr. World* XII. 140 (*heading*) Dynamo current interference with telephone systems and means of relief. 1899 *Electrician* 17 Nov. 106/2 Before beginning the experiments, Mr. Marconi wrote to the Commission stating that he had an instrument which would render interference practically impossible. 1902 *Windsor Mag.* May 720/2 Two messages were sent, one in English and one in French. Both were received at the same time on the same wire at Poole..without the least interference. 1926 *Encycl. Brit.* I. 459/1 The atmosphere is nearly always filled with vagrant radio waves which enter the receiving set, producing noises called 'interference'. 1932 R. W. Hallows *Finding Foreign Stations* xiv. 113 Other kinds of apparatus which are apt to radiate interference are refrigerators..and flashing signs. 1943 A. L. Albert *Fund. Telephony* xiii. 314 When two or more telephone circuits parallel each other, electric energy may be transferred from one to the other, causing..inductive interference. 1962 J. H. & P. J. Reyner *Radio Communication* viii. 311 The interference is conducted by the mains.. to the point where the receiver is located. 1964 R. F. Ficchi *Electr. Interference* ii. 9 Interference is an electrical disturbance created by equipment in one part of a system which is carried into equipment in another part of the system, causing malfunctioning of the latter part. 1966 *B.B.C. Handbk.* 133 On the television screen the interference is seen as patterns of lines, white flashes or bands of light. 1967 E. L. Gruenberg *Handbk. Telemetry & Remote Control* xi. 11 Some of the remedies for interference are the use of shielded line between signal source and input section of the data-acquisition equipment, [etc.].
b. *Engin.* (i) The collision of the tips of the teeth of one gear-wheel with the flanks of those of the mating wheel which occurs if the teeth are not cut to a suitable profile.
1914 A. E. Ingham *Gearing* i. 32 If, however, pinions having a low number of teeth are constructed, they are much more undercut below the base line than is consistent with strength or with tooth contact. This 'undercut' or 'interference' is clearly shown in Fig. 11. 1926 Bradford & Eaton *Machine Design* viii. 149 Contact will have taken place between the tip of the driven tooth and the radial flank of the driving-gear tooth. Since the latter is not the conjugate involute of the former the two curves will not run together and interference takes place. 1948 Parkinson & Dawney *Gears* iv. 39 (1) Involute interference..is avoided by making the whole working profile of involute form, (2) tip interference..is avoided by making the diameter of the spur pinion a sufficient amount smaller than that of the internal gear. 1966 G. W. Michalec *Precision Gearing* xi. 591 It is important to avoid even the most isolated interference points because they cause wear that results in rapid degradation of precision quality.
(ii) The amount by which the external dimension of a part exceeds the internal dimension of the part into which it has to fit.
[1919: see *interference fit* below.] 1930 L. S. Marks *Mech. Engineers' Handbk.* (ed. 3) 896 In Table 47 is given a summary of the allowances, allowances plus tolerance, and average interferences for the various classes of fits, as recommended tentatively by the A.S.A. Interference here denotes negative allowance. 1950 T. Nuttall *Nat. Cert. Workshop Technol.* xv. 124 The force required to pass in the shaft will be much greater with the maximum interference; for this reason the modern tendency is to specify very close limits for both hole and shaft when interference fits are required. 1969 M. Haslehurst *Manuf. Technol.* xiii. 320 This gives a maximum interference of 0·056 mm and a minimum interference of 0·025 mm.
c. *Genetics.* The action of one cross-over in reducing or increasing (orig. only reducing) the chance of a second cross-over occurring along the same chromosome, the effect being generally proportional to distance.
1916 H. J. Muller in *Amer. Naturalist* L. 288 In a sense, then, the occurrence of one crossing-over interferes with the coincident occurrence of another crossing-over in the same pair of chromosomes, and I have accordingly termed this phenomenon 'interference'. 1969 G. W. Burns *Sci. Genetics* vii. 115 Interference appears to be unequal in different parts of a chromosome... In general, interference appears to be greatest near the centromere and at the ends of a chromosome.
d. *Aeronaut.* (See quot. 1940.)
1932 *Gloss. Aeronaut. Terms* (*B.S.I.*) (Proofs) III. 3 *Interference*, the aerodynamic influence of two or more bodies on one another. 1940 *Chambers's Techn. Dict.* 452/2 *Interference*, the aerodynamic influence of one body upon another. Usually, the head resistance, or drag, of two bodies placed close together will be greater than the total of their separate drags, because of interference.
e. *Biol.* and *Med.* The action of a virus of one kind in inhibiting a virus of another kind in the same host.
1937 *Jrnl. Path. & Bacteriol.* XLIV. 420 In plant viruses the phenomenon of interference is only seen in connection with those viruses that are generically related. 1970 Passmore & Robson *Compan. Med. Stud.* II. xviii. 108/2 If the viruses are inoculated at different times, the second may not replicate. This is known as interference.
f. *Philol.* (See quot. 1953.) Also *attrib.*
1940 *Language* XVI. iii. 219 Thus, in describing the

difficulty of pronouncing foreign sounds (5), there is no mention of the interference of the speaker's native phonemic habits. 1953 U. Weinreich *Lang. in Contact* i. 1 Those instances of deviation from the norms of either language which occur in the speech of bilinguals as a result of their familiarity with more than one language..will be referred to as interference phenomena. *Ibid.* 11 In speech, interference is like sand carried by a stream; in language, it is the sedimented sand deposited on the bottom of a lake. 1962 W. F. Mackey in J. A. Fishman *Readings Sociol. of Lang.* (1968) 569 The foregoing characteristics of degree, function, and alternation determine the interference of one language with another in the speech of bilinguals. Interference is the use of features belonging to one language while speaking or writing another. *Ibid.* 570 In the speech of bilinguals the pattern and amount of interference is not the same at all times and under all circumstances... Interference also varies with the style of discourse used—descriptive, narrative, conversational, etc. 1964 E. Palmer tr. *Martinet's Elem. Gen. Ling.* v. 160 Interference..may have the result of increasing the range of phoneme variation. 1965 *Amer. Speech* XL. 63 Galinsky deals less with regular interference phenomena than with occasional borrowings of language-conscious individuals. 1972 J. L. Dillard *Black English* i. 36 *Interference* is the term for the influence of one's native language on a language acquired later.
5. *interference pattern*; (sense 4* a) **interference-free** *a.*, not causing or not affected by interference; † **interference preventer**, an apparatus for reducing interference at a radio receiver; **interference suppressor**, an electrical device designed to prevent or reduce the production of interference by the apparatus to which it is fitted; (sense 4* b (ii)) **interference fit** *Engin.*, a fit between two mating parts for which, within the specified tolerances, there is always an interference between them.
1919 *Engineer* 23 May 511/2 The following three classes of fit would be needed, *i.e.*, running fits, transition fits, and interference fits. 1973 A. Parrish *Mech. Engineer's Ref. Bk.* vii. 31 The magnitude of the interference fit will depend upon the conditions required, i.e. axial, torsional or radial holding ability. 1950 *Engineering* 3 Feb. 140/2 A range of waterproofed and interference-free [electrical] components is being produced. 1965 *B.B.C. Handbk.* 49 Reception in the overcrowded medium- and long-wave bands continues to be difficult in many areas, in marked contrast to the interference-free reception which VHF can provide. 1933 *Discovery* May 151/2 As a typical example, I need only mention the interference patterns which appear when light from one source can travel to a screen along two different paths. 1973 *Nature* 12 Oct. 297/1 The only consistent difference between cross-over and non-cross-over events, apart from the presence or absence of an exchange of homologous segments of chromatids, is in their interference pattern. 1905 Interference preventer [see *ATMOSPHERICS *sb. pl.*]. 1914 R. Stanley *Text-bk. Wireless Telegr.* xix. 292 Marconi patented an interference preventer which involved the use of two aerials and a rotating machine. 1951 *Gloss. Terms Plastics Industry* (*B.S.I.*) 41 Interference suppressor. 1966 *B.B.C. Handbk.* 133 It has for some years been compulsory for all new vehicles and stationary engines using spark ignition to have interference suppressors fitted.

interferer. Add: **b.** *U.S. Football.* One who interposes between a runner and a tackler.
1897 *Encycl. Sport* I. 424/2 Interference..once established,..the query was immediately raised of how much aid the interferers could give the runner. 1922 D. Canfield *Rough-Hewn* xxvi. 243 Neale could see Rogers rock a second, undecided, on tip-toe; side-step an interferer; and then shoot his body like a projectile into the play.

interfering, *ppl. a.* Add: **b.** That causes or constitutes interference (sense 4* a).
1914 R. Stanley *Text-bk. Wireless Telegr.* xix. 292 If the International Rules are duly observed an interfering station should be one which considers itself out of range. 1954 E. Molloy *Radio & Television Engineers' Ref. Bk.* xxxiii. 10 As with all forms of interference, the effect will largely depend upon the ratio of the levels of the interfering signal to the picture signal.

interferogram (intəɹfiə·rogræm). [f. INTERFER(E *v.* + -O + -GRAM.] A pattern formed by the interference of radiation, esp. one represented in a photograph or diagram.
1921 *Trans. Optical Soc.* XXII. 185 Fig. 12 represents an interferogram as seen on one of the lens interferometers described, the thick lines representing dark bands. 1949 *Jrnl. R. Aeronaut. Soc.* LIII. 637/2 The interferograms give the density variation across the mixing zone. 1959 *Physics of Fluids* II. 165 (*caption to photograph*) Monochromatic interferograms taken simultaneously at λ = 4122 Å and 5463 Å of an ionizing shock wave in argon... The fringes move upward with increasing refractivity. 1965 E. B. Brown *Mod. Optics* viii. 440 The intensity plots..are referred to as interferograms, and the interferometers used in this fashion are often called interferometer modulators. 1966 D. G. Brandon *Mod. Techniques Metallogr.* 27 For first-order interference at λ = 5,500 Å with a high refractive index oil between the plates..the cone angle α that can be tolerated is..about 25°, corresponding to a 0·65 *NA* objective, and critical illumination can be used to form the interferogram.

interferometer. Substitute for def.: Any instrument in which the interference of waves (e.g. of light) from a common source is employed to make precise measurements of (linear or angular) length or displacement in terms of the wavelength. (Add earlier and later examples.)
1897 *Physical Rev.* IV. 480 An application of the Michelson 'interferometer' to the measurement of the small linear movements of the electrometer disk. 1897 *Phil. Mag.* XLIV. 91 To these two methods of measuring angular motions we must now add a third, the interferometer method, first suggested by Michelson. 1921 *Discovery* July 181/2 Betelgeux, the size of which was measured in December last at Mount Wilson by interferometer methods, has a diameter of 273,000,000 miles. 1932 Hardy & Perrin *Princ. Optics* xxviii. 592 These interferometers are often used in the analysis of a mixture of gases to determine the proportion in which the components are present. 1957 *New Scientist* 15 Aug. 14/1 At the Cambridge radio-observatory four aerials with a total area of one acre are placed at the corners of a rectangle to form a double interferometer which was used in 1954 to plot the positions of nearly 2,000 radio stars. 1969 C. A. Ronan *Invisible Astron.* vi. 83 For mapping radio sources..the radio interferometer has become widely adopted. This is based on the same principles as that of the stellar interferometer.., and at its simplest uses two antennae separated by as large a distance as is convenient, the signals being fed into an electronic mixing unit and then passed to the radio telescope receiver.

interferometric (i:ntəɹferome·trik), *a.* [f. prec. + -IC, after *barometric*, etc.] Of or pertaining to interferometry; employing or of the nature of an interferometer.
1932 Hardy & Perrin *Princ. Optics* xvii. 378 Lens-bench methods [of measuring aberrations of lenses], the Hartmann method, and the interferometric method yield this information in different ways. 1951 *Engineering* 18 May 589/1 Dr. J. S. Courtney-Pratt has used interferometric methods to study the uniformity and thickness of thin films adsorbed on solid surfaces. 1954 *Oxf. Univ. Gas.* 15 June 1036/1 Spectra for the interferometric determination of wavelengths. 1957 *New Scientist* 24 Oct. 11 An interferometric radio telescope consists of two separate aerials each joined to the same receiver whose output measures, in effect, the difference in the distances from the source to each of the aerials. 1974 *Times* 12 Jan. 14/7 He carried interferometric methods to an extreme degree of precision.
Hence **i:nterferome·trically** *adv.*, by means of interferometry.
1959 *Physics of Fluids* II. 166/2 Photomultiplier observations through 4122-Å and 4515-Å interference filters show the appearance of peak luminosity to be closely associated with the attainment of interferometrically observable ionization. 1972 *McGraw-Hill Yearbk. Sci. & Technol.* 388/2 Along with the recent development of stable lasers..have come interferometrically ruled gratings practically free of 'ghosts'. 1973 *Physics Bull.* May 307/3 The longitudinal and transverse curvatures of a plate subjected to a bending moment are determined interferometrically.

interferometry (i:ntəɹfĕrǫ·mĕtri). [INTERFEROMETER: see -METRY.] The action or art of measuring interference phenomena; the study and use of the interferometer.
1911 C. Barus in *Publ. Carnegie Inst. Washington* CXLIX. I. p. iii, I came across a principle of interferometry which seemed of sufficient importance to justify special investigation. 1946 *Nature* 12 Oct. 519/2 It is possible to evaluate an approximate crystal lattice spacing with visible light waves by virtue of multiple beam interferometry. 1950 *Endeavour* IX. 196 Multiple-beam interferometry offers a delicate method of detecting extremely small imperfections in surfaces. 1962 *Punch* 29 Aug. 298/2 The distinguished custodian of Jodrell Bank is well up in the cepheid variables, interferometry and the Laplace hypothesis. 1971 *Physics Bull.* July 397/2 Several of these laboratories..are now using laser interferometry for calibration purposes.

interferon (intəɹfiə·rǫn). *Biol.* [f. INTERFER(E *v.* + *-on.*] A protein released by an animal cell, usu. in response to the entry of a virus, which has the property of inhibiting further development of viruses of any kind in the animal (or in others of the same species).
1957 Isaacs & Lindenmann in *Proc. R. Soc.* B. CXLVII. 263 To distinguish it from the heated influenza virus we have called the newly released interfering agent 'interferon'. 1961 *New Scientist* 13 July 81/1 It looks as if ..interferon is capable of blocking the multiplication of virtually all the animal viruses that have been tested. 1961 *Lancet* 23 Sept. 680/2 Recently there has been growing evidence that interferon may play an important part in recovery from virus infections. 1963 *Ann. Reg. 1962* 393 The investigation of Interferon, a chemical substance produced in men and animals and believed to act as the body's first line of defence against a wide range of virus infections, passed a critical stage in May 1962. 1968 B. D. Davis et al. *Princ. Microbiol. & Immunol.* IV. xxvi. 789 Purified interferons from various sources consist of small proteins usually stable at low pH... Interferons are not virus-specific but cell-specific in both their production and their effects. 1970 Passmore & Robson *Compan. Med. Stud.* II. xviii. 109/2 Interferon has many of the properties of an ideal antiviral agent but its present clinical applications are limited by difficulties associated with the production of large quantities of purified human interferon and the assessment of its activity in man.

inter-fertile, -fertility: see *INTER- *pref.* 2 a.

inter-fibre, interfile *v.*: see *INTER- *pref.* 5, 1 a.

interfi·nger, *v. Geol.* [INTER- 1 b.] *intr.* Of strata, sediments, etc.: to interdigitate (*with*). So **interfi·ngering** *ppl. a.*
1921 A. W. Grabau *Textbk. Geol.* II. xxxvi. 441 (*caption*) The source of the black mud is the land on the south from which it was repeatedly washed into the sea, interfingering with clastics derived from the east. 1960 *New Scientist* 23 June 1609/1 Detailed geological mapping of interfingering and overlapping lava flows. 1965 G. J. Williams *Econ. Geol. N.Z.* xiii. 205/2 The granitic facies of the Hawks Crag Breccia nearby and the interfingering facies containing both granite and greywacke components were quickly found to contain uranium minerals. 1971 *Nature* 28 May 247/1 The lower part of this lacustrine sequence interfingers with about 20 m of fluvio-deltaic sediments.

inter-firm, *a.* [INTER- 5.] Carried on between two or more business firms.
1949 *Collier's Year Bk.* 132 The rulings..disapproved secondary boycotts affecting interfirm activities. 1961 *Times* 2 May 2/5 The growing volume of interfirm comparison work. 1964 *Economist* 21 Mar. 1067/1 Confidential interfirm comparisons. 1967 *Ibid.* 7 Jan. 24/3 What sort of mergers, and what sort of inter-firm co-operation, does Britain want? 1972 *Times* 31 Oct. 25/2 'Interfirm comparison' is one of those rare animals, a management technique which the British have managed to sell to the Americans.

Interflora (intəɹflōə·ră). Name of the Florists' Telegraph Delivery Association, an international agency which organizes the delivery of flowers to order; also the trade name of the British branch of this association.
1949 *Sell's Directory of Registered Telegraphic Addresses* 795/2 Taylor William (florist), .. Worthing, Sussex. Interflora Worthing. 1951 G. Lewis in *Mod. Florist* xxiii. 229 The four large groups..are closely connected and compose the overall organisation Interflora, set up in 1946 to control and develop the interchange of overseas orders. 1960 L. Johns in *Retail Florist's Handbk.* xxii. 250 The clearing house at Interflora headquarters is an impressive sight. 1966 *Guardian* 4 Oct. 8/2, 22 million Interflora orders were sent world-wide last year. 1971 N. Freeling *Over High Side* I. 51 The ladies of Belgrave Square had sent long emotional telegrams and Interflora wreaths. 1972 F. Warner *Maquettes* 30, I arrange bouquets for Interflora.

interfluve (i·ntəɹflūv). [Back-formation from next.] A region lying between (the valleys of) adjacent watercourses, esp. one between the valleys of a dissected upland.
1902 in Webster Suppl. 1913 *Bull. Geol. Soc. Amer.* XXIV. 206 In an early stage of the new cycle the fault-line scarp will be highest near the incised valleys of transverse streams, and it may remain for a time undeveloped on the interfluves. 1937 *Geogr. Jrnl.* LXXXIX. 356 A cross-section of a typical portion of deltaic country from river to interfluve. 1956 D. L. Linton *Sheffield* 29 Here are some eighty or more square miles where dips are gentle and the grits build broad tabular interfluves swathed in peat and separating deep valleys with benched sides. 1968 R. W. Fairbridge *Encycl. Geomorphol.* 559 *Interfluve*,.. Many geomorphologists today use the term for the interstream area to imply a discrete landscape or geomorphic unit, composed of uni- or polycyclic slope facets. Interfluve almost always appears in a phrase explicitly or implicitly denoting its dissection.

interfluvial (intəɹflū·viăl), *a.* [f. INTER- 4 a + FLUVIAL *a.*] Situated between (the valleys of) adjacent watercourses.
1830 *New Monthly Mag.* III. *Hist. Reg.* Jan. 6/2 Returning from Bagdad across the interfluvial country, he took up his quarters at Hillah. 1903 *Sci. Amer. Suppl.* 14 Feb. 22679/1 A deposit of the flooded rivers during a stage of abundant ice melting, with considerable redistribution over the interfluvial upland areas by winds. 1933 *Antiquity* VII. 27 Beech Bottom and the Verulamium Devil's Dyke were relics of a transverse *limes* covering the whole tract of interfluvial country. 1971 [see *HUM *sb.*[4]].

inter-follicular: see *INTER- *pref.* 4 a.

intergalactic (intəɹgălæ·ktik), *a.* [INTER- 4 a.] Situated between the galaxies; of, pertaining to, or occupying the regions between galaxies.
1928 *Nature* 6 Oct. 556/1 In other words, in interstellar or intergalactic space. 1930 R. A. Millikan *Sci. & New Civilization* iv. 106 We think that the atom-building processes that give rise to the observed cosmic rays can take place only under the extreme conditions of temperature and pressure existing in interstellar, or intergalactic, space. 1953 J. S. Huxley *Evolution in Action* i. 13 The whole of space intergalactic as well as interstellar. 1969 *Times* 29 May 12/8 Extremely far away even by intergalactic standards. 1971 *Nature* 29 Jan. 305/2 At these energies both the interstellar and the intergalactic gas should become gradually opaque.

inter-generational: see *INTER- *pref.* 4 c.

intergene·ric, *a.* [INTER- 4 c.] Formed or obtained from (individuals of) different genera.
1921 *Genetics* VI. 380 Intergeneric crosses between genera only distantly related. 1926 *Jrnl. Agric. Res.* XXXIII. 101 (*heading*) Intergeneric hybrids in Aegilops, Triticum, and Secale. 1946 *Nature* 10 Aug. 204/2 The successful production of hybrid plants from interspecific, intergeneric and even interfamily crosses. 1973 W. J.

Bean *Trees & Shrubs Hardy in Brit. Isles* (ed. 8) II. 103 It is difficult to agree..that these plants [*sc.* varieties of *Erica cinerea*] are intergeneric hybrids between the bell-heather and *Calluna vulgaris*.

intergenic: see *inter- *pref.* 6.

interglacial, *a.* Add: Also *absol.*, an interglacial period.
1922 *Bull. Geol. Soc. Amer.* XXXIII. 421 In the terraces corresponding to the First Interglacial of the Alps he holds that there is no Scandinavian material. **1939** G. Clark *Archaeol. & Society* v. 137 An extra long gap, which..would be equivalent to the Mindel-Riss inter-glacial. **1970** I. Cornwall *Ice Ages* iii. 69 This down-cutting formed the bench..on which the deposits of the following Interglacial were laid down.

Interglossa (intəɹglǫ·să). [f. Inter- + Gr. γλῶσσα tongue.] An artificial auxiliary language devised by Lancelot Hogben (b. 1895).
1943 L. Hogben *Interglossa* 7 The author of *Interglossa* does not flatter himself with the hope that it will ever become the common language of international communication. **1946** H. Jacob *On Choice of Common Lang.* iii. 36 The potential extent of the Interglossa vocabulary will be many times greater than the number of terms contained in its vocabulary. **1954** Pei & Gaynor *Dict. Ling.* 103 *Interglossa*, an artificial language proposed by Hogben, based on Greek and Latin roots with a system of syntax resembling that of Chinese.

interglyph, inter-governmental: see *inter- *pref.* 3 a, 4 c.

intergranular, *a.* Delete *Anat.* and add to def.: occurring between granules or grains; intercrystalline. (Further examples.)
1932 *Jrnl. Iron & Steel Inst.* CXXV. 680 Intergranular corrosion of a chromium-nickel steel. **1946** *Nature* 24 Aug. 275/1 It is suggested that the migrations responsible for metasomatism occur partly by ionic diffusion through the crystal lattices and partly by molecular and/or ionic diffusion through the interstices ('intergranular film' of Wegmann) between the minerals. **1962** *Science Survey* III. 329 The final failure of the metal is essentially mechanical and it usually occurs by the propagation, through the body of the metal, of intercrystalline (or intergranular) cracks. **1971** *Nature* 30 July 327/1 It is quite likely that fluids have been able to flow through these rocks by means of both intergranular and fracture channels.

intergroup (i·ntəɹgrūp), *a.* [Inter- 5.] Situated, distributed, carried on, etc., between groups. Also as *sb.* So **intergrou·ping** *vbl. sb.* and (as a back-formation) **intergrou·p** *v.*
1883 W. Robinson *Eng. Flower Garden* p. cx/2 Breadth of mass and intergrouping. **1931** *Economist* 11 Apr. 783/1 The formation of an 'intergroup' for mutual benefit in connection with the coming contest at the polls. **1936** *Discovery* May 160/1 Inter-group relations..are cemented by an elaborate exchange of gifts. **1943** Wyndham Lewis *Let.* 17 Aug. (1963) 365 The inter-group and inter-racial quarrels seem to grow in number. **1965** *Math. in Biol. & Med.* (Med. Res. Council) iii. 87 (*caption*) The centres are at points representing the average intergroup distance. **1970** G. F. Newman *Sir, You Bastard* i. 17 The fifteen recruits..were moved and taught en masse, intergrouped with students further advanced and those who followed them in. **1972** *Accountant* 17 Aug. 191/2 Inter-group dividends may thus be paid gross.

interim, *adv.*, *sb.*, and *adj.* Add: **B.** *sb.* **5.** An interim dividend.
1930 *Daily Express* 6 Nov. 14/2 An interim of 5 per cent. actual was declared in May. **1935** *Economist* 14 Sept. 527/2 Associated Portland Cement had decided to pay the first interim in its history, at the rate of 5 per cent. **1964** *Financial Times* 10 Feb. 8/1 On Wednesday an interim is expected from Triplex Holdings.
C. *adj.* (Further examples.)
1869 *Bradshaw's Railway Manual* XXI. 247 The interim dividend of gross receipts. **1905** *Daily Chron.* 21 June 4/3 The Committee should..be invited to present an interim Report. **1950** T. S. Eliot *Cocktail Party* iii. 145 We have just drawn up an interim report. **1968** J. Lock *Lady Policeman* xviii. 149 A four-week interim order was made so that enquiries could continue. **1972** *Accountant* 17 Aug. 197/1 What would happen if that same company deferred the interim dividend.

‖ **Interimsethik** (i·ntərimzē:tik). *Theol.* Also in anglicized forms **interim-ethic, interim ethic.** [G. (Schweitzer, 1901, 1906) *interims-*, comb. form of *interim*, provisional, temporary + *ethik* ethics, ethical values, principles.] The moral principles laid down by Jesus, interpreted as formulated for the guidance of men expecting the imminent end of the world; hence, a code of behaviour for use in a specific, temporary situation.
1910 W. Montgomery tr. *Schweitzer's Quest Historical Jesus* xix. 352 What this repentance, supplementary to the law, the special ethic of the interval before the coming of the kingdom (*Interimsethik*) is,..He explains in the Sermon on the Mount. [**1915** A. Hall *Jesus & Christianity in 20th Cent.* iii. 43 Schweitzer..perceives the difficulty, and endeavours to overcome it by describing this side of the message and character of Jesus as 'the ethic of the interim'.] **1918** *Encycl. Relig. & Ethics* X. 733/1 Many NT students argue that the ethics of Jesus is conditional, an *Interimsethik*, and was proclaimed in indissoluble connexion with the eschatological expectation of a state of perfect blessedness to be supernaturally brought about. **1946** R. Knox tr. *Epistles & Gospels* 270 But Matthew makes it more plain than Luke that we are not merely dealing with what Schweitzer called an *interimsethik*, a scale of values only appropriate to a world which is shortly to go up in smoke. **1947** G. Seaver *Albert Schweitzer* ii. xiii. 183 Our Lord's teaching was an *interim-ethic*, that is to say it was conditioned by His conviction of the nearness of the supernatural Kingdom. **1960** N. Sykes *Sixty Yrs. Since* 7 It was in the light of this eschatology that all the ethical teaching of the Gospels must be understood..and the whole was an *Interimsethik*. **1961** *Times Lit. Suppl.* 17 Feb. p. x/3 He also points out..that their advocacy of toleration was intended only as an *Interimsethik*, until reconciliation..had been attained. *Ibid.* p. xii/3 Schweitzer's thorough-going eschatology and his consequent estimate of the ethical teaching of Jesus as *Interimsethik*. **1964** M. Ratter *Schweitzer* 116 The teaching of Jesus is seen to be an 'Interim Ethic'.

interindividual (-indivi·diu,ăl), *a.* [Inter- 5.] Subsisting, carried on, taking place, or forming a communication between, individuals. So **interindivi·dually** *a.*
1922 Joyce *Ulysses* 651 The converse domain of interindividual relations. **1951** T. Parsons et al. in Parsons & Shils *Toward Gen. Theory Action* i. i. 24 These selections cannot be inter-individually random in a social system. **1953** N. Tinbergen *Herring Gull's World* xvi. 122 The female's posture is identical with the submissive posture which the species has developed for other kinds of inter-individual conflict. **1965** *Philos. Rev.* LXXIV. 181 According to Kant interindividual inferences concerning existence are impossible by analytic means. **1971** *Jrnl. Gen. Psychol.* Oct. 175 The actual conditioning procedures should be introduced only after interindividual differences have been reduced to a minimum.

inter-influence: see *inter- *pref.* 1 b, 2 a.

interionic (intəɹəi,ǫ·nik), *a.* [Inter- 4 a.] Existing or occurring between ions.
1903 *Nature* 19 Nov. 65 The interionic forces..probably exert an effect even at dilutions at which the inter-molecular forces are negligible. **1929** *Jrnl. Chem. Soc.* 1487 The influence of interionic attractions. **1965** Phillips & Williams *Inorg. Chem.* I. v. 153 The observed interionic distances and the sums of the radii for the alkali halides with the NaCl structure are given in Table 5. vii.

interior, *a.* and *sb.* Add: **A.** *adj.* **3.** Special collocations, as **interior monologue** (cf. F. *monologue intérieur*), a form of writing in which the inner thoughts of a character are presented; **interior spring** (also **interior-sprung**) **mattress**, one with coiled springs within.
1922 tr. V. Larbaud in *Criterion* I. 103 It is of course especially in the interior monologues..that sexual instinct and erotic revery emerge. **1933** Joyce *Let.* 17 Mar. (1966) III. 270 The other [is] going to Rome to lecture on the Interior Monologue. **1952** A. Wilson *Emile Zola* iii. 66 The Impressionist approach which he [*sc.* Zola] used could have led to a development of the interior monologue, as it did for Tolstoy. **1971** *Guardian* 27 May 8/7 Eight ward-inmates of an old people's home describe themselves in interior monologue that rambles on. **1959** *Spectator* 9 Oct. 498/2 A fridge, interior spring mattresses, even a bath. **1948** Binnie & Boxall *Housecraft* (ed. 4) xvii. 197 Interior sprung beds..may occasionally be turned from side, to side or from end to end. **1951** *Good Housek. Home Encycl.* 187/2 Interior-sprung mattresses do not require such frequent turning. **1961** *Countryman* LVIII. 601 The stretcher felt like an interior-sprung mattress.
B. *sb.* **1. c.** (Earlier and later examples.) Also, in a theatre, a 'set' consisting of the inside of a building or room.
1829 H. Foote *Compan. to Theatres* 57 A few interiors, two or three streets, and about the same number of country views, would last as stock scenery for several seasons. **1858** Geo. Eliot *Jrnl.* 20 May in J. W. Cross *George Eliot's Life* (1885) II. viii. 40 The two interiors of Westminster Abbey by Ainmueller admirable. **1898** W. Archer *Theatr. 'World' 1897* 180 Mr. and Mrs. Alfred Wigan at the Olympic, made great strides towards realism in the dressing of modern plays and setting of everyday 'interiors'. **1916** J. R. Towse *Sixty Yrs. Theater* ii. 23 There were no elaborate and costly interiors, no enclosed box scenes, flats and wings were shifted before the eyes of the spectators. **1966** J. Potts *Footsteps on Stairs* (1967) i. 9 She does the most divine interiors.
d. The internal parts of the body, esp. the digestive system. *colloq.*
1835 Dickens *Let.* 4 Nov. (1965) I. 87 A..Pill—which..is performing such singular evolutions in my interior. **1906** Joyce *Let.* 18 Oct. (1966) II. 183 It [*sc.* wine] had not the least effect upon..'my interiors'. **1922** —— *Ulysses* 622 Mr Bloom..noticed when he stood up that he had two flasks of presumably ship's rum sticking one out of each pocket for the private consumption of his burning interior.

interior decoration (intīᵊ·riəɹ dekərēˈ·ʃən). [f. Interior *sb.* 1 c + Decoration 1.] The planned co-ordination for artistic effect of colours and furniture, etc., in a room or building. So **interior de·corator**, one who practises interior decoration. Also **interior-de·corate** *v. trans.*, to paint and furnish (a room or the inside of a building) in accordance with an artistic design; *interior-decorated* ppl. adj.; *interior decorating* vbl. sb. Also *attrib.* and *fig.* (see also quot. 1926).
1807 T. Hope (*title*) Household furniture and interior decoration. **1861** C. M. Yonge *Young Step-Mother* vii. 83 She was..too fond of out-of-door occupation, to regard interior decoration as one of the domestic graces. **1867** D. R. Hay (*title*) The interior decorator, being the laws of harmonious coloring adapted to interior decorations. **1906** *Dress* Oct. p. xxvi/1, I am a student of interior decoration—have had splendid success in this direction. It will pay you to consult me as to your plans for home beautifying. **1921** W. S. Maugham *Circle* II. 51 Arnold should have been an interior decorator. **1926** Maines & Grant *Wise-Crack Dict.* 9/2 *Interior decorator*, bartender. **1930** E. Ferber *Cimarron* xxiii. 373 Fascinating little..interior decorating shops. **1933** *Burlington Mag.* Aug. p. xv/1 Arranging flowers for interior decoration. **1935** *Discovery* Nov. 326/2 Ideal for interior decoration work. **1935** S. Lewis *It can't happen Here* ii. 24 The bar-room had been professionally interior-decorated by a young New York gentleman. **1943** F. L. Wright *Autobiogr.* (1945) iii. 160 John was to regale the inner man, interior-decorate him. **1950** E. H. Gombrich *Story of Art* iv. 77 The painters and interior decorators of Pompeii obviously drew freely on the stock of inventions made by the great Hellenistic artists. **1960** *House & Garden* July 4/3, I wonder if there are any courses in Interior Decorating that can be studied at home. I am thinking of starting a small interior decorating business. **1967** E. Short *Embroidery & Fabric Collage* i. 18 Texture is a factor which plays a very important part in modern architecture and interior decoration. **1972** L. P. Bachmann *Ultimate Act* i. 9 The library..was as interior-decorated as the rest of the house.

interior design (intīᵊ·riəɹ dīzəi·n). [f. Interior *sb.* 1 c + Design *sb.* 7.] The design of the interior of a building, including wall-paper, furniture, fittings, etc., according to artistic and architectural criteria. So **interior designer**, one whose business is to plan such interiors. Cf. prec. entry.
1927 T. P. Bennett *Archit. Design in Concrete* 13 The effect of this adjustment of proportion and detail upon interior design is..in some ways more fundamental than the effect of the new proportions upon exterior designs. **1938** *Decorative Art* p. xxxii, The favourable attention of all leading Interior Designers can be secured..through a single publication, 'Interior Design and Decoration'. **1957** *Encycl. Brit.* XII. pl. x (*caption*) Living room by Ain, Johnson and Day, showing relationship of interior and exterior design. **1962** H. Stephenson *Design & Decoration in Home* 7 There is more to interior design than just putting furniture in a room and paint on a wall. **1967** E. Short *Embroidery & Fabric Collage* iii. 64 Good interior designers sometimes go as far as having carpets specially designed and manufactured. **1972** *Guardian* 17 May 9/4 She turned to interior design.

interiority. (Later examples.)
1890 W. James *Princ. Psychol.* II. xvii. 43 It is surely subjectivity and interiority which are the notions *latest* acquired by the human mind. **1934** E. Bowen *Cat Jumps* 51 Voices came out from some dark interiority. **1941** *Theology* XLII. 156 The characteristic of the new period was, as Hegel put it, interiority. **1967** *Listener* 26 Oct. 552/1 Alan Bates as Gabriel Oak suffers..from Schlesinger's reluctance to suggest, as Hardy might put it, interiority. **1973** *Times Lit. Suppl.* 2 Nov. 1348/4 For all its imaginative ambitiousness, the volume lacks a certain human interiority.

interiorize (intīᵊ·riərəiz), *v.* [f. Interior *a.* + -ize.] *trans.* To connect with the soul, as distinguished from the body; also, to locate within the mind.
1906 *Academy* 20 Oct. 392/2 To 'interiorise' the struggle, to place it on the stage of the soul, with eternity for background. **1916** Stanford & Forsyth *Hist. Mus.* xvi. 329 The second [feature in American life] is the interiorizing and democratic habit-of-mind which partly connotes the term *Americanism*. **1934** *Mind* XLIII. 89 In so far as habits of co-operation have convinced the child at a later age of the necessity of not lying, rules will become comprehensible and interiorised. **1937** G. W. Allport *Personality* (1938) p. viii, From this point of view culture is relevant only when it has become *interiorized* within the person as a set of personal ideals, attitudes, and traits. **1971** *Jrnl. Gen. Psychol.* Apr. 206 The child interiorizes what he already understands.
Hence **interioriza·tion**.
1941 *Theology* XLII. 156 To discover anew the meaning of authority *in* their immanent freedom, by making it itself immanent within them by a process of interiorization. **1956** *Scottish Jrnl. Theol.* IX. 74 Hence he is unable to rise to the thought of suffering as the gift of divine love, as a sacramental medium through the purifying effect of which man attains a deeper realisation and interiorisation of God. **1961** J. N. Findlay *Values & Intentions* i. 40 Our talk about thoughts, decisions, etc., is always largely an 'interiorization' of our talk about words. **1966** L. Jones in A. Chapman *New Black Voices* (1972) 465 The Black man must seek a Black politics, an ordering of the world that is beneficial to his culture, to his interiorization and judgment of the world. **1971** *Jrnl. Gen. Psychol.* Apr. 206 Since these interiorizations are not sufficient for the child, he gradually proceeds to representations based upon more logical..principles of operation.

interkinesis (intəɹkəinī·sis). *Cytology.* [ad. F. *intercinèse* (V. Grégoire 1905, in *La Cellule* XXII. 226): see Inter- 2 b and *kinesis.] A stage which sometimes intervenes between

the first and second divisions of meiosis; also, any stage between mitoses.
1906 *Jrnl. R. Microsc. Soc.* 283 The daughter-chromosomes..preserve their autonomy during the interkinesis. **1925** E. B. Wilson *Cell* (ed. 3) i. 28 The central body..retains its morphological identity during the interkinesis or vegetative (non-mitotic) condition of the cell. **1965** Penny & Waern *Biol.* xlii. 775 In some cases the chromosomes decondense as in ordinary mitosis, and there is a brief pause before the second division commences. This stage, if it occurs, is called interkinesis.
Hence **interkine·tic** *a.*
1927 *Protoplasma* II. 189 The interkinetic nuclear substance is devoid of any structural element. **1931** J. Gray *Text-bk. Exper. Cytol.* viii. 141 A 'resting' or interkinetic nucleus in the living condition is usually optically homogeneous. **1960** L. Picken *Organization of Cells* iv. 101 (*heading*) The interphasic, interkinetic, 'resting', or non-mitotic nucleus.

interlace, *v.* Add: **6.** *Television.* (*trans.*) To present (scanning lines) so that alternate lines of a picture form one sequence and are followed by the intervening lines in a second sequence; to present (dots) similarly so that several fields of regularly spaced dots go to form each picture. Also, to combine (two or more fields), or form (a picture or raster), in this way. Freq. as pa. pple. Cf. also *interlaced *ppl. a.*
1927 M. Latour *Brit. Pat.* 267,513, The elements of the image transmitted by each system AB are..the ones within the others, or interlacing each other. **1936** [see *frame *sb.* 12 c]. **1955** D. G. Fink *Color Television Standards* iii. 92 Dot interlace, in which minute dots..of different primary colors, produced adjacent to each other during the color-scanning process, are interlaced in various repeated and prearranged sequences. **1966** G. H. Hutson *Television Receiver Theory* I. xii. 187 If these conditions are met the resulting raster must be interlaced. **1967** [see *field *sb.* 16 d]. **1972** *Sci. Amer.* Sept. 132/2 Every other line is scanned in just under a sixtieth of a second and the missing lines are interlaced in the next sixtieth of a second.

interlace (i·ntəɹlēˈs), *sb.* [f. the vb.] The action or result of interlacing.
1904 Goodchild & Tweney *Technol. & Sci. Dict.* 312/2 *Interlace.* This relates to the crossing of warp and weft, the order of the interlacing in a weave prescribing the structure of the cloth. **1923** *Daily Mail* 19 Mar. 1 The upturned brim has fancy straw interlace, giving a ribbon effect. **1936** A. W. Clapham *Romanesque Archit.* iii. 61 The acanthus-scroll..commonly has a stem composed of three strands, a trick of the carver which was probably inherited from the universal use of the triple strand interlace in Italy at an earlier date. **1948** N. Gray in *Papers Brit. Sch. Rome* XVI. 116 The cross is carved with a foliage pattern on one side..and an interlace on the other. **1973** *Country Life* 29 Nov. 1761/1 Strap-work and interlace patterns on plain velvet.
b. *spec.* in *Television* (see *interlace *v.* 6).
1936 O. S. Puckle tr. *M. von Ardenne's Television Reception* i. 5 The line component and the frame component of scanning are regularly recurrent, the interlace being derived from the fractional relationship between line and frame frequencies. **1937** *Electronics* June 15/2 At the end of each half-frame or 'interlace', the frame synchronizing impulses are imposed in a similar manner. **1961** *Listener* 2 Nov. 725/3 The television service reopened on June 7, 1946, using the pre-war system (405-lines, 25 pictures per second with 2:1 interlace, positive modulation and AM sound). **1966** G. H. Hutson *Television Receiver Theory* I. xii. 192 The alternate scanning field is ½ line late in starting. This causes very poor interlace.

interlaced, *ppl. a.* (in Dict. s.v. Interlace *v.*). Add: Also stressed *i·nterlaced*. **b.** *Television.* Applied to scanning in which the lines or the dots of the picture are interlaced (see *interlace *v.* 6), so that each is built up from two or more fields; composed or combined in this way.
1935 R. W. Hutchinson *Television Up-to-Date* v. 131 It has been suggested that a system of interlaced scanning, as it is called, should be adopted: in this, lines 1, 3, 5, 7, etc., would first be scanned, and then the intermediate ones, 2, 4, 6, 8, and so on. **1936** O. S. Puckle tr. *M. von Ardenne's Television Reception* i. 21 With the resulting 50 pictures per second, the flicker of the interlaced picture is almost entirely avoided. **1953** H. A. Chinn *Television Broadcasting* i. 5 In order to conserve bandwidth without sacrificing freedom from flicker, the standard television system employs a system of interlaced scanning. **1961** *New Scientist* 26 Jan. 199/3 The British system transmits 25 complete pictures per second, forming each from 405 horizontal lines laid down in two interlaced sequences of 202½ lines each.

interlacing, *vbl. sb.* (Further examples.)
1927 *Jrnl. Genetics* XVIII. 182 There is interlacing of the chromatids. **1935** R. W. Hutchinson *Television Up-to-Date* v. 132 Both Baird and E.M.I. have experimented with interlacing. **1936** *Proc. IRE* XXIV. 573 The difference between the amplitude of the alternate discharges is somewhat critical for perfect interlacing. **1974** J. Robinson *Penguin Bk. Sewing* xii. 328/2 Interlacing..consists of horizontal stitches interlaced with the herring bone stitch.

interlacustrine (intəɹlǎkᵊ·strin), *a.* [f. Inter- 4 a + Lacustrine *a.*] Lying between lakes.
1900 *Geogr. Jrnl.* Feb. 179 This region forms part of the great interlacustrine plateau. **1958** E. Winter in Middleton & Tait *Tribes without Rulers* 158 The other Interlacustrine kingdoms of which the best known is Buganda.

interlaminar, *a.* Add: **2.** Situated or occurring between the reinforcing layers or components of a laminate or composite.
1963 *Symposium Standards for Filament-Wound Reinforced Plastics, 1962* (Amer. Soc. Testing Materials STP 327) 17 In this method, a bending test is conducted with a short span to induce failure in horizontal ('interlaminar') shear. **1964** *Plastics Inst. Trans. & Jrnl.* XXXII. 292/2 Creep effects in the composite thus become limited principally to the resin interlaminar planes. **1966** K. L. Lowenstein in L. Holliday *Composite Materials* v. 200 The consensus of opinion is that failure [of composites] occurs by interlaminar shear, *i.e.* fracture of resin between reinforcing fibres followed, as a secondary step, by the fracture of individual fibres themselves. **1971** *New Scientist* 8 July 70/1 Composites of [carbon] fibres in resin having the expected high tensile strength and modulus..had a low interlaminar shear strength of only 2500 lb/sq in.

interlanguage (i·ntəɹlæŋgwėdʒ), *sb.* [Inter- 2 a.] An artificial auxiliary language. Also *attrib.*
1927 E. S. Pankhurst *Delphos* vii. 86 The Interlanguage cannot be the creation of Governments. **1928** O. Jespersen *Internat. Lang.* i. 45 The Delegation and the Ido academy have left their indelible mark on the interlanguage movement. **1929** T. C. Macaulay in *S.P.E. Tract* (1930) xxxiv. 462 Interlanguage will have no idiomatic tradition of its own. **1960** E. Delavenay *Introd. Machine Transl.* iv. 47 Georges Mounin rightly distinguishes between pseudolanguages—of which Esperanto is the classic example—intended to be *speakable*, and interlanguages, designed for use as auxiliary languages, such as the *interlingua* of Peano or that of Gode and Blair.

interla·nguage, *a.* [Inter- 5.] Between or relating to two languages.
1953 U. Weinreich *Lang. in Contact* i. 8 Interference resulting from such inter-language identification. **1964** E. A. Nida *Towards Sci. Transl.* vii. 147 In an attempt to describe these interlanguage and intercultural factors, we must reckon with differences of time.

interlay (i·ntəɹlēˈ), *sb.* [Back-formation f. the vb.] That which is intercalated; esp. in *Printing*.
1901 *Westm. Gaz.* 10 Oct. 2/2 A delicate Chantilly lace mounted over cream satin, with an interlay of cream chiffon. **1940** *Chambers's Techn. Dict.* 453/2 *Interlay*, paper inserted between a printing plate and its mount in order to raise the plate to type height. **1948** H. Missingham *Student's Guide Commercial Art* ii. 82 *Interlay*, paper placed between the mount and the metal plate to raise those portions of the block, representing usually the dark tones and solids.

interlayer (i·ntəɹlēˈ:əɹ), *sb.* and *a.* [f. Inter- + Layer *sb.*] **A.** *sb.* [Inter- 2 b.] A layer situated between two other layers.
1936 *Brit. Pat.* 453,578, The polymer..is..dried in vacuo and may be mixed with a suitable amount of an appropriate plasticiser..compressed to form a block either of rectangular or of cylindrical shape, and..cut into sheets ready for use as a laminated safety glass interlayer. **1960** *Nature* 16 July 262/1 The laboratory introduction of aluminium interlayers into montmorillonite and vermiculite produced clays that resembled the chloritized soil clays in their 14-Å. spacing. **1962** L. S. Sasieni *Princ. & Pract. Optical Dispensing* xiii. 329 The three laminae are joined under heat and pressure, and if one or both of the layers of glass are broken by impact..the pieces of glass remain adherent to the interlayer. **1972** *Soil Sci.* CXIII. 172/2 It was anticipated that the removal of interlayer liquid in K-saturated illite might be similar to montmorillonite and vermiculite, where entire interlayers are usually lost abruptly.
B. *adj.* [Inter- 5.] Situated or occurring between two layers.
1956 *Nature* 4 Feb. 239/2 A flake of magnesium vermiculite, which has been partially dehydrated to the phase containing single sheets of interlayer water molecules. **1970** *Proc. Soil Sci. Soc. Amer.* XXXIV. 201/1 Interlayer swelling of the soil clays.

inter-lend, *v.*: see *inter- *pref.* 1 b.

interlevel (i·ntəɹlevěl). *Linguistics.* [Inter- 2 a.] A 'level' of language serving to relate other linguistic levels (see quots.).
1961 M. A. K. Halliday in *Word* XVII. 269 Context is an interlevel in a different sense, since it relates language to something that is not language. **1963** R. M. W. Dixon *Ling. Sci. & Logic* i. 23 An interlevel will itself contain theories which will perform the necessary level-relating. **1964** R. H. Robins *Gen. Ling.* p. xix, Both treat phonology as an 'interlevel' serving to relate statements made at the level of grammar with those made at the level of phonetics. **1966** J. Ellis in C. E. Bazell *In Memory of J. R. Firth* 79 This classification of levels is represented by the diagram showing the levels of subst... form and situation and interlevels of phonology/graphology and context.

inter-library: see *inter- *pref.* 5.

interline (intəɹləi·n), *a.* [f. Inter- 5 + Line *sb.*²] Of or pertaining to transport: using more than one route, service, etc.
1946 T. C. Bigham *Transportation* iii. 96 Service may be local or interline. **1957** *Encycl. Brit.* XVIII. 940 (*heading*) Joint or interline rates. **1969** *Jane's Freight Containers 1968–69* 48/1 The New York Port District may be likened to the centre of a web of gleaming steel rails..of ten long-haul railroads which, through interline connections, serve the entire nation, Canada and Mexico. **1970** *Daily Tel.* 30 June 24 Logically the first flight after the official connecting time would be the correct interline connection.

interlinearly, *adv.* Delete *rare* and add later examples.
1906 *Hibbert Jrnl.* Apr. 682 A distinction that is indeed repeatedly suggested interlinearly in Mr Mallock's pages. **1955** C. E. Wright in *Bald's Leechbook* Pref. 11 One manuscript, the later of the two, is a volume of small format..(glossed interlinearly).

Interlingua (intəɹli·ŋgwă). Also **interlingua.** [f. Inter- + L. *lingua* tongue.] An artificially devised international language; *spec.* one promoted by the International Auxiliary Language Association of New York.
Several invented languages have been so named.
1922 A. L. Guérard *Short Hist. Internat. Lang. Movement* v. 127 Those which, like Neutral, Ido, Interlingua, present themselves as the collective work of some 'Academy' bear..the mark of one master mind. **1927** E. S. Pankhurst *Delphos* iv. 38 'Interlingua is the standard of the insurrection against the routine of red tape and the tyranny of the ancient grammarian', thus wrote Kerchoffs, the first director of the Volapük Academy, in 1886. Giuseppe Peano..was presently to translate those words into a language scheme, and to adopt Interlingua as its title. **1928** O. Jespersen *Internat. Lang.* i. 45 This 'interlingua' is now employed in the publication of *Academia pro Interlingua*. **1953** J. B. Carroll *Study of Lang.* iv. 127 A recent attempt to rationalize an artificial language by making maximal use of elements common to the most widely used natural languages is Interlingua, the work of the International Auxiliary Language Association of New York. **1955** *Sci. News Let.* 29 Jan. 79/2 Interlingua is a language composed of elements common to Spanish, Portuguese, French, Italian and other Romance languages. *Ibid.* 22 Oct. 258/2 Interlingua is no overnight creation of one linguist or even one group of linguists. No one sat down and theorized as to what an international language should be. Instead three decades ago, in 1924, an ambassador-to-be, a chemist, several radio engineers, several educators, editors and linguistic experts started a long and detailed inquiry into what an international language should be. **1956** J. Whatmough *Lang.* iv. 59 An interlingua needs more than expletives and exclamations. **1960** [see *interlanguage *sb.*].

interlingual, *a.* Delete *rare* and add later examples. So **interli·ngually** *adv.*
In quot. 1931 *spec.* = of or relating to an artificial interlanguage (cf. *interlanguage *sb.*).
1931 *Mod. Lang. Notes* XLVI. 18 Those who..have paid little attention to the interlingual movement. **1941** P. B. Gove *Imaginary Voy. Prose Fiction* i. i. 8 The term is in some degree interlingual; occasionally to German writers it seems so apposite that it is not translated. **1951** W. Empson *Struct. Complex Words* xxi. 397 The confusion of translation equivalents is so great that many students have been warned against the interlingual dictionary. **1958** J. Berry in J. A. Fishman *Readings Sociol. of Lang.* (1968) 741 Is it..preferable to follow the orthographic practice of the trade language or that of an interlingual notational inventory such as The Africa Script? **1964** *Language* XL. 243 A minimum of inter-lingual interference. **1966** *Publ. Amer. Dial. Soc.* xlvi. 5 Whenever two interlingually identified forms are similar in sound and..meaning,..we have a homologous diamorph.

interlinguist (-li·ŋgwist). [Inter- 2 a.] One versed in or an adherent of an interlanguage or interlanguages.
1928 O. Jespersen *Internat. Lang.* i. 12 What then we interlinguists are thinking of, is..what another inventor of an artificial language, Bollack, took as his motto: The second language to everybody. **1934** *N. & Q.* 8 Sept. 168/1 The true interlinguist..contends that a suitable auxiliary language must be an efficient medium for the communication of thought internationally. **1947** H. Jacob *Planned Auxiliary Lang.* i. i. 31 Leopold Einstein, a well-known interlinguist.

interlinguistic, *a.* Delete *rare* and add: **2.** Of or relating to an interlanguage; between or relating to two languages.
1947 H. Jacob *Planned Auxiliary Lang.* iii. xiv. 131 For the purposes of interlinguistic discussion they have been termed the naturalistic school and the autonomistic school. **1962** P. S. Ray in J. A. Fishman *Readings Sociol. of Lang.* (1968) 756 We might speak, instead of 'closure' and 'opening', of intra-linguistic and inter-linguistic uniformity.

interlinguistics (-liŋgwi·stiks). [Inter- 2 a.] The study of the relationships of two or more languages, e.g. for the purpose of devising an interlanguage. Hence **interlinguisti·cian** = *interlinguist.
1931 O. Jespersen in H. N. Shenton et al. *Internat. Communication* iii. 95 A new science is developing, Interlinguistics. **1934** *N. & Q.* 8 Sept. 168/1 The new or coming science of 'interlinguistics' is based upon certain fundamental principles of its own. **1938** *Encycl. Brit. Bk. of Yr.* 673/2 In recent years, interlinguisticians have been feeling their way to an average, as it were, of existing languages. **1947** H. Jacob *Planned Auxiliary Lang.* i. iii. 70 In interlinguistics the term *pleonastic endings* has been used to describe specific grammatical terminations. **1953** M. Weinreich in J. A. Fishman *Readings Sociol. of*

Lang. (1968) 394 As the raw material..is drawn from the most divergent sources: Persian.., Arabic, Slavic, Greek, several Romance languages, Teutonic, Jewish interlinguistics should prove the comparativist's delight.

interlining, *sb.* (Earlier and later examples.)
1881 C. C. Harrison *Woman's Handiwork* i. 76 Lay the work upon the interlining of canton flannel, and turn the edges down. **1959** [see *iron-on *a.*]. **1974** J. Robinson *Penguin Bk. Sewing* v. 133 Interlining..used in tailoring..consists of the introduction of a section of tailor's canvas, Vilene, [etc.]..to stiffen the fabric.

interlink, *v.* Add: So **interli·nkage.**
1904 *Westm. Gaz.* 3 Dec. 16/3 The phenomena to be seen in the living being, their inter-linkage, their apparent adaptation to an end. **1930** *Times* (Empire Press No.) 31 May p. xi/4 An Empire broadcasting system... A total problem that includes both interlinkage with the United States and [etc.]. **1957** V. W. Turner *Schism & Continuity in Afr. Soc.* x. 301 The..values shared by all Ndembu are prominently displayed..in the ritual association of those who have suffered regardless of their kinship or other interlinkages.

interlocal, *a.* (Example.)
1920 A. C. Pigou *Econ. of Welfare* ii. vi. 171 So soon as people become thoroughly familiarised with town-planning, local patriotism and inter-local emulation will make resort to external pressure from the central Government no longer necessary.

interlock, *v.* Add: **4.** *Cinemat.* To connect (the electric motors of cameras or the like) electrically in such a way that they rotate in synchronism with one another.
1928 *Trans. Soc. Motion Picture Engin.* XII. 704 It has been necessary..to develop a motor drive equipment which will satisfactorily interlock the camera and the recording machine... It is essential that the interlock should hold..during acceleration and deceleration. **1931** B. Brown *Talking Pictures* ix. 206 Where we have cameras working in conjunction with sound recorders..there is absolute necessity for both devices to be interlocked or driven together, so that sound and photograph are always exactly in phase. **1953** L. J. Wheeler *Princ. Cinematogr.* ii. 68 When all was ready to take the scene the camera was interlocked with the sound recorder so that, on starting up, both camera and recorder would rotate in synchronism.

interlock, *sb.* Restrict *rare* to senses in Dict. and add: **2. a.** *Cinemat.* Synchronism between two or more electric motors (e.g. in a camera and in sound-recording apparatus); also, the mechanism by which this is effected. Freq. *attrib.*
1928 [see *interlock *v.* 4]. **1931** B. Brown *Talking Pictures* ix. 206 Gaumont coupled the armature of his projector motor, and using a common power supply, obtained synchronism. To-day we use a similar type of arrangement known as the 'electrical interlock'. **1938** *Motion Picture Sound Engin.* (Acad. Motion Pict. Arts & Sci.) viii. 118 Another commonly used driving system employs the Selsyn type of motor and is known as the interlock system. This is virtually an electrical gear system, whereby all the motors connected together on several separated units will start together, come up to speed at the same rate, and continue to run at identical speeds. *Ibid.* 119 Field rheostats for each motor are located in one place and all speeds are adjusted to maintain interlock at the estimated camera and recorder loads. **1949** Frayne & Wolfe *Elem. Sound Recording* xxiv. 477 Because the three-phase interlock windings are interconnected with the d-c windings voltage appears on the interlock circuit as soon as the d-c leads are excited. **1962** E. L. Levitan *Animation Technique* i. 70/1 The process whereby the picture reel and the sound track are run and projected at the same time is called the interlock.
b. A mechanism for preventing a set of operations from being performed in any but the prescribed sequence.
1934 in Webster. **1945** *Rev. Sci. Instruments* XVI. 57/2 There are two mechanical interlocks on the controls... The selector switch is locked so that it cannot be moved unless the Variac is set to zero. **1955** *Archit. Rev.* CXVII. 142/3 The most important piece of equipment associated with lift doors is the inter-lock, an electro-mechanical device which ensures that the lift cannot move until both car and landing doors are locked in position. **1958** *Engineering* 28 Feb. 261/3 To prevent incorrect operation, electrical interlocks are provided to ensure that the hopper can only be tilted when all the pressing rams are clear and the box is open..and that the ejection door cannot be closed until the final pressing ram has been withdrawn. **1958** *Newnes Compl. Amat. Photogr.* 54 It is common nowadays for even simple cameras to have a shutter-film wind interlock which prevents blank negatives or double exposures. **1963** B. Fozard *Instrumentation Nucl. Reactors* xiii. 170 Interlocks must be fitted to ensure that the chambers cannot be inserted under high flux conditions.
B. *attrib.* or *adj.* Esp. designating woven material in which the stitches are woven together.
1928 *Daily Mail* 25 July 3/6 (Advt.), The merits of Meridian Interlock Underwear. **1935** *Economist* 23 Nov. 1003/1 Considerable progress..was made with the development of spun yarns for..pyjama cloths and for underwear fabric manufactured on circular interlock knitting machines. **1969** *Sears Catal.* Spring/Summer 26 Cardigan sweater interlock knit of Orlon acrylic.

interlocutor¹. Add: **c.** The compère in a troupe of nigger minstrels; the man in the middle of the minstrel line who questions the end men.
1880 E. James *Amat. Negro Minstrel's Guide* 2 Interlocutor or Middle Man, in the Center. **1884** [see Banjoist]. **1957** W. C. Handy *Father of Blues* xxi. 276 Henry Troy acted as the interlocutor, with Tom Fletcher and Laurence Deas as end men.

interlot, *v.*: see *inter- *pref.* 1 a.

interludial (intəɹliū·diăl), *a.* [f. Interlude *sb.* + -ial.] Of, pertaining to, or of the nature of an interlude.
1884 *Encycl. Brit.* XVII. 94/1 Admitted for interludial purposes in a fabrication styled *intermezzo* that was played between the acts of a serious composition, comedy became [etc.]. **1922** S. Grew *Art of Player-Piano* 75 The interludial figure is extended to lead into a *forte*.

inter-marginal: see *inter- *pref.* 4 a.

intermat (intəɹmæ·t), *v.* [Inter- 1 b.] *trans.* and *intr.* To mat together.
1904 Goodchild & Tweney *Technol. & Sci. Dict.* 312/2 *Intermat* (Textile Manufac.), the term applied to the felting or shrinking of cloths, the fibres intermatting or felting together. **1927** *Daily Express* 18 Apr. 3/7 As the hair grows, it is worked into a kind of felt by intermatting it.

intermède. **2.** Now current in alien form *intermède* (æ̃ntɛrmęd).
1887 *Gentl. Mag.* June 540 The singularly appropriate *intermède* arranged by Beaumarchais for performance between the acts of his 'Eugenie'. **1931** *Times Lit. Suppl.* 6 Aug. 606/2 The inclusion between the first and second acts of an *intermède* of song and dance. **1970** *Oxf. Compan. Mus.* (ed. 10) 517/2 It was as *intermezzo* or *intermède* that the comic opera grew up.

intermedial, *a.* and *sb.* Delete †*Obs.* and add later example in sense A. *adj.* 1.
1942 *Mind* LI. 80 Part II is a systematic and richly illustrated description of generic and special forms, of inter-medial and intra-medial forms.

intermediate, *a.* and *sb.* Add: **A.** *adj.* **d.** (Further examples.) Also *intermediate education, school.*
1842 E. Lazarus *Let.* 19 July in N. E. Eliason *Tarheel Talk* (1956) 278 There are the primary & the intermediate schools, & the high-school. **1882** W. D. Hay *Brighter Britain* I. ii. 57 It doesn't matter twopence *how* you go out, whether saloon, intermediate, or steerage, so far as your future prospects are concerned. **1886** Kipling *Plain Tales from Hills* (1888) 120 The four constables saw him safe to Umritsar in an 'intermediate' compartment. **1889** *Act* 52–54 *Vict.* c. 40 (*title*) An Act to promote Intermediate Education in Wales..sect. 1. This Act may be cited for all purposes as the Welsh Intermediate Education Act, 1889. *Ibid.* sect. 17 The expression 'intermediate education' means a course of education which does not consist chiefly of elementary instruction..but which includes instruction in Latin, Greek, the Welsh and English language and literature,..mathematics, natural and applied science. **1893** *Harper's Mag.* Apr. 806 Oh, she was a rose half-budded, in the intermediate school, And her face and form I studied twice as much as task or rule. **1945** C. V. Good *Dict. Educ.* 223 Intermediate school: a school that enrolls pupils in intermediate grades, usually comprising the fourth, fifth, and sixth years of schoolwork. **1974** *Times* 1 Apr. (Yorkshire & Humberside Suppl.) p. i/2 Yorkshire and Humberside is classified as an 'intermediate area'. As such, while enjoying the benefits of financial inducements available to incoming and expanding industry, it does not rank for the benefits available in the other two types of aided regions.
2. Specific techn. uses.
a. *Petrol.* Of a rock: having a silicate content that falls between that of the acidic and that of the basic rocks (cf. Acidic *a.* 2, Basic *a.* 2 b); often *spec.* having a silicate content between 52 and 66 per cent by weight.
1888 J. J. H. Teall *Brit. Petrogr.* viii. 253 The basic rocks shade into the intermediate rocks, and these again into the acid rocks, in the most gradual manner. **1892** F. H. Hatch *Text-bk. Petrol.* (ed. 2) vi. 107 In respect to the percentage of silica, igneous rocks fall naturally into four groups, viz.:—(1) An acid group with 65–80% of silica... (2) An intermediate group with 55–70% of silica... (3) A basic group with 45–60% of silica... (4) An ultra-basic group with silica between 35 and 50%. **1909** *Ibid.* (ed. 5) iii. i. 152 Arranged in the order of their silica contents, the plutonic rocks can be divided into three groups: 1. Acid, with silica contents above 66 per cent. 2. Intermediate, with silica contents between 66 and 52 per cent; and 3. Basic, with silica contents below 52. **1939** A. Johannsen *Descr. Petrogr. Ign. Rocks* (ed. 2) I. 181 (*gloss.*) *Intermediate rocks*, rocks intermediate between the 'acid' and 'basic' groups. Syn. Neutral rocks, mediosilicic. Cf. Acid. **1968** B. Bayly *Introd. Petrol.* vi. 53 The second system is chemical rather than mineralogical, being based on the weight percentage of SiO_2 in the rock; thus if SiO_2 percentage is: over 66, rock is acid; 52–66, rock is intermediate; 45–52, rock is basic; under 45, rock is ultrabasic.
b. *Nuclear Physics.* Applied to neutrons with less energy than fast neutrons but more than thermal neutrons, and also to nuclear reactors in which such neutrons are the chief cause of fission.
1947 *Rep. U.S. Atomic Energy Commission* A-4315 (*title*) A multi-group method for computing critica masses of intermediate piles. **1949** *Nucleonics* Dec. 41/1 Intermediate piles may operate with neutrons at any energy level between thermal and fission or even at several different energy levels. **1956** Glasstone *Princ. Nucl. Reactor Engin.* i. 15 In nuclear reactor work, the term *fast neutrons* is applied to neutrons having energies of about 0·1 Mev, i.e., 10^5 ev, or more. Those with energies from 10^5 ev down to 1 ev are called intermediate neutrons. **1959** L. F. Curtiss *Introd. Neutron Physics* i. 18 Less information has been accumulated about intermediate neutrons than about neutrons of lower energies because of [the] difficulty of finding efficient detectors. **1966** *McGraw-Hill Encycl. Sci. & Technol.* XI. 358/1 An example of an intermediate reactor is the first propulsion reactor for the submarine USS *Seawolf*. The fuel core consisted of enriched uranium with beryllium as a moderator.
3. Specific collocations: **intermediate boson:** see *boson; **intermediate frequency** *Electronics*, the frequency to which an incoming carrier wave is converted by the frequency changer of a superheterodyne receiver; abbrev. **I.F.*; **intermediate host** *Zool.*, an organism infected by a parasitic animal which then goes on to complete its life cycle in another host; **intermediate-range**, used *attrib.* of a ballistic missile of medium range (less than 'intercontinental').
1924 *Proc. IRE* XII. 540 Three intermediate frequency amplifiers. **1947** D. G. Fink *Radar Engin.* x. 504 The [radar receiver] system which avoids the foregoing difficulties is the superheterodyne, which introduces an initial change from radio frequency to a lower frequency (intermediate frequency) followed by a high-gain amplification at this frequency. **1968** B. P. Lathi *Communication Syst.* iii. 202 The advantage of conversion to an intermediate frequency is that to receive different stations it is necessary to tune only the first stage (and the local oscillator). All of the amplification is achieved at a constant intermediate frequency and needs no tuning. **1878** *Jrnl. R. Microsc. Soc.* I. 377 The ultimate form assumed by the larvae whilst still within the body of the intermediate host. **1892** [see Host *sb.*² 3]. **1901** *Practitioner* Mar. 273 It is parasitic in man and in a certain genus of mosquito (Anopheles); the former is its intermediate host and the latter its definitive host. **1925** A. D. Imms *Gen. Textbk. Ent.* iii. 365 The latter issue from the galls and are divisible into winged *gallicolæ migrantes* (migrantes), which fly to the intermediate host, and *gallicolæ non-migrantes* which remain on the spruce and give rise to further fundatrices. **1971** E. R. & G. A. Noble *Parasitol.* (ed. 3) xxiv. 528/1 As a generalization, there is less host specificity when there are two intermediate hosts than when only one is employed. **1956** *Newsweek* 30 Jan. 27/1 Developing a 1,500-mile intermediate-range ballistic missile (IRBM) is now largely a question of 'straightforward engineering'. **1957** *Economist* 30 Nov. 774/2 The Polaris, the intermediate range ballistic missile to be launched from submarines. **1959** *Listener* 18 June 1053/1 Intermediate-range ballistic missiles—in other words, medium-range rockets.
B. *sb.* **1. c.** *Chem.* and *Biochem.* A compound which after being produced by one reaction participates in another; *esp.* one manufactured from naturally occurring materials for use in the synthesis of dyes, plastics, or other substances.
1919 E. de B. Barnett *Coal Tar Dyes* i. ii. 29 Aniline.—This is the most important intermediate, and is invariably manufactured by the reduction of nitrobenzole. *Ibid.* 31 Factories preparing their own intermediates. **1938** *Nature* 30 July 203/2 Mr. F. P. Garvan's appreciation of the dependence of the United States on Germany for dyes, intermediates, photographic chemicals, medicinals, etc., led him to organize the Chemical Foundation. **1953** *Nature* 27 June 1160/2 (*heading*) Occurrence of hydroxylamine in lake waters as an intermediate in bacterial reduction of nitrate. **1961** *Times* 30 May (I.C.I. Suppl.) p. xvi/1 All of them are 'intermediates'—the raw materials of other products such as Terylene or plastics. **1962** J. Hine *Physical Org. Chem.* (ed. 2) vii. 163 In most of the S_N1 reactions that have been studied kinetically the carbonium ion is a very reactive intermediate that is rapidly transformed into the final product. **1962** [see *homocysteine].

intermediateness. (Examples.)
1854 Geo. Eliot tr. *Feuerbach's Essence Christianity* xii. 123 Thoughts of intermediateness and dependence. **1909** W. R. Sorley *Interpretation of Evolution* 28 The characteristics of life, indeed,..have a certain appearance of intermediateness. **1958** P. A. M. Dirac *Princ. Quantum Mech.* (ed. 4) i. 13 The probability of a particular result for the state formed by superposition is not always intermediate between those for the original states..so there are restrictions on the 'intermediateness' of a state formed by superposition.

intermedin (intəɹmī·din). *Physiol.* Also **-ine.** [a. G. *intermedin* (Zondek & Krohn 1932, in *Klin. Wochenschr.* 5 Mar. 406/1), f. mod.L. (*pars*) *intermed*(*ia* the intermediate part (of the hypophysis) + -in¹.] = **melanocyte-stimulating hormone*.
1932 *Q. Cumulative Index Medicus* XI. 900/1 Red coloring of European minnow as test of intermedin from pars intermedia. **1948** G. H. Parker *Animal Colour Changes* vi. 194 The intermedine system in amphibians is assumed to operate in the following manner. **1963** M. Fingerman *Control Chromatophores* vi. 103 Two general types of intermedin have been found. **1964** E. J. W. Barrington in Pincus & Thimann *Hormones* IV. v. 327 The particular hormone involved, formerly known as intermedin, or B-substance, but now more usually termed melanocyte-stimulating hormone, or MSH.

intermedio-lateral (intəɹmī:diolæ·təɹăl), *a.* *Anat.* Also **intermediolateral.** [ad. mod.L. *intermedio-lateralis*, f. mod.L. *intermedio-*, comb. form of L. *intermedius* intermediate + L. *lateralis* lateral.] Both intermediate and lateral: applied *spec.* to the tract of nerve cells which constitutes the lateral grey column of the spinal cord.
[**1859** J. L. Clarke in *Phil. Trans. R. Soc.* CXLIX. 446 This tract..was first pointed out by myself in 1851. I shall call it, on account of its position, the *tractus intermedio-lateralis*.] **1875** *Encycl. Brit.* I. 866/2 Lockhart Clarke has described an intermedio-lateral group of nerve cells. **1906** *Fifth Rep. Carnegie Trust Scotland* 20 Pathology of the intermedio-lateral tract of the spinal cord. **1972** J. Minckler *Introd. Neuroscience* xxii. 382 The preganglionic fibers arise in the intermediolateral column of cells of the ventral horn from segments T1 to L2.

intermesh (intəɹme·ʃ), *v.* [f. Inter- 1 b + Mesh *v.* 3 b.] *intr.* Of gears, etc.: to mesh or interlock with one another. Also *fig.*
1909 in Webster. **1928** *Daily Tel.* 27 Mar. 7 (Advt.), The light yarns are vibrated to cause them to intermesh in the fabric. **1948** M. J. Herskovits *Man & his Works* xiv. 215 To achieve some expression of the unities in culture by indicating how trait and complex and pattern..intermesh, as the gears of some machine to constitute a smoothly running, effectively functioning whole. **1955** J. G. Davis *Dict. Dairying* (ed. 2) 153 The cylinders have square-cut threads which intermesh. **1957** G. E. Hutchinson *Treat. Limnol.* I. iii. 201 These chains form an intermeshed net and run in any direction in the free liquid. **1971** *Flying* Apr. 92/3 Medical records..are not yet intermeshed with airman records. **1971** T. F. Mitchell in *Archivum Linguisticum* II. 39 It should perhaps be said again that the aspects of meaning we are distinguishing intermesh and often meet in one text, in one sentence, even in one word or syllable.

intermetallic (intəɹmitæ·lik), *a.* Also inter-metallic (with hyphen). [f. Inter- 4 c + Metallic *a.* and *sb.*] Applied to compounds formed from two or more metals. Hence as *sb.*, an intermetallic compound.
1900 *Rep. Brit. Assoc. Adv. Sci. 1900* 131 Most students of alloys are now convinced that they often contain definite chemical compounds, yet these 'intermetallic' compounds are still passed over in silence by..books on descriptive chemistry. **1923** U. R. Evans *Metals & Metallic Compounds* I. 198 Inter-metallic compounds are rarely—if ever—formed between two metals belonging to the same group of the periodic table. **1937** *Chem. & Industry* XV. 677/1 A consideration of the formulæ of typical intermetallic compounds..shows at once that their formation is not governed by the simple rules of valency. **1956** *Nature* 10 Mar. 487/1 Its analysis is complicated by the presence of..intermetallics. **1962** Simpson & Richards *Junction Transistors* iii. 36 E.g. InP and GaAs among the intermetallic compounds. **1967** *New Scientist* 27 Apr. 209/2 Intermetallics are compounds of two or more metals, and are commonly found in alloys. **1972** *Sci. Amer.* Aug. 46/3 They have successfully synthesized intermetallic compounds in which the hydrogen is held in the form of hydrides.

intermezzo. **1.** (Earlier example.)
1771 [see *bad *a.* 1 c].

inter-micellar: see *inter- *pref.* 4 a.

interministerial, *a.* Add: **2.** Involving the participation of two or more ministers or of the representatives of two or more departments of state; constituted from members of different departments of state.
1937 *Times* 30 Dec. 10/1 After a day of almost incessant inter-Ministerial consultation..the Government decided to take a firm line with the strikers. **1967** *Economist* 29 Apr. 477/1 A secret report on the labour market by the commissioner-general for national planning, M. Francois-Xavier Ortoli, is to be discussed by a special interministerial committee within the next 10 days. **1970** *Nature* 8 Aug. 544/2 In most countries..there is an inter-ministerial committee for science and technology.

intermission¹. Add: **2.** *spec.* = the interval between the parts of a play, film, concert, etc. Chiefly *U.S.*
1927 *N.Y. World* 24 July, *Intermission*, interval. **1933** H. Footner *Ring of Eyes* xv. 104 When the curtain descended for the intermission, [etc.]. **1955** Keepnews & Grauer *P. . Hist. Jazz* xvi. 197 Nick Rongetti..loved to join the intermission pianist. **1956** B. Holiday *Lady sings Blues* (1973) iii. 37 In between ups,..there was Garland Wilson at the piano for intermission. **1961** *Listener* 7 Dec. 1002/2 *Rocco* does have an intermission, but when a film is [etc.].

intermitotic (intəɹmaitǫ·tik), *a.* *Cytology.* Also inter-mitotic. [f. Inter- 4 b + Mitotic *a.*] Occurring or existing between mitoses; capable of dividing again. Also *absol.*, a cell which is capable of dividing again.
1942 E. V. Cowdry *Probl. Ageing* (ed. 2) xxiv. 627 The daughter epidermal cells that retain a basal location..are comparatively undifferentiated and may be called vegetative intermitotics. *Ibid.*, The cells that are formed from vegetative intermitotics, and begin to differentiate, may also be intermitotic with lives extending from one mitosis to the next. Their lives are, however, different insofar that they exhibit increasing specialization. Consequently they may be designated differentiating intermitotics. **1948** W. Andrew tr. *E. D. P. de Robertis's Gen. Cytol.* xii. 324 Cowdry has classified cells into four groups on the basis of their degree of differentiation and ability to divide. The first group, vegetative intermitotic cells, comprises those undifferentiated cells which have the capacity to multiply continually. **1962** L. von Sallmann et al. in A. Pirie *Lens Metabolism Rel. Cataract* 449 The intermitotic, or turnover, time was estimated as 19 days for the equatorial zone..of the lens epithelium of young rats. **1964** G. H. Haggis et al. *Introd. Molecular Biol.* i. 13 The appearance of the inter-mitotic nucleus in living cells. **1968** M. W. Strickberger *Genetics* ii. 13 As compared to the period of active mitosis..the interphase or 'intermitotic' period of most cells is usually many times longer.

intermittency. Add: **2.** Special Comb.: **intermittency effect** *Photogr.*, the difference in the density of an emulsion when it is exposed intermittently from that resulting from a continuous exposure (the total light received being the same in both cases).
1907 Sheppard & Mees *Investigations Theory Photogr. Process* ii. vi. 223 The intermittency effects are all in good agreement with this view, since the reverse reaction would have the greater effect, (*a*) the longer the pause, and (*b*) the more numerous the pauses, a pause being the period of no illumination. **1958** H. Baines *Sci. of Photogr.* xiv. 172 Another phenomenon now known to be closely connected with reciprocity failure is the intermittency effect.

intermittent, *a.* (*sb.*). Add: **A.** *adj.* **1. a.** *intermittent claudication*: see *claudication.
b. *intermittent movement* (see quot. 1959); *intermittent sterilization*, a microbiological procedure which accomplishes sterilization without using the high temperatures required to kill spores outright, and which involves alternately maintaining the materials to be sterilized at a temperature high enough to kill vegetative cells and at a much lower temperature during which germination of spores occurs (producing new vegetative cells to be killed during the next high-temperature period).
1893 tr. *W. Migula's Introd. Pract. Bacteriol.* ii. 41 The test-tubes containing the blood serum may be now subjected to 'fractional or intermittent sterilisation', by exposing them for an hour a day for eight days to a temperature of 58°C. **1959** W. S. Sharps *Dict. Cinematogr.* 104/1 *Intermittent movement*, the term used for the method of film transport in a camera, projector or printer, whereby the film is moved intermittently and only exposed to light when stationary. **1969** S. T. Lyles *Biol. Microorganisms* v. 130 Intermittent sterilization may be accomplished by boiling or steaming in the autoclave at O pressure.

intermodal (intəɹmōᵘ·dăl), *a.* Also inter-modal (with hyphen). [f. Inter- + Mod(e *sb.* 4 + -al.] Of the conveyance of goods: making use of differing modes of transport during the journey between the place of dispatch and the destination. Hence **intermo·dally** *adv.*
1963 A. H. Norling *Future U.S. Transportation Needs* (NASA Rep. N-64-25006) VIII-4 The elimination of the wasteful, duplicative handling now required in the intermodal transfer of goods will call for the increased use of containerized shipments. **1967** *Containerization* (McKinsey & Co.) 27 Low-cost intermodal transfer makes it economic to switch modes to take advantage of the lowest cost alternative. **1968** *Economist* 14 Sept. p. xxxiv, The sea-going container will fit nicely into the big aircraft, and hey presto we have real 'intermodal' transport. **1969** *Jane's Freight Containers 1968–69* 3/2 This subject is part of the legal issues arising in connection with combined or inter-modal transport in general. **1970** *Commercial Motor* 25 Sept. 133/1 The exhibition is by no means limited to container traffic and equipment, but will deal with all the intermodal systems currently available. **1970** *Times* 16 Sept. (Road Haulage Suppl.) p. vii/8 By the mid 1970s, more than 1 m. containers will be circulating intermodally on an international basis. **1971** Johnson & Garnett *Econ. of Containerisation* i. 11 It [*sc.* containerization] has been standardised, making it intermodal: i.e. a container can be carried by almost any mode of transport and easily transhipped between modes. **1972** *Times* 29 Sept. 20/6 The association gives a warning that attempts to take over small ports are likely to result in 'some quite extraordinary distortions of the inter-modal transport patterns'.

intermodulation (i:ntəɹmǫdiulēˈ·ʃən). *Electr. Engin.* [f. Inter- 2 a + Modulation.] Modulation of the sinusoidal components of a signal or wave-form by one another, a kind of distortion that produces new components with frequencies equal to the sums and differences of those already present. Freq. *attrib.*, esp. in *intermodulation distortion*.
1931 *Inst. Radio Engin. Year Bk.* 50 Intermodulation is the production, in a non-linear circuit element, of frequencies corresponding to the sums and differences of the fundamentals and harmonics of two or more frequencies which are transmitted to that element. **1940** *Chambers's Techn. Dict.* 454/1 Intermodulation distortion. **1950** K. Henney *Radio Engin. Handbk.* (ed. 4) xvii. 854 These frequencies resulting from intermodulation are generally inharmonic or discordant and also likely to fall in the middle or upper portion of the audio range, both of which conditions make them quite objectionable. **1958** *Engineering* 31 Jan. 160/2 If two musical tones are applied, both appear at the output with all their combination tones (intermodulation products). **1964** R. F. Ficchi *Electr. Interference* iii. 24 The mixing of at least two signals in a nonlinear device which causes intermodulation distortion and results in the generation of harmonics should not be permitted. **1971** *Sci. Amer.* Sept. 178/1 (Advt.), That's 250 watts..over the whole power-pushin' listening range. Right through from 20 Hz to 20,000 kHz [*sic*] with total harmonic and intermodulation distortion at less than 0·1%!

intermolecular, *a.* Add: Hence **intermole·cularly** *adv.*, between molecules.
1936, 1971 [see *intramolecular *a.*]. **1973** *Nature* 13 Apr. 473/2 In a condensed phase, initially formed vibrational excited 'states' are coupled intra- and intermolecularly..to a quasi-continuum of other vibrational 'states'.

intermontane (intəɹmǫ·ntēˈn), *a.* [f. Inter- 4 a + Montane *a.*] Situated or lying between mountains. Also **i·ntermont, intermou·ntain** *adjs.*
1807 J. Mease *Geol. Acct. U.S.* 59 Whatever of saline..the soil of the upland contains, is thus floated or rolled along to the low lands, and constitutes with proportional diversity and mixture, the intermontane soil. **1828** Webster, Intermontane. **1900** *Congress. Rec.* 24 Jan. 1147/2 Polygamy..was taught by the Mormon Church throughout the whole of the intermountain country as a religious duty. **1901** *Science* 8 Mar. 396/1 The erosion of narrow valleys in the floor of the weak-rock intermont peneplains. **1902** *Encycl. Brit.* XXV. 362/2 The intermontane basins and the piedmontese plains that slope eastward from the Rocky Mountains in middle latitudes are treeless. **1903** *Science* 12 June 950/1 The study of the anthropology of the Indian races in this intermountain region. **1911** J. L. Myres *Dawn of History* iv. 91 We are..concerned..with the intermont plains and upland valleys which sustained the old Medes and Persians. **1920** L. V. Pirsson *Textbk. Geol.* (ed. 2) I. 269 Intermontane Plateaus. **1927** *Glasgow Herald* 21 May 8 This fungus..attacks both the green or Oregon and the blue or Colorado Douglas [fir], as well as the intermountain variety. **1936** *Geogr. Jrnl.* LXXXVII. 27 Coarse detritus which came to rest in the intermont basins. **1959** Wooldridge & Morgan *Outl. Geomorphol.* (ed. 2) xx. 285 Plains are here in a much less dominant rôle; we have an intermont-basin landscape as compared with a true Inselberge landscape. **1962** G. MacEwan *Blazing Old Cattle Trail* xxv. 167 Ranching of western Canada's prairie, foothill and intermountain ranges was just two or three decades old. **1965** G. J. Williams *Econ. Geol. N.Z.* xviii. 297/1 The Tertiary-filled intermontane basin is 50 miles long and averages 15 miles in width.

inter-morainic, -mutule: see *inter- *pref.* 4 a, 3 a.

intern. **B.** *sb.* For *U.S.* read Chiefly *U.S.* Also **interne.** Now usu., a recent medical graduate who is working under supervision in a hospital (and often living there) as part of his training, prior to entering general practice or becoming a resident. (Broadly equivalent to a houseman in Great Britain.) Also *transf.*, used of individuals in other professions (esp. teaching) who are receiving practical experience under supervision. (Examples.)
Freq. pronounced with stress on first syllable.
[**1699** M. Lister *Journey to Paris* (ed. 3) 74 Monsieur du Pes Surgeon Interne of the Hôtel-dieu.] **1879** in Webster Suppl. **1889** *Kansas City* (Missouri) *Times & Star* 16 Oct., Convalescent women and young nurses are given too much freedom with the internes. **1890** W. James *Princ. Psychol.* I. x. 385 His brother, Dr. Jules Janet..was *interne* at the Salpêtrière Hospital. **1914** M. Gerry *Masks of Love* 123 The young interne..had alarmed them. **1938** *Amer. Speech* XIII. 228/1 *Resident* ranks over *intern* by reason of experience, but both are medical school graduates studying in the hospital. **1955** W. Gaddis *Recognitions* i. i. 41 Physicians, technicians, and internes X-rayed the boy from every possible angle. **1961** J. Heller *Catch-22* (1962) xviii. 175 In the morning a helpful young English intern popped in to ask him about his liver. **1963** A. Beales *Educ. under Penalty* i. ii. 18 In the field of lay education the loss caused by the dissolution was less than formerly supposed, since the monasteries had long ceased, most of them, to teach any but interns. **1967** Mrs. L. B. Johnson *White House Diary* 14 Mar. (1970) 497 One intern was giving remedial reading. **1969** *Eugene* (Oregon) *Register-Guard* 3 Dec. 1A/3 Representing Springfield teachers are Rita Castleberry, Lee Elementary School instructor; Roy Van Horn, principal at Mt. Vernon Elementary School, and Moore, who is intern teacher supervisor at Lee Elementary School. **1970** *Toronto Daily Star* 24 Sept. 31/6 (Advt.), Broderick Crawford..gives young intern Steve Brooks some food for thought. **1972** *Nature* 4 Feb. 291/2 Doctor of medicine and intern of the hospitals of Lyon, in 1908 Professor Lacassagne entered the Laboratory of Histology to work under the direction of Claudius Regaud.
Hence **inte·rnship, interneship,** the position or station of an intern; the period of such a position. Chiefly *U.S.*
1904 *Jrnl. Amer. Med. Assoc.* 13 Aug. 469/2 From one to two years of what might be called practical apprenticeship is the privilege sought by the earnest [medical] student. To gain this privilege he..delays for the period of his interneship his start in the world and his establishment

in his profession. 1924 *Scribner's Mag.* Feb. 183/1 My father, who had the practice..told me the story in professional confidence... I was at that time just finishing my internship. 1934 A. Woollcott *While Rome Burns* 46 The fiery young doctor, in the days of his interneship, had already tasted the experience of spending two months in the lockup. 1938 *Internships & Residences in N.Y. City, 1934–1937* p. xxvii, An internship is a period of service as a member of the hospital staff while residing in the hospital and receiving a period of formal education subsequent to graduation from medical school. 1948 *Training of a Doctor* (B.M.A.) xxxii. 127 This prescribed period of compulsory pre-registration house-appointments corresponds to the post known in America as an 'internship', and the period is called 'the intern year'. 1957 Kendall & Selvin in R. K. Merton *Student-Physician* 155 Nearly every medical student is required to take an internship when he has completed medical school, regardless of his plans for his later career. 1968 *Globe & Mail* (Toronto) 17 Feb. 50 (Advt.), The internship plan for the training of elementary school teachers. 1968 *New Scientist* 14 Nov. 388/2 One suggestion is very similar to the principle of our sandwich systems, under the more anaesthetised name of an 'internship'. 1971 *Black Scholar* June 62/1 (Advt.), An individually tailored, accelerated career program which begins with a two-year internship and leads to positions of increasing responsibility in the developing countries. 1971 *Nature* 1 Oct. 301/1 The money will be used to establish internships in federally funded laboratories for some 420 young unemployed scientists and engineers who hold advanced degrees.

intern, *v.* Add: **4.** (Usu. pronounced with stress on first syllable.) *intr.* To act as an intern. *U.S.*

1933 S. Kingsley *Men in White* I. i. 24 You interned here? 1969 *Eugene* (Oregon) *Register-Guard* 3 Dec. 5D/1 He..interned at Cook County Hospital in Chicago for one year. 1971 'D. Shannon' *Murder with Love* (1972) v. 83 Harlow interned at the General... He had the makings of a very fine surgeon.

internal, *a.* and *sb.* Add: **A.** *adj.* **1. a.** *internal object*: in *Psychoanalysis*, the inward image formed of an object invested with the emotional energy which would normally have been expended on the object itself.

1940 *Internat. Jrnl. Psycho-Anal.* XXI. 280 (*title*) Temper tantrums in early childhood in their relation to internal objects. 1949 *Brit. Jrnl. Med. Psychol.* XXII. 13 Thus we see the internal object represents also the child himself. 1973 H. Segal *Introd. Work M. Klein* (new ed.) ii. 12 These internal objects are not 'objects' situated in the body or the psyche: like Freud, Melanie Klein is describing unconscious phantasies which people have about what they contain.

2. a. *spec.* in *Philos.*, as *internal property, relation*: a property or relation which belongs essentially to an object or proposition.

1883 F. H. Bradley *Princ. Logic* 432 As the material supplied is in each case different, so again the product is not the same. In one case the whole precedes and is followed by its internal relations. 1920 G. E. Moore in *Proc. Aristotelian Soc.* XX. 40 (*title*) External and internal relations. *Ibid.* 41 And I shall maintain that, if we give to the assertion that a relation is 'internal' the meaning which this proposition would give to it, then..*some* relations are 'internal'. 1922 tr. *Wittgenstein's Tractatus* 4. 122 Instead of property of the structure I also say 'internal property'; instead of relation of structures 'internal relation'. 1937 A. Smeaton tr. *Carnap's Logical Syntax of Lang.* v. § 79. 304 The numerous discussions and controversies about external and internal properties and relations. 1959 R. Wollheim *F. H. Bradley* iii. 92 The most important of them all, the doctrine of Internal Relations. 1966 *Amer. Philos. Q.* III. 45/1 This forces one to acknowledge that a thing changes when what it is related to changes. In short one is involved with *internal relations*.

b. *internal revenue*: revenue derived from duties and taxes imposed on domestic trade and commerce; inland revenue. *U.S.*

1796 *Deb. Congress U.S.* 26 Feb. 379 Mr. W. Smith moved the order of the day on the report of the Committee of Ways and Means on the Internal Revenue. 1873 *Newton Kansan* 2 Jan. 2/1 The most important bills ..are those reducing the internal revenue force. 1959 *Chambers's Encycl.* VII. 417/1 In the United States collection is made by the Bureau of Internal Revenue. 1972 *Lebende Sprachen* XVII. 34/1 US internal revenue—BE inland revenue.

5. Special collocations: **internal clock**, a person's innate sense of time; = **biological clock*; **internal-combustion**, used *attrib.* to designate any engine in which combustion of the fuel takes place inside it in the chamber where the force is developed (or a part continuous with it); also *fig.*; **internal conversion** *Physics*, (*a*) the process whereby the whole energy of a gamma-ray photon emitted by a nucleus is given up to an orbital electron, causing its emission from the atom; (*b*) (see quot. 1972²); **internal energy**, the energy possessed by a physical system in consequence of the positions and relative motions and interactions of its component parts: a function of its state (usu. of undefined absolute magnitude) such that any change in the function is equal to the sum of the heat absorbed by the system and the work done on it; **internal friction**, resistance to the deformation or flow of a substance that occurs inside it with the production of heat, and arises from the relative motion of adjacent parts; **internal medicine** = Medicine *sb.*¹ 1 (in the 'more restricted sense'); **internal phloem** (see quot. 1933); **internal pressure**, the pressure which exists within a fluid arising from intermolecular attraction; **internal secretion** *Physiol.* [tr. F. *sécrétion interne* (C. Bernard *Leçons sur les Propriétés physiol. des Liquides de l'Organisme* (1859) II. xvii. 408)], any secretion that is delivered into the internal environment of the body, esp. into the blood stream; *spec.* a hormone; also, the process of secreting in this manner; **internal stress**, stress that arises inside a substance (e.g. as a result of differential heating of different parts) and is not imposed from without; **internal wave**, any transverse wave occurring within a fluid either along the interface between layers of different density or within a layer having a vertical density gradient.

1960 I. Jefferies *Dignity & Purity* vi. 123, I suppose he works according to his own internal clock. 1974 *Guardian* 28 Jan. 8/3 If a programme is on too late for our internal clocks, we would almost rather it were not shown at all. 1884 H. C. F. Jenkin in *Heat in its Mech. Applications* (Inst. Civil Engin.) (1885) 105 The internal combustion engine..is really the fore-runner of the gas-engine. 1921 W. H. Berry *Mod. Motor Car Pract.* i. 14 In effect liquid fuel made the internal combustion motor practicable for locomotive purposes, its essential qualities being that it can be easily stored and carried, and be readily converted into an explosive gas ignitable within the cylinder. 1949 D. G. Shepherd *Introd. Gas Turbine* i. 1 The gas turbine, in common with other forms of internal combustion engines, converts heat into work by a cycle using a gas as the working medium, the processes being compression, addition of heat and expansion, and requiring continuous flow of the gas during these changes of state. 1951 M. McLuhan *Mech. Bride* 113/2 What Kipling was to the aggressive British imperialists, these ads are to our domestic economy. They act as a sort of firing spark in the internal combustion engine. 1967 E. Delmar-Morgan *Maintenance Inboard Engines* i. 19 Within the broad category of the internal-combustion engine there are the spark-ignition engine (petrol and paraffin) and the compression ignition (diesel). 1927 *Proc. R. Soc.* A. CXVI. 491 (*heading*) The internal conversion of γ-rays. 1927 *Proc. Cambr. Philos. Soc.* 718 The homogeneous groups in the β-ray spectra are due to the conversion of the γ-rays, and the intensities of these groups depend not only on the intensities of these γ-rays but also on the extent to which they are converted, that is the magnitude of the internal conversion coefficient. 1949 P. Pringsheim *Fluorescence & Phosphorescence* iii. 272 Since no light emission accompanies the transition from the higher electronic state to the emitting state, the transition must correspond to a process of internal conversion. 1955 R. D. Evans *Atomic Nucleus* i. 23 The second general class of nuclear transitions which invariably result in X-ray-emission spectra is the internal-conversion transitions... Internal conversion..often predominates over γ-ray emission if the nuclear excitation energy is small and the angular-momentum change is large. 1959 *Q. Rev.* XIII. 5 The strong fluorescence indicates that internal conversion does not take place as often as would be expected. 1972 H. A. Enge et al. *Introd. Atomic Physics* xii. 386 Internal conversion competes with gamma emission, and the ratio of the probabilities of the two processes depends strongly upon the multipolarity. 1972 C. H. J. Wells *Introd. Molecular Photochem.* iii. 34 Internal conversion (ic) is the term given to the radiationless process whereby a molecule transfers from one electronic state to another electronic state of the same multiplicity. 1887 *Encycl. Brit.* XXII. 480/1 Since the gas had neither gained nor lost heat, and had done no work, its internal energy was the same at the end as at the beginning of the experiment. 1921 A. W. Judge *Automobile & Aircraft Engines* ii. 60 The internal energy of a given quantity of gas depends only upon its temperature. 1927 H. S. Taylor *Elem. Physical Chem.* ii. 64 The internal energies of the gas in the two states may include the energy resultant from such factors as motion and position of the molecules, molecular attraction, intramolecular forces, intra-atomic vibrations, chemical and other unknown forces. We therefore note that the absolute magnitudes..are not ascertainable. The change of internal energy, ΔU, is definite and measurable. 1966 *McGraw-Hill Encycl. Sci. & Technol.* VII. 209/1 The change in internal energy is fixed by the initial and final states, and is independent of the path by which the change in state is accomplished. 1972 A. L. Ruoff *Introd. Materials Sci.* iv. 159 The change in internal energy associated with stress, electric and magnetic fields is due to the work done on the material. 1860 *Phil. Mag.* XIX. 20 The internal friction of gases. 1875 *Encycl. Brit.* III. 39/2 The compressed gas expanded to twice its volume, and the work of expansion..was soon converted into heat by the internal friction of the gas. 1922 Glazebrook *Dict. Appl. Physics* I. 351/1 The characteristics of the internal friction of fluids when the general motion is eddying or turbulent. 1931 [see *internal wave* below]. 1958 *Jrnl. Iron & Steel Inst.* CXC. 93/2 Stress relaxation across grain boundaries and phase interfaces, an important source of internal friction, is considered. [1894 Gould *Dict. Med.* 735/1 On the European continent it is customary to divide medicine into internal and external, the former implying the restricted sense of the term, or the study of diseases of internal organs.] 1904 Stedman *Dunglison's Dict. Med. Sci.* (ed. 23) 583/2 *I*[*nternal*] *medicine*, that branch of medicine which has to do with diseases of the body not amenable to operative treatment; medicine as contrasted with surgery. 1968 Talso & Remenchik (*title*) Internal medicine, based on mechanisms of disease. 1933 *Tropical Woods* XXXVI. 3 Internal Phloem.—Primary phloem internal to the primary xylem. (To replace *Intraxylary Phloem.*) 1953 K. Esau *Plant Anat.* xii. 268 In angiosperms the internal phloem is initiated somewhat later than the external. 1967 S. Broido-Altman tr. *Fahn's Plant Anat.* xi. 167 The internal phloem may be present as separate strands on the border of the pith. 1911 *Trans. Faraday Soc.* VII. 94 It is therefore immaterial..whether we fix our attention on the internal pressure or on the thermal pressure, which acts in the opposite sense. 1940 Glasstone *Text-bk. Physical Chem.* vii. 472 The internal pressure of naphthalene is about 3600 atm., so that the forces of cohesion in liquids are evidently very large. 1973 A. W. Adamson *Textbk. Physical Chem.* iv. 151 The internal pressure is zero for an ideal gas and for most real gases it is small compared to *P* [*sc.* the measured pressure]. 1895 *Brit. Med. Jrnl.* 10 Aug. 341/1 On the other hand, some secreted materials are not poured out upon an external surface at all, but are returned to the blood. These may be termed internal secretions. 1921 I. G. Cobb *Organs Internal Secretion* (ed. 3) i. 19 The term hormone is in more or less general use to denote an internal secretion. 1924 A. Lipschütz *Internal Secretions Sex Glands* p. vi, Prof. Starling suggested the term 'hormone' for the active principles of those internal secretions which act as chemical messengers. 1926 J. S. Huxley *Ess. Pop. Sci.* 203 Claude Bernard introduced physiology to the general idea of internal secretion. 1941 R. G. Hoskins *Endocrinol.* i. 20 Usually, however, formal credit for first demonstrating the process of internal secretion is given to Claude Bernard, who in 1848 reported the discovery that sugar stored in the liver in the form of glycogen is discharged as dextrose directly into the blood rather than through the ducts of the organ after the fashion of an ordinary secreting gland. The term 'internal secretion' is no longer applied to substances like sugar, but is restricted to those having a more specific regulatory function. 1966 W. S. Hoar *Gen. & Compar. Physiol.* ii. 34 As a matter of fact, the first internal secretion conclusively demonstrated was the hormone secretion produced by the wall of the gut. 1904 Goodchild & Tweney *Technol. & Sci. Dict.* 313/1 Internal stress. 1906 *Jrnl. Iron & Steel Inst.* LXXII. 608 (*heading*) Internal stresses and strains in iron and steel. 1923 Glazebrook *Dict. Appl. Physics* V. 344/1 Carpenter and Edwards argue that in quenching steel these internal stresses cause internal straining of the metal. 1950 *Jrnl. Iron & Steel Inst.* CLXIV. 166/2 In a partially transformed steel, the volume changes accompanying transformation presumably set up a system of internal stresses. 1931 *Rapports et Proc.-Verb. des Réunions Conseil Perm. Internat. Explor. Mer* LXXVI. 5 (*heading*) On internal waves. *Ibid.* 10 Free internal waves must abate owing to internal friction and eddy-viscosity. 1966 *McGraw-Hill Encycl. Sci. & Technol.* XIV. 417/2 Internal waves have been found in the atmosphere as lee waves (waves in the wind stream down-wind from a mountain) and as waves propagated along an inversion layer. 1966 R. W. Fairbridge *Encycl. Oceanogr.* 402 Internal waves are subsurface waves found between layers of different density or within layers where vertical density gradients are present. They can exist in any stratified fluid and can be caused by flow over an irregular bottom, atmospheric disturbances, tidal forces, and shear flow.

internalization. (Later examples.)

1942 *Internat. Jrnl. Psycho-Anal.* XXIII. 8 (*title*) A contribution to the problem of sublimation and its relation to processes of internalization. 1952 Gerth & Martindale tr. *Weber's Anc. Judaism* iv. xii. 328 For Jeremiah this internalization..went hand in hand with the formation of quite modest external hopes. 1954 T. Grygier *Oppression* i. i. 8 Civilization takes certain measures to check it [*sc.* aggression]. The most important is the introjection or internalization of aggressiveness. 1964 M. Argyle *Psychol. & Social Probl.* i. 20 Such people often behave differently when away from the group—they are showing overt compliance without internalization, and we should only speak of learning taking place when there is also internalization. 1970 P. Kelvin *Bases of Social Behaviour* ix. 281 'Internalization' here refers to the process and condition whereby the norms of the group or society become the individual's own norms, whether of behaviour or of values. 1971 *Sci. Amer.* Sept. 192/3 One method of fostering the incorporation of technology assessment into their decisional processes would involve the internalization of costs. 1973 *Nature* 20 Apr. 488/1 The most economical way of controlling both pollution and congestion in cities must depend on what the economists call an internalization of the externalities —making the motorist bear not merely the costs of operating his vehicle but the costs of what he does to other people.

internalize, *v.* Add: *spec.* **a.** *Psychol.* To transfer to a subjectively formed image (the emotions connected with some object) (see also *introject *v.*); to adopt or incorporate as one's own (the values, etc., of a social group).

1942 *Internat. Jrnl. Psycho-Anal.* XXIII. 15/1 The atonement for guilt is here carried out by internalizing the attacked external object. 1950 Bettelheim & Sylvester in P. Greenacre et al. *Psychoanal. Study Child* V. 330 While their own actions are motivated by 'what will get by'..they demand that he himself be guided by 'what is right'... The child who tries to internalize these discrepant demands is confronted by a seemingly insoluble conflict. 1958 M. Argyle *Relig. Behaviour* v. 42 The suggestion that Protestants internalize the super-ego to a greater extent. 1960 Koestler *Lotus & Robot* ii. viii. 203 The chances are that his aggressive impulses.. will be internalized and deflected against himself. 1968 J. M. Ziman *Public Knowl.* v. 77 He must internalize the scientific attitude so that he cannot even conceive of, say, ..recording the epoch of an eclipse by reference to the age of the reigning monarch. 1971 *Black Scholar* June 7/1 Some [black women] even internalize white society's low regard for black men.

b. *Linguistics.* To acquire knowledge of (a

set of rules in a given language): used esp. of the language-learning procedures of a child.

1965 [implied in quots. for *internalized *ppl. a.*]. 1968 Chomsky & Halle *Sound Pattern Eng.* 3 The person who has acquired knowledge of a language has internalized a system of rules. 1971 D. Crystal *Ling.* iii. 104 We have mastered ('internalized' is a word often used here) a technique for breaking each new sentence up.

c. *Econ.* To incorporate (costs) as part of the internal cost structure, esp. 'social' costs, resulting from the use or manufacture of a particular product.

1971 *Sci. Amer.* Aug. 44/3 The first is the alteration of the ground rules and incentives under which the market operates by such devices as taxes, subsidies, and judicial actions to internalize (i.e., make explicit) social costs. *Ibid.* Sept. 194/2 A consideration of means whereby the legal system can internalize costs that power companies have been allowed to treat as external and therefore to exclude from their cost-benefit calculations.

internalized (intǝ̄·ɹnăləizd), *ppl. a.* [f. the vb.] Made internal, acquired or mastered internally. Also in other senses of the vb.

1932 A. Strachey tr. *Klein's Psycho-Anal. Children* vii. 169 The boy could not maintain this alliance with a cruel external super-ego against his id and his internalized objects. 1942 *Internat. Jrnl. Psycho-Anal.* XXIII. 9/2 The connection between her phantasies about internalized objects and her artistic productivity. 1959 *Psychiatry* XXII. 290 The successfully treated patient comes to be his own critic, using in relation to himself evaluations previously consensually validated with the therapist and hence perpetuating the relationship in an internalized form. 1964 M. Argyle *Psychol. & Social Probl.* iv. 54 There is considerable evidence that the stronger the forces which are restraining aggression—whether external fear of punishment or internalized anxiety about aggression—then the more remote the displaced aggression will be. 1965 N. Chomsky *Aspects of Theory of Syntax* 21 Few hearers may be aware..that their internalized grammar ..provides at least three structural descriptions for this sentence. 1965 H. A. Gleason *Ling. & Eng. Gram.* 476 This means strengthened competence in the student's internalized grammar of literary English. 1968 Chomsky & Halle *Sound Pattern Eng.* 3 New utterances..are formed and interpreted by the same grammar, the same internalized system of rules. 1971 *Sci. Amer.* Sept. 194/2 The company may merely pay the internalized costs (whether they are in the form of a judgment for damages, a tax or a negotiated payment), absorbing the payment or passing it on to customers.

international, *a.* (*sb.*) Add: **A.** *adj.* **1. a.** (Later examples.)

1948 *Written Statement U.K.* 11 Dec. in *I.C.J. Pleadings* I. v. 25 Does the United Nations possess international personality? 1959 R. A. Graham *Vatican Diplomacy* vii. 185 After 1870... Could the papacy..be properly regarded as a member of the international community of the law of nations? This was tantamount to asking whether the papacy had 'international personality'. 1963 T. Tullett *Inside Interpol* ii. 22 What is, in actual fact, an 'international criminal'? The definition of this type of wrongdoer is not based on any legal concept.. but simply on practical convenience. 1970 D. W. Greig *Internat. Law* iii. 73 The brief statement that an international person is an entity having the power of independent action on the international plane has the obvious advantage of including not only states, but also communities like 'protected states'. 1971 *Mod. Law Rev.* XXXIV. vi. 599 A country may be sovereign in the sense that it satisfies the international law criteria for independent statehood and is recognized as independent by the international community. *Ibid.* 613 The sovereignty of the United Kingdom as an international person has been abridged..a..formidable camel for a court to swallow. 1974 *Times* 15 Feb. 14/1 Renegotiation of our membership of the European Community..will add yet another uncertainty to international relations.

c. Applied to various units of physical quantities (now mostly obsolete) agreed upon by different nations, and in some cases forming part of the formal name of a unit, so as to distinguish it from a unit numerically similar but differently defined; *international system* (*of units*) (usu. with capital initials) [tr. F. *système international d'unités*], a system of physical units (together with a set of prefixes indicating multiplication or division by a power of ten) based on the metre, kilogramme, second, ampere, kelvin, candela, and mole as independent basic units, with each of the derived units defined in terms of these without any multiplying factor.

Quot. 1932 represents an isolated and fortuitous use of *international system of units*, unrelated to the later Système International.

1893 *Electrician* 29 Sept. 578/2 To distinguish these units as now defined from the definitions given by previous meetings or Congresses, they are denominated the 'international' ohm, the 'international' ampere, &c. 1908 *Nature* 29 Oct. 678/2 The international ohm is the resistance offered to an unvarying electric current by a column of mercury at the temperature of melting ice, 14·4521 grams in mass, of a constant cross-sectional area, and of a length of 106·300 centimetres. *Ibid.* 679/1 The difference between the ohm and the international ohm remains a matter for experiment. 1909 *Engineering* 17 Sept. 397/2 The British Committee..now have the honour to submit..that the name 'International Candle' be adopted. 1924 A. W. Smith *Electr. Measurements* (ed. 2) i. 3 In practical measurements the electron current is measured in international amperes. 1932 *Phil. Mag.* XIV. 292 The international system of [electrical] units differs but little from the practical system, and the two may be taken as identical for the present purpose. 1934 *Brit. Inst. Radiol. Year Bk.* 39 The International Unit of X-radiation shall be called the 'Röntgen'. 1937 *Discovery* Sept. 285/2 Last month, the first international conference on acoustics met in Paris... Thanks in particular to the conciliatory attitude of the delegates of Germany and the United States the 'phon' and the 'decibel' become international units. 1938 G. P. Harnwell *Princ. Electr. & Electromagn.* 602 It is evident that 1 international volt is equal to 1·00043 absolute volts, 1 international joule is equal to 1·00034 absolute or mechanical joules, etc. [1957 *Proc.-Verb. du Comité Internat. Poids et Mesures, 1956* 81 Après discussion, la Commission adopte la dénomination de 'Système International d'Unités'.] 1957 *Nature* 21 Dec. 1388/1, 1 A. becomes, by definition, equal to 10^{-10} m. exactly. It may be recalled that the value λ_R = 6438·4696 A. was adopted as the spectroscopic reference standard in 1907 and thereby became the means of defining the international angstrom—the unit that has since served for all spectroscopic measurements of wave-length. 1961 *Ibid.* 21 Jan. 196/2 Among other scientific resolutions adopted by the Conference [*sc.* the eleventh General Conference of Weights and Measures] are the following... Approval of a detailed list of the international system of units (designation *SI*) founded on six basic units (metre, kilogram, second, ampere, degree Kelvin and candela). 1967 A. J. Lissaman *Metrology* i. 5 The U.K., U.S.A., Canada, Australia, New Zealand and South Africa have agreed to adopt an international yard based upon the metre. 1971 P. Vigoureux *Units & Standards for Electromagn.* ii. 11 The hour, the day and the year, so long established and in such common use, will undoubtedly outlive the International System.

2. Special collocations: **International Brigade**, a body of volunteers, raised internationally by foreign Communist parties, although open to non-Communists, with the purpose of fighting for the Republic in the Spanish Civil War of 1936–39; also *transf.*; hence **International Brigader**; **international code**, a code of signals by which seamen of all nations can hold communication at sea; **international copyright**, the protection of literary and artistic property by international agreement, particularly the Berne Convention of 1885, which led to the foundation of the International Copyright Union, and the Universal Copyright Convention of 1952; **International Court of Justice**, a judicial court of the United Nations which replaced the Cour Permanente de Justice in 1945; **international date-line**, the date-line (see Date *sb.*² 8) in the Pacific Ocean; **international driving licence, permit**, a licence allowing the holder to drive a specified class of vehicle in foreign countries; **international Gothic**, name given to a style of Gothic art which spread across western Europe in the late 14th and early 15th centuries; also called *international style*; **International Monetary Fund**, an organization having a monetary pool on which member nations can draw, established in 1945 to promote international trade and stabilization of currencies; **international orange**, a bright orange colour, visible from a great distance; **International Phonetic Alphabet**, a set of phonetic symbols for international use, introduced in the late 19th century by the International Phonetic Association: constructed on the basis of the Roman and Greek alphabets with the addition of some special symbols and diacritical marks; **international style**, name given to a naturalistic style of twentieth-century architecture associated esp. with Walter Gropius (1883–1969) and his associates; also (*rare*) = **international Gothic*; **international unit**, (*a*) *Physics* (see sense *A. c above); (*b*) *Biol.* and *Med.*, a unit of activity or potency of sera, hormones, vitamins, etc., defined individually for each substance in terms of the activity of a standard quantity or preparation.

1937 'G. Orwell' *Let.* 9 May in *Coll. Ess.* (1968) I. 267 Owing partly to an accident I joined the POUM militia instead of the International Brigade, which..meant that I have never seen the Madrid front. 1968 K. Martin *Editor* x. 214 The International Brigade..came from every part of Europe and America. 1968 *N.Y. Rev. Books* 11 Apr. 42/3 Genuine American patriots must.. work for the downfall of American imperialism even if this means joining international brigades should the Vietnamese request them. 1972 *N.Y. Times* 3 Nov. 2/2 Well-known as a writer, Mr. Popovic studied philosophy in Paris before World War II and later fought in the International Brigades during the Spanish Civil War. 1949 Koestler *Promise & Fulfilment* vi. 57 This same argument..has served the Soviet Government as a constant pretext for refusing to admit..International Brigaders.. into Soviet territory. 1885 H. Paasch *From Keel to Truck* 161/1 Signal,..international code of signals. 1961 F. H. Burgess *Dict. Sailing* 121 *International code of signals*, the system of signals adopted for use between ships and shore stations of all nations, with simplicity of use in coding and decoding in any language. 1838 *Act* 1 & 2 *Vict.* c. 59 An act for securing to Authors, in certain Cases, the Benefit of International Copyright. 1844 *Act* 7 & 8 *Vict.* c. 12. § 1 Which Act [*sc.* 1 & 2 Vict. c. 59] is herein-after, for the sake of Perspicuity, designated as 'the International Copyright Act'. 1894 in H. Paasch *From Keel to Truck* (ed. 2) (title-page), Protected by International Copyright. 1901 *Empire Rev.* May 364 A book is subject to national copyright..and also to international copyright in the several countries admitting that right. 1942 *Times Lit. Suppl.* 14 Nov. 559/3 The last International Copyright Convention, signed at Berne in 1908, afforded a very considerable degree of protection to the literary and artistic property belonging to the nationals of those countries which signed it. 1971 E. P. Skone James *Copinger & Skone James on Copyright* (ed. 11) xxiv. 435 International copyright is concerned with treaties or conventions between nations requiring their signatories to respect, in their own countries, the copyright of nationals of other signatories. 1945 *Times* 10 Sept. 2/5 The nomination of candidates to the posts of Judge of the International Court of Justice. 1972 *Mod. Law Rev.* XXXV. i. 52 In the North Sea Continental Shelf Cases the International Court of Justice had the opportunity of considering this phenomenon. 1973 *Times* 15 May 6/4 Mr van der Stoel is in favour of increasing the powers of the International Court of Justice at The Hague so that countries which accept its jurisdiction in principle cannot, as Iceland has done recently and France now threatens to do, disregard the court's competence to judge a case which may turn out wrongly for them. 1910 *Encycl. Brit.* II. 134/1 The so-called 'International Date Line'..is..practically only due to American initiative. 1957 Von Engeln & Netschert *Gen. Geogr.* iv. 71/2 The international date line does not coincide with 180° meridian..but zigzags to give portions of the same political territories the same date. 1966 Mrs. L. B. Johnson *White House Diary* 2 Nov. (1970) 453 Sometime in the dark hours over the Pacific we had crossed the International Dateline. 1931 *Automobile Assoc. Foreign Touring Guide* 19 The International Driving Permit acts as a driving licence in all the countries recognising the International Certificate for Motor Vehicles. 1966 R. E. Pickering *Himself Again* x. 67 Do you have an international driving licence? 1966 A. A. Randall *Flashpoint* ii. 41 The two men book a passage on the Car Sleeper Express from Boulogne to Narbonne... You bring the Car Temporary Exportation Certificate, International Driving Permits and Insurance Green Card. 1951 M. Davies *Nat. Gallery Catal. Earlier Italian Schools* 280 They [*sc.* works by Master of the Bambino Vispo] are somewhat related to Lorenzo Monaco in style, and are even more markedly in the 'International Gothic' current. 1959 P. & L. Murray *Dict. Art & Artists* 199 Masolino may have worked under Ghiberti on the First Baptistry Doors..and this would explain his normal International Gothic style. 1961 H. E. Scott tr. *Chiarelli's European Painting 15th Cent.* 10 In international Gothic painting there is a contrast between a growing interest in naturalism..and..a transference of natural forms into abstract figures which are frankly decorative and heraldic. 1970 *Oxf. Compan. Art* 584/1 International Gothic..first distinguished by Louis Courajod in 1892. *Ibid.* 584/2 International Gothic was nurtured in the cosmopolitan courts of France and Burgundy from the soil of the French aristocratic court style. 1944 H. Morgenthau in *U.N. Monetary & Financial Conference* p. iii, The International Monetary Fund agreed upon at Bretton Woods will help remedy this situation [*sc.* economic tactics which contribute to world-wide depression and war]. 1955 *Times* 4 July 13/4 The International Monetary Fund has agreed to a change in the par value of the Nicaraguan cordoba from five to seven a United States dollar. 1957 *Economist* 21 Dec. 1075/2 The British Government this week decided to extend for 12 months its $738,530,000 standby credit with the International Monetary Fund. 1958 *Colour Index* (Soc. Dyers & Colourists) (ed. 2) IV. 4323 International Orange 2221... C.I. Pigment Orange 21. 1967 'G. Carr' *Lewker in Tirol* vi. 85 The cagoule is in the colour we call 'international orange'... It is the colour that can best be seen from a distance, so that climbers in difficulties can quickly be found and rescued. 1969 *New Yorker* 12 Apr. 110/3 The dial is coated with a luminous orange paint called International Orange, a psychedelic color that is glaringly visible against the milky-white experiment. 1898 *Mod. Q. Lang. & Lit.* I. i. 69/2 It is of the utmost importance that, for the more advanced students of languages, we should adopt an international phonetic alphabet. 1912 *Princ. Internat. Phonetic Assoc.* 1 In 1888, after consulting the opinion of its members, the Association drew up an *International Phonetic Alphabet*, by means of which the pronunciation of any language may be accurately represented. 1962 A. C. Gimson *Introd. Pronunc. Eng.* iv. 35 The International Phonetic Alphabet..provides symbols to denote the sound types occurring in languages. 1964 J. Downing *Initial Teaching Alphabet* p. ix, Some educational reformers..favour an alternative type of 'rational orthography' or 'systematized notation' such as the International Phonetic Alphabet. 1932 Hitchcock & Johnson *Internat. Style* i. 20 There is, first, a new conception of architecture as volume rather than mass. Secondly, regularity rather than axial symmetry serves as the chief means of ordering design. These two principles, with a third proscribing arbitrary applied decoration, mark the productions of the international style. 1937 *Archit. Rev.* LXXXI. 133/2 This natural phenomenon, specifically variegated in each of these regions, constitutes a severe impediment in considering the rational mode of design as quickly resulting in a so-called 'International Style'. 1937 *Time* 8 Feb. 32/1 Walter Gropius, one of the founders of the concrete-pipe-and-plate-glass school of architectural modernism known as the 'International Style'. 1950 E. H. Gombrich *Story of Art* xiii. 178 Until round about 1400, art in different parts of Europe had developed on similar lines... The style of the Gothic painters and sculptors of that period is known as the International Style. 1922 *Internat. Conf. Standardisation of Sera & Serological Tests 1921* (League of Nations) 6 The Sub-Committee for investigating methods for testing the potency of anti-diphtheritic and anti-tetanic sera considers it both possible and

desirable to fix for both these sera an antitoxin unit which could be generally accepted and acknowledged as an International Unit. 1944 L. E. H. Whitby *Med. Bacteriol.* (ed. 4) xxv. 234 Staphylococcus: an 'international unit' is the specific antitoxic activity contained in 0·5 mg. of dry standard serum preserved at the National Institute of Medical Research, London. 1961 *Brit. Med. Dict.* 1500/2 *International unit of male hormone*, the specific activity contained in 0·1 milligram of pure androsterone. ..*International unit of vitamin C*, the antiscorbutic activity contained in 0·05 mg. of pure ascorbic acid. 1970 *Daily Tel.* 12 Oct. 3/2 The Medical Research Council now requires that all influenza vaccines should contain 600 of the new international units of virus to each human dose.

B. *sb.* **a.** (Later examples.) Also, an international contest.

1890 *North British Daily Mail* 7 Apr. 6/6 Their goal was again and again besieged in a way never before seen at the close of an international. 1905 A. Conan Doyle *Return of Sherlock Holmes* 330 The defeat of the Light Blues may be entirely attributed to the unfortunate absence of the crack International, Godfrey Staunton, whose want was felt at every instant of the game. 1942 *R.A.F. Jrnl.* 30 May 33 It was like..Wembley or Hampden Park after an international. 1959 *Times* 21 Sept. 3/6 The British Lions were given but a faint hope of defeating the world champion All Blacks in the fourth and final international of their tour at Eden Park yesterday. 1971 *Leader* (Durban) 7 May 1/2 Efforts were being made for the Mauritian team to play a series of games including 'internationals' against South African Indian, Coloured and African elevens. 1973 *Times* 7 Feb. 15/4 It is no wonder, she feels, that so many former internationals never stay on in the sport.

b. (Earlier and later examples.) Also **Internationale.** Spec. **First International**, founded in London by Karl Marx in 1864 for promoting the joint political action of working classes in all countries, and dissolved in Philadelphia in 1876; **Second International**, an organization founded at Paris in 1889 to celebrate the 100th anniversary of the French Revolution; **Third International**, founded at Moscow in 1919 by delegates from twelve countries to promote communism and support the Russian Revolution, and dissolved in 1943; also called **Communist International** (abbrev. *Comintern); **Fourth International**, founded in 1936 by followers of Trotsky.

1871 *Observer* 9 Apr. 6/4 Most of the decrees are the work of illiterate members of the *Internationale*; they consist of candid attacks on the rights of property,.. such as you may expect from vulgar Socialists. 1871 G. M. Hopkins *Let.* 2 Aug. (1935) 27, I feel inclined to begin by asking whether you are secretary to the International... I am always thinking about the Communist future. 1919 Communist International [see *communist 1 b]. 1919 G. B. Shaw *Matter with Ireland* (1962) 193 The number of branches of The International formed in France in the 1860s by the police agents of Napoleon III must have filled the Castle with envy. 1920 B. Russell *Pract. & Theory Bolshevism* 6, I do not believe that the methods of the Third International can lead to the desired goal. 1935 C. Isherwood *Mr. Norris changes Trains* vi. 89 My association with..the representatives of the Third International..have even excited favourable comment in certain quarters in Moscow. 1937 E. Snow *Red Star over China* iv. 154 In 1920, M. Martin, an energetic and persuasive representative of the Third International.. came to Shanghai. 1950 Theimer & Campbell *Encycl. World Politics* 430/1 Trotsky's adherents in a number of countries, recruited from discontented left-wing communists, tried to found a Fourth International. *Ibid.* 430/2 The Fourth International, a shadowy organization which held conferences in Paris in 1936 and 1948. 1951 W. Pickles in *Political Q.* XXII. 335 June 30th, 1951..was the fourth occasion in less than 100 years on which a Socialist International has been founded..and the longest-lived of these ancestors, the Second International of 1889, survived for only 25 years. 1959 *Chambers's Encycl.* VII. 684/1 The first International spread rapidly, sections being founded in France, Belgium, Germany, Switzerland, Italy, Spain and other countries. 1967 N. McInnes in R. P. Arnot et al. *Impact Russ. Revolution* 85 The Communist International having been disbanded by Stalin in 1943, the only socialist international in existence was the Fourth, demonstrating again that Trotskyism was the last refuge of several of socialism's discarded illusions. 1968 K. Martin *Editor* xi. 234 'I tell you,' he [*sc.* Trotsky in 1937] said, 'that in three to five years from now the Fourth International will be a great force in the world.' 1969 *Listener* 27 Mar. 429/1 If the ILP had joined the Communist International, as its French and German equivalents did, and as the Scottish divisional council wanted it to, a powerful Communist Party might have come into existence on British soil. 1972 *Guardian* 10 Mar. 3/2 The Fourth International (FI) arose after the disintegration under Stalin of the Third International.

c. *pl.* International bonds.

1909 *Westm. Gaz.* 17 Sept. 12/2 Internationals featureless.

d. *The International* = next.

1912 *Songs for Socialists* (Fabian Soc.) 2 The International. 1928 A. Huxley *Point Counter Point* xxiii. 413 Organized singing of the International.

Internationale (-næʃiǒnā·l, ‖ æ̃ntɛrnasyonal). [Fr. (sc. *chanson* song).] A revolutionary hymn composed by Eugène Pottier in 1871 and adopted by French socialists and subsequently by others. Also *fig.*

[1912 H. M. Hyndman *Further Reminisc.* 124 We all left Paris regretfully, with the strains of 'L'Internationale' ringing in our ears.] 1920 S. Lewis *Main St.* i. 9 A Russian Jewess who sang the Internationale. 1933 *Morning Post* 30 Mar. 11/3 Communists..waved red flags and sang the 'Internationale'. 1938 I. Goldberg *Wonder of Words* p. vii, This *internationale* of poverty in a West-End slum. 1949 I. Deutscher *Stalin* xii. 491 The Internationale, the hymn of the Labour movement of the world. 1974 *Times* 7 Mar. 16/8 Cardew had intended to end the programme with the *Internationale.*

internationalism. (Earlier and later examples.)

1851 M. Tupper in D. Hudson *Martin Tupper* (1949) x. 121 [President Fillmore..received me] very kindly and cordially, and avowing himself a great lover of my works and my Internationalism. 1955 *Times* 17 May 10/7 Internationalism seems to have become by now an accepted thing even in trades far less easily internationalized than ours. 1971 *World Archaeol.* III. 226 In Childe's view, it was..the Iron Curtain which abruptly severed this age-old tradition of internationalism. 1973 *Times* 16 Apr. 4/6 No city has done more than Coventry since the war to further the cause of internationalism.

internationalist. Add: (Further examples.)

1916 A. Huxley *Let.* 19 Mar. (1969) 94, I get extraordinarily irritated with some of these Internationalists, who conscientiously object. 1955 H. Hodgkinson *Doubletalk* 60 In practice it is enough to discover what the USSR considers her own interests to be to show to the 'proletarian internationalist' his own. 1965 M. Bradbury *Stepping Westward* i. 47 They were avaricious internationalists, evidently, their legs turned nutmeg by a sun that had come to find them daily in different places. 1973 *Listener* 17 May 636/1 A little-known Maoist organisation known as the Internationalists, founded in Vancouver in March 1963.

c. *attrib.* or as *adj.*

1941 J. S. Huxley *Uniqueness of Man* xiv. 288 The nationalistic element in socialized religion will be subordinated or adjusted to the internationalist. 1955 Koestler *Trail of Dinosaur* 190 Socialism has lost its claim to represent the internationalist trend of humanity.

Hence **internationali·stic** *a.*

1930 A. Flexner *Universities* i. 13 This present-day world, compounded of tradition, good and bad, racial mixtures, nationalistic and internationalistic strivings. 1973 M. Truman *Harry S. Truman* x. 204 Arthur Vandenberg, leader of the internationalistic Republicans, fulminated against what was happening in Poland.

inter-nebular: see *inter- *pref.* 4 a.

internee (intǝ̄ɹnī·). [f. Intern *v.* 2 + -ee¹.] One who is interned; an interned person.

1918 *Rep. Detenus & Internees Bengal* 2 in *Parl. Papers* (Cd. 9198) VIII. 106 The cases of the State prisoners and internees. 1920 *Observer* 5 Dec. 13/4 Lorries with military guards..removed the internees to the camp. 1941 Wodehouse *Let.* 13 Aug. in *Performing Flea* (1953) 179 We are elderly internees, most of us with corns and swollen joints. 1972 *Daily Tel.* 4 Mar. 32/6 A policy to end imprisonment without trial by a phased programme of releases of every detainee and internee.

interneuron (intǝɹniū̂ǝ·rǫn). *Physiol.* Also **interneurone.** [f. Inter(nuncial *a.* + Neuron.] Any of the neurons which transmit nerve impulses from receptor neurons to effector neurons; an internuncial neuron.

1939 R. Lorente de Nó in *Jrnl. Neurophysiol.* II. 402 Following the example of Prof. Gerard the simple term 'interneuron' will henceforth be used instead of the rather cumbersome designation 'internuncial neuron'. 1963 S. Ochs in E. E. Selkurt *Physiol.* v. 104 (*caption*) Sensory fibers terminate upon interneurons within the spinal cord which, by multiple branches, engage a large number of motoneurons. 1971 *Jrnl. Exper. Biol.* LV. 727 The DCMD (descending contralateral movement detector) neurone is a monocular visual interneurone..which sends a large axon down the contralateral connective to the thoracic ganglia. 1971 Villee & Dethier *Biol. Princ. & Proc.* xx. 619 Interneurons are those that connect two or more neurons. They usually lie wholly within the central nervous system.

interneuronal (intǝɹniūǝrōᵘ·năl), *a.* *Physiol.* [Partly f. prec. + -al, partly f. Inter- 4 a + *neuronal *a.*] **a.** Of, pertaining to, or affecting an interneuron. **b.** Occurring or existing between neurons.

1948 E. Gardner *Fund. Neurol.* viii. 174 The afferent tracts of the cord evolve from such interneuronal pools. 1952 J. D. Hardy et al. *Pain Sensations* i. 15 Lorente de Nó..inferred that all interneuronal connections fall into two categories. 1959 *Jrnl. Amer. Chem. Soc.* LXXXI. 4347/1 In quite a few of the compounds analgesic activity is accompanied by marked interneuronal blocking action. 1969 *New Scientist* 6 Feb. 278/1 The *specificity* of this interneuronal communication, every neuron making contact with other, often carefully specified, neurons.

internist (intǝ̄·ɹnist). *U.S.* [f. **intern*(*al medicine* + -ist.] A general physician; also, a specialist in internal medicine.

1904 *Science* 29 Apr. 696/1 Many internists ('general physicians') of experience and authority. 1910 [see *hypoparathyroid *a.*]. 1918 *Nation* (N.Y.) Feb. 170/1 A maternity bureau composed of expert obstetricians and competent internists. 1961 *Brit. Jrnl. Clin. Pract.* June 510/2 The role of the internist is different. He grasps intuitively the whole of the patient—the 'gestalt' —and combines the various organic findings. 1970 *Sci. Amer.* Mar. 60/3 The British surgeon is a consultant who sees only patients referred to him by a general practitioner or internist. 1972 *Science* 5 May 489/1 The use of antidepressants by general practitioners, internists and psychiatrists.

internment. Add: **b.** *attrib.*, as internment camp, a detention camp for prisoners of war and aliens.

1916 *Daily Colonist* (Victoria, B.C.) 26 July 1/3 Brandon, Man. July 25.—The internment camp here is to be closed at the end of this month. 1941 *Star* 10 Apr. 8/3 A revolt at an internment camp at Vernet les Bains.

internuclear, *a.* Add: **2.** *Physics.* Existing between atomic nuclei.

1946 *Nature* 17 Aug. 224/1 The carbon chain must then be coiled to the same extent if normal internuclear distances are to be maintained. 1949 Koestler *Insight & Outlook* xi. 156 The integrative tendencies in the various forms of attracting or binding forces: internuclear, gravitational, electromagnetic, and so on. 1962 *Science Survey* III. 36 Typical known values for atomic masses and approximate values for typical inter-nuclear distances.

3. *Phonetics.* Situated between nuclei.

1961 Y. Olsson *On Syntax Eng. Verb* ii. 23 In pre-, inter-, and post-nuclear sections, 4/S-units and 4/0-units may occur in various combinations.

inter-nucleon: see *inter- *pref.* 5.

interoceanic, *a.* Add: (Earlier U.S. example.) So **intero·cean** *a.*

1850 *Clayton–Bulwer Treaty* 19 Apr. 8 Interoceanic communication. 1913 *Chambers's Jrnl.* Jan. 22/2 The isthmus ..became a centre of attraction for interocean transit. 1969 *Jane's Freight Containers 1968–69* 93/1 Parcel Tankers, Inc. Agent: Interocean Steamship Corp.

interoceptor (i:ntǝrose·ptǫ̈ɹ). *Physiol.* [perh. f. Inter(ior *a.* and *sb.* + -o + Re)ceptor, after *exteroceptor.] Any sensory receptor which receives stimuli arising within the body, or *spec.* within the viscera. So **intero·ce·ptive** *a.*

1906 C. S. Sherrington *Integrative Action Nervous Syst.* ix. 317 This [inner] surface of the animal may be termed the intero-ceptive. *Ibid.* 352 Spinal shock hardly at all affects the nervous reactions of the intero-ceptors (visceral system). 1930 L. T. Troland *Princ. Psychophysiol.* II. iv. xiii. 14 The visceral or interoceptive systems include those for the feelings of hunger, thirst, nausea. 1940 Fraenkel & Gunn *Orientation of Animals* (1961) iii. 33 The interoceptors in the alimentary canal which are responsible for belly-aches. 1971 D. J. Aidley *Physiol. Excitable Cells* xv. 307 Exteroceptors are sensitive to stimuli originating outside the body, and interoceptors are excited by stimuli inside the body. 1972 C. H. Hockman *Limbic Syst. Mechanisms & Autonomic Function* viii. 128/2 Intracellular recording in awake, sitting squirrel monkeys have shown a differential effect of interoceptive and exteroceptive inputs on the excitability of hippocampal neurons.

inter-office: see *inter- *pref.* 5.

interoperable (i:ntǝrǫ·pǝrăb'l), *a.* [f. Inter- 2 a + Operable *a.*] Able to operate in conjunction.

1969 *Nature* 29 Nov. 839/1 Skynet is inter-operable with the IDCSP system and is intended to be inter-operable with its successor now in view. 1970 *Sci. Jrnl.* Mar. 13 In other respects the two nations' forces are working in parallel and Skynet [the Royal Navy's military satellite communications system] is compatible and interoperable with the US Defence Communication Satellite system. 1970 *Fremdsprachen* XIV. 220/1 Sometime in the mid-1970s the communications..will not only be interoperable among themselves, but with United States, Canadian and Australian services also.

inter-palpebral, -particle, -party: see *inter- *pref.* 6, 5, 5.

inter partes: see *inter L. prep.

interpellate, *v.* Add: Also and now more usually with pronunc. (intǝ̄·ɹpĕlēⁱt).

inter-perceptual: see *inter- *pref.* 4 c.

interpermeate (intǝɹpǝ̄·ɹmi,ēⁱt), *v.* [Inter- 1 b.] To pervade or penetrate reciprocally. Hence **interpermea·tion.**

1909 in *Cent. Dict.* Suppl. 1926 *Blackw. Mag.* Oct. 518/2 It [*sc.* the Vital Principle] came into being by the interpermeation of air and water. 1954 *Jrnl. Theol. Stud.* V. 54 It is surely more intelligible to postulate a personal unity in which persons coinhere and interpermeate in a way unknown to us on earth.

interpersonal (intǝɹpǝ̄·ɹsǝnăl), *a.* *Psychol.* [f. Inter- 4 c + Personal *a.* 5.] Between persons; *spec.* a term introduced by H. S. Sullivan (1892–1949) to describe behaviour between people in any encounter. So **interpersona·lity**, the state or fact of being interpersonal; **interpe·rsonally** *adv.*, in an interpersonal manner.

1842 [see Inter- 4 c]. 1938 H. S. Sullivan in *Psychiatry* I. 121 Psychiatry seeks to discover and formulate the laws of human personality... Its peculiar field is the study of interpersonal phenomena. Personality is made mani-

fest in interpersonal situations, and not otherwise. It is to the elucidation of interpersonal relations, therefore, that psychiatry applies itself. 1946 C. Morris *Signs, Lang. & Behavior* 39 (*heading*) The interpersonality of the language sign. 1947 P. L. Harriman *Dict. Psychol.* 184 *Interpersonal relationships*, the reciprocal influences which persons exert upon one another in primary (face-to-face) social groups. 1949 M. Mead *Male & Female* x. 212 The specific interpersonality of the sexual act. *Ibid.* ii. 39 Using the understanding not interpersonally but personally. 1959 H. Bonner *Group Dynamics* ii. iii. 90 An important feature of interpersonal aggression is that it is usually accompanied by the members' insight into their own behavior. 1962 U. Weinreich in Householder & Saporta *Probl. in Lexicogr.* 43 Lexicography as a descriptive (rather than a normative) discipline must also take the criterion of interpersonality seriously. 1963 *Dissertation Abstr.* XXIII. 3484/1 He was impulsive, emotionally and interpersonally immature, and not emancipated from his parents. 1964 M. Argyle *Psychol. & Social Probl.* xv. 183 Sociological field studies of neighbourhoods, factories and families can disclose facts about inter-personal behaviour and relationships. 1971 M. Cook (*title*) Interpersonal perception. *Ibid.* 11 That area of social psychology—variously called interpersonal perception, person perception and impression formation. 1972 *Jrnl. Social Psychol.* LXXXVI. 177 The 15-year-old boy..was involved in interpersonal conflicts with his stepfather.

interphase (i·ntǝɹfēⁱz), *sb.* and *a.* [f. Inter- + Phase.] **A.** *sb.* [Inter- 2 b.] **1.** *Cytology.* [a. G. *interphase* (H. Lundegårdh 1912, in *Arch. f. mikrosk. Anat.* LXXX. 27).] A stage in the cycle of nuclear division which intervenes between one mitosis and the next; also, a stage between the first and second division of meiosis. Also *attrib.* or as *adj.*

1925 E. B. Wilson *Cell* (ed. 3) 1134 *Interkinesis*... (*Interphase* of Lundegårdh, 1912). 1932 C. D. Darlington *Recent Adv. Cytol.* iv. 90 The chromosomes may form two daughter-nuclei at the poles and pass into a resting stage, the interphase. 1961 *Lancet* 26 Aug. 488/1 Any X chromosomes in excess may produce a sex-chromatin body in the interphase nucleus. 1968 M. W. Strickberger *Genetics* ii. 13 As compared to the period of active mitosis..the interphase or 'intermitotic' period of most cells is usually many times longer. *Ibid.* 21 The chromosomes pass into a short interphase before the second meiotic division begins.

2. *Physical Chem.* The region between two phases in which the properties are significantly different from the bulk properties of either phase.

1933 *Phil. Mag.* XVI. 849 In recent years attention has been concentrated once more on the electro-chemistry of the interphase between mercury and an aqueous solution containing electrolytes. *Ibid.* 864 The surface tension of a liquid–liquid interphase is defined in such a way as to be applicable to any desired boundary in an interphase of finite thickness. 1955 E. J. Pryor *Introd. Mineral Dressing* xvii. 428 In the solid–fluid interphase, forces of each phase meet and merge through a transition stage. 1970 Bockris & Reddy *Mod. Electrochem.* I. i. 2 An interface formed by two phases..is an apparent surface because..there is a *region* in which there is a continuous transition from the properties of one phase to the properties of the other. If one aims to refer specifically to this three-dimensional transition region, then it is more appropriate to use the term interphase. 1972 Bockris & Dražić *Electro-Chem. Sci.* ii. 22 An interphase is a less sharply definable region, which will range over at least two molecular diameters, but may extend over thousands of Ångströms.

B. *adj.* [Inter- 5.] **1.** Occurring or existing between two phases or states of matter.

1933 *Jrnl. Chem. Physics* I. 8/1 The equilization [*sic*] of spreading force *f* in different phases within one field may take place in 3 ways: (a) Interphase mobility by which adatoms cross the phase boundary [etc.]. 1959 *Times* 22 Jan. 2/5 (Advt.), The work will include calculation of interphase reactions in which water or steam is normally one phase. 1967 B. P. Nicolsky et al. in G. Eisenman *Glass Electrodes for Hydrogen* vi. 175 We shall agree to regard the interphase potential as the electrical potential of the solid phase as compared with the liquid one.

2. *Electr. Engin.* Applied to an inductor used to connect two or more polyphase rectifiers so that they operate in parallel with the current divided between them.

1935 *Gen. Electric Rev.* XXXVIII. 499/1 Polyphase rectifiers usually have an interphase transformer which serves to divide the anodes into two or more groups which operate in parallel. 1952 Erikson & Bryant *Electr. Engin.* xxiv. 431 The presence of the interphase transformer is necessary to allow conduction in two tubes ..at any given time. 1970 J. Shepherd et al. *Higher Electr. Engin.* (ed. 2) xxv. 804 By the addition of an interphase reactor to the simple 6-phase connexion it is possible to obtain rectifier action with the smoothness of normal 6-phase rectification and the utilization factor of 3-phase rectification.

interphasic (intǝɹfēⁱ·zik), *a.* *Cytology.* [f. prec. + -ic.] Of or pertaining to interphase.

1948 W. Andrew tr. *E. D. P. de Robertis's Gen. Cytol.* vii. 134 Every cell passes..through two periods: one being interphasic or metabolic and the other being mitotic, or a period of division. 1960 [see *interkinetic *a.*].

i:nterpheno·menon. *Physics.* Pl. **-phenomena.** [Inter- 2 b.] Reichenbach's name for a phenomenon that cannot (even in principle) be inferred or demonstrated straightforwardly from observations in accordance with the laws of classical physics. Hence used for any unobserved phenomenon.

1944 H. Reichenbach *Philos. Found. Quantum Mech.* i. § 6. 21 We..shall consider as unobservable all those occurrences which happen between the coincidences, such as the movement of an electron, or of a light ray from its source to a collision with matter. We call this class of occurrences the interphenomena. Occurrences of this kind are introduced by inferential chains of a much more complicated sort; they are constructed in the form of an interpolation within the world of phenomena, and we can therefore consider the distinction between phenomena and interphenomena as the quantum mechanical analogue of the distinction between observed and unobserved things. 1956 —— *Direction of Time* v. § 25. 217 The terms 'particle' and 'wave' both belong to the language of interphenomena. They assert something about what happens between the localized phenomena. 1956 E. H. Hutten *Lang. Mod. Physics* v. 195 In classical physics..the unobserved phenomena are supposed to follow the same laws as the observed phenomena. The idealised character of measurement in classical theory allows us to speak about these interphenomena. *Ibid.*, The uncertainty principle makes it impossible to introduce interphenomena.

interphone (i·ntǝɹfōᵘn). orig. *U.S.* [f. Inter- 2 a + Phone *sb.*²] An intercommunication system whereby telephones are used to connect points within a small area, as a building, aeroplane, etc.

1942 *Time* 20 Apr. 25/2 During the same fracas, the pilot inquired over the [Flying] Fortress interphone: 'Are you firing at the enemy?' 1958 J. Cannan *And be a Villain* v. 118 Pelly was speaking on his interphone. 1961 'B. Wells' *Day Earth caught Fire* viii. 119 He was urgently dialling his inter-phone. 1962 *Punch* 5 Dec. 816/1 You'll need a Home Interphone, to communicate with the party in the other rooms.

interplanar: see *inter- *pref.* 6.

interplane (i·ntǝɹplēⁱn), *a.* *Aeronaut.* [f. Inter- 5 + Plane *sb.*³] Situated between or connecting the upper and lower 'planes' or wings of a biplane.

1916 H. Barber *Aeroplane Speaks* 96 Interplane struts ..have to keep the lifting surfaces or 'planes' apart. 1920 *Blackw. Mag.* Nov. 565/1 Great lanky beasts with double interplane engines, heavy bombers of freak aspect. 1969 K. Munson *Pioneer Aircraft 1903–14* 152/2 The rigid construction of the wings needed no tie-wires, the sole bracing consisting of two I-type interplane struts on each side.

interplanetary, *a.* Add: Also, existing between planets or pertaining to travel between planets.

1897 *Pearson's Mag.* July 116/1 Cylinder followed cylinder in its interplanetary flight. 1901 *Daily News* 7 Jan. 5/1 Dreams of inter-planetary communication. 1931 D. Lasser *Conquest of Space* 13 Ignorance and prejudice.. surround the 'interplanetary rocket' question. 1936 'J. Beynon' *Planet Plane* iii. 28 Mr. Curtance is going to try to win the Keuntz Prize for the inter-planetary flight. 1943 C. S. Lewis *Perelandra* vi. 91 In obscure works of 'scientification', in little Interplanetary Societies and Rocketry Clubs. 1950 *Jrnl. Brit. Interplanetary Soc.* IX. 300 Should be able to take the Interplanetary Age, when it comes, without hysteria. 1957 *Economist* 2 Nov. 377 The artificial satellite circling the earth was to be a proud reminder of the distance covered along the road from the wooden plough to the inter-planetary rocket. 1960 *Analog Science Fact/Fiction* Oct. 30/1 Maybe they were bravely defending their own planet by hunting down an interplanetary raider. 1962 [see *cosmic *a.* 3 b].

interplant (intǝɹplɑ·nt), *v.* [f. Inter- 1 a + Plant *v.*] *trans.* To plant (land) so that it is occupied by a mixture of plant species; to plant (a specified crop) *with* another crop. Hence **interpla·nting** *vbl. sb.* So **i·nterplant** *sb.*, a plant growing among others of different species.

1927 *Daily Tel.* 11 May 3/4 Included in the 1,442 acres of rubber were 98 acres which were interplanted with coffee. 1942 *R.A.F. Jrnl.* 30 May 27 Fill every inch of space by inter-planting one crop with another. This principle has already been illustrated by interplanting early potatoes with winter greens. 1950 *N.Z. Jrnl. Agric.* LXXX. 539/1 Other interplants are chiefly peaches. 1966 Webster & Wilson *Agric. in Tropics* x. 230 Interplanting with suitable crops, in a manner which maintains fertility and avoids competition, may be profitable. 1969 G. Wrigley *Tropical Agric.* (ed. 2) ii. 119 On the fertile soils of East Africa bananas are interplanted with coffee. 1971 *Countryman* LXXVI. iv. 109 Interplant them with dark purple tulips, or white splashed with purple, to give spring interest.

interplical: see *inter- *pref.* 6.

interpluvial (intǝɹplū̆·viăl), *a.* and *sb.* [f. Inter- 4 b + Pluvial *a.*] **A.** *adj.* Designating any of the periods of generally drier conditions in equatorial latitudes during the geological past (esp. the Pleistocene) which alternated with pluvial periods in a cycle which may be correlated with or related to the better-known cycle of interglacial and glacial periods in more polar latitudes. **B.** *sb.* An interpluvial period.

1907 *Bull. Geol. Soc. Amer.* XVIII. 362 The time when a thesis occurs is a thesial epoch, which may be described as interglacial, or interfluvial, and so forth, when the agents which modify the earth's surface are considered, or as intervegetal, interpluvial, and so forth, when other phases are considered. 1931 L. S. B. Leakey *Stone Age Cultures Kenya Colony* ii. 13 During the dry inter-pluvial periods..the lakes in the Rift Valley dried up. *Ibid.*, One would expect that the prehistoric tribes moved to the high land..during the inter-pluvials. 1940 *Geogr. Jrnl.* XCVI. 333 There have been very long wet and dry phases, the so-called pluvials and interpluvials. 1957 G. E. Hutchinson *Treat. Limnol.* I. i. 133 Such deposits are supposed to be the result of water-borne material, washed toward the pan in pluvial times, becoming cemented during the following interpluvial. *Ibid.*, Interpluvial periods of deflation. 1963 D. W. & E. E. Humphries tr. *Termier's Erosion & Sedimentation* i. 9 In Equatorial Zones it is now suspected that the pluvial phases were hot times and interpluvial (dry) phases were equivalent to the glacial epochs. 1973 *Nature* 16 Mar. 187/1 It seems.. that the tectonic movement which caused the tilting of the Ubeidiya Formation occurred some time within the Mindel-Riss Interpluvial and is younger than 640,000 yr.

Interpol (i·ntǝɹpǫl). [abbrev. of *Inter*national *pol*ice.] The International Criminal Police Commission (founded in 1923), with headquarters in Paris.

1952 Söderman & O'Connell *Mod. Criminal Investigation* (ed. 4) ii. 24 All national bureaus of the ICPC have adopted the telegraphic address of 'Interpol'. 1955 *Times* 7 July 7/5 They have issued the description of a man of about 45..whom they would like to interview. This has been circulated through the international police organization, Interpol, since it is thought likely that the man may have gone abroad. 1957 *Times* 24 June 10/2 Wire-tapping was one of the subjects discussed at the Lisbon meeting of the International Criminal Police Commission (Interpol) which ended yesterday. 1958 *Daily Mail* 31 July 1/5 British police last night sent a call to Interpol, the international police organisation, for help in solving the riddle of a British airman whose father believes he is still alive 15 years after he was shot down over Germany. 1964 J. Cramer *World's Police* vi. 442 The word 'Interpol' was originally the telegraphic address of the I.C.P.O. but it became so well known that it is now part of the official title of the organization.

interpolability (intǝ̄:ɹpǒlăbi·lĭti). [f. Interpolab(le *a.* + -ility.] The state or quality of being interpolable.

1938 *Nature* 12 Nov. 852/2 Attention is paid everywhere to interpolability, and the volume ends with a number of tables of interpolation coefficients. 1971 T. F. Mitchell in *Archivum Linguisticum* II. 46 Other points of contrast with the other examples include the interpolability before *off* of the 'adverbial' forms *straight* or *right*, the substitutability of *away* for *off*, [etc.].

interpolant (intǝ̄·ɹpǒlănt). *Math.* [f. L. *interpolant-em*, pres. pple. of *interpolāre* (see Interpolate *v.*), or Interpol(ate *v.* + -ant¹.] A value or expression (given or calculated) used in finding some other value by interpolation.

1920 *Tracts for Computers* ii. 17 Forward difference formulae, central difference formulae and Lagrangian formulae, when the interpolants are spaced equally apart are really different aspects of the same process, i.e. running a parabola of the $(n-1)$th order through n points. 1946 *Phil. Mag.* XXXVII. 260 A function value interpolated in this way may indeed be more accurate than any of the eight values used to obtain it. For in a perfect table the rounding-off errors of the eight interpolants have uniform probability distributions between −0·5 and 0·5 in units of the last decimal, and it may be shown that the probability of the resulting error of the interpolate is very nearly normally distributed with standard deviation 0·12. 1965 A. Ralston *First Course Numerical Analysis* iii. 57 An important advantage of finite-difference interpolation formulas..would seem to be the property..that enables a term to be added to them merely by adding one tabular point and computing an additional row of differences... This enables us to generate a sequence of interpolants each involving one more tabular point than the previous one.

interpolate, *v.* Add: **3. c.** (With the words spoken as object.) To interpose orally.

1881 Mrs. J. H. Riddell *Senior Partner* I. vii. 151 'I have not a word to say against Effie,' began..Robert... 'If ye had, ye'd best not say it before me,' interpolated his father. 1908 *Smart Set* June 142/1 'I suppose I may claim some eminence as physicians go.' 'Top notch of the whole bunch, dad!' interpolated Keene. 1922 Joyce *Ulysses* 614 What year would that be about? Mr Bloom interpolated.

6. (Further examples.) Also *absol.* or *intr.*, to use or perform interpolation. Also *fig.*

1888 *Encycl. Brit.* XXIX. 8/2 The reciprocal of a number of five figures is therefore taken out at once, and two more figures may be interpolated for as in logarithms. 1905 [see *extrapolate *v.* 2 b]. 1928 *Monthly Notices R. Astron. Soc.* LXXXVIII. 506 The Lagrangian method of interpolating to fixed sub-divisions of an interval. 1968 Fox & Mayers *Computing Methods for Scientists & Engineers* i. 11 Interpolating for $y(2{\cdot}5)$, from given values at $x = 2$ and $x = 3$, the formula gives $y(2{\cdot}5) = 22y(3)/35 + 18y(2)/35$.

interpolate (intǝ̄·ɹpǒlĕt), *sb.* *Math.* [f. L. *interpolāt-us*, pa. pple. of *interpolāre* (see

INTERPOLATE *v.*): see -ATE[1] b.] A value arrived at by interpolation.

1920 *Tracts for Computers* II. 17 As a rule the interpolation formulae work, but once in a while bitter experience forces us up against cases in which increasing the number of differences..is quite ineffectual as a method of obtaining accurate interpolates. **1932** *Proc. Edin. Math. Soc.* III. 56 For example, a linear interpolate $u_{0.683}$ can be computed from u_0 and u_1 as $u_{0.683} = 0.317u_0 + 0.683u_1$. **1946** [see *INTERPOLANT]. **1956** F. B. HILDEBRAND *Introd. Numerical Analysis* ii. 35 If more reliable interpolates are desired, it is clearly necessary to make use of more information than that consisting merely of tabulated values (ordinates) of a function, corresponding to two successive abscissas.

interpolation. Add: **3. b.** (Later examples.) Freq. *attrib.*

1872 [see *EXTRAPOLATION]. **1888** *Encycl. Brit.* XXIII. 13/1 All tables of proportional parts may be regarded as interpolation tables. **1928** *Monthly Notices R. Astron. Soc.* LXXXVIII. 511 The interpolation formula known as Bessel's, but really discovered by Newton. **1968** FOX & MAYERS *Computing Methods for Scientists & Engineers* viii. 147 Moreover the 'interpolation coefficients'..are available in tabular form,..and our interpolation process is reasonably straightforward. **1970** O. DOPPING *Computers & Data Processing* xvi. 256 The computer calculates a first approximation to the desired address by interpolation between the known minimum and maximum values of the argument.

interpolator. Add: **2.** A mechanical contrivance for securing correct retransmission from a submarine cable of any consecutive letter-elements having the same sign.

1902 *Encycl. Brit.* XXXIII. 226/2 The special form of curb sender mentioned, termed the 'Interpolator'. **1958** *Economist* 1 Nov. 425 (Advt.), The Interpolator is the essential part of the mechanism used to re-form signals automatically at the end of their journey through a long cable, before they are sent into another.

3. *Engin.* A device or apparatus which, when fed with a set of datum points defining a curve, produces a continuous output for guiding a tool over the curve.

1953 *IRE Trans. Industr. Electronics* I. 29 The interpolator output provides a continuous angular rotation whose instantaneous angular position represents the radius vector of the machine tool's cutter locus. **1961** S. FIFER *Analogue Computation* IV. xxx. 1279 The function of the analogue interpolator is to compute a continuous curve from the discrete set of digital information and to position a mechanical lathe accordingly. **1973** tr. *W. Simon's Numerical Control Machine Tools* ii. 43 The cost of the control equipment located on the machine tool itself is considerably reduced, since the machine no longer has its own interpolator.

interpolatory (intǝ·ɹpǫ̆lǎtǝri), *a.* [f. INTERPOLAT(E *v.* + -ORY[2].] Serving to interpolate.

1946 *Mind* LV. 201, I would suggest approaching the matter from the Quantum Analogue angle and boldly taking the line that the latter were Interpolatory expressions. **1965** W. S. ALLEN *Vox Latina* iii. 72 The interpolatory nature of the '*gnus*' passage.

interpole (i·ntǝɹpǝ̄ul). *Electr.* [INTER- 2 b.] In an electric motor or generator having a commutator, each of a set of auxiliary poles situated between the main poles and connected in series with the armature, the function of which is to facilitate commutation of the current by cancelling the induced e.m.f. in the coils that tends to hinder it.

1907 H. H. NORRIS *Introd. Study Electr. Engin.* x. 280 The function of the commutating or 'inter' pole is to produce a reversing field for the coil undergoing commutation. **1927** H. H. BROUGHTON *Electr. Winders* iii. 82 Sparkless commutation can be obtained by means of interpoles alone. **1964** R. F. FICCHI *Electrical Interference* vii. 108 If interpoles or compensating windings are introduced in the design, the commutation process will usually become more efficient and interference will be minimized.

interpolymer (i·ntǝɹpǫ:limǝɹ). *Chem.* [INTER- 2 b.] (See quot. 1966.)

1936 *Chem. Abstr.* XXX. 6478/2 Mixts. of polymers or interpolymers of these compds. **1950** *Thorpe's Dict. Appl. Chem.* (ed. 4) X. 87/2 The product is not a mixture of polymers A-A-A-A-A-A-...with B-B-B-B-B-B-..., but an interpolymer, e.g. A-B-A-A-A-B-A-A-..., the order in which the units are arranged not being readily determinable. **1955** *Jrnl. Polymer Sci.* XVIII. 129 A general method of producing graft and block interpolymers between elastomers is therefore indicated. **1966** M. L. MILLER *Struct. Polymers* ix. 426 Polymers made up of molecules containing two or more types of monomer units are called copolymers or interpolymers... The terms 'copolymer' and 'interpolymer' are essentially equivalent but the term 'interpolymer' is more restrictive because it implies an intimate distribution of monomer units in the molecule while the term 'copolymer' has no such implication.

Hence **interpo·lymerize** *v. trans.* and *intr.*, to combine so as to form an interpolymer; **i:nterpolymeriza·tion.**

1950 *Thorpe's Dict. Appl. Chem.* (ed. 4) X. 87/2 Vinyl acetate is interpolymerised with vinyl chloride to give plastics which are lower softening and more soluble than polyvinyl chloride. *Ibid.* 88/1 There are monomers.. which do not give high molecular-weight polymers by themselves but may readily interpolymerise with a second component... Interpolymerisation may even be effected in cases where neither component is separately polymerisable.

interpopulational: see *INTER- *pref.* 6.

interpose, *v.* Add: **1. c.** *Chess.* To move (a man) so as to obstruct the line of action of an opposing piece, esp. when the latter is giving check. Also *absol.*, or with the interposed man as subject.

1761 E. HOYLE *Ess. Game of Chess* 2, 6 B. The Queen gives Check. W. The Bishop interposes. **1765** R. LAMBE *Hist. Chess* 107 If the B. Bp. instead of retreating, checks your K. you must interpose your Bp. **1808** J. H. SARRATT *Treat. Game of Chess* I. 49 If he interpose his King's Knight, you must take his Rook with your Queen. **1844** W. LEWIS *Treat. Game of Chess* 19 Sometimes a piece or pawn, which before was out of play, may be interposed, and force the piece that checks to retire. **1861** *Chambers's Encycl.* II. 799/2 The king..must either move out of check or interpose some one of his subjects, unless the checking piece can be captured. **1876** *Encycl. Brit.* V. 593/1 If..the king, being thus in check, cannot move to another square.. and there is no piece or pawn which can be interposed, and the checking piece or pawn cannot be taken, then it is 'checkmate'. **1950** *Hoyle's Games Modernized* (ed. 20) 366 The player may..interpose one of his own men between the King and the attacking piece.

interpret, *v.* Add: **1. d.** To obtain significant information from (a photograph), used esp. of aerial photographs taken for military purposes.

1950 A. LEE *Soviet Air Force* vii. 154 Photographs, developed in mobile front-line vans, were interpreted and the main points signalled..to army and corps headquarters. **1958** C. B. SMITH *Evidence in Camera* i. 21 Although the art of interpreting air photographs had been developed quite a long way in the First World War, in the summer of 1939 there was precisely one experienced interpreter.

interpretability. (Earlier and later examples.)

1854 G. BOOLE *Investigation Laws of Tht.* 67 The same conditions of interpretability. **1956** E. H. HUTTEN *Lang. Mod. Physics* ii. 64 It is not translatability but interpretability in terms of an empiricist language that can establish meaning.

interpretant (intǝ·ɹprǐtǎnt). *Philos.* [f. INTERPRET *v.* + -ANT[1].] Peirce's term for the effect of a proposition, or sign-series, upon its interpreter, the person who understands it; thus, the meaning, in one sense.

c **1905** C. S. PEIRCE *Coll. Papers* (1931) I. III. 147 Now the relation of every sign to its object and interpretant is plainly a triad. *a* **1910** W. JAMES *Some Probl. Philos.* (1911) iv. 71 The interpretants are then substituted for the sensations, which thus get rationally conceived. **1939** *Mind* XLVIII. 480 The class of interpretations of the kind described in L will be called the *interpretant* of that sign. **1946** C. MORRIS *Signs, Lang. & Behavior* i. 17 The disposition of an interpreter to respond, because of the sign,..will be called an interpretant. **1947** H. REICHENBACH *Elem. Symbolic Logic* vii. 280 These psychological processes are sometimes called the *interpretant* of a sentence. **1964** *Language* XL. 234 The relationship between linguistic signs and their users, the relationship which Morris calls the 'interpretant'.

interpretation. Add: **1. c.** The technique of obtaining information from a photograph, esp. an aerial photograph. Cf. *INTERPRET *v.* 1 d.

1948 CARLING & ROSS *Brit. Surg. Pract.* V. 276 The interpretation of the pyelogram may be difficult. **1958** C. B. SMITH *Evidence in Camera* i. 21 Interpretation was supposed to be a job that any Station Intelligence Officer could be trained for in a week or two.

5. *attrib.*: **interpretation clause,** a clause in an Act of Parliament which defines the meaning of certain terms for the purposes of the Act.

1897 *Bouvier's Law Dict.* I. 1107/2. **1971** *Halsbury's Statutes Eng.* (ed. 3) XXXII. 363 *Interpretation sections.* Most modern Acts contain an interpretation clause which states the meanings which particular expressions used in the Act in question are to bear or include.

interpretatively, *adv.* **a.** Delete *rare* and add later examples.

1950 E. E. EVANS-PRITCHARD *Ess. Social Anthropol.* (1962) i. 22 He then lives the experiences over again critically and interpretatively in the conceptual categories and values of his own culture. **1969** *Daily Tel.* 17 Dec. 10/8 It was already obvious that we were to be treated to playing which was free of all technical cares, and which developed interpretatively with disciplined spontaneity.

interpretativeness (intǝ·ɹprǐtǎtivnės). [f. INTERPRETATIVE *a.* + -NESS.] The quality or condition of being interpretative.

1932 *Mind* XLI. 216 The several associations of 'meaning' with intellectual interpretativeness, value-carrying, and purposefulness are being confounded. **1957** D. L. BOLINGER in *Publ. Amer. Dial. Soc.* XXVIII. 26 The interpretativeness may be indicated merely by accent.

interpreter. Add: **1. d.** One who interprets (sense *1 d) photographs.

1950 A. LEE *Soviet Air Force* ii. 55 Large flying units.. had photographic interpreter specialists. **1958** [see *INTERPRET *v.* 1 d].

5. *Computers.* **a.** A machine that prints on a punched card fed into it the alphabetic or numerical equivalent of the pattern of holes.

1949 E. C. BERKELEY *Giant Brains* iv. 47 The chief IBM punch-card machines are: the key punch, the verifier, the sorter, the interpreter, the reproducer, the collator, the multiplying punch, the calculating punch, and the tabulator. **1970** O. DOPPING *Computers & Data Processing* iii. 54 If the cards have no text, a special machine, an 'interpreter', can be used to read the cards and print their contents on the top line.

b. An interpretive routine (see *INTERPRETIVE *a.* b).

1954 *Computers & Automation* Dec. 15/2 *Interpreter*, an executive routine which, as the computation progresses, translates a stored program expressed in some machine-like pseudo-code into machine code and performs the indicated operations, by means of subroutines, as they are translated. **1958** GOTLIEB & HUME *High-Speed Data Processing* xiv. 291 If a complete translation of the program is prepared *before* the initiation of its operation the system is a compiler; if the pseudo-code is retranslated each time it is used during operation, the system is an interpreter. **1964** [see *INTERPRETIVE *a.* b].

interpretive, *a.* Delete *rare* and add later examples.

1953 D. CECIL in W. de la Mare *Private View* p. vi, In his critical essays, Mr. de la Mare had not the scope for his interpretive genius to display itself so freely. **1955** *Bull. Atomic Sci.* June 218/3 There will be some 200 to 300 supplementary and interpretive books on atomic energy provided by American publishers. **1966** J. J. KATZ *Philos. of Lang.* iv. 111 Both the phonological and semantic components are, therefore, purely interpretive. **1966** MEIER & ELLIOTT (*title*) From plantation to ghetto: an interpretive history of American negroes. **1969** A. B. CALLOW (*title*) American urban history: an interpretive reader with commentaries. **1970** A. CAMERON et al. *Computers & O.E. Concordances* 107, I will admit that what follows will be interpretive. **1971** *World Archaeol.* III. 197 The field of interpretive functions which archaeologists call upon their pottery collections to sustain. **1973** *Studies in Eng. Lit.: Eng. Number* (Tokyo) 65 This is a highly controversial analysis, and has been much criticized by the so-called Interpretive Semanticists.

b. *Computers.* Applied to a routine that executes a source program by translating each instruction into machine language and then executing it (by means of subroutines) before proceeding to the next instruction; of or pertaining to the operation of such a routine.

1951 M. V. WILKES et al. *Preparation of Programs for an Electronic Computer* I. iv. 34 There are in the library a number of subroutines which, when called in, execute series of operations according to sets of parameters in the store... These subroutines are usually called in by the method used for the closed type, the parameter following the orders which call in the routine. The routines do, however, form a distinct class, and have been labelled 'interpretive'. **1957** W. D. BELL *Managem. Guide Electronic Computers* viii. 148 The availability of already established libraries is an important consideration in selecting a data system. And then there are 'interpretive routines', 'pseudo coding', 'compilers', and many other specialized programming methods and techniques. **1964** F. L. WESTWATER *Electronic Computers* ix. 143 As the number of subroutines available increased..whole programs could be written without any machine code being necessary. The coded orders were read in and a special routine (called the interpretive program or 'interpreter') selected each required subroutine in turn. **1970** O. DOPPING *Computers & Data Processing* xix. 306 Nowadays, the interpretive systems are used mainly for simulating a computer *B* using another computer *A* with other programming characteristics.

interproximal: see *INTER- *pref.* 6.

interpulse (i·ntǝɹpɒls), *a.* and *sb.* Also **inter-pulse.** [f. INTER- + PULSE *sb.*[1]] **A.** *adj.* [INTER- 5.] Existing or occurring between one pulse and the next.

1948 G. N. GLASOE in Glasoe & Lebacqz *Pulse Generators* iv. 119 For such a succession of pulses, it is generally desired that the interpulse intervals, as well as the pulses, be of controlled duration. **1962** SIMPSON & RICHARDS *Physical Princ. Junction Transistors* xvii. 444 The interpulse period must be long enough to allow C_1 to replenish its charge before another sweep begins. **1966** *Jrnl. Exper. Psychol.* LXXI. 881/1 As the interpulse intervals increase the threshold for the gap monotonically increases. **1971** *Jrnl. Gen. Psychol.* LXXXIV. 86 It is possible that pairs of photic pulses of the proper durations and appropriate interpulse intervals would provide impairment of visual resolution. **1971** *Sci. Amer.* Jan. 50/1 The amplitude of the pulses from the objects was observed to be very uneven, but the interpulse spacing was quite regular.

B. *sb. Astr.* [INTER- 3.] A weaker pulse occurring between the main pulses of radiation from some pulsars.

1969 *Nature* 8 Feb. 525/2 The interpulse trails the main pulse by 14·0 ms. **1971** *New Scientist* 1 July 6/2 In many pulsars..there is a weak 'interpulse' roughly halfway between main pulses.

interpupillary: see *INTER- *pref.* 6.

interquartile (intǝɹkwǭ·ɹtǝil), *a. Statistics.* [f. INTER- 5 + QUARTILE *sb.*] Situated between the first and third quartiles of a distribution.

1882 F. GALTON in *Rep. Brit. Assoc. Adv. Sci. 1881* 245 This gave the upper and lower 'quartile' values, and consequently the 'interquartile' range (which is equal to twice the 'probable error'). **1944** *Mind* LIII. 175 Does a statistical description in terms of mean, mean deviation, interquartile range, describe well enough the individuals we have examined? **1952** *New Biol.* XII. 23 The interquartile range will be about 7 per cent of the median. That is to say, if we take 101 teeth, and arrange them in order for the particular length measured, the difference in length between the 26th and the 76th will be about 7 per cent of the length of the 51st.

inter-racial, -racialism: see *INTER- *pref.* 4 c, 2 a.

interrelate, *v.* Delete *rare* and add later examples. Also *intr.*

1964 R. JOHNSTON in *Oceanogr. & Marine Biol.* II. 100 Because the conductivity-chlorinity relationship has been shown to be somewhat variable, one cannot interrelate density, chlorinity and conductivity. **1964** M. MCLUHAN *Understanding Media* (1967) II. xiv. 149 Money becomes..the principal means of interrelating the ever more specialist activities of literate society. **1973** *Times Lit. Suppl.* 23 Mar. 320/3 The essays do not interrelate very adequately.

interrenal, *a.* (Further examples.)

1902 *Nature* 18 Sept. 516 On the interrenal bodies of Plagiostomes, by M. Ed. Grynfeltt. **1929** *Physiol. Abstr.* XIII. 530 The interrenal organ is essential to life in Selachians. **1940** *Chambers's Techn. Dict.* 455/2 *Interrenal body*, in selachian Fish, a ductless gland which lies between the kidneys and corresponds to the cortex of the suprarenal gland of higher Vertebrates. **1965** LEE & KNOWLES *Animal Hormones* iv. 69 The term adrenal cortex can be retained for all the vertebrate classes, but in the fish in view of the anatomical position the synonym interrenal may be preferred.

inter-responsive: see *INTER- *pref.* 2 a.

interrogate, *v.* Add: **4.** *trans.* **a.** To cause (a transponder, or a vehicle or craft fitted with one) to transmit a signal, usually coded to give information about the device or its surroundings, by transmitting a triggering signal to it.

1945 R. WATSON-WATT in *Nature* 15 Sept. 319/2 The 'H' and 'Oboe' systems, in which the primary radio pulses 'interrogating' the mobile craft automatically release from it a series of reinforced, coded, and conveniently frequency-displaced reply pulses. **1951** *Engineering* 3 Aug. 151/3 The aircraft would carry a transponder which, when interrogated would give the aircraft's identity and possibly other information such as height. **1953** *Electronic Engin.* XXV. 416/1 As the balloon ascends into the upper air, its airborne transponder is interrogated from the ground station by radar pulses. **1960** *Electronics Weekly* 23 Nov. 20/1 It..allows the control stations to interrogate each repeater for checking its operational condition. **1961** *Observer* 19 Feb. 11/1 The Russians have said they will 'interrogate' the rocket every five days. **1970** *Sci. Jrnl.* Apr. 52/1 Orbiting satellites interrogate the moving platforms..as they pass over them, using a coded VHF signal, and receive back the data measured by sensors carried by the platforms.

b. To cause (a computer memory or memory element) to give a signal that corresponds to or reveals information contained in it.

1964 T. W. MCRAE *Impact of Computers on Accounting* i. 7 A system whereby one can interrogate any record (or group of records) directly without having to process all of the previous records is called a random access storage system. **1971** *Nature* 19 Mar. 156/2 It is now possible, in principle, for the user to interrogate the computer store directly. **1971** *Publishers' Weekly* 9 Aug. 25/3 The customer's file will be interrogated to determine his credit status. **1972** *Nature* 31 Mar. 206/1 The memories are interrogated (read) by much lower voltage signals.

interrogatee. (Later example.)

1964 D. F. GALOUYE *Counterfeit World* i. 11 'I won't answer any questions,' Siskin said... The pollster frowned. 'You must. You are an officially registered interrogatee.'

interrogating (inte·rǭgēitiŋ), *ppl. a.* [f. INTERROGAT(E *v.* + -ING[2].] Of a radio signal: intended to cause a transponder to emit a signal.

1946 *Jrnl. Inst. Electr. Engin.* XCIII. IIIA. 332/1 In the earliest radar beacon systems the responder replied on the same frequency as the radar interrogating pulse. **1953** *Electronic Engin.* XXV. 416/1 The interrogating pulses are received by the balloon-borne unit.

interrogation. Add: **2. b.** *note of interrogation*: also *fig.*

1889 W. B. YEATS *Let.* 7 May (1954) 125, I have no theories about her. She is simply a note of interrogation. **1897** G. B. SHAW *Our Theatres in Nineties* (1932) III. 130 The ending of the play is not on the face of it particularly tragic: the alleged 'note of interrogation' is a sentimental fancy. **1931** H. G. WELLS *Work, Wealth & Happiness of Mankind* (1932) xv. 747 The note of interrogation which is born in the nature of every human being has been released. **1944** —— *'42 to '44* 116 No battle is a foregone conclusion, and so I must end with a note of interrogation. Will this thing be done now, rapidly?

3. The process of obtaining a signal from a transponder by transmitting an appropriate 'interrogating' signal to it; also, an interrogating signal.

1946 *Jrnl. Inst. Electr. Engin.* XCIII. IIIA. 487/2 The earlier varieties of I.F.F. all used direct interrogation. *Ibid.*, Each display..showed one echo fixed in position for each aircraft, due to its station's own interrogation. **1947** L. N. RIDENOUR *Radar Syst. Engin.* viii. 258 The interrogations were at 10 cm, the replies at 1·5 m. **1959** K. HENNEY *Radio Engin. Handbk.* (ed. 5) xix. 112 The moving craft sends out the interrogation and the beacon returns the reply. **1966** *Guardian* 19 Aug. 16/6 The new Interrogation Recording and Location System..to be embodied in future Nimbus satellites. **1970** *Sci. Jrnl.* Apr. 52/2 Direct interrogation of the package is made from the satellite using normal VHF interrogation telecommunications.

4. *attrib.* (sense 1), as *interrogation centre, room.*

1973 *Freedom* 2 June 3/1 Always there are the Chicago-style shootings from passing cars, the churches bombed or burned, the homes gutted by fire and watching over it all the army with its own State-sponsored violence, the interrogation centre, the beatings-up, [etc.]. **1944** *Sat. Even. Post* 15 July 19, I went into the interrogation rooms with these pilots and listened to the laconic reports they were relating to the Intelligence officers. **1971** 'O. BLEECK' *Thief who painted Sunlight* (1972) ix. 79 We were in one of those small, brown interrogation rooms at Homicide South. **1972** H. C. RAE *Shooting Gallery* iii. 142 Surely, sir, you've *seen* them round here, in the Interrogation Room, at the desk?

interrogator[1]. Add: **2.** A radio or radar transmitter designed to transmit interrogating signals to a transponder; also (more fully *interrogator-responsor, -responder*), one that in addition receives the resulting signals from the transponder.

1945 *Nature* 15 Sept. 324/1 Pulses from an airborne or shipborne interrogator. **1945** *Electronic Engin.* XVII. 735/3 Aircraft and other moving vehicles could now carry small questioning transmitters ('interrogators'). **1946** *Jrnl. Inst. Electr. Engin.* XCIII. IIIA. 331/2 (*heading*) Introduction of an independent interrogator-responder system. **1947** A. ROBERTS *Radar Beacons* ii. 28 An interrogator-responsor can be so designed as to be useful as a radar. **1962** *Aeroplane & Commercial Aviation News* 29 Mar. 27/1 The aerial system..was delivered in February and the interrogator-responsor is now being shipped. An airborne transponder is also being obtained.

in terrorem: see IN *Lat. prep.* in Dict. and Suppl.

interrupt (intěrɒ·pt), *sb. Computers.* [f. the vb.] The action (usu. automatic) of interrupting the execution of a program as a result of the need for the immediate execution of another program, after which the original program is automatically resumed.

1957 *Computers & Automation* VI. 17/2 The interrupt feature of the 1103A will no doubt be embodied in many future computers. **1967** *Technology Week* 23 Jan. 11/1 (Advt.), Sigma 5 can deal with foreground real-time interrupts in 6 microseconds. **1969** P. B. JORDAIN *Condensed Computer Encycl.* 257 Some interrupts are initiated by the computer operator when he wants to give a command, or by a remote user who wishes to request service. **1970** O. DOPPING *Computers & Data Processing* vi. 106 In older computers, the response to an error discovered by the built-in checks was usually to stop... In modern machines, the response usually is an interrupt.. which automatically gives a branch to a special routine for analyzing and, if possible, correcting the error. *Ibid.* ix. 126 During these time intervals, the computer can run a background program which is not a real-time job. When a transaction has been input, perhaps over a telecommunication line, an interrupt signal causes the CPU to hand over control to the real-time program. When the transaction is dealt with.., a return jump to background takes place.

interrupter, -or. Add: **b.** (Earlier U.S. example.)

1851 C. CIST *Cincinnati* 302 It has been represented that Prof. Locke had merely invented a new species of 'electrical interrupter'.

c. *attrib.* **interrupter gear,** a timing device attached to machine-guns in aeroplanes to prevent the discharge of bullets when the propeller is in the line of fire. Also *transf.*

1928 C. F. S. GAMBLE *Story N. Sea Air Station* x. 150 The Fökker monoplane, which was equipped with a mechanical interrupter gear. **1932** *Discovery* Oct. 323/2 A very great advance..has been made in reducing the attrition of fabrics during washing, by an invention of the Research Association, called the Interrupter Gear. **1940** *Flight* 4 Jan. 7/1 That machine was..fitted with a rather crude but serviceable type of interrupter gear.

interruptible, *a.* Delete *rare* and add later examples. Also **interruptable.**

1939 H. NICOLSON *Diary* 4 Feb. (1966) 390 Compared with a film, it [sc. television] is a bleary..interruptible thing. **1961** R. B. LONG *Sentence & its Parts* xvii. 389 *Awhile*, of course, is readily interruptible: *I waited a long while* is common alongside *I waited awhile.* **1961** *Economist* 2 Dec. 910 A supply of gas that is interruptible. **1966** *Ibid.* 14 May 734/1 So customers prepared to accept 'interruptable' electricity and have their power cut at peak hours can buy electricity very cheap indeed—possibly too cheap. **1971** *Daily Tel.* 4 Mar. 14/3 Supplies will be on an interruptible basis, so that a cold-weather surge in domestic demand for gas might be met by cutting off supplies to industrial users such as Tunnel.

intersection. Add: **2.** *spec.* = CROSS-ROAD 2. Chiefly *N. Amer.*

a **1864** [in Dict.]. **1931** W. FAULKNER *Sanctuary* xxiv. 231 He whipped the car in and out of traffic.., shooting recklessly across inter-sections. **1953** W. BURROUGHS *Junkie* (1972) ii. 31, I had to pull over to the side of a road and wait until the weed wore off. I could not tell how far away anything was or when to turn or put on the brakes for an intersection. **1970** *Rand Daily Mail* 28 Feb. 7/4 In the same way it does not matter that South Africans should say..'intersection' for cross roads. **1970** *Globe & Mail* (Toronto) 25 Sept. 2/4 Two girls, 14 and 15 years old, were killed last night when struck by a car at the intersection of Park and Cherry Streets.

3. b. *Logic* and *Math.* The set which comprises all the elements common to two or more given sets, and no others; also, the operation of forming such a set.

1909 in WEBSTER. **1941** BIRKHOFF & MACLANE *Survey Mod. Algebra* xi. 313 Intersection and union are related to each other and to inclusion by a fundamental principle of *Consistency*: The three conditions $X \leq Y$, $X \cap Y = X$, and $X \cup Y = Y$, are mutually equivalent. **1950** W. FELLER *Introd. Probability* I. i. 14 In the theory of probability we can describe the event *AB* in words as the simultaneous occurrence of *A* and *B*. In standard mathematical terminology *AB* is called the (logical) intersection of *A* and *B*. **1952** F. B. FITCH *Symbolic Logic* iv. 101 For any two classes F and G there is a class [F ∩ G] known as the intersection of F with G and having as members just those things which belong to both F and G. **1963** ANDERSON & HALL *Sets, Sequences, & Mappings* iii. 50 A point is in the intersection of a collection of sets iff it is a member of every set in the collection. **1968** E. T. COPSON *Metric Spaces* i. 9 Let *A*, *B*, *C* be subsets of a given set *E*. Then the operations of union, intersection and complementation have the following properties: (*a*) $A \cup A' = E$, $A \cap A' = \emptyset$. [Etc.]

intersegment. Restrict *rare* to sense in Dict. and add: **2.** *Zool.* In certain animals, for example earthworms or caterpillars, the part of the body between two segments.

1901 *Proc. Zool. Soc.* 211 Papillæ with the arrangement just described were only found in one example [of the earthworm, *Benhamia gambiana*]; in another the intersegment xv./xvi. bore but a single median papilla.

intersegmental, *a.* (in Dict. s.v. INTER- *pref.* 4 a). Add: (Examples.) Used esp. in *Zool.* and *Linguistics.*

1894 *Jrnl. R. Microsc. Soc.* 545 Miss Platt describes.. three longitudinal ectodermic ridges, which become connected by intersegmental transverse ridges. **1906** J. B. SMITH *Explanation Terms Entomol.* 69 Intersegmental: = interarticular. **1925** A. D. IMMS *Gen. Textbk. Entomol.* I. 11 The flexible infolded portion of the cuticle between adjacent segments is the intersegmental membrane. **1952** *New Biol.* XIII. 86 The abdomen when empty is flat (hence the name 'mahogany flat') and the intersegmental membranes are telescoped under the segments. **1964** *Language* XL. 220 One could speak of this intersegmental dependency as a form of redundancy. **1970** I. M. MACKERRAS in *Insects of Australia* (Commonwealth Sci. & Industr. Res. Organization) i. 5/2 When sclerotization develops [in insect skeletons], it does so in dorsal and ventral plates extending from just anterior to each intersegmental groove. **1972** *Language* XLVIII. 46 In intersegmental variation, different values for a feature are distributed in different contexts.

Hence **intersegme·ntally** *adv.*

1909 *Cent. Dict.* Suppl., Intersegmentally. **1941** JOHANNSEN & BUTT *Embryol. of Insects & Myriapods* xv. 242 At about the 112-hour stage the neuroblasts of the median cord..at first lying intersegmentally shift forward. **1957** S. POTTER *Mod. Ling.* v. 105 Sentences may be described..intersegmentally in respect of the prosodeme of syllabic transition or juncture.

intersensory, inter-sensory (intǝɹse·nsŏri), *a.* [f. INTER- 4 + SENSORY *a.*] Registered by two or more senses, e.g. sight and hearing. Hence **intersenso·rial** *a.*, **interse·nsorily** *adv.*

1933 *Proc. Brit. Acad.* XIX. 71 This [protocol-] language is inter-subjective and inter-sensory. **1934** M. BLACK tr. *Carnap's Unity of Sci.* 62 Physical determinations are valid inter-sensorily. **1935** *Brit. Jrnl. Psychol.* XXV. 468 A perception is produced by sensations of different modalities, which often create intersensorial patterns. **1964** M. CRITCHLEY *Developmental Dyslexia* xiii. 79 The evolution of behaviour can be 'conceptualised' as the process of development of intersensory patterning. **1973** *Jrnl. Genetic Psychol.* Sept. 115 (*title*) Effects of training on intersensory communication by three- and five-year olds.

intersertal (intǝɹsǝ̄·ɹtǎl), *a. Petrogr.* [a. G. *intersertal* (F. Zirkel *Untersuch. über Basalt.* (1870) iv. 111), f. L. *intersert-*, ppl. stem of *interserere* (see INTERSERT *v.*): see -AL.] Applied to the texture of igneous rocks containing a relatively small proportion of glass, augite, or other minerals in interstices between feldspar laths.

1893 A. GEIKIE *Text Bk. Geol.* (ed. 3) II. vii. 168 Under the name of Tholeite some interesting augite-andesites have been described, in which the felspar prisms form a network filled in with granular augite and interstitial matter (intersertal structure). **1916** A. JOHANNSEN tr. *Weinschenk's Fund. Princ. Petrol.* x. 199 Among the basic melaphyres there occurs a texture analogous to the ophitic, namely the intersertal. **1962** READ & WATSON *Introd. Geol.* vii. 412 Pyroxenes often include both augite and pigeonite; they occupy the spaces between the feldspars and are sometimes granular giving the intersertal texture. **1972** *Science* 2 June 977/1 Sample 14310 is an intersertal plagioclase-rich basalt.

inter-service(s) (intǝɹsǝ̄·ɹvis, -sǝ̄·ɹvisėz), *a.* [INTER- 5 + *SERVICE[1].] Existing or constituted between, or common to, the armed services.

1942 *Aircraft Recognition* Sept. 2, I am very pleased to write a preface to the first edition of the *Inter-Services Journal on Aircraft Recognition* which is being launched by my Department. **1946** *Daily Tel.* 24 Sept., Fortunately, experience since 1939 has destroyed many inter-Service prejudices without modifying the legitimate pride each Service takes in its corporate spirit. **1958** 'P. BRYANT' *Two Hours to Doom* 74 In the few weeks he had been President he had already experienced one major inter-service row and three or four minor squabbles. **1958** *Times* 19 May 11/5 Coordinating purchases by the formation of inter-service supply committees. *Ibid.*, Inter-service differences revolve largely round the allocation of finance and resources.

intersex (i·ntǝɹseks). *Biol.* [ad. G. *intersexe* (R. Goldschmidt 1915, in *Biol. Centralblatt* XXXV. 566): see INTER- 2 b and SEX *sb.*] In a dioecious species, an abnormal form or individual having characteristics of both sexes; the condition of being of this type.

[**1920** K. MAYER (*title*) The intersexes, the history of similisexualism as a problem in social life.] **1917** R. GOLDSCHMIDT in *Endocrinology* I. 437 We have proposed the use of the terms intersexes, intersexual, intersexuality instead of sex-intergrades. **1920** *Science* 26 Mar. 326 The intersex gene itself has been found not to be sex-linked. **1930** G. R. DE BEER *Embryol. & Evolution* iii. 22 Animals which have switched over from a period of development in one sex to a period of development in the other are called intersexes. **1955** T. DOBZHANSKY *Evolution, Genetics & Man* xi. 262 This group consists of ordinary diploid females, diploid males, triploid females, and three new sexual types—intersexes, superfemales, and supermales. **1966** *Nature* 21 Jan. 144/1 The birds were found to be six males, ten females and one intersex. **1962** J. B. ASHLEY *Human Intersex* xi. 190 There is recognized interest in the problems of intersex particularly from the viewpoint of my own discipline. *Ibid.*, Gonadal tumours present a difficult problem in the intersex patient. **1971** *Daily Tel.* (Colour Suppl.) 10 Dec. 18/2 There are four main criteria involved in diagnosing and assigning the true sex of an individual: the chromosome sex, the sex of the gonads, the sex of the genitals and the psychological sex. Dr Armstrong has defined intersex as the state in which any of these fail to agree.

So **i·ntersexed** *a.* = *INTERSEXUAL *a.* 2.

1939 *Ann. Reg.* 1938 372 The production of feminized males and intersexed rats by antenatal administration of oestrogens. **1968** R. J. STOLLER *Sex & Gender* ii. 17 [*heading*] The intersexed patient with 'normal' gender identity.

intersexual (intǝɹse·ksiŭăl), *a.* [f. INTER- 4 c + SEXUAL *a.*] **1.** Existing between the sexes.

a **1866** J. GROTE *Treat. Moral Ideals* (1876) viii. 108 The intersexual feeling which belongs to all animals. **1889** in H. Ellis *Stud. Psychol. Sex* I. 183 This is quite as true of intersexual..love. **1904** G. S. HALL *Adolescence* II. iii. 223 A subtle but potent intersexual influence is among the strongest factors of all adolescent sport.

2. *Biol.* Typified by or having both male and female characteristics; having some characteristics proper to the other sex. Also *absol.* as *sb.*, an intersexual individual.

1916 [see *INTERSEXUALITY]. **1917** R. GOLDSCHMIDT in *Endocrinology* I. 438 With the male of race *B*, all would-be females with his high grade intersexuals, almost transformed into males. **1940** J. S. HUXLEY *Ess. Pop. Sci.* 51 Intersexual males—i.e. animals which have started as males and been forced to finish their development as females. **1962** D. J. B. ASHLEY *Human Intersex* xii. 269 When the diagnosis of an intersexual state is deferred. **1969** *Punch* 10 Nov. 691/2 The intersexual goats are sterile. **1970** *Guardian* 7 Jan. 7/3 With..intersexuals, the problem begins at birth, for in rare cases it is uncertain from the genital organs whether a baby is a boy or a girl.

intersexuality (i:ntǝɹseksiŭæ·lĭti). *Biol.* [ad. G. *intersexualität* (R. Goldschmidt 1915, in *Biol. Centralblatt* XXXV. 566): see INTER- 2 b and SEXUALITY.] The state or condition of being intersexual; intersexual character.

1916 R. GOLDSCHMIDT in *Amer. Naturalist* L. 708 It seems advisable to use another term for these forms, which in general represent a definite step between the two sexes. The phenomenon shall therefore be called intersexuality. Further experiments now proved that intersexuality segregates, F_2 giving normal and intersexual animals. **1930** G. R. DE BEER *Embryol. & Evolution* iii. 22 A series of degrees of intersexuality can be established from very slight to complete sex-reversal. **1965** *Punch* 10 Nov. 691 Goat-breeders have tried to maximise hornlessness..but in so doing they have unwittingly increased the frequency of the gene for intersexuality. **1966** L. L. MOORE *Sex Chromatin* xiv. 229 Intersexuality occurs about as frequently in cattle, swine and goats as it does in humans.

inter-species: see *INTER- *pref.* 5.

interspecific, *a.* Add: Also, formed or obtained from (individuals of) different species. (Further examples.)

1913 *Amer. Naturalist* XLVII. 243 Interspecific hybrids seldom show the typical Mendelian behavior. **1946** [see *INTERGENERIC *a.*]. **1968** H. HARRIS *Nucleus & Cytoplasm* v. 93 If somatic cells derived from the two species of animal are fused together with inactivated virus, the resultant interspecific hybrids not only multiply, but actually overgrow the two parent types.

Hence **interspeci·fically** *adv.*, between species, from one species to another.

1970 *Nature* 10 Oct. 189/2 The number of cellular enzymes known to differ interspecifically, for example, between mouse and man, is increasing rapidly.

interstadial (intǝɹstē·diăl), *sb.* and *a. Geol.* [a. G. *interstadial*, ad. F. *interstadiaire* adj. (A. Penck et al. 1894, in *Bull. de la Soc. d. Sci. nat. de Neuchâtel* XXII. 81): see INTER- and *STADIAL.] **A.** *sb.* A minor period of ice retreat during a glacial period, less pronounced than an interglacial. **B.** *adj.* Pertaining to or characteristic of such a period.

1914 W. B. WRIGHT *Quaternary Ice Age* vii. 156 It is, of course, extremely difficult to distinguish between interstadial and interglacial. This can only be done with certainty when the flora of the deposit in question indicates a condition of things too severe to be regarded as interstadial. **1922** *Bull. Geol. Soc. Amer.* XXXIII. 421 Baltic interstadial time (= Achen recession). **1946** L. D. STAMP *Britain's Struct.* xiv. 163 Arkell attributes the Aurignacian buried Channel period to an 'interstadial' period in the midst of the Fourth or Final Glaciation rather than to the Third Interglacial. **1953** *Proc. Prehist. Soc.* XIX. 181 The additional nodes at about 5 m. may correspond to the Epi-Monastirian beach of the First Interstadial of the Last Glaciation. **1968** ERICSON & WOLLIN *Ever Changing Sea* iv. 119 We may be living during a short interval of mild climate, or interstadial, within the last ice age. **1969** *Proc. Geol. Soc.* Aug. 155 Includes Upton Warren interstadial complex. **1973** *Times Lit. Suppl.* 1 June 607/3 (Advt.), The evidence for high interstadial sea levels.

i-stage, *a.* [INTER- 5.] Situated or occurring between successive stages of an apparatus.

1950 R. F. YATES *ABC of Television* v. 89 This inter-stage shielding is further insurance against audio-frequency regeneration. **1953** A. HUND *High Frequency Measurement* viii. 255 This happens..when inter-stage transformers are used in amplifiers. **1958** W. R. BARRETT et al. in H. W. Cremer *Chem. Engin. Pract.* V. 125 The interstage relief valves protect the compressor against excessive interstage pressure. **1970** V. O. STOKES *Radio Transmitters* xi. 147 The most suitable method of interstage coupling when driving a final amplifier in a grounded-cathode arrangement is a π circuit. **1972** *Physics Bull.* Oct. 583/1 In the manufactured version of this pump the interstage pressure is raised to about 0·1 Torr.

interstate, *a.* For *U.S.* read orig. *U.S.* Change def. to read: Lying, extending, or carried on between independent states, or between states belonging to a Union, Federation, etc. Also as *sb.*, a road between states. (Add further examples.)

1887 *Statutes at Large U.S.A.* XXIV. 383 A Commission is hereby created and established to be known as the Inter-State Commerce Commission. **1928** *Manch. Guardian Weekly* 10 Aug. 75/1 The problem of inter-State legal aid was bound to throw light upon the larger question of international legal aid. **1943** WYNDHAM LEWIS *Let.* 21 Aug. (1963) 367, I hate to think how much inter-state competition.. will probably cost me. **1953** M. OPPLER *Murder must Wait* iv. 20 Such matters as State boundaries [in Australia] and inter-State jealousies wouldn't register with Bony. **1968** *Globe & Mail* (Toronto) 3 Feb. 33/5 Here in Georgia where Interstate 75 slices through the heart of a peanut, cotton and general farming area. **1971** *Ibid.* 12 Feb. 8a/3 The U.S. Food and Drug Administration proposes to ban the interstate shipment of products containing carbon tetrachloride because of health hazards. **1969** *Jane's Freight Containers 1968–69* 115/1 During 1967, the Interstate Commerce Commission approved the merger of the company with the Chicago Great Western Railway. **1969** *Sydney Morning Herald* 24 May 35/4 (Advt.), Interstate carrier requires haulier. **1971** *Sunday Australian* 8 Aug. 8/3 Builders and interstate truckdrivers, always high on the bankruptcy lists, continue to provide more than their share of bankruptcies. **1971** D. E. WESTLAKE *I gave at the Office* (1972) 84 The shipments..traveled via truck on Interstate 40 and US 64 to a small town near Little Rock. **1972** *Times of India* 28 Nov. 10/7 Referring to inter-state disputes, Dr. Gajendragadkar urged the government to set up an inter-state council under Art. 263 to deal with such disputes instead of attempting to solve them on a political plane. **1973** J. M. WHITE *Garden Game* 173 The five-hour drive along Interstate 10 from Phoenix to Los Angeles. **1973** *Sun-Herald* (Sydney) 26 Aug. 73/4 As NSW champion she would have been the first player picked in the State team for the interstate series.

inter-sta·tion, *a.* [INTER- 5.] Occurring (in a radio) between two stations or tuning positions.

1940 *Chambers's Techn. Dict.* 456/1 *Inter-station interference*, interference which arises from the presence of another transmitter on the same (or on an adjacent) wavelength as that to which a receiver is tuned. **1946** C. A. QUARRINGTON *Mod. Pract. Radio & Television* I. xxvii. 225 Another unpleasant inter-station noise is experienced when tuning off one station and when tuning on to the next. **1970** J. EARL *Tuners & Amplifiers* i. 24 Other features found in tuner-amplifiers..are automatic monostereo switching..and an inter-station muting device which in the 'on' position deadens the audio channels on weak signals..and on noise signals between stations, thus making it possible to tune over the f.m. band without the annoyance of a high background noise, the tuner audio channels only being activated when a transmission of a useable strength is tuned.

interstellar, *a.* Add: (Further examples.) Also, relating to matter or to travel in interstellar space; occurring in such regions.

1925 A. S. EDDINGTON *Internal Constitution of Stars* xiii. 377 Fixed calcium lines..are formed either in an envelope surrounding the whole system or during the passage of the light through interstellar space. *Ibid.* 379 The question..arises whether, granting that there is a cloud extending through interstellar space, the fixed lines are produced uniformly during the transit of the light from the star to the earth. **1930** *Astrophysical Jrnl.* LXXII. 199 A star at a distance of 100 parsecs from the sun should show an interstellar K line of intensity 2. **1936** S. ROSSELAND *Theoret. Astrophysics* xxiii. 243 An intensive study of interstellar absorption is a necessary feature of stellar spectroscopy. **1937** *Physical Rev.* LII. 1075/1 If..the third hypothesis..seems to point toward a physical relationship between space reddening (presumably caused by interstellar dust) and the formation of the interstellar line. **1937** WHIPPLE & GREENSTEIN in *Proc. Nat. Acad. Sci.* XXIII. 178 The energy density in the galactic nucleus will be considerably increased, resulting in an increased temperature and ionization of the interstellar particles. **1946** A. HUXLEY *Let.* 18 Mar. (1969) 539 People either wanted to talk about themselves, their job and its issues, or else they couldn't stick to the point at issue, but wandered off into interstellar space. **1950** [see *COSMONAUTICS]. **1960** O. STRUVE *Stellar Evolution* ii. 106 All indications are that interstellar clouds..gradually condense into protostars which later form real stars. **1962** LOVELL & CLEGG *Radio Astron.* ix. 141 The current belief about the remaining mass is that it is concentrated in the dust and gas which permeates interstellar space. By terrestrial standards this interstellar material is highly attenuated. **1963** *Mag. Fantasy & Sci. Fiction* I. 95 The cruiser *Ilker* had just gone into her interstellar overdrive beyond the orbit of Pluto. **1966** *McGraw-Hill Encycl. Sci. & Technol.* VII. 221/1 Interstellar lines can be distinguished from the stellar atmospheric lines by their extreme narrowness. **1969** *Times* 11 Jan. 4/7 Measurements of the radio signals from sulphur hydride may be a valuable check of estimates of the amount of sulphur in interstellar space. **1973** *Times* 28 Dec. 10/8 Amino acids ..are formed when three of the chemicals known to be present in the interstellar clouds..are mixed in a flask and subjected to an electric spark.

inter-sterile: see *INTER- *pref.* 5.

interstice. Add: **1. b.** *Physics.* The space between adjacent atoms or ions in a crystal lattice. Cf. *INTERSTITIAL *a.* 2 c.

1933 [see *INTERSTITIAL *a.* 2 c]. **1940** F. SEITZ *Mod. Theory Solids* xiv. 496 The interstices of metals forming substitutional alloys usually are much smaller than the atomic size. **1944** SNEED & MAYNARD *Gen. Inorg. Chem.* xlvii. 1108 The interstices in a metal lattice are small, hence only the smallest metalloid atoms, C, N, B and H, can form such compounds and solid solutions. **1972** GROSS & WEBSTER *Princ. Phys. Chem.* xix. 339 The holes or interstices formed between the layers of spheres in close-packed structures are of two different types, (a) tetrahedral interstices and (b) octahedral interstices.

interstitial, *a.* Add: **2. c.** *Physics.* Situated between the normally occupied points of a crystal lattice. Cf. sense *5.

1938 *Trans. Faraday Soc.* XXXIV. 851 Motion..is possible by jumps of ions from one interstitial position to the neighbouring one...An interstitial ion takes the place of a neighbouring lattice ion. **1940** F. SEITZ *Mod. Theory Solids* xiv. 496 Semi-conductors, such as zinc oxide, that have interstitial atoms. **1966** C. R. TOTTLE *Sci. Engin. Materi* vi. 92 There may be interstitial atoms present in a lattice as impurities.

5. Containing atoms or ions in interstitial positions; cf. sense *2 c.

1933 W. H. & W. L. BRAGG *Crystalline State* I. vii. 158 The atoms of H, N, C, or B are placed in the interstices of the metallic structures... Such structures are called *Einlagerungsstrukturen* by Hägg, and will be referred to here as 'interstitial structures'... All hydrides and nitrides [of transition elements], and some carbides, fall within the first category of interstitial compounds. **1948** *Acta Crystallogr.* I. 189/1 A large number of metals form metal-like phases with hydrogen, boron, carbon, nitrogen, and occasionally oxygen... Following suggestions of Hägg, these phases are generally regarded as interstitial solutions of the light elements in the metals. **1966** B. L. SHAW *Inorg. Hydrides* xi. 101 It is now known that the metal lattice often does change when the hydrogen is absorbed so that the term 'interstitial hydride' is not a good one.

B. *sb. Physics.* An interstitial atom or ion.

1961 *Physical Rev.* CXXIV. 669/2 If the interstitial is displaced slightly from its position in configuration 'B' toward surrounding crowdion, 'body-centred' or 'A' interstitial positions the calculations show that the interstitial returns to its position in configuration 'B'. **1969** [see *DISLOCATION 1 e]. **1972** *Physics Bull.* July 400/1 At the temperatures relevant to the operation of fast reactors both vacancies and interstitials, produced as a consequence of radiation damage, are able to migrate under thermal activation.

interstitialcy (intǝɹsti·fǎlsi). *Physics.* [f. INTERSTITIAL *a.* + -CY, after *vacancy*.] A kind of imperfection in a crystal lattice, characterized by an interstitial atom able to displace an atom from an adjacent lattice position so that it becomes the interstitial in its turn, able to displace another atom.

1950 F. SEITZ in *Acta Crystallogr.* III. 361/1 The lattice imperfection which is the inverse counterpart of a lattice vacancy is an interstitial atom which moves by jumping into a normal lattice site and forces the atom that is there into a neighboring interstitial site... We shall term it an interstitialcy. *Ibid.* 361/2 The interstitial atom associated with a passing interstitialcy at any point in its migration may be either an *A* atom or a *B* atom. **1956** *Physical Rev.* CI. 1210/2 The effects discussed here may be useful in distinguishing experimentally between interstitialcy and vacancy migration in other crystals. **1967** F. C. BROWN *Physics of Solids* x. 315 In the silver halides the interstitial ion moves by an interstitialcy mechanism. **1968** J. R. MANNING *Diffusion Kinetics for Atoms in Crystals* i. 6 Two jumps of the interstitialcy are required to move an atom from one lattice site to another.

interstitially, *adv.* (Examples in *Biol.* and *Physics.*)

1913 F. H. HUMPHRIS *Electro-Therapeutics for Practitioners* xvi. 115 Antiseptic chemicals..interstitially diffused throughout the growth. **1938** *Trans. Faraday Soc.* XXXIV. 851 Some ions exist interstitially within the normal lattice. **1965** *Nuclear Instruments & Methods* XXXVIII. 158/2 Interstitially located impurities. **1972** *Sci. Amer.* Nov. 45/1 In silver halide many cations wander interstitially through the lattice.

interstrain, inter-strain (i·ntǝɹstrēin), *a. Biol.* [f. INTER- 5 + STRAIN *sb.*[1]] Occurring or existing between one strain (STRAIN *sb.*[1] 7) and another; formed by crossing two strains.

1950 *Amer. Jrnl. Bot.* XXXVII. 631/1 Exceptions to the relatively ready inter-strain crossability among the strains hybridized were 109 × 105 and 109 × 110. **1955** *New Biol.* XIX. 52 Livesay's interstrain F_1 hybrids should be genetically uniform. **1965** *Jrnl. Gen. Microbiol.* XLI. 7 The progeny of inter-strain crosses.

inter-stream: see *INTER- *pref.* 5.

intersubjective (intǝɹsǝbdʒe·ktiv), *a. Philos.* [INTER- 4 c.] Existing between conscious minds. Hence **intersubje·ctively** *adv.*, in an intersubjective manner; **intersubjecti·vity,** the fact or state of being intersubjective.

1899 J. WARD *Naturalism & Agnosticism* II. 170 It seems to depend upon three elements or conditions which are consequences of intersubjective intercourse. **1925** J. E. TURNER *Theory Direct Realism* iv. 44 Intersubjective communication. **1934** M. BLACK tr. *Carnap's Unity of Sci.* iii. 42 It will be proved..that the physical language is inter-subjective and can serve as a *universal* language. *Ibid.* iii. 51 In such a case p will be said to have sense (for those persons) inter-subjectively. **1938** *Mind* XLVII. 133 Propositions about private experiences are intersubjectively understood. **1938** C. MORRIS *Found. Theory of Signs* § 12. 125 The thesis of the potential intersubjectivity of every meaning. **1945** K. R. POPPER *Open Soc.* II. xxiii. 205 Scientific objectivity can be described as the inter-subjectivity of scientific method. **1945** *Psychol. Rev.* LII. 258/2 Concepts which are to be of value to the factual sciences must be definable by operations which are..intersubjective and repeatable. **1956** *Essays in Crit.* VI. 10 The validation for the new myth philosophy ..plunges, in the vast reservoir of racial unconsciousness, for an intersubjective base for universality. **1957** C. LA DRIÈRE in N. Frye *Sound & Poetry* 97 The acknowledged intersubjectivity of meaning is a kind of objectivity. **1963** R. CARNAP in P. A. Schilpp *Philos. R. Carnap* 52 One of the most important advantages of the physicalistic language is its intersubjectivity, i.e., the fact that the events described in this language are in principle observable by all users of the language. **1967** W. SELLARS *Philos. Perspectives* xv. 371 Universals are..identities not only with respect to their many instances, but also with respect to the many minds which think in terms of them... This inter-subjective and inter-linguistic character must be accounted for by any adequate theory of abstract entities.

intersubstitutability, -substitutable, -territorial, -testamental: see *INTER- *pref.* 2 a, 2 a, 4 c, 4 b.

interti·llage. *U.S.* [INTER- 1.] Intercropping. So **interti·lled** *a.*

1912 L. A. MOREHOUSE in *Bull. Bureau Plant Industry, U.S. Dept. Agric.* No. 260. 55 The alfalfa roots give some trouble at times, especially with the first cultivations of an intertilled crop, but this difficulty does not seem to be encountered when the cantaloupe or the cucumber is selected as the initial crop in the rotation. *Ibid.* 60 Regular intertillage can not be omitted without incurring losses. **1914** *Rep. Brit. Assoc. Adv. Sci.* 270 The intertilled crop has in many sections largely taken the place of fallow. **1973** *Times Lit. Suppl.* 21 Dec. 1572/3 The schemes of the Altai region to replace fallow and grass with intertilled crops (sugar-beet, maize, and fodder beans).

intertonic: see *INTER- *pref.* 6.

intertrade (i·ntǝɹtrēid). [INTER- 2 a.] Reciprocal trade. Also *attrib.*

1915 E. CARPENTER *Healing of Nations* xi. 168 Merchants and dealers came in and effected the exchange, and so an intertrade has sprung up. **1929** *Daily Tel.* 15 Jan. 6/4 The variations in the curve of consumer demand are much less than the curve of inter-trade demand.

intertranslatable (intǝɹtrɑnslē̆i·tǎb'l), *a.* [INTER- 2 a.] Capable of being translated from one language to another and vice versa. Also *transf.* Hence **intertranslatabi·lity.**

a **1866** J. GROTE in *Jrnl. Philol.* (1872) IV. 58 We assume that all languages are pretty accurately intertranslatable. **1951** *Mind* LX. 93 There is thus no genuine intertranslateability. **1960** W. V. O. QUINE *Word & Object* ii. 76 An illusion that our so readily intertranslatable sentences are diverse verbal embodiments of some intercultural proposition or meaning. **1969** *Canad. Jrnl. Linguistics* XV. 47 From these (and other) approaches, therefore, we are forcefully reminded that linguistics now has more facets than ever—and that we must be on guard to maintain a degree of integration and inter-translatability among its schools.

intertrial, -tubercular: see *INTER- *pref.* 5, 6.

Intertype (i·ntǝɹtǝip). [f. name of the *International Type*setting Machine Company, which manufactured the first machines of this type.] The proprietary name of a composing machine which produces type in whole lines rather than individual letters; cf. *LINOTYPE, *MONOTYPE.

1913 *Official Gaz.* (U.S. Patent Office) 1 Apr. 242/2 International Typesetting Machine Company, New York, N.Y. Intertype—Type Casting, Setting, and Composing Machines. **1916** LEGROS & GRANT *Typogr. Printing-Surfaces* 437 *The Intertype*..closely resembles the two-letter, single-magazine Linotype... The constructors have given special attention to speed of operating... It appears to be virtually a copy of American model 5 Linotype. **1932** *Year Bk. Printing Dept. North-Western Polytechnic 1931–2* (Intertype Section, *subtitle*) Examples of Intertype composition composed by students in the Intertype Classes, and printed direct from the slugs. **1946** A. MONKMAN in H. Whetton *Pract. Printing & Binding* iii. 35/2 The Intertype... These [machines] are designed to produce a wide variety of composition, in the form of slugs. **1959** *Times* 14 Jan. 12/4 The great bulk of typesetting is achieved..by keyboard-operated Monotype, Linotype, and Intertype. **1967** V. STRAUSS *Printing Industry* ii. 74/1 The Intertype casts lines of type or slugs instead of individual types.

inter-union, -university: see *INTER- *pref.* 5.

interurban, *a.* Add: (Later examples.)

1898 *Engineering Mag.* May 163/1 A finely-illustrated description of an English inter-urban street railway. **1901** *Westm. Gaz.* 15 Feb. 4/3 The Minister of the Interior [in Spain] has prohibited inter-urban telephonic communication. **1903** *Ibid.* 14 May 16/3 In the provision of means of inter-urban transport, London was originally..not inferior to any other city. **1962** *Coast to Coast 1961–62* 72 The cameras zoomed to a middle-aged man alighting from an inter-urban Hovercraft. **1971** *Selling Today* Sept. 4/3 Inter-urban motorways.

b. Ellipt. as *sb.*, an interurban railway or train. *U.S.*

1912 *Out West* Mar. 204/1 There is no more beautiful road..than the Los Angeles interurban through Azusa. **1948** *Daily Ardmoreite* (Ardmore, Okla.) 11 Apr. 1/6 Fifty persons were hurt..when two interurbans smashed head-on.

interval, *sb.* Add: **2. c.** *Physics.* A quantity *ds*, invariant under the Lorentz transformation, that represents the separation of two events in space-time and is defined by $ds^2 = dx^2 + dy^2 + dz^2 - c^2dt^2$ (or by the negative of the right-hand side), where *dx*, *dy*, *dz*, and *dt* are the differences in the space and time coordinates of the events and *c* is the speed of light.

1918 A. S. EDDINGTON *Rep. Relativity Theory Gravitation* ii. 15 In the four-dimensional continuum the interval δs between two point-events..is unaffected by any rotation of the axes, and is therefore invariant for all observers. **1929** W. C. D. DAMPIER-WHETHAM *Hist. Sci.* ix. 422 Just as the distance between two points in the continuous space of Euclidean geometry is the same however measured, so, in the new continuum of space-time, two events may be said to be separated by an 'interval', involving both space and time. **1952** C. MØLLER *Theory of Relativity* iv. 99 This expression for the line element or the interval defines the geometry in (3+1)-space. **1959** J. AHARONI *Special Theory of Relativity* i. 25 When the interval is time-like it is always possible to find a system of coordinates in which the events appear in the same place.

6*. *Math.* **a.** A range between one number and another; *spec.* that between successive values of the argument in a mathematical table.

1838 *Penny Cycl.* XII. 508/2 The smaller the tabular interval, the more correctly will a given number of differences serve to make the interpolation. **1911** *Encycl. Brit.* XIX. 862/2 The actual calculation of the number of primes in a given interval may be effected by a formula constructed and used by D. F. E. Meissel. **1928** [see *INTERPOLATE *v.* 6]. **1946** *Nature* 12 Oct. 504/1 The basic sequence is the integration procedure for one interval of the integration, which is a sequence of operations starting from the initial values for that interval and giving final values which become the initial values for the next interval. **1968** FOX & MAYERS *Computing Methods for Scientists & Engineers* i. 8 If $1{\cdot}5 \leq x \leq 2{\cdot}5$, then $y = x^{10}$ is in the approximate interval $57{\cdot}6 \leq y \leq 9536{\cdot}7$. **1974** *Nature* 26 Apr. 739/1 The grain temperature..is determined by the emissivity between 4 and 15 μm. We have carried out calculations..over this spectral interval.

b. A set composed of all the numbers between two given numbers, which may be either included in the set (a *closed interval*) or excluded (an *open interval*); an analogously defined subset of any partially ordered set.

The distinction made in quot. 1949 is not usual.

1902 *Encycl. Brit.* XXVIII. 545/2 This domain may be an 'interval', i.e., it may consist of two terminal numbers, all the numbers between them and no others. **1949** S. LEFSCHETZ *Introd. Topology* i. 27 If the real line *L* is parameterized by means of a parameter *u*, then an interval is a set: $a < u < b$, and a segment is a set $a \leq u \leq b$, $a < b$. **1962** B. H. ARNOLD *Intuitive Concepts Elem. Topology* viii. 164 It is easy to see that a set $A \subset R$ is an interval if and only if it contains all points which lie between any two of its members. **1965** A. ABIAN *Theory of Sets* iv. 184 Let $(P, \leq)$ be a partially ordered set. For every two elements *a* and *b* of *P*, the set of all elements of *P* such that $a \leq x \leq b$ is called a closed interval and is denoted by $[a, b]$, and the set of all elements *x* of *P* such that $a < x < b$ is called an open interval and is denoted by (a, b).

7. (sense 1) *interval music, prose*; also **interval running,** in *Athletics*, a method of training by running set distances at pre-determined speeds (opp. *FARTLEK); so **interval training**; (sense 2) **interval signal** *Broadcasting* (see quot. 1941).

1951 *Catal. of Exhibits, South Bank Exhib., Festival of Britain* 176 Recordings for Interval Music. **1967** *Daily Tel.* 12 May 20/6 The unchanging interval-music. **1970** *Listener* 8 Jan. 60/2 The material..provided interval prose of just the right emotional tone between two parts of a good concert. **1957** *Oxf. Pocket Bk. Athletic Training* (ed. 2) 22 *Interval running*,..a series of runs over a particular distance each in a certain time..linked up by jogging between each. **1932** *B.B.C. Year Bk.* 373 The sudden failure of the programme..is likely to cause the listener to think his set has become faulty. The radiation of the interval signal relieves all anxieties on this account. **1941** *B.B.C. Gloss. Broadcasting Terms* 16 *Interval Signal*, particular sequence of sounds used by a broadcasting organization to fill short intervals between programmes, and to enable listeners to identify its transmissions. **1962** *Times* 9 Apr. 5/7 Oxford's application of interval training, on the other hand, may have been at fault.

intervallary (intǝɹvæ·lǎri), *a. rare.* [See INTERVALLIC *a.* and -ARY[2].] = INTERVALLIC *a.*

c **1864** G. M. HOPKINS *Jrnls. & Papers* (1959) 76 The division then is of abrupt and gradual,..of intervallary and chromatic..and qualitative beauty. **1889** *London Med. Recorder* 20 Aug. 310/2 Herr Pfeiffer recommends in the intervallary treatment of gout the Wiesbaden baths.

intervalometer (intǝɹvălǫ·mītǝɹ). *Photogr.* [f. INTERVAL *sb.* + -OMETER.] An attachment for a camera that enables photographs to be taken automatically at set intervals.

1933 *Discovery* Feb. 60/1 The photographer works out the intervals that he must allow between exposures, so as to secure a 60 per cent. overlap, and sets the required interval on the intervalometer on the remote control. **1971** *Amat. Photographer* 13 Jan. 51/3 Time-lapse devices known as Intervalometers (repeating timers giving various ranges of delays between 0·2 sec. and 10 min.). With these the camera can be left unattended for indefinite periods (for motion analysis of such things as plant growth, etc.).

inter-valve, -varietal, -veinal: see *INTER- *pref.* 5, 6, 6.

intervening, *ppl. a.* (in Dict. s.v. INTERVENE *v.*). *spec.* in *Psychol.*, *intervening variable*, a factor, such as individual memory, desire, or habit, which may affect the results of psychological tests or experiments in a way which is hard to predict.

1935 E. C. TOLMAN in *Philos. of Sci.* II. 365 The nature of this their resulting behavior is determined by a set of intervening variables to be conceived as lying in the organism... The molar behaviorist seeks to state the intervening variables as specific types of behavior-readiness. **1951** *Mind* LX. 50 The methodological device of the 'intervening variable' (drive, habit, demand, etc.). **1963** A. PAP *Introd. Philos. of Sci.* v. xx. 383 Human behavior also has mental determinants, such factors as memories, expectations, desires, [etc.]. But since these 'intervening variables' were supposed to be inaccessible to scientific investigation, the tendency developed to interpret them as certain *dispositions* to overt behavior.

intervention. Add: **1. b.** *Law.* The action of one, not originally a party, who intervenes in a suit.

1860 *Act* 23 & 24 *Vict.* c. 144 § 7 The said Proctor.. may..intervene in the Suit, alleging such Case of Collusion..and it shall be lawful for the Court to order the Costs..arising from such Intervention, to be paid by the Parties. **1864** G. BROWNE *Treat. Princ. & Pract. Court for Divorce & Matrimonial Causes* 152 There are two kinds of intervention, one 'by any person' under the first branch of the section, the second by the Queen's proctor under the latter branch... It appears that at any time before a decree for dissolution of marriage is made absolute, it is competent for one of the public to intervene.

Ibid. 153 The Court will not act on an intervention, when satisfied that it is made at the instance of the respondent or co-respondent. **1883** [see Intervener^1]. **1952** *Stroud's Judicial Dict.* (ed. 3) II. 1500 Intervention in divorce proceedings (generally by the King's Proctor) is for (*a*) collusion, or (*b*) suppression of a material fact.

Hence **interve·ntionism**, the principle or policy of intervening, esp. in international and economic affairs; **interve·ntionist**, also, one who favours a doctrine of intervention; one who favours intervention with the course of a disease on medical grounds (*Cent. Dict.*); also as *adj.*

1915 *Morning Post* 19 Apr. 8/4 Interventionist, and, in some places, neutralist meetings were held yesterday at Milan. **1921** *Glasgow Herald* 22 July 7 M. Tchitcherin sees in this fact another 'interventionist' manœuvre. **1923** *Ibid.* 29 Mar. 5 The methods of interventionism..are contrary to the spirit of cooperation. **1930** *Times Lit. Suppl.* 14 Aug. 648/2 The 'interventionist' conception of miracle has passed out of fashion. **1940** *Economist* 6 July 2/2 It was not primarily because of this advocacy of support for Britain that Mr Willkie was nominated...We cannot go further than to say that Mr Willkie's interventionism did not prevent his success. Moreover, there are severe limitations on interventionism even of the Willkie type... For example, there is the almost universal qualification that aid must stop 'short of war'. **1945** K. R. Popper *Open Soc.* II. xvii. 117 We must demand that *laissez-faire capitalism* give way to an *economic interventionism*. *Ibid. Notes* 318, I suggest using the name *laissez-faire capitalism* for that period which Marx analysed..and the name *interventionism* for our own period. The name 'interventionism' could indeed cover the three main types of social engineering in our time: the collectivist interventionism of Russia; the democratic interventionism of Sweden and the 'Smaller Democracies' and the New Deal in America. **1962** *Listener* 20 Dec. 1040/1 The long-standing tradition [*sc.* in France] of an active, confident, and interventionist civil service. **1967** *Economist* 30 Sept. 1162/1 The new interventionism is an economic theory that came into vogue with some Labour intellectuals about three years ago, and with some core of justification. Its kernel was the entirely true argument that the governments of some countries with successful postwar economic records (e.g. France, Japan, Italy) have had greater scope for influencing the course of major industrial investment than have the less successful governments of postwar Britain. **1969** *Observer* 26 Jan. 8/4 America has never swung back..to isolationism and Russia has never regressed to all-out revolutionary interventionism. **1970** *Times* 2 July 8/3 It will not be easy to persuade the country to prefer government interventionism to the freedom which is Mr. Heath's aim. **1971** *Physics Bull.* June 261/2 Interventionists such as Joan of Arc, Ralph Nader, Ghandi [*sic*] and Sir Alan Herbert. **1973** *Financial Times* 28 Feb. 27/2 The contrast between the philosophy of the Conservative Government that believed so earnestly in the miraculous powers of laissez-faire in 1970 and the interventionist Conservative Government of today.

interview, *sb.* Add: **1. c.** Similarly in broadcasting.

1956 B. Paulu *Brit. Broadcasting* vii. 176 The BBC prefers straight talks to interviews, believing that talks are more apt to be carefully worked out. **1965** *Listener* 18 Feb. 260/2 The right interview..has to be used in the right place with the right person if the programme is to be craftsmanlike. **1974** *Radio Times* 21 Feb. 5/5 The background to my interviews is firmly in my head, though I still face the interviewer's perennial problem of..when to change the topic.

4. *interview room* (in a police station or prison).

1967 E. Grierson *Crime of one's Own* xvi. 135 Donald was called..to the interview room on the ground floor [of the prison]. **1969** J. Gardner *Compl. State of Death* v. 76 Interview room. Usual. Stone walls and glossy paint. Table. Three chairs. **1974** J. Wainwright *Evidence I shall give* xxxiii. 188 He..left the Murder Room and found a telephone in one of the Interview Rooms, where he couldn't be overheard.

interview, *v.*^{2} Add: Now *spec.*, to talk with or question (a person) for a programme broadcast on radio or television.

1933 *Radio Times* 14 Apr. 96/1 Three speakers..will interview experts in design before the microphone on behalf of listeners. **1957** G. Harding *Along my Line* xvi. 159 One or two of the guests whom I had agreed to interview had gone to the trouble to notify their friends and relations across the width and breadth of Canada. **1974** *Radio Times* 21 Feb. 41/2 Robin Day interviewing those making the news.

interviewee. (Later examples.)

1959 F. Grisewood *My Story of B.B.C.* xii. 190 The laborious business of discovering exactly how many.. interviewees watched or listened to each of more than a hundred programmes a day. **1965** *Listener* 20 May 756/2 It [*sc.* a radio programme] seemed to have caught some of the town-hall atmosphere, tersely described by a teenage interviewee as 'dead'. **1970** *Nature* 8 Aug. 641/1 Hammerton might contemplate the exact nature of the questions he asked his interviewees. **1973** *Guardian* 17 Apr. 15/1 The interviewee usually remembers the one-to-one relationship with the reporter.

interviewer. Add to *spec.* sense: also, a person employed by a broadcasting organization to perform a similar function.

1939 *Radio Times* 25 Aug. 20/2 'Come and be televised.' Interviewer, Elizabeth Cowell. **1941** *B.B.C. Gloss. Broadcasting Terms* 16 *Interviewer*, person whose role is to interrogate a broadcaster at the microphone and to elicit his story. **1969** *Times* 24 Nov. 17/3 Praise the Lord for a television interviewer who does not ram his personality down our throats. **1974** [see *interview *sb.* 1 c].

interviewing, *vbl. sb.* (Later examples.)

1949 *Radio Times* 15 July 9/2 As compère, Brian Reece returns to the job in which he first showed his flair for radio—that of interviewing, 'gagging', and knitting a programme together. **1960** *20th Cent.* May 458 There is no substitute for direct contact through intimate interviewing. **1968** *Guardian* 26 Apr. 9/1, I wish I had some interviewing officers who could interview.

inter-village, -villous: see *inter- *pref.* 5, 4 a.

Intervision (i·ntəɹviʒən). [f. *Inter*national tele*vision*.] (See quot. 1962^2.)

1961 *Listener* 2 Nov. 729/2 Other countries in the Intervision (East European) network. **1962** *Ibid.* 20 Dec. 1042/2 It [*sc.* the speech] was also carried by Intervision for viewers in Finland, Denmark, and Sweden. **1962** *B.B.C. Handbk.* 25 International co-operation between BBC engineers, Eurovision, and its Eastern counterpart, Intervision.

inter vivos: see *inter L. prep.

intervocal, *a.* Add: So **intervoca·lically** *adv.*

1950 D. Jones *Phoneme* 22 Double k occurs only intervocalically. **1964** R. H. Robins *Gen. Ling.* viii. 327 Cockney speakers often have glottalized stops intervocalically. **1972** *Language* XLVIII. 465 Whatever analysis is adopted, the patterning of these segments is admittedly anomalous. They occur only intervocalically.

inter-war, *a.* [Inter- 5.] Of a period occurring between wars; *spec.* of the period 1919 to 1939 between the two world wars.

1939 O. Lancaster *Homes Sweet Homes* 74 The interwar period through which we have just passed. **1944** W. Temple *Church looks Forward* v. 39 There must be no slipping back into the self-seeking and self-indulgence of the interwar years. **1959** *Encounter* Aug. 47/1 In the inter-war period [in Hungary] a limited number of Government and Opposition newspapers could be printed. **1973** *Listener* 25 Jan. 104/3 *Mein Kampf* is a vital document of the inter-war period.

interweft (i·ntəɹweft). *rare.* [f. Inter- 2 a + Weft *sb.*^{1}] = Interweftage.

1927 R. Fry *Cézanne* 80 In nature such a scene gives an effect of a confused interweft. **1939** —— *Last Lectures* 19 We begin to yield ourselves to the rhythmical movements of Botticelli's linear design, to its mazy interweft of curves.

interxylary: see *inter- *pref.* 6.

interzonal (intəɹzōᵘ·năl), *a.* [Inter- 4 a, c.] Existing or carried on between zones.

1881 E. L. Mark in *Bull. Mus. Compar. Zool. Harvard* VI. ii. 198 When seen lengthwise of the spindle, the numerous..thickenings appear arranged..in the form of a ring... Between the two zones of thickenings are stretched delicate nearly parallel threads, which I shall designate as interzonal filaments. **1920** L. Doncaster *Introd. Study Cytol.* iii. 32 As the chromosomes travel to the poles, the interzonal spindle-fibres are seen still running continuously from pole to pole. **1956** *Ann. Reg. 1955* 229 In interzonal trade [in Germany] the volume reached little more than half the target figure for the year. **1959** *Ann. Reg. 1958* 263 Dr Adenauer..warned the East Germans that interzonal trade..would meet with difficulties if there were any interference with West Berlin's supplies.

inter-zone: see *inter- *pref.* 5.

intestinal, *a.* **1.** *fig.* (Later examples.)

1945 *N.Y. Times Mag.* 21 Oct. 18/4 (*caption*) Edgar Bergen—'That guy has a lot of intestinal fortitude.' Charlie McCarthy—'I know a quicker way to say that.' **1961** J. S. Mayer *Restorative Art* (ed. 4) 21 This is the time for intestinal fortitude and determination.

in-thing: see *in *a.* 2.

‖ **intichiuma** (intitʃi,ū·mā). Pl. intichiuma. [Native name.] Sacred ceremonies performed by some Central Australian Aborigines with the purpose of increasing the totemic plants or animals, and thus ensuring a good food supply.

1899 Spencer & Gillon *Native Tribes Cent. Austral.* vi. 170 It sometimes happens that the members of the totem, such as..the rain or water totem, will hold their *Intichiuma* when there has been a long drought and water is badly needed. **1911** J. G. Frazer *Golden Bough* (ed. 3) I. iii. 85 The general result supposed to be accomplished by these magical totemic ceremonies, or *intichiuma*, as the Arunta call them, is that of supplying the tribe with food and other necessaries. **1950** J. Strachey tr. *Freud's Totem & Taboo* iv. 139 The *intichiuma* ceremonies of the Central Australian tribes.

intilted (i·ntiltėd), *ppl. a.* [In *adv.* 11 b.] Tilted inwards.

1940 F. Smythe *Adventures Mountaineer* viii. 117 At the base of the buttress lay a big intilted granite slab. **1956** M. Stewart *Wildfire at Midnight* x. 87 He climbed.. easily, making for the next stance, which was an in-tilted ledge some fifteen feet above him.

intimacy. 1. b. Delete 'illicit' and add further examples.

1906 B. Webb in S. Hynes *Edwardian Turn of Mind* (1968) iv. 114 Friendship between particular men and women..is practically impossible..without physical intimacy... There remains the question whether, with all the perturbation caused by such intimacies, you would have any brain left to think with? **1907** *Westm. Gaz.* 14 Dec. 11/2 She stayed the night with Wood at his father's house. ..Intimacy took place on that occasion. **1963** A. Heron *Towards Quaker View of Sex* 71 *Intimacy*, close friendship, but also as a synonym for sexual intercourse.

intimal (i·ntimăl), *a.* *Biol.* [f. Intim(a + -al.] Of the intima.

1907 *Buck's Handbk. Med. Sci.* (rev. ed.) II. 98/2 A ligated vessel may be occluded by intimal proliferation without thrombosis taking place. **1961** *Lancet* 22 July 187/1 Numerous small systemic arteries can be seen and some of these show intimal hypertrophy.

intimate, *a.* **3. b.** Delete 'illicit' and add later examples.

1926 R. Macaulay *Crewe Train* ii. vi. 139 Some of them were..what newspapers call intimate together, without having undergone marriage. **1963** 'E. McBain' *Ten Plus One* (1964) xiv. 162 Do you mean that you and the other two girls were *intimate* with these boys? **1969** *Times* 15 Nov. 3/2, I ripped her dress off. She was lying on her face. I was intimate with her.

d. Used allusively of women's underclothing.

1904 P. Gibbon *Souls in Bondage* i. 5 Clothes hung on lines in all directions, intimate linen flapped in the wind. **1970** *Globe & Mail* (Toronto) 28 Sept. 32/4 (Advt.), Next week we'll be highlighting panti hose and the week after it will be Intimate Apparel week in our Foundations Departments. **1973** *Tucson* (Arizona) *Daily Citizen* 22 Aug. 10 (Advt.), Intimate Apparel, mall level.

e. Of a theatrical performance, esp. a revue: that aims at establishing familiar and friendly relations with the audience. Also of a theatre itself.

1915 H. K. Moderwell *Theatre To-Day* xvi. 309 The [Manchester Repertory] theatre happens to work mostly with the modern 'intimate' or 'realistic' play, and so is enabled to get along with one company of actors, albeit a large one. **1919** A. Hornblow *Hist. Theatre Amer.* II. xxxi. 343 The new method is to build a smaller house, or *théâtre intime*, allowing of an auditorium with limited capacity so that no seat will be very far from the stage. Among these theatres may be mentioned..the Maxine Elliott, one of the first of the intimate theatres. **1929** *N.Y. Times* 1 May 28/5 The Little Show. An intimate revue in two acts and twenty-seven scenes. **1930** *Nation* (N.Y.) 24 Sept. 331/1 The place and the popularity of the intimate music show is assured. People are delighted if it sounds like an impromptu affair. **1948** *Penguin Music Mag.* VI. 51 A series of intimate opera to be given at La Scala with a small audience seated on the stage. **1952** Granville *Dict. Theatr. Terms* 102 *Intimate revue*, a smart, topical revue played in a small (*intimate*) theatre. **1959** *Times* 22 Jan. 3/2 Intimate revue, at a glance, appears to be the theatre's gift to television. **1961** A. Berkman *Singers' Gloss. Show Business* 52 The Intimate Position of the head is that in which both the face and the eyes are directed squarely toward the other person. **1974** *Times* 27 Aug. 8/5 That quiet British archness which put the phrase Intimate Revue into the language.

‖ **intime** (æntīm), *a.* Examples of the French word revived in modern English use.

1857 W. Bagehot in *National Rev.* V. 411 The real, rougeless, *intime* Flicflac we know not. **1947** *Ballet Ann.* I. 73 *Ballet intime* bears the same relation to *Ballet Russe* as chamber music bears to a symphony. **1963** J. Wiesenfarth *Henry James* iii. 60 Another carriage ride introduces Maisie to an intime relationship between her father and her governess. **1968** W. Garner *Deep, Deep Freeze* vii. 87 She..patted the sofa, 'Come and sit here. It is more *intime*.' **1973** *New York* 26 Mar. 21/2 Intime restaurant with continental cuisine.

intimism (i·ntimiz'm). Also ‖ **intimisme**. [ad. F. *intimisme*.] (See quot. 1959.)

1903 *Daily Chron.* 19 Dec. 3/2 The great result of the emancipation of French art from Italian influence is what he [*sc.* C. Mauclair] calls 'intimism'. **1959** P. & L. Murray *Dict. Art & Artists* 164 *Intimisme*, a form of Impressionist technique applied to the depiction of everyday life in domestic interiors rather than to landscape. The work of Bonnard and Vuillard is usually meant. **1960** *Times* 16 Feb. 14/7 To paint a group of television viewers would be a caricature of 'intimisme'.

Hence **i·ntimist**, ‖ **i·ntimiste** *a.*, relating to intimism. Also *transf.* Also as *sb.*, a painter following the principles of intimism.

1903 P. G. Konody tr. *Mauclair's French Impressionists* ix. 196 Simon Bussy is decidedly the most personal of that young generation of 'Intimists' who seem to have retained the best principles of the Impressionist masters. **1937** *Times* 7 Oct. 12/3 In artistic slang they [*sc.* Bonnard and Vuillard] are grouped together as 'intimists', and the name serves well enough to describe..their habit of dwelling upon their material so as to extract from it the last possibilities in subtle colour relationships. **1959** *Listener* 21 May 898/1 The modified impressionism of Vuillard, a style especially suitable for such *intimiste* themes. **1967** *Ibid.* 18 May 654/1 Károly Ferenczy's 'Sunny Morning' of 1905 and the *intimiste* paintings of about the same period by..an associate of Bonnard in Paris. **1968** *Ibid.* 23 May 680/1 Jiri Menzel has everywhere been praised, like other Czech directors, as *intimiste*. **1973** *Times* 20 Mar. 9/7 The gallery shows a mixture of contemporary art and regular *Paris–Londres* exhibitions together with minor impressionists and intimists.

into, *prep.* Add: **23.** Interested or involved in; knowledgeable about. *colloq.*

1969 *Rolling Stone* 28 Jan. 19/1, I tend to like the stuff the rock groups are doing because they're creative and original, and that's something I'm very much into. **1969** *Down Beat* 20 Mar. 17/1 She is a Libra, for those of you who are into that. **1969** *It* 4–17 July 15/3 He was basically into being a hustler, which he was very, very good at. **1971** *Ink* 12 June 19/1 This should have been the highlight of the evening, but the audience just wasn't into it. **1971** *New Yorker* 11 Sept. 48 First I was into Zen, then I was into peace, then I was into love, then I was into freedom, then I was into religion. Now I'm into money. **1973** *Listener* 15 Feb. 209/1 Margaret is 'into' astrology, and consults the *I-Ching* each morning.

B. *adj.* *Math.* Used to designate a mapping (of one set 'into' another (Into *prep.* 6 a)) that is not necessarily 'onto'.

1949 S. Lefschetz *Introd. Topology* 215 (Index), Into transformation. **1958** K. S. Miller *Elem. Mod. Abstr. Algebra* i. 21 Since the mapping is into, there may exist elements in $\bar{S}$ which have no preimage in *S*. **1968** E. T. Corson *Metric Spaces* vii. 85 Every 'onto' mapping is 'into' but not all 'into' mappings are 'onto'.

intolerant, *a.* Add: **1. b.** *Ecol.* Of trees or other plants: unable to flourish in deep shade.

1898 G. Pinchot *Adirondack Spruce* 5 The Poplar, a tree very intolerant of shade..rapidly takes possession of the soil. *Ibid.* 22 If the intolerant species can get the start, being often rapid of growth, they may hold their position by growing above the other trees about them. **1929** Weaver & Clements *Plant Ecol.* xiv. 320 The leaves of intolerant trees can not make food in weak, diffuse light. **1952** P. W. Richards *Tropical Rain Forest* iii. 42 Of the young trees in this patch of undergrowth the more light-demanding (intolerant) species respond more quickly than the shade-bearing (tolerant) species. **1965** G. L. Clarke *Elem. Ecol.* (rev. ed.) vi. 233 Plants that require strong illumination and will not survive or develop in reduced light are referred to as intolerant species.

‖ **intombi** (intǫ·mbi). [Xhosa, Zulu *i-ntombi* maiden.] (See quot. 1913.)

1809 R. Collins *Jrnl.* in D. Moodie *Record: Papers Native Tribes S. Afr.* (1842) v. 46 Cattle are never given for a Tumbee, but her father or brother is supplied with assagays by her keeper. **1833** S. Kay *Trav. Caffraria* xviii. 470 'That,' said he, 'contains the body of an *intombi* (young woman) who was killed by lightning from heaven, about two years ago.' **1855** J. W. Colenso *Ten Weeks in Natal* 26 There is another special one [*sc.* reason] for the young men wishing to go home from time to time—namely, to make acquaintance with the *intombies*, or young women, whom they will one day acquire for wives. **1913** C. Pettman *Africanderisms* 228 *Intombi*,..a girl or young unmarried woman. **1947** L. Hastings *Dragons are Extra* vi. 118 The little girls, the intombis, were busy with bunches of twigs tidying up the huts.

intonation^1. Add: **4.** (Later examples.)

1935 M. Schubiger *Role of Intonation in Spoken Eng.* 2 Word-order can remain unaltered, and then the different intonation, the rising instead of the falling tune, is the sole bearer of the interrogative relation. **1965** W. S. Allen *Vox Latina* 6 It is important to distinguish tone from intonation. The former refers to the pitch-patterns operative within individual words, whereas 'intonation' refers to the pitch-pattern operative over the whole clause or sentence.

5. *attrib.* and *Comb.*, as *intonation change, morpheme*; **intonation contour**, a succession of levels of pitch extending over an utterance; **intonation curve**, the rising and falling of pitch within an utterance; **intonation pattern**, a pattern of variations in pitch; **intonation phoneme** = *intoneme; **intonation tune** (see quot. 1964); **intonation turn**, the point, usually at a prominent part of an utterance, at which the intonation rises or falls.

1964 C. Barber *Present-Day Eng.* iii. 50 There are intonation-changes inside the syllable which require a certain length of vowel to manifest themselves. **1946** K. L. Pike *Intonation Amer. Eng.* iii. 20 All speakers of the language use basic pitch sequences in similar ways under similar circumstances. These abstracted characteristic sentence melodies may be called *Intonation Contours.* **1960** [see *attitudinal *a.*]. **1964** R. A. Hall *Introd. Ling.* xix. 114 An intonation contour does not..make any difference in the 'dictionary meaning' of an utterance. **1970** Intonation contour [see *contour *sb.* 1 e]. **1936** H. Mulder *Cognition & Volition in Lang.* ii. 65 Wishes, commands, and questions introduced by interrogative pronouns, can be communicated on the same intonation-curve as plain statements. **1965** *Language* XLI. 498 In refusing to consider intonation-curves as subject to division into significant units, Martinet alleges that every modification..of a melodic curve brings with it a corresponding modification of meaning. **1953** *Internat. Jrnl. Amer. Ling.* XIX. ii. Suppl. 29 Hjelmslev requested a metalinguistic analysis of the English intonation morphemes which Smith had demonstrated earlier so as to make clear the difference between differential meaning and the meaning in general of the intonation patterns. **1966** G. N. Leech *Eng. in Advertising* ii. 18 The relationship of apposition between elements is marked in speech by tone-concord, or equivalent intonation patterns on each element. **1971** D. Crystal *Ling.* Interlude 134 A noun phrase may not have any separate intonation pattern at all. **1948** Intonation phoneme [see *intoneme]. **1934** J. J. Hogan *Outl. Eng. Philol.* i. v. 31 Its [*sc.* a verse's] accompanying intonation-tune suffers modification by the regular tune which is what the line always retains of its character as a sentence or clause. **1964** R. H. Robins *Gen. Ling.* iv. 148 Intonations or intonation *tunes*, as they are often called, are regular sequences of pitch differences coextensive with a whole sentence or with successive parts thereof, and constituting an essential feature of normal spoken utterance. **1935** M. Schubiger *Role of Intonation in Spoken Eng.* 9 If the psychological predicate consists of several words the most important gets the intonation turn.

Hence **intona·tional** *a.*, relating to intonation; **intona·tionally** *adv.*, in an intonational manner.

1895 J. Osgood in *Forum* June 503 The misused intonational 'twist', technically noted as the circumflex inflection. **1949** E. A. Nida *Morphol.* (ed. 2) 62 In English the sentence-final glides which follow the last intonationally stressed syllable constitute morphemes. **1952** *Trans. Philol. Soc.* 91 Differences of intonational relationship between stem and ending. **1957** *Publ. Amer. Dial. Soc.* xxviii. 6 We might start intonationally with Qs [*sc.* questions] classed as upmoving and downmoving. **1958** C. F. Hockett *Course in Mod. Ling.* 45 Certain types of speech..show..a total loss of intonational contrasts. **1964** R. H. Robins *Gen. Ling.* iii. 112 The different ways in which pitch differences are exploited intonationally and tonally. *Ibid.* iv. 149 In English..stressed syllables carry more intonational weight than unstressed syllables. **1971** D. Crystal *Ling.* 133 The intonational movement over the noun phrase as a whole must be indicated.

intoneme (intōᵘ·nīm). [Short for **intonation phoneme*: see *-eme.] An intonation pattern that contributes to the meaning of an utterance.

1948 K. L. Pike *Tone Lang.* v. 60 One may choose to call the key pitches *intonation phonemes* or *intonemes*. **1965** *Language* XLI. 498 Intonation-curves..can nevertheless be classified into contrasting units (call them 'tonemes' or 'intonemes' if you like). **1967** *Word* XXIII. 473 To turn again to my son's efforts to communicate, it was possible to isolate at the age of four months two intonemes, the one for desiderative and the other for deictic purposes. **1968** *Language* XLIV. 82 These two intonemes operate in the high range in the intonation system.

intorsion. Add: Also *spec.* in *Ophthalm.* (see quots.).

1899 A. Duane tr. *Fuchs's Text-bk. Ophthalm.* (ed. 2) xiv. 578 It [*sc.* the superior rectus muscle] also rolls the eye in such a way that the upper extremity of its vertical meridian is inclined inward (intorsion). **1964** [see *extorsion].

in toto: see In *L. prep.* in Dict. and Suppl.

Intourist (i:ntūᵊ·rist). [Russ. *Inturist*, abbrev. of *inostránnyĭ turist* foreign tourist.] Name of the State Travel Bureau of the U.S.S.R. Also *attrib.*

1932 *Intourist's Pocket Guide to Soviet Union* 581 The State Joint-Stock Company for Foreign Tourists—*Intourist*—was established in 1929 jointly by the People's Commissariat for Foreign and Domestic Trade, the People's Commissariat for Ways and Communications and the Soviet Mercantile Fleet. **1935** N. Mitchison *We have been Warned* ii. 194 All arrangements in the competent hands of the Intourist Office, who undertook all Russian travel. **1940** H. G. Wells *Babes in Darkling Wood* iii. ii. 251, I suppose they were packing off most of the Intourist people. **1949** F. Maclean *Eastern Approaches* i. iii. 35, I walked into the local branch of Intourist and informed the seedy little Armenian clerk behind the counter that I wished to book a passage across the Caspian to Central Asia. **1952** M. McCarthy *Groves of Academe* (1953) ix. 169 He took you through that session like a regular Intourist guide with a party of dumb fellow-travellers. **1958** J. Gunther *Inside Russia Today* i. 29 You go to any travel agency that has an arrangement with Intourist, the official Russian agency..and apply for a tourist visa. **1973** T. Allbeury *Choice of Enemies* xix. 95 The failed ballerinas seconded to Intourist as easy lays for Western salesmen.

in-town. Restrict *Sc.* to sense in Dict. and add: **2.** *adj.* and *adv.* Within (the central part of) a town.

1817 J. Keats *Let.* 16 May (1958) I. 147, I am glad to hear of Mr T's health and of the Wellfare of the In-town-stayers. **1941** W. Stevens *Let.* 25 Mar. (1967) 388 Today, as I walked in-town, I heard..song sparrows. **1958** S. Spender *Engaged in Writing* i. 24 A concourse of canals that formed a minute intown harbour. **1967** *Boston Herald* 1 Apr. 22/1 Three intown congregations..will join in worship here at 11 a.m. Sunday.

intra (i·ntră), *prep.* [L., = within.] **1.** In phr. *intra vires*, within the powers or legal authority (*of* a person, etc.). Cf. Ultra *prep.* 1.

1877 S. Brice *Treat. Doctrine Ultra Vires* (ed. 2) ii. ii. 56 It will be best to retain the secondary use of the term Ultra Vires, and..to deal only with those proceedings which are bad in reference to matters admitted to be Intra Vires in both meanings. **1884** *Law Times* LXXVIII. 110 (Stanford), If this were *intra vires*, the other securities which they had accepted were not *bonâ fide* ones. **1930** A. Palmer *Company Secretarial Pract.* 185 Excess borrowing *intra vires* of the company but *ultra vires* of the directors may be ratified by ordinary resolution. **1966** *Rep. Comm. Inquiry Univ. Oxf.* I. 256 We have already explained..the methods by which this power may be kept *intra vires*. **1970** *Internat. & Compar. Law Q.* 4th Ser. XIX. ii. 244 It would seem that, at least in New Zealand municipal law, if the Antarctica Act is *intra vires*, the Dependency includes the high seas south of 60° S.

2. intra vitam, during life; while still living. Freq. *attrib.*, = *intravital *a.*

1881 *Bull. Mus. Compar. Zool. Harvard* VI. ii. 264 Whether the intranuclear network was present as a structure *intra vitam*. **1898** *Jrnl. Microsc. Soc.* 133 Herr A. M. Przesmycki has made a long series of experiments on *intra-vitam* staining. **1965** *Acta Med. Scand.* CLXXVIII. 155 (*heading*) Intra-vitam diagnosis of oxalosis.

intra-, *prefix.* Add: **1. intra-amnio·tic**, taking place, situated, or administered within the amnion; so **intra-amnio·tically** *adv.*, within the amnion; **intra-arterial** (further examples); also, administered into an artery; **intra-arte·rially** *adv.*, (by injection) into an artery; **intra-arti·cular**, situated within or passing into a joint of the body; **intracapi·llary**, existing within a blood capillary; **intraca·rdially** *adv.*, into the heart; **intra-cerebral** (earlier and later examples); **intrace·rebrally** *adv.*, in or into the cerebrum; **intraciste·rnal**, occurring within or (of an injection) administered into a cistern of the body, esp. one in the brain; hence **intraciste·rnally** *adv.*; **intracoa·stal**, situated close to the coast; **intraco·ronal** *Dentistry*, placed or performed within the crown of a tooth; hence **intraco·ronally** *adv.*; **intraco·rtical**, situated or occurring within the cortex of the brain; **intracru·stal** *Geol.*, situated within the earth's crust; **intracry·stalline** *Min.*, occurring within a crystal; **intracu·ltural**, occurring within a culture; hence **intracu·lturally** *adv.*; **intracuta·neous** = *intradermal* (below); hence **intracuta·neously** *adv.*; **intracytopla·smic**, situated or occurring within the cytoplasm of a cell; **intra-departme·ntal**, done or occurring within a department; **intrade·rmal, -de·rmic**, situated or applied within the skin; hence **intrade·rmally** *adv.*; **intradu·ctal**, situated or applied within a duct (of a breast); **intradu·ral**, situated or performed within the dura mater; hence **intradu·rally**; **intraepide·rmal, -epide·rmic**, situated or occurring within the epidermis; **intra-Europe·an**, occurring or carried on within Europe; **intra-experie·ntial**, within experience; **intrafasci·cular** *Bot.*, situated within a vascular bundle; **intraforma·tional** *Geol.*, formed or occurring within a geological formation; **intrafu·sal** [L. *fusus* spindle], situated or occurring within a muscle spindle; **intraga·stric**, applied, existing, or situated within the stomach; hence **intraga·strically** *adv.*, into the stomach; **intragene·ric**, occurring or existing within a genus or between individuals of a single genus; **intrage·nic** [*genic *a.*], occurring within a gene; **intragla·cial** = *englacial *a.*; also, lying upon or within, or being, the terrain formerly occupied by a glacier or ice-sheet; **intragla·ndular**, existing or carried out within a gland; **intraglute·al**, administered into the gluteal muscles; **intra-governme·ntal**, occurring within the institutions or branches of a government; **intra-lenti·cular** *Ophthalm.*, situated within the lens of the eye; **intra-li·ngual**, (*a*) *Med.*, situated or occurring in the substance of the tongue (*New Syd. Soc. Lex. Med.* 1888); (*b*) of communication, etc.: within a given language; within the bounds of language; **intra-lingui·stic** = **intralingual* (*b*); **intralu·minal**, existing within a lumen, esp. that of the intestine; **intrama·mmary**, existing or applied within a breast; **intra-me·ntal**, occurring within the mind; so **intra-menta·lity**; **intramorphe·mic**, occurring or existing within a morpheme; **intramuscular** (further examples); also, administered into a muscle; hence **intramu·scularly** *adv.*, (by injection) into a muscle; **intrana·sally** *adv.*, through, in, or into the nose; **intrana·tal**, taking place at the time of birth; **intra-na·tional**, occurring or carried on within a nation-state; **intraneu·ral**, situated or occurring within a nerve; **intranucleolar** (-niuklīōᵘ·lăɹ, -niūklī·ŏlăɹ), situated or occurring within a nucleolus; **intra-orga·nic, -organi·smal, -orga·ni·smic**, within an organism; **intra-pe·rsonal**, occurring inside a person's mind or character; **intra-psy·chic, -psy·chical**, occurring or existing within the psyche; **intrapulmo·nic** = *intrapulmonary* (in Dict.); **intra-ra·cial**, within, or occurring within, a race; **intrare·gional**, occurring within a region; **intrasegme·ntal** *Zool.* and *Linguistics*, occurring within a 'segment'; **intrase·minal** *Bot.*, occurring or existing within a seed; **intra-soma·tic** [Gr. σῶμα body], situated or occurring within the body; **intraspi·nally** *adv.*, within the spinal cord; **intra-subje·ctive**, of a reaction, response, etc., which occurs within a person; **intratesti·cular**, existing or carried out within, or administered into, a testicle; hence **i:ntratesti·cularly** *adv.*; **intrathecal**, (*b*) going into or occurring within the spinal theca; hence **intrathe·cally** *adv.*; **intratra·cheally** *adv.*, within the trachea; **intra-ty·pical**, occurring within one type; **intravaginal**, (*b*) *Bot.*, within the sheath of a leaf; **intravari·etal**, occurring or existing within a variety (Variety 6 b), or between individuals of the same variety; **intravehi·cular**, of, pertaining to, or used within a space vehicle; **intrave·rbal**, within a word; **intrave·rtebral**, situated within a vertebra; hence **intrave·rtebrally** *adv.*

1960 *Biol. Abstr.* XXXV. 1865/2 (*heading*) The lipids in the coating layer of the epidermis during the intra-amniotic life. **1973** *Nature* 26 Jan. 280/1 The observed periodicity in response to the intra-amniotic injection of PGF$_{2\alpha}$ must..be related to endocrine and/or other rhythmic metabolic changes. **1961** *Lancet* 5 Aug. 279/1 If the endocrine condition of one uterine horn is altered experimentally (..by injecting progesterone intra-amniotically into one horn) the two horns deliver at different times. **1973** *Nature* 26 Jan. 280/1 Intra-amniotically injected PGF$_{2\alpha}$ appears to be slowly transferred from the amniotic compartment. **1946** *Ibid.* 17 Aug. 238/1 Buchtal and Kahlson have pointed out that the close intra-arterial injection of 5 μgm. acetylcholine after introduction of adenosine triphosphate increases the intensity and duration of the mechanical response of muscle. **1962** *Lancet* 29 Dec. 1338/1 Substances given by intra-arterial infusion. **1938** *Coll. Papers Mayo Clinic* XXIX. 533 Histamine phosphate..and acetyl β-methylcholine..were injected intra-arterially. **1964** W. G. Smith *Allergy & Tissue Metabolism* ii. 26 Doses of antigen given intra-arterially or intravenously. **1890** Billings *Med. Dict.* I. 711/1 Intra-articular. **1908** *Practitioner* Apr. 516 There was much intra-articular effusion into both knee-joints. **1961** *Lancet* 29 July 266/2 Intra-articular cartilage is avascular. **1879** *Jrnl. Physiol.* II. 336 The relation which exists between the intracapillary pressure and the degree of dilation of these delicate tubes. **1961** *Lancet* 16 Sept. 664/2 A higher intracapillary hydrostatic pressure is presumably related to a greater leakage of protein and fluid. **1917** *Jrnl. Immunol.* II. 141 Römer..injected guinea pigs sensitized to horse serum intracardially with diphtheria antitoxin. **1958** *Immunology* I. 104 Injections were made intracardially. **1881** G. Sigerson tr. *Charcot's Lect. Dis. Nervous System* II. 281 An intra-cerebral focus of hæmorrhage. **1890** W. James *Princ. Psychol.* I. iii. 100 He found very regularly an immediate deflection of the galvanometer, indicating an abrupt alteration of the intracerebral temperature. **1964** S. Duke-Elder *Parsons' Dis. Eye* (ed. 14) xxxiii. 530 They spread slowly within the sheaths, and death is due to intracerebral extension. **1910** *Jrnl. Exper. Med.* XII. 253 Inoculated intracerebrally with spinal cord. **1937** *Jrnl. Path. & Bacteriol.* XLIV. 418 The strain of fowl pest virus..was only very slightly pathogenic for mice when inoculated intraperitoneally or even intracerebrally. **1970** *European Jrnl. Cancer* VI. 173/1 Inbred Swiss/Ry-female mice were inoculated intracerebrally with Ehrlich carcinoma. **1932** Dorland & Miller *Med. Dict.* (ed. 16) 642/1 *Intracisternal*, within a cistern, especially the cisterna magna. **1958** *Technology* Jan. 384/3 By intracisternal injection *nor*-morphine was rather more active than morphine. **1964** G. H. Haggis et al. *Introd. Molecular Biol.* v. 141 One hour after the first meal following starvation..the endoplasmic reticular cavities of the basal zone..are distended and contain small granules (intracisternal granules). **1934** *Physiol. Abstr.* XIX. 481 Histamine.., acetylcholine, and padutin have no influence upon the blood pressure when injected intracisternally. **1971** *Nature* 5 Mar. 54/1 Intracisternally injected radioactive noradrenaline. **1928** *Daily Tel.* 26 June 10/6 Houston['s]..hinterland is shortly to be widened by the construction of the intra-coastal canal from New Orleans. **1964** *Times Rev. Industry* Mar. 86/1 Between Jacksonville and Palatka..there is a navigable channel in the St. Johns River forming part of the Atlantic Intracoastal waterways. **1972** *Countryman* Winter 41 Fifty miles along the intracoastal canal to the Aransas wildlife sanctuary [in Texas]. **1940** S. D. Tylman *Theory & Pract. Crown & Bridge Prosthesis* xix. 177 Retainers are classified into three types: The first is the intracoronal, or inlays... As the name indicates, the prepared cavity and its cast retainer lie largely within the body of the coronal portion of the tooth. **1963** C. R. Cowell et al. *Inlays, Crowns & Bridges* ii. 4 A gold inlay is an intracoronal restoration for a vital tooth. *Ibid.* vi. 60 Tooth substance may have been lost to such an extent that a restoration cannot be retained intracoronally. **1890** W. James *Princ. Psychol.* II. xix. 128 The normal forward irradiation of intra-cortical excitement through association-paths is checked. **1970** *Jrnl. Physiol.* CCX. 57P (*heading*) Relation of movements induced by intracortical stimulation to receptive fields of points in the perirolandic and parietal cortex of the monkey. **1933** Intracrustal [see *bysmalithic *a.*]. **1921** *Sci. Papers U.S. Bureau of Standards* XVI. 215 The fracture of normal material is, in general, intracrystalline; that is, it consists of a break across the grains rather than of a separation between them. **1955** *Soil Sci.* LXXX. 425 The intracrystalline swelling of montmorillonite. **1973** *Nature* 3 Aug. 277/1 The increase in intracrystalline slip..may lead to the elongation of the recrystallized grains. **1937** R. H. Lowie *Hist. Ethnol. Theory* (1938) xiii. 237 As soon as functionalism is reduced to what it is—a worthy programme for ascertaining what intracultural bonds may exist—the neglect of other methods appears as solely a matter of personal preference. **1956** Garvin & Mathiot in J. A. Fishman *Readings Social. of Lang.* (1968) 366 In intra-cultural terms, different segments of a speech community can be compared as to the degree to which the standard language has penetrated them, just as different subcultures of the same culture can be compared in terms of different degrees of penetration by urban elements. **1956** Lenneberg & Roberts *Language of Experience* 4 Not all hypotheses can be verified intra-culturally. **1972** *Jrnl. Social Psychol.* LXXXVII. 13 The study of values both intraculturally and cross-culturally has a long, and..distinguished history. **1885** M. Hay tr. *H. von Ziemssen's Handbk. Gen. Therapeutics* II. 391 We may characterise all such procedures, in contradistinction to the epidermic method, as the endermic or intracutaneous administration of remedies. **1905** Hyde & Montgomery *Pract. Treat. Dis. Skin* (ed. 7) 96 Hypodermatic and intracutaneous injections. **1956** D. M. Pillsbury et al. *Dermatol.* xv. 139 Scratch or intracutaneous tests are routinely employed to detect the presence of skin-sensitizing antibodies. **1925** *Jrnl. Immunol.* X. 729, 0·1 cc. was the amount intracutaneously injected in each case. **1927** *Proc. Soc. Exper. Biol. & Med.* XXV. 97 An injection..was given intracutaneously. **1961** *Jrnl. Amer. Med. Assoc.* 29 July 279/2 Alum-precipitated toxoids were injected intracutaneously. **1916** *Jrnl. R. Microsc. Soc.* 307 The chromatic margin of the undulating membrane represents an intracytoplasmic posteriorly-directed flagellum. **1971** *Biol. Abstr.* LII. 2029/1 In other organs..secretory granules may undergo intracytoplasmic lysis without destruction of their limiting membrane. **1961** P. Fleming *Bayonets to Lhasa* xxiii. 291 The fruits of intradepartmental research. **1967** A. Battersby *Network Analysis* (ed. 2) iii. 42 The single activity..might well be a summary of an intra-departmental arrow diagram. **1900** Dorland *Med. Dict.* 327/2 Intradermal. **1946** *Nature* 31 Aug. 311/2 The intradermal test [for tuberculosis] is conceded to be eminently satisfactory in cattle. **1964** W. G. Smith *Allergy & Tissue Metabolism* i. 8 A patient who developed a local reaction to an intradermal injection. **1926** *Amer. Year Bk. 1925* 951/1 The time required for the disappearance of intradermally injected salt solution. **1937** *Jrnl. Path. & Bacteriol.* XLIV. 410 Four rhesus monkeys were inoculated intradermally on the inner side of the left thigh with 0·2 c.c. of a 20 per cent. suspension of mouse brain. **1962** *Lancet* 26 May 1107/2 Sarcoid tissue..was..inoculated intradermally into 4 patients with suspected sarcoidosis. **1888** *Syd. Soc. Lex.* III, Intradermic. **1966** *Amer. Jrnl. Vet. Res.* XXVII. 541/1 The cervical region of cattle is the most responsive site for applying intradermic tuberculin tests. **1953** L. C. de Leborgne tr. *R. A. Leborgne's Breast in Roent. Diagn.* i. 14 When contrast mammography shows intraductal lesions..we collect the liquid that has been injected, by softly expressing the breast (intraductal rinse). **1961** *Lancet* 29 July 241/2 Prolactin activity was detected and assayed semiquantitatively by the localised lactogenic response of the mammary gland of the pseudo-pregnant rabbit to intraductal injections. **1971** *Amer. Jrnl. Obstetr. & Gynecol.* CX. 505/1 In our study of 20 intraductal papillomas of the breast, the similarity of these lesions to the papillary hidradenoma of the vulva was striking. **1890** Billings *Med. Dict.* I. 711/1 Intradural. **1901** J. Collins in Hektoen & Riesman *Text-bk. Path.* xi. 618 Intradural tumors arise either from the inner surface of the dura or the pia. **1950** *Jrnl. Neurosurg.* VII. 1 Intradural granulomas are very rare and may be intramedullary or extramedullary. **1971** *Ibid.* XXXIV. 378/1 We therefore think that the intradural approach..is not more hazardous than the extradural approach. **1944** *Brit. Jrnl. Ophthalm.* XXVIII. 328 Some surgeons again approach the ganglion intradurally. **1960** *Cleveland Clinic Q.* XXVII. 198 When this extradural treatment fails..we have injected corticosteroids intradurally by lumbar puncture. **1971** *Jrnl. Neurosurg.* XXXIV. 378/2 No facial palsy occurred in more than 1000 patients, most of whom were operated on intradurally. **1904** F. P. Foster *Appleton's Med. Dict.* 1161 Intraepidermal. **1951** J. J. & W. D. Eller *Tumors Skin* (ed. 2) vii. 336 (*heading*) Intra-epidermal and superficial carcinomatous changes. **1971** *Dermatologica* CXLII. 29 (*heading*) On benign intra-epidermal follicular acanthomas. **1904** F. P. Foster *Appleton's Med. Dict.* 1161 Intraepidermic. **1910** *Practitioner* June 871 An intra-epidermic abscess. **1949** *Time* 11 Apr. 39 They had launched the Intra-European Payments plan also as 'the little ECA'. **1962** H. O. Beecheno *Introd. Business Stud.* i. 4 The European countries felt that there was no further danger of another intra-European War. **1967** *Guardian* 15 May 6/6 Intra-European bilateral contacts continue to proliferate benignly. **1973** *Nature* 16 Mar. 150/2 Designed for intra-European communication the satellite will handle telephone, telegram, telex and television channels. **1895** W. James *Meaning of Truth* (1909) ii. 45 Mental images..are one phenomenal fact; the tigers are another; and their pointing to the tigers is a perfectly commonplace intra-experiential relation. **1909** W. M. Urban *Valuation* vi. 188 For Ehrenfels the real test of the rationality of the desire is not an intra-experiential test. **1900** B. D. Jackson *Gloss. Bot. Terms* 137/1 Intrafascicular (*fasciculus*, a bundle), within a bundle. **1914** M. Drummond tr. *Haberlandt's Physiol. Plant. Anat.* ii. 97 So-called fascicular or intrafascicular cambium forms a strip which extends tangentially right across the bundle. **1917** *Ann. Bot.* XXXI. 45 The existence of this vestigial, intrafascicular cambium indicates that Monocotyledons have been derived from a dicotyledonous stock. **1960** W. B. R. Laidlaw *Guide Brit. Hardwoods* 228 *Interfascicular*, between bundles..(cp. intrafascicular—cambium inside a bundle). **1894** C. D. Walcott in *Bull. Geol. Soc. Amer.* V. 192 An intra-formational conglomerate is one formed within a geologic formation of material derived from and deposited within that formation. **1938** Hatch & Rastall *Petrol. Sedimentary Rocks* (ed. 3) iv. 76 Remarkable beds of intraformational breccia and conglomerate are found in some limestone and dolomite formations. **1940** E. S. Hills *Outl. Structural Geol.* i. 15 When a slumped mass slides down on to undisturbed sediments it may later be covered by younger deposits. Severely disturbed beds will then be found between undisturbed strata, an arrangement that is known as intraformational contortion or corrugation. **1963** D. W. & E. E. Humphries tr. *Termier's Erosion & Sedimentation* x. 194 False unconformities and intraformational unconformities can often be explained by the effect of 'creep'. **1894** C. S. Sherrington in *Jrnl. Physiol.* XVII. 240 Its own contained muscle-fibres, or as they may be termed the intrafusal muscle-fibres, always however run parallel with the long axis of the spindle itself. **1905** J. S. Ferguson *Normal Histol.* ix. 136 The bundle of intrafusal muscle fibres is again surrounded by a delicate axial sheath of connective tissue. **1968** Passmore & Robson *Compan. Med. Stud.* I. xxiv. 12/2 A muscle spindle..contains a few short and very slender striated muscle cells; these are the intrafusal fibres which contrast with the main mass of extrafusal fibres. The intrafusal fibres receive motor nerve terminals. **1970** F. J. Schulte in U. Stave *Physiol. Prenatal Period* II. xxv. 805 In kittens, the intrafusal activity..seems to be less tonic than in adult cats. **1900** Dorland *Med. Dict.* 327/2 Intragastric. **1903** G. Herschell *Man. Intragastric Technique* vii. 107 The intragastric needle-douche..may be described as the application of fine jets of fluid under considerable pressure to the interior of the stomach. **1926** J. A. Ryle *Gastric Function* i. i. 15 When the stomach is filled and the intragastric tension thereby raised, the tension..quickly falls again. **1972** *Nature* 21 Apr. 385/2 Contractions of the body of the stomach of an anaesthetized rat measured from pressure changes in an intragastric balloon. **1959** *Proc. Amer. Assoc. Cancer Res.* III. 63/2 (*heading*) Relationship of dose of intragastrically administered methylcholanthrene to incidence of breast cancer in rats. **1971** *Nature* 16 Apr. 461/1 Propane sultone administered intragastrically at two dose levels gave rise chiefly to gliomas. **1947** *Biol. Abstr.* XXI. 2107/2 The role of intra- and intergeneric hybridization in the breeding of cultivated plants. **1971** *Nature* 22 Oct. 526/1 They approach the limits of intrageneric relationship when compared with other fish taxa. **1937** *Ibid.* 30 Oct. 760/2 These are changes of 'balance', and rank with intra-genic changes and position changes as one of the three effective means of variation. **1971** *Ibid.* 10 Dec. 337/1 One or more amino-acids have been deleted, probably during intragenic crossing over. **1895** J. D. Dana *Man. Geol.* (ed. 4) 957 Nearly all transported debris of the glacier was confined at first to its lower part... It was intraglacial, as now in Greenland. *Ibid.*, The term englacial..is not here adopted because it is half Greek. Intraglacial accords with Latin usage. **1896** *Amer. Geologist* XVIII. 153 Intraglacial..is here used for drift deposited in the field occupied by the ice, in contradistinction to extraglacial drift which has come to rest on ground not actually covered by the ice when it was deposited. According to Prof. Dana's use of this term, however, it would be synonymous with englacial, as Prof. Chamberlin designates the drift enclosed in the lower part of the ice-sheet. **1898** *Q. Jrnl. Geol. Soc.* LIV. 205 The débris-bands and intraglacial material with which the lower part of the glacier is so richly charged. **1966** *Geogr. Abstr.* A. 287 In the intraglacial zone are found: forms of marginal dissection, numerous eskers and kames. **1892** *Amer. Jrnl. Med. Sci.* CIII. 466 The author believes that in circumscribed growths, in the future, the only operation that will be admissible will be the bloodless intra-glandular method. **1909** *Practitioner* Nov. 682 Intraglandular enucleation.. should be performed. **1969** *Cancer* XXIV. 765/2 The intraglandular dissemination of thyroid cancer was seen in a very high proportion in follicular adenocarcinoma. **1940** Becker & Obermayer *Mod. Dermatol. & Syphilol.* xlviii. 798/1 Intragluteal injection should be made in the upper outer quadrant of the buttock. **1961** *Lancet* 29 July 268/2 They were given 125 units of chorionic gonadotrophin daily by deep intragluteal injection. **1964** P. Worsley in I. L. Horowitz *New Sociol.* 388 The institutional norms..intra-governmental and intra-social. **1967** *Time* 10 Feb. 58 An ugly intragovernmental feud over the creation of an electronics and broadcasting giant. **1944** *Amer. Jrnl. Ophthalm.* XXVII. 1427/1 A case of intralenticular foreign body is presented. **1962** Harris & Gruber in A. Pirie *Lens Metabolism Rel. Cataract* 383 If.. there are capsular and intralenticular barriers to diffusion ..localized accumulation of fluid could occur. **1937** O. Jespersen *Analytic Syntax* ii. xxix. 109 In the case of ..the German or Latin masculine the correspondence with the extralingual quality 'male sex' is far from being pure: here the extralingual and the intralingual are inextricably mingled. *Ibid.*, When the preterit is used for 'the shifted present time' in indirect speech this cannot in the same way be said to refer to something outside the linguistic expression: it is intralingual. **1956** J. Whatmough *Lang.* 63 For intralingual purposes (i.e. communication within a single language) every language is a good language for those who actually use it. **1964** E. A. Nida *Toward Sci. Transl.* i. 3 'Intralingual' translation..consists essentially in rewording something within the same language. **1965** *Language* XLI. 505 Meaning refers to the intralingual relations contracted by linguistic units. **1937** J. R. Firth *Tongues of Men* iii. 34 For the vast majority of mankind clicks are extra-linguistic sounds, but for..Zulus they are intra-linguistic—that is to say, they form part of grammatical words. **1945** *Mind* LIV. 149 Morris' 'syntactics'.. deals only with relations among expressions in a language and thus with an entirely intra-linguistic subject-matter. **1962** U. Weinreich in Householder & Saporta *Probl. Lexicogr.* 35 Over-schematized though it may be, ad hoc intralinguistic considerations suggest that 'c_1 and c_3' should have been considered a single condition. **1936** Stedman *Med. Dict.* (ed. 13) 555/2 *Intraluminal*, intratubal. **1943** *Jrnl. Clin. Invest.* XXII. 225/1 The clinical importance of intra-luminal pressure measurements in hollow viscera is well exemplified by the syndromes of hyper- and hypotension in the cardiovascular system. **1961** *Lancet* 30 Sept. 738/2 This flora may have utilised a relatively large part of the intraluminal nutrients during their logarithmic rate of growth. **1971** *Gut* XII. 268 Intraluminal pressure measurements may be a useful guide for the treatment of achalasia by pneumatic dilatations. **1892** *Amer. Jrnl. Med. Sci.* CIV. 96 (*heading*) The treatment of intra-mammary abscesses. **1960** *Farmer & Stockbreeder* 8 Mar. 107/1 His vet bill was too high... The balance was for drugs—almost entirely expensive intramammary antibiotics. **1971** *Jrnl. Endocrinol.* LI. p. xiii, The myoepithelial cells of the mammary gland contract in response to oxytocin and this results in a rise in intramammary pressure. **1904** *Jrnl. Philos., Psychol., & Sci. Methods* I. 300 The intra-mental and the extra-mental objects..differ only in position and in relational context. **1935** *Mind* XLIV. 356 Purely intra-mental causation. **1958** Intra-mental [see *extra-mental* (*extra- 1)]. **1946** *Mind* LV. 372 Notwithstanding his insistence on the intramentality of objects, he keeps 'what is in the mind "intirely distinct" from the mind itself'. **1962** A. Tietze in Householder & Saporta *Probl. Lexicogr.* 272 It has no intramorphemic changes. **1973** A. H. Sommerstein *Sound Pattern Anc. Greek* ii. 22 Words containing intramorphemic clusters /pm bm pʰm/..all, historically, underwent assimilation. **1946** *Nature* 17 Aug. 242/2 The present data ..cover some twenty-six treatment schedules with intramuscular penicillin. **1962** Luntz & Wright in A. Pirie *Lens Metabolism Rel. Cataract* 317 These persons,..when given a desensitizing course of intramuscular injections.., rapidly improved. **1909** *Practitioner* Dec. 871 Digalen.. can be used by the mouth, by the rectum, intravenously, intramuscularly, and hypodermically. **1933** *Discovery* Jan. 4/2 This could be safely injected intramuscularly and intravenously. **1961** *Ann. N.Y. Acad. Sci.* XCIV. 917 A carcinoma had been transplanted intramuscularly, intraperitoneally, or into the lungs. **1933** *Jrnl. Amer. Med. Assoc.* 24 June 2014/1 These agents were given intranasally. **1961** *Lancet* 23 Sept. 680/2 Swiss white mice.. were infected intranasally with..egg cultures of influenza viruses. **1971** *Infection & Immunity* IV. 738 Statolon.. when instilled intranasally (IN) protects mice infected with lethal doses of influenza virus. **1902** *Encycl. Brit.* XXXI. 304/1 The causes of the high death-rate among infants, whether due to ante-natal, intra-natal, or neo-natal conditions, come under..observation. **1922** *Rep. Public Health & Med. Subjects* (Ministry of Health) No. 7. 8 Death during labour—intra-natal fœtal death. **1963** D. G. W. Clyne *Textbk. Gynaecol. & Obstetr.* xxxvii. 875 The intranatal care of the premature infant has been summarized. **1923** G. O'Brien *Ess. Econ. Effects Reformation* iv. 179 It is an institution at once intranational and international. **1940** A. Huxley *Let.* 15 Dec. (1969) 464 Two satisfactory alternatives, either voluntary international and intranational co-operation, or balance of power. **1901** *Buck's Handbk. Med. Sci.* (rev. ed.) II. 110/1 Foci of degeneration in the nerves were found to correspond to nodules upon intraneural arterial branches. **1936** H. Mulder *Cognition & Volition in Lang.* iii. 118 A bit of intraneural behavior in the brain of the man who coined it. **1954** Zachary & Roaf in H. J. Seddon *Peripheral Nerve Injuries* ii. 73 Complete palsies have been caused by the accidental intraneural injection of a noxious substance. **1879** *Jrnl. R. Microsc. Soc.* II. 138 Finally, the ordinary intranucleolar network is produced, marking the completion of the division process and the entrance of the nuclei into a state of quiescence. **1970** *Jrnl. Cell Biol.* XLV. 584 (*heading*) Aberrant intranucleolar maturation of ribosomal precursors in the absence of protein synthesis. **1894** A. C. Fraser in Locke *Hum. Und.* II. iv. xi. 327 An odd and inadequate illustration; for 'writing' and 'moving the hand', although intraorganic, as much need to have their reality indicated as the sight of black or white does. **1962** S. K. Langer *Philos. Sk.* i. 9 Sentience is a phase of vital process itself, a strictly intraorganic phase, i.e., an appearance which is presented only within the organism in which the activity occurs. **1941** J. S. Huxley *Uniqueness of Man* xi. 241 Analogy may very readily mislead. Weismann sought to apply this same analogy of intra-organismal struggle and selection to the units of heredity; but the analogy happens not to hold good. **1952** *Brit. Jrnl. Psychol.* XLIII. 245 The production of a neurosis in a cat by a method that is of special interest in that a difficult discrimination of intraorganismal cues was involved. **1955** Intraorganismic [see *extra-organismic* (*extra- 1)]. **1909** W. M. Urban *Valuation* x. 286 The terms of estimation are..wholly intra-personal, within the ego. **1935** G. K. Zipf *Psycho-Biol. of Lang.* (1936) 297 The occasions of inter-personal conflict are often occasions for intra-personal conflict as well. **1955** R. Jakobson in H. Werner *On Expressive Lang.* 79 A competition of both devices is manifest in any symbolic process, either intrapersonal or social. **1917** C. R. Payne tr. *Pfister's Psychoanal. Method* viii. 169 The sleeping state lowers the power of the intra-psychic censor. **1954** J. A. C. Brown *Social Psychol. of Industry* ix. 252 Frustration, whether intrapsychic or due to external factors. **1935** *Mind* XLIV. 215 The interplay of conflicting intra-psychical forces. **1923** *Amer. Jrnl. Physiol.* LXV. 229 Blood pressure changes are..complicated by the increased intrapulmonic pressure interfering with filling of the heart. **1968** *Biol. Abstr.* XLIX. 2076/2 Excess intrapulmonic pressure in the abdominal and thoracic cavities. **1903** *Biometrika* Feb. 152 The interracial correlation of the mean numbers of stamens and pistils is very much greater than the mean intraracial correlation between stamens and pistils. **1922** Joyce *Ulysses* 717 Intraracial inhibition. **1957** *Antiquity* XXXI. 196 These two groups [of early Man] are in effect intraracial variants. **1964** *Ann. Reg. 1963* 204 Intra-regional trade increased. **1966** *B.B.C. Handbk. 1966* 51 These intra-regional services are directed..to the region as a whole. **1909** Webster, Intrasegmental. **1940** *Chambers's Techn. Dict.* 456/2 *Intrasegmental*, said of vertebrae which arise by the fusion of the cranial and caudal elements of the same somite. **1953** C. E. Bazell *Ling. Form* 3 It is to be noted however that the intra-segmental range of a suprasegmental phoneme has normally no distinctive relevance. **1972** *Language* XLVIII. 46 In intra-segmental variation, as has been seen, different values for a feature are distributed over different portions of one and the same segment. **1895** S. H. Vines *Students' Text-bk. Bot.* II. iii. 440 The intra-seminal stage includes the whole of the development which the embryo undergoes during the conversion of the ovule into the ripe seed. **1902** *Nature* 3 Apr. 519/1 Seeds from infected plants are entirely free from disease, thereby proving fairly conclusively the impossibility of intra-seminal sources of infection. **1930** *Ann. Bot.* XLIV. 772 The germination of the seed [of *Garrya elliptica*] is reported as 'intra-seminal'. **1932** H. H. Price *Perception* viii. 257 The revival of kinaesthetic and other intra-somatic data..is..far less common. **1938** [see *extrasomatic *a.*]. **1939** *Mind* XLVIII. 517 His treatment of intrasomatic sense-perception is equally precise. **1948** *Mind* LVII. 246 Any 'map' which may be elaborated at this level has a purely 'intrasomatic' significance. **1906** *Jrnl. Physiol.* XXXIV. 31 The reflex arcs from separate areas of the receptive field are closely knit together intraspinally. **1970** *Biol. Abstr.* LI. 7775/1 (*heading*) Clinical trials with the use of 5% phenol intraspinally in the treatment of cancer pain. **1914** Myers & Valentine in *Brit. Jrnl. Psychol.* VII. i. 72 Bullough's..physiological aspect has here been extended to include not merely the sensory effects and the changes in feeling..but also the experiences of self-activity which the sounds may produce in the subject; in consequence, the word 'intra-subjective' will be substituted for this aspect. **1921** E. Bullough *Ibid.* XII. i. 86 After the experiments of C. S. Myers..I would adopt the term 'intra-subjective' proposed by him, in place of 'physiological', which..was found by him to be too narrow, when applied to musical experiences. **1955** *Times* 26 May 3/4 Bruckner was a late romantic and as such was concerned with intra-subjective emotion as the material of his art. His fondling of an idea, then dropping it, at a silent pause and picking up another involves no incongruity. **1959** W. V. Quine in R. A. Brower *On Translation* ii. 159 Intrasubjective synonymy ..is intrasubjective in that the synonyms are joined for each subject by sameness of stimulus meaning for him;.. intrasubjective synonymy is in principle just as objective, just as discoverable by the outside linguist, as is translation. **1888** *Syd. Soc. Lex.*, Intratesticular. **1921** *Brain* XLIV. 168 In case XXIX an intratesticular injection of 3 c.c. [was given]. **1958** *Immunology* I. 4 For intratesticular grafting, the right testis was delivered through the smallest possible incision in the scrotum. **1970** *Sci. Jrnl.* June 68/2 The spermatozoa..are transferred through an uninterrupted system of ducts which originate from a complex network of intratesticular ducts. **1942** *Cancer Res.* II. 288 After the blood examinations had been concluded, the animals were inoculated intratesticularly with the Brown-Pearce tumor. **1971** *Lipids* VI. 706/1 Fatty acid synthesis was studied in testes of young and adult rats either injected intratesticularly or incubated with 1-^{14}C acetate. **1921** *Brain* XLIV. 168 In Case III an intrathecal injection of 1·5 c.c. cerebrospinal fluid was given. **1960** P. W. Daykin *Vet. Appl. Pharmacol. & Therapeutics* iv. 36 Administration by the intrathecal route involves the penetration of the dura mater, and the route is not frequently used in veterinary practice. **1971** *Jrnl. Path.* CIV. 141 An intrathecal lipoma of the filum terminale..was found in a 2-yr-old female infant dying from extensive burns. **1928** L. E. H. Whitby *Med. Bacteriol.* xxiii. 236 The most popular method is to inject 500 units intrathecally immediately the diagnosis is made. **1972** *Lancet* 30 Dec. 1401/2 An account of the toxicity of antifolate drugs given intrathecally. **1923** *Jrnl. Exper. Med.* XXXVII. 793 (*caption*) Protocols of rabbits treated intratracheally and intraperitoneally with killed cultures of hog-cholera bacillus. **1930** *Jrnl. Laboratory & Clin. Med.* XVI. 87 (*heading*) A new apparatus for administering volatile anesthetics intratracheally without tracheotomy. **1973** *Nature* 16 Mar. 203/2 Turkey poults were inoculated intratracheally. **1934** *Mind* XLIII. 74 An intra-typical variable is a variable all of whose arguments are of one type. **1959** P. F. Strawson *Individuals* i. i. 53 The cases in which this intra-typical identification is possible are severely restricted. For they require that the parties to an identifying reference should be operating with one and the same type-homogeneous referential framework. **1900** I. B. Balfour tr. *Goebel's Organogr. Plants* II. 359 The intravaginal squamules are organs which secrete mucilage for the protection of the bud. **1951** McLean & Ivimey-Cook *Textbk. Theoret. Bot.* I. xxii. 993 The first [leaf structures] are called intravaginal scales because they appear in the axil within the leaf sheath or vagina of certain Monocotyledons... They take the form of small tooth-like scales. **1916** *Mem. N.Y. Bot. Garden* VI. 352 The marked self-sterility of individual plants, the intra-varietal sterility, and the cases of inter-varietal sterility are not due to 'any inherent weakness of either ovaries or pollen grains'. **1970** *Euphytica* XIX. 382 (*heading*) Intra-varietal variation of yield in two varieties of *Lolium perenne* L. **1969** *Encycl. Sci. Suppl.* (Grolier) 329 The intravehicular space suit consists of: fecal containment subsystem, constant wear garment, [etc.]. **1970** N. Armstrong et al. *First on Moon* iii. 62 There were two kinds of suits... Mike Collins had the lighter 'intra-vehicular' version. **1909** Webster, Intraverbal. **1953** C. E. Bazell *Ling. Form* 87 On the level of the sememe itself, all such facts as congruence, rection, extra-verbal as opposed to intra-verbal formation, are excluded. **1957** B. F. Skinner in Saporta & Bastian *Psycholinguistics* (1961) 229/2 A comparable minimal repertoire was found to be lacking in intraverbal behavior. **1855** R. G. Mayne *Expos. Lex. Med. Sci.* (1860) 535/2 Intra-vertebral chord. **1896** *Phil. Trans. R. Soc.* B. CLXXXVII. 10 One of the last changes of importance is the appearance of intravertebral cartilage. *Ibid.* 12 In all Amphibia and Amniota each spinal nerve lies intravertebrally, *i.e.* issues behind the dorsal arch of its vertebra.

2. intra-class, -party, -state, occurring within a class, political party, state. Also **intracloud**, within a cloud; **intra-day**, occurring within one day; **intra-list** *Psychol.*, occurring between lists (of words, numbers, etc.) within a test situation; **intra-sentence**, occurring within a sentence.

1950 Intra-class [see *inter-class* (*inter- 5)]. **1971** *Jrnl. Gen. Psychol.* Apr. 306 The intraclass correlation was significant but small. **1970** *Sci. News* 28 Mar. 320 Zeroing in on intracloud lightning. **1973** J. L. Marshall *Lightning Protection* iii. 31 The higher frequency radiation is dominant in intracloud discharges. **1972** *Daily Tel.* 11 Nov. 17/4 The previous intra-day high for this most illustrious of all stock market indices was 1,001·11 which was reached more than six-and-a-half years ago on Feb. 9, 1966. **1972** *Korea Times* 16 Nov. 3/5 The Dow topped 1,000 on an intraday basis—a compilation of the day's highs for all component issues. **1942** *Jrnl. Exper. Psychol.* XXX. 185 (*heading*) Intra-list generalization as a factor in verbal learning. **1958** G. A. Miller in Saporta & Bastian *Psycholinguistics* (1961) 213/2 The redundant strings show greater intralist similarity. **1970** *Jrnl. Gen. Psychol.* Oct. 256 The last letter being thus the only characteristic by which the entire series was to be differentiated..would reduce the level of intralist interference. **1971** *Ibid.* Apr. 194 Overt intralist intrusions were scored as if stimuli had been clustered for both groups. **1923** *Glasgow Herald* 24 Apr. 8/4 Mr Harding proposed to the Senate that the United States should accept membership of the Court of International Justice at The Hague... Present indications are that he will refuse to retreat, and will make participation in the Court the issue of the preliminary intra-party campaign. **1963** *Economist* 20 July 257/1 This is nothing less than an intraparty revolt. **1969** *Sunday Statesman* (Calcutta) 27 July 12/3 The intra-party quarrel of the SSP deepens. **1969** *Computers & Humanities* IV. 129 Investigation of intra-sentence parallelism in present-day American prose. **1903** E. Johnson *Amer. Railway Transportation* 370 In 1886 the Supreme Court in the Wabash decision..limited the authority of the State strictly to the intrastate traffic and excluded that moving from one State to another. **1969** *Jane's Freight Containers 1968–69* 37/2, 101 inter- and intra-state motor freight lines serving the Port of Toledo. **1971** M. Tak *Truck Talk* 90 *Intrastate shipping*, any transportation of goods that has its origin and destination within the same state.

intra-amniotic to **intra-articular:** see *intra- *pref.* 1.

i:ntra-ato·mic, *a.* [Intra- 1.] Occurring or existing within an, or the, atom.

1904 *Nature* 16 June 151/2 Enormous orbital velocities due to intra-atomic rearrangement. **1946** *Nature* 23 Nov. 726/1 For 'chemical process' is suggested the definition: 'a process in which a product is formed by a re-arrangement or re-distribution of atoms of chemical elements present in the starting materials, or by intra-atomic change'. **1955** *Bull. Atomic Sci.* Mar. 92/1 The possibility of releasing very large reserves of intra-atomic energy appeared suddenly a few months after the discovery by Hahn and Strassmann of the neutron fission of uranium atoms. **1971** B. A. Lengyel *Lasers* (ed. 2) iii. 212 The general laws of emission and absorption of radiation are as valid for recombinations [of electrons and holes in semiconductors] as they are for ordinary intra-atomic transitions.

intra-capillary to **intracortical:** see *intra- *pref.* 1.

intra-class, intracloud: see *intra- *pref.* 2.

intracranial, *a.* Add: Hence **intracra·nially** *adv.*, in or within the cranium.

1908 *Practitioner* Aug. 296 Rabbits were inoculated intracranially with the fluid. **1971** *Biol. Abstr.* LII. 7922/2 A large stick remained intracranially for a long time without clinical neurological deficit.

intracrustal to **intracytoplasmic:** see *intra- *pref.* 1.

intrada (intrā·dă). *Mus.* [Modified f. It. *intrata*, older form of *entrata* entry, prelude.] An introduction or prelude; = Entrée 3 b.

1740 J. Grassineau *Mus. Dict.* 113 *Intrada*, an entry, much the same as prelude, or overture. **1801** Busby *Dict. Mus.*, *Intrada*, the old Italian name for an opera overture, or prelude. **1883** Grove *Dict. Mus.* III. 756/1 In Purcell's suites, for instance, which date from the last 10 or 20 years of the 17th century, besides the Allemande and Courante..in one case the group also comprises a Sarabande, Cebell, Minuet, Riggadoon, Intrade, and March. **1954** *Grove's Dict. Mus.* (ed. 5) IV. 525/1 These intradas usually form part of a suite of dances. The term is seldom found after the middle of the 17th century, being replaced by the French term *entrée*, though Bach, Mozart..and Beethoven..still used it.

intra-day: see *intra- *pref.* 2.

intra-departmental to **intragovernmental:** see *intra- *pref.* 1.

intragroup (i·ntrăgrūp), *a.* [Intra- 2.] Existing or occurring within a group or between the members of a group.

1918 *Genetics* III. 477 Intra-group competition. **1952** C. P. Blacker *Eugenics* iv. 90 If we owe as much to the social instincts and co-operativeness of our forebears as to their competitive success in a struggle for existence, how far, we may ask, is the continued strife between man and man necessary for the further development of the race? Diverse answers were given by the controversialists of the period, and discussion turned much on the effects of two kinds of struggle—that *inside* social groups and that *between* social groups—or, as they were called *intra*-group and *inter*-group struggle. **1958** W. J. H. Sprott *Human Groups* ix. 141 The organized group displayed much more intra-group aggression than did the unorganized group. **1963** C. S. Johnson in M. M. Grossack *Mental Health & Segregation* 44 For the Negro youth, the tensions inherent in the intra-group class struggle become extremely important. **1965** *Math. in Biol. & Med.* (*Med. Res. Council*) III. 86 The degree of variation within the groups was estimated by the average intragroup distance. **1972** *Accountant* 23 Mar. 384/1 The accounting practice persists whereby we defer profits on all intra-group sales and chalk up another defeat for reality.

intralenticular to **intra-mentality:** see *intra- *pref.* 1.

intra-list: see *intra- *pref.* 2.

intramolecular, *a.* Add: Hence **intramole·cularly** *adv.*, within a molecule.

1936 G. M. Kline *Organic Plastics* (U.S. Bureau of Standards Circular C411) 6 Condensations may take place intramolecularly, as in the case of the formation of phthalic anhydride from phthalic acid, or intermolecularly. **1962** Walker & Straw *Spectroscopy* II. 181 Dipole attachment occurs between the intramolecularly hydrogen bonded and the solvent molecule. **1964** N. G. Clark *Mod. Org. Chem.* x. 195 It undergoes a Cannizzaro reaction with sodium hydroxide; this occurs intramolecularly, i.e. within the same molecule. **1971** *Biochemistry* (Easton, Pa.) X. 925/1 Is the S-S polylysine intramolecularly or intermolecularly cross-linked?

intramorphemic: see *intra- *pref.* 1.

intramurally (-miūə·rāli), *adv.* [Intra- 1.] Within the walls or boundaries; inside a particular community, institution, etc.

1927 [see *extramurally *adv.*]. **1952** *Facilities for Advanced Study Univ. Oxf.* 29 Research studies in all branches of agricultural economics have been conducted both 'in the field', extra-murally, and intra-murally by means of records. **1968** *Sat. Rev.* (U.S.) 2 Nov. 32 Older families, with their own special and intramurally recognized ways of walking, talking and thinking.

intramuscular to **intraneural:** see *intra- *pref.* 1.

intra·nsigeance. *rare.* [Fr.] = Intransigence.

1899 J. W. Mackail *Life W. Morris* II. 291 Socialism.. from extreme intransigeance..had swung back to something approaching opportunism. **1909** *Daily Chron.* 20 July 3/2 The loves of Jeanne-Jeannette and the young man from Montpellier are threatened by the intransigeance of their respective ancestors.

intransigeantly (intrɑ·nsidʒāntli), *adv.* [f. F. *intransigeant* (see Intransigent *a.* and *sb.*) + -ly².] Uncompromisingly.

1921 *Contemp. Rev.* Sept. 331 The peasants are intransigeantly anti-Karlist. **1925** *Glasgow Herald* 17 Mar. 8 The advocates of an intransigeantly nationalist policy.

in-tra·nsit, intransit, *a.* [f. *in *prep.* 17* + Transit *sb.*] Of or pertaining to people, goods, etc., that are in transit; being in transit.

1918 Johnson & Huebner *Princ. Ocean Transportation* III. xx. 309 The granting of in-transit privileges on all-rail routes. **1951** *Manila Daily Bulletin* 26 Mar. 1-P, The ship had seven intransit passengers and around 1,741 tons of transit cargo. **1967** V. S. Naipaul *Mimic Men* III. vii. 279 There were no aeroplanes..that day... Sixteen intransit hours awaited me. **1969** *Jane's Freight Containers 1968–69* 102/2 The in-transit time between UK/Europe and the Far East could be considerably reduced by transporting containers by rail across the North American continent. **1971** *Jamaican Weekly Gleaner* 17 Nov. 38/4 Miss Milford Rohrbaugh, 53, an American tourist slipped and fell in the intransit lounge of the Palisadoes Airport yesterday on her way to board an aircraft to Miami, Florida.

intransitable, *a.* (U.S. and later examples.)

1838 'Texian' *Mexico v. Texas* 9 In that singular region of Mexico..there extends..a desert..so utterly devoid of water and vegetation as to be intransitable. **1897** *Geogr. Jrnl.* X. 64 A road along the coast would become lost in intransitable gorges of the coast range of mountains.

intransitive, *a.* Add: **3.** *Logic.* Of a relation: such that if the relation holds between a first and a second item, and also between the second and third, it cannot hold (or more widely, does not hold) between the first and the third.

1870 C. S. Peirce in *Mem. Amer. Acad.* (1873) IX. 369 Repeating relatives may be divided (after De Morgan) into those whose products into themselves are contained under themselves, and those of which this is not true. The former are well named by De Morgan transitive, the latter intransitive. **1881** J. Venn *Symbolic Logic* xix. 403 Relations..may be divided into those which are 'transitive' and those which are 'intransitive'. **1903** B. Russell *Princ. Math.* xxvi. 218 Relations which do not possess the second property I shall call not transitive; those which possess the property that *xRy*, *yRz* always exclude *xRz* I shall call intransitive. All these cases may be illustrated from human relationships... *Spouse* is symmetrical but intransitive;..*father* is both asymmetrical and intransitive. **1930** L. S. Stebbing *Mod. Introd. Logic* vii. 113 Symmetry and transitiveness..are independent, so that relations can be symmetrical and either transitive or intransitive; asymmetrical and either transitive or intransitive. **1964** E. Bach *Introd. Transformational Gram.* vii. 155 *Equal, Smaller* are transitive; *Father* is an intransitive relation; *Friend* is non-transitive if you are my friend and have friends who are not my friends.

4. *Math.* Of a group: not transitive (see Transitive *a.* (*sb.*) 6 and quot. 1889).

1889 *Amer. Jrnl. Math.* XI. 195 If a substitution-group *I* is intransitive, the letters upon which it operates can be distributed into 'systems of intransitivity', $x_1, x_2 \ldots$; $y_1, y_2 \ldots$; $z_1, z_2 \ldots$ such that the substitutions of *I* interchange among each other only the letters $x_1, x_2 \ldots$; the letters $y_1, y_2 \ldots$; the letters $z_1, z_2 \ldots$, and so on, and connect transitively the letters of each system. **1940** D. E. Littlewood *Theory Group Characters* iii. 42 In an intransitive group the symbols are divided into transitive sets, the symbols of each set being permuted amongst themselves. **1971** Powell & Higman *Finite Simple Groups* vii. 234 If the union of these is Ω, they form a fixed trio, and if not, the group is intransitive.

intransitivity (i:ntrɑnsiti·viti). [f. Intransitiv(e *a.* + -ity.] The property or quality of being intransitive (in any sense).

1889 [see *intransitive *a.* 4]. **1933** Chapman & Henle *Fund. Logic* 3 We have chosen four logical properties of certain relations—transitivity and intransitivity, symmetry and asymmetry. **1950** W. V. Quine *Methods of Logic* (1952) 177 A dyadic relative term is called..intransitive..according as it fulfils:..(x)(y)(z) (Fxy.Fyz.⊃–Fxz) (intransitivity). **1971** *Archivum Linguisticum* II. 104 Aspect, as well as transitivity, intransitivity, expression of cause, wish, etc., are secondary developments and more recent than the use of *s* to denote tense or mood.

intransitivize (intrɑ·nsitivəiz), *v.* *Gram.* [f. Intransitiv(e *a.* + -ize.] *trans.* To make intransitive. Chiefly as **intra·nsitivizing** *ppl. a.*

1949 E. A. Nida *Morphol.* (ed. 2) iii. 68 In Tzeltal there is an intransitivizing verbal infix -h-. **1964** *Language* XL. 76 The intransitivizing stem formative -di.

intransmissibility (i:ntrɑnsmisibi·liti). [f. Intransmissible *a.* + -ity.] The state or quality of being intransmissible.

1913 H. Goudy in P. Vinogradoff *Ess. Legal Hist.* 225 The passive intransmissibility of actions of Debt and Account.

intransparency (intrɑnspēə·rĕnsi, -pæ·r-). [f. Intransparent *a.*: see -ency.] The quality of being opaque; also, an instance of this.

1902 *Encycl. Brit.* XXXI. 570/1 This intransparency caused by a mere infiltration generally clears away in the course of time. *Ibid.*, Centrally placed intransparencies, which cover the pupil, are relatively the most disturbing.

intranucleolar to **intra-personal:** see *intra- *pref.* 1.

intra-party: see *intra- *pref.* 2.

intrapluvial (intrāplū·viăl), *a.* and *sb.* [f. Intra- 1 + Pluvial *a.*] **A.** *adj.* Of, pertaining to, or designating relatively short, drier periods (less marked than interpluvials) that may have occurred during pluvials. **B.** *sb.* An intrapluvial period.

1934 E. J. Wayland in *Jrnl. R. Anthrop. Inst.* LXIV. 344 In each of them [*sc.* pluvials] there is a break, or intrapluvial period. **1939** T. P. O'Brien *Prehist. Uganda* i. 9 The evidence in support of the intrapluvial in Pluvial I was provided by local, though not intense, soil reddening, selenite beds within Pluvial I deposits and by talus accumulations. **1940** *Geogr. Jrnl.* XCVI. 333 There have been very long wet and dry phases, the so-called pluvials and interpluvials; there have been shorter intrapluvials and epipluvials; and there have been short-period oscillations. ..The intrapluvial oscillations are measured in terms of one hundred years. **1946** F. E. Zeuner *Dating Past* viii. 247 The succession of pluvials with intrapluvials and separated by interpluvials, however, appears at present to be somewhat uncertain.

intrapolation (intræ:pŏlēi·ʃən). [f. Intra- 3 + *-polation*, after *extrapolation*.] = Interpolation 3; also more widely, (an) inference within the scope or framework of what is known.

1923 C. D. Broad *Sci. Thought* xi. 428 Like all extrapolations, this argument is weaker than an intrapolation. **1949** Koestler *Insight & Outlook* xxv. 352 The listener is forced to complete its pattern himself, by a process of intrapolation and extrapolation. **1956** E. H. Hutten *Lang. Mod. Physics* v. 195 This intrapolation between observations is not, as such, required as support for the theory. **1969** D. F. Horrobin *Sci. is God* iv. 49 This is known as intrapolation because..the process operates only within the range..actually measured.

So **intra·polate** *v.* = Interpolate *v.* 6 (in Dict. and Suppl.).

1956 E. H. Hutten *Lang. Mod. Physics* v. 189 In classical mechanics we are able to intrapolate between observations and to say that, between the space-time points where it is observed, the particle follows a definite path.

intra-psychic to **intrasomatic:** see *intra- *pref.* 1.

intra-sentence: see *intra- *pref.* 2.

intraspe·cies, *a.* [Intra- 2.] = next.

1927 *Techn. Bull. N.Y. State Agric. Exper. Stat.* No. 127. 5 During the past 18 years many intra- and interspecies crosses of *Prunus* have been made. **1953** *New Biol.* XIV. 9 Intraspecies fighting is different from interspecies fighting. **1963** *Language* XXXIX. 461 The whistles, squeals, chirps, clicks, rasps, and other noises of marine mammals have suggested three areas of inquiry: orientation by echolocation, intraspecies communication, and interspecies communication. **1973** *Sci. Amer.* Aug. 29/1 This interspecies difference contrasts with remarkable intraspecies similarity.

intraspeci·fic, *a.* [f. Intra- 1 + Specific *a.*, as adj. from *species*.] Produced, occurring, or existing within a (taxonomic) species or between individuals of a single species.

1919 *Genetics* IV. 501 (*heading*) An analysis of certain cases of intraspecific sterility. **1929** *Hereditas* XIII. 185 Partial sterility in intra- and interspecific crosses. **1937** A. Huxley *Ends & Means* xiv. 265 Progress is dependent on the preponderance of intra-specific co-operation over intra-specific competition. **1953** N. Tinbergen *Herring Gull's World* x. 88 Species are known in which the reproductive season brings much intraspecific strife. **1966** M. Latzke tr. *Lorenz's On Aggression* iii. 18 The survival value of inter-specific fights is much more evident than that of intra-specific contests. **1968** R. D. Martin tr. *Wickler's Mimicry* xvi. 227, I have given the term intraspecific mimicry to this form of self-imitation within a species. **1969** *Daily Tel.* 13 Feb. 22/4 Mankind's allegedly instinctive affection for weaponry and intraspecific killing has been a selling subject.

intraspinally to **intra-typical:** see *intra- *pref.* 1.

intra-state: see *intra- *pref.* 2.

intra-uterine, *a.* Add: Also **intrauterine.** Applied *spec.* to contraceptive devices for placing in the uterus. (Further examples.)

1931 F. W. S. Browne tr. *T. H. van de Velde's Fertility & Sterility in Marriage* xiv. 354 The intra-uterine appliances were based on the same principle as the vaginal; but they soon completely changed their mechanism and structure. **1969** N. W. Pirie *Food Resources* 15 The contraceptive ideal is a treatment of some sort that makes one or other sex infertile until it is reversed. The closest approach to this ideal so far is the intrauterine device. **1972** *Guardian* 10 July 9/2 Asking, at a Family Planning Association branch, for an intra-uterine device.

intravaginal, -varietal: see *intra- *pref.* 1.

intravascular, *a.* Add: Hence **intrava·scularly** *adv.*, within the vascular system.

1906 [see *anti-¹ 7 c]. **1971** *Nature* 5 Feb. 412/2 For preservation, dog brain was perfused intravascularly with 3% glutaraldehyde for 5–20 min.

intravehicular: see *intra- *pref.* 1.

intravenous, *a.* Add: Also as *sb.*, an intravenous injection or feeding.

1960 J. G. Ballard in D. Knight *100 Yrs. Sci. Fiction* (1969) 355 Took me half an hour to steady myself enough for an intravenous. **1970** *New Yorker* 21 Nov. 66/2 You have intravenouses going in both your hands.

intraventricular, *a.* Add: Hence **intraventri·cularly** *adv.*, into or within a ventricle.

1951 *Year Bk. Drug Therapy* 218 Trial of streptomycin intraventricularly..seems indicated. **1955** *Brit. Jrnl. Pharmacol.* X. 373/1 Bulbocapnine, when injected intraventricularly, produces its characteristic effects in doses a fraction of those effective on subcutaneous injection. **1961** *Lancet* 22 July 178/1 The sodium salt of polymyxin B methane sulphonic acid..is well tolerated intraventricularly. **1972** *Science* 29 Sept. 1213/2, 6-HDA administered intraventricularly can produce several aspects of the classical lateral hypothalamic syndrome.

intraverbal: see *intra- *pref.* 1.

intraversable, *a.* Add: (Later example.) Also **intraversible.**

1900 W. James *Let.* 26 Sept. (1920) II. 137 It must seem strange to you that the way from the mind to the pen should be as intraversable as it has been in this case of mine. **1923** R. F. Horton *Mystical Quest Christ* III. xxiv. 255 The vast universe seems so homeless, so comfortless; its spaces of darkness so intraversible.

intravertebral, -vertebrally: see *intra- *pref.* 1.

intravital (intrăvəi·tăl), *a.* *Biol.* [f. Intra- 1 + Vital *a.*, perh. suggested by *intra vitam* (see *intra *prep.* 2).] Performed on, applied to, or occurring in something alive.

1890 in Billings *Med. Dict.* 712/1. **1922** *Proc. R. Soc.* B. XCIV. 138 On intra-vital staining by polychrome methylene blue, some of these granules are stained blue, others red. **1938** *Biochem. Jrnl.* XXXII. 381 The distribution of the lyochromes in the kidneys of frogs and rats was studied by intravital microscopy. **1963** *Lancet* 19 Jan. 152/2 The technique of intravital staining of the motor end plate..is of limited diagnostic value.

Hence **intravi·tally** *adv.*, during life, in a living organism.

1930 *Biol. Abstr.* IV. 3299/3 (Index), Iron lactate, storage intravitally. **1937** *Acta Med. Scand.* XCI. 350 It has been demonstrated intravitally in retinal vessels in cases of stasis provoked by pressure on the eye.

intra vitam: see *intra *prep.* 2.

in-tray (i·ntrēi). [Cf. In *adv.* 12 a.] In an office, etc.: a tray for incoming correspondence and other papers; = *in-basket. Cf. *out-tray.

Sometimes written as two separate words with *in* regarded adjectivally.

1941 *Punch* 25 June 602/2, I found the first one lying quietly in my IN tray (or IN wire-basket). **1943** W. S. Churchill *Second World War* (1951) IV. 852 A habit of having secret papers placed in 'In' or 'Out' trays on the desks of important officers. **1958** *New Statesman* 25 Jan. 112/3, I hope some copies reach the In-trays of the Indian officials who refused to support him at Oxford. **1968** P. McKellar *Experience & Behaviour* vi. 163 In an office, some item waiting to be dealt with in the older man's in-tray may seem to him to wait only a short time. **1970** *New Scientist* 30 Apr. 239/1 The issue of industrial noise is finding its way to the political in-trays.

intrazonal (intrăzōu·năl), *a.* *Soil Sci.* [a. F. *intrazonal* (N. Sibirtsev 1897, in *Compt. Rend. de la VII*^e^ *Session, Congr. géol. internat.* (1899) II. III. v. 83): see Intra- 1 and Zonal *a.*] Designating any soil which occurs within a major soil zone but differs from the characteristic soil of that zone owing to the overriding influence of relief, parent material, or other local factor.

[**1908** *Jrnl. Agric. Sci.* III. 84 In all these zones also certain interzonal soils are sure to occur, which owe their peculiarities to special conditions of topography alone. *Ibid.*, Such soils constitute the six..interzonal types enumerated by Sibirtzev.] **1927** C. F. Marbut tr. *Glinka's Great Soil Groups* 28 Intrazonal soils originate, according to Sibirceff, where local soil forming forces predominate over the general or zonal forces. **1946** Lutz & Chandler *Forest Soils* xi. 385 Intrazonal soils have more or less well-defined characteristics which reflect the dominating influence of some local factor, such as relief, parent material, or age, over the normal effect of climate and vegetation. **1965** B. T. Bunting *Geogr. Soils* i. 18 The soils of the world have been considered as belonging to one of the three major orders—zonal, azonal and intrazonal—each divisible into many groups.

intrigue, *v.* Add: **5.** *trans.* To excite the curiosity or interest of; to interest so as to puzzle or fascinate. Also *absol.* (A modern gallicism.)

1894 *Month* May 122 The publishers often become so intrigued by these claims of authorship, that we find them at times passing by the matter altogether. **1896** [in Dict., sense 1]. **1900** *Westm. Gaz.* 5 Dec. 2/2 We do agree most heartily, but the observation intrigues us not a little. **1909** H. G. Wells *Ann Veronica* iii. 78 The New Woman and the New Girl intrigue me profoundly. **1918** A. Quiller-Couch *Stud. in Lit.* 1st Ser. 147 These theological poets and preachers of the seventeenth century..were intrigued ..by man's lapse from a state of innocence. **1924** W. M. Raine *Troubled Waters* xxi. 225 The conspiracy she proposed intrigued his interest. **1957** Partridge *English gone Wrong* i. 9 Such words as..'to be *intrigued*' for 'deeply or much interested'..have degenerated from definite sense to indefinite nonsense.

intriguing, *ppl. a.* Add: Also, in sense of prec. Hence **intri·guingly** *adv.*

1909 *Daily Chron.* 29 Apr. 3/2 A brisk, intriguing, and entertaining story. **1920** *Isis* (Oxf.) 27 Oct. 2/1 Edited.. by three members of Oriel..with a longish and intriguing introduction by Mr. John Masefield. **1922** *Blackw. Mag.* June 778/2 The line of alders on the far bank was intriguingly punctuated with squatting figures. **1935** W. S. Maugham *Don Fernando* x. 190, I would say boldly then that no great artist is more intriguing than El Greco. **1970** *Daily Tel.* (Colour Suppl.) 30 Oct. 15/1 She was a mine of intriguingly useless information. **1974** *Observer* 10 Feb. 32/8 Even more intriguing than the sociology of fashion is its psychology.

intrinsic, *a.* Add: **3. d.** *intrinsic factor*, a substance (perhaps a mucoprotein) which is secreted in the gastric juice and makes possible the absorption by the body of vitamin B_{12} ('extrinsic factor').

1930 [see *extrinsic *a.* 3 c]. **1961** *Lancet* 26 Aug. 483/2 Vitamin-B_{12} deficiency through lack of intrinsic factor (I.F.), as in pernicious anæmia, has stimulated efforts to purify and isolate I.F. **1965** A. Doscherholmen *Stud. Metabolism Vitamin B_{12}* 4 The intrinsic factor has not yet been isolated in pure form.., but it is believed to be a mucoprotein or mucopolypeptide... The purpose of the intrinsic factor is to bring about the absorption from the food, by some mechanism still unknown.., of the small amount of cyanocobalamin needed.

e. *Physics.* Of a semiconductor: owing its electrical conductivity to thermally excited electrons from the principal substance present, rather than to electrons from impurity atoms. Hence applied to conduction that arises in this way.

1933 R. H. Fowler in *Proc. R. Soc.* A. CXL. 507 Semiconductors without impurities owe their conductivity and other electrical properties to thermal excitation of electrons from band 2 to band 1. These we shall refer to as intrinsic semi-conductors. **1945** *Jrnl. Appl. Physics* XVI. 562/2 The atoms of the bulk material hold their valence electrons at low temperatures but become thermally ionized at elevated temperatures. An electronic conductivity of this type is called intrinsic. **1948** Torrey & Whitmer *Crystal Rectifiers* iii. 47 Intrinsic semiconduction occurs in materials that have a band structure similar to that of insulators..but with the difference that the gap in energy between the highest filled band and the lowest empty band is relatively small. **1962** Simpson & Richards *Physical Princ. Junction Transistors* viii. 167 The region near the collector is practically intrinsic and under proper operating conditions the transition region (depletion region) of the collector barrier occupies the whole of it. **1966** C. R. Tottle *Sci. Engin. Materials* ii. 46 Intrinsic semiconductors..are insulators below a given temperature and conductors with a negative temperature coefficient above it. Silicon and germanium are examples of this type of semiconductor.

intro, intro., colloq. abbrev. of Introduction.

1923 *Daily Mail* 17 July 12 (Advt.), It's extraordinary the number of fellows who..write to us..thanking me for the intro! **1928** *Melody Maker* Feb. 195/3 The sixth and eighth bars of the intro. worried me a little. **1929** J. B. Priestley *Good Companions* III. ii. 515 You'd never see him if you hadn't an intro, but when you do see him, 's'business. **1949** L. Feather *Inside Be-Bop* ii. 17 The intro and coda of *Ko-ko*. **1964** *Listener* 17 Dec. 987/2 The intro film's symbols..appear to be so dully strange that they induce an uneasy feeling. **1968** *Blues Unlimited* Dec. 6 Al Smith compered. Not very imaginatively, in fact he started off two intros word for word the same. **1973** J. Jones *Touch of Danger* xx. 116 Fred Tarkoff..gave me a letter of intro to her. **1974** *Melody Maker* 13 Apr. 50/2 There are guitar intros and solos.

introducing, *vbl. sb.* (In Dict. s.v. Introduce *v.*) Add: Also *attrib.*, as † **introducing house** (see quots.). *Obs.* exc. *Hist.*

1846 *Swell's Night Guide* 41 French introducing houses. These accommodation cribs have become so numerous, that it requires some tact and *nous* to discover them. **1857** W. Acton *Prostitution* vii. 97 The establishments of certain procuresses..vulgarly called 'introducing houses'..are worth notice as the leading centres of the more select circles of prostitution here. **1861** Mayhew *Lond. Labour* (1862) Extra vol. 214/1 Under this head [*sc.* brothels] we must include introducing houses, where the women do not reside, but merely use the house as a place of resort in the daytime. **1955** C. Pearl *Girl with Swansdown Seat* ii. 36 The Victorian 'introducing house' where the pleasant ceremony of introducing wealthy amateurs to willing girls was carried out with dignity and delicacy.

introduction. Add: **1. d.** The issuing of new shares by a company not directly to the public but through the medium of the Stock Exchange.

1929 *Economist* 27 July 175/1 The Stock Exchange 'introduction' (as distinct from the issue by prospectus or offer for sale) will fill an increasingly important *rôle*. **1966** P. A. S. Taylor *New Dict. Econ.* 152 *Introduction*, the offer of a new issue to the public, not directly but through the Stock Exchange... This method does away with part of normal new issue procedure, but necessitates that there should be a large body of shareholders, securities available to start the market, and no large 'deal' involved in the marketing of the securities. **1970** G. D. Newbould *Business Finance* III. 193 Two less common techniques are the public issue and the introduction... The *introduction* is unique, since there is no formal issue of shares... The introduction can be used only where there are a sufficient number of shareholders to ensure a market when quotation is granted. **1971** J. Bates *Financing Small Business* (ed. 2) vi. 95 The Stock Exchange introduction is effectively an application by the issuing house on behalf of the shareholders for shares to be quoted on a Stock Exchange.

5. b. (Earlier and later examples.)

1801 C. Wilmot *Let.* 13 Dec. in *Irish Peer* (1920) 14 Mr. Holcroft, to whom Lady Mount Cashell had a letter of introduction. **1816** Jane Austen *Emma* II. xiv. 267 The idea of her being indebted to Mrs. Elton for what was called an *introduction*—of her going into public under the auspices of a friend of Mrs. Elton's. **1885** A. Edwardes *Girton Girl* III. xv. 267 It was well for her, she said, to.. look at Newnham and Girton from without, before delivering her letters of introduction. **1949** *Radio Times* 15 July 6/3 The BBC's New York office has given me introductions to the broadcasting people in Montreal. **1969** L. Hellman *Unfinished Woman* vii. 81, I had sent off a few letters of introduction, but..they weren't answered.

c. (Earlier example.)

1808 Jane Austen *Let.* 26 June (1952) 200 They have nice weather for their introduction to the Island.

introessive (-e·siv), *a.* *Gram.* [f. L. *intrō* within + *esse* to be + -ive.] Designating the case which expresses 'motion into'.

1903 *Amer. Anthropologist* Jan.–Mar. 13 Besides a general locative some of the most frequently occurring are inessive, superessive, introessive, ablative, and terminative.

introgress (introgre·s), *v.* *Biol.* [Back-formation from next.] *intr.* To be transferred by introgression *into* another species. So **intro·gre·ssed** *ppl. a.*, (*a*) transferred in this way; (*b*) produced as a result of introgression; **introgre·ssing** *ppl. a.*

1958 F. C. Elliott *Plant Breeding & Cytogenetics* ix. 267 Genes from one species introgress into another. **1963** E. Mayr *Animal Species & Evolution* vi. 124 Introgressing hybrid swarms. *Ibid.*, They are introgressing hybrid populations between *D. middendorfiana*, *D. schoedleri*, and *D. pulex*. **1965** D. Zohary in Baker & Stebbins *Genetics Colonizing Species* 415 Introgressed types are quickly fixed by self-pollination. **1968** J. A. Serra *Mod. Genetics* III. xx. 198 This introduction of alien loci or introgressed loci may be of as much interest as the introduction of entire alien chromosomes. **1972** *Science* 22 Sept. 1076/2 Where the introgressed maize at Tehuacán came from is not known.

introgre·ssion. [See s.v. Intro-.] **1.** (In Dict. s.v. Intro-.)

2. *Biol.* The transfer of a small amount of genetic material from one (usu. plant) species to another as a result of hybridization between them and repeated back-crossing.

1938 Anderson & Hubricht in *Amer. Jrnl. Bot.* XXV. 399 We conclude that there is a strong introgression of *T. canaliculata* into *T. occidentalis*. **1950** G. L. Stebbins *Variation & Evolution in Plants* vii. 265 In this instance, therefore, frequent interspecific hybridization is not accompanied by introgression of genes across the barrier formed by partial sterility of the F_1 hybrid. **1963** Lewis & John *Chromosome Marker* IV. iv. 373 Low sterility and open pollination will favour introgression. **1971** V. Grant *Plant Speciation* iv. 52 If the introgression goes far enough, it may obliterate the morphological and ecological distinctions between the original species.

introgressive (introgre·siv), *a.* *Biol.* [f. Introgress(ion + -ive.] Characterized by, bringing about, or resulting from introgression; *introgressive hybridization*, introgression; hybridization leading to introgression.

1938 Anderson & Hubricht in *Amer. Jrnl. Bot.* XXV. 396 We have therefore given it a distinctive name, introgressive hybridization. *Ibid.* 398 The introgressive effect of *T. canaliculata*. **1950** G. L. Stebbins *Variation & Evolution in Plants* vii. 263 The habitat provides an ecological niche for the establishment of the introgressive types. **1968** J. A. Serra *Mod. Genetics* III. xx. 199 The introgressive introduction of some loci of *Nicotiana plumbaginifolia* into *N. tabacum*. **1970** *Watsonia* VIII. 85 (*heading*) Introgressive hybridisation between British annual *Senecio* species. **1971** V. Grant *Plant Speciation* iv. 51 Such bridging populations of hybrid origin fall into two general classes: hybrid swarms and introgressive populations. **1971** G. L. Stebbins *Chromosomal Evolution Higher Plants* v. 149 The phenomenon of introgressive hybridization or introgression is a sequence of three processes: hybridization, back crossing, and natural selection of back cross derivatives in a habitat where they are superior to either of the original parents.

introject (introdʒe·kt), *v.* *Psychol.* [Back-formation f. next.] *trans.* To incorporate an inward image of (an external object, or the values and attitudes of others) into oneself. Cf. *introjection 2, 3. Hence **introje·ctive** *a.*, that is introjected; **introje·cted** *ppl. a.*

1925 J. Riviere et al. tr. *Freud's Coll. Papers* IV. 78 The objects presenting themselves, are absorbed by the ego into itself, 'introjected'. **1932** *Brit. Jrnl. Psychol.* Oct. 156 Between the ages of 1 and 2½ his [the child's] mentality alternates between an introjective psychotic pattern..and a projective psychotic pattern. **1935** *Internat. Jrnl. Psycho-Anal.* XVI. 145 From the beginning the ego introjects objects 'good' and 'bad', for both of which the mother's breast is the prototype—for good objects when the child obtains it and for bad when it fails him. **1937** 'C. Caudwell' *Illusion & Reality* viii. 158 A blue rose, which was in the *speaker's* perceptual world,..has been formed in the common perceptual world and introjected into the hearer's perceptual world. **1952** W. J. H. Sprott *Social Psychol.* ix. 169 We are familiar with a theory that we 'introject' our version of our parents. **1962** *Sci. & Psychoanal.* V. 74 Second, the problem was said to be related to the *cathexis* of a deceased sister now existing as an 'introjected object'. **1964** Gould & Kolb *Dict. Social Sci.* 353/2 It is more probable that as a result of the powerful oral medium of the initial object-relations, identification may assume a more or less oral-cannibalistic, incorporative, and therefore introjective terminology. **1967** M. Argyle *Psychol. Interpersonal Behaviour* vii. 123 Children introject their parents' love and admiration of themselves. If they are never loved they will come to reject themselves and suffer from low self-esteem in later life.

introjection (introdʒe·kʃən). [Cf. Intro-. In senses 2 and 3, ad. G. *introjektion*.] **1.** (In Dict. s.v. Intro-.)

2. *Philos.* A theory whereby external objects are images of elements within the consciousness of the individual. Hence **introje·ctionism,** belief in a theory of introjection, **introje·ctionist** *a.*, pertaining to introjection.

1899 J. Ward *Naturalism & Agnosticism* II. xvi. 172 The term 'introjection' we owe..to the late Richard Avenarius of Zurich. *Ibid.*, Thus while my environment is an external world for me, his experience is for me an internal world in him. This is introjection. **1903** A. E. Taylor *Elem. Metaphysics* II. i. 81 Subjectivism is thus the last step in the development of the fallacy which begins with what Avenarius calls 'introjection'. *Ibid.* IV. i. 304 To translate it into the introjectionist psychology. **1912** *Mind* XXI. 10 The theory appears..to be really a piece of lingering introjectionism. **1931** G. F. Stout *Mind & Matter* 291 Ward finds this in a supposed process ..which, following Avenarius, he calls Introjection.

3. a. *Psycho-analysis.* A term used by S. Ferenczi (1909 *Jahrb. f. Psychoanalyt. Forschungen* I. 422–57) to denote the forming of a subjective image of an object and the transfer to it of emotional energy previously given to the object itself.

1916 E. Jones tr. *Ferenczi's Contrib. Psycho-Anal.* ii. 40 One might give to this process, in contrast to projection, the name of *Introjection*. **1917** C. R. Payne tr. *Pfister's Psychoanal. Method* xii. 387 In the projection, one feels subjective processes producing discomfort as influences of the outer world; in the introjection, inversely, processes of the outer world as one's own. **1922** J. Strachey tr. *Freud's Group Psychol.* 65 First, identification is the original form of emotional tie with an object; secondly, in a regressive way it becomes a substitute for a libidinal tie, as it were by means of the introjection of the object into the ego. **1946** *Mind* LV. 83 This growth of the super-ego has four main features, 'narcissism', 'introjection', 'nemesism', and 'sado-masochism'. **1963** *Listener* 7 Mar. 431/2 The ego has incorporated (by means of introjection ..) certain fantasy figures acquired in infancy (such as the image of the breast).

b. *Psychol.* The forming of an inward image of the attitudes, values, and expectations of people or groups by whom one is anxious to be accepted.

1931 J. C. Flügel in W. Rose *Outl. Mod. Knowl.* ix. 384 An introjection into the self of the earliest external moral forces, i.e. the moral attitudes and precepts of parents. **1955** M. Klein *New Directions in Psycho-Anal.* i. 21 External and internal situations are always interdependent, since introjection and projection operate side by side from the beginning of life. **1962** *Listener* 21 June 1055/2 When a child's parents have habitually reacted to his behaviour in a certain way—for instance, being shocked if he cheats—then after a time the child begins to react in this way towards himself. This is the process sometimes described as 'introjection'. **1967** M. Argyle *Psychol. Interpersonal Behaviour* vii. 121 A process of *introjection*, whereby children adopt the perceptions, attitudes, and reactions to themselves of parents and others.

intropunitive (intropiū·nitiv), *a.* *Psychol.* Also **intrapunitive.** [f. L. *intrō* inwardly (cf. Intro-) + Punitive *a.*] Blaming oneself rather than other people or events; of or pertaining to an unreasonable feeling of responsibility for frustrations or the like. Contrasted with *extrapunitive *a.* and *impunitive *a.*

1938 S. Rosenzweig in H. A. Murray *Explorations in Personality* vi. 587 He may react with emotions of guilt and remorse and tend to condemn himself as the blameworthy object. This type of reaction may be termed 'intropunitive'. **1954** G. W. Allport *Nature of Prejudice* xxvii. 437 This inwardness and ability to know and to laugh at oneself make for the intropunitive tendency that we examined... Self-blame takes the place of projected external blame. **1958** [see *impunitive *a.*]. **1965** B. I. Murstein *Handbk. Projective Techniques* xxxii. 575 These criteria included diagnostic council ratings.., Rosenzweig Extrapunitive and Intrapunitive Scores, and the self-ratings of the subject. **1969** [see *extrapunitive *a.*].

Hence **intropu·nitiveness,** the condition of being intropunitive.

1943 *Psychol. Abstr.* XXVII. 99/2 Nonhypnotizability is associated with other defense mechanisms.., and with other reactions to frustration, such as intropunitiveness and extrapunitiveness. **1958** M. Argyle *Relig. Behaviour* xii. 161 There is no evidence concerning the intropunitiveness or private religious activities of sect members, but the above three findings confirm the application of the reduction of guilt theory to sects. **1969** M. D. Vernon *Human Motivation* ix. 145 One of the most frequently occurring types of intropunitiveness is anxiety.

introscope (i·ntroskōup). [f. Intro- + -scope.] An instrument designed to be inserted into tubes so as to permit a visual examination of their interiors, and provided with a light source and some kind of optical system.

1937 *Nature* 27 Feb. 380/2 Charles Baker showed an 'introscope', an instrument for inspecting the interiors of boiler tubes, ship shaftings, oxygen bottles and aeroplane spars, etc. By means of this instrument, it is possible to illuminate and examine microscopically surfaces which cannot easily be inspected in other circumstances. **1958** *Ann. Rep. Chief Insp. Factories for 1957* (Cmnd. 521) 27 Entry into the reactor vessel itself will not be possible. Accordingly,..great attention is being given to the developments of introscopes..which will allow remote inspection of internal surfaces. **1962** *Punch* 18 Apr. 604/2 There are men who spend their best years bending pipes; ..others who peer inside them, with the aid of introscopes and boroscopes capable of seeing round four or five corners. **1973** A. Parrish *Mech. Engineer's Ref. Bk.* viii. 3 Introscopes, Endoscopes, Borescopes, etc. (The Trade name depending on the manufacturer) are forms of rigid, narrow, long industrial telescopes which introduce light and permit visual examination through small apertures e.g. down a small bore tube. They range from 2–50 mm in diameter and, in special cases, may be made from small sections up to 50 m long.

introspectible (introspe·ktĭb'l), *a.* Also **-able.** [f. Introspect *v.* + -ible.] Of a thought, sensation, experience, or other mental phenomenon: capable of being examined by introspection.

1925 C. D. Broad *Mind & its Place* ix. 419 How little of this..is introspected or is introspectible! **1937** *Mind* XLVI. 22 We must include, under the psychological responses which the words tend to produce, not only immediately introspectable experiences, but *dispositions* to react in a given way with appropriate stimuli. **1940** *Philosophy* Jan. 10 Show me the impression, the sensible or introspectible datum.., from which your general symbol derives its meaning. **1959** A. J. Ayer *Logical Positivism* 17 The prevailing view was that these [elementary] statements referred to the subject's introspectible or sensory experiences. **1971** A. Quinton in A. Bullock *20th Cent.* 257/2 The introspectible facts of mental life.

introspectionism (introspe·kʃəniz'm). *Psychol.* [f. Introspection + -ism.] Introspective psychology (see *introspective *a.*); also, more generally, = Introspection 2.

1922 R. S. Woodworth *Psychol.* ii. 21 Without caring to attach ourselves exclusively to either introspectionism or behaviourism. **1931** *Psyche* Jan. 68 Straightforward introspectionism, on the whole, judged by the contents of this book, has very little status in modern psychology. **1965** N. Chomsky *Aspects of Theory of Syntax* 193 Introspectionism—should one make use of introspective data in the attempt to ascertain the properties of these underlying systems? **1967** *Lancet* 11 Nov. 1050/1 Behaviourism..arose as a reaction to the even less fruitful introspectionism of the late 19th century.

introspectionist. Add: **c.** *attrib.* or as *adj.*

1934 in Webster. **1949** Koestler *Insight & Outlook* xiii. 184 Behaviourism has rendered a service to science by its puritan intolerance towards introspectionist debauch.

introspectioni·stic, *a.* = prec.

1943 *Mind* LII. 133 The methodologically correct use..of introspectionistic terms..is not impugned.

introspective, *a.* Add: *introspective psychology*, psychology based on introspection and on the direct observation of one's own mental states.

1878 W. James in R. B. Perry *Tht. & Char. W. James* (1935) II. LIII. 29 Those whose highest flights are articles in the *Popular Science Monthly* will talk of the exploded superstitions of introspective psychology. **1931** R. S. Woodworth *Contemp. Schools Psychol.* ii. 17 What we do find..is..more precise formulation of the aim of introspective psychology. *a* **1942** B. Malinowski *Sci. Theory of Culture* (1944) vii. 71 Whether we use introspective psychology, and say that understanding means identification of the mental processes, or whether, as behaviourists, we affirm that his response to the integral stimulus of the situation follows lines familiar to us from our own experiences, does not change the argument profoundly. **1951** E. E. Evans-Pritchard *Social Anthropol.* iii. 44 Other anthropologists were later left in a similar way in the fashion of introspective psychology.

introsusception. 2. (Earlier example.)

1786 J. C. Lettsom (*title*) The history of an extraordinary introsusception... With an account of the dissection. By..Whately.

‖ **introuvable** (æ̃ntruvāb'l), *a.* [Fr.] Unfindable, undiscoverable; *spec.* of books. Also as *sb.*

1824 *Edin. Rev.* Mar. 2 We are by no means of opinion that it [*sc.* a correct standard of national prosperity] is really *introuvable*. **1856** *Newsp. & Gen. Reader's Compan.* II. § 1436 Give me a mere *annonce* of any thing, that can tell me of your *introuvable* friend. **1895** H. B. Forman in Nicoll & Wise *Lit. Anecdotes 19th Cent.* I. 67 The almost introuvable tract *Prothanasia and other Poems*. **1963** *Times Lit. Suppl.* 15 Feb. 116/3 A potential *introuvable* to future collectors.

introversion. Add: **1. b.** The tendency to turn psychic energy inwards and to withdraw from the external world; opp. *extraversion 2, *extroversion 3.

1912 *Psychol. Bull.* IX. 159 So that when in later life there occurs an introversion (in the sense of Jung), it consists of a harking back to regressive, reminiscent, infantile material. **1915**, etc. [see *extraversion 2]. **1935** C. G. Jung *Analytical Psychol.* (1968) ii. 41 The psychological mechanism of introversion of the conscious mind into the deeper layers of the unconscious psyche. **1955** *Sci. News Let.* 19 Mar. 185/2 Patients with this disease are at times completely withdrawn from the world around them and give the picture of the very extreme of introversion. **1964** M. Argyle *Psychol. & Social Probl.* vi. 75 Eysenck has suggested the three dimensions of neuroticism, psychoticism and introversion–extraversion.

introversive. Add: **1. c.** Characterized by introversion (sense *1 b).

1923, 1932 [see *extroversion 3]. **1970** *Jrnl. Gen. Psychol.* July 65 Rorschach writes:..all of these findings indicate a more introversive type. **1972** *Daily Tel.* (Colour Suppl.) 25 Aug. 31/1 Indoor flying is one of the world's most esoteric and introversive hobbies.

introvert, *sb.* Add: **2.** *Psychol.* A person characterized by introversion; a withdrawn or reserved person; opp. *extrovert *sb.* Also *attrib.* and as *adj.* Also **i·ntrovertish** *a.*, said of such a person, his activities, etc.

1918, etc. [see *extrovert *sb.* (and *a.*)]. **1925** C. Fox *Educational Psychol.* 254 The introvert abstracts from the object and deals with it by concepts concentrating upon the inner world of thought. **1934** *Brit. Jrnl. Psychol.* July 26 They were noticeably more introvert, schizoid and desurgent in temperament. **1946** R. P. Basler in W. S. Knickerbocker *20th Cent. English* III. 392 In the snugness of introvertish isolation, there is always time, an eternity for continual deception and indecision. **1955** L. Langstroth *Struct. of Ego* vii. 82 This question of the relative strength of the social and biological selves suggests at once Jung's broad division of personalities into two main types: the introvert and the extrovert. **1957** H. J. Eysenck *Dynamics Anxiety & Hysteria* vi. 213 The introvert, as we have seen, is socialized. **1960** *Encounter* XV. 47 The introvert-intellectual is the hero of several of Buchan's works. **1967** M. Argyle *Psychol. Interpersonal Behaviour* iii. 50 Experiments with schoolchildren show that introverts respond better to praise.

introverted, *ppl. a.* **1.** Add to def.: esp. in *Psychol.*

1915 [see *extraversion 2]. **1916** C. E. Long tr. *Jung's Coll. Papers Analytical Psychol.* 348 The introverted type is characterised by the fact that his libido is turned towards his own personality to a certain extent. **1923** [see *extrovert *sb.* (and *a.*)]. **1957** H. J. Eysenck *Dynamics Anxiety & Hysteria* i. 31 We may take dysthymics on the one hand, and hysterics and psychopaths on the other, as examples of our introverted and extroverted groups. **1968** C. Rycroft *Critical Dict. Psychoanal.* 48 There is a tendency to equate 'introverted' with 'withdrawn' or 'schizoid'. **1974** *Country Life* 17 Jan. 76/3 Dorothy Osborne... This witty and introverted girl.

intrude, *v.* **5.** For † *Obs. rare* substitute: *Obs. rare* except in *Geol.* To be forced or thrust into. (Add later examples.)

1925 J. Joly *Surface-Hist. Earth* viii. 130 Both these [series], together with a series mainly of basaltic eruptives .., were finally intruded by vast uprisings of granites (largely batholithic). **1932** W. H. Emmons et al. *Geol.* xii. 361 This sill intrudes limestone, is nearly flat lying, and in general is almost parallel to the beds it intrudes. **1955** *Econ. Geol.* L. 715 Four locations where sills, dikes, and stocks intrude or cut off the phosphate-bearing beds. **1957** *Mineral. Mag.* XXXI. 588 This block is intruded by three stock-like masses of fine-grained granite. **1966** C. O. Dunbar *Earth* iv. 61 Remnants of the oldest sedimentary formations were intensively deformed and intruded by the underlying granite. **1971** I. G. Gass et al. *Understanding Earth* ii. 48/2 A small igneous intrusion.. which intruded a fossiliferous sediment.

intruder. Add: **3.** An aeroplane (or its pilot) that invades the enemy's aerodromes to interfere with his operations. Also *attrib.*, as *intruder attack, raid.*

1941 *Aeroplane Spotter* 9 Oct. 174 *The Intruder*. 'Night intrusion' is the name of one type of operation on which Douglas Havocs are earning distinction. The Havocs fly out over enemy aerodromes when the night bombers are returning and shoot them down over their home stations. *Ibid.* 6 Nov. 205 Messerschmitt Me 110 two-motor fighters are now being used with Ju 88s for night defence over Germany and 'intruder' work over Great Britain. **1943** *Times* 22 Dec. 4/5 R.C.A.F. intruders destroyed two enemy aircraft. **1944** *Times* 6 Nov. 4/4 Night fighters and intruder aircraft of Bomber Command supported the bombers. **1966** M. R. D. Foot *SOE in France* iv. 82 But 418 was an ordinary Canadian intruder squadron.

intrusion. Add: **1. b.** (Earlier example.)

1839 R. I. Murchison *Silurian Syst.* v. 78 This intrusion having taken place on a line of ancient volcanic eruption, the origin of which cannot be understood without a previous acquaintance with the history of the Silurian System, the account of this new red trap dyke is necessarily deferred.

3. *spec.* in contexts of Journalism.

1958 *Spectator* 18 July 110/3 Newspaper intrusion into private lives. **1960** *New Statesman* 15 Oct. 556/2 The intrusion and impertinence of some of the gossip writers.

intrusive, *a.* Add: **B.** *sb.* *Geol.* An intrusive rock or rock mass.

1895 A. Harker *Petrol.* vii. 87 Acid intrusives. The acid intrusive rocks embrace a considerable range of varieties. **1925** N. E. Odell in E. F. Norton *Fight for Everest, 1924* 300 Yet the character of the former [*sc.* the limestone series] may be entirely due to its proximity to the hard crystalline rocks and its alteration brought about by pressure against them, if not also by their igneous intrusives. **1962** W. T. Huang *Petrol.* iii. 51 It [*sc.* magma] may be solidified at great depth, forming such large intrusives as batholiths, stocks, and lopoliths. **1968** J. Gilluly et al. *Princ. Geol.* (ed. 3) xviii. 435 (*caption*) Devil's Tower, Wyoming, probably a volcanic plug, but perhaps part of a roofed intrusive.

intuit, *v.* Add: **2.** (Later example.)

1968 J. C. Holmes *Nothing more to Declare* 105 You had to be able to intuit on the bias, to hear music *being* music.

b. (Later examples.)

1926 A. Huxley *Two or three Graces* 85 You intuit things that aren't there at all. **1968** *Times* 13 Jan. 20/3 We may intuit his reality, but we cannot share it. **1972** *Times Lit. Suppl.* 18 Feb. 178/1 What Johnson intuited and asserted in 1765, that of the Folios only the first had any textual authority, was demonstrated by Malone in 1790.

intuitable (intiū·ităb'l), *a.* [f. Intuit *v.* + -able.] That can be known by intuition.

1887 G. T. Ladd tr. *Lotze's Outl. Logic* § 64 'Abstraction' makes the content of the concept intuitable only as a whole. **1904** J. H. Muirhead in J. E. Hand *Ideals of Sci. & Faith* 91 If the mere absence of intuitable continuity were the only difficulty in the way of assimilation, we might ignore it. **1932** H. H. Price *Perception* vi. 168 The intuitable characteristics of the sense-datum now sensed. **1971** R. Harrod *Sociol.* 51 In the case of lying, there is no question of its having a certain innate intuitable quality.

intuited (intiū·ited), *ppl. a.* [f. Intuit *v.* + -ed¹.] Arrived at or known by intuition.

1886 A. Weir *Hist. Basis Mod. Europe* xii. 481 The mathematical sciences..drew their conclusions from intuited figures and series. **1890** W. James *Princ. Psychol.* I. 630 Meanwhile, the specious present, the intuited duration, stands permanent, like the rainbow on the waterfall, with its own quality unchanged by the events that stream through it. **1924** H. J. W. Hetherington *Life & Lett. Sir H. Jones* I. 64 Even intuited truth is not reality. **1967** *Listener* 4 May 593/1 Some felt or intuited absolute ethical standard. **1973** S. Henderson *Understanding New Black Poetry* 10 Fidelity to the observed or intuited truth of the Black Experience in the United States.

intuition. 5. c. (Earlier example.)

1796 F. A. Nitsch *Gen. View Kant's Princ. concerning Man* 75 Those ideas which immediately arise in consequence of our external sense being affected are external perceptions or external intuitions.

intuitionism. Add: **3.** *Math.* The theory put forward by L. E. J. Brouwer (1908) that mathematics is founded on extra-linguistic constructs based on pure intuition (in the Kantian sense, cf. Intuition 5 c); that space geometry is reducible to arithmetic and that therefore the law of the excluded middle, applying to finite classes, might not be valid for infinite classes. Cf. *formalism 3.

1913 tr. L. E. J. Brouwer in *Bull. Amer. Math. Soc.* XXV. 86 From the present point of view of intuitionism.. all mathematical sets of units..can be developed out of the basal intuition, and this can only be done by combining a finite number of times the two operations: 'to create a finite ordinal number' and 'to create the infinite ordinal number ω'. **1940** [see *formalism 3 a]. **1941** Courant & Robbins *What is Math.?* i. 87 Some distinguished mathematicians have recently advocated the more or less complete banishment from mathematics of all non-constructive proofs... The school of 'intuitionism', which has adopted this program, has met with strong resistance. **1959** E. W. Beth *Found. Math.* xv. 413 One of the most spectacular features in Brouwer's intuitionism is..his rejection of the unrestricted application of the principle of the excluded third in mathematical reasoning. **1965** Kleene & Vesley *Found. Intuitionistic Math.* i. 1 Modern intuitionism, founded by Brouwer, constitutes a vigorous manifestation of the constructive tendency. **1973** *Sci. Amer.* Mar. 103/2 Three modern schools of mathematical thought: logicism, formalism and intuitionism.

intuitionist. Add: **1.** *attrib.* (Earlier example.)

1872 Mill *Exam. Hamilton's Philos.* (ed. 4) xiv. 339 This..is the staple of the Intuitionist argument.

2. (Later examples.) Used esp. in *Math.* Cf. *intuitionism 3. Also *attrib.*

1913 *Bull. Amer. Math. Soc.* XXV. 86 The intuitionist can never feel assured of the exactness of a mathematical theory by such guarantees as the proof of its being non-contradictory, the possibility of defining its concepts by a finite number of words or the practical certainty that it will never lead to a misunderstanding in human relations. **1926** *Proc. London Math. Soc.* 2nd Ser. XXV. 339 Apart from formalism, there are two main general attitudes to the foundation of mathematics: that of intuitionists or finitists..and that of the logicians. **1933** M. Black *Nature of Math.* 195 The intuitionist recognises only the existence of denumerable sets. **1941** [see *formalist 5]. **1952** R. M. Hare *Lang. Morals* iii. 30 The word 'good' is treated in the fashion that many intuitionists have treated it. **1959** E. W. Beth *Found. Math.* xv. 421 The intuitionist..can point to the edifice of intuitionistic mathematics which has been built alongside classical mathematics. **1967** S. C. Kleene *Math. Logic* § 36. 196 To prove an existence statement ∃xA(*x*), an intuitionist insists that it be shown how to find an *x* such that A(*x*). **1973** *Sci. Amer.* Mar. 103/2 As for intuitionists, they have in effect returned to the Pythagorean position that the natural numbers must be accepted without further analysis as the foundation of mathematics.

b. *attrib.* or as *adj.* Of or pertaining to intuitionism.

1926 [see *formalist B]. **1933** M. Black *Nature of Math.* 11 Intuitionist doctrines require the larger part of mathematics to be rewritten. *Ibid.*, The intuitionists.. are beginning to produce an intuitionist formal logic. **1960** S. Körner *Philos. of Math.* vi. 131 The intuitionist logic is a *post factum* record of the principles of reasoning which have been employed in mathematical constructions. *Ibid.*, *Every* intuitionist proposition *p*, whether or not the (intuitionist-)negation occurs in it, is the record of a construction. **1970** A. Kino et al. *Intuitionism & Proof Theory* 19 That is the general way of accepting-on-faith in the domains of religion and philosophy, and in traditional intuitionist or constructivist mathematics also.

intuitionistic *a.* (later examples); also **intuitioni·stically** *adv.*, in an intuitionistic manner.

1940 *Mathematical Rev.* Nov. 323/1 The treatment is 'intuitionistic' in the sense that it is purely algebraic, involving reference to order, absolute value, boundedness, etc., but not to limiting processes. **1942** D. D. Runes *Dict. Philos.* 150 The resulting disjunction becomes intuitionistically acceptable. **1944** *Mathematical Rev.* Sept. 198/2 The author proves intuitionistically seven theorems concerning the full product of a finite or denumerable number of virtually ordered sets, generalizing some results of Brouwer concerning the full products of sets of integers. **1945** E. T. Bell *Devel. Math.* (ed. 2) xxiii. 560 The 'objects' with which intuitionistic mathematics is concerned are said to be immediately apprehended in thought. **1946** *Nature* 7 Sept. 323/1 Nevertheless, if ethics is to be scientific in this sense, some possible theories, particularly those commonly called intuitionistic, are excluded. **1957** *Encycl. Brit.* XV. 82B/1 Most theorems of classical arithmetic can be established intuitionistically. **1962** B. Meltzer tr. *Gödel's On formally Undecidable Propositions* 60 The following is demonstrated in an intuitionistically unobjectionable way. **1965** Kleene & Vesley (*title*) The foundations of intuitionistic mathematics. **1967** S. C. Kleene *Math. Logic* § 44. 257 The consistency proof by a truth definition can even be managed intuitionistically. **1971** R. Schock *Quasi-Connectives* vi. 58 The resulting quasi-connectives satisfy all the axioms of the intuitionistic sentential calculus, but not the principle of the excluded middle.

intuitive, *a.* Add: **3. b.** (Earlier example.)

1833 Mill *Let.* 5 July (1910) I. ii. 54, I conceive that most of the highest truths are..intuitive; that is, they need neither explanation nor proof, but if not known before are assented to as soon as stated.

6. (Earlier example.)

1852 Mill in *Westm. Rev.* LVIII. 362 If it is alleged that the intuitive school require, as an authority for the feeling, that it should *in fact* be universal, we deny it.

B. *sb.* One who works by intuition.

1907 *Westm. Gaz.* 28 Nov. 2/1 Poincaré divides mathematicians into analysts and geometers—i.e., into logicians and 'intuitives'. **1927** A. Huxley *Proper Stud.* 207 Intellectuals and intuitives.

intuitively, *adv.* Add: **3.** *Logic.* By, or to, unaided reflection; without the use of any technique of logic.

1942 J. C. Cooley *Primer of Formal Logic* vi. 218 Most of the statements excluded by the theory of types would probably be avoided intuitively because they do not make sensible English sentences.

intuitivism. 1. (Earlier example.)

a **1866** J. Grote *Exam. Utilitarian Philos.* (1870) i. 21 That doctrine, hostile to utilitarianism, to which he has given the name of 'intuitivism'.

intuitivist, *a.* (Earlier example.)

a **1866** J. Grote *Exam. Utilitarian Philos.* (1870) x. 168 The..description of conscience which Mr. Mill gives.. seems to me, if anything is, intuitivist.

inturn, *sb.* Add: **4.** *Curling.* An inward turn of the elbow made in delivering a stone.

1890 [see *out-turn c]. **1897** *Encycl. Sport* I. 262/1 The inturn is made when the curl is to be toward the right. **1923** G. Rae *Langsyne in Braefoot* iv. 42, I want the in-turn, an' I want ye here. Dinna lie back for ony sake. **1969** R. Welsh *Beginner's Guide Curling* xi. 83 Almost all beginners are taught first to play the in-turn.

inturned, *ppl. a.* (Later examples.)

1906 R. H. Benson *Queen's Tragedy* III. iv. 367 She.. touched the palms of her hands with her in-turned fingers. **1923** D. H. Lawrence *Birds, Beasts & Flowers* 6 The fig-fruit: Involved, Inturned, The flowering all inward and womb-fibrilled. **1967** *Antiquaries Jrnl.* XLVII. 256 The ditch as now defined with its right-angle turn near the western corner of the Arbour and its possible inturned entrance is best interpreted as the defence or boundary work of a Belgic *oppidum*.

intussuscept, *v.* (Earlier example.)

1802 *Med. & Physical Jrnl.* VII. 36 The part of the intestine inflated by Mr. Muir, was an intus-suscepted portion.

intussusception. Add: **3. a.** (Earlier and later examples.) Also, the mass of intestine involved in this.

1802 *Med. & Physical Jrnl.* VII. 36 Intestinal intussusceptions vary much in their extent, situation, and other circumstances. **1838** *Guy's Hosp. Rep.* III. 332 There were four intus-susceptions of the small intestines. **1960, 1970** [see *intussusceptum]. **1970** H. M. Spiro *Clin. Gastroenterol.* xxii. 356/1 At laparotomy, the intussusception is easily recognized and reduced or..resected.

intussusceptum (i:ntŏs,sŏse·ptŏm). *Med.* [mod.L., neut. sing. of *intussusceptus*: see Intussuscept *v.*] The middle and the innermost tubes of intestine (taken together) of the three present in an intussusception; also (less commonly), the innermost tube alone.

1857 Dunglison *Dict. Med. Sci.* (rev. ed.) 507/2 It [sc. *intussusceptio*] means the introduction of one part of the intestinal canal—*intussusceptum*—into another, which serves it as a sort of vagina or sheath. Generally, it is the upper part of the small intestine, which is received into the lower—*intussuscipiens*—when the intussusception is said to be progressive. **1884** F. Treves *Intestinal Obstructions* viii. 168 The external of the three layers is known as the intussuscipiens, the sheath, or the receiving layer... The innermost cylinder is known as the entering layer..and the middle one as the returning layer... Taken together, these two layers form the intussusceptum. **1960** Jones & Gummer *Clin. Gastroenterol.* iii. 93 An intussusception when fully developed is made up of three layers of intestine. The innermost or entering portion of the gut is called the intussusceptum... The middle or returning layer..is referred to as the resceptum and the outermost or ensheathing layer is the intussuscipiens. **1965** Lumsden & Truelove *Radiol. Digestive Syst.* xv. 497 The barium may penetrate between the intussuscipiens and the intussusceptum, thus outlining the apex of the intussusceptum. **1970** H. M. Spiro *Clin. Gastroenterol.* xxii. 353/2 Intussusception is the telescoping of the wall of one segment of the bowel into the adjacent distal portion. ..In formal circles, the invaginated segment is known as the intussusceptum, the receiving segment is the intussuscipiens.

intussuscipiens (i:ntŏs,sŏsi·pienz). *Med.* [mod.L., pres. pple. of *intussuscipere*: cf. prec.] The outermost of the three tubes of intestine which are present in an intussusception, and into which the other two are inserted.

1857, etc. [see prec.].

Inuit, var. *Innuit.

inulase (i·niulēiz, -s). *Biochem.* [f. Inul(in + *-ase.] An enzyme which hydrolyses inulin to fructose, found esp. in some fungi.

1893 J. R. Green in *Ann. Bot.* VII. 90 This transformation has been found to be due to a special enzyme, to which the name inulase may be given. It is a different body from diastase, for it has no action upon starch-paste. **1931** E. C. Miller *Plant Physiol.* xi. 628 Inulase may occur with diastase but has no action whatsoever upon starch. **1965** Florkin & Stotz *Comprehensive Biochem.* (ed. 2) XIII. 136 Systematic name: β-2, 1-Fructan fructanohydrolase. Recommended trivial name: Inulase.

inundatal (i·nŏndēitāl), *a.* *Ecol.* [irreg. f. Inundate *v.* + -al.] Of plants: growing in areas subject to flooding.

1847 H. C. Watson *Cybele Britannica* I. 65 Inundatal. Plants of places liable to be inundated in wet weather, but often dry in summer. **1897** G. C. Druce *Flora Berkshire* 575 Water Whorl Grass... Native. Inundatal. On the muddy margins of ditches, ponds, &c. **1926** *Nat. Hist. Oxford District* 88 Paludal flora (including uliginal and inundatal species).

inurn, *v.* Add: **inu·rnment,** the process of placing the ashes of a cremated body into an urn.

1934 *Amer. Speech* IX. 317/1 Olivet Memorial Park provides every service for Entombments, Inurnments, Interments. **1948** E. Waugh *Loved One* 36 Normal disposal is by inhumement, entombment, inurnment or immurement, but many people..prefer insarcophagusment.

in utero, in vacuo: see *in *Lat. prep.*

invalid. Add: **B.** *sb.* **3.** (Further examples.)

1822 M. Edgeworth *Let.* 28 May (1971) 402 Her own invalid breakfast as she called it, a glass of Seltzer-water and milk! **1847** W. M. Thackeray *Vanity Fair* (1848) xli. 375 Sir Pitt's invalid-chair was wheeled away into a tool-house in the garden. **1859** Dickens in *N.Y. Ledger* 27 Aug. 5/5 The hand-carriage was spinning away ..at a most indecorous pace for an invalid vehicle. **1861** Mrs. Beeton *Bk. Househ. Managem.* 893 (*heading*) Invalid cookery... A few rules to be observed in cooking for invalids. **1862** *Illustr. London News* 1 Nov. 473/1 An elongated invalid-chair is shown which is capable of being arranged as an ordinary easy-chair and of being extended into a camp bedstead. **1873** C. M. Yonge *Pillars of House* I. iii. 38 Wilmet could..do invalid cookery. **1876** —— *Three Brides* II. xiv. 259 He diverged to the invalid-carriage he had secured. **1880** Invalid furniture [see *carrying-chair* (*carrying *vbl. sb.* 4)]. **1902** 'Mark Twain' *Let.* 23 Oct. in C. Clemens *Mark Twain* (1932) 95 We brought Mrs. Clemens through successfully in an invalid car. **1911** *Daily Colonist* (Victoria, B.C.) 27 Apr. 6/1 (Advt.), Wiese & Brohn's Oporto Invalid Port, best on the market. **1934** T. S. Eliot *Rock* I. 15 And political religion is like invalid port: you calls it a medicine but it's soon just a 'abit. **1953** 'N. Blake' *Dreadful Hollow* I. ii. 28 She's got an electric invalid-carriage. **1967** *Guardian* 12 June 6/4 As an Oxford undergraduate with a muscular dystrophy, I am grateful for a Ministry of Health invalid tricycle. **1972** K. Bonfiglioli *Don't point that Thing at Me* i. 1 Invalid Port of an unbelievable nastiness. **1972** *Guardian* 4 Sept. 11/8 The middle-aged spina bifida sufferer who, after having saved to buy an invalid car, lost his job.

invalidity. Add: **3. b.** *attrib.*

1906 *Daily Chron.* 18 Nov. 3 A State scheme of invalidity insurance would lead to the destruction of the successful voluntary system. **1907** *Ibid.* 15 July 3/2 [Germany] The invalidity pensions go to all those of any age who are unfitted for relief. **1908** *Ibid.* 16 June 5/7 There is a great deal to be said for the German invalidity scheme. **1969** *Daily Tel.* 25 July 23 The new 'invalidity pension' will be introduced as part of the proposed earnings related retirement pension scheme..due to come into force in 1972. *Ibid.*, The new invalidity pension is designed to take over after a person has been away from work through illness for more than 28 weeks.

invalidy (i·nvălĭdi), *a.* *colloq.* [f. Invalid *sb.* + -y¹.] Of the nature of an invalid.

1894 Mrs. H. Ward *Marcella* III. 9 She's fussy, you know, and invalidy, and has to be wrapped up in shawls. **1912** 'R. Dehan' *Between two Thieves* 626 We invaded the Crimea with a weakly, invalidy, or crippled army.

Invar (i·nvāɪ). Also **invar.** [a. F. *invar* (M. Thury *Nouveau Pendule Compensateur* (1897) 4), f. *invar*(*iable* invariable.] The proprietary name of an alloy of iron or steel (about 64%) and nickel (about 36%), which has a very small coefficient of expansion.

1902 *Encycl. Brit.* XXXIII. 805/2 For ordinary standards of length Guillaume's alloy (invar) of nickel (35·7 per cent.) and steel (64·3 per cent.) is used, as it is a metal that can be highly polished, and is capable of receiving fine graduations. **1928** J. E. Haswell *Horology* iii. 29 The more recent discovery of the nickel-steel alloy, 'Invar', by Dr. C. E. Guillaume has, however, to a considerable extent revolutionised compensated pendulums. **1929** J. A. Ratcliffe *Physical Princ. Wireless* iii. 50 The oscillating system..is not an electrical circuit, but is a tuning-fork made of 'invar' metal, whose frequency remains very constant under all conditions. **1955** [see *gravimeter 2]. **1962** J. G. Tweeddale *Metall. Princ. for Engineers* iv. 106 Nickel may be used to reduce the coefficient of thermal expansion of steel, 36% reducing the coefficient practically to zero at room temperature when the carbon is 0·2% and manganese 0·5% (this alloy is sold in this country under trade names such as 'Invar' and 'Nilo 36'). **1971** R. J. P. Wilson *Land Surveying* viii. 134 Invar tapes are more delicate than steel tapes and cannot stand up to everyday survey use. Also the degree of accuracy they can provide is unnecessary on most survey work. However, they are invaluable on precise base measurements, all the British bases being measured with 100-ft invar tapes.

invariance. Add: Hence applied to a similar property with respect to any transformation or operation. Also *attrib.*

1910 [see *Galilean a.² b]. 1919 A. N. Whitehead Enquiry Princ. Nat. Knowl. 41 This invariance, with these formulae for transformation, does not extend to Maxwell's equations for the electromagnetic field. 1941 Courant & Robbins What is Math.? iii. 159 A particular consequence of the invariance of angle under inversion is that two circles or lines that are orthogonal, i.e. that intersect at right angles, remain orthogonal after an inversion. 1969 Observer 13 Apr. 1 These rules [sc. conservation laws of physics] are based on much more fundamental ideas about nature, called invariance principles. Ibid. 2/6 The conservation of energy is a direct result of the principle of 'time invariance'. This says that the laws of nature do not change with time.

2. *gen.* The property of remaining unaltered or of being the same in different circumstances; an instance of this, an invariant.

1939 Mind XLVIII. 50 This (heuristic) principle of ethical invariance is what I wish to oppose to the principle of ethical relativity. 1956 Jakobson & Halle Fund. of Lang. iii. 28 The study of invariances within the phonemic pattern of one language must be supplemented by a search for universal invariances in the phonemic patterning of language. 1962 F. I. Ordway et al. Basic Astronautics xiii. 525 Drugs..might also minimize the psychological problems of space flight by reducing the tensions associated with sensory deprivation or invariance. 1964 Language XL. 202 Concepts concerning the invariance of linguistic units. 1973 [see *invariant a. a].

invariant, *a.* and *sb.* Add: **A.** *adj.* (Further examples.)

1957 G. E. Hutchinson Treat. Limnol. I. ix. 634 The relatively invariant climatic conditions of the equatorial regions. 1964 M. Argyle Psychol. & Social Probl. i. 14 These so-called 'first-order factors' are not a satisfactory final solution, since they will reflect the numbers of tests of different kinds which have been used, and fail to yield a stable or 'invariant' set of factors. 1966 [see sense B. 2 below]. 1973 Sci. Amer. Feb. 26/1 The locations of nerve cells, the trajectories of nerve fibers and the spatial arrays of synaptic connections are invariant in all individuals of the same species. This invariance is termed neuronal specificity.

b. *Physical Chem.* Having no degrees of freedom (see *freedom 10 b).

1899 R. A. Lehfeldt Text-bk. Physical Chem. v. 208 Such systems may conveniently be called invariant, univariant, divariant, &c., according as they possess no, one, two, &c., degrees of freedom. 1923 A. C. D. Rivett Phase Rule i. 25 When $F = 0$ the system is said to be invariant, since none of its variables may be altered at will without destroying the system in the sense of altering the number of coexisting phases. 1971 F. A. Bettelheim Exper. Physical Chem. xxvii. 259 Phase diagrams of ternary systems that are plotted on a triple co-ordinate graph have two important features: an invariant point, O, and the solubility curves.

c. *Math.* and *Physics.* Unchanged by a specified transformation or operation. Const. *under.*

1908 H. Hilton Introd. Theory Groups of Finite Order v. 62 If every element of a group G transforms an element g of G into itself, so that g is permutable with every element of G, g is called a normal, self-conjugate, or invariant element of G... Similarly, if every element of G transforms a subgroup H into itself, H is called a normal, self-conjugate, or invariant subgroup of G. 1914 L. Silberstein Theory of Relativity iv. 111 The principle of relativity excludes all such laws as are not invariant with respect to the Lorentz transformation. 1919 A. N. Whitehead Enquiry Princ. Nat. Knowl. 39 They [sc. Newton's equations] are invariant for the spatio-temporal transformations from one such set to another within the Newtonian group. 1941 Birkhoff & MacLane Survey Mod. Algebra vi. 153 A subgroup S of a group G is normal (in G) if and only if it is invariant under all inner automorphisms of G (i.e., contains with any element all its conjugates). 1955 W. Pauli Niels Bohr 34 The theory is invariant with respect to space or time reflection separately. 1968 [see *covariant B].

B. *sb.* Also *attrib.*

1908 J. E. Wright Invariants of Quadratic Differential Forms Pref., The aim of this tract is to give..an account of the invariant theory connected with a single quadratic differential form. 1940 J. L. Coolidge Hist. Geom. Methods II. ii. 156 The invariant idea, thus launched, was eagerly seized on, especially..by the great twin brethren, Cayley and Sylvester.

b. Any quantity or expression which is invariant under a specified transformation or operation.

1908 H. Hilton Introd. Theory Groups of Finite Order vii. 99 An expression is an invariant of [a substitution-group] G if it is not altered when we perform on it every one of a set of substitutions which generate G. 1914 L. Silberstein Theory of Relativity iv. 112 It may be expressed shortly by saying that $x^2 + y^2 + z^2 - c^2t^2$..is a relativistic invariant. 1956 E. M. Patterson Topology i. 11 Such entities are called topological invariants, because they are the same for all topologically equivalent spaces. 1959, 1967 [see *covariant].

2. *gen.* An invariant property or feature.

1939 Mind XLVIII. 39 There is a widespread view that the sole invariant of morals is their sociological function to secure the preservation and welfare of a social group. 1960 E. Delavenay Introd. Machine Transl. iii. 28 In 1949 Weaver pointed out that..one discovers statistical invariants, as found in cryptography.., semantic invariants,..and logical invariants. 1966 J. J. Katz Philos. of Lang. ii. 9 The study of language in general provides us with generalizations expressing the invariant features of language which we may particularize as the requirement that an empirically correct description of a natural language represents such invariants.

invasion. Add: **1. c.** *Path.* The spreading of pathogenic microorganisms or malignant cells that are already in the body to new sites.

1891 F. P. Foster Med. Dict. III. 2011/2 Invasion, the process by which a particular organ or part becomes affected with disease or parasites existing elsewhere in the organism. 1892 G. M. Sternberg Man. Bacteriol. III. i. 221 The invasion of the blood which occurs in anthrax and in various forms of septicæmia in the lower animals, induced by subcutaneous inoculation with pure cultures of certain pathogenic bacteria, does not generally immediately follow the inoculation. 1924 R. Muir Text-bk. Path. xii. 332 Incision of the capsule may permit of renewed invasion by the tumor. 1949 H. T. Karsner Human Path. (ed. 7) 312/2 Although certain normal cells may infiltrate into neighboring structures,..yet infiltration and invasion are properties of neoplastic cells. 1965 T. F. Nealon Managem. Patient with Cancer i. 13/2 Extension or invasion is the spread of a cancer by direct involvement of surrounding structures, lymphatics, or blood vessels.

2. Esp. in phr. *invasion of privacy.*

1890 Harvard Law Rev. 15 Dec. 198 The common law recognizes and upholds a principle applicable to cases of invasion of privacy. 1912 Kipling Divers. Creatures (1917) 2 Any complaint of invasion of privacy needs immediate investigation. 1967 H. McCloy Further Side of Fear i. 3 He would have no motive for coming by stealth ... Besides, it was impossible to associate Mr. Erskine with any invasion of privacy.

4. *Ecology.* The spread of a plant or animal population into an area formerly free of the species concerned.

1905 F. E. Clements Res. Methods Ecol. iv. 210 By invasion is understood the movement of plants from an area of a certain character into one of a different character, and their colonization in the latter. 1932 Fuller & Conard tr. Braun-Blanquet's Plant Sociol. xiii. 308 The colonization of new unoccupied land by plant disseminules may be either by centrifugal or by marginal invasion. 1940 H. F. Witherby et al. Handbk. Brit. Birds I. 31 The slender-billed nutcracker... Not confined to conifer woods in its 'invasions' of Europe. 1958 C. Elton Ecol. of Invasions vi. 111 We have to accept the proposition that invasions of animals and plants and their parasites—and our parasites—will continue. 1964 Gould & Kolb Dict. Social Sci. 354/1 The process which occurs when one kind of population begins to occupy a territory (or an occupational niche) already occupied by another, or it increases its rate of occupancy... In human ecology invasion is restricted to this..meaning.

5. *attrib.*

1915 Mrs. Belloc Lowndes Let. 10 Mar. (1971) 57 No, I do not believe in either Zeppelins or an invasion... There is an invasion scare but I don't believe in that. 1923 W. S. Churchill World Crisis (1938) I. xix. 404 An invasion scare took a firm hold of the military and naval authorities. 1941 Times (Weekly ed.) 15 Oct., One of the heaviest attacks of the war was made on the invasion coast. 1942 R.A.F. Jrnl. 3 Oct. 8 The men..who would fly tomorrow's dawn patrol along the invasion coast. Ibid. 18 A tradition which was..renewed in France in 1939..; over the 'invasion ports' of Northern France in 1940 and 1941.

invasive, *a.* Add: **2. c.** *Path.* Of, exhibiting, or characterized by invasiveness.

1926 H. T. Karsner Human Path. xii. 332 It is probable that nearly all tumors begin originally as a somewhat invasive growth... The invasive character depends very largely upon the multiplication of the cells in the margin of the tumor. 1948 R. A. Willis Path. Tumours ix. 148 While proliferation is an important factor in invasive growth, it is certainly not the only or even the most essential one. 1970 [see *invasiveness]. 1971 Brit. Med. Bull. XXVII. 4/2 The attempt is made to relate the number of cases of carcinoma-in-situ entering a population to the number leaving it to become cases of invasive cancer.

invasiveness (invēi·zivnės). *Path.* [f. Invasive *a.* + -ness.] The ability of pathogenic microorganisms or malignant cells that are already in the body to spread to new sites.

1937 R. W. Fairbrother Text-bk. Med. Bacteriol. vii. 69 The term invasiveness is sometimes erroneously used as synonymous with virulence; it, however, applies strictly to the power of the organism to invade the tissues of the host. 1949 Cancer Res. IX. 559/1 Evidence was found indicating that the invasiveness of the cancer cells depended on their lessened adhesiveness. 1967 R. A. Willis Path. Tumours (ed. 4) ix. 149 The factors determining invasiveness are still uncertain. Continued excessive proliferation is one of them. 1970 Passmore & Robson Compan. Med. Stud. II. xviii. 28/1 Virulence may involve invasiveness or toxigenicity; some pathogenic organisms are typically invasive, e.g. the causative organisms of typhoid fever.

inveigle, *v.* Now also with pronunc. (invī·g'l).

inventory, *sb.* Add: **2. b.** *spec.* in *Linguistics.*

1945–49 Acta Linguistica V. 88 The graphemes become manifested in concrete letters and letter-attributes. ..These items make up what may be called the graphic inventory of the given language, which has..its counterpart in the phonic inventory of the same language. 1954 U. Weinreich in Word X. 394 One thing is certain: In the study of language contact and interference..a clear picture of differences in inventory is a prerequisite. Ibid. 395 We are not told whether in the phoneme inventory of Southeastern American English, the /æy/ of pass does or does not correspond as an inventory item to the /æ/ of other varieties. 1971 B. Mafeni in J. Spencer Eng. Lang. W. Afr. 109 It is possible that a number of nasalised vowel phonemes may be included in the vowel inventories of some of the conservative varieties of Nigerian Pidgin spoken in certain parts of the country.

3. b. *spec.* The quantity of material, etc., in use or held in stock in an installation at any one time. Also *attrib.*

1955 Proc. Internat. Conf. Peaceful Uses Atomic Energy (United Nations) III. 4/2 The inventory is the total of all separated fissile fuel supplied to the reactor or system other than produced internally in the reactor, less the amount assignable to the steady rate of make-up. 1963 Chem. Engineer 292/1 Returns..may range from inventory statements to quality control rights. Ibid., The problem of maintaining minimum inventories at minimum cost to the company. 1970 New Scientist 25 June 628/2 Freezing also wins over evaporating as the 'heat inventory' is far lower—the latent heat of fusion is far less than the latent heat of evaporation. 1970 Supervisor XXI. 65/2 Vehicles are built for inventory in slack sales periods in anticipation of the peak sales periods. Production hours can thus be stored as physical inventory.

4. (Further *attrib.* examples.)

1906 Daily Chron. 1 Feb. 5/1 An attempt to enforce the inventory clause of the new separation law was responsible for serious conflict between the police and the people in many quarters of Paris to-day. 1927 Carr-Saunders & Jones Survey Social Struct. Eng. & Wales 107 There is a distinction between what we may call the 'going concern' and the 'inventory' methods of estimating wealth. 1943 N. J. Silberling Dynamics of Business xviii. 443 Manufacturing and mining industries are now able to operate more freely on a rational inventory-valuation method that unquestionably will introduce greater stability in income statements. Ibid. xvi. 367 Speculation in raw materials or merchandise or what is known as 'inventory write-up'. 1951 A. H. Hansen Business Cycles & National Income II. ix. 122 What is happening..to inventory accumulation..throws light on how business activity, employment, and income are likely to unfold. 1958 Spectator 7 Feb. 186/1 One would expect the present cyclical-type recession [in the U.S.A.] to last longer than the previous inventory-type recession of twelve to sixteen months. 1962 A. Battersby Guide to Stock Control p. vii, Not only do excessive stocks immobilize our capital resources: they can generate the so-called 'inventory recession'. 1969 J. Argenti Managem. Techniques 137 That aspect of inventory control concerned with how large a consignment of any given item one puts into stock at a time.

inventory, *v.* Add: **2.** *intr.* and *trans.* To amount to or be worth (so much) on an inventory.

1902 G. H. Lorimer Lett. Merchant ix. 113 The last time I saw her, she inventoried about $10,000 as she stood. 1905 Springfield (Mass.) Weekly Republ. 20 Oct. 12 The late Senator Platt left an estate which inventories at $20,880.

inverse, *a.* and *sb.* Add: **A.** *adj.* **5.** *inverse spelling,* an unetymological spelling based on the spelling of another word containing an element that is no longer pronounced, e.g. *limb* from OE. *lim* after *lamb* (from OE. *lamb*).

1933 L. Bloomfield Lang. xvii. 294 So-called inverse spellings tell the same story... When we find the word deleite..spelled delight, then we may be sure that the [x] was no longer spoken in words like light. 1956 N. E. Eliason Tarheel Talk v. 194 Occasionally inverse spellings occur, for example forks for folks,..and polk for pork.

6. *Cryst.* Designating a spinel structure, $B[AB]O_4$, in which half the B (trivalent) cations are in tetrahedral holes and the A (bivalent) cations together with the other half of the B cations are in octahedral holes in the array of oxide ions (in contrast to the normal structure $A[B_2]O_4$).

[1947 Verwey & Heilmann in Jrnl. Chem. Physics XV. 175/2 For the sake of convenience, the first arrangement will be indicated as characteristic for 'normal spinels', the Barth and Posnjak arrangement as that of 'inversed spinels'.] 1957 Jrnl. Physics & Chem. Solids III. 313 These energies..may be sufficient to determine whether a given spinel shall be normal or inverse. 1970 R. G. Burns Mineral. Applications Crystal Field Theory vi. 112 Ni^{2+} and Cu^{2+} have a strong tendency to form inverse spinels. Ions such as Fe^{3+} and Mn^{2+}, which have zero octahedral site preference energies, form both normal and inverse spinels.

B. *sb.* **3.** *Rouge et Noir.* The section at the end of the table in which are placed bets wagering that the colour of the card that wins the coup will not be the same as that first dealt for a colour.

1850 Bohn's Hand-bk. Games 343 If the player..be determined to try his luck on the inverse, he must place his money on a yellow circle, or rather a collection of circles, situated at the extremity of the table. Ibid., The punters place on the Rouge, the Noir, the Couleur, or the Inverse, the sum they wish to risk. 1909 [see *couleur 2]. 1950 Hoyle's Games Modernized (ed. 20) 291 The tailleur never mentions the words 'Black' or 'Inverse', but always says that Red wins or Red loses, and that the colour wins or the colour loses.

4. *Math.* An element which, when combined with a given element by a given operation, produces the identity element for that operation.

E.g. the inverse of any number with respect to multiplication is the reciprocal of that number, and with respect to addition the negative of it.

1900 Ann. Math. II. 48 For every element T_κ exists an element (denote it by T_κ^{-1}) such that $T_\kappa T_\kappa^{-1} = T_\kappa^{-1}T_\kappa = 1$. T_κ^{-1} is called the inverse of T_κ. 1905 Trans. Amer. Math. Soc. VI. 187 The identity, 1, of this group, is

called the zero-element of the field, and is denoted by 0; while the inverse of an element a is here called the negative of a, and is denoted by −a. 1951 N. Jacobson Lect. Abstr. Algebra I. i. 22 If the operation in 𝔊 is denoted as +, we denote the identity as 0. The inverse of a if it exists is written as −a. 1953 Birkhoff & MacLane Survey Mod. Algebra (ed. 2) vi. 122 Hence $\phi\psi = I$ and ψ is a right-inverse of ϕ. 1965 Patterson & Rutherford Elem. Abstr. Algebra i. 19 Let ∧ be the closed binary operation in R defined by $x \wedge y = x + y - 2x^2y^2$... Show..that every..non-zero element x of R such that $x > -\frac{1}{2}$ has two inverses but that if $x < -\frac{1}{2}$, then x has no inverse.

inversion. Add: **2. f.** *Meteorol.* In full, *temperature inversion.* An increase of temperature with height in part of the atmosphere (the reverse of the usual situation); a layer of air having such a temperature gradient; also, more widely, an analogous deviation from the normal temperature gradient in bodies of water.

1902 Sci. Abstr. V. 285 There are instances of temperature inversion, ascribed to insulation of the clouds. 1903 Sci. Abstr. A. VI. 491 Temperature inversions were observed..mostly between altitudes 200 and 1,500 m. 1906 W. Marriott Hints to Meteorol. Observers (ed. 6) 66/2 Inversion of temperature, a warmer stratum of air above a colder one. 1928 D. Brunt Meteorol. vi. 59 An inversion is most readily produced during a clear night in winter. 1957 G. E. Hutchinson Treat. Limnol. I. vii. 487 Slight temperature inversions of the order of 0·1° to 0·2°C often occurred in the temperature curves at the bottom of two Austrian lakes. 1969 Courier-Mail (Brisbane) 16 Sept. 2/3 New York's skyscraper skyline is now badly clouded ..only when the city is trapped under what is called an 'inversion'. 1971 Nature 26 Feb. 583/2 It is only a decade since old people in their hundreds were killed off in northern cities when sulphur dioxide and other noisome products were trapped beneath inversion layers.

g. *Philol.* *inversion-compound,* a compound place-name in which the second element is a personal name, or a word designating a person, in the genitive (or genitive-equivalent) case, as the property-name *Kuikobba* (in Kirkwall), f. ON. *kvi Kobba* sheep-pen of Kobbi.

1918 E. Ekwall Scandinavians & Celts in N.-W. Eng. i. 15 In inversion-compounds the second element is a necessary ingredient, which can be absent only owing to ellipsis. 1963 C. Matras in Brown & Foote Early Eng. & Norse Stud. xii. 148 'Inversion-compounds' can serve as a criterion for classifying a settlement as Norwegian (as distinct from Danish).

h. *Biol.* A reversal of the order of the genes in a chromosome segment as compared with the corresponding segment of a normal homologous chromosome; a chromosome segment exhibiting such a reversal.

1921 A. H. Sturtevant in Proc. Nat. Acad. Sci. VII. 236 The simple inversion of a section of a normal chromosome. 1939 C. H. Waddington Introd. Mod. Genetics iv. 93 Pairing may be more nearly complete in flies heterozygous for very long inversions. 1964 G. H. Haggis et al. Introd. Molecular Biol. x. 258 The term mutation embraces stable chromosomal variations..including..the turning of an interstitial segment back to front (inversion). 1969 G. W. Burns Sci. Genetics xii. 221 Inversions act chiefly as so-called crossover suppressors. 1972 W. V. Brown Textbk. Cytogenetics xiv. 204/2 In most interbreeding populations inversions are eliminated because of the numerous types of inviability they produce. Ibid. 205/1 Drosophila pseudoobscura is unusual for the number of different inversions within especially the third chromosome of the species.

i. *Electr.* The conversion of direct current into alternating current: the opposite of rectification.

1926 L. B. W. Jolley A.C. Rectification (ed. 2) xix. 443 The problem of inversion, viz. the conversion from direct to alternating current. 1964 New Scientist 2 Apr. 25/1 Thyristors can also be used to convert d.c. to a.c.—a process known as 'inversion'. 1967 F. G. Spreadbury Electr. Inverters i. 2 The choice of a current converter for inversion purposes will, of course, depend on the magnitude of the output voltage, current and frequency required.

j. *Telecommunications.* Reversal of the order of the component frequencies of a signal: cf. *invert *v.* 2 g.

1930 Engineering 14 Nov. 625/3 Though the simple inversion..is satisfactory on short waves, it is not so effective on long waves, since, as only one side band is present, it is immaterial whether the transmission is inverted or not, provided the oscillator at the receiving end injects a local carrier of the correct frequency. 1933 K. Henney Radio Engin. Handbk. x. 273 To insure secrecy of transmission, frequency inversion or frequency scrambling methods are used. 1967 D. H. Hamsher Communication Syst. Engin. Handbk. x. 23 A reduction in interfering energy of about 3 db is realized by frequency inversion.

k. *Computers.* The conversion of either of the two binary digits or signals into the other; negation.

1955 R. K. Richards Arithmetic Operations in Digital Computers ii. 32 The 'not' operation, or inversion, is symbolized by a block labeled with the letter I. 1970 O. Dopping Computers & Data Processing ii. 39 The number that is somewhat inappropriately called the 1-complement of x is obtained simply by inversion, i.e. by substituting zeroes for all ones and vice versa. 1972 B. H. Vassos Analog & Digital Electronics for Scientists viii. 207 There are two other basic gates.., called NOR and NAND, respectively. They cannot be implemented by simple diode logic, because inversion is required, which can only be produced by an active device.

l. *Physics.* In full *population inversion.* A transposition of the relative numbers of atoms or molecules occupying certain energy levels.

1961 Physical Rev. Lett. VI. 106/1 Population inversions are achieved between several Ne levels by means of excitation transfer. 1968 E. L. Steele Optical Lasers in Electronics iv. 135 In the limit when total inversion occurs the ground state is entirely empty. 1973 Sci. Amer. Feb. 92/3 The condition of having more atoms in the upper state is called a population inversion (because it goes against the normal processes of nature, which tend to keep more electrons at lower energies than at higher energies).

3. c. *Math.* The process of finding a function $g(y)$ which either (*a*) yields a variable x when its argument is a given function $y = f(x)$ of that variable, or else (*b*) yields a given function when transformed by a given transformation.

1880 Encycl. Brit. XIII. 66/2 We have mentioned.. the problem of inversion which leads to elliptic functions, viz., that if $u = F(\kappa, \phi)$, then $\phi = \mathrm{am}\ u$. 1934 Trans. Amer. Math. Soc. XXXVI. 107 $f(x) = \int_0^\infty e^{-xt}d\alpha(t)$... By the inversion of the integral we mean the determination of the function $\alpha(t)$ in terms of the function $f(x)$. 1962 D. R. Cox Renewal Theory i. 11 Suppose that we have calculated the Laplace transform $k^*(s)$ of an as yet unknown function $k(x)$. The problem of finding $k(x)$ from $k^*(s)$ is called the inversion problem. Ibid. 13 We shall commonly find that, although we can find a quite simple expression for the Laplace transform, $k^*(s)$, of the function $k(x)$ in which we are interested, the inversion cannot be done explicitly in simple terms. 1968 P. A. P. Moran Introd. Probability Theory vi. 250 $\phi(t)$ [$= \int_{-\infty}^{\infty} e^{itx}dF(x)$] is uniquely determined by $F(x)$. We shall show that the reverse is true by obtaining an explicit expression, the inversion formula, for $F(x)$ in terms of $\phi(t)$.

5. b. In full, *Walden inversion* [tr. G. *Walden'sche umkehrung* (E. Fischer 1906, in *Ber. d. Deut. Chem. Ges.* XXXIX. 2895), named after P. von *Walden* (1863–1957), Latvian chemist]. Originally, the reversal of the direction of optical rotation observed in certain substitution reactions. Now interpreted as a change of configuration (from D to L or vice versa) occurring when a reactant enters along the axis of the bond between a central atom and the leaving group and causes the other substituents on the central atom to pass through a plane perpendicular to this axis; hence extended to other substitution reactions in which there is such a reversal of configuration, regardless of whether the molecule is optically active or the direction of activity reversed.

1899 Jrnl. Chem. Soc. LXXVI. II. 540 If..the sign of the rotatory power alone is considered, then optical inversion occurs in the action of phosphorus pentachloride and pentabromide..and the only 'normal' action in which no inversion occurs is the hydrolysis by means of silver oxide. 1911 Ann. Rep. Progr. Chem. VIII. 64 Most cases of the Walden inversion have been observed in transformations of α-substituted acids and their derivatives. 1937 F. C. Whitmore Org. Chem. xxi. 480 In the Walden inversion, the process takes place on a single carbon atom instead of with a system of several atoms. 1962 E. L. Eliel Stereochem. Carbon Compds. xiii. 375 In most cases the covalent attachment occurs from the side opposite to the one from which X had departed..and the stereochemical course is therefore predominantly inversion. 1964 D. A. Shirley Org. Chem. ix. 202 The molecule of methyl bromide is said to undergo inversion in its conversion to methyl alcohol. 1966 Morrison & Boyd Org. Chem. (ed. 2) xiv. 473 Paul Walden..discovered the phenomenon of inversion in 1896 when he encountered one of the exceptional reactions in which inversion [of configuration] does not take place. 1968 J. March Adv. Org. Chem. x. 254 The Walden inversion has been found at a primary carbon atom..and at a sulfur (in sulfoxides), a silicon, and a phosphorus atom.

6*. *Physical Chem.* A transformation of a substance, esp. an enantiotropic one, from one solid form to another; **inversion temperature,** the temperature at which the two forms can coexist in equilibrium.

1903 M. H. Fischer tr. Cohen's Physical Chem. vii. 111 This temperature, above which the Glauber's salt is transformed into the anhydride, is designated the inversion temperature of the Glauber's salt. 1904 A. Findlay Phase Rule iii. 34 If the vapour phase [of sulphur] is absent and the system maintained under a constant pressure..there will also be a definite temperature at which the two solid forms are in equilibrium... This temperature..is known as the transition temperature or inversion temperature. 1928 Jrnl. Physical Chem. XXXII. 1205 When quartz is subjected to a uniform hydrostatic pressure..the temperature of its high–low inversion is raised. 1947 R. H. Bogue Chem. Portland Cement vii. 125 The α–β inversion temperature of C_2S was found to be 1456°. 1966 W. A. Deer et al. Introd. Rock-forming Min. 129 It [sc. pigeonite] then later inverts to an orthorhombic pyroxene and the inversion is accompanied by the exsolution of a second generation of augite lamellae.

8. c. *Anat.* (See quots.)

1869 G. V. Ellis Demonstrations Anat. (ed. 6) ix. 762 In inversion the great toe is adducted, the inner border of the foot is shortened, and is raised from the ground so that the sole looks inwards, whilst the outer border is depressed. 1902 D. Hepburn in D. J. Cunningham Text-bk. Anat. 304 By inversion we mean the raising of the inner border of the foot so that the sole looks inwards, while the toes are depressed towards the ground. 1971 D. L. Kelley Kinesiology vi. 75 Inversion lifts the medial border of the foot to turn the sole inward.

10. In full, *sexual inversion.* Homosexuality (see also quot. 1958).

1895 A. Beardsley Let. 15 May (1971) 85 [To André Raffalovich] Your study of inversion is I think quite brilliant. 1897 H. Ellis Stud. Psychol. Sex I. ii. 27 In Italy, also, Ritti, Tamassia, Lombroso, and others began to study these phenomena, and it seems to have been in Italy that the convenient term 'sexual inversion' was first used. When the matter was taken up in France, the same term was used. 1901 J. A. Godfrey Sci. Sex v. 206 Sexual inversion—that is, the turning-in of the sex instinct towards individuals of the same sex—is an abnormal phenomenon. 1927 Scots Observer 1 Oct. 15/3 It will help to approach the problems of inversion with knowledge and charity. 1958 Amer. Jrnl. Orthopsychiatry XXVIII. 424 Many workers fail to distinguish between homosexuality and sexual inversion, or more accurately, sex-role inversion. Freud..himself..equated the two terms. Ibid., The following distinction is offered: homosexuality refers to sexual activity or the desire for such activity between two members of the same sex, while the criterion of inversion is a personality in which a person's thinking, feeling, and acting are typical of the opposite sex. 1965 J. Marmor (title) Sexual inversion.

11. Special Comb.: **inversion temperature,** (*a*) *Physics,* the temperature (for any particular gas) at which the Joule–Thomson effect changes sign, so that the gas is neither heated nor cooled when allowed to expand without doing any work; (*b*) (see sense 6* above).

1902 Phil. Mag. III. 536 It had been deduced by Wilkowski..by assuming the thermodynamic coincidence of the inversion temperatures for hydrogen and for air. 1940 S. Glasstone Text-bk. Physical Chem. iv. 285 The inversion temperature, as derived from the van der Waals equation, should be twice the Boyle point... The observed inversion temperature for hydrogen is about 190°K. 1971 Nature 20 Aug. 519/1 For every gas, however, there is at a given pressure an inversion temperature above which throttled expansion results in a temperature increase. (The inversion temperature decreases with increasing pressure.)

invert, *v.* Add: **2. g.** *Telecommunications.* To subject (a signal) to a heterodyning process that reverses the order of the component frequencies (either completely or in a restricted range) prior to modulation for transmission, those at the two extremes being interchanged.

1930 Engineering 14 Nov. 625/1 The risk of important conversations being overheard..has still further been reduced by using a complex heterodyning process, which inverts and mixes the signals. Ibid. 626/1 The original speech-frequency range is divided into a number of bands. ..These bands can then be inverted as explained above, and re-arranged in transposed bands. 1933 K. Henney Radio Engin. Handbk. x. 274 To obtain an intelligible signal at the receiving end, it is necessary to again invert the modulation frequencies by the reverse process. 1966 McGraw-Hill Encycl. Sci. & Technol. X. 624/2 The first method employs equipment for inverting speech to make overseas telephone conversations unintelligible to the casual listener.

5*. *intr.* Of a substance: to undergo inversion.

1887 [see *caramel v.]. 1933 Amer. Jrnl. Sci. XXV. 284 It would..be possible for β-solid solutions to invert to the γ-form. 1966 [see *inversion 6*].

9. *Geom.* (*intr.*) To be transformed by inversion *into.*

1865 R. Townsend Chapters on Mod. Geom. II. xxiv. 384 Every two circles invert into two whose radii have a constant ratio from every point on any third circle coaxal with themselves. 1916 J. L. Coolidge Treat. Circle & Sphere i. 22 Points within the circle of inversion other than the centre will invert into points without, points without will always invert into points within. 1966 J. H. Cadwell Topics in Recreational Math. v. 42 Hence the locus..is a circle centre C_1. We note that C_1 is not the inverse of C, i.e. centres do not invert into each other.

invert, *sb.* Add: **2.** *Psychol.* One whose sex instincts are inverted. (Cf. *inversion 10; *inverted *ppl. a.* 3 c.)

1897 H. Ellis Stud. Psychol. Sex I. 12 Caesar was proud of his physical beauty, and like many modern inverts he was accustomed carefully to shave his skin. Ibid. 144 The sexual invert is specially liable to suffer from a high degree of neurasthenia. 1911 R. W. Chambers Common Law i. 29 This world is full of pale, enraptured artists;..full of unwashed little inverts. 1957 L. Durrell Justine II. 96 At least the invert escapes this fearful struggle to give oneself to another. 1957 Observer 1 Sept. 11/5 For once, the hero is not wrongly but rightly accused: he is an irrevocable invert. 1971 R. Reisner Graffiti (1974) viii. 115 The inverts (a word preferred by homosexuals to perverts) attempt to win converts.

invert, *a.* Add: Also *ellipt.*

1910 Encycl. Brit. IV. 508/2 This method is more suited to the preparation of invert in the brewery itself than the acid process. 1940 H. L. Hind Brewing II. xxii. 545 The sugars must be pale in colour and carefully selected in accordance with the flavour required in the

beer. No. 1 and No. 2 inverts and other sugars of somewhat similar character are suitable. 1971 J. S. Hough et al. Malting & Brewing Sci. xi. 296 The solid invert can be added directly to the copper or dissolved in liquor before addition.

b. invert soap [tr. G. *invertseife* (Kuhn & Bielig 1940, in *Ber. d. Deut. Chem. Ges.* LXXIII. 1080)]. A soap whose surface-active ion is a cation (rather than the more usual anion): a cationic detergent.

1941 Chem. Abstr. XXXV. 3596 K. and B. intended to study the mechanism of the bactericidal action of Zephirol (mixt. of alkyldimethylbenzylammonium chlorides) and other quaternary ammonium, sulfonium and phosphonium salts by means of expts. on the interaction of these 'invert soaps'..with proteins, chromoproteids, ferments, symplexes and genes. 1947 Conant & Blatt Chem. Org. Compounds (ed. 3) x. 213 Another group of detergents consists of the invert soaps... These are quaternary ammonium salts containing at least one large alkyl group. 1966 Smith & Cristol Org. Chem. xl. 765 Cationic detergents or invert soaps are used mainly for germicidal properties.

invertase (i·nvɜːt-, invɜ̄·tēiz, -s). *Biochem.* [f. Invert(in + *-ase.] = Invertin.

1887 Jrnl. Chem. Soc. LI. 60 Solutions of invertase and diastase were prepared from Mr. O'Sullivan's specimens. 1949 [see *ficin]. 1964 N. G. Clark Mod. Org. Chem. xvi. 330 Hydrolysis is also brought about by the enzyme, invertase. 1970 Sci. Jrnl. Mar. 19/3 A 1 ml sample of this water is then treated with the enzyme invertase, which converts any of the sugar sucrose..into a mixture of glucose and fructose.

inverted, *ppl. a.* Add: **1. c.** (Later examples.)

1902 [see *cacuminal a.]. 1918, 1934 [see *cerebral a. 2].

3. c. *Psychol.* spec. of the sex instincts.

1897 H. Ellis Stud. Psychol. Sex I. 16 The painter Bazzi seems to have been radically inverted. Ibid. 156 Social opinion is most amply adequate to deal with the manifestations of inverted sexuality. 1958 Amer. Jrnl. Orthopsychiatry XXVIII. 428 Inverted females would be expected to show..a relatively complete identification with the masculine role. 1974 Times Lit. Suppl. 1 Feb. 109 A telling parody of inverted sexist trends in our own society.

9. Special collocations: *inverted comma* (see also Comma 4), *engine* (see quot. 1961), *loop, pleat;* **inverted snob,** one who dislikes, or avoids contact or association with, the upper classes; one who tries to appear to be a member of, or sympathetic to, the lower classes; so *inverted snobbery, snobbism;* **inverted spelling** = *inverse spelling* (*inverse *a.* 5).

1789 J. Robertson Ess. Punctuation 130 Two inverted commas are generally placed at the beginning of a phrase or a passage, which is quoted or transcribed from some author, in his own words. 1927 R. B. McKerrow Introd. Bibliogr. 316 Inverted commas were, until late in the seventeenth century, frequently used at the beginnings of lines to call attention to sententious remarks... They were not especially associated with quotations until the eighteenth century. 1933 Week-End Rev. 28 Oct. 439/1 'Dunky Fitlow'! What a name! One could hardly ask for it without embarrassment, without qualifying it by inverted commas in the voice. 1956 A. S. C. Ross in M. Black Importance of Lang. (1962) 95 It is non-U to place the name of a house in inverted commas. 1963 Auden Dyer's Hand 520 One has a sense, and nowhere more strongly than in the songs [of Twelfth Night], of there being inverted commas around the 'fun'. [1885 List of Subscribers, Classified (United Telephone Co.) (ed. 6) 236 High-class..stationary engines. Beam, Compound Condensing, Inverted, Vertical.] 1933 Meccano Mag. Feb. 109/2 It is fitted with three de Havilland 'Gipsy III' inverted engines. 1961 Webster, Inverted engine, an engine whose crankshaft is above the cylinders. 1940 Chambers's Techn. Dict. 458/1 Inverted loop, a manœuvre of an aeroplane consisting of a complete revolution about a lateral axis, with the normally upper surface of the machine on the outside of the path of the loop. Must be commenced while flying inverted. 1915 T. Eaton & Co. Catal. Spring & Summer Suppl. 1/2 The smart patch pockets are finished with stitched inverted pleats. 1964 McCall's Sewing ii. 30/1 Inverted pleat, two side pleats which turn toward each other. 1971 'D. Halliday' Dolly & Doctor Bird iii. 29 For golf, I have always worn an Orkney tweed skirt with a low inverted pleat at the back. 1943 N. Marsh Colour Scheme vi. 99 Don't call me Mr. Bell. I'm afraid you're an inverted snob. 1955 Koestler Trail of Dinosaur II. 87 Only an inverted snob will pretend..that 'the aristocracy'..is devoid of value. 1966 Guardian 14 May 7/6 But, gosh, don't be an inverted snob! Subculture is fun! 1883 R. L. Stevenson Silverado Squatters 176 His book..was a capital instance of the Penny Messalina school of literature; and there arose from it..a rank atmosphere of..sickening, inverted snobbery. 1930 R. Lehmann Note in Music III. 108 Inverted-snobbery complex, Clare would have called it. 1937 L. Bromfield Rains Came I. xxxviii. 159 It was that eternal, inverted snobbery of his, that hatred of anyone born with the things he had never achieved. 1958 Times 25 Jan. 7/3 Inverted snobbery, we are told, is rampant in Cambridge. 1971 Lancet 23 Apr. 918/2 The curious inverted snobbery of the use of 'Mr.' by surgeons (even if they have an M.D. or a PH.D. as well as their M.B.),..with all its regional and specialty variations, is well worth the attention of a sociologist. 1939 A. Thirkell Before Lunch vi. 160 The inverted snobbism that at once overcame her against her will. 1958 G. L. Brook Hist. Eng. Lang. v. 102 Another contributory cause of the confusion of Modern English spelling is to be found in what are called inverted spellings. 1970 B. M. H. Strang Hist. English II. v. 291 Inverted spellings, e.g. y for e in districts suspected to be y-less, are as revealing as direct e for y spellings.

inverter. Restrict *rare* to sense in Dict. and add: **2.** That which inverts, or produces inversion. **a.** *Electr.* Also **invertor.** Any apparatus which converts direct current into alternating current.

1926 L. B. W. Jolley A.C. Rectification (ed. 2) xix. 444 In an inverter such as is shown in Fig. 324 a direct current generator G is supplying a steady current to two three-electrode valves and the primary of a transformer. 1947 P. Kemp Alternating Current Electr. Engin. (ed. 7) xxviii. 498 Inverters are used for regenerative braking on electric railways. 1961 Listener 9 Nov. 773/1 The valve serves not only as a rectifier..but—at the other end of the direct current link—as an invertor, turning direct current into alternating current. 1962 A. Shepard in Into Orbit 103 Inverters are used to convert the DC current which comes from the batteries into the AC current required to power some of the systems in the booster. 1970 J. Shepherd Higher Electr. Engin. (ed. 2) xxv. 817 D.C. links (with a rectifier at one end and an invertor at the other end of the link) are now commonly used to connect two a.c. power distribution systems together. 1973 Wireless World June 27 (Advt.), Models 107A and B are precision built inverters providing 240 volts a.c. from 12 and 24 volt battery systems.

b. *Telecommunications.* A device that inverts a signal (see *invert *v.* 2 g).

1930 Engineering 14 Nov. 625/3 The inverter is usually inserted at the point where the transmitting and receiving signals are combined.

c. *Computers.* A device which converts either of the two binary digits or signals into the other.

1955 R. K. Richards Arithmetic Operations in Digital Computers ii. 32 If the input to an inverter is A, the output is Ā. 1960 T. C. Bartee Digital Computer Fund. iv. 63 Because of its high input resistance, low output resistance, and voltage- and current-gain characteristics, the inverter is one of the most commonly used transistor configurations in digital systems. 1972 Millman & Halkias Integrated Electronics vi. 164 The output of an inverter is relatively more positive if and only if the input is relatively less positive.

invertible, *a.*² Add: **2.** *Math.* Of an element of a set: having an inverse in the set (*spec.* an inverse for multiplication).

1956 C. Chevalley Fund. Concepts Algebra ii. 27 Let G be the set of all invertible elements of a monoid A. 1963 G. D. Mostow et al. Fund. Struct. Algebra vi. 141 Invertible elements are divisible only by invertible elements. Ibid. xii. 341 An element a is called invertible if there is an element a′ in A such that $aa' = a'a = e$. 1965 Patterson & Rutherford Elem. Abstract Algebra iv. 122 An element $x \in R$ is said to be invertible if it has an inverse with respect to multiplication: that is, if there exists y such that $xy = e = yx$.

Hence **invertibi·lity.**

1963 H. B. Curry Found. Math. Logic v. 249 Ketonen ..showed the invertibility of his classical propositional rules. 1973 J. J. Zeman Modal Logic iii. 50 The number of rules we must examine for invertibility.

invertor. Add: **2.** *Anat.* A muscle which turns a part (as the foot) inwards.

1903 [see *evertor]. 1967 Rowe & Wheble Conc. Textbk. Anat. & Physiol. (ed. 2) v. 258 The invertors of the foot are:—Tibialis posterior [and] Tibialis anterior. 1969 MacConaill & Basmajian Muscles & Movements xiii. 254 The tibialis posterior is an invertor in non-weightbearing movements of the foot.

3. *Electr.* var. *inverter 2 a.

invest, *v.* Add: **2. c.** To embed in or surround *with* investment (sense *2 b).

1892 C. Hunter Man. Dental Laboratory viii. 107 We.. apply the little plaster-mould to the face of the model and settle the tooth in it, cement tooth to auxiliary plate, then carefully remove from denture, invest and solder. 1946 Foundry Aug. 85/1 Precision casting is a somewhat abused term often employed to designate the lost wax investment casting process... Properly, the phrase investment casting is the most suitable for use in defining the lost wax investment casting process since it is characterized primarily by the use of a molding material which completely enrobes or invests an expendable pattern. 1964 S. Crawford Basic Engin. Processes xii. 255 This casting process involves the use of a heat-disposable wax pattern which is invested with refractory material forming a mould (or shell).

9. c. To lay out money in betting on a horse race, or in football pools, etc.

1951 'H. Cecil' Painswick Line ix. 107 He went to the £5 tote windows and invested (as they euphemistically call it) £100 on Maiden Aunt. 1958 Punch 27 Aug. 265/1 Your skill can put you on top of the world when it's invested with ——'s [Pools]. 1973 Times 21 Apr. 12/1 In bookie parlance, one does not bet on a horse; one invests.

investible, *a.* [f. Invest *v.* + -ible.] = Investable *a.*

1931 Economist 18 July 110/2 Central Banks..should take steps..to counteract any tendency which their own nationals may show either to keep their investible resources excessively liquid or to undertake excessive long-term commitments. 1968 Daily Tel. 21 Nov. 3 These funds form a pool of investible money.

investigational (investigēi·ʃənăl), *a.* [f. Investigation + -al.] Of or pertaining to investigation.

1905 Science 29 Sept. 387/1 Investigational apparatus of great importance. 1930 Aberdeen Press & Jrnl. 4 Sept. 7/3 An unprecedented demand for botanical specialists to fill investigational and advisory posts. 1946 Nature 12 Oct. 522/1 This would be sufficient for the investigational purposes now envisaged by the Board. 1971 Mod. Law Rev. XXXIV. VI. 651 What..is to be the future of the social sciences as part of the normal decisional or investigational equipment of lawyers?

investigative, *a.* (Later examples.)

1925 [see *flair¹ 2]. 1955 Sci. News Let. 16 Apr. 249/2 The pathologists can and will assume their part in this investigative obligation and I'm sure that practicing physicians will do likewise. 1972 Village Voice (N.Y.) 1 June 37/1 What is vitally needed is more investigative reporting on this growing national phenomenon. 1975 Times 24 Jan. 14/5 Some suggest that I should turn my investigative powers to other vanishing drinks.

investigator. (Later examples.)

1971 Nature 15 Oct. 469/2 Other investigators who have scanned the North Polar spur and portions of other galactic spurs..at large solar zenith angles have not observed any enhancement at soft X-ray energies. 1972 Daily Tel. 21 Jan. 7/1 Department of Employment investigators, only 30 for the whole country, are too few to catch offenders.

investigatory, *a.* (Later examples.)

1966 Punch 23 Feb. 276/3 During his editorship, Sam Campbell, as he was known to most of Fleet Street, added more than one and a half million paying readers to his paper by sustained crusading investigatory journalism of a kind unequalled since W. T. Stead. 1969 Listener 27 Mar. 436/2 World in Action has an aggressive investigatory policy. 1969 Daily Tel. 31 Oct. 1/4 The judges said they believed it was desirable to have the inquest kept secret to 'protect its integrity, investigatory character and effectiveness'. Ibid., Inquest proceedings are not accusatory and..they should be regarded as investigatory.

investment. Add: **2. b.** Refractory material which can be used to embed or surround an object and then is allowed to harden, so that soldering can be carried out (in *Dentistry*) or a mould made from it; freq. *attrib.*, as *investment material;* **investment casting,** a technique for making small, accurate castings from alloys having high melting points, the mould being made by investing a pattern of wax or similar material that can be removed from the investment by melting it.

1892 C. Hunter Man. Dental Laboratory viii. 115 If the new tooth must be soldered, an hour-and-a-half, from the time the case is ready to be put in the investment, is not too much for its safe accomplishment. 1942 Iron Age 9 July 39 Production of castings..by a modern modification of the lost-wax process is described herein. This method utilizes a refractory investment which makes possible the casting of stainless and other alloy steels. 1946 Investment casting [see *invest v. 2 c]. 1947 J. C. Rich Materials & Methods Sculpture vi. 140 For the solid casting of small bronze statuettes, the Greeks employed a simple technique. The work was modeled in wax and incased in a containing negative mold composed of sand and earth-clay. When this investment material was dry, an opening was made and the entire mass was heated until the wax ran off. 1956 J. N. Anderson Appl. Dental Materials xvii. 191 A wax pattern may be expanded directly by using warm water to mix the investment. 1963 C. R. Cowell et al. Inlays, Crowns & Bridges ii. 3 Modern inlay waxes and investment techniques make dimensional control possible. 1964 S. Crawford Basic Engin. Processes xii. 255 Precision casting, investment casting, and the lost-wax process, are various names given to the process by which small intricate castings can be produced to a high grade of dimensional accuracy and surface finish in materials which prove difficult or impossible to cast and subsequently machine by the more traditional methods... The principle of the process..has been in use in the jewellery and dental trades for many years.

5. b. *attrib.* (Further examples.) Also **investment currency,** the currency resulting from the sale of foreign securities or used for their purchase when such transactions are controlled and channelled through a market separate from the market in foreign exchange; **investment trust,** a trust (Trust *sb.* 7 a) whose business is the investment of money; also *attrib.*

1885 List of Subscribers, Classified (United Telephone Co.) (ed. 6) 213 American Investment Trust Co. 1924 B. D. Nash Investment Banking in Eng. 3 The function of investment banking is to gather together the savings and surplus capital of individuals and companies and to turn this accumulation into the hands of others for use in the construction of fixed plants. 1924 L. M. Speaker Investment Trust 3 The investment trust in its characteristic form is an institution mainly of British origin. It is essentially a corporation, the business of which consists chiefly of judicious investment of its capital. 1929 Observer 17 Nov. 2/3 Among the many high-class securities which have had to be sold to meet the demand for money caused by the depressed conditions have been the junior stocks of the large investment trust companies. 1933 B. Ellinger This Money Business iv. 33 There are a number of firms and brokers connected with trust and investment companies. 1935 Economist 5 Jan. 23/1 The extent of the..movement may be appreciated from

the following table of representative values and yields based on the Actuaries' Investment Index. 1951 R. W. Jones Thomson's Dict. Banking (ed. 10) 346/1 Investment Ledger. A separate account is opened in this ledger for each different investment. 1961 'E. Lathen' Banking on Death (1962) iii. 21 Robichaux had..surprised his friends and relations by becoming an astute and competent investment banker. 1963 Listener 31 Jan. 191/1 Bankers have looked with jaundiced or favourable eye on the investment bank. 1965 D. Greenwald et al. McGraw-Hill Dict. Mod. Econ. 274 The first real investment trust was the Foreign and Colonial Trust, which was established in London in 1868. Its purpose was to give small investors the same advantage, diversification, that large capitalists had. 1967 Spectator 21 July 85/3 Investment trusts are closed portfolios. Selling or buying shares in such trusts neither increases nor decreases the size of the portfolio. 1969 Times 30 Apr. 30/6 Investment companies qualify, so do unit trusts, and a company is eligible for relief if the interest would have been eligible when payable by an individual. 1969 Times 5 May (Suppl.) p. vi/5 Now the insurance companies have joined the fight for this part of the investment dollar. 1971 Mod. Law Rev. XXXI. VI. 699 The second chapter..contains a detailed practical account of the work of American investment bankers. 1972 D. Lees Zodiac 117 He's among the top six investment consultants in the world today. 1973 Country Life 12 July 120/3 It is necessary to buy what is known as investment currency, which is a pool of funds created by the sale of British investments abroad. 1973 Times 17 Dec. 14/3 Investment funds..went too much into property or overseas, too little into productive industry.

investor. Add: **2. b.** One who bets on a horse race, or in football pools, etc. Cf. *invest *v.* 9 c.

1958 [see *banker² 5].

inviable (invəi·ăb'l), *a.* *Biol.* [f. In-³ + Viable *a.*¹] Unable to survive; unable to germinate, grow or develop; unable to perform its proper biological role.

1918 Genetics III. 476 The disadvantage of so many individuals being rendered sterile or inviable. 1946 Nature 12 Oct. 520/2 Among 986 seeds collected in 1941 from chimeral plants there were found six triploid ones, three of which proved inviable. 1948 Mind LVII. 297 He found that within one 'species' of frog..ranging from lat. 45° N. to 27° N. in the United States, representatives of the northern and southern ends of the range gave inviable hybrids, though parents not too far apart geographically gave normal development. 1955 New Biol. XIX. 21 More usually an unbalanced and inviable nucleus is formed [from an irregular mitosis] in which some chromosomes are represented twice, others once or not at all. 1956 Nature 31 Mar. 626/2 Damage arising from inviable recombinations of broken chromosome ends. 1962 Lancet 29 Dec. 1384/2 Since the O-sperm, if functional, leads to the Turner zygote which is in great deficit in the general population, it seems likely that O-sperm are inviable and that the Turners are due to a phenomenon other than fertilisation by O-sperm.

So **inviabi·lity,** the state or condition of being inviable.

1918 Genetics III. 614/2 (Index), Inviability. 1929 Encycl. Brit. XI. 495/1 The inviability of most gametic and zygotic combinations. 1972 [see *inversion 2 h].

in vino veritas: see *in *Lat. prep.*

inviscid, *a.* Add: **2.** Not possessing viscosity.

1913 Phil. Mag. XXVI. 1001 (heading) On the stability of the laminar motion of an inviscid fluid. 1930 Flight 3 Jan. 29/1 The practically useful theory of Prandtl comes from considering air as frictionless or inviscid. 1938 L. M. Milne-Thomson Theoret. Hydrodynamics 1 An inviscid fluid is a continuous substance which will yield instantly to any shearing stress however small. 1971 Nature 6 Aug. 427/1 The problem of the plane transonic flow of an inviscid gas.

invisible, *a.* Add: **1. c.** *invisible college* (further examples).

1962 D. J. de S. Price Science since Babylon (new ed.) v. 99 Collaborative work now exceeds the single-author paper, and the device of prepublication duplicated sheets circulated to the new Invisible Colleges has begun to trespass upon the traditional functions of the printed paper in a published journal. Ibid., The new Invisible Colleges, rapidly growing up in all the most hard-pressed sections of the scholarly research front, might well be the subject of an interesting sociological study. 1967 Garvey & Griffith in De Reuck & Knight Communication in Sci. 25 Once an author submits his manuscript the dissemination of this piece of work is small and usually only to members of his invisible college. 1971 Halsey & Trow Brit. Academics xiii. 362 Here is evidence of the 'invisible college' having its greatest importance.

d. *invisible exports, imports:* those items which do not appear in returns of exports and imports for which payment has to be accepted from or made to a foreign country, such as shipping services, insurance, profits on foreign investment, money spent by visitors from a country with a different currency, etc. Also *invisible earnings, trader, transaction,* etc.

1911 C. G. Robertson Eng. under Hanoverians II. iv. 344 Prior to the Industrial Revolution the seaborne and carrying trades, with their invisible exports, are an expanding source of wealth, but are not indispensable. 1919 J. A. Todd Mech. Exchange (ed. 2) xiv. 174 An invisible export is something which enables a country to import goods without paying for them directly by the export of other goods, and conversely, an invisible import is something which makes it necessary for a country to export goods without receiving payment directly in other goods. 1923 Westm. Gaz. 24 Feb., Depressed shipping means less insurance, less banking, and a smaller earning in freights —all the things we mean when we speak of our invisible exports. 1935 Economist 12 Oct. 712/2 Among 'invisible' traders with Italy, the shipping lines with Mediterranean routes would suffer some inconvenience. 1957 Encycl. Brit. VIII. 991/2 Invisible exports are of vital importance to countries whose limited resources..make them dependent on large imports. Italy's high percentage is due to foreign tourists' expenditure and to a steady flow of remittances from Italians living abroad. 1958 Economist 18 Oct. 256/1 The..government deficit on invisible account. 1963 Ann. Reg. 1962 479 Net earnings on invisible transactions improved somewhat. 1969 Times 27 Nov. 6/6 The rapid growth of invisible earnings. 1971 Daily Tel. 24 Apr. 2 The works of songwriters and composers are now one of Britain's major 'invisible' exports.

e. *invisible man:* used esp. with direct or implied allusion to H. G. Wells's novel *The Invisible Man* (1897).

1911 Chesterton Innocence of Father Brown v. 120 (title) The invisible man. 1940 'G. Orwell' in Horizon Mar. 191 The one theme that is really new is the scientific one. Death-rays, Martians, invisible men, robots. 1959 N. Polsky in N. Mailer Advts. for Myself (1961) 313 Even in the world of the hipster the Negro remains essentially what Ralph Ellison called him—an invisible man. 1963 D. B. Hughes Expendable Man (1964) v. 139 He's safe only so long as he's the invisible man. I'm going to have to..find out who he is. 1966 J. Wainwright Evil Intent 123 He isn't The Invisible Man. Somebody must have seen him. 1973 E. Berckman Victorian Album 144 Dancey'll work out how to be the invisible man, it's just up his street.

f. *invisible mending:* repair of material, clothing, etc., so carefully executed that little or no sign of the repair can be seen. So *invisible mender,* one who undertakes such repairs; *invisibly-mend* v. trans.

1921 Dict. Occup. Terms (1927) § 419 Invisible mender, ..repairs tears, by hand, using stoating, finedrawing, or rentering process, according to kind of tear and material. 1931 W. Holtby Poor Caroline i. 33 She complained of the price of invisible mending. 1937 E. Price Enter—Jane i. 1 Until you have paid the invisible-mending bill..you shan't have your threepence a week pocket-money! 1959 G. Freeman Jack would be Gent. i. 16 Sitting in the shop window invisibly-mending nylons with a little hook. 1969 D. Francis Enquiry xv. 201 My coat would cost a fortune at the invisible menders. 1970 —— Rat Race xv. 199 My coat was soaked... Have to get it cleaned, and the slit invisibly mended.

3. (Further examples.)

1834 E. E. Perkins Lady's Shopping Manual 103 Invisible Wire for Lace and Nett Caps. 1881 Queen 12 Mar. (Advt.), Invisible fringe nets made of hair. 1895 Montgomery Ward Catal. 87/1 Invisible Hairpins, about 50 in a box. 1897 Sears, Roebuck Catal. 322/2 Invisible Hair Pins made of Good Wire. 1901 Invisible stitch [see Stoat v.]. 1964 McCall's Sewing ii. 30/1 Invisible stitch, used for hems and attaching facings and interfacings in tailoring. 1966 J. S. Cox Illustr. Dict. Hairdressing 83/2 Invisible pins, very fine hairpins which are difficult to see when placed in the hairdress. 1967 J. Caird Murder Scholastic xiv. 190 Mabel Glossop had worn her thick, beautiful white hair in a bun, over which she put a fine 'invisible' net. 1973 Times 19 Oct. 3/1 (Advt.), Without these foils such modern marvels as 'invisible' hearing aids or heart pacemakers simply could not exist.

B. *sb.* **3.** Usu. in *pl.* Invisible exports and imports. Cf. sense *1 d of the adj.

1958 Economist 18 Oct. 256/1 Net earnings from commercial services ranked as invisibles also rose. 1962 H. O. Beecheno Introd. Business Stud. xv. 143 These 'invisibles' arise mainly from services which we supply to other countries. 1964 New Statesman 3 Apr. 514/1 Angola and Mozambique provide one-third of all Portuguese exports, valued at £50 million, and this sum does not include invisibles from tourism, shipping and railways. 1973 Daily Tel. 13 July 19 The average deficit for the latest three months, excluding invisibles, is £22 million lower than in May at £133 million.

invita Minerva (invəi·tă minɜ̄·rvă), *advb. phr.* [L., = 'Minerva (the goddess of wisdom) unwilling'.] When one is not in the vein or mood; without inspiration.

1584 R. Scot Discov. Witchcr. XII. iii. 219 It should be vnto them (Inuita Minerua) to banket or danse with Minerua. 1626 Bacon Sylva § 292 That nothing bee done Inuitâ Minerua, but Secundum Genium. 1848 Geo. Eliot Let. Feb. (1954) I. 250, I have tired myself with trying to write cleverly 'invita Minerva'. 1954 'M. Cost' Invitation from Minerva 126 It is always a mistake to do anything invita Minerva... Briefly, against the grain.

invitation. Add: **2. b.** *Bridge.* (See quot. 1964.)

1928 A. E. M. Foster Auction Bridge for All I. i. 22 There still lurks amongst a number of players the notion that a bid of a minor suit by their partner is what they call 'an invitation to a No Trumper'. Again and again I have heard said: 'I responded to your invitation, partner.' 1958 Listener 23 Oct. 669/3 With his excellent controls it is easy for West to accept the slam invitation. 1964 Official Encycl. Bridge 265/1 Invitation, a bid which encourages the bidder's partner to continue to game or slam, but gives him the option of passing if he has no reserve values in terms of high-card strength or distribution.

3. (Further examples.)

c1855 in M. Johnson Amer. Advertising, 1800–1900 (1960), Invitation notices, and every arrangement for funerals personally attended to with correctness and dispatch. 1902 A. Bennett Grand Babylon Hotel vi. 71 The invitation-list..contained no reference to any such person. 1931 A. E. M. Foster Auction Bridge made Clear 124 An original bid of Two of a suit is by them reserved as a 'Demand' or 'Invitation' bid. That is, a demand or invitation to partner to show his quick tricks. Ibid. 125 Z has..a hand of over seven tricks but under eleven—the 'Invitation Area'.. Z bid 'Two Hearts' as an 'Invitation'. 1968 Radio Times 28 Nov. 27/1 Invitation concert... Recorded before an invited audience.

invitational (invitēi·ʃənăl), *a.* [f. Invitation + -al.] Characterized by invitation. Also as *sb.*, an invitational tournament (? only *N. Amer.*).

1922 Chambers's Jrnl. Aug. 543/1 Philip walked the room's length with invitational pauses. 1957 Publ. Amer. Dial. Soc. XXVIII. 58 Somebody like to help me?..someone in the context..would sound more factual..than invitational. 1958 Listener 23 Oct. 669/3 A cue bid of Five Hearts would be too strongly invitational. 1964 Official Encycl. Bridge 265/1 Invitational bid, bid indicating strong game prospects, which requests partner to continue if he has some reserve strength. 1964 N. Squire Bidding at Bridge ii. 17 Your partner..will give an invitational raise to 2NT. 1968 Globe & Mail (Toronto) 3 Feb. 35/5 Yesterday Sinclair's mixed rink of his wife, Betty, Don Maclellan and Mildred Maclellan advanced to the quarter-finals of the Thornhill mixed invitational tournament. Ibid. (headline), 2 Thornhill rinks survive club's mixed invitational. 1972 Evening Telegram (St. John's, Newfoundland) 27 June 11/1 John Hamilton won the June Invitational on his first try Monday, despite a desperate charge by two Corner Brook golfers on the final day of the tournament.

invite, *v.* Add: **1. a.** Also, *to invite in:* to ask (a person) to come into one's house.

1875 Harper's Mag. Aug. 417 She found a house where she was invited in. 1888 Mrs. H. Ward R. Elsmere II. III. xxii. 213 It was evidently the Squire's purpose to come in, so Robert invited him in. 1899 Skeel & Brearley King Washington 37 Being occupied in a nice calculation whether or not her breakfast menu would stand the strain of an extra appetite should Mr. Ettrick..invite the captain in. 1974 'M. Underwood' Pinch of Snuff xi. 94 'I doubt whether we need a warrant,' Sergeant Ellis remarked. 'After all, he invited us in.'

invite, *sb.* Add: **1.** (Later examples.) Now usu. with stress on first syllable.

1937 Times 28 Dec. 7/5 This little servant girl, who.. believed that she would some day receive an 'invite' to a royal ball and marry the prince, [etc.]. 1968 Listener 13 June 770/1 Is it just an invite from the colonel for a working week-end? 1970 G. F. Newman Sir, You Bastard v. 143 The four detectives didn't await an invite into the house.

invitee. Add: **b.** *spec.* (See quot. 1913.)

1913 Law Rep. King's Bench I. 410 The duty of the owner or occupier to use care..is..never very definitely measured... More care, though not much, is owed to a licensee—more again to an invitee. The latter term is reserved for those who are invited into the premises by the owner or occupier for some purpose of business or of material interest. Those who are invited as guests..are not in law invitees but licensees. 1953 Times 14 Nov. 2/7 Their Lordships now held that the plaintiff was himself to blame for the accident, and that he was not an 'invitee' because the business on which he was engaged was not one in which the defendant had a common interest. 1965 Mod. Law Rev. XXVIII. v. 519 A finding that a workman was sciens was sufficient to defeat his claim as an invitee, even if he was not volens. 1971 N.Y. Law Jrnl. 23 Nov. 19/3 In England the distinction between licensees and invitees have [sic] been abolished by statute.

in vitro, in vivo: see *in *Lat. prep.*

invoice, *v.* Add: (Further examples.) Also, to send or submit an invoice to (a person).

Rarely occurs in written English, but common in speech. The contextual sense in quot. 1939 is not certain.

1939 Joyce Finnegans Wake (1964) III. 623 You invoiced him last Eatster so he ought to give us hockockles and everything. 1956 Invoicing Methods (Brit. Inst. Managem.) xv. 124 An opaque signal is placed over the ledger folio on the customer record in order to prevent subsequent orders from the customer being inadvertently stamped with the ledger folio as a sign that they may be invoiced in the ordinary way. 1972 Daily Tel. 12 July 11 (Advt.), I enclose my remittance of £6 for the first ingot, and agree to pay for each subsequent ingot upon being invoiced on a monthly prepayment basis.

involatile, *a.* **2.** (Further examples.)

1962 Simpson & Richards Physical Princ. Junction Transistors xii. 263 When wetted with an involatile (silicone) oil the mica may introduce a resistance of some 0·5°C/watt. 1965 Phillips & Williams Inorg. Chem. I. xvii. 623 Many of the compounds of this section are very high-melting solids, while many others are relatively involatile solids or oils.

involute (i·nvŏliūt), *v.* [Back-formation from Involuted *a.*] *intr.* **a.** 'To return to a normal condition' (*Cent. Dict.* Suppl.). **b.** To undergo involution (sense 4). Hence **i·nvoluting** *ppl. a.*

1904 Buck's Handbk. Med. Sci. (rev. ed.) VII. 782/2 A circular scaly pink patch that spreads peripherally with a pinkish border, and clears up or involutes in the central portion. 1910 Practitioner July 106 The uterus had involuted normally. 1968 Amer. Jrnl. Obstetr. & Gynecol. CII. 33/1 Deeper arteries and veins contract and are compressed from without by the involuting muscle mass of the uterus. 1971 Jrnl. Insect Physiol. XVII. 857 These..

glands reach the peak of their metabolic and synthetic activity shortly before rapidly involuting, leaving only a small remnant.

involuted, *a.* Add: **1. b.** *fig.*

a **1910** 'Mark Twain' *Speeches* (1910) 290 Whatever moral..you put into a speech,..gets diffused among those involuted sentences. **1972** *Times Lit. Suppl.* 22 Dec. 1552/1 Clothed in orthodoxy, that could be no more than an involuted way of saying that God is love.

involution. Add: **6. c.** A function or transformation that is equal to its inverse.

1916 E. Kasner in *Amer. Jrnl. Math.* XXXVIII. 177 It is easy to determine all regular transformations of period 2. In the direct type $Z = f(z)$ the functional equation is $f(f(z)) \equiv z$, that is, $f^2 = 1$; in the reverse type $Z = f(z_0)$ the functional equation is $f(f_0(z)) \equiv z$, that is, $ff_0 = 1$, where f_0 denotes the series whose coefficients are the conjugates of the coefficients of series *f*. We shall call a transformation of the former type (excluding the identical transformation) a conformal involution, and one of the latter type a conformal symmetry. **1969** F. M. Hall *Introd. Abstr. Algebra* II. ii. 31 If θ is a 1–1 correspondence between elements of *A* and itself such that $\theta = \theta^{-1}$, then θ is said to be an involution.

involutional (invŏliū·ʃənăl), *a.* *Psychol.* [f. Involution + -al.] Of or pertaining to the bodily change of involution (sense 4), or to mental disturbances associated with this change.

1910 *Rev. Neurol. & Psychiatry* VIII. 8 I refer to the work of Dreyfus, who, after reviewing Kraepelin's own cases of involutional melancholia, concludes that the involutional depression is a mixed form of manic-depressive insanity. **1934** P. Bottome *Private Worlds* 111 It's not an ordinary melancholic case. She hasn't had any attacks before; and it is not involutional; she is only thirty. **1945** E. Davidoff in O. J. Kaplan *Mental Disorders Later Life* viii. 189 There may be a qualitative as well as a quantitative difference between the nonpsychotic involutional syndrome and the involutional psychosis. **1950** J. Zinkin tr. *Bleuler's Dementia Praecox* II. 242 Kraepelin was the first to draw attention to the high incidence of cases who develop an apparently common melancholia during the involutional period. **1965** Foulds & Caine *Personality & Personal Illness* II. iv. 67 We do not have much difficulty in understanding what is meant by an involutional melancholia with obsessional features. **1968** *Amer. Jrnl. Obstetr. & Gynecol.* CII. 29/1 Pronounced involutional atrophy occurred in the..myometrium at 7 weeks post partum. **1969** *Biochem. Jrnl.* CXII. 641/1 Oestradiol might not have a differential effect on two separate involutionary processes, but might just decrease the water content of the treated uteri.

involutionary (invŏliū·ʃənări), *a.* [f. Involution + -ary[1].] Characterized by involution; retrograde.

1920 *Discovery* Nov. 338/2 Our conceptions of psychical ..'regression'.., the backward or involutionary path of mental processes to more infantile conditions. **1942** *Mind* LI. 146 The infinite series of causes to which *Ethices I, xxviii* [of Spinoza] refers is not a temporal regress to an impossible 'first' cause, but the involutionary sequence of eternal causes to a necessary First Cause.

involutory (invŏliū·tŏri), *a.* *Math.* [f. L. type *involūtōri-us: see Involutorial *a.* and -ory[2].] That is an involution (sense *6 c).

1941 Birkhoff & MacLane *Survey Mod. Algebra* viii. 203 The correspondence $A \leftrightarrow A'$ therefore preserves sums and inverts the order of products, so is sometimes called an anti-automorphism. Since $(A')' = A$, this anti-automorphism is called 'involutory'. **1971** J. H. Conway in Powell & Higman *Finite Simple Groups* vii. 217 The *p* objects permuted by $L_2(p)$..can in each case be taken as *p* involutory permutations of the set Ω. **1972** F. J. Budden *Fascination of Groups* viii. 81, 2-groups always arise as subgroups of larger groups whenever there is an 'involutory' element present, that is, an element whose square is equal to the identity.

invulnerability. (Later examples.)

1900 *Daily News* 4 Sept. 6/1 The superstition of their [*sc.* the Boxers'] invulnerability to cannon shot... The Empress-Dowager..is said to have been deceived by the invulnerability superstition. **1962** *Listener* 29 Mar. 540/1 'Invulnerability' suggests, first of all, protecting your missiles under concrete, or hiding missiles and bombers and submarines by having them moving about. But 'invulnerability' also includes having too many missile sites for your enemy to be able to knock out even with a perfectly executed first strike; it includes having a mixed deterrent of bombers and land-base missiles and sea-going missiles, and it includes the case where your enemy knows that you are sure to have enough warning to launch your missiles and get your bombers aloft. This business of warning introduces serious ambiguities in the idea of 'invulnerability'. **1972** *Sci. Amer.* July 14/2 As a result of this possibility each country will be concerned to maintain the invulnerability of its submarine-based strategic missiles, which are essentially immune to attack from land-based weapons.

invulnerable, *a.* **1.** (Later examples.)

1962 *Listener* 29 Mar. 540/1 It is essential, isn't it, to have what are called 'invulnerable strategic forces'? *Ibid.* 554/1 The speediest possible development of 'invulnerable' nuclear retaliatory power. *Ibid.*, The 'increasingly invulnerable' missile.

inward, *adv.* **3.** (Further examples.)

1850 J. G. Whittier *Songs of Labor* 59 Still dreamed my inward-turning eye. **1890** W. James *Princ. Psychol.* I. x. 320 The more utterly 'selfish' I am in this primitive way, the more blindly absorbed my thought will be in the objects and impulses of my lusts, and the more devoid of any inward looking glance. **1910** Kipling *Rewards & Fairies* p. x, These shall cleanse and purify Webbed and inward-turning eye. **1946** Koestler *Thieves in Night* 207 She sipped her dry Martini with an inward-turned look. **1961** A. Miller *Misfits* xii. 131 His eyes are sightless, inward-looking. **1963** *Times* 28 Jan. 5/2 Yet Schumann's poetry has its feet on German earth, and it was possible to feel that Mr. Richter's presentation of it was just a little too disembodied and wraithlike, his interpretation a little too inward-looking. **1968** *Guardian* 15 Apr. 9/6 Pressure from inward-looking, anti-national groups.

in-winter (i·nwintəɹ), *v.* [In- *pref.*[1]] *trans.* To protect (animals, particularly sheep) by keeping them indoors during severe weather and providing food for them. So **in-wi·ntered** *ppl. a.*, **in-wi·ntering** *sb.*

1961 *2nd Rep. Hill Farming Res. Organisation 1958–61* 25 Hoggs have been in-wintered experimentally in improvised cattle courts... In-wintered Blackface hoggs.. thrived normally on diets of hay... Indoor wintering is quite satisfactory. **1962** *Times* 26 Nov. 17/6 The in-wintering of in-lamb ewes. **1965** Cooper & Thomas *Profitable Sheep Farming* viii. 69 The main interest today ..is the inwintering of breeding ewes on lowland farms. *Ibid.* 70 A hundred six and seven-crop Mule ewes..were inwintered in a Dutch barn. **1971** *Farmer & Stockbreeder* 16 Feb. 25/3 In-wintered ewes should also be dosed two weeks after being brought in.

inwit. **1, 2.** Used as a conscious archaism by some modern writers.

In Joyce's *Ulysses* adopted from Dan Michel's title *Ayenbite of Inwyt* (1340); so also in quots. 1967 and 1968. Cf. Ayenbite.

1922 R. Bridges *Coll. Ess.* (1928) III. 68 If..such good old English words as *inwit* and *wanhope* should be rehabilitated (and they have been pushing up their heads for thirty years), we should gain a great deal. **1922** Joyce *Ulysses* 17 They wash and tub and scrub. Agenbite of inwit. Conscience. **1955** E. Pound *Classic Anthol.* III. 150 Designing in his heart felicity From inwit to his act moved ever so straight He got in sovereignty the whole Quadrate. *Ibid.* 174 There is no light in your conscience And your acts shed, therefore, no light In your inwit. **1967** *Punch* 12 July 72/1 There's this dedicated, totally frenetic fumble under the skin of the times' underbelly.. for the agenbite of inwit. **1968** *Listener* 28 Mar. 411/3 Very probably Bond fans will be able to turn a blind eye to the bites and agenbites of new-Bond's *inwit*.

in-word, -work: see *in *a.* 2 b, *prep.* 10 a.

inworker. Add: **2.** One who works on the premises of a shop or factory.

1909 *Fabian News* XX. 75/1 A minimum wage for both inworkers and outworkers.

‖ **inyanga** (inya·ŋā). *S. Afr.* Also **enyanga, inyanger.** [Zulu.] A medicine-man (see also quot. 1826).

1826 N. Isaacs *Jrnl.* 1 Apr. in *Trav. E. Afr.* (1936) I. vi. 82 The 'inyangers', or water doctors, arrived to take us across the river. *Ibid.* 83 The inyangers from their great muscular power, and experience in their occupation, kept us above the water. **1862** G. H. Mason *Zululand* xiv. 180 An 'Enyanga', as the said professor is called, is a formidable foe; merely as being master of the most deadly poisons in the world. **1946** *Archit. Rev.* C. 22/1 Before the first stick or stone is cleared the headman has the new site approved and then treated by an inyanga—a specialist in various types of native medicine, a man usually labelled by the ethnocentric European as 'magician'. **1954** H. Gibbs *Background to Bitterness* II. ii. 66 He [*sc.* Dingaan] decided to spare Owen and a few American missionaries in Natal—the next morning his inyangas told Owen they would be spared. **1970** *Golden City Post* 28 June, Millionaire inyanga Sethuntsa Khotso, who last year threw the bones and successfully tipped Naval Escort, says that Golden Jewel will win the Durban July next Saturday.

inyoite (i·nyōuɔit). *Min.* [f. the name of *Inyo* County, California + -ite[1].] A hydrated calcium borate, $Ca_2B_6O_{11}.13H_2O$, found as colourless rhombic crystals.

1914 W. T. Schaller in *Jrnl. Washington Acad. Sci.* IV. 355 Inyoite and Meyerhofferite, two new Calcium Borates from Death Valley, Inyo County, California, belong to the Colemanite series. **1921** *Bull. Geol. Survey Canada* XXXII. 2 The crystals of inyoite from Hillsborough are remarkably well developed and clear. **1959** *Acta Crystallogr.* XII. 162 Polyions of inyoite are connected to one another and to neighboring water molecules by bonding through calcium ions and by hydrogen bonds.

io[2] (ɔi·ōu). [mod.L., the specific name of the insect, a. Gr. Ἰώ, daughter of the river god Inachus.] In full, *io moth.* A large yellow North American moth, *Automeris io*, distinguished by prominent ocelli on the hind wings.

[**1869** *Canad. Entomologist* II. 20 These caterpillars.. proved to be the very familiar larvae of what is commonly known as the Io Emperor-moth.] **1870** *Canada Farmer* 15 Sept. 336/2 The beautiful, pale yellowish-green caterpillar..is the larva of the handsome Io-moth. **1873** C. V. Riley in *5th Ann. Rep. Missouri State Entom.* 133 The Io Moth..is one of our most beautiful moths, receiving its name from two conspicuous eye-spots on the hind wings, in allusion to the ancient Greek heroine, Io. **1912** G. Stratton-Porter *Moths of Limberlost* viii. 207 Mr. Eisen presented me with a pair of *Hyperchiria Io*... Because the Io was yellow, I wanted it. **1954** Borror & DeLong *Introd. Study Insects* i. 6 A few caterpillars, such as the saddleback and the larva of the io moth, have stinging hairs.

iod-. Add: **iodargyrite** (examples); see quot. 1971; **iode·mbolite** *Min.* [Embolite], the name now given to Iodobromite.

1868 A. Ramsay *Rudiments of Mineral.* v. 143 Iodargyrite readily fuses, colours the flame red, and yields a globule of silver when reduced. **1950** *Thorpe's Dict. Appl. Chem.* (ed. 4) X. 766/1 Silver iodide, AgI, occurs in Chili, Peru, Mexico, and Spain as the mineral iodargyrite in citron-yellow hexagonal crystals. **1971** *Mineral. Mag.* XXXVIII. 104 Recommendations of the Commission [on New Minerals and Mineral Names of the International Mineralogical Association] on minerals for which more than one name is in common use... Iodargyrite, not iodyrite. **1902** Prior & Spencer in *Mineral. Mag.* XIII. 176 Since the name [*sc.* iodobromite] is a particularly misleading one..we propose to refer to the varieties of the cerargyrite group which contain all three halogens as iodiferous embolites or briefly iodembolites. **1944** *Mineral. Abstr.* IX. 59 At Maikan, near Pavlodar, iodembolite and embolite occur in native sulphur.

iodimetry (ɔi,ŏdi·mĕtri). *Chem.* [f. Iodi(ne *sb.* + -metry.] The titrimetric analysis of an oxidizing or reducing agent using the iodine/iodide redox system; *spec.* the quantitative analysis of a solution of a reducing agent by titration with a standard solution of iodine. Cf. *iodometry.

1897 *Jrnl. Chem. Soc.* LXXII. II. 342 (*heading*) Barium thiosulphate as basis for iodimetry. **1907** Lincoln & Walton *Exerc. Elem. Quantitative Chem. Analysis* 78 The methods of determination in iodimetry may be divided into three general classes: 1. The titration of..reducing agents. 2...oxidizing agents... 3. Free chlorine. **1939** A. I. Vogel *Text-bk. Quantitative Inorg. Analysis* iii. 401 Iodimetry covers titrations *with* a standard solution of iodine. Iodometry deals with the titration *of* iodine liberated in chemical reactions. **1960** R. J. Winterton in C. L. & D. W. Wilson *Comprehensive Analytical Chem.* IB. vii. 249 Iodine..may be used as a standard solution and titrated into the reducing solution, or it may be liberated from potassium iodide or some other compound of iodine by an oxidising agent in solution, and titrated with thiosulphate solution...The use of standard iodine is called direct iodimetry, and the other method indirect iodimetry.

So **iodime·tric** *a.*, of or pertaining to iodimetry; **iodime·trically** *adv.*, by means of iodimetry.

1887 *Jrnl. Chem. Soc.* LII. 997 The iodimetric method serves well for the assay of commercial sodium sulphide. *Ibid.* 998 Ferrous and ferric oxides, in hydrochloric acid solution, can be determined iodimetrically. **1931** Jenkins & DuMez *Quantitative Pharmaceutical Chem.* viii. 123 Iodimetric methods include some of the most exact processes of volumetric analysis, because..the presence of one part of iodine in several million parts of solution can be recognised by means of starch indicator solution. **1963** Skoog & West *Fund. Analytical Chem.* xx. 458 The first [category] is made up of procedures that use a standard solution of iodine to titrate easily oxidized substances. These are termed direct or iodimetric methods and have rather limited applicability since iodine is a relatively weak oxidizing agent. **1969** H. T. Evans tr. *Hägg's Gen. & Inorg. Chem.* xxi. 534 Since many oxidizing agents oxidize iodide ion to iodine, such substances can also be determined iodimetrically.

iodinate (ɔi·ŏdinēit), *v.* [f. Iodin(e *sb.* + -ate *suffix*[3].] *trans.* **a.** To introduce an iodine atom into (a compound or molecule) in place of a hydrogen atom. **b.** To treat with iodine. Hence **i·odinated** *ppl. a.*, **i·odinating** *vbl. sb.*

1908 *Jrnl. Chem. Soc.* XCIV. 778, *m*-Nitroaniline is readily iodinated by adding iodine..to a solution of the base. **1917** *Jrnl. Amer. Chem. Soc.* XXXIX. 444 The xylenols are not so readily iodinated as the lower members. **1921** *Biochem. Jrnl.* XV. 320 The monochloride of iodine has been used before as an iodinating agent. **1926** *Chem. Abstr.* XX. 596 (*heading*) Iodinated pyrrole derivatives. **1948** *New Biol.* IV. 138 Similar improvements in the yields of cows can be obtained by injecting the thyroid hormone—thyroxine—or by feeding the closely related substances known as iodinated proteins. **1964** N. G. Clark *Mod. Org. Chem.* xiv. 275 They are then all iodinated three times on the methyl group. **1968** *Canad. Jrnl. Physiol. & Pharmacol.* XLVI. 449 The effects of increasing or decreasing the endogenous secretion of thyroid-stimulating hormone on the iodinating activity of the rat thyroid gland were investigated. **1969** *Nature* 23 Aug. 778/2 Thomas [etc.]..have measured..the uptake of radioactive iodine in twenty-two adults before and after swimming in an iodinated pool.

iodination (ɔi,ŏdinē·ʃən). *Chem.* [f. Iodin(e *sb.* + -ation.] **a.** The replacement of a hydrogen atom in a compound or molecule by an iodine atom.

1873 *Chem. News* 26 Sept. 167/2 (*heading*) On the chlorination and iodination of anthracen. **1919** *Jrnl. Chem. Soc.* XLI. 293 Iodobenzene on iodination with the required quantity of iodine and nitric acid gives *p*-diiodobenzene. **1971** *Jrnl. Chem. Educ.* XLVIII. 508/1 Iodination by molecular iodine occurs only with phenols and amines.

b. Addition of iodine (e.g. to a water supply).

1966 *Jrnl. Clin. Endocrinol. & Metabolism* XVI. 620/1 Prior to iodination of the water. **1969** *Arch. Environmental Health* XIX. 127/1 Iodination of a public water supply has proven to be technically and commercially feasible.

iodine, *sb.* Add: **2. d.** Other *attrib.* uses, as **iodine number, value** [tr. G. *jodzahl* (A. Hübl 1884, in *Dingler's polytechn. Jrnl.* CCLIII. 287)], the proportion of unsaturated matter present in a substance as measured by the number of grammes of iodine which can be taken up by 100 grammes of the substance; **iodine scarlet,** mercuric iodide, HgI_2, a brilliant red powder.

1885 *Analyst* X. 123 About everything on the list which might be used to adulterate or be substituted for butter, gave iodine numbers so far removed from that for genuine butter that no difficulty would occur. **1969** J. R. Holum *Introd. Org. & Biol. Chem.* ix. 302 Oleic acid has an iodine number of 90, linoleic acid 181, and linolenic acid 274. **1835** G. Field *Chromatography* x. 94 Iodine scarlet is a new pigment of a most vivid and beautiful scarlet colour, exceeding the brilliancy of vermilion. **1885** *Encycl. Brit.* XIX. 87/2 From mercury combined with iodine is prepared a pigment of unequalled vivacity and brilliance, Iodine Scarlet, but unfortunately as fugitive as it is bright. **1969** R. Mayer *Dict. Art Terms & Techniques* 198/1 *Iodine Scarlet*, mercuric iodide; a dangerously poisonous inorganic pigment of the most brilliant scarlet hue. It is useless as a paint pigment because it fades to a pale yellow. **1898** *Analyst* XXIII. 241 Purified allyl alcohol has a theoretical iodine value of 435. **1921** *Biochem. Jrnl.* XV. 319 Resins also show great variations in iodine values according to the conditions of the determination. **1953** J. Davidsohn et al. *Soap Manuf.* I. xviii. 365 In order to produce a hard soap, a fat charge with low iodine value and low saponification value should be selected.

iodine *v.* (examples other than in *Photogr.*).

1929 C. C. Martindale *Risen Sun* 173 His grooms.. were iodining his abrasions. **1936** A. Huxley *Eyeless in Gaza* li. 571 How I regret those cretins one used to see in Switzerland when I was a child! They've iodined them out of existence now.

i·odinized, *ppl. a.* [f. Iodine *sb.*: see -ize.] Of a material: treated or impregnated with iodine.

1919 *Chem. Abstr.* XIII. 2254 (*heading*) Iodinized emulsion. **1962** *Jrnl. Laboratory & Clin. Med.* LIX. 118 (*heading*) Effects of iodinized contrast media upon electrophoretic mobilities of blood proteins. **1963** *Neurology* XIII. 492/1 All the water-soluble iodinized contrast media currently used in diagnostic procedures..have been reported to cause side effects and complications.

iodipin (ɔi,ōu·dipin). *Pharm.* [ad. G. *jodipin*, f. *jod* iodine + L. *adip-*, *adeps* fat: see -in[1].] A liquid obtained by treating sesame oil with iodine, formerly used in treating syphilis and scrofula and as a contrast medium in radiography.

1899 *Brit. Med. Jrnl. Epitome* 18 Nov. 81/1 Iodipin, apart from its uses as a test is, as an iodine preparation, also therapeutically active. **1907** Wood & Bache *Dispensatory U.S.A.* (ed. 19) 1532/1 Iodipin.—A yellow, oily fluid, of a purely oleaginous taste. **1930** *Biol. Abstr.* IV. 2568/2 Injection of a medicinal iodine compound ('Iodipin') into the uterine cavity promises to yield clinical diagnostic results of some value by outlining the uterine cavity and the Fallopian tubes. **1940** H. A. McGuigan *Appl. Pharmacol.* 190 Iodipin is iodized sesame oil. The dose of 10 per cent iodipin is 4 to 8 Gm.

iodization (ɔi:ŏdəizē·ʃən). [f. Iodiz(e *v.* + -ation.] The process or practice of iodizing; the addition of iodine or an iodine compound to a substance.

1909 *Jrnl. Chem. Soc.* XCVI. II. 919 The species specificity of a protein is considerably modified as regards its biological action by iodisation. **1956** *Nature* 24 Mar. 562/2 Iodine metabolism and the iodization of food will first be considered. **1971** *Sci. Amer.* June 97/1 This aspect of thyroid physiology, together with increasing reports of severe iodine toxicity..elicited strenuous opposition to the iodization of table salt.

iodized, *ppl. a.* (Later example.)

1933 *Radio Times* 14 Apr. 112/2 Finest iodised table salt.

iodo-. Add: In combination stressed *i·odo-* or *io·do-*. **a.** **iodoacetic** *a.*, now the usual form of *iodacetic* adj. (s.v. Iod-); (examples); hence **iodoa·cetate,** an ester or salt of iodoacetic acid; **iodopro·tein,** any protein containing iodine; **iodopyracet** (-pɔiə·răset) *U.S.* [f. di-*iodo-pyr*idone-*acet*ic], the diethanolamine salt, $C_{11}H_{16}N_2O_5I_2$, of 3,5-di-iodo-4-pyridone-*N*-acetic acid, used in radiography as a contrast medium, principally for intravenous urography and for measuring renal plasma flow; called *diodone* in the U.K.; **iodopy·rin(e** *Pharm.*, a crystalline iodinated derivative, $C_{11}H_{11}N_2OI$, of antipyrin, used as an antipyretic; **iodothy·rin** = *thyro-iodine* (s.v. Thyro-). **b.** Delete **iodometric, iodometry** and see *iodometry; **i·odo-, io·dophil(e** [-phil, -phile], **iodophi·lic** *adjs.*, readily stained by iodine.

1902 *Jrnl. Chem. Soc.* LXXXII. I. 585 By the condensation of ethyl iodoacetate with citraldehyde, a mixture of substances is apparently obtained. **1931** *Times Lit. Suppl.* 9 July 550/3 Sodium iodoacetate prevents a similar breakdown of methylglyoxal by its inhibition of the enzyme glyoxalase. **1964** W. G. Smith *Allergy & Tissue Metabolism* ii. 25 Inhibition of histamine and SRS-A release by antigen was noted with iodoacetate. **1888** *Syd. Soc. Lex.*, *Iodo-acetic acid*, see Iodacetic acid. **1948** *Jrnl. Biol. Chem.* CLXXVI. 88 The irreversible toxic action of the inhibitor toward yeast cells is proportional to the concentration of undissociated iodoacetic acid. **1962** *Lancet* 15 Dec. 1275/1 Although iodoacetic acid and acetoacetate might have a different action on the metabolism of the heart muscle, both seem to inhibit contractions. **1902** *Encycl. Brit.* XXXII. 817/2 The Myxobolidæ.., which have an iodophile vacuole. **1927** Thaysen & Bunker *Microbiol. Cellulose* viii. 218 Henneberg's statement that most of the cellulose-decomposing organisms found in the intestine are iodophil..disagrees with all previous observations. **1942** *Nature* 23 May 582/1 Both the rumen in ruminant and the cæcum in non-ruminant Herbivora support an abundant iodophile microflora: that is, an association of taxonomically diverse species exhibiting, in consequence of the decomposition within them of bacterial starch or granulose, the common characteristic of giving a blue colour with iodine. **1948** *Jrnl. Bacteriol.* LV. 197 In the majority of instances iodophilic colonies were frequent on the sucrose agar plates containing penicillin. **1971** Hawker & Linton *Micro-Organisms* xv. 543 Cellulose decomposition is associated with iodophilic bacteria (i.e. organisms which reveal starch-like substances when stained with iodine). **1909** *Jrnl. Chem. Soc.* XCVI. II. 919 The behaviour of iodoproteins as regards the formation of specific antisubstances was investigated. **1951** *Adv. Protein Chem.* VI. 253 (*heading*) Natural and artificial iodoproteins. **1963** *Lancet* 5 Jan. 12/1 Very rare examples of absent T_3 suppression in patients not having hyperthyroidism have been recorded when a thyroid nodule with autonomous iodine uptake synthesises abnormal iodoproteins. **1947** *Dispensatory U.S.A.* (ed. 24) 584/2 The iodopyracet injection is warmed to body temperature and 20 cc..., containing 7 Gm..of the drug, is injected slowly intravenously. **1951** A. Grollman *Pharmacol. & Therapeutics* xxxii. 751 Iodopyracet may be injected into the aorta in order to visualize the renal arteries; or into the brachial vein to visualize the chambers of the heart in the diagnosis of congenital cardiac lesions. **1971** *Amer. Jrnl. Obstetr. & Gynecol.* CX. 566/1 Following injection of radioactive iodopyracet into the fetal circulation radioactivity moved equally rapidly into amniotic fluid in preparations involving nephrectomized or normal fetuses. **1892** *Brit. Med. Jrnl.* 2 Jan. 22/1 Iodopyrin. This product is a chemical compound of iodine with antipyrin, in which an atom of hydrogen is replaced by iodine. It occurs in colourless crystals. **1909** H. A. Hare et al. *National Stand. Dispensatory* (ed. 2) 201 Iodopyrine presents the combined physiological action of iodine and an antipyretic, and has been recommended in bronchial asthma and tertiary syphilis. **1961** L. Martin *Clinical Endocrinol.* (ed. 3) iii. 68 There are also certain other therapeutic substances which will produce a simple goitre... These include..iodopyrine (..found in Felsol powders which are used for asthma). **1897** *Jrnl. Chem. Soc.* LXXII. II. 330 The animal was taking..diet to which was added either fresh thyroid or iodothyrin. **1909** H. A. Hare et al. *National Stand. Dispensatory* (ed. 2) 855 Iodothyrine is a white or yellowish-white powder, having a sweet taste, and is prepared..by boiling fresh thyroid glands..with dilute sulphuric acid. **1932** *Discovery* Mar. 96/2 The flattened gilled water newt called the Axolotl [is] changed into a smaller differently shaped land newt in the course of a few weeks by feeding it with thyroid. In this case the active substance can be extracted from the gland and chemically analysed; it is termed iodothyrin.

iodometry (ɔiŏdǫ·mĕtri). *Chem.* [f. Iodo- + -metry.] The titrimetric analysis of an oxidizing or reducing agent using the iodine/iodide redox system; *spec.* the quantitative analysis of a solution of an oxidizing agent by the addition of excess iodide followed by titration of the iodine so liberated with thiosulphate or arsenite solution. Cf. *iodimetry.

1883 *Chem. News* 6 Apr. 166/1 (*heading*) Preparation of a durable starch solution for iodometry. **1916** *Chem. Abstr.* X. 730 (*heading*) Differential iodometry. I. Determination of periodates, iodates, bromates and chlorates in the presence of each other. **1928** *Ibid.* XXII. 2901 Reaction between iodic and hydriodic acids in very dilute solution and the titration of the liberated iodine with thiosulfate... An investigation of reactions involved in iodometry. **1939** [see *iodimetry]. **1973** *Chem. Abstr.* LXXVIII. 23524 Iodometry can be employed for detn. of small amts. of Zn, Cu, Cd, Co and Hg.

So **iodome·tric** *a.*, of or pertaining to iodometry; **iodome·trically** *adv.*, by means of iodometry.

1856 *Q. Jrnl. Chem. Soc.* VIII. 194 We have employed the much more accurate and convenient iodometric method, which..gives a degree of accuracy attainable by very few analytical processes. **1891** *Jrnl. Chem. Soc.* LX. 614 The facility with which pure potassium iodate is prepared renders it an admirable basis for iodometric analysis. **1928** A. W. Wellings *Volumetric Analysis* v. 102 Nitrites may also be estimated iodometrically. **1957** G. E. Hutchinson *Treat. Limnol.* I. xiii. 769 The excess chromate, equivalent to the original sulfate, is determined iodometrically. **1960** *Limnology & Oceanogr.* V. 343 (*heading*) A note on a stabilized starch indicator for use in iodometric and iodimetric determinations. **1971** G. D. Christian *Analytical Chem.* xv. 238 Sodium thiosulfate solution is standardized iodometrically against a pure oxidizing agent such as $K_2Cr_2O_7$, KIO_3, $KBrO_3$, or metallic copper.

iodonium (ɔi,ŏdōu·niə̆m). *Chem.* [f. Iod- + *-onium.] The name of cations of the type $RR'I_2^+$ where R and R′ are (different or the same) alkyl or aryl radicals, or part of a ring. Usu. *attrib.* or as a formative element.

1894 *Jrnl. Chem. Soc.* LXVI. I. 242 Diphenyliodonium iodide, $I·IPh_2$. *Ibid.* 243 Phenyl, which does not form ammonium and sulphonium bases, forms iodonium bases. **1922** J. Mellor *Comprehensive Treat. Inorg. & Theoret. Chem.* II. 108 The iodonium bases and salts resemble those of lead and silver but particularly those of thallium. **1950** N. V. Sidgwick *Chem. Elements* II. 1257 The free iodonium bases $Ar_2I·OH$ are only known in solution, being made from the halides with silver oxide. **1962** P. J. & B. Durrant *Introd. Adv. Inorg. Chem.* xxiii. 931 Diphenyliodonium hydroxide $[(C_6H_5)_2I]OH$. **1966** *Chem. Abstr.* LXV. 8872 (*heading*) Chemistry of heterocyclic iodonium compounds.

iodophor (ɔi,ōu·dǫf-, ɔi·ŏdǫfǭɹ). [f. Iodo- + *-phor.] Any substance in which iodine is combined with a surface-active agent to render it more soluble and chemically stable in aqueous solution, and so more suitable for use (in solution) as a disinfectant.

1952 Terry & Shelanski in *Mod. Sanitation & Building Maintenance* IV. 62/2 This combination of iodine with a carrier is called a halophor (or more specifically it could be referred to as an iodophor). **1962** *Lancet* 22 Dec. 1330/2 A convenient way of using iodine would be to employ an iodophor, which, incidentally, would add a little cleaning power to the water. **1963** *Surg., Gynecol. & Obstetr.* CXVI. 363 The iodophor solutions were all pleasant to use. They did not irritate the skin, eyes, or nasal passages. **1967** *Biol. Abstr.* XLVIII. 10833/1 The percentage dissociation of the iodine complex in 16 commercial iodophors at use dilution was calculated from the partition coefficients. **1972** *Dairy Sci. Abstr.* XXXIV. 205/2 The development and use of foamless iodophors, particularly for cleaning milking equipment..and in combatting mastitis when used as a teat dip.

iodoso(-) (ɔiŏdōu·so), used in *Chem.* as comb. form of Iodous *a.* to indicate the presence of an IO— group in a compound, as in **iodo:sobe·nzene,** C_6H_5IO, a yellow amorphous powder which disproportionates to iodobenzene and iodoxybenzene, and explodes when heated. Also as quasi-*adj.*

1892 *Jrnl. Chem. Soc.* LXII. 1460 On heating the solution for some seconds, cooling, and pouring into water, a precipitate of iodosobenzoic acid, $C_7H_5IO_3$, separates. **1893** Iodosobenzene [see next]. **1893** *Jrnl. Chem. Soc.* LXIV. I. 508 All attempts to obtain iodoso-derivatives from meta- and para-iodobenzoic acid have been without success. **1935** *Ibid.* 1671 Other reactions of iodoso-compounds point to the vulnerability of their I-O link to addition. **1936** *Chem. Abstr.* XXX. 6662 (*heading*) Use of iodosobenzene in developers and emulsions. **1950** N. V. Sidgwick *Chem. Elements* II. 1249 The iodoso-compounds are greyish-white, amorphous substances, with a very characteristic penetrating 'iodoso smell'. **1968** Iodosobenzene [see next].

iodoxy(-) (ɔiŏdǫ·ksi), *prefix.* [f. Iod- + Oxy-[2].] Used in chemical names to indicate the presence of the radical IO_2— (or, formerly, IO—), as in **i:odoxybe·nzene,** a compound, $C_6H_5IO_2$, forming colourless needles and obtained by disproportionation of iodosobenzene. Also as quasi-*adj.*

1865 *Jrnl. Chem. Soc.* XVIII. 308 A combustion made of the pure acid gave numbers leading to the formula of iodoxybenzoic acid $C_7H_5IO_4$. **1889** *Ibid.* LVI. 1150 The authors are inclined to regard it as an iodoxydiiodobenzene, $C_6H_3I_2·OI$. **1893** *Ibid.* LXIV. I. 506 Iodoxybenzene, $C_6H_5IO_2$, when treated with hydrochloric acid in aqueous solution, is converted into iodosobenzene hydrochloride. **1905** *Jrnl. Soc. Chem. Industry* 31 Jan. 104/1 (*heading*) Iodoxy compound (*p*-iodoxyphenol ester): and process of making same. **1935** *Jrnl. Chem. Soc.* 1669 The present summary of several of our investigations, dealing particularly with the iodoxy-group IO_2. **1950** N. V. Sidgwick *Chem. Elements* II. 1251 The iodoxy-compounds hold their oxygen more firmly than the iodoso, and they are not such strong oxidizing agents. **1968** R. O. C. Norman *Princ. Org. Synthesis* xvi. 535 Iodoxybenzene may be obtained in over 90% yield by steam-distilling iodosobenzene to remove the iodobenzene formed by the disproportionation.

ion. Add: (Further examples.) In modern use, any individual atom, molecule, or group having a net electric charge (either positive or negative), whether in an electrolytic solution or not.

In the def. in Dict. *element* is to be understood in the sense 'component substance' (Element *sb.* 2), not 'chemical element'.

1834 W. Whewell *Let. to Faraday* 5 May in I. Todhunter *William Whewell* (1876) II. 182 For the two together you might use the term *ions*. **1834** Faraday in *Phil. Trans. R. Soc.* CXXIV. 79 Finally, I require a term to express those bodies which can pass to the electrodes... I propose to distinguish these bodies by calling those anions which go to the anode of the decomposing body; and those passing to the cathode, cations; and when I have occasion to speak of these together, I shall call them ions. [*Note*] Since this paper was read, I have changed some of the terms which were first proposed. *Ibid.* 112 A body decomposable directly by the electric current, i.e. an electrolyte, must consist of two ions. *Ibid.*, Compound ions are not necessarily composed of electro-chemical equivalents of simple ions. For instance, sulphuric acid,

boracic acid, phosphoric acid, are ions, but not electrolytes, i.e. not composed of electro-chemical equivalents of simple ions. **1856** W. A. Miller *Elem. Chem.* II. xviii. 1110 When a binary compound, such as a fused chloride, ..is submitted to electrolysis, the ions or components of the compound are separated at the respective electrodes in equivalent proportions. **1879** *Encycl. Brit.* VIII. 107/1 Sodium acetate and silver chloride are therefore electrolytes of which Ag, Cl, Na, $C_2H_3O_2$ are the respective ions. **1896** W. R. Whitney tr. *Le Blanc's Elem. Electrochem.* iii. 60 Only those substances conduct which are at least partly dissociated, and therefore the conductivity is due to the dissociated parts; to the latter, which were called by him the 'ions', Arrhenius ascribed electric charges. **1896** Rutherford & Thomson in *Phil. Mag.* XLII. 405 We have made..experiments with the view to seeing whether there is any polarization when a current of electricity passes through a gas; we have not, however, been able to satisfy ourselves of the existence of this effect. The absence of polarization implies, however, that the ions are able to give up their charges to the metal electrodes. **1899** Rutherford in *Ibid.* XLVII. 112 The theory has been put forward that the rays in passing through the gas produce positively and negatively charged particles in the gas, and that the number produced per second depends on the intensity of the radiation and the pressure... The term ion was given to them from analogy with electrolytic conduction, but in using the term it is not assumed that the ion is necessarily of atomic dimensions; it may be a multiple or submultiple of the atom. **1927** N. V. Sidgwick *Electronic Theory of Valency* vi. 91 In a crystal like calcium carbonate we find the same kind of relation between the calcium ion and the CO_3 ion, but a different one for the constituent atoms of the CO_3 group;..this may be taken as evidence that the calcium and the CO_3 are themselves ions, but that the atoms of the CO_3 group are covalently linked to one another. **1962** P. J. & B. Durrant *Introd. Adv. Inorg. Chem.* xii. 346 An ionic crystal is one in which the units of crystal structure are the ions of a salt. **1967** *New Scientist* 30 Nov. 531/1 The normal electrode separation is less than 1 mm, so there is a very strong electric field which ionizes the gas atoms, giving ions and electrons.

2. Special Comb.: **ion beam,** a current of ions moving in a fixed direction; **ion bombardment,** the process of bombarding a surface with ions (usu. of an inert gas), so breaking up the surface, used to remove impurities; hence **ion-bombarded** *a.*; **ion burn,** the damaging of the phosphor of a cathode-ray tube by negatively ionized gas molecules produced by the electron beam and focused on to the screen; also, an ion spot so produced; **ion chamber,** an ionization chamber; **ion drive,** (*a*) = *ion propulsion*; (*b*) = *ion engine*; **ion engine,** a rocket engine that employs ion propulsion; **ion etching,** the controlled removal of extremely thin layers of material from the surface of an object by the use of an ion beam; **ion gun,** a device in which ions are produced (usu. by the ionization of a gas) and emitted in a beam; **ion implantation,** the implantation of ions in a crystalline material (see *implantation 7); **ion pair,** (*a*) a pair of oppositely charged ions held together in a solution by electrostatic attraction; (*b*) a negative ion (or an electron) and a positive ion formed from a neutral atom or molecule by the action of radiation; **ion propulsion,** a mode of rocket propulsion in which thrust is produced by the ejection of ions produced inside the engine and accelerated by an electric field; **ion rocket,** (*a*) = *ion engine*; (*b*) a rocket in which an ion engine is the means of propulsion; **ion source,** a device for producing ions, *spec.* an ion gun; **ion spot,** (*a*) a dark spot in the middle of the screen of a cathode-ray tube where the phosphor is damaged as a result of ion burn; (*b*) a white spot in a television picture produced as a spurious signal when ionized gas molecules strike the target of a television camera tube; **ion trap,** a device designed to catch ions; *spec.* one in a cathode-ray tube or television camera tube that prevents ionized molecules from reaching the screen or the target and causing an ion spot. Also *ion exchange.

1932 *Physical Rev.* XL. 33 Intense high speed ion beams. **1951** *Jrnl. Brit. Interplanetary Soc.* X. 253 The acceleration of a space ship by an ion beam seems to offer no particular difficulties. **1970** *New Scientist* 5 Feb. 256/1 The ion beam, projected at a small area of the sample, consists of heavy, positively charged ions of inert gas, which remove atoms from the specimen's surface layers. **1959** *Jrnl. Chem. Physics* XXX. 926/2 The relative ease of removal of oxygen can therefore not be explained by the assumption that the ion-bombarded and annealed surface is partially oxygen contaminated. **1930** *Rev. Mod. Physics* II. 186 'Sputtering', or disintegration of an electrode subjected to positive ion bombardment is a well known and often troublesome phenomenon. **1952** *Trans. Faraday Soc.* XLVIII. 747 As a general procedure for cleaning surfaces, the inert gas ion-bombardment has some advantages: its main disadvantage is that there is sputtering of the metal on to the walls of the vessel. **1960** *Jrnl. Appl. Physics* XXXI. 1516 (*title*) Ion-bombardment etching of synthetic fibers. **1954** E. Molloy *Radio & Television Engineers' Ref. Bk.* xxiv. 27 The third technique of preventing ion burn is by protecting the screen with a layer of aluminium. **1956** [see *ion spot below]. **1963** J. R. Davies *Understanding Television* ii. 62 Ion burns normally show up as brown circles an inch or so in diameter, the discoloration being greatest at the centre of the burn. In some cathode ray tubes, protection against ion burn is achieved by mounting part of the electron gun assembly at an angle, and..applying across the tube neck a fixed magnetic field which causes electrons only to be deflected across the screen. **1955** *Bull. Atomic Sci.* June 213/3 An ion chamber, however, is a device which directly measures the dose to the air volume it encloses. **1962** F. I. Ordway et al. *Basic Astronautics* iv. 121 Solar radiation intensities in the Lyman-alpha experiment were measured by a photo-sensitive ion chamber. **1958** C. C. Adams et al. *Space Flight* xiv. 346 Both rubidium and cesium have been considered as the propellant for the ion drive. **1960** *Aeroplane* XCVIII. 776/2 Ion-drive cannot be used in propelling space-vehicles from the Earth's surface because of their inherently low thrust. **1962** F. I. Ordway et al. *Basic Astronautics* x. 424 The three basic elements of the ion drive are the emitter, the accelerator, and the beam neutralizer. **1960** *Aeroplane* XCVIII. 776/2 (*heading*) Experimental ion engine. **1961** [see *ion rocket* below]. **1971** *Jrnl. Brit. Interplanetary Soc.* XXIV. 573 The use of pulsed plasma thrusters rather than ion engines for attitude control has..become a distinct possibility. **1965** *Chem. Abstr.* LXII. 3491 (*heading*) Ion etching: an effective method for the elimination of foreign layers in ultra vacuum. **1968** *Times* 13 Nov. 16/1 The inside appearance of a red blood cell has been revealed for the first time by the novel combination of two physical techniques, scanning electron microscopy and ion etching... Ion etching has so far been used chiefly to study the structure of metals. The specimen is bombarded with a stream of high energy ions and the inner structure of the material is revealed as successive layers are stripped off. **1970** *New Scientist* 5 Feb. 256/2 It has been possible to use ion etching to penetrate the cuticular layers of insects. **1952** *Jrnl. Brit. Interplanetary Society* XI. 179 Any rocket system requiring the conversion of electric to kinetic energy will require one or more ion guns. **1957** *Physical Rev.* CVII. 642/1 The performance of the spectrometers was tested with an electron gun and with an ion gun. **1967** *New Scientist* 30 Nov. 531/1 The apparatus..consists of two ion sources (ion guns) which direct beams of ions at shallow angles on to the centres of the faces of a disc-like sample of ceramic. **1965** *Nuclear Instruments & Methods* XXXVIII. 169 (*heading*) Doping of silicon by ion implantation. **1970** *Times* 23 Jan. 27/8 This makes a low-temperature method of doping the silicon, such as ion implantation, an attractive approach to..flexible circuit fabrication. **1970** *New Scientist* 15 Oct. Suppl. 16/1 In ion implantation, the necessary impurities are accelerated by an electric field to an energy sufficient to embed them into the silicon to the depth required. **1972** *Physics Bull.* Oct. 612/3 Ion implantation can be helpful in understanding already known damage centres by careful choice of bombarding isotope. **1933** *Jrnl. Amer. Chem. Soc.* LV. 477 The changes in the properties of the solvent..caused by the presence of undissociated (non-conducting) ion pairs will be neglected. **1941** *Proc. 7th Internat. Congr. Genetics* 246 At 0·01 r. per min. there is on the average only one ion pair produced in a sperm nucleus every 27 min. and since the ion pairs produced by γ-rays are rather far apart, it is now more difficult than ever to avoid the conclusion that individual mutations arise from individual ionizations and that there is no threshold intensity. **1963** B. Fozard *Instrumentation Nucl. Reactors* ii. 13 When a charged particle of high energy is introduced into the sensitive volume of a gas ionisation detector it undergoes large numbers of ionising and exciting collisions... In suitable conditions the whole of this energy is expended within the sensitive volume. In such cases the total number of ion pairs produced is a direct measure of the particle energy. **1964** Black & Wagner *Dynamic Path.* xi. 231 The different types of ionizing radiation produce the same fundamental change in matter, that is, the ejection of planetary electrons from atoms or molecules, leading to the formation of ion pairs. **1957** *Jrnl. Brit. Interplanetary Soc.* XVI. 233 A system which may ultimately utilize power from fusion is that of ion (or plasma) propulsion. *Ibid.*, Much serious attention is being given to the ion-propulsion problem in the U.S.A. **1966** *McGraw-Hill Encycl. Sci. & Technol.* VII. 245/2 The space charge represents one of the serious obstacles of any ion-propulsion system. **1949** *Jrnl. Brit. Interplanetary Soc.* VIII. 64 These techniques form the basis for what we shall here call the 'ion rocket', the jet of which might more correctly be termed an exhaust beam. **1951** *Ibid.* X. 248 The use of ion rockets as a means for propelling vehicles between satellite stations. **1953** *Ann. Reg. 1952* 406 The 'most feasible project' was described as consisting of space stations circling the most important planets with ion rockets plying between them. **1961** *Flight* LXXIX. 330/2 The earliest ion rockets to be considered for spaceflight will have a thrust of one-tenth of a pound and will be powered by 30kW SNAP-8 nuclear reactors after first-stage launch by Atlas-Centaur or Saturn C-1 boosters. Higher-powered ion engines (up to one megawatt) will also be considered. **1971** *Nature* 6 Aug. 357/1 What was called space technology—the design and operation of solar cells, ion rockets and the like. **1955** *Gloss. Terms Radiology (B.S.I.)* 43 *Ion source*, a device in which gas ions are produced, focused and accelerated, and emitted as a narrow beam. **1966** *McGraw-Hill Encycl. Sci. & Technol.* VII. 245/1 The ionizer or ion source converts the propellant from its original stored form to a system of charged corpuscles. **1967** Ion source [see *ion gun* above]. **1940** D. G. Fink *Princ. Television Engin.* viii. 341 If magneto-static focusing is used, the ion spot is spread over a much larger area (often the full area of the screen) and is much less troublesome, whether electric or magnetic deflection is used. **1953** Amos & Birkinshaw *Television Engin.* I. v. 93 Spurious signals in the form of a white ion spot are eliminated by use of an ion-trap mesh situated close to the target. **1953** H. A. Chinn *Television Broadcasting* ii. 69 Occasionally, a white spot..may be observed in the center of the picture. Such a spot, especially if it is visible on the monitor with the camera lens capped, is probably an ion spot. **1956** M. Slurzberg et al. *Essent. Television* x. 375 A concentrated bombardment of the fluorescent screen by the heavy ions produces a small brown circle in the centre of the picture screen, called an ion burn or ion spot. **1905** Bragg & Kleeman in *Phil. Mag.* X. 321 An arrangement which we find to be of great importance is the ion-trap which is placed under the gauze of the ionization chamber. **1940** D. G. Fink *Princ. Television Engin.* viii. 342 The ion spot may..be eliminated by an ingenious construction in the electron gun known as an 'ion trap'. **1953** Ion-trap [see *ion spot* above]. **1954** E. Molloy *Radio & Television Engineers' Ref. Bk.* xxiv. 27 The successful operation of..ion traps depends upon the different paths followed by the electrons and negative ions under the action of magnetic fields. **1967** Wharton & Howorth *Princ. Television Reception* iv. 59 A small permanent magnet known as the ion trap magnet was mounted on the neck of the tube to deflect the electrons so that they travelled axially and landed on the screen.

-ion, *suffix*[2], the ending of Inion[1], proposed by P. Broca, 1875 (in *Bull. de la Soc. d'Anthrop. de Paris* X. 346), as a suffix for forming the names of other craniometric points (as *gnathion, *gonion).

-ion, *suffix*[3], the word Ion added to abbreviated forms of the names of elements and radicals to form the names of their ions, as *chlorion*, *hydrion, *nitrion*, *sodion*.

-ion, *suffix*[4]. (See quot. and Alliance *sb.* 6.)

1930 F. R. Bharucha tr. *Braun-Blanquet's Vocab. Plant Sociol.* 23 For designation of Alliances, the suffix '-ion' is added to the radical of the name of one of the principal associations of the alliance. Ex. *Ammophilion-Genisteto-Vaccinion.*

ionamine (ɔi,ǫ·nămīn). *Dye Chem.* Also **Ionamine.** [f. Ion + Amine.] Any of various [(arylazo)arylamino] methanesulphonates, $X·N_2·Y·NRCH_2SO_3Na$ (where X and Y are (substituted) benzene nuclei), formerly used as specific dyes for acetate silk.

1922 *Glasgow Herald* 24 Apr. 10 The 'ionamines' will have nothing to do with this homely fibre [*sc.* cotton]. They will only dye silk, either real or artificial. **1923** Green & Saunders in *Jrnl. Soc. Dyers & Colourists* XXXIX. 12/1 Those Ionamines which are derived from primary amido compounds or contain free amido groups are capable of being diazotised upon the fibre, and in combination with various developers produce a wide range of shades from orange to scarlet, red, maroon, violet, blue and black. **1946** S. R. & E. R. Trotman *Bleaching, Dyeing & Chem. Technol. Textile Fibres* (ed. 2) 474 The chief advantages of the ionamines are (i) ready solubility, and (ii) miscibility with direct dyestuffs. **1971** R. L. M. Allen *Colour Chem.* i. 12 Systems in which the solubilising groups are eliminated by hydrolysis during or after dyeing have also been used in dyes such as the Ionamine (British Dyestuffs Corp.) and Neocotone..ranges, but these have now been superseded.

ionene (ɔi·ŏnīn). *Chem.* [ad. G. *ionen* (Tiemann & Krüger 1893, in *Ber. d. Deut. Chem. Ges.* XXVI. 2693), f. Gr. ἴον violet: see -ene.] 1,1,6-Trimethyltetralin, $C_{13}H_{18}$; a bicyclic oily liquid obtained on elimination of a water molecule from α- or β-ionone.

1894 *Jrnl. Chem. Soc.* LXVI. I. 82 When heated with hydriodic acid and phosphorus, it [*sc.* ionone] loses water, and yields ionene, $C_{13}H_{18}$, which boils at 106–107° (10 mm.)... It resembles irene very closely. **1933** *Jrnl. Amer. Chem. Soc.* LV. 4680 That ionene is a true 1,1,6-trimethyltetralin is evidenced by its smooth sulphonation and nitration, and by the oxidation products obtained from ionene. **1960** *Jrnl. Chem. Soc.* 3128 *Ionene.*—This hydrocarbon was prepared in 53% yield from β-ionone.

ion exchange. The interchange of ions of like charge between an insoluble solid and a solution in contact with it. Freq. *attrib.* (usu. hyphenated), as **ion-exchange resin,** any synthetic resin or polymer suitable for use as an ion exchanger, characterized in general by a cross-linked molecular network which allows the penetration of solvent and ions (causing swelling) and has ionized or ionizable groups weakly attached to it.

1923 *Chem. Abstr.* XVII. 578 (*heading*) The ion exchange between the blood corpuscles and the serum. **1943** *Industr. & Engin. Chem.* Aug. 858 (*heading*) Ion exchange resins. New tools for process industries. **1946** *Jrnl. Amer. Leather Chemists' Assoc.* XLI. 555 The present work is an attempt to explore the possibilities of utilizing ion exchange resins in the analysis of some chromium liquors of various types. **1949** [see *ion exchanger* below]. **1950** *Sci. Amer.* Nov. 51/3 Ion exchange has provided a tool for fascinating studies of the constituents of cell nuclei—nucleic acids and nucleotides—which are so similar to one another that they are nearly impossible to separate by conventional means. **1951** J. E. Maudru in R. A. McGinnis *Beet-Sugar Technol.* x. 289 Ion exchange..is another type of juice purification which has recently been applied to the beet-sugar industry. This process consists of passing dilute juice..through beds of active synthetic resinous material, which removes ionized impurities. **1952** *Sci. News* XXVI. 83 Ion-exchange resins are usually marketed in the form of coarse or spherical granules, about 1–2 millimetres in diameter... The major use of these resins is in the complete removal of salts from water by ion-exchange. **1963** *Guardian* 28 Feb. 18/1 The Government had decided to establish an experimental pilot plant to examine the problems in the removal of strontium 90 from milk by an ion-exchange process. **1964** *Oceanogr. & Marine Biol.* II. 149 The amino acid compositions of phytoplankton and pure cultures of phytoplanktonic species have been determined by paper chromatography and by ion-exchange chromatography.

Also **ion exchanger,** a solid involved or used in ion exchange; also, an apparatus for effecting ion exchange.

1941 *Jrnl. Franklin Inst.* CCXXXII. 317 It is possible ..that the characteristics of an ion exchanger are conditioned by the nature of the ion with which it was last saturated. **1949** H. F. Walton in F. C. Nachod *Ion Exchange* 4 In the materials which find practical use as ion exchangers, such as the synthetic aluminosilicates, synthetic resins, or sulfuric acid-treated coals, nearly all of the ion exchange takes place in the interior of the granules, which have a gel structure, something like a sponge on the molecular scale. **1950** *Sci. Amer.* Nov. 50/2 During the war compact little ion exchangers were used in life-raft emergency kits for making drinking water from sea water. **1960** R. Kunin *Elem. Ion Exchange* iii. 24 Only the zeolite minerals have a sufficiently satisfactory combination of the above-mentioned characteristics..to permit their use as ion exchangers on a commercial scale.

Ionic, *a.*[1] and *sb.* Add: **B.** *sb.* **4.** *Typogr.* A type face distinguished by prominent serifs and a high degree of legibility.

1842 H. Caslon *Specimen of Printing Types* in *Two Centuries of Typefounding* (Caslon Letter Foundry) (1920) 71 (*caption*) Diamond two-line Ionic. **1934** A. F. Johnson *Type Designs* viii. 205 Ionic in some cases appears to be only another name for Egyptian. **1954** *Archit. Rev.* CXVI. 119/1 Ionic, or Clarendon, is familiar to all readers of *The Architectural Review* as a type face. It can also be pleasing and useful as an architectural letter. **1970** W. P. Jaspert et al. *Encycl. Type Faces* (ed. 4) 121 The first Ionic was a bold face cut by Caslon and shown 1842... It has been revived as a suitable newspaper type... Linotype Ionic was introduced in 1926 in the *Newark Evening News*.

ionic, *a.*[2] Add to def.: composed of or containing ions; that is an ion. (Further examples.)

1913 *Q. Rev.* July 122 A knowledge of the total mass of water precipitated by the expansion enabled Mr. Wilson..to estimate the number of ionic nuclei required to form the cloud. **1914** J. J. Thomson in *Phil. Mag.* XXVII. 761 Molecules of this type, which I shall call ionic molecules. **1936** *Discovery* June 197/2 A description of ionic clouds formed in electrolytes is given. **1936** *Jrnl. R. Aeronaut. Soc.* XL. 594 Experiments on ionic crystals have determined the weakening effect of minute cracks. **1962** [see *ion 1]. **1966** C. R. Tottle *Sci. Engin. Materials* ii. 33 Magnesium..will form an ionic compound with two chlorine atoms, $MgCl_2$, by donating its two 3*s* electrons, one to each of two chlorine atoms. The magnesium will be doubly charged (positive) and will therefore take up a position in the crystal lattice in which for each magnesium atom there are two chlorine atoms.

b. Brought about by, employing, or depending on ions; applied *spec.* to an electrovalent bond.

1907 *Brit. Med. Jrnl.* 14 Sept. 631/1 The study of ionic medication—the subject of Professor Leduc's paper—has been in many respects the most fruitful of modern times. **1910** E. R. Morton *Essent. Med. Electr. & Radiogr.* (ed. 2) v. 120 The advantages of the ionic method are very obvious... We can introduce a drug exactly where it is required. **1930** [see *heteropolar *a.* 3]. **1938** R. W. Lawson tr. *Hevesy & Paneth's Man. Radioactivity* (ed. 2) xxiv. 261 The phenomenon of the 'ionic wind'... When the air between the plates of an ionization chamber is ionized, then on applying an electric field the air between the plates is set in motion. **1939** L. Pauling *Nature Chem. Bond* i. 4 We describe the interactions in this crystal by saying that each ion forms ionic bonds with its six neighbors, these bonds combining all of the ions in the crystal into one giant molecule. **1958** C. C. Adams et al. *Space Flight* ix. 236 When atomic energy, ionic drive, and light beams have been harnessed for propulsion, combined perhaps with anti-gravity, flight to the stars may be possible. **1966** C. R. Tottle *Sci. Engin. Materials* vi. 127 In solid materials ionic conduction is strongly structure-sensitive, particularly with respect to impurities and flaws or defects in the lattice. **1972** *Physics Bull.* Nov. 651/3 The atomic bonding within the network is partly covalent, partly ionic.

ionically (ɔi,ǫ·nikăli), *adv.* [f. Ionic *a.*[2]: see -ically.] By means of ions or an ionic bond; as regards or in terms of ions.

1912 *Marconigraph* II. 340/1 The author's own observations on the influence of ordinary cloud on signals proved that the ionically turbulent belt was above the ordinary cloud level. **1925** H. H. U. Cross *Electro-Therapy & Ionic Medication* vi. 172 The efficacy of introducing drugs ionically depends..upon the fact that a drug introduced by ionic medication does not quickly pass out of the tissues which surround the point of introduction. **1965** Phillips & Williams *Inorg. Chem.* I. xv. 567 The electrical conductivity of liquid SO_2 has been commonly interpreted as due to the doubly-charged ions SO^{2+} and SO_3^{2-}, which would mean that a typical acid-base reaction would be written ionically as $SO^{2+} + SO_3^{2-} \rightarrow 2SO_2$. **1972** *Nature* 4 Feb. 262/1 The passage of dyes and small molecules between excitable cells which are ionically coupled has also been observed.

ionicity (ɔi,ŏni·siti). *Chem.* [f. Ionic *a.*[2]: see -icity.] Ionic character (in a chemical bond or a crystal).

1946 *Nature* 26 Oct. 592/2 The absence of a similar fall in the C—O bond-breaking energy passing from methyl to ethyl alcohol can be satisfactorily explained in terms of the increased ionicity of the C—O bond in ethyl alcohol relative to C—O in methyl alcohol. **1952** C. A. Coulson *Valence* v. 123 The asymmetry of charge, i.e. the polar character of the bond, sometimes called the degree of ionicity. **1960** *Nature* 13 Aug. 590/1 In discussions..regarding bonding in inorganic semiconducting (or insulating) crystals there seems to be some confusion between bond ionicity on one hand and charge separation effects due to electronegativity differences on the other. **1966** *Bull. Amer. Physical Soc.* XI. 187/1 The large hyperfine parameter for the NaCl structure is consistent with the expected greater ionicity.

ionium (ɔi,ōu·niə̆m). *Chem.* [f. Ion + -ium: see quot. 1907.] A radioactive isotope of thorium with mass number 230, produced by the α-decay of uranium 234. Symbols Io, ^{230}Th.

1907 B. B. Boltwood in *Amer. Jrnl. Sci.* XXIV. 372 The name 'Ionium' is proposed for this new substance... This name is believed to be appropriate because of the ionizing action which it possesses in common with the other elements which emit α-radiations. **1930** *Engineering* 4 Apr. 461/2 The elements ionium and thorium were actually found to have the same chemical properties, though differing in mass. **1954** K. Rankama *Isotope Geol.* lxxxiii. 410 The age of deep-sea sediments younger than approximately $0·5 \times 10^6$ years may be determined from their ^{230}Th (ionium) content. A plot of the ionium content against the depth in these sediments reveals the decay of the 80 000-year ^{230}Th, and from this graph the age of the sediment at a given depth may be calculated... This is the ionium method for dating deep-ocean sediments. **1970** *Geochim. et Cosmochim. Acta* XXXIV. 389 (*heading*) A system for detection of ionium, thorium and protactinium to date deep-sea cores.

ionizable (ɔi·ŏnəizăb'l), *a.* [f. Ioniz(e *v.*[2] + -able.] Capable of being ionized.

1907 *Jrnl. Chem. Soc.* XCII. II. 560 In the haloid salts all the halogen is in an ionisable state. **1946** *Nature* 7 Sept. 325/2 One notable change in this edition is that the 1940 table of 'available' (ionizable) iron contents has been dropped. **1968** R. F. Steiner *Life Chem.* ii. 27 Each amino acid in the free state contains at least two ionizable sites. **1972** *Exper. Cell Res.* LXX. 122/1 The pI of the cells would..give information about the nature of ionisable groups present on the surface.

ionization[2] (ɔi:ŏnəizē·ʃən). [f. Ion + -ization.] **1.** The state of being ionized, or the process of ionizing.

1891 G. F. Fitzgerald in *Rep. Brit. Assoc. Adv. Sci. 1890* 327, I object to the term dissociation as applied to the ions in an electrolyte,... I would..appeal to both sides to adopt some neutral term such as 'ionisation' to express the state of ions in electrolytes. **1898** *Nature* 8 Dec. 142/1 The measurements of the ionisation produced by Röntgen rays in fourteen gases showed that the ionisation was connected with the chemical composition in a very simple manner. **1908** *Westm. Gaz.* 13 Mar. 2/1 Paulsen urges that the aurora is due to an immense ionisation of the upper layers of the air. **1926** A. S. Eddington *Internal Constitution of Stars* i. 10 In the actual conditions of a star the ionisation is not quite complete. **1938** R. W. Lawson tr. *Hevesy & Paneth's Man. Radioactivity* (ed. 2) i. 12 This kind of ionization of a gas, in which the ions initially produced gain sufficient energy by their acceleration in the electric field to produce fresh ions by collision, is called 'ionization by collision'. **1947** Glasstone *Elem. Physical Chem.* xiii. 422 It is essential to distinguish between 'dissociation' and 'ionization'... The term 'ionization' applies to the total number of ions, irrespective of whether they are free or are held in ion-pairs. **1962** *Newnes Conc. Encycl. Electr. Engin.* 161/2 The ionization of an atom by the removal of an outermost electron (or electrons) requires a definite amount of energy.

2. *Med.* = *cataphoresis a.

1908 *Practitioner* June 785 The modern zinc mercury ionisation treatment is a very definite way of directly applying powerful drugs to a diseased area. **1909** [see *iontophoresis]. **1934** E. P. Cumberbatch *Lect. Med. Electr.* iii. 32 Ionization is a form of treatment in which ions possessing therapeutic properties are made to migrate into the body by the agency of electro-motive force. **1944** E. B. Clayton *Electrotherapy* xi. 171 The patient should receive the ionisation reclining on a couch. **1960** B. Savage *Pract. Electrotherapy* iv. 63 Histamine ionization is used whenever there is need for a marked increase in local circulation.

3. *Comb.*: **ionization chamber,** an instrument for measuring the intensity of ionizing radiation by collecting and measuring the charge on the ions which the radiation produces in a volume of gas; **ionization constant** *Physical Chem.* = *dissociation constant* (*dissociation 2); **ionization current,** an electric current arising out of the movement, under the influence of an electric field, of ions and electrons produced in a gas; **ionization energy** = **ionization potential*; **ionization gauge,** an instrument for measuring the pressure in an evacuated vessel by ionizing the residual gas and measuring the resulting ionization current; **ionization potential,** the potential difference through which an electron must be accelerated in an electron impact experiment, or the energy required, to remove an electron in its lowest energy state from an atom or molecule of a gas.

1904 *Phil. Mag.* VIII. 721 In Rutherford's experiment ..the radioactive material was scattered over the floor of the ionization chamber. **1919** J. A. Crowther *Ions, Electrons & Ionizing Radiations* ii. 15 The gas under investigation is contained in a metal box which is connected to earth... The box and its electrodes forms what is known as an ionization chamber. **1945** *Electronic Engin.* XVII. 405 These experiments were carried out by means of ionisation chambers, like most of the early investigations on cosmic rays. **1966** *McGraw-Hill Encycl. Sci. & Technol.* VI. 92/1 If the voltage is too low, the electron avalanche never builds up, and the [Geiger] counter operates only as an ionization chamber (a device which gives the total ionization produced in the gas) or proportional counter, in which the output pulses are much smaller. **1904** T. S. Moore in R. A. Lehfeldt *Electro-Chem.* I. ii. 101 For binary electrolytes, in comparing the ionisation-constants, we are comparing the concentration necessary to produce a given degree of dissociation. **1924** J. R. Partington in H. S. Taylor *Treat. Physical Chem.* I. xi. 538 The ionization constant of water..increases very rapidly at lower temperatures, passing through a maximum about 218°. **1972** Moeller & O'Connor *Ions in Aqueous Systems* iv. 74 Inasmuch as numerical values for ionization constants are available for many weak acids and bases, it is possible to calculate ion concentrations in solutions containing these substances under a variety of conditions. **1902** *Phil. Mag.* IV. 375 The time taken to pass over 100 divisions of the scale is taken by a stop-watch. The rate of movement is a measure of the ionization-current between the plates. **1956** Hine & Brownell *Radiation Dosimetry* i. 18 The determination of dose in roentgens requires the measurement of an ionization current under saturation conditions. **1928** F. J. Fuchs tr. *Gerlach's Matter, Electr., Energy* v. 69 On account of its large diameter and high ionization energy.., this singly charged helium atom appropriates to itself very rapidly a second electron. **1940** Glasstone *Text-bk. Physical Chem.* i. 52 The ionization energy is the work required to remove an electron from its lowest level to infinity. **1973** *Sci. Amer.* Feb. 91/1 Each kind of atom has a different ionization energy depending on the number of protons, neutrons and electrons it has. **1934** *Physical Rev.* XLV. 611/2 The pressure within the vacuum chamber as recorded by an ionization gauge..is often less than 10^{-6} mm Hg. **1966** *McGraw-Hill Encycl. Sci. & Technol.* VII. 253/2 In another type of ionization gage, the gas is ionized by high-energy alpha particles emitted by a radioactive source such as radium. **1914** *Phil. Mag.* XXVIII. 753 (*heading*) Note on the ionization potential of mercury vapour. **1927** *Physical Rev.* XXIX. 287 The second ionization potential of lithium, the amount of work required in order to remove one further electron from the singly ionized lithium atom, Li^+, in the normal state 1S, has not been directly determined by the method of electron impact. **1935** J. N. Friend *Text-bk. Physical Chem.* II. viii. 362 Upon raising the P.D. still more, gaseous ionisation occurs, the lowest potential producing this being known as the ionisation potential. **1950** W. J. Moore *Physical Chem.* x. 262 The alkali metals have low ionization potentials; the inert gases, high ionization potentials. **1972** C. E. Brion in A. Maccoll *Mass Spectrometry* iii. 61 PES [*sc.* photoelectron spectroscopy] is a detector of direct photoionisation and as a result has proved to be the best and most prolific source of both inner and outer ionisation potentials.

ionize, *v.*[2] Add to def.: to produce ions in (a substance or medium). Also *absol.* (Further examples.)

1901 B. Blount *Pract. Electro-Chem.* i. 16 Solutions of moderate strength..behave as if a portion of the molecules were ionised and a portion were present as ordinary molecules. **1915** *Proc. R. Soc.* XCI. 485 The minimum energy required to ionise an atom of mercury is that acquired by an electron in passing through a fall of potential of 4·9 volts. **1916** *Physical Rev.* VIII. 386 None of the electrons emitted by the cathode are able to ionize until they have moved a fraction V_0/V of the distance toward the anode. **1927** N. V. Sidgwick *Electronic Theory of Valency* i. 10 Whenever an atom is ionized, whether by chemical combination, or by exposure under suitable conditions to the action of light, heat, or electricity, it gains or loses one or more electrons. **1947** *Sci. News* IV. 55 Through the action of various radiations the gas in this region is ionised, *i.e.*, split into electrically positive or negative particles. **1956** Hine & Brownell *Radiation Dosimetry* i. 2 Electromagnetic radiations, having energies above a few kev, ionize by virtue of the secondary electrons released when they are absorbed. **1963** C. A. McDowell *Mass Spectrometry* xii. 507 Electrons with sufficient energy can ionize molecules by the following process.

2. *intr.* To dissociate into ions; to become converted (wholly or partly) into ions.

1904 T. S. Moore in R. A. Lehfeldt *Electro-Chem.* ii. 130 Those di-acid bases which ionise in stages show the same relation between the first and second ionisation as the dibasic acids. **1930** Field & Weill *Electro-Plating* vii. 92 Copper sulphate ionises as follows: $CuSO_4 \rightleftharpoons Cu^{++} + SO_4^{--}$. **1966** *McGraw-Hill Encycl. Sci. & Technol.* VI. 60/1 At..the breakdown voltage, the nearly conducting gas ionizes and becomes a good conductor, and a self-sustained discharge is established. **1968** R. O. C. Norman *Princ. Org. Synthesis* ii. 63 There is an increase in stabilization energy when acetic acid ionizes.

3. *Med.* (*trans.*) **a.** To introduce (a substance) *into* tissue by means of medical ionization. **b.** To treat by medical ionization.

1909 *Lancet* 13 Mar. 756/2 If a person ionises cocaine into himself..the skin becomes anæsthetic. *Ibid.*, I have placed under the microscope a piece of skin ionised with calcium. **1913** H. L. Jones *Ionic Medication* iii. 84 They [*sc.* corns] should have a thorough preliminary soaking.., and then may be ionized. **1936** J. N. Dyson *Pract. Ionization* iv. 51 For certain reasons the ulcer was ionized in two parts. **1949** *Brit. Jrnl. Physical Med.* XII. 144/2 Penicillin was 'ionized' into the skin covering the paronychia.

Hence **i·onized** *ppl. a.*; **i·onizing** *vbl. sb.* and *ppl. a.*; † *ionizing potential* = *ionization potential* (s.v. *IONIZATION² 3); *ionizing radiation*, radiation which produces ionization in matter through which it passes.

1899 *Nature* 30 Nov. 114/1 The only ionising inorganic solvents hitherto found in addition to water are nitric acid and liquefied ammonia. **1902** *Phil. Mag.* IV. 704 The ionization observed in gases may be due..to the emission of an ionizing radiation from the walls of the containing vessel. **1914** *Proc. R. Soc.* A. XC. 398 It appeared to be of interest to measure the ionising potential for negative corpuscles under experimental conditions somewhat different from those previously adopted. **1919** J. A. CROWTHER (*title*) Ions, electrons, and ionizing radiations. *Ibid.* v. 52 We have already seen..that the current through an ionized gas increases with increasing electric field up to a certain maximum value known as the saturation current. **1924** S. DUSHMAN in H. S. Taylor *Treat. Physical Chem.* II. xvi. 1110 This accounts for the production of arcs in gases at voltages below the ionizing potential. **1935** *Brit. Jrnl. Radiol.* VIII. 479 The Hospital Physicist is required to measure ionising radiations covering the vast range of intensities from many röntgens per second down to comparatively weak intensities. **1935** *Discovery* Mar. 76/2 It was clear that the sun is the ionising agent. **1937** *Ibid.* Jan. 5/2 (*caption*) The transmitter sends out 50 pulses per second, which after travelling to the ionised layers are reflected and received by the receiver. **1956** A. H. COMPTON *Atomic Quest* 42 He had perfected to a high degree the Geiger counters that have become so valuable for measuring ionizing radiations. **1969** R. & E. BRECHER *Rays* xxvi. 416 Of special interest during the 1960's were the studies of what happens when ionizing radiation is absorbed by human cells. **1970** D. W. TURNER et al. *Molecular Photoelectron Spectroscopy* i. 2 This instrument possessed the advantage of..intimate juxtaposition of ionizing region and analyser entrance slit.

ionizer (əi·ŏnəizəɹ). [f. prec. + -ER¹.] That which produces ionization.

1901 *Smithsonian Contrib. Knowl.* No. 1309. 27 Instead of a single tube, two or more ionizers..are used simultaneously, each of them being charged with phosphorus. **1905** *Phil. Mag.* X. 207 An α particle is nearly twice as efficient an ionizer as the electron at its maximum efficiency. **1933** *Discovery* Apr. 107/2 This recoiling nucleus, on the other hand, is an efficient ionizer. **1967** S. W. TROMP et al. *Biometeorol.* II. II. 1025 Unavoidable by-production of chemical compounds, and poor knowledge of ion distribution around emitting electrodes, have largely hampered the design and practical use of electrical ionizers. *Ibid.*, Sufficient ion concentration can be achieved in the centre of a room by inconspicuous ionizers.

iono- (əi·ŏno-, əi,ǫ·no-), used as comb. form of (*a*) ION, (*b*) *IONOSPHERE.

ionogen (əi·ŏnodʒėn, əi,ǫ·nodʒėn). *Physical Chem.* [f. *IONO- + -GEN 1.] Any compound which exists as ions when dissolved in a solvent.

1906 A. SMITH *Introd. Gen. Inorg. Chem.* xvii. 296 The substances of the three classes which alone are ionized may be designated ionogens... The electrolytic property of ions is only one amongst many special properties of electrolytes, and the majority of these properties are chemical and have nothing to do with electrolysis. Hence we have preferred the more general word 'ionogen'. **1922** *Chem. Abstr.* XVI. 1890 (*heading*) Nature of the ionogen linkage. **1935** J. N. FRIEND *Text-bk. Physical Chem.* II. vii. 217 The undissociated molecules and the ions of an ionogen in aqueous solution constitute a system in equilibrium. **1943** *Gloss. Terms Electr. Engin.* (*B.S.I.*) 91 The term 'ionogen' has been suggested as an alternative to the term 'electrolyte' when used with this meaning [*sc.* a substance which, when dissolved in a specified solvent, produces a conducting medium].

ionogenic (əi,ǫ:nodʒe·nik), *a. Chem.* [f. *IONO- + *-GENIC.] † **a.** [ad. G. *ionogen* (O. Hinsberg 1911, in *Jrnl. f. prakt. Chem.* LXXIV. 179).] Of an atom or radical: promoting ionization elsewhere in the molecule of which it forms part. *Obs.*

1912 *Chem. Abstr.* VI. 351 (*heading*) Ionogenic atomic groups and atoms.

b. Capable of being ionized chemically.

1922 J. LOEB *Proteins & Theory Colloidal Behavior* ii. 34 If we wish to prepare gelatin or casein free from ionogenic impurities, we must bring these proteins..to the isoelectric point and then wash them. **1955** *Federation Proc.* XIV. 735/1 The electrophoretic mobility of a given protein is a function of the net electric charge and hence of the number of ionogenic groups. **1967** *Jrnl. Cell. Physiol.* LXIX. 287/1 It appears likely..that ionogenic groups of RNA, i.e. phosphate groups, are present at the electrokinetic surfaces of the two cell types. **1972** *Materials & Technol.* V. x. 275 Of the water-soluble surface-active agents.., some undergo ionization when dissolved in water whereas others do not ionize. On this basis, they are divided into two broad groups, namely ionogenic, or ion-forming, and non-ionogenic.

ionogram (əi,ǫ·nogræm). [f. *IONO- + -GRAM.] **1.** A record of radio pulses received by an ionosonde following their reflection by the ionosphere.

1955 *Sci. Amer.* Sept. 128/2 To make an ionosonde record (ionogram) the transmitter and receiver are tuned rapidly through a range of frequencies. The echoes received are displayed on a cathode ray oscilloscope and photographed. **1963** *Times* 15 Feb. 7/7 Some 70,000 'ionograms'—graphs of frequency against depth of reflection below the satellite—have so far been obtained and are of good quality. **1967** *Encycl. Dict. Physics* Suppl. II. 240 The resulting 'ionogram', showing time delay as a function of frequency, can be interpreted as electron density as a function of 'virtual height' in the atmosphere.

2. *Chem.* The result of an ionographic separation, usually a series of spots or bands on the support medium.

1955 *Federation Proc.* XIV. 736/2 To determine the location of the various zones on an ionogram occupied by the individual components of a mixture, a number of procedures have been developed. **1963** *Proc. 17th World Veterinary Congr.* II. 1107 The ionogram of serum represents the electrolytic compounds of extracellular fluid in dogs. **1970** *Biol. Abstr.* LI. 354/1 Pre- and post-operational ionograms should be examined, in patients with a funnel-shaped deformation of the thorax.

ionography (əi,ŏnǫ·grăfi). *Chem.* [f. *IONO- + -GRAPHY.] The migration of ions or charged colloidal particles in a buffer solution held on a support (usu. filter paper) under the influence of an electric field, esp. as used to separate the components of a mixture; = *electrochromatography* (s.v. *ELECTRO-).

1950 H. J. MCDONALD et al. in *Science* 25 Aug. 228/1 It is suggested that the term *ionography* be used to describe the technique. *Ibid.* 228/2 The isoelectric point of proteins can be determined, and..amino acids and proteins can be separated by ionography. **1955** *Federation Proc.* XIV. 736/2 Although paper has been used as the stabilizer in over 95% of the published articles on ionography, the use of starch has been found to have certain advantages in special cases. **1957** *Texas Rep. Biol. & Med.* XV. 235 An investigation into the lipoprotein levels of serum from patients was undertaken, using the technique of ionography.

Hence **ionogra·phic** *a.*, of or pertaining to ionography.

1955 *Federation Proc.* XIV. 733/1 The term 'ionographic apparatus' will refer to the instrument used in carrying out ionographic separations. **1968** *Compar. Biochem. & Physiol.* XXV. 727 The ionographic properties of the hemoglobins from ten species of *Peromyscus* have been compared.

ionomer (əi,ǫ·nŏməɹ). [f. *IONO- + *-MER.] Any of a class of thermoplastics in which there is ionic bonding between the polymer chains.

1964 *Mod. Plastics* Sept. 98/1 E. I. du Pont de Nemours & Co. Inc. has finally announced the introduction of an exciting new transparent thermoplastic it calls 'ionomer'. *Ibid.* 98/2 The tensile strength of many grades of ionomers is higher than that of any commercial polyolefin. **1967** H. F. MARK et al. *Encycl. Polymer Sci. & Technol.* VI. 395 Another class of copolymers with exceptional properties are the ionomers, which are essentially the metal salts of ethylene-acrylic acid copolymers. The resulting ionic bonding makes these materials a special class. **1967** *Times Rev. Industry* May 76/2 Among the newer plastics materials offered for use in packaging are ionomer and ethylene/vinyl acetate (EVA) films for meat and other retail produce.

ionone (əi·ŏnōᵘn). *Chem.* [ad. G. *ionon* (Tiemann & Krüger 1893, in *Ber. d. Deut. Chem. Ges.* XXVI. 2693), f. Gr. ἴον violet + -ONE.] **a.** Either of two liquids (*α-ionone* and *β-ionone*) that are used in perfumery for their strong odour of violets and are isomeric cyclic ketones, $(CH_3)_3C_6H_6CH{:}CH{\cdot}CO{\cdot}CH_3$, differing only in the position of the double bond in the C_6H_6 ring.

Ionone is a proprietary name in the U.S.

1894 *Jrnl. Chem. Soc.* LXVI. I. 82 When pseudoionone is heated with dilute sulphuric acid and a little glycerol, it is converted into the isomeric ionone... This substance boils at 126–128° at 12 mm. **1900** *Ibid.* LXXVIII. II. 375 In order to separate and detect the α- and β-ionone, the crude ketones are boiled with a little alcohol and an aqueous solution of sodium sulphite. **1929** *Encycl. Brit.* VIII. 721/1 Ionone, derived from citral isolated from oil of lemongrass,..is the basis of all violet scents, and perhaps the most important of all synthetic perfumes. **1943** *Jrnl. Amer. Chem. Soc.* LXV. 2062/2 The resolution of *dl*-α-ionone was accomplished by means of *l*-menthydrazone. **1957** T. MOORE *Vitamin A* x. 100 The synthesis of vitamin A from β-ionone obviously requires the formation of a longer side chain to the trimethylcyclohexenyl ring. **1971** *New Scientist* 6 May 351/2 The pleasant-smelling ionones are breakdown products of carotenoids.

b. *ψ*-**ionone** (*pseudoionone*), $C_{13}H_{20}O$, a third isomer which can be cyclized to form α- and β-ionones.

1897 *Jrnl. Chem. Soc.* LXXII. I. 538 Ionone is..easily prepared by the action of sulphuric acid on ψ-ionone, the latter substance being produced from geranaldehyde by condensation with acetone. **1969** W. TEMPLETON *Introd. Chem. Terpenoids & Steroids* iii. 46 Similarly ψ-ionone.., obtained by condensation of citral with acetone, cyclizes under acidic conditions to α- and β-ionones, which are found in Nature.

ionophone (əi,ǫ·nofōᵘn). [f. *IONO- + *-PHONE.] A type of loudspeaker using the vibrations of a volume of ionized air in place of those of a diaphragm.

1955 *Sci. News Let.* 23 July 63/3 The vibrating pocket of air in the new loud speaker, called the Ionophone, will create the sound without this setback, the designers say. **1962** A. NISBETT *Technique Sound Studio* 257 *Ionophone*, a type of loudspeaker which has no moving parts.

ionophore (əi,ǫ·nofōᵊɹ). *Biol.* [f. *IONO- + -PHORE.] An agent which is able to transport ions across a lipid membrane in a cell. So **ionopho·rous** *a.*

1967 B. C. PRESSMAN et al. in *Proc. Nat. Acad. Sci.* LVIII. 1954 It is our conclusion that both the valinomycin and the nigericin classes of antibiotics induce alkali ion permeability in mitochondrial and other systems by carrying ions across lipid barriers as lipid-soluble complexes. Accordingly we propose to classify them generically as ionophores or ionophorous agents. **1970** *Jrnl. Biol. Chem.* CCXLV. 5606 (*heading*) Effect of nucleotides on the transport of alkali metal cations catalyzed by ionophorous antibiotics across mitochondrial membranes. **1971** *Nature* 5 Nov. 12/2 Cyclic peptides such as gramicidin-S or alamethecin which can act as ionophores ..in membranes.

ionophoresis (əi,ǫ:nofŏrī·sis). *Biochem.* [f. *IONO- + *-PHORESIS.] The migration of ions in solution under the influence of an electric field, esp. as used to separate the components of a mixture. Hence **io:nophore·tic** *a.*

1945 *Adv. Protein Chem.* II. 31 We propose to use *ionophoresis* to describe processes concerned with the movement in an electric field of relatively small ions, *electrophoresis* for movement of large molecules and particles. *Ibid.* (*heading*), Ionophoretic methods. **1949** ABRAHAM & HEATLEY in H. W. Florey et al. *Antibiotics* I. ii. 106 An apparatus in which ionophoretic separation takes place in a slab of silica jelly, the substances under investigation being inlaid in a gutter at right angles to the length of the slab. **1956** *Nature* 25 Feb. 393/2 A number of primary aliphatic amines can readily be separated.. by ionophoresis on paper at *p*H 7 using collidine acetate buffer, and the spots revealed with ninhydrin. **1967** *Biol. Abstr.* XLVIII. 895/2 (*heading*) Detection of metal cations in soil by chromatographic and ionophoretic techniques. **1971** *European Jrnl. Biochem.* XIX. 125/1 Ionophoresis in the second dimension was carried out on an 85 × 45 cm sheet of DEAE-cellulose paper..in 7% formic acid.

ionosonde (əi,ǫ·nosǫnd). [f. *IONO- + *SONDE.] An instrument for obtaining information about the ionosphere by transmitting a succession of radio pulses into it at different frequencies and recording their echoes.

1955 *Sci. Amer.* Sept. 128/2 The instrument used to explore the ionosphere is called an ionosonde. **1962** *Flight International* LXXXI. 681/2 The 95 (US) tons of water ballast was released by explosion of the vehicle at a height of about 65 miles 160 sec after lift-off, and a large white cloud of ice particles was formed... Observations were made by ground-based cameras, aircraft at various altitudes, ground radars and ionosondes. **1971** *New Scientist* 18 Mar. 598/2 Reflections of a swept-frequency pulse of radio waves by the ionosphere—the ionosonde technique—provide information about the distribution and density of its constituent electrons. **1972** *Science* 5 May 464/1 Lightweight ionosondes have been placed in satellites.

ionosphere (əi,ǫ·nŏsfiᵊɹ). [f. *IONO- + *-SPHERE.] A region of the outer atmosphere, beginning at a height of 50–80 km. (30–50 miles), which contains many ions and free electrons and is capable of reflecting radio waves; also, a corresponding region above other planets.

1926 R. A. WATSON-WATT *Let.* 8 Nov. in *Nature* (1969) 13 Dec. 1096/1 We have in quite recent years seen the universal adoption of the term 'stratosphere'..and.. the companion term 'troposphere'... The term 'ionosphere', for the region in which the main characteristic is large scale ionisation with considerable mean free paths, appears appropriate as an addition to this series. **1929** —— in *Q. Jrnl. R. Meteorol. Soc.* LV. 278, I have suggested the name ionosphere to make the systematic group troposphere, stratosphere, ionosphere, but meanwhile the term 'upper conducting layers' seems to hold the field. **1932** E. V. APPLETON in *Jrnl. Inst. Electr. Engin.* LXXI. 642 (*heading*) Wireless studies of the ionosphere. [*Note*] This is a convenient term, suggested by Mr R. A. Watson Watt, connoting the ionized regions of the upper atmosphere. **1934** *Times* 23 Feb. 20/2 Beyond lies the ionosphere, and of the layers in this we learn from the paths of radio-waves. **1947** *Sci. News* IV. 55 Sir Edward Appleton, whom one might call the Master of the Ionosphere, distinguishes in it the following layers: the complex F-layer 250 kilometres (150 miles) above the earth (also called the Appleton layer), the E- or Heaviside layer at about 100 km., and further down the D layer which is little noticeable in normal times, but whose ionisation is considerably increased as long as a solar flare is visible. **1955** *Times* 15 June 5/5 The ionosphere, a belt of gases ionized.. by ultra violet rays from the sun and lying between 70km. and 300–400km. above the earth, makes medium and short wave radio propagation possible by reflecting signals back to earth. **1959** DAVIES & PALMER *Radio Stud. Universe* iv. 66 This is difficult to explain if Jupiter is surrounded by an ionosphere similar to the Earth's. **1973** *Physics Bull.* Feb. 93/3 Spacecraft observations of the ionospheres of Venus and Mars.

ionospheric (əi,ǫ:nosfe·rik), *a.* [f. prec. + -IC.] Of, pertaining to, or involving the ionosphere.

1933 *Proc. Physical Soc.* XLV. 673 (*heading*) On two

methods of ionospheric investigation. **1937** *Discovery* Jan. 3/2 (*heading*) The base hut in Brandy Bay..with the wireless and ionospheric research masts behind. **1955** *Sci. News Let.* 28 May 339 Most of the radio frequency power in the new method is lost, but some is scattered by the lower part of the E region... This portion is received hundreds of miles away by high-gain antennas aimed at the exact spot. The system is called either forward scatter, or FPIS, for forward propagation by ionospheric scatter. **1958** *Times* 8 Nov. 4/7 Assurances that the ionospheric forward-scatter radio station..will not represent a radiation hazard to the public have now been given officially. **1966** *Electronics* 3 Oct. 181 Boelkow has designed a fixed-wing rocket—intended for ionospheric studies—that can be recovered from 48 miles.

Hence **io:nosphe·rically** *adv.*, by the ionosphere.

1955 *Proc. IRE* XLIII. 1174 Ionospherically propagated waves. **1971** *Nature Physical Sci.* 5 Apr. 125/1 There is no fading such as occurs in ionospherically reflected sky waves.

ionospherist (əi,ǫ:nosfe·rist). *rare.* [f. as prec. + -IST 4 b.] One who studies the ionosphere. Also **io:nosphe·ricist**, in the same sense.

1933 *Proc. R. Soc.* A. CXLI. 715, I should be most disappointed if time permitted only those whom I might call the professional ionospherists to take part in the discussion. **1955** *Jrnl. Brit. Interplanetary Soc.* XIV. 19 Moving concentrations of electrons account for the other main type of irregularity [in the ionosphere]; ionospherists seem never to grow tired of studying them. **1971** *Nature* 1 Jan. 15/3 The angle of arrival of radio waves is determined by ionosphericists using spaced receivers.

ionotropy (əiŏnǫ·trŏpi). *Chem.* [f. *IONO- + Gr. -τροπ-ία turning (f. τρέπειν to turn) + -Y³.] **1.** Tautomerism or tautomeric change, regarded as occurring through the detachment of an ion from the molecule concerned followed by a redistribution of the charge in the resulting molecular ion and reattachment of the ion to a different part of it. *rare.*

[**1926** T. M. LOWRY in *Inst. Internat. de Chimie Solvay: Deuxième Conseil de Chimie..1925..Rapports et Discussions* 151 Les transformations dans lesquelles des radicaux autres que l'hydrogène se déplacent peuvent.. être formulées comme comportant une *migration d'ions* et peuvent être désignées sous le terme d'*ionotropie*.] **1928** *Ann. Rep. Progr. Chem. 1927* XXIV. 107 Lowry has also used the word 'ionotropy' (Report of the Second Solvay Conference 1925, p. 182). **1955** *Chem. Abstr.* XLIX. 40 Reactions of acids and bases can be described by assuming the fundamental reaction of 'ionotropy'.

2. [ad. G. *ionotropie* (H. Thiele 1947, in *Naturwiss.* XXXIV. 123/2.] The ordering of particles in a gel that results when an electrolyte is added to a colloidal suspension.

1949 *Chem. Abstr.* XLIII. 4539 Usually the sign of the ionotropy is the same as that of flow anisotropy. **1958** *Jrnl. Physical Chem.* LXII. 1277/1 The phenomenon of ionotropy can be simply demonstrated by placing a thin layer of 2% suspension of salt-free Wyoming bentonite on a glass slide and then placing a crystal of sodium chloride on the layer of suspension.

Hence **ionotro·pic** *a.*, exhibiting ionotropy (sense *2).

1944 *Handbk. Descr.* in *Chem. & Chem. Engin.* (U.S. Manpower Comm.) Nov. 66 Polarity, ionotropic change and conjugation. **1954** H. THIELE in *Discussions Faraday Soc.* XVIII. 295 Since ionotropic gels are formed by chemical or electrical forces the degree of order decreases with decreasing dissociation of the ionic groups of the colloid particles. **1969** *Jrnl. Biomed. Materials Res.* III. 432 The ordered ionotropic structure of the fibers of the polyelectrolyte directs the mineralization and calcification in the same manner as in biological tissue.

iontophoresis (əi,ǫn:tofŏrī·sis). *Med.* [f. Gr. ἰοντ-, ἰόν pr. pple. of ἰέναι to go + -O + *-PHORESIS.] = *CATAPHORESIS a.

1909 *Lancet* 13 Mar. 756/1 By medical ionisation, or, as it is called abroad..'iontophoresis', we mean the introduction of ions of different sorts into the tissues of the human body. **1937** *New Eng. Jrnl. Med.* 5 Aug. 202/2 Iontophoresis is a term used for denoting the act of driving negatively or positively charged ions into the subcutaneous tissues by means of a galvanic or direct current of electricity. **1949** A. HUXLEY *Let.* 26 Feb. (1969) 591 Dr Gustav Erlanger..has developed a method for treating the eyes by iontophoresis and gets remarkable results. **1971** *Jrnl. Physiol.* CCXV. 199 DNP and other metabolic inhibitors were tested on cortical neurones by iontophoresis from micropipettes.

Hence **io:ntophore·tic** *a.*, of, pertaining to, or employing iontophoresis; **io:ntophore·tically** *adv.*, by means of iontophoresis.

1937 *New Eng. Jrnl. Med.* 5 Aug. 204/1 The changes were not of sufficient degree to make us feel that the iontophoretic administration of acetyl β-methylcholine produced constant changes in the blood-sugar level. **1947** ABRAMOWITSCH & NEOUSSIKINE *Treatm. by Ion Transfer* iii. 68 Novocain was introduced iontophoretically. **1955** *Jrnl. Investigative Dermatol.* XXV. 223 It seemed logical to first find out whether or not the thorium X..could be introduced into a test material by the galvanic iontophoretic current. **1964** D. M. GORDON *Med. Managem. Ocular Dis.* iv. 20 Many of the systemic preparations can be..administered iontophoretically. **1971** *Jrnl. Cell. Physiol.* LXXVII. 339/2 Iontophoretic application of materials was accomplished by connecting a variable voltage source with a micropipette containing the material. **1972** *Science* 5 May 515/1 The ACh was applied iontophoretically by a micropipette containing a 4M solution of ACh chloride.

Iowan (əi·ŏwăn), *sb.* and *a.* Also formerly **Iowaian** (from the pronunciation *Ioway* of *Iowa*, which is still heard). [f. *Iowa*, name of one of the United States of America, formerly of a tribe of Indians inhabiting Iowa and Minnesota.] **A.** *sb.* A native or inhabitant of Iowa.

c **1848** W. WHITMAN in *Amer. Speech* (1961) XXXVI. 297 Iowans *Gophers*. **1856** N. H. PARKER *Iowa as it Is* 56 Our ferry is busy all hours in passing over the large canvas-backed wagons, densely populated with becoming Iowaians. **1928** *Glasgow Herald* 6 Apr. 12 Iowans..used to console themselves by telling the world..that their largest city had the greatest consumption of ice-cream per head of population. **1964** MRS. L. B. JOHNSON *White House Diary* 21 Nov. (1970) 208 Just enough places for the Iowans and the kinfolks.

B. *adj.* Of, pertaining to, or designating what was formerly considered the fourth Pleistocene glaciation of North America, but which is now considered the earliest phase of the Wisconsin glaciation. Also *absol.*, the Iowan glaciation or the deposits it produced.

1894 in J. Geikie *Gt. Ice Age* (ed. 3) xlii. 756 It may.. be the equivalent of the next later formation—the East-Iowan. *Ibid.* 760 The East-Iowan till-sheet..is not usually bordered by any definite terminal moraine. *Ibid.*, The designation 'East-Iowan formation' is chosen because it has been most carefully worked out by Mr. McGee in north-eastern Iowa. **1896**, etc. [see *ILLINOIAN *a.*]. **1968** R. W. FAIRBRIDGE *Encycl. Geomorphol.* 292/2 In the Wisconsin phase, the Iowan Lobe was considerably smaller.

ipecacuanha. 5. *ipecacuanha lozenge* (examples).

1847 F. A. KEMBLE *Let.* 7 Mar. in *Rec. Later Life* (1882) III. 173, I..am swallowing ipecacuanha lozenges by the gross. **1926–7** *Army & Navy Stores Catal.* p. xlix, Ipecacuanha lozenges.

ipid (i·pid, əi·pid), *sb.* and *a.* [a. mod.L. *Ipidæ*, f. Gr. ἰπ-, ἴψ woodworm.] **a.** *sb.* A bark-beetle of the family Ipidæ, which is now included in the Curculionidæ. **b.** *adj.* Of or pertaining to a beetle of this type. Cf. *SCOLYTID.

1866 E. C. RYE *Brit. Beetles* xii. 100 The Ipides have a single lobe to the maxillæ. **1922** W. M. WHEELER *Soc. Life Insects* 40 The females of the Ipid ambrosia beetles carry the fungus in the fore part of the stomach.

ipiti (i·piti). *S. Afr.* Also **impiti, ipiete.** [Zulu *i-mpiti*.] The blue duiker, *Cephalophus monticola*, the smallest South African antelope.

1836 R. M. MARTIN *Hist. S. Afr.* I. iv. 138 They [*sc.* the 'Caffres'] display considerable taste in the arrangement of their dress, particularly for the head, which is covered by a turban made of the skin of the '*ipiete*', a species of antelope. **1878** T. J. LUCAS *Camp Life & Sport S. Afr.* viii. 101 Immediately around Pietermaritzberg the game is comparatively scarce, consisting principally of the small Ipite Bok, a graceful little antelope. **1879** R. J. ATCHERLEY *Trip to Boerland* 26, I shot a few small buck known as *impiti*..not much larger than a hare. **1905** D. BLACKBURN *Richard Hartley* xiii. 244 The beautiful little ipiti, no bigger than a toy-terrier and quite as sprightly and alert. **1926** R. LYDEKKER *Game Animals Afr.* (ed. 2) 143 (*heading*) Blue Buck or Blue Duiker... Ipiti, Zulu. **1934** B. I. BUCHANAN *Pioneer Days Natal* ii. 33 Twice we saw some of the fairy bucks—the darling little ipiti—which immediately won our hearts.

ipoh (ī·po). Also **Ipoh.** [Malay.] The upas tree, *Antiaris toxicaria*, or a creeping shrub, *Strychnos ignatii*, both of which are native to SE. Asia and have a poisonous sap; also, the poison itself.

1779 T. FORREST *Voy. New Guinea* I. xviii. 369 They [*sc.* the Sooloos] are acquainted with a subtle poison called Ippoo, the juice of a tree, in which they dip small darts. **1820** J. CRAWFURD *Hist. Indian Archipelago* I. IV. 468 The Malays call this last [shrub] *Ipoh*. **1836** J. LOW *Diss. Soil & Agric. Penang* iv. 206 Many arrows tipped with Ipoh were given to me by the different tribes in Perak. **1839** T. J. NEWBOLD *Pol. & Statistical Acct. Straits of Malacca* I. vii. 444 The Upas tree of the Javanese, or the Ipoh of the Malays, is found though rarely in the forests. *Ibid.* II. xii. 211, I had the greatest difficulty to prevail on one of this wild race to part with..a small quantity of the ipoh poison. **1907** *Q. Rev.* July 190 The poison for their arrows is obtained from the ipoh or upas tree. **1958** J. SLIMMING *Temiar Jungle* v. 78 The poison with which the darts are smeared is a mixture of the saps from the Ipoh tree and the Ipoh creeper. **1965** C. SHUTTLEWORTH *Malayan Safari* ii. 27 Perhaps the most dangerous poison is found in the sap of the ipoh tree from which the aborigines obtain poison for their blowpipe darts.

iproniazid (əiprǫnəi·ăzid). *Pharm.* [f. *i*(*so*)-*pro*(*pyl* (s.v. ISO- b) + *ISO)NIAZID (iproniazid being the isopropyl derivative of isoniazid).] A derivative of isoniazid which was used for a time (usually as the phosphate, a white crystalline powder) in the treatment of tuberculosis and as an antidepressant; 1-isonicotinyl-2-isopropyl hydrazine, $(CH_3)_2CH{\cdot}NH{\cdot}NH{\cdot}CO{\cdot}C_5H_4N$.

1953 *Amer. Rev. Tuberculosis & Pulmonary Dis.* LXVII. 214 A definite pattern of withdrawal has been observed following discontinuance of isoniazid and iproniazid therapy. **1960** *Times* 4 Mar. 15/7 Yet another of the psychotropic drugs now being used, iproniazid, reverses the effect of reserpine. **1968** W. C. BOWMAN et al. *Textbk. Pharmacol.* xxiii. 623 The antitubercular drug iproniazid ..possesses marked antidepressant activity and inhibits MAO [*sc.* monoamine oxidase]... Iproniazid may cause a number of side-effects, the most serious being acute hepatic necrosis which is occasionally fatal. Because of this, the use of iproniazid as an antidepressant has been discontinued. **1971** *Brit. Med. Bull.* XXVII. 28/2 Hypertensive attacks were described in patients who were given iproniazid for tuberculosis.

ipsilateral (ipsilæ·tĕrăl), *a.* Formerly also **ipse-, ipsolateral.** [*ipselateral* f. L. *ipse* self + LATERAL *a.*, after CONTRALATERAL *a.*; *ipsi-*, *ipsolateral* app. formed by alteration.] Belonging to or occurring on the same side of the body; connecting two parts on the same side.

1907 *Jrnl. Physiol.* XXXVI. 187 On then eliciting the ipselateral 'flexion-reflex' this latter will..appear in the full diphasic form. **1910** *Ibid.* XL. 30 This reflex contraction is readily seen to be inhibited by stimulation of the ipsilateral musculo-cutaneous nerve. **1911** STEDMAN *Med. Dict.* 438/1 Ipsolateral. **1917** *Jrnl. Physiol.* LI. 405 The ipsilateral hind-foot is brought forward to scratch the stimulated part. **1962** J. W. PATTERSON et al. in A. Pirie *Lens Metabolism Rel. Cataract* 413 The initial cataract was contralateral in 9 animals, ipsilateral in 4 animals, and in 3 animals it developed simultaneously in both eyes. **1972** *Sci. Amer.* Aug. 84/3 The rabbit, with only a tiny binocular portion in its visual field, has a very small number of ipsilateral, or uncrossed, fibers in the optic chiasm.

Hence **ipsila·terally** *adv.*, on the same or on one side of the body.

1950 *Merck's Man. Materia Med.* (ed. 8) 300 It [*sc.* the pain] may be referred ipsilaterally to the ear or to an adjacent tooth. **1962** J. W. PATTERSON et al. in A. Pirie *Lens Metabolism Rel. Cataract* 413, 14 developed cataracts contralaterally and 2 ipsilaterally. **1968** A. SOULAIRAC et al. *Pain* 39 A number of fibers which ascend ipsilaterally to their origin.

ipsissima verba (ipsi·simă vǝ̄·ɹbă). [L.] The precise words used by a writer or speaker.

1807 SOUTHEY *Lett.* (1856) II. 40 Last night I was in too much haste to look for the *ipsissima verba* of Fuller. **1834** *Edin. Rev.* Apr. 151 We..shall, therefore,..treat the reader to our author's *ipsissima verba*. **1886** *Athenæum* 13 Nov. 630/1 An assurance that the extracts contain the *ipsissima verba* of the poet would be..valuable. **1888** W. JAMES *Let.* 2 Jan. in R. B. Perry *Tht. & Char. W. James* (1935) I. 403 Your *ipsissima verba* of latest date were what I required. **1931** *Times Lit. Suppl.* 28 May 416/2 An undergraduate named Brauer,..who.. seems to have taken down the *ipsissima verba* of Kant with almost impeccable accuracy. **1937** *Mind* XLVI. 302 Mr. Paton's interpretations..where they exceed Kant's *ipsissima verba*..seem to be careful and reasonable. **1961** A. HUXLEY *Let.* 21 Feb. (1969) 904 D. H. L[awrence]'s *ipsissima verba* are: 'The hot stinging centrality of the goose on the cold shifting flux of mud and waters.' **1972** J. A. W. BENNETT *Piers Plowman* p. ix, Some of my notes may retain vestiges of his *ipsissima verba*.

ipso jure (i·pso dʒūᵊ·ri), *advb. phr.* [L.] By the operation of the law itself.

1909 in WEBSTER. **1913** *Act* 3 & 4 *Geo. V* c. 20 § 97 The act and warrant of confirmation in favour of the trustee shall ipso jure transfer to and vest in him..the whole property of the debtor.

-ir-. *Chem.* [See quot.] A formative element in the names of three-membered heterocyclic ring systems. A systematic nomenclature is generated by appending to *-ir-* a prefix (or prefixes) representing the hetero-atom(s), and one of the suffixes -ANE 2, -ENE, *-IDINE c, or *-INE⁵, to indicate saturation or unsaturation and the presence or absence of a nitrogen atom in the ring. (Cf. *-IRANE, *-IRENE, *-IRIDINE, *-IRINE.)

1928 A. M. PATTERSON in *Jrnl. Amer. Chem. Soc.* L. 3079 With a precedent already set for *et* (four members, compare '*tetra*') and with *ol* (five members) and *in* (six members) already confirmed by usage, it would only be necessary to supply three syllables more to cover all but rare cases. These might be *ir* (from '*tri*', reversed) for three members. *Ibid.*, Examples of the resulting names and combining forms are: oxirene, oxirane (ethylene oxide).

-irane. *Chem.* [See *-IR-.] A suffix used systematically to form the names of saturated heterocyclic monocyclic compounds having a three-membered ring containing no nitrogen atom. Cf. *-IR-.

Iranian, *a.* and *sb.* Add: **B.** *sb.* **2.** The Iranian language.

1885 *Encycl. Brit.* XVIII. 655/2 Afghan..is at bottom a pure Iranian language, not merely intermediate between Iranian and Indian. **1898** R. BROWN *Semitic Influence Hellenic Mythol.* III. iii. 89 Its truth does not depend upon the fact that Iranian and Greek are two dialects of an

original common speech. **1952** G. SARTON *Hist. Sci.* I. xii. 312 Scythian was probably a form of Iranian, the northwestern branch of it. **1966** *Chambers's Encycl.* VII. 705/2 The nations speaking Iranian are divided linguistically into two main groups.

So **Ira·nicize, Ira·nize** *v. trans.*, to make Iranian in character, etc.

1899 W. Z. RIPLEY *Races of Europe* xv. 420 The Azerbeidjians..have become much Iranized by contact with the dolichocephalic peoples of this region. **1945** *Jrnl. R. Anthrop. Inst.* LXXV. 74/1 There is some reason to believe that the Daylamites were of pre-Iranian origin, but by the tenth century they were iranicized, although not entirely assimilated to the dominant race. **1948** D. DIRINGER *Alphabet* 316 They [*sc.* Uighurs]..may conveniently be called Iranized Turks. **1961** L. F. BROSNAHAN *Sounds of Lang.* viii. 177 A palatalisation..occurs in most of the Turco-Tartar languages..but not in..the iranised Uzbek dialects.

Iraqi (irā·ki). Also **Iraki.** [Arab.: f. *'irāq* (see def.) + *-ī* adj. suffix.] A native or inhabitant of Iraq, a republic in the Persian Gulf, formerly (before 23 Aug. 1921) known as Mesopotamia: a kingdom from 1921 until the assassination of Faisal II in 1958 when it became a republic; the capital is Baghdad. Also as *adj.*, of or belonging to Iraq or its inhabitants. Also (now *Obs.?*) **Ira·q(u)ian** *sb.* and *a.*

1777 J. RICHARDSON *Dict. Persian, Arabic & Eng.* I. p. xliv/2 They [*sc.* Arabians] have three principal *perdes* or modes; called the *Isphahani*, which appears to be the original Persian melody; the *Iraki* or the Babylonish; and the *Hejazi* or the Arabian. **1824** J. MORIER *Adventures Hajji Baba* II. xi. 163 'Untie the string of your trowsers,' said an old Irâki. **1923** *Glasgow Herald* 9 Apr. 10 The Iraqians declare that it took the British all their time to win. **1924** *Blackw. Mag.* Mar. 345/1 The administration is now in the hands of Iraqis, assisted by a mere handful of British advisers. **1927** *Observer* 7 Aug. 13/5 An addition of Iraqi or Syrian territory to the Turkish Republic would constitute a danger to his State. *Ibid.*, The Iraqi movement towards Westernisation. **1939** L. H. GRAY *Foundations of Language* 364 [Arabic] was divided into several dialects, of which only that of Mekkah has survived, this being the parent of a large number of modern vernaculars, notably Arabian.., Irāqian (Baghdādh, Mōṣul, Mardīn), [etc.]. **1958** *Economist* 26 July 267 Among Iraqis in general there were..widespread desires to be Arab and to be neutral. *Ibid.* 268 No Iraqi statement has been made concerning withdrawal from the [Baghdad] pact. **1965** *Listener* 17 June 900/1 Zamal, an Iraki..was ..expected to be the life and soul of a diplomatic party. **1973** *Times* 16 Apr. 14/1 Such a flexing of Iraqi muscles would (if taken too far) provide the Shah with ample pretext to cut the Iraqis down to size.

So **Ira·qize** *v. trans.*, to make Iraqi in character, etc.; hence **Iraqiza·tion.**

1959 *Economist* 23 May 751/1 'Iraqisation'..aims at a system in which 'no job is done by a foreigner if an Iraqi can do it as well'. *Ibid.* 752/1 The company is prepared to help in its exploitation on behalf of a fully 'Iraqised' industry. **1961** *Ann. Reg. 1960* 311 Prolonged conversations took place..on..increased 'iraqization' of senior staff. **1964** *Ann. Reg. 1963* 312 Already 633 commercial agencies..had 'Iraqized' themselves by allocating 51 per cent of their shareholding to Iraqis.

Ireland (əiᵊ·ɹlănd). The name of a classical scholarship at Oxford University founded in 1825 by John *Ireland*, D.D. (1761–1842), of Oriel College, Dean of Westminster.

1861 J. A. SYMONDS *Let.* 13 Mar. (1967) I. 282 We hope to secure the Ireland too this Term. If we do not, we shall be in a poor way. **1877** O. WILDE *Let.* Mar. (1962) 32, I have been in for 'the Ireland' and of course lost it: on six weeks' reading I could not expect to get a prize for which men work two and three years. **1951** M. KENNEDY *Lucy Carmichael* II. 80 He really is clever; he got the Ireland or the Hertford, I forget which, at Oxford. **1953** E. BARKER *Age & Youth* II. iii. 317 A year later, when I tried my luck for the Ireland, the king of classical scholarships, I had less confidence. **1972** *Oxf. Univ. Cal. 1972–73* 216 Dean Ireland's Scholarship... Value: £120. Awarded annually in Michaelmas Term after an examination... The examination is the same as that for the Craven Scholarships and the person elected to the Ireland Scholarship is, if not already a Craven Scholar, elected to the first Craven Scholarship.

irene (əiᵊ·rīn). *Chem.* [ad. G. *iren* (Tiemann & Krüger 1893, in *Ber. d. Deut. Chem. Ges.* XXVI. 2682), f. G. *ir-on* *IRONE + *-en* -ENE.] A colourless, liquid, bicyclic hydrocarbon, $C_{14}H_{20}$, obtained by the dehydration of irone.

1894 *Jrnl. Chem. Soc.* LXVI. I. 81 When treated with hydriodic acid and phosphorus, irone loses a molecule of water, and forms irene, $C_{13}H_{18}$, which is a colourless oil boiling at 113–115° (9 mm.). **1938** *Jrnl. Amer. Chem. Soc.* LX. 933 The 1, 1, 2, 6-tetramethyltetralin structure for irene..has been supported by the synthesis of this tetralin from *m*-bromotoluene.

-irene. *Chem.* [See *-IR-.] A suffix used systematically to form the names of unsaturated heterocyclic monocyclic compounds having a three-membered ring containing no nitrogen atom. Cf. *-IR-.

iretol (əiᵊ·rėtǫl). *Chem.* [a. G. *iretol* (De Laire & Tiemann 1893, in *Ber. d. Deut. Chem. Ges.* XXVI. 2015), f. G. *ir-igenin* *IRIGENIN: see -ET and -OL.] 2,4,6-Trihydroxyanisole, $C_7H_8O_4$, a crystalline compound obtained by alkaline hydrolysis of irigenin.

1894 *Jrnl. Chem. Soc.* LXVI. I. 48 Iretol, $C_6H_2(OH)_3{\cdot}OMe$,..is ultimately obtained in white needles melting at 186°. **1937** *Jrnl. Amer. Chem. Soc.* LIX. 933/1 Iretol is stable when pure and dry, but readily turns dark if exposed to moisture and oxygen. **1960** *Tetrahedron Let.* v. 8 Iridonitrile and iretol (Hoesch reaction) yielded the deoxybenzoin (I).

Irgun (iɹgu·n). [mod.Heb., organization; in full *Irgun Zvai Leumi* national military organization (also used).] Name of a militant right-wing Zionist organization. Hence **Irgu·nist**, a member of this organization.

1946 KOESTLER *Thieves in Night* 243 The new organisation, 'Irgun', was numerically smaller and organised on the conspiratorial lines of a terrorist underground movement. **1947** *Ann. Reg. 1946* 299 The Irgun Zvai Leumi and the Stern Group had worked with the Hagana High Command in certain of these operations. **1949** KOESTLER *Promise & Fulfilment* II. v. 260 A woman of sixty-five, who escorted her grandson, an Irgunist wanted by the police, across Tel Aviv to a safe hiding place. **1959** *Chambers's Encycl.* XII. 398/1 The immediate post-war years saw such Jewish organizations as the *Irgun Zvai Leumi* and the Stern Gang undertake planned terrorist activities against the mandatory power. **1968** P. DURST *Badge of Infamy* iii. 26 Eichmann..was kidnapped in South America by the Irgun and brought to trial by the State of Israel. **1970** *Observer* 12 Sept. 9/6 Leila Khaled is inspired by the same fanatical devotion..as the Irgun Zvai Leumi and the Stern Group, who fought against the British and the Arabs to establish the State of Israel. **1972** *Guardian* 11 Jan. 11/1 The news agency's stringer was a secret member of the Irgun. **1972** *Times* 23 Sept. 5/1 The district Court here refused a police request to remand Mr Amihai Paglin, the key strategist of the Irgun Zvai Leumi during British rule in Palestine.

Irian (i·riăn), *a.* [Native name.] Of or pertaining to Irian Barat or West Irian, formerly Dutch New Guinea or Netherlands New Guinea, since 1963 a province of Indonesia. Hence **Iriane·se**, a native or inhabitant of this province; also *collect.*

1950 tr. *Rep. Comm. New Guinea* (*Irian*) III. iv. 96 The Netherlands promise that it is prepared to listen to the voice of the original Irian people. **1958** *Times* 9 Sept. 9/3 'West Irian action', meaning the 'liberation' of west New Guinea from the Dutch. **1967** *Guardian* 17 Nov. 12/7 To talk of self-determination for West Irian by 1969 is, of course, unrealistic... To most of the Irianese.. considerations of this sort are academic. **1969** *New Yorker* 31 May 44/2 Suharto..rose steadily... His outstanding military performance was as a co-ordinator of the West Irian campaign in 1961–62.

iridal, *a.* Add: **2.** *Med.* Of or pertaining to the iris of the eye; = *IRIDIC *a.*²

1830 W. MACKENZIE *Pract. Treat. Dis. Eye* xix. 770 One anatomist supposes that he has traced branches from the ciliary or iridal nerves, where they lie between the sclerotica and choroid. **1879** *Brain* I. 8 In addition to..absence of iridal susceptibility to influence from the other eye, we have..failure of accommodation. **1965** *Ophthalmologica* CXLIX. 35 Although the iridal tumour revealed no sign of malignancy, it metastasized to the ipsilateral cervical lymph nodes prior to the enucleation of the eyeball.

iridesce (iride·s), *v. rare.* [Back-formation from IRIDESCENT *a.*] *intr.* To exhibit iridescence; to shine in an iridescent manner.

1884 E. COUES *Key to N. Amer. Birds* (ed. 2) III. i. 427 General plumage of metallic lustre, iridescing dark green on most parts. **1905** J. LONDON *Jacket* (1915) 48 Sun-flashed water where coral-growths iridesced from profounds of turquoise deeps.

iridiagnosis (əiᵊ:ridəiăgnōᵘ·sis). [f. IRIS *sb.* 4 a + DIAGNOSIS.] Diagnosis of disease from observation of the iris of the eye; iris diagnosis.

1918 F. W. COLLINS in P. J. Thiel *Dis. diagnosed by Observations of Eye* xvi. 76 See our works under preparation:..'Pathology, Iridiagnosis and Treatment', by F. W. Collins..and Charles F. Haverin. **1919** H. LINDLAHR (*title*) Iridiagnosis and other diagnostic methods. **1928** *Sunday Express* 8 Apr. 10/7, I had never come across iridiagnosis before.

iridial (əiri·diăl), *a. Med.* [Irreg. f. L. *īrid-* (see IRID) + -IAL.] = *IRIDAL *a.* 2, *IRIDIC *a.*²

1911 STEDMAN *Med. Dict.* 438/2 *Iridial, iridian*, iridal. **1920** I. F. & W. D. HENDERSON *Dict. Sci. Terms* 156/2 *Iridial angle*, the filtration angle of the eye; an angular recess at the anterior surface of the attached margin of the eye. **1960** *Anatomical Rec.* CXXXVI. 345/1 The iridial muscles of reptiles.

iridic (əiri·dik), *a.*² *Med.* [f. as prec. + -IC.] Of or pertaining to the iris of the eye; = *IRIDAL *a.* 2.

1891 M. FOSTER *Textbk. Physiol.* (ed. 5) IV. iii. 717 On the inside the curved circumferential portion of the cornea makes a blunt angle, 'iridic angle', with the outer edge of the more or less horizontal iris. **1907** *Ophthalmoscope* V. 378 A true fistulous passage lined by iridic pigmented epithelium. **1964** S. DUKE-ELDER *Parsons' Dis. Eye* (ed. 14) i. 6 The third portion of the muscle is composed of a few tenuous iridic fibres..finding insertion in the root of the iris.

iridin (əi·ridin). Also **iridine.** [f. as prec. + -IN¹.] **1.** *Pharm.* A substance obtained from the rhizome of the blue flag, *Iris versicolor*, and formerly used as a hepatic stimulant.

1879 *Pharmaceutical Jrnl. & Trans.* IX. 988/2 Iridin is obtained from the fresh rhizome of *Iris versicolor*. It is usually mixed with liquorice, or some other absorbent powder. **1907** *Yesterday's Shopping* (1969) 504/1 Iridin Compound. **1921** *Oxf. Index Therap.* 992 *Iridin*. Dose 1 to 5 grains..in pill... Cholagogue and diuretic in its action.

2. *Chem.* [a. G. *iridin* (De Laire & Tiemann 1893, in *Ber. d. Deut. Chem. Ges.* XXVI. 2011).] 5, 7, 3′-Trihydroxy-6, 4′, 5′-trimethoxyisoflavone 7-glucoside, $C_{24}H_{26}O_{13}$, a crystalline compound occurring in the rhizome (orris-root) of the Florentine iris (*Iris florentina*).

1894 *Jrnl. Chem. Soc.* LXVI. I. 47 The dried roots of the violet (*Iris florentina*) on extraction with alcohol, yield a glucoside iridin, $C_{24}H_{26}O_{13}$; it forms slender, white needles. **1928** *Ibid.* 1027 An alcoholic solution of iridin develops with a trace of ferric chloride a somewhat intense, dull reddish-violet colour. **1966** *Chem. Abstr.* LXIV. 15826 (*heading*) Final structure elucidation and complete synthesis of iridine.

3. *Chem.* [a. G. *iridin* (K. Felix et al. 1951, in *Zeitschr. f. physiol. Chem.* CCLXXXVII. 226).] A protamine or mixture of protamines found in the heads of spermatozoa of the rainbow trout, *Salmo gairdneri* (= *S. irideus*).

1953 *Chem. Abstr.* XLVII. 10018 Fresh spermatozoa heads of *Salmo iridis* [*sic*] and..of herring were shown to consist of nucleoprotamine..[which] contained the protamines iridin and clupein. **1969** *Internat. Jrnl. Protein Res.* I. 221/1 Iridine from rainbow-trout..has been separated by column chromatography..into three main fractions.

-iridine. *Chem.* [See *-IR-.] A suffix used systematically to form the names of saturated monocyclic compounds having a three-membered ring which includes a nitrogen atom. Cf. *-IR-.

iridium. Add: *Comb.*, as *iridium-pointed* adj.

1897 *Sears, Roebuck Catal.* 433 Gold filled and ebony telescopic Holder, with best quality iridium pointed pens. **1908** *Daily Chron.* 27 Feb. 7/3 A gold-iridium-pointed nib. **1926–7** *Army & Navy Stores Catal.* 425 Fountain pens. Fitted with 14-ct. gold nib, iridium-pointed.

irido-. Add: Also before a vowel **irid-**. **iridencleisis** (-klǝi·sis) [Gr. ἐγκλείειν to shut up], the trapping of a portion of the iris in an incision of the cornea; **iridople·gia** [*-PLEGIA], paralysis of the iris.

1855 R. G. MAYNE *Expos. Lex. Med. Sci.* (1860) 540/2 *Iridencleisis*, old term for strangulation of a prolapsed portion of the iris, between the lips of an incision in the cornea. **1879** *Arch. Ophthalm.* VIII. 497 (*heading*) A case of severe iritis and glaucoma, following iridencleisis. **1964** D. M. GORDON *Med. Managem. Ocular Dis.* xviii. 335 Chandler prefers peripheral iridectomy in acute cases in which the pressure is rapidly brought under control. If the pressure has been elevated for 36 or more hours.., he prefers an iridencleisis. [**1854** W. MACKENZIE *Pract. Treat. Dis. Eye* (ed. 4) xxii. 882 (*heading*) Myosis... Iridoplegia pupillam contrahens.] **1878** *Brain* I. 8 Iridoplegia is a term applicable wherever both circular and radiating fibres of the iris are paralysed, and the pupil fails wholly to respond to the stimulus of light. **1970** E. ZAGORA *Eye Injuries* i. 12/1 Occasionally traumatic iridoplegia and cycloplegia occur as a result of concussion of the eye.

iridocyte (əiᵊ·ridosəit). *Zool.* [a. Fr. *iridocyte* (G. Pouchet 1876, in *Jrnl. de l'Anat. & de la Physiol.* XII. 45), f. Gr. ἰριδ-, IRIS *sb.* + κύτος a hollow, cell.] A cell which refracts light to cause iridescence, found in the skin of fishes, cephalopods, and certain other animals.

1893 CUNNINGHAM & MACMUNN in *Phil. Trans. R. Soc.* B. CLXXXIV. 767 The chief features of the iridocytes are their regularity of outline, and their great reflecting power. **1923** *Glasgow Herald* 23 June 4 The silveriness [of plaice] is due to the reflection of light from minute spangles of a waste-product called guanin, which accumulates in certain skin-cells called iridocytes. **1940** *Proc. Zool. Soc.* A. CX. 19 Throughout the epithelial tissues of the animal [*sc. Sepia officinalis*], lying below the chromatophores, is a layer of immobile reflector cells, the iridocytes or iridophores. **1960** FOX & VEVERS *Nature of Animal Colours* x. 150 In scales the guanine crystals may cause iridescence by interference of light and the chromatophores are then called iridocytes.

irigenin (əiridʒe·nin). *Chem.* [a. G. *irigenin* (De Laire & Tiemann 1893, in *Ber. d. Deut. Chem. Ges.* XXVI. 2011), f. G. *iri-din* *IRIDIN 2: see *GENIN.] A pale yellow crystalline flavonoid, $C_{18}H_{16}O_8$, obtained on acid hydrolysis of iridin, of which it is the aglycone.

1894 *Jrnl. Chem. Soc.* LXVI. I. 47 Iridin undergoes hydrolysis when heated with dilute sulphuric acid at 80–100°, yielding *d*-glucose and irigenin... This substance

crystallises in rhombohedra, melting at 186°. **1928** *Ibid.* 1925 These results prove that irigenin..is consequently 5:7:3′-trihydroxy-6:4′:5′-trimethoxy*iso*flavone. **1960** *Tetrahedron Let.* v. 9 The synthetic irigenin..was identical with the natural product.

-irine. *Chem.* [See *-IR-.] A suffix used systematically to form the names of unsaturated monocyclic compounds having a three-membered ring which includes a nitrogen atom. Cf. *-IR-.

irinite (i·rinəit). *Min.* [ad. Russ. *irinit* (Borodin & Kazakova 1954, in *Doklady Akad. Nauk S.S.S.R.* XCVII. 725), f. the name of *Irin*-a Dmitrievna Borneman-Starynkevich, Russian geochemist: see -ITE¹.] An oxide-hydroxide of sodium, cerium, thorium, titanium, and niobium occurring as red-brown crystals in the Khibiny Massif, U.S.S.R., and belonging to the perovskite group of minerals.

1955 *Mineral. Abstr.* XII. 462 Irinite occurs as combination of cube and octahedron from 0·5 to 1 cm. in size... Formula ABX_3, namely $(Na, Ce, Th)_{1-x}(Ti, Nb)(O_{3-x}, (OH)_x)$ with deficiency of A cations. **1962** W. A. DEER et al. *Rock-forming Min.* V. 51 Irinite, described as a new mineral of the perovskite group..has a high thorium and rare earth content.

iris, *sb.* Add: **4. c.** *Photogr.* = *iris-diaphragm*; also *attrib.* Also **iris-in** *sb.*, **iris-out** *sb.* (see quot. 1959): both used as vbs. Also *transf.* and *fig.*

1911 C. N. BENNETT et al. *Handbk. Kinematogr.* I. iv. 28 In the form of lens attached to kinematograph cameras, alteration of diaphragm is effected by the movement of a ring or pin on the lens mount which causes the 'iris' inside to open and close like the iris of a cat's eye, except that the hole in the middle always remains circular in shape. **1929** I. MONTAGU tr. *Pudovkin's On Film Technique* App. 191 The author gives..the iris-in and iris-out, mentioning what is called the fade only as a variant. **1934** WEBSTER, *Iris*, to operate the iris of a camera so as to fade (a picture). With *in* or *out*. **1958** *Spectator* 6 June 730/3 He can iris himself in and out of the scene at will. **1959** W. S. SHARPS *Dict. Cinematogr.* 104/2 *Iris-in*, the film wipe, in which the image viewed progressively disappears and at the same time is replaced by another, moving from the centre of the frame outwards in the form of a circle. *Ibid.*, *Iris-out*, the film wipe, in which the image viewed progressively disappears and is replaced by another from the outside of the frame, moving inwards in the form of a circle. **1961** K. REISZ *Technique Film Editing* (ed. 9) iii. 246 An iris may on occasion introduce or close a shot in a more telling way than a fade. **1962** *B.B.C. Handbook* 119 The control of the zooming, focusing, and iris setting, of a television camera. **1966** WODEHOUSE *Plum Pie* vii. 177 After a terrific struggle the hood called it a day and irised out. **1968** BLISH & KNIGHT *Torrent of Faces* II. vii. 137 Fongaváro glided silently through the Rest Stop door, which irised shut as silently behind him. **1969** GISH & PINCHOT *Lilian Gish* xi. 145 The scene of Sherman's march to the sea opened with an iris shot—a small area. *Ibid.*, Slowly the iris opened wider to reveal a great panorama. **1972** WODEHOUSE *Pearls, Girls, & Monty Bodkin* vii. 102 When a bunch of flatfeet burst in with their uncouth cry of 'Everybody keep their seats, please,' the thing to do is to iris out unobtrusively through the kitchen.

7. (sense 5) *iris blue, green*; **iris diagnosis** = *IRIDIAGNOSIS; **iris-diaphragm** (earlier examples).

1908 F. TREVES *Cradle of Deep* x. 57 A West Indian island..rising aloft from an iris-blue sea. **1921** *Glasgow Herald* 17 Feb. 9 Iris diagnosis and its relation to true health. **1938** H. ORBELL (*title*) The science of iris-diagnosis. **1867** *Trans. R. Microsc. Soc.* XV. 74 (*heading*) Iris diaphragm proving the circular form whether expanding or contracting. *a* **1877** KNIGHT *Dict. Mech.* II. 1195/2 *Iris-diaphragm*, a contractile diaphragm, simulating the action of the natural iris. **1875** E. SPON *Workshop Receipts* 95/1 *Iris green*, a pigment prepared by grinding the juice of the petals of the blue flag with quicklime. It is very fugitive. **1880** *Encycl. Brit.* XIII. 276/2 From the flowers of *Iris florentina* a pigment—the 'verdelis', 'vert d'iris', or iris-green, formerly used by miniature painters—was prepared by maceration.

Irish, *a.* and *sb.* Add: **A.** *adj.* **2.** *Irish butter, guipure, linen, poplin, tweed, whiskey* (examples).

1741 RICHARDSON *Pamela* IV. xiii. 75 A Piece of Irish or Scotish Linen. **1751** [see POPLIN¹ *attrib.*]. **1780** A. YOUNG *Tour in Ireland* I. 276 The salt for the fish trade [comes] from Rochelle: for butter english and irish. **1785** J. WEDGWOOD *Let.* 3 Oct. (1965) 285 Irish linens in the British market. **1798** C. MORDAUNT *Let.* July in E. Hamilton *Mordaunts* (1965) x. 243, I hope our conduct may gain us credit for discipline, but am terribly afraid of the cheap Irish whiskey. **1805** *Times* 6 Nov. 1/2 (Advt.), The Public supplied, as usual, with pieces of Irish linen, at the wholesale prices, at the Irish Linen Company's, No. 4, Bloomsbury-square, near Hart-street. **1813** JANE AUSTEN *Let.* 15 Sept. (1932) II. 321 Very pretty English poplins at 4*s.* 3*d.*; Irish, ditto at 6*s.* **1828** M. R. MITFORD *Our Village* III. 178 Crockery ware was piled on one side of her door-way, Dutch cheese and Irish butter encumbered the other. **1851** *Illustr. Catal. Gt. Exhib.* III. 516/1 Dowlas is a strong kind of Irish linen, for shirting. *Ibid.* 561/1 Laces: Royal Irish guipure; Irish appliqué. **1855** [see SCOTCH *a.* 2 a]. **1861** MRS. BEETON *Bk. Househ. Managem.* 808 Irish butter sold in London is all salted, but is generally good. **1865** F. B. PALLISER *Hist. Lace* xxxv. 416 Irish Brussels is made at Clones,...Irish guipure at Carrickmacross, in the same Co. **1879** M. E. BRADDON *Vixen* III. 319 She wore Irish poplin, and Irish lace, Irish stockings, and Irish linen. **1892–3** *T. Eaton & Co. Catal.* Fall & Winter 33/1 Boys' Overcoats... In Scotch, English and Irish tweeds. **1895** *Army & Navy Co-op. Soc. Price List* 15 Sept. 1094 Irish poplin. *Ibid.* 1099 Irish lawn... Irish diapers. **1896** *Illustr. London News* 25 Feb. (jacket) 4/3 (Advt.), A three-garment suit for boys... In Scotch and Irish tweeds. **1907** *Yesterday's Shopping* (1969) 12/1 Butter..Irish (finest Creamery)—lb. 1/2. **1909** Irish tweed [see *DONEGAL]. **1966** MRS. L. B. JOHNSON *White House Diary* 17 Mar. (1970) 373 Irish whiskey for St. Patrick's Day, the first time it's been served in the White House, I'll bet, and not a soul wanted tea! **1968** J. IRONSIDE *Fashion Alphabet* 233 *Irish linen*, a very fine light-weight linen woven of Irish flax. *Ibid.*, The linen industry in Ireland now, however, produces all weights of linen and linen mixtures..which are collectively and erroneously referred to as 'Irish' linen. **1968** L. ROSTEN *Joys of Yiddish* p. xiii, Yiddish phrasing and overtones are found in, say, the way an Irish whiskey advertises itself. **1970** J. FLEMING *Young Man, I think you're Dying* viii. 106 There was an Irish tweed jacket for Joe and a toy leprechaun. **1973** P. GEDDES *Ottawa Allegation* viii. 102 In a department store on the Mall she stood and fingered Irish linens.

b. *Irish potato, setter* (see the sbs.); **Irish deer**, the extinct giant deer, *Megaceros giganteus*, whose remains have been found in Ireland and other parts of Europe; **Irish elk** = *Irish deer* (examples); **Irish terrier**, a large wire-haired terrier, with a sandy or reddish-coloured coat; **Irish wolfhound**, a large, rough-coated hound, often grey in colour.

1697 T. MOLYNEUX in *Phil. Trans. R. Soc.* XIX. 505 We shall not have the least Reason to question but these vastly large Irish Deer and the American Moose, were certainly one and the same sort of Animal. **1825** S. HIBBERT in *Edin. Jrnl. Sci.* III. 15 The Irish Elk attracts no small share of attention. **1844** *Rep. Brit. Assoc. Adv. Sci.* 237 The most remarkable of the unquestionably extinct species of the Cervine family is that which is commonly called the Irish Elk. *Ibid.* 238 Mr. Parkinson refers the beams of two antlers found in the till at Walton in Essex, on account of their large size, to the great Irish Deer. **1857** C. KINGSLEY *Two Yrs. Ago* II. viii. 269 Round the burleaf-bed dances a rough, white Irish terrier. **1880** G. A. GRAHAM in H. Dalziel *Brit. Dogs* iii. 34 The Irish wolf-hound, being used for both the capture and despatch of the wolf, it would necessarily have been of greyhound conformation, besides being of enormous power. **1892** H. N. HUTCHINSON *Extinct Monsters* xvi. 224 The 'Great Irish Elk', as it is generally called, deserves special notice. *Ibid.*, The term 'Elk' is misleading, for it is not an elk (*Alces*) at all, but a true *Cervus* (stag). It should be called 'the Great Irish Deer'. **1897** *Encycl. Sport* I. 323/1 It is just twenty years since the Irish terrier first obtained recognition in the Kennel Club Stud Book. **1908** A. J. DAWSON *Finn* vi. 59 Finn had won two special prizes; one, a medal offered by the Irish Wolfhound Club..and another..for the biggest Irish Wolfhound in the Show. **1933** A. S. ROMER *Vertebr. Paleontol.* xviii. 356 Among the more interesting forms was the gigantic 'Irish elk', *Megaceros*, with the largest antlers of any known deer. **1947** J. STEVENSON-HAMILTON *Wild Life S. Afr.* viii. 68 A plucky Irish terrier diverted the pig's attention by attacking it from behind. **1964** G. K. WHITEHEAD *Deer Gt. Brit. & Ireland* xxx. 435 The giant deer—variously called great fallow deer and, quite erroneously, the Irish elk—was undoubtedly the finest deer that has ever inhabited Great Britain. **1969** E. H. HART *Encycl. Dog Breeds* 313 The Irish Wolfhound is remarkable in combining power and swiftness with keen sight. *Ibid.* 521 Origin of the Irish Terrier is a subject likely to provide debate as long as interest in the breed continues.

c. Irish American, an American of Irish origin; as *adj.* (with hyphen), of or pertaining to such a person, or to the Irish community, in the United States; **Irish apricot** *humorous*, a potato; **Irish articles**, articles of belief drawn up by Archbishop Ussher in 1615; **Irish-Australian** *a.*, of or pertaining to an Australian of Irish origin; (as two words) such a person; **Irish bridge**, an open stone drain carrying water across a road (see quots.); **Irish coffee** = *Gaelic coffee*; **Irish confetti** *slang*, stones, bricks, etc., esp. when used as weapons; **Irish crochet** = **Irish lace* (*a*); *also attrib.*, as *Irish crochet lace, point*; **Irish Free State**, (from 1921) name of the independent democratic State of Southern Ireland (since 1937 called Eire, and since 1949 the Republic of Ireland); **Irish green**, Connemara marble; **Irish Guards** [GUARD *sb.* 8], an infantry regiment of the British Army formed to signalize the bravery of Irish troops in the Boer War of 1899–1902; **Irish harp** = CLAIRSCHACH; **Irish horse** *slang*, salted meat; **Irish hurricane** *Naut. slang*, see quots.; **Irish Ireland** *a.*, designating a movement to arouse the interest of all Irish people in their own country; **Irish lace**, (*a*) a type of lace that resembles crochet; (*b*) see quot. 1957; **Irish martingale** (see quot. 1958); **Irish pennant** *Naut. slang*, see quot. 1962; **Irish point**, a kind of lace made in Ireland; **Irish Society**, a society founded in the reign of James I to have jurisdiction over the new Protestant settlement in Ulster; **Irish stitch** (earlier and later examples); **Irish Sweep** (or **Sweepstake**), a sweepstake organized by Irish hospitals on the results of English horse-races, esp. the Derby and the Grand National Steeplechase.

The examples are arranged for convenient reference in alphabetical order.

1832 *New-England Mag.* June 490 Irish-American Literature. **1836** T. POWER *Impressions Amer.* I. 185 The accent of the Irish American..differs [little] from that of the settler of a year. **1891** KIPLING *Life's Handicap* 197 My Irish American-Jew boy. **1902** *Irish Rosary* Jan. 77/1 The Irish-American Press should urge upon the Irish in America a sense of their duty in this matter. **1957** P. KEMP *Mine were of Trouble* x. 197 The famous Irish-American plastic surgeon, Eastman Sheean. **1971** *Guardian* 27 Oct. 12/6 The Irish-Americans are interested in Ireland. **1972** T. P. MCMAHON *Issue of Bishop's Blood* (1973) xiii. 188 The nylon-curtain Irish Americans of the Seventies. **1785** GROSE *Dict. Vulgar T.*, *Irish apricots*, potatoes; it is a common joke against the Irish vessels to say they are loaded with fruit and timber, that is, potatoes and broomsticks. **1846** *Swell's Night Guide* 122/2 *Irish apricots*, potatoes. **1973** *Times* 15 Feb. 14/3 Like the Norfolk Capon (a red herring), the Irish Apricot (a potato)..the Welsh Rabbit is a time-honoured joke. **1877** P. SCHAFF *Hist. Creeds of Christendom* I. 664 The Irish Articles are one hundred and four in number. **1967** D. T. KAUFFMAN *Dict. Relig. Terms* 253/2 *Irish articles*, the Calvinistic statement of faith, adopted in 1615 by the Irish Episcopal Church in the form of 104 articles, which became part of the basis of the Westminster Confession. **1907** *Westm. Gaz.* 17 Sept. 1/3 The..Irish-Australian baronet. **1957** P. KEMP *Mine were of Trouble* vi. 106 Peter Lawler, a hard-bitten little Irish-Australian. **1923** W. L. STRANGE *Indian Engin.* xlvi. 205 For lower-class roads..stream crossings may be substituted... The cheapest form is a level crossing... The next higher type is a road dam, or 'Irish bridge', for which two parallel curtain walls are constructed across the stream bed and continued for a short distance as flanks up the approaches; between them is the roadway: the downstream wall should be securely founded. **1969** 'M. INNES' *Family Affair* xv. 167 You cross the river by an Irish bridge... It's just a bridge, but built under the water instead of over it... It's really a reliable sort of ford. **1950** *Social & Personal* Dec. 57/1, I am..drinking Irish coffee, which.. is a mixture of very, very good Power's whiskey and very, very bad coffee. I am deciding that my next drink will be an Irish coffee without the coffee. **1959** D. O'NEILL *Life has no Price* vii. 102 Donal ordered himself an Irish coffee. **1966** 'S. FORBES' *Terror touches Me* iv. 37 Everyone ..had Irish coffee after dinner... He spooned brown sugar into glasses... Over the sugar he poured dollops of Irish whiskey... Next came steaming dark coffee..and.. cream..poured slowly over the back of a spoon. **1969** *Observer* (Colour Suppl.) 30 Nov. 10/1, 17/6 each... The Irish Coffee glass. **1935** A. J. POLLOCK *Underworld Speaks* 62/1 *Irish confetti*, bricks. **1939** G. KERSH *I got References* xii. 161, I learned the use of Irish Confetti, or Brickbats, at a tender age. **1966** *Observer* 19 June 40/1 An American friend in Amsterdam, describing last week's riots there, said: 'There's just a lot of Irish confetti around.' **1966** F. SHAW et al. *Lern Yerself Scouse* 57 *A cargo uv Irish confetti*, a cargo of stone chippings. **1881** C. C. HARRISON *Woman's Handiwork* III. 217 A deep fall of large-patterned Irish crochet lace. **1900** Irish crochet point [see POINT *sb.*¹ A. 31]. **1932** D. C. MINTER *Mod. Needlecraft* 103/1 A fine linen thread is most suitable for Irish crochet. **1965** *Daily Express* 14 Apr. 8/3 Linen..covers her drawing room walls (held down by strips of Irish crochet). **1969** *New Yorker* 31 May 92/2 The attached vestees are done in oyster bolillo lace, which is rather like heavy Irish crochet. **1922** *Act* 13 *Geo. V* c. 2 An Act to make such provisions as are consequential on or incidental to the establishment of the Irish Free State. **1929** *Encycl. Brit.* XII. 628/1 The Irish Free State, with the status of a British Dominion, came officially into being on January 15, 1922. **1937** V. BARTLETT *This is my Life* x. 144 When the Irish Free State was admitted [to the League of Nations]..President Cosgrave made his opening speech in Gaelic. **1959** *Chambers's Encycl.* V. 49/1 The constitution is considered by the *dáil* to apply to the whole of Ireland..but, 'pending the reintegration of the national territory' to have effect only in the area formerly known as the Irish Free State. **1883** *Encycl. Brit.* XV. 529/1 The 'Irish green' of architects is a similar rock from Connemara in western Galway. **1886** Irish green [see *CONNEMARA]. **1902** *Encycl. Brit.* XXXIII. 684/1 The Queen.. issued an order..for a new regiment of Irish Guards to be constituted. **1923** KIPLING (*title*) The Irish Guards in the Great War. **1967** A. FARRAR-HOCKLEY *Death of Army* v. 179 General Landon had sent him the 2nd Grenadier Guards and the Irish Guards. **1611** G. VADIANUS in *Coryat's Crudities* sig. l2 Torn is an Irish Harpe, whose heart-strings tune As fancies wrest doth straine or slacke his cord. *a* **1700** EVELYN *Diary* an. 1654 (1955) III. 92 My old acquaintance & most incomparable player on the Irish-Harp, Mr. Clarke. **1797** *Encycl. Brit.* VIII. 326/1 There are among us two sorts of this instrument, *viz.* the Welch harp..and the Irish harp. **1879** GROVE *Dict. Mus.* I. 686/1 The beautiful form of the more modern Irish harp is well known from its representation in the royal coat of arms. **1969** *Guardian* 16 Sept. 11/3 They only need to play Irish harps..to complete the absurd romanticism of Free Belfast. **1973** *Country Life* 29 Mar. 861/3 A finished harp costs at least £1,300. Happily.. small Irish harps..made almost entirely of wood..can be bought for £130. They serve as a sort of apprenticeship for would be serious harp players. **1748** SMOLLETT *R. Random* I. xxxiii. 291 Our provision consisted of putrid salt beef, to which the sailors gave the name of Irish horse. **1886** H. BAUMANN *Londinismen* 83/2 *Irish horse*, Pökelfleisch, s. salt horse. **1929** F. C. BOWEN *Sea Slang* 72 *Irish hurricane*, a flat calm with drizzling rain. **1962** *John o' London's* 14 June 571/2 An Irish hurricane is a flat calm. **1904** W. B. YEATS in *Daily Chron.* 18 Mar. 3/4, I went..to tell the Irish of America of what we call the Irish Ireland movement. **1854** C. M. YONGE *Heartsease*

I. xiv. 336 She was..prettily dressed with some Irish lace. **1880** L. Higgin *Handbk. Embroidery* v. 51 Tambour work..is now almost confined to the manufacture of what is known as Irish or Limerick lace..made as net..with a tambour or crochet hook. **1881** C. C. Harrison *Woman's Handiwork* 1. 94 Irish lace, made of flax-thread with a ground-work of crochet. **1895** *Army & Navy Co-op. Soc. Price List* 1126/1 (*heading*) Real Lace Sets..Irish..per set 4/6 to 20/0. **1907** E. Wharton *Fruit of Tree* II. ix. 139 Let me lend you my dress with the Irish lace. *a* **1929** L. Troubridge *Life amongst Troubridges* (1966) iii. 16 We all had new poplin dresses with Irish lace collars. **1957** M. B. Picken *Fashion Dict.* 201/2 *Irish lace*, variety of laces made in Ireland. Best known are crochet, net embroideries of Limerick, and Carrickmacross cut work or Irish guipure. **1937** P. Rodzianko *Mod. Horsemanship* iii. 94 An ordinary Irish martingale has two rings that are connected with a leather strap, of about three inches long. This..prevents the reins going over his [the horse's] head as it catches him under the jaws. **1946** M. C. Self *Horseman's Encycl.* 272 *Irish martingale*. This is a short strap with a ring at each end. The reins of the snaffle bridle run through the rings and the strap slide up to about six inches from the bit. It gives additional control and prevents the snaffle being pulled too far through the animal's mouth. **1958** J. Hislop *From Start to Finish* 172 *Irish martingale*, two rings joined by a leather strap, about six inches long, through which the reins are passed. It is the type of martingale most used on racecourses and is more usually called 'rings', sometimes 'spectacles'. **1883** W. C. Russell *Sailors' Lang.* 73 *Irish pennants*, fag-ends of rope, rope-yarns, etc., flying about. **1910** D. W. Bone *Brassbounder* i. 14 'Irish pennants' fluttering wildly on spar and rigging tell of scamped work of those whose names are not on our 'Articles'. **1962** A. G. Course *Dict. Naut. Terms* 108 *Irish Pennants or Pendants*, untidy ends of ropes, rope yarns, etc., flying in the wind. It is said that it was originally the name given to a flag with a torn or frayed fly or end. **1865** F. B. Palliser *Hist. Lace* xxxv. 416 The fabric flourishes, and is known by the name of 'Irish', or 'Curragh point'. **1882** Caulfeild & Saward *Dict. Needlework* 272/1 Irish Point can be worked entirely as old Brussels needle point. **1613** in *Hist. Narr. Irish Soc.* (1916) 163 After all which done information was given by the Governor and Assistants of the Irish Society, that all the monies formerly levied towards that charge is altogether issued. **1846** T. MacNevin *Confiscation of Ulster* vii. 214 The Irish Society is a type and symbolical representation of English rule in Ireland from the beginning. **1877** *Encycl. Brit.* VI. 224/2 The separate estates are still held to be under the paramount jurisdiction of the Irish Society. **1634** J. Taylor *Needles Excellency* (ed. 10) sig. A2 Fisher stitch, Irish-stitch, and Queen stitch. **1738** C. Fiennes *Journeys* (1947) 364 A 'seatee of Irish stitch'. . . '8 Irish stitch coushons.' **1932** D. C. Minter *Mod. Needlecraft* 10/1 Intergradating one stitch and colour with another, as is possible with Irish stitch. **1931** F. Lenwood *Why all this Fuss about 'Sweeps'?* 10 How unsocial it all is may be seen from the excitement over the Irish Sweep and the Derby of 1931. **1933** W. S. Maugham *Sheppey* I. 12 Did you have a ticket for the Irish Sweep? **1937** G. Greene *Coll. Ess.* (1969) iv. 425 The great muted chromium shadows wait..the novelist's Irish sweep: money for no thought, for the banal situation and the inhuman romance. **1963** 'G. Bagby' *Murder's Little Helper* (1964) vi. 52 When you people come around, it's never been to tell one of my roomers that he's won the Irish Sweepstakes. **1965** N. Gulbenkian *Pantaraxia* viii. 152 Although I was not a gambler, I did buy a ticket in the first Irish Sweep. . . One afternoon..the..commissionaire came in..and he said..'You've drawn a horse in the Irish Sweep.' **1974** A. Ross *Bradford Business* 48 A wad of Irish Sweepstake tickets.

4. *spec.* Used of seemingly contradictory statements. (See also *Irish hurricane* s.v. sense *A. 2 c.)

1820 H. Brougham *Let.* 5 Feb. in H. Maxwell *Creevey Papers* (1903) I. 297 Your advice has been followed by anticipation (to speak Irish). **1838** Geo. Eliot *Let.* 18 Aug. (1954) I. 6 Isaac and I went alone (that seems rather Irish), and staid only a week. **1843** F. A. Kemble *Let.* 25 Aug. in *Rec. Later Life* (1882) III. 36 We are going out of town,..to-morrow at half-past six in the morning, and it is now past midnight, and I have every mortal and immortal thing to pack, with my own single pair of hands, which is Irish, Lord bless us! **1857** Mrs. Gaskell *Let.* 7 Dec. (1966) 491 The lecture was not (to me) so very interesting, being a sort of recapitulation of what he was going to say (if that's not Irish). **1892** C. H. Fretwell *Anc. Mariner* 94, I had what sailors call 'an Irish rise', becoming second officer after being for a time commanding officer. **1903** H. C. Rowland *Sea Scamps* 4, I was promptly addressed as 'lieutenant', which struck me as being rather an Irish promotion, having once previously served as major. **1926** J. S. Huxley *Essays Pop. Sci.* 121 To be Irish, the longer it lives, the sooner it ought to die. **1937** A. Upfield *Mr. Jelly's Business* (1938) iii. 28 He doesn't seem to mind me courting his daughter, but he doesn't give me a chance to do any courting. That's Irish, but it's a fact. **1970** R. Hill *Clubbable Woman* vi. 192 'Marcus wouldn't dare to tell a lie like that unless it was true!' 'Irish,' said Pascoe.

B. *sb.* **1. a.** Also, *black Irish*, Irish of Mediterranean appearance.

1888 Kipling *Soldiers Three* (1890) 82 Those are the Black Oirish an' 'tis they that bring dishgrace upon the name av Oireland. **1953** K. Tennant *Joyful Condemned* v. 46 His fleshy hooked nose..suggested Jewish blood, but he claimed he was black Irish. **1961** J. B. Priestley *Saturn over Water* xiv. 201 He was a black Irish type, with centuries of rebelliousness behind him. **1962** *Guardian* 18 July 5/5 That haunted 'Black Irish' face [of Eugene O'Neill]. **1970** K. Giles *Death in Church* vii. 177 Have you ever seen the black Irish?

4. (Earlier and later examples.)

1784 Cowper *Let.* 21 Mar. in *Corr.* (1904) II. 181 Your mother wishes you to buy for her ten yards and a half of yard-wide Irish, from two shillings to two shillings and sixpence per yard. **1834** E. E. Perkins *Lady's Shopping Manual* 63 The regard to time and other circumstances which has been recommended in choosing Irishes, should be observed in the purchase of all linens. **1889** J. K. Jerome *Three Men in Boat* ii. 29 Harris..proposed that we should go out and have a smile, saying that he had found a place..where you could really get a drop of Irish worth drinking. **1914** [see *Apollinaris]. **1972** P. Ruell *Red Christmas* vi. 58 Irish. I drink Irish. Not this muck.

5. Temper; passion. orig. *U.S.* and *dial.*

1834 D. Crockett *Narr. Life* iv. 30 Her Irish was up too high to do any thing with her. **1860** Bartlett *Dict. Amer.* (ed. 3) 217 My friends say that my Irish is getting up, meaning, I am getting angry. **1877** F. Ross et al. *Gloss. Words Holderness* 80/1 *Iry*; *Irish*, E. and N., passion; anger; rage; fury. **1933** Partridge *Words, Words, Words!* I. 9 Both *Irish* and the colloquial *Paddy* are used for anger. **1949** R. Harvey *Curtain Time* vii. 73 But George's Irish was up. **1972** *Evening Telegram* (St. John's, Newfoundland) 23 June 1/4 'I got my Irish up,' he said, 'and here's a man that's going to fight back.'

Irisher (ai³·riʃəɪ). *colloq.* [f. Irish *a.* + -er¹.] A person of Irish origin, an Irishman.

1807 C. Schultz *Let.* 15 July in *Trav. on Inland Voy.* (1810) I. 8 The inhabitants are mostly of German descent, and still, in a great measure, retain their national prejudices, and consider all who do not speak their own language either as Yankees or Irishers. **1832** J. Barrington *Personal Sk. Own Times* (ed. 2) III. 35 What would the poor Irishers have done in owld times? **1854** Mrs. Gaskell *North & South* (1855) II. iii. 35 This indignation was tempered..by contempt for 'them Irishers'. **1882** W. D. Hay *Brighter Britain!* II. 283 A big, red-headed Irisher. **1956** H. Gold *Man who was not with It* (1965) xi. 90 I'm a Wop Irisher from Boston. **1973** M. Mackintosh *King & Two Queens* xii. 180 I'm only a no-account Irisher, but I like to pay my debts.

Irishly. Delete *rare* and add later examples.

1907 *Daily Chron.* 23 May 6/4 That solitary English engineer—who is, to speak Irishly, of course, a Scotchman—holds the Nile in fee. **1915** W. Owen *Let.* 4 Apr. (1967) 329 She is..badly off; and though spending irishly, spends every penny. **1929** E. Bowen *Joining Charles* 64 Grizelda and Doris were best in the Irish Jig; so saucy, quite Irishly saucy. **1959** R. Graves *Coll. Poems* 290 How Irishly you sacrifice Love to pity, pity to ill-humour.

Irishman. Add: **2.** In full, *wild Irishman*. A thorny New Zealand shrub, *Discaria toumatou*; also called Matagouri and Tumatakuru.

1857 C. W. Richmond in *Richmond–Atkinson Papers* (1960) I. vi. 317 Besides the grasses there is flax, *tutu* and wild Irishman. **1860** S. Butler in H. F. Jones *Samuel Butler* (1919) I. 80 There is a large quantity of Irishman (the name given to a thorny shrub which, in the back country, attains to a considerable size)...A glorious lurid flare marks the ignition of an Irishman. **1883** [see Tumata-kuru]. **1896** *Australasian* 28 Aug. 407/5 It seems uncivil to a whole nation—another injustice to Ireland—to call a bramble a wild Irishman. **1941** O. Duff *N.Z. Now* i. 1 If a wild Irishman is grown in a hothouse it loses its spines and develops soft leaves. **1966** G. W. Turner *Eng. Lang. Austral. & N.Z.* viii. 168 The shrub matogowrie, called Irishman in early writings, is especially common in the South Island where Maori influence is least strong and anglicization is most likely.

3. Irishman's hurricane *Naut. slang*, a dead calm (see also quot. 1961); cf. *Irish hurricane* (*Irish *a.* 2 c); **Irishman's promotion, rise,** reduced wages; cf. *Irish *a.* 4 (quots. 1892, 1903).

1827 J. F. Cooper *Red Rover* III. v. 107 There was an Irishman's hurricane, right up and down, for a day. **1873** 'Vanderdecken' *Yachts & Yachting* xxxi. 264 She is like a hurrah's nest, or a billyboy caught in an Irishman's hurricane! **1889** Barrère & Leland *Dict. Slang* I. 488/1 *Irishman's rise*,..wages reduced. **1902** C. J. C. Hyne *Mr. Horrocks, Purser* ii. 27 I've a sort of memory that you got Irishman's promotion for a bit of a mistake just recently. **1915** *Truth* 25 Aug. 295/2 The utmost the surveyor might expect from most corporations would be an Irishman's rise. **1957** A. MacNab *Bulls of Iberia* xiii. 140 He..asked for a job as a picador. His offer was accepted immediately—on condition that he would stand guarantee for any horses that might be killed under him! This would have been an Irishman's rise, even in those days when they used worn-out horses from the U.S.A. at five dollars a head. **1961** F. H. Burgess *Dict. Sailing* 121 *Irishman's hurricane*, a slight drizzle in a calm. **1972** *Times* 29 Sept. 15/4 For many low-paid workers with children, an extra £2 a week may be no more than an 'Irishman's rise'.

Irishness. (Later examples.)

1902 *Daily Chron.* 12 Sept. 3/3 Mr. Cobb's 'man of sentiment' is an Irishman with a broad brogue, who out-Herods Herod in his Irishness. **1930** H. V. Morton *In Search of Ireland* vi. 119 At times this Irishness is a rather pleasant melancholy. **1972** 'R. Crawford' *Whip Hand* I. iv. 18 What made Mullins' [a public house] special was its authentic uncompromising Irishness. **1973** *Times* 5 Dec. 18/1 The contemplation of a common Irishness between north and south.

Irishy (ai³·riʃi), *a.* [f. Irish *a.* and *sb.* + -y¹.] Like the Irish, somewhat Irish.

1884 'Mark Twain' *Lett. to Publishers* (1967) 174 The boy's mouth is a trifle more Irishy than necessary. **1913** D. H. Lawrence *Let.* 1 Feb. (1962) I. 182 A reaction against Shaw and Galsworthy and Barker and Irishy (except Synge) people.

irisin (ai·risin). *Chem.* [a. G. *irisin* (O. Wallach 1886, in *Ann. d. Chem.* CCXXXIV. 374), f. G. *iris* Iris *sb.* + *-in* -in¹.] A fructan present in the rhizome of the yellow iris, *Iris pseudacorus*.

1887 *Jrnl. Chem. Soc.* LII. 1. 26 The rhizome of the water lily, *Iris pseudacorus*, contains a peculiar carbohydrate, called 'irisin' by the author. Irisin..closely resembles inulin. **1925** *Chem. Abstr.* XIX. 2813 Irisin..by the diffusion method shows a mol. wt. of 10,300. **1946** *Ibid.* XL. 550 Of the fructosans, irisin is the most stable to dry heat. **1953** *Q. Jrnl. Exper. Physiol.* XXXVIII. 9 Intravenous administration of low molecular weight grass levan and irisin is unaccompanied by vasomotor and respiratory disturbances.

iris-in, iris-out: see *Iris *sb.* 4 c.

irk, var. *Erk.

‖ **iroha** (*iro*·ha). Also **irofa, irova.** [Jap., named from the first three syllables *i, ro*, and *ha* or *fa*.] The Japanese kana or syllabary.

1845 *Encycl. Metrop.* XX. 482/2 They..formed a collection of 47 syllables, comprehending all the sounds which are found in their language...This syllabarium, or alphabet, is called *irofa*, from its three first elements. **1868** J. J. Hoffman *Japanese Gram.* 9 To facilitate the learning of the Japanese sounds or syllables, they have been so arranged as to compose a couple of sentences, and as these begin with the word *Irova*, that name has been given to the Japanese alphabet. **1890** B. H. Chamberlain *Things Japanese* 379 The order of the I-ro-ha bears witness to the Buddhist belief of the father of Japanese writing. **1903** C. Noss tr. *Lange's Text-bk. Colloquial Japanese* Introd. p. xvi, There is another arrangement of the syllabary called *iroha*...This is in the form of a stanza of poetry giving expression to Buddhistic sentiment. **1937** I. Nitobé *Lect. Japan* xi. 177 Our language is polysyllabic and is expressed in forty-seven sounds... We call the whole collection a syllab[a]ry instead of an alphabet. In our own language it is called *i-ro-ha* from the first three syllables. **1967** R. A. Miller *Japanese Lang.* iii. 127 The *kana* symbols have at various times been arranged into mnemonic word lists...One such is the *ame tsuchi* list... A somewhat later example..is the *iroha* list..thought today to have been the work of the priest Kūya (903–972) or of the priest Senkan (918–983).

iroko (irōᵘ·ko). [Yoruba.] A hardwood tree of the genus *Chlorophora*, either *C. excelsa*, which is found across the central part of Africa, or *C. regia*, which grows in the western region; also, the timber from these trees, sometimes called West African or Nigerian teak.

1890 A. Millson in *Kew Bull.* 240 The only trees of unusual bulk [in Yorubaland] are the cotton trees and an occasional 'Iroko' tree. **1894** H. M. Ward *Laslett's Timber & Timber Trees* (ed. 2) xxv. 307 Mention may also be made of Iroko, a very valuable and handsome building and cabinet wood. **1939** Eggeling & Harris in L. Chalk et al. *Forest Trees & Timbers Brit. Empire* IV. 84 The harder, more durable 'Brown Iroko' is probably *C. excelsa*; the softer, less durable 'Yellow Iroko' *C. regia*. **1955** *Times* 18 July 9/7 The good stout hull, framed with English oak, planked with iroko—a type of teak from Africa—and copper fastened, shows not the slightest sign of movement or decay. **1964** R. W. J. Keay et al. *Nigerian Trees* II. 188 This [sc. *Chlorophora excelsa*] is a very common large forest tree with dark green foliage and distinctive leaves. It is widely known by the Yoruba name Iroko. **1967** W. Soyinka *Kongi's Harvest* 58 When a squirrel Seeks sanctuary up the *iroko* tree The hunter's chase is ended. **1974** *Habitat Catal.* 76/3 Work top; warm honey coloured Iroko, a West African hardwood similar to teak.

iron, *sb.*¹ Add: **1. d.** *Geol.* Any meteorite which contains a high proportion of iron.

[**1826** *Phil. Mag.* LXVII. 4 (*heading*) Falls of meteoric stones and masses of iron.] **1842** *Amer. Jrnl. Sci. & Arts* XLIII. 358 The imbedded grains of olivin in the Pallas iron of Siberia, and the Otampa iron of South America. **1868** *Geol. Mag.* V. 75 The bodies which are comprised under the general name of meteorites have long since been arranged under two great divisions, the irons and the stones. **1920** *Mineral. Mag.* XIX. 56 In this scheme [of the author], meteorites are divided into four classes, viz. Irons, Stony-irons, Chondritic Stones, and Non-chondritic Stones. **1962** B. Mason *Meteorites* ix. 130 In total mass the iron meteorites far outweigh the stones, since all large meteorites are irons, and the average mass of an iron is much greater than that of a stone. **1971** I. G. Gass et al. *Understanding Earth* viii. 116/1 The basic division [of meteorites] into irons, stony-irons, and stones is simple and straightforward.

4. g. *slang.* Money. Cf. *iron-man 1 c and d.

1785 Grose *Dict. Vulgar T.*, *Iron*, money in general. **1906** E. Pugh *Spoilers* i. 5 The iron you're goin' to give me. **1966** C. Rougvie *Gredos Reckoning* iii. 50 He was earning a bit of iron.

h. *pl.* Iron supports to correct bow-legs, etc.

1838 Dickens *Nickleby* (1839) viii. 67 Children..with irons upon their limbs, boys of stunted growth. **1884** W. Pye *Surg. Handicraft* xxv. 319 Wooden splints are..preferable to 'irons'. **1927** W. E. Collinson *Contemp. Eng.* 56 We could see..deformities due to rickets or injuries and the remedies e.g. irons to correct bow-legs.

i. (Usu. in *pl.*) A stirrup. Cf. Stirrup-iron 1.

1894 *Country Gentleman's Catal.* 173 Saddles..with girths, stirrup leathers and irons, complete. **1907** *Yesterday's Shopping* (1969) 304 Gentleman's spring-side safety irons, with Prussian sides. **1955** *Times* 30 June 3/7 He bumped Gawthorpe badly, causing Nevett to stand up quickly in his irons. **1963** E. H. Edwards *Saddlery* xix. 145 There are two main variations of the basic iron which are the Bent Top iron..and the Kournakoff. **1969** D. M.

Goodall tr. *Müller's Pocket Dict. Horseman's Terms* 70 *Stirrup/irons*, der Steigbügel.

j. *pl.* Eating utensils. *dial.* and *slang.*

1905 *Eng. Dial. Dict.* Suppl., *Irons*... Cum. Knife and fork, in phr. *to be a good fist with one's irons*, to have a good appetite. **1943** in Hunt & Pringle *Service Slang* 40. **1946** J. Irving *Royal Navalese* 97 *Irons* (*eating irons*), the sailor's name for his knife, fork and spoon.

k. *slang.* An old motor vehicle.

1935 *Sun* (N.Y.) 19 Feb. 28/1 'Iron' is the dealer's name for an obsolete [automobile]. **1961** J. Stroud *Touch & Go* xi. 105 'This iron of yours——' began Frank. **1963** *Amer. Speech* XXXVIII. 42 *Iron*, an old truck. **1967** M. Reynolds *After Some Tomorrow* 9 Well, it would mean being able to maintain a decent hovercar rather than the..four wheel iron he was currently driving.

l. Used as a form of currency in Sierra Leone.

1936 G. Greene *Journey without Maps* I. iii. 64 One could speculate in irons: the rate that day was twenty for fourpence.

m. *slang.* A jemmy used in housebreaking.

1941 'V. Davis' *Phenomena in Crime* xix. 251 The bishop, cane, iron, or stick. **1962** *John o' London's* 25 Jan. 82/1 Tools for breaking into other people's premises are *irons*.

8*. = *corrugated iron. Austral.* and *N.Z.*

1924 'R. Daly' *Outpost* iii. 28 The Residency was a large iron-and-weatherboard bungalow. **1944** D. Stewart in D.M. Davin *N.Z. Short Stories* (1953) 270 We sat for hours..listening to the rain hammering on the iron roof. **1948** V. Palmer *Golconda* x. 73 Her banter usually glanced off Neda like hail from an iron roof. **1956** G. Bowen *Wool Away!* (ed. 2) x. 115 Building paper should be used under the iron above the shearing board.

8.** Ellipt. form of *iron hoof*, rhyming slang for 'poof', a homosexual.

1936 J. Curtis *Gilt Kid* viii. 79 You gets into bed and goes straight off to kip, never touched me you didn't, you great iron. **1938** J. Phelan *Lifer* iv. 39 Harry had a young iron an' Painter butted in on him. **1961** Partridge *Adventuring among Words* xii. 58 Gorblimey, 'e's an *iron*, did'n yeh know?

8*.** *Theatr. slang.* Ellipt. form of *iron curtain 1.

1951 R. Southern in *Oxf. Compan. Theatre* 171/2 Another curtain in the proscenium opening is the Safety or Fireproof Curtain, sometimes nicknamed the Iron. **1952** Granville *Dict. Theatr. Terms* 102 Iron's down. **1967** N. Marsh *Death at Dolphin* v. 112 'I'll take the Iron up and you can see Jeremy Jones's set for the first act.' He..sent up the elegantly painted fireproof curtain.

11. a. (sense 1 c) *iron pill, tablet, tonic.* **b.** *iron-containing* adj. **c.** *iron-stained* (examples). **d.** *iron-coloured* (later examples), *-like* (later examples), *-red* (examples).

1849 D. G. Rossetti *Let.* 27 Sept. (1965) I. 61 The iron-coloured sea. **1909** S. W. Bushell *Chinese Art* (ed. 2) II. viii. 26 Bowls and cups with iron-coloured feet and brown mouths. **1901** *Brit. Med. Jrnl.* 23 Nov. 1540/1 It also revealed a yellowish-brown iron-containing substance within the primitive nuclei of red globules. **1926** *Jrnl. Biol. Chem.* LXX. 474 The amounts of iron-containing supplements to be fed. **1946** *Nature* 12 Oct. 516/2 The intensification due to the formation of the iron-containing complex would increase the slope of the density/concentration curves at the lower concentrations of molybdenum. **1908** *Daily Chron.* 17 Sept. 6/3 The discipline is as iron-like as ever. **1963** *Times* 25 Feb. 5/7 Shepherds with Byzantine, iron-like faces protect their flocks against the wolves. **1879** *Family Physician* 809 This condition, known as *chlorosis* or *green sickness*, is readily controlled by the use of iron... The systematic use of the iron pills is almost invariably attended with the most satisfactory results. **1912** *More Secret Remedies* (B.M.A.) 203 The pills were..a form of Blaud's pill, somewhat weaker than the official iron pill. **1820** D. Wordsworth *Jrnl.* 11 Aug. (1941) II. 125 Its crags of grey and iron-red hues. **1909** S. W. Bushell *Chinese Art* (ed. 2) II. viii. 29 Coral or iron-red (*fan hung*). **1974** *Country Life* 3/10 Jan. 10/1 Asil game cocks..head and breast enamelled with iron-red flecks over a wash of rouge-de-fer. **1876** Meredith in *Fortn. Rev.* 1 June 829 A shape in stone, Sword-hacked and iron-stained. **1915** E. R. Lankester *Diversions of Naturalist* vii. 63 A few only iron-stained and yellow. **1957** *Brit. Nat. Formulary* (B.M.A.) (ed. 4) 147 *Ferrous Carbonate Tablets*, B.P.C. Synonyms: Blaud's tablets: iron tablets. **1933** E. C. Pearce *Short Encycl. for Nurses* 24 Good nourishing food,..combined with the use of iron tonics, is all that is necessary. **1973** J. Porter *It's Murder with Dover* ix. 88 Getting her daughter to take a dose of iron tonic.

12. iron bacterium, any of various bacteria, found esp. in fresh water, which are capable of oxidizing ferrous salts to ferric hydroxide (perhaps obtaining energy thereby) and storing the end product in their structure; **iron-binding** *a.*, able to combine chemically with iron; *sb.*, combination with iron; also *attrib.*; **iron buff,** hydrated ferric oxide used as a dye for cotton by impregnating the cloth with a soluble iron salt, passing it through an alkali solution, and oxidizing; **iron deficiency,** insufficient iron in an organism or in its food; also *attrib.*, so *iron-deficient* adj.; **iron-fall** (earlier example); **iron gang** *Austral.*, a gang of prisoners working in irons; **iron loss** *Electr.* = *core-loss* (*Core *sb.*¹ 15); **iron maiden,** an instrument of torture consisting of a coffin-shaped box lined inside with iron spikes, inside which the victim is confined; also *transf.* and *fig.*; also *iron maid*; **iron mask,** a mask, supposedly made of iron, worn by a political prisoner in France at the time of Louis XIV who died in the Bastille in 1703 and whose identity is disputed; hence used as the name of the prisoner himself; also *fig.*; **iron mountain,** a mountain rich in iron ore; **iron play** *Golf*, a specified manner of playing with irons (sense 4 g); so *iron player*; **iron ration,** (*a*) (usu. *pl.*) an emergency ration of tinned food, esp. as provided in the armed services; (*b*) various extended and *fig.* uses; **iron shot** *Golf*, a shot made with an iron; **iron virgin** = *iron maiden*.

1888 *Jrnl. R. Microsc. Soc.* 786 Bacteria which assume a rust-coloured hue were denominated iron-bacteria by Ehrenberg. *Ibid.*, The oxidizing power of the cells of iron-bacteria must be extremely great. **1919** D. Ellis (*title*) Iron bacteria. **1945** *Science* 23 Nov. 533/1 It seems unreasonable..to conclude that an organism is an iron bacterium because of that it is developing in an iron bacterium unless there are far greater quantities of ferric hydrate than cell substance in the accumulated materials resulting from bacterial growth. **1955** K. V. Thimann *Life of Bacteria* xxi. 598 The iron bacteria are of two types, unicellular and multicellular. **1946** *Science* 12 Oct. 340/1 (*heading*) An iron-binding component in human blood plasma. [illegible] **1954** *Arch. Biochem.* XX. 170 (*heading*) Carbon dioxide and oxygen in complex formation with iron and siderophilin, the iron-binding component of human plasma. *Ibid.* 173 (*heading*) On the mode of iron binding by siderophilin. [illegible] An automated method has been developed for determining serum iron-binding capacity. **1902** *Encycl. Brit.* XXVI. 524/1 Iron Buff is produced by impregnating the cotton with a solution of ferrous sulphate, squeezing, passing into sodium hydrate or carbonate solution, and finally exposing to air. **1925** S. R. & E. R. Trotman *Bleaching, Dyeing & Chem. Technol. Textile Fibres* xxxii. 315 Iron buffs are fast to light, washing and alkalis, but are sensitive to acids. **1971** R. J. Adrian [illegible] might deduce correctly that iron buff would not necessarily produce anaemia, and that it was expected to last for the life of the textile. **1923** *Biochem. Jrnl.* XVII. 203 The result was that the symptoms and effects of iron deficiency as stated appeared in the pigs. **1929** *Trans. & Proc. N.Z. Inst.* [illegible] The theory..that iron deficiency in the pastures was the cause of 'Bush Sickness', was finally adopted. **1956** *Nature* 18 Jan. 336/2 Metal-induced iron-deficiency in crop plants. **1971** *Brit. Med. Jrnl.* 15 Jan. 6/2 The detection of iron deficiency anaemia by measuring the haemoglobin content of the blood. *Ibid.* 12/1 Iron deficiency is believed to be common in Great Britain in adult women of all ages. **1932** *Biol. Abstr.* VI. 792/1 'Salt sick' of cattle on certain iron-deficient sandy and residual soils has proved to be a nutritional anemia due to deficiency of Fe [illegible]. **1956** *Nature* 14 Jan. 93/1 In iron-deficient plants there is observed an increase of the soluble forms of nitrogen, with a simultaneous decrease of its protein forms. **1846** *Amer. Jrnl. Sci.* 2nd Ser. II. 186 This is the way of the heavy iron-falls (Croatia, 1752, and Tennessee, 1835) as set off against that of all the stones,..a ratio approximating that of one (for irons) to twenty (for stones). **1840** *Sydney Gaz.* 8 Feb. in Stewart & Keesing *Old Bush Songs* (1957) 29 I'll tell the Mahers, MacNamaras and McCartys All about iron gangs and road parties. **1848** H. W. Haygarth *Recoll. Bush Life Austral.* ix. 93 Men chained together [illegible] 'iron gang'. **1945** Baker *Austral. Lang.* ii. 44 A bullock wagon taking supplies to men in an iron gang. **1894** *Electrician* 14 Dec. 190/2 In the case of certain transformers specially tested for the purpose no true increase in the iron loss takes place. **1931** A. W. Hull [illegible] (*heading*) [illegible] iron losses. **1958** L. H. Fisher *Theory & Design Magnetic Amplifiers* xii. 366 Excessive iron losses cause reduced gain. **1957** E. E. Cummings *Let.* 10 Feb. (1969) 211 Later or sooner I always glimpse a miserably twisting dwarf...waiting & the 'iron maiden'. **1895** *Brewer's Dict. Phr. & Fable* (new ed.) 662/2 *Iron Maiden of Nuremberg*, a box big enough to admit a man, with folding doors, the whole studded with sharp iron spikes. When the doors were pressed to these spikes were forced into the body of the victim... (German, *Eiserne Jungfrau*.) **1958** J. Baldwin in [illegible] *Black Short Story Anthology* [illegible] he did not hear himself; thinking into what an iron maiden of love and hatred he had placed her [illegible]. **1964** F. I. Ordway et al. *Basic Astronautics* xii. 465 Submerged in this water-filled 'iron maiden', a subject was able to withstand 31 G for a period of 5 sec. **1972** *Daily Tel.* 11 Oct. 1/7 The 'callous and inhuman' treatment of a Moscow Jewish scientist said to have been incarcerated in an 'iron maiden'... He had been held..in a cell measuring 3 ft by 18 inches... The walls of the cell were covered with spikes which prevented him leaning or sitting down. **1752** tr. *Voltaire's Age of Louis XIV* II. xxv. 8 [illegible] the marquis de Feuillade..has told me, that when his father-in-law [*sc.* Chamillard] was dying, he conjured him..to tell him who this person was, who had been known by no other name than that of *the man with the iron mask*. Chamillard answered him, it was the secret of state, which he had sworn never to reveal it. **1826** G. A. Ellis (*title*) The true history of the state prisoner commonly called the Iron Mask, extracted from documents in the French archives. **1841** C. Fox *Jrnl.* 20 Apr. (1972) 104 Sterling..believes that Kasper Häuser was an imposter. The Iron Mask much more fascinating, but unluckily there was no prince in Europe missing at the time. **1858** Geo. Eliot *Let.* 21 Jan. (1954) II. 424 The iron mask of my incognito seems quite painful in forbidding me to tell Dickens how thoroughly his generous impulse has been appreciated. **1869** 'Mark Twain' *Innoc. Abr.* (1870) xi. 75 They showed us the noisome cell where the celebrated 'Iron Mask'—that ill-starred brother of a hard-hearted king of France—was confined. **1893** H. L. Williams tr. *Dumas's Man in Iron Mask* xliii. 332 D'Artagnan..asked himself..why the Iron Mask had thrown the silver plate at the feet of Raoul? **1959** *Oxf. Compan. French Lit.* 460/2 So persistent was the rumour that Monmouth had survived his supposed execution in 1685 that Voltaire thought it necessary to deny..that he was the Man in the Iron Mask. **1968** M. Guybon tr. *Solzhenitsyn's First Circle* x. 52 At all times the privileged prisoner's cell was in semi-darkness. .. The other prisoners nicknamed him 'The Man in the Iron Mask'. No one knew his real name. **1838** *Boston Weekly Mag.* 24 Nov. 91/1 Having visited the Iron Mountain in Missouri..I am happy to add my testimony..respecting the remarkable deposites of iron ores. **1846** *Sci. Amer.* 12 Dec. 90/1 The new blast furnace at the iron mountain is again in blast. **1887** *Encycl. Brit.* XXII. 638/2 [Sumatra]. Iron is not unfrequent, and magnetic iron is obtained at the 'Iron Mountain' near Fort van der Capellen. **1969** *Times* 21 Nov. 27/3 The most poignant part of the trip was a visit to Hamersley [in Western Australia] to see Mount Tom Price, the 'iron mountain'. **1892** Iron play [see Putting *vbl. sb.*² 1]. **1973** *Country Life* 21 June 1806/3 An historic exhibition of iron play by a master of the game [*sc.* golf]. **1909** *Westm. Gaz.* 22 Feb. 12/2 He was also a most accomplished iron player. **1876** Voyle & Stevenson *Mil. Dict.* (ed. 3) 20/2 The ordinary iron rations for two days should be 2 lbs. preserved meat and 2 lbs. biscuits, supplemented in such manner as circumstances admit. **1896** Farmer & Henley *Slang* IV. 16/1 *Iron-rations* (nautical), tinned meat; specifically boiled salt-beef. **1915** 'I. Hay' *First Hundred Thousand* xvi. 215 A haversack, occupied by his 'iron ration'—an emergency meal of the tinned variety, which must never on any account be opened except by order of the C.O. **1918** E. S. Farrow *Dict. Mil. Terms* 318 Fritz is getting his iron rations. **1925** Fraser & Gibbons *Soldier & Sailor Words* 128 'Iron rations' was in the War also a colloquial expression in speaking of a hot shell-fire, *e.g.*, 'Jerry is letting them have it, lots of iron rations flying about!' **1951** L. MacNeice tr. *Goethe's Faust* I. 61 O believe *me*, who have been chewing These iron rations many a thousand year. **1970** R. Lowell *Notebk.* 235 The new painting has to live on iron rations. **1973** *Daily Tel.* 25 Apr. 36/6 When the boys left on their expedition they took with them only one day's supply of food, and intended to pick up more provisions during their journey. They..included in their packs 'iron rations' of chocolate, raisins and Kendal mint cake. **1909** *Westm. Gaz.* 28 May 12/3 Maxwell..had made a splendid iron shot. *c* **1895** B. Stoker *Squaw* in *B. Stoker Bedside Compan.* (1973) 123 When we got back to the chamber we found Hutcheson still opposite the Iron Virgin. **1906** J. Huneker *Melomaniacs* 273 You remember the summer I spent at Nuremberg digging up the old legend, and the numberless times I visited the torture chamber where stands the real Iron Virgin, her interior studded with horrid spikes? **1973** C. Osborne in B. Stoker *B. Stoker Bedside Compan.* 11 'The Squaw'..involves a torture device which Bram had examined in Nuremburg, a contraption known as the 'Iron Virgin'.

b. iron-monticellite [tr. G. *eisenmonticellit* (C. Doelter *Handb. d. Mineralchem.* (1914) II. I. 499)], a silicate of calcium and iron, $CaFeSiO_4$, analogous to monticellite, found as a constituent of slag and more recently as a natural mineral (kirschsteinite); **iron pan** (earlier and later examples).

1937 *Mineral. Mag.* XXIV. 613 Iron-monticellite. **1950** *Jrnl. Amer. Ceramic Soc.* XXXIII. 164/2 Iron-monticellite ($CaFeSiO_4$) is a compound that forms an unbroken series of solid solutions with fayalite (Fe_2SiO_4). **1957** *Mineral. Mag.* XXXI. 698 (*heading*) Kirschsteinite, a natural analogue to synthetic iron monticellite, from the Belgian Congo. **1840** *Outl. Flemish Husbandry* in *Brit. Husbandry* III. II. ii. 12 Between the sand and the loam, an indurated crust of earth cemented by carbonate of iron, which is well known to all improvers of poor sands by the name of the iron pan. **1949** *Antiquity* XXIII. 35 A thick layer of the same blue clay..was incorporated in the body of the cairn between two layers of stones: it had been trampled down and iron-pan had formed on it. **1961** *Listener* 12 Oct. 559/1 The soil is a 'podsol', with its contrasting dark, humic, white-leached, and iron-pan layers.

iron, *a.* Add: **3. c.** *iron hand*: in var. phrases with *velvet glove* indicative of firmness or inflexibility combined with apparent softness or gentleness. Also, formerly in Australia, the closure (see quots. 1876, 1883).

1850 T. Carlyle *Latter-Day Pamph.* ii. 8 Soft of speech and manner, yet with an inflexible rigour of command..'iron hand in a velvet glove', as Napoleon defined it. **1876** *Victorian Hansard* 20 Jan. 2002 They [*sc.* the Government] have dealt with the Opposition with a velvet glove; but the iron hand is beneath, and they shall feel it. **1883** G. W. Rusden *Hist. Austral.* III. 406 The *clôture*, or the 'iron hand', as McCulloch's resolution was called, was adopted in Victoria, for one session. **1925** Wodehouse *Carry on, Jeeves!* i. 14 You have to keep these fellows in their place, don't you know. You have to work the good old iron-hand-in-the-velvet-glove wheeze. **1941** S. Wood *Murder of Novelist* (1946) xvi. 123 She..runs the town..with an iron hand. And no nonsense about a velvet glove. **1942** R. A. J. Walling *Corpse with Eerie Eye* vi. 177 The velvet glove and the iron hand are outmoded. **1963** *Times* 2 Feb. 7/2 Mr. Winston Field, who has so far shown only the velvet glove since he was elected Prime Minister of Southern Rhodesia in December, today revealed the iron hand expected of Rhodesian Front Government when he announced a Bill to introduce the death penalty by hanging for various offences, including the throwing of petrol bombs.

4. a. *iron-jawed, -knobbed, -minded, -railed, -studded, -willed* (examples), adjs.

1883 'Mark Twain' *Life on Mississippi* iii. 45 I'm the old original iron-jawed, brass-mounted..corpse-maker from the wilds of Arkansaw! **1926** E. Hemingway *Torrents of Spring* (1933) i. 26 A short, iron-jawed man. **1895** B. M. Croker *Village Tales* 196 Shut it out by merely closing to the street an iron-knobbed wooden door. **1949** Blunden *After Bombing* 41 At the iron-knobbed

church door. **1897** J. L. Allen *Choir Invisible* xii. 168 Fighting it all over in his foolish, iron-minded way. **1944** Blunden *Shells by Stream* 9 Squadrons of gem-eyed hobby-horses [illegible]. **1968** *Punch* 28 Feb. 323/2 [illegible] iron-minded [illegible] Yorkshire. **1883** 'Mark Twain' *Life on Mississippi* xliv. 442 Lone, iron-railed verandas running along the several stories. **1904** A. Wynne [illegible] It was [illegible] an iron-railed podium. **1903** [illegible] May 30 In the wall in front of him was an iron-studded door. **1848** J. R. Lowell in *National Anti-Slavery Standard* 9 Nov. 96/1 In that far isle, whence, iron-willed, The new-world's sires their bark unmoored. **1951** M. McLuhan *Mech. Bride* [illegible] 'Mother' is a flint-eyed, iron-willed [illegible].

c. iron cap = Iron hat 2; **iron chink** [*Chink *sb.*²] (see quot. 1942); **Iron Cross** [G. *das eiserne kreuz*], a German and Austrian decoration awarded for distinguished services in war (founded by Frederick William III of Prussia in 1813, to reward those who served in the wars against Napoleon, and later revived by William I in 1870); **Iron Duke,** a name for the first Duke of Wellington (1769–1852); **Iron Guard,** an anti-Semitic, Fascist, terrorist Roumanian political party developed from the Legion of the Archangel Michael, founded by C. Z. Codreanu (?1899–1938) in 1927; **iron gum(-tree)** *Austral.*, one of several Queensland species of *Eucalyptus* which have particularly strong wood; **iron horse** (earlier and later examples); **iron jubilee,** the seventieth anniversary of an event; **iron law** (of wages), the law or idea that wages tend to sink to the level of mere subsistence; **iron mike** *slang*, a familiar name for the automatic steering device of a ship; **iron oak,** for def. (s.v. sense 12 of the sb.) read: any of several oaks with particularly durable wood, as *Quercus cerris*, *Q. stellata*, etc.; add examples; **iron paper,** extremely thin sheet-iron.

1911 *Chambers's Jrnl.* Mar. 166/1 The indication of a deposit of pyrites is the appearance of an outcrop of oxide of iron more or less honeycombed. This is called the 'iron cap', or in Cornwall 'gossan'. **1913** *Hadden's Guide to Western Canada* 11 In the salmon canneries the introduction of a machine called the 'Iron Chink', for cleaning and cutting the fish has made a great economy in the cost of labor. [illegible] The 'iron chink' cuts off the heads, tails, and fins, dresses the fish at the rate of [illegible] per hour. **1942** [illegible] The fish [illegible] to the so-called 'iron chink', a machine which automatically severs the head and eviscerates the fish. In the early days of the [salmon] industry, this work was performed by Chinese labor, hence the name 'iron chink' for the machine. **1961** *Vancouver Sun* 5 Apr. 24/1 (Advt.), Fishing company requires qualified iron chink operator. **1871** *Monthly Packet* July [illegible] Two prints which appeared during the autumn of 1870 in the *Illustrated London News*, one of the Crown Prince distributing the Iron Cross. **1902** *Encycl. Brit.* XXXI. [illegible] The Austrian Iron Cross, founded by Napoleon I. as King of Italy in 1805, [illegible]. **1914** *Punch* 11 Nov. [illegible] The Iron Cross. (For German loafers.) [illegible] My friends would rather have me back with pox than dead with the Iron Cross! [illegible] The Wisdom of Wellington; or maxims of the Iron Duke. [illegible] (*heading*) The Iron Duke. **1957** *Encycl. Brit.* XXIII. [illegible] As a diplomatist the 'Iron Duke'..was no match for the 'Iron Tsar'. **1965** W. R. Benét *Reader's Encycl.* (ed. 2) [illegible] Wellington, 1st duke of, Arthur Wellesley.. Known as the Iron Duke. **1933** *Times* 16 Nov. 13/2 It is reported that the new Government has decided to dissolve the Iron Guard organization..Anti-semitic and pro-Fascist and Hitlerist. **1934** *Ann. Reg.* 1933 [illegible] was assassinated..by a student member of the Iron Guard. **1942** L. B. Namier *Conflicts* 46 The Iron Guard, which was indebted to Germany for much of its income and of its revolutionary *élan*, indulged in the extremest forms of anti-Semitism, [illegible]. **1971** W. Laqueur *Dict. Politics* 438 After mounting Iron Guard atrocities and massacres of Gentiles as well as Jews, Antonescu repressed the Iron Guards with German consent. **1889** F. von Mueller *Eucalyptographia* I. s.v. *Eucalyptus Raveretiana*, Vernacularly it passes in the districts of its growth as 'Grey Gum-tree' and 'Iron Gum-tree'. **1888** F. M. Bailey *Queensland Woods* 63 *E[ucalyptus] Raveretiana*.. Thozet's Box or Iron Gum-tree. A large tree with a scaly bark persistent on the trunk. **1919** R. T. Baker *Hardwoods Austral.* 273 *Eucalyptus Raveretiana*, F. von M. 'Thozet's Box' or 'Iron Gum Tree'..close-grained, very hard, and tough; valuable for building purposes. **1957** *N.Z. Timber Jrnl.* Sept. 61/1 Irongum: Eucalyptus spp. Australia. The best known is E. maculata Hook, or spotted gum. Weight 64 lbs. per cubic foot. **1970** N. Hall et al. *Forest Trees Austral.* 46 Spotted iron gum (Qld) *Eucalyptus maculata*. **1840** D. March *Yankee Land* 23 There were noble steeds in the days of old... But the iron horse, there were none like him! **1846** *Congress. Globe* 6 Feb. 323/1 The iron horse..must be whistling on the wind,..vomiting fire and smoke. **1958** C. Achebe *Things Fall Apart* xvi. 129 Stories about these strange men [*sc.* missionaries] had grown since one of them had been killed in Abame and his iron horse tied to the sacred silk-cotton tree. **1967** C. O. Skinner *Madame Sarah* viii. 171 The engineer returned to his iron horse and the train started. **1903** *Westm. Gaz.* 6 Feb. 10/1 Pope Leo XIII. will celebrate..during the present year..his 'Iron Jubilee' as a priest—he was ordained seventy years ago. **1896** R. H. I. Palgrave *Dict. Pol. Econ.* II. 568/1 He [*sc.* Lassalle] dwelt on what he called the iron or 'brazen law' (*ehernes Gesetz*) of wages, already laid down by Turgot and Ricardo. **1907** J. S. Nicholson in *Cambr. Mod. Hist.* X. [illegible] Ricardo..was credited with the 'iron law of wages' as the one side and the theories of the continuous growth of rent and the 'unearned increment' on the other. **1913** *Pitman's Commercial Encycl.* IV. 1662/2 The 'iron law'..[illegible] that the general rate of wages constantly tended to fall to the mere subsistence limit. **1966** A. Gilpin *Dict. Econ. Terms* 119 *Subsistence theory*. Of French origin, this 'iron law' of wages asserted that if wages rose above subsistence level an increase in population would inevitably follow, thus forcing wages down again to subsistence level. [illegible]

iron, *v.* Add: **3.** *fig.* (Further examples of *iron out*.)

1903 *Springfield* (Mass.) *Weekly Republ.* 31 Mar. 8 The differences between Chairman Flaherty..and Col. William A. Gaston are in a fair way to be amicably ironed out. [illegible]

iron-bark. Add: (Later examples.)

1909 A. E. Mack *Bush Calendar* 4 The blue-gums, iron-barks, and turpentines gave way to scribbly gums and banksias. [illegible]

2. Passing into *adj.*: hard, unyielding. *Austral.*

1888 'R. Boldrewood' *Robbery under Arms* I. vi. 85 I always thought he was ironbark outside and in. **1945** Baker *Austral. Lang.* 90 From countrymen in general, *ironbark*, unyielding.

iron-blue. Add: **c.** *sb.* The pigment Prussian blue (Prussian *a.* 2).

1930 A. C. Harrison *Manuf. Lakes & Precipitated Pigments* ix. 124 The term 'iron blue'..comprises all types and shades of 'Prussian blues'. [illegible] Iron blue is probably the most important blue pigment used by the paint industry. [illegible] (available in a number of shades, such as milori blue, bronze blue, prussian blue, etc.) is a chemical compound of iron.

iron-cased, *a.* Add: Also in other uses.

1901 L. M. Waterhouse *Conduit Wiring* 50 The Simplex iron-cased distributing boards. **1906** *Westm. Gaz.* 29 Aug. 10/1 His tubular iron-cased telegraph. **1921** *Wireless World* 25 June 200/1 [illegible] are fixed audio-frequency transformers, iron-cased and with iron cores.

ironclad, iron-clad, *a.* and *sb.* Add: **A.** *adj.* **1. b.** Applied to electrical apparatus.

1903 *Encycl. Brit.* XXVII. 584/2 The two-poled ironclad type [of field-magnet], so called from the exciting coil being more or less encased by the iron yoke. **1910** *Hawkins' Electr. Dict.*, *Iron-clad dynamo*, a dynamo having an iron-clad armature. **1927** W. Wilson *Electr. Control Gear* xvi. 226 There are..reasonably cheap ironclad types of starting equipment on the market. **1939** Reed & Corcoran *Electr. Engin.* [illegible] The [illegible] solenoid can be made more effective by providing it with a stationary iron core which occupies about 20 or 30 per cent of the axial length.

2. *fig.* Also, of plants, able to withstand cold and frost. *U.S.*

1872 *Rep. Vermont Board Agric.* I. 54 Currants and gooseberries are iron clad as regards climate. **1882** *Rep. Maine Board Agric.* XXVI. 336 It is an early winter fruit, the tree not perfectly 'iron clad', notwithstanding its origin, yet hardy enough for most places.

3. (Earlier and later examples.)

1866 *Congress. Globe* 14 Feb. 835/1 Traitors never would be troubled with the 'iron-clad oath', for they never would have a chance to take it. **1868** *Harper's Mag.* Sept. 484/2 [He] was as well reconstructed as a man might be who could not take the iron-clad oath. **1873** J. Miller *Life amongst Modocs* xxvi. 304 Some hard, iron-clad oaths and then shot after shot. **1911** H. S. Harrison *Queed* x. 114 He insisted on doing it after an ironclad schedule. **1930** *Observer* 1 June 11 The rationing system..is not so complete or ironclad as those which prevailed in various European countries during the war. **1951** E. Kefauver *Crime in America* (1952) xix. 223 The operations of Murder, Inc., finally came to light when..police obtained ironclad evidence against a gangster known as Abe (Kid Twist) Reles.

B. *sb.* **2.** *transf.* and *fig.*

1867 J. N. Edwards *Shelby* 483 The West Pointers were the iron-clads in our wooden navy. **1875** 'Mark Twain' *Sk. New & Old* (1900) 157 After the Tortoises came another long train of ironclads—stately and spacious Mud Turtles. **1889** —— *Connecticut Yankee* xxxix. 456 Things began to look serious to the iron-clads [*sc.* knights in armour]. **1892** O. F. Whitney *Hist. Utah* I. 547 The ravages of the 'iron-clads' [i.e. grasshoppers] were widespread and far-reaching. **1974** *Country Life* 28 Mar. 752/1 Those plants that are always a risk, climatewise, but which do offer rewards..such as the hardy ironclads can never quite match. **1974** *Oxford Times* 19 Apr. 3/3 'Iron clads' are the..term for day-old Chelsea buns.

iron curtain. [f. Iron *sb.*¹ + Curtain *sb.*¹] **1.** In a theatre, a curtain of iron which can be lowered between the stage and the auditorium in order to prevent passage or communication, or for protection.

1794 *Times* 13 Mar. 3/2 Besides other precautions, an iron curtain has been contrived, which, on such occasion [of fire], would compleatly prevent all communication between the audience and stage. **1829** H. Foote *Compan. to Theatres* 30 As a precaution against fire, an iron curtain was constructed, so as to let down in a moment of danger and separate the audience from the stage. **1891** Scott & Howard *Life E. L. Blanchard* II. 557 And provision against fire is made [in the Prince of Wales' Theatre] by an iron curtain. **1908** Wodehouse & Westbrook *Globe by the Way Bk.* 51/2 'The iron curtain,' she gasped, and exerting her full strength, held the actor under the descending sheet of metal.

2. *fig.* Any impenetrable barrier.

After 1946 regarded as a transf. use of sense *2 b.

1819 Earl of Munster *Jrnl. Route across India* 1817–18 iv. 58 On the 19th November we crossed the river Betwah, and as if an iron curtain had dropt between us and the avenging angel, the deaths diminished. **1904** H. G. Wells *Food of Gods* III. iv. 274 It became evident that Redwood had still imperfectly apprehended the fact that an iron curtain had dropped between him and the outer world. **1915** G. W. Crile *Mechanistic View War & Peace* iv. 69 Suppose that Mexico were a rich, cultured, and brave nation of forty million with a deep-rooted grievance, and an iron curtain at its frontier. [**1916** E. Howard *Potsdam Princes* 250 The war, of course, has made them bitter enemies, and when reminded of her German relations, Queen Elizabeth of Belgium is reported to have said that between her and her people in Bavaria a curtain of iron had fallen.] **1939** J. Gloag *Word Warfare* xi. 113 In an international crisis Germany can be cut off from the world by an iron curtain of censorship. **1946** J. C. Davies *Episcopal Acts Welsh Dioceses 1066–1272* I. 43 Within this iron curtain in the south the princes of Deheubarth maintained a precarious independence until the middle of the thirteenth century. **1959** *Listener* 29 Jan. 198/1 By tact and determination she gradually broke down the iron curtain and mobilized the women into a united Conservative front. **1962** *Daily Tel.* 15 Aug. 13/2 There was no 'Iron Curtain' between them and the consumer. **1967** *Freedomways* VII. 119 Since I don't want the United States to appear like an 'Iron Curtain' to the Vietnamese, I think it would be good for several Vietnamese women to visit this country.

b. *spec.* (usually with initial capitals). A barrier to the passage of information, etc., at the limit of the sphere of influence of the Soviet Union. Cf. **bamboo curtain* (*Bamboo *sb.* 2), *Curtain *sb.*¹ 3 c. Also in extended use (as quot. 1924).

The *locus classicus* is quot. 1946.

1920 Mrs. P. Snowden *Through Bolshevik Russia* ii. 32 We were behind the 'iron curtain' at last! **1924** Lord D'Abernon *Diary* 14 Sept. in *Ambassador of Peace* (1930) III. iv. 101 Stresemann considered that it was essential for the Rhineland to be frankly part of Germany, also for Danzig to be reincorporated. Without this there could be no permanent peace. I put forward my view of the reciprocal iron curtain or strip of inviolable territory as a protection. [**1945** *Times* 23 Feb. 3/4 (tr. Paul Josef Goebbels, German Minister for Propaganda, from *Das Reich*) If the German people lay down their arms, the whole of eastern and south-eastern Europe, together with the Reich, would come under Russian occupation. Behind an iron screen [*ein eiserner Vorhang*] mass butcheries of peoples would begin.] **1945** *Times* 3 May 4/5 (speech by the German Foreign Minister, Schwerin von Krosigk) In the East the iron curtain behind which, unseen by the eyes of the world, the work of destruction goes on is moving steadily forward. **1945** Sir St. V.

Troubridge in *Sunday Empire News* 21 Oct. 2/2 A Curtain Across Europe... Yet at present an iron curtain of silence has descended, cutting off the Russian zone from the Western Allies. **1946** W. S. Churchill in *Times* 6 Mar. 6/1 (Address at Westminster College, Fulton, U.S.A., 5 March) From Stettin, in the Baltic, to Trieste, in the Adriatic, an iron curtain has descended across the Continent. **1953** *Encounter* Oct. 58/1 If they live behind the Iron Curtain they can do none of these things—for, while the Communists agree that knowledge is power, they are persuaded that they are already in essential possession of both. **1953** W. S. Churchill *Second World War* (1954) VI. xxxiv. 498 In these same days I also sent what may be called the 'Iron Curtain' telegram to President Truman.. *Prime Minister to President Truman* 12 May 45..An iron curtain is drawn down upon their front. We do not know what is going on behind. **1971** *Times Lit. Suppl.* 31 Dec. 1621/5 The theory of convergence informs us that societies on both sides of the Iron Curtain are conditioned by similar forces in all essential respects, whatever the differences in kind or degree of individual liberty enjoyed by their members. **1972** *Guardian* 23 Aug. 7/5 The iron curtain that fell on Anglo-Soviet cultural deals when 105 Soviet diplomats were expelled from Britain last year has been lifted. **1973** J. M. White *Gooden Game* 184 One of Harrison's local contacts, an experienced Iron Curtain operator.

irone (ai³·rōᵘn). *Chem.* [ad. G. *iron* (Tiemann & Krüger 1893, in *Ber. d. Deut. Chem. Ges.* XXVI. 2679), f. G. *ir-is* Iris *sb.* 5: see -one.] Orig., an essential oil obtained from orris-root; hence, any of the three isomers (α-, β-, and γ-*irone*) of which this is composed, which are cyclic ketones, $(CH_3)_4C_6H_5{\cdot}CH{:}CH{\cdot}CO{\cdot}CH_3$, that differ only in the position of the double bond in the C_6H_5 ring and are methyl derivatives of the ionones.

1894 *Jrnl. Chem. Soc.* LXVI. I. 80 The readily-volatile fraction [of iris root] contains the irone or fragrant oil. *Ibid.*, Irone..is an oil which is scarcely soluble in water. **1934** *Chem. Abstr.* XXVIII. 4053 Analyses of irone..establish the compn. $C_{14}H_{22}O$. **1948** *Ann. Rep. Progr. Chem.* XLIV. 148 While α- and γ-irone possess a fresh violet odour (the 'irone' odour), the β-isomer has an ionone smell. **1952** Simonsen & Owen *Terpenes* (ed. 2) III. 497 It is now recognised that irone is 6-methylionone, and that it can exist in the α-, β-, and γ-forms..each of which has several stereoisomeric modifications. *Ibid.* 503 γ-Irone is transformed into a mixture of α- and β-irones when treated with dilute sulphuric acid. **1963** *Chem. Abstr.* LVIII. 1500 Irones are used in the manuf. of perfumes; and they can be synthesized from dextro-α-pinene.

ironer. 1. (Earlier examples.)

1773 *South Carolina Gaz.* 26 July, To be sold..a complete Washer and Ironer, and remarkable for doing up Ladies fine Cloaths. **1820** *Barbados Mercury & Bridge-Town Gaz.* 16 Sept. 4/3 (Advt.), Also for sale, a family of Negroes, consisting of a woman, a complete washer and ironer, [etc.].

ironing, *vbl. sb.* Add: **1.** Further *attrib.* uses, as *ironing blanket, board, machine, room, stool*; *ironing table* (later examples).

1810 E. Weeton *Let.* 25 Feb. in *Jrnl. of Governess* (1969) I. 232, I flew into the servants' hall for the ironing blanket. **1854** Mrs. Gaskell *North & South* (1855) I. iv. 44 Dixon had complained that the ironing-blanket had been burnt again. **1843** *New Mirror* 8 Apr. 4/2 Swapped away for a wash-bench or an ironing-board. **1876** H. E. Scudder *Dwellers in Five-Sisters Court* iii. 52 Nicholas will have to carry the ironing-board for her. **1928–9** *T. Eaton & Co. Catal.* Fall & Winter 391/4 Ironing Board Cover..padded with wadding..fits any regulation-sized board. **1969** E. H. Pinto *Treen* 151 Ironing boards..are found in a variety of shapes and sizes, for skirts,..trousers,..and sleeves. **1972** *Guardian* 30 June 7/7 This superb new Milium Ironing Board Cover..is scorch resistant. **1972** *Sci. Amer.* Dec. 48/3 The development of fine glass yarns..has brought glass fibers into such products as bedspreads and covers for ironing boards. *a* **1877** Knight *Dict. Mech.* II. 1203/2 *Ironing-machine*, one for ironing clothes, etc. **1962** *Which?* June 170/1 We should like to hear from members who have bought a dishwashing machine or an ironing machine during 1960 or 1961. **1894** E. Banks *Campaigns of Curiosity* 195 The one large room ..should be divided into..three apartments; wash-house, sorting-room, and ironing-room. **1906** *Westm. Gaz.* 5 July 2/1 The hooded fireplace of the ironing-room. **1962** *Guardian* 3 Dec. 4/7 Mothers come..to enjoy working in the little ironing room. **1878** Mrs. Stowe *Poganuc People* xxv. 268 Will seized her off the ironing stool and, perching her on his shoulder, danced round the table. **1911** F. B. Jack *Woman's Bk.* 273/1 The Ironing Table.—This must be of a good size, strong, and steady, and of a convenient height for working at. *Ibid.* 273/2 In addition to the ironing table, a smaller one for placing the work on will be found a great convenience if space permits. **1959** *Sears, Roebuck Catal.* Spring & Summer 966/2 Ironing Table Designed for stand-up and sit-down ironing. **1974** *J. Fraser Catal.* 739/1 Ironing surface 36″ × 13″ All-metal ironing table..£4.50.

ironize, *v.*¹ Delete †*Obs.* and add later examples.

1906 H. Bland *Lett. to Daughter* 116 Does one satirise, ironise, slate, bully-rag, and squirt verbal vitriol at the thing one loves? **1933** Partridge *Words, Words, Words!* III. 184 The tendency either to ironize or to belittle one's fears, sufferings and discomforts. **1969** *Daily Tel.* 13 Feb. 22/4 There is also the ironist, ironising himself.

iron lung. 1. A kind of respirator for giving prolonged artificial respiration mechanically, consisting of a metal case that fits over the patient's chest or trunk with an air-tight aperture for the neck (and limbs), so that air may be forced into and out of the lungs by producing rhythmic variations in the air pressure in the case. orig. *U.S.*

1932 *N.Y. Times* 3 Oct. 19/5 Surgeons revealed today that Birdsall Sweet, 14, of Beacon has just rounded out one year in the artificial respirator or 'iron lung' in Vassar Brothers hospital, Poughkeepsie. **1934** E. Podolsky *Medicine marches On* v. xvi. 140 Dr. Drinker's machine was called into service. The patient was placed in the machine that was to do his breathing for him. The iron lung forced air into his paralyzed chest and drew it out again. **1938** *Encycl. Brit. Bk. of Yr.* 406/2 The new type of respirator or iron lung..covers only the upper half of the body and does not involve the insertion of the complete person into the apparatus. **1938** *Times* 12 Aug. 14/4 Eric Baker, aged 11, died at Braintree..from infantile paralysis. He was taken ill on Monday and was placed in an iron lung. **1948** *Electronic Engin.* XX. 4/1 In the modern 'iron lung' bellows are used which take upon them the work usually done by the muscles of chest and diaphragm. *a* **1963** S. Plath *Ariel* (1965) 78 No fingers to grip, no tongue, My god the iron lung That loves me, pumps My two Dust bags in and out.

2. *slang.* A Nissen hut.

1943 Hunt & Pringle *Service Slang* 40 *Iron lung*, Barrage Balloon boys' phrase for Nissen hut.

iron-man. Add: **1. c.** *U.S. slang.* A dollar.

1908 H. Green *Maison de Shine* 45 A feller who shells out his six iron men every week. **1926** *Flynn's* 16 Jan. 639/1 Still I can flash a toad-skin now and then, and have a few iron men planted where th' berries grow. **1932** J. Dos Passos *1919* 52 He still had more'n fifty iron men. **1945** *This Week Mag.* 21 Apr. 15/2 When I'd given the boys back their dough I had a nice little profit of two hundred iron men. **1970** E. R. Johnson *God Keepers* (1971) vi. 62 An ounce should bring a street pusher about two thousand iron men.

d. *slang* (orig. *Austral.*). A pound. Cf. *iron *sb.*¹ 4 g.

1959 Baker *Drum* 119 *Iron man*, a £1 note. **1973** J. Leasor *Host of Extras* i. 16 A nut..to whom I'd sold a Bean Tourer—not the most exciting car..but what did he expect for five hundred iron men? **1974** J. Wainwright *Evidence I shall Give* xxviii. 141 Ten thousand iron men... We're talking bank-notes.

ironmongery. Add: **1. c.** *slang.* Firearms.

1902 Kipling *Traffics & Discov.* (1904) 11 They'd all stand to their horses and pile on the ironmongery, and washers, and typewriters, and..they'd sail out..lying down and firing. **1905** A. H. Lewis *Sunset Trail* (1914) x. 160 All men have their delicate side, and it was Mr. Allison's to regard the open wearing of one's iron-mongery as bad form. **1942** E. Waugh *Put out More Flags* 275 Their 'o98 stores arrived; a vast profusion of ironmongery which..included Alastair's mortar. **1960** 'W. Haggard' *Closed Circuit* xv. 177 He wasn't armed. He had never carried ironmongery. **1973** J. Wainwright *Devil you Don't* 112 Shove it. You are only here for the ride. If you hadn't been so damned handy with the ironmongery—.

d. Also in other applications (see quots.).

1892 *Century Mag.* Dec. 218/2 The broad veranda.. had the customary array of..hanging lamps set with bosses of colored glass, and much ironmongery in spirals and curlicues. **1934** H. G. Wells *Exper. Autobiogr.* II. viii. 568, I tried to make him [*sc.* Gissing] a cyclist...'Get on to your ironmongery,' said I. **1955** *Times* 4 Aug. 8/7 A pike which was already the subject of legend..had broken anglers' tackle seven times..appeared to thrive on a diet of triangles, and snap-tackles, plug-baits, five-inch spoons, and other ironmongery. **1958** *Engineering* 4 Apr. 425/1 No good starting with ironmongery—computers—got to start with systems analysis. **1967** *Guardian* 5 May 6/7 With the 'attached ironmongery', there is a disturbing lack of coordination... The lighting standards are not inelegant, but there are too many of them... Guard-rails, crash barriers, gantries, and direction signs..have the appearance of rushed afterthoughts. **1971** D. Haston in C. Bonington *Annapurna South Face* xvii. 212, I started off on the chimney. At this time we were badly short of rope and ironmongery. I had about three hundred feet of rope, four pegs and about six karabiners.

i-ron-o-n, *a.* [f. Iron *v.* 3 + On *adv.*] Such as can be affixed to the surface of a fabric by ironing.

1959 *Times* 12 Jan. 11/5 Woven cotton iron-on interlining. **1966** *Olney Amsden & Sons Ltd. Price List* 35 Iron-on Sheet Markers. **1967** E. Short *Embroidery & Fabric Collage* iv. 94 (*caption*) The fabric, previously backed with an iron-on interlining, is cut with a razor blade and pressed. **1970** *Guardian* 10 Sept. 13/3 Sewing on name tapes is a tedious job. Iron-on tapes can cut down the sewing.

iron-sand. 1. (Earlier and later examples.)

1805 D. McClure *Diary* (1899) 29 The soil abounds in iron sand Ore. **1862** W. K. Hulke *Let.* 9 Aug. in *Richmond–Atkinson Papers* (1960) I. 781 Yesterday the battle about the iron-sand leases came off. **1963** *Times* 6 Feb. (N.Z. Suppl.) p. iv/5 New methods have been developed to process the ironsands of the North Island beaches.

ironweed. (Earlier and later U.S. examples.)

1819 D. Thomas *Trav. Western Country* 231 The *iron-weed*, which I first saw above Pittsburgh, extends on clayey lands all the way to the Wabash. **1963** Gleason & Cronquist *Man. Vasc. Plants Northeastern U.S.* 746 *Vernonia* Schreb. Ironweed... Our spp. bloom in late summer and fall.

Iroquoian (i-rŏkwoi,ăn, -koi-), *a.* and *sb.* [f. *Iroquoi(s *a.* and *sb.* + -an.] **A.** *adj.* Of or pertaining to the Iroquois, or to the language family defined in sense B a below. **B.** *sb.* **a.** A language family which includes the Iroquois, Huron, Cherokee, and several lesser-known American Indian languages. **b.** A member of this linguistic group.

1697 L. Le Compte *Mem. Journey through China* I. v. 124 Iroquian Mathematicians, or Learned Alkonkins. **1888** J. C. Pilling (*title*) Bibliography of the Iroquoian languages. *Ibid.* Pref. p. v, To the Iroquoian perhaps belongs the honor of being the first of our American families of languages to be placed upon record. **1906** *Rep. Brit. Assoc. Adv. Sci.* 679 The Iroquoian tribes of North America possess a word which exactly expresses this potentiality. **1917** W. K. Moorehead *Stone Ornaments Indians* 170 A surface find in the Iroquoian area in New York is no sure indication that the artifact is Iroquoian. **1933** L. Bloomfield *Lang.* iv. 72 The Iroquoian family was spoken in a district surrounded by Algonquian; it includes..the Huron (or Wyandot) language, and the languages of the Iroquois type..; in a detached region to the south Cherokee was spoken. **1934** D. Jenness *Indians of Canada* (ed. 2) xix. 288 When Jacques Cartier sailed up the St. Lawrence river in 1535 he found Iroquoians cultivating the land and controlling the country around the present site of Montreal. **1959** E. Tunis *Indians* 19/2 Remnants of the temple-mound culture survived into historical times among the Muskhogean Indians of the Southeast and there are broad hints of a connection between the burial-mound people and the Iroquoians. **1968** F. G. Lounsbury in J. A. Fishman *Readings Sociol. of Lang.* (1968) 51 The writer has done intensive field work on a language—Cayuga, one of the languages of the Iroquoian family. **1969** *Observer* (Colour Suppl.) 25 May 53/1 Various groups that spoke dialects of the Iroquoian language.

Iroquois (i-rŏkwoi, -koi *U.S.*, i·rŏkwa *Canad.*), *a.* and *sb.* [Fr., from some Algonquian language, perh. Montagnais (G. Day in E. Tooker *Iroquois Culture* (1967), 57–61).] **A.** *adj.* Of or pertaining to an American Indian group of peoples (the 'Five Nations') encountered esp. in Ontario and central and northern New York State, or to the languages of this group; = *Iroquoian *a.* **B.** *sb.* **a.** A member of this group (see *Five Nations* (*five *a.* and *sb.* C. 2)); also as collect. sing., this confederacy. **b.** More widely = *Iroquoian *sb.* **c.** The languages of this group.

1666 in *Documents Colonial Hist. New-York* (1853) III. 134 The Irocquois Indians should not cōmit any Act of hostilety. **1677** Locke tr. *Nicole's Ess.* (1828) 71 There needs not many other thoughts to make up the complete idea of an Iroquois. **1705** [see *Algonquin, -kin *sb.* and *a.*]. **1710** Shaftesbury *Advice to Author* 179 Historys of Incas or Iroquois, written by Friars and Missionarys. **1756** [see *Huron]. **1791** [see *Eskimo *a.* 1]. **1851** *Harper's Mag.* Aug. 390/2 He fancied he heard her mutter in Iroquois one word—'revenged'! **1852** [see *Five Nations* (*five *a.* and *sb.* C. 2)]. **1866** G. M. Wallace tr. *Beethoven's Lett.* II. 91, I must also strive to make some future provision for him [his nephew]; being neither Indians nor Iroquois, who, as we know, leave everything to Providence, whereas we consider a pauper's existence to be a very sad one. **1881** [see Gyneocracy]. **1922** D. Canfield *Rough-Hewn* (1923) i, Neale was silent as an Iroquois. *Ibid.* xxv, His Iroquois mask of insensibility. **1933** [see *Cayuga]. **1959** [see *Five Nations* (*five *a.* and *sb.* C. 2)]. **1965** *Canad. Jrnl. Ling.* Spring 135 Iroquois speakers remain in New York State;..others remain in the Carolinas (Cherokee). **1969** *Listener* 12 June 834/1 It is curious to think of Winston Churchill having Red Indian blood, but his mother was one-eighth Iroquois.

irradiance. Add: **2.** The flux of radiant energy per unit area, esp. an area normal to the direction of travel through a medium.

1956 A. Hollaender *Radiation Biol.* III. iii. 129 'Irradiance' is the intensity term applicable to the interception of radiant energy by objects and is power per unit area. Frequently 'intensity' is used loosely as a substitute for 'irradiance'. *Ibid.* 136 The U.S. Weather Bureau..still uses the calorie per minute per square centimeter for the specification of solar irradiance. **1966** *McGraw-Hill Encycl. Sci. & Technol.* XII. 105/2 At a depth of 64 m where the irradiance is 0·5 watt/m²..approximately 0·04 watt of radiant power is absorbed by every cubic meter of sea water. **1969** *Physics Bull.* Oct. 409/2 Outside the earth's atmosphere..the annual mean irradiance of the sun is about 270 W m⁻² on a plane parallel to the earth's surface at latitude 52° N. **1973** *Nature* 9 Feb. 402/1 The total irradiance of the unattenuated light..was approximately $4{\cdot}5 \times 10^3\ \mu W/cm^2$.

irradiate, *v.* Add: **1. d.** To expose to the action of some kind of radiation (other than visible light, as X-rays, ultra-violet radiation, or neutrons).

1901 *N.Y. Med. Jrnl.* 16 Nov. 909/1 The inguinal tumor was removed..and now the inguinal area is also irradiated. **1927** *Jrnl. Biol. Chem.* LXXIII. 383 Cholesterol solutions were also irradiated by the γ-rays from radium emanation. **1952** Cook & Duncan *Mod. Radiochem. Pract.* v. 203 To obtain the maximum yield of a radio-element by use of a laboratory neutron source it is often of advantage to irradiate an aqueous solution of the absorbing element. **1957** *Technology* Mar. 14/2 Food can be preserved for long periods if irradiated. **1963** Bowen & Gibbons *Radioactivation Analysis* ii. 5 When a material is bombarded or irradiated by the nuclear particles produced in a nuclear reactor, particle accelerator, or other suitable source, some of the atoms present in the sample will interact with the bombarding particles. **1964** M. PYKE *Food Sci. & Technol.* viii. 153 By irradiating frozen whole egg with from 0·1 to 1·0 Mrad, it is possible to destroy certain pathogenic micro-organisms. **1971** *Nature* 4 June 317/1 When the spores were initially irradiated with γ-rays, the germination rate increased remarkably. **1973** *Sci. Amer.* Apr. 71/1 The diffusion constant can now be determined by irradiating the silicon with slow neutrons and measuring the energy of the alpha particles that emerge.

irradiated, *ppl. a.* Add: **1. c.** Exposed to the action of some kind of radiation (see prec., 1 d).

1915 COLWELL & RUSS *Radium, X-Rays & Living Cell* iii. 117 The nuclear changes observed in the development of such irradiated ova have been investigated. **1931** *Times* 13 May 17/4 A cheap and effective means of obtaining Vitamin D has been made available in the form of irradiated ergosterol. **1957** BENEDICT & PIGFORD *Nucl. Chem. Engin.* i. 17 The most important neutron-absorbing and long-lived fission products in irradiated uranium are listed in Table 1.2. **1968** *Observer* 11 May 8/1 Japanese doctors could draw upon a wealth of medical information gained by systematic examinations of men, women and children who had survived the blast, heat and radiation at Hiroshima and Nagasaki. But..knowledge about the treatment of irradiated individuals was woefully inadequate. **1970** *New Scientist* 6 Aug. 284/1 Most countries have banned the sale of irradiated food... However, in the UK the Minister of Health can exempt a particular food..if evidence is submitted to show that irradiation is harmless.

irradiation. Add: **3. b.** *Photogr.* The scattering of light by silver halide crystals in a photographic emulsion causing diffuseness of the image obtained on development.

1924 L. P. CLERC *Ilford Man. Process Work* xi. 53 The effect of irradiation, evidently, is the more marked,..as the exposure is longer. **1940** C. I. JACOBSON *Developing* ii. 43 If the exposure is longer, the then light is scattered so that it spreads beyond the area protected by the metal, and hence irradiation takes place. **1968** H. BAINES in E. S. BOMBACK *Sci. Photogr.* (ed. 2) 1. 169 This scatter from one crystal to others is known as 'irradiation'.

9. Exposure to the action of some kind of radiation (other than visible light, as X-rays, ultra-violet radiation, or neutrons); the (or an) action or process of irradiating something. Also, radiation allowed to be incident upon something.

1901 *N.Y. Med. Jrnl.* 16 Nov. 908/1 Up to today irradiation has been done seven times (on the same patient). *Ibid.* 909/1 If a strong effect is desired, intense irradiation must naturally be employed. **1915** COLWELL & RUSS *Radium, X-Rays & Living Cell* iii. 116 Radium, like X rays, does not effect the immediate death of the cell; specimens subjected to three days' continuous irradiation still underwent division. **1935** *Practitioner* Libr. *Med. & Surg.* VII. v. 158 Ultraviolet irradiation of the skin is effective in preventing or curing rickets. *Ibid.* 159 Short exposures of thin films of milk to ultraviolet irradiation. **1958** B. J. M. HARRISON *Textbk. Roentgenology* iii. 52 If he moves out of position the irradiation falls on the protected covers and not upon the patient. **1951** *Jrnl. Sci. Instruments* XXVIII. 191/1 The neutron irradiation of small quantities of material in the pile is often carried out in aluminium foil 'envelopes' or in silica capsules. **1953** CARTER & MERRITT in Smith & Werner *Med. Treatm.* xx. 433/1 Daily shortwave diathermy in combination with intra-red irradiation twice applied to the lumbar area may be of value. **1953** *Cold Spring Harbor Symp. Quant. Biol.* XVIII. 101/2 Irradiation of cultures of lysogenic *Bacillus megatherium* with ultraviolet light greatly increased the proportion of bacteria producing phage. **1957** *Times* 3 Sept. 9/2 Therapeutic irradiation of the pelvic region would certainly involve considerable risk to an embryo in the direct beam. **1972** *Physics Bull.* July 398 The damage produced during irradiations with 10 MeV C ions and 48 MeV Ni ions has been normalized to that produced by 4 MeV protons where we can make reasonably accurate estimates of the number of displaced atoms.

irradiator. (Later example: cf. *IRRADIATE *v.* 1 d.)

1971 *Nature* 12 Mar. 120/2 Corneas were first irradiated for 15 s in a γ-irradiator..delivering 1·2 × 10⁴ rad/h.

irrationalist. Add: Also *attrib.* or as *adj.*

1867 H. M. CECIL (*title*) Pseudo-philosophy at the end of the nineteenth century: an irrationalist trio: Kidd, Drummond, Balfour. **1910** W. JAMES *Mem. & Stud.* (1911) xv. 392 Listen for a moment to such irrationalist deliverances on his part as these.

irrationalistic (iræʃənăli·stik), *a.* [f. IRRATIONAL *a.* + -ISTIC.] Characterized by irrationalism; contrary to reason; illogical.

1910 W. JAMES *Mem. & Stud.* (1911) xv. 400, I spoke a while ago of its being an 'irrationalistic' philosophy in its latest phase. **1912** *Q. Rev.* Oct. 364 This brings us to the fundamental difference between the standpoints of history and science, which the theology called 'irrationalistic' appears to have overlooked. **1920** A. S. PRINGLE-PATTISON *Idea of God* (ed. 2) 64, I have dwelt in the latter part of this lecture on the tendency to slip into an anti-intellectualistic, and even irrationalistic, mode of statement in expressing the principle of value. **1934** A. C. EWING *Idealism* v. 250 The irrationalistic and excessively pluralistic tendencies of the present day. **1963** R. WELLEK in N. Frye *Romanticism Reconsidered* 121 He singles out the most irrationalistic writers.

irreconciliation. Delete † *Obs.* and add later examples.

1906 *Daily Chron.* 1 Oct. 5/6 Where..brotherly love and charity [have long been] enemies sworn to irreconciliation. **1927** *Brit. Weekly* 24 Mar. 639/1 Science has its confusions and irreconciliations no less than religion.

irredeemable, *a.* (*sb.*) Add: **1.** Also, *irredeemable debenture* (see quot. 1965).

1837 *Daily News* 3 July 4/5 £600,000 in Four-and-a-Half per Cent. Irredeemable Mortgage Debenture stock at £108. **1965** PERRY & RYDER *Thomson's Dict. Banking* (ed. 11) 327/1 *Irredeemable debenture*, a debenture which does not contain any provision for repayment of the principal money. Even if irredeemable, it falls to be paid upon the company going into liquidation.

b. (Earlier U.S. example.)

1837 H. WEBSTER in *Niles' Weekly Reg.* 6 May 155/3, I abhor paper; that is to say irredeemable paper, paper that may not be converted into gold or silver at the will of the holder.

B. sb. Restrict † *Obs.* to sense in Dict. and add: **b.** Anything that is irredeemable; *spec.* an irredeemable debenture.

1904 *Daily Chron.* 6 Feb. 3/2 The redemption of the irredeemable by woman's sweet and subtle influence the author has sapped us. **1953** *Economist* 20 Aug. 514/1 Prices of most stocks at their lowest for twenty years, with irredeemables offering flat yields ranging up to..4½ per cent. **1967** *Ibid.* 18 Nov. 785/2 The main effect..would eventually be felt by the long end of the market, especially by the irredeemables. **1973** *Daily Tel.* 24 Nov. 27/4 Most of the irredeemables return over 11 p.c. on income.

‖ **irredenta** (irĕde·ntă). [It. (see IRREDENTIST).] A region containing people who are ethnically related to the inhabitants of one state but are politically subject to another. Also, post-positively, as *adj.*

1914 *Everybody's Mag.* Sept. 333 These million Germans along the Baltic..constitute a Germania Irredenta which is one of the things that Kaiser Wilhelm is after. **1916** *Times Hist. War* 31 Oct. 407/1 Rumanian interests would naturally be first in the region. **1926** *Times Lit. Suppl.* ... Italy's..irredenta..by the coming into being of the Central European bloc but for the fact that between them loomed the question of Rumania's irredenta. *Ibid.* 408/2 She [sc. Rumania] had no territorial Irredenta to secure. **1934** A. HUXLEY *Beyond Mexique Bay* 73 British Honduras still is regarded by the Guatemalans as an *irredenta*. **1938** H. NICOLSON *Diary* 22 Aug. (1966) 356 Even if Germany absorbed a portion of the Ukraine, how would it benefit her? It would create an *irredenta* and perpetuate Russo-German enmity. **1967** *Punch* 1 Feb. 141/2 They annexed Tibet, but they had argued themselves into the belief that this was ancient Chinese territory, China irredenta.

irredentism. Add: Also in extended use of any policy of seeking the recovery and reunion to one country of a region or regions for the time being subject to another country.

1922 *Encycl. Brit.* XXX. 314/1 Up to the World War there was actually no articulate irredentism among the Austrian Poles. **1933** *Times Lit. Suppl.* 5 May 319/1 Never was 'irredentism' so rampant as it is today; and so far as Germans and Magyars are concerned there does not seem much prospect of their reshaping their minds. **1961** *Listener* 21 Dec. 1057/1 The young African states are..learning that they are no more immune than the wicked old nations from the evils of frontier disputes, irredentism, and even ideological differences. **1973** *Ibid.* 20 Dec. 845/3 The seeds of Japanese irredentism, already latent, will begin to sprout.

irredentist. Add: Also in extended use (see prec.).

1919 G. B. SHAW *Peace Conf. Hints* vi. 75 The French and Italian Jingos have been..making no secret of their determination to annex parts of the Rhineland and the Austrian Tyrol..without regard to the irredentist movements which must follow such annexations. **1920** (see *GRAB *v.* 2 b). **1958** G. MIKES *East is East* 18 Naturally, I instinctively sympathized with Hungary and became a staunch irredentist at the age of six or so. Later..some subversive older friends assured me that irredentism was wrong. **1961** *Listener* 21 Sept. 411/1 The extremist parties [in the German Federal Republic] representing communist, neo-nazi, and irredentist tendencies. **1972** D. DAKIN *Unification of Greece* v. 73 The bolder irredentists in Greece had taken the field.

irreducibly, *adv.* (Later example.)

1923 C. D. BROAD *Sci. Thought* x. 368 The temporal relations..are really irreducibly triadic.

irreducible, *a.* (Later examples.)

1922 JOYCE *Ulysses* 673 What anthem did Bloom chant partially in anticipation of that multiple, ethnically irreducible consummation? **1964** *English Studies* XLV. 98 He described the scientist's labour as a hammering out of truth in the teeth of irreducible stubborn facts.

irredundant (irĭdʌ·ndănt), *a. Math.* [f. IR-² + REDUNDANT *a.*] Containing no redundant elements.

1925 A. CHURCH in *Trans. Amer. Math. Soc.* XXVII. 318 A set of postulates is irredundant if the postulates are independent and no one of them can be weakened with respect to the set. **1957** *IBM Jrnl. Res. & Devel.* I. 175/2 To be irredundant the statement has to involve the complete list of reasons which are necessary and sufficient to make this prime implicant dispensable. **1965** R. E. MILLER *Switching Theory* I. 195 An 'irredundant cover' of a complex has the property that if any cube is eliminated from the cover, the resulting set of cubes is no longer a cover. **1966** *Math. Rev.* XXXI. 37/1 A factorization $a = a_1 \ldots a_n$ of a into simple factors is irredundant if no product $a_1 a_2 \ldots a_{i-1} a_{i+1} \ldots a_n$, $p > 0$, is simple.

Hence **irredu·ndance, irredu·ndancy** *sbs.*, the property of being irredundant.

1925 *Trans. Amer. Math. Soc.* XXVII. 320 (*heading*) A criterion for irredundance. **1952** *Proc. Amer. Catholic Philos. Assoc.* XXVI. 112, I fulfills the irredundancy requirement. **1960** *IRE Trans. Electronic Computers* IX. 246/2 Comparing all the resulting implications for irredundancy.

irreflexive, *a.* Delete *rare* and add: *Logic* and *Math.* Of a relation which no postulate can have to itself. Hence **irrefle·xiveness, irreflexi·vity**, the quality or property of being irreflexive. See also *NON-REFLEXIVE *a.*, *REFLEXIVE *a.*

1933 *Mind* XLII. 36, *a* < *b*. ⊃. *a* ≠ *b*. gives..the property of irreflexiveness. **1937** A. SMEATON tr. *Carnap's Logical Syntax of Lang.* iv. § 63. 234 A relation F is called irreflexive when no object has this relation to itself. **1943** J. C. COOLEY *Primer of Formal Logic* 157 A relation, *f*, is said to be *irreflexive* when: (*x*) ~*f*ᵡ. **1952** S. C. KLEENE *Introd. Metamath.* vii. 188 Connexity, irreflexiveness, asymmetry. **1954** I. M. COPI *Symbolic Logic* v. 143 An irreflexive relation is one which no individual has to itself. **1963** W. V. QUINE *Set Theory* § 21. 147 Or, better, stipulate the irreflexivity and derive the asymmetry. **1964** E. BACH *Introd. Transformational Gram.* vii. 156 No item precedes itself (irreflexivity). **1967** R. A. GEORGE tr. *Carnap's Logical Struct. of World* § 11. 22 A relation is called..irreflexive if it is irreflexive and transitive (and hence asymmetrical) and connected.

irregardless, *a.* and *adv.* Chiefly *N. Amer.* [Prob. blend of *irrespective* and *regardless*.] In non-standard or humorous use: regardless.

1912 in WENTWORTH *Amer. Dial. Dict.* **1923** *Lit. Digest* 17 Feb. 76 Is there such a word as irregardless in the English language? **1934** in WEBSTER (labelled Errone. or Humorous, U.S.). **1938** T. KUHN *Assigned to Adventure* xxi. 170, I made a grand entrance and suffered immediate and complete obliteration, except on the pay-roll, which functioned automatically to present me with a three-figure cheque every week, 'irregardless', as Hollywood says. **1939** C. MORLEY *Kitty Foyle* xxvii. 267 But she can take things in her stride, irregardless what's happened. **1955** *Publ. Amer. Dial. Soc.* xxiv. 19, I don't think like other people do and irregardless of money or how little the dope would cost me [etc.]. **1970** *Current Trends in Linguistics* X. 590 She tells the pastor that he should please quit using the word 'irregardless' in his sermons as there is no such word. **1972** M. MCCRAY *Any who left Well Enough* iv. 96 The sun poured down on Purity irregardless of the fact that it received no welcome.

irregular, *a.* Add: **II. 8*.** *Astr.* **a.** Of a galaxy: having an irregular shape and lacking any axis of symmetry or central nucleus. Also *ellipt.* as *sb.*

1811 W. HERSCHEL in *Phil. Trans. R. Soc.* CI. 296 By calling the figure of a nebula irregular, it must be understood that I saw no particular dimension of it sufficiently marked to deserve the name of length. **1875** *Encycl. Brit.* II. 811/1 Among the varieties of form may be noted spiral, elliptic, and ring nebulæ, double nebulæ, and irregular nebulæ. **1928** J. H. JEANS *Astron. & Cosmogony* i. 28 The irregular nebulae show the bright line spectrum which is characteristic of a transparent gas. **1936** E. HUBBLE *Realm of Nebula* ii. 47 About half of the irregulars form a homogeneous group, in which the Magellanic Clouds are typical examples. **1959** *Listener* 31 Dec. 1152/1 There are a few galaxies that do not fit conveniently into this classification of spirals, ellipticals, and irregulars. **1965** J. MULLER *Handbk. Astron.* xxiv. 259 Irregular galaxies seem to be primarily late developers; spirals are in the prime of life, while elliptical galaxies are bankrupt.

b. Of a variable star: fluctuating in brightness in a way that lacks any definite rhythm.

1903 A. M. CLERKE *Probl. Astrophysics* xxiii. 363 (*heading*) Peculiar and irregular variables. **1935** F. HOYLE *Frontiers Astron.* xi. 190 Certain irregular variables are of exceptional interest. **1967** D. H. MENZEL et al. *Survey Universe* xxvi. 551 Truly irregular variables, like R Coronae Borealis, suffer brightness changes in abrupt and unpredictable fashion.

irremeable, *a.* (Later example.)

1974 *Encounter* Feb. 54/1 The subject of correctness in language is now tending to be lost in an irremeable labyrinth.

irrepealability. (Earlier U.S. example.)

1802 *Deb. Congress U.S.* 18 Feb. 592 The character of irrepealability was not exclusively attached to this law.

irrepressible, *a.* (*sb.*) (Earlier example.)

1811 JANE AUSTEN *Sense & Sens.* III. ix. 298 His was an involuntary confidence, an irrepressible effusion.

irreption. Delete † *Obs.* and add later examples.

1966 G. W. S. FRIEDRICHSEN *Gothic Version of Gospels* 190 Previous to this there had been casual but continued irreptions from the Old Latin. *Ibid.* 249 The Gothic reading could..be explained as a corruption due to the irreption of some parallel or reminiscent passage. **1974** *Encounter* Feb. 54/1 A protection against casual and deplorable irreptions creeping into the language.

irretention. (Later example.)

1963 V. NABOKOV *Gift* ii. 85 All this sick irretention of electric light.

irreticence. [IR-².] The condition of being irreticent. With *an* and *plural*: an instance of this.

1919 V. WOOLF *Night & Day* xvi. 211 Rodney might begin to talk about his feelings, and irreticence is apt to be extremely painful. **a 1941** —— *Captain's Death Bed* (1950) 111 Those irreticences and hyperboles which the voice of the speaker corrects in talk.

irreticent, *a.* (Later example.)

1931 V. WOOLF *Let.* to Young Poet 6 Therefore you could afford to be intimate, irreticent, indiscreet in the extreme.

irreversible, *a.* Add: **3.** *Physical Chem.* Of a colloid or colloidal system: incapable of being changed from a gelatinous state into a sol by a reversal of the treatment which turns the sol into a gel or gelatinous precipitate. Of a change of state: characterized by this property.

[**1899** *Jrnl. Physiol.* XXIV. 180 (*heading*) Colloidal mixtures which form irreversible molecular aggregates when they pass into the gel state.] **1900** *Proc. R. Soc.* LXVI. 110 A large number of colloidal solutions..belong to the class of irreversible colloidal mixtures. **1915** M. H. FISCHER tr. *Ostwald's Handbk. Colloid-Chem.* 40 When a change in the state of a colloid may be reversed by reversing the conditions which brought that change about, it is said to be 'reversible'. Thus when a colloid which has been precipitated by a salt goes back into solution on removal of the salt, the colloid change is said to be 'reversible'. On the other hand, if this does not occur it is 'irreversible'. **1930** J. C. WARE *Chem. Colloidal State* ix. 204 When a reversible colloid is evaporated to dryness and later stirred into the fluid which constituted the external phase, a very complete dispersion will again result. With an irreversible colloid, a suspension will not result by mixing with the solvent but one of the regular methods for the preparation of the colloidal state must be applied. **1930** *Engineering* 18 July 61/1 Gels which cannot be converted into sols are 'irreversible'. **1959** K. J. MYSELS *Introd. Colloid Chem.* iv. 81 These different behaviors are quite generally called reversible and irreversible flocculations.

irrigate, *v.* Add: **4.** *intr.* To drink; to take a drink. *slang* (chiefly *U.S.*).

1856 'J. PHOENIX' *Phœnixiana* 104 [He] was invited by the urbane proprietor to irrigate. **a 1880** in Thornton *Amer. Gloss.* (1912) II. 475 'Stranger, do you ever irrigate?' 'No, I thank you, sir, I do not.' **1905** A. ADAMS *Outlet* xxi. 296 Spondulier..called every one up to the bar to irrigate. **1911** E. M. CLOWES *On Wallaby* viii. 202 There was even a further decrease in drunkenness, people having no money, I suppose, for what our hero [sc. Victoria] they call 'irrigating'.

irrigation. Add: **4.** *irrigation canal, ditch*; **irrigation-wheel**, a wheel by means of which land is irrigated.

1910 *Encycl. Brit.* XIV. 841/1 When a river partakes of the nature of a torrent, ..it is impossible to construct a system of irrigation canals without very costly engineering works. **1925** A. FRY *Lit.* 24 May (1972) II. 536 A vast flat plain perfectly green with corn,..and all run by irrigation canals from the Ebro. **1870** *Gov. Hunt Own Story* (MS.) 1 Irrigation ditches were made for us as much of the land as could be covered with water. **1902** O. WISTER *Virginian* vi. 77 One of the irrigation ditches ran under the fence from the hay-field to supply the house with water. **1958** [see *ACEQUIA]. **1974** J. THOMSON *Long Revenge* vi. 67 The only possible cover was that provided by the tall reeds that grew in the deeper irrigation ditches. **1864** J. M. DA COSTA *Med. Diagn.* iv. 280 These statements are not intended to be final. They are but a very short summary of the results of a large number of observations which I have had an opportunity of making on these cases of 'irritable heart', and which elsewhere..will be laid before the profession. **1922** [see *cardiological adj.* s.v. *CARDIO-]. **1973** COOK & HORWITZ *Cardiac & Vascular Dis.* I. xxiv. 600/1 This condition has been known by many names, such as irritable heart, soldier's heart, disordered action of the heart, functional heart disease, effort syndrome, and neurocirculatory asthenia.

irritable, *a.* Add: **2. b.** *irritable heart*, a syndrome characterized by shortness of breath on exertion, palpitation, fatigue, chest pain, and dizziness, and believed to be psychosomatic; 'soldier's heart'.

1864 J. M. DA COSTA *Med. Diagn.* iv. 280 These statements are not intended to be final.

irrotationality (i:rotēʃənæ·liti). *Physics.* [f. IRROTATIONAL *a.* + -ITY.] The property or state of being irrotational.

1960 *McGraw-Hill Encycl. Sci. & Technol.* VII. 402/1 For an incompressible fluid, the continuity equation..is div v = 0; hence, combining this relation with irrotationality gives Laplace's equation, div (grad ϕ) = 0. **1967** *Progress Aeronaut. Sci.* VIII. 71 The velocity field..is determined by the equation of continuity and the condition of irrotationality.

irrumation (irumē·ʃən). [f. ppl. stem of L. *irrumāre* to practise fellatio on or L. *irrumātiō, -ōnis* the action of the verb.] (See quot. 1901.) Also **i·rrumate** *v. trans.*, to practise irrumation on (a person); **i·rrumator**, one who practises irrumation.

1887 L. C. SMITHERS tr. *Forberg's Man. Classical Erotology* iii. 72 To put the member in erection into some one's mouth is called to irrumate, a word, which in its proper sense means to give the breast. *Ibid.* 90 Irrumators are less feared by married men. *Ibid.* 97 Examples, in his *Sayings*,..does not deny that in his time the obscene practice of irrumation was still known. **1888** tr. *Priapeia* 39 If thou shalt attempt a third theft,..I will sodomise and irrumate thee. **1901** A. ALLINSON tr. *Rosenbaum's Plague of Lust* II. xxi. 3 Very much more abominable and repulsive still is the habit of Irrumation (..to erect the penis and insert it into the mouth of another person). *Ibid.* 20 The *irrumator*..takes the *fellator* between his opened thighs. **1947** F. L. HARRISON *Dict. Psychol.* 187 *Irrumation*, the act of obtaining an orgasm by the mouth; fellatio or penilingus.

irruption. Add: *spec.* An abrupt local increase in the numbers of a species of animals.

1912 W. E. CLARKE *Stud. Bird Migration* II. xxi. 112 During the remarkable irruption of Crossbills from the Continent in the summer of 1909, Fair Isle received many of the visitors. **1936** A. L. THOMSON *Bird Migration* ii. 42 Apart from all the categories of annual movements, there are movements which occur at irregular intervals in the form of invasions or irruptions... In the spring of certain years the birds have 'irrupted' in large numbers. **1968** *New Scientist* 21 Nov. 425/1 The majority of migrants [sc. butterflies] reaching Britain are regularly seen within a few days of an irruption in their home territory.

¶ Confused with ERUPTION. Delete *Obs.* and add later U.S. examples.

1883 'MARK TWAIN' *Life on Mississippi* 244 A filament-obliteration irruption of gorgeous flame. **1892** —— *Amer. Claimant* 62 A volcanic irruption.

is. (Properly s.v. *BE *v.*) Infinite vars. of the expression (*a*) — *is a* — *is a* —, on the model of Gertrude Stein's line (see quot. 1922).

1922 G. STEIN *Geogr. & Plays* 187 Rose is a rose is a rose is a rose. Loveliness extreme. **1968** *Listener* 5 Sept. 292/1 Trevelyan is Trevelyan is Trevelyan, and film censorship will endure for as long as he keeps on fighting. **1970** *Guardian* 2 Apr. 10/1 There is only one art form common to all sorts and conditions of people: the poster..a hoarding is a hoarding is a hoarding. **1971** 'L. BLACK' *Death has Green Fingers* xiv. 158 Let me adapt the quotation. 'A crook is a crook is a crook.' **1974** *Times* 3 Jan. 15/1 As Miss Gertrude Stein would have said: 'A union is a union is a union.'

is (iz), *sb.* [f. the vb.] That which exists, that which is; the fact or quality of existence.

1897 F. THOMPSON *New Poems* 164 Could I face firm the Is, and with Tobit trust Heaven. **1949** *North Amer. Rev.* Apr. 507 She is not is Has been, she is as is. **1951** R. K. MERTON *Social Theory & Social Structure* Introd. 37 The blueprint of his culture and society, the 'should-be' rather than the 'is'. **1958** E. PEPLER *Eng. Relig. Heritage* iv. viii. 300 The man is conscious that he is, and in comparison to the Is of God, this realization is itself the greatest sorrow.

I's: see *ISE, I'SE b.

Isabella. **2. b.** (U.S. examples of absol. use.)

1846 *Knickerbocker* XXVII. 419 A snaky looking vine..from which glorious bunches of Catawbas and Isabellas were to depend. **1864** [see *CONCORD *sb.*² 5]. **1949** *Amer. Photography* Apr. 244/1 *Vitis labrusca*..has furnished the catawba,..the Concord,..the Isabella, and others.

isacoustic: see *ISO-.

is all. (Properly s.v. *ALL *a.* 8 g.) *U.S.* and *Canad. colloq. phr.*, a shortened form of *that is all*.

1954 G. BREWER *Killer is Loose* iii. 27 You didn't see the boss, is all. **1967** W. WRIGHT *Shadows don't Bleed* vii. 126 I'm not married. Paula looks after the house for me, is all. **1969** C. HIMES *Blind Man with Pistol* xiv. 152, I help you look, just don't call me nigger is all.

isallobar (əis,æ·lobăr). *Meteorol.* [f. Gr. ἴσ-ος equal (ISO-) + ἄλλο-ς other (*ALLO-) + βάρ-ος weight, after *isobar*.] A line (imaginary or on a map) connecting points at which the barometric pressure has changed by an equal amount during a specified time. Hence **isallo·baric** *a.*

1911 N. SHAW *Forecasting Weather* xv. 337 Dr. Nils Ekholm, of Stockholm,..uses charts of isallobars. *Ibid.*, Similar groups of isallobaric areas may be traced on many occasions. **1934** D. BRUNT *Physical & Dynamical Meteorol.* ix. 189 Brunt and Douglas found that the isallobaric component of wind frequently amounted to as much as 5 metres/sec. **1959** S. L. HESS *Introd. Theoret. Meteorol.* xiv. 226 This solution is for the case of isobars and isallobars (lines of constant pressure tendency) which are oriented east-west. **1962** *Course Elem. Meteorol.* (Meteorol. Office) iii. 25 By analogy with isobars we have isallobaric highs and lows, showing centres of rising and falling pressure respectively.

isanomalous, -anomaly: see *ISO-.

isat-. Add: **isa·togen**, any compound containing the bicyclic group $C_8H_4NO_2$ found in isatin; hence **isatoge·nic** *a.*

1882 *Jrnl. Chem. Soc.* XLII. 198 Isatogenic acid is very unstable, and cannot be obtained either from the above salt, or from free orthonitropropiolic acid, as it is immediately converted into isatin. *Ibid.* 620 To ascertain the nature of the isatogen-group, experiments were made on its derivatives. **1898** ROSCOE & SCHORLEMMER *Treat. Chem.* III. v. 262 When the solution is poured into water, the isatogenic acid decomposes instantaneously into carbon dioxide and isatin. **1916** *Chem. Abstr.* X. 1334 The isatogens are typical quinones, forming a new class of o-quinoid compds., in which the group N:O plays the role of a CO group. **1948** WEYGAND & BAUMANN in *Diseren's Chem. Technol. Dyeing & Printing* I. i. 26 The formation of indigo can be explained by an intermediate rearrangement of the o-nitrophenyl propiolic acid into isatogenic acid... The latter product is reduced to indigo by splitting off CO_2. **1965** *Jrnl. Pharmacy & Pharmacol.* XVII. 736 The isatogens..were all active against Gram-positive organisms. **1970** W. A. REMERS in W. J. Houlihan *Chem. Heterocyclic Compounds* XXV. i. 19 Isatogens..also undergo addition to the 1,2-double bond.

isatoic (aisătō·ik), *a. Chem.* [See below.] **1.** *isatoic acid* [tr. G. *isatosäure* (H. Kolbe 1884, in *Jrnl. f. prakt. Chem.* XXX. 85): see ISAT-]. † **a.** = *isatoic anhydride* (below). *Obs.* **b.** *N*-Carboxyanthranilic acid, $C_6H_4(COOH)(NH\cdot COOH)$, known only as its derivatives.

1884 *Jrnl. Chem. Soc.* XLVI. 1. 939 'Isatoic acid' is thus not an acid, but the anhydride of a dicarboxylic acid, COOH·C_6H_4·NH·COOH. The author calls this dibasic acid isatic acid, and the anhydride, previously known as isatoic acid, he terms isatoic anhydride. **1944** *Jrnl. Org. Chem.* IX. 55 Isatoic anhydride is a convenient reagent for certain anthraniloylations. **1972** *Jrnl. Pharmaceutical Sci.* LX. 1252/1 The condensation of isatoic anhydride with primary amines was found to yield the corresponding substituted anthranilamides.

2. *isatoic anhydride* [after G. *isatosäureanhydrid* (E. Erdmann 1889, in *Ber. d. Deut. Chem. Ges.* XXXII. 2163)]. The anhydride,

$$C_6H_4\begin{matrix}\text{CO·O·CO·NH,}\\\text{|}\end{matrix}$$

of isatoic acid, obtained by oxidation of isatin.

1899 *Jrnl. Chem. Soc.* LXXVI. 1. 939 'Isatoic acid' is thus not an acid... [see above]. **1930** *Chem. Abstr.* XXIV. 4793 Isatoic acid and its derivs. are obtained by treating cong acid solns. of aromatic o-aminocarboxylic acids with $COCl_2$.

Isaurian (aisǭ·riăn), *sb.* and *a.* [f. *Isauria* (see below) + -AN.] **A.** *sb.* A native or inhabitant of Isauria, an ancient country in Asia Minor, between Cilicia and Phrygia; *spec.* applied to a line of emperors of the Eastern Roman Empire. **B.** *adj.* Of or belonging to Isauria, or to the emperors thus called.

1776 GIBBON *Decl. & F.* I. x. 285 In the heart of the Roman monarchy, the Isaurians were converted into a nation of wild barbarians. **1843** *Penny Cycl.* XXVII. 770/2 The increasing power of the brothers and other Isaurian friends of Zeno. **1880** *Encycl. Brit.* XI. 114/2 The Emperors of this time were those of the Isaurian, Armenian, and Amorian dynasties. **1904** W. M. RAMSAY *Lett. Seven Churches* xviii. 210 The Empire of Rome has been..transformed into a Roman-Asiatic Empire, on whose throne sat successively Phrygians, Isaurians, Cappadocians, and Armenians. **1957** J. M. HUSSEY *Byzantine World* ii. 29 The Isaurians were administrators as well as soldiers.

‖ **isba** (izbā·). Also isbah, izba. [ad. Russ. *izbá* (related to STOVE *sb.*¹).] A Russian hut or log-house.

1784 J. KING in Cook *Voy. Pacific Ocean* III. 374 These houses consist of three distinct sorts, jourts, balagans, and log-houses, called here isbas. **1823** R. PINKERTON *Russia* 24 These simple articles compose the whole..furniture of a Russian isba. **1883** *Harper's Mag.* Jan. 251/1 Here we left, if they wanted new isbahs—alias log huts—would get from the priests..to venerate their petition. **1894** *Daily News* 22 Jan. 6/7 The meanest Kirghis yourt was more artistically decorated than any grimy, unventilated isba. **1943** E. M. ALMEDINGEN *Frossia* iii. 89, I was twenty-one, married, had my own isba, owned my own livestock. **1962** *Observer* 20 May 21/1 We lived in a tiny isba..of log cabin, by the roadside.

‖ **isblink** (ī·sblink). Also 8–9 **eis-blink, ice-blink**. [Sw. (and mod. Dan.) *isblink*, etc.: see ICEBLINK.] = ICEBLINK 2.

1796 *Kis-blink* [see ICEBLINK 2]. **1870** *Q. Jrnl. Sci. Soc.* XXVI. 679 Here the 'Isblink', or 'ice glance', of the Danes (i.e. the projecting glacier—though English seamen use the word iceblink in a totally different sense, meaning thereby the 'loom' of ice at a distance), projects boldly out to sea for more than a mile. **1880** *Encycl. Brit.* XI. 167/2 If..the sea is shallow, the glacier will protrude for a considerable distance, as in the case of the Isblink. **1957** J. K. CHARLESWORTH *Quaternary Era* I. iv. 72 For this type [sc. continental ice-sheet] there is almost universal assent; it constitutes O. Nordenskiöld's 'continental glaciers', Hobbs' 'ice-cap type', and the 'inland ice' of Drygalski, Ferrar and Gourdon. Early writers, following H. Rink (1851), named it *Isblink* from the peculiar light seen over the inland ice when approached from the sea.

ischæmia, -emia. Add: Hence **ischæ·mically** *adv.*, by, or as a result of, ischæmia.

1967 *Urologia Internat.* XXII. 381 (*heading*) Electron-microscopic investigation of the action of mannitol in ischemically-damaged rat kidneys. **1970** *Nature* 27 June 1272/1 In 1963 and 1964..a high attrition rate resulted from the use of ischaemically damaged grafts..and

absence of tissue typing facilities. **1972** *Ibid.* 21 Jan. 171/2 We have tried to develop a simple perfusing system with better protection of the organ..and eventually to test this system with ischaemically damaged kidneys.

ischio-. Add: **ischiocaverno·sus** *Anat.* [mod. L. (ad. F. *ischio-caverneux* (J.-B. Winslow *Expos. anat. de la Struct. du Corps Humain* (1732) 571/1)), f. L. *cavernōsus* full of hollows, in *corpus cavernosum* (see CORPUS 2)], either of a pair of small perineal muscles, each of which arises partly from the ischial tuberosity and partly from the ramus of the ischium and is inserted into the crus of the penis (or clitoris) and perhaps helps to maintain erection.

1733 G. DOUGLAS tr. *Winslow's Anat. Expos. Struct. Human Body* II. viii. 198 The first two muscles are commonly termed erectores, but might be more properly named ischio-cavernosi. **1867** *Quain's Elem. Anat.* (ed. 7) I. 264 The ischio-cavernosus, or erector penis muscle, embracing the crus penis, arises from the inner part of the tuber ischii, behind the extremity of the crus penis, and from the pubic arch along the inner and outer sides of the crus. **1967** G. M. WYBURN et al. *Conc. Anat.* i. 17/2 Each crus is covered by the ischiocavernosus muscle.

Ise, I'se. Add: (Later example.)

c **1863** T. TAYLOR in M. R. Booth *Eng. Plays of 19th Cent.* (1969) II. 91 *Waiter.* Beg pardon sir, it's for No. 1. *Brierly.* I'se No. 1.

b. *spec.* in the United States. Also **I's.**

Freq. in Black English writings.

1852 W. L. G. SMITH *Life at South* iii. 51 I'se tinking wha' jolly time we will hab on Saturday arternoon, down under ole elm trees, on bank ob de riber. **1875** S. & C. LANIER in *Scribner's Monthly* June 240 I'se pow'ful skeered; but neversomeless I ain't gwine run away. *Ibid.*, I's like a word dat somebody done said, and den forgotten. **1898** P. L. DUNBAR *Folks from Dixie* 32 No, suh, I's a Babtist myse'f. *Ibid.* 49 I's mighty fond o' fishin', myse'f. **1902** J. D. CORROTHERS *Black Cat Club: Negro Humor & Folklore* i. 14 I'se a genamum, mase'f. *Ibid.* iii. 43, I writes a good han' and I's done read de dictionary. *Ibid.* iv. 56 De kine o' dahkey I'se talkin' 'bout am de feller whut's done bumped his head up ag'inst some college 'tel he cain't talk nothin' but Greek an' Latin, an' cuss you in Trinogometry. **1922** T. W. TALLEY *Negro Folk Rhymes* 135 I'se a bird o' one fedder, w'en it comes to you. **1927** A. P. RANDOLPH in A. Dundes *Mother Wit* (1973) 201/2 Wait a minute, son. I's wid you. *Ibid.*, Suppose des white folks find out I'se jined dis Brotherhood? *Ibid.*, Suppose des white folks ask me whether I'se a member? **1940** J. STREET in *Sat. Even. Post* 6 Jan. 32/2 I knows Ise outfigger'd. I knows Ise whupped. **1967** C. HIMES *Black on Black* (1973) 133 'I'se tired as you are,' she said evilly. *Ibid.* 136 Ise already free.

isenthalpic, isentropic, -ally: see *ISO-.

isepiptesis, isopiptesis (əise:pipti·sis, əisopipti·sis). *rare.* [a. G. *isepiptese* (A. T. von Middendorff 1855, in *Mém. Acad. Imp. Sci. St. Pétersbourg, 6e série, Sci. nat.* VIII. 8), f. Gr. ἐπίπτησις flying down upon.] A line (either imaginary or on a map) connecting points which migrating birds reach at the same time. Hence **isepipte·sial** *a.*

1875 A. NEWTON in *Encycl. Brit.* III. 768/1 His [*sc.* Middendorff's] chief object has been to trace what he has termed the isepipteses... Assuming that the advance is directly across the isepiptesial lines..the whole course of the migration is thus most accurately made known. **1926** A. L. THOMSON *Probl. Bird-Migration* viii. 132 His [*sc.* Middendorff's] method was to plot out 'isepipteses', or isochronal lines, joining up all localities corresponding with each other as to their average dates for the arrival of particular summer visitors. [**1962** C. D. SHERMAN tr. *Dorst's Migrations of Birds* viii. 233 He [*sc.* Middendorff] plotted these data on a map, linked the localities which migrants reach on the same date and thus obtained a line he called the isopiptes..(this line was later named the isochronal line).]

Isfahan: see *ISPAHAN.

-ish¹. Add: **4.** Added to names of hours of the day or numbers of years to denote: round about, somewhere near (the time or period of) (prob. after *earlyish, latish*).

1916 'PETER' *Trench Yarns* ix. 110 'What time shall I come?' 'Elevenish,' Sam replied. **1925** B. TRAVERS *Mischief* xiv. 209, I shall be going to Shady Nook at about tenish. **1930** J. B. PRIESTLEY *Angel Pavement* iii. 106 Eightish then, next Tuesday, eh? **1941** *Britannia & Eve* Sept. 15/3 Lady Regan was probably thirty. Sir Gerald looked fifty-five-ish. **1950** *Sat. Rev. Lit.* 28 Jan., The ninetyish, gentle Chandler at the reception desk. **1955** E. HYAMS *Slaughterhouse Informer* xiv. 220 'We'll make a party of it.' 'Sixish?' **1967** B. NORMAN *Matter of Mandrake* xx. 174 Will you be in your room about sevenish? **1971** P. PURSER *Holy Father's Navy* iii. 19 Thirtyish furniture in pale, shabby wood. **1972** C. FREMLIN *Appointment with Yesterday* iv. 24 This anxious thirty-five-ish person.

ishan (ī·ʃān). [Pers., = hill, landmark.] A prehistoric mound in Iraq.

1921 *Blackw. Mag.* June 708/1 They had just moved their home to an ishan or mound only some ten or fifteen minutes from the river. **1927** 'FULANAIN' *Haji Rikkan* i. 8 Facing us was a group of the strange mounds or *ishans* which here and there, in the marshes of Southern 'Iraq, stand out high, or seeming high in that vast watery expanse. **1964** W. THESIGER *Marsh Arabs* xix. 173 We passed a bare black mound..known to the Madan as Ishan Waqif, or Standing Island.

Ishihara (iʃihā·rā). *Ophthalm.* The name of Shinobu *Ishihara* (1879–1963), Japanese ophthalmologist, used *attrib.* with reference to a test for colour-blindness (the *Ishihara test*) devised by him in 1917, in which the subject is asked to name the numbers in (or, for illiterates, to distinguish pathways through) a series of printed plates (*Ishihara plates*), in each of which the numbers or pathways are formed of coloured spots in a background of spots of a different colour or colours; so *Ishihara-blind* adj., *method*, etc.

1924 *Amer. Jrnl. Physiol. Optics* V. 269 (*heading*) The Ishihara test for color blindness. **1944** *Amer. Jrnl. Physiol.* CXL. 578 The color vision was tested by the Ishihara method. *Ibid.* 579 Twelve of the 'Ishihara blind' students were trichromats. *Ibid.*, The Ishihara test gave a correct diagnosis. **1968** C. BEARD et al. *Symposium Surg. & Med. Managem. Congenital Anomalies Eye* xvi. 423 In the Ishihara plates the colors of the test symbols and of the background are of such saturation, hue, and brightness that they are regularly confused by either the deutan or protan or both. *Ibid.* 425 The tracing plates in the Ishihara test..are useful for young children and illiterates.

ishikawaite (iʃikā·wă,əit). *Min.* Also † **ishikawite**. [ad. Jap. *ishikawaishi* (K. Kimura 1922, in *Jrnl. Geol. Soc. Tokyo* XXIX. 320), f. *Ishikawa*, the name of a district in Honshu, Japan + *ishi* stone, mineral: see -ITE¹.] A black oxide of various metals, perhaps (U, Fe, Yt, Ce) (Nb, Ta)O_4.

1922 *Jrnl. Geol. Soc. Tokyo* XXIX. No. 347 (contents list), On ishikawite, a new mineral from Ishikawa District. **1923** *Mineral. Abstr.* II. 380 For ishikawaite..new axes are suggested. **1944** C. PALACHE et al. *Dana's Syst. Min.* (ed. 7) I. 766 *Ishikawaite*... The tabular crystals {100} are supposedly orthorhombic.

isidium. Add: [ad. mod.L. generic name *Isidium* (E. Acharius *Lichenographiae Suecicae Prodromus* (1798) 87), formerly used to include all lichens bearing isidia.] Hence **isi·dial** *a.*, of or pertaining to an isidium; **isi·diate, isi·dioid** *adjs.*, bearing isidia.

1921 A. L. SMITH *Lichens* iii. 149 In the genus [sc. *Isidium*] were included the more densely isidioid states of various crustaceous species. *Ibid.* 150 The centre of the isidial tuft [of *Umbilicaria pustulata*] may fall out. **1959** U. K. DUNCAN *Guide to Study of Lichens* 7 *P*[*armelia*] *crinita*... Resembles an isidiate form of *P. trichotera*. **1962** *Lichenologist* II. 3 *P*[*armelia*] *reddenda* is a pseudocyphellate species with peculiar granular or granular-isidiate outgrowths on the upper surface. **1967** M. E. HALE *Biol. Lichens* i. 22 It is not apparent that isidiate species hold any advantage over non-isidiate species. **1970** *Lichenologist* IV. 216 Further development of the isidial initials leads to a definite dorsi-ventral organization.

Islam. Now freq. with pronunc. (izlā·m).

Islamic, *a.* (Earlier example.)

1791 W. ENFIELD *Hist. Philos.* II. 244 Avenpace..applied it to the illustration of the Islamic system of theology.

Islamitic, *a.* (Earlier example.)

1791 W. ENFIELD *Hist. Philos.* II. 240 Al-Ashari..applied an extensive knowledge of the Peripatetic philosophy to the explanation of the Islamitic law.

island, *sb.* Add: **1. d.** In specific elliptical uses for some particular island or islands, as the Isle of Wight, the Hebrides, some islands in the western Pacific. Also, by further extension, for a specific prison on an island.

1814 JANE AUSTEN *Mansf. Park* I. ii. 34 She thinks of nothing but the Isle of Wight, and she calls it *the Island*, as if there were no other island in the world. **1817** KEATS *Let.* 17 Apr. (1931) I. 19, I intend to walk over the Island east—West—North South. **1852** C. M. YONGE *Two Guardians* xiii. 239 Suppose I was to take him to Marchmont's grouse shooting place in Scotland, and about among the Highlands and Islands. **1896** CONRAD *Outcast of Islands* I. ii. 15 There was not a white man in the islands, from Palembang to Ternate, from Ombawa to Palawan, that did not know Captain Tom and his lucky craft. **1901** *N.E.D.* s.v. *Isle*, The Isle of Wight is commonly referred to as 'the island'. **1902** *Captain* VII. 141 We used to gather the niggers in from all round the islands [*sc.* Pacific Islands]. *a* **1911** D. G. PHILLIPS *Susan Lenox* (1917) II. vii. 185 He was caught, did a year on the Island before his 'pull' could get him out. **1930** V. PALMER *Men are Human* xxiii. 205 He was tormented by sporadic impulses to scrap his responsibilities and go off to the [Pacific] Islands. **1935** A. J. POLLOCK *Underworld Speaks* 62/2 *Island*, Portland prison, England; Blackwell's Island, N.Y. **1939** J. PHELAN *In Can* iii. 28 He's bin on the Moor and the Island an' in the Ville, but I ain't never heard as he was in *Eton*. **1968** R. C. GALWAY *Assignment Gaolbreak* viii. 71 You're going straight to the island, via the cells at the Old Bailey. **1974** *Times* 9 Mar. 3/1 It was here in the Isle of Wight that the Conservatives last week suffered their biggest electoral disaster... For the past 50 years the island..had been considered a Tory fortress.

2. (Further examples.)

1776 A. HENRY *Trav. & Adv. Canada* (1901) xi. 282 The country was one uninterrupted plain,..a frozen sea, of which the little coppices were the islands. **1834** *Visit to Texas* iv. 41 These groves are called islands, from the striking resemblance they present to small tracts of land surrounded by water. **1843** *Amer. Pioneer* II. 283 An island of timber. **1853** F. W. THOMAS *John Randolph* 61 Islands—that is, great clumps of trees, covering sometimes many acres, appearing just like many islands in an outstretched ocean. **1902** S. E. WHITE *Blazed Trail* ix. 63 The pine there grew thick on isolated 'islands' of not more than an acre or so in extent,—little knolls rising from the level of a marsh. **1930** *19th Cent.* Dec. 713 Now, the drawback of this plan, from the Zionist point of view, is that it will prevent land purchase for the meantime and the growth of the Jewish 'islands' in the country. **1962** A. FRY *Ranch on Cariboo* iv. 42 The islands were small patches of pine and spruce timber on little rises of high ground that occurred here and there in the several hundred acres of the meadow. **1974** *Country Life* 21 Feb. 350/1 A churchyard..is..perhaps a well-wooded 'island' in an agricultural countryside.

b. *island* (or *Island*) *of Langerhans* = *islet of Langerhans* (*ISLET 2 b); also *ellipt.*

1899 *Jrnl. Exper. Med.* IV. 285 Pigment is frequently abundant in the cells composing the intertubular cell-groups or islands of Langerhans. **1900** *Bull. Johns Hopkins Hosp.* XI. 205 (*heading*) On the histology of the islands of Langerhans of the pancreas. *Ibid.* 207 When the fœtal pancreas is affected by congenital syphillis, the islands..retain their continuity with the secreting structures. **1951** A. GROLLMAN *Pharmacol. & Therapeutics* xxvi. 571 Banting and Best..obtained a preparation which was named insulin, since it is derived from the Islands of Langerhans in the pancreas, and not from the general parenchyma of the gland. **1962** W. H. HOLLINSHEAD *Textbk. Anat.* ix. 140/2 The endocrine tissue of the pancreas, the pancreatic islands (islets) or islands of Langerhans, consists of small groups of cells scattered among the more numerous acini.

c. = REFUGE *sb.* 3 c.

1869 *Spectator* 12 June 695/1 We have already 'refuges', or 'islands', or whatever they are, in most crossings. **1878** *Social Notes* 10 Aug. 358/1 It is only very lately that 'islands'—those necessary havens of refuge—have been placed at the most dangerous portions of the boulevards. **1899** *Daily Tel.* 31 Jan. 6/6 The statue being situated on an 'island', a certain amount of skirmishing was necessary in order to reach it. **1926** C. SIDGWICK *Sack & Sugar* xi. 131, I took Gerda's arm, and was nearly at the island, when the bus swept round a corner and was on us. **1930** L. COOPER *Ship of Truth* ii. 178 He stood on an island in the middle and saw the traffic sweep past him. **1956** D. GASCOYNE *Night Thoughts* 26 Street-crossing islands stand becalmed. **1970** P. LAURIE *Scotland Yard* iii. 72 The cart collided with a concrete bollard and finished up on an island. **1972** *Daily Tel.* 20 Jan. 17/7 The gang lifted a grill on a Shaftesbury Avenue island to gain access to the inspection tunnel.

d. A small isolated ridge or structure between the lines in finger-prints.

1891 *Proc. R. Soc.* XLIX. 545 Any one well-marked characteristic of a minute kind, such as an island, or enclosure, or a couple of adjacent bifurcations. **1930** E. WALLACE *White Face* xii. 183 Before we start discussing whorls, islands and circles..what is this? **1950** *Gross's Criminal Investigation* (ed. 4) v. 127 On Fig. 4 are marked some of the common types of ridge characteristics:—'A' is an enclosure or lake... 'F' is a short independent ridge or island.

e. = **speech island*.

1882 A. J. ELLIS in *Trans. Philol. Soc.* 30 The maps being thus arranged, coloured lines are drawn on them marking boundaries, which sometimes unite and form islands. **1892** *Dialect Notes* I. iv. 225 One of his own students was thinking of a linguistic island in the Tennessee mountains as a field for future work. **1923** A. L. KROEBER *Anthropol.* v. 105 This explains the numerous survivals and 'islands' of speech. **1934** H. KURATH in *Proc. Amer. Philos. Soc.* LXXIV. 239 R-islands still exist in eastern New England.

f. A piece of furniture, in a private house or in a museum, library, etc., surrounded by unoccupied floor space. Freq. *attrib.*

1932 *Museums Jrnl.* June 127 In the vertical island-cases with different displays on opposite sides. **1960** *Guardian* 1 Mar. 3/5 Living and dining space planned round a large island range and barbecue grill. *Ibid.*, Peter Jones and Heal's both show island fireplaces. **1960** *Oxf. Univ. Gaz.* 4 Mar. 806/2 A new island bookcase has been acquired for the library. **1960** *House & Garden* Aug. 65/2 The cooking island screens a small corner used for informal meals. **1968** *Globe & Mail* (Toronto) 13 Feb. 30/1 (Advt.), Huge kitchen with built-ins and double sink plus island sink for children. **1972** *House & Garden* Dec.–Jan. 77 Country-style kitchen..has a central butcher-block island with built-in hot-plates.

g. The superstructure of a ship, esp. an aircraft carrier.

1937 *Jane's Fighting Ships* 497 Adding 2½ feet to the beam..to balance the island superstructure. **1964** *New Scientist* 2 July 22 (*caption*) The 'island' which supports a giant radar scanner on HMS *Hermes*.

3. a. *island fortress, race* (i.e. the British).

1898 H. NEWBOLT (*title*) The island race. **1902** BELLOC *Path to Rome* 51 Some kinds of men begin talking of Dogged Determination, Bull-dog pluck, the stubborn spirit of the Island race, and so forth. **1942** *R.A.F. Jrnl.* 18 Apr. 28 In the strongly defended island fortress of Corregidor. **1958** *Spectator* 14 Feb. 199/2, I could wish that the Island Race as a whole were a little more discriminating in its drooling. **1966** *Listener* 26 May 771/1 Aware of Mr Menuhin's devotion to his adopted country, one felt it matched by the island-race stateliness of Sir Adrian.

4. island arc, any arcuate chain of islands located and aligned in relation to an orogenic belt and characteristically having a deep

trench on the convex side; **Island Carib**, (*a*) the Carib people of the Lesser Antilles; (*b*) the language of this people; **island-hill, -mountain**, a hill or mountain rising out of a plain; **island-hop** *v. intr.* (of the U.S. army in the Pacific during the war of 1941–45), to recapture Japanese-occupied islands one after another; also *transf.*; chiefly in *vbl. sb.*; **island-mountain**: see *island-hill* above; **island plot, site**, a plot of land on a building site surrounded by streets or open spaces; **island-refuge** = *2 c; **island site**: see *island plot* above; **island-universe**, add: [app. tr. G. *weltinsel* (von Humboldt), though the term has been attributed to Sir William Herschel] (earlier and later examples).

1906 H. B. C. & W. J. SOLLAS tr. *Suess's Face of Earth* II. iv. 207 It would be sound geology to draw the western boundary of the Pacific Ocean outside the island arcs from Kamchatka through Japan. **1971** I. G. GASS et al. *Understanding Earth* xix. 271/1 The zone meets the surface close to the line of the deep ocean trench and dips away beneath the island-arc. **1938** D. TAYLOR in *U.S. Bureau Amer. Ethnol. Bull.* No. 119. 140 The most typical product of the Island Carib is..the dugout canoe. **1951** —— *Black Carib of Brit. Honduras* 41 The Black Carib of Central America speak a dialect of Island Carib—of the language, that is to say, spoken by the native Indian inhabitants of the 'Caribee' islands at the time of Breton's stay among them (1635–1653). **1968** *Encycl. Brit.* XXIII. 433/1 These island Carib [of Dominica] have not retained as much of their aboriginal language and culture [as the Black Carib]. **1969** *Word* XXV. 276 In Island-Carib, the plural is employed only with reference to animate..beings. **1839** H. T. DE LA BECHE *Rep. Geol. Cornwall, Devon & W. Somerset* i. 26 The lower island-hills of Pawlet and Chedzoy..rise out of the plain near Bridgewater. **1907** Island-hill [see *INSELBERG]. **1944** *Sat. Even. Post* 28 Oct. 98/3 American air power won the battles for Attu, Kwajalein and Tarawa, and has made possible island hopping. **1946** *Sat. Rev. Lit.* 23 Feb. 33/1 Cant takes us along on the island-hopping campaigns. **1955** in *Amer. Speech* (1956) XXXI. 85 Did you island-hop or did you take the plane directly? **1971** P. DRISCOLL *White Lie Assignment* vii. 61 The caiques are built for coasting and island-hopping. They can't take much rough weather. **1972** *Guardian* 11 Aug. 1/1 The airlift by small, island-hopping aircraft. **1906** *Daily Chron.* 31 Aug. 4/4 They [*sc.* the Malvern Hills] lie precisely north by south, moored like some great island-mountain to the westward of the central plain of England. **1913** *Geogr. Jrnl.* XLII. 149 The fantastic peaks and domes of the rocky island-mountains. **1941** F. H. LAHEE *Field Geol.* (ed. 4) xi. 360 A similar residual type in arid regions is the island mountain, or island mount (German, Inselberg). **1908** *Daily Chron.* 20 Apr. 3/5 On this 'island' plot of land has been erected a building which is certainly an adornment to Great Portland-street. **1922** F. MUIRHEAD *London & Environs* (ed. 2) 8 A busy street should be crossed only at a point where an 'island-refuge' is provided in the middle. **1907** *Westm. Gaz.* 20 Sept. 10/1 Australia and the Strand 'Island Site'. **1936** C. ROUSE *Old Towns* i. 17 The market house or town hall was often built on an island site in those wide streets. **1972** *Accountant* 17 Aug. 206/1 Two important island sites with main frontages to Great Portland Street have been assembled over the course of many years. [**1845** A. VON HUMBOLDT *Kosmos* I. 93 Unter den vielen selbstleuchtenden ihren Ort verändernden Sonnen ..welche unsre Weltinsel bilden.] **1867** A. J. DAVIS *Stellar Key to Summer Land* vi. 32 The expression 'Island Universe' was suggested by the immense distance of the fixed stars from our Sun and Planets; giving the impression that our Solar System occupies an isolated position in the boundless ocean of space. **1887** R. A. PROCTOR *Other Suns than Ours* i. 1 Our 'island universe', as Humboldt poetically called the stellar system. *Ibid.* 11 The results which Sir W. Herschel published in 1817 and 1818 justify the belief that..large numbers of the nebulæ must be regarded as external galaxies. This grand conception fascinated..some who, like Humboldt,..had understood and appreciated the work of the great observer. The idea of 'island universes' strewn throughout the ocean of space impressed the world. **1928** J. H. JEANS *Astron. & Cosmogony* i. 19 These figures amply show that these nebulae and star-clouds are quite outside our system of stars; they constitute what Herschel described as 'island-universes' distinct from the universe which contains our sun. **1959** DAVIES & PALMER *Radio Stud. Universe* i. 2 Many..nebulae were discovered. However, it was not until the 1920's that the problem of their distance and spacing was unravelled by Hubble, who showed that they were very distant island universes (galaxies) of stars many of which were similar to our Milky Way system.

Isle of Wight (əiləv,wəi·t). The name of an island off the coast of Hampshire, used *attrib.* to designate a disease of bees first found there in 1904, caused by the parasitic mite *Acarapis woodsi*; also called *acarine disease*.

[**1907** A. D. IMMS in *Jrnl. Board Agric.* XIV. 129 (*title*) Report on a disease of bees in the Isle of Wight.] **1909** W. MALDEN in *Jrnl. Board Agric.* XV. 815 In commencing my observations on the Isle of Wight disease, I first endeavoured to ascertain whether, by careful dissections, any characteristic lesions could be discovered in the internal anatomy of the diseased bees. **1932** E. B. WEDMORE *Man. Beekeeping* xvii. 401 In 1920 Miss E. J. Harvey observed a mite in the tracheæ of bees suffering from 'Isle of Wight' disease. **1965** M. HOYT *World of Bees* x. 171 Acarapis is sometimes called Isle of Wight disease.

islet. Add: **2.** Also = *ISLAND *sb.* 2 d.

1921 *Discovery* Oct. 257/2 A dozen guiding marks, consisting of the beginnings of lines, bifurcations and islets.

b. *islet* (or *Islet*) *of Langerhans* [tr. F. *îlot de Langerhans* (E. Laguesse 1893, in *Compt. rend. hebdom. d. Séances et Mém. de la Soc. de Biol.* XLV. 819): named after Paul *Langerhans* (1847–1888), German anatomist, who, in 1869, first described such islets], any of numerous highly vascular islets of tissue in the pancreas, composed of light-staining cells of two principal types, one of which secretes insulin and the other glucagon; also *ellipt.*

1896 *Jrnl. R. Microsc. Soc.* 35 Besides the solid buds which form the first 'islets of Langerhans', they give rise to numerous hollow buds. **1904** *Proc. R. Soc.* LXXIII. 84 The degeneration or absence of the islets in diabetes. **1910** *Practitioner* Jan. 32 Many other pathological conditions.. have been ascribed to abnormalities of other ductless glands; e.g...glycosuria to the loss of the internal secretion formed by the islets of Langerhans in the pancreas. **1968** *Times* 27 Nov. 9/3 Those parts of the organ [*sc.* the pancreas] concerned with carbohydrate metabolism—known as the islets of Langerhans. **1973** *Nature* 23 Mar. 259 Yields of up to 350 Islets per rat pancreas have been achieved using this method.

3. (*attrib.* examples of sense 2 b.)

1914, etc. Islet-tissue [see *INSULIN 1]. **1927** HALDANE & HUXLEY *Animal Biol.* xii. 279 Islet cell of human pancreas. **1962** *Lancet* 12 May 1003/2 The characteristics of the Zollinger–Ellison syndrome are extreme gastric hypersecretion, intractable peptic ulceration..and either hyperplasia or tumour of the islet-cells of the pancreas. **1963** E. J. W. BARRINGTON *Introd. Gen. & Compar. Endocrinol.* iii. 43 We have evidence that islet tissue arose very early in vertebrate evolution. **1965** LEE & KNOWLES *Animal Hormones* vii. 113 Occasionally in man a tumour of the islet cell occurs and there is excessive secretion of insulin.

islomania (əi:lomē·niă). [f. ISL(AND *sb.* + -O + -MANIA.] A passion or craze for islands.

1962 *Listener* 25 Oct. 693/1, I suffer from acute islomania, and was therefore specially interested in 'Let's Imagine'..which investigated the Channel Island of Herm. **1971** D. CONOVER *One Man's Island* 110 Islomania runs in my blood. I would rather talk about islands than eat, would rather—and often do—think about islands than sleep. **1972** *Times* 4 May 11/6 Psychologists have a word for it—islomania—an insatiable attraction to islands.

ism, *quasi-sb.* (Earlier and later examples.)

1680 'HERACLIO DEMOCRITUS' *Vision of Purgatory* 46 He was the great Hieroglyphick of Jesuitism, Puritanism, Quaquerism, and of all Isms from Schism. **1756** *Monthly Rev.* XIV. 359 Arianism, Socinianism, Arminianism, or any other *ism*. *a* **1773** [see GERMANISM 1]. **1928** G. B. SHAW *Intelligent Woman's Guide Socialism* lxxxiii. 447 The proletarian Isms are very much alike. **1944** J. S. HUXLEY *On Living in Revolution* iii. 29 Democracy could become more dynamic than Fascism or Communism or any other ism or ideology. **1963** F. W. FREY in L. W. Pye *Communications & Political Devel.* xvii. 299 Movement towards..'Communism', or 'totalitarianism', or..whatever one's preferred 'ism'..happens to be. **1968** S. C. HUTCHISON *Hist. R. Acad.* xvii. 183 He saw no place in art for abstractions and 'isms' and had a very low opinion of their adherents. **1974** *Listener* 14 Feb. 220/1 Impressionism became the most successful 'ism' in the history of art.

-ismus, *suffix*, repr. G. *-ismus* or L. *-ismus* (see -ISM), used similarly to -ISM, indicating a typical condition or typical conduct, in nouns formed from proper names, or indicating a system or principle, as in *historismus, mysticismus, Sherlockismus, snobismus*. Freq. with ironical or pejorative overtones.

1912 R. A. KNOX in *Blue Bk.* (Oxf.) July 132 There is a special kind of epigram, known as the Sherlockismus, of which the indefatigable Ratzegger has collected no less than 173 instances. The following may serve as examples: 'Let me call your attention to the curious incident of the dog in the night-time.' 'The dog did nothing at all in the night-time.' 'That was the curious incident,' said Sherlock Holmes. And again: 'I was following you, of course.' 'Following me? I saw nobody.' 'That is what you must expect to see when I am following you,' said Sherlock Holmes. **1948** *Scottish Jrnl. Theol.* I. 142 This kind of unhistorical thinking is neither Biblical nor Christian and.. is allied to the speculations of Philo... *This* is certainly not the way to escape from the errors of *Historismus*. **1952** *Spectator* 14 Mar. 344/2 A sincere and compelling protest against *mysticismus* in power politics. **1959** H. KENNER *Invisible Poet: T. S. Eliot* (1960) v. 241 His exchange with the Third Tempter opens with the figure of rhetoric the late Rev. Ronald Knox christened the Sherlockismus: *Third Tempter*: I am an unexpected visitor. *Thomas*: I expected you. **1961** *Daily Tel.* 23 June 18/5 Of all the practitioners of this new cult of 'Sherlockismus', Mr Starrett is..far the most interesting. **1971** A. QUINTON in A. Bullock *20th Cent.* 260/2 Massive studies of *Historismus* by the historians Troeltsch and Meinecke. **1974** *Times Lit. Suppl.* 1 Feb. 113/2 Perhaps he has also done more than he reckoned to put an end to the tiresome excesses of Sherlockismus. **1974** V. GIELGUD *In Such a Night* xx. 172 His acquaintances accused him of Victorian *snobismus*.

isness. Delete *nonce-wd.* and add earlier and later examples.

1865 J. H. STIRLING *Secret of Hegel* II. III. i. 4 *Seyn*, in Germany, often in Hegel himself, means the abstraction of sensuous Isness. **1884** 'SCOTUS NOVANTICUS' *Metaphysica Nova et Vetusta* VI. 146 The moment of being or is-ness yields identity (A = A); and as this is-ness is given *in concreto* as a determined somewhat which *is*, we have the category of Essence as derivative category from Being. **1918** [see *GOOD A. *adj.* 14 c]. **1942** *Mind* LI. 257 Any one..who sets out along the *via negativa* in this spirit is confessing in the very act that 'is' can never be tortured into 'isness'. **1955** A. HUXLEY *Genius & Goddess* 46 The girl is who she is. Some of her isness spills over and impregnates the entire universe. **1965** L. R. HUBBARD *Scientology Abridged Dict.*, *Is-ness*, one of the four conditions of existence. It is an apparency of existence brought about by the continuous alteration of an As-isness. This is called, when agreed upon, reality.

Isnik (izni·k). The name of a town in Asian Turkey, the classical Nicæa (see NICENE *a.* (and *sb.*)), used *attrib.* to denote pottery or tiles made there, or imitations thereof, from the fifteenth to the seventeenth centuries, characterized by the use of brilliant pigments.

[**1909** F. R. MARTIN in *Burlington Mag.* Aug. 270/2 The splendid blue and white bowls, commonly ascribed to Kutaya, which I am convinced come from Isnik.] **1932** R. L. HOBSON *Guide Islamic Pott. Near East* III. 87 There were doubtless potters at work in other towns in the sixteenth century,..and it is highly probable that they would adapt their wares to the prevailing Turkish taste as expressed by the Isnik pottery. **1939** A. LANE *Guide Coll. Tiles* (V. & A. Mus.) ii. 16 The pure white ground and tense drawing so characteristic of the later Isnik tiles. **1957** —— *Later Islamic Pott.* iii. 60 It will probably never be possible to stop dealers and collectors calling the later Isnik wares 'Rhodian', and the nickname is at any rate a convenient label for the whole class in which the 'scaling-wax red' appears. **1966** J. FOWLES *Magus* xv. 88 A triangular cabinet full of pale-blue and green Isnik ware. **1972** *Daily Tel.* 5 Dec. 12/5 A large Isnik pottery dish of the late 16th century was bought by Eskenazi for ..£8,000 at Sotheby's yesterday.

iso-. Add: **isacoustic** *a.*, (*b*) *Seismology*, applied to a line (imaginary or on a map) connecting places where an equal percentage of observers heard the sound of an earthquake; **isano·malous** *a.*, (of a line) isabnormal; (of a map) depicting such lines; **isanomaly**, add: also **isonomaly**; also used with reference to other kinds of anomaly (see quots.); **isentha·lpic** *a.*, of or denoting equal enthalpy; **isentropic** *a.* (earlier and later examples); also, taking place at constant entropy, involving no change in entropy; *sb.* (examples); hence **isentro·pically** *adv.*, without a change in entropy; **i:soacce·ntual** *a.*, of verse in which the syllables are of equal length and accent; **iso-agglu·tinate** *v. trans.* (also *absol.*), to cause isoagglutination (of); chiefly in **isoagglu·tinating** *vbl. sb.* and *ppl. a.*; **i:soagglutina·tion** *Immunol.* [a. G. *isoagglutination* (A. Klein 1902, in *Wiener klin. Wochenschr.* XV. 415/1)], agglutination of cells of an individual by a substance obtained from another individual of the same species; so **i:soagglu·tinative** *a.*, pertaining to or causing isoagglutination; **i:so-agglu·tinin** *Immunol.*, an agglutinin that agglutinates cells of other individuals of the same species as that in which it is found; **i:soagglu·tinogen** *Immunol.*, a substance that elicits or reacts with an isoagglutinin; **iso-allele** (əiso,æ·līl) *Genetics*, an allele indistinguishable from another allele in its effect on the phenotype except when special techniques are employed; hence **isoalle·lic** *a.*; **isoa·ntibody** *Immunol.*, an antibody elicited by an isoantigen; **isoa·ntigen** *Immunol.*, an antigen in one individual which is capable of eliciting antibody formation only in other, genetically different, individuals of the same species; so **isoantige·nic** *a.*; **i·sobase** *Geol.* [ad. Sw. *isobas* (G. De Geer 1890, in *Geol. Fören. i Stockholm Forhandl.* XII. 72), f. Gr. βάσις stepping, step], a line (either imaginary or on a map) connecting points on the earth which have undergone equal amounts of uplift (or more rarely depression) over a period of geological time; hence **isoba·sic** *a.*; **i·sobath** *sb.* [Gr. βάθος depth], a line (either imaginary or on a map) joining places where water has equal depth; an underwater contour; **isoba·thic** *a.*; **isoca·loric** *a.*, of equal calorific value; hence **isocalo·rically** *adv.*, in a way that leaves the calorific value unchanged; **isoche·mical** *a. Geol.*, taking place with or characterized by constant chemical composition; hence **isoche·mically** *adv.*, without a change in chemical composition; **i·sochlor** (-klǭɹ) [CHLOR(INE *sb.*], a line (imaginary or on a map) connecting points where the concentration of chlorine in the surface water is the same; **isochro·mosome** *Cytology*, an abnormal chromosome having a pair of identical arms; **isoco·lloid** *Chem.* [ad. G. *isokolloid* (W. Ostwald *Grundriss* *der Kolloidchemie* (ed. 2, 1911) I. iv. 128)], a colloidal solution in which the disperse phase and the dispersion medium are chemically identical (as ice dispersed in water) or chemically related (as a polymer dispersed in its monomer); **isoco·rtex** *Anat.* [mod.L. (C. & O. Vogt 1919, in *Jrnl. f. Psychol. u. Neurol.* XXV. 293)] = *NEOPALLIUM; **isocy·clic** *a. Chem.* = *homocyclic* adj. s.v. *HOMO-; **iso-e·cho** *a.*, applied to a line on a radar display along which the echo signal (from clouds and the like) has the same strength, and to devices and techniques relating to such lines (as in the detection of rain and atmospheric turbulence); **isoelectric,** delete def. in Dict. and see *ISOELECTRIC *a.*; **isoenerge·tic** *a.*, having, denoting, or giving rise to equal amounts of energy; taking place at constant energy; **i·sogel** *Chem.*, an isocolloid in the form of a gel; **i·sograd** *Petrol.* [GRAD(E *sb.*], a line or surface (either imaginary or in a diagram) joining points where the rock originated under the same conditions of pressure and temperature, as indicated by similarity of facies or metamorphic grade; hence **i·sograde, isogra·dic** *adjs.*, that is an isograd; of the same facies or metamorphic grade; **i·sogyre,** a thick band or bow of darkness seen crossing the coloured fringes in an interference pattern; **i:sohæmagglutina·tion** (also **-hem-**) *Immunol.*, isoagglutination of red blood cells; **i:sohæmagglu·tinin** (also **-hem-**) *Immunol.*, an isoagglutinin which agglutinates red blood cells; **i:sohæmoly·sin** (also **-hem-**) *Immunol.*, a hæmolysin that lyses red blood cells of other individuals of the same species as that in which it is found; **isohaline** (-hē·ləin) *sb. Oceanogr.* [Gr. ἅλινος of salt], a line (imaginary or on a chart), or an imaginary surface, connecting points which have the same salinity; *a.*, connecting such points; also, of a constant salinity throughout; **i·sohel** [Gr. ἥλιος sun], a line (imaginary or on a map) connecting points having the same amount or duration of sunshine; so **isohelic** (-hī·lik), *a.*; **i·sohyet** (-həiĭt) [Gr. ὑετ-ός rain], = *isohyetal* sb. s.v. Iso-; **isohyetal** *a.* and *sb.* (examples); **isoimmu·ne** *a.*, of, producing, or exhibiting isoimmunization; **i:soimmuniza·tion** (an instance of) the development of an isoantibody in an individual against an antigen derived from another individual of the same species; **isokine·tic** *a.*, characterized by no disturbance to the speed and direction of a fluid when it is withdrawn as a sample from a flow; hence **isokine·tically** *adv.*, in an isokinetic manner, i.e. without causing such a disturbance; **isole·cithal** *a.* = *homolecithal* adj. (s.v. *HOMO-); **i·solex** *Linguistics*, a line connecting places in which there is uniformity of vocabulary; hence **isole·xic** *a.*; also **isole·ctic** *a.*; **i·soline** = *ISOPLETH 1; **isoly·sin** *Immunol.* [a. G. *isolysin* (Ehrlich & Morgenroth 1900, in *Berliner klin. Wochenschr.* XXXVII. 455/1)] = *isohæmolysin* above; **isomagne·tic** *a.*, denoting a line (either imaginary or on a map) connecting places which have the same value of a particular parameter of the earth's magnetic field, and a map or chart showing such lines; also as *sb.*, an isomagnetic line; **isonomaly,** var. *isanomaly* (in Dict. and Suppl. s.v. Iso-); **isonu·clear** *a. Chem.* = *HOMONUCLEAR *a.* b; **iso-osmo·tic** *a. Physiol.* = *isosmotic* adj. (below); **isopiestic** *a.* (substitute for def.:) representing, having, or characterized by equal or constant pressure; also as *sb.*, a line in a diagram representing states of equal pressure; (examples); **i·sopor** [Gr. πόρ-ος passage, way], a line (either imaginary or on a map) connecting points at which equal annual changes in some parameter of the earth's magnetic field are observed; hence **isopo·ric** *a.*; **isosmotic** (əisǫzmǫ·tik) *a. Physiol.*, of or having the same osmotic pressure; const. *with*; **i·sostich** (-stik) *Biochem.* [Gr. στίχ-ος line (of poetry)], each of two or more fractions of a polynucleotide that contain the same number of nucleotides; **isostru·ctural** *a. Min.*, having the same or similar crystal structure; const. *with*; **isosylla·bic** *a. Linguistics*, of a metrical structure in which the syllables are of the same length; **isosynta·ctic** *a.* (see quot.); **isosynta·gmic** *a. Linguistics* (see quot. 1954); **i·sotach** [Gr. ταχ-ύς swift], a line on a chart or diagram connecting points where the speed of something, esp. the wind, is the same; **isote·niscope** [irreg. f. TEN(SION *sb.* + -I- + -SCOPE], an instrument for measuring the vapour pressure of a liquid over a range of temperatures, consisting of a bulb for containing the liquid attached to one arm of a U-tube, which is used as a manometer to show when the pressure applied to the other arm is equal to the vapour pressure; **i·sovol** [VOL(ATILE *sb.* and *a.*], a line (either imaginary or on a map) joining places at which the coal has the same ratio of fixed to volatile carbon.

1900 C. DAVISON in *Phil. Mag.* XLIX. 43 An isacoustic line may be defined as a line which passes through all places in which the percentage of persons who hear the sound is the same. **1938** L. D. LEET *Pract. Seismol.* viii. 282 Curves passing through the places at which equal percentages of the observers heard the earthquake sound were drawn in 1899, and Davison, followed by Knott and de Montessus, called them isacoustic lines. **1900** *Geogr. Jrnl.* XV. 662 Maps of isotherms and isanomalous lines for January and July. **1943** G. T. TREWARTHA *Introd. Climate* (ed. 2) i. 56/2 If lines, called isanomals, are drawn on a world map, joining places of equal thermal anomaly, an isanomalous map is the result. **1967** R. W. FAIRBRIDGE *Encycl. Atmospheric Sci.* 985/1 Isanomalous lines drawn on a map..reflect regional distortions of the world patterns. **1930** *Meteorol. Gloss.* (Meteorol. Office) (ed. 2) 109 *Isanomaly.* This word..is used of lines joining all points on a map or chart having equal anomalies, or differences from normal, of a particular meteorological element. **1931** F. H. LAHEE *Field Geol.* (ed. 3) xxiii. 674 Equal anomalies may be connected by flowing lines called isonomalies, or isanomalies, which express in gammas the local variations from the average total magnetic intensities in the area. **1962** F. I. ORDWAY et al. *Basic Astronautics* v. 199 Lines of regional vertical intensity are constructed, as are isonomaly charts expressing in gammas the local variations in the average total magnetic intensity of a given region. **1967** R. W. FAIRBRIDGE *Encycl. Atmospheric Sci.* 507/2 *Isanomaly*, lines or contours of equal anomalies or departures from normal (often used with gravity anomalies, cf. *Isogal*). **1925** *Proc. Amer. Acad. Arts & Sci.* LX. 581 The isenthalpic curves themselves contain valuable information. **1937** M. W. ZEMANSKY *Heat & Thermodynamics* xiv. 245 (*caption*) Isenthalpic states of a gas. *Ibid.* 246 The numerical value of the slope of an isenthalpic curve on a *t–P* diagram at any point is called the Joule-Kelvin coefficient. **1973** J. S. TURTON *Macroscopic Thermodynamics* iv. 81 The process undergone by the gas or vapour in passing through the constriction cannot be represented by an isenthalpic curve. **1873** J. W. GIBBS in *Trans. Connecticut Acad. Arts & Sci.* II. 311 If, however, we..call that quantity entropy..it seems natural to..call the lines in which this quantity has a constant value isentropic. *Ibid.* 327 Although the inclination of the isentropics is independent of the quantity of gas under consideration, the rate of increase of η will vary with this quantity. **1923** LEWIS & RANDALL *Thermodynamics* xii. 137 In such an isentropic compression there will ordinarily be a change in temperature. **1951** C. L. BROWN *Basic Thermodynamics* vi. 88 Two additional relations..relate temperature and volume and pressure and volume for an isentropic. **1956** G. C. MCVITTIE *Gen. Relativity & Cosmol.* vii. 123 An equation determining the function *F* in isentropic flow. **1951** C. L. BROWN *Basic Thermodynamics* vi. 89 The gas expands isentropically and is exhausted at 15 psia. **1972** *Nature* 15 Sept. 139/2 The pressure applied to an implosion system does *PdV* work generating kinetic energy which is converted near isentropically to internal energy concentrated in the compressed volume. **1956** H. WHITEHALL in *Kenyon Rev.* 420 There is..in much of Milton, isochronic counterpointed with isoaccentual rhythm. In some poems of Dylan Thomas, we find a most elaborate counterpoint of isoaccentual..and, apparently, isotonic rhythms. **1957** N. FRYE *Sound & Poetry* 143 Isoaccentual, or, as it is often called, isosyllabic rhythm weights with stress. **1904** *Alienist & Neurologist* XXV. 386 The serum from such blood was also isoagglutinating. **1910** *Johns Hopkins Hosp. Bull.* XXI. 70/1 Human beings can be divided into four groups according to the ability of their serum to cause isoagglutination and of their corpuscles to be isoagglutinated. **1921** *Biol. Bull.* XL. 18 The iso-agglutinating power of the egg-water. **1967** D. M. WEIR *Handbk. Exper. Immunol.* iv. 87 The isoagglutinating activity found in this sedimentation region of human serum. **1907** *Jrnl. Med. Res.* XVII. 338 Human bloods may be separated into three rather definite groups as regards isoagglutination. **1921** *Biol. Bull.* XL. 17 In iso-agglutination round solid masses of agglutinated spermatozoa form in a few seconds. **1927** Iso-agglutination [see *A 7]. **1970** *Exper. Cell Res.* LIX. 37 (*heading*) Jelly coat substances of sea urchin eggs. I. Sperm isoagglutination and sialopolysaccharide in the jelly. **1902** *Science* 28 Nov. 858/1 The isoagglutinative and isolytic properties of human serums in health and in disease. **1911** *Jrnl. Exper. Med.* XIII. 537 Isoagglutinative serum is active at a considerable dilution. **1903** DORLAND *Med. Dict.* (ed. 3) 350/1 iso-agglutinin. **1907** [see *isohæmagglutination* below]. **1971** J. A. BELLANTI *Immunol.* iii. 78 All mature individuals possess antibody in their serum, the so-called 'naturally occurring' isoantibodies (isoagglutinins) directed against the antigenic determinant absent from their own erythrocytes. **1926** *Jrnl. Immunol.* XI. 240 The main human iso-agglutinogens A and B are to be detected not only by human serum but also by the sera of animals. **1972** W. E. HAESLER *Immunohematol.* i. 2 Immunohematology deals with hemagglutinogens (isoagglutinogens, immunogens, antigens) that are a natural phenomenon in human beings. **1943** STERN & SCHAEFFER in *Proc. Nat. Acad. Sci.* XXIX. 361 Different alleles indistinguishable except by special tests will be called isoalleles. **1970** *Sci. Amer.* Mar. 104 These slightly different forms of the same gene that perform the same function are called isoalleles. **1944** *Genetics* XXIX. 485 These crosses established the iso-allelic if not identical nature of all seven *pyd* mutants. **1961** *Lancet* 29 July 262/2 An isochromosome carries, on its two symmetrical arms, duplicate gene loci each influencing the same character. These arms can be both isologous and isoallelic. **1919** L. & H. HIRSCHFELD in *Lancet* 18 Oct. 676/1 The antibodies produced within the species which we call isoantibodies..act..only against the differences between the blood of the animal which provides the blood for injection and that of the recipient. **1971** Isoantibody [see *isoagglutinin* above]. **1936** *Jrnl. Immunol.* XXX. 445 (*heading*) Procedure for the determination of isoantigens in saliva. **1971** J. A. BELLANTI *Immunol.* iii. 76 Homologous antigens or isoantigens (alloantigens) are those genetically controlled antigenic determinants which distinguish one individual of a given species from another. **1938** *Jrnl. Path. & Bacteriol.* XLVI. 249 Iso-antigenic factors present in the grafted tissue and absent in the host are capable of eliciting a response which results in the destruction of the graft. **1971** J. A. BELLANTI *Immunol.* iii. 80 New human isoantigenic specificities. **1892** G. DE GEER in *Proc. Boston Soc. Nat. Hist.* XXV. 457 To get a general view of the warping of land..I have used the graphic method of Mr. G. K. Gilbert..and have connected with lines of equal deformation, or as I have called them isobases, such points of the limit as were uplifted to the same height. **1957** J. K. CHARLESWORTH *Quaternary Era* II. xlv. 1289 (*caption*) Map of the sea's greatest area in Baltoscandia (black), with areas of most important glacier-lakes (shaded) and isobases of uplift in metres. **1969** BENNISON & WRIGHT *Geol. Hist. Brit. Isles* xvi. 366 It is still not possible to construct accurate isobases (lines joining points of equal uplift) for Britain, as has been done in the case of Scandinavia. **1932** E. G. WOODS *Baltic Region* x. 121 A glance at such a map with the isobasic lines indicated, shows the late-glacial sea at about its maximum development. **1895** *Rep. Sci. Results Voy. H.M.S. Challenger, Summary* I. 55 Bauche..is..considered as the first to make use of isobaths for the sea. **1938** *Jrnl. Marine Res.* I. 138 (*caption*) Isobaths (light broken lines) for every ten meters depth. **1956** *Trans. Amer. Microsc. Soc.* LXXV. 335 Miner's map is our primary source of that shore line..just as it is for the contained isobaths. The Commission..has in no way modified the lakes except to build a bathing beach and boat docks. **1972** *Nature* 4 Feb. 253/2 Thus Corsica makes a good fit with France along the 1,000-metre isobath. **1895** *Rep. Sci. Results Voy. H.M.S. Challenger, Summary* I. 50 These isobathic curves are intended to show that certain elevations of the sea-bottom correspond with the orography of the neighbouring land. **1957** *Encycl. Brit.* VIII. 743/2 The isobathic chart of the Severn estuary..shows a progressive deepening seaward by means of V-shaped lines which become blunter westward. **1922** *Experiment Station Rec.* Sept. 370 The sugar tolerance of the diabetic patient was..more markedly lowered by protein than by isocaloric amounts of fat. **1956** *Biol. Abstr.* XXX. 2208/1 Male Wistar rats were divided into 3 groups receiving an isocaloric diet. **1971** *Jrnl. Gen. Psychol.* LXXXV. 155 Decreasing the amount of protein in the diet, while holding it isocaloric through addition of carbohydrate, resulted in an increase in the 'excitatory process' of rats. **1973** *Lancet* 2 June 1201/2 A programme of work on an isolated group of healthy young men, using isocaloric substitution of glucose syrup..for dietary sucrose. **1971** *Jrnl. Gen. Psychol.* LXXXV. 156 Diets in which carbohydrate and protein were interchanged isocalorically. **1972** *Science* 19 May 795/1 The animals fed alcohol received the identical diet except that ethanol..isocalorically replaced carbohydrate. **1937** *Trans. R. Soc. Edin.* LIX. 218 Correlation between rocks of the same bulk-composition metamorphosed under different physical controls—i.e. isochemical correlation. **1951** TURNER & VERHOOGEN *Ign. & Metamorphic Petrol.* xv. 369 Metamorphism may be considered as commonly approaching, though seldom attaining, the nature of an isochemical change. **1952** T. F. W. BARTH *Theoret. Petrol.* IV. xii. 356 In geological discussions the fact that sediments at the very incipience of metamorphism regularly change their chemical composition has often been neglected. However, these changes are not to be neglected... Isochemical regional metamorphism *senso strictu* [sic]..does not exist. **1969** W. D. JOHNS tr. *Correns's Introd. Mineral.* ix. 298 Transformations in an isochemical system depend on the temperature and pressure to which the system has been subjected. **1964** J. CHALLINOR *Dict. Geol.* (ed. 2) 133/2 A rock changing its mineral composition isochemically remains a closed 'system'. **1973** *Nature* 23 Mar. 243/2 The thickness of halite salt that could be precipitated isochemically from one basinful of Mediterranean waters..is..only about 20 m. **1890** *Rep. on Water Supply & Sewerage Pt. 1: Examinations of Water Supplies & Inland Waters* (Massachusetts State Board of Health) 679 In the accompanying map of normal chlorine of Massachusetts, the points of like normal chlorine have been connected by lines which we will call isochlors. **1943** *Proc. R. Irish Acad.* XLVIII. B. 157 The geochemical data of chlorine are considered with respect to river and surface waters, and equations developed relating the distance of any particular isochlor from the sea coast. **1957** G. E. HUTCHINSON *Treat. Limnol.* I. viii. 545 In Europe a greater proportion of rain is derived from air which has moved in perpendicular to the isochlors than is likely to be the case in North America. **1939** C. D. DARLINGTON in *Jrnl. Genetics* XXXVII. 357 The attached-*X* chromosome has two exactly similar arms united at the centromere. It is what we may call an iso-chromosome. **1972** W. V. BROWN *Textbk. Cytogenetics* xix. 268 The only likely isochromosomes found in human beings are of the long arm of a G-group chromosome, probably No. 21, and of the X chromosome. **1915** M. H. FISCHER tr. *W. Ostwald's Handbk. Colloid-Chem.* iv. 103 We shall term these structures in which disperse phase and dispersion means are chemically isomeric, isocolloids. **1946** J. ALEXANDER *Colloid Chem., Theoret. & Appl.* VI. xxiii. 531 Wolfgang Ostwald considers high-boiling petroleum fractions as isocolloids, in which the dispersed phase and the dispersion medium possess the same or similar chemical constitution. **1934** *Biol. Abstr.* VIII. 1476/1 Myelin reduction, at least in the iso-cortex, is discontinuous. **1937** BEST & TAYLOR *Physiol. Basis Med. Pract.* lxiii. 1418 The laminated cortex, which in man constitutes the remaining eleven-

twelfths [of the cortical area] and in animals is a much smaller fraction of the whole, is called the isocortex. **1951** K. S. LASHLEY in L. A. Jeffress *Cerebral Mechanisms in Behaviour* 132 In the rat, I have removed..practically every other part of the isocortex without disturbing visual perception or memory. **1970** *Developmental Biol.* XXII. 575 Fetal mouse cerebral isocortex from normal animals ..was dissociated and aggregated. **1900** E. F. SMITH tr. *V. von Richer's Org. Chem.* (ed. 3) II. 435 The carbocyclic substances..belong to the class of isocyclic compounds which consist of rings of atoms of one and the same element. **1932** [see *homocyclic* adj. s.v. *HOMO-]. **1951** I. L. FINAR *Org. Chem.* I. 667 The subject matter is divided into four main divisions: (i) Alicyclic compounds. (ii) Isocyclic compounds. (iii) Heterocyclic compounds. (iv) Natural products. **1951** *Jrnl. Meteorol.* VIII. 274/1 The distribution of echo intensity from all points in the two-dimensional cut through a storm may be plotted by use of the contour-mapping techniques suggested by Langille and Gunn or Atlas. The latter techniques produce a contour chart showing isopleths of reflected power throughout the area of the storm. (These isopleths will be referred to as 'isoecho lines' or 'power contours'.) **1959** L. J. BATTAN *Radar Meteorol.* xv. 141 (*heading*) Isoecho contouring. *Ibid.*, If range differences are negligible, the isoecho contour corresponds to a line of equal cloud reflectivity and equal rain intensity. **1961** *Aeroplane* CI. 573/2 For general weather observation the C.R. 353 has a 12-in. PPI unit incorporating iso-echo circuits which enable the operator to make cloud density determinations. **1969** *New Scientist* 4 Dec. 509/1 A..height-finding radar system has been modified to observe cloud and rain up to 160 miles out to sea..and to plot contours of constant rain intensity (iso-echo contours). **1899** *Ann. Rep. Board of Regents Smithsonian Inst., 1897–8* 543 Two weights of different aliments for which these numerical values are the same are said to be..isoenergetic weights. **1937** *Proc. R. Soc.* A. CLXI. 259 The mean free path in paraffin wax of the iso-energetic neutrons obtained by bombarding heavy hydrogen with deuterium ions has been measured. **1937** M. W. ZEMANSKY *Heat & Thermodynamics* xiv. 281 Show that the slope of an isoenergetic curve on a T–V diagram is equal to [etc.]. **1962** *Jrnl. Aerospace Sci.* XXIX. 400/2 During the first phase of re-entry, the motion is isoenergetic and no significant heating or deceleration problem will exist. **1936** *Trans. Faraday Soc.* XXXII. 124 As a consequence of the..growth of the micelles the resin passes from its (assumed) isogel state..into the infusible *C* stage. **1937** *Jrnl. R. Aeronaut. Soc.* XLI. 531 Phenolformaldehyde resins (at least in their initial stages of condensation) are 'isogels'. **1950** ROBITSCHEK & LEWIN *Phenolic Resins* iv. 56 As condensation proceeds, crosslinking takes place between some of the largely spherical bigger molecules (micelles) leading to a structure which.. can be likened to a sponge and is termed an isogel. **1924** C. E. TILLEY in *Geol. Mag.* LXI. 169 In the terminology suggested above, this line may be said to be an Isograd... In reality an isograd is the intersection of an inclined isograd surface with the earth's surface. **1956** E. W. HEINRICH *Microscopic Petrogr.* vi. 173 By joining points on a map that mark the initial appearance of each of the diagnostic minerals, mineral isograds (biotite isograd, sillimanite isograd, etc.) may be defined. **1971** I. G. GASS et al. *Understanding Earth* i. 36/2 The garnet isograd for example is the surface (line on the map) separating the garnet-bearing rocks of the garnet zone (high-grade) from the garnet-free rocks of the biotite zone (low grade). **1924** C. E. TILLEY in *Geol. Mag.* LXI. 168 Isograde rocks are those which have originated under closely similar physical conditions of temperature and pressure. **1966** *McGraw-Hill Encycl. Sci. & Technol.* VIII. 298/1 Rocks within the same zone [of metamorphism] may be called..isograde. **1924** C. E. TILLEY in *Geol. Mag.* LXI. 168 Rocks which belong to the same facies can be said to be in the same metamorphic grade, and can be referred to by the terms which I now suggest as isofacial or isogradic. **1926** G. W. TYRRELL *Princ. Petrol.* xv. 259 In the green-schist facies, a chlorite-quartz-muscovite-schist is isogradic with a green schist composed of chlorite, epidote, and albite. **1968** F. J. TURNER *Metamorphic Petrol.* viii. 376 On this model the isogradic surfaces near the heat source have a reversed dip. **1902** MANN & MILLIKAN tr. *Drude's Theory of Optics* II. II. iii. 354 The whole field of view is now.. traversed by a black curve, the so-called principal isogyre. **1922** N. H. & A. N. WINCHELL *Elem. Optical Mineral.* (ed. 2) I. xviii. 168 The biaxial optic axis interference figure differs from the uniaxial optic axis interference figure most clearly in the fact that it has only one isogyre instead of two. It also differs..in the fact that the single isogyre is not fixed in position, nor constantly straight, when the crystal is rotated. *Ibid.* 172 As in uniaxial crystals, the isogyres are the locus of all points at which the light emerges with its vibration planes parallel with the planes of the nicols. **1964** HARTSHORNE & STUART *Pract. Optical Crystallogr.* v. 203 These interference bands are symmetrically arranged around the optic axis (or axes)... In addition to these bands there are dark 'brushes' or isogyres. **1907** *Jrnl. Med. Res.* XVII. 321 Earlier observers of human isohemagglutination asserted that isoagglutinins occurred only in the serum of pathological states. **1940** *Amer. Jrnl. Physiol.* CXXXI. 205 Breed, sex and age of animals seemed to have no influence on the occurrence of natural isohemagglutination. **1907** *Jrnl. Med. Res.* XVII. 334 Such an experiment, in the case of human isohemagglutinins, does more to prove the strict specificity of each 'bound agglutinin'. **1971** J. A. BELLANTI *Immunol.* ii. 44 It is known that isohemagglutinins—the antibody to blood groups—..may develop as a result of exposure to enteric bacilli, containing blood grouplike substances in their structure. **1905** GOULD *Dict. New Med. Terms* 318/2 Isohemolysin. **1916** Isohemolysin [see *blood group* s.v. *BLOOD *sb.* 19]. **1972** W. E. HAESLER *Immunohematol.* i. 2 Hemolysis is most frequently observed in the detection of the group A and B isohemolysins and the Le[a] antibodies. **1902** *Encycl. Brit.* XXXI. 404/2 South of the Tropic of Capricorn the isohalines run nearly east and west. **1964** *Oceanogr. & Marine Biol.* II. 71 In winter the fiord water becomes isothermal and isohaline with average temperature and salinity of −1·76°C and 32·75‰. *Ibid.* 375 (*caption*) The increase in salinity in recent decades has altered..the course of the isohalines given in the chart. **1968** G. NEUMANN *Ocean Currents* iv. 129 The slope of isobaric surfaces..is small when compared to the slope of isothermal, isohaline (surfaces of equal salinity) and isopycnal surfaces. **1904** GOODCHILD & TWENEY *Technol. & Sci. Dict.* 323/1 *Isohels*, lines connecting places having the same amount of sunshine. **1931** A. A. MILLER *Climatol.* 22 The deviation of sunshine (shown on maps by lines of equal duration known as isohels). **1968** J. GENTILLI *Sun, Climate & Life* (1971) xiii. 141/2 These lines are called isopleths as a general term, but..more specifically..isobars (equal pressure), isotherms (equal temperature),..isohels (equal sunshine),..etc. **1897** *Geogr. Jrnl.* X. 306 König.. has found sufficient material for a first attempt to draw 'isohelic' lines for Western Europe. **1899** *Nature* 21 Dec. 172/2 Isobars and isohyets indicating monthly and annual distribution of barometric pressure and rainfall. **1911** C. E. W. BEAN *'Dreadnought' of Darling* xv. 144 The carpet which covers a country within the ten-inch isohyet (rainfall line) is a pretty ticklish thing to play with. **1967** M. J. COE *Ecol. Alpine Zone Mt. Kenya* 63 The main peak area has a rainfall of between 30" and 40" per annum, the 30" Isohyet being displaced slightly to the South-west of the peaks. **1889** *Cent. Dict.*, *Isohyetal, a.* and *n.* **1895** T. RUSSELL *Meteorol.* vii. 141 A graphic representation of quantity of rainfall by lines through places having equal depths of rainfall are 'isohyetals'. **1909** *British Rainfall 1908* 140 The isohyetal lines are drawn about the 18th of the following month. **1923** *Glasgow Herald* 1 Feb. 6 Most of these features are..illustrated by maps exhibiting them..by isohyetal lines. **1962** W. STEGNER *Wolf Willow* IV. iv. 281 She knew nothing about minimal annual rainfall, distribution of precipitation, isohyetal lines. **1938** *Jrnl. Path. & Bacteriol.* XLVII. 234 Fleisher found evidence of iso-immune reactions working with grafts of renal tissue in the guinea-pig. **1967** D. M. WEIR *Handbk. Exper. Immunol.* XXX. 991 Isoantigenic differences, though usually detected by isoimmune sera, may in some cases also be picked up with antisera produced by immunization across a species gap. **1969** B. PIROFSKY *Autoimmunization* xxi. 479/1 The bulk of knowledge concerning the nature and effect of immune interactions on the erythrocytes has been derived from heteroimmune and isoimmune studies. **1939** *Jrnl. Amer. Med. Assoc.* 8 July 126/1 It would seem to resemble agglutinins resulting from iso-immunization following repeated transfusions. **1971** J. A. BELLANTI *Immunol.* iii. 77 Alternatively, isoimmunization can occur during the course of pregnancy when fetal cells..or proteins gain access to the maternal circulation. **1971** *Nature* 29 Oct. 608/1 The human foetus is at greater risk from maternal isoimmunizations than the bovid foetus. **1958** *Engineering* 22 Aug. 230/1 Regular production has..commenced..of the isokinetic sampling apparatus developed by the British Iron and Steel Research Association. **1959** *Brit. Jrnl. Appl. Physics* X. 26/1 The dust concentration calculated from the sample will be correct provided the gas sample is drawn into the nozzle at the same velocity as that of the gas stream. This is known as isokinetic sampling. *Ibid.* 27/2 In the extreme case of isokinetic sampling the nozzle becomes effectively non-existent. **1967** *Ann. Occupational Hygiene* X. 77 Brass nozzles were fitted around the filter retaining ring in order to obtain isokinetic samples at windspeeds from 5·4 m.p.h. to 30 m.p.h. *Ibid.*, 3-in.-long nozzles were attached to the filter holder to allow isokinetic flow into the inlet. **1958** *Engineering* 22 Aug. 230/1 The stainless steel probe faces directly into the stream of dusty gas and a sample is withdrawn isokinetically, that is, it flows into the nozzle in the same direction and with the same velocity as the local undisturbed gas stream. **1972** *Science* 16 June 1232/3 We sampled the suspended fly ash isokinetically at several locations across the outlet duct of the electrostatic precipitator. **1926** JORDAN & KINDRED *Textbk. Embryol.* v. 31 Since the small amount of yolk is evenly distributed throughout the cytoplasm, it [*sc.* the human egg] may also properly be called an isolecithal egg. **1940** L. H. HYMAN *Invertebrates* I. v. 256 When the yolk is slight in amount, it is also more or less evenly dispersed; such eggs are variously termed isolecithal, alecithal, or homolecithal. **1972** P. A. MEGLITSCH *Invertebr. Zool.* (ed. 2) iv. 81/2 Isolecithal ova usually cleave in a characteristic manner. **1921** *Jrnl. Eng. & Gmc. Philol.* XX. 183 The finding and fixing of the isolectic lines is a task of word geography. *Ibid.* 182 There is danger in delay if certain phases of dialect life are to be recorded at all, and an accelerated pace in registering them would be advisable, especially in establishing the boundary lines of present dialects, with their maze of isophones, isomorphs, isolexes, and isotaxes, i.e., lines connecting places of identical or nearly identical sounds, forms, words, and syntactical peculiarities. **1963** *Amer. Speech* XXXVIII. 127 The regional diversity and the complicated grid of isolexes in the Northern counties. *Ibid.* 128 Any attempt to correlate the isolexes with demographic factors. **1926** *Germanic Rev.* I. iv. 285 The isolexic lines of this word..present a hopeless tangle. **1939** L. H. GRAY *Foundations of Lang.* ii. 26 These lines will be isophonic, isotonic, isomorphic, isosyntagmic, or isolexic according as they indicate identical sounds, tones, inflexions, syntax, or vocabulary. **1954** PEI & GAYNOR *Dict. Ling.* 107 *Isolexic lines*, lines on a linguistic map, indicating the approximate boundaries of the speech-areas in which a uniformity in the vocabulary of the speakers and in their use of words can be observed. **1944** V. CONRAD *Methods in Climatol.* xiii. 167 Isolines..are fully analogous to contour lines, or to the equipotential lines used in physics... Closed isolines surrounding a region indicate that this is either depressed or elevated. **1961** G. T. TREWARTHA *Earth's Problem Climates* xvi. 237/2 In summer..the isolines of rainfall frequency show a strong zonal arrangement. **1969** *Nature* 29 Nov. 903/1 Vertical sections show that isolines slope steeply to the surface [of the sea] during upwelling, but the slope of the isoline varies with the parameter chosen. **1970** *Biol. Abstr.* LI. 11534/1 The resulting pattern of the isolines can be readily related to the distribution of a particular organism. **1901** DORLAND *Med. Dict.* (ed. 2) 333/2 Isolysin. **1910** *Jrnl. Hygiene* X. 186 The injection of goats' blood into other goats resulted as a rule in the formation of isolysins. **1969** L. H. CRISP *Clin. Immunol. & Allergy* (ed. 2) xliv. 433/1 Isohemolysins, isolysins, or antibodies capable of lysis of erythrocytes may be found in the serum in paroxysmal hemoglobinuria. **1898** J. MILNE *Seismol.* xii. 225 Slight changes in the isomagnetics of a district. **1899** *Nature* 6 July 236/2 An opportunity will thus be afforded..to obtain some idea of the accuracy with which the isomagnetic lines can be determined. **1940** CHAPMAN & BARTELS *Geomagnetism* I. iii. 96 The lines are called isomagnetic lines, and a chart in which the distribution of a magnetic element is thus indicated..is called an isomagnetic chart. **1967** E. H. VESTINE in Matsushita & Campbell *Physics of Geomagnetic Phenomena* I. II. ii. 185 The isomagnetic lines for declination *D* or variation of the compass are also called 'isogonic' lines. **1900** E. F. SMITH tr. *V. von Richter's Org. Chem.* (ed. 3) II. 390 Isonuclear substitution products with adjacent substituents show in general the same deportment as the ortho-substitution products of benzene. **1951** I. L. FINAR *Org. Chem.* I. xxix. 586 Introduction of a second substituent can give rise to homonuclear (isonuclear) substitution..or to heteronuclear substitution. **1908** *Jrnl. Exper. Med.* X. 137 An iso-osmotic physiologically balanced solution. **1971** *Biochem. Jrnl.* CXXI. 261 Protein-polysaccharides of knee-joint cartilage of 9-month-old pigs were extracted sequentially with neutral iso-osmotic sodium acetate. **1873** J. W. GIBBS in *Trans. Connecticut Acad. Arts & Sci.* II. 311 In the same way we may conceive of lines of equal pressure... These lines we may also call..isopiestic. *Ibid.* 313 To prove that the ratio is independent of the shape of the circuit, let us suppose the area..divided up by an infinite number of isometrics..with equal differences of volume *dv*, and an infinite number of isopiestics..with equal differences of pressure *dp*. **1902** *Encycl. Brit.* XXXIII. 283/2 The isothermals..coincide with the isopiestics for a saturated vapour in presence of its liquid. **1940** GLASSTONE *Text-bk. Physical Chem.* ix. 622 If two vessels containing different solutes in the same solvent are placed side by side in a closed space, vapor will distil from the solution of higher vapor pressure and condense in the one having the lower pressure until, when equilibrium is attained, both solutions are exerting the same pressure, that is to say they are isopiestic. **1966** R. JOEL *Basic Engin. Thermodynamics* i. 80 The pressure remains constant throughout the process. It is often referred to as an isobaric or isopiestic process. **1931** *Compt. Rend. de l'Assemb. de Stockholm 1930* (Union Géod. et Géophys. Internat., Sect. de Magn. et Électr. Terr.) 284 In that year the zero-isopor (the line dividing easterly and westerly change) crossed central Siberia. **1963** J. A. JACOBS *Earth's Core* v. 53 Considerable changes take place in the general distribution of isopors even within 20 years. **1931** *Compt. Rend. de l'Assemb. de Stockholm 1930* (Union Géod. et Géophys. Internat., Sect. de Magn. et Électr. Terr.) 280 A consideration of the most recent published results of secular-variation observations..has revealed some very interesting and important conditions governing isoporic movements. **1940** CHAPMAN & BARTELS *Geomagnetism* I. iii. 114 The rate of the secular variation in each element at any epoch..can..be represented by..isoporic charts, a term proposed by Harradon. **1973** M. W. MCELHINNY *Palaeomagnetism & Plate Tectonics* i. 6 Isoporic foci are not permanent, but grow and decay, their lifetime being of the order of 100 years, during which they move on the earth's surface in a somewhat irregular fashion. **1895** C. S. PALMER tr. *Nernst's Theoret. Chem.* I. v. 121 The investigation of solutions having the same osmotic pressure, viz. the so-called is-osmotic solutions. **1905** W. H. HOWELL *Text-bk. Physiol.* 884 A 0·95 per cent. solution of NaCl is isotonic or isosmotic with mammalian serum. **1967** *Oceanogr. & Marine Biol.* V. 383 In life the muscle cells are probably isosmotic with the interstitial fluid and the plasma. **1964** SHAPIRO & CHARGAFF in *Biochim. & Biophys. Acta* XCI. 263 There exist procedures permitting the separation of the hydrolysates [of DNA] into a series of fractions, each comprising the equinumerant oligonucleotide runs of a given length. .. It may be convenient to refer to such a size group as an isostich. **1970** *Nature* 26 Sept. 1296/1 The tracts were fractionated into isostichs by chromatography on DEAE-cellulose. **1971** F. VON DER HARR et al. in Cantoni & Davies *Procedures Nucleic Acid Res.* II. 682 The isostichs can be further separated into their components, differing in base composition, by paper chromatography. **1906** *Jrnl. Chem. Soc.* LXXXIX. II. 1129 It is conceivable that in the case of two 'isostructural' substances the actual size of the structural unit may be of the greatest importance. **1965** PHILLIPS & WILLIAMS *Inorg. Chem.* I. xvi. 576 Tellurium is only known in one form, isostructural with grey Se. **1922** JOYCE *Ulysses* 307 The intricate alliterative and isosyllabic rules of the Welsh englyn. **1943** *Jrnl. Theol. Stud.* XLIV. 51 The isosyllabic metre of the Greek homilies of Ephraem, typical of Syriac poetry. **1957** Isosyllabic [see *isoaccentual* adj. above]. **1956** H. WHITEHALL in *Kenyon Rev.* XVIII. III. 420 Of the non-syllabic rhythms, the first, found typically in Old Testament Hebrew verse and in some, though not all, 'free verse' is *isosyntactic*—the recurrent factor is repetition of the same syntactic construction, usually a phrase or clause, in strictly parallel sequences. **1939** Isosyntagmic [see *isolexic* adj. above]. **1954** PEI & GAYNOR *Dict. Ling.* 107 *Isosyntagmic lines*, lines on a linguistic map, indicating the approximate boundaries of the speech-areas in which a uniformity of syntax can be observed. **1957** Isosyntagmic [see *ISOMORPHIC *a.* 5]. **1947** *Mineral. Abstr.* X. 159 Isogyres and isotaches (curves of equal velocity) are plotted on a stereographic net. **1955** W. J. SAUCIER *Princ. Meteorol. Analysis* x. 304/1 The wind field is analyzed by drawing streamlines and isotachs, which give, respectively, the course of flow and its speed. **1970** *Nature* 11 Apr. 133/2 (*caption*) Geostrophic isotachs (in knots) at 500 mbar for 00 GMT November 14, 1968. **1910** SMITH & MENZIES in *Jrnl. Amer. Chem. Soc.* XXXII. 1420 The purpose of the apparatus being to show when two pressures have become equal, the arrangement may be called an isoteniscope. **1960** *Jrnl. Chem. Education* XXXVII. 533/1 Livingstone suggests the inclusion of a thermometer well in the isoteniscope bulb so that temperature equilibrium can be ascertained at the time of the pressure measurements. **1915** D. WHITE in *Jrnl. Washington Acad. Sci.* V. 198 Lines were then drawn through the points of equal fixed carbon (or volatile matter). Such lines,..which I have termed 'isovols', are drawn to mark each 5 per cent increase in the fixed carbon in the pure coal. **1923** *Glasgow Herald* 11 June 7 The isovols for the Hutton seam take the form of a number of rings with a common centre to the north-west of Durham. **1928** E. R. LILLEY *Geol. Petroleum & Nat. Gas* v. 113 The greater part of the oil of Pennsylvania is produced from pools lying between the isovols.. of 55 and 60%. **1968** MURCHISON & WESTOLL *Coal* xv. 370 Fig. 15 shows lines of equal magnetic vertical intensity and the isovols of the Wealden coals.

b. Add to note:

Iso- was formerly regarded as a separable prefix and printed in italics (often with a hyphen). The International Union of Pure and Applied Chemistry recommendation is that *iso-* should always be directly attached to the remainder of the parent name (and be printed in ordinary type).

Where the isomerism occurs within an alkane or alkyl radical, the prefix *iso-* is used to form the name of the isomer having a $(CH_3)_2CH$— group at the end of an otherwise straight chain: so $CH_3(CH_2)_3CH_3$ (normal or *n-*) pentane, $(CH_3)_2CHCH_2CH_3$ isopentane, $(CH_3)_3CCH_3$ neopentane.

i:soallo·xazine [ALLOX(AN + *AZINE], the hypothetical tricyclic parent compound, $C_{10}H_6N_4O_2$, of the flavins, which has a structure formed of fused benzene, pyrazine, and pyrimidine nuclei and is known only as substituted derivatives; **isoa·myl alcohol** = *isopentyl alcohol*; **isobo·rneol** [a. G. *isoborneol* (Bertram & Walbaum 1894, in *Jrnl. f. prakt. Chem.* XLIX. 1)], a crystalline bicyclic alcohol, $C_{10}H_{18}O$, which is a stereoisomer of borneol and like it yields camphor on oxidation; **isobu·tane,** 2-methylpropane, $(CH_3)_2\cdot CH\cdot CH_3$, a gaseous hydrocarbon used as a fuel; **isobu·tene** = *isobutylene*; **isobu·tyl,** the radical $(CH_3)_2CH\cdot CH_2$—, 2-methylpropyl, as in **isobutyl alcohol,** $(CH_3)_2CH\cdot CH_2OH$, a primary alcohol which is a colourless liquid and occurs in fusel oils; **isobu·tylene** († -en), 2-methylpropylene, $(CH_3)_2C{:}CH_2$, an easily liquefied gas used in the manufacture of butyl rubber; **isobu·tyrate,** any of the esters of isobutyric acid, several of which are used as flavourings and in perfumery; **isobuty·ric acid** [tr. G. *isobuttersäure* (H. Kolbe 1864, in *Zeitschr. f. Chem. und Pharm.* VII. 33)], a liquid carboxylic acid, $(CH_3)_2CH\cdot COOH$, found in many plants and also obtained by oxidation of isobutyl alcohol; **isoci·trate,** the anion, or an ester or salt, of isocitric acid; **isoci·tric acid** [tr. G. *isocitronsäure* (F. Rochleder 1869, in *Jrnl. f. prakt. Chem.* CVI. 320)], 1-hydroxypropane-1,2,3-tricarboxylic acid, $HOOC\cdot CH(OH)CH(COOH)CH_2COOH$, which occurs in blackberry juice and is formed in the Krebs cycle by dehydration of citric acid to *cis*-aconitic acid followed by rehydration; **isocy·anate** [a. F. *isocyanate* (F.-S. Cloëz *Rech. sur Éthers Cyaniques* (Thesis, 1866) 18)], any of the class of compounds containing the group —N:C:O, some of which are used in making polyurethane resins; **isocya·nic acid,** the acid HN:C:O, which exists in equilibrium with cyanic acid (HO·CN); **isocy·anide,** any member of the class of compounds having the formula R—NC (where R is an alkyl, aryl, etc., radical), which in general are poisonous liquids with a strong unpleasant odour; also called carbylamines, isonitriles; **isoeugenol** (əisǫ,yū·dʒĕnǫl) [a. G. *isoeugenol* (Tiemann & Kraaz 1882, in *Ber. d. Deut. Chem. Ges.* XV. 2067)], an aromatic liquid that occurs in ylang-ylang and other essential oils, is produced commercially from eugenol, and has been used in the manufacture of vanillin and in perfumery; 2-methoxy-4-propenylphenol, $CH_3CH{:}CH\cdot C_6H_3(OCH_3)OH$; **isofla·vone,** (*a*) the crystalline tricyclic ketone 3-phenylbenzo-4-pyrone, $C_{15}H_{10}O_2$; (*b*) any of the derivatives of this compound, which occur (often as glycosides) in many plants; **isoli·chenin** [a. G. *isolichenin* (F. Beilstein *Handbuch d. Org. Chem.* (1881) I. xxxix. 602)], a water-soluble starch occurring in lichens which yields glucose on hydrolysis; **isoma·ltose** [a. G. *isomaltose* (E. Fischer 1890, in *Ber. d. Deut. Chem. Ges.* XXII. 3688)], 6-*O*-α-D-glucopyranosyl-D-glucose, $C_6H_{11}O_5\cdot O\cdot C_6H_{11}O_5$, a syrupy disaccharide formed by the action of acid on glucose; **i:sonicoti·nic acid** [tr. G. *isonicotinsäure* (Weidel & Russo 1883, in *Sitzungsber. d. K. Akad. d. Wissen.* (Math.-Nat. Classe) LXXXVI. II. 1172)], pyridine-4-carboxylic acid, $(C_5H_4N)COOH$, a crystalline compound used in the synthesis of isoniazid; **isonicotinic (acid) hydrazide** = *ISONIAZID; **isoni·trile** = *isocyanide*; **isopa·raffin,** any branched-chain paraffin, *spec.* one containing the isopropyl group, $(CH_3)_2CH$—, attached to an otherwise straight chain; **isope·ntane,** 2-methylbutane, $(CH_3)_2CHCH_2\cdot CH_3$, a volatile liquid hydrocarbon present in petroleum; **isope·ntyl,** the radical $(CH_3)_2\cdot CHCH_2CH_2$—, as in **isopentyl alcohol,** a liquid primary alcohol, $(CH_3)_2CH_2CH_2OH$, which has a disagreeable odour and is obtained from fusel oil; **isophtha·late,** a salt or ester of isophthalic acid; **isophtha·lic acid,** benzene-*m*-dicarboxylic acid, $C_6H_4(COOH)_2$, a crystalline compound made by the oxidation of *m*-xylene and used in the manufacture of polyester and alkyd resins; **isopro·panol** = *isopropyl alcohol*; **isopro·penyl,** the radical $CH_2{:}C(CH_3)$—; **isopro·pyl,** the radical $(CH_3)_2CH$—, as in **isopropyl alcohol,** a liquid secondary alcohol, $CH_3CHOH\cdot CH_3$, made by the hydration of propylene and widely used as a solvent and in the production of acetone; **isopropy·lidene,** the bivalent radical $(CH_3)_2C$=, frequently introduced into compounds by reaction with acetone; **isoqui·noline** [ad. F. *isoquinoléine* (Hoogewerff & Van Dorp 1885, in *Rec. des Trav. chim. des Pays-Bas* IV. 128)], a low-melting, crystalline, bicyclic compound, C_9H_7N, found in coal tar and forming the nucleus of many alkaloids; **isovalera·ldehyde,** 2-methylbutyraldehyde, $(CH_3)_2CH\cdot CH_2\cdot CHO$, a liquid which occurs in peppermint, sandalwood, eucalyptus, and other oils; **isova·lerate,** a salt or ester of isovaleric acid; **isovaleria·nic (acid)** = *isovaleric acid*; **isovale·ric acid,** 2-methylbutyric acid, $(CH_3)_2\cdot CHCH_2COOH$, a liquid with a disagreeable odour found free in valerian root and as esters in porpoise and dolphin oils; **iso·xazole** [ad. G. *isoxazol* (A. Hantzsch 1888, in *Ann. d. Chem.* CCXLIX. 3)], (*a*) a liquid heterocyclic compound,

$$\underbrace{CH{:}CH\cdot O\cdot N{:}CH}$$

with a penetrating odour; (*b*) a derivative of this compound.

1936 *Chem. Abstr.* XXX. 4512 Isoalloxazines with a substituent, e.g., alkyl, cycloalkyl or aryl, in the 9-position are prepd. by condensing *N*-monosubstituted aromatic *o*-diamines with alloxan. **1953** FRUTON & SIMMONDS *Gen. Biochem.* xiii. 320 In riboflavin, a sugar residue, D-ribitol, is attached to a nitrogen atom of a heterocyclic nucleus, termed an isoalloxazine ring. **1968** I. L. FINAR *Org. Chem.* (ed. 4) II. 556 It appears that isoalloxazine, the tautomer of alloxazine, does not exist as such; only when the hydrogen atom is substituted is the isoalloxazine form retained. **1886** *Jrnl. Chem. Soc.* XLIX. 770 The value of the ratio [of vapour pressures] of isobutyl and isoamyl alcohol is practically a constant. **1927** *Chem. Abstr.* XXI. 985 Isoamyl alc. is considered as the mother substance of most of the compds. constituting oil of lavender. **1970** *New Phytologist* LXIX. 557 Isovaleric acid and isoamyl alcohol have been identified as metabolites of *Agaricus bisporus*. **1894** *Chem. News* 30 Mar. 156/2 Isoborneol..crystallises out of petroleum ether in thin, feathery leaflets. **1951** P. Z. BEDOUKIAN *Perfumery Synthetics & Isolates* 93 Iso-borneol and its esters, particularly the acetate, are employed in many types of industrial perfumes—for example, in sprays. **1972** G. D. SARGENT in Olah & Schleyer *Carbonium Ions* III. xxiv. 1122 On reduction with lithium aluminium hydride, camphor (34) gives isoborneol (35) in high yield. **1876** *Jrnl. Chem. Soc.* XXIX. 540 When isobutane is heated to 250° with iodine trichloride it gives the same products as propane. **1936** [see *CALOR]. **1959** *Times Rev. Industry* Aug. 98/2 Other processes..include alkylation, in which iso-butane is reacted with olefins to produce high octane material for aviation gasoline and high quality motor spirit. **1876** *Jrnl. Chem. Soc.* XXX. 397 Isobutene combines readily with hypochlorous acid. **1964** N. G. CLARK *Mod. Org. Chem.* viii. 141 t-Butyl alcohol..is obtained from a 2-methylpropene [isobutene] using 65 per cent sulphuric acid. **1870** *Chem. News* 21 Jan. 34/2 (*heading*) Conversion of isobutyl-alcohol into tertiary pseudobutyl-alcohol. **1873** [in Dict.]. **1964** D. A. SHIRLEY *Org. Chem.* x. 251 Isobutyl alcohol..is manufactured by a modification of the carbon monoxide and hydrogen method for synthesis of methanol. **1966** *Nomencl. Org. Chem.* (I.U.P.A.C.) (ed. 2) A. 8 The following names are retained for the unsubstituted radicals only: Isopropyl..Isobutyl..Isopentyl [etc.]. **1872** *Chem. News* 29 Nov. 265/2 Chlorhydric acid behaves with isobutylen in the same manner as iodhydric acid, the result being the formation of a tertiary chloride of butyl. **1913** J. B. COHEN *Org. Chem. Adv. Students* II. ii. 119 Isobutylene when heated with strong sulphuric acid yields a mixture of isomeric diisobutylenes. **1951** *Economist* 29 Dec. 1599/2 It should produce butyl rubber based almost entirely on iso-butylene. **1969** R. F. LANG tr. *Henglein's Chem. Technol.* 576 Cracking gases consist of ethylene, propylene and isobutylene... Isobutylene yields on polymerization a dimer which can be hydrogenated to iso-octane. **1873** *Jrnl. Chem. Soc.* XXVI. 55 Ethylic isobutyrate boils at 113°. **1928** *Chem. Abstr.* XXII. 2809 A table of 23 butyrates and isobutyrates, giving their name, odor and specific use in perfume. **1973** *Proc. Soc. Exper. Biol. & Med.* CXLII. 595/1 Sucrose acetate isobutyrate.. is employed as a flavor-suspending agent in the manufacture of soft drinks. **1871** *Jrnl. Chem. Soc.* XXIV. 126 Isobutyric acid..is scarcely attacked by a mixture of potassium dichromate and dilute sulphuric acid. **1881** ROSCOE & SCHORLEMMER *Treat. Chem.* III. I. 599 Isobutyric acid is found in the free state in the flowers of the *Arnica montana*, as well as in the carob bean, and amongst the acids of croton oil. **1970** *Exper. Parasitol.* XXVII. 408 The branched chain acids, isobutyric and isovaleric, are also excreted by *A. caninum*. **1925** *Jrnl. Amer. Chem. Soc.* XLVII. 572 The ethyl isocitrate obtained from the first lot of blackberries was levorotatory. **1952** *Biochem. Jrnl.* LII. 528/2 The observations are in agreement with Martius's earlier assumption..that *cis*-aconitate is an intermediate in the conversion of citrate into *iso*citrate. **1971** *Jrnl. Biol. Chem.* CCXLV. 4807/1 Studies are reported..which indicate that there is a specific carrier system for the transport of citrate and isocitrate in mitochondria. **1869** *Chem. News* 11 June 287/1 The isocitric acid may be obtained pure..and then exhibits a crystalline mass. **1930** *Jrnl. Amer. Chem. Soc.* LII. 2928 As there are two asymmetric carbon atoms in isocitric acid, and no meso form is possible, four optically active forms and two racemic forms of the acid may exist. **1968** R. F. STEINER *Life Chem.* xii. 219 The reversible transformation of citric acid to *cis*-aconitic acid, and of the latter to isocitric acid, is catalyzed by a single enzyme, aconitase. **1872** *Jrnl. Chem. Soc.* XXV. 446 The first members of the group of compounds, now generally described as the isocyanates and isocyanurates, were discovered by Wurtz in the ethyl and methyl series. **1877** [in Dict.]. **1944** S. J. SMITH *Princ. Org. Chem.* xiv. 312 The alkyl isocyanates are liquids with a powerful stifling odour. **1961** *Times* 30 May (I.C.I. Suppl.) p. xxii (Advt.), Italians need isocyanates for lightweight rigid and flexible polyurethane foams. **1963** in *Amer. Speech* (1964) XXXIX. 146 A dummy man made of material resembling human flesh which has a base of isocyanate rubber. **1891** *Jrnl. Chem. Soc.* LX. I. 282 Hydrocyanic acid and a small quantity of isocyanic acid are evolved. **1919** *Jrnl. Amer. Chem. Soc.* XLI. 381 The reaction between isocyanic acid and benzylideneaniline leads directly to a four-membered cyclic-urea. **1973** J. J. LAGOWSKI *Mod. Inorg. Chem.* xi. 349 Isocyanic acid (mp −86·8°, bp 23·5°) is formed when cyanuric acid is passed through a hot tube; the product reverts to cyanuric acid spontaneously. **1877** Isocyanide [in Dict.]. **1881** ROSCOE & SCHORLEMMER *Treat. Chem.* III. I. 162 Cyanides of the alcohol radicals. These bodies are formed when an alcoholic iodide is heated with silver cyanide... The compounds obtained in this way are usually termed isocyanides or carbamines. **1907** *Daily Chron.* 12 Dec. 5/5 Someone noiselessly discharged several squibs of isocyanide, and two ladies in the audience fled. **1928** *Sunday Dispatch* 16 Dec. 13/5 The new gas, cacodyl isocyanide, which..was..so terrible and destructive..in its effect on life. **1964** N. G. CLARK *Mod. Org. Chem.* xiii. 265 The isocyanides are of no practical value, with the possible exception of the Carbylamine Reaction. However, the elucidation of their structure..has provided an interesting chapter in the development of organic chemistry. **1883** *Jrnl. Chem. Soc.* XLIV. 201 By distilling it [*sc.* homoferulic acid] with lime a body is obtained which is isomeric with eugenol, and termed isoeugenol. **1891** *Jrnl. Soc. Chem. Industry* 31 Oct. 854/1 Iso-eugenol..is prepared from eugenol, or the Essence of Cloves, by heating it with caustic potash in amyl alcohol for 16–24 hours. **1936** A. HUXLEY *Eyeless in Gaza* xviii. 243 That's one of the reasons why your scent costs you so much. The poor..have to be content with plain iso-eugenol. **1965** *Chem. Abstr.* LXII. 6815 Eugenol and isoeugenol were estd. in the smoke at 4 and 14·7 mg./1000 Turkish tobacco cigarets, resp. **1925** *Jrnl. Chem. Soc.* CXXVII. 1981 The occurrence of derivatives of 3-phenylchromone (*iso*-flavone) has not yet been definitely proved. **1948** *Proc. Indian Acad. Sci.* A. XXVII. 36 Hydroxy isoflavones are more toxic than the corresponding flavones. **1951** *Ann. Rev. Biochem.* XX. 508 The..occurrence of the isoflavone (prunetin) along with its isomeric flavone (genkwanin) is a rare example of such association. **1965** T. SWAIN in Pridham & Swain *Biosynthetic Pathways Higher Plants* 33 Isoflavones..are common in other members of the Leguminosae. **1898** *Amer. Jrnl. Physiol.* I. 455 The unusual behavior of isolichenin towards amylolytic enzymes—the formation of dextrins without sugars—recalls the formation (from glycogen) of dystropo-dextrin. **1934** *Chem. Abstr.* XXVIII. 2375 Isolichenin was proved in some varieties of the lichens produced in Japan, such as *Alectoria ochroleuca*... Isolichenin closely resembles amylose. **1967** M. E. HALE *Biol. Lichens* viii. 103 Isolichenin, the rarer of the two major lichen starches, is distinguished by a positive iodine test and consists of D-glucose residues with α-1,3 and α-1,4 glucosidic linkages. **1891** *Jrnl. Chem. Soc.* LX. I. 413 The author has obtained from glucose a new glucobiose, which from its properties is doubtless constituted like maltose, and is hence called isomaltose. **1892** *Jrnl. Soc. Chem. Industry* 30 July 627/2 Iso-maltose is an important constituent of beer and forms 25–30 per cent of beer extract. **1956** *New Biol.* XXI. 12 Maltose appears after two days [during malting], and then maltatriose and isomaltose, as a result of the degradation of starch. **1883** *Jrnl. Chem. Soc.* XLIV. 484 In the form of sulphate it [*sc.* γ-dipyridyl] is easily oxidised by potassium permanganate, yielding pyridine-monocarboxylic or isonicotinic acid... This acid..forms a white crystalline mass melting at 307°. **1952** *Biol. Abstr.* XXVI. 35208 Isonicotinic acid hydrazide..is effective against tuberculosis in the mouse. **1956** *Nature* 25 Feb. 367/2 An active programme of leprosy work at Singapore included a chemotherapeutic trial of isonicotinic hydrazide. **1961** *Biol. Abstr.* XXXVI. 2299/1 (*heading*) Antituberculous activity of isonicotinic acid derivatives in vitro. **1972** *Biochem. & Biophys. Res. Communications* XLVIII. 58 (*heading*) In vitro inhibition of tRNA and protein methylation by nicotinamide and isonicotinic acid hydrazide. **1871** *Jrnl. Chem. Soc.* XXIV. 137 This body [*sc.* a urea], when heated, is resolved into triethylphosphine sulphide and the isonitrile of the allyl series. **1915** R. H. A. PLIMMER *Pract. Org. & Biochem.* 61 To the dilute solution of chloroform in water is added some alcoholic sodium hydroxide and a drop of aniline and the mixture heated. Phenyl isonitrile or carbylamine is formed. **1965** *Chem. Communications* May 181/1 The ability of isonitriles to act as bridging groups has now been demonstrated by the preparation of the iron complex (I). **1971** GREEN & HOFFMANN in I. Ugi *Isonitrile Chem.* i. 1 The term isonitriles is used for the

general class of compounds, whereas the term isocyanide is used for specific designations (e.g., ethyl isocyanide). **1876** *Phil. Mag.* I. 206 The dimethylated and trimethylated paraffins have been distinguished for some time past as normal and isoparaffins respectively. **1889** G. M'GOWAN tr. *Bernthsen's Text-bk. Org. Chem.* i. 43 Iso-paraffins, in which one assumes a single branching in the molecule. **1939** GRUENER & LANKELMA *Introd. Org. Chem.* vi. 85 A tertiary alcohol is obtained only in the case of branching carbon chains, or 'isoparaffins'. **1969** R. F. LANG tr. *Henglein's Chem. Technol.* xxi. 576 The synthesis of antiknock isoparaffins for aviation gasoline is achieved also by chemical means. **1876** *Encycl. Brit.* V. 558/2 Isopentane..is formed by the dehydration of amyl alcohol by means of zinc chloride. **1943** V. A. KALICHEVSKY *Amazing Petroleum Industry* iv. 55 Natural gasoline contains certain quantities of a hydrocarbon known as isopentane which is a valuable component of high-grade gasolines. **1964** ROBERTS & CASERIO *Basic Princ. Org. Chem.* iii. 92 The chlorination of isopentane at 300° gives all four possible monosubstitution products. **1876** *Phil. Mag.* I. 217 One of the two conceivable isopentyl alcohols would be derivable in this way from pseudobutyl alcohol. **1970** H. E. NURSTEN in A. C. Hulme *Biochem. Fruits* I. x. 247 Hultin and Proctor..had already found..isopentyl alcohol to be significant as regards the rank odour of overripe fruit. **1886** E. F. SMITH tr. *V. von Richter's Chem. Carbon Compounds* 566 It [*sc.* the barium salt] is not precipitated by barium chloride from a solution of ammonium isophthalate. **1968** A. L. WADDAMS *Chemicals from Petrol.* (ed. 2) xii. 182 Isophthalates have many properties in common with the phthalate esters so that the two are in competition to some extent. As the isophthalates are the more expensive their use is limited to more specialized applications. **1870** *Chem. News* 22 Apr. 191/2 (*heading*) On isophthalic acid and some of its derivatives. **1914** H. T. CLARKE *Introd. Study Org. Chem.* xxxiii. 399 Isophthalic acid..differs from phthalic acid in being incapable of forming an anhydride or an imide. **1968** A. L. WADDAMS *Chemicals from Petrol.* (ed. 2) xii. 182 The production of isophthalic acid in the U.S.A. is about 30,000 long tons a year. Its major use is in unsaturated polyester resins (38 per cent of the total). 31 per cent of consumption is for alkyd resins. **1945** *Chem. Abstr.* XXXIX. 349 Isopropanol ..in blood and body fluids can be detd. iodometrically. **1956** *Nature* 11 Feb. 271/1 Fractions of ribonucleic acid were hydrolysed..to mononucleotides, which were separated by paper chromatography in isopropanol-water-ammonia. **1972** P. WISEMAN *Introd. Industr. Org. Chem.* vi. 218 Acetone is made by the dehydrogenation of isopropanol. **1885** *Jrnl. Chem. Soc.* XLVIII. 645 On heating isopropenyl carbinol with a small quantity of acid, isobutaldehyde is formed. **1950** R. C. FUSON *Adv. Org. Chem.* xv. 347 The conversion of isopropenyl acetate to acetylacetone. **1965** *Nomencl. Org. Chem.* (I.U.P.A.C.) C. 239 Isopropenyl (replacing 1-methylvinyl) (unsubstituted only). **1866** *Jrnl. Chem. Soc.* XIX. 487 We know that isopropyl compounds do not yield propionic acid by oxidation. **1872** *Ibid.* XXV. 237 The production of isopropyl alcohol, instead of the normal alcohol, by the decomposition of normal propylamine nitrite. **1888** [in Dict.]. **1934** H. HILER *Notes Technique Painting* v. 288 Isopropyl alcohol or petrohol is one of the latest solvents used. It will dissolve most varnishes. **1948** *Economist* 31 July 193/1 American production of isopropyl alcohol (now the principal source of acetone) began in small quantities about fifteen years ago. **1955** H. R. DOWNS *Chem. Living Cells* xii. 410 Administration of doubly labeled isovaleric acid gives rise to cholesterol in which it appears that the carbons of the isopropyl group of that acid have been incorporated as a unit. **1966** [see *isobutyl* above]. **1970** PASSMORE & ROBSON *Compan. Med. Stud.* II. xviii. 40/2 Ethyl alcohol, or better isopropyl alcohol, are used for rapid skin disinfection and are probably the best substances for this purpose. **1900** E. F. SMITH tr. *V. von Richter's Org. Chem.* (ed. 3) II. 495 The condensation of the same oxime in the presence of ketones or aldehydes gives rise to isopropylidene and benzylidene methyl isoxazolons, $(C_4H_3NO_2){:}C(CH_3)_2$ [etc.]. **1932** H. PRINGSHEIM *Chem. Monosaccharides & Polysaccharides* ii. 31 Into the hexoses and pentoses there can be introduced two isopropylidene remainders. **1967** R. J. MCILROY *Introd. Carbohydrate Chem.* v. 52 This displacement of the ring has led to the employment of isopropylidene derivatives in the synthesis of reference compounds of the furanose type. **1886** *Jrnl. Chem. Soc.* L. 78 By adding concentrated sulphuric acid to an alcoholic solution of crude quinoline from coal-tar, the sulphates of quinoline and isoquinoline, C_9H_7N, are precipitated. **1932** I. D. GARRARD *Introd. Org. Chem.* xii. 180 Morphine is one of the alkaloids obtained from opium. It is a derivative of isoquinoline. **1960** R. M. ACHESON *Introd. Chem. Heterocyclic Compounds* vi. 231 Many alkaloids (*e.g.* papaverine) contain either the aromatic, or the reduced, isoquinoline system. **1972** N. L. ALLINGER et al. *Org. Chem.* xxviii. 746 Isoquinoline may be synthesized from benzaldehyde by a cyclization reaction known as the Pomeranz–Fritsch synthesis. **1883** *Jrnl. Chem. Soc.* XLIII. 86 The same remark applies to the polymeride obtained..from isovaleraldehyde. **1946** *Chem. Abstr.* XL. 6757 The oil examd., obtained in 1·2% yield by direct steam distn... of *Lavandula delphinensis* plants..had..isovaleraldehyde 0·02[%]. **1970** *Jrnl. Econ. Ent.* LXIII. 1819/1 Bioassay results with 37 terpenoids and related plant constituents indicate that..menthone, isovaleraldehyde, and linalool were among the most attractive to *Anthomus grandis* Boheman. **1882** *Jrnl. Chem. Soc.* XLII. 30 Fraction 3..yielded impure isopentyl isovalerate. **1888** *Ibid.* LIV. 251 Silver isovalerate. **1963** *Chem. Abstr.* LVIII. 4974 Other compds. identified [in nutmeg oil] were.. isoeugenol..and menthyl isovalerate. **1894** *Chem. News* 9 Feb. 66/1 (*heading*) Condensation of isovalerianic aldehyd with ordinary acetone. **1927** *Chem. Abstr.* XXI. 985 By oxidation iso-AmOH gives the aldehyde and isovalerianic acid, all 3 of which have been found in oil of lavender. **1971** *Angewandte Parasitol.* XII. 107 The pure attractants ..valerianic acid and iso-valerianic acid had only a little attraction [for synanthropic flies]. **1882** *Jrnl. Chem. Soc.* XLII. 162 Of the lower homologues of isocaproic acid which contain the isopropyl group, isovaleric acid alone yields an acid similar to the above on oxidation. **1934** *Biochem. Jrnl.* XXXVIII. 401 The porpoise and dolphin depôt fats are unique in containing large amounts of isovaleric acid. **1950** J. BONNER *Plant Biochem.* xxv. 393 Fig. 25-3 gives an example of a fractionation conducted on oil of peppermint... Acetaldehyde, acetone, isovaleric acid, and isoamyl alcohol first distil over. **1960** K. S. MARKLEY *Fatty Acids* (ed. 2) II. ii. 55 Isovaleric acid has been reported to occur in the free state in large amounts in valerian root; in lesser amounts in the oils of pineapple and lavender, and among the volatile acids of mutton tallow.., and in the rumen of the sheep. **1891** *Jrnl. Chem. Soc.* LIX. 410 (*heading*) Formation of isoxazoles. **1946** A. A. MORTON *Chem. Heterocyclic Compounds* xiv. 421 No naturally occurring isoxazole compounds are known. They are often obtained in the course of laboratory work with nitroso and isonitroso compounds. **1960** R. M. ACHESON *Introd. Chem. Heterocyclic Compounds* vii. 272 Isoxazole itself is obtainable from propargyl aldehyde and hydroxylamine; 1,3-dicarbonyl compounds also give isoxazoles with hydroxylamine in a very general synthesis.

isoaccentual, isoagglutination (and derivs.), **isoagglutinin, isoagglutinogen, isoallele, isoallelic, isoantibody, isoantigen, -genic:** see *ISO-. **isoalloxazine, isoamyl alcohol:** see *ISO- b.

isobar. Add: **1. b.** A line in a diagram that represents states or conditions of equal pressure.

1892 P. ALEXANDER *Treat. Thermodynamics* v. 39 An isobar is a constant pressure thermogram. **1924** A. E. HILL in H. S. Taylor *Treat. Physical Chem.* I. ix. 383 The isobar *ab* drawn at 1 atmosphere pressure intersects the curve *BO* at −78·3°. **1969** S. M. BLINDER *Adv. Physical Chem.* viii. 144 In Fig. 8.2, the same data are represented as a series of isobars, plots of *V* versus *t* for fixed *P*.

2. *Physics.* Each of two or more nuclides which have the same mass number but different atomic numbers (and so are different elements). Orig. † **isobare** (*obs.*).

1918 A. W. STEWART in *Phil. Mag.* XXVI. 331 These elements [*sc.* mesothorium and radiothorium] differ completely from one another in chemical character; but they all possess the same atomic weight. For this reason the name isobares..is here suggested for them. **1928** M. STEEL *Physical Chem. & Biophysics* ii. 37 Any product of radioactive change due to the loss of a beta particle..is an isobar of the parent element. **1946** *Electronic Engin.* XVIII. 88/2 Neptunium and plutonium are isobares (same atomic mass but different atomic number). **1952** *Sci. News* XXIII. 37 These nuclei are called isobares. **1954** K. RANKAMA *Isotope Geol.* xxv. 302 The isotope ^{40}K is the active nuclide in the series of the three neighbouring isobars, ^{40}A—^{40}K—^{40}Ca. **1955** R. D. EVANS *Atomic Nucleus* iii. 99 The chemical properties of isobars are generally dissimilar, but their nuclear properties tend to present many parallel features. **1966** PHILLIPS & WILLIAMS *Inorg. Chem.* II. xxxv. 621 We shall use the symbol *Z* for the number of protons in a nucleus (i.e. the atomic number) and *N* for the number of neutrons, so that $A = Z + N$, where *A* is the mass number or 'rounded' atomic weight. All nuclei of the same *Z* are called isotopes, of the same *N* isotones, and of the same *A* isobars.

isobare, obs. var. *ISOBAR 2.

isobaric, *a.* Add: **1. b.** Occurring at or pertaining to a constant pressure.

1903 A. OGG tr. *Planck's Treat. Thermodynamics* I. i. 7 (*heading*) Behaviour under constant pressure (isopiestic or isobaric changes). **1933** D. J. MARTIN *Introd. Thermodynamics for Chemists* xiii. 315 The isobaric heat of adsorption corresponds to the heat of a reaction in a condensed system, the amount adsorbed changes while the pressure remains practically constant. **1937** P. S. EPSTEIN *Textbk. Thermodynamics* iii. 46 The heat function acquires a particular importance in the so-called isobaric process, i.e. a process which takes place without change of pressure. **1966** [see *isopiestic* adj. s.v. *ISO-].

2. *Physics.* Of, pertaining to, or being isobars (sense *2); *isobaric spin* = *ISOSPIN. Const. *with*.

1919 *Nature* 18 Sept. 61/2 Elements can be regarded as divisible into three classes:—(1) Isotopic elements, each set of which have different atomic weights but identical chemical properties; (2) isobaric elements which have identical atomic weights but different chemical properties; and (3) normal elements which differ..both in atomic weights and chemical properties. **1933** F. W. ASTON *Mass-Spectra & Isotopes* xii. 158 Spectra were obtained which showed four strong isotopes and two very weak ones, one of which is isobaric with W^{186}. **1948** GLASSTONE *Textbk. Physical Chem.* (ed. 2) ii. 171 With the development of the study of artificial radioactivity it has become apparent that there are more than seventy pairs of such isobaric isotopes, now known as isomeric nuclei. **1953** D. R. INGLIS in *Rev. Mod. Physics* XXV. 395/2 One finds the states listed according to the usual quantum numbers *L* and *S* of the (*LS*)-coupling scheme and also according to the isobaric spin quantum number *T* (or isotopic spin as it has been less aptly called since it was named long ago by Wigner, who agrees to this renaming of his child). **1971** *Physics Bull.* Mar. 137/1 Isobaric spin was first introduced into nuclear interactions by Heisenberg purely as a mathematical convenience.

3. *Med.* Of, pertaining to, or designating a solution for spinal anæsthesia having the same density as the cerebro-spinal fluid.

1930, 1946 [see *HYPERBARIC *a.* a]. **1947** J. ADRIANI *Techniques & Procedures Anesthesia* v. 222 A 2½% solution of procaine in distilled water is nearly isobaric. **1962** —— *Chem. & Phys. Anaesthesia* (ed. 2) xxxi. 653/1 The results obtained with isobaric techniques are unpredictable and variable. **1971** P. C. LUND *Princ. & Pract. Spinal Anesthesia* vii. 371 Isobaric solutions have not been popular..because of the difficulty of achieving predictable levels of anesthesia. *Ibid.*, An adaptation of isobaric spinal anesthesia for anorectal surgery was introduced by Jacoby et al. in 1965.

B. *sb.* = *ISOBAR 1 b.

1903 *Engineer* 24 July 83/3 The isobarics of evaporation happen to be isothermals. **1937** M. W. ZEMANSKY *Heat & Thermodynamics* iv. 56 The series of short isobarics and isovolumics from *i* to *f* and the continuous curve from *i* to *f* represent other possibilities.

isobarically (əisobæ·rikăli), *adv.* [f. prec. + -LY².] At constant pressure, without a change in pressure.

1951 R. B. MONTGOMERY in D. E. Kerr *Propagation Short Radio Waves* iii. 186 Wet-bulb Temperature.— This may be defined as the temperature at which saturation would occur if the air were cooled isobarically and adiabatically by means of contact with a water surface.

isobase, isobasic, isobath *sb.*: see *ISO-.

isobestic, var. *ISOSBESTIC *a.*

isoborneol, isobutane, isobutene, isobutyl, isobutylene, isobutyrate, isobutyric acid: see *ISO- b. **isocaloric, -calorically:** see *ISO-.

isocarboxazid (əi:sokāɹbǫ·ksăzid). *Pharm.* [f. *isocarbox-* (by rearrangement of part of the chemical name: see quot. 1959) + *HYDR)AZID(E.] A whitish powder that is a hydrazine derivative, $C_6H_5CH_2NH\cdot NHCO\cdot C_3HNO\cdot CH_3$, and is used as an antidepressant.

1959 *Diseases Nervous Syst.* XX. 269/1 An analog of iproniazid, isocarboxazid.., has been used in pharmacological studies which show it to be a more potent amine oxidase inhibitor than its progenitor. Its chemical formula is 1-benzyl-2-(5-methyl-3-isoxazolylcarbonyl)-hydrazine. **1963** *Jrnl. Amer. Med. Assoc.* 16 Mar. 952/1 It is the opinion of the Council that isocarboxazid seems to be effective in the depressed phases of anxiety and manic-depressive states as well as in certain involutional, obsessive, and disassociative reactions. **1965** J. POLLITT *Depression & its Treatment* iv. 56 The monoamine oxidase inhibitors include several members, among them phenelzine (Nardil), iproniazid (Marsilid), isocarboxazid (Marplan).

isocentre (əi·sosentəɹ). Also (*U.S.*) **-center.** [f. Iso- + CENTRE *sb.*] In aerial photography, the point at which the bisector of the angle between the optical axis of the camera and the vertical at the inner nodal point of the lens meets the plane of the camera plate or film (or the corresponding object point where the bisector meets the ground).

1931 C. M. HOTINE *Surveying from Air Photogr.* v. 63 We may, therefore, consider the plate isocentre as the centre of tilt distortions, in that images are displaced radially from the isocentre from the positions they would occupy if the photograph had not been tilted. **1950** L. G. TROREY *Handbk. Aerial Mapping* i. 3 Any photograph is angle true with respect to the isocentre for all points in the plane of the isocentre. **1970** J. A. HOWARD *Aerial Photo-Ecology* ix. 102 The isocentre and nadir are very important due to their relation to tilt and radial displacement.

isochar (əi·sokāɹ). [f. Iso- + CHAR(ACTER *sb.* rendering G. *isopsepher* (W. Rothmaler 1938, in *Beih. Rep. Spec. Nov. Reg. Veg.* C. 90), f. Gr. ψῆφος number + ἦρος sign.] A line (imaginary or on a map) linking areas containing plants showing similar numbers of distinguishing characteristics.

1963 DAVIS & HEYWOOD *Princ. Angiosperm Taxon.* ix. 317 The second procedure involves the drawing of *Isopsepheren*, a term which we can anglicise to *isochars*. These are based on the largest possible number of contrasting character expressions found in the taxa being studied... We are grateful to Dr. W. T. Stearn for suggesting this term.

isochemical, -chemically, isochlor: see *ISO-.

isochromatic, *a.* Add: **1.** Substitute for def.: Of a single colour or tint: applied to a fringe in an interference pattern obtained with birefringent material, such as a biaxial crystal or material used in photoelastic experiments (when such a fringe corresponds to points where the difference between the principal stresses is the same). Also, representing or depicting such fringes. (Further examples.)

1931 COKER & FILON *Treat. Photo-Elasticity* iii. 248 The integrated tint remains the same over the whole of this locus, and it is for this reason that such lines are called lines of equal tint or isochromatic lines. **1932** HARDY & PERRIN *Princ. Optics* xxix. 617 Although the values of *p* and *q* themselves can be computed from the isoclinic and isochromatic patterns, the operations are extremely tedious. **1966** *McGraw-Hill Encycl. Sci. & Technol.* X. 150/1 Utilizing equations for the difference of the principal stresses from the isochromatic fringe orders, the stresses may be found by solving the two equations for the two principal stresses.

2. (Examples.)

1884 *Philadelphia Photographer* Oct. 315/1 (*heading*) Isochromatic gelatin plates. **1885** *Jrnl. Franklin Inst.* CXIX. 368 It was..the only truly isochromatic process ever discovered. Dr. Vogel's new process was not only no better in any respect, but the plates were insensitive to scarlet and ruby-red. *Ibid.* 371 Truly isochromatic photography. **1903** A. M. Clerke *Probl. Astrophysics* II. iii. 291 D. appeared conspicuously on Professor Campbell's isochromatic plates. **1904** *Westm. Gaz.* 19 Nov. 16/2 Not much has been heard of late about isochromatic plates, and it is to be feared that among amateurs their use is not on the increase. Generally speaking, the more recent advances in orthochromatic photography have been in the direction of increasing rather than lessening the difficulties. **1932** *Discovery* Sept. 294/1 The extension of the sensitivity of photographic emulsions..has given rise to three..types of colour sensitive material. The first type includes materials in which the sensitivity has been extended to cover the green; such materials are generally known as 'orthochromatic' or 'isochromatic'. **1955** H. & A. Gernsheim *Hist. Photogr.* xxii. 268 Vogel and others had transformed the hitherto colour-blind emulsion into one which was more accurately sensitive for most colours—i.e. the so-called iso- or orthochromatic plates.

B. *sb.* An isochromatic fringe or line.

1924 *Rep. Brit. Assoc. Adv. Sci.* 1923 314 The disc carried a network of reference lines and the appearances were projected on a screen, upon which the isoclinics and isochromatics were traced with a pencil. **1948** M. M. Frocht *Photoelasticity* II. iv. 139 The fringes or isochromatics..all pass through the points of application of the loads. **1958** Condon & Odishaw *Handbk. Physics* III. vi. 86/2 With a white light source the stress patterns consist of colored bands, called isochromatics, which form in the order of yellow, red, and green followed by similar cycles.

isochromosome: see *iso-.

isochron (ai·sokrǫn), *a.* and *sb.* Also **isochrone** (-krǫun). [f. Gr. ἰσόχρον-ος (see Isochronal *a.*).] **A.** *adj.* (In form *isochrone*) = Isochronous *a.* 1697, etc. (in Dict. s.v. Isochronal *a.*).

B. *sb.* †**1.** An isochronal line. *Obs.*

a **1774** (in Dict. s.v. Isochronal *a.*).

2. A line (imaginary or on a map) connecting points at which a particular event occurs or occurred at the same time.

1881 F. Galton in *Proc. R. Geogr. Soc.* III. 658 Along the coast of West Africa..the ports are regularly served by steamers that touch at every one of them..and which consequently occupy more than forty days to reach even the mouth of the Congo, whereas steamers occasionally sail direct to one or other of those ports in considerably shorter time than these small steamers. This particular difficulty is met and explained by the sea isochrones, which in this case do not conform to those of the land. **1900** *Cent. Dict. Suppl.*, *Isochrone*, *n.*, a line connecting points at which the same events occur simultaneously. Thus the isochrone of travel is the line connecting points attainable by a person riding or an army marching from a given center forward during a given interval of time; the phenological isochrone, the line connecting points at which plants of any species attain simultaneously the same stage of development. **1948** *Antiquity* XXII. 114 While all competent authorities will agree that the practice of producing food..must have spread in some such way as this map shows, there will be differences of opinion... We expect for instance that the bulge made by the isochrones to include Anau and no more is largely artificial. **1955** W. J. Saucier *Princ. Meteorol. Analysis* xii. 389/1 If the weather occurs along a line, successive positions (isochrones) will be curves on the map. **1966** *Nature* 24 Mar. 571/1 The plots of radio blackout distribution in the North American and North Atlantic region..agree with the isochrons of the 'morning' maximum of magnetic disturbance in the Arctic region. **1970** *Ibid.* 17 Jan. 224 The map depicts..the course of withdrawal of the ice sheet from its greatest extent some 18,000 years ago... The detailed isochrons, separated in places by as little as 100 years, graphically depict the north-south corridor that had opened up to the east of the Rocky Mountains about 7,000 years ago.

3. A line (imaginary or on a map) connecting points at which some chosen time interval has the same value.

1940 C. A. Heiland *Geophysical Explor.* ix. 548 Adjusted times are plotted against the location of depth points; points with equal time differences are connected by isochrons which, barring velocity variations and steep dips, give a true picture of the depth contours of the structure. **1945** *Electronic Engin.* XVII. 713/2 Sets of lines can be drawn, joining all the points having the same time-differences; and it has been agreed to call these lines 'isochrones', analogous to the 'isobars' of a weather-map. In general, these 'isochrone' lines are hyperbolae. **1952** F. H. Lahee *Field Geol.* (ed. 5) xxiii. 779 On the assumption that velocity values [of seismic waves] are essentially constant and that reflecting horizons are continuous over a given area, the differences in arrival time from two such horizons can be plotted at each station and then lines of equal time difference (isochrons) can be drawn to produce an isochron map. **1968** *Jrnl. Brit. Interplanetary Soc.* XXI. 340 We have expressed the duration of the voyage in terms of *q*, *ε* and *n* only. In our diagram we can now draw lines of equal duration, which we call isochrones. In Fig. 9 are shown the isochrones for voyages from the Earth to Venus in days.

4. In the isotopic dating of rock, a straight line whose gradient is taken to represent the time since the isotopic content of a sample was fixed (e.g. by crystallization), and obtained by plotting the ratio of the amount of a radiogenic isotope to that of a non-radiogenic isotope against a corresponding ratio for a second radiogenic isotope and the same non-radiogenic one in two or more samples having the same history but different ratios.

1953 F. G. Houtermans in *Nuovo Cimento* X. 1624 By dividing (3b) by (3a) the equations of 'isochrones' are obtained... These are a number of straight lines, intersecting at the point a_0, β_0 corresponding to the isotopic constitution of 'primeval lead' at the time t_0. **1963** K. Rankama *Progr. Isotope Geol.* xxxvii. 543 When the $^{206}Pb/^{204}Pb$ ratio was plotted against the $^{207}Pb/^{204}Pb$ ratio, the slope of the isochron yielded the age 4·55 Gy for meteoritic matter, and the isotopic constitution of rock lead fell close to the isochron. **1969** Bennison & Wright *Geol. Hist. Brit. Isles* iii. 41 This data can also be presented as isochrons..the slope of a whole-rock isochron being proportional to the age of initial crystallization and the slope of the mineral isochron to that of the metamorphism. **1972** *Nature* 23 June 500/1 Bodinger..has carried out extensive radiometric dating on illitic sedimentary rocks..; he produced eight separate total-rock Rb-Sr isochrons from seventy-two samples.

5. A line (imaginary or on a map) connecting points on the sea-floor formed at the same time.

1968 *New Scientist* 30 May 452/2 American workers hope eventually to produce a complete 'isochron' (lines of equal age) map of the world's oceans. **1972** *Nature* 8 Dec. 329/2 The discharge curve for Iceland was constructed..by extrapolating seafloor spreading isochrons from the ocean floor immediately southwest of the aseismic ridge.

isochronal, *a.* Add: **2.** Of a line: connecting points at which a particular event occurs or occurred at the same time. Of a diagram: depicting such lines. Also as *sb.*, = *isochron 2.

1928 [see *isepiptesis, isopiptesis]. **1937** D. Kennedy tr. *Imamura's Theoret. & Appl. Seismol.* iv. 41 If a line is drawn through places where a certain phase of the earthquake motion..appears simultaneously..a line resembling an isoseismal is obtained... It is the coseismal, or the isochronal of some recent workers. **1948** R. B. Hounsfield *Traffic Surveys* vi. 39 Another type of diagram used in planning is the 'isochronal' diagram. *Ibid.* 40 (*caption*) Isochronal diagram showing accessibility of different areas. **1962** C. D. Sherman tr. *Dorst's Migrations of Birds* vii. 236 The influence of temperature, revealed in this parallelism between isochronal lines and isotherms, may be less apparent because of other factors which also govern migration.

isochronic, *a.* (in Dict. s.v. Isochronal *a.*). Add: **2.** = *isochronal *a.* 2; *spec.* (see quots. 1881, 1959).

1881 F. Galton in *Proc. R. Geogr. Soc.* III. 657 By 'isochronic' passage-charts, I mean charts constructed to show the extreme distances that can be traversed in 'equal times' from a common starting point. *Ibid.* 658 Isochronic maps might be..constructed for Continental travel or for home excursions. **1948** *Antiquity* XXII. 114 The second [map]..covers the Old World and shows the spread of the food-producing economy from five possible independent centres... The date at which food-production first appeared is shown by isochronic lines. These are an ingenious invention which show 'equal dates' exactly as contours show equal heights and isobars equal pressures. **1959** L. M. Harrod *Librarians' Gloss.* (ed. 2) 155 *Isochronic map*, one which shows possible progress of travel in all directions from a given centre in certain specified time intervals.

3. *Prosody.* Equal in metrical length.

1956 H. Whitehall in *Kenyon Rev.* XVIII. iii. 418 Unlike such 'syllable-timed' languages as Spanish, English is 'stress-timed' or *isochronic*. **1959** *PMLA* LXXIV. 587/1 Mr. Whitehall distinguishes a type of 'rhythm' which he calls the isochronic: it 'depends on equal time lapses between primary stresses'.

isochronism. Add: **2.** *Prosody.* The character or property of being isochronous.

1954 Wellek & Warren *Theory of Lit.* 166 The artistic rhythm of prose..must not reach an apparent isochronism (that is, a regularity of time intervals between stresses). **1956** H. Whitehall in *Kenyon Rev.* XVIII. iii. 418 Isochronism is produced not only by accelerating and crushing together the syllables between primary stresses but also by temporizing devices—the pauses. **1959** *PMLA* LXXIV. 587/1 If isochronism were a general principle, or even an approximate principle, of all English speech, it would clearly be a different thing from meter. **1966** M. Pei *Gloss. Ling. Terminol.* 133 *Isochronism*, a term applied to verse in which the amount of time between two primary stresses tends to be the same, irrespective of the amount of material between them. **1973** *Studies in Eng. Lit.: Eng. Number* (Tokyo) 31 Hopkins has nowhere upheld a principle of absolute isochronism.

isochronous, *a.* Add: *spec.* in *Prosody*, equal in metrical length.

1784–1822 [in Dict.]. **1857** C. Patmore in *North Brit. Rev.* XXVII. 149 A metre which, totally abandoning the element of natural syllabic quantity, takes the isochronous bar for the metrical integer. **1942** J. C. Pope *Rhythm of Beowulf* 9 Isochronous measures are the rule..and it is easy to produce them in *Beowulf* by means of limited quantitative variation. **1948** *Mod. Philology* XLVI. 75 There is..no reason to suppose that, *if* the *Beowulf* was chanted to a real musical accompaniment, the lines were therefore delivered in isochronous groups. **1971** *Times Lit. Suppl.* 1 Oct. 1179/3 Its technique of isochronous rhythm—a metrical sequence which remains constant for a given part, though the pitch relationships change—is comparable with the Oriental tala.

2. *Palæont.* [ad. G. *isochron* (E. Mojsisovics *Die Cephalopoden der Hallstätter Kalke* (1893) II. 5).] Originating or formed at the same period.

1895 [see *homomorphy]. **1913** [see *heterochronous *a.* b]. **1952** R. C. Moore et al. *Invertebr. Fossils* vi. 218/1 Such contemporaneous or near-contemporaneous forms he [sc. Buckman] designated as isochronous homœomorphs.

isochrony (ai·sǫ-krǫni). [f. as Isochronism, after *synchrony*, etc.] Isochronism; the character or property of being isochronous.

1953 *Word* Apr. 3 [The] tendency toward word isochrony whereby every simple word gets two moras either in one long syllable or in two short ones. **1961** *Brno Studies in English* III. 48 In his [sc. A. Martinet's] opinion, 12th century English was characterized by..isochrony, i.e. the state of things resulting from the elimination of vocalic quantity as a phonematic feature. **1961** *Rev. Eng. Stud.* XII. 342 There exist all sorts of musical rhythms very different from the isochrony which has dominated European music for so long a time. **1966** J. C. Pope *Rhythm of Beowulf* (rev. ed.) p. x, Isochrony and initial rests are vital, in my opinion, for the achievement of an adequate sense of order in opposition to the extraordinary variety of syllabic patterns in the verses. **1973** *Word* 1966 XXII. 5 It is a whole chapter of the history of isochrony, the process through which the quantitative pattern of Proto-Indo-European was reorganized in most of the languages of that family.

isocitrate, isocitric: see *iso-.

isoclasite (aisǫklǣ·sait, -zait). *Min.* [f. G. *isoklas* (F. Sandberger 1870, in *Jrnl. f. prakt. Chem.* II. 125), f. Gr. κλάσ-ις fracture + -ite¹.] A colourless or white hydrated phosphate and hydroxide of calcium, $Ca_2(PO_4)(OH).2H_2O$, known from a single locality in Bohemia.

1872 G. J. Brush in J. D. Dana *Syst. Min.* (ed. 5) App. I. 7 Isoclasite. **1955** M. H. Hey *Index Min. Species* (ed. 2) 421 Isoclasite.

isoclinally, *adv.* *Geol.* [f. Isoclinal *a.* + -ly².] In the manner of isoclinal strata (see Isoclinal *a.* 2).

1936 *Bull. Geol. Soc. Amer.* XLVII. 720 Straight, seemingly undeformed layers of marble enclose isoclinally folded and dismembered fragments of stronger rocks. **1970** *Nature* 19 Sept. 1252 Greenstones..and associated serpentinites and stratiform basic complexes have been isoclinally folded on NNW-trending axes.

isoclinic, *a.* and *sb.* Add: **A.** *adj.* **2.** Corresponding to or depicting the locus of points in a body where each of the principal stresses is in some fixed direction.

1913 Filon & Coker in *Rep. Brit. Assoc. Adv. Sci.* 1912 203 The lines of principal stress are parallel to the axes of the Nicols... These may be called the lines of equal inclination or isoclinic lines. **1939** *Jrnl. Appl. Physics* X. 254/1 The direction of the stresses are taken from the isoclinic sketch. **1966** *McGraw-Hill Encycl. Sci. & Technol.* X. 130/1 Isoclinic fringes are a different set of interference patterns made by using white light, removing the quarter wave plates and rotating the polarizer and analyzer a fixed number of degrees. These fringes represent lines making known angles with the principal planes of stress.

B. *sb.* **2.** An isoclinic line or curve.

1924 L. N. G. Filon in *Rep. Brit. Assoc. Adv. Sci.* 1923 352 ..through which passes the isoclinic of parameter φ. *Ibid.* 353 The isoclinics are usually well-defined brushes, of which the direction, at any point, can be observed with considerable accuracy. **1948** M. M. Frocht *Photoelasticity* II. 147 All isoclinics above the X axis pass through the point of application of the downward load and those below the X axis pass through the point of application of the upward load. *Ibid.* 150 The isoclinics are all horizontal where they cross the horizontal boundary of the disk. **1958** Condon & Odishaw *Handbk. Physics* III. vi. 86/2 Isoclinic lines will be exhibited against colored backgrounds.

isocolloid: see *iso-.

isocolon. (Earlier and later examples.)

1550 R. Sherry *Treat. Schemes & Tropes* sig. Dv, Isocolon. Compare, euen or equall, is when the oracion hath in it the partes of the whyche we speake before (Articulus, Dialyton), and that they be made of euen number of syllables: but thys equalitie must not stand by numbrynge of them, but by perceyvyng of it in the mynde. **1941** *English Studies* XXIII. 16 When we start looking for Pettie's isocolon we do find an average of 8·5 words, but there is much more irregularity than in Lyly. **1962** [see *agnomination 2].

isocortex, isocyclic: see *iso-. **isocyanate, isocyanic, isocyanide:** see *iso- b.

isodose (ai·sodǭus). [f. Iso- + Dose *sb.*] An imaginary line or surface, or a graphical representation of one, connecting points, esp. points in the body, that receive equal doses of radiation; now always used *attrib.*, esp. of such lines and surfaces and of diagrams depicting them.

1922 H. Schmitz tr. *Kroenig & Friedrich's Princ. Physics & Biol. Radiation Therapy* 249 To render a graphical presentation of the distribution of the dose in radiated tissue one proceeded with the conception that the equal intensity curves of like doses in a body, which we may term 'isodoses', are spheres which surround the center of the preparation concentrically. *Ibid.*, The isodoses of such strong capsules as are used in deep therapy must deviate from the circular or cylindrical forms. **1923** O. Glasser in *Amer. Jrnl. Roentgenology* X. 405 (*heading*) Isodose charts. *Ibid.* 405/2 The curves are called 'isodoses', a name I gave first to these curves in connection with radium five years ago. **1939** *Brit. Jrnl. Radiol.* XII. 263/1 The isodose curves..of complex solids whose surfaces are 'isodose surfaces'. **1950** *Rev. Sci. Instruments* XXI. 363/2 The automatic isodose recorder..satisfies the demand for tracing radiation fields with a good accuracy in a short length of time. *Ibid.* 365/1 Both the speed of the isodose tracing and the accuracy..are quite adequate. **1956** Hine & Brownell *Radiation Dosimetry* xix. 571 (*caption*) Isodose contours for cancer of the esophagus obtained by combining six 6 × 15-cm fields using Co^{60}. **1966** R. D. Cadle *Particles in Atmosphere & Space* iv. 104 Fallout patterns are maps consisting of families of isodose rate lines or contours. **1968** *Brit. Med. Bull.* XXIV. 242/1 The computer produced a mass of figures..which were easily converted with an x-y plotter to..isodose distribution form, with which all radiotherapists are familiar.

isodrin (ai·sodrin). [f. Iso- b + *al)drin.] An insecticide that is a stereoisomer of aldrin, $C_{12}H_8Cl_6$.

1953 *Rev. Appl. Entomol.* XLI. 144 Isodrin is defined by the Committee as 1,2,3,4,10,10-hexachloro-1,4,4a,5,8,8a-hexahydro-1,4,5,8-endo-endo-dimethanonaphthalene. **1955** *Jrnl. Hort. Sci.* XXX. 181 Isodrin and endrin are more persistent than dieldrin, but show great promise in their versatility of action; control of such completely different insects as aphids and cut-worms has been obtained by their use. **1971** *Jrnl. Agric. & Food Chem.* XIX. 5/1 Homogenates prepared from the excised roots of bean seedlings..oxidized isodrin, producing a compound corresponding chromatographically to endrin ketone.

isoelectric (aisǫ,ile·ktrik), *a.* [f. Iso- + Electric *a.*] **1.** Equal in electrical potential; containing or indicating no potential difference.

1877 [in Dict. s.v. Iso-]. **1901** J. H. Raymond *Human Physiol.* (ed. 2) 445 A normal muscle in a condition of rest is iso-electric—i.e., it is 'equally electric throughout', and hence has no electric current; the same is true of dead muscle. **1940** Schery & Boyd *Clin. Electrocardiogr.* i. 7 The two waves are separated from each other by an isoelectric line. **1972** T. P. Foote in R. G. Sanderson *Cardiac Patient* iv. 149 The electrocardiographic tracing is flat or isoelectric during this event.

2. (Composed of particles) having no net electric charge; equal as regards electric charge; chiefly in *isoelectric point*, the point (usually *p*H value) at which an amphoteric molecule or a colloidal particle is electrically neutral in a solution.

1900 W. B. Hardy in *Proc. R. Soc.* LXVI. 112 It is clear that there exists some point at which the particles and the fluid in which they are immersed are iso-electric. This iso-electric point is found to be one of great importance. As it is neared, the stability of the hydrosol diminishes until, at the iso-electric point, it vanishes, and coagulation or precipitation occurs. **1922** J. Loeb *Proteins & Theory Colloidal Behavior* i. 6 The conception of the 'isoelectric point' of proteins was introduced before its chemical meaning was recognized and it attracted attention because it was supposed to coincide with the precipitation of colloids. *Ibid.* iv. 42 When an acid, e.g., HCl, is added to isoelectric gelatin (or any other isoelectric protein), an equilibrium is established between free HCl, protein chloride, and non-ionogenic (or isoelectric) protein. **1946** P. H. Mitchell *Textbk. Biochem.* iv. 107 Crystallization of a protein is usually carried out at its isoelectric point. **1958** Packer & Vaughan *Mod. Approach Org. Chem.* xiii. 433 In a solution at its isoelectric point, amino acid molecules will not migrate in the electric field created by the introduction of a cathode and an anode. **1966** *Acta Chem. Scand.* XX. 821 The peptides isoelectric between pH 5·0 and 6·5 proved to possess poor carrier ampholyte properties.

b. Carried out or occurring at the isoelectric point.

1961 *Acta Chem. Scand.* XV. 326 Isoelectric analysis and fractionation by electric transport is based on sending a direct current through a system of electrolytes such that the pH increases gradually from anode to cathode... Proteins and other ampholytes will..collect in a region where the local pH is identical with the isoelectric point of the ampholyte. **1971** *European Jrnl. Biochem.* XXI. 110/1 One might expect the 'isoelectric coagulation'..to represent an isoelectric fractionation based on unaltered native molecules.

Hence **isoele·ctrically** *adv.* *Biochem.*, by making use of the different isoelectric points of the components of a mixture (in order to separate them).

1966 *Acta Chem. Scand.* XX. 834 Viruses with an average molecular weight of 20×10^6 can be expected to be isoelectrically resolvable only about three times more effectively than myoglobins. **1970** *Zeitschr. Klin. Chem. Klin. Biochem.* VIII. 383 Iron-free transferrin was separated isoelectrically into 2 components with isoelectric points at pH 5·8 and 5·4.

isoelectronic (aisǫ,ilektrǫ·nik), *a.* *Chem.* and *Physics.* [f. Iso- + *electron² + -ic.] (Composed of atoms or molecules) having the same number of electrons. Const. *with*.

1928 *Chem. Rev.* V. 155 Vertical lines represent atoms having the same number of external electrons ('isoelectronic systems'). **1929** *Physical Rev.* XXXIII. 538 The succeeding elements calcium, scandium, titanium, vanadium, etc., are made isoelectronic (that is, having the same number of electrons) with potassium by removing one electron from calcium, two electrons from scandium, etc. **1946** *Nature* 5 Oct. 480/1 In particular, they have assigned the line at 1,400 cm.⁻¹ to NO_2^+, comparison with the isoelectronic molecule CO_2 having shown that a polarized Raman frequency would be expected to appear in this region. **1964** L. Candler *Atomic Spectra* (ed. 2) ix. 138 Ions which have only one electron outside an inert gas shell and are therefore isoelectronic with the alkalis. **1971** *Internat. Jrnl. Quantum Chem.* V. 335 Wave functions of the ²S (ground state), ²P and ²P states for the lithium isoelectronic sequence have been obtained.

isoenergetic: see *iso-.

isoenzyme (ai·so,e·nzaim). *Biochem.* Also **iso-enzyme**. [f. Iso- + *enzyme.] One of two or more chemically different forms of an enzyme (see quot. 1968). Cf. *isozyme.

1960 *New England Jrnl. Med.* 15 Sept. 531 (*heading*) Isoenzymes in human serum. **1964** *Oceanogr. & Marine Biol.* II. 120 Species- and tissue-specific forms of enzymes (iso-enzymes) have been reported in animal and vegetable tissues. **1968** Latner & Skillen *Isoenzymes in Biol. & Med.* i. 1 It is now well recognized that a large number of enzymes exist in multiple forms. This applies not only to tissues and tissue extracts but also to enzyme proteins which have been isolated in the crystalline state and are really mixtures. Isoenzymes are examples of these multiple forms. Precise definition of the word 'isoenzyme' is, however, rather difficult. Different tissues of the same individual or even of different species may possess closely similar enzymes, which are not really isoenzymes. For the time being, most authorities believe that a broad definition such as 'different proteins with similar enzymatic activity' best suits the current state of our knowledge... It is customary, for the most part, to limit this definition to multiple enzymes obtained from one tissue of one individual animal or plant or possibly a small organ, or a culture of a unicellular organism... An exception would be the major multiple forms of human alkaline phosphatase. They are nevertheless referred to as isoenzymes. **1970** *Nature* 30 May 862/1 The enzyme glucose-6-phosphate dehydrogenase..exists in several different forms (isoenzymes) in the human erythrocyte... The structural differences between the isoenzymes lead to different enzymatic activities.

Hence **isoenzy·mic** *a.*

1968 *Brit. Med. Bull.* XXIV. 222/2 Chemical differences between individuals have been much studied..for example..the occurrence of isoenzymic variants of a number of well-known enzymes.

isoetes (aisǫu·ĭtīz, isǫ-ĭ·tīz). [mod.L. (Linnæus *Skånska Resa* (1751) 417), f. Gr. ἰσοετής, f. ἴσος equal + ἔτος year: the plants are evergreen.] An aquatic plant of the genus so called; = *quill-wort* (s.v. Quill *sb.*¹ 8 b).

1886 *Encycl. Brit.* XX. 431/2 An Isoetes plant was produced on the leaf instead of a sporangium. **1910** *Rep. Brit. Assoc. Adv. Sci.* 784 The whole axis of the *Isoetes* plant can be compared with that of *Lepidodendron*. **1965** R. F. Scagel et al. *Evolutionary Survey Plant Kingdom* xix. 375/1 One of the most distinctive vascular plants, *Isoetes*..consists of a thick mass of tubular to strap-shaped leaves borne in a rosette on a very short stem.

isoeugenol, isoflavone: see *iso- b.

isoflor (ai·sǫflǫ̂ə). [f. Iso- + Flor(a render-ing G. *isoporie* (W. Rothmaler 1938, in *Beih. Rep. Spec. Nov. Reg. Veg.* C. 89), perh. f. Gr. πορεία load.] A line (imaginary or on a map) linking areas containing equal numbers of plant species.

1944 S. A. Cain *Found. Plant Geogr.* xii. 163 Isoflors, lines delimiting regions with equal numbers of species (within the circle of affinity), can be drawn for the generic area as a whole. **1960** N. Polunin *Introd. Plant Geogr.* vii. 208 Sometimes a fair one [sc. indication of a centre of origin] may be given by isoflors, which are lines delimiting regions supporting equal numbers of species. **1963** Davis & Heywood *Princ. Angiosperm Taxon.* ix. 317 In the construction of isoflors, it is not the distribution of the individual species that is important, but only the number of species occurring together at any one point. In order to produce the isoflor map, a grid is drawn on the map and the number of taxa noted which occur in each quadrat.

isogam (ai·sǫgæm). [app. f. Iso- + Gam(ma).] A line (imaginary or on a map) connecting points where the acceleration due to gravity has the same value. Freq. *attrib.*

1928 *Science* 13 July 37 Isogams, surface or line of equal gravitative attraction. **1931** F. H. Lahee *Field Geol.* (ed. 3) xxiii. 661 *Isogam* is the name applied to lines of equal value of relative or absolute gravity. Isogam maps..are used to picture the variation of gravity. **1940** *Geogr. Jrnl.* XCV. 135 The gravity pendulum..cannot produce a survey of gravity in which isogams at 2 milligal intervals can be drawn with confidence. **1954** *Geophysical Suppl. Monthly Notices R. Astron. Soc.* VI. 180 The easiest approach to the problem..is to prepare a chart of density anomalies..and next an isogam chart of the gravity effect of these density anomalies.

2. [cf. *gamma I c (iv).] An isodynamic line.

1940 L. L. Nettleton *Geophysical Prospecting for Oil* viii. 163 Contours, or lines of equal magnetic intensity, are commonly called 'isogams'. **1965** G. J. Williams *Econ. Geol. N.Z.* xix. 354/2 A recent magnetic survey revealed a group of high values corresponding fairly well with the inferred position of the anticline; the trend of the isogams and of the high axis is almost parallel to that of the axis.

isogamy. Add: (Later examples.) **isogamete, isogamous** *a.* (earlier and later examples).

1889 Bennett & Murray *Handbk. Cryptogamic Bot.* 272 The only known sexual mode of reproduction [in the Confervoideae] is an isogamous one between two masses of protoplasm. **1935** F. E. Fritsch *Struct. & Reprod. Algae* I. 43 Isogamy (i.e. the fusion of morphologically identical gametes) is very usually..combined with an absence of differentiated organs for the production of the sexual cells or gametes. *Ibid.* 46 Isogametes are commonly positively phototactic. **1938** G. M. Smith *Cryptogamic Bot.* I. ii. 21 All green algae in which both gametes are nonflagellated..are isogamous. **1952** C. J. Alexopoulos *Introd. Mycol.* i. 17 We use the terms isogametangia and isogametes..to designate gametangia and gametes which are morphologically indistinguishable. **1964** E. J. H. Corner *Life of Plants* vi. 85 The process is called isogamy or the union of outwardly similar gametes. **1968** Bell & Woodcock *Diversity of Green Plants* ii. 31 The gametes [of *Stigeoclonium*] are noticeably smaller than the zoospores, and copulation is isogamous.

isogel: see *iso-.

isogeneic (aisodʒenī·ik, -ēi·ik), *a.* *Immunol.* [f. Iso- + Gr. γενε-ά race, stock + -ic.] = *syngeneic *a.*

1963 Humphrey & White *Immunol. for Students of Med.* xi. 359 When grafts are made..from one animal to another isogeneic animal they are 'isografts'. *Ibid.* 365 Since surgeons rarely have the opportunity of working with isogeneic patients, successful grafting has been largely limited to autografts. **1973** *Nature* 23 Mar. 259/2 An animal strain in which inbred immunologically isogeneic lines are available, making possible transplantation studies uncomplicated by rejection problems.

isogenic (aisodʒe·nik, aisodʒī·nik), *a.* *Biol.* [ad. G. *isogen* (W. Johannsen *Elem. d. exakten Erblichkeitslehre* (ed. 2, 1913) xii. 208), f. Iso- + *gen(e + -ic.] Having the same genotype (*genotype *sb.*²).

1933 *Biochem. Jrnl.* XXVII. 6 In all the work discussed in this paper we have, unless otherwise stated, compared only isogenic animals (animals of the same sex and from the same litter). **1943** *Proc. Nat. Acad. Sci.* XXIX. 36 Three homozygous stocks were obtained, which were isogenic except for the fourth chromosome. **1950** *Jrnl. Exper. Zool.* CXIII. 123 To obtain a culture isogenic with the recessive parent, employ the usual backcross procedure. **1956** *Nature* 7 Jan. 42/2 Males of a wild-type ('Canton-S') stock of *Drosophila melanogaster*, which had been made isogenic some time before the experiments were crossed to homozygous *ci ey^R* females. **1970** *Ibid.* 7 Feb. 557/1 This method..is especially suitable for characters in which the distributions of the isogenic parental strains are not clearly dichotomous. **1973** *Ibid.* 9 Feb. 383/2 The relatively uniform appearance and behaviour of the aggressive isolates suggest that these are highly isogenic, possibly with a recent origin from a single clone.

isogloss (ai·soglǫs). [a. G. *isogloss* (A. Bielenstein *Die Grenzen des Lettischen Volksstammes* (1892) 397, f. Iso- + Gloss *sb.*¹] In Linguistic Geography, the boundary of an area of local concentration or dominance of a significant feature (as of vocabulary or pronunciation). Also, a line plotted on a map indicating the area in which such a feature is concentrated or is dominant. Hence **isoglo·ssic** *a.*

1925 O. Jespersen *Mankind, Nation & Individual* iii. 41 As a rule we find that the frontiers for one phenomenon do not exactly coincide with the frontiers for another, so that the 'isoglosses' (as such frontiers of single phenomena are called) sometimes agree, sometimes run somewhat parallel with one another, but fairly often cross one another in the most distracting manner. **1927** L. Bloomfield in *Mod. Philology* Nov. 220 The boundaries of successive linguistic changes (isogloss lines) do not coincide... Isoglosses may be bundled along barriers to communication; if the barrier (e.g., a political boundary) is removed, the isoglossic bundle may be overlaid by later linguistic development. **1933** —— *Lang.* iii. 51 Within a dialect area, we can draw lines between places which differ as to any feature of language. Such lines are called isoglosses. **1949** H. Kurath *Word Geogr. Eastern U.S.* 12 In figure 2 the focus of each of the more distinctive sub-areas of the Eastern States is set off by a single word line or isogloss. **1953** J. B. Carroll *Study of Lang.* ii. 59 The purpose of linguistic geography is to make a detailed analysis of local linguistic variations... Efforts are made to establish isoglosses, which may be described as cartographic lines separating the geographical localities which show a difference with respect to a specific linguistic item. **1955** *Publ. Amer. Dialect Soc.* xxiii. 40 When it [sc. a form] is found within a sharply defined area, one may draw a line called an isogloss, to indicate the boundaries of the area in which this form occurs. **1958** *Ibid.* xxx. 3 But a number of lexical terms do not occur in geographical patterns corresponding to the isogloss divisions for known Northern and Midland features. **1964** H. B. Allen in A. A. Marckwardt *Stud. Lang. & Ling. in*

Honor C. C. Fries 305 The major isogloss bundle is shown on the map by the 1–1 line. **1968** *Amer. Speech* XLIII. 185 It is separated from other dialects of Yiddish by a bundle of grammatical, phonological, and lexical isoglosses. **1968** W. S. Allen *Vox Graeca* i. 12 The ᾱ of pure Attic is part of an isogloss having its probable point of origin in Boeotian. **1972** H. Kurath *Stud. Area Ling.* 60 The location of some representative isoglossic lines..is shown in Figure 21.

Hence also **isoglo·ttal, -glottic** *adjs.*

1932 *Missouri Alumnus* Apr. 232/1 The American Council of Learned Societies is financing a 'Dialect Atlas of the United States and Canada', and already the workers are making a survey of New England. They have a 'work-sheet' of 800 questions which bring to light the notable speech variations, and from the information so gathered they construct maps with 'iso-glottal' lines like those on a topographic map. **1939** L. H. Gray *Foundations of Lang.* ii. 26 Such lines are termed isoglottic lines or isographs. **1954** *Word* X. 375 The location of different isoglottic lines did not coincide in one linguistic area.

isogonic, *a.*² Read: *Biol.* [ad. Fr. *isogonique* (A. Pézard 1918, in *Bull. Biol. de la France et de la Belgique* LII. 24).] Of an organ, growing at the same rate as its parent body. So **iso·gony**, growth of this kind.

1924 J. S. Huxley in *Nature* 20 Dec. 895/1 Pézard..has styled the growth of such an organ heterogonic..as opposed to isogonic. **1932** —— *Probl. Relative Growth* i. ii. 8 An organ which..is growing at the same rate as the body..must be styled isogonic. *Ibid.* iii. 38 It is justifiable to regard isogony as a special case of heterogony, with growth-coefficiency unity. **1945** Richards & Kavanagh in Clark & Medawar *Ess. Growth & Form* 219 Isogonic growth is a special case of isotropic growth in which the specific growth-rate in length is the same at every point throughout the organism.

isograd, isograde, isogradic: see *iso-.

isograft (ai·sogrɑft), *sb.* *Med.* and *Biol.* [f. Iso- + Graft *sb.*¹] **a.** = *homograft. Now *rare* or *obs.*

1909 *Jrnl. Exper. Med.* XI. 194 Three isografts were placed. **1919** J. S. Davis *Plastic Surg.* iv. 50, I have thought for some time that the success or failure of isografts may be dependent on the similarity or dissimilarity of blood groups of the host and donor. **1942** J. P. Webster in F. Christopher *Textbk. Surg.* (ed. 3) xxxv. 1587 Successful free corneal isografts have been reported with maintenance of transparency. **1950** F. Smith *Plastic & Reconstruction Surg.* i. 27 Homograft, homologous graft and isograft are synonymous terms indicating tissue transplanted from one to another person of the same species.

b. A graft taken from an identical twin of the recipient or from an animal of the same inbred strain.

1958 *Immunology* I. 1 Adrenal cortical grafts transplanted between members of an inbred strain of mice ('isografts') are held to be successful when they empower adrenalectomised mice to subsist on a diet low in NaCl. *Ibid.* 6 Homografts transplanted to the brain and testis were significantly inferior to isografts. **1963** Humphrey & White *Immunol. for Students of Med.* xi. 359 Isografts usually also take permanently, although even within a pure line of mice male skin may be rejected by a female recipient, by virtue of the fact that the Y chromosome can carry sex-linked genetic differences. **1970** *Microvascular Res.* II. 91 The vascular dilation and petechiae, which were observed in the skin isografts in animals treated with cortisol, were similar in the allograft group.

isograft (ai·sogrɑft), *v.* *Med.* and *Biol.* [f. Iso- + Graft *v.*¹ or f. prec.] *trans.* **a.** To transplant from one individual to another of the same species; = *homotransplant *v.* **b.** To transplant between identical twins or animals of the same inbred strain. So **i·sografted** *ppl. a.*, **i·sografting** *vbl. sb.*

1909 *Jrnl. Exper. Med.* XI. 194 The following day the isografting was made. **1919** J. S. Davis *Plastic Surg.* iv. 50 Care must be taken when isografting is contemplated, not to transmit disease to a healthy person. **1959** *Proc. Soc. Exper. Biol. & Med.* CII. 651/1 (*caption*) Light-dark synchronized activity rhythm whether or not pituitaries were isografted. **1962** *Dissertation Abstr.* XXII. 2829/1 The sex of the recipient animal had no effect on the survival of isografted fetal tissue. **1970** *Zeitschr. für Anat. & Entwickl.-Gesch.* CXXXII. 318, 92 mammary duct-segments..were excised..and microdissected for isografting into the fourth mammary gland-free fat pads of the 46 female hosts.

isograph (ai·sogrɑf). [f. Iso- + -graph.] **1.** A drawing instrument (see quots.).

1838 *Civil Engin. & Arch. Jrnl.* I. 349/1 These are a set of triangular rulers, drawing paper, with the lines and ellipses described upon it, and the isograph which consists of a number of rulers, made of brass or ivory, the fiducial edge of each being an inch apart, and parallel to each other. **1909** Webster, *Isograph*, an instrument consisting of two short straightedges connected by a large circular joint marked with angular degrees, used with a T square on a drawing board and combining the functions of a protractor and a set square. **1926–7** *Army & Navy Stores Catal.* 456/1 Isograph. 12 in. boxwood, brass joint.

2. *Linguistics.* (See quots.)

1939 L. H. Gray *Foundations of Lang.* ii. 26 Certain features of any one of the dialects will be common to some of the rest..so that it will be possible to draw upon a map lines indicating at least the approximate boundaries of these features... Such lines are termed isoglottic lines or isographs. **1954** Pei & Gaynor *Dict. Ling.* 106 *Isograph*, any line on a linguistic map, indicating a uniformity in the use of sounds, vocabulary, syntax, inflexion, etc.

isogyre, isohæmagglutination, isohæmagglutinin, isohæmolysin, isohaline, isohel, isohelic: see *iso-.

isohydric (aisohai·drik), *a.* [ad. G. *isohydrisch* (S. Arrhenius 1887, in *Ann. d. Physik u. Chem.* XXX. 54): see Iso- and Hydric *a.*] **a.** *Physical Chem.* Having the same hydrogen ion concentration; maintaining the same hydrogen ion concentration after mixing; also used with reference to other ions.

1887 *Jrnl. Chem. Soc.* LII. i. 415 The molecular conductivity of a mixture of butyric and acetic acid solutions in any proportions is always the sum of the molecular conductivities of the constituent solutions. Each electrolyte behaves as if the other were absent. Such solutions are termed by the author [sc. S. Arrhenius] 'isohydric'. **1899** J. Walker *Introd. Physical Chem.* xxv. 288 Let there be prepared isohydric solutions of the different salts, NaCl being made isohydric with NaBr, by getting the sodium ions of the same concentration in both solutions. **1930** Glasstone *Electrochem. Solutions* viii. 151 Solutions of acids which do not change their ionization on mixing are said to be isohydric with one another. **1952** J. E. Ricci *Hydrogen Ion Concentration* ii. 26 We shall here examine the conditions for the validity of the theorem of isohydric solutions, that isohydric solutions (solutions of the same *H*) mix without change of *H*.

b. *Physiol.* Occurring without causing any change in the *p*H of the blood: applied to the reactions by which carbon dioxide is removed from the tissues and taken up by the blood.

1920 *Proc. Soc. Exper. Biol. & Med.* XVII. 181 Curves ..have been obtained, showing an isohydric shift of base between hemoglobin and the other constituents of the blood. **1946** J. F. Fulton *Howell's Textbk. Physiol.* (ed. 15) xxxix. 887 This series of chemical reactions in the erythrocyte has been designated as the isohydric cycle because the uptake of CO_2 and the release of O_2 is accomplished without the production of an excess of H^+. **1954** A. White et al. *Princ. Biochem.* xxvi. 691 The isohydric shift entails formation of about 0·7 meq. of bicarbonate for each millimole of oxygen which dissociates from oxyhemoglobin. **1970** R. W. McGilvery *Biochem.* xxv. 612 This action of the Bohr effect to permit the blood to take up CO_2 without a change in pH is known as the isohydric carriage of CO_2.

isohyet, isohyetal, isoimmune, isoimmunization: see *iso-.

isoionic (aiso,ai,ǫ·nik), *a.* *Physical Chem.* and *Biochem.* [f. Iso- + Ionic *a.*²] Of a solute or solution: giving rise to or containing no non-colloidal ions other than those formed by dissociation of the solvent; *isoionic point*, †*reaction*, the point (usually *p*H value) at which the average number of hydrogen ions attached to the basic groups of solute molecules is equal to the average number dissociated from the acidic groups.

This is the 'theoretical' definition of *isoionic point*; for the 'practical' definitions, see quot. 1959.

1926 Linderstrøm-Lang & Lund in *Compt. Rend. Lab. Carlsberg* XVI. v. 22 We will therefore define isoionic reaction as the value of pa_H, pa_H^0 at which h [sc. the specific hydrogen ionisation of the ampholyte] is 0. **1934** *Biochem. Jrnl.* XXVIII. 1257 At 0°, the isoionic point of crystallised haemoglobin was at p_H 7·6. **1943** J. D. Edsall in Cohn & Edsall *Proteins* xx. 446 If the protein binds no other ions than protons, the isoionic point may correspond to the isoelectric point of the protein... If the protein combines with other ions also, the isoelectric and isoionic points are different. **1949** *Jrnl. Physical & Colloid Chem.* LIII. 88 Operationally we may define the isoionic material as the limit approached by successful electrodialysis. *Ibid.* 95 We may..calculate the change in pH of a solution which is not isoionic. **1959** Linderstrøm-Lang & Nielsen in M. Bier *Electrophoresis* I. ii. 63 According to the first definition the isoionic point is the pH of the protein solution which does not change on the addition of more isoionic protein. According to the second definition the isoionic point is the pH of a solution of the isoionic protein in water, or in a solution which does not produce H^+ or OH^- ions when dissolved in water alone. Thus a mixture of proteins may be isoionic. **1969** Ottaway & Irvine tr. *Netter's Theoret. Biochem.* vi. 192 Amino acids and proteins are usually characterised by their isoelectric or isoionic point.

isokinetic, -kinetically: see *iso-.

isokite (isǭu·kait). *Min.* [f. *Isok-a*, name of a small town in Zambia: see -ite¹.] A white, buff, or pinkish phosphate and fluoride of calcium and magnesium, $CaMgPO_4F$.

1955 Deans & McConnell in *Mineral. Mag.* XXX. 681 The mineral, for which the name isokite is now proposed, occurs in the carbonatite plug which forms Nkumbwa Hill (lat. 10°10′ S. long. 32°51′ E), 15 miles east of Isoka (pronounced Isōka) in Northern Rhodesia. *Ibid.* 686 The analysis leaves no doubt that isokite is essentially $CaMgPO_4F$. **1968** I. Kostov *Mineral.* II. 463 Tilasite and isokite are monoclinic (*C*2/*c*), isostructural with sphene.

isolable, *a.* (Later examples.)

1926 A. R. Lord *Princ. Pol.* v. 147 Democracy is no isolable, dead element in the composition of social and political humanity. **1965** W. Lamb *Posture & Gesture* vii. 97 We can demonstrate, however, that there is a distinctly isolable quality of children's physical behaviour.

isolatable (aisolǣi·tăb'l), *a.* = Isolable *a.*

1936 *Chem. Abstr.* XXX. 6354 G[rundmann] has attempted to apply the reaction to certain substituted phenols in order to obtain aliphatic polyene polycarboxylic acids, *o*-Coumaric acid gave, as the only isolatable product, *I* in very small amt. **1949** E. A. Nida *Morphol.* (ed. 2) ii. 60 The morphemes *cran-*, *rasp-*, and *cray-* are isolatable because the elements *berry* and *fish* occur in isolation or in other combinations. **1957** N. Frye *Anat. Criticism* 198 Romance, like Comedy, has six isolatable phases. **1964** R. H. Robins *Gen. Ling.* vii. 277 The bound morpheme..is likely to be a much less semantically isolatable unit. **1971** *Jrnl. Gen. Psychol.* LXXXIV. 169 Personality is the name for something that is not 'time-bound': that is, it is not an event that occurs in a limited period of time and is not isolatable in the way other categories are.

Hence **i:solatabi·lity**.

1949 E. A. Nida *Morphol.* (ed. 2) 59 On the basis of the first condition of isolatability we may identify as morphemes such forms as *boy*, *cow*..since it is possible to utter..these forms in isolation.

isolate, *a.* (*sb.*) Add: (Later examples.)

1923 D. H. Lawrence *Kangaroo* vii. 151 In the visible world I am alone, an isolate instance. **1956** R. Redfield *Peasant Soc. & Cult.* 8 Little isolate societies. *a* **1963** S. Plath *Ariel* (1965) 26 These are the isolate, slow faults That kill, that kill, that kill. **1967** T. Gunn *Touch* 33 Drops are isolate on leaves. **1973** *Archivum Linguisticum* IV. 14 'Am' is a bound form, a pronoun which only occurs as an object and never as an isolate form.

B. *sb.* Add to def.: Esp. something abstracted from its normal context for study. (Further examples.)

1934 *Nature* 8 Dec. 889/2 The method of science to search for useful isolates may easily lead the scientific worker to overlook the reactions of his social environment on his own scientific work. **1937** D. J. B. Hawkins *Causality & Implication* iii. 57 Perhaps it will be best to use the term *isolate*, meaning what is *isolated* in thought but referring specifically to the factual element which is thus isolated. *Isolate* is an appropriate name for any conceptual object in itself, whether it be a simple character or a complex of characters. **1950** J. E. L. Farradane in *Jrnl. Documentation* June 87 An item of knowledge will thus be an object..or an abstract..which is clearly and, at its own level of complexity, uniquely definable, as far as may be possible. Any other item would in reality be composed of two or more concepts, leading to logical confusions. Let us call these items, as defined, isolates. **1951** S. F. Nadel *Found. Social Anthropol.* v. 75 In this sense no legitimate isolate can be discovered more basic than that of a standardized pattern of behaviour rendered unitary and relatively self-contained. **1956** R. Redfield *Peasant Society & Culture* 7 The primitive isolate, the community that is a whole all by itself,..became the model of research. **1958** *Antiquity* XXXII. 148 Cultural isolates from archaeological material. **1961** *Encounter* May 74 'Homosexuality' is a false isolate, a term covering a number of conditions. **1964** C. D. Needham *Organizing Knowl. in Libraries* vii. 71 The isolates now need grouping so that those which are related are proximate. **1969** A. C. Foskett *Subject Approach to Information* I. v. 56 Copper as a topic taken out of context is an isolate, but if we place it in a facet in a particular basic class we can refer to it as a focus. **1972** *Jrnl. Social Psychol.* LXXXVI. 109 Ss reported that the isolate reduced the homogeneity.

b. *Perfumery.* A compound which is isolated in a more or less pure condition from a natural essential oil for use in perfumery.

1923 W. A. Poucher *Perfumes & Cosmetics* iii. 214 It is usual to combine both with a natural isolate of rose odour such as geraniol. **1949** R. W. Moncrieff *Chem. Perfumery Materials* I. ii. 29 Most of the alcohols used in the perfume industry are isolates rather than synthetics. **1957** E. Sagarin *Cosmetics* xxxiii. 743 Of the chemical bodies used in perfumery, the isolates from plant oils bear the strongest resemblance to the plant materials themselves.

c. *Biol.* A group of like micro-organisms obtained by isolation or culturing for study or experiment; *esp.* a pure culture.

1931 W. B. Brierley in *Ann. Appl. Biol.* XVIII. 421 The procedure I adopt in my *Botrytis* work is as follows. Each separate pure culture made by direct isolation from fresh material, whether a number of cultures are made from a single lesion or from one or more host plants, I term an Isolate. If the first culture direct from the diseased tissue contains, as is very often the case, two or more types, the pure or single-spore isolations from this mixture and not the first impure culture are the isolates. Each isolate is an individual line and sub-cultures are merely duplicates or replicates of that isolate or line. The isolate is the nearest equivalent to Lotsy's 'species'. **1949** H. W. Florey et al. *Antibiotics* I. i. 66 Many surveys for antibiotic activity have been performed on type-culture collections of fungi and on new isolates. **1958** *New Biol.* XXVII. 63 It seems to be no easier to establish experimental infections with isolates of *Candida albicans* derived from epidemic outbreaks than with isolates from any other source. **1971** *E. Afr. Standard* (Nairobi) 13 Apr. 9/8 Identification of isolates and antisera received from associate laboratories in Senegal..continued. **1972** *Nature* 17 Mar. 122/1 In an investigation of this increased severity, isolates of *Ceratocystis ulmi* were made from infected trees in the outbreak regions.

d. *Soc. Psychol.* A person who, either from choice or through separation or rejection, is isolated from normal social interaction; also occas. an animal separated from its kind.

1942 *Psychol. Bull.* XXXIX. 458 Differences in interpersonal capacity for participation with others, differences which are revealed when the personalities of isolates and leaders are studied. **1953** J. L. Moreno *Who shall Survive?* I. 100 A rough classification of the position of the individual in the groups was possible—the isolates, the pairs, and the bunch that clung to the leader. **1963** T. & P. Morris *Pentonville* vii. 174 The retreatist is difficult to detect because he *is* an isolate. **1966** T. Pynchon *Crying of Lot 49* v. 113 Nobody knows anybody else's name... We're isolates, Arnold. Meetings would destroy the whole point of it. **1968** *Observer* (Colour Suppl.) 14 Apr. 36/1 Woolly monkeys..who..live on as 'pets' become desocialised, isolates. **1969** *Sunday Times* (Colour Suppl.) 16 Mar. 23/4 Social isolates..often become so careless of their own welfare that they become undernourished. **1970** *New Scientist* 14 May 319/1 In previous attempts at social rehabilitation, normal monkeys of the same age as the isolates were placed in the cage with them. But the isolate ignored its visitor.

e. *Biol.* A group of plants or animals which has developed characteristics distinct from those of the parent species through the operation of an **isolating mechanism*.

1948 G. Dahlberg in *Adv. Genetics* II. 92 Recessive mutations will come to the fore more quickly in a population built up of small isolates than in one composed of large ones. **1967** M. E. Hale *Biol. Lichens* iv. 62 The form and appearance of colonies of different mycobionts and even different isolates of the same mycobiont vary considerably. **1969** E. Mayr *Princ. Systematic Zoology* iii. 49 Isolates are frequently of sufficient difference to merit subspecies rank.

f. *Linguistics.* A word, or words, or other linguistic feature(s) abstracted from context for special study. Also, a word or short phrase that functions as a clause.

1949 *Trans. Philol. Soc. 1948* 128 For the purpose of distinguishing prosodic systems from phonematic systems, words will be my principal isolates. **1961** R. B. Long *Sentence & its Parts* i. 20 Isolates sometimes take adjunct modifiers, much as nucleuses do. Adjuncts are italicized in the following sentences. Thanks *very much*. Good night, *Marian*. **1965** W. S. Allen *Vox Latina* 7 In French, stress is a feature of the word only as an isolate (in which case it falls on the final syllable). **1972** Hartmann & Stork *Dict. Lang. & Ling.* 119/2 *Isolate*, (*a*) a single word functioning as a clause.., (*b*) a term used occasionally as an alternative to segment.

isolate, *v.* **4.** Add: Also *absol.*

1888 Mrs. H. Ward *R. Elsmere* II. iii. xxi. 178 Three cases of diphtheria..I must go for..a nurse, and we must isolate and make a fight for it.

isolated, *a.* Add: *isolated pawn*: see quots.

1842 C. Pearson *Chess Exemplified* 27 An isolated pawn is one that has no comrade on the same or either adjoining file, so that he requires the support of a Piece. **1847** H. Staunton *Chess-Player's Handbk.* 23 A Pawn which stands alone, without the support or protection of other Pawns, is termed an isolated pawn. **1950** S. Tartakover in R. N. Coles *Chess-Player's Week-End Bk.* 153 An isolated pawn spreads gloom all over the chess-board. **1957** Cunnington & Du Mont *Chess Traps & Stratagems* II. 75 An isolated pawn is normally weak, a pawn supported by its neighbours is a strong asset.

isolating (ai·sǒlēⁱtiŋ), *ppl. a.* [f. Isolate *v.* + -ing².] **1.** *Linguistics.* Designating languages (e.g. Vietnamese) in which (for the most part) words do not vary in form according to their grammatical functions in sentences, as contrasted with agglutinating and inflecting languages.

1860 [see Inflexional, inflectional *a.*]. **1861** Max Müller *Lect. Sci. Lang.* 1st Ser. viii. 274 Languages belonging to this first or Radical Stage, have sometimes been called Monosyllabic or Isolating. **1868** —— *Sel. Ess. Lang.* (1881) I. 44 We find it repeated again and again in most works on Comparative Philology, that Chinese belongs to the isolating class, the Turanian languages to the combinatory, the Aryan and Semitic to the inflectional. **1885** *Encycl. Brit.* XVIII. 774/2 Such languages, constituting the small minority of human tongues, are wont to be called 'isolating', i.e., using each element by itself, in its integral form. **1921** E. Sapir *Lang.* vi. 133 More justifiable [than the traditional classification of linguistic types as i) isolating, ii) agglutinative, iii) inflective] would be a classification according to the formal processes most typically developed in the language. Those languages that always identify the word with the radical element would be set off as an 'isolating' group against such as either affix modifying elements..or possess the power to change the significance of the radical element by internal changes. **1953** C. E. Bazell *Ling. Form* i. 2 In languages of so-called 'isolating' structure the syllable may often be regarded as the minimal unit having independent morphological relevance.

2. *Biol.* **isolating barrier, mechanism,** a geographical, ecological, seasonal, physiological, or other factor which limits or prevents interbreeding between groups of plants or animals.

1913 W. Bateson *Probl. Genetics* vi. 119 In one remarkable case the season of appearance plainly acts as the isolating barrier. **1937** T. Dobzhansky in *Amer. Naturalist* LXXI. 405 The expression 'isolating mechanisms' seems to be a convenient general name for all the mechanisms hindering or preventing the interbreeding of racial complexes or species. **1973** I. H. Herskowitz *Princ. Genetics* xxxvi. 564 It is expected..that natural selection would favor isolating mechanisms that operate prior to mating.

isolation. Add: **2. a.** *Psychol.* and *Sociol.* The separation of a person or thing from its normal environment or context, either for purposes of experiment and study or as a result of its being, for some reason, set apart. Also *attrib.* or as *adj.*

1890 C. L. Morgan *Animal Life & Intelligence* viii. 322 We may call the process by which we select a certain quality, and consider it by itself to the neglect of other qualities, *isolation*. **1902** *Amer. Jrnl. Sociol.* VIII. 37 Thus isolation, apparently confined to a single person, consisting in the negation of sociality, is really a phenomenon of very positive sociological significance. **1934** *Ibid.* XL. 157 The hypothesis is that the cause of schizophrenia is isolation of the person. **1950** K. H. Wolff tr. *Simmel's Sociol.* iii. 119 Isolation thus is a relation which is lodged within an individual but which exists between him and a certain group or group life in general. **1961** D. O. Hebb in P. Solomon et al. *Sensory Deprivation* ii. 7 The isolation procedure seems to be contributing to more effective interrelations between psychiatry and psychology. **1964** Gould & Kolb *Dict. Social Sci.* 355/2 Isolation is regarded as one of the dynamic variables in the failure to acquire personality. **1969** Zigler & Child in Lindzey & Aronson *Handbk. Social Psychol.* (ed. 2) III. xxiv. 523 That early isolation increases later aggression is an especially interesting phenomenon which has also been found in mice..and monkeys. **1970** G. A. & A. G. Theodorson *Mod. Dict. Sociol.* 216 The prolonged isolation of an individual from satisfying social..involvement with others usually leads to or is a result of a mental disorder. **1971** *Jrnl. Gen. Psychol.* LXXXV. 107 Isolation fails to enhance total list acquisition. *Ibid.* 157 Operant tasks were performed in an isolation chamber. **1972** *Jrnl. Social Psychol.* LXXXVI. 106 The results of the experiment showed an isolation effect to the name 'Cecil'.

b. *Psychoanal.* A defence mechanism whereby a particular wish or thought loses emotional significance by being isolated from its normal context.

1926 *Brit. Jrnl. Med. Psychol.* VI. 125 In obsessional neurosis the isolation is given magical motor reinforcement—motor isolation is a guarantee for rupture of thought connections. **1937** tr. *Freud's Gen. Sel. Works* 280 Our attention has..been drawn to a process of 'isolation' (whose technique cannot as yet be elucidated) which has direct symptomatic manifestations of its own. **1946** O. Fenichel *Psychoanal. Theory of Neurosis* ii. ix. 155 Another mechanism of defense prevalent in compulsion neuroses and of very general significance for psychopathology is isolation. **1951** P. M. Symonds *Ego & Self* xii. 181 The compulsive neurotic may use the mechanism of isolation in which a portion of his personality is walled off through lack of feeling. **1964** H. Hartmann *Ess. Ego Psychol.* i. iii. 48 A tendency toward isolation ('good' things must not be contaminated with 'bad' things [etc.]). **1970** P. Chodoff in H. S. Abram *Psychol. Aspects Stress* 54 Isolation of affect [among concentration camp inmates], which could be so extreme as to involve a kind of emotional anesthesia, seemed to have functioned particularly to protect the ego.

3. *Biol.* The limitation or prevention of interbreeding between groups of plants or animals by geographical, ecological, seasonal, or other factors, leading to the development of new species or varieties.

[**1859** Darwin *Origin of Species* iv. 105 Isolation, by checking immigration and consequently competition, will give time for a new variety to be improved at a slow rate.] **1913** W. Bateson *Probl. Genetics* vi. 119 The distinctness of the two forms [of the moth *Tephrosia bistortata*] in the places where they co-exist is maintained by the seasonal isolation. **1929** *Biol. Abstr.* III. 1621/1 If foreign hereditary elements are mixed in a population, correlations will be established, partly through polymery, partly through isolation. **1937** T. Dobzhansky *Genetics & Origin of Species* viii. 230 The mechanisms that prevent the interbreeding of groups of individuals, and consequently engender isolation, are remarkably diversified. **1973** I. H. Herskowitz *Princ. Genetics* xxxvi. 563 Although cross breeding may occur naturally or experimentally between closely related species, each maintains its unique gene pool via reproductive isolation.

isolationism (aisǒlēⁱ·ʃəniz'm). [f. Isolation + -ism.] The policy of seeking (political or national) isolation: with special reference to the U.S.A. Also *transf.*

1922 *19th Cent.* Nov. 731 Her isolationism..discovered that the strain of a formidable advance against freedom was more than it could bear. **1930** *Headway* June 112/2 Add to this the fact that half the people..who have emigrated to America in the last generation or so are Europeans who have left Europe because they wanted to get away from Europe, and the secret of America's 'isolationism' is very largely explained. **1931** *Time & Tide* Suppl. 4 July, However much an instructed minority in America might be in favour of the abandonment of isolationism and a larger co-operation with Europe. **1934** [see *autarky b]. **1953** *Manch. Guardian Weekly* 27 Aug. 1 He was now ready to say that 'unilateralism is the new face of isolationism'. **1955** *Bull. Atomic Sci.* Oct. 274 Apparently, the period of enforced scientific isolationism imposed on them for several years has not destroyed this tradition. **1956** K. Clark *Nude* v. 167 The rigid isolationism of the Parthenon metopes. **1969** *Nature* 8 Feb. 524/2 One of the lamentable consequences of intellectual isolation is isolationism—the unconscious fear of exposing the inadequacy of one's achievement to one's fellow scientists. **1973** *Guardian* 25 Apr. 11/1 The difference..between the old isolationism and the new was that whereas the United States felt itself too good for the world in the 1930s, it now felt it was not good enough.

isolationist. Add: Also *transf.*

1901 *Rep. Brit. Assoc. Adv. Sci.* 676 This way of accounting for progress in one or more directions may prove as inadequate as the one suggested by isolationists. **1920** *Times* 31 Oct. 16/1 If a grave crisis were ever to arise on the western borders of Russia, the isolationists might be swept off their feet. **1962** [see *comparatist]. **1969** *Guardian* 6 Feb. 10 A teacher cannot be an isolationist. He has to go out and meet the people whose opinions matter in education.

b. *attrib.* or as *adj.*

1921 *Glasgow Herald* 21 Apr. 8 Regarding the future policy of the United States...The isolationist attitude.. is..much less obvious. **1922** *Ibid.* 30 Dec. 7 Senator Borah has been regarded as the foremost advocate..of the isolationist policy. **1930** *New Statesman* 26 Apr. 69/1 When Mr. Hoover signed last year the prospect was good, but the prolonged bickerings in London have stiffened the isolationist sentiment of the interior. **1958** *Punch* 1 Jan. 75/1 They go further and speculate about the future of an isolationist Britain. **1958** M. West *Second Victory* iii. 56 You know what these mountain people are —irredentist, isolationist, intolerant of foreigners and officials alike.

Hence **isolationi·stic** *a.*; **isolationi·stically** *adv.*

1943 E. W. Hall in *Mind* LII. 232, I shall speak of them as the 'primitivistic', the 'isolationistic', and the 'semantical' usages, respectively. **1964** *Economist* 7 Nov. 567 To be..isolationistically independent and..internationally dominant.

isolative (ai·sǒlēⁱtiv), *a.* [f. Isolate *v.* + -ive.] **a.** *Phonology.* Of a sound-change: taking place without reference to neighbouring sounds: opp. *combinative *a.* 3.

1888 H. Sweet *Hist. Eng. Sounds* 17 Isolative changes are those which affect a sound without any reference to its surroundings. *Ibid.* 26 Isolative change of s into ʃ is regular in Gm initial *s* followed by a cons., as in *schwan*, *stein*. **1906** H. C. Wyld *Hist. Study Mother Tongue* iv. 73 Sound changes are conveniently divided into two main classes: *Isolative Changes*, which take place independent of other neighbouring sounds in the word or sentence, and uninfluenced by them; and *Combinative Changes*. **1972** Hartmann & Stork *Dict. Lang. & Ling.* 212/2 A sound change is said to be..isolative, autonomous, spontaneous, sporadic when it is in no way dependent on its environment but occurs in all positions in which the sound in question occurs.

b. *gen.* Tending to isolate.

1957 *Antiquity* XXXI. 189 Natural selection and isolative mechanisms. **1968** G. Jones *Hist. Vikings* ii. i. 69 The axe-resisting, isolative forests of central Sweden.

‖ **isolato** (isolā·to). [It.] An isolated person, an outcast.

The spelling in quot. 1950 is *erron.*

1851 H. Melville *Moby Dick* I. xxvi. 191 They were nearly all Islanders in the *Pequod*, *Isolatos* too, I call such, not acknowledging the common continent of men,..yet ..what a set these Isolatos were! **1950** Auden *Enchafèd Flood* i. 35 What it feels like to be such an isolatoe, who ..is left standing there alone in the wide waste. **1958** *New Statesman* 6 Sept. 322/2 What most of them failed to see..was that the Negro was also, primarily, an American and faced many of the same problems as the displaced isolato, the white American.

isolecithal, isolectic: see *iso-.

Isolette (aisole·t). *N. Amer.* Also **isolette**. [f. Isol(ation + -ette.] The proprietary name of a type of infants' incubator.

1949 *Official Gaz.* (U.S. Patent Office) 7 June 45/2 [Trade-mark] No. 542,499 Air Shields, Inc., Hatboro, Pa... Isolette. For Infant Incubators. **1950** *Collier's* 1 Apr. 71/2 Through the isolettes the baby can be seen at all times, yet need never be moved or exposed. **1951** *Trade Marks Jrnl.* 21 Feb. 182/1 Isolette... Incubators for the care of infants. Airshields, Inc.., Hatboro,..Pennsylvania, United States of America; Manufacturers. **1962** *Jrnl. Pediatrics* Aug. 306/1 Four model C 77 Isolette incubators were used. Two of these were equipped with an Infant Servo-Controller. **1970** *Daily Colonist* (Victoria, B.C.) 27 Feb. 23/2 Mrs. Margaret Kienast..got her first look at her brood [sc. quintuplets] last Wednesday when she was brought from her room to the isolette in the centre's babies hospital.

isoleucine (aisolⁱū·sīn). *Biochem.* [ad. G. *isoleucin* (F. Ehrlich 1903, in *Zeitschr. Ver. d. Deut. Zucker-Ind.* LIII. 821), f. *iso-* Iso- b + *leucin* Leucin.] An amino-acid of which the dextrorotatory or L-form is an essential nutrient and a general constituent of proteins; 1-amino-2-methylvaleric acid, $CH_3CH_2CH(CH_3)CH(NH_2)COOH$.

1903 *Jrnl. Chem. Soc.* LXXXIV. i. 796 The author has isolated leucine and a new compound, *d*-isoleucine, isomeric with it. *d-iso*Leucine crystallises in shining rods or leaflets. **1936** *Jrnl. Biol. Chem.* CXIV. 89 Leucine and isoleucine have been isolated in considerable quantities by extracting the dried, defatted mycelium of *Aspergillus syndowi* with acetone. **1962** *Lancet* 6 Jan. 26/1 They demonstrated a marked elevation of four amino-acids in the blood and urine—namely, the three branched-chain aminoacids (leucine, isoleucine, and valine) and also methionine.

isolex, -lexic, isoline: see *ISO-. **isolichenin:** see *ISO- b.

isologue (əi·sǫlǫg). *Org. Chem.* Also (*U.S.*) **isolog.** [a. F. *isologue* (C. Gerhardt *Traité de Chim. Org.* (1853) I. II. 127): see Iso- and -LOGUE.] Each of two or more isologous compounds.

1889 in *Cent. Dict.* **1924** *Chem. Abstr.* XVIII. 2158 Preparation of the thiopene isolog of cocaine. **1949** G. B. Bachman *Org. Chem.* xi. 110 Propanoic, propenoic, and propynoic acids are isologs of each other. **1965** *Jrnl. Med. Chem.* VIII. 846/1 While the former compound has a pK_a of 7·7, that of its selenium isolog was found to be 4·68 ± 0·05. **1970** *Tetrahedron* XXVI. 2151 (*heading*) NMR studies on the conformation of acetylcholine isologues.

isologous, *a.* Add: [First formed as F. *isologue* (C. Gerhardt *Traité de Chim. Org.* (1853) I. II. 127).]

b. Applied to each of two or more chemically similar compounds having some difference in composition other than a multiple of CH_2; now usu. applied *spec.* to compounds which have different atoms of the same valency at some position(s) in the molecule but are otherwise of identical molecular structure.

1884 *Jrnl. Chem. Soc.* XLVI. 12 In the author's results differences of Br_2 in isologous compounds, as allyl and dibromopropyl alcohols, correspond with differences of specific volume varying from 49·96 to 59·14. **1931** F. J. Moore *Hist. Chem.* (ed. 2) xiv. 163 By isologous compounds Gerhardt understood substances of analogous function like acetic and benzoic acids whose formulæ showed some other difference than CH_2. **1959** *Jrnl. Amer. Chem. Soc.* LXXXI. 6272/1 (*heading*) Partition of isologous oxygen, sulfur and selenium compounds between buffers and organic solvents. **1964** *Jrnl. Med. Chem.* VII. 229/1 Selenoacyl compounds undergo aminolysis much more readily than isologous thioacyl compounds.

2. *Med.* and *Biol.* Genetically identical, esp. with respect to immunological factors; derived from another individual that is genetically identical or belongs to the same inbred strain; carried out between such individuals.

1955 *Jrnl. Nat. Cancer Inst.* XV. 1023 Lorenz, Congdon and Uphoff have increased the 30-day survival time of mice..by the injection of isologous, homologous and, in some cases, heterologous bone marrow. **1959** *Folia Biologica* V. 24/1 An isologous combination..fulfils the conditions for reciprocal tolerance between the host and the haematopoietic tissue graft. **1961** [see *isoallelic* adj. s.v. *ISO-]. **1962** *Cancer Res.* XXII. 947 Sera from mice which had been rendered resistant against the isologous transplantation of Gross virus-induced lymphomas.. were found to be cytotoxic. *Ibid.*, Admixture of lymph node cells from isologous mice..inhibited the outgrowth. **1971** *Nature* 1 Oct. 310/2 Human antibodies..against diphtheria toxoid likewise disappear rapidly when injected into isologous human skin.

isolux (əi·sǫlʊks), *a.* Also **iso-lux.** [f. Iso- + L. *lux* light.] = *ISOPHOTAL *a.*

1911 A. P. Trotter *Illumination* iv. 47 (*heading*) Contour lines of equal illumination or iso-lux curves, due to two lights at a distance apart equal to four times their height. **1926** J. W. T. Walsh *Photometry* iv. 99 An iso-lux diagram may be used to obtain a mental conception of the effect of a given system of light sources of known distribution. **1934** *Archit. Rev.* LXXV. 66 (*caption*) Isolux diagram. This should be read similar to a contour map, the rings representing foot candle intensities on a plane 2 ft. 6 in. above the floor. **1967** E. Chambers *Photolitho-Offset* xi. 164 With the glass screen the unit dots grow by exposure to a graded pattern formed by diffraction effects and..their contour shapes follow the isolux lines of equal intensity.

isolysin, isomagnetic: see *ISO-. **isomaltose:** see *ISO- b.

isomer. Add: **2.** *Physics.* A nucleus having the same atomic number and mass number but different radioactive properties, as a result of being in a different, long-lived, energy state from which a transition is inhibited; *esp.* one in a metastable excited state rather than the ground state. Also called *nuclear isomer*.

1934 *Physical Rev.* XLV. 729/2 The introduction of isomers may be of help for the removal of existing contradictions in the estimation of neutronic mass from different nuclear reactions. **1950** Glasstone *Sourcebk. Atomic Energy* x. 278/1 The second class consists of genetically related isomers,..in which the metastable state decays to the ground state with a definite half life. **1955** R. D. Evans *Atomic Nucleus* iii. 97 Isobaric isotopes with distinguishable nuclear properties are called isomers. **1968** F. B. Morinigo tr. *H. von Buttlar's Nucl. Physics* xiv. 478 If the half-life is larger than 1 msec, the state is called a nuclear isomer (the choice of 1 msec for this definition is arbitrary).

isomerase (əisǫ·mərēiz, -s). *Biochem.* [f. Isomer + *-ASE.] Any enzyme which brings about an isomerization reaction; orig. applied to two particular enzymes (see quots. 1943, 1944).

1943 I. Banga in *Stud. Inst. Med. Chem. Univ. Szeged* III. 67 When ADP had been incubated with isomerase, the product was split by actomyosin which suggested that isomerase had changed the molecular structure of ADP. It was for this reason that the name isomerase was given to the new protein. **1944** *Jrnl. Biol. Chem.* CLVI. 109 Triose phosphate isomerase (or more briefly isomerase) is an enzyme..which catalyzes the reaction glyceraldehyde phosphate ⇌ dihydroxyacetone phosphate. **1953** *Ibid.* CCI. 83 *Escherichia coli* contains an adaptive isomerase catalyzing the equilibrium, D-arabinose ⇌ D-ribulose. **1963** C. H. Doering tr. *Karlson's Introd. Mod. Biochem.* ix. 173 Two enzyme systems are involved in the biosynthesis of uroporphyrin III from porphobilinogen: a deaminase and an isomerase. **1972** J. R. Whitaker *Princ. Enzymol. for Food Scientists* xiv. 417 The 'isomerases' are enzymes which bring about an isomerization of the substrate. The name is formed on the order 'substrate prefix-isomerase'. The prefix indicates the type of isomerization involved, e.g., 'maleate *cis-trans*-isomerase'.

isomeric, *a.* Add: **1. b.** *Physics.* Of, pertaining to, or designating nuclear isomers; *isomeric transition*, a radioactive transition from a metastable state to a lower energy state of the same nuclide.

1934 *Physical Rev.* XLV. 729/1 Another consequence of the introduction of negative protons is the possibility of the existence of isomeric nuclei, that is, nuclei with the same charge and mass but different internal structure. **1937** *Ibid.* LI. 1011 The 18 minute, 4·2 hour, and 35 hour activities must all belong to Br^{80} and Br^{82}, i.e. one of these isotopes appears to exist in two isomeric forms, from which it decays with different periods. **1950** Glasstone *Sourcebk. Atomic Energy* x. 276/1 Over seventy examples of nuclear isomerism have been discovered, and the phenomenon of decay by isomeric transition, that is, by the spontaneous conversion of one nuclear isomer into another, has been elucidated. In most instances there are isomeric pairs only, but triple isomerism has been observed in a few cases. **1955** R. D. Evans *Atomic Nucleus* vi. 231 By the usual definition, an isomeric level is one whose half-period is 'measurably' long. **1970** Marmier & Sheldon *Physics of Nuclei & Particles* II. xv. 1276 Those nuclei whose ground-state configuration is modified by the preferential-filling rule would be expected to display low-lying excited states whose configuration corresponds to that of an unmodified ground state; because of the large spin change occasioned by this effect, the excited state is usually an isomeric state.

isomerism. Add: **b.** *Physics.* The fact or condition of being nuclear isomers.

1938 R. W. Lawson tr. *Hevesy & Paneth's Man. Radioactivity* (ed. 2) x. 124 This isotope of silver..has two half-value periods, 24·5 m. and 8·2 d., the former being associated with positron emission, and the latter with the emission of β- and γ-rays... Here we are confronted with the phenomenon of nuclear isomerism. **1950** [see *ISOMERIC *a.* 1 b].

isomerization (əisǫ:mĕrəizē¹·ʃən). *Chem.* [f. *ISOMERIZ(E *v.* + -ATION.] The conversion of a compound into an isomer of itself.

1891 *Chem. News* 27 Nov. 274/1 The isomerisation is complete at 140°–170°. **1946** *Industr. Chemist* XXII. 45/1 Since 1944 Tide Water Associated Oil Company, California, has operated a liquid-phase pentane isomerisation plant of the Shell type. **1971** *Sci. Amer.* Dec. 48/3 Isomerization increases the compactness of gasoline molecules, thereby improving the fuel's antiknock quality.

isomerize (əisǫ·mĕrəiz), *v.* *Chem.* [f. Isomer + -IZE 1.] *trans.* and *intr.* To change *into* or *to* an isomer (of the original substance).

1891 *Chem. News* 27 Nov. 274/1 Monosubstituted acetylenes with primary alkyl radicles are isomerised to bisubstituted acetylenes. **1903** *Notices Proc. R. Inst.* (1906) XVII. 100 It is isomerised into the quinonoid form by the addition of water. **1920** *Biochem. Jrnl.* XIV. 185 The free oxonium base (III) which more or less easily isomerises into the colourless pseudo-base (II). **1924** *Jrnl. Chem. Soc.* CXXV. II. 2545 These cycloids isomerise to hydroxyamidines at the moment of their formation. **1955** Pines & Mavity in B. T. Brooks et al. *Chem. Petroleum Hydrocarbons* III. xxxix. 44 Incapable of isomerizing to a six membered ring compound, cyclopentane showed quite low reactivity in the presence of aluminium chloride. **1971** *Sci. Amer.* Dec. 48/3 The straight-chain saturated hydrocarbons can be catalytically isomerized to more compact structures.

Hence **iso·merized** *ppl. a.*, having undergone isomerization; **iso·merizing** *ppl. a.* and *vbl. sb.*

1908 *Chem. Abstr.* II. 1970 The velocity and the limits of the isomerizing power of a solvent are independent of each other. **1941** *Jrnl. Amer. Chem. Soc.* LXIII. 519 The question arises whether these catalysts have any isomerizing properties. **1942** *Canad. Chem. & Process Industries* XXVI. 637/2 The most outstanding recent development allied with fatty acids lies in the field of synthetic drying oils. Of these, two are outstanding—dehydrated castor oil, and the isomerized oils with conjugated double bond structure. **1954** Isomerized [see *CYCLIZE *v.*]. **1963** *Agric. & Biol. Chem.* XXVII. A4/1 (*heading*) Identification and cultural conditions of glucose isomerizing bacteria. **1965** Nylén & Sunderland *Mod. Surface Coatings* iii. 99 Isomerized linseed oil dries more rapidly than the normal oil.

isometric, *a.* Add: **4.** Delete the reference to Isotonic *a.* 2 and substitute: [ad. G. *isometrisch* (A. Fick *Mech. Arbeit u. Wärmeentwickelung bei d. Muskelthätigkeit* (1882) vii. 112).] Of, pertaining to, or designating muscular action in which tension is developed but appreciable shortening of the muscle is prevented. (Examples.)

1891 A. D. Waller *Introd. Human Physiol.* ix. 330 If a muscle contracts against a large resistance..so that it can shorten very little, the curve described by a lever attached to it is termed 'isometric'. **1895** *Proc. R. Soc.* LVII. 423 The shortening of the muscle is prevented by resistance (isometric contraction of Fick). **1900** [see Isotonic *a.* 2]. **1920** *Jrnl. Physiol.* LIV. 85 During the development of tension in an isometric contraction both heat and potential energy are being produced by the muscle. **1951** [see *ISOTONIC *a.* 3]. **1969** *New Yorker* 20 Dec. 43/2 He neither drinks nor smokes, and he devotes an hour a day to yoga and isometric exercises. **1971** [see *ISOTONIC *a.* 3]. **1973** *Sci. Amer.* Mar. 83 Whereas a machine only performs mechanical work when a force moves through a distance, muscles consume energy when they are in tension but not moving (doing what is sometimes called 'isometric' work).

5. *Physics.* Indicating, or taking place under, conditions of constant volume.

1873 J. W. Gibbs in *Trans. Connecticut Acad. Arts & Sci.* II. 311 The points associated with states of equal volume will form lines, which..we may also call isometric. **1912** G. A. Goodenough *Princ. Thermodynamics* (ed. 2) ii. 22 Lines of constant volume [are called] isometric lines. **1957** V. M. Faires *Thermodynamics* (ed. 3) v. 69 For a reversible steady flow isometric process..we have the energy equation from equation (7A).

6. *Biol.* = *ISOGONIC *a.*[2]

1950 J. S. Huxley in *Proc. R. Soc.* B. CXXXVII. 467 In an isometric organ we find no change in the proportions of its parts with increase in its absolute size.

7. *Math.* That is an isometry; related by an isometry. Const. *to*.

1952 C. C. Krieger tr. *Sierpinski's Gen. Topology* vi. 100 Two metric spaces consisting of the same elements but with different metrics may be isometric. **1959** L. F. Boron tr. *M. A. Naimark's Normed Rings* i. 38 Two metric spaces X, X' are said to be isometric if there exists an isometric mapping of X onto X'. **1966** J. H. Cadwell *Topics in Recreational Math.* xi. 117 A pattern is characterised by the set of isometric motions that bring it into self-coincidence, and this set forms a discrete group. **1968** E. T. Copson *Metric Spaces* iv. 53 If the points **a**, **b** of A^* correspond to the points a and b of M, we have proved that $\rho^*(\mathbf{a}, \mathbf{b}) = \rho(a, b)$, so that A^* is isometric to M.

B. *sb.* **1.** *Physics.* A line in a diagram that corresponds to or represents states of equal volume.

1873 [see *isopiestic* adj. s.v. *ISO-]. **1936** *Industr. & Engin. Chem.* Feb. 261/2 Not only are the isometrics of the pure substances themselves substantially straight, but the same is also true for mixtures. **1963** Obert & Gaggioli *Thermodynamics* (ed. 2) x. 215 For the real gas, the isometrics are displaced from the origin and are either straight or slightly curved.

2. *pl.* A system of stationary physical exercises in which muscles are exercised isometrically by pitting one against another or against an unyielding object. orig. *U.S.*

1962 *Scholastic Coach* Nov. 31/1 This concludes a series of three comprehensive articles on isometrics. **1964** *Life* 17 Apr. 47/1 In 1921 Atlas began preaching the muscle-building system he called Dynamic Tension, which pits one muscle against another. Now variations of his system, in which muscles struggle against immobile objects, are the latest U.S. exercise fad. Scientists lump the systems under one fancy word, isometrics. **1970** *New Scientist* 20 Aug. 365/1 Isometrics, but recently hailed as the key to the good life, have also now been condemned for dangerously rocketing the blood pressure.

isometrically, *adv.* (in Dict. s.v. Isometrical *a.*). Add: **2.** *Physiol.* Under isometric conditions (see *ISOMETRIC *a.* 4).

1920 *Jrnl. Physiol.* LIV. 84 When a muscle is stimulated isometrically it passes gradually into a new elastic condition. **1945** *Amer. Jrnl. Physiol.* CXLIV. 477 Strings were tied to the tendons and the contractions were recorded isometrically under an initial tension of 50 grams. **1973** *Nature* 23 Feb. 537/2 The muscle twitch was measured isometrically.

3. *Math.* By means of or in the manner of an isometry.

1959 L. F. Boron tr. *Naimark's Normed Rings* ii. 181 Two normed rings R and R' are said to be isometrically isomorphic if there exists an isometric isomorphism of R onto R'. **1971** E. C. Dade in Powell & Higman *Finite Simple Groups* viii. 318 The map $\phi \to \phi^*$ sends x_+ isometrically into x_-.

isometry (əisǫ·mĭtri). [ad. Gr. ἰσομετρία equality of measure (f. μέτρος measure): see -Y[3].] **1.** *Math.* A one-to-one transformation of one metric space into another that preserves the distances or metrics between each pair of points.

1941 Birkhoff & MacLane *Survey Mod. Algebra* vi. 128 An obvious example is furnished by the symmetries of the cube. Geometrically speaking, these are the one-one transformations which preserve distances on the cube. They are known as 'isometries', and are 48 in number. **1965** S. Lang *Algebra* xiv. 356 If σ is a linear isomorphism, and is metric, then we say that σ is an isometry. **1966** J. H. Cadwell *Topics in Recreational Math.* xi. 113 A fundamental property of isometries is that any two carried out in succession define a third. Thus if isometry U takes figure F into figure F', while V takes F' into F'', their combined effect, taken in this order, is an isometry carrying F into F''.

2. *Biol.* = *ISOGONY (s.v. *ISOGONIC *a.*[2]).

1950 J. S. Huxley in *Proc. R. Soc.* B. CXXXVII. 465 Isometry is used of the special case when the organ grows at the same rate as the body.

isomorph. Add: **2.** *Linguistics.* A line in a linguistic atlas connecting places exhibiting identical or nearly identical morphological forms; a morphological isogloss.

1921 E. C. Roedder in *Jrnl. Eng. & Germ. Philol.* XX. 182 The boundary lines of present dialects, with their maze of..isomorphs,..and isotaxes, i.e., lines connecting places of identical or nearly identical..forms,..and syntactical peculiarities. **1926** *Germanic Rev.* I. iv. 303 Like the larger dialects, they would have to be portioned off against each other by isophones, isomorphs, isolexes, and isotaxes, i.e., lines of agreement in sound, form, word (meaning), and syntax. **1937** J. Orr tr. *Iordan's Introd. Romance Ling.* iii. 215 Val d'Ajol alone offers certain well-marked individual features... A group of 'isophones' and 'isomorphs' passes exactly through the point where the two valleys..investigated join each other.

isomorphic, *a.* **2.** (Also *Logic* and *Linguistics.*) For 'groups' read 'groups or other sets' and add to def.: related by an isomorphism; that is an isomorphism. Const. *to*, *with*. (Earlier and later examples.)

1892 F. N. Cole tr. *Netto's Theory of Substitutions* iv. 83 If to every substitution of G correspond q substitutions of 𝔊, and to every substitution of 𝔊 p substitutions of G, then G and 𝔊 are said to be $(p\text{-}q)$-fold isomorphic, or if p and q are not specified, manifold isomorphic. If $p = q = 1$, the groups are said to be simply isomorphic. **1900** *Ann. Math.* II. 48 Two abstract groups G, H which are 1–1 isomorphic are evidently, as far as structure goes, not distinct. **1934** J. A. Eldridge *Physical Basis of Things* xxviii. 372 The scientist studies the infinite variety of nature's processes, acquaints himself with nature's group of operations, and then attempts to find a representation, in models or symbols, which is isomorphic therewith. **1937** A. Smeaton tr. *Carnap's Logical Syntax of Lang.* iv. 224 If S_1 is reversibly transformable in S_2 in respect of symbols, then S_1 and S_2 are called isomorphic. **1952** H. B. Veatch *Intentional Logic* i. iii. 36 Whether it is necessary and proper that the structure of logical entities should correspond to, or be isomorphic with, or be similar to the structure of the facts which they are supposed to signify or represent. **1962** C. O. Frake in J. A. Fishman *Readings Sociol. of Lang.* (1968) 437 The items and arrangements of a structural description of the language code need not be isomorphic with the categories and propositions of the message. **1965** Patterson & Rutherford *Elem. Abstr. Algebra* ii. 51 In other words, G_2 is isomorphic to G_1 if G_1 is isomorphic to G_2. **1968** J. Lyons *Introd. Theoret. Ling.* ii. 55 To the degree that the meanings of one language can be brought into one-to-one correspondence with those of another we will say that the two languages are semantically isomorphic (have the same semantic structure). **1971** *Nature* 22 Jan. 233/2 We can imagine the state of the whole system described by the states $S_1, S_2, \ldots S_n$ of its S_n components. These values will determine a unique point in the n dimensional phase-space S such that there is an isomorphic correspondence between points in the space and the states of the system. **1973** J. Hintikka *Logic, Lang.-Games & Information* ii. 28 In order for anything to be an isomorphic representation of anything else, they must both have a certain structure.

3. *Biol.* Of the same or an analogous form.

1888 *Nature* 20 Dec. 180/1 Dicholophus..has assumed peculiar Raptorial characters isomorphic with those of Gypogeranus, which is a true bird of prey.

4. *Bot.* In algæ and certain fungi, designating a type of alternation of generations in which the two forms are morphologically similar.

1935 F. E. Fritsch *Struct. & Reprod. Algae* I. 52 Isomorphic (homologous) alternation occurs in each of the three largest classes of the Algae. **1951** M. O. P. Iyengar in G. M. Smith *Man. Phycology* iii. 59 In the third type of life-cycle [of Chlorophyta], there is an alternation of two generations both externally similar... This kind of alternation has been called by Fritsch isomorphic alternation. *Ibid.* 60 Isomorphic types very probably originated from haploid types. **1964** E. J. H. Corner *Life of Plants* vi. 93 Their [*sc.* slender brown seaweeds'] life-cycle is an isomorphic alternation of diploid sporophyte with haploid gametophyte. **1964** J. S. Karling *Synchytrium* iii. 51 In 1905 Lowenthal described fusion of two motile isomorphic cells in *S*[*ynchytrium*] *taraxaci* and illustrated what appears to be a binucleate zygote. **1970** Raven & Curtis *Biol. Plants* vi. 438/2 In *Ulva*, the alternation is isomorphic—that is, the mature sporophytes are morphologically identical to mature gametophytes.

5. *Linguistics.* Similar in morphological structure, having similar morphological forms.

1950 *Language Learning* III. iii. 93 The presentation of such items seems to be indicated in terms of lists the members of which are isomorphic (i.e. show similar differences between morpheme alternants). **1954** Pei & Gaynor *Dict. Ling.* 107 *Isomorphic lines*, lines on a linguistic map, indicating the approximate boundaries of the speech-areas in which a uniformity in grammatical forms, inflections, etc. can be observed. **1957** S. Potter *Mod. Ling.* vi. 134 On more detailed [linguistic] maps we might construct different kinds of lines called isophonic, isotonic, isomorphic, or isosyntagmic, which would record differentiating features of sound, tone, word-form and clause-structure respectively.

Hence **isomo·rphically** *adv.*, by an isomorphism (sense 2); in an isomorphic manner.

1935 *Amer. Jrnl. Math.* LVII. 434 By means of $P_i \to P^*_i$ we thus map the algebra 𝔄 reduced modulo $(1-u)$ isomorphically upon itself. **1960** Brown & Gilman in J. A. Fishman *Readings Sociol. of Lang.* (1968) 266 There are a large number of expressions of subordination which are patterned isomorphically with T [*sc.* a familiar pronoun such as *tu*] and V [*sc.* a polite pronoun such as *vous*]. **1971** E. C. Dade in Powell & Higman *Finite Simple Groups* viii. 258 R_i sends the algebra A_i isomorphically onto $\mathrm{Hom}_F(I_i, I_i)$ for all $i = 1, \ldots, k$.

isomorphism. 2. (Also *Logic* and *Linguistics.*) For 'groups' read 'groups or other sets' and add to def.: An exact correspondence as regards the number of constituent elements and the relations between them; *spec.* a one-to-one homomorphism. (Examples.)

1892 F. N. Cole tr. *Netto's Theory of Substitutions* iv. 83 The correspondence of two groups as just defined is called isomorphism. **1897** W. Burnside *Theory of Groups of Finite Order* xi. 222 A correspondence between the operations of a group, such that to every operation S there corresponds a single operation S', while to the product ST of two operations there corresponds the product $S'T'$ of the corresponding operations, is said to define an isomorphism of the group itself. **1941** W. V. Quine in P. A. Schilpp *Philos. A. N. Whitehead* 160 The theory of relation numbers is the general theory of isomorphism, i.e., of structural identity among relations. **1949** Hutten & Reichenbach tr. *H. Reichenbach's Theory of Probability* vi. § 41. 207 By showing that both the frequency interpretation and the geometrical interpretation satisfy the axioms of the formal system of probability..we have demonstrated the isomorphism, or structural identity, of the two interpretations. **1956** E. M. Patterson *Topology* iv. 81 An isomorphism ϕ between two groups G_1, G_2 is a one–one transformation ϕ: $G_1 \to G_2$ of G_1 onto G_2 which preserves the group operation. **1962** *Anthropol. Ling.* June 26 An analysis of the special uses that an individual may make of other dialects..may tell us something about the range of prestige variants and the relative isomorphism of prestige scales that are involved in inter-dialect interchange. **1963** J. Lyons *Structural Semantics* ii. 17 In so far as semantic and distributional criteria converge on the establishment of the same [grammatical] units, this is due to the admitted partial isomorphism between the 'expression-plane' and the 'content-plane' of language. **1965** Patterson & Rutherford *Elem. Abstr. Algebra* iii. 77 An isomorphism of a ring R_1 onto a ring R_2 is a one–one mapping f of R_1 onto R_2 such that, for all $x_1, x_2 \in R_1$, we have $f(x_1 + x_2) = f(x_1) + f(x_2)$, $f(x_1x_2) = f(x_1)f(x_2)$... In a similar manner, we define isomorphisms between integral domains and fields. **1971** *Sci. Amer.* Aug. 96/3 Boole.. noticed that there was a considerable similarity (in fact a mathematical isomorphism) between the rules that govern the use in logic of the connectives 'and', 'or' and 'not', and the operations on sets of, respectively, intersection, union and complementation. **1973** J. Hintikka *Logic, Lang.-Games & Information* ii. 42 The basic idea of Wittgenstein's picture theory is the idea of an isomorphism obtaining between language and reality.., an isomorphism which can be established by any correlation or mapping.

3. *Biol.* A similarity of appearance displayed by organisms having different genotypes.

1902 *Encycl. Brit.* XXVI. 255/2 A few fundamental characters are better indications of the affinities of given groups of birds than a great number of agreements if these can be shown to be cases of isomorphism or heterophyletic, convergent analogy. **1920** I. F. & W. D. Henderson *Dict. Sci. Terms* 158/1 *Isomorphism*, apparent similarity of individuals of different race or species.

4. *Psychol.* The correspondence assumed to exist between mental perception and physiological processes.

1930 W. Köhler *Gestalt Psychol.* ii. 46 But I should have ever so much difficulty in trying to relate definite experiences to definite processes so long as I failed to assume one specific relationship between the two orders, viz., that of *congruence or isomorphism in their systematic properties*. This principle is sometimes formulated more explicitly in a number of 'psycho-physical axioms'. **1937** R. H. Thouless *Gen. & Social Psychol.* (ed. 2) xii. 241 One of the boldest of the Gestalt speculations has been that of the 'isomorphism' of brain processes and mental processes. **1951** R. B. MacLeod in Rohrer & Sherif *Social Psychol. at Crossroads* 223 In the doctrine of isomorphism, however, it is asserted that the all-important parallel to perceptual organization is to be found not in the patterning of stimulus processes but rather in the immediately underlying organization of brain activities. **1970** H. C. Shands *Semiotic Approaches to Psychiatry* xviii. 290 The evidence suggests strongly that at every level there is an isomorphism between physiological and psychological mechanisms [in cancer patients] when both are abstractly considered.

isomorphously (əisomǭ·ɹfəsli), *adv.* *Min.* [f. Isomorphous *a.* + -LY[2].] In such a way as to produce isomorphous substances.

1901 *Mineral. Mag.* XIII. 45 Miersite..mixes isomorphously with marshite (CuI) on the one hand, and on the other forms intimate intergrowths with iodyrite. **1906** *Jrnl. Chem. Soc.* LXXXIX. II. 1148 Chlorine and fluorine sometimes replace each other isomorphously. **1928** *Amer. Mineralogist* XIII. 52 $AlAlO_3$ may enter $CaMgSi_2O_6$.. isomorphously to as much as 25 per cent. **1947** *Q. Rev.* I. 254 When it is found that F′ can replace OH′ isomorphously, as in the apatites..this can be taken as strong presumptive evidence that the OH..is ionic. **1963** *Geochemistry* 1173 Manganese..is associated with the bivalent iron and calcium which it replaces isomorphously in the rock-forming minerals.

isoniazid (əisonəi·ăzid). *Pharm.* [f. *isoni*(*cotinic acid* (s.v. *ISO- b) + *HYDR)AZID(E.] A white or colourless crystalline compound, $C_5H_4N{\cdot}CO{\cdot}NH{\cdot}NH_2$, which has bacteriostatic properties against mycobacteria and is used esp. in the treatment of tuberculosis.

1953 *Jrnl. Pharmacol. & Exper. Therap.* CVII. 219 Isonicotinic acid hydrazide (Nydrazid), generically designated as isoniazid, has been shown to be an effective chemotherapeutic agent against experimental..and human tuberculosis. **1962** *Lancet* 29 Dec. 1364/2 In the second phase, since the bacilli are no longer multiplying rapidly and resistant mutants are therefore unlikely to appear, a single drug only—isoniazid—need be continued to kill off those remaining. **1970** Passmore & Robson *Compan. Med. Stud.* II. xx. 32/2 Isoniazid (isonicotinic acid hydrazine INH) is the most powerful and effective antituberculous drug. **1973** Turner & Richens *Clin. Pharmacol.* xiv. 186 Isoniazid can interfere with the metabolism of pyridoxine and can cause peripheral neuritis when prolonged courses are given.

isonicotinic, isonitrile: see *ISO- b. **isonomaly:** see *ISO-.

isonomia (əisonō̆u·miă). = Isonomy.

1853 E. Creasy *Rise & Progress Eng. Constitution* xiii. 198 There is no part of our constitution so admirable as this equality of civil rights, this *isonomia*, which the philosophers of ancient Greece only hoped to find in democratical government. **1926** *Contemp. Rev.* Oct. 485 It was the *name* 'isonomia' that so commended a democracy.

isonuclear: see *ISO-.

isooctane (əiso,ǫ·ktēin). Also **iso-octane.** [f. Iso- + Octane.] † **a.** 2-Methylheptane, $CH_3{\cdot}(CH_2)_4{\cdot}CH(CH_3){\cdot}CH_3$, a liquid hydrocarbon that occurs in petroleum. *Obs.*

1909 L. Clarke in *Jrnl. Amer. Chem. Soc.* XXXI. 107 In this paper are described the preparation and properties of iso-octane or 2-methyl heptane... Iso-octane is the ninth hydrocarbon to be prepared in the series C_8H_{18}. **1911** *Ibid.* XXXIII. 524 2-Methylheptane (or iso-octane) has the boiling point 116°. [**1966** *McGraw-Hill Encycl. Sci. & Technol.* I. 249/1 Although the name isooctane correctly designates 2-methylheptane, it should be avoided because of the unfortunate use of the misnomer 'isooctane' in the petroleum industry to represent 2,2,4-trimethylpentane.]

b. 2,2,4-Trimethylpentane, $(CH_3)_3C{\cdot}CH_2{\cdot}CH(CH_3){\cdot}CH_3$, a colourless liquid hydrocarbon which is used in aviation fuels and as a solvent and which, because of its good antiknock properties, is taken as a standard in the determination of octane numbers (being assigned the number 100).

This use of the name does not conform to the convention for using *iso-* described s.v. *ISO- b.

1932 *Bureau of Standards Jrnl. Res.* (U.S.) IX. 269 As a criterion for the purity of commercial 'iso-octane', it is desirable to have a reliable value for the freezing point. **1938** *Jrnl. Chem. Soc.* 239 *iso*Octane..is now produced commercially for blending purposes from *iso*butylene via diisobutylene. **1946** *Industr. Chemist* XXII. 519/1 This small yield of *iso*octane was the backbone of the 87 octane no. finished aviation grade produced commercially before the war. **1954** D. M. Desoutter *All about Aircraft* vii. 120/2 A fuel of 130 octane is one which allows a standard test engine to give 30 per cent more power than it would when using pure iso-octane. **1962** *Biochim. & Biophys. Acta* LXI. 467 It occurred to us that the use of an organic solvent to detach lipid and lipoprotein from rat-liver microsomes might be possible, and this note presents results of the successful use of isooctane. **1971** R. J. & J. S. Fessenden *Basis Org. Chem.* ii. 36 At one time, the branched-chain alkane erroneously named isooctane was the best antiknock fuel known. It was given an octane number of 100.

iso-osmotic: see *ISO-.

isopach (əi·sopæk). [f. Iso- + Gr. παχ-ύς thick.] **1.** *Geol.* = *ISOPACHYTE.

1918 *Private Let. U.S. Geol. Survey* (G. & C. Merriam Co. files) 10 May, *Isopach, isopachous*—noun and adjective, referring to lines on a map that shows equal thicknesses of beds, e.g. coal beds. **1925** *Bull. Amer. Assoc. Petroleum Geologists* IX. 890 The term 'isopach' is used to designate lines through points of equal thickness. *Ibid.*, The isopachs swing from a general east-and-west direction to the north. **1957** *Bull. Geol. Survey Gt. Brit.* XII. 55 The total stratigraphical thickness in feet at each locality was plotted and isopachs drawn through points of equal thickness. **1969** *Nature* 8 Nov. 536/2 The measured thickness of the pyroclastic deposits at various localities and the proposed isopachs are shown in Fig. 2. The isopachs generally extend south-eastwards from the eruptive centres, indicating distribution under the influence of a prevailing north-westerly wind.

2. = *ISOPACHIC *sb.*

1936 *Jrnl. R. Aeronaut. Soc.* XL. 55 An isopach is defined..as lines along which the sum of the principal stresses is constant. **1962** J. C. Jaeger *Elasticity, Fracture & Flow* (ed. 2) iv. 182 Isopachs are curves of constant mean stress $\sigma_1 + \sigma_2$.

isopachic (əisopæ·kik), *a.* and *sb.* [f. as prec. + -IC.] **A.** *adj.* Corresponding to or depicting the locus of points in a body where the sum of the principal stresses has the same value.

1931 Coker & Filon *Treat. Photo-Elasticity* ii. 178 By observations with the lateral extensometer we can..obtain..the curves for which $P+Q$ has an equal value. These may be briefly referred to as Isopachic curves, or curves of equal thickness. **1950** M. Hetényi *Handbk. Exper. Stress Analysis* xvii. 910 If the isochromatic and isopachic lines for a given model and loading are known, it becomes..simple..to evaluate the two principal stresses at any point by direct addition and subtraction of the equivalent values. **1972** *Phys. Bull.* May 276/3 The use of double exposure holography to obtain isopachic–isochromatic fringe patterns in the reconstructed image has also been reported.

B. *sb.* An isopachic line or curve.

1931 Coker & Filon *Treat. Photo-Elasticity* ii. 179 The lines of equal P are..obtained by taking successive intersections of isochromatics and isopachics, the parameters of the isochromatics going up one at a time, those of the isopachics going down one at a time; or conversely. **1958** Condon & Odishaw *Handbk. Physics* III. vi. 88/1 Curves of constant $p+q$ are called isopachics.

isopachous (əisopæ·kəs, əisǫ·pǎkəs), *a.* *Geol.* [f. as prec. + -OUS.] Depicting or pertaining to isopachs (sense *1); that is an isopach.

1913 *Bull. U.S. Geol. Survey* No. 537. 89 (*caption*) Sketch map showing lines along which a coal bed is of equal thickness (isopachous lines), drawn for use in valuation of coal land. **1938** *Bull. Amer. Assoc. Petroleum Geologists* XXII. 425 (*heading*) Thinning of Devonian shown by isopachous lines. **1940** *Ibid.* XXIV. 2151 (*heading*) Isopachous studies in Michigan. **1968** Harbaugh & Merriam *Computer Applications in Stratigr. Analysis* v. 87/1 A thickness map (also termed an isopachous map)..portrays the interval between two stratigraphic horizons.

isopachyte (əisopæ·kəit). *Geol.* [f. Iso- + Gr. παχύτ-ης thickness.] A line on a map or diagram joining points below which a particular stratum or group of strata has the same thickness.

1912 *Compt. Rend. XI Congrès Géol. Internat.* I. 250 A series of maps was prepared..and on them were drawn isopachytes, or lines marking points of equal thickness. **1946** *Nature* 13 July 49/2 The detailed stratigraphy of the Coal Measures, shown by isopachyte maps and many other data, gives an insight into the development of this sedimentary series. **1969** Bennison & Wright *Geol. Hist. Brit. Isles* iv. 77 Isopachytes for the Cambrian as a whole ..must be interpreted with caution.., but they provide some indication of the form of the Cambrian geosyncline in Wales.

isoparaffin, isopentane, isopentyl: see *ISO- b.

isophane (əi·sofē¹n), **isophene** (-fīn). [f. as next; cf. Phen-, pheno-.] A line (imaginary or on a map) linking places in which seasonal biological phenomena (the flowering of plants, etc.) occur at the same time. Hence **isopha·nal, isophe·nal** *adjs.*

1918 A. D. Hopkins in *U.S. Dept. Agric. Monthly Weather Rev. Suppl.* IX. 8 (*caption*) Isophenal map of North America. *Ibid.* 9/1 Taking base maps of North America and of the major and minor political divisions, parallel lines (designated as isophanes) are drawn on them to define, according to the bioclimatic law, theoretical lines and zones of equal phenomena as to time of occurrence and equal bioclimatic conditions, at the same level. **1929** V. E. Shelford *Lab. & Field Ecol.* i. 15 An isophenal map was prepared with isophenes drawn through equal-event dates at the same altitude. **1931** *Trans. Entomol. Soc. Lond.* LXXIX. 183 Local conditions.. cause considerable departures in the actual dates of seasonal events from those indicated by isophanes. *Ibid.*, Instead of basing the isophanal map on degrees of latitude and longitude, he [*sc.* Znamenskii] made use of the mean annual isotherms. **1947** R. F. Daubenmire *Plants & Environment* iv. 219 From phenologic data maps can be drawn with lines (isophenes) connecting locations where plants are in the same stage of development at the same time. **1963** E. Mayr *Animal Species & Evolution* xiii. 362 This north-south cline is crossed by the isophenes, lines connecting all populations with the same phenotype.

isophane (əi·sofē¹n), *a.* *Pharm.* [f. Iso- + Gr. -φανης showing, appearing (f. φαίνειν to show, cause to appear).] **a.** Designating that ratio of protamine to insulin which, in a solution made by mixing solutions of the individual substances, gives rise to equal turbidity in two equal samples taken after a precipitate has been allowed to form when one sample has sufficient insulin added to it to precipitate all the protamine in it and the other a sufficient amount of protamine to precipitate all the insulin.

1946 Krayenbühl & Rosenberg in *Rep. Steno Memorial Hosp.* I. 65 The proportion of protamine to insulin must be nearly isophane. At isophane proportions the formation of crystals is rapid. **1950** *Federal Register* (U.S.) 2 Nov. 7363/2 The isophane ratio shall be expressed as milligrams of protamine per 100 U.S.P. Units of insulin. **1955** *Dispensatory U.S.A.* (ed. 25) 686/2 The most useful of these mixtures, for many patients, proved to be the 2:1 mixture of regular insulin and protamine zinc insulin. In this mixture protamine is present to the extent of about 0·5 mg. per 100 units of insulin, which is the combining proportion, or so-called isophane ratio, of protamine and insulin.

b. Applied to a crystalline mixture of insulin and protamine in the isophane ratio with zinc, which has longer-lasting effects than pure insulin. Also *ellipt.* as *sb.*

1954 A. Grollman *Pharmacol. & Therapeutics* (ed. 2) xxvi. 606 (*table*) NPH Insulin (Isophane). **1955** *Dispensatory U.S.A.* (ed. 25) 686/2 In preparing Isophane Insulin Injection, sufficient insulin is used to provide either 40 or 80 U.S.P. Insulin Units for each ml. of the Injection. **1955** *Brit. Med. Jrnl.* 19 Feb. 478/2 Having used isophane.. since November, 1953, and in more than 60 diabetics, I find it easily the most satisfactory one-shot-a-day insulin available. **1966** *Lancet* 24 Dec. 1389/2 A 20-year-old diabetic patient, who was poorly controlled on 45 units isophane insulin in the morning and 35 units isophane insulin in the evening, was admitted in diabetic ketoacidosis. **1968** Pond & Oakley in W. G. Oakley et al. *Clin. Diabetes* xxiii. 602 A comparable method of control by a single morning injection is the use of a mixture of isophane (NPH) and SI.

isophone (əi·sofōun). *Linguistics.* [f. Iso- + Phone *sb.*[1]] A phonetic isogloss; also, a phonetic feature shared by speakers in contiguous areas. Hence **isopho·nic** *a.*

1921 E. C. Roedder in *Jrnl. Eng. & Germ. Philol.* XX. 182 The boundary lines of present dialects, with their maze of isophones,..and isotaxes, i.e., lines connecting places of identical or nearly identical sounds,..and syntactical peculiarities. **1926** *Germanic Rev.* I. iv. 286 A map of France with all isophonic lines would show the wildest maze and confusion. **1926** [see *ISOMORPH 2]. **1932** W. L. Graff *Lang.* 364 There were isoglosses and isophones which contained the germs of later developments. **1935** *Univ. Mich. Publ. Lang. & Lit.* XIII. 23 We believe ..that these isophonic lines do reflect roughly at least approximate boundaries for eleven of the most important characteristics of the regional dialects of Middle English as they existed about the middle of the fourteenth century. *Ibid.* 32 The physiographical boundaries are an interpretation of the isophones and are therefore subject to the modifications that additional documentary evidence may bring. In some cases these modifications will result in a closer approximation of the isophones to the corresponding physiographical boundaries. **1936** *Trans. Philol. Soc.* 1936 79 It certainly raises the question as to how far the delimitation of dialects and dialect features in Middle English by means of 'isophones' can really be reliable. **1937** *Year's Work Eng. Stud.* 1935 42 The country marked out by 'isophonic' lines into ten areas, each of which presents a more or less distinct complex of dialect-characteristics. **1955** W. S. Avis in *Amer. Speech* XXX. 7 In a very loose way this boundary approximates an *r*-isophone. **1970** *English Studies* LI. 445 The new and very clear ā/ǭ isophone there [*sc.* in Lincolnshire] is an important result of the investigation.

isophot, var. *ISOPHOTE.

isophotal (əisofōu·tăl), *a.* [f. as next + -AL.] Applied to an isophote and to a diagram depicting isophotes. Also as *sb.*, = *ISOPHOTE.

1904 D. Burnett in *Trans. Amer. Inst. Electr. Engin.* XX. 74 The impression they convey would much more nearly accord with what actually occurs if the curves showed not the intensity of light in the various directions, but the lines along which the lighting is equal at any point. Such curves may be called 'isophotals', and while they are of the same general shape as intensity curves, their dimensions are as the square root of the radii. *Ibid.*, The tendency of enclosing an arc.., either in the interior gas globe or more particularly in the outer diffusing globe, is to make such an isophotal curve approach more nearly a circle. **1930** *Astrophysical Jrnl.* LXXI. 365 The total luminosity of the entire nebula out to the distance *r/a* is found by integrating successive elliptical rings on the assumption that the isophotal contours are all similar ellipses. **1964** *Jrnl. Geophysical Res.* LXIX. 487/1 These regions were characterized by systems of isophotal contour lines, sometimes occupying virtually the whole observed area. *Ibid.* 487/2 Isophotal sky maps. **1964** A. Cox *Syst. Optical Design* ix. 367 The contour lines are 'isophotals' mapping areas of equal intensity.

isophote (əi·sofōut). Also **-phot** (-fǫt). [f. Iso- + Gr. φῶς, φωτ- light.] A line (imaginary or in a diagram) connecting points where the brightness or the illumination is the same. Also *transf.* (of radiation other than light).

1909 *Cent. Dict. Suppl.*, Isophote. **1937** *Nature* 2 Oct. 577/1 The successive isophotes of the corona round the sun are nearly circular. **1957** G. E. Hutchinson *Treat. Limnol.* I. vi. 373 (*caption*) Isophots in 1000 cal. cm.$^{-2}$ received during April, May, June, and July on the Scandinavian Peninsula. **1959** Davies & Palmer *Radio Stud. Universe* iv. 47 In some cases the radio isophotes are far more extensive than their optical counterparts. **1966** E. Harrison-Jones in Hewitt & Vause *Lamps & Lighting* xxix. 466 In planning general window lighting of this kind..isophot diagrams are used to determine direct illumination. **1973** *Physics Bull.* July 417 (*caption*) Intensity map of elliptical galaxies NGC 4278 and 4283, showing the outermost faint isophotes.

isophotic (əisofōu·tik), *a.* [f. as prec. + -IC.] = *ISOPHOTAL *a.*

1931 *Nature* 14 Mar. 418/2 Diagrams are given, showing the isophotic lines in the two [Magellanic] clouds. **1935** E. A. Milne *Relativity, Gravitation & World-Struct.* xiv. 279 The circular shape of the observed isophotic contours near a nebular nucleus. **1971** *Jrnl. Astron. Soc. Canada* LXV. 251 (*heading*) Monochromatic photographs and isophotic contours of planetary nebulae.

isophthalate, isophthalic: see *ISO- b. **isopiestic:** see *ISO-.

isoplasty (əi·sǫplæsti). *Med.* and *Biol.* [f. Iso- + -PLASTY.] = *HOMOPLASTY, *HOMOTRANSPLANTATION. Hence **isopla·stic** *a.* = *HOMOPLASTIC *a.* 2.

1923 Isoplasty, -plastic [see *HOMOGRAFTING *vbl. sb.*]. **1929** *Ann. Surg.* XC. 926 Homoplasty, homeoplasty and isoplasty mean tissue transplantation from one individual to another of the same species.

isopleth (əi·sopleþ). [ad. Gr. ἰσοπληθ-ής equal in quantity, f. πλῆθος multitude, quantity.] **1.** [ad. G. *isoplethe* (C. A. Vogler *Anleitung zum Entwerfen graph. Tafeln* (1877) i. 7).] A line (either imaginary or on a map or diagram) connecting points for which some chosen quantity has the same value, the points (if on a diagram) being defined by two variables of which one is usually distance and the other either distance or time.

1909 in *Cent. Dict. Suppl.* **1911** N. Shaw *Forecasting Weather* 4 Isobars are isopleths of pressure, isotherms are isopleths of temperature, and so on. **1935** *Geogr. Jrnl.* LXXXV. 145 Isopleths of population density are drawn. **1950** Conrad & Pollak *Methods in Climatol.* (ed. 2) iii. 69 The curves z = const., called isopleths, thus show the relationship of the two variables x and y only for selected values of the third variable z, which is constant for each curve. **1957** G. E. Hutchinson *Treat. Limnol.* I. ix. 630 Three days later a still more striking deflection of the oxygen isopleths downward at the upper end of the basin was noted. **1971** *Sci. Amer.* Sept. 92 (*caption*) Mean radiation of the earth is portrayed by isopleths..that give the net radiation in terms of calories per square centimeter per minute.

2. *Physical Chem.* A line or surface (in a diagram) joining points that represent mixtures of the same composition.

1924 G. Edgar in H. S. Taylor *Treat. Physical Chem.* I. ix. 400 Assume a complex containing 60 per cent phenol to be warmed from 10° to 70°; the process can be followed by drawing a line of equal concentrations, *ab*, called an isopleth. **1950** W. J. Moore *Physical Chem.* vi. 136 Let us follow the sequence of events as the pressure is gradually reduced along the line of constant composition, or isopleth, *xx′*. **1951** J. E. Ricci *Phase Rule* iii. 59 The representation of the melting point region may be completed by a constant composition section, the isopleth shown in Fig. 3–12.

Hence **i·soplethal** (-plī·þăl) *a.*, carried out or occurring at constant composition.

1924 G. Edgar in H. S. Taylor *Treat. Physical Chem.* I. ix. 411 The general diagram..may be used to illustrate the effect of isothermal and isoplethal changes. **1951** J. E. Ricci *Phase Rule* ix. 189 Isoplethal or Synthetic Methods. These involve the measurement of the temperature of phase transition or of the curve of a miscibility gap, upon complexes of known composition.

isopor, isoporic: see *ISO-.

isoprenaline (əisopre·nălīn). *Pharm.* [f. the chemical name N-*iso*pro*pyl*nor*adrenaline*.] A sympathomimetic amine, $C_6H_3(OH)_2CH(OH)CH_2NHCH(CH_3)_2$, that is a derivative of adrenaline and is used (usu. as the hydrochloride or sulphate, both whitish, bitter-tasting powders) either sublingually or in an aerosol for the relief of bronchial asthma and pulmonary emphysema.

This is the name in the *British Pharmacopœia*: cf. *ISOPROTERENOL.

1951 *Brit. Jrnl. Pharmacol.* VI. 295 The amines used were *l*-adrenaline..and *dl*-isopropyl-*nor*adrenaline hydrochloride (isoprenaline). **1965** *Lancet* 19 June 1301/2, 75 patients with symptoms due to heart-block were treated with a long-acting formulation of isoprenaline. **1966** Dunlop & Alstead *Textbk. Med. Treatm.* (ed. 10) 716 Isoprenaline given by inhalation is effective within a few minutes and, although the relief is relatively short-lived, this form of treatment has a useful place in the treatment of episodic asthma. **1971** *Brit. Med. Bull.* XXVII. 27/1 Death might be caused by the excessive use of aerosol inhalers containing..isoprenaline.

isoprene (əi·soprīn). [f. Iso-, perh. + Pr(opyl)ene.] **a.** 2-Methyl-1,3-butadiene, $CH_2{:}C(CH_3)CH{:}CH_2$, a colourless liquid obtained by the destructive distillation of rubber and from petroleum and used in the manufacture of certain synthetic rubbers.

1860 C. G. Williams in *Phil. Trans. R. Soc.* CL. 244, I have given the substance thus examined the name of isoprene. *Ibid.* 249 Isoprene combines explosively with bromine. *Ibid.* 255 The isolation of isoprene, $C^{10}H^8$. **1867** Bloxam *Chem.* 481 Heated in a retort, caoutchouc is decomposed into several hydrocarbons, one of which, called isoprene, boils at about 100°F. **1906** *Westm. Gaz.* 8 Sept. 10/2 In forming caoutchouc from isoprene, the great problem is already practically solved. **1957** *Technology* July 176/3 By the use of Ziegler catalysts..Goodrich-Gulf were able to synthesize from isoprene a polymer almost identical with natural rubber. **1972** P. Wiseman *Introd. Industr. Org. Chem.* vii. 253 More important than high molecular weight homopolymers of isobutene are copolymers with 1·5 to 4·5% of isoprene, known as butyl rubber.

b. *attrib.*, as **isoprene rule**, the rule that the carbon skeleton of a terpene is made up of isoprene units linked together; **isoprene unit**, the arrangement of five carbon atoms found in the isoprene molecule (the single or double nature of the bonds between them being disregarded).

1926 C. K. Ingold in *Proc. Leeds Philos. & Lit. Soc.* (*Sci. Sect.*) I. 11 It is known..that the carbon skeletons of all terpenes of established constitution can empirically be built up from the skeleton of isoprene, the common product of thermal degradation of almost all terpenes. Divisibility into isoprene units may therefore be regarded as a *necessary* condition to be satisfied by the structure of any plant-synthesised terpene product. **1931** J. L. Simonsen *Terpenes* I. p. xii, An apparent exception to the 'isoprene rule', the hydrocarbon, sylvestrene, has been shown not to be a plant product. **1942** Fuson & Snyder *Org. Chem.* xxvii. 358 Caoutchouc is composed of the isoprene unit repeated over and over. *Ibid.*, The occurrence of isoprene units is so general that investigators of the structures of natural products generally consider the most probable formula of an unknown compound to be that which contains the maximum number of isoprene units. **1959** L. Ruzicka in *Proc. Chem. Soc.* 343/2 Up to 1921, the idea that the higher terpenes are structurally composed of isoprene units was not used as an accepted working hypothesis. **1967** *Nature* 18 Nov. 642/2 The occurrence of a series of hydrocarbons, with structures based on the C_5H_8 isoprene unit, has been invoked as evidence for life-forms in Pre-Cambrian times. **1971** J. D. Roberts et al. *Org. Chem.* xxix. 779 The empirical isoprene rule resulted from Ruzicka's observation that the majority of the terpene families could be considered as arising from head-to-tail combinations of isoprene units.

isoprenoid (əi·sǒprĭnoid, əisoprī·noid), *a.* and *sb.* *Chem.* [f. prec. + -OID.] **A.** *adj.* Containing or designating the isoprene unit; having a molecular structure composed of such units. **B.** *sb.* An isoprenoid compound.

1945 *Chem. Abstr.* XXXIX. 5887/1 Other isoprenoid hydrocarbons, as isoprene, diterpenes, triterpenes, sesquiterpenes, are either entered at their trivial names or named systematically. **1958** *Proc. Nat. Acad. Sci.* XLIV. 167 Squalene synthesis from branched-chain subunits will be treated as a two stage process, namely (a) the condensation of three 'isoprenoid' units to a sesquiterpene (C_{15}) [etc.]. **1965** *New Scientist* 24 June 867/2 The branched-chain compounds called isoprenoids, which have a branch every fourth carbon atom. **1971** *Nature* 24 Dec. 449/1 Pristane (2,6,10,14-tetramethylpentadecane) and phytane (2,6,10,14-tetramethylhexadecane), isoprenoid hydrocarbons, which probably arise from diagenetic degradation of algal chlorophylls, were present [in the sediment] in trace amounts. **1972** *Science* 9 June 1121/2 After 2 years, isoprenoids..and alicyclic and aromatic hydrocarbons remained prominent in the polluted sediments.

isopropanol, isopropenyl, isopropyl, isopropylidene: see *ISO- b.

isoproterenol (əi:soprotĕrī·nǫl). *Pharm.* [f. the chemical name N-*iso*pro*pyl*ar*terenol*.] The name in the *United States Pharmacopeia* for *ISOPRENALINE.

1957 *Jrnl. Pharmacol. & Exper. Therap.* CXIX. 253 Isoproterenol..produces vasodilation and a marked increase in heart contractile force. **1961** *Lancet* 9 Sept. 574/1 Isoproterenol hydrochloride..was administered intravenously, and the nodal pacemaker gradually increased in rate with sudden re-emergence of ventricular fibrillation. **1972** R. Haiat et al. in J. Han *Cardiac Arrhythmias* v. 91 Although some authors reported satisfactory medical management of chronic heart block, long-term drug therapy with isoproterenol is often unreliable and even hazardous.

isopycnal (əisopi·knăl). *Oceanogr.* [f. Iso- + Gr. πυκν-ός dense + -AL.] A line (imaginary or on a chart) or an imaginary surface connecting points which have the same density. Also *attrib.* or as *adj.*

1927 *Geofysiske Publikasjoner* IV. II. 18 We may therefore assume the chief cause of the variations in the depth of the isopycnals to have been oscillations of the water-masses. **1962** C. A. M. King *Oceanogr. for Geographers* iv. 119 A line joining points of equal density is called an 'isopycnal'. **1968** G. Neumann *Ocean Currents* iv. 129 The slope of isobaric surfaces..is small when compared to the slope of isothermal, isohaline (surfaces of equal salinity) and isopycnal surfaces. *Ibid.* 168 In the case of frictionless ocean currents and without mixing a non-accelerated current must always be parallel to the isobars and to the isopycnals.

isopycnic (əisopi·knik), *a.* and *sb.* [f. as prec. + -AL.] **A.** *adj.* (Connecting points) of the same density or of constant density; also, in *Biochem.*, used with reference to ultracentrifugal separative techniques which rely on differences in density between the components of a mixture.

1890 Webster *s.v.*, An isopycnic line or surface. **1910** V. Bjerknes et al. *Dynamic Meteorol. & Hydrogr.* I. iii. 39 We may represent the distribution of mass by drawing surfaces of constant value of the density, or isopycnic surfaces. **1930** N. Shaw *Man. Meteorol.* III. vi. 262 (*caption*) The isothermic, isobaric, isopycnic lines. **1955** *Exper. Cell Res.* IX. 457 Isopycnic gradient centrifugation for the stratification and separation of marine egg halves and quarters was first described by the Harveys. **1962** *Aeroplane* CII. 179/1 The isopycnic level at about 25,000 ft., known to meteorologists for its relative constancy, is being suggested as an accurate reference level for shifting safely from pressure instrument to density altimeter set according to the standard atmosphere. **1964** *Jrnl. Appl. Meteorol.* III. 292/2 A thick isopycnic zone might reasonably be said to exist there [in the equatorial region] from about 5–14 km. **1964** C. J. O. R. & P. Morris *Separation Methods in Biochem.* xxi. 807 On the application of the centrifugal field each particle migrates to a position in the gradient corresponding to its own density (the isopycnic point). Particles initially situated below their isopycnic level float upwards to it. *Ibid.* 809 The apparent density..found from isopycnic measurements. **1971** *Nature* 29 Jan. 299/3 Preparing material entirely uncontaminated by cytoplasmic ribosomes, by isolating the mitochondria by isopycnic centrifugation through sucrose gradients.

B. *sb.* *Meteorol.* A line (imaginary or on a map) or an imaginary surface connecting points at which the density is the same.

1890 Webster, *Isopycnic* n., a line or surface passing through those points in a medium, at which the density is the same. **1924** *Q. Jrnl. R. Meteorol. Soc.* L. 32 In all probability the concentration of the isopycnics takes place over the antarctic continent. **1930** N. Shaw *Man. Meteorol.* III. vi. 260 (*table*) Isobars..Isotherms..Isopycnics.

isopycnosis (əi:sopiknōu·sis). *Cytology.* Also **-pyknosis.** [f. Iso + *HETERO)PYCNOSIS.] The character or condition of having such a degree of condensation and staining as is typical of the majority of chromosomes or chromosomal regions within a particular nucleus. Hence **i:sopycno·tic** *a.*

1950 G. Östergren in *Hereditas* XXXVI. 511 Heteropycnosis is a term in current use in chromosome studies... It might be useful to have a term for the opposite effect to that called heteropycnosis. Since such a term might actually have a function to fulfil, I suggest that the term isopycnosis be introduced. *Ibid.* 512 That a chromosome region is isopycnotic should then mean simply that it does not differ from the majority of the chromosome regions of the cell in its appearance. **1962** *Lancet* 29 Dec. 1384/1 In female meiosis the two X chromosomes may show complete or partial isopyknosis according to the type of pairing. **1967** U. Mittwoch *Sex Chromosomes* xii. 242 Both X-chromosomes of the female become isopycnotic [during gametogenesis].

isoquinoline: see *ISO- b.

isorhythm (əi·soriđ'm). Also **isorr-.** [f. Iso- + Gr. ῥυθμ-ός measured motion.] The rhythmic structure of isorhythmic music.

1954 *Grove's Dict. Mus.* IV. 551/1 *Isorhythm*, the main feature of the French motet in the 14th and early 15th centuries. **1959** *Times* 13 Feb. 13/5 The hockets and isorrhythms of the Middle Ages. **1963** *Listener* 17 Jan. 141/1 A set of variations using the medieval technique of isorhythm (the repetition throughout of a certain rhythmic pattern, in the manner of a passacaglia).

isorrhythmic, *a.* Add: Now more usually **isorhythmic. 3.** *Mus.* 'A modern musicological term applied to fourteenth-century choral works in which the tenor *canto fermo* (or sometimes an upper part) is many times repeated as to its rhythmic features, the pitch of the notes, however, being varied each time it appears' (*Oxf. Compan. Mus.* ed. 9). Also in more general use.

1954 *Grove's Dict. Mus.* IV. 551/1 The isorhythmic motet came to life shortly before 1316. **1957** *Listener* 26 Dec. 1086/1 An aria..on an isorrhythmic bass, that is to say, a 'ground' whose recurrencies begin on different beats in the bar. **1962** *Ibid.* 20 Sept. 453/1 Whether the demonstration is of isorhythmic patterns, classical thematic structures or rotations of a twelve-note succession, it rarely does more than rationalize a unity already intuitively experienced. **1963** *Medium Ævum* XXXII. 151 The elaborate iso-rhythmic setting of *O potores exquisiti* from the Carmina Burana. **1972** *Daily Tel.* 23 Feb. 11/7 To pinpoint some isorhythmic features when a given sequence of durations, though not of pitches, is reiterated in one or more parts. **1972** *Composer & Conductor* Aug. 6/1 Friedrich Ludwig proposed the term Isorhythmic to designate identical rhythmic patterns of different melodies in medieval motets.

isosbestic (əiso,sbe·stik), *a.* *Physical Chem.* Also erron. **isobestic.** [ad. G. *isosbestisch* (A. Thiel et al. 1924, in *Fortschr. d. Chem.* XVIII. 116), f. Gr. ἴσος equal + σβεστ-ός extinguished (f. σβεννύναι to quench, extinguish): see -IC.] *isosbestic point*: a wavelength at which the absorption of light by a liquid remains constant as the acidity varies or, more generally, as the state of equilibrium between two interconvertible substances or states shifts.

1925 *Chem. Abstr.* XIX. 2180 Photometric measurements of the absorption spectra were plotted for equal degrees of acidity and these curves intersected in general in the 'isobetic' point [*sic*: rendering G. *isosbestischer Punkt*.]. **1943** W. R. Brode *Chem. Spectroscopy* (ed. 2) ix. 249 (*caption*) Spectrophotometric determination of hydrogen-ion concentration; mixed indicator (methyl red + bromothymol blue). (Note isosbestic points at 467 and 500 mμ.) **1954** *Trans. Faraday Soc.* L. 802 The ionized form exhibits absorption at a longer wave-length (251 mμ) than the non-ionized (239 mμ) and there is an isobestic point at 245 mμ. **1958** Meites & Thomas *Adv. Analytical Chem.* viii. 277 The spectra of a number of solutions having equal formal concentrations of bromthymol blue but different pH values will all intersect at 501 mμ. Such a point is known as an isosbestic point, and the appearance of such a point on a family of spectra is a necessary (but not quite sufficient) criterion of the presence of two and only two forms of the absorbing substance in equilibrium with each other. **1960** *Jrnl. Biol. Chem.* CCXXXV. 1026/2 Each of the absorption maxima and isosbestic points is essentially the same for normal as well as for sickle cells and homozygous C hemoglobins. **1962** R. E. Dodd *Chem. Spectroscopy* v. 299 The absence of an isobestic point indicates a more complex system. **1971** F. A. Bettelheim *Exper. Physical Chem.* xiii. 144 The absence of an isobestic point is definite proof of the presence of more than two absorbing species. **1972** *Nature* 24 Mar. 140/2 Control measurements are made at an isosbestic point for the rhodopsin and the product.

¶ Other anomalous forms (with quot. 1939 cf. quot. 1943 above).

1939 W. R. Brode *Chem. Spectroscopy* ix. 206 There will be a point..where the extinction curves should have a common value at any *p*H concentration. Such a point is known as an isobastique point. **1949** A. C. Candler *Pract. Spectroscopy* iv. 90 As one absorption curve fades, another develops, while the absorption at an intermediate wavelength, called by Brode the isobastic point, remains unaltered. *Ibid.* 91 The indicator should be of standard strength, but slight variations may be corrected if the absorption at both the centre of the absorption band and at the isobastic point are measured.

isosmotic: see *ISO-.

isospin (əi·sospin). *Physics.* [Contraction of **isotopic spin*, **isobaric spin*.] A vector quantity associated with elementary particles and atomic nuclei which is used to give mathematical expression to the fact that the strong interaction is independent of electric charge; its quantum number (symbol T or I, $T = 0, \pm\frac{1}{2}, \pm 1, \pm\frac{3}{2}$, etc.) is assigned on the basis of there being $2T+1$ particles in a charge multiplet, each with the same value of the quantum number T but differing in the value of the third component of isospin T_3 (or T_z) according to the charge of each particle, with the result that these can be treated as different states of a single particle.

1963 S. Tolansky *Introd. Atomic Physics* (ed. 5) xxiii. 395 The decay of a particle is controlled by a quantity T which has been called the isotopic spin vector or more simply, the isospin. **1965** H. Muirhead *Physics Elem. Particles* i. 14 The isospin of a system is formally similar to angular momentum but is linked to the charge states of the system. **1970** *Physics Bull.* Jan. 22/1 These hadrons and their compounds (the atomic nuclei) appeared as charge multiplets, the members of which had essentially the same mass and the same strong interactions, but different charge values, these being the $(2I+1)$ values $(\frac{1}{2}B - I)e, (\frac{1}{2}B - I + 1)e, \ldots (\frac{1}{2}B + I)e$ where B is the baryon number..and I (called the isospin) is characteristic of the multiplet. **1971** P. E. Hodgson *Nucl. Reactions* xix. 575 The isospin T of a nucleus is the sum of the isospins t_i of its constituent nucleons... The isospins t of the nucleons have the same mathematical properties as the spin vectors σ, that is $t^2\chi = t(t+1)\chi$, $t_z\chi = \pm\frac{1}{2}\chi$, where $t_z = +\frac{1}{2}$ refers to neutrons and $t_z = -\frac{1}{2}$ to protons. Here the z component refers to isospin space and not to ordinary space.

isostasy. Also **isostacy** (*rare*). Add to def.: The condition thought to exist within the earth's crust of approximate hydrostatic equilibrium between portions of different density, the land masses being supported by underlying denser material that yields or flows under their weight and those parts of them that reach to a greater height also extending to a greater depth, any (large) part slowly rising (or falling) if matter is removed from (or added to) its surface. (Earlier and later examples.)

1889 C. E. Dutton in *Bull. Philos. Soc. Washington* XI. 53 For this condition of equilibrium of figure, to which gravitation tends to reduce a planetary body, irrespective of whether it is homogeneous or not, I propose the name isostasy... We may also use the corresponding adjective isostatic. **1893** *Ann. Rep. U.S. Geol. Survey 1891–2* II. 280 The condition of isostasy prevailing in the earth's mass demanded that compensation should be made to the continental area for the load taken from it. **1914** [see *ASTHENOSPHERE]. **1923** *Discovery* Dec. 315/2 The suggestion which has gained most favour, and seems to fit in with many observations, is the theory of Isostacy, which contends that all visible land is composed of the lighter parts of the Earth's crust and is floating upon the components of greater density. **1944** A. Holmes *Princ. Physical Geol.* iii. 34 It must..be clearly realized that isostasy is only a state of balance; it is not a force or a geological agent. It is the disturbance of isostasy by denudation and deposition, earth movements and igneous activity, that brings into play the gravitational forces that restore isostasy. **1950** *Antiquity* XXIV. 43 The dating of these sites involves the post-Algonquin changes in Great Lakes water-levels through isostasy and drainage. **1960** *New Scientist* 19 May 1278/2 The gravity variations are generally consistent with the theory of isostasy, according to which mountains 'float' in hydrostatic equilibrium in the dense mantle as do icebergs in the sea.

isostatic, *a.* Substitute for def.: Pertaining to, produced by, or characterized by isostasy. (Examples.)

1889 [see *ISOSTASY]. **1893** *Ann. Rep. U.S. Geol. Survey 1891–2* II. 280 During the period of sedimentation, which ultimately set up isostatic adjustment, there had been continuous shrinkage of a nucleus cooling beneath the accumulating strata. **1927** Peake & Fleure *Apes &*

Men v. 80 There may have been a slight compensating, or as it is called isostatic, uprise in Denmark and other regions around the margin of the ice sheet. **1937** *Proc. Prehist. Soc.* III. 181 The isostatic emergence of the land from the sea since the last glacial maximum. **1944** A. Holmes *Princ. Physical Geol.* xi. 289 While the crust was being thus unloaded by denudation, slow isostatic uplift must have been continuously in progress. **1955** *Antiquity* XXIX. 181 The complex eustatic and isostatic movements which determine the part played by land movements on one hand and changes of sea-level on the other. **1960** [see *isostatically adv. a]. **1960** B. W. Sparks *Geomorphol.* xiv. 311 Large ice sheets caused an isostatic depression of the areas they occupied and a subsequent rise as the ice melted away.

2. Performed under or involving conditions in which equal pressure is applied from all directions.

1957 *Ceramic News* Apr. 20 (*heading*) Unique 'isostatic process' marks manufacture of Coors famous high density grinding media. **1965** Hove & Riley *Ceramics for Adv. Technologies* iii. 79 In isostatic pressing, the powder material is compacted under uniform pressure. **1967** M. Chandler *Ceramics in Mod. World* vi. 163 (*caption*) New ceramic materials demand new forming techniques, such as this isostatic press for making ceramic spark plug insulators.

isostatically (aisostæ·tikăli), *adv.* [f. prec.: see -ICALLY.] **a.** *Geol.* As regards isostasy; by, or as a result of, isostatic forces.

1901 *Geogr. Jrnl.* XVIII. 517 The elevation of the land caused an ice-sheet to form gradually over it until the surface was depressed, isostatically, by the weight of accumulated ice. **1924** J. G. A. Skerl tr. *Wegener's Orig. Continents & Oceans* xi. 160 It [sc. the map] shows us immediately the mass-defect under mountain chains through which the latter are isostatically compensated. **1957** G. E. Hutchinson *Treat. Limnol.* I. i. 8 The rebound of the southern part of the recently deglaciated Scandinavia, which rose isostatically more rapidly than the postglacial eustatic rise in sea level. **1960** *New Scientist* 19 May 1278/3 Considerable local departures from isostatic equilibrium are known, but an area forming a considerable fraction of a continent is in general found to be isostatically compensated. **1969** *Nature* 21 June 1120/1 Continents and continental margins respond isostatically to pressure variations of ten bars or less.

b. By (means of) pressure applied equally from all directions.

1960 *Times Rev. Industry* Feb. 24/2 Silicon nitride.. powder may be formed to the required shape by..slip casting, or pressing either in steel dies, or isostatically in a thin walled rubber sac. **1972** *McGraw-Hill Yearbk. Sci. & Technol.* 332/2 The table shows a comparison of some of the properties achieved by hot isostatically pressed alloys compared to those of cast products of the same composition.

isoster: see *ISOSTERE 2.

isostere (ai·sostiə̄r). [f. Iso- + Gr. στερε-ός solid.] **1. a.** [ad. G. *isoster* (A. R. v. Miller-Hauenfels *Theoret. Meteorol.* (1883) i. 2).] A line or surface (either imaginary or on a map or diagram) connecting points where some substance has an equal specific volume.

1900 V. Bjerknes in *U.S. Dept. Agric. Monthly Weather Rev.* XXVIII. 436/1 The upper isosteres surround the whole earth, whereas the lower ones intersect the earth's surface along the isosteric curves. **1944** H. R. Byers *Gen. Meteorol.* ix. 217 The lines of constant specific volume, called isosteres, slope upward toward the pole. **1957** G. E. Hutchinson *Treat. Limnol.* I. v. 287 It is probable that in all cases in stratified lakes the deep water currents do not represent steady-state circulation systems maintained by a more or less constant wind stress, but rather continually changing readjustments to the changing position of the isosteres.

b. *Physical Chem.* [ad. G. *isoster* (H. Freundlich *Kapillarchemie* (1909) ii. 102).] A line on a graph showing the pressure of a gas required to produce a given amount of adsorption at different temperatures.

1919 *Proc. R. Soc.* XCVI. 289 It is..difficult to obtain observations of an adsorption isostere over a very wide stretch of temperature. **1940** Glasstone *Text-bk. Physical Chem.* xiv. 1173 The isostere for log *p* against 1/*T* is linear, as it should be if the van't Hoff equation is applicable. **1970** A. J. B. Robertson *Catalysis Gas Reactions by Metals* ii. 22 Such isosteres can be derived from isotherms at different temperatures.

2. Also **isoster.** *Chem.* Each of two or more isosteric molecules or ions (see *ISOSTERIC *a.* 4).

1919 I. Langmuir in *Jrnl. Amer. Chem. Soc.* XLI. 1544 Compounds showing a relationship to one another like that between carbon dioxide and nitrous oxide will be called isosteric compounds or isosteres. **1924** J. A. Cranston *Struct. Matter* viii. 160 The tabulation of isosteres is of special value in the study of the crystalline forms of compounds. **1958** A. Albert in *Current Trends in Heterocyclic Chem.* (Chem. Soc.) viii. 64 Thiazole and pyridine are another pair of isosteres said to resemble one another closely in properties. **1968** *European Jrnl. Cancer* IV. 222 The carcinogenic activity of 10 heteropolycyclic substances, of which 7 were sulfur-containing isosters, and 3 were isosters.

isosteric (aisoste·rik), *a.* [f. as prec. + -IC.] **1.** (In Dict. s.v. Iso-.)

2. Indicating equal specific volume.

1900 V. Bjerknes in *U.S. Dept. Agric. Monthly Weather Rev.* XXVIII. 436/1 This distribution can be expressed with the help of surfaces of equal specific volume, or isosteric surfaces. **1934** D. Brunt *Physical & Dynamical Meteorol.* viii. 166 Any isosteric surface is also isopycnic. **1957** G. E. Hutchinson *Treat. Limnol.* I. v. 263 When isosteric surfaces or surfaces of equal density or specific volume do not correspond to the equigeopotential or level surfaces, a current must be flowing.

3. *Physical Chem.* Of a heat of adsorption: corresponding to a constant amount of adsorbed material as the pressure and temperature vary (equilibrium being maintained).

1918 *Proc. R. Soc. Edin.* XXXVIII. 31 Freundlich..considers an 'isosteric' heat of adsorption where *a* [= number of mols. of gas adsorbed per grm. adsorbent, adsorbing at *p*, *T*] is constant, *p* and *T* variables... But it is difficult to see how we can have an isosteric heat of adsorption per mol. adsorbed, for under the defined conditions nothing is adsorbed. We have merely a readjustment of *p* and *T*. **1933** D. J. Martin *Introd. Thermodynamics for Chemists* xiii. 314 The isosteric heat of adsorption is such that the amount adsorbed is kept constant while the pressure and temperature are varied. **1970** *Biopolymers* IX. 1531 A comparison between calorimetric and isosteric heats of sorption..will indicate the degree of irreversibility of the sorption process.

4. *Chem.* Having the same number of valence electrons arranged in a similar manner.

1919 [see *ISOSTERE 2]. **1921** F. H. Loring *Atomic Theories* xv. 134 Methane is isosteric with the ammonium ion. **1938** *Jrnl. Amer. Chem. Soc.* LX. 2628/1 In a few cases, isosteric compounds are interchangeable in serological reactions. **1971** *Arch. Biochem. & Biophysics* CXLIII. 252 These overall studies support the concept that the 1,4-cyclohexadienyl group is an effective isosteric replacement for a phenyl group.

isosterism (aisostiə·riz'm). *Chem.* [f. as prec. + -ISM.] The condition of being isosteric (sense *4).

1919 I. Langmuir in *Jrnl. Amer. Chem. Soc.* XLI. 1545 The isosterism of the cyanate and trinitride ions applies of course also to compounds derived from them. **1935** J. N. Friend *Text-bk. Physical Chem.* II. ix. 384 (*caption*) Isosterism of CO_2 and N_2O. **1963** *Pharmaceutica Acta Helv.* XXXVIII. 706 The two most important postulates for isosterism were there: similarity in shape and polarity.

isostich, isostructural, isosyllabic, isosyntactic, isosyntagmic, isotach: see *ISO-.

isotactic (aisotæ·ktik), *a.* *Chem.* [f. Iso- + Gr. τακτ-ός arranged, ordered + -IC, as ad. It. *isotattico* (G. Natta 1955, in *Atti d. Accad. Naz. d. Lincei, Mem.* (Classe d. Sci. fis.) 8th Ser. IV. II. 69), f. Attic τάττ-ειν (= Gr. τάσσειν) to arrange.] Having or designating a polymeric structure in which all the repeating units have the same stereochemical configuration.

1955 *Chem. Abstr.* XLIX. 12912 The designation 'isotactic' is applied to chain polymers contg. alternating CH_2 and asym. C groups of identical configuration. **1963** *Times* 11 June 19/1 The entry into production of the Brindisi plant..considerably increased output of the isotactic polypropylene plastic ('Moplen'), now becoming popular in many countries throughout the world. **1964** *New Scientist* 23 Jan. 225 These arrangements, known as 'isotactic' and 'syndiotactic' configurations.., have now been realised with most of the common vinyl monomers. **1972** *Physics Bull.* Nov. 666/1 Isotactic polypropylene is a useful commercial plastic noted for its high strength/weight ratio.

Hence **isotacti·city**, the quality of being isotactic.

1959 Natta & Danusso in *Jrnl. Polymer Sci.* XXXIV. 7 All the arrangements which may compose the isotacticity must be only repetitive arrangements. **1968** *Ibid.* A-2. VI. 1495 The isotacticity and the crystallizability of the original sample remain unchanged after degradation.

isoteniscope: see *ISO-.

isotherm. Add: Also, a similar line (either imaginary or in a diagram) connecting points other than on the earth's surface. (Further example.)

1973 *Nature* 16 Feb. 446/2 A pressure-sensing float and a barrier driven by a motor, each fencing the film and connected to a recorder.., made it possible..to obtain isotherms of films.

b. = *ISOTHERMAL *sb.* b.

1895 C. S. Palmer tr. *Nernst's Theoret. Chem.* II. ii. 192 The behaviour of gaseous and liquid carbon dioxide was studied very exactly by Andrews for those temperatures the isotherms of which are plotted. **1924** A. E. Hill in H. S. Taylor *Treat. Physical Chem.* I. ix. 383 The result of changing the pressure at constant temperature can be deduced by following a line drawn parallel to the pressure axis, and called an isotherm, or isothermal line. **1961** M. Tribus *Thermostatics & Thermodynamics* viii. 205 There is shown a line of constant temperature which is tangent to the top of the vapor–liquid region. This line represents the critical isotherm, since it represents the temperature above which only one phase exists. **1971** F. A. Bettelheim *Exper. Physical Chem.* xli. 378 In the graphic representation of the adsorption isotherm, one usually plots the amount of component 1 adsorbed on the surface of the solid per gram (or per unit surface area) of the solid against the equilibrium composition of the solution.

isothermal, *a.* and *sb.* Add: **A.** *adj.* **1. c.** Applied to a line in a diagram that represents states or conditions of equal temperature.

1873 J. W. Gibbs in *Trans. Connecticut Acad. Arts & Sci.* II. 311 In the same way we may conceive of lines..of equal temperature... These lines we may also call..isothermal. **1922** Glazebrook *Dict. Appl. Physics* I. 930/1 When a perfect gas expands isothermally *PV* is constant, and hence its isothermal line on the pressure–volume diagram, for any assigned temperature, is a rectangular hyperbola. **1949** F. Tyler *Intermediate Heat* vii. 146 The relative slopes of the isothermal and adiabatic curves passing through the point P_1..may be obtained as follows.

2. Occurring at a constant temperature; pertaining to or involving a constancy of temperature with time.

1887 *Encycl. Brit.* XXII. 481/1 During isothermal expansion a gas must take in an amount of heat just equal to the work it does. **1937** M. W. Zemansky *Heat & Thermodynamics* xiii. 231 The isothermal compressibility of a gas may be calculated from an empiric equation expressing the dependence of *V* upon *P* at constant temperature. **1957** *Encycl. Brit.* VIII. 123/2 Under ordinary conditions the adiabatic constants for a solid substance are practically the same as the corresponding isothermal constants. **1967** A. H. Cottrell *Introd. Metall.* xx. 373 Decisive progress [in the heat-treatment of steel] came only after the development of the technique of isothermal transformation..in which specimens are quenched into a bath of molten lead or salt at some pre-determined temperature and the course of their transformation at this fixed temperature is then determined.

3. Of the same temperature throughout.

1909 *Rep. Brit. Assoc. Adv. Sci.* 1908 591 Permit me to open the discussion on the Isothermal Layer, and the inversions of temperature which are found there. **1912** H. N. Dickson *Climate & Weather* iii. 77 At heights greater than about nine miles the temperature..remains nearly constant at about −70°F at all levels... In this 'isothermal layer'..there would seem to be..little movement. **1951** J. A. Hynek *Astrophysics* xiv. 662 It is possible that at this stage..the star develops a small isothermal core at the center. **1964** *Oceanogr. & Marine Biol.* II. 71 In winter the fiord water becomes isothermal and isohaline with average temperature and salinity of −1·76°C and 32·75‰.

B. *sb.* **b.** A line in a diagram that is isothermal (sense *A. 1 c).

1873 J. W. Gibbs in *Trans. Connecticut Acad. Arts & Sci.* II. 323 In that part of any diagram which represents a mixture of vapor and liquid, the isopiestics and isothermals will be identical, as the pressure is determined by the temperature alone. **1879** *Amer. Jrnl. Sci.* XVIII. 463 In Diagram A, are drawn Isothermals, curves of equal temperature, in which the abscissæ are wave lengths, the ordinates intensities. **1937** P. S. Epstein *Textbk. Thermodynamics* iii. 49 The slope of the adiabatic ($dp/dv = -\gamma p/v$) is always steeper than the slope of the isothermal ($dp/dv = -p/v$) passing through the same point. **1968** Wallace & Linning *Basic Engin. Thermodynamics* 458 At very low pressures all isothermals tend towards a value of the compressibility function equal to unity, i.e. towards perfect gas behaviour.

isothermally (aisoþɔ̄·rmăli), *adv.* [f. prec. + -LY².] At a constant temperature, without a change in temperature.

1886 *Proc. R. Soc. Edin.* XIII. 79 Steam might be condensed isothermally to supersaturation without condensing. **1897** [in Dict. after Isothermal *a.* and *sb.*]. **1922** [see *ISOTHERMAL *a.* 1 c]. **1952** *New Biol.* XII. 75 In a system such as this the simultaneous dismantling and assembly can be described as 'energy-linked'; and since no heat is lost or gained the changes occur isothermally, that is, at constant temperature. **1967** A. H. Cottrell *Introd. Metall.* xx. 374 Over the range in which the austenite decomposes isothermally the time of transformation follows a C-curve like that of Fig. 20.2.

isothermic (aisoþɔ̄·rmik), *a.* [f. Iso- + Gr. θέρμ-η heat, θερμ-ός hot + -IC.] **a.** = Isothermal *a.* 1 (in Dict. and Suppl.).

1879 *Amer. Jrnl. Sci.* CXVIII. 466 Such an error, in order to produce such an effect in all the isothermic curves, must be looked for either in my comparison of the sun's spectrum with that of the platinum wire, or in Lamansky's measurements of the sun's heat. **1881** *Proc. R. Geogr. Soc.* III. 657 Fall of temperature as we proceed in different directions is shown by means of isothermic lines. **1930** [see *ISOPYCNIC *a.*].

b. = *ISOTHERMAL *a.* 2.

1936 *Jrnl. Iron & Steel Inst.* CXXXIV. 265 A, A certain lowering of the tensile strength and limit of proportionality is revealed between certain temperature limits after isothermic treatment. **1951** R. A. Dutcher et al. *Introd. Agric. Biochem.* xix. 383 Hydrolyses are characterized by no perceptible heat changes and..are known as isothermic reactions. **1965** B. E. Freeman tr. *Vandel's Biospeleol.* xxx. 474 The humus of these mountains possesses an isothermic climate.

c. = *ISOTHERMAL *a.* 3.

1963 S. W. Tromp *Med. Biometeorol.* i. 22 Above the troposphere lies a region known as stratosphere..divided into three layers: a lower cold isothermic layer 12–35 km above sea level with fairly constant temperatures..a warm layer..and a very cold upper layer.

isotone (ai·sotōᵘn). *Physics.* [a. F. *isotone* (K. Guggenheimer 1934, in *Jrnl. de Physique et le Radium* V. 253), coined by replacing the *p* of *isotope* (the initial letter of *proton*) by *n* for *neutron*.] Each of two or more nuclides having the same number of neutrons (but usually different numbers of protons).

1934 *Sci. Abstr.* A. XXXVII. 734 It is proposed that elements having the same number of neutrons, but different atomic numbers shall be termed 'isotones'. **1952** *Sci. News* XXIII. 39 There is an exceptionally large number of nuclei all containing 20 protons (isotopes of calcium)... Nuclei containing 20, 50 or 82 neutrons also have a large number of isotones. **1966** [see *ISOBAR 2]. **1972** *Physics Bull.* Mar. 148/3 The isotone shift differs from the isotope shifts in that the former measures the energy difference of atoms when protons are added to the nucleus with the number of neutrons remaining the same, whereas for the latter the role of protons and neutrons are reversed. The study of isotope as well as isotone shifts reveal the nuclear shell structure.

isotonic, *a.* **2.** Delete '(See quot. 1900.)' and substitute: [ad. G. *isotonisch* (H. de Vries 1884, in *Jahrb. f. wiss. Bot.* XIV. 427).] Of, pertaining to, or (of a solution) having the same osmotic pressure as some particular solution (usually that in a cell, or a body fluid). Const. *with*. (Earlier and later examples.)

Quot. 1900 in Dict. belongs to sense 3 below.

1895 [see *HYPERTONIC *a.* 2]. **1902** Poynting & Thomson *Text-bk. Physics: Properties of Matter* xvi. 190 A series of solutions can be prepared which are isotonic with each other. **1936** A. P. Mathews *Princ. Biochem.* xxxiv. 364 Practically all the secretions of the body, except the urine, as for example the bile, the pancreatic juice, the saliva and so on, have the same osmotic pressure as the blood. They are isotonic with the blood. **1951** H. Davson *Textbk. Gen. Physiol.* vii. 159 For marine eggs the isotonic concentration will generally be approximately 0·5 M salts or 1 M non-electrolyte. *Ibid.* 158 In isotonic glycerol solution the cell is unstable because no osmotic equilibrium is possible. **1969** J. H. Green *Basic Clin. Physiol.* vi. 37/1 The sodium chloride is present in plasma to the extent of 0·9 g. per 100 ml. A solution containing this amount of sodium chloride in water is termed normal, isotonic, or physiological saline, and it has the same electrolyte strength as blood. Such a solution..could be run into the veins of a patient, whereas pure distilled water would destroy the red blood cells by the process of haemolysis. **1972** *Nature* 17 Mar. 117/2 Each animal received an intraperitoneal injection of ethanol (4 g kg^{-1}, 25% v/v solution in isotonic saline).

3. *Physiol.* [ad. G. *isotonisch* (A. Fick *Mech. Arbeit u. Wärmeentwickelung bei d. Muskelthätigkeit* (1882) vii. 112).] Of, pertaining to, or designating muscular action in which the muscle contracts more or less freely against a small, constant resistance.

1891 A. D. Waller *Introd. Human Physiol.* ix. 330 If a muscle contracts against a small and constant resistance, so as to be extended by a constant force during its contraction, the curve described by a light lever attached to it is termed 'isotonic'. **1900** [in Dict., sense 2]. **1939** *Jrnl. Physiol.* XCVI. 63 The relation between applied force and speed of isotonic shortening was studied in frog and tortoise muscle. **1951** H. Davson *Textbk. Gen. Physiol.* xvii. 481 A record of an isometric twitch is thus a record of changes in tension of the muscle, whilst that for an isotonic twitch is a record of changes in length. **1971** A. C. Guyton *Basic Human Physiol.* vii. 77/1 In comparing the rapidity of contraction of different types of muscles, isometric recordings..are usually used instead of isotonic recordings, because the duration of an isotonic recording is almost as dependent on the inertia of the recording system as upon the contraction itself.

isotonically (aisotǫ·nikăli), *adv.* *Physiol.* [f. prec.: see -ICALLY.] Under isotonic conditions (see *ISOTONIC *a.* 3).

1953 A. Szent-Györgyi *Chem. Physiol. Contraction in Body & Heart Muscle* ix. 52 Fig. 8..sums up the experience..on a widely varied material, registered both isotonically and isometrically, excited in different ways. **1970** *Nature* 16 May 656/1 The muscular activity was recorded isotonically on a kymograph. **1971** A. C. Guyton *Basic Human Physiol.* vii. 77/2 Muscles can contract both isometrically and isotonically in the body, but most contractions are actually a mixture of the two.

isotonicity (ai:sotǫni·siti). *Physiol.* [f. as prec. + -ITY.] The property or state of being isotonic (sense *2); equality of osmotic pressure; also, degree of osmotic pressure (of the blood).

1896 T. L. Stedman *20th Cent. Pract.* VII. 284 After twelve or twenty-four hours the solution which is exactly isotonic is determined..and the isotonicity of the blood corpuscles..determined at the same time. *Ibid.*, The isotonicity of the corpuscles is 0·45. **1906** *Amer. Jrnl. Physiol.* XV. 366 Every degree of dilution from isotonicity down to distilled water acts as a stimulus to the heart ganglion of Limulus. **1909** R. J. M. Buchanan *Blood in Health & Dis.* ix. 182 The isotonicity of the plasma is rather below normal. **1939** A. Krogh *Osmotic Regulation Aquatic Animals* 196 The isotonicity normally present between tissue cells and the surrounding medium need not mean, however, that the sum of free ions on both sides of the cellular membrane is the same. *Ibid.* 66 Schwabe (1933)..observed a definite hypotonicity for [the crab] *Pachygrapsus marmoratus* amounting to 57mM., while he confirmed the absolute isotonicity for *Portunus corrugatus*. **1951** [see *HYPERTONIC *a.* 2]. **1974** *Nature* 4 Jan. 50/1 The concentration of calcium was varied from 0 to 2·5 mM (isotonicity was kept constant by appropriate adjustments in the concentration of NaCl).

isotope (ai·sotōᵘp). [f. Iso- + Gr. τόπ-ος place.] **1.** A variety of a chemical element (strictly, of one particular element) which is distinguished from the other varieties of the element by a different mass number but shares the same atomic number and chemical properties (and so occupies the same position in the periodic table); freq. used to denote any individual variety without reference to identity of atomic numbers (see quot. 1947 and cf. *NUCLIDE).

1913 F. Soddy in *Nature* 4 Dec. 400/1 The same algebraic sum of the positive and negative charges in the nucleus, when the arithmetical sum is different, gives what I call 'isotopes' or 'isotopic elements', because they occupy the same place in the periodic table. They are chemically identical, and save only as regards the relatively few physical properties which depend upon atomic mass directly, physically identical also. **1915** *Rep. Brit. Assoc. Adv. Sci.* 1914 301 Sir E. Rutherford (replying) said that the chemical inseparability of certain isotopes was, indeed, derived from experiments with small quantities, but the methods used were very delicate. **1923** *Glasgow Herald* 7 Apr. 7 Another entirely revolutionary factor in chemical theory has been the discovery that the majority of the chemical elements are mixtures of isotopes. **1927** N. V. Sidgwick *Electronic Theory of Valency* i. 11 What Soddy has called an isotope—an element of the same atomic number but a different atomic weight. *Ibid.*, The one element of which the isotopes can be obtained in any quantity in a state of approximate purity..is lead. **1942** J. D. Stranathan *'Particles' of Mod. Physics* v. 205 Hertz, employing gaseous diffusion at low pressure through a special porous material, succeeded in producing a real separation for the isotopes of Ne. *Ibid.* 207 By this method the isotopes H^2, Li^7, C^{13} and N^{15} have been produced in significant concentration. **1947** *Amer. Jrnl. Physics* XV. 356/2 There is at present no word in the English language to express the concept of a particular species of atom, differing from all others in the constitution of its nucleus... In recent years the word *isotope* has come into use for this purpose, less by design than by default. According to its definition [by Soddy], this general usage is incorrect, since *isotope* properly refers to a species of a particular and designated element, and emphasizes its relationship to other isotopes of that element. It is analogous to such words as *brother* and *colleague*. **1959** *New Biol.* XXX. 96 Radioactive isotopes behave chemically in identical fashion with the normal isotopes, but can be distinguished by means of their radioactivity. **1962** *Nature* 19 May 621/2 There are about 300 naturally occurring isotopes, and several hundred additional isotopes can be prepared artificially. **1963** D. W. & E. E. Humphries tr. *Termier's Erosion & Sedimentation* i. 4 The isotopes O^{16}, O^{17} and O^{18} of oxygen are in different proportions in atmospheric and sea water, the latter being rich in heavy isotopes as a result of differential evaporation. **1966** *McGraw-Hill Encycl. Sci. & Technol.* VII. 291/2, 20 elements possess no isotopes; each of these consists of one type of atom only. **1966** C. R. Tottle *Sci. Engin. Materials* i. 23 Hydrogen has three isotopes, the two heavier ones being known as deuterium (originally called heavy hydrogen), with one neutron per atom, and tritium, with two neutrons per atom.

b. In the usage of biologists and biochemists: a less common, usually radioactive, isotope of an element as used in tracer or other studies, in contradistinction to the common, naturally occurring isotope of the element; freq. without *the*.

1945 *Jrnl. Biol. Chem.* CLIX. 697 Isotopic analyses were usually made on these inorganic salts, thus avoiding the dilution of the isotope [sc. C^{13}] by non-isotopic carbon. **1954** A. White et al. *Princ. Biochem.* xiv. 307 By degradation of the product and determination of the distribution of isotope among its atoms, one may often procure information as to the mechanism of the transformation. **1954** Cantarow & Schepartz *Biochem.* xiv. 352 The assay of stable isotopes is based upon the difference in atomic weight between the normal element and the isotope used. **1971** Mazur & Harrow *Textbk. Biochem.* (ed. 10) iv. 99 Synthesis of physiologically important compounds, containing isotopes in place of the normally occurring elements, furnishes the worker with a 'tag' that can easily be followed. **1972** *Science* 2 June 1032/1 Swiss albino mice were labeled with ^{45}Ca by injection with 10 μc of isotope 1 day after birth.

2. Special combs.: **isotope dilution,** a diminution of the concentration of one isotope (or isotopically labelled compound) by the addition or presence of another isotope of the same element (or of the unlabelled compound); *esp.* as a technique for measuring the amount of an element or compound in a system by introducing a known amount of a different isotope (or a labelled compound) and then measuring its concentration in a sample withdrawn from the system after mixing; **isotope effect,** a variation in some physical or chemical characteristic between one isotope of an element and another; **isotope shift,** a small difference in the wavelength of corresponding spectral lines of different isotopes of an element owing to the different masses and charge distributions of their nuclei.

1940 Rittenberg & Foster in *Jrnl. Biol. Chem.* CXXXIII. 737 (*heading*) A new procedure for quantitative analysis by isotope dilution. *Ibid.* 744 A new method (isotope dilution procedure) for the analysis of complex mixtures is described. **1956** *Nature* 28 Jan. 159/1 Isotope-dilution analysis, a technique which is now being increasingly applied in geology, agriculture, metallurgy and biology. **1971** *Metabolism* XX. 1099 (*heading*) Kinetics of potassium distribution in man using isotope dilution and whole-body counting. **1972** *Chemico-Biol. Interactions* IV. 103 This finding suggested that the decreases in specific activity were a result of isotope dilution rather than inhibition of protein synthesis. **1923** *Science* 31 Aug. 164/2 (*heading*) The vibrational isotope effect in the band spectrum of boron nitride. **1956** *Nature* 28 Jan. 159/2 Most properties of solids depend on the atomic masses, and the study of isotope effects can be a powerful means of testing the validity of theories. **1967** A. H. Cottrell *Introd. Metall.* xxiv. 503 The importance of the lattice ions in superconductivity is shown by the isotope effect. Different isotopes of mercury (also tin) are found to have different critical temperatures, proportional to $M^{-\frac{1}{2}}$, where *M* is the atomic mass of the isotope concerned. **1968** A. White et al. *Princ. Biochem.* (ed. 4) xiii. 288 Distinct isotope effects, e.g. differential rates of reaction between normal and tritiated compounds, have been observed. **1932** *Physical Rev.* XLII. 350 It appears that this determination fits in with the magnitude of the isotope shift. **1958** *Oxf. Mag.* 8 May 418/1 Work on high resolution optical spectroscopy is now mainly directed towards the study of 'isotope shifts'. **1971** *Physics Bull.* Oct. 583/2 Isotope shifts occur when the energy levels of different isotopes are slightly shifted with respect to one another, resulting in an apparent splitting of the 'line' if two or more isotopes are present in the source.

isotopic (aisǫtǫ·pik), *a.* [f. prec. + -IC.] **a.** Of, pertaining to, or being an isotope or isotopes of an element; *isotopic number*, the number of neutrons in a nucleus minus the number of protons.

Before Soddy coined the word in 1913, *isotopic* had previously (unknown to him) been used in a different sense (see quot. 1904), but this use did not gain currency.

[**1904** Cohen & Miller in *Jrnl. Chem. Soc.* LXXXV. 1624 Supposing the bromine in the ortho- and meta-positions to retard oxidation more than the chlorine in these positions, the effect should become apparent when the isotopic dichloro-, chlorobromo-, and dibromo-toluenes are oxidised together. [*Note*] We propose..to employ the word 'isotopic'..in place of the rather awkward expression 'similarly substituted'.] **1913** [see *ISOTOPE 1]. **1919** F. Soddy in *Jrnl. Chem. Soc.* CXV. 18 The chemistry of actinium has been enormously simplified by the discovery that mesothorium-2 is isotopic with it, for the latter may be used as an indicator to show in what way the actinium distributes itself after any chemical treatment. **1920** *Glasgow Herald* 7 Oct. 8 It had now been found that certain elements which hitherto had been stated to have atomic weights fractional were mixtures of isotopes, and such isotopic elements had atomic weights as whole numbers on the oxygen scale. **1921** W. D. Harkins in *Nature* 14 Apr. 202/2 Let us specify the atoms of this important class as those of isotopic number 0. Then the isotopes of magnesium of atomic weights 24, 25, and 26 will have isotopic numbers 0, 1, and 2. *Ibid.*, The isotopic number *n* is the number which, when added to twice the atomic number, gives the atomic weight. **1933** F. W. Aston *Mass-Spectra & Isotopes* xiv. 176 Atoms of isotopic number 0, such as C^{12}, O^{16}, Mg^{24}, etc., predominate in a marked degree. Abundance is much less for isotopic number 1 such as F^{19}, Na^{23}, Al^{27}, decreases again for 2:—N^{14}, Ne^{22}, and becomes practically zero for 3:—Cl^{37}. **1942** J. D. Stranathan *'Particles' of Mod. Physics* v. 200 Considerable isotopic data has been obtained from studies of hyperfine structure. **1954** *Sci. News* XXXIV. 51 The next simplest use of the mass spectrometer is to measure the quantity of one isotope of an element compared with another—the isotopic ratio as it is called. **1956** A. H. Compton *Atomic Quest* i. 52 The isotopic separation of U-235 from U-238. **1957** G. E. Hutchinson *Treat. Limnol.* I. viii. 549 A study of the isotopic constitution of the sulfur in the sulfate of rain. **1971** J. Z. Young *Introd. Study Man* v. 82 Many of the elements exist in isotopic forms, which break down spontaneously.

b. Of, pertaining to, or designating isospin; orig. and chiefly in *isotopic spin* = *ISOSPIN.

1937 E. Wigner in *Physical Rev.* LI. 106/2 W. Heisenberg..considered protons and neutrons as different states of the same particle. Heisenberg introduced a variable τ which we shall call the isotopic spin, the value −1 of this variable can be assigned to the proton state of the particle, the value +1 to the neutron state. *Ibid.* 107/1 The Pauli principle requires that the wave function $\Psi(r_1s_1\tau_1, r_2s_2\tau_2, \ldots, r_ns_n\tau_n)$ be antisymmetric with respect to the simultaneous interchange of Cartesian, spin and isotopic spin variables of any pair of heavy particles. This fact is quite analogous to the similar statement for ordinary spin. *Ibid.* 117/1 Terms with different ζ components of the isotopic spin have the same energy in approximation 2. These are, of course, terms of different isobaric nuclei, and a total isotopic spin *T* will be a term with the same binding energy for all nuclei with isotopic numbers from −*T* to *T*. **1953** M. Gell-Mann in *Physical Rev.* XCII. 833/2 Let us suppose that the new unstable particles are fermions with integral isotopic spin and bosons with half-integral isotopic spin. For example, the V_1 particles may form an isotopic triplet, consisting of V_1^+, V_1^0, and V_1^-. The τ^+ and V_4^0 may form an isotopic doublet, which we may call τ^+ and τ^0. **1963** K. W. Ford *World of Elem. Particles* viii. 235 For a group of nucleons, a rotation of the total isotopic-spin vector in *I* space corresponds to changing from one nucleus to another without changing the total number of nucleons. *Ibid.*, New particles..were found to come in closely linked groups, like proton and neutron. There were three pions which formed a triplet and could be described as three states of a single pion with one unit of isotopic spin. **1965** *Listener* 2 Sept. 332/1 Abstract quantities known as 'isotopic spin' and 'hypercharge'. **1969** D. H. Wilkinson *Isospin* i. 3 It is clear that Wigner gave the isotopic spin that name because it is a vector whose *z*-projection, $\frac{1}{2}(N-Z)$, distinguished one isotope from another along an isobaric multiplet. In fact Wigner often called T_z itself the isotopic spin and *T* 'the total isotopic spin quantum number'. Others felt that a stronger case could be made to call the new quantum

number the isobaric spin because its magnitude was constant along an isobaric multiplet.

c. Containing or being a less common or special isotope, e.g. as a label. Cf. *ISOTOPE 1 b.

1942 J. D. Stranathan *'Particles' of Mod. Physics* v. 209 Heavy water..is one of the few compounds which can be produced in practically pure isotopic form. **1953** Fruton & Simmonds *Gen. Biochem.* xxviii. 641 The administered isotopic glycine had 'mixed' with nonisotopic glycine already present in the body of the animal. **1954** A. White et al. *Princ. Biochem.* xiv. 306 The isotopic material may be administered and, from the subsequent analyses for the appropriate isotope in the various tissues.., the distribution of the isotopic atom may be ascertained. **1958** *Oxf. Univ. Gaz.* 5 June 1127/1 Studies on intermediary metabolism in whole animals using isotopic carbon.

d. Employing or depending on isotopes; obtained by such methods.

1956 A. H. Compton *Atomic Quest* v. 326 As of 1956, three hundred American companies are using isotopic thickness gauges. **1962** *Newnes Conc. Encycl. Nucl. Energy* 365/2 Another example of the application of the isotopic tracer technique..is furnished by the use of radioactive iron to investigate the economy of haemoglobin iron. **1969** Bennison & Wright *Geol. Hist. Brit. Isles* ii. 26 While isotopic dating does provide a system of dating of this kind, giving dates in millions of years, it is too imprecise to be relevant to this problem. **1971** I. G. Gass et al. *Understanding the Earth* ii. 42/1 Laboratory techniques for determining so-called radioisotope, isotopic or radiometric ages of rocks (which used to be erroneously called absolute ages) are extremely complex.

isotopically (aisǫtǫ·pikăli), *adv.* [f. prec.: see -ICALLY.] **a.** As regards isotopes or isotopic constitution.

1933 in O.E.D. Suppl. **1940** *Jrnl. Biol. Chem.* CXXXIII. 737 A deuteropalmitic acid is added to a mixture of isotopically normal fatty acids. **1960** *New Scientist* 14 July 136/3 For an age determination it is not sufficient only to analyse the mineral for uranium, thorium and total lead, but the lead must also be analysed isotopically. **1963** G. Troup *Masers & Lasers* (ed. 2) viii. 136 The electrons were localized on donor (Phosphorus) atoms in isotopically pure silicon of atomic weight 28 which has no nuclear magnetic moment. **1970** *Nature* 23 May 738/2 The solar wind may actually deposit isotopically heavier carbon on the lunar surface. **1970** D. W. Tenquist et al. *University Optics* II. v. 190 Isotopically pure elements must therefore be used.

b. By means of isotopes or isotopic methods.

1954 *New Biol.* XVI. 74 The behaviour of isotopically labelled molecules. **1959** *Times* 15 May 3/3 (Advt.), Vacancies for Chemists to work on the following projects: ..c) the synthesis of isotopically labelled compounds. **1964** G. H. Haggis et al. *Introd. Molecular Biol.* ii. 33 Many steps in the metabolic pathways of cells have been discovered or confirmed using isotopically labelled compounds. **1969** Bennison & Wright *Geol. Hist. Brit. Isles* xii. 272 The rarity of isotopically datable rocks in the Triassic System of Britain and north-western Europe presents a further problem.

isotopy (aisǫ·tǫpi). [f. *ISOTOP(IC *a.* + -Y³.] The fact or condition of being isotopic.

1919 F. Soddy in *Jrnl. Chem. Soc.* CXV. 18 Its [sc. actinium's] definite location in the periodic table, by virtue of its isotopy with mesothorium-2. **1925** J. Joly *Surface-Hist. Earth* ix. 150 Isotopy is not peculiar to the uranium and thorium groups of elements. **1926** R. W. Lawson tr. *Hevesy & Paneth's Man. Radioactivity* xvi. 125 (*heading*) Isotopy and the displacement laws in the light of atomic constitution. **1966** *McGraw-Hill Encycl. Sci. & Technol.* VII. 291/2 The occurrence and degree of isotopy among the 83 elements that are found in nature in significant amounts have been ascertained by means of mass spectrometer studies.

Also **i·sotopism** (in the same sense).

1914 F. Soddy *Chem. Radio-Elements* II. 17 The new data are the isotopism of mesothorium II and actinium, and of radio-actinium and thorium. **1938** *Chem. Abstr.* XXXII. 8925 Evidence is offered that the discovery of the existence of isotopism in nonradioactive elements is another confirmation of O.'s nuclear law of 4.

isotransplant (aisotra·nsplant). *Med.* and *Biol.* [f. Iso- + Transplant *sb.*] A piece of tissue transplanted from one individual to another of the same inbred strain.

1953 G. D. Snell in Homburger & Fishman *Physiopath. Cancer* xiv. 358 There exists in the case of isotransplants a possibility for genetic difference between tumor and host that is not found in autotransplants. **1966** *Nature* 23 Apr. 429/2 There was a significantly higher number of acceptances of tumour isotransplants in mice which were pre-treated with 'Cytoxan'. **1971** *Radiology* XCIX. 187/1 The autotransplants were irradiated when they reached an average of approximately 1 cm diameter, significantly larger than the isotransplants.

Hence **isotranspla·nted** *ppl. a.*

1967 *Jrnl. Nat. Cancer Inst.* XXXIX. 1 (*heading*) Morphologic studies of lymphoid tissues during the growth of an isotransplanted mouse tumor. **1970** *Biol. Abstr.* LI. 3130/2 (*heading*) Role of the immune system in the growth of an isotransplanted urethan-induced lymphosarcoma.

isotransplantation (ai:sotransplantēⁱ·ʃən). *Med.* and *Biol.* [f. Iso- + Transplantation.] † **1.** = *HOMOTRANSPLANTATION. *Obs.*

1909 *Jrnl. Exper. Med.* XI. 194 Of 35 isotransplantations in 17 dogs.., thirty-two parathyroids were either absorbed or necrotic.

2. The operation of transplanting tissue from one individual to another of the same inbred strain.

1962 *Cancer Res.* XXII. 955 A state of relative resistance can be built up in highly inbred skin-compatible C3H mice against the isotransplantation of lymphomas recently induced in the same strain. **1963** *Exper. Cell Res.* XXXII. 618 (*heading*) Resistance against isotransplantation of mouse tumors induced by Rous sarcoma virus. **1966** *Cancer Res.* XXVI. 127/2 Spontaneous loss.., induced loss.., and even gain of antigens..can occur on isotransplantation.

isotron (ai·sotrǫn). [f. Iso- (repr. *isotope*) + *-TRON.] A machine for separating isotopes emitted as ions from an extended source by accelerating them by means of a varying electric field, which causes ions of like mass to bunch together, and applying a transverse radio-frequency field synchronized with the arrival of the bunches, so that the ions are deflected by an amount dependent on their mass.

1945 H. D. Smyth *Gen. Acct. Devel. Atomic Energy Mil. Purposes* xi. 112 The objective..shifted to the effecting of large-scale separation of uranium isotopes by electromagnetic methods... Of the many electromagnetic schemes suggested, three soon were recognized as being the most promising: the 'calutron' mass separator, the magnetron-type separator.., and the 'isotron' method of 'bunching' a beam of ions. *Ibid.* 118 The device which resulted from Wilson's idea was given the deliberately meaningless name 'isotron'. **1946** *Electronic Engin.* XVIII. 153/3 Another method, dispensing altogether with a magnetic field, makes use of the klystron bunching principle and the device has been called the isotron. *Ibid.*, It is understood from publications on the subject that the isotron method has not yet been fully developed.

isotropic, *a.* Add: Also **isotro·pically** *adv.*, equally in all directions.

1885 *Electrician* 24 Jan. 220/1 If μ = constant, all space equally magnetisable isotropically, then B is the same multiple of H everywhere. **1946** *Nature* 5 Oct. 483/1 The crystal structure of silver is cubic and it expands isotropically. **1970** G. K. Woodgate *Elem. Atomic Struct.* vi. 98 The electrons are moving freely, with no forces acting on them, and their translational momenta can be taken to be directed isotropically in momentum space.

isotype (ai·sotaip). [f. Iso- + Gr. τύπ-ος Type *sb.*¹; cf. Gr. ἰσότυπος shaped alike.] † **1.** *Biol.* (In Dict. s.v. Iso-.) *Obs.*

1881 [in Dict. s.v. Iso-]. **1905** *Bull. U.S. Nat. Museum* No. 53. 16 Isotypes (equal or like and form): Forms common to different countries; e.g. the Shrews are isotypes in Europe and North America.

2. *Min.* Any mineral which is isotypic with another; an assemblage of minerals of which all the members are isotypic with one another.

1901 *Mineral. Mag.* XIII. 52 These minerals are placed by F. Rinne in the 'magnesium type' of his isotypes. **1943** *Science* 22 Jan. 99/1 A more complex type formula is necessary to cover the phosphates, arsenates and hydrates which are now recognized as isotypes. **1947** *Mineral. Abstr.* X. 52 With substitution of Si by other elements various isotypes are illustrated.

3. *Bot.* A duplicate of the *HOLOTYPE.

1919 F. W. Pennell in *Torreya* XIX. 13 To meet this need [for a term for a duplicate type] I suggest the term isotype. **1943** *Rhodora* XLV. 485 If a type specimen is lost or unavailable an isotype is in general much more important than a paratype for the interpretation of a name. **1966** *Internat. Code Bot. Nomencl.* ii. 19 An isotype is any duplicate (part of a single gathering made by a collector at one time) of the holotype.

4. Also **Isotype.** [f. *I*nternational *s*ystem *o*f *ty*pographic *p*icture *e*ducation.] The name of an international picture language devised by O. Neurath (1882–1945): used esp. to display statistical information in a convenient visual form.

1936 O. Neurath *Internat. Picture Lang.* 17 We have made one international picture language..into which statements may be put from all the normal languages of the earth. We have given it the name 'isotype'. **1945** G. Williams *Women & Work* 10 Isotype charts..illustrate and elaborate the argument. **1948** *Sunday Times* 15 Feb. 2/6 A record in powerful and sometimes horrifying images of misery, interrupted only by a few isotypes. **1950** Neurath & Lauwerys *Living in Early Times* 1 These Isotype symbols..are always kept as simple and clear as possible. **1966** *Chambers's Encycl.* IX. 792/2 He [sc. O. Neurath] developed the 'isotype', a system of visual linguistics... His isotype films, e.g. *World of Plenty*, attracted wide attention. **1966** M. Pei *Gloss. Ling. Terminology* 135 *Isotype*, a system of writing that uses non-phonetic symbols of universal significance, designed as a medium of education, but also to convey information in ready visual form (the picture of one soldier to indicate an army division).

isotypic (aisoti·pik), *a.* [f. prec. + -IC.] **1.** (In Dict. s.v. Iso-.)

2. *Min.* [ad. G. *isotyp* (F. Rinne 1894, in *Neues Jahrb. f. Mineral., Geol. u. Palaeontol.* I. 55).] Having the same or similar crystal structure (i.e. isostructural) and, with some authors, having in addition analogous chemical formulæ; exhibiting such similarity. Const. *with*.

1929 H. L. Bowman *Miers's Mineral.* (ed. 2) 251 Substances which thus have the same type of composition and resemble one another in their angles and structure have been described by F. Rinne as 'isotypic'. **1942** *Mineral. Abstr.* VIII. 220 It [sc. berlinite] is therefore identical with artificial $AlPO_4$..and isomorphous with $AlAsO_4$.., which are isotypic with quartz. **1943** *Amer. Mineralogist* XXVIII. 598 Na_2BeF_4 was shown by them to be isotypic with γ-Ca_2SiO_4. **1943** *Science* 22 Jan. 98/2 Investigations..have revealed new isomorphic and isotypic relationships among silicates and phosphates. **1959** W. H. Dennen *Princ. Mineral.* iii. 104 Isotypic minerals are those with both isostructural characteristics and chemical similarity. *Ibid.*, The members of an isomorphous series are isotypic, but..isotypic minerals are not necessarily isomorphous. **1959** W. F. de Jong *Gen. Crystallogr.* 182 If the structures of two compounds are entirely or nearly equal, and if the ratios of the dimensions do not differ much, the structures are called isotypic. NaCl and PbS form isotypic structures. **1968** Mason & Berry *Elem. Mineral.* (new ed.) iii. 83 Sometimes the terms isostructural or isotypic are used instead of isomorphous.

isotypy (ai·sotaipi). *Min.* [ad. G. *isotypie* (F. Rinne 1894, in *Neues Jahrb. f. Mineral., Geol. u. Palaeontol.* I. 55): see *ISOTYPE and -Y³.] The character or state of being isotypic (sense *2); isotypic relationship. Also **isoty·pism** (in the same sense).

1938 *Mineral. Abstr.* VII. 92 A list of similar isotypism between other silicates and phosphates (or arsenates) is given. **1942** *Chem. Abstr.* XXXVI. 694 (*heading*) Isotypy between phosphates of general composition $MLiPO_4$ and silicates of the olivine-monticellite series. **1943** *Science* 22 Jan. 98/2 The isotypism of $AlPO_4$ and SiO_2. **1943** *Amer. Mineralogist* XXVIII. 598 O'Daniel and Tscheischwili consider the isotypy of Ca_2SiO_4 and Na_2BeF_4 as limited to the low-temperature modifications. **1959** W. F. de Jong *Gen. Crystallogr.* 182 (*heading*) Isotypism. **1967** Isotypy [see *FLINKITE].

isovaleraldehyde, isovalerate, isovalerianic, isovaleric, isoxazole: see *ISO- b. **isovol**: see *ISO-.

isozyme (ai·sozaim). *Biochem.* [f. Iso- + *(EN)ZYME.] = *ISOENZYME.

1959 Markert & Møller in *Proc. Nat. Acad. Sci.* XLV. 753 We propose, therefore, to use the term *isozyme* to describe the different molecular forms in which proteins may exist with the same enzymatic activity. **1968** T. Wieland in Thoai & Roche *Homologous Enzymes & Biochem. Evolution* iii. 3 The second group of multiple forms consists of the so called 'isozymes'. They may be defined as the multiple molecular forms of an enzyme existing within a single organism. *Ibid.* 9 Isozymes are multiple forms of enzymes, which occur in one cell or—less strictly defined—in one organ of an organism. **1968** *New Scientist* 21 Mar. 649/2 LDH exists in five very slightly different forms... Each version is known as an isozyme of LDH, and each will catalyze the final glycolysis step.

Hence **isozy·mic** *a.*

1959 Markert & Møller in *Proc. Nat. Acad. Sci.* XLV. 756 The relative enzymatic activity disposed in each isozymic band. **1973** *Nature* 3 Aug. 261/2 The genes for isozymic forms of various enzymes are often not linked to each other.

ispaghul (i·spagūl). Also **ishabgul, ispagool, ispughool, isubgol.** [Hind., a. Pers. *asp* horse + *gol* ear, in allusion to the shape of the leaves or seeds.] A plantain, esp. *Plantago ovata*, native to India and Persia, the dried seeds of which are used medicinally.

1810 J. Fleming in *Asiatick Researches* XI. 174 *Plantago Ispaghul* (Roxb. MS.) Ispaghul, H[industani]... Ispughool [Sanskrit]. **1820** W. Roxburgh et al. *Flora Indica* I. 404 *Ispagool*, the Hindee and Persian name, and that by which it is most generally known in Bengal and on the coast of Coromandel. **1880** Bentley & Trimen *Medicinal Plants* III. § 211 Ispaghūl seeds have long been highly valued in India and other parts of the East for their cooling and demulcent properties. **1889** G. S. Boulger *Uses of Plants* II. 111 The mucilaginous seeds of *Plantago decumbens*, Forsk., Ispaghūl, or Spogel, are used in India as a demulcent drink, especially in dysentery. **1931** M. Grieve *Mod. Herbal* II. 643/1 Plantain, Ispaghul. *Ibid.* 643/2 The seeds of the Indian species, *Plantago Amplexicaulis*, are sold in the bazaars as Ispaghula. **1953** B. Mukerji *Indian Pharmaceutical Codex* I. 124 Commercial samples of ishabgul consist of the seeds of *P. arenaria* Waldst., *P. lanceolata* Linn., *P. major* Linn. besides *P. ovata* Forsk. and *P. psyllium* Linn. **1969** *Wealth of India* (Council Sci. & Industr. Res. India) VIII. 151/1 The efficacy of isubgol is due entirely to the large quantity of mucilage present in the husk.

Ispahan (i·spăhān). Also **Isfahan.** The name of a province and town in West Central Persia, used *attrib.* and *ellipt.* to designate a type of hand-woven rug, the most distinguished of which were produced there in the 16th century.

1931 A. U. Dilley *Oriental Rugs & Carpets* ii. 32 From the succeeding court came, during the sixteenth century, the 'Ispahan' rugs so greatly esteemed by collectors. *Ibid.* iii. 61 A fourth group of superior rugs, distinguished by pattern of palmette and now called Ispahan. **1960** H. Hayward *Antique Coll.* 149/2 *Ispahan carpets*, Persian carpets or rugs, often employing medallion or vase designs..in a wide range of colours, the composition being very well balanced. **1969** 'W. Haggard' *Doubtful Disciple*

xv. 166 The dealers call them Isfahans and..know they're Heratis. **1969** B. Weil *Dossier IX* v. 38 A carpet that..Asher could see was an Isfahan. **1970** V. Canning *Great Affair* iv. 50 He took the heavy Passavant carpet from the floor..added a smaller Ispahan rug.

‖ **ispravnik** (ispra·vnik). *Hist.* Pl. **ispravniki, ispravniks.** [Russ., lit. 'executor'.] A chief of police in a rural district in Czarist Russia.

1886 *Encycl. Brit.* XXI. 70/2 The organs of the central government..are..the stanovoys and ispravniks (chiefs of the police) in the districts. **1898** J. Y. Simpson *Side-Lights on Siberia* 90 The ispravnik of the district happened to be in the village. **1906** *Daily Chron.* 24 Aug. 5/6 The police informed the ispravnik (chief of rural police). **1911** *Encycl. Brit.* XXIII. 876/1 As organs of the central government there are further, the *ispravniki*, chiefs of police in the districts into which the governments are divided. **1952** H. Altschuler tr. *Gorky's Artamonovs* i. 156 The ispravnik shook a threatening finger, and shouted: 'Hush, fool!' **1967** H. Seton-Watson *Russ. Empire* I. i. 20 The chief executive officer at the uezd level was the *ispravnik*, who was elected by the nobility, and presided over the local court which was an administrative rather than a judicial body.

Israel. Add: **3.** An independent Jewish State established in 1948 in the country formerly called Palestine. Also *attrib.*

1948 *Manch. Guardian* 15 May 5/1 The Jews yesterday proclaimed in Tel Aviv the new State of Israel. **1957** L. F. R. Williams *State of Israel* 13 The Israel Government has been kindness personified in helping me with my researches. *Ibid.* 17 The modern state of Israel occupies part of territory which in ancient times formed a link between the civilizations of Egypt, Asia Minor and Mesopotamia. **1958** *Economist* 26 July 300/2 The western armies cannot stay in the Middle East indefinitely, while Israel must live in the region forever. **1972** *Encycl. Judaica* I. 12 According to official government usage, Israel (and not Israeli) is the adjective relating to Israel: Israeli is a citizen (or permanent resident) of the State of Israel. **1972** *Times* 24 Mar. 10/4 Israel has decided to recall its military experts from Uganda. **1974** *Times* 17 Apr. 14/7 Israel newspaper correspondents were given preferential treatment. *Ibid.* 14/8 The ban remained for Israel-based reporters... The Israel government has followed a curious policy.

Hence **Isra·eli** *sb.*, (*a*) a native or inhabitant of the State of Israel; (*b*) the language spoken by the inhabitants of the State of Israel (more usually called *Modern Hebrew*). Also **Israeli** *a.*, relating to the State of Israel, its inhabitants, etc.

1948 *Daily Tel.* 23 June 1/1 A coastal vessel of the Israeli Navy stood by. **1948** *N.Y. Times Mag.* 5 Dec. 4/3 Shortly after the proclamation of the new State of Israel.. Moshe Shertok announced that its citizens would be called Israelis. **1949** Koestler *Promise & Fulfilment* II. ii. 222 His English is good with a specific Israeli-*sabra* accent. **1949** H. Nicolson *Diary* 25 July (1968) 173 George Weidenfeld..has been asked by [Chaim] Weizmann to become his personal assistant. The problem is whether he should become an Israeli and get a definite job in the Israeli civil service. **1958** *Economist* 26 July 300/1 The Israeli government has asked Britain to order another course for the airflight to Jordan. **1964** R. MacGregor-Hastie *Pope Paul VI* x. 215 The Pope greeted Dr. Shazar in Israeli—'Shalom, Shalom!' **1967** *Listener* 8 June 741/1 The Israelis, with their small country and just over two million population, can never conquer the Arab States. **1972** [see 3 above].

iss (is), *adv.* *Dial.* var. Yes *adv.* (*sb.*²); also used to represent an inaccurate pronunciation of 'yes' by a speaker whose native language is not English.

1795 'P. Pindar' *Royal Visit to Exeter* II. 17 Iss, iss, he'll do the feat. **1829** G. Griffin *Collegians* (ed. 2) I. xii. 259 'Erra, no!' 'Iss, dear knows.' **1853** Mrs. Gaskell *Ruth* I. vi. 154 One servant-girl..was wearied out of what little English she had knowledge of..and could only answer, 'Iss, indeed, ma'am,' to any question. **1890** A. Gissing *Village Hampden* I. i. 19 Iss, Miss—but 'ere her be. *a* **1966** C. S. Forester *Hornblower & Crisis* (1967) vii. 71 'You boy!..D'you want to earn a shilling?' 'Iss, that do I.'

issuance. (Earlier and later examples.)

1863 *National Almanac & Ann. Rec.* 545/2 A proclamation of neutrality..was issued by Victoria, Queen of England... A brigade of British Volunteers..enlisted in the North..disbanded after its issuance. **1905** *Westm. Gaz.* 11 Jan. 1/3 The new canon insists on at least a year's delay after the issuance of a decree of divorce before any marriage can take place. **1931** G. T. Clark *Leland Stanford* vii. 214 The Supreme Court rendered another decision..to compel the issuance of the bonds voted by the people of San Francisco. **1967** *Boston Sunday Herald* VI. 5/1 (Advt.), You save 25% on the issuance charge. **1972** *Science* 12 May 640/2 Overzealous concern with style and format of writing imposed by the issuance, first in 1909, of the Survey's style manual. **1972** *N.Y. Law Jrnl.* 22 Aug. 12/1 A brother..filed objections and cross-moved for issuance of letters to him.

issue, *sb.* Add: **13*.** *the (whole) issue*: everything, the lot. *colloq.*

1919 W. H. Downing *Digger Dial.* 29 *Issue*, 1. A portion; 2. 'to get one's issue'—to be killed; 3. 'to get the whole issue of a shell'—to be struck bodily by a shell. **1925** Fraser & Gibbons *Soldier & Sailor Words* 129 Not a soul got back; the whole issue were done in last night. **1930** Brophy & Partridge *Songs & Slang 1914–18* 131 *The issue* was also used for 'the whole lot', e.g. 'There's no rum tonight. The sergeant's snaffled the issue.' **1941** Baker *Dict. Austral. Slang* 38 *Issue, all*, everything, the lot. **1960** K. Amis *Take Girl like You* i. 10 He put a metal flint-scratcher into the mouth of the geyser and went clicking away... 'Now the water... Funny. The whole issue should light up now.' **1966** 'L. Lane' *ABZ of Scouse* 53 Yer've buggered up ther 'ole issue.

14. b. (Of books and periodicals: examples.)

1833 *Penny Mag.* Monthly Suppl. Nov.–Dec. 511/1, 12,000 copies of each number (the quantity required for the first issue of the volume [sc. of the *Penny Magazine*]) will have been delivered to two book-binders. **1895** B. Wood in *Brontë Soc. Trans.* I. i. 4 Many of the editions of the Brontë Works are merely reprints of previous issues. **1910** C. Shorter in E. Brontë *Compl. Poems* p. vi, It is a curious irony of circumstance that this little volume [sc. the Brontë *Poems*]..now sells..for more money than the whole issue cost Charlotte Brontë and her sisters when they had it published at their own expense. **1929** J. L. Young *Bks. from MS. to Bookseller* viii. 93 Quite a considerable proportion of the issue of every book is given away for one purpose or another. **1954** *Willing's Press Guide* p. vii, With this issue..we announce important policy developments... As from this edition special supplements will be issued. **1962** M. Toase *Guide Current Brit. Periodicals* p. viii, Information given in the entries... Date of first issue. **1974** *Leisure Painter & Craftsman* Aug. (verso front cover), We have invited Mr. Harry Richardson to outline, in this issue, his..approach to the problem.

(ii) *Bibliogr.* In bibliogr. classification and description, a subdivision of an edition (or of an impression [see *IMPRESSION *sb.* 3 d] of an edition), denoting a distinct form, planned and put on sale by the publishers, of the edition (or impression) sheets; a new issue is normally indicated by the provision of a new title-leaf, with or without other changes.

In some cases the precise application of *issue* in this sense remains subject to discussion.

1928 M. Sadleir *Trollope: a Bibliogr.* p. xiii, It would be well to distinguish clearly between an 'issue' and a 'binding-up'. An 'issue' represents an order from the publishers for a definite effort in publication... It is a *piece of publishing*, and reflects some definite intention in the publishers' mind. A 'binding-up', on the other hand, is a mere replenishment of the stock in a publisher's warehouse. **1949** F. Bowers *Princ. Bibliogr. Descr.* ii. 41 *Issue* is included within *edition* and itself includes only *state*. *Ibid.* 78 Removed from consideration [sc. of issue] are all alterations made during continuous printing of the original sheets, as well as alterations made after public sale had begun which are largely for the purpose of constructing an 'ideal copy'. We must take it as a fundamental assumption that, except in the most uncommon circumstances, sheets will not be re-issued without a change of title-page... It is impossible to set up any standards for issue [sc. for hand-printed books] which have any likelihood of uniform and logical application unless the title-page is taken as the prime evidence. *Ibid.* xi. 403 Sadleir's correct refusal to admit as issues those bindings-up exhibiting small variants in the binding not the result of a publisher's order helps sweep away the majority of the ridiculously conceived modern 'issues' but still leaves various ambiguities. *Ibid.* 419 Alterations to the text [sc. of 19th- and 20th-century books] are a cause of re-issue even if the title-page is unaffected, providing they go beyond the standards of *ideal copy* and were not made during the course of printing the impression affected... At the publisher's request, the text of D. H. Lawrence's.. *White Peacock* (1911) was altered in certain respects by the substitution of two cancellans leaves..causing re-issue... In *The White Peacock*..the second issue is found in a second state. **1952** J. Carter *ABC for Bk.-Collectors* 108 Since differences of issue are bibliographically tidier and more straightforward than differences of state, and since the term falls much more pleasingly on the priority-conscious ear, a good many undeterminable cases have been, and no doubt will continue to be, given the benefit of the doubt. **1960** G. A. Glaister *Gloss. Bk.* 200/1 Before the term 'issue' can first be applied, copies of the work without any of the changes now involved must have already been published. Bibliographers then speak of a 'first issue' and a 'second issue' of the 'first edition'. The term is less precisely used in the book trade. **1969** E. W. Padwick *Bibliogr. Method* xvi. 206 Both *issue* and *state* are included within *impression*... Basically, *issue* should be regarded as an act of intent by the publisher to effect some change in the yet unbound copies of an impression after the publication of some copies has taken place. **1972** P. Gaskell *New Introd. Bibliogr.* 317 The term 'edition' has always been used in the trade for 'impression' or 'issue' as well as for edition in the bibliographical sense; a book that is advertised as a 'new edition'..may be.. simply a reissue of the original sheets with a new title-page.

c. An item or amount of something given out or distributed. orig. *U.S.*

1861 *Regulations Army U.S.* 283 His descriptive list.. on which the surgeon shall enter all payments, stoppages, and issues of clothing to him in hospital. **1881** *Rep. Indian Affairs* (U.S.) 10 They agreed to go as soon as the issue of beef..had been made. **1882** *Cassell's Family Mag.* June 399/1 He is also responsible that the proper issues of bread and meat are made to the men. **1899** T. W. Hall *Tales* 109 Then our..aching bodies are loaded down with a further issue of ammunition. **1911** H. Quick *Yellowstone Nights* xii. 305 'You represent the Elkins interests in the matter of supplying for the issue do you not?' says he. *Ibid.* 321 She hove in sight of the issue. **1919** *Athenæum* 8 Aug. 727/2 Anything supplied by the Army was an 'issue'. **1940** 'Gun Buster' *Return via Dunkirk* I. ii. 20 His beard..had all the luxuriance of a ten-days unhampered growth... Some of our gunners began to give themselves a morning shave... His face..said..'Do as you please. I'm sticking to mine. I'm not likely to get a better issue.' **1942** E. E. Dale *Cow Country* 163 They [sc. Amerindians] could not subsist upon the present issue of eighty thousand pounds a week.

15. (sense 11) *issue roll*; (sense *14 c) *issue boot, cigarette, day, house, mess tin, shoe.*

1927 *Daily Express* 4 Oct. 3 Men..running up and down perpendicular 4-inch steel stairs in issue boots without arriving in hospital. **1925** Fraser & Gibbons *Soldier & Sailor Words* 129 An 'Issue' cigarette..was a ration cigarette, in contradistinction to one bought at the Canteen. **1874** R. Glisan *Jrnl. Army Life* xxxi. 447 They gave the white physicians much annoyance by coming for medicine only on issue or ration day. **1878** *Rep. Indian Affairs* (U.S.) 39 Other mechanics are putting up new store and issue-houses. **1911** H. Quick *Yellowstone Nights* xii. 316 The way we..hit the trail f'r the Issue House was a high-class piece o' teamin'. **1944** *Living off Land* vii. 145 The Australian issue mess tin is light, durable and well-suited to bush cooking. **1886** *Encycl. Brit.* XX. 312/1 The judgment rolls pass through three stages—first, they are plea rolls; then, when the parties join issue, issue rolls; and lastly,..judgment rolls. **1946** *R.A.F. Jrnl.* May 155, I put on a pair of R.A.F. issue shoes which I had grabbed.

issue, *v.* Add: **9.** (Of books and periodicals: earlier and later examples.)

1833 *Penny Mag.* Monthly Suppl. Oct.–Nov. 472/1 Twenty million 'Penny Magazines' have been issued from the commencement. **1876** 'Mark Twain' *Lett. to Publishers* (1967) 95 It is going to rush you too tight to do your canvassing and issue 'Tom' [sc. *Tom Sawyer*] in the middle of April isn't it? **1954** [see *ISSUE *sb.* 14 b]. **1966** (*title*) The English catalogue of books..1963–1965, giving..the size, price, date of publication, and publisher of books issued in the United Kingdom.

11. To give things out to (a person); to supply (a person) *with*. (Cf. *ISSUE *sb.* 14 c.)

1925 T. G. Bruce in E. F. Norton *Fight for Everest, 1924* 344 Every man in the Expedition should be issued with one blanket either in Kalimpong or Phari. **1927** *Daily Tel.* 27 Sept. 8/6 Infantry battalions were issued with two weighted dummies apiece. **1928** *Sunday Express* 18 Mar. 3/2 The extraordinary experience of being twice in a year issued by the same bank with a faulty £1 Treasury note. **1929** F. A. Pottle *Stretchers* (1930) 37 Before we were issued our heavy trench shoes. **1953** *Listener* 6 Aug. 224/2 Then, the idiom 'issued with': 'He was issued with' a rifle, and a packet of cigarettes, or what not. I suppose this horror has come to stay. It is undeniably convenient. **1961** *Oxford Times* 15 Dec. 14/9 He stated that he had not been issued with a licence previously when, in fact, one had been issued by Surrey County Council earlier that year.

issuing, *vbl. sb.* Add: **3. issuing house**: see quot. 1965.

1929 *Issuing House Year Bk.* 3 The requirements of an Issuing House in the production of a Prospectus are more exacting than for almost any other work the printer is called upon to undertake. **1932** *Daily Tel.* 8 Oct. 2/3 Regarding the continued default on City of Riga 4½ p.c. Bonds,..what is inexplicable is that the bondholders have shown such apathy and negligence in this matter respecting the negotiations between the issuing house and the City of Riga. **1965** Perry & Ryder *Thomson's Dict. Banking* (ed. 11) 317/2 The main function of an issuing house is to obtain from the public capital for the expansion of existing companies, for new public companies, or for private companies being converted into public concerns.

‖ **istana** (istā·na). Also **astana, astanah.** [Mal. *istana*, ad. Skr. *ā-sthāna* place, site, assembly.] In Malay kingdoms, a ruler's palace.

1839 T. J. Newbold *Pol. & Statistical Acct. Straits of Malacca* II. viii. 90 He married Tuanku Itam..and proceeded to his astánah in Srimenanti. **1907** F. A. Swettenham *Brit. Malaya* xiii. 315 The Sultan's *astána* or palace ..surrounds on three sides a court of sand. **1927** R. J. H. Sidney *In Brit. Malaya Today* xxiii. 271 His [sc. the Sultan's] *Istana* at Klang is filled with work done by his own hand. **1965** *Festival Malaysia 1965: Calendar of Events* 6/1 The Investiture Ceremony held in the afternoon at the Istana in Sri-Menanti is followed by a Garden Party in the evening. **1972** *Sunday Times* (Kuala Lumpur) 30 Apr. 5/2 The Yang di-Pertuan Agong today presented awards to 76 people at an investiture ceremony at the Istana Negara. **1972** *Straits Times* (Malaysia ed.) 24 Nov. 20/6 This Istana in the village of Bandar was the residence of Sultan Abdul Samad in the middle of the nineteenth century.

Istrian (i·striăn), *a.* and *sb.* [f. *Istria*, a peninsula near the head of the Adriatic sea; see -AN.] **A.** *adj.* **1.** Of or belonging to Istria.

1607 Topsell *Four-f. Beasts* 288 The Istrian Horsses are of good able feete. **1881** E. A. Freeman *Sk. Subject Lands Venice* 98 The Istrian shore has lost its beauty. **1920** *Glasgow Herald* 16 Apr. 8 The 'conversation' broke down solely on the question of the Istrian frontier. **1967** C. Seton-Watson *Italy from Liberalism to Fascism* xi. 426 A new and energetic Minister of War, General Vittorio Zupelli, of Istrian origin, took the job of military preparation systematically in hand.

2. Special Comb.: **Istrian marble, stone,** a fine limestone resembling marble; also *ellipt.*; **Istrian pointer,** a hunting dog (in Yugoslavia).

1611 Coryat *Crudities* 163 Stately pillers made partly of white stone, and partly of Istrian marble. **1947** J. C. Rich *Materials & Methods Sculpture* viii. 228 Istrian marble, which is found on the Istrian peninsula and on some of the Dalmatian islands, is a large-grained, buff marble that compares favorably with the Greek marbles. **1971** *Country Life* 18 Feb. 16F Suppl., A most important 17th century Italian Istrian marble wall fountain. **1948** A. Lokar in B. Vesey-Fitzgerald *Bk. Dog* II. 536 The Istrski Brak (Istrian Pointer) has a wide range in Yugoslavia. **1888** *Encycl. Brit.* XXIV. 149/2 This Istrian stone has for most architectural purposes all the beauty of the finest white marble. **1962** *Listener* 27 Dec. 1105/3 The sculptor is not above counterpointing the folds of a drapery or the positioning of a hand with the natural pressure lines in a fine piece of green Istrian. **1972** *Country Life* 6 Jan. 16/1 This colonnade..culminates in an early baroque Italian fountain of Istrian stone.

B. *sb.* A native or inhabitant of Istria.

1880 *Encycl. Brit.* XIII. 433/2 The Istrians..were only subdued by the Romans in 177 B.C. after two wars. **1974** *She* Jan. 28/2 The Istrians are hospitable and friendly.

it, *pron.* Add: **1. d.** Sexual intercourse. Now *slang* or *colloq.* Cf. *DO *v.* 16 b.

1611 COTGRAVE *Dict.*, *Fretiller*..to..lust to be at it. **1896** FARMER *Vocab. Amatoria* 118/1 *Faire*, to copulate; 'to do it'. **1922** JOYCE *Ulysses* 747 Gardner said no man could look at my mouth and teeth smiling like that and not think of it. **1923** T. WOLFE *Lett.* (1956) 45, I have been reading the *Amores* of Ovid this morning. It is beautiful Latin and beautiful poetry—although it is altogether concerned with two topics: How am I going to get it and How fine it was when you let me have it. **1941** H. G. WELLS *You can't be too Careful* III. xi. 161 Edward Albert knew.. of venereal disease, clumsy 'precautions' and the repulsive aspects of the overwhelming desire for 'It'. **1949** N. MITFORD *Love in Cold Climate* I. xi. 119, I was lugged off to their secret meeting-place..to be asked what IT was like. **1972** F. WARNER *Maquettes* 16 He doesn't even know I'm overdue. And he hasn't had it for a week.

e. In emphatic predicative use: the actual or very thing required or expected; that beyond which one cannot go; the *ne plus ultra*; the acme. (In 20th-c. use from U.S.)

a **1834** C. LAMB *Dramatic Ess.* (1891) 52 Lovegrove.. revived the character..and made it sufficiently grotesque; but Dodd was *it*, as it came out of Nature's hands. **1896** ADE *Artie* i. 4, I didn't do a thing but push my face in there about eight o'clock last night, and I was 'it' from the start. **1900** *Dialect Notes* II. I. 42 Did he know his Greek? I should say so. He was *it*. **1904** F. LYNDE *Grafters* xxx. 397 Mrs. Hepzibah..thinks you are It. **1906** *Daily Chron.* 5 Mar. 6/6 There is in America a curious use of the word 'it' conveyed by emphasis. Pre-eminently Roosevelt is 'it'. Next after Roosevelt an American would say 'Shaw is it'. **1915** 'I. HAY' *First Hundred Thousand* xx. 307 You can't go anywhere in London without running up against him. He is It. **1916** 'TAFFRAIL' *Pincher Martin* vii. 111 On board his ship he had a very poor time; but ashore he was absolutely *it*, so far as the ladies were concerned. **1923** D. H. LAWRENCE *Birds, Beasts & Flowers* 206 Red Men still stick themselves over with bits of his fluff, And feel absolutely *it*. **1930** H. M. SMITH *Inspector Frost in City* III. v. 103, I have some new plus-fours which are 'it'. **1963** P. WILLMOTT *Evolution of Community* i. 8 People were making themselves out to be something they weren't... They thought they were it.

f. In children's games, the player who has the task of catching or touching the others. Also *transf.* and *fig.*

1842 R. CHAMBERS *Pop. Rhymes Scotl.* 62/2 The tig usually catches and touches some one upon the crown before all are *in*—otherwise he has to be *it* for another game. **1888** [see *COUNT *v.* 15 b]. **1923** KIPLING *Land & Sea Tales* 279 As the sides are chosen and all submit To the chance of the lot that shall make them 'It'. **1949** J. B. PRIESTLEY *Delight* 137 The boy who was 'it' retrieved the can and replaced it in the circle. **1950** C. S. LEWIS *Lion, Witch & Wardrobe* iii. 30 They decided to play hide-and-seek. Susan was 'It' and..the others scattered to hide. **1969** I. & P. OPIE *Children's Games* i. 18 They do whatever they can to avoid being..the one who, as they express it, is..'it'. **1970** G. JACKSON *Let.* 23 Mar. in *Soledad Brother* (1971) 188 It's us against them, hide and seek. They're always it and getting caught means getting dusted. **1974** S. GULLIVER *Vulcan Bulletins* 111 'I'm not helping to get him knocked off to suit..the CIA.' 'That's too bad, Lee,' said Selby quietly, 'because you're it.'

g. 'Sex appeal.'

1904 KIPLING *Traffics & Discoveries* 352 'Tisn't beauty, so to speak, nor good talk necessarily. It's just It. Some women'll stay in a man's memory if they once walk down a street. **1927** E. GLYN *'It'* i. 10 He had that nameless charm, with a strong magnetism which can only be called 'It'. **1930** G. B. STERN *Mosaic* III. i. 205 The Viennese composer made his first awed acquaintance with the words pep, kick, body-urge, sex-appeal, a hundred-per-cent, stuff, spin it along, put it over, and It. **1932** *Bystander* 23 Mar. 546 A film star who has proved to producers and film public alike that she is blessed with that undefinable quality called 'It'. **1972** L. P. BACHMANN *Ultimate Act* i. 16 She really had 'It', as it was called.

2. e. *this is it* colloq. phr.: used when something previously spoken about or foreboded has come to pass or is about to happen.

[**1908** G. MURRAY tr. *Aristophanes' Frogs* I. iii. 34 That's it, sir. These are the Initiated Rejoicing somewhere, just as he told us.] **1942** *Newsweek* 27 July 23/3 Finucane tried to settle the rocking plane onto the water. It hit the waves, then sank like a rock. Just before, Paddy spoke into the two-way radio: 'This is it, chaps.' **1959** *Sunday Times* 5 Apr. 15/7 He heard the sound of countless aircraft overhead. This is it, he thought.

f. *that's it* (and similar colloq. phrases): there is no more to it than that.

1966 P. WILLMOTT *Adolescent Boys E. London* v. 88 You got no encouragement. They'd look at the report at the end of term and that was about it. *Ibid.* 92 We just sat about..doing half-witted things. You only had to find a weak teacher and that was it. **1968** *Listener* 31 Oct. 574/3 Really I think the Brummie likes to stay at home. And work. And shop in Birmingham. Holiday in Majorca —and that's it. **1968** *New Yorker* 2 Nov. 163 They are excellent musicians (except for one of the girls, whose function is obscure; she shakes a tambourine now and then, but that's about it). **1972** *Observer* (Colour Suppl.) 13 Feb. 18/1 Adoption agencies are wary..of any people who want to adopt for any reason other than that they love children, that they have homes and that there are children who need homes. To put it briefly, parents are parents and that's it.

3. f. (Later examples.)

1840 K. H. DIGBY *Mores Catholici* x. vii. 171 In Saxon histories... Thus it says. **1894** G. F. X. GRIFFITH tr. *C. Fouard's St. Paul* xv. 352 From the sequel, as it reads in the Acts, it would seem [etc.]. **1902** H. K. MANN *Lives of Popes* I. II. 234 'In mense Junius Indictione ii,' or x., as by mistake it reads in the *Chronicle*. **1932** CHESTERTON *Chaucer* iii. 108 Chaucer was a man for whom the world teemed with quiet fun, as it says in the comic opera. **1955** tr. *William of Rubruck's Journey* in C. Dawson *Mongol Mission* xxxvii. 212 In Isaias it says that they fled into the land of Ararat.

g. (Later examples of *were it not* or *had it not been for, if it were not*, etc., *for*.)

1710 SWIFT *Jrnl. to Stella* 30 Sept. (1948) I. 47 They may talk of the *you know what*; but, gad, if it had not been for that I should never have been able to get the access I have had. *a* **1732** J. GAY *Fables* (1738) II. xiii. 118 Were it not for this cursed show'r, The park had whil'd away an hour. **1736** J. BUTLER *Analogy of Relig.* II. v. 200 Assistance which they would have had no Occasion for, had it not been for their Misconduct. **1780** *Mirror* No. 102 The misapplication of the term is so completely ridiculous, as to be beneath contempt, were it not for the mischief that I am convinced has been occasioned by it. **1864** G. MEREDITH *Emilia in England* II. v. 69, I feel better already, if it weren't for my legs. **1974** *Times Lit. Suppl.* 5 Apr. 375/2 Disaster that would have been total had it not been for the conventional crime-writer's beginning and end.

It, it (it), *sb.* See *gin and it* s.v. *GIN *sb.*² 2 b.

ita (i·tă). Substitute for entry ITA-PALM: Also **eta, ite.** [f. Arawak *ite.*] In full, *ita palm.* The tropical South American fan-palm, *Mauritia flexuosa*, or the drink made from its fermented sap; cf. *ETA².

1845 *Encycl. Metrop.* XX. 6/2 The Eta, a smaller kind of this [cabbage] palm, furnishes nuts. **1860** MAYNE REID *Odd People* 360 The *ità* is a true palm-tree. **1866** [in Dict. s.v. ITA-PALM]. **1904** W. H. HUDSON *Green Mansions* xxii. 292 Even the Ita palm and mountain glory..had lost all grace and beauty. **1922** W. E. ROTH tr. *Schomburgk's Trav. Brit. Guiana* I. 150 There was..a considerable supply of a rarer drink, the Ite, manufactured by the Indians from the juice of *Mauritia flexuosa*. **1957** M. SWAN *Brit. Guiana* x. 152 An outcrop of sandy soil has produced a cluster of ite palms, those 'trees of life' found all over the coastal areas.

‖ **itai-itai** (i·tɔi,i·tɔi). [Jap., lit. 'ouch-ouch'.] A disease, first reported in Japan in 1955, caused by the ingestion of cadmium and characterized by severe pain particularly in the back.

1969 *Keio Jrnl. Med.* (Tokyo) XVIII. 181 (*title*) Causation of ouch-ouch disease (Itai-Itai byō). **1970** *Biol. Abstr.* LI. 4256/1 Rice straw samples used were 2 kinds of stubble grown in the 'Itai-itai' disease epidemic district. **1973** *Biol. Conservation* V. 143 *Itai-itai* originated in the prefectures of Toyama and Gumma, in north-western Japan. The cause of the affliction, however, is world-wide: rivers poisoned by effluence from smelters.

Italian, *a.* and *sb.* Add: **A.** *adj.* **4.** *Italian clover, paper, vermouth*; **Italian cypress,** *Cupressus sempervirens* var. *stricta*; **Italian earth,** the colour sienna; **Italian garden,** a formal garden, characterized by clipped trees, box-edged beds of flowers, paved paths, statues, fountains, etc.; often arranged in terraces linked by steps and balustrades; **Italian paste,** the paste from which macaroni and vermicelli are made; **Italian pink** = **Dutch pink*; **Italian quilting** (see quots.); **Italian stitch,** a form of CROSS-STITCH *sb.*

1840 C. DEWEY in Dewey & Emerson *Rep. Herbaceous Plants & Quadrupeds Mass.* 66 *Trifolium incarnatum*, Italian clover. **1908** *Animal Managem.* 109 'Valerian', 'Italian' or 'Crimson clover', commonly called 'Trifolium'. **1838** J. C. LOUDON *Arboretum* IV. 2464 The common, or evergreen, Cypress... Synonymes..the Italian Cypress. **1923** L. H. BAILEY *Cultivated Evergreens* II. 208 Italian Cypress... Much planted since ancient times in southern Europe particularly in its columnar form. *Ibid.*, Columnar Italian C[ypress]... With erect branches, forming a narrow, columnar head. The classical cypress of the Greek and Roman writers. **1969** T. H. EVERETT *Living Trees of World* 35/1 The Italian cypress is really a horticultural form of venerable ancestry, its exact origin unknown. **1854** F. W. FAIRHOLT *Dict. Terms Art* 260/1 *Italian earth*, a pigment known as burnt Italian earth. **1897** *Sears, Roebuck Catal.* 361/2 Pastel Crayons..Burnt Sienna..Italian Earth..Purple Brown. **1969** R. MAYER *Dict. Art Terms & Techniques* 200 *Italian earth*, an old name for sienna. [**1822** J. C. LOUDON *Encycl. Gardening* I. 16 He [*sc.* Volkman] considers the Italian gardens as inferior to those of France in point of superb alleys, lofty clipt hedges, and cabinets of verdure.] **1883** W. ROBINSON *Eng. Flower Garden* p. vi/2 It has been affirmed that none but an Italian garden would have suited South Kensington. **1928** L. ARCHER-HIND tr. *Gothein's Hist. Garden Art* II. xvi. 329 We feel the resemblance to the parterre of the Doria Pamfili when we walk through an 'Italian garden' at an English country seat. **1942** A. E. W. MASON *Musk & Amber* i. 10 The Italian garden.., an oblong of grass paths and glowing flower beds, of box trees and hedges, of stone seats..and..a ridiculous charming little temple with open pillars. **1961** G. MASSON *Italian Gardens* 274 Within the space of two hundred and fifty years, Italian gardens had been introduced into France, developed and expanded until they represented a national style that became the model for Europe, and then via Spain returned to their point of departure, the Neapolitan Realm, as a foreign innovation. **1924** A. HUXLEY *Let.* 25 Feb. (1969) 228 The best form, I think, would be something small, cheap and pretty. Covers of Italian paper or something of the kind. **1930** *Times Lit. Suppl.* 6 Feb. 108/4 A large variety of Italian papers. **1845** E. ACTON *Mod. Cookery* i. 4 All the ingredients used for soups should be fresh,..particularly Italian pastes of every kind (maccaroni, vermicelli, &c.). **1907** *Army & Navy Stores Catal.* 1246/2 Italian paste, for soups. **1957** *Encycl. Brit.* XIV. 544/2 *Macaroni*... The same substance in different forms is also known as *vermicelli*, *pasta* or Italian pastes, *spaghetti*, *taglioni*, *fanti*, etc. **1835** Italian pink [see **English pink*]. **1934** H. HILER *Notes Technique Painting* ii. 111 *Italian pink*, *quercitron lake*, etc., organic pigments prepared from Turkish or Avignon berries, quercitron bark, etc. **1971** *Country Life* 10 June 1428/3 This leads into the north-facing hall, which has been painted an Italian pink as a background to full-length portraits. **1937** E. HAKE *Eng. Quilting* iii. 16 Italian quilting..was as prevalent in England as in any other European country during the seventeenth and eighteenth centuries. **1955** *Oxf. Jun. Encycl.* XI. 323/1 Italian quilting consists of two layers of cloth sewn together in a design built up entirely of parallel lines. A padding of soft wool or piping cord is then threaded between the narrow channels. **1967** E. SHORT *Embroidery & Fabric Collage* ii. 47 In Italian quilting the design is worked entirely in parallel lines of running or back stitch, which are then padded by inserting a thick wool from the back. **1882** Italian stitch [see *HOLBEIN]. **1913** M. K. GIFFORD *Needlework* xvii. 262 *Italian stitch* can be worked either open or close. The latter makes a very solid filling. **1957** M. B. PICKEN *Fashion Dict.* 185/2 *Italian-stitch*, running stitch done twice on the same line. **1896** *T. W. Stapleton & Co. Wine List* Dec., Vermouth, Italian—36/-. **1925** TOYE & ADAIR *Drinks Long & Short* 12 Three and a half glasses of gin, one and a half of Italian Vermouth. **1967** A. LICHINE *Encycl. Wines* 541/1 Vermouth was certainly being made in Italy in the seventeenth century, and now it is produced all over the world, and the two main types are 'French' and 'Italian'.

B. *sb.* **5.** Ellipt. for *Italian cloth.* Also *attrib.*

1897 *Sears, Roebuck Catal.* 177/1 Fine Italian lining in fancy figured effects. **1900** *T. Eaton & Co. Catal.* Midwinter Sale 13/2 Ladies' black boucle curl cloth jackets.., lined throughout with mercerized Italian. **1907** *Yesterday's Shopping* (1969) 743/1 Jackets..in serges and cloths, lined Italian. **1960** *Textile Terms & Definitions* (Textile Inst.) (ed. 4) 82 *Italian*,..a cloth of 5-end sateen weave with a lustrous finish, used chiefly as a lining material.

6. = **Italian vermouth.* Chiefly in phr. *gin and Italian.* Cf. *GIN *sb.*² 2 b.

1929 J. B. PRIESTLEY *Good Companions* II. i. 264 Couldn't you take some cocktails—gin and Italian or sherry and bitters or something—upstairs to those people. **1957** G. BELLAIRS *Death in High Provence* vii. 80 Two long Italians with some ice and perrier, please.

italic, *a.* and *sb.* Add: **A.** *adj.* **4.** Pertaining to the older Latin version of the Bible known as *Vetus Itala*.

1861 C. D. GINSBURG tr. *Coheleth* App. I. 501 The Old Italic Version forms the basis of the one on which St. Jerome wrote the *Commentarium ad Paulam et Eustochium*. **1957** *Oxf. Dict. Chr. Ch.* 981/1 It has been generally supposed that there are two main types—the 'Italic' (represented by the MSS. *f*..and *q*..) and the 'European'.

B. *sb.* **2.** Add to def.: Also, a modern adaptation of the old Italic hand. Hence **ita·licist,** one who favours or practises this style.

1955 W. BLUNT *Handwriting* 9 Many firms..produce fountain-pens designed for Italic. **1956** *Jrnl. Educ.* July 304/1 Having myself been lambasted more than once by the Italicists because I dared to qualify my praise of their handwriting. *Ibid.* 304/2, I know of *no* school where italic is given a disproportionate share of the timetable. **1963** A. FAIRBANK *How to teach Italic Hand* 14 The following remarks relate to pen-and-ink italic.

Italo-. Add: **Italo-Celtic, Italo-Keltic,** a postulated common parent language of Italian and Celtic; also as *adj.*, of, pertaining to, or characteristic of this language; **I·talophi:l, -phi:le** *a.*, friendly to Italy or to what is Italian; *sb.*, one who is Italophile; **I·talopho:be,** one affected with **I:talopho·bia,** intense dislike or fear of Italy.

1877 J. RHYS *Lect. Welsh Philol.* i. 5 Some subdivide the Southern division into an Italo-Celtic and a Hellenic group, while others prefer to suppose a Celtic and a Greco-Italic group. **1888** J. WRIGHT tr. *Brugmann's Elem. Compar. Gram. Indo-Gmc. Lang.* I. 3 The Italo-Keltic hypothesis has perhaps the best prospect of attaining a greater degree of probability in the future. **1906** *Daily Chron.* 28 Mar. 5/4 The appointment of the Italophile reactionary Miushkovitch Ministry. **1920** G. E. BUCKLE *Life Disraeli* V. iii. 130 Protestant and Italophil England rejoiced. **1921** *Contemp. Rev.* Oct. 494 Some Italophobe Germans. **1922** *Ibid.* Sept. 302 The official state of hostility between the Vatican and Italy and the periodical protests against it..maintain in the Catholic masses a feeling of Italophobia. **1927** *Scots Observer* 26 Mar. 12/5 Italy found it easier to buy Ahmed Zogu, turn him into an Italophil. **1932** W. L. GRAFF *Lang.* 375 Italo-Celtic was probably once a common group of closely related dialects. **1933** L. BLOOMFIELD *Lang.* iv. 62 Some scholars believe that Italic and Celtic are connected.., so as to form an

Italo-Celtic sub-group within the Indo-European family. **1958** N. GORDIMER *World of Strangers* 20 She..was a passionate Italophile, scattering her speech with *cara mia*'s. **1966** E. P. HAMP in Birnbaum & Puhvel *Anc. Indo-European Dial.* 116 Germanic and..Italo-Keltic have *-yo/i*, almost leveled out for both verb types. **1967** C. SETON-WATSON *Italy from Liberalism to Fascism* x. 378 The Italophil Prime Minister of Bulgaria, Gueshov, offered an alliance. **1973** *Times Lit. Suppl.* 5 Oct. 1151/1 Such an experienced Italophile.

ita-palm: see *ITA.

itatartaric (itătaɹtæ·rik), *a. Chem.* [tr. G. *itaweinsäure* itatartaric acid (T. Wilm 1867, in *Ann. d. Chem. u. Pharm.* CXLI. 33), f. *ita-* (in *itaconsäure* itaconic acid (see ITACONIC *a.*)) + *weinsäure* tartaric acid.] *itatartaric acid*: dihydroxyitaconic acid, $CH_2OH\cdot C(OH)(COOH)\cdot CH_2COOH$. Hence **itata·rtrate,** a salt of this acid.

1872 WATTS *Dict. Chem.* Suppl. 762 Pure itatartaric acid is amorphous, vitreous, smells like honey when gently heated, deliquesces in the air, dissolves easily in alcohol, and does not volatilise perceptibly at 100° with aqueous vapour. *Ibid.*, Calcium itatartrate..forms crystalline masses sparingly soluble in water. **1945** *Jrnl. Biol. Chem.* CLXI. 739 (*heading*) Itatartaric acid, a metabolic product of an ultraviolet-induced mutant of *Aspergillus terreus*.

itch, *v.*¹ **3.** Delete †*Obs.* and add later examples. Also *refl.* and *fig.*

1900 J. LONDON *Let.* 16 June (1966) 107 It is a fascinating subject. It has itched me for long, and it is often all I can do to keep away from writing on it. **1922** JOYCE *Ulysses* 748 My hole is itching me always when I think of him. **1947** *Penguin New Writing* XXIX. 12 With long sensuous strokes he smoothed a patina of paint down the chairlegs, then itched with fussing dabs the corners and underneath. **1951** R. CAMPBELL *Light on Dark Horse* vi. 99 The thick super-salty water of the Mediterranean, which tires and itches the naked eye. **1951** L. MACNEICE tr. *Goethe's Faust* II. I. 171 The dice already itch me in my pocket. **1954** S. BECKETT *Waiting for Godot* II. 46 Then I can keep it [*sc.* a hat]. Mine irked me... How shall I say? ..It itched me. **1973** *Welcomat* (Philadelphia) 10 Oct. 4/2 The sticker that itches her most is the one that says: 'School's Open. Drive Carefully.'

ite (ɔit). [The suffix -ITE¹ used as an independent word: cf. ISM *quasi-sb.*] A person or thing that is or may be designated by a sb. in *-ite*.

1852 *Blackw. Mag.* Aug. 260/1 The right honourable gentleman has shown that he is neither a Derby-*ite* nor a Russell-*ite*. Then what *ite* are you? **1906** *Westm. Gaz.* 1 Dec. 9/2 A big factory for explosives, holding dynamite, ballistite, cordite..Heaven knows what 'ites'—sufficient to wreck half the world. **1926** R. W. HUTCHINSON *First Course Wireless* viii. 138 Most of the 'ites' on the market.. are galena subjected to various treatments.

ite: see *ITA.

-ite, *suffix*¹. Add: **2. b.** Also used more widely in *tektite*, and hence in the names of tektites from different regions (as *australite, indochinite*).

item, *adv.* and *sb.* Add: **B.** *sb.* **2.** (Further examples.)

1961 *Lancet* 12 Aug. 358/2 Questions [set]..included items that the undergraduate who had read something more..than an introductory text could be expected to answer. **1970** *Jrnl. Gen. Psychol.* LXXXII. 63 A person may agree with an item and its opposite because he believes statements worded in the dogmatic (authoritarian) direction but agrees with a reversed item because of the high social desirability of this item.

c. *Computers.* Any quantity of data treated as a unit, such as a field, a group of fields, or a record.

1954 *Computers & Automation* May 17/2 *Item*, a set of one or more fields containing related information. **1958** *Computer Jrnl.* I. 71/1 Let us call the units which are to be sorted items. *Ibid.* 71/2 If every item of data has a unique key, complete sorting will result in each place holding no more than one item. *Ibid.* 72/2 A typical item in commercial data is an 80-column punched card. **1964** A. LYTEL *Fund. Data Processing* (1965) iv. 108 Information is stored in variable-length memory areas called fields... Consecutive fields can be combined to form a larger unit of information called an item. Grouping fields to form an item simplifies the manipulation of related-data fields, and minimizes the number of instruction executions required to move consecutive fields within the main memory. **1967** B. S. WALKER *Introd. Computer Engin.* vi. 157 The field or item is typically a group of letters or numbers, in association, to mean a name, or reference number, or a heading of some kind. **1971** *Computers & Humanities* VI. 67 Each item file is composed of an open-ended sequence of variable length records called 'items', every item being the description of one entity.

d. A member of a set of linguistic units.

1954 *Word* X. 230 'Items'..are either morphemes or sequences of morphemes, but still one has to contend with the independent status of order, constructions, and hierarchical structure. Even so, there is a clear difference between taking some phonemic material as 'root' (= item) and some as 'marker' of processes. **1964** R. A. HALL *Introd. Ling.* vi. 34 The first group [of linguistic analysts] wish to limit their description strictly to an enumeration of items and the arrangements or sequences in which they are found... The other group..take into account the passage of time when the observer moves from one part of his material to another... This approach is based on a listing of items involved and of 'processes' which the items 'undergo'. *Ibid.* 35 In some respects, the item-and-process (IP) approach is closer to our traditional type of grammatical description than is the item-and-arrangement (IA) approach. **1964** M. A. K. HALLIDAY et al. *Ling. Sci.* ii. 24 In English..'chair' is a lexical item: it operates as an item in open set choices. 'the', 'chair', the '-s' in 'chairs', 'in case'..are grammatical items. **1970** *Canad. Jrnl. Linguistics* XV. 95 This system grew out of a union of the item-and-process approach to linguistics with automata theory. **1971** T. F. MITCHELL in *Archivum Linguisticum* II. 64 In the English-speaking world of linguists little more than ten short years ago, the talk was of grammatical models labelled 'Item and Arrangement', 'Item and Process', and 'Word and Paradigm'.

C. *attrib.* and *Comb.*

1859 E. H. N. PATTERSON in L. Hafen *Overland Routes to Gold Fields* (1942) 68 This..is one of those cases, probably, that will remain a mystery only to be solved when the great item book of the recording angel shall be opened to justify the final sentence. **1961** *Lancet* 12 Aug. 359/2 Item analysis showed that there were fairly large differences between the groups on a few questions, and these were seen to be due to specific differences in teaching. **1970** *Jrnl. Gen. Psychol.* LXXXII. 166 Item-item and item-test correlations were computed. **1972** *Jrnl. Social Psychol.* LXXXVI. 221 An item analysis technique was used to select those items that discriminated between the high and low scores by 20 percentage points.

iterate, *v.* Add: **4.** *intr. Math.* To employ iteration; to make repeated use of a formula by substituting in it each time the result of the previous application.

1953 A. S. HOUSEHOLDER *Princ. Numerical Analysis* ii. 45 Since the 'approximation' x_0 with which one may start an iteration does not necessarily need to be close, it is sometimes advantageous to..start with an arbitrary x_0, perhaps $x_0 = 0$, and iterate until the approach is sufficiently close. **1957** L. FOX *Numerical Solution Two-Point Boundary Probl.* iv. 81 As an example..of an uncommon but convenient iterative procedure consider the solution of the differential equation $y'' - (1/x^4 - 2/x^3)y + 1/x^3 = 0$... We iterate according to the scheme $y_{n+1} = x^2(1+x^2y_n'')/(1-2x)$, with $y_0 = 0$.

iterate (i·tĕrĕt), *sb. Math.* [f. the vb., or ad. L. *iterāt-us*, pa. pple. of *iterāre* (see ITERATE *v.*): see -ATE¹.] A quantity arrived at by iteration.

1941 R. J. SCHMIDT in *Phil. Mag.* XXXII. 370 Denote the *n*th approximation to the value of any unknown x_r by $x_r^{(n)}$. To solve these [simultaneous] equations by the method of successive approximations, we assume approximations $x_2^{(0)}, x_3^{(0)},\ldots x_m^{(0)}$ to the unknowns $x_2, x_3,\ldots x_m$. Making use of these values we then use the first equation to find $x_1^{(1)}$, the first approximation to x_1. As the approximations will not always converge to the value of the unknowns, we shall, in future, call the numbers $x_s^{(r)}$ iterates. **1956** F. B. HILDEBRAND *Introd. Numerical Analysis* x. 449 The error t_k in the *k*th iterate will be approximately the *square* of that in the preceding iterate, and will be of opposite sign. **1968** FOX & MAYERS *Computing Methods for Scientists & Engineers* i. 10 In the days of machine saturation it is tempting to take 25 iterates even though 250 are really necessary. **1968** N. BOURBAKI *Theory of Sets* iii. 186 The mapping f^n is called the *n*th iterate of the mapping *f*.

iteration. **1. b.** Add: *esp.* The repeated application of a formula devised to provide a closer approximation to the solution of a given equation when an approximate solution is substituted in the formula, so that a series of successively closer approximations may be obtained; a single application of such a formula; also, the formula itself.

1924 WHITTAKER & ROBINSON *Calculus of Observations* vi. 79 In 1674 a method depending on a new principle, the principle of iteration, was communicated in a letter from Gregory to Collins. **1941** *Phil. Mag.* XXXII. 374 After a few iterations the values of e_r much less than unity will hardly affect the results. **1960** G. N. LANCE *Numerical Methods for High Speed Computers* i. 8 Whichever criterion is used to determine the end of the iteration, it is clear that the orders to evaluate $f(x_r)$ and $f(x_{r+1})$ are identical except that x_{r+1} is used instead of x_r. This kind of modification is made extremely simple on high-speed computers. **1968** E. T. COPSON *Metric Spaces* viii. 115 The Newton–Raphson iteration $x_{n+1} = x_n - f(x_n)/f'(x_n)$ for solving the equation $f(x) = 0$.

iterative, *a.* Add: **1. c.** *Math.* Of the nature of or employing iteration.

1924 WHITTAKER & ROBINSON *Calculus of Observations* vi. 81 A pleasing characteristic of iterative processes may be observed..that a mistake in the performance of the numerical work does not invalidate the whole calculation. **1943** *Phil. Mag.* XXXIV. 409 Iterative methods have found favour with computers, despite an outward semblance of clumsiness which masks their solid advantages from the casual critic. **1949** *Proc. Cambr. Philos. Soc.* XLV. 230 Iterative methods seem very suitable, in principle, for application in large automatic calculating machines. **1953** A. S. HOUSEHOLDER *Princ. Numerical Analysis* ii. 44 Generally speaking, an iterative method for solving an equation or set of equations is a rule for operating upon an approximate solution x_p in order to obtain an improved solution x_{p+1}, and such that the sequence $\{x_q\}$ so defined has the solution *x* as its limit. **1968** E. T. COPSON *Metric Spaces* viii. 115 The iterative process $x_{n+1} = f(x_n)$ leads to a solution of the equation $x = f(x)$ when the mapping of the real line into itself is a contraction mapping.

B. *sb. Philol.* **a.** An iterative verb or aspect. **b.** A word expressing repetition of an action, sound, etc.

1853, 1884 [see *ASPECT *sb.* 9 b]. **1934** PRIEBSCH & COLLINSON *German Lang.* I. i. 13 Formation of distributive numerals in Latin by adding *-no* to the iteratives. *Ibid.* II. iii. 225 Verbs in *-ern*. Some are..iteratives.. e.g. *flattern*. **1961** F. G. CASSIDY *Jamaica Talk* iv. 69 In Standard English one finds three kinds of iteratives: the simple ones like *hush-hush*..; those with vowel gradation like *ding-dong*..; and the rhyming ones like *handy-dandy*.

ithel (i·þel). Also **athel, athleh, atl, ithil.** [Local Arab.] A tamarisk, *Tamarix aphylla*, bearing panicles of pink flowers and minute leaves, native to western Asia and north-east Africa.

1838 W. AINSWORTH *Res. Assyria* 125 The common tamarisk of the country, the Athleh or Atle of Sonnini, is the Tamarix orientalis of Forskahl. **1875** *Encycl. Brit.* II. 236/2 Of plants there is an endless variety [in Arabia]. ..The tamarisk or 'Talh', the southern larch or 'Ithel', the chestnut, the sycamore, and several other trees. **1881** A. BLUNT *Pilgrimage to Nejd* I. 84 The Ithel, a tree grown in every village of Central Arabia. *Ibid.* 85 The roof was of ithel beams. **1924** *Blackw. Mag.* Mar. 351/1 Mackintosh..was watching them from the shelter of a clump of *ithil* bushes. **1949** L. H. BAILEY *Man. Cultivated Plants* (rev. ed.) 678 *T[amarix] aphylla*... Athel Tamarisk. Shrub or small tree... Useful as a windbreak in desert regions. **1963** F. VON BREITENBACH *Indigenous Trees Ethiopia* (ed. 2) 150 *Tamarix aphylla*... Atl (Arabic)... Leaves reduced to a minute triangular tooth on a sheathing base.

ithomiid, ithomiine, *sb.* and *a.* (iþōᵘ·mi,id, -īn). [f. mod.L. family and subfamily names *Ithomiidæ*, *Ithomiinæ*, f. the generic name *Ithomia* (J. Hübner *Verzeichniss bekannter Schmetterlinge* (1816) 9), f. Gk. ἰθύς straight + ὦμος shoulder.] A tropical Central or South American butterfly belonging to the family Ithomiidæ, or this group treated as the subfamily Ithomiinæ of the family Nymphalidæ; of or pertaining to an insect of this kind.

1899 D. SHARP in *Cambr. Nat. Hist.* VI. vi. 346 The Ithomiides are peculiar to tropical America. **1912** G. B. LONGSTAFF *Butterfly-Hunting* vi. 312 The hour was earlier and the Ithomiines were not so closely packed. **1930** *Proc. Entomol. Soc. London* V. 91 An Ithomiine butterfly and its Heliconine mimic taken flying together in NW. Peru. **1972** BROWN & HEINEMAN *Jamaica & its Butterflies* 97 The number of ithomiids found in Central America is limited. *Ibid.* 99 *Greta* is among the most advanced in structure of all ithomiid genera.

I-thou (ɔi·ðau). [tr. G. *ich-du* (M. Buber (1923) *Ich und Du* 9), f. I *pers. pron., 1st sing. nom.* + THOU *pers. pron., 2nd sing. nom.*] Used *attrib.* of a personal relationship between man and God. Also *transf.*

1937 R. G. SMITH tr. *Buber's I and Thou* I. 3 The primary words are not isolated words, but combined words. The one primary word is the combination *I-Thou*. The other primary word is the combination *I-It*. *Ibid.*, If *Thou* is said, the *I* of the combination *I-Thou* is said along with it. **1958** *Church Times* 14 Feb. 10/3 Am I really prepared to obey God, in the utter loneliness of the 'I-Thou' relationship, even if it means the actual hatred of other people? **1961** *English Studies* Oct. 323 But Dr Esch also stresses the differences: while the other of the love-poetry is an equal partner, his God is *totaliter aliter*, which makes for a completely different I-Thou relation. **1967** C. DAVIS *Question of Conscience* 223 We may first take the I-Thou relationship of deep personal commitment. **1967** *Guardian* 19 May 9/1 When any chorus stands up to sing..it is so much more an I-Thou affair than opera.

ithyphallic, *a.* and *sb.* Add: So **ithypha·llus,** an erect phallus.

1889 in *Cent. Dict.* **1967** *Listener* 28 Sept. 401/2 The fathers..exhibit their ithyphalluses. **1968** *Punch* 27 Mar. 469/3 The spike (Laius's sword) Jocasta steadily presses through her womb is replaced by a golden ithyphallus ten foot tall.

Iti (ɔi·tɔi), *sb.* and *a.* Also **Itie, Ity.** [Dim. of ITALIAN *a.* and *sb.*] Used with disparaging overtones: **A.** *sb.* An Italian. **B.** *adj.* Italian. Cf. *EYETIE *sb.* and *a.*

1941 R. MOORE in Michie & Graebner *Lights of Freedom* x. 130 Ity planes were circling overhead. *Ibid.* 131 How surprised the Ities would be to see us. **1942** *E. Afr. Ann.* 1941–2 109/1 The Itis ran away with it all, fancy even taking the macaroni, poor devils. **1944** A. JACOB *Traveller's War* xiv. 246 We thought you were Iti. *Ibid.* xviii. 278 The petulant Itie or the solid Boche. **1947** [see *BRUSH *sb.*⁵]. **1959** G. JENKINS *Twist of Sand* iv. 68 Those Itie destroyers will have to come mighty close. **1965** *Economist* 4 Dec. 1100/3 With all these Ities and squareheads and Greeks around, it might be worth remembering. **1973** *Times* 1 Jan. 14/4 I'm going to be a German, an Iti, a Dutchman.

itis (ɔi·tis), *sb.* [The suffix -ITIS used as an independent word.] A bodily condition, affection, or disease that is or may be described or designated by a word ending in *-itis*.

[**1896**: cf. Dict. s.v. -ITIS.] **1909** *Practitioner* Nov. 706 It must be remembered that the complaint referred to [*sc.* mucous colitis] is not, strictly speaking, an *itis* at all.

-itis. Add: In irregular trivial use applied to a state of mind or tendency fancifully regarded as a disease.

1903 ASQUITH in *Westm. Gaz.* 19 Oct. 5/1 All the people were suffering from a new disease—the disease of fiscalitis. **1906** *Ibid.* 27 Apr. 4/2 Several members of Parliament are suffering from a slight attack of Suffragitis. **1912** *Q. Rev.* Oct. 504 Cricket has just suffered from so severe an attack of 'testitis' as to render it highly improbable [etc.]. **1944** F. CLUNE *Red Heart* 68 Those were the days when the nor'-west of New South Wales was agog with bushranger-itis. **1945** W. S. CHURCHILL *Victory* (1946) 186 It was impossible to go on in a state of 'electionitis' all through the summer and autumn. **1969** *Sunday Express* 28 Dec. 24/3 As the year wears on, politicians' electionitis will have more influence on events than central bankers' views.

itisket, itasket (iti·skĕt ita·skĕt). A vocal utterance in verses accompanying any of several children's games, esp. 'Drop the handkerchief'.

1926 D. LA SALLE *Play Activities Elem. Sch.* II. 71 Itisket, itasket, a green and yellow basket, I lost a letter to my love and on the way I found it... (Drop and pick up the handkerchief.) **1969** R. D. ABRAHAMS *Jump-Rope Rhymes* p. xviii, This game [*sc.* drop the handkerchief] is found with a number of accompanying chants, including 'Itisket, Itasket'.

-itol (itǫl). *Chem.* [f. -IT(E¹ 4 + -OL 1.] A suffix used to form the names of polyhydric alcohols other than di- or trihydric alcohols.

Such compounds were formerly given names terminated by *-ite*, as in *dulcite*, *inosite*, *mannite*. These names were later modified by the addition of *-ol* to express their alcoholic nature, giving e.g. *dulcitol*, *inositol*, *mannitol*. Hence in mod. use *-itol* has become an independent suffix, as in *hexitol*.

I told you so. [TELL *v.* 8.] *Phr.* used to remind the person addressed that he has previously been warned that his actions would incur misfortune. As *sb.*, a person who uses this expression or adopts this attitude; such an expression or attitude; used *attrib.* (as *I-told-you-so*) to denote such an attitude. Also used as a kind of quasi-vb.

1609 [see TELL *v.* 8]. **1823** BYRON *Don Juan* XIV. I. 3 Sadder than owl-songs or the midnight blast, Is that portentous phrase, 'I told you so'. **1919** W. DE MORGAN *Old Madhouse* xxviii. 435 Perhaps I'm only I-told-you-soing. **1926** WHITEMAN & MCBRIDE *Jazz* iii. 49, I really did debate whether I hadn't better give up and let the I-told-you-so's, who said jazz would bring me to no good end, have it their own way. **1930** J. DOS PASSOS *42nd Parallel* IV. 299 Alice had an Itoldyouso manner. **1954** W. FAULKNER *Fable* (1955) 43 His I-told-you-so to the elders. **1959** A. LEJEUNE *Crowded & Dangerous* xi. 125 She'll.. put on that disapproving I-told-you-so look.

itsy-bitsy (i·tsi,bi·tsi), *a. colloq.* [Baby-form of LITTLE *a.* + *BITSY *a.*: see -SY.] Small, (charmingly) insubstantial, tiny; also, used disparagingly: arty-crafty, twee. So **i·tsy-bi·tsiness.** Cf. *ITTY-BITTY *a.*

1938 I. GOLDBERG *Wonder of Words* viii. 162 Itsy-bitsy (little bit). **1939** R. CHANDLER *Trouble is my Business* (1950) 23 The same clerk was nuzzling at the same itsy-bitsy moustache. **1953** P. JONES in *Plays of Year* IX. 567 You should be allowed to revolve radiantly at some cocktail party..a dry martini in one hand and an itsy-bitsy little thing on a stick in the other. **1957** P. WILDEBLOOD *Main Chance* 197 An itsy-bitsy tot of vodka and a teeny-weeny tranquilliser. **1958** *Observer* 15 June 11/2 The rather sentimental and itsy-bitsy patterns that used to be considered suitable for wear by the young. **1967** *House & Garden* Mar. 74/1 The Peter Pan statue..is the embodiment of all sentimental itsy-bitsiness. **1972** 'H. HOWARD' *Nice Day for Funeral* iii. 45 If Frankie was here he'd break you into itsy-bitsy pieces. **1972** *Lady* 29 June 1052/1 Country accessories are also bold, far less itsy-bitsy than many town-worn ones.

itty (i·ti), *a. colloq.* Also **ittie.** [Baby-form of LITTLE *a.* + -Y *suffix*⁶.] Used hypocoristically for 'little' (chiefly in reference to babies or small domestic animals).

1798 JANE AUSTEN *Let.* 27 Oct. (1932) I. 10 My dear itty Dordy's remembrance of me is very pleasing to me. **1853** MRS. GASKELL *Cranford* vii. 132 Come down stairs with me, poor ittie doggie, and it shall have its tea. **1938** D. RUNYON *Furthermore* viii. 159 He..starts whispering, 'There, there, there, my itty oddleums.' **1964** *Guardian* 30 Oct. 13/3 Now, ah reckon Lady Bird an' ah will git ahselves an itty bit o' sleep.

i·tty-bi·tty, *a.* = *ITSY-BITSY *a.*

[**1938** *Amer. Speech* XIII. 314/1 Itty bitty kitty from the city, a country yokel who tries to be a cat.] **1940** R. CHANDLER *Farewell, my Lovely* xxxiii. 253 Itty-bitty frame houses on the wrong side of town. **1959** H. HOBSON *Mission House Murder* xiii. 89 No big swallows now. Itty-bitty sips—and take your time. **1968** W. GARNER *Deep, Deep Freeze* v. 62 I'll move every ittybitty piece on that goddammed fancy board. **1969** L. KENNEDY *Very Lovely People* i. 35, I felt, here I am in this itty-bitty tropical village, a tremendous long ways from anywhere.

Ity, var. *ITI *sb.* and *a.*

-ium, *suffix.* Add: **b.** Used to form the names of various protonated, mostly organic, bases, as *anilinium, benzenium, ethenium, flavylium, guanidinium, hydrazinium, imidazolium, pyrylium.* Cf. *-ONIUM.

This usage of the suffix derives from AMMONIUM. For rules governing the application of the suffix see *Nomenclature of Organic Chemistry* and *Nomenclature of Inorganic Chemistry*, published by the International Union of Pure and Applied Chemistry.

i-umlaut: see *I I. 2 b.

Ivan (ɔi·văn, īvā·n). [Russ., = John.] Used for: a Russian, esp. a Russian soldier (as typical of the Russian army).

[**1870** *Brewer's Dict. Phr. & Fable* 448/1 *Ivanovitch*, a lazy, good-natured person, the national impersonation of the Russians as a people, as *John Bull* is of the English. **1890** WEBSTER, *Ivan Ivanovitch*, an ideal personification of the typical Russian or of the Russian people;—used as 'John Bull' is used for the typical Englishman.] **1925** FRASER & GIBBONS *Soldier & Sailor Words* 129 *Ivan*, the everyday name in the Russian Army, at any rate down to 1916, for a private soldier, equivalent to our 'Tommy Atkins'. **1959** M. CROSLAND tr. *Rovan's Germany* 51 The Russian 'Ivan' is the brutal sub-man and the giant with the kind, noble and spontaneous heart..[to] the German spectator. **1968** 'B. MATHER' *Springers* xii. 128 We'd knocked off quite a few of their side so far, and even dedicated Ivans could be expected to show a little exacerbation under the circumstances. **1971** C. EGLETON *Last Post for Partisan* xvii. 174 So long as the Ivan kept on coming he wasn't worried. **1972** *Guardian* 8 Sept. 12/4 A situation in which Ivan continues to come a lot cheaper than GI Joe.

ivi (ī·vi). Also **eevie, ifi, ihi.** [Fijian *ivi*, Samoan *ifi*.] The Tahitian chestnut, *Inocarpus fagiferus* (*I. edulis*), a leguminous evergreen tree bearing spikes of white or yellow flowers and dark red, edible fruit.

1862 B. SEEMANN *Viti* xvi. 318 The Ivi, or Tahitian chestnut,..is one of the common trees [in Fiji]. **1874** LINDLEY & MOORE *Treas. Bot.* II. Suppl. 1308/1 Ivi (Feejee). *Inocarpus edulis.* **1881** C. F. CUMMING *At Home in Fiji* I. 275 A group of eevie trees appears like one gigantic mass of lovely trailing foliage. **1888** W. HILLEBRAND *Flora Hawaiian Islands* 109 Here [*sc.* among the Caesalpinieae] also must be given a place to the anomalous *Inocarpus edulis*, Forst., or Tahitian Chestnut, the Ivi or Mapé. **1894** B. THOMSON *S. Sea Yarns* 7 He repaired to the mainland to consult a rival oracle named *Na-ivi* (the ivi-tree). **1935** *B. P. Bishop Mus. Bull.* (Honolulu) CXXX. 119 The native names [of *Inocarpus edulis*] are *ihi* or *mape* in Nukuhiva and Hivaoa of the Marquesas. **1964** C. S. BELSHAW (*title*) Under the ivi tree. Society and economic growth in rural Fiji. **1970** W. R. SYKES *Contrib. Flora Niue* 156 No ifi trees were seen growing in places other than those connected with man's activities.

Ivicene (i·visīn), *sb.* and *a.* [f. *Ibiza, Iviça*, one of the Balearic Islands + Sp. *-eño*, -ENE.] A type of hound, said to be descended from ancient Egyptian hunting dogs, native to the island of Ibiza and characterized by large, pointed, pricked ears and white, fawn, or reddish-brown colouring; also as *adj.*; also called **Ibizan* or *Balearic hound*.

1929 *Morning Post* 5 Feb. 15/3 The Ivicene Described. *Ibid.*, The Ivicene dog, a breed which has never before been seen in this country. **1929** *Dog World* 8 Feb. 912/1 Pedro of Chardia, Ivecine [*sic*], Greyhound type, wants time and should do well, rare and undeveloped at the moment. **1948** C. L. B. HUBBARD *Dogs in Brit.* III. xv. 124 The correct name of Cà Eivissenne is loosely pronounced as 'Ivicene', which led to considerable publicity in Britain about 1930, giving the dog the breed name of Ivicene. **1964** E. F. DAGLISH tr. *Schneider-Leyer's Dogs of World* 165 (*heading*) Balearic Hound or Ivicene.

-ivity (in Dict. s.v. -IVE *suffix*). Add: (See quot. 1895.)

1885 O. HEAVISIDE in *Electrician* 4 Sept. 311/1 Thus, 'specific resistance' may well be called 'resistivity', and specific conductance 'conductivity', referring to the unit volume. Resistivity is the reciprocal of conductivity, and resistance of conductance. **1895** *Rep. Brit. Assoc. Adv. Sci.* 197 That the termination *-ance* be used in general for words expressing the properties of a definite body or piece of matter; *e.g.*, resistance, conductance, inductance, permeance, reluctance, &c.; and that the termination *-ivity* or *-ility* or the like be used for words expressing the specific properties of a material; *e.g.*, conductivity, resistivity, inductivity, refractivity, permeability, &c.

-ivore, usual comb. form of the suffix *-VORE.

ivoried, *a.* (Later examples.)

1893 *19th Cent.* Nov. 843 On thy bare and ivoried shoulder. **1928** HARDY *Coll. Poems* 156, I borrowed deep to carve the screen And raise the ivoried Rood. **1945** W. DE LA MARE *Burning-Glass* 76 Tipped arrow, ivoried bow, and rain-soaked quiver.

-ivorous, usual comb. form of the suffix -VOROUS.

ivory. Add: **5. d.** *collect. sing.* and *pl.* The keys of a piano or similar instrument. *colloq.*

1818 KEATS *Let.* 18 Dec. (1958) II. 13 She plays the Music without one sensation but the feel of the ivory at her fingers. **1854** THACKERAY *Newcomes* I. xi. 114 It is a wonder how any fingers can move over the jingling ivory so quickly as Miss Cann's. **1918** [see *JAZZ *v.* 2]. **1940** *S.P.E. Tract* LV. 196 Tickle the ivories. **1962** *Times* 10 Feb. 4/2 'Ivory-tickling' has become an outmoded and faintly derogatory description of piano-playing. **1974** *Times* 15 Feb. 14/7 Its cover portrays the Prime Minister, seated at the organ, tinkling one lot of ivories and flashing the other lot.

10. ivory-bill (earlier U.S. examples); **ivory board,** a kind of pasteboard with both surfaces smooth; **ivory dome** *U.S. slang*, a stupid person; **ivory (nut) palm,** a South American palm of the genus *Phytelephas*, or a Micronesian one of the genus *Metroxylon*, both of which bear nuts yielding vegetable ivory; **ivory plum** *U.S.*, the wintergreen, *Gaultheria procumbens*, or the creeping snowberry, *Chiogenes hispidula*, or their fruit; **ivory-wood** *Austral.*, the tree *Siphonodon australe*, or its timber, which is used for drawing-instruments, etc.

1787 *Ellicott Almanac 1788* (Winchester, Virginia) sig. B2, The land fowls [of Kentucky] are turkeys, pheasants, ..the perraquet, ivory-bill, woodcock, and the great owl. **1872** E. COUES *Key to N. Amer. Birds* 191 The ivory-bill and the flicker stand nearly at extremes of the family. **1926** *Paper Terminol.* (Spalding & Hodge) 14 *Ivory boards*, superfine cardboard highly finished by means of bees-waxed rolls. **1962** F. T. DAY *Introd. to Paper* iv. 46 A large variety of boards is produced by the paper maker, in grades ranging from high class Bristol and Ivory boards to the cheaper kind of Triplex board. **1923** Ivory dome [see *DOME *sb.* 4 d]. **1844** W. PURDIE *Let.* 14 Dec. in *Compan. Bot. Mag.* (1847) LXXIII. 11 The *Phytelephas* (Vegetable Ivory Palm) is procurable at Sta. Martha. *Ibid.* 12, I think of ascending the Magdalena, and myself collecting growing plants and seeds of the Ivory Palm. **1871** C. KINGSLEY *At Last* I. viii. 303 The ripe fruit [of the Moriche palm] contains first a rich pulpy nut, and at last a hard cone, something like that of the vegetable ivory palm. **1916** *Jrnl. Bombay Nat. Hist. Soc.* XXIV. 682 Names of the Tree [*sc. Phytelephas macrocarpa*]. English: Large-fruited Ivory Plant, Ivory Palm, Ivory nut Palm, Vegetable Ivory Plant. **1951** J. H. KRAEMER *Trees W. Pacific Region* 10 In this genus [*sc. Metroxylon*] are the sago palms and the ivory-nut palms. **1966** E. J. H. CORNER *Nat. Hist. Palms* xiii. 315 This anomalous species is the Caroline ivory-nut palm, *M[etroxylon] amicorum* of Micronesia. **1828** J. NEAL *Rachel Dyer* 55 The more brilliant ivory-plumbs or clustered bunch-berries rattled among the withered herbage. **1891** *Jrnl. Amer. Folk-Lore* IV. 149 *Gaultheria procumbens* seems to have an almost endless variety of epithets... The berries are called Ivory Plums. **1892** *Ibid.* V. 99 *Chiogenes serpyllifolia*, ivory plums. **1887** *Colonial & Indian Exhib. Rep. Col. Sect.* 429 Ivory-wood. **1888** F. M. BAILEY *Queensland Woods* 29 Ivory-wood. A tall tree with straight erect stem, the bark of a light colour... Found in the dense scrubs both north and south in Queensland, and also in New South Wales. **1932** R. H. ANDERSON *Trees New South Wales* 145 Ivorywood..is sometimes known as Native Guava. **1965** *Austral. Encycl.* II. 310/1 Though widely spread in coastal New South Wales and Queensland, the ivorywood is a comparatively rare tree.

ivory tower. [tr. F. *tour d'ivoire* (see below).] A condition of seclusion or separation from the world; in general, protection or shelter from the harsh realities of life. Also (with hyphen) *attrib.* Hence *ivory-towered* adj., *ivory-toweredness*; *ivory-towerish* adj., somewhat ivory-towered; *ivory-towerism, -towerist.*

[**1837** SAINTE-BEUVE *Pensées d'Août, à M. Villemain* 152 Et Vigny, plus secret, Comme en sa tour d'ivoire, avant midi rentrait.] **1911** BRERETON & ROTHWELL tr. *Bergson's Laughter* iii. 135 Each member [of society] must be ever attentive to his social surroundings..he must avoid shutting himself up in his own peculiar character as a philosopher in his ivory tower. *a* **1916** H. JAMES *Ivory Tower* (1917) II. iii. 142 Doesn't living in an ivory tower just mean the most distinguished retirement? **1922** H. CRANE *Let.* 10 Dec. (1965) 108, I have grown accustomed to an 'ivory tower' sort of existence. **1936** E. POUND *Let.* Jan. (1971) 277 Ivory tower aesthetes. **1938** R. G. COLLINGWOOD *Princ. Art* vi. 120 The tendency was for each artist to construct an ivory tower of his own: to live, that is to say, in a world of his own devising. **1940** H. G. WELLS *New World Order* § 9. 133 We want a Minister of Education who can..electrify and rejuvenate old dons or put them away in ivory towers, and stimulate the younger ones. **1945** A. HUXLEY *Let.* 2 Apr. (1969) 518 Between ivory-towerism and art for art's sake on the one hand and direct political action on the other lies the alternative of spirituality. **1947** J. HAYWARD *Prose Lit. since 1939* 46 If [literature] fails in this task it will be reduced to the status of an art pursued for art's sake by isolated groups of writers, segregated from the world in their ivory towers and 'private worlds'. **1953** G. VANN *Water & Fire* iii. 50 That ivory-tower æstheticism which averts its gaze from the squalors of humanity. **1954** 'N. BLAKE' *Whisper in Gloom* vii. 94 I'm going to plunge you into reality, my little Ivory-Towerist. **1959** *20th Cent.* Nov. 401 British governments..have been badly informed..and Britain's ivory-towered embassies may have to bear some of the blame. **1963** M. MCCARTHY *Group* vi. 120 We called you the Ivory Tower group. Aloof from the battle. **1963** *Daily Tel.* 12 Oct. 8/7 Pity the poor parson!..If he eschews all worldly contact, he's accused of being ivory-towerish and out of touch. **1963** *Economist* 26 Oct. 355/1 Every don..however attached to

academic ivory-toweredness. **1967** P. NOKES *Professional Task in Welfare Pract.* vii. 113 When I began teaching at the Prison Staff College..I soon became aware of a well established tradition that what was taught there was 'ivory-towered'. **1968** J. J. C. SMART *Between Sci. & Philos.* 17 It would be unwise to think that philosophy is exclusively a subject for inhabitants of ivory towers. **1972** *Science* 19 May 769/3 New realities which make it impossible for them to think and perform in such ivory-tower isolation.

ivy, *sb.* Add: **1. c.** *U.S.* = *poison ivy* (see sense 2).

1788 J. MAY *Jrnl.* 9 June (1873) 65, I have been clearing land for eight days, and now begin to feel the effects of poison—from ivy, doubtless. **1849** F. PARKMAN *Calif. & Oregon Trail* xiii. 205 In the morning Shaw found himself poisoned by ivy.

ivy-berry. Add: (Later example.)

1971 *Country Life* 25 Mar. 667/3 The woodpigeons.. had been on the ivy berries and the clover fields.

b. *U.S.* The wintergreen or checkerberry, *Gaultheria procumbens*.

1840 *Southern Lit. Messenger* VI. 518/2 There were the fringed polygula, the buttercup, wild geranium, bunch-plum, ivy-berry. **1892** *Jrnl. Amer. Folk-Lore* V. 100 *Gaultheria procumbens*, ivy-berry.

ivy-leaf. Add: **b.** *attrib.* = IVY-LEAVED *a.*

1909 *Daily Chron.* 5 June 9/5 Ivy-leaf geraniums can be depended on to produce a long succession of blooms. **1939** E. BOWEN *Coll. Impressions* (1950) 64 A window-box gay with pink ivy-leaf geraniums. **1957** *Encycl. Brit.* X. 205/2 The ivy-leaf geranium, derived from *P[elargonium] peltatum*, has given rise to an important class of both double- and single-flowered forms adapted especially for pot culture, hanging baskets, [etc.].

Ivy League. Name given to a group of long-established eastern U.S. universities; also *attrib.* of the social and intellectual prestige or other characteristics of these universities, or of, relating to, or characteristic of the members or former members of these universities. So **Ivy Leaguer,** a member or former member of an Ivy League university.

[**1933** S. WOODWARD in *N.Y. Herald Tribune* 16 Oct. 18/1 The fates which govern [football] play among the ivy colleges and academic boiler-factories alike seem to be going around the circuit.] **1939** *Princeton Alumni Weekly* 29 Sept., The 'Ivy League' is something which does not exist and is simply a term which has been increasingly used in recent years by sports writers, applied rather loosely to a group of eastern colleges. **1943** K. P. KEMPTON in *Sat. Even. Post* 22 May 14 (*title*) Ivy Leaguer. **1949** S. WOODWARD *Sports Page* xii. 132 For many years the colleges seemed to feel that there was something not quite nice about the press. It was fashionable, especially in the hallowed precincts of the Eastern Ivy League, to snub and bamboozle the sports writers. **1951** J. D. SALINGER *Catcher in Rye* xii. 103 My father wants me to go to Yale, or maybe Princeton, but I swear, I wouldn't go to one of those Ivy League colleges. *Ibid.* xvii. 133 The jerk had one of those very phoney, Ivy League voices, one of those very tired, snobby voices. **1959** *Listener* 12 Feb. 283/1 Hemingway's sense of Ivy-League social life and its complex snobberies. **1962** 'S. RANSOME' *Without Trace* v. 49 He could drop his natural guttersnipe talk and sound like an Ivy Leaguer whenever it suited him. **1965** *Times Lit. Suppl.* 25 Nov. 1053/1 James Purdy's Ivy-league rapist. **1970** *Daily Tel.* 28 Apr. 4/6 Rising costs are driving away middle-income students from some of America's Ivy League universities, turning the campuses into places for the poor, supported by scholarships, and the rich, admissions officers say. *Ibid.*, The Ivy League colleges are Brown, Columbia, Cornell, Dartmouth, Harvard, Pennsylvania, Princeton and Yale. **1973** S. ALSOP *Stay of Execution* (1974) II. 171 Waspish Ivy Leaguers like Roosevelt or Welles.

ixodid (iksōᵘ·did), *sb.* Pl. **ixodides.** [f. mod.L. family name *Ixodidæ*, f. the generic name *Ixodes* (P. A. Latreille 1795, in *Magazin encyclopédique* IV. 15), f. Gk. ἰξώδης sticky.] A tick of the family Ixodidæ. Also *attrib.* or as *adj.*, of or pertaining to a tick of this type.

1911 G. H. F. NUTTALL et al. *Ticks* II. 105 Leach called the Acarina *Monomerosomata*, and divided them into 11 families, one of which, Ixodides, included *Argas*, *Ixodes* and *Europoda*. *Ibid.* 113 The Argasid and Ixodid ticks were recognized as possessing very distinctive features. **1935** T. H. SAVORY *Arachnida* xiii. 146 The Ixodides possess a soft-skinned opisthosoma capable of great distension necessary for the alternate gorging and fasting that is inseparable from the life of a tick. **1952** BAKER & WHARTON *Introd. Acarology* iv. 144 The majority of ixodids usually parasitize different hosts in the immature and mature stages. *Ibid.*, Heavy infestation with ixodid ticks can cause anemia in domestic animals. **1965** B. E. FREEMAN tr. *Vandel's Biospeleol.* xv. 256 An ixodid parasitic on bats has been found frequently in Europe, Asia and Africa.

ixora (iksōᵊ·rā). [mod.L. (Linnæus in J. Burmann *Thesaurus Zeylanicus* (1737) 125), ad. *Iswara*, name of a Hindu divinity, = Skr. *īśvarā* lord, master: so named because the flowers of some species are used as votive offerings.] An evergreen shrub or small tree of the genus so called, mostly native to tropical Africa and Asia and bearing clusters of white or brightly-coloured flowers.

1816 *Bot. Reg.* II. 154 (*heading*) Large-flowered scarlet Ixora. **1846** J. LINDLEY *Veget. Kingd.* 764 The fragrance or beauty..of the Gardenias, Hindsias, Posoquerias, Ixoras..&c. is unsurpassed. **1871** C. KINGSLEY *At Last* I. vii. 231 The air was..almost too heavy with the fragrance of the 'white Ixora'. **1926** D. H. CAMPBELL *Outl. Plant Geogr.* vi. 214 In the upper forest [of Sarawak], a handsome Ixora..was abundant. **1961** *Amateur Gardening* 21 Oct. Suppl. 29/3 For those with plenty of heat in the greenhouse the ixoras are well worth growing.

Iyyar (ī·yāɹ). Also **Iyar, Jiar, Jyar, Yiar.** [ad. Heb. *iyyār*.] The post-exilic name of one of the Jewish months, being the eighth of the civil and second of the ecclesiastical year; its pre-exilic name was *Zif*.

1737 W. WHISTON tr. *Josephus' Works* 910 The Romans began to raise their banks on the twelfth day of the month Artemisius (Jyar). **1738** A. CRUDEN *Compl. Concordance Holy Scriptures* s.v. *month*, *Zif*..is the second month of the holy year, and..answers to that which afterwards had the name *Jiar*, or *April*. **1902** *Encycl. Brit.* XXVI. 43/2 In 746 B.C. Calah rebelled, and on the 13th of Iyyar (April), in the following year, Pulu or Pul..seized the throne. **1962** *New Jewish Encycl.* 229/2 Most of the *Sefirah* days occur during the month of Iyar. **1973** *Jewish Chron.* 9 Feb. 29/2 On May 7 this year the State of Israel will be 25 years old. (The Hebrew date is the fifth day of the month of Iyar.)

Izarra (izā·ra). Also **Izzara.** [Basque, lit. 'star'.] A liqueur from the Pyrenees, brandy-based and flavoured with herbs.

1926 E. HEMINGWAY *Sun also Rises* xix. 243 The waiter recommended a Basque liqueur called Izzara.. made of the flowers of the Pyrenees... It looked like hair-oil and smelled like Italian *strega*. **1965** R. POSTGATE *Plain Man's Guide to Wine* (ed. 2) ix. 149 *Name* Izarra. *Flavour* Wild herbs. *Origin* France. **1971** *Good Food Guide* 397 Grapefruit sorbet with Izarra.

izba, var. *ISBA.

Izod (ɔi·zǫd). The name of E. G. *Izod* (fl. 1903), British engineer, used *attrib.* with reference to a kind of impact test devised by him in which a notched specimen fixed at one end is broken by a blow from a pendulum hammer and the energy absorbed determined by the decrease in the swing of the pendulum.

1904 *Proc. Inst. Mech. Engin.* IV. 1225 A single-blow test of the Izod type. **1922** GLAZEBROOK *Dict. Applied Physics* I. 206/1 The brittleness in steel resulting from a high percentage of sulphur and phosphorus is well marked by the Izod test. *Ibid.*, The form of notch selected, and sometimes called the Izod notch, is a 45° vee, 2 mm. deep, with a root radius of 0·25 mm. **1939** E. C. ROLLASON *Metall. for Engineers* i. 11 As a material with a low Izod figure offers a poor resistance to the development of a crack, a low Izod value indicates that in service there will be a greater chance of final failure before the crack is discovered. **1954** A. R. BAILEY *Text-bk. Metall.* xiii. 457 The Izod test..is standard in Great Britain. **1968** D. R. CLIFFE *Technical Metallurgy* vii. 152 The Izod machine is a cantilever type with the knife-edge of the hammer striking the specimen at the horizontal at a point 22 mm above the plane of gripping.

‖ **izzat** (i·zɒt). Also **izzut.** [Urdu, ad. Arab. *'izzah* glory.] Honour, reputation, credit, prestige.

1857 H. LAWRENCE *Let.* 26 Feb. in Edwardes & Merivale *Life Sir H. Lawrence* (1872) II. xviii. 279 Man can but die once, and if I die in Oude, after having saved some poor fellow's hearths, or skins, or *izzut* (reputation), I shall have no reason for discontent. **1893** KIPLING *Many Inventions* 207 Thou hast done great wrong, and altogether lost thy *izzat* and thy reputation. **1922** *Blackw. Mag.* Feb. 201/1 *Izzat*, too, generally prescribes that he should be an hour or two late. **1924** E. M. FORSTER *Passage to India* iii. 31 The educated native..[is] trying to increase his izzat—in plain Anglo-Saxon, to score. **1953** —— *Hill of Devi* 27 In every remark and gesture, does not the Indian prince either decrease his own 'izzat' or that of his interlocutor? **1968** J. BARZUN *Amer. University* (1969) viii. 252 We saw how much faculties want presidents 'prestigious'. I suggest that instead of prestige we use the Hindi word *izzat* and see how absurd we are: 'Has he *izzat*? Have we enough *izzat* in the house? Our friends across the way are getting ahead of us in *izzat*.' Newspapers could then rank institutions *izzat*-wise.

J

J

J. Add: **II. 6. b.** In *Electr.* j is used (in place of the mathematical symbol *i*, which is used for electric current) to represent $\sqrt{(-1)}$, or an angular displacement of 90°.

1893 *Electrician* 15 Sept. 522/2 The next Paper was by Mr. C. P. Steinmetz on 'Complex Quantities and their Application in Electrical Engineering'. This Paper contained a novel method of the treatment of alternating currents... The device is to introduce the letter *j* in the expression of the sine curve, at first simply as a distinguishing index without mathematical meaning, and afterward as $j = \sqrt{(-1)}$. **1907** Franklin & Esty *Elem. Electr. Engin.* II. v. 90 It can be shown..by ordinary trigonometry that the components of a line whose length is $I\sqrt{R^2+X^2}$ and which is θ degrees ahead of *I*, where θ is the angle whose tangent is *X/R*, are $e_1 = Ri_1 - Xi_{11}$ and $e_{11} = Xi_1 + Ri_{11}$. Therefore the term j^2Xi_{11}..must be equal to $-Xi_{11}$, or, in other words, we must have: $j^2 = -1$. **1945** 'Electr. Engineer' *Ref. Bk.* I. 67 Considering *j* as a vector-operator, it is a means whereby a length in one direction is rotated anticlockwise by 90°. **1961** R. B. Angus *Electr. Engin. Fund.* xiii. 341 Since $j^2A = -A$.., then $j^2 = -1$ or $j = \sqrt{(-1)}$. *Ibid.* 343 The *j* is always placed to the left in the product. This practice helps avoid the overlooking of a *j*-term when real and *j*-terms are being grouped.

c. In *Physics* J and j represent quantized angular momentum.

j was first used by A. Sommerfeld (in *Ann. d. Physik* (1923) LXX. 33), in place of n_j, which had been used by W. Heisenberg (in *Zeitschr. f. Physik* (1922) VIII. 274) in place of Sommerfeld's original symbol n_i, which he introduced (in *Atombau und Spektrallinien* (ed. 3, 1922) vi. 446 and *Zeitschr. f. Physik* (1922) VIII. 269) as an empirical 'inner' quantum number (G. *innere Quantenzahl*) in describing the Zeeman effect. As this was interpreted in terms of later theory, *j* became the quantum number of the total angular momentum of an electron, corresponding to the resultant of its spin *s* and orbital momentum *l*; *J* is used similarly for an assemblage of electrons, and in molecular spectroscopy is the quantum number for angular momentum due to rotation of a molecule as a whole. Apart from these specific uses, j and J are often used as symbols for a general angular momentum.

jj-coupling, an approximation used in the quantum theory of the atom when the spin-orbit interaction of individual electrons is large compared with the remaining electrostatic interaction between one electron and another, so that spin and orbital angular momenta may be coupled to give a resultant *j* for each electron, and the resultants in turn coupled to give the total angular momentum *J* of the electrons. Cf. *LS-coupling.

1924 *Phil. Mag.* XLVIII. 720 The X-ray atomic levels may be conveniently classified by means of three quantum numbers—*n* (total), *k* (azimuthal), and *j* (inner). **1929** *jj*-coupling [see *COUPLING *vbl. sb.* 6 f (ii)]. **1930** *Physical Rev.* XXXVI. 613 The resultant angular momentum of the molecule (quantum number *J*). **1955** Townes & Schawlow *Microwave Spectroscopy* i. 5 As *J* increases and the molecule rotates faster, it stretches. **1957** M. E. Rose *Elem. Theory Angular Momentum* i. 11 Consider the case of two spins (or angular momenta) j_1 and j_2. **1970** G. K. Woodgate *Elem. Atomic Struct.* vii. 140 In the lower-lying configurations of neutral atoms the *j-j* coupling approximation is not often found, in other words the conditions for small electrostatic interaction are not often satisfied, even in heavy elements.

III. J, joule; J.A., Judge Advocate; Justice of Appeal; J.C.R., Junior Common Room; j.n.d., just noticeable difference; J.P. (examples).

1951 *Symbols, Signs & Abbreviations* (R. Soc.) 14 Joule. ..J. **1967** W. H. Hayt *Engin. Electromagn.* (ed. 2) ii. 32 The volt..having the label of joules per coulomb (J/C) or newton-meters per coulomb.., we shall..measure electric field intensity in..volts per meter (V/m). **1881** D. Jones *Notes Mil. Law* p. vii, J.A., Judge Advocate. *Ibid.* v. 67 The J.A. and prosecutor must not be the same person. **1918** E. S. Farrow *Dict. Mil. Terms* p. x, J.A., Judge Advocate. **1883** *Wharton's Law Lexicon* (ed. 7) 432 J.A., Justice of Appeal. **1972** *Mod. Law Rev.* XXXV. 1. 45 The dictum already referred to of Watermeyer J.A. came at the end of his judgment. **1892** *Isis* (Oxf.) 8 June 35/2 Only the Pres. of the J.C.R. and the Captain of the Eleven retain their equanimity. **1914** C. Mackenzie *Sinister St.* II. III. viii. 656 The editor rejected the frivolous attentions of his audience, and left the J.C.R. **1968** *Listener* 1 Aug. 147/1 The Student Council received the support..of the vast majority of JCRs. **1929** *Encycl. Brit.* VII. 420/1 The just noticeable difference, often called the 'j.n.d.', between the stimuli of two sensations. **1948** *Sci. News* VII. 14 Rubbers could be compared with one another with about three times as small a j.n.d. as that found for bitumens. **1732** *Calendar State Papers Amer. & W. Indies* (1939) 48 A dispute arising between two of his men on shore, Francis Squib and Jacob Taverner, J.P's in that harbour, put them in the stocks. **1869** *Bradshaw's Railway Manual* XXI. 328 Directors..Sir Benjamin Morris, J.P. **1972** P. Johnson *Offshore Islanders* v. 291 The Local Government Act of 1888..replaced the old Quarter Sessions of JPs by democratically elected County Councils.

jab, *v.* Add: **d.** *trans.* To give (a person) a stabbing blow with the fist. Also *fig.*

1901 R. Fitzsimmons *Physical Culture & Self-Defense* 114 Jab him, if you can, with your left. **1915** E. Corri *30 Yrs. Boxing Referee* 38 Time and again he jabbed and patted Smith cleverly on the nose with his left hand. **1959** *Amer. Speech* XXXIV. 155 One may be shafted or jabbed by the opposite sex, a professor..or anyone else for any real or imagined injury.

e. *trans.* and *intr.* To inject or inoculate (a person) with a hypodermic needle; to use (a hypodermic needle) to make an injection. So **ja·bbing** *vbl. sb. slang* (orig. *U.S.*).

1926 *Flynn's* 16 Jan. 638/2 Some stiffs uses mud but coke don't need any jabbin', cookin', or flops. You just hit it an' go. **1938** *Amer. Speech* XIII. 186/1 To jab, to take drugs hypodermically. **1948** Partridge *Dict. Forces' Slang* 100 Be jabbed, to be inoculated or vaccinated. **1956** S. Longstreet *Real Jazz* xviii. 114 Not all jazz-players smoke marijuana or opium, or sniff snow or jab a vein. **1968** J. R. Ackerley *My Father & Myself* xiii. 146 Dr Wadd..dashed in with a hypodermic syringe of digitalis and jabbed it so hastily, though successfully, into the back of one of his hands that it raised a large lump. **1968** 'L. Black' *Outbreak* ix. 82 Smith-road primary got jabbed. Why not our school?

jab, *sb.* Add: **b.** An injection with a hypodermic needle. *slang* (orig. *U.S.*).

1914 Jackson & Hellyer *Vocab. Criminal Slang* 48 Jab, current amongst morphine and cocaine fiends. A hypodermic injection. **1959** *Punch* 13 May 658/3 Resenting the hypodermic jab intended for the bullock. **1972** G. Durrell *Catch me a Colobus* i. 11 Can't you give me a jab of something to keep me going? **1973** *Times* 17 Apr. (Liberia Suppl.) p. ii/2 The visitor must..take precautions and submit to a variety of jabs.

c. A radio signal of momentary duration. *colloq.*

1932 *News Chron.* 23 Sept. 10/7 Another film I saw was of the flight of a jab of wireless energy on its journey from the East End to King's College, Strand. **1945** *Electronic Engin.* XVII. 679/3 For this purpose they devised a radio transmitter which sent out very short pulses, or jabs, of radio energy.

jabal, var. *JEBEL.

jabbers, jabers (dʒæ·bəɪz, dʒēi·bəɪz). Also **jap(p)ers.** Corruption of *Jesus*, used expletively (see *BEJAB(B)ERS *int.*).

1821, etc. [see *BEJAB(B)ERS *int.*]. **1934** A. P. Herbert *Holy Deadlock* 311 'Be Jabers!' said the girl, 'is that the way you feel?' **1970** *Alberta Hist. Rev.* XVIII. III. 18/2 'Did you hit that man?' The gunner replied, 'Be japers, I ate what's left of him.'

jabberwock (dʒæ·bəɪwǫk). The name of the fabulous monster in Lewis Carroll's poem *Jabberwocky*. Hence in allusive and extended uses, esp. 'incoherent or nonsensical expression'. So **ja·bberwocky**, invented language, meaningless language, nonsensical behaviour; also as *adj.*, nonsensical, meaningless, topsy-turvy. Also **ja·bberwock(y)** *v. intr.*, to write, speak, etc., in jabberwocky style.

1871 'L. Carroll' *Through Looking-Glass* i. 22 The Jabberwock, with eyes of flame, Came whiffling through the tulgey wood. **1902** J. Buchan *Watcher by Threshold* iii. 38 It was the strangest jumble of vowels and consonants I had ever met... It was some maniac talking Jabberwock to himself. **1908** *Daily Chron.* 10 Apr. 4/7 Those exceptional modern folk who write with equal ease in the ordinary left-to-right manner and in 'Jabberwocky' fashion..right to left. **1917** [see *CHRISTMAS-TREE 2]. **1926** *Glasgow Herald* 7 Oct. 5 From 'Jabberwocky' it is but a short step to the old-fashioned nursery-rhyme. **1931** E. Wilson *Axel's Castle* vi. 227 The dreaming mind does not usually speak—and when it does, it is more likely to express itself in the looking-glass language of 'Jabberwocky' than in anything resembling ordinary speech. **1939** *Times* 25 Feb. 15/5 It is all very Jabberwocky, and so far the writers of the movement [surrealism] have the advantage of the artists. **1953** H. Miller *Plexus* (1963) ii. 57 Realizing in a short time that I was not in the least interested in all this jabberwocky, and thinking of Mona waiting for me to lunch with her, I suddenly interrupted him. **1959** *Listener* 29 Jan. 226/1 His [sc. Skelton's] tendency to jabberwock (like Pound or Eliot, Skelton was a polylingual versifier). **1963** *Guardian* 22 July 5/5 He jabberwockied, pulling furiously at his ear-lobe as he talked. **1964** A. Swinson *Six Minutes to Sunset* vi. 126 Sometimes, to confuse the issue,..he would indulge in his own subtle form of jabberwocky. **1970** J. Fleming *Young Man, I think you're Dying* xii. 164 He was going to wear her down, intimidate her, tame and train her to obey, and a whole lot of other jabberwocky which he couldn't..remember now. **1972** *Collector's Guide* Aug. 12/1 (Advt.), Worcester first period teapot stand 'Jabberwocky pattern'.

jabiru. Also *Austral.* **jaberoo, jabiroo.** Add later examples of the Australian name for the black-necked stork, *Xenorhynchus asiaticus*, which is closely related to the tropical American *Jabiru mycteria*.

1912 A. Searcy *By Flood & Field* viii. 60 There were also Jaberoo, Spoonbill, Ibis and Spur-wing Plover. **1943** W. E. Harney *Taboo* 158 The jaberoo struts along its sands. **1946** I. L. Idriess *In Crocodile Land* xxxiii. 237 The call of the jabiru. **1952** *Coast to Coast 1951* 211 He had ridden up quietly, and the big, black-and-white, dark-blue-necked jabiroo, pacing slowly on its red-yellow legs at the further end of the lagoon, had not flown away. **1965** G. McInnes *Road to Gundagai* viii. 125 The brick walls dissolved into..polygonum swamps over which the long legged jabiroo flew creaking on its way. **1965** *Austral. Encycl.* V. 114/1 Sometimes the jabiru is quite solitary, hence perhaps the term 'policeman-bird', by which it is known in North Queensland.

Jablochkoff (yā·blǫχkǫf). The name of Paul Jablochkoff (1847–94), Russian physicist, used *attrib.* and in the possessive to denote an electric arc lamp invented by him (now obsolete) in which carbon rod electrodes were placed side by side and separated by an insulating material such as plaster of Paris, which gradually melted as the electrodes burned away.

1877 *Jrnl. Franklin Inst.* CIV. 295 Jablochkoff's electric candles burn for two or three hours without interruption. **1884** J. E. H. Gordon *Pract. Treat. Electr. Lighting* vii. 99 The lighting of the Avenue de l'Opera in Paris in 1878 by Jablochkoff candles first demonstrated the possibility of street lighting by electricity. **1894** R. M. Walmsley *Electr. Current* xii. 508 Because of the difference in the rate of burning of the positive and negative carbons in the ordinary arc, the Jablochkoff candle has to be supplied with alternate currents. **1922** S. G. Starling *Electricity* vi. 85 In that year [sc. 1876] Jablochkoff employed two parallel carbon rods, separated by an insulating material. ..These were called Jablochkoff's candles. They soon gave place to more convenient arc lamps.

jaboticaba (dʒabǫtikā·bā). Also **jabuticaba.** [Tupi.] An evergreen Brazilian tree, *Myrciaria* (or *Eugenia*) *cauliflora*, of the family Myrtaceæ, which bears clusters of white flowers and purple fruits directly on the trunk and branches; also, the fruit of this tree.

1824 H. E. Lloyd tr. *Spix & Martius's Trav. Brazil* II. III. ii. 85 A light and agreeable wine is also prepared from the fruits of the jabuticaba (*Myrtus cauliflora*). **1862** *Chambers's Encycl.* IV. 161/1 *E*[*ugenia*] *cauliflora*, a Brazilian species,..yields a very fine fruit of a black colour, about the size of a greengage plum, called the Jabuticaba. **1931** B. Miall tr. *Guenther's Naturalist in Brazil* iv. 79 In the Jaboticaba, a myrtle-tree, even the trunk looks as though beaded with the blue-black fruits, about the size of a damson, that seem to be nailed on to the bark. **1974** F. N. Howes *Dict. Useful Plants* 132 Jaboticaba (Braz[il]) *Myrciaria*, esp. *M. cauliflora*, (Braz. grape), fr[uit] ed[ible], used for wine and jelly.

jacal (hākā·l). Also **hackal, jackal(l), jucal.** [Mexican Sp., ad. Nahuatl *xacalli*.] A hut constructed of erect poles or stakes filled in with wattle and mud, a type common in Mexico and the south-western United States; an adobe house; also, the material or method used in building such a hut.

1838 'Texian' *Mexico v. Texas* 249 It was a little Jacal, or cabin, built with large unburnt bricks, called adobes, in the language of the country. **1844** J. J. Webb *Adventures in Santa Fe Trade* (1931) 104 In a valley..where the herders had a temporary corrall and jacal made of bushes laid upon poles. *Ibid.* 105 The jacal was full, packed so thick it was impossible to count them. **1850** J. W. Audubon *Western Jrnl.* (1906) 230 We..saw a comfortable (for this country), log and jacal built house. **1897** *16th Ann. Rep. U.S. Bureau Amer. Ethnol. 1894–95* 108 This method is known to the Mexicans as 'jacal', and much used by them. It consists of a row of sticks or thin poles set vertically in the ground and heavily plastered with mud. **1900** R. B. C. Graham *Thirteen Stories* 127 A straw-thatched jacal. **1947** *Chicago Sun Bk. Week* 8 June 2/2 In the back alleys..are to be found..primitive jacals.

J-acid. *Dye Chem.* A crystalline dye intermediate, 2-amino-5-naphthol-7-sulphonic acid, $C_{10}H_9NO_4S$, used in coupling reactions with diazotized amines to form direct azo dyes.

1914 F. W. Atack tr. *A. Wahl's Manuf. Org. Dyestuffs* vii. 66 The third [sc. compound] is obtained by alkaline fusion of 2-naphthylamine-5:7-disulphonic acid, and is called 'J acid'. **1917** Fort & Lloyd *Chem. Dyestuffs* xvi. 159 There are certain peculiarities about J-acid and its azo derivatives. Thus coupled with diazobenzene derivatives cotton-dyeing properties are absent, but with diazonaphthalene derivatives are strongly marked. **1937** *Chem. Abstr.* XXXI. 3281 Absorption by cotton cellulose of J acid..was studied. **1971** R. L. M. Allen *Colour Chem.* v. 57 About 35 monoazo direct dyes are in use, most of them containing either a thiazole or a J acid residue.

jacinth. Add: **1. e.** = *HYACINTH 3 b.

1854 L. A. Meall *Moubray's Treat. Poultry* 288

Jacinth,..slaty-blue, and pied on back and wings with white.

jacitara (dʒæsitā·rā). [Tupi.] In full, *jacitara palm*. A prickly climbing palm, *Desmoncus macroacanthus* or *D. orthacanthos*, native to the Amazon region.

1853 A. R. Wallace *Palm Trees of Amazon* 74 The 'jacitara' never loses its hold, and it is only by deliberately extracting its fangs that the intruder can expect to depart unhurt. **1860** Mayne Reid *Odd People* 52 The grated pulp [of manioc]..is afterwards put into a long elastic cylinder-shaped basket or net, of the bark of the 'jacitara' palm. **1863** H. W. Bates *Naturalist on River Amazons* I. ii. 48 There is even a climbing genus of palms (*Desmoncus*), the species of which are called, in the Tupi language, Jacitára. *Ibid.* vii. 322 When the rolls [of tobacco] are sufficiently well pressed they are bound round with narrow thongs of remarkable toughness, cut from the bark of the climbing Jacitára palm tree. **1878** *Chambers's Encycl.* Suppl. X. 579/1 Jacitara palm, (*Desmoncus macroacanthus*), a palm found in the forests of the low lands of the Amazon district in South America. **1927** R. R. Gates *Botanist in Amazon Valley* vii. 167 Among climbers collected here was a small climbing palm called jacitara (*Desmoncus* sp.). The slender stem bears sharp spines in pairs, projecting backwards. **1931** B. Miall tr. *Guenther's Naturalist in Brazil* v. 92 In Brazil there is another climbing palm, the Jacytára. I found it in Pernambuco, and was quite intimidated by its armament, which has earned it the name of 'the terrible'.

Jack, *sb.*¹ Add: **1. d.** Used as a form of address to an unknown person. *colloq.* (orig. *U.S.*).

1889 Barrère & Leland *Dict. Slang* I. 490/2 Jack (American), it is common among schoolboys in Philadelphia to address a stranger as Jack. **1933** Partridge *Words, Words, Words!* i. 71 Jack is still very frequent among 'the common people' when, in addressing a stranger, one wishes to avoid the abruptness caused by omitting the unknown name. **1943** *N.Y. Times* 9 May II. 5/4 Jack, that man had them rolling in the aisles. **1966** S. Kelly in F. Shaw et al. *Lern Yerself Scouse* 76 Dawn taps yer winder [sc. window]..another day before yiz [sc. you], Jack. **1970** C. Major *Dict. Afro-Amer. Slang* 70 Jack, term of address by one male to another.

2. d. A policeman or detective; a military policeman. Cf. *JOHN 1 c. *slang.*

1889 Clarkson & Richardson *Police!* xxiii. 320 A policeman, a fly, Jack,..crusher, peeler. **1899** *Birmingham Daily Mail* 1 Nov., A couple of men who were in plain clothes in the tap-room of a public-house, and were suspected by the 'gaffer' of being 'Jacks'. **1919** W. H. Downing *Digger Dial.* 29 Jack, a military policeman. **1930** *Bulletin* (Sydney) 1 Jan. 11/2 Blue..looked up and saw two Jacks waiting. 'Where are you going?' demanded one M.P. **1941** *Argus* (Melbourne) *Week-End Mag.* 15 Nov. 1/4 Jacks, military police. **1946** F. Sargeson *That Summer* 102 We all had to stand there with a crowd of jacks in plain clothes standing round. **1967** J. Gardner *Madrigal* viii. 199 You're not going to believe it. I haven't told all to the jacks naturally. **1971** J. Wainwright *Dig Grave* 45 These county coppers..couldn't get their minds unhooked from the words 'New Scotland Yard'—as if every jack in the Metropolitan Police District worked from there.

e. Slang phr. *on one's jack* = on one's own, alone (short for *on one's Jack Jones*: see 34 e below).

1931 'G. Orwell' *Coll. Ess.* (1968) I. 71 *Jack, on his*: on his own. **1935** —— *Clergyman's Daughter* iii. 197 Michael went off on his jack an' left me wid de bloody baby. **1936** [see *GRASS *sb.*¹ 11*]. **1968** M. Woodhouse *Rock Baby* ix. 93 You're off on your jack then? **1973** R. Parkes *Guardians* x. 193, I thought I could go sneaking in there all on my jack and bring out the evidence.

3. b. Phr. *Jack ashore*: see quot. 1909. *slang.*

[**1875** R. Rowe (*title*) Jack afloat and ashore.] **1909** J. R. Ware *Passing Eng.* 158/2 *Jack ashore*, Jack elevated—practically drunk, and larky. **1920** Conrad *Victory* (1921) Author's note p. xviii, It was long after the sea-chapter of my life had been closed, but it is difficult to discard completely the characteristics of half a lifetime, and it was in something of the Jack-ashore spirit that I dropped a five-franc piece into the saucebox. **1970** E. McGirr *Death pays Wages* iv. 90 Jack Ashore does not check bills.

c. Colloq. phr. *I'm all right, Jack*: a saying indicating selfish complacency on the part of the speaker.

1910 D. W. Bone *Brassbounder* iii. 37 It's 'Damn you Jack—I'm all right!' with you chaps. **1919** F. Ninette *Tiddley Sailors* 26 They dodged as much work as possible and generally assumed the manner 'I'm all right Jack'. **1958** A. Hackney (*title*) Private life or I'm all right Jack. **1960** *News Chron.* 2 May 4/6 This 'I'm all right, Jack' attitude towards such relatives is deplorable. **1970** *Times* 17 Feb. 3/2 He adopted an 'I'm all right, Jack' attitude in leaving his convoy. **1971** John & Humphry *Because they're Black* (1972) x. 105 Right now it is, as I said before, dog eat dog... I'm all right, Jack, damn you.

4. b. Colloq. shortening of *lumberjack* (Lumber *sb.*¹ 4). *N. Amer.*

c **1900** in F. Rickaby *Ballads & Songs Shanty-Boy* (1926) 97 Every jack's a cant-hook man... They do some heavy loggin'. **1913** *Collier's* 18 Jan. 21/1 The breaking up of the lumber camps and the streaming southward of thousands of 'jacks'. **1947** *Sat. Even. Post* 8 Mar. 20/1 The red-bearded jack came on again, head low and shielded. **1961** W. E. Greening *Ottawa* 101 The jacks who felled the trees and the workers who stripped them were called *piqueteurs*. **1973** L. Gutteridge *Killer Pine* iv. 46, I had to fire thirty Jacks last fall... Lumberjacks.

5. b. *California jack* (examples). Also *Californian jack.*

1882 J. W. Steele *Frontier Army Sk.* (1883) 50 Here is the down-east Yankee..turning his native cunning to account at poker and California jack. **1893** W. Raleigh *Let.* 28 July (1926) I. 177 His losses at Californian Jack..were my gains. **1921** C. E. Mulford *Bar-20 Three* iii. 39 For two hours they sat and played California jack in plain sight of the street.

15. d. Substitute for def.: A socket or receptacle having one or more pairs of terminals and designed so that insertion of a suitable plug enables a device to be quickly introduced into a circuit. (Examples.)

1891 J. Poole *Pract. Telephone Handbk.* vii. 128 The effect of inserting a plug in one of the jacks is that the end of the plug lifts the line spring R from pin Y. **1905** Preece & Sivewright *Telegraphy* (new ed.) xi. 233 The home sections and the multiple panels are made up of 'jacks', mounted in strips. *Ibid.* 234 The 'jacks' consist of two springs of unequal length and a collar or socket. **1926** J. L. Pritchard *Broadcast Reception* xi. 185 The last three jacks have filament controlling contacts by which the insertion of the plug automatically lights the filaments. **1970** *Toronto Daily Star* 24 Sept. 38/2 (Advt.), Lloyd's AM/FM digital clock radio... Features..earphone jack and solid-state circuitry. **1971** R. Thomas *Backup Men* xiii. 119 Is there another jack in this room?.. Can you get another phone and plug it in?

18. b. = Jack-stone; usu. *pl.*; also, a game played with these (see also quot. 1908).

1900 S. R. Crockett *Little Anna Mark* xii. 97 Playing at quoits, tops, marbles, tic-tac-toe, jacks, knucklebones. **1908** *Dialect Notes* III. 323 *Jack*,..a piece of metal with five tines or protuberances, used in the game of jacks. *Jacks*, an indoor catching game played with small five-tined metal pieces. **1922** A. C. Sies *Spontaneous & Supervised Play* xix. 293 'Jacks' is a game in which throwing and catching are not the main centers of interest; rather are the attention and interest focussed on what is done between catches. **1960** H. Miller *Nexus* (1964) i. 11 Do you know how to skate?..Did you ever play jacks?

19. c. Money. *slang* (orig. *U.S.*).

1890 M. Townsend *Index U.S.A.* 427 The..verbal wealth of the United States language is illustrated in an inquiry for a loan of money; by using any of the following words in conjunction with the inquiry, Have you any.. Jacks, [etc.]. **1920** *Collier's* 28 Aug. 33/2 The fans which paid their jack to see a fight would be gypped. **1932** J. Dos Passos *1919* 24 This way every bastardly tourist with a little jack thinks he can hire you. **1945** 'N. Shute' *Most Secret* 111, I hadn't that much jack.... I worked a passage home. **1960** A. Prior in *Pick of Today's Short Stories* XI. 184, I asked him..to think of the new suits he could get..when the jack came in.

d. *slang.* Also **jacks, jax.** Five pounds; a five-pound note. Cf. *Jack's alive* s.v. sense *36.

1958 F. Norman *Bang to Rights* iii. 150 I'll bet you a jacks that I nick you. **1968** *Gloss. Brit. Argot*, Jack, five pounds. **1968** *Guardian* 13 Apr. 7/3 'That one,' says the dealer from Islington, 'that one we know she died in; so it'll cost you a jax.'..Five quid for a shroud; cheap at the price. **1970** G. F. Newman *Sir, You Bastard* viii. 230 'Couldn't lend me a Jack's, Terry, could you?' 'Sure.' He gave the DS a fiver. **1972** K. Royce *Miniatures Frame* v. 64 From under a pottery sugar jar..protruded two jacks.

25. (Examples.) Also *jack boat*.

1891 *Rep. U.S. Comm. Fish & Fisheries 1887* App. VI. 529 The jack varies from 5 to 15 tons; is schooner-rigged, carrying three sails as a rule. **1895** *St. Nicholas* Apr. 448/2 The gashers [were] dashing in and out among the punts and jacks (stoutly built two-stickers larger than gashers). **1908** *Daily Chron.* 8 Jan. 3/3 A little jack (much the same in rig as our North Sea smacks). **1937** *Beaver* June 29/2 Bill had a nice little jack (small fishing schooner, in this case with outriggers on the quarters instead of booms). **1951** *Maine Coast Fisherman* Oct. 26 A typical Newfoundland jack seen in the cruise to the Bras D'Or Lakes. **1954** *9th Census of Canada 1951* IX. Table 4 Jack boats... Bateaux 'Jack'. **1965** *National Fisherman* Mar. 24/2 The 'jack schooners' or 'jack boats' (so-called in Cape Breton), or 'two-spar boats' (as they are known in Newfoundland) were 40' to 50' from stemhead to taffrail. They were gaff-rigged on both masts and they usually carried a longish bowsprit. **1969** H. Horwood *Newfoundland* xx. 157 The sons of men who had built windjammers were confined to building trap skiffs and jack boats.

26*. A tablet of heroin. *slang.*

1967 M. M. Glatt et al. *Drug Scene* 115 *Jack*, heroin tablet. **1971** R. Busby *Deadlock* xii. 177 He's been cranking up on horse [sc. heroin]. His last jack is wearing off, and he's grovelling on the floor for another pill.

29. c. *Austral.* A laughing jackass, a kookaburra. Cf. *JACKO, *JACKY 1 c.

1898 Morris *Austral Eng.* 216/1 The bird is generally called only a *Jackass*, and this is becoming contracted into the simple abbreviation of *Jack*. **1934** *Bulletin* (Sydney) 14 Feb. 26/2 Jack came to the conclusion that it was, as tucker, a washout, and departed. **1949** *Geogr. Mag.* Feb. 374/1 Hence such names for the kookaburra as *laughing jackass*, *jack*.

33. a. jack-hammer, jackhammer, a portable rock-drill worked by compressed air; **jack-ladder,** (*b*) = *JACK-CHAIN 2; **jack-plug** *Electr.*, a single-pronged plug for use with a jack (sense *15 d); **jack-pot** (earlier and later examples); also, any large prize, as from a lottery or a gambling machine; often, a prize that accumulates until it is won; *to hit the jack-pot*: to win such a prize; to have an extraordinary stroke of luck; (*b*) (see quot. 1914); **jack-roller** (see quots.); so *jack-roll* v. trans., *jack-rolling*; **jack shaft,** any of various kinds of auxiliary or intermediate shafts which are driven by another shaft or by a set of gears, esp. in locomotives and motor vehicles (see quots.); **jack-socket** *Electr.* = *JACK *sb.*¹ 15 d.

1930 *Engineering* 24 Oct. 511/2 A battery of about 45 jackhammers..drilled two lines of holes. **1936** *Economist* 12 Oct. 714/1 While this may be possible with machine drills in development faces, it is difficult at the moment to see how such appliances could be used with mobile jackhammers in stoping. **1971** *Daily Tel.* 2 Aug. 20/5 A specially-designed percussion drill similar to a jack-hammer. **1972** *Southerly* XXXII. 102 He works a ship and has about the delicacy of a jackhammer. **1886** *Encycl. Brit.* XXI. 345/1 From the rear end of the mill..a 'jack ladder' is constructed of heavy timber. **1929** *Ibid.* XIV. 482/1 An endless spiked conveyor known as a jack ladder. **1931** Moyer & Wostrel *Radio Handbk.* 874 (Index), Jack plugs. **1953** W. MacLanachan *Television & Radar Encycl.* 103/2 *Jack and Jack plug*, a socket with two or more contacts..into which a jack plug with corresponding contacts can be inserted. **1973** Jack plug [see *jack socket* below]. **1881** *Harvard Lampoon* 6 Apr. 40/2 Poker-playing is not to be learned in one evening, and Jack Pots are often a snake in the grass. **1884** *Virginia* (Nev.) *Chron.* 1 Oct. 3/3 Old Bill [the warden] just lays back until there is a good jack-pot of trout in hand, and then he makes a bold bluff and walks off with it. **1886** 'M. Kershaw' *Colonial Facts & Fictions* 229 They call this [growing poker] pool a Jack Pot. **1887** *Grip* (Toronto) 21 May 10/2 What was written of..was a jack-pot. **1902** L. McKee *Land of Nome* 123 On the occasion of his getting into a 'jack-pot' (some trouble) he had hunted me for legal advice. **1914** Jackson & Hellyer *Vocab. Criminal Slang* 48 *Jackpot*, a dilemma; a difficult strait; a retribution; trouble; an arrest. **1923** G. Ade *Let.* 24 Oct. (1973) 97, I..have been rather interested to learn that the Governor did not show you any degree of gratitude for your work in organizing the jack-pot. **1944** *Newsweek* 25 Dec. 67/1 The 'Vick's Vaper' had indeed hit the jack-pot. **1949** *Radio Times* 15 July 6/3 We saw our first American audience-participation show. The prizes included a diamond wrist watch... The jackpot was 1,250 dollars! **1956** N. Streatfeild *Judith* ii. 150 Those sort of weak good looks..quite often hit the jackpot. **1959** *Maclean's Mag.* 4 July 34/3 Canada House receives SOS messages from 'distressed Canadians', the official designation for those who get themselves into various jackpots. **1962** *Sunday Times* Suppl. 10 June 10 There is always the chance that one or other number or artist will hit the jackpot. **1963** *Listener* 28 Mar. 568/3 Cabinet Ministers are hauled out in front of the cameras and asked increasingly impertinent leading questions. A week or two ago Mr Butler copped one of these jackpots from Robert Mackenzie: did he, or did he not, want to be Prime Minister? **1967** *Times* 18 Dec. 5/6 A jackpot may be opened by any player who holds a pair of jacks or a higher ranking hand. **1970** *New Scientist* 14 May 341/1 Rolls-Royce landed a jack-pot order from the USA for the supply of aircraft engines. **1973** 'M. Innes' *Appleby's Answer* i. 10 It was nice to be fairly widely read... It would be even nicer if one day she contrived to hit the jackpot. **1923** N. Anderson *Hobo* i. 5 The 'jack roller',..the man who robs his fellows, while they are drunk or asleep. *Ibid.* iv. 51 'Jack rolling' may be anything from picking a man's pocket in a crowd to robbing him while he is drunk or asleep. **1926** *Flynn's* 16 Jan. 638/1 Jack rollin' th' workstiffs was like takin' candy from th' kids. **1930** C. R. Shaw *Jack-Roller* vii. 85, I was in a predicament, for I had no money, and you can't enjoy life without dough. My buddy, being an old 'jack-roller', suggested 'jack-rolling' as a way out of the delima [*sic*]. **1955** *Publ. Amer. Dial. Soc.* xxiv. 171 Thieves who specialize..on lumberjacks or other seasonal workers who get paid off in a lump sum at the end of the season are called *jack rollers*. **1896** T. Thornley *Cotton Spinning Calculations* ii. 70 The jack shaft of the frame makes 298 revolutions per minute. **1901** J. Black *Illustr. Carpenter & Builder Ser.: Scaffolding* 72 The electric motor..with jack-shaft and friction drive. **1907** R. B. Whitman *Motor-Car Princ.* ix. 146 The jack shaft..is a shaft passing across the car, and bearing on its ends the sprockets by which the wheels are driven. **1925** A. W. Judge *Mechanism of Car* vii. 105 The object of the differential is to enable either jack shaft to rotate at a different speed to the other..whilst transmitting the drive. **1936** W. Staniar *Mech. Power Transmission Handbk.* v. 144 The jackshaft is used for ratio purposes and also to break up long center distances. *Ibid.* 145 Jackshafts. Location.—Either between head and line shafts, or between line- and countershafts. **1940** *Chambers's Techn. Dict.* 465/2 *Jack shaft*, an intermediate shaft used in locomotives having collective drive; the jack shaft is geared to the motor shaft and carries cranks which drive the coupling rods on the driving wheels. **1950** *Engineering* 23 June 699/2 The final drive is through reduction gearing to a jackshaft which is fitted with pressed-on balanced cranks; thence the drive is by rods to the crankpins of the middle coupled wheels [of the Diesel locomotive]. **1962** *Diesel Traction* (Brit. Railways Board) 266 *Jackshaft*, a shaft with cranks at each end mounted across the frames for driving the road wheels through connecting rods. **1966** *McGraw-Hill Encycl. Sci. & Technol.* XII. 240/2 A countershaft, especially when used as an auxiliary shaft between two other shafts, is termed a jack shaft. **1970** J. Earl *Tuners & Amplifiers* i. 14 Modern amplifiers are often equipped with a jack socket wired to accommodate stereo headphones. **1973** *Radio & Electronics Constructor* XXVI. 700/2 A suitable circuit for a miniature 2·5 mm. or 3·5 mm. switched jack socket is given in Fig. 4, and this causes the speaker to be silenced when the jack plug is inserted.

34. b. (Further examples.)

1903 [see *DOGIE, DOGY]. **1942** G. Mitchell *Laurels are Poison* xvii. 181 It'll be all over College before you can say Jack Robinson. **1956** N. Mitford *Noblesse Oblige* 43 That picture will appear at Christies before you can say Jack Robinson, though there is no necessity whatever for such a sale.

c. *Jack Scott* = *Jock Scott* (see Jock¹ 1).

1908 R. Bagot *Anthony Cuthbert* xxix. 378 'There are some grilse come up. What do you think would be a good fly to use?'..'I should put on a small Jack Scott.' **1915** J. Webster *Dear Enemy* 167 He..got out his case of fishing-flies, and gallantly presented Betsy and me with ..a 'Jack Scott'..to make hat-pins.

d. *Jack Johnson* [from the name of a noted American Negro boxer, whose nickname was 'The Big Smoke'], = *BLACK MARIA 2.

1914 *Illustr. London News* 10 Oct. 504/1 The German 'Jack Johnson' siege-guns. *Ibid.* 505 The gigantic projectile which on bursting makes the black smoke called 'Jack Johnson'. **1919** [see *BLACK MARIA 2]. **1962** J. B. Priestley *Margin Released* II. iii. 101 The German heavy batteries..dropped 'Jack Johnsons' among us.

e. *Jack Jones*: rhyming slang for 'alone'; usu. in phr. *on one's Jack Jones*: on one's own; alone. Cf. sense 2 e above.

1925 Fraser & Gibbons *Soldier & Sailor Words* 130 Jack Jones, alone. **1935** 'G. Orwell' *Clergyman's Daughter* ii. 156 A good night's kip all alone... All on your Jack Jones. **1958** P. Scott *Mark of Warrior* II. 168 You're on your jack jones again. What do you do? **1972** A. Draper *Death Penalty* xx. 134 You're on your Jack Jones. Ben's deserted you.

35. Jack Dusty *Naut. slang* (see quots.); **Jack Mormon** *U.S.*, a non-Mormon on friendly terms with Mormons; also, a nominal or backsliding Mormon; **Jack Shalloo, Shilloo** *Naut. slang* (see quots.); **Jack Strop** *Naut. slang* (see quots.).

c **1931** W. N. T. Beckett *Few Naval Customs* 18 The junior member of the Paymaster's Victualling staff is known as The Dusty Boy or Jack Dusty. **1938** 'Giraldus' *Merry Matloe Again* 263 *Jack Dusty*, a ship's steward's assistant, *i.e.* any member of the supply branch below the rating of a petty officer. **1974** P. Wright *Lang. Brit. Industry* x. 85 In the Second World War, destroyers had a *Jack Dusty*, a supply assistant who kept ledgers listing all the stocks. **1845** *Quincy* (Ill.) *Whig* 30 Oct. 2/1 Jack Mormons, and sympathizers abroad may croak and groan over the poor Mormons. **1846** Jack-mormon [see *BIG-HEAD 3 b]. **1890** *Congress. Rec.* 2 Apr. 2941/2 In our country we have a *genus homo* called 'Jack-Mormon',..a class of individuals who do not belong to the Mormon church,..yet who are ever found doing the bidding of Mormon priests. **1947** *Time* 21 July 21/1 The number of backsliding 'jack-Mormons' is increasing. **1904** E. P. Statham *Story of 'Britannia'* iv. 70 This lad [sc. an officer cadet]..was already rather a favourite, being of the breezy type, which sailors call a 'Jack-shilloo'. **1929** F. C. Bowen *Sea Slang* 73 *Jack Shalloo*, a braggart according to its old naval meaning. Nowadays it is applied to a happy-go-lucky, careless officer and a slack ship is called a Jack Shalloo ship. **1962** Granville *Dict. Sailors' Slang* 65/1 *Jack Shalloo*, naval officer whose aim in life is to be popular with the men. A corruption of *John Chellew*, an officer who earned himself the title Popularity Jack. **1945** Baker *Austral. Lang.* viii. 162 *Jack Strop*, a new recruit who tries to pass himself off as an old hand. **1946** J. Irving *Royal Navalese* 98 *Jack Strop*, the Mess Deck's sobriquet for a conceited, opinionated sort of man.

36. *Jack in the basket* Naut., a type of warning beacon (see quot. *a* 1865); *Jack of the dust* (examples); *Jack's alive*, (*b*) rhyming slang for 'five'; also, a five-pound note; *Jack the Painter*, a kind of acrid green tea formerly used in the Australian bush; *Jack the Ripper*, popular name for a murderer of women in London in 1888, who mutilated the bodies of his victims; also used *allusively*.

a **1865** Smyth *Sailor's Word-Bk.* (1867) 407 *Jack in the basket*, a sort of wooden cap or basket on the top of a pole, to mark a sand-bank or hidden danger. **1921** *Yachting Monthly* Mar. 299/1 How comfortingly the cocoa boiled on an even keel at dawn with Jack-in-the-basket in sight! **1941** *Beaver* Sept. 38/1 Jack-in-the-Basket. A beacon. The Moose River was well buoyed, and the many shoals were marked with high poles surmounted with long wicker baskets, or broom heads of willows. **1878** *Detroit Free Press* (Suppl.) 12 Jan. 2/4 Forward, on the gun-deck, the Paymaster's Steward, and his assistant, the Jack of the Dust, were serving out small stores. **1882** J. W. Danenhower *Narr. 'Jeannette'* ii. 41 He was doing duty at the time as paymaster's yeoman, or 'Jack of the Dust'. **1931** Brophy & Partridge *Songs & Slang 1914–18* (ed. 3) 322 *Jacks alive*, the number 5, especially at House. **1938** F. D. Sharpe *Sharpe of Flying Squad* 331 *Jack's Alive*, a five-pound note. **1852** G. C. Mundy *Our Antipodes* I. x. 329 Another notorious ration tea of the bush is called 'Jack the Painter'. This is a *very* green tea indeed. **1878** *Australian* I. 418 The billy wins, and 'Jack the Painter' tea Steams on the hob, from aught like fragrance free. **1945** Baker *Austral. Lang.* iv. 83 *Jack the painter*..from the stain left round the drinker's mouth or in the billy (at least, that is the approved explanation). **1890** *Pall Mall Gaz.* 7 Mar. 5/1 A Jack the Ripper outrage at Moscow. **1902** *To-Day* XXXV. 99/2 Now we know 'oo Jack the Ripper was! **1919** C. P. Thompson *Cocktails* 17 If only the officer would let him have a whack at her over the open sights, he'd do the Jack-the-ripper act on her in half a tick. **1958** Hayward & Harari tr. *Pasternak's Dr. Zhivago* II. ix. 268, I expected to see a bashi-bazook or a revolutionary Jack the Ripper, but he was neither. **1959** 'H. Carmichael' *Stranglehold* i. 13, I had to obtain a Home Office permit. And in case you still think I'm Jack the Ripper, here it is. **1972** A. E. Lindop *Journey into Stone* (1973) ii. 21 There's a *lousy* fog... It's a Jack the Ripper's paradise.

37. Jack Russell (terrier), a small terrier named after John Russell (1795–1883), the so-called 'sporting parson'; **jack salmon,** read: *U.S.*, a large freshwater fish, *Stizostedion vitreum*, also called walleyed pike; **jack-sharp,** a northern dialect name for the Stickleback; **jack-spaniard** (earlier and later examples).

1907 R. Leighton *New Bk. Dog* xxix. 317/1 In another decade or so the neglected Sealy Ham Terrier,..and the almost forgotten Jack Russell strain, may have claimed a due recompense for their long neglect. *Ibid.* 318/1 'I have kept the Jack Russell type of terrier for nearly twenty years,' says Mr. Reginald Bates, 'and have used them for fox and badger digging.' **1931** *Times Lit. Suppl.* 13 Aug. 620/4 Perhaps the most popular is the 'Jack Russell' which..may now be considered a breed. **1965** B. Vesey-Fitzgerald *Dog Owner's Encycl.* 125 Jack Russell terrier. A small working terrier, named (for no very sound reason) after a West Country parson [sc. the Rev. John Russell of Devonshire] who, a century ago, was renowned for his working terriers. Any small Hunt terrier may be called a Jack Russell... There is no distinct type and the breed is not recognised by the Kennel Club. **1973** 'I. Drummond' *Jaws of Watchdog* i. 13 A little tubby man..with..the look of an overfed Jack Russell terrier. **1871** *Game Laws* (Pennsylvania) in *Fur, Fin & Feather* (1872) 122 The species commonly known as Susquehanna salmon, pike, perch, jack salmon..shall henceforth not be taken..during their spawning time. **1920** *Outing* (U.S.) May 118/2 We always hope..to catch a few jack salmon. **1811** A. Clarke *Jrnl.* 12 June in *Acct. Life A. Clarke* (1833) II. 261, I went into the grounds where I had often sported, read, talked, searched for birds' nests, and caught jack-sharps. **1876** I. Banks *Manchester Man* I. v. 81 He mun larn to tak' care on himsel' th' next time hé marlocks among th' Jack-sharps. **1901** *Westm. Gaz.* 24 Apr. 8/1 A boy of ten was attempting to catch jacksharps when he fell from the embankment into the stream... [South-country readers will be interested to learn that jacksharps are the little fishes known to them as sticklebacks or tittlebats.] **1925** J. T. Jenkins *Fishes Brit. Isles* 124 There are numerous local names for this fish [sc. the three-spined stickleback], such as Jack Sharp, Prickleback, [etc.]. **1974** P. Wright *Lang. Brit. Industry* vii. 63 The worker pursuing his week-end hobby of angling may find round a pond *cockies* or *jacksharps*, 'sticklebacks' to the uninitiated. **1833** *Chambers's Edin. Jrnl.* 21 Sept. 269/3 The jack-spaniard may be called the wasp of the West Indies; it is twice as large as a British wasp. **1867** J. K. Lord *At Home in Wilderness* xviii. 299 A hornet, called by the packers a 'Jack-Spaniard',..builds a circular paper nest. **1938** *Trans. R. Entomol. Soc.* LXXXVII. 181 *Polistes* Latreille. Four species of this genus were collected in Trinidad. *P. canadensis* (L.). Typical form. 'Jack Spaniard'. Common throughout all settled areas. **1963** D. M. Douglass *Saba's Treasure* i. 14 You could as well discover the particular jack-spaniard that stung you a week ago.

38. jack bean, a sub-tropical, climbing, leguminous plant of the genus *Canavalia*, esp. *C. ensiformis*; **jack-go-to-bed-at-noon** (examples); **Jack-in-the-bush,** (*a*) (example); (*b*) = Jack-in-the-green 1; **jack-in-the-hedge** = *Jack-by-the-hedge*, the hedge-garlic, *Alliaria petiolata*; **jack-in-the-pulpit** (earlier and later examples); **jack oak** (earlier and later examples); **jack-pine,** any of several species of pine, esp. the Banksian pine, *Pinus banksiana*.

1885 'C. E. Craddock' *Prophet Gt. Smoky Mts.* xv. 280 He sat upon the cabin porch beneath the yellow gourds and the purple blooms of the Jack-bean. **1951** *Dict. Gardening* (R. Hort. Soc.) I. 382/1 Jack Bean, Chickasaw Lima Bean... The young beans (about 4 to 6 in. long) are sometimes used like Scarlet Runners; the ripe seeds are said to be used as a substitute for coffee. **1853** A. Pratt *Wild Flowers* II. 47 The leaves and stems are often the only portions of the plant [sc. Yellow Goat's Beard] to attract the eye of the wanderer, for the flower is closed by mid-day. Several of its country names refer to this peculiarity, as Noontide, and Jack-go-to-bed-at-noon. **1931** M. Grieve *Mod. Herbal* I. 360/1 The Goat's Beard opens its blossoms at daybreak and closes them before noon, except in cloudy weather, hence its old country name of 'Noon-flower' and 'Jack-go-to-bed-at-noon'. **1951** W. de la Mare *Winged Chariot* 9 How punctual they!.. As testifies 'Jack-go-to-bed-at-noon'. **1807** R. Southey *Lett. from Eng.* I. xiii. 146 A more extraordinary figure is sometimes in company, whom they call Jack-in-the-Bush; as the name indicates, nothing but bush is to be seen, except the feet which dance under it. The man stands in a frame-work which is supported upon his shoulders, and is completely covered with the boughs of a thick and short-branched shrub. **1872** F. Kilvert *Diary* 10 May (1939) II. 196 He [sc. the sawyer] said wild garlic, called Jack-in-the-Bush, is a famous pot herb. **1875** Jack-in-the-hedge [in Dict.]. **1925** W. de la Mare *Broomsticks* 312 Young spring flowers—primroses, violets, jack-in-the-hedge, stitchwort—in palest blossom starred the banks. **1941** A. L. Rowse *Poems of Decade* 111 The many and various scents of the flowers, Cuckoo-pint, cow-parsley and jack-in-the-hedge. **1837** H. Martineau *Society in Amer.* I. ii. 211 Fine specimens of Jack-in-the pulpit, and the moccasin-flower. **1906** *N.Y. Even. Post* (Suppl.) 16 June 2 In these woods I made acquaintance with Jack-in-the-pulpits, or, as the English call them, 'Lords and Ladies'. **1949** *Nature Mag.* Apr. 178 A few of these, like Indian turnip or jack-in-the-pulpit, cowslip and milkweed, may be considered mildly inedible. **1972** T. McHugh *Time of Buffalo* viii. 91 Soups were popular, some brewed with..buffalo meat, berries, fat, and the roots of jack-in-the-pulpits. **1816** U. Brown *Jrnl.* in *Maryland Hist. Mag.* (1915) X. 266 Jack Oaks and other Scrub Wood. **1836** D. B. Edward *Hist. Texas* iv. 68 The post-oak and Jack-oak are considered in Texas as every man's property. **1901** Duncan & Scott *Hist. Allen & Woodson Counties, Kansas* 581 The 'jack oak hills' have been fenced. **1883** G. O. Shields *Rustlings in Rockies* xxxi. 285 This [country] is now grown up with scattering dwarf pines or, as the settlers call them, jack-pines. **1925** *Chambers's Jrnl.* June 381/2 The jack-pine grows low and twisted. **1957** J. Kerouac *On Road* (1968) II. ix. 55, I..looked up and saw jackpines in the moon. **1965** G. McInnes *Road to Gundagai* xii. 210, I saw, moving majestically towards me and looming over the head of the crowd like a Douglas fir among jackpine, a prematurely grey giant about nine feet tall. **1969** T. H. Everett *Living Trees of World* 51/1 The range of the jack pine (*P. banksiana*), the most northern of eastern American pines, extends south from near the Arctic Circle to New York and Minnesota.

jack, *sb.*⁴ Add: (Further examples.)

1820 J. Crawfurd *Hist. Indian Archipelago* I. iv. 122 Of the Jack fruit..two species occur in the Indian islands. ..The Jack is highly nutritious. **1824** H. E. Lloyd tr. *Spix & Martius's Trav. Brazil* I. ii. i. 175 Brazil is indebted to the intercourse of the Portuguese with the East Indies..for the excellent fruits of the jaca, the mango, and the jambos. **1839** T. J. Newbold *Pol. & Stat. Acct. Straits of Malacca* ii. 53 In the valley grow various fruit-trees, such as..the jack. **1919** *Nature* 25 Sept. 78/2 The native of the country is content with the fruit that is easily produced there and is already well known (in this case the durian, mango, sapodilla, mangosteen, jak, etc.). **1931** B. Miall tr. *Guenther's Naturalist in Brazil* iv. 80 The fruits of the Jaca, as large as a man's head, seem to grow directly from the trunk and boughs. **1969** *Oxf. Bk. Food Plants* 114/2 Jak or Jack Fruit (*Artocarpus integrifolia*) is a related species with enormous fruits which can weigh up to 70 lb. each. In spite of their very strong odour, they are relished especially in Asia and may be eaten cooked or raw.

b. (Further examples.)

1869 A. R. Wallace *Malay Archipelago* I. xvi. 362 There were also great numbers of a wild Jack-fruit tree (Artocarpus), which bore abundance of large reticulated fruit. **1900** W. W. Skeat *Malay Magic* vi. 563 To dream about eating jack-fruit (*nangka*)..is an indication of great trouble impending. **1908** E. J. Banfield *Confessions of Beachcomber* I. i. 43 We have..Jack fruit..in plenty. **1920** *Outward Bound* Oct. 44/1 The shameless stranger..issued from the saffron sunset, somewhere behind the spiky line of the mission jack fruit trees. **1921** E. M. Forster *Let.* 4 June in *Hill of Devi* (1953) 92 The Jack Fruits are ripening a little—they are extraordinary, with crocodile scales. **1947** *E. Afr. Ann. 1946–7* 23/2 Doors, mostly of local jack-fruit tree wood, are still being carved by Swahili craftsmen. **1962** *Housewife* (Ceylon) Apr. 23 The rich green foliage of mature plaintain and jak trees. **1966** D. Forbes *Heart of Malaya* ii. 27 A Malay kampong stands..half hidden by an assortment of fruit trees—durian, jack-fruit, mangosteen and rambutan. **1967** Singha & Massey *Indian Dances* xx. 175 It is said that he once had a vision in which Krishna asked him to carve his image from the wood of a certain jack-fruit tree. **1972** A. Amin tr. *Ahmad's No Harvest but Thorn* ii. 13 The handle of his *parang*, which was made of the heart-wood of a jack-fruit tree.

jack, *v.*¹ Add: **1.** (Further examples.)

1971 J. D. MacDonald *Seven* (1974) iii. 46 When we decided to give up the apple stand, I said it might make a nice little cabin. My husband Ralph jacked it up and put it on a flatbed wagon and tractored it up through the west orchard. **1973** *Country Life* 18 Oct. 1190/1, I enjoy reminiscing over early motoring days... Jack up the back tyre to ensure easier starting.

b. *transf.* and *fig.* To raise, increase; to force or bolster *up*. *colloq.* (orig. *U.S.*).

1904 *N.Y. Tribune* 8 May 10 The management thought it saw a chance to jack up rents, and made a sudden announcement of a raise. **1939** J. P. Marquand *Wickford Point* x. 126 She had jacked up the price on the table d'hôte ten francs. **1959** *Economist* 7 Feb. 504/1 At his first trial, Cho Bong-Am got only five years, but a second trial jacked this up to capital punishment. **1964** *Ann. Reg. 1963* 44 Mr Heath unveiled the plans to jack up the punctured local economies. **1971** *Daily Tel.* 26 July 15/7 Reinvestment would then jack up earnings per share and hence the value of the equity.

c. To arrange, organize, fix *up*; to put right, spruce *up*. *N.Z. slang.*

1942 *2 N.Z.E.F. Times* 12 Oct. 6/5 *Jack-up*, a term meaning to achieve the apparently impossible or to bring out of chaos; to arrange, inveigle, wangle or bolster up (a) Any transport (b) any alibi (c) leave (d) ED pay—usually without the prior knowledge or consent of authority. **1943** *Ibid.* 16 Aug. 6/5 Some of the pubs could do with a jacking up... Not enough service and civility. **1944** J. H. Fullarton *Troop Target* xxvi. 187 I've jacked up a hot snack for the end of the shoot. **1945** E. G. Webber *Johnny Enzed in Middle East* 13/3 May take a year to jack it up again. **1956** D. M. Davin *Sullen Bell* I. iii. 24 I've jacked it up to stay the night with a friend of mine. **1960** B. Crump *Good Keen Man* 59 Harry decided to try his hand at jacking up a home brew. **1963** *Weekly News* (Auckland) 8 May 39/3 We jacked it up between ourselves that one evening we'd get Honey's gloves. **1971** *N.Z. Listener* 22 Mar. 13/1 I'll see you right at a boardin' place until you get jacked up.

4. Slang phr. *to jack in*: to abandon, leave, give up, stop. Freq. in phr. *to jack it in*.

1948 A. Baron *From City from Plough* i. 12 'What's your ol' woman do to you, Charlie?' 'Jacked me in for a civvy. I got home; no one there, no furniture, nothing.' **1958** F. Norman *Bang to Rights* iii. 77 There was only fifteen of us on the hunger strike, I suppose the others must have got hungrey anyway they jacked it in. **1963** 'A. Garve' *Sea Monks* v. 130, I ain't goin' to jack it all in now for Chris or anyone. **1972** K. Royce *Miniatures Frame* ix. 123 I'm beginning to wonder if it's worth it... Let me jack it in. **1973** *Times* 31 May 10/7 Private landlords jack-in the shaky business of letting.

5. *to jack off* (intr.): (*a*) to go away, depart; (*b*) to masturbate. *slang.*

1935 'G. Orwell' *Clergyman's Daughter* ii. 109 Flo and Charlie would probably 'jack off' if they got the chance of a lift. *a* **1950** —— *Coll. Ess.* (1968) I. 71 *Jack off, to*, to go away. **1959** W. Burroughs *Naked Lunch* 74 He plummets from the eyeless lighthouse, kissing and jacking off in face of the black mirror. **1971** R. A. Carter *Manhattan Primitive* (1972) xxiv. 237 You miserable little queer... You can jack off in Llewellyn's best hat for all I care. **1972** V. Ferdinand in A. Chapman *New Black Voices* 472 We might as well be jacking off, masturbating for our own self-gratification.

jack (dʒæk), *a. Austral. slang.* [f. Jack *v.*¹ 3.] Tired *of* (something or someone); bored.

1901 E. Dyson *Gold Stealers* iv. 41 Oh, well, Twitter's jack of it, an' I don't think it's much fun. **1944** J. Devanny *By Tropic Sea & Jungle* xviii. 155 Too much of it makes you jack of it quick. **1959** Baker *Drum* vi. 50 He was clearly a bore and they were jack of him. **1969** *Coast to Coast 1967–8* 4 He was willing to bet she'd get jack of it.

ja·ckal, *v.* [f. the sb.] *intr.* To play the jackal (see Jackal 2); to do subordinate work or drudgery.

1900 Kipling in *Daily Mail* 21 Apr. 4/5 For three months she had jackalled behind the army..and in that time had carried over thirteen hundred sick and wounded. **1914** A. Dobson *18th Cent. Stud.* 204 Johnson..lost many of the papers lent to him by Percy. Malone, who jackalled for him, lost others. **1940** V. Woolf *Writer's Diary* 6 Apr. (1953) 331, I..brooded quietly till the tyre punctured: we had to jackal in mid-road.

jackal(l), varr. *JACAL.

jackaroo, jackeroo. Add: (Earlier and later examples.) Now esp. a cadet or novice on a sheep-station or cattle-station.

1878 'R. Boldrewood' *Ups & Downs* vii. 72 If these here fences is to be run up all along the river, any Jackaroo can go stock-keeping. **1911** C. E. W. Bean *'Dreadnought' of Darling* xi. 99 In the bachelors' quarters there will probably live..one or two 'jackeroos'—young Australians, or sometimes young Englishmen, learning the work of a sheep run by taking an ordinary part in it. **1918** *Chambers's Jrnl.* Apr. 267/2 The tracker's methods are noted in the following true story of a lost 'jackeroo' (a 'new chum' learning station work). **1936** I. L. Idriess *Cattle King* xxxii. 287 Some of the blacks told me he was coming with a couple of young jackeroos. **1956** 'N. Shute' *Beyond Black Stump* ii. 45 At the age of sixteen David had gone as a jackeroo upon a sheep station, to learn the business. **1962** *Coast to Coast 1961–62* 28 Coorabin's Royal Hotel had afforded her plenty of practice in turning on a stare that caused the invitations of commercial travellers, boundary-riders and jackaroos to trail off before they were half spoken. **1968** *TV Times* (Austral.) 28 Aug. 30/1 Jackaroo. 45 min. documentary. The jackaroo is..an apprentice to the rural industry, and in theory is in training to become a manager or owner. **1969** *West Australian* 5 July 58/2 (Advt.), 2 Jackaroos or Junior Stockmen with riding experience 16/18 years for West Kimberley Cattle Station.

jackaroo, jackeroo *v.* (earlier and later examples).

1878 'R. Boldrewood' *Ups & Downs* xix. 239 A year or two more Jackerooing would only mean the consumption of so many more figs of negro-head. **1911** C. E. W. Bean *'Dreadnought' of Darling* xxxiv. 302 A houseful of bachelors—three or four young fellows jackerooing (that is to say, learning colonial experience) under a bachelor manager. **1936** A. Russell *Gone Nomad* ii. 12 My graduation in jackerooing, or, as I usually call this period of my life, my 'pack-mule and damper days', had begun. **1967** *Coast to Coast 1965–6* 161 Young and old..jackerooed for thirty bob and tucker.

jackass, *sb.* Add: **4. b.** (Examples.)

1891 in H. Patterson *Illustr. Naut. Dict.* **1907** A. T. Mahan *From Sail to Steam* iv. 96 The absurd-sounding, but legitimate, message to have the jackasses put in the hawse-holes. **1918** F. Riesenberg *Under Sail* ii. 29 'Jackasses' were then bowsed into the hawse holes *for fair*. **1948** R. de Kerchove *Internat. Maritime Dict.* 368/1 Jackasses are a most effective method of making hawseholes watertight in ships using stocked anchors.

5. jackass barque, a sailing ship having the same sails as a barquentine but rigged in a manner varying in some respect from the orthodox barquentine rig; **jackass brandy** *U.S. slang*, home-made brandy; **jackass brig** (examples); **jackass-rigged** *a.*, substitute for def.: of a schooner, having three masts with square sails set on the foremast and having no main topmast; add further example; **jackass schooner,** a schooner which is jackass-rigged.

1861 *Mitchell's Maritime Reg.* 890 On the 10th instant there was launched from the building-yard of Messrs. J. and J. Hall, at Arbroath, a beautifully modelled jackass barque, named the Princess Alice, of 190 tons N.N.M. **1923** F. C. Bowen *Ships for All* ii. 39 The sails of a barquentine and of a jackass barque are precisely the same. **1969** B. Landström *Sailing Ships* 178 It sometimes occurred that a ship was rigged, for example, with a fore-and-aft sail on the lower mainmast and square sails on the top and topgallant masts... In England all such types of pseudo-barques were called *jackass barques*. **1920** *Federal Reporter* (U.S.) CCLXXVIII. 42 Intoxicating liquors, to wit, one pint bottle of jackass brandy. **1921** *Dialect Notes* V. 109 *Jackass brandy*, a home-made brandy with a powerful 'kick'. **1923** *San Francisco Examiner* 18 Feb. 16/7 A still in operation and a stock of jackass brandy close by. **1883** R. B. Dixon *Fore & Aft* 191 The potatoes would have to be paid for before that 'jackass brig' could sail. **1887** G. Davis *Recoll. Sea Wanderer's Life* 231 She is what is called a jackass brig. She has one mast square rigged and the other is schooner rigged, with topsail. **1926** *Sea Breezes* VIII. 214 There was a jackass brig, no main course, main yard high up, rigged like a topsail schooner on main mast. **1898** A. Ansted *Dict. Sea Terms* 239 There is another class of trading schooner, with three masts... When it sets square sails on the foremast it is sometimes called *jackass* rigged. **1929** H. W. Smyth *Mast & Sail* (new ed.) 520 *Jackass schooner*, still used occasionally of a schooner without main-topmast. **1951** H. Benham *Down Tops'l* v. 71 'Jackass' schooners..were so called because they were square rigged forward like a tops'l schooner, but had a ketch's mizzen. **1961** F. H. Burgess *Dict. Sailing* 122 A 'jackass schooner' has no main topmast.

jack boat: see *JACK *sb.*¹ 25.

jack-boot, jackboot. Add: Also, those worn by German soldiers during the Nazi regime.

1942 *R.A.F. Jrnl.* 2 May 36 Either the British way shall survive or the Nazi jackboot and whip shall take its place. **1955** *Times* 13 June 8/5 There were jackboots to be seen, iron crosses were pinned to the pockets of a few green shirts.

b. *fig.* Military oppression; rough bullying tactics. Also *attrib.*

1768 J. Mordaunt *Let.* 10 May in E. Hamilton *Mordaunts* (1965) ix. 215 Some insults were offer'd to some of the Lords & Commons in their Coaches as they went down with a Cry of Wilkes & Liberty & no Jack Boot. **1910** *Westm. Gaz.* 14 Feb. 5/1 A large numerical majority..excluded from power and honour by a mere jack-boot minority. **1968** *Sat. Rev.* 26 Oct. 36 The writers' trials and the jackboot on Prague have made it quite clear that Brezhnev and Kosygin's views on liberalism are akin to Stalin's. **1968** 'R. Simons' *Death on Display* vii. 105 He also objected strongly to what he called your jackboot methods when you interviewed Mrs Hurd. **1970** *Times* 18 Mar. 26 This..attempt by the Government to bludgeon this Bill through with jackboot tactics.

Hence **ja·ck-booted** *a.*

1846 R. Ford *Gatherings from Spain* xxii. 300 The clumsy look of a French jackbooted postilion. **1939** 'N. Blake' *Smiler with Knife* iv. 60 Silent, jack-booted watchers standing outside frightened houses. **1946** *R.A.F. Jrnl.* May 175 The once-jackbooted Germans. **1972** J. Rossiter *Rope for General Dietz* ii. 28 He stood..with his jackbooted legs apart..as if to underline what they'd done in the name of Nazi justice.

jack-chain. Add: **2.** *Logging.* (See quot. 1905.)

1905 *Terms Forestry & Logging* (U.S. Dept. Agric. Bureau Forestry) 40 *Jack chain*, an endless spiked chain, which moves logs from one point to another, usually from the mill pond into the sawmill. **1957** in *Brit. Commonwealth Forest Terminol.* II. 97.

jacker. Add: **b.** *jacker-off, -up*: see quots. 1921.

1881 in *Instructions to Census Clerks* (1885) 70. **1904** *Westm. Gaz.* 28 Apr. 4/1 It was in the lace factory that the lad was set to work as a 'jacker-off'. **1921** *Dict. Occup. Terms* (1927) § 399 *Jacker-off* (lace); takes off from bobbins, waste lengths of unused threads, and winds them on to large wooden bobbins, using a small winding machine. *Ibid.* § 688 *Jacker-up* (lead pencil making); places a number of glued pencils together in a clamp, and screws down clamp to make glueing secure; removes pencils when clamping is complete. **1924** J. Marchant *Dr. John Clifford* i. 5 Three-fourths of the children were jackers-off or 'piecers'.

jackeroo: see Jackaroo in Dict. and Suppl.

jacket, *sb.* Add: **1. f.** See also Dust *v.*¹ 6 b in Dict. and Suppl.

†g. *Mil. colloq.* (See quots.) *Obs.*

1898 *Geogr. Jrnl.* May 556 Lieut. Tanner obtained his 'jacket', and was the *beau ideal* of a horse-artillery officer. **1908** *Westm. Gaz.* 15 Oct. 5/3 Until 1895..a 'jacket'—i.e., a post in the Royal Field Horse Artillery—might be given to an officer of Field Artillery or of Garrison Artillery. **1909** J. R. Ware *Passing Eng.* 158/2 *Jacket* (Military), a soldier who wears a jacket (chiefly cavalry or horse artillery). **1925** Fraser & Gibbons *Soldier & Sailor Words* 130 *Jacket, to get the*: colloquial for an appointment to the Royal Horse Artillery.

2. b. See also *dust-jacket* (*DUST *sb.*¹ 8 e).

d. *Ordnance.* A coil or cylinder of wrought iron or steel placed around the barrel of a gun to strengthen or protect it.

1876 *Engineering* XXI. 17 This improvement consists in the addition of a steel jacket to the body of the gun from the breech to beyond the trunnions. **1888** O. E. Michaelis tr. *Monthaye's Krupp & De Bange* ii. 24 The tube..is encircled by a single band or jacket (Mantel, in German), shrunk on. **1902** *Kynoch Jrnl.* Apr.–May 79/2 A second gun..having a jacket of cast steel. **1972** *Internat. Defense Rev.* Feb. 61 Sheathing the gun tube in an insulating jacket of cloth, plastic material or aluminum plate.

4. jacket crown *Dentistry*, a crown (*CROWN *sb.* 28 b), often of porcelain or vinyl, fitted over an existing natural crown, which is usually ground down to receive it; **jacket potato,** a potato cooked in its jacket, i.e. unpeeled.

1903 H. J. Goslee *Princ. & Pract. Crowning Teeth* viii. 129 The so-called jacket crown is often a most useful style of construction. **1963** [see *bridge-work* (*BRIDGE *sb.*¹ 11 b)]. **1966** L. Deighton *Billion-Dollar Brain* xix. 197 One of my jacket crowns is loose. [**1902** Wright *Eng. Dial. Dict.* (s.v. *jacket*), *Jackutty-taters*, potatoes boiled with their skins on.] **1928** Galsworthy *Swan Song* I. v. 37 A young woman was handing him 'jacket' potatoes. **1967** M. Summerton *Memory of Darkness* iv. 56 You'll fare better on Rosie's stew and my baked jacket potatoes.

jacket, *v.* Add: **1. c.** To enclose (a person) in a strait-jacket.

1856 C. Reade *Never too Late* I. xv. 276 He found himself surrounded, jacketed, strapped, and collared. **1905** J. London *Jacket* (1915) 52 They told me plainly that they would jacket me to death if I did not confess.

Jackfield (dʒæ·kfīld). The name of a village in Shropshire, used *attrib.* to denote a kind of black-glazed pottery of a type manufactured there in the 18th century. Also *ellipt.* as *sb.*

1904 A. H. Church *Eng. Earthenware* (rev. ed.) xv. 115 Collectors of English earthenware are generally inclined to assign the black so-called 'Jackfield ware' to Whieldon. Possibly the black glaze..may have been added at Jackfield. **1967** *Sunday Times* 5 Mar. 47/1 Leeds pottery dates from about 1750, and though the creamware is the commonest product, the factory also produced some rather nasty shiny black earthenware called 'Jackfield'. **1968** *Canad. Antiques Collector* Oct. 5/2 A smaller exquisitely modelled Jackfield ware jug.

jack-fruit: see Jack *sb.*⁴ b (in Dict. and Suppl.).

jack-knife, *sb.* Add: **1.** (Earlier Amer. example.)

1711 *Official Rec. Springfield, Mass.* (1898–9) IX. 39 One Dozen of Jack Knives, at six pence the knife.

3. *Swimming.* In full, *jack-knife dive*. A kind of dive executed by first doubling up and then straightening the body before entering the water.

1922 *Country Life* (U.S.) July 60/3 All variety dives fall into four main groups—somersaults, twists, gainers, jack-knives. **1928** *Radio Times* 11 May 274/2 Doing jack-knife and swallow dives from incredible heights. **1942** J. D. Carr *Seat of Scornful* xiv. 191 That's not a jack-knife, you ass... A jack-knife dive is where you bend double and touch your toes in mid-air, and then straighten out before you hit the water. **1956** J. Symons *Paper Chase* x. 60 His long legs drawn up like those of a jack-knife-diving swimmer.

4. The accidental folding up of an articulated lorry.

1966 *Times* 29 Sept. 11/6 A 'jack-knife' is an ugly complaint of 'artics' as these articulated monsters [sc. lorry and trailer] are known familiarly in the trade. The trailer and lorry fold up like a Boy Scout's jack-knife.

jack-knife, *v.* Add: (*a*) (Earlier U.S. example.) (*b*) (Earlier and later examples.) *spec.* Of the sections of an articulated lorry: in an accident, to fold together like a jack-knife. (*c*) To do a jack-knife dive. So **ja·ck-knifing** *vbl. sb.*

1806 *Balance* (Hudson, N.Y.) 22 July 228 (Th.), A sailor..Jacknifed (as he termed it) the poor creature [sc. a cat] in several places about the head. **1888** *Century Mag.* June 251/2 The practice..of dodging shots, 'jack-knifing' under fire. **1920** T. S. Eliot *Ara Vos Prec* 22 The sickle motion from the thighs Jackknifes upward at the knees Then straightens down from heel to hip. **1949** *Sun* (Baltimore) 4 Aug. 1/8 An automobile crashed into a tractor-trailer truck that jackknifed in a driving rain. **1955** T. Sterling *Evil of Day* v. 50 He..jack-knifed into a chintz armchair. **1958** *Times* 12 Apr. 7/7 Nobody envies Joe Cree his new 'artic,' for there is always the danger that, on ice, the rear will swing round, or jack-knife. **1964** [see *FIN *v.* 4]. **1968** *New Scientist* 14 Mar. 573/3 Jack-knifing accidents to articulated vehicles are all too common... The vehicle folds up—jack-knifes—at the kingpin, which is the hinge between the two parts. **1971** *Rand Daily Mail* 27 Mar. 8/2 This was particularly so in the case of drivers, where jackknifing of the upper torso onto the steering wheel rim could lead to serious injuries. **1971** *New Scientist* 12 Aug. 359/2 Commercial vehicles..need them [sc. antilock brakes] most urgently because of their widely varying states of load and their tendency to jack-knife if articulated.

jack-leg, jackleg (dʒæ·kleg), *a.* and *sb. U.S. colloq.* and *dial.* [f. Jack *sb.*¹ + *-leg* as in Black-leg 2, 3.] Used as a term of contempt or depreciation: **A.** *adj.* Incompetent, unskilled; unscrupulous, dishonest. Freq. used of lawyers and preachers. **B.** *sb.* An incompetent or unskilled or unprincipled person.

1850 *Amer. Rev. Mag.* XI. 465/2 A party of some twenty of the most notorious rode up, headed by what is there [sc. in Texas] known as a 'jack-leg' lawyer. **1853** 'P. Paxton' *Stray Yankee in Texas* xiii. 137 A sorter jack-leg lawyer. *Ibid.* xxviii. 284 In the Texan vocabulary, all men who have a mere inkling of any trade or profession are called 'jack-legs'. *Ibid.*, These men were 'jack-leg' carpenters. **1891** *Harper's Mag.* June 160/1 Once I was called a jack-leg and shyster. **1902** W. N. Harben *Abner Daniel* ii. 16 The Atlanta jack-leg lawyer is akin to the Tompkins family some way. **1943** R. Ottley *New World A-Coming* 86 The cultists were augmented by a number of herb doctors, clairvoyants, and 'jackleg' preachers. **1958** P. Oliver in P. Gammond *Decca Bk. Jazz* i. 19 The wandering evangelists, and 'Jack-leg' preachers. 1974 Amer. Speech 1971 XLVI. 70 One innovation possibly attributable to population shift is jackleg preacher, which Carlson heard from a black informant in Roxbury.

So **ja·ck-legged** a.

1839 Congress. Globe App. 127 A set of jack-legged pettifogging lawyers. 1892 Congress. Rec. 27 May 4777/1 He goes away, and a jack-legged [army] officer could do nothing except to mark him as a deserter.

Jacko (dʒæ·ko). Austral. slang. [f. *JACK sb.^{1} 29 c + *-O^2.] A kookaburra (= *JACK sb.^{1} 29 c).

1941 BAKER Dict. Austral. Slang 38 Jacko, a kookaburra. 1942 C. BARRETT On Wallaby iv. 80 Were they only having a close-up view, having mistaken the moving figure among the dunes for a Jacko?

Jack-o'-lantern, sb. 3. (Examples.)

1837 HAWTHORNE Twice-Told Tales 222 Hide it [sc. the great carbuncle] under thy cloak, say'st thou? Why, it will gleam through the holes, and make thee look like a jack-o'-lantern! 1959 I. & P. OPIE Lore & Lang. Schoolch. xii. 269 As soon as it is dark on Hallowe'en they take the lighted 'Jack-o-lanterns' and put them on their gate-posts.

jack-pot: see JACK sb.^{1} 33 a (in Dict. and Suppl.).

jack-rabbit. Add: (Earlier and later examples.) Also attrib. and fig.

1863 N. S. KEITH Let. 24 Aug. in Colorado Mag. (1940) XVII. 69 We saw wolves, buffalos, antelopes, jack-rabbits, prairie-dogs innumerable, deer, and birds of various sizes. 1906 Chambers's Jrnl. July 538/1 For miles one may ride without seeing a living thing larger than a jack-rabbit. 1929 W. FAULKNER Sartoris III. iv. 206 The mules flapped their jack-rabbit ears. 1961 'E. LATHEN' Banking on Death (1962) vii. 57 He was thrown backward by a jack-rabbit start from a stop sign. 1962 Amer. Speech XXXVII. 269 Jack rabbit, a motorist who is proficient at watching the cross-street traffic light; when it turns yellow, he starts up and is into the intersection before the light in front of him has turned green. 1963 D. P. MANNIX All Creatures Great & Small xi. 180 We saw our first live jackrabbit just at dawn while crossing the plains of Nebraska. A big, white-tailed jack with black-and-white squares like signal flags on his long ears bolted across the road. 1972 Guardian 16 Dec. 10/1 You surely have hit the jack-rabbit on the head with a fire-iron.

Jacksonian (dʒæksōu·niăn), a.^{1} [See -IAN.] Pertaining to or characteristic of Andrew Jackson (1767–1845), seventh president of the United States, a prominent leader of the Democratic party. Also as sb., a follower of Jackson. Hence **Jackso·nianism.**

1824 Amer. Sentinel (Georgetown, Ky.) 18 Oct. 3/1 At Mount-sterling..they collected together six Jacksonians. 1824 Commentator (Frankfort, Ky.) 23 Oct. 3/2 The old Jacksonian aristocratic leaven of the Adams faction. 1906 W. CHURCHILL Coniston v. 51 He..preached the word of Jacksonian Democracy in all the farmhouses round about. Ibid. 57 The conscientious Jacksonians who were misguided enough to believe in such a ticket. 1929 Encycl. Brit. I. 156/2 Up to this point Adams's career had been almost uniformly successful, but his presidency (1825–29) was in most respects a failure, owing to the virulent opposition of the Jacksonians. Ibid. IV. 585/1 Calhoun,..during the remainder of the Jackson regime, was a severe critic of Jacksonianism. 1966 AUDEN About House 16 A Proustian snob or a sound Jacksonian Democrat. 1973 New Yorker 28 Apr. 146/2 Douglas's creed was Jacksonianism, which to him meant a United States expanding over the whole continent and demonstrating that a democracy could not only survive but prosper.

Jacksonian (dʒæksōu·niăn), a.^{2} Med. [f. the name of John Hughlings Jackson (1834–1911), English physician and neurologist: see -IAN.] Designating a form of epilepsy (and the associated convulsions, fits, etc.) in which the seizures are usually confined to one side of the body (in which case consciousness is retained) and begin at one site (usually a digit or the angle of the mouth), progressing from there to neighbouring parts.

1877 Brit. & Foreign Med.-Chirurg. Rev. LIX. 51 The epithet 'Jacksonian' is even being introduced in France by Charcot to be applied to epilepsy or convulsions of a localised and partial character. 1878 Glasgow Med. Jrnl. X. 92 Dr Robertson showed a patient who suffers from Jacksonian Epilepsy. 1878 Brit. Med. Jrnl. 28 Dec. 959/1 (heading) Jacksonian convulsions. 1890 Jrnl. Nerv. & Mental Dis. XVII. 56 The Jacksonian spasm confined chiefly to the left hand and arm. 1933 W. R. BRAIN Dis. Nervous Syst. iii. 224 Jacksonian attacks may occur at long intervals, or with great frequency, even up to several hundreds a day—serial Jacksonian epilepsy. 1960 W. G. LENNOX Epilepsy I. vii. 212 Most authorities, including Penfield.., use the term Jacksonian to cover all forms of local motor seizures. 1969 H. H. JASPER et al. Basic Mechanisms Epilepsies xix. 517/1 Particular interest was centered in Jacksonian seizures.

jacksy (dʒæ·ksi). slang. Also **jacksie, jacksy-pardo, -pardy, jaxey, jaxie.** [f. JACK sb.^{1} + -SY.] The posterior, backside, arse.

1896 FARMER & HENLEY Slang IV. 33/1 Jacksy-pardy, the posteriors. 1943 HUNT & PRINGLE Service Slang 40 Jacksie, service slang for 'rear', 'tail', or 'bottom'. 1959 K. WATERHOUSE Billy Liar v. 78 Why don't you tell the boring little man to stick the job up his jacksy? 1963 Sunday Times 15 Sept. 29/8 Tonbridge boys many years ago said 'a root on the Jaxie (or Jacksie)' for 'a kick in the pants'. 1966 B. NAUGHTON Alfie xxxvi. 206 She's sitting there on her jacksie, reading one of those colour things out of a newspaper. 1970 A. DRAPER Swansong for Rare Bird i. 9 The amount of love in our house you could stick up a dog's jacksie and he wouldn't even yelp.

jack-up (dʒæ·k,ʊp). [f. JACK v.^{1} 1.] In full, jack-up rig. A type of drilling rig for use in an offshore oil-field the legs of which are lowered to the sea bed from the operating platform.

1965 Oil & Petroleum Year Bk. 511 (Advt.), Santa Fe's offshore operations now include drilling services from fixed platforms, jack-ups and the semi-submersible Blue Water No. 2. 1967 Ocean Industry Dec. 11/1 Husky's New Jack-up Rig Designed for North Sea Husky Oils new $10-million jack-up Gulftide was christened October 6. 1970 New Scientist 29 Oct. 219/2 In the North Sea.. of the dozen or so mobile rigs usually operating the most common types are the semi-submersible and the jack-up. Ibid., Jack-ups are..very vulnerable when jacking up or down to move location. 1972 Times 23 Nov. 23/2 The discovery was made by the new jack-up rig Zapata Nordic which is under long term contract to Shell.

Jacky. Add: **1.** (Further examples in sense 'a sailor'.)

1893 Funk's Stand. Dict., Jacky, a sailor. 1909 Daily Chron. 1 Oct. 1/6 The place of honour..was given to the British 'jackies', who were easily the most popular.

b. Also Jacky-Jacky. A white man's name for an Aboriginal. Austral. slang.

1928 'BRENT OF BIN BIN' Up Country ii. 35 He was.. going to..outdo Jacky-Jacky as a swell. 1944 W. E. HARNEY Taboo (ed. 3) 87 Such was the view of the boss—not how good a cattle-man you were, but how cheap and good were your methods with the 'jackies'—the name given to the natives out there. 1969 Sun-Herald (Sydney) 13 July 52/1 'Jackie Jackie' is dead. There is a new and growing spirit in the Aboriginal community.

c. Austral. A kookaburra. Cf. *JACK sb.^{1} 29 c.

1934 Bulletin (Sydney) 11 July 21/2 Jacky left hurriedly and didn't return for his dinner. Ibid. 17 Oct. 20/4 Cockatoos are the latest-feeders among the bush birds, bar the jacky.

3. Jacky (or **Jackie**) **Howe** (see quot. 1965); also **Jimmy Howe** Austral. and N.Z. slang; **Jacky Winter** Austral., the brown flycatcher, Micræca fascinans.

1930 Bulletin (Sydney) 9 Apr. 19/2 It took nine bars of soap to wash his 'Jacky Howe' flannel. 1937 E. HILL Great Austral. Loneliness xl. 301 A weird figure in a flour-bag, Jimmie Howe, with shoes of raw-hide laced with pandanus, Leng carried no luggage. 1948 V. PALMER Golconda xi. 85 In his sleeveless Jackie Howe he swaggered around showing the punctures on his burly arm. 1965 N.Z. Listener 26 Feb. 15/2 Jackie Howe, the familiar navy or black woollen singlet worn by Australian and New Zealand shearers and bushmen. It is named after Jackie Howe, an Australian shearer who in 1892 established a world shearing record by shearing 321 Merinoes with hand shears in 8 hr 40 min. 1898 MORRIS Austral Eng. 218/2 Jacky Winter..the vernacular name in New South Wales of the Brown Flycatcher, Micræca fascinans, a common little bird about Sydney. 1911 LUCAS & LE SOUEF Birds Austral. 272 Jacky Winter is indeed seldom molested by even the thoughtless schoolboys. 1936 F. D. DAVISON Children of Dark People 5 Jacky winters..flitted among the bushes. 1969 A. BELL Common Austral. Birds (ed. 2) 22 Post Boy, Jacky Winter, Spinks—half-a-dozen unofficial names suggest the friendly feeling towards this quiet, confidential, contented bird.

jackyard. (Earlier and later examples.)

1873 'VANDERDECKEN' Yachts & Yachting 186, I have never seen a jack yard used with a jib-headed gaff-topsail. 1931 Rudder Mar. 82/2 The topsail might be termed half a club topsail, having a jack yard or club at the luff only.

Jacob. Add: **4.** [f. Genesis 30:40 A.V. 'Jacob did separate the lambs, and set the faces of the flocks toward the ring-straked'.] A variety of two- or four-horned piebald sheep, believed to have been introduced from Spain in the eighteenth century, and since used as an ornamental park breed.

1913 H. J. ELWES Guide Primitive Breeds of Sheep 30 'Spanish' or Piebald Sheep...These sheep are called by various names—'Syrian', 'Persian', 'Zulu', 'Barbary', 'Jacobs', and 'Spotted'. 1970 Observer (Colour Suppl.) 26 Apr. 35/2 The black and white Jacobs..are another rare breed of sheep. 1972 Kent Life July 28/2 Ten years ago the number of Jacob Sheep in Britain could be counted in dozens. Today the National Agricultural Centre calculates that there are more than 3,000 sheep in 132 flocks. 1973 Times 14 Apr. 14/5 Most horned breeds are hill breeds. Exceptions are Dorset..and Jacob.

jacobæa (dʒækǒbī·ă). [mod.L. (R. Dodoens in Trium Priorum de Historia Stirpium (1553) 15), perh. f. G. popular name of the plant S. Jacobs kraut.] **1.** The ragwort, Senecio jacobæa, formerly called St. James's wort, or a related purple-flowered species from South Africa, S. elegans.

[1578 H. LYTE tr. Dodoens's Niewe Herball 69 Jacobea. S. James' worte. Jacobea marina. S. James' worte of the Sea...The first kinde of S. James worte, hath long, browne, red, crested or straked stalkes. 1728 R. BRADLEY Dictionarium Botanicum I. s.v., Jacobæa, in English Ragwort, is of different Sorts. Ibid., James-wort, or Ragwort, is Jacobæa, which see.] 1789 W. AITON Hortus Kewensis III. 193 Elegant Groundsel, or Purple Jacobea. Nat[ive] of the Cape of Good Hope. 1884 W. MILLER Dict. Eng. Names Plants 69/2 Jacobæa, Purple, Senecio elegans. c 1903 E. T. COOK Cent. Bk. Gardening 27/1 The botanical name of the Jacobaea is Senecio elegans. 1972 W. T. STEARN in A. W. Smith's Gardener's Dict. Plant Names (rev. ed.) 371/3 Jacobea. Senecio jacobaea.

2. jacobæa lily: see JACOBEAN a. 2 b.

1752 P. MILLER Gardeners Dict. (ed. 6) s.v. Amaryllis, The third Sort, which is commonly called Jacobæa-lily, is now become pretty common. 1760 [see JACOBEAN a. 2 b]. 1789 W. AITON Hortus Kewensis I. 416 Jacobea Lily. Nat[ive] of South America. Cult[ivated] 1658, in the Oxford garden. 1864 L. H. GRINDON Brit. & Garden Bot. 644 Perhaps the commonest [Amaryllid] is the 'Jacobæa-lily', Amaryllis formosissima, easily told by its dark hue. 1962 Jrnl. Roy. Hort. Soc. LXXXVII. 284 (title) The Jacobaea Lily—Sprekelia formosissima.

Jacobean, a. Add: **1. b.** In the furniture trade, designating wood of the colour of dark oak, or the colour itself; also denoting furniture made in mock-Jacobean style.

1918 Heal & Son Catal. 28 Jacobean refectory table in dark oak. 1928 Daily Mail 31 July 1/2 It can be obtained in Light Brown or Jacobean coloured solid oak. 1930 Daily Express 8 Sept. 2 This fine Chest is..finished Jacobean colour. 1974 Times 8 Apr. 13/3 Philips can provide you with a colour television set in a Jacobean chest. Ibid. 18 May 5/5 (Advt.), Reproduction styling in Jacobean oak, walnut, white or ivory and gilt finishes.

3. Of or pertaining to Henry James (1843–1916), American novelist and critic.

1906 M. BEERBOHM Around Theatres (1953) II. 442, I cannot imagine two minds..more divergent than the Shavian and the Jacobean. Mr James must excuse my invention of this adjective. 1932 Q. D. LEAVIS Fiction & Reading Public III. iii. 264 Those interested will even find a telegram in Jacobean English in The Great Good Place. 1958 Times 6 Mar. 13/5 The masterly Jacobean answer to this insult is laid up in the Lubbock volumes of James's letters.

Jacobethan (dʒækǒbī·þăn), a. [Blend of JACOBEAN a. and ELIZABETHAN a.] Of design: that displays a combination of the Elizabethan and Jacobean styles. Also transf. and ellipt. as sb.

1933 J. BETJEMAN Ghastly Good Taste iv. 53 The style in which Gothic predominates may be called, inaccurately enough, Elizabethan, and the style in which the classical predominates over the Gothic, equally inaccurately, may be called Jacobean. To save the time of those who do not wish to distinguish between these periods of architectural uncertainty, I will henceforward use the term 'Jacobethan'. Ibid. 54 To me the appeal of Jacobethan is indeed remote... Jacobethan architecture may be ugly, but it is never dull. 1945 Archit. Rev. Nov. 124/3 Westcombe Park Road..shows an early tendency towards those ornamental features which long afterwards gave the names of 'sham Tudor' and 'Jacobethan' to a rather pathetic phase in domestic design. 1966 Listener 22 Dec. 928/1 The hacienda-Jacobethan garage with which the inhabitants of Sompting are so rightfully delighted. 1969 J. GROSS Rise & Fall Man of Lett. i. 13 Archaisms, hand-picked quotations, artful Jacobethan echoes. 1974 Times Lit. Suppl. 7 June 602/4 The specimen quoted is a pseudo-archaic piece of 'Jacobethan' prose, modelled on the style of the King James Bible.

Jacob Evertsen (ya·kǫb ĕ·vɑːtsən). S. Afr. Also **Jacob Evertson, jacopever.** [Name of a 17th-c. Dutch sea-captain.] Any of several marine food fishes distinguished by reddish skin and large eyes, esp. Sebastichthys capensis.

1727 J. G. SCHEUCHZER tr. Kæmpfer's Hist. Japan I. I. xi. 136 Ara is what the Dutch in the Indies call Jacobs Evertz. 1798 S. H. WILCOCKE tr. Stavorinus's Voy. E. Indies II. II. iii. 352 There is likewise, it is said, a large fish near the pier-head at Amboyna, to which the name of Jacob Evertsen has been given. 1801 J. BARROW Acct. Trav. S. Afr. I. i. 31 The Scorpæna Capensis, here called Jacob Evertson, is a firm, dry fish, but not very commonly used. 1853 L. PAPPE Synopsis Edible Fishes Cape of Good Hope 14 Sebastes Capensis... Called Jacob Evertsen, after a Dutch Captain, remarkable for a red face and large projecting eyes. 1927 Ann. S. Afr. Mus. XXI. 908 Sebastichthys capensis (Gmel.). Jacob Evertson; Jacopever... Red, shading to orange below; several silvery-white or pinkish irregular spots on sides. Ibid. 910 Sebastosemus capensis (G. and v. B.). Spiny Jacopever. 1973 Farmer's Weekly (S. Afr.) 18 Apr. 102 In the very early days of the Colony, the Dutch East India Company had as captain of one of its ships a certain Jacob Evert. When the local fishermen caught a fish with a red face, bulging eyes and thick lips they named it jacopever.

Jacobite, sb.^{4} and a.^{2} Add: **B.** adj. **2. b.** Of glass or pottery: bearing inscriptions and emblems which indicate Jacobite sympathies.

1936 Burlington Mag. Oct. p. xxiii/1 There are also many specimens of engraved glasses including Jacobite specimens with their symbolic references. Ibid. 175/1 A series of Jacobite glasses. 1957 MANKOWITZ & HAGGAR Conc. Encycl. Eng. Pott. & Porc. 117/2 Jacobite pottery. 1960 H. HAYWARD Antique Coll. 150 Jacobite glasses, propaganda glasses bearing emblems and mottoes of a cryptic character associated with the Jacobite cause. 1970 Canad. Antiques Collector Oct. 29/2 The emblems to be found on these Jacobite glasses include a rose..and the Latin word 'Fiat'.

Jacobite (dʒæ·kǒbəit), sb.^{5} [f. as JACOBITE sb.^{4}] An admirer of Henry James (1843–1916). Cf. *JACOBEAN a. 3.

1909 BEERBOHM Around Theatres (1953) II. 541 There, in those six last words, is quintessence of Mr James; and the sound of them sent innumerable little vibrations through the heart of every good Jacobite in the audience. 1961 L. AUCHINCLOSS Reflections of Jacobite p. vii, I have called myself a Jacobite because so much of my lifetime's reading has been over the shoulder of Henry James.

Jacob's ladder. Add: **5.** An elevator consisting of a series of bucket-shaped receptacles fixed upon an endless chain.

1845 G. DODD Brit. Manuf. 5th Ser. ii. 31 The hops are raised to the boiler by a contrivance something like the buckets of a dredging-machine; it is called a 'Jacob's ladder'. 1853 Househ. Words VII. 491/1 The malt.. being precipitated up a curious contrivance called a 'Jacob's ladder'. 1860 Ure's Dict. Arts (ed. 5) II. 589 It [sc. the bloom] is squeezed four times before it leaves the rolls and falls upon the Jacob's ladder. 1884 W. H. GREENWOOD Steel & Iron xvi. 303 The puddled ball.. falling from the bottom shoot of the machine on to a Jacob's ladder or other elevator.

Jacobson (dʒē·kəbsən). The name of Ludwig Levin Jacobson (1783–1843), Danish anatomist and physician, used in the possessive or with of-adjunct to designate structures investigated by him, as **Jacobson's nerve,** the tympanic nerve, a branch of the glosso-pharyngeal (ninth cranial) nerve; **Jacobson's organ,** an organ which is only vestigial in adult man but is a well-developed olfactory organ in many vertebrates, notably snakes and lizards, occurring as (one of) a pair of sacs or tubes in communication with the mouth (or in some cases the nose).

1836–9 R. B. TODD Cycl. Anat. & Physiol. II. 495/2 The most important of these is a small branch which proceeds from the ganglion into the tympanum (ramus tympanicus nervi glosso-pharyngei; nerve of Jacobson). 1860 GRAY Anat. (ed. 2) 524 The tympanic branch (Jacobson's nerve), arises from the petrous ganglion. 1871 T. H. HUXLEY Man. Anat. Vertebrated Animals ii. 72 In the latter case they are the canals of Stenson... Glandular diverticula of the mucous membrane, supplied with nervous filaments from both the olfactory and the fifth pair, may open into these canals. They are called..the 'organs of Jacobson'. 1888 W. R. GOWERS Man. Dis. Nervous Syst. II. 212 The gangliform enlargement also receives a twig from the nerve (small petrosal) which connects the otic ganglion, through the nerve of Jacobson, with the glosso-pharyngeal. 1896 KIRKALDY & POLLARD tr. Boas's Text Bk. Zool. 333 In some Reptiles and most Mammals, there is a peculiar saccular or tubular paired organ, Jacobson's organ, in close connection with the olfactory apparatus. 1951 C. K. WEICHERT Anat. Chordates vi. 254 Although the function of Jacobson's organ is obscure, it is believed to aid in the recognition of food, since it is best developed in animals which hold food in their mouths. 1954 T. L. PEELE Neuroanat. Basis Clin. Neurol. ix. 217/2 By way of the tympanic branch of the glossopharyngeal (Jacobson's nerve) sensory fibers are supplied to the medial surface of the ear drum, to the tympanic cavity, to the eustachian tube in part, and to mastoid cells. 1965 R. & D. MORRIS Men & Snakes viii. 195 Snakes, like lizards, possess a..Jacobsen's [sic] Organ. This is a scent-sensitive pair of pits lying in the roof of the mouth.

Hence † **Jacobso·nian** a. (obs.).

1878 A. MACALISTER Introd. Syst. Zool. & Morphol. Vertebr. Animals iii. 16 In connexion with this 5th pair [of nerves] there may be six separate ganglia:..5th, Cloquet's, on the Jacobsonian organ. 1893 GUNN & HENSMAN in H. Morris Treat. Human Anat. VII. 928 In the septal cartilage above the opening of Stenson's canal there is a small pouch which presents a minute opening below. This is the representative of the Jacobsonian organ. A strip of cartilage underneath this..is known as the Jacobsonian cartilage.

Jacqueminot (ʒa·kmino). [Name of the Vicomte J. F. Jacqueminot (1787–1865), French soldier.] In full Général Jacqueminot. A red-flowered, hybrid perpetual variety of rose; also, formerly used for a colour resembling that of the flower.

1857 T. RIVERS Rose Amateur's Guide (ed. 6) 98 We must not forget to rank under our crimson flag General Jacqueminot, which..is so glorious in its colour. 1881 C. C. HARRISON Woman's Handiwork I. 12 Contrast the works of art of quondam florists..with the sumptuous modern assembling of Maréchal Neil [sic] or glowing Jacqueminot roses. 1893 W. ROBINSON Eng. Flower Garden (ed. 3) 644/2 General Jacqueminot and many other H.Ps. do not usually bloom after the month of August. 1895 Montgomery Ward Catal. 6/3 French crepon... Colors: Beige,..heliotrope, magenta or jacqueminot. 1908 Daily Chron. 10 Mar. 3/2 English roses have ..arrived.., and include the beautiful Jacqueminot. 1920 E. WHARTON Age of Innocence I. ix. 69 He tried to analyse the trick, to find a clue to it in..the fact that only two Jacqueminot roses (of which nobody ever bought less than a dozen) had been placed in the slender vase. 1955 C. C. HURST in G. S. Thomas Old Shrub Roses ix. 87 In 1852 'Jules Margottin' and the famous 'Général Jacqueminot' arrived... 'The General', with its brilliant scarlet-crimson flowers and damask fragrance was a fertile tetraploid.

jacu (dʒăkū·). [Pg., f. Tupi jacú.] A large, turkey-like, South American game bird of the genus Penelope, esp. P. marail or P. jacquacu; cf. GUAN.

1824 H. E. LLOYD tr. Spix & Martius's Trav. Brazil II. IV. ii. 227 Several species of wood-hens, particularly the pretty Jacú (Penelope Marail, leucoptera). 1876 Encycl. Brit. IV. 227/2 The gallinaceous jacús, the hoccus, and different kinds of pigeons, haunt the woods [of Brazil]. 1933 P. FLEMING Brazilian Adv. I. xvi. 137 There were a lot of the birds called jacú, a kind of stringy, dowdy pheasant with subfusc plumage. 1964 A. L. THOMSON New Dict. Birds 175/2 There are 12 species of guans, some of them with the names 'jacu' and 'camata'.

jacutinga (dʒækiuti·ŋgă). Also 9 **jaco-.** [f. Pg. jacutinga (formerly jacú-tinga), Brazilian name of a kind of guan (jacu) (probably the black-fronted piping guan, Pipile jacutinga) whose plumage the ore is said to resemble.] A name given to various kinds of soft gold-bearing iron ore found in Brazil (see quot. 1963).

1846 Trans. R. Geol. Soc. Cornwall VI. 227 In both mines [sc. Itabira and Santa Anna, in Brazil] the directions and inclinations of the gold-bearing beds conform to the configuration of the neighbouring mountain, as well as to the structure of the contiguous rock;—a circumstance of common occurrence in jacotinga formations. 1851 Edin. New Philos. Jrnl. L. 61 These [strata] are followed by the Jacotinga, the principal auriferous rock, which is for the most part composed of specular iron-ore and oxide [printed oixide] of manganese. 1869 R. F. BURTON Explor. Highlands Brazil I. 301 The mysterious Jacutinga. The name is evidently derived from the well-known Penelope called Jácu-tinga (P. Leucoptera) from the white spots upon its crested head and blue-black wings. This substance of iron-black, with metallic lustre, sparkles in the sun with silvery mica... The constituents are micaceous iron schist and friable quartz mixed with specular iron, oxide of manganese, and fragments of talc. 1908 J. M. MACLAREN Gold 649 With the itabirite are associated thin beds..of sandy micaceous and limonitic iron-ore containing yellowish talc and earthy oxides of manganese. These beds are generally friable, and appear to be a decomposition product of itabirite. The rock is locally known as jacutinga. 1934 BAIN & READ Ores & Industries S. Amer. vi. 112 In places this [sc. itabirite] contains narrow veins of native gold, or of gold-bearing quartz, giving rise to what are known as jacutinga ores, which are peculiar to Minas Geraes. 1963 Prof. Papers U.S. Geol. Survey No. 341-C. 104/2 The term 'jacutinga' is now used indiscriminately for soft high-grade hematite, soft itabirite, or both, depending entirely upon the locality within the Quadrilátero Ferrífero. In short, it has only a local, not a general meaning and should not be used in scientific description without precise definition.

‖ **jadam** (dʒa·dam). [Malay.] A type of silver or brass niello ware from the Malay Peninsula and Sumatra: used esp. for decorating belt-buckles.

1907 F. SWETTENHAM Brit. Malaya vii. 138 The most original and artistic work of all is called chútam or jádam, and it was made originally in the Province of Ligor. 1908 L. WRAY in A. Wright 20th Cent. Impressions Brit. Malaya 244/1 The art of enamelling is also known to the Malays. The ware is called Jadam, which is equivalent to niello in England. 1910 C. W. HARRISON Illustr. Guide Federated Malay States II. 166 Jadam is the fashion in Rembau where the women wear large belt-buckles of it called pinding, but it is also made in boxes of all shapes. It may be of silver or of brass filled with enamel. 1953 C. A. GIBSON-HILL Malay Arts & Crafts § Silverwork, A form of silverware called Jadam, in which the hollows of the pattern were filled in with black enamel to give a smooth surface, used to be made in the western Malay states, principally in Negri Sembilan. The art is said to have come from Sumatra... In Malaya jadam work was used mostly for the production of decorated waist buckles.

jade, sb.^{2} Add: **1. c.** A colour resembling that of jade; jade-green. Also attrib.

1921 H. WALPOLE Young Enchanted IV. iv. 391 The faint jade of the fading light. 1926 M. LEINSTER Dew on Leaf iii. 42 The jade rabbit (moon) nibbles the clouds. 1928 Manch. Guardian Weekly 31 Aug. 175/3 A faint breeze blowing in from a North Sea of misty jade. 1972 Guardian 5 Dec. 11/2 Toga dress..in..midnight blue, jade, red, sapphire.

2. b. jade-coloured adj.

1868 G. M. HOPKINS Jrnl. 19 July (1959) 178 The Aar sallow and jade-coloured. 1926 A. HUXLEY Essays New & Old 17 The brown or jade-coloured water.

jadeite. Add: Hence **jadeitic** (dʒēidi·tik) a., approximating to jadeite in composition.

1965 Prof. Papers U.S. Geol. Survey No. 525. 25 (heading) Composition of jadeitic pyroxene from the California metagraywackes. 1971 I. G. GASS et al. Understanding Earth viii. 116/1 A notable feature of the overall mineralogy of meteorites is the absence of phases, such as..jadeitic pyroxenes, indicative of high pressure.

jadoo (dʒā·dū). Also **jadu.** [Hind. jādū enchantment.] Magic, conjuring. Comb. **ja·doo-wa:llah** [WALLAH], a Hindu conjurer.

1886 KIPLING Plain Tales from Hills (1888) 126 If there was any jadoo afoot. 1890 Q. Rev. July 244 The Indian conjurers, or Jadoo-walla. 1924 J. A. TYSON Barge of Haunted Lives iv. 93 These took me before a jaboowallah [sic], who..had performed some of his tricks before me at Rajiid.

‖ **j'adoube** (ʒadūb). Chess. [Fr., = I adjust.] An expression used when a player wishes to touch a chessman without making a move.

1808 J. H. SARRATT Treat. Game of Chess I. 3 If a player touch one of his adversary's pieces, without saying 'J'adoube', he may be compelled to take it. 1847 H. STAUNTON Chess-Player's Handbk. 36 A Piece or Pawn touched must be played, unless at the moment of touching it the player say 'J'adoube', or words to that effect. 1967 Chess 11 Dec. 99/2 You may adjust a piece if you previously warn your adversary you are going to do so by saying 'j'adoube'.

Jaeger^1 (yēi·gəɹ). The proprietary name of an all-wool clothing material manufactured originally by Dr. [Gustav] Jaeger's Sanitary Woollen System Co. Ltd.; also, a garment made of this material. Also attrib. and fig.

1887 G. B. SHAW Let. 8 Feb. (1965) 163 Seeing me arrive, clad in an irresistible new Jaeger samite. 1893 K. SANBORN Truthful Woman in S. California 121, I really suffered during a drive, although encased in the heaviest of Jaeger flannels. 1905 CHESTERTON Heretics x. 140 Those who talk to us with interfering eloquence about Jaeger and the pores of the skin. 1925 Trade Marks Jrnl. 7 Oct. 2200 Jaeger... Cloths and stuffs of wool, worsted or hair. The Jaeger Company, Limited,.. London,..merchants. 1932 THORNE SMITH Bishop's Jaegers (1934) 2 If they could not be called things of beauty, these brave long jaegers of the Bishop's, they did ..represent the highest expression of the drawers-maker's craftsmanship. 1936 R. LEHMANN Weather in Streets I. i. 12 A grubby jaeger shroud lay over the first suburbs. 1942 N. BALCHIN Darkness falls from Air vii. 127 The inspector was wearing a roll-necked Jaegar [sic] pull-over. 1973 A. MORICE Death & Dutiful Daughter v. 53 Betsy's old Jaeger dressing-gown.

Jaeger^2 (yēi·gəɹ). The name of E. R. Jaeger von Jastthal (1818–1884), Austrian ophthalmologist, used in the possessive (or occas. attrib.) to designate a series of short passages printed in type-faces of different sizes and used for testing visual acuity at reading distances.

1869 Trans. Amer. Ophthalm. Soc., 4th and 5th Ann. Meeting 68 In Jaeger's Schrift-Scalen the numbering is altogether arbitrary and irregular. 1884 G. HARTRIDGE Refraction Eye iii. 45 Snellen and Jaeger's are the types most commonly in use. 1897 NORRIS & OLIVER Syst. Dis. Eye II. 28 Jaeger's Schriftscalen consisted of a very complete set of reading tests printed in several languages. 1907 J. H. PARSONS Dis. Eye ix. 161 Jaeger's near test types..are simply the ordinary printers' founts of type, from the smallest upwards (nonpareil, minion, etc.). 1962 H. C. WESTON Sight, Light & Work (ed. 2) viii. 224 Another series of test types has been selected by the British Faculty of Ophthalmologists and either this or the Jaeger series is readily obtainable.

Jaffa (dʒæ·fă). The modern name of Joppa, a port in Israel, used attrib. or absol. to designate an oval, thick-skinned variety of orange first cultivated near Jaffa, and later introduced to other parts of Israel and suitable regions elsewhere.

1881 A. H. MANVILLE in T. W. Moore Treat. & Handbk. Orange Culture in Florida (ed. 2) 110 Jaffa and other recently imported varieties have not been fruited long enough in the State to determine their qualities. 1909 Westm. Gaz. 20 Apr. 7/2 The practice of 'faking' oranges by boiling and greasing them and selling them as Jaffas. a 1916 'SAKI' Toys of Peace (1919) 134 We have some very fine Jaffa oranges. 1943 WEBBER & BATCHELOR Citrus Industry I. v. 511 In Florida, however, the Jaffa develops a fruit of very good quality. 1962 J. FLEMING When I grow Rich xix. 207 Oven-hot rolls, ice-cold Jaffa juice, fragrant hot coffee. 1973 'D. JORDAN' Nile Green xx. 76 The orange-juice they serve..in the Beirut St Georges isn't Israeli but I'm sure they'd produce the Jaffas if you insisted. 1974 W. GARNER Big Enough Wreath iv. 43 El Al's flying in fresh-picked Jaffa oranges. Ibid., Miss Cleverley's..face showed no more emotion than a Jaffa.

jag, sb.^{2} Add: **1. c.** (Earlier and later examples.) For U.S. read dial. and colloq. Also, a drinking bout; the state or a period of being drunk.

1678 J. RAY Coll. Eng. Proverbs (ed. 2) 87 Proverbiall Periphrases of one drunk... He has a jagg or load. 1872 J. GLYDE Norfolk Garland I. 149 He has got his jag, i.e., as much drink as he can fairly carry. 1894 [see *GIN v.^{3}]. 1895 N.Y. Dramatic News 26 Oct. 7/2 An ability to acquire a 'jag' in a wonderfully short space of time and with a single drink. 1904 [see *HANG-OVER 2]. 1921 E. WALLACE Law Four Just Men iv. 112 He had been on a jag the night before and had finished up in what he called an opium house. 1924 [see *CROOK v.^{1} 1 d]. 1934 WODEHOUSE Right Ho, Jeeves xix. 250, I took the whole thing as a great compliment, proud to feel that any drink from my cellars could have produced such a majestic jag. 1966 Listener 28 Apr. 619/1 Sid Chaplin's Saturday Saga, the account of two miners on a memorable jag.

d. transf. and fig. A period of indulgence in a particular pastime, emotion, interest, etc.; = FIT sb.^{2} 4 a; freq. with defining word prefixed, as crying jag; spec. (see quot. 1946). colloq. (orig. U.S.).

1913 J. LONDON Valley of Moon (1914) I. xv. 119 'Aw, it's only one of his cryin' jags,' Mary said. 1924 P. MARKS Plastic Age xix. 213 One had a 'crying jag'. Ibid. xxii. 254 A girl got a 'laughing jag' and shrieked with idiotic laughter. 1933 S. HOWARD Alien Corn III. 97 Isn't seventy-one fifty cheap for the jag I've got to-night? 1945 S. LEWIS Cass Timberlane (1946) xlix. 347 Now you're beginning to get over your love-jag, maybe you can see that Jinny is as..tricky and grabbing as a monkey. 1946 MEZZROW & WOLFE Really Blues 375 Jag, a state of extreme stimulation, produced by marihuana or some other stimulant. 1958 Spectator 4 July 15/2 The British public are on an enormous clean-clothes jag. 1972 New Yorker 26 Aug. 38/3 A neurotic habit..may be overt, like a temper tantrum or a crying jag. 1973 Times Lit. Suppl. 8 June 631/3 The Kennedy years..launched the Americans on a jag of hope and fear.

Jag (dʒæg), sb.^{3} Colloq. abbrev. of Jaguar (the proprietary name of a make of motor car). See also gin-and-Jag s.v. *GIN sb.^{2} 2 b.

1959 J. DRUMMOND Black Unicorn xxi. 146 'He will meet us opposite the clock-tower in Point Road. I gave him a little word-picture of the Jag.' The Jaguar was a long cream drop-head. 1962 Times Lit. Suppl. 19 Oct. 805/4 Stephen, the boss's son, with his Jag, to Wilf, the miner's son, with his typewriter. 1968 J. FLEMING Kill or Cure vii. 88 People with lots of money, living in the Jag belt. 1974 T. ALLBEURY Snowball x. 55 They've bought a car. A Jag—second-hand.

Jagatai (dʒægătā·i). [The native name of Turkestan, f. Jagatai, a son of Jenghiz Khan, who inherited it.] The branch of the Turkic group of languages spoken in Turkestan. Also attrib. Also **Jagata·ic.** Hence **Jagata·ian** a., pertaining to Turkestan or the dialects spoken there.

1843 Penny Cycl. XXV. 406/1 Jagatái, [spoken] in the greater part of independent Turkistán... The Jagatái language has a valuable literature. 1867 Chambers's Encycl. IX. 589/2 The former [sc. Eastern Turkish] is mainly represented by the Uigur (Jagatai). Ibid. 590/1 The Eastern [Turkish] or Jagataian [literature]. 1908 T. G. TUCKER Introd. Nat. Hist. Lang. 134 (d) Uiguric, including Uigur proper, Jagataic, and Turkoman, the dialects spoken in the parts of Turkestan not occupied by the Kirghiz. 1929 Encycl. Brit. XXII. 920/1 The Uzbeg (Jagatai Turkish) tongue. 1962 A. TIETZE in Householder & Saporta Probl. Lexicogr. 266 Jagatai and other non-Ottoman Turkic languages.

jagatī (dʒɒ·gātī). Also **gagatī.** [Skr.] A Vedic metre of twelve syllables.

1843 [see *GĀYATRĪ]. 1869 MAX MÜLLER tr. Rig-Veda-Sanhita p. cxxx, I maintain by no means that this was the actual origin of Gagatī metres... Theories..would wish us to look upon the hendecasyllabic Trishṭubh as originally a dodecasyllabic Gagatī, only deprived of its tail. 1937 M. D. SHASTRI Ṛgveda-Prātiśākhya III. 137 All that is light by nature is related to light syllables, and a Jagatī, one should know, has light syllables. 1969 Language XLV. 251 A cadence of four or five syllables depending on whether the pāda is in triṣṭubh meter (typically 11 syllables) or jagatī (12 syllables).

jagged (dʒægd), a.^{2} slang (chiefly U.S.). [f. *JAG sb.^{2} 1 c + -ED^2.] Drunk, intoxicated.

1737 Pennsylvania Gaz. 6–13 Jan. 1/3 He's Jagg'd. 1902 Telegram (Winnipeg) 20 Aug. 7/4 Miller was pretty well jagged. 1904 'O. HENRY' in N.Y. World Mag. 1 May 8/2 What I want is a masterful man that slugs you when he's jagged, and hugs you when he ain't jagged. 1956 Amer. Speech XXXI. 279 Jagged, adj. In the sense of intoxicated..[it] is probably not of American origin.

b. Intoxicated by, or under the influence of, drugs.

1938 C. HIMES Black on Black (1973) 175 She made him smoke pot and when he got jagged..she put him out on the street. 1973 BOYD & PARKES Dark Number xiii. 151 Solange is—was—God help her, a heroin addict. When we first met, she was all jagged up. She was a reject on the junk heap.

‖ **jagt** (yakt). [a. Da. jagt (cf. Norw., Sw. jakt) YACHT sb.] In Scandinavia, a small single-masted coastal vessel, rigged either with square sails or as a cutter or sloop.

1861 J. LAMONT Seasons with Sea-Horses ii. 22 We saw two small vessels, which we made out to be a brig and a sloop, or 'jagt', at some distance amongst the ice. 1906 H. W. SMYTH Mast & Sail iii. 44 Square-rigged Nordland jaegts which formerly were so characteristic a feature of the Norwegian coastline. Ibid. 51 The old Nordland 'jaegt' which formerly did so much of the coast-wise trade of Norway. 1925 A. MOORE Last Days Mast & Sail vi. 181 Amundsen's ship, the Gjoa, in which he made the North-West Passage, was described in the English papers as a sloop, and she was one according to the old definition; but in the list of the Bureau Veritas for 1907 she is called a Jagt. 1971 Mariner's Mirror LVII. 152 In Schleswig-Holstein the Jachten doubtless were of the classical Danish jagt design in hull and rig; carvel-built keel craft with transom stern and outside rudder, convex stem, and pole-masted gaff sloop rig.

jaguarondi. Add: Also **jaguarundi, yaguarundi.**

1906, 1955 [see *EYRA]. 1959 A. S. LEOPOLD Wildlife Mexico 482 Little is known of the jaguarundi. 1964 L. S. CRANDALL Managem. Wild Mammals in Captivity 368 The jaguarundi is a shy, secretive creature.

jai alai (hɔi·lɔi, hɔi·ālɔi, hɔi,ālɔi·). [Sp., f. Basque jai festival + alai merry.] = PELOTA.

1910 I. A. WRIGHT Cuba i. 7 (caption) Ball Players on the Court, General Leonard Wood, a Jai Alai enthusiast among them. 1923 W. STEVENS Let. 4 Feb. (1967) 234 In the evening I went to see a game of jai alai, the Spanish national game. 1947 M. LOWRY Under Volcano i. 9 Its jai-alai courts are grass-grown and deserted. 1972 Times 8 Aug. (Asian Suppl.) p. vii/8 Do not forget to see a game of jai-alai [in the Philippines]. 1973 Times 28 May (Macao Suppl.) p. iii/3 The fast and furious Basque ball and racquet game of Jai Alai (or Pelota Basque in Portuguese) is the latest of Macau's diversions for the visitor. The opulent new Jai Alai Centre on the Outer Harbour offers the last word in de luxe comfort for fans or gamblers betting on the matches, plus nightclub, restaurant and bar facilities.

‖ **Jai Hind** (dʒɔi hind), int. [Hindi, f. jai long live! + Hind India.] In India, a salutation: used in exchange of greetings, at a public meeting, etc.

1948 A. MOOREHEAD Rage of Vulture v. 76 The Indians ..raised their cry of 'Jai Hind', and it really meant 'Expel the British'. And finally they have expelled us. They go on crying 'Jai Hind', and now it means exactly what it says—'Long Live India'. 1969 Commerce (Bombay) 26 July 170/3 We remain committed to the freedom and progress of the people of this great country. Jai Hind.

jail, gaol, sb. Add: **2. jail-bait** slang (orig. U.S.), a girl who is under the legal age of consent; **jail-break** orig. U.S., the act of escaping from a jail.

1934 J. T. FARRELL Calico Shoes 48 She's not hard on the eyes but she's jail bait. 1957 J. BRAINE Room at Top xxiv. 198 I'm not interested in little girls. Particularly not in jail-bait like that one. 1972 A. DRAPER Death Penalty vi. 45 She looks young enough to be jail bait. 1910 J. HART Vigilante Girl xix. 266 Hamlin did not yet know of the jail-break. 1952 J. STEINBECK East of Eden 440 Not with her holding that jail-break over him. 1973 E. HYAMS Final Agenda ii. 24 He..led a jail-break of seventeen political prisoners.

Jaipur (dʒɔi·pūəɹ). The name of a former Indian native state and its capital city, now capital of the State of Rajasthan, used attrib. to designate products of this State. Hence **Jaipuri,** the dialect of this region.

1889 KIPLING in Macm. Mag. Dec. 152/1 The cedar sliding doors were fitted with hasps of translucent Jaipur enamel. 1901 Jrnl. R. Asiatic Soc. 787 Sixteen real dialects spoken over the area in which Rājasthānī is a vernacular..fall into four main groups, which may be called Mēwātī, Mālwī, Jaipurī, and Mārwārī. Ibid. 788 Jaipurī may be taken as representing the dialects of Eastern Rājputānā, as far east as Gwāliōr. 1931 A. U. DILLEY Oriental Rugs & Carpets Pl. 37 (caption), Agra copy of Jaipur plant rug. 1957 Encycl. Brit. XVIII. 957/2 Rajasthani language..has several dialects, the principal of which are Jaipuri, Marwari, Mewati and Malvi. 1963 Listener 28 Feb. 365/2 Then there is Jaipur work.

jak, var. JACK sb.^{4} (in Dict. and Suppl.).

jake (dʒēik), sb.^{1} U.S. colloq. [Prob. the personal name Jake, abbrev. of Jacob.] A rustic lout or simpleton: usually country jake.

a 1854, etc. [see country jake s.v. *COUNTRY 16]. 1884 G. W. PECK Peck's Boss Bk. 68 A masher, like many of the Jakes of the present day. 1915 Dialect Notes IV. 199 He's no jake even though he did come from a Nebraska farm. 1941 H. S. TRUMAN Let. 5 Oct. in M. Truman Harry S. Truman (1973) viii. 142 You'd think I was Cicero or Cato. But I'm not. Just a country jake who works at the job.

jake (dʒēik), sb.^{2} slang (orig. U.S.). [Abbrev. form of *Jamaica (ginger).] An alcoholic beverage made from Jamaica ginger. **b.** Methylated spirits used as an alcoholic drink.

1926 [see *HIP a.]. 1932 Fortn. Rev. Mar. 324 Over twenty-five per cent. are 'jake' or 'feke' drinkers... They drink methylated spirits. 1935 H. NEVILLE Sneak Thief on Road II. 162 That's pure meths... They call it Blue Billy or jake, and it's known among all tramps and kip-houses and wherever men have empty lives. 1939 J. STEINBECK Grapes of Wrath x. 131 He would drink jake or whisky until he was a shaken paralytic.

jake (dʒēik), a. slang (orig. U.S.). [Origin obscure.] Excellent, admirable, fine, 'O.K.'

1914 JACKSON & HELLYER Vocab. Criminal Slang 48 As an adjective 'jake' means good; satisfactory; acceptable; all-right. 1921 P. & T. CASEY in Adventure (U.S.) 18 May 40/2 Well, if it is Jerrold, everything's jake. 1924 P. MARKS Plastic Age xxii. 247 She said the whole college seemed jake to her. 1924 WODEHOUSE Bill the Conqueror vi. 130 Everything was jake with Horace. 1930 [see *BY prep. 33 e]. 1943 2 N.Z.E.F. Times 10 May 5/1 She'll be Jake on the counter. 1947 D. M. DAVIN Gorse blooms Pale 190 We'll just give [the tea] a minute to draw and she'll be jake. 1958 R. FRANCE Race 100 'Will you be all right..if I lie down in the wheelhouse?'..'I'm jake. You have a rest.' 1958 'A. GILBERT' Death against Clock viii. 105 'If it's about the election we vote conservative here.' 'Jake by me,' said Mr. Crook politely. 1967 Southerly XXVII. 149 'Is she [sc. a train] on time?'.. 'She's jake tonight mate.' 1970 N.Z. Listener 12 Oct. 12/4 Long as there's plenty of beer, she'll be jake.

So **jake(a)loo, jakerloo** a. Austral. and N.Z. slang, in the same sense.

1919 W. H. DOWNING Digger Dial. 29 Jake-aloo. 1936 N. MARSH Death in Ecstasy xvii. 211 It'll all come out what the Australians call 'jakealoo'. 1936 'R. HYDE' Passport to Hell xi. 174 Jakeloo, Starkie; she's a little beauty, clean through my arm. 1938 X. HERBERT Capricornia xii. 169 'Lambkin, you're not wounded, are you?'..'Na—ow! I'm jakerloo.' 'You're what?' she demanded, looking scared. 'Jakerloo Mum, jakerloo.' 'What—not a disease, my darling?' 'Na—ow—that's French for "I'm good-o".' 1965 G. MCINNES Road to Gundagai viii. 123 Jakeloo! Let's have the names then, and the addresses.

jakes. 1. (Later examples.)

1913 L. WOOLF Village in Jungle iv. 54 The headman's brother is to marry a sweeper of jakes! 1922 JOYCE Ulysses 68 He kicked open the crazy door of the jakes. 1969 Listener 26 June 902/3 He is at his best when not occupied with symbols..but concerned to tell how the keeper of an 'underground jakes' mistakes a police stool-pigeon for a real poof.

jakkalsbessie (ya·kălsbe:si). S. Afr. [Afrikaans, f. jakhals jackal + bessie berry.] Either of two trees, the evergreen Diospyros mespiliformis or Sideroxylon inerme, or their fruit. Also attrib.

1854 L. PAPPE Silva Capensis 22 Sideroxylon inerme. Lin. (Melkhout.),.. The fruit (Jackalsbesjes) are edible. 1917 R. MARLOTH Dict. Common Names of Plants 42 Jakkalsbessie. Diospyros mespiliformis, but also Sideroxylon inerme (milkwood). 1932 WATT & BREYER-BRANDWIJK Medicinal & Poisonous Plants S. Afr. 137 The Zulus take an infusion of the bark of Sideroxylon inerme L., White milkwood, Wit melkhout, Jakkalsbessie..to dispel bad dreams. 1953 [see *GEELHOUT]. 1963 S. CLOETE Rags of Glory 377 Here and there a giant tree, a baobab, jakkals-bessie, or fig, stood out—a blob dropped on the carpet. 1969 T. H. EVERETT Living Trees of World 285/1 A fine African member of the genus [Diospyros] is the jakkalsbessie, West African ebony or Transvaal ebony (D. mespiliformis), which becomes 70 feet tall with a trunk diameter of 3 feet. 1973 PALMER & PITMAN Trees of Southern Africa III. 1795 The bar at Leydsdorp is made out of a solid piece of jakkalsbessie timber.

Jakun (dʒakū·n). [Native name.] An aboriginal people of the southern part of the Malay peninsula; a member of this people; also, their language. Also attrib. or as adj.

1839 T. J. NEWBOLD Pol. & Stat. Acct. Straits of Malacca I. vii. 421 The Jakuns do not differ materially from the Malay in colour or physiognomy. 1883 Encycl. Brit. XV. 323/2 The aborigines..are divided into a great many tribes, of which the best known are the Jakuns, widespread in the south. 1906 SKEAT & BLAGDEN Pagan Races Malay Peninsula I. II. v. 235 An old Jakun, who was singularly free from superstition. Ibid. II. III. vi. 196 In the Semang tribes the office of chief medicine-man appears to be generally combined with that of chief, but amongst the Sakai and Jakun these offices are sometimes separated. Ibid. IV. i. 405 It may be that before their decay, the other Jakun dialects resembled it [sc. Kenaboi] more than they do now... Kenaboi must be regarded either as the best specimen of Jakun recorded or else as not being Jakun at all. 1935 Discovery Sept. 262/2 The Jakun, straight haired and akin to the Malay, occupy Southern Johore. 1947 R. O. WINSTEDT Malays 14 A Jakun (or Proto-Malay) marriage ceremony..requires the groom to walk or run after his bride three or seven times round a hillock. 1958 Listener 13 Nov. 793/3 Ethnically, they [sc. the aborigines of Malaya] were divided into Jakun (proto-Malays), Negritos, and a group that was primarily Caucasoid. 1972 A. AMIN tr. Ahmad's No Harvest but Thorn iii. 18 If the jakuns come out what can we say?

jāl, var. *JOL.

jalapinolic (dʒæ:lăpinǫ·lik), a. Chem. [tr. G. jalappinolsäure jalapinolic acid (W. Mayer 1854, in Ann. d. Chem. u. Pharm. XCII. 128), f. jalappin (now jalapin) JALAPIN + -ol -OL: see -IC.] jalapinolic acid: the dextrorotatory form of 11-hydroxyhexadecanoic acid, $CH_3(CH_2)_4CHOH(CH_2)_9COOH$, a crystalline derivative of palmitic acid obtained by hydrolysis of jalapin.

1855 Chem. Gaz. 15 Mar. 115 The oxidation of convolvulinolic and jalapinolic acids by nitric acid. 1928 Jrnl. Amer. Chem. Soc. L. 1749 The resin from Orizaba root yields on hydrolysis an hydroxyhexadecanoic acid, jalapinolic acid. 1964 Phytochemistry III. 289 The aglycons most frequently obtained from convolvulaceous glycosidic acids have been (+)-11-hydroxyhexadecanoic acid (jalapinolic acid) and convolvulinolic acid.

‖ **jaleo** (hale·o). [Sp., lit. 'halloo'.] A lively Andalusian dance, or the clapping that accompanies it.

1865 H. O'SHEA Guide to Spain p. xlv, The dances differ in each province... Andalucia is the land of the jaleo de Jerez. 1893 Funk's Stand. Dict., Jaleo, a vivacious Spanish dance. 1967 'LA MERI' Spanish Dancing (ed. 2) v. 69 Probably because of the centuries of dancing to a percussive sound, the jaleo accompanying the dance has developed into an art in itself.

jalopy (dʒălǫ·pi). colloq. (orig. U.S.). Also † **gillopy, jalapa, jollopy; jallopy, jaloppi(e), -y.** [Origin unknown.] A battered old motor vehicle; also, an old aeroplane.

1929 HOSTETTER & BEESLEY It's a Racket! 229 Jaloppi, a cheap make of automobile; an automobile fit only for junking. 1936 J. STEINBECK In Dubious Battle vi. 90 Mac and Jim circled the buildings and went to the ancient Ford touring car. 'Get in, Jim. You drive the gillopy.' 1937 Time 17 May 35/1 They announced they would burn..used cars at the fair grounds... The huge pile of jalopies was touched off while firemen..looked on. 1938 P. GALLICO in Sat. Even. Post 8 Oct. 8/3 He..made a living out of the Red Arrow Hangar down at one end of the field, teaching beginners how to hoist a couple of training jalopies around the field without killing themselves. 1941 AUDEN New Year Let. III. 68 And in jalopies there migrates A rootless tribe from windblown states. 1951 WODEHOUSE Old Reliable xxi. 233 This afternoon we'll go out in my jalopy and start pricing ministers. 1955 M. E. B. BANKS Commando Climber vi. 114 Perhaps a succession of broken down jalopies has impaired my faith in the internal combustion engine. 1973 A. HUNTER Gently French ii. 22 He'd get in the jalopy beside him, start trying to pressure him.

jalousie. (Earlier and later examples.)

1766 DUCHESS OF NORTHUMBERLAND Diary 23 Oct. (1926) 76, Rows of Seats with Jalousies in Front that they [sc. the women] may not be seen. 1961 I. FLEMING Thunderball xxiv. 254 Inside the small room, the jalousies threw bands of light and shadow over the bed. 1974 K. BENTON Craig & Tunisian Tangle v. 47 Tall windows shielded against the sun by wooden jalousies.

jalpaite (dʒæ·lpă,əit). Min. [ad. G. jalpait (A. Breithaupt 1858, in Berg- und hüttenmännische Zeitung 17 Mar. 85/2), f. Jalpa, the name of a locality in Mexico (probably the town of that name in southern Zacatecas): see -ITE^1.] A sulphide of silver and copper, Ag_3CuS_2, of a light metallic-grey colour when freshly fractured.

1868 J. D. DANA Syst. Min. (ed. 5) ii. 39 Jalpaite is a cupriferous silver-glance from Jalpa, Mexico. 1925 Mineral. Abstr. II. 519 Jalpaite.., as narrow veins in hornstone, gave analyses agreeing closely with the formula $3Ag_2S.Cu_2S$. 1968 Amer. Mineralogist LIII. 1539 Jalpaite from Silver Plume and Boulder County, Colorado probably formed at some temperature below 117°C.

jam, sb.^{1} Add: **1.** (Further examples.) Also, spec. in logging, an accumulation of logs in a river. Also attrib.

1805 Deb. Congress U.S. 7 Apr. 1076 Its overflowing [is] occasioned by a jam of timber choking the river. 1836 Bytown (Ottawa) Gaz. 9 June 4/3 A canoe with nine men..were engaged in taking some timber in a jam at the head of Colton's shoots. 1905 Terms Forestry & Logging (U.S. Dept. Agric. Bureau Forestry) 40 Jam, to break a, to start in motion logs which have been jammed. 1910 S. E. WHITE Rules of Game I. xii. 69 'Where's the drive, doctor?' asked the lumberman. 'This is the jam camp,' replied the cook. 'The jam's upstream a mile or so.' 1929 Encycl. Brit. XIV. 482/1 A log jam in the Montreal river, Ontario, Canada. 1955 Times 31 May 4/3 From all around the capital came reports of traffic jams. 1968 R. M. PATTERSON Finlay's River 32 So they made a risky crossing of the Parsnip [River] on a jam, wondering as they did so whether the ice-bridge over the deepest water would not give way beneath them. 1971 Daily Tel. (Colour Suppl.) 22 Oct. 22/1 There would be fewer frayed tempers and thus far fewer accidents—not to mention fewer jams.

b. Also fig., an awkward or difficult situation; trouble; = FIX sb. 1; freq. in phr. in a jam. colloq. (orig. U.S.).

1914 San Francisco Call 26 Oct. 7, I knew we'd get in a jam coming here. 1926 Clues Nov. 159/1, I think some one single-duked us, but if so I'll shiv the heel. There'll be plenty of jam. 1927 WODEHOUSE Small Bachelor vi. 93 'I've gone and got myself into the devil of a jam.' 'A position of embarrassment?' 'You said it!' 1938 R. D. FINLAYSON Brown Man's Burden 81 Henare would give his whole-hearted sympathy and his last shilling to any-one in a bit of a jam. 1950 [see *CLEANER 2]. 1958 New Statesman 12 Apr. 459/3 He knew instinctively that in a jam it was not done to let down one's own side.

2. Jamming (of broadcasts, devices, etc.), or an instance of this. Hence **ja·m-proof** a., proof against jamming. Cf. *JAM v.^{1} 3 c.

1914 P. VAUX Sea-Salt & Cordite 129, I don't like this wireless jam! 1927 W. E. COLLINSON Contemp. Eng. 113 The trouble caused by jams, atmospherics, and howlings. 1964 Ann. Reg. 1963 185 It was said..that the needles in orbit round the earth could provide an inexpensive and jam-proof global communications system. 1972 Sci. Amer. June 17/1 These communications must be jam-proof; the potential attacker cannot be allowed to hope that a communications failure might prevent a retaliatory strike.

3. (This sense may belong to JAM sb.^{2}) colloq. Jazz or similar music simultaneously extemporized by a number of performers; a period of playing such music. Freq. attrib., esp. as **jam session,** a gathering of musicians to improvise jazz; also transf. and fig.

1929 Melody Maker Jan. 75/3 There are many variations on this rhythm..which make excellent breaks—or 'jams' as they now call them when they are taken by the whole band, the word 'break' being used only when it is intended to signify that it is played by one instrument or a section moving together or unaccompanied. 1933 Fortune Aug. 90/1 The jazz musicians' jam sessions where the players vie with one another in hot solos. 1935 Swing Music July 120/2 The best Chicagoans very often had 'jam' sessions. 1935 Vanity Fair (N.Y.) Nov. 71/3 Extremely hot ensemble improvisations are jams. 1937 Amer. Speech XII. 46/2 A jam band depends entirely on improvisation, using no written music. 1944 Theology XLVII. 278 This contemporary jam session gives enormous pleasure to the participants. But we [sc. the Church of England] have had little enough success in charming the ear of the nation to the extent of persuading it to come and join the band. 1949 Chicago Daily News 25 Mar. 33/2 One of his ambitions reportedly was to sit in on a jam session with some of our jazz musicians. 1959 R. GANT World in Jug 116 Everyone saw back to hear Mitch give a muted chorus which had them roaring again as we went into a final jam. 1967 'LA MERI' Spanish Dancing (ed. 2) vi. 78 Martinez called bulerias 'the Cachucha of the gitanos', while Argentinita described it as a 'flamenco jam-session'. 1969 S. GREENLEE Spook who sat by Door xx. 170 He..moved to the stereo. 'Let's see if I can remember the jams you dig.' 1972 Jazz & Blues Feb. 18/3 Several musicians told me how much they enjoyed the jam sessions.

jam, sb.^{2} Add: **b.** (Further examples.) Colloq. phrases: to have (or like, want) jam on it: to have, etc., something exceedingly pleasant or easy; jam tomorrow: something pleasant promised or expected for the future, esp. something that one never receives; money for jam: see *MONEY sb. 6 h.

1871 'L. CARROLL' Through Looking-Glass v. 94 The rule is, jam to-morrow and jam yesterday—but never jam to-day. 1874 HOTTEN Slang Dict. 268 Real jam, a sporting phrase, meaning anything exceptionally good. 1919 Athenæum 8 Aug. 727/2 'Having jam on it' (i.e. something nice and easy, a 'cushy' job). 1925 FRASER & GIBBONS Soldier & Sailor Words 130 'You want jam on it', i.e., You expect too much. 1936 J. CURTIS Gilt Kid 2, You want jam on it, you do. 1939 A. HUXLEY After Many a Summer II. iii. 201 The entire capital outlay had already been amortized, so that everything from now on would be pure jam. 1946 J. IRVING Royal Navalese 9, The ironical suggestion made to a sailor already 'moaning' about his job—'Do you want jam on it?' 1951 'J. WYNDHAM' Day of Triffids xii. 225 Just put the Americans into the jam-tomorrow-pie-in-the-sky department awhile. 1962 Listener 29 Nov. 925/1 Dr Leavis sees C. P. Snow as a gross materialist concerned only with jam tomorrow. 1970 Times 21 Feb. 6/8 Freedman says he can break even during the 10 weeks, with the jam to follow in the summer. 1972 Daily Tel. 30 Mar. 22/6 Ultramar has ever been the 'jam tomorrow' stock par excellence, with not a penny paid out in dividends. Instead, shareholders get scrip issues. 1973 G. MITCHELL Murder of Busy Lizzie I. 14 'I think Greece might be a very good idea—later on.' 'Never jam today!' muttered Margaret. 1974 O. MANNING Rain Forest I. vi. 87 Hugh.. was free to leave at six... Pedley..said: 'You've got jam on it: walking home in the sunset.'

c. jam-making (examples), -pudding, -tin; **jam-buttie, -butty,** a butty (*BUTTY^2) spread with jam; **jam-jar,** (a) a jar designed for holding jam; (b) rhyming slang for 'motor-car' (see also quot. 1943).

1927, 1965 Jam-buttie, butty [see *BUTTY^2]. 1970 Times 29 Jan. 9/8 You could have knocked us all down with a jam buttie when she [sc. Gracie Fields] first took up with those foreigners. 1972 Observer (Colour Suppl.) 16 Jan. 17/4, I am sluggish and sapped of energy and living on an occasional 'jam butty'. 1895 Army & Navy Co-op. Soc. Price List 785 [Cut glass] Jam Jar. 1902 M. BARNES-GRUNDY Thames Camp iv. 67 Jane went on with her jam-jar trap [for wasps]. 1934 P. ALLINGHAM Cheapjack xiii. 163 Have you got a jam-jar—a car? 1943 C. H. WARD-JACKSON It's a Piece of Cake 38 Jam jars, armoured cars. 1962 R. COOK Crust on its Uppers i. 23 Parking this dreadful great orange-and-cream jamjar..slap under a no-parking sign. 1967 N. FREELING Strike Out 81 A few brushes in a jam-jar. 1908 G. JEKYLL Children & Gardens ii. 12 In the kitchen the children..learn the elements of even more serious cookery, such as jam making. 1968 P. JENNINGS Living Village 122 Most wives buy cakes and preserves, a few still do their own baking and jam-making. 1841 THACKERAY Gt. Hoggarty Diamond (1849) ix. 100 My dear wife..vowed she would cook all the best dishes herself (especially jam pudding, of which..I am very fond). 1906 E. DYSON Fact'ry 'Ands xii. 161 Gets 'is quid a week..solderin' jam-tins. 1956 Coast to Coast 1955–6 59 He had one of the jam-tins in his hand.

2. Affected manners; self-importance; freq. in phr. to lay (or put) on jam. Austral. slang.

1882 Sydney Slang Dict. 5/1 Jam (putting on), assuming fast airs of importance. 1901 'M. FRANKLIN' My Brilliant Career (1966) xxvi. 159 People who knew how to conduct themselves properly, and who paid one every attention without a bit of fear of being twitted with 'laying the jam on'. 1924 LAWRENCE & SKINNER Boy in Bush 46 Don't y' get sidey..puttin' on jam an' suchlike. 1945 BAKER Austral. Lang. vi. 119 Terms like..jam and quiver, connoting 'side' or affectation.

Jam, sb.^{4} Substitute for def.: A hereditary title of certain princes and noblemen in Sind, Kutch, and Saurashtra. Add earlier and later examples.

1727 A. HAMILTON New Acct. E. Indies I. xi. 115 The Jams to the Eastward, who being Borderers, are much given to Thieving, and they rob all whom they are able to master. 1913 A. G. GARDINER Pillars of Society 293 And so, 'hats off' to the Jam Sahib—the prince of a little State, but the king of a great game. 1958 L. F. R. WILLIAMS Black Hills 70 Certain chiefs..whose original early title of Jam is by tradition associated with the mighty Iranian monarch Jamshed.

jam, v.^{1} Add: **3. c.** trans. To cause interference with (radio or radar signals) so as to render them unintelligible or useless, esp. deliberately; to prevent reception of (a transmitter or station) by such means. Also transf.

1914 P. VAUX Sea-Salt & Cordite 46 Communications became regularly jammed. Ibid. 47 We'll stop this jamming, wherever it's coming from. 1914 Wireless World July 246/1 Electricity in our language..is not 'juice'; neither is radio interference 'jamming'. 1920 Discovery Apr. 116/2 When the reception of a message is thus interfered with by other messages being sent at the same time, the message is said to be 'jammed'. Ibid., The jamming of a message may also be caused by stray ether disturbances in the atmosphere itself. 1920 Telegraph & Telephone Jrnl. VI. 165/1 As the number of aeroplanes multiplied 'jambing'—the great drawback of wireless—became more acute. 'Jambing'..refers to the general mix up which results from the reception of two or more sets of signals at once in the same instrument. 1926 E. F. SPANNER Naviators x. 124 The Admiral had answered the Japanese C.-in-C. by sending out jamming signals immediately the British scout had been driven down. 1939 War Illustr. 7 Oct. 126/2 Gramophone records of pledges given by Hitler in his public speeches have been broadcast from France—and jammed by the Germans! 1947 Amer. Speech XXII. 154/1 Allied bombers jammed (rendered ineffective) German radar equipment by dropping quantities of metal foil when over enemy targets. 1947 Jrnl. R. Aeronaut. Soc. LI. 432/2 We developed a jamming screen for the purpose of blinding the enemy's early warning system and so preventing him from obtaining information of our approach. 1955 Times 18 Aug. 9/2 In 1933 the Vienna transmitters were put on to jam Nazi attacks on the Dolfuss Government from the Munich transmitters. Between 1934 and the outbreak of the war the device was copied wholesale. 1959 Ann. Reg. 1958 235 Jamming of Western broadcasts continued. 1970 Daily Tel. 16 Apr. 1 The Post Office is jamming broadcasts by the pirate radio ship North Sea International. 1971 Sci. Amer. June 132/2 Any sonar can be jammed, and clever moving jammers would pretty surely beat the art of beam shaping. 1971 New Scientist 2 Sept. 536/3 In 1942 they investigated the severe jamming of army radar stations, and concluded that radio waves of amazing intensity are emitted by the Sun.

5. b. To apply or put (a brake) on violently.

1925 Morris Owner's Manual 11 Jambing on the brakes at the last moment.

6. intr. To play in a 'jam' or 'jam session' (see *JAM sb.^{1} 3); to extemporize. Also trans., to improvise (a tune, etc.). colloq. (orig. U.S.).

1935 Stage Sept. 46/2 Jam, to improvise hot music, usually in groups. 1936 Delineator Nov. 11/2 He just comes on in here once in a while because he likes to jam. 1951 E. PAUL Springtime in Paris xi. 203 Pierre Braslavsky could sit in anywhere old-school jazzmen are jamming. 1955 L. FEATHER Encycl. Jazz 132 He was seen in the Norman Granz film, Jamming the Blues. 1958 R. HORRICKS in P. Gammond Decca Bk. Jazz ix. 115 This became an important factor in Kansas City jamming. Ibid. 117 The legendary Art Tatum loved to jam with the resident jazz musicians. 1960 Melody Maker 31 Dec. 5/3 They just wanted me to jam a blues for the fourth number. 1971 It 2–16 June 19/1 They've been jamming together at a studio in Greenwich Village.

jam, v.^{2} Add: **2.** To make into jam. Hence **jammed** ppl. a.^{2}; **ja·mming** vbl. sb.^{2}

1854 THOREAU Walden 256 The cranberries,..destined to be jammed. 1905 Daily Chron. 2 Dec. 4/4 Apples, pears, plums, berries, &c. (fresh or dried, or jammed, or tinned, or bottled). 1949 Hansard Commons 16 May 12 'Jamming sugar' is a term very frequently used by housewives. 1969 Islander (Victoria, B.C.) 21 Sept. 8/1 In many kitchens there is jamming, jelly and pickle making, perking.

jam (dʒæm), adv. and a. orig. U.S. Also **jamb.** [f. JAM v.^{1}] **A.** adv. **1.** Closely; in close contact or with firm pressure. Often with up (against).

1825 J. NEAL Bro. Jonathan II. 52 He had been sitting, for two or three hours,..'jam up' in a back seat. 1842 American Pioneer I. 184 The next moment the sloop ran jamb against it. 1852 MRS. STOWE Uncle Tom's Cabin I. 49 'It'll stand, if it only keeps jam up agin de wall!' said Mose. 1932 Kansas City (Missouri) Star 24 May 18 His Chevrolet..[ran] jam up against a house.

2. jam up: thoroughly, perfectly, excellently; right up to; so jam-full adj.: packed full, completely filled; jam-packed adj.: tightly packed; closely crowded or squeezed together; hence (as a back-formation) jam-pack v. trans., to pack tightly, fill.

1835 D. CROCKETT Acct. Col. Crockett's Tour 192 [Andrew Jackson] went jam up for war; but the cabinet got him down to half heat. 1846 Congress. Globe 22 May 852 Their notion is that we go jam up to 54° 40', and the Russians come jam down to the same. 1858 S. A. HAMMETT Piney Woods Tavern xiv. 146 The regular stage was jam full, and there was an extra put on, and that was jam full, and a leetle more. 1866 C. H. SMITH Bill Arp 61 Linton played his part of the programme jam up. 1893 G. B. SHAW Let. 27 Apr. (1965) 392 Friday & Saturday are jam full. 1925 R. LARDNER in Liberty 28 Mar. 5/1 This place is jam-packed Saturdays, from four o'clock on. 1928 WODEHOUSE Money for Nothing v. 96 How can you be poor, when that gallery place you showed us round yesterday is jam full of pictures worth a fortune an inch? 1936 F. CLUNE Roaming round Darling xxi. 214 Eventually we were jampacked in, with the ladies alternately sitting on the Poet's knees. 1938 State Jrnl. (Lincoln, Nebr.) 5 May 14 The foursome finally got a chance to try the floor of the ballroom before the crowd, which later jampacked it, got there. 1947 J. BERTRAM Shadow of War 262 It was jam-packed with neglected cargo. 1958 Archit. Rev. CXXIV. 383 In surprising and welcome contrast to the jam-packed streets. 1970 N. ARMSTRONG et al. First on Moon xiii. 322 They passed the rock boxes through to me, and I handled them as if they were absolutely jam-packed with rare jewels.

B. *adj.* Usu. *jam-up.* Excellent, perfect; thorough. *colloq.*

1832 *Boston Transcript* 6 Aug. 1/1 Do you like jam spruce beer, Miss? **1839** F. Trollope *Dom. Manners Amer.* (ed. 5) 270, I must have everything jam. *Ibid.* 273 That's a jam gal. **1841** *Southern Lit. Messenger* VII. 54/2 Introduced him to the 'jam-up little company' in his command. **1855** T. C. Haliburton *Nat. & Human Nat.* II. ix. 261 In Paradise..connubial bliss, I allot was rael [*sic*] jam up. **1946** Mezzrow & Wolfe *Really Blues* i. 4, I got my first chance to play in a real man-size band, with jam-up instruments.

jama² (dʒā·mă). Colloq. abbrev. of *pyjama.* Usu. *pl.*

1960 *Galaxy Mag.* Feb. 120/2, I groped under the bed for the paper bag that had my jamas in it. **1969** E. Gébler *Shall I eat you Now?* 59 She pulled her skirt on up over her pyjama trousers, and rolled the jama legs up. **1969** *Guardian* 26 Aug. 7/1 Pyjamas..in the language of lingerie..are now called 'jamas.

jama, jamabandi, varr. *JUMMA, **jummabundi.*

Jamaica. Add: *Jamaica bark, bilberry, cherry*: examples; *Jamaica shorts*; **Jamaica ginger,** white ginger (see Ginger *sb.* 1).

1801 T. Dancer *Med. Assistant* 363 *Jamaica bark*... These [species] are indigenous species of the Jesuits or Peruvian bark. **1811** W. J. Titford *Sk. Hortus Bot. Amer.* 60 Jamaica bilberry, or Whortle Berry. **1942** A. L. Simon *Conc. Encycl. Gastron.* V. 40/2 Jamaica Bilberry. The edible berry of a Jamaica mountain shrub. *Ibid.*, Jamaica Cherry. A West Indian Fig, globose and no bigger than a cherry. **1946** Jamaica cherry [see *ceriman]. **1818** *Public Ledger* LVIII. 4/5, 20 Bags and 10 Casks Jamaica Ginger. **1870** Jamaica ginger [see Ginger *sb.* 1]. **1920** *T. Eaton & Co. Catal.* Spring & Summer 367/3 Ginger, Jamaica, powdered. **1959** *Sears, Roebuck Catal.* Spring/Summer 333/3 Jamaica shorts. One pocket. **1972** *Evening Telegram* (St. John's, Newfoundland) 23 June 2/6 (Advt.), 500 only Ladies Jamaica Shorts.

b. *ellipt.* for *Jamaica coffee, rum,* etc.

1899 *T. Eaton & Co. Catal.* Spring & Summer 171/3 Coffees...Finest Santos and Jamaica with chicory. **1910** S. R. Crockett *Dew of their Youth* ii. xviii. 142 My grandfather got his ale, of the sort just then beginning to be made—called 'Jamaica', because a quantity of the cheap sugar refuse from the hogsheads was used in its production. **1938** S. V. Benét *Thirteen o'Clock* 206 He bought his Jamaica personal and in the jug. **1964** *Women's Wear Daily* 30 Nov. 36 From short-shorts to slacks—with Jamaicas, Nassaus, Bermudas, knee pants in between. **1969** *Sears Catal.* Spring/Summer 41D Package of 3 Jamaicas [*sc.* shorts]. (Above-the-knee length.)

Jamaican (dʒămē¹·kăn), *sb.* and *a.* [f. Jamaica + -an.] **A.** *sb.* A native or inhabitant of Jamaica; the form of English spoken there.

1693 *Truest & Largest Acct. Earthquake in Jamaica* 23 God curb'd their Malice, restrain'd their Power, and gave the Jamaicans a Signal Victory over them. **1770** W. Guthrie *New Geogr. Gram.* 613 The Jamaicans were undoubtedly very numerous, until reduced by earthquakes. **1902** J. D. Hooker in L. Huxley *Life J. D. Hooker* (1918) II. 408 The Jamaicans do not deserve the sacrifice England is making in respect of its fruit trade. **1960** R. B. Le Page *Jamaican Creole* x. 116 'English with a Jamaican accent'..one could well call Standard Jamaican. *Ibid.* 120 The grammar of Standard Jamaican is not very greatly different from that of Standard English. **1970** *Sunday Times* (Colour Suppl.) 6 Dec. 39/1 Everyone's English accent..had lapsed comfortably into Jamaican during the flight. **1971** I. F. Hancock in J. Spencer *Eng. Lang. W. Afr.* 114 *Jamaican Creole*: spoken in one form or another by about one million Jamaicans.

b. A Jamaican cigar.

1964 I. Fleming *You only live Twice* ii. 27 The best of the Jamaicans are quite up to the Havanas these days. They've got the outer leaf right at last. **1971** 'A. York' *Infiltrator* x. 144 Lucinda carefully pierced his cigar... 'I like these stogies... Better than your Jamaicans.'

B. *adj.* Of or pertaining to Jamaica or the Jamaicans.

1881 *Handbk. Jamaica* iv. 131 The Calipeva or 'Jamaica Salmon'..ranks among three specially Jamaican dainties. **1907** W. Jekyll (*title*) Jamaican song and story. **1913** [see *gros Michel* s.v. *gros a.]. **1955** *Caribbean Q.* IV. ii. 125, 4 Jamaican plums, hard round sugar sweets, a roast corn. **1971** [see the *sb.*, above]. **1972** 'B. Graeme' *Tomorrow's Yesterday* xiv. 145 He..pushed a box of cigars across the table. 'Do you smoke these? They are not too bad. Jamaican.'

Hence **Jamai·canism,** a Jamaican word or idiom.

1961 [see *chinchy a.]. **1963** *Amer. Speech* XXXVIII. 136 The cultural and historical setting in which these Jamaicanisms have developed. **1967** F. G. Cassidy in *Amer. Speech* XLII. 190 (*title*) Some new light on old Jamaicanisms.

jaman. Add: (Further examples of the forms *jamoon* and *jamun.*) Also *attrib.*

1914 *Indian Forester* XL. 268 *Eugenia Jambolana*, Lam. Jam, Jamun, H[indi]. **1958** J. Carew *Black Midas* x. 206 Get you mango,..soursop, starapple and sweet jamoon. **1965** 'Lauchmonen' *Old Thom's Harvest* x. 133 Everybody make their cake and ginger-beer and jamoon and rice wine. **1969** S. M. Sadeek *Windswept & Other Stories* 32 Rance lay under the jamoon tree.

jambalaya (dʒæmbălē¹·ă). orig. *U.S.* Also **jambalayah, jambolaya.** [Louisiana Fr., f. Provençal *jambalaia.*] A dish composed of rice together with shrimps, chicken, turkey, etc. Also *fig.*

1872 *New Orleans Times* 28 June, Those who brought victuals, such as gumbo, jambalaya, etc., all began eating and drinking. **1905** 'O. Henry' in *Munsey's Mag.* July 467/2 Terrapines,..jambolaya, and canvas-covered ducks. **1916** *Dialect Notes* IV. 269 The show was a regular jambalaya of stunts. **1949** B. A. Botkin *Treas. S. Folklore* iv. i. 552 Louisianians [grow lyrical] over the superiorities of the Cajun and Creole cuisine—gombo, jambalaya, bouillabaisse. **1961** *Listener* 14 Dec. 1050/2 *Jambalaya*..is based on a creole mixture of ham chunks, prawns, and rice, highly flavoured and simmered in chicken stock. **1973** L. Hellman *Pentimento* (1974) 78 The dinner was wonderful: jambalaya, racoon stew, and wild duck.

jambone (dʒæ·mbōᵘn). *Euchre.* ? *Obs.* (See quots.)

1864 W. B. Dick *Amer. Hoyle* (1866) 83 A party who plays Jambone plays a lone hand with his cards exposed upon the table. **1886** *Euchre: how to play It* 42 A Jambone is to play a lone hand with the cards exposed on the table, and to give to that adversary who is entitled to the lead, or whose first play it is, the privilege of calling one of the exposed cards to the first trick played, or if the jambone player has the lead, to call upon him to lead any one of the exposed cards.

jamboree. For *U.S. slang* read 'orig. *U.S. slang*'. Add earlier and later examples.

1868 *N.Y. Herald* 10 July 8/3 The Seventh regiment has gone on a jamboree to Norwich, Connecticut. **1955** H. Spring *These Lovers fled Away* iii. 90 There was some jamboree or other at the Assembly Rooms. **1959** *Economist* 27 June 1144/1 This is sufficiently important, and difficult, to warrant a top-level western conference this year—not merely another Atlantic jamboree. **1960** A. Huxley *Let.* 17 July (1969) 893 Meet me in Boston with the Microbus and drive me..to Hanover, where you might stay for all or part of the Jamboree (at which I am to receive an honorary degree). **1973** G. Scott *Water Horse* (1974) v. 33 Even the most respectable souls milled around the countryside..holding all sorts of jamborees.

b. *Euchre.* A lone hand containing the five highest cards. ? *Obs.*

1886 *Euchre: how to play It* 45 *Jamboree* signifies the combination of the five highest cards, as, for example, the two Bowers, Ace, King, and Queen of trumps in one hand, which entitles the holder to count sixteen points. The holder of such a hand simply announces the fact, as no play is necessary; but should he play the hand as a Jambone, he can count only eight points, whereas he could count sixteen if he announced it as a Jamboree.

c. The name given to the 1920 International Rally of Boy Scouts, and now applied to any large scout rally. Also *attrib.*

1919 *Times* 17 Oct. 9/6 The Council of the Boy Scouts Association announce that a 'Jamboree' will be held at Olympia, for about eight days next August. **1931** *Mag. Univ. Students' Union* Apr. 2 The Jamboree spirit was marvellous. **1955** *Times* 20 Aug. 9/3 The four-year cycle for jamborees is being broken in order to commemorate in 1957 the fiftieth anniversary of scouting.

jambu: see Jambo.

James Bond (dʒē¹·mz bǫ·nd). The name of the hero of a series of novels by the British writer Ian Fleming (1908–64), used allusively (freq. *attrib.* or as *adj.*) of adventurous, sophisticated men resembling the hero, or of situations similar to those in the novels. So **James Bo·ndish** *a.*

1966 *Economist* 19 Mar. 1115/1 There was nothing James Bondish about his arrival, for Mr Smith's external relations department was informed well in advance. **1967** *Guardian* 23 Dec. 14/2 The escape itself was James Bond with the merest trace of Keystone Cops. **1968** *Ibid.* 23 Feb. 7/4 This fast James Bond stuff doesn't blend happily with the more realistic comedy. **1968** K. Bird *Smash Glass Image* iv. 54, I should love to deck myself out as a two-fisted James Bond because that would impress you. **1970** *Motoring Which?* Apr. 43/2 Aston Martin DB5—very powerful James Bond car. **1972** G. Lyall *Blame the Dead* vii. 47 People who..go running around France playing James Bond with unlicensed Walthers.

Jamesian (dʒē¹·mziăn), *a.* (*sb.*) [f. *James* + -ian.] **1.** Of or pertaining to the American philosopher and psychologist William *James* (1842–1910) or his works. Also as *sb.*, a follower or admirer of William James.

1875 C. Wright *Let.* 18 July in R. B. Perry *Tht. & Char. W. James* (1935) I. 532 He rather attracts me by the Jamesian traits. **1935** R. B. Perry *Tht. & Char. W. James* II. 668 While there are very few pure Jamesians, in the sense of direct descent, the world is full of mixed Jamesians, who acknowledge their common relationship to him without feeling any bond with one another. **1955** Koestler *Trail of Dinosaur* 249 The Jamesian view that a transcendental faith was a biological necessity for man. **1964** *Amer. Philos. Q.* I. 115/1 Elements in the German phenomenology of the 1930's..are clearly repugnant to the post-Jamesian American philosophical temper.

2. Of or pertaining to the American (later naturalized British) writer Henry *James* (1843–1916) or his works. Also as *sb.*, a follower or admirer of Henry James.

1905 *Daily Chron.* 7 Sept. 3/1 The plot..is Jamesian..in its tenuity. **1954** *Essays in Crit.* IV. 371, I am not a good enough Jamesian to decide the other claim. **1958** *Listener* 17 July 98/1 For Jamesians, it is the first thrilling scent of the great chase. **1972** J. Symons *Bloody Murder* xvi. 235 The motivations of his characters seem at times to be of Jamesian complexity.

James–Lange (dʒē¹·mz,lɑ̈·ŋə). *Psychol.* The names of W. *James* (1840–1900) and C. G. *Lange* (1834–1900), used *attrib.* to designate a theory propounded by each of them separately that the response to an emotional stimulus is, in the first place, an organic reaction rather than a mental awareness of the emotion.

1909 H. R. Marshall *Consciousness* I. App. A. 109 It is not correct to say either that the emotions cause the instinctive reaction..nor to say that the instinctive reactions cause the emotions, as the common statement of the James–Lange theory implies. **1914** M. Prince *Unconscious* xiv. 423 The James–Lange theory is disregarded here as untenable. **1918** E. Jones *Papers on Psycho-Anal.* (ed. 2) xxvii. 485 The occurrence of the abortive anxiety attacks..stands in direct conflict with the James–Lange hypothesis. **1933** G. Murphy *Gen. Psychol.* v. 75 Such evidences might even lead to a partial confirmation of the James–Lange view. **1949** D. O. Hebb *Organization of Behavior* x. 237 Just such an inconsistency of thought has led to an endless, and pointless, debate on the James–Lange theory of emotion. **1970** H. C. Shands *Semiotic Approaches to Psychiatry* xx. 307 The James–Lange notion is an interpretation of a communicational process involving feedback from one's own behavior.

Jameson (dʒæ·misǫ̆n, dʒē¹-, dʒi-). The proprietary name of a brand of Irish whiskey. Also, a drink of this whiskey.

[**1877** *Trade Marks Jrnl.* 28 Feb. 568/1 W. Jameson & Co. Distillers. Dublin. William Robertson, on behalf of self and partners, William Jameson and James Jameson, trading as Wm. Jameson and Co.,..Dublin; whisky distillers.] **1922** Joyce *Ulysses* 176 We'll take two of your small Jamesons after that. **1948** W. Stevens *Let.* 7 Sept. (1967) 613 Perhaps if I could have that bottle of Jameson..on the reading stand, I could really get somewhere. **1965** *Trade Marks Jrnl.* 24 Mar. 414/1 Jameson..Irish whiskey..John Jameson & Son Limited, ..Dublin.

Jamie Green (dʒē¹·mi grīn). *Naut.* [Proper name, orig. unknown.] The name of a type of sail found on tea-clippers. Cf. *Jimmy Green* (*Jimmy 8).

1866 Capt. Keay *Jrnl.* 20 June in B. Lubbock *China Clippers* (1914) App. H. p. xxi, About 5.30 a.m., all staysails and fore-topmast and topgallant stunsails set and Jamie Green. **1927** G. Bradford *Gloss. Sea Terms* 109/2 *Jamie Green*, a sail set beneath the bowsprit and jib-boom of a tea clipper. The halyard hauled the sail to the end of the jib-boom and the tack to the lower end of the martingale boom. **1934** P. Mitchell *Deep Water* I. xxiii. 184 They..had..Jamie Greens for under the jib boom, ringtails, watersails, and an extra flying jib. **1961** F. H. Burgess *Dict. Sailing* 123 *Jamie Green*, a sail set under the jib-boom, to the dolphin striker.

jammer (dʒæ·məɹ). [f. Jam *v.*¹ + -er¹.] A transmitter used for jamming.

1947 *Jrnl. R. Aeronaut. Soc.* LI. 428/1 We could..dispose our limited number of transmitters along the tracks usually followed by the beam-flying enemy bombers and site our relatively low-powered jammers where they would produce their greatest effect. **1957** *B.B.C. Handbk.* 39 Jamming by the Communist authorities is more efficient than the wartime efforts of the Germans and Italians, but except in certain centres of population where local jammers are used, it is not completely effective. **1965** H. Kahn *On Escalation* vii. 141 The Soviets might even use shipboard jammers. **1971** [see *jam *v.*¹ 3 c].

jammy (dʒæ·mi), *a.* [f. Jam *sb.*² + -y¹.] Covered with jam, sticky. Also *fig.* (*colloq.*), excellent; very lucky or profitable; easy, 'soft'. Hence **ja·mminess.**

1853 D. G. Rossetti *Let.* 2 Nov. (1965) I. 161 The frame for my water-colour..is..jammy, nobby, stunning, jolly, splendacious. **1895** *Punch* 12 Oct. 180/1 The way as that Sam chewed the rag was just jammy **1899** Kipling *Stalky & Co.* 228 Jam for the Sixth! Jam for us! Either way it's jammy! **1908** 'I. Hay' *Right Stuff* xi. 205 We had disposed of grouse sandwiches..and jammy scones. **1915** D. O. Barnett *Lett.* 170 If I get a 'jammy' one as it is called, I shall be back pretty soon. **1920** *Chambers's Jrnl.* X. 862/2 She was aroused by the..jammy caresses of her blue-eyed nephews. **1929** W. Deeping *Roper's Row* 179 There was a jamminess about these meals and about the ladies' fingers. **1932** A. J. Worrall *Eng. Idioms* 32, I don't know any one like him. He's jammy. **1955** M. Allingham *Beckoning Lady* x. 137 The twins, jammy-faced and excited. **1973** 'Trevanian' *Loo Sanction* (1974) 130 'I almost always win. Isn't that odd?' The Sergeant regarded the slim body... 'I'd say you were bloody jammy.'

jam-packed, *a.*: see *jam *adv.* 2.

jams (dʒæmz), *sb. pl.* [Shortened from Pyjamas, pajamas *sb. pl.*] A garment, derived from Pyjamas, worn as leisure-wear, and *spec.* as a type of swimming-trunks.

1966 *Telegraph* (Brisbane) 3 Feb. 18 It's the season of the hubba hubba jams—and that's..a patio fashion stopper. It is snug..just-above-the-knee jams worn with a covered camisole top... Jams can be worn with a bikini top or a frilly feminine flip-top style. **1968** *N.Y. Times* 22 Jan. 36 There will also be ascots (which can double as belts), walking shorts, swim trunks and surfers' 'jams', knee-length trunks with drawstring waists. **1970** *New Yorker* 14 Mar. 34/3 He wore His-'n-Hers flowered at-home jams. **1971** *Telegraph* (Brisbane) 29 Dec. 31 (Advt.), Boys' cotton floral swim jams... Cord tie at waist. Hip pocket detail.

jam-tree. Also **jamwood.** = *raspberry jam tree* (s.v. Raspberry 4). *Austral.*

1934 T. Wood *Cobbers* vii. 91 Jam trees grow here [*sc.* in Western Australia], and blackboy, that strange fascination. **1941** *Coast to Coast* 46 The wheat..rippled up the hillside and lost itself..in the shadow of a jam-tree belt. **1947** *Coast to Coast 1946* 202 Old Tom Caseley had just finished whittling the fifty-seventh link in his jamwood chain when he died. **1962** A. Upfield *Will of Tribe* iv. 39 He..dismounted and neck-roped the horse to a stout desert jamwood. **1965** *Austral. Encycl.* IX. 219/2 Myall wood has a peculiar violet-like fragrance, while jamwood is reminiscent of raspberries.

jam-up, *sb. colloq.* [Jam *sb.*¹ + Up *adv.*¹] = Jam *sb.*¹ 1 (in Dict. and Suppl.).

1941 J. M. Cain *Mildred Pierce* xi. 229 To forestall the possibility of another jam-up. **1961** *Time* 14 Apr. 105/3 The jam-up at the post-office when the social security checks arrive. **1970** *Globe & Mail* (Toronto) 28 Sept. 22/4 Larry Foubert of Weston ended up on the infield grass after a jam-up at the north end of the track. **1973** 'D. Jordan' *Nile Green* xi. 46 There was the usual jam-up..and the auto horns blared in frustration.

janapa (dʒæ·năpă). Also **-um.** [Tamil.] = Sunn.

1851 *Illustr. Catal. Gt. Exhib.* iv. 882/2 *Crotolaria juncea*..is cultivated in most parts of India for its fibre, which..is called *sun* and *sunnee* in different parts of India, but, in the Madras peninsula, *janapum.* **1866** Lindley & Moore *Treas. Bot.* II. 635/1 *Janapa*, an Indian name for Sunn Hemp.

Jane, jane² (dʒē¹n). *slang* (orig. *U.S.*). [Female Christian name.] A woman, girl, girlfriend.

1906 *Dialect Notes* III. 142 'It's the magazine over yonder with a red Jane on it.' 'Going to take your Jane to the show?' **1916** C. J. Dennis *Moods of Ginger Mick* 79 She's like some fat ole Jane 'oo loves to smile. **1923** [see *dame 2 c]. **1924** P. Marks *Plastic Age* 149, I met a bunch of janes down at Bar Harbour. **1929** 'G. Daviot' *Man in Queue* iii. 30 He has a new 'jane'. He probably wants money. **1941** *Coast to Coast* 129, I was trailing round with a high-class little jane in those days. **1958** 'A. Gilbert' *Death against Clock* x. 145 She didn't see that it could, and for once I agree with a jane. **1967** E. S. Gardner *Case of Queenly Contestant* (1973) xiii. 150 'Who was this jane? Anybody I know?' 'No one you know... She had been a nurse in San Francisco.'

Janeite (dʒē¹·nəit). Also **Jane-ite, Janite.** [f. the Christian name of *Jane* Austen (1775–1817), novelist + -ite¹.] = *Austenite¹.

1896 G. Saintsbury *Hist. 19th Cent. Lit.* 129 It did not apparently occur to this critic that he (or she) was in the first place paying Miss Austen an extraordinarily high compliment—a compliment almost greater than the most enthusiastic 'Janites' have ventured. **1924** Kipling *Debits & Credits* (1926) 147 (*title*) The Janeites. **1927** *Observer* 5 June 6/3 Clearly he is *not* a Janeite. **1928** *Daily Tel.* 7 Aug. 3/3 The best page of her book by far is her spirited defence of Jane Austen, which will endear her to the 'Jane-ites' for all time. **1947** W. J. Blyton in *Eng. Lang. & Lit.* vii. 111/1 Men as masculine as Scott and Kipling have been Jane-ites and have been enthralled by her sly humour and fidelity to reality. **1972** N. Freeling *Lone Silence* ii. 149 Arlette was not a Janeite, and..it was accidentally that she let slip..that Piet used to call her 'Bates'.

jang(g)a, varr. *jonga.

janitor. Add: **1. b.** A caretaker of a building, esp. a school, who has charge of the cleaning, heating, etc., of it. Also *attrib.*

1708 C. Morthland *Acct. Govt. Church of Scotl.* 18 When a Student enters the University [of Edinburgh] he must pay Half a Crown to the Porter or Janitor and Seven Pence to his Servant. **1835** Dr. Neill in J. J. Audubon *Ornith. Biogr.* III. 313 A cock-heron..had been..kept for some weeks in a cellar in the old College, and then presented to me by the late Mr John Wilson, the janitor. **1868** *Mich. Agric. Rep.* VII. 22 Expenses for wood, furniture, janitor work, &c. **1884** H. Butterworth *Zigzag Journeys Western States* 50 He was employed merely as janitor at Yule [school]. **1952** *Manch. Guardian Weekly* 4 Dec. 13 The Civil Service now protects janitors and chauffeurs. **1954** *Press & Jrnl.* (Aberdeen) 7 Apr., The Education Committee..had recommended that a janitor be appointed, and that he be responsible for all janitorial and cleaning work at the school, including maintenance of the gardens. **1961** *Evening Standard* 12 Sept. 19/1 Janitor required for building in W.1. area. **1968** *Globe & Mail* (Toronto) 13 Feb. 31/6 (Advt.), Prestige offices, full janitor services.

janizary, janissary. Add: **4. janizary music** [G. *janitscharenmusik*] = *Turkish music* (Turkish *a.* 2 b); **janizary pedal,** a pedal attached to some old pianofortes, etc., having an arrangement of drums and cymbals connected with it, by which a sound as of martial music was produced.

1888 F. Moscheles tr. *Mendelssohn's Lett. to I. & C. Moscheles* 54 He must have a cradle song with drums and trumpets and janissary music. **1896** A. J. Hipkins *Descr. & Hist. Pianoforte* 106 Drum and triangle (for Janissary music). **1922** J. Riviere tr. *Freud's Introd. Lect. Psycho-Anal.* 75 The little bells, shaken violently, begin their familiar janizary music. **1900** *Pall Mall Gaz.* 21 May 4/2 Even Mozart condescended to employ the 'Janissary pedal' in one or two of his sonatas.

‖ **janken** (dʒa·ŋken). [Jap.] A Japanese children's game played with the hands. Now principally used in sport to decide ends, etc.

1936 E. K. Venables *Behind Smile in Real Japan* viii. 269 The ceremony which is the Oriental equivalent of tossing up... *Jan-ken-poh*, as it is called, is used in all such cases of decision by chance. **1964** *Japan* (Jap. Nat. Commission for Unesco) 831/2 *Janken* is the only form of the game which remains today as a means to decide the dealer in card games, the server in a match, and so on. **1967** D. & E. T. Riesman *Conversations in Japan* 157 *Janken* (games of three: paper, scissors, and stone). **1972** *Nat. Geographic* CXLI. 689/2 There was the ritual with children of *jan ken pon*, a game in which fist and fingers represent paper, rock, or scissors.

jankers (dʒæ·ŋkəɹz). *Services' slang.* [Origin unknown.] Punishment for defaulters; the defaulters themselves; the cells in which they are placed. Also *attrib.* (occas. in *sing.* form).

1916 J. N. Hall *Kitchener's Mob* 35 The 'jankers' or defaulters' squad was always rather large. **1919** *Athenæum* 25 July 664/2 The advent of the Royal Navy Division introduced to the Army the sailor's slang word 'jankers', the equivalent of the soldier's 'clink', punishment cells. *Ibid.* 8 Aug. 727/2 When doing C.B. or 'time' he [*sc.* the soldier] was doing 'jankers' or 'Paddy Doyle'. *a* **1935** T. E. Lawrence *Mint* (1950) ii. xxii. 160 A week before my last jankers. **1936** F. Richards *Old-Soldier Sahib* ii. 54, I was now a defaulter, or 'on jankers' as the troops called it. **1946** *Penguin New Writing* XXVII. 72, I stepped into the hall of B.H.Q. over two janker-wallahs. *Ibid.* 73 He broke off to bawl out the jankermen. **1960** T. Rattigan *Ross* i. ii. 20 None of your lip, Parsons, now—unless you want a dose of jankers. **1965** J. Porter *Dover Two* ix. 122, I pulled her leg about it a bit, you know, said something about having her put on jankers if she was late again. **1971** *Sunday Mail Mag.* (Brisbane) 25 July 6/2 Jankers can be painful. It usually means confined to barracks and menial tasks. *Ibid.*, My first jankers was for causing a fire.

jann (dʒān). Also **jan.** [a. Arab. *jānn* demon.] = Jinn *sb.*

1777 J. Richardson *Dict. Persian, Arabic & Eng.* I. 667 That race of creatures called by the Arabians *Jan* or *Jinn.* **1891** E. S. Hartland *Sci. Fairy Tales* x. 256 Hasan is favoured with the sight of 'ten virgins'... He fell madly in love with the chief damsel, who turns out to be a daughter of a King of the Jann. **1931** A. Wilson in E. S. Stevens *Folk Tales Iraq* p. xiv, Stories in which *jānn*, or fairy-folk, don at will the appearance of birds.

janney (dʒæ·ni). *Newfoundland.* Also **janny, jenny, johnny.** [Prob. var. Johnny, Johnnie.] A Christmas mummer in Newfoundland. Hence as *v. intr.*, to act as a janney; to dress *up* as a janney.

1896 *Jrnl. Amer. Folklore* IX. 36 *Old teaks* and *jannies*, boys and men who turn out in various disguises and carry on various pranks during the Christmas holidays. **1925** *Dialect Notes* V. 335 *Johnnies*, Christmas mummers. Also *jennies.* **1964** *Canad. Jrnl. Linguistics* Fall 44 At Christmas time it was usual for people to..'janny up'—dress up as mummers. **1969** J. D. A. Widdowson in Halpert & Story *Christmas Mumming in Newfoundland* 218 Janneys 'janney up', but mummers 'rig out' or 'dress up'.

January. Add: **2.** *attrib.*, as *January sale.*

1896 in G. Eley *Ruined Maid* (1970) 24/1 A friend of mine..bought at a January sale six yards of brocade in pale electric blue. **1936** *Discovery* Oct. 320/2 A rush..which must have equalled the January sales of our own days. **1967** E. G. Cousins *Death in Quiet Place* xiv. 177 Mollie's getting so used to Town that she's threatening to stay on over the January Sales.

Janus. Add: **2.** Designating materials with a double facing, or things having a two-way action, as *Janus-beaver, -cloth, -cord, -lock.*

1851 *Illustr. Catal. Gt. Exhib.* iii. 486/2 Fur Janus beaver. *Ibid.* v. 1468/2 Janus locks. *a* **1877** Knight *Dict. Mech.* II. 1210/2 *Janus-cloth*, a fabric having each side dressed, and different colors on the respective sides. Such fabric is used for reversible garments. **1881** in A. Adburgham *Shops & Shopping* (1964) vi. 64 Mourning materials such as Victoria Cords, Janus Cords, Cashmere, and so on. **1960** Cunnington & Beard *Dict. Eng. Costume* 259/2 *Janus cord*, a black rep of wool and cotton, the fine cord showing equally on both sides. Much used for mourning.

b. *Dye Chem.* Applied to a group of basic azo dyes that contain a quaternary ammonium group, often with safranine as the diazo component, and are capable of dyeing cotton directly; **Janus** (also **janus**) **green** (**B**), either of the two basic dyes obtained by coupling diazotized safranine or dimethylsafranine with dimethylaniline, of little importance as dyes but extensively used as biological stains.

1898 *Jrnl. Soc. Dyers & Colourists* XIV. 146 The first eight patterns..were dyed with Janus Colours on Union linings. *Ibid.* (Table), Janus Black I. Janus Blue R... Janus Green G. Janus Green B. **1926** *Stain Technol.* I. 35 The use of janus green to stain mitochondria has long been known. While using it to study the mitochondria in *Trichomonas buccalis* it was found to stain flagella also. **1928** C. E. Mullin *Acetate Silk* x. 142 Some of the Janus dyes are also applicable to acetate silk, however most of them give rather dull shades. **1949** N. G. Heatley in H. W. Florey et al. *Antibiotics* I. iii. 118 The test organism was *Welchii perfringens* strain HA, the indicator wss janus green. **1971** D. R. Baer in K. Venkataraman *Chem. Synthetic Dyes* IV. iv. 165 Derivatives of (*m*-aminophenyl)trimethylammonium chloride..formed part of the Janus range of colors which were of limited use on rayon, silk-wool unions, and on cotton. **1971** E. Gurr *Synthetic Dyes* 128 There are apparently two types of Janus green B, one from the USA and the other of European origin. *Ibid.* 129 Both types of Janus green are employed as oxidation-reduction indicators.

Jap, *sb.* Add: (Later examples.) Also as *adj.*; spec. *Jap silk* = *habutai.

As *sb.* and *adj.* the word *Jap* has strong derogatory connotations and is now falling into disuse.

1895 *Montgomery Ward Catal.* 37/3 Ladies' short silk waists, made of plain colored Habutai Jap silk. **1900** in *American Mail Order Fashions* (1961) 18 Jap braid edge. **1902** *To-Day* 14 May 95/1 Printed Jap silks are lovely. **1914** *Dialect Notes* IV. 136 This is a Jap Kimona. **1921** [see *chow *sb.* 1]. **1936** *Times* 3 Jan. 10/3 The most jealously limited edition of an unacted Swinburne drama, 50 copies only on jap vellum. **1940** [see *Dago, dago]. **1944** *Living off Land* viii. 156 The Jap pearlers have landed on his coasts. **1951** *Good Housek. Home Encycl.* 251/2 Many attractive natural silk fabrics..jap and spun silks and ninons. **1959** [see *Jew *sb.* 3 a]. **1970** G. F. Newman *Sir, You Bastard* viii. 211 Nice little tape-recorder... Snazzy Jap job.

jap (dʒæp), *v. U.S. slang.* [f. Jap *sb.*] *trans.* To make a sneak attack on; also, to queer the pitch of (a person).

In restricted use.

1957 *New Yorker* 21 Sept. 135/1 'They japped us,' a third boy said, meaning that the Cherubs had taken them by surprise. **1958** H. E. Salisbury *Shook-up Generation* (1959) ii. 29 An uncertain area where one side or another may at any sudden moment 'jap' an unwary alien. **1971** D. E. Westlake *I gave at the Office* 170 Joe would hate me forever and would probably jap me with Mr. Clarebridge.

Japan, *sb.* Add: **6. Japan anemone,** one of several varieties or hybrids of *Anemone hupehensis*, bearing large pink or white flowers; **Japan camphor** = *tub-camphor* (see Tub *sb.* 10); **Japan cedar** = *cryptomeria; **Japan current** = Kuroshiwo; **Japan lacquer (tree)** (see Lacquer *sb.* 2 b, 4); **Japan lily,** any of several species of *Lilium* native to Japan, esp. *L. japonicum*; **Japan paper** = **Japanese paper*; **Japan pepper** = *Japanese pepper* (see Pepper *sb.* 3); **Japan quince** (see *japonica); **Japan rose,** a name once used for the camellia; later = **Japanese rose*; **Japan varnish (tree)** = *varnish sumach* (see Varnish *sb.*¹ 5); = **Japan lacquer*; **Japan wax** = **Japanese wax.*

1847 *Curtis's Bot. Mag.* LXXIII. 4341 (*heading*) Japan Anemone. **1870** W. Robinson *Wild Garden* i. 28 The Japan Anemone and A. hybrida..grow so strongly that they will take care of themselves. **1882** [in Dict.]. **1908** G. Jekyll *Colour in Flower Garden* ix. 81 The pink colourings are the wide-headed *Sedum spectabile*, pink Japan Anemone and a few pale pink Gladioli. **1882** R. Bentley *Man. Bot.* (ed. 4) 642 Commercial camphor is derived entirely from the island of Formosa and Japan,..the latter [being known] as *Japan* or *Dutch Camphor.* **1852** Japan cedar [see *cryptomeria]. **1900** M. Thorn in W. D. Drury *Bk. Gardening* xi. 488 *C. japonica* (Japan Cedar) is a charming tree. **1865** D. Page *Handbk. Geol. Terms* (ed. 2) 263 *Japan current*, that branch of the equatorial current of the Pacific which trends northward along the Japan coasts. **1885** [see Kuroshiwo]. **1936** Russell & Yonge *Seas* (ed. 2) x. 231 The Japan Current or 'Kuro Shiwo' corresponding to our Gulf Stream. **1835** W. J. Hooker *Compan. Bot. Mag.* I. 268/1 The so much celebrated Japan lacquer or varnish. **1880** C. E. Bessey *Bot.* xx. 535 Japan Lacquer, so much used by the Japanese in the manufacture of many wares, is obtained..from *Rhus vernicifera*, and probably other species. **1884** W. Miller *Dict. Eng. Names Plants* 241/1 Japan Lacquer-, or Varnish-, Tree. **1911** *Encycl. Brit.* XXVI. 70/2 *Rhus vernicifera* is the Japan lacquer or varnish-tree. **1813** *Curtis's Bot. Mag.* XXXVIII. 1591 (*heading*) White one-flowered Japan Lily. **1854** C. M. Yonge *Heartsease* I. iv. 56 You should go and look at the Japan lilies. **1625** Purchas *Pilgrimes* III. ii. v. 339 Nothing gave him such content as two Bookes of Iapon paper, smooth and hard bound in Europæan manner. **1914** *Photo-Era* Feb. 102/1 There are certain tricks to produce a soft print from a sharp negative, one of which is to put a piece of Japan paper between the plate and the sensitive paper in the printing-frame. **1974** *Country Life* 17 Jan. 75/1 The exquisite *Le Bain* of 1905, a dry point (an impression from the edition of 27 or 29 on Japan paper). **1866** Japan pepper [see Pepper *sb.* 3]. **1914** W. J. Bean *Trees & Shrubs Hardy in Brit. Isles* II. 692 *Z*[*anthoxylum*] *piperitum*, De Candolle. Japan Pepper... The seeds when ground are used by the Japanese as pepper. **1850** W. Howitt *Year-bk. Country* 24 In gardens..come forth the vernal crocus, various hellebores, the Japan quince. **1916** L. H. Bailey *Pruning-Manual* 222 *Chænomeles japonica* (Japan quince) **1789** W. Aiton *Hortus Kewensis* II. 460 Camellia. ..Japan Rose. **1793** B. Edwards *Hist. Brit. Colonies W. Indies* I. 204 *Camellia japonica*, Japan Rose. **1895** W. Robinson *Eng. Flower Garden* (ed. 4) 732/1 It [*Rosa acicularis*] has a showy fruit, which differs from that of the Japan Rose, for, instead of being roundish and smooth, it is long and Pear-shaped. [**1727** J. G. Scheuchzer tr. *Kæmpfer's Hist. Japan* I. i. ix. 114 The *Urusi* or *Varnish-Tree* is another of the noblest and most useful Trees of this country.] **1789** J. Belknap *Let.* 13 Mar. in W. Parker & J. P. Cutler *Life M. Cutler* (1888) II. 252, I have sent for the seeds of the Japan varnish tree. **1843** *Penny Cycl.* XXVI. 147/2 The Japan varnish of Kæmpfer and Thunberg is Rhus vernix. **1889** W. Robinson *Eng. Flower Garden* (ed. 2) 788/2 Others in cultivation..include *R*[*hus*] *vernicifera*, or Japan Varnish Tree. **1859** Japan wax [see *Japanese wax* s.v. *Japanese a. b]. **1887** *Colonial & Indian Exhib. Rep. Col. Sect.* 275 Myrtle wax..which, like Japan wax, is rather a fat than a true wax. **1969** R. Mayer *Dict. Art Terms & Techniques* 201/2 Japan wax is sometimes called vegetable wax.

Japanese, *a.* and *sb.* Add: **A.** *adj.* **b.** Special collocations: **Japanese anemone** = **Japan anemone*; **Japanese ape** = **Japanese monkey*; **Japanese artichoke** = **Chinese artichoke*; **Japanese beetle,** a scarabæid beetle, *Popillia japonica*, which has become a pest of foliage and grasses in eastern North America; **Japanese camphor** = **Japan camphor*; **Japanese cedar** = *cryptomeria; **Japanese cherry,** an ornamental flowering tree belonging to a variety or hybrid of several species of *Prunus* native to Japan; **Japanese current** = Kuroshiwo; **Japanese flower,** a piece of coloured paper which unfolds like a flower when placed in water; **Japanese garden,** a garden in which clipped shrubs, water, bridges, rocks, stepping-stones, raked gravel, stone lanterns, etc., are used in a formal design, without masses of bright colour; **Japanese gold thread** (see quot. 1880); **Japanese iris,** a variety of *Iris kæmpferi* or *I. lævigata*; **Japanese lantern** = *Chinese-lantern* (see Chinese *a.* 2); **Japanese larch,** *Larix leptolepis*, which was introduced to Britain in 1861; **Japanese lily** = **Japan lily*; **Japanese macaque** = **Japanese monkey*; **Japanese maple,** a variety of *Acer palmatum* or *A. japonicum*, cultivated esp. for its decorative foliage; **Japanese medlar** = Loquat; **Japanese monkey,** a large monkey, *Macaca fuscata*, which is native to Japan; **Japanese pagoda tree,** *Sophora japonica*, the scholar tree; **Japanese paper,** paper made by hand, originally and chiefly in Japan, from the bark of the mulberry-tree; **Japanese pepper** (see Pepper *sb.* 3); **Japanese print,** a coloured print made in Japan from a wood-block; **Japanese quince** = *japonica; **Japanese rose,** any of several species of *Rosa* native to Japan, esp. *R. rugosa*; **Japanese screen,** an embroidered screen made in Japan; **Japanese silk** = **Jap silk*; **Japanese spaniel,** a breed of small, black-and-white or brown-and-white, long-coated dog; **Japanese stitch** (see quot. 1880); **Japanese tissue (paper),** a type of strong thin transparent paper; **Japanese vellum** (see quot. 1923); **Japanese waltzing mouse,** a mutant of *Mus musculus bactrianus*, a house mouse native to Central and Eastern Asia; also **Japanese waltzer; Japanese wax,** a yellow wax obtained from the berries of certain plants of the genus *Rhus*; **Japanese wolf,** *Canis lupus hodophylax*, a subspecies of the common wolf.

c **1908** E. J. Cook *Century Bk. Gardening* 71/1 The Japanese Anemone..was originally seen only in its pink-blossomed form. **1913** C. Mackenzie *Sinister St.* I. i. vi. 87 All along the paths were masses of flowers, phloxes and early Michaelmas daisies and Japanese anemones. **1969** H. R. Fletcher *Story R. Hort. Soc.* x. 151 All the 'Japanese anemones'..are now grouped under the hybrid name of *A.* x *elegans.* **1883** *List Animals* (Zool. Soc.) (ed. 8) 22 *Macacus speciosus*... Japanese Ape. **1966** R. & D. Morris *Men & Apes* i. 18 The famous three wise monkeys, See-no-evil, Hear-no-evil and Speak-no-evil..are based on the Japanese ape (*Macaca fuscata*), a delightful monkey with a short, stumpy tail and a bright pink face that flushes scarlet when the animal is sexually active. **1905** tr. *Vegetable Garden* (Vilmorin-Andrieux et Cie) 671 Chinese or Japanese Artichoke... These rhizomes ..are white, watery, and tender. **1970** Simon & Howe *Dict. Gastron.* 40/2 Artichokes, Japanese or Chinese. These are small tubers which one writer describes as looking like 'petrified worms'. **1919** *Rev. Appl. Entomol.* A. VII. 101 Eradication work in connection with a Japanese beetle (*Popillia japonica*) has been systematically undertaken. **1922** *Jrnl. Econ. Entomol.* XV. 303 An efficient contact spray will no doubt have considerable application in the control of the Japanese beetle at this time. **1936** *Discovery* XVII. 36/1 The Japanese Beetle (*Popillia japonica*)..accidentally introduced into America in the larval stage in a shipment of Japanese iris in 1916, has now invaded the eastern coastal states. **1972** Swan & Papp *Common Insects N. Amer.* xx. 431 The Asiatic Garden Beetle..is similar in habits to the Japanese beetle ..but it flies only at night. **1727** J. G. Scheuchzer tr. *Kæmpfer's Hist. Japan* I. 179 Japanese boil'd Camphire may be had for one single Catti of the true Bornean Camphire. *Ibid.*, The Japanese Camphire-tree. **1880** *Encycl. Brit.* XIII. 574/2 *Cryptomeria* (Japanese cedar). **1954** *New Biol.* XVI. 97 The Japanese Cedar or Sugi, a tree found in China and Japan and, in the latter country, an important and abundant timber tree. **1901** L. H. Bailey *Cycl. Amer. Hort.* III. 1452/2 Japanese Flowering Cherry. **1913** W. P. Wright *Garden Trees & Shrubs* xxvi. 215 The double Japanese Cherries..have beautiful rosy flowers. **1925** *Jrnl. R. Hort. Soc.* L. 73 (*title*) Notes on Japanese cherries. **1951** *Dict. Gardening* (R. Hort. Soc.) II. 1085/2 It is only comparatively recently that the Japanese Cherries have been widely planted. **1972** G. Chadbund *Flowering Cherries* i. 16 As far as we know none of the upright garden varieties of Japanese cherries were introduced into the western world until 1894. **1926** *Daily Colonist* (Victoria, B.C.) 16 Jan. 1/3 Similar conditions..gave rise to the belief that the Japanese current had changed its course. **1972** *Islander* (Victoria, B.C.) 26 Mar. 16/1 The fickle Japanese current which sweeps in a circular motion across the Pacific. **1917** N. Douglas *South Wind* xxii. 271 Those Japanese flowers..those paper flowers, I mean, which we used to put in our finger-bowls... They look like shrivelled specks of cardboard. But in the water they begin..to unfold themselves into unexpected patterns of flowers of all colours. **1968** D. Hopkinson *Incense-Tree* iv. 42 A Japanese flower in a glass of water slowly uncurling to reveal its coloured pattern. **1863** R. Alcock *Capital of Tycoon* I. iv. 103 We ..gained a fine suite of apartments looking on to as beautiful a specimen of Japanese garden..as can well be conceived. **1902** C. H. Townsend in G. Brown *European & Japanese Gardens* 162 The composition of the Japanese garden depends chiefly upon the arrangement of its trees, boulders, paths, streams, bridges and other artificial structures. It is, least of all, a flower garden, and is probably best understood when regarded as a reduced copy of the scenery of a country—conveying the impression produced by a picture. **1912** Mrs. B. Taylor *Japanese Gardens* iv. 53 Seldom does a Japanese garden lack water, or the appearance of water, in its scenery. **1920** W. J. Locke *House of Baltazar* i. 13 The Japanese garden with its pond of great water-lilies and fairy bridge across. **1957** M. G. Sims tr. *Yoshida's Gardens of Japan* i. 7 Much in the Japanese garden is merely symbolical of nature. *Ibid.* 9 The Japanese garden is monochrome, the European polychrome. **1971** S. Eliovson *Gardening the Japanese Way* 26 Another misconception is that Japanese gardens are composed only of sand and stone. **1972** T. Ito (*title*) The Japanese garden—an approach to nature. **1880** L. Higgin *Handbk. Embroidery* i. 8 '*Japanese gold thread*', which has the advantage of never tarnishing, is..made of gilt paper twisted round cotton thread. **1883** W. Robinson *Eng. Flower Garden* 155/2 *I*[*ris*] *Kæmpferi* (Japanese Iris). The large number of varieties in cultivation under this name have sprung from *I. lævigata* and *I. setosa.* **1900** L. H. Bailey *Cycl. Amer. Hort.* II. 822/1 There are few handsomer flowers than good forms of the white Japanese Iris. **1936** *Discovery* XVII. 86/2 An iris garden with special emphasis on Japanese Iris. **1895** *Brit. Warehouseman* Feb. 26/1 Japanese lanterns and Kakemonos (wall-pictures) are shown. **1901** *Daily Colonist* (Victoria, B.C.) 2 Oct. 10/2 There, added to the effect of the bunting drapings and clusters of flags, three long strings of Japanese lanterns stretch from the roof. **1966** G. Baxt *Queer Kind of Death* (1967) xii. 174 The garden will be festooned with Japanese lanterns. **1861** *Gardeners' Chron.* 12 Jan. 23/1 The Japanese larch, *A*[*bies*] *leptolepis* of Zuccarini, is represented with cones four times larger than those sent home by Mr. Veitch. **1914** W. J. Bean *Trees & Shrubs Hardy in Brit. Isles* II. 8 The Japanese larch has been almost, although not wholly, immune from the attacks of larch canker. **1957** M. Hadfield *Brit. Trees* 46 In Britain the European and Japanese larches have been planted extensively. **1870** J. C. Patteson *Let.* 21 Dec. in C. M. Yonge *Life J. C. Patteson* (1874) II. xii. 488, I have *such* Japanese lilies making ready to put forth their splendours. **1943** R. Godden *Rungli-Rungliot* 45 What else is there in the garden? Wild coffee flowers, roses, Japanese lilies. **1894** H. O. Forbes *Hand-bk. Primates* II. 14 Nothing is known of the habits of the Japanese Macaque. **1967** J. R. & P. H. Napier *Handbk. Living Primates* 405 Japanese macaque. Yellowish-brown shaggy fur. **1898** W. Robinson *English Flower Garden* (ed. 6) 379/1 The varieties of the Japanese Maple (*A. palmatum*) and its numerous forms..have been found of much interest for the garden. **1904** *Jrnl. R. Hort. Soc.* XXIX. 328 The popularly known 'Japanese Maples' are varieties of the two species *Acer palmatum* and *Acer japonicum.* **1973** C. Lloyd *Foliage Plants* x. 172 The beautiful but slow growing Japanese maple of the golden foliage is the corner piece at the back. **1866** Lindley & Moore *Treas. Bot.* I. 462/1 The Loquat, or Japanese Medlar..is a native of Japan and the southern parts of China. **1950** G. Brenan *Face of Spain* iv. 80 The Japanese medlars with their fish-shaped leaves and thick snake-like branches. **1972** A. F. Simmons *Growing Unusual Fruit* 179 It [*sc.* the loquat] arrived in the Mediterranean area in the nineteenth century, under the name of the Japanese medlar. **1872** *Proc. Zool. Soc.* 780 (*heading*) Observations on the Macaques.—IV. The Japanese Monkey. **1932** S. Zuckerman *Social Life Monkeys* xix. 310 The Japanese monkeys are confined together. **1961–2** *Primates* (Inuyama, Japan) III. ii. 3 The Japanese monkey is an endemic species which usually inhabits the thick forests of the mountains which cover the greater part of the Japanese Islands. **1924** L. H. Bailey *Man. Cultivated Plants* 413 Japanese Pagoda-Tree. Round-headed deciduous tree attaining 60 ft., with spreading branches. **1973** *Times* 20 Feb. 16/3 The 152-year-old scholar tree, or Japanese pagoda tree, at Oxford University Botanic Garden is being felled because it is dying. **1727** J. G. Scheuchzer tr. *Kæmpfer's Hist. Japan* II. App. ii. 25 The Japanese paper is very tight and strong, and will bear being twisted into ropes. **1822** F. Shoberl tr. *Titsingh's Illustr. Japan* ii. 319 Two hundred and thirty-four different flowers, painted with great truth on thin Japanese paper. **1877** *Trans. Asiatic Soc. Japan* V. 77 The kites are constructed of Japanese paper which is both thin and strong. **1905** F. H. Collins *Author & Printer* 190/2 *Japanese paper*, hand-made in Japan with vellum surface. Used for proofs of etchings and engravings. **1958** J. R. Biggs *Woodcuts* 90 The best Japanese papers are made from the fibres of the mulberry tree. *a* **1963** S. Plath *Ariel* (1965) 59 My head a moon Of Japanese paper. **1861** R. Bentley *Man. Bot.* ii. 503 The fruit of *X*[*anthoxylum*] *piperitum* is employed by the Chinese and Japanese as a condiment... It is commonly termed in commerce, Japanese Pepper. **1972** Y. Lovelock *Vegetable Bk.* iii. 344 The most important among these [members of the genus *Zanthoxylum*] include Chinese pepper (*Z. bungei*) and Japanese pepper (*Z. aromaticum*). *c* **1895** A. Beardsley *Lett.* (1971) 98 All the books I have left behind are at your disposal. Also a set of erotic Japanese prints. *a* **1922** T. S. Eliot *Waste Land Drafts* (1971) 33 line 140 A touch of art is given by the false Japanese print, purchased in Oxford Street. **1972** *Country Life* 5 Oct. 805/1 William Burges was collecting Japanese prints in the 1850s. **1900** L. H. Bailey *Cycl. Amer. Hort.* I. 427/1 Rarer kinds..are grafted in the greenhouse in early spring, on stock of the Japanese or common Quince. **1972** A. F. Simmons *Growing Unusual Fruit* 152 A species known often as *Cydonia sinensis* and classed with the Japanese quince or japonica ..does not, however, belong to the same genus. **1883** W. Robinson *Eng. Flower Garden* 244/1 The palm for hardiness and decorativeness in exposed situations must be given to another Japanese Rose (Rosa rugosa). **1922** T. G. W. Henslow *Rose Encycl.* xiv. 176 The Japanese Rose (*R. Rugosa*). These roses are gaining in popularity every day. **1956** B. Park *Collins Guide to Roses* xi. 196 *R*[*osa*] *rugosa.* The Ramanas Rose. The Japanese Rose. The typical form has deep purplish-pink single flowers. **1872** D. G. Rossetti *Let.* 18 Dec. (1967) III. 1108 If you could look in at Hewitt's one day, would you see what Japanese screens he has, and what he wants for them? **1881** C. C. Harrison *Woman's Handiwork* iii. 151 A Japanese screen in the house is a liberal education to the follower of art-needlework. **1935** C. Isherwood *Mr. Norris changes Trains* xv. 254 In addition to the etchings and the Japanese screen, Arthur gave her three flasks of perfume. **1973** *Country Life* 22 Nov. 1691 A few delectable painted Japanese screens..averaged about £750 each. **1873** *Young Englishwoman* May 258 How to clean a white Japanese silk, which has got soiled in the wearing. **1895** *Montgomery Ward Catal.* 79/3 Ladies' Japanese Silk Chemisette. [**1863** R. Alcock *Capital of Tycoon* I. xv. 309 First I am to find a pair of well-bred Japanese dogs, 'with eyes like saucers, no nose, the tongue hanging out at the side, too large for the mouth, and white and tan if possible'.] **1880** H. Dalziel *Brit. Dogs* iii. 444 At the New York Dog Show..they [*sc.* Japanese pugs] were classed as Japanese spaniels. **1894** R. B. Lee *Hist. & Descr. Mod. Dogs Gt. Brit. & Ireland* (*Non-Sporting Division*) xiii. 302 These little dogs are now called and identified as Japanese spaniels because they are supposed to have originally been brought from Japan. **1948** C. L. B. Hubbard *Dogs in Brit.* 253 The native Japanese spaniel is quite distinct from the Pekingese of China. **1971** F. Hamilton *World Encycl. Dogs* 524 One such rarity is what is known in its country of origin, Japan, as the Chin; in Britain as the Japanese; and in the Americas as the Japanese Spaniel. **1880** L. Higgin *Handbk. Embroidery* v. 51 *Japanese Stitch* is a modification of stem..taking very long stitches, and then bringing the needle back to within a short distance of the first starting-place. **1900** *Knowledge* 1 Dec. 285/1 Japanese tissue paper used by dentists. **1936** *Discovery* May 157/2 Paste a good quality Japanese tissue on to both sides of the document. **1888** C. T. Jacobi *Printers' Vocab.* 68 *Japanese vellum paper*, thick hand-made paper with a vellum surface manufactured in Japan. **1923** H. A. Maddox *Dict. Stationery* 41 *Japanese Vellum*, a stout toned printing or cover paper with smooth surface and of exceptional strength made from long Japanese fibre by natural methods. **1952** J. Carter *ABC for Bk.-Collectors* 109 Japanese vellum is a very costly paper, hand-made in Japan from the inner bark of the mulberry tree. **1902** *Biometrika* II. 101 (*title*) Note on the results of crossing Japanese waltzing mice with European albino races. **1904** [see Waltzing *vbl. sb.* and *ppl. a.*, Waltzer b]. **1943** H. Grüneberg *Genetics Mouse* iv. 50 Most stocks of Japanese waltzing mice are homozygous for it [*sc.* the gene for 'recessive' or piebald spotting]. *Ibid.* 85 Japanese waltzers differ slightly, but significantly, from albino mice in their temperature of choice. **1964** G. Durrell *Menagerie Manor* vii. 141 Rich people who do nothing all day long but revolve from one cocktail party to another, like a set of Japanese waltzing mice. **1859** L. Oliphant *Narr. Earl of Elgin's Mission China & Japan* II. 257 Hitherto the most successful cargo brought to this country from Japan has been one of Japanese wax. Mr. Simmonds..gives the following account of Japan wax:—'Rhus succedanea, the species which furnishes the Japan wax, has long been grown in our greenhouses, having been introduced from China nearly a century ago.' **1951** Japanese wax [see *heneicosane]. **1878** *Proc. Zool. Soc.* 788 Judging from the present specimen the Japanese Wolf, although nearly allied to *Canis lupus*, would seem to be a distinct species, to be recognized by its smaller size and shorter legs. **1968** R. & A. Fiennes *Nat. Hist. Dog* 160 Japanese wolves are very like the common northern wolves, but they are smaller.

Japaneseness (dʒæpănī·znes). [f. Japanese *a.* + -ness.] The quality or state of being Japanese, or of displaying Japanese characteristics.

1965 *New Statesman* 19 Mar. 462/2 There is no point in trying to skirt the sheer Japaneseness of Ozu's film. **1967** D. & E. T. Riesman *Conversations in Japan* 8 The wife being the one to bear the burden of Japaneseness.

Japanesery (dʒæpănī·zəri). [f. Japanese *a.* and *sb.* + -ery, after F. *japonaiserie* (see *japonaiserie).] Japanese characteristics or fashion; also *pl.*, Japanese ornaments, knick-knacks, etc.

1885 *Daily News* 30 Apr. 4/8 The 'Mikado' may even bring in a little agreeable Japanesery. **1894** W. J. Locke *At Gate of Samaria* ix. 103 Cheap Japaneseries that had lent it the suggestion of artistic atmosphere the girl of eighteen had craved. **1906** E. Nesbit *Man & Maid* viii. 179 Bright, picturesque cushions and screens and Japaneseries. **1929** Blunden *Near & Far* Pref., Those, however, who go from England to Japan without succumbing first to Japanesery will find that there is no great gulf between the old experiences and the new. **1958** *Spectator* 14 Feb. 204/2 One very soon starts longing for a bit of simple Japanesery without all that *obtrusive* local detail.

Japanesey, Japanesy (dʒæpănī·zi), *a.* [f. Japanese *a.* and *sb.* + -y¹.] Having or inclining to a Japanese character.

1890 B. H. Chamberlain *Things Japanese* 144 Criticism is not at all a 'Japanesey' thing. **1891** S. J. Duncan *Amer. Girl in London* 55 Her parlour was Japanesy, too, in places. **1901** 'C. Holland' *Mousmé* xxiii. 328 He..has referred to..their figures as '*petite*' and Japanesy. **1923** E. F. Wyatt *Invis. Gods* i. ii. 17 The Japanesy shadows of the black scrub pine. **1925** *Glasgow Herald* 8 May 10 High-art, Japanesy tenements. **1941** *Penguin New Writing* X. 23 The stained Japanesey walls. **1971** *Country Life* 16 Dec. 1720/3 A trifle too fussy, too obviously Japanesey?

Japanned, *ppl. a.* Add: **3.** *Japanned leather* (see quot. *a* 1877); *Japanned peacock, peafowl,* the black-winged peafowl, *Pavo cristatus* mut. *nigripennis.*

1814 R. Heron *Jrnl.* in *Proc. Zool. Soc.* (1835) 54 The hens..would not suffer a japanned peacock to touch them. **1851** *Illustr. Catal. Gt. Exhib.* iv. 1252/1 Japanned leather, grained calf-skin for boots and shoes, and trimmings. **1855** J. C. Morton *Cycl. Agric.* II. 698/2 The Japanned pea-fowl, as it ought to be styled, instead of the *Japan* pea-fowl. *a* **1877** Knight *Dict. Mech.* II. 1211/1 *Japanned leather*, leather treated with several coats of Japan varnish and dried in a stove. **1894** A. Newton *Dict. Birds* iii. 699 The 'japanned' Peacock, often erroneously named the Japanese or Japan Peacock,..has received the name of *P*[*avo*] *nigripennis*, as though it were a distinct species. *Ibid.* 701 The 'japanned' bird is not known to exist anywhere as a wild race.

Japano- (dʒæ·pănǫ), used as combining form of Japanese, esp. in adjs. meaning 'belonging to Japan (and some other country)'. Also **Ja·panophile,** a lover of Japan or the Japanese.

1904 *Daily Chron.* 10 May 5/2 The Japano-Russian war. **1905** D. Sladen *Playing Game* ii. iii. 185 Under the influence of the Japanophile Jevons. **1906** *Daily Chron.* 21 May 7/5 The Japano-Korean treaty. **1910** *Westm. Gaz.* 3 Feb. 3/1 The Japano-Chinese war of 1894–5. **1973** *Record* (Oxf. Univ. Press) xviii. 5/1 We..made our way to..Tokyo where TK greeted us in his Japano-Oxford accent and bore us off to the Palace Hotel.

Japlish (dʒæ·pliʃ). [f. Jap(anese *sb.* + Eng)lish *sb.*] A blend of Japanese and English spoken in Japan: either the Japanese language freely interlarded with English expressions or the English language spoken in an unidiomatic way by a Japanese speaker. Also *attrib.* or *adj.*

1960 *N.Y. Times* 25 Dec. 3/5 (*caption*) New traffic instructions, written in 'Japlish', cause linguistic bewilderment. **1963** *Harper's Mag.* Jan. 54 A great many Japanese speak English nowadays (or at least 'Japlish', as the American colony calls it), and their words are usually understandable. **1966** *Time* 22 July 30 Japanese sometimes sounds like Japlish: *masukomi* for mass communications, *terebi* for TV, *demo* for demonstration and the inevitable baseballisms *pray bollu, storiku* and *hitto.* **1970** *Times* 26 Nov. 12 A word of warning to tourists and others: the Japlish veneer can be deceptive. The Japanese may use more English words, but they still think like Japanese. **1970** *Bull. Inst. Res. Lang. Teaching* (Tokyo) CCXCIII. 13 There are, moreover, hosts of 'Japlish'..words and expressions, such as '*old miss*'..(meaning 'old maid' or 'spinster').

‖ **japonaiserie** (ʒaponęzri). [Fr.] = *Japanesery.

1896 S. R. Crockett *Cleg Kelly* xxvii. 195 The little alcove..was cobwebbed with the latest artistic Japonaiseries of the period. **1902** *Westm. Gaz.* 11 Aug. 12/2 His earlier passion for Gothic art had been succeeded by one for *japonaiseries.* **1905** W. Stevens *Let.* 21 Aug. (1967) 84 The japonaiserie of the pines. **1966** *New Statesman* 8 July 64/2 This..enthralling piece of contemporary *japonaiserie* [*sc.* a film] circles round a gentle young wife.

japonica (dʒăpǫ·nikă). [mod.L., fem. of *japonicus* pertaining to Japan.] A name used for various plants originally native to Japan, esp. formerly the camellia (*C. japonica*); now usually designating a spring-flowering, deciduous shrub of the genus *Chænomeles*, esp. *C. japonica*, the Japanese quince, or its fruit.

1819 Keats *Let.* 13 Mar. (1931) II. 309, I would put it [*sc.* a globe of gold-fish] before a handsome painted window and shade it all round with myrtles and Japonicas. **1851** H. Melville *Moby Dick* I. xli. 299 Whiteness refiningly enhances beauty,..as in marbles, japonicas, and pearls. **1859** A. Van Buren *Jottings of Sojourn in South* 134 That richest and sweetest blossomed of tropical shrubs—the japonica—that never blossoms only in winter. **1910** 'O. Henry' *Strictly Business* (1917) ix. 91 You put me in mind of a japonica in a window. **1933** A. Osborn *Shrubs & Trees for Garden* xxxv. 318 *C*[*ydonia*] *japonica* (Japanese Quince), of which there are varieties with flowers of many shades of white, yellow, pink and red, is often known as 'Japonica'. **1963** W. Blunt *Of Flowers & Village* 46 The japonica has had no less than five different names during my lifetime. **1969** *Listener* 9 Jan. 45/2 The Japonica is strangled with convolvulus.

b. The fruit of the Asian plant *Zizyphus jujuba* (*Z. sinensis*); cf. Jujube.

1874 Lindley & Moore *Treas. Bot.* II. Suppl. 1308/1 Japonica. A market name for the fruits of *Zizyphus sinensis*, which are occasionally sold in Covent Garden as dessert fruits. **1887** R. Bentley *Man. Bot.* (ed. 5) ii. 517 Another Japanese species, *Z*[*izyphus*] *sinensis*, yields the fruits known as Japonicas.

Hence **japo·nicadom** *U.S.* (see quot. 1860). *Obs.*

1851 A. O. Hall *Manhattaner* 123 The general society of New Orleans is still in a chaotic state, and she has no located, acknowledged empire of Japonicadom. **1860** Bartlett *Dict. Amer.* (ed. 3), *Japonicadom*, a word invented by N. P. Willis to denote the upper classes of society.

jar, *sb.*¹ Add: **V. 9.** *attrib.* and *Comb.*, as **jar ramming** *Founding* = *jolt ramming* (s.v. *jolt *sb.* 5).

1909 *Iron Age* LXXXIV. 1165/1 The working foundryman has carried this development, especially as applied to what are known as jolt or jar ramming machines, far beyond what was anticipated. **1912** *Jrnl. Iron & Steel Inst.* LXXXVI. 546 A jar-ramming moulding-machine with a roll-over device, which also lowers the mould away from the pattern, is described. **1934** Laing & Rolfe *Man. Foundry Pract.* v. 102 (*heading*) Jolt- or jar-ramming.

jar, *sb.*² Add: † **2. b.** *Electr.* A unit of capacity (see quot. 1920¹). *Obs.*

[**1834** W. S. Harris in *Rep. Brit. Assoc. Adv. Sci. 1833* 387 The unit of measure consists of a small electrical jar, having a discharging electrometer. **1889** A. W. Poyser *Magn. & Electr.* xiii. 139 Harris's unit jar.—This instrument is used for measuring the charge given to a Leyden jar.] **1920** *Admiralty Handbk. Wireless Telegr.* iii. 68, 1 farad = ..9 × 10⁸ (nine hundred million) 'jars'. *Ibid.*, The jar is a Service unit, and is very useful when dealing with small capacities. *Ibid.* viii. 245 Provide a condenser composed of two elements of 100 jars each. **1932** *Admiralty Handbk. Wireless Telegr. 1931* vi. 338 The low reactance of even minute capacities and the easy shunt paths they provide..may best be realised by giving a comparative table of the reactances of a capacity of 1 jar and an inductance of 100 microhenries at various frequencies.

c. A drink (of beer, etc.). *colloq.*

1925 S. O'Casey *Juno & Paycock* i, in *Two Plays* 42 *Boyle.* An' now, Mr. Bentham, you'll have to have a wet. *Bentham.* A wet? *Boyle.* A wet—a jar—a boul! **1941** Baker *Dict. Austral. Slang* 38 *Jar*, a pint or 'handle' of beer. **1961** C. Willock *Death in Covert* i. 8 'Have a jar, Goss,' he said, and poured him at least three fingers of whisky. **1966** P. Moloney *Plea for Mersey* 56 Whan lads frae Scotty Road gang far, Untae the boozer for a jar. **1969** V. Canning *Queen's Pawn* iii. 41 Hot morning. Care for a jar? They keep good beer. **1972** *Observer* 26 Nov. 26/4 The painter, Raymond Piper, took us for a jar at his local. **1973** *New Society* 6 Sept. 563/1 A great place to meet old friends and make new ones, to knock back the jars and sit gossiping into the early hours.

jar (dʒāɹ), *v.*² [f. Jar *sb.*²] *trans.* To preserve (fruit) in a jar; to bottle. Also *transf.*

1747 H. Glasse *Art of Cookery* xviii. 152 (*heading*) To jar Cherries. **1962** *Guardian* 24 Dec. 4/3 There's no point in jarring it away. You have to buy clothes with anything you get.

‖ **jarabe** (hărā·be). Also **jarave.** [Amer. Sp., a. Sp. *jarabe* syrup.] One of several Mexican pair dances in which the man dances the zapateado steps.

1834 A. Pike *Prose Sk. & Poems* 103 In the jarabes, or singular dances of the fandango, her first partner was always Rafael. **1903** C. Lumholtz *Unknown Mexico* II. xxii. 382 Professional male and female dancers, engaged as special attractions, execute the national dance, *jarave.* **1932** H. Crane *Let.* 12 Apr. (1965) 408 Someone dances a *jarabe*, a dance that is all vibrant gristle. **1964** *Spectator* 3 Apr. 450/1 The typical dance [of Mexico], the *Jarabe*, slightly Spanish in origin.

jardinière. Add: **2.** *Cookery.* (See quot. 1877). So *jardinière soup*, vegetable soup.

1841 Thackeray in *Fraser's Mag.* June 723/1 They..served us..Jardinière cutlets (particularly seedy). **1846** A. Soyer *Gastronomic Regenerator* 40 [Sauces.] Jardinière. **1846** C. E. Francatelli *Mod. Cook* 234 Fill the centre of the *entrée* with a *jardinière* of vegetables. **1877** *Cassell's Dict. Cookery* 338/1 Jardinière.—This is a garnish made of cooked vegetables, which gives its name to the dish with which it is served. Thus, fillet of beef à la jardinière, mutton à la jardinière, goose à la jardinière, simply mean fillet of beef, mutton, and goose served with a garnish à la jardinière. *Ibid.*, Jardinière Soup. **1907** G. A. Escoffier *Guide Mod. Cookery* 357 Prepare the fillet as directed under 'Filet de Bœuf Jardinière'. Set it on a long dish and surround it with a *Macédoine* garnish. The latter comprises the same ingredients as the '*Jardinière*'. *Ibid.* 512 *Sauté* the *suprêmes* in butter. Dish and surround with small heaps of vegetables, arranged very neatly, as explained in the case of the *Jardinière* garnish. **1969** R. & D. De Sola *Dict. Cooking* 128/1 Vegetables served in a savory sauce or soup, usually labelled *à la jardinière.*

jargoneer (dʒāɹgənīᵊ·ɹ). [f. Jargon *sb.*¹ + -eer.] One who uses jargon.

1913 A. Quiller-Couch *On Art of Writing* (1916) 90 A Jargoneer would have said that 'among the beneficent qualities of sleep its capacity for withdrawing the human consciousness from the contemplation of immediate circumstances may perhaps be accounted not the least remarkable'. **1923** *19th Cent.* Nov. 786 Your true jargoneer must have at least two languages in the same word. **1947** I. Brown *Say Word* 92 Among the jargoneers of our time..the advertisers of women's garments have a notable place. **1973** *N.Y. Law Jrnl.* 4 Sept. 5/1 Another miserable heritage of Watergate has been the bilge loosed on the public by..psychopathic jargoneers..and belly-achers.

jarlite (yā·ɹləit). *Min.* [f. the name of C. F. *Jarl*, 20th-century Danish mining official: see -ite¹.] A colourless to greyish-white fluoride of sodium, strontium, and aluminium, $NaSr_3Al_3F_{16}$, found as monoclinic crystals at Ivigtut, Greenland.

1933 R. Bøgvad in *Meddelelser om Grønland* XCII. viii. 3 The mineral has been named jarlite, after Mr. C. F. Jarl, by whose courtesy I have been enabled to carry out the present investigations. **1949** *Amer. Mineralogist* XXXIV. 386 Jarlite..occurs as colorless, flat crystals about 1 mm. in size, often fused in the shape of a fan and growing with thomsenolite in vugs in the cryolite.

jarool (dʒărū·l). Also **jarul.** [Hind., a. Bengali *jarūl.*] A deciduous tree, *Lagerstrœmia speciosa*, of the family Lythraceæ, which is native to tropical Asia and bears large panicles of purple flowers; also, the wood of this tree.

1850 J. D. Hooker *Himalayan Jrnls.* (1854) II. xxxi. 327 These forests are frequented by timber-cutters, who fell jarool (*Lagerstrœmia Reginæ*), a magnificent tree with red wood, which, though soft, is durable under water, and therefore in universal use for boat-building. **1889** G. S. Boulger *Uses of Plants* vii. 188 Among numerous proposed teak substitutes we can..mention..the Jarúl. **1922** J. S. Gamble *Man. Indian Timbers* (ed. 2) 373 The Jarúl tree is never likely to be important beyond its use for local requirements. **1962** *Wealth of India* (Council Sci. & Industr. Res. India) VI. 24 Jarul is a constructional timber of considerable commercial importance, particularly in North-East India.

jarovization (yæːrovəizē¹·ʃən). Also **iarovization.** [ad. Russ. *yarovizátsiya*: see *vernalization.] = *vernalization.

[**1933** Whyte & Hudson in *Bull. Imperial Bureau Plant Genetics* No. 9. 3 The Russian word 'Jarovizatzia', referred to in German publications as 'Jarowisation', has here been translated by the latinized equivalent 'vernalization' in consultation with the School of Slavonic Studies, London, and Miss M. V. Cytovich.] **1934** *Biol. Abstr.* VIII. 1655/2 Discussion is given of..the process of vernalization (or iarovization) by means of which an early development of the reproductive phase of many plants..was obtained. **1946** *Nature* 19 Oct. 548/1 Vol. 1, No. 1 [of *Genetica Agraria*] contains papers on such subjects as ..Jarovization of the potato and resistance to rust in wheat.

jas, var. *jazz *sb.*

jasbo, var. *jazzbo.

jasmine. Add: **3. jasmine tea,** tea perfumed with jasmine; also *ellipt.*

1933 N. Waln *House of Exile* i. iii. 44 We finished our meal with..cups of jasmine tea..served without milk or sugar. **1967** V. C. Clinton-Baddeley *Death's Bright Dart* 64 Neither Lapsang..nor Jasmine, was what he wanted. **1967** S. Knight *Window on Shanghai* 119 Just broke off to..make myself a cup of jasmine tea which smells delicious. **1972** *Korea Herald* 17 Nov. 5/1 Between sips of jasmine tea they haggle.

jaspé (dʒɑ·spe, ‖ ʒaspe). Also **jaspe.** [Fr., pa. pple. of *jasper* to marble.] Resembling jasper; mottled or variegated.

1851 *Illustr. Catal. Gt. Exhib.* iii. 500/2 Printed and embroidered 'jaspé' cashmere. **1908** *Ladies' Field* 24 Oct. 318/1 The Ghiberti damask has a *jaspé* ground. **1923** *Weekly Dispatch* 11 Feb. 16 Harmonising shades of Pink, Blue, Mauve, Brown and Green on Cream, Jaspe, Black or Grey backgrounds. **1931** *Daily Express* 23 Sept. 4/1 New clasp in Imitation Jaspe Shell. **1961** J. I. M. Stewart *Man who won Pools* 36 Crossing the jaspé lino. **1968** J. Arnold *Shell Bk. Country Crafts* 274 Products of horn are obtainable in four colours:—a mottled horn called 'Jaspé', [etc.].

jasper (dʒæ·spəɹ), *sb.*² *U.S. colloq.* [Male Christian name.] A person, fellow: usu. with contemptuous overtones; *spec.* a rustic simpleton, 'hick'. Also (with capital initial) used as a nickname (see quots. 1929, 1952).

1896 H. M. Blossom *Checkers* x. 229 After supper.. I went over to the only shanty in the place that looked like a store, and opened the door. There were a lot of 'Jaspers' sitting around the stove, chewing Tobacco and swapping lies. **1914** 'B. M. Bower' *Flying U Ranch* 174 Some uh you boys help me rope him—like him and that other jasper over there done to Andy. **1929** T. Gordon *Born to Be* 236 Zigaboo, Dingo, Jasper, nicknames for Ethiopians. **1952** Granville *Dict. Theatr. Terms* 103 *Jasper*, the traditional name for the villain of the piece in melodrama. **1963** 'M. Corrigan' *Why do Women —?* xxiii. 173 If that dark jasper calls on you again, try and keep him here. **1970** *New Yorker* 17 Oct. 40/3 What's with those jaspers?

jasperoid, *a.* Add: (Further example.)

1965 G. J. Williams *Econ. Geol. N.Z.* xii. 176/1 The lode..lies immediately alongside and to the south-east of a belt of banded jasperoid chert.

B. *sb. Petrol.* A rock in which silica, in the form of fine-grained quartz or chalcedony, has replaced some of the original constituents (usually the carbonate of limestone).

1898 J. E. SPURR in *Monogr. U.S. Geol. Survey* XXXI. 219 He [*sc.* the writer] has become impressed with the widespread occurrence of this variety of quartz, which arises from the replacement of some original rock, ordinarily limestone or dolomite, by silica from circulating waters; and there seems to be need of some term which may specifically indicate it. For this use the word 'jasperoid' is suggested. Jasperoid may then be defined as a rock consisting essentially of cryptocrystalline, chalcedonic, or phenocrystalline silica, which has formed by the replacement of some other material, ordinarily calcite or dolomite. **1928** W. LINDGREN *Mineral Deposits* (ed. 3) xiii. 207 Silicification of limestone, argillaceous shale and rhyolite is..a very common process taking place frequently with preservation of texture. The quartz will usually be fine-grained. Silicified limestones are called jasperoids. **1944** *Proc. Prehist. Soc.* X. 12 The cherts and jasperoids probably came from the conglomerates at Bankrom. **1968** *Prof. Papers U.S. Geol. Survey* No. 600-B. 112/1 Sixty-eight samples of jasperoid associated with base- and precious-metal ore deposits in 25 western mining districts..were analysed for gold.

jass, var. *JAZZ *sb.*

jassid (dʒæ·sid), *a.* and *sb.* [f. mod.L. family name *Jassidæ* f. the generic name *Iassus* (*Jassus*) (J. C. Fabricius *Systema Rhyngotorum* (1803) 85), f. L. *Iāsus* a town in Asia Minor.] **A.** *adj.* Of or pertaining to a homopterous insect of the family Jassidæ, sometimes considered equivalent to the Cicadellidæ, and including several pests of cereals, fodder crops, etc. **B.** *sb.* A leaf-hopper, a small, jumping insect of this kind, which attacks the leaves of plants.

1892 *Trans. Amer. Entomol. Soc.* XIX. 307 Mr. Howard Ewarts Weed..has brought to light many new and interesting members of the little-known Jassid fauna of the 'Mississippi Bottoms'. **1895** *Insect Life* VII. 323 The half-clothed jassid (*Eutettix seminudus* Say) was often taken on the stalk of the cotton, and was observed to feed upon the juices of the plant. **1918** *Agric. Gaz. New South Wales* 568 (*title*) The apple-leaf Jassid (*Empoasca australis*). **1926** *Proc. U.S. Nat. Museum* LXVIII. 1 (*title*) Revision of the American leaf hoppers of the Jassid genus *Typhlocybe*. **1956** *Nature* 11 Feb. 282/2 The influence of sprayed areas in reducing populations of jassids infesting adjacent unsprayed cotton was elucidated recently.

Jat[1] (dʒāt). Also 7 **Jett, Jutt,** 8 **Jaut.** [Hindi *Jāṭ*.] A member of an Indian tribe settled in the Punjab, Sind, and North-West Provinces. Also *attrib.*

1622 in W. Foster *Eng. Factories India 1622–3* (1908) 90 [There] goeth with the carts 27 Jutts, etc. for their safer passadge. *Ibid.* 111 A Jett whome some tymes you have [app]roved off for trusty. **1787** C. HAMILTON *Hist. Relation of Rohilla Afgans in N. Provinces of Hindostan* 104 By similar means he procured the assistance of Soorâj Mull with a number of Jâts. **1797** *Encycl. Brit.* VIII. 529/1 The Jauts, or Jats. **1845** *Encycl. Metrop.* XXIII. 781/2 The Ják'har, Shiváɡ'hs and Punyás..are all of Jat origin. **1880** *Encycl. Brit.* XIII. 507/2 Early Jât settlements on the shores of the Persian gulf. **1901** *North Amer. Rev.* Feb. 301 The Jats are the most important people in the Punjab. **1946** J. H. HUTTON *Caste in India* I. iv. 33 The Jats who form perhaps the most important element in the population of the Punjab may be either Hindu or Sikh or Muslim, though..the typical Punjab Jat is probably a Sikh. The Jat is a typical yeoman, devoted to agriculture and..takes a lower social level than the Rajput. *Ibid.* 34 The Sikh of Rajput or Jat origin. **1956** SINGH & SETI in Aiyappan & Ratnam *Soc. in India* xxii. 241 The Sardars..have been converted from the higher castes of Hindus and Jats. **1964** A. SWINSON *Six Minutes to Sunset* iii. 49 Some retired Jat and Sikh police officers. **1972** *Nat. Geographic* Oct. 535/1 Each of the three basic divisions of the Sikhs—Jats, non-Jats, and untouchables—lives behind boundaries that the others cross with difficulty.

jat[2] (dʒāt). Also **jāt, jāti.** [Hind. *jāt*.] A caste, tribe, sect.

[**1855** H. H. WILSON *Gloss. Judicial & Revenue Terms* 234 *Ját*, H., and in most dialects; corruptly, *Jaut*... Caste, clan, tribe, occupation, kind, sort.] **1873** E. BALFOUR *Cycl. India* (ed. 2) III. 151/2 *Jat* or Jet or Jut or Zat, pronounced thus variously in different parts of India, means a race, a tribe, a clan, a manner, a kind. **1894** M. DYAN *All in Man's Keeping* I. i. 5 Are they not all one jât or caste? **1909** M. DIVER *Candles in Wind* viii. 80 She's another 'jāt' [*note*, class] from us altogether. **1931** N. K. DUTT *Orig. & Growth Caste in India* i. 6 Many castes or jatis were produced by a series of crosses..between members of the four varnas. **1960** A. C. MAYER *Caste & Kinship in Cent. India* III. viii. 152 *Jat*, the general word for species, is the most commonly used term for 'caste'... You ask a man..'what is your *jat*'. **1971** *Illustr. Weekly India* 4 Apr. 11/3 Besides, they have helped in removing a number of malpractices in the community like child marriage, polygamy and in settling *Jati* disputes.

jat' (yatʸ), *sb.*[3] Also **ěti, iet, jat, jet', yat,** etc. [ad. OSl. *jatĭ*; cf. quot. 1964.] The name of the characters Ѣ and ѣ of the Slavonic Glagolitic and Cyrillic alphabets; the sound represented by these characters, or the Common Slavonic sound from which it developed.

1763 W. MASSEY *Origin & Progress of Letters* (tables facing 114) Yat..iet. **1883** I. TAYLOR *Alphabet* II. 196 (*table*) Yet. **1887** M. GASTER *Ilchester Lectures on Greeko-Slavonic Lit.* 214 (*table*) ěti. **1950** *Slavistična Revija* III. 257 In both alphabets *jat'* as a letter is distinct from the other vowel-symbols. **1955** R. JAKOBSON *Slavic Languages* (ed. 2) 14, *ě* (called *jat'*, the reflex of *ē* and *oi*). **1964** M. SAMILOV *Phoneme jat' in Slavic* 11 One of the main reasons for disagreement is the extreme diversity of the reflexes of *jat'* (as the Glagolitic and Cyrillic letter for the sound *ě has been called)... The old name for the Glagolitic Ѣ and Cyrillic ѣ seems to have been *ěd'* or *jad'* 'food'. **1965** G. Y. SHEVELOV *Prehist. of Slavic* xi. 164 Long *ēa* has traditionally been denoted in Slavistics by *ě* (and known as *jat'*).

Jataka (dʒā·tăkā). [Skr. *jātaka* engendered by, born under, f. *jātá*, pa. pple. of *jan* to produce.] In Buddhist literature, a story of one or other of the former births of the Buddha; also, the name of the Pāli collection of these stories. Also *attrib.*

1828 *Asiatic Researches* XVI. 427 *Játaka*, treat of the actions of former births. **1861** V. FAUSBÖLL *Five Jātakas* Pref., We.., in the Jâtaka, meet with some of the Comical stories that are well known all over Europe under different names. **1876** *Encycl. Brit.* IV. 430/1 Jātaka stories..containing..the oldest known versions of many of the nursery songs, and fairy tales, and comic stories, and fables, which are the common property of Europe in the present day. **1876** J. FERGUSSON *Hist. Indian & Eastern Archit.* I. 88 Bas-reliefs..representing some scene or legend of the time, and..inscribed..with the title of the jataka or legend. **1951** G. DE *Significance of Jātakas* p. xii, Prof. Rhys Davids..has observed..that a typical Jātaka is one which has..an introductory episode,..the story of the past being the Jātaka proper in prose. **1956** M. WICKRAMASINGHE *Buddhist Jataka Stories & Russ. Novel* p. ix, The linguistic, religious and sociological aspects of the Jataka Stories have attracted the attention of Western Orientalists. *Ibid.* i. 3 There are two kinds of narratives in the Jataka Book.

jatha (dʒā·tā). *India.* Also **jathā.** [Hind. *jathā*.] An armed or organized band, *spec.* of Sikhs.

1922 *Glasgow Herald* 9 Sept. 9 There is no waning in Sikh fanaticism, and organised jathas, or companies, come forward increasingly. A jatha of Sikh women wearing daggers has been formed. **1924** *Ibid.* 31 Mar. 11 A continuous procession of martyr jathas from Amritsar to Jaito. **1964** H. SINGH *Heritage of Sikhs* xviii. 156 A Jatha of 150 Sikhs came to make obeisance at the Gurdwara. **1964** A. SWINSON *Six Minutes to Sunset* x. 181 The Sikh leaders [in 1947] sat at the feet of Master Tara Singh and listened as he roused them to violence. His plan was to form murder gangs known as *jathas*. **1966** K. SINGH *Hist. Sikhs* II. IV. xiii. 203 *Jathās* of 100 Akalis each were formed. They first took an oath..to remain non-violent.

jati, jâti (dʒā·ti). [Skr. *jāti*.] (See quots.)

1891 C. R. DAY *Mus. & Mus. Instruments S. India* iii. 36 The different degrees of time are termed Tálas, of which there are seven, each being sub-divided into five 'jâtis', or kinds; so that there are in use no less than thirty-five distinct measures. **1914** A. H. F. STRANGWAYS *Mus. Hindostan* iv. 112 There are seven modes (*mūrchanā*) of each of the two *grāmas*, i.e. fourteen in all; but only seven of these fourteen are in practical use under the name of *jātis*, species. **1967** SINGHA & MASSEY *Indian Dances* ii. 42 The adavus are combined in the tirmanas with complex rhythm patterns known as *jatis*. Jatis are a combination of long and short syllables. **1968** A. DANIÉLOU *Rāga-s of N. Indian Mus.* ii. 55 The classes obtained by grouping the modes according to the number of their notes are sometimes known as *jāti-s*.

jāti, var. *DJATI, *JAT[2].

jato (dʒēⁱ·to). *Aeronaut.* orig. *U.S.* Also **Jato, JATO.** [f. the initial letters of *j*et-*a*ssisted *t*ake-*o*ff.] **a.** A take-off assisted by a jato unit. **b.** An auxiliary, usually detachable, unit of one or more jet engines (usually rockets) for providing temporary extra thrust when an aircraft takes off. Usu. *attrib.*, esp. in *jato unit*.

1944 *Sci. News Let.* 7 Oct. 229/3 Of particular value on the restricted area of carrier flight decks, JATO, as jet-assisted take-offs are known in the Navy, will also be extremely useful for lifting heavily-laden flying boats from the water. **1950** J. V. CASAMASSA *Jet Aircraft Power Syst.* v. 94 The application of the 1,000-lb thrust of a single JATO rocket motor to an aircraft traveling at a take-off speed of 115 mph at sea level is equivalent to an additional 395 engine-brake-horsepower for the duration of the thrust. **1951** COGGINS & PRATT *Rockets, Jets, Guided Missiles & Space Ships* iv. 35/1 When the Jato unit burned out the pilot could push a release lever and drop the empty cylinder. *Ibid.* 35/2 Jatos were being used to help Navy fighters go almost straight up into the air. **1962** F. I. ORDWAY et al. *Basic Astronautics* ii. 21 During World War II he worked for the U.S. Navy on propellant pumps, jatos, and liquid propelled engines. **1965** C. N. VAN DEVENTER *Introd. Gen. Aeronaut.* vi. 125/1 JATO (jet-assisted takeoff) is sometimes used to accelerate to the required flying speed. **1967** *Propulsion* (Amat. Rocket Assoc. U.S.) ii. 39 Jato units permit aircraft to take off with a heavier payload or to utilize unusually short airfields. **1973** W. T. GUNSTON *Bombers of West* i. 36 We flew an F-80 jet at Muroc.., and with it also experienced JATO take-offs.

jatoba (dʒatŏbā·). [Tupi.] = COURBARIL.

1890 BILLINGS *Med. Dict.* I. 726/1 *Jatoba*, resin of *Hymenæa courbaril*. **1933** P. FLEMING *Brazilian Adv.* II. ii. 191 We landed on a beach strewn with *jatobá* nuts, of which we subsequently made a dish as nauseous as any I have ever tasted.

jaul, var. *JOL.

jaundy, var. *JAUNTY *sb.*

jaune, *a.* Add: **b.** In the names of pigments: *jaune brill(i)ant*, cadmium yellow (alone or in a mixture with white lead); *jaune jonquille* (see quot. 1960).

1851 H. WATTS tr. *Gmelin's Hand-bk. Chem.* V. 57 Sulphide of Cadmium.—Found native in the form of Greenockite. Prepared as a pigment known by the name of *Jaune brillant*. **1895** *Montgomery Ward Catal.* 252/3 Artists tube oil colours...Jaune Brilliant. **1910** B. RACKHAM *Bk. of Porc.* iv. 56 On a ground of yellow (*jaune jonquille*) are rococo-bordered panels. **1924** F. W. WEBER *Artists' Pigments* 30 Mixtures of Cadmium Yellow and Lead White are offered under the name *Jaune Brilliant*. **1948** KIRK & OTHMER *Encycl. Chem. Technol.* II. 736 Cadmium sulfide (cadmium yellow, jaune brilliant). **1960** R. G. HAGGAR *Conc. Encycl. Cont. Pott. & Porc.* 234/2 *Jaune jonquille*, a yellow ground colour introduced at Sèvres in 1753. It is of great intensity and beauty and somewhat different from the shade of yellow employed at Meissen.

Jaune Desprez (ʒōn depre). The name of Monsieur *Desprez*, nineteenth-century French horticulturist, used to designate a variety of yellow climbing rose developed by him about 1830.

1837 T. RIVERS *Rose Amateur's Guide* 82 Jaune Desprez, or the new French Yellow Noisette, is a well-known and much esteemed rose... This was originated by M. Desprez about seven years since. **1867** H. KINGSLEY *Silcote of Silcotes* I. x. 93 Clinging to the house itself, hung the deep dark porch..with festoons of Jaune d'Esprez, and Dundee Rambler. **1869** S. R. HOLE *Bk. about Roses* ix. 134 Jaune Desprez, Noisette.—Phoebus, what a name!..Yellow Desprez, moreover, is not yellow, but buff or fawn colour, deliciously fragrant. **1955** C. C. HURST in G. S. Thomas *Old Shrub Roses* ix. 82 In 1830 'Lamarque' and 'Jaune Desprez' appeared...yellowish Noisettes which, selfed, gave Yellow Climbing Teas.

jaunty (dʒǭ·nti), *sb. Naut. slang.* Also **jaundy, jonty.** [Said to be a sailor's corruption of GENDARME.] The master-at-arms on board ship.

1902 KIPLING *Traffics & Discov.* (1904) 197 The other jaunty is now pursuin' us on his lily feet. **1903** 'L. YEXLEY' *Grog Time Yarns* 3 The Chief of the Police—the Master-at-Arms—is always referred to as the 'Jonty'. **1909** J. R. WARE *Passing Eng.* 159/2 *Jaundy*, master-at-arms. **1927** *Blackw. Mag.* Oct. 457/1 Mounting the rope ladder in that awful sea..proved a bit of a task to the unaccustomed Jonties. **1928** *Weekly Dispatch* 27 May 14 The sailor spun a yarn that would make the hardest-hearted jonty (master-at-arms) weep.

Java. Add: **Java ape-man** = **Java man* (below); **Java canvas,** a loosely-woven linen cloth with an even mesh used in embroidery; **Java man,** the fossil hominid, *Homo erectus* (formerly *Pithecanthropus erectus*) whose remains were first found by E. Dubois in Java in 1891; cf. PITHECANTHROPE b.

1931 STUBBS & BLIGH *Sixty Centuries Health & Physick* 3 Pithecanthropus, the Java ape-man. **1872** *Omen* 26 Oct. 342/2 Antimacassar (Java canvas). **1878** *Cassell's Family Mag.* 494/2 The materials used for these curtains are many—velvet, .., Java canvass, .. and serges. **1882** CAULFEILD & SAWARD *Dict. Needlework* 278/1 *Java canvas*, a close make of canvas, having an appearance of being plaited, and made in many sizes and degrees of fineness. *Ibid.*, *Java canvas work*, this Embroidery is named from the material upon which it is worked and is used for mats, work cases, music cases. **1932** D. C. MINTER *Mod. Needlecraft* 246 Two 4" squares of Java canvas. [**1896** E. DUBOIS in *Nature* 16 Jan. 247/1, I could find no place for the fossil Javanese form, which I consider as intermediate between Man and the Anthropoid apes.] **1911** A. KEITH *Anc. Types of Man* xiv. 136 One is compelled to believe that the human brain had attained a greater size than that of the Java man at the end of the Pliocene Period. **1945** *Anthropol. Papers Amer. Mus. Nat. Hist.* XL. 62/2 Considering..the complexity of the problem of Java man..I deem it best to put aside the Sangiran Mandible of 1939 for the present. **1965** M. H. DAY *Guide to Fossil Man* 224 The features of this femur..suggest strongly that Java man was capable of standing and walking erectly.

b. *ellipt.* Coffee from Java; also (*slang*) any sort of coffee.

1850 L. H. GARRARD *Wah-to-Yah* xiii. 169 Partaking of the nectar-like Java. **1926** J. BLACK *You can't Win* vi. 67 We went back to the fire and discussed breakfast. 'Nothing but Java,' said the bum that had the coffee. **1945** G. MILLAR *Maquis* ii. 29 Later on, to wake you up a bit.., we'll give you a nice cup of Java, strong and hot. **1956** H. GOLD *Man who was not with It* (1965) xxv. 235 Lots a guys come in for chatter and java, friend.

javaite (dʒā·vă,əit). *Geol.* [f. JAVA + -ITE[1], rendering Du. *javaniet* (see etym. of *JAVANITE).] Any tektite from the tektite field of Java.

1938 *Mineral. Abstr.* VII. 78 The tektites ('javaites') from Solo in central Java..have the form of balls or

drops. **1961** *Sci. Amer.* Nov. 63/1 Those [tektites] from Java, Indo-China and the Philippines are designated javaites, indochinites and philippinites respectively. **1964** *Geochim. & Cosmochim. Acta* XXVIII. 761 Rhenium and osmium abundances have been determined in two australites, two javaites, one philippinite, two indochinites, two moldavites and one bediasite.

Javan, *a.* and *sb.* (Further examples.)

1817 [see *BANTENG, BANTING]. **1892** in G. T. Wrench *Restoration of Peasantries* (1939) v. 59 The Javans have escaped the fatal gift of proprietary right which has been the ruin of so many tens of thousands of our peasantry in India. **1893** W. B. WORSFOLD *Visit to Java* x. 172 The average size of the Javan coffee plantation is from 400 to 500 acres. **1910** *Encycl. Brit.* VII. 795/2 The Javans are perhaps unique in their distinct and graceful gestures of the hands and fingers [in dancing]. **1939** G. T. WRENCH *Restoration of Peasantries* v. 58 They [*sc.* the Dutch] made no effort to impose the methods of their civilization upon the Javans for their good. *Ibid.*, They stood aside from indifference and lack of interest in their Javan subjects. **1962** *Times* 20 Dec. 9/6 Some such as..the Javan Rhinoceros are on the verge of extinction. **1964** R. PERRY *World of Tiger* xiii. 192 The reputations of some Javan tigers as man-eaters have served to scare poachers away from the haunts of the lesser one-horned rhino. **1965** R. MCKIE *Company of Animals* xiii. 184 Two species of rhinoceros..have lived in the jungles of Malaya for thousands of years. One, known as the Javan, has one horn on its snout. **1966** C. A. W. GUGGISBERG *S.O.S. Rhino* v. 125 The 300 to 400 bantengs (*Bibos javanicus*) and the numerous Javan deer (*Rusa timorensis*)..frequent the more open parts of the reserve.

‖ **javanais** (ʒavane). [Fr.] A form of French *argot* or slang in which *av* or *va* is introduced after each syllable of a word.

1925 P. RADIN tr. *Vendryes's Lang.* 255 French children often employ *javanais* in the school. **1939** L. H. GRAY *Foundations of Lang.* 31 The relatively recent French *loucherbème* and *javanais*.

Javanese, *a.* and *sb.* Add: Now normally with pronunc. (dʒāv-). (Further examples.)

1836 N. WISEMAN *Twelve Lect. Sci. & Revealed Relig.* I. i. 47 Among these languages [*sc.* the vernacular Indo-Chinese languages on the continent] he [*sc.* J. Leyden] reckons the Bugis, Javanese, Malayu,..and others. *Ibid.* 48 Javanese..is peculiarly deficient in grammatical forms. **1847** F. A. KEMBLE *Let.* 23 June in *Rec. Later Life* (1882) III. 204 Patagonians, Javanese, from the Cordilleras, from Peru..—the flower tribes of the whole earth. **1869** *Month* June 612 Javanese servants and porters. **1933** [see *FORMOSAN *a.* and *sb.*]. **1961** P. KEMP *Alms for Oblivion* v. 85 The peoples of Indonesia, many of whom differed widely in language, religion and culture from the Moslem Javanese. *Ibid.* 86 The full fury of Javanese chauvinism was reserved for the Eurasians. **1964** R. PERRY *World of Tiger* ii. 26 A Javanese tiger..carried off a girl. **1969** *Word* XXV. 323 All masculine names are either first names (in Javanese terminology *nama alit*, little names) or second names. **1973** D. MAY *Laughter in Djakarta* i. 7 He peered into the warm Javanese night from the betjak—the bicycle rickshaw—in which he was riding. *Ibid.* 18 The other, Subekto, evidently a Javanese like Sumitro, laughed softly.

javanicin (dʒăvæ·nisin, dʒāvănəi·sin). *Pharm.* [f. mod.L. *javanic-um* (f. JAVA), specific epithet (see def.) + -IN[1].] A red, crystalline, bicyclic compound, $C_{15}H_{14}O_6$, isolated from the fungus *Fusarium javanicum* and having antibacterial properties.

1946 H. R. V. ARNSTEIN et al. in *Nature* 16 Mar. 333/2 One pigment formed red laths with a coppery lustre, m.p. 208° (decomp.)... It had the molecular formula $C_{15}H_{14}O_6$, and as it does not seem to have been encountered before, the name 'javanicin' is proposed for it. **1949** ABRAHAM & FLOREY in H. W. Florey et al. *Antibiotics* I. vii. 306 Germination of lettuce seeds was entirely suppressed by a concentration of 1 in 20,000 of javanicin. **1967** E. PARYSKI tr. *T. Korzybski's Antibiotics* II. iii. 1320 Gram-negative micro-organisms..are very slightly sensitive to javanicin.

javanite (dʒā·vănəit). *Geol.* [f. JAVAN *a.* and *sb.* + -ITE[1]; before coining the word in Eng. von Koenigswald had previously coined its Du. equivalent, *javaniet*, in *Natuurkundig Tijdschrift voor Nederlandsch-Indië* (1936) XCVI. 285.] = *JAVAITE.

1957 G. H. R. VON KOENIGSWALD in *Proc. Kon. Nederl. Akad. van Wetensch.* B. LX. 371 Here for the first time we will give a survey of the types we can distinguish among the javanese tektites, for which the name 'Javanites' is proposed. *Ibid.* 380 The Javanites (nom. nov.) from Java. **1963** J. A. O'KEEFE *Tektites* i. 1 Accepted varieties [of tektites] are..javanites from Java. **1964** *Geochim. & Cosmochim. Acta* XXVIII. 793 Forty-two tektites and two 'amerikanites' were..analysed for major elements. Included were 7 moldavites, 1 bediasite, 2 javanites, 15 philippinites, 12 indochinites and 5 australites.

javelin, *sb.* Add: **4. javelin-throwing,** the throwing of a javelin as an athletic field event; also *ellipt.* as *javelin*.

1902 *Daily Chron.* 7 Apr. 3/1 Professors in a university to teach javelin-throwing. **1906** *Westm. Gaz.* 3 May 10/2 Very pretty was the javelin-throwing, the long thin spear being launched high into the air. **1958** *Times* 20 Aug. 2/6 Mrs. Zatopkova..won the women's javelin with a new European record of 183 ft. 9½ in. **1964** J. J. WALSH *Understanding Paraplegia* xix. 128 Throwing and catching a medicine ball, javelin-throwing, and shot-putting, were all of use. **1974** *Country Life* 14 Feb. 292/1 The big break-through..came from young Charles Clover in the javelin as he threw 278 ft 7¼ in.

Javelle (ʒave·l). Also **Javel, javelle.** [ad. *Javel*, name of the village near (now a suburb of) Paris where the solution was first made as a bleach.] *eau de Javelle* (also *water of Javelle*, *Javelle('s) water*): an aqueous solution containing potassium hypochlorite and used as a bleach or a disinfectant; also applied to a similar solution of sodium hypochlorite, which has largely replaced it in modern use. Cf. *LABARRAQUE.

1807 J. A. CHAPTAL *Chem. Arts & Manuf.* III. v. 98 What is known at Paris by the name of *eau de javelle*, is oxy-muriatic acid, combined with an alkali. It is used for taking out the stains of fruit, &c. from linen. **1815** S. PARKES *Chem. Ess.* IV. xii. 57 Some manufacturers at Javelle near Paris announced..that they had discovered a particular liquor which they called the Lye of Javelle, having the property of bleaching cloth by a few hours immersion. This composition..was found to be..a solution of the oxy-muriate of potash. *Ibid.* 61 The manufacturers of Javelle,..having been disappointed in their commercial prospects at home, came over to England, and settled at Liverpool for the purpose of manufacturing the solution of oxy-muriate of potash, which they proposed to sell to the English bleachers in bottles, and which they still denominated the Liquor de Javelle. **1875** *Ure's Dict. Arts* (ed. 7) I. 787 When a weak solution of caustic potash or soda is saturated with chlorine, it affords a bleaching liquor, still used by some bleachers and calico-printers for delicate processes. The chloride of potash is known as Water of Javelle, and the chloride of soda as Labarraque's Liquor. These so-called chlorides are now generally considered to be mixtures or compounds of chlorides and hypochlorites. **1888** J. P. REMINGTON *Pract. Pharmacy* xxxii. 422 Solution of chlorinated soda..is sometimes substituted for *Eau de Javelle* (Javelle's water), a French preparation made with potassium carbonate instead of sodium carbonate. **1892** A. BROTHERS *Photogr.* IV. 260 As a reducing agent for negatives which are too dense and for removing the last traces of sodium thiosulphate, the following solution, which forms *eau de javelle*, may be used:—Chloride of lime 2 ounces, Potassium carbonate 4 ounces, Water 40 ounces. **1928** *Daily Express* 19 Dec. 5 Javelle water, made by dissolving half a pound of washing soda in a quart of cold water, adding four ounces of bleaching powder. **1951** A. GROLLMAN *Pharmacol. & Therapeutics* xxv. 517 Compounds which give off chlorine have long been used as disinfectants, e.g., potassium hypochlorite (Javelle water), alkaline sodium hypochlorite (Labarraque's solution). **1958** L. DURRELL *Balthazar* vi. 117 It's all right..and smells fashionably of armpits and *eau de javel*. **1967** *Martindale's Extra Pharmacopoeia* (ed. 25) 314/2 Eau de Javel (Fr[ench] P[harmacopoeia]) is a concentrated solution of sodium hypochlorite. **1969** B. WEIL *Dossier IX* iii. 22 The bathroom with shower, redolent of *Eau de Javel*.

jaw, *sb.*[1] Add: **1. b.** The parts of certain invertebrates used for the ingestion of food.

1870 H. A. NICHOLSON *Man. Zool.* 163 The Medicinal Leech (*Sanguisuga officinalis*)..has its mouth furnished with three crescentic jaws. **1877** T. H. HUXLEY *Man. Anat. Invertebr. Animals* i. 56 In the Arthropoda, what are usually termed jaws are modified limbs. **1902** *Encycl. Brit.* XXV. 696/1 The jaws of Peripatus are formed by the axis or corm itself. **1932** BORRADAILE & POTTS *Invertebrata* xviii. 567 The lantern [of echinoids] consists of five composite jaws, each clasping a tooth. **1971** J. E. SMITH et al. *Invertebr. Panorama* iv. 55 The gut [of leeches] has a muscularized blood-sucking pharynx often armed with piercing jaws.

6. (Further examples.) Also, a long talk, incessant chatter.

1842 F. J. FURNIVALL in *F. J. Furnivall: Personal Rec.* (1911) p. xi, Had a jaw with Young, which ended, as it began, in nothing. **1846** *Swell's Night Guide* 123/1 *Jaw*, abusive language. **1861** D. G. ROSSETTI *Lett.* (1965) II. 387 We would go to a theatre afterwards or else have a jaw here. **1916** G. B. SHAW *Pygmalion* I. 115 Come with me now and lets have a jaw over some supper. **1964** *Guardian* 2 Mar. 7/6 So before the show starts the promoter gives me a bit of a jaw. **1968** J. R. ACKERLEY *My Father & Myself* viii. 79 He invited the two of us into the billiard-room of Grafton House..for a 'jaw'. **1972** *Times Lit. Suppl.* 14 Apr. 420/3 Without these things, committee work is just endless jaw and empty substitute.

7. *jaw-line, -muscle, -opening*; **jaw clutch,** a claw clutch or a dog clutch; **jaw-jerk** *Med.*, a jerk (JERK *sb.*[1] 2 b (*a*)) of the lower jaw elicited by a downward blow on it when the mouth is open; **jaw-piece,** read: † (*a*) = JOWPY, JOPY; (*b*) (see quot. 1886); **jaw-process** = *GNATHOBASE; **jaw-smith, jawsmith** *U.S. slang*, a talkative person; esp. a loud-mouthed demagogue.

1893 LANGMAID & GAISFORD *Elem. Less. Steam Machinery* vi. 62 A common form of this fitting is the jaw clutch. **1907** Jaw clutch [see *change-speed* (*CHANGE *sb.* 12 a)]. **1911** *Encycl. Brit.* XVIII. 927/1 A clutch of this description can be made to engage without difficulty, there being no fixed positions or steps such as one associates with the ordinary jaw-clutch. **1936** W. STANIAR *Mech. Power Transmission Handbk.* vii. 267 Jaw Clutches.—This type of clutch is employed for moderate and heavy rough driving... It consists of a square or spiral jaw portion which is keyed to the driving shaft and a sleeve portion equipped with square or spiral jaws into which the driving portion can be engaged. With this type of clutch..the pick-up is instantaneous, resulting therefore in shock. **1966** *McGraw-Hill Encycl. Sci. & Technol.* III. 224/1 Although square jaw clutches are the strongest and most elementary to construct, the difficulty of engagement limits their use... A modification..to permit more convenient engagement and to provide a more gradual movement of the mating faces toward each other produces the spiral jaw clutch. **1886** A. DE WATTEVILLE in *Brain* VIII. 518 It does not appear to be generally known that a 'jaw-jerk' can be readily elicited...The phenomenon is clearly of the same nature as that of the 'knee-jerk', and is due to the sudden stretching of the masseter and other muscles of mastication. Hence the name I have ventured to give to it, in preference to the longer and less accurate term mandibular (or masseteric) tendon-reaction (or reflex). **1968** PASSMORE & ROBSON *Compan. Med. Stud.* I. xxiv. 12/2 This jerk [*sc.* the knee jerk] is one of a whole family of tendon jerks... They include the Achilles tendon or ankle jerk..the biceps and triceps jerks.., and the masseter-temporalis or jaw jerk. **1936** J. CURTIS *Gilt Kid* xiv. 144 Perhaps his jaw-line was a little tenser. **1971** *Chatelaine* Aug. 41/2 Tweezing stragglers over nose bridge and applying tawny blusher from brows to jawline livened her skin and slimmed her face. **1890** W. JAMES *Princ. Psychol.* I. x. 301 In *effort* of any sort, contractions of the jaw-muscles and of those of respiration are added to those of the brow and glottis. **1929** W. FAULKNER *Sound & Fury* 140 My jaw-muscles getting numb. **1958** E. FISCHER-JORGENSEN in Saporta & Bastian *Psycholinguistics* (1961) 131/2 Slow jaw-opening might rather be combined with a fortis–lenis difference. **1886** WILLIS & CLARK *Archit. Hist. Univ. Cambr.* I. 283 A 'jaw-piece' or triangular piece of wood..interposed between [the principal] itself and the spars forming the roof. **1881** Jaw process [see *GNATHOBASE]. **1902** *Encycl. Brit.* XXV. 697/1 The usual uni-ramose limb found in the various classes of Arthropoda..varies as to the presence or absence of the jaw-process. **1887** *Chicago Tribune* 13 May 5/2 George Schilling, Socialist and jawsmith. **1910** Jawsmith [see *HOT AIR 2]. **1942** BERREY & VAN DEN BARK *Amer. Thes. Slang* § 422 Talker,..*jawsmith* (esp. a public speaker).

‖ **jawan** (dʒăwā·n). Also **juwan.** [a. Urdu *jawān*.] An Indian soldier.

1839 M. TAYLOR *Confessions of Thug* I. i. 6 Chumpa was busy cooking and the Juwans were all out of the way. **1923** *Blackw. Mag.* July 22/1 The cool deliberately-aimed fire of the jawans lashed them each time. **1962** *Listener* 29 Nov. 895/2 For the public, the cult of the Indian soldier, the Jawan, remains unshaken. **1965** P. ROBINSON *Pakistani Agent* i. 9 A *jawan* in a dark-green uniform..was resting upright against the wall of the cargo shed, with his rifle propped up next to him. **1969** *Amrita Bazar Patrika* 5 Aug. 1/1 Ten persons, including five Security Force personnel, were killed and three jawans were wounded in two incidents of ambush in far-flung areas of the State in the last 48 hours. **1971** *Sunday Times* 13 June 12/2 One of the *jawans* (privates) crouched in the back of the Toyota Land Cruiser called out sharply: 'There's a man running, Sahib.'

jawbation: see JOBATION (in Dict. and Suppl.).

jaw-bone, jawbone. Add: **2.** An animal's jaw-bone used as a musical instrument; also, castanets or a jew's harp.

[**1790** W. BECKFORD *Descr. Acct. Jamaica* II. 387 Their musical instruments..consist of..the jaw-bone of an animal, from which is produced a harsh and disagreeable sound.] **1844** in C. Cist *Cincinnati Misc.* (1845) I. 14/2 Fowler..found the truant..at a dance house..playing the *jaw bones* or Castanets. **1952** B. ULANOV *Hist. Jazz in Amer.* (1958) v. 46 By the end of the 1880s New Orleans Negro musicians were no longer playing jawbones, hide-covered casks, or bamboo tubes. **1970** P. OLIVER *Savannah Syncopators* 109 *Jawbone*, jawbone of a mule, ass, cow or other domestic animal used as a rattle. A North American plantation instrument. The jawbone was also struck or played with a nail or length of iron.

3. Credit. *N. Amer.* (orig. *Canadian*) *slang.*

1862 *Times* 21 Oct. 9/4 Individuals who, in digger's parlance, live on jawbone (credit). **1885** A. S. HILL *From Home to Home* 413 His ready money gone, he has nothing to live on but 'jawbone', *i.e.* credit. **1941** J. SMILEY *Hash House Lingo* 33 *Jawbone*, credit. **1970** *New Yorker* 31 Oct. 130/3 A young Canadian..started this film on a small grant..and apparently finished it on jawbone and by deferring processing costs. **1971** A. P. MCINNES *Dunlevy* 54 No jaw-bone credit is allowed and all bets must be matched with goods.

Hence **jawboning** *U.S. slang*, name applied to a policy, first associated with the administration of President Lyndon Johnson (1963–1969), of urging management and union leaders to adopt a policy of restraint in wage and price negotiations. Also in extended use. Also (as a back-formation) **jaw-bone** *v.*

1966 *N.Y. Times* 2 Jan. IV. 2 Every price increase that happens to catch the public's eye must be 'jawboned' to death by the Government. **1969** *Time* 19 Sept. 32 Since June, Feather has been jawboning his union chiefs on the virtues of labor discipline on the shop floor. **1969** *Time* 10 Oct. 57 As for jawboning, Nixon's Republican advisers consider it unfair and almost immoral to single out individual companies or industries. **1970** *Daily Tel.* 19 June 15 Policy will almost certainly concentrate on 'jawboning', the American tactic of trying belatedly to talk both sides of industry out of outlandish wage and price increases. **1970** *Harper's Mag.* Mar. 48 Lecturing business and labor on their responsibilities to hold down prices and wages—jawboning as it was called in the Johnson Administration—has been foregone. **1970** *Times* 7 Nov. 7/1 He criticized the Nixon Administration's decision on coming to office to drop the practice of 'jawboning', or presidential persuasion, on the prices and wages front.

Jawi (dʒā·wi). [Malay.] Formerly, the Malay vernacular; now, the Malay language written in Arabic script.

1808 *Asiatick Researches* X. iii. 164 The *Basa Jawi* or written language of composition, is nearly the same in all dialects]. **1812** W. MARSDEN *Gram. Malay Lang.* p. xiv, *Jâwi* or *bhâsa jâwi* is..employed in writings to denote the vernacular language of the Malays, especially that of books, as distinguished from all foreign languages. **1920** R. O. WINSTEDT *Colloquial Malay* (ed. 2) 138 The Malay word spelt in Arabic characters is a puzzle to be solved by the possession of a large vocabulary. And it is impossible to draw up a logical system of *Jawi* spelling. **1964** M. TAIB BIN OSMAN in Wang Gungwu *Malaysia* III. xv. 210 Traditional literature as known today is the literature written in the Perso-Arabic script known as *Jawi*. **1972** *Straits Times* (Malaysian ed.) 24 Nov. 21/1 A tombstone found in the same area has inscriptions in Jawi, stating that it was erected in 1292.

jaw-jaw (dʒǭ·dʒǭ), *v.* [Redupl. form of JAW *v.*[1]] *intr.* To talk in a tedious manner or at great length.

Quot. 1831 is an isolated early example.

1831 M. EDGEWORTH *Let.* 14 Apr. (1971) 523 Mrs. Sotheby is *jaw-jawing* in the drawing room to that poor victim Fanny. **1954** *Times* 28 June 8/1 [W. S. Churchill at a White House lunch] Eden's two words are pretty good words. To jaw-jaw is always better than to war-war. **1969** *Manch. Guardian Weekly* 13 Dec. 12 The novelty of the reaction to the latest call for a European Summit is not in any changed assessment of the super-Powers' intentions, but in the feeling that it is time at last to start jaw-jawing.

jaw-jaw (dʒǭ·dʒǭ), *sb.* [Redupl. form of JAW *sb.*[1] 6.] Talking, often with the implication of lengthy and sterile discussion.

1958 *Times* 31 Jan. 6/7 He [*sc.* Mr. H. Macmillan] believed, in the words of Sir Winston Churchill, that 'jaw jaw is better than war war'. **1960** *Economist* 22 Oct. 326/3 The easy going jaw-jaw policy introduced to Lancaster House by Mr Macleod has been successful enough so far. **1963** *Daily Tel.* 7 Aug. 10/2 We seem to be entering an era of 'jaw-jaw'. **1969** *Listener* 10 July 59/2 Here..is one idea..which might actually enliven the jaw-jaw on Radio-4.

jawl, var. *JOL.

jawless, *a.* (Later examples.)

1968 *Times* 19 Dec. 4/8 Lampreys, like hagfish, are surviving members of the jawless fishes, the first group of vertebrates to evolve. **1972** *Sci. Amer.* Nov. 61/3 Primitive jawless ostracoderms belonging to the orders Anaspida and Thelodonti.

jaw's harp, jaws harp, varr. JEWS' HARP.

1880 GROVE *Dict. Mus.* II. 34/1 *Jew's-harp*, possibly a corruption of Jaw's-harp. **1927** *Melody Maker* Aug. 811/1 When I tune my 'Jaws Harp' as a 'Jujulele', I do not seem to get the *full* banjolic effect. **1958** P. OLIVER in P. Gammond *Decca Bk. Jazz* i. 23 Cheap guitars, jaw's harps and harmonicas became available in greater numbers in the 'nineties.

jax: see *JACK *sb.*[1] 19 d; **jaxey, jaxie,** varr. *JACKSY.

jay. Add: **1. b. blue jay,** read: (*a*), a North American jay, *Cyanocitta cristata* (earlier and later examples); (*b*) = ROLLER *sb.*[2] 1.

(*a*) **1709** *Gleanings Anc. Rec. Bristol, R.I.* 18 Mar. in *Narragansett Hist. Reg.* (1885) III. III. 211 The same order shall extend to the killing of blew Jawes [*sic*]. **1731** M. CATESBY *Nat. Hist. Carolina* I. 15 The Blew Jay is full as big, or bigger than a Starling. **1792** J. BELKNAP *Hist. New Hampshire* III. 173 The blue jay, the wood pecker and the partridge..are then seen flying. **1886** *Harper's Mag.* Nov. 877/2 The bell note of the blue-jay comes up from some mysterious haunt. **1961** O. L. AUSTIN *Birds of World* 226/2 Essentially a woodland species fond of open forest, the Blue Jay has become a common resident of the parks and suburbs of most North American cities. (*b*) **1878** T. J. LUCAS *Camp Life & Sport S. Afr.* vi. 83 Conspicuous among them [*sc.* the birds] were..the beautiful blue jay, and the Kaffir finch. **1896** H. A. BRYDEN *Tales S. Afr.* viii. 185 Please don't forget the blue jay [*fn.* the 'roller' is usually called 'blue jay' by colonists] feathers. **1911** *Encycl. Brit.* XV. 298/1 The birds known as blue jays in India and Africa are rollers. **1964** A. L. THOMSON *New Dict. Birds* 411/2 'Blue Jay' is a misnomer in India for *Coracias benghalensis*.

3. d. Also *attrib.* or as *adj.*, dull, unsophisticated; inferior, poor. *U.S. colloq.*

1889 *Daily Even. Bulletin* (San Francisco) 13 July 1/6 Smith has a poor opinion..of St. Joseph, which he alludes to as a 'jay' town of the worst description. **1891** H. C. BUNNER *Short Sixes* 91 'T ain't neuralogy, you jay pill-box, she's *cooked!* **1900** ADE *More Fables* (1902) 185 It was a Shame to String these Jay Amateurs. *a* **1911** D. G. PHILLIPS *Susan Lenox* (1917) II. ii. 23 Gee, what awful jay things we work off on them, sometimes! They can't see the dress for the figure. **1916** H. L. WILSON *Somewhere in Red Gap* viii. 348 Them jay New York newspapers would fall for it. **1942** BERREY & VAN DEN BARK *Amer. Thes. Slang* § 21/10 Small; insignificant,..*jay*. *Ibid.* § 30/4 Poor; mean; contemptible,..*jay*. *Ibid.* § 45/3 Small country town; 'hick' town,..*jay town*.

jay-bird. (Earlier and later Amer. examples.)

1661 *Early Rec. Dedham, Mass.* (1894) IV. 41 En Dani Fisher is creditor to the Towne for his sonne catching of Jaybirds. **1832** J. P. KENNEDY *Swallow Barn* II. iii. 55 A scream of jay-birds heard at intervals. **1896** J. C. HARRIS *Sister Jane* 84, I hear a flutter in the chaney-berry tree and look up and see a jaybird. **1972** D. DELMAN *Sudden Death* (1973) iii. 74 The corpus was naked as a jaybird.

Jaycee (dʒēⁱ·sī·). orig. *U.S.* [f. J + C, initial letters of *junior chamber*.] A colloquial name for a member of a junior chamber of commerce. Also **Jaycee-ette, Jaycette,** a female member of such an organization. Also *attrib.*

1946 in *Amer. Speech* XXI. 295. **1952** *Ibid.* XXVII. 75 *Jaycee-ettes* are the wives of Jaycees—members of the United States Junior Chamber of Commerce. **1969** *Sun-Herald* (Sydney) 13 July 77/1 Blue and white—Jaycee colours—was the theme for decorations at Menzies Sydney Hotel when the Combined Sydney Metropolitan Jaycees held their 2nd annual ball last night. **1971** *Illustr. Weekly India* 25 Apr. 20/3 Through his association with the Jaycees he met Paru, a fellow-Jaycee, whom he married. **1972** *Malay Mail* 25 May 2/3 (*headline*) Four Malaysian Jaycees for Asian youth voyage. **1973** *Sunday Advocate-News* (Barbados) 25 Feb. 14/1 Mrs. Hazel Ann Edwards..[was] named the 1972 Jaycette of the Year... Mrs. Edwards..received her award at the annual Jaycees of Barbados Awards dinner. **1974** *Ibid.* 3 Mar. 5/7 Yesterday Mr. Mann toured projects of the Jaycees and met members of the State Board, Bridgetown Jaycees and Jaycettes at Viamede.

jay-walker (dʒēⁱ·wǭkəɹ). orig. *U.S.* [f. JAY 3 d + WALKER *sb.*[1]] A pedestrian who crosses a street without regard to traffic regulations. Hence **jay-walk** *v. intr.*; **jay-walking** *vbl. sb.*

1917 *Harper's Mag.* June 70/2 The Bostonian..has reduced 'a pedestrian who crosses streets in disregard of traffic signals' to the compact *jaywalker*. **1919** S. LEWIS *Free Air* 257 He had..been cursed by a policeman for jaywalking. **1933** *Bulletin* (Sydney) 1 Nov. 23/4 His car brushed plaintiff, who was jay-walking. **1937** *Times* 25 Jan. 8/2 In many streets like Oxford Street, for instance, the jaywalker wanders complacently in the very middle of the roadway as if it was a country lane. **1957** C. BROOKE-ROSE *Lang. of Love* 15 She jay-walked through the traffic-jam of St. Giles, vaguely hoping to be run over. **1970** P. LAURIE *Scotland Yard* v. 123 A quarter of all London's accidents involve pedestrians—we have no effective jay-walking law. **1972** *Police Rev.* 24 Nov. 1528/3 Realising his mistake, [he] pulls back quickly, narrowly missing the jay-walker. **1973** *Scotsman* 7 Aug. 8/3 Although there are penalties for jay-walking they do not seem to be much needed. Indeed a friend of mine from another part of Canada once jay-walked in the city I refer to and said that she would never do so again because she had been so embarrassed when all the traffic stopped for her.

jazz (dʒæz), *sb.* orig. *U.S. slang.* Also † **jas, jascz, jass, jasz, jaz.** [Origin unknown: see quots. for some of the many suggested derivations. Cf. *JAZZBO.] **1.** A kind of ragtime dance (see quot. 1919[2]); hence, the kind of music to which this is danced; (the usual sense) a type of music originating among American Negroes, characterized by its use of improvisation, syncopated phrasing, a regular or forceful rhythm, often in common time, and a 'swinging' quality (see quots.); loosely, syncopated dance-music.

Connection with Amer. Eng. *jasm* 'energy, enthusiasm' (see Mathews *Dict. Amer.* s.v.) cannot be demonstrated.

1909 C. STEWART *Uncle Josh in Society* (*gramophone-record*), One lady asked me if I danced the jazz. **1913** *Bulletin* (San Francisco) 6 Mar. 16 The team which speeded into town this morning comes pretty close to representing the pick of the army. Its members have trained on ragtime and 'jazz'. **1917** *Sun* (N.Y.) 5 Aug. III. 3/6 Variously spelled Jas, Jass, Jaz, Jazz, Jasz and Jascz. The word is African in origin. It is common on the Gold Coast of Africa and in the hinterland of Cape Coast Castle. *Ibid.* 3/7 Jazz is based on the savage musician's wonderful gift for progressive retarding and acceleration guided by his sense of 'swing'. **1918** *Era* 11 Sept. 21 John Lester's Frisco Five. The Jollities of 'Jazz'. **1919** *Punch* 12 Mar. 193/1 'Whitehall,' says a society organ, 'has succumbed to the Jazz, the Fox-trot and the Bunny-hug.' **1919** 'MONSIEUR PIERRE' *How to Jazz* 7 The Jazz is a three-step dance done to four-beat time. The three steps fall on the first three beats of the bar, the third being prolonged to last two beats, namely, the third and fourth. There are three distinct movements, which may be described as the Straight Jazz, the Side Jazz and the Jazz-Roll. **1920** V. LINDSAY *Daniel Jazz* 1 Daniel was the chief hired man of the land. He stirred up the jazz in the palace band. **1921** W. LE QUEUX *Secret Telephone* 48, I was thoroughly enjoying a delightful jazz with the child. **1922** C. ENGEL *Discords Mingled* (1931) 147 Jazz is rag-time, *plus* 'blues', *plus* orchestral polyphony; it is the combination, in the popular music current, of melody, rhythm, harmony, and counterpoint. **1925** *Amer. Mercury* Sept. 7 According to tradition, jazz has taken its name from Jasbo Brown, an itinerant Negro player along the Mississippi, and later, in Chicago cabarets. **1927** [see *BEETHOVENIZED *ppl. a.*]. **1928** GALSWORTHY *Swan Song* I. iv. 26 'The faster you can move your legs, the more you think you're dancing.'..'You don't like jazz?' queried the young lady. 'I do not,' said Soames. **1930** AUDEN *Poems* 41 In a hot room with the sagging melody of jazz. **1933** *Fortune* Aug. 47/2 Their use of 'jazz' includes both Duke Ellington's Afric brass and Rudy Vallée crooning *I'm a Dreamer, Aren't We All?* **1934** S. R. NELSON *All about Jazz* i. 23 It has been suggested that jas, jass, jaz, jazz, jasz, or jascz were originally part of the patois of the Negro in his native Africa. **1937** *Amer. Speech* XII. 180 Mr. Preston Jackson, negro trombonist of note, says that the 'Creole Jazz Band' was in New Orleans in 1911. Almost as early as that was the 'Original Dixieland Jass Band' which was the first of the lot to have the name printed on a phonograph record, 1917...The word 'Jass' was a verb of the negro patois meaning 'to excite' with an erotic and rhythmic connotation. Later becoming pronounced 'Jazz', it was used attributively to describe bands which by the intensity of their rhythm produced excitement. **1950** *N.Y. Times* 30 June 21 Dr. Bender..was stumped by the word 'jazz'. In..three years..he..tracked it to the West Coast of Africa, the contact point for the slave trade with colonial America. He said that the word meant 'hurry up' in the native tongue, and was first applied in the Creole dialect to mean 'speed up' in the syncopated music in New Orleans. **1952** B. ULANOV *Hist. Jazz in Amer.* viii. 82 When..[Brown's] band came to Chicago, directly from New Orleans, the word 'jass' had a semi-sordid sexual connotation. Chicago Musicians Union officials..thought that labeling this group a jass or jazz band would be a very successful smear. But..the term caught on, and Brown's Dixieland Band became Brown's Dixieland Jass Band. **1955** L. FEATHER *Encycl. Jazz* 50 Improvisation has always been the life blood of jazz. There are critics today who still claim that true jazz cannot be written down. This is not literally true, but it is true that orchestrated jazz, if it is to remain jazz, must retain the same rhythmic feeling, the same concept of phrasing, that is inherent in improvised jazz... Just as the three basic elements of music as a whole are melody, harmony and rhythm, the three additional elements essential to jazz may be said to be syncopation, improvisation and inspiration. **1955** J. JONES in Shapiro & Hentoff *Hear Me Talkin' to Ya* 358 What is jazz? The closest thing I can get to saying what jazz is, is when you play what you feel. All jazz musicians express themselves through their instruments and they express the types of person they are, the experiences they've had during the day, during the night before, during their lives. D. BRUBECK *Ibid.* 361 When there is not complete freedom of the soloist, it ceases to be jazz... If it's spontaneous, it's going to be rough, not clean, but it's going to have the spirit which is the essence of jazz. **1956** M. STEARNS *Story of Jazz* (1957) xxii. 282 We may define jazz tentatively as a semi-improvisational American music distinguished by an immediacy of communication, an expressiveness characteristic of the free use of the human voice, and a complex flowing rhythm; it is the result of a three-hundred-years' blending in the United States of the European and West African musical traditions. **1968** A. DANKWORTH *Jazz* 1 Most jazz is in the form of melodic or rhythmic variations upon a theme. The theme is usually a twelve-bar blues melody, the chorus of a popular dance-tune, or a specially composed theme. **1970** *Melody Maker* 12 Sept. 35/1 The essential motive of all jazz, at least through the end of the 1950s, was to relate to the audience through a steadily pulsing improvisational concept—in other words, to swing. Today we find young jazzmen effecting radical alterations in this objective. Much of the jazz presented by today's innovators avoids the free-flowing 4/4 or 3/4 essence in favour of a beat that is often heavier though not necessarily cruder.

† **b.** A piece of jazz music. *U.S. Obs.*

1920 *Harvey's Weekly* 24 July 14/2 That isn't a keynote; it's a jazz. **1921** *Ladies' Home Jrnl.* Jan. 50 All the latest popular hits..all this season's jolliest jazzes.

c. *spec.* A passage of improvised music in a jazz performance.

1926 [see *BREAK *sb.*[1] 9 c].

2. *transf.* Energy, excitement, 'pep'; restlessness, excitability.

1913 *Bulletin* (San Francisco) 6 Mar. 16 What is the 'jazz'? Why it's a little of that 'old life', the 'gin-i-ker', the 'pep', otherwise known as the enthusiasalum. **1922** *Dialect Notes* V. 142 *Jazz*,..animation, animal good spirits. 'She's just full of jazz.' **1923** L. J. VANCE *Baroque* vi. 34 Only about enough heroin to give every man, woman and child in N'York the jazz for a week. **1924** GALSWORTHY *White Monkey* II. iii. 145 With all the jazz there is about, she'd appreciate somebody restful. **1928** 'J. SUTHERLAND' *Knot* xii. 163 'What is really the matter?' she asked. 'You look extraordinarily queer, and you ought to be full of jazz.'

3. Meaningless or empty talk, nonsense, rot, 'rubbish'; unnecessary ornamentation; anything unpleasant or disagreeable.

1918 *Dialect Notes* V. 25 *Jazz*, talk; 'gas'. College students. **1930** E. POUND *XXX Cantos* vii. 27 'Toc' sphinxes, sham-Memphis columns, And beneath the jazz a cortex, a stiffness or stillness, Shell of the older house. **1936** *Harper's Mag.* Apr. 567 The word jazz has been used to describe every disagreeable phenomenon since the year 1916, when it came into common use. **1944** *Metronome* Apr. 22 Some of them [*sc.* swing musicians] use the noun 'jazz' to denote corn, especially those who are opposed to the Dixieland type of music and sum it up derogatorily with the word 'jazz'. **1953** D. WALLOP *Night Light* III. 153 What do you call that jazz, alpaca or something? **1958** 'E. MCBAIN' *Killer's Choice* (1960) iii. 31 'How was school today, darling?' 'Oh, the same old jazz,' Monica said. **1962** M. BARRETT *Return of Cornish Sailor* x. 129 All this jazz about your poppa and..his old ship, all that stuff. **1969** C. F. BURKE *God is Beautiful, Man* (1970) 20, I asked one of the young men if he understood what had been read from the Bible. His response was that he 'didn't get that jazz'. **1971** B. MALAMUD *Tenants* 165, I read all about that formalism jazz in the library and it's bullshit.

b. Colloq. phr. *and all that jazz*: and all that sort of thing; et cetera.

1959 F. ASTAIRE *Steps in Time* (1960) i. 3 But it's nice to hear, 'How does the old boy do it...why isn't he falling apart?' And all that jazz. **1960** *Punch* 9 Mar. 345/1 Politics, world affairs, film stars' babies and all that jazz, the things that the adult world seems obsessed with, do not interest us at all. **1968** B. TURNER *Sex Trap* ix. 69 Always been a good girl and all that jazz, but a bit stage-

struck. **1972** J. PORTER *Meddler & her Murder* x. 132 Come to identify the body..and all that jazz.

4. *slang.* Sexual intercourse. Cf. *JAZZING *vbl. sb.* 2.

[**1924** *Étude* Sept. 595/3 If the truth were known about the origin of the word 'Jazz' it would never be mentioned in polite society... The vulgar word 'Jazz' was in general currency in those dance halls thirty years or more ago.] **1927** *Jrnl. Abnormal & Social Psychol.* XXII. I. 14 The word *jazz*,.. Used both as a verb and as a noun to denote the sex act,..has long been common vulgarity among Negroes in the South, and it is very likely from this usage that the term 'jazz music' was derived. **1950** A. LOMAX *Mister Jelly Roll* (1952) 47 Winding Boy is a bit on the vulgar side. Let's see—how could I put it—means a fellow that makes good jazz with the women.

5. *attrib.* and *Comb.*, as *jazz ballet, band, banjo, club, cult, dance, -dancing, drum, -drummer, -drumming, fan, festival, joint, king, -land, -lover, music, musician, opera, orchestra, -player, queen, record, scene, -singer, song, tune*; *jazz-conscious, mad* adjs.; *jazz-loving, -minded, -oriented, -struck, -tinged* ppl. adjs.; **jazz age** (freq. with capital initials), the era of jazz; *spec.* (see quot. 1959); **jazz baby,** a girl who is interested in jazz; a flapper; **jazz poem,** a poem that is read aloud to the accompaniment of jazz; so *jazz poetry*; **jazz-rock,** music that has the characteristics of both jazz and 'rock' music. Also *JAZZMAN.

1922 F. SCOTT FITZGERALD (*title*) Tales of the jazz age. **1926** WHITEMAN & MCBRIDE *Jazz* vii. 137 The jazz age has been the subject of profound and careful condemnation. **1959** T. GRIFFITH *Waist-High Culture* (1960) iii. 31 In the years between the Armistice and the stock-market crash, came the period we used to call..the Jazz Age. **1920** F. SCOTT FITZGERALD in *Metropolitan* Oct. 65/2 Evylyn and Ben followed singing a drowsy song about a Jazz baby. **1964** M. MCLUHAN *Understanding Media* (1967) II. xxxi. 348 Baseball..will always remain a symbol of the era of the hot mommas, jazz babies..and the fast buck. **1961** WEBSTER, Jazz ballet. **1972** *Times* 16 May 14/8 We have a jazz ballet by a Canadian choreographer, set to music that uses a string quartet and a rock quartet. **1916** *Ragtime Rev.* Oct. 6/3 The 'Jaz' bands that are so popular at the present time. **1916** *Variety* 27 Oct. 12/4 The Jazz Band is composed of three or more instruments and seldom plays regulated music. **1917** *Era* 20 Aug. 20 Holborn Empire... Frank Powell and The Magleys and the Jazz Band. **1919**, etc. [see *DIXIE[2] 1 c]. **1956** M. STEARNS *Story of Jazz* (1957) vii. 71 The 101 Ranch, a cabaret which employed many jazz-bands, was particularly famous. **1923** Jazz banjo [see *BANJOLIN]. **1917** *Spiker* 25 Dec. 10/3 H. Williams, Rickard, Putney, Short, Hermann and Duncan were the Jazz Club entertainment committee. **1958** *New Statesman* 25 Jan. 102/3 Our native music..still flourishes nightly in the jazz-clubs, though best in those where musicians..like to drop in for a little drinking, gossiping, watching the dancers..and perhaps sitting in with the band. **1968** A. DIMENT *Great Spy Race* vii. 114 Past the pub with built-in Jazz Club. **1956** M. STEARNS *Story of Jazz* (1957) xxiii. 286 With the arrival of jazz-conscious American troops, the murmur of interest grew to a rhythmic roar. **1933** *Fortune* Aug. 47/3 The jazz cult is apathetic to nine-tenths of modern dance music. **1919** *Punch* 30 Apr. 333/3 An early bather was seen executing the Jazz-dance on the beach at Ventnor on Easter Monday. **1963** *Spectator* 27 Dec. 852/1 In America the jazz-dance..has a validity as..a pop-art expression of one side of the national culture. *Ibid.*, The so-called jazz-dancing which has insidiously crept into our ballet repertory. **1919** *Observer* 16 Mar. 14/4 There has been a good deal of curiosity concerning the origin of the term 'Jazz'. Authorities on Jazz dancing say it is a word used by niggers to denote a scramble. **1922** *Encycl. Brit.* XXX. 796/1 The music..consists of various combinations, the most common of which perhaps is:—piano, violin, alto or tenor saxophone, banjo and jazz-drum. **1925** *Chambers's Encycl.* VI. 302/2 The jazz-drummer, a sort of one-man band, provides the characteristic feature of jazz, which is noise. **1930** *Melody Maker* Feb. 123/2 Murray Pilcer—Britain's first 'jazz' drummer and still going strong. **1956** B. EDWARDS in S. Traill *Play that Music* vi. 59 There have been five major stages in jazz-drumming during the last three and a half decades. **1958** D. HALPERIN in P. Gammond *Decca Bk. Jazz* xx. 241 Calling the young man..a jazz-fan would be off-centre; he is, rather, a jazz convert. **1961** *Observer* 16 July 5/5 He caught on, first with his fellow college intellectuals and jazz fans (whom he still calls 'my people') and then with a wider audience. **1959** 'F. NEWTON' *Jazz Scene* xi. 184 'Jazz festivals'—in Newport, Conn., in Nice, Cannes, San Remo and other European holiday resorts. **1970** *New Statesman* 9 Oct. 454/3 The crowd was very similar..to the audience that came to the Beaulieu jazz festivals. **1942** BERREY & VAN DEN BARK *Amer. Thes. Slang* § 366/4 Dance hall,..*jazz joint*. *Ibid.* § 576/27 *Jazz King*, Paul Whiteman, jazz orchestra leader. **1920** *Quill* Sept. 6 They're the best jazz kings in the jazzy world. **1958** *Daily Herald* 24 Mar. 2/4 Gladys Hampton, wife of jazz-king Lionel, arrived in London yesterday. **1942** BERREY & VAN DEN BARK *Amer. Thes. Slang* § 578/2 *Jazzland*, the world of jazz. **1947** *Penguin Music Mag.* May 30 'You can't make a gentleman out of jazz'—a perfectly true statement, and one which all jazz-lovers will applaud. **1974** *Guardian* 22 Mar. 14/6 The European jazz lover can go to Harlem or New Orleans. **1956** M. STEARNS *Story of Jazz* (1957) xvii. 201 Jazz-loving record buyers wore out the grooves. **1924** 'J. SUTHERLAND' *Circle of Stars* xxvi. 273 Far more dangerous to a silly, headstrong child than the jazz-mad boys she chose as habitual companions. **1955** L. FEATHER *Encycl. Jazz* vii. 141 A style that was found palatable by many non-jazz-minded people. **1917** *Sun* (N.Y.) 5 Aug. III. 3/7 Jazz music is the delirium tremens of syncopation. **1922** W. J. LOCKE *Tale of Triona* v. 51 The crash of jazz music welcomed them. **1941** B. SCHULBERG *What makes Sammy Run?* iii. 46 It made me realize again how true jazz music was, how it echoed everything that was churning inside us. **1973** *Listener* 19 Apr. 522/2 The excitement in jazz music is usually concerned with nerve. **1917** *Sun* (N.Y.) 5 Aug. III. 3/7 The jazz musicians and their auditors have the most rhythmic aggressiveness. **1969** Jazz musician [see *BIT *sb.*[2] 4 i]. **1924** *N.Y. Times* 18 Nov. 23/1 Irving Berlin, Jerome Kern and George Gershwin might submit..a jazz opera to..the Metropolitan Opera House. **1970** *New Yorker* 29 Aug. 22/2 Recently, I [*sc.* Rolf Liebermann] commissioned a jazz opera, because I think that is a way to make contact with..young people. **1916** *Variety* 27 Oct. 12/4 The low cost makes it possible for all the smaller places to carry their Jazz orchestra. **1925** *Scribner's Mag.* July 45/1 The faint echo of a jazz orchestra in the background. **1955** KEEPNEWS & GRAUER *Pict. Hist. Jazz* xv. 155/2 The personnel were primarily jazz-oriented musicians. **1918** *Quill* June 31 (Advt.), Music and Entertainment by the original Jackson Jazz Players. **1958** *New Statesman* 25 Jan. 102/3 Jazz-players and promoters..are so much more difficult to handle than the good old-fashioned pit and palais musicians. **1960** *Guardian* 21 Nov. 7/7 A 'jazz poem' read at a recital of modern poetry and jazz. **1959** *Listener* 26 Mar. 567/3 In the current craze for jazz poetry a mistaken attempt is made to bend the verse to the last of the music. **1922** *Canad. Mag.* June 96 What will they make of a literature which is garnished with references to 'snuggle puppers' and 'jazz queens' and 'flappers' and 'face dancers'? **1923** H. CRANE *Let.* 5 Dec. (1965) 159 We had a Victrola... Lots of jazz records, etc. **1970** *Americana Ann.* 578 The year [*sc.* 1969] was flooded with such new combinations as jazz-rock, folk-rock, and country-rock. **1974** *Down Beat* 18 July 26/3 Here are two exploratory jazz-rock albums. **1959** 'F. NEWTON' (*title*) The jazz scene. **1969** Jazz scene [see *BIT *sb.*[2] 4 i]. **1927** (*film-title*) The jazz singer. **1929** A. HUXLEY *Do what you Will* 57 He is employed as a jazz-singer on the music-hall stage. **1923** H. CRANE *Let.* 9 May (1965) 133 Marvelous jazz songs, jokes, etc. **1947** R. DE TOLEDANO *Frontiers of Jazz* vii. 82 The jazz-struck kids who are today the core of the non-commercial white bands. **1955** L. FEATHER *Encycl. Jazz* vii. 227 A honey-toned, jazz-tinged, original song stylist. **1918** F. HUNT *Blown in by Draft* vi. 144 This, he felt deep in his heart, might be a fighting army, but it was a jazz tune army. **1926** [see *CLASSIC *sb.* 1 c].

6. *attrib.*, passing into *adj.* Of grotesque or fantastic design, marked by vivid or riotous colouring; also, lively, sophisticated, unconventional.

1919 *Punch* 7 May 357 Jazz stockings are the latest thing. **1919** *Current Opinion* Aug. 98/3 Boston is only slightly Jazz. **1922** *Glasgow Herald* 14 Dec. 5 He has some justification for using this jazz language. **1923** *Daily Mail* 5 May 8 Jazz patterns in dress. **1927** R. H. WILENSKI *Mod. Movement in Art* III. 165 The 'jazz' curtains and sunshades..and the prevalence of bright tints in the theatre. **1928** E. WEEKLEY *Eng. Lang.* 76 The rather jazz-patterned idiom which is now spoken. **1930** J. COLLIER *His Monkey Wife* xiv. 198 Drawing a jazz silk dressing-gown about her shoulders, she went to the bathroom. **1938** S. M. BESSIE (*title*) Jazz journalism: the story of the tabloid newspapers. **1957** H. CROOME *Forgotten Place* 15 A jazz-patterned carpet on the floor.

jazz (dʒæz), *v.* orig. *U.S. slang.* [Cf. prec.] **1.** *trans.* To speed or liven *up*; to render more colourful, 'modern', or sensational; to excite.

1917 *Sun* (N.Y.) 5 Aug. III. 3/6 In the old plantation days when the slaves were having one of their rare holidays and the fun languished some West Coast African would cry out, 'Jaz her up', and this would be the cue for fast and furious fun... Curiously enough the phrase 'Jaz her up' is a common one to-day in vaudeville and on the circus lot. **1919** *Amer. Mag.* Nov. 69/1 For ways that is dark and tricks which is vain, the daughters of Eve is peculiar, to jazz up a line of Bret Harte's. **1923** *Daily Mail* 27 Mar. 8 My colour scheme is rather fetching, don't you think? X—a famous artist—jazzed it up for me. **1923** WODEHOUSE *Inimitable Jeeves* xv. 195 It's rather too late to alter the thing [*sc.* a little fairy play] entirely, but at least I can jazz it up. **1926** [see *GAS *sb.*[2]]. **1959** *Times Lit. Suppl.* 2 Oct. 557/3 The 'honesty' and the depth of the rejection here carry no conviction and the attempt to jazz it up leads Mr. Sillitoe where we would expect it to lead him—to bluster and sentimentality. **1967** *Surfabout* IV. III. 33/2, I could hardly sleep at all; man, I was jazzed just listening to the hissing swells as they smoothly broke through the night. **1974** *Guardian* 13 June 10/5 He..jazzes the mixture up with a series of film-makers' clichés that one can only describe as stylised film school.

b. To play (music, or an instrument) in the style of jazz. Freq. const. *up*.

1919 E. SCOTT *All about Latest Dances* 76 The nigger bands at home 'jazz' a tune; that is to say, they slur the notes, they syncopate, and each instrument puts in a lot of little fancy bits on its own. **1922** C. SANDBURG *Slabs of Sunburnt West* 6 Listen while they jazz the classics. **1934** *Hound & Horn* VII. 599 The saxophone..can be as 'hot' as the clarinet when it is 'jazzed up'. **1934** C. LAMBERT *Music Ho!* II. 74 A Frenchman or an Italian might have felt some embarrassment about jazzing up the classics. **1965** *Listener* 20 May 738/2 He had jazzed up Weill's music in the modern American manner.

2. *intr.* To play jazz; to dance to jazz music. Hence *transf.*, to move in a grotesque or fantastic manner; to behave wildly (see also quot. 1918[2]).

1918 F. HUNT *Blown in by Draft* vi. 143 Ole Hen Sauser..started, walking up and down the black and white ivories until he had the brown box rocking and swaying and jazzing like eight electric pianos. **1918** *Dialect Notes* V. 25 *To jazz*. 1. To talk to kill time. 2. To walk about to kill time. Rare. 'I *jazzed* around all forenoon.' **1919** E. SCOTT *All about Latest Dances* 80 When the band is 'jazzing' along. **1919** *Punch* 23 Apr. 318/1 She did not ask whether I could jazz, mainly, I think, because I had already danced with her. **1922** *Dialect Notes* V. 142 You mustn't expect to pass your quizzes if you keep jazzing around like this. **1923** *Daily Mail* 18 Apr. 8 There are a good many present-day books that just give the reader a view of the protagonists jazzing across the pages in a vivid pattern of action, passion or crime. **1934** S. SPENDER *Vienna* ii. 23 Where radio crazily jazzes. **1966** 'J. HACKSTON' *Father clears Out* 19 Chester..waltzed, jazzed, did Catherine wheels, [etc.].

3. *trans.* and *intr.* To have sexual intercourse (with). *slang.*

1927 [see *JAZZ *sb.* 4]. **1929** T. WOLFE *Look Homeward, Angel* (1930) II. xiv. 176 Jazz 'em all you like,..but get the money. **1930** J. T. FARRELL in *This Quarter* July–Sept. 193 'She's cute. I jazzed her too,' O'Keefe said. **1931** G. IRWIN *Amer. Tramp & Underworld Slang* 109 *Jazz*,..to have intercourse. **1948** H. MACLENNAN *Precipice* (1949) I. 81 My sister was being jazzed by half the neighbourhood cats by the time she was fifteen. **1968** B. FOSTER *Changing Eng. Lang.* ii. 114 The original verb 'jazz' denoting the human male's most important generic activity (itself probably an African word).

jazzbo (dʒæ·zbo). *U.S. slang.* Also **jasbo, jazz-bo.** [Origin unknown; perh. a corruption of the name *Jasper*; cf. *JAZZ *sb.*] (See quots.)

1917 *Sun* (N.Y.) 5 Aug. III. 3/7 'Jasbo' is a form of the word [*sc.* jazz] common in the varieties, meaning the same as 'hokum', or low comedy verging on vulgarity. **1919** *Quill* June 9 Have you heard Jazzbo the chocolate syncopated Hobohemian at the Moulin Rouge Cave? **1923** H. C. WITWER *Fighting Blood* ix. 272, I merely commence to stutter an apology, when the old jazzbo [= fellow] shuts me off kind of angrily. **1923** *N.Y. Times* 9 Sept. VII. 2/1 *Jazzbo*, bladder and slapstick comedy. **1926** WHITEMAN & MCBRIDE *Jazz* v. 122 Sousa..says jazz slid into our vocabulary by way of the vaudeville stage, where at the end of a performance, all the acts came back on the stage to give a rousing, boisterous *finale* called a 'jazzbo'. **1942** BERREY & VAN DEN BARK *Amer. Thes. Slang* § 438/1 Dissolute person,..*jazz-bo*. *Ibid.* § 583/18 *Jazz-bo*, jazzbo, a negro performer, esp. in a minstrel show. *Ibid.* § 868/3 *Jazzbo*, a colored American soldier. **1944** [see *JIGABOO]. **1957** J. KEROUAC *On Road* (1958) II. i. 113 He dodged a mule wagon; in it sat an old Negro plodding along... He slowed down the car for all of us to turn and look at the old jazzbo moaning along.

jazzed (dʒæzd), *ppl. a.* [f. *JAZZ *v.* + -ED[1].] Played in the style of jazz; hence, enlivened; made more colourful, 'modern', or sensational. Freq. with *up*.

1919 E. SCOTT *All about Latest Dances* 75 Certain steps and movements already in practice may be more or less adaptable to 'jazzed' music. **1926** *Bulletin* 9 June 5 Some of our own jazzed thoroughfares. **1929** *Musical Times* Feb. 129 The music is jazzed-up, restless stuff. **1958** E. BORNEMAN in P. Gammond *Decca Bk. Jazz* xxi. 270 The real Cuban rumba..had next to nothing to do with the jazzed-up songs which had passed under the name 'rumba' in the U.S. **1969** *Listener* 16 Jan. 85/2 Denouncing his inaugural lecture at the University of Malaya (or rather, a tendentious and jazzed-up version of part of the lecture which had appeared in the local Singapore press). **1972** *Country Life* 7 Dec. 1587 (Advt.), Jazzed up drinks aren't quite our style.

jazzer (dʒæ·zəɹ). orig. *U.S. slang.* [f. *JAZZ *v.* + -ER[1].] One who plays or dances to jazz; a jazz fan.

1919 *Current Opinion* Aug. 99/3 The 'klaxon' in particular..as one of the Jazzers explains,..reminds them that they have an automobile. **1922** *Public Opinion* 5 May 418/2 The son of an agricultural labourer has won the second prize as the best jazzer in the village. **1927** *Musical Times* Nov. 978 In the hands of jazzers, syncopation is a ruthless and mechanical defiance of strict time. **1928** A. HUXLEY *Point Counter Point* xxiii. 418 A real complete human being. Not a newspaper reader, not a jazzer, not a radio fan. **1935** *Punch* 22 May 620/2 D'ye think the earliest jazzer Who coined the ragtime strain Was the ill-starred Belshazzar, Or was it Tubal Cain? **1973** *Melody Maker* 31 Mar. 18 Most of the musical action for jazzers is, then, in the studios, away from the public gaze.

jazzetry (dʒæ·zĕtri). [f. *JAZZ *sb.* + PO)ETRY.] The reading aloud of poetry accompanied by jazz; a recital of poetry accompanied by jazz.

1959 *Daily Mail* 17 Feb. 4 A hybrid art-form..known variously as jazzetry, jazz-poetry readings or..just J. & P. **1959** C. LOGUE in *Encounter* June 86 A jazzetry, using two narrative poems of mine and some short epigrammatic fables will have been given at the Royal Court. **1963** *Spectator* 26 July 113 A new art form had been developed called (by Lindsey Anderson) 'jazzetry' and consisting of poetry read to a jazz accompaniment.

jazzify (dʒæ·zifəi), *v.* [f. *JAZZ *sb.* + -IFY.] *trans.* To render jazzy; = *JAZZ *v.* 1 a and b. So **ja:zzifica·tion**; **ja·zzified** *ppl. a.*

1922 [see *crazy flying* s.v. *CRAZY *a.* 5]. **1927** *Daily Express* 9 Nov. 9 In 'Hit the Deck',..British bluejackets kneel down, and, with arms uplifted to heaven, jazzify a negro spiritual. **1928** *Daily Tel.* 28 Feb. 15/1 We had already seen our musical taste jazzified, and our British standards of art and life were being jazzified by foreign films. **1958** E. BORNEMAN in P. Gammond *Decca Bk. Jazz* xxi. 270 The mambo, of course, was part of the gradual 'jazzification' of Cuban folk-music that had begun with the so-called 'rumba'.

jazzily, jazziness: see *JAZZY *a.*

jazzing (dʒæ·ziŋ), *vbl. sb.* (and *ppl. a.*). orig. *U.S. slang.* [f. *JAZZ *v.* + -ING^{1,2}.] **1.** The playing of jazz music; jazz dancing. Also *attrib.* or as *ppl. adj.*

1918 *Red Cross Mag.* Oct. 53/1 When I was singing this, one poor fellow..endeavored to keep time to my jazzing by wiggling his toes! **1919** *Lit. Digest* 26 Apr. 28/1 The negro loves anything that is peculiar in music, and this 'jazzing' appeals to him strongly. **1920** P. GIBBS *Realities of War* VIII. v. 437 There was an epidemic of dancing, Jazzing, card-playing, theatre-going. **1920** *Glasgow Herald* 26 Feb. 9 This business woman of 1920 has nothing of the jazzing featherhead about her appearance. **1920** *Chambers's Jrnl.* 28 Aug. 617/1 Good jazzing partners were scarce. **1922** CHESTERTON *Ballad St. Barbara* 75 Of earth's other tributes are plenty to choose, Tobacco and petrol and Jazzing and Jews. **1928** D. L. SAYERS *Unpleasantness at Bellona Club* vii. 80 They had a much better time than they have now, with all this jazzing and short skirts and pretending to have careers. **1938** *Life* 26 Dec. 52/2 Whiteman's new-fangled jazzing of the classics.

2. Sexual intercourse. *slang.*

1958 MURTAGH & HARRIS *Cast First Stone* xiv. 205 She asked if I wanted to do a little jazzing... I said, 'How much?' 'Two dollars,' she said.

jazzist (dʒæ·zist). orig. *U.S.* [f. *JAZZ *sb.* + -IST.] = *JAZZER.

1926 WHITEMAN & MCBRIDE *Jazz* viii. 181 Jazzists chuckle over lowbrows who say they can't abide classical music and highbrows who squirm when they hear jazz. **1927** *Radio Times* 1 July 1/1 Whether the jazzist ever listens to oratorio with an appreciative ear is possibly a doubtful point. **1941** *New Yorker* 1 Mar. 44/2 Basie makes effective use of tone-shading, a technique which some of our noisier jazzists might do well to cultivate.

jazzman, jazz man (dʒæ·zmæn). orig. *U.S.* [f. *JAZZ *sb.* + MAN *sb.*^{1}] A man who plays jazz.

1926 *Amer. Mercury* Apr. 392/1 Alfredo Casella..bore public testimony in writing that the American jazz men have invented effects that he and his colleagues never dreamt of. **1935** *Vanity Fair* (N.Y.) Nov. 38/1 Our jazzmen have had no attention. **1939** W. HOBSON *Amer. Jazz Mus.* 30 Traditionalists accuse the jazz players..and the jazz men accuse the traditionalists. **1970** [see *JAZZ *sb.* 1]. **1972** J. L. DILLARD *Black English* vi. 262 New Orleans jazzmen.

jazzophile (dʒæ·zofəil). *U.S.* [f. *JAZZ *sb.* + -O + -PHIL, -PHILE.] A devotee of jazz.

1941 *Jazz Information* Nov. 24/2 Many hardshelled jazzophiles. **1959** *Look* 21 July 76/2 But 3,000 brave jazzophiles sat patiently through three hours of liquefied jazz.

jazzy (dʒæ·zi), *a.* orig. *U.S.* [f. *JAZZ *sb.* + -Y^1.] Pertaining to or resembling jazz; characterized by jazz; spirited, lively, exciting; vivid, gaudy. Hence in more pejorative senses: 'corny', false, phoney.

1919 *Quill* Apr. 8 Gil Boag promises the Jazziest dance of the season. **1920** *Collier's* 13 Mar. 57/2 Bergstrom's two-piece orchestra was in the throes of its jazziest foxtrot number. **1922** *Dialect Notes* V. 142 She's the jazziest girl in Pem East. **1924** GALSWORTHY *White Monkey* I. ix. 76 'Whom do you think to meet him, besides Alison?' 'Nothing jazzy.' *Ibid.* III. iii. 238, I should like to change my bedroom curtains to blue... The present curtains really are too jazzy. **1925** *Chambers's Jrnl.* 466/2 To sing some jazzy stuff called 'Alexander's Rag Time Band'. **1925** D. H. LAWRENCE *Refl. Death Porcupine* 82 Inside it, the worms will jig the same jazzy dances. **1928** 'J. SUTHERLAND' *Knot* xv. 204, I may be frivolous and modern and jazzy and all the things you clever people hate. **1934** [see *CORNINESS]. **1937** *Amer. Speech* XII. 46/2 *Jazzy*, outmoded, showy, ostentatious style of playing. **1944** *Metronome* Apr. 22 Most musicians use the adjective 'jazzy' to denote 'corny'. **1957** *Sunday Mail* (Glasgow) 10 Feb. 11 Poor ole-fashioned jazz is almost a dirty word with the kids. So Jazzy means phoney or false. **1959** *Times Lit. Suppl.* 2 Oct. 556/3 The jazzy ebullience of the United States seems curiously out of date. **1961** C. MCCULLERS *Clock without Hands* vii. 149 Hard as it was for Jester to make up jazzy hurtful remarks, he was learning to do it. **1963** N. & J. KANTROWITZ in A. Dundes *Mother Wit* (1973) 351 *Jazzy motherfucker*..describes someone fluent, glib, animated. **1967** E. SHORT *Embroidery & Fabric Collage* ii. 54 Beach clothes can take bold areas of really 'jazzy' colours. **1967** *Melody Maker* 27 May 10/3 The material he works over on eleven tracks isn't the jazziest ever. **1971** *Homes & Gardens* Aug. 32 Jazzy colours are confined to the bathrooms and kitchen, where they make a vibrant contrast to the other rooms. **1973** B. BROADFOOT *Ten Lost Years* xxvii. 311 He bought a new car, the jazziest in Calgary.

Hence **ja·zzily** *adv.*; **ja·zziness.**

1921 J. C. LINCOLN *Galusha the Magnificent* xv. 253 They danced jazzily in the hotel parlor and on the porches. **1927** *Melody Maker* Sept. 847/2 Quite as full of jazziness of song and dance as it is now. **1928** *Ibid.* Feb. 183/2 The slow tempo and complete absence of 'jazziness' is the ideal treatment for a melody number. **1928** *Gramophone* VI. 300/1 With a musically artistic legato rhythmic swing and not 'jazzily'. **1951** *Archit. Rev.* CIX. 220/2 The freedom of handling, the faith in elementary cubic forms, the occasional jazziness of detail. **1959** *Encounter* Nov. 60/1 When the ballets met with resistance, it often hinged on this question of jazziness. **1968** D. E. ALLEN *Brit. Tastes* v. 126 These jazzily unnerving designs and patterns.

J-curve: see *J-SHAPED *a.*

jean. Add: **2. b.** (Earlier and later examples.) Now usu. close-fitting trousers of this (or other) material. See also *blue-jean* s.v. *BLUE *a.* 13.

1843 R. S. SURTEES *Handley Cross* I. xiii. 276 Septimus arrived flourishin' his cambric, with his white jeans strapped under his chammy leather opera boots. **1846** S. F. SMITH *Theatr. Apprenticeship* 48 My friend in the jeans and white hat. **1873** 'MARK TWAIN' & WARNER *Gilded Age* (1874) i. 19 They were dressed in homespun 'jeans', blue or yellow. **1904** G. V. HOBART *Jim Hickey* ii. 35 Wouldn't we be a nice pair of turtles to stand around with coin in our jeans and see a nice girl like Amy getting the ice? **1923** J. GALSWORTHY *Captures* 62 He wore, not white ducks, but blue jeans. **1936** WODEHOUSE *Laughing Gas* xvii. 187 No doubt this fiend in butler's shape was even now on his way east with the stuff in his jeans, gone beyond recall. **1957** *Times* 12 Nov. (Canada Suppl.) p. xv/4 For miles and miles of suburban area you will rarely see a young woman out of blue jeans, shorts or slim-jim pants during the day. **1958** *Economist* 11 Jan. 94/1 Girls in tight jeans and dazzle socks. **1958** *Daily Herald* 24 Mar. 2/6 In bright red jeans and thick nautical sweater, secretary Fiona stood on the deck of her Thames-side barge. **1969** I. & P. OPIE *Children's Games* xi. 317 Little girls, dressed in T-shirts and jeans.

c. *jean boot, trousers* (examples); **jean-age** [cf. *TEEN-AGE *a.*], the age at which a young person is likely to wear jeans; so *jean-aged* adj.; *jean-ager.*

1960 *Guardian* 15 Sept. 9/1 A full-house audience, ranging from the peerage to the jean-age. **1967** *Observer* (Colour Suppl.) 30 Apr. 34/2 (*heading*) Up to jean-age. *Ibid.*, The most comprehensive analysis yet of the difficult jean-age years. **1962** *Times Lit. Suppl.* 18 May 359/2 Mr. Kerouac is..far more than the mere apostle of jean-aged, teen-aged..American youth. **1959** *Design* Oct. 29/2 Pony-tailed jean-agers. **1961** *Times Lit. Suppl.* 1 Dec. p. iv (*heading*) Tales for jean-agers. **1849** THACKERAY *Pendennis* I. xxiv. 231 His jean-boots, with tips of shiny leather. **1860** *Observer* 16 Sept. 3/6 The disinterested and brave liberator of Italy, in his red shirt, in a dirty pair of jean trowsers, and worn-out boots. *c* **1871** J. ALBERY *Dramatic Works* (1939) I. 234 He..wears..brown, soiled jean trousers.

Hence **jeaned** *a.*, clad in jean trousers.

1970 'E. LINDALL' *Gathering of Eagles* iii. 30 Van Jordan stood up and looked down at him, legs wide and hands on her jeaned hips. **1971** D. WALLIS *Bad Luck Girl* I. i. 8 She forked out a length of jeaned thigh to fall against his own.

jean(n)ette (dʒăne·t). Also **jean(n)et.** [f. JEAN + -ETTE.] A name for various types of material resembling jean (see quot. 1950).

1785 *Daily Universal Reg.* 1 Jan. 4/2 Half-ell printed jeanets. **1862** *Illustr. Catal. Internat. Exhib., Industr. Dept., Brit. Div.* II. No. 3667 Spinners and manufacturers of jeannettes. **1882** CAULFEILD & SAWARD *Dict. Needlework* 278/2 *Jeanette*, a variety of jean, coarser in quality, yet not so closely woven. Some Jeanettes are twilled. **1950** '*Mercury*' *Dict. Textile Terms* 286 *Jeanettes*, a cloth similar to jean, but usually finer yarns... There is very little difference between jeans and jeanettes. Many makers make no distinction at all. *Jeannet*, a strong coarse fabric, made with cotton warp and wool weft in a twill weave, especially for working clothes for use in cold weather.

jebel (dʒe·bel). Also **djebel, djibel, gebel, jabal.** [a. colloq. Arab. *jebel*, classical Arab. *jabal* mountain.] A hill or mountain; freq. in specific names.

[**1600** J. PORY tr. *Leo Africanus' Geogr. Hist. Afr.* II. 58 (*heading*) Of the mountaine of Iron, commonly called Gebelelhadih. **1834** *Penny Cycl.* II. 210/1 The mountainous region in the interior is distinguished by the appellation of Jabal, 'the Hills'.] **1844** A. W. KINGLAKE *Eothen* xxvii. 387 For a part of two days I wound under the base of the snow-crowned Djibel el Sheik. **1848** H. MARTINEAU *Eastern Life* I. iv. 53 It [*sc.* the causeway] extends from the river bank to the town, and thence on to the Djebel (mountain) with many limbs from this main trunk. **1920** *Blackw. Mag.* Nov. 667/1 An enormous waste of rocky jebels and broad sandy plains. **1924** *Ibid.* Oct. 562/1 The mountains or gebels, to give them their local name, are honeycombed with caves. **1927** *Chambers's Jrnl.* 535/1 A low-lying bank of white mist wound itself round the gaunt jabal. **1942** F. STARK *Lett. from Syria* 186 He thought I was referring to the Intelligence in the Jebel Druse! **1968** *New Scientist* 30 May 447/3 Swarms of desert locusts have whirled up like djinns from the moist jebels on both sides of the Red Sea. **1970** *Guardian* 3 Aug. 3/1 The annual monsoon..hits the Jebel, the mountain range behind Salala.

jeep (dʒīp), *sb.* orig. *U.S.* [f. the initials *G.P.* (dʒī pī) 'general purpose', prob. influenced by the name 'Eugene the Jeep', a creature of amazing resource and power, first introduced into the cartoon strip 'Popeye' on 16 March 1936 by his creator E. C. Segar.] A small, sturdy, four-wheel-drive army vehicle, used chiefly for reconnaissance; a similar vehicle in non-military use; hence (*colloq.*) any vehicle. Also *attrib.* and *Comb.*

For an earlier application in 1937 of the term to a commercial motor vehicle, see **1944** *Amer. N. & Q.* May 26/2 and June 43/1.

1941 *Amer. Speech* XVI. 166/2 *Jeep*, a term applied to bantam cars, and occasionally to other motor vehicles; in the Air Corps, the Link trainer; in the Armored Force, the 1½ ton command car. **1941** J. DANIELS *Tar Heels* 47 Beer wagons moved on the road with the brown jeeps of soldiers and marines. **1942** *News Chron.* 7 Apr. 4/5 The light armoured car which hauls the gun and carries the gun crew is called a jeep. **1942** *R.A.F. Jrnl.* 3 Oct. 24 The Canadians..put their dollars on a Jeep. There were a few of these contraptions beetling about the countryside. **1943** *Archit. Rev.* XCIV. 52 The Jeep is no family car, but it has points that private-car designers might well study. **1945** *Trade Marks Jrnl.* 22 Aug. 423/2 Jeep... Motor cars, and parts thereof..Willys-Overland Motors Inc.., Toledo, State of Ohio, United States of America; manufacturers. **1946** *R.A.F. Jrnl.* May 174 The men and machines, drivers and Jeeps..have finished their job. **1950** G. GREENE *Third Man* iv. 37 A jeep came tearing round the corner and bowled him over. **1952** MORIN & SMITH tr. *Herzog's Annapurna* ii. 30 The jeep now took us along a dusty stony road. **1954** D. DODGE *Lights of Skaro* vi. 213 We were picked up by a jeep-load of soldiers. **1955** *Sci. News Let.* 18 June 396/3 The cosmetic preparations women use to remove superfluous hair might help speed recovery from jeep disease. This is the slow-healing painful infection at the base of the spine which thousands of World War II soldiers developed from riding in jeeps. Doctors called it pilonidal sinus. **1958** *Punch* 22 Jan. 141/1 We flew over more icebergs and then landed at Gander, Newfoundland, where people in woolly tartan jackets drove around in jeeps. **1960** *Guardian* 25 Feb. 9/4 The arrival of jeep-loads of armed police. **1966** F. SHAW et al. *Lern Yerself Scouse* 64 Here comes the police jeep. **1967** WODEHOUSE *Company for Henry* ix. 155 Repeated inquiries as to whether his darned jeep couldn't do more than three miles an hour. **1971** *Country Life* 19 Aug. 431/1 There are no paths, pony tracks or jeep roads to this inaccessible spot. **1973** *Daily Mirror* 24 Jan. 15 On the US side, ADSIDS (air delivered seismic intrusion detectors) dropped by plane, could tell a Communist jeep from a truck. **1973** J. LEASOR *Host of Extras* vii. 127 The Jeep driver started his engine... The Jeep shuddered.

jeep (dʒīp), *v.* orig. *U.S.* [f. the sb.] *intr.* To travel by jeep. So **jee·ping** *vbl. sb.*

1942 *Time* 28 Sept. 57 Yanks: drinking English tea; jeeping in the Middle East. **1945** *Picture Post* 14 Apr. 7 Sometimes, it took us hours to jeep a few miles. **1959** *Times Lit. Suppl.* 27 Feb. 115/4 Jeeping across the mountains on sheep tracks, ingeniously harrying the Germans. **1966** M. R. D. FOOT *SOE in France* xii. 406 An attempt to harass the impending German retreat by jeeping and ambushing.

jeepable (dʒī·păb'l), *a.* [f. *JEEP *sb.* + -ABLE.] Negotiable by jeep.

1944 *Daily Tel.* 12 July 4/6 A new word is now being used on the maps of the South-East Asia Command. It designates tracks impassable to vehicles other than jeeps. These are marked as 'jeepable'. **1946** G. HANLEY *Monsoon Victory* vii. 71 The infantry battalions took off their jackets and started to lay thousands of logs across the mud in an endeavour to make the road even 'Jeepable'. **1960** D. MORAES *Gone Away* 195 There are two passes into Tibet: the Jelep La and the more strategically important Nathu La, to which a jeepable road has been constructed from Gangtok. **1967** *Guardian* 19 Jan. 9/4 Now at least many roads are jeepable.

jeepers (dʒī·pəɪz), *int. slang* (orig. *U.S.*). Also **jeepers-creepers** (-krī·pəɪz). [Corruption of JESUS.] = *JEEZ(E *int.*

1929 W. D. EDMONDS *Rome Haul* ii. 24 Jeepers! A cat would n't stand no show at all. *Ibid.* 30 Spinning swore. 'I'll bet that's right. Jeepers Cripus! How can they expect us to help a marshal if he don't let us know who he is?' **1930** *Amer. Speech* VI. 98 *Jeepers*, an expression of astonishment. 'By Jeepers!' **1937** *Sat. Even. Post* 13 Feb. 13/2 Jeepers Creepers! Where are you going to find a couple of goats and a red wagon on Christmas Eve in three hours? *Ibid.* 82/2 Jeepers, what a story! **1938** MERCER & WARREN (*song-title*) Jeepers creepers. **1939** A. HUXLEY *After Many a Summer* II. ix. 265 'Jeepers Creepers!' he said to himself, remembering the expression on Mr. Stoyte's face. **1940** 'N. BLAKE' *Malice in Wonderland* I. vii. 94 'I think you're very pretty.' 'Jeepers-creepers! Put some conviction into your voice, my pet.' **1959** C. MACINNES *Absolute Beginners* 57, I put my head around the door, and jeepers-creepers, nearly had a fit. **1972** J. AIKEN *Butterfly Picnic* vii. 129 Jeepers, Fernand, we had a bullfight... It was real great!

jeepney (dʒī·pni). *Philippine Islands.* [f. *JEEP *sb.* + *JIT)NEY.] A jitney bus converted from a jeep.

1961 in WEBSTER. **1963** *New Yorker* 8 June 137 A flock of jeepneys—United States Army surplus jeeps the Filipinos have converted into buses. **1972** *Asian* 26 Nov.–2 Dec., The Philippine *jeepney*, the jeep bus (which derives its name from the New Orleans *jitney*, or five cent ride..).

jeerga, var. *JIRGA.

Jeeves (dʒīvz). The name of a character in the novels of P. G. Wodehouse represented as the perfect valet, used allusively. Hence **Jee·vesian, Jeeves-like** *adjs.*, resembling Jeeves.

1930 E. WAUGH *Labels* iii. 46 The stewards..maintained a Jeeves-like standard of courtesy and efficiency. **1952** 'R. GORDON' *Doctor in House* xiii. 149 I'll..get my evening clothes out, and appear as the Jeeves of the chafing dish. **1957** M. SUMMERTON *Sunset Hour* i. 7 That dummy Jeeves we've hired for the evening. **1962** *Observer* 24 June 24/8 A periphrastic and Jeevesian repetition. **1970** *Times* 8 Apr. 18/3 (Advt.), Jeeves required to attend the needs of wealthy bachelor. **1972** *Listener* 27 Jan. 125/2 His Jeeves-like valet. **1972** E. ROUTLEY *Puritan Pleasures of Detective Story* xii. 140 Wimsey is buried in a trench and rescued by his batman. The batman becomes his Jeeves. In Bunter there is about 70 per cent Jeeves.

Jeez(e (dʒīz), *int. slang* (orig. *U.S.*). Also **Geez(e), Jese, Jez,** and with lower-case initial. [Corruption of JESUS.] = *GEE *int.*^{2}

1923 G. EMERY in A. H. Quinn *Contemp. Amer. Plays* 252 Gee-z—it's a cold night, I'll tell the world. **1930** J. DOS PASSOS *42nd Parallel* I. 73 Jez we got to get us women. **1932** L. GOLDING *Magnolia St.* III. vi. 546 'Jeeze!' he said. 'It's going to be a swell party!' **1934** J. BROPHY *Waterfront* i. 17, I wish you wouldn't say, Jeez. **1937** N. MARSH *Vintage Murder* xxi. 238 'Aw, Geeze!' said Wade disgustedly. 'What a case!' **1939** I. BAIRD *Waste Heritage* III. 40 He muttered, 'Jese, Eddy, I don't know whether I ought to apologise to you or turn you over to the cops.' **1946** 'S. RUSSELL' *To Bed with Grand Music* iii. 45 Jeez, I certainly am proud to be seen out with you. **1965** D. LODGE *Brit. Mus. is falling Down* x. 164 Geeze, was that you? What were you doing up there? **1968** 'A. HAILEY' *Airport* I. iii. 43 At an adjoining table, a woman said loudly, 'Geez! Lookit the time!' **1970** *Private Eye* 2 Jan. 12 Jeez, that's nice of you to say so. **1971** *Melody Maker* 18 Dec. 30/6 Jeez, this travels.

jeff (dʒef), *sb.*^{2} Also **Jeff Davis.** [f. *Jefferson Davis* (1808–89), president of the Confederate States 1861–5.] A derogatory term for a man, usu. a 'hick' or a bore; esp. used by American Blacks of white men. Also *attrib.*, as *jeff artist, hat.*

1870 O. LOGAN *Before Footlights* 202, I thought perhaps they imagined I was a female Jeff Davis, and were going to make a '*charge a la bayonette*' instanter. **1917** E. E. CUMMINGS *Let.* 4 June (1969) 26, I escaped repairing with the bums, mutts and Jeffs. **1938** C. CALLOWAY *Hi De Ho* 16 *Jeff*, a pest, a bore, an icky. **1946** MEZZROW & WOLFE *Really Blues* (1957) 375 *Jeff Davis*, an unenlightened person, a hick from down South; sometimes shortened to *jeff*. **1952** BERREY & VAN DEN BARK *Amer. Thes. Slang* (ed. 2) (1954) 391/3 *Jeff Davis, jeff*, a Southern 'hick'. **1969** *Publ. Amer. Dial. Soc.* LI. 29 Names used exclusively by Negroes..jeff, jeffer, jeff davis, jeff artist. **1970** C. MAJOR *Dict. Afro-Amer. Slang* 70 *Jeff*,..a white person;..a dull person; a horrible square. **1973** *Black World* Apr. 57 He wears a jeff hat and a light raincoat.

jeff, *v.* Add: (Earlier and later examples.) Hence **je·ffing** *vbl. sb.*

1837 *Baltimore Commercial Transcript* 7 Nov. 2/1 (Th.), We move that the printers of the U.S. divide off in halves, and 'jeff' to see which shall go to digging ditches or picking stone coal for a living. **1841** W. SAVAGE *Dict. Art of Printing* 428 *Jeff*. See *Throw*. **1875** J. SOUTHWARD *Dict. Typogr.* (ed. 2) 58 *Jeffing*, throwing with quads... One of ..[the party interested] takes up the quads, shakes them ..and throws them..after the manner of throwing dice, when the number of quads with the nicks appearing uppermost are counted,..the highest thrower being the winner. **1884** J. GOULD *Letter-Press Printer* (ed. 3) 166/1 *Jeff*, to throw for a choice with quadrats instead of dice. **1892** A. POWELL *Southward's Pract. Printing* (ed. 4) lxv. 577 In the old companionship system, the fat [*sc.* easy work] is distributed by 'jeffing', or 'throwing quads'. **1942** BERREY & VAN DEN BARK *Amer. Thes. Slang* § 526 *Jeff*, to play dice with em quadrates. **1947** E. HOWE *London Compositor* 24 A custom [*sc.* playing at quadrats] known in the nineteenth century as 'Jeffing'. **1967** F. J. M. WIJNEKUS *Elsevier's Dict. Printing & Allied Industr.* 177/2 *Jeffing*, gambling with nine one-em quadrats, i.e., to throw quards [*sic*] like dice, using the nick side, appearing uppermost, representing one and the other sides blanks. It is a very old custom, but now almost entirely out of practice.

Jeffersonian, *a.* and *sb.* Add: **A.** *adj.* (Earlier and later examples.)

1800 *Connecticut Courant* 6 Oct. 1/3 The Jeffersonian party boast, that the increase of Jacobinism has been great in the Northern part of the United States, within the last year. **1806** *Balance* (Hudson, N.Y.) V. 35/3 Occlusion.. is a Jeffersonian word. **1813** *Niles' Reg.* IV. Suppl. 65/1 This is true Jeffersonian, Madisonian, democratic economy. **1838** *U.S. Mag & Democratic Rev.* Jan. 145 Jeffersonian republicanism. **1972** *Science* 16 June 1223/3 The group takes a Jeffersonian view of the virtues of rotation in office and recommends that a committee member's service be limited to a 3-year term. **1972** *Listener* 21 Dec. 858 They didn't say: 'Yessir, Mr Jefferson, we are going to conquer the North-West for Jeffersonian democracy.'

B. *sb.* (Earlier and later examples.)

1799 *Spectator* (N.Y.) 3 Apr. 1/1 Rouse, ye insurgents, rioters, refugees, and deserters! Ye friends to liberty and equality... Ye Jeffersonians, Gallatonians, Nicholites. **1803** *Fredericktown* (Maryland) *Herald* 30 Apr. 3/3 A thorough going Jeffersonian. **1948** *Antioch Rev.* Spring 10 American Federalists openly sympathized with British Tories and American Jeffersonians with French Girondists.

Jeffrey (dʒe·fri). The name of John *Jeffrey* (d. ?1854), British plant-collector, used *attrib.* or in the possessive to designate *Pinus jeffreyi*, a large pine with a spreading head of drooping branches, collected by him in California in 1852.

1858 G. GORDON *Pinetum* 198 *Pinus Jeffreyii*, Hort. Jeffrey's Pine... A noble tree, growing 150 feet high. **1908** N. L. BRITTON *N. Amer. Trees* 25 Black Pine, *Pinus Jeffreyi* A. Murray, also known as Jeffrey's pine..occurs on dry volcanic mountains from southern Oregon through California. **1931** *Discovery* XII. 92/1 The insect [*sc.* the pandora moth] attacks only pines, its principal hosts being western yellow pine (*Pinus ponderosa*) and Jeffrey pine (*P. jeffreyi*). **1942** N. A. BOWERS *Cone-bearing Trees Pacific Coast* 107 As compared with the Yellow Pine, the Jeffrey cones are much denser; they are shaped like the old-fashioned beehive. **1965** *Listener* 20 May 742/3 There are a good many specimen trees of considerable age—Jeffrey's pine, the Bishop pine. **1967** N. T. MIROV *Genus Pinus* iii. 159 *Pinus jeffreyi*, commonly known as 'Jeffrey pine', has been considered by some botanists as a variety of *P. ponderosa*, while other specialists maintain it to be an independent species.

Jehoshaphat (dʒīhǫ·ʃăfæt, -săfæt). orig. *U.S.* Also **Jehosaphat,** etc. A biblical name (2 Sam. viii. 16, etc.) used interjectionally as a mild expletive. Freq. *jumping Jehoshaphat.*

1857 S. A. HAMMETT *Sam Slick in Texas* xxiv. 161 'Jehosophat!'.. 'Easy over the stones, Joe,' ses I. **1866** MAYNE REID *Headless Horseman* vi. 39 Geehosofat! what a putty beest it air! *Ibid.* xviii. 100 By the jumpin' Geehosofat, what a gurl she air sure enuf! **1876** [see GREAT *a.* 12 c]. **1898** J. D. BRAYSHAW *Slum Silhouettes* 123 'Oh, Jehosephat!' cried the old man, with a chuckle. **1935** G. HEYER *Death in Stocks* iv. 41 'Great jumping Jehoshaphat!' he exclaimed. 'Who did it?' **1972** *Guardian* 16 Dec. 10/1 Why Jeehosophat, Houston, you surely have hit the jack-rabbit on the head.

Jehovah. Add: **2.** *Jehovah's Witness*, a member of a fundamentalist millenary sect, the Watchtower Bible and Tract Society, founded *c* 1879 (under the name 'International Bible Students') by Charles Taze Russell (1852–1916), which rejects institutional religion and refuses to acknowledge the claims of the State when these are in conflict with the principles of the sect. Occas. *Jehovah Witness.*

[**1932** J. F. RUTHERFORD *Health & Life* (Watchtower Bible & Tract Soc.) 7 This commandment is written, in Isaiah 62:10, directed to Jehovah's faithful witnesses.] **1933** M. S. CZATT (*title*) The International Bible Students, Jehovah's Witnesses. **1941** H. G. WELLS *You can't be too Careful* IV. i. 223 A single declared Fascist or Communist or Jehovah's Witness or Single Taxer. **1944** G. B. SHAW *Everybody's Pol. What's What?* v. 46 May he choose..a tribal idol as the sect called Jehovah's Witnesses now do? **1955** W. GADDIS *Recognitions* II. i. 305 It's like Jehovah's Witnesses when you sit down at a table there, everybody comes over. **1961** *Times* 5 May 20/6 A rhesus-positive baby whose father, a Jehovah Witness, refused to consent to a blood transfusion. **1962** *Lancet* 7 Apr. 747/2 Jehovah Witnesses will usually accept their own blood at operation, though they would not take that of a donor. **1973** *Guardian* 14 Mar. 7/4 A mother who divorced her husband because of his adultery is to lose her children because of her religious beliefs. She is a Jehovah's Witness. *Ibid.*, To bring up the children under the Jehovah Witness cult would tend to isolate them.

jejuno-. Add: **jeju:noi·leum** [prob. ad. F. *jéjuno-iléon* (J. Cruveilhier *Traité d'Anat. path.* (1849) I. 718)], the small intestine exclusive of the duodenum; the jejunum and ileum considered together; **jeju:nojejuno·stomy** [*-STOMY], the operation of joining, and creating a passage between, two parts of the jejunum so that the intervening part is bypassed; **jejuno·stomy** [*-STOMY], the operation of attaching the jejunum to the abdominal wall and making an opening, through which the patient may be fed, directly into the jejunum from the exterior; also, the opening thus made.

1876 DUNGLISON *Dict. Med. Sci.* (rev. ed.) 561/2 Jejunoileum. **1955** R. T. SHACKELFORD *Bickham-Callander Surg. Alimentary Tract* II. vii. 999 The jejunoileum..is coiled in a complicated fashion. **1925** DORLAND *Med. Dict.* (ed. 13) 595/2 Jejunojejunostomy. **1955** R. T. SHACKELFORD *Bickham-Callander Surg. Alimentary Tract* I. ii. 330 Most surgeons do not perform jejunojejunostomy routinely with antecolic gastrojejunostomy. **1885** *Brit. Med. Jrnl.* 5 Dec. 1063/2 For the operation of jejunostomy, as he [*sc.* C. H. Golding-Bird] termed the one that he detailed, he claimed that..it was..the best palliative operation for pyloric cancer. **1948** T. H. SOMERVELL *Surg. Stomach & Duodenum* xxvii. 500 The patient is well nursed, and nourished through the jejunostomy. **1971** CAREY & ALBERTIN *Ellison's Atlas Surg. Stomach & Duodenum* ix. 125 Even though parenteral hyperalimentation has decreased the need for jejunostomy, it remains a valuable procedure. *Ibid.* 130 Feedings through the jejunostomy should not be attempted until the return of normal bowel activity.

Jekyll (dʒe·kil, dʒe·k'l). The name of the hero of R. L. Stevenson's story, 'Strange Case of Dr. Jekyll and Mr. Hyde' (published 1886), who appears as a benevolent and respectable character under the name of *Jekyll* and the opposite under the name of *Hyde*: used allusively in reference to opposite sides of a person's character or to persons or things of a dual character, alternately good and evil. Also *attrib.* (Cf. *HYDE.)

?**1885** F. STEVENSON *Let.* in J. Pope-Hennessy *R. L. Stevenson* (1974) ix. 179 While one side represents an angel, the devil must have posed for another... Plainly Jekyll and Hyde. **1887** *Puck* (U.S.) XXII. 188 Is that you, Livingston?.. No, m' dearsh, it'sh Doct' Hyde. Mist' Jekyll didn't..g-g' out t'night! **1902** *Daily Chron.* 22 May 3/4 While the left lobe is the Jekyll of the intellect, the right, on occasion at least, is apt to play the part of Hyde. **1905** *Strand Mag.* Apr. 455/2 Meeting a young and winsome feminine counterpart of Dr. Jekyll and Mr. Hyde in real life is a very pleasant, if novel, experience. **1915** 'I. HAY' *First Hundred Thousand* xiv. 187 When he is good he is very good indeed, and when he is bad he is horrid. He is either Jekyll or Hyde. *Ibid.*, But we encountered surprisingly few Hydes. Nearly all were Jekylls —Jekylls of the most competent and courteous type. **1929** W. J. LOCKE *Ancestor Jorico* xviii. 253 Suppose it pleased him to lead a Jekyll and Hyde sort of life? **1931** *Times Lit. Suppl.* 2 July 522/1 Turner was a case of Jekyll and Hyde in real life and oscillated continuously between the Victorian respectability of Bloomsbury..and the Rabelaisian society of the London Docks. **1945** A. J. P. TAYLOR *Course German Hist.* 146 A Jekyll and Hyde policy, the bureaucrat Jekylls confident until too late that they could always shake off the Pan-German Hyde at their convenience. **1971** L. P. DAVIES *Shadow Before* xi. 128 We are all a mixture of good and evil, Jekyll and Hyde, if you like.

jeldi, var. *JILDI.

jell, *v.* For *U.S. colloq.* read orig. *U.S. colloq.* (Add further examples.) Also *fig.*, to take definite or satisfactory shape; = CRYSTALLIZE *v.* 5. Cf. *GEL *v.* Hence **jelled** *ppl. a.*

1869 L. M. ALCOTT *Little Women* II. 60 The jelly won't jell. **1902** *Fortn. Rev.* June 1021 (*heading*) Why a nerve tends to 'jell'. **1908** *Daily Chron.* 20 Mar. 3/3 [He] remarked of his countrywomen's minds that they 'didn't jell'; but he possibly, and mistakenly, thought he was talking American. **1937** *Maclean's Mag.* 15 Apr. 17/3 Davis shook his head, but the look of innocent disclaimer in his cherubic eyes didn't quite jell. **1956** K. FARRELL *Cost of Living* 138 'Not going well?' 'Hardly going at all. Even the cat book doesn't quite jell.' **1958** K. AMIS *I like it Here* ix. 113 His uncertainty..now felt more or less permanently jelled. **1958** *Observer* 18 May 10/5 Let jell in cool larder. **1959** P. H. JOHNSON *Unspeakable Skipton* iv. 26 They did the music by itself, later, at the Wigmore Hall, but it didn't jell. **1970** *Times* 5 Dec. 21/1 The present Parliament is only two parliamentary months old.., and although it will jell in time, it has not yet done so. **1972** *Times Lit. Suppl.* 12 Jan. 45/5 Somehow his case against RTZ as a sort of Hydra does not quite jell. **1972** *Village Voice* (N.Y.) 1 June 24/3 These detective novels were written very fast, read extremely well, and, as Himes told me: 'They worked. They jelled.'

2. *trans.* = JELLIFY *v.* 1. Also *fig.*, to give shape to; to make clear and definite.

1905 *Dialect Notes* III. 62 *Jell*, make or turn into jelly. **1935** *Forres Gaz.* 6 Nov. 4/5 To jell (to firm jelly). **1941** A. J. CRONIN *Keys of Kingdom* (1942) ii. 20 The tea was delicious, the scones and bannocks home-made, the preserves jelled by Elizabeth's own hands. **1948** *Newsweek* 10 May 58/3 The studio also ordered that no scripts be bought unless it was certain they could be jelled onto film. **1968** J. M. ZIMAN *Public Knowl.* v. 91 The course work is too rich a diet, and the knowledge it contains has been jelled too soon.

jell (dʒel), *sb.* orig. *U.S.* [f. the vb.] A jelly or gel.

1870 'F. FERN' *Ginger-Snaps* 262 My excellent country friends put up pounds and quarts of 'jell' every fall. **1951** *Good Housek. Home Encycl.* 423/1 Allow it to boil briskly, without stirring, until a jell is obtained on testing. **1959** *Listener* 11 June 1043/1 The sticky jells given by other starches such as tapioca and arrowroot.

jellaba(h), jellibee, varr. *GALABIYA. Cf. JELAB.

1904 A. E. W. MASON *Truants* xxv. 233 A black jellaba and cap, such as the Jews must wear in Morocco. **1922** JOYCE *Ulysses* 742 Longbearded jews in their jellibees. **1964** *Punch* 19 Feb. 278/3 'He had always a bottle of whisky under his jellaba,' Si Mohammed went on. **1969** *Daily Tel.* (Colour Suppl.) 24 Jan. 17/2 A girl called Susy in a jellabah like John Hanson in *Desert Song.*

jellied, *a.* Add: **3.** Coated with jelly; made or cooked inside jelly.

1895 'M. RONALD' *Century Cook Bk.* iv. 171 Jellied veal ..a good cold dish to use with salad. *Ibid.* v. 182 Jellied boned chicken. A braised boned chicken may be..jellied as follows. **1907** *Daily Chron.* 6 June 5/5 'Jellied eel! 'Ave a plate; lovally jelly,' shouts a third. **1908** *Ibid.* 6 July 3/4 She..knows the secrets of jellied eels. **1911** *Daily Colonist* (Victoria, B.C.) 26 Apr. 10/5 (Advt.), Jellied Ham per lb. 40¢. **1960** *Good Housek. Cookery Bk.* (rev. ed.) 84/2 Jellied eels. *Ibid.* 157/2 Jellied sheep's tongues. *Ibid.* 524/1 Jellied eggs.

Jell-o, Jello (dʒe·lo). Chiefly *N. Amer.* [f. JELLY *sb.*] The proprietary name of a powder used to make a fruit-flavoured gelatin dessert; loosely (with lower-case initial), jelly.

1934 *Trade Marks Jrnl.* 11 July 907 *Jell-o*. Brand of Jelly Powder... Jell-o can be placed in the refrigerator as soon as dissolved... The Jell-o Company of Canada Limited..Montreal, Canada; Manufacturers. **1936** *Amer. Speech* XI. 40 *Nervous pudding, Shivering Liz*, or *Shimmy* for jello..describe neatly. **1945** *New Yorker* 1 Dec. 36/2 Carola..learned to prefer salt fish and bean curd..to Jello and cereal. **1961** *Guardian* 29 Mar. 12/4 All the side dishes of salad and jello were there. **1971** *Black Scholar* Sept. 44/2 Her firm young breasts quivering like a dish of molded jello. **1972** *Sat. Rev.* (U.S.) 27 May 7/2 Herb Edelman's well-turned contribution as the agent with a heart of jello.

jelly, *sb.*^{1} Add: **1. c.** A table-jelly.

1728 E. SMITH *Compleat Housewife* (ed. 2) 146 *To make Riben Jelly*.. run the Jelly into little high Glasses..one Colour must be thorough cold before you put another on.. colour red with Cochineal, green with Spinage..and sometimes the Jelly by it self. **1845** E. ACTON *Mod. Cookery* (ed. 2) xx. 438 A great variety of..excellent jellies for the table may be made with clarified isinglass, clear syrup, and the juice of..fresh fruit. **1851** *Illustr. Catal. Gt. Exhib.* III. 650/1 Moulds for jellies, cakes, &c. **1916** *Punch* 6 Dec. 394/2 He shook all over like a badly-set jelly. **1930** C. MACKENZIE *April Fools* vii. 152 Mr. Wenlow, balanced like a pale jelly on the edge of a chair in the drawing-room. **1974** *Radio Times* 4 Apr. 42/4 (Advt.), Rowntrees strawberry flavour jelly.

2. *glycerin(e) jelly*, any of various mixtures of glycerol and gelatin, principally used as mounting media in microscopy; cf. JELLY *sb.* 2 d; *royal jelly*, the secretion produced by honey bees to feed the larvae of the colony, esp. those that will become queens.

1817 KIRBY & SPENCE *Introd. Entomol.* II. xix. 130 They will select one or more to be educated as queens; which..being fed with royal jelly for not more than two days..will come forth complete queens. **1859** *Q. Jrnl. Microsc. Sci.* VII. 257 The bottle of glycerine jelly is put into a cup of hot water, until liquefied. **1880** *Amer. Monthly Microsc. Jrnl.* I. 208/1, I have used such a medicine dropper to hold and apply glycerin jelly, with great satisfaction. **1886** F. R. CHESHIRE *Bees* I. vi. 82 In the case of the queen larva..that secretion, commonly, though, as I hold, erroneously, called royal jelly, is added unstintingly. **1954** C. G. BUTLER *World of Honeybee* iv. 46 It has come to be believed that any female honeybee larva..that is fed exclusively on royal jelly always develops into a queen bee. **1958** J. R. BAKER *Princ. Biol. Microtechnique* xiii. 255 Preparations mounted in glycerine-jelly, balsam, or other commonly-used media. **1973** L. HELLMAN *Pentimento* (1974) 230 He was always having mysterious operations... He took royal jelly.

e. *slang.* A pretty girl; a girl-friend.

1889 BARRÈRE & LELAND *Dict. Slang* I. 496/1 *Jelly*, or *All jelly*, a buxom, good-looking girl. **1931** W. FAULKNER *Sanctuary* iii. 23 Gowan goes to Oxford a lot... He's got a jelly there. He takes her to the dances. *Ibid.* iv. 36 Don't think I spent last night with a couple of your barber-shop jellies for nothing.

f. A gelatinous contraceptive substance.

1931 F. W. S. BROWNE tr. *T. H. van de Velde's Fertility & Sterility in Marriage* III. xiv. 348 The most important chemical contraceptives are the lubricant jellies. **1935** E. F. GRIFFITH *Mod. Marriage* iv. 84 There are numberless chemical substances, made up either in the form of tablets or jellies... These are introduced into the vagina and are intended to kill the sperms. **1937** —— *Voluntary Parenthood* iii. 46 Many other devices have been invented to carry the chemical, such as jellies, ointments, foaming tablets and foaming jellies... All these solubles, jellies and tablets are unsatisfactory and if used by themselves are liable to fail in a high percentage of cases. **1943** in T. H. van de Velde *Ideal Marriage* (1947) 148 Arrangements have now been made for the manufacture of Dr. Van de Velde's Jellies ('Eugam'): Lubricant, Contraceptive and Proconceptive. **1949** *New Gould Med. Dict.* 529/1 *Contraceptive jelly*, any one of a number of viscous substances introduced into the vagina to prevent conception. **1970** *Which? Contraceptives Suppl.* (ed. 3) 58 (*heading*) Spermicidal creams, jellies and pastes. **1972** *Guardian* 9 June 5/5 Her doctor..prescribed a diaphragm and a contraceptive jelly but the jelly was seized by the [Irish] customs authorities.

4. *jelly-boned* adj.; **jelly baby,** a soft gelatinous sweet in the shape of a baby; **jelly bean** orig. *U.S.*, (*a*) a bean-shaped sweet with a gelatinous centre and a hard sugar coating; (*b*) *slang*, an unpleasant, weak, or dishonest person; *spec.* a pimp; **jelly-belly,** a fat person; hence **jelly-bellied** *a.*; **jelly paint,** a non-drip paint with the consistency of jelly; **jelly powder,** (*a*) a kind of explosive (see quot. *a* 1884); (*b*) a crystalline powder used in the preparation of table-jellies; **jelly roll** *U.S.*, a cylindrical cake containing jelly or jam; freq. in *transf. slang* senses: (*a*) a lover; (*b*) sexual intercourse; (*c*) the female genitalia or vagina.

1945 DYLAN THOMAS in *Listener* 20 Dec. 734/2 A bag of moist and many-coloured jelly-babies. **1950** 'R. CROMPTON' *William—the Bold* i. 13 'Jelly babyth are nithe, too,' said the small shrill voice behind them. **1972** *Guardian* 27 Dec. 9/1 A lady going round Europe buying..jelly babies. **1905** *Chicago Daily News* 5 July 11/5 Jelly beans, assorted, per lb., 9 c. **1919** *Dialect Notes* V. 65 Mary is such a jelly-bean that she never gets her lessons. **1923** WODEHOUSE *Adv. Sally* 223 What's the idea, you jelly bean? **1929** W. FAULKNER *Sound & Fury* 202 Are you hiding out in the woods with one of those damn slick-headed jellybeans? **1935** A. J. POLLOCK *Underworld Speaks* 63/2 *Jelly bean*, a pimp (ellyjay eanbay). **1940** C. MCCULLERS *Heart is Lonely Hunter* (1943) I. iii. 32 She..took from her shirt pocket a blue-coloured jelly bean. **1972** M. J. BOSSE *Incident at Naha* i. 56, I went into the kitchen for jelly beans, which taste good after you've been blowing grass. **1899** KIPLING *Stalky & Co.* 214 He was..a Flopshus Cad, an Outrageous Stinker, a Jelly-bellied Flag-flapper (this was Stalky's contribution). **1938** *Times Lit. Suppl.* 26 Nov. 753/2 The most jelly-bellied of them is Nabucet, a flowery and despicable humbug. **1950** M. LOWRY *Let.* 23 June (1967) 213 She [*sc.* Russia]..lives in a state of constant 'war effort', with its attendant..'jelly-bellied flag-flapping'. **1896** FARMER & HENLEY *Slang* IV. 44/1 *Jelly-belly*, a fat man or woman. **1903** [see *EH *int.* 3]. **1935** L. A. G. STRONG *Tuesday Afternoon* 88 If ever I want a ginger-chinned jelly-belly's advice,..I'll ask for it. **1912** D. H. LAWRENCE *Let.* 3 July (1962) I. 134 Curse the blasted, jelly-boned swines. **1961** *Spectator* 17 Nov. 698 It is so spineless and 'jelly-boned'. **1958** *Listener* 28 Aug. 323/1 If anyone were to ask me what has been the most interesting new development for the do-it-yourself painter over the last year or two I think I would say the dripless, or jelly, paints... Some complain that the jelly paints do not cover up the surface underneath. *a* **1884** KNIGHT *Dict. Mech.* Suppl. 511/2 *Jelly powder*, so called from its resemblance to calf's-foot jelly. It consists of 94% or 95% of nitro-glycerine and 5% or 6% collodion cotton, so mixed as to assume a gelatinous form. **1895** *Army & Navy Co-op Soc. Price List* 16 Table Jelly powder..in packets. **1921** *Daily Colonist* (Victoria, B.C.) 6 Apr. 6/1 Lipton's and Shirriff's Jelly Powders, per packet..10¢. **1895** 'M. RONALD' *Century Cook Bk.* xxi. 468 *Jelly Rolls*. Make a layer of Genoese..press it through a pastry bag in lines onto the tins... Before it has had time to cool, cut off the hard edges, spread it with..any jelly or jam, and roll it up evenly; then roll it in paper and tie, so it will cool in a round, even shape. **1914** W. C. HANDY *St. Louis Blues* (song), I'm most wile 'bout mah Jelly-Roll. **1919** S. & C. WILLIAMS (*song-title*) I ain't gonna give nobody none o' this jelly roll. **1927** *Jrnl. Abnormal & Social Psychol.* XXII. 13 By far the most common of these terms is *jelly roll*. As used by the lower class Negro it stands for vagina, or the female genitalia in general, and sometimes for sexual intercourse. *Ibid.* 14 Angels in heaven do the sweet jelly roll. **1929** T. WOLFE *Look Homeward, Angel* (1930) xxii. 324 'What yo' want?' she asked softly. 'Jelly roll?' **1964** *Amer. Folk Music Occasional* 1. 12 Negro blues where women are sweet food (*biscuit-roller..jelly roll baker*). **1970** G. GREER *Female Eunuch* 265 If a woman is food, her sex organ is for consumption also, in the form of..*cake-* or *jelly-roll*. **1971** B. MALAMUD *Tenants* 205 Irene Lost Queen I miss To be between Your Jelly Roll. **1974** *Amer. Speech 1971* XLVI. 79 Jelly-filled doughnut, *bismarck, jelly roll*.

jelly (dʒe·li), *sb.*^{2} *slang.* Also **gelly.** [Shortening of the pronunciation of GELIGNITE, influenced by its jelly-like appearance.] Gelignite.

1941 in BAKER *Dict. Austral. Slang.* **1948** D. L. G. MUNDY *There's Gold in them Hills* x. 128 Put a charge of 'gelly' under it. **1955** [see *CREEP *v.* 5 b]. **1960** *Observer* 24 Jan. 5/2 A hut where they knew gelly was kept. *Ibid.* 5/3 There was always the gelly and oxy-acetylene if necessary. **1971** *Guardian* 28 Aug. 1/1 Stolen 'gelly' found.

jelly-fish. 2. b. (Further examples.)

1910 [see *GABFEST]. **1928** [see *FACE *v.* 6]. **1966** C. ACHEBE *Man of People* viii. 88 Max's unfortunate..presentation of me as a kind of pitiable jellyfish.

jellygraph (dʒe·ligraf). [f. JELLY *sb.* + -GRAPH.] An appliance used for multiplying copies of writing, etc., of which the essential part is a sheet of jelly. Also *attrib.* Hence **je·llygraph** *v. trans.*, to copy with a jellygraph; **je·llygraphed** *ppl. a.*

1900 H. G. WELLS *Love & Mr. Lewisham* xxv. 241 A letter of atrociously jellygraphed advices from Messrs. Danks & Wimborne. **1902** WODEHOUSE *Pothunters* xii. 182 On Sunday we jellygraph it—it'll have to be a jellygraphed number this time. **1904** *Sat. Rev.* 9 Jan. 40 It is better 'jellygraphing' questions for some one else's form than [etc.]. **1919** *Brit. Jrnl. Photogr. Alm.* 615/1 Jellygraph mixture for enlarging easel. **1936** 'G. ORWELL' *Keep Aspidistra Flying* iii. 58 They ran an unofficial monthly paper..duplicated with a jellygraph. **1972** W. A. PANTIN *Oxf. Life* iv. 52 We can trace the belated triumph of the typewriter..over the older methods of manuscript, jellygraph, and printed fly-sheet.

jelutong (dʒelū·tǫŋ). Also 9 **jolo-, julu-;** 20 **jela-, jelo-.** [Malay.] A Malaysian tree of the genus *Dyera*, esp. *D. costulata*, which produces a latex when tapped; the latex or the light-coloured wood from a tree of this kind.

1836 J. LOW *Diss. Soil & Agric. Penang* iv. 205 *Julutong*—very white. These woods are used chiefly by undertakers. **1885** *Spons' Mechanics' Own Bk.* 164 Jolotong... Well adapted for patterns and mouldings, excellent for carving purposes. **1900** W. W. SKEAT *Malay Magic* 205 Other haunted trees..are the Jawi-jawi, the Jĕlotong, and Bĕrombong. **1904, 1927** [see *GUTTA^2 2]. **1940** E. J. H. CORNER *Wayside Trees Malaya* I. 144 The *Jelutong* is deciduous partly or wholly. **1947** H. BARRON *Mod. Rubber Chem.* (ed. 2) iii. 27 Jelutong is obtained by tapping. **1970** *Timber Trades Jrnl.* 21 Mar. 54/1 Few sales of jelutong or kapur have been made, but meranti is said to be selling fairly well.

‖ **je m'en fiche** (ʒəmaṅfiʃ), *phr.* [Fr.] I couldn't care less; I don't care at all. Hence **je-m'en-fich(e)ism(e),** indifference.

1889 E. DOWSON *Let.* 10 Mar. (1967) 48 As for *esprit*—je m'en fiche—there never was a woman spirituelle before she was thirty. **1902** W. JAMES *Varieties Relig. Experience* ii. 36 *Je m'en fiche* is the vulgar French equivalent for our English ejaculation 'Who cares?' And the happy term *je m'en fichisme* recently has been invented to designate the systematic determination not to take anything in life too solemnly. **1905** A. BENNETT *Sacred & Profane Love* III. iii. 251 'Oh!' with a disdainful gesture. '*Je m'en fiche*. Let him go.' **1916** A. HUXLEY *Let.* Dec. (1969) 117 Old Birrell came down..professing an almost total je-m'en-ficheisme about the war. **1921** GALSWORTHY *To Let* II. ii. 139 Why want to know anything of that 'small' mystery—*Je m'en fiche*, as Profond says? **1923** A. HUXLEY *Antic Hay* v. 69 His divining eyes pierced through the veil of cynical *je-m'en-fichisme* to the bruised heart beneath. **1968** W. GARNER *Deep, Deep Freeze* ix. 240 As they say in this part of the world, *je m'en fiche*. **1970** R. HAUGHTON *Love* i. 25 The school of Anglo-Saxon 'je m'en fiche-ism'.

‖ **je m'en fous** (ʒəmaṅfu), *phr.* [Fr.] = prec. Hence **je-m'en-foutism(e).**

1918 A. BENNETT *Pretty Lady* xxi. 134 'But, madame, it is raining terribly.' '*Je m'en fous*. Run for a taxi.' **1936** 'G. ORWELL' *Keep Aspidistra Flying* x. 269 At the bottom of all his feelings there was a sulkiness, a je m'en fous in the face of the world. **1954** *Landfall* VIII. I. 27 The Sydneysider's traditional lounging *jemenfoutism*. **1959** *Guardian* 5 Nov. 6 France..has gone from despair and j'menfoutisme to optimism. **1964** *Ibid.* 4 Dec. 11/3 There is a species of *je m'en foutisme* in Godard that I like. **1966** *Ibid.* 9 Dec. 7/3 Happiness, they appear to say, is square, happiness is corny—and *je m'en fous*.

Jemima, jemima (dʒĕmai·mă). [Female Christian name.] **1.** A made-up tie. Also *attrib.*

1899 SOMERVILLE & 'ROSS' *Some Experiences Irish R.M.* v. 97 We indulged in..'Jemima' ties with diagonal stripes. **1920** *Glasgow Herald* 3 Apr. 4, I have never learned the knack of fixing a dress tie, and I have not the moral courage to wear a jemima.

2. *pl.* Elastic-sided boots; the British name for *Congress boots.*

1902 *M.A.P.* 29 Mar. 323/1, I spoke of Mr. Chamberlain's having fallen from sartorial grace to the extent of wearing 'Jemimas'. **1906** *Westm. Gaz.* 31 Dec. 3/1 A pair of well-preserved 'jemimas'. They are a kind of footgear the immortal Teufelsdröckh himself might have worn, unless he had a weakness for bluchers. **1927** *Glasgow Herald* 24 Aug. 8 The old-fashioned, long obsolete elastic-sided boots, known for some obscure reason as 'Jemimas'. **1961** *Times* 2 Oct. 13/4 The..Dame would at last be caught in the mangle out of whose rollers protruded the struggling feet in the supposedly ludicrous Jemimas.

Jemmy O'Goblin, var. *Jimmy O'Goblin* (s.v. *JIMMY 5).

1889 in BARRÈRE & LELAND *Dict. Slang.*

Jena (yē·nă). The name of a city in East Germany, used *attrib.* to designate glass made there, which originated in the experiments of Ernst Abbe and Otto Schott and became famous for its high quality and the special kinds that were developed.

1892 *Work* IV. 145/2 The new Schott Jena glass. **1902** J. D. & A. EVERETT tr. *Hovestadt's Jena Glass* i. 20 The introduction of Jena glass into practical optics was initiated by Abbe, who was now enabled, with the help of the technical resources of Zeiss' optical works, to realise his long-cherished plans for the improvement of the microscope. *Ibid.* vii. 189 Results for 29 different Jena glasses. **1902** [see *ANASTIGMAT]. **1908** W. ROSENHAIN *Glass Manuf.* i. 7 Another instance of these refractory glasses is to be found in the Jena special thermometer glasses..; the best of these glasses show little or no plasticity at temperatures approaching 500°C. **1941** *Jrnl. Inst. Petroleum* XXVII. 434 The solution was sucked off via a Jena-glass immersion filter. **1954** G. W. MOREY *Properties Glass* (ed. 2) iii. 78 The first Jena 'Geräte' glass, made before 1910.., differed from the ordinary soda-lime glass in the substitution of magnesia and zinc oxide for lime, and in having a small content of B_2O_3. Later..the soda content was reduced and the magnesia largely removed, with increase in the zinc oxide, alumina and boric oxide.

Jennerian (dʒenɪə·riăn), *a. Med.* [f. the name of Edward *Jenner* (1749–1823), English physician, who in 1796 vaccinated a subject with cow-pox against small-pox and thereby laid the foundations of vaccination in medicine and of the science of immunology: see -IAN.] Of, pertaining to, or commemorating Jenner; made by or following the methods of Jenner.

1801 *Med. & Physical Jrnl.* VI. 5 The Jennerian Inoculation is universally adopted by the medical gentlemen of this town and neighbourhood. **1805** W. ROWLEY *Cow-Pox Inoculation No Security against Small-Pox* p. v, The names of..those who formed the principals in the Royal Jennerian Institution, etc. are omitted. **1824** MILL *Prefaces to Liberty* (1959) 72 The great spiritual physicians who would vaccinate the nation with hypocrisy to prevent the eruption of infidelity, are not acting on a true Jennerian analogy. **1911** G. B. SHAW *Doctor's Dilemma* p. lxvi, To this day the law which prescribes Jennerian vaccination is carried out with an anti-Jennerian inoculation because the public would have it so. **1951** W. R. LEFANU *Bio-Bibliogr. E. Jenner* ii. 59 The text was included in Jones' and Crookshank's Jennerian collections.

jer (yer). Also **ier, yeer, yer(r).** Pl. **jers, jery.** [ad. OSl. *jerŭ* and *jerĭ*, Russ. *er* (pl. *ěrȳ*) and *er'*.] The name of either member of the pairs of characters ѫ, ѫ and ъ, ь of the Slavonic Glagolitic and Cyrillic alphabets, sometimes distinguished according to their Slavonic names as **jerŭ** (**yerŭ, jer**) and **jerĭ** (**yerĭ, jer'**) respectively; one or both of the sounds represented by these characters, or the Common Slavonic sounds from which they developed. Also *attrib.* (**Jerek, Yerek** in quots. 1861, 1883, referring to the second member of each pair, are from the deriv. seen in Russ. *ěrik*.)

1763 W. MASSEY *Origin & Progress of Letters* (tables facing 114) Yerr... Yeer... ier. **1861** *Grammatography* Based on the German compilation of F. Ballhorn 58 (table Jer... Jerek. *Ibid.* 61 (table) Yerr... Yer. **1883** I. TAYLOR *Alphabet* II. 196 (table) Yer... Yerek. **1949** *Archivum Linguisticum* I. 158 The left-to-right characters, viz. *jestŭ, kako, onŭ, ukŭ*, and the left-inclined, *jerŭ* and *jerĭ*.. are significant. **1949** ENTWISTLE & MORISON *Russian & Slavonic Languages* 66 Peter the Great's Russian 'civil alphabet'..retained the two *jers* which had no longer an alphabetical value. *Ibid.* 82 In Russian the *jers* continued to be distinguished as back/front vowels, and so developed into the open vowels *o/e* respectively. **1952** *Word* VIII. 323 Leskien..thought that the spellings indicated a change in the quality of the consonant preceding the jer-letter. *Ibid.* 325 The category of jer-phonemes had been sharply reduced as to occurrence. **1955** H. G. LUNT *Old Church Slavonic Gram.* 30 A jer is weak in a syllable followed directly by a non-reduced vowel or pause... A jer is strong only in a syllable directly before a weak jer... When the jers were lost as a category, the weak jers simply disappeared, and the strong jers were replaced by another vowel. **1969** *Language* XLV. 558 The rule specifying sequences of obstruents as fricative + stop was eliminated by the time the jers were lost.

Jeremiah (dʒerīmɔi·ă). The name of a Hebrew prophet (see JEREMIAD), used allusively to denote a person given to lamentation or woeful complaining.

1781 GIBBON *Decl. & Fall* III. 620 The vague and tedious lamentations of the British Jeremiah [*sc.* Gildas]. **1902** *Daily Chron.* 15 Oct. 3/1 This talk about the ratepayers only came from municipal Jeremiahs. **1905** *Ibid.* 1 Sept. 5/7 The Jeremiahs have been on the rampage; the dismal and the doleful would-be experts [etc.]. **1928** *Daily Express* 23 Feb. 3/5 There are always Jeremiahs who go about saying that we have never had such bad times. **1928** *Observer* 22 July 16/3 The Socialists are..bound to be confirmed Jeremiahs by the necessity of their propaganda. **1963** *Times* 22 Apr. 8/5 Mr. Selwyn Lloyd..wanted to see Young Conservatives 'rise up in protest against the Jeremiahs, defeatists, pessimists, denigrators', [etc.]. **1973** *Listener* 15 Nov. 655/1, I am not going to try to play the role of prophet, least of all Jeremiah.

Jeremianic (dʒe:rīmɔi,æ·nik), *a.* [f. *JEREMIAH, after *Messianic*.] Of or pertaining to the prophet Jeremiah or the book of the Old Testament which bears his name.

1880 *Encycl. Brit.* XIII. 628/1 Brought into its present form by a captivity prophet, working on a Jeremianic basis. **1891** T. K. CHEYNE *Orig. Psalter* 151 A group of literary works which we may call Jeremianic. **1921** J. MOFFATT *Approach to New Testament* ii. 62 The Jeremianic prediction of the new covenant has been fulfilled in Christianity.

jerk, *sb.*^{1} Add: **2. b.** (*a*) (Examples of the unqualified use.)

1895 G. N. STEWART *Man. Physiol.* xii. 625 The interval which elapses between the tap and the jerk ($\frac{1}{25}$ to $\frac{1}{20}$ second) is distinctly shorter than the reflex time of the extremely rapid lid-reflex. **1936** M. G. EGGLETON *Muscular Exercise* viii. 181 The jerk is equally readily obtained if the skin has been de-sensitized by application of a local anæsthetic. **1968** PASSMORE & ROBSON *Compan. Med. Stud.* I. xxiv. 14/1 The presence of a normal jerk shows that the receptors are sensitive..and the responding muscle is in order.

d. Colloq. phr. *physical jerks*, physical or gymnastic exercises.

1919 [see sense 2 e below]. **1930** E. RAYMOND *Jesting Army* I. i. 7 It was the parade for 'Physical Jerks'. *Ibid.* 14 Now the whole family party must go out into the garden to do physical jerks. **1966** A. SACHS *Jail Diary* ii. 24 In the afternoon, I am busy doing physical jerks. *Ibid.* 25 My physical jerks period.

e. Colloq. phr. *to put a jerk in it*, to act vigorously, smartly, or quickly.

1919 *Athenæum* 25 July 664/2 'Physical jerks' dates from war-time, as does also the admonition 'put a jerk in it', which is the equivalent of the ante-bellum 'jump to it'. **1921** N. KENT *Quest M. Harland* II. viii. 241 'I like to see young people enthusiastic. Put a jerk in it, can't you?' 'A—a *what*?' stammered Anthea, tottering. 'Put a jerk in it,' repeated Roger. **1939** C. DAY LEWIS *Child of Misfortune* III. ii. 271 Put a jerk in it. I'm meeting my boy at the second house at the Royal. **1974** 'J. ROSS' *Burning of Billy Toober* xv. 147 If you put a jerk into it, you'll probably have something for me by lunchtime.

f. The name of a dance characterized by jerking movements.

1966 *N.Y. Times Mag.* 9 Jan. 106/2 There is the Watusi, basically a side-to-side stumble, the Shake, and the Jerk—whose movements come as no surprise to old fans of burlesque. **1969** N. COHN *AWopBopaLooBop* (1970) ix. 85 Dance-crazes bossed pop right up until the Beatles broke... The Jerk and the Block. **1972** T. KOCHMAN *Rappin' & Stylin' Out* 161 Names for dances (Rock and Roll,..Jerk, and Black Power Stomp) all embody kinetic elements.

5. *slang* (orig. *U.S.*). Someone of little or no account; a fool, a stupid person. Cf. *JERKWATER b.

1935 A. J. POLLOCK *Underworld Speaks* 63/2 *Jerk*, a boob; chump; a sucker. **1938** *New Republic* 7 Sept. 129/1 A jerk not only bores you but pats you on the shoulder as he does so. **1945** *Daily Express* 11 Sept. 2/4 See this lighter? A dying Jerry gave it to me. I gave the jerk a smoke from my last cigarette. **1950** [see *brown-noser* s.v. *BROWN *a.* 7]. **1956** L. MCINTOSH *Oxford Folly* 85 Julian sounds a dismal little jerk when you sum him up like that. **1958** *Listener* 15 May 802/1 If..the sponsors get eight letters saying that their comedian is an idiot, or a foul-mouthed jerk, they're terrified. **1971** J. BALL *First Team* (1972) xxiv. 382 '*I say* you're a nigger!'.. 'And I say that you're a goddamned jerk.'

6. *Physics.* Rate of change of acceleration (with respect to time).

1955 J. S. BEGGS *Mechanism* iv. 122 Since the forces to produce accelerations must arise from strains in the materials of the system, the rate of change of acceleration, or jerk, is important. **1964** A. G. FADELL *Calculus* vii. 195 In the automobile industry the concept of a jerk is used as a 'comfort index'. *Ibid.*, Free fall has an ideal comfort index, namely zero of jerk. **1973** *Nature Physical Sci.* 19 Feb. 140/1 Large values of jerk (third derivative of the position of *m*) can occur if *M* is sufficiently large.

jerk, *v.*^{1} Add: **7.** *trans.* To serve (soda, beer, etc.) at a soda-fountain, bar, etc. Cf. **soda-jerker*. *U.S. colloq.*

1883 G. W. PECK *Peck's Bad Boy* xiii. 126 Well, I must go down to the sweetened wind factory, and jerk soda! **1884** J. MILLER *Memorie & Rime* 20 They stared at me, but went on jerking beer behind the counter. **1935** *Amer. Mercury* May 102/1 They had spent their tender years jerking sodas. **1949** WODEHOUSE *Uncle Dynamite* iv. 54, I also jerked soda.

8. *to jerk off* (trans. and intr.), to masturbate. *slang.*

1937 in PARTRIDGE *Dict. Slang.* **1947** A. BERNSTEIN *Home is the Hunted* 152 Big enough for a bush and jerking off like crazy, disconcerted and embarrassed by the riches of manhood flooding through your bones, veins, and gizzard. **1969** P. ROTH *Portnoy's Complaint* 177 She will jerk off one guy, but only with his pants on. **1971** B. MALAMUD *Tenants* 202 The mother..dies unattended, of malnutrition, as Herbert jerks off in the hall toilet.

jerkin^1. Add: (Later examples.) In modern use, usu. a sleeveless jacket or waistcoat (see quots.).

1935 E. WEEKLEY *Something about Words* 44 Until the War one vaguely associates *jerkin* with the costume of Robin Hood, but after the War 'army jerkins' were to be had cheap. **1957** M. B. PICKEN *Fashion Dict.* 187/2 *Jerkin*, jacket, short coat, or doublet, sometimes of leather. Occasionally, waistcoat without sleeves. Also straight woolen pull-over. **1963** *Austral. T.V. Times* 18 Apr. 10/1 *Jerkin*, a sort of zipper jacket. **1974** *Guardian* 4 Apr. 11/2 It is legal to shoot clouts, which used to be the jerkins of longbowmen hung on a stick.

jerking, *vbl. sb.* Add: **b.** The action of *JERK *v.*^{1} 8. Also with *off.*

1889 BARRÈRE & LELAND *Dict. Slang* I. 497/2 *Jerking*, masturbation. **1960** E. L. WALLANT *Human Season* ix. 94 Another thing bothers me is your jerking off in bed at night. **1969** F. NORMAN *Banana Boy* 108 'Ninety-nine change hands,' he would laugh when he came into the kitchen after a hard night's jerking-off. **1972** *Screw* 12 June 12/1 Such sexual behavior as homosexuality, jerking off, [etc.].

jerk-line (dʒə·ɪk,ləin). *N. Amer.* [f. JERK *v.*^{1} 2.] A rope used in place of reins to guide a team of horses, etc. Also *attrib.*

1888 M. GRIGSBY *Smoked Yank* xix. 156 The driver rode on the nigh wheel mule, and drove the leader with a jerk-line. **1907** S. E. WHITE *Arizona Nights* III. iv. 287, I bet that Sang would get a wiggle on him... if he had a woman ahold of his jerk line. **1910** J. HART *Vigilante Girl* x. 140 This train of animals was driven by a 'jerk line' instead of reins. **1937** J. STEINBECK *Of Mice & Men* ii. 61 He was a jerkline skinner..capable of driving ten, sixteen, even twenty mules with a single line to the leaders. **1960** A. DOWNS *Wagon Road North* 43 Sometimes there were so many horses or mules hitched to a freight wagon that the use of reins was impractical. Under these circumstances a 'jerk-line' was used. This was a single line connected to the bridles of the lead horses.

jerk-off (dʒə̄·ɪk,ǫf), *a.* [f. vbl. phr. *to jerk off* (*JERK *v.*^{1} 8).] Erotic; encouraging masturbation.

c **1957** in N. Mailer *Advts. for Myself* (1961) 310 They turn their backs..on..the grandeurs of the past, restrict their horizon to..tripe..jazz magazines and jerk-off magazines. **1965** *New Society* 30 Dec. 21/1 It would prove Screeches was not a 'jerk-off' press but was serious.

jerk-off (dʒə̄·ɪk,ǫf), *sb.* [Perh. f. prec. but cf. *JERK *sb.*^{1} 5.] = *JERK *sb.*^{1} 5.

1968–70 *Current Slang* (Univ. S. Dakota) III–IV. 75 *Jerk off*,..a rustic; a simpleton. **1972** J. MILLS *Report to Commissioner* 145 I'm sitting there alone..feeling like the biggest jerk-off in the history of the world. **1972** R. A. WILSON *Playboy's Bk. Forbidden Words* 172 A jerk or jerk-off is a fool. **1973** W. SHEED *People will always be Kind* I. xiv. 182 You know perfectly well that the jerk-offs do all the talking at meetings.

jerkwater (dʒə̄·ɪk,wǭ:təɪ). *U.S.* [f. JERK *v.*^{1}] In full, *jerkwater train.* A train on a branch railway (see quot. 1945). Also *attrib.*

1878 F. H. HART *Sazerac Lying Club* 16, I wish I may be run over by a two-horse jerk-water if there was a sage-hen in sight. **1905** *Dialect Notes* III. 84 The St. Paul branch is a jerkwater railroad. *Ibid.*, *Jerkwater* (*train*), train on a branch railway. 'Has the jerkwater come in yet?' **1909** *Sat. Even. Post* 15 May 9/3 The farther along Thorpe got in the list the more disgusted he became with the prospect of living on jerk-water trains. **1920** *Bulletin* 22 June 10/1 This Oriental who, with perfect self-possession..descended from the jerk-water train carrying a modern suitcase. **1926** J. BLACK *You can't Win* xx. 303, I followed the pay-roll aboard the jerkwater train that carried it to the waiting miners. **1941** *Sun* (Baltimore) 7 Mar. 12/7 In the early days of railroads the small boilers of the locomotives required frequent refilling, and water tanks were very few. Every train crew carried a leather bucket on a long rope with which they 'jerked water' from the streams along their track. As locomotives increased in size the small 'jerk-water' engines were relegated to branch-line service. Today no train crew carries a bucket, but the name 'jerk water' still sticks and has become part of our national heritage of American slang. **1945** J. L. MARSHALL *Santa Fe, Railroad that built an Empire* 68 The Santa Fe was the Jerkwater Line—because train crews, when the water got low, often had to stop by a creek, form a bucket brigade and jerk water from the stream to fill the tender tank.

b. Used *attrib.* or as *adj.* in sense 'small, insignificant, inferior'. *colloq.*

1897 *Chicago Tribune* 25 July 15/2 John J. Ingalls regards the Swiss Mission as a jerkwater job, and would not take it if it were offered to him. **1911** H. S. HARRISON *Queed* xviii. 225 The spring found West stronger and more contented with his lot as president of a jerkwater college. **1923** J. DOS PASSOS *Streets of Night* 130 Perkinville, a little jerkwater town back in South Dakota. **1936** *Jrnl. Genetic Psychol.* XLIX. 492 It was one of those jerkwater towns that have one lawyer, one drug store and no traffic cops. **1950** BLESH & JANIS *They all played Ragtime* (1958) 3 Vaudeville teams—from the jerkwater acts to specialists like Ben Harney. **1970** R. LOCKRIDGE *Twice Retired* (1971) v. 68 It won't be easy for him to get another job if he's fired... Maybe at some jerkwater college at half what he's getting now.

jerrican, jerrycan (dʒe·rikæn). Also **jerry can, jerry-can.** [f. *JERRY *sb.*^{2} + CAN *sb.*^{1}] A five-gallon (usu. metal) container for petrol, water, etc., of a type first used in Germany and later adopted by the Allied forces in the war of 1939–45.

1943 *Hutchinson's Pict. Hist. War* 17 Feb.–11 May 258 Mules carrying 'jerricans' to British troops... Jerricans are a special type of petrol container for transporting water. **1944** [see *JERRY *sb.*^{2}]. **1944** *Times* 25 Nov. 8/3 The Germans had a very efficient five-gallon petrol can. The Eighth Army captured some of the cans. They were sent back to England, and the British started manufacturing them. They were called jerricans. **1955** M. E. B. BANKS *Commando Climber* v. 87 We counted our burden of spuds a privilege in comparison to the jerrycan of red wine which was foisted on to one unfortunate. **1958** *Punch* 9 July 60/2 It [*sc.* sea-water] is collected *each week*..and rushed to Paris in white plastic jerricans. **1969** *New Yorker* 4 Oct. 55/1 Foreigners flying into Biafra now bring their own food and, if the pilot permits it, their own gasoline in jerry cans. **1972** D. HASTON *In High Places* x. 109 Water is the only problem: you have to bring big jerry-cans in from Cassis.

Jerry (dʒe·ri), *sb.*^{2} *colloq.* (orig. *Mil. slang*). [Prob. alteration of GERMAN *a.*^{2} and *sb.*^{2}, perh. infl. by JERRY *sb.*] A German; *spec.* a German soldier; a German aircraft; also, the Germans or German soldiers collectively. Also *attrib.* or as *adj.* Cf. *FRITZ^1.

1919 J. B. MORTON *Barber of Putney* ii, There was three Jerries waiting for 'im to get tired and chuck it. **1925** FRASER & GIBBONS *Soldier & Sailor Words* 131 *Jerry over*, 'Lights out!' The word passed along the lines at the Front at night on the nearing overhead of an enemy aeroplane. *Ibid.*, *Jerry up*, a warning call on the approach of a German aeroplane. **1929** E. W. SPRINGS *Above Bright Blue Sky* 272 If you have many chaps like him, it won't take long to chase Jerry back to the Rhine. **1931** W. V. TILSLEY *Other Ranks* 8 The way they referred to the Germans—almost affectionately. Old Fritz, or Old Jerry! Might be an ally! **1941** *Southern Daily Echo* (Southampton) 26 Mar., Last time the enemy was generally called the *Hun* by the people at home, and *Jerry* by the soldiers. The latter is the term which remains in use in the present war. **1942** *Tee Emm* (Air Ministry) II. 58 Net result: all square—instead of one Jerry down. **1943** [see *BAG *sb.* 17 (*c*)]. **1944** G. NETHERWOOD *Desert Squadron* xii. 118 The well known Jerry boat, the canvas and leather affair, was soon put into active service by our men and also the petrol container known as the 'Jerrycan'. **1955** J. THOMAS *No Banners* xx. 198, I thought you were a Jerry, trailing me. **1961** W. VAUGHAN-THOMAS *Anzio* viii. 185 They almost felt a sympathy for the Jerries under that merciless rain of explosions. **1972** *Daily Mail* 4 May 3/3 Give us a Jerry paper, love... There's a German bloke on top wants one.

jerry (dʒe·ri), *sb.*^{3} *slang.* [Prob. abbrev. of JEROBOAM.] A chamber-pot.

[**1827** W. MAGINN *Whitehall* II. iv. 140 The naval officer ..came into the Clarendon for a Jerry [= jeroboam] of punch. **1850** *Sessions Papers Cent. Criminal Court* (Surrey cases) May 124, I went into the *jerry* [= water-closet], but they had got there before me.] **1859** HOTTEN *Dict. Slang* 53 *Jerry*, a chamber utensil. **1932** J. CARY *Aissa Saved* xvii. 96 A thin handsome young man carrying a tin jerry in his hand and a broken kettle among the tatters on his back. **1939** 'G. ORWELL' *Coming Up for Air* IV. vi. 271 A bed not yet made and a jerry under the bed. **1968** *Canad. Antiques Collector* Dec. 10/1 Young English ladies and gentlemen were beginning to find it offensive to have the old man keep a jerry in the side-board.

jerry (dʒe·ri), *sb.*^{4} *Austral.* and *N.Z. slang.* [Cf. *JERRY *a.*^{2} and *v.*] Phr. *to take a jerry* (*to*): to investigate and understand (something), to 'tumble' *to* (something).

1919 W. H. DOWNING *Digger Dial.* 30 *Take a jerry*, change (for the better) one's course of conduct. **1937** PARTRIDGE *Dict. Slang* 437/1 *Jerry*,..a recognition, discovery, 'tumble'. **1969** *Landfall* XXIII. 328 It was time this country—ah! Took a jerry to itself. Ha ha.

jerry (dʒe·ri), *a.*² *U.S. slang.* [Origin unknown.] Phr. *to be* (or *get*) *jerry* (*on, on to, to*): to be aware (of); to be 'wise' (to); to understand.

1908 K. McGaffey *Sorrows of Show Girl* 200 She accepted the attentions of the comedian which his wife was not supposed to be jerry to. **1921** *Adventure* (U.S.) 18 May 25/1 I've got a strong hunch that thousand bucks is all stowed away, neat as pie, in the pendulum box o' that clock. I'm wise, Kid; I'm jerry. **1926** *Flynn's* 16 Jan. 639/1, I know that th' fly was jerry because he gave me th' once over as I was comin' out. **1942** Berry & Van den Bark *Amer. Thes. Slang* § 149/7 Know; be aware of, *be hep, — jerry, — on, — onto, — wise to. Ibid.* § 149/12 Knowing; cognizant; aware of,..*jerry*.

jerry (dʒe·ri), *v. slang* (chiefly *Austral.* and *N.Z.*). [Cf. *jerry *sb.*⁴] *intr.* To understand, realize; to 'tumble' *to* something.

1917 *Digger* 4/3 The excuse was so full of Mer(r)it that the officer failed to 'Jerry' to it. **1918** *Chrons. N.Z.E.F.* 21 June 221/1 Unless the sergeant jerries to your lurk. **1925** Fraser & Gibbons *Soldier & Sailor Words* 131 *Jerry, to*, to understand, *e.g.*, 'Do you Jerry it, man?' **1959** G. Slatter *Gun in my Hand* viii. 91 Tried to cut me out with me sheila. Hadn't jerried to it before.

jerry-built, *a.* Add: Also *fig.*

1901 *Daily Chron.* 13 Aug. 3/2 In an age of jerry-built books it is refreshing to come across a volume that has taken forty years to compile. **1903** *Ibid.* 20 Feb. 3/2 Fiction, he said, was now jerry-built. **1933** J. Baillie *And Life Everlasting* (1934) i. 15 That great cataclysm undoubtedly came as a rude shock to many jerry-built philosophies of life.

jerrycan: see *jerrican.

Jersey¹. Add: **1.** (Further examples.) Also used of fine machine-knitted fabric generally.

1881 *Queen* 8 Jan. (Advt.), Boy's jersey suit. **1938** D. Baker *Young Man with Horn* I. iv. 38 His brother Henry..was selling jersey knits. **1959** *Observer* 22 Mar. 18/4 Some lovely short evening dresses were in silk jersey and had to be bought therefore, a little surprisingly, in the jersey wear department. **1966** *Economist* 9 Apr. 175/1 Acrylics (woolly, fluffy fibres used in knitwear mostly) and polyesters (mostly used in wool mixes, now bulked in jersey-knits) are bound to follow the same course as nylon. **1969** N. Freeling *Tsing-Boum* vi. 36 A nearly new jersey cocktail dress with the boutique label of a couturier. **1972** *Sci. Amer.* Dec. 4/2 The fabric is a plain knitted jersey... It was knitted..of a spun, staple yarn.

4. Also *attrib.* or as *adj.*

1842 *Guide to Island of Jersey* v. 94 The English reader need scarcely be told in what great estimation the Jersey Cows and Heifers are held... The Jersey cow is small and slender in its make, with short crumpled horns. **1845** *Jrnl. R. Agric. Soc.* V. 47 The Jersey cow is a singularly docile and gentle animal. **1875** [see *Alderney]. **1964** G. Durrell *Menagerie Manor* ii. 35 You have vivid mental pictures of an escaped tiger stalking your pedigree herd of Jersey cows.

5. (sense 4) *Jersey cream, milk*; **Jersey elm**, *Ulmus stricta* var. *sarniensis*, a variety of elm of more erect growth than the parent species; **Jersey lily**, a nickname for Li(l)lie or Lily Langtry (1852–1929), an actress born in Jersey.

1895 *Montgomery Ward Catal.* 110/1 Williams' Medicated Jersey Cream Toilet Soap; contains pure, rich cream from Jersey Cows. **1971** 'J. J. Marric' *Gideon's Art* iv. 35 The butler..offered more of the delicious apricot and peach flan, more of the rich Jersey cream. **1838** J. C. Loudon *Arboretum* III. 1376 The Jersey Elm: is a free-growing variety, differing very little from the species. **1914** W. J. Bean *Trees & Shrubs Hardy in Brit. Isles* II. 620 A yellow-leaved form of the Jersey elm originated in the nurseries of Messrs Dickson at Chester in 1900. **1957** M. Hadfield *Brit. Trees* 242 The Jersey Elm is not infrequently seen planted as a street or ornamental tree, but is believed to be indigenous only to the Channel Islands. **1971** *Country Life* 8 Apr. 823/1 It should be widely known that the Jersey elm (*Ulmus stricta sarniensis*) is not immune to elm disease. **1882** W. Hamilton *Aesthetic Movement* 103 *La bella Donna della mia Mente* exists, but she is not the Jersey Lily, though I have grovelled at her feet. **1895** M. Beerbohm in *Yellow Bk.* IV. 275 To have strained my eyes for a glimpse of the Jersey Lily. **1930** W. S. Maugham *Cakes & Ale* xii. 145 She had..a fringe on the forehead with a bun at the nape of the neck as you may see in old photographs of the Jersey Lily. **1972** *Times* 3 July 14/8 The Millais portrait ..showed her [*sc.* Lily Langtry] holding a lily in one hand which earned her the nickname 'Jersey Lily'. **1881** *Jrnl. R. Agric. Soc.* 2nd Ser. XVII. 234 Jersey milk is considered too rich [for a calf]. **1970** E. McGirr *Death pays Wages* vi. 115 Her own breakfast..was Jersey milk and sliced banana.

Jersey² (dʒə̄·ɪzi). *U.S.* = *New Jersey*, the name of the state situated south of New York, used *attrib.* and *Comb.* to denote people or things coming from or associated with New Jersey, as *Jersey girl, maid*; *Jersey-built, -made* adjs.; **Jersey blue**, (*a*) a colonial New Jersey soldier (so called from his blue uniform); (*b*) a native or inhabitant of New Jersey; (*c*) a breed of chicken; **Jersey justice**, strict or severe justice; **Jersey lightning** *colloq.*, a strong kind of apple-jack, peach-brandy, or whisky; **Jersey pine**, the scrub pine, *Pinus virginiana*; **Jersey tea**, the red-root, *Ceanothus americanus*, the leaves of which were used as a substitute for tea; **Jersey wag(g)on**, a light carriage formerly used in New Jersey.

1758 C. Rea *Jrnl.* 16 July in *Essex Inst. Hist. Coll.* (1881) XVIII. 110 Excepting ye Yorkers and Jersy Blews all ye Provincials didn't loose more than 100 men. **1849** J. F. Cooper *Sea Lions* I. i. 3 Distinctions..exist between the eastern and the western man,..the Buckeye or Wolverine, and the Jersey Blue. **1850** D. J. Browne *Amer. Poultry Yard* 77 The Jersey-Blue Fowl..is another large mongrel of a bluish cast. **1942** C. Weygandt *Plenty of Pennsylvania* 277 We have no native breeds of fowls in Pennsylvania as New Jersey had its Jersey Blues. **1829** R. C. Sands *Writings* (1834) II. 121 Trim Jersey-built wagons. **1770** *Boston Gaz.* 7 May (Th.), A likely active Jersey girl. **1903** *N.Y. Tribune* 18 Oct. 8 Even with a faithful judge..'Jersey justice' did not shine as brilliantly as usual. **1948** *Collier's* 20 Nov. 78/3 Pellechia is having dealings with Jersey justice. **1852** *Alta California* (San Francisco) 23 Aug. 2/5 The rumsellers dealt out Jersey lightning by the gallon. **1872** G. P. Burnham *Mem. U.S. Secret Service* p. vi, *Jersey lightning*, a peculiar New Jersey drink; 'blue ruin'. **1970** *Observer* 19 Apr. 9/4 This urbane and sophisticated man came to believe that after repeal Jersey Lightning would capture the fancy of the whole country, and become a standard national drink. **1778** *Boston Gaz.* 25 Aug. 373 Handy, light, Jersey made waggon. **1713** *Boston News-Let.* 5 Oct. 2/2 A Jersey Maids times for four years and an half very good Servant to be disposed of. **1743** J. F. Gronovius *Flora Virginica* 191 The common Jersay-Pine. **1908** N. L. Britton *N. Amer. Trees* 46 Jersey Pine—*Pinus virginiana* Miller... This tree grows in poor rocky or sandy soil from southern New York to Indiana. **1971** *Country Life* 23 Dec. 1773/1 A singular collection of conifers, some species such as the Jersey pine (*Pinus virginiana*) being planted in considerable numbers. **1759** P. Miller *Gardeners Dict.* (ed. 7) s.v. Ceanothus, New Holland Dogwood with female cornel leaves commonly called New Jersey Thea grows naturally in most Parts of North America, from whence great Plenty of the Seeds have been of late Years brought to Europe, by the Title of New Jersey Thea, where I have been informed the inhabitants dry the leaves of this Shrub to use as Thea. **1808** H. Muhlenburg *Let.* 5 July in E. Rowland *Life W. Dunbar* (1930) 203 Red Root or Jersey Tea we call the Ceanothus americanus a little shrub with white Flowers, cordate leaves—Some times the sanguinaria is called so. **1870** *Amer. Naturalist* IV. 583 The *Ceanothus*, or Jersey tea, is a frequent inhabitant of the prairies. **1948** *Green Bay* (Wisconsin) *Press-Gaz.* 13 July 11/4 Jersey tea, the shrub whose leaves were the 'tea' used by the colonists after the Boston tea party. **1811** R. Sutcliff *Travels N. Amer.* p. viii, Jersey waggons..are made very light, hung on springs with leather braces, and travel very pleasantly. **1835** J. J. Audubon *Ornith. Biogr.* III. 606 Fishermen gunners passed daily..with Jersey wagons, laden with fish, fowls, and other provisions. **1944** *Sat. Rev.* (U.S.) 14 Oct. 24/2 The vagabond Reverend Mason Weems 'bumping along in his Jersey wagon'.

Jerseyman (dʒə̄·ɪzimæ̆n), *sb.*¹ Now *rare.* A native or inhabitant of New Jersey, U.S.A.

1679 *Rec. Early Hist. Boston* (1881) VII. 58 Thomas Begretia [entertained] at James Wardens, Jersiman. **1839** *Southern Lit. Messenger* V. 800/2 A Jerseyman is pre-eminently calculated to make a good traveller. **1873** C. G. Leland *Egypt. Sk.-Bk.* 45 The last number of the *Anglo-American* contains the names of half-a-dozen as veritable Jerseymen as ever drank apple-jack. **1878** *Harper's Mag.* 318/2 'Pretty hard times,' said the Jerseyman; 'but I want three hundred dollars in cash.' **1949** *Hist. & Philos. Soc. Ohio Bull.* Apr. 74 They were especially obnoxious to the Pennsylvanians and Jerseymen.

Jerseyman (dʒə̄·ɪzimæ̆n), *sb.*² A native or inhabitant of Jersey (one of the Channel Islands).

1825 J. C. Loudon *Encycl. Agric.* 1129/2 The treasure highest in a Jersey-man's estimation, is his cow. **1842** *Guide to Island of Jersey* x. 144 There is an independence of character in a Jerseyman. **1973** A. Grey *Some put their Trust in Chariots* v. 23 Jerseymen turn every statement into a question with that little word 'Ay?'

Jerusalem. Add: **Jerusalem cherry** = Winter cherry 1 b.

1884 W. Miller *Dict. Eng. Names Plants* 251 *Solanum Pseudo-Capsicum.* Jerusalem Cherry, Winter-Cherry Capsicum. **1887** G. Nicholson *Illustr. Dict. Gardening* III. 455/1 *S*[*olanum*] *Pseudo-capsicum* (false Capsicum). Jerusalem Cherry. A few handsome hybrids have been raised from this species. **1902** L. H. Bailey *Cycl. Amer. Hort.* IV. 1679/1 [*Solanum*] *Pseudo-Capsicum*, Linn. Jerusalem Cherry... An old-fashioned plant..grown for its showy berry-like fruits, which persist a long time. **1969** *Northwest* (*Sunday Oregonian Mag.*) 14 Dec. 19/3 Christmas potted plants are just as important a part of the scene. Poinsettias, cyclamens, chrysanthemums, Christmas peppers, Jerusalem cherries, foliage plants, dish gardens, and others are among them.

2. *fig.* An ideal or heavenly city; *spec.* = City 3 b; freq. *the new Jerusalem.*

1382, etc. [see City 3 b]. **1535** [see New *a.* 5 c]. **1601** *Song of Mary Mother of Christ* 38 Ierusalem my happy home, when shall I come to thee. **1804–8** W. Blake *Milton* Pref., in *Compl. Writings* (1972) 481, I will not cease from Mental Fight, Nor shall my Sword sleep in my hand Till we have built Jerusalem In England's green & pleasant Land. **1933** J. Buchan *Prince of Captivity* I. ii. 70 For the first time Adam told another of..his childhood home and its lonely peace... 'Ah,' he said, 'it is as I guessed. We have each our Jerusalem.' **1959** A. Huxley *Let.* 10 Jan. (1969) 863 The nature of these visions is often paradisal and the descriptions of them remind one irresistibly of the description of the New Jerusalem in the Apocalypse. **1968** *Listener* 21 Mar. 377/3 Both men indeed looked forward to the same new Jerusalem—a society in which the state would wither away, bureaucracy would be no more.

3. Used as an exclamation, usu. of surprise.

1840 *Spirit of Times* 8 Aug. 276/2 By Jerusalem! **1868** Reade & Boucicault *Foul Play* III. xvii. 199 'What is your name, when you are ashore?' 'Robert Penfold. The Reverend Robert Penfold.' 'The Reverend!—Jerusalem.' **1872** [see *geewhillikins *int.*]. **1898** J. D. Brayshaw *Slum Silhouettes* 179 Jee-roosalem! You can't stand there; the police won't allow it. **1914** Chesterton *Wisdom of Father Brown* i. 19 'Jerusalem!' ejaculated Brown suddenly. **1968** J. F. Straker *SIN & Johnny Inch* i. 19 He consulted his watch. 'Jerusalem! It's nearly three hours.' **1974** P. Lovesey *Invitation to Dynamite Party* ii. 27 'He kept..buying me whisky.' Jerusalem! Cribb blinked. 'You were on spirits that night, then?'

jes, jes' (dʒes), colloq. shortening of Just *adv.*; freq. in American (esp. Black English) writings. Cf. *jest.

1851 Mrs. Stowe *Uncle Tom's Cabin* (1852) I. iv. 41 Missis let Sally try to make some cake t'other day, jes to *larn* her, she said. **1886** in H. Baumann *Londinismen* 85/2. **1895** *Southern Workman* XXIV. 15/3 Ef I was starvin' an' had jes one ginger-cake, I would give you half. **1925** Odum & Johnson *Negro & his Songs* ii. 39 Ef you want to see old Satan run, Jes fire off dat gospel gun. **1929** *Amer. Mercury* XVIII. 48/1 This big sargent wus jes' too mean to live. **1938** R. E. Bass in B. A. Botkin *Treas. S. Folklore* (1949) III. i. 457 He jes rared back and 'lowed, 'I ain't never told a lie in my life.' **1962** *Jrnl. Amer. Folklore* LXXV. 311 They thinks that science has solved jes' about ever'thin. **1966** *Massachusetts Rev.* VII. 659 He be sittin' theah at crop time, jes' afigurin' an' areckonin'. **1969** 'J. Morris' *Fever Grass* vi. 65 You jes talked a whole lotta sense.

Jesse² (dʒe·si). *U.S. slang.* (? *Obs.*). Also **jesse, jessie, jessy.** [Perh. derived from a jocular interpretation of 'There shall come a rod out of the stem of Jesse' (Isa. xi. 1).] *To give* (*a person*) *Jesse*: to treat or handle severely; to beat or scold soundly. Similarly *to catch* or *get Jesse.*

1839 *Spirit of Times* 19 Oct. 396/3 If I thought he had been shot and creesed in that savigerous sorta fashion, I'd give him jessy with my butcher knife. **1840** *Daily Even. Transcript* (Boston) 12 Feb. 1/1 If any of you ever come to Saco, I kalkilate you'll get *jesse*. **1846** *Spirit of Times* 4 July 223/3 One of the combatants 'caught Jessie'. **1846** D. Corcoran *Pickings* 126 Threatening to give Miss Martin 'jessy' when she would next meet her. **1865** A. H. Stephens *Diary* 29 Sept. (1910) 518 While I thought I was giving you Jesse on hearts, you were giving me fits on spades. **1946** *Amer. Speech* XXI. 153/1 Thornton's latest citation for *give him Jesse* is from 1865... In February, 1946, I heard the expression used in a game of bridge by a player from Sidney, Nebraska, when her partner was ruffing an opponent's suit.

Jessie (dʒe·si). *colloq.* Also **Jessy**, and with lower-case initial. [Female proper name.] A cowardly or effeminate man; a male homosexual.

1923 G. Blake *Mince Collop Close* i. 20 He was a big Jessie, ...but she liked him. **1938** [see *go-round]. **1958** K. Amis *I like it Here* ii. 31 Darling, you really don't have to convince me that you're not a jessie. **1958** M. Dickens *Man Overboard* iv. 59 Don't listen to those timid old Jessies at Southampton. **1971** G. Sims *Deadhand* II. ii. 88 Duff had been scathing about 'soft jessies who couldn't get their fat heads down'. **1973** D. Lees *Rape of Quiet Town* xix. 153 The implication that J. Plummer Esquire was a soft jessie—because he had a rotten objection to getting himself killed.

jest (dʒest), colloq. and dial. var. Just *adv.* Cf. *jes, jes'.

1815 D. Humphreys *Yankey in Eng.* I. 22 I'm rather in a strait, jest now. **1890** Kipling *Barrack-Room Ballads* (1892) 19 Jest send in your Chief an' surrender. **1896** in G. F. Northall *Warwickshire Word-Bk.* 119. **1908** A. J. Dawson *Finn* xix. 293 Jest you remember, my boy, that where I sleeps I breakfast. **1971** M. Babson *Cover-up Story* x. 112 Hell, it was jest a thought. **1973** *Black World* June 63 Jest git on da good foot.

Jesuit, *sb.* Add: **1.** (Further examples.)

For note in Dict. read:

The object of the Society of Jesus was to support the Roman Church in its struggle with the 16th c. Reformers and to propagate the faith among the heathen. Hated and feared by Protestants, the Order, with its authoritarian constitution and its principle of total obedience to papal commands, became suspect to many in Roman Catholic countries too—more especially when Jesuit schools and confessionals came to exercise great influence on rulers and high society. By their enemies, the Jesuits were accused of teaching that the end justifies the means, and the lax principles of casuistry put forward by a few of their moralists were ascribed to the Order as a whole, thus giving rise, not only in English but in French and other languages, to sense 2, and to the opprobrious sense attached to *Jesuitical, Jesuitry*, and other derivatives.

1846 W. F. Hook *Church Dict.* (ed. 5) 491 The Jesuits assume neither the name, quality, nor way of living, of monks. They call themselves an *order of priests*... The end of their institution is the salvation of souls: they preach, instruct youth, read lectures, and dispute and write against heretics. **1913** G. P. Gooch *Hist. & Historians in 19th Cent.* xxvi. 530 Renan sharply castigates the futility of the Priestly Code and the sterile scholasticism of its commentators. Nehemiah is described as the first Jesuit, who turned Jerusalem into a tomb. **1914** [see *bovo]. **1932** E. Bevan *Christianity* ix. 191 Jesuits were trained by a severe discipline, not to live in retirement from the world, but to mingle with the world in order to conquer it for the Church. **1934** H. H. Gowen *Hist. Relig.* xxxix. 598 The Jesuits as a body, by their splendid training, their broad-minded knowledge of human nature, and by their extraordinary personal devotion, did much to win for the Roman Churches territory far larger in area than..had been lost. **1939** [see *iconology 1]. **1953** J. E. Neale *Elizabeth I & her Parliaments* vii. i. 370 The Jesuits came, as it were, direct from England's capital enemy, the Pope. **1959** L. Hanke *Aristotle & Amer. Indians* viii. 108 The efforts of another Jesuit, Antonio Vieira, in the seventeenth century to protect the natives of Brazil. **1972** J. P. Kenyon *Popish Plot* i. 22 The Jesuits..aimed to draw England into the revitalized Church Universal of the Counter-Reformation.

2. (Further examples.) Also *fig. depreciatory.*

[**1852** Thackeray *Esmond* I. v. 99 Father Holt wore more suits of clothes than one. All Jesuits do. You know what deceivers we are, Harry.] **1855** C. Kingsley *Westward Ho!* III. ii. 34 Eustace is a man no longer; he is become a thing, a tool, a Jesuit. **1856** J. W. Carlyle *Jrnl.* 11 Apr. in *Lett. & Memorials* (1883) II. 271 'I'll tell you what to do,' said this Jesuit of a baker; 'Go and join the Methodists' chapel for six months; make yourself agreeable to them, and you'll soon have friends that will help you in your object.' *c* **1879** E. Dickinson *Poems* (1955) III. 1015 The Jesuit of Orchards He cheats as he enchants. **1907** G. B. Shaw *Major Barbara* III. 285 Charles Lomax: you are a fool. Adolphus Cusins: you are a Jesuit. Stephen: you are a prig. Barbara: you are a lunatic. **1923** D. L. Sayers *Whose Body?* ii. 40 Gentlemen, we are not Jesuits, we are straightforward Englishmen. You cannot ask a British-born jury to convict any man on the authority of a probable opinion. **1947** V. S. Pritchett in *Horizon* May 241 Rubashov and Gletkin are a sad pair of Jesuits consumed and dulled as human beings by their casuistry. **1948** D. Shub *Lenin* vii. 152 In July 1916, Viacheslav Menzhinsky, later chief of the Soviet secret police,..wrote: Lenin is a political Jesuit who over the course of many years has molded Marxism to his aims of the moment.

4. a. *adj.* (Further examples.)

1874 J. R. Green *Short Hist. Eng. People* vii. 402 The torture and death of the Jesuit martyrs sent a thrill of horror through the whole Catholic Church. **1922** Joyce *Ulysses* 220 Father Conmee..thought ..of the book that might be written about jesuit houses. **1939** *Times Lit. Suppl.* 14 Jan. 20 One of the most heroic members of one of the most heroic bodies in the history of the world, the Jesuit mission to the Hurons. **1950** *Chambers's Encycl.* VIII. 273/1 In 1872 supervision of schools was reserved exclusively for the government of Prussia and the Jesuit Order was banned from Germany altogether. **1953** J. E. Neale *Elizabeth I & her Parliaments* vii. i. 370 The Jesuit mission to England led by two distinguished and contrasting men, Parsons and Campion. **1956** *Atlantic Monthly* Nov. 43/2 It was a form of answer well known to the examiners—the famous Jesuit equivocation. **1966** D. Johnson *France & Dreyfus Affair* xiii. 220 The tighter the organisation of the group, Assumptionist or Jesuit, the greater the hostility towards Dreyfus. **1972** J. P. Kenyon *Popish Plot* vi. 182 The suspicion that he had designs on the family estate, which should have descended to his Jesuit brother, now in Newgate.

c. Used *attrib.* to designate a type of Chinese 18th-century export porcelain decorated with religious pictures copied from European designs.

1882 W. W. Old *Indo-European Porc.* 5 Dealers as a rule calling it [*sc.* Indo-European porcelain] 'Jesuit china', a general impression has prevailed that it was the work of the converts to the early Romish missions in China and Japan. **1898** W. G. Gulland *Chinese Porc.* 12 Christianity has left little mark on the ceramics of China;..few pieces display biblical subjects or Christian emblems, and such are known as 'Jesuit China'. **1900** F. Litchfield *Pott. & Porc.* vii. 114 This is called 'Jesuit china', because it is said that it was painted to the order of..the Jesuit missionaries. **1927** W. B. Honey *Guide Later Chinese Porc.* viii. 67 'Jesuit china' was..probably copied from designs supplied by the merchants. **1952** M. Anderson *Story Chinese Porc.* 46 Flat-edged plates with biblical subjects painted in a grey-black are the usual features of these 'Jesuit' pieces. **1962** D. Imber tr. *Beurdeley's Porc. E. India Co.* 141/2 The Roman Catholic creed was foremost in Europe at the time, but the great reformers, Calvin and Luther, are also represented on different plates and servers. The British Museum has a plate representing John of Leyden, the leader of the Anabaptists. It will be clear, then, that the title *Jesuit porcelain* cannot be maintained in the light of the facts.

Jesuitical, *a.* Add: **1.** (Further examples.)

1848 *Jesuitism in Church* iv. 41 Jesuitical principles are by no means confined to those who openly adhere to the rule of Loyola. *c* **1905** H. A. Henderson *Shall we tolerate Jesuits?* vii. 24 An influx of Jesuitical religious Orders. **1932** G. F.-H. Berkeley *Italy in Making* I. xvi. 238 The head of a Jesuitical society, the *Amicizia cattolica*. **1943** H. J. Muller *Sci. & Crit.* 195 It has therefore incorporated the Jesuitical principle that the end justifies the means. **1947** C. J. Cadoux *Philip of Spain & Netherlands* vii. 147 The need of money for..launching his Jesuitical seminaries..caused the pope great financial difficulty.

2. (Further examples.) Often used in sense 'hair-splitting', keenly analytical.

1829 *Jesuitism & Methodism* I. viii. 119 She trusted..to preserve Cordelia from being perverted, by Jesuitical craft, from the true light. **1925** T. Dreiser *Amer. Trag.* (1926) II. III. xxvi. 327 'Hesitating fatally but not criminally at the one time in his life when he should not have hesitated'—a really strong if jesuitical plea which was not without its merits and its weight. **1932** E. Bevan *Christianity* ix. 192 Two things especially brought odium upon the name of Jesuit. One is the suspicion of vast subterranean intrigue carried on to gain worldly power..; and the other is the belief that the Jesuits cultivate an immoral casuistry—whence the adjective 'jesuitical'—and especially teach that a good end justifies any kind of means. **1950** *Nation* (N.Y.) 16 Dec. 595 If they insist on debating jesuitical verbal formulas instead of effective and enforceable agreements..then no worth-while end would be served by attempting to bargain with them. **1971** M. Hastings *Jesuit Child* I. i. 14 People only call a man jesuitical when they are beaten in an argument. **1972** *New Yorker* 29 Apr. 112/2 The film ['Loot'] is made up of people who regard themselves as lone wolves with a unique hold on sense who have to put up with the Jesuitical twitterings of everybody else. **1974** *Daily Tel.* 17 Dec. 12 An argument of such Jesuitical subtlety that one would have thought it could impress no one of moderate common sense or sanity.

jesuitically, *adv.* (Later examples.)

1946 *Protestant* (Boston, Mass.) Oct.–Nov. 48 Yet so jesuitically has the argument of 'discrimination' been presented that it has taken in even some well-intentioned progressive citizens. **1962** *Punch* 7 Feb. 253/3 'It's an ancient highway, a bit of old England,' he pleads jesuitically, caring nothing for the means so that his end be achieved. **1974** I. Murdoch *Sacred & Profane Love Machine* 240 Monty in black linen jacket and white shirt, his dark hair well combed and neat, his black shoes ludicrously well polished, was looking his most jesuitically untouchable.

Jesuitism. 1, 2. (Further examples.)

1749 J. Wesley *Jrnl.* 2 June (1912) III. 404 A gentlewoman informed me that Dr. B had averred to her..that it [*sc.* Methodism] was all Jesuitism at the bottom. Alas for poor Dr. B.! **1829** *Jesuitism & Methodism* I. xi. 187 Had he been present,..spite of Jesuitism and priestly pride, it is probable his secret would have escaped him. **1897** J. McCabe *Twelve Yrs. in Monastery* ix. 187, I have met very many priests who quite accept the Protestant Alliance version of Jesuitism. **1901** W. C. Copeland *Empire's Greatest Danger* is. 58 How can one prove that the origin of this particular 'genius' of Jesuitism belongs to Mussulman monasticism? *c* **1905** H. A. Henderson *Shall we tolerate Jesuits?* viii. 32 Jesuitism is too deadly a danger for us as a Protestant nation to tolerate. **1913** J. McCabe *Candid Hist. Jesuits* vii. 167 (*heading*) The first century of Jesuitism. **1956** *Time* 2 Jan. 58/2 Her plans to give 'freedom' to Bavaria were blocked by what she called 'the cloven foot of Jesuitism'. **1970** *Guardian* 25 July 7/2 Both Catholics and Communists suffered from Jesuitism, from a belief that the ends justified the means, and the Church would be willing to come to agreement when the terms suited its books.

Jesuitry. 1. (Further examples.)

1901 W. C. Copeland *Empire's Greatest Danger* vi. 40 Belgium cleared its borders of Jesuitry in 1818. **1911** *Encycl. Brit.* XV. 340/1 A certain characteristic, which soon began to manifest itself in an impatience of episcopal control, showed that the quality of 'Jesuitry', usually associated with the Society, was singularly lacking in their dealings with opponents. **1944** *Atlantic Monthly* Nov. 75/1 Humanism considered as an intellectual discipline-for-discipline's-sake..is a prime specific for such ills as bigotry and puritanism and jesuitry and vulgarity.. and the complacency of the bourgeois mind. **1951** R. Hall *Short Hist. Ital. Lit.* 261 The words *Jesuitry* and *Jesuitical* have become proverbial in reference to dishonestly subtle dialectics and hypocritical condonement of evil practice. **1972** *Times Lit. Suppl.* 22 Dec. 1551/1 Casuistry like Jesuitry is a perfectly respectable word, but it has acquired a sinister connotation.

Jesus. Add: **1. b.** Used as (or as part of) an oath or as a strong exclamation of surprise, disbelief, dismay, or the like; also in various phrases, as *by Jesus, Jesus* (*H.*) *Christ, Jesus wept.* Cf. *gee *int.*², *Jeez(e *int.*

1377, 1676 [in Dict., sense 1]. **1592** Shakes. *Rom. & Jul.* II. iv. 31 Iesu a very good blade, a very tall man. *Ibid.* v. 29 Iesu what hast? can you not stay a while? **1753** Gray *Long Story* in *Six Poems* 22 Jesu-Maria! Madam Bridget, Why, what can the Viscountess mean? **1922** Joyce *Ulysses* 38 Jesus! If I fell over a cliff. *Ibid.* 39 Yes, sir. No, sir. Jesus wept: and no wonder, by Christ. *Ibid.* 131 By Jesus, she had the foot and mouth disease and no mistake! **1923** *Dialect Notes* V. 212 *Jesus Christ*, an expletive or exclamation common to both men and women and considered by neither as in any wise profane. **1924** *Ibid.* 264 Jesus Christ, Jesus H. Christ, holy jumping Jesus Christ. **1936** S. Sassoon *Sherston's Progress* IV. iii. 273 Someone gasping by, carrying a bag of rations—'Jesus, ain't we there yet?' **1943** 'C. Dickson' *She died a Lady* xvi. 135 Have we heard about him?.. Jesus H. Christ! **1966** A. La Bern *Goodbye Piccadilly* vi. 66 It's you she's describing, your clothes, everything. Oh, Jesus wept! **1968** B. Healey *Murder without Crime* iii. 59 Jesus! It's murder out there. **1970** T. Lewis *Jack's Return Home* 223 'Jesus Christ!' he says softly. 'Jesus H. Christ.' **1974** I. Murdoch *Sacred & Profane Love Machine* 94 He's so spineless... He just wants to be let off and I let him off. Jesus wept!

3. b. Used *attrib.* or *Comb.* to designate a fervently evangelical type of Christian, or a group of such people, or things characteristic of them.

1970 *Catholic Worker* Feb. 2/3 He still spoke contemptuously of Jesus-shouters. **1970** *N.Y. Times* 22 Feb. 64 They gladly accept the name Jesus Freaks and have long hair and other marks of hip culture. **1971** *New Scientist* 3 June 588/1 Its Jesus shops, symbols of the brand of freaked-out Christianity that has replaced Flower Power as a culture. **1971** *Listener* 9 Sept. 324/1 The Jesus people, offering religion as a substitute for the drug cult among young people, arrived from America. **1971** *Daily Colonist* (Victoria, B.C.) 12 Dec. 4/1 Certainly the sect seems to have outfreaked all its competitors in the 'Jesus Freak' movement. **1972** *Sat. Rev.* (U.S.) 6 May 58/2 (*heading*) Meditations on the Jesus Movement. *Ibid.*, The Jesus revolution is simply the most recent, popular, and obvious expression of that need. *Ibid.*, The Jesus-cults..offer the young..an instantaneous and push-button forgiveness. **1972** *Awake!* 8 Nov. 3/1 Also called 'Jesus freaks', or 'Street Christians', they put up Jesus posters, wear Jesus buttons, and emblazon their car bumpers with stickers that say: 'Honk, if you love Jesus.' **1973** *Times* 3 Mar. 6/3 Dr Ramsey said: 'I welcome Jesus Freaks as a genuine religious movement.'

jet, *sb.*³ Add: **III. 4. c.** *Astr.* (i) A thin, well-defined stream of luminous material extending from the nucleus in the head of a comet.

1866 W. Lockyer tr. *Guillemin's Heavens* (ed. 2) 293 If we look at the drawings..of the comet of 1862..we shall be astonished at the rapidity of the changes of position and form of the luminous jets which successively were emitted from the nucleus... M. Chacornac was able to distinguish the formation of thirteen..jets, similar to jets of steam. **1888** C. A. Young *Textbk. Gen. Astron.* (1889) xvii. 404 In the case of a very brilliant comet, its head is often veined by short jets of light which appear to be continually emitted by the nucleus. **1931** *Publ. Lick Observatory* (Univ. Calif.) XVII. 481 Secondary nuclei were found showing all the properties of the primary nucleus, namely, halos, jets, and streamers. **1966** *McGraw-Hill Encycl. Sci. & Technol.* III. 314/1 At smaller distances [from the sun] (less than 0·5 AU) there may be profuse emission of material but distinguishable features, such as jets, are seldom observed.

(ii) A spicule or similar structure in the solar chromosphere.

1948 *Astrophysical Jrnl.* CVIII. 130 An interpretation of the chromospheric spicules as a system of superthermic jets is presented. **1951** *Monthly Notices R. Astron. Soc.* CXI. 630 The solar chromosphere exhibits an intricate fimbriate structure that consists of very small prominences in contact with the photosphere. For brevity, these chromospheric details will be called 'jets'. *Ibid.*, 'Spicule', introduced by W. O. Roberts.., apparently refers to the largest of the jets that appear in the Sun's polar regions. **1953** G. P. Kuiper *Sun* v. 212 The chromosphere seems composed of a more or less homogeneous layer, from which emerge fine streaks or spikes... Roberts has introduced the name spicules for these fine details, Lyot and Mohler call them jets.

(iii) A narrow strip extending radially outwards from some galaxies and quasars and usually differing from the rest of the galaxy or quasar in the radiation it emits.

1954 *Astrophysical Jrnl.* CXIX. 221 NGC 4486 has a unique peculiarity which has been known for a long time. In the center of the nebula..is a straight jet, extending from the nucleus in position angle 290°... Several strong condensations are in the outer parts of the jet, which extends about 20″ from the nucleus and has an average width of about 2″. **1966** *Nature* 13 Aug. 698/2 The synchrotron radiation of the jet of 3C 273 is detectable optically but not in the radio wave-lengths. **1967** *Internat. Astron. Union Symposium* XXXI. 442 A well-known feature of M87 is a jet, about 1000 pc long, emitting polarized light via the synchrotron mechanism. **1972** *Physics Bull.* Apr. 202/2 An example is galaxy M87, the famous 'jet' galaxy showing a high velocity jet of matter flying from its centre.

d. = *jet stream.

1953 *Q. Jrnl. R. Meteorol. Soc.* LXXIX. 236 The wind components perpendicular to this axis were evaluated on each sounding at 50-mb intervals below and above the level of the jet. **1957** *Ibid.* LXXXIII. 222 The local and shallow nature of the near-discontinuity surface compared to the upper-temperature gradients in the jet certainly suggests a minor role for the front in the overall mechanics of the waves. **1963** tr. *E. R. Reiter's Jet-Stream Meteorol.* iv. 167 (*heading*) The structure of the frontal zone in the region of the polar-front jet. **1968** *Jrnl. Atmospheric Sci.* XXV. 1169/1 An equatorial jet can be driven by a convergence of momentum flux. **1970** *Nature* 17 Jan. 254/2 Thermal convection on a global scale in the lower atmosphere of Jupiter redistributes angular momentum so as to produce a well defined westerly jet at high level astride the equator... On Earth, equatorial jets are found in the lower stratosphere and in the oceans.

5. (Further examples.)

1887 *Encycl. Brit.* XXII. 500/2 The oil is injected in the form of a spray..by a steam jet arranged in such a way that air will be drawn into the furnace along with the petroleum. **1901** *Motor-Car World* II. 42/1 Sometimes the jet gets stopped up, causing the engine to cease working. **1929** Newton & Steeds *Motor Vehicle* vii. 108 [In a Diesel engine] the jet must also be so disposed and directed that a stream of liquid is not likely to impinge on the cylinder wall or piston, where rapid carbonization would occur. **1943** A. P. Fraas *Aircraft Power Plants* vii. 116 The simple carburetor has been used to meter fuel to gasoline engines from the time they first came into use. Essentially, it consists of two orifices in parallel. The first, the air venturi, meters the air flow. The second, the fuel jet, meters the fuel flow. **1963** C. Campbell *Sports Car Engine* vi. 97 This is a simple auxiliary carburettor with an air jet and a fuel jet feeding into a passage leading to the engine side of the throttle plate.

c. [Ellipt. use.] A jet plane or a jet engine.

1944 *Flight* 10 Feb. 153/2 The advantages of the jet are so great that I am sure their development will be rapid. **1944** *Collier's* 22 Apr. 13/3 The jet..is capable of faster flight at low altitudes than any airplane with conventional engines and propeller. **1948** 'N. Shute' *No Highway* i. 7 The Mark I model..had radial engines, though now they all have jets. **1957** *Economist* 31 Aug. 697/1 In military air weapons, the jet is now giving way to the rocket motor. **1973** *Observer* 14 Jan. 7/2 The enormous capacity of the latest generation big jets can be filled only by promotional and concessionary fares.

9. *jet speed, velocity*; **jet boat** *N.Z.* (see quot. 1968); **jet engine**, an engine utilizing jet propulsion to provide forward thrust, esp. an aircraft engine that takes in air and ejects hot compressed air and exhaust gases; so *jet-engined* adj.; **jet flap** *Aeronaut.*, a jet of gas from a jet engine ejected downward as a sheet through a slot in the trailing edge of an aircraft wing, so as to act as a flap and increase the lift; **jet injector** *Med.*, an instrument for giving injections without breaking the skin by means of a fine jet of fluid forced through it under pressure; **jet motor** = *jet engine* above; **jet pipe**, a pipe or duct from which a jet of fluid is expelled; *spec.* the exhaust duct of a jet engine; **jet turbine**, a turbojet engine. Also *jet stream.

1963 J. Hamilton (*title*) White water, the Colorado jet boat expedition, 1960. **1967** A. & D. Reid *Paddle Wheels on Wanganui* p. x, Canoes, rafts and jetboats notwithstanding, there can be no equal to the experience of shooting a long twisting rapid in..a Wanganui riverboat in its heyday. **1968** *Times* 31 Aug. 7/5 The jet boat is..a simple enough contraption—a strong glass fibre hull..; a powerful and reliable inboard engine; and..a highly developed water pump. Water is sucked in through a flat grating..and projected violently out the back to supply propulsion and steering. **1970** *N.Z. Listener* 27 Feb. 1 (*caption*) The experimental jetboat on the Waimakariri River, 1951. **1975** *N.Z. News* 8 Jan. 9/3 Another development of consequence in New Zealand has been the invention of the jet boat... Propelled and steered by its water jet, the [Sir William] Hamilton jet boat can make 180-degree turns in its own length. **1943** *Jrnl. R. Aeronaut. Soc.* XLVII. 413 In general, the jet engine performance is given not in h.p. but in kg. of thrust. **1950** *Sci. News* XV. 72 The ordinary jet engine uses oxygen from the air, but the rocket carries its own oxygen supply with it in some form. **1952** A. Y. Bramble *Air-Plane Flight* x. 156 There are, at present, three types of air-plane jet 'engine'—(*a*) rocket, (*b*) turbo-jet, (*c*) athodyd. **1969** *Jet Engine* (Rolls-Royce Ltd.) (ed. 3) i. 4/1 Although a rocket engine..is a jet engine, it has one major difference in that it does not use atmospheric air as the propulsive fluid stream. **1952** *Lancet* 15 Nov. 967/2 In jet-engined aeroplanes the high sonic frequencies predominate. **1960** *Guide Civil Land Aerodrome Lighting* (*B.S.I.*) 30 Where a runway is used by large jet-engined aircraft, blister type fittings are advisable as elevated fittings can be blown over by the jet efflux. **1958** *Spectator* 7 Feb. 160/3 The vast amount of British work on blown flaps, jet flaps, jet lift and boundary layer control. **1963** *Engineering* 5 Apr. 493/1 The principle of the jet flap was..patented in 1952 by the National Gas Turbine Establishment and is a technique whereby the power of jet engines installed for propulsion can also be used greatly to increase the wing lift at low speeds. **1947** *Current Res. Anesthesia & Analgesia* XXVI. 223 Figure 2 shows position of the jet injector for anesthetizing the skin preparatory to insertion of lumbar puncture needle. **1965** *Brit. Med. Jrnl.* 25 Dec. 1541/1 High-pressure jet injectors..throw the material to be injected with considerable force through a very fine nozzle. **1973** *Lancet* 28 Apr. 927/2 Contacts of the early cases were vaccinated by intradermal injection, but two jet injectors were ordered for the vaccination of the 5 and 6-year-old children. **1944** *Discovery* Nov. 346/2 This is the Whittle aircraft, with a gas-turbine jet-motor. **1950** J. V. Casamassa *Jet Aircraft Power Syst.* v. 89 The Aerojet motor is a true rocket, since its propellent charge contains both the fuel and the oxygen necessary to burn it. Consequently, this jet motor delivers its rated power independent of altitude or atmospheric conditions. **1946** *Jrnl. R. Aeronaut. Soc.* L. 317/1 After compression in the compressor, the air is circulated through a radiator or heat exchanger situated in the jet pipe of the engine. **1966** *McGraw-Hill Encycl. Sci. & Technol.* X. 227/2 Jetting consists of displacing the soil at the pile tip by means of a quantity of water or air through an internal or external jet pipe, which also lubricates the sides of the pile as it rises to escape. **1966** D. Stinton *Anat. Aeroplane* vii. 126 The thrust of a turbojet is frequently augmented by burning additional fuel in the jetpipe, thus utilizing unburnt air. **1934** *Aircraft Engineering* VI. 170/2 At the maximum pressure end the difference between the static pressure inside the [wind] tunnel and atmospheric at a jet speed of 200 ft./sec. is about 45 lb./sq. ft. **1945** *Sci. Amer.* June 365 Jet turbines have extremely low oil consumption ratings. **1949** *Punch* 14 Sept. 286/3 Emerging miraculously from its dive, the monster..flashes low across the runway, the whiffling roar of its jet-turbine manfully pursuing. **1935** in *Aeronaut. Jrnl.* (1970) LXXIV. 130/2 The jet velocity resulting from a useful heat drop of 250 units is..2320 ft/s. **1966** *McGraw-Hill Encycl. Sci. & Technol.* XIV. 159/1 Under these conditions, afterburning to an exhaust temperature of 3460°R provides an increase in jet velocity of..1·5.

IV. 10. Combs. in which *jet* represents 'jet engine', as *jet efflux, fuel*; *jet-assisted, -powered* adjs.; esp. in the designations of aircraft (and occas. other forms of transport) powered by jet engines, as *jet aeroplane, aircraft, air liner, airplane, bomber, fighter, lifter, liner, plane, tanker, trainer.* Also **jet lift**, lift (vertical thrust) provided by a jet engine. (For *jet engine, motor*, etc., see sense *9.) Also *jet propulsion.

1951 'J. Wyndham' *Day of Triffids* ii. 36 Also I must buy an aeroplane—a jet aeroplane, very fast. **1944** *Flight* 3 Feb. 130/1 Now about the jet aircraft. Instead of the jets coming out of the tail, shape the aircraft like an orange pip and have several jets coming out of the shoulders, pressing air against and all round the fuselage. **1954** *Economist* 11 Sept. 12/2 On multi-engined jet aircraft the horizontal tail must also be placed away from the jet blast. **1959** *Daily Tel.* 23 Feb. 11/8, I have seen films of man's experience of weightlessness in a jet aircraft. **1970** *Times* 25 Feb. 8/6 The first Whittle jet was tested three years later [*sc.* in 1937].., and the first British jet aircraft flew in 1941. **1947** *Jrnl. R. Aeronaut. Soc.* LI. 178/2 With regard to the comfort of jet air liners, he wondered whether people were being misled. **1959** *Daily Tel.* 2 Mar. 16/4 The new large jet airliners are probably as fast, and can fly as high, as most jet bombers now in service. **1944** *Collier's* 22 Apr. 13/1 Few standard conveniences for upper stratosphere flying had yet been built into the jet airplane. **1944** Jet-assisted [see *jato]. **1957** *Jane's Fighting Ships 1957–8* 413 The missile is using jet-assisted rocket bottles to launch it. **1952** *Oxf. Jun. Encycl.* X. 56/2 One of the first tactical jet bombers to go into service was the English Electric *Canberra*. **1965** H. Kahn *On Escalation* x. 200 In a jet-bomber and ballistic-missile age, events go so fast. **1960** Jet efflux [see *jet-engined* adj. in *9]. **1966** D. Stinton *Anat. Aeroplane* 244 The position of the fin was an initial argument against mounting the engine above the boom, because of the hot jet efflux playing on the surfaces. **1944** *Sat. Even. Post* 6 May 20/2 The British had flown a jet plane successfully, and now the USAAF proposed to develop a twin-engined jet fighter of its own. **1955** *Ann. Reg. 1954* 149 An outstanding item in Canadian mutual aid within N.A.T.O...was the transfer of 164 Sabre jet fighters to Greece and Turkey. **1964** M. McLuhan *Understanding Media* II. xxvii. 272 The two pilots of one Canadian jet fighter. **1953** *Ann. Reg. 1952* 164 Agreement was also reached on financing the first half of the..programme for air-fields, communications, and jet-fuel supplies. **1966** *McGraw-Hill Encycl. Sci. & Technol.* I. 178/2 Volatility is the most important consideration in the selection of jet fuels; combustion qualities are of secondary concern. **1958** Jet lift [see *jet flap* in *9]. **1968** *New Scientist* 25 July 167/1 Jet lift, as a means of personal mobility, has been a tempting idea since Rolls-Royce began using the Flying Bedstead about 15 years ago... This invites speculation as to whether paratroopers, deployed from a transport aeroplane, might subsequently be gathered back into it with the help of their individual jet-lift belts. **1954** P. Masefield in *Listener* 30 Sept. 511/2 The commercial jet-lifters are yet to come. **1967** *Times Rev. Industry* Feb. 68/2 As long as a decade ago, technically competent enthusiasts were saying that another form of vertical take-off and landing (VTOL) transport, the jet-lifter, was just round the historical corner. **1949** *Birmingham* (Alabama) *News-Age-Herald* 13 Nov. A. 18/3 But there is much work yet to be done before the combination of the jetliner and the helibus can be fully utilized. **1961** A. Miller *Misfits* i. 12 A great jet liner roars over, flying quite low. **1970** *New Scientist* 23 Apr. 172/2 The cruising speed would be competitive with current jetliners at Mach 0·93. **1971** *Daily Tel.* (Colour Suppl.) 30 July 25 This frenetic era of automated jetliners and impersonal airports. **1944** G. Smith *Gas Turbines & Jet Propulsion for Aircraft* (ed. 3) 104 He believed the jet plane would be used in the near future. **1944** *War Illustr.* 10 Nov. 404/3 The first enemy jet-plane to fall in Allied lines was shot down over Nijmegen on October 5 by six R.A.F. Spitfires; it was a Me 262. **1964** M. McLuhan *Understanding Media* II. xviii. 177 The effect of..jet-plane speeds. **1956** W. A. Heflin *U.S. Air Force Dict.* 281/1 *Jet-powered*, powered by one or more jet engines. **1957** *Encycl. Brit.* I. 246/2 Aeroplanes are either propeller driven or jet powered. **1959** *Daily Tel.* 2 Mar. 16/4 The United States Air Force already has a large number, probably about 400, Boeing KC-135 jet tankers. **1967** *Times* 28 Feb. (Canada Suppl.) 27 Jet tankers are also needed for long distance inflight refuelling of the CF5. **1959** *Daily Tel.* 23 Feb. 11/8 The first such experiments were conducted by a pilot and a doctor in a T-33 jet trainer.

11. In numerous combinations (in which *jet* represents 'jet plane') relating to locomotion by means of jet aircraft, as *jet route, transport, transportation, travel, traveller*; *jet-borne* adj. Also **jet age**, the age or era of travel by jet aircraft; **jet hop**, a short or rapid flight by a jet aircraft; also as *v. intr.*; **jet lag**, the delayed effects (esp. temporal disorientation) suffered by a person after a long flight on a (jet) aircraft; **jetport** [after *airport], an airport served by jet aircraft; **jet set**, a smart set of wealthy people who conduct business by jet travel, or who make frequent journeys, e.g. to holiday resorts, by jet aircraft; also in weakened senses; so **jet-setter.**

1952 R. Walker (*title*) Jet age. **1953** *Sci. News Let.* 21 Feb. 122/1 A weird-looking, skin-tight 'space suit' now clothes the jet-age test pilot. **1958** *New Statesman* 21 June 804/2 The uppermost deck of this air-pier or jet-age jetty is an open public area—'waving base' in official terminology—that gives one a grandstand or pierhead view of the comings and goings. **1963** *PMLA* LXXVIII. I. p. v, Air France ground hostesses threaten to work in civilian clothes unless their employer comes up with jet age fashions. **1971** *Guardian* 12 June 12/3 You can cleanse yourself immediately of most jet-age pollution in the..local Turkish bath. **1968** *N.Y. Times* 21 Apr. IV. 1 Johnson..climbed aboard his jetborne White House for the flight back to his Texas ranch. **1963** *Times* 26 Feb. 8/6 A certified real Orient just a jet-hop away. **1969** J. Mander *Static Soc.* i. 27 As he jet-hops from country to country, it is the similarities that strike him. **1969** 'A. Cade' *Turn up Stone* i. 23 The long journey, jet-lag and the heat had given him a headache. **1971** *Times* 16 Apr. 7 (Advt.), A team of scientists has researched the phenomenon of 'jet lag' thoroughly. The disorientation of a jet flight has been proved to upset..decision-making. **1972** *Sat. Rev.* (U.S.) 10 June 15/2 The news of Nixon—booked into Salzburg to fight jet lag for a weekend on his way to the summit meeting in Moscow. **1973** *Observer* 3 June 29/1 They had the dazed, sleep-walking look of people still confused by jet-lag. **1961** *N.Y. Times* 1 June 37/1 (*caption*) New study backs jetport in Morris. **1972** *Fortune* Jan. 40D/3 (Advt.), A fourth major jetport to serve the New York metropolitan area. **1970** *Times* 27 Feb. 3/8 (*heading*) Windsor on jet route. **1951** *San Francisco Examiner* 5 Aug. 5/1 You're strictly jet set.. if you stake your claim in the dunes..*never* descend to ocean level except for a quick dunk. **1956** *N.Y. Times Mag.* 4 Nov. 14 This is the Soviet 'Jet Set', an element of the younger generation... The term was originated by a young member of a foreign embassy staff in Moscow and refers to the Soviet youth who are attracted by things foreign. **1963** *Amer. Speech* XXXVIII. 206 *Jet set* (analogous to *smart set*), rhyming slang term referring to the sophisticated, well-to-do skiers who go abroad by jet just for skiing vacations in the best known resorts. **1964** *Sat. Rev.* (U.S.) 10 Oct. 70/1 The Jet Set..has rediscovered St. Tropez. **1966** *Daily Tel.* 24 Oct. 13/6 Clothes for the jet set, for you who seek the sun in winter. **1970** *Americana Ann.* 694 His campaign managers created a new image of him [*sc.* Pierre Trudeau] as the youthful, debonaire, 'with-it' man of the jet-set age. **1972** *House & Garden* Dec.–Jan. 126/2 Even in the jet-set hotels..local dishes are presented. **1965** *N.Y. Times* 13 June 17 (*caption*) Jet-setter Mrs...and model..prepare for ABC's fall special, 'The Wild, Wild East'. **1966** *Time* 9 Dec. 88/3 In swarmed the jet-setters (Gloria Guinness, Lee Radziwill, Count and Countess Rudolfo Crespi). **1973** *Daily Tel.* 22 Sept. 19/8 The crowd..naturally included show business celebrities and jet-setters who had paid £40 for court-side seats. **1949** *Flight* 29 Sept. 438/2 A large jet transport might take five years to develop to the production stage. **1961** L. Mumford *City in Hist.* xvi. 505 Jet transportation brings an area twelve hundred miles away as near as one sixty miles distant today. **1962** *Daily Tel.* 13 June 11/1 What with jet travel, the Common Market and [etc.]. **1964** M. McLuhan *Understanding Media* II. x. 94 The jet traveler..might just as well be in a cocktail lounge.

jet, *v.*² Add: **9.** *trans. Building.* To loosen and remove (sand, gravel, etc.), or to sink (a pile), by the technique of jetting (see *jetting *vbl. sb.*² 4).

1956 H. L. Nichols *Mod. Techniques Excavation* v. 58/2 In peat, the points are jetted down and sand pumped in the hole around them. *Ibid.* vi. 25/2 If the cover should be left off, and the vertical pipe filled with dirt.., it may be jetted out by the use of an engine-driven water pump.

jet (dʒet), *v.*³ [f. Jet *sb.*³] *intr.* To travel by jet plane. Also *trans.* (and *refl.*), to convey by jet plane or jet engine. So **je·tted** *ppl. a.*

1946 *All Hands* June 50 A Martin Marauder was jetted into the air for the scientist-spectators. **1949** *Sat. Rev.* (U.S.) 8 Jan. 17, I rather think Captain Osborne will be the first of us to go jetting to the moon. **1951** A. C. Clarke *Sands of Mars* vii. 81 They jetted themselves slowly out across the surface of Deimos. **1965** F. Pohl *Alternating Currents* (1966) 96 We can't jet home through normal space because we don't have the fuel. **1959** *Time* 23 Mar. 20/3 Jetting home to Moscow..Krushchev exuded confidence. **1962** *Daily Tel.* 13 June 11/1 There's no rest for the 'jetted' British businessman. **1966** *Guardian* 7 June 12/6 In the last few days, Brown has jetted over an area the size of Europe. **1968** *Daily Tel.* 28 Sept. 9/6 (Advt.), Clarksons jet you to top resorts like Alpbach, Auffach, [etc.]. **1971** *Radio Times* 21 Oct. 71/3 Perhaps I'd been lucky to catch him in today, before he jetted off to Tokyo or the Bahamas? **1973** *Daily Tel.* (Colour Suppl.) 10 Aug. 7/2 But Beer now jets between Santiago and his Surrey home.

jetavator (dʒe·tăvēⁱtəɪ). *Astronautics.* Also **jetevator.** [f. (irreg., in the case of *jetavator*) Jet *sb.*³ + El)evator.] A ring-shaped deflector surrounding the exit nozzle of a rocket engine which can be swivelled into the exhaust gases to divert them and so alter the direction of the thrust.

1960 *Aeroplane* XCVIII. 176/2 The missile is powered by a solid-propellent Aerojet rocket motor fitted with 'jetavator' deflectors of the kind developed for Polaris. **1961** *Flight* LXXIX. 13/2 The use of swivelling nozzles is claimed to be lighter than the Polaris system of fixed nozzles plus jetevators, and undoubtedly minimizes drag. **1967** *Propulsion* (Amat. Rocket Assoc. U.S.) ii. 45 A jetavator consists of a ring surrounding the exhaust-nozzle exit. The ring is mounted so that it can be moved into the jet exhaust as needed.

‖ **jeté** (ʒəte). Also erron. **jété, jetée, jetté.** [F. *jeté*, pa. pple. (sc. *pas* step) of *jeter* to throw.] A ballet-step, having a wide variety of forms, in which a spring is made from one foot to land on the other. So **jeté en tournant** (aṅ tʊrnaṅ), a leap executed with a turning movement. Cf. *grand jeté.

1830 R. Barton tr. *Blasis' Code of Terpsichore* II. vi. 78 *Entrechats* are generally begun with an *assemblé, coupé*, or *jetté*. **1877** [see *contretemps 3]. **1927** *Dancing Times* Apr. 9 All dancers know that the coupé cannot be done alone, as the position of the foot which is being 'cut-away' depends entirely upon the step which follows the coupé. I am therefore taking the 'coupé-dessous', followed in the first instance by a 'posé', and afterwards the 'coupé-dessous' followed by a 'jeté'. **1930** Craske & Beaumont *Theory & Pract. Allegro in Classical Ballet* 95 Execute a *Jeté en Tournant*, that is:—With a slight spring, change the position of the feet so that the *right* foot is *sur le cou de pied devant*. **1953** *Ballet Ann.* VII. 59 She took off with a wonderful *jeté* and landed only a few inches from his nose. **1958** *Observer* 28 Sept. 18/5 A special salute is due to Yelle Bettencourt's splendid *jétés*. **1961** *Times* 28 Dec. 12/7 M. Jean-Paul Andreani is quite electrifying with his *jétés en tournant*.

jetevator, var. *jetavator.

jeton (dʒe·tən). [See Jetton.] **1.** = Jetton.

1933 H. G. Wells *Shape of Things to Come* II. § 11. 229 Today our museums contain hundreds of thousands of specimens of these improvised European coins of lead, nickel, tin and all sorts of alloys, *jetons* or checks of wood. **1969** R. C. Bell *Board & Table Games* II. x. 138 Many of these casting-counters, or *jetons*, simulated coins, and cause difficulty to collectors of medieval money.

2. (ʒetoṅ). A metal disc used, chiefly in France, instead of a coin for insertion in a public telephone box. Also *attrib.*

1942 E. Paul *Narrow St.* xi. 82 In order for a client to use the phone he had to buy from her a metal disc or *jeton*. **1957** *Times* 16 Oct. 11/7 The French *jeton* system for calls from public telephone boxes. *Jetons* were introduced in France because of frequent changes in the local call tariff (15 changes in 20 years) and in the diameter and material of the coinage. **1962** A. Williams *Long Run South* VI. ii. 193 An old crone sold him a jeton and he dialled the number of the house in the Viale Piemonte. **1972** *Times* 4 Jan. 14/8 (*heading*) Use the nickel sixpence as a jeton [for telephone coin boxes].

jet propulsion. [Jet *sb.*³] The ejection of a usu. high-speed jet of gas (or liquid) as a source of propulsive power, esp. for aircraft.

1867 *Ann. Rep. Aëronaut. Soc.* 50 In spite of the costly experiments..made to revive the system of jet propulsion, it was clear to him that the screw propeller must be much superior. **1877** C. B. Mansfield *Aerial Navigation* II. xiii. 453 One of the great advantages attending the system of pump or jet propulsion in the air is that, however the force be generated, the direction in which it is applied may be instantly altered, without shifting the position of the actual instrument of motion. **1886** *Ann. Rep. Aëronaut. Soc.* 67 (*heading*) Jet propulsion for aëronautical purposes. **1892** J. H. Cotterill *Appl. Mech.* (ed. 3) 572 The efficiency of jet propulsion [of ships] can hardly be estimated as greater than ·33. **1920** *Chambers's Jrnl.* June 415/1 In the case of life-boats liable to ground on sandbanks..jet-propulsion has obvious advantages. **1935** Balmer & Wylie *After Worlds Collide* iv. 75 The most energetic members of the colony were working upon a small metal jet-propulsion [air]ship. **1944** *Time* 24 Jan. 66 This..drawing illustrates..the operation of a propellerless, jet-propulsion airplane. **1945** *Daily Express* 4 June 4/8 America's car manufacturers are secretly experimenting to produce the first jet-propelled car, using exhaust gases from the present car to boost power by jet propulsion. **1953** J. Y. Cousteau *Silent World* xi. 106 The octopus was downright terrified... It made off by slow jet propulsion, exuding spurts of its famous ink. **1962** F. I. Ordway et al. *Basic Astronautics* x. 392 Jet propulsion may be divided into two categories: ducted or air-breathing engines and rocket engines.

So **jet-propelled** *a.*, having or employing a means of jet propulsion; also *fig.*, very fast, frenzied; also (as a back-formation) **jet-propel** *v. trans.* (chiefly *fig.*).

[**1867** J. Bourne *Treat. Screw Propeller* (new ed.) p. cxix/3 Johnson's jet propeller.] **1877** W. H. White *Man. Naval Archit.* xiv. 612 Jet-propelled vessels, when moving ahead at full speed, derive their steering power from the reaction of the water in the wake upon the rudder. **1904** R. Kennedy *Mod. Engines & Power Generators* I. ii. 104 The Viper and Vixen, twin screw vessels of about the same tonnage..were built as fair competitors with the jet propelled Waterwitch. **1922** *Flight* XIV. 276/2 Many of the inventions imagined by Jules Verne..have already been realised, so who shall say that the design for a jet-propelled monoplane seaplane described by M. Maurice Armende..will not materialise—some day? **1936** *Daily Sketch* 17 Nov. 13/4 In visualising a jet-propelled aeroplane of the future, the mind immediately thinks of a large hole in the tail, from which would come a blast of whatever gases were being released. **1944** A. Huxley *Let.* 9 July (1969) 507 One can be safe in betting that, within ten years, there will be rockets, or jet-propelled flying bombs,..capable of flying any distance up to five thousand miles. **1945** F. Whittle in *Proc. Inst. Mech. Engin.* CLII. 422/2 At the beginning of 1940 the Air Ministry.. began to work on the assumption that there was a good chance of getting jet propelled fighter aircraft into production for use in the war. **1947** *Sci. News* IV. 35 When Sir Malcolm Campbell makes his next attempt on the world's speed record, he will be using a true 'jet-propelled' speedboat. **1949** *Sat. Rev.* (U.S.) 25 June 17/2 It was that courage..that jet-propelled her onward and upward to stardom and marriage to Florenz Ziegfeld, Jr. **1950** in *Amer. Speech* (1956) XXXI. 285 They have learned to jet-propel bodies, but they haven't taken the first step in jet-propelling the human spirit. **1954** 'N. Blake' *Whisper in Gloom* I. i. 15 If the jet-propelled craft he had built failed to jet-propel itself. **1958** A. Huxley *Let.* 22 June (1969) 850 Time, as one advances in life, seems to become jet propelled. **1959** *Encounter* Aug. 35 He was lit up and jet-propelled by a sort of crazy, electric frenzy. **1960** *Times* 6 Aug. 9/5 On the same disc..is Mendelssohn's concerto which Heifetz carries by assault. He lays into the triplets of the first movement.., plays the slow movement with a fair amount of elbow-power, and jet-propels the fairies through the last. **1962** *Economist* 19 May 672/1 Dr Nkrumah's phrase about achieving a 'jet-propelled' rate of economic growth. **1969** *Daily Tel.* 13 Feb. 22/8 The book conveys the period of music-making in the 1920s and 1930s and its air of domesticity far removed from today's jet-propelled bustle.

jet stream. [f. Jet *sb.*³ + Stream *sb.*] **a.** A fast-moving, relatively narrow stream of fluid that is present as a current in an atmosphere or ocean; *spec.* in *Meteorol.*, a strong wind confined to a narrow region of the atmosphere, esp. one in the upper troposphere at middle latitudes that blows in an approximately horizontal direction, predominantly from west to east.

1947 *Bull. Amer. Meteorol. Soc.* XXVIII. 255 During periods of reasonably straight west wind circulation over the North American continent, there exists normally, at levels between 5 km and 15 km above sea level, a fairly narrow zone of extremely strong west wind circulation (jet stream), reaching its maximum intensity and sharpness at the tropopause level. **1950** *Time* 29 May 70/2 Six miles up, where the air is thin and cold, a fearful wind zigzags round the earth at 200 m.p.h. Meteorologists call it the 'jet stream'. **1957** G. E. Hutchinson *Treat. Limnol.* I. iv. 225 The jet-stream mechanisms believed to maintain circulation in the region of the westerlies. **1963** tr. *E. R. Reiter's Jet-Stream Meteorology* iv. 271 (*heading*) Jet streams in the oceans. *Ibid.*, The current does not spread out..but it remains concentrated in a narrow band of high velocities—the oceanic jet stream. **1971** *New Scientist* 8 Apr. 90/2 In the tropopause..there is a meandering hemispheric band of high speed westerly winds, called the jet streams.

b. A jet of fluid (Jet *sb.*³ 4), esp. one ejected by a jet engine.

1955 *Sci. News Let.* 5 Mar. 160/3 Graphite lubricant is propelled in a jet-stream reaching many often inaccessible trouble spots. **1955** *Ibid.* 26 Mar. 200/1 These air whirlpools are not caused by either the propeller backwash or jet stream, but by the action of the plane's wings. **1959** S. N. Samburoff tr. *Feodosiev & Siniarev's Introd. Rocket Technol.* vii. 232 The presence of the jet stream.. introduces certain specific peculiarities into rocket aerodynamics. **1962** H. E. Newell *Express to Stars* iii. 29 It takes a force to expel material into a jet stream..and by Newton's law of reaction there must be a reaction force.

‖ **jettatura** (yetătūə·rā). [ad. It. *iettatura.*] The evil eye (see Evil *a.* 6); bad luck. So **jettato·re** [ad. It. *iettatore*], a person who brings bad luck.

1855 Mrs. Gaskell *Accursed Race* in *Househ. Words* 25 Aug. 78/2 Their glance, if you meet it, is the jettatura, or evil eye. **1882** C. M. Yonge *Unknown to Hist.* II. iii. 34 'Tis not only the *jettatura* wherewith the Queen Mother used to reproach me. Men need but bear me good will, and misery overtakes them. **1892** A. Lang *Bks. & Bookmen* (new ed.) 122 The superstitious might have been excused for crediting him with the gift of *jettatura*,—of the evil eye. **1921** *Glasgow Herald* 29 July 6 This simple remedy is much in use throughout Italy to-day as an antidote to the evil power of the Jettatore.

jetted (dʒe·ted), *a.*² *Tailoring.* [app. f. Jet *v.*²] Of a pocket: having no flap, but an outside seam on either edge, called the jetting.

1923 *Daily Mail* 23 Apr. 8 The skirt pockets, which are finished in jetted fashion. **1928** [see *flapless *a.*]. **1933** J. E. Liberty *Pract. Tailoring* v. 38 *Jettings.* These are strips of material about 9 in. long and 2 in. wide. They.. are made up in two forms, flat jettings,..and as a narrow piping.

jetting, *vbl. sb.*² Add: **2. b.** (See quot. 1957.)

1941 *Nature* 5 Apr. 422/2 Work in Britain has been concentrated mainly on chemical means of protecting the host by the use of dips. Jetting has been tried in Scotland, but protective dusts have not been experimented with on a large scale. **1957** *New Biol.* XXII. 99 Mr A. J. Gillespie, a woolgrower in Queensland, suggested jetting. This consisted of spraying the breech of the sheep with a jet of fluid delivered at a pressure of up to 120 lb per square inch from a nozzle whose aperture varied between 3/64 and 5/64 of an inch. A solution of sodium arsenite was the most popular jetting fluid.

4. *Building.* The loosening and removal of sand, gravel, etc., by directing jets of water or compressed air on to it, esp. so as to make a hole for pile-driving; the sinking of a pile by this means.

1942 *Amer. Speech* XVII. 280/1 *Jetting*, the act of pushing aside the mud or sand at the foot of a pile by a jet of compressed air or water. **1948** D. W. Taylor *Fund. Soil Mech.* xx. 647 Jetting often aids greatly in getting piles to the desired final grade. **1951** G. P. Tschebotarioff *Soil Mech.* xv. 445 It is not always possible to drive piles.. through a compact layer of sand without serious damage to the piles... Jetting of the piles has to be used then. **1956** H. L. Nichols *Mod. Techniques Excavation* v. 60/1 Jetting with high pressure water, or less commonly, compressed air, is used in making deep narrow holes for setting piles, installing vertical drains, obtaining soil samples, and for various other purposes. **1966** *McGraw-Hill Encycl. Sci. & Technol.* X. 228/1 Open-type piles are usually cleared of soil by jetting.

je·tting, *vbl. sb.*³ [f. *jet *v.*³ + -ing¹.] Travelling in a jet plane.

1966 in Webster. **1971** *Guardian* 9 Sept. 11/3 It is work which involves much jetting about,..lecture tours across the States, visits to North Vietnam.

jettison, *v.* Add: **b.** To release or drop from an aircraft or spacecraft in flight; *spec.* to drop (a bomb) intentionally from an aircraft elsewhere than over an assigned target.

1934 [implied in *jettisonable *a.*]. **1942** D. M. Crook *Spitfire Pilot* 66 Enemy bombers showed a much greater tendency to jettison their bombs. **1942** *R.A.F. Jrnl.* 18 Apr. 14 'C for Charlie's' pilot jettisoned his guns rather too near the wife of a senior officer who happened to be below. *Ibid.* 34 The crew went to their ditching stations. ..No fuel was jettisoned. **1946** Taylor & Allward *Spitfire* 101/1 The hood..could be jettisoned. **1955** *Times* 10 Aug. 8/7 The world's fastest aircraft..was jettisoned by its mother aircraft after an explosion at 30,000 ft. **1974** *Flight International* 29 Aug. (Suppl.) 11/1 Surface-to-air missiles often use a booster to produce a high initial acceleration and push the missile clear of its launcher, the booster then being jettisoned.

2. *intr.* To drop off, or fall away from, an aircraft or spacecraft in flight.

1962 J. Glenn in *Into Orbit* 138 The escape tower, which we would not need once we got above the atmosphere, would jettison when the boosters dropped off and lighten the load even more.

jettisonable (dʒe·tisənăb'l), *a.* [f. Jettison *v.* + -able.] Capable of being jettisoned from an aircraft or spacecraft; designed to be readily detachable in flight.

1934 *Shell Aviation News* No. 34. 18/1 As a result, the machine can be used for a variety of purposes... It is fitted with a 'jettisonable' fuel tank riveted with duralumin. **1946** *Aeroplane Spotter* 14 Dec. 290/1 The pilot and test engineer are housed in a conical nose section which is jettisonable and contains a decelerating parachute in case of emergency. **1956** *Spaceflight* I. 24/1 The second stage ..is a new rocket; the final stage and the satellite package are enclosed within its jettisonable nose. **1961** E. Brown *Wings on my Sleeve* x. 113 The undercarriage consisted of a retractable skid mounted on the jettisonable two-wheel chassis which was automatically dropped when the skid was retracted on take-off. **1963** A. Smith *Throw out Two Hands* v. 52 The balloon commander is all,..the passenger is nothing more than jettisonable ballast.

jettisoning, *vbl. sb.* (*Fig.* and *Aeronaut.* examples.)

1950 *Engineering* 10 Feb. 158/1 Landing operations.. could take the form of 'jettisoning' operations. **1957** J. S. Huxley *Relig. without Revelation* ix. 209 This wholesale jettisoning.

jeu. Add: **c. jeu de paume** (ʒö də pōm) [Fr., lit. 'game of palm (of the hand)']: tennis (not lawn-tennis); also, a tennis court; also *fig.*

1789 A. Young *Jrnl.* 20 June in *Trav. France* (1792) I. 115 The resolution..was to assemble instantly at the *Jeu de paume*, and there the whole assembly took a solemn oath. **1880** Geo. Eliot *Let.* 7 Oct. (1956) VII. 329 Johnnie gets a game of real tennis—jeu de paume—every day. **1910** *Encycl. Brit.* X. 450/1 Fives and racquets are probably both descended from the *jeu de paume*, of which they are simplified forms. **1939** A. Toynbee *Study of Hist.* IV. 161 Warfare now is no longer just a *jeu de paume* among a party of kings. **1940** E. Pound *Cantos* lviii. 74 Playing at jeu de paume and escrime. **1970** *Guardian* 3 June 9/2 The story that the *jeu de paume* has been played by members of the opposing army with a living infant as the ball, goes back hundreds..of years.

d. jeu de règle (ʒö də rɛ̄gl') [Fr., lit. 'game of rule']: in the game of *Écarté* (see quot. 1963).

1850 *Bohn's Hand-bk. Games* 261 On this principle all '*Jeux de Regles*' are played without changing (although there be a few which can scarcely reckon in their favor 2 to 1). **1934** *Neuphilol. Mitt.* XXXV. III/IV. 132 Écarté.. *jeu-de-règle*, hand which should be played without taking cards. **1963** G. F. Hervey *Handbk. Card Games* 61 There are certain stock hands, called *jeux de règle*, holding which a player should play and not propose, and equally refuse the opponent's proposal.

e. jeu de société (ʒö də sosiete) [Fr., lit. 'game of society']: (freq. in *pl.*) a game or amusement at a party.

1827 *Edin. Rev.* XLVI. XCII. 382 To these pantomimes succeeded ballets, and such *jeux de société* as 'La Peur'.. a sort of dumb show. **1854** Thackeray *Newcomes* I. xxviii. 271 These little diversions and *jeux de société* can go on anywhere. **1932** H. Nicolson *Diary* 1 Jan. (1966) 104 He..had been kept up doing *jeux de société* till 8 am. **1963** A. Zaina in B. Sewell *Two Friends* 78 The two duologues..were obviously destined to be a *jeu de société*... Perhaps it was performed at one of the Raffalovich parties.

jeuk skei: see *jukskei.

‖ **jeune fille** (ʒön fīy). [Fr.] A young girl. Used *attrib.* or as *adj.*: characteristic of an *ingénue.*

1802 H. Eckersall *Jrnl.* 4 July in Malthus *Trav. Diaries* (1966) 296 Les jeunes filles who were sitting near us were much diverted but did not look as if they would take his advice. *a* **1855** C. Brontë *Professor* (1857) I. xii. 191, I took these sketches in the second-class school-room of Mdlle. Reuter's establishment, where about a hundred specimens of the genus 'jeune fille' collected together, offered a fertile variety of subject. **1888** E. Dowson *Let.* 13 Nov. (1967) 19 It is not les filles de l'opéra or horizontales that I protest against—it is les jeunes filles de société. **1892** [see *enfant gâté]. **1903** H. James *Ambassadors* VI. xiv. 189 It will really be a chance for you..to see the *jeune fille*—I mean the type—as she actually is. **1926** C. Beaton *Diary* in *Wandering Yrs.* (1961) 110 Miss Wilberforce plays safe in a *jeune fille* frock. **1928** D. H. Lawrence *Phoenix II* (1968) 519 Talking to an intelligent girl, the famous '*jeune fille*'..you find..that she, the innocent girl in question, is flinging all sorts of fierce questions at your head. **1946** 'S. Russell' *To Bed with Grand Music* ix. 115, I always think romance is rather *jeune fille*. **1974** *Times* 30 Apr. 8/6 The dresses are ravishing—*jeune fille* dresses of the right sort, drifting and romantic.

‖ **jeune premier** (ʒön prəmye). [Fr., lit. 'first young man'.] An actor who plays the part of the principal lover or young hero. So **jeune première** (prəmyɛ̄r), the performer of the corresponding female part.

1852 *Blackw. Mag.* Nov. 600/2 The *prima donna* and *jeune premiere* of the troop. **1877** *Sat. Rev.* 24 Nov. 662/2 What the *jeune premier* would necessarily be when acting the part of a ruined country gentleman. **1888** *Athenæum* 3 Nov. 588/2 Theology always plays a part, albeit in the form of the *jeune premier*, the handsome curate with Broad Church instead of agnostic views. **1896** Mrs. H. Ward *Sir G. Tressady* xiii. 283 Ancoats always seems to me the *jeune premier* in his own play. **1902** *Sat. Rev.* 13 Sept. 329/2 [He] plays him with no more intelligence than would suffice for the part of a quite ordinary jeune premier. **1924** 'R. Crompton' *William—the Fourth* iv. 71 William..has eloped with a *jeune première* and a bear-skin. An entire Christmas pantomime is searching the village for him. **1946** Mrs. Belloc Lowndes *Let.* 1 Jan. (1971) 269 Lord Rosebery..was the *jeune premier* of that generation. **1958** L. Durrell *Mountolive* xv. 272 Mountolive was now irradiated by an appalling sense of futility as he sat (like some ageing *jeune premier*) and listened to the torrent of Nur's excuses. **1960** *Guardian* 20 May 8/4 He..became a highly successful jeune premier in light comedy and revue.

‖ **jeunesse** (ʒönɛs). [Fr.] Young people; the young.

1781 H. Walpole *Let.* 25 July (1904) XII. 31 All the *jeunesse* strolled about the garden. We ancients..retired from the dew. **1858** Trollope *Doctor Thorne* II. xv. 309 'La jeunesse' was beginning to get a lesson; experience.. sometimes comes early in life. **1884** H. James in *Atlantic Monthly* Sept. 310/2 In the early days of October when the whole *jeunesse* of the country is going back to school. **1947** *Horizon* Oct. 11 As if the wonderful *jeunesse* of America were suddenly to retain their idealism.

‖ **jeunesse dorée** (ʒönɛs dore). [Fr., lit. = gilded youth.] Orig. applied in France to the group of fashionable counter-revolutionaries formed after the fall of Robespierre; now *gen.*, young people of wealth and fashion.

1830 W. Hazlitt *Life Napoleon Buonaparte* I. viii. 383 The volatile genius of this people [*sc.* the French] ..decked out the youth of the city [*sc.* Paris] (*La Jeunesse Dorée*) in the Chouan uniform. **1836** *Edin. Rev.* LXIII. 460 The indefatigable opponents of the ancient rudeness, the *jeunesse dorée* of Germany. **1837** [see Gilt *ppl. a.* 2]. **1845** *Encycl. Metrop.* XIII. 374/1 These young men, who were commonly known as the 'Jeunesse dorée', no longer permitted the Jacobins to hold assemblies in the public places. *Ibid.*, The Jeunesse dorée and the Thermidorians had on their side the tradesmen. **1886** *Athenæum* 11 Sept. 329/2 We shall not envy the *jeunesse dorée* of the period these so-called sports. **1888** *N. & Q.* 7th Ser. V. 190/1 *Jeunesse dorée* answers, perhaps, rather to Disraeli's expression of 'curled darlings' than to 'dandy'. **1910** D. Schwann *Bk. Bachelor* 47 Mason..received the guests, who were the fine flower of dramatic and critical Bohemia, with a sprinkling of the *jeunesse dorée* of Society and high finance. **1939** N. Monsarrat *This is Schoolroom* i. i. 28 One of the favourite children, the *jeunesse dorée*. **1965** *New Statesman* 9 Apr. 568/2 The *jeunesse dorée* would have to go out and work for their living. **1969** J. Mander *Static Soc.* v. 134 Those Country Clubs where the *jeunesse dorée* of Lima, Caracas and Bogota cavort. **1974** D. Smith *Look back with Love* x. 90 The love-life of the Old Trafford *jeunesse dorée*.

Jew, *sb.* **1.** Read: A person of Hebrew descent; one whose religion is Judaism; an Israelite. (Add further examples.)

Orig. a Hebrew of the kingdom of Judah, as opposed to those of the ten tribes of Israel; later, any Israelite who adhered to the worship of Jehovah as conducted at Jerusalem. Applied comparatively rarely to the ancient nation before the exile (cf. Hebrew *sb.* 1), but the commonest name for contemporary or modern representatives of this group, now spread throughout the world. The word 'Jew' is also applied to groups, e.g. the Falashas in Ethiopia, not ethnically related to persons of the main European groups, the Ashkenazim and the Sephardim.

1710 etc. [see *Falasha]. **1940** Auden *Another Time* 116 He [*sc.* Sigmund Freud] Was taken away from his old interest To go back to the earth in London, An important Jew who died in exile. **1956** I. Murdoch *Flight from Enchanter* ix. 126 'Of course, you realize that I could rescue you with my little finger,' said Mrs Wingfield. 'I'm as rich as a Jew!' **1968** L. Rosten *Joys of Yiddish* 142 Relentless persecution of Jews, century after century, in nation after nation, left a legacy of bitter sayings: 'Dos ken nor a goy.' ('That, only a *goy* is capable of doing'). **1970** R. D. Abrahams *Positively Black* iii. 76 The Englishman is arrogant and overbearing, the American is a check-writing millionaire who doesn't mind the cost, the Jew tries to push down the entry price into heaven. **1970** *Times* 28 Jan. 10/4 At the heart of the matter lies the rabbinical definition of a Jew: a person born of a Jewish mother, or a person who has converted to Judaism according to rabbinical law. **1971** B. Malamud *Tenants* 50 The Jews got to keep us bloods stayin weak. **1974** J. R. Baker *Race* xiv. 232 In various parts of the world today there are communities that practise the Jewish faith in one form or another, but are ethnically distinct from the Jews of Europe and North America. *Ibid.* 234 From the traditional religious point of view, a Jew was a person born of a Jewish mother, but this formula suffers from the defect that the defined word is included in adjectival form in the definition. The same flaw occurs in part of the new definition enacted by the Israeli Parliament in..1970, according to which a person is a Jew if he or she is the offspring of a Jewish mother or has been converted to the Jewish faith by the Orthodox Rabbinate or by the Rabbis of the Jewish Reform Movement or by the Rabbis of the Jewish Conservative Movement.

c. *Black Jew* (see quot. 1967); also = *Falasha.

1807 C. Buchanan *Jrnl.* 4 Feb. in *Christian Res. in Asia* (1811) 192 The resident Jews are divided into two classes, called the Jerusalem or *White* Jews; and the Ancient or *Black* Jews. The White Jews reside at this place [*sc.* Cochin]. The Black Jews have also a Synagogue here; but the great body of that tribe inhabit towns in the interior of the province. *a* **1817** T. Dwight *Trav. New-Eng.* (1823) III. 174 The black Jews in Hindostan. **1822** *Imperial Mag.* IV. 358 A copy of the Hebrew Pentateuch ..found in one of the Black Jews' Synagogues, at Cochin. **1843** J. C. Maitland *Lett. from Madras* xviii. 178, I told him about the first preachers, the Black Jews, the Syrian Christians, &c. **1892** G. M. Rae *Syrian Church in India* x. 150 These black Jews are converts to the faith from among the people of the land. **1907** I. Zangwill *Ghetto Comedies* 155 The black Jews..surrounded by all those millions of Hindoos. **1930** H. Norden *Africa's Last Empire* 185 The black Jews among whom he works. **1964** [see *Falasha]. **1967** D. T. Kauffman *Dict. Relig. Terms* 77/1 *Black Jews*, in India, term applied to brown-skinned Jews to distinguish them from a group known as 'White Jews'. Sometimes used also for Negro Jewish groups. **1974** J. R. Baker *Race* xiv. 232 The Falasha or 'black Jews' of Ethiopia are members of the Aethiopid subrace, a hybrid taxon.

d. A ship's tailor. Hence also **jewing** *vbl. sb.* and *ppl. a.* *Naut. slang.*

1916 *Chambers's Jrnl.* May 278/2 They [*sc.* ships' tailors] were still known as 'jews'. *Ibid.*, The term 'jewing', as sewing is still called. **1945** 'Tackline' *Holiday Sailor* x. 102 There was the 'Jewing-bloke', who undertook tailoring repairs. *Ibid.*, The 'Jewing-bloke' had a rather ancient Singer sewing machine, bought when ashore at Alexandria with..pay in his pocket. **1946** J. Irving *Royal Navalese* 100 A sailor-tailor is known as a 'Jew'. **1962** Granville *Dict. Sailors' Slang* 66/1 *Jewing firm*, ship's tailoring 'firm' run by one or more ratings who repair and make clothing.

2. Read: *transf.* and *offensive.* As a name of opprobrium: *spec.* applied to a grasping or extortionate person (whether Jewish or not) who drives hard bargains. (Add further examples.)

In medieval England, Jews, though engaged in many pursuits, were particularly familiar as money-lenders, their activities being publicly regulated for them by the Crown, whose protégés they were. In private, Christians also practised money-lending, though forbidden to do so by Canon Law. Thus the name of Jew came to be associated in the popular mind with usury and any extortionate practices that might be supposed to accompany it, and gained an opprobrious sense.

1846 *Swell's Night Guide* 123/1 *Jew*, an overreaching fellow. *c* **1861** E. Dickinson *Poems* (1955) I. 160 'Twould be 'a Bargain' for a *Jew*! Say—may I have it—Sir? **1906** J. M. Synge *Lett. to Molly* (1971) 31 What have I done that you should write to me as if I was a dunning Jew? **1920** T. S. Eliot *Ara Vos Prec* 14 The jew is underneath the lot. Money in furs. **1931** T. R. G. Lyell *Slang* 428 Why waste your time asking him for a subscription? He's a perfect Jew where money's concerned. **1944** *Britannica Bk. of Year* 693 In March 1943 there were tirades from Bangkok radio against the 'Jews of Siam' (probably Chinese), who were accused of profiteering. **1952** G. Bone *Came to Oxf.* xi. 34 There is a curious fallacy, rather wide-spread, that a borrower of money is an innocent and hapless person, while a lender is a shark, a harpy, a 'Jew'. **1964** H. Brotz *Black Jews of Harlem* iii. 54 Occasionally the Black Jews forget they are Jews when complaining about the fact that 'the Jews' own all or most of Harlem!

b. A pedlar.

In this use not depreciatory.

1803 G. Colman *John Bull* iii. ii. 32 Here is two poets, and a poll-parrot, the best image the Jew had over his head, over the mantle-piece. **1963** 'E. McBain' *Ten Plus One* (1964) iv. 42 There was a guy who used to come around to the door selling stuff, and my mother called him 'The Jew'... For her, 'Jew' was synonymous with pedlar. **1970** J. H. Gray *Boy from Winnipeg* 43 For us, however, 'Jew' was just another generic word that often included the peddlers who were Greek or Italian. When we scrounged bottles it was to sell to 'the Jew', who was anybody that came along buying junk.

3. a. *Jew boy* (examples), *girl, man.*

Such expressions now mainly in offensive use but not originally opprobrious.

1796 P. Colquhoun *Treat. Police of Metropolis* (ed. 3) vi. 125 Jew Boys..go out every morning loaded with counterfeit Copper, which they exchange for bad Silver, to be afterwards coloured anew, and again put into circulation. **1817** M. Edgeworth *Harrington* iii. 45 Mowbray easily engaged me to join him against the Jew boy; and a zealous partizan against Jacob I became. **1873** Trollope *Eustace Diamonds* II. liii. 361 You used to be very wicked, and say he was once a Jew-boy in the streets. **1929** D. H. Lawrence *Let.* 10 Oct. (1962) II. 1208, I do hate John's Jewish nasal sort of style—so uglily moral. ..Spring doesn't only come for the moral Jew-boys—for them perhaps least. **1948** J. Baldwin in *Commentary* Oct. 334/2 Jules Weissman, a Jewboy, had got the room for me. **1954** Jewboy [see *ikey *sb.* and *a.*]. **1959** N. Mailer *Advts. for Myself* (1961) 50 Jewboy, blond Jewboy Wexler perched by the cellar window, tackling Japs with machine-gun bullets. **1968** *Daily Mail* 9 Feb. 3/3 Angry viewers rang the BBC last night to complain about an 'anti-Semitic' remark on the TV programme *Softly, Softly*. In last night's episode..the detective tells the man: 'You always were a great one for putting things in your wife's name, Bob, just like a Jewboy heading for bankruptcy.' **1972** *Observer* 7 May, Mrs Lane Fox dismisses what she calls the country set, who call their children 'the brats', talk about 'thrashing them into shape', support Enoch Powell and still refer to 'jew boys'. **1974** *New Society* 3 Jan. 11/2 A car's desirability also creates the opposite reaction, in that envy is easily turned to resentment and aggression towards, for instance, the 'jewboy', the 'poser', the 'toffee nose' and the 'business classes' who sport expensive and powerful cars. **1796** E. Wynne *Diary* 11 Dec. (1937) II. ix. 139 Before having ever seen us she declared that we were all excepting the youngest, like little crows and Jew girls. **1930** E. Pound *XXX Cantos* x. 45 Wives, jew-girls, nuns. **1971** B. Malamud *Tenants* 50 Jewgirls are the best whores. **1905** Joyce *Let.* 29 Oct. (1966) II. 127 For a Jewman it's better than having to bathe. **1922** —— *Ulysses* 336 I'll brain that bloody jewman for using the holy name. **1938** W. B. Yeats *John Kinsella's Lament* in *London Mercury* Dec. 114 Though stiff to strike a bargain Like an old Jew man.

c. Jew-baiter (later examples); **Jew-baiting** *sb.* and *a.* (later examples); **Jew plum** = Otaheite apple; **Jew Tongo,** a language spoken among Bush Negroes in Surinam, possessing a structure largely derived from West African languages and a vocabulary largely derived from English.

1907 I. Zangwill *Ghetto Comedies* 85 She's honest... She won't fall back on the old Jew-baiter. **1945** W. S. Churchill *Victory* (1946) 145 Julius Streicher, most notorious of Jew-baiters, was captured by the Americans. **1960** C. Day Lewis *Buried Day* vi. 116 The same herd instinct that produces Teddy Boys, Jew-baiters and Ku-Klux-Klansmen. **1974** G. Mitchell *Javelin for Jonah* xi. 133 Benjy was unlucky enough to fall foul at school of a ring of young Jew-baiters. **1922** Joyce *Ulysses* 202 Shylock chimes with the jewbaiting that followed the hanging. **1939** *Ann. Reg. 1938* 203 The brutalities began on April 23, and it was clear that the scheme of Jew-baiting had been worked out in readiness for the 'Anschluss'. **1969** J. Mander *Static Soc.* iii. 99 The American, however ugly, is no Jew-baiting *Gauleiter*. **1913** W. Harris *Notes Fruit & Veg. in Jamaica* 18 The Jew Plum..was introduced to Jamaica in 1782 and again in 1792. **1920** W. Popenoe *Man. Tropical & Subtropical Fruits* iv. 156 Jew-plum is another name for the ambarelle, used in Jamaica. **1971** *Caribbean Q.* XVII. ii. 14 Different name, same referent..golden apple / Jew plum / pomme-citerre. **1933** L. Bloomfield *Lang.* xxvi. 474 Two creolized forms of English are spoken in Suriname (Dutch Guiana). One of these.., more divergent from ordinary types of English, is known as Jew-Tongo. **1968** W. J. Samarin in J. A. Fishman *Readings Sociol. of Lang.* 666 Amerindian pidgins... Saramakan (Jew Tongo, Ningre-Tongo).

Jew, jew, *v.* Add: (Further examples.) Also, to drive a hard bargain. Also *intr.*, to haggle. Phr. *to jew down*, to beat down in price; also *transf.*

These uses are now considered to be offensive.

1824 C. Harding *Diary* 29 Apr. in *Sketch* (1929) 75 He is a country clergyman; and, from his Jewing disposition, I should judge he had more taste in tithes than pictures. **1825** *Constitutional Adv.* (Frankfort, Kentucky) 15 Dec. 3/1 We hope, for the honour and character of the state, that neither the legislature nor the people, will Jew the items of expence. **1833** L. Dow *Dealings of God* (1849) 189 If they [*sc.* the Jews] will *Jew* people, they cannot flourish among Yankees, who are said to '*out-jew*' them in trading. **1848** W. Bagley *Let.* 14 Mar. in N. E. Eliason *Tarheel Talk* (1956) 279, I Jewed old Galloway down to 1·50 for ploughs. **1851** H. Mayhew *London Labour* I. 368/1 Some of the ladies in the squares..sets to work Jewing away as hard as they can, pricing up their own things, and downcrying yourn. **1870** *Congress. Globe* 7 July 5340/1 This bill supposes that Congress..is ready to commence jewing down the pay of its General. **1872** *Chicago Tribune* 14 Oct. 8/2 The prices [for lodging] asked vary—the lodger being generally asked as much as it is thought he will give. If he jews, he will get it for comparatively little. **1883** G. M. Hopkins *Let.* 6 Dec. (1938) 195 You will I know say..that Jew is a reproach because the Jews have corrupted their race and nature, so that it is their vices and their free acts we stigmatise when we call cheating 'jewing'—and that you mean that Disraeli in 1871 overreached and jewed his constituents. **1897** [see *Highland *a.* 2 b]. **1908** *Dialect Notes* III. 324 *Jew*, to beat down the price. 'I tried to jew him, but he wouldn't jew.' **1926** *Market Growers Jrnl.* 1 July 3, I make my retail prices about half way between grocery store wholesale and retail prices, and do not stand for any 'jewing' down. **1937** *Scribner's Mag.* Apr. 25 Thought we might get the divorce a little under fifty dollars. Maybe we might jew the young man down. **1939** A. Powell *What's become of Waring* v. 140 Then we can meet again and jew each other down. **1946** W. G. Hammond *Remembrance of Amherst* 121 Both here and at the mountain top we were unmercifully jewed for all the refreshments. **1947** L. Z. Hobson *Gentleman's Agreement* i. 9 Now she was describing the large new house she and Dick wanted to buy. 'Did you close the sale on the old place?' Mrs. Green asked. 'Not yet. That cheap Pat Curran keeps trying to Jew us down.' **1968** L. Rosten *Joys of Yiddish* 142 Just as some Gentiles use 'Jew' as a contemptuous synonym for too-shrewd, sly bargaining ('He tried to Jew the price down,' is about as unappetizing an idiom as I know), so some Jews use *goy* in a pejorative sense. **1970** R. Lowell *Notebk.* 69 This embankment, jewed—No, yankeed—by the highways down to a grassy lip. **1971** R. Thomas *Backup Men* xxi. 184, I say how much and he says this much and I say it's not enough so we jew around with each other until we make a price. **1972** *Harper's Mag.* May 83 Jew the fruitman down for his last Christmas tree. **1972** *New Society* 11 May 301 1, I got jewed down..over the cheap offer.

jewel, *sb.* Add: **3.** *jewels of the crown*, a rhetorical phrase for the colonies of the British Empire. *temporary.*

1901 P. Manson in *Daily Mail Year Bk.* 101, 10 or 15 years hence that region [*sc.* West Africa] would be regarded as one of the richest jewels in the crown of England. **1931** *N. & Q.* 5 Sept. 166/2 Those irritations against tyranny and stupidity which lost us those jewels of the Crown [*sc.* the American colonies].

5. d. jewel-weed, read: *U.S.*, either of two annual herbs, *Impatiens capensis*, with orange flowers, or *I. pallida*, with yellow flowers, native to eastern and central North America.

1818 A. Eaton *Man. Bot.* (ed. 2) 283 *Impatiens*.. *nolitangere* (jewel-weed, touch-me-not). **1869** J. G. Fuller *Uncle John's Flower-Gatherers* 223, I have seen.. the wild jewel-weed in our meadow. **1884** [in Dict.]. **1907** *St. Nicholas* July 842/2 A sudden thrust of the lantern into this clump of jewel weed near our path, produces a shower of miniature drops. **1951** R. Frost *Coll. Poems* 143 It will be found Either to have gone groping underground..Or flourished and come up as jewel-weed. **1968** *Herb Grower Mag.* (Falls Village, Connecticut) XXI. 11. 23 Jewelweed is our favorite [name] for the plant which makes the spotted-touch-me-nots..droopy, poised to defy gravity even on leaves tipped downwards, and cold rain, dew or even tap water that falls upon the foliage.

jeweller, -eler. Add: *jeweller's* (or *jewellers'*) *rouge*, a fine preparation of ferric oxide used as rouge.

1839 [see Rouge *sb.*¹ 2]. **1886** *Encycl. Brit.* XXI. 13/2 Jeweller's rouge for polishing gold and silver plate is a fine red oxide of iron prepared by calcination from sulphate of iron. **1947** J. C. Rich *Materials & Methods Sculpture* vii. 210 For the final stages of polishing, a chamois skin can be used to apply a fine buffing powder or jeweler's rouge.

Jewess. (Later examples.)

1922 Joyce *Ulysses* 756 Are you not a Jewess?.. On purpose of my mother... I.. mother..why she...Shaw..crat..we..**1970** J... Rumanian beauty?' 'I think..she is a typical little Jewess.' [illegible]

Jewessy (dʒū·ɛsi), *a. depreciatory.* [f. Jewess + -y¹.] Resembling or characteristic of a Jewess.

1930 J. B. Priestley *Angel Pavement* iv. 151 And there were two or three [girls] worth looking at, the flashy young Jewessy type.

jew-fish. (Earlier and later examples.)

1679 T. Trapham *Discourse State of Health in Jamaica* 65 The Jew fish conceived to be one of the first three of our most worthy Fish. **1968** J. E. Randall *Caribbean Reef Fishes* 60 Jewfish, *Epinephelus itajara*... Tropical western Atlantic and eastern Pacific... Highly esteemed as food. **1972** *Nature* 7 Apr. 266/2 The groupers (some species are known as jewfishes and hinds) are well represented in American warmer seas.

jewhil(l)iken(s), -kin(s), varr. *Gewhilli-kins.

Jewish, *a.* Add: **1.** (Later examples.)

1925 P. Gibbs *Unchanging Quest* xiv. 109 These peasants think the Duma will..kill all the *Koulaks*, or Jewish moneylenders. **1935** [see *Amerindian a.]. **1941** *Times* 24 Feb. 102/3 Alfred Rosenberg last fortnight opened in Frankfurt an Institute for Research into the Jewish Problem.. in the world dealing exclusively with the Jewish problem'. **1963** A. Powell *Question of Upbringing* iii. 96 'Jew—and have you seen it? A Jewish old clothes man would think twice about wearing it. **1957** *Ovf. Dict. Chr. Ch.* 1093/2 The Jewish problem arose at a later date than in Spain, and it was not until 1536 that, under the influence of the civil power, the Inquisition was set up. **1968** L. Rosten *Joys of Yiddish* xxi It is important to note that the idea of respect for others and the values of a pluralistic society form an old, integral part of Judaism and Jewish tradition. **1970** I. Shur *Memoirs* vi. 113 In February we submitted a memorandum to the Chief Rabbi, in which we asked that Palestine should be placed under such political, administrative, and economic conditions as will ensure the establishment therein of the Jewish national home.' **1970** *Times* 24 Jan. 7/1 It is now declared that under the law of Israel a person may be of Jewish nationality although not of Jewish religion. **1972** *Jewish Chron.* 4 Feb. 14/3 Though the PLO leaders seek to identify a distinction between the Palestinian Arabs and these oriental Jews whom they call 'Jewish Arabs'. **1973** *Daily Mail* 4 Oct. 4/5 The strain of anguish and uncertainty shows in the face of this Jewish woman as she waits with her grandson following a bomb alert before going to Schoenau transit camp in Austria. **1974** *Times* 18 Feb. 2/4 *Publishers Weekly* 18 Feb. 12/2, I do not see why writers should be singled out as ethnic identities when the habit of so designating individuals has been abandoned generally. I suppose it is because of the new use of the term 'Jewish writers'. But those are writers—I believe—who concentrate on the experience of being Jewish in America.

Jewy (dʒū·i), *a. depreciatory.* Also **Jewey.** [f. Jew *sb.* + -y¹.] Resembling or characteristic of a Jew or the Jews; having the characteristics attributed to Jewish people. Also as *sb.*

1904 Fowler & Felkin *Kate of Kate Hall* xvii. 194 They [*sc.* the boots] are as like as two peas, and both in my mind, a bit Jewy. **1914** E. Pugh *Cockney at Home* 67 You're a dirty, bloomin' lyke, Jewey. **1919** W. N. P. Barbellion *Jrnl. Disappointed Man* 154 That beard, Jewy, secretive, petty creature, J. M. W. Turner. **1922** Joyce *Ulysses* 287 Jesus, I had to laugh at the little jewy getting his shirt out. *Ibid.* 299 I'm told those Jewies does have a sort of a queer odour coming off them for dogs. **1930** J. B. Priestley *Angel Pavement* iv. 184 A neat dark Jewy sort of chap.

Jeyes(') fluid (dʒēiz). [f. *Jeyes*, the name of the manufacturing company.] The proprietary name of a disinfectant consisting of a saponified solution of phenols, resins, and other ingredients. Also *Jeyes(') disinfecting fluid.*

1888 *Trade Marks Jrnl.* 18 July 995/2 [In] sickness always use Jeyes' disinfectants.. Jeyes' Sanitary Compounds Company, Limited, ..London, ..manufacturers]. **1900** *Encycl. Medica* III. 31 Cresol or methyl phenol with its derivatives forms the chief constituent of.. 'Jeyes' Fluid' and a host of similar mixtures. **1911** *T. Eaton & Co. Catal.* Spring & Summer 173/2 Jeyes' Disinfecting Fluid. **1933** D. L. Sayers *Murder must Advertise* iv. 72 'It's the boy that goes round with the disinfectant, sir.' 'Ah, of course! Spray with Sanfect and you're safe.' 'That's right, sir, except that they use Jeyes' Fluid.' **1957** S. Lockett *Clin. Toxicol.* xiv. 490 Phenolic substances and cresols are contained in a number of disinfectants, e.g... Izal, Jeyes' Disinfecting Fluid and Creolin. For sanitary purposes varying mixtures with soap and alkali are used. **1973** *Times* 19 June 1/3 Jeyes Fluid will rise by a maximum of 20 per cent.

jhoom, jhum, varr. Joom, jūm.

1886 Yule & Burnell *Hobson-Jobson* 351/2 Jhoom. **1895** W. R. Fisher *Schlich's Man. Forestry* IV. 350 Jhuming, or the thorough burning of brushwood on the soil. *Ibid.* 343 In *jhums*, or cultivations on forest clearings, where the branches and undergrowth are burned. **1925** J. H. Hutton *Angami Nagas* ii. 72 Good jhum land, cleared once in twelve or fifteen years. *Ibid.*, The Lhotas, Semas, Aos, and trans-Dikhu and trans-Tizu tribes cultivate only by 'jhuming'. **1927** *Blackw. Mag.* June 816/1 Crops are grown by a simple method known as 'Jhoom'. *Ibid.*, A fresh patch of jungle is chosen.. and the 'Jhuming' process repeated. **1936** *Nature* 5 Sept. 408/1 A fairer form, who.. practised a rude form of agriculture, comparable to jhuming, but not terracing or systematic irrigation. **1937** *Times Lit. Suppl.* 20 Feb. 123/2 The low-caste Indian addicted to jhum cultivation. **1946** Ali & Lambert *Assam* 16 One of two tribes have adopted terrace cultivation; but the others depend on the *taungya* or *jhuming* system, which is very wasteful.

jhula (dʒū·la). *India.* Also **joolah.** [Hind., Hindi *jhūlā* swing, swing-rope.] A simple swinging rope-bridge used in the Himalayas.

1830 A. S. H. Mountain *Mem. & Lett.* (1857) 114 Our chief object (according to the Sutlej) was to swing on a Joolah bridge. **1844** J. C. Stocqueler *Hand-bk. India* 458 The rustic bridge is superfluous.. by planks, formed by ropes stretched across, constituting a species of loose parapet. **1923** *Blackw. Mag.* Aug. 259/2 A *jhula* or swinging rope-bridge.

jiao (dʒau). Also **chiao.** [Chinese.] A unit of currency and coin of China.

1949 *Whitaker's Almanack* 953/2 By Presidential Decree of August 19, 1948, the Chinese National Dollar currency ..was replaced by a new currency named the *Gold Yuan*. ..The subsidiary coinage consists of *fen* (cents) and *Chiao* (tenths of a yuan). **1962** R. A. G. Carson *Coins* 544 The coinage of Nationalist China in Formosa, issued since 1949 in various values of the chiao in aluminium and bronze. **1971** *Whitaker's Almanack* 980 *China*—*Renminbi* or Yuan of 10 Jiao or 100 Fen. **1973** *Times* 21 Mar. (China Trade Suppl.) p. i/3 The subsidiary units are jiao (10 jiao = 1 yuan).

jib, *sb.*³ Add: **3.** jib topsail (see quot. 1961).

1860 in Smyth *Sailor's Word-Bk.* (1867) 411. **1866** Capt. Kray *Jrnl.* in B. Lubbock *China Clippers* (1914) App. H. p. xxii, Again lost the spare flying jib-boom and jib topsail set on fore-royal stay sail up. **1892** *Rudder* Sept. 202 The Princess had her jib topsail hoisted. **1927** G. Bradford *Gloss. Sea Terms* 92/2 *Jib-topsail*, a light jib set outside of the outer jib and above them. **1961** Burgess *Dict. Sailing* 123 *Jib topsail*, a light triangular sail, set above the jib; it is hanked to the topmast stay and sometimes called 'jib o' jib'.

jib (dʒib), *sb.*⁴ *slang.* Also **gib, jibb.** [Origin unknown.] A first-year student at Trinity College, Dublin.

Unknown to the present generation of TCD students (1973).

1827 in J. E. Walsh *Trinity College in 19th Cent.* (1902) 21 About a hundred young jibbs..forced the gates and sallied out into the streets. **1840** C. J. Lever *Confessions H. Lorrequer* xiii. 99 During all this melée tournament, I perceived that the worthy jib as he would be called in the parlance of Trinity, Mr. Cudmore, remained perfectly silent. **1902** W. M. Dixon *Trinity College, Dublin* vi. 137 It was a proud thing for a 'gib' to present himself to a new friend, and the worthy sizar..to take care of [illegible] greatness of the College!' **1922** Joyce *Ulysses* 160 And the Trinity jibs in their mortarboards.

jibbah. (Earlier and later examples.)

1848 J. Richardson *Trav. Sahara* I. xiv. 386 Feel glad I took the advice of the Governor of Ghadames, and purchased a quantity of warm woollen clothing, heik, bornouse, and jibbah. **1922** [see *abaya].

jibber, *sb.* Add: **b.** One who jibs.

1936 F. D. Davison *Children of Dark People* x. 147 He said you'd got lost and he'd been sent out by the elders of the tribe to look for you; the old jibber! **1961** F. H. Burgess *Dict. Sailing* 123 *Jibber*, one who has second thoughts about attempting something, and then refuses to make the effort.

jibber, *v.* Add: Also **jibber-jabber** *sb.* and *v.*

1922 A. Haddon *Green Room Gossip* ix. 240 The jibber-jabber was entertaining, not because the utterances were those of ordinary human beings, but because they were the voice of Shaw. **1945** L. Shelly *Jive Talk Dict.* 26 *Jibber jabber*, senseless talk. **1948** *Bulletin* (Philadelphia) 23 Mar. 24/5 Time for Congress to quit jibber-jabbering.

jibe, *v.* For '*U.S.*' read 'Chiefly *U.S.*' Add later examples.

1959 [see *gyrate]. **1970** *New Yorker* 28 Nov. 101/1, I didn't jibe with my mother as a personality, but there was no argument. I could relate to. **1973** *Word* 1970 XXVI. 36 The first does not jibe with the patterning of meanings.

‖ **jicara** (hī·kārā). Also **jicaro.** [Amer. Sp. *jícara, jícaro*, ad. Nahuatl *xicalli*.] A Central American name for the *calabash-tree* (s.v. Calabash 7) or its fruit (Calabash 2).

1859 J. Froebel *Seven Years' Travel in Central Amer.* v. 91 On the market place, there..were interesting and highly creditable articles of Indian skill and industry are offered for sale..gourds and *jicaras* (cups and basins made of calabashes) ornamented with reliefs. **1892** C. F. Lummis *Tramp across the Continent* 245 Each bore upon her head a big, flaring basket—the rush *chiquihuite* of home make, or the elegantly woven Apache *jicara*. **1924** *Chambers's Jrnl.* Sept. 615/1 They saw a woman carrying a jicara gourd. **1927** D. H. Lawrence *Mornings in Mexico* 73 He bargained for a carved jicara. **1949** Record & Hess *Timbers of New World* 78/2 The shells [of calabash fruits] are used by the natives for making cups, dishes, and other household utensils, and some of them are ornately carved or painted... Common names: Calabash ..jícaro. **1964** Little & Wadsworth *Common Trees Puerto Rico & Virgin Islands* 490/2 Calabash-tree... Widely planted in Puerto Rico and through the tropics for the fruits... Other common names..jícara, jícaro (Central America).

Jicarilla (hikarī·lyä). Also **Jiccarilla.** [Mexican Sp. (dim. of prec.).] In full, *Jicarilla Apache.* An Apache people in New Mexico and nearby states; a member of this people; also, the name of their language. Also *attrib.* or as *adj.*

1850 *Ex. Doc. 31st U.S. Congress 1 Sess. Senate* No. 64. 57 The vocabulary is distinctly shown the 35 [illegible] character of the language of the Navajos and of the Ticorillas [*sic*] branch of the Apache. **1858** D. C. Peters *Life Kit Carson* xv. 416 The Jiccarillas..had broken out in open defiance of the authorities. **1871** *Republican Rev.* (Albuquerque, New Mexico) 1 Apr. 2/1 News also came on Sunday of the murder of Francisco, the lawyer of the Jicarilla Apache tribe. **1911** *Anthropol. Papers Amer. Mus. Nat. Hist.* VIII. 8 A large number of texts was secured from Cas Mira, a Jicarilla now about seventy years of age and nearly blind. *Ibid.*, This shift..has taken place in the Lipan, Jicarilla, and Kiowa-Apache. **1964** J. Adair *Navajo & Pueblo Silversmiths* vi. 97 The Jicarilla Apache move down to the southern edge of their reservation in the winter. **1971** *Observer* 24 Jan. 32/5 The land-rich Jicarilla tribe in New Mexico. *Ibid.*, The Jicarillas didn't ask for an anti-Western with a trendy social message.

jiffle, *v.* Add: (Later examples.) Also as *sb.*

1963 S. Marshall *Exper. in Educ.* iii. 93, I have usually been aware of a discreet rustling (and jiffling) among my audience, and..little note-books and pencils have appeared. **1968** G. J. Barrett *Guilty, be Damned* ix. 104 Fenton's stomach squirmed, and he jiffled in his seat. **1971** B. W. Aldiss *Soldier Erect* 13 Ann was winding our gramophone,..and Sylvia was standing beside her, jiffling to the tunes. *Ibid.*, I prowled round the perimeter of the little conversations, attracted by Sylvia's jiffle.

jiffy. Add: **2.** *Comb.* The proprietary name *Jiffy* in *Jiffy* (**book**) **bag**, a type of large envelope padded to protect the contents; also *fig.*; also *ellipt.* as **Jiffy**; *Jiffy pot*, a type of small pot in which seeds may be sown (see quot. 1972); also **jiffy-quick** *adv.*, in a jiffy.

1956 *Library Jrnl.* 1 June 1384/1 Book handling equipment... Mailing bags: Jiffy. *Tranfile*. *Ibid.* 1385/1 Jiffy book bag. **1968** *New Scientist* 10 Oct. 77/1 Dr. Pratt has recently been using 'jiffy bags' for his cube—padded things that mail paper envelopes used for mailing books. **1969** *Trade Marks Jrnl.* 2 July 1084/2 Jiffy... Padded paper bags, cushioned paper pads and padded paper in sheet or roll form, all being wrapping and packaging materials for industrial use, for protecting goods in transit. Jiffy Packaging Company Limited,.. Winsford, Cheshire; manufacturers. **1973** 'D. Halliday' *Dolly & Starry Bird* xvi. 249 Maurice..had backed discreetly into a Jiffy bag of deliberate ignorance. **1957** *Trade Marks Jrnl.* 4 Dec. 1278/1 Jiffapot [*sic*]... Hormonal peat in the form of propagating pots for plants. **1959** *Greenhouse* IV. iv. 243 (Advt.), These tomato plants grew up side by side, 'B' in a standard clay pot; 'A' in a Jiffy Pot... Exhibit 'A' was ahead... Ideal for root cuttings too—hyacinths start evenly in Jiffy Pots. **1965** G. V. Williams *Econ. Geol. N.Z.* xvi. 256/2 It is of interest to note that 25 million 'jiffy pots' are used annually for horticultural purposes in New Zealand. **1968** *Trade Marks Jrnl.* 27 Mar. 484/2 Jiffypots... Flower, planting and transplanting pots made of peat. Odd Smaaberg Melvold and Leif Fraas Koxvold, trading as Me-Kox Industri Melvold & Koxvold, Østre Akers Vei 210 Grorud, near Oslo, Norway; manufacturers and merchants. **1972** *Guardian* 16 Dec. 12/7 Jiffy pots are ..made of peat and wood pulp, through which plant roots can grow. **1927** *Ladies' Home Jrnl.* Dec. 34/3 A waxed surface that..you can dust up jiffy-quick promises speck-and-span floors with but little trouble.

jig, *sb.*¹ Add: **1.** *Irish jig* (earlier example). Also as *v. intr.*

1780 A. Young *Tour in Ireland* ii. xvii. 75 The Irish jig, which they can dance with a most [illegible] expression. **1919** [see *fox-trot *v.*].

5. (Earlier U.S. and later examples of the *jig is up*). Also, *the jig is over.*

1777 *Maryland Jrnl.* 17 June (Th.), Mr. John Miller came in and said, 'The jig is over with us.' **1800** *Aurora* (Philadelphia) 17 Dec. (Th.), As the Baltimore paper says, 'The Jigg's up, Paddy.' **1923** E. Wallace *Missing Million* xii. 100 It was almost like the last spiteful act of a man who knew the jig was up. **1961** Wodehouse *Service with a Smile* (1962) ix. 134 You're in the soup, Miss Briggs. The gaff has been blown, and the jig is up. **1965** *New Yorker* 18 Sept. 56 O.K., Frankie, the jig's up! **1973** *Nature* 15 Feb. 420/3 The weight of opinion seems to be that the jig is up for the map's supporters.

6. b. (Earlier and later examples.)

1839 *Ex. Doc. 31st U.S. Congress 1 Sess. House* No. 5. III. 479 Assay and analysis of the washed metals from the jigs at the Boston and Pittsburg Company's mine. **1953** T. R. Mitchell in *Symposium Recent Devel. Mineral Dressing* (Inst. Mining & Metall.) 263 The jig is by no means obsolete and in the United States, indeed, it is finding increased use for the treatment of those fractions which are too fine for economical concentration by dense media.

c. (Earlier U.S. example.)

1846 *Knickerbocker* XXVII. 513 See that your jigs are in perfect order, for if we do get hold of 'em, our lines and hooks will have to take it, I guess, for a spell.

e. A device for accurately guiding and positioning a drill or other tool in relation to the workpiece, or for positioning the parts of an object during assembly, and used when a large number of similar articles have to be made with high precision; = Template¹ 2 b.

1894 W. L. Lineham *Text-bk. Mech. Engin.* vi. 274 Jigs are an extension of the template principle. Instead of thin plates, castings of an inch or so in thickness are used, supplied with holes where needed, the object being to guide the drill to its proper place on the work without the necessity of lining-out. **1903** W. H. van Dervoort *Mod. Machine Shop Tools* xxvii. 410 Jigs are manufacturing tools of, as a rule, high first cost and their economy depends very largely on the number of pieces to be drilled. **1912** R. W. A. Brewer *Motor Car Construction* ii. 13 Modern competition has made jig work absolutely essential. **1942** B. A. Shields *Prine. Flight* iii. 91 The complete fuselage is built in a jig. **1947** Bryant & Dickinson *Jigs & Fixtures for Mass Production* i. 4 In the machine shop, a jig is usually an appliance which guides a cutting tool... In the automotive industry, a jig is a work-holding device wherein all positions for assembly or fabrication operations are preplaced. **1967** M. Chandler *Ceramics in Mod. World* iv. 127 (*caption*) Assembling a large post insulator in a jig.

f. *Dyeing* = *Jigger *sb.*¹ 5 n.

1942 Whittaker & Wilcock *Dyeing with Coal-Tar Dyestuffs* (ed. 4) iv. 67 The jig or jigger is a machine designed for dyeing piece goods at full width. **1963** Meitner & Kertess tr. *Schmidlin's Preparation & Dyeing Synthetic Fibres* xii. 108 High-temperature pressure jigs are mainly used for heavier fabrics.

8. jig borer, (*a*) a machine for drilling holes in or machining the surfaces of a component (esp. a jig (sense *6 e)), usually having a vertical spindle mounted above a table which can be accurately positioned relative to the spindle; (*b*) (see quot. 1972); so **jig-boring** *vbl. sb.*; hence (as a back-formation) **jig-bore** *v. trans.*, to drill (a hole) by means of a jig borer; **jig box**, the box or sieve of a jig (sense 6 b); **jig-brow** (earlier example); **jig button,** a steel bush used for accurate positioning of a jig plate when making jigs on a lathe; **jigman,** one who works an ore-dressing jigger; **jig plate,** (a part of) a jig consisting of a steel plate which carries the bushes which guide the drill; *jig-time, colloq.* (chiefly *U.S.*) in phrase *in jig-time* expressing a very short space of time.

1930 C. B. Cole *Tool Making* 258 The bushing plate is made from cold-rolled steel, and this is laid out carefully and the hole jig-bored for the drill bushing. **1967** A. J. Lineham *Metrology* vi. 71 The holes in the plate would be jig-bored prior to the fitting of the bushes and the centre distances would need to be checked, both after jig-boring, and after the fitting of the drill bushes. **1932** Gwiazdowski & Lord *Econ. Tool Engin.* xiii. 189 A Swiss firm.. developed a large jig borer that derives its accuracy from.. its lead screws. **1941** W. C. Durney *Machine Shop Pract.* v. 149 In the majority of modern engineering establishments, manufacturing jigs and fixtures in any quantity, these pieces of vital auxiliary apparatus are usually machined up in jig borers. **1959** *Times* 5 Oct. (Switzerland Suppl.) p. viii/3 Over one thousand jig-borers are installed in the United Kingdom to-day. **1972** *Classification of Occupations & Directory Occupational Titles* (Department of Employment) III. 290/2 *Jig borer*, [one who] sets up and operates a jig boring machine to drill and bore holes in workpieces to extra fine limits of accuracy. **1932** Gwiazdowski & Lord *Econ. Tool Engin.* xiii. 183 (*heading*) Jig-boring methods. **1938** H. J. Davies *Precision Workshop Methods* vii. 115 A large proportion of the time occupied in jig-boring..is taken up in the initial setting up of the discs or buttons on the work. **1970** W. F. Patton *Mod. Manuf.* vii. 137 Jig-boring machines are not production machines but toolmaking equipment for the accurate location and drilling of holes. **1902** *Encycl. Brit.* XXXI. 371/2 The pulsating current is obtained by placing a vertical longitudinal partition.. extending part of the way down to the bottom of the jig box. **1951** A. F. Taggart *Elem. Ore Dressing* x. 190 The supporting reactions of the relatively rigid screen and sidewalls of the jig box are highly variable. **1859** [see *or sb.*⁵ 5 b]. **1932** Gwiazdowski & Lord *Econ. Tool Engin.* xiii. 184 (*caption*) Toolmaker's jig buttons. **1964** S. Crawford *Basic Engin. Processes* x. 134 Fig. 19 (*a*) shows a sectional view of a jig-button, a small cylindrical steel bush accurately ground on the circumference and end faces, the hole being the larger in diameter than the retaining screw. These buttons can be set by end measurement. **1809** *Ex. Doc. 31st U.S. Congress 1 Sess. House* No. 5. III. 479 The jigman. **1932** Gwiazdowski & Lord *Econ. Tool Engin.* xiii. 183 *Jig plates*.. jig plates and generally attends to jig (?) ..have centres located by bored holes in the jig plate. **1963** *Rep. Sci. Comm. Occup. Terms* (1927) 18/1 *Jigger, jigman*; controls by lever, tappet..one of the jig boxes. **1966** *Gloss. Terms Motor Veh.* (B.S.I.)..The drill bushes..are attached to the jig plate by screws passing through loosely fitting holes. **1970** W. J. Patton *Mod. Manuf.* v. 79 Drill bushings are inserted into a jig plate and secured to the jig plate. **1918** *Dial. Notes* V. 29 *In jig time*. **1940** *Amer. Speech* XV. 341 Kate has about four more of 'em licked to a standstill in jigtime. **1946** H. Lewis *Plant Somewhere* iv. ..Get your jig time. **1947** J. Steinbeck *Wayward Bus* ..in jig time..

jig (dʒig), *sb.*² *U.S. coarse slang.* Also **jigg(s).** [Origin unknown, but perhaps the same word as prec.] A Negro.

Like *jigaboo*, a term that gives offence.

1924 F. J. Wilstach *Slang Dict.* (Typescript in N.Y. Public Libr.), *Jiggs*, Negro actor. **1927** K. Nicholson *Barker* iii. i. 128 You go along and give 'em a hand, son. Nat Brody's there and a crew of jigs. **1931** *Amer. Mercury* Nov. 352/2 *Jig*, a Negro.—*Jigaband*, the sideshow band. **1932** J. T. Farrell *Young Lonigan* iii. 113 Janitor's jobs went to jiggs, and Hunkeys, and Polacks, anyway. **1933** *Fortune* Aug. 47/1 A couple of jigs got on the bus with a dogbouse. **1935** E. Hemingway *Green Hills Afr.* (1936) ii. vi. 169 This jig we call Othello falls in love with this girl. **1939** *New Yorker* 23 Oct. 24/3 He said a jig band..could be a big success in Paris why not a kike in love with any jig. **1963** *Ebony* ..jigs ..were getting a new deal—to cause a big kick..an *Amer. Dialect* ..the jig that..

jig, *v.* Add: **9.** *trans.* To provide or equip with jigs (sense *6 e). Also *absol.*

1900 *Machinery* (N.Y.) Dec. 119/1 There are many other considerations..which cannot be overlooked when the question of 'To jig or not to jig' arises. **1927** *Observer* 16 Oct. 20 A sound financial scheme always includes writing off the heavy cost of jigging and tooling up a factory to manufacture a given type during the first years of its production. **1957** *Times* 23 Aug. 3/6 There was long idle floor space jigged and tooled to produce six Britannia fuselages a month.

jigaboo (dʒi·gābū). *U.S. coarse slang.* Also **jiggabo, jigabo, zigabo,** etc. [Related to *jig *sb.*² after Bugaboo.] A Negro.

Like *jig sb.*², a term that gives offence.

1909 Weston & Barnes *I've got Rings on my Fingers* (song), So come to your na-bob, and next Pat-rick's Day Be Mis-tress Mum-bo Jum-bo Jij-ji-boo J. O'Shea. **1920** *Sat. Even. Post* 13 Apr. 34/4 Jigaboo (underworld). **1929** D. Gordon *Born to be* 238 Jigaboo, Dinge, Jasper, Tar-kin... **1935** *Amer. Mercury* Dec. 456/1 He broad's saxophone the jigaboo hop tries to make her. **1935** J. T. Farrell *Studs Lonigan* iii. 87 Yes, but the pupils are all jigaboos, and the parents..very poor now. **1935** W. R. Riddle ..jigaboo at a [illegible]. **1944** ..I will take Follow You and that zigaboo lock of yours to the Gold Vase for mine. **1945** W. W. Burnett *High Sierra* vi. 55 'Yeah,' said Red, 'I thank the jigaboo has got your cabin all set.' **1944** *Amer. Speech* XIX. 174 Such vulgar synonyms for Negro as..jazbo, jigaboo (with the variants, jijiboo, jiggerboo, zigaboo, zigabo). **1961** J. H. Griffin *Black like Me* (1962) 57 The Negro..hearing himself referred to as nigger, coon, jigaboo! **1970** L. Sanders *Anderson Tapes* ix. 160 The tall one..he's a jigaboo. **1973** *Washington Post* 31 Mar. 6/6 'All that I tell Black there is a bunch of boos'—short for 'jigaboos', a derogatory term for blacks.

jig-a-jig. Add: Also **jig-jig.** Used in sense 'sexual intercourse'; also as *vb.*, to copulate. *slang.*

1896 Farmer & Henley *Slang* IV. 54/2 *Jig-a-jig*, to copulate. **1932** L. Golding *Magnolia St.* xix. 488 'Jig-a-jig, Johnny? Very nice!' they said. **1939** C. Isherwood *Dog beneath Skin* i. v. 58 Come wiv me. Good Jig-a-jig. **1949** A. J. Pollock *Underworld Speaks* 64/1 Jig-a-jig, a lewd act. **1948** G. Greene *Heart of Matter* i. i. 'Captain want a jig jig, my sister pretty girl school-teacher, captain want jig jig.' **1953** A. Baron *Human Kind* xvii. 124 He put his hand on her knee. 'You like jig-a-jig?' **1966** 'E. Lindall' *Time too Soon* (1967) iv. 49 This woman's master has jig-a-jig with his blood sister. **1971** B. W. Aldiss *Soldier Erect* 125 'Hello, sweetheart. You like jig-jig?' 'That's the idea. Let's look at you first.'..She said something—... All we had in common was the word, the call-sign, 'jig-jig'.

jigger, *sb.*¹ Add: **1. b.** In full, *jigger coat.* A woman's short loosely-fitting jacket.

1957 M. B. Picken *Fashion Dict.* 190/2 *Jigger coat*, short semi-tailored, informal coat. **1966** *Olney Amsden & Sons Ltd. Price List* 30, 50 numbers Pinarettes, Aprons and Jiggers. **1968** J. Ironside *Fashion Alphabet* 36 *Jigger*, a jacket popular in the 1930s—loose, finger-tip length, often with a tuxedo front. **1974** *Times* 12 Feb. 11/7, I bought this little mink jigger..was a coming look.

c. *N.Z.* (See quot. 1971.)

1961 *Countryman* LVIII. iii. 500 The axeman has to insert at least two 'jiggers' or steps. **1966** *Wanganui* (N.Z.) *Photo News* 4 June 43/2 (*caption*) Champion axeman Sonny Bolstad is watched by the Queen Mother as he competes in a jigger chop. **1971** F. C. Ford-Robertson *Terminol. Forest Sci.* 252/2 *Jigger(-board)* (New Zealand) ..a short board or plank, its end notched into the bole, on which the cutter stands so as to enable him to fell the tree at a level not reachable from the ground.

d. Also, in ice-fishing. *Canad.*

1946 *Beaver* June 17/1 The jigger is a wooden plank with a slot in the middle through which a wooden arm controlled by a metal lever, moves. **1972** D. Pryde *Nunaga* i. 16 The ice is eight feet thick on the lakes and it's almost impossible to set a net without a jigger. We had a jigger in here once and showed the Eskimos how to use it to string a net under the ice, but no one ever bothered to make one.

5. Other specific applications:

l. Any small mechanical contrivance; a 'thingummy'. *U.S. colloq.* **m.** *Golf.* A short iron-headed club used for approaching shots. **n.** *Dyeing.* A device for dyeing piece goods by passing them back and forth through a dye-bath over a set of rollers. **o.** A bicycle or small motor vehicle or hand-car. †**p.** *Radio.* A high-frequency transformer used in early spark transmitters to couple the aerial circuit to the circuit in which the oscillations were generated; an oscillation transformer. *Obs.* **q.** A light vehicle, esp. one that moves on rails. *dial.* and *N.Z.* (see *E.D.D.*, sense 4). **r.** = ouija. **s.** A device for administering electric shocks (see quot. 1973).

l. 1874 Hotten *Slang Dict.* 203 Jigger has many meanings, the word being applied to any small mechanical contrivance. **1926** *Amer. Speech* I. 628/2 The term jigger has long been used of small mechanical devices... In America, jigger is often used as an indefinite name, not undignified, of the same order as thingumbob. **1944** H. Wentworth *Amer. Dial. Dict.* 328 *Jigger*, thingema-bob. **m. 1893** H. G. Hutchinson *Golfing* 21 The learner will probably do better..to employ..—supposing that he finds he cannot play the short approaches with sufficiently good loft off an ordinary iron—a much-laid-back approaching cleek. On some links these are a great deal used, under the name of jiggers. **1931** *Punch* 1 July 717/2 *Fully-equipped Visitor*. This looks a weird sort of hole. What on earth does one take here? *Local player*. One takes a jigger, if that's all one has. **1970** H. Taylor *Golf Dict.*, *Jigger*, an iron club, of value in all kinds of golfing situations. **n. 1893** E. Knecht et al. *Man. Dyeing* II. ix. 694 The modern jigger consists of a wooden or cast-iron dye vessel heated by steam and provided with water supply and waste-pipe. In the vessel are three rollers..at the top and two..at the bottom, which guide the pieces in their passage through the dye-liquor. **1915** T. Beacall et al. *Dyestuffs & Coal-Tar Products* iv. 82 The machine most frequently met with in the dyeing of cotton cloth is the jigger. In this machine the cloth in full width is passed through the dye liquor several times over guide rollers. **1963** Meitner & Kertess tr. *Schmidlin's Preparation & Dyeing Synthetic Fibres* xi. 90 Although a discontinuous machine the automatic jigger is very suitable for de-sizing, boiling-off, bleaching and dyeing of heavy fabrics sensitive to creasing. **o. 1897** H. G. Wells in *Humours of Cycling* 7 'Pretty jigger!' said the Bounder... 'Nice-looking machine you've got.' **1930** 'E. Bramah' *Little Flutter* iv. 52 My little jigger is no good for a job like this. **1958** *Globe Mag.* (Toronto) 9 Aug. 18/1 A jigger carrying eight men came bouncing around the mountains and ran smack into a moose. **1973** *Courier-Mail* (Brisbane) 28 July 17/11 It takes a little time, too, to absorb the antique public school language... Jigger—a bicycle. **p. 1902** *Encycl. Brit.* XXXIII. 230/2 The plugs of the sensitive tube..are joined to the terminals of the secondary circuit S'S' of a small transformer, called a 'jigger'. **1906** J. A. Fleming *Princ. Electr. Wave Telegr.* vii. 437 If the oscillation transformer, or jigger, is not wound to suit the wave length employed, so far from being a benefit, it prevents any signals being received at all. **1924** P. J. Risdon *Wireless* xii. 116 The oscillating current in the primary circuit induces, through the 'jigger' coils, as they were called, another current of the same frequency in the aerial circuit. **1937** in 'Decibel' *Wireless Terms Explained* 41. **q. 1904** 'G. B. Lancaster' *Sons o' Men* 158 Two men sat on the little iron jigger that straddled the wooden tram-line. **1918** *N.Z.E.F. Chrons.* 8 Nov. 179/1 At last I commenced the second stage—this time on a 'jigger', a frame on two rubber-tyred wheels which holds the swagger. **1949** E. de Mauny *Huntsman in Career* 162 Chancey..went off on the jigger down the narrow track into the bush each morning. **r. 1916** O. Lodge *Raymond* 186 Jigger. (A kind of ouija.) **s. 1972** *Sunday Sun* (Brisbane) 26 Nov. 1/2 Battery operated jiggers are being used on mentally retarded children..to bring them into line. The electric shock treatment is followed by..lollies if they behave. **1973** *Sunday Mail Mag.* (Brisbane) 25 Feb. 14/1 Occasionally, a blue spark would flash forth as a recalcitrant beast was touched with the 'jigger' (a battery-operated device

carried over the shoulder and imparting an electric shock through an insulated rod held in the hand).

6. c. A passage between or at the back of houses; a back entry or alley. *Merseyside.*

1902 in *Eng. Dial. Dict.* **1966** [see *knee-trembler* s.v. *knee *sb.* 13]. **1966** P. Moloney *Plea for Mersey* 21 'A seen a scuffer up a jigger wid a rozzer' ('I saw two policemen in the side entry'). **1967** A. Henri in *Penguin Mod. Poets* X. 16 A Polish gunman..collapses down a back jigger.

7. b. (Earlier examples.) Also, a small glass or metal cup, a measure used in mixing cocktails; the contents of such a glass or measure.

1836 W. O'Bryan *Narr. Trav. U.S.* 107 These canal labourers have a boy to supply them with Whiskey, called a *Jiggar boss*, who goes on the canal and carries a half gill (half noggin) of Whiskey to every man sixteen times a day! **1870** J. H. B. Nowland *Early Reminisc. Indianapolis* 361 By jiggers was meant a small cup of whisky, say about a gill; they had cups made on purpose for this use. **1879** *N.Y. Herald* 21 Nov. 8/2 A jigger..is a conical metal cup in which to mix fancy drinks. **1946** E. Holding *Innocent Mrs. Duff* 17 On a shelf there was a fine array of bottles, with jiggers of two sizes, swizzle sticks, glass mixers. **1946** 'P. Quentin' *Puzzle for Fiends* (1947) viii. 62 A jigger of liquor clutched between thumb and first finger. **1952** S. Kaufmann *Philanderer* (1953) vii. 114 What I came out to ask you is, do you have any gin in the house and could you spare me a jigger? **1971** R. Dentry *Encounter at Kharmel* ix. 148 A baker's dozen perversions for a jigger of vodka.

8. (senses 5 a, b) *jigger-boy*; (sense *1 c) **jigger-board** *N.Z.* = *Jigger 1 c; **jigger-saw** = Jig-saw *sb.*

1944 R. Gilberd in *N.Z. New Writing* III. 55 We would have given you..nerves to stand the narrow insecurity of the jigger-board. **1961** B. Crump *Hang on a Minute* 42 During the next few weeks Jack learned about scarfing, backing, limbing, deeing, sniping, jigger-boards, platforms, toms, strops, drives, triggers, and saw and axe sharpening. **1963** N. Hilliard *Piece of Land* 176 The best thing of the day to watch, the three-tier jigger-board chop. **1869** *Good Words* 1 Mar. 172/2 A plaster-of-Paris cast is placed on a disc which a handle-turning 'jigger-boy' causes to revolve. **1921** Jigger boy [see *jiggerer]. **1888** *Lockwood's Dict. Mech. Engin.* 197 *Jigger saw*, or *jig saw*. **1957** *N.Z. Timber Jrnl.* Sept. 61/1 *Jigger saw*, a reciprocating saw. A fret or scroll or jig saw, used for pierced and tracery work.

jigger, *sb.*² Also **jigger flea** (examples). The Latin name of the insect is now *Tunga penetrans.*

1947 H. Vaughan-Williams *Visit to Lobengula* xxvii. 176 Arthur's feet suffered badly from jigger fleas—horrid little tiny insects that burrow under the skin and lay a bunch of eggs there in a capsule. **1953** *New Biol.* XIV. 120 *Tunga penetrans*, the tropical jigger flea of man, is the best-known of these [burrowing fleas]. **1962** Gordon & Lavoipierre *Entomol. for Students of Med.* xxxv. 217 Both the male and the female jigger flea are blood-suckers.

jigger, *v.*¹ Add: **2.** orig. *pass.*, usu. with *up*: to be tired out, exhausted; so, to be 'done for', devitalized. Also *actively*: to break, destroy, ruin. *dial.* and *slang.*

1862 C. C. Robinson *Dial. Leeds* 332 *Jigger'd up*, Av tramp'd a matter o' fotty mile to-daay, an' am fair jigger'd up. *a* **1865** Smyth *Sailor's Word-Bk.* (1867) 412 *Jiggered-up*, done up; tired out. **1885** B. Brierley *Ab-o' th-Yate in Yankeeland* v. 42 A generation or two would see it jiggered up if it wurno' for th' fresh blood ut's bein sent into it. **1895** 'G. Mortimer' *Like Stars that Fall* xii. 167 Bates will jigger us if he can... I wouldn't trust that fellow. **1896** *Yorks. Weekly Post* 6 June 6/8 T'chap wor reight jiggered up. **1923** *Daily Mail* 13 June 12 I've 'jiggered' up my Rolls-Royce. **1949** E. de Mauny *Huntsman in Career* 150 He jiggered up his ankle last Saturday. **1969** *Telegraph* (Brisbane) 19 May 8/1 The firing pin's jiggered and the sights are sloppy.

jigger, *v.*² Add: Also **ji·ggering** *ppl. a.* and intensive *adv.*

1903 *Daily Chron.* 14 Sept. 3/3 Once you've made up your mind, as you may say, about a young man, you've got to be jiggerin' well careful you don't go and lose him. *a* **1950** X. Herbert in Murdoch & Drake-Brockman *Austral. Short Stories* (1951) 301 Take the lot. Take the rintin' jiggerin' lot!

ji·gger, *v.*⁴ *Pottery.* [f. Jigger *sb.*¹ 5 a.] *trans.* To shape with a jigger.

1931 W. H. Warburton *Hist. Trade Union Organization in Potteries* xi. 208, I will try and get the price you want for this article, but you must remember that this.. is being jiggered by a firm in the next town at a much less price. **1967** M. Chandler *Ceramics in Mod. World* ii. 63 Shaping methods..include..throwing, jollying or jiggering, plastic pressing, and extrusion.

ji·gger, *v.*⁵ *Book-binding.* [? f. Jigger *sb.*¹] *trans.* To rub (a tool) backwards and forwards along a line or other impression in a leather binding, in order to polish it. Hence **ji·ggering** *vbl. sb.*

1880 J. W. Zaehnsdorf *Art of Bookbinding* xxii. 114 The lines impressed on the back must now have their gloss given to them. This is done by *giggering* the pallets over them. **1901** D. Cockerell *Bookbinding* xv. 224, I have found that a tool guided by a straight-edge, and 'jiggered' backwards and forwards, makes by far the best lines for blind-tool work. **1946** E. Diehl *Bookbinding* II. xxiii. 352 It [*sc.* the fillet] is then pushed over the line a few times with a 'jiggering' motion, until the line is polished. **1951** L. Town *Bookbinding by Hand* x. 229 The tool can be put down again and rocked slightly. This gives a polished surface as well as a darkened one, and is known as 'jiggering'. **1963** B. C. Middleton *Hist. Eng. Craft Bookbinding Technique* xii. 167 The tools are usually rocked or jiggered to produce a polish.

jiggerer (dʒi·gərəɹ). [f. Jigger *sb.*¹ 5 + -er¹.] One who uses or works with a jigger (in various trades).

1881 *Instructions to Census Clerks* (1885) 84 Coal Miner... Loader, Jiggerer, Trammer, Hooker-on. *Ibid.* 88 China, Porcelain, Manufacture:..Jigger or Gigger Turner, Jiggerer. **1921** *Dict. Occup. Terms* (1927) § 043 *Jigger, jigger boy, jiggerer*..attaches or detaches tubs, singly or in pairs, to or from endless rope, by placing rope in fork of 'jigger', or double crook, in socket at one side or end of tubs. *Ibid.* § 105 *Jigger, jiggerer*, makes flat-ware such as plates, saucers; presses bat of clay on top of mould, already revolving on vertical spindle. **1931** W. H. Warburton *Hist. Trade Union Organization in Potteries* xi. 205 By this scheme his hollow-ware pressers would employ female assistants in the same way as did his hollow-ware jiggerers and 'jolliers'.

jiggery-pokery (dʒi·gəri,pō^u·kəri). *colloq.* [Cf. Sc. *joukery-pawkery* (see Joukery, jookery b).] Deceitful or dishonest 'manipulation'; hocus-pocus, humbug.

1893 in Dartnell & Goddard *Gloss. Words Wiltshire* 86. **1926** E. F. Spanner *Naviators* ix. 104, I thought.. it was some more jiggery-pokery to keep down the expenditure this financial year. **1943** *Mind* LII. 304, I share with Gray the feeling that there's some *jiggery pokery* here and that what you are doing is not what one tends to feel you are doing. **1973** G. Mitchell *Murder of Busy Lizzie* ii. 23 Business reasons could make any alliance respectable..so long as there was no jiggery-pokery.

jigging, *vbl. sb.* **2.** (Examples in the sense of *jig *v.* 9.)

1903 W. H. van Dervoort *Mod. Machine Shop Tools* xxvii. 410 No class of work in the manufacturing shop presents as many possibilities for jigging as does the work handled in the drilling machine. **1945** Lea & Simons *Machining of Steel* xii. 107 A fraction of a minute per part saved in large-scale production by better jigging soon outweighs any appreciable resultant increase in the cost of the jig.

jiggoty, var. Jiggety *a.*

1876 H. E. Scudder *Dwellers in Five-Sisters Court* ii. 30 Mr. le Clear appeared and received the jiggoty Miss Pix's welcome in a smiling and well-bred manner.

‖ **jigotai** (dʒigotai·). *Judo.* [Jap.] In Judo, a defensive posture.

1950 E. J. Harrison *Judo* 102 *Jigotai*, self-defensive posture. **1957** Takagaki & Sharp *Techniques Judo* i. ii. 12 *Jigo-tai*..is assumed by spreading your legs wide apart and bending your knees to lower your body.

jig-saw, *sb.* Add: For '*U.S.*' read 'orig. *U.S.*' Also **jigsaw.** (Later example.) Also *attrib.*, of a type of architectural decoration using fretwork patterns.

1892 Kipling *Lett. of Travel* (1920) 21 The jig-saw days, when it behoved respectability to use unlovely turned rails and pierced gable-ends. **1928** E. O'Neill *Strange Interlude* iii. 86 The room is one of those big, misproportioned dining rooms that are found in the large, jigsaw country houses scattered around the country as a result of the rural taste for grandeur in the eighties. **1966** M. M. Pegler *Dict. Interior Design* 245 *Jigsaw detail*, a cutout or fretwork design made with a jigsaw. It was used for the enhancement of buildings of the mid and late 19th century. The bargework was often made with a jigsaw. The 'gingerbread' or 'steamboat Gothic', late Victorian period was jigsaw work in its most aggravated form. **1968** J. Arnold *Shell Bk. Country Crafts* 274 Salad-forks, for example, are cut without previous marking by a hand jig-saw.

b. In full, *jig-saw puzzle.* A puzzle formed by cutting into small irregular pieces (orig. with a jig-saw) a picture mounted on a sheet of wood, cardboard, or the like. (Now the usual sense.) So *jig-saw map.* Also *transf.* and *fig.*

1909 *Daily Mirror* 17 Aug. 4/4 A jigsaw map of England. *Ibid.*, These jigsaw geography puzzles should be introduced into all the Council schools in London. **1910** *Punch* 9 Mar. 172 (*caption*) What if the jig-saw epidemic spreads? **1915** *Morning Post* 15 Apr. 2/4 A kind of verbal jig-saw. **1919** E. Shackleton *South* i. 11 Pack-ice might be described as a gigantic and interminable jigsaw-puzzle devised by nature. **1935** W. S. Maugham *Don Fernando* x. 213 The various particulars fit like the pieces of a jig-saw puzzle. **1947** *People* 22 June 4/3 How can this jigsaw be pieced together? Many of the facts are now known as a result of most painstaking police inquiries. **1955** A. Huxley *Let.* 25 Sept. (1969) 766 All this jigsaw work entailed in shaping a play for stage production is extremely boring. **1964** M. Critchley *Developmental Dyslexia* ix. 57 Constructional tasks which embrace spatial concepts include the assembling of jigsaw puzzles, a game which may not be easy for some of these dyslexics. **1972** *Oxford Times* 11 Aug. 3 As the excavation proceeds more and more tiny pieces of the archaeological jigsaw puzzle will be discovered. **1974** G. Markstein *Cooler* xl. 149 Sylvia was turning into the little jigsaw piece that often remained the hardest one to find.

jig-saw *v.*, also, to fit together the pieces of a jig-saw puzzle. Freq. *fig.*

1938 *Times* 2 Feb. 13/4 Jig-sawing is one of the few pastimes in which bludgeoning methods definitely do not pay. **1963** *Harper's Bazaar* Jan. 29/3 It taxed all Miss Molesworth's expertise to jigsaw the requirements into a pleasing ensemble. **1966** *Punch* 18 May 720 I've often cut his articles into line sentences, mixed them up, and tried to jigsaw them together, a most difficult thing to do until you've caught the drift of his mind. **1967** *Listener* 9 Feb. 207/3 The interviews through which Lowry's character was jigsawed together were wholly fascinating. **1973** *Guardian* 27 Feb. 11/5 We..jig-sawed our bits together into one consecutive piece, typing on maddening French machines.

jildi (dʒi·ldi). *Mil. slang* (orig. *Anglo-Indian*). Also **jeldi, jildy, juldie,** and other varr. [ad. Hind. *jaldī* quickness.] Haste, as in phrases *on the jildi*, in a hurry, and to *do* or *move a jildi*. So **jildi** *a.*, quick; **jildi, juldily** *adv.*, quickly; **jildi** *v. trans.* and *absol.*, to hurry.

1890 Kipling *Barrack-Room Ballads* (1892) 24 You put some *juldee* in it Or I'll *marrow* you this minute. **1919** W. H. Downing *Digger Dial.* 59 *Jeldi*, hurry; run. *Ibid.* 30 *Jeldy* (Hind.), quickly. **1926** *Scots Observer* 30 Oct. 21/1 Wullie..would have seen..that the rabbits were moving juldie into the thistles and long grass as he came along. Juldie! What's that? Hindustani for quick. **1929** [see *fuck *v.* 3]. **1930** R. Blaker *Medal without Bar* vii. 47 Come on. We'll catch 'im if we jildi. **1930** Brophy & Partridge *Songs & Slang 1914–18* 132 *Jildi*, quick, look sharp, hurry. Also used in the phrase 'on the jildi', e.g. 'Get them bags filled on the jildi.' **1948** Partridge *Dict. Forces' Slang* 102 *Jildi*, hurry! Also *get a jildi on*. (Indian Army.) *Ibid.*, *Do a jildi move*, to beat a hasty retreat. General Army slang... Among tank men: to take evasive action. **1957** M. K. Joseph *I'll soldier no More* (1960) 14 Hey, Antonio, where's me rooty? And make it juldy, see? **1972** J. Brown *Chancer* i. 12, I went out jilty..and there he was.

jill (dʒil), *v.* [Presumably var. of Gill *v.*²: cf. quot. 1855 s.v.] *intr.* Of a boat: to move *about*, to move *around*; to idle around.

1955 *Times* 18 July 9/6 In the early morning light Falmouth bay looked as lovely as ever, with its rounded green hills and little fishing boats jilling about under sail off the Manacles. **1956** W. Golding *Pincher Martin* i. 18 Survivors, a raft, the whaler, the dinghy, wreckage may be jilling about only a swell or two away hidden in the mist and waiting for rescue with at least bully and perhaps a tot. **1964** F. Chichester *Lonely Sea & Sky* xxix. 303 After breakfast I took over for an hour or two until we had cleared the point, when I gratefully lowered all sail and *Gipsy Moth* jilled along. **1972** *Guardian* 23 Sept. 5/4 Bosun dinghies jilling around the windless Medway... Picture by Peter Johns.

jillaroo (dʒilarū·). *Austral.* Also **jilleroo.** [Jocular formation from Jackaroo *sb.*] A female station-hand. So as *v. intr.*, to work as a jillaroo.

1945 Baker *Austral. Lang.* iii. 62 The past few years have given us a variant [of jackeroo] in *jillaroo*, a female station-hand.., especially used during World War No. 2 for a Land Girl. **1945** *Salt* 26 Feb. 16/1 Miss Garraty is our new 'jilleroo'. **1964** *Pix* 22 Aug. 41 Dr. Davies said *Pix* should have called Miss Lukis a jenniroo or jennyroo. That might be all right for West Australians but the word used in the Eastern States is jillaroo. **1969** *Courier-Mail* (Brisbane) 15 Aug. 7/3 A part-time model, jilleroo and cattle judge was named the top under-21 Rural Youth member yesterday. **1970** *Sunday Mail Mag.* (Brisbane) 28 June 6/2 Isabel has been jillarooing all over Australia for the last four years.

jillion (dʒi·lyən). orig. *U.S.* [Fanciful formation after Billion, Million.] Very many, a great many.

1942 Berrey & Van den Bark *Amer. Thes. Slang* § 18/6 *Indefinite number*, jillion. **1945** L. Shelly *Jive Talk Dict.*, *Jillion*, a lot of people or money. **1950** *Time* 13 Feb. 10/1 After Spindletop, in the superlatives of the oilfields, came a jillion jackpots—roaring booms at Electra, Ranger, Burkburnett, Desdemona and Mexia proved that oil was where you found it. **1957** *New Yorker* 2 Nov. 141/1 'Oh, I've got a jillion,' he replied. 'A jillion ties and no wife. I have ties from many parts of the world.' **1963** 'R. L. Pike' *Mute Witness* (1965) vi. 100, I bet I've hit about a jillion travel agencies today. **1971** *Physics Bull.* Nov. 682/3 Gamesters with 'game theories' of a jillion different maliciously fixable kinds.

jills (dʒilz). *slang.* [Shelta.] Used with a possessive pronoun: *my jills* = 'I', *his jills* = 'he', etc.

1906 E. Dyson *Fact'ry 'Ands* ix. 117 'They thort his jills had done er get,' said the packer. **1940** *N. & Q.* 15 June 421/1 In the current slang of the Variety profession and other Bohemian circles, 'jills', coupled with a possessive pronoun, stands for 'I', 'you', 'he', etc., according to the possessive pronoun prefixed... So 'my jills', 'your jills', 'his jills'.

jim (dʒim). *Austral. colloq.* [f. *Jimmy O'Goblin.*] A pound, a pound note.

1906 E. Dyson *Fact'ry 'Ands* xvi. 214 He was tearin' ratty t' raise another jim. **1919** V. Marshall *World Living Dead* in Baker *Austral. Lang.* (1945) xvii. 310 Half a jim (10s.). **1959** Baker *Drum* 120 *Jim*, the sum of £1.

Jim-crow. Substitute for entry:
Jim Crow¹, Jim-crow, jim crow (dʒi·m,krōᵘ·). [From the refrain of a popular old Negro song, 'Wheel about and turn about and jump Jim Crow'.]

Orig. *U.S.*, but the main current senses (1 f, and its *attrib.* and verbal uses, with *Jim Crowism*) are used throughout the English-speaking world, although chiefly in U.S. contexts.

1. The name of a Negro plantation song of the early nineteenth century; also, a stage presentation of a song and dance first performed by Thomas D. Rice (1808–60) and subsequently by other actors dressed as 'nigger minstrels'.

c **1832** T. D. Rice (*title*) Jim Crow. Celebrated comic song or ballad. **1832** *Amer. Sentinel* (Philadelphia) 11 Sept. 3/1 (Advt.), Mr. Rice will appear and sing Jim Crow. **1835** *Vade Mecum* (Philadelphia) 28 Mar. 2/7 'Ditanti Palpita', 'Jim Crow', 'Old Hundred', with two or three waltzes played in *different* keys usually form the *Hotchpotchiana* of their delicious entertainment. **1837** *New Yorker* 16 Dec. 610/2 The impulse of despair must have tempted them to strike up 'Jim Crow'. **1841** Thackeray in *Britannia* 15 May 315/4 The organ-man.. struck up two beautiful melodies, viz., 'Getting up Stairs', and 'Jim Crow'. **1926** *N.Y. Times* 26 Dec. vii. 8/2 From 'Old Jim Crow' to 'Black Bottom', the negro dances..are African in inspiration.

b. Phr. *to jump Jim Crow*: to execute the dance that was part of a theatrical (or street) performance of Jim Crow; to jump about. Also *fig.*, to change one's political principles, to desert one's party.

1833 *Sk. & Eccentr. D. Crockett* 41 You nebber get to Heben till you jump Jim Crow. **1836** *Louisville* (Kentucky) *Jrnl.* 16 Sept., A Mr. Collier of Virginia has 'jumped Jim Crow'. **1840** *Log Cabin Song-Bk.* 38 Fo he's the man to jump Jim Crow, And prove that black is white. **1857** *Observer* 12 Apr. 2/4 A street clown once told him (Mr. Mayhew) that..he jumped 'Jim Crow' for twelve hours in the mud and wet of the streets, and he carried home..the sum of 15d. **1922** Galsworthy *Windows* i. 19 Not much balance about us. We just run about and jump Jim Crow.

c. A Negro character in the Jim Crow song; T. D. Rice or another performer of the Jim Crow act; in England, a street performer of this type of act (see b).

1835 *Vade Mecum* (Philadelphia) 24 Jan. 3/7 Jim Crow is in the town, about to 'wheel about' for the edification of the Brandywine. Daddy Rice will surprise them. **1851** H. Mayhew *London Labour* I. 4/1 The street-actors—as clowns, 'Billy Barlows', 'Jim Crows', and others. **1861** *Ibid.* III. 121/1 A few minutes afterwards I saw this man dressed as Jim Crow, with his face blackened, dancing and singing in the streets as if he was the lightest-hearted fellow in all London. **1867** *Atlantic Monthly* Nov. 608/2 As a national or 'race' illustration, behind the footlights, might not 'Jim Crow' and a black face tickle the fancy of pit and circle?

d. A turncoat. (Cf. the *fig.* sense of b.)

1837 *New Yorker* 16 Dec. 610/2 An engraving of the veritable Jim Crow is to be seen in every print-shop, with the exception that the face of Lord Lyndhurst usurps that of Rice, his lordship being placed in that peculiar attitude which the Liberals denominate 'turning about—wheeling about' from political consistency and common sense. **1840** J. Romilly *Diary* 13 Nov. (1967) 204 The blackgds in the gallery hooted & called him Jim Crow.

e. A Negro. *depreciatory.*

1838 'Uncle Sam' in *Bentley's Misc.* IV. 582 Don't be standing there like the wooden Jim Crow at the blacking maker's store. **1841** H. Playfair *Hugo Playfair Papers* I. 3 A portmanteau and carpet-bag..were snatched up by one of the hundreds of nigger-porters, or Jim Crows, who swarm at the many landing-places to *help* passengers. **1948** *Sat. Rev.* (U.S.) 27 Mar. 36/1 Jim Crow works at the depot.

f. Racial discrimination, *spec.* against Negroes in the U.S. More usually *attrib.*, or as *Jim Crowism* (below).

1943 R. Ottley *New World A-Coming* 69 Negro soldiers had suffered all forms of Jim Crow, humiliation, discrimination, slander, and even violence at the hands of the white civilian population. **1946** J. H. Burma in A. Dundes *Mother Wit* (1973) 625/2 To the Negro any joke is particularly humorous if it shows Jim Crow 'backfiring' on a Southerner. **1958** J. Asman in P. Gammond *Decca Bk. Jazz* xiv. 166 The difficulties facing any studio recordings by Negro artists are almost insurmountable in the Southern States, where Jim Crow is predominant, even to the present time. **1969** *N.Y. Times* 16 Jan. 40/5 Those of them who are young and loud want segregated colleges... It's Jim Crow when we want it... Why don't they think of it as James Eagle? **1971** *Black Scholar* June 4/1 The historical literature..suggests that Jim Crow was directed more at the black male than the black female. **1973** A. Dundes *Mother Wit* 231 Since white southerners obviously spoke in dialect..this practice was little more than another insidious form of Jim Crow. **1973** *Freedomways* XIII. 30 One hundred years of frustration and battle have not resulted in victory over Jim Crow and racism.

2. In various technical senses. (Dict. sense 1, *jim-crow planing-machine* in Dict. sense 2.)

3. *attrib.*, as (sense 1 a) *Jim Crow song*; (sense 1 c) *Jim Crow boots, dance, hat*; (sense 1 e) *Jim Crow line*; (sense 1 f) *Jim Crow bill, car, college, conditions, law, school, town.* Also (with possessive case) **Jim Crow's nose** = *John Crow('s) nose* (*John Crow).

1835 *Knickerbocker* V. 47 Some jolly slaves..were waiting to take us into a ferry-boat, which they rowed, singing some Jim Crow song. **1842** *Liberator* (Boston) 21 Jan. 10/1 It is this spirit that compels the colored man to ..ride in the 'Jim-Crow car'. **1847** *Chicago Jrnl.* 7 Oct., We do not mean *Jim Crow* dances and poor songs worse sung. **1851** G. S. Cooper *Jrnl. Expedition Auckland to Taranaki* 58 A man in a common shooting jacket, a Jim Crow hat, trowsers rather the worse for wear and a pair of moustaches. **1851** Mrs. Stowe *Uncle Tom's Cabin* (1852) II. xx. 33, I thought she was rather a funny specimen in the Jim Crow line. **1866** Lindley & Moore *Treas. Bot.* II. 638/1 Jim Crow's nose. A West Indian name for *Phyllocoryne.* **1887** C. B. George *40 Yrs. on Rail* viii. 160 An educated colored man..found, on going from Boston to Salem, his home, that he must ride in the Jim Crow car. **1902** A. H. Lewis *Wolfville Days* xvi. 235 An' whyever don't you-all wear leather chapps that a-way, instead of them jimcrow boots an' trousers? **1903** *Sun* (N.Y.) 29 Nov. 7 The members of the committee have arranged with the parents of negro children to send them all to the Jim Crow school, thus entirely separating the white and negro pupils. **1904** *Nation* (N.Y.) 17 Mar. 202 The Jim Crow bills now before the Maryland Legislature. **1904** *Richmond* (Virginia) *Times-Despatch* 25 May 10 Violating the Jim Crow law by allowing negroes to ride in the same car with whites. **1926** A. Niles in W. C. Handy *Blues* 20 'Jim Crow' songs with syncopated airs..were current long before the Civil War. **1931** W. Faulkner *Sanctuary* xix. 167 It was full too, the door between it and the jim crow car swinging open. **1949** L. Feather *Inside Be-Bop* i. 3 Big band jazz had been played by Negro orchestras, frequently under Jim Crow conditions. **1957** W. C. Handy *Father of Blues* xiv. 195 Having spent much time in Jim Crow towns, I was under the illusion that these Negro musicians would jump at the chance to patronize one of their own publishers. **1960** *New Left Rev.* Sept.–Oct. 39/2 The Uncle Tom presidents of the captive Jim Crow colleges. **1971** *Black World* Mar. 75/1 Black people will continue to see themselves under jim crow conditions. **1973** A. Dundes *Mother Wit* 397 Riding on Jim Crow cars could literally make a Negro sick.

So **Jim Crow** *v. trans.*, to segregate persons ethnically, to discriminate against (Negroes or other non-whites); so **Jim Crowing** *vbl. sb.*; **Jim Crowism,** the institution of segregation, the practice of 'racial' discrimination; also, the act of deserting one's political party.

1837 *N.Y. Mirror* 7 Oct. 118/1 Then, to counterbalance this good, you have entailed upon those British islands the curse of Jim Crowism. **1841** *Times* 21 June 5/6 His propensities to what they call 'Jim Crowism' in politics. **1921** *United Free Ch. Miss. Rec.* Jan. 2/1 [S. Africa] The picture he gives of the 'jimcrowing' and ostracizing of the natives in public places and the working of the pass-laws up-country is dark enough. **1923** *Nation* (N.Y.) 15 Aug. 155 But they are not 'jim crowed'. **1925** *Amer. Mercury* Jan. 87/2 In his celebrated Atlanta speech he justified all the forms of Jim-Crowism. **1932** E. Wilson *Devil take Hindmost* iii. 17 The Negroes..are discriminated against by a general policy of Jim Crowism. **1942** C. Himes *Black on Black* (1973) 216 Out of the bootings and the lynchings and the jim-crowings..will come our strength. **1948** *Sat. Rev.* (U.S.) 24 July 16/3 It is to his eternal credit that he ripped through the Jim Crowism of our national game by giving a fine Negro athlete a chance to play in organized baseball. **1955** L. Feather *Encycl. Jazz* i. 22 The Negro musician was Jim Crowed from the day he first became aware of music. **1969** C. Himes *Blind Man with Pistol* i. 16 It was obviously a jim-crowed convent, and no one ever dreamed that white Catholics would act any different from anyone else who was white. **1971** *Rand Daily Mail* (Johannesburg) 4 Dec. 4/2 Since I visited South Africa, the Government has opened the door to dozens of prominent Blacks and ensured that they were not Jim Crowed. (Jim Crowed means treated like Blacks in the old segregationist United States South.)

Jim Crow² (dʒi·m,krōᵘ·). *colloq.* [Coined by W. S. Churchill; cf. Crow *sb.*¹ 8 and Crow's nest.] A roof-top spotter of enemy aircraft; also his look-out post (see also quot. 1943).

1940 W. S. Churchill *Into Battle* (1941) 278 Our plan must be to use the siren..as an alert and not as an alarm, and to have a system of highly trained what I may call Jim Crows or look-out men, who will give the alarm when immediate danger is expected at any point. **1941** *Battle of Britain Aug.–Oct. 1940* (Ministry of Information) 21 Except when roof-watchers—the Prime Minister's 'Jim Crows'—signalled that danger was imminent, life went on as usual and still does. **1943** Hunt & Pringle *Service Slang* 41 *Jim Crow*, normally used to describe the corps of roof-spotters guarding our large buildings. Now taken into service slang to denote the man on watch when 'unofficial business', such as cards, is being transacted. **1952** R. Sherbrooke-Walker *Khaki & Blue* v. 43 An anti-aircraft post, with a good view of the sky, had already been installed on top of one of the forts. To convert this into 'Jim Crow', it was turned into a post de luxe.

ji·m-da·ndy, *sb.* and *a.* *U.S. colloq.* [Cf. Dandy *sb.*¹ (and *a.*).] **A.** *sb.* An excellent person or thing. **B.** *adj.* Remarkably fine, outstanding.

1887 *Courier-Jrnl.* (Louisville, Kentucky) 12 Jan. 2/1 Dear Sir: Though a stranger to you (yet a Democrat), let me say you are a 'Jim Dandy'. **1888** *Inter-Ocean* (Chicago) 14 Feb. (Farmer), George C. Ball came upon the floor yesterday arrayed in a jim-dandy suit of clothes. **1902** O. Wister *Virginian* xxvii. 346 He must have been a jim-dandy of a boy. **1904** W. N. Harben *Georgians* ix. 87 Right thar I baked pies—open-top jim dandies, too. **1912** *Collier's* 21 Dec. 23/2 Prince Albert rolls up the jim dandiest cigarette. **1919** H. L. Wilson *Ma Pettengill* iv. 119, I bet you made a jim-dandy good report. **1941** J. Stuart *Men of Mountains* 154 He said the blue-tick hound was a jim-dandy possum dog. **1953** *New Yorker* 7 Feb. 92/2 When one suffers from headaches,..Anacin is the jim-dandy remedy. **1963** *Spectator* 12 July 45 The new GPO tower building is a jim-dandy, it excites and pleases me. **1972** B. Garfield *Line of Succession* (1974) ii. 128 Your voting record on foreign affairs was fine, jim-dandy.

ji·m-hi·ckey. [Cf. *hickey *sb.*] = prec. *sb.*

1895 S. Crane *Red Badge of Courage* xxi. 207 Th' lieutenant, he ses: 'Ahem! ahem! he is, indeed, a very good man.' **1907** J. Masefield *Tarpaulin Muster* xii. 131 Them topsails had a good look along the yard..or there was a jim hickey of a stink raised.

jiminy. Substitute for entry:
jiminy, jimminy, now the usual form of Gemini (sense 4). Used esp. in phrases *jiminy Christmas* (see *Christmas *sb.* 1 c) and *jiminy cricket.*

1803 G. Colman *John Bull* i. i. 5 *Den.* A customer... *Mrs. B.* Jemmeny! and so there is. c **1816** [see Crimine, -iny *int.*]. **1848** in *Amer. Speech* (1935) X. 40 *By Jiminy Cricket*, an exclamation of surprise. **1863** [see Gemini 4]. **1890** *Dialect Notes* I. 49 Jimminy Cripes! and Jimminy Christmas! are forms of oaths overheard. **1894** J. T. Moore *Songs & Stories from Tennessee* (1897) 18 Jiminy! But didn't we hab a good time on de road? **1897** [see *Christmas *sb.* 1 c]. **1930** J. Dos Passos *42nd Parallel* ii. 147 Jimminy criskets [*sic*], what I couldn't do to a watermelon, Joe. **1934** W. Saroyan *Daring Young Man* (1935) 123 He was telling her. Jiminy crickets. **1940** N. Marsh *Surfeit of Lampreys* (1941) x. 139 'I am completely baffled.' 'Jiminy cricket!'

jim-jam. 1. Delete † *Obs.* and add later example.

1890 Kipling *Phantom 'Rickshaw* 85 There was half a dozen big stone idols. Dravot he goes to the biggest.. and says: 'That's all right..all these old jim-jams are my friends.'

3. b. *pl.* The fidgets, jitteriness; a fit of depression. *colloq.*

1896 'Mark Twain' in *Harper's Mag.* Sept. 537 They gave me the jimjams and the fantods and caked up what brains I had. **1904** *Strand Mag.* Dec. 770/2 By Gosh, look there. Enough to give a fellow the jim-jams, isn't it? **1926** Galsworthy *Silver Spoon* i. vi. 44 Who was the old buffer?..he gave me 'the jim-jams'. **1931** C. Mackenzie *Buttercups & Daisies* xiii. 169, I reckon if you put him in the upper circle at *Charley's Aunt* he'd give half the audience the jim-jams and upset the whole piece. **1946** K. Tennant *Lost Haven* (1947) xix. 332 'There is someone out there,' he said. 'Or else I'm giving myself the jim-jams.' **1961** R. M. Patterson *Buffalo Head* vi. 223 The sides gave me the jim-jams merely to look at them: on the south, sliding rock and then a drop of 2000 feet down.

jimmies, *sb. pl. colloq.* = Jim-jam 3 and *3 b.

1900 J. C. Harris *On Wing of Occasions* 42 Take 'im to the hospital, Tim; 'tis the only way to clear the jimmies from his head. **1921** A. Mason *Flying Bo'sun* xxvi. 216 Riley,..you drank too much Scotch last night; be careful that you don't get the Jimmies and jump overboard. **1928** *New Yorker* 22 Dec. 18/3 Him popping down the chimney —well, frankly, it gives me the jimmies. **1945** Baker *Austral. Lang.* x. 199 *Britts up, to have the*, to be alarmed. From 'to have the wind up' and 'have the sh-ts', by rhyming slang on the name of the former lightweight boxing champion of the world, Jimmy Britt, who was on vaudeville tour in Australia during World War No. 1. *To have the jimmies* is an extension. **1961** P. White *Riders in Chariot* xi. 415 She was not accustomed to see the grey light sprawling on an empty bed; it gave her the jimmies.

jimmy, dial. and colloq. pronunciation of Jemmy *sb.* Add: Now the more usual form in the U.S. (Examples.) Also as *v. trans.*, to open with a jemmy.

1848 G. W. M. Reynolds *Mysteries of London* IV. cxcv. 369/1, I have got my own clasp-knife..and a small jimmey. **1854** *Alta California* (San Francisco) 23 Feb., Officer Powers,..upon examining the lock, found it had been broken open with a 'jimmy'. **1893** J. Hawthorne *Confessions of Convict* iii. 49 We took the safe..and carried it..to the basement... We jimmied it open in no time. *Ibid.* xi. 172, I have drilled holes in large safes so accurately that the bolts could be 'jimmied' without leaving a mark. **1904** G. H. Lorimer *Old Gorgon Graham* viii. 159 You can't break a big merchant with a jimmy and a stick of dynamite. **1905** *N.Y. Even. Post* 22 Dec. 3 The thieves jimmied the front door. **1922** R. Parrish *Case & Girl* xxxii. 247 Finally we jimmied open the back door of this garage. **1973** 'E. McBain' *Let's hear It* xiii. 194 The patrolman..was examining a door and jamb for jimmy marks. **1973** *Sat. Rev. Society* (U.S.) May 42/2 Any attempt to jimmy the doors, hood, or trunk will cause the horn to begin sounding. **1973** P. B. Austin tr. *Sjöwall & Wahlöö's Locked Room* v. 24 The door was equipped with a jimmy-proof lock. *Ibid.* xxiii. 186 [He] had brought with him every thinkable jimmy and other tool for opening the door.

Jimmy (dʒi·mi). [A male personal name, pet-form and familiar equivalent of the name James.] In various transferred senses:

1. In full, **Jimmy Grant.** Rhyming slang for *immigrant* or *emigrant.* *Austral.*, *N.Z.*, and *S. Afr.*

1845 E. J. Wakefield *Adventure N.Z.* I. xi. 337 The profound contempt which the whaler expresses for the 'lubber of a *jimmy-grant*', as he calls the emigrant. **1850** *McLean Papers* VIII. 177 (MS.), I consider Davy has done a foolish thing in selling his farm... I am glad it has not as usual fallen into the hands of 'Jimmies', usurpers of the soil. **1859** H. Kingsley *Recoll. G. Hamlyn* II. ix. 154 'What are these men that we are going to see?' 'Why one,' said Lee, 'is a young Jimmy (I beg your pardon, sir, an emigrant), the other two are old prisoners.' **1878** A. Aylward *Transvaal of To-Day* xi. 216, I was a raw emigrant, and still what Natalians call a '*Jimmy*'. **1922** *Daily Mail* 11 Dec. 8 With his wife and child he had just come over as a 'Jimmie-grant'. **1948** F. Irvine-Smith *Streets of my City* ii. 32 At the close of 1840, there were 2,500 settlers, or in whaler parlance, 'Jimmy Grants', upon its shores.

2. Jimmy Ducks, Jimmy Dux. Also **Jemmy Ducks.** A sailor who had charge of the live-stock carried on merchant ships to serve as food on long voyages. *Naut. slang* (*Obs.* exc. *Hist.*).

1849 H. Melville *Redburn* I. ix. 97 He in the rudest kind of manner laughed aloud in my face, and called me a 'Jimmy Dux'. **1850** —— *White Jacket* I. iii. 12 These fellows are all Jimmy Duxes—sorry chaps, who never put foot in ratlin, or venture above the bulwarks. **1890** R. C. Leslie *Old Sea Wings* xiii. 177 The ship's butcher and his mate, 'Jemmy Ducks', formed an important part of the economy of our old East Indiamen. *Ibid.* 182 A gay rooster, after an exciting chase round the decks by Jemmy Ducks, escaped overboard. **1928** J. Mason *Before Mast in Sailing Ships* 13 One of our men had deserted an Orient Liner in Sydney, and had been a 'Jimmy Ducks' on that vessel. **1938** W. E. Dexter *Rope-Yarns* v. 32 'Sails' or 'Jimmy Ducks' was the first one to get round.

3. Jimmy Low *Austral.*, a name for red mahogany, *Eucalyptus resinifera.*

1887 *Colonial & Indian Exhib. Rep. Col. Sect.* 428 Jimmy Low is usually a large tree, yielding timber of a rich red colour. **1888** F. M. Bailey *Queensland Woods* 65 'Jimmy Low'. Forest Mahogany of N.S.W. Usually a very large tree with a rough, reddish, fibrous bark. **1904** J. H. Maiden *Forest Flora N.S.W.* I. 67 In Queensland it [sc. *Eucalyptus resinifera*] is often called 'Jimmy Low', after the late Mr. James Low, of Maroochie River, a locality for some of the finest specimens in that State. **1945** Baker *Austral. Lang.* xii. 215 Among popular names for various trees noted by Morris were Jemmy Donnelly, Jimmy Low and Roger Gough.

4. Jimmy (also **Johnny**) **Wood(s, Woodser,** a solitary drinker; a drink taken on one's own; also *transf.*; so **Jimmy Woods** *nonce vb.*, to drink alone. *Austral.* and *N.Z. slang.*

The usual form is *Jimmy Woodser.*

1892 B. H. Boake in *Bulletin* (Sydney) 7 May 15/1 At the thought the heart beats quicker Than an old Bohemian's should... Bah! I'll go and have a liquor With the genial 'Jimmy Wood'. **1898** *Ibid.* 17 Dec. Red Page/2 The use of Christian names in this form of slang seems to have originated the Australian..*Jimmy-Woodser*, a solitary drinker. **1900** H. Lawson *Verses Pop. & Humorous* 67 The old Jimmy Woodser comes into the bar, Unwelcomed, unnoticed, unknown. **1928** J. Devanny *Dawn Beloved* ii. xxxiv. 307 Duke preferred to drink alone.., Jimmy Wood-sing, as the miners called it. **1930** *Bulletin* (Sydney) 19 Feb. 51/4 'You have your holiday.' 'Oh, no,' she told him..'I'm getting too old for Jimmy Woodsers, thanks.' **1933** *Press* (Christchurch, N.Z.) 28 Oct. 17/7 *Jimmy* or *Johnny Woodser*, slang. A drink by yourself. It is a common expression up-country in New Zealand. **1942** *2 N.Z.E.F. Times* 21 Dec. 18/2 You'll find me lonesome in a Naafi, a-drinkin' to me sins, A-sippin' like a Jimmy Woodser. **1945** Baker *Austral. Lang.* ix. 171 A *Jimmy Woodser* is not only a lone drinker, but also a drink consumed by such a person. **1957** D. Niland *Call me when Cross turns Over* i. 9 I'm a real Jimmy Woodser now. On my own. The first bird on the family tree and the last.

5. In full, **Jimmy O'Goblin** (also with lower-case initials). Rhyming slang for 'sovereign', twenty shillings. Cf. *goblin², *Jemmy O'Goblin. *slang.*

1899 A. E. W. Mason *Miranda of Balcony* xv. 206, I want one thousand jimmies per annum. **1931** T. H. Dey *Leaves from Bookmaker's Bk.* xi. 180 Here's another story of a lost 'Jimmy o' Goblin'. **1932** D. L. Sayers *Have his Carcase* xi. 128 Three hundred golden sovereigns... Three hundred round, golden jimmy o' goblins. **1934** E. Waugh *Handful of Dust* ii. 35 He had won five Jimmy-o-goblins at ten to three at Chester. **1956** C. Smith *Deadly Reaper* xiv. 108 Her first husband left her half a million. Yes, sir, five hundred thousand jimmy o'goblins. **1959** *Spectator* 3 July 5/2 The proposed sale of the Watford firm S. G. Brown Ltd., at present owned by the Admiralty..is a most unlikely haunt of businessmen, if the thirty-eight million jimmy-o'goblins they were wasting on obsolete warships a month or so ago is anything to go by. **1967** C. Watson *Lonelyheart 4122* xv. 149 'The money's paid in—five hundred nice shiny Jimmy O'goblins.' (Dear God! Where had she last come across that one? Sapper? Henty?) **1973** *Times* 28 June 16/2 He..had made a profit of some six million jimmy-o-goblins.

6. In full, **Jimmy the One.** First Lieutenant. *Naut. slang.* Also in various other applications (see quots.).

1916 'Taffrail' *Pincher Martin* viii. 147 Th' Bloke, an' Jimmy the One, an' most o' th' other officers made a bit too. **1935** Wodehouse *Luck of Bodkins* viii. 80 'By rights I ought to just go to Jimmy the One.'..'The chief steward, sir?' **1945** [see *flannel *sb.* 1 f]. **1953** E. Hyams *Gentian Violet* v. 76 He became second in command of a very old destroyer and as such he was perhaps the only Jimmy-the-One in the Royal Navy who was not detested by the ship's company. **1962** 'E. Peters' *Funeral of Figaro* i. 10 He was..used to being Number One or Jimmy the One [in a theatre]. **1970** *Guardian* 19 Aug. 16/8 Smith told Petty Officer David Lewis, 'We are going to have a sit-in and give the "Jimmy" a hard time.'

7. dismal Jimmy: see *dismal *a.* 7.

8. Jimmy Green = *Jamie Green; a sail under the jib-boom.

1913 E. K. Chatterton *Ships & Ways of Other Days* xii. 265 The reader will remember we called attention some time back to those spritsails which seem so curious to us moderns. They were also known as 'water sails' and as 'Jimmy Greens', both appellations being due, obviously, to the unhappy knack they possessed of scooping up the sea. **1933** *Sea Breezes* XVII. 186 A 'Jimmy Green' was set along the bowsprit and jibboom under the head sails. **1944** J. Masefield *New Chum* 155 Once one came by with an odd triangular yardless topsail, then, very rare, but now in use again. We were told that it would be called a Jimmy Green if set above a royal.

9. In full, **Jimmy Riddle.** Rhyming slang for *piddle.*

1937 in Partridge *Dict. Slang.* **1959** R. Fuller *Ruined Boys* i. 10 Come and have a Jimmy Riddle. **1964** J. Symons *End of Solomon Grundy* i. ii. 34, I must do a Jimmy Riddle before I go. **1966** 'L. Lane' *ABZ of Scouse* 56 Said of a person with a weak bladder: 'e's got ther jimmy riddles. **1971** D. Clark *Sick to Death* vii. 154 Mrs D. was in there having a jimmy.

jimsonweed. Also jim(p)son, jimpsonweed. = Jamestown-weed.

1812 *Cramer's Almanac 1813* (Pittsburgh) 26 James'-town weed..also known by the name of Jimson, and Thorn-Apple. **1842** *Amer. Pioneer* I. 314 She said her principal dressings were made of..the leaves of stramonium, or 'jimson'. **1880, 1892** [see Jamestown-weed]. **1911** C. Harris *Eve's Second Husband* xiv. 299 Clumps of sweet jimson and borders of balsam and pinks. **1930** R. Bass in A. Dundes *Mother Wit* (1973) 382 He cures.. headaches with a poultice of jimson-weeds. **1934** *Amer. Ballads & Folk Songs* 23 Mike was pilin' ties near the ditch by the road Out among the jimpson where the boys ain't mowed. **1943** R. Peattie *Great Smokies* 118 Jimson root for ulcers. **1954** C. J. Hylander *Macmillan Wild Flower Book* 360 Jimsonweed is a stout plant, growing to a height of five feet, with ovate lobed leaves and white or violet, trumpet-shaped flowers up to four inches in length.

jimswinger (dʒi·m,swiŋəɹ). *Southern U.S.* Also **jim swinger, jim-swinger.** [Origin unknown.] In full, *jimswinger coat.* A swallow-tailed coat, esp. a frock-coat.

1890 *Dialect Notes* I. 389 *Jim-swinger*, long-tailed coat, especially a 'Prince Albert'. **1898** P. L. Dunbar *Folks from Dixie* 52 He walked to church, flanked on one side by Aunt Caroline..and on the other by her husband stately in the magnificence of an antiquated 'Jim-swinger'. **1912** H. & W. B. Crumpton *Adv. Two Alabama Boys* 78, I was a tall, slim, awkward lad, about eighteen years old, thin as a match, pale as a ghost and had on a long Jim Swinger. **1933** C. Miller *Lamb in Bosom* v. 47 If you were a lawyer, ..you wore a jimswinger coat. **1950** *Publ. Amer. Dial. Soc.* xiv. 75 *Cut-away and Jim swinger*, formal clothes. Suwannee backwoods. **1951** E. Paul *Springtime in Paris* i. 9 She wore indoors and out a strictly tailored swallow-tail coat, the kind American Negroes call a 'Jimswinger'. **1972** *News & Observer* (Raleigh, North Carolina) 30 Dec. 4/1 Jim-swinger coat..and brogans have almost vanished.

Jina (dʒi·nă). Also **Gina.** [Skr.: see Jain.] Title given to Mahavira or to any other of the twenty-four Jain tirthankars or to one of the five Dhyani Buddhas; a sculptured representation of such a saint.

1807 *Asiatick Researches* IX. 303 The first chapter [of a vocabulary of synonymous terms, by an author of the Jaina sect] begins with the synonyma of a *Jina* or deified saint. **1833** *Ibid.* XVII. 250 Notwithstanding the sameness of the general character and identity of generic attributes, the twenty-four *Jinas* are distinguished from each other in colour, stature, and longevity. Two of them are red, two white, two blue, two black, the rest are of a golden hue, or a yellowish brown. The other two peculiarities..observe a series of decrement from *Rishabha*, the first *Jina*, who was five hundred poles in stature, and lived 8,400,000 great years to Mahávína the 24th, who had degenerated to the size of man, and was not more than forty years on earth. **1875** M. Williams *Indian Wisdom* vii. 129 Twenty-four Jinas or 'perfect saints' raised to the rank of gods have appeared in the present Avasarpini cycle. **1884** *Sacred Bks. East* XXII. i. 201 The Venerable One had become an Arhat and Gina, he was..omniscient. **1901** V. A. Smith *Jain Stūpa of Mathurā* xviii. 46 The Jina shown in the plate is Pārśvanāth. **1915** A. M. Stevenson *Heart of Jainism* iii. 39 Mahāvīra now added to his titles those of *Jina* (or Conqueror of the Eight Karma, the great enemies), from which Jainism derives its name, [etc.]. **1924** B. Bhattacharya *Indian Buddhist Iconogr.* i. 2 The Jinas (victorious ones) are Vairocana, Ratnasambhava, Amitābha, Amoghasiddhi and Akṣobhya, whose colours respectively are white, yellow, red, green and blue. **1971** *Illustr. Weekly India* 11 Apr. 8/1 (*caption*) Risabha and Vardhamana are the most honoured among the twenty-four Tirthankaras and the Jain scriptures narrate in detail the life of these two Jinas.

Jindyworobak (dʒi:ndiwǫ·robæk). [Austral. Aboriginal word.] A member of a group founded in 1938 by R. C. Ingamells (1913–55) to promote Australianism in literature, art, etc. Also *attrib.* or as *adj.*

1938 (*title*) Jindyworobak anthology. **1958** *N. & Q.* Oct. 459/1 Roland Robinson, the Jindyworobak, is a perspicuous observer of natural beauty. **1959** *Times Lit. Suppl.* 2 Jan. 9/1 This book demonstrates that, while in the direct Jindyworobak line, he is no fanatic.

jing, *sb.*² Add non-*Sc.* examples.

1809 [see *good *a.* 19]. **1881** 'Mark Twain' *Let.* 16 Dec. (1917) I. 412 By jings! the postman will be here in a minute. **1925** T. Dreiser *Amer. Tragedy* ii. iii. 184 Well, by jing, if it ain't Tom. **1941** Baker *Dict. Austral. Slang* 39 *By jings!*, an exclamation, derived from 'by jingo!'

jingle, *sb.* Add: **2. b.** *Austral. slang.* Money. Cf. *jingle-boy.*

1906 E. Dyson *Fact'ry 'Ands* viii. 99 Ther Elder dug in 'n' brought up er 'andful iv jingle. **1941** Baker *Dict. Austral. Slang* 39 *Jingle*, money.

3. b. A short verse or song in a radio or television commercial or in general advertising.

1930 A. Flexner *Universities* i. xxv. 165 Let the psychologists study advertising..in order to understand what takes place when a jingle like 'not a cough in a carload' persuades a nation to buy a new brand of cigarettes. **1949** *Life* 28 Mar. 36/1 She abhors the jingle's suggestion that she be taken home and squeezed. **1959** *Punch* 10 June 769/1 He says I can't possibly get on in the jingles business without going to Oxford. *Ibid.* 17 June 815/2 Channel 9 already gets into hot water when its 'natural breaks' happen to clip a speech in mid-sentence: its life would not be worth living if it saw fit.. to substitute a jingle for the heavyweight's knock-out hook. **1968** *Listener* 26 Sept. 421/3 Certainly those hideous jingles could go: it made good sense for Radio London or Radio Caroline, as new, commercial stations, to tell us that they were wonderful, but the BBC could spare itself that reassurance. **1972** D. Ramsay *Little Murder Music* 76 Colby was working a jingle date. *Ibid.*, The jingle, a singing commercial for a detergent, was being recorded.

4. Also in Cornwall.

1874 A. I. Thackeray *Let.* 12 July (1924) viii. 158 You come to a most detestable little object called Bude.. and then you..come home in a little thing called a Jingle. **1906** *Daily Chron.* 10 Mar. 4/4 When I asked one of the drivers how to reach the Cornish border, he offered to convey me..in a jingle-cart. **1924** C. Mackenzie *Heavenly Ladder* ii. 34 He saw the black-coated train toiling up Pendhu hill,..some leading the ponies in the jingles. **1942** A. L. Rowse *Cornish Childhood* vii. 189, I was sent to put the donkey into the shay or jingle.

jingled (dʒi·ŋg'ld), *ppl. a.* *U.S. slang* (now *rare*). [f. Jingle *v.* + -ed¹.] Intoxicated, fuddled.

1908 G. H. Lorimer *Jack Spurlock* xii. 315 Old Mrs. Corliss was purple with pleasure at having so plausible a pretext for getting comfortably jingled. **1917** [see *blotto *a.*] **1942** in Berrey & Van den Bark *Amer. Thes. Slang* § 106/7.

jingling, *ppl. a.* Add: **jingling Johnny,** (*a*) *slang* = *Chinese pavilion*; (*b*) *Austral.* and *N.Z. slang*, one who shears sheep by hand; *pl.* hand shears.

a. 1904 H. G. Farmer *Mem. R. Artillery Band* ii. 51 An instrument known as the 'Jingling Johnnie', and tambourines were employed. **1920** G. B. Shaw *How to become Mus. Critic* (1960) 311 Every scorer of ballets could scatter pearls from the *pavillon chinois* (alias Jingling Johnny) over the plush and cotton velvet of his harmonies. **1970** *Times* 24 Aug. 20/5 (Advt.), French Pavillon Chinois (jingling Johnny) for sale. Recently restored. £100 o.n.o.

b. 1934 L. G. D. Acland in *Press* (Christchurch, N.Z.) 20 Jan. 15/7 *Jingling Johnnies*, old time slang term for hand shearers. **1941** Baker *Dict. Austral. Slang* 39 *Jingling johnnies*, hand shears. **1945** [see *dagger *sb.*² a]. **1965** J. S. Gunn *Terminol. Shearing Industry* i. 33 *Jingling Johnny*, originally a swagman or bagman but in many districts this was also another name for a hand-shearer.

jink, *sb.*¹ Add: **1.** Used esp. of a tricky turn in Rugby Football, or in Aeronautics. Also *transf.* and *fig.*

a **1914** J. E. Raphael *Mod. Rugby Football* (1918) 122 [Poulton Palmer's] 'jink' is all by itself in modern-day Rugger. **1921** E. H. D. Sewell *Rugby Football* vi. 126 The defender..moves to *his* left as the dummy pass is made, thus making wider the gap which the attacker widens still further by his jink to the defender's right. **1943** Hunt & Pringle *Service Slang* 41 *Jinks*, quick turns in the air, a form of aerobatics and of avoiding action. **1959** V. Fuchs *Antarctic Adv.* xii. 168 Suddenly, ..below them, they spotted vehicle tracks in the snow. From the outward flight they remembered that except for one 'jink' these led in a straight line from South Ice—and the 'jink' was only about a quarter of a mile from the station. **1969** P. Dickinson *Pride of Heroes* 164 A jink in his train of thought made Pibble wonder who the next heir was.

3. high jinks. Hence **high-ji·nking.**

1891 *Pall Mall Gaz.* 22 Oct. 2/1 On board we were very jovial and had much high jinking. **1904** *Daily Chron.* 13 July 8/1 It is evening—eight o'clock—and the festival is at its very top notch of high-jinking.

jink, *v.*¹ For 'Chiefly *Sc.*' read 'orig. *Sc.*' Add later examples of senses 1 and 2, esp. in Rugby Football and Aeronautics; cf. *jink *sb.*¹ 1. Also *to jink one's way*, to advance by means of jinks.

a **1914** J. E. Raphael *Mod. Rugby Football* (1918) 103 This is a method which..should not be copied by the ordinary performer—except perhaps when 'jinking'. **1927** Wakefield & Marshall *Rugger* 93 A curious jinking side-step. **1932** *Daily Tel.* 19 Mar. 17/2, I can see him jinking his way past our mid-field players. **1940** *Aeroplane* 30 Aug. 235/2 One example of Air Ministry verbosity is the continual use of the phrase 'took evasive action', instead of saying 'dodged' or 'jinked'. **1942** *R.A.F. Jrnl.* 18 Apr. 3 You boost and you dive and you jink. **1942** E. Waugh *Put out More Flags* iii. 243 If they come in now from the rear the cars may jink round and give the other companies a chance to get out. **1944** *R.A.F. Jrnl.* Aug. 288 The aircraft crossed in front of them, jinking steadily at high speed. **1959** V. Fuchs *Antarctic Adv.* xii. 169 The visibility was about a hundred yards, but where the tracks jinked sharply right, he closed the throttle. **1961** R. Jeffries *Evidence of Accused* i. 14 A rabbit..jinked away under a rhododendron bush. **1963** I. Fleming *On H.M. Secret Service* xvii. 189 Bond.. put on all the speed he could, crouching low and jinking occasionally to spoil the man's aim. **1969** G. Macbeth *War Quartet* 35 Untailed, I jinked, Flipped over in a half-roll. **1973** *Times* 1 Jan. 17/2 Hales scored two tries, but I only saw the first of them, and a good one it was as he jinked inside his man.

jinker². Add: (Later examples.) Also, in *Trotting*, a sulky.

1916 J. B. Cooper *Coo-oo-ee* i. 1 Often the wheels of the jinkers bogged in a soak on the track. **1941** Baker *Dict. Austral. Slang* 39 *Jinker*, a trotting sulky. **1941** *Coast to Coast* 185 Then would come a decision, a harnessing of the old horse to the jinker and a trip to the township. **1966** 'J. Hackston' *Father clears Out* 15 We hated the horse that was tied to the man's jinker.

jinker, *v.* *Austral. trans.* To manipulate with a jinker (see Jinker²).

1903 R. Bedford *True Eyes & Whirlwind* 240 Waiting for a fine day to jinker those trees out of the bush.

jinny. 1. (Earlier example of *jinny-road.*)

1877 [see *dip *sb.* 5 b].

jintawan (dʒintā·wăn). [Malay.] A kind of caoutchouc derived from *Urceola elastica*, a woody climbing plant of the family Apocynaceæ; also, the plant itself.

1851 *Illustr. Catal. Gt. Exhib.* iv. 877/2 Raw caoutchouc from Assam, Singapore (*Urceola elastica*, the Jintawan of the Malays). **1853** Ure *Dict. Arts* (ed. 4) I. 984 The said apparatus may be used also for purifying caoutchouc and jintawan. **1880** *Encycl. Brit.* XII. 817/1 Besides the orange ..we have the rambutan,..the jintawan,..and the jambosa.

jinx (dʒiŋks). orig. *U.S.* Also **ginks, jinks.** [app. f. Jynx 2.] A person or thing that brings bad luck or exercises evil influence; a hoodoo, a Jonah. Also *attrib.* and *Comb.*

1911 *Chicago Daily News* 19 Sept. 6/3 Dave Shean and 'Peaches' Graham..have not escaped the jinx that has been following the champions. **1919** *Oxf. Mag.* 7 Mar., Will some one remove the jinx? On Friday, February 28, we lost to Oriel and Merton by 3 goals to nil. On Saturday, March 1,..we lost to Queen's by 1–0. **1926** Anderson & Stallings *What Price Glory* iii, in *Three American Plays* 88 This town is a jinx for me. **1927** A. Miller *Colfax Bk.-Plate* xiv. 172, I thrust that jinx of a book back into the lowest left-hand drawer. **1931** *Collier's* 26 Sept. 57/3 They say he is a jinx to them and they cannot beat him in any manner, shape, or form. **1932** J. T. Farrell *Young Lonigan* iv. 157 Young Corbett who was born with a horse shoe in his hands..put a jinx on Terrible Terry. **1948** Menjou & Musselman *It took Nine Tailors* 96 There has never been such a jinx picture as that one. **1955** E. Pound *Classic Anthol.* iii. 179 Jinx on the remnants everywhere. **1958** *Times Lit. Suppl.* 24 Jan. 41/4 The family jinx, in this case dark strangers who erupt into the bourgeois marriages of three generations, leaving behind an illegitimate child and the torturing memory of a bliss that might have been. **1962** *Times* 19 Feb. 3/7 A jinx-ridden Thomas somehow contrived to miss the kick. **1969** *Daily Colonist* (Victoria, B.C.) 30 Nov. 8/4 A man and two boys went fishing Friday, and Mary went along. The boat capsized, and apparently she alone survived. 'I'm either a jinx, or very lucky,' she says. **1972** 'E. Lathen' *Murder without Icing* (1973) xxiii. 198 I'm beginning to think that damned team is a jinx... It's been a bad-luck team from the beginning.

So **jinx** *v. trans.* (freq. *pass.*) to cast a spell on, to bring bad luck upon; **jinxed** *ppl. a.*

[**1912** C. Mathewson *Pitching in a Pinch* xi. 244 He outjinxed our champion jinx killer.] **1917** *Amer. Mag.* Apr. 43/1 What do you mean—humming love songs when their darn pitcher is forcing in runs? You jinxed my ball club. **1934** J. T. Farrell *Young Manhood* iv. 61, I know they ain't loaded. But use these ones. Them damn things is jinxed! **1946** E. O'Neill *Iceman Cometh* (1947) iii. 147 I's an ole gamblin' man and I knows bad luck when I feels it!.. But it's white man's bad luck. He can't jinx me! **1956** D. Barnham *One Man's Window* 26 He told me he was 'jinxed'. He explained that on his last three trips [in a fighter aircraft] he had been in trouble. **1962** K. Orvis *Damned & Destroyed* xi. 73 I'm jinxed. No fix. All blew up. **1972** G. Durrell *Catch me a Colobus* vii. 152 I'm damned if I'm going to take up all my animals and myself and my wife in a jinxed plane. No, I'm afraid I'll have to get another charter flight.

jipijapa (hī·pi,hā·pă). Also **jipi-japa, jippa-jappa, jippi-jappa.** [Name of a town in Ecuador.] **a.** The palm-like tree, *Carludovica palmata*, native to tropical America, or the fibre produced from its leaves. **b.** A Panama hat. Also *attrib.*

1858 P. L. Simmonds *Dict. Trade Products* 273/2 (s.v. *Panama-hats*), In Central America where they are made, the palm is called Jipijapa. **1877** *Encycl. Brit.* VI. 155/1 Straw hats, usually known as jipijapa or Panamá hats. **1901** *Amer. Anthropologist* III. 206 Ecuador is the real home of the hats wrongly designated under the name of 'panama'... Everywhere in Latin America the hat is known under the name of *jipijapa*, in honor of the city where its manufacture was first started. **1901** *Daily Chron.* 7 Feb. 4/6 The Jippi-Jappa plant, known sometimes as 'the broom thatch'. **1953** V. Bell in *Caribbean Anthol. Short Stories* 72 He could almost see her there in her usual seat at the back of the church in her white linen dress and her jippi-jappa hat. **1961** F. G. Cassidy *Jamaica Talk* vi. 115 The best known Jamaica-made hat is the *jipi-japa*..from the name of the town..in Ecuador. **1965** I. Fleming *Man with Golden Gun* iv. 51 There were only two other passengers..Cubans, perhaps, with jippa-jappa luggage. **1972** D. Sale *Love Bite* iv. 52 Shops crammed with liqueur, tweeds, English doeskin, jippa-jappa straw goods.

jipper (dʒi·pəɹ). *dial.* Also **gipper.** Gravy; dripping; stew. So as *v. trans.*, to baste.

1822 Scott *Fortunes of Nigel* III. vi. 176 This man Gregory is not fit to jipper a joint with him. **1886** W. H. Long *Dict. Isle of Wight Dial.* 31 Thee'st fat all the jipper out of the pudden. **1896** Farmer & Henley *Slang* IV. 59/2 *Jipper* (nautical), gravy. **1900** *N. & Q.* 9th Ser. V. 295 The *chef de cuisine* was an old navy pensioner, and his instructions were: 'Mind you jipper them [*sc.* thrushes] well.' From him I learned to call gravy 'jipper', and bread and dripping 'bread-and-jipper'. **1904** *Daily Chron.* 12 Aug. 3/7 The mysteries of Irish stew, or 'gipper', as it is mysteriously termed in camp. **1909** H. G. Wells *Tono-Bungay* ii. iii. 195 They [*sc.* shavings] might be anything. Soak 'em in jipper,—Xylo-tobacco!

jippo². = *gippo².

1929 *Papers Mich. Acad. Sci., Arts & Lett.* X. 303 *Jippo*, fat. **1931** W. V. Tilsley *Other Ranks* 119 Every morning, a good thick slice of bread and toasted cheese or bacon. Plenty of jippo (if you were pally with the orderly man). **1937** D. Jones *In Parenthesis* v. 118 He's with his longe ladel To test the jippo.

jirga (dʒiə·ɹgă). Also **jeerga, jirgah.** [Pushtu.] An assembly or council of the headmen of Pathan or Baluchi tribes.

1843 F. Sale *Jrnl. Disasters Afghanistan* p. xii, *Jeerga*, An assembly or council—a diet. **1894** M. Dyan *All in Man's Keeping* I. iv. 75 All large issues had to be referred to his Jeerga. **1906** *Westm. Gaz.* 14 May 3/2 They can.. mobilise as easily as a Pathan jirga. **1908** *Ibid.* 24 Feb. 7/1 Afridis other than Zakka Khels are collecting at Chora for a jirgah. **1926** *Chambers's Jrnl.* Apr. 293/2 He eyed the elders of Pir Mahommed, assembled in *jirga.* **1935** *Pathfinder* July 125/1 The 'jirga', that is the council of the headmen of each tribe, would meet us at each new boundary..to..see if they would let us pass or not. **1967** A. Swinson *N.-W. Frontier* xv. 326 This triumph seemed to expend the tribesmen's energy and soon the jirgas came in to sue for peace. **1972** *Times of India* 28 Nov. 11/5 The paper quoted Mr. Bhutto as telling a tribal jirga at Landikotal..that unfortunately Pakistan did not accept this offer.

jism (dʒiz'm). *slang* (orig. *U.S.*). Also **chism, gism, jizz,** etc. [Origin unknown.] **a.** Energy, strength. **b.** Semen, sperm.

In sense b often regarded as a taboo-word.

1842 *Spirit of Times* 29 Oct. 409/3 At the drawgate Spicer tried it on again, but his horse was knocked up— 'the gism' and the starch was effectively taken out of him by the long and desperate struggles he had been obliged to maintain. **1899** B. W. Green *Word-bk. Virginia Folk-Speech* 85 *Chism, chissum*, seminal fluid. **1937** J. Weidman *I can get it for you Wholesale!* xxxii. 314 'Step on it, will you?' 'Sure...' 'All right,..but put a little jism into it, will you?' **1942** C. Morley *Thorofare* (1943) xxxvii. 137 One of the same fields that the Major said had had the jizzum leached out of it by tobacco. **1955** T. Sterling *Evil of Day* vii. 76 The man had more jisum than he'd counted on. **1959** W. Burroughs *Naked Lunch* 90 The Moslems must have blood and jissom... See, see where Christ's blood streams in the spermament. **1967** S. Beckett *Stories & Texts for Nothing* 85 A week will be ample, a week in spring, that puts the jizz in you. **1968** J. Updike *Couples* iv. 311 Georgene would wash herself before and after. Said his jizz ran down her leg, too much of it. **1969** P. Roth *Portnoy's Complaint* 132 You've got to..walk round downtown Newark dripping gissum down your forehead. **1972** *Screw* 12 June 35/3 At last I felt my gism rushing up like electricity and I..felt the love bolt burst out of my cock into her vacuum-sucking mouth.

jist (dʒist), colloq. and dial. var. Just *adv.* Cf. *jest.

c **1820** in *Amer. Speech* (1941) XVI. 157/1 *Jist*, just. **1835** A. B. Longstreet *Georgia Scenes* 29 His eyes don't look like it; but he *jist as live go agin* the house with you, or in a ditch, as anyhow. **1839** [see *lop *v.*² 2 b]. **1848** Trollope *Kellys & O'Kellys* II. iii. 69 You must do more than jist ask her. **1931** D. L. Sayers *Five Red Herrings* xxv. 283 'I was jist thinkin', then,' said Duncan, '..whether it was not, after all, possible.' **1941** *Sat. Even. Post* 10 May 36/2 Jist taste a grain o' patience. *Ibid.* 112/3, I jist saw a quare thing.

jit¹ (dʒit), abbrev. *jitney 1.

1913 *Awgwan* (Univ. of Nebraska) 3 June 10 We went to the second jit show. **1915** G. Bronson-Howard *God's Man* iv. iv. 264 Keeping tab on you and knowing how much you snatch to the last jit. **1926** J. Black *You can't Win* xx. 318 A 'jit', as the Southern negro affectionately calls his nickel. **1931** J. T. Farrell in *New Rev.* (Paris) I. 24 Say Vinc lemme take a nickel will yuh?.. You'll lemme take a jit won'tcha?..I'll give yuh the jitney back with a nickel interest. **1946** B. Treadwell *Big Bk. of Swing* 124/2 Jit, five-cent piece.

jit² (dʒit). *U.S. slang. depreciatory.* [Origin unknown.] A Negro.

1931 G. Irwin *Amer. Tramp & Underworld Slang* 110 *Jit*... A negro, or, more usually, a negress, and seemingly a term of derision. **1936** Mencken *Amer. Lang.* (ed. 4) 296 For Negro..jit. **1942** Berrey & Van den Bark *Amer. Thes. Slang* § 385/14 *Negro*, jit.

jitney (dʒi·tni). *N. Amer.* [Origin unknown.] **1.** A five-cent piece, a nickel. *slang.*

1903 *Cincinnati Enquirer* 2 May 11/5 [In St. Louis] a 'crown guy' is a policeman, a 'gitney' is a nickel, and 'mug's landing' is the Union Station. **1915** *Nation* (N.Y.) 4 Feb. 142/1 The word 'jitney'..is the Jewish slang term for a nickel. **1916** *Chambers's Jrnl.* June 400/1 Five cents ..is the charge for any distance, and as the colloquial name for this coin is a 'jitney', this form of traffic has become known as 'jitney competition'. **1931** 'D. Stiff' *Milk & Honey Route* 177 Beer or wine at a jitney a throw. **1947** W. Saroyan *Jim Dandy* i. i. 11 Call that money? A jitney? A nickel?

2. In full *jitney bus, omnibus.* An omnibus or other motor vehicle which carries passengers for a fare, orig. five cents. So, on account of the low fare or the poor quality of these buses, used *attrib.* to denote anything cheap, improvised, or ramshackle.

1914 *Let.* 28 Nov. in *Nation* (N.Y.) (1915) 14 Jan. 50/3 This autumn automobiles, mostly of the Ford variety, have begun in competition with the street cars in this city [*sc.* Los Angeles]. The newspapers call them 'Jitney 'buses'. **1915** *N.Y. Even. Post* 16 Apr., The jitney wears out the streets and should contribute to their repair. **1916** *Daily Colonist* (Victoria, B.C.) 6 July 4/5 The daily jitney service between Nanaimo and South Wellington has been discontinued. Hereafter cars will run on Saturdays only and the fare has been raised from 25 cents to 50 cents. **1916** H. L. Wilson *Somewhere in Red Gap* ii. 59 He..sells these jitney pianos and phonographs and truck like that. **1917** E. Frost *Lett.* (1972) 13, I hired McGrath, the jitney driver, to bring us home. **1919** M. A. von Arnim *Christopher & Columbus* xxxi. 400 He had come in the jitney omnibus to the nearest point. **1920** [see *Daiquiri]. **1921** *Daily Colonist* (Victoria, B.C.) 25 Mar. 2/4 The decision of the Provincial Legislature to prohibit the sale of near beer will affect 58 jitney bars in this city, according to City Licence Inspector Charles Jones. **1923** F. Parsons *Everybody's Business* 215 All the jitney operators on the line. **1925** *Amer. Speech* I. 152/1 That bastard word 'jitney' is still used in outlying places, where a 'jitney dance' means a nickel dance. **1933** *N.Y. Herald Tribune* 5 Dec. 17/3 We refer to the jitney economists, the boys who play the money tunes only by ear. **1946** E. O'Neill *Iceman Cometh* (1947) i. 36 He never worries in hard times because there's always old friends from the days when he was a jitney Tammany politician. **1947** E. A. McCourt *Music at Close* 108 He would go to the jitney dance held in the big open space behind the town hall unless an outdoor floor had been set up. **1967–8** *Bahamas Handbk. & Businessmen's Ann.* (ed. 7) 197 There are jitneys on New Providence..which travel over unscheduled and not necessarily prescribed routes picking up and dropping off passengers. **1973** *Philadelphia Inquirer* (Today Suppl.) 7 Oct. 8/2 From the museum a jitney carries visitors to the Mills.

jito (dʒī·to). Also **gito.** [Jap.] In the Japanese feudal system: a military land steward (see also quot. 1974).

1845 *Encycl. Metrop.* XX. 472/2 A military commander ..appoints the *gitos*, or revenue officers,..who are stationed in every township to levy the portion of the produce claimed by the Crown. **1902** L. Hearn *Kottō* viii. 65 The Jitō of the district, with a hunting party. **1974** *Encycl. Brit. Macropædia* X. 64/2 In 1185..Yoritomo appointed military governors (*shugo*) in all the provinces and military stewards (*jitō*) in both public and private landed estates... The *jitō* collected taxes, supervised the management of landed estates, and maintained public order.

jitter (dʒi·təɹ), *sb. colloq.* [Origin unknown.] **1.** *pl.* (usu. *the jitters*). Extreme nervousness, nervous incapacity; a state of emotional and (often) physical tension; agitation.

1929 P. Sturges *Strictly Dishonorable* ii. 123 *Isabelle.* Willie's got the jitters—. *Judge.* Jitters? *Isabelle.* You know, he makes faces all the time. **1930** *N.Y. Press* 2 Apr. 6/4 The game is played only after the mugs and wenches have taken on too much gin and they arrive at the state of jitters, a disease known among the common herd as heebie jeebies. **1931** *Charlottesville* (Virginia) *Progress* 23 Mar. 12/8 Swift moving elevators and roller coasters also give her the jitters. **1931** *Harper's Mag.* Mar. 420/2 How much of a price did we pay next day for a not very good party? How many kinds of hangovers are there? How much do you have to drink to get 'jitters'? I know my hostess called me up to say she had the jitters and her husband an awful hangover. **1932** E. Wilson (*title*) The American jitters. **1933** *Passing Show* 15 July 14/1 Oh Heck, tell some photographer I can't be photographed. The very sight of a camera nowadays gives me the jitters. **1934** *Redbook Mag.* June 38/2 You begin to have the 'jitters'. Your placid disposition has given way to irritability, sarcasm and dissatisfaction. **1936** 'P. Quentin' *Puzzle for Fools* vi. 44 My old jitters had returned and I had started to shake. **1937** *Daily Herald* 12 Jan. 15/5 All this chatter about rough play has been to give players..the jitters. **1939** F. M. Ford *Let.* 23 May (1965) 324 All the publishers here have the jitters so badly that they won't look at anything new at all. **1945** *Univ. of Colorado Stud.* Ser. B. II. iv. 42 The cures of General Semantics are not limited, fortunately, to those suffering from platform jitters. **1957** *Economist* 7 Sept. 759/1 The recession jitters now afflicting significant numbers of economists..seem to be better founded on economic evidence than did the others. **1971** B. W. Aldiss *Soldier Erect* 229 The signal came, the machine-gun fire stopped... Geordie was next to me, not showing a sign of his earlier jitters.

b. jitter party *Mil. slang* (see quot. 1948).

1948 Partridge *Dict. Forces' Slang 1939–45* 102 *Jitter party*, a party of Japanese sneaking around the perimeter of a camp and trying to cause panic with strange noises and grenades. **1956** W. Slim *Defeat into Victory* xviii. 430 Only a few jitter parties prowled about the perimeter. **1962** *Times* 16 Oct. 14/6 Still less was anyone to return the unaimed small-arms fire so often indulged in after nightfall by Japanese 'jitter-parties'.

2. Chiefly *Electronics.* Random or irregular variation in the shape or timing of a regularly repeated pulse; the resulting unsteadiness of an image in a cathode-ray tube. Also in *Cinemat.*, jerkiness of the picture.

1943 *Electronic Engin.* XVI. 55/2 Absence of 'jitter' or hum derived from fluctuations of the H.T. supply potential. **1954** *Ibid.* XXVI. 37/2 V_4..failed to fire properly.. and caused considerable 'jitter' in the output pulse length. **1959** Halas & Manvell *Technique Film Animation* xix. 231 The main danger of rostrum tracks and pans is strobing, or jitter. *Ibid.* 339 *Jitter*, uncontrolled movement on the screen caused by faulty animation, tracing or camera work. **1960** J. D. Haigh *Radiolocation Techniques* v. 76 If the displayed picture is to be..free from 'jitter' echoes must appear in exactly the same position at each recurrence of the time-base. **1966** *Jrnl. Acoustical Soc. Amer.* XXXIX. 920 The principal experimental parameters are the mean interval between pulses and root-mean-square deviation or jitter about this interval. **1966** R. J. Ross *Television Film Engin.* i. 5 Very slight errors in registration cause vertical or horizontal unsteadiness in screen presentations, commonly known as jitter or weave.

So **ji·tteriness; ji·ttery** *a.*, nervy, jumpy, upset, 'on edge'; also *transf.*

1931 H. Crane *Let.* 1 May (1965) 369 I'm too jittery to write a straight sentence. **1936** 'P. Quentin' *Puzzle for Fools* iv. 26 A slight return of jitteriness. **1937** *Daily Mirror* 27 Feb. 10/4 Stop being so jittery. What if we have missed the bus? There will be another later. **1941** *Penguin New Writing* X. 23 Jittery light sprang on the screen. **1946** *Mind* LV. 137 There is the effect of their [*sc.* the Wittgensteinians'] therapy upon ordinary philosophers. Under its attack, the latter tend, in general, to become 'jittery' and to retreat into worried silence. **1950** H. E. Bates *Scarlet Sword* ii. 17 An atmosphere of tense and growing jitteriness. **1957** *Listener* 12 Dec. 968/2 An impression of weakness and jitteriness. **1963** A. Ross *Australia 63* viii. 153 Barrington made 33, in his more jittery manner, before flicking at an outswinger and being caught at slip. **1970** B. W. Aldiss *Moment of Eclipse* 47 Nobody gets jittery down there? **1972** *Jrnl. Social Psychol.* LXXXVIII. 279 Clutched-up, jittery, stirred-up, fearful.

jitter (dʒi·təɹ), *v. colloq.* [Origin unknown.] **1.** *intr.* To move in an agitated manner; to exhibit alarm, to act in a nervous way, to get the 'jitters'.

1931 A. M. Mackenzie *Cypress in Moonlight* iv. iv. 226 Her lace cap jittered tremulously till her earrings rattled. **1932** *Brevities* (N.Y.) 5 Dec. 16 (*heading*) Jittering junkies sniff sexy joy flakes. **1936** M. Allingham *Flowers for Judge* xxi. 301 You've been jittering around the Continent like an agitated tourist. **1938** *N. & Q.* 5 Mar. 172/1 My daily paper this morning points out that the total of the workless is higher, and adds, 'But don't jitter!' **1959** 'P. Quentin' *Shadow of Guilt* vii. 67 My thoughts jittered around Chuck. **1960** 'S. Harvester' *Chinese Hammer* vi. 64 There was a funny atmosphere about him. He jittered. **1966** *Electronics* 17 Oct. 109 If the error signal during the measurement period is too low, the antenna may jitter. **1970** *New Yorker* 28 Feb. 73/1 He sat there quietly, taking part in the dinner, but I knew that under the table his leg was jittering wildly.

2. *trans.* To propel by nervous energy; to fluster.

1932 E. Hemingway *Death in Afternoon* ii. 18 His effort to be statuesque while his feet jittered him away out of danger was very funny to the crowd. **1951** C. Simak *Time & Again* (1956) xvi. 78 He knew that I would catch and he thinks he can jitter me.

3. *trans.* To subject (a series of pulses, or some characteristic of it) to rapid variation. So **ji·ttered** *ppl. a.*

1960 J. D. Haigh *Radiolocation Techniques* xvii. 251 In order that an aircraft might be able to identify the responses to its own transmissions its pulse recurrence frequency was 'jittered' automatically. **1966** *Jrnl. Acoustical Soc. Amer.* XXXIX. 920/2 Jittering the interval between pulses might also eliminate perceptual differences. *Ibid.* 922/1 Assumptions of ideal distributions were used to obtain power-density spectra of jittered pulse-train signals.

jitterbug (dʒi·təɹbʌg), *sb.* orig. *U.S.* Also **jitter bug, jitter-bug.** [f. *jitter *v.* + *bug *sb.*² 4a.] **1.** A jittery or nervous person; an alarmist; an attack of the jitters. Also as *adj.*

1934 C. Calloway et al. (*song title*) Jitter bug. **1934** —— in *Song Hits Mag.* (1939) Nov. 19 They're four little jitter bugs. He has the jitters ev'ry morn, That's why jitter sauce was born. **1939** *Times* 30 Jan. 13/5 It [*sc.* the Government] is the only body capable of giving the information which would largely dispel the apprehensions of the 'Jitter-bugs', who, though perhaps few in number, are vocal and often influential. **1941** N. Marsh *Death & Dancing Footman* (1942) iv. 65 One man will keep his head in a crisis where another will go jitter-bug. *Ibid.* 76 Nick had a jitter-bug and wanted to make off. **1944** [see *bomb-happiness*]. **1966** E. H. Jones *Margery Fry* xiv. 193 Sir Samuel Hoare denounced the 'jitterbugs' who feared war... Five days after..German troops moved unresisted into Czechoslovakia.

2. A jazz musician; a devotee of jazz; a person who dances the jitterbug.

1937 *Down Beat* Feb. 9/4 (*heading*) 'Jives of the jitterbugs.' High and low-down on the swing men. **1938** *Manch. Guardian Weekly* 2 Sept. 188/3 A 'jitterbug' is a person keen on swing. **1939** *Times* 27 Jan. 7/5, I am told that in the U.S.A. there is a class of people who sit listening in hysterical excitement to what is called 'hot-music' and waiting for the final crash. Americans in their forcible language call them the 'Jitter-bugs'. There are many people in Europe to-day who seem to be behaving in much the same way. **1942** *Dancing Times* May 411/2 Equally pleasing to Jitterbugs and some rhythm dancers will be 'Russian Salad'. **1951** E. Paul *Springtime in Paris* xii. 218 On the corner, the Nest of Vipers nightly harboured jitterbugs and Lindy hoppers whose antics shamed the praying mantis.

3. A dance, popular esp. in the early 1940s, performed chiefly to boogie-woogie and swing music, and consisting of a few standardized steps augmented by much improvisation.

1939 R. Chandler in *Sat. Even. Post* 14 Oct. 11/1 This jitterbug music gives me the backdrop of a beer flat. I like something with roses in it. **1943** *Dancing Times* Sept. 560/1 The wildest Jitterbug yet is danced by Dorothy Lamour. **1969** N. Cohn *AWopBopaLooBop* (1970) ix. 81 With old favourites like the jitterbug and the jive.. the girls spun like tops and everyone got fast flashes of knicker. **1971** *Daily Tel.* 1 Nov. 9/1 To pop music..she performed movements reminiscent of the jitterbug of a generation ago.

ji·tterbug, *v.* orig. *U.S.* [See prec., sense 3.] *intr.* To dance the jitterbug.

1939 *Amer. Mag.* Sept. 160 Susy Shag..begins thinking seriously of marrying the guy she's been jitterbugging with. **1949** N. Marsh *Swing, Brother, Swing* v. 77 D'you jitter-bug? **1951** J. D. Salinger *Catcher in Rye* x. 80 The band was starting a fast one. She started jitterbugging with me. **1955** C. Fox in *Jazzbook 1955* 4 They jitter-bugged down the aisles to the exhilarating music, horrifying their parents and bringing worried frowns to the faces of the jazz purists of those days. **1972** *Jazz & Blues* Nov. 5/1 They jitterbugged to 'One O'Clock Jump' and stomped their feet to 'Maple Leaf Rag'. **1973** *Washington Post* 5 Apr. B.2 The happiest part..is a salute to the big bands of the '40s, with Miss Adams in a snood and Candoli jitterbugging in a zoot suit.

Hence **ji·tterbugging** *vbl. sb.* and *ppl. a.*

1938 *Call-Bulletin* (San Francisco) 8 Dec. 17/1 'Jitter-bugging is just a phase,' he said... 'It isn't music. This town wants real music and nothing else will do.' **1939** *San Francisco Examiner* 17 Sept. 2/2, I love ballroom dancing and I usually do just that, but at times I find it a job to express my joy of living by doing a little jitter-bugging. **1940** Graves & Hodge *Long Week-End* xxii. 390 Jitterbugging had then [1939] just come over from the United States..an ecstatic mode of dancing to fast swing music in which the two partners could perform absolutely any..acrobatic feat they liked. **1942** A. P. Jephcott *Girls growing Up* iii. 63 The slick, apparently assured, smoking, jitterbugging town girl of sixteen still regards friendliness as the all-important feature of any group with which she intends to associate. **1952** M. McCarthy *Groves of Academe* (1953) ii. 20 He did not dance, but..vigorously nodded with approval when a jitter-bugging pair desisted. **1973** G. Durrell *Beasts in my Belfry* v. 105 The only folk dances I have ever witnessed were danced by elderly, aesthetic ladies with fringes and strings of beads and they were nothing like the gnus' wild jitterbugging.

jiu-jitsu, -jutsu, varr. *ju-jitsu *sb.*

jiva (dʒī·vă). *Hindu* and *Jain Philos.* [Skr. *jīvá* living being, life, the highest personal principle of life.] Life, the soul, the self; the vital principle.

1807 *Asiatick Researches* IX. 290 The *Jainas* conceive the soul (*Jíva*) to have been eternally united to a very subtil material body. **1885** E. Balfour *Cycl. India* (ed. 3) II. 442/2 *Jiva.* Sansk. Life, the soul. **1887** W. J. Wilkins *Mod. Hinduism* viii. 101 All beings possessed of jiva are of two kinds: those who can move and those who cannot. **1915** A. M. Stevenson *Heart of Jainism* 176 As long as the jīva or ātmā is fettered by karma, so long must it undergo rebirth, and it must be remembered that karma is acquired through good as well as through evil actions. **1951** H. Zimmer *Philosophies of India* iii. i. 229 Jainism regards the life-monad (*jiva*) as pervading the whole organism; the body constitutes, as it were, its garb; the life-monad is the body's animating principle. *Ibid.* 277 Some vegetables, such as trees, are provided with a collectivity of jīvas. **1971** *Times Weekender* (Ceylon) 3 Oct. 4/3 As the macrocosmic reality it is called Braman and its manifestation through human beings is the microcosmic reality known as the atman or soul or jiva.

Jivaro (hī·văro). [ad. Sp. *jíbaro*, f. a native name.] An Indian people of Ecuador and Peru; a member of this people; also, the language of this people or a group of related languages of which this is the principal. Also *attrib.* or as *adj.* So **Ji·varan, Jivaro·an** *adjs.*

1862 *Intellectual Observer* Mar. 134 On the eastern side of the Republic of Ecuador..live a tribe of Indians called Jívaros. *Ibid.* 135 It is in the forests among these rivers that the Jívaro Indians now make their homes. **1902** *Encycl. Brit.* XXV. 374/1 The families of..South America are..Jivaroan, [etc.]. **1927** K. G. Grubb *Lowland Indians of Amazonia* v. 77 The Jivaro live in communal houses divided into numerous sub-tribes. *Ibid.*, The Jivaro language has been attested by a variety of documents. **1934** Webster, Jivaran. **1957** *Encycl. Brit.* VII. 941/1 The Cayapa-Colorados and the Jivaro [languages] still survive owing to their existence in inaccessible regions. *Ibid.* XIII. 69/2 *Jivaran*, an independent linguistic stock of South American Indians, so called from the best known tribe, the Jivaros. The Jivaran tribes live in eastern Ecuador and the adjacent portions of Peru. **1962** N. Maxwell *Witch-Doctor's Apprentice* i. 3 That was my first long jungle trip and I was lucky enough to get to a tribe of Jivaros who had never seen a white woman before. **1969** *Times* 24 Jan. 6/7 Jivaro give the seeds to refractory children so that during their visions the ancestral spirits may come to admonish them.

jive (dʒəiv), *sb.* *slang* (orig. *U.S.*). [Origin unknown.] **1.** Talk or conversation; *spec.* talk that is misleading, untrue, empty, or pretentious; hence, anything false, worthless, or unpleasant; vaguely, 'stuff'; = *jazz *sb.* 3 a.

1928 R. Fisher *Walls of Jericho* 301 *Jive*, pursuit in love or any device thereof. Usually flattery with intent to win. **1929** T. Gordon *Born to Be* 236 *Jive*, a misleading remark. **1932** Muse & Arlen *Way down South* 50 Thus the enamoured customer completed his meal, without ever having taken his eyes off that tantalizing brown, with her suave Birmingham jive. **1935** *Swing Music* Autumn 55/2 Maybe you think that that is all jive. You are wrong if you do. It is the way I felt about these new records. **1946** Mezzrow & Wolfe *Really Blues* iii. 37, I used to hear a lot of medical jive. *Ibid.* 375 *Jive n.*, confusing doubletalk, pretentious conversation, anything false or phony. *Jive that makes it drip*, clouds that produce rain. **1954** L. Armstrong *Satchmo* x. 150 There was lots of just plain common shooting and cutting. But.. that jive didn't faze me at all. *Ibid.* xii. 193, I bought a lot of cheap jive at the five and ten cents store to give to the kids. **1956** M. Stearns *Story of Jazz* (1957) v. 50 The attitude of several modern jazzmen, born and bred in the South, is striking: 'This hoodoo jive is nowhere,' they say, 'but man, watch out!' **1960** in P. Oliver *Blues fell this Morning* vii. 197 I'm evil and mean and funny, so don't come back with that line of jive. **1972** M. J. Bosse *Incident at Naha* iii. 152 Maybe some of his Christian sentiments sound corny today, but..he had cut through a lot of the jive of his own time, and he had, like, the balls to fight injustice. **1973** *Black World* Oct. 36/2 Everything that we do must be aimed toward the total liberation, unification and empowerment of Afrika... Anything short of that is jive.

2. Jazz, esp. a type of fast, lively jazz; 'swing'.

1928 (*title of gramophone record by Cow Cow Davenport*) State Street Jive. **1937** *New Yorker* 17 Apr. 31/3 The music of hot bands..is referred to as *swing* or *jive*, of which, in turn, there are several kinds. **1939** *San Francisco News Let.* 1 Sept. 12/2 Fats Waller..is the King of Jive and gets off some fine stuff. **1946** *N. & Q.* 13 July 20/1 Mr. Mitchell Parish, the American song-writer,.. told me that he uses *jive* to describe syncopated music played noisily, and (usually) fast, with great emphasis on rhythm. **1959** 'F. Newton' *Jazz Scene* i. 12 In Sophia-town and the rest of the South African ghettoes the 'jive bands' play what is patently jazz. **1960** *Down Beat* 9 June 15 Regarding the word jive, Wilson said, it is nothing more than an obsolete slang term for jazz.

b. Lively and uninhibited dancing to dance-music or jazz; *spec.* 'jitterbugging'.

1943 *Dancing Times* Dec. 117/1 The rhythm of the Jive is not an entirely new one. **1957** C. MacInnes *City of Spades* i. iv. 24 I'll teach you..bop steps, and jive, and all. **1958** *Listener* 20 Nov. 848/1 Jive and tribal dancing. **1969** H. Horwood *Newfoundland* x. 69 The jive..is still the universal dance of..outport youngsters.

3. A variety of American English associated with the Harlem area of New York; slang used by American Blacks, or by jazz musicians and their followers. Also *attrib.*, as *jive talk.*

1938 C. Calloway *Hi De Ho* 16 *Jive.* 1. Harlemese speech or lingo. 2. To kid along, to blarney, to give a girl a line. **1943** *Time* 26 July 56/2 A jive-talk glossary that is strictly Dracula has been put out by Parents' Institute. **1944** D. Burley (*title*) Original handbook of Harlem jive. E. Conrad in *Ibid.* 5 Jive is one more contribution of Negro America to the United States. *Ibid.* 6 Jive talk may have been originally a kind of 'Pig Latin' that the slaves talked with each other, a code—when they were in the presence of whites. **1960** *Time & Tide* 24 Dec. 1599/2 Jive-talk is nothing new. It goes back at least to the thirties when for the first time a brand of jazz, swing, grew to be a cult. Jive was originally the patois of Harlem, not jazz musicians' slang; but with time the distinction was lost. **1965** *Economist* 4 Sept. 888/2 Some common American jive-words (nappy, funky) are left out [of the *Penguin English Dictionary*]. **1971** *Black World* June 92/2 All the rest of that jive talk about white liberals and Rhett Butler is part of another conversation, Sam. **1971** *Melody Maker* 13 Nov. 31/1 That is if you forget the usual jive phrases that whittle their way into his conversation. **1973** *Times Lit. Suppl.* 1 June 604/4 A narrative tone which frequently coincides with the fast, obscene jive-talk of his characters.

4. Marijuana, or a cigarette containing it.

1938 *Call-Bulletin* (San Francisco) 19 Mar., The cigarettes are variously called sticks, reefers, tea gyves, Mary Anns and goofy butts. **1952** *N.Y. Times* 29 Apr. 25 So Diane smoked jive, pod, and tea. **1955** *U.S. Senate Hearings* (1956) viii. 4168 'Sticks', 'reefers', 'jive sticks'. **1963** 'D. Rutherford' *Creeping Flesh* ii. 124 'Jive' originally meant marijuana. **1972** *Lancet* 16 Sept. 565/1 She was convinced that only in the institution could she 'make it without jive', for she invariably used heroin whenever she was sent home.

jive (dʒəiv), *v.* *slang* (orig. *U.S.*). Also **gieve.** [Cf. prec.] **1. a.** *trans.* To mislead, to deceive, to 'kid'; to taunt or sneer at. Also *intr.*, to talk jive, to talk nonsense, to act foolishly.

1928 L. Armstrong (*title of gramophone record*) Don't jive me. **1929** W. Thurman *Blacker the Berry* 128 But I jived her along, so she ditched him, and gave me her address. **1934** *Amer. Speech* IX. 26/2 *Gieve*.., to mislead with words; to take into one's confidence. *Ibid.* 27/1 *Jive*, see *gieve*. **1938** *Ibid.* XIII. 317/1 *To jive around*,..'to fool around'. **1939** J. Dollard in A. Dundes *Mother Wit* (1973) 281/2 Willy kept 'jiving' him until Jimmy finally left. **1944** D. Burley *Orig. Handbk. Harlem Jive* 71 Jive is a distortion of that staid, old respectable English word 'jibe'... In the sense in which it came into use among Negroes in Chicago about the year 1921, it meant to taunt, to scoff, to sneer. *Ibid.*, A highly effective manner of talking about each other's ancestors and hereditary traits..called 'Jiving' someone. **1946** Mezzrow & Wolfe *Really Blues* vi. 70 Monkey wasn't jiving about that bartender. **1950** A. Lomax *Mr. Jelly Roll* (1952) iv. 170, I..jived the expressman to hank my trunks to the station by telling him my money was uptown. **1969** J. McPherson in A. Chapman *New Black Voices* (1972) 162, I don't need no money. Nobody's jiving me. I'm jiving them. You know I can still pull in a hundred in tips in one trip [as a waiter on a train]. **1973** *Black World* Mar. 57 Lawd, don't jive Miz Jackson,..Ride on King Jesus!

b. *intr.* To make sense; to fit in. *U.S.* Cf. Jibe *v.*

1943 *Amer. Speech* XVIII. 153/2 *Doesn't jive*, doesn't make sense. **1955** W. Gaddis *Recognitions* II. i. 308 His analyst says he's in love with her for all the neurotic reasons in the book. It don't jive, man. **1973** *To our Returned Prisoners of War* (Office of U.S. Secretary of Defense) 7 *Jive*, verb meaning fit in, go with, to make sense.

2. *intr.* **a.** To play 'jive' (*jive *sb.* 2 a).

c **1938** N. E. Williams *His Hi De Highness of Ho De Ho* 35/2 'Jiving', meaning to improvise. **1942** Berrey & Van den Bark *Amer. Thes. Slang* § 579/9 Play 'hot jazz'; 'swing',..*jive*. **1947** Auden *Age of Anxiety* (1948) ii. 44 The juke-box jives rejoicing madly.

b. To dance the 'jive' (*jive *sb.* 2 b).

1939 *San Francisco News Let.* 1 Sept. 12/2 If you should dance to the rhythms of either gentleman you will be jiving. **1957** *Observer* 13 Oct. 3/4 Young people from the East End and the West End came there [the Humphrey Lyttelton Club] to jive or listen. **1958** *New Statesman* 4 Jan. 10/2 A couple began a little hesitantly to jive.

Hence **ji·ver**, one who jives; **ji·ving** *vbl. sb.* and *ppl. a.*

1936 *N.Y. World-Telegram* 6 Oct. 16/1 High jiving—tall, if you know what I mean, talk. *c* **1938** [see 2 above]. **1939** Blind Boy Fuller (*song title*) Jivin' Big Bill Blues. **1939** *San Francisco Examiner* 18 Aug. 3 (*heading*) Jiving deluxe. **1943** *N.Y. Times* 9 May II. 5/7 I'm a jivin jitterbug. **1943** *Gramophone* Aug. 47/2 Lawd, don't cut out 'Jazz'! I'll write you as many letters as Robert Mackenzie likes, but..don't forget the 'jiver', sir, don't forget the 'jiver'! **1944** S. J. Perelman *Crazy like Fox* (1945) 163 A jiving, hot-hosing jitterbug. **1947** M. H. Boulware *Jive & Slang of Students in Negro Colleges*, *Jiver*,..a flatterer. **1951** 'A. Garve' *Murder in Moscow* iii. 46 Her daughter..had won a prize for jiving at some South London *palais*. **1959** *Spectator* 12 June 856/1 One..finds the jazz virtuoso Stephane Grapelly performing to an admiring crowd of jivers. **1973** *Black World* Mar. 85 Jiving, bopping, napping, signifying, sounding—all modes of Afro-American expression—seek to affirm the vitality of the Black American experience. *Ibid.* May 84/1 He comes down hard on white racists, but he also attacks Black 'jivers' who seek to exploit their brethren under the guise of Blackness.

jive (dʒaɪv), *a.* *U.S. slang.* [f. the sb.] Used, chiefly by American Blacks, in the primary sense 'not acting correctly' but with a wide range of connotations ('pretentious', 'deceitful', etc.).

[**1959** *Esquire* Nov. 70 *Jive*, to fool, to kid. The adjective is bogus.] **1971** E. E. Landy *Underground Dict.* 112 *Jive*,..not acting right; doing something wrong—e.g. *You never showed up. You're a jive dude.* **1973** E. Bullins *Theme is Blackness* 131 Kiss ma ass ya jive mathafukker! **1973** *Black World* June 61 Huh? Awh, Sistuh, u sho is jive. *Ibid.* 79/1 The hero..is 'hip', but not 'jive'. *Ibid.* Sept. 53, I been confused, fucked up, scared, phony And jive To a whole/lot of people.

jive-ass (dʒaɪ·væs, -ɑ̄s). *U.S. slang.* Also **jiveass.** [f. *jive *sb.* + Ass (= Arse *sb.*).] **a.** A person who loves fun or excitement. **b.** A deceitful or pretentious person. Freq. *attrib.*

A word of fluid meaning and application.

1964 R. S. Gold *Jazz Lexicon* 169 Jive-ass motherfucker. **1969** C. Brown in A. Chapman *New Black Voices* (1972) 181 'You jiveass nigger,' Reb said, laughing. 'No, I'm telling the truth.' *Ibid.* 184 He became a hustler, a jiveass, a jazz player. **1970** *New York* 16 Nov. 42/1 This happens to be a jive-ass *af-flu*-ent neighborhood. **1973** *Black World* Aug. 55/2 Silly-ass so-called hangups of jive-ass white folks. *Ibid.* 67/1 Jive-ass jivers. Stupid. Looking clever.

jivey (dʒaɪ·vi), *a.* *slang* (chiefly *U.S.*). Also **jivy.** [f. *jive *sb.* + -y[1].] Jazzy, 'swinging', lively (see also quot. 1954); also, misleading, phoney, pretentious.

1944 [in the popular song *Mairzy Doats* by M. Drake, A. Hoffman, and J. Livingston]. **1953** Berrey & Van den Bark *Amer. Thes. Slang* (1954) § 149/13 'Knowing'; 'hep' (manifesting knowledge of what is proper or fashionable)..*jivey*. **1955** Keepnews & Grauer *Pict. Hist. Jazz* xv. 187 Stuff Smith, one of the jiviest. *Ibid.* xix. 248 Bassist Slam Steuart was..primarily a jivey Swing man who hummed solos loudly as he played. **1972** M. J. Bosse *Incident at Naha* ii. 85 I'm not sure I would have accepted that sort of jivey explanation, but Mrs. Halliday did. **1972** *Jazz & Blues* Dec. 28/1 This coupling finds him in jivey mood.

jiwan (dʒiwā·n). [Skr. *jīvana* a son; cf. Hindi *jawān*.] = *jawan; also, an Indian youth.

1914 Kipling *New Army* (1915) 49 The trouble in India was that all the young men—the mere *jiwans*—wanted to come out at once. **1924** *Blackw. Mag.* Aug. 227/1, I met the Gurkha officer in the back of beyond with two or three of his *jiwans*. *Ibid.* 232/2, I ruled over as jolly a crowd of students as a Principal of a College can hope to collect, north-country lads—Sikh, Muhammadans, and Hindus... At cricket..we had an English professional..who coached them... He became attached to my *jiwans*.

Jixi, Jixie (dʒɪ·ksi). *temporary.* [f. *Jix*, nickname of *Joynson-Hicks* (see below) + *-i*, after Taxi.] A two-seater taxi-cab licensed in 1926 while Sir William Joynson-Hicks (1865–1932) was Home Secretary.

1926 *Westm. Gaz.* 7 Apr., Jixi is the name given by the Westminster Gazette, and now used by everyone, to 2-seater taxicabs. **1926** *Punch* 21 Apr. 438/2 The Home Secretary explained to Sir F. Meyer that the police have licensed two types of 'Jixi'. **1927** *Observer* 3 Apr. 9 The first 'Jixie' or two-seater taxicab will probably be seen on the streets of London at the end of the present week.

jizz (dʒɪz). [Etym. unknown.] The characteristic impression given by an animal or plant.

Guise *sb.* 5 is coincident in sense but the phonetic relationship remains unexplained and the two words may therefore be unrelated.

1922 T. A. Coward *Bird Haunts & Nature Memories* 141 A West Coast Irishman was familiar with the wild creatures which dwelt on or visited his rocks and shores; at a glance he could name them, usually correctly, but if asked how he knew them would reply 'By their "jizz".' What is jizz?..We have not coined it, but how wide its use in Ireland is we cannot say... Jizz may be applied to or possessed by any animate and some inanimate objects, yet we cannot clearly define it. A single character may supply it, or it may be the combination of many. *Ibid.* 143 Jizz, of course, is not confined to birds. The small mammal and the plant alike have jizz. **1950** *Brit. Birds* XLIII. 29 Miss Quick obviously looks at her birds more than once and does so with an artist's eye for those peculiarities of shape, outline and stance which give a species its 'jizz'. **1960** *Times* 14 June 14/7 The boy could name any bat at a glance..(an indefinable accomplishment which the late T. A. Coward..once described as 'jizz' when speaking of bird identification). **1966** D. McClintock *Compan. Flowers* ix. 117, I know only too well the problem of trying to express what there is in a plant that enables me, or you, to tell it from another at sight. The word I use for these intangible characteristics, that defy being put into words, is jizz.

jizz, var. *jism.

‖ **jnana** (dʒənā·nă). *Hinduism.* Also **jnyana.** [Skr. *jñāna*, f. *jñā* to know.] Spiritual knowledge, as a means of salvation.

1827 *Trans. R. Asiatic Soc.* I. 576 *Vāsudēva*, or *Bhagavat*,..has six especial attributes,..1st. Knowledge (jnyana), or acquaintance with everything animate or inanimate constituting the universe. **1875** M. Williams *Indian Wisdom* iii. 59 Jnāna, 'knowledge of universal truth'. **1916** T. A. Gopinatha Rao *Elem. Hindu Iconogr.* II. i. 273 Siva was seated facing south when he taught the *rishis yōga* and *jñāna*. **1945** A. Huxley *Let.* 10 Apr. (1969) 520 There is gnosis or jnana after death only for those who have chosen to pay the price..of gnosis during life. **1971** *Illustr. Weekly India* 11 Apr. 15/4 The essence of the world is religion..the essence of religion is knowledge (*jnana*).

b. *Comb.*, as **jnana-marga** [Skr. *mārga* road, path], the way to salvation through spiritual knowledge or asceticism.

1877 M. Williams *Hinduism* i. 11 Popular Hindûism..though supposed to accept this creed as the way of true knowledge (*Jnāna-mārga*)..adds to it two other..ways. **1883** —— *Relig. Thought & Life in India* iii. 63 The way of love and faith (bhakti-mārga)..superseded the other two ways of salvation—knowledge and works (jñāna-mārga and karma-mārga). **1937** [see *bhakti-marga]. **1945** A. Huxley *Let.* 21 Jan. (1969) 513 The three types of religion enumerated in the *Bhagavad Gita*, (karma marga, the way of works or action; Bhakti marga, the way of devotion; and jnana marga, the way of knowledge) correspond precisely to the three extremes of somatotonic, viscerotonic and cerebrotonic temperament. **1963** *Listener* 28 Mar. 550/2 In earliest times, in pre-Buddhist texts such as the first *Upanisads* is described what later became known as *Jnana-marga*, or the way of knowledge, in which the realization of the identity of the individual self or soul with the Universal, the absolute, Brahma, was sought.

‖ **jnani** (dʒənā·ni). *Hinduism.* [Skr., f. *jnana.] A worshipper or devotee of *jnana-marga*.

1885 E. Balfour *Cycl. India* (ed. 3) II. 442/2 *Jnani*, or divine Buddhas, are five. **1916** T. A. Gopinatha Rao *Elem. Hindu Iconogr.* III. i. 242 The silent *jñānis* destroying the threefold bond..behold the sacred and are filled with bliss. **1962** Brahmacharini Usha *Ramakrishna-Vedanta Wordbk.*, *Jnani*. 1. One who follows the path of knowledge and discrimination to reach the impersonal Reality; a non-dualist. 2. A knower of Brahman. **1969** *Indian Music Jrnl.* V. Suppl., Arunagiri completed the song, got cured of his ailments and became a Jñānī and a yōgēśvara.

Joachimite (dʒōᵘ·ākimait). *Ch. Hist.* [f. the name of *Joachim*, abbot of Fiore in Calabria (*c* 1132–1202) + -ite[1].] A heretical follower of the Italian mystic, Joachim of Fiore. Also **Jo·achimist, Jo·achist, Jo·achite, Jo·achitist**; so **Jo·achimism, Jo·achism.**

1797 *Encycl. Brit.* IX. 290/2 *Joachimites*, in church history, the disciples of Joachim, a Cistertian monk. **1842** K. H. Digby *Mores Catholici* xi. xiii. 478 The execrable book, entitled the eternal Gospel,..whose adherents were termed Joachimites, as it was ascribed to Joachim, were again condemned by the Council of Arles. **1874** J. H. Blunt *Dict. Sects* 242/1 As early as 1260 the Council of Arles spoke of the Joachitist error as spreading like a cancer. **1906** G. G. Coulton *From St. Francis to Dante* 104 The Pope..insisted on his resignation, partly on account of his Joachism. *Ibid.* 108 A great orator and a great Joachite. **1913** E. G. Gardner *Dante & Mystics* vi. 192 The Joachists went considerably beyond what Joachim himself had taught. **1913** A. G. F. Howell *S. Bernardino of Siena* i. 10 They cultivated the Joachite literature. **1926** E. Hutton *Franciscans in Eng.* ix. 141 A sect of Joachists had arisen among the 'spiritual' party within the Franciscan Order. **1929** E. C. Messenger tr. *Guiraud's Mediæval Inquisition* v. 143 In 1247, a friar with Joachimist tendencies, John of Parma was elected minister general of the Order. *Ibid.*, Peter John Olivi..expounded a modified Joachimism which was henceforth the doctrine of the Spirituals. **1931** H. Bett *Joachim of Flora* v. 112 These Joachites maintain that the future age of the Holy Spirit shall be enlightened with a more perfect law. **1932** *Speculum* VII. 260 These works..beside throwing some lights on several problems of the Joachite tradition, were..useful in awakening more interest in the subject. *Ibid.* 267 Grundmann's work is..epoch-making in the studies on Joachim and Joachism. **1947** H. R. Williamson *Arrow & Sword* x. 158 The Joachists..had taken their master's ideas to their logical conclusion. *Ibid.*, It is possible that the *Divine Comedy*..is the classic Joachist apologia.

joanna (dʒoæ·nă). Also **joano, johanna,** and other varr. Rhyming slang for 'piano'.

1846 *Swell's Night Guide* 34 As a sort of whipper-in, music is provided—viz.,..a joano. **1912** A. N. Lyons *Clara* iii. 28 All the time a party down below was bangin' out a song called 'Bedelia' on a old johanner. **1923** J. Manchon *Le Slang* 168 *Johannah, johanner*, un piano. **1961** S. Price *Just for Record* ix. 89 He gets up from the joanna-stool. **1972** *Listener* 20 July 92/2 The old Jo-anna intrudes its amateurish thumpings.

joaquinite (wā-, wokī·nait, dʒō·ākwinait). *Min.* [f. *Joaquin*, the name of a ridge in San Benito County, California: see -ite[1].] A honey-yellow to brown silicate of sodium, barium, iron, and titanium that occurs as small, isolated, orthorhombic crystals.

1909 G. D. Louderback in *Bull. Dept. Geol. Univ. Calif.* V. 376 Joaquinite... A honey yellow or light brown substance in small generally individual crystals or crystal grains..which is believed to be a new mineral. **1932** *Amer. Mineralogist* XVII. 308 (*heading*) The chemical nature of joaquinite. *Ibid.* 312 There are, therefore, 4 molecules of composition $NaBa(Ti,Fe)_3Si_4O_{15}$ in the unit cell of joaquinite. **1967** *Ibid.* LII. 1762 An apparently new rare-earth mineral of the joaquinite group occurs in nephelin-syenite pegmatites of Ilimaussaq alkaline massif (S. Greenland). Optical and X-ray properties are the same, as for standard joaquinite $NaBaTi_2Si_4O_{12}$ from California, but the chemical analysis is quite different... The formula is $NaBa_2Fe^{2+}Ce_2Ti_2Si_8O_{26}(OH)$.

joar, var. *johar.

job, *sb.*[2] Add: **1. c.** (Earlier and later examples.)

1795 *Let. of Compositors of London* in E. Howe *London Compositor* (1947) i. 76 That all jobs, not exceeding one sheet, be paid at the rate of six pence per thousand. **1841** W. Savage *Dict. Art of Printing* 428 Any thing which printed does not exceed a sheet, is termed a Job, and is paid for extra to the compositor, because there is no return of furniture or of letter: he has generally to put up fresh cases, and has some additional trouble in getting the right letter, and in making up the furniture. **1960** P. M. Handover *Printing in London* vii. 183 In general, however, both the Bagford and Johnson Collections [*sc.* of ephemera] reveal that during the eighteenth century the letterpress printers made little effort to extend the range of jobs.

3. *jobs for the boys*: see *boy *sb.*[1] 6 d.

4. a. *spec.* A paid position of employment, a situation (sense 6 b).

1858 in *Amer. Speech* (1965) XL. 130 But when he gets a good fat job For dat am all he cares. *a* **1861** T. Winthrop *Edwin Brothertoft* (1862) 38, I will find you a fat job and plenty of pickings! **1883** [see Fat *a.* 9 b]. **1931** *Daily Express* 13 Oct. 5/2 When I got my job as social secretary..I got a large salary and lived in luxury. **1940** G. D. H. & M. Cole *Counterpoint Murder* xv. 237 He was in the same job up to about a week ago. **1974** *New Statesman* 17 May 701/3 After the Tory general election defeat of 1964, the outgoing Chancellor of the Exchequer..told a newspaper gossip writer: I shall have to get a job and earn some money.

b. *bad job* (earlier and later examples); *job of work*: a task, piece of work; *to have a job* (*to*): to have difficulty (in doing something); similarly, *to be a job*: to present difficulties;

also, (*to be*) *the devil's* (*own*) *job*; *to get on with the job* [Get *v.* 63 f], to proceed with one's work, to continue with one's affairs; *just the job* (colloq.): exactly what is wanted, the very thing.

1557 in A. Feuillerat *Documents Revels Court Ed. VI & Q. Mary* (1914) 236 Doinge certen Iobbes of woorke. **1862** Borrow *Wild Wales* I. xix. 210 The minister of the parish..had frequently entered into argument with him, but quite unsuccessfully and had at last given up the matter, as a bad job. **1873** Trollope *Eustace Diamonds* I. xix. 252 Arthur did not go on the search, because he had a job of work to do. *Ibid.* III. lxxii. 251 The barrister who will have the cross-examining of her..will have a job of work on his hand. **1878** [see *coast *v.* 14]. **1887** G. H. Devol *Forty Yrs. a Gambler* 267 For ten or fifteen years during my early life, the sporting men of the South tried to find a man to whip me, but they couldn't do it, and finally gave it up as a bad job. **1907** J. M. Synge *Tinker's Wedding* I. 1 It's the divil's job making a ring. **1922** F. H. Burnett *Robin* xviii. 154 'You propose to suggest that she shall marry *you*?'..'Yes. It will be the devil's own job..she has abhorred me all her life.' **1928** M. Walsh *While Rivers Run* xx. 279 'A sound job of work!' boasted Alistair. 'We have arriven.' **1931** A. Huxley *Let.* 24 Aug. (1969) 351 It has been a job writing the book and I'm glad it's done. **1931** 'P. Williams' *Word of To-morrow* IV. xvi. 263 Tramps..who wouldn't do an honest job of work not if it was offered them. **1935** *Punch* 21 Aug. 208/3, I had a job to find the meter, and I can't think why they want to hide it behind a table. **1941** *London Opinion* June 48 (*caption*) I'll have a job explaining things to my wife. **1942** 'A. Bridge' *Frontier Passage* viii. 131 Why the hell couldn't all the extremists..allow sensible people..to get on with the job?—the job being to live the good life. **1943** Hunt & Pringle *Service Slang* 42 *Just the job*, if you see anything that you like, whether it is something in a shop window or a new billet, this is what you say, meaning of course that it suits you all right. **1944** R. Lehmann *Ballad & Source* 55 My riding lessons had been given up as a bad job. **1959** 'J. Ross' *Boy in Grey Overcoat* ii. 21, I thought..she'd be just the job. **1960** H. Pinter *Dumb Waiter* 148 Just the job. We should have used it before. **1966** *Listener* 1 Sept. 300/2, I have had quite a job to get a copy [of a book]. **1969** *Ottawa Commons Debates* 24 July 11570/2 If the opposition accords so little respect to parliament that it defines governing as evil, that getting on with the job is regarded as tyranny, then I am happy to be given the chance to join issue. **1973** E. Page *Fortnight by Sea* ix. 98 If Mrs Barratt could possibly see her way to letting us stay on..it really would be just the job.

c. A consignment of goods to be sold cheaply as bargains, a job lot.

1858 *Illustr. News World* I. 257/3 Butcher's meat, the week's gathering, to be sold by the job. **1905** *Daily Chron.* 18 Nov. 3/7 As soon as a girl can do a corset, which is at all passable, even if we have to put it into the 'jobs'—that is, lots for selling cheap—she can earn much more.

d. *on the job*: (*a*) hard at work, busy (also in extended senses, committing a crime, engaged in sexual intercourse, etc.); (*b*) (of a racehorse) out to win and well backed; (*c*) = *by the job* (in Dict., sense 1 d); (*d*) used as *attrib. phr.* (hyphenated): done or occurring while a person is at work.

1882 W. Burnot *Old Mother Goose* iii. 12 He will have to hide his nob. Come along, we're on the job. *a* **1889** in Barrère & Leland *Dict. Slang* (1889) I. 502/1 Trainers and jockeys..very easily gathered whether a particular horse..was 'out for an airing' or was on the job. **1890** Lloyd George *Let.* 7 Aug. (1973) 32, I am on the job Saturday or Monday. **1891** *Licensed Victuallers' Gaz.* 23 Jan. (Farmer & Henley), There was a long wrangle over the choice of referee, for no one cared to occupy that thankless post when the Lambs were on the job. **1892** E. J. Milliken *'Arry Ballads* 3 'Arry is fair on the job. **1901** *To-Day* 22 Aug. 115/1 'This cook..is very good.' 'She is, but she is only here on the job.' **1909** *Post & Paddock* 22 Nov. 1/3 Their denunciations of horses 'not trying', being 'out for an airing', or 'not on the job' on every occasion when their speculations go wrong. **1914** G. Atherton *Perch of Devil* I. iv. 22 She was 'on the job' every minute until the cottage was 'on wheels'. **1922** Joyce *Ulysses* 466 Mother Slipperslapper. (*Familiarly.*) She's on the job herself tonight with the vet. **1922** *N. & Q.* 9 Sept. 206/2 To be 'on the job' is for a horse to be 'busy', to be 'out', *i.e.*, backed and trying. **1937** *Burlington Mag.* June p. xxvi/1 The slippery pickpocket is depicted 'on the job'. **1947** [see *biotechnology]. **1958** *Times* 24 Mar. (Careers in Industry Suppl.) p. vi/4 Everyone should receive the training appropriate to his particular aptitude—which may range from the on-the-job training to a university course. **1966** 'L. Lane' *ABZ of Scouse* 78 *On ther job*: engaged in sexual intercourse. **1971** *Optometry Today* 17 On-the-job accidents and injuries could be appreciably reduced if every worker had maximum visual efficiency for the task at hand. **1972** *Daily Tel.* (Colour Suppl.) 16 Nov. 94/3 'Why the hell did you play Eric Clapton's *Easy Now*?..Didn't you realise it was all about some guy on the job?' And I said, 'Yeah. How many songs aren't?' **1973** *Amer. Speech 1969* XLIV. 243 This approach..falls short of giving the reader an accurate idea of the large role that railroad language plays in on-the-job communication.

e. A commission to back a racehorse; so, a horse on which such bets are placed.

1889 Barrère & Leland *Dict. Slang* I. 502/1 'He has got the job', he has the putting on of the stable money. **1907** *Favourite* 16 Nov. 9/2 Elfin Revel was a big starting-price job for the Croxteth Plate. **1911** *Turf* 10 Oct. 1/1, I am not now referring to s.p. jobs.

f. *colloq.* A term of wide application, often with suggestions of excellence, to describe something, esp. something manufactured (as a motor vehicle, aircraft, etc.); also *joc.* of persons, esp. a pretty girl.

1927 D. Hammett in *Black Mask* May 14/2 She's a tough little job who was probably fired for dropping her chewing gum in the soup the last place she worked. **1928** *Daily Mail* 7 May 6/4 [U.S. slang] *A job*, always used when a particular aeroplane is mentioned. **1930** 'A. Armstrong' *Taxi* xii. 164 The fare is 'the rider' or 'the job'. **1930** *Engineering* 18 July 79/3 The motorship press would hail with delight the figures given..as showing the fuel cost of Diesel engines to be almost half that of any of the steam jobs. **1938** *Harper's Mag.* Jan. 141/2 There was an antiquated high-wing monoplane job. **1939** *Ottawa Jrnl.* 25 July 7/3 He was just about to sit down to a good breakfast in Texas opposite a young woman who was a first class job. **1942** Berrey & Van den Bark *Amer. Thes. Slang* § 81.2 *Automobile*,..job. **1942** *R.A.F. Jrnl.* 30 May 23 The bath was a limited affair: a quick sponge job from a gallon or so of warm water. **1942** *Gen* 1 Sept. 12/2 A 'ropey job' is likely as not to be a blonde who proved uncollaborative. **1946** Brickhill & Norton *Escape to Danger* v. 50 A rather imposing moustache. It was one of those bushy black jobs. **1948** C. Day Lewis *Otterbury Incident* i. 2 It [*sc.* a chronometer] was one of those super Swiss jobs. **1964** 'A. Gilbert' *Knock, knock, who's There?* i. 9 You make good contacts in a pub... There was a job called the Admiral Box where I'd never been. **1972** G. Lyall *Blame the Dead* xvii. 117 The only desk was Steen's own oiled-teak job. **1973** *Daily Tel.* 22 June 9 (Advt.), Its suspension's almost identical to that of a £2,966 Mercedes 220. (Not to mention a few other equally pricey jobs.)

5. *good job, bad job*: (later examples).

1923 *Spectator* 9 June 972/2 The novel ends, therefore, with Derek making the best of a bad job. **1930** J. B. Priestley *Angel Pavement* ii. 65 Girls *are* a bit silly..and it's a good job for the men they are. **1942** *Punch* 9 Sept. 215 (*caption*) It's a good job that pig-headed old fool of a farmer ignored our instructions, or we'd have been running on ploughed land. **1950** T. S. Eliot *Cocktail Party* II. 111 *Edward*: Lavinia, we must make the best of a bad job. That is what he means. *Reilly*:..The best of a bad job is all any of us make of it.

7. *attrib.* and *Comb.* (now chiefly in sense *4), as *job assessment, assignment, centre, content, -counselling, definition, description, discrimination, displacement, enlargement, enrichment, evaluation, -hungry* adj., *hygiene, insecurity, mobility, opportunity, placement, reservation, -rich* adj., *rotation, satisfaction, security, -seeking, situation, specification, structuring, study*. Also **job analysis**, analysis of the essential factors of a particular piece of work and the necessary qualifications of the person who is to perform it; so *job analyst*; **job-buyer**, one who buys job lots; **job case** (*Printing*), a type case used in job printing with boxes for both upper- and lower-case types; **Job Corps** *U.S.*, 'an organization..that operates rural conservation camps and urban training centers for poor youths' (Random House Dict.); **job fount, font** (*Printing*), see quots. 1888, 1894; **job-hopping**, the act or practice of changing from one job to another; so **job-hop** *v. intr.*; **job-hopper**; **job house** (*Printing*), = *job-office*; **job-office** (*Printing*) (examples); **job press** (*Printing*), a small press designed for job-printing; **job-print, -printing**, the printing of job-work (see sense 1 c); **job-printer** (examples); **job sheet**, a sheet on which are recorded details of a job that has been done; **job shop**, (*a*) a workshop where small pieces of work are performed; (*b*) = *job-office*; (*c*) (see quot. 1972); **job-type** (examples); **job-work**, (*b*) the printing of jobs (sense 1 c).

1923 J. D. Hackett in *Managem. Engin.* May, *Job analysis*, the determination of the essential factors in a specific kind of work and of the qualifications of a worker necessary for its competent performance. **1929** *Encycl. Brit.* XIII. 78/2 This 'questionnaire method' has received severe condemnation from scientists and should be used sparingly by investigators in the field of job analysis. **1962** H. C. Weston *Light, Sight & Work* (ed. 2) viii. 239 This is subsequently fitted to a job visual standards profile based on the results of job-analysis. **1946** R. H. Warnhoff *Automobile Accessory Industry* 18 *Job analyst.* Performs detailed job analysis and job evaluation studies. **1971** *State Service* Nov. 340/2 The job analyst..assesses the post. **1959** *Gloss. Terms Work Study (B.S.I.)* 26 *Job assessment*, the process of ascertaining the relative value of a job by examination of the job analysis and job description. **1964** G. L. Cohen *What's Wrong with Hospitals?* i. 17 Through job-assignment, nurses gain expertise but also become frustrated. **1903** *Daily Chron.* 7 Oct. 10/1 E. H...described as a 'Job buyer'. **1894** *Amer. Dict. Printing & Bookmaking* 308/2 *Job case*, a lower case condensed into two-thirds of its usual width, half of the boxes in the upper case being placed by its side. **1946** A. Monkman in H. Whetton *Pract. Printing & Binding* ii. 20/1 The two best double cases are the California Job Case..and the Improved Double Case. **1972** *Jrnl. Printing Hist. Soc.* 1971 VII. 39 By 1860 the full size job case was generally available as a printers' supplier's stock item on both sides of the Atlantic. **1972** *Times* 8 Sept. 15/1 Forget about the Government employment exchanges..think about the bright, new offices with a new image and a sign outside saying job centre. **1973** *Guardian* 23 May 6/4 Britain's first Job Centre will be opened this afternoon. Job Centres are the modern version of what have..been called employment exchanges or..labour exchanges. **1969** J. Argenti *Managem. Techniques* 153 The use of a computer to analyse trends in job content. **1965** Mrs. L. B. Johnson *White House Diary* 3 Feb. (1970) 234 Dr. Otis Singletary..director of the Job Corps..says that applications to join are coming in about six thousand a week. **1965** *Economist* 4 Sept. 884/1 Over forty of the Job Corps centres are already in operation. **1970** *Times* 1 June 2/6 The separation of benefit from job counselling. **1967** A. Battersby *Network Analysis* (ed. 2) iv. 68 It will always be more convenient to go on using existing job-definitions for which data on performance and cost are already known. **1951** J. M. Fraser *Psychol.* II. xi. 126 Once we have drawn up a complete and realistic description of a job we can work out from it a specification of the kind of person who is likely to do it satisfactorily. It is very important that..each quality laid down in it should be linked with some aspect of the actual duties shown in the job description. **1972** *Accountant* 28 Sept. 385/1 Essential records are..those for employee turnover, performance appraisals, dynamic job descriptions/specifications. **1970** G. Greer *Female Eunuch* 297 A protest against job discrimination. **1964** Job displacement [see *across *prep.* 2 c]. **1954** *Time* 12 Apr., Job enlargement (a phrase coined by Yale's Human Relations Expert Charles R. Walker) was formally born in 1943. On a trip through International Business Machine Corp.'s plant at Endicott, N.Y., I.B.M. President (now Chairman) Thomas J. Watson spotted a young woman standing idly by a milling machine. She explained that she was waiting for the 'machine inspector' (there was a 'setup man', whose task it was to adjust the machine for each new operation, and an inspector, who okayed the adjustment). Actually, she was able to adjust the machine herself, but it was against plant rules to do so. **1966** T. Lupton *Managem. & Social Sci.* ii. 32 It might well be possible to look at individual jobs to see whether it is possible to make them less rigidly paced and repetitive and not lose output, by 'job enlargement' for example. **1972** *Listener* 7 Sept. 301/1 A growing interest in the idea of work-structuring or 'job enrichment'. **1957** *Encycl. Brit.* XXIII. 272/1 Where job evaluation wage rate systems prevail jobs are analyzed and defined in terms of basic attributes and requirements. **1973** *Times* 17 Jan. 13/3 Some techniques that until recently were virtually unheard of in the [hotel] industry are now beginning to make their appearance, such as job evaluation and formal performance appraisal. **1888** C. T. Jacobi *Printers' Vocab.* 68 *Job fount*, a small fount of type used for displaying purposes. **1894** *Amer. Dict. Printing & Bookmaking* 308/2 *Job font*, a small font of type used for display, distinct from a book font. **1972** *Jrnl. Printing Hist. Soc.* 1971 VII. 39 A more efficient use of the full size case in accommodating the caps, figures and points only job fonts came with the devising of the treble case. **1970** *Time* 19 Jan. 43 He job-hopped, serving briefly as operating vice president of Servo-mechanisms Inc. and later organizing his own law firm in Los Angeles. **1967** *Time* 13 Oct. 63 Boyden's prospects are rarely aware that Boyden is aware of them as potential job hoppers. **1953** *Britannica Bk. of Year* 639/2 *Job-hopping*, a policy of changing jobs so as to place a higher value on one's services. **1972** *Times* 22 Sept. 22/6 Increased job-hopping tendencies among young male graduates. **1825** T. C. Hansard *Typographia* 700 But, that it [*sc.* a kind of printing machine] could be introduced into a book-work house, or even a job-house, to execute all the variety of sizes of works and papers, was a thing which I could never believe. **1841** W. Savage *Dict. Art of Printing* 428 *Job house*, a printing office, the general run of business in which is the printing of Jobs, namely, cards, shop bills,..play bills,..and all other work of a similar description. **1888** C. T. Jacobi *Printers' Vocab.* 68 *Job house*, a term applied to printing offices distinct from book or newspaper offices. **1894** *Amer. Dict. Printing & Bookmaking* 309/1 *Job house*, a term applied in England to printing-offices where the chief kind of work done is in jobs. **1946** Job-hungry [see *gadget 2]. **1966** T. Lupton *Managem. & Social Sci.* iii. 63 Good environmental conditions—so called 'job hygiene', i.e. good welfare facilities, meals, lighting, heating, good mates. **1959** *Listener* 25 June 1095/2 The very notion of a job insecurity..is alien to the Russians. **1966** I. Jefferies *House-Surgeon* iii. 30 Rosalind..was used to even more job-mobility than most people would regard as desirable. **1972** *Accountant* 5 Oct. 441/2 To cope with the problems presented by job mobility it was proposed to concentrate employment records in ten computer centres. **1882** T. MacKellar *Amer. Printer* (ed. 13) 301 A useful thing to have in a job office is Ame's Paper and Card Scale. **1894** *Amer. Dict. Printing & Bookmaking* 309/2 *Job office*, a printing-house where the chief work is in jobs. **1971** *Wall St. Jrnl.* 2 July IV. 1/1 While thousands of 1971 graduates pound the pavements in pursuit of jobs, an exclusive little group here has found alluring job opportunities popping up on all sides. **1963** *New Society* 7 Nov. 7/2 The private employment agencies, who handle a lucrative job-placement business. **1894** *Amer. Dict. Printing & Bookmaking* 309/2 *Job press*, a press on which job-work is done. It is, however, in practice usually limited to the treadle-machines which do not print a sheet larger than 14 or 15 by 20 [inches]. **1936** Greenhood & Gentry *Chronol. Bks. & Printing* (rev. ed.) 111, 1850..Gordon of New York is awarded a patent for first 'job' press. **1971** R. W. & E. W. Polk *Pract. Printing* (ed. 7) xv. 115 Most of the small presses for general job work are of the platen type. These presses are often classified as job presses. **1921** C. E. Mulford *Bar-20 Three* vi. 77 A hard-riding courier, relaying twice, carried the work of the job-print towards Mesquite. **1884** J. Gould *Letter-Press Printer* (ed. 3) 132 For job-printers the small platen machines are invaluable. **1960** P. M. Handover *Printing in London* vii. 194 Without the lavish production of publicity material fashionable since the second world war many job printers would be unable to meet the necessary cost of new machines and the rising cost of labour and rent. **1825** *New Lisbon* (Ohio) *Patriot* 29 Oct. 1/4 Job Printing, Neatly and expeditiously done at this office. **1859** *Abridgments of Specifications relating to Printing* (Patent Office) 365 The main object of this invention is to carry on 'job printing' without the necessity of employing skilled labour. **1892** A. Powell *Southward's Pract. Printing* (ed. 4) xxix. 266 Bookwork is almost always executed in black ink; while in job printing any colour is permissible. **1924** *Southward's Mod. Printing* (ed. 5) I. Job Design Suppl. 2 (following p. 336), Fine book and job printing a speciality. **1960** P. M. Handover *Printing in London* viii. 196 As job printing in London followed a course separate from that of periodicals, so the course of book printing was equally distinct. **1969** *Times* 19 Nov. 30/2 Some form of job reservation [in South Africa] imposed either by legislation or custom will persist for some time to come. **1971** *Sunday Express* (Johannesburg) 28 Mar. 4/3 Sir Francis de Guingand..said he found job reservation difficult to understand. **1972** *Sat. Rev.* (U.S.) 6 May 39/1 Industrialized and job-rich Morgan City. **1963** R. Stewart *Reality of Management* II. vii. 119 The companies which use job rotation as a conscious policy are likely to have a general policy of moving people in their early years and later to practise selective job rotation. **1955** Brayfield & Crockett in *Psychol. Bull.* LII. 397/1 We have not attempted to define such terms as job satisfaction or morale. Instead, we have found it necessary to assume that the measuring operations define the variables involved. **1972** *Accountant* 21 Sept. 346/1 Prinny..would attend the first course to assess its suitability in terms of personal job-satisfaction. **1959** *Listener* 25 June 1095/1 A number of other attitudes towards job security. **1972** *Sat. Rev.* (U.S.) 6 May 39/1 Job-seeking itself can be costly. **1970** P. Laurie *Scotland Yard* v. 116 The ambulance crew which we met on the way are in Sister's office drinking tea and filling out their job-sheets. **1974** 'J. Ross' *Burning of Billy Toober* xvi. 149 Completing job sheets and questionnaires. **1851** H. Melville *Moby Dick* III. xl. 233 When I kept my job-shop in the Vineyard. **1963** *Times* 15 Feb. 7/1 The west coast and international editions of the *New York Times* are printed in 'job shops', commercial printing offices working on a contract basis. **1967** *Electronics* 6 Mar. 28 (Advt.), Whether you have an in-plant plating operation or a job shop, we'll help you select a process. **1972** *Times* 7 June 27/5 The Butts Centre..will feature a job shop—the officially approved title of a room full of advertising boards where people can browse. **1951** J. M. Fraser *Psychol.* II. xi. 120 We are confronted then with two variables, the job and the individual... Our task is to match up one with the other, to find the individual who will fit neatly into the job situation. **1923** J. D. Hackett in *Managem. Engin.* May, *Job specification*, a record of the essential factors in a given piece of work and of the human qualifications necessary for its performance. **1970** *Times* 11 May 9/1 The company has begun the groundwork for job structuring. **1951** J. M. Fraser *Psychol.* II. xi. 128 The job-study provides a complete factual description of the job and the conditions in which it is carried out. **1882** T. MacKellar *Amer. Printer* (ed. 13) 304 Capitals and lower-case job types should not be laid together in the same boxes. **1888** Job type [see *book type* (*book *sb.* 18)]. **1892** A. Powell *Southward's Pract. Printing* (ed. 4) viii. 53 Job types are laid in upper and lower cases, in double cases, or in half cases, according to the extent of the founts. **1832** *Reg. Deb. Congress U.S.* 5 May 2766 The occasional advertising and job-work for the Government. **1859** *Abridgments of Specifications relating to Printing* (Patent Office) 390 This invention relates to an arrangement of machinery intended chiefly for printing 'job work' with great expedition. **1892** A. Powell *Southward's Pract. Printing* (ed. 4) xxix. 266 It is obvious that there is hardly any limit to the modes in which job work may be executed. **1948** *Words into Type* 543 *Job work*, miscellaneous printing—all except newspaper, magazine, or book work. **1966** Berry & Poole *Ann. Printing* 256/1 The extravagant style of American job work.

Job, *sb.*[4] Add: **2. Job's cat, turkey** *U.S. joc.*, used as types of poverty.

1854 S. Smith *Way down East* 184, I should rather be as poor as Job's cat all the days of my life. **1824** *The Microscope* 22 May 42/2 We have seen fit to say 'the patience of Job's turkey', instead of the common phrase, 'as patient as Job'. **1871** E. Eggleston *Hoosier Schoolmaster* (1872) iv. 22 But laws! don't I remember when he was poorer nor Job's turkey! **1951** *Publ. Amer. Dial. Soc.* xv. 58 Poor as Job's turkey.

job, *v.*[2] Add: **6.** *to job off*: to sell goods at very low prices.

1903 *Lett. that bring Business* 68 We have had some very unpleasant experiences in the past through our goods being held on consignment for months, and then jobbed off at suicidal prices. **1936** *Economist* 22 Feb. 400/1 Motor spirit..has been comparatively easy to sell, while heavier oils have either been 'jobbed off' or used for 'cracking' into lighter oils. **1955** *Times* 22 Aug. 7/6 The bottom has dropped out of the rice market and hard decisions are being taken about jobbing off as cattle fodder a million tons of surplus rice.

c. *to job backwards*: to engage retrospectively in calculations, e.g. of profits on Stock Exchange transactions, that presume knowledge of subsequent events. Freq. *transf.*

1919 D. Lloyd George *Let.* 8 July in A. J. P. Taylor *My Darling Pussy* (1975) 27 The election was muddled but it is no use jobbing backwards. **1931** *Economist* 21 Mar. 621/1 Calculations based on 'jobbing backwards' on a Fixed Trust holding are altogether illusive. **1939** *War Illustr.* 18 Nov. p. ii/1, I notice that in his [*sc.* Lloyd George's] writings about this later, and quite possibly greater, War he shows a sad inclination to 'job backwards', as they say on the Stock Exchange. By jobbing backwards I have no difficulty in proving how I could have been worth £20,000,000, whereas if I die, or get bombed, before this war finishes I shall figure for ever so much less than that, having lost a considerable fortune by declining values of stocks and shares. **1959** *Times* 4 Sept. 11/4 Did he [*sc.* the Prime Minister] not actually say 'We never job backwards'? This is a Stock Market term meaning, in this context, 'We let bygones be bygones.' **1959** *Observer* 11 Oct. 15/3 Meanwhile, jobbing backwards, how much was one inclined to overestimate the effect of those excellently produced Labour Party telecasts? **1968** J. M. Ziman *Public Knowl.* iii. 31 All too often it becomes an exercise in 'jobbing backwards': it tells us how we ought to have derived our result if only we had known the answer before we began.

10. To cheat; to betray; = *frame *v.* 10. *slang* (orig. *U.S.*).

1903 A. H. Lewis *Boss* viii. 100 Twelve honest dullards who called themselves a jury, despite his protestations that he was 'being jobbed', instantly declared him guilty. **1904** *McClure's Mag.* Nov. 64/1 Now she was coming back, swearing she'd been 'jobbed', the judge had been bought, and the jury corrupted. **1926** J. Black *You can't Win* xxiii. 353 It has always been a question with me where this framing and jobbing started; whether the defense originally began it..or whether it was the other way round. *Ibid.* 366, I was in the district attorney's office..and I know you got 'jobbed'. I'll take your case for nothing. **1972** C. Drummond *Death at Bar* v. 110 Funny you not minding Alwyn jobbin' your mum, not to mention your lawful wedded hubby. **1972** J. Philips *Vanishing Senator* (1973) I. ii. 11 'Peter is troubled by the possibility that Jeremy Lloyd was jobbed by the Justice Department.'..'You think Lloyd is innocent?' **1973** K. Giles *File on Death* ii. 28 You want to watch or they'll job you on that.

jobation. Add: (Later examples of *jawbation*.) Also, a long discussion.

1916 'Taffrail' *Pincher Martin* v. 75 Well, after a lot of jawbation we got him into the boat. **1925** P. Gibbs *Unchanging Quest* vii. 51, I used to watch Katherine while all this jawbation was in progress. **1938** H. G. Wells *Apropos of Dolores* iv. 227 They aren't happy until the hand's been played [*sc.* at bridge] and the jawbation begins.

jobbing, *vbl. sb.*[2] Add: **1.** (*Spec. printing*), the printing of jobs (Job *sb.*[2] 1 c.); also *attrib.* and *Comb.* as Job *sb.*[2] 7, *jobbing case, fount, office, press, printer, printing, type, work*.

1841 W. Savage *Dict. Art of Printing* 428 Jobbing is an extensive business in London. **1861** [in Dict.]. **1872** *Printers' Register* Apr. 114/1 The difficulty often experienced in laying Jobbing Founts. **1884** J. Gould *Letter-Press Printer* (ed. 3) 78 Every compositor should endeavour to make himself acquainted with the composition of miscellaneous work, as it is a great source of annoyance to find that a man who is put on to assist cannot earn his wages, but is in fact a hindrance to others in a jobbing office. **1888** C. T. Jacobi *Printers' Vocab.* 68 *Jobbing cases*, double cases made with upper and lower in one. They are sometimes made treble. **1892** A. Powell *Southward's Pract. Printing* (ed. 4) vi. 41 Jobbing Type is so called because it is used for 'jobs', i.e., for work like cards, circulars, letter headings, and advertisements. **1924** *Southward's Mod. Printing* (ed. 5) I. ii. 9 (*caption*) Small jobbing office. *Ibid.* v. 37 The ordinary arrangement causes the 49 boxes, in seven rows, for the capitals, to be much too small for jobbing founts. For this reason, double cases are now being made with only five or six rows of boxes. *Ibid.* li. 328 Jobbing work is a term applied to every kind of printing except book-work and newspaper work. **1946** A. Monkman in H. Whetton *Pract. Printing & Binding* vii. 75/1 There is still a wide variety of work available to the general or jobbing printer. **1960** P. M. Handover *Printing in London* vii. 172 (*heading*) Jobbing. *Ibid.* vii. 173 The range of jobbing is so great that it would be possible to confine the following pages to a single subject, such as printed games. **1960** G. A. Glaister *Gloss. Bk.* 204/1 *Jobbing types*, types used for jobbing printing. **1966** Berry & Poole *Ann. Printing* 236/2 This [*sc.* Gordon's 'Alligator' Press 1851] was a jobbing press with a platen standing fixed at an angle of 45 degrees. *Ibid.* 248/2, 1870..A collection of 'artistic' jobbing work..the work of Oscar H. Harpel..was published in Cincinnati. *Ibid.* 253/2, 1880..The [Printers' International Specimen] Exchange was commenced with the object of fostering good jobbing printing. **1972** *Jrnl. Printing Hist. Soc.* 1971 VII. 38 *Double case, improved double case, improved jobbing case*, British terms for full size job cases which put the lower case letters in the left two thirds and the capitals in the right third of the case. *Ibid.*, Most printers both in America and in Britain were forced to add more and more of the new letters to their stores of type in order to remain competitive in the jobbing printing business. *Ibid.*, J. L. Ringwalt [and others]..offer succinct but accurate accounts of the forces that brought about the design and production of these new jobbing types. **1973** *Univ. of Stirling Press Room* (Univ. of Stirling Library), The Press Room contains a large assortment of printing types, jobbing cases, galleys, chases, composing sticks, rules and leads. **1973** *Times Lit. Suppl.* 7 Dec. 1500/3 These principles underlie the construction of most of the presses now operating, from jobbing presses to the largest newspaper machine.

3. (Earlier example.)

1780 A. Young *Tour in Ireland* II. ix. 41 Lists and tables of the names of all persons who have obtained presentments,..should be given freely by the jurymen, to all their acquaintance, that every man might know, to whose carelessness or jobbing, the public wes indebted for bad roads.

jobbing-master. (Later example.)

1960 G. E. Evans *House in Furrow* xv. 193 A jobbing-master at The Rampant Horse, kept about six horses for use in broughams and so on.

jo·b-hunter. *colloq.* [f. Job *sb.*[2] + Hunter 1 b.] One who seeks employment. So **jo·b-hunting** *vbl. sb.* and *ppl. a.*, and (as back-formations) **jo·b-hunt** *v.* and *sb.*

1928 *Daily Tel.* 13 Nov. 9/3 His abrupt departure without returning to Washington will serve as a notice to job-hunters..that he intends to make up his own mind about all important appointments. **1930** H. Crane *Let.* 16 Mar. (1965) 349 I've really never known so discouraging a time job-hunting. **1946** *Vogue* June 88/2 Hardly anyone's prepared to job-hunt till they've had their demobilisation leave. **1956** W. H. Whyte *Organization Man* (1957) ix. 113 Job-hunting seniors. **1965** H. Waugh *Girl on Run* (1966) x. 63 If the idea falls through, then you can job hunt. **1970** *Physics Bull.* Mar. 106/1, I launched into my own 'job hunt': the only way to see the full picture was to go out and see industry for myself. **1972** M. Jones *Life on Dole* II. v. 120 Every day, they..study the advertisements. Then they go job-hunting. **1973** *Sci. Amer.* Apr. 63/2 (Advt.), How to job hunt for a 20% to 50% salary increase.

jobless, *a.* Restrict *rare* to sense in Dict. and add: **2.** Out of work, unemployed. Also as *sb. collect.* So **jo·blessness**, the state of being out of work.

1923 *Public Opinion* 30 Mar. 304/1 He means not the fear of foremen so much as the fear of joblessness. **1923** *Glasgow Herald* 25 Oct. 7 The demand that would ensue for land users would mean jobs for jobless men. **1937** *Times Lit. Suppl.* 16 Oct. 755/2 The later episodes showing the graduates of the college being set free into a jobless world full of depression are not so agreeable. **1958** *Economist* 1 Nov. 415/1 One-sixth to one-fourth of the labour force has been jobless for several months. **1964** *Economist* 11 Jan. 110/1 The realities behind joblessness. **1970** *Times* 7 Sept. 16 (*headline*) Jobless total must go higher says Paish. **1971** *Nature* 8 Oct. 369/2, 4·4 per cent of the engineers who do not possess a degree of any kind were jobless in the summer.

Jo block, colloq. abbrev. of *Johansson block* (see *Johansson). Also *fig.*

1936 H. D. Burghardt *Machine Tool Operation* (ed. 2) I. xi. 274 Size blocks may be used also instead of the pin gauge for spacing the holes. The most accurate of these gauges are the Johansson gauge blocks or 'Jo-blocks', which measure accurately in millionths of an inch. **1942** F. H. Colvin *Gages* viii. 96 Standard measuring blocks, once known largely as 'Jo' blocks from the name of the first maker, Johannsen. **1957** R. A. Heinlein *Door into Summer* (1960) ii. 41, I had known Ricky half her life and if there ever was a human being honest as a Jo block, Ricky was she. **1966** G. W. Michalec *Precision Gearing* xi. 564 Using a very accurate input for positioning the carriage, such as master Jo-blocks.., the output can be observed and plotted.

jo·bmongering, *vbl. sb. rare.* [f. Jobmonger + -ing[1].] The action or practice of a jobmonger.

1901 *Daily Chron.* 18 Dec. 5/1 Has Tammany no synonym? Or what about jobmongering? **1927** *Glasgow Herald* 24 Mar. 4 The intriguing and jobmongering of the Base.

jobster (dʒɒ·bstəɹ). [f. Job *sb.*[2] + -ster.] = Jobber[2].

1892 W. W. Greener *Breech-Loader* 117 If unable to send [a damaged gun] to the makers, avoid advertising jobsters. **1897** *N.Y. Times* 15 Nov., The Hawaiian jobsters are astir again... They seem to feel sure of the administration. **1901** *Westm. Gaz.* 7 May 2/2 All the jobsters, speculators, South African financiers, all the coal and steel owners, who in 1899 cheered on the war. **1913** R. M. La Follette *Autobiogr.* 167 He was not in favor of the spoilsman or the jobster. **1964** R. M. Buck *Grim Truth about Fluoridation* iv. 23 (*heading*) Federal jobsters disagree.

Jocism (dʒōᵘ·siz'm). [ad. F. *Jocisme*, acronym from the initial letters of *Jeunesse Ouvrière Chrétienne*, Christian working youth, set up by Joseph Cardijn in Belgium in 1924, and subsequently extended in Europe: see -ism.] An organization which aimed at spreading Christianity amongst working people. So **Jo·cist** *sb.* and *a.*

1935 *Catholic Worker* Nov. 3/2 It is necessary to have study circles exclusively for the young wage-earners... Let us see how the Jocists (Catholic working youth) run theirs. **1939** *Theology* XXXIX. 432 Jocism has as its final end the conquest of the entire working class, the whole world, for Christ and His Church. *Ibid.* 437 The apex of the Jocist organization. **1951** *Scottish Jrnl. Theol.* IV. 160 This theory, sometimes derided by its opponents as 'Jocist romanticism', has been the main support, particularly in Anglo-Catholic circles, of much Christian social doctrine. **1960** *Rev. Politics* XXII. 339 From the beginning Mounier was against any political movement by a specifically confessional group (such as the Jocist movement).

Jock[1]. Add: **1. b.** A Scottish (or † northern English) sailor; a Scottish soldier or a member of a Scottish regiment; any Scotsman. Freq. as a nickname. *slang.*

1788 Grose *Dict. Vulgar T.* (ed. 2), *Jock*,..a jeering appellation for a north-country seaman, particularly a collier. *a* **1865** Smyth *Sailor's Word-Bk.* (1867) 413 *Jocks*, Scotch seamen. **1914** R. Hodder *Brit. Regiments* 17 'The Jocks.' The origin of this name for the Scots Guards is obvious. **1918** H. Matthews in Murdoch & Drake-Brockman *Austral. Short Stories* (1951) 242 And he had to admit that some of the Jocks and Tommies from France were fine fellows. **1925** Fraser & Gibbons *Soldier & Sailor Words* 132 *Jocks*, Scotsmen in general. Men of a Highland regiment. **1930** Blunden *De Bello Germanico* 17 An interminable stream of displeased and ejaculatory Jocks repeat the act. **1952** 'J. Tey' *Singing Sands* vii. 101 Mr. Mackay had been in North Africa with the Jocks. **1965** *New Statesman* 16 Apr. 606/3 Why can't the Jocks support their team without dressing up like that? **1968** *Scottish Field* Feb. 35/1 All the infantry officers..attached tremendous importance to the Scottishness of their regiments. Kilts, trews, bonnets, pipe bands..have helped

enormously to make the Jock the man he is. *Ibid.* 37/2 The Scots Guards could be described as typical Jocks. **1970** G. M. Fraser *General danced at Dawn* 46 Who knows your Jocks aren't my matelots?

jock[2]. Add: Also as colloq. abbrev. of *disc-jockey.

1952 in Wentworth & Flexner *Dict. Amer. Slang* (1960) 294/1 Already the jukes and jocks are dinning our ears with Christmas songs. **1972** *Islander* (Victoria, B.C.) 9 Jan. 14/4 Jocks..are pretty much a thing of the past.

jock[3] (dʒɒk). *coarse slang.* [Origin unknown; perh. f. an old slang word *jockum, -am* penis (Farmer & Henley).] The genitals of a man (or † of a woman). So † **jock-gagger**, a man living upon the earnings of a prostitute (*Obs.*).

a **1790** H. T. Potter *New Dict. Cant & Flash* (1795) 36 *Jock*, private parts of a man or woman. **1809** G. Andrewes *Dict. Slang & Cant*, *Jock-gagger*, a sort of fellows who live on the prostitution of their wives, &c. **1846** *Swell's Night Guide* 123/1 *Jock*, man's privates. **1960** J. Cross *Backward Sex* iii. 73 Sprigs clattering on the floor, knees, jocks, backsides and shouting as everybody dressed. **1966** 'L. Lane' *ABZ of Scouse* 56 *Jocks*, testicles. **1972** *Dict. Contemp. & Colloq. Usage* (Eng.-Lang. Inst. Amer.) 17/1 *Jock*,..*vulgar*, the penis.

jock[4] (dʒɒk). *dial.* and *slang.* [Origin unknown.] Food. (See also *E.D.D.*)

1879 *Yorkshireman's Comic Ann.* 33 Monny a shift he wor put to to get jock eniff. **1881** B. Preston *Dial. & Other Poems* 3 An' bumps 'em dahn i' t'corner chair, An' gloars reyt hard at t'jock. **1966** H. Sheppard *Dict. Railway Slang* (ed. 2) 7 *Jock*, food. **1974** P. Wright *Lang. Brit. Industry* vi. 59 Food becomes..*jock*.., and contrasts oddly with officialese.

jock[5] (dʒɒk). *N. Amer. slang.* **1.** Abbrev. of *jock-strap 1.

1952 B. Malamud *Natural* (1963) 68 He located his jock, with two red apples in it, swinging from a cord. **1973** W. McCarthy *Detail* ii. 87 He found the Beretta..as well as the jock strap. He quickly took off his trousers, put on the jock.

2. Abbrev. of *jock-strap 2.

1963 [see *jock-strap 2]. **1968** *N.Y. Times* 12 May iv. 13 An obstacle to such trust is the attitude of some students and faculty members who, for example, smear all anti-strike students with the blanket label of 'jocks'. **1969** C. Davidson in Cockburn & Blackburn *Student Power* 351 The administration will try by a whole range of 'divide and rule' tactics such as fostering the 'Greek-Independent Split', sexual double standards, intellectuals *vs* 'jocks', [etc.]. **1970** *Globe Mag.* (Toronto) 26 Sept. 2/1 On the sundeck are the clubbers, the sweats, the jocks. **1972** *Time* 2 Oct. 41/2 Rocks for jocks, elementary geology course popular among athletes at Pennsylvania.

jocker (dʒɒ·kəɹ). *N. Amer. slang.* [f. *jock[3] + -er[1].] **a.** A tramp who is accompanied by a youth who begs for him or acts as his catamite. **b.** A male homosexual.

1893 'J. Flynt' in *Cent. Mag.* Nov. 107/2 Subject to the whims and passions of various 'jockers', or protectors. **1913** 'A No. 1' *Trail of Tramps* 72 A simple-faced chap..played the deaf and dumb game, for which purpose his jocker had forced him to learn the sign language. **1923** [see *gunsel 1]. **1931** 'D. Stiff' *Milk & Honey Route* iii. 34 While it is true that the punk serves the jocker..it is not accurate to call this exploitation. *Ibid.* xiv. 161 Whenever a man travels around with a lad he is apt to be labeled a 'jocker' or a 'wolf'. **1935** A. J. Pollock *Underworld Speaks* 64/2 *Jocker*, one who practices sodomy (ockerjay). **1950** H. E. Goldin *Dict. Amer. Underworld Lingo* 111/1 *Jocker*, an active pederast. **1972** B. Rodgers *Queens' Vernacular* 156 *Jocker* is synonymous with wolf: his role in prison parodies the husband's relationship in a marriage. **1972** J. Wambaugh *Blue Knight* (1973) iv. 54 Roxie hustles the guys who want a queen, and the kid goes after the ones who want a jocker. This jocker would probably become a queen himself.

jockette (dʒɒke·t). [f. Jock[2] + -ette.] Occas. used for: a female jockey.

1969 *N.Y. Times* 17 Jan. 21/3 The male riders, determined not to let the so-called jockettes compete against them, continue to bellow 'Vive la Différence!' **1972** *Sunday Express* 11 June 16/4 That leaves Mr. Rippon about as much chance of displacing him as jockette Miss Meriel Tufnell has of dislodging Lester Piggott from his chosen mount in the St. Leger. **1973** *Times* 17 Oct. 13/3 We have now reached the stage when theoretically a 'jockette' could ride in the 1975 Derby.

jockey, *sb.* Add: **5. b.** *transf.* A driver of a motor vehicle. Freq. preceded by a *sb.* used *attrib.* So *garage jockey*, a garage attendant. Chiefly *N. Amer.*

1912 *Collier's* 28 Sept. 11/2 Some are, so to speak, 'gentlemen jockeys', and own, enter, and drive their own cars for the fun of the thing. **1929** D. Runyon in *Hearst's International* Nov. 72/2 Jerking me into the cab and telling the jockey to go to the Penn Station. **1936** *Daily Herald* 5 Aug. 8/4 Here is a short list of busmen's slang phrases:..*Jockey* (*Driver*). **1942** Berrey & Van den Bark *Amer. Thes. Slang* § 723/2 *Automobile racer*, auto or buzzer jockey,..speed jockey,..suicide jockey. *Ibid.* § 723/4 *Motorcycle racer*, broadsider, jockey,..motor jockey. *Ibid.* § 765/2 *Commercial driver* (*bus, taxicab, truck*),..jockey, motor jockey. *Ibid.* § 765/5 *Truck driver*, truck jockey *or* spinner. ..*Spec.* juice jockey, a gasoline-truck driver; grunt-and-squeal jockey, a stock hauler;..suicide jockey, a nitro-glycerine hauler. **1945** *Amer. Speech* XX. 148/1 *Jeep jockey*, truck driver. **1954** *Chicago Tribune* 20 July A. 17/1 Trolley jockey gets even with slowpokes. **1968** *Drive* Spring 113 The driver of any heavy-load vehicle is known as the pilot or jockey. **1970** *Globe & Mail* (Toronto) 25 Sept. 37/3 (Advt.), Executive 2 bedroom, 2 baths professionally decorated, overlooking rose garden, pool treed ravine. Doorman and car jockey. **1972** *Nat. Geographic* Aug. 209/1 Sweat-streaked truck jockey. **1973** J. Cleary *Ransom* i. 33 There had been no trouble with the garage jockey. He had been lounging in a chair in the tiny office.

c. = *disc-jockey.

1963 B. W. Aldiss *Airs of Earth* 69 We stayed with the jocular jockey, hoping to catch a news bulletin, as I drove south. **1971** *Daily Tel.* 5 Jan. 10/1 The same 30 turns. Different jockeys. Different stations, but no real choice. **1972** P. Black *Biggest Aspidistra* iii. iv. 179 *Housewives' Choice*, a favourite shop-window for records and jockeys from 1946 to 1967.

7. [Shortening of *jockey-sleeve*.] *Fashion.* (See quot. 1960.)

1896 *Godey's Mag.* Feb. 211/1 Jockeys cut in fanciful shape are set into the shoulders of many of the sleeves. **1960** Cunnington & Beard *Dict. Eng. Costume* 118/2 *Jockey*,..a flat trimming applied over the outer part of the shoulder of a dress and having the lower border free. **1963** A. Gernsheim *Fashion & Reality* i. 27 By 1841 tight sleeves were..often headed by a *mancheron* (later called jockey).

8. a. (*b*) *jockey-back, -leg*: applied to a style of boot.

1909 *Boot Catal.*, Gentleman's brown calf lace, whole golosh, jockey back. **1862** *Illustr. Catal. Internat. Exhib., Industr. Dept., Brit. Div.* II. No. 4658, Skins, kips, fronts, shoe legs, jockey legs, cordovan, grained calf.

9. jockey briefs, shorts, short under-drawers for men; **jockey club**, (*b*) a toilet-water releasing chiefly rose and jasmine scents; **jockey spider** *Austral.*, a venomous black spider, *Latrodectus hasseltii*, the female of which is distinguished by a red stripe on the upper side of its abdomen; also = Katipo; also ellipt. *jockey*; **jockey-stick** *U.S.* (see quots.); **jockey strap** = *jock-strap 1 *Obs.*; **jockey-wheel**, (*b*) a small adjustable wheel at the nose of a caravan.

1966 J. Gardner *Amber Nine* ix. 130 He stripped to his jockey briefs. **1855** G. W. S. Piesse *Art of Perfumery* 122 Jockey Club Bouquet. **1859** in Bartlett *Dict. Amer.* (1860) 396, I..used cologne, hair oil, and scented my handkerchief with 'jockey-club'. **1907** *Yesterday's Shopping* (1969) 521/1 Highly concentrated Essences..Jockey Club, Maiglöckchen, [etc.]. **1941** W. A. Poucher *Perfumes, Cosmetics & Soaps* (ed. 6) II. vii. 281 Jockey Club..Bergamot oil..Jasmin..Rose..Tuberose...Mace oil... Civet extract. **1973** *Sat. Rev. World* (U.S.) 6 Nov. 29/2 The following..of our own exclusive scents..are presently available: 1 Jockey Club, 2 Number Six, [etc.]. **1951** G. Marx *Let.* 18 Jan. (1967) 194, I remember in 1932 telling Truman to stick to the haberdashery business, that there was a fortune in shirts and jockey shorts. **1968** J. R. Ackerley *My Father & Myself* 210, I took to wearing tight jockey shorts. **1971** B. Malamud *Tenants* 210 The bridegroom, in a smoked raffia skirt from waist to knees over his jockey shorts. **1933** *Bulletin* (Sydney) 15 Nov. 20/4 *Latrodectus*, which we call the 'redback' or 'jockey'. **1936** K. C. McKeown *Spider Wonders Austral.* xi. 155 In the Jockey Spider the red stripe is a direct and unmistakable warning. **1942** C. Barrett *On Wallaby* ii. 30 An infant bitten by a 'jockey'..died six hours later. **1944** *Living off Land* vi. 129 The red-back spider, or katipo, also called the 'jockey' spider, definitely is harmful. **1965** *Austral. Encycl.* VIII. 236/2 The best-known Australian member [of the comb-footed group of spiders] is the red-back, red-spot, or jockey spider, *Latrodectus hasseltii*, with a distribution from Arabia to New Guinea, the whole of Australia, the Pacific Islands, Hawaii and New Zealand; in the last-named country it is called 'katipo' (night-stinger). **1887** E. B. Custer *Tenting on Plains* xii. 352 [In driving a prairie schooner] a small hickory stick, about five feet long, called the jockey-stick, not unlike a rake-handle, is stretched between a pilot [mule] and his mate. **1968** R. F. Adams *Western Words* (rev. ed.), *Jockey stick*, in freighting, a hickory stick,..stretched between a pilot mule or horse and its mate. **1896** *Crescent* (Brooklyn, N.Y.) 1 Dec. 33/1 (Advt.), Suspensories, Jockey Straps. **1909** *Spalding's Athletic Library* (N.Y.) Group XV. No. 333 (Advt.), Bike Jockey Strap Suspensory. **1952** *Motor Manual* (ed. 34) xiii. 248 Car and caravan can be connected or disconnected in a matter of a few seconds. To help this operation, the forward end of the van chassis is provided with a jockey wheel to take the forward weight. **1966** *Caravanning* ('Know the Game' Series) 12 The nose weight of a caravan can vary from a few pounds to over a hundredweight. On most vans there is a telescopic jockey wheel to facilitate raising the nose.

jockey, *v.* Add: **1. d.** Freq. in phr. *to jockey for position*, to try to gain an advantageous position (in a race, contest, etc.).

1908 *Daily Chron.* 16 July 8/1 The fastest time was that by J. Matthews..but that counts for little in a cycle race owing to the..jockeying..for position in the final sprint. **1955** *Times* 27 July 6/7 In Alberta when there was no jury, congestion was caused by lawyers jockeying for position in order to appear before the right judge. **1969** Auden *City without Walls* 105 They're jockeying for position round the first bend.

jockeying, *vbl. sb.* **2.** (Earlier Amer. example.)

1779 G. Williams *Let.* 28 Feb. in *Essex Inst. Hist. Coll.* (1907) XLIII. 202 The Makers of Money sent it to there servants to purchas goods and thay knew it was to be out of circulation in a few months. It would be called by some Jockeing.

Jocko. (Earlier examples.)

1777 P. Thicknesse *Year's Journey* II. xl. 68 My monkey..rode postilion upon my sturdy horse... Jocko put whole towns in motion. **1778** *Ibid.* (ed. 2) II. xl. 106, I have seen an animal of the Jocko kind, who, when chained to a spot, contrive to get his food, which was out of his reach, by an address which many human creatures would have perished for want of abilities to put in practice.

jock-strap (dʒɒ·kˌstræp). [f. *jock[3] + Strap *sb.*] **1.** A supporter or protector for the male genitals, worn esp. by sportsmen; also (occas.), a *cache-sexe.

1897 *U.S. Patent Office Index* 27 Bennett, Charles F., Chicago, Ill. Combined jock-strap and suspensory. **1911** *T. Eaton & Co. Catal.* Spring & Summer 179/2 Athletic Support. Sharp & Smith's Bike-Jock Strap, being made entirely of elastic. **1929** T. Wolfe *Look Homeward, Angel* (1930) II. xiv. 191 The checking of overcoats, evening wraps, jock-straps, and jewellery. **1933** E. E. Cummings *Eimi* 52 Do I seriously..seem to see your human mind squatting in your magic jockstrap, freely watching Hitchy Goomy Gitchie Koo (the sickest medicine-man of them all) turn your ailments into formulas? **1935** *New Statesman* 14 Sept. 357/1 (Advt.), The Linia Belt, including the Linia Jock Strap. **1936** H. Miller *Black Spring* 247 The night is cold but the queen is naked save for a jock-strap. **1947** 'A. P. Gaskell' *Big Game* 15, I had to undo my pants and look to see whether I'd put on my jock-strap. **1954** *Dancing Times* Mar. 400 Articles for Sale. Jock-straps for either sex in pink, white or black. **1960** *Guardian* 23 Sept. 26/7 A Nubian chauffeur in a leopard-skin jockstrap. **1969** *Daily Tel.* 7 Mar. 21/1 It is the first play in which I remember to have encountered an actor with a jockstrap which squeaks when pushed.

2. An athletic (as distinguished from an æsthetic or intellectual) man (esp. one at a university); a 'hearty'. *N. Amer. slang.*

1956 *Amer. Speech* XXXI. 192 A healthy, athletic young man addicted to sports may wince under the pointed *jockstrap marine*. **1963** *Ibid.* XXXVIII. [illegible] A college athlete: jock (81), animal (32)... The full synecdoche itself, jockstrap, was also reported once.

Hence **jo·ck-strapped** *a.*; **jo·ck-strapper.**

1960 *Spectator* 23 Sept. 432 Even prints biographical notes on its jock-strapped heroes. **1967** N. Mailer *Cannibals & Christians* i. 14 Scranton had none of the heft of a political jockstrapper like Goldwater.

jodhpurs (dʒɒ·dpʊəz, dʒōᵘ·d-), *sb. pl.* Also **jodhpor, jodpor(e, judhpur**, etc. [Name of a town and district in Rajasthan, a state in north-western India.] **1. a.** Riding-breeches reaching to the ankle, combining breeches and gaiters. Also *attrib.* in *sing.* **b.** Indian trousers cut loosely at the top but close-fitting below the knee.

1899 G. W. Steevens *In India* iv. 28 The Jodhpur riding-breeches—breeches and gaiters all in one piece, as full as you like above the knee, fitting tight below it without a single button or strap..are on the way to be world-famous. **1913** E. M. Forster *Let.* 1 Jan. in *Hill of Devi* (1953) 21 My legs were clad in Jodpores made of white muslin. **1925** *Vogue* 1 Sept. 66 Judhpur. **1927** *Daily Express* 14 Nov. 5/2 These ski-ing suits are made with a plain, well-cut coat and jodhpurs or trousers. **1928** *Ibid.* 2 May 13 Finding some new jodhpors in his room, [he] could not resist trying them on. **1931** *Times* 7 May 17/5 Princess Elizabeth of York is shown dressed for a ride with her fair curls shining and wearing a yellow jumper pulled down over her jodhpurs. **1952** R. S. Summerhays *Elements of Riding* (ed. 3) xv. 87 Jodhpurs ..are less expensive than breeches. **1957** *Encycl. Brit.* XIII. 80/2 The Jodhpur style of riding breeches (as is seen, made of cotton, as normal walking garb in northern India) comprises breeches and gaiters in one. **1969** G. Greene *Travels with my Aunt* i. iii. 19 Men..in tweed coats..split horsily behind, gathered round a girl in jodhpurs. **1973** 'B. Mather' *Snowline* xvii. 203 The Chitralis..dress in a distinctive style—jodhpurs, silk frock-coats and small embroidered skull-caps.

2. jodhpur boot, an ankle-high boot, orig. as worn with jodhpurs.

1939 E. Hemingway *Fifth Column* ii. iii. 61 Dorothy Bridges, wearing..a tweed skirt, wool stockings and jodhpur boots. **1972** *Police Rev.* 10 Nov. 1472/1 (Advt.), Angel (Reg. Brand) Best Jodhpur Boot.

Hence **jo·dhpured** *a.*, wearing jodhpurs.

1969 *Daily Tel.* 8 Sept. 10/5 Small, pig-tailed Thelwell girls in voluminous riding-macs that flap round the bottoms of their plump, jodhpured legs. **1971** D. Francis *Bonecrack* vii. 94, I watched the trim jodhpured figure walk off.

Jōdo (dʒōᵘ·do). Also **Jō-do.** [Jap., lit. 'purified land'.] **a.** A Japanese Buddhist sect which teaches salvation through absolute faith in the Buddha Amida and constant repetition of a formulaic prayer invoking his name. **b.** One of the heavens of Buddhist faith; *spec.* the Western Paradise, where the Buddha Amida resides. Also *attrib.* or as *adj.*

1727 J. G. Scheuchzer tr. *Kæmpfer's Hist. Japan* I. iv. iii. 287 Zealous persons, chiefly the followers of the Sect of Siodo. **1886** B. Nanjio *Short Hist. Twelve Jap. Buddhist Sects* ix. (*title*) The Jō-do-shū, or Pure Land sect. **1895** W. E. Griffis *Relig. Japan* ix. 268 The Japanese technical term, '*tariki*', or relying upon the strength of another, renouncing all idea of *ji-riki* or self-power is the substance of the Jō-dō doctrine. **1911** B. H. Chamberlain *Jap. Poetry* iv. 215 *Jōdo*, literally, 'the Pure Land', is one of the Buddhist heavens fabled to exist in the West. **1938** D. T. Suzuki *Zen Buddhism & its Influence on Jap. Culture* i. iii. 37 The Jōdo appeals naturally more to plebeian requirements because of the simpleness of its faith and teaching. **1970** J. W. Hall *Japan from Prehist. to Mod. Times* vi. 74 In the tenth century new and more accessible teachings began to gain currency among the aristocracy. Among these was the worship of Amida, the Buddha of the Pure Land (Jōdō) or Western Paradise.

jods (dʒɒdz, dʒəʊdz), *colloq.* abbrev. of *JODHPURS *sb. pl.* Also (in *sing.*) *attrib.*
1959 J. Verney *Friday's Tunnel* xxvii. 258, I had nothing on but a sopping shirt and jods. *Ibid.* 259 The thought of riding reminded me of gob-stoppers. I found three left in my jod pocket. **1960** E. H. Clements *Honey for Marshal* ii. 43 Better take her jods! Your children do ride, don't they?

Joe, *sb.*² Add: **1. b.** phr. *not for Joe (Joseph)*, by no means, not on any account.
1844 C. Selby *London by Night* (1886) ii. i. 9/1 *Jack.* Who's to pay? *Ned.* Whichever you please. *Jack.* Oh! in that case you may as well settle it. *Ned.* Not for Joseph! You asked me to tea. *c* **1867** *Broadside Ballad* in Farmer & Henley *Slang* (1896) IV. 76/2 Not for Joe... Not for Joseph, if he knows it. **1928** Galsworthy *Swan Song* ii. xiii. 219 Not if he knew it—not for Joe.
c. *int.* (See quots. 1855¹ and 1862.) So as *v. trans.* (See quot. 1861.) *Austral.* and *N.Z. slang.*
1855 W. Howitt *Land, Labour & Gold, or Two Yrs. in Victoria* I. xxii. 400 The well-known cry of 'Joe! Joe!'..which means..one of the myrmidons of Charley Joe, as they familiarly style Mr. [Charles Joseph] La Trobe [Governor of Victoria]. **1855** C. R. Thatcher in Stewart & Keesing *Old Bush Songs* (1957) 111 Should a body 'Joe' a body For having on a hat? **1861** T. McCombie *Austral. Sk.* 135 To 'Joey' or 'Joe' a person on the diggings, or anywhere else in Australia, is grossly to insult and ridicule him. **1862** E. Hodder *Memories N.Z. Life* 188 As [the diggers] descried us approaching on the rocks, a simultaneous cry of 'Joe! Joe!' was raised. This is a popular cry on the New Zealand diggings and is used to hail any 'new chums' who may appear. It had its origin at the Australian diggings, where licenses were granted to all who held claims... When the police came upon the ground, to inspect licences, the cry of 'Joe!' was raised. **1871** C. L. Money *Knocking about in N.Z.* vii. 103 The word 'Joe' expresses derision usually bestowed on new chums on the diggings, or any man acting, or dressing, or speaking in any way considered as outré by the diggers themselves. *Ibid.*, Among the first to 'Joe' me at the beginning. **1917** 'H. H. Richardson' *Fortunes R. Mahony* i. i. 11 An odd figure..crying at the top of her voice: 'Joe, boys!—Joe, Joe, Joey!'
5. *colloq.* A fellow, 'guy', chap; (in certain countries) an American. Cf. *G.I. Joe, *Holy Joe.
1846 *Swell's Night Guide* 123/1 *Joe*, an imaginary person, nobody, as Who do those things belong to? Joe. **1906** E. Dyson *Fact'ry 'Ands* viii. 92 Why, man, it's meat 'n' beer t' them Joes what go in fer bringin' ther wanderers 'ome. **1932** *Amer. Speech* June 333 *Joe*, term used to designate anyone whose real name is unknown. When used with a place or profession 'Joe' indicates a perfect example of the type connected with that place or profession. **1945** 'A. Boucher' in M. & G. Gordon *Pride of Felons* (1964) 80 The customers are mostly elderly Italian businessmen who are good Joes. **1947** *Amer. Speech* XXII. 55/1 *Joe*, name given by the natives [in the Pacific] to any American. **1952** S. Selvon *Brighter Sun* iv. 55 In Trinidad... All Americans..are known as 'Joes'. **1957** J. Osborne *Entertainer* viii. 71 While everyone else is sitting on their hands you're the Joe at the back cheering. **1962** E. Lacy *Freeloaders* vii. 133 Way I see it, Gil is an American... We joes have to stick together. **1969** M. Pugh *Last Place Left* xvi. 110 A few of his men were Irish moleskin joes from the hydro-electric operation. **1972** J. Burmeister *Running Scared* xii. 169 Be a good Joe and take the pills. Please. **1973** *Publishers Weekly* 26 Mar. 65/2 The average Joe probably thinks that cyclists..are eccentric folk.
b. Joe College. 'A college boy; *esp.* one devoted to amusement' (Webster, 1961).
1932 *Amer. Speech* June 333 'Joe College' is a perfect specimen of the college man. Often used with 'himself'. **1964** *Ibid.* XXXIX. 193 The net effect of the publications on college slang has been to encourage the image of perky Joe College.
c. Joe Bloggs, Joe Blow (*U.S.*), **Joe Do(a)kes:** names applied to a hypothetical average or ordinary man.
c **1941** Kendall & 'Viney' *Dict. Army–Navy Slang*, Joe Blow...means any soldier. **1943** *Amer. Speech* XVIII. 109 *Joe Doakes*, generic term for all ball players. **1945** L. Lane *How to become a Comedian* x. 92 He would ask the conductor if he had seen 'Joe Dokes' last night. **1956** B. Holiday *Lady sings Blues* (1973) ix. 90 But just let me walk out of the club one night with a young white boy of my age, whether it was John Roosevelt, the President's son, or Joe Blow. **1968** *Jazz Monthly* 15/1 All these items are essentially jazz-tinged versions of Joe Doakes's favourite melodies. **1969** *Guardian* 6 Mar. 7/2 LSD can be taken by Joe Bloggs on a lump of sugar. **1970** C. Major *Dict. Afro-Amer. Slang* 71 *Joe Blow*,..horn-blowing musician..came to mean any male person. **1971** *Daily Tel.* 27 July 13/5 In too many cases these forms arrive on the desk of a busy executive who concludes that Joe Bloggs down the corridor must have signed the order. **1973** K. Giles *File on Death* iv. 98 Joe Bloggs, the honest garage mechanic.
d. Joe Soap: name applied to a 'dumb' person, a mug; also, more generally, a quite ordinary person, any person.
1943 Hunt & Pringle *Service Slang* 41 *Joe Soap*, the 'dumb' or not so intelligent members of the forces. The men who are 'over-willing' and therefore the usual 'stooges'. **1966** 'L. Lane' *ABZ of Scouse* 56 *Joe Soap*, whozit, whatzisname. When asked by the police to account for the possession of a stolen article the answer often is *I got guv it by Joe Soap.* **1968** 'B. Mather' *Springers* xi. 111, I sat there like Joe Soap on guard; feeling guilty if I dozed off. **1969** *Guardian* 31 Mar. 15/8 Socialists have become..over-eager to find out what Joe Soap is doing in order to tell him not to do it. **1972** J. Brown *Chancer* ix. 118 Who do you think I am, moosh? Joe Soap?
6. Joe Blake. *Austral. rhyming slang.* A snake.
1927 M. Terry *Through Land of Promise* ix. 123 I'll bet you what you like there are Joe Blakes in this camp. **1934** *Bulletin* (Sydney) 22 Aug. 20/2 At Bob's prompting Billy sneaked up quietly and quickly behind Joe Blake, and, seizing the tail firmly, gave a terrific swing round his head. **1970** *Sunday Mail Mag.* (Brisbane) 11 Jan. 3/1 We've camped..with the Joe Blakes, the goannas, the flies, and 4000 skinny jumbucks.
7. A French Canadian. *Canad. slang.*
1963 R. I. McDavid *Mencken's Amer. Lang.* 368 Frog is not applied in Canada to a French Canadian, who is a Canuck or simply Joe, and prefers to be called a Habitant. **1966** *Globe & Mail* (Toronto) 19 Apr. 6/6 Their waspish counterparts in Quebec always refer to 'pea-soupers' or 'Joes'. The word 'Frog' in that connection went out of fashion 50 years ago.

joe (dʒəʊ), *sb.*³ *U.S. colloq.* [Origin unknown.] Coffee.
1941 J. Smiley *Hash House Lingo* 34 *Joe*, coffee. **1944** K. D. McCracken *Baby Flat-Top* 87 Quartermasters..are inexhaustible in furnishing the Officer of the Deck with anything from the true wind and the relative humidity to a flashlight and a pot of Joe. **1953** C. M. Kornbluth *Syndic* (1964) viii. 80 Get me a mug of joe, sailor... Go get the coffee. **1963** 'E. McBain' *Ten plus One* (1964) v. 48 'Would you like some coffee?' Carella asked. 'Is there some?' 'Sure... Can we get two cups of joe?'

joe-pye weed. *N. Amer.* Also **joe pye.** [See quot. 1893.] Either of two tall, perennial herbs, *Eupatorium purpureum* or *E. maculatum*, belonging to the family Compositæ, and bearing clusters of purplish flowers.
1818 A. Eaton *Man. Bot.* (ed. 2) ii. 245 *Eupatorium..purpureum* (purple thoroughwort, or joe-pye). *Ibid.*, *Eupatorium..verticillatum* (joe-pye's weed). **1866** Lindley & Moore *Treas. Bot.* II. 638/1 Joe-pye weed. An American name for *Eupatorium purpureum*. **1893** F. T. Dana *Wild Flowers* iii. 210 Joe-Pye-Weed. Trumpet-Weed. *Eupatorium purpureum*... 'Joe Pye' is said to have been the name of an Indian who cured typhus fever in New England by means of this plant. **1895** K. D. Wiggin *Village Watch-Tower* 42 Purple asters and gay Joe Pye waved their colors by the road-side. **1903** E. C. Waltz *Pa Gladden* vii. 144 In other Septembers the slope ..was lovely with..joepye-weed. **1949** *Nature Mag.* Apr. 187/2 Later the cowslips are replaced by purple loosestrife and that butterfly magnet, Joe-pye weed. **1954** C. J. Hylander *Macmillan Wild Flower Bk.* 412 Joe-Pye-weed is found from New England south to North Carolina and New Mexico.

joes (dʒəʊz), *sb. pl. Austral. slang.* [Origin unknown.] Depression of spirits, the blues.
1916 C. J. Dennis *Moods of Ginger Mick* 27 'E's got the joes reel bad. *a* **1938** —— in *Penguin Bk. Austral. Ballads* (1964) 235 It gimme Joes to sit an' watch them two! **1957** V. Palmer *Seedtime* xx. 138 What I saw in the sugar country gave me the joes.

joey¹. Restrict † *Obs.* to sense in Dict. and add: **2.** A threepenny bit. *slang.*
1936 'G. Orwell' *Keep Aspidistra Flying* i. 7 A Joey. He took out the miserable little threepenny-bit. **1945** Baker *Austral. Lang.* v. 109 3*d.*,..joey. **1965** *Australasian Post* 4 Mar. 46 *Threepence*, 'joe' or 'joey', though these..two words are now rarely heard. (A 'joey' was originally London slang for a fourpenny bit; and later was transferred to the English threepence). **1966** F. Shaw et al. *Lern Yerself Scouse* 33 *Joey*, threepence.

Joey³ (dʒəʊ·i). *colloq.* [Familiar abbrev. of the name of the clown *Joseph* Grimaldi (1779–1837).] A clown.
1896 G. B. Shaw *Our Theatres in Nineties* (1932) II. 105 Its [*sc.* philosophic comedy's] common Joeys with red-hot poker and sausages. **1926** *Amer. Speech* Feb. 283/2 I'm through with bein' a joey. Gettin' too old, thought I would never troupe again. **1962** D. H. Laurence in G. B. Shaw *Platform & Pulpit* p. xiii, Thus was created a paradox of the public simultaneously decrying Shaw for playing a 'Joey' rôle and seeking desperately to set a jester's cap upon his head. **1973** J. Wainwright *Devil you Don't* 121 A pensioner—gnarled hand gripping a walking-stick, grey hair peeping from beneath an old-fashioned flat cap, pullover, frayed jacket and baggy trousers—like a Joey, without make-up.

jog, *sb.*¹ Add: **2. b.** (Further examples in sense 'a slow measured trot or run'.)
1948 *Oxf. Pocket Bk. Athletic Training* iii. 32 A very slow jog, where the runner lands flat-footed with a slight jarring action. **1969** [see *jog *v.* 4].

jog, *sb.*² Add: **3.** *Cryst.* A step in a dislocation where it passes from one atomic plane to another.
1951 N. F. Mott in *Proc. Physical Soc.* LXIV. B. 733 Supposing one of the expanding dislocation loops..cuts a screw dislocation, pictured as perpendicular to the plane of the paper. This will normally happen several times in the expansion of a loop in a real crystal. The loop will then necessarily contain what we call a 'jog', i.e. a point where the dislocation jumps from one slip plane to an adjacent one. **1955** *Rep. Conf. Defects Crystalline Solids 1954* (Physical Soc.) 391 If the section of the dislocation between the two parallel planes has a length of the order of the atomic distance, it is called a dislocation jog. **1960** *New Scientist* 6 Oct. 915/3 There is a theory based on the strain fields round the dislocations and the way they attract and repel each other, and a theory based on the way they get kinky ('jogs' is the technical word) when they cut through each other. **1966** C. R. Tottle *Sci. Engin. Materials* iv. 98 If the jog climbs, by moving into another slip plane, point defects are created, but the screw dislocation containing the jog can continue to slip.

jog, *v.* Add: **4.** Later examples in sense 'to run at a gentle pace' (esp. as part of a 'keep-fit' schedule).
1968 'E. V. Cunningham' *Margie* ix. 156 Fenton, who was jogging in place to keep his circulation up, explained that they were in a local elevator. **1969** *Age* (Melbourne) 24 May 17/4 Latest to join the 'jog' set is Sir Reginald Sholl, who was seen jogging around Fawkner Park. **1970** N. Armstrong et al. *First on Moon* ii. 37 There would be little time..to read or jog on the beach.
c. (Later example with *along*.)
1847 A. Brontë *Agnes Grey* xxii. 325 They're jogging along as usual, I suppose.

jogah (dʒəʊ·gä). *slang.* Also **jogar.** [Origin unknown.] = *busker².
1928 *Radio Times* 2 Nov. 302/1 'Varda the polone,' murmurs a 'jogah'. **1952** Granville *Dict. Theatr. Terms* 104 *Jogar*, a street-singer, or queue entertainer.

jogee, jogi, varr. Yogi.
1864 J. A. Grant *Walk across Afr.* 317 Like mad 'jogees' or devotees. **1899** *Folk-Lore* X. 394 The struggle for local supremacy between a Musalman saint and his rival and counterpart, a Hindu *jogi*. **1922** *Chambers's Jrnl.* 29 Apr. 343/2 Other *jogis* there are, with heavy iron rings in their ears.

‖ **joget** (dʒǫ·gĕt). Also *erron.* **jogget.** [Mal. *joget*.] A Malay popular dance, in which a couple improvises to the accompanying music. Also *attrib.*, and *transf.*, the place where such dancing occurs.
1895 F. A. Swettenham *Malay Sk.* vii. 45 Dancing girls..perform what is called the 'Jôget'—a real dance with an accompaniment of something like real music, though the orchestral instruments are very rude indeed. **1900** W. W. Skeat *Malay Magic* vi. 513 The *Joget*, a kind of dramatic and symbolical dance. **1910** R. J. Wilkinson *Papers on Malay Subjects: Life & Customs* iii. 22 The *ronggeng*, *gambok* and *joget* come from Java. **1959** P. Brown *As Far as Singapore* vi. 125 'It is the Malayan *jogget* dancing.'.. It was rather like an Oriental version of jiving, with the boy and girl facing each other and executing intricate steps to the rhythmic music, but never actually touching each other. **1963** J. Kirkup *Tropic Temper* xv. 163 The place where the dancing is done is called a joget. **1966** D. Forbes *Heart of Malaya* vii. 89 We were shown..an entertainment at court, providing the occasion for more music and for *joget* dances. **1972** M. Sheppard *Taman Indera* 92 (*caption*) The Trengganu palace *Joget* dancers in an 18th century dance. **1972** *Straits Times* (Malaysian ed.) 23 Nov. 4/4 Among the dances to be performed are..Joget.

jogged, *a.* Restrict ? *Obs.* to sense in Dict. and add: **b.** *Cryst.* Having a jog or jogs.
1962 *Phil. Mag.* VII. 83 Jogged dislocations can advance only if the jogs glide conservatively along the Burgers vector direction or by producing lines of point defects. **1973** *Nature* 3 Aug. 276/2 Bowed, looped and jogged dislocations.

jogger. Add: (Examples of 'one who jogs at a gentle pace for physical exercise'.)
1968 *Times* 24 May (N.Z. Suppl.) p. viii/5 The caption refers to the runner as 'a jogger out for his morning run'. In New Zealand the noun jogger is an acceptable..word because so many people..run for fitness. **1968** *Chicago Tribune* 9 July 1. 12/1 Joggers have become an almost familiar sight thruout America in the last year. **1969** *Age* (Melbourne) 24 May 17/4 Ken Myer is a well-known jogger around the Toorak area.

jogging, *vbl. sb.* (Later examples.)
1948 K. S. Duncan *Oxf. Pocket Bk. Athletic Training* iii. 33 The runner should start any piece of jogging with the slow jog style. **1968** *Courier-Mail* (Brisbane) 4 May 4/6 Jogging enthusiasts say that it improves posture and waistline. **1972** *Publishers Weekly* 3 Apr. 64/1 His whole-hearted commitment to jogging..on five mile jaunts around New York's Central Park.

joggling-table (s.v. Joggle *v.*¹). Add: (U.S. example); (*b*) *Printing* (see quot. 1947).
1849 *Ex. Doc. 31st U.S. Congress 1 Sess. House* No. 5. i. 435 It is probable that a set of joggling and sleeping tables will be added to the washing machinery. **1947** *Jrnl. R. Aeronaut. Soc.* LI. 319/1 Offset lithographic process... The mat surface is obtained by placing the template on a joggling table... When the table is joggled or agitated by a lever.., the constant scouring action.. produces a delicate mat surface.

jog-trot, *v.* **1.** (Later example.)
1968 *Courier-Mail* (Brisbane) 4 May 4/6 The latest American keep-fit craze—jogging. The idea is to jog-trot for 50 yards, then walk 50 yards and repeat this sequence until you are tired.

johachidolite (dʒohæ·tʃi-dŏləit). *Min.* [f. *Jōhachidō*, Japanese name for the village of Sāngpal-dong in Kilchu Co., North Hamgyong Province, Korea: see -lite.] A colourless hydrous fluoborate of sodium, calcium, and aluminium, $Na_4Ca_8Al_8B_{16}O_{25}(OH, F)_{20}$.
1942 Iwase & Saito in *Sci. Papers Inst. Physical & Chem. Res.* (Tokyo) XXXIX. 302 Investigation on chemical composition and determination of some physical constants convincingly demonstrate that the blue fluorescent mineral here studied is a new mineral species for which we would like to present the name jōhachidōlite after Jōhachidō district, where we have come across it. **1968** *Amer. Mineralogist* LIII. 2082 (*heading*) Johachidolite, a revised chemical formula.

Johannean, *a.* (Earlier example.)
1847 W. Smith tr. *Fichte's Characteristics Present Age* 98 The Johannean Jesus knows no other God than the true God.

Johannes, Joannes. (Earlier example.)
1758 *Essex Inst. Hist. Coll.* (1881) XVIII. 102, I this day delivered one Johannes to Major Gage.

Johannisberger. Delete (?) *erron.* -berg and add later examples of the form *Johannisberg*.
1853 S. Whiting *Mem. Stomach* 112 From the Johannisberg, with the golden seal..down to the poverty-stricken Marsala. **1888** *Century Mag.* Apr. 875/1 This brief but finished composition stood in the same relation to the usual 'magazine story' that a glass of Johannisberg occupies to a draught of table d'hôte *vin ordinaire*. **1958** A. L. Simon *Dict. Wines* 93/2 Plain Johannisberg..is never one of the best wines of Johannisberg. **1967** A. Lichine *Encycl. Wines* 437/1 The estate-bottled (*Original-abfüllung*) Schloss Johannisberg is divided into three grades of quality and price.

johannsenite (dʒohæ·nsənəit). *Min.* [f. the name of Albert *Johannsen* (1871–1962), American geologist: see -ite¹.] A brown, greyish, or greenish silicate of calcium and manganese, $CaMnSi_2O_6$, which often occurs with partial substitution of ferrous iron for manganese.
1932 *Amer. Mineralogist* XVII. 575 The Mineralogical Society of America... List of papers to be presented Wednesday, December 28, 1932...W. T. Schaller.—Johannsenite, a new manganese pyroxene. **1963** W. A. Deer et al. *Rock-Forming Min.* II. 75 Johannsenite, the manganese analogue of diopside and hedenbergite, occurs in metasomatized limestones, and is found less frequently as a vein mineral.

Johansson (yohæ·nsən). Also erron. **Johannson, Johannsen.** The name of Carl E. *Johansson*, 20th-century Swedish armaments inspector, used *attrib.* to designate a type of steel block originated by him which is made with flat, parallel faces and of a designated length to a high degree of accuracy, and forms one of a set of different sizes used in making up standard lengths.
1918 D. T. Hamilton *Gages* i. 3 (*heading*) Johansson reference blocks. **1920** *Proc. Inst. Mech. Engin.* Nov. 1102 A gauge of known length is made up by wringing together a series of Johansson gauges. **1932** Hardy & Perrin *Princ. Optics* xxviii. 577 A set of Johansson gauges used in conjunction with a plane test glass..can be used for the extremely accurate measurement of thickness. **1935** O. W. Boston *Engin. Shop Pract.* II. ix. 448 So-called master gage blocks with very hard, smooth, flat, and parallel surfaces in the form of Johansson gages..which are known as Swedish gages as they were made in Sweden, but which are now owned and made by the Ford Motor Company. **1941** A. W. Judge *Aircraft Engines* II. iv. 141 The inside walls..receive a micro-finish..guaranteed to be accurate within 2 micro-inches..longitudinally; this corresponds to the surface finish of Johansson gauge blocks. **1943** S. H. Lebowitz *Pre-Service Course Machine Sci.* vi. 149 When two adjoining surfaces of solids are sufficiently smooth and plane, they cohere from relatively slight pressure. This fact can be demonstrated with Johansson gage blocks. **1968** J. A. Rankin *Workshop Processes & Materials* II. ix. 156 Slip gauges, often called Johannsen gauges after their originator, are rectangular blocks of steel with a cross-section of about 1⅜ in. by ⅜ in. which are hardened before being finished to size.

johar (dʒəʊ·haɪ). Also **jauhar, joar.** [ad. Hindi *jauhar*, f. Skr. *jatu-griha* a house built of combustible materials.] The sacrificial burning of Rajput women to avoid their being captured by the enemy.
1802 C. James *New Mil. Dict.*, *Joar*, a general massacre of the women and children, which is sometimes performed by the Hindoos, when they find they cannot prevent the enemy from taking the town. **1907** *Westm. Gaz.* 27 Nov. 2/1 The last siege of Chitor, terminated by the greatest of the Johars. **1919** V. A. Smith *Oxf. Hist. India* 350 The women were immolated on funeral pyres to save them from dishonour, a dread rite known as *jauhar*, and usually practised by Rājpūts when hard pressed. **1965** *Handbk. for Travellers India* (ed. 20) p. xlviii, Three times Chitor suffered the horrors of sack. Time after time, when all hope was lost, the fatal *johar* was commanded. The women committed their bodies to the flames, and the men, arrayed in bridal robes of saffron, sallied out and died fighting.

johl, var. *jol.

John. Add: **1. c.** A policeman; (less commonly) a detective. In full, **johndarm** (dʒǫ·ndāɪm) [ad. F. *gendarme*]. Also with suffixed quasi-surname, as *John Dunn* (Austral.), *John Hop* (Austral. and N.Z.), *John Law* (U.S.). Cf. *jonnop. [Austral., N.Z., and U.S. *john* perh. shortened directly from *John Hop* (rhyming slang for Cop *sb.*²) and *John Law*.] *depreciatory.*
1633 [in Dict., sense 1 b]. **1858** S. A. Hammett *Piney Woods Tavern* vi. 63 He larnt that time to let the John Darmes hunt thar own varmint. **1901** *Westm. Gaz.* 18 Sept. 8/2 'George Johns are sure to visit the old girl to see if anyone has got to her... Brake the cab then shift it again as the John will be sure to tell the porter of that.'.. Detective-sergeant Stevens said the word 'Johns' on the paper signified 'detectives'. **1905** *Munsey's Mag.* July 466/2 Somebody hollers that the johndarms are coming. **1906** E. Dyson *Fact'ry 'Ands* viii. 99 'Twas near two hours afore I see His Whiskers steamin' back, 'n' he had er John in tow... Ther policeman was fer me. **1907** J. London in *Cosmopolitan* May 17/2 A lot of my brother hoboes had been gathered in by John Law. **1918** *Chrons. N.Z.E.F.* 19 July 280 In view of the fact that the camp 'John Hops' were off duty. *Ibid.* 22 Nov. 206 'Never mind about that,' said the John—and dumped him in the cell. **1925** [see *fink *sb.*²]. *c* **1926** 'Mixer' *Transport Workers' Song Bk.* 94 'Tis us..The wily 'John Hop' loves to track. **1933** *Bulletin* (Sydney) 31 May 12/3 Even on beat duty a John Hop..can cogitate. **1939** H. Hodge *Cab, Sir?* xv. 222 A policeman is the usual cockney 'Grass'..Or sometimes 'Johndarm'—thus proving we know French. **1940** R. Chandler *Farewell, my Lovely* xxxix. 302 'The johns tied me to it?' 'I don't know.' **1941** Baker *Dict. Austral. Slang* 39 *John Dunns*, policemen. **1941** W. G. Davis *Phenomena in Crime* xix. 254 Bogies, busies, gendarmes, johns...Detectives. **1961** P. White *Riders in Chariot* xi. 380 'I'll fetch the johns in the mornin'!' she shrieked. 'Layin' into a white woman!' **1965** G. McInnes *Road to Gundagai* xi. 187 Then we would climb the steps, leaving our bicycles chained to a lamp post (courtesy of the John Hops). **1965** M. Shadbolt *Among Cinders* xxi. 202 Bill Halloran... He's the local john. **1968** K. Weatherly *Roo Shooter* 23, I didn't tell the John to sweep your cell out.
d. (With lower-case initial.) A lavatory, water-closet. *slang* (chiefly *U.S.*).
[**1735** *Harvard Laws* in W. Bentinck-Smith *Harvard Bk.* (1953) 146 No freshman shall mingo against the College wall or go into the fellows' cuzjohn.] **1932** *Amer. Speech* VII. 333 *John, johnny*, a lavatory. **1946** 'J. Evans' *Halo in Blood* xvi. 181, I..made a brief visit to the john. **1959** C. MacInnes *Absolute Beginners* 54 'You poor old bastard,' I said to the Hoplite, as he sat there on my john. **1972** *Last Whole Earth Catalog* (Portola Inst.) 247/3 Every time you take a dump or a leak in a standard john, you flush five gallons of water out with your piddle. **1973** *Black World* June 19 They gave me my Status Symbol The key to the white Locked John.
e. In full, *John Chinaman*. A Chinaman; the Chinese collectively. *depreciatory.*
1818 B. Hall *Acct. Voy. Corea* ii. 92 This interpreter is called 'John' by all the parties. **1826** *Ibid.* (rev. ed.) i. 37 The seamen..not caring whether John Chinaman, as they called him, understood them or not. **1834** *Amer. Railroad Jrnl.* III. 189/1 They are required to..ascertain the height of John Chinaman in a breath. **1853** *Alta California* (San Francisco) 20 Apr. 2/2 The May Adams brought 118 'Johns' from the terrestrial kingdom of heaven. **1854** C. A. Corbyn *Sydney Revels of Bacchus, Cupid & Momus* 125 They came to the determination of giving wretched John Chinaman such a dose that prussic acid is fool to it. They committed John for trial at the Quarter Sessions. **1858** *Brit. Colonist* (Victoria, B.C.) 27 Dec. 4/3 The Johns had a high time, drinking brandy and eating fried hog. **1865** *Dunedin Punch* 16 Sept. 119 John Chinaman is to immigrate here. **1869** J. R. Browne *Adv. Apache Country* 308 John Chinamen, with long tails rolled up on the backs of their heads, running distractedly through the crowd in search of their lost bundles. **1872** [see *Chinaman 2]. **1921** R. Watson *Spoilers of Valley* 310 John always was a better truck farmer anyway. **1933** *Bulletin* (Sydney) 8 Mar. 10 Old John Chinaman, patient, hardworking, scrupulously honest. **1970** *Brewer's Dict. Phr. & Fable* (rev. ed.) 591/2 *John Chinaman*, a Chinaman or the Chinese as a people.
f. A ponce; the client of a prostitute. *slang* (orig. *U.S.*).
a **1911** D. G. Phillips *Susan Lenox* (1917) II. vi. 154 A John's a sucker—a fellow that keeps a girl. **1928** [see *creep *v.* 5 b]. **1953** W. Burroughs *Junkie* (1972) ii. 27 Mary was describing the techniques she used to get money from the 'Johns' who formed her principal source of revenue. **1967** M. M. Glatt et al. *Drug Scene* 117 *John*, prostitute's customer. **1972** *New York* 24 Apr. 38/3 Many working girls, when they are new in the city, spend at least a few months with a madam to meet the better johns. **1973** *Times* 22 Mar. 8/7 The customers (Johns or tricks) are the usual solitary, bored, out-of-town..men.
g. Abbrev. of *John Thomas (*b*).
1934 *Neuphilol. Mitt.* XXXV. 130 Here [at public-school] his first linguistic experience will be with mumfordish and swear-words (e.g...*john* 'penis'..). **1948** D. Ballantyne *Cunninghams* ii. xvi. 241 How often did the nurse find him with his old john lying limply? **1972** C. Murry *Private View* i. 33 The tip of old John brushed against the inside of my thigh.
4. John Barleycorn: see Barley-corn 1 b; **John Birch Society** (see *Bircher); hence **John Bircher** = *Bircher; **John boat** (also **Jon boat**) *U.S.*, a small, flat-bottomed boat for use principally on inland waterways; **John Citizen,** the ordinary man (esp. considered as a member of the community); **John Collins,** a Collins (see *Collins²) made with a base of gin or whisky; a *Tom Collins*; **John Doe,** also in *U.S. Law*; (*b*) name given to an ordinary or typical citizen (see also quot. 1942); **John Down,** a Newfoundland name for the fulmar; **John Hancock** *U.S. colloq.*, a signature; **John Henry** *U.S. local slang* = prec.; **John Innes** (**compost**), one of a group of composts prepared according to formulæ developed at the John Innes Horticultural Institution in the late 1930s; **John Q. Public** *U.S. colloq.*, the general public, or a member of this; **John Roscoe,** a gun (*U.S. slang*); see also *roscoe; **John Thomas,** (*b*) *slang*, the penis.
1961 John Bircher [see *Bircher]. **1961** *Listener* 20 Apr. 684/1 The patriots of the right—and they seem to be mushrooming belligerent societies round the country, as witness the sudden, and alarmingly widespread, rise of the John Birch Society. **1963** *Ibid.* 21 Feb. 350/2 The line taken by the interviewer, whose questions might have been drafted by the John Birch Society, so puritanical did they sound. **1967** *Boston Sunday Herald* 30 Apr. 1. 1/1 A strange hybrid bird is hovering..over the town of Lincoln. Is it a dove? Is it a hawk? No, it's a John Bircher who may be somebody's pigeon. **1973** W. McCarthy *Detail* iii. 201 We have covered every John Bircher and nut that has sworn to kill the President. **1905** *N.Y. Even. Post* (Sat. Suppl.) 2 Sept. 3/1 Two men came down the Mississippi in an Illinois Jon-boat, paddling slowly with rough-whittled boards. **1917** H. Kephart *Camping & Woodcraft* II. 134 We..hit upon what we conceived to be a brilliant scheme for transporting a gallon of whiskey inconspicuously in our John-boat. **1921** *Rudder* Feb. 7 Their scope of understanding..ranges from the 150-foot schooner to the ten-foot 'john boat'. **1965** A. J. McClane *Standard Fishing Encycl.* 441/2 The original john boat.. was made of clear pine boards 20 feet long and 18 inches wide. **1924** H. H. Curran (*title*) John Citizen's job. **1931** T. E. Gregory in W. Rose *Outl. Mod. Knowl.* 626 If John Citizen buys a house out of his own savings, [etc.]. **1937** *Evening News* 12 Mar. 8/1 No Government department has yet thought of issuing a pamphlet extolling the virtues of John Citizen, who, as taxpayer, every year takes a stronger weight and never complains. **1958** *Daily Tel.* 8 Aug. 11/4 When all is quiet John Citizen says what a great humanitarian he is. **1965** Mrs. L. B. Johnson *White House Diary* 13 July (1970) 296, I would depend upon that man if I were John Citizen looking on from my living room. **1865** *Australasian* 25 Feb. 8/7 That most angelic of drinks for a hot climate—a John Collins... Take a bottle of sodawater, a wineglassful of gin, a lump of sugar, a piece of ice, with a slice of lemon—mix well together. **1913** R. Brooke *Let.* 3 Aug. in *Coll. Poems* (1918) p. lxxxvi, I believe I could do a deal in Real Estate, ..over a John Collins, with a clean-shaven Yankee. **1928** T. M. Healy *Lett. & Leaders* i. x. 142 Two barmen who.. served me with a 'John Collins' [in U.S.]. **1962** John Collins [see *Collins²]. [**1825** O'Hara Family' (*title*) Tales, containing..John Doe.] **1852** Dickens *Bleak Ho.* (1853) xx. 195 It is reported at the public offices that his father was John Doe. **1900** (*title*) The bridge manual..by John Doe [pseudonym of Francis Reginald Roe]. **1928** M. H. Weseen *Crowell's Dict. Eng. Gram.* 355 John Doe. Properly used in legal actions as the name of a fictitious plaintiff. Richard Roe is the corresponding name of a fictitious defendant. By extension these names are used loosely as substitutes for real names in other affairs. **1932** John Doe [see *Jones 1]. **1935** M. M. Atwater *Murder in Midsummer* xxi. 196 Serve on them one of the John Doe warrants with which you have been provided. **1938** S. Robertson in *For Men Only* Feb. 50 (*heading*) John Doe.. crosses the Styx. **1940** L. E. Frailey *How to write Better Business Lett.* ii. 14 You start thinking: 'Here's a salesman, John Doe.' **1941** *N.Y. Times* 13 Mar. 25/2 (*title*) Meet John Doe, screen play by Robert Riskin...'John Doe'.. Gary Cooper. **1942** Berrey & Van den Bark *Amer. Thes. Slang* § 442/3 *One whose name is not known*,..Joe Doe, Joe Zilch, John Doe, John Henry, John Smith. **1957** B. & C. Evans *Dict. Contemp. Amer. Usage* 259/2 John Doe is a fictitious person in legal proceedings, usually the plaintiff. The corresponding fictitious defendant is Richard Roe. **1972** *N.Y. Law Jrnl.* 31 Oct. 15/4 USA v. Jose Valenzuela-Correa,...John Doe. **1973** *Washington Post* 18 Apr. iii. 1/8 They didn't include any small guys struggling down here. I think they should have had some John Doe on there. **1852** *Arctic Miscellanies* 10 On the banks of Newfoundland, where this bird is known by the name of 'John Down', it attends the fishing vessels for the offal of the cod fish. **1893** A. Newton *Dict. Birds* ii. 471 John-Down, the name given to the Fulmar by Newfoundland fishermen. **1917** T. G. Pearson *Birds of America* I. 80 Fulmar. *Fulmarus glacialis glacialis*... [Also called] Molly Hawk; John Down; Sea Horse. **1957** W. L. McAtee *Folk-Names Canad. Birds* 4 Fulmar... John Down (Sailor's name, significance unknown. Nfld.). [**1846** Holmes in R. W. Griswold *Passages from Corr.* (1898) 221 Avoiding..the pretentious boldness of John Hancock..I subscribe myself Yours very truly.] **1903** Ade *People you Know* 150 After he got through filling in the Blank Spaces with his John Hancock, he didn't have a Window to hoist or a Fence to lean on. **1937** *N. & Q.* 6 Mar. 178/1 There is also a popular phrase for signature, 'John Hancock', as in, 'Go ahead. Put down your John Hancock.' **1937** *Greensboro* (N. Carolina) *News* 23 Dec., Lindberg's *John Hancock* is worth ten to twenty-five dollars on the market. **1972** *Listener* 7 Dec. 784/1 Even today an American handing you a contract is apt to say: 'And now if you will just give us your John Hancock.' **1914** *Dialect Notes* IV. 109 *John Henry* or *John Hancock*, autograph. **1951** *Publ. Amer. Dial. Soc.* xv. 56 Put your John Henry on that. **1974** T. Barling *Shooter Man* iii. 20 Sign your John Henry there... Your name is Balkin. You'd better get used to it. **1939** *Ann. Rep. John Innes Hort. Inst. 1938* XXIX. 17 A great many enquiries about the John Innes composts have been received. The John Innes Base Fertiliser is now being widely used. **1939** Lawrence & Newell *Seed & Potting Composts* 9 These composts are the outcome

of experiments to meet the practical needs of cultivation at the John Innes Horticultural Institution. *Ibid.* 10 In the John Innes composts the gardener has soil mixtures which, properly used, always give excellent results. **1951** *Good Housek. Home Encycl.* 94/2 The compost for..practically all house plants should be that known as 'John Innes'. **1959** *Listener* 12 Nov. 850/1 The standard John Innes compost is excellent. **1970** *Nature* 25 July 377/1 These results show that fertilizer is crucial for satisfactory growth, particularly slow acting types such as John Innes base. **1973** *Country Life* 29 Mar. 864/3 Acidanthera corms are being potted..in 7 or 8 in. pots of John Innes No. 2 compost. **1937** *N. & Q.* 6 Mar. 177/2 'John Citizen'..is not so frequent in American usage as 'John Q. Public'... It is probably a play on the name of an early president, John Q(uincy) Adams. **1938** S. Chase *Tyranny of Words* xiv. 175 There is no 'public'... Calling it 'John Q. Public' does not help. **1972** M. Kaye *Lively Game of Death* ix. 49 My main concern at that moment was to play John Q. Public by phoning the cops. **1938** D. Runyon *Furthermore* iii. 54 The joke is, I will not be asleep in the sack, and my hands will not be tied, and in each of my hands I will have a John Roscoe, so..I pop out blasting away. **1973** A. S. Neill *Neill! Neill! Orange Peel!* (rev. ed.) ii. 130 The USA.., where anyone can carry a gun, or, to be topical, should I say a Betsy or a John Roscoe? **1879–80** *Pearl* (1970) 76 As around her fair form I a firm hold took, And John Thomas I silently buried. **1928** D. H. Lawrence *Lady Chatterley* xiv. 253 'John Thomas! John Thomas!' and she quickly kissed the soft penis. **1957** L. Durrell *Justine* 249 She had neatly tied his dress-tie to his John Thomas, a perfect bow. **1972** *Times Lit. Suppl.* 7 July 783/3 The grotesquely coy accounts of sex, during which Tony tells us that his 'John Thomas' was 'up and raring to go'.

John Canoe (dʒɒn kănū·). *W. Indies* (and †*U.S.*). Also (with various local pronunciations) **John Connú, Johnkannau, joncanoe, jonkanoo,** etc. [f. a W. African lang.] **a.** The chief dancer, or one of several dancers, in a Christmas celebration. **b.** Any of various masks or structures worn or carried on the head by the dancer at Christmas time. **c.** The celebration itself.
See also Cassidy and Le Page *Dict. Jamaican Eng.* s.v.
1774 E. Long *Hist. Jamaica* II. iii. iii. 424 The masquerader..dances at every door, bellowing out *John Connú!* **1816** M. G. Lewis *Jrnl. W. India Proprietor* (1834) 51 The John-Canoe is a Merry-Andrew..bearing upon his head a kind of pasteboard house-boat. **1825** R. Bickell *W. Indies as they Are* 214 The crowds of Slaves..making John Canoe, as they term it, according to the customs of Africa. **1826** A. Barclay *Pract. View Slavery W. Indies* 11 One or two Joncanoe-men, smart youths, fantastically dressed. **1826** C. R. Williams *Tour through Jamaica* 25 This [model of a] house is called the Jonkanoo. **1844** W. Bagley *Let.* 18 Dec. in N. E. Eliason *Tarheel Talk* (1956) 279 On Christmas day..four young fellows blacked themselves & dressed up in negro clothes ..& the fiddler would play & the rest dance. They acted the part of 'John Cunners' very well. **1861** H. Jacobs *Incidents Life Slave Girl* 179 Every child rises early on Christmas morning to see the Johnkannaus. **1929** M. Beckwith *Black Roadways* ix. 155 The instruments which I saw used in the John Canoe performance [were] identical with those accompanying the myal dance. **1952** M. Kerr *Personality & Conflict in Jamaica* 144 These dances..are usually called the John Canoe Dance. **1961** F. G. Cassidy *Jamaica Talk* xii. 256 The John Canoe dancing, recently revived, has had a complex history, and the meaning of the name is uncertain. **1962** S. Wynter *Hills of Hebron* ii. 36 The mask on the face of a 'junkonoo' dancer, striped black and white with slits for eyes and huge white teeth. **1966** *Punch* 5 Jan. 27/2 The wild revelry of the Junkanoo Parade. **1972** E. Hargreaves *Fair Green Weed* xii. 172 John Kanoos...They're rather like old-fashioned mummers they go round at Christmas time.

John Crow (dʒɒn krōᵘ). [Reduction, in folk pronunciation, of *carrion crow*.] In Jamaica, a name for the red-headed turkey vulture, *Cathartes aura*. Hence **John Crow('s) nose,** a parasitic plant, *Scybalium jamaicense*, which bears bright red flowers.
Entered in Dict. s.v. John 4 without examples.
1826 C. R. Williams *Tour through Jamaica* 82 The dead carcass of a mule, on which a score of john-crows were holding an inquest. **1864** A. H. R. Grisebach *Flora Brit. W. Indian Islands* 785 John-Crow's-nose. **1893** A. Newton *Dict. Birds* ii. 470 In Jamaica, within a few years, the John-Crow, though there protected by human law, has been nearly extirpated by the introduction of the Mongoose. **1963** Robertson & Gooding *Bot. for Caribbean* xvi. 122 In the rain-forests of the West Indies, several members of the family Balanophoraceae occur as parasites on the roots of trees. An example is the John Crow Nose (*Scybalium jamaicense*), which occurs in Jamaica, Cuba and Puerto Rico. **1963** A. L. Thomson *New Dict. Birds* 867/1 The Turkey Vulture *Cathartes aura* (also called 'Turkey Buzzard' and 'John Crow' in the United States) is regarded as a pest on the Peruvian guano islands. **1972** E. Hargreaves *Fair Green Weed* viii. 88 The john-crows hadn't yet caught sight of him, and given warning by their insistent, wheeling flight.

Johne (yō·nə). The name of Heinrich Albert *Johne* (1839–1910), German veterinary surgeon, used in the possessive to designate **Johne's bacillus,** *Mycobacterium johnei*, first recognized by him in 1895, which causes **Johne's disease,** an infectious enteritis of cattle and sheep, characterized by diarrhœa and progressive emaciation.
1907 J. M. Fadyean in *Jrnl. Compar. Path. & Therapeutics* XX. 53, I have therefore ventured to suggest that in this country the disease might in future be known as Johne's disease. Similarly, the causal organism would be known as the bacillus of Johne's disease. **1910** *Proc. R. Soc.* B. LXXXIII. 158, I have also succeeded in isolating and growing the acid-fast bacillus found in the intestine of cows in Jöhne's [*sic*] disease... Jöhne's bacillus grows somewhat more easily than Hansen's lepra bacillus. **1913** Twort & Ingram *Monogr. Johne's Dis.* iii. 29 'Johne's disease' should be the name given to the disease of sheep produced by an infection of Johne's bacillus in the gut and mesenteric glands. **1943** *Jrnl. Compar. Path. & Therapeutics* LIII. 140 Lesions could be produced with various saprophytic acid fast bacilli and killed cultures of Johne or tubercle bacilli suspended in saline. **1945** *Ibid.* LV. 41 (*title*) Ovine paratuberculosis (Johne's disease of sheep). **1972** *Jrnl. Compar. Path.* LXXXII. 333 (*title*) *Corynebacterium renale* as a cause of reactions to the complement fixation test for Johne's disease.

Johnny, Johnnie. Add: **1.** (Further examples.)
1865 R. H. Kellogg *Life & Death in Rebel Prisons* 194 The 'cheekiest' thing that had been done by the 'Johnnies' ..was an attempt to secure the services of our men as artillerists. **1900** [see *crumpet 4 b]. **1930** [see *cool *v.* 3 e]. **1960** M. Spark *Bachelors* i. 9 He felt himself to be, not the amiable johnnie he had by then..affected to be. **1965** A. Nicol *Truly Married Woman* 42 Why are you always trying to be fair, you Johnnies?
b. [Cf. *John 1 c.] A policeman. Also *Johnny Darby*, *Johnny Hop*. *slang.*
a **1852** H. Mayhew *London Labour* (1861) II. 154/1 The 'Johnnys' on the water are always on the look out, and..we has to cut our lucky. **1886** *Graphic* 30 Jan. 130/1 Constables used to be known as 'Johnny Darbies', said to be a corruption of the French *gensdarmes*, and they are still occasionally called 'Johnnies'. **1898** W. T. Goodge in M. Davitt *Life & Progress Australasia* xxxv. 192 A policeman is a 'johnny', Or a 'copman' or a 'trap'. **1923** D. H. Lawrence *Kangaroo* xvi. 357 Somers knew that Johnny Hops was Australian for a policeman. **1935** H. Neville *Sneak Thief on Road* 347 Johnny, a policeman. A Lincoln johnny, a stupid and impudent policeman.
c. (*a*) = *John 1 e; (*b*) a soldier of the Indian Army (before 1947); (*c*) a Gurkha; (*d*) = *Johnny Turk*; (*e*) an Arab; (*f*) an onion-seller from Brittany.
1857 T. B. Gunn *Physiology of N.Y. Boarding Houses* 275 He's seed the *Johnnies* goin' into that there doorway next block. **1858** *Leisure Hour* 27 May 326/1 Sepoys..known as Johnnys. **1888** Kipling *Wee Willie Winkie* (1889) 103 The Highlander..turning to a Gurkha, said, 'Hya, Johnny!' **1897** I. Scott *How I stole over 10,000 Sheep in Austral. & N.Z.* 46 His name was Lim Hung Ching...this Johnny knowing he would have no show in such a scotch town if he sent in his tender with his own name,..signed it Angus McPherson. **1916** *Anzac Book* 50 What should we at Anzac have done without 'Johnnie' and his sturdy little mules? **1925** Fraser & Gibbons *Soldier & Sailor Words* 132 *Johnny*, a Turk. (As a Service nickname, dating from the Crimean War.) **1948** Partridge *Dict. Forces' Slang* 103 *Johnny*, soldiers' word for Arab. It rebounded, as Arabs also used the name for British soldiers. **1960** *Guardian* 8 Mar. 8/7 Any time now we shall have the Johnnies [from Brittany] coming around in small vans. **1967** J. Caird *Murder Scholastic* xi. 130 She heard the knock at the door... It could be a tradesman.. or Onion Johnny, or a tramp. **1969** S. Mays *Fall out Officers* xxii. 173 'Now we are all the fighting brothers,' said Johnny with delight.
2. b. Read: A sailor's name for a penguin, *Pygoscelis papua*, the Gentoo (*sb.*²). (Earlier and later examples.)
1879 H. N. Moseley *Notes by Naturalist on 'Challenger'* vii. 175 One [sort] was a penguin called by the sealers the 'Johnny' (*Pygoscelis tæniata*), the 'Gentoo' of the Falklands. **1893** A. Newton *Dict. Birds* ii. 471 Johnny, the South-Sea sealers' name for a Penguin, *Pygoscelis papua* or *tæniata*, one of the widely-distributed species. **1954** W. B. Alexander *Birds of Ocean* (ed. 2) 155 At the Falkland Islands it [*sc.* the gentoo penguin] is usually known as the 'Johnny'.
3. *Johnny Armstrong* joc. slang (see quot. 1962); *Johnny-come-lately* orig. *U.S.*: (*a*) a newcomer; (*b*) = *Johnny Raw*; (*c*) *fig.* and *attrib.*; *Johnny Crapaud* [i.e. toad], also *Johnny Crapeau*: (*a*) nickname for a Frenchman or French Canadian; (*b*) (*Obs.*) the French nation; *Johnny Head-in(-the)-Air*: nickname for a man with 'his head in the clouds', unconscious of his surroundings; *Johnny Magorey* (see quots.); *Johnny Newcome*: (*a*) = *Johnny Raw*; (*b*) a newcomer of any kind; *Johnny-on-the-spot*: a person who is available when needed, or 'at the psychological moment'; quasi-*adj.*, present, in readiness; *Johnny Reb*: a *U.S.* (Northern) name for a 'rebel' or Confederate soldier in the American Civil War; *Johnny Turk*: a Turkish soldier; any Turk; *Johnny Woodser* = *Jimmy Woodser* (*Jimmy 4).
1922 *N. & Q.* 9 Sept. 206/2 *Johnny Armstrong*, the action of 'pulling' or restraining a horse. **1962** Granville *Dict. Sailors' Slang* 66/2 *Johnny Armstrong*, any hard work involving pulling or hauling. **1839** C. F. Briggs *Adv. H. Franco* I. 249 'But it's Johnny Comelately, aint it, you?' said a young mizzen topman. **1924** 'R. Daly' *Outpost* xiv. 139 He may be an old barbarian, but he's entitled to more consideration than these Johnny-come-lately's who cruise along the coast after trade. **1933** *Press* (Christchurch, N.Z.) 28 Oct. 17/7 *Johnny-come-lately*, nickname for a cowboy or any newly-joined hand or recent immigrant. **1946** M. Shulman *Zebra Derby* iii. 22 Postwar planning in these United States was no Johnny-come-lately. **1952** E. Coxhead *Play Toward* iii. 88 The Midlands are..all Johnny-come-latelys who coined money out of the war. **1953** *Amer. Scholar* XXIII. 17 The excessive power and renown of many Johnny-come-lately anti-Communists. **1972** *Listener* 16 Nov. 671/2 Here [in Utah] man himself is a Johnny-come-lately. [**1818** 'A. Burton' *Adv. Johnny Newcome* iii. 131 Jean Crappeau's mighty stout, He surely means to fight it out.] **1834** W. N. Glascock *Naval Sketch-Bk.* 2nd Ser. II. 137 Mister Bull and Johnny Crappo..have..suddenly taken for each other..a fit of affection. **1839** *Spirit of Times* 31 Aug. 312/1 Poor Johnny Crapeau has been the favorite mark for every shaft of ridicule. **1840** J. F. Cooper *Pathfinder* II. vi. 201 We are no Johnny Crapauds to hide ourselves behind a..fort on account of a puff of wind. **1851** J. F. W. Johnston *Notes N. Amer.* II. 6 These Ayrshire emigrants appear to be shrewd enough to buy out Johnny Crapaud. **1887** *Gentl. Mag.* Feb. 135 Those vessels went armed, too, as befitted the majesty of the bunting under which old Dance had gloriously licked Johnny Crapeau. **1915** G. Parker *Money Master* xxiii. 323 Yet, raging as he was, and ready to take the Johnny Crapaud..by the throat, he was not yet sure that Jean Jacques was not armed. **1919** W. A. Fraser *Bulldog Carney* 70 Them Johnnie Crapeaus from Quebec. **1965** G. R. Stevens *Incompleat Canadian* 142 'Johnny Crapeau', esteemed alike for his good temper..and ebullient spirits. **1848** Johnny Head-in-Air [see *head *sb.* 39*]. **1913** C. Mackenzie *Sinister St.* I. i. ii. 23 Much of his morning walk was passed in a dream... Nanny used to jeer at him, calling him Little Johnny Head-in-Air. **1929** C. Day Lewis *Transitional Poem* i. 14 What happier place For Johnny Head-in-Air, Who never would hear Time mumbling at the base? **1942** [see *head *sb.* 39*]. **1958** B. Nichols *Sweet & Twenties* xvi. 225 Yet he was not a Johnny-Head-in-the-Air; he had a shrewd grasp of the contemporary scene. **1870** G. M. Hopkins *Jrnl.* 4 Apr. (1959) 198 The following Irish expressions..Johnny Magoreys (seeds of the hip). **1922** Joyce *Ulysses* 172 Poisonous berries. Johnny Magories. **1815** (*title*) The military adventures of Johnny Newcome, with an account of his campaigns on the Peninsula and in Pall Mall. **1818** 'A. Burton' (*title*) The adventures of Johnny Newcome in the navy; a poem, in four cantos. **1833** *United Service Jrnl.* XI. i. 59 On first landing at Port Royal, a Johnny Newcome (as all strangers are there called)..will at once imagine himself transported into the community of Bedlamites. **1839** Barham *Ingol. Leg.* (1840) 1st Ser. 308 Now to young 'Johnny Newcome' she seems to confine hers, Neglecting the poor little dear out at dry-nurse. **1865** J. H. A. Bone *Petroleum & Petroleum Wells* 40 The Johnny Newcomes had to fight their way back to the bar, and deposit seventy-five cents for the bit of blue pasteboard. **1961** F. H. Burgess *Dict. Sailing* 125 *Johnny Newcome*,..a new hand or a landsman, unused to the sea. **1896** Johnny-on-the-spot [see *case *sb.*¹ 1 d]. **1902** 'D. Dix' *Fables of Elite* 43 Because they had never Tried it, they thought they were Johnny on the Spot, and could do anything. **1914** Wodehouse *Man Upstairs* 102 'I must be close at hand. I must be—what's your expression?—'Johnny-on-the-spot.' **1916** *Daily Colonist* (Victoria, B.C.) 23 July 7/5 We have our eyes on some splendid billets, and will be 'johnny on the spot' when the present residents move out. **1944** E. Bennett-Bremner *Front-Line Airline* (1945) viii. 50 Qantas Empire Airways have been called upon to conduct searches for missing aircraft, and it was only natural, therefore, that being 'Johnny-on-the-spot' they should be asked to join in when aircraft went missing. **1961** J. McCabe *Mr. Laurel & Mr. Hardy* (1962) ii. 57 Suddenly they needed a fat boy for a comedy sequence. He was Johnny-on-the-spot and he was hired at $5 a day. **1968** *Globe & Mail* (Toronto) 13 Jan. 42/7 He..was johnny-on-the-spot to rap in Mike Walton's rebound. **1865** *Nation* (N.Y.) I. 584 They said he was a Johnny Reb. **1948** Johnny Reb [see *grey, gray *sb.* 1 c]. **1919** *Mr. Punch's Hist. Gt. War* 24 Now it is the turn of 'Johnny Turk', who has had his knock on the Suez Canal. **1972** E. Ambler *Levanter* ii. 22 'Johnny Turk is a gentleman,' he used to say. **1933** Johnny Woodser [see *Jimmy 4].
4. (With lower-case initial.) = *John 1 d. *slang* (chiefly *U.S.*).
1932 [see *John 1 d]. **1934** J. O'Hara *Appointment in Samarra* (1935) iv. 98 Kitty Hofman came in the johnny. **1946** C. S. Archer *China Servant* ii. 32 He left a smell of stale tobacco where-ever he went. Even in the johnny. **1969** L. Greenbaum *Out of Shape* (1970) xviii. 136 You're not going to find me in aisle four next to the toilet paper and the johnny mop. **1971** D. Conover *One Man's Island* 65 Why, oh, why, do little boys (and big ones) rush to a johnny when nature provides opportunity everywhere?
5. johnny-jump-up *N. Amer.*, a name used for several kinds of wild or cultivated pansy or violet; **Johnny penguin** = *Johnny 2 b.
1842 *Knickerbocker* XIX. 115 Mr. Ketchup had now kissed little Chip and stuck a johnny-jump-up in his cap. **1872** E. Eggleston *End of World* i. 12 Julia..was hoeing a bed in which she meant to plant some johnny-jump-ups. **1929** F. A. Pottle *Stretchers* (1930) iii. 48 Under foot the ground teemed with lupine and phlox and those large scentless violets which the natives call 'Johnny-jump-ups'. **1972** T. A. Bulman *Kamloops Cattlemen* xii. 75 Green grass, buttercups and Johnny-jump-ups were much in evidence. **1879** H. N. Moseley *Notes by Naturalist on 'Challenger'* viii. 189 The whole beach of Christmas Harbour [in Kerguelen's Land] was covered with droves of the Johnny Penguin. **1882** *Encycl. Brit.* XIV. 49/1 All parts of the coast [of Kerguelen's Land] and even the lower slopes are covered with penguins of various species, mainly the Johnny penguin..rock-hopper..and king penguin. **1968** Shackleton & Stokes *Birds Atlantic Ocean* 20/2 Affectionately known as 'Johnny' penguin, the gentoo is the only penguin with white on top of its head.

Johnny-cake. Add: **a.** Also *W. Indies*, a scone or dumpling (cf. quot. 1831 in Dict.). (Earlier and later examples.)
1739 *S. Carolina Gaz.* 22 Dec. 4/2 (Advt.) (Th. Suppl.), New Iron Plates to cook Johnny Cakes or gridel bread on. **1893** in C. Sullivan *Jamaica Cookery Bk.* **1952** S. Selvon *Brighter Sun* vi. 109 Over there, nastiness and poverty, a tin cup of weak tea and a johnny cake or a roti. **1972** J. Hewitt *N.Y. Times Heritage Cook Bk.* 96 Massachusetts johnnycakes usually have wheat flour and corn meal in them.

Johnson (dʒɒ·nsən). A common surname, used in *low slang* to designate: **a.** The penis. Also *Jim Johnson*.
1863 W. B. Cheadle *Jrnl. Trip across Canada* (1931) 108 Neck frozen. Face ditto; thighs ditto; Johnson ditto, & sphincter vesicae partially paralyzed. **1972** *Screw* 12 June 10/3 So I went to take my turn with the hopes of somehow getting my *Jim Johnson* wet.
b. A man who is kept by a prostitute or prostitutes, a ponce.
1960 'J. Ashford' *Counsel for Defence* v. 66 You got Legs off... He was a Johnson.

Johnson bar (dʒɒ·nsən bāɪ). *U.S.* [Origin unknown.] A long heavy lever used to reverse the motion of a steam locomotive. Also *transf.* (see quot. 1971).
1930 *Railroad Man's Mag.* II. 471/1 *Johnson bar*, reverse lever on a locomotive. **1943** *Railroad Mag.* XXXIV. 58/1 He jerked two short ones from his chime, eased the Johnson bar down to the first corner and dropped his left hand on the throttle. **1959** P. Ransome-Wallis *Conc. Encycl. World Railway Locomotives* 508 *Johnson bar*, reversing lever. **1971** M. Tak *Truck Talk* 93 *Johnson bar*, the emergency brake handle.

Johnsoniana (dʒɒnsōᵘniǣ·na, -ā·nā). [f. *Johnson* (see Johnsonian *a.* and *sb.*) + *-iana.] Sayings, etc., of or about Dr. Samuel Johnson; matters connected with Dr. Johnson.
1776 (*title*) Johnsoniana; or, a collection of Bon Mots, etc., by Dr. Johnson and others. **1909** *Daily Chron.* 14 Sept. 3/2 [Mr. Reade's] latest compilation of Johnsoniana. **1928** *Daily Tel.* 12 June 17/3 Miss Anna Seward's delightful Johnsoniana. **1952** *Bull. John Rylands Libr.* XXXV. 211 (*title*) Johnsoniana from the Bagshawe muniments.

Johnson noise. *Electronics.* [Named after John Bertrand *Johnson* (b. 1887), naturalized U.S. physicist, who published an account of it in 1928.] Electrical noise caused by the random thermal motion of conduction electrons.
1931 *Proc. IRE* XIX. 1407 Finally..an irreducible minimum of noise is encountered, commonly referred to as 'Johnson' or circuit noise. **1947** [see *flicker noise* (*flicker *sb.*³ 4)]. **1958** Condon & Odishaw *Handbk. Physics* v. iii. 31/1 This voltage fluctuation, called Johnson noise, which places a limit on the ultimate sensitivity of amplifiers, provides a method for measuring thermodynamic temperature. **1973** *Physics Bull.* Oct. 596/2 Any electrical resistor exhibits a Johnson noise voltage, with a mean square amplitude proportional to absolute temperature.

johnstrupite (yɒ·nstrupəit, dʒɒ·nstrŭpəit). *Min.* [ad. G. *johnstrupit* (W. C. Brögger 1890, in *Zeitschr. f. Krystallogr. u. Mineral.* XVI. 75), f. the name of J. F. *Johnstrup* (1818–1894), Danish geologist: see -ite¹.] A brownish green hydrous silicate and fluoride of several metals.
1890 *Jrnl. Chem. Soc.* LVIII. 1077 In addition to the minerals previously known, the author [*sc.* Brögger] has found..johnstrupite. **1906** J. P. Iddings *Rock Minerals* ii. 458 Mosandrite and johnstrupite occur in nephelite-syenites and closely related rocks, chiefly in the region of the Langesund fjord, Norway; also in the aphanitic and lava equivalents in other localities. **1959** *Mineral. Abstr.* XIV. 105/1 Calculations of the unit cell contents..give the following formulae:..johnstrupite $(Ca, Na, Ce)_{12}(Ti, Mg, Al, Zr)_3Si_8O_{31}H_2F_{4-5}$... The authors conclude that mosandrite and johnstrupite are varieties of the same species. **1965** *Ibid.* XVII. 287/1 Selective concentration of B..has been observed in specimens of..mosandrite and johnstrupite.

Johnswort. = *St. John's-wort* (John 5).
1753 J. Hempstead *Diary* (1901) 611 Wee pulld up the yellow Blossoms (alias Johnsworth) in the upper end & back Side the Lot. **1819** *Mass. Hist. Soc. Coll.* 2nd Ser. VIII. 170 In July the lover of plants is gratified with..two species of pyrala, the small geranium, several species of hypericum or John's wort. **1874** *Rep. Vermont Board Agric.* II. 390 It is very desirable that the dairyman's pastures should be well stocked with good nutritious grasses, not wild grass of the low boggy pasture, or johnswort, or daisies, but June or blue grass and clover.

‖ **joie de vivre** (ʒwa də vīvr). [Fr., = joy of living.] A feeling of healthy enjoyment of life; exuberance, high spirits.
1889 E. Dowson *Let.* 1 Apr. (1967) 58, I do not suppose 'la joie de vivre' will be revealed to me any more at Limehouse than in the Temple. **1901** 'L. Malet' *Hist. R. Calmady* iii. v. 204 The hungry all-compelling *joie de vivre* which is begotten whensoever youth thus seeks and finds youth. **1907** *Westm. Gaz.* 5 Feb. 4/2 The new *joie de vivre* of motoring. **1917** S. McKenna *Sonia* vi. 263 It was only when the twanging banjos changed to rag-time that the majority of our neighbours sheepishly unbent and put forth an assumption of *joie de vivre*. **1925** A. P. Herbert *Laughing Ann* 82 The simple mind and manly air, Not Brains so much as Breeding, With *joie de vivre* and *savoir faire*, Are constantly succeeding. **1935** *Discovery* Jan. 28/2 Song is an outlet for any of a bird's emotions.., but chiefly *joie de vivre*. **1944** *Burlington Mag.* Sept. 216/2 Many delightful travel impressions have also been noted by him in spirited pen and wash drawings, glistening with *joie-de-vivre*. **1955** *Times* 16 Aug. 5/3 She is excitable, but rather in the sense of joie de vivre. Children play with her, ride on her back, twist her tail and do anything that might aggravate a ferocious dog. **1964** R. Church *Voyage Home* i. 10 Suffering acutely from the effort to ignore the bursts of *joie-de-vivre* from the corridors. **1974** I. Murdoch *Sacred & Profane Love Machine* 96 He had laughed at Sophie's spiteful jokes, of which the sheer *joie de vivre* had seemed to lessen the bite.

join, *sb.* Add: Also *join-up*.
1945 A. Lunn *Third Day* v. 47 If we are expected to accept the hypothesis of a second-century compiler who worked his heterogeneous materials into an artistic unity with..skill we may ask why the join-ups between the 'we' sections and the rest of the *Acts* are so inartistic. **1969** *Sun* 22 July 1/2 On the join-up, Eagle and Columbia ended their separate existences [as space vehicles] and became Apollo-11 again.

join, *v.*¹ **6. b.** Delete † *Obs.* and add later examples.
1882 R. L. Stevenson *New Arabian Nights* I. 95 You join yourselves to persons of condition..for no other purpose than to escape the consequences of your crimes. **1904** S. J. Weyman *Abbess of Vlaye* ii. 41 Had I known of what sort they were to whom I was joining myself.
15. d. *to join up*: to enlist in the army.
1916 'Boyd Cable' *Action Front* 5 Just joined up to get a finger in the fighting? **1922** D. H. Lawrence *England, my England* 40 Egbert went and joined up immediately as a private soldier. **1934** J. T. Farrell *Young Manhood* i. 20 He stepped up to a beefy-faced sergeant... 'We came to join up.' **1943** J. B. Priestley *Daylight on Saturday* xxxviii. 301, I wish you'd leave here—join up or something.
22. *to join the ladies*: to go into the room to which the ladies have retired after dinner.
1848 *Punch* XIV. 204/1 (*caption*) Walter, I think you had better join the ladies. **1921** J. M. Barrie (*title*) Shall we join the ladies? **1956** N. Marsh *Off with his Head* (1957) x. 212 'Join the ladies?' Dr. Otterley suggested, and they did so. **1974** N. Freeling *Dressing of Diamond* 95 And that, thought Richard grinning, will be denounced as 'sexual fascism'... And shall we now join the ladies?
23. *if you can't beat them, join them*: a semi-proverbial assertion applied to a situation where a person crosses to another side or party because he is unable to defeat them by opposition alone.
1955 [see *epistemological *a.*]. **1970** *Times* 26 Nov. 12/3 Gale will also be applying for a ticket to the Parliamentary lobby for himself: 'If you can't beat them, join them.'

joiner, *sb.* Add: **1. b.** One who makes a habit of joining societies, etc. Cf. Join *v.*¹ 15 b. *colloq.* (orig. *U.S.*).
1890 *Ann Arbor* (Mich.) *Reg.* 13 Mar., Ypsilanti is a good place for 'jiners'. There are..100 societies and organizations that a person can join. **1923** R. Macaulay *Told by Idiot* 36 That was Stanley's headlong manner of entering into movements. She was a great and impetuous joiner. **1957** *Times Lit. Suppl.* 15 Nov. 690/2 He is not a 'joiner', that [is] he resists the temptation to cash in on stock fashionable attitudes. **1967** W. Murray *Sweet Ride* vii. 109 I'm not a joiner, you know? I mean, all they do is party, party, party all the time. **1971** *Guardian* 6 July 8/5 You could not see Bogart as a marcher or a singer or a joiner—he was strictly solo. **1973** C. Mullard *Black Brit.* iii. vii. 78 Sadly, this meant that people who were not 'joiners', but who might have been active in race relations, were not considered.

joint, *sb.* Add: **4. c.** *Bookbinding.* The flexible cloth or leather which forms the junction between the spine and the sides of the binding of a book; also the projection along the edge of this junction.
1835 J. Hannett *Bibliopegia* 104 The volume being laid upon the table or press, with the head towards the workman and the upper board open, the guard or false end paper must be removed and all other substances cleared out of the joint with the folder. **1861** *Chambers's Encycl.* II. 226/2 Coming to his hands flat and solid, and with its joints well formed. **1894** *Amer. Dict. Printing & Bookmaking* 313/2 *Joints*, the projection formed in backing to admit the millboards. The leather or cloth placed from the projection to the millboard is called a joint. **1901** D. Cockerell *Bookbinding* xii. 165 Ensure that there is enough leather in the turn-in of the joint to allow the cover to open freely. **1951** L. Town *Bookbinding by Hand* xvi. 203 If a slight amount of moisture is still present the leather will bed itself very neatly into position in the joint. **1967** V. Strauss *Printing Industry* x. 673/2 Backing does not give the book a new back but provides the joints of the back.
11. b. *flat joint, set joint, strong joint*: see the adjs. in Suppl.
14. More generally, a place; a house. (Earlier and later examples.)
1821 *Real Life in Ireland* xvii. 199, I had my education at the boarding-school of Phelim Firebrass..; and when I slipt the joint, and fang'd the arm, he strengthened the sinews. **1877** *Sessions Papers Cent. Criminal Court* 25 Oct. 631 The *joints*—that means the offices where the swindle was carried on—that is a cant word. **1880** *Weekly Times* 4 Jan. 8/3 They soon found him a 'joint' to do... Doing a 'joint' means effecting a burglarious entrance. **1904** *Sun* (N.Y.) 6 Mar. 7/4 Of course, there are no saloons in Kansas; no one would dream of calling them by that name. They are all 'joints', whether the drinks are passed over a polished counter by a white aproned attendant, or shoved through a hole in the wall by a dirty fist. **1905** 'H. McHugh' *You can search Me* 20, I took Clara J. to the St. Regis to dinner... It's a swell joint, all right. **1912** *Maclean's Mag.* Sept. 69/2 Mr. Kelley, to whom few streets were unfamiliar, knew the place exteriorly as a 'Dago joint'. *a* **1922** T. S. Eliot *Waste Land Drafts* (1971) 59 line 50 So the men..thought Of home, and dollars, and the pleasant violin At Marm Brown's joint, and the girls and gin. **1925** H. L. Foster *Trop. Tramp with Tourists* 32 Been to Havana? Good Lord! I've been to that damned joint with Cook, Clarke, Frank, Raymond-Whitcomb, and the American Express. **1934** Wodehouse *Right Ho, Jeeves* xiv. 167 Hanging out in a joint called Kingham Manor. **1935** A. Squire *Sing Sing Doctor* v. 73 In cities the market for brothels, gambling joints, & narcotic dens is better. **1946** F. Sargeson *That Summer* 55, I found a joint that was kept by a Mrs. Clegg. **1953** G. Lamming *In Castle of my Skin* xiv. 282, I see one or two things change round this joint... I mean the village. **1959** 'A. Gilbert' *Death takes Wife* xv. 198 Put down money for a joint you didn't frisk in advance. **1959** 'M. M. Kaye' *House of Shade* iii. 31 He turns the joint upside down until he finds it. **1974** *Listener* 13 June 766/2 A rather pokey, smokey little jazz joint in San Francisco.
b. *Fairground slang.* A stall, tent, etc., in a circus or fair; a concession stand. orig. *U.S.*
1927 *Amer. Speech* June 414 A carnival concession is known as a joint or store. **1931** *Amer. Mercury* Nov. 353/1 *Joint*, any concession stand, or novelty spindle. **1934** P. Allingham *Cheapjack* ii. 16 'You can build up yonder... Where's your joint?'.. It took some time to discover that by a 'joint' he meant my tent. **1968** D. Braithwaite *Fairground Archit.* 22 The tober is a composite of many elements—roundabouts, booths, joints, and transport vehicles. *Ibid.* 68 Joints fall into roughly three categories—round ones, generically termed hoop-las, side-stuff,..and casual stalls for vending.
c. A marijuana cigarette; also, hypodermic equipment used by drug addicts.
1935 A. J. Pollock *Underworld Speaks* 65/1 *Joint*, a complete hypodermic outfit consisting of syringe and needles (ointjay). **1938** *Amer. Speech* XIII. 186/1 *Joint*. The hollow needle, or a substitute... The hypodermic outfit including all accessories... The opium smoker's outfit complete. **1952** G. Mandel *Flee Angry Strangers* 171 You got a couple of joints to take along?.. I know I'll want to get on. Take some pod, Dinch. **1967** M. M. Glatt et al. *Drug Scene* i. 5 In Britain, cannabis is..almost always smoked in the form of a cigarette which is referred to as a smoke, joint or reefer. **1970** *Times* 28 Apr. 10/8 (*caption*) Please fasten your seat belts and extinguish your joints. **1971** *Black Scholar* Sept. 33/1 When the shower stopped he lit two joints and went to the bathroom. **1972** *Daily Tel.* 3 Apr. 8 The making of the joint seemed to be as much a part of the ritual as smoking it.
d. Prison. *U.S. slang.*
1953 W. Burroughs *Junkie* (1972) xii. 124 He said even the best thieves spend most of their time in the joint. **1969** C. F. Burke *God is Beautiful, Man* (1970) 27 She made things so rotten for him that the king threw him in the joint. **1972** J. Wambaugh *Blue Knight* (1973) i. 28 He was a no-good asshole and belonged in the joint.
15. (sense 5) *joint-face, -filling, -surface*; **joint bolt,** a bolt for holding together the two parts of a joint; *spec.* (see quots. *a* 1884, 1964); **joint box,** a junction-box, esp. one designed to be filled with an insulating material; **joint mouse** *Med.* [tr. G. *gelenkmaus* (see quot. 1886)], a loose fragment (as of cartilage or bone) floating in the cavity of a joint; usu. *pl.*; **joint-plane** *Geol.*, a plane in rock in which a joint exists or is liable to form; also, an exposed surface that was once such a plane; **joint set** *Geol.*, a group of parallel joints; **joint system** *Geol.*, a group of two or more intersecting joint sets; **joint-vetch,** a leguminous plant of the genus *Æschynomene*, so called from its jointed seed pods.
1844 H. Stephens *Bk. Farm* II. 75 The top bar..swells out in the middle, where it is perforated for the joint-bolt of the lever. *a* **1884** Knight *Dict. Mech.* Suppl. 514/2 *Joint bolt*, a bolt used for fastening two timbers, one endwise to the other... Used commonly as a fastening for a bed-rail to the bed-post. **1964** J. S. Scott *Dict. Building* 159 *Handrail bolt* or *joint bolt*, a bolt threaded at both ends. **1901** *Chambers's Jrnl.* Dec. 845/2 A new form of joint-box for forming connections. **1907** J. F. C. Snell *Distrib. Electr. Energy* v. 269 Joint boxes must be used to connect lengths of cables or conductors equivalent to, or larger than 7–16 S.W.G. **1929** G. W. Stubbings *Underground Cable Systems* v. 50 A cast-iron joint box must provide a compound-tight chamber for the filling compound. **1966** J. F. Whitfield *Electr. Installations & Regulations* vi. 120 There is some saving in cable but extra man-hours are needed for joint-box connections. **1912** *Jrnl. Geol.* XX. 76 The cliff face is in many parts composed of projecting and re-entrant angles formed by the joint faces of large area meeting in obtuse angles. **1931** *Times Lit. Suppl.* 11 June 466/2 A block of which the bounding joint-faces slope downwards and towards each other will be squeezed upwards. **1961** *Amer. Jrnl. Sci.* CCLIX. 502 The convergence of plumose and radial structures toward the center of the joint face strongly suggests that joints are fractures initiated at a point. **1916** F. H. Lahee *Field Geol.* viii. 223 By their shape and relations to the surface of unconformity, these joint fillings may indicate something of the original character and arrangement of the fractures in which they were formed. **1965** G. J. Williams *Econ. Geol. N.Z.* xiv. 230/1 Soft white

veinlets and joint-fillings of crystalline laumontite are abundant in greywackes throughout New Zealand. **1886** H. Marsh *Dis. Joints* xv. 185 On account of the manner in which these bodies change their site, and slip out of reach, the Germans have suggestively called them 'joint-mice' (*gelenk-mäuse*). **1920** R. Stockman *Rheumatism & Arthritis* ix. 115 'Joint-mice', if present, can be felt, and slip from under the fingers in a very characteristic fashion. **1952** E. F. Traut *Rheumatic Dis.* ix. 194 A detached portion of the internal meniscus constitutes a loose body or joint mouse. **1961** R. D. Baker *Essent. Path.* xxi. 560 (*caption*) Aching, weakness and locking, inability to extend left knee, begun 11 years previously, following which joint mice were removed on two occasions. **1855** J. Phillips *Man. Geol.* ii. 44 The cleavage and joint planes in these beds are not parallel to the general cleavage. **1905** J. Geikie *Struct. & Field Geol.* x. 144 Common household coal..is divided by three sets of planes disposed at right angles to each other: namely (a) planes of bedding, and (b) and (c) joint-planes. **1944** A. Holmes *Princ. Physical Geol.* vi. 76 The joint pattern may also control the course of rivers, the joint planes themselves commonly forming the walls of steep-sided gorges and canyons. **1970** R. J. Small *Study of Landforms* iv. 122 The spectacular cliffs on the east side of the A'Chir ridge coincide with near-vertical joint-planes. **1931** C. M. Nevin *Princ. Struct. Geol.* v. 144 That it [sc. tension] is the effective stress which caused the actual break has not been proven for dip and strike joint sets. **1942** *Bull. Geol. Soc. Amer.* LIII. 392 The relationships of the joint sets to other structures indicate their age. **1965** F. C. Badgley *Struct. & Tectonic Princ.* iv. 117/1 A close relationship exists between these lineament patterns and regional joint sets. **1943** *Bull. Geol. Soc. Amer.* LIII. 396 On most of the more planar joint surfaces feather-like or flamelike markings..are found. **1961** *Amer. Jrnl. Sci.* CCLIX. 493 Freshly exposed joint surfaces commonly are marked with faint ridges that form plume-like or radial patterns. **1973** E. E. Wahlstrom *Tunneling in Rock* v. 102 Joint surfaces that are smooth..commonly display slickensides and contain crushed materials. **1929** C. R. Longwell *Pirsson's Textbk. Geol.* (ed. 3) I. xii. 315 In many places there are two prominent sets of joints, approximately at right angles to each other and each very nearly vertical. Such a combination of two or more intersecting sets constitutes a joint system. **1952** Von Engeln & Caster *Geology* xiv. 191 Joint systems are commonly much more distinctively and conspicuously developed in sedimentary strata than in other classes of rocks. **1973** E. E. Wahlstrom *Tunneling in Rock* v. 99 Regional geological studies often reveal a systematic geometrical relationship between joint systems and faults and folds. **1829** J. C. Loudon *Encycl. Plants* 1284 *Anthrodebium*..Joint-Vetch. **1884** W. Miller *Dict. Eng. Names Plants* 156/1 *Æschynomene..hispida.* Sensitive Joint-Vetch.

joint, *a.* Add: **2.** (Further examples.)

1828 P. Bingham *Rep. Court Common Pleas* IV. 70 [Addison v. Jarvis 1827]. The Plaintiff and Defendant..must..be both considered as joint tort feasors, and the present action is nothing but an attempt by one tort feasor to recover contribution from another. **1886** [see Tortfeasor]. **1908** J. C. Miles in E. Jenks *Digest Eng. Civil Law* ii. i. 337 Persons are joint tort-feasors when one aids, counsels, or joins the other in the commission of a tort. **1920** H. Crane *Let.* 28 Jan. (1965) 32, I am engaged at Mencken and Nathan..the joint authors. **1925** J. A. Holden *Bookman's Guide* 69 Joint author, a person who writes a book in collaboration with one or more associates. **1935** *Act* 25 & 26 *Geo. V* c. 30 § 6 Any tort-feasor liable in respect of..damage may recover contribution from any other tort-feasor who is..liable in respect of the same damage, whether as a joint tort-feasor or otherwise. **1957** J. G. Fleming *Law of Torts* xxviii. 744 The common law..does attach some significance to the distinction between 'joint' and 'several' tortfeasors. **1967** *Anglo-Amer. Catal. Rules: Brit. Text* 267 *Joint author*, a person who collaborates with one or more associates to produce a work in which the contribution of each is not separable from that of the others. **1971** *Mod. Law Rev.* XXXIV. vi. 674 The accessory after the fact has become, perhaps more appropriately, something more like the joint-tortfeasor.

b. *joint family*, a type of extended family in which married children share the family home, living under the authority of the head of the family. Also *attrib.*

1876 W. K. Sullivan in *Encycl. Brit.* V. 799/2 Beside the 'joint and undivided family' there was another kind of family which we might call the 'joint family'. This was a partnership composed of three or four members of a sept whose individual wealth was not sufficient to qualify each of them to be an aire, but whose joint wealth qualified one of the co-partners as head of the joint family to be one. **1889** C. N. Starcke *Primitive Family* ii. i. 139 Polyandry belongs to the category of facts which have to do with the ordinary family communism, and especially with the joint family group. **1937** *Jrnl. R. Anthrop. Inst.* LXVII. 137 Members of the owner's joint-family and lineage. **1953** J. H. Weakland in Mead & Métraux *Study of Culture at Distance* x. v. 423 Dissent among the brothers of a Chinese joint family is concealed from the public. **1957** *Encycl. Brit.* XII. 235/2 The striking feature of Hindu society and Hindu law is the joint family. It is the form, no doubt, in which the Aryan patriarchal family has survived. **1968** G. D. Mitchell *Dict. Sociol.* 77 Extended family is sometimes used not merely to include but as a synonym for joint family. It is more useful to restrict this last term to a form of family which has a number of distinctive characteristics: co-residence, commensality and often some common family cult. **1971** *Illustr. Weekly India* 11 Apr. 23/2 (*caption*) The house was full of Tagore relatives who all lived together as a traditional Bengali joint-family.

3. (Further examples.)

1778 *N.Y. Laws* 27 Mar., The joint Committee..[shall] canvas and estimate the votes. **1869** *Bradshaw's Railway Manual* XXI. 80 The companies should enter into a joint purse agreement with the Irish North Western. **1884** G. C. Brodrick *Hist. Univ. Oxf.* xviii. 213 A vote of censure on Dr. Hampden..was defeated in Convocation by the Proctors' joint-veto. **1909** Webster, Joint sitting. **1936** T. S. Eliot *Coll. Poems 1909–1935* 138 The fetchers and joint-makers and smiths Have appointed a joint committee to protest against the reduction of orders. **1937** *Discovery* May 139/2 The formal invitation of the Indian Science Congress Association for the British Association to send a party to hold a joint session in India. **1942** N. Balchin *Darkness falls from Air* x. 184 You'll want some money... Use the joint account. **1952** H. Nicolson *King George V* ix. 137 It was agreed..that if..an irreconcilable conflict arose between the House of Commons and the House of Lords, the matter should be settled by a Joint Sitting of both Houses. **1964** W. Duff *Indian Hist. Brit. Columbia* I. 73 Of great and growing importance in recent years has been the development of 'joint' or 'integrated' school. **1969** H. I. Ansoff *Corporate Strategy* ii. 16 Joint-venture opportunities. **1966** *Rep. Comm. Inquiry Univ. Oxf.* I. 83 A statutory 'joint' committee of Council, the General Board and the Conference of Colleges. **1967** T. Lupton *Managem. & Social Sci.* iii. 67 The term Joint Consultation is usually used to describe the formal machinery through which the managers and the managed, in a firm..discuss their common problems. **1972** *Wodehouse Pearls, Girls, & Monty Bodkin* iv. 50 He and his wife have a joint account, and he can't draw a cheque without her approval. **1973** *Oxf. Univ. Gaz.* 16 May 866/1 A universal system of joint appointments is in my view unnecessary.

b. *joint denial*: the negation of each of two or more stated propositions; 'neither..nor..'.

1940 W. V. Quine *Math. Logic* i. 45 A joint denial..e.g. 'Neither is Jones away nor is Smith ill',..is true just in case its components are both false. *Ibid.* 49 The possibility of denial, conjunction, and alternation in terms of joint denial was first pointed out by Sheffer in 1913. **1954** I. M. Copi *Symbolic Logic* viii. 256 The other operator which suffices for a functionally complete logic is that of 'joint denial'. **1960** G. Kemeny *Philos. of Math.* ii. 40 All truth functions..can be introduced by definition if we start either (a) with the single notion of alternative denial..., or else (b) with the single notion of joint denial.

joint, *v.* Add: **1. d.** (Earlier U.S. example.)

1835 *Niles' Reg.* IX. 36/1 The power is given by one or two horses, which with a man and a boy can dress and joint..these staves necessary for one hundred barrels.

4. *intr.* To form joints.

1772 *Carroll Papers* in *Maryland Hist. Mag.* (1919) XIV. 287, I am apprehensive it will be too thick and joint if the weather proves warme. **1904** *Topeka* (*Kansas*) *Daily Capital* 1 June 8 Wheat has not done well, though it is jointing now.

jointed, *a.* Add: **1.** (Examples in *Geol.*: cf. Joint *sb.* 5.)

1821 J. Macculloch *Geol. Classification of Rocks* viii. 129 In a few instances, from the extreme shortness of the prisms, the columnar passes to a tabular, or a lamellar and jointed structure. **1865** D. T. Ansted *Gt. Stone Bk.* xi. 139 The harder kinds of sand-rock are generally jointed. **1968** R. W. Fairbridge *Encycl. Geomorphol.* 489/2 The contrast between granite that is widely jointed and that which is closely jointed and generally shattered is easily seen in many cases.

3. *Bot.* Having or appearing to have joints; separating readily at the joints; as a specific vernacular name.

1597 Jointed Glasswort [see Glasswort]. **1793** T. Martyn *Lang. Bot.* s.v., Jointed..Applied to the root..to the stem or culm..to the pod, siliqua or loment, when one knot is placed above the top of another. **1821** S. F. Gray *Nat. Arrangement Brit. Plants* II. 160 Leaves ..knotty, jointed, or smooth. **1839** J. Lindley *Sch. Bot.* 4 If a stem is swelled at the part where the leaves grow, and capable of being snapped across, or apparently so, it is called *articulated* or *jointed*, as in Stellaria Holostea, and Grasses. **1850** W. H. Harvey *Man. Brit. Algæ* 52 Sea weeds... Jointed..Polysiphonia etc. **1967** C. P. Pettman *Africanderisms* 254 Jointed cactus, *Opuntia pusilla*, a dangerous weed. **1971** J. E. Lousley *Flora Isles of Scilly* 268 [*Juncus*] *articulatus* L. Jointed Rush. Native.

jointer², **1.** *a.* (Further examples.)

1885 J. Richards *Wood-Working Machinery* 147 The first and leading tools are bench planes, a set of which should consist of one 26-inch jointer..one 24-inch jointer ..one 22-inch foreplane [etc.]. **1900** C. G. Wheeler *Woodworking for Beginners* xv. 448 The jointer is more accurate for making a surface level and true, or for shooting the edges of boards. **1953** W. Coventry *Woodwork Tools* ii. 48 The trying plane..will be used for finishing the jointing, unless there is much jointing to do in hardwood, when it will pay to follow with a steel panel and jointer plane. **1958** G. H. Love *Theory & Pract. Woodwork* ii. 28 Working on the assumption that the longer the sole, the longer the plane required to true it, we have the jointer, a plane which is even longer than the trying plane. **1959** *Handyman & Home Mechanic* I. 36/2 Steel trying planes, often known as jointers, and are similar to steel jack planes except in length.

joint-grass. (U.S. examples.)

1851 W. G. Simms *Partisan* 55 Rebellion grows like joint-grass when it once takes root. **1894** J. M. Coulter *Bot. W. Texas* III. 499 Joint grass... Moist places throughout Texas and across the continent.

jointless, *a.* Add: **b.** In one piece; without a seam or joint of any kind.

1909 *Prosp. Rubber-Tanned Leather Co.* 2 May, Automobile tyres in seamless and jointless bands. **1921** *Dict. Occup. Terms* (1927) § 571 'Composition' or jointless floors.

joint stock, joint-stock. Add: **2.** *attrib.* *joint-stock bank*, *-company* (earlier and later examples). Also *joint-stock-companyship*.

1776 Joint-stock company [see Company *sb.* 7]. **1843** *Ainsworth's Mag.* IV. 110 It is a pleasant privilege of the honourable corps..that a sort of joint-stock-companyship prevails among them. **1853** Mrs. Gaskell *Cranford* xiii. 225 Her poor opinion of joint-stock banks in general. **1964** Gould & Kolb *Dict. Social Sci.* 350/1 The modern business corporation or joint stock company originated in the 18th century. **1969** *Lebende Sprachen* XIV. 97/2 Clearing arrangements have been agreed with the joint stock banks, so that money can be readily transferred from a giro account to a bank account. **1972** *Accountant* 5 Oct. 413/1 Sweden's national pension fund, it has been recommended, should be free to invest approximately 1 per cent of its resources in selected joint-stock companies.

joint-worm. 2. Substitute for def.: The larval form of several species of chalcid flies belonging to the genus *Harmolita* (or *Tetramesa*), which form galls near joints on grain stems, causing them to bend. (Examples.)

1852 *Southern Planter* (Richmond, Virginia) July 197/2 Joint worm has injured White Flint most, Early Purple Straw next, and Mediterranean least. **1863** A. Fitch in *Cultivator* VIII. 321/1, I first observe, lying upon the infested stalk, the insect to which you allude in your letter, as perhaps having been hatched from the joint-worm. **1882** T. W. Harris *Treat. Insects Injurious to Vegetation* (ed. 2) 443 The ravages of the joint-worm in the wheat-fields of Virginia are said to have been first observed in Albemarle county, about four or five years ago. **1882** *Illinois Entomol. Rep.* XI. 63 We have obtained another specimen of this species from a gall in a stalk, produced evidently by the regular joint-worm. **1918** *U.S. Dept. Agric. Farmers' Bull.* No. 1006. 2 This joint-worm can be controlled in Virginia, Tennessee, and Kentucky by plowing down wheat stubble deeply after harvest. **1972** Swan & Papp *Common Insects N. Amer.* xxi. 537 The Rye Jointworm, *Tetramesa secale*, occurs in eastern U.S., westward to Utah and Alberta.

Jokari (dʒǫkā·ri). The proprietary name of a game played with bat and ball.

The name is registered for a number of games, etc. (see 1953).

1953 *N.Y. Times* 2 Aug. 43/1. (Advt.) Jokari requires little space. You use paddles and a ball that's fastened to the Jokari center control box on a long rubber cable. **1960** 'A. Garve' *Golden Deed* xxiv. 181 A tall, plump man..was playing Jokari with an equally plump youth on the grass outside the door... Mellanby waited till the ball had come to rest. **1962** R. W. Aldiss *Primal Urge* iii. 34 That body creature who..played Jokari from a sitting position. **1963** L. Fleming *On H.M. Secret Service* ii. 22 Two golden girls in exciting bikinis packed up the game of Jokari they had been so decoratively playing. **1973** *Trade Marks Jrnl.* 4 Apr. 664/1 Jokari... Games (other than ordinary playing cards), playthings, gymnastic and sporting articles (other than clothing), and parts..of all the aforesaid goods. R. & J. Travis Limited, Fenbright Works,..Berkhamsted, Hertfordshire; manufacturers.

joke, *sb.* Add: **1.** *practical joke* (earlier and later examples); *a joke is a joke*: a joke is not to be taken seriously (freq. with the implication that the matter referred to is too serious for jokes); *joke over*: the joke is finished (usu. implying that the speaker is not amused by the words or behaviour of the person addressed).

1804 M. Wilmot *Let.* 4 June in *Russ. Jrnls.* (1934) i. 104 Such are the practical Jokes of the Great. **1824** J. Hogg *Private Mem. Justified Sinner* 356 Some of your practical jokes on strangers or honest folks. **1837** Dickens *Sk. Boz* 2nd Ser. 125 A joke's a joke; and even practical jests are very capital in their way, if you can only get the other party to see the fun. **1838** T. C. Haliburton *Clockmaker* 2nd Ser. xvii. 253 A joke is a joke, but that's no joke. **1849** [see Practical a. 1]. **1892** [see *bear-fight*]. **1930** R. Gore-Browne *By Way of Confession* vi. 52 'Goes too far, 'e does!' agreed an elder with a righteous look. **1935** *Punch* 10 Dec. 626/2 'A joke's a joke'... 'And politics is politics.' The resort came from a skinny-necked man. **1942** [see *boo v.*]. **1953** E. O'Neill *Long Day's Journey* (1956) i. 25 No, Mary, a joke is a joke, but—. **1961** Partridge *Dict. Slang Suppl.* 1153/2 Joke over! ..when do we laugh? **1965** G. M. Wilson *Devil's Shal* ix. 112 'Telephone!' He laughed again. 'All right, joke over,' said Lovick, nettled. **1967** V. Williams *Laughter* 21 It's nice to find a joke about a crack on the head. **1970** *Times Lit. Suppl.* 180 All right. Joke over. Now just what is it you think you're up to? **1970** 'J. Lucas' *Nk. Fearfully Hot Night* 180 All right. He was always interested in Raft. **1969** 'B. Mather' *Midnight* Highlight beasts, practical jokes and all the fun of the dormitory.

5. *b. black joke* (see quot. 1796). Also *coal-black joke*. *slang.*

The c 1710 example in Dict. should be dated not earlier than 1734 (*Amer. Speech* (1961) XXXVIII. 152).

1729 C. Coffey *Beggar's Wedding* I. iv. 27 (*heading*) Hunter with Musick. Air X. Coal-black Joak. **1748** Smollett *R. Random* II. lviii. 186 He whistled one part and hummed another of Black Joke. **1796** Grose *Dict. Vulgar T.* (ed. 3) *Black joke*, a popular tune to a song, having for the burden, 'Her black joke and belly so white:' figuratively the black joke signifies the monosyllable. **1808** S. W. Ryley *Itinerant* I. iv. 91 A blind fiddler, mounted on a three footed stool, rasped away very seriously the black Joke. c **1835** [in Dict., sense 1]. **1970** P. O'Brian *Master & Commander* ix. 252 He decided..to sit it out until the drum beat to quarters,..humming the Black Joke.

4. *joke book, shop.*

1951 J. Steinbeck *Burning Bright* 23 His malformed wisdom, his pool-hall, locker-room, joke-book wisdom.

1973 T. Tobin *Lett. of George Ade* 174 This rare pamphlet is in the Ford joke book tradition. **1947** K. P. Kempton *Short Story* 189 At a joke shop in town he had bought one of those spark-plug bombs with the idea of hitching it to a friend's jalopy. **1951** J. M. Fraser *Psychol.* iii. 28 Any 'joke-shop' will furnish examples of similar bits of apparatus designed to suggest a misleading total situation. **1959** V. C. Clinton-Baddeley *Death's Bright Dart* 122 Sinister in the silence gleamed the window of the Joke Shop. **1973** E. Page *Fortnight by Sea* ix. 101 One of those joke shops, you know, rubber poached eggs, exploding cigars.

joke, *v.* Add: **1. b.** *you are* (or *have got to be*) *joking*, etc.: in phrases indicative of incredulity. Cf. also *have v. 7 d.

1907 G. B. Shaw *John Bull's Other Island* ii. 36 Youre joking, Mr. Keegan: I'm sure yar. **1967** Partridge *Dict. Slang Suppl.* 1164/2 You're, *be*, *etc. joking*, *of course*. A c.p. of modified optimism. **1967** N. Lucas *C.I.D.* xiii. 195 'Norman, would you be interested in meeting Charlie de Silva?' 'Probably,' I replied, 'but where do I have to go, Paris, New York?..' 'You're joking. He's skint.' **1968** M. Stewart *Wind off Small Isles* i. 26 'If I've got to turn on sand—' 'You could reverse up.' 'You've got to be joking.' **1972** *Times* 29 Sept. 15/5 'We can consider ourselves fortunate in the character of the Prime Minister.' In the parlance of the day, you must be joking, mate.

joker. Add: **1.** (Earlier and later examples of *practical joker*.)

1810 G. Colman *Random Rec.* II. vi. 205 It convinced me that he was no practical joker. **1899** Beerbohm *More* 70 Severe of Licensed practical jokers. **1926** A. Conan Doyle *Hist. Spiritualism* I. vii. 153 The contemptible crew of practical jokers and ill-natured researchers who visited her found her a ready victim. **1950** T. S. Eliot *Cocktail Party* ii. 104 Are you a devil Or merely a lunatic practical joker?

2. esp. *Austral.* and *N.Z.*

1868 *Auckland Punch* No. 2. 7/1 If Louis Napoleon could be prevailed to bring over his Zouaves, and join the mounted marines, together with some Indian jokers. **1881** H. Nisbet *Bail Up!* x. 92 The driver of the coach was the joker that wanted to charge me a shilling. **1895** [see *stake* *v.* 5 b]. **1941** *Coast to Coast 1941* 97 We'd get tired of tramping around the mines listening to well-fed jokers say 'No!' **1949** [see *cobber* *v.* 6 b]. **1951** *Landfall* V. 21 There are about a dozen brush in the place and about five times as many jokers. **1963** *Australian Post* 14 Mar. 51/2 Mr Pingleton turned out to be a tall, rather thin, pleasant-looking joker with big feet. **1970** N.Z. *Listener* 12 Oct. 15/1 Their benevolent gaff's even harder for a decent-spirited joker to accept.

3. b. Also in fig. phr. *joker in the pack*: a person whose behaviour is unexpected or unpredictable.

1963 *Times* 25 Feb. (Canada Suppl.) p. ii/2 Mr. Caouette as a joker in the pack; they guide the future of the country's 26 federal seats reflects a morbid situation. **1973** G. Sims *Hunters Point* vi. 47 Fred Wheeler may be the joker in the pack. He might have got Dave involved in something wild. **1974** *Times* 9 May 4/1 Although the Government has a majority of four votes over the Conservatives, there are 37 'jokers in the pack'.

4. A clause unobtrusively inserted in a legislative enactment and affecting its operation in a way not immediately apparent. Also *transf.* of a clause, etc., in a contract, administrative order, etc., which frustrates its intention or puts one of the parties at a disadvantage; *fig.*, a drawback, a 'snag', a trick. *U.S.*

1904 *N.Y. Even. Post* 11 May 1 There are all nervous over the possibility that there may be a hitherto unperceived joker in the present bill. **1906** *Ibid.* 30 Apr. 6 The Malby 'joker' to the Adirondack Reserve bill. **1914** S. H. Adams *Clarion* 241 Even her single friend engaged the joker in the contract. **1928** *Daily Express* 17 July 8/1 The surtax was slipped into the Finance Act of 1927 very much as a 'joker' is occasionally insinuated into an American Tariff Act—that is to say, surreptitiously, without anybody except those in the know being aware of the significance of what was happening. **1935** A. J. Pollock *Underworld Speaks* 63/1 Joker (*nigger in the woodpile*), hidden meaning or clause. **1951** *Pol. Sci. Q.* laws, investigations, vice elements, court connections, political appointments, newspaper articles, speeches or contracts arranged for..so that the looters in on the in will be financially benefited. **1952** E. Paul *Narrow St.* xiv. 200 Early in 1926 Painlevé introduced a bill to reorganize the French army from thirty-two to twenty divisions. To this was attached a joker, increasing the term of universal compulsory military service from one year to eighteen months. **1947** *Harper's Mag.* Nov. 441/1 The order also contains this crucial joker: The charges shall be stated as specifically and completely as security conditions permit. **1953** *Congress. Rec.* XCIX. App. A2592/2 A postal rate increase bill..had within it a 'joker' which seriously affected the nonprofit publications of organized labor in all parts of America.

jokery. Delete † and add later example.

1970 *Daily Tel.* 21 May 6/5, I could find very little to interest one laugh..although it has all the ingredients of current jokery.

jokey: see *joky a.

jokiness (dʒōᵘ·kinės). [f. Joky *a.* + -ness.] A joking style or manner.

1869 D. G. Rossetti *Let.* 26 Aug. (1965) II. 720 His letters remind one by their ponderous jokiness of Holman Hunt. **1966** *Punch* 26 Jan. 138/3 Very short chapters and much coy jokiness about methods of fiction add to the Shandean charm. **1971** *Daily Tel.* (Colour Suppl.) 8 Jan. 6/2 The jokiness at a first night of an amateur theatrical company. **1974** *Times* 11 May 9/8 The self-conscious jokiness and triviality that is the bane of newscasting on British television.

joking, *vbl. sb.* Add: (Earlier examples of *practical joking*.) Phr. *joking apart* (Apart *adv.* 5 b).

1745 S. Croker *Let.* 18 July in D. Garrick *Private Corr.* (1831) I. 35 But joking apart, I beg till you come that we may consult together. **1841** J. Ellitson *Let.* 4 Aug. in Dickens *Lett.* (1969) II. 345 Joking apart, you make me ashamed with your over fondness of the milk of kindness. **1853** Mrs. Gaskell *Cranford* xi. 201 The captain of the school in the art of practical joking. **1860** Practical joking [see Brit. *sb.* 9]. **1918** C. Mackenzie *Early Life Sylvia Scarlett* II. i. 256 No, joking apart, I think it would be a great fun. **1931** G. 'Trevor' *Murder at School* xii. 234 Joking apart, you had some ingenious ideas. **1973** *Times* 17 Nov. 11/8 Joking apart, this is the kind of show difficult to write about without..spoiling the jokes.

b. *joking relationship* (Anthropol.), a relationship of familiarity between specific persons which is sanctioned in certain tribal groups.

1920 R. H. Lowie *Primitive Soc.* (1921) v. 95 Of a licensed familiarity with the Crow and Hidatsa. **1933** E. E. Evans-Pritchard *Ess. Social Anthropol.* (1962) vii. 151 One way in which intimacy and equality are expressed is a joking relationship (that of blood-brothers) is by each publicly insulting the other, a custom commonly described by chronologists as a 'joking relationship'. *Ibid.* 152 A 'joking relationship' may grow up between two clans. **1958** A. R. Radcliffe-Brown *Method in Social Anthropol.* i. v. 119 The expression of opposition between the societies may take various forms. One is the institution to which anthropologists have given the not very satisfactory name of 'the joking relationship'. **1964** Gould & Kolb *Dict. Social Sci.* 356/1 A joking relationship is a relationship between two persons (sometimes between two groups) in which one is by custom permitted (and in some cases obliged) to tease or make fun of the other, who must take no offence... Obscenity.. and the taking of property are common forms.

jokist. (Earlier example.)

1873 'Mark Twain' & Warner *Gilded Age* xxxvi. 331 And here is The Jokist's Own Treasury.

joky, *a.* Add: Freq. jokey. Also, subject to jokes, ridiculous. (Later examples.)

1964 'N. Blake' *Sad Variety* viii. 149 'Leake,' I said, 'your mind's as jokey as your clothes.' **1970** *Homes & Gardens* May 145/1 Despite the jokey reputation that middle-class British hotels enjoy, they compare very well indeed for comfort with their European and US counterparts. **1972** *Times Lit. Suppl.* 1 Sept. 1014/4 Crops up again in less jokey contexts. **1973** *Listener* 4 Oct. 461/1 The jokey and toy-like elements..so many contexts. **1974** *Times Lit. Suppl.* 11 Oct. 1142/4 The use of anachronism..is now a joky, selfconscious device.

‖ **jol** (dʒǫl). Also **djöl, jäl, jaul, jawl, johl, jöl.** [Arab.] The local name (often with capital initial) for a barren, much dissected, limestone plateau south of Wadi Hadhramaut in the Arabian peninsula; also applied (often with capital initial) to different regions within this plateau, and (with lower-case initial) to the individual blocks of tableland into which the whole plateau is dissected.

By some writers the plateau itself is designated by the plural form.

1904 D. G. Hogarth *Penetration of Arabia* ii. ix. 217 The guides conducted Hirsch through the cultivated coastal district..up to a bare down-land, 'Jol' rising to six thousand five hundred feet. *Ibid.* 219 At the head of this [sc. a valley] he found himself once more on the downs of the Jol, a wild, arid, broken country. **1921** *Handbk. Arabia* (Admiralty, Naval Intelligence Div.) I. vii. 218 Physically the Hadhramaut may be divided into four main horizontal belts:..(b) a broad belt of downs or plateaux (*jöl*) diversified by a few isolated peaks. **1932** Van der Meulen & Von Wissmann *Hadramaut* iii. 23 Our way now led over the *djöl* as far as Wâdi Dô'an. The *djöl* consists of vast table-lands of reddish-brown limestone. *Ibid.* xvi. 210 This *djöl* differed little in character from that which we had crossed on our way to Wadi Dô'an. **1936** F. Stark *Southern Gates Arabia* ix. 87 The Jöl has usually been dismissed by travellers as a piece of flat featureless, a plateau where heat and cold are alike unbearable. **1940** — *Winter in Arabia* iii. 14 Our lorry ..began to climb long broken ridges that lead to the tilted plateau of the jöl. *Ibid.*, Jöl—waterless steppe plateau. **1945** *Antiquity* XIX. 189, I expect that these stone-age people lived very much the life that the Beduins of the northern jols of the Hadhramaut on the edge of the Rub'al Khali live today. **1946** *Western Arabia & Red Sea* (Geogr. Handbk. Ser. B.R. 527, Admiralty, Naval Intelligence Div.) ii. 32 To the north the land rises to a 3,000 feet above the sea and the desert, but to the south altitudes of 6,000 feet and more are attained. These barren *jöl* are dissected into detached blocks of tableland by an extraordinarily intricate network of canyons and valleys. **1958** *Geogr. Jrnl.* CXXIV. 165 This is Wadi Hadramaut..which is enclosed by the precipices of the limestone strata of a wide and barren table land, the Jaul. **1961** *U.S. Board on Geogr. Names, Gazetteer no. 54: Arabian Peninsula* p. vi, Jawl..plateau, plain. **1966** J. Lunt *Barren Rocks Aden* ix. 139 We were now on the *johl*, as the Arabs call the mountainous plateau which lies between the coast and the Wadi Hadhramaut, and between the Wadi Hadhramaut and the Empty Quarter. **1966** W. C. Brice *S.-W. Asia* xii. 253 Further inland, in the districts known as the Jols, the limestone surface..is scored with a maze of ravines.

‖ **jolie laide** (ʒoli lęd). [Fr., fem. sing. of *joli* pretty + *laid* ugly.] A woman or girl who is attractive in spite of not being pretty. Occas. *joli laid* (*m.*), applied to a man.

1894 Lady Monkswell *Diary* 12 May in E. C. F. Collier *Victorian Diarist* (1944) I. 245 Mrs. Flower is a 'jolie laide' of about 50, rather charming. **1949** N. Mitford *Love in Cold Climate* i. viii. 87 One was a beauty or a *jolie-laide* and that was that. **1960** D. Holman-Hunt *My Grandmothers & I* i. 6 The lazy young ladies who came to stay were called *jolies laides*. **1961** *Times* 10 Oct. 16/1 The arrival in the house..of a jolie-laide innocent looking for the Y.W.C.A. **1966** *Observer* 20 Mar. 25/4 He has the *joli-laid* face of a boxer. **1972** *Times* 22 June 12/5 Sexual teases like jolie-laide Israeli Dvora. **1973** R. Rendell *Some lie & Some Die* xi. 95 He had an ugly attractive face, *joli laid*.

jolley (dʒǫ·li). *Pottery.* Also **jolly.** [Of unknown origin.] A variety of jigger (see Jigger *sb.*¹ 5 a). So **jo-lleying, jo-llying** *vbl. sb.*, the act of using a jolley; **jo-llied** *ppl. a.*, manufactured by jolley; **jo-llier¹, jollyer,** one who makes pottery by means of a jolley.

1881 *Instructions to Census Clerks* (1885) 88 Earthenware manufacture..Jollier. Jolly Maker. **1891** *Pop. Sci. Monthly* Dec. 168 A 'jolly' (consisting of a plaster-mould, wheel), the thing is a revolving mold [etc.]. **1901** W. P. Rix tr. *Bourry's Treat. Ceramic Industries* iv. 182 In jollying, the body is compressed, and forced to become stretched out by means of a profile, having the outline of the article to be made. *Ibid.* 183 The heads of jolleys are arranged so as to receive the moulds, the placing and removal of which must be done very rapidly. **1913** A. B. Searle *Clays & Clay Products* xv. 118 As plates and saucers are always shaped with the profile..the profile forms the 'bottom' of the plate. The machines used for this work are known as *jiggers* and *jolleys*. **1921** *Dict. Occup. Terms* (1927) § 165 *Bowl maker*,..a jollier who makes bowls by jollying process. *Ibid.*, *Cup maker*, a jollier who makes cups on a single or double cup jolley. **1931** J. B. Priestley *Eng. Journey* 213 [It Bennett] had been either a master potter on one hand, or a 'thrower' or 'potter' on the other. *Ibid.* 220 If you work at the stuff [sc. potter's clay] inside a mould, you are, I believe, either 'jollying' or 'jiggering'. **1951** *Electronic Engin.* XXIII. 337 Automatic jollying machines have now been developed for the production of plates, cups and saucers. **1966** H. Powell *Beginner's Bk. Pott.* I. i. 14 A good investment for a factory would cut-out lay to the jigger, or jolley (no one seems to know which is which). **1971** N. French *Industr. Ceramics: Tableware* iii. 26 The extreme precision and expense of mould-making normally offsets the advantages of jollying.

jo-llier². *U.S.* [f. Jolly *v.* 2 c.] One who 'jollies' or flatters others; a jovial or sociable person.

1896 *Ade Artie* ix. 78 He's one o' the biggest jolliers that ever come over the hills. **1901** Merwin & Webster *Calumet 'K'* i. 23 Oh, he's a good-looking young chap... He's a great jollier. **1904** *Sun* 155 Nobody, he said, 'is a great jollier'. **1905** *N.Y. Even. Post* 12 Oct. 2 He was talkative, and as the attendants say, 'quite a jollier'. **1946** W. A. White *Autobiog.* 443, I was errand boy, peace-maker, jollier, fixer, horse-trader.

jollo (dʒǫ·lō). *Austral. slang.* [app. f. Joll(i-fication + Austral. termination *-o*¹.] A party, esp. one at which liquor is drunk. Cf. *jolly *sb.*⁴

1934 *Bulletin* (Sydney) 25 July 46/3 Most..came to the wedding jollo to drink the beer and smoke the bride's new fawn coat-and-skirt. **1966** Baker *Austral. Lang.* (ed. 2) xi. 230 Australians have a fair selection of terms to describe drinking and drinking bouts, such as..jollo [etc.].

Jollof, Jolof, var. *Wolof.

jollop (dʒǫ·ləp), *sb.*² *slang.* [See Jalap *sb.*] **a.** A purgative, a medicine. **b.** Strong liquor, or a drink of this.

1920 *Conferre Pier* Aug. 250 We may imagine them [sc. smugglers c 1820] stowing away their precious booty in caches, then a jollop of brandy all round. **1955** D. Niland *Shiralee* 146 He nutted out some jollop for her dog. **1961** C. Willock *Death in Covert* ii. 32 'Tell 'em up at the house to bring out the jollop.' The keeper uttered this in a tone that made it quite clear that he considered serving refreshment something completely outside his duties. **1966** 'L. Lane' *ABZ of Scouse* 57 If yer don't wallop yer jollop yer'll get ther beezers in yer belly.

jollopy, var. *jalopy.

jolly, *a.* Add: **5. b.** *jolly roger*: see Roger² 4. **13. b.** Also *ironically*.

1916 Galsworthy *Sheaf* i. 13 'Jolly for my new coat!' I said.

C. **jolly-tail** *Austral.*, a small fresh-water fish of the genus *Galaxias*, esp. *G. attenuatus*.

1892 P. L. Simmonds *Commercial Dict. Trade Products* (rev. ed.) Suppl. 463/2 *Jolly-tail*, a small fresh-water fish of Australia..highly esteemed as a delicacy for the table. There are several species. **1898** E. E. Morris *Austral Eng.* 224/1 Jolly-tail, *n.* a Tasmanian name for the larger

variety of the fish *Galaxias attenuatus*, Jenyns, and other species of *Galaxias*. **1951** T. C. Roughley *Fish Austral.* (rev. ed.) 156 It [sc. whitebait] is composed mainly of the young fry of small fish called minnows or jollytails (*Galaxias attenuatus*). **1965** *Austral. Encycl.* V. 141/2 Jollytails, small freshwater fishes (*Galaxias*) rarely attaining a length of 8 inches.

Jolly (dʒǫ·li, yǫ·li), *sb.*³ The name of P. von *Jolly* (1809–1884), German physicist, used *attrib.* († or in the possessive) to denote a balance invented by him, used esp. in determining the specific gravities of minerals, in which the elongation of a helical spring when a body is hung on it indicates the weight of the body.

1882 A. Geikie *Text-bk. Geol.* ii. ii. 93 Jolly's spring balance is a simple and serviceable instrument. **1906** J. P. Iddings *Rock Minerals* ii. 92 By means of a Jolly balance or spring..the specific gravity of a crystal may be found with approximate accuracy. **1964** J. Sinkankas *Mineral. for Amateurs* vii. 187 In the homemade Jolly balance shown.., a specimen attached by thread to the lower hook of the spring carries it down a certain distance which can be measured.

jolly, *sb.*⁴ *colloq.* Short for Jollification; so, a thrill of enjoyment or excitement, as in phr. *to get one's jollies*. Also *jollyo, jolly-up*. Cf. *jollo.

1905 in *Dialect Notes* (1908) III. 325 Justice Brewer's jolly. **1907** F. H. Burnett *Shuttle* xxiii. 237 If you can give 'em a jolly and make 'em laugh, they'll listen. **1920** *Spectator* 4 Dec. 740/1 Every age must be allowed an occasional 'jolly'. **1921** Galsworthy *To Let* ii. iv. 154 Come and have a 'jolly' with us. **1924** M. Newman *Consummation* iv. xv. 197 Troops fed to the teeth with.. relentless routine, broken only by the occasional horror of a 'show' (what Bossy called a 'jolly'). **1927** *Amer. Speech* II. 277/1 *Jolly-up*, informal dance. **1932** E. Waugh *Black Mischief* iv. 140 Why can't the silly mutt go off home and leave us to have a jolly up. **1957** M. Shulman *Rally round the Flag, Boys!* (1958) ix. 100 If she wasn't so goddam busy..then he wouldn't be thinking about getting his jollies elsewhere! **1962** in Wentworth & Flexner *Dict. Amer. Slang* Suppl. (1967) 686/1 The owner of this place gets his jollies by walking around most of the day in a Sioux war bonnet. **1963** *Daily Tel.* 3 Oct. 20/2 Many parents rejoice inwardly at..their sons being immured under a régime of spartan rigour, but if the compensating jolly-up in the holidays is carried beyond a certain point the whole exercise becomes..unrealistic. **1966** D. Francis *Flying Finish* v. 64 You couldn't just go up alone for an afternoon's jolly in an airliner. **1968** *Surfer Mag.* Jan. 18/2 The announcer acted like this is where all of the surfers go after dark to get their jollies. **1970** V. C. Clinton-Baddeley *No Case for Police* iv. 78, I had gone for a jollyo to one of the rather swell hotels. **1971** 'W. Haggard' *Bitter Harvest* xiv. 145 It would be a splendid wedding, the sort of big jolly Charles Russell enjoyed.

jolly, *v.* Add: **2. a.** (Later example.)

1924 H. de Selincourt *Cricket Match* vii. 219 Their main effort seems not to be..jollied out for a depressing total.

c. For '*U.S.*' read 'orig. *U.S.*' (Earlier and later examples.) Also with *impers.* obj.

1890 H. Palmer *Stories of Base Ball Field* 81, I jollied him along as strong as I could. **1901** *Daily Colonist* (Victoria, B.C.) 19 Oct. 2/5 It is now asserted that the message was an artful device of the astute Mr. Bratnober, who wanted to keep Mr. Macdonald 'jollied up' until the time should come to dispense with his services. **1908** G. H. Lorimer *Jack Spurlock* v. 91 Our customers expect the boys to have a little snap and jolly their grub along. **1929** E. Wilson *I thought of Daisy* iv. 253, I thought that I'd jolly him along a little. **1938** E. Waugh *Scoop* ii. i. 124 We've got to..make contacts, dig up some news sources, jolly up the locals a bit. **1943** J. B. Priestley *Daylight on Saturday* xi. 68 He had seen himself arranging sports and entertainments..and generally jollying everybody along. **1958** 'A. Gilbert' *Death against Clock* 10 Her clothes were varying shades of brown, jollied up with a purple scarf. **1959** *Times* 28 July 11/2 Mr. Maurice Browning..jollies things along briskly enough, with enthusiastic aid from the audience. **1973** H. McCloy *Change of Heart* iv. 40 He protested, he argued, he even tried to jolly them along. They only became bolder.

jolt, *v.* Add: **2. b.** To startle, to surprise. Cf. *jolt *sb.* 2 b.

1872 'Mark Twain' *Roughing It* (1873) ii. 27 She would launch a slap at him that would have jolted a cow. **1875** —— in *Atlantic Monthly* Feb. 219, I said I didn't know. 'Don't know?' His manner jolted me. **1919** H. Crane *Let.* 27 Dec. (1965) 28 Yes, the last word will jolt you. **1972** *Guardian* 23 Dec. 17/2 Those mega-million pound takeover bids which jolt the City.

jolt, *sb.* Add: **2. b.** *fig.* A surprise; a shock which disturbs one's mental composure.

1884 'Mark Twain' *Huck. Finn* v. 30, I was scared now,..but in a minute I see I was mistaken. That is, after the first jolt,..he being so unexpected. **1905** D. G. Phillips *Plum Tree* 3 I'd like to give him a jolt. **1924** H. T. Lowe-Porter tr. *Mann's Buddenbrooks* I. vi. 208 Oh, no! I know they gave you a jolt yesterday—a very, very stimulating jolt.

c. A blow in boxing. Also in phr. *to pass a jolt*, to deliver a blow. Also *fig.*

1908 S. E. White *Riverman* xvii. 160 Murphy blocked, ducked, and kept away, occasionally delivering a jolt as opportunity offered. **1912** [see *hay-maker 4]. **1916** C. J. Dennis *Songs Sentimental Bloke* 124 *Jolt, to pass a*, to deliver a short, sharp blow. **1950** J. Dempsey *Championship Fighting* vii. 26 Best of all the punches is the 'stepping straight jolt'. **1954** F. C. Avis *Boxing Reference Dict.* 60 *Jolt*, a kind of jab punch that brings up short an advancing opponent.

4. a. A drink of liquor. *slang* (chiefly *U.S.*).

1904 *McClure's Mag.* Mar. 560/2, I stopped at a blacksmith's shop..and had my arm dressed and a big jolt of whiskey. **1920** F. Scott Fitzgerald *This Side of Paradise* (1921) ii. iv. 252 We'll take you to some secluded nook and give you a wee jolt of Bourbon. **1935** G. Blunden *No More Reality* xxiii. 344 'Take another jolt, sport,' said Clarrie with a grin. **1957** A. MacNab *Bulls of Iberia* xii. 125 'You've been drinking.'..'I shoved in a couple of jolts on the way here.' **1959** T. Griffith *Waist-High Culture* (1960) 231 Jolts of whiskey or vodka. **1973** R. Thomas *If you can't be Good* (1974) xvi. 145 She took two green plastic glasses... I poured a generous jolt into both of them.

b. A prison sentence. *slang* (orig. *U.S.*).

1912 D. Lowrie *My Life in Prison* ii. 17 A professional 'pete' man had..returned exultingly to the jail with a six-year 'jolt'. **1926** J. Black *You can't Win* xii. 197 He was in good spirits and condition after 'stopping his jolt' in the stir and anxious to start 'rooting'. **1928** [see *book *sb.* 9 d]. **1936** 'D. Hume' *Meet Dragon* ix. 96 They are only too ready to turn King's evidence,..you'd take a very stiff jolt.

c. = *bang *sb.*³; a quantity of a drug in the form of a cigarette, tablet, etc. *slang* (chiefly *U.S.*).

1916 T. Burke *Limehouse Nights* 19 A little later he would take a jolt of opium at the place at the corner of Formosa Street. **1926** J. Black *You can't Win* xii. 162 He wouldn't give us a jolt if we had the horrors... Given a sufficient quantity of hop, no fiend is ever at a loss for a sound reason for taking a jolt of it. **1929** D. Hammett *Dain Curse* (1930) xxi. 233 You can take your jolt in front of me. I won't blush. **1955** *U.S. Senate Hearings* (1956) VIII. 4164 Terms used in the traffic pertaining to the alkaloid morphine are as follows:..jolt,..a dose. **1970** K. Platt *Pushbutton Butterfly* (1971) vi. 58 Her LSD cap would cost about two dollars and fifty cents for the jolt.

5. *attrib.* and *Comb.*, as **jolt ramming** *Founding*, a method of packing the sand around a pattern in which the moulding box, pattern, and sand are repeatedly lifted by machine and allowed to fall; freq. *attrib.*; **jolt-squeeze** *Founding*, simultaneous or successive jolting of a moulding box and 'squeezing' of the sand in it (i.e. application of pressure at the top), as a means of packing the sand around a pattern; usu. *attrib.*

1909 *Iron Age* LXXXIV. 1165/1 Today we have pneumatic jolt-ramming machines in successful service with lifting capacities from 10 to 15 tons. *Ibid.* 1165/2 How to adapt our foundry methods to this new principle of jolt-ramming green sand molds. **1926** *Jrnl. Iron & Steel Inst.* CXIII. 568 The whole of the mould and core are rammed on a Mumford jolt ramming machine. **1950** J. S. Campbell *Casting & Forming Processes* xii. 104 Jolt ramming packs the lower portions of the sand next to the pattern best. **1931** *Jrnl. Iron & Steel Inst.* CXXIII. 602 (*heading*) A novel combination jolt-squeeze moulding machine. **1955** Heine & Rosenthal *Princ. Metal Casting* iv. 53 Match-plate molding using jolt-squeeze machines is perhaps the simplest method of speeding up the molding of small castings. **1971** W. B. Parkes *Clay-Bonded Foundry Sand* viii. 235 For most moulds, all that is needed is a simultaneous jolt-squeeze of a few seconds.

jolter, *sb.*¹ (Further example.)

1843 *Knickerbocker* XXI. 39 The traveller has but to express a wish to visit a distant plantation, and his.. luggage is placed in the donkeyed jolter.

jo·ltiness. [f. Jolty *a.* + -ness.] The condition of being jolty.

1891 'L. Malet' *Wages of Sin* iv. iii. 78 Oh! the joltiness of this conversational road. **1905** *Westm. Gaz.* 9 May 4/2 The existing motor-'buses..with their perpetual pulling up, their joltiness, and malodorousness.

Joly (dʒǫ·li). The name of John *Joly* (1857–1933), Irish physicist, used *attrib.* and in the possessive to denote apparatus and processes devised by him, as **Joly('s) method, process** *Photogr.*, a process of colour photography using a screen (*Joly screen*) ruled with a repeating sequence of adjacent orange, blue-green, and blue lines placed in front of the plate in the camera and a screen similarly ruled with red, green, and blue lines placed in contact with the transparency when it is being projected; **Joly's steam calorimeter**, a device for determining the specific heat of a substance by measuring the weight of steam that condenses on a known mass of the substance in raising its temperature to that of the steam.

1894 T. Preston *Theory of Heat* v. 236 (*heading*) Joly's steam calorimeter. **1902** *Encycl. Brit.* XXXI. 686/1 (*heading*) Joly's process. **1906** E. J. Wall tr. *König's Natural-Color Photogr.* 20 In 1904 Professor Wood improved his process and applied it to positives obtained by the Joly process... Gratings were ruled with three sets of lines in bands corresponding to the width of the red, green, and blue lines of the Joly screen. **1909** S. E. Bottomley *Photogr. in Princ. & Pract.* xxvi. 207 At the time the Joly process was put on the market, a suitable panchromatic emulsion was not obtainable, and to this cause alone must be ascribed its non-success. **1938** W. J. Sparrow *Heat* vii. 139 Joly's steam calorimeter may be used (*a*) to find the specific heat of a solid.., (*b*) to determine the specific heat of a gas at constant volume. **1940** *Chambers's Techn. Dict.* 469/1 *Joly screen*, a colour mosaic consisting of ruled lines. **1958** H. Baines *Sci. of Photogr.* xx. 246 Fig. 6 illustrates the reproduction of a red colour by the Joly and similar methods... The Joly experiments showed the possibility of colour photography using a mosaic screen. **1963** A. E. E. McKenzie *Second Course Heat* ii. 32 Joly's steam calorimeter measures heat in terms of the mass of steam condensed on a body.

Jomon (dʒōᵘ·mǫn). Also **Jōmon.** [Jap.] Used *attrib.* in *Jomon pottery* to denote a kind of very early hand-made Japanese pottery; hence applied to the early neolithic or pre-neolithic culture which is characterized by this pottery.

1946 G. B. Sansom *Japan* (rev. ed.) i. 1 Two main types of neolithic culture are distinguished. One is known as the Jōmon ('rope-pattern') type, because the pottery which characterises it was made by coiling or has a coil as conventional decoration. **1957** *Encycl. Brit.* XII. 901/1 The Jomon is classified as Neolithic. *Ibid.*, The Jomon ('cord-pattern') cultures have handmade pottery with patterns formed by rolling or pressing a cord-wrapped stick..into the still wet clay. **1960** B. Leach *Potter in Japan* vii. 156 The earliest Jomon pots are long pre-Ainu and are estimated to be upwards of 6,000 years old. **1970** J. Kirkup *Japan behind Fan* 27 Rare Haniwa and Jomon figures.

Jon (dʒǫn), abbrev. *Jonathan 3.

1931 *Daily Tel.* 21 May 5/1 Jons 10s 6d–12s 6d.

Jonathan. Add: **3.** A red-skinned variety of dessert apple, first introduced in the United States.

1831 J. Buel in *Gardeners' Mag.* VII. 239 Besides the Newton, I know of none except the Aesopus, the Pownal, the White and the New, or Jonathan. **1831** *Catal. Fruits in Garden of Hort. Soc.* (ed. 2) 18 Apples... Jonathan. **1845** A. J. Downing *Fruits Amer.* 113 The Jonathan is a very beautiful dessert apple... The original tree of this new sort is growing on the farm of Mr. Philip Rick, of Kingston, New York. **1879** *Chicago Tribune* 3 May 10/3 Our best winter apples are..Jeniton, Jonathan, Red Canada, Wythe. **1924** *Glasgow Herald* 23 Dec. 5 In 1922 the price realised for Jonathans did not pay the cost of packing. **1932** [see *delicious *a.* 2 b]. **1950** Smock & Neubert *Apples & Apple Products* ix. 233 Jonathan spot occurs primarily on Jonathan apples. **1959** [see *delicious *a.* 2 b]. **1963** A. Lubbock *Austral. Roundabout* 68 The crimson-cheeked Jonathan apples.

Jones (dʒōᵘnz). **1.** One of the commonest British family names, used esp. in the plural to designate one's neighbours or social equals, as in phrase *to keep up with the Joneses*: see *keep *v.* 57 j; now usu. with allusion to this phrase.

1879 E. J. Simmons *Mem. Station Master* (1974) vi. 83 There is a considerable amount of importance attached to this public place of meeting—the railway station. The Jones's who don't associate with the Robinsons, meet there. Mr Jones would not like the station master to touch his cap to the Robinsons, and pass him without notice. **1932** E. C. Harwood *Cause & Control of Business Cycle* x. 124 Why..does John Doe choose to speculate on margin?.. An ages-old desire to get something for nothing; keeping up with the Joneses; [etc.]. **1955** J. Cannan *Long Shadows* vii. 106 On a higher scale you're simply outdoing the Joneses. **1959** 'L. Gibb' (*title*) The Joneses: how to keep up with them. **1959** *Times* 18 Sept. 14/5 By addressing Mr. and Mrs. Jones through The Reader's Digest, you thus exert a selling leverage [etc.]..that's Jonesmanship. **1962** *Friend* 29 June 801/1 What we decide will be affected by whether we think morals are something stuck on to people by outside pressure (law, custom, tradition, and 'what the Joneses do') or whether they are an outcome of our nature. **1967** *Guardian* 3 Jan. 4/7, I pretended not to hear..but in the end you follow the Jones. **1967** G. Household *Courtesy of Death* 38 Dunton..avoided any nonsense of competing with Joneses.

2. *slang.* A drug addict's habit.

1968 *Sun Mag.* (Baltimore) 13 Oct. 19/4 Soon you're out to keep from getting the Jones. **1970** C. Major *Dict. Afro-Amer. Slang* 71 *Jones*, a fixation; drug habit; compulsive attachment. **1971** *Black World* Mar. 54/1, I don't have a long jones. I ain't been on it too long. **1971** E. E. Landy *Underground Dict.* 113 *Jones*, the habit of a drug addict—eg. *His jones is heavy.* **1974** *Publishers Weekly* 12 Aug. 50/2 Knows the reality of Detroit's heroin sub-culture as few of those who are not 'Jones men' do. ('Jones' stands for both heroin and the habit.)

Jonesian (dʒōᵘ·nziăn), *a.* [f. name of Daniel *Jones* (1881–1967), English phonetician + -ian.] Used to designate the phonetic system or the system of classifying phonemes adopted by Daniel Jones. Hence as *sb.*, an adherent or follower of Jones.

1951 *Language* XXVII. 334 Neither 'pure' phonetics nor Jonesian phonemics stands the ghost of a chance. **1953** A. Martinet in *Anthropol. Today* 578 A Jonesian origin of the Bloomfieldian phoneme might account for the opinion. **1964** Y. R. Chao in D. Abercrombie et al. *Daniel Jones* 40 Linguists trained under the Jonesian cardinal vowels. **1965** *Language* XLI. 308 Linguists who are neither Firthians nor Jonesians. **1969** *Ibid.* XLV. 109 A reaction against Jonesian and Bloomfieldian phonemics.

jong¹ (yǫŋ). *S. Afr.* [Afrikaans *jong* a coloured servant, f. Du. *jong* a young person.] Formerly, a young male slave; now a coloured male servant; also used as a familiar term of address to any person.

1615 W. Peyton in R. Raven-Hart *Before Van Riebeeck* (1967) 71 One of the Condemned menn (with twoe other of the Peppercornes companye) caryed awaye her pinasse the next night, at which instant twoe yeongers of my Shipps companye allsoe conserted to carry awaye my boat. **1812** A. Plumptre tr. *Lichtenstein's Trav. S. Afr.* I. viii. 119 A Hottentot..takes it extremely amiss if he is addressed by the words *Pay* or *Jonge*, as the slaves are. **1846** J. C. Brown tr. *Arbousset & Daumas's Narr. Tour N.-E. of Cape Good Hope* xxiii. 253 Tied his *jong*, or young bushman slave, to the wheel of his wagon, where he was severely flogged. **1886** G. A. Farini *Through Kalahari Desert* xvii. 279 These slaves were called 'yungs' or 'boys'. **1912** *East London* (Cape Province) *Dispatch* 13 Feb. 3 (Pettman), Presently a couple of jongs came along with dainty cigarettes in their mouths. **1926** E. Lewis *Mantis* i. v. 88 And then, another time (so Dan wrote to his father), he'd make you feel a piccanin yourself, a proper little *jong* that'd never been beyond the dorp. **1939** S. Cloete *Watch for Dawn* v. 65 Where is your whip, jong? **1953** J. Collin-Smith *Locusts & Wild Honey* ii. iv. 155, I..shouted into the house for the 'jong', the coloured houseboy, to bring me coffee. **1956** A. G. McRae *Hill called Grazing* vii. 55 Now look here, *jong*, I've got all these pigs yere on my farm, man. *Ibid.*, Now, if I can't get a permit to sell them, man *jong*, I'll have to drive them up on to the main road. **1960** D. Lytton *Goddam White Man* i. 10 The sergeant at the police station says—'Watch it jong.' **1973** *Sunday Tribune* (Durban) 1 Apr. 20 Jong, I've had enough of this!

jong² (dʒǫŋ). Also **dzong.** [ad. Tibetan *rdzoṅ* fortress.] A Tibetan building (also, a territorial and administrative division) constituting a prefecture, freq. also serving as a fortress, a monastery, or both. Hence **dzo·ngpōn, jo·ngpen, jong-pon,** a prefect; **jong-nyer,** a sub-prefect.

[**1888** *Encycl. Brit.* XXIII. 340/2 A valley to the south-east..contains the towns of Pena-jong and Gyangtse-jong.] **1904** *Times* 11 May 5/1 The Tibetans have strongly fortified the jong. **1904** *Daily Chron.* 21 May 5/4 The collection and equipment of the local levies are conducted by various Jongpen. **1911** *Encycl. Brit.* XXVI. 922/1 The administrative subdivisions of the Lhasa country.. are called *jong*, or 'prefecture', each of which is under the rule of two *jong-pon*. *Ibid.*, There are 123 sub-prefectures under *jong-nyer*. **1921** *Glasgow Herald* 13 July 9 The Jongpen of the district rode out to meet us with a few followers. **1938** F. S. Chapman *Lhasa* iii. 33 The *Dzongpön* (fort commander) of Phari came in for tea. *Ibid.* iv. 49 The dzong is..on the summit of a volcano-like rock six or seven hundred feet high. *Ibid.* v. 83 He is..magistrate, or dzongpön, of a district called Purang. **1960** 'S. Harvester' *Chinese Hammer* xi. 99 Phari Dzong, Hog Fort, a large square grey fortress. **1960** *Times* 9 June 15/6 The castle-monastery (the *dzong*) which rears high above the floor of the wide valley. **1966** *Times* 17 June 13/5 The great *dzong*..rears up on the side of the valley, inward-leaning walls around a keep.

jonga (dʒǫ·ŋgă). Also **jangga.** [f. Doualla-Bakweri (Cameroons) *njaŋga* crayfish.] In Jamaica, a small freshwater prawn, *Macrobrachium jamaicensis*.

See Cassidy & Le Page *Dict. Jamaican Eng.* s.v. *jangga* for oral uses of the word.

1893 C. Sullivan *Jamaica Cookery Book* 4 Jonga Soup. *Ibid.* 8 Jongas are a kind of small cray-fish which are often found in our mountain rivers. **1929** M. W. Beckwith *Black Roadways* 51 Fresh-water shrimps, called *jonga*, from the mountain streams. **1967** *Amer. Speech* XLII. 191 *Jangga*, the river prawn *Macrobrachium jamaicense* [sic]. This word is usually found in print as *jonga*, an unfortunate misrepresentation which has probably led would-be explainers astray. Its actual pronunciation among the folk may always be respelled *jangga*. **1969** 'J. Morris' *Fever Grass* iii. 31 They're even canning *jongas* from the head-waters of the Martha Brae. **1971** E. Jones in J. Figueroa *Caribbean Voices* I. 4 Up in de hills, where de streams are cool, An' mullet an' janga swim in de pool.

joning (dʒōᵘ·niŋ), *vbl. sb.* [Origin unknown.] A name given by American Blacks to a game characterized by the exchange of insults; 'playing the dozens'.

1970 H. E. Roberts *Third Ear* 12/2 *Signifying*,.. language behavior that makes direct or indirect implications of baiting or boasting, the essence of which is making fun of another's appearance, relatives or situation. Variations include joning, playing the dozens. **1971** C. M. Kernan in A. Dundes *Mother Wit* (1973) 312 The verbal insult game known in other regions as..'Joning'. **1972** J. L. Dillard *Black English* vi. 260 There are, of course, some characteristic topics, speech styles, and modes of discourse. One of the more interesting is called *jonin'* by the Washington community—*sounding* or *the dozens* elsewhere.

jonnop (dʒǫ·nəp). *Austral. slang.* [Contraction of *John Hop* (*John 1 c).] A policeman.

1938 X. Herbert *Capricornia* xxiv. 353 But even the jonnops don't treat the Binghis anything like they used to. **1963** A. Lubbock *Austral. Roundabout* 45 He's not a bad sort for a jonnop. **1966** Baker *Austral. Lang.* vii. 142 England was originally responsible for the word cop, which Australia elaborated into the now-obsolete copman and copperman and which, by a process of rhyme, gave us the popular Australianisms john hop, jonnop and hop.

Jonsonian (dʒǫnsōᵘ·niăn), *a.* [f. the name of Ben *Jonson* (?1573–1637), English dramatist + -ian.] Of, pertaining to, or characteristic of Jonson or his works. So **Jonso·nianly** *adv.*, in the manner of Jonson or his works.

1886 J. A. Symonds *Ben Jonson* 153 The most truly Jonsonian of all these places..was the Old Devil Tavern at Temple Bar. **1909** G. G. Greenwood *In re Shakespeare* ii. 76, I here leave the Jonsonian riddle. **1928** C. J. Sisson *Eliz. Dramatists* iv. 43 The Jonsonian comedy of humour. **1931** *Times Lit. Suppl.* 21 May 405/1 But the 'laws of dramatic art' have already ceased to look Aristotelian or Jonsonian. **1947** *Partisan Rev.* Mar.–Apr. 176 They are monsters in the Jonsonian manner; a human trait has been carried in them to the point of inhumanity. **1948** F. R. Leavis *Great Tradition* v. 235 Bounderby..remains Jonsonianly consistent in his last testament and death. **1959** *Times* 26 Nov. 6/5 Mr. Raymond Raikes's production took pleasure in the Jonsonian tumult. **1973** M. R. Booth *Eng. Plays of 19th Cent.* IV. 2 The father.. may be a traditional Jonsonian 'humours' character with a dominant eccentricity.

jonty, var. *jaunty *sb.*

Jonval (ʒo·ṅval). *Engin.* The name of Nicolas Joseph *Jonval*, nineteenth-century Frenchman, used *attrib.* and *absol.* to denote an obsolete kind of reaction turbine (see quots.), patented by him in 1841.

1851 *Jrnl. Franklin Inst.* LII. 419 The advantage of the Jonval Turbine over the over-shot wheel, using the same amount of water, is consequently seven per cent. **1894** W. J. Lineham *Textbk. Mech. Engin.* xi. 724 The Jonval turbine, like the Fourneyron, is a pressure turbine; but while the latter works best above tail water, the Jonval is always drowned or else connected to tail water by a 'suction' tube. **1909** H. Louis *Dressing of Minerals* xii. 491 For lower falls turbines are best employed; a twin Jonval parallel-flow turbine..offers many advantages, amongst which may be reckoned the fact that the turbine itself may be situated 10 to 15 feet above the level of the tailwater. **1957** *Encycl. Brit.* XXII. 580/2 In..the Fourneyron, the guide vanes were inside the runner, and the water flowed outward. This was followed by the Jonval turbine, in which the guide vanes are above the runner and the water flows axially into and through the wheel. Both types are now obsolete.

jook, var. *juke *sb.*

Jordan² (dʒǭ·ɹdăn). The name of a river in Palestine, the crossing of which is used (after Num. 33:51) in pietistic language to symbolize death.

1684 Bunyan *Pilgr.* ii. 222 Mr. Stand-fast..went down to the River..he said..my Foot is fixed upon that, upon which the Feet of the Priests..stood while Israel went over this Jordan. **1707** I. Watts *Hymns & Spiritual Songs* ii. No. lxv. 139 Could we but climb where Moses stood, And view the Landskip o're, Not Jordan's Stream, nor Death's cold Flood Should fright us from the Shore. **1772** W. Williams *Hymn, 'Guide me, O Thou Great Jehovah'*, When I tread the Verge of Jordan. **1786** S. Stennett (*title of hymn*), On Jordan's stormy banks I stand. c **1871** in J. W. Johnson *Bk. Amer. Negro Spirituals* (1925) 63, I look'd over Jordan, an' what did I see... A band of angels comin' after me. a **1890** in Barrère & Leland *Dict. Slang* (1890) II. 368/1 (s.v. *T'other*), And I saw a mighty charret a comin'.., To take us to de odder side of Jordan... Jordan am a hard road to trabble. **1944** D. Van de Voort in B. A. Botkin *Treas. Amer. Folklore* v. 682 Everybody thought that when Wiley's pappy died he'd never cross Jordan because the Hairy Man would be there waiting for him.

Jordan³ (dʒǭ·ɹdăn). *Math.* [Name of Marie Ennemond Camille *Jordan* (1838–1922), French mathematician.] *Jordan curve*, any curve that is topologically equivalent to a circle, i.e. is closed and does not cross itself; so *Jordan('s) (curve) theorem*, the theorem that any Jordan curve in a plane divides the plane into just two distinct regions having the curve as their common boundary.

1900 W. F. Osgood in *Trans. Amer. Math. Soc.* I. 310 By a Jordan curve is meant a curve of the general class of continuous curves without multiple points, considered by Jordan, *Cours d'Analyse*, vol. I, 2d edition, 1893, p. 90. **1919** *Ann. Math.* XXI. 180 (*heading*) A proof of Jordan's theorem about a simple closed curve. **1939** M. H. A. Newman *Elem. Topology of Plane Sets of Points* vi. 132 Theorem 2·3. (Jordan's Theorem for a polygon.) A simple polygon determines two domains, of each of which it is the frontier. **1947** Courant & Robbins *What is Math.?* (ed. 4) v. 246 The Jordan curve theorem is quite simple to prove for the reasonably well-behaved curves, such as polygons or curves with continuously turning tangents, which occur in most important problems. **1965** S. Barr *Exper. Topology* i. 13 A Jordan curve can be drawn on the side of the torus and still divide it into two, but not if it circles, or goes through the hole.

Jordanian (dʒǭɹdē·niăn), *a.* and *sb.* [f. *Jordan* + -ian.] **A.** *adj.* Of or pertaining to the Hashemite kingdom of the Jordan (see *Jordan²), formerly Transjordan. **B.** *sb.* A native or inhabitant of Jordan. Hence **Jo·rdanize** *v. trans.*, to render Jordanian in character; **Jo:rdaniza·tion.**

1950 *Encycl. Brit. Bk. of Year* 371/1 On Aug. 26 the Jordanian minister in London announced that a firm of British irrigation engineers..were now engaged in drafting practical schemes of irrigation for the Jordan valley. **1953** *Daily Tel.* 10 Mar. 6/5 The refugees, sometimes aided by Jordanian..officials, 'infiltrate' into Israel. **1955** *Times* 12 May 11/7 As totally inaccessible to Jordanians as Siberia or Tibet. **1959** *Chambers's Encycl.* XIII. 745/1 There is..a national guard, to which all Jordanians between 20 and 40 years of age are liable to be called for service. **1960** A. Ramati *Rebel against Light* v. 110 Then the Jordanian guns opened fire again, but this time at another part of the city. **1962** *Oxf. Univ. Gaz.* 9 Mar. 776/2 The British School in Jordanian Jerusalem, where a further lecture was given. **1969** *Listener* 6 Mar. 305/1 The dignity, pride, reticence..of the Jordanian and Palestinian Arab. **1970** *Times* 13 May 11/2 Israel hawks have often spoken of the 'Jordanization' of Lebanon. **1972** *Times* 27 June 9/1 They asserted that the main aim of the 'Israeli aggressors' was to 'Jordanize' Lebanon. **1972** *Guardian* 11 Oct. 4/2 Jordanians, like Israelis, wanted.. peace, the newscaster assured viewers.

Jordanite². [f. the proper name *Jordan* (see below) + -ite¹.] A believer in the doctrines of Jordan, a 20th-century Jamaican preacher who found followers in Guyana.

1934 E. Waugh *Ninety-two Days* ii. 43 The Jordanites are one of the many queer sects that flourish among negroes. They derive their name not..from the river, but from a recently deceased Mr. Jordan from Jamaica. **1958** J. Carew *Black Midas* iv. 60 Processions of torch bearing Jordanites passed by the house. **1965** 'Lauchmonen' *Old Thom's Harvest* ix. 113 Brother Polo..was done up in flowing white cotton gown like the Jordanite sect wear.

Jordanon (dʒǭ·ɹdănǫn). [f. the name of Alexis *Jordan* (1814–1897), French botanist.] = *microspecies.

1916 J. P. Lotsy *Evolution by Means of Hybridization* i. 23 Forms from which we know nothing else than that they breed true to type, may consequently not be designated as species but must receive another name; as such I propose *Jordanons*. **1963** Davis & Heywood *Princ. Angiosperm Taxon.* xi. 378 The formation of numbers of morphologically uniform populations, all easily separable from each other and eligible for consideration as distinct taxa..is well known to the practising taxonomist who often refers to the populations as microspecies or Jordanons after the nineteenth-century French botanist who described so many breeding lines as distinct 'true' species.

jordisite (dʒǭ·ɹdisəit, yǭ·ɹd-). *Min.* [ad. G. *jordisit* (F. Cornu 1909, in *Zeitschr. f. Chem. u. Ind. d. Kolloide* IV. 190/2), f. the name of E. F. A. *Jordis* (1868–1917), German chemist: see -ite¹.] A black amorphous sulphide of molybdenum, MoS_2.

1910 *Mineral. Mag.* XV. 423 Jordisite... A black, powdery, colloidal form of molybdenum sulphide. **1951** *Amer. Mineralogist* XXXVI. 614 The present study.. indicates that Cornu was correct in his statement that ilsemannite is derived from a black colloidal molybdenum sulfide, and although Cornu did nothing to establish the name 'jordisite', the occurrence in Oregon should confirm the name as a valid one for the mineraloid. **1966** *Prof. Papers U.S. Geol. Survey* No. 550-B. 120/1 Much of the jordisite occurs in medium-grained to coarse-grained sandstone.

joree (dʒǫrī·). *U.S.* [Echoic, from the call of the bird.] = *ground robin* (*ground *sb.* 18 b).

1884 J. C. Harris *Mingo* 179 We seem to agree, Brother Brannum, like the jay-bird and the joree—one in the tree and t'other on the ground. **1938** M. K. Rawlings *Yearling* iv. 35 Jorees flew from the denseness. **1955** *Sci. News Let.* 23 Apr. 271/2 Other names by which he [sc. the towhee or ground robin] is known include swamp robin, joree, bush-bird and turkey sparrow.

jornada. Add: **2.** Also **journada, -ado.** (Earlier examples of forms.)

1828 in *Missouri Hist. Rev.* (1914) VIII. 190 At 4 p.m. we entered *Jornada*. **1844** J. J. Webb *Adv. Santa Fe Trade* 156 Stopping over for a few hours..to prepare for the journey of fifty miles to the Arkansas, without water, [we] started into the *jornada*. *Ibid.* 119 Whether..to.. travel a longer distance and through two *jornadas*. **1845** J. C. Frémont *Rep. Exploring Expedition* 260 The caravans sometimes continue below to the end of the river, from which there is a very long *jornada* of perhaps sixty miles. **1859** [in Dict.].

‖ **joro** (dʒōᵘ·ro). [Jap. *jorō, jōro*, a prostitute.] In Japan: a prostitute.

1884 [see *hinin]. **1891** A. M. Bacon *Jap. Girls & Women* x. 289 Below the gëisha in respectability stands the jōrō, or licensed prostitute. **1899** J. London *Let.* 22 Feb. (1966) 17 My last posed foto was taken in sailor costume with a Joro girl in Yokohama. **1959** N. Mailer *Advts. for Myself* (1961) 109 The Americans..did not want geisha girls. They wanted a *joro*, a common whore.

‖ **jōruri** (dʒōᵘ·ruri). [Jap., f. the name of Lady *Jōruri*, whose story was one of the most popular subjects for recitation.] **1.** A type of dramatic recitation to musical accompaniment, associated with the Japanese puppet theatre. Also *attrib.* or as *adj.*

1890 B. H. Chamberlain *Things Japanese* 342 The plays given at these [kabuki] theatres originated..partly in marionette dances accompanied by explanatory songs, called *jōruri* or *gidaiyū*. **1911** *Encycl. Brit.* XV. 170/1 The Jōruri is a dramatic ballad, sung or recited to the accompaniment of the *samisen* and in unison with the movements of puppets. **1965** W. Swaan *Jap. Lantern* xv. 175 *Joruri* singing or chanting or recitation derives from an ancient form of metrical story-telling analogous to that of the Greeks and the mediaeval bards.

2. In full, *ningyo-jōruri* [Jap. *ningyo* doll]. The Japanese puppet drama.

1950 *Chambers's Encycl.* VIII. 60/2 The *jōruri* (puppet play) arose from recitations..which were combined in the early 17th century with marionette performances. **1954** F. Bowers *Jap. Theatre* i. 30 The word *Joruri* applies specifically to musical dramas which developed from this style of chanting, and which retain for their accompaniment a combination singer-narrator and one or more samisen players. **1959** E. Ernst *Three Jap. Plays* 36 It is not clear when the puppets were added to jōruri, but it is thought that puppet *jōruri* performances were given in 1596 in the city of Kyoto. **1966** C. J. Dunn *Early Jap. Puppet Drama* 4 The name *bunraku* has widely displaced *jōruri* as the name of the puppet drama.

Joseph. Add: **4.** A violin made by Joseph Guarnieri del Gesù (1698–1744). Cf. *Guarnerius.

1875 G. Hart *Violin* vi. 95 How is this 'Joseph', unaccustomed to elbow his legitimate namesakes in the world of fiddles, to maintain the character he has assumed? **1879** Grove *Dict. Mus.* I. 637/2 The value of a good 'Joseph' now varies from £150 to £400. **1968** *Encycl. Brit.* X. 986/2 It was not until Paganini played on the 'Joseph' that the taste of amateurs turned from the sweetness of the Amati and the Stradivari violins in favour of the more robust tone of the Giuseppe Guarneri.

josephinite (dʒōu·zėfinait, dʒōuzėfī·nait). *Min.* [f. *Josephine*, the name of the county in south-western Oregon where it was first found: see -ite[1].] The terrestrial (as opposed to meteoric) alloy of nickel and iron, having about 67 to 77 per cent of nickel.

1892 W. H. Melville in *Amer. Jrnl. Sci.* CXLIII. 509 (*heading*) Josephinite; a new nickel-iron. *Ibid.* 514 The placer gravel, in which josephinite is found. **1905** *Ibid.* CLXIX. 415 It is seen from a glance at the analyses of the alloys from the five different localities, that there is a certain uniformity in composition, but that they are not a definite compound of iron and nickel is evident, nor would this be expected... It seems unfortunate that so rare a substance should have received three distinct names, awaruite, josephinite, and souesite, and it is urged that awaruite, which has priority, should alone be used. **1950** *Mineral. Mag.* XXIX. 390 The occurrences of native nickel-iron regarded as magmatic secretions from peridotites (awaruite, josephinite, souesite, &c.) are certainly formations of extremely low temperatures.

Josephite[1] (dʒōu·zėfait). [f. the name *Joseph* (see below) + -ite[1].] A member of either of two orders of St. Joseph, the Priests of the Mission of St. Joseph (founded *c* 1640), or a teaching institute founded in 1817 by Constant-Guillaume van Crombrugghe (1789–1865). Also *attrib.* or as *adj.*

1846 in N. French *Works* I. p. lvi, At the end of the street, there is a large building yet occupied as a school, and now held by the Josephites. **1897** Addis & Arnold *Cath. Dict.* (ed. 5) 531/2 *Josephites.* Two communities bear, or have borne, this name. The first was founded by Jacques Cretenet at Lyons... At the Revolution it was suppressed. The second.., founded in 1817 at Grammont in Belgium. **1898** C. H. Bowden *Simple Dict. for Catholics* 18 *Josephites*, a teaching institute..for the education of the commercial and industrial classes. **1973** *Daily Colonist* (Victoria, B.C.) 28 June 43/1 The 50-year-old former Josephite priest was parolled last December.

Josephite[2] (dʒōu·zėfait). Also **Josephine.** [f. the name of St. *Joseph* (1439–1515), Abbot of Volokolamsk, a Russian zealot.] A member of an ascetic and caesaro-papist party formed among Russian Orthodox monks in the sixteenth century. Hence **Josephism**[2], **Josephitism**, the doctrines of this party.

1944 B. H. Sumner *Survey Russ. Hist.* iv. 187 The Josephines won the day in the church council of 1503. **1946** G. P. Fedotov *Russ. Relig. Mind* p. xii, The study of the conflict between the Josephites and the Trans-Volgans led at least one of our scholars to deeper inquiry. **1947** R. M. French tr. *Berdyaev's Russ. Idea* i. 7 Fedotov explains this as due to the fatal influence of 'Josephism' which has distorted the portrait of Christ among the Russian people, so that the Russian people wants to take shelter from the frightful God of Joseph Volotsky behind Mother Earth. **1950** G. P. Fedotov *Treas. Russ. Spirituality* p. xiii, Josephitism degenerated into static ritualism. **1953** K. S. Latourette *Hist. Christianity* xxvii. 618 Probably it was inevitable that the Josephites and the Non-Possessors should clash. *Ibid.* xl. 905 The Josephites..believed in the possession of property by the monasteries and the Church. They were ascetic, approved works of charity, and stressed ritual... The Josephites believed in the maintenance of a close tie between Church and State. **1963** T. Ware *Orthodox Church* vi. 117 Russia needed both the Josephite and the Transvolgian forms of monasticism.

Josephson (dʒōu·zėfsən). *Physics.* The name of Brian David *Josephson* (b. 1940), British physicist, used *attrib.* to denote an effect predicted by him in 1962, whereby an electric current (*Josephson current*) can flow from one superconducting metal to another with no potential difference between them if they are separated by a sufficiently thin layer of an insulator (owing to tunnelling by coherent pairs of electrons), the application of a potential difference causing the current to oscillate with a frequency equal to the voltage multiplied by $2e/h$ (e = the electronic charge, h = Planck's constant). So **Josephson junction,** a metal–insulator–metal junction exhibiting this effect.

1963 Anderson & Rowell in *Physical Rev. Lett.* X. 231/1 There are..four experimental points suggesting that this is indeed the Josephson effect. **1966** *Physical Rev.* CXLI. 366/2 The Josephson junction behaves as a parametric inductance element in which the voltage is proportional to the frequency. **1966** *McGraw-Hill Encycl. Sci. & Technol.* XIII. 301/2 If a very small voltage difference is maintained between the two sides of the barrier..the Josephson current should oscillate. Its frequency should be 4·836 × 10[14] cycles per second per volt of potential difference. **1969** J. E. Mercerau in R. D. Parks *Superconductivity* I. viii. 399 Such structures of two superconductors nearly in contact (weakly connected) are now called Josephson junctions... The device provides a unique opportunity to study the coupling of electromagnetic and quantum waves. **1972** *Physics Bull.* Aug. 457/1 The Josephson effects have led to new types of galvanometer, magnetometer and computer memory elements, to new methods of harmonic mixing of two frequencies and of harmonic generation, and to simple oscillators.

josh, *v.* Add: Also **joss.** For '*U.S. slang*' read '*slang* (orig. *U.S.*)'. (Earlier and later examples.)

1852 *Lantern* (N.Y.) I. 199/2 The squint eyed chap's been jossin' ye. **1921** Wodehouse *Indiscretions of Archie* vi. 54 Are you trying to josh me? **1927** J. Devanny *Old Savage* 30 They jossed him in the usual way about his proverbial amours. **1940** L. A. G. Strong *Sun on Water* 58 Though every now and then she'd remember and clap a hand to her mouth, we'd soon josh her out of it. **1950** C. M. Kornbluth in D. Knight *100 Yrs. Sci. Fiction* (1969) 249 A humorous sergeant, the Mindworm was pleased to note, joshed the loafer out of his temper. **1972** P. H. Kocher *Master of Middle-Earth* (1973) v. 107 When Pippin and Merry are reunited with their comrades.. Gimli joshes them over and over as 'truants' who had to be rescued.

2. *intr.* To indulge in banter or ridicule.

1845 *St. Louis Reveille* 19 Apr. 2/4 Look out in future, and if you must *Josh*, why, give a *private* one. **1887** F. Francis *Saddle & Mocassin* 185 He..liked nothing better than to..chin and josh [*note*, chat and joke] with them in his funereal fashion. **1905** *Amer. Illustr. Mag.* Dec. 214, I was jus' joshin', mother, 'cause I 'spect all your plans are made. **1958** J. G. MacGregor *North-West of 16* xiv. 195 Each, according to his outlook on life, grumbled or whined, laughed or 'joshed'. **1963** *Guardian* 25 Jan. 9/4 The bearded Alberts tootle and joss with goon-like ebullience. **1966** J. Dos Passos *Best Times* (1968) i. 13 I'd huddle in acute misery while he joshed with the waiters.

Hence **jo·sher, jo·shing** *vbl. sb.* and *ppl. a.*, **jo·shingly** *adv.*

1864 in *Ohio Arch. & Hist. Q.* LII. 175 The Bay was rough; thirty minutes out and the boys began to get sick. There was a good deal of joshing. **1899** F. Norris *McTeague* iv. 57 What a josher was this Marcus! Sure, you never could tell what he would do next. **1909** R. A. Wason *Happy Hawkins* xxii. 263 Dick was smilin' now.. an' makin' funny, joshin' remarks. **1921** *Daily Colonist* (Victoria, B.C.) 25 Mar. 4/3 Indifference is blazingly conspicuous by the small audiences that attend pre-election meetings; also the joshing manner in which the electors speak to and about the candidates. **1924** Wodehouse *Bill the Conqueror* xvii. 273 Miss Stryker formally stamped him with the seal of her approval as 'a good kid'. And..it is but a step from being a good kid to being..a great old josher. **1948** V. Palmer *Golconda* xi. 85 It was only Donovan's energy and easy joshing that carried the day with them. **1957** —— *Seedtime* 122 Donovan had said joshingly to the..conductor..'Well, how's this strike of yours going?' **1959** J. Thurber *Years with Ross* ii. 21 The intramural joshing turns up everywhere in the crumbling documents. **1966** H. Marriott *Cariboo Cowboy* viii. 75 This P.G.E., or 'Please Go Easy' as some smart joshers called it, started its way at Squamish on Howe Sound. **1970** B. Spock *Decent & Indecent* 18 There has been a trend in social gatherings to substitute loud joshing and playful insults for conversation. **1972** P. Dickinson *Lizard in Cup* xi. 172 Buck maintained his peculiar brand of joshing bonhomie.

josh (dʒǫʃ), *sb. U.S. slang.* [f. the vb.] A piece of banter or badinage; a good-natured or bantering joke. Also as *adj.*, ridiculous.

1878 F. H. Hart *Sazerac Lying Club* 57 Be there anything in this..or aint it only one of them 'joshes' they get up in the *Reveille* sometimes? **1896** Ade *Artie* iii. 30 That ain't no josh, neither. **1908** G. H. Lorimer *Jack Spurlock* iii. 40 First, I sat there chuckling, but by and by I began to forget the josh end of it I had joined [the union] for, and to remember my own grievances against the house. **1948** *Sat. Rev.* (U.S.) 12 June 19/1 We found him tired-eyed and peaked,..not a man for josh and chatter. **1959** J. Thurber *Years with Ross* ii. 20, I shall spare you the *New Yorker*'s prospectus, drawn up in..1924, except for.. 'There will be a personal mention column... This will contain some josh.'

Joshua[1] (dʒǫ·ʃuă). [f. the Christian name of Sir *Joshua* Reynolds (1723–92), English painter.] In full, *Sir Joshua.* **a.** A portrait painted by Reynolds. **b.** A woman resembling those depicted in the portraits of Reynolds.

[**1866** A. J. Munby *Diary* 6 July in D. Hudson *Munby* (1972) 227 One [girl] with a rose in her bosom..looked like a masterpiece of Sir Joshua.] **1875** G. H. Lewes *Diary* 8 July in Geo. Eliot *Lett.* (1956) VI. 155 Lady Clementina Mitford—quite a Gainsborough or Sir Joshua —charmed me with her beauty. **1972** 'M. Innes' *Open House* xi. 102 It's a poor class of thing to my eye—me 'aving been in places with Joshuas and the like on the walls.

Joshua[2] (dʒǫ·ʃuă). *U.S.* [prob. f. the name of the Old Testament leader of the Israelites, in allusion to the branching shape of the tree, compared with that of Joshua brandishing a spear: see Josh. 8:18.] In full **Joshua palm, tree,** or **yucca.** A small evergreen tree, *Yucca brevifolia*, bearing clustered white flowers and found in western, desert regions.

1867 W. H. Jackson *Diary* 15 Feb. (1959) I. 134 Sage brush is used for cooking and the cactus or Joshuas, as I hear them called, for other fires. **1897** G. B. Sudworth *Nomencl. Arborescent Flora U.S.* 106 *Yucca arborescens*... Joshua Yucca... Common names... The Joshua (Utah). Joshua Tree (Utah, Ariz., N. Mex.). **1946** E. S. Gardner *D.A. Breaks Seal* 3 Joshua palms, thrusting up grotesque spine-covered arms, made the scenery resemble some fantastic reconstruction of life on another planet. **1948** A. Huxley *Ape & Essence* (1949) 12 A strangely gesticulating Joshua-tree, rough-barked, or furred with dry prickles. **1955** *Sci. Amer.* Apr. 69 (*caption*) Joshua trees (Yucca brevifolia) grow among the fantastic rock formations of the Joshua Tree National Monument in southern California east of Los Angeles. These plants are characteristic of high steppe-desert vegetation in the southwestern U.S. **1964** Gleason & Cronquist *Nat. Geogr. Plants* xxii. 408 Some of the most interesting plants of the Sonoran Floristic Province are giant Joshua (*Yucca brevifolia*) [etc.].

joss. Add: Also, (*colloq.*) luck: *bad, good, joss.*

1913 *Chambers's Jrnl.* Aug. 590/2 This comprises a good 'joss' for the voyage, and may be likened to a system of 'blessing'. **1915** Kipling *Fringes of Fleet* 48 Mines are all Joss. You either hit 'em or you don't. **1948** Partridge *Dict. Forces' Slang* 104 Bad joss your leave being stopped.

b. joss-man, a priest of a Chinese religion; *slang*, of any other religion (see also quot. 1948); **joss-pidgin,** a religious ceremony; **joss-pidgin-man,** a minister of religion; **joss-stick** (earlier and later examples).

1845 J. R. Peters *Misc. Remarks upon Chinese* 111 Every one has a shrine and Jos, or representation of one, before which a jos stick is kept continually burning. **1876** C. G. Leland *Pidgin-Eng. Sing-Song* 43 Burnee joss-stick, talkee plitty. **1879** I. L. Bird *Let.* 4 Jan. in *Golden Chersonese* (1883) 49 There is a recess outside each shop, and at dusk the joss-sticks..fill the city with the fragrance of incense. **1886** Yule & Burnell *Hobson-Jobson* 354/1 Joss-house-man or Joss-pidgin-man is a priest, or a missionary. *a* **1889** *Mary Coe* in Barrère & Leland *Dict. Slang* (1889) I. 508/1 Allo tim he make joss-pidgin, Wat you fan-kwei cālly 'ligion. *a* **1889** *The Rebel Pig* in *Ibid.* I. 508/1 When dey talkey pig look all-samee like he joss-pidgin-man. **1913** *Chambers's Jrnl.* Aug. 590/2 A missionary is known as a European 'joss-man'. **1926** M. Leinster *Dew on Leaf* iv. 45 He do joss-pidgin. **1948** Partridge *Dict. Forces' Slang* 104 *Jossman*, a term used on China-side for Plymouth gin... (From the picture of the monk on the bottle.) All holy men are 'jossmen' to the Chinese. **1964** *Navy News* Nov. 12/4, I was watch aboard and tried to get a sub, but no joy. I asked the Jossman if I could go ashore, and he told me to go to... **1970** *Islander* (Victoria, B.C.) 17 May 5/2 The jossman picked up hammer and nails and, with great care, fixed the eyes, exactly 15 feet from the bow [of a junk]. **1974** *Times* 15 Aug. (India Suppl.) p. iv/9 The sandalwood-based joss sticks.. were produced to the tune of 5,000 tonnes last year.

joss[2] (dʒǫs). *dial.*, *Austral.*, and *N.Z.* [Eng. dial., of unknown origin.] = Boss *sb.*[6]

c **1860** in H. Maxwell *Evening Memories* (1932) xviii. 323 Come where the boss is a deuce of a joss, Come to the pub next door! *a* **1876** E. Leigh *Gloss. Words Dial. Cheshire* (1877) 111 *Joss*, a foreman. Used in Macclesfield. **1877** F. Ross et al. *Gloss. Words Holderness* 81/2 *Joss*, a head man; a superior. *a* **1922** H. Lawson in *Penguin Bk. Austral. Ballads* (1964) 149 Must I turn aside from my destined way For a task your Joss would find me? **1948** V. Palmer *Golconda* iv. 28 'Then why don't you go to the Golconda Mining Company?' asked Donovan slyly. 'Tilburg Kloss is there, a big joss from down south.' **1955–6** —— in *Coast to Coast* 36 A big joss among the young bucks and gins.

joss, var. *josh *v.*

josser (dʒǫ·səɹ). *slang.* [f. Joss + -er[1].] **1.** A clergyman or minister of religion, 'padre'. *Austral.*

1887 J. Farrell *How he Died* 22 The reverend josser.. hammering the pulpit. **1889** Barrère & Leland *Dict. Slang* I. 507/1 *Josser*,..a priest... Australian slang designated those who ministered in them [*sc.* joss-houses] *jossers*, and then extended this term it had created to mean ministers of any religion. **1941** Baker *Dict. Austral. Slang* 40 *Josser*, a parson. **1973** G. Rose *Clear Road to Archangel* iii. 35 The old josser, all black robe and beard and upside-down hat and silver cross, addressed himself to me. In German.

2. a. A simpleton; a soft or silly fellow. So, in flippant or contemptuous use, a fellow, an (old) chap (see also quots. 1933 and 1946).

1886 *Broadside Ballad*, 'I took it On' in Farmer & Henley *Slang* (1896) IV. 77/1, I took it on, Of course I was a josser. **1890** *Punch* 22 Feb. 85/2 These 'Equality' jossers would spile it; if arf their reforms they can carry. **1894** Wilkins & Vivian *Green Bay Tree* I. vii. 176 The josser next me, who had won his money. **1901** *Westm. Gaz.* 30 Mar. 8/2 The old josser asked me where the fire was, and I said: 'Jolly well go and find out yourself, you old bounder!' **1902** Strong & Osbourne *Memories of Vailima* (1903) 61 Though he had known and liked Mr. Stevenson all this time, it was only the other day..that it came over him all of a heap—'*he's* the josser that wrote *Treasure Island*'. **1904** 'G. B. Lancaster' *Sons o' Men* 184 There's a big josser at Gatefield—Hunt, I s'pose. He gits what he wants. **1911** G. B. Shaw *Getting Married* 244 *Reginald*: Boxer is rather a fine old josser in his way. **1933** E. Seago *Circus Company* iii. 29 One of their number would speak of a subject not meant for a 'gajo's' ears, to be checked immediately..'Nante palari before the josher [*sic*] cul.' *Ibid.* 294 *Josser*, an outsider. **1936** F. Clune *Roaming round Darling* vi. 53 Meanwhile, a hard-boiled josser, with beard and gun complete, has seated himself on a post to see we didn't get away with his wire fence (for a tow-line). **1938** 'R. Hyde' *Nor Years Condemn* xv. 270 Some josser with a necktie tells 'em he's got a job waiting. **1946** G. Tyrwhitt-Drake *Eng. Circus & Fair Ground* vi. 67 A 'josser' is an amateur, or in fact any one of the public who does not belong to the profession.

b. *attrib.*

1891 *Daily News* 29 June 2/4 Any 'josser' policeman would be enabled..to pry into their show. **1893** *Standard* 29 Jan. 2 (Farmer & Henley), Now suppose we are on the road..and we meet a josser policeman?

‖ **jota** (hō·tă). [Sp.] A Spanish folk dance in ¾ or ⅜ time. Also, the music of this dance.

1846 A. Robinson *Life in California* iii. 23 Singing or whistling the air of some favorite '*jota*'. **1850** E. G. Buffum *Six Months in Gold Mines* 143 Often have I seen little girls, scarce six years of age, flying through a *cotillon*..or dancing with great skill their favourite *jotah* or *jarabe*. **1902** *Encycl. Brit.* XXVII. 374/2 The *Jota* is the national dance of Aragon. **1912** J. E. C. Flitch *Mod. Dancing & Dancers* xiii. 195 The Spanish dance is intensely national. The snapping of the castanets, the short and insolent skirt, the exciting rhythm of the music, do not alone suffice for the performance of the *jota* or *fandango*, as some foreign artists would appear to suppose. **1922** J. Hergesheimer *Bright Shawl* (1923) 12 The air.. was Liszt's Spanish Rhapsody. The accent of its measure, the jota, was at once perceptible and immaterial. **1960** *Spectator* 16 Dec. 988 Antonio's own solo numbers have vitality, and the company's final *jota* is attractive. **1967** 'La Meri' *Spanish Dancing* (ed. 2) ii. 36 It is said that the Jota, 'the fastest dance in the world', is less a dance than an endurance contest.

jotter (s.v. Jot *v.*[2]). Add: **2.** A small pad or writing-book used for jotting down notes, memoranda, etc.; a memorandum book.

1882 Ogilvie *Imp. Dict.*, *Jotter*,..the book in which notes or memoranda are made. **1916** A. S. Neill *Dominie's Log* xii. 132 More attention..should be paid to neatness of method and penmanship in copybooks and jotters. **1930** *Times Educ. Suppl.* 29 Nov. 487/2 The present practice is to supply to all primary schools a sufficient supply of 'Second' readers and jotters sufficient for free issue to all pupils. **1968** M. Woodhouse *Rock Baby* vii. 75 A scribbled list on a desk jotter.

‖ **joual** (ʒwāl). *Canada.* [Dialectal Canad. Fr., ad. F. *cheval* horse.] 'Uneducated or dialectal Canadian French considered as debased or inferior by educated French Canadians, characterized by regional pronunciations, non-standard grammar, and often, especially in cities, by numerous English words and syntactical arrangements' (*Dict. Canadianisms*).

[**1959** *Le Devoir* (Montreal) 21 Oct. 4/6 Faut-il expliquer ce que c'est que *parler joual*?... Tout y passe: les syllabes mangées, le vocabulaire tronqué ou élargi, toujours dans le même sens, les phrases qui boitent, la vulgarité virile, la voix qui fait de son mieux pour être canaille.] **1962** M. Chapin tr. *Impertinences of Brother Anonymous* 27 Our pupils talk joual, write joual, and don't want to talk or write any other way. Joual is their language. *Ibid.* 30 To be understood, I often must have recourse to one or another joual expression. **1963** *Maclean's Mag.* 16 Nov. 54/3 [I] have less trouble with workers' *joual* than with Montreal cabdrivers. **1965** *New Statesman* 10 Dec. 931/1 The uneducated [in Quebec] often speak a *patois* which is not even a genuine dialect so much as a lazy slurring—it is known as *joual*, which reproduces the pronunciation of *cheval*. **1972** *Evening Telegram* (St. John's, Newfoundland) 5 Aug. 36/7 Michel Tremblay's play, Les Belles-Soeurs, done in the joual. **1972** *Islander* (Victoria, B.C.) 15 Oct. 15/1 The French is mere tokenism, too; a few words or sentences of high-school français and joual sandwiched between paragraphs of English. **1973** *N.Y. Times* 22 Apr. 14/3 There are six million Canadians whose native language is French, but much of what they say would be baffling to a Parisian because many of them speak a patois known as '*Joual*', its name drawn from the way some rural people in Quebec pronounce the word 'cheval' or horse. **1975** *Canadian Forum* (Toronto) Oct. 7/2 In some of his books he lapses into *joual*.

jouk, var. *juke *sb.*

Joule. Add: The pronunc. (dʒūl) is now usual for the name of the unit and the physicist.

Although some people of this name call themselves (dʒaul) and others (dʒōul) (D. Jones *Everyman's Eng. Pronouncing Dict.* (ed. 11, 1956), G. M. Miller *BBC Pronouncing Dict. British Names* (1971)), it is almost certain that J. P. Joule (and some at least of his relatives) used (dʒūl). For evidence on this point see *Nature* (1943) CLII. 354, 418, 479, 602.

Add to def.: Later adopted as the M.K.S. unit of energy and work in non-electrical contexts also; it is equal to 10 million ergs, and is now incorporated in the International System of Units, in which it is given the equivalent definition of the work done when the point of application of a force of one newton is displaced through a distance of one metre in the direction of the force. Usually written **joule.** (Further examples.)

1886 J. A. Fleming *Short Lect. to Electr. Artisans* v. 87 An amount of work equal to 10 million ergs is called one joule. Hence one foot-pound = 1·356 joules, or one joule = ·7373 foot-pound. **1889** *Rep. Brit. Assoc. Adv. Sci. 1888* 56 It was also agreed to adopt the name 'Joule' for 10[7] C.G.S. units of work. Thus a Joule is equal to 10[7] ergs. It is the work done in one second by a power of one Watt. **1893** *Operator & Electr. World* 2 Sept. 175/1 The several governments represented by the delegates of this International Congress of Electricians..hereby recommended to formally adopt as legal units of electrical measure the following... As a unit of work the joule, which is 10[7] units of work in a C.G.S. system. **1896** *Rep. Brit. Assoc. Adv. Sci. 1896* 154 The amount of heat required to raise the temperature of 1 gramme of water 1° C. of the scale of the hydrogen thermometer, at a mean temperature which may be taken as 10° C. of that thermometer, is 4·2 Joules. **1923** N. Shaw *Forecasting Weather* (ed. 2) i. 24 Observations for nine stations within greater London, comprising measures of atmospheric pollution at Richmond and South Kensington, total solar radiation at South Kensington in joules, and its average and maximum rate in milliwatts. **1949** *Nature* 19 Mar. 428/1 It was also agreed [by the ninth International Conference of Weights and Measures] that the unit of heat should be the 'Joule'. **1951** *Engineering* 11 May 577/1 The international units, as realised in the various national laboratories, differed by small amounts, but it has been agreed that the mean international joule of the various laboratories is equal to 1·00019 absolute joules. *Ibid.* 577/3 If the value of the 15-deg. calorie is taken as 4·1855 absolute joules, a B.Th.U. is equal to 1054·54 absolute joules. **1963** [see *calorie]. **1970** *Daily Tel.* 4 May 3 Adoption of the metric system and loss of the calorie as a unit of energy may increase the hazard of obesity, Prof. J. B. M. Coppock, an expert on nutrition, said yesterday... There was a danger of losing the calorie for another unit, the joule, which was little understood outside physics and engineering... It would be years before the public..could adjust to a new jargon.

2. Used *attrib.* and in the possessive as the personal name to designate quantities, effects, etc., discovered by Joule or arising out of his work: **Joule** († or **Joule's**) **effect,** (*a*) the heating effect of an electric current flowing through a resistance (described by *Joule's law* (*a*)); (*b*) a change in the linear dimensions (esp. length) of a body when it is subjected to a magnetic field; **Joule's equivalent** (examples); **Joule heating,** the heating that occurs when an electric current flows through a resistance; **Joule's law,** (*a*) the law that the heat produced by an electric current i flowing through a resistance R for a time t is proportional to i^2Rt; (*b*) a law concerning ideal gases, now usually given in the form: the internal energy of a given mass of a gas depends only on its temperature.

¶ See also *Joule–Kelvin, *Joule–Thomson.

1879 *Encycl. Brit.* VIII. 57/2 In general the Peltier effect is..mixed up with Joule's effect, and makes itself felt by producing a disturbance at the junction. **1914** *Physical Rev.* IV. 499 If the change in length of *AC* is measured as the field, *H*, is varied, we have the ordinary longitudinal Joule effect, whereas if *DB* or the length normal to the page is measured we have the transverse Joule effect. **1931** S. R. Williams *Magn. Phenomena* ii. 108 In the Joule effect the changes in length are..very small. *Ibid.* 111 (*heading*) Transverse Joule effect—change in dimensions normal to the magnetic field. **1933** J. E. Phillips *Intermediate Magn. & Electr.* xiii. 303 The Peltier effect..is proportional to the first power of the current, while the Joule effect varies as the square of the current. **1939** L. F. Bates *Mod. Magn.* x. 303 The existence of a Joule effect in non-ferromagnetic substances was first proved by Kapitza. **1966** F. Brailsford *Physical Princ. Magn.* vi. 143 Additional to the high-field volume change is the Joule effect involving changes in linear dimensions in relatively low fields, which in most cases may be 10 to 100 times greater than those due to the volume magnetostriction. **1853** W. J. M. Rankine in *Trans. R. Soc. Edin.* XX. 192 This last quantity..is.. one-ninth of the value according to Mr Joule's equivalent. W. Thomson in *Ibid.* 298 In terms of Joule's mechanical equivalent of heat. **1912** H. A. Perkins *Introd. Gen. Thermodynamics* ii. 19 The numerical constant connecting heat energy..and mechanical energy..is known as 'Joule's equivalent'. **1958** A. J. Woodall *Heat* v. 99 Joule's equivalent is, of course, expressible in different units. **1929** J. A. Ratcliffe *Physical Princ. Wireless* iv. 57 The known forms of A.C. galvanometer, which depend on Joule heating. **1971** *Physics Bull.* Apr. 207/3 Some of the power generated is wasted by joule heating within the element. **1855** W. Thomson in *Proc. Philos. Soc. Glasgow* III. 286 We thus have an equation between the diminution of the electrical energy in any infinitely small time, and the expression according to Joule's law for the heat generated in the same time in the discharges multiplied by the mechanical equivalent of the thermal unit. **1879** *Encycl. Brit.* VIII. 57/1 The current of the battery heats the pile in part uniformly according to Joule's law. **1887** *Ibid.* XXII. 480/1 Joule's law... When a gas expands without doing external work, and without taking in or giving out heat, its temperature does not change... We must therefore conclude that the internal energy of a given mass of a gas depends only on its temperature, and not upon its temperature or volume. **1945** R. H. Frazier *Elem. Electr.-Circuit Theory* i. 21 Ohm's law and Joule's law are interdependent; either can be derived from the other. **1951** C. L. Brown *Basic Thermodynamics* vi. 83 For a perfect gas the internal energy is a function of temperature only. This statement, known as Joule's law after its formulator, can be expressed symbolically as $u = f(T)$.

joulean, Joulean (dʒū·liăn), *a. Physics.* Chiefly *U.S.* [f. prec. + -an.] Of or pertaining to Joule heating.

1891 *Electrician* 27 Feb. 507/2 The waste in Joulean heating. **1903** C. O. Mailloux tr. *Boy de la Tour's Induction Motor* p. xii, Joulean effect. This term is used ..to denote the loss of power (watts), and the heating, due to ohmic resistance. **1937** M. W. Zemansky *Heat & Thermodynamics* xiv. 268 A phantom experiment in which the two irreversible phenomena, namely, Joulean dissipation and heat conduction are absent. **1952** A. Hund *Short-Wave Radiation Phenomena* I. i. 3 The electric field component along the conductor surface accounts for the joulean heat loss in the conductor. *Ibid.* II. ix. 1196 This energy is dissipated in the thin skin as joulean heat. **1960** E. M. & E. W. Pugh *Princ. Electr. & Magn.* vi. 179 The joulean heat produced in the windings of an electromagnetic generator.

Joule–Kelvin (dʒūl ke·lvin). [See next.] = next.

1909 *Cent. Dict.* Suppl. s.v. *Effect*, Joule-Kelvin effect. **1916** I. B. Hart *Student's Heat* xvi. 320 The liquefaction of hydrogen is achieved as a result of the fact that at 200 atmospheres and a temperature of −200° C. the Joule-Kelvin effect changes from a heating to a cooling. **1937** M. W. Zemansky *Heat & Thermodynamics* xiv. 246 One of the most important commercial applications of the Joule-Kelvin effect involves the transition of a liquid at room temperature and at high pressure (15 to 20 atm) into a mixture of liquid and vapor at a temperature about 30°C colder. **1965** W. C. Reynolds *Thermodynamics* viii. 216 Show that the Joule-Kelvin coefficient for a perfect gas is zero.

Joule–Thomson (dʒūl tǫ·msən). The names of James Prescott *Joule* (1818–89), English physicist, and Sir William *Thomson*, Lord Kelvin (see *Kelvin), used *attrib.* with reference to an effect discovered jointly by them, viz. the change of temperature of a gas that occurs when it expands through a porous plug or a throttle without doing external work, the gas being heated or cooled according as it is initially above or below its inversion temperature (which is above room temperature for most gases).

1899 *Proc. Physical Soc.* XVI. 454 He desires to indicate the relation which must exist between the formula assigned to the Joule-Thomson effect, considered as a function of the temperature, and the particular form adopted for the characteristic equation of a gas. *Ibid.* 464 The Joule-Thomson results..were necessary for the establishment of the thermodynamic scale [of temperature]. **1930** *Engineering* 7 Feb. 163/3 The first commercial pure oxygen which Carl Linde put on the market in 1902 was prepared by the application of the Joule-Thomson effect. **1957** G. E. Hutchinson *Treat. Limnol.* I. iii. 207 The amount of change in temperature per unit change of pressure is known as the Joule-Thomson coefficient. **1966** D. G. Brandon *Mod. Techniques Metallogr.* 221 Miniaturized liquid-hydrogen refrigerators based on Joule–Thomson cooling are now available commercially.

jounce, *v.* Add: **1.** (Later U.S. examples.)

1967 C. O. Skinner *Madame Sarah* viii. 171 The train ..swayed, rocked, jounced and hustled a couple of passengers from their seats. **1969** *New Yorker* 12 Apr. 118/3 The drill, which is a percussive one, jounces up and down. **1971** D. E. Westlake *I gave at the Office* (1972) 55 The two trucks jouncing off along the narrow dirt road through the swamp.

2. (Later U.S. examples.)

1902 H. L. Wilson *Spenders* xiv. 148 Then I jounced Hank. **1910** *N.Y. Even. Post* 4 Aug. (Th.), The raft was jounced about so severely that it broke its anchorages. **1972** *Time* 17 Apr. 40/3 The rover's seat belts have been redesigned to anchor passengers more comfortably during the jouncing ride in the moon's weak gravity.

jour. Add: **2.** (Earlier and later examples.)

1865 F. B. Palliser *Hist. Lace* xiii. 181 Twined among them appear a variety of 'jours', filled up with patterns of endless variety, the whole wreathed and garlanded like the decoration of a theatre. **1953** M. Powys *Lace & Lace-Making* iv. 27 Binche, Modern Belgian Lace... As a filling in the center is the Fond à la Mariage or honey-comb, mode or jour.

3. jour de fête = *fête-day.*

1806 C. Wilmot *Let.* 21 Oct. in *Russ. Jrnls.* (1934) II. 234 The gift of their Wives on such a *Jour de Fête* or *Jour de Nom.* **1864** Mrs. Gaskell *French Life* in *Fraser's Mag.* Apr. 439/2 The gowns they wear on jours de fêtes. **1869** C. Schreiber *Jrnl.* 17 May (1911) I. 4 Jour de Fête, no shops open.

jour[2] (dʒɜ̄r), U.S. colloq. abbrev. of Journeyman.

1835 *Gent's. Vade-Mecum* (Philadelphia) 27 June 3/1 The *jours* are in the habit of *spouting* their work from one week to another. **1854** B. P. Shillaber *Life & Sayings Mrs. Partington* 146 'I wouldn't be so bothered about *my* meals,' said a jour printer to a brother typo. **1868** E. E. Hale *If, Yes & Perhaps* 35, I stopped at one or two cabinet-makers, and talked with the 'jours' about work. **1898** *Milwaukee Sentinel* 16 Jan. II. 2/7 Where the hundreds of old time-honored 'jours'..have gone to, no one seems able to decide.

journal, *sb.* Add: **11. b.** (senses 2 and 4) **journal-letter,** a letter written as a diary. (Cf. sense 11 a, quot. 1742.)

1756 J. Hanway *Jrnl. Eight Days Journey from Portsmouth* III. 9 You see I have begun my Journal Letters, with the solemnity of a dedication. **1869** L. M. Alcott *Little Women* II. x. 141, I shall keep a journal-letter, and send it once a week. **1906** *Daily Chron.* 26 Oct. 3/3 Fanny's journal-letters to her dear 'Daddy Crisp'..are delightful and vivid effusions. **1964** *Listener* 17 Dec. 983/2 Sir Edward Marsh's long journal-letters..reveal that he was often silly.

journey, *sb.* Add: **3. e.** The travelling of a vehicle along a certain route between two fixed points and at a stated time.

1851 *Illustr. London News* 25 Oct. 526/1 If they.. obtained..12 passengers at 2d. each per journey, the profit would be 19s. 3d. per diem. **1878** *Porcupine* XX. 507/2 The conductor..shouts, 'Journey's end.' **1908** *Daily Chron.* 4 Jan. 1/7 London Motor Bus Strike... The company is determined to insist on the journey system of payment. **1909** *Westm. Gaz.* 8 Sept. 2/1 The journey-time to Glasgow is 8½ h. **1954** *Gloss. Highway Engin. Terms* (*B.S.I.*) 55 *Journey time*, the overall time taken to travel between two specified points on a route, excluding the times of any stoppages other than those due to interruptions of traffic.

10. c. A set of trams in a colliery.

1883 W. S. Gresley *Gloss. Terms Coal Mining* 144 *Journey*, a train or set of trams all coupled together running upon an engine plane. **1896** Mrs. H. Ward *Sir G. Tressady* xxiv. 553 The 'journey' of trucks..was standing laden in the entrance of the mine. **1901** *Daily Chron.* 8 Nov. 11/3 He was caught by the 'journey' and killed. **1921** *Dict. Occup. Terms* (1927) § 043 *Journey rider*,.. rides on trams or tubs on haulage planes. **1967** *Gloss. Mining Terms* (*B.S.I.*) x. 14 *Train* (*journey, set, trip*), a number of tubs or cars coupled together.

11. journey-money, read: travelling expenses; (later examples); **journey-pride** *dial.*, excitement or alarm occasioned by the prospect of travelling; so **journey-proud** *a.*

1914 'I. Hay' *Lighter Side School Life* iv. 95 You've been a long time getting your journey-money. **1936** 'R. Hyde' *Check to your King* vii. 84 Vigneti is parcelled off to Guadeloupe, supplied by his Sovereign Chief with a thousand francs journey-money. **1960** G. E. Evans *Horse in Furrow* v. 69 Wagoners and horsemen on long journeys..even where they had adequate journey-money, ..often preferred to sleep out. **1902** *Eng. Dial. Dict.*, *Journey-proud*, excited like children, at the prospect of a journey. **1908** *Daily Chron.* 5 Nov. 4/7 In Cheshire,.. a village good-wife, describing her farm-labourer husband's first visit to Manchester, declared that he was 'that journey-proud that he couldn't eat a bite o' breakfast'. **1938** *Times* 21 Dec. 10/4 'Journey-pride'..will be familiar to your west-country readers. The adjective is still more useful, for 'feeling-upset-physically-and-mentally-with-anticipatory-excitement-and/or-anxiety' can all be expressed by 'journey-proud'. **1956** *Sunday Times* 3 June 2/6 The lengthy German phrase for holiday anxiety... People who suffered from it used to be described in Yorkshire..as being journey proud.

journeyman. Add: **3. b.** = *impulse dial* (*impulse *sb.* 5 b).

1904 Goodchild & Tweney *Technol. & Sci. Dict.* 384/1 *Master clock*, the timepiece controlling and actuating by electricity a series of dial works, or 'journeymen', at different points in the circuit. **1923** [see *impulse clock* (*impulse *sb.* 5 b)]. **1938** J. W. Player *Britten's Watch & Clock Maker's Handbk.* (ed. 14) 159 Secondary clocks, sometimes called impulse clocks, dial works, or journeyman clocks, are simple constructions.

jowler[2]. *Merseyside slang.* = *jigger *sb.*[1] 6 c.

1961 Partridge *Dict. Slang* Suppl. 1154/1 *Jowler*, a lane between back-to-back houses: Liverpool. **1966** S. Kelly in F. Shaw et al. *Lern Yerself Scouse* 80 When I was young half of me time was spent Up jowlers.

joy, *sb.* Add: **1. g.** *colloq.* Result, satisfaction, success. Esp. with negative, and freq. *ironical.*

1945 *Tee Emm* (Air Ministry) V. 53 There's even less joy in sending us the money. **1945** C. H. Ward-Jackson *Piece of Cake* (ed. 2) 40 *Joy*, satisfaction. Thus, 'Johnnie took the new kite up this morning—had bags of joy', or 'no joy at all'. **1946** Brickhill & Norton *Escape to Danger* xxxiii. 294 At 9.15 the workers had been down nearly forty minutes and still 'no joy'. **1961** S. Price *Just for Record* ii. 17, I..tried to get a taxi. No joy, so back into the studio. **1961** H. R. Williamson *Wicked Pack Cards* ix. 94 Did you get any joy at the picture gallery? **1971** D. Bagley *Freedom Trap* vii. 147 He reported, 'No joy!' *Ibid.* viii. 178 'Any joy there?' She looked up. 'There's not much more than I told you last night.' **1972** R. Fiennes *Ice Fall in Norway* vi. 86 It was becoming late—we tried to locate Patrick's position again, but without joy. **1973** *Scotsman* 7 Aug. 8/2 Parking the car in this bay we started to look for a path and a break in the barbed wire—again with no joy.

10. *joy-night*; **joy-firing:** delete (*nonce-wd.*) and add later example in sense: the firing of celebratory shots (cf. Feu de joie 2); **joy-flight,** an aerial joy-ride; so **joy-flying; joy-house** *slang*, a brothel; **joy juice** *U.S. slang*, alcoholic drink; **joy-plank,** a plank leading from the stage to the audience in a theatre, for the use of performers; **joy-popper** *slang* (orig. *U.S.*), an occasional taker of illegal drugs; hence [back-formations] **joy-pop,** (an inhalation or injection of) a drug; **joy-pop** *v. intr.*, **joy-popping** *vbl. sb.*; **joy-stick** *slang*, the control-lever of an aeroplane; the controls of another vehicle; also *attrib.*, *transf.*, and *fig.*; **joy-wheel,** a form of amusement consisting of either (*a*) a gigantic wheel-shaped structure, as on a fairground, on which passengers are carried in cars rotating round the axis, or (*b*) (see quot. 1954).

1926 T. E. Lawrence *Seven Pillars* (1935) cxvii. 635 The shooting he heard was joy-firing. **1923** *Daily Mail* 7 Aug. 8/2 The 'joy flights' in three-seater Avros, at 5s. a time. **1928** *Daily Express* 3 July 12 Strict regulations were made against joy-flights during the war. **1940** R. Chandler *Farewell, my Lovely* iii. 24, I ain't been in a joy house in twenty years. **1970** 'B. Mather' *Break in Line* iii. 43 All right—so you're a sailor in a joy-house with a sore foot. **1960** Wentworth & Flexner *Dict. Amer. Slang* 297/2 *Joy-juice*, liquor. **1974** *Black World* Mar. 56/2 He could hear the others as in a dream, laughing, telling dirty jokes, playing cards and swizzling joy-juice. **1925** T. Dreiser *Amer. Tragedy* I. I. viii. 53 Being invited by them to a joy-night supper—a 'blow-out' as they termed it..he decided to go. **1928** *Daily Mail* 7 Aug. 12/7 It was a 'joy night', although many people were still unable to believe that they could..buy a packet of cigarettes openly. **1924** *Illustr. London News* 27 Dec. 1265/2 The picture of the Grand Ballet at Florence in 1616.. shows a method of presentation which was in vogue here in Revues a year or two ago, and is still continued in the Cabarets; performers leaving the stage by means of steps and 'joy-planks'. **1970** J. B. Priestley *Edwardians* III. 247 (*caption*) Shirley Kellogg leading the chorus along the joy plank in *Hullo Ragtime!* at the London Hippodrome in 1912. **1939** *Detective Fiction Weekly* 18 Mar. 59/1 If you should happen to hear anybody speaking of a sueypow or a joy-pop or of going out the lemon bowl,.. bring him right here. **1951** *Time* 26 Feb. 24/3 A sniff of heroin is a 'snort of horse', and an injection under the skin a 'joy pop'. **1954** Beckhardt & Brown *Violators* viii. 238 Every now and then he would 'joy-pop' (take an occasional injection) but he thought he could avoid the 'hook' (addiction) by 'spacing his shots'. **1962** K. Orvis *Damned & Destroyed* v. 41, I take a joy-pop once in a while. *Ibid.* vii. 51 So you're handling a bit of hot stuff as well as joy-popping? **1964** D. Warner *Death of Dreamer* I. i. 8 Each junkie is taking an average of twenty joy-pops a day. The joy-pops are sold in one-grain packets, called decks by the junkies. **1936** *Amer. Speech* XI. 123/1 *Joy-popper*, a person, not a confirmed addict, who indulges in an occasional shot of dope. However, joy-popping is usually the beginning of a permanent addiction. If the joy-popper has trouble establishing the desire and pleasure from indulging it, he is called a student. **1949** N. Algren *Man with Golden Arm* I. 24 They called those using the stuff only occasionally 'joy-poppers' and wished them all great joy. For the 'joy-poppers' had no intention of becoming addicts in the true sense. **1972** J. Brown *Chancer* ii. 30 The weekend ravers and joy-poppers..for whom smoke and amphetamines alone were not enough. **1910** R. Loraine *Diary* 9 Apr. in W. Loraine *Robert Loraine* (1938) vi. 105 In order that he shall not blunder inadvertently into the air, the central lever—otherwise the *cloche*, or joy-stick is tied well forward. **1916** Joy-stick [see *control lever]. **1932** Auden *Orators* II. 52 *Joystick*—Pivot of power And responder to pressure And grip for the glove. **1936** *Amer. Speech* XI. 123/1 *Joy stick*, an opium pipe. **1948** Partridge *Dict. Forces' Slang* 104 *Joysticks*, the two levers by which the steering of tanks and some other tracked vehicles was controlled. **1950** G. Barker *News of World* 44 My love, my love, lift up your joystick hand. Dismiss the dividing Grief. **1952** A. Tustin *Automatic & Manual Control* 467 Both hands held at approximately elbow height a joystick that could be rotated or deflected about a universal coupling, to control the spot movement in elevation and traverse. *Ibid.* 468 The joystick controlled the movements of the spotlight. **1964** S. Duke-Elder *Parsons' Dis. Eye* (ed. 14) x. 98 (*caption*) Joystick lever for horizontal course and fine adjustments. **1967** *Times Rev. Industry* Mar. 43/3 The system is simply operated by a keyboard and joystick—no computer programming knowledge is needed. **1969** G. Macbeth *War Quartet* 34 Easing the joy-sticks to their mid-riffs. **1911** *Oxford Times* 9 Sept. 10/6 A new form of amusement to Oxford, known as the 'Joy Wheel'. **1942** 'M. Innes' *Daffodil Affair* I. i. 7 Perhaps twenty times it passed to and fro, as if outside some great joy-wheel were oscillating idly in a derelict amusement park. **1954** *Engineer* 27 Aug. 282/2 A once-popular novelty, now obsolete, was the 'Joywheel', or 'Devil's Disc'. This consisted of a power-driven spinning disc, slightly domed and having a smooth surface. It was surrounded by a stationary padded circular platform, which in turn was surrounded by a padded wall. Riders sat on the disc while it was stationary and, as it accelerated, were eventually thrown off against the padding. **1968** D. Braithwaite *Fairground Archit.* 65 A panoramic 'Joy Wheel' using kinetoscope effects to create the illusion of a race between motor-car and train.

Joycean (dʒoi·siăn), *a.* and *sb.* Also **Joycian.** **A.** *adj.* Of, pertaining to, or characteristic of the Irish writer James *Joyce* (1882–1941), or his work. **B.** *sb.* An admirer or follower of Joyce.

1927 *New Republic* 20 July 236 Joycean passages and bursts of purple lyricism. **1932** *Times Lit. Suppl.* 26 May 386/3 The modified 'Joycean' language is clever and expressive. **1935** *Discovery* Dec. 378/2 The verbatim description of one 'disturbed case' by another, with its progressive lapse into Joycian language. **1938** Partridge *World of Words* vi. 161 The Joyceans are artificial, but, except at the cost of a highly gymnastic cerebration, unintelligible. **1953** M. Lowry *Lett.* (1967) 330 Even Bernard de Voto had to interpolate that he was 'a good Joycean—he hoped'. **1965** *Times Lit. Suppl.* 11 Mar. 199/3 This particular 'deviation' has a rather Joycean flavour to it, as it combines the meaning and the effect of exorbitance in one word. **1966** *Listener* 2 June 805/3 The hero's name, Strumienski, would delight any Joycean: it is derived from 'stream'. **1971** *Ibid.* 16 Sept. 381/2 There was a fine Joycian slide into a muck wake. **1971** *Daily Tel.* (Colour Suppl.) 3 Dec. 9/2 The lecturer..led his summer school audience down the howling avenues of Joycean puns.

joyful, *a.* Add: **4.** *o* (or *oh*) *be joyful*, an alcoholic drink. *slang.*

1823 P. Egan *Grose's Dict. Vulgar T.* (rev. ed.), *O be joyful*, good liquor; brandy. *Sea term.* **1830** *Greensborough* (N. Carolina) *Patriot* 4 Aug. 4/2 They didn't come to, till the old woman and her darter poured some o be joy full down their throates. **1846** L. Crawford *Hist. White Mts.* iv. 45, I was loaded..with a plenty of what some call..'O-be-joyful'. **1865** *London Jrnl.* 8 Apr. 222/3 Like a great many other clever fellows, he was too much addicted to the 'O be joyful!' **1934** T. S. Eliot *Rock* ii. 77 Well, boys, what do you say to a pint of Oh be Joyful? **1968** *Listener* 21 Mar. 369/3 It's a lively scene on this part of the Oldham Road: some very jolly pubs, Dutton's O Be Joyful ales in little bottles.

joy-ride (dʒoi·raid), *sb. colloq.* (orig. *U.S.*). [f. Joy *sb.* + Ride *sb.*[1]] A pleasure trip in a motor car, aeroplane, etc., often without the permission of the owner of the vehicle. Also *transf.* and *fig.* Hence **joy–ride** *v. intr.*, to go for a joy-ride; *trans.*, to convey (as) on a joy-ride; **joy–rider,** one who goes on a joy-ride; **joy–riding** *vbl. sb.*; also *attrib.*

1908 W. S. Blunt *Diary* 5 Aug. (1920) II. 217 Joy-riders..light-hearted Londoner folk concerned with nothing but their own pleasure. **1909** *N.Y. Even. Post* (Semi-weekly ed.) 15 July 2 [The] Acting Mayor vetoed the ordinance passed last week to prevent city officers from taking 'joy rides'. *Ibid.* 2 Sept. 8 'This was no hap-hazard expedition,' he said, 'no intensified Arctic joy ride, undertaken on nerve.' **1910** *National Police Gaz.* 2 July 3/4 As a rule there isn't much chance of joy riding in a taxi. **1913** *Aeroplane* 17 Apr. 455/2 The pilots of machines already qualified naturally took no risks of damaging their mounts by 'joy riding'. **1913** *Collier's* 27 Dec. 9 (*heading*) With the joy-riding fleet. **1915** D. O. Barnett *Let.* 6 July 206 We joined forces and captured a motor-ambulance which joy-rided us back here. **1915** *Amer. Mag.* Oct. 28/2 You take your dago friend 'n' joy-ride with the gang all night. **1918** H. Hayens *Lords of Air* 41 Unfortunately I was not joy-riding, but bent on serious business. **1919** *Aeroplane* 5 Feb. 555/1 There are two million people in this country who wish for what is vulgarly called a joy-ride in an aeroplane. *Ibid.* 555/2 Joy-riders will not be taken aloft one at a time but in half-dozens and dozens, in big flying boats..and multiple-engined land machines. **1922** F. Harris *My Life & Loves* I. ix. 187 'Aren't you tired?'..'No,' I replied, 'I feel no fatigue, indeed, I feel the better for our joy ride!' **1922** Joyce *Ulysses* 497 You once nobble that, congregation, and a buck joy ride to heaven becomes a back number. **1925** *Flynn's* 14 Feb. 664/2 *Joy-ride*, intoxication resulting from use of drugs. **1925** *Times* 26 Sept., For the third year..no fatal accident involving passengers in other flying for hire, mainly 'joy ride' flights. **1930** E. Pound *XXX Cantos* ix. 36 With three pages of secret instructions To the effect: 'Did he think the campaign was a joy-ride?' **1952** C. Day Lewis tr. *Virgil's Aeneid* XII. 280 Around them..The hostile formations are pressing..While you're joy-riding over these empty fields. **1972** J. Brown *Chancer* xii. 161 That's what you need now, he says—need a joy-ride, man... No danger neither, he says, on the joy-pop. **1972** *N.Y. Law Jrnl.* 10 Oct. 20/6 The annotation..dealing with elements of the offense defined in 'joyriding' statutes fails to disclose any case in which the person using a rented vehicle with the consent of the custodian has been held guilty of unauthorized use. **1973** *Scottish Sunday Express* 5 Aug. 7/8 A man who drove away two cars for a 'joy ride' was fined £75.

J-shaped (dʒē[i]·,ʃē[i]pt), *a.* Having the shape of the letter J; used *esp.* of a graph or of a variable expressed by it. Also **J-curve,** a J-shaped curve.

1911 G. U. Yule *Introd. Theory Statistics* II. vi. 98 The extremely asymmetrical, or 'J-shaped', distribution, the class-frequencies running up to a maximum at one end of the range. **1927** H. L. Rietz *Math. Statistics* iii. 57 Type III would be J-shaped if γα were negative. **1939** *Amer. Jrnl. Sociol.* XLIV. 975 He [*sc.* Professor Allport] advances the 'J-curve hypothesis of conforming behavior'. **1950** S. A. Stouffer *Measurement & Prediction* i. 41 In general, as predicted, a J-shaped or U-shaped curve was obtained with attitude data. **1968** P. McKellar *Experience & Behaviour* xi. 304 If..the situation is changed by the introduction of a traffic light we have J-curve behaviour. **1973** *Times* 7 May 16/4 The notion of the post-devaluation 'J-curve', according to which an initial deterioration in the balance of payments is followed by a more-than-offsetting improvement as the volume of

exports responds over time to the gain in price competitiveness.

Ju (dʒū). [ad. *Ju Chou*, the name of a town in Honan, China.] Used to designate pottery with buff body and blue-green glaze produced in Ju Chou in the 12th century.

1906 W. Burton *Porcelain* vi. 57 Dr. Bushell mentions a fine piece of *Ju-Yao* in his possession, which, however, has a delicate greenish glaze with no trace of blue. **1936** *Burlington Mag.* 4/1 The original Ju ware was made at the potteries at Ju Chou. **1949** G. Savage *Ceramics for Collector* ii. 44 The kiln was in operation only for about twenty years, and at one time the '*ying ch'ing*' type was identified with Ju ware. **1969** *Times* 9 Dec. (Taiwan Suppl.) p. vii/2 The 24 items of Ju ware are the best of the porcelain. **1972** *Collector's Guide* June 95 Ju ware, only identified as recently as 1936 by Sir Percival David, is much rarer than Ting.

jube[3] (dʒūb), abbrev. of Jujube 2.

1937 Partridge *Dict. Slang* 445/2 *Jube*, a coll. abbrev. of *jujube* (the lozenge). **1967** E. & M. A. Radford *No Reason for Murder* iii. 16 He bought..bars of chocolate and jubes. **1970** J. Cleary *Helga's Web* vii. 118 Do you have any menthol jubes?

jube[4] (dʒūb). [Persian.] An open water-course in Iranian cities.

1948 in *Punch's Almanack* 2. **1953** A. Smith *Blind White Fish in Persia* iii. 48 By every pavement ran the jube, a stream of water which had doubtless been clean at the top of the town but did not remain so for long. **1959** T. Griffith *Waist-High Culture* (1960) ix. 115 Tehran, its streets vivid with sidewalk hawkers, the flow of filth in the jubes.

jubilee, *sb.* Add: **5. c.** A Negro folk-song of an optimistic and joyful kind, often having a religious basis; freq. *attrib.*, esp. *jubilee singer, song.*

1872 *N.Y. Tribune* 17 Feb. 3/6 Unique Vocal Concert by the Jubilee Singers. *c* **1872** T. F. Seward (*title*) Jubilee songs as sung by the Jubilee Singers, of Fisk University. **1873** G. D. Pike *Jubilee Singers* 163 The excellent rendering of the Jubilee Band is made more effective..by the comparison of their former state of slavery..with the present prospects and hopes of their race. **1922** *Jrnl. Amer. Folk-Lore* XXXV. 248 Every Time I Feel the Spirit. This is one of the most thrilling of the later jubilee songs. **1946** S. H. Holbrook *Lost Men Amer. Hist.* 133 The Chautauqua offered no such strong meat as the Lyceum, but went in for bell ringers, jubilee singers, preachers..and assorted stuffed shirts. **1949** B. A. Botkin *Treas. S. Folklore* v. i. 701 Besides his 'jubilee' and 'sorrow songs', the Negro has his 'sinful' songs. **1956** M. Stearns *Story of Jazz* (1957) xii. 133 The next step in this blending, which produced both ring-shout and spiritual, is the jubilee. Jubilees are both cheerful and rhythmic, usually announcing some sort of good news. **1961** J. Jahn in A. Dundes *Mother Wit* (1973) 99/1 The hymns of Christian European origin used by the missions are Africanized, producing jubilees. **1968** P. Oliver *Screening Blues* Introd. 7 Blues recordings accounted for nearly half the total output in the 'twenties whereas religious issues, including those by solo evangelists and jubilee groups, totalled only a fraction over a sixth of the number. *Ibid.* ii. 81 The gospel, or jubilee, quartets performed all their songs in harmonized style which smoothed out the differences between the blues, popular and gospel song forms.

jubjub (dʒɒ·bdʒɒb). [Invented by 'Lewis Carroll' (C. L. Dodgson); perh. a portmanteau word or formed after such representations of bird-cry as *jug-jug* (Jug *sb.*[2]).] An imaginary bird of a ferocious, desperate and occasionally charitable nature, noted for its excellence when cooked.

1871 'L. Carroll' *Through Looking-Glass* i. 22 Beware the Jubjub bird, and shun The frumious Bandersnatch. **1876** —— *Hunting of Snark* 44 Should we meet with a Jubjub, that desperate bird, We shall need all our strength for the job! *Ibid.* 58 The song of the Jubjub recurred to their minds, And cemented their friendship for ever! **1962** M. Gardner in 'L. Carroll' *Annotated Snark* 104 Their common fear of the *Jubjub Bird* is interesting. Philologically..Jubjub is a 'portmanteau bird', compounded of *jabber* and *jujube*... It flashed across me that the Jubjub was Society itself.

jucal, var. *jacal.

Judæan, Judean (dʒudī·ăn), *a.* and *sb.* [f. L. *Jūdæus*, a. G. Ἰουδαῖος, f. Ἰουδαία, f. Ἰούδας, ad. Heb. *Jehūdah* Judah, name of a son of Jacob.] **A.** *adj.* Of or pertaining to Judæa or southern Palestine. **B.** *sb.* A native or inhabitant of this region.

In Shakes. *Oth.* v. ii. 347 the first Folio has the doubtful reading *Iudean*; the other Folios and the Quartos have *Indian*.

1652 Heylin *Cosmogr.* III. 89 Accaron, on the South of Gath, of great wealth and power, and one that held out notably against the Danites and Judæans. **1832** J. Bell *Syst. Geogr.* IV. 206 Judean Mountains. **1852** H. W. Dulcken tr. *Pfeiffer's Visit Holy Land* 103 The foreground of the picture is formed by the Judæan mountains. **1880** *Encycl. Brit.* XIII. 410/1 A Judæan, Amos of Tekoa. **1922** A. E. Garvie *Beloved Disciple* x. 210 The Synoptic record is incomplete as regards the Judaean ministry. **1931** *Times Lit. Suppl.* 5 Nov. 853/1 A preliminary survey of Judaean sites. **1955** R. W. Millar tr. *Daniel-Rops's Jesus in his Time* 57 Eleven of the disciples were Galileans, the one Judean was Judas. **1965** M. Spark *Mandelbaum Gate* ii. 22 She..had driven northward through the Judean hills to Galilee. **1973** *Times* 3 Dec. 17/6 David Ben-Gurion..took a leading part..in recruiting for the Judaean battalions of the Royal Fusiliers.

Judæo-, Judeo- (dʒudī·o), used as comb. form of L. *Judæus* *Judæan, Judean *a.* and *sb.*, designating persons or things pertaining to Judæa and hence (more widely) to the Jews; often = Jewish (and); **Judaeo-German,** Yiddish.

1823 *Christian Observer* App. 828/1 The New Testament in German-Hebrew, and Judeo-Polish. **1851** *Illustr. Catal. Gt. Exhib.* III. 552/1 Judæo Spanish, Old Testament. *Ibid.* 552/2 Judæo Arabic, four books of New Testament. **1863** *Chambers's Encycl.* V. 712/2 The number of Judæo-Greek fragments..which have survived. *Ibid.* 721/2 Numerous authors wrote in Hebrew,..and Judæo-German. **1899** *Lit. Guide* 1 Oct. 146/1 The total abandonment of the Judæo-Christian 'continuity' theory. **1900** tr. *J. Deniker's Races of Man* 424 Particular kinds of jargon, the most common of which is the Judeo-German. **1906** *Westm. Gaz.* 17 Aug. 10/2 The Judæo-Spanish world of the Levant. **1910** *Ibid.* 12 Mar. 2/1 'The Judæo-Masonic and Protestant coalition' which now governs France. **1910** *Encycl. Brit.* VI. 494/1 The Clementine literature throws light upon a very obscure phase of Christian development, that of Judaeo-Christianity. **1937** *Times Lit. Suppl.* 16 Jan. 37/2 Irrelevances about the world hegemony of Judaeo-Masonry. **1939** *New English Weekly* 27 July 237/2 The Judaeo-Christian scheme of morals. **1939** L. H. Gray *Foundations of Language* 349 The two chief forms of creolised languages of the Teutonic group, in addition to those already noted for English.., are *Afrikaans*..; and *Yiddish*, or *Judaeo-German*, based upon a Franconian German dialect of the fourteenth century with many Hebrew words, and spoken by Jewish communities in Lithuania, Poland, Russia, and parts of Rumania as well as by Jewish emigrants from those areas. **1948** D. Diringer *Alphabet* II. iv. 267 Judezmo or Judæo-Spanish, called also Ladino, contains many Hebrew words, but is principally based on Old Spanish or Castilian. **1952** Koestler *Arrow in Blue* xxv. 234 Immune against German chauvinism through a hereditary judeo-cosmopolitan touch. **1957** N. Frye *Anat. Criticism* 145 The appearance of the Judaeo-Christian deity in fire. **1960** *Encounter* Mar. 34/2 The religious zeal of Judeo-Christianity. **1964** *Language* XL. 282 Mediaeval Judaeo-French. **1974** R. A. Hall *External Hist. Romance Languages* 24 Judaeo-Spanish or 'Ladino'. *Ibid.* 30 Judaeo-Italian, now dying out, has been attested in some medieval documents, and in scattered remnants in modern Italian dialects.

Judahite (dʒū·dă,ǝit), *a.* and *sb.* [f. *Judah* name of an ancient Hebrew tribe and kingdom + -ite[1].] **A.** *adj.* Of or pertaining to the tribe or the kingdom of Judah. **B.** *sb.* A member of the tribe, or an inhabitant of the kingdom, of Judah.

1899 J. Hastings *Dict. Bible* II. 792/2 Five Judahite clans were in the course of time formed. **1901** R. L. Ottley *Short Hist. Hebrews* vi. 137 In the early stages of the conquest we find the Kenites coalescing with the Judahites. **1932** T. H. Robinson *Hist. Israel* I. II. viii. 170 In Judges xv...Judahites are introduced into the story of Samson. *Ibid.*, An old Judahite tradition of the capture of Hebron..is transferred to Caleb. **1950** H. H. Rowley *From Joseph to Joshua* i. 5 We find a Levite.. who regarded Benjamites equally with Judahites as associated peoples. **1971** *Catholic Biblical Q.* July 394, 5 and 8a describe a Judahite defeat.

Judas. Add: **4.** *Judas kiss* (earlier and later examples); **Judas goat,** an animal used to lead others to destruction; also *transf.*; **Judas priest** *int.*, a euphemism for *Jesus Christ* in an oath.

1941 *Amer. Speech* XVI. 236/1 Sheep are led to the shackling pen by a Tony or Judas goat. **1964** *Punch* 19 Feb. 283/2 Irene only used Billy as a 'judas goat' to catch Des. **1972** D. Anthony *Blood on Harvest Moon* xxv. 220 Not dead. Not me as the Judas goat. *c* **1400** R. Lavynham *Litil Tretys* (1956) 11 Hate of herte is whan a man spekyth litil & menyth moche malyce..& at þe laste wt a iudas kesse schewith a love y fayned. **1860** Geo. Eliot *Mill on Floss* III. vi. i. 15 A woman who was loving and thoughtful for other women, not giving them Judas-kisses with eyes askance. **1942** 'G. Orwell' *Diary* 28 Sept. in *Coll. Ess.* (1968) II. 449 A special prayer 'for the people of Stalingrad'—the Judas kiss. **1947** E. Sitwell *Shadow of Cain* 18 When the last Judas-kiss Has died upon the cheek of the Starved Man Christ. **1973** R. Lewis *Blood Money* ix. 146 Candour shone from his eyes, as insincere as a Judas kiss. **1914** 'High Jinks, Jr.' *Choice Slang* 13 *Judas Priest*, an exclamation of surprise. **1922** S. Lewis *Babbitt* ix. 126 Judas Priest, I could write poetry myself if I had a whole year for it. **1972** J. D. Buchanan *Professional* xi. 123 Judas Priest! What the hell are you saying?

judder (dʒɒ·dəɹ), *v.* [Imit.; cf. Shudder *v.*] *intr.* To shake violently, esp. of the mechanism in cars, cameras, etc.; also of the voice in singing, to oscillate between greater and less intensity. Hence **ju·ddering** *vbl. sb.* and *ppl. a.*

1931 A. J. Cronin *Hatter's Castle* I. xii. 213 The train entered this tunnel. It entered slowly,..juddering in every bolt and rivet of its frame as the hurricane assaulted [it]. **1933** Auden in M. Roberts *New Country* 197 His [*sc.* a motorcyclist's] back wheel juddered; he was off with a roar. **1958** 'P. Bryant' *Two Hours to Doom* 122 He felt the onset of compressibility vibration... The juddering stopped. **1960** *New Scientist* 21 Apr. 1015/2 The drum must run perfectly true and not 'judder' at its circumference more than 1/1,000 inch. **1962** *Times* 28 Feb. 14/6 All day long the school has been juddering and shuddering in the hysterical blasts of north-west wind. **1965** *Punch* 3 Feb. 175/1 As the train juddered a few yards back towards Westminster, he flung a huge arm round my shoulder to steady himself. **1970** *Times* 1 May 2/7 Driver Gwynn started the engine..with much.. cranking and juddering. **1970** *Motoring Which?* July 85/1 During our track tests the front brakes vibrated and juddered when used hard. **1971** B. W. Aldiss *Soldier Erect* 120 Beggars with heroic deformities lay juddering in the gutter. **1973** *Weekly News* (Glasgow) 11 Aug. 3/3 A sustained campaign has been under way on behalf of crippled people who are provided with those juddering little three-wheeled cars to get about in.

judder (dʒɒ·dəɹ), *sb.* [f. prec.] An instance or state of juddering.

1935 *Practical Motorist* 26 Oct. 993/1 (*heading*) Curing clutch judder. **1946** F. C. Field-Hyde *Vocal Vibrato, Tremolo & Judder* iii. 10 Judder consists of marked, rapid changes in intensity during the emission of tone, due to involuntary variations..of the vocal tension. **1952** *Times* 18 Dec., A pronounced judder was felt..and in spite of two corrective movements of the control column [of the aircraft] the judder continued. **1956** *Sunday Times* 28 Oct. 13/3 There is a little 'judder' from the steering wheel at speed, but it is not worrying. **1958** *Times Rev. Industry* May 31/1 The coupling operates with the smoothness and absence of jerk or judder found in hydraulic devices. **1960** O. Skilbeck *ABC of Film & TV*, *Judder*, violent vertical unsteadiness of picture due to faulty camera motion. **1972** *Daily Tel.* 12 Apr. 13/5 It is not difficult to get a touch of wheelspin and axle judder by fierce acceleration in the lower gears.

juddite (dʒɒ·dait). *Min.* [f. the name of John Wesley *Judd* (1840–1916), English geologist: see -ite[1].] A mineral of the amphibole group (see quots.).

1908 L. L. Fermor in *Rec. Geol. Survey India* XXXVII. 212 The optical characters of the mineral leave no doubt that it is a new variety of the amphibole group; I propose to name it juddite in honour of Professor J. W. Judd, F.R.S., and as a respectful tribute from a former student. **1963** W. A. Deer et al. *Rock-Forming Min.* II. 369 Although Bilgrami (1955) considered it appropriate to classify the manganese-rich alkali amphibole juddite.. with the richterite group, it has been included here in the eckermannite-arfvedsonite series. The mineral has a high content of Na..and high Fe^{+3}, as well as high Mn, and the formula $(Na_{2.5}Ca_{0.5})(Mg_{3.5}Mn_{1.0}Fe^{+3}_{1.0})Si_{7.5}Al_{0.5}O_{22}(OH)_2$ is typical of the eckermannite-arfvedsonite group: it is suggested that the mineral may be adequately characterized by the name manganarfvedsonite. **1970** *Mineral. Mag.* XXXVII. 708 It may thus be concluded that the juddites generally refer to (manganoan) magnesioriebeckite compositions.

‖ **Judenhetze** (yū·dənhetsə). [G., = Jew-baiting.] Systematic persecution of the Jews.

1882 *19th Cent.* Aug. 254 Those forces which Europe has confessed are too powerful for it to deal with, and which have led to persecution in Russia and to *Judenhetze* in Germany.

‖ **Judenrat** (yū·dənrat). [G., = Jewish council.] A council representing a Jewish community in a locality controlled by the Germans during the war of 1939–45.

1950 J. Hersey *Wall* I. i. 20 The *Judenrat* had been formed a month earlier, to the day. The Germans told us then that the *Kehilla*, or Community Council, would henceforth become the *Judenrat* and would be the instrument of German authority over the Jews. **1959** M. Levin *Eva* 20 In the very first days after the Germans came, they had ordered the Jews to select a Judenrat. **1960** H. Agar *Saving Remnant* v. 146 In the Ghettos of Eastern Europe the Germans invariably set up a *Judenrat*: a Council of Jews to transmit orders to the community. **1967** *Sat. Rev.* (U.S.) 15 Apr. 20/2 Krüger sent his blood-spattered uniform and boots to the *Judenrat*, the community council.

‖ **judenrein** (yū·dənrain), *a.* [G., = free of Jews.] Of a society or organization: without Jewish members, out of which Jews have been expelled.

1942 *New Republic* 21 Dec. 817 It is not merely central and eastern Europe which are being 'purged', or rendered 'Judenrein', as the Nazis like to say. **1951** H. Arendt *Origins of Totalitarianism* i. 4 When Hitler came to power, the German banks were already almost *judenrein*. **1960** H. Agar *Saving Remnant* vii. 188 A country which was becoming almost Hitlerianly *Judenrein*. **1969** C. Roth *Short Hist. Jewish People* (rev. ed.) vi. xxxi. 444 The survivors..were rounded up and sent to the death camps, Warsaw being now *judenrein*.

judge, *sb.* Add: **1. c.** Phrases, *as grave, sober, as a judge.*

1650 Cowley *Guardian* II. ii. sig. B3/2 If I look not as grave as a Judge upon the bench, let me be hanged for't. **1734** Fielding *Don Quixote in England* III. xiv. 57, I am as sober as a Judge. **1842** Dickens *Let.* 28 Mar. (1938) I. 423, I remained as grave as a judge. **1856** [in Dict., sense 1 a]. **1866** Mayne Reid *Headless Horseman* xliv. 210 I'm as sober as a judge. **1889** [see Grave *a.*[1] 3].

d. *Judges' Rules*: (see quot. 1965[2]).

1925 *Criminal Appeal Rep.* XVIII. 47 The Judges' Rules for the guidance of the Police discussed. **1931** D. L. Sayers *Five Red Herrings* vii. 80 Between the Judges' Rules..and his anxiety to pull off a coup, he felt his position to be a difficult one. **1965** J. Porter *Dover Two* iv. 54 And don't start quoting the Judges' Rules to me... The only time you need bother about the Judges' Rules is when the accused person's likely to know more about 'em than you do. **1965** *Stone's Justices' Manual* (ed. 97) I. 368 The Judges' Rules, made by Her Majesty's judges of the Queen's Bench Division, are concerned with the admissibility in evidence against a person of answers, oral or written, given by that person to questions asked by police officers and of statements made by that person. **1973** 'M. Innes' *Appleby's Answer* xix. 167 Miss Pringle wondered whether..she would..perhaps receive some caution required by what are called Judges' Rules.

10. **judge-made** *a.* (earlier and later examples).

1824 J. S. Mill in *Westm. Rev.* I. 510 The common judge-made definition of a public libel, is, any thing which tends to bring the constituted authorities into hatred and contempt. **1965** *Mod. Law Rev.* XXVIII. v. 510 The clash between democracy and judge-made law.

judgement, judgment. Add: **13.** **judgement sample** *Statistics* (see quot.).

1947 W. E. Deming in *Jrnl. Marketing* Oct. 145/1 *Judgment-samples*, wherein the biases and sampling errors cannot be calculated from the sample, but instead must be settled by judgment.

judgemental, judgmental (dʒʌdʒme·ntăl), *a.* [f. Judgement, judgment + -al.] Involving the exercise of judgement; inclined to make moral judgements.

1909 W. M. Urban *Valuation* ii. 40 Whether exclusively judgmental or not is a question to be determined. **1952** S. Kauffmann *Philanderer* (1953) ix. 154 'Russell,' she said with a queer grin that was friendly and yet vaguely judgemental, 'you puzzle me.' **1964** *Philos. Rev.* LXXIII. 373 There is also..a 'judgmental' constituent of perceptual experience. **1965** Hall & Howes *Church in Social Work* xv. 246 A criticism sometimes made of moral welfare workers..is that they are 'judgmental' in the approach to clients. **1969** *Sci. Jrnl.* Feb. 64/1 Jenny was described rather well and the more quantitative structure analysis (versus content analysis by judgmental methods) yielded greater amounts of information about this woman. **1972** *Jrnl. Social Psychol.* LXXXVI. 13 Conforming behavior was produced..with the use of a two-choice judgmental task. **1973** *Times* 16 July 13/4 As one who is entirely unconvinced about the usefulness of boycotts—of any kind—and a little suspicious of judgmental attitudes to South Africa [etc.].

So **judg(e)me·ntally** *adv.*

1971 J. B. Carroll et al. *Word Frequency Bk.* p. xli, If the citations..are gathered in a number of separate judgmentally biased ways, the significance of each..must be assessed individually.

‖ **judicatum** (dʒūdikēⁱ·tǝm, -ātǝm). *Philos. rare.* Pl. **-ata.** [L. *jūdicātum* judgement, pa.pple. of *jūdicāre* to judge.] (See quots.)

1913 *Mind* XXII. 15 As I use the term, the proposition is what the logicians call the import of the judgment or proposition. It is the *propositum* or *judicatum*. I do not use it as equivalent either to the act of judging or the verbal sentence. **1935** *Mind* XLIV. 365 A judgement *to the effect that* A is B seems to be just a judgement (act of judging) whose *object* or *judicatum* is that A is B. **1936** H. H. Price *Truth & Corrigibility* 17 The relation is between *judicata* or *judicabilia*, or—as some call them—'propositions'.

judo (dʒū·do). Also formerly **jiudo, ju-do.** [Jap., f. *ju* gentleness, ad. Chinese *jou* soft, gentle + Jap. *dō*, ad. Chinese *tao* way.] A refined form of ju-jitsu introduced in 1882 by Dr. Jigoro Kano, using principles of movement and balance, and practised as a sport or form of physical exercise. Also *attrib.*

1889 *Trans. Asiatic Soc. Japan* XVI. 192 The art of Jiujutsu, from which the present Jiudo..has sprung up. *Ibid.* 204 In Judo, which is an investigation of the laws by which one may gain by yielding, *practice* is made subservient to the *theory*. **1892** *Trans. & Proc. Japan Soc.* I. 9 It is due to the study of *Jū-do* that the Japanese police..are so skilful in seizing malefactors. **1905** Hancock & Higashi *Compl. Kano Jiu-Jitsu* p. xi, *Jiudo* is the term selected by Professor Kano as describing his system more accurately than *jiu-jitsu* does. **1921** *Glasgow Herald* 1 Jan. 9/1 This 'Judo' is practised all over Japan. **1931** E. V. Gatenby in *Studies in Eng. Lit.* (Tokyo) XI. 515 There is at least one jûdô society in London. **1953** *Encounter* Oct. 24/1 She ties it in front like a judo jacket. **1958** *Radio Times* 7 Feb. 9/4 A judo club. **1971** R. Busby *Deadlock* xiii. 193 A pair of judo pyjamas. **1972** *Oxford Mail* 1 Aug. 10/4 Judo is a way of learning to control yourself and your opponent. *Ibid.*, The first judo club in England was founded in 1918 in London.

Hence **ju-dogi**, the costume worn for judo; **ju-doka**, one who practises, or is expert in, judo.

1952 *Time* 22 Dec. 40/2 France, center of the European cult, now has 150,000 judo wrestlers (called judoka). **1954** E. Dominy *Teach yourself Judo* 190 *Judogi*,..Judo costume. **1961** *New Statesman* 22 Sept. 402/2 The thrower's..right hand lifts the lapels of the Uke's judogi—his canvas wrestling-jacket. **1964** *Sunday Mail Mag.* (Brisbane) 1 Mar. 7 The only major expense in Judo is £3 for the uniform, which is known as a judogi. *Ibid.*, The examiners..award various gradings to the different judoka. **1974** *Times* 9 Jan. 14/5 With tough agility, however, the British *judoka* emerged from a tight corner.

judoist (dʒū·doist). [f. prec. + -ist.] An expert in judo; one who practises judo.

1950 *Chambers's Encycl.* XIV. 757/1 Whatever the attacker's movement the judoist yields to it in order to lure him into a weakened stance or position. **1966** *Times* 29 Nov. 4 Mr. Gleeson [is] one of the most experienced judoists in this country, and national coach to the British Judo Association. **1969** S. Greenlee *Spook who sat by Door* iii. 26 Soo thinks him one of the finest natural judoists of his experience.

judy. Add: Now somewhat less disparaging than indicated in Dict. (Further examples.)

Often used colloq. simply to mean 'a girl, woman'.

1929 *Papers Mich. Acad. Sci. Arts & Lett.* X. 303/1 *Judies*, girls. **1932** Kipling *Limits & Renewals* 244 'We began to bustle our people over the bows before she went to pieces. You'll admit Paul was a help there, Red?' 'I dare say he herded the old judies well enough.' **1934** G. B. Shaw *Too True to be Good* III. 83 *The Sergeant*... [*He kisses her*] How does that feel, Judy? **1938** J. Masefield *Dead Ned* 284 You'd ought to be careful with the judy. It's best to keep in her good books. **1944** *Penguin New Writing* XIX. 105 I'm off the beer, but I could use a judy for a battering-ram against any wall in creation. **1963** [see *gear sb.* 5 e]. **1966** F. Shaw et al. *Lern Yerself Scouse* 25 Me judy, me tart, me gerl. My lady-friend; my fiancée; my wife. **1973** *Guardian* 31 May 13/7 During a strike a man whose judy is working is obviously better off than the man with a wife and three kids about the house.

jug, *sb.*² Add: **1. c.** A jug used as an instrument in a jazz band. So *jug-blower*, etc. See *jug band* (sense 3 below).

1946 R. Blesh *Shining Trumpets* (1949) v. 104 Exotic instruments may be utilized as well, such as harmonica, kazoo, jug, washboard, wood blocks and musical saw. **1956** M. Stearns *Story of Jazz* (1957) xiv. 157 He didn't even get a chance to team up with washboard beaters, jug blowers, kazoo players, tub thumpers, or alley fiddlers. **1960** *20th Cent.* Dec. 556 The hillbilly form..is played on ..the twelve-string guitar, the jug, the jew's harp. **1964** *Amer. Folk Music Occasional* I. 95 Horses of many different colors run loose in this album, the common denominator being the use by all of a jug, which (blown like a coke bottle) produces rich, booming sound, able to take the bass part. **1968** *Blues Unlimited* Nov. 8 In Memphis, we recorded Dewey Corley, who used to blow jug with the Memphis Jug Band.

2. (Earlier U.S. example.)

1815–16 *Niles' Reg.* IX. Suppl. 190/1 A full grown villain, who with an accomplice, were shortly after safely lodged in the jug.

b. A bank. *slang.*

1845 *National Police Gaz.* 15 Nov. 97/3 Jim Morgan..disdained no branch of business, from 'craking a jug' (entering a bank) to picking a pocket. **1862** *Cornh. Mag.* Nov. 648 It is all in single pennifs on the England jug... It is in 5*l* notes on the Bank of England. **1904** [see *box sb.*² 3 i]. **1935** G. Ingram *Cockney Cavalcade* viii. 121 Give me time to go to the 'jug'. **1960** *Observer* 24 Jan. 5/1 If a villain had seriously suggested screwing a jug (breaking into a bank).

3. **jug and bottle**, used *attrib.* of the bar of a public house at which alcoholic liquors are sold for consumption off the premises; **jug band**, a jazz band in which jugs (sense 1 c above) are used; **jug handle**, the handle of a jug; also *attrib.* and *fig.*, shaped like a jug handle; hence **jug-handled** *a.*, (*a*) *lit.* placed on one side, as the handle of a jug; (*b*) *fig.* (*U.S.*) unilateral, one-sided, unbalanced.

1894 G. Moore *Esther Waters* xxx. 236 The public entrance and the jug and bottle entrance were in a side street. *Ibid.* xlii. 327 Journeyman was surprised to see Ketley sitting quite composedly in the jug and bottle bar. **1909** *Daily Chron.* 31 Mar. 1/3 A 'jug and bottle' department..does not come within the definition of an open bar. **1932** L. Golding *Magnolia St.* I. ix. 144 She got her pint from the Jug and Bottle Department. **1953** *Word for Word: Encycl. Beer* (Whitbread & Co.) 11/2 *Jug-and-bottle bar*, specially reserved for the purchase of drinks for consumption off the premises; only to be found in older pubs. **1946** R. Blesh *Shining Trumpets* (1949) xi. 253 The southern 'jug' band typical of Tennessee and Mississippi. **1970** P. Oliver *Savannah Syncopators* 63 The recordings of some of the jug bands. **1970** *Western Folklore* XXIX. 229 The..'gutbucket'..is generally played in ensembles such as jug and skiffle bands. **1846** S. F. Smith *Theatr. Apprenticeship* 118 Not perceiving the entire justice of this arrangement, it being somewhat on the jug-handle principle, all on one side. **1900** E. Glyn *Visits of Elizabeth* 245 She has a jug-handle chignon. **1955** M. E. B. Banks *Commando Climber* iii. 38 A final wall, almost vertical but amply provided with the largest of jug-handles, remained. **1961** L. Mumford *City in Hist.* xvi. 506 To ensure the continuous flow of traffic,..immense clover leaves and jug handles are designed. **1967** R. J. Serling *President's Plane is Missing* (1968) ii. 27 He was one of those homely men whose virile masculinity masked such features as a big nose and jug-handle ears. **1970** A. Blackshaw *Mountaineering* (rev. ed.) vi. 163 A large incut hold (a 'jug-handle') in good rock is the most secure of all holds. **1881** *Congress. Rec.* 8 Dec. 60/2 English reciprocity in pleasure travel,..like their often proposed commercial reciprocity, is comparatively jug-handled. **1904** *Boston Herald* 28 Sept. 6 The trade between Canada and the United States is..jug-handled.

jug (dʒʌg), *sb.*⁵ Slang abbrev. of Juggins.

1914 D. H. Lawrence *Let.* 18 Dec. (1962) I. 299 But he is a jug... Don't bother anymore. **1956** H. Gold *Man who was not with It* (1965) xi. 96 It was nice to find the born jug of the continent and to decide even about him: he knows what he wants.

Jug (dʒŭg, yŭg), *a.* and *sb.*⁶ Abbrev. of *Jugoslav(ian.

1949 V. Gielgud *Fall of Sparrow* xvii. 168 I've been down among the Jugs during this last week or two. **1958** P. Kemp *No Colours or Crest* x. 218 If Cairo's relations with the Jug Partisans are really so important,..why the hell didn't they warn us off before? **1961** R. B. Amos *Wasp in Web* vii. 74 A Jug friend of mine was standing next to me. **1967** L. Forrester *Girl called Fathom* iv. 33 Most..are based here all the time..the nursing sister, Radik the Jug.

jug (dʒʌg), *sb.*⁷ *West Indies.* Also **jug-jug.** [Origin unkn.] A savoury Barbadian dish served esp. at Christmas (see quots.).

1945 E. P. Clark *West Indian Cookery* xix. 80 *Jug Jug*... Clean, cut up, and season the beef and pork... Stew the pork.., then add beef and peas and stew..until peas are soft. **1957** F. A. Collymore *Notes for Gloss. Barbadian Dial.* (ed. 2) 49 *Jug, jug-jug*, a famous Christmas dish. **1958** B. Hamilton *Too Much of Water* iv. 74 A real Barbadian breakfast.., sweet potatoes, an' jug-jug, an' okras. **1970** M. Slater *Caribbean Cooking* 142 'Jug' is a Bajan Christmas speciality, served with turkey. Jug-Jug came to Barbados..via the Scottish 17th Century exiles, in an attempt to produce something resembling haggis. **1973** *Advocate-News* (Barbados) 25 Dec. 4/6 The specialities of the Yuletide dinner—the turkey, the ham, the plum pudding; the pudding and souse; the peas and rice; the jug-jug; and even caviar.

jug, *v.*¹ Add: **4.** *intr.* To fish with a bait attached to a floating jug (see *jug-fishing* s.v. Jug *sb.*² 3). *U.S.*

1872 *Kansas Mag.* Feb. 178 Jugging for catfish in the chutes of the Missouri and the Kaw. **1947** *Life* 15 Sept. 155 The boys go jugging for catfish. They tie their fishing lines to jugs and haul them in when the jug bobs in the water.

‖ **juge** (ʒüʒ). [Fr.] A judge. *juge d'instruction*, an examining magistrate, a police magistrate.

1882 *Standard* 25 Dec. 3/2 After giving their names and addresses they were permitted to retire, but were informed that they would be called up for examination by a *juge d'instruction*. **1911** Chesterton *Innocence of Father Brown* i. 2 How he turned the *juge d'instruction* upside down and stood him on his head, 'to clear his mind'. **1964** *Ann. Reg. 1963* 51 One of Lord Shawcross's solutions was recourse to an examining magistrate, on the lines of a continental *juge d'instruction*. **1973** M. Catto *Sam Casanova* ii. 22 You're going to have a *juge d'instruction* on your back.

‖ **Jugendstil** (yū·gǝnt,ʃtīl). [G., f. *jugend* youth (the name of a German magazine started in 1896) + *stil* style.] The German equivalent of *art nouveau*. Also *attrib.* (occas. with lower-case initial).

1928 *Minutes of G. & C. Merriam Co. Editorial Board* III. 707 Note the expression *Jugend Stil*, meaning nouveau art. **1950** B. S. Myers *Mod. Art in Making* 254 In the works of the *Jugendstil* painters in Germany.., those ideas found their expression. **1958** M. L. Wolf *Dict. Painting* 148 *Jugendstil*, in Germany, the name applied to the *art nouveau* movement, popular from 1890 to 1905, and affecting all of the visual arts. **1960** *Times* 4 May 16/6 The individual stature of the two leading figures of the *jugendstil* period represented in this exhibition. **1963** *Listener* 24 Jan. 160/2 The Jugendstil was much more exuberant, much more demonstratively independent of any past style, and much more violent. **1967** A. Wilson *No Laughing Matter* III. 369 'Not our happiest sort of architecture, I'm afraid.' 'No? I suppose it's the usual Jugendstil of its period.' **1972** *Guardian* 16 Mar. 14/4 In Cracow there's a political cabaret 80 years old, with hair-raising original *Jugendstil* décor. **1974** *Times Lit. Suppl.* 11 Oct. 1116/2 Sheer frustration may have thrown up a Jugendstil *Parsifal*.

jugful. (Earlier U.S. and further examples.)

1831 *Boston Transcript* 14 Nov. 2/1 'Vote on your side!' says another: 'Not by a jug full.' **1865** Dickens *Mut. Fr.* II. iv. vii. 217 Mr. Riderhood..fetched his jug-full of water. **1967** Wodehouse *Company for Henry* xii. 220 Did your heart melt? Not by a jugful. You gave him the sleeve across the windpipe and kicked him out. **1971** —— *Much Obliged, Jeeves* vi. 57 'Then why is Tuppy short of cash? Didn't he inherit them?' 'Not by a jugful.'

juggernaut, *sb.* Add: **3.** A large heavy vehicle; *spec.* a heavy lorry.

In quot. 1927 prob. just an ordinary motor car.

1841 Thackeray *Second Funeral of Napoleon* iii. 58 Fancy, then,..the body landed at day-break..and transferred to the car; and fancy the car, a huge Juggernaut of a machine. **1927** W. G. M. Dobie *Game-Bag & Creel* 20 How gladly would I have him killed or shot, That blind, fat fool who drove the Juggernaut! **1946** G. Tyrwhitt-Drake *Eng. Circus & Fair Ground* xvii. 198 In 1919 I bid £900 for one of these juggernauts [*sc.* a traction engine] to use for hauling my circus equipment. **1969** *Evening Star* (Ipswich) 28 Apr. 16/2 Experienced colleagues of mine are concerned about container lorries—the 30-ton juggernauts—which are completely disregarding speed limits. **1972** *Guardian* 30 Nov. 32/6 (*headline*) No entry for juggernauts. *Ibid.*, The juggernaut lorry will be kept out. **1972** *Oxford Mail* 11 Dec. 3/3 A plan to banish the juggernaut lorry from many Oxfordshire village roads is being prepared. **1973** *Press & Jrnl.* (Aberdeen) 7 Aug. 8/6 (*caption*) Sunshine and shade in the Church Square, Ballater. Continental juggernauts, on their way to the west coast, would certainly change this peaceful scene. **1973** *Daily Tel.* 10 Aug. 3, French police were last night interviewing 27 Asians—17 Pakistanis and 10 Indians found in Calais under a false floor in a juggernaut bound for Britain.

ju·ghead. *slang* (chiefly *U.S.*). [f. Jug *sb.*² + Head *sb.*] **a.** A mule; also, a stubborn horse. **b.** A foolish or stupid person; also, as a mere term of abuse.

1926 L. H. Nason *Chevrons* iii. 86 'Unload everything,' he says. Hear that, you jugheads? At four-thirty! **1936** D. McCarthy *Lang. Mosshorn, Jughead*, a horse which seemingly has no brains. **1942** *Sat. Even. Post* 3 Oct. 36/1 Unloose dat jughead jaw! **1966** *Publ. Amer. Dial. Soc. 1964* XLII. 18 Hardtail, a derogatory name for a mule. Others were *jarhead* and *jughead*. **1971** *Illustr. Weekly India* 25 Apr. 67/5 Jughead, what took you so long?

juglet (dʒʌ·glėt). *Archæol.* [f. Jug *sb.*² + -let.] A small vessel of jug shape.

1932 *Discovery* Sept. 305/1 The small juglets were commonly made in two halves and imperfectly joined together, though no mark was visible on the outer surface. **1942** *Antiquity* XVI. 187 The juglet with elongated stump base is not among the earliest types of Bronze Age pottery. **1957** K. Kenyon *Digging up Jericho* 98 Out of a total of 251 vessels..111 are juglets with a round base, a wide mouth and a rather high loop-handle.

juglone (dʒū·-, dʒʌ·glōᵘn). *Chem.* Also † **juglon.** [ad. G. *juglon* (C. Reischauer 1877, in *Ber. d. Deut. Chem. Ges.* X. 1542), f. L. *iūglans* walnut: see -one.] 5-Hydroxy-1,4-naphthoquinone, $C_{10}H_6O_3$, a reddish-yellow crystalline compound obtained from the green shells of walnuts and having fungicidal properties.

1878 *Jrnl. Chem. Soc.* XXXIV. 233 Juglone (Nucin)... This body, prepared from the green shells of walnuts (*Juglans regia*), has been analysed by the author, who assigns to it the empirical formula, $C_{36}H_{12}O_{14}$. **1887** *Chem. News* 9 Nov. 214/1 'Juglon'. An extended series of analytical results with regard to this body. **1892** Roscoe & Schorlemmer *Treat. Chem.* III. vi. 138 About 150 grms. of juglone are thus obtained from rather more than 100 kilos. of walnut-shells. **1925** *Phytopathology* XV. 783 Juglone..is an irritant causing violent sneezing and has been found..to be especially valuable as a medicant for skin diseases. **1949** E. P. Abraham in H. W. Florey et al. *Antibiotics* I. xiv. 619 Brissemoret and Michaud (1917) reported the use of juglone in clinical cases of eczema, impetigo, and other skin diseases. **1971** R. H. Thomson *Naturally Occurring Quinones* (ed. 2) iv. 222 Finely powdered juglone is an effective sternutator and like many other quinones has weak fungicidal and bactericidal properties.

Jugoslav, Jugoslavian: see *Yugoslav, etc.

jugular, *a.* and *sb.* Add: Now usually pronounced (dʒʌ·giŭlăɹ).

jugulo- (dʒū·-, dʒʌ·giŭlo), combining form of Jugular *a.* and *sb.*, Jugulum, in a few anatomical terms, as **ju:gulodiga·stric** *a.*, designating a pair of lymph nodes (sometimes also including several smaller associated nodes) in the neck (see quots.); **ju:gulo-omohy·oid** *a.*, designating either of a pair of lymph nodes in the neck (see quots.). See also Jugulo-cephalic *a.*

1920 Jamieson & Dobson in *Brit. Jrnl. Surg.* VIII. 80 Examined from in front, certain members are particularly conspicuous.., and merit special mention: these are a large gland and a variable number of smaller nodes lying in a triangle formed by the posterior belly of the digastric, common facial vein, and internal jugular vein. These may be called the jugulodigastric glands. **1967** G. M. Wyburn et al. *Conc. Anat.* iv. 120/1 The lymph node at the angle of the jaw, the jugulodigastric, drains the tonsil. **1972** C. R. & T. S. Leeson *Human Struct.* xxix. 151/1 Jugulodigastric. This node (or nodes) lies where the deep chain is crossed by the posterior belly of the digastric muscle. **1920** Jamieson & Dobson in *Brit. Jrnl. Surg.* VIII. 80 One gland..situated on or just above the omohyoid, frequently presents at the anterior border of the sternomastoid... From its connection with all parts of the tongue we have to refer to it frequently, and suggest the name jugulo-omohyoid gland. **1963** Tobias & Arnold *Man's Anat.* II. xv. 120 The jugulo-omohyoid node lies on the internal jugular vein just above the tendon of the omohyoid.

jugum. Add: **2. a.** *Zool.* In certain brachiopods with hinged shells, a process of the dorsal valve. **b.** *Ent.* In certain Lepidoptera, a lobe on the fore wing which is able to link it to the hind wing. Hence **ju·gal, ju·gate** *adjs.*

1888 Rolleston & Jackson *Forms Animal Life* (ed. 2) 693 One valve may have depression or sinuses to which correspond elevations or juga on the other. **1893** J. H. Comstock in *Wilder Quarter-Century Bk.* 43 There are two distinct ways of uniting the two wings of each side in the Lepidoptera; they may be united by a frenulum or..by a jugum. *Ibid.* 84 In this genus [*sc. Hepialis*]..they [*sc.* the wings] are joined by a membranous lobe extending back from near the base of the inner margin of the fore wing. To this lobe I have applied the name jugum. **1895** J. H. & A. Comstock *Man. Study Insects* xviii. 215 This projecting lobe is named the jugum or yoke; and the moths possessing this organ are termed the Jugatæ or the Jugate Lepidoptera. **1952** R. C. Moore et al. *Invertebr. Fossils* vi. 213/1 A variously shaped crossbar between the spiralia is called [the] jugum; it has considerable importance in classifying some groups of spine-bearing brachiopods. **1957** Richards & Davies *Imms's Gen. Textbk. Ent.* (ed. 9) 538 Wing-coupling apparatus of jugate type, the jugal lobe elongate and resting upon the hind wing. **1969** R. F. Chapman *Insects* x. 175 The Hepialidae have a strong jugal lobe which lies beneath the costal margin of the hind wing so that this is held between the jugum and the rest of the fore wing.

juice, *sb.* Add: **1. b.** Also more generally, alcoholic liquor. *U.S. slang.*

1932 *Evening Sun* (Baltimore) 9 Dec. 31/4 *Juice*, whisky. **1940** D. Ellington in *Swing* May 10/3 Everybody in our band at that time was a juice-hound, juice meaning any kind of firewater. **1956** B. Holiday *Lady sings Blues* (1973) xix. 157 There was no place I could work in New York—not if they sold juice there. **1961** R. Russell *Sound* 22 'Nuthin' at all like juice, either,' Hassan said. 'No hangover.' **1971** *Harper's Mag.* May 83 But they need their juice, for their kind of tension would not be relieved by the head-lightening stuff, they need the down-deep sleep of the intelligence that comes with liquor.

c. (*a*) The liquor from the sugar cane; (*b*) this made ready for evaporation.

(*a*) **1697** *Phil. Trans. R. Soc.* XIX. 381 The Juice of the Cane. **1784** P. H. Maty in *New Rev.* Sept. 194 To..cut the cane,..to have the juice expressed, and boiled into sugar. **1812** J. Taylor *Arbores Mirabiles* 39 The season continues..about six weeks, when the juice is found to be too thin and poor to make sugar. **1830** G. R. Porter *Nature & Properties Sugar Cane* 17 The cane contains three sorts of juice, one aqueous, another saccharine, and the third mucous.

(*b*) **1839** Ure *Dict. Arts* 1202 Where canes grow on a calcareous marly soil, in a favourable season the saccharine matter gets so thoroughly elaborated, and the glutinous mucilage so completely condensed, that a clear juice and a fine sugar may be obtained without the use of lime. **1887** *Encycl. Brit.* XXII. 626/1 Wetzel's pan,..and similar devices for the efficient evaporation of juice..are also in use.

d. Electricity, electric current. *slang.*

1896 *Boston Herald* 25 Dec. 4/5 Now we know what a blessing the trolley is—when the juice isn't turned off. **1903** *Electrical Engineer* 28 Aug. 327/2 The first he asked, a councillor Whose town had got the juice. **1916** 'Taffrail' *Pincher Martin* xiii. 238 Call her up by wireless... Don't make our name, but use all the juice you can, so that they'll think we're very close. **1928** U. Sinclair *Boston* (1929) xxiv. 724 The juice was turned off, and Vanzetti was officially pronounced dead. **1934** J. M. Cain *Postman always rings Twice* ii. 18 They got neon signs, they show up better, and they don't burn as much juice. **1966** H. Sheppard *Dict. Railway Slang* (ed. 2) 9 *On the juice*, running on electrified lines—particularly LTE.

e. Petrol. *slang.*

1909 *Installation News* III. 52/2 We are not faced with a threepenny tax on each gallon of 'juice'. **1918** E. M. Roberts *Flying Fighter* 281 Then I discovered that the tank was nearly empty. That meant that I would have to go in search of 'juice'. **1959** *N.Z. Listener* 12 June 20/4 'Turn the juice on!' He felt sheepish as he twisted the key. **1968** K. Weatherly *Roo Shooter* 56 The Rover had him worried. If she ran out of juice..he had to walk in. **1973** *Nation Rev.* (Melbourne) 24–30 Aug. 1399/6 'Tis cheaper to slow down—you use less juice then, be it petroleum or gastric.

f. A drug or drugs. *slang.*

1957 [see *gang-bang* s.v. *gang sb.*¹ 12]. **1972** H. C. Rae *Shooting Gallery* iii. 187, I wasn't interested in him. I mean, when you shoot juice, you lose the other thing.

4. c. Political influence (exercised by or on behalf of criminals); money paid to obtain immunity from prosecution, or lent at a usurious rate of interest, or the interest thus extorted; money acquired by corruption, gambling, or threats. Also *attrib.* *U.S.*

1935 A. J. Pollock *Underworld Speaks* 65/1 *Juice*, corrupt influence (shake-down) for protection to operate unlawfully. **1951** [see *ice sb.* 4 e]. **1961** *Chicago Daily Tribune* 12 Aug. 1 William ('Action') Jackson..a 'juice man' (loan collector) for syndicate hoodlum bosses. **1962** A. Buchwald *How Much is that in Dollars?* 75 'Well, use some juice,' Mr Cahn said. 'Juice' is a Hollywood expression which means influence. **1963** P. Wyden *Hired Killers* xii. 196 'Juice'—usurious interest of up to twenty per cent—was known to fester at the root of some of these assassinations. The juice racket has been flourishing for decades. **1964** [see *Cosa Nostra]. **1968** *N.Y. Times* 9 June 1, 29 At least two murders and perhaps more have been connected to the loan shark, or 'juice' racket, as it is called here, as well as beatings and threats. **1969** *Time* 11 July 24 This Las Vegas is..a venal demimonde in which the greatest compliment that can be paid a man is to say that he has 'juice' (influence in the right places). **1970** E. R. Johnson *God Keepers* (1971) xiv. 146 Vito Lucchese was involved in the case and..he had a certain amount of juice around the city. **1971** *Ink* 12 June 14/2 His high-paid whizkid managers weren't whizzing too well so he went after some extra juice.

6. **juice-head** *slang*, an alcoholic; **juice-joint** *N. Amer. slang*, a bar, club, or stall serving either alcoholic or non-alcoholic liquor.

1955 S. Whitmore *Solo* 247 The juiceheads..got so fractured [i.e., drunk] that they wouldn't show up for a date. **1967** *New Yorker* 9 Sept. 41 If anybody wanted to get stoned the guy who owned the pad made them go up on the roof. Juice-heads drank Red Mountain. **1969** A. H. Cain *Young People & Drugs* 159 *Juice head*, one whose hang-up is booze; an alcoholic. **1927** K. Nicholson *Barker* 149 *Juice joint*, soft drink stand. **1932** *Evening Sun* (Baltimore) 9 Dec. 31/4 *Juice-joint*, speakeasy. **1958** G. Lea *Somewhere there's Music* iv. 35 Six lonely nights a week in a juice joint. **1960** Wentworth & Flexner *Dict. Amer. Slang* 298/2 *Juice-joint*. 1. A soft-drink tent, stand, booth, or concession. Carnival and circus use. 2. A speakeasy; a bar or nightclub. Orig. 1920 use. **1970** C. Major *Dict. Afro-Amer. Slang* 72 *Juice joint*, 'tavern, bar, cabaret.

juice, *v.* Add: **2.** To animate, liven *up*, inspire. *slang.*

1964 *Time* 23 Oct. 61 A thing like that can really juice you up. **1972** J. Mills *Report to Commissioner* 259 The departmental surgeon asked Jackson if he wanted him to give Lockley a shot of something, he meant juice him up a little, keep him from passing out.

juiced, *a.* Add: **2.** Drunk. Also const. *up*. *slang.*

1946 C. Himes *Black on Black* 260 She was an old wino used to come there every night and get juiced up. **1955** D. W. Maurer in *Publ. Amer. Dial. Soc.* XXIV. x. 170 Men sufficiently *juiced up* to be robbed without much interference. **1956** B. Holiday *Lady sings Blues* (1973) ii. 17 The doctor said if her man had even come to enough to raise the window and let in some air he could have saved her. But he was too juiced even for that. **1968** A. Young in A. Chapman *New Black Voices* (1972) 149 Chicken Hawk and Wine, well-juiced, eased quietly up the back steps. **1969** *Time* 24 Oct. 70 Later Marvin apologized: 'I'm sorry I was so rotten this afternoon. I was a little juiced.' **1971** 'S. Ransome' *Trap 6* (1972) xiii. 147 He was sitting at the bar brooding over a drink—not making any trouble, not getting juiced up.

juice-harp. Corrupted form of Jews' harp.

1942 Berrey & Van den Bark *Amer. Thes. Slang* § 577/22 Jews'-harp, jaw-harp, juice-harp. **1950** Blesh & Janis *They all played Ragtime* (1958) vi. 109 The childhood band he led, consisting of two tin whistles, jew's-harp ('juice-harp'), and triangle, was a country rag band. **1959** M. Stearns in M. T. Williams *Art of Jazz* (1960) ii. 9 Sonny Terry listened to his father's harmonica and 'juice' harp. **1968** *Publ. Amer. Dial. Soc.* XLIX. 15 The decline of *juice harp* and *jew's harp* and the students' use of *mouth harp*.

juicer (dʒū·sǝɹ). [f. Juice *sb.* + -er¹.] **1.** An electrician. *slang.* (Cf. *juice *sb.* 1 d.)

1928 *Amer. Speech* III. 366 'Juicers' have a most juicy vocabulary of their own. Thanks to them Hollywood has 'kliegs' (klieg lights). **1934** *Tit-Bits* 31 Mar. 12/2 Here are some more studio terms. 'Juicers' are electricians. **1950** B. Schulberg *Disenchanted* (1951) i. 11 Assistant directors, cutters, bit players, second cameramen, sound men, juicers, grips, special effects men, Hollywood's exclusive proletariat. **1957** V. J. Kehoe *Technique Film & Television Make-up* i. 18 He directs the..juicers to place the lights in the most effective positions.

2. An appliance used to extract juice from fruit and vegetables.

1938 *Cooperative Distributors Catal.* in *Amer. Speech* (1942) XVII. 272/1 A well constructed juicer for oranges. **1945** *Archit. Rev.* XCVII. 152 (*caption*) The 'Dazey' super juicer..in aluminium. **1961** *Listener* 31 Aug. 331/2 Self-contained grinders, liquidizers, juicers and meat mincers are also made. **1972** *New York* 1 May 27/2 Juicers—used to extract the juice of fresh vegetables and fruits.

3. An alcoholic. *U.S. slang.*

1967 *New Yorker* 13 May 40 The difference between the juicers and the heads is that the juicers sometimes get a little obstreperous but the heads go sit quietly in a corner someplace. **1970** *Ibid.* 28 Feb. 31/1 'Forget him, he's catatonic.' 'Forget him, he's a juicer.'

juicily (dʒū·sili), *adv. slang.* [f. Juicy *a.* + -ly².] **a.** Excellently, vigorously, well. **b.** Suggestively.

1916 E. F. Benson *David Blaize* vi. 118 [He] hit it juicily to square leg. **1927** Wodehouse in *Sunday Express* 18 Oct. 9 Abstemious cove though I am as a general thing ..on this occasion..I had been doing myself rather juicily. **1969** *Daily Tel.* 3 Dec. 13/8 The hand-outs asked juicily whether the girls who took on these jobs were really on the first step to the harem and white slavery.

juicy, *a.* Add: **2. c.** Suggestive, esp. in a sexual way; piquant, racy, sensational. *colloq.*

1883 J. Greenwood *Odd People* 59 'Let me play you a tune, then,' said the frightened lad... 'All right, then. Play us something juicy,' exclaimed the ruffian. **1908** Kipling *Lett. of Travel* (1920) 153 They interpolated no juicy anecdotes of murder or theft among their acquaintance. **1920** S. Lewis *Main St.* ix. 102 The gang..gathered in a snickering knot to listen to the 'juicy stories'. **1923** A. Bennett *Riceyman Steps* I. ii. 4 Accounts of bloody crimes and juicy sexual irregularities. **1929** [see *duck soup* s.v. *duck sb.*¹ 12]. **1953** [see *gang-bang* s.v. *gang sb.*¹ 12]. **1958** B. Nichols *Sweet & Twenties* 106 Extracting what he was pleased to call 'the juicy bits'. **1965** 'O. Mills' *Dusty Death* xix. 177 'Aren't you going to give us a look at it?' 'Not this one, no. Far, far too juicy.' **1972** J. McClure *Caterpillar Cop* xiii. 218 His parents sound the sort who'd run a mile before they'd say the word 'sex'. They wouldn't have any juicy books in the house either.

d. Excellent, vigorous, first-rate; serious; profitable.

1916 E. F. Benson *David Blaize* vi. 119 It didn't often happen that the first ball of an innings was slogged for six. Juicy hit, too! **1922** G. Ade *Let.* 8 Mar. (1973) 80 The author who has only a few stories ready to market can see no prospect of juicy returns. **1934** *Time & Tide* 8 Sept. 1122/2 But still, with this juicy price in prospect, the shrewd professionals are hesitant. **1948** Partridge *Dict. Forces' Slang* 104 *Juicy*, (of targets) easy, or well worth the trouble of destroying. **1962** *Times* 18 Oct. 6/3, I have been thinking of some very juicy ways of how we can use the South Carriage Drive in one of the royal parks.

ju-jitsu (dʒū·dʒi·tsu), *sb.* Also formerly **jiu-jitsu, -jutsu, ju-jitzu, -jutsu;** and as one unhyphenated word. [ad. Jap. *jūjutsu*, f.

jū (Chinese *jou*: see *judo) + *jutsu* (Chinese *shu, shut* art, science).] A Japanese system of wrestling and physical training, characterized by the use of certain techniques and holds to overcome an adversary. Cf. *judo. Also *attrib.*

1875 *Japan Mail* 10 Mar. 133/1 *Jiu-jitsu* (wrestling) is also taught, but not much practised by gentlemen. **1889** [see *judo]. **1891** L. Hearn *Let.* Nov. in E. Bisland *Life & Lett. L. Hearn* (1906) II. 70 A building in which jū-jutsu is taught by Mr. Kano. **1893** —— *Let.* 13 Oct. (1910) 183, I am working out an essay—a philosophical essay on 'Jiujutsu'. **1895** J. Inouye *Wrestlers & Wrestling* 3 These methods were adopted and extensively practised by *Samurai*, and were finally developed into what is now known as *Jujitsu*. **1905** *Daily Chron.* 21 Feb. 7/4 Their gymnasium is often visited by jiu-jitsu wrestlers. **1910** H. G. Wells *Hist. Mr. Polly* vii. 238 A combination of something romantic called 'Ju-jitsu' and..the 'Police Grip'. **1925** N. Venner *Imperfect Impostor* xvi, Jos Polkins ..enwrapped him in a benevolent jiu jitsu grip that left him powerless to move. **1946** R. Benedict *Chrysanthemum & Sword* (1947) xii. 277 He perfects himself in jujitsu or sword play.

fig. **1906** R. Whiteing *Ring in New* xxix. 206 To lay him flat on his back by a sort of intellectual jiu-jitsu. **1928** F. Romer *Numbers Up!* 11 'Revenge?'..'nothing of the kind. I shall merely practise Moral Jiu-jitsu.' **1942** Wyndham Lewis *Let.* 15 July (1963) 334 There are omens of promise—in much that Mr. Roosevelt has done, or in the splendid massive jujitzu of the Russians. **1965** K. Briggs in Battiscombe & Laski *Chaplet for C. Yonge* 25 Practising a kind of spiritual ju-jitsu, in which by falling with the misfortune you overcome it.

Hence **ju-ji·tsu** *v. trans.*, to overcome by means of ju-jitsu; **ju-ji·tsian, ju-ji·tsuist**, one who teaches or practises ju-jitsu.

1905 D. Sladen *Playing Game* II. iv. 202 The wiry little Japanese having Jujitsu'd the three biggest men on the Russian flag-ship. **1905** Hancock & Higashi *Compl. Kano Jiu-Jitsu* p. v, These famous *jiu-jitsians*, Hoshino and Tsutsumi. **1905** *Westm. Gaz.* 23 Nov. 4/2 He..issued ..a challenge to all jiu-jitsuists of the world. **1928** *Observer* 4 Mar. 15/2 They seem to me to put up no fight at all, and to be very easily ju-jitsued by the Japanese servants.

ju-ju² (dʒū·dʒū). *slang.* [redupl. form of *mari)ju(ana.] A marijuana cigarette. Also *attrib.*

1940 R. Chandler *Farewell, my Lovely* xi. 83 'I knew a guy once who smoked ju-jus,' she said. 'Three high balls and three sticks of tea and it took a pipe wrench to get him off the chandelier.' **1963** N. Freeling *Because of Cats* x. 163 'He had juju cigarettes too; like Russians, with a big mouth piece and pretty loose...' 'The jujus are—you feel very clever.'

juke (dʒūk), *sb. slang* (orig. *U.S.*). Also **jook, jouk.** [Prob. f. Gullah *juke, joog* disorderly, wicked, of W. Afr. origin; cf. Wolof *dzug* to live wickedly.] **1.** A roadhouse or brothel; *spec.* a cheap roadside establishment providing food and drinks, and music for dancing. In full *juke-house, juke-joint*.

1935 Z. N. Hurston *Mules & Men* I. iii. 82 They talked and told strong stories of Ella, Wall, East Coast Mary,..and lesser jook lights around whom the glory of Polk County surged. **1936** *Scribner's Mag.* Dec. 27/2 Jim's daddy owned the General Store and a nigger jook. **1937** in *Florida Rev.* (1938) Spring 28/1 Back yonder a 'juke' was a place, usually a shack somewhere off the road, where a field negro could go for a snort of moonshine. *Ibid.*, There were negro juke-joints as far back as I can remember. **1941** J. Faulkner *Men Working* ii. 39 The glow from the lights of the jook house at the lake appeared above the trees. **1956** S. Longstreet *Real Jazz* xviii. 151 *Juke* from juke box came from juke house—which was once a whorehouse. **1958** P. Oliver in P. Gammond *Decca Bk. Jazz* i. 23 The crude, wood-frame dance-halls called 'jooks' (or jukes). **1964** *Amer. Folk Music Occasional* I. 93 You go into the bars and the juke joints and you ask around. **1968** *Blues Unlimited* Nov. 6 Now Ike told Dave that he cut Elmore's 'Dust my broom' at a Canton juke-joint. **1971** *Black World* June 72/2 Had done sent Lueta and Carol Ann to every juke joint in Greenwood askin bout you.

2. **juke-box** (occas. **juke-organ**), a machine that automatically plays selected gramophone records when a coin is inserted; also ellipt. as *juke*. Also *attrib.*, *transf.*, and *fig.*

1937 in *Florida Rev.* (1938) Spring 25/3 The screeching of the 'jook' organ. **1939** *Time* 27 Nov. 56/2 Glenn Miller attributes his crescendo to the 'juke-box', which retails recorded music at 5 ¢ a shot in bars, restaurants and small roadside dance joints. *Ibid.* 25 Dec. 3/1 To the Florida Man such an instrument is a jook organ. **1942** D. Powell *Time to be Born* (1943) x. 241 Corinne put a quarter in the juke-box to play 'Let's Be Buddies' five times. **1944** Auden *For Time Being* (1945) 65 War has become Like a juke-box tune that we dare not stop. **1947** *Gramophone* Dec. 95/1 It is the Petrillo thesis that records which are played in juke-boxes and broadcast over the radio..are monsters which have threatened the musicians' very existence. **1947** [see *jive v.* 2 a]. **1954** *Archit. Rev.* CXVI. 92/2 The stupefying juke-box façade of the Wertheim project. **1959** C. MacInnes *Absolute Beginners* 83 See all the kids jam-packed in there beside the jukes? **1961** L. Mumford *City in Hist.* viii. 224 Pennsylvania Station in New York retains this noble quality—or did until that structure was converted..into a vast jukebox. **1968** J. Winearls *Mod. Dance* (ed. 2) vi. 143 There emerges a constant from which the immediate must not depart. Our teenage Juke Box Juries demonstrate this ably. **1973** C. Bonington *Next Horizon* xvii. 244 A coffee bar for teenagers, complete with juke box. **1973** *Nation Rev.* (Melbourne) 31 Aug. 1453/4 The America they traverse, of fibro suburbs and prison farms and jukebox bars, is..correctly and compassionately observed.

juke (dʒūk), *v. slang* (orig. *U.S.*). Also **jook, jouk.** [Cf. prec.] *intr.* To dance, esp. at a juke-joint or to the music of a juke-box (see also quot. 1958). So **ju·king** *vbl. sb.*

1933 W. Roland (*title of disc*) Jookit Jookit. **1937** C. R. Cooper *Here's to Crime* ix. 190 In the 'jukin' joints' there is, of course, the prime requisite of liquor. **1941** *Amer. Speech* XVI. 319/2 'Let's jouk' is an invitation to dance, but 'let's go joukin'' is a request for a date. **1958** T. Williams *Orpheus Descending* I. 28 I'd like to go out jooking with you tonight... That's where you get in a car and drink a little and drive a little and stop and dance a little to a juke box. **1960** *20th Cent.* Aug. 144 But living, in these terms, is reduced to jooking. **1967** *Daily Tel.* 15 May 12/8 To juke also came to mean to dance and to go pub-crawling.

jukskei (yō·kskēⁱ). *S. Afr.* Also **jeuk skei, jukschei.** Pl. **jukskeie, jukskeis.** [Afrikaans, f. *juk* yoke + *skei* pin, Skey *sb.*²; cf. *yoke-skey*.] **1.** A yoke-skey.

1822 W. J. Burchell *Trav. S. Afr.* I. viii. 151 The yokes are straight, and pierced with two pair of mortices to receive the *jukschei* which fits in loosely, and answer to what in English husbandry are called the *bows*. **1871** Lord & Baines *Shifts & Expedients Camp Life* ix. 452 Near each end are two mortices..through which to pass the 'jeuk skeis', or yoke keys, which keep it in place on the neck of the ox. **1971** *Evening Post Mag.* (Port Elizabeth) 27 Feb. 2 A stinkwood yoke, with 'jukskeis', serves as a hatstand.

2. A quoits-like game; the bottle-shaped 'quoit' used in this game.

[**1934** C. P. Swart *Africanderisms* (M.A. Thesis, S. Afr.), *Jukskeigooi*, a game, very similar to quoits, played by the Boers, a skey being used instead of the customary circular ring.] **1942** *Cape Times* 10 Nov. 4/2 Saturday's Jukskei results were as expected. **1947** *Ibid.* 21 Apr. 7 The jukskeis were brought..by Mr. Tom Naude. **1950** S. de Wet *Hour of Breath* ix. 69 In the late afternoon we used to play Jukskei on the river bank below our houses. **1956** *Cape Times* 26 Jan. 1/1 A few minutes after throwing about six practice *jukskeie* at the Newlands circus grounds.. Mr. P. A. Meiring..dropped dead. **1971** *Progress* (Cape Town) May 2/4 No one has challenged us at jukskei.

juldie, var. *jildi.

julienite (dʒū·liĕnǝit, yūli,ēⁱ·nǝit). *Min.* [ad. Flemish *juliĕniet* (A. Schoep 1928, in *Natuurwet. Tijdschr.* X. 58), f. the name of Henri *Julien* (d. 1920), Belgian geologist: see -ite¹.] A hydrous thiocyanate of sodium and cobalt, $Na_2Co(SCN)_4.8H_2O$, that occurs as minute blue needles and has been made artificially; orig. wrongly regarded as a cobalt chloronitrate.

1928 *Mineral. Mag.* XXI. 567 Julienite... Hydrated chloro-nitrate of cobalt as minute blue needles, presumably hexagonal and isomorphous with buttgenbachite: from Katanga. **1954** *Mineral. Abstr.* XII. 337 Needles of julienite.., prepared by digesting Na-Co nitrite with an aqueous solution of Na sulphocyanide, are tetragonal. **1969** I. Kostov *Mineral.* 117 The interesting mineral julienite..is obtained during the exogenetic decomposition of primary cobalt sulphides.

julienne. Add: **2.** Applied *attrib.* or as *adj.* to vegetables cut into small thin strips.

1889 A. B. Marshall *Cookery Bk.* ii. 31 *Julienne Garnish.*—Peel and cut the vegetables, such as carrots, turnips, etc., into strips about an inch long. **1906** A. Filippini *Internat. Cook Bk.* 228 Cut with a sharp knife—if no julienne-shaped potato cutter is at hand—into even julienne, match-like strips. **1920** [see *cornflakes* sb. pl. s.v. *corn sb.*¹ 11]. **1971** *Salads* (Cordon Bleu Cookery School) 100/2 Wash, peel and slice celeriac, cut across into julienne strips (about 1½–2 inches long).

Juliet (dʒū·liĕt). [Female personal name (F. *Juliette*, It. *Giulietta*), dim. of *Julia*.] *Juliet cap* (see quot. 1957).

1909 *Westm. Gaz.* 9 Feb. 8/3 Their Juliet caps were composed of violets. **1930** *Daily Tel.* 7 Apr. 7/6 The 'Juliet' cap idea is to be found in the little theatre hats worn abroad. **1957** M. B. Picken *Fashion Dict.* 49/2 *Juliet cap*, small, round cap of wide, open mesh, usually decorated with pearls or other jewels, similar to that worn on the stage by Shakespeare's Juliet. Worn chiefly for evening. **1973** *Times* 15 Nov. 6/3 The bridesmaid..wore a pinafore dress and a jewelled Juliet cap.

jum, júm: see Joom.

jumar (dʒū·măɹ). *Mountaineering.* [Swiss name.] A clip which when attached to a fixed rope automatically tightens when weight is applied and relaxes when it is removed, thus facilitating the climbing of the rope; also, a climb using jumars. Hence as *v. intr.*, to climb with the aid of jumars; **ju·maring** *vbl. sb.*, the action of so doing.

1966 *Climbers' Club Jrnl.* 77 Somebody spotted him dangling from his Jumars, half-way down one of the fixed ropes. **1968** P. Crew *Encycl. Dict. Mountaineering* 73/1 Jumars are the most effective device for prusiking, but they have the disadvantage of not working well on iced ropes. **1969** *Sunday Times* (Colour Suppl.) 16 Feb. 38/3, I dreamed that I was on the ropes setting off for the summit by myself and I was jumaring all night without getting anywhere. **1971** C. Bonington *Annapurna South Face* viii. 90 Once the ropes are in place, the climbers no longer move roped together, but simply clip onto the rope, using on the upward sections a device called a jumar clamp. This is a metal handle, with a knurled lever which is fitted over the rope. It works on a ratchet principle, sliding easily up the rope, but biting into it when put under tension. *Ibid.* xi. 125 This was the first time that this type of free jumaring had ever been necessary at this altitude in the Himalayas. **1971** D. Haston in *Ibid.* xvii. 214 The sun was two hours away from the gully when we started the upward jumar. **1972** —— *In High Places* 2 Onward, outward, creeping on jumars and rimed ropes. It is cold... Dachstein-mitted hands freeze in jumar clutch. *Ibid.* ix. 104 The hook fitted, a quick move, and I was up. Mick jumared and I pushed on again. **1973** C. Bonington *Next Horizon* ix. 141 Later, I learned that an essential precaution for any jumaring is to tie a knot in the rope, so that if the jumars do slip on the rope, you don't slide straight off the end.

jumbal, jumble. (Later examples.)

1892 *Mrs. Beeton's Bk. Househ. Managem.* xl. 1125 California jumbles... sugar,..butter,..flour,..grated lemon-peel..whites of 4 eggs. **1923** Chesterton *Fancies versus Fads* ii. 21 It involved eating jumbles (a brown flexible cake now almost gone from us). **1974** *Daily Tel.* 29 Apr. 16/7 Miss Lane mentions the absence from the glossary of such terms as cracknells, jumbells, trencher-bread, etc.

jumbee, jumbi, jumbie, varr. *jumby.

jumble, *sb.*¹ Add: **1. b.** *collect. sing.* Articles for a jumble-sale; also, a jumble-sale or sales. *colloq.*

1931 *Times* 16 Mar. 1/3 Maternity Hospital, holding annual Jumble Sale.—Please deluge us with jumble. **1932** *Daily Tel.* 17 Mar. 1/2 Do please help us with our Easter Jumble on March 18th by sending anything saleable, old or new. **1962** [see *fête sb.* 1 b]. **1966** *Listener* 20 Oct. 570/1 This feat of administration, this orgy of jumble and whist. **1973** J. Burrows *Like Evening Gone* ii. 27 When did the scouts have their jumble? I'd have thought every gloryhole..was empty.

Jumble (dʒʌ·mb'l), *sb.*² *slang.* [Corruption of John Bull.] A black man's nickname for a white man. Also *attrib.* or as *adj.*

1957 C. MacInnes *City of Spades* I. iii. 17 'You're a Jumble, man... That's what we call you... It's cheeky, perhaps, but not so very insulting.' 'May I enquire how it is spelt?' 'J-o-h-n-b-u-l-l.' '..But pronounced as you pronounce it?' 'Yes: Jumble.' **1957** *Listener* 12 Sept. 402/1 Jumble, a happy corruption of John Bull, is the Englishman's nickname in the mouths of the thousands of Africans and West Indians who have flocked to London since the war. *Ibid.*, The Jumble capital. *Ibid.*, An alien and uncomprehending Jumble world. **1961** M. Dickens *Heart of London* II. 190 Get all you can out of the Jumbles. *Ibid.* III. 294 He feeling his way about the Jumbles, he got no time to worry about Trinidad.

ju·mble-bead. [Alteration of *jumby-bead* (see *jumby b), prob. after *mumble*.] The particoloured seed of the jequirity.

1855 R. G. Mayne *Expos. Lex. Med. Sci.* (1860) 554/2 *Jumble beads*, an irreverent name for the seeds of the *Abrus precatorius*, from the purpose they are applied to in forming rosaries. **1887** [see Jequirity]. **1951** A. Nelson *Medical Bot.* xxiii. 456 The seeds [of *Abrus precatorius*] may be referred to as Abrus, Jequirity, Jumble beads or Prayer beads, and are known to contain two poisonous principles.

jumbo. Add: **1. b.** (Earlier and later examples.) Also *Comb.*, as **jumboburger** *U.S.*, a large hamburger; **jumbo jet**, a large jet aeroplane with a seating capacity of several hundred passengers; **jumbo-size(d)** *a.*, of a large size.

1897 *Sears, Roebuck Catal.* 12/2 Peaches, Jumbo California, halves. **1916** *Amer. City* Apr. 373/1 Large jumbo peanuts were bought, instead of culls. **1940** S. Lewis *Bethel Merriday* i. 19 The three children..supped on a jumbo malted milk. **1949** *Sat. Even. Post* 2 Apr. 116/3 Davis..is a jumbo-sized (six feet two, 202 pounds) man of forty-six. **1958** *Daily Express* 1 Apr. 1/4 You and your jumbo Martinis!! **1959** *Observer* 8 Nov. 3/5 We can expect the kind of hamburger proliferation that exists in America, with beefburgers, eggburgers, cheeseburgers, jumbo burgers, superior 'Hamburger Heavens'. **1960** *Guardian* 9 Sept. 8/4 The jumbo-sized plastic bone. **1964** *Punch* 15 July 79/3 The silly girl spends most of her time eating jumboburgers. **1964** *Economist* 19 Dec. 1317/2 Some..airlines do not expect..an American supersonic airliner ever to go into production. Jumbo-jets, with 500 seats and two decks,..look much cheaper to fly. **1965** *Teacher's World* 18 June 5/4 The shelves of the local supermarket are full of 'king-sized', 'giant economy' and 'jumbo' packages. **1965** *Business Weekly* 25 Dec. 31 Nosing into the Jumbo jet race: Lockheed, Boeing and Douglas. **1968** 'R. Raine' *Night of Hawk* xxvii. 126 The meal was avocado salad, a jumbo-size steak, and a mountain of ice cream. **1969** *Listener* 13 Nov. 683/3 They can replace two flights on conventional aircraft with a single jumbo jet flight. **1970** *Drum* (E. Afr. ed.) Feb. 31/3 This airport will be capable of handling jumbo jets and is ideally situated in the heart of the safari country. **1972** J. Rossiter *Rope*

for General Dietz iii. 34 The shrill rasping of a jumbo-size cicada..woke me. **1973** *Sun* 18 Jan. 6 The Prime Minister handled his jumbo Press conference amid the splendour of Lancaster House with poise and style.

c. *Engin.* Any of various large types of equipment used in drilling, lifting, dumping, etc.

1908 *Sat. Even. Post* 7 Nov. 11/3 Nearest the portal was the 'jumbo', a great, movable platform through which the 'muckers' dumped their barrow-loads to cars beneath. **1909** Webster, *Jumbo*,..a large traveling carriage for transporting excavated material, as in tunnel driving. **1950** *Engineering* 10 Mar. 264/3 The method is well illustrated in the Stockholm tunnel, using a truck-mounted jumbo. **1951** *Ibid.* 12 Jan. 34/2 The concrete was placed from a travelling frame or 'jumbo', as it is termed in American practice. This straddled the pipe and travelled on rails laid in an excavated trench.

d. (See quot. 1948.)

1912 *Outing* Aug. 629/2 The pilot schooners when lying at their stations, the sail carried being jumbo or forestaysail, foresail, and reefed mainsail. **1916** F. W. Wallace *Shack Locker* (1922) i. 22 Slack off that jumbo an' the foresheet! **1927** *Canad. Fisherman* 141/2 A gigantic sea boarded the vessel and broke about fifteen feet up the jumbo stay. **1932** J. Barbour *48 Days Adrift* vii. 94 'Hard, down hard,' I cried; 'haul over the jumbo sheet. If we are not careful we will all be lost this time.' **1948** R. de Kerchove *Internat. Maritime Dict.* 377/1 *Jumbo* (U.S.). 1. The forestaysail on fore-and-aft-rigged vessels. 2. A triangular sail which sets, point downward, on the foreyard of a square-rigged vessel or a topsail schooner in place of the regular foresail. **1956** A. F. Loomis *'Hotspur' Story* x. 117 The weather looked a little threatening and I set the jumbo and storm jib.

e. Short for *jumbo jet.

1966 *New Statesman* 22 Apr. 591/2 The competitors..need only a small handful of jumbos..on the popular long-distance routes. *Ibid.*, A rival jumbo designed by Douglas. **1971** *Daily Tel.* (Colour Suppl.) 8 Jan. 15/4 A collision between two jumbos on the approaches to Heathrow could produce a deathroll of thousands. **1972** H. Osborne *Pay-Day* III. xiii. 154 They can't get him on a plane before tonight. They want someone to go down on the Jumbo today. **1973** *Guardian* 12 Mar. 10/4 Jumbo-loads of Soviet immigrants. **1974** *Observer* (Colour Suppl.) 29 Sept. 65/1 The visitors who disembarked from our jumbo at Kai Tak Airport.

jumboize (dʒʌ·mbo,ǝiz), *v.* [f. Jumbo + -ize.] *trans.* To enlarge a ship, esp. a tanker, by inserting a new middle section between the bow and stern. So **ju·mboized** *ppl. a.*; **ju·mboizer**; **ju·mboizing** *vbl. sb.*

1956 *N.Y. Times* 23 Dec. S11/7 (*heading*) Concern to build ten and 'jumboize' nine. *Ibid.*, 'Jumboizing' consists of building an entire new center section for a tanker. The existing vessel is then cut off at the bow and stern which are later joined to the new section. **1957** *Sun* (Baltimore) 4 Apr. 16 We're 'jumboizing' present tankers. *Ibid.*, First completed 'jumboized' oil tanker. **1957** *New Yorker* 14 Sept. 33/3 In language controls, the trend is toward bastardization, if you ask us, and we wish the National Association of Manufacturers would call in all the jumboizers and miniaturizationists and bang their heads together. **1961** *Economist* 28 Jan. 366/1 Conversions and jumboizing within 56 days. *Ibid.*, The Schlieker yard had 'jumboized' on the spot the T2 tanker *Kaposia*. **1973** *Daily Colonist* (Victoria, B.C.) 10 Feb. 25/2 Ferries on this route currently do not have the double kitchens of the jumboized ships on the Shwartz Bay–Tsawassen service.

jumbuck. Add: (Earlier and later examples.) Also *N.Z.*

1824 W. Walker *Let.* 26 Jan. in W. S. Ramson *Austral. Eng.* (1966) vi. 107 They smacked their lips and stroked their breasts, 'boodjerry patta! murry boodjerry!—fat as jimbuck!!' i.e. good food, very good, fat as mutton. **1871** C. L. Money *Knocking about in N.Z.* i. 7 A pivot..shot the unsuspecting 'jumbucks' into the water below. **1896** H. Tichborne *Noqu Talanoa: Stories S. Seas* 95 Our girls very often possess fathers who own considerable stretches of 'jumbuck' property. **1933** L. Acland in *Press* (Christchurch, N.Z.) 28 Oct. 17/7 *Jumbuck*, slang for sheep..has always been in common use here. **1934** [see *guts v.]. **1967** *Sunday Mail Mag.* (Brisbane) 16 Apr. 2 The swagman who shoved a jumbuck in his tucker bag was stealing a sheep.

jumby (dʒʌ·mbi). Chiefly *West Indies.* Also **jumbee, jumbi, jumbie, zumbi.** [ad. Kongo *zumbi* fetish.] A ghost or evil spirit among American and West Indian Negroes. Cf. *zombie.

1871 C. Kingsley *At Last* II. x. 56 Out of the mud comes up—not jumbies, but—a multitude of small stones. **1876** R. F. Burton *Two Trips Gorilla Land* II. 124 There was no danger of the Zumbi, or ghost. **1887** W. H. Bentley *Dict. & Gram. Kongo Lang.* 505 The fetish Zumbi is supposed to bring good luck with it. It sometimes consists of a bundle of charms, at others it is an image, or even an animal. **1891** 'J. Evelyn' *Baffled Vengeance* iv. 60 The 'jumbies' (evil spirits) that haunted the ill-omened spot. **1894** 'A. Spinner' *Study in Colour* iv. 47 To tremble over the terrible tales of the Jumbi and Duppies, that..terrify belated travellers. **1918** C. W. Beebe *Jungle Peace* (1919) vii. 138 Sam had formerly been a warden in the Georgetown jail, and rumour had it that he left because he saw 'jumbies' in the court where one hundred and nine men had been hanged. **1951** E. A. Mittelholzer *Shadows move among Them* I. ii. 18 A pencil drawing of an ugly face—the face of a jumbie as imagined by Berton. *Ibid.* II. ix. 239 Believe in fairies and ghosts and jumbies and goblins. **1955** *Caribbean Q.* IV. I. 32 Caribs..delight in telling as a matter of fact how they met a certain 'jumby' in the road after dark. **1973** *Daily Tel.* (Colour Suppl.) 21 Sept. 25/3 To comb your hair at night is an open invitation for the Jumbies to call. (Jumbies are West Indian ghosts.)

b. *attrib.*, as **jumby-bead**, the hard seed of any of several West Indian plants; **jumby-bean** = *jumby-tree*; **jumby-bird**, a bird of ill omen, esp. an owl; **jumby-tree**, any of several West Indian trees, esp. the lead tree, *Leucæna glauca*.

1802 H. Swinburne *Let.* 17 Feb. in *Courts of Europe* (1841) II. 339, I gathered to-day, a handful of Jumbee beads. **1871** C. Kingsley *At Last* II. xiii. 197 Hedges of dwarf Erythrina, dotted with red jumby beads. *Ibid.* xvi. 276 The scarlet flowers of the Jumby-bead bush. **1934** J. Rhys *Voy. in Dark* I. vi. 66 The niggers say that jumbie-beads are lucky, don't they? **1956** *Caribbean Q.* IV. III. 198 They also sold charms made of 'Jumby beads' (*abrus precatorius* and allied species). **1920** Britton & Millspaugh *Bahama Flora* 162 *Leucaena glauca*... Probably native of continental tropical America. Jumbie Bean. **1871** C. Kingsley *At Last* II. xiii. 206 The obnoxious bird was not an owl, but a large goat-sucker, a Nycteribius, I believe, who goes by the name of jumby-bird among the English Negros. **1893** A. Newton *Dict. Birds* II. 471 Jumby-bird, a Negro name for almost any kind that is of bad omen, but especially for an Owl. **1923** E. Ingersoll *Birds in Legend* 168 The 'jumbie-bird', or 'big witch', of the West Indian region..is the dead-black ani, a kind of cuckoo. **1960** J. Bond *Birds W. Indies* 120 Barn Owl *Tyto alba*. Local names: Owl; Screech Owl; Night Owl; Death Owl; Death Bird; Jumbie Bird. **1928** M. Summers *Vampire* 265 The occult silk-cotton tree (*bombax ceiba*, often known as the Devil's tree or Jumbie tree).

Jumeau (dʒū·mo). [Proprietary name.] The name of Pierre François and Émile *Jumeau*, 19th-century French doll-makers, used to designate a doll manufactured at their establishment; freq. *attrib.*

[**1888** *Official Gaz.* (U.S. Patent Off.) 27 Nov. 950/2 Dolls. Emile Jumeau, Paris, France. Application filed Oct. 8, 1888. Used since 1840. 'The words "Bébé Jumeau".'] **1897** *Pearson's Mag.* July 60/1 The Bébé Jumeau is known all over the world. *Ibid.* 65/1 The Jumeau dolls serve two purposes, the one being to amuse children and the other to instruct ladies..in the latest phases of Parisian fashion. **1951** E. St. George *Dolls of Three Centuries* iii. 27 The Jumeau doll is still being made... The present day Jumeau bears little or no resemblance to the old, collector's Jumeau. **1962** G. White *Dolls of World* 192 By 1862, the Jumeau dolls were quite famous; their well-modelled heads, waxen complexions, life-like eyes and movable joints attracted much attention. **1972** *National Observer* (U.S.) 27 May 9/4 (Advt.), Antique dolls, mint condition dressed and wigged. Guaranteed no reproductions. German or French, many rare French dolls, Jumeaus all sizes. **1973** *Times* 31 July 23/3 (*caption*) A late 19th century Jumeau doll, papier mâché with bisque head, height 20½ inches.

jumma (dʒʌ·mă). Also **jama, jummah.** [ad. Hind. *jama* collection, amount, account, a. Arab. *jamaʿ* total, aggregate.] During British rule in India, the assessment for land revenue from an estate or division of country. So **jummabu·ndi** [Pers.-Arab. *jamaʿbandī* (Hind. *bandī*, a. Pers. *bandi* a tie, band; Skr. *bandh* bind)], the settlement of the revenues; the document recording this settlement.

1772 H. Verelst *View of Bengal* 214 The rents of the province, according to the Jummabundy, or rent-roll... The amount..was taken at a considerable discount from the former Jumma or valuation. **1800** *Asiatic Ann. Reg., Proc. Parl.* 38/1 The collections on the current jumma have fallen short 16,875*l*. **1845** *Encycl. Metrop.* XXI. 672/1 Jumma-bundee Customs. **1851** *Illustr. Catal. Gt. Exhib.* IV. 926/2 Model of a Jamma Bundi. Collector making the annual jambundi. **1858** J. B. Norton *Topics for Indian Statesmen* 269 When he reached the station he found the magistrate absent on jumabundy. **1892** B. H. Baden-Powell *Land-Syst. Brit. India* I. I. i. 24 To say that the 'jama' of village A. is Rs. 300 means that the Government Land Revenue demand on the village as a whole is Rs. 300 each year. *Ibid.* II. IV. i. 563 The 'jamabandī' is perhaps the most generally useful of all the papers.

jump, *sb.*¹ Add: **1. c.** A descent on a parachute.

1922 *Encycl. Brit.* XXX. 14/2 The parachute..is of little use unless the jump is made over 200 ft. from the ground. **1935** C. G. Burge *Compl. Bk. Aviation* 490/1 Jumps have been made from heights up to 30,000 ft., the descent taking about 40 min. **1970** *Times* 9 Dec. 16/1 The man..made his astonishing parachute jump into allied territory.

d. A journey, trip. *slang* (orig. *U.S.*).

1923 N. Anderson *Hobo* v. 83 He likes to tell of making 'big jumps' on passenger trains as from the coast to Chicago in five days, or from Chicago to Kansas City or Omaha in one day. **1932** E. Smith *Satan's Circus* 17 The performers..amused themselves, during the tedium of long 'jumps', by making him sing to them. **1956** B. Holiday *Lady sings Blues* (1973) viii. 77 We were playing big towns and little towns, proms and fairs. A six-hundred-mile jump overnight was standard. **1967** C. O. Skinner *Madame Sarah* xii. 268 On sleeper jumps after the star had got to bed, it was Pitou's nightly duty to hear her read the lines of a role.

e. Jazz music with a strong beat; a jazz tune with a strong rhythm. Freq. *attrib.* orig. *U.S.*

1937 *Metronome* Sept. 32/2 Count Basie, *John's Idea*; *One O'Clock Jump*. **1943** R. Blesh *This is Jazz* 30 You have left only the intolerable monotony of 'jump' (riff) phrases played over and over. **1946** Mezzrow & Wolfe *Really Blues* xvii. 325 This mechanical swing-band age of jump, organ-grinder riffs, mop-mop and rip-bop. **1946** R. Blesh *Shining Trumpets* (1949) xii. 279 Jump-swing, that aptly named music which cannot be danced to but must be jumped to. *Ibid.* 282 The small jump bands that are constantly forming and disbanding. *Ibid.*, Exploiting jump rhythms. **1955** C. Fox in A. J. McCarthy *Jazzbook 1955* 6 Ellington..produced jazz in the 'twenties, 'swing' during the 'thirties, 'jump' in the early 'forties. **1971** *Melody Maker* 9 Oct. 17/4 In a way, his band was a 1940's jump band with amplifiers. **1972** *Jazz & Blues* Sept. 10/3 The West Coast 'jump' style adopted by artists like Roy Milton and Joe Liggins. *Ibid.* 11/3 'Jump' instrumentals.

f. An act of copulation; sexual intercourse. *slang.*

1934 J. O'Hara *Appointment in Samarra* (1935) vii. 212 Then you get cockeyed and take her out for a quick jump and ruin the whole works. **1970** G. Greer *Female Eunuch* 249 A wank was as good as a jump in those days.

5. b. *Contract Bridge.* A bid higher than is necessary in the suit concerned. Also *attrib.*

1927 M. C. Work *Contract Bridge* (1928) 24 One more trick than would be required in Bridge to justify a raise or jump. *Ibid.* 56 With Ace-King-Queen..make a jump denial by bidding three. **1931** E. Culbertson *Contract Bridge at Glance* 11 A jump bid in a new suit by Opening bidder, after a minimum response by partner, is a Forcing Re-bid. **1933, 1959, 1970** [see *forcing ppl. a.*]. **1970** *Globe & Mail* (Toronto) 26 Sept. 51/4 The jump to three diamonds shows 13 to 16 points. **1973** *Country Life* 27 Dec. 2184/1 The jump bid in the opponents' suit is to show that he really has them.

7. *on the jump* (earlier and later examples); also (*b*) abruptly; swiftly; (*c*) in a nervous condition; *all of a jump* (orig. *U.S.*), in a jumpy or nervous state; (*at a*) *full jump* (*U.S.*), at full speed; *at one jump* (*U.S.*), in one go; *at the jump* = *at the first jump*; *for* (or *on*) *the* (*high*) *jump*, *for the jumps*, up for trial, on a charge for misdemeanour; due for punishment, *spec.* hanging; *to get* (or *have*) *the jump on* (orig. *U.S.*), to gain a lead on, get an advantage over (someone); *one jump ahead*, one step in front *of* (someone or something); just avoiding a pursuer or the like (*lit.* and *fig.*).

1825 J. Neal *Bro. Jonathan* II. 291 What's the matter with you,—all of a jump! **1854** M. J. Holmes *Tempest & Sunshine* i. 12 What you ridin' Prince full jump down the pike for? **1859** *Southern Lit. Messenger* XXVIII. 143, I run down stream, an I meets Bill on the jump. **1870** De B. R. Keim *Sheridan's Troopers* vi. 39 The irate quadruped made for our party, coming at a 'full jump'. **1884** 'Mark Twain' *Huck. Finn* xviii. 162 My nigger had a monstrous easy time..but Buck's was on the jump most of the time. **1896** Ade *Artie* xvi. 147, I put up a holler right at the jump. **1899** 'Mark Twain' in *Century Mag.* Nov. 76/1 It was my idea to spread [a name] all over the world, now, at this one jump. **1905** J. C. Lincoln *Partners of Tide* vi. 106 When one of us three says, 'Nickerson, do thus and so,' you *do* it, and do it on the jump. Don't stop to think 'bout it. **1912** Ade *Knocking Neighbors* 123 Rufus was sinfully Rich, but nevertheless Detestable, because his Family had drilled into him the low-down Habit of getting the Jump on the Other Fellow. **1912** F. M. Hueffer *Panel* III. i. 289 That elderly gentleman was exceedingly 'on the jump', as nervous as a man well could be. **1914** 'High Jinks Jr.' *Choice Slang* 21 Only about three jumps ahead of a young conniption. **1919** *Athenæum* 1 Aug. 695/2 'He's for the high jump' is a favourite expression meaning that someone is to be charged before his company or commanding officer. *Ibid.* 8 Aug. 727/2 'For the jumps' (up for trial). **1921** C. Mulford *Bar-20 Three* xviii. 230 Hurrying men pulled thick planks from the pile..and hauled them, on the jump, to windows and doors. **1922** E. O'Neill *Anna Christie* (1923) 29, I didn't go wrong all at one jump. **1925** Fraser & Gibbons *Soldier & Sailor Words* 119 *High jump, on the*, a term used of a man entered on a 'Crime sheet', and for trial for a military offence; the suggestion being that the accused would need to jump very high to get over the trouble. **1936** E. Ambler *Dark Frontier* xi. 173 If we fall down on this job..it's me for the high jump. **1936** G. Greene *Gun for Sale* i. 23 He sounded all of a jump. **1940** Wodehouse *Eggs, Beans & Crumpets* 103 If ever I saw a baby that looked like something that was one jump ahead of the police..it is this baby of Bingo's. Definitely the criminal type. **1942** 'N. Shute' *Pied Piper* 247 I'm for the high jump. They got the goods on me all right. **1956** 'A. Gilbert' *And Death came Too* xiv. 145, I can't afford to act for someone who's going to be found guilty. And.. it looks to me remarkably likely Mrs. Appleyard is going to find herself for the high jump. **1960** D. Lytton *Goddam White Man* xi. 183 He thinks he has the jump on us. **1963** J. Prescot *Case for Hearing* viii. 123 All of the accused are for the high jump. **1971** M. Sinclair *Sonntag* ii. 14 Someone is for the jump, I can tell you. Misinformed, that's what I was. **1972** *Real Estate Rev.* Winter 22/1 Each of these new developers hopes to get the jump on the other by adding more square footage to the units and giving more in amenities. **1972** J. Philips *Vanishing Senator* (1973) I. iv. 37 Get over here on the jump... Step on it, will you? **1973** *Sun* 18 Jan. 6 That would allow the Government to permit wage rises to keep one jump ahead of prices.

8. A robbery (see quots.). *slang.*

1777 in Partridge *Dict. Underworld* (1949) 374/1 *The jump*... The dusk of the evening is the time allotted for this, as it prevents any one at a distance from observing what passes; a great number of rogues then gets lurking about, taking advantage of the unpardonable neglect of others; every window they come near that has no light in, they open, if it happens not to be fastened; they then take what is most valuable out of that room, and very often go into others in the same house. 1781 G. Parker *View of Soc.* II. i. 140 As soon as they have completed this robbery, the *Jumper* descends...The *Jump* being thus completed, they sheer off immediately. 1788 Grose *Dict. Vulgar T.* (ed. 2), The jump, or dining room jump; a species of robbery effected by ascending a ladder placed by a sham lamp-lighter, against the house intended to be robbed. It is so called, because, should the lamp-lighter be put to flight, the thief who ascended the ladder has no means of escape but that of jumping down. 1901 'Linesman' *Words by Eyewitness* (1902) 293 They are sure to see a 'jump' in everything, even in concessions. *Note*: South African euphemism for a robbery.

9. *Comb.* (sense *1 c), as **jump boot**, a parachutist's boot; **jump-master**, a man in charge of parachutists; **jump-sack** *slang*, a parachute; **jump-suit, jump suit** orig. *U.S.*, a parachutist's one-piece garment; also, a similar garment worn by other people.

1948 *Amer. Speech* XXIII. 319 *Jump boots*, paratroopers' shoes. 1972 *Daily Colonist* (Victoria, B.C.) 4 Aug. 1/6 She ran to her husband's prostrate body, unlaced his jump boots and pulled them off. 1942 *Look* 3 Nov. 43/2 (*caption*) The jumpmaster..cries, 'Stand in the door!' and the men crowd forward, waiting for the electrifying order: 'Jump!' 1970 N. Armstrong et al. *First on Moon* xiv. 353 It's like riding an airplane, getting ready to jump. Anything could go wrong. Something could happen to the airplane.., the jumpmaster. 1973 *Daily Colonist* (Victoria, B.C.) 20 May 30/5 She was standing on the wing of the airplane hanging on and waiting for the jumpmaster to say 'go'. 1942 'B. J. Ellan' *Spitfire!* p. x, A parachute is called a *brolly* or a *jumpsack*. 1948 *Amer. Speech* XXIII. 319 *Jump suit*, uniform worn when jumping from airplane. 1965 *Guardian* 7 May 10/1 Rayon linen jump-suit with turn-up trousers and Orlon fish-net midriff. 1965 H. Kane *Devil to Pay* (1966) xxii. 129 Nora was slender and graceful in a crisp white narrow-legged jump suit. 1969 *New Yorker* 30 Aug. 73/1 Three parachutists in jump suits. 1971 *Black World* Apr. 38/2 We worry over horsepower (no pun intended) and power steering, fashionable jump suits and Afro haircuts. 1972 *Time* 17 Apr. 58/2 He..sews conservation patches all over his jumpsuit.

jump, *v.* Add: **1. c.** *to jump out of one's skin*: see Skin *sb.* 5 f.

d. Colloq. phr. *to jump* (or *go* (*and*) *jump*) *in the lake*: to go away and cease being a nuisance; usu. *imp.* as a contemptuous dismissal.

1912 [see *go *v.* 32 a]. 1937 E. S. Gardner *Case of Lame Canary* vii. 67 Suppose she tells us to go jump in the lake? 1946 H. Croome *Faithless Mirror* ix. 97 'I'm here to stop that particular change.' 'You and what ten other fellows? Go jump in the lake.' 1966 M. Woodhouse *Tree Frog* xviii. 129 There was no real reason why I shouldn't have told Andy to jump in the lake as soon as he'd got us through Customs at Heathrow. 1968 [see *fruit *sb.* 2 e]. 1974 D. Gray *Dead Give Away* xxii. 202 She smelt pot in his room... He destroyed the evidence, and told her to jump in the lake.

e. *to jump to the eye*(*s*) [tr. F. *sauter aux yeux*]: to be noticed; to be obvious or prominent.

1926 Fowler *Mod. Eng. Usage* 311/1 *Jump to the eye*(*s*) is a bad Gallicism. 1929 G. Goodwin *Conversations with G. Moore* xxvii. 174 The fact that the Banquo scene in 'Macbeth'—a scene which jumps to the eye—was overlooked, encourages me, obliges me, to think that no one reads Shakespeare. 1931 M. D. George *England in Transition* iii. 59 Things jump to the eyes of the reader of this passage which have yet been ignored.

f. *to jump rope*: to skip with a skipping-rope. Cf. *jump-rope* s.v. *jump-. *N. Amer.*

1934 in Webster. 1961 *Western Folklore* July 179 If only two children are jumping rope, one end of the rope may be tied to a tree, the other end being turned by one of the children. *Ibid.* 193/1 When she died she told me this, When I jump rope I always miss. 1972 *Nat. Geogr.* Sept. 414 When they aren't shooting marbles or jumping rope, the youngsters lurch about on..stilts.

g. To jump out of an aeroplane.

1935 C. G. Burge *Compl. Bk. Aviation* 490/1 After jumping and pulling the release cord the parachutist finds that the parachute opens fully in about 1½ sec. 1942 [see *jump-master* s.v. *jump *sb.*[1] 9]. 1969 A. White *Long Drop* 220 Ben jumped. His parachute failed to open.

h. Of jazz or similar music: to have a strong or exciting rhythm; to 'swing'; so of a place, esp. a place of entertainment: to pulsate with activity; to be full of excitement or enjoyment. *colloq.* (orig. *U.S.*).

c 1938 N. E. Williams *His Hi de Highness of Ho de Ho* 16 *The joint is jumping*, the place is lively, the club is leaping with fun. 1943 H. A. Smith *Life in Putty Knife Factory* vi. 89 He then called up a couple of his friends.. and they came, and before long the joint was jumpin'. 1944 *Needle* July 23/2 The jumping-jive Harlem musicians who think that to obtain any semblance to rhythmic excitement they must leave the theme and become lost altogether. 1946 Mezzrow & Wolfe *Really Blues* (1957) vi. 71 Indiana Harbor was small but it jumped like mad. 1946 F. Stacy in Rosenthal & Zachary *Jazzways* 49/2 The meaning of a 'jump tune' should be clear enough from the term itself; literally, it jumps. 1959 'F. Newton' *Jazz Scene* v. 86 Nobody minded what was played so long as it 'jumped'. 1968 J. Sangster *Foreign Exchange* i. 31 The place was really jumping. It took me three minutes to locate the bar through the smoke haze. 1972 *Jazz & Blues* Sept. 12/1 We should give some mention to the jumping instrumentals which Fats and the band were committing to wax during the early 50's.

i. *to jump up*. To dance the 'jump-up' (*jump-up 2). *West Indies.*

1959 'M. Underwood' *Arm of Law* xiv. 165 Glad to see you enjoying carnival. But why aren't you jumping up? 1968 C. Nicole *Self Lovers* v. 71 'Alex!..I don't suppose you'd care to jump up.'..'I'd love to.' He took her in his arms. The tempo changed to a calypso beat. 1973 *Sunday Advocate-News* (Barbados) 21 Jan. 6/6 Barbadians will have the opportunity to 'jump up' in real carnival fashion at the Barbados Cruising Club's annual carnival dance which will take place at Culloden Farm on Saturday, March 3.

4. b. Also *to jump on*.

1887 *Lantern* (New Orleans) 1 Oct. 2/1 The idea of two big chaps jumping on one man. 1891 [in Dict.]. 1917 D. Canfield *Understood Betsy* (1922) viii. 153 If you had to live the way he does, you'd be dirty!..And then you go and jump on him! 1939 I. Baird *Waste Heritage* xxi. 292 I'm sorry, Eddy, I didn't mean to jump on you that way. 1973 'M. Innes' *Appleby's Answer* xvi. 138 She jumped on the butler for misunderstanding something about the drinks.

c. *to jump down one's throat*: see Throat *sb.* 3 a.

d. *to jump to it*: to make an energetic start upon something; to take prompt action; usu. *imp.* Also occas. *to jump to* = to obey readily. *colloq.* (orig. *Mil.*).

[1886 F. T. Elworthy *West Somerset Word-Bk.* 390 *Jump*, to readily accept an offer. 'Not her hab'm? Let-n ax o' her, that's all; I tell ee her'd jump to un.'] 1917 W. Owen *Let.* 12 Feb. (1967) 434 He does nothing off his own bat, and doesn't always 'jump to' my orders! 1919 [see *jerk *sb.*[1] 2 e]. 1929 *Morning Post* 13 July 16 He does not know whether the service will come to his fore- or his back-hand; but he is ready to 'jump to it', whatever happens. 1956 J. Masters *Bugles & Tiger* xiv. 178 A P. & O. run like a warship, where the passengers would do as they were told and jump to it, and like it. 1974 M. Babson *Stalking Lamb* xvi. 121 When you hear my signal—jump to it!

6. b. Also, to leave (a place or thing) suddenly; *spec.* of a seaman: to desert (his ship) before his contract expires. orig. *U.S.*

1875 J. Miller *First Fam'lies Sierras* vii. 47 Even the head man of the company..jumped a first-class poker game..to come in and weigh out dust. 1883 *American* VI. 40 This evasion of imperative duty affords impunity to the men, if they jump the boat on the route. 1921 C. E. Mulford *Bar-20 Three* vii. 88 I'm admittin' I'm walkin' soft, an' ready to jump th' country right quick. 1923 R. D. Paine *Comrades of Rolling Ocean* xiv. 260, I told you about jumping the town because I had stove up a limousine. 1939 G. Greene *Lawless Roads* 302 He thought perhaps he'd jump the ship at Lisbon—but..he was carried remorselessly on. 1957 'N. Shute' *On Beach* iv. 131 Most of them would probably jump ship.

d. *to jump the bite* (Dentistry): to correct a faulty occlusion or 'bite', esp. one due to a retracted mandible, by bringing the mandible forward as a whole.

1880 [see *bite *sb.* 1 f]. 1901 Smale & Colyer *Dis. & Injuries Teeth* (ed. 2) iv. 158 If..the patient can be made to acquire the permanent habit of bringing the mandible forward so as to make the teeth articulate normally, the bite will have been 'jumped'. 1951 J. M. Schweitzer *Oral Rehabilitation* xxxv. 830 Nearly 80 years ago Class II, Division 1 (Angle), cases were treated by 'jumping the bite'... An attempt was made to reposition the mandible in an anteroposterior as well as a vertical and lateral direction.

e. *Contract Bridge.* To raise (a bid) higher than necessary in the suit concerned. Also *intr.*

1927 M. C. Work *Contract Bridge* (1928) 33 If the partner jump, it must be with three cards of a suit. *Ibid.* 55 Cases of one No Trump jumped to two, and two of a Major jumped to three. 1929 —— *Complete Contract Bridge* i. 7 His proper procedure may be to shift to another declaration, or it may be to jump the original bid. 1963 G. F. Hervey *Handbk. Card Games* 142 If responder has a count of 12 points he can jump straight to three No-Trumps.

8. c. (Earlier examples.)

1836 *Southern Rose* 10 Dec. 57/3 The boys were ordered to stick close to the dogs, and if they jumped the buck to catch him. 1839 *Southern Lit. Messenger* V. 377/1, I would go, but I am a going to jump mullet to-night. 1874 J. W. Long *Amer. Wild-Fowl Shooting* 205 The most successful method of hunting ducks is identical with..'jumping them up' along the creeks.

10. *to jump one's bail*: for *U.S. slang* read orig. *U.S. slang*, and add earlier and later examples. Freq. *to jump bail*.

1859 G. W. Matsell *Vocabulum* 47 *Jumped his bail*, run away from his bail. 1872 G. P. Burnham *Memoirs U.S. Secret Service* 55 Pete's friend Fred Biebusch had hid himself, after jumping his bail. 1911 L. J. Vance *Cynthia* 177 He's jumped bail on a bigamy indictment. 1973 M. Russell *Double Hit* xxii. 165, I shan't jump bail. They'll see me..back in court. 1974 *Guardian* 25 Jan. 24/1 [He] was given a three years' sentence in his absence, after he had jumped bail.

b. To drive past (traffic lights) when they indicate that one should stop. Also *transf.* orig. *U.S.*

1938 *Words* Mar. 44/2 *Jump*, v.t., to anticipate (the go signal of a traffic director). 1958 *Listener* 6 Nov. 731/1 Cutting in, jumping the lights, blind corners at sixty,.. they're things I'd never dream of doing. 1961 J. Barlow *Term of Trial* II. ii. 160 She stared at the conflict of traffic. 'Good God!' she protested... 'They jump the lights!' 1970 J. Porter *Rather Common Sort of Crime* iv. 42 She jumped a red light..it was a damned silly place to have traffic lights. 1973 *Daily Tel.* 5 Jan. 2/7 The driver of the local train..said he had jumped a red signal light.

c. *to jump the gun*: see *gun *sb.* 6 e; *to jump the queue*: to go unfairly to or near the front of a queue of people; to push forward out of one's turn; also *fig.*, to gain an unfair advantage or preferential treatment.

1947 *Hansard Commons* 9 Dec. 951 There is no local authority who can clear these camps by allowing the people in them to jump the queues. 1955 L. P. Hartley *Perfect Woman* viii. 82 He distrusted the quality of imagination; it was a rogue quality that jumped the queue. 1955 *Times* 27 June 8/2 The Port of London Authority gave permission for the ship to 'jump the queue' of other vessels waiting in the river for berths. 1958 Hayward & Harari tr. *Pasternak's Dr. Zhivago* II. ix. 273 There was always a queue in the street... Of course I didn't try to jump the queue, I didn't say I was his wife. 1958 P. Townsend in N. Mackenzie *Conviction* 118 Choosing whether to dodge some taxes..or jump the queue at the hospital. 1973 'M. Innes' *Appleby's Answer* v. 49 One of the women makes a gesture, indicating that you should jump the queue.

jump-. Add: **jump-ball, jump ball** *Basketball*, a ball thrown between two opposing players by the referee; **jump-cut, jump cut** *Cinemat.* and *Television* (see quot. 1953); also *transf.*, *attrib.* and as *v. trans.* and *intr.*; **jump-jet, jump jet**, a vertical take-off/landing jet aircraft; **jump jockey** *Horse racing*, a jockey who rides in steeplechases; **jump-rope, jump rope** (chiefly *N. Amer.*), a skipping-rope; **jump-seat, jump seat**, (*b*) a folding seat in a motor car; also *transf.*; **jump-shot, jump shot**, (*a*) *Billiards*, etc., a shot which causes the ball to jump; (*b*) *Basketball* (see quot. 1961); **jump-spark, jump spark**, a spark produced by the application of a potential difference to two electrical conductors separated by a narrow gap; usu. *attrib.*, designating devices or methods employing this; **jump take-off** *Aeronaut.*, a vertical take-off; **jump-turn, jump turn**, a turn made while jumping; spec. in *Skiing*.

1924 W. E. Meanwell *Sci. of Basket Ball* 62 During scrimmage watch the tip-off formations and also those for jump ball and from out of bounds. 1939 Jourdet & Hashagen *Mod. Basketball* xi. 63 Cover your man well on all jump balls. If you are jumping someone else's man, make sure he covers yours. 1969 *Eugene* (Oregon) *Register-Guard* 3 Dec. 3D/1 The game had started on a technical foul when the Beavers were awarded a free throw instead of the usual jump ball. 1953 K. Reisz *Technique Film Editing* 280 *Jump cut*, cut which breaks continuity of time by jumping forward from one part of an action to another obviously separated from the first by an interval of time. 1962 *Listener* 9 Aug. 223/1 The eye-jerking, ear-jarring jump cuts which result when the commercials are removed from imported American programmes. 1962 *Punch* 19 Sept. 428/1 Harsh jump-cuts that might almost be breaks in the film. 1964 *Observer* 12 July 25/5 A jump-cut speeded-up sequence mostly shot from a helicopter. 1965 *Time* 18 June 80 He recklessly jump-cuts from scene to scene, using gimmicky transitions. 1966 *Punch* 6 July 26/2 The restless, jump-cutting style is sometimes disconcerting—one takes a second or two to realise that an expected bridging passage has been waived. 1968 P. Dickinson *Skin Deep* vii. 130 The soft lines of the black visage jump-cut into wary maturity. 1972 *Times Lit. Suppl.* 9 June 649/4 A series of frenzied incidents..a matter of jump-cuts and unfinished sentences suggestive of a painful collaboration between Ken Kesey and Ford Madox Ford. 1974 *Ibid.* 14 June 629/4 The reader adjusts soon enough to the breakneck jump-cuts of the first few pages. 1975 *New Yorker* 20 Jan. 79/1 Once Trintignant takes over as the filmmaker, the movie loses its playful movie-within-a-movie spirit, and the technique, which had been a sprinting, jump-cutting shorthand that didn't take itself too seriously, turns glassy smooth. 1964 *Sunday Times* 12 Jan. 1/4 The Ministry of Defence delays ordering the revolutionary Hawker P. 1154 'jump jet' fighter. 1970 *New Scientist* 19 Feb. 362/1 The trials that the RAF and RN are now conducting on the employment at sea of the Harrier 'jump-jet'. 1973 *Guardian* 18 Apr. 24/5 A command cruiser designed to carry jump-jet aircraft. 1970 J. Leach *Rider on Stand* ix. 89 Jump jockeys are a devil-may-care bunch. They accept the hazards of their profession in a happy-go-lucky manner. 1972 *Times* 29 Nov. 1/1 Michael Eddery, the jump jockey,..had his right leg amputated. 1973 *Scotsman* 7 Aug. 15/6 Barry Brogan's dispute with the Jockey Club over renewal of his jump jockey's licence ended amicably. [1805 G. McIndoe *Poems & Songs* 40 At three year auld he crys for whips,..And guns, and girrs, and jumpin'-rapes.] 1834 Emerson *Jrnl.* (1964) IV. 359 Thus is one reminded of the children's prayers who in confessing their sins, say, 'Yes, I did take the jumprope from Mary.' 1869 L. M. Alcott *Little Women* II. x. 147 Mr. Bhaer down on his hands and knees..Kitty leading him with a jump-rope. 1969 R. D. Abrahams (*title*) Jump-rope rhymes. 1973 *Islander* (Victoria, B.C.) 8 Apr. 8/1 Spring brings activity—jacks and jump ropes pop up with the flowers. 1931 *Automotive Abstr.* Aug. 238/2 Treatment of the jump-seat explains unique problems presented by this accessory. 1963 Mrs. L. B. Johnson *White House Diary* 24 Nov. (1970) 8 We all got into the same limousine—Mrs. Kennedy and Lyndon in the back seat, the Attorney General and I in the jump seats. 1972 *Guardian* 18 Sept. 14/1, I was sitting in the jump seat... We were racing through the city. 1973 *Black Panther* 6 Oct. 10/1 Only the jump seat for stewardesses was behind him. 1909 P. A. Vaile

Mod. Golf 92 This shot has its exact counterpart on the billiard table in the useful jump-shot. 1966 Miller & Thorp *Croquet* 174 *Jump shot*, a shot in which the ball is struck so that it leaves the ground. 1961 J. S. Salak *Dict. Amer. Sports* 250 *Jump shot* (basketball), a shot taken with both feet off the floor. It can be made with one or two hands with the one-handed shot in general use in the National Basketball Association. 1969 *New Yorker* 14 June 79/1 You go through Harlem and you'll see kids less than five feet tall with pretty good jump shots and hook shots. 1974 *Anderson* (S. Carolina) *Independent* 20 Apr. 8A/1 New York's Julius Erving, known as Dr. J., drilled the 20-foot jump shot that gave the Nets their 3-0 margin in the series with an 89-87 victory at Louisville on Wednesday. 1908 J. H. Adams in Onker & Baker *Harper's How to understand Electr. Work* 340/2 *Jump-spark*, a disruptive spark excited between two conducting surfaces in distinction from a spark excited by a rubbing contact. 1911 *Daily Colonist* (Victoria, B.C.) 20 Apr. 8/1 A regal Marine Engine... Jump spark or make-and-break ignition. 1922 A. F. Collins *Bk. Wireless Telegraph & Telephone* i. 6 The spark-coil, or induction coil,..is used to change the battery current into a current of high pressure to make jump sparks. 1938 A. W. Judge *Automobile Electr. Maintenance* ii. 27 A brass plate at the end of the arm..passes very close to the brass contacts..as it rotates, so that a spark leaps across the small air gap. This is known as the jump spark method. 1963 Bird & Hutton-Stott *Veteran Motor Car* 8 'Jump-spark' ignition in American usage generally, though not always, referred to high tension coil and battery apparatus with a mechanical contact-breaker and non-trembling coil, but in English usage, at one time, 'jump-spark ignition' meant any form of H.T. ignition with spark-gaps or 'intensifiers' included in the circuit. 1939 *Jrnl. R. Aeronaut. Soc.* XLIII. 62 A sustaining rotor for a gyroplane of the 'jump take-off' type. 1924 *Tourist* Winter Sports No. 12/1 *Jump turn*, a method of changing direction or stopping. 1949 Shurr & Yocom *Mod. Dance* v. 147 Add a jump in place after each landing, before executing jump-turn movement in air. 1972 M. Yorke *Silent Witness* ii. 13 He did a quick jump turn and took the narrow track, running fast along the twisting *piste*.

jumped, *ppl. a.* Add: **b.** *jumped-up*: that has newly or suddenly risen in status or importance (often with an implication of conceit or arrogance). Also *transf.*

1835 'T. Treddlehoyle' *Bairnsla Ann.* 35 (E.D.D.), A bit ov a jumpt up dress-macker, wot reckans ta be t' biggest beauty it taan. 1867 E. Waugh *Tufts of Heather* 23 What a stark, starin', jumped-up foo aw wur to send tho up theer! 1895 *Punch* 24 Aug. 93 You jumped-up, cheap, Coventry bagman. 1919 J. C. Snaith *Love Lane* xxxiv. 189 Democracy. Between you and me, Gert, it's mainly a name for a lot of jumped-up ignoramuses. 1934 J. B. Priestley *Eng. Journey* 380 It has flourished as the big city in the minds of men for generations. It is no mere jumped-up conglomeration of factories, warehouses and dormitories. 1942 L. A. G. Strong *Slocombe Dies* xxvii. 127 The better class despise me as a jumped-up chap with too good a conceit of himself. 1972 J. Wilson *Hide & Seek* i. 19 That jumped-up tarty little madam who couldn't even keep her own husband. 1973 J. Wainwright *Pride of Pigs* 177 He didn't like being talked to.. as if he was some moss-green recruit..by a jumped-up C.I.D. clown from headquarters.

jumper, *sb.*[1] Add: **1. b.** A ticket-inspector or ticket-collector. *slang.*

1900 [in Dict., sense 1]. 1906 *Daily Chron.* 24 July 3/7 It was not a fact that unless the 'jumpers'—travelling ticket inspectors—made a certain number of reports they were discharged. 1931 *Evening Express* (Aberdeen) 4 Apr., It is not at all uncommon for a 'jumper' to find that fifty per cent. of the occupants of a second class compartment have only third class tickets. 1937 *Daily Express* 21 Jan. 3/4 If you use a second [class carriage] with a 'third' ticket, watch for the 'jumpers', ready to pounce and demand excess. 1966 H. Sheppard *Dict. Railway Slang* (ed. 2) 7 *Jumper*, travelling ticket collector.

c. *Basketball.* A jump-ball or jump-shot; a player of such a ball or shot.

1937 F. C. Allen *Better Basketball* ii. xiii. 182 Any jumper must keep his eyes fixed upon the ball until it is tapped. He must always play the ball and not the other jumper. *Ibid.* 188 Many jumpers are taught illegally to jump sooner than their opponent in order to get above him. 1958 F. McGuire *Offensive Basketball* ii. 75 Getting possession of the ball depends upon a number of items which are more or less related. First comes the leaping and timing ability of the jumper. *Ibid.* 112 The two-hand overhead jump shot is made in the same manner as the one-hand jumper except that the ball is carried above the head instead of over the shoulder. 1969 Z. Hollander *Mod. Encycl. Basketball* 121/2 Lucus..could also score on jumpers from the corner. 1969 *Eugene* (Oregon) *Register-Guard* 3 Dec. 3D/4 The Vikings took the lead on Snider's free throw with 46 seconds left, but Steve Halberg hit a 15-foot jumper to put the Irish back on top and North couldn't come up with an equalizer.

3. b. In full, *jumper ant*. An Australian ant of the genus *Myrmecia*.

1907 W. W. Froggatt *Austral. Insects* 92 The 'Jumper', *Myrmecia albo-cincta*..is one of the smaller species, about ½ an inch in length. *Ibid.* 436/2 (*index*) Jumper ant. 1926 R. J. Tillyard *Insects Austral. & N.Z.* xxii. 287 The genus *Myrmecia*..contains the huge Bull-dog Ants and the smaller Jumpers..which swarm out of their nests and advance to the attack in a series of jumps or springs. 1970 E. F. Reek in *Insects of Australia* (Commonwealth Sci. & Industr. Res. Organization) xxxvii. 956/1 Forms such as the bulldog or jumper ants..feed largely on nectar and honey-dew as adults.

6. d. For *Telegraphy* read *Electr.* and add examples.

1906 T. E. Herbert *Telegraphy* xviii. 586 When any cross is necessary, the cross-connecting or 'jumper' wires between the vertical and horizontal sides of the frame are altered, so avoiding the necessity for disturbing the cabling. 1931 Moyer & Wostrel *Radio Handbk.* xi. 560 A temporary jumper may be used to close the circuit. 1948 *Aircraft Power Plants* (Northrop Aeronaut. Inst.) ix. 216/2 Test the switch by placing heavy jumpers across the terminals. In other words, close the circuit through the switch with temporary conductors. 1967 *Electronics* 6 Mar. 282/3 The mode selector..includes a 'battery' position that enables checking the condition of the battery without removing it or connecting jumpers. 1972 G. H. Reed *Refrigeration* xiii. 120 A single wire 'jumper' lead.. is useful both for by-passing faulty controls or for incorporating a capacitor in the test cord.

7. Also *Canad.*

1834 J. Langton *Let.* 2 Feb. in *Early Days Upper Canada* (1926) 81 A jumper..is a most admirable conveyance and most properly called a jumper... It sticks at nothing; wherever the horses can scramble the jumper can leap after them. 1902 A. C. Laut *Story of Trapper* xv. 221 The rutted marks of a 'jumper' sleigh cut the hard crust. 1903 B. W. Carr-Harris *White Chief of Ottawa* 119 They had not gone far when the Indian drew their attention to the tracks of a jumper in the snow. 1941 *Beaver* June 28 We loaded twelve hundred pounds of freight into a canoe, besides the dogs and a jumper sleigh. 1964 E. C. Guillet *Pioneer Days Upper Canada* 74 Early settlers from the vicinity of Meaford and Owen Sound brought their grists in home-made sleighs called jumpers, which were hauled by oxen. 1971 J. McDougall *Parsons on Plains* i. 5 Then, in winter, with our little white pony and jumper, we would make the same trips.

jumper, *sb.*[2] Add: **3.** = Jersey 3 a; also, a loose-fitting blouse worn over a skirt; (see also quot. 1968).

1908 *Sears, Roebuck Catal.* 1149/4 The jumper is made in surplice effect. 1908 *Dialect Notes* III. 326 *Jumpers*, a one-piece garment for children to play in, 'rompers'. 1909 *Public Ledger* (Philadelphia) 24 June 7/6 One-piece & jumper styles. 1909 *Westm. Gaz.* 7 Aug. 15/2 For smaller girls the jumper still holds its own. 1925 W. Deeping *Sorrell & Son* i. 13 The modiste had received a consignment of silk 'jumpers'. She was unpacking them and hanging them up on the stands in her showroom where they glowed brilliantly like jewels in a case. 1928 Galsworthy *Swan Song* II. ix. 181 He came on Anne herself, without a hat, sitting on a gate, her hands in the pockets of her jumper. 1930 *N. & Q.* 14 June 431/1 Some five years ago the fashion-mongers gave the name of jumper to the knitted blouses ladies had been wearing under the name of sports coats. 1945 *Wales* IV. vi. 44 He turned up the cuff of his jumper and showed her the word 'Sue' tattooed with a border of foliage on his forearm. 1965 *Australian* 13 Apr. 5 She also prefers casual clothes like the jumper and skirt she is wearing here. 1968 J. Ironside *Fashion Alphabet* 61 *Jump-suit.* This is an abbreviation of 'jumpers', another name for rompers (i.e. top and bloomers in one) worn by children.

b. *U.S.* A pinafore dress. Also *jumper dress*.

1939 M. B. Picken *Lang. Fashion* 84/3 *Jumper-dress*, sleeveless, one-piece garment worn with guimpe. 1967 *Boston Sunday Herald* (Mag.) 16 Apr. 6/1 (Advt.), Wear as a jumper over blouses. 1971 *New Yorker* 11 Dec. 3 (Advt.), Wear a jumper to dinner!

c. *Comb.*, as **jumper suit**, (*a*) a pinafore dress; (*b*) a woman's suit consisting of a jumper and skirt made of the same material, freq. wool.

1908 *Sears, Roebuck Catal.* 1149/1 An unusually pretty jumper suit made of soft striped taffeta silk. 1925 *Times* 29 Dec. 7/6 Sports stockinette jumper suits. 1931 E. Raymond *Mary Leith* III. ii. 225 Mary was in a jumper suit of primrose silk. 1973 *Country Life* 2 Aug. 335/2 Soft jumper-suits in fine printed wools.

jumper, *v.*[2] (s.v. Jumper *sb.*[1]). Add: **2.** *Electr.* To connect by means of a jumper (sense 6 d).

1929 *Post Office Electr. Engineers' Jrnl.* XXII. 79/1 From the cable terminal tag blocks all lines are jumpered *via* protecting apparatus to the 'line' tag blocks of the test boards. 1968 T. Howard *Black Light* xxi. 183 He made no attempt to force the locked ignition, but simply 'jumpered' the ignition wiring so that it by-passed the locked switch.

jumping, *vbl. sb.* Add: **b.** **jumping-board**, a spring-board; also *fig.*; **jumping jockey** = *jump jockey* s.v. *jump-; **jumping-off board** = **jumping-board*; **jumping-off ground, place** (earlier and later examples); also *transf.* and *fig.*; (*b*) *N. Amer.*, a place regarded as being the farthest limit of civilization or settlement; a very remote place; the extreme limit of the earth; also *fig.*; (*c*) a starting-point for aircraft or the like; so *jumping-off point, spot*; **jumping-pole**, a long pole used in jumping long distances or in making pole-vaults; **jumping-wire**, on a submarine: see quot. 1974.

1878 H. H. Jackson *Bits Trav. at Home* 53 There are public gardens..with little ponds, and boats, and targets, and jumping-boards. 1909 *Athenæum* 21 Aug. 218/2 A jumping-board for the imagination to spring from. 1947 W. Bebbington *Rogues go Racing* xviii. 115 There are some [jockeys] who are known to me as habitual gamblers. Particularly is this so with certain of our 'jumping' jockeys. 1914 *Eng. Rev.* Sept. 237 Salonika..was to be the German jumping-off board to Asia Minor. 1931 *Musical Times* 1 June 497/2 His studies abroad had given him a stock of admirably nurtured gifts, but no jumping-off board such as that offered by a career in an English institution. 1934 R. Macaulay *Going Abroad* xi. 82 That's abso*lutely* the best jumping-off ground for the new life. 1959 P. Moyes *Dead Men don't Ski* vi. 74 Tangiers is a convenient jumping-off ground. 1826 T. Flint *Recoll.* 366 Being, as they phrase it, the 'jumping off place', it is necessarily the resort of desperate, wicked, and strange creatures who wish to fly away from poverty, infamy, and the laws. 1834 H. M. Brackenridge *Recoll.* x. 111, I had no jumping off or jumping up place, like those who prepare their exordium and perorations, and leave the body of the speech to take care of itself. 1847 W. I. Paulding in J. K. & W. I. Paulding *Amer. Comedies* 197, I *have* hunted all over them parts, almost clean out to the jumping off place of creation. 1899 B. Tarkington *Gentleman from Indiana* xv. 266 He had come to a jumping-off place in his life—why had they not let him jump? 1909 F. Ash *Trip to Mars* xvii. 131 A narrow platform which had been erected as a 'jumping-off place' for fliers. 1922 *Encycl. Brit.* XXX. 14/2 The Governments demanded that their aeroplanes should be transported in crates, or towed with folded wings to their jumping-off places. 1930 G. B. Shaw *Apple Cart* I. 37 Today the nation would be equally amazed if a man of his ability thought it worth his while to prefer the woolsack even to the stool of an office boy as a jumping-off place for his ambition. 1953 F. Stark *Coast of Incense* 242 The way to carry out an adventure is to organize the jumping-off place as near to its borders as possible. 1964 D. Jenness *Eskimo Admin.* II. 14 Archdeacon Stuck described Herschel Island during the whaling period as 'the world's last jumping-off place, where no law existed and no writs ran'. 1927 R. H. Wilenski *Mod. Movement in Art* I. 13 An emotional reaction as the sole jumping-off point. 1958 G. Lascelles in P. Gammond *Decca Bk. Jazz* viii. 100 It is not unnatural.. for New York to have been the proving ground and the jumping-off point for a new sort of music. 1909 *Daily Chron.* 8 Sept. 1/4 To reach the neighbourhood of Cape Columbia.., his elected jumping-off spot for the Pole. 1966 *Beautiful Brit. Columbia* Spring 23/1 Prince Rupert ..is a jumping-off spot for the Queen Charlotte Islands. 1873 L. Troubridge *Life amongst Troubridges* (1966) vi. 47 We..jumped loads of ditches, and when we came to a very large one we made a bridge of our jumping poles. 1972 *Listener* 31 Aug. 274/2 We had jumping-poles and we jumped from one rock to another. 1919 *Jane's Fighting Ships* 318 Jumping wires were added to French submarines. 1940 'N. Shute' *Landfall* iii. 73 'Did you notice how many jumping-wires she had?' 'That's the wire that runs from bow to stern over the conning-tower, isn't it?' 'That's right. Did she have one or two?' 1974 G. Jenkins *Bridge of Magpies* xv. 223 Her jumping-wire—the thick cable designed to slice through undersea objects like mine moorings—which runs from bow to stern via the conning-tower.

jumping, *ppl. a.* Add: **b. jumping deer**, substitute for def.: either of two North American animals, the pronghorn, *Antilocapra americana*, or the mule deer, *Odocoileus hemionus* (examples); **jumping-shrew** (examples).

1806 A. Henry *Jrnl.* 14 July in E. Coues *New Light Hist. Greater Northwest* (1897) I. ix. 305 Herds of cabbrie or jumping deer were always in sight. 1831 R. Cox *Adv. Columbia River* II. 364 The jumping-deer, or chevreuil, ..frequent the vicinity of the mountains in considerable numbers. 1908 J. W. Tyrrell *Across Sub-Arctics of Canada* (ed. 3) xxi. 243 Jumping Deer are found in more or less abundance throughout the timbered country about southern parts of the [Hudson] Bay. 1936 D. McCowan *Animals Canad. Rockies* xxxi. 265 The Mule deer is most common... In some parts of Canada the animal is called Jumping deer, this from its well known habit of progressing when alarmed in a series of immense leaps and bounds. 1961 R. P. Hobson *Rancher takes Wife* vii. 108 There were the tiny little white-tailed jumping deer that would make about four meals for one man. 1900 H. A. Bryden *Animals Afr.* ii. 16 The typical Cape jumping shrew has a long, proboscis-like nose, large ears, long, thin hind legs, which enable him to take enormous leaps for his size, and a long, rat-like tail. 1920 F. W. Fitzsimons *Nat. Hist. S. Afr. Mammals* IV. 2 There are several species or kinds of Jumping or Elephant Shrews inhabiting South Africa. 1971 D. J. Potgieter et al. *Animal Life S. Afr.* 346/2 The elephant-shrews or jumping shrews (*Macroscelidea*) are insect-eaters, and the whole order is confined to Africa.

c. jumping-bean (examples); (*b*) a toy consisting of a small bean-shaped capsule containing a weight such as a lead ball which causes it to move unaided down a sloping surface; **jumping-seed** (examples).

1889 *Cent. Dict.*, Jumping-bean. 1896 *Chambers's Jrnl.* 18 Apr. 249 A new botanical curiosity..has lately been brought into notice in England under the name of 'A Jumping Bean'. 1910 *Boy's Own Paper* 15 Jan. 256 *Tommy* (who has been watching the jumping beans for some time): 'Oi'm waitin' to see them sticks walk.' 1972 F. Warner *Maquettes* 14 Along they go, like jumping beans from a toy factory. 1972 Swan & Papp *Common Insects N. Amer.* 312 The wriggling larva of an olethreutid moth, *Laspeyresia saltitans*, is the activator of the Mexican 'jumping bean', the seed of a species of *Croton*. 1876 *Field & Forest* II. 53 These so-called jumping seeds received from California. 1889 *Wesley Naturalist* III. 22 Those are the only 'jumping seeds' of which I had even heard until I met with these of Natal.

jump-off (dʒʌ·mpɒf). [f. *to jump off*.] **1.** A precipitous descent; a place from which a person must jump. *U.S. colloq.*

1873 [see *bed *sb.* 12 f]. 1884 C. Phillipps-Wolley *Trottings of Tenderfoot* v. 129 The broad stem of a fallen giant gives you 150 feet of splendid wooden road; but.. you find you have been gradually ascending, and now stand on what the Americans would call a 'jump off'. 1909 R. A. Wason *Happy Hawkins* ii. 26 The lantern shed a splash o' light on the shelf, but the jump-off looked like the mouth o' the pit.

b. The start of a military operation or campaign. *U.S. slang.*

1918 [see *H-Hour* s.v. *H. III]. 1944 *Daily Progress* (Charlottesville, Va.) 12 May 1/8 The Fifth and Eighth armies have launched the greatest drive in Mediterranean warfare in the jump-off of Allied spring offensives. 1945 *Sun* (Baltimore) 23 Feb. 1/6 (*heading*) Canadians find going toughest since jumpoff.

2. *Aeronaut.* A vertical take-off.

In quot. 1969 'jump off point' = **jumping-off point*.

1939 *Jrnl. R. Aeronaut. Soc.* XLIII. 110 (*heading*) Possibilities of the jump-off autogyro. *Ibid.*, The available kinetic energy in the rotor system for 'jump-off' is directly proportional to the weight of the blades and to the square of the rotational speed of the rotor. 1969 *Daily Tel.* 16 Jan. 1/6 A space station in permanent orbit round the earth..could..be used as a 'jump off' point for travel to the moon.

3. *Show-jumping.* An additional round to resolve or determine a tie.

1947 H. Disston *Equestionnaire* (rev. ed.) 77 In the event of a jump off over triple bars, how is the obstacle altered? 1954 P. Smythe *Jump for Joy* vii. 113 The final jump-off included a high wall with a pole on top. 1969 *Times* 10 May 6/5 Alan Oliver..justified..the odds laid on him in the..competition yesterday. Riding Pitz Palu, who was made favourite at 7–2 and shortened to 2–1 for the jumpoff, he..made the running from start to finish.

jump-up (dʒʌ·mpʌp). *colloq.* [f. *to jump up*.] **1.** An escarpment. *local Austral.*

1927 M. Terry *Through Land of Promise* 85 We had been looking at the 'jump-up' marking the extremity of the Barkly Tableland. 1969 'A. Garve' *Boomerang* i. 32 There's a sharpish rise from the plain to the tableland, with a steep edge—what we call the 'jump-up'.

2. An informal West Indian dance.

1955 *Caribbean Q.* IV. ii. 102 Children and adults dancing in the shuffling manner of the Trinidad Carnival 'jump-up'. 1959 'M. Underwood' *Arm of Law* xiv. 164 Some might be found doing a traditional mid-day jump-up. 1965 'Lauchmonen' *Old Thom's Harvest* vii. 96 A few people..gathered around the calypsonian... Saul was doing a real jump-up in the centre. 1971 *Sunday Times* 13 June 9 You have some rum and a bloody good lunch, then you have a jump-up, and a proper shindig. 1973 *Advocate-News* (Barbados) 17 Feb. 6/1 With the carnival fever in the air, the students at the Cave Hill campus will stage their pre-carnival jump-up and calypso tent at the campus tomorrow, beginning at 5 p.m.

jumpy, *a.* **2. a.** Add to def.: nervous, apprehensive. (Add further examples.)

1918 *Wine, Women & War* (1926) 10 No trip for anybody with jumpy nerves. 1935 [see *gooey *a.*]. 1957 *Sat. Even. Post* 21 Sept. 94 One of our pals in the night-club business is jumpy and he needs a bodyguard tonight. 1974 G. Markstein *Cooler* xlvi. 167 She was jumpy about the blackout too... She *is* on edge, he decided.

jun. (Examples.)

1708 [see Junior *a.* 1]. 1837 Dickens *Sk. Boz* 2nd Ser. 222 Mr. Green, sen., and his noble companion entered one car, and Mr. Green, jun., and *his* companion the other. 1922 Joyce *Ulysses* 391 Dixon jun., scholar of my lady of Mercy. 1955 *Times* 8 July 6/7 On June 8, continued counsel, Thomas Foote, jun., saw a bank official carrying out two bags of silver to a car. 1974 *Country Life* 11 Apr. (Suppl.) 66 John Frederick Herring Jun. A farmyard scene.

junction, *sb.* Add: **1. c.** In Jespersen's terminology, a group of words consisting of a primary word and an adjunct (*adjunct *sb.* 5 b).

1924 O. Jespersen *Philos. Gram.* vii. 97 If..we compare the combination *a furiously barking dog*..with *the dog barks furiously*..there is a fundamental difference between them, which calls for separate terms for the two kinds of combination: we shall call the former kind *junction*, and the latter *nexus*. *Ibid.* viii. 115 In a junction a secondary element (an adjunct) is joined to a primary word as a label or distinguishing mark. 1935 *Jrnl. Eng. & Germ. Philol.* XXXIV. 415 Two entirely different classifications are involved: (1) an assignment of importance within the frame of a sentence..(2) a scrutiny of subordination within a group ('junction'). 1966 M. Pei *Gloss. Ling. Terminol.* 136 *Junction*, a grammatical unit formed by qualified and qualifying terms (*the red barn*).

2. b. *Electronics.* A transition zone in a semiconductor between two regions of different conductivity type (usually *n*-type and *p*-type).

1949 W. Shockley in *Bell Syst. Techn. Jrnl.* XXVIII. 435 Silicon and germanium may be either *n*-type or *p*-type semiconductors... If, in a single sample, there is a transition from one type to the other, a rectifying photosensitive *p-n* junction is formed. *Ibid.* 436 We shall use the word *junction* to include all the material near the transition region in which significant contributions to the rectification process occur. 1959 R. A. Smith *Semiconductors* xii. 444 *p-n* junctions are generally much more stable mechanically than fine metal point contacts and the modern tendency is to use them whenever possible. 1962 Simpson & Richards *Physical Princ. Junction Transistors* iii. 43 If the change from n^+-type to *n*-type is sufficiently gradual, electrons diffusing from the n^+-type material will recombine with holes before reaching the *n*-type material and the n^+-*n* junction will be non-rectifying. 1965 Burford & Verner *Semiconductor Junctions & Devices* vii. 91 All junctions in semiconductors are inherently rectifying. To make such a *p-n* junction into an operable rectifier, we merely attach leads to the *p* and *n* regions and protect the active element..by suitable encapsulation.

4. junction-box, a closed, rigid box or casing used to enclose and protect the junctions of electric wires or cables; **junction diode** *Electronics*, a diode consisting essentially of a piece of semiconductor containing a rectifying *p-n* junction; **junction rectifier** *Electronics* = **junction diode*; **junction transistor** *Electronics*, a transistor consisting essentially of a piece of semiconductor containing two (or more) junctions that divide it into three (or more) regions.

1885 E. S. Farrow *Mil. Encycl.* II. 147/1 In submarine mining, when it is necessary to employ a multiple cable, a junction-box is used to facilitate the connection of the several separate wires diverging from the extremities of such a cable. 1934 *Archit. Rev.* LXXV. 141/3 Junction boxes are arranged at close intervals all over the floor before the blocks or floorboards are put down. 1958 M. Dickens *Man Overboard* ii. 28 He tripped over a large metal junction-box, where several thick cables met in a writhing tangle. 1972 *Police Rev.* 10 Nov. 1453/3 A security van..crashes into an electric junction box at the side of the road. 1952 *Proc. IRE* XL. 1348/1 This paper describes..a new type of silicon diode, namely, the *p-n* junction diode prepared by alloying. 1970 J. Earl *Tuners & Amplifiers* ii. 28 Very few tuners are now being made with valves. The vast majority employ semiconductor devices, and of these many use transistors and junction diodes, but the trend is also towards the use of ICs. 1951 *Physical Rev.* LXXXI. 475/1 The holes move mainly under the influence of diffusion in a manner similar to that discussed in connection with carriers injected across the junction in a *p-n* junction rectifier. 1962 Simpson & Richards *Physical Princ. Junction Transistors* iii. 35 While it is possible to produce junction rectifiers and transistors from many different semiconductors, the devices in successful commercial production are.. made from either germanium or silicon. 1949 W. Shockley in *Bell Syst. Techn. Jrnl.* XXVIII. 435 (*heading*) The theory of *p-n* junctions in semiconductors and *p-n* junction transistors. 1959 R. A. Smith *Semiconductors* xii. 449 The first type of transistor to be used was the point-contact transistor, but this has been almost entirely replaced by the junction transistor. 1962 Simpson & Richards *Physical Princ. Junction Transistors* i. 1 The extraordinary technological growth that has taken place since that time [*sc.* 1948] has established the junction transistor as a device of major engineering and economic importance.

junction (dʒʌ·ŋkʃən), *v.* [f. the sb.] *intr.* To form a junction; to join *with* or *on to*.

1904 *Electrical Investments* IV. 771/2 Railway companies whose lines junctioned with each other did not always give either the passenger or goods traffic the advantages that the physical junctions rendered possible. 1909 R. A. Wason *Happy Hawkins* xxvii, Deuced if I ever could see where your trail could have junctioned onto the Clarenden family. 1936 I. L. Idriess *Cattle King* xii. 105 This line..gradually draws in towards the Diamantina until it junctions with it here, just above the South Australian border. 1959 *Tararua* XIII. 47 New Zealanders and Australians occasionally use the verb *junction with* of rivers, though why this is necessary when there is the verb *join* is hard to say.

junctive (dʒʌ·ŋktiv), *a.* *poet. rare.* [ad. L. *junctiv-us*, f. *junct-*, pa. ppl. stem of *jungĕre* to join.] Having the quality of joining.

1898 Hardy *Wessex Poems* 28 So may I live no junctive law fulfilling, And my heart's table bear no woman's name.

juncture. Add: **2. c.** *Linguistics.* The transition between two linguistic segments or between an utterance and preceding or following silence; the phonetic feature that marks such a transition. Also *attrib.*

[1934 Priebsch & Collinson *German Lang.* iii. 210 When a stem-vowel or declensional suffix in its crude form occurs in the first components of compound substantives and adjectives, it is called the 'Fugenvokal', i.e. juncture vowel.] 1941 *Language* XVII. 224 Those [phonemes] that relate to the way in which utterances begin and end.. we call juncture phonemes... A logical order of exposition ..will begin with the juncture phenomena... The present study..will deal with junctures, stresses, [etc.]. *Ibid.* 225 The transition from a pause preceding an isolated utterance to the first segmental phoneme, and from the last segmental phoneme to the following pause, we call open juncture. 1942 Bloch & Trager *Outl. Ling. Analysis* ii. 35 Phenomena relating to the way in which sounds are joined together are summarized under the term *juncture*. 1946 E. A. Nida *Morphol.* 94 When two items are combined, there are potentially several different types of junctures, or seams, at the point of contact. 1957 S. Potter *Mod. Ling.* iii. 73 No less elusive than intonation.. are the related features of *juncture* and *pause*. Where precisely does one syllable end and another begin? 1969 *English Studies* L. 292 They regard juncture as a special type of phoneme, neither segmental nor prosodic, causing sub-signemic changes in the environment. 1972 R. Wardhaugh *Introd. Linguistics* 64 We can say that such words as *nitrate*, *night rate*, and *Nye trait* require the postulation of a juncture phoneme to show the difference.

Hence **ju·nctural** *a.*, **ju·ncturally** *adv.*

1942 *Language* XVIII. 14 A suprasegmental phoneme is junctural if each member phone has a determining starting-point (or a determining end-point). 1964 [see *demarcative *a.*]. 1965 *Language* XLI. 499 A separate phoneme..(which is subject to frequent loss, mainly but not wholly juncturally conditioned). 1966 W. S. Allen in C. E. Bazell *In Memory of J. R. Firth* 11 This ..would have been contrary to Greek junctural principles (being characteristic of close and not open juncture).

June, *sb.* Add: **2. June-berry** (earlier and later examples); also used for other trees or shrubs of the genus *Amelanchier*, or their fruit; **June-bug** (earlier and later examples); **June grass** (examples); **June Week**, at Durham University, the last week of the summer term, Commemoration week.

1832 D. J. Browne *Sylva Amer.* 217 The wood of the June berry is of a pure white. 1854 Mayne Reid *Young Voyageurs* 356 The berries..are known as..'June-berries' [or] 'service-berries'. 1914 I. Cowie *Company of Adventurers* 327 A grizzly bear [was] found among the saskatoon (Juneberry) bushes. 1928 J. E. LeRossignol *Beauport Road* 274 The chief attraction..was along the fences and hedgerows where, in season, were strawberries, June berries..dew berries. 1969 R. C. Hosie *Native Trees Canada* (ed. 7) 234/1 Mountain Juneberry..and Saskatoon-berry..occasionally become trees. 1829 in *Amer. Speech* (1965) XL. 132 *Jim Crow*... Dere's possum up de gumtree, An Raccoon in de hollow, Wake Shakes for June Bugs Stole my half a dollar. 1836 *Congress. Globe* 5 May 349/2 They hopped upon it, to use a homely phrase, like a duck on a June-bug. 1906 W. Churchill *Coniston* xv. 189 June-bugs hummed in at the high windows. 1972 L. E. Chadwick tr. *Linsenmaier's Insects of World* 162/3 (*caption*) An American June beetle, the June bug, having completed development but still in its underground cavity. 1855 *Trans. Mich. Agric. Soc.* VI. 160 A stiff June grass sod plat. 1919 *Maine, my State* (Maine Writers Research Club) 336 How fair her fields when June-grass waves! 1970 *Alberta Hist. Rev.* Winter 1/2 Prairie wool, a nutritious blend of..June grass..and other grasses. 1889 *Durham Univ. Jrnl.* IX. 1 The end of last term was signalised by what was called by some 'Commemoration' and by others 'the June Week'. 1900 *Ibid.* XIV. 229 Those who have visited the race-course during the June Week.

june (dʒūn), *v.* *U.S. colloq.* and *dial.* [? f. the sb.] **a.** *intr.* To move in a lively fashion, hurry; to be restless or aimless; to wander *around*. **b.** *trans.* To drive briskly. *rare.*

1869 *Overland Monthly* III. 127 A trig, smirk little horse is a 'lace-horse', and he often has to 'june' or 'quill'. 1892 *Dialect Notes* I. 230 *June-in'*, running fast. 'She came a-june-in'.' 1895 W. C. Gore in *Inlander* Nov. 61 *June around*, to be busy but not accomplish anything. 1903 A. Adams *Log of Cowboy* xiv. 228 To june a herd of cattle across in this manner would have been shameful. 1948 *Amer. Speech* XXIII. 305/1 You stay here and I'll go down and june around awhile.

jungar, var. Jangar.

‖ **Junggrammatiker** (yu·ŋˌgramā:tikəɹ), *sb. pl. Philol.* [G.] A name given to members of a late 19th-century school of historical linguistics who held that phonetic changes (sound laws) operated without exceptions. The name was accepted by the persons concerned and by others, and has been anglicized as 'neogrammarians'. Hence **junggrammatisch** *a.*

1922 O. Jespersen *Language* iv. 93 The 'blind' operation of phonetic laws became the chief tenet of a new school of 'young-grammarians' or 'junggrammatiker' (Brugmann, Delbrück, Osthoff, Paul, and others). 1936 J. R. Kantor *Objective Psychol. Gram.* viii. 108 The *Junggrammatiker*..believed themselves to have discovered absolute phonetic laws. 1936 *Language* XII. 58 There is another point of view,..an organic development..of the essential core of truth in the Junggrammatiker doctrine. 1938 *Year's Work Eng. Stud.* 1936 30 Wilhelm Havers.. well summarizes the history of the changing attitude towards 'sound-laws' since the first confident days of the *Ausnahmslosigkeit* of the *Junggrammatiker*. 1953 J. B. Carroll *Study of Lang.* ii. 50 In opposition to the neo-grammarians (*Junggrammatiker*), as Leskien, Osthoff, and Brugmann came to be called, Schuchardt (1885), Curtius, and Ascoli pointed to what seemed to be exceptions to phonetic laws. 1958 A. S. C. Ross *Etym.* 8 It is certainly quite impossible for anyone to understand Laryngeal Theory without being thoroughly familiar with junggrammatisch Ablaut. 1965 *Language* XLI. 187 Brugmann.. accepted the originally humorous epithet *Junggrammatiker* and used it as a rallying cry.

Jungian (yu·ŋiăn), *a.* [See -ian.] Of or pertaining to Dr. Carl Gustav *Jung* (1875–1961), the Swiss leader of the school of analytic psychology, or his teaching. Also *sb.*, a follower or adherent of Jung. Hence **Ju·ngianism**, the teaching or system of Jung; a characteristic specimen of this.

1933 D. C. Daking (*title*) Jungian psychology and modern spiritual thought. 1942 K. W. Bash tr. *Jacobi's Psychol. C. G. Jung* iii. 59 Jungian psychotherapy is no analytical procedure... It is..a 'way of healing'. 1947 *Downside Rev.* 35 For the Jungian the whole position is altered... The historical causality of complexes is not denied, and the methods of releasing these complexes, as discovered by Freud, are recognized. 1956 *Essays in Crit.* VI. 417 How we can now avoid Jungianism is a difficult problem. 1958 *Times Lit. Suppl.* 1 Aug. 438/5 The church has a resident Jungian psychiatrist. 1959 *Ibid.* 20 Mar. 164/5 To non-Jungians..this will seem merely another Jungian book. 1964 M. McLuhan *Understanding Media* (1967) II. xx. 207 Myth and Jungian archetypes.

jungle, *sb.* Add: **2. a.** (Further examples.) Also, a place of bewildering complexity or confusion; a place where the 'law of the

jungle' prevails; a scene of ruthless competition, struggle, or exploitation; esp. with qualification, as **blackboard jungle* in schools, *asphalt jungle, concrete jungle* in cities.

1906 U. Sinclair (*title*) The Jungle. 1920 Ade *Hand-Made Fables* 83 After the newly arrived Delegate from the Asphalt Jungles had read a Telegram..he..sauntered back to the Bureau of Information. 1924 A. D. Sedgwick *Little French Girl* II. vi. 150 The jungle itself was part of the order, since the *demimondaine* was taken as much for granted as the *femme du monde*. 1949 W. R. Burnett (*title*) The asphalt jungle. 1954 [see *blackboard]. 1956 'E. McBain' *Cop Hater* (1958) viii. 70 Their front page.. shouted 'The Police Jungle—What Goes On In Our Precincts.' 1958 [see *blackboard]. 1969 D. Morris *Human Zoo* 8 The city is not a concrete jungle, it is a human zoo. 1971 *Sunday Times* 30 May 31/5 Namier.. fitted especially ill in the academic jungle. 1971 *Times* 17 July 5/2 New York seemed to me infernal... By night the streets become concrete jungles, their occupants hysteric, terrified of predators. 1972 *Guardian* 14 Feb. 10/5 The Minister lit up some lurid corners of the taxation jungle. 1974 *Black World* Jan. 38 The Waikiki jungle is kind of a—you might call it a ghetto surrounded by high-rise buildings in Waik!ki.

b. *pl.* Shares in West African concerns. Also *attrib.* ? *Obs.*

1904 *Daily Chron.* 2 Dec. 1/7 Kaffirs weakened, but Jungles moved upward. 1906 *Ibid.* 9 Feb. 2/3 Jungle shares were..firm. 1908 *Westm. Gaz.* 10 Dec. 15/4 A Jungle Dividend.

c. A camp for hoboes, tramps, or the like. Also *attrib.* *slang* (orig. *U.S.*).

[1908 C. Johnson *Highways & Byways Pacific Coast* 215 My companions spoke of the grove they were in as the 'Hoboes Jungle'.] 1914 *Sat. Even. Post* 4 Apr. 10/3 It followed the two along the tracks and into the jungle. *Ibid.* 11/3 Frisco Red slouched into the jungle. 1915 *N.Y. World Mag.* 9 May 14 *Jungle buzzard*, a tramp who sneaks around hobo or tramp camps to get a free meal. *Ibid.*, *Jungle court*, a make-believe court held in woods by hoboes. 1923 N. Anderson *Hobo* ii. 21 Most 'jungle buzzards', men who linger in the jungles from season to season, take an interest in the running of things. 1926 J. Black *You can't Win* vi. 65 'This is a pretty snide jungle,' he said, 'no cans.' *Ibid.* 82 There was a grand jungle by a small, clean river where they boiled up their verminous clothes. 1971 *Islander* (Victoria, B.C.) 4 Apr. 12/1 During the depression in the 1930s gangs of youths ranged across the country, riding the rails and sleeping in jungles, and caused us concern.

3. *jungle craft*.

1942 *R.A.F. Jrnl.* 27 June 24 Even an expert can make mistakes in jungle craft. 1946 W. S. Churchill *Secret Session Speeches* 59 The Japanese armies..having added their jungle-craft..have established themselves..in the whole of these wide regions.

b. jungle-bashing *slang* [*bashing *vbl. sb.* 3], movement through a jungle, esp. by soldiers; so *jungle-basher*; **jungle bunny** orig. *U.S.*, a term used by some white people to designate Negroes, Australian Aborigines, etc.; **jungle green**, a dark green colour; clothes of this colour; also *attrib.*; **jungle gym** (formerly a registered trade mark in the U.S.), a type of climbing frame; **jungle juice** *slang*, alcoholic liquor, esp. liquor that is either very powerful or that has been prepared illicitly or amateurishly; also *transf.*; **jungle law**, the 'law of the jungle' (see *law *sb.*[1] 16 c); **jungle rot** *slang*, name given to a tropical skin disease; **jungle war**, a war fought in jungle; also *fig.*; so *jungle warfare*.

1963 *Times* 24 May 14/6 All the poor 'jungle-bashers' could offer by way of city reminiscence was the egregious Calcutta. 1954 V. Bartlett *Rep. from Malaya* iii. 46 A man does an average of 700 hours 'jungle-bashing' before he kills a Communist. 1969 J. M. Gullick *Malaysia* ii. 113 British, Malay and other Commonwealth troops spent many weary hours on patrol, 'jungle-bashing' as they called it, with the object of contacting terrorists. 1966 *Publ. Amer. Dial. Soc.* 1964 xlii. 27 Both middle-aged informants giving *jungle bunny*.. work with adolescents. 1968–70 *Current Slang* (Univ. S. Dakota) III–IV. 76 *Junglebunny*, n. Negro (derogatory). 1973 *Sunday Times* (Colour Suppl.) 10 June 51/3 Australians in the Territory can be grossly insensitive to the pride of the local people, using terms like 'jungle bunnies'. 1974 *New Society* 14 Mar. 627/2 White South Africans who wanted to gamble, buy *Playboy*..and go to bed with a 'jungle bunny'. 1946 *Nature* 14 Sept. 386/2 Land Army hose, sea-boot stockings, R.A.F. socks and jungle-green pullovers also came under the scheme. 1947 Jungle green [see *cobber *sb.*[1]]. 1973 D. Lees *Rape of Quiet Town* iii. 53 A commanding figure in jungle green with a Lüger pistol in his hand. 1923 *Official Gaz.* (U.S. Patent Off.) 30 Jan. 844/2 Junglegym, Inc., Chicago, Ill. Filed Nov. 14, 1921. Junglegym... Playground Apparatus, in Particular Climbing Frames. 1925 *Playground* Mar. 721 (Advt.), 22 Units—Now in the New York City Playgrounds... Junglegym is six years old this spring. 1929 L. F. Zwarg *Study of Hist. Apparatus Physical Educ.* i. 81 Many odd contrivances [of physical education apparatus] of former years have disappeared entirely, others have from time to time been rediscovered or reinvented. The climbing tower (jungle gym) and the teeter ladder, are examples of this. 1931 *Recreation* May 97 The low climbing device (which is known as the Junglegym). 1951 W. van Hagen et al. *Physical Educ. Elem. Sch.* v. 93 Monkey rings... Manufactured under various names, such as climbing trees, junglegyms, climbing towers, castle towers, and climbing maze. 1963 Barnard & Lauwerys *Handbk. Brit. Educ. Terms* 115 *Jungle gym*, a simple gymnastic apparatus on which children in an infant school can climb or swing as part of their free activity curriculum. 1967 J. Redgate *Killing Season* (1968) ii. vii. 104 Through the kitchen window he could see the children laughing and wrestling with each other inside their jungle gym. 1973 *Washington Post* 3 Oct. B1/4 (*heading*) Recreation 1973: Everything from jungle gyms to the Bataca bromb. 1945 Baker *Austral. Lang.* viii. 157 *Jungle juice*, any alcoholic beverage concocted by servicemen in the tropics. *Ibid.* 158 *Jungle juice*, poor quality petrol. 1958 R. Stow *To Islands* i. 19 The cartoons.. about going troppo and drinking jungle juice. 1960 *News Chron.* 9 Mar. 7/4 The draught cider and gin they drink in the West of England and call 'jungle juice'. 1967 O. Norton *Now lying Dead* vi. 99 Oh, I know what our ale can do! Jungle-juice, as the lads call it. 1894 Kipling *Jungle Bk.* 63 One of the beauties of Jungle Law is that punishment settles all scores. 1957 M. Kennedy *Heroes of Clone* III. vi. 204 It was awkward having to explain jungle law to someone who had never..emerged from a well-kept shrubbery. 1971 *Daily Nation* (Nairobi) 10 Apr. 13/2 The Obote regime had turned the country into 'a political jungle ruled by jungle law' whereby some people earned their living by putting others into prison. 1944 *Amer. N. & Q.* Mar. 183/1 Can somebody identify a tropical disease called 'jungle rot'? Is it a new name for an old illness? 1945 Jungle rot [see *crud 2 b]. 1958 *Times* 28 Oct. 4/5 (*headline*) N.A.L.G.O. fear 'jungle war' —arbitration move opposed. 1955 E. Waugh *Officers & Gentlemen* I. vi. 70 They put me in charge of a jungle warfare school. 1972 D. Bloodworth *Any Number can Play* viii. 61 He had forgotten more about jungle warfare than a fellow like that would learn in a lifetime. *Ibid.* xix. 194 It has..a big tangle of forest and swamp for jungle-warfare training.

c. Passing into *adj.* = characteristic of the jungle; savage, untamed; *spec.* designating a style of jazz music characterized by primitive sounds redolent of the jungle.

1908 A. Noyes *William Morris* 118 Torn by the savage jungle-cries of the elemental passions. 1909 *Daily Chron.* 22 Jan. 3/3 These wild poems of fierce jungle-passion and horror. 1935 *Vanity Fair* (N.Y.) Nov. 71/3 The savagery of their rhythm calls forth the terms '*shake music*' and '*jungle music*'. 1955 Keepnews & Grauer *Pict. Hist. Jazz* xiii. 141 Cootie Williams..produced a fine, muted 'jungle' sound. 1955 L. Feather *Encycl. Jazz* vii. 133 Early Ellington orchestral characteristics included the use of what he originally called 'jungle style' effects, through the use of plunger mutes. 1957 *Times Lit. Suppl.* 25 Oct. 637/2 A clearly truthful account of the lives and jungle-fights of those cold-hearted career-women who make fortunes from knocking pounds of unnecessary weight off sad fat ladies without love. 1972 *Jazz & Blues* Feb. 20/2 Duke's 'jungle' sounds.

jungle (dʒʌ·ŋg'l), *v.* [f. *jungle *sb.* 2 c.] *intr.* To prepare a meal at a hoboes' camp; to form such a camp; to join forces with another person. Usu. with *up*.

1922 J. Tully *Emmett Lawler* 252 The fire was built in the improvised furnace, and water was carried from the brook. They returned laden with meat and eggs, potatoes, and coffee... The method is called 'jungling up' by tramps. 1924 'Digit' *Confessions 20th Cent. Hobo* 12 *Jungle up*, bivouac in the weeds and clean up generally. 1926 J. Black *You can't Win* vi. 70 You're welcome to travel with me, kid, if you want to jungle-up for a month or two. 1931 U. Ledoux *Mr. Zero's Scrapbk.: Ho-bo-ho Medley No. 1* 11 Hoboes and Yeggs never mix and jungle in separate camps. 1937 J. Steinbeck *Of Mice & Men* i. 8 Tramps who come wearily down from the highway in the evening to jungle-up near water.

jungli (dʒʌ·ŋgli), *a.* and *sb.* [f. Jungle + -*i*, adj. suffix as in *Hindi*, etc.] **A.** *adj.* = Jungly *a.* 2; (see also quot. 1927[1]). **B.** *sb.* An inhabitant of the jungle.

1920 *Blackw. Mag.* Oct. 463/1 Just oneself with half a dozen of one's men and some jungli villagers. 1927 *Chambers's Jrnl.* 29 Jan. 138/2 Already he ceases to be jungli. *Note*, Wild and boorish, a clodhopper or uneducated peasant. 1927 *Blackw. Mag.* Mar. 290/1 His crew of two junglis managed to make him understand. 1928 *Ibid.* Jan. 1/2 A system of small flying columns kept the junglis from the inertia which breeds mischief.

Junian, *a.* Delete *rare* and add later examples.

1963 *Times Lit. Suppl.* 25 Jan. 67/1 Candidates for the Junian crown of laurels. 1964 *Language* XL. 87 A word count of the Junian material.

2. Of or pertaining to Francis Junius (1589–1677), philologist and antiquary.

1826 J. J. Conybeare in W. D. Conybeare *Illustr. Anglo-Saxon Poetry* 197 The Junian Cædmon. 1840 J. Petheram *Hist. Sk. Progress Anglo-Saxon Lit. in England* 73 The copy used was the Junian transcript in the Bodleian. 1892 S. A. Brooke *Hist. Early Eng. Lit.* II. 67 Archbishop Ussher..found this manuscript and gave it to Francis Dujon, a scholar of Leyden, who is known in literature as Junius, and from whom the manuscript derives its name of the *Junian Caedmon*. 1897 F. A. Blackburn in *Anglia* XIX. 91 The same device is found in other manuscripts, for example in the Genesis and the Exodus of the Junian MS. in the Bodleian. 1914 *PMLA* XXIX. 146 This book was printed at Dort from the famous and beautiful Junian types representing the Gothic and Anglo-Saxon alphabets, which Junius later presented to the University of Oxford.

junior, *a.* (*sb.*) Add: **3. b.** Designating something intended for children or young people; also applied to a product, device, etc., that is smaller than the normal size.

1860 (*title*) The junior atlas, for schools; fourteen maps selected from the college atlas. 1884 *Chambers's Hist. Readers* (*title*) Junior English history. 1941 *Tennessean* (Nashville) 12 Aug. 9 (Advt.), Handyhot junior electric washer. 1948 (*title*) Oxford junior encyclopaedia. 1948 *Tennessean Mag.* (Nashville) 7 Nov. 23 The idea that 'Junior is a size, not an age', has been plugged with rather half-hearted vigor for several years..in clothes in the 9 to 17 size range. 1967 L. B. Archer in Wills & Yearsley *Handbk. Managem. Technol.* 125 It will be an important design consideration to know whether the product is to be presented as one of a family of different products.. and/or one of a family of similar products (standard, de luxe, junior, and portable models?). 1967 M. Drabble *Jerusalem the Golden* vii. 172 She had been in the afternoon to the chemist's to buy some Junior Aspirin. 1972 *Practical Motorist* Oct. 212/1 A full-size hacksaw won't fit into the average tool box, nor will it work in tight corners. A 'junior' frame saw is a useful back-up, since it's small enough to travel with any tool kit.

5. Special collocations: *junior college* (in *U.S.*), 'a college, operating as a separate institution or as part of a standard college, which does not offer courses more advanced than those of the sophomore year' (D.A.E.); also, a similar institution in Britain and elsewhere; *junior high* (*school*) (*N. Amer.*), a school intermediate between elementary school and high school; *junior miss* (orig. *U.S.*), a young teen-aged girl; = Miss *sb.*[2] 4; also *attrib.*; *junior school*, (*a*) in the state educational system, a school for children aged roughly between 7 and 11; a primary school; (*b*) the lower forms of some fee-paying schools; *junior service*, the Army; *junior stock* (see quot. 1914); *junior technical school*, a school providing a technical and secondary education for boys.

1899 *Univ. Chicago Reg. 1898–99* 37/1 The Faculties of the Schools of Arts, Literature, and Science have been organized as follows: (1) The Faculty of the Junior Colleges; [etc.]. 1949 *Manch. Guardian Weekly* 7 Apr. 8 You will not learn what are the ambitions of the students at a junior college. 1957 *Encycl. Brit.* XX. 258/1 Some schools extended secondary education upward by offering two years of additional work of 'junior college' type. 1963 *Higher Educ.: Rep. Comm. under Ld. Robbins 1961–3* 148 in *Parl. Papers 1962–3* (Cmnd. 2154) XI. 639 Other witnesses..advocated the creation of separate junior or preparatory colleges to undertake the later stages of sixth form work and the first year of university work. 1971 *English Studies* LII. 569 The action..takes place in a classroom of a Southern Californian junior college. 1909 *Ann. Rep. Bd. Educ.* (Columbus, Ohio) 168 The Board has declared itself in favour of the Junior High School System. 1929 *Encycl. Brit.* XX. 258/1 The junior high school has also been a vehicle for innovations in teaching methods. 1948 *Daily Ardmoreite* (Ardmore, Okla.) 12 Oct. 10/1 They met in the ninth grade in junior high. 1968 *Globe & Mail* (Toronto) 13 Feb. 30/2 (Advt.), Convenient to public, junior high and separate schools. *Ibid.* 33/6 (Advt.), Experienced Junior High, social studies and science teacher required. 1927 *Vogue* 15 Jan. 106/2 Junior Misses' Frock. 1950 M. Allingham *Take Two at Bedtime* 17, I was still wearing the junior-miss dresses I had had at school. 1965 *Harper's Bazaar* May 75 Both dresses £1 19s. 11d. by Marks & Spencer Junior Miss. 1871 *Minutes of School Board for London* I. 156 Public elementary day schools are conveniently classified into infant schools, for children below seven years of age; junior schools, for children between seven and ten years of age; and senior schools, for older children. 1902 *Captain* VII. 221/1 Workington passed out of the Junior school. 1971 *Times Educ. Suppl.* 5 Feb. 41/4 (Advt.), Small groups of immigrant pupils in Junior Schools, who need additional language instruction. 1915 E. Wallace *Man who bought London* viii. 81 She had a son in the army, and she bore the junior service a grudge in consequence. 1914 H. Halford *Dict. Stock Market Terms* 50 *Junior stocks*, ordinary and deferred stocks ranking for dividend after debentures and preference stocks. 1932 *Daily Tel.* 8 Oct. 2/4 The current quotations of the junior stocks remove the likelihood of an issue in that form. 1929 *Encycl. Brit.* VII. 988/2 Its lower grades have shown a considerable increase, whether in junior technical schools, art schools or evening classes. 1931 *Education Outlook* June 183/1 Its pupils [*sc.* of the new senior school] are distinguished from their contemporaries in grammar schools, modern schools, and junior technical schools.

B. *sb.* **a.** More generally (chiefly *U.S.*), a child, esp. a young boy. Freq. with capital initial.

1946 *Sun* (Baltimore) 14 Dec., Lest the joy of Christmas be marred by..Junior's nipping his pal's arm with an arrow from his archery set, [etc.]. 1951 O. Nash *Family Reunion* 31 But you take ingenuous Junior, and it's just a radio to him. 1968 *Globe & Mail* (Toronto) 13 Jan. 25/6 (Advt.), Enclosed is cheque..for..adults..and.. juniors. 1970 G. Greer *Female Eunuch* 230 If junior finds out that his parents are going out, he'll scream.

2. A barrister who has not taken silk; a junior barrister.

1837 [in Dict., sense B. *sb.* b]. 1842 Dickens *Amer. Notes* I. iii. 127 The counsel who interrogated the witness ..was alone and had no 'junior'. 1872 G. H. Lewes *Let.* 5 Jan. in Geo. Eliot *Lett.* (1956) V. 234 We had..Bowen (the junior in the Tichborne case on whom Coleridge mainly relies),..and had..lots of fun. 1958 S. Hyland *Who goes Hang?* xlix. 260 Oliver Passmore K.C., M.P. And there was a 'junior' with him called Mortimer. 1972 'W. Haggard' *Protectors* iii. 26 This barrister was..a strong Junior at the criminal bar, and he'd defended Martiny's friend.

junk, *sb.*[2] Add: **1. d.** (Earlier and later

examples.) Also, second-hand or discarded articles of little or no use or value; rubbish.

1842 *Congress. Globe* 23 Feb. 261 Champagne was charged for under the head of 'old junk'. 1913 V. Steer *Romance of Cinema* 30 The life of a film is very short. It is 'first run' to-day and 'junk' a few short weeks hence. 1924 Galsworthy *White Monkey* i. v. 33 His 'junk', however, was not devoid of the taste and luxury which overflows from the greater houses of England. 1935 A. Christie *Death in Clouds* xi. 118, I have my collection ..that all connoisseurs know—and also I have—well, frankly, Messieurs, let us call it junk! 1940 *Punch* 3 Apr. 386/2 A great deal of ecclesiastical junk which the Vatican ..would have obvious difficulty in dispersing. 1952 R. Finlayson *Schooner came to Atia* 112 The boat was found battered to junk in a pool. 1974 *Woman* 4 May 5/1 Collecting junk for creative work is a way of life at school.

e. Any narcotic drug, esp. heroin; also, such drugs collectively. Also *attrib.* *slang* (orig. *U.S.*).

1925 *Writer's Monthly* June 487/1 *Junk*, dope. 1930 *Liberty* 5 July 24/1 When he has the junk in him there is no telling what he'll do. 1933 'J. Spenser' *Limey* iii. 37 You shouldda seen him when he was only half light with a shot o' the junk (dope) he used. 1937 B. Reitman *Sister of Road* (1941) vii. 74 She was full of junk, which she had received in her mail. 1938 [see *connect *v.* 5 b]. 1953 W. Burroughs *Junkie* (1972) vii. 66 It doesn't take the manager long to spot a bookie or a junk-pusher. *Ibid.* x. 108 You cannot escape from junk-sickness any more than you can escape from junk-kick after a shot. *Ibid.* xii. 120 Lupita got her start with one gram of junk and built up from there to a monopoly of the junk business in Mexico City. 1957 P. Frank *Seven Days to Never* vii. 204 Pretty slick way to carry your junk... You're a junkie, aren't you? 1960 J. Gelber *Connection* i. 25 You cats ought to smoke pot instead of using junk. It would make you more agreeable. 1963 A. Trocchi *Cain's Book* 73 One must go where the junk is and one is never certain where the junk is, never sure that where the junk is is not the anteroom of the penitentiary. 1966 *Listener* 17 Mar. 401/2 Burroughs's position is explicit: junk—*i.e.*, all narcotic drugs, as opposed to hallucinogenic ones like hashish and mescalin—are purely evil. 1972 J. Brown *Chancer* ix. 85 You do anything for junk... Cheat. Lie. Steal.

5. (sense 1 d) *junk car, cart, collector, -heap, merchant, -pile, -room, -stall, store, -yard*; **junk art** (see quot.); so *junk artist*; **junk jewellery** = *costume jewellery*; **junk mail** *N. Amer.*, circulars, advertisements, etc., sent by post to a large number of addresses; **junk-playground** = *adventure playground*; **junk sculpture** = *junk art*; so *junk sculptor*; **junk-shop** (further examples).

1966 *Britannica Bk. of Year* (U.S.) 807/1 *Junk art*, three-dimensional art made from discarded material (as of metal, mortar, glass, or wood); junk artist, *n.* 1954 *Amer. Speech* XXIX. 99 *Junk car* or *junker*, a car in poor condition that has been made to look like a hot rod. Usually used sarcastically of any car ready for the junk yard. 1967 *Boston Sunday Herald* 26 Mar. ii. 9/2 Thirty-five junk cars hauled away. 1879 *Scribner's Monthly* May 34/1 These piers..are too narrow even for the circulation of a junk-cart. 1934 Webster, Junk collector. 1956 J. M. Mogey *Family & Neighbourhood* 10 Junk-collectors' yards. 1906 *Westm. Gaz.* 26 Oct. 2/1 He [*sc.* Hearst]..took hold of a junk-heap relic of Pacific-coast journalism called the *Examiner*. 1917 R. L. Alsaker *Eating for Health* ii. xiii. 195 You and I..have to conform to the laws of nature, or else we are thrown into the junk heap. 1953 C. Day Lewis *Italian Visit* iii. 34 It would take three life-times to cover The glorious junk-heap. 1973 *Guardian* 23 Mar. 12/4 In Plymouth once they raided the local junk-heap to supplement the props which they had brought with them. 1939 *Sun* (Baltimore) 8 May 17 The Duchess of Kent..envies American women their smart 'junk jewelry' (costume stuff). 1960 *Sunday Express* 21 Feb. 14/3 American women have..loaded themselves with so much 'junk' jewellery they jangle as they walk. 1954 *Reader's Digest* Oct. 37/1 The argument for junk mail is that sorting requires little time—the postman simply delivers one throwaway to each home. 1967 *Economist* 23 Dec. 1227/1 These high charges are necessary to subsidise third class bulk mail, often called 'junk' mail, which includes millions of unsolicited advertisements sent out to lists of names as well as mail order catalogues. 1972 *Edmonton* (Alberta) *Jrnl.* 26 June 5/5 Within the first few months of their baby's life, the Duncans had received more than 1,000 pieces of junk mail. 1901 *Westm. Gaz.* 8 July 3/2 Twenty tons of unsold copies of a well-known cheap magazine were sold for waste-paper to junk merchants. 1880 *Harper's Mag.* June 67/1 The junk pile in the barn is invaded, and the rusty plough abstracted. 1912 J. H. Moore *Ethics & Educ.* 10 They should be sent without sighs or lamentations to the junk-pile. 1957 *Economist* 5 Oct. 28/2 That lively social experiment the Grimsby 'shanty town', a junk-playground which has evolved into a children's community. 1936 M. Allingham *Flowers for Judge* xiv. 207 They didn't keep anything valuable there. It's a sort of junk-room. 1966 *Time* 14 Jan. E1 Long before the *tachiste* painter or the junk sculptor, the American Indian shaped art from sand, bone, feathers—whatever he had at hand. 1969 *Harper's Mag.* Sept. 14 Students interested in musicology, junk sculpture, the Theater of the Absurd. 1970 *Times* 7 Mar. p. iv/6 He [*sc.* Brion Gysin] had time on his hands and became one of the pioneers of junk sculpture. 1841 'Dow, Jr.' *Short Patent Sermons* 77 Trash, that wouldn't fetch two cents in the market of heaven, and but a trifle more in the junk-shops of hell. 1879 *Scribner's Monthly* May 33/2 New York is bordered with rat-holes and rotten cribs, gin-mills and junk-shops. 1881 C. C. Harrison *Woman's Handiwork* iii. 171 She confessed to having purchased from a junk-shop the charming little gilt-framed oblong mirror. 1939 C. Day Lewis *Child of Misfortune* i. v. 70 Lady Gresham's mind was dark, dusty and furtive as a junk-shop. 1951 M. Kennedy *Lucy Carmichael* iii. i. 157, I saw it..in a tray in a junk shop. 1969 *Listener* 17 Apr. 533/1 The Metropolitan Museum in New York was a really dismal junk shop 40 years ago. 1962 *John o' London's* 11 Jan. 40/3 Oddities off junk-stalls. 1882 J. D. McCabe *New York* 583 They..sell it to the junk and rag stores. 1908 Kipling *Lett. of Travel* (1920) 197 We went to look at a marine junk-store. 1880 G. W. Cable *Grandissimes* 192 You may still..see one [*sc.* a villa] standing..among..junk-yards, and long-shoremen's hovels. 1952 W. R. Burnett *Vanity Row* (1953) vi. 50 All about loomed gas tanks, warehouses, junk yards. 1953 *Manch. Guardian Weekly* 27 Aug. 7 Suffocating in a junkyard of chairs. 1966 T. Pynchon *Crying of Lot 49* i. 14 If it had been an outright junkyard, probably he could have stuck things out, made a career. 1974 M. Hastings *Dragon Island* ix. 77 The only reason this floating junkyard's still in service is that the Chinese perishers that own her can't even give her away.

junk, *v.* Add: **2.** To treat as junk or rubbish; to discard, abandon; to 'scrap'.

1916 B. Hall *Diary* 11 Oct. in Hall & Niles *One Man's War* (1929) 196 When he got home his ship was a complete wreck. It will be junked. 1922 H. Titus *Timber* xxxii. 281 Perhaps he had friends..who are junking their mills now and getting ready to move. 1930 *Time & Tide* 20 Sept. 1164 Jugo-Slavia will not disband a soldier, scrap a gunboat, or junk a gun while Italy menaces her. 1945 [see *cow-country* s.v. *cow *sb.*¹ 7]. *a* 1963 S. Plath *Crossing Water* (1971) 33 The roses in the Toby jug Gave up the ghost last night... You should have junked them before they died. 1972 *Listener* 21 Dec. 860/2 The free-wheeling teams..might have to junk their wagons and stumble on foot down the western slopes of the Sierra. 1973 K. Giles *File on Death* vi. 149, I dare say you can pick up junked bits of weapons. 1973 *Times* 21 Mar. 9/4 The story of a man who junked his education and went after the short-term riches of sport.

junkanoo, junkonoo: see *John Canoe.

Junker¹. Add: Also *transf.*

1914 G. B. Shaw *What I really wrote about War* (1931) 25 The Junker is by no means peculiar to Prussia... Lord Cromer is a Junker. 1916 *Ibid.* 159 British Junker stupidity. 1919 W. H. Downing *Digger Dial.* 30 *Junker*, a superior staff-officer. 1920 E. Antonelli *Bolshevist Russia* i. ii. 32 A delegation of 'Junkers' (pupils of the Military Schools) appeared before Kerensky. *Ibid.* 35 Some Junker and regimental delegations..received no reply.

junker² (dʒɒ·ŋkəɪ). *Austral.* and *N.Z.* = Jinker².

1888 P. W. Barlow *Kaipara* xiii. 94 He mounted the 'junker'... The unhappy [horse] gave a feeble shake with one hind leg. 1924 Lawrence & Skinner *Boy in Bush* 236 'What's a junker, Tom?' 'A low, four-wheeled log hauler, with a long pole.'

junker³ (dʒɒ·ŋkəɪ). *U.S. slang.* [f. *Junk *sb.*² 1 e + -er¹.] A drug-addict; a drug-peddler.

1922 E. F. Murphy *Black Candle* ii. xvii. 276 One must ..be known as a 'junker' or addict to make the purchase. 1930 *Detective Fiction Weekly* 15 Nov. 473/2 He got the poppy gum from the smugglers and turned it over to the chemists at a good profit, getting back half in cash and half in drugs, which his junkers peddled. 1930 C. R. Shaw *Jack Roller* xii. 161 Next to me in the hospital was Herbie, a junker, who was taking the cure. 1949 'J. Evans' *Halo in Brass* (1951) iv. 29 No slim-waisted junker with a snapbrim hat and a deck of nose candy for sale to the right guy.

junket, *sb.* Add: **4.** Also *transf.* and *fig.*; *spec.* (see quot. 1886). Chiefly *U.S.*

1886 *Detroit Free Press* 4 Sept. 4/2 The term 'junket' in America is generally applied to a trip taken by an American official at the expense of the government. 1946 R. Blesh *Shining Trumpets* (1949) vii. 162 The first recording junket of modern times. 1954 Koestler *Invis. Writing* xxxi. 326 Bloomsbury and Greenwich Village went on a revolutionary junket. 1966 *Telegraph* (Brisbane) 13 Apr. 51/1 United States delegates to the Inter-Parliamentary Union conference in Canberra are upset that their trip has been described as a junket. 1966 *Sunday Times* 11 Dec. 3/2 A week here as a member of a gambling junket. 1973 *Black World* Apr. 96 On a junket to L.A. and New York looking for scripts. 1973 *Times* 18 Aug. 14/1 The only way I could get to see the countries about whose politics I would write so knowledgeably was to get myself attached to groups of travel writers on facility trips, or what the Americans call junkets.

junketeer (dʒɒːŋkĕtīə·ɪ). orig. *U.S.* [f. Junket *sb.* + -eer.] = Junketer.

1939 *Amer. Speech* XIV. 237/1 Junketeer is a term used in the press in these days. 1963 *Economist* 1 June 902/2 The 'junketeers' need not submit any receipts or documents. 1969 *Daily Tel.* 21 Feb. 19/6 Most of the junketeers have been Americans with personal credit ratings of 2,000 dollars upwards.

junkie (dʒɒ·ŋki). *slang* (orig. *U.S.*). Also **junkey, junky.** [f. *Junk *sb.*² 1 e + -ie.] A drug-addict; also occas., a drug-peddler. Also *attrib.* or as *adj.* (chiefly in form *junky*).

1923 [see *gun *sb.* 3 e]. 1929 *Flynn's* 2 Feb. 125/2 He's a confirmed 'junkie' and, of course, he'll get hold of it if he can. 1930 *Time & Tide* 20 Sept. 1181 He became a recognised junky and earned his living as a useful subordinate member of a drug-traffic gang. 1930 *Detective Fiction Weekly* 20 Dec. 362 (*heading*) Undercover man Murray follows the junkey trail into the greatest city's street of evil. 1949 N. Algren *Man with Golden Arm* 25 You're not a student any more... Junkie—you're hooked. 1951 *N.Y. Times* (City ed.) 27 June 19/5 An addict who sells narcotics, for which he usually gets his own daily supply, is called a 'junkie', he explained, and they hang out in taverns and cafes until the police raid them. 1959 W. Burroughs *Naked Lunch* 13 He would suck the juice right out of every junky he ran down. *Ibid.* 53 He spoke in his dead, junky whisper. 1959 C. MacInnes *Absolute Beginners* 68 If you have a friend who's a junkie..you soon discover there's no point whatever discussing his addiction. 1969 *Daily Tel.* (Colour Suppl.) 16 May 33/2 Combined with my appetite-killing pills and my tranquillisers I was possibly the junkiest man on the [ski] slopes. 1970 *Times* 22 Sept. 10/4 Over-prescribing by the notorious junkie doctors. 1971 *Daily Tel.* 18 June 17/3 They plan to tour villages showing a film about 'junkies' and giving lectures to youngsters about the dangers of drugs. 1972 J. Brown *Chancer* iv. 59 Lacerated hands, the hands of junkies, scarred where needles had searched for veins. 1973 *Black World* May 47 Talk about how to get more money, how to get educated, how to have scientists for children rather than junkies.

junkman². For *U.S.* read orig. *U.S.*, and add earlier and later examples.

1872 E. Crapsey *Nether Side N.Y.* 39 The shop of a junkman selected..as the purchaser of the plunder. 1879 *Scribner's Mag.* May 37/2 He is a licensed junkman; he holds a license for running a boat and buying and selling old refuse articles of any kind—a kind of water-ragman. 1963 H. Garner in R. Weaver *Canad. Short Stories* (1968) 2nd Ser. 38 The two of them helped the junkman load his truck. 1972 R. K. Smith *Ransom* iv. 150 The old pawnbroker watched the junkman.

junky (dʒɒ·ŋki), *a.* [f. Junk *sb.*² + -y¹.] **1.** Worthless, valueless, rubbishy.

1946 'G. Orwell' *Coll. Ess.* (1968) IV. 92 The kind of junky books..that accumulate in the bottoms of cupboards. 1966 *Punch* 27 July p. vii, Dealing in junky but odd bric-à-brac and more expensive furniture. 1972 *N.Y. Times* 3 Nov. 16/8 'Those are junky schools,' she declared. 'They have riots up there every day.' 1974 *New Yorker* 25 Feb. 62/3 Not necessarily cheap junk, however. In the past couple of years, an extraordinary demand for junky furniture seems to have arisen in London.

2. See *junkie.

junr. (Examples.)

1813 Jane Austen *Let.* 16 Sept. (1952) 326 The Letter you forwarded from Edwd Junr has been duly received. 1819 M. Edgeworth *Let. c* 1 Jan. (1971) 154 Mr. Brooke Boothby (Junr.) and Mr. Henry Vernon dined here.

‖ **junshi** (dʒu·nʃi). *Hist.* [Jap.] In Japan, suicide at the death of one's lord, self-immolation.

1871 A. B. Mitford *Tales Old Japan* II. 57 The ancient Japanese custom of *Junshi*; that is to say 'dying with the master'. 1904 L. Hearn *Japan: Attempt at Interpretation* iv. 47 With the rise of the military power there gradually came into existence another custom of *junshi*, or following one's lord in death,—suicide by the sword.

jupati (dʒū·-, hū·pāti). [Pg., a. Tupi.] A Brazilian palm, *Raphia tædigera*, bearing large leaves whose long stalks are used locally as a building material.

1856 B. Seemann *Popular Hist. Palms* 331 *Raphia tædigera* Mart., the Jupati of the Lingoa Geral, is one of the many noble Palms which grow on the rich alluvium of the Amazon. 1863 H. W. Bates *Naturalist on River Amazons* I. v. 196 The moon now broke forth and lighted up..the leaves of monstrous Jupatí palms which arched over the creek. 1908 in R. Spruce *Notes of Botanist on Amazon & Andes* II. 521 Glossary of native names... Jupati. *Rhaphia* [*sic*] *tædigera*. A short-stemmed but noble palm with immense leaves. 1972 Y. Lovelock *Vegetable Bk.* 302 Palms are to be found spanning the equatorial belt around the world. They include..the Japanese [*sic*] jupati palm.

Jupiter. Add: **1. b.** *Jupiter Pluvius*, Jupiter as the dispenser of rain; hence used trivially in reference to a fall or storm of rain.

1864 G. A. Sala *Quite Alone* I. ii. 39 'Take my advice, and..borrow somebody else's umbrella.'.. 'Are you, too, ready for the wrath of Jupiter Pluvius?' 1874 G. H. West *Rugby Union Football Ann.* 62 But 'Jupiter Pluvius' and the Fates were against it. 1922 Joyce *Ulysses* 598 It cleared up after the recent visitation of Jupiter Pluvius.

Jura (dʒūə·rā, ‖ ʒūra). [Name of the range of mountains on the borders of France and Switzerland that gave its name to the Jurassic system; in this use prob. a. Ger. *Jura*, *Jura-Jurassic* (sb. and adj.).] Jurassic rocks or strata; the Jurassic system. Also *attrib.*, = Jurassic *a.*

Lower or *Black Jura* = Lias 2; *Middle* or *Brown Jura* = Dogger² 2; *Upper* or *White Jura* = Malm *sb.* 1 a.

1829 *Proc. Geol. Soc.* I. 94 On the descent to Sospello [near Nice] are found, in a regular descending series, greensand, Jura, oolitic (or younger Alpine) limestone, lias, red-marl, and older Alpine limestone or dolomite. 1851 *Q. Jrnl. Geol. Soc.* VII. ii. 42 (*heading*) On the comparison of the German Jura formation with those of France and England. *Ibid.* 46 In the Jura, or in the lower land of Berne, where the 'black Jura' is developed, there is no great extension in width. *Ibid.* 47 The Marne, Seine,.. Indre, &c., all of which flow westward, descend from the Lias to the 'brown Jura', and thence to the 'white'; whilst in the German 'Jura' the rivers flow from the heights of the 'white Jura' through the 'brown' and the

'black'. *Ibid.* 85 The 'Jura'-formations in northern Asia. 1893 P. Lake tr. *E. Kayser's Text Bk. Compar. Geol.* 241 In England the Jura forms a broad zone striking N.N.E. 1895 *Jrnl. Geol.* III. 380 During all this time the Upper Jura of South America..retained its central European character. 1900 *Jrnl. Geol.* VIII. 216 The rocks of the region represent..remnants of formations belonging to the Algonkian, Cambrian, Silurian, Carboniferous, Jura-trias, Cretaceous, and Eocene periods. 1929 *Bull. Geol. Soc. Amer.* XL. 176 (*heading*) Delimitation of Jura and Trias in British Columbia. 1955 G. G. Woodford tr. *M. Gignoux's Stratigr. Geol.* vii. 314 In describing the Swabian Jura, Quenstedt and Leopold von Buch distinguished three successive groups of rocks, to which they gave the names Black Jura, Brown Jura, and White Jura, in accordance with their appearance in outcrops. These three groups correspond approximately to what we now call the Lower Jurassic or Lias, the Middle Jurassic or Dogger and the Upper Jurassic or Malm.

‖ **Jurançon** (ʒūrañsoñ). [Name of a commune in the Basses Pyrénées.] The white wine produced in Jurançon.

[1833 C. Redding *Hist. Mod. Wines* v. 138 The first growths of Jurançon and Gan bring two hundred francs the hectolitre.] 1920 A. L. Simon *Blood of Grape* ix. 243 Fine wines..from the Mediterranean Coast and the Pyrénées (..*Rivesaltes, Jurancon*, etc., etc.). 1926 P. M. Shand *Bk. Wines* v. 64 Like all fine wines, Jurançon owes its quality to one supreme and informing vine. 1969 L. Orizet *Wine Bk.* 144 The production seldom exceeds 400,000 gallons of genuine white Jurançon a year, and a few gallons of red wine.

Jurassic, *a.* Add to def.: Also applied to the period itself and to flora and fauna found in Jurassic formations. (Earlier and additional examples.)

1831 *Proc. Geol. Soc.* I. 241 The Jurassic and Alpine limestones. 1847 *Q. Jrnl. Geol. Soc.* III. 117 A succession of deposits..formed during the Jurassic period. 1851 *Ibid.* VII. 179 But for the 'plant-beds' at Gristhorpe, Cloughton, Kiburn, and Whitby, we should know little of the ancient vegetation of the Jurassic period. These localities have supplied, in fact, the types of the Jurassic Flora. 1909 Chamberlin & Salisbury *Geol.: Shorter Course* xxiv. 712 (*caption*) A group of Jurassic ammonites. 1938 L. D. Stamp *Physical Geogr. & Geol.* xiii. 207 At the close of the Jurassic period the sea retreated and left a great lake in south-eastern England. 1968 D. A. Robson *Sci. Geol.* ix. 242 The succession is reminiscent of that of the Carboniferous Coal Measures, except for its distinctive Jurassic flora.

b. *absol.* as *sb.* The Jurassic system or the Jurassic period.

1831 *Proc. Geol. Soc.* I. 241 Chalk does not exist in the Carpathians, nor could the author recognise it at Cracow, the limestone of which he refers to the Upper Jurassic. 1873 [in Dict.]. 1902 A. J. Jukes-Browne *Student's Handbk. Stratigr. Geol.* xiv. 331 The Upper Jurassic, again, is an argillaceous series. 1938 A. K. Wells *Outl. Hist. Geol.* xv. 153 The Lower Jurassic is co-extensive with the Lias, a formation..splendidly exposed in parts of the English coast that enjoy a measure of popularity as seaside resorts. 1956 W. J. Arkell *Jurassic Geol. of World* i. 3 In the popular imagination the Jurassic is the period of great marine reptiles and flying dragons. 1973 *Nature* 13 July 92/1 Rifting probably started in the Triassic and by the Jurassic a reasonably large Tethyan Ocean was in existence.

juriballi (yūəribæ·li). Also **euri-, youraballi.** [Arawak (Makuchi).] Any of several trees belonging to the family Meliaceæ, especially a species of *Trichilia*, the bark of which was formerly used as a febrifuge; also, the bark itself.

1834 J. Hancock in *Trans. Medico-Botanical Soc.* 1832–33 36 (*title*) Remarks on the Juribali, or Euribali. 1846 J. Lindley *Veget. Kingd.* 462 Juriballi bark, a Demerara product..is described as being a potent bitter and astringent, far superior to Peruvian bark in fevers of a typhoid and malignant nature. 1851 *Illustr. Catal. Gt. Exhib.* iv. 982/1 Youraballi. 1903 *Imperial Inst. Techn. Rep.* 285 The Crabwood and Euriballi would have shown fair results.

jurimetrics (dʒūərime·triks). [f. L. *juris*, gen. of *jūs* law + *-metrics*, as in *biometrics*, *econometrics*.] The use of scientific methods in the study of legal matters. So **jurimetri·cian, jurime·tricist**, a student of, or expert in, jurimetrics.

1949 L. Loevinger in *Minnesota Law Rev.* XXXIII. 483 The next step forward in the long path of man's progress must be from jurisprudence (which is mere speculation about law) to *jurimetrics*—which is the scientific investigation of legal problems. 1964 *Jrnl. Politics* XXVI. 915 Professor Spaeth's view of what jurimetrics may hope to accomplish. 1966 *Sci. Amer.* Sept. 296 Being, in this enterprise, jurimetricists and not legal historians, they chose the molds of analytic rather than historical jurisprudence for the ordering of their materials. *Ibid.* 295 The fruit of the union of jurisprudence and social science has inevitably been christened 'jurimetrics'. 1970 *Encycl. Sci. Suppl.* (Grolier) 287 Polimetricians, psychometricians, jurimetricians are all rapidly proliferating species of a genus of mathematically minded scholars.

jurist. 1. For † *Obs.* read *Obs.* exc. *U.S.*, and add further examples. Also, a judge.

1905 E. B. Holt tr. *H. Münsterberg's Americans* iv. 88 Sixty-one of them [*sc.* members of the Senate] were jurists... As to the jurists, they are not men who are still active as attorneys or judges. 1931 W. G. McAdoo *Crowded Years* iii. 41 A well-known jurist at that time was Judge Trewhitt. 1936 *S. P. E. Tract* xlv. 188 Even more important is the divergence between the English and American uses of *jurist*, which is not restricted in the United States to the meaning of an expert in the science of law. It is commonly applied to any one who has obtained the qualifications required for legal practice. 1973 N. W. Schur *British Self-Taught* 211 In America *jurist* is synonymous with *judge*. Unfortunately not all *jurists* (in the American sense) are *jurists* (in the English sense).

jury, *sb.* Add: **6.** *jury service, system*; **jury-fixer** *U.S.*, one who bribes or otherwise illegally influences a jury or juror; so *jury-fixing*; **jury-woman**, (*b*) a female juror.

1882 *Washington Post* 18 Mar. (Th.), There might be some scope in the proceedings before the Grand Jury for a 'jury fixer'. 1931 *Blue Valley Farmer* (Okla. City) 24 Dec. 1/6 Fill the town with secret service men to catch the jury fixers. 1887 *Library Mag.* Apr. 531/2 Bribery and jury-fixing would speedily disappear. 1946 C. McWilliams *Southern California Country* 245 The long and sordid aftermath, involving jury-fixing, bribery, and murder. 1955 *Radio Times* 22 Apr. 28/3 Jury Service is often regarded as a tiresome duty. 1973 E. McGirr *Bardel's Murder* i. 26, I got caught for jury service, six horrible days of it. 1875 W. Stubbs *Constitutional Hist. Eng.* I. xiii. 611 Many writers of authority have maintained that the entire jury system is indigenous in England. 1974 *Times* 2 May 18/6 The acquittals..demonstrate the strength of a jury system which acquits when there is reasonable doubt. 1927 *Daily Tel.* 24 May 17/6 The manner in which the recalcitrant jurywoman is eventually brought round is not altogether convincing. 1962 *Punch* 21 Nov. 733/2 One of the two jury-women who made up our twelve disappeared completely during the last day's hearing.

jus, jus', colloq. and dial. shortening of Just *adv.*

1801 T. Tenney *Female Quixotism* II. xiv. 145 He kill us all, one after toder, jus as easy as we kill chickens. 1884 J. C. Egerton *Sussex Folk* iii. 33, I jus should be glad if you could get rid an 'em for me. 1935 Z. N. Hurston *Mules & Men* (1970) i. i. 28 Jus' go wid Gabriel. 1973 *Black World* July 56/2 We jus' laid on the corner and watched. *Ibid.* Aug. 55/1 He walked over to the Village an' dug a backdoor with a lock jus' beggin' to be picked.

‖ **jus cogens** (dʒɒs kōᵘ·genz). [L., compelling law.] A principle of international law which cannot be set aside by agreement or acquiescence. So, in modern use, as laid down by the Vienna Convention on the Law of Treaties (1969), 'a peremptory norm of general international law'. (See quots.)

1895 W. H. Rattigan *Private Internat. Law* vi. 163 A *jus cogens* regulating rights of this description is a product of modern commercial and intellectual activity, and finds no place, for instance, in the jurisprudence of the Roman Empire. 1937 *Amer. Jrnl. Internat. Law* XXXI. 571 The answer to this question depends on the preliminary question, whether general international law contains rules which have the character of *jus cogens*. 1945 M. Wolff *Private Internat. Law* iii. 168 Savigny..clearly showed that the rules of an absolute, imperative character (the *ius cogens*) to be found in any legal system are of two kinds. There are first of all those rules 'that are enacted merely for the sake of persons who are the possessors of rights', such as the laws limiting the capacity to act on account of age or sex, or laws concerning the transfer of property; and secondly the rules that are not made solely for the benefit of single individuals but rest on moral grounds, or on the 'public interest'... *ius cogens privatorum pactis mutari non potest*. 1957 G. Schwarzenberger *Internat. Law* (ed. 3) I. xx. 352 In this respect, the rules governing the principle of the freedom of the seas are not *jus cogens* which as such is unalterable, but as much *jus dispositivum* as any of the other rules of international law. 1960 F. A. Mann in *Brit. Year Bk. Internat. Law* 1959 45 Lord McNair must allow the *jus cogens* of the proper law to override the general principles which are merely incorporated into it. 1965 *Texas Law Rev.* XLIII. 455 The problem of international *jus cogens* can be stated in a simple question: Are there rules of international law which, by consent, individual subjects of international law may not modify? 1969 *Vienna Convention on Law of Treaties* 18 in *Parl. Papers* 1968–9 (Cmnd. 4140) LV. 395 Treaties conflicting with a peremptory norm of general international law (jus cogens). A treaty is void if, at the time of its conclusion, it conflicts with a peremptory norm of general international law. For the purposes of the present Convention, a peremptory norm of general international law is a norm accepted and recognized by the international community of States as a whole as a norm from which no derogation is permitted and which can be modified only by a subsequent norm of general international law having the same character. 1971 M. Akehurst *Mod. Introd. Internat. Law* (ed. 2) iii. 60 The technical name now given to the basic principles of international law, which states are not allowed to contract out of, is 'peremptory norms of general international law', otherwise known as *jus cogens*.

‖ **jus gentium** (dʒɒs dʒe·nʃiɒm). [L.] = *law of nations* (see Law *sb.*¹ 4 c).

1548 Hooper *Declar. Commandm.* iii. 31 They shuld observe the commune lawes vsyd among all people whiche is callid ius gentium. 1682 Evelyn *Let.* 19 Sept. in *Diary & Corresp.* (1906) 666 The right of passes, and petitions thereupon, were formed upon another part of the *Jus Gentium*, than our pretended dominion of the seas. 1771 'Junius' *Lett.* (1772) II. lxi. 285 Any law that contradicts or excludes the common law of England; whether it be *canon, civil, jus gentium*, or *levitical*. 1839 *Penny Cycl.* XIII. 361/2 According to their [*sc.* Roman lawyers'] phraseology,..*jus gentium* consists of those rules of law which are common to all nations. 1856 Bouvier *Law Dict.* (ed. 6) I. 685 Among the Romans by *jus civile* was understood the civil law, in contradistinction to the public law, or *jus gentium*. 1880 *Encycl. Brit.* XIII. 191/1 The ambiguity of the phrase *jus gentium* enabled the early founders of international law to apply the principles of the *jus naturæ* to the conduct of states *inter se* in a way of which there is no example in the Roman law-books. 1959 Jowitt *Dict. Eng. Law* II. 1035/1 *Jus gentium*, the law of nations.

jusi (hū·si). Also **husi, jussi.** [a. Sp. *jusi*, ad. Tagalog *husi*.] A delicate fibrous fabric woven in the Philippine Islands.

1851 *Illustr. Catal. Gt. Exhib.* iv. 1344/1 Piece of 'jusi', and a shawl of 'jusi'. *Ibid.*, Pieces of striped jusi dresses. 1902 *Encycl. Brit.* XXXI. 667/1 Beautiful fabrics called 'piña' and 'jusi', the former woven of pineapple-leaf fibre and the latter of this fibre mixed with silk.

‖ **jus primæ noctis** (dʒɒs prəi·mi nǫ·ktis). [L., right of the first night.] = *droit du seigneur* (see *droit¹ 1 b).

1887 F. K. Wischnewetzky tr. *Engels's Condition of Working-Class in England in 1844* 99 Factory servitude.. confers the *jus primæ noctis* upon the master... If the master is mean enough,..his mill is also his harem. 1911 *Encycl. Brit.* XV. 593/1 *Jus primae noctis* or *droit du seigneur*, a custom alleged to have existed in medieval Europe, giving the overlord a right to the virginity of his vassals' daughters on their wedding-night... The *jus*, it seems, is a myth, invented no earlier than the 16th or 17th century. 1923 A. Huxley *Antic Hay* xviii. 250 It was on a splendid subject—the 'Jus Primæ Noctis, or Droit du Seigneur'. 1933 'G. Orwell' *Down & Out* xxxiv. 256 He had heard..of..the *jus primae noctis* (he believed it had really existed). 1970 G. Greer *Female Eunuch* 228 No serf, writhing under the law of *jus primae noctis*,..ever had it worse. 1971 *Black Scholar* Dec. 13/1 The integration of rape into the sparsely furnished legitimate social life of the slaves harks back to the feudal 'right of the first night', the *jus primae noctis*.

‖ **jusqu'au bout** (ʒüskobu), *adv. phr.* [Fr.] To the end, to the conclusion, to completion; to the bitter end; *spec.* in the war of 1914–18, used in the context of carrying on fighting until a conclusive victory has been gained. Hence **jusqu'auboutisme**, the policy of carrying on to the bitter end; **jusqu'auboutist(e** (also without apostrophe), an advocate of *jusqu'auboutisme*.

1917 A. Huxley *Let.* 30 Sept. (1969) 134 Are the French socialists going to make much fuss at Bordeaux? I think and hope we are all Jusquauboutistes here. 1918 G. B. Shaw *What I really wrote about War* (1931) 280 In Constantinople it will be a matter of fighting *jusqu'au bout*. *Ibid.* 282, I am a Jusqu'auboutist. I do not want this war to be compromised as long as it will be possible for any of the belligerent Powers afterwards to pretend that if it had only gone on for another year it would have won. 1933 A. Huxley *Let.* 13 Aug. (1969) 372, I have never more passionately felt the need of using reason jusqu'au bout. 1968 *Listener* 4 July 12/1 The rhetoric of the klieg-lit European student leaders spouting mocking *jusqu'auboutisme* like post-Marxist Beatles.

Jussiæan (dʒɒsi,ī·ăn), **Jussieu(e)an** (dʒɒsiyū·-ăn), *a.* Also **Jussiean.** [f. mod.L. *Jussiæus*, or its origin, the French surname *Jussieu* + -an.] Of or pertaining to Bernard de Jussieu (1699–1777) and his nephew Antoine Laurent de Jussieu (1748–1836), or to the natural system of botanical classification devised by them.

1824 J. C. Loudon *Encycl. Gardening* (ed. 2) 47 All the hardy plants..arranged in groups, according to the Jussieuean system. 1857 A. Henfrey *Elem. Bot.* § 392 The Jussieuean System. 1865 G. Bentham *Handbk. Brit. Flora* p. viii, The so-called Linnæan or Jussiæan systems. 1876 *Encycl. Brit.* IV. 81/1 It [*sc.* the Linnæan method] was superseded by the Jussiean method.

just, joust, *sb.*¹ Now usu. spelt joust and pronounced (dʒaust).

just, joust, *v.*¹ Now usu. spelt joust and pronounced (dʒaust).

just, *adv.* Add: **1. c.** *just so* (earlier and later examples); also, (*b*) in the required or appropriate manner; (*c*) very close or friendly; (*d*) neatly and tidily; also as *adj.*; *just-so story*, a story which purports to explain the origin of something; a myth.

1751 Richardson in Johnson *Rambler* No. 97. ¶ 24 When I courted and married my Lætitia, then a blooming Beauty, every Thing passed just so! 1794 *Massachusetts Spy* 3 Sept. (Th.), A few years ago, every body supposed that if people did not behave just so, they ought to be punished. 1824 'A. Singleton' *Lett.* 18 Their *just so* garb, which, when adopted, was the court costume of the time, makes them [*sc.* Quakers] appear like antediluvians. 1831 J. Constable *Let.* 4 Dec. (1966) IV. 361 Was it *now* in its first state, I would rejoice to publish it *just so*. 1886 *Lantern* (New Orleans) 15 Sept. 4/2 It looks as though the contractors and the aldermen are just so, and fully understand each other. 1887 Parish & Shaw *Dict. Kentish Dial.* 85 He's not a bad master, but he will have everything done just-so; and you wunt please him

without everything is just-so. 1902 Kipling (*title*) Just so stories. 1922 J. Strachey tr. *Freud's Group Psychol.* x. 90 This is only a hypothesis..a 'Just-So Story'. 1930 *Times Lit. Suppl.* 27 Feb. 155/2 The charm of an album of Just-so stories, such as proves that even the humblest members of the human race possess..the saving grace of vision. 1952 L. MacNeice *Ten Burnt Offerings* 76 Reposed on a Sunday lap in the just-so room. 1960 R. Pound *Selfridge* i. 8 He always looked as if he had just come out of the bandbox. His mother..kept him just so. 1969 E. Bishop *Compl. Poems* 198 A raccoon..was the executioner. He was very fastidious and did everything just so. 1969 R. A. Noblett *Stavin' Chain* 5 Now Stavin' Chain wuz a man just so: When he got good whiskey, he would gurgle it slow. 1973 *Times Lit. Suppl.* 25 May 576/2 At first sight *My Country* is another of these just-so stories.

g. Delete 'in *U.S.*' and add later U.K. examples.

1960 R. A. Knox *Occasional Sermons* 227 Many who value the name of Christian still find it reasonable to believe that he did just that. 1971 *You* Sept. 24/1 Doctors and researchers are spending their energies and our money on finding out just how the female body ticks. 1974 *Times Lit. Suppl.* 8 Mar. 243/5 One wonders just how biased a view we develop of the human ecology of tropical Africa.

6. Delete 'Chiefly *Sc.* and *dial.*' and add further examples.

1937 A. Huxley *Ends & Means* xii. 199 In the abstract this scheme seems good enough; but in practice it just doesn't work. 1962 *Listener* 18 Jan. 135/1 The functionalist revolution just doesn't seem to have reached architectural photography. *Ibid.* 25 Oct. 694/2 It is just not true that cultivated people are able to pronounce equally on films, books, paintings, the theatre. 1972 *Newsweek* 10 Jan. 34/1 Now most businessmen think they just have to have a guard.

b. Delete 'in *Sc.* and *north. dial.*' and add further examples.

c 1863 T. Taylor in M. R. Booth *Eng. Plays of 19th Cent.* (1969) II. 119 Ain't it a bore, just! *c* 1875 'Brenda' *Froggy's Little Brother* (new ed.) iv. 41 'Now, haven't we 'ad a supper *just*?' exclaimed Froggy with satisfaction. 1894 Kipling *Let.* 28 July in C. E. Carrington *Rudyard Kipling* (1955) ix. 217 Won't New York be hot—just! 1903 A. Bennett *Leonora* viii. 228 'He's a good dancer.' 'I should think he was!..Isn't he just, mother?' 1904 E. Nesbit *Phoenix & Carpet* v. 94 'Luv us!' said Ike, 'ain't it been taught its schoolin', just!' 1930 J. B. Priestley *Angel Pavement* i. 16 She let herself go all right, didn't she just! 1943 K. Tennant *Ride on Stranger* (1968) iv. 36 'I don't believe you'd do that anyway yourself. Just grab money.' 'Wouldn't I, just,' her mentor said exultantly.

c. *just too bad*, unfortunate but inevitable.

1935 H. L. Ickes *Secret Diary* (1953) I. 270 If, in the course of his investigation, Glavis runs across your brother or my son, that will be just too bad, but it will be in the line of his duty. 1938 Auden & Isherwood *On Frontier* ii. 75 If some one's mistaken or lying or mad, Or if we're defeated, it will be just too bad. 1962 J. F. Powers *Morte d'Urban* ix. 199 Well, isn't that just too bad? 1962 'S. Woods' *Bloody Instructions* viii. 87, I admit you'll come in for some rough handling, and that's just too bad.

7. c. Also *S. Afr.* [tr. Afrikaans *netnou*].

1939 'D. Rame' *Wine of Good Hope* i. iii. 40 'Well, eat then,' said Lowell. 'I'll come just now.' 1953 N. Gordimer *Lying Days* ii. ix. 92 'Well,' I said, 'I'll open it just now—.' 1966 A. Sachs *Jail Diary* xvi. 143 'Would you mind switching off the light after you lock up.' 'The men on cell duty will do that just now.'

‖ **juste milieu** (ʒüst' milyö). [Fr., lit. 'the right mean'.] The happy medium, the golden mean; judicious moderation, esp. in politics.

1833 Mill *Lett.* (1910) I. 76 The Enfantin portion.. have become *juste milieu* men in politics. 1866 Mrs. Gaskell *Wives & Daughters* II. xxvii. 279 Now I hope that man in the garden is the *juste milieu*,—I'm that myself, for I don't think I'm vicious, and I know I'm not virtuous. 1875 J. G. Saxe *Leisure-Day Rhymes* 261 For me, the *juste milieu* I seek; I fain would leave alone The girl who rudely slaps my cheek Or volunteers her own! 1882 E. W. Hamilton *Diary* 13 June (1972) I. 289, I suppose between the extremes there must be a *juste milieu*, but as to where it is I am at sea. 1936 E. Sitwell *Victoria of Eng.* iii. 48 It was the *juste milieu* that ruled her life. 1945 A. J. P. Taylor *Course of German Hist.* 13 One looks in vain in their history for a *juste milieu*, for common sense.

Justicialism (dʒɒsti·ʃăliz'm). Also in Sp. form **Justicialismo** (hustisiali·zmo), and with small initial. [f. Sp. *justicia* justice + -al + *-ismo* -ism.] The name given by Juan Domingo Perón (1895–1974), President of Argentina (1946–55 and 1973–4), to his political doctrine: a combination of Fascism and socialism. Cf. *Peronism. So **Justi·cialist**(a *adj.*

1949 tr. *J. Perón's Speech at Opening 83rd Parl. Session Nat. Congr.* 7 The year that has gone by will be recorded in the Argentine history as the year of the 'justicialist' Constitution. [1950 R. A. Mende (*title*) El Justicialismo.] 1952 *Time* 12 May 28/3 If we take advantage of this historic moment, we shall impose Justicialism on the world and the coming century will be Justicialist. 1953 G. I. Blanksten *Perón's Argentina* xii. 276 The current label [for Perón's system] is *Justicialismo*. 1955 G. Pendle *Argentina* vi. 107 The basis of 'justicialist philosophy' was never more clearly expressed, perhaps, than in the following presidential assertion: 'For us there is nothing fixed and nothing to deny. We are anti-Communist because Communists are sectarians, and anti-capitalist because capitalists are sectarians. Our Third Position is not a central position. It is an ideological position which is in the centre, on the right or on the left, according to specific circumstances.' 1971 A. Hennessy in A. Bullock *20th Cent.* 120/2 In spite of..exaggerated claims for the ideology of 'justicialism' as a Third Position —neither capitalism nor communism—Peronism's impact outside Argentina was limited. 1972 *Buenos Aires Herald* 3 Feb. 9/2 A Justicialista Party executive committee member. 1972 *Times* 17 Nov. 16/2 Justicialismo, a vague word described as a middle way between communism and capitalism. *Ibid.* 16 Dec. 4/5 The candidate will be a Justicialist civilian.

jut, *sb.*² Add: **3.** *Comb.*, as **jut-jawed** *a.*, having a jutting jaw.

1943 *Commonweal* (N.Y.) 11 June 195 The company sergeant-major, a tough, jut-jawed, red-faced, bull-necked veteran of the last war. 1952 T. Pyles *Words & Ways Amer. Eng.* v. 94 The dour, thin-lipped, jut-jawed righteousness of his [*sc.* N. Webster's] later portraits. 1959 *Daily Mail* 21 Mar. 6/4 That jut-jawed Marshal of the R.A.F.

jutia, var. *hutia.

jutka (dʒɒ·tkā). [f. Hindi *jhaṭka* a jerk, jolt, lurch.] In southern India, a light two-wheeled vehicle drawn by a horse.

1886 Yule & Burnell *Hobson-Jobson* 362/2 *Jutka*,.. the native cab of Madras, and of Mofussil towns in that Presidency; a conveyance only to be characterised by the epithet *ramshackle*... It consists of a sort of box with venetian windows, on two wheels, and drawn by a miserable pony. It is entered by a door at the back. 1907 B. M. Croker *Company's Servant* xxxi. 318 She..got her baggage on a jutka and drove away. 1927 *Scots Observer* 16 Apr. 3/2 All afternoon jutkas and carts..are arriving. 1947 R. K. Narayan *Astrologer's Day* 4 *Jutka* drivers swore at their horses. 1961 K. Nagarajan *Chron. Kedaram* 117 It was so reminiscent of a *jutka* pony.

jutty, *a.* Delete *rare*⁻¹ and add later examples.

1868 G. M. Hopkins *Jrnl.* 19 July (1959) 178 Lying in jutty bends. 1922 A. S. M. Hutchinson *This Freedom* ii. 23 He swung round and pushed his dark face and jutty nose into the face of Bolas. 1937 G. M. Young *Daylight & Champaign* 93 Driving away the birds who nest in the jutty frieze.

juve (dʒūv). Colloq. abbrev. of *juvenile lead. Also *attrib.*

1935 *Variety* 17 Apr., Jones is a good-looking juve and possesses a corking tenor. 1968 P. Loraine *Dead Men of Sestos* viii. 115 I'm the young tenor, the juve, the one with the tennis racket. 1974 *Observer* 25 Aug. 22/3 The cherished tradition which demands that the worst thing about the Hammer horror [film]s shall be their juve leads.

Juvenal (dʒū·vənăl), *sb.*² Anglicized form of the cognomen of the Roman satirist Decimus Junius *Juvenalis*, used gen. to designate a satirist.

1592 Greene *Groats W. Wit* sig. F1, With thee I ioyne yong Iuuenall [*sc.* Nashe], that byting Satirist. 1693 Dryden tr. *Juvenal's Satires* p. vii, I might find in France, a living Horace and a Juvenal, in the person of the admirable Boileau. 1841 I. D'Israeli *Amenities of Lit.* III. 132 Jonson, the Juvenal of our drama. 1885 *Brewer's Dict. Phrase & Fable* (ed. 17) 469/1 The English Juvenal. John Oldham... The Juvenal of Painters. William Hogarth. 1902 *Daily Chron.* 20 Feb. 3/2 The art of satire is dead in England... The Juvenals of Fleet-street are no more.

juvenile, *a.* and *sb.* Add: **A.** *adj.* **1.** *spec.* Designating young offenders against the law, or the offences committed by them; esp. in *juvenile delinquency, delinquent*; also *juvenile adult*, a person below the legal age of responsibility and above a certain minimum age, who is held to be punishable for breaking the law (the term was discontinued by the Family Law Reform Act of 1969).

1816 *Rep. Soc. investigating Causes Increase in Juvenile Delinquency* 5 It was found that Juvenile Delinquency existed in the metropolis to a very alarming extent. 1817 *Observer* 14 Sept. 1/3 Your Committee have anxiously sought for information as to the number of juvenile delinquents who are annually committed to the different prisons in the metropolis;..the greater part of these Juvenile Offenders..are mixed indiscriminately with old offenders of all ages. 1837 Dickens *Let. c* 2 Oct. (1965) I. 315 Many thanks for your statistical Magazine, which contains some tables concerning juvenile delinquency. 1837 —— *O. Twist* (1838) I. xix. 321 Then the Juvenile Delinquent Society comes, and takes the boy away. 1847, 1854 [see Offender]. 1902 [see *Borstal, borstal]. 1916 *Lancet* 2 Feb. 365/1 (*title*) Juvenile crime. 1917 C. Leeson (*title*) The child and the war, being notes on juvenile delinquency. 1926 *Encycl. Brit.* I. 411/1 In 1894 two public inquiries into the administration of prisons and of Home Office schools arrived..at the same.. conclusion, viz: 'that the age 16–21 was the dangerous age; that we must concentrate on that; on the incipient criminal' or, as he was officially christened, the juvenile adult. 1958 *New Statesman* 25 Oct. 551/1 Some years ago when the current crop of juvenile delinquents were being labelled in the press as cosh-boys, I had written an article that contained interviews with some live specimens, hand-picked for me by an underworld acquaintance. 1959 Jowitt *Dict. Eng. Law* II. 1041/1 *Juvenile adult*, a person not less than sixteen and not more than twenty-one. 1964 M. Argyle *Psychol. & Social Probl.* v. 59 Juvenile delinquency is one of our most pressing social problems. *Ibid.* xv. 187 The rate of juvenile crime has risen rapidly during the decade, particularly since 1955.

b. *juvenile lead* = Juvenile *sb.* 2; *spec.* an actor who plays the leading youthful part in a play, etc.; the rôle so played. So *juvenile leading*.

1870 [see *heavy *a.*¹ 21]. 1885 J. K. Jerome *On Stage* (1891) xi. 102 Juvenile Lead's opinion is that the stage manager is a fool. 1897 G. B. Shaw *Our Theatres in Nineties* (1932) III. 210 This is not human nature or dramatic character; it is juvenile lead, first old man, heavy lead, heavy father. 1910 M. Beerbohm *Around Theatres* (1924) II. 460 The 'ingénue' and 'juvenile lead' of old-fashioned commercial drama. 1946 G. Millar *Horned Pigeon* i. 7 The part..might lead to great things, even to juvenile leads. 1973 J. Porter *It's Murder with Dover* xvi. 160 MacGregor flashed his juvenile lead smile.

2. *juvenile court*, a court of law for the trial of young offenders.

1899 *Illinois Laws* 132 A special court room, to be designated as the juvenile court room, shall be provided ..and the court may, for convenience, be called the 'Juvenile Court'. 1908 *Act* 8 *Edw. VII* c. 67 § 111 A court of summary jurisdiction when hearing charges against children..shall..sit either in a different building ..or on different days..., and a court of summary jurisdiction so sitting is in this Act referred to as a juvenile court. 1944 *Ann. Reg.* 1943 381 More attention than usual was focussed on the work of the Juvenile Courts. 1972 *Daily Tel.* 5 May 13/1 The Scots..abolished juvenile courts and replaced them with a system of children's panels.

3. *Geol.* [tr. G. *juvenil* (E. Suess 1902, in *Verh. d. Ges. deutsch. Naturf. u. Ärzte* 141).] Originating within the earth (or another planet) and brought to the surface for the first time.

1907 *Econ. Geol.* II. 266 Many mineral springs may be of magmatic origin, but since their starting points are inaccessible they can be proved to be juvenile only by showing that they cannot be meteoric. 1909 H. B. C. & W. J. Sollas tr. *Suess' Face of Earth* IV. xv. 549 [The hot springs] of Carlsbad..bring yearly to surface a million kilogrammes of juvenile salt. *Ibid.* 559 We must assume that the juvenile gases are originally liberated beneath the Sal mantle. 1944 *Bull. Geol. Soc. Amer.* LV. 1375 Clearly one cannot hold that the volcano was kept alive merely by free juvenile gas rising from an abyssolithic injection. 1955 [see *connate *a.* 5]. 1973 *Sci. Amer.* Jan. 56/2 One speculation is that deep permafrost is involved, associated perhaps with the arrival near the surface of juvenile water preceding and accompanying the rise of molten rock near the surface of the planet during the volcanic episode apparent to the west.

4. Special collocations: **juvenile foliage, leaf**, a type of foliage characteristic of the immature stages of certain trees, shrubs, or woody climbers, differing in shape, colour, etc., from the adult form; **juvenile hormone** *Ent.*, the hormone that controls the development of larval characteristics in insects; **juvenile wood**, an inner core of wood in a tree, distinguished by particularly small cells.

1957 M. Hadfield *Brit. Trees* 96 The juvenile foliage [of *Cryptomeria*] is spreading, with flatter and softer leaves than the adult. 1971 T. T. Kozlowski *Growth & Devel. Trees* I. iii. 95 Plants derived from needle-leaved cuttings [of *Chamæcyparis pisifera*] retained juvenile foliage if the source tree did so. 1940 V. B. Wigglesworth in *Jrnl. Exper. Biol.* XVII. 221 In previous papers the 'inhibitory hormone' was so called because in its presence the production of imaginal characters at moulting is suppressed. But in view of its probable mode of action through the activation of the nymphal system at the expense of the imaginal, it might be preferable to refer to this hormone as the 'nymphal' or 'juvenile' hormone. 1965 Lee & Knowles *Animal Hormones* xiii. 161 Normal development depends on changes in the relative amounts of ecdysone and the juvenile hormone which are available to the tissues. 1967 *New Scientist* 20 Apr. 154/1 Juvenile hormone..is necessary for the normal growth of immature insects—caterpillars and grubs, for example. 1970 *Daily Tel.* 16 Nov. 6/1 Many South American plants contain similar substances to insect juvenile hormone. 1910 L. Cockayne *N.Z. Plants* iv. 60 After a few weeks its [*sc.* a veronica's] new growth will be of the juvenile form, and juvenile and adult leaves will be on the plant at the same time. 1946 A. B. Jackson *Identification of Conifers* 2 In many cases the juvenile leaves differ in form, attachment or arrangement from those on the adult tree. 1956 F. W. Jane *Struct. Wood* ix. 191 Juvenile wood often has cells of smaller dimensions, often much smaller, than those of the trunk. 1971 T. T. Kozlowski *Growth & Devel. Trees* I. iii. 109 The wood in the region of the pith, which is formed early ..is termed juvenile (sometimes called core or pith) wood.

B. *sb.* **2.** (Later examples.)

1898 G. B. Shaw *Our Theatres in Nineties* (1932) III. 307 Ferdinand Gadd, the leading juvenile of The Wells. 1933 P. Godfrey *Back-Stage* iv. 46 His legs are too short or too long..for him to be a successful male juvenile. *Ibid.* vi. 72 Dramatists, like stage juveniles, are considered young until they are past the age of forty. 1973 *Times* 17 Nov. 11/5 I'm going to be your juvenile next season.

3. A book written for children. Freq. *pl.* Also *attrib.*

1849 *Mother Goose in Hieroglyphics* (1963) (Advt.), Pictures from the history of the Swiss... A very instructive and entertaining Juvenile, designed for children from ten to fifteen years of age. 1908 *Daily Chron.* 27 Nov. 3/5 What would John Newbery say if he were to.. see his old shop..filled with this season's 'juveniles'? 1930 *Publishers' Weekly* 5 July 28 We announce 10 juveniles. 1947 *Times Lit. Suppl.* 15 Nov. 593/2 (Advt.), The exacting and critical Juvenile public in this country.

juvenilia (dʒūvĭni·liă), *sb. pl.* [L., neut. pl. of *juvenilis* Juvenile *a.*] Literary or artistic works produced in the author's youth (freq. as a title of such works collected). Also *transf.*

1622 G. Wither (*title*) Ivvenilia: a collection of those poemes which were heretofore imprinted and written by George Wither. 1633 Donne (*title*) Iuuenilia: or certaine paradoxes and problemes. 1693 Dryden tr. *Juvenal's Satires* p. ix, His *Juvenilia*, or Verses written in his Youth. *a* 1849 H. Coleridge *Ess. & Marginalia* (1851) II. 265 Whatever effect these juvenilia may have produced at the time, they are quite worthless now. 1896 in Tennyson *Works* 2 Juvenilia. 1929 *Sunday Dispatch* 13 Jan. 10/5 Not that I belong to the school which would trace, in these innocent juvenilia of our nation, an anthropological or historical origin. 1952 *Brontë Soc. Trans.* XII. 126 Prose and verse by Charlotte and Branwell [in M. Christian's 'Census of Brontë Manuscripts in the United States'] classified as unpublished are mostly in the *Juvenilia* category. 1971 W. Gérin in C. Brontë *Five Novelettes* 7 Charlotte Brontë wrote the following novelettes between the years 1836 and 1839, from the time when she had just turned twenty until her twenty-third year. They cannot, therefore, be technically reckoned as belonging to her juvenilia. 1973 *Times Lit. Suppl.* 23 Nov. 1453/1 It is always difficult to judge the literary merit of juvenilia.

juvescence (dʒūve·sĕns). *rare.* [irreg. f. Juve(nile *a.* and *sb.*: see Adolescence.] The state of becoming young, juvenescence.

1920 T. S. Eliot *Ara Vos Prec* 11 In the juvescence of the year Came Christ the tiger. 1948 S. Spender in *Time & Tide* 10 Jan. 33/1 That kissing of steel furies Preparing a world's childless juvescence.

juvey, var. *juvie.

juvia (dʒū·-, hū·viă). [Amer. Sp.] The Brazil-nut. Also *attrib.*

c 1840 W. Rhind *Hist. Vegetable Kingdom* xxxix. 387 The triangular grains which the shell of the juvia incloses, are known in commerce under the name of Brazil nuts. 1852 T. Ross tr. *A. von Humboldt's Personal Narr. Trav. Amer.* II. xxiii. 390 Juvia-trees, which furnish the triangular nuts called in Europe the almonds of the Amazon, or Brazil-nuts. 1858 W. Baird *Cycl. Nat. Sci.* 69/1 The natives are very fond of this nut, and celebrate the harvest of the *Juvia* with great rejoicings. 1860 Mayne Reid *Odd People* 142 The splendid fruits of the *Bertholetia excelsa*, or juvia-tree, known in Europe as 'Brazil nuts'.

juvie (dʒū·vi). *U.S. slang.* Also **juvey.** [Colloq. shortening of Juvenile *a.* and *sb.*] A juvenile or juvenile delinquent; also, a detention centre or a court for juvenile delinquents.

1941 J. Smiley *Hash House Lingo* 34 *Juvie*, child. 1966 *Time* 2 Dec. 52/3 Los Angeles County police went after the 'juvies' (minors under 18), began carting them off by the busload. 1967 *New Yorker* 25 Feb. 128/3 But the teacher at juvey said, 'You have to finish it.' 1967 E. B. Nickerson *Kayaks to Arctic* vii. 58, I wondered..if we would have to bail them out of the Canadian equivalent of 'Juvies'. 1970 P. Stadley *Autumn of Hunter* (1971) viii. 115 Just where would you take me, little juvie? To a drive-in movie?

juxta-. Add: **ju:xta-arti·cular** *a. Anat.*, situated near a joint.

1900 in Dorland *Med. Dict.* 1910 Castellani & Chalmers *Man. Trop. Med.* lviii. 1137 (*heading*) Juxta-articular nodules. 1940 *Q. Bull. Northwestern Univ. Med. Sch.* XIV. 270/1 Juxta-articular nodes are non-inflammatory, painless, fibrous, subcutaneous growths.

juxtaglomerular (dʒɒːkstăglǫme·rŭlăɪ), *a. Anat.* [f. Juxta- + Glomerular *a.*] Situated next to a glomerulus of the kidney; *juxtaglomerular apparatus*, a structure variously considered to comprise (i) a juxtaglomerular body alone, (ii) one of these bodies and a *macula densa*, or (iii) (also *juxtaglomerular complex*) one of each of these, together with lacis cells and the afferent and efferent arterioles of a glomerulus; *juxtaglomerular body*, a mass of tissue scattered along the wall of the afferent arteriole of a glomerulus, composed of cells (*juxtaglomerular cells*) with conspicuous cytoplasmic granules, believed to be the site of renin secretion.

1935 *Biol. Abstr.* IX. 2181/2 The neuro-myo-arterial juxtaglomerular segments of the kidney. 1939 *Proc. Soc. Exper. Biol. & Med.* XLII. 227 The media of these vessels [*sc.* the interlobular and glomerular arteries] is composed not only of ordinary smooth muscle cells but also of larger, more afibrillar and probably less contractile cells. In the kidney these cells are found all along the arterial vascular tree and accumulate in groups at the vascular poles of the glomeruli to form the 'juxta glomerular apparatus' or 'polkissen'. 1942 J. F. A. McManus in *Lancet* 3 Oct. 394/2 (*heading*) The juxtaglomerular complex. *Ibid.*, There are an increasing number of references ..to a structure termed the 'juxtaglomerular apparatus'. By this term we mean a group of apparently specialised structures in relation to, and including, the afferent and efferent arterioles of the glomerulus. The term 'complex' is used here in preference to Goormaghtigh's term 'apparatus', since he restricts the meaning to the vascular and perivascular arteriolar structures. *Ibid.* 395/1 Ruyter in 1925 saw granular, afibrillar cells in the juxtaglomerular portion of the afferent arteriole in the mouse. 1952 A. C. Allen *Kidney* ii. 32/2 The macula densa in combination with the juxtaglomerular body is referred to as the juxtaglomerular apparatus. 1958 *Amer. Jrnl. Path.* XXXIV. 863 (*heading*) The juxtaglomerular cells in man. 1968 Page & McCubbin *Renal Hypertension* i. 40 (*heading*) Structure of the juxtaglomerular complex. 1968 E. L. Becker *Struct. Basis Renal Dis.* i. 34 The juxtaglomerular apparatus..is considered to consist of four portions: the specialized portion of the afferent arteriole containing characteristic granules; the first portion of the distal convoluted tubule, or macula densa; the cushion of cells at the vascular pole, continuous with the mesangium, termed the polkissen 'lacis' cells; and the efferent arteriole.

juxtaposition. Add: † **b.** *spec.* in *Cryst.* Contactual union between twinned crystals; *juxtaposition twin*, a composite crystal of two (or more) crystals joined along a plane; a contact twin. *Obs.*

1883 *Encycl. Brit.* XVI. 366/1 In aragonite the crystals are partly interpenetrating, and partly merely in juxtaposition. 1890 G. H. Williams *Elem. Crystallogr.* ix. 185 Two individual crystals in twinning position are usually, though by no means always, united in a plane, which may or may not coincide with the twinning plane. Twins of this sort are called contact or juxtaposition twins. 1911 A. E. H. Tutton *Crystallogr.* xxvi. 420 In the case of 'juxtaposition twins' the plane of union, whether the twin plane or not, is known as the 'plane of composition'. 1917 F. M. Jaeger *Lect. Princ. Symmetry* vii. 171 The classification of twins into such as are produced by juxtaposition or by penetration, may have certain advantages from a practical standpoint.

juxtapositive (dʒɒkstăpǫ·zĭtiv), *a.* *Gram.* [f. Juxtaposit(ion + -ive.] The designation of a case expressing juxtaposition.

1880 A. H. Sayce *Introd. Sci. Lang.* I. v. 370 Steinthal, however, goes on to divide his formless languages into 'juxta-positive' and 'compositive'. 1890 A. S. Gatschet *Klamath Indians* i. 490 Juxtapositive case in -tana.

K

K. Add: **3. b.** In *Physics k* (or *K*) is the symbol of thermal conductivity. [Introduced by J. B. J. Fourier, 1822.]

[1822 J. B. J. Fourier *Théorie anal. de la Chaleur* i. 54 Nous avons choisi ce même coëfficient K, qui entre dans la seconde équation, pour la mesure de la conducibilité spécifique de chaque substance.] 1850 in *Trans. R. Soc. Edin.* (1864) XXIII. 137 The specific heat of the metal being known, we can convert this amount of heat or flux across *x* into absolute measure; for the Flux is = – K *dv/dx* and *dv/dx* is known... Thus every experiment becomes an independent means of finding K. 1880 *Encycl. Brit.* XI. 579/2 Let *k* be the thermal conductivity of the substance and *c* its thermal capacity per unit bulk. 1947 *Sci. News* IV. 147 With glass..the heat conductivity (*k*) is 0·002. *Ibid.* 148 Steel (*k* = 0·10). 1969 *Jane's Freight Containers 1968–69* 239/1 The k-value according to choice of insulating material is about 0·4 to 0·5 kcal/m² h°C.

c. *Physics.* The designation of one of the strongest Fraunhofer lines, situated in the extreme violet at a wavelength of 3934 Å and due to absorption by calcium ions.

1879 *Proc. R. Soc.* XXVIII. 367 The calcium line with wave-length 4226..appears more or less expanded with a dark line in the middle..; the remaining bright lines of calcium are also frequently seen in the like condition, but sometimes the dark line appears in the middle of K (the more refrangible of Fraunhofer's lines H), when there is none in the middle of H. 1897 *Ibid.* LXI. 437 The H and K lines have become thin and defined. 1967 R. G. Giovanelli in J. N. Xanthakis *Solar Physics* xii. 353 The Balmer lines and the *H* and *K* lines of ionized calcium are..strong Fraunhofer absorption lines.

d. In *Physics* and *Chem.* *k* is the symbol of Boltzmann's constant.

1901 *Sci. Abstr.* IV. 230 For a comparison of his own reasoning with that of Boltzmann on gas molecules, the author deduces from *k* an estimate (6·175 × 10²²) for the number of molecules in the gramme molecule of any element. 1915, etc. [see *Boltzmann]. 1962 W. B. Thompson *Introd. Plasma Physics* ii. 17 A temperature of 11,600°K is needed to give an energy *kT* of 1 eV so that the mean kinetic energy of a molecule, $\frac{3}{2}kT$, reaches 1 eV only when *T* = 7,730°K.

e. In *Physics K* is used to designate the series of X-ray emission lines of shorter wavelength obtained by exciting the atoms of any particular element (cf. *L 6* a); these arise from electron transitions to the innermost, lowest-energy atomic orbit, of principal quantum number 1, which is thus termed the *K-shell*, and electrons in this shell *K-electrons*. **K(-electron) capture,** the capture by an atomic nucleus of one of the *K*-electrons.

1911 C. G. Barkla in *Phil. Mag.* XXII. 406 It is seen that the radiations fall into two distinct series, here denoted by the letters K and L. [*Note*] Previously denoted by letters B and A... The letters K and L are, however, preferable, as it is highly probable that series of radiations both more absorbable and more penetrating exist. 1923 H. L. Brose tr. *Sommerfeld's Atomic Struct. & Spectral Lines* iii. 144 If the excitation occurs through the agency of cathode rays, it is easy to imagine that the tearing-off of the 'K-electron' is effected by the impact of a cathode-ray particle that has penetrated into the atom. 1923 E. N. da C. Andrade *Struct. Atom* vi. 100 Moseley identified in the *K* series the two lines which he called α and β. ..In the *L* series he identified five lines. 1938 L. B. Loeb *Atomic Struct.* iii. 83 A tube with 1,470 volt electrons may excite *K* x-rays of Al, while it takes 60,600 volt electrons to excite the *K* x-rays of tungsten. 1946 H. Semat *Introd. Atomic Physics* (ed. 2) viii. 342 Probably the most clear-cut example of *K*-electron capture is the radioactive disintegration of vanadium, $_{23}V^{49}$, into titanium, $_{22}Ti^{49}$, with the capture of a *K* electron by the vanadium nucleus to form a titanium atom in the *K* state. 1951 J. Dougall tr. *Born's Atomic Physics* (ed. 5) vii. 220 *K*-capture should therefore compete with β-decay. 1970 Hurst & Turner *Elem. Radiation Physics* ii. 24 A few nuclei..capture an atomic electron from outside the nucleus, most often from the K-shell, and emit a neutrino.

f. *Physics* and *Chem.* In the old quantum theory *k* is the azimuthal or subordinate quantum number (introduced by N. Bohr 1920, in *Zeitschr. f. Physik* II. 445), which determines the shape of electronic orbits of the same *n*; (now superseded by the quantum number *l*). In molecular spectroscopy *K* is a quantum number which in diatomic and linear molecules represents the total angular momentum apart from electronic spin (now usu. replaced by *N*), and in polyatomic molecules represents the component of the total momentum about an axis of symmetry.

1922 A. D. Udden tr. *Bohr's Theory of Spectra* II. iii. 44 The perturbations are periodic, so that we may assume that to each energy value of a stationary state of the unperturbed system there belongs a series of discrete energy values of a whole number *k*. *Ibid.* III. iii. 85 Where it is necessary to differentiate between orbits corresponding to various values of the quantum number *k*, a central orbit, characterized by given values of the quantum numbers *n* and *k*, will be referred to as an n_k orbit. 1930 R. S. Mulliken in *Physical Rev.* XXXVI. 613 In [Hund's] case *b*, *Λh*/2π and the nuclear angular momentum combine to give a quantised resultant... For the corresponding quantum number..the designation *K* is now recommended. The possible values of *K* are *Λ*, *Λ*+1, *Λ*+2,.. There is usually a small magnetic field in the molecule parallel to *K*, so that *K* and *S* form a resultant *J*. 1934 H. L. Brose tr. *Sommerfeld's Atomic Struct. & Spectral Lines* (ed. 3) ii. 115 In wave mechanics the azimuthal quantum number, our n_ϕ or Bohr's *k*.., becomes replaced by the quantity *l* = n_ϕ – 1, *l* = 0, 1, 2... 1961 Powell & Crasemann *Quantum Mech.* i. 24 A perturbation of the force, such as might be produced by the presence of other electrons, has the effect of removing the degeneracy, so that states with the same value of *n* but different values of *k* have different energies. 1962 P. J. & B. Durrant *Introd. Adv. Inorg. Chem.* vii. 226 Paschen-Back effect [for diatomic molecules] (strong magnetic field)... **K** and **S** are not coupled together but are coupled directly to the field. 1966 C. N. Banwell *Fund. Molecular Spectroscopy* iii. 94 Parallel Vibrations [of Symmetric Top Molecules]. Here the selection rule is: *Δv* = ± 1, *ΔJ* = 0, ± 1, *ΔK* = 0.

g. *Psychol.* The letter chosen to represent the spatial factor, or aptitude for remembering form and structure, in some ability-tests.

1935 *Brit. Jrnl. Psychol.* Monogr. Suppl. xx. vii. 65 In order to distinguish these eight tests from the rest of the table they may for convenience be called the *K* tests. *Ibid.* 75 Therefore there is in them [*sc.* specific correlations], over and above '*g*', *one group factor*; this we name the *K* factor. 1944 L. L. Thurstone *Factorial Study Perception* iii. 117 It is quite likely that the factor *K* is determined by experimental dependence. 1950 Spearman & Jones *Human Ability* xii. 132 For the first time, the spatial test does, in some degree, measure K. 1969 P. E. Vernon *Intelligence & Cultural Environment* ix. 59 Embedded Figures and the Kohs Block test are good measures of the *k* factor of British psychologists..and this is much the same as Thurstone's original *S* (spatial) factor.

h. [From its use as an abbrev. for *kilo-*.] In connection with *Computers* K or k is used to represent 1,000 (or 1,024: see quot. 1970).

1966 P. D. Reynolds *Computer ABC* 54 The internal storage of computers is commonly arranged..to hold a quantity of data which is some power of 2, for example, 4096 characters, bytes or words, which is 2¹². The convention is to refer to this number as 4K. 64K..amounts to 65,536(2¹⁶). 1967 Cox & Grose *Organiz. Bibliogr. Rec. by Computer* II. 2 It seemed desirable..wherever possible to ignore the limitations of the computer available to us (a KDF 9 with a store size of 16 K 48-bit words). 1970 O. Dopping *Computers & Data Processing* ii. 35 Sometimes, a 'K' is used for a number which is either 1,000 or 1,024, depending on whether the context calls for integral powers of 10 or 2. If we say that a certain computer has a memory capacity of 4 K words, then, this means either 4,000 or 4,096 words, depending on whether the computer in question has a decimal or binary address system. However, the usage is a little loose: the number 2¹⁶ = 65,536 is written either as 64 K or 65 K. 1971 *Daily Tel.* 21 July 20 (Advt.), I.B.M. programmers... Sal. from £1,600 to £2·4k. 1971 *New Scientist* 9 Sept. 569/1 Typically, a minicomputer has a minimum memory of between one and eight K words.

4. b. K.H.B. *colloq.* (see quot. 1925); K.O.S.B., King's Own Scottish Borderers. **c.** K used *colloq.* for *knighthood*; also K.B.E., Knight Commander of the Order of the British Empire; K.B.S., Knight of the Blessed Sacrament; K.C.B. (examples); K.C.V.O., Knight Commander of the Victorian Order; K.G. (examples); K.S.G., Knight of the Order of St. Gregory the Great. **e.** Also many other abbreviations of names of units in which *k* represents *kilo-*, as kcal., kcal, kilocalorie(s); kHz, kilohertz; kVA (also kv.-a., etc.), kilovolt-ampere(s); kW (also K.W., kw., etc.), kilowatt(s); kWh, kilowatt-hour(s). **f.** In miscellaneous abbreviations, as °K, K, (degree) Kelvin (see *Kelvin, kelvin 3 b); K.E., k.e., kinetic energy; K.G.B. [Komitet Gosudarstvennoĭ Bezopasnosti] = Committee of State Security (U.S.S.R.); K.i.H., K.I.H., Kaisar-i-Hind; K.K.K., Ku Klux Klan; K.L., Kuala Lumpur; K.L.M. [Koninklijke Luchtvaart Maatschappij], Royal Dutch Airlines; K.M.T., Kuomintang; K.O., k.o. = knock(ed) out (cf. *kayo *v.* and *sb.*); K.P. (*U.S.*), kitchen police(man); K.P.D. [Kommunistische Partei Deutschlands], German Communist Party; k.p.h., kilometres per hour; K *ration* [f. the initial letter of the surname of Ancel Keys (1904–), American physiologist], a package of concentrated food; KWIC = *key-word-in-context* (*key *sb.*¹ 17).

1911 *Physical Rev.* XXXIII. 226 The value of the ice point..was taken as 273·2°K. 1937 M. W. Zemansky *Heat & Thermodynamics* xiii. 232 The isothermal compressibility of copper is plotted against the Kelvin temperature in Fig. 72. Above about 100°K the rise..is approximately linear. 1959 *Sci. News* LI. 11 The lowest temperature that can conveniently be obtained by evaporating helium under reduced pressure is about 1°K. 1970 *Nature* 10 Oct. 144/2 The coolest spectrum recorded over the Antarctic plateau indicates a surface temperature of 190 K (–83° C). 1972 *Amat. Photographer* 12 Jan. 38 Household Bulbs...150 watts is usually the maximum, with a colour temperature of around 2,700 to 2,900°K. 1910 *Lett. Lord Kilbracken & Gen. Godley* (*c* 1932) 13 Long may you live to wear your honours, and I hope that 'the coming K' will not be long deferred. 1966 J. Betjeman *High & Low* 70 That very near miss for an All Souls' Fellowship, The recent compensation of a 'K'. 1968 *Listener* 13 June 770/3 A 'K' isn't certain any more, even if you're a civil servant. 1973 *Times* 24 Aug. 12/8 There might not have been much merit in a political knighthood, but there was no harm in it... The 'K', when it came, was a boon to the Member's wife, and a blessing to the Member himself. 1952 'W. Cooper' *Struggles of Albert Woods* IV. i. 206 They tell me your Principal got a plain K and was hoping for K.B.E. 1968 *Listener* 12 Dec. 787/3 He was given a KBE for his efforts. 1916 *Let.* 12 June in *Knights of Blessed Sacrament* (Catholic Truth Soc.) (1918) 14, I have started the K.B.S. among my men here. 1923 A. O'Connor *Knight in Palestine* ii. 11 The K.B.S. and another Catholic made ready the altar. 1954 R. T. Sanderson *Introd. Chem.* iv. 43 Larger quantities of heat are measured in kilocalories, which are 1000 calories each. They are sometimes abbreviated as Cal with a capital C; more commonly they are simply kcal. 1964 N. G. Clark *Mod. Org. Chem.* iii. 29 The energy required to bring about this rotation is seen to be 5 kcal, a relatively trivial amount capable of being supplied by the thermal motion of neighbouring molecules. 1849 Thackeray *Three Sailors* in S. Bevan *Sand & Canvas* xxv. 341 There's the British fleet a riding at anchor, With Admiral Napier, K.C.B. 1880 E. W. Hamilton *Diary* 29 Aug. (1972) I. 40 Loch, the Governor of the Isle of Man, is to be a KCB. 1904 *Westm. Gaz.* 8 Jan. 6/2 Lord Lawrence used to speak of England's aggressive policy in India as the 'K.C.B.' mania. 1972 *Times* 6 May 1/4 Sir James..was appointed KCB in the New Year Honours. 1897 *Whitaker's Almanack* 108 The Royal Victorian Order. Instituted 21st April, 1896... Knights Grand Cross. G.C.V.O... Knights Commanders. K.C.V.O. 1968 *Listener* 29 Aug. 278/1 His sufferings..were not assuaged by a KCVO. 1888 A. Aveling *Mech. & Exper. Sci.: Mech.* xiii. 137, *k.e.* of the two masses after collision = *mv*²/2. 1909 Jackson & Roberts *First Dynamics* 88 The gain of K.E. equals arithmetically the work done by the forces. 1965 Van Wylen & Sonntag *Fund. Classical Thermodynamics* v. 84 *E* = Internal energy + Kinetic energy + Potential energy or *E* = *U* + KE + PE. 1876 Trollope *Prime Minister* IV. iv. 52 (*heading*) The new K.G. 1880 E. W. Hamilton *Diary* 23 Aug. (1972) I. 37 A Garter is vacant by death of Lord Stratford de Redcliffe ..Lord Palmerston is a precedent for a commoner receiving a KG. 1904 K.G. [see *gunner 1 c]. 1972 *Whitaker's Almanack 1973* 463 The Viscount Montgomery of Alamein K.G., G.C.B., D.S.O. 1960 *Analog Science Fact/Fiction* Oct. 122/2 The KGB was once again checking on every foreigner. 1966 J. Porter *Sour Cream* v. 58 The K.G.B. is willing to shell out a small fortune in roubles for me, dead or alive. 1972 K. Benton *Spy in Chancery* i. 8 The Russian who's made the approach..isn't the type of KGB operative one would expect to make a run for it. 1916 'Taffrail' *Pincher Martin* iv. 62 He was a K.H.B., and they were not sorry to be rid of his presence. 1925 Fraser & Gibbons *Soldier & Sailor Words* 134 *A K.H.B.*: a King's Hard Bargain. A worthless or incorrigible fellow. (Old Service term.) 1955 *Proc. IRE* XLIII. 880/3 A happy solution would be more widespread use of the term 'hertz', meaning cycle-per-second. Thus the units of frequency would be hertz (or hz), khz, and Mhz. 1974 *Electronics* 26 Dec. 48E A portable a-m signal generator..covers 85 kHz to 100 MHz. *a* 1912 W. T. Rogers *Dict. Abbrev.* (1913) 108/2 K.I.H., Kaiser-i-Hind (Emperor of India). 1942 Partridge *Dict. Abbrev.* K.i.H. or K.I.H., the Kaisar-i-Hind Medal; for useful service in India. 1973 *Daily Tel.* 20 Sept. 36/7 Alice Headwards-Hunter, K.I.H., F.R.C.S.E., D.C.H., formerly of Calcutta. 1872 in W. L. Fleming *Documentary Hist. Reconstruction* (1907) II. 132 We advanced upon the supposed K.K.K.'s. 1877 J. M. Beard *K.K.K. Sk.* 35 The horses of the raid were.. furnished with all those *cap-a-pie* appointments of K.K.K. regalia. 1952 *N.Y. Times* 1 Aug. 16/2 For conspiracy to flog a Negro woman, the so-called 'Imperial Wizard' of the local KKK has been given..four years. 1970 G. Jackson *Let.* Apr. in *Soledad Brother* (1971) 48 I've already mentioned that most of them are K.K.K. types. 1961 'G. Black' *Suddenly, at Singapore* ix. 129 From there we could have been going to K.L. or anywhere in Malaya. 1973 *Observer* 7 Oct. 36/6 Here is the capital, Kuala Lumpur, which the old Malay hands called KL and now everybody does. 1933 *Meccano Mag.* Mar. 193/1 A pilot flying on the K.L.M. route to Batavia. 1968 *Listener* 28 Nov. 704/3 KLM suggest you come to Amsterdam just to see the airport. 1959 *Times Lit. Suppl.* 8 May 270/4 Two months later Chiang struck again by excluding Communists from any higher posts in the K.M.T. 1969 J. M. Gullick *Malaysia* ii. 85 At one time the Chinese middle class gave its support to the Kuomintang, an effective if externally orientated nationalist movement, but the KMT was ground between..British restrictions.. and..communist penetration of the Chinese working class. 1972 'M. Hebden' *Killer for Chairman* I. ix. 115, I last saw you in Canton... There was one of the K.M.T.

generals still hiding there. 1922 T. Burke *London Spy* 209 As a youth the ring attracted him... A few k.o.'s put an end to that. 1923 H. Cox *Dogs & I* xxii. 209 The Field Spaniel has received the 'K.O.' and taken the count! 1927 *Observer* 25 Dec. 12/6 His record..includes a k.o. victory over Paul Berlenbach. *Ibid.*, Knut Hansen, who k.o. Phil Scott in the first round. 1928 *Daily Express* 25 June 17/7 Young Stanley..was then k.o. by a right swing to the jaw. 1951 'J. Wyndham' *Day of Triffids* viii. 142 Coker an' another chap was giving them the k.o. as they tripped. 1971 *Weekend World* (Johannesburg) 9 May 1/2 Morodi said he was not upset by the defeat—the first k.o. he has suffered in 70 fights. 1909 *Who's Who* p. xi/1 K.O.S.B. King's Own Scottish Borderers. 1914 F. W. Spicer *Diary* in P. Young *Brit. Army* (1967) xv. 203 On our right was the 13th Infantry Brigade, with the 2/K.O.S.Bs. joining up with our right Company. 1924 *Cricketer Ann. 1923–4* 82 The band and pipers of the K.O.S.B.'s. 1964 'T. Carew' *Vanished Army* II. 128 The K.O.Y.L.I. saw strange faces in their depleted ranks—men of the Suffolks, K.O.S.B. and Manchesters. 1917 D. C. Falls *Army & Navy Information* 84 K.P., Kitchen Police. A mild form of punishment. 1921 J. Dos Passos *3 Soldiers* I. 10 The men..filed by the great tin buckets at the door, out of which meat and potatoes were splashed into each plate by a sweating K.P. in blue denims. 1929 [see *kitchen police* s.v. *kitchen *sb.* 7]. 1956 B. Holiday *Lady sings Blues* (1973) xviii. 143 After all the big personnel experts got together to figure out a job that was right for a city girl like me, I was cast for the part of Cinderella of Cottage No. 6. This was nothing but a fairy-tale name for permanent K.P. 1973 *Publishers Weekly* 25 June 70/2 For an uncertain spell he struggles through the miserable childhood of 'Norma Jean' in a welter of secondhand conjectures—Mailer doing biographical KP. 1922 *Encycl. Brit.* XXXI. 280/2 The violent agitation conducted by the central committee of the K.P.D. in Berlin. 1935 C. Isherwood *Mr Norris changes Trains* vi. 98, I am not a member of the Communist Party..I merely sympathize with the attitude of the K.P.D. to certain non-political problems. 1964 *New Statesman* 28 Feb. 332/1 Relations with Germany—and with the German Communist Party, the KPD—stood at the centre of the picture. 1966 G. B. Mair *Kisses from Satan* ix. 106 They were doing over a hundred k.p.h. on a snaky road. 1972 W. Garner *Ditto, Brother Rat!* xxiii. 172 The speedo needle crept past the 150 kph mark. 1944 *R.A.F. Jrnl.* Aug. 283 We get American 'K' rations, with a few extras. 1967 'T. Carew' *Korea* ii. 19 A K ration consisted of a tin of compressed meat hash, coffee, powdered milk, a fruit bar, cigarettes, chewing gum, and toilet paper. 1909 Webster, K.S.G. 1905 S. P. Thompson *Dynamo-Electr. Machinery* (ed. 7) II. iii. 173 An 8-pole, 60 KVA three-phase generator. 1930 *Engineering* 14 Mar. 355/1 Supplies to the villages will be..through pole transformers with capacities of 50, 20 and 10 kv.-a. 1959 *B.S.I. News* June 9/1 The standard applies to power transformers, reactors and earthing transformers having windings insulated with four different classes of insulating material, with single-phase ratings of 1kVA and above or polyphase ratings of 2kVA or above. 1905 A. H. Bate *Princ. Electr. Power* iii. 34 Electrical power can thus be expressed in either of three units, namely:—The watt, equal to 1 volt multiplied by 1 ampere. The kilowatt (K.W.), equal to 1000 watts. And the electrical horse-power. 1930 *Engineering* 23 May 667/1 The maximum load in the area during 1928–29 was 26,059 kw., [etc.]. 1959 *Chambers's Encycl.* V. 93/2 Central power stations for public electricity supply may range in capacity from the comparatively small size of 40,000 kW to the very large capacity of 500,000 kW or more. 1930 *Engineering* 28 Feb. 299/2 The best yearly record has been reduced to as low as 12,500 B.Th.U. per kw.-h. sent out. 1963 *Times* 3 June 12/1 The Minister expressed his belief that in 1968–70 30,000m. kWh of a total electricity production of 260,000m. kWh will be produced by France's atomic plants. 1959 H. P. Luhn in *IBM Corporation ASDD Rep.* RC-127 (*title*) Keyword-in-context index for technical literature (KWIC index). 1967 Cox & Grose *Organiz. Bibliogr. Rec. by Computer* VI. 154 It has been decided not to produce a KWIC index at this stage. 1969 *Computers & Humanities* III. 166 An obvious prerequisite for this kind of dictionary construction is large key-word-in-context (KWIC) lists drawing on large samples of the language.

ka (kā), *sb.* Also **kaa.** The name given by the ancient Egyptians to a spiritual part of a human being or a god which survived after death and could reside in a statue of the dead person.

a 1892 Tennyson in A. G. Weld *Glimpses Tennyson* (1903) 119, I believe that beside our material body we possess an immaterial body, something like what the ancient Egyptians called the Ka. 1905 E. F. Benson *Image in Sand* i. 11 Somebody's Ka—his ghost, you know, or his astral body. 1923 *Glasgow Herald* 22 Feb. 4 The Princess has a Ka, or better self. 1952 J. M. White *Anc. Egypt* 40 The Ka lived in the tomb with the mummy. 1968 V. Ions *Egyptian Mythol.* 123 The *ka*, the vital principle of a man or of a god. 1972 *Daily Tel.* (Colour Suppl.) 3 Mar. 29/4 The Ancient Egyptians, whose civilisation goes back earlier than 3,000 BC, believed that the *ka* or spirit of a man could only survive if his body was preserved.

kaaba, kaabeh, varr. Caaba.

Kababish (kābā·biʃ). Also 8 **Cubbabeesh,** 9 **Kobabeish.** [Arab. *kabābīš*, pl. of *kabbāši*.] A nomadic Arab people of the northern Kordofan and Dongola provinces of the Sudan. Also *attrib.*

1790 J. Bruce *Trav.* IV. viii. x. 515 Lately the Beni Gerar, Beni Faisara, and Cubba-beesh, have expelled the ancient Arabs of Bahiouda. *a* 1817 J. L. Burckhardt *Trav. Nubia* (1822) 438 The Djerar, Kobabeish, and Feysara live to the north and north-east. 1861 J. Petherick *Egypt, Soudan & Cent. Afr.* xvii. 282 The entire Kababish nomade tribe inhabiting the northern confines of the province emigrated to Darfour. 1891 F. R. Wingate *Mahdiism & Egyptian Sudan* ix. 288 The Kababish, who lived in the desert west of Dongolee. 1898 Alford & Sword *Egyptian Soudan* I. ii. 26 The earliest and most important revolt against the Khalifa was the rising of the Kababish Arabs. 1918 C. G. & B. Z. Seligman in *Harvard Afr. Stud.* II. 105 The Kabâbîsh constitute the richest and most powerful of the Arab tribes of the Anglo-Egyptian Sudan. 1954 H. A. MacMichael *Sudan* vii. 74 Little tribal authority survived except among those nomadic tribes, such as the Kababish. 1960 A. Tibble *With Gordon in Sudan* ii. 48 Those camel-owning nomads whom Gordon would meet between Dongola and Metemma were the Kababish.

kabad(d)i (kăbā·di). [Tamil.] A game popular in northern India and Pakistan played between two teams of nine boys or young men (see quot. 1935). Also *attrib.*

1935 W. M. Ryburn *School Organization* 278–80 Kabaddi... 3. Each team consists of nine players... 5. The members of each team remain in their respective semi-circles... 10. A player scores a point for his team if he succeeds in getting back to his semicircle after touching some opponent (with the hand only) or after pushing some opponent out of his semicircle provided that he holds his breath all the time. He will say that word 'kabaddi' over and over to show that he is holding his breath... An attempt by a player to touch an opponent is known as a kabaddi. 1943 Moraes & Stimson *Introd. India* (ed. 3) 171 Among Indian games *kabadi* (*hu-tu-tu*) and *kho-kho* are prominent (they are too complicated to describe here). 1965 E. Linton *World in Grain of Sand* x. 172 There was volley ball, tug of war and 'kabadi'. Kabadi, the Indian wrestling game, is fascinating to watch. 1969 *Femina* (Bombay) 26 Dec. 45/1 The versatile collegians of Bombay lifted the Inter-Varsity table tennis, kabaddi and swimming titles. *Ibid.* 45/4 (*caption*) Rajkumari Jain..during the inter-varsity kabaddi tourney. 1971 *Illustr. Weekly India* 11 Apr. 51/1 (*caption*) A variety of Indian games (like *khokho* and *kabaddi*) find play here. 1974 *Dawn* 17 Apr. 4/3 Special attention was being paid to the popularization of indigenous games like Kabaddi and even a National Kabaddi Federation had been formed.

Kabaka (kăbā·kā). The Bantu title which was given to the ruler of the province of Buganda in Uganda. Hence **Kaba·kaship,** the office of Kabaka.

1878 H. M. Stanley *Through Dark Continent* I. 189 General jack-of-all-trades for the *Kabaka*. 1925 J. W. Mackail *J. L. Strachan-Davidson* 90 The choice of an English tutor for the eight-year-old Kabaka of Uganda. 1954 *Ann. Reg. 1953* 123 The U.K. Government had withdrawn recognition from the Kabaka of Buganda. 1960 *Economist* 8 Oct. 136/2 The public esteem for the institution of the Kabakaship in Buganda. 1966 *Ibid.* 9 July 139/1 Mr Obote seems to intend to maintain the Kabakaship. *Ibid.* 139/2 Any descendant of Mutesa I is eligible for selection as Kabaka. 1971 *Sunday Nation* (Nairobi) 11 Apr. 6/2 Mutesa was their last Kabaka and..Uganda is to stay a Republic.

kabane (kăbā·ne). Also **kabané.** [Jap.] In ancient Japan, a series or system of titles of rank.

1890 B. H. Chamberlain *Things Japanese* 252 The *kabane* or *sei*, a very ancient and aristocratic sort of family name, but now so widely diffused as to include several surnames in the narrower sense of the word. 1904 L. Hearn *Japan: Attempt at Interpretation* xii. 260 Caste would not seem to have developed any very rigid structure in Japan; and there were early tendencies to a confusion of the kabanè. 1931 G. B. Sansom *Japan* I. ii. 39 An integral part of the clan system..was the system of titles, or *kabane*. 1970 J. W. Hall *Japan from Prehist. to Mod. Times* v. 36 A more precise set of titles of rank (*kabane*) had also been evolved.

kabaragoya (kabaragōᵘ·yā). Also **kabragoya.** [Etym. unknown.] The water monitor, *Varanus salvator*, a large lizard found in south-eastern Asia.

1681 R. Knox *Ceylon* I. vii. 30 There is a Creature here called *Kobberaguion*, resembling an Alligator. The biggest may be five or six foot long, speckled black and white. 1892 C. F. G. Cumming *Two Happy Years in Ceylon* (ed. 4) II. xx. 176 A gigantic lizard, or rather iguana, of a greenish-grey colour, with yellow stripes and spots, called by the natives kabragoya, awoke from its midday sleep. ..The kabragoya is amphibious. 1960 H. W. Parker tr. *Mertens's World of Amphibians & Reptiles* x. 181 In south-eastern Asia the large Kabara Goya (*Varanus salvator*), despite its timidity, prefers to be associated with man.

Kabardian (kăbā·ɹdiăn), *sb.* and *a.* Also **Kabard, Kabardan, Kabardin(e), Kabardinian.** [f. Russ. *Kabarda* (place-name) + -ian.] **A.** *sb.* **a.** A member of one of the peoples inhabiting the Kabardino-Balkarian Republic in the northern Caucasus, related ethnically to the Circassians, of Caucasian race but non-Indo-European in language. **b.** The north-western Caucasian language of this people.

1824 *Encycl. Brit.* III. 166/2 The most remarkable are the Kabardines,..the Kisti, the Ossetes. 1888 *Ibid.* XXIII. 514/2 Kabards, Circassians, Osses, and Karapapakhs. 1931 G. C. Wheeler tr. *Nansen's Through Caucasus* 43 The Kabardians live on the north side of the Caucasus along the Terek. 1956 J. Lotz in Saporta & Bastian *Psycholinguistics* (1961) 10/1 In Kabardian, also of the Caucasus. 1960 A. H. Kuipers *Phoneme & Morpheme in Kabardian* 8 The Kabardians differed from their Western relatives in that they formed a well-developed feudal ceremony. 1965 D. Fidlon tr. *Trunov's Trip N. Caucasus* 11 'The place swarms with birds,' the Kabardinian continued.

B. *adj.* Of, pertaining to, or characteristic of the Kabardians or their language.

1902 [see *Ingush]. 1940 J. F. Baddeley *Rugged Flanks Caucasus* II. xxii. 220 This ceremonial observance is..a point of honour with the Kabardán nobility. 1950 D. Jones *Phoneme* 222 The Kabardian language of the Caucasus. 1965 D. Fidlon tr. *Trunov's Trip N. Caucasus* 14 She was a graceful Kabardinian girl with a healthy suntan.

kabeljou (kā·bəlyəu, kǫ·bəlyəu, ‖ ka·bəlyōᵘ). *S. Afr.* Also **cabaljao, cabeliau, kabbeljou, kabeljaauw, -jauw, -jouw, kobeljauw, -jouw** and, abbreviated, *kob² [Afrikaans, = Du. *kabeljauw* (see Cabilliau).] A large marine food fish, *Johnius hololepidotus*, of the family Sciænidæ.

1731 [see Cabilliau]. 1838 J. E. Alexander *Expedition Interior Afr.* II. iv. 83 We got a great prize in a stranded cabaljao, fifty pounds weight, like a huge salmon. 1912 J. T. Cunningham *Reptiles, Amphibia, Fishes* 277 The name kabeljaauw [= cod]..in South Africa has been transferred to a fish of a very different species,..the maigre, *Sciaena aquila*. 1913 C. Pettman *Africanderisms* 271 Kobeljauw or Kabeljauw. 1930 C. L. Biden *Sea-Angling Fishes of Cape* 109 The kabeljou is often mistaken for the geelbek. 1950 *Cape Times* 17 Nov. 13/8 Last week-end kobeljouw catches were kept down by the difficulty of casting into the teeth of the wind. 1952 *Ibid.* 7 Nov. 3/5 Fishing-boats brought home the first few kabbeljou of the season. 1959 *Ibid.* 16 Nov 2/5 A..visitor..landed a 94-pound Kabeljouw in the Klein Rivier lagoon. 1973 *Farmer's Weekly* (S. Afr.) 13 June 102 The species being tagged include grunter and white steenbras, elf (shad), kob (kabeljauw or Cape Salmon in Natal) haarders (mullet) and leervis (garrick).

Kabistan (kæbIstā·n). Also **Cabistan.** Erron. for *Kubistan* after the district of Kuba in Azerbaijan, used *attrib.* or *absol.* to designate a finely woven, short napped rug or carpet with intricate geometric design made in that area; also called *Kuba.

1900 J. K. Mumford *Oriental Rugs* ix. 111 In the Kabistans the weft and sometimes the warp is of cotton cord. 1904 M. B. Langton *How to know Oriental Rugs* iii. 117 The Cabistan rug is a most desirable, moderate-priced rug, of artistic worth. 1913 G. G. Lewis *Pract. Bk. Oriental Rugs* (rev. ed.) x. 128 S Forms..are very common in the Caucasian fabrics, especially in the Kabistans and Shirvans. 1931 A. U. Dilley *Oriental Rugs & Carpets* Pl. 52 (*caption*) Kabistan or Hila. Geometric adaptation of Persian pattern. 1964 *Sunday Times* (Colour Suppl.) 19 Jan. 24 Cabistan. Similar to Shirvan rugs in weave and colour. 1970 J. Franses *European & Oriental Rugs* 44 (*caption*) Antique Kabistan rug. 1972 [see *Kuba].

kabouri (kabu·ri). *Guyana.* Also **kaboura, kaburi.** [Origin unknown.] A fly of the genus *Simulium*.

1899 J. Rodway *In Guiana Wilds* xi. 161 The great pest was the kaburi, which raised a blister in every spot where its venomous proboscis was inserted. 1917 W. G. White in C. W. Beebe et al. *Trop. Wild Life Brit. Guiana* I. xxxiv. 486 No one can know the district without knowing the kabouri fly... It is a blood-sucker, which marks one as with fine pocks... I am told that there are two species of kabourie; *Simulium guianense*, and *Simulium amazonicum*. 1954 G. Durrell *Three Singles to Adventure* iv. 92 We were fiercely attacked by great numbers of tiny black flies a little larger than a pin-head but with a bite that was out of all proportion to their size...'They're kaboura flies...' The kabouras continued their assault on us.

‖ **kabouter** (kābəu·təɹ). [Du.] A gnome, goblin, or dwarf; *transf.*, used for a member of an anti-establishment political movement in Holland.

1961 T. Henrot *Belgium* 182 Each region has its familiar spirits..the dwarves kindly and quarrelsome—*Nûtons* in the grottoes of Namur, *Kabouters* on Campine's heaths, *Sotáis* of the Amblève. 1970 *Guardian* 6 June 11/4 Remember the Provos, Amsterdam's pacifist rebels..? Their successors, the Kaboute [*sic*] or Elves, have gone one better. They campaigned in the nude for this week's city council elections. 1970 *Daily Tel.* 30 June 16 *World in Action* (ITV) introduced the Kabouter, the gnomes of Amsterdam, an anti-establishment movement who wear pointed hats..and, to general astonishment, have won five seats on the city council. 1971 *Guardian* 13 Apr. 11/4 The Kabouters—the Dutch street action group. 1973 *Reader's Digest* Apr. 97/1 What city would have a political movement called the *kabouters*, or pixies?

Kabuki (kăbū·ki, ‖ kabuki). Also with small initial. [Jap., f. *ka* song + *bu* dance + *ki* art, skill.] A traditional and popular form of Japanese drama which employs highly stylized singing, miming, and dancing in addition to acting, and in which (since *c* 1650) all the parts are played by males. Also *attrib.* Hence **Kabukie·sque** *a.*, in the style or manner of the Kabuki theatre.

1899 W. G. Aston *Hist. Jap. Lit.* VI. iii. 288 Kabuki theatres, which had men for actors, had been established there before the middle of the seventeenth century. 1928 *Daily Tel.* 4 Dec. 8/4 The Kabuki affords freedom for old and favourite plays, for new ones on Western lines, and for adaptations of Western drama. 1951 *Oxf. Compan. Theatre* 411 The present day Japanese theatre takes three distinct, although related forms, the *Nō* or lyrical drama, *Ningyō-shibai* or marionettes, and *Kabuki*, the popular theatre. 1954 F. Bowers *Jap. Theatre* vii. 224 Another woman who lies down to offer herself as a substitute for the married woman—a postwar Kabukiesque 'substitution'. 1960 B. Leach *Potter in Japan* viii. 190 The merchant class with its popular arts of the Kabuki theatre and the colour print. 1970 *Oxf. Compan. Art* 1171 Favourite subjects were theatre scenes, which began to appear along with the development of the popular *Kabuki* theatre in the 17th c. 1972 *Nat. Geographic* Sept. 378 Man in maiden's guise charms theater-goers in the classical drama known as Kabuki. 1972 *Mainichi Daily News* (Japan) 6 Nov. 3/5 Collection and sale of kabuki dolls.

Kabuli (kăbū·li), *a.* and *sb.* Also **Cabuli, Kabulian. A.** *adj.* Of, pertaining to, or characteristic of the city or province of Kabul in Afghanistan.

1887 Kipling *Plain Tales from Hills* (1888) 173 Baskets of fruit and pistachio nuts and Cabuli grapes. 1893 —— *Day's Work* (1898) 3 A little switch-tailed Kabuli pony. 1909 *Westm. Gaz.* 2 Feb. 5/1 These Cabuli rooks do not visit us every winter. 1966 P. King *Afghanistan* 64 (*caption*) Kabuli carpet seller.

B. *sb.* A person or animal from Kabul; also, the Iranian language of some of the people of this area or its script.

1895 Kipling *Day's Work* (1898) 188 You're to take one of Sir Jim's horses. There's a gray Cabuli here. 1909 M. Diver *Candles in Wind* xi. 112 He urged the Kabuli forward. 1948 D. Diringer *Alphabet* 301 The Kharoshthi, which is called also Bactrian..Kabulian..and so forth. 1964 H. H. Paper tr. *Shafeev's Short Gram. Outl. Pashto* 1 Until 1936 the official language of Afghanistan was Kabuli, one of the dialects of Tajik.

kaburi, var. *kabouri.

Kabyle (kăboi·l). Also **Kabail, Kabile, Kabyl.** [Arab. *kabā'il*, pl. of *kabīla* tribe.] A member of a group of Berber peoples inhabiting northern Algeria and Tunisia; these peoples collectively; also, the Berber (Hamitic) language of these peoples. Also *attrib.* *Kabyle dog* (see quot. 1945).

1738 T. Shaw *Trav. Barbary* i. 8 The several Portions and Districts..are chiefly known and distinguished by the particular Names of the Kabyles..or African Families, who respectively possess them. 1818 E. Blaquière tr. *Pananti's Narr. Residence Algiers* 181 The Bedouins are divided into many scattered tribes, called *Kabiles*, and vulgarly *Nege*. 1861 *Chambers's Encycl.* II. 44/1 In Algeria, where they [*sc.* Berbers] usually are termed Kabyles, they are yet unconquered by the French. 1868 L. Wingfield *Under Palms in Algeria & Tunis* I. iv. 84 He told us many things about the 'interieur' of Kabyle life. 1882 F. W. Newman *Libyan Vocab.* 3 The Libyan language..was but one, according to St. Augustine. Now there are at least four, the Kabail in Algeria, the Shilha in the mountains of Morocco, the Tuarik..beyond the Atlas, and the Ghadamsi at Ghadames. *Ibid.* 38 Kabail verbs and verbals, including adjectives. 1900 A. Wilkin *Among Berbers of Algeria* 178 Cheek by jowl the villages of Frenchmen and Kabyles stand. 1900 *Knowledge* 1 Aug. 173/2 The various Kabyle tribes. 1924 *Glasgow Herald* 14 Aug. 7 The Kabyles are agricultural Berbers living in the uplands of Morocco and Algeria. 1945 C. L. B. Hubbard *Observer's Bk. Dogs* 176 Kabyle Dog..a sheepdog native to the Kabyle Mountains. 1963 *Guardian* 9 Oct. 11/1 President Ben Bella..does not speak Kabyle and many Kabyles do not understand Arabic. 1969 E. Gellner *Saints of Atlas* i. 14 A Kabyle worker in France..has also kept his Berber speech. 1974 *Times* 30 May 10/5 Sons are what Kabyl husbands demand... Sterility and daughters alike are no good, and Kabyl women know it.

Kach(ch)eri, var. Cutcherry, cutchery.

1903 *Oxf. Mag.* 11 Feb. 208/1 The Kachcheri..is the centre of official life in the province. 1913 L. Woolf *Village in Jungle* vii. 179 He was standing..frightened, on the Kachcheri verandah. 1926 *United Free Ch. Miss. Rec.* Sept. 391/2, I can't have the Kacheri turned into a pawnshop.

kach(h)a, var. Cutcha *a.*

1920 *Blackw. Mag.* Oct. 921/1 My friend transported me over three extra miles of 'kacha' road. 1975 *Bangladesh Times* 20 July 2/5 A good number of kacha houses were damaged.

Kachin (kătʃi·n). [Burmese.] A member of a Tibeto-Burman ethnic group inhabiting the mountainous regions of north-east Burma; its language group, which includes Chingpaw, Atsi, Maru, etc. Also *attrib.*

1892 W. R. Winston *Four Yrs. Upper Burma* 102 In the north of Burma..are found the Kachins, a warlike hill people. 1903 G. A. Grierson *Linguistic Survey of India* III. ii. (*title*) Specimens of the Bodo, Nāgā, and Kachin groups. 1906 O. Hanson (*title*) A dictionary of the Kachin language. 1925 *Blackw. Mag.* Dec. 864/1 The reiterated word accented in Kachin is most descriptive. 1934 'G. Orwell' *Burmese Days* xiv. 211 Producing two green pigeons from his Kachin bag. 1962 [see *Chin *sb.*²]. 1963 C. Maxwell-Lefroy *Land & People of Burma* vii. 47 Warm green tea,..shyly handed to us by a pretty Kachin girl. 1970 F. S. V. Donnison *Burma* ii. 47 The Kachins are nine-tenths Animists and the rest Christians.

kachina (kătʃī·nă). Also **cachina, katc(h)ina,** and with capital initial. [Hopi *qacina* supernatural.] In North American Pueblo Indian mythology, one of the deified ancestral spirits which periodically visit the pueblos, to bring rain, etc. Also in *Comb.*, as *kachina dance*, a dance performed at annual ceremonies by masked and elaborately costumed men impersonating the kachinas whom they seek to invoke; *kachina doll*, a doll representing a kachina, given to children at the annual ceremonies by the kachina dancers.

1888 S. Wallace *Land of Pueblos* 47 The *cachina* dance, which they celebrate at certain seasons of the year. 1907 F. W. Hodge *Handbk. Amer. Indians* I. 638/1 Kachina. A term applied by the Hopi to 'supernatural beings impersonated by men wearing masks'. 1910 G. W. James *Grand Canyon* 122 For use in the katchina dances, katchina baskets are made. 1948 *Seattle Sunday Times* (Mag.) 26 Sept. 3/4 The masked dancers who impersonate deities, or katchinas, in the great rain rites and rituals. 1959 E. Tunis *Indians* 126/2 He saw the sacred kachinas, and he was given dolls dressed like them to play with. 1968 N. Benchley *Welcome to Xanadu* ii. 29 Harry was a Hopi Indian, and..he sold silverware and pottery and kachina dolls. 1972 *Sat. Rev.* (U.S.) 3 June 33/2 The high-voltage transmission towers..march across the desert like giant skeletal Kachinas—Hopi ceremonial figures—carrying the white man's electricity.

kack-handed, var. *cack-handed *a.*

1955 M. Allingham *Beckoning Lady* vi. 91, I never met such a kack-handed jackass in all my born days. 1964 S. Jepson *Fear in Wind* i. 17 A series of kack-handed manœuvres..designed to keep one end of a stick on the ground.

kackle, var. Cackle *v.*² *Naut.*

1883 *Man. Seamanship for Boys' Training Ships R. Navy* (Admiralty) (1886) 128 The cable is then served, or, as is termed, kackled with 2½-in. rounding, for the distance of 9 ft. from the eye.

Kadet, var. *Cadet².

Kadiak, var. *Kodiak.

‖ **kadin** (kā·din). Also **kadine.** [Turk.; the form *kadine* is prob. through Fr.] A lady of the Sultan's harem.

1843 *Penny Cycl.* XXV. 394/2 The women of the harem are divided into five classes:—1, 'Kâdin', or 'wives of the Sultan', in number from four to seven. 1896 *Westm. Gaz.* 16 Sept. 1/3 He seldom notices any woman in his harem except the chief kadine. 1937 *Times Lit. Suppl.* 16 Jan. 36/2 He [*sc.* the Sultan] was content with four *kadins*.

kadir, var. *khadar.

kadish (ka·diʃ). Also 9 **gedish.** [Arab. *kadīš*, f. O.Turk. *igdiş* a cross-breed, f. *igid-* to feed, rear; cf. Turk. *idiş* gelding.] An Arabian horse that is not a thoroughbred; a cross-bred horse, a nag, a gelding.

a 1817 J. L. Burckhardt *Trav. Syria & Holy Land* (1822) 295 A man with two or three is esteemed wealthy; and such a one has probably two camels, perhaps a mare, or at least a Gedish (a gelding), or a couple of asses. 1879 A. Blunt *Let.* 6 May in Lady Wentworth *Authentic Arabian Horse* (1945) 26 They have made my horses vulgar 'Kadishes'. 1924 *Blackw. Mag.* Mar. 346/2 Others less generous or more cautious sent *kadish* horses, only fit for the plough. 1945 Lady Wentworth *Authentic Arabian Horse* i. 34 All Oriental potentates regarded the Arabian as the only Horse and the rest as 'Kadishes', a word which exactly corresponds to the word 'cur' in English. It was the excellence of these Arabians..whose fame earned for the country-bred Kadishes a reflected radiance. 1970 A. Dent tr. *Schiele's Arab Horse in Europe* 24 If there were the tiniest lacuna or unknown quantity in its ancestry, it was no longer regarded as 'asil' and was demoted to the status of 'kadish'.

‖ **kadkhoda** (kadkō·dā). Also **kadkhuda,** and with capital initial. [Pers. *kadkhudā*.] The head man of an Iranian village.

1934 F. Stark *Valleys of Assassins* 56 We..were welcomed by the *kadkhuda* and a dozen tribesmen or so. 1953 A. Smith *Blind White Fish in Persia* v. 80 The Kadkhoda, the bailiff of the village, blows out his lamp and retires to bed. 1963 *Times* 6 Feb. 12/6 The *kadkhoda*, or head man.

KADU, Kadu (kā·du). [f. the initials of the words *K*enya *A*frican *D*emocratic *U*nion.] The name of a Kenyan political party.

1962 *Listener* 12 Apr. 643/3 The two main African parties, Kadu and Kanu. 1964 *Ann. Reg. 1963* 107 During the campaign rifts appeared in KADU.

kæmmererite, obs. var. *kämmererite.

kaempferol (kæ–, ke·mpfĕrǫl). *Chem.* Also **kampherol** (kæ·mfĕrǫl). [f. mod.L. *Kaempferia*, generic name of the plant (*K. galanga*) from which it was first obtained, f. the name of Engelbert *Kaempfer* (1651–1716), Ger. physician and traveller: see -ol.] A yellow crystalline flavonoid, $C_{15}H_{10}O_6$, that occurs alone and in glycosides in various plants; 3,4′,5,7-tetrahydroxyflavone.

1897 H. M. Gordin *On Crystallised Substances contained in Galangal Root* (Dissertation, Univ. of Berne) 37 Demethylised kampherid or kampherol as we agreed to call it crystallises from alcohol in light yellow needles. It contains one molecule of water of crystallisation. 1900 *Proc. Chem. Soc.* XVI. 183 The colouring matter has properties resembling those assigned to kampherol. 1923 *Nature* 17 Nov. 747/2 Four different species of Acacia.. have been examined... The water soluble yellow pigment was a glucoside of kæmpferol. 1963 *Brit. Pharmaceutical Codex* 718 The drug [*sc.* senna leaf] also contains rhein, aloe-emodin, kæmpferol and isorhamnetin in the free state and combined as glycosides. 1969 Kirk & Othmer *Encycl. Chem. Technol.* (ed. 2) XIX. 743 China tea, *C. sinensis*, contains relatively large amounts of triglycosides of quercetin and kaempferol.

kaersutite (kĕə·ɹsu·təit). *Min.* [ad. Da. *kaersutit* (J. Lorenzen 1884, in *Meddelelser om Grønland* VII. 27), f. *Kaersut*, name of the locality (on the shore of Umanak Fiord on the west coast of Greenland) where it was first found: see -ite¹.] A dark brown or black kind of hornblende characterized by a high titanium content and occurring as a constituent of many volcanic rocks.

1886 *Jrnl. Chem. Soc.* L. 519 Analyses of minerals from Greenland... Kærsutite, colour black, streak brown. 1939 *Trans. & Proc. R. Soc. N.Z.* LXIX. 305 It appears that in the Dunedin district brown monoclinic amphiboles occur very generally in the only slightly alkaline basalts and trachybasalts of the first and second volcanic phases, and prove to be basaltic hornblendes which in the two samples analysed have just sufficient content of TiO_2 to permit their being classed as kaersutite. 1966 W. A. Deer et al. *Introd. Rock-Forming Min.* II. 177 Kaersutite is a typical constituent of alkaline volcanic rocks, and occurs as phenocrysts in trachybasalts, trachyandesites, trachytes..and alkali rhyolites; in the more silica-rich rocks it occurs also as a groundmass constituent. 1968 *Mineral. Mag.* XXXVI. 1001 Aoki (1963) considered an amphibole from a trachyte with 4·36% TiO_2 (0·49 Ti) as not a typical kaersutite... It is clear that any lower limit of titanium for kaersutite must be an arbitrary one.

Hence **kaersuti·tic** *a.*, of the nature of or resembling kaersutite.

1968 *Mineral. Mag.* XXXVI. 1001 The data..strongly suggest that upper mantle amphibole is probably a kaersutitic hornblende.

‖ **kafenion** (kæfĕnī·ən). [Gr. καφενεῖον.] A Greek coffee-house or café.

In the Greek word the final -*n* is often dropped in the spoken form (thus quot. 1939). The Gk. pl. form is καφενεῖα.

1939 E. Ambler *Mask of Dimitrios* iv. 63 It took another week, a week of waiting, of sitting in *kafenios*, of being introduced to thirsty gentlemen with connections in the municipal offices. 1964 *Punch* 24 June 925/2 They overflowed the local kafenion. 1967 J. Eastwood *Little Dragon from Peking* xiii. 130 The kind of reception accorded to a woman unaccompanied in, say, a male-dominated, male-preserve Kafenion in rural Greece. 1974 'M. Yorke' *Mortal Remains* I. vii. 24 There was a Kafenion on the water-front, so he sat at a table..and ordered coffee.

‖ **kaffeeklat(s)ch** (ka·feklatʃ). Also with hyphen or as two words, and with capital initial(s). [G., f. *kaffee* coffee + *klatsch* gossip.] Gossip over coffee cups; a coffee party; cf. *coffee klatsch* (s.v. *coffee *sb.* 5 b). Hence **kaffeeklatscher** *sb.*, **kaffeeklatsching** *vbl. sb.* Cf. *klatsch.

1888 in A. Randall-Diehl *2000 Words & Definitions.* 1903 *Current Opinion* Aug. 205/2 She usually operates as an amateur, appearing at *Kaffeeklatsches*. 1906 S. Ford *Shorty McCabe* 111 He let it out one day after we'd had our little kaffee klatch with the gloves. 1911 *International* (N.Y.) July 35/2 Theatrical *Kaffeeklatsch* has now absorbed the space where once the redoubtable Charles Edward fought his battles. 1919 F. Hurst *Humoresque* 322 They're a darn sight better than the wads of respectability I see waddlin' in here to swap *Kaffee Klatsches* with you! 1936 H. Miller *Black Spring* 134 It's the hour of the kaffee-klatchers sitting around the family table. 1956 W. H. Whyte *Organization Man* (1957) xxii. 286 Dot will be *Kaffee-klatsching* and sun-bathing with the girls. 1958 M. West *Second Victory* i. 5 They came in summer to take the waters, to sit on the terrace for *Kaffeeklatsch*. 1969 R. Lockridge *Murder in False Face* ix. 112 If [he]..wants a morning kaffee-klatch it's all right with me. 1972 J. Williams *Home Fronts* xiii. 233 The traditional friendly *Kaffeeklatsch*—the afternoon coffee party with friends.

Kaffir. Add: **1.** (Earlier example.)

1790 J. Bruce *Trav.* IV. viii. ix. 497 Why did not you tell those black Kafrs..to stay a little longer.

2. (Further examples.) Also, usu. disparagingly, with reference to any black African. Also *transf.*, as a term of opprobrium, a white man who associates with or is thought to favour black Africans.

1792 E. Riou tr. *J. van Reenen's Jrnl. Journey from Cape Good Hope* 22 We saw several Kaffers. 1852 Godlonton & Irving *Narr. Kaffir War* III. xv. 180 The other teachers ..who could speak Kaffir. 1926 S. G. Millin *S. Africans* 209 In the old days..men, thrusting their ancestry, their traditions..completely behind them, became what people sometimes call in South Africa 'white Kafirs'. They

merged themselves with the natives, stayed for ever with the wives they had bought and with their African children. 1949 [see *Hotnot]. 1949 *Cape Argus* 9 July 3/5 'Did you think he was a perfectly reliable person to give information to?'—'I would have given the statement to a Kaffir if someone had sent a Kaffir along.' 1959 *New Statesman* 2 May 62/3 How, for instance, does one describe negroid South Africans? The early missionary word 'kaffir', meaning heathen, has become a term of abuse. 1960 *Cape Times* 6 Sept. 7 A mob which swore at the police, called them 'white Kafirs', and hurled bottles at them. 1961 L. van der Post *Heart of Hunter* I. iii. 62 Kaffir is the term used by Europeans to describe all black people in Africa irrespective of their race and origin. 1967 [see *coolie, cooly 2 b]. 1973 *Deb. Senate S. Afr.* 17 May 2777 When we..were young people the word 'kaffir' meant nothing more than to indicate a Black man... It has deteriorated to such an extent that it offends people with a dark coloured skin and..we try to avoid it. *Ibid.* 2798, I have heard people when I visit a farm call out 'Kaffir' and a wife appears, and he says 'my kaffir, prepare food for us' ..; but, if I called my friend the hon. Senator.., 'You are a kaffir', then it has another meaning.

3. Substitute for def.: Usu. **Kafir.** A member of a people inhabiting the Hindu Kush mountains of north-east Afghanistan; **Kafir harp,** a primitive harp with four or five strings used by this people.

1961 A. Baines *Mus. Instruments* 43 In the Kafir harp the lower end of the bow reappears above the skin.

4. Kaffir beer, an alcoholic beverage brewed from Kaffir corn by the black inhabitants of S. Africa; **Kaffirboetie** (kæ·fəɹbūti) *S. Afr.* [partial tr. Afrikaans *Kafferboetie*, f. *Kaffer* Kaffir + *boetie* little brother], an opprobrious term for a Negrophil; **Kaffir-boom** (earlier and later examples); **Kaffir bread** (earlier and later examples); **Kaffir Circus** *Stock Exchange slang*, the body of brokers who operate in 'Kaffirs', or the place where they operate; **Kaffir corn** (earlier and later examples); **Kaffir crane,** a name formerly used for the crowned crane, *Balearica pavonina regulorum*, which is grey with a tuft of black feathers on top of its head; **Kaffir finch, fink,** the red bishop-bird, *Pyromelana oryx*, or a closely related bird of the sub-family Ploceinæ; **Kaffirland,** the land of the Kaffirs; **Kaffir lily,** a herb of the family Iridaceæ, *Schizostylis coccinea*, bearing spikes of gladiolus-like flowers; also = *clivia; **Kaffir (water-)melon,** either of two species of melon, *Citrullus caffer* or *C. vulgaris*; **Kaffir orange,** a shrub or small tree of the genus *Strychnos*, esp. *S. pungens*, or its fruit; **Kaffir piano,** a S. African marimba or xylophone; **Kaffir pot,** an iron cooking-pot usu. on three short metal legs; **Kaffir tea** (examples); **Kaffir('s) tree** (earlier example); **Kaffir truck** *S. Afr.*, term applied to small miscellaneous general goods for barter or sale.

1837 R. B. Hulley in F. Owen *Diary* (1926) 174 About a hundred pots filled with Kaffir beer were brought and placed before the..men. 1905 *Transvaal Agric. Jrnl.* Jan. 314 Kaffir beer, which..is not..a bad drink for natives. 1952 L. Marquard *Peoples & Policies S. Afr.* iv. 101 Africans may..drink kaffir beer. This is a traditional African drink, brewed by African women from fermented kaffir corn and containing a maximum of 2 per cent of alcohol. 1968 M. Pyke *Food & Society* iv. 43 It is a very great error..to assume..that this Kaffir-beer is simply and solely an intoxicating drink. 1939 R. F. A. Hoernlé *S. Afr. Native Policy* p. vii, For a member of the White group to be concerned about the impact of white domination on the non-European population of the Union..is to earn for himself the title of 'negrophilist', *kafir-boetie*, or—most scathing of all—'liberal'. 1942 P. Abrahams *Dark Testament* I. xiii. 71 One's got to live. I can't let the other fellows call me a 'Kaffir boetie'. 1947 A. Keppel-Jones *When Smuts Goes* ii. 20 The disgruntled factions of trade unionists and liberals, Indians and 'Kafferboeties', seemed very small fry. 1958 N. Gordimer *World of Strangers* v. 115 You must be Communist or Anti-Communist, Nationalist or Kaffirboetie. 1965 *Punch* 24 Feb. 272/3 Multi-racialists, educationalists, Kafir-boeties. 1827 G. Thompson *Trav. & Adv. S. Afr.* I. i. ii. 18 The stakes of this fence, consisting chiefly of Caffer-boom (*Erythrina Caffra*) which grows abundantly in the neighbourhood, had in numerous instances struck root. 1949 *Cape Argus* 15 Oct. 4/5 The Alexandria forests, red with giant kaffir-booms. 1953 J. Packer *Apes & Ivory* xv. 160 The red flowers of the *kaffir-booms* flamed in the hard clear light. 1964 A. Rothmann *Elephant Shrew* 19 There was a big clump of wild bananas in this camp, and also two large, spreading kaffirbooms. 1801 J. Barrow *Acct. Trav. S. Afr.* I. iii. 189 The *zamia cycadis*, or Kaffer's bread-tree, growing on the plains. 1958 L. G. Green *S. Afr. Beachcomber* 14 Beyond the Buffalo River lies the Wild Coast, with the frangipane and kaffir-bread trees growing down to the beaches. 1896 M. Donovan *Kaffir Circus* 96 A big boom is on in the Kaffir Circus, and Laure's shares are worth £15,000. 1901 C. Duguid *How to read Money Article* 121 The market in which they are dealt in the Stock Exchange is often called the 'Kaffir Circus'. Term does not comprise Rhodesians. 1902 *Encycl. Brit.* XXXII. 865/1 At first..the 'Kaffre circus', as it was called, was regarded with contempt by the older *habitués* of the Stock Exchange. 1928 *Daily Chron.* 9 Aug. 8/6 Otherwise the Kaffir Circus presented a very idle appearance. 1785 G. Forster tr. *Sparrman's Voy. Cape Good Hope* II. x. 10 The kind of corn which they sow, is.. known to yield abundantly. The colonists call it caffer-corn. 1792 Kaffir corn [see *black bean* (*black *a.* 19)]. 1954 R. St. John *Through Malan's Afr.* i. 14 Indian millet is called Kaffir corn. 1973 *Farmer's Weekly* (S. Afr.) 13 June 3 (Advt.), Prevent fallen kaffir corn and other crops from being double cut. 1834 A. Smith *Diary* 15 Nov. (1939) I. 137 Saw several Caffer cranes. 1853 F. Fleming *Kaffraria* iii. 68 The handsomest of these [birds] to be met with is the Kaffir Crane..a species of the *Anthropoides-Pavonia* or Crowned-Demoiselle. 1906 W. L. Sclater *Birds S. Afr.* IV. 279 *Bugeranus carunculatus*. Wattled Crane..sometimes 'Kaffir Crane' of Colonists. *Ibid.* 284 *Balearica regulorum*. Crowned Crane. ..'Kaffir Crane' of some. 1822 W. J. Burchell *Trav. S. Afr.* I. i. 20 In the aviary I saw..the *Kaffers Fink*. 1834 [see *fink *sb.*¹]. 1844 J. Backhouse *Narr. Visit Mauritius & S. Afr.* xiv. 202 The Caffer Finch of this part of the country is *Ploceus spilonotus*. 1897, 1908 [see *fink *sb.*¹]. 1931 Kaffir fink [see *bishop-bird]. 1973 A. P. Brink *Birds* 4 You're so cocky. What are you—a kaffir finch? 1821 E. Blount *Notes on Cape Good Hope* 137 A poet of great respectability..was ready to invoke the muse of Kaffer-land. 1853 *Househ. Words* 11 June 338/1 Let us.. see what the noble savage does in Zulu Kaffirland. 1900 W. D. Drury *Bk. Gardening* x. 348 *Schizostylis coccinea* (Crimson Flag; Kaffir Lily) is a lovely iridaceous subject with bright crimson gladiolus-like spikes of flower. 1946 M. Free *All about House Plants* xii. 94 *Clivia miniata*, Kafir Lily. Give only enough water to keep leaves from wilting. 1951 *Dict. Gardening* (R. Hort. Soc.) IV. 1904/1 S[*chizostylis*] *coccinea*. Crimson Flag; Kaffir Lily. 1970 M. Allan *Tom's Weeds* ii. 27 A feature of Number 1 greenhouse was the imantophyllum or Kaffir lily, renamed clivea by John Lindley in honour of the Duchess of Northumberland (of Syon House) who was a member of the Clive family. 1886 G. A. Farini *Through Kalahari Desert* xii. 199 Close by were a lot of young gourds growing which Kert said were Kaffir melons; they were quite unlike our English watermelons; nor were they like a pumpkin. 1950 *Cape Times* 1 June 7/6 The National Parks Board has authorized expenditure on kafir melons with which to feed the elephants in summer-time. 1859 R. J. Mann *Colony of Natal* viii. 159 The 'Kafir orange' of the sea-coast-bush is a 'strychnos' and has strychnine in its seeds. 1907 T. R. Sim *Forests & Forest Flora Cape Good Hope* 274 *Strychnos spinosa* (Kafir Orange..). An evergreen shrub 8–10 feet high, seldom a small tree... Fruit size of an orange, or larger, with rind green when young, yellow when ripe, hard shell, and numerous flat seeds lying in acidulous edible pulp. 1932 Watt & Breyer-Brandwijk *Medicinal & Poisonous Plants S. Afr.* 140 The pulp of the fruit of *Strychnos pungens* Solered., Wild orange, Kaffir orange..is acidulous from the presence of citric acid, and is very refreshing. 1952 S. Cloete *Curve & Tusk* (1953) xiii. 112 There were patches where the marsala or kaffir orange grew, its round, hard-shelled fruit a favourite dish of the baboon and kudu. 1891 Monteiro *Delagoa Bay* 253 (Pettman), The song had a rapidly played accompaniment on the Kaffir piano. 1897 J. Bryce *Impressions S. Afr.* xiv. 251 The so-called 'Kaffir piano', made of pieces of iron of unequal length fastened side by side in a frame. 1931 J. Mockford *Khama* xxiii. 157 To the throb and wail of these kafir pianos the big-bodied, lusty mine-boys dance freely in two long lines. 1948 H. V. Morton *In Search of S. Afr.* x. 311 A native band was thrumming on 'Kaffir pianos', instruments like large xylophones. 1896 H. A. Bryden *Tales of South Africa* 260 The *kaptein* ..persuaded the *vrouw* to follow his own example, and roast wild duck or a joint of springbok in a Kaffir pot. 1922 S. G. Millin *Adam's Rest* III. x. 254 Over the fire stood a big black tripod Kaffir-pot. 1959 A. Fullerton *Yellow Ford* xiii. 177, I use a kaffirpot, a three-legged thing made of cast iron. 1851 J. J. Freeman *Tour S. Afr.* xv. 362 One kind hearted woman..prepared a Kaffir meal for us—a pot of sour-milk, some Kaffir corn bread and some Kaffir tea. 1899 G. Russell *Hist. Old Durban* 96 An indigenous herb both nutritive and refreshing, which is known to us as *Kaffir tea*. 1949 L. G. Green *In Land of Afternoon* 55 Bush tea is popular in the fashionable cafes of the United States. They call it 'Kaffir tea' over there. 1792 E. Riou tr. *J. van Reenen's Jrnl. Journey from Cape Good Hope* 38 We interred the body of our friend, under a large kaffer-tree standing alone. 1855 G. H. Mason *Life with Zulus* 133 (Pettman), This portion of South Africa is dependent entirely on the P. M. Berg traders for..*caffre truck*. 1900 J. Robinson *Life Time in S. Afr.* 279 (Pettman), Glass, beads, knives, scissors, needles,..small looking-glasses—such are the chief staples of *Kaffir truck*. 1948 E. Rosenthal *Afr. Switzerland: Basutoland* vii. 83 Basuto, who crowd a quaint kiosk loaded with what is called 'Kaffir Truck' in South Africa—bangles, beads, mirrors, combs, and the like. 1951 D. Lessing *This was Old Chief's Country* ix. 198 He had gone into town and was down among the kaffir-truck shops buying a supply of aprons for his houseboys. 1832 *Graham's Town* (Cape Province) *Jrnl.* 1 June 92 All the Caffers..were dispatched forward, that they might get Caffer water melons to make soup. 1838 W. H. Harvey *Genera S. Afr. Plants* 105 The water-melon of which two colonial species *C. Caffer* (Kaffir-water-melon) and *C. amarus*..are described. 1932 Watt & Breyer-Brandwijk *Medicinal & Poisonous Plants S. Afr.* 180 *Citrullus vulgaris* Schrad is known as Water-melon, Wild water-melon, Kaffir water-melon.

B. *adj. S. Afr. slang.* Bad, unreliable.

1934 'N. Giles' *Ridge of White Waters* II. vii. 266 'Another kaffir bargain!' said Sir Alfred wearily. 1961 *Spectator* 14 July 53 'That was a real Kaffir shot.'.. This ..was the first time I had come across Kaffir, *adj.*: bad, clumsy, inferior..etc.

Kaffrarian, var. *Caffrarian *a.*

Kafkaesque (kæfkă,e·sk), *a.* Also with hyphen. [See -esque.] Of or relating to the Austrian writer Franz *Kafka* (1883–1924) or his writings; resembling the state of affairs or a state of mind described by Kafka. Hence **Kafkae·squely** *adv.*

1947 *New Yorker* 4 Jan. 61/1 Warned, he said, by a Kafka-esque nightmare of blind alleys. 1954 Koestler *Invis. Writing* x. 120 Long before the Moscow purges revealed that weird, Kafka-esque pattern to the incredulous world. 1958 *Spectator* 24 Jan. 114/2 An authentic Kafkaesque atmosphere of despair and horror. 1958 E. Dundy *Dud Avocado* I. viii. 147 Postcards and wires to the Paris Embassy were all Kafkaesquely re-routed to that powerful Man in Charge. 1963 *Times* 23 May 6/7 Kafkaesque in its grip and pitiless in its exposition of the cruellest of tortures, that of hope. 1972 *Newsweek* 10 Jan. 51/2 The Kafkaesque self-abnegation of the infamous 'show trials' [in Russia].

Also **Ka·fka** *sb.* used *attrib.*, **Ka·fkan, Ka·fkaish, Ka·fkian** *adjs.* = *Kafkaesque *a.*

1936 M. Lowry *Let.* (1967) 11 This is the perfect Kafka situation. 1951 S. Spender *World within World* v. 272 They became more Kafkaish than ever. 1959 *N. & Q.* Oct. 381/1 A re-statement of the Kafkan anguish. 1962 tr. *J. L. Borges's Labyrinths* (1970) 234 The moving object and the arrow and Achilles are the first Kafkian characters in literature. 1962 *Guardian* 26 Sept. 8/6, I..had wondered if the whole project would turn out to be a Kafka nightmare. 1966 *New Statesman* 25 March 437/1 All the Kafkan stuff..gone seedy and suicidal in a back-street rooming-house. 1971 N. Freeling *Over High Side* II. 82 So little of what one did made any sense. One lived in a Kafka world.

kaftan, var. Caftan (in Dict. and Suppl.) (further examples).

1965 [see *caftan 2]. 1967 *Daily Mail* 10 Aug. 4/7 I'd like to see men in this country wearing kaftans—those long cotton robes—to relax at home. 1969 *Daily Nation* (Nairobi) 31 Oct. 19/2 It comprises a Kaftan, slit on either side to the knees to reveal pencil slim hipster pants. 1972 J. Wilson *Hide & Seek* ii. 37 Her wedding dress..had been an all enveloping Kaftan because she'd been hugely pregnant.

kafuffle, var. Curfuffle *sb.*, *gefuffle, *kerfuffle.

1946 F. Sargeson *That Summer* 94, I bet it [*sc.* the domestic row] ended up in a good old kafuffle.

kagg: see *cag *sb.*³

‖ **kagura** (kā·gurā). [Jap.] A sacred dance performed at Shinto festivals, one of the oldest dances of Japan; also, one performed at a village shrine on a festive day.

1884 Satow & Hawes *Handbk. for Travellers Cent. & N. Japan* (ed. 2) 63 At some temples young girls fill the office of priestess, but their duties do not appear to extend beyond the performance of the pantomimic dances known as *Kagura*, [etc.]. 1899 W. G. Aston *Hist. Jap. Lit.* v. iii. 197 The drama in Japan was in its beginnings closely associated with religion. Its immediate parent was the Kagura, a pantomimic dance, which is performed at this day to the sound of fife and drum at Shinto festivals. 1936 K. Nohara *True Face of Japan* v. 191 Dances and festival plays were performed in front of Shinto shrines, which were called *Kagura*, or 'Joys of the Gods'. 1946 R. Benedict *Chrysanthemum & Sword* (1947) v. 90 Watching wrestling matches or exorcism or *kagura* dances, which are liberally enlivened by clowns. 1966 P. S. Buck *People of Japan* (1968) xii. 132 At special times the Shinto priests perform their own religious dance, the *kagura*.

‖ **Kahal** (kā·hal). [Heb. *kāhal* assembly, community.] One of the former localized Jewish communities in Europe; also, the governing body of such a community.

1901 *Daily Chron.* 14 June 3/4 The power of the *Kahal* —the court of the congregation. 1907 I. Zangwill *Ghetto Comedies* 342 The very Rabbi was petrified; the elders of the *Kahal* stood dumb. 1916 H. Sacher *Zionism & Jewish Future* 19 In every Kahal (community) many youths..studied. *Ibid.* 20 A Kahal of fifty families. 1937 Wyndham Lewis *Blasting & Bombarding* v. v. 280, I took no further interest in this cowboy songster, said to be a young sprig of the Kahal. 1971 *Encycl. Judaica* V. Gloss., *Kahal*, Jewish congregation; among Ashkenazim, *Kehillah*.

kahawai (kā·wai, ka·hawai). Also **kawai.** [Maori.] A perciform, marine, food fish, *Arripis trutta*, found in shoals in New Zealand and south-eastern Australian waters.

1838 J. S. Polack *New Zealand* I. 322 The *kâhâwai*, or colourless salmon. 1845 E. J. Wakefield *Adventure N.Z.* I. 92 A shoal of *kawai* came into that part of the bay. The *kawai* has somewhat of the habits of the salmon. 1849 W. T. Power *Sketches in N.Z.* ix. 76 The kawai is not unlike the salmon in size and shape, and, like it, comes up the rivers in shoals in the spring. 1870 R. Taylor *Te Ika a Maui* (ed. 2) 623 The Kahawai..is one of the most abundant, and is called mackerel by the settlers. 1927 *Daily Express* 26 Feb. 1 The Duchess returned to the Renown with seventeen 'schnapper' and one 'kahawai'. 1962 *Antiquity* XXXVI. 272 Fish-hooks.. another type of lure with barbed point notched for lashing (the *kahawai* lure). 1962 G. W. Jackson *N.Z. Beach & Boating Bk.* viii. 82 The bait is..a bundle of sprats or a whole kahawai. 1963 *Evening Post* (Wellington, N.Z.) 30 Nov., Stripping fillets of kahawai and baracouta we cut them into fish shapes and the reaction was immediate, the big fellows accepting them as readily as they had the rock cod earlier.

kahika (kā·ˌikā). [Maori.] = Kahikatea.

1921 H. Guthrie-Smith *Tutira* x. 71 Close to this orchard grew..three tall white pines, survivors of the kahika grove, from which the flat had probably taken its name. 1949 P. Buck *Coming of Maori* (1950) IV. i. 450

She [sc. Hinewaoriki] gave birth to twins in the form of the *kahika* and *matai* trees.

kahili (kāhī·li). [Hawaiian.] A feather standard, mounted on a tall pole, symbolic of royalty in Hawaii and used on ceremonial occasions.

1866 'Mark Twain' *Lett. from Hawaii* (1967) 180 A dozen or more of these gaudy kahilis were upheld by pallbearers. **1883** C. F. Gordon Cumming *Fire Fountains* I. 35 At the door of the mausoleum are placed tall *kahilis*, honorific symbols, which to irreverent foreign eyes are suggestive of gigantic feather-brushes, or rather bottle-brushes. **1915** W. A. Bryan *Nat. Hist. Hawaii* 81 On the other hand is a small kahili with ivory and tortoise shell handle. **1937** D. & H. Teilhet *Feather Cloak Murders* xv. 267 The Baron was next to find two rotted kahilis, ancient feather standards. **1948** Kuykendall & Day *Hawaii* xi. 108 In the shadow of somber *kahilis* (royal standards) his ministers and his subjects marched past. *Ibid.* xvi. 166 He stood beside..the *kahili*, symbolic of Hawaiian chieftainship.

Kahn (kān). *Med.* The name of Reuben Leon *Kahn* (b. 1887), Lithuanian-born U.S. bacteriologist, used *attrib.* and *absol.* to designate a diagnostic test for syphilis devised by him in 1922, in which serum or spinal fluid that has been inactivated by heating is shaken with a suspension of antigen obtained from beef heart and the mixture examined for flocculation (usually after a period of incubation).

1922 *Jrnl. Amer. Med. Assoc.* 9 Sept. 874/2 The clinical applications of the Kahn precipitation test compares favorably in sensitiveness with the standard Wassermann reaction. *Ibid.* 873/1 In the serums examined from patients with late bone and joint involvement, the Kahn reaction again compares very favorably with the two Wassermann reactions. **1950** R. R. Willcox *Text-bk. Venereal Dis.* ix. 115 As a verification test the Kahn test performed at different temperatures..has not proved entirely satisfactory. **1953** *Med. Ann.* LXXI. 7 Positive Wassermann and Kahn reactions may be given, for a time, by the serum of patients recovering from glandular fever. **1964** King & Nicol *Venereal Diseases* vii. 96 Of the many flocculation tests available, the Kahn test has been the most widely used, but in recent years the Price precipitation test (PPR) has increased in popularity in Great Britain.

kahuna (kahū·nă). [Hawaiian.] **a.** A Hawaiian priest or minister; an expert or wise man. **b.** *Surfing.* (With capital initial.) A term adopted to designate a 'god' of surfing.

1886 H. H. Gowen *Let.* 6 Dec. in *Paradise of Pacific* (Hawaii) (1892) viii. 85 The Kahunas advised him to starve off the calamity by getting rid of the white power. **1915** W. A. Bryan *Nat. Hist. Hawaii* 54 A numerous class of more irregular priests or Kahunas, that were little more than sorcerers. **1920** *Nature* 15 July 628/1 A much longer paper..deals with the functions of the *Kahunas*, the priesthood called the Order of Sorcery'. The word in varying form (*kahuna*, *tahunga*, *tauna*) is used throughout the Eastern Pacific to denote possessed of varying degrees of wisdom from priesthood to sorcery. **1948** Kuykendall & Day *Hawaii* i. 8 The *kahunas* (priests, doctors, sorcerers, navigators, and experts in various other lines) comprised a class closely associated with the chief. **1962** *Austral. Women's Weekly* (Suppl.) 24 Oct. 3/2 Kahuna..the god of the Californian and Hawaiian board-riders. **1970** *Studies in English* (Univ. Cape Town) I. 25 The word 'kahuna' has been personified into *Kahuna*, the god of surfing.

kai (kai). *N.Z.* [Maori.] Food, victuals.

[**1838** J. S. Polack *New Zealand* I. 289 There is a much larger variety of this esculent [sc. potato] called *kai pakeha*, or white man's food.] **1846** E. J. Wakefield *Adventure N.Z.* I. 265 The determination of the natives not to move till all the *kai* was exhausted. **1925** Fraser & Gibbons *Soldier & Sailor Words* 134 *Kai*, food. (A Maori word, used among the New Zealand troops in the War.) **1927** T. E. Donne *The Maori, Past & Present* 95 I bring in to share for tree day, but no *kai* (food). **1952** E. Finlayson *Schooner came to Atia* iii. 17, I can take Tua smokes and good kai. **1970** *N.Z. Listener* 12 Oct. 12/1 Some kai would go nicely now. Empty bellies do things to people.

So (in reduplicated form) **kaikai** (kai·kai), food; feasting; a feast.

[**1807** J. Savage *Some Acct. N.Z.* xi. 75 *Kiki*..food.] **1845** E. J. Wakefield *Adventure N.Z.* I. 79 He explained..that there would be much *kai kai* or feasting. **1894** Stevenson & Osbourne *Ebb-Tide* I. iv. 60 There shall be no growling about the kaikai, which will be above allowance. **1901** A. C. Haddon *Head-Hunters* 39 One afternoon some of us went to a *kaikai*, or feast. **1941** Baker *N.Z. Slang* 26 In early records the 'pidgin' forms *kaikai* or *kiki* are often discovered. **1969** *Coast to Coast 1967–68* 48 No, she didn't say 'tucker'. The kanakas said kai-kai.

|| **kaid.** Add: (Earlier and later examples.) Also **kayed, kaïd.** Hence **kai·dship.**

1816 'Ali Bey' *Trav.* I. 5, I handed it [sc. my passport] to the captain, who ordered that no one should come on shore, and went away to shew my passport to the *Kaid*, or Governor. *a* **1817** J. L. Burckhardt *Trav. Nubia* (1819) 364 The Shikh of the tribe is never the commander..of the armed parties, which the tribe sends out against an enemy. He may join the expedition, but the command of it is in the Kayed, or leader, a dignity which is always hereditary in the same family. **1920** *Glasgow Herald* 23 Sept. 6 Kaid of Tangier. **1920** *Blackw. Mag.* Dec. 742/2 His half-brother was already nominated to the kaidship. **1925** *Ibid.* Nov. 622/2 Presently the Kayed will appear, walking with his chief villagers.

kaik, var. *KAINGA.

kaikomako (ka,ikomā·ko). *N.Z.* [Maori.] A New Zealand tree, *Pennantia corymbosa*, which bears panicles of fragrant white flowers.

1832 G. Bennett in *London Med. Gaz.* 22 Sept. 794/2 (*heading*) Kaiko-mako tree of the natives of New Zealand, ..This tree..attains the elevation of twenty-five to thirty feet...The wood of the..Kaiko-mako, is only used by the natives for procuring fire. **1882** W. D. Hay *Brighter Britain!* II. 198 The Kaikomako..will be much cultivated as a garden ornament. **1910** L. Cockayne *N.Z. Plants* iii. 33 *Pennantia corymbosa* (the kaikomako) vies in its purity with any bridal flower. **1963** Poole & Adams *Trees & Shrubs N.Z.* 122 *Pennantia corymbosa*...Kaikomako. Tree reaching 12m... Flowers small, dioecious, fragrant.

|| **kain** (ka·in, kain). [Malay.] Cloth, a piece of cloth. Usu. with defining word following (see quots.).

1783 W. Marsden *Hist. Sumatra* 44 The *cayen sarrong* is not unlike a Scots highlander's plaid, being a piece of party coloured cloth about six or eight feet long, and three or four wide, sewed together at the ends. **1848** H. Low *Sarawak* v. 143 Their dress..consists of the kain tape, or cloth, which has been described as a wide sack open at both ends. **1910** C. W. Harrison *Illustr. Guide Federated Malay States* ii. 106 The great majority of sarongs are of cotton cloth known as *kain plekat*. **1919** P. Mijer *Batiks* ... 'Kains' can be bought as cheaply as a dollar each. **1947** R. O. Winstedt *Malaya* viii. 148 The obsolescent 'time' pattern (*kain limau*)..reminds one of Indian designs. *Ibid.*, Batani, Pahang and Selangor produce cloths (*kain telepoh*) gilded by a technique practised also in the Punjab. **1958** H. Forster *Flowering Lotus* ii. 25 Formal Javanese wear, for gentlemen as for ladies, was the *kain batik*. This was a simple length of cotton cloth, decorated with an elaborate pattern. **1963** J. Kirkup *Tropic Temper* iv. 41 There are many ways of folding the big, starched kerchief in a form of head-dress...On the East Coast it is called *kain satangan*. *Ibid.* xv. 167 Round their waists they wore vivid red or purple *kain songket* which is a silver or gold-threaded sarong reaching down to the knee and tied at the hip in an especially intricate knot. **1967** F. Mullally *Prince* ... 39 Indonesians in brown-moulding *kebaya* and batik kains. **1971** *Carry Singapore in your Pocket* (Singapore Tourist Promotion Board) (ed. 3) 53 *Kain songket*, spun silk woven on cottage handlooms..prized for its dazzling beauty of gleaming motifs in gold or silver. **1972** M. Sheppard *Taman Indera* 117 *Kain lepas*, unsewn sarong length, often of gold thread silk. *Ibid.* 118 (*caption*) Part of a length of *Kain Limar*, from Kelantan. Fine silk cloth with a rose red warp and a mosaic pattern. **1972** *Sunday Times* (Kuala Lumpur) 25 June 7/4 The Queen was given a kain kebat—a dress resembling the sarong.

kainga (kā,i-ngă, || kaiŋa). *N.Z.* Also (*South Island*) **kaik.** [Maori.] A place of residence; a settlement, village.

1820 *Gram. & Vocab. Lang. N.Z.* (Church Missionary Soc.) 157 *Kainga*,..A place of residence, a home, &c. **1838** J. S. Polack *New Zealand* I. 66 These animals were a part of the *kainga* or village, of which they formed part. **1879** [see *gum-digging* (*gum sb.*²)]. **1884** *Maoriland* 84 The drive may be continued from Porirua to the Maori kaik. **1904** G. B. Lancaster *Sons o' Men* 56 He had..fallen foul of many native kaingas where the pakeha was unwelcome. **1905** W. B. *Where White Man Treads* 181 And so to-day; when you see small square potato patches dot the landscape near his kaingas, this meagre husbandry is no sign of improvidential laziness. **1926** J. Cowan *Trav. N.Z.* I. 114 They are places for the artist, these out-of-the-way kaingas. **1938** R. D. Finlayson *Brown Man's Burden* 49 The Maori just sits in his kainga and takes what comes to him. **1944** *Mod. Jam. Dict.* (Whitcombe & Tombs) 229 *Kaik, Kainga*,..A Maori village; 'kaik' in the South Island only. **1967** A. & D. Reed *Paddle Wheels on Wanganui* 12 The excitement of the riverside Maoris as the ship passed their lonely kaingas.

kainite. Add: Also **kainit.**

1877, 1882 Kainit [in Dict.]. **1950** *Engineering* 4 Aug. 101/2 Among the interesting and unusual cargoes which had been shipped from time to time, there were some which had been described, respectively, as praff, shooks, kainit, and thiolith. **1971** *Farmers Weekly* 19 Mar. 38/3 This week we have been dressing feet ground with kainit, a cheap source of potash and salt.

kainosite (kai-, kəi·nǝsait), the usual spelling of *CENOSITE.

1895 [see *cenosite]. **1925** *Mineral. Mag.* XX. 356 Unless obviously in error, species names should be accepted as nearly as possible the same form as that given by the first author...As an example, take the two names Kainosite and Kainite, originally given in Swedish and German in the forms Kainosit and Kainit respectively, and both derived from the Greek καινός, new. The first was altered by Dana (1892) to Cenosite, which is scarcely recognizable even in English. **1968** I. Kostov *Mineral.* 306 Nordite and kainosite are orthorhombic.

kairoline (kai·rolin, -in). *Chem.* [ad. G. *kairolin* (Hoffmann & Konigs 1883, in *Ber. d. Deut. Chem. Ges.* XVI. 740), f. *kaïrin* KAIRINE with insertion of *-ol* -OL.] 1,2,3,4-Tetrahydro-*N*-methylquinoline: an oily liquid, $C_{10}H_{13}N$, with antipyretic properties.

1883 *Pharmaceutical Jrnl.* XIV. 384 The chinolinmethylhydride (kairoline) of Konigs and Hoffmann..[was] tried and also found to have antipyretic properties. Kairoline is built up in precisely the same way as kairine, except that one atom of H is replaced by HO. **1953** *Jrnl. Amer. Chem. Soc.* LXXV. 3030 Following the procedure of Gilman and Banner kairoline was prepared from freshly distilled 1,2,3,4-tetrahydroquinoline..and freshly distilled dimethyl sulfate.

|| **kairos** (koi·rǫs). [Gr. καιρός right or proper time.] Fullness of time; the propitious moment for the performance of an action or the coming into being of a new state.

1936 E. L. Talbot tr. *Tillich's Interpretation of Hist.* II. ii. 129 We call this fulfilled moment, the moment of time approaching us as fate and decision, *Kairos*. In doing this we take up a word that was, to be sure, created by the Greek linguistic sense, but attained the deeper meaning of fullness of time, of decisive time, only in the thinking of early Christianity and its historical consciousness. **1939** V. A. Demant *Religious Prospect* viii. 220 A teaching that all man can know is how to respond to the Unconditioned at each moment of decision, which he calls the Kairos. **1948** J. L. Adams tr. *Tillich's Protestant Era* (1951) i. iii. 47 Every kairos is..implicitly..an actualization of the unique kairos, the appearance of the Christ. *Ibid.* 48 We are convinced that today a kairos, an epochal moment of history, is visible. **1963** Auden *Dyer's Hand* III. 140 The Greek notion of *Kairos*, the propitious moment for doing something, contained the seed of the notion of punctuality.

Kaiser. Add: **Kai·ser-ate, Kai·serdom** = KAISERSHIP; **Kai·serish** *a.*; **Kai·serism**, absolutism as exhibited in the rule of the German emperor; **Kai·serist**, an adherent of the absolutist political system of the German emperor, *esp.* that of Wilhelm II (ruled 1888–1918); so **Kaiseri·stic** *a.*; **Kaisership** (earlier U.S. example).

1848 J. R. Lowell *Fable for Critics* 23 Two dozen of Italy's exiles who shoot us his Kaisership daily. **1882** R. Adamson *Fichte* 81 Even the shadowy bond which seemed to unite the German States had been dissolved by the Austrian emperor's renunciation of the Kaiserate. **1905** *Westm. Gaz.* 27 Apr. 2/2 We confess to finding his speech..distinctly Kaiserish. **1914** C. Brereton *Who is Responsible?* 101 In order to smash and pulverize Kaiserdom and all that it stands for in the world. **1914** T. Roosevelt in *N.Y. Herald* 5 Sept. 8/2 The American people will countenance nothing that resembles Kaiserism. **1915** *Morning Post* 13 Feb. 6/7 The Revolutionaries, who declare that 'Kaiserism' is as deadly a form of 'Absolutism' as any that can be conceived. **1919** *New Appeal* (Girard, Kansas) 18 Jan. 4/6 The Tribune's editorials..have generally been in a Kaiseristic vein. **1920** R. Crowin *Timber Wolves* viii. 121 A more flagrant example of business Kaiserism never happened. **1920** *Glasgow Herald* 10 June 4 The curious relations between Kaiserists and anarchists. **1923** *Ibid.* 13 May 7 The Kaiserist system of political autocracy. **1973** *Times Lit. Suppl.* 27 Oct. 1272/5 A brilliant imperial stroke in the direction of world liberty against Kaiserism.

2. *Comb.* **Kaiser moustache** (also **Kaiser Bill moustache, Kaiser Wilhelm moustache**): see quot. 1966; **Kaiser's war**, the war of 1914–18.

1938 J. Cary *Castle Corner* 108 His small, reddish moustache was curled up at the ends. It was not a Kaiser moustache. **1946** G. Millar *Horned Pigeon* xvi. 215 The clientèle favoured cropped hair and either Hitler or Kaiser Wilhelm moustaches. **1952** M. Innes *Private View* iv. 70 Absolutely trustworthy—my batman..in the Kaiser's war. **1957** R. Kemp *Colours or Crest* (1960) ix. 205 The one appointed as our personal bodyguard was a bibulous, red-faced fellow with an enormous 'Kaiser Bill' moustache. **1963** Bird & Hutton-Stott *Veteran Motor Car* 16 Before the Kaiser's war famous firms made cars in Scotland. **1966** J. S. Cox *Illustr. Dict. Hairdressing* 86/1 *Kaiser Moustache*, a moustache of which the ends are turned up in the manner of William II, Emperor of Germany, the Kaiser. **1971** A. Price *Alamut Ambush* x. 118 The war the wheelwright was remembering was the Kaiser's, not Hitler's. **1974** A. Ross *Bradford Business* 20 The man..was tall..with a shock of white hair and a Kaiser Bill moustache.

|| **kaitaka** (kaitā·kă). *N.Z.* [Maori.] A flaxen cloak worn by Maoris.

1882 W. D. Hay *Brighter Britain!* II. 148 The kaitaka, a toga with a silky gloss and texture, was very highly esteemed. **1884** M. Martin *Our Maoris* vi. 84 The kaitaka, made of the finest flax and ornamented by a handsome border. **1949** P. Buck *Coming of Maori* (1950) II. v. 173 The plain cloaks with taniko borders divide into two classes. 1. *Parawai* or *Kaitaka*... 2. *Paepaeroa*.

|| **kajang** (kā·dʒæŋ). [Malay *kajang*.] Matting made from the leaves of palms or pandanus. Also **cajang, kadjan, kedgang.**

1821 J. Leyden tr. *Malay Annals* viii. 261 Raja Ahmed..flew a huge kite, as big as a cajang, (or tent fold covering) screen. **1839** T. J. Newbold *Pol. & Statistical Acct. Straits of Malacca* I. vi. 369 The Rhio gambier is often adulterated with sago, and rendered heavier by the Chinese purposely packing it in baskets lined with wet cajangs. **1845** J. Brooke *Jrnl.* 1 Feb. in H. Keppel *Expedition Borneo* (1846) II. vi. 131 The number of fallen trees..obliged us to quit our boat, and remove all the kajang covers. **1848** F. S. Marryat *Borneo* 63 The Malay war-boat..is built of timber at the lower part, the upper is of bamboo, rattan, and kedgang (the dried leaf of the Nepa palm). **1901** A. C. Haddon *Head-Hunters* 299 The roof was covered with *kajangs* from the boats. **1904** E. H. Giglioli tr. *Beccari's Wanderings Gt. Forests Borneo* 223 Sampans have generally a roofing of 'kadjan', a sort of matting made with palm or pandanus leaves. **1922** *Chambers's Jrnl.* 503/1 Under the *kajang* (native rush matting) covering. **1959** 'M. Derby' *Tigress* iv. 161 The *kajang*-maker, who wove waterproof strips of matting of screw-pine leaves for roofing boats and carts.

Kajar (kā·dʒā:). Also **Kadjar, Qajar.** [Pers. *ḳājār*.] A member of a northern Iranian people of Turcoman origin, who formed the ruling dynasty of Persia from 1794 to 1925. Also *attrib.*

1911 *Encycl. Brit.* XV. 651/2 Branches of the royal Afshâr and Kājār tribes of Turki descent. **1902** P. M. Sykes *Ten Thousand Miles in Persia* 121 Baluchis call all Persians Gajar, a corruption of Kâjâr, the reigning dynasty. **1932** A. T. Wilson *Persia* v. 103 Under the Qajars, roads and caravanserais were allowed to decay. **1966** R. Sarkis ... *Shah of Iran* xxiii. 232 Power sprang originally from their association with the Qajar monarchy. **1973** *Country Life* 31 May 1548/1 Those pictures are of a kind made in large numbers, mostly at Isfahan and Shiraz, throughout the Qajar period.

kajat, var. *KIAAT.

kajatenhout: see *KIAAT.

|| **kakahi** (kā·kāhi). *N.Z.* [Maori.] The freshwater mussel, *Hyridella menziesii*.

1921 H. Guthrie-Smith *Tutira* viii. 55 The shallows of the lake were paved with mussel-beds—kakahi. **1949** P. Buck *Coming of Maori* (1950) II. viii. 235 The rake was lowered from a canoe on to the shoals with beds of the fresh-water clam termed *kakahi*. **1962** *Post-Primary School Bull.* XV. i. 19 Kakahi are to be found embedded in the muddy and sandy bottoms of rivers and lakes. **1966** *Encycl. N.Z.* II. 616/1 These bivalve shellfish are..the kakahi of the Maoris.

|| **kakaho** (kā·kaho). *N.Z.* [Maori.] **a.** The dry flower-stalk of the *toetoe raupo* (*Arundo conspicua*), used for thatching and the making of mats and cloaks. **b.** From its use for cloaks, a general term for clothing.

1843 J. Earle *Narr. Residence N.Z.* (1966) 59 They were clothed in mats, called Ka-ka-hoos. **1936** *Punch* 14 Oct. 443/1 Pingao and Kakaho, which are used in tuku-tuku work. **1949** P. Buck *Coming of Maori* (1950) II. vi. 126 The thin stalks of kakaho were used..to make reeds but without cross stitch designs. *Ibid.* v. 166 The general term for clothing is *kakahu* or *kahu* but the different types of garments have received specific names. **1974** *Nat. Geographic* Aug. 209/1 Mrs. Emily Schuster, ... of women's work, showed me how girls put together a *kakahu*, the magnificent feathered cape for ceremonial occasions.

kaki. (Earlier and later examples.)

1727 J. G. Scheuchzer tr. *Kæmpfer's Hist. Japan* I. I. 116 There are three different sorts of Fig-trees growing in Japan. One is call'd Kaki, if otherwise it may be called a Fig-tree, it differing from it in several particulars. **1795** tr. *C. P. Thunberg's Trav. Europe, Afr. & Asia* (ed. 2) III. 61 Another cause [of diarrhœa] supervened, viz. the excessive eating of the fruit of the Kaki (*Diospyros kaki*) which was at this time ripe. *Ibid.* IV. 38 For the desert [sic], they have kaki-figs. **1892** F. T. Piggott *Garden of Japan* 43 The golden clusters of tiny flowers of *Diospyros kaki* give promise of a rich harvest of luscious fruit in the autumn—the Kaki loved of the Japanese. **1920** W. Popenoe *Man. Tropical & Subtropical Fruits* xii. 375 (*heading*) The Kaki or Japanese Persimmon. *Ibid.* 376 The kaki is a deciduous tree growing up to 40 feet in height. *Ibid.* 355 From Japan the kaki has been carried around the world. **1936** T. Nohara *True Face of Japan* v. 162 The fruiterer with apples, mandarines and kaki fruits. **1951** *Dict. Gardening* (R. Hort. Soc.) II. 1098/1 D[iospyros] Kaki. Kakee or Chinese Persimmon. *Ibid.* III. 1098/1 [see *Diospyros*] **1962** J. Ohwi *Flora Japan* ... *Diospyros kaki* Thunb....Kaki-no-ki.

Kakiemon (kā·ki-emǫn). [f. the name of Sakaida *Kakiemon*, a 17th-c. Japanese potter.] A Japanese porcelain first made by Kakiemon at Arita, characterized by asymmetrical designs, large areas of undecorated porcelain, and the use of iron red enamel, with blue, green, and yellow enamels as fuller; also, any porcelain in the style of Kakiemon, which was widely imitated in Europe. Also *attrib.*

1890 J. L. Bowes *Jap. Pott.* 171 (*heading*) Arita wares, Kakiyemon wares... A tea bowl. **1902** W. G. Gulland *Chinese Porc.* II. 321 The shapes and decoration..never were appreciated..by them with the exception of the Kakiemon porcelain. **1906** R. L. Hobson *Porcelain* xx. 187 The enamel painting was largely in the Kakiemon style, ..a pattern consisting of one or two birds..and a spray of bamboo or plum. **1932** W. A. Thorpe tr. *Schmidt's Porc. as Art* 20 The decoration which best kept the equivalence of white ground and coloured design was the type known as Kakiemon. **1965** Fisher & Savage in J. Wedgwood *Sel. Lett.* viii. 151 The Chelsea mark rarely appears on their extremely close copies of Japanese Kakiemon porcelain. **1970** *Oxf. Compan. Art* 610 For refinement of shape, material, and decoration these Kakiemon wares fully equal their Chinese counterparts. **1970** *Times* 11 Mar. 12/6 Ten years ago Meissen copies of Kakiemon wares fetched about ten times the Japanese originals.

|| **kakke** (kæ·ke). Also **kakké.** [Jap., f. *kyaku, kaku* leg + *ki, ke* illness, disease.] The Japanese name for beriberi.

1874 *Boston Med. & Surg. Jrnl.* XC. 361 We have received a late number of the *Japan Mail*, containing an interesting description..of a species of endemic disease, known as *kak-ke*, peculiar to the Islands of Japan. **1893** A. M. Bacon *Jap. Interior* ix. 153 Had I died of kakke the year before last, there would have been no help, would there? **1906** *Practitioner* Nov. 695 In her previous wars, Japan saw her armies practically prostrate with beri-beri or kakke. **1930** A. C. Reed *Trop. Med. in U.S.* vii. 311 This [belief] would relate wet or edematous beriberi..to the 'kakke' of the Labrador fisherman..and certain forms of ship beriberi. **1951** E. R. Whitmore in R. B. H. Gradwohl *Clin. Trop. Med.* lxi. 1333 Kakké is mentioned in a Chinese pamphlet of the second century B.C...; it is recorded as occurring in Japan in the ninth century.

kakoto·pia. [f. Kako- + Eu)topia, U)topia.] = *DYSTOPIA. Cf. CACOTOPIA.

1915 P. Geddes *Cities in Evolution* ii. 74 The material alternatives of real economics, which these obsessions of money economics have too long obfuscating, are broadly two, and each is towards realising an ideal, a Utopia. These are the paleotechnic and the neotechnic—Kakotopia and Eutopia respectively. **1970** *New Yorker* 10 Oct. 100/3, I use 'kakotopia'..as the opposite of 'utopia', to describe a misplanned and ugly urbanized technological kakotopia. **1970** L. Mumford *Pentagon of Power* 49 This nightmarish conclusion..has been a recurrent theme of later technological kakotopias.

kakur (kā·kuə). Also **kakar, karkur.** [prob. f. Hindi *kākar*.] The muntjak or barking deer, *Muntiacus muntjak*.

1876 A. A. Kinloch *Large Game Shooting* II. 26 The Kakur is one of the smallest Deer, not being much more than eighteen inches in height. **1887** J. M. Brown *Shikar Sk.* 254 The harsh roar of a kakur rang out close to us. **1925** A. G. Arbuthnot in L. Buttard *Big Game Hunting* 141 The flesh of kakur is excellent. **1960** J. Corbett *Man-Eaters of Kumaon* (ed. 2) 185, I went to sleep..a kakur started barking in the scrub jungle behind my tent. **1966** R. Perry *World of Tiger* iv. 56 Occasionally in India, and more frequently in Burma and Malaya, the little kakar, the muntjac or barking deer, is preyed upon.

Kalá, var. *KULLAH.

kala azar (kā·lă-, kæ·lă,ăzā·ɹ). *Path.* Also with hyphen. [Assamese, f. *kālā* black + *āzār* disease.] A febrile disease of tropical and subtropical regions caused by the protozoan *Leishmania donovani* and transmitted by sand flies of the genus *Phlebotomus*; usually associated with emaciation, enlargement of the spleen and liver, and often bronzing of the skin. Also known as visceral leishmaniasis.

1883 J. J. Clarke in *Ann. Sanitary Rep. Assam 1882* 36 As far back as 1869, the attention of administrative officers in Assam became directed to a peculiar disorder (called 'Kala Azar', the 'Black Disease', from the singular bronzing of the skin so often observed with it), the ravages of which decimated..numerous villages in the district of the Gáro Hills. **1908** L. Rogers *Fevers in Tropics* i. 31 Kala-azar is the epidemic manifestation of a fever, endemic in extensive areas of India, which has spread slowly for thirty years up the Assam valley. **1930** Rogers & Megaw *Trop. Med.* ii. 81 The name kala-azar ..was in use in Bengal and Assam long before the nature of the disease was known. **1932** Gaiger & Davies *Vet. Path. & Bacteriol.* xxiii. 346 Visceral leishmaniasis is known in human beings as kala-azar, and is found frequently in dogs.., especially on the shores of the Mediterranean. **1966** 'Han Suyin' *Mortal Flower* i. 35 The German pension..was run by Mrs Apelt, whose husband was dying of kala-azar acquired in Manchuria. **1970** Passmore & Robson *Compan. Med. Stud.* II. xix. 8 *Leishmania donovani* causes kala azar or visceral leishmaniasis in man and animals, and the disease occurs in certain parts of all continents except Australia.

kalanchoe (kæ:lănkō^u·i). [Fr. (M. Adanson *Familles des Plantes* (1763) II. 247), ult. f. Chinese.] A sub-shrub of the genus so called, belonging to the family Crassulaceæ, native to Africa and southern Asia, and distinguished by succulent leaves and red, pink, or white flowers borne in terminal panicles; often cultivated as a house or greenhouse plant.

1830 J. C. Loudon *Hortus Britannicus* 160 *Kalanchoe* Adan. Kalanchoe. (Chinese name). **1864** *Curtis's Bot. Mag.* XC. 5460 (*caption*) Large-flowered Kalanchoe. **1915** L. H. Bailey *Stand. Cycl. Hort.* III. 1732/2 Any number of kalanchoës may appear in the collections of fanciers. **1942** *E. Afr. Ann. 1941–2* 44/1 Kalanchoes, with their large heads of four-petalled flowers are very showy. **1951** *Dict. Gardening* (R. Hort. Soc.) III. 1098/1 Kalanchoes require a good, well-drained soil. **1970** M. Allan *Tom's Weeds* xix. 150 There were plenty of interesting plants.. [including] the pretty little kalanchoe which Rochford's are now popularizing.

kalashi, kalashy, kalas(s)i, varr. *KHALASI.

|| **kalashnikov** (kălæ·ʃnikǫf, kălā·ʃnikǫf). [Russ.] The name of a type of rifle or submachine gun made in the U.S.S.R. Also *attrib.*

1970 *N.Y. Times* 30 Oct. 41 A ragtag group of *fedayeen* bearing *kalashnikovs*, hand grenades and often Pepsi-Cola bottles, swarms around the headquarters area. **1971** E. Luttwak *Dict. Mod. War* 19/2 AK-47 (Avtomat Kalashnikov). Soviet rifle (includes AKM and RPK). **1972** *Times* 12 Jan. 10 They consist of forays across the border by from 40 to 100 men armed with Russian mortars, rockets, recoilless guns and kalashnikov automatics. **1973** *Times* 11 Apr. 1/8 He ran to get his *kalashnikov* (a Russian assault weapon) but when he returned, the Israelis had burst through the door.

kale. Add: **2*.** *N. Amer. slang.* Money.

1912 J. Sandilands *Western Canad. Dict.* 26 Kale, money, or wealth. **1922** [see *funfest* s.v. *fun sb.* 3 b]. **1926** *Flynn's* 16 Jan. 638/1 The kale is cut up an th' biggest corner goes to th' brains. **1927** *Daily Express* 23 Sept. 1 Enough 'kale' (prize-fighters' name for money) has been received..to assure the promoters a profit of approximately £100,000. **1946** B. Treadwell *Big Bk. of Swing* 124/2 *Kale*, paper money.

kalei·doscope, *v.* [f. the sb.] To present the appearance of a brightly coloured and constantly changing pattern; to cause to come together or coalesce with pleasing results. Hence **kalei·doscoping** *ppl. a.* and *vbl. sb.*

1891 *Daily News* 5 Mar. 5/3 The spectators in the gallery cheered heartily when some particularly effective kaleidoscoping of colours happened amongst the dancers on the floor below. **1894** *Ibid.* 1 Feb. 5/1 If the ladies and gentlemen so industriously kaleidoscoping their new frocks, they might do something better on these carnival nights than play at devils and clowns. **1920** *Literature* 14 July 25/1 In 'Isis'..Villiers kaleidoscoped together in his memory and imagination what he had read in many Oriental and medieval books. **1933** *Discovery* July 218/2 The sitting-rooms, parlour, drawing-room, morning room, study, library, ballroom and so on have all been kaleidoscoped into the living room. **1971** *Guardian* 4 Jan. 9/2 These days of kaleidoscoping time.

kalgan (kæ·lgăn). [f. *Kalgan*, name of the capital of Chahar Province, China.] Used *attrib.* or *absol.* to designate a fur obtained from the kalgan lamb.

[**1930** M. Bachrach *Fur* iii. 35 Kalgan in Chihli, northern China, is a junction city on the Kiachta–Peiping route, and furs offered from that city are taken in the districts just mentioned.] **1960** *Guardian* 22 Apr. 8/3 Furs..never seen in London before. Among them were..Kalgan lamb gilt. **1970** *Ibid.* 18 Nov. 13/3 A wild mottled grey suede coat with snowy white Kalgan. **1972** *Times* 28 July 10/4 (*caption*) Tank top and battle jacket lined in white kalgan lamb.

Kali² (kā·li). *Hinduism.* Also **Cali.** [Skr. *kālī* fem. of *kāla* black, dark; also taken as fem. of *Kāla* time (as destroyer), one of the names of Siva.] The name given to the Hindu mother-goddess Devi, consort of Siva, in her most terrible form as goddess of destruction and death, when she is depicted as black-skinned, smeared with blood, and wearing a necklace of skulls and a girdle of snakes.

1798 W. C. Blaquire tr. in *Asiatick Researches* V. 369 Let the sacrificer say *Hrang, hring.* Ca'li, Ca'li, O horrid-toothed goddess; eat, cut, destroy all the malignant. **1810** E. Moor *Hindu Pantheon* 145 Of the many names of the goddess..those of Parvati, Bhavani, Durga, Kali, and Devi..are the most common. **1832** C. Coleman *Mythol. Hindus* vii. 92 Kali is also called the goddess of cemeteries, under which form she is described dancing with the infant Siva in her arms. **1882** W. J. Wilkins *Hindu Mythol.* vii. 257 Kāli (the black woman), or, as she is more commonly called Kāli Mā, the black mother, with the aid of Chandi, slew Raktavija, the principal leader of the giant's army. **1917** A. Coomaraswamy in Coomaraswamy & Duggirala tr. *Mirror of Gesture* 8 Kālī..dances in the burning ground..to signify the heart of the devotee made empty by renunciation. **1933** E. A. Payne *Śāktas* ii. 13 Hibiscus flowers have become a favourite present to Kālī, probably because they are the colour of blood. **1952** S. Selvon *Brighter Sun* vii. 130 Don't know Indian people haveam own god?..To pray Kali for rain? **1971** 'G. Black' *Time for Pirates* i. 9 They were Moslems, which meant that there shouldn't have been a place in their faith for a worship of Kali, the Destroyer.

kalicine (kæ·lisīn). *Min.* [a. F. *kalicine* (F. Pisani 1865, in *Compt. Rend.* LX. 919), irreg. f. *kali*, mod.L. *kalium* (see Kali 2), old names for potassium + *c* + *-ine* -INE⁵.] Potassium bicarbonate, $KHCO_3$.

1892 E. S. Dana *Dana's Syst. Min.* (ed. 6) 294 Kalicine. ..Potassium bicarbonate. **1922** J. W. Mellor *Comprehensive Treat. Inorg. & Theoret. Chem.* II. xx. 774 The corresponding potassium hydrocarbonate, $KHCO_3$, was reported by F. Pisani to occur at Chypis (Canton Wallis) as a mineral which he called kalicine or kalicinite. **1965** *Chem. Abstr.* LXIII. 1514 Investigations on various blast furnaces showed the destructive effects of alkali metals, Zn, Pb, and C on the refractory shaft lining... Alkali metal and Zn compds...form a no. of minerals, such as kalicine, kalsilite,..and zincite. **1968** I. Kostov *Mineral.* 530 Kalicine and teschemacherite occur as finely crystalline white masses.

kalicinite (kæ̌li·sinəit). *Min.* [f. prec. + -ITE¹.] = prec.

1922 *Mineral. Mag.* XIX. 343 Kalicinite...Variant of Kalicine..for monosymmetric $HKCO_3$. **1971** *Mineral. Rec.* II. 130/1 Buetschliite, kalicinite, carbonaceous material, and quartz are constituents of fused wood-ash clinkers formed as a result of the burning of a dead, but still standing chestnut-oak tree near Long Shop, Montgomery County, Virginia.

|| **kalimba** (kăli·mbă). [Bantu.] A musical instrument played with the thumbs, consisting of metal strips along a small, hollow piece of wood.

1968 *Sat. Rev.* (U.S.) 26 Oct. 89 Buckley blends unusual 'noises' into the music: clinks, kalimba, calliope, gunfire, and an odd assortment of rhythm instruments. **1971** *Ink* 12 June 19/2 Toni Brown wrote most of the songs, sings lead on a couple of them, and plays keyboards..and something called a kalimba.

kalistrontite (kælistrǫ·ntəit). *Min.* [ad. Russ. *kalistrontsit* (M. L. Voronova 1962, in *Zap. Vsesoyuz. Mineral. Obshch.* XCI. 712), f. *káli-ĭ* potassium (cogn. w. Kali) + *strónts-iĭ* Strontium + *-it* -ITE¹.] A sulphate of potassium and strontium, $K_2Sr(SO_4)_2$, found as colourless hexagonal crystals.

1963 *Mineral. Abstr.* XVI. 183/2 A new potassium and strontium sulphate found in saline anhydrite rocks from a borehole near the village of Alshtan, Bashkir, A.S.S.R., is named kalistrontite. **1968** I. Kostov *Mineral.* 504 Kalistrontite is trigonal, isostructural with palmierite $K_2Pb(SO_4)_2$.

kallidin (kæ·lĭdin). *Biochem.* [a. G. *kallidin* (E. Werle 1948, in *Angew. Chem.* LX. A. 53), f. G. *kall-ikrein* *KALLIKREIN + *pept-id* *PEPTIDE: see -IN¹.] † **a.** A supposed hypotensive peptide, released from a globulin by the enzyme kallikrein, which stimulates the uterus and intestine; later shown to be a mixture of two such peptides, bradykinin and kallidin (sense *b). *Obs.*

1950 *Chem. Abstr.* XLIV. 5476 Kallidin (previously called DK substance) is a low-mol.-wt. peptide. **1959** *New Scientist* 16 Apr. 856/1 When its properties were studied, bradykinin showed itself to be extremely similar to the peptide kallidin, whose existence was first demonstrated in Germany between the wars.

b. (Also *kallidin II*.) The peptide having the same sequence of constituent amino-acids as bradykinin except for an additional lysergic acid residue at the hydrogen end.

1961 Pierce & Webster in *Biochem. & Biophys. Research Communications* V. 353 The present report describes the isolation of two kallidins from the incubation of human urinary kallikrein with acid-treated human plasma. *Ibid.* 354 Two peaks of activity [on a chromatograph] were obtained. These were designated kallidins I and II. *Ibid.* 356 These data indicate that kallidin I is identical with bradykinin and that kallidin II is a decapeptide. **1965** *Jrnl. Physiol.* CLXXVI. 1 Bradykinin and kallidin II are respectively, a nona- and a deca-peptide, which occur in man and other species... Kallidin II (hereafter referred to as kallidin), which is converted from kallidinogen by the enzyme kallikrein, appears to be the immediate precursor of bradykinin. **1970** Passmore & Robson *Compan. Med. Stud.* II. xvii. 1 The name kallidin now refers specifically to lysyl-bradykinin.

kallikrein (kæ·likri,in, -krīn). *Biochem.* Also † **callicrein.** [a. G. *kallikrein* (H. Kraut et al. 1930, in *Zeitschr. f. physiol. Chem.* CLXXXIX. 99), f. Gr. καλλίκρε-ας sweetbread (f. καλλι-, comb. form of κάλλος beauty + κρέας flesh): see -IN¹.] **a.** An enzyme found in the human pancreas and elsewhere in the body, which releases kallidin from a plasma precursor and has been used therapeutically as a vasodilator and hypotensive agent.

1930 *Chem. Abstr.* XXIV. 4541 The name *callicrein*, taken from a Greek synonym for pancreas, is proposed for this circulatory hormone. **1967** *Martindale's Extra Pharmacopoeia* (ed. 25) 1440 Kallikrein is a dilator of the peripheral blood-vessels... It has been used in the treatment of vasospastic circulatory disorders, including.. chilblains. **1956** Robson & Keele *Recent Adv. Pharmacol.* (ed. 2) xiv. 480 Kallidin is formed by the proteolytic action of a widely distributed (in saliva, serum, urine, pancreas, etc.) enzyme called Kallikrein, which under the name Padutin has been used therapeutically.

b. Any enzyme which liberates a kinin from a protein.

1966 M. E. Webster in E. G. Erdös *Hypotensive Peptides* 650 A kallikrein is defined as an endogenous enzyme which rapidly and specifically liberates a kinin from kininogen... In view of the many known chemical and physical differences between the kallikreins, each kallikrein should be identified by species and source, e.g. hog pancreatic kallikrein. **1969** *Physiol. Rev.* XLIX. 510 The kallikreins (kininogenases) are a group of enzymes... They have marked actions on blood vessels and smooth muscles in vivo that..are due to the rapid enzymatic cleavage of a specific substrate, an α_2 globulin present in plasma and lymph.

ka·llipyg. *rare.* [Cf. Callipygian *a.*] A person with finely developed buttocks.

a **1913** F. Rolfe *Desire & Pursuit of Whole* (1934) xvii. 178 Some bulgy kallipyg with swung skirts and cardboard waist.

kallitype (kæ·litəip). *Photogr.* [f. Gr. καλλι-, comb. form of κάλλος beauty + -TYPE.] A disused photographic printing-process using either paper coated with a ferric salt and silver nitrate which is developed in a solution of borax and Rochelle salt, or paper sensitized with a ferric salt and developed in silver nitrate solution.

1890 *Brit. Jrnl. Photogr.* 23 May 335/2 We may here state that we have received..two specimens of kallitype. **1941** *Ibid.* LXXXVIII. 445 Owing to the present scarcity of printing material it seemed worth while canvassing some of the old abandoned processes of photographic printing; the most probably useful seemed to be Kallitype. This was widely used before the days of P.O.P. and bromide. **1965** J. Kosar *Light-Sensitive Syst.* i. 39 In the past, silver-iron printing papers were employed also for producing continuous-tone prints. This process was named 'Kallitype'; its attractive feature was the possibility of toning the prints with toning baths.

kalloscope (kæ·loskō^up). *Disused.* [Irreg. f. Gr. κάλλος beauty + -SCOPE.] A type of Stereoscope.

1901 *Daily Chron.* 10 Aug. 5/6 The suppression of what are known as kallascopes [*sic*], stereoscopes and similar machines. **1902** *Ibid.* 3 May 4/5 The 'Automatic Kalloscope'..which attracted the attention of the police.

kalmia (kæ·lmiă). [mod.L. (Linnæus *Nova Plantarum Genera* (1751) no. 1079), f. the name of Pehr *Kalm* (1716–1779), Swedish botanist + -IA¹.] An evergreen shrub of the genus so called, belonging to the family Ericaceæ, native to North America, and bearing clusters of pink or white flowers; also called mountain laurel or calico-bush.

1765 J. Bartram *Diary* 6 Sept. in *Trans. Amer. Philos. Soc.* (1942) XXXIII. 24/1 Here grows most northward trees here [*sic*] except..white pine & our 3 calmias. **1785** H. Marshall *Arbustrum Americanum* 72 Narrow leaved Kalmia..delights in moist or swampy places. **1838** *Boston Weekly Mag.* 22 Sept. 17/2 The *Rhodora* is followed in succession by the *Honeysuckles*, the *Kalmias* or *Laurels*, the *Azalea*. **1878** R. T. Cooke *Happy Dodd* 347 With.. glittering clusters of Kalmia leaves..she adorned all the rooms. **1900** L. H. Bailey *Cycl. Amer. Hort.* II. 854/1 The Kalmias thrive well in a sandy, peaty or loamy soil. **1955** *Sci. News Let.* 21 May 334/2 Kalmia, or mountain laurel, is a most attractive plant at any time, for its dark shining leaves are evergreen...Its clusters of closed starflowers, pink..are things for poets to write sonnets about.

Kalmuck (kæ·lmɒk). Also 8 **-muc,** 9 **-muk, -myk,** 7–9 **Calmuc(k.** [Russ. *kalmy̆k*.] **1. a.** A member of a Mongolian people living on the north-west shores of the Caspian Sea. Also *attrib.* or as *adj.*

1613 Purchas *Pilgrimage* iv. xiii. 358 Master Ienkinson mentioneth a Nation liuing among the Tartars, called Kings; which are also Gentiles, as are also the Kirgessen.. and the Colmackes, which worship the sunne. **1617** *Ibid.* (ed. 3) iv. xv. 482 There are some..which are not Mahumetans, nor shaue their haire of their heads after the Tartarian manner; and therefore they call them Calmuck or Pagans. **1757** J. Dyer *Fleece* iv. 126 The Cossac there, The Calmuc, and Mungalian, round the bales In crouds resort. **1783** W. Tooke tr. *Georgi's Russia* IV. 121 The dwellings of the heathenish and christian Kalmucs. *Ibid.* 125 The Kalmuc hords on the Volga. **1822** Byron *Don Juan* vii. lviii. 273 Suwarrow, who was standing in his shirt Before a company of Calmucks. **1882** R. L. Stevenson *New Arabian Nights* (ed. 2) II. 66 His broken nose and high cheekbones gave him somewhat the air of a Kalmuck. **1902** *Encycl. Brit.* XXX. 8/2 Kalmyk, or Kalmuck Steppe, a territory or reservation belonging to the Kalmyks. **1903** Ld. R. Gower *Rec. & Reminisc.* 430 A man..with a rather Kalmuk-featured face and white curly hair. **1963** V. Nabokov *Gift* ii. 114 The compact, sturdy Kalmuk ponies walk in single file forming echelons. **1972** J. Poyer *Chinese Agenda* (1973) xiii. 180 A caravan of Kalmuck traders. *Ibid.* 181 These Kalmucks are strictly traders.

b. The language of this people, belonging to the Ural-Altaic group.

1883 *Encycl. Brit.* XVI. 750/2 The Kalmuk and East Mongolian dialects do not differ much...In Kalmuk..the guttural can only be traced through the lengthening of the syllable. **1947** [see *Buriat].

2. (With small initial letter.) A kind of shaggy cloth, resembling bearskin (see also quot. 1940).

1860 S. Jubb *Hist. Shoddy-Trade* 40 A cloth called calmucks..has..replaced 'short ends'. **1940** *Chambers's Techn. Dict.* 128/1 *Calmuc*, a coarse type of wool, from the Khirghiz district, Central Asia.

Hence **Kalmu·ckian** *a.*

1727 J. G. Scheuchzer tr. *Kæmpfer's Hist. Japan* I. I. vi. 90 The Prince of the Calmuckian Tartars.

|| **kalokagathia** (kæ:lokægæ·þiă, kē^i:lo-). [Gr. καλοκαγαθία, f. καλοκάγαθος = καλὸς κἀγαθός for καλὸς καὶ ἀγαθός beautiful and good (the perfect character).] Nobility and goodness of character.

1921 tr. *W. Rathenau's New Soc.* x. 102 The Greeks.. adopted as their highest law..that impulse of the will which they called *Kalokagathia*. **1930** N. Mitchison in *Time & Tide* 14 June 773/1 The formal kalokagathia of that incredible time.

kalsilite (kæ·lsiləit). *Min.* [f. the letters *KAlSi* in its chemical formula + -LITE.] A rare silicate of potassium and aluminium, $KAlSiO_4$, that is chemically and physically similar to nepheline and occurs in some potassium-rich lavas.

1942 A. Holmes in *Mineral. Mag.* XXVI. 198 Mr. Bannister's results, supplemented by micro-chemical analyses made by Dr. M. H. Hey, are recorded in the communication which follows this paper. The new data indicate that the mineral is..a hitherto unrecognized polymorph of $KAlSiO_4$ for which the appropriately mnemonic name kalsilite is proposed. **1942** Bannister & Hey in *Ibid.* 221 It is proposed to name the new mineral kalsilite after its composition $KAlSiO_4$. **1967** *Nature* 24 June 1322/1 Kalsilite ($KAlSiO_4$) occurs as a major constituent of a large sedimentary xenolith metamorphosed to the sanidinite facies within the gabbro of Brome Mountain, Quebec. Kalsilite is extremely rare and has previously been found only as an igneous mineral in three volcanic areas. **1972** M. H. Battey *Mineral. for Students* II. 292/2 The ratio of K to Na in nepheline varies and there is a series towards kalsilite $KAlSiO_4$ which is a rare mineral found in certain potassic lavas on the Uganda–Congo border.

kalsomine (kæ·lsoməin). For 'erron.' read 'orig.' Add examples. Also as *v. trans.* and *intr.*, to whitewash with kalsomine. Hence **ka·lsomined** *ppl. a.*, **ka·lsominer, ka·lsomining** *vbl. sb.*

1840 *Athenæum* 20 June 502 Kalsomine. **1858** W. A. Butler *Two Millions* 42 From lowest basement up to topmost attic, The whole was gorgeous, glaring and prismatic; Pannelled and kalsomined. **1883** *Harper's Mag.* Mar. 503/2 Paint and kalsomine can not be counted upon. **1884** H. G. Carleton *Thompson St. Poker Club* 20 An extensive kalsomining contract. **1888** *Pall Mall Gaz.* 3 Mar. 11/1 Over face, arms, neck,..and bosom she spreads a coat of liquid white... In plain words, she, as it were, kalsomines herself. **1891** H. C. Bunner *Zadoc Pine* 166 White kalsomined bedrooms. **1893** K. A. Sanborn *Truthful Woman S. California* 81 Those who feel an unctuous joy in painting the lily, kalsomining the calla, and adding perfumes to the violet. **1904** 'O. Henry' *Cabbages & Kings* xiv. 248 Let me kalsomine you a little mental sketch to consider. **1916** H. L. Wilson *Somewhere in Red Gap* iv. 128 He was a painter and grainer and kalsominer and paperhanger. *Ibid.* 135 He..began to paper and paint and grain and kalsomine. **1924** *Spectator* 1 Nov. 640 These walls can be whitewashed or covered with vines on the outside and kalsomined within. **1936** L. C. Douglas *White Banners* iii. 57 The specifications for improvements: a new sink, repair of the hot-water machine, kalsomining in the kitchen. **1945** B. MacDonald *Egg & I* (1946) 49 We laid new floors; put in windows; kalsomined the walls.

Kalydor (kæ·lidō^ə1). Also **kalydor.** (The proprietary name of) a type of skin tonic of which almond oil forms the basis.

1824 *Advt.* in C. W. Cunnington *Feminine Attitudes 19th Cent.* (1935) 309 The Kalydor..a never-failing specific for all cutaneous deformities. **1828** tr. *L. af Holberg's Journey to World under Ground* 239 The fourth attended with a bottle of kalydor, for the improvement of the complexion. **1861** [see Amandin(e)]. **1876** *Trade Marks Jrnl.* 13 Dec. 993 Rowland's Kalydor for improving and beautifying the complexion. Eradicates all cutaneous eruptions... Henry Edward Rowland and George William Rowland, trading as Alexander Rowland and Sons,.. Hatton Gardens, Middlesex. **1901** C'tess C. *Beauty's Aids* 242 (Advt.), Rowland's Kalydor..successfully opposes the attacks of the hot summer sun or damp chilly weather..allays all smarting irritations..removes freckles, tan, sunburn, prickly heat..imparts a luxuriant beauty to the complexion, and arrays the neck, hands, and arms in matchless whiteness. **1907** *Yesterday's Shopping* (1969) 539/2 Rowland's Kalydor bot. 2/0. **1939–40** *Army & Navy Stores Catal.* 438/3 Rowland's Kalydor—bot. 2/6.

kamachili (kāmātʃi·li). Also **camanchile, guamachil, kamachile, kamachilis.** [Tagalog: see quot. 1923.] A tree, *Pithecolobium dulce*, of the family Leguminosæ, native to tropical America and naturalized in the Philippines, having edible pods and bark that yields a yellow dye. Also *attrib.*

1866 Lindley & Moore *Treas. Bot.* II. 898/2 *P[ithecolobium] dulce*..produces cylindrical irregularly swollen pods, curled at the top, containing a sweet edible pulp, which the Mexicans, who call the tree Guamachil, boil and eat. **1903** E. D. Merrill *Dict. Plant Names Philippine Islands* 44 Camachilis, T[agalog]. Pithecolobium dulce Benth. **1915** *Philippine Jrnl. Sci.* A. X. 353 The price of air-dried camanchile bark has risen. **1923** E. D. Merrill *Enumeration Philippine Flowering Plants* II. 243 *Pithecolobium dulce*... Throughout the Philippines... Introduced from Mexico; now pantropic. Local names:.. kamachili (Tag.); kamachilis (Tag.)..all corruptions or modifications of the Aztec *kwamochitl*. **1937** *Nature* 16 Oct. 687/1 Several trees growing in the Philippine Islands yield liquors suitable for tanning purposes. The betel nut, *Areca Catechu*..and kamachile, *Pithecolobium dulce*, are.. the most important trees. **1954** W. H. Brown *Useful Plants of Philippines* II. 154 Camanchile bark is used almost exclusively by Filipino tanners.

kamagraph (kæ·măgrɑf). [f. next.] **a.** A painting reproduced by kamagraphy. **b.** A printing press which produces kamagraphs.

1967 *Time* 23 June 49 Each kamagraph looks as though the artist had painted it by hand. **1968** *Collier's Encycl. Year Book* 131 Max Ernst, the well-known dada and surrealist painter; Edouard Pignon, a French abstractionist; and the late René Magritte, the extraordinary Belgian surrealist who died this year, have all executed special work for the kamagraph.

Hence **kama·grapher, kamagra·phic** *a.*

1970 *Britannica Bk. of Year* (U.S.) 798/2 *Kamagraphy*, a process for making multiple copies of a painting produced by an artist on a specially treated canvas in which the copies retain the texture of the brushstrokes of the original but the original is destroyed in the process; *kamagraph, kamagrapher*; *kamagraphic, adj.*

kamagraphy (kămæ·grăfi). [ad. F. *kamagraphie*.] A process for making copies of original paintings, using a special press and treated canvas, which reproduces exactly the colour and texture of the brushstrokes.

1967 *Time* 23 June 49 Kamagraphy faithfully produces 250 perfect copies of a painting on a special press, destroying the original in the process. **1968** *Collier's Encycl. Year Book* 131 A French process called kamagraphy has been developed by engineer André Cocard, with the backing of art collector and vintner Alexis Lichine. **1970** [see prec.].

kamahi (kā·māhī, kā·məi). *N.Z.* Also **kaamahi, karmahi, karmai.** [Maori.] A forest tree, *Weinmannia racemosa*, belonging to the family Cunoniaceæ, and bearing racemes of small, cream flowers.

1867 J. D. Hooker *Handbk. N.Z. Flora* II. 765/2 Karmahi, *Hector*. *Weinmannia silvicola* and *racemosa*. **1868** J. Hector in *Trans. N.Z. Inst.* I. iii. Essay 4 The flat land and low spurs are covered with the common species of Pines and Birch, such as Rimu, Totara, Weinmannia (Karmahi), and Fagus (Tawai). **1868** J. Buchanan in *Ibid.* Essay 37 Towai, or Karmai (*Weinmannia racemosa*). A beautiful large tree, especially when in flower. **1899** T. Kirk *Students' Flora N.Z.* 140 *W[einmannia] racemosa*, Linn. f. A shrub or large tree, often from 70 ft.–90 ft. high... *Kamahi*. **1953** *Landfall* VII. 122 He had played his violin under a kamahi tree. **1963** B. Pearson *Coal Flat* iv. 59 In a little clearing in the kaamahi and fuchsia and young bush..was Mrs Seldom's little house. **1966** *Weekly News* (Auckland) 22 June 44 The kamahi present on this charming little hill is a hardy tree.

Kamakura (kā·măkū^ərā). The name of a town in central Japan, used *attrib.* to designate the art of the period (1192–1333) during which Kamakura was the seat of government.

[**1890** B. H. Chamberlain *Things Japanese* 227 The grandest example of..colossal bronze-casting is the *Daibutsu* (literally, 'great Buddha') at Kamakura.] **1902** F. Brinkley *Oriental Series: Japan* VII. iii. 110 Nearly four hundred years may be regarded as the Kamakura epoch from the point of view of the sculptor's art, and may also be regarded..as the final era of vigorous originality in religious sculpture. **1912** E. F. Fenollosa *Epochs Chinese & Jap. Art* I. ix. 197 Several phases of Kamakura art..went on parallel to the main stream of secular makimono painting. **1952** L. Warner *Enduring Art Japan* iv. 52 Kamakura scrolls seemed to epitomise the other changes. **1970** *Oxf. Compan. Art* 606/2 Works of the spiritual stature of the Tempyō sculptures or the Kamakura portraits were isolated incidents rarely attained.

Kamares (kămā·rīz). Also **Kamarais.** [Gr. Καμάραις, name of a cave-sanctuary of the Minoans on Mt. Ida in Crete, where the pottery was first found.] A type of Minoan pottery from the Middle Bronze Age, characterized by the use of red, white, and yellow ornaments on a black ground, depicting abstract or stylized plant designs. Also *attrib.*

1895 *Proc. Soc. Antiquaries London* XV. 356 The red.. on very thin black-glazed ware, is exactly of the Kamárais tint, while the drawing has the Kamárais touch. *Ibid.*, We may consider that the Kamárais pottery began at least as early as 2300 B.C. **1902** *Encycl. Brit.* XXXI. 56/1 This ware, known as 'Kamáres', from a cave near a village on the south-east of Mount Ida. **1948** A. Lane *Greek Pott.* iv. 22 The 'Kamares' style named after a cave in Crete where many examples were found. **1949** W. F. Albright *Archaeol. of Palestine* v. 93 There was nothing in Palestine like the delicate Kamares ware of Middle Bronze Crete, which was in great demand in Egypt. **1960** T. Burton-Brown *Early Medit. Migrations* i. 1 Certain classes of pottery..were believed to indicate direct contacts between the peoples of Egypt and those of the Aegaean area. Amongst these are sherds of polychrome painted fabrics, some of which may be either Kamares wares from Crete, or related wares. **1970** *Oxf. Compan. Art* 725/2 The 'egg-shell' Kamares ware..bears some of the very best of Minoan decoration, with some floral motifs possibly derived from geometric forms rather than copied from nature.

kamarezite (kămā·rəzəit). *Min.* [ad. G. *kamarezit* (K. H. E. G. Busz 1893, in *Verh. d. naturhist. Vereins* L. *Sitzungsber. d. naturwiss. Sekt.* 84) f. *Kamareza*, name of the place in Greece where it was first found: see -ITE¹.] A grass-green basic sulphate of copper, originally thought to be $Cu_3(SO_4)(OH)_4.6H_2O$, but later found to be identical with brochantite, $Cu_4SO_4(OH)_6$.

1895 *Jrnl. Chem. Soc.* LXVIII. II. 506 Kamarezite... This new mineral from Kamareza, Laurium, Greece, is grass-green and shows a crystalline structure. **1951** C. Palache et al. *Dana's Syst. Min.* (ed. 7) II. xxxi. 588 *Kamarezite*...Orthorhombic (?). In minute crystals. **1965** *Amer. Mineral.* L. 1456 The optical, physical, crystallographic, and chemical evidence that have been presented ..is (*sic*) strong support for the contention that the mineral described as kamarezite by Busz (1893, 1895) is brochantite. *Ibid.* 1457 We recommended that kamarezite be removed from the list of accepted mineral species; this recommendation has been accepted by..the Commission on New Minerals and Mineral Names.

kamassi (kama·si). *S. Afr.* Also **camassie, kamasse, kamassie.** [Afrikaans *kammassie*, f. native name.] A South African evergreen tree, *Gonioma kamassi*, of the family Apocynaceæ, or its hard yellow wood. Also *attrib.*

1793 tr. *C. P. Thunberg's Trav. Europe, Afr., & Asia* II. 110 Camassie wood (Camassie-hout), is merely a shrub, and consequently produces small pieces only, which serve for veneering. **1814** R. B. Fisher *Importance of Cape Good Hope* 84 The kamasse, a sort of bark, being the rhind or shavings of the tree of that name. **1907** T. R. Sim *Forests & Forest Flora Cape Good Hope* 323 The Knysna export under the name of Boxwood was all, or mostly, Kamassi-wood, without any Boxwood. **1924** Record & Mell *Timbers Tropical Amer.* 506 One of the two South African woods known to the world trade is the so-called Knysna or Kamassi boxwood. **1935** L. Chalk et al. *Forest Trees & Timbers Brit. Empire* III. 15 Kamassi attains about 40 ft. in height and generally a maximum girth of 2 to 3 ft. *Ibid.* 17 Kamassi is one of the two timbers exported from South Africa regularly in small quantities. **1951** *Dict. Gardening* (R. Hort. Soc.) II. 908/2 *G[onioma] Kamassi*. Evergreen shrub... Yields the hard Kamassi wood of S. Africa. **1973** Palmer & Pitman *Trees S. Afr.* III. 1905 Kamassi occurs in numbers in the Midland forests of the Cape. *Ibid.* 1906 It is exported in small quantities, as is Cape box for which kamassie is often mistaken.

Kama Sutra (kā:mă sū·tră). Also (as one word) **Kamasutra.** [Skr. *kāma* love + *sūtra* (see Sutra).] The title of an ancient Sanskrit treatise on the art of love and sexual techniques; hence used allusively.

1883 Burton & Arbuthnot *Kama Sutra of Vatsyayana* 3 In the present publication it is proposed to give a complete translation of what is considered the standard work on love in Sanscrit literature, and which is called the 'Vatsyayana Kama Sutra', or Aphorisms on Love, by Vatsyayana. *Ibid.* iii. 24 Man should study the Kama Sutra and the arts and sciences subordinate thereto... Even young maids should study this Kama Sutra. **1915** *Encycl. Relig. & Ethics* VIII. 450/1 The *Kāmasūtra* permits love matches generally. **1960** 'S. Harvester' *Chinese Hammer* xix. 172 A large illustration of many-armed embrace whose postures owed more to the Hindu *Kamasutra* than to a depiction of a Tibetan *Dukor* with his bejewelled *shakti*. **1961** C. Willock *Death in Covert* iii. 73 Individual pillars were decorated with some rather questionable designs. Whynne..said: 'Don't worry. We shall be building an outdoor Espresso Bar..round the most Kama Sutra of those.' **1964** *Listener* 26 Nov. 848/3 For its delineation of the almost innumerable techniques for handling a locomotive, *The Train* must be accounted the *Kama Sutra* of the permanent way. **1970** J. Boland *Big Job* xii. 100 The things I've taught that girl. Proper little Kama Sutra, I am. **1972** R. Quilty *Tenth Session* 93 He'll be bursting in any minute. Kama Sutra lips—ready for the last waltz.

Kamba (kæ·mbă). [Bantu.] A Bantu-speaking people of central Kenya, related ethnically to the Kikuyu; a member of this people; their language.

1885 J. T. Last *Gram. Kamba Lang.* 14 Probably the cardinal numbers are seldom used by the Kambas beyond *iyana*, 100. *Ibid.* 31 The simplest form of the verb is used in Kamba, as in Swahili and English, for the second person singular of the imperative. **1938** W. M. Hailey *Afr. Survey* ii. 24 Among the Kikuyu and Kamba, a whole age-grade of boys is initiated at one time. **1959** B. Bernardi *Mugwe* i. 2 They share this tradition with the other bordering Bantu peoples, the Kikuyu,..the Mbere and the Kamba. **1963** *Times* 31 May 10/2 Mr. Sagini is a Kisii, Mr. Mwanyumba (Works) is a Taita and Mr. Mwendwa (Labour) is a Kamba.

Kamba, var. *Khamba.

Kamchadal (kæ·mtʃădæl). Also **Kamtchat(ka)dale, Kamt(s)chadale.** [Russ.] **a.** A member of a Mongoloid people inhabiting the Kamchatka peninsula on the Pacific coast of Siberia. **b.** The language of this people.

1764 J. Grieve tr. *Krasheninnikov's Hist. Kamtschatka* III. iii. 175 Thus it appears likely, that the Kamtschadales lived formerly in Mungalia beyond the river Amur, and made one people with the Mungals. **1790** tr. *J. B. B. de Lesseps's Trav. Kamtschatka* I. 134 It is observed..that Kamtschadales of either sex, do not live longer than Russians. **1824** J. D. Cochrane *Narr. Journey Russia & Siberian Tartary* (ed. 2) II. xi. 43 The number of real Kamtchatdales who retain their ancient usages is small. **1855** *Eng. Cycl.: Nat. Hist.* III. 557 The Kalan of the Kamtchatkadales. **1871** E. B. Tylor *Primitive Culture* I. iii. 98 This spiritualistic belief among the Kamchadals is, no doubt, the key to their superstition as to rescuing drowning men. **1909** *Westm. Gaz.* 28 Aug. 10/2 The Kamchadals, one of the numerous tribes inhabiting the North of Siberia. **1933** L. Bloomfield *Lang.* iv. 70 The Hyperborean family..consists of Chukchee..Koryak..and Kamchadal. **1937** R. H. Lowie *Hist. Ethnol. Theory* ii. 15 The Kamchadal cooked meat in wooden troughs filled with water into which they threw heated rocks. **1956** [see *Gilyak].

Kamchatkan (kæmtʃæ·tkăn), *sb.* and *a.* Also **Kamschatkan, Kamskatchan, Kamt(s)chatkan.** [f. *Kamchatka* (place-name); see -AN.] **A.** *sb.* A person from the peninsula of Kamchatka in Siberia; often = prec. **B.** *adj.* Of or pertaining to Kamchatka.

1797 *Encycl. Brit.* IX. 429/2 The southern Kamtchatkans commonly build their villages in thick woods. **1833** W. L. Mackenzie *Sk. Canada & U.S.* 25 Even the wild Greenlander, the grim Kamschatkan, and the desolate Siberian love their barren wastes. **1865** Dickens *Mut. Fr.* II. iv. v. 201 [She] sometimes might have issued her directions to equal purpose in the Kamskatchan language. **1871** *Month* May–June 552 The marvellous fertility and floweriness of a Kamchatkan summer. **1888** *Athenæum* 3 Mar. 270/3 An Eskimo offshoot, though mixed with Tuski or Kamtchatkan blood. **1890** J. G. Frazer *Golden Bough* II. iii. 110 It was a principle with the Kamtchatkans never to kill a land or sea animal without first making excuses to it. **1917** W. M. Salter *Nietzsche* 263 The Kamschatkans required that snow should never be scraped off with a knife. **1937** *Times* 30 Dec. 9/3 By a Convention signed in 1928 Japan was given extensive fishing rights in Russian waters off the Kamchatkan and other eastern coasts.

kameel (kămī·l). *S. Afr.* [Afrikaans, a. Du. *kameel* camel.] The giraffe, *Giraffa camelopardalis*.

1839 W. C. Harris *Wild Sports S. Afr.* 373 *Camelopardalis Giraffa*, The Giraffe. Kameel of the Cape Colonists. **1896** H. A. Bryden *Tales S. Afr.* 70 As..we wanted meat, I rammed the spurs in and galloped headlong for the kameels. **1900** W. L. Sclater *Fauna S. Afr.* I. 264 The name giraffe..is practically unknown in South Africa where the term 'kameel' is always used. **1925** F. C. Slater *Shining River* 234 Kameel—The Southern giraffe, formerly found throughout the country north of the Orange River.

kameeldoorn (kămī·ldu^ə1n). *S. Afr.* Also **kameeldoring.** [Afrikaans, ad. Du. *kameel* camel + *doorn* thorn.] = *camel-thorn* (*b*) (s.v. *Camel *sb.* 5).

1822 W. J. Burchell *Trav. S. Afr.* I. xviii. 453 A large solitary tree of *Kameel-doorn* (Camel-thorn, or the tree on which, generally, the Camelopardalis browses), the first I had seen of the species, was standing here. **1896** H. A. Bryden *Tales S. Afr.* 44 Groves of giraffe acacia (*kameel doorn*), through which still wander freely in these pathless, waterless solitudes the tall giraffe. **1937** S. Cloete *Turning Wheels* 53 Passing the big group of kameel-doorns that were a landmark, they turned slightly west. **1948** [see *Gompaauw, gompauw]. **1957** *Cape Times* (Mag. Section) 20 July 2/7 Ghani got out lazily.. stationing himself beneath a shady kameeldoring overlooking the hotel yard. **1972** *S. Afr. Garden & Home* Oct. 33 The striking crimson-breasted shrike..seen perched on a *kameel-doring*.

kameez(e, var. Camise, camiss.

1966 J. & R. Godden *Two under Indian Sun* iii. 73 A coat and trousers instead of dhoti and kameeze. **1971** *Femina* (Bombay) 2 Apr. 51/2 An off-shoot of the same idea is a two-piece unit, basically a kameez and gharara, but the kameez-sleeves being of the same material as the gharara, extra colourful, while the body of the kameez itself is a plain dark self-colour. **1972** 'E. Peters' *Death to Landlords!* i. 21 She had taken to the *shalwar* and *kameez* of the Punjabi women.

kamerad (kæ·mĕrăd, || kamĕrā·t). [G., ad. F. *camarade* Comrade.] Comrade, companion: the exclamation used as an appeal for quarter by a German-speaking soldier on surrendering. Hence jocularly as *v. intr.*, to say 'kamerad', express one's wish to surrender.

1914 *Illustr. London News* 10 Oct. 497 How the enemy surrenders, saying, 'Kamerad..Pardon!' **1916** 'Boyd Cable' *Action Front* 63 'Nein, nein!' answered Ainsley. 'You kamarade—sie kamarade.' The other, in somewhat voluble gutturals, insisted that Ainsley must 'kamarade', otherwise surrender. **1917** P. MacGill *Brown Brethren* vii. 105 'Kamerad! Kamerad!' they whined, their arms shaking as if stricken with palsy. **1917** *Times Hist. War* XIV. 199/2 Then Gardener shouted to the others..'You're late. Everybody else has Kameraded.' **1918** *Daily News* 21 Sept. 5/2 When our men came down the steps of the dugout the card-players perfunctorily held up their hands and 'Kameraded'. **1923** *Westm. Gaz.* 3 July, Sir W. Joynson-Hicks cried 'Kamerad' at once. He tried to let himself down lightly by saying that he had expected a unanimous acceptance. **1930** Kipling *Limits & Renewals* (1932) 259 'Kamerad, Bull! I'll come in,' said Loftie. Vaughan's hands had gone up first. **1973** C. Egleton *Seven Days to Killing* viii. 89 He threw his weapon aside, held up his hands and yelled Kamerad.

kami. Add: (Earlier and later examples.) Also **kami-dana** = *god-shelf* s.v. God *sb.* 16 *a.*

1616 R. Cocks *Diary* (1883) I. 131 Micarna Camme Samme, the Emperours sonns sonne. **1663** R. Manley tr. *Caron & Schouten's True Descr. Kingdoms Japan & Siam* 115 *Owarny Cammy Samma*, the old Emperors Brother. **1876** [see *god-shelf*]. **1904** L. Hearn *Japan: Attempt at Interpretation* viii. 150 The domestic god-shelf —*Kamidana*. **1931** G. B. Sansom *Japan* I. iii. 46 At one end of the scale the Sun Goddess, that Heaven-Shining-Great-August Deity is a *kami*, and at the other mud and sand and even vermin are *kami*. **1965** W. Swaan *Jap. Lantern* iii. 33 Mirrors are sacred objects associated with *kami* (spirits). **1970** J. W. Hall *Japan from Prehist. to Mod. Times* iv. 32 Often translated as 'god', 'deity', or 'spirit', *kami* can best be described as localized spiritual forces. **1972** *Guardian* 23 Sept. 10/3 He spends twenty minutes in personal prayer to the 'kami' of the shrine, who in this case are the Emperor Meiji, the monarch who presided over the modernisation of Japan in the late nineteenth century, and his consort, Empress Shoken.

kamik (kæ·mik). Also **kammik.** [Eskimo.] A long boot of sealskin worn by the Eskimos.

1891 L. GIBSON *Jrnl.* 15 Aug. in R. E. Peary *Northward over Gt. Ice* (1898) I. iii. 109 The [Eskimo] woman made us a pair of Kamiks. 1900 *Scribner's Mag.* Sept. 297/1 Seal-skin kammiks, or top boots. 1910 R. E. PEARY *North Pole* xiv. 128 The *kamiks*, or boots, of seal-skin, soled with the heavier skin of the square-flipper seal. 1922 *Chambers's Jrnl.* 425/1 Untying the upper part of his *kamik*, or long boot. 1933 J. BUCHAN *Prince of Captivity* iii. 106 Their reindeer-skin kamiks had been worn into holes. 1945 D. LEECHMAN *Eskimo Summer* 29 Nearly all of them [*sc.* Eskimo girls] were wearing sealskin *kamiks*, but one or two girls actually had on silk stockings and shoes with medium high heels. 1969 *Daily Tel.* (Colour Suppl.) 18 Apr. 11/4 Eskimo wives stitching parkas and *kamiks*, or sealskin boots, which they trade at the local co-op store.

‖ **kamikaze** (kæmīkā·zi). Also with capital initial. [Jap., 'divine wind', f. *kami* god, KAMI + *kaze* wind.]

The word was originally used in Jap. lore with reference to the supposed divine wind which blew on a night in August 1281, destroying the navy of the invading Mongols.

A. *sb.* **1.** 'The wind of the gods' (see small-type note above).

1896 L. HEARN *Kokoro* x. 137 That mighty wind still called *Kami-kazé*,—'the Wind of the Gods', by which the fleets of Kublai Khan were given to the abyss. *Ibid.*, But ..the Kami-kazé did not come. 1970 J. W. HALL *Japan* vii. 93 The 'divine wind' (*kamikaze*) which Japan's protective *kami* had generated against its enemies.

2. One of the Japanese airmen who in the war of 1939–45 made deliberate suicidal crashes into enemy targets (usu. ships). **b.** An aircraft, usu. loaded with explosives, used in such an attack. Also *transf.*

1945 *Newsweek* 27 Aug. 25 As a British task force was hoisting victory pennants a Kamikaze darted out of the clouds toward the ship. 1952 *Time* 22 Dec. 17/1 No land-based bomber—including the Japanese *Kamikaze*—has ever sunk a U.S. carrier while the carrier was traveling in a task group. 1954 *Time* 4 Jan. 67/1 Fleets of Kamikazes plunged out of the sky, their suicidal pilots aiming their bomb loads at the destroyers. 1959 *Sunday Times* 5 Apr. 13/5 The Kamikaze hit the bridge, killing thirty and wounding eighty-seven. 1964 *Sun-Herald* (Sydney) 21 June 28/2 Three Australian guards and more than 200 kamikazes died. 1971 *Observer* 28 Nov. 3/6 The stand of the *kamikazes* means that in any critical division the Government is assured of a working majority.

3. *Surfing.* (See quots.)

1963 *Pix* 28 Sept. 62/3 Kamikaze: riding the nose with the hands in cross across chest. 1967 J. SEVERSON *Great Surfing* Gloss., *Kamikaze*, a planned wipe-out; taken on purpose with no hope of saving the board or avoiding the swim. 1970 *Studies in English* (Univ. of Cape Town) I. 32 A *kamikaze* occurs when the surfer takes a wipe-out fair on the nose of his surfboard.

B. *adj.* **1.** Of, pertaining to, or characteristic of a *kamikaze* (sense 2, above).

1946 *Chem. & Engin. News* XXIV. 1030/2 The Army and Navy..provided protective [smoke] screens against the Kamikaze attack of the Japanese. 1954 P. K. KEMP *Fleet Air Arm* 203 A Kamikaze, or suicide, plane dived into the base of H.M.S. *Indefatigable*'s island. 1956 A. H. COMPTON *Atomic Quest* iv. 225 Japan's one great new weapon was her 'kamikaze' planes, loaded with bombs and guided to their targets by heroic suicide pilots. 1960 *Spectator* 3 June 803 With the suicidal self confidence of kamikaze pilots ramming an aircraft-carrier. 1966 *New Scientist* 11 Aug. 305/3 After the mobilization of the bed-bugs for guard duties in Vietnam comes news of *kamikaze* porpoises. 1974 *Illustr. London News* Feb. 25/3 Newspaper speculation that *kamikaze* dolphins, with explosives strapped to them, had been trained to ram and destroy enemy craft.

2. *transf.* and *fig.* Reckless, dangerous, or potentially self-destructive (*lit.* and *fig.*).

a 1963 S. PLATH *Ariel* (1965) 23, I have taken a pill to kill The thin Papery feeling. Saboteur, Kamikaze man. 1963 *Punch* 16 Jan. 81/3 One of the *Kamikaze* apes at the RAF's Central Ape School. 1964 J. H. ROBERTS *Q Document* (1965) viii. 198 The Ginza was crowded... The kamikaze cabs did not seem to be affected..by the condition of the streets. They followed the same erratic courses through the staggered lines of more cautious drivers. 1966 L. COHEN *Beautiful Losers* (1970) I. 92 Kamikaze insects splashed against the glass. 1967 *Telegraph* (Brisbane) 5 Apr. 8/1 No one is too anxious to be a 'Kamikaze kid', and take on a seat without hopes. 1968 *Evening Standard* 29 Aug. 13/3 He developed a contempt for the kamikaze liberals who prefer glorious defeat to sensible accommodation. 1974 D. SEAMAN *Bomb that could Lip-Read* xi. 88 The Royal Army Ordnance Corps—not normally looked upon as a kamikaze outfit—supplies two such [bomb disposal] units, the only ones in the whole British Army.

Kamilaroi (kămi·lăroi). [Austral. Aboriginal.] A group of Australian Aboriginal peoples living between the Gwydir and Lachlan rivers in New South Wales; also, their language.

1856 W. RIDLEY (*title*) Kamilaroi and other Australian languages. 1877 L. H. MORGAN *Anc. Society* II. i. 51 The Kamilaroi are divided into six gentes, standing with reference to the right of marriage, in two divisions. 1911 J. G. FRAZER *Golden Bough: Magic Art* (ed. 3) I. iii. 101 The Cammeroi of whom Collins speaks are no doubt the tribe now better known as the Kamilaroi. 1952 A. G. MITCHELL in *Chambers's Shorter Eng. Dict.* Suppl., *Kamilaroi*,..one of the New South Wales aboriginal tribes. 1972 *Talanya* I. 21 The Wiradjuri language..met Kamilaroi on the north.

‖ **kamish** (kămi·ʃ). [ad. Russ. *kamýsh* reed.] The common reed, *Phragmites communis*.

1902 *Westm. Gaz.* 25 Jan. 3/1 As we advanced the mountainous country changed gradually into desert, and the desert again into steppes overgrown with kamish, or reeds, where good water could be dug out almost anywhere. 1964 R. PERRY *World of Tiger* i. 2 The vast beds of kamish reeds..which stretch for miles from the slow flood of the Kuban to the Persian shore of the Caspian.

kämmererite (ke·m-, kæ·mərərəit). *Min.* Also † **kæm-, kam-.** [ad. Sw. *kæmmererit* (N. Nordenskiöld 1842, in *Acta Soc. Sci. Fennicæ* I. 486), f. the name of August Alexander *Kämmerer* (1789–1858), Prussian surveyor of mines: see -ITE[1].] A mineral of the chlorite group that is a chromiferous variety of pennine and occurs as soft, flaky, pale violet crystals.

1854 J. D. DANA *Syst. Min.* (ed. 4) II. 292 Kæmmererite occurs in hexagonal prisms, of a reddish violet color... Found with chromic iron at Bissersk, Siberia; also at Texas, Lancaster Co., Pennsylvania. 1926 *Amer. Jrnl. Sci.* CCXI. 284 Chlorite includes as end members.. (7) $H_4Mg_5Cr_2SiO_9$ (kämmererite or Kr). 1928 *Q. Jrnl. Geol. Soc.* LXXXIII. 647 The most extensively-developed chromiferous silicate in the Shetlands is the chrome-chlorite (kämmererite). 1958 *Amer. Mineralogist* XLIII. 954 Chromium substitutes into both the octahedral and tetrahedral positions of chlorite. Above 2 per cent chromic oxide content, the former are termed kammererite, and the latter kotschubeite.

Kampa, var. *KHAMBA; **kampherol,** var. *KAEMPFEROL.

kampong. Add: Also **kampung.**

1972 *Straits Times* (Malaysian ed.) 23 Nov. 13/7 A kampung worker..was sentenced to 18 months' jail today. *Ibid.* 24 Nov. 19/8 (*heading*) Elephant visits keep a whole kampung on edge.

Kamschatkan, Kamskatchan, varr. *KAMCHATKAN *sb.* and *a.*

Kamtchat(ka)dale, Kamt(s)chadale, varr. *KAMCHADAL.

Kamt(s)chatkan, varr. *KAMCHATKAN *sb.* and *a.*

kana (kā·nă). Also 8 **canna, kanno.** [Jap.] Japanese syllabic writing, the chief varieties of which are *HIRAGANA and *KATAKANA.

1727 J. G. SCHEUCHZER tr. *Kæmpfer's Hist. Japan* I. I. iv. 68 The Names of the Provinces..are only in their *Canna*, or common Writing. *Ibid.* IV. iv. 305 Publish'd in the vulgar characters, call'd *Kanno.* 1874 *Trans. Asiatic Soc. Japan* I. 104 The invention of the Japanese syllabic *kana* ten centuries ago. 1879 *Ibid.* VII. 101 It is supposed that he is responsible for the *kana* readings given by the side of the Chinese text. *Ibid.* 230 The *kana* in the Kozhiki and Nihongi are the earliest examples of the use of Chinese characters by the Japanese as phonetic symbols. 1931 H. O. YARDLEY *Amer. Black Chamber* 194 Every time I..saw this benevolent-faced whiskered old missionary as he puzzled over Japanese words, *kana* and code groups. 1965 W. SWAAN *Jap. Lantern* vi. 74 Each of the symbols, known as *kana*, represented either a vowel or a syllable. 1968 *Encycl. Brit.* XII. 882 These syllabaries or *kana*, originally consisting of about 50 syllables, greatly stimulated the development of literature. 1973 *Physics Bull.* May 279/2 Japanese is normally written, typed or printed in a mixture of Chinese and Japanese Kana characters. The simplest set of these, the 'education set', already contains 881 Chinese characters and the 'daily use' and 'standard' sets contain 1850 and 2669 characters respectively. The dictionary set amounts to between 10 000 and 15 000 characters.

kanae (ka·nəi). *N.Z.* Also **kanai.** [Maori.] A grey mullet, *Mugil cephalus*, found in New Zealand waters.

1820 *Gram. & Vocab. Lang. N.Z.* (Church Missionary Soc.) 158 *Kanáe*,..the mullet fish. 1838 J. S. POLACK *New Zealand* I. 322 Some deep banks lie off the east coast, on which the *kanai*, or mullet..abound. 1860 A. S. ATKINSON *Jrnl.* 25 Dec. in *Richmond–Atkinson Papers* (1960) I. 671 Magnificent breakfast of kanae & oysters. 1888 in E. E. Morris *Austral Eng.* (1898) 229/1 The months of December, January, and February in each year are here prescribed a close season for the fish of the species of the mugil known as mullet or kanae. 1966 *Encycl. N.Z.* II. 600/2 Mullet, grey (*Mugil cephalus*), or kanæ of the Maoris, is an excellent food fish, rich in fat and protein, and especially suitable for smoking.

kanaff, var. *KENAF.

‖ **kanaima** (kănai·mă). [Native name.] The name given by the Indians of Guyana to an evil avenging spirit.

1951 E. MITTELHOLZER *Shadows move among Them* I. i. 8 The Genie might lure you away—or a *kanaima*. *Ibid.* xiii. 115 A *kanaima* bit me... He's a terrible Indian man who stalks through the jungle looking for people to attack.

Kanaka. Add to def.: Also, the Hawaiian language. *Obs.*

1866 'MARK TWAIN' *Lett. from Hawaii* (1967) 68, *k* and *t* are the same in the Kanaka alphabet. *Ibid.*, A white chief clerk..handed the document to Bill Ragsdale.. who translated and clattered it off in Kanaka.

Hence **Kanakaland,** Queensland; **Kanakalander,** an inhabitant of Queensland. *Obs.*

1945 BAKER *Austral. Lang.* x. 186 Kanakalanders... Used during the closing decades of the last century when many Pacific island natives were imported. *Ibid.*, 187 Queensland: *Bananaland, Kanakaland* (now obsolete) and *the Nigger State* (now obsolete).

Kanam (kā·nă·m). The name of a place in Kenya, on the south shore of the Kavirondo gulf of Lake Victoria, used *attrib.* in **Kanam jaw, man, mandible,** to designate the fossil hominid remains found there by L. S. B. Leakey in 1932.

1933 L. S. B. LEAKEY in *Man* XXXIII. 200 (*title*) The status of the Kanam mandible and the Kanjera skulls. 1935 —— *Stone Age Races Kenya* ii. 11 Further examination of the Kanam mandible has led me to separate it from the species *Homo sapiens* and to give it a new specific name *Homo kanamensis*. *Ibid.* 148/2 (*index*) Kanam Man. 1952 M. R. SAHNI *Man in Evolution* x. 246 If the evidence of the Kanam jaw from East Africa were relied upon, it would take us back to the Lower Pleistocene. 1965 M. H. DAY *Guide to Fossil Man* 145 Kanam man. *Ibid.* 147 Originally Keith thought..that the Kanam mandible was evidence of the early development of a modern type of man... Later Keith said that the small front teeth of the Kanam jaw suggested a closer relationship to the australopithecines than to modern man. 1968 P. V. TOBIAS in G. Kurth *Evolution & Hominisation* (ed. 2) 180 A re-assessment of the morphology of the Kanam mandibular fragments convinced me that...the specimen clearly contains a number of archaic features which relate it most closely to the Upper Pleistocene mandible from Dire-Dawa, in the Harrar district of Ethiopia.

kanamycin (kænămai·sin). *Pharm.* [f. mod.L. *kanamyc-eticus*, specific epithet (see def.) + -IN[1]; cf. *-MYCIN.] (One of) a mixture of antibiotics, chemically related to neomycin, which are produced by the bacterium *Streptomyces kanamyceticus* and are effective against a wide range of bacteria.

1957 H. UMEZAWA et al. in *Jrnl. Antibiotics* (Tokyo) A. X. 188 An antibiotic was isolated from a streptomyces which was assigned to a new species *S. kanamyceticus, n. sp.* Okami et Umezawa... This antibiotic was named kanamycin. 1958 *Ann. N.Y. Acad. Sci.* LXXVI. 27 Paper chromatography of kanamycin preparations revealed a second antibiotic, designated kanamycin B, which has been isolated and characterized. 1958 *Observer* 3 Aug. 9/4 A new antibiotic, kanamycin, discovered and developed by Professor Umezawa and his colleagues..of Tokyo, is now being marketed in the United States. It is active against staphylococcus... It also appears to be highly effective in acute gonorrhea in males. 1963 *Brit. Pharmaceutical Codex* 414 Toxic effects occur with sufficient frequency to make kanamycin useful only when the infecting organisms are resistant to other antibiotics. 1969 H. SMITH *Antibiotics in Clin. Pract.* vii. 97 Chemically, kanamycin is related to neomycin, consisting of two amino sugars linked to a deoxystreptamine. This polybasic water soluble antibiotic in commercial preparations consists..of kanamycin A and B... It is a white crystalline powder dissolving easily in water. 1969 KIRK & OTHMER *Encycl. Chem. Technol.* (ed. 2) 42 The kanamycins are produced by selected strains of *Streptomyces kanamyceticus*.

Kanarese (kænărī·z), *a.* and *sb.* Also **Canarese.** [f. *Kanara* + -ESE.] **A.** *adj.* Of or pertaining to Kanara in western India, or its people. **B.** *sb.* **a.** A native of Kanara. **b.** The language of Kanara, belonging to the Tamulic class of the Dravidian family, closely allied to Telugu; also called Karnata, and now generally and officially Kannada.

1838 KRISHNAMACHARYA (*title*) A grammar of the modern Canarese language. 1847 *Jrnl. Asiatic Soc. Bengal* XVI. II. 1142 The same effect is observable even in Telugu and Canarese. 1856 [see *KARNATA *a.* and *sb.*]. 1875 *Encycl. Brit.* III. 513/2 Marāthī and Kanarese are both spoken. 1880 *Ibid.* XII. 428/2 In the different parts of the [Hyderabad] territory the Marathī, the Kanarese, and Telugu languages are spoken. 1920 *Publ. Opinion* 26 Nov. 521/2 Instruction in the following tongues, Hindustani, Kaffir, Kanarese. 1921 *Q. Rev.* Oct. 328 That Indian Kings were deified after death is placed beyond doubt by a Kanarese inscription. 1939 L. H. GRAY *Foundations of Lang.* 30 The languages of South India (Tamil, Telugu, Kanarese, etc.). 1969 *Enactment* (Delhi) Nov. 12/1, I do not hurl abuses and acid bulbs on my neighbours because they speak..Kanarese and I speak Konkani. 1971 *Hindustan Times Weekly Rev.* (New Delhi) 4 Apr. p. iv/5 (Advt.), Please send me a free set of recipes in..Malayalam/Gujarati/Marathi/Kanarese.

‖ **kanat** (kanā·t). Also **kanát, kanaut,** etc. [Pers., a. Arab. *ḳanāt*.] A gently sloping underground channel or tunnel, usu. *spec.* in Persia, leading water from the interior of a hill to a village in the valley below, and provided at regular intervals with a series of vertical shafts communicating with the surface of the ground to assist in its construction and maintenance.

1855 *Q. Jrnl. Geol. Soc.* XI. I. 252 Subterraneous canals called Konáts, for irrigation derived from the river, have been cut by Persian perseverance for miles through the gravel at a great depth below the surface. Their course is traceable by the heaps of pebbles thrown out at regular intervals through walls. 1861 *Jrnl. R. Geogr. Soc.* XXXI. 37 By means of a subterranean passage—the excavated earth being thrown up, forming these cannauts..—water

is brought sometimes 5 and 6 miles across the plain. 1865 *Ibid.* XXXV. 180 The Afladj is not a province of itself, but, as its name denotes, is that portion of Dowasser which is watered by kanaats, or underground water-ducts. 1874 *Ibid.* XLIV. 185 For irrigation the plains and valleys depend on the mountains, and at the base of these are 'kanats', or underground channels. 1894 G. BELL *Safar Nameh Persian Pict.* 81 A kanat which is carrying water to many gardens. 1902 P. M. SYKES *Ten Thousand Miles in Persia* iv. 44 A heavy shower or a sandstorm frequently choking up the *kanát*. 1902 EARL OF RONALDSHAY *Sport & Politics under Eastern Sky* 364 Our road took us along the *karez* or *kanat* which brought water from the mountains at the head of the plain. 1968 *Encycl. Brit.* X. 949 Teheran and Marrakesh are among the many modern cities whose water is supplied by kanats.

Kandh, var. *KHOND.

Kandyan (kæ·ndiăn), *a.* and *sb.* Also **Kandian.** [f. *Kandy, Candy*, in Sri Lanka; see -AN.] **A.** *adj.* Of, pertaining to, or characteristic of the town or kingdom of Kandy, or of its inhabitants. **B.** *sb.* A native or inhabitant of Kandy; the language of the Kandyans.

1849 in T. Skinner *Fifty Yrs. Ceylon* (1891) 220 Robberies and bloodshed became familiar to the Kandyan. 1883 J. FERGUSON *Ceylon in 1883* 129 The Kandyan Buddhist temples. *Ibid.* 138 Of nothing is the elephant so much afraid as of fire, and with nothing will a Kandyan approach a wild elephant so readily. 1891 T. SKINNER *Fifty Yrs. Ceylon* 30 My raw untaught Kandians. 1892 C. F. G. CUMMING *Two Happy Years in Ceylon* I. 255 Most of the chiefs who attended the reception could talk more or less English, but the ladies were as deficient therein as we were in Kandyan. 1923 A. GIBSON *Cinnamon & Frangipanni* 23 Some of the real old Kandyan brass. 1933 P. FLEMING *Brazilian Adventure* ii. 20 He spent all his.. money on a fruitless search..for the buried treasure of the Kandyan Kings. 1942 H. A. J. HULUGALLE *Ceylon* 10 The up-country Sinhalese are loosely called Kandyans. 1956 R. PIERIS *Sinhalese Social Organization* 3 The Kandyan kingdom maintained its independence under the kings of Kandy from Vimala Dharma Sūrya I (reg., A.D. 1591–1604) to the deposition of Śrī Vikrama Rājasiṃha by the British in 1815. 1969 *Femina* (Bombay) 26 Dec. 5/4 She also excels in the Kandyan dances of Ceylon. 1971 *Weekend* (Ceylon) 12 Sept. 4/1 (Advt.), Father Burgher Mother Kandyan Goigama Buddhist seek partner for their pretty educated homely daughter.

Kanesian (kănī·ʒən). [Irreg. f. *Kanesh*, ancient city of Asia Minor + -IAN.] The principal dialect of Hittite, also called Kaneshite.

1921 [see *NESITE]. 1928 C. DAWSON *Age of Gods* xiii. 302 The official language of the [Hittite] empire has been named by its discoverers Nashili or Kanesian. 1950 H. L. LORIMER *Homer & Monuments* i. 4 Lydian, the language of the Lemnian inscription, and Etruscan are all three closely interrelated and belong, like Kanesian, to that early (though post-Luvian) stage of I.-E.

kang, var. CANGUE, CANG.

1928 H. LAMB *Genghis Khan* 27 Targoutai..commanded that a *kang* be put upon him—a wooden yoke resting on the shoulders and holding the wrists of a captive prisoned at both ends.

kanga[1] (kā·ŋă). *N.Z.* Also **kaanga.** [Maori.] Indian corn, *Zea mays.*

1843 E. DIEFFENBACH *Trav. in N.Z.* II. III. 366 *Kanga*, corn, maize. 1905 W. B. *Where White Man Treads* 2 Plantations of..Kaanga (Indian corn). 1949 P. BUCK *Coming of Maori* (1950) II. i. 111 Indian corn.. received the name of *Kanga*, the Maori form of the word corn... A unique method of utilizing it as food was evolved... The unhusked cobs were placed in fenced enclosures in still water where the grain became soft. 1959 *Economic Botany* XIII. 321/2 The vernacular name covering all maize is *kaanga*.

kanga[2] (kæ·ŋgă). *Austral. colloq.* [Shortened form of *kangaroo*.] = KANGAROO *sb.* 1.

[1926 A. A. MILNE *Winnie-the-Pooh* vii. 91 Christopher Robin..said that a Kanga was Generally Regarded as One of the Fiercer Animals. *Ibid.* 93 Kanga never takes her eye off Baby Roo, except when he's safely buttoned up in her pocket.] 1942 C. BARRETT *On Wallaby* iv. 73 'Leaves a kanga cold,' Tim declared after seeing a jerboa in action. 1969 *Sun* (Melbourne) 12 July 9/3 (*heading*) Kangas: all are to blame.

kanga, var. *KHANGA.

‖ **kangany** (kăngā·ni). Also **canganeme, cangany, kangani.** [f. Tamil *kaṅkāṇi*, f. *kaṇ* eye + *kāṇ* to see.] An overseer or headman of a gang of local labourers in Sri Lanka, southern India, and Malaya.

1817 'PHILALETHES' *Hist. Ceylon* lvi. 324 *Canganeme*. This officer musters the people of the village, and calls them together when there is any work to be done. *Ibid.* 336 Canganys, corporals under the aratsches. 1886 R. W. JENKINS *Ceylon in Fifties & Eighties* 78 Kangani, head of a gang of coolies. 1903 *Westm. Gaz.* 30 May 4/3 The kanganies (head coolies)..say if their coolies are not given work they must remove them to another estate. 1923 *Glasgow Herald* 21 Apr. 4/1 When the tasks are all filled the tappers go to work..under the native overseers or Kanganies. 1926 *Blackw. Mag.* Apr. 507/1 Several Tamil headmen—Kanganies—are sent to India with recruiting licenses. Each Kangany will go to his own village in South India. 1964 K. G. TREGONNING *Hist. Mod. Malaya* 199 As indentured labour diminished in importance..it was replaced by the *kangany* system, whereby a foreman or senior labourer from the estate was sent back to India, empowered to recruit in his old village... The *kangany* system was abolished in 1938.

kangaroo, *sb.* Add: **3. g.** Applied to a form of Parliamentary closure by which some amendments are selected for discussion and others excluded.

1913 *Q. Rev.* Apr. 551 The 'kangaroo' or selection by the Chairman of Committee of the amendments to be discussed. 1927 [see *guillotine resolution* s.v. *GUILLOTINE *sb.* 4].

h. A system of containerized freight transportation by railway in which a loaded road trailer complete with wheels is carried on a flat rail car; also called 'piggyback'.

1967 *Guardian* 3 July 6/3 On the European continent.. there has recently been a very rapid increase in 'Kangaroo', the system of piggyback for road trailers and semi-trailers developed by French railways. 1969 *Jane's Freight Containers 1968–69* p. iii/2 Rolling stock and terminals designed for containers should be owned or operated, or..TOFC ('piggyback' or 'kangaroo') equivalent should be owned or operated. *Ibid.* 178/1 Vehicles available: T.I.R. flat and kangaroo trailers.

4. b. kangaroo closure (see *3 g); **kangaroo court** orig. *U.S.*, an improperly constituted court having no legal standing, e.g. one held by strikers, mutineers, prisoners, etc.; **kangaroo justice,** the trying of a person by an unauthorized court, as a kangaroo court; also, the decision of such a court, taken with a disregard for normal legal procedures and criteria; **kangaroo mouse** (earlier examples); **kangaroo paw,** an Australian herb belonging to the genus *Anigozanthos* of the family Hæmodoraceæ; **kangaroo ship** (see quot.); **kangaroo-shoot,** a hunting expedition to shoot kangaroos; hence **kangaroo-shooter, -shooting.**

1930 *Times Educ. Suppl.* 22 Nov. p. i/1 It will be necessary to further restrict the rights of private members of the House of Commons by use of what is known as 'The Kangaroo Closure'. 1853 'P. PAXTON' *Stray Yankee in Texas* 205 By a unanimous vote, Judge G—— was elected to the bench and the 'Mestang' or 'Kangaroo Court' regularly organized. 1895 *Harper's Mag.* Apr. 718/2 The most interesting of these impromptu clubs is the one called in the vernacular the 'Kangaroo Court'. It is found almost entirely in county jails. 1931 'DEAN STIFF' *Milk & Honey Route* 209 *Kangaroo court*, mock court held in jail for the purpose of forcing new prisoners to divide their money. 1935 A. J. POLLOCK *Underworld Speaks* 66/1 *Kangaroo Court*, a jail tribunal comprised of inmates which collects money from prisoners awaiting trial to supply the needy with tobacco, food and a few luxuries—its decision regarding disputes is final. 1966 *Times* 14 Mar. 10/1 Shop stewards at Theale are to meet tomorrow to consider paying back the sums levied by a kangaroo court. 1971 *Times* 20 Jan. 15/3 Citizens who live in the riotous areas [of N. Ireland] deserve protection from.. kangaroo courts. 1973 C. MULLARD *Black Brit.* III. vii. 81 Such practices are surely more like those of a kangaroo court than those that the Race Relations Board should encourage. 1909 *Daily Chron.* 15 Jan. 6/7 It seems to me to be something like Kangaroo justice. 1966 *Oxford Mail* 11 Mar. 1/6 The unconstitutional strike at B.M.C. Service, Cowley, and the 'kangaroo justice' to which seven men were subjected. 1867 *Amer. Naturalist* I. 394 They are known in the vernacular as 'Kangaroo' or 'Jumping' Rats and Mice, and are entirely confined to Transmississippian regions. 1875 J. MILLER *First Fam'lies Sierras* (1876) xxx. 243 Wood-rats, kangaroo-mice..had gone into winter-quarters under the great logs. 1902 *Western Austral. Year-Bk. 1900–01* I. II. ix 304 Some of the most remarkable flowers in the flora of Western Australia [are].. Kangaroo Paws, of which there are nine species altogether. 1949 D. WALKER *We went to Austral.* 184 With the kangaroo paw it is the stalk that is scarlet and the blending of the colours peculiar. 1966 *Times* 11 Nov. (W. Austral. Suppl.) p. iv/2 The red-and-green kangaroo paw (*Anigosanthus manglesii*, the state's floral emblem) is a barbaric cluster of rich green-and-gold, paw-like flowers on a regal three-foot stem of deep scarlet, yet it is only one of nine species known to exist in the state. 1919 H. JENKINS *John Dene of Toronto* (1920) ii. 32 'A "mother",' he explained, 'is a kangaroo-ship, a dry-dock ship for salvage and repair of submarines.' 1933 *Bulletin* (Sydney) 11 Oct. 11/3 The royal pair had been participating in a kangaroo-shoot. 1902 J. H. M. ABBOTT *Tommy Cornstalk* i. 11 Indeed, it is doubtful whether there is any better shot in the world than the kangaroo-shooter. 1963 A. LUBBOCK *Austral. Roundabout* 15 The kangaroo-shooters go out at night in cars, and the kangaroos..are shot down. 1888 A. C. GUNTER *Mr. Potter* vi. 80 The Australian has been accustomed to kangaroo-shooting.

Hence **kangaroo-er,** one who hunts kangaroos.

1909 in *Cent. Dict.* Suppl. 1936 A. RUSSELL *Gone Nomad* viii. 63 That night we hobbled out at a kangarooer's camp.

Kang-Hai, Kang-He, Kang-Hi, erron. forms of next.

1910 *Encycl. Brit.* V. 747/1 On the finely prepared K'ang-hi wares much more striking and brilliant colour effects were obtained. 1911 in C. Schreiber *Jrnls.* II. (plate facing p. 16), Powdered blue..bottle of the Kang-He period. 1966 S. MORROW *Moonlighters* xvi. 166 A.. game of hide-and-seek which had toppled an irreplaceable K'ang Hai pot from his mother's desk.

K'ang-Hsi (kæŋʃī). [Royal name of Hsüan-Yeh, emperor of China 1661–1722.] Used *attrib.* with ref. to the Chinese pottery and porcelain of the latter half of the seventeenth century and the first quarter of the eighteenth, notable for very fine blue-and-white wares and the development of *famille verte* and *famille noire* enamels.

1906 S. W. BUSHELL *Chinese Art* II. viii. 33 The use of cobalt as a ground wash, foreshadowing the greater triumphs of the coming K'ang Hsi epoch. 1934 *Burlington Mag.* Mar. p. xv/1 The K'ang Hsi figures of *Kuan-yin* ..are especially remarkable. 1937 E. LINKLATER *Juan in China* xii. 213 A small flower-painted black vase. 'K'ang-hsi,' she said. 1943 D. WELCH *Maiden Voy.* xxii. 188 The K'ang-Hsi blue and white which the Dean was buying for his grey-walled room at home. 1965 D. TORR *Diplomatic Cover* vi. 103 Powder blue Chinese vases, a Sèvres biscuit group, a Nevers faience jug and the precious little Kang Hsi figures.

‖ **kangri** (ka·ŋgri). [Hindi *kaṅgṛī*; cf. Kashmiri *kangurü*.] A small wicker-covered clay-lined pot filled with glowing charcoal, carried esp. by Kashmiris next to the skin to warm the air beneath the clothing.

1911 *Allbutt's Syst. Med.* (ed. 2) IX. 591 Scars due to burns (compare the kangri carcinoma of Thibet). 1912 *Med. Ann.* 197 A form of cancer..in the natives of Kashmir,..due to their habit of carrying a portable fire-basket (kangri) beneath their clothes in contact with the abdominal skin. 1956 LD. AMULREE in A. Pryce-Jones *New Outl. Mod. Knowl.* 214 Among the Indians it has been shown that cancer of the skin is associated with the use of kangri heaters applied to the abdomen. 1968 *Daily Tel.* (Colour Suppl.) 20 Dec. 27 (*caption*) She keeps warm with a kangri, a pot filled with hot embers. 1972 *Times* 5 Aug. 11/1 If it is cold..a Kashmiri may offer the visitor his kangri... The kangri is a small wicker basket with a metal pan filled with glowing coals.

kanickanick, etc. = KINNIKINIC.

1839 J. K. TOWNSEND *Narr. Rocky Mts.* ii. 146 He smokes the article called *kanikanik*,—a mixture of tobacco and the dried leaves of the poke plant. 1847 C. LANMAN *Summer in Wilderness* xiv. 87 A bag of ka-nick-a-nick and tobacco was circulated and a cloud of fragrant smoke ascended to the sky.

Kanjar (kæ·ndʒaɹ). A generic term for certain small gypsy communities which wander about India.

1875 *Encycl. Brit.* III. 508/1 The aboriginal tribes consist of the Bhars, Cherus, Dhāngars, Kanjhārs, Kharwārs, Kols. 1885 *Ibid.* XVIII. 72/1 The Nats and Kanjars wander like gipsies over the country. 1916 R. V. RUSSELL *Tribes & Castes Cent. Provinces India* III. 333 The Kanjars and Berias are the typical gipsy castes of India. 1924 *Chambers's Jrnl.* Aug. 497/2 Dulloc..was nothing better than an outcast Kanjar. 1931 E. A. H. BLUNT *Caste System of N. India* 150 All Kanjars, however, are not criminals; many are poor and fairly respectable hunters and *shikaris*.

‖ **kanji** (ka·ndʒi). [Jap., f. *kan* Chinese + *ji* letter, character.] **a.** The corpus of borrowed and adapted Chinese ideographs which forms the principal part of the Japanese writing system. Cf. *KANA. **b.** Any one of these ideographs. Used esp. *attrib.*

1920 W. M. MCGOVERN *Colloq. Japanese* 7 The *Honji* or *Kanji* consist of the ideographs taken over from China. 1960 *New Scientist* 21 Apr. 1014/3 The Japanese newspaper contains roughly 50 per cent. kanji or Chinese characters, 40 per cent. kana letters or phonetic signs, and 10 per cent. Arabic figures and other signs. 1964 M. CRITCHLEY *Developmental Dyslexia* iv. 14 More difficulty was experienced in reading the syllabary Kana script than the ideographic Kanji symbols of Chinese origin. 1965 W. SWAAN *Jap. Lantern* vi. 74 The immense prestige of the *kanji* (Chinese characters) proved irresistible. 1972 *Mainichi Daily News* (Japan) 6 Nov. 17/4 (Advt.), Plastic Kanji Cards... All the 1,900 symbolic characters now used in Japanese newspapers and magazines, in Plastic Cards. *Ibid.*, An easy method to learn the 1,900 Chinese-Japanese characters (Kanji) now used.

kankerbos, kankerbossie (ka·ŋkəɹbɒs, -bǫ:si). *S. Afr.* [Afrikaans.] = *cancer bush* (*CANCER *sb.* 5).

1913 C. PETTMAN *Africanderisms* 248 Kanker boschje.—*Sutherlandia frutescens, R. Br.* is so named in the Riversdale district. 1931 *Farming in S. Afr.* Sept. 216 They preferred by far the 'klein kankerbos'. 1949 L. G. GREEN *In Land of Afternoon* 50 There is the kankerbos, which has failed to provide a cure for cancer. 1953 *Cape Times* 14 July 2/7 These [*sc.* existing plantings] will be followed by the introduction of more woody species, such as bitou, blombos, kankerbossie and waxberrie.

Kannada (kæ·nădă). Now the official and the more usual name for *KANARESE *sb.* b.

1856 [see *KARNATA *a.* and *sb.*]. 1877 A. HOVELACQUE *Sci. of Lang.* 78 The *Kanarese*, or *Kannada*, occupies the north Dravidian district. 1942 MORAES & STIMSON *Introd. India* ii. 16 Other widely spoken languages are the southern group—Tamil, Telugu, Kannada and Malayalam. 1959 *Clarendonian* June 46 New editions of the *First Aid* textbook have recently been published in Arabic, Greek, Turkish, Chinese, Gujarati, Kannada, and Bengali.

1969 *Filmfare* (Bombay) 1 Aug. 23/4 There is friction between distributors, exhibitors and the State Government regarding exhibition of Kannada films in Bangalore.

‖ **káns** (kāns). *India.* [a. Hindi *kans*, f. Skr. *kāśa*.] A large grass, *Saccharum spontaneum.*

1874 E. F. T. ATKINSON et al. *Statistical Acct. N.-W. Provinces India* I. 89 The very destructive weed *káns*.. yields a good coarse grass for thatching. 1884 *Encycl. Brit.* XVII. 234/2 Cultivators dare not leave their lands fallow, even for a single year, for the ground would be immediately occupied by rank *káns* grass. 1918 R. N. PARKER *Forest Flora Punjab* 535 *Saccharum spontaneum* ..Vern. *káns, káhi, kán.* Throughout the plains of the Punjab.

Kansa (kæ·nză). Also **Kansas, Kanzas,** etc. [Native name.] A Siouan Indian people formerly of Kansas and now in Oklahoma; also known as *KAW; a member of this people; the name of their language. Also *attrib.* or as *adj.*

1722 D. COXE *Descr. Carolana* i. 11 The Southerly of these two Rivers, is that of the Ousoutiwy upon which dwell..the Kansæ, Mintou, Erabacha and others. 1806 WILKINSON in Z. M. Pike *Acct. Expeditions Sources Mississippi* (1810) 108 You may attach to this deputation ..the same number of Kanses chiefs. 1847 D. COYNER *Lost Trappers* (1859) 41 They were informed of the trade made by Captain Williams and the chief of the Kansas village. 1848 E. BRYANT *California* 41, I asked him if he was a Kansas. 1907 F. W. HODGE *Handbk. Amer. Indians* I. 654/1 The Kansa figured but slightly in the history of the country until after the beginning of the 19th century. 1933 L. BLOOMFIELD *Lang.* iv. 72 The Siouan family includes many languages, such as..Kansa. 1957 *Encycl. Brit.* XX. 714/1 The principal Siouan tribes were..in the west, Dakota and Assiniboin,..tribes speaking Dhegiha, viz., Omaha, Ponca, Kansas, [etc.].

Kansan (kæ·nzăn), *sb.* and *a.* [f. *Kansas*, the name of one of the United States: see -AN.] **A.** *sb.* A native or inhabitant of the State of Kansas; = *KANSIAN.

1868 J. N. HOLLOWAY *Hist. Kansas* 574 At a favorable opportunity the Kansans seized the Ruffians,..and held them for trial. 1924 M. CRETCHER (*title*) The Kansan. 1964 *Amer. Folk Music Occasional* I. 40 Now I am teamed up with another singer. He's a Kansan like myself. 1973 *Times* 9 Nov. 12/1 The editor of the *Journal* of Salina, Kansas..suggests Kansans are longing for Mr Nixon to make a last-minute comeback.

B. *adj. Geol.* Of, pertaining to, or designating the second Pleistocene glaciation of North America, which was probably contemporary with the Mindel glaciation in the Alps. Also *absol.*, the Kansan glaciation or the deposits it produced.

1894 in J. Geikie *Gt. Ice Age* (ed. 3) xlii. 755 While the formation in Kansas is subordinately divisible, it appears to be an essential unity, and therefore the name *Kansan formation* is selected..as a convenient designation of the outermost drift sheet. 1896, 1934, 1957 [see *ILLINOIAN *a.*].

Kansas City. [City in Missouri, U.S.A.] Used *attrib.* with ref. to a style of big-band jazz which evolved in Kansas City in the 1930s.

1946 *Hollywood Q.* July 449 We have..Kansas City jazz. 1955 L. FEATHER *Encycl. Jazz* i. 22 Kansas City jazz was brewing in the bands of Benny Moten and Andy Kirk. 1958 P. GAMMOND *Decca Bk. Jazz* ix. 116 A further facet of Kansas City jazz: the singularity of the beat which it evolved to underline the inspiration of its soloists, and of its principal orchestral exponent, the Count Basie band. 1959 M. T. WILLIAMS *Art of Jazz* (1960) xviii. 188 Charlie Parker is an exponent of the Kansas City style in general and Lester Young in particular.

† **Kansian** (kæ·nziăn). *Obs.* [f. as *KANSAN *sb.* + -IAN.] = *KANSAN *sb.*

1855 W. WHITMAN *Leaves of Grass* 58 Not only the free Utahan, Kansian, or Arkansian. 1873 J. H. BEADLE *Undevel. West* 214 It was laid out by a town company of ambitious Kansians. 1878 —— *Western Wilds* ix. 133 Discount sixty per cent when a Kansian talks about snakes. 1879 W. WHITMAN *Specimen Days* (1882–3) 141 We found a train ready and a crowd of hospitable Kansians to take us on to Lawrence.

Kantean *a.*, var. KANTIAN *a.*

‖ **kantele** (ka·ntĭli). [Finn.] A form of zither used in Finland and Karelia.

1921 *Glasgow Herald* 24 Nov. 8 Eastern Karelia, where the kantele players still survive, is the home of the most characteristic Finnish music. 1960 G. TAYLOR *Mortlake* III. ix. 272 The kantele..was a large wooden box, pentagonal and tapering at one end. The five strings were stretched the length of the instrument. 1969 *Listener* 24 July 124/1 The best bits were musical: a lovely Lapp pentatonic chant, a zither-like instrument called a *kantele*, and, of course, liberal doses of Sibelius.

kantharos: see *CANTHARUS.

Kantian. Add: Also **Kantean.** **A.** *adj.* (Earlier examples.)

1796 F. A. NITSCH *Gen. View Kant's Princ. concerning Man* 1, I venture to address the Learned and Philosophers, on the Kantean principles. 1798 A. F. M. WILLICH *Elem. Critical Philos.* 22 Plattner..has employed rational scepticism against the Kantian system. 1811 H. C. ROBINSON *Diary* 29 Mar. (1967) 3, I doubt the concurrence of this explanation of a general idea with the Kantian theory.

B. *sb.* (Earlier examples.)

1799 COLERIDGE *Notebooks* (1957) I. 390 Brown was no Kantian & probably held nothing but high degrees of Probability possible. 1805 J. MACKINTOSH in R. J. Mackintosh *Mem. Life Sir J. Mackintosh* (1836) I. 260, I own to you that I am not a whit more near being a Kantian than I was before.

KANU, Kanu (kā·nu). [f. the initials of the words *K*enya *A*frican *N*ational *U*nion.] The name of a Kenyan political party.

1960 *Times* 6 June 6/5 Kenyatta was recently elected as President of Kanu. 1961 *Economist* 17 June 1244/1 The divisional and provincial boards..are manned by officials and trusty Africans of the old school. KANU branches have asked..to have representation on these boards. 1962 [see *KADU, KADU]. 1969 *Reporter* (Nairobi) 13 June 5/2 The KANU Government had been sustained by the faith and confidence of the people. 1971 *Daily Nation* (Nairobi) 10 Apr. 2/3 The people of Kisumu District are fully behind him, the Government and Kanu.

Kanuck, var. *CANUCK.

kanuka (ka·nukă). *N.Z.* [Maori.] A small, white-flowered, evergreen tree, *Leptospermum ericoides*; also called white tea-tree.

1906 T. F. CHEESEMAN *Man. N.Z. Flora* 161 *L[eptospermum] ericoides*... Abundant from the North Cape to the Bluff... Kanuka; Maru. 1929 W. MARTIN *N.Z. Nature Bk.* II. 84 The endemic Kanuka or Tea-tree.. is restricted to low levels. 1963 *Weekly News* (Auckland) 26 June 28 (*caption*) Kanuka plants being lined out in rows in the outdoor section of the nursery. 1966 *Ibid.* 1 June 16/3 Smoke is drifting, blue and hazy, through the kanukas above the campfire. 1966 *Encycl. N.Z.* II. 406/2 Kanuka grows into a small tree up to 40 or more feet high.

Kanuri (kănū·ri). [Native name.] A group of Negroid peoples living in the region of Lake Chad, in north-eastern Nigeria; their language, which belongs to the Central Saharan group.

1876 *Encycl. Brit.* IV. 61/1 The leading people of the country, called Bornuese or Kanuri, present a perfect specimen of the negro form and features. 1888 *Ibid.* XXIII. 334 Barth on linguistic grounds grouped them with the Kanuri of Bornu, who are undoubtedly Negroes. 1932 W. L. GRAFF *Lang.* 434 Kanuri, of the Nilo-Chadian group. 1959 *Listener* 29 Oct. 740/1 The..Kanuri of Nigeria. 1966 J. H. GREENBERG *Languages of Africa* (ed. 2) 132 In agreement with Eastern Sudanic Kanuri has a *tə-* prefix for the causative in a few verbs of the 'strong' conjugation.

‖ **kanzu** (ka·nzū). [Swahili.] A long white cotton or linen robe as worn by East African men.

1902 *Westm. Gaz.* 24 Nov. 2/1 Dressed in a white kanzu, or long shirt, with a tweed coat over it,..he was an attractive little figure. 1920 *Chambers's Jrnl.* Sept. 572/2 He wears..the white linen Kanzu or long robe rather like a nightgown. 1966 B. KIMENYE *Kalasanda Revisited* 17 He was wearing a kanzu in deference, no doubt, to the Sabbath, instead of his week-day costume. 1970 *Times* 20 May 7/1 He was wearing a kanzu (cotton robe) over his shirt.

kaoliang (kē·o,lyæŋ). [Chinese; lit. 'high grain'.] The Indian millet, *Sorghum vulgare.*

1904 *Westm. Gaz.* 5 Sept. 7/1 The kao-ling (millet). 1909 *Official Hist. Russo-Jap. War* (ed. 2) I. vi. 56 The advanced guard..erected screens of *kao-liang* and trees at every point. 1923 *Chambers's Jrnl.* 40/2, I found the god of rain..glaring at me from the middle of a parched kaoliang patch. 1928 *Brit. Chem. Abstracts* B. 443 A relatively easy-bleaching soda pulp can be obtained from the stalk of 'Kaoliang'. 1945 R. HARGREAVES *Enemy at Gate* 265 The field of fire was invariably obscured by meadows of high-standing *kao-ling*. 1958 A. N. STEWARD *Man. Vasc. Plants Lower Yangtze Valley* 485 *Sorghum vulgare*..(Kao Liang; Tall Millet)... Cultivated in warm countries.

kaolinitic (kē͡ŏlini·tik), *a. Min.* [f. KAOLINIT(E + -IC.] Of the nature of or containing kaolinite.

1885 *Nature* 8 Oct. 559/1 The interstitial dusty, siliceous, and kaolinitic paste has only crystallised in part. 1940 *Mineral. Abstr.* VII. 427 The formation of kaolinitic-nontronitic clays is primarily due to the hydrothermal processes. 1965 G. J. WILLIAMS *Econ. Geol. N.Z.* xx. 364/2 Clay from well-weathered greywacke, transported and deposited in fresh water is kaolinitic and of fine grain size. 1971 *Nature* 6 Aug. 371/2 The thin, curious, kaolinitic bands called tonsteins discovered more recently in the coalfields of Western Europe.

kaon (kē͡i·ǫn). *Nuclear Physics.* Also † **kayon.** [f. *ka(y)*- (repr. the pronunc. of the letter *K* in *K(-)MESON, *K(-)PARTICLE) + *-ON[1].] Any of a group of mesons which have masses several times those of the pions and non-zero hypercharge, and on decaying usually produce two or three pions or else a muon and a neutrino.

1958 *Phil. Mag.* III. 330 Writing $m_{K^+}, m_{\pi^+}, m_{\pi^0}, m_\mu$, for the kayon, charged pion, neutral pion, and muon masses. 1958 *Proc. 2nd U.N. Internat. Conf. Peaceful Uses Atomic Energy* XXX. 44/2 When the newly produced kaons disintegrate, only half of them should exhibit the short lifetime 10^{-10} sec. 1960 *Rev. Mod. Physics* XXXII. 479/1 Kayons were clearly observed at least as early as 1947. 1960 *McGraw-Hill Encycl. Sci. & Technol.* IV. 543 *K*-meson (kaon). 1963 K. W. FORD *World of Elem. Particles* vi. 180 The weak interactions.. act in their own leisurely fashion and bring about the kaon decay after about 10^{-10} sec. 1968 *New Scientist* 18 Jan. 144/2 It has generally been supposed that cosmic-ray muons are the product of the radioactive decay of the mesons known as pions and kaons. 1972 G. L. WICK *Elem. Particles* iv. 77 Both the lambda and the kaon decay into strongly interacting particles, but the lifetimes are very long.

Hence **kao·nic** *a.*, of or pertaining to a kaon, or an atom having a kaon orbiting the nucleus.

1965 *McGraw-Hill Yearbk. Sci. & Technol.* 181/1 (In table), Kaonic state. 1969 *Physical Rev. Lett.* XXII. 1238/1 The ultimate ability of kaonic x rays to yield information on the nuclear-size parameters requires the accurate determination of the kaon-nucleus potential. 1972 *Physics Bull.* Mar. 149/3 The study of kaonic atoms may provide us with information about the extreme surface region of the nucleus. 1972 *Sci. Amer.* Nov. 106/3 Increasing the charge of the nucleus shrinks the kaonic orbits and brings the kaons closer to the nucleus.

‖ **kapa,** another form of TAPA.

1909 in *Cent. Dict.* Suppl. 1913 R. BROOKE *Let.* 28 Oct. (1968) 521 Their women held up pieces of 'kapa'—which is bark beaten into a stiff cloth, and covered with a brown pattern.

‖ **kapai** (kæ·pəi), *a.* and *adv. N.Z.* Also **carpi, ka pai.** [Maori *ka pai.*] Good, fine; also as an exclamation of pleasure or approval.

1836 W. B. MARSHALL *Personal Narr. Two Visits N.Z.* II. 256 'Kapai! Good!' being the only vocable by which satisfaction at the receipt of kindness is communicable. 1840 E. HOBSON *Let.* 29 June in A. Drummond *Married & gone to N.Z.* (1960) 50 My dear husband..is on excellent terms with the natives, who call him the Carpi (good) Governor. 1861 *Taranaki Punch* 2 Jan. 2/1 The pudding turned out *kapai* until we came to cut it. 1918 *N.Z.E.F. Chrons.* 27 Feb. 37/1 The evening was simply 'kapai'. 1933 F. E. BAUME *Half-Caste* v. 55 You're looking kapai, Ngaire. 1938 R. D. FINLAYSON *Brown Man's Burden* 30 They can do South Sea hulas... Ka pai! they're very good. 1960 N. HILLIARD *Maori Girl* II. xiii. 153 '*Kapai!*' shouted Henry.

‖ **kaparrang, -ring** (kăpă·răŋ, -riŋ). *S. Afr.* Also **caparran, kaproen.** [ad. Jav. *gamparan.*] A wooden sandal worn by the Cape Malays.

1867 M. KOLLISCH *Mussulman Population Cape Good Hope* 23 Both sexes in some instances carefully ignoring the use of shoes, rather preferring clogs, called *kaparrans* (which is a small piece of wood with two slips joined underneath, and a wooden knob on the upper side). 1870 *Cape Monthly Mag.* Aug. 109 A Malay beauty..clatters upon '*caparrans*' (a species of wooden buskin). 1911 *State* Dec. 596 (Pettman), The old coloured woman walking carefully in *kaproens*. 1953 *Cape Argus* (Mag. Section) 28 Feb. 3/6 Malays, with 'Kapparings' on their feet,..were our fish vendors. 1972 I. D. DU PLESSIS *Cape Malays* (ed. 3) iii. 33 Kaparrings (probably from the Javanese gamparan: wooden sandals with a knot to push between the big and second toes) are still in use in the Mosques.

‖ **Kapenaar** (kā·pənār). *S. Afr.* Also **Kaapenaar.** [Afrikaans *kapenaar*, f. *kaap* Cape + *-enaar* pers. suffix.] **1.** An inhabitant of Cape Town or of the Cape Peninsula and its environs.

1834 *Cape of Good Hope Lit. Gaz.* IV. 180 (Pettman), The Capenaars have..attempted to justify the holding of human flesh in bondage by appeals to Scripture. 1902 *Dowey's Early Annals of Kobstad* 99 (Pettman), He was a Kaapenaar. 1946 *Spotlight* (Johannesburg) 13 Dec. 2b Kapenaars..feel that the Cabinet is really learning something of agricultural economy. 1959 *Cape Times* 9 Mar. 10/5 A statement more pleasing to..all true Kapenaars than that the long battle for the preservation of the building has been won. 1973 *Ibid.* 22 Jan. 5 Is it the mountain which makes Kaapenaars so allergic to change?

2. A large, silver-coloured, marine, food fish, *Argyrozona* (or *Polystegamus*) *argyrozona*, of the family Denticidæ.

1902 J. D. F. GILCHRIST in *Trans. S. Afr. Philos. Soc.* XL. 223 At Port Elizabeth it is called not Silver-fish, but Kapenaar. 1913 W. W. THOMPSON *Sea Fisheries of Cape Colony* 154 *Dentex argyrozona*... Silver-fish; Kaapenaar (Port Elizabeth). 1947 K. H. BARNARD *Pictorial Guide S. Afr. Fishes* 157 Silverfish, Kapenaar (*Polystegamus argyrozona*)... Table Bay to Natal, down to 70 fathoms. One of the best known, and economically important South African fishes. As well as being sold fresh, it is cured and smoked and sold as 'haddock'. 1962 *Pretoria News* 5 Jan. 4/2 The Japanese come for what they call red bream, or 'red fish', which includes the red roman, 74, kaapenaar and silver fish.

kapok. Add: Now usu. with pronunc. (kē͡i·pǫk). Substitute for def.: A large tropical tree, *Ceiba casearia*; silk cotton, the fibre produced from the soft covering of the seeds within its fruit, used to stuff mattresses, cushions, etc. Also *attrib.* (Earlier and later examples.)

1735 T. SALMON *Mod. Hist.* XXVII. vi. 186 There is also [in Guinea] the Capot Tree, that bears a sort of Cotton. 1795 tr. *C. P. Thunberg's Trav. Europe, Afr. & Asia* II. 284 The cotton which encloses the seed in the capsule, is called Kapock, and is not used for spinning, but for making mattrasses, bolsters, and pillows. 1907 *Yesterday's Shopping* (1969) 692/2 Life belts. 'Kapok' pillow belt. 1940 E. J. H. CORNER *Wayside Trees Malaya* I. 436 The Kapok is grown mainly for the wool obtained from the fruits. 1958 J. SLIMMING *Temiar Jungle* v. 77 Chabok..raised his blow-pipe, quickly inserting a dart into the mouthpiece and packing the end with a small twist of raw kapok. 1963 J. KIRKUP *Tropic Temper* 33 It had a good firm mattress stuffed with locally-grown fresh kapok. *Ibid.* 127 The kapok tree's gaunt appearance with its thin branches stuck out at right-angles to the dead-straight trunk, has earned it the nickname of P.W.D. tree. (Because people think it resembles a telegraph pole erected by the Public Works Department.) 1966 D. FORBES *Heart of Malaya* vi. 66 The kapok tree, called the midnight horror, with its long seed pods hanging down like drooping fingers, and the fan palm were common enough. 1974 L. DEIGHTON *Spy Story* xx. 209 We stood around..wearing kapok-lined white snow-suits.

kappa (kæ·pă). [Gr. κάππα.] **1.** The tenth letter of the Greek alphabet, *K*, κ.

c 1400 [see LAMBDA 1]. 1746 T. NUGENT tr. *De Port Royal's New Method of Learning Gk. Tongue* I. ii. 3 The Greeks have 24 Letters, whose Figure, Name, and Power are as followeth:..Thêta..Iôta..Cappa..Lambda [etc.]. *Ibid.* xi. 41 Κάππα, Kappa, from the Hebrew *Cap* or *Caph*, or rather from the ancient *Kappa.* 1791 R. P. KNIGHT *Analytical Ess. on Gk. Alphabet* i. 5 After the invention of the Kappa, the simple Gamma seems to have fallen into disuse in some dialects. *Ibid.*, The harshest and most emphatical palatial consonant..is the Kappa. 1871 [see *CHAPTER *sb.* 5 b]. 1948 D. DIRINGER *Alphabet* II. viii. 453 The *qoph*, which expresses the Semitic emphatic *k*, was adopted [*sc.* into the Greek alphabet] as *koppa*, differentiated from *kappa*.

2. *Biol.* An agent in some strains of *Paramecium aurelia* that confers on cells possessing it the property of producing a substance toxic to *Paramecium* cells lacking it, and exists as small cytoplasmic particles (*kappa particles*) capable of reproducing independently of the cell containing them and of infecting other *Paramecium* cells; these particles collectively. Freq. *attrib.*

1945 T. M. SONNEBORN in *Amer. Naturalist* LXXIX. 319 The alternative characters, killer and non-killer, are determined by a pair of allelic genes, K and k, and a cytoplasmic factor, which may be called kappa. Clones are killers only when both the dominant gene K and the cytoplasmic factor kappa are present; they are non-killers when kappa is absent, regardless of genic constitution. *Ibid.* 332 How can the unequal division of kappa at fission be accounted for? 1947 *Genetics* XXXII. 106 Under conditions in which the concentration of kappa decreases, two methods may be used to find the number of kappa particles per original killer cell. 1951 G. H. BOURNE *Cytol. & Cell Physiol.* (ed. 2) ii. 91 The factor kappa can reproduce itself in the cytoplasm but only if the gene *K* is present in the nucleus. 1965 PEACOCKE & DRYSDALE *Molecular Basis Heredity* ii. 7 These so-called kappa particles are probably best regarded as self-determining invaders and not as part of the normal cell. 1969 J. R. PREER in *Res. Protozool.* 174 Kappa itself has been shown..to respire and utilize glucose and sucrose *in vitro.* The presence in kappa of the biochemical apparatus required for such a sophisticated metabolism clearly eliminates the possibility that it is a virus. *Ibid.* 183 The kappas of stocks 7 and 51.

‖ **kappie** (ka·pi). *S. Afr.* Also 9 **cappie, kap(p)je.** [Afrikaans, = Du. *kapje* dim. of *kap* hood.] A sun-bonnet or coal-scuttle bonnet.

1834 A. SMITH *Diary* (1939) I. 92 All that a farmer's daughter can do or ever does is to make cappies. 1871 J. MACKENZIE *Ten Yrs. North of Orange River* iv. 62 Her taste as to colours and shapes in kapjes, handkerchiefs and dresses. 1883 'R. IRON' *Story Afr. Farm* I. ii. 26 Em took off her big brown Kappje and began..to fan her red face with it. 1902 *Daily Chron.* 18 Mar. 3/4 Women, in big, flapping kapjes, were at work on household business. 1939 S. CLOETE *Watch for Dawn* iii. 47 In her anger she had pulled her kappie from her head. 1966 E. PALMER *Plains of Camdeboo* xviii. 295 In burst Hannie Rafferty, her blue kappie askew, her round face red. 1974 *Africana Notes & News* Sept. 93 The strong contrast of her black kappie against the white kappie of the woman at her side.

‖ **kapu** (ka·pu), *a.* and *sb.* [Hawaiian.] = TABOO, TABU *a.* and *sb.*

1933 *Ancient Hawaiian Civilization* (Bishop Museum, Honolulu) 35 The kapu was the ancient social and religious law of Hawaii... The word itself can best be translated 'forbidden'. The king and all his family, his clothes, and his possessions, were kapu because of the ali'i's sacredness. All the rules and prohibitions established by the ali'i..were called the 'kapu system'. *Ibid.* 71 These prayers and the feast removed the kapu on the new dwelling. 1954 J. SHERIDAN in J. Macdonald *Lethal Sex* (1959) 151 That room is *kapu*. Definitely not the background for the sort of shindig his wife throws. 1968 *Awake!* 8 May 21/1 Under this [Polynesian] pagan religious system there was a set of taboos called *kapu*,.. that controlled almost every action of every person.

kapur (ka·pŭɹ). [Malay.] A large dipterocarp timber tree of the genus *Dryobalanops*, esp. *D. aromatica*, native to Malaya, Sumatra, and Borneo; also, the wood of a tree of this kind.

1935 I. H. BURKILL *Dict. Econ. Products Malay Peninsula* I. 863 There is probably a total of more than 500,000 acres of 'kapur' forest in the Peninsula. 1940 E. J. H. CORNER *Wayside Trees Malaya* I. 212 The *Kapur* may form almost pure forests, and it is at once distinguished by its beautiful, open, fine-leafed crown. 1956 *Handbk. Hardwoods* (Forest Prod. Res. Lab.) 125 Kapur grows to a height often exceeding 200 ft. 1963 *Engineering* 25 Oct. 525/2 Beams composed of four species, western hemlock, European redwood, jarrah and kapur. 1972 *Timber Trades Jrnl.* 13 May 39/2 Of the lesser species merbau and kapur are readily available.

kaput (kăpu·t), *a.* (in pred. use). *slang.* Also **kaputt.** [ad. G. *kaputt*, f. Fr. (*être*) *capot* (to be) without tricks in the card-game of piquet; see CAPOT *sb.*[1]] **a.** Finished, worn out; dead or destroyed. **b.** Rendered useless or unable to function.

1895 W. M. CONWAY *Alps from End to End* iii. 59 The thing would then go *wie's Donnerwetter* and the man would be *kaput* at once. [1914 DUCHESS OF SUTHERLAND *Six Weeks at the War* p. xiv, When Prussian military despotism is 'Kaputt'—to use the Germans' favourite word about their foe, meaning 'done'.] 1924 *Glasgow Herald* 11 Dec. 7 The intellectual consciousness is kaput. *a* 1930 D. H. LAWRENCE *Last Poems* (1932) 286 So self-willed, self-centred, self-conscious people die The death of nothingness, worn-out machines, kaput! 1955 E. POUND *Classic Anthol.* I. 19 North gate, sorrow's edge, Purse kaput, nothing to pledge. 1959 J. BRAINE *Vodi* iv. 63 He could now see that his whole life was *kaputt*. 1972 *New Yorker* 9 Sept. 30 It's a real American tragedy —Wunderkind at twenty, Übermensch at thirty, kaputt at forty. 1975 J. SYMONS *Three Pipe Problem* xix. 220 Sherlock Holmes is finished. Finito. Kaput.

‖ **kar, Kar** (kāɹ). *Physical Geogr.* Pl. **kare, kars;** also *erron.* **kar, karen.** [G.] = CIRQUE 2, *CWM.

1893 *Geogr. Jrnl.* I. 352 The direct erosive action of ice in forming 'glacier-pots' (Karen) and inland lakes. 1905 *Jrnl. Geol.* XIII. 2 Above the shoulders the valley slopes are far from being regular; often they form cirque-like niches, at the bottoms of which little tarns occur. These are the *Kar* of the Alps, the 'corries' of Scotland. *Ibid.* 16 Now and then we find here mountains with a single corrie or *Kar*. *Ibid.*, There is a sharp crest line which separates the corries or *Kare* of opposite sides. 1957 G. E. HUTCHINSON *Treat. Limnol.* I. i. 59 Such amphitheaters are called *cirques* in the French-speaking parts of the Alps, *Kars* in the German-speaking regions, *cwms* in Wales, and *corries* in Scotland. All four terms have achieved some degree of international usage, but the first seems to have been the most widely employed. 1963 D. W. & E. E. HUMPHRIES tr. *Termier's Erosion & Sedimentation* v. 128 The heads of the fjords formed in this way are called *botn, dalbotn* or *saekkedaler*..and differ from the true glacial cirque or *fjeldbotn*. They are almost the equivalent of the *trogschluss* or *kar* of the Alps.

Karabagh (kæ·răbā). Also **Carabagh.** The name of a region in the Soviet republic of Azerbaijan used *attrib.* or *ellipt.* to denote a thick knotted carpet or rug (orig.) made there, usually with a floral but occas. an animal pattern.

1900 J. K. MUMFORD *Oriental Rugs* ix. 118 The production of Karabaghs..has of late been pushed forward without stint. 1962 C. W. JACOBSEN *Oriental Rugs* 227 Karabagh Rugs...Also spelled Carabagh...Where made: In the southern part of what used to be called Caucasia, now the Russian States of Armenia, Azerbaijan, and Nakichevan. 1973 *Country Life* 6 Dec. (Suppl.) 40g Karabagh rug... Caucasus dated 1876.

karabiner (kærăbī·nəɹ). *Mountaineering.* Also **carabiner,** *erron.* **karibiner.** [Shortened form of G. *karabiner-haken* spring-hook.] A coupling device consisting of a metal oval or D-shaped link with a gate protected against accidental opening. Cf. *KRAB.

1932 *Amer. Alpine Jrnl.* 526 (*caption*) Safety snap (carabiner). 1933 G. D. ABRAHAM *Mod. Mountaineering* x. 182 Light pitons..are used and karibiners. The latter are special oval-shaped rings with a hinge or swivel somewhat like that found on the end of a watch chain. 1942 K. A. HENDERSON *Amer. Alpine Club's Handbk. Amer. Mountaineering* vi. 124 *Snap-rings*, sometimes called by their German name, *karabiner*, are used to fasten the rope to the piton. They come in both oval and pear-shaped form. 1946 J. E. Q. BARFORD *Climbing in Brit.* ii. 25 These are called karibiners in Germany..and consist of an oval steel ring with a spring loaded hinged link on one side. 1959 W. H. MURRAY *Five Frontiers* iv. 93 Round my waist I tied a loop of rope, clipped on a karabiner (a steel ring with a spring clip, used in rock-climbing). 1965 *New Scientist* 22 July 205/3 The karabiner is basically a hook which is closed by a pivoted arm that may or may not snap home on the latch. 1972 D. HASTON *In High Places* i. 7 Oval metal snaplinks, called carabiners. 1973 C. BONINGTON *Next Horizon* xii. 169 A jerk —you drop three inches... But you're alive! The knot in the sling attached to the karabiner in your waist harness had jammed on the gate of the karabiner, and had then freed itself, letting you drop those few inches.

‖ **karaburan** (kærăbiū͡ə·răn). [Turk., f. *kara* black + *buran* whirlwind.] A hot dusty wind in Central Asia.

1903 J. T. BEALBY tr. *Hedin's Cent. Asia & Tibet* I. xx. 331 The first real *kara-buran*, or 'black tempest',.. came..early... Drift-sand and dust swept down the Tarim and made the air so thick that we were unable to see the steep dune which faced us. 1931 A. A. MILLER *Climatol.* xiii. 243 Strong winds from almost any point of the compass spring up by day, carrying clouds of dust and sand... The *karaburan* is of this type, blowing strongly from the north-east in the Tarim basin. 1949 W. MOORE *Dict. Geogr.* 90 The sand blown along by the karaburan is one of the principal causes of the changes in the courses of rivers through the desert.

Karadagh, Kara Dagh (kæ·rădā). The name of a range of mountains in north-western Iran, used *attrib.* or *ellipt.* to denote a thick woollen carpet or rug (orig.) made there and knotted with either a geometrical or a floral pattern.

1900 J. K. MUMFORD *Oriental Rugs* xi. 180 The Kara Dagh weavings are not often seen in market. 1962 C. W. JACOBSEN *Oriental Rugs* 228 If there are any rugs coming to Ardebil or Tabriz as Karadaghs, it will be news to most buyers in Tabriz. What little information is available describes these as being very much like the Caucasian Karabaghs, thick rugs with both geometric and floral designs... The new Ardebil rugs are successors to the Karadaghs.

karaji, var. *KORADJI.

karaka. (Earlier and later examples.)

1834 G. BENNETT *Wanderings New South Wales* I. xvii. 336 The Karaka tree, (*Corynocarpus lævigata*,) of New Zealand, was in thriving condition, having reached the elevation of from six to nearly fourteen feet, and borne fruit. 1905 W. B. *Where White Man Treads* 16 And of nuts—the karaka, with its coating of soft yellow pulp. 1921 H. GUTHRIE-SMITH *Tutira* xii. 102 Single plants of karaka (*Corynocarpus lævigatus*) grew also. 1938 R. D. FINLAYSON *Brown Man's Burden* 77 Herding the cows down by the karaka trees. 1949 E. DE MAUNY *Huntsman in Career* 72 He could see the red karaka berry. 1957 J. FRAME *Owls do Cry* II. xv. 67 Green as karaka leaf.

Karakalpak, Kara-Kalpak (ka·răkălpa·k), *a.* and *sb.* [Kirghiz, f. *kara* black + *kalpak* cap.] **A.** *adj.* Of or pertaining to a Turkic people inhabiting a region south of the Aral Sea. **B.** *sb.* **a.** This Turkic people; a member of this people. **b.** (Also **Kara-Kalpaki.**) The Turkic language of the Karakalpak people.

1832 [see *KAZAKH]. 1865 J. & R. MICHELL tr. *Valikhanof's Russians Cent. Asia* ii. 24 The Mission was met by four deputies... These were the Karakalpak Prince Istleu [etc.]. 1875 G. R. ABERIGH-MACKAY *Notes W. Turkistan* 79 Karakalpaks offer allegiance to Russia..1723. 1882 *Encycl. Brit.* XIV. 94/1 The Kipchaks, forming a connecting link between the nomad and settled Turki peoples of Ferghana and Bokhara, and the Kara-Kalpaks on the south-east side of the Aral Sea. 1888 *Encycl. Brit.* XXIII. 661/2 Tatar dialects (..Kumi, Karatchai, Kara-Kalpaki). 1959 E. H. CARR *Socialism in One Country* II. xx. 270 The Kara-Kalpak autonomous region of the Kazakh autonomous SSR. 1970 *Encycl. Brit.* XXII. 399/1 Kara-Kalpak, which is hardly more than a Kazakh dialect, is used..by the Kara-Kalpaks living in Afghanistan.

‖ **karakia** (karăkī·ă). *N.Z.* Pl. **-kia, -kias.** [Maori.] An incantation (see quot. 1949).

1832 H. WILLIAMS *Jrnl.* 6 Jan. in H. Carleton *H. Life Williams* (1874) I. 111 If they should be where they cannot land, everyone immediately ceases talking, and they commence *karakia* [their incantations]. 1843 E. DIEFFENBACH *Trav. N.Z.* II. I. vii. 114 The 'Karakia' (prayers).. are most powerful when coming from a priest who is distinguished by high birth. 1862 A. S. ATKINSON *Jrnl.* 10 Sept. in *Richmond-Atkinson Papers* (1960) I. xiii. 791 We got there just after the flag had been hoisted & while the karakia were being chanted. 1874 J. C. JOHNSTONE *Maoria* 191 *Karakia*, incantations. In their use of this word the Maoris had no idea of prayer, and it is a mistake to attach that meaning to Karakia. 1905 W. B. *Where White Man Treads* 38 [The Maori's] 'karakia' (incantations) were invocations to his gods to preserve him from the Unknown. 1921 H. GUTHRIE-SMITH *Tutira* x. 70 Preparing herself..by the recitation of proper *karakias*—incantations. 1949 P. BUCK *Coming of Maori* (1950) IV. iii, The priests established oral communication with their gods by means of *Karakia*. A *Karakia* may be defined as a formula of words which was chanted to obtain benefit or avert trouble... They cover a range which exceeds the bounds of religion. It is therefore impossible for one English word to cover adequately all the meanings of *Karakia*. All *Karakia* are chants but there are a number of chants,..which are not *Karakia*... Probably incantation is the nearest in general meaning. 1959 TINDALE & LINDSAY *Rangatira* iv. 41 The karakia ceremony in which the canoe said farewell to the stump on which it had grown. 1968 *Landfall* XXII. 255 Underneath, Back, back The old life seeths, Sound of old chants, Of karakias In my skin.

Hence as *v. trans.*, to put a spell on (a person or object) by chanting or reciting *karakia*; also *intr.*, to chant or recite *karakia*.

1833 H. WILLIAMS *Let.* in H. Carleton *Life H. Williams* (1874) I. 134 His old superstition was too strong, though he did not submit to be *karakia'd* previous to dissolution. 1836 J. A. WILSON *Jrnl.* in *Missionary Life & Work N.Z.* (1889) III. 42 They invited me to hold afternoon prayers.. I reminded them of the cruel and bloody war they had long waged... They..again pressed me to *karakia*. 1874 H. CARLETON *Life H. Williams* I. 41 The natives said he had '*Karakia'd*' us—a term they apply to our religious worship.

Kara-Kirghiz (ka·rǎ,kiəɹgī·z). [Native name, f. *kara* black + *KIRGHIZ.] = *KIRGHIZ *sb.* and *a.*

1879 *Encycl. Brit.* IX. 85/2 The nomads are mainly Kipchaks and Kara Kirghiz. **1959** E. H. CARR *Socialism in One Country* II. xx. 268 The Kirgiz (here called Kara-Kirgiz) population was to form a new autonomous region within the RSFSR. **1967** D. S. PARLETT *Short Dict. Lang.* 73 *Kirgiz* (Kara-Kirgiz).., apparently one of the oldest (or oldest attested) Turkic tribes. **1971** *Whitaker's Almanack 1972* 966 In 1924, a Kara-Kirghiz Autonomous Province was formed within the R.S.F.S.R.

karakul (kæ·rǎkŭl). Also **caracul(e), carakul, karacul.** [Russ., f. *Karakul*, name of a province and lake in Bokhara, where the breed originated.] **a.** A breed of sheep with coarse wiry fur; a sheep of this breed. **b.** The glossy curled coat of a young karakul lamb, valued as fur. Also *attrib.*, as *karakul cloth*, a kind of cloth made in imitation of karakul.

1853 M. ARNOLD *Sohrab & Rustum* in *Poems* 10 And on his head he plac'd his sheep-skin cap, Black, glossy, curl'd, the fleece of Kara-Kul. **1894** *Westm. Gaz.* 20 Sept. 3/3 Here is fashion's forecast for the winter season:—Blue the leading colour... Caracule the popular fur. *Ibid.* 4 Oct. 3/3 The most striking of this season's productions is caracule-cloth, which closely resembles the fur of that name. **1894** *Queen* 27 Oct. 735/3 A 'Caracule' plush, which simulates that fashionable fur. **1895** *Army & Navy Co-op. Soc. Price List* 1095 Astrachans, &c... Black Caracul. **1898** *Daily News* 15 Oct. 6/4 A tight-fitting caracul with revers of chinchilla. **1913** *Illustr. Technical World* XIX. 700/1 The Karakul is a desert sheep native to Bokhara, Central Asia. **1929** *Daily Express* 26 Jan. 5/3 The smarter coats are generally collared with a flat fur. These include astrachan, krimma, and caracul, in black, beige, or grey. **1930** *Economist* 4 Jan. 10/2 There have also been established in the prairie provinces karakul sheep farms, from which astrakhan and broadtail are secured. **1948** L. G. GREEN *To River's End* viii. 88 On both sides of the river the karakul lamb is flourishing. *Ibid.* x. 111 Glossy black karakul skins drying on their frames. **1949** *Amer. Speech* XXIV. 95 Lambs from China or the interior of Asia possess pelts distinguished by a flat, open, wavy curl and give their name to a common variety of skin very common in the industry called *karacul*, also spelled *carakul*. **1957** V. NABOKOV *Pnin* v. 134 The warm rose-red silk lining of her karakul muff. **1960** *Times* 31 May (S. Afr. Suppl.) p. vi/7 Wool, of course, is a major export and is well supported by karakul pelts.

karakurt (ka·rǎkūəɹt). [Turki, f. *kara* black + *kurt* wolf.] A venomous, black spider, *Latrodectus tredecimguttatus*, found in southern Russia and eastern Europe; = MAL-MIGNATTE.

[**1909** A. PETRUNKEVITCH in *Sci. Amer.* 22 May 395/3 Like other species of *Latrodectus*, *Karakurt* does not show much inclination to attack.] **1932** E. NIELSEN *Biol. Spiders* I. i. 43 One of the most poisonous spiders is the Russian *Karakurt* (*Lathrodectus lugubris*), which causes great damage in the steppes of South Russia in the hot months of the summer.

karamat, var. *KRAMAT.

Karamojo (kærǎmōu·dʒo). Also **Karamojong.** [Native name.] **a.** A Nilotic people of north-eastern Uganda; a member of this people. **b.** The language of this people.

1911 *Encycl. Brit.* XXVII. 559/1 (*heading*) Nile negroes (Aluru,..Turkana and Karamojo). **1959** A. MOOREHEAD *No Room in Ark* vii. 137 The Karamojong is a warrior and a herdsman of cattle. *Ibid.* 144 The Karamojong warrior wears the kind of accessories a well-dressed western woman might have, but without the clothes. **1961** WEBSTER, *Karamojong*, a Nilotic language of the Karamojong people. **1964** C. WILLOCK *Enormous Zoo* i. 3 In the north-east [of Uganda] lay the Karamoja,..a semi-desert peopled by the Karamojong, a group of nomadic, pastoralist tribes. **1966** J. H. GREENBERG *Lang. Afr.* (ed. 2) v. 86 The Eastern Sudanic family has the following..branches: ..Nilotic:..Eastern: Dodoth, Karamojong, Teso. *Ibid.*, We may compare Nilo-Hamitic..Karamojong *ano*, Kakwa *an* with Nilotic Shilluk *an*, Anuak *ana*.

karana (kǎrǎ·nǎ). *India.* [Skr. *kárana* doing, making, (hence) position, posture.] One of the 108 basic postures in Indian dance, details of which were set out in the *Natya Sastra* by the sage Bharata Muni, traditionally after instruction from the god Siva, lord of the dance.

1936 B. V. N. NAIDU et al. tr. *Bharata's Tandava Lakṣanam* ii. 19, I shall now enumerate..the Karaṇas and Rēcakas... A Karaṇa in dance is the co-ordination of movements of the hands and feet. **1948** G. VENKATACHALAM *Dance in India* 132 *Karana*, a fundamental pose. **1956** P. BANERJI *Dance of India* (ed. 5) iv. 51 The Karanas are single postures, the special feature of them being that the left hand is generally put on the breast while the right hand follows the movements of the feet. *Ibid.* 52 A Karana is the source and origin of all the movements. **1965** E. BHAVNANI *Dance in India* iii. 17 In the Nataraja temple at Chidambaram..there are carvings of the 108 *Karanas* or basic dance postures. **1967** SINGHA & MASSEY *Indian Dances* ii. 41 Just as in Bharata's time the basic unit of dance was the karana, so it is generally agreed that today this unit is the *adavu*, which seems to have evolved from the karana. **1969** *Weekly Mail* (Madras) 26 July 10/4 Karanas can be classified as those which are anukaranas (imitations)..and those which are abstract meant only for aesthetic (internal) pleasure.

Karankawa (kǎræ·ŋkǎwǎ). Also **Carancahua, Carancoway, Carankoua.** [Native name.] **A.** *sb.* An Indian people of the Gulf coast of Texas; a member of this people; also, their language. **B.** *adj.* Of or pertaining to this people. Also **Kara·nkawan.**

1806 J. SIBLEY in *Message from President of U.S., communicating Discoveries made in exploring the Missouri by Captains Lewis & Clark* 72 Carankouas, live on an island, or peninsula, in the bay of St. Bernard. **1823** W. B. DEWEES *Lett. from Early Settler Texas* (1852) 30 During their absence the Carancoway Indians attacked the vessel. **1891** *7th Ann. Rep. U.S. Bureau Amer. Ethnol.* *1885–6* 82 (*heading*) Karankawan [linguistic] family. (*text*) The Karankawa formerly dwelt upon the Texan coast. **1907** F. W. HODGE *Handbk. Amer. Indians* I. 657/1 Karankawa... The signification of the name has not been ascertained. **1948** *True* May 126/2, I hear the Carancahuas been putting on war paint down south of here. **1957** *Encycl. Brit.* XII. 203 A/1 Powell classification (1891)..Language family..Karankawan (Extinct). *Ibid.* 203 B/2 (*heading*) Sapir arrangement (1929). (*text*) Coahuiltecan *a.* Tonkawa. *b.* Coahuilteco. *c.* Karankawa. **1966** C. F. & F. M. VOEGELIN *Map N. Amer. Indian Langs.* (*caption*) Language isolates and families with undetermined phylum affiliations... Karankawa Language Isolate.

karanteen (kæ·rāntīn). *S. Afr.* Also **-ine.** Either of two marine fishes of the family Sparidae, *Crenidens crenidens*, or the larger *bamboo-fish* (*BAMBOO *sb.* 2), *Sarpa salpa*, also known, in Natal, as the striped karanteen.

1905 *Natal Mercury Pictorial* 334 (Pettman), The fish pictured today is a Karantine. It is a local species. **1913** [see *bamboo-fish*]. **1930** C. L. BIDEN *Sea-Angling Fishes of Cape* 62 Mackerel, mullet, sardine, and bamboo-fish (Natal karanteen) are the best lures. **1947** K. H. BARNARD *Pict. Guide S. Afr. Fishes* 152 White Karanteen (*Crenidens crenidens*)... Silvery, greenish or bluish above, narrow dark longitudinal stripes. *Ibid.* 156 Striped Karanteen (Natal) (*Sarpa salpa*)... Silvery, greenish or bluish above, yellow-orange longitudinal stripes. **1953** J. L. B. SMITH *Sea Fishes S. Afr.* (rev. ed.) 274 *Sarpa salpa*... Striped Karanteen, Bamboo Fish (Natal)..is said to occur right round Africa. *Ibid.* 275 *Crenidens crenidens*... Karanteen... Comes from Indian waters, extends to Durban. **1970** *Albany Mercury* (S. Afr.) 29 Jan. 15 Smaller fish—such as mullet and karanteen—making up his [*sc.* the leerfish's] staple diet.

karat. Delete obs. and add U.S. examples.

1901 *Manufacturing Jeweler* 28 Feb. 244 (Advt.), Our rose gold, green gold, Roman and karat gold..are well known and universally used. **1903** 'O. HENRY' in *Ainslee's Mag.* Mar. 125/1 There stood..a first-class looking little man, with a four-karat diamond on his finger. **1945** *Jewelers' Dict.* 124/2 In U.S. usage *karat* designates the proportion of fine gold in an alloy; while *carat* is applied to the weight of a stone. **1972** *Newsweek* 10 Jan. 15/1 (Advt.), *Cross* pens and pencils are recognized as fine writing instruments. Available in lustrous chrome, sterling silver, gold filled and solid fourteen karat gold, they are purchased with pride. **1972** *Sci. Amer.* Dec. 8/1 (Advt.), Exclusive, extra-large, 18-karat gold point with etched facing assures super-smooth writing action.

karate (kǎrǎ·ti), *sb.* [Jap., lit. 'empty hand'.] A Japanese system of unarmed combat in which hands and feet are used as weapons. Also *attrib.*, esp. **karate chop**, a sharp slanting blow with the hand.

1955 E. J. HARRISON *Fighting Spirit Japan* (ed. 2) vii. 74 Karate resembles both jujutsu and judo. *Ibid.*, A single karate technique..is capable of inflicting fatal injury upon its victim. **1962** *Movie* Dec. 35/3 The interest of the karate techniques employed. **1964** *Guardian* 11 Jan. 5/1 Her unsporting habit of dispatching people of both sexes with a carefully rehearsed Karate blow. **1964** I. FLEMING *You only live Twice* x. 127 Your judo and karate are special skills requiring years of practice. **1966** J. PORTER *Sour Cream* ix. 126 She probably knew the lot: unarmed combat, judo, karate. **1970** *New Yorker* 5 Dec. 49 'I'm Larry Taylor,' a breathless, sharp-featured young man said, offering a karate-chop handshake to Jay Steffy. **1971** *Ink* 12 June 17/4 He floored the guard with a karate chop.

Hence ‖ **karateka** (kǎrǎ·tikǎ), an exponent or devotee of karate, a karate expert.

1966 *New Scientist* 7 July 8/1 Karatekas, those fearsome exponents of the Japanese technique of self-defence called Karate..often display their prowess by breaking..bricks with their bare hands. **1972** *Straits Times* (Malaysian ed.) 26 Sept. 3/2 Some 700 karatekas from the Police Reserve Unit..will be attending. **1972** D. LEES *Zodiac* 122 The other guy was making like a karateka. **1973** [see *KATA].

karate (kǎrǎ·ti), *v.* [f. prec.] *trans.* To strike or beat with karate blows. Also **karate-chop** *v.*, to strike with a karate chop.

1966 T. PYNCHON *Crying of Lot 49* v. 134 'I'm unarmed. You can frisk me.' 'While you karate-chop me in the spine, no thank you.' **1968** *New Yorker* 14 Sept. 129 A wolf was bugging me, so I..karated him, and called the fuzz. **1970** *Time* 11 May 62 The wife..can karate-chop hell out of her husband.

karaya (kǎrai·ǎ). [ad. Hind. *karāl*, *karāyal* resin.] In full, *karaya gum*, *gum karaya.* A gum exuded by the Indian tree *Sterculia urens* when the bark is pierced: used industrially, esp. as a substitute for gum tragacanth.

1893 G. WATT *Dict. Econ. Products of India* VI. III. 365 The gum [of *Sterculia urens*], under the name of *karai-gond*, is largely used in Bombay in the manufacture of native sweetmeats (*Dymock*). **1916** *Sci. Amer. Suppl.* 16 Dec. 393/2 (*heading*) What is karaya gum? **1918** *Jrnl. Amer. Pharmaceutical Assoc.* VII. 789 Karaya gum is used extensively in India as a substitute for tragacanth in the preparation of sweetmeats, and also locally as a demulcent in the treatment of throat affections. **1947** C. L. MANTELL *Water-Soluble Gums* iii. 50 Karaya gum has been used in the United States since the latter part of the 19th century, but the large scale use in the United States dates from World War I, when the price of tragacanth was high. Karaya is frequently sold as tragacanth. **1954** KIRK & OTHMER *Encycl. Chem. Technol.* XIII. 883 Gums derived from Irish and Iceland moss, karaya, algin, and quince seeds have received limited applications in textile finishes. **1962** M. G. DENAVARRE *Chem. & Manuf. Cosmetics* (ed. 2) II. 119 Although karaya has sometimes been used as a tragacanth adulterant, its acidity, odor, color, negative starch reaction, and ability to swell in 60% alcohol will quickly differentiate it from tragacanth. Gum karaya has found its widest cosmetic use in making finger-waving concentrates for subsequent dilution to proper consistency with water. **1974** *Daily Colonist* (Victoria, B.C.) 20 Oct. 2/3 Mrs. Wallace decided to experiment with Karaya vegetable gum powder made from the extract from a tree grown in India, in an effort to promote healing of Nunn's wound.

karbi (kǎ·ɹbi). *Austral.* [Aboriginal name.] A small, dark, stingless, native bee, *Trigona carbonaria.*

1884 H. J. HOCKINGS in *Trans. Entomol. Soc. London* 149 Of these stingless bees of Australia two varieties only have come under my immediate observation...'Karbi' or 'Keelar' and 'Kootchar' are the names given to them by the natives. *Ibid.* 150 'Karbi' gather but little honey. **1932** *Victorian Naturalist* (Melbourne) XLVIII. x. 185 The aborigines were familiar with several species, and.. *Trigona cassiæ* Ckll. is known as 'Koochee', and *Trigona carbonaria* Smith, as 'Koobee', or 'Karbi'. **1948** *Bull. Amer. Mus. Nat. Hist.* XC. 22/1 The spiral staircase type of nest was recorded by Hockings..in the case of an Australian *Trigona* known as 'karbi' or 'keelar' that he believed to be *carbonaria* F. Smith.

karee, var. KARREE (in Dict. and Suppl.).

karela (ka·rēlǎ). Also **kareli, kurilla.** [Hindi.] The balsam pear or bitter gourd, *Momordica charantia*, a climbing plant of the family Cucurbitaceæ, native to south-east Asia and tropical Africa, and cultivated in India for its edible, rough-skinned, yellow or white fruit, which is lemon-shaped and may be as much as eight inches long.

1839 J. W. MASTERS in *Trans. Agric. & Hort. Soc. India* III. 200 *Karilla*, annual, propagated by seeds. **1881** J. A. MURRAY *Plants & Drugs of Sind* 41 (*heading*) Momordica balsamina... The Balsam Apple. Vernacular—Karelo-Jangro, *Sind.* Fruit ovoid, smooth,..tapering at both ends. Eaten in pickle. **1893** G. WATT *Dict. Econ. Products India* V. 256 The rainy season kind, called *kareli*, has rather smaller fruits and is more esteemed than the hot-weather variety, known in some districts under the name of *karela*. **1895** KIPLING *Second Jungle Bk.* 81 The Karela, the vine that bears the bitter wild gourd. **1962** J. F. DASTUR *Medicinal Plants India & Pakistan* (ed. 2) 112 (*heading*) *Momordica charantia* Linn. Family: Cucurbitaceae. *Local names*: Karela, karvel. *English names*: African Cucumber, Bitter Gourd. **1969** *Sunday Statesman* (Calcutta) 27 July 4 Vegetable vendors have lao sag at 30 paise a bundle... Tinda, french beans and karela are all Re 1 a kg.

Karelian (kǎrī·liǎn), *sb.* and *a.* Also **Carelian.** [f. *Karelia*, name of a region in Eastern Finland and of a republic in the adjoining parts of the U.S.S.R.] **A.** *sb.* **a.** A native or inhabitant of Karelia. **b.** The Finno-Ugric language of this people.

1855 R. SEARS *Illustr. Descr. Russ. Empire* iii. 94 The district of the government traversed by this canal is inhabited by a tribe of Carelians. **1879** *Encycl. Brit.* VIII. 700/1 Finnic or Ugrian represented by (*a*) Finnish proper or Suonic, (*b*) Karelian. **1882** *Ibid.* XIV. 307/2 The Karelians were pressing on the Eastern Lapps. **1921** *Glasgow Herald* 21 Nov. 10 The Karelians want..to open ..peace negotiations. **1931** K. M. STEWART-MURRAY *Conscription of a People* vii. 75 The conscription of 30,000 Karelians for lumbering early in 1931. **1935** HUXLEY & HADDON *We Europeans* vii. 220 The east [of Finland is inhabited] by Karelians. **1942** K. W. DEUTSCH in J. A. Fishman *Readings Sociol. of Lang.* (1968) 600 Seven of these [smaller nations] reached some form of statehood between 1900 and 1941: Albanians, Irish, Byelo-Russians, Karelians, Moldavians—the three last-named only as 'Union Republics' within the federal framework of the U.S.S.R., [etc.]. **1961** L. F. BROSNAHAN *Sounds of Lang.* viii. 177 In Finno-Ugrian it [*sc.* palatalisation opposition] is present in eastern dialects of Finnish and Estonian, southern Karelian, and in all the languages of this family which occur to the east. **1967** E.-L. WUORIO *Midsummer Lokki* iii. 93 Kalle Koli, Finnish war-hero, Karelian, ex-gentleman farmer, and now an evacuee.

B. *adj.* Of or relating to Karelia, the Karelians, or their language.

1879 *Encycl. Brit.* IX. 218/2 It was Torkel Knutson who conquered and connected the Karelian Finlanders in 1293. **1920** *Glasgow Herald* 29 Oct. 9/2 The Karelian population in the governments of Archangel and Olonets will enjoy the right to national self-determination. **1954** *Chambers's World Gazetteer* 357/1 The great Finnish epic

Kalevala is of Karelian origin. **1967** E.-L. WUORIO *Midsummer Lokki* iii. 72 The heavy Karelian accent made it difficult for Luke to understand. **1969** E. H. PINTO *Treen* 331 Karelian birch cigar cases..enjoyed quite a vogue in the first 30 years of this century. **1974** *Sci. Amer.* Sept. 101/1 After World War II some 2·7 million Sudeten Germans were transferred to Germany and 415,000 Karelian Finns were moved to Finland.

karelianite (kǎrī·liǎnəit). *Min.* [f. prec. + -ITE[1].] An oxide of vanadium, V_2O_3.

1963 J. V. P. LONG et al. in *Amer. Mineralogist* XLVIII. 40 Electron microprobe analyses and the crystallographic data obtained suggest the existence of a new vanadium mineral, V_2O_3, isostructural with hematite, eskolaite and corundum. The name karelianite, after the region of Finland in which the Outokumpu mine is situated, is proposed for this mineral, the name to be applied to the pure end member. **1967** *Mineral. Abstr.* XVIII. 282/2 Three principal minerals compose the dominantly vanadiferous ores [at Mounana, Gabon]—karelianite, montroseite, and roscoelite. Karelianite, rarely more than 0·5 mm in size, occurs as grains in the grit cement or as blades forming 60° stars in quartz.

Karen (kǎre·n), *sb.* and *a.* Also † **Carayner, Carian(er), Carianner.** [f. Burmese *ka-reng* wild, dirty, low-caste man.] **A.** *sb.* **1.** One of a group of non-Burmese Mongoloid tribes scattered throughout Burma, esp. to the east; a member of one of these tribes.

c **1759** in A. Dalrymple *Oriental Repertory* (1793) I. 100 This Country contains two Nations, the Bûraghmahns, and Peguers... There is another People in this Country called Carianners, whiter than either. **1800** M. SYMES *Acct. Embassy to Kingdom of Ava* vi. 207 Carayners, or Carianers..inhabit different parts of the country. **1833** W. TANDY tr. *Sangermano's Descr. Burmese Empire* vi. 34 The Carian, a good and peaceable people. **1833** BENNETT *Let.* 11 Nov. in F. Mason *Karen Apostle* (1847) v. 45 Four of the Karens were yesterday baptized. **1885** [see *crow-flight* s.v. *CROW *sb.*[1] 11]. **1910** *Jrnl. R. Soc. Arts* LVIII. 701/2 The Karen undoubtedly had his original home in China. **1922** W. G. WHITE *Sea Gypsies of Malaya* i. 18 Upon the 'backbones' of Burma live the Karens... The Karens are..monolatrists, if not monotheists. **1937** R. H. LOWIE *Hist. Ethnological Theory* (1938) vii. 84 He [*sc.* Tylor] definitely ascribes the latter view only to Algonkians, Fijians, and Karens. **1962** [see *CHIN *sb.*[2]]. **1965** B. SWEET-ESCOTT *Baker St. Irreg.* viii. 236 Hitherto we had been compelled to discourage widespread guerrilla activity among the loyal tribes such as the Karens. **1972** *Nat. Geographic* Feb. 270/1, I knew that a handful of Karens had moved into these hills about 125 years ago and now far outnumbered the native Lua.

2. The language of this people.

1861 J. WADE *Karen Vernacular Gram.* p. iii, Without this knowledge no one can determine, with confidence, what is good..Karen, and what is mere colloquial jargon. **1871** C. M. YONGE *Pioneers & Founders* vi. 161 She tried to learn Karen, but never had time, and..she could always have an interpreter. **1887** D. M. SMEATON *Loyal Karens Burma* ii. 73 Reduplication of words in Karen conveys an adverbial signification. **1961** *Dental Practitioner* XI. 369/1 Indeed, 'PAT' has now been programmed to produce phrases in English, Swedish, Polish, and Karen (a language of Burma). **1970** M. PEREIRA *Pigeon's Blood* vii. 81, I can speak ten dialects of Chinese, but not a word of Karen or Shan.

B. *adj.* Of or relating to the Karens or their language.

1839 [see *FINE-TOOTH *a.*]. **1885** A. R. COLQUHOUN *Burma & Burmans* i. 11 The White Karen chief of Western Karennee. **1887** D. M. SMEATON *Loyal Karens Burma* ii. 73 The Karen language is monosyllabic. **1892** W. R. WINSTON *Four Yrs. Upper Burma* x. 99, I have.. listened in Upper Burma to..glees, choruses and solos, rendered by Karen young men and maidens. **1948** D. DIRINGER *Alphabet* vii. 412 The Karen character is a modern adaptation of the Burmese script to the Karen tongue. **1959** *Listener* 30 Apr. 746/1 The Karen rebels. **1972** *Nat. Geographic* Feb. 270/1 We found our Karen neighbors bore their troubles in a cheery, resilient, and generally relaxed way.

‖ **karez** (kǎ·rez). Also **kareze.** [Pers. (whence Pushtu) *kārēz.*] In Afghanistan and Baluchistan = *KANAT.

1875 *Encycl. Brit.* I. 232/1 The water of the *kárez*, or subterranean canals. **1880** *Ibid.* XIII. 836/2 Irrigation by 'karez' is also largely resorted to. **1902** [see *KANAT]. **1920** *Blackw. Mag.* Feb. 246/2 Then I planned a *kareze*, a subterranean water-cut. **1924** *Glasgow Herald* 29 May 9 The Persian husbandmen had even begun to repair and refit their wonderful 'karezes'. **1969** B. L. C. JOHNSON *South Asia* iv. 95/1 A karez system consists of a near horizontal tunnel driven from the level of the cultivable land near the centre of an intermontane basin to intersect the water-table in the gravelly detritus which constitutes the fans along the mountain foot... At the lower end of the tunnel emerges a small stream of water which is led to the fields. An important advantage of the karez is that the water travels most of its way underground, and so with the minimum risk of loss by evaporation.

‖ **karezza** (kǎre·tsǎ). Also **carezza.** [It. *carezza* caress.] Sexual union in which ejaculation or complete orgasm is avoided (see quot. 1896); = *coitus reservatus.*

1896 A. B. STOCKHAM *Karezza* ii. 22 Karezza consummates marriage in such a manner that through the power of will, and loving thoughts, the final crisis is not reached, but a complete control by both husband and wife is maintained throughout the entire relation. **1953** [see *coitus reservatus* s.v. *COITUS c]. **1956** A. HUXLEY *Adonis & Alphabet* 276 He advocated Male Continence and what Dr Stockham was later to call *Karezza*. **1970** W. W. ROBSON *Mod. Eng. Lit.* v. 105 Late in his life.. he [*sc.* Aldous Huxley] speculated optimistically on the benefits of a technique of sexual intercourse known as *carezza*. **1970** B. WALKER *Sex & Supernatural* ix. 83 In time 'karezza' was believed to bestow extraordinary.. blessings.

Karitane (kæritǎ·ni). *N.Z.* [f. *Karitane*, a township in the South Island of New Zealand.] Used *attrib.*, of or pertaining to the system of ante- and post-natal care for mothers and babies initiated by Sir F. Truby King (1858–1938).

1913 F. T. KING *Feeding & Care of Baby* 43 What has been achieved..on a relatively small scale at the Karitane Hospital is reflected enormously magnified in the district work. **1917** W. WRENCH *Let.* in M. King *Truby King* (1948) xxiv. 234, I am convinced that our greatest need in this country is a Training Centre on Karitane lines. **1930** *Bulletin* (Sydney) 26 Mar. 42/3 The Karitane Training Centre in Nelson-street, Woolahra, has appointed Sister May Richardson as its matron. **1945** R. M. BURDON *N.Z. Notables* 2nd Ser. iv. 62 [Truby King] had long wished to build a proper Karitane Hospital in Wellington where, as yet, there was only a mothercraft home. **1947** 'A. P. GASKELL' *Big Game* 122 Margaret had started training as a Karitane nurse. **1948** M. KING *Truby King* xv. 107 The word 'Karitane' has become famous in connection with Truby King Mothercraft work; but its origin is far removed from babies. The Huriawa or Karitane Peninsula lies at the south end of the wide Waikonaiti Bay. **1958** *N.Z. News* 11 Mar. 3/1 As an example, he [*sc.* Sir Truby King] successfully treated thirteen neglected infants in his own home, and ultimately six model 'Karitane' homes were founded throughout the Dominion. **1966** *Encycl. N.Z.* II. 222/2 A nurse was needed for the first Karitane baby.

Karlowitz(er), varr. *CARLOWITZ.

karma. Add: Also with pronunc. (kǎ·ɹmǎ). (Further examples.) Also in Hinduism.

Latterly adopted by Western popular 'meditative' groups.

1827 H. T. COLEBROOKE in *Trans. R. Asiatic Soc.* I. 111 The next head in Canáde's arrangement, after quality, is action (*carme*). **1830** —— *Ibid.* II. 38 Questions most recondite, which are agitated by theologians, have engaged the attention of the *védántins* likewise..such as.. efficacy of works (*carman*). **1881** A. P. SINNETT *Occult World* 132 Every thought of man upon being evolved passes into the inner world... It survives as an active intelligence..so man is continually peopling his current in space with a world of his own... The Buddhist calls this his 'Shandba'; the Hindu gives it the name of 'Karma'. **1899** L. HEARN *In Ghostly Japan* iii. 28 The destruction of Karma by virtuous effort is likened to the burning of incense by a pure flame. **1915** A. M. STEVENSON *Heart of Jainism* viii. 175 While Hindus think of karma as formless (*amūrta*), Jaina believe karma to have shape. **1948** N. MICKLEM *Relig.* ii. 45 Thus the doctrine of *karma* (*literally* 'action') or transmigration is intimately associated with the philosophy of the *Upanishads*. **1962** T. C. LETHBRIDGE *Witches* x. 133 The individual's choice of a future earthly body is limited however by what is known as the 'Law of Karma'. **1969** *Surfer* IX. vi. 38 May your destiny always be flavoured with good—good karma in this life. *Ibid.* 62 This is a bad Karma contest. **1971** *Nat. Geographic* Mar. 321/2 Officials, too, are subject to the law of karma—that sooner or later every action brings its retribution, in this existence or in one to come. **1971** *Rolling Stone* 24 June 32/2 John Sebastian's recording career has been plagued with an unusually bad karma. **1972** *Last Whole Earth Catalog* (Portola Inst.) 407/2 Because the karma is a little slower when you're not stoned, but it's the same karma and it works the same way.

Hence **ka·rmic** *a.*, pertaining to, relating to, or concerning *karma.*

1883 A. P. SINNETT *Esoteric Buddhism* xii. 195 Thus, on a careful examination of the matter the Karmic law.. will be seen not only to reconcile itself to the sense of justice, but to constitute the only imaginable method of natural action that would do this. **1931** L. H. MYERS *Prince Jali* vi. 63 Would not a good Buddhist see in her only a link in the long karmic chain? **1949** S. BARBANELL *Silver Birch Speaks* xiii. 142 One limit imposed, and always imposed, is what I would call the karmic debt of the patient, the relationship between mind and body determined by that individual's spiritual growth and attainment. **1974** *Publishers Weekly* 29 Apr. 47/2 Brodsky begins with an introduction to current genetic theory and proceeds to relate that to Eastern karmic understanding.

‖ **karmadharaya** (kaɹmǎdǎ·rǎyǎ). *Linguistics.* [Skr., f. KARMA + *dhāraya* holding, bearing.] A compound in which the first member describes the second, as *highway* (adjective + noun), *steamboat* (attrib. noun + noun).

1846 M. WILLIAMS *Elem. Gram. Sanscrit Lang.* ix. 158 Native grammarians class compound nouns under five heads:..The 3d, *Karmadháraya*, or those composed of an adjective and substantive. **1933** L. BLOOMFIELD *Lang.* xiv. 235 The Hindus found it convenient to set off..a special class of syntactic attribute-and-head compounds (*Karmadharaya*), such as *blackbird*. **1957** S. POTTER *Mod. Ling.* iv. 92 That important type of compound called *karmadharaya* by Indian grammarians is represented by E *blackbird*, consisting of attribute + substantive.

‖ **karma-marga** (kǎ:ɹmǎ,mǎ·ɹgǎ). *Hinduism.* [Skr., f. KARMA + *mārga* road, path.] A strict following of Hindu precepts as a means of attaining a better life in one's next incarnation; the way of works or action (contrasted with *bhakti-marga* and *jnana-marga*).

1877 M. WILLIAMS *Hinduism* i. 11 Belief in the efficacy of works, penances, and austerities,..is the *Karma-mârga*, 'way of works'. **1883** [see *JNANA b]. **1937** [see *BHAKTI b]. **1945** [see JNANA b].

Kármán (kǎ·ɹmǎn, -ǎn). Also **Karman.** [Name of Theodore von *Kármán* (1881–1963), Hungarian-born physicist and aeronautical engineer who investigated the phenomenon.] *Kármán (vortex) street*, or *Kármán street of vortices*: a vortex street in which the vortices of one line are situated opposite points midway between those of the other line, an arrangement which is stable in certain conditions.

1928 *Proc. R. Soc.* A. CXX. 34 (*heading*) The characteristics of the Karman vortex street in a channel of finite breadth. **1929** *Phil. Trans. R. Soc.* A. CCXXVIII. 275 The following investigations deal with a Kármán street of vortices, or unsymmetrical double row, in a channel of finite width. **1936** *Proc. R. Soc.* A. CLIV. 68 Some very fine photographs of Kármán streets have been taken..in the water tank at the Aeronautical Research Laboratory of the Imperial College. **1949** O. G. SUTTON *Sci. of Flight* iii. 64 In certain conditions the eddies detach themselves alternately from either edge of an obstacle placed across stream and as they break away they form in the wake the pattern known as the Kármán vortex street. **1956** A. A. TOWNSEND *Struct. Turbulent Shear Flow* vii. 149 The eddies occur alternately in each half of the wake (as in a Kármán vortex street).

Karman, var. *KIRMAN.

‖ **karma-yoga** (kǎ:ɹmǎ,yōu·gǎ). *Hinduism.* [Skr., f. KARMA + YOGA.] The attainment of perfection through disinterested action. So **ka:rma-yo·gi**, an adherent or devotee of *karma-yoga.*

1896 'VIVEKÂNANDA' *Addresses Vedânta Philos.* I. 19 The Karma Yogi..is the man who understands that the highest ideal is non-resistance, but who also knows that it is the highest manifestation of power. *Ibid.* 55 Karma Yoga is the Yoga of work, to reach the goal of perfection by means of work. **1960** KOESTLER *Lotus & Robot* I. i. 36 Vinoba Bhave could be described as a Karma-Yogi—a person who seeks fulfilment by action. **1962** A. HUXLEY *Island* vi. 78 It's a real yoga... As good as raja yoga, or Karma yoga, or bhakti yoga. **1969** *Indo-Asian Culture* Oct. 42 Gandhiji's philosophy of action is *karmayoga*.

karmic: see *KARMA.

Karnata (kaɹnǎ·tǎ), *a.* and *sb.* Also **Carnataca, Carnati(c), Karnatak(a), Karnatakam, Karnatic, Kurnata.** [f. *Karnata* (Karnataka), a region of south-west India; also var. of *Carnatic* (Karnatak), the name given, under British rule, to a region of southern India in the presidency of Madras.] **A.** *adj.* Of or pertaining to Karnata or the Carnatic. **B.** *sb.* **a.** The music of southern India, purer in form and more ancient than the Hindustani music of the north. † **b.** *Obs.* The native language of Karnata and the Carnatic; = *KANARESE *sb.* b.

1792 W. JONES in *Asiatick Researches* III. 82 On the formulas exhibited by Mirzakhan I have less reliance; but, since he professes to give them from *Sanscrit* authorities, it seemed proper to transcribe them: (*heading*) Dipaca:..*Netta*:..*Cédari*:..*Carnati.* **1801** H. T. COLEBROOKE *Ibid.* VII. 227 Carnáta, or Cánara, is the ancient language of Carnátaca. **1814** W. CAREY *Gram. Telinga Lang.* p. i, The languages of the south of India, i.e. The Telinga, Kurnata, Tamul, Malayala, and Cingalese. **1820** J. MCKERRELL *Gram. Carnataca Lang.* p. i, The Carnátaca Language..is nearly the universal Language of all the dominions of the late Tippoo Sultan. **1855** H. H. WILSON *Gloss. Judicial & Revenue Terms* p. ix, In Karnáta, *kul* is a payer of government revenue. **1856** R. CALDWELL *Compar. Gram. Dravidian Lang.* 6 The Kannadi or Karnâtaka..is spoken throughout the plateau of Mysore. ..'Karnâtaka'..is defined to mean primarily 'a species of dramatic music', or 'comedy': it is used secondarily in Telugu as an adjective to signify 'native', 'aboriginal' ..it then became the common designation of the Telugu and Canarese, or 'native' languages: and, finally, was restricted still further, and became the distinctive appellation of the Canarese alone. **1892** A. M. CHINNASWAMI MUDALIYAR *Oriental Mus. in European Notation* 9 The unfettered use of accidentals in European Music is regarded by the Karnâta musician as due to want of principle and system. *Ibid.* 12 The primary distinction is into two classes, Mârga (celestial) and Dêsi (terrestrial); the latter is now broadly divided into Hindustani and Karnâṭa. **1914** A. H. F. STRANGWAYS *Mus. Hindostan* 15 The Carnatic system frankly ignores the niceties..of intonation. **1920** E. CLEMENTS *Rāgas of Tanjore* 43/2 The Chaturdandi Prakāshikā..is said to be the principal treatise followed by modern Karnatic musicians. **1929** E. ROSENTHAL *Story Indian Mus.* i. 10 There is a wide diversity between the classification of *rágas* in the northern or Hindustani system, and in the southern or Carnatic system. **1948** G. VENKATACHALAM *Dance in India* x. 90 Carnatic music had its origin in the Telugu country. **1952** P. SAMBAMOORTHY *Dict. S. Indian Mus. & Musicians* p. vi, A few pictures are from the valuable collection of musical instruments in the Central College of Karnataka

Music,..Madras. **1968** *Jrnl. Musical Acad. Madras* XXXIX. 49 Is the wealth of Karnataka music preserved and handed down to posterity? **1971** *Shankar's Weekly* (Delhi) 4 Apr. 25/1 There is much more honesty..in a Jon Higgins submitting himself to the discipline of Karnatak music wholeheartedly. **1972** P. HOLROYDE *Indian Mus.* iii. 84 The Karnatakam remained uninfluenced... South Indian music, properly known as Carnatic, and South Indian dance, survived therefore in more-or-less pure form.

Karnaugh (kǎ·ɹnǭ). [The name of Maurice *Karnaugh* (b. 1924), U.S. physicist, who published an account of the diagram in 1953.] *Karnaugh map* or *diagram*: a diagram that consists of a rectangular array of squares each representing a different combination of the variables of a Boolean function (used, e.g., to find by inspection a simpler equivalent function).

1958 M. PHISTER *Logical Design Digital Computers* 405 (Index), Karnaugh map. **1960** W. C. IRWIN *Digital Computer Princ.* xiv. 76 The most elementary Karnaugh map represents a function of two variables. **1968** HILL & PETERSON *Introd. Switching Theory* vi. 87 The Karnaugh map is essentially a diagrammatic form of truth table, and..the Venn diagram concepts of union and intersection of areas aid us in setting up or interpreting a Karnaugh map. **1970** O. DOPPING *Computers & Data Processing* i. 28 When the number of sets is large, the overall view is improved by using a kind of modified Venn diagrams called Karnaugh diagrams or Veitch diagrams. **1973** *Math. Teaching* Sept. 48/1 The Karnaugh map takes the place of a truth table.

karo (kǎ·ro). *N.Z.* [Maori.] An evergreen shrub or small tree, *Pittosporum crassifolium*, which bears crimson flowers.

1853 J. D. HOOKER *Bot. Antarctic Voy.: Flora Novæ-Zelandiæ* I. 23 *Pittosporum cornifolium*... Nat[ive] name, 'Karo'. **1868** W. COLENSO in *Trans. N.Z. Inst.* I. III. Essay. 5 Of our shrubs and smaller timber trees, several are of strikingly beautiful growth, or blossom, or foliage; and are often seen to advantage in some clear glade..and, on the sea-coast,..the Karo, *Pittosporum crassifolium*. **1928** COCKAYNE & TURNER *Trees N.Z.* 104 *Pittosporum crassifolium*... Karo. A small tree, 15–30 ft. high, or frequently a dense shrub. **1950** *N.Z. Jrnl. Agric.* Dec. 533/1 The native karo (*Pittosporum crassifolium*) will thrive in heavy shade and will stand strong salt winds and much drier soil conditions than boxthorn, but it requires permanent double fencing to protect it from stock. **1963** J. T. SALMON *N.Z. Flowers & Plants in Colour* 53 The rich sweet scent of karo flowers fills the air of a calm evening when the tree is in flower.

Karok (kǎrǫ·k). Also † **Cahroc, Kahruk.** [f. Karok *kâruk* upstream.] **a.** An Indian people of the Klamath river valley in north-western California. **b.** The language of this people. Also *attrib.*

1851 G. GIBBS *Jrnl.* 12 Oct. in H. R. Schoolcraft *Hist. & Stat. Information Indian Tribes* (1853) III. 151 They do not seem to have any generic appellation for themselves, but apply the terms 'Kahruk', up and 'Youruk', down, to all who live above or below themselves. **1872** *Overland Monthly* Apr. 328/2 The Cahrocs are probably the finest tribe of Indians in California. *Ibid.* 330/2 The Cahroc language, though rich in its vocabulary, is said to contain no expression for 'virtue'. **1877** *Contrib. N. Amer. Ethnol.* III. 32 The Karok language is said by those acquainted with it to be copious, sonorous, and rich in new combinations. **1903** G. W. JAMES *Indian Basketry* (ed. 3) 53 The Karoks (often spelled Cahrocs) are a fine, vigorous people. **1913** [see *HOKAN]. **1921** E. SAPIR *Lang.* ix. 220 'Hokan' languages (Shasta, Karok). **1940** *Oregon Guide* 34 The southern part of Oregon was occupied by..two 'spill-overs' from California—the Shastas and Karoks. **1962** J. J. GUMPERZ in J. A. Fishman *Readings Sociol. of Lang.* (1968) 467 Certain Californian Indian tribes (the Yurok, Karok and Hupa). **1971** *Language* XLVII. 830 The Karok shift suggests consonant frequency raising in addition to palatalization.

karoro (ka·roro). *N.Z.* [Maori.] The southern black-backed gull, *Larus dominicanus.*

1861 A. S. ATKINSON *Jrnl.* 18 Jan. in *Richmond-Atkinson Papers* (1960) I. 680 Saw another noteworthy sight..a young karoro just out of the nest, running along in a curious human way all by himself. **1888** W. L. BULLER *Hist. Birds N.Z.* (ed. 2) II. 47 *Larus dominicanus.* (Southern Black-backed Gull.)..Native names. Karoro; the young bird distinguished as Ngoiro, Koiro, and Punua. **1930** W. R. B. OLIVER *N.Z. Birds* 260 Black-backed Gull. Karoro..[is] one of the most common and conspicuous birds on our coasts. **1966** R. A. FALLA et al. *Field Guide to Birds N.Z.* 154 Southern black-backed gull *Larus dominicanus*... Maori name: Karoro.

karper, var. *KURPER.

karpinskyite (kǎɹpi·nski,əit). *Min.* [ad. Russ. *karpinskiit* (L. L. Shilin 1956, in *Doklady Akademii Nauk SSSR* CVII. 737), f. the name of A. P. *Karpinsky* (1847–1936), Russ. geologist: see -ITE[1].] A hydrous alumino-silicate of sodium, beryllium, zinc, and magnesium, $Na_2(Be,Zn,Mg)Al_2Si_6O_{16}(OH)_2$, occurring as radial aggregates of white, needle-shaped crystals.

1956 *Doklady Akademii Nauk SSSR* CVII. 628 (*table of contents*) Karpinskyite—a new mineral. **1958** *Mineral. Mag.* XXXI. 963 Karpinskyite. **1968** I. KOSTOV *Mineral.* 406 Leifite..is similar to cancrinite, and so also is karpinskyite.

karree. Also karee (now the usual form). Substitute for def.: Either of two South African trees of the genus *Rhus*, *R. lancia* or *R. viminalis*, of the family Anacardiaceæ. (Earlier and later examples.)

1815 A. PLUMPTRE tr. *Lichtenstein's Trav. S. Afr.* II. xliv. 223 Mimosas,..willows, and *karree bushes*. Among the latter the colonists include several sorts of *rhus*. **1834** A. G. BAIN *Let.* 18 Dec. in *Jrnls.* (1949) 151 It was actually two large karee trees with large pools of rain-water around them. **1898** W. C. SCULLY *Vendetta of Desert* xviii. 177 They quickened their paces so as to reach a long, low ridge dotted with *karee* bushes. **1934** L. VAN DER POST *In a Province* viii. 110 Noon beats on tattered parasols of karee-thorn over them. **1939** S. CLOETE *Watch for Dawn* xxviii. 418 Behind a thick clump of karee boom a Kaffir squatted motionless. **1958** L. VAN DER POST *Lost World of Kalahari* i. 15 In the rivers and streams he constructed traps beautifully woven out of reeds and buttressed with young karee wood or hardekool. **1966** E. PALMER *Plains of Camdeboo* xvii. 287 Along the river-beds across the plains are thorn trees and karees. *Ibid.* 288 The karee trees, with their round crowns of yellow-green drooping foliage, were a welcome sight to the travellers.

‖ **karren, Karren** (ka·rən), *sb. pl.* *Geomorphol.* [G.] The furrows, fissures, or grikes of a *karrenfeld*; also, = *karrenfelder.*

1894 *Geogr. Jrnl.* III. 322 The chief features of such limestone regions are those known as *karren, dolinen,* blind valleys, and *poljen.* The *karren*, or surfaces composed of blocks of limestone separated by narrow fissures, are dealt with very briefly. **1898** *Ibid.* XII. 90 Karren are peculiar fissures in the rocky surface of some mountainous regions. **1902** LD. AVEBURY *Scenery of Eng.* xiii. 437 In calcareous districts the surface is sometimes quite bare and intersected by furrows... Such districts are known on the Continent as 'lapiées' or 'karren'. **1924** *Geogr. Rev.* XIV. 27 Lapiés are found at all altitudes from sea level to lofty mountain summits. They were first observed and described in the limestone Alps in Switzerland, where in the cantons of German speech they are called *Karren* or *Schratten.* **1960** B. W. SPARKS *Geomorphol.* vii. 155 The surface of the limestone..is often conspicuously furrowed and fretted (clints, grykes, lapiés, rascles, Schratten, Karren). **1971** J. N. JENNINGS *Karst* iv. 41 Very many *Karren* in central and western Europe must be regarded as stripped.

‖ **karrenfeld, Karrenfeld** (ka·rənfelt, -feld). *Geomorphol.* Pl. **-felder, -felds.** [G., f. *karren* (see prec.) + *feld* field.] An area or landscape, usu. of limestone bare of soil, which has been eroded by solution of the rock giving an extremely dissected surface with conspicuous furrows and fissures, often separated by knife-like ridges.

1885 A. GEIKIE *Text-bk. Geol.* (ed. 2) III. ii. 322 Limestones frequently assume a remarkable channelled rugose surface, with projecting knobs, ridges and pinnacles especially developed in high bare tracts of ground. (Karrenfelder.) **1922** *Geol. Mag.* LIX. 394 In Switzerland, notably in the canton Glarus, these bare surfaces known as *Karrenfelder* have been the subject of detailed studies by Heim. **1923** *Nature* 9 June 787/1 These have originated in guano, which gathered in the hollows of a 'karrenfeld', worn out of upraised coral-limestone. **1948** C. A. COTTON *Landscape* (ed. 2) xxiii. 445 It has been suggested that all karrenfelds are such surfaces stripped of soil in recent times. **1957** G. E. HUTCHINSON *Treat. Limnol.* I. i. 100 In some cases a *karrenfeld*, consisting of small pinnacles separated by crevices, may form.

Karri-Kot: see *CARRY-COT.

‖ **karrozzin** (kǎrǫ·tsin). Also **carozzi, carrozzi, karrozin.** [Maltese, f. It. *carrozza* carriage.] A horse-drawn cab used in Malta. Also *attrib.*

1926 E. SHEPHERD *Malta & Me* vii. 53, I..preferred the dreary carozzi. *Ibid.* xvi. 129 The roads leading to Valletta are black from early morning with carozzis. **1943** *Epic of Malta* 25 (*caption*) The picture above shows a 'carrozzi' (Maltese cab) at its usual..stand. *Ibid.* 89 (*caption*) In the picture below Maltese ponies are seen sheltering beside their 'carozzis' (cabs). **1960** *Sunday Express* 6 Nov. 19/4 You can ride in a horse-drawn cab (called a *karrozin*). **1964** G. BUTLER *Coffin in Malta* iii. 66 The flower-sellers and the karrozzin drivers in the square. **1968** *Clarendonian* XXII. 269 One could get up into town by a lift up the cliff face for 2½d...or by karrozzin, the Maltese name for the colourful horse and buggy. **1972** *Daily Tel.* (Colour Suppl.) 21 Jan. 21 (Advt.), They're horse-drawn 'karrozzins', and each one is so beautifully painted you get the feeling all the owners are in earnest competition. Hire one, and go for a ride through Malta's tiny winding streets, at a good steady trot.

karrusel (karuse·l). *Horology.* Also **karussell.** [Da., = *CAROUSEL 2.] (See quot. 1962.)

1892 B. BONNIKSEN *Brit. Pat. No. 21,421* 24 Nov., The drawing accompanying this following description, is an illustration of what I have named a position-equalizing-karrusel (hereinafter termed' karrusel' the word 'karrusel' means the same as a 'roundabout') its purpose is to make the balance of..any..portable timepiece, turn round itself in any given time. **1929** G. H. BAILIE *Watches* xix. 347 A different type of tourbillon was invented by Bonniksen of Coventry in 1892 and called by him a Karussell. **1962** E. BRUTON *Dict. Clocks & Watches* 98 *Karrusel*, arrangement similar to the tourbillon in which the escapement revolves every 52½ minutes.

karsey, var. *KARZY.

Karshuni, var. *GARSHUNI.

karst, Karst (kǎɹst). **1.** The name (*the Karst*, G. *der Karst* (= Serbo-Croat *Kras*)) of a high barren limestone region south of Ljubljana in N.W. Yugoslavia that has given its name to a kind of topography typified there (see 2); used *attrib.* in *Geomorphol.* (now usu. with small *k*) to designate similar regions and scenery, features, and phenomena associated with them, etc.; **karst land, karstland,** karstic land; a karstic region.

1894 *Geogr. Jrnl.* III. 321 Under the general designation *Karst-phenomena*, physical geographers in Germany include a variety of land-surface features, all characteristic of limestone regions, which, when the features in question are present, are known as 'Karst-regions'. *Ibid.* 509 A monotonous limestone plateau almost without water and vegetation, and occasionally exhibiting the very worst Karst features. **1898** Karst-land [see *BLIND *a.* (and *adv.*) 11 c]. **1903** *Westm. Gaz.* 10 Feb. 3/1 The latter [*sc.* Herzegovina], although fertile in parts and well cultivated, is a Karst country, warm and southern. **1908** H. B. C. & W. J. SOLLAS tr. *Suess's Face of Earth* III. vi. 231 In the Shan states of Burma, several of the coulisses which approach from the north and north-east disappear beneath a karst-like plateau of Palaeozoic limestone. **1909** CHAMBERLIN & SALISBURY *Geol.: Shorter Course* xxiii. 689 The limestone and dolomite are much more resistant than the associated shales, and as a result, erosion has developed a distinctive topography (Karst topography) at several points in the southern Alps. **1921** *Geogr. Rev.* XI. 594 In a region where karst topography is fully developed the water circulates almost entirely underground. *Ibid.* 631 To Penck and others we owe something for their development of the idea of the karst cycle. **1922** *Geol. Mag.* LIX. 394 Karst phenomena are not infrequent in the Alpine limestone districts. **1924** *Glasgow Herald* 25 Aug. 4 Karakul sheep were also introduced..in the Austrian 'karst land', in the mountains between Croatia and the Adriatic. **1932** W. H. EMMONS et al. *Geol.* v. 80 In the United States similar topography is developed in limestone areas in central Tennessee and Kentucky and is referred to as Karst topography. **1939** BAILEY & WEIR *Introd. Geol.* xxxi. 183 Karst land is rare in Scotland. **1954** W. D. THORNBURY *Princ. Geomorphol.* xiii. 349 Whether there exists a distinct cycle of land-form evolution in limestone terrains which we may designate as a karst cycle or whether what has been so designated is better considered as the karst phase of a fluvial cycle is a disputed question. **1957** G. E. HUTCHINSON *Treat. Limnol.* I. i. 107 Along the southwestern shore of the lakes where the karst topography is best developed, there are not only many islands but also very deep holes. **1958** *Geogr. Jrnl.* CXXIV. 184 (*heading*) The karstlands of Jamaica. **1963** 'M. ALBRAND' *Call from Austria* xvii. 145 He might have fallen into one of those *karst* holes you can't spot until it's too late. **1963** D. W. & E. E. HUMPHRIES tr. *Termier's Erosion & Sedimentation* xiv. 316 A karst-eroded surface. **1971** *Guardian* 5 June 9/2 This is Italy's deep south, karst country, hot and dry, split with canyons.

2. (With small *k.*) A kind of topography of which the Yugoslavian Karst is typical, found in areas of readily dissolved rock (usually limestone) and predominantly underground drainage and marked by numerous abrupt ridges, fissures, sink-holes, and caverns; a region dominated by this kind of topography.

In quot. 1902 the Karst in Yugoslavia is referred to.

1902 *Geogr. Jrnl.* XX. 429 The uvala is a large, broad sinking in the karst with uneven floor. **1916** H. F. CLELAND *Geol.* iii. 72 Karst is used as a descriptive term for any limestone region which has been etched and eroded by water into a rough surface. **1922** *Geol. Mag.* LIX. 394 The south of France where the karst attains its greatest development in Western Europe. **1937** WOOLDRIDGE & MORGAN *Physical Basis Geogr.* xix. 289 The surface of a well-developed karst has lost all semblance of normal water-modelled forms. It is a stone desert, a chaos of pits, elongated hollows, and ridges. **1958** *Geogr. Jrnl.* CXXIV. 192 Considerable areas of degraded karst occur in northern and central Jamaica. **1963** D. W. & E. E. HUMPHRIES tr. *Termier's Erosion & Sedimentation* vi. 148 Bauxites occur very frequently on karsts. **1966** J. C. PUGH in G. H. Dury *Ess. Geomorphol.* 135 Some workers prefer to regard cockpit country and tower karst as typical of karst in general, holding that in the past too much emphasis has been placed on solution and on associated collapse of passages, and too little on surface features of solution which are not restricted to tropical regions. **1968** R. W. FAIRBRIDGE *Encycl. Geomorphol.* 682/2 The terrace surface is a deeply etched karst ('karrenfeld'), with pinnacles 15 feet high, alternating with deep crevices, partly filled with red soil. **1972** J. ROGLIĆ in Herak & Stringfield *Karst* i. 6 Following Cvijić, he considered the doline a basic feature ('Leitform') of karst.

Hence **ka·rstic** *a.*, of or characteristic of karst; that is (a) karst.

1925 *Geogr. Rev.* XV. 72 The poljes have been formed by karstic and fluvial erosion. *Ibid.* 140 They occur in volcanic and karstic terrains. **1933** *Geogr. Jrnl.* LXXXI. 275 Karstic drainage occurs in limestone, dolomite, gypsum. **1957** G. E. HUTCHINSON *Treat. Limnol.* I. i. 19 In common with the other large lakes of the region [*sc.* the Balkans], both show some karstic features. **1963** D. W. & E. E. HUMPHRIES tr. *Termier's Erosion & Sedimentation* vi. 148 The karstic form of the limestone has thus acted

as a trap for the transported sediment. **1970** R. J. SMALL *Study of Landforms* iv. 152 Enclosed depressions in karstic areas are usually attributed either to (i) slow downward development by solution processes..or to (ii) collapse of rock above an underground passage or cavern.

karstification (kǎ:ɹstifikē·ʃən). *Geomorphol.* [f. prec. + -IFICATION.] Development of karst or karstic features; alteration into karst. So **ka·rstify** *v. trans.*, to subject to karstification; **ka·rstified, ka·rstifying** *ppl. adjs.*

1958 *Geogr. Jrnl.* CXXIV. 186 'Karstification' of the White Limestones began..after the mid-Miocene movements. **1968** R. W. FAIRBRIDGE *Encycl. Geomorphol.* 582/2 For a karst to develop fully or for an area to become karstified, the region must possess the following set of features. **1972** *Science* 12 May 664/2 The book is a bit thin on chemistry, considering that karstification is mainly a chemical process. **1972** F. DARÁNYI in Herak & Stringfield *Karst* viii. 283 Karst surfaces in Hungary occupy an area of some 3,000 km². The contributions of the various epochs are, however, very different, being dependent on stratigraphic position, thickness, tectonic setting, and karstifying agents. *Ibid.*, On the present-day surface of the western half of the Mecsek, the 400 m thick Anisian sequence has been karstified.

karsting (kǎ·ɹstiŋ), *vbl. sb.* *Geomorphol.* [f. as prec. + -ING[1].] = *KARSTIFICATION.

1921 *Geogr. Rev.* XI. 597 In youth, the surface of the land is still principally drained by the rivers which flow on the former surface of the land before the limestone subject to karsting was laid bare. **1956** D. L. LINTON *Sheffield* ii. 27 Our knowledge of the 'karsting' of the limestone plateau of the southern Pennines is still elementary.

karsto·logy. *Geomorphol.* [f. as prec. + -OLOGY.] The study of karst.

1968 R. W. FAIRBRIDGE *Encycl. Geomorphol.* 560/2 There are furthermore several national and international organizations that promote congresses and publications in Speleology and Karstology. **1972** HERAK & STRINGFIELD *Karst* i. 2 The classical connection of karstology with the Dinaric karst was not accidental.

kart. Restrict † *obs.* to sense in Dict. and add: **2.** = *GO-KART. Hence **ka·rting** *vbl. sb.*, the sport of driving or racing a kart.

1959 *Motor* 9 Sept. 111/1 The whole affair may well seem reprehensible to those who feel that an injustice has been wrought on the children or that karting is too juvenile for adults, but after trying a kart, although one must admit that they are a joke, there is no doubt that they are one of the very best jokes to come out of the U.S.A. since Thurber's dogs. After 10 minutes' karting the cynic is invariably asking where he can get one. **1961** *Sunday Express* 1 Jan. 15/3 '*Karting clothes*' the newest fad. **1964** K. WHEELER *Sport* 123 The newest motor sport is Karting. In the late 1950s the first little midget cars were seen on the tracks... Now karting is much faster, and an officially recognised sport with its own trophies and its own champions... A kart can cost as little as £80. **1968** *Times* 2 Aug. 13/2 Her tiny 100 c.c. machine is capable of 80 m.p.h. She has been driving since she was 13, the minimum age for kart racing in this country.

‖ **karuna** (kǎrū·nǎ). *Buddhism.* [Skr. *karuṇā* charity, compassion.] Loving compassion, as that sought and attained by a *Bodhisattva.*

1850 R. S. HARDY *Eastern Monachism* xx. 246 When we see any object in distress, we feel kampáwīma, agitation, in the mind; and from this arises karuná, pity or compassion. **1960** A. HUXLEY *Let.* 17 July (1969) 893 Then no me any more and a kind of *sat chit ananda*, at one moment without *karuna* or charity. **1962** —— *Island* ix. 152 A fair chance..of eventual *prajnaparamita* and *karuna*, eventual wisdom and compassion.

karyo-. Add: **karyo·gamy** [*-GAMY], fusion of cell nuclei; **ka·ryomere** [Gr. μέρος part], a vesicular chromosome enclosed in a nuclear membrane of its own, such as forms at telophase in the division of some cells; **karyoplasma·tic, -pla·smic** *adjs.*, of or pertaining to karyoplasm; **karyothe·ca** *rare*, the nuclear membrane.

1891 *Jrnl. R. Microsc. Soc.* 49 In *Hydatina*, as in some Hymenoptera, there is established between arrhenotoky (parthenogenetic production of males) and fecundating karyogamy, a relation so necessary that the second is impossible without the first. **1901** G. N. CALKINS *Protozoa* iii. 97 It is quite possible that many cases of so-called conjugation are only instances of plastogamy, or fusion of the cell-body, and are not followed by union of the nuclei (karyogamy), as in fertilization. **1970** J. WEBSTER *Introd. Fungi* 187 Karyogamy (i.e. nuclear fusion) occurs within certain of the binucleate cells. **1912** *Jrnl. Acad. Nat. Sci. Philadelphia* XV. 525 The most general results of increased temperature are:..(4) Formation of numerous karyomeres from these scattered chromosomes; indeed by slight increase of temperature almost every chromosome may be caused to remain distinct from every other one, and to give rise to a separate chromosomal vesicle. **1934** L. W. SHARP *Introd. Cytol.* (ed. 3) x. 136 Of considerable interest are those nuclei in which every chromosome of the telophase group forms an individual vesicle, or karyomere. In some cases the karyomeres may eventually fuse partially or completely, but in others they remain separate although in contact, forming what is virtually a group of small nuclei containing one chromosome each. *Ibid.* 146 The limits of the several chromosomes remain visible through this stage [*sc.* between mitoses] in certain nuclei; in extreme cases the nucleus is virtually a group of separate elementary nuclei, or karyomeres. **1969** BROWN & BERTKE *Textbk. Cytol.* xvii. 318/2 Karyomeres are rather like micronuclei except that they are normal and become associated to form a 'compound' nucleus. **1920** L. DONCASTER *Introd. Study Cytol.* ii. 16 Hertwig regards a disturbance of the normal karyo-plasmatic ratio as the immediate cause of cell-division, and supposes that the unequal rate of growth of nucleus and cytoplasm brings about a condition of 'karyo-plasmatic strain' leading to cell-division and a consequent restoration of the normal ratio. **1909** *Cent. Dict. Suppl.*, Karyoplasmic. **1924** E. V. COWDRY *Gen. Cytol.* VI. 351 Hertwig's karyoplasmic relation hypothesis. **1925** E. B. WILSON *Cell* (ed. 3) iii. 237 These various facts show on how precarious a basis rest theories of senescence and rejuvenescence which refer these processes to changes in the karyoplasmic ratio. **1948** R. A. R. GRESSON *Essent. Gen. Cytol.* i. 2 It has been found that a quantitative relationship exists between nuclear mass and cytoplasmic mass; this is known as the karyoplasmic ratio. **1966** D. M. KRAMSCH tr. *Grundmann's Gen. Cytol.* ii. 66 The optical appearance of the karyoplasmic area is homogeneous and frequently seems to be empty. **1896** E. B. WILSON *Cell* 337 *Karyotheca*, the nuclear membrane. **1948** W. ANDREW tr. *E. D. P. de Robertis's Gen. Cytol.* iii. 45 Between the two parts of this heterogeneous system.., there is found the karyotheca or nuclear membrane.

karyogram (kæ·riogræm). *Biol.* [f. KARYO- + -GRAM.] **a.** A karyotype or idiogram.

In quot. 1952 (by a Jap. writer) perh. an error for *karyotype*, which is used throughout the paper (in sense *1 a).

1952 *Cytologia* XVII. 311 (*heading*) Karyogram studies in birds. **1965** *Amer. Jrnl. Bot.* LII. 968 (*caption*) Karyogram from untreated apical and nodal cells..showing chromosome length greater than that in pretreated antheridial filament cells. **1971** *Nature* 9 Apr. 368/1 (*caption*) A typical karyogram of Chinese hamster strain *wg 3 IMP*⁻: twenty-two chromosomes, all but one (bottom right) distinguishable from those of mouse strain *3T3 TK*⁻.

b. A diagram in which each chromosome of a set is represented by a point positioned with respect to a vertical and a horizontal axis, these axes corresponding to two numerical characteristics of the chromosomes.

1960 K. PATAU in *Amer. Jrnl. Human Genetics* XII. 255 A chromosome characterized by two quantities is best represented by a point in a two-dimensional co-ordinate system. The choice of these quantities—total length and arm index or two arm lengths, is in principle irrelevant... Let the percent length, l, of the longer arm serve as abscissa and the percent length, s, of the shorter arm as ordinate. The scatter diagram obtained by plotting in this manner all chromosomes of the given complement will henceforth be referred to as a 'karyogram'. **1966** STEWART & KILLEAN in Darlington & Lewis *Chromosomes Today* I. 211 In the production of karyograms the basic data are taken from a photographic enlargement of the chromosome plate to be analysed. **1969** *Nature* 22 Nov. 801/2 The results of a chromosome analysis can be displayed in several ways. The methods commonly used are the karyotype..and the karyogram, in which each chromosome is represented by a point in a bivariate coordinate system.

karyology (kæriǫ·lŏdʒi). *Biol.* [f. KARYO- + -LOGY.] **a.** The distinctive or characteristic features (esp. as regards chromosomes) of a particular cell nucleus, or of the nuclei of a particular species, strain, etc. **b.** The study of cell nuclei, esp. the chromosomes they contain.

1895 *Ann. Bot.* IX. 631 Wager ('89) has carefully investigated the karyology of *Peronospora parasitica*. **1932** *Symbolae Bot. Upsalienses* I. 150 The chromosomes have individuality or genetic continuity from one cell generation to another. This conception..may be considered as the foundation of all comparative karyology. **1948** W. ANDREW tr. *E. D. P. de Robertis's Gen. Cytol.* vii. 134 There was considerable progress in karyology, which is a branch of cytology dealing with the nucleus or karyosome, to the detriment of the study of the cytoplasm or cytosome. **1965** *Jrnl. Nat. Cancer Inst.* (U.S.) XXXV. 766/1 This..indicates a significant difference between the general karyology of the normal cell and of the tumor cell population. **1970** *Cytologia* XXXV. 294 (*heading*) Karyology of *Sequoia sempervirens*: karyotype and accessory chromosomes. **1971** *Daily Colonist* (Victoria, B.C.) 24 Dec. 4/2 Karyology, or the study of the number, shape, and size of chromosomes, is Dr. Wiens' specialty.

Hence **karyolo·gic** (chiefly *U.S.*), **-lo·gical** *adjs.*, of or pertaining to karyology; **karyolo·gically** *adv.*, as regards or in relation to karyology.

1927 *Trudy po Prikladnoi Botanike i Selektsii* XVII. III. 64 Presence, shape and size of the satellite are characters of much importance in karyological systematics. *Ibid.* 65 The plates given..represent races which are karyologically most sharply different. **1929** *Cytologia* I. 76 (*heading*) Karyological studies in *Hemerocallis*. **1935** *Experiment Station Rec.* LXXII. 754 Karyologic and genetic studies with Fragaria. **1962** *Exper. Cell Res.* XXVI. 434 (*heading*) Karyologic studies on polyoma virus induced mouse tumors. **1962** *Lancet* 27 Jan. 219/1 Karyologically the cells with 47 chromosomes were typically mongoloid, and those with 46 chromosomes were normal. **1970** *Nature* 11 July 169/1 The karyological properties of MRC-5 cells conform to those required of a diploid cell of human origin to be used for producing viral vaccines intended for human use. **1974** *Ibid.* 5 Apr. 504/2 A unique possibility of investigating RSV production on a permanent cell line which is well defined karyologically.

karyosome. Substitute for def. (in Dict. s.v. KARYO-): **1.** [ad. G. *karyosoma* (M. Ogata 1883, in *Arch. f. Anat. u. Physiol.* (Physiol. Abt.) 414).] **a.** A body of chromatin in a nucleus resembling a nucleolus but distinguished from the 'true' nucleolus or plasmosome. **b.** Any densely staining central body of a nucleus.

1889 [see *hyalosome* s.v. *HYALO-]. **1890** [in Dict. s.v. KARYO-, where for *Soc.* read *Sci.* in the title]. **1896** E. B. WILSON *Cell* i. 24 The bodies known by this name [*sc.* nucleolus] are of at least two different kinds. The first of these, the so-called true nucleoli or plasmosomes.., are of spherical form... Those of the other form, the 'net-knots' (Netzknoten), or karyosomes, are either spherical or irregular in form, stain like the chromatin, and appear to be no more than thickened portions of the chromatic network. **1901** G. N. CALKINS *Protozoa* v. 146 In most cases they [*sc.* the nuclei of Sporozoa] consist of a firm and resisting membrane containing a single large chromatin reservoir or karyosome. **1920** L. DONCASTER *Introd. Study Cytol.* ii. 18 The karyosome or chromatic nucleolus is a mass of chromatin of varying size... It appears to serve as a reservoir of chromatin from which the chromosomes..may draw part at least of their supply when nuclear division is approaching. **1926** G. N. CALKINS *Biol. Protozoa* ii. 60 The centrally placed intranuclear body is generally described under the name karyosome, a term which has been so widely used by students of the Protozoa and for so many obviously different structures that it is practically synonymous with endosome or Binnenkörper. **1948** W. ANDREW tr. *E. D. P. de Robertis's Gen. Cytol.* iii. 59 In the fixed nucleus one can distinguish:..(4) in some nuclei there are found larger and denser flakes of chromatin situated in the chromonemata, the chromocenters or karyosomes, also called false nucleoli or chromatin nucleoli. **1961** MACKINNON & HAWES *Introd. Study Protozoa* i. 12 The term karyosome (the endosome of some authors) is here used descriptively as the name of any conspicuous, deeply staining body lying in the nuclear sap, without regard to its constitution. Undoubtedly some karyosomes are really nucleoli (*Trichomonas vaginalis*) and they disappear during mitosis; others play an important part in that process (*Naegleria*) and some are Feulgen positive (*Trichomonas sanguisugae*). **1969** BROWN & BERTKE *Textbk. Cytol.* xvii. 320/1 The terms 'karyosome', 'endosome', and 'central body' are applied to nucleolus-like nuclear organelles that are permanent structures... Karyosomes, etc. are probably permanent nucleoli that undergo division and are found in numerous algae, protozoa, and some fungi.

† **2.** = NUCLEUS *sb.* 6 a. *Obs.*

1894 S. WATASÉ in *Biol. Lect. Marine Biol. Lab. Wood's Holl 1893* 84 An animal cell may be described as composed of two sharply distinct organs: the cell body (cytosome), and the nucleus (caryosome). **1948** [see *KARYOLOGY].

karyotin (kæ·riotin). *Biol.* Also **-ine.** [a. G. *karyotin* (H. Lundegård 1910, in *Svensk bot. Tidskr.* IV. 177), f. *karyo-* KARYO- + *-tin* (after *chromatin*).] The stainable material of a cell nucleus; the substance of which the nuclear reticulum is composed.

1925 E. B. WILSON *Cell* (ed. 3) i. 90 Such considerations led Lundegårdh ('10) to propose that the term 'chromatin' be replaced by 'karyotin' (caryotin), the substance thus designated appearing in either a basichromatic or an oxyphilic phase. **1934** L. W. SHARP *Introd. Cytol.* (ed. 3) iii. 55 Only future research can decide whether karyotin ('chromatin' in the wide sense) is a true chemical compound. **1948** W. ANDREW tr. *E. D. P. de Robertis's Gen. Cytol.* vii. 138 Other authors admit the existence of two distinct phases: the karyolymph, which is considered as a colloid sufficiently stable to be precipitated only with acids and fixatives, and the karyotin, a more labile and complex colloid dispersed in the karyolymph, which would precipitate with great facility. **1966** D. H. KRAMSCH tr. *Grundmann's Gen. Cytol.* ii. 66 The nucleus was..regarded as a solution, a sol in which little droplets, the so-called karyotine droplets, representing a somewhat more solid 'phase', were dispersed.

karyotype (kæ·riotəip), *sb.* *Biol.* and *Med.* [ad. Russ. *kariotip* (L. N. Delone (Delaunay) 1922, in *Vestnik Tiflisskogo botanicheskogo Sada* 2nd Ser. I. 49): see KARYO- and -TYPE.

Orig. coined by Delaunay in sense 2, but according to quot. 1931 it was later coined independently (again in Russ.) by Lewitsky (1924), in sense 1 a.]

1. a. The chromosomal constitution of a cell (and hence of an individual, species, etc.) as determined by the number, size, shape, etc., of the chromosomes (usually, as observed at metaphase during cell division).

1929 *Amer. Jrnl. Bot.* XVI. 415 In studying the karyotypes of different varieties of *Hyacinthus orientalis* which have chromosomes with permanent secondary constrictions, de Mol (1927) found correlation between the number of chromosomes with constrictions and the number of nucleoli. **1931** G. A. LEWITSKY in *Trudy po Prikladnoi Botanike i Selektsii* XXVII. I. 221, I have proposed myself, independently from Delaunay, the same term 'Karyotype', but merely for designation of nuclear peculiarities of a given organism or systematical unit. **1934** L. W. SHARP *Introd. Cytol.* (ed. 3) ix. 128 The diagrammatic representation of a karyotype..is called an idiogram. *Ibid.* 129 Groups of related genera, as well as the related species of a genus, often have the same general karyotype. **1956** *Nature* 18 Feb. 336/1 In species of *Hemerocallis*, etc., different individuals of the same species differ in the karyotypes of the normal cells, even if these do not differ in chromosome number. **1957** C. P. SWANSON *Cytol. & Cytogenetics* xiii. 449 It is from a close study of related species ..that the evolution of the karyotype has been to a limited extent unravelled. **1957**, etc. [see *IDIOGRAM]. **1961** *Lancet* 29 July 263/2 The mother..and the father..were also examined by fibroblast tissue culture and their karyotypes were apparently normal. **1964** M. HARRIS *Cell Culture & Somatic Variation* iv. 239 In these materials, normal chromosome complements were reported to predominate, with abnormal karyotypes appearing ordinarily only in tumors of large size or after transplantation. **1971** *Nature* 31 Dec. 506/3 It should now be possible, in theory at least, to eradicate Down's syndrome (mongolism) by recognizing its karyotype in foetal cells, and aborting the foetuses concerned.

b. A systematized representation of the chromosomes of a cell or cells, esp. a photographic one (cf. *IDIOGRAM).

1950 in WEBSTER Add. **1960** *Lancet* 14 May 1063/1 In contemporary publications the terms, karyotype and idiogram, have often been used indiscriminately. We would recommend that the term, *karyotype*, should be applied to a systematised array of the chromosomes of a single cell prepared either by drawing or by photography, with the extension in meaning that the chromosomes of a single cell can typify the chromosomes of an individual or even a species. **1969** *Nature* 22 Nov. 801/2 The results of a chromosome analysis can be displayed in several ways. The methods commonly used are the karyotype, 'a systematized array of the chromosomes of a single cell', and the karyogram. **1970** *Sci. Jrnl.* June 76/3 The best way we know of detecting abnormalities is to cut out each chromosome from a photomicrograph and pair them up like a jigsaw. This array, known as a karyotype, is constructed according to an international convention. **1970** J. D. BURKE *Cell Biol.* ix. 267 (*caption*) The mitotic metaphase chromosomes of a somatic cell of a male, arranged in a karyotype. **1973** *Lancet* 24 Feb. 420/1 Strictly speaking the actual pictures are karyotypes, and an idiogram is a diagram of the chromosome state of an individual.

† **2.** [After the original meaning of the Russ.] A group of species having similar karyotypes (sense *1 a). *Obs. rare.*

1931 *Trudy po Prikladnoĭ Botanike i Selektsii* XXVII. I. 221 [*Referring to the work of L. N. Delaunay.*] The variations from species to species within the indicated genera, or 'karyotypes' are, as it were, quantitative. In most cases it is but the length of the arms of the chromosomes..which undergoes variation. **1932** H. G. BRUUN in *Symbolae Bot. Upsalienses* I. 111 The species [of *Primula*] can be divided according to their nuclear constitution into different cytological types..called 'karyotypes'. *Ibid.* 117 Allied species as a rule also belong to the same karyotype. *Ibid.* 195 In agreement with this [statement of Delaunay's], 'Karyotypus' has been defined as a taxonomical conception, for which reason its use is made impossible in discussing the relation between taxonomy and cytology. ..In this work..the unconditional term 'cytological type' has had to be employed instead. But there exists an obvious need of a term to include all the species with nuclei of similar type, independent of taxonomical considerations, and in this sense I suggest that the term 'karyotype' be used... The karyotype should, therefore, be founded exclusively on chromosome-morphology.

karyotype (kæ·riotəip), *v. Biol.* and *Med.* [f. prec. sb.] *trans.* To determine or investigate the karyotype of.

1963 *Lancet* 24 Aug. 417/1 The more cells examined (karyotyped) the greater the chance of encountering neoplastic cells. **1971** *Nature* 11 June 387/1 Ninety-seven of these modal cells were karyotyped according to the Denver–London system. **1972** *Lancet* 29 July 213/2 Newborn infants who were suspected of having an anomalous cytogenetic constitution of any kind have been photographed, X-rayed, and karyotyped.

Hence **ka·ryotyped** *ppl. a.*, **ka·ryotyping** *vbl. sb.*

1963 *Lancet* 24 Aug. 417/1 We have undertaken..the karyotyping of cells obtained by lumbar puncture. **1966** *New Scientist* 3 Nov. 217/1 Foetal karyotyping would not, of course, reveal all genetic defects. It would show up gross chromosome abnormalities. **1971** *Nature* 2 July 25/1 In the karyotyped cells, all the normal chromosomes could be separated into pairs on the basis of their distinctive patterns of fluorescence.

karyotypic (kærioti·pik), *a. Biol.* and *Med.* [f. as prec. + -IC.] Of or pertaining to a karyotype. Also **karyoty·pical** *a.*, in the same sense.

1931 *Trudy po Prikladnoĭ Botanike i Selektsii* XXVII. I. 236 The lowest type of karyotypical variations are certainly the polyploid multiplication of sets. **1959** *Cytologia* XXIV. 390 No work has yet been carried out to find out how far karyotypic changes..have contributed to the origin of these varieties. **1963** *Canad. Jrnl. Genetics & Cytol.* V. 132 The idiogram can also be considered as a standard for detecting potential karyotypical alterations. **1964** M. HARRIS *Cell Culture & Somatic Variation* iv. 201 Various neoplasms among a series of related tumors were found to share these features, although individual differences were found in the detailed karyotypic patterns. **1968** *New Scientist* 2 May 219/2 Karyotypic analysis, or biochemical tests for the presence of sex-linked genes.., might be employed as alternative methods of sexing.

Hence **karyoty·pically** *adv.*, as regards karyotype.

1965 *Canad. Jrnl. Genetics & Cytol.* VII. 358 Karyotypically abnormal cells. **1972** *Science* 23 June 1333/1 Differences in..chromosome lengths between two karyotypically divergent groups of *Peromyscus maniculatus* are taken as evidence for an addition–deletion mechanism of chromosomal variation. **1972** *Nature* 8 Sept. 88/2 Several HAT-resistant clones were karyotypically male.

karzy (kā·ızi). *slang.* Also **carsey, carsy, karsey, karzey.** [Corruption of It. *casa* house.] = WATER-CLOSET.

1961 PARTRIDGE *Dict. Slang* Suppl. 1029/1 *Carsey*,..a w.c. **1965** *Daily Mail* 2 Oct. 5/4 Where do you spend a penny? (a) Toilet..(d) Karzy. **1966** D. FRANCIS *Flying Finish* ix. 118, I was in the cockpit most of the time... I went aft to the karzy once. **1967** J. BURKE *Till Death us do Part* v. 84 Have you seen the carsy? Just a bucket with a seat on top. **1968** T. E. B. CLARKE *Trail of Serpent* xiii. 122 You made a real thorough search? Everywhere? Outhouses, karzey, the lot? **1969** K. GILES *Death cracks Bottle* iv. 38 Apart from a working pee none of my ladies nor me got out of here. There's only one door to the carsey. **1970** G. F. NEWMAN *Sir, You Bastard* 262 Visits to the karsey.

kasbah (kæ·zbā). Also **casbah, cashbah, cassaubah, kasba, kasbar.** [ad. F. *casbah*, f. N. Afr. Arab. dial. *ḳaṣba* fortress.] **a.** A North African castle or fortress. **b.** The Arab quarter surrounding a castle or fortress in a North African town, esp. that of Algiers.

1738 T. SHAW *Trav. Barbary & the Levant* 313 They made an unsuccessful Attempt upon the Government, by endeavouring to seize upon the *Cassaubah*. **1844** J. H. DRUMMOND HAY *Western Barbary* 113 We paced with measured steps to the *Kasba*, or citadel, wherein is situated the '*Dar-al-Kebeer*', the residence of the governor. **1857** H. BARTH *Travels* I. 147 The little kasbah, which is never wanting in any of these towns, was in tolerable condition. **1867** 'OUIDA' *Under Two Flags* II. i. 12 Singing her refrain,..as she bounded over the picturesque desolation of the Cashbah. **1890** G. W. HARRIS *Pract. Guide Algiers* I. 41 This Casbah, in the good old days of Algerian predominance, was a magnificent palace fitted with all the luxury and refinement of the epoch. **1902** *Westm. Gaz.* 9 Dec. 3/1 Algiers.., with its quaint Kasbah, or native quarter, its Citadel,..and its interesting suburbs. **1930** E. WAUGH *Labels* viii. 189 Five Scots people..were caught by a very shady guide who took them up to the Kasbar in a taxi-cab. **1958** *Times Lit. Suppl.* 11 July 393/4 The kasbah country south of the Atlas *is* on the tourist route, and well organized. **1961** *Times* 6 May 9/7 In Algiers the Casbah is entirely without romance. **1969** R. LANDAU *Kasbas S. Morocco* 11 The first kasbas (qasbas, more precisely) were built by the Almohad dynasty in the thirteenth century; they were not simply fortresses, but were the walled-in section of the ruler's capital, a complex with palace, main mosque, . . and various state offices. **1971** *Country Life* 28 Oct. 1120 The kasbahs, or mud fortresses, of the pre-Saharan oases in southern Morocco are virtually unique as a form of defensive architecture.

‖ **kasha**[1] (kæ·ʃā). Also **casha.** [Russ.] **1.** A gruel or porridge made from cooked buckwheat or other meals or cereals.

1808 M. WILMOT *Jrnl.* 5 July in *Russ. Jrnls.* (1934) III. 356 Their Casha is very like *Stirabout*, & this is a favourite dish. **1903** [see *BLINTZE]. **1958** HAYWARD & HARARI tr. *Pasternak's Dr. Zhivago* II. ix. 270 I'll get Uncle Yury to stay to dinner and take the kasha out of the oven. **1961** N. FROUD et al. tr. *Montagné's Larousse Gastronomique* 554 Kasha is the Russian for cooked buckwheat. **1966** N. BEHN *Kremlin Let.* xiv. 141 Breakfast..consisted of one small bowl of kasha..the [Russian] equivalent to American hot cereals. **1971** *New Statesman* 1 Jan. 8/3 Rents are low in Russia..and the basic necessities—bread, potatoes and *kasha* flour—are not dear. **1973** *Times* 3 Feb. 13/5 You can try the kneidlach soup (with matzo-meal dumplings), the kasha (buckwheat) and the tzimmes.

2. A beige colour resembling that of buckwheat groats.

1957 M. MCCARTHY *Memories Catholic Girlhood* viii. 200 She had an outfit made..in a new colour called 'kasha'. **1971** *Guardian* 19 Jan. 9/3 Principal colours are navy, 'Kasha' (a Russian buckwheat porridge beige), and 'smoke'.

Kasha[2] (kæ·ʃā). Also **kasha.** The proprietary name, originated by Rodier, a French textile manufacturer, of a soft napped fabric made from wool and hair. Also in various *Combs.* (see quots.). Also applied to a cotton lining material.

1920 *Queen* 10 Apr. 466 Dress with very wide skirt in pale green kasha. **1923** *Daily Mail* 12 Feb. 15 White Kasha cloth. **1926** *Queen* 17 Feb. 10 A modified Inverness coat made by Lelong in the new kasha with the slightly spongy surface—kashatoile. **1926** G. G. DENNY *Fabrics* (ed. 2) II. 53 *Kasha*, (a) fine, soft, napped wool dress fabric originated with Rodier Freres, Paris. (b) Cotton plain weave napped on reverse side, for linings. **1928** *Observer* 4 Mar. 20/4 The couturiers..give prominence to jersey, crêpella, Kashatoil, Kashangora, and a host of other materials. **1941** R. STOUT *Red Threads* i. 6 He must have the natural kasha, the one with nubs, by tomorrow. **1942** G. G. DENNY *Fabrics* (ed. 5) I. 37 *Kasha*, fine, soft napped wool dress fabric with crosswise streaked effect in dark hairs... Similar textures in wool and cotton now made in United States called Kasha. **1967** *Boston Sunday Globe* 23 Apr. (Advt. Suppl.), Roomy enough to sleep 2 or separate into two single full size sleeping bags. Top, bottom and snap-on canopy made of cotton duck, with kasha flannel lining. **1968** J. IRONSIDE *Fashion Alphabet* 234 *Kasha*, very soft pale beige fabric made from goat-hair and wool.

Kashan (kāʃā·n). Also **Keshan.** The name of a province and town in central Iran, used (freq. *attrib.*) to designate a finely woven rug, usu. of wool or silk, made there.

1905 M. C. RIPLEY *Oriental Rug Bk.* 305 Kashan rugs. **1920** C. J. D. MAY *How to identify Persian Rugs* v. 61 *Sarouks*... The student may regard these merely as a slightly inferior grade of Kashan. **1931** A. U. DILLEY *Oriental Rugs & Carpets* iv. 122 Kashan rugs, always important, were so far unrecognized as a separate weaving thirty years ago that no mention of them was made in the first modern rug books. **1960** H. HAYWARD *Antique Collecting* 157/1 *Kashan carpets*, Persian carpets in which medallion and prayer designs predominate, woven in wool or silk. **1962** *Times* 12 June 20/1 (Advt.), A beautiful and rare carved silk Keshan rug...a superb natural Keshan carpet. **1962** C. W. JACOBSEN *Oriental Rugs* II. 230 In the 20 nearby villages..Kashans are woven which are equal to those made in the city itself. **1963** L. DEIGHTON *Horse under Water* xxxviii. 150 I'm leaving before I vomit over your beautiful Kashan carpet. **1967** J. MORGAN *Involved* 126 Thomas looked up from an all-silk Kashan rug he was lying on.

Kashgai (kæ·ʃgəi). Also **Kashkai, Qashgai, Qashqai.** A Turkic-speaking people living around Shiraz in Persia; a member of this people. Also *attrib.*

1885 *Encycl. Brit.* XVIII. 625/1 The Kashgais, or those wandering semi-Turkish tribes brought down from Turkestan to the neighbourhood of Shiráz, have the credit of possessing good steeds. **1921** *Glasgow Herald* 19 April 5 Among the most important questions were the relations with the Kashgais and other nomad tribes. The Kashgais were about 135,000 strong. **1937** *Sunday Times* 29 Aug. 7/5 The region is occupied by scattered and turbulent Qashgais upon whom the Tehran Government has only very recently imposed its authority. **1957** H. H. VREELAND *Iran* ix. 105 The Qashqa'i tribesmen are important in this area. *Ibid.* x. 128 The Qashqa'i have firmly resisted conscription, and no tribal members have yet been taken into the army.

b. Used *attrib.* or *absol.* to designate a type of nomad rug made by them.

1922 KENDRICK & TATTERSALL *Hand-Woven Carpets* 196 Kashkai carpets. **1931** A. U. DILLEY *Oriental Rugs & Carpets* iv. 95 Both Afshar and Kashgai rugs are Caucasian nomad transplants. **1962** C. W. JACOBSEN *Oriental Rugs* II. 257 Plate 104 is a typical design found in many Qashqai (Mecca Shiraz) but no two designs are exactly alike. *Ibid.* III. 403 The correct technical name is 'Qashqai Prayer Rug'.

Kashgar (kā·ʃgāɹ). **1.** A language or dialect of the central Turkic or Turco-Tatar group of Altaic languages, spoken in Kashgar. Also *attrib.* or as *adj.*

1875 R. B. SHAW *Sk. Turki Lang.* p. xv, An examination of the Yarkánd and Káshghar dialect accounts for them in another way. **1918** G. W. HUNTER *Examples Turki Dial.* III. 1 Under this section is included Kashgar Turki and Kirghese Turki as they have much in common... Kashgar Turki is spoken by the Sart..people of Chinese Turkestan. **1948** D. DIRINGER *Alphabet* 568 Kashgar Turkish, spoken between the T'ien Shan mountains and northern Tibet. **1954** PEI & GAYNOR *Dict. Ling.* 113 *Kashgar*, an Asiatic language; member of the Central Turkic group of the Altaic sub-family of the Ural-Altaic family of languages.

2. The name of a city and district of Sinkiang-Uighur (formerly East Turkestan), used *attrib.* or *absol.* to designate a type of Turkoman carpet.

1900 J. K. MUMFORD *Oriental Rugs* (1901) 273 Kashgar rugs. **1931** A. U. DILLEY *Oriental Rugs & Carpets* x. 224 Kashgar and Yarkand rugs..are attributed more or less arbitrarily. *Ibid.* pl. 69 (*caption*) Kashgar, Disc-Star-Seal Pattern. **1962** C. W. JACOBSEN *Oriental Rugs* II. 232 Kashgars are more like a Chinese rug than a Bokhara. None are available, and I have seen a few rugs that were called Kashgar but there is not enough definite information on these rugs... We seldom come across the name.

Kashmir (kæ·ʃmiəɹ, kæʃmiə·ɹ). [var. CASHMERE.] **1.** Used *attrib.* in the sense: of or pertaining to Kashmir. Also as *sb.*, a native or inhabitant of Kashmir. Hence **Kashmi·rian** *sb.*

1876 C. F. G. CUMMING *From Hebrides to Himalayas* II. 309 Kashmerians, Persians, Paharis, Hindus of every possible sect. **1882** F. M. CRAWFORD *Let.* 21 July in M. H. Elliott *My Cousin F. M. Crawford* (1934) vi. 136 The Tibetans have no postage stamps, but the Kashmirs have. **1887** J. M. BROWN *Shikar Sk.* 237 The Kashmir stag—cervus Cashmeriensis. **1925** A. G. ARBUTHNOT in G. Burrard *Big Game Hunting* 126 A very old Kashmir shikari. **1971** R. RUSSELL tr. *Ahmad's Shore & Wave* i. 15 Courtiers..who had been..patiently gazing at the Kashmir carvings on the wooden columns.

2. Also **Cashmere.** Used *attrib.* of a Caucasian pileless carpet, characterized by the many loose yarn ends on the back resembling a Cashmere shawl. Also *ellipt.*

1900 J. K. MUMFORD *Oriental Rugs* (1901) ix. 119 It is the shaggy ends of the colored nap-yarns, left loose at the back of these rugs, which has given them the name of 'Kashmir'... The true name of the so-called 'Kashmir' rugs is Shemakha, derived from the city where they are marketed. **1904** M. B. LANGTON *How to know Oriental Rugs* iii. 127 The Cashmere or Soumak rugs are not from the Vale of Cashmere, in India,..but are made by the nomad tribes in and about Shemakha, the old capital of Shirvan. **1931** A. U. DILLEY *Oriental Rugs & Carpets* vii. 179 Soumak rugs, formerly called Kashmir,..derive their true name from the town of Shemakha. **1962** C. W. JACOBSEN *Oriental Rugs* II. 297 Rugs are generally called Soumak or Kashmirs... The distinguishing feature of this rug is the loose ends of stitch yarn at the back of the rug, about the same effect as we find in the Cashmere shawl.

Kashmiri (kæʃmiə·ri), *a.* and *sb.* Also **Kashmiree.** [f. prec. + *-I.] **A.** *adj.* Of or pertaining to Kashmir, a state in the western Himalayas. **B.** *sb.* **a.** A native or inhabitant of Kashmir. **b.** The Dardic language spoken in Kashmir.

1880 *Encycl. Brit.* XIII. 821/2 The language distinct from Turki, Persian, Hindi, and Kashmiri. **1884** KIPLING *Let.* in C. E. Carrington *Rudyard Kipling* (1955) iv. 56 There came out a Cashmiri girl that Moore might have raved over. **1891** 'L. MALET' *Wages of Sin* III. v. vii. 56 The Kashmiree beauties. **1901** KIPLING *Kim* i. 33 A smooth-faced Kashmiri pundit. **1936** P. FLEMING *News from Tartary* VII. xi. 375 Kashmiri herdsmen camped in grubby tents. **1953** R. GODDEN *Kingfishers catch Fire* xiii. 159 She had to speak to them in Urdu and they probably only knew Kashmiri. **1955** *Times* 27 June 6/3 It is therefore surprising that so little attention is given to the political welfare of the four million Kashmiris. **1971** *Shankar's Weekly* (Delhi) 4 Apr. 5/3 He was taken for a Kashmiri priest. **1974** 'S. HARVESTER' *Forgotten Road* i. 11 He spoke..Hindi and Kashmiri, Tamil and Gujerati and Konkani among the many tongues of India.

Kashrut (kāʃrū·t). Also **Kashres, Kashrus, Kashruth,** and with small initial. [Heb., = fitness, legitimacy (in religion), f. *kāshēr* KOSHER *a.*] The body of Jewish religious laws relating to the fitness of food, and also of persons and objects; the observance of these laws. Also *attrib.*

1907 *Daily Chron.* 7 Dec. 4/7 The word 'Kashruth'.. denotes the dietary laws as laid down by Moses, together with..commentaries and explanations thereof made by the Rabbis in the intervening centuries... 'Questions of Kashruth' concern themselves even with the utensils used in the preparation and serving of the food... [They] can only be decided by the Rabbi. **1953** *Jewish Chron. Trav. Guide* 12 In this group come those establishments in regard to the kashrut of which no information is to hand, or which, while not insisting on strict observance of kashrut, have a special interest for Jewish clients. *Ibid.* 43 The following..are under the official supervision of the Beth Din and Kashrus Commission. **1962** *New Jewish Encycl.* 264 Most Jewish communities the world over set up Kashrut regulations and provide..for the supervision of the slaughtering and distribution of meat. **1970** L. M. FEINSILVER *Taste of Yiddish* 242 *Kashres*, the system of kashruth observance, or 'keeping kosher'... Though the Reform movement originally rejected the binding force of kashruth, some Reform Jews today 'keep kosher'. **1973** *Jewish Chron.* 19 Jan. 8/1 The freedom and the high standards of kashrut which we in Britain enjoy should be shared equally by the Jewish communities in the rest of the EEC. **1973** *Times* 3 Feb. 13/4 The laws of Kashrus, which govern Jewish cooking and give us the label 'Kosher Food'.

Kashube (kāʃū·b). Also **Kashub, Kaszube.** [f. *Kashubia* (Pol. *Kaszuby*), a region of Poland west and north-west of Gdansk.] **a.** A member of the Slavonic people inhabiting Kashubia. **b.** The Slavonic language spoken in this region. Also *attrib.* or as *adj.* So **Cassu·bian, Kashu·bian, Kashu·bish, Kassu·bian** *sbs.* and *adjs.*

1893 W. R. MORFILL *Poland* i. 13 The language of the Kashubes differs in some interesting points from the Polish, having a fluctuating accent..and more nasal sounds. **1919** A. B. BOSWELL *Poland & the Poles* 14 The original Pomeranians were absorbed by German colonists. But in the region west of the Vistula there still dwells a tribe called the Kaszubes who are descended from them... This region of Pomerania..is known to the Poles as the Kaszubian Switzerland. *Ibid.* 26 Lower Polish dialects.. Kaszubian. **1934** PRIEBSCH & COLLINSON *German Lang.* I. i. 11 The Cassubian and almost extinct Slovinzian (brought by Lorentz under the collective name Pomoranian). **1935** *Times Lit. Suppl.* 15 Aug. 506/2 A Slav language, called sometimes Slovincian and sometimes Cassubian. **1936** *Discovery* Mar. 95/1 The Cassubians are an ancient and peculiar tribe who live on the seashore on both sides of the German-Polish frontier line. **1950** A. P. COUDY in *Cambr. Hist. Poland* i. 9 From the linguistic point of view Slovinzish and Kashubish belong to the Polish group. **1950** [see *LECHITIC *sb.* and *a.*]. **1955** *Archivum Linguisticum* VII. 133 The accent is free in North Kashubian. **1957** *Encycl. Brit.* XIII. 293 *Kashubes*, a Slavonic people living in the northwest of Poland. *Ibid.* In Kashube, as against Polish, all vowels can be nasal instead of *a* and *e* only. *Ibid.* XVIII. 152/1 Linguistically.. two local Pomeranian dialects remained until the 20th century, the Slovince (Slowinski) and the Cassubian (Kaszubski). **1972** W. B. LOCKWOOD *Panorama Indo-Europ. Lang.* 158 The present territory of these Pomeranian Slavs, Kashubs as they call themselves, comprises no more than the north-eastern tip from Lake Leba to the southern outskirts of Gdynja (Gdingen). Its southern border is ill-defined, being followed by a broad band of transitional dialects, basically Kashubian, but already highly polonised.

kasidah, var. *QASIDA.

kas-kas (kɒs,kɒs). *Jamaican.* Also **cuss-cuss, kass-kass, kos-kos, kus-kus.** [f. Twi *kasákàsa* to dispute.] In folk usage: a dispute, quarrel. Pronunciation perh. influenced by Eng. *cuss* var. *curse.*

1873 C. J. G. RAMPINI *Lett. from Jamaica* 176 Cuss-cuss (calling names) no bore hole in my skin. **1943** L. BENNETT *Jamaican Humour in Dial.* 15 Dat marga gal Wingy Want put me eena kus-kus An big lian story. **1950** —— et al. *Anancy Stories & Dial. Verse* 33 Anancy never like fe se' two people live neutral, so him start fe carry lie and story between dem, and start big kaskas. **1961** F. G. CASSIDY *Jamaica Talk* iii. 29 When Jamaicans become angry and indulge in a *kas-kas*, the lilt is quite lost and the imprecations come pelting in a high-pitched volley. **1971** *Jamaican Weekly Gleaner* 3 Nov. 5/1 She's..fed up of kass-kass with customers.

kasolite (kæ·so-, kēi·zoləit). *Min.* [a. F. *kasolite* (A. Schoep 1921, in *Compt. Rend.* CLXXIII. 1476), f. *Kasolo*, the name of a locality (prob. near Shinkolobwe, west of Likasi) in Katanga province, Zaïre: see -ITE[1].] A yellow, rather soft, hydrous silicate of lead and uranium, $Pb(UO_2)(SiO_4).H_2O$.

1922 *Mineral. Mag.* XIX. 343 Kasolite... Hydrated silicate of uranium and lead, $3PbO.3UO_3.3SiO_2.4H_2O$, forming ochre-yellow, monoclinic crystals. **1958** E. W. HEINRICH *Mineral. & Geol. Radioactive Raw Materials* viii. 294 Kasolite occurs preferentially in dolomitic-graphitic shale.

kassidar, var. *KHASSADAR.

Kassite (kæ·səit), *sb.* and *a.* Also **Cossæan, Kasshi, Kossæan.** [Native name.] **A.** *sb.* A member of an Elamite people from the central range of the Zagros mountains, who ruled Babylon from the 18th to the 12th century B.C.; also, their language. **B.** *adj.* Of or pertaining to the Kassites.

1888 Z. A. RAGOZIN *Assyria* (ed. 2) ix. 300 The next.. expedition, against the very warlike and turbulent mountain tribes of the *Kasshi* (*Cossæans* of classical writers), is of some interest because of the details we are given concerning that most rugged region of the Zagros range. **1894** A. H. SAYCE *Primer of Assyriology* iii. 47 Babylonia was conquered by Kassite princes who ruled over it for 576 years and nine months (B.C. 1806–1229). **1898** C. R. CONDER *Hittites & their Lang.* ii. 41 The Kassites thus became dependent on Assyria. **1902** *Encycl. Brit.* XXVI. 43/1 Babylonia was conquered by Kassites or Kossæans from the mountain of Elam, under Kandis or Gaddas (in 1800 B.C.), who established a dynasty which lasted for 576 years and nine months. **1909** *Daily Chron.* 14 Jan. 4/4 There is a letter to Kadashman-Targu, the Kassite King of Babylon, as to the appointment of a successor. **1928** [see *GUTIAN *sb.* and *a.*]. **1934** A. TOYNBEE *Study of Hist.* I. 116 The Kingdom of the Sea-Land..had been annexed to the barbarian 'successor-state', subsequently established by the Kassites at Babylon, at the turn of the eighteenth and seventeenth centuries B.C. **1938** T. FISH in E. I. J. Rosenthal *Judaism & Christianity* III. ii. 31 In the second millennium Babylonia was Semitic, though the rulers and the court at Babylon were non-Semitic (Kassite) for several centuries. **1939** L. H. GRAY *Found. Lang.* xii. 380 Attempts have been made to connect the language [*sc.* Elamite] with Altaic,..with Kassite, and with Carian... The language of the *Kassites* (or *Cossaeans*) ..is known only from a scanty glossary. **1964** G. ROUX *Ancient Iraq* vi. 79 The Sumerians..and..the Amorites, Kassites, Assyrians and Chaldaeans who, after them, ruled in succession over Mesopotamia.

Kassubian: see *KASHUBE.

kastura (kastū·rā). Also **kasturi, kustoorah.** [Hindi *kastūrī*.] The Himalayan musk deer, *Moschus moschiferus.*

1837 T. HUTTON in *Jrnl. Asiatic Soc. Bengal* VI. 936 The Kastúra, or musk deer of these hills is to be found in the deep forest shades of Mahássú throughout the year. **1867** T. C. JERDON *Mammals India* 267 *Moschus moschiferus*... *Kastúrá*, H.— *Rous* or *Roos* and *Kasturé*, Kashmir. **1884** R. A. STERNDALE *Nat. Hist. Mammalia India & Ceylon* 494 The Musk Deer. Native names—*Kastura*, Hindi;.. *Kasturé*, in Kashmir. **1893** [see MUSK-DEER]. **1904** F. G. AFLALO *Sportsman's Bk. India* 186 Local Names of various Game... Musk deer. Kustoorah. **1951** ELLERMAN & MORRISON-SCOTT *Checklist Palaearctic & Indian Mammals* 353 *Moschus moschiferus* Linnaeus, 1758. Musk Deer (Kastura). **1955** P. BAUER *Kanchenjunga Challenge* IV. ii. 184 The Kasturi is a [Himalayan] musk deer and on this account it has been almost exterminated.

‖ **kata** (ka·tā). [Jap.] A system of basic exercises or formal practice used to teach and improve the execution of Judo techniques, devised by Prof. Jigoro Kano (1860–1938).

1954 E. DOMINY *Teach Yourself Judo* 190 *Kata*, a prearranged series of movements performed for the purpose of demonstration. **1956** C. YERKOW *Judo Katas* i. 14 This training is called *kata* and means form-practice, both for stand-up techniques and in mat-work. **1961** *New Statesman* 22 Sept. 402/2 Kata is the general technique of posture and balance. **1970** G. JACKSON *Let.* in *Soledad Brother* (1971) 247 It's never bothered me too much before, the sex thing. I would do my exercises and the hundreds of katas, stay busy with something. **1973** *Express* (Trinidad & Tobago) 27 Apr. 31/3 Over 500 karatekas will be competing for titles in..katas (imaginary combat).

kata, shortened f. *KATATHERMOMETER.

katabatic (kætăbæ·tik), *a. Meteorol.* [f. Gr. καταβατ-ός descending (f. καταβαίνειν to go down) + -IC (or ad. Gr. καταβατικ-ός affording a means of descent).] Of a wind: blowing down a slope, or from an elevated region to a lower one, esp. when caused by the effect of gravity on air cooled by the underlying ground. Cf. *ANABATIC *a.* 2.

1918 *Meteorol. Gloss.* (Meteorol. Office) 182 A local cold wind is called Katabatic if it is caused by the gravitation of cold air off high ground. **1920** W. J. HUMPHREYS *Physics of Air* vii. 111 Where the valley is long and rather steep..the down-flowing air current may attain the velocity of a gale and become a veritable aerial torrent. This drainage flow is known..as the mountain breeze, or mountain wind; also canyon wind, katabatic wind, and gravity wind. **1936** *Geogr. Jrnl.* LXXXVII. 433 The extremely cold katabatic winds blowing off the high Greenland ice-cap..lowered the temperature. **1954** W. D. THORNBURY *Princ. Geomorphol.* xiv. 362 The radially outflowing winds which Hobbs took as evidence of the existence of a permanent anticyclone over the interior of the Greenland ice cap are really katabatic winds or cold air draining downslope under the influence of gravity. **1967** R. W. FAIRBRIDGE *Encycl. Atmospheric Sci.* 1153/1 Foehn... This is a gusty katabatic wind which crosses the Alps..and is characterized by dryness and warmth.

Hence **kataba·tically** *adv.*, as a result of downward motion.

1967 R. W. FAIRBRIDGE *Encycl. Atmospheric Sci.* 1152/2 They consist of more or less cool, dry, continental air, nearly always of anticyclonic source, at times katabatically warmed.

‖ **katabothron.** Add: Forms with initial *k* and medial *v* are most usual (with varying endings: see note below). Substitute for etym.: [a. mod.Gr. καταβόθρα swallow-hole, f. κατά down + βόθρος hole.

Katavothra (or *-bothra*) is the correct sing. form, with plurals *-ai* (*-æ*) and *-es* (corresponding respectively to the mod.Gr. pl. forms καταβόθραι and καταβόθρες). *Katavothre* is an erron. sing. formation from *katavothres*; the sing. ending *-on* (pl. *-a*, *-ons*) corresponds to nothing in mod. or ancient Gr. and prob. arose as a result of mistaking καταβόθρα for a neut. pl.]

(Further examples.)

1892 A. J. JUKES-BROWNE *Student's Handbk. Physical Geol.* (ed. 2) xi. 197 The torrents of the Morea are usually charged with reddish mud, sand, and pebbles, when they enter the *katavothra*, but are pure and limpid when they flow out again. **1892** *Proc. R. Geogr. Soc.* XIV. 466 (*heading*) The katavothræ of the Morea. *Ibid.* 467 All these basins are drained by underground channels, to the entrance of which the name katavothra is given. **1937** *Geogr. Jrnl.* XC. 448 Since the water was drawn off, the mouths of twenty-five large katavothrai..have appeared in the sides of the surrounding hills. **1957** G. E. HUTCHINSON *Treat. Limnol.* I. i. 107 Though these variations [in level] can in part depend on rainfall, they are largely independent of it and may be attributed to the silting of the katavothrai or sinks which drain the lake. **1970** *Water-Supply Paper U.S. Geol. Survey* No. 1899-K. 11/2 *Katavothron*, a closed depression or swallow hole. **1971** J. N. JENNINGS *Karst* v. 74 The French name *estavelle* is commonly used for these alternating orifices but they are also well known by the Greek *katavothre.*

katakana (kætăkā·nă). Also 8 **kattakanna,** 9 **katagana.** [Jap., f. *kata* side + *kana* *KANA.] One of the two varieties of the Japanese syllabic writing, the characters of which are more angular than the hiragana, derived from abbreviated forms of Chinese ideographs of the corresponding sounds, and used chiefly in scientific and official documents and in spelling out foreign words adopted into the Japanese language. Cf. *HIRAGANA.

1727 J. G. SCHEUCHZER tr. *Kæmpfer's Hist. Japan* II. v. xiv. 590 The other was a map of the whole world, of their own making, in an oval form, and mark'd with the Japanese *Kattakanna* characters. **1822** F. SHOBERL tr. *Titsingh's Illustr. Japan* 194 These works, published in the learned language, *Gago*, with the *kata-kana*, or women's letters, have been re-printed expressly for them. **1859** A. STEINMETZ *Japan & her People* I. vii. 305 *Katagana* is very simple, each sound having one invariable representative. **1861** G. SMITH *Ten Weeks in Japan* xi. 173 A copy of St. Luke's gospel in the Katagana Japanese character published at Hong-kong. **1880, 1928** [see *HIRAGANA]. **1970** *Jrnl. Gen. Psychol.* LXXXII. 40 The transmission of the Katakana syllabary by electrical signals applied to the skin. **1973** *Physics Bull.* May 280/3 We have also extended the process by including in addition to the 881 Chinese characters, the 50 Japanese Kata-Kana and 50 Hira-Gana characters and 10 numerals.

katalase, var. *CATALASE.

katamorphism (kætămǭ·ɹfiz'm). *Petrol.* [f. Gr. κατά down + μορφή form: see -ISM.] Alteration of rocks (usually at or near the earth's surface) characterized by the formation of chemically simpler minerals from more complex ones. Hence **katamo·rphic** *a.*

1904 C. R. VAN HISE in *Monogr. U.S. Geol. Survey* XLVII. 43 The geological factor which in this treatise will serve as the primary basis for a classification of metamorphism is the dominant factor of depth. On this basis metamorphism will be classified into (1) alterations in the zone of katamorphism and (2) alterations in the zone of anamorphism... The zone of katamorphism may be defined as the zone in which the alterations of rocks result in the production of simple compounds from more complex ones. *Ibid.* 162 Katamorphic zone. **1916** F. H. LAHEE *Field Geol.* ix. 230 Any kind of alteration that any rock has undergone, whether katamorphic or anamorphic, comes under the head of metamorphism, but there are many geologists who prefer to restrict the meaning of the more general term so that it does not

include weathering. **1946** *Amer. Mineralogist* XXXI. 288 It is very well known that oxidation is one of the important chemical processes of katamorphism, especially of weathering.

‖ **katana** (kătā·nă). Also 7 **cattan.** [Jap.] A long single-edged sword of the Japanese samurai.

1613 J. SARIS *Jrnl.* 11 June in *Voy. Japan* (1900) 79 Either of them had two Cattans or swords of that Countrey by his side. **1615** R. COCKS *Diary* 17 June (1883) I. 10, I delivered Mr. Richard Wickham the rich *cattan* he left in my custody at his departure towardes Siam. **1874** *Trans. Asiatic Soc. Japan* 57 The word 'sword' is invariably rendered by the Japanese word '*ken*', which signifies a long, straight, double-edged sword, as opposed to the '*katana*', of modern times, which has but a single edge, and is slightly curved towards the point. **1890** B. H. CHAMBERLAIN *Things Japanese* 328 The Japanese sword of ancient days (the *tsurugi*) was a straight double-edged heavy weapon some three feet long... That of medieval and modern times (the *katana*) is lighter, shorter, has but a single edge, and is slightly curved towards the point. **1906** *Macm. Mag.* Apr. 457 An escort of sturdy little Japanese armed with service rifles and the keen-bladed *katana.* **1959** R. KIRKBRIDE *Tamiko* xvii. 138 He'd cut both our heads off with honourable katana. **1963** *Art of Armourer: Exhib. Armour, Swords & Firearms* (V. & A. Mus.) 97 Blade of the long sword (*katana*) signed.., and dated in the 6th year of the period Yeishō (1509).

katathermo·meter (kæ:tă-). Also **kata thermometer, kata-thermometer.** [f. KATA- + THERMOMETER.] An alcohol-in-glass thermometer with an enlarged bulb and restricted scale, used for determining the cooling power of ambient air by measuring the time taken for its temperature to fall from one fixed value to another. Also shortened to **ka·ta.**

1914 *Rep. Brit. Assoc. Adv. Sci. 1913* 673 (*heading*) The katathermometer. By Professor Leonard Hill. **1915** L. HILL et al. in *Phil. Trans. R. Soc.* B. CCVII. 185 The katathermometer..is an instrument designed primarily for the measurement of its own rate of cooling when its temperature approximates to that of the human body. **1915** *Ibid.* 191 The heat lost from the kata at body temperature. **1930** W. G. KENDREW *Climate* xxx. 189 The conditions of a perspiring body may be imitated by surrounding the bulb of the kata-thermometer with wet muslin. **1936** *Discovery* Sept. 280/1 When the air conditions of the bakery were tested, the 'Katathermometer' revealed most unsuitable conditions. **1938** *Jrnl. Amer. Med. Assoc.* 18 Nov. 1649/2 The time required for the alcohol meniscus to fall from the 100 to the 95 degree mark is then observed, and the 'kata factor' marked on the stem..is divided by this time in seconds, giving the 'cooling power' of the air. **1948** W. N. WITHERIDGE in F. A. Patty *Industr. Hygiene & Toxicol.* I. x. 346 The kata thermometer has been calibrated as an air-velocity instrument. **1963** HERTIG & BELDING in C. M. Herzfeld *Temperature* III. III. xxxii. 348 The thermoanemometer.. and the katathermometer are representative of instruments whose readings are proportional to the rate of heat loss.

Hence **ka:tathermome·tric** *a.*

1923 *Med. Res. Council Special Rep. Ser.* No. 73. 90 (*heading*) A kata-thermometric comparison of methods of heating and ventilation.

katavothra, -vothre, -vothron: see KATABOTHRON in Dict. and Suppl.

kate(h)ina, varr. *KACHINA.

Kate Greenaway (kēit grī·nāwēi). The name of Kate (Catherine) Greenaway (1846–1901), English artist and illustrator of children's books, used *attrib.* and *absol.* to designate the style of children's clothing modelled on her drawings.

1902 *Little Folks* 153/1 Mothers began to dress their little boys and girls in what they came to call 'Kate Greenaway dresses'. **1907** 'E. GODFREY' *Eng. Children in Olden Time* xvi. 262 The Little Female Academy is supposed to exist about 1770, so by that time the pretty little 'Kate Greeneway' [*sic*] garments prevailed. **1940** BEERBOHM *Lett. to R. Turner* (1964) App. B. 294 The Preraphaelite influence, with the Kate Greenaway influence thrown in. **1957** M. B. PICKEN *Fashion Dict.* 193 (*caption*) Kate Greenaway..Coat, 1890's. **1959** *Times* 21 Sept. 12/3 The child bridesmaid..wore a long Kate Greenaway dress. **1960** S. KALE *Fire Escape* v. 27 Priscilla was naturally chosen to take part in a scene calling for Kate Greenaway fashions. **1965** M. SHARP *Sun in Scorpio* III. xxv. 131 Elspet was peddling lavender-bags in her Kate Greenaway. *Ibid.* 130 'That's me being bridesmaid in Kate Greenaway,' pointed out Elspet. **1966** J. S. COX *Illustr. Dict. Hairdressing* 86 *Kate Greenaway style*, hairdresses for young children inspired by the hair fashions depicted in the coloured illustrations by Kate Greenaway published between 1879–1895. **1968** J. IRONSIDE *Fashion Alphabet* 140 *Kate Greenaway*, a child's bonnet of the style illustrated by Kate Greenaway, similar to those worn in the Empire period and with a frill round the face. Still worn by small bridesmaids sometimes.

‖ **κατ' ἐξοχήν** (kăte·ksokīn, -kēin). More usual form of KATEXOKEN. [Gr. phr.] Pre-eminently, par excellence.

1588 A. FRAUNCE *Lawiers Logike* sig. ¶¶1v The Romayne Lawe, which Iustinian calleth the Cyuill law κατ' ἐξοχήν, (as Homer is called the Poet). **1698** J. SERGEANT *Non Ultra* (in *Monist* XXXIX, 1929) 605. § 21 Which Propositions being..most fully and Properly such we do therefore, κατ' ἐξοχήν, call Identical. **1841** MILL *Let.* 1 Mar. in *Works* (1963) XIII. 466 Poetry κατ' ἐξοχήν or poet's poetry as opposed to everybody's poetry. **1865** —— *Exam. Hamilton's Philos.* xx. 402 If any general theory of the sufficiency of Evidence and the legitimacy of Generalization be possible, this must be Logic κατ' ἐξοχήν. **1879** W. JAMES *Coll. Ess. & Rev.* (1920) 88 Schopenhauer ..says that Intuition..'is knowledge κατ' ἐξοχήν'.

kathak (kătā·k). [Skr., = professional story-teller.] **a.** A North Indian caste of story-tellers and musicians; a member of this caste. **b.** A North Indian classical dance composed of passages of mime alternating with passages of dance.

1931 E. A. H. BLUNT *Caste System of N. India* 244 Kathak.—These religious troubadours carefully preserve their ancient ballads, and allow nobody to tamper with them. **1941** 'LA MERI' *Gesture Lang. Hindu Dance* 17 Pantomime is relatively unimportant in Kathak, virtuosity being entirely in the feet. **1957** G. B. L. WILSON *Dict. Ballet* 158 *Kathak*, one of the four main forms of Indian dancing. **1959** *Marg* XII. iv. 10 Thakur Prasad, a Kathak, migrated to Lucknow about the beginning of the 19th century and became the Court Dancer of Wajid Ali Shah. *Ibid.* 12 The contribution which the Jaipur gharana has made to the preservation and spread of the Kathak dance is very considerable. To this school goes the credit of having given us some of our best Kathaks. **1967** SINGHA & MASSEY *Indian Dances* xv. 125 The word Kathak, story-teller, derives from 'katha' which means story. **1969** *Amrita Bazar Patrika* 5 Aug. 4/6 The group-dances were generally imposing based on the kathak, and soft touch of Manipuri and choreographer Namita Chatterjee intelligently introduced the various folk-dances—Rajasthani, Manipuri, Garba, Lotus to avoid the spell of monotony. **1969** *Cultural News from India* Nov. 25 To observe the sixtieth birth-day of Ustad Dabir Khan a two-hour Saraswat Veena and Surbahar recital was given... Munawar Ali Khan (vocal), Ustad Bahadur Khan (sarod), Bandana Sen (Kathak dance) and Keramatullah (tabla) were among those who took part in the programme. **1971** *E. Afr. Standard* (Nairobi) 10 Apr. 6/7 Apart from the film songs, there was a delightful Kathak group dance by three very young pupils of Uma Devi.

Kathakali (kătākă·li). Also **Kadhakali.** [Malayalam *kathakali* drama, f. *katha* story (Skr. *kathā*) + *kali* play.] A South Indian dance-drama based on Hindu literature, and characterized by its stylized costume and make-up, and frequent use of mime.

1900 T. K. GOPAL PANIKKAR *Malabar & its Folk* v. 74 Malabar Drama... Our drama..assumes various forms of which Krishnattom and Ramanattom are the principal ones. The latter is usually called *Kadhakali* and constitutes our drama proper... Our drama is altogether a dumb-show in which the actors never utter a word but do everything by signs and gestures. **1933** R. K. YAJNIK *Indian Theatre* iii. 61 The 'Kathākali' of the Malabar district..fairly gives an idea of what a country drama is like. **1967** *Spectator* 18 Aug. 193/3 At the Saville..we are seeing Kathakali for the first time in its pure form. Dating from the sixteenth century, indigenous to Malabar, it is a dance-drama drawn from the great Indian epics. **1969** *Cultural News from India* Nov. 13 Kanak has devoted her whole heart and soul to the study of Kathakali and defied tradition in the fact that females are traditionally debarred from dancing the genre. **1969** R. SHANKAR *My Music* iii. 63/2 The noble style of the dance drama, *Kathakali.* **1974** H. R. F. KEATING *Bats fly Up* i. 10 The measured-to-a-fingersbreadth gestures of a Kathakali dancer.

katharevousa (kæ:þăre·vūsă). Also **katharevoussa, katharevusa.** [mod.Gr. καθαρεύουσα, fem. of καθαρεύων, pres. pple. of Anc.Gr. καθαρεύειν to be pure, f. καθαρός pure.] The purist form of Modern Greek; the 'official' language as opposed to the spoken and literary *DEMOTIC *a.* 1 b.

1912 S. ANGUS tr. *Thumb's Handbk. Mod. Gr. Vernacular* p. xi, The term 'modern Greek'..designates *two* forms of language—first, the living language..; and, secondly, the literary language, the καθαρεύουσα. **1956** J. PRING *Compan. Greece* 31 Katharevusa is used not only in official and technical matters, but also in public notices, shop signs and the news columns of the press. **1959** *Times Lit. Suppl.* 15 May 288/4 The literary language was..the *katharevousa*, that artificial pseudo-antique speech. **1969** R. BROWNING *Med. & Mod. Gr.* 150 *Katharevousa*, the learned, archaising form of Modern Greek. **1969** 'E. LATHEN' *When in Greece* vi. 63 His Greek was..a shade pedantic. He used the precise formulations of the *Katharevoussa.* **1974** C. SPENCER *How the Greeks kidnapped Mrs Nixon* viii. 53 The Greek Prime Minister was..using, as was customary, the official language, *Katharevousa.*

katharometer (kæþărǭ·mītəɹ). [f. Gr. καθαρός pure: see -METER.] An instrument for determining the concentration of one gas in another by comparing the rate of heat loss of an electrically heated wire in the mixture with that in the second gas alone.

1917 G. A. SHAKESPEAR in *Rep. & Mem. Advisory Comm. Aeronaut.* No. 317. 3 The following is a brief account of a permeability tester..designed for the rapid testing of balloon and airship fabrics... At the centre of the lower part a katharometer (an instrument for measuring directly the percentage of hydrogen in the air) is fixed. **1945** G. R. NOAKES *Text-bk. Heat* ix. 354 The conductivity of hydrogen is about seven times as great as that of air under similar conditions, and this has been made the basis of an instrument called the 'katharometer' (or purity tester) originally designed by Shakespear and Daynes for detecting the leakage of hydrogen through balloon fabrics. **1961** *Engineering* 2 June 780/1 One [detector] widely adopted is the katharometer which measures thermal conductivity... It has a zero or constant response so long as carrier gas alone is flowing, but produces a signal each time one of the components emerges from the column. **1971** D. W. GRANT *Gas-Liquid Chromatogr.* vi. 116 The katharometer is normally situated in a separately heated air oven, next to the column oven.

kathete (kæ·þīt), anglicized f. *kathetus*, CATHETUS.

1912 G. KAPP *Electr.* viii. 210 The well-known Pythagorean axiom that the sum of the squares of the kathetes in a rectangular triangle is equal to the square of the hypotenuse.

Kathi (kā·ði). Also **kathi.** [Malay.] A judge in Islamic law, who also functions as a registrar of Muslim marriages, divorces, etc.

1947 R. O. WINSTEDT *Malays* 117 The Chief Kathi of Kelantan fined 14 men and 2 women $15 each, with imprisonment in default. **1963** J. KIRKUP *Tropic Temper* 31 In religious matters the Kathi or Muslim religious authority exercises magisterial powers. **1967** W. R. ROFF *Origins Malay Nationalism* iii. 73 In June 1884 the Council decided to appoint a State Kathi, 'to decide disputes involving Muhammadan Law and Custom', and the appointment of assistant kathis for the districts was ratified in the following year. **1972** M. SHEPPARD *Taman Indera* 96 He [*sc.* a *Mufti*] can, if requested, conduct a marriage, but this is normally undertaken by a minor official called *Kathi.* *Ibid.* 102 *Kathi*, a Registrar of Muslim marriages. **1972** *Straits Times* (Malaysian ed.) 25 Nov. 13/2 A former school servant, who declared before a kathi that he was a bachelor and married a second time, was jailed for a day and fined $500 or three months' jail yesterday.

‖ **kati** (kæ·ti). Also **katti.** Var. CATTY.

1727 J. G. SCHEUCHZER tr. *Kæmpfer's Hist. Japan* I. IV. viii. 367 Camphire of Baros, a *Katti*, or 1 pound and a qr. *à* 33 *Siumome*, or *Thails.* **1820** J. CRAWFURD *Hist. Indian Archipelago* I. III. i. 273 One hundred katis make a pikul, or 133⅓ lbs. avoirdupois. **1900** W. W. SKEAT *Malay Magic* v. 214 Prices varying according to the quality from $15 to $40 per *katti.* **1947** R. O. WINSTEDT *Malays* 126 One buffalo costing a *kati* of silver. **1969** J. M. GULLICK *Malaysia* v. 232 The miners..are expected to recover 6 *katis* (8 lbs.) of tin ore from each cubic yard of soil. **1972** *Straits Times* (Malaysian ed.) 25 Nov. 13/2 Chicken cost $1·20 per kati.

kationoid, var. *CATIONOID *a.*

katipo. Substitute for def.: A large, black, venomous New Zealand spider, *Latrodectus katipo*, closely related to the Australian *jockey spider* (*JOCKEY *sb.* 9) and the American *black widow* (*BLACK *a.* 19). Add earlier and later examples.

1843 E. DIEFFENBACH *Trav. N.Z.* II. III. ix. 366/2 Katipo—a black spider on the seashore, regarded as poisonous. **1915** *Chambers's Jrnl.* May 319/1 The katipo spider..is the Dominion's [*sc.* New Zealand's] one poisonous creature. **1934** *Bulletin* (Sydney) 4 July 21/4, I gathered katipo by the dozen on the sandhills at the entrance of the Wanganui. **1942** C. BARRETT *On the Wallaby* ii. 29 The katipo or red-backed spider enjoys a very wide distribution in Australia. **1963** *Evening Post* (Wellington, N.Z.) 20 Dec., The katipo spider..is black with a red stripe on its back.

katjiepiering (ka:tɪ'ipi·riŋ). *S. Afr.* Also **catjepiring, katjepeering, katjiepering.** [Afrikaans, f. Malay *katja-piring, kachapiring*, the Cape jasmine, *Gardenia jasminoides.*] A South African evergreen shrub of the genus *Gardenia*, esp. *G. thunbergia*, belonging to the family Rubiaceæ and bearing large, fragrant, white or yellow flowers.

1793 tr. *C. P. Thunberg's Trav. Europe, Afr. & Asia* II. 111 Wild Catjepiring (*Gardenia Thunbergia*) is a hard and strong kind of wood, and on this account used for clubs. **1869** W. G. ATHERSTONE in R. Noble *Cape & its People* 373 The wagons..stand on a bed of wild flowers..fragrant clematis—the 'traveller's joy',—vying in sweetness with the wild 'katjepeering'. **1910** D. FAIRBRIDGE *That which hath Been* 269 Friends bring the first daphne of the year, the richest purple violets, the sweetest katjiepierings, until the house is heavy with the perfume of flowers. **1949** *Cape Argus* 3 Dec. 18/5 Gardenias (Katjiepering), flowering, 4s. 6d. each. **1972** *Stand. Encycl. S. Afr.* V. 120/1 Gardenia. Katjiepiering. Genus of evergreen shrubs..which bear sweetly scented white flowers.

katonkel (kătǭ·ŋkəl). *S. Afr.* Also **katonker, katunka, katunker.** [Afrikaans, f. Malay *kentangkai* a kind of fish.] Either of two marine game fishes: (*a*) *Scomberomorus commersoni*, of the family Scomberomoridæ, which may be six feet long, occurs in the Indo-Pacific ocean, and is also called barracuda; (*b*) *Sarda sarda*, of the family Scombridæ, a much smaller fish found in the Atlantic, and also called bonito.

1853 L. PAPPE *Synopsis Edible Fishes Cape Good Hope* 26 Stromateus Capensis. Mihi. N. Sp. (*Katunker*)... A good table-fish, but not common. It is caught with the hook and net, chiefly East of Table Bay. **1893** H. A. BRYDEN *Gun & Camera S. Afr.* 449 Many of the Cape fish are endowed with the quaintest Dutch names. Here are a few of them: Kabeljouw,..katunka, elftvisch. **1930** C. L. BIDEN *Sea-Angling Fishes of Cape* vii. 144 The word 'Katonkel' has been corrupted from what was known by the Port Elizabeth Malays as 'katunker' or 'katonker' which originated from the original Malay word 'kentangkai', a kind of sea-fish. **1959** *Cape Times* 18 Feb. 2/4 This is the first time in eight years that so many barracuda (katonkels) have been caught at one time. **1974** *Eastern Province Herald* (S. Afr.) 1 Aug. 21 Katankel were a regular summer fishing feature of Port Elizabeth when the harbour breakwater was open to anglers.

katoptrite (kătǭ·ptrəit). *Min.* Also **catoptrite.** [ad. G. *katoptrit* (G. Flink 1917, in *Geol. För. Förh.* XXXIX. 432), f. Gr. κάτοπτρ-ον mirror (alluding to its brilliant lustre): see -ITE[1].] A silico-antimonate of bivalent manganese (partly replaced by magnesium and bivalent iron) and aluminium (partly replaced by trivalent iron) occurring as black, lustrous, monoclinic crystals.

1917 *Chem. Abstr.* XI. 2650 Catoptrite, a new mineral from Nordmarken. **1919** *Jrnl. Chem. Soc.* CXVI. II. 112 Katoptrite occurs as tabular crystals and irregular lumps with magnetite in granular limestone in the Brattfors mine at Nordmark. **1951** C. PALACHE et al. *Dana's Syst. Min.* (ed. 7) II. 1029 Catoptrite. **1966** *Amer. Mineralogist* LI. 1495 The analysis of Flink (1917)..seems to be essentially correct, even considering the peculiar composition of catoptrite. The analysis computes to very nearly $(Mn_{.88},Mg_{.10},Fe_{.04})_{14}{}^{2+}(Al_{.83},Fe_{.17})_4{}^{3+}Sb_2Si_2O_{29}$. **1968** I. KOSTOV *Mineral.* 328 Katoptrite and parwelite are monoclinic.

Kat stitch (kæt stitʃ). [f. *Kat*, abbrev. of Katharine, in ref. to Katharine (Catherine) of Aragon who was thought to have invented the stitch: see STITCH *sb.*[1] 9.] In lace-making, a stitch which forms a star-shaped ground net.

1919 T. WRIGHT *Romance of Lace Pillow* ix. 84 Downton, five miles from Salisbury, produced..a lace with a net..which is similar to Bucks Point, and the Kat Stitch finds favour with the workers. *Ibid.* xiv. 194 Catherine the Saint became confused with Katharine the Queen—that is to say, Katharine of Aragon..of Kat Stitch fame. **1931** M. MAIDMENT *Man. Hand-Made Bobbin Lace Work* x. 158 *Bucks Kat Stitch or Wire Ground.* This stitch is used as a ground net or filling, and it is sometimes varied by the use of plaits. **1960** H. HAYWARD *Antique Coll.* 156/2 *Kat stitch*, a term used in lace-making to describe a *fond chant* ground but frequently found in Bedfordshire lace and given legendary association with Katherine of Aragon.

‖ **katsuo** (ka·tswo). [Jap.] = BONITO, *Katsuwonus pelamis*, an important food fish in Japan, whether fresh or dried. So **ka·tsuobushi,** a dried quarter of this fish.

1727 J. G. SCHEUCHZER tr. *Kæmpfer's Hist. Japan* I. I. 136 The best sort of *Katsuwo* fish is caught about Gotho. **1884** tr. *J. J. Rein's Japan* I. vii. 194 Most conspicuous is the common bonito or Katsu-uwo (Thynnus pelamys), one of the most important and most valued fishes of Japan. **1891** A. M. BACON *Jap. Girls & Women* i. 5 Sometimes a box of eggs, or a peculiar kind of dried fish, called *Katsuobushi*, is sent with this present. **1899** L. HEARN *In Ghostly Japan* xiv. 227 The Yaidzu-fishing-industry, which supplies dried *katsuo* (bonito) to all parts of the Empire. **1965** W. SWAAN *Jap. Lantern* iv. 51 Sticks of *katsuobushi*, a form of *bonito*, a species of striped tunny. **1969** *Guardian* 16 July 16/4 One staple [Japanese food]..is dried bonito (*Katsuobashi*).

‖ **katsura** (katsu·rā). [Jap.] A type of wig worn mainly by Japanese women.

[**1894** L. HEARN *Glimpses Unfamiliar Japan* II. xviii. 421 As soon as the girl becomes old enough to go to a female public day-school, her hair is dressed in the pretty, simple style called katsurashita.] **1908** N. G. MUNRO *Prehistoric Japan* xiii. 567 The word *Katsura* means a vine, such as the Ainu use on certain occasions for personal decoration. In the middle ages it was applied to artificial hair, which meaning is still retained; the evidence scarcely justifies the conviction that wigs were worn by the prehistoric Yamato. **1970** J. KIRKUP *Japan behind Fan* ii. 63, I watched..a display of graceful dances by girls in kimono,..*obi* (belt) and black-lacquered *katsura*, or wig.

‖ **katsuramono** (katsu:rāmǭ-no). Also **kazuramono.** [Jap., f. prec. + *mono* piece, play.] One of the categories of Japanese Noh plays in which the chief character is female and the theme romantic. It is usually presented third in the sequence of five plays of different categories which makes up a performance of Noh. Also *ellipt.* **katsura.**

1916 FENOLLOSA & POUND *Noh* 15 Kazura, or Onnamono, 'wig-pieces', or pieces for females, come third. Many think that any Kazura will do, but it must be a 'female Kazura', for after battle comes peace. **1932** B. L. SUZUKI *Nōgaku* 19 A romantic play (*jo* or *katsuramono*), in which the chief character is a woman and the chief motive love. **1948** *Introd. Classic Jap. Lit.* 141 The regular *kazuramono* contains a *jonomai*, a dance consisting of five movements and a prelude. **1964** W. G. RAFFÉ *Dict. Dance* 353/2 They [*sc.* Noh plays] range.. from romantic themes (*katsuramono*) to farce (*kyogen*). **1965** W. SWAAN *Jap. Lantern* xiii. 153 *Katsura-mono* or wig-plays consisting chiefly of posturing by a woman.

Kattern (kæ·tɜːn). Also **Cathern, Cattern.** [Corruption of Catherine, in ref. to St. Catherine of Alexandria, the patron saint of spinners, who was martyred in A.D. 307.] Used in the possessive in **Kattern's day,** 25 November, the feast day of St. Catherine, which was formerly celebrated by lace-makers in the Midlands (see also quot. 1849). Also *ellipt.* **Kattern('s).** Hence **ka·tterner; ka·tter(n)ing** *vbl. sb.*

[**1521** in H. Ellis *Brand's Pop. Antiq.* (1813) I. 322 Mem. that reste in the hands of the wyffe of John Kelyoke and John Atye, 4 merkes, the yere of ower Lorde God 1521, of Sent Kateryn mony.] **1730** C. LAMOTTE *Ess. Poetry & Painting* ii. 126 St. Catherine..is held in so much Veneration in the Church of Rome...her Holiday is observed, not in Popish Countries only, but even in many Places in this Nation, young Women meeting on the 25th of November, and making merry together, which they call Catherning. **1849** H. ELLIS *Brand's Pop. Antiq.* (rev. ed.) I. 413 Until within a very recent period, it was the custom of the dean and chapter of Worcester, yearly, on St. Catharine's Day,..to distribute amongst the inhabitants of the college precincts a rich compound of wine, spices, &c., which was..called the Cattern or Catharine bowl. **1862** *N. & Q.* 17 May 387/2 In Buckinghamshire, on Cattern Day..these hard-working people hold merry-makings, and eat a sort of cakes they call 'wigs'. **1865** F. B. PALLISER *Hist. Lace* xxx. 352 To this very day..the lace-makers still hold 'Cattern's day', the 25th Nov., as the holiday of their craft. **1875** W. D. PARISH *Dict. Sussex Dial.* 25 *Catterning*, to go catterning is to go round begging for apples and beer for a festival on St. Catherine's Day, and singing,—'Cattern' and Clemen' be here, here, here, Give us your apples and give us your beer.' **1899** A. M. SHARP *Point & Pillow Lace* vii. 171 Till well within the present century the name-day of the kind but most unhappy lady [*sc.* Catherine of Aragon], St. Catherine's Day,..was annually kept as a treat-day for young lace-makers,..and called 'Kattern's Day'. **1919** T. WRIGHT *Romance of Lace Pillow* xiv. 194 In some parts of Northants, Bucks and Beds, the leading festival [for lace-makers] was Catterns (St. Catharine's Day), Nov. 25th—St. Catharine being the patron saint of the spinners. **1942** W. ROSE *Good Neighbours* xv. 131 The mid-winter turning of the days, with its joyous celebrations, of which the visit of the 'Katterners' or 'Mummers'—as they were called—was an announcement and a beginning. **1950** *Bedfordshire Mag.* II. xv. 254 How many Bedfordshire people have kept Kattern? Yet St. Catherine's day was once an important occasion to pillow-lace makers. **1959** *Times* 4 Dec. 19/1 English 'luck-visits'..take their names from the Christian usage that overlies pagan observance—as souling, katterning (St. Catherine), [etc.].

katti: see *KATI.

katun (kā·tūn). [Maya.] A period of twenty years, each with 360 days, in the calendar of the Mayan Indians.

1902 *Amer. Anthropologist* Jan.–Mar. 135 Moreover, in A3 we find the Katun sign with the number 1, which may be a declaration that the date is in a first Katun or beginning Katun, for I can see no reason why the beginning Cycle, Katun, Tun, Uinal and Kin should not have been called the first. **1934** A. HUXLEY *Beyond Mexique Bay* 212 Each stela marks the close of one of the shorter of the chronological periods, in terms of which they [*sc.* the Mayas] reckoned their position in endless duration—the close of a Katun of 7200 days. **1950** *Caribbean Q.* II. II. 28 Each altar is in fact a huge day glyph marking the end of a Katun (7,200 days). *Ibid.*, The ancient inhabitants..set up beside the altars stelae bearing Long Count glyphs including the altars' Katun numbers, thus setting the Katun dates into their correct positions. **1968** D. BAGLEY *Vivero Let.* vii. 179 It's a stele—a Mayan date-stone. In a given community they erected a stele every katun—that's a period of nearly twenty years.

katydid. Substitute for first part of def.: A large longhorn grasshopper of the family Tettigoniidæ. (Add earlier and later examples.) See also *CATYDID.

[**1751** J. BARTRAM *Observations Pensilvania to Ontario* 70 It was fair and pleasant, and the great green grasshopper began to sing (*Catedidist*) these were the first I observed this year.] **1784** J. F. D. SMYTH *Tour U.S.A.* II. 243 They are named by the inhabitants here *katy did's*, from their note, which is loud and strong, bearing a striking resemblance to those words. **1886** *Outing* (U.S.) IX. 106/2 Soon the chiding katydids mingled their voices with the rush of the foaming river. **1909** *Springfield* (Mass.) *Weekly Republ.* 16 Sept. 1 All around the globe people are like katydids, saying he did and he didn't in an endless reiteration. **1935** M. MOORE *Sel. Poems* 64 The small tuft of fronds or katydid legs above each eye. **1942** E. O. ESSIG *College Entomology* viii. 95 The name 'katydid' has gradually replaced all others. It originated in the United States, having been derived from characteristic stridulatory sounds produced by the males of certain green species, notably *Pterophylla camellifolia*.. whose note, the loudest of all species of the eastern states, simulates 'Katy did, Katy she did'. **1957** L. EISELEY *Immense Journey* 25 The skilled listener can distinguish man's noise from the katydid's rhythmic assertion. **1972** SWAN & PAPP *Common Insects N. Amer.* iii. 74 Katydids are predominantly green, have exceedingly long antennae, are more often heard than seen.

Katyusha (katiū·ʃă). [Russ.] A Russian rocket launcher.

1955 M. REIFER *Dict. New Words* 116/1 *Katyusha*, Soviet counterpart of the bazooka, an anti-tank weapon employing rockets. **1970** *Guardian* 28 Jan. 10/6 To talk of demilitarisation in the days of Katyushas is obsolete. *Ibid.* 2 June 2/5 Arab guerillas fired Russian-made Katyusha rockets into the Northern Israeli town of Beisan. **1972** E. AMBLER *Levanter* vi. 170, I remembered what Barlev had told me about the 120-mm. Katyusha rocket: fifty kilo warhead, range of about eleven kilometres.

katzenjammer (kæ·tsĕndʒæ:məɹ). *U.S. colloq.* [G., f. *katzen* (comb. form of *katze* cat) + *jammer* distress, wailing.] **a.** A hangover, or a symptom of one.

1849 *Ex. Doc. 31st U.S. Congress 1 Sess. House* No. 5. III. 733 Some of Mr. Hale's men had kept up a drunken frolic all night, general kakenjammer [*sic*], therefore all day. **1877** R. J. BURDETTE *Rise & Fall of Mustache* 291 This 'Centennial Cordial and American Indian Aboriginal Invigorator'..has positively no equal for the cure of.. katzenjammer. **1948** *Life* 5 Apr. 111/2 Attempting to drink himself to death on..vodka and champagne... The result one of the most colossal *Katzenjammers* ever recorded. He was in bed for a week. **1965** P. DE VRIES *Let Me count the Ways* ix. 125 The symptoms classic to hangover persist. Dizziness, nausea, headache—you know that katzenjammer just above the eyes.

b. *transf.* and *fig.* An unpleasant aftermath or reaction; depression, 'blues'; clamour, uproar.

1897 W. W. COOK in *Yellow Kid* 8 May 26/2 He has a deplorable habit of constantly looking for something. Either he is the relict of some individual who was brought up in a sawmill or else he is suffering from a bad attack of katzenjammer. **1900** W. JAMES *Let.* 8 June in R. B. Perry *Tht. & Char. W. James* (1935) II. 198, I am afraid of what the French people may do during the *Katzenjammer* which will inevitably succeed the Exhibition. **1922** W. STEVENS *Let.* 24 Aug. (1967) 228 Nothing has survived the subsequent katzenjammer. **1949** I. DEUTSCHER *Stalin* x. 406 Amid the *Katzenjammer* which befell them after 1933, most leaders of the German left were only too eager to explain away their own failure. **1960** B. KEATON *Wonderful World of Slapstick* (1967) iv. 74 The Katzenjammer spirit of the other acts on the bill inspired them to contribute new plot turns.

c. *Katzenjammer Kids* (or *Children*), mischievous, naughty children; *enfants terribles.* So called from the title of a comic strip, first drawn by Rudolph Dirks in 1897 for the *New York Journal*, featuring Hans and Fritz, two incorrigible children. Also *attrib.*

1897 R. DIRKS in *N.Y. Jrnl.* 12 Dec. 8 (*comic strip title*) Ach, Those Katzenjammer Kids! **1910** M. G. PEDRICK in *Good Housekeeping* (N.Y.) May 625/2 The children are engrossed with the mishaps of Happy Hooligan, Smarty, Gaston or the 'demoniacal ingenuity of the Katzenjammer Kids'. **1947** W. STEVENS *Let.* June (1967) 558 The noisy Katzenjammer children..were upstairs saying their prayers. **1962** *Times Lit. Suppl.* 13 July 511/3 He has no difficulty in matching the varied dialects of the Greek, drawing as required on such rich sources as mint-julep Southern, broad Brooklynese or Katzenjammerkids German.

‖ **kau kau** (kau kau). [Native name.] In New Guinea, the sweet potato.

1937 *Official Handbk. New Guinea* v. 435 In Morobe District, taro, sweet potatoes (*kau-kau*), sugar-cane and bananas furnish the main food, supplemented by yam, coco-nut and sago in certain areas. **1957** M. WEST *Kundu* xiv. 166 They returned laden with taro and kau-kau and sugar-cane. **1964** *Economist* 30 May 985/1 Sago and *kau kau* provide a basic if poor quality diet [in Papua].

‖ **kavadi** (kā·vădi). [ad. Tamil *kāvaṭi*.] A decorated arch carried on the shoulders as an act of penance, esp. by Hindus in Malaysia.

1954 V. BARTLETT *Rep. from Malaya* ii. 27 A 'kavadi'.. is a gaily-decorated but heavy wooden harness normally supported on one shoulder. **1961** *Times* 8 Mar. 14/6 [Ceylon] On their shoulders the traditional kāvadis. **1963** J. KIRKUP *Tropic Temper* 254 Blocks of pure camphor for burning..before the holy men and penitents carrying kavadis. **1966** S. WAVELL et al. *Trances* 148 Many are poor and walk long distances carrying upon their shoulders a heavy superstructure known as *kavadi* bearing flowers and fruit and peacock feathers to place before the image of Subramaniam. **1970** S. ARASARATNAM *Indians in Malaysia & Singapore* vi. 171 Popular forms of devotion..as practised in South India have persisted... Among the most significant of these is the carrying of the *kavadi*, a large wooden decorated arch, as an act of penance. **1972** *Straits Times* (Malaysian ed.) 27 Nov. 26/1 (Advt.), Devotees of Lord Subramaniam carry penitent 'kavadis'. **1973** *Observer* 7 Oct. 36/6 The penitents..climb 272 steps to the place where the Lord's image is installed, their bodies enclosed in the kavadi, a wooden frame decorated with feathers and supported by long metal rods which are hooked into their flesh.

kavir (kăviə·ɹ). Also **kevir.** [Pers.] A salt-desert, or more rarely a saline swamp, in Persia; terrain of this type; also *spec.* (*the Kavir*), the great central salt-desert of Persia, more commonly called the Dasht-i-kavir.

1881 *Proc. R. Geogr. Soc.* III. 515 The road..across the Kavir or Great Salt Desert was very difficult. *Ibid.* 517 Soon after passing Illahabad a small piece of *kavir* or salt desert is passed. *Ibid.* 518 There are various sorts of kavir, depending upon the soil and the amount of salt. **1896** *Geogr. Jrnl.* VII. 34 These bushes grow..along the borders of the swampy part of the salt desert, which is known as the Kavir or Dasht-i-Kavir. *Ibid.* 166 The greater portion of the tract consists of *kavir*, or sandy soil strongly impregnated with salt. **1902** P. M. Sykes *Ten Thousand Miles in Persia* iii. 33 And that Kavir is applied to every saline swamp in the whole blighted expanse. **1963** D. W. & E. E. Humphries tr. *Termier's Erosion & Sedimentation* v. 118 As a result of evaporation, some closed basins are floored by a saline crust which is called *sebkha* in North Africa.. *solonchak* in the region of the Caspian, and *kevirs* in Transcaspia.

Kavirondo (kā·virǫ·ndo). Also Kaverond(a), Kavirond. [Native name.] **a.** The name of two Kenyan peoples, one Nilotic (cf. *Luo *sb.* and *a.*), one Bantu (also called Wa-Kavirondo); a member of these peoples; any of the Nilotic and Bantu languages spoken by these peoples. Also *attrib.*

1870 *Jrnl. R. Geogr. Soc.* XL. 308 At Kaverond.. there are villages. The people of this place are called Wa-Kaverond. They are the same as the Wa-Kosova, only a different tribe or clan. The language is one. **1873** C. New *Life E. Afr.* xxiii. 468 Captain Speke gives only a few words of the Gani dialect.. and these are the very words which are used by the Wakavirondo for the same things. *Ibid.* 526 A Table showing the variations in the dialects and languages spoken by some of the tribes.. Kisuaheli.. Masai, Kavirondo. **1882** *Proc. R. Geogr. Soc.* New Ser. IV. 743 The town of the Kavirondo chief Sendege. *Ibid.* 744 Mr. Wakefield's vocabulary of the Kavirondo language clearly shows that this tribe does not belong to the Bantu family... Two islands lie off Kisumo... Both are cultivated by Kavirondo. **1895** J. Thomson *Through Masai Land* (ed. 3) xi. 475 Their shields are of all shapes and sizes, though the characteristic Kavirondo form is enormous in dimensions and weight. *Ibid.* 478 We picked the bones of fat Kavirondo fowls. *Ibid.* 485 The Wa-kavirondo are apparently a homogeneous race.. Yet.. there were two totally distinct languages. The inhabitants of..Lower Kavirondo..speak a language resembling..that spoken by the Nile tribes, while those of Upper Kavirondo speak a Bantu dialect. **1902** C. W. Hobley *Eastern Uganda* (Ethnol. Survey) i. 11 No Kavirondo marries in his own clan, and the degeneracy due to inbreeding is obviated. *Ibid.* vi. 88 There are many striking resemblances between the Nyanza language and the Bantu language of Kavirondo. **1921** *Manual on E. Afr.* (Church Missionary Soc.) 13 Leaving the coast.., we come to the Nilotic tribes—the Masai.. and Nilotic Kavirondo. The greatest tribe of all in point of numbers is the Kavirondo, of which there are about 400,000; but it is divided by language into two groups, Nilotic and Bantu. **1950** Huntingford & Bell *E. Afr. Background* (ed. 2) xiv. 111 Kavirondo, this is properly a name of the Nilotic Luo people in Kenya. According to one native explanation, it was applied by the people of one side of Kavirondo Gulf to the people of the other side as a term of abuse, derived from a Luo word *rondo* 'to deceive'... From this restricted area it has become applied to (1) The Nilotic Luo as a whole;..(3) the Bantu inhabitants of this area. **1956** *Linguistic Survey Northern Bantu Borderland* (Internat. Afr. Inst.) I. iii. 129 The northern corner of the 'Bantu Kavirondo pocket'. *Ibid.* 130 A note on 'Bantu Kavirondo'. This name refers to a pocket of Bantu-speaking peoples between the..'Nilotic Kavirondo'..and the Jopadhola and speakers of Nandi dialects.

b. Kavirondo crane = *Kaffir crane* (*Kaffir 4).

1928 *Daily Express* 31 July 4 The handsomest [bird] is the Kavirondo or golden-crested crane, kept as a pet by some native tribes in Kenya, a gorgeous stork-like bird plumaged in browns, blues, greys, and gold. **1938** F. J. Jackson *Birds Kenya Colony* I. 317 The East African Crowned Crane, known locally as the Straw-crested, Golden-crested, and sometimes the Kavirondo crane, is found throughout Kenya colony... Why it ever became known as the Kavirondo crane is a mystery.

Kaw (kǫ). *U.S.* Also Caw, Kah. Another name for the *Kansa. Also *attrib.* or as *adj.*

1804 Lewis & Clark in E. G. Thwaites *Orig. Jrnls. Lewis & Clark Exped.* (1905) VI. 84 (in list) 5. Kanzas; Karsea; Kah. **1823** W. Becknell in *Missouri Intelligencer* 22 Apr. 2/5 We..shaped our course over the high land which separates the waters of that and the Caw rivers. Among the Caw Indians were we treated hospitably. **1844** J. Gregg *Commerce Prairies* I. 42 It was either a Kaw, or else a straggler of the Kaws (or Kansas Indians). **1930** E. Ferber *Cimarron* xviii. 260 The Oklahoma Territory and the Indian Territory, with an Indian population of..various tribes.. Kaws, Choctaws, Seminoles. **1942** *R.A.F. Jrnl.* 16 May 18 The festivities..were planned by Mose Bellmard, chief of the Kaw Indian tribe.

kawai, var. *kahawai.

kawaka (kā·wăkă). *N.Z.* [Maori.] A New Zealand cedar, *Libocedrus plumosa.*

1832 G. Bennett in *London Med. Gaz.* 7 Jan. 506/2 A tree of the Natural Family *Coniferæ*, collected without..flower or fruit: it is named Kawaka by the natives of New Zealand, attaining the height of from 30 to 50 feet... The natives informed me that it derived the name Kawaka from the branches growing out regularly on each side of the tree. **1855** R. Taylor *Te Ika a Maui* 440 *Kawaka, koaka (dacrydium plumosum).* This tree grows in large quantities on the central plains; the wood is of a very dark red grain, and is said to be as durable as the *totara*. **1906** T. F. Cheeseman *Man. N.Z. Flora* 647 *Kawaka; New Zealand Arbor-vitæ.* Wood dark-red, beautifully grained, said to be durable, but on account of its scarcity little used. **1966** *Encycl. N.Z.* II. 208/2 Kawaka occurs in lowland forest from Northland to the centre of the North Island and again in the north-west tip of the South Island.

kawa-kawa[1] (ka·wăka·wă). [Maori.] **1.** A shrub or small tree, *Macropiper excelsum*, of the family Piperaceæ, native to New Zealand and neighbouring islands; also called pepper-tree.

1850 J. Greenwood *Journey to Taupo* 30 A most refreshing light beverage made from the leaves of the Kawa-kawa tree. **1910** L. Cockayne *N.Z. Plants* vi. 80 On the Little Barrier, at the foot of the cliffs, it [*sc.* a member of the gourd family] is abundant, scrambling over the kawa-kawa. **1938** R. Finlayson *Brown Man's Burden* 47 The wreaths of bitter kawakawa around their heads were not more bitter than their tears of grief. **1949** P. Buck *Coming of Maori* (1950) iii. vi. 407 Hot infusions of leaves such as the kawakawa. **1966** *Encycl. N.Z.* II. 785/1 Numerous other berries were eaten raw, especially by children... Examples are kahikatea..and kawakawa.

2. *N.Z.* A variety of Greenstone 2.

1880 *Encycl. Brit.* XIII. 540/2 The green jade-like stones which are known to the Maories as *kawa-kawa* and *tangiwai* do not appear to be either jade or jadeite. **1909** *Q. Jrnl. Geol. Soc.* LXV. 368 The variety [of greenstone], however, which is almost exclusively used by the lapidary and jeweller, is the common green kawakawa. **1965** G. J. Williams *Econ. Geol. N.Z.* x. 156/2 Kawakawa..was named from its resemblance to the leaf of a shrub; its colour is dark green in various shades including spinach-green, seaweed-green, and olive-green.

kawakawa[2] (ka:văka·vă). [Hawaiian.] The little tuna, *Euthynnus yaito.*

1887 H. H. Gowen *Lat. in Paradise of Pacific* (1892) xi. 130 The kawakawa, a large fish tasting somewhat like mackerel. **1965** W. A. Bryan *Nat. Hist. Hawaii* xxvii. 363 The little tunny or kawakawa is of a size and shape like the bonito. **1944** S. W. Tinker *Hawaiian Fishes* 156 The kawakawa is dark blue in color above and almost silvery beneath.

kay (kē), *int.* [Representation of the sound of the letter *K.*] = *O.K. used as *int.*

1950 I. Jefferies *Thirteen Days* v. 66 'How about a quick half-hour now?' I said. 'Kay.' **1968** S. Challis *Death on Quiet Beach* v. 72 'Kay. We'll check on that.' **1973** J. Wainwright *Night is Time to Die* 189 'And sling him in the cells,' added the D.D.I... 'Kay.'

kaya (ka·ya). [Jap.] A Japanese evergreen tree, *Torreya nucifera*, of the family Taxaceæ, with large seeds which contain oil; also, the wood of this tree.

[**1727** J. G. Scheuchzer tr. *Kæmpfer's Hist. Japan* I. ii. x. 129 All of the Oils expressed of the seeds of these several plants, only that of the *Sesamum* and *Kai*, are made use of in the kitchen.] **1889** J. J. Rein *Industries Japan* I. iii. 257 Kaya-no-abura, Kaya-oil, is manufactured by the Japanese from the seeds of *Torreya nucifera*... The Kaya resembles our yew. It is found in most cases not as underwood, scattered like brush in mountain forests; seldom as a tree. In autumn the plant is laden with nuts, which are good to eat, although having a resinous after-taste. **1894** C. S. Sargent *Notes Forest Flora Japan* 76 The Kaya should be cultivated wherever the climate permits it to display its beauty. **1923** Dallimore & Jackson *Handbk. Coniferæ* 75 *Torreya nucifera*, Siebold & Zuccarini. Kaya... Wood lustrous yellow to pale brown, durable under water, used for chests, boxes, cabinets, furniture, vessels, toys, fine appearance. **1965** J. Ohwi *Flora Japan* 110/1 Kaya. Glabrous tree with spreading branches. **1969** C. L. Bell *Board & Table Games* II. iii. 61 The best boards are made of a species of yew, called 'Kaya' (*Torreya Nucifera*).

kayak[1]. Add: **b.** Any canoe developed from the Eskimo kayak, used for touring or sport.

1933 A. B. Blake *Canoeing for Beginners* i. 10 Construction of kayaks... This craft is a very seaworthy little piece of work... It is about 15 feet in length. **1946** P. W. Blandford *Canoeing To-Day* i. 4 It is common to refer to a folding craft as a canoe and a rigid craft as a kayak, although neither is built on the Eskimo lines. *Ibid.* 62 Kayak, originally the Eskimo craft, but now generally applied to any rigid canvas-covered decked canoe. **1962** — *Canoes & Canoeing* i. 13 What the European calls a canoe the American calls a kayak, and what the American calls a canoe the European usually qualifies as a Canadian canoe, keeping the word kayak for the special slim craft based on the Eskimo pattern. **1966** J. Samson in B. C. Skilling *Canoeing Complete* i. 24 About 1840 the first copies of Greenland kayaks appeared in Europe... After 1865 canoeing began to rise as a new kind of sport. The Scot, McGregor, made his sensational voyages in kayaks of his own design.

c. *attrib.* and *Comb.*

1963 *Internat. Jrnl. Social Psychiatry* IX. 1. 19 Kayak-angst (kayak-phobia, kayak dizziness) is well known throughout all districts of West Greenland... Kayak-angst is scarcely mentioned in English written accounts, with the exception of brief references in Freuchen, Birket-Smith and a few others. **1965** B. C. Skilling in *Canoe Camper* 10 In the kayak world it was accepted that to have complete control the canoe and man must become as one. **1964** *Slalom & White Water Course* (Ontario Voyageurs Kayak Club) i. 4 Modern kayak-paddling technique is a combination of the classical style and of elements adopted from the Canadian canoe. **1964** N. Hunt *Adventures in Canoeing* i. 13 He feels his individual make-up is in line with these exciting aspects of kayak sport. *Ibid.* ii. 24 A lightweight, strong and watertight hull, easy to make and no more expensive than a soft-skinned kayak kit. *Ibid.* iii. 37 Free your legs from the knee grips by pressing the thighs towards the kayak floor. *Ibid.* vii. 89 A broken kayak paddle.

Hence **kay·ak** *v. intr.*, to travel by kayak; **kayaker, kayaking** *vbl. sb.* (further examples); also *attrib.*; **kay·akist**, one who paddles a kayak.

1875 H. Rink *Tales & Trad. Eskimo* I. 294 Another day, when he was kayaking along the coast, he remarked some loose pieces of ice. *Ibid.* 295 His wife repeated the tale of his misfortunes to every kayaker on his return home. *Ibid.* 329 The following day the father kayaked the same way past the cape, and came in sight of the tents. **1960** *Spectator* 2 Sept. 340 She's training with the world's top kayakists. **1963** R. George in *Canoe Venture* 5 The Australian and American founders, who swore that they would never raise the status of kayak to canoe even though it was joining a canoeing club of kayakists. *Ibid.* 21 The paddlers sat on their wickerwork seats and watched with disdain the gyrations of the new kayakers. **1964** *Slalom & White Water Course* (Ontario Voyageurs Kayak Club) 17 His name is probably not mentioned in Czech kayaking literature. *Ibid.* 22 Slalom competitions are considered the 'matriculation' of the kayaking fraternity. **1966** B. C. Skilling *Canoeing Complete* ii. 11 The knowledge and experience which he and other members of his parties gained in kayaking with the Eskimos. *Ibid.* x. 176 It will be found that few kayakers can paddle with equal facility on both sides of their canoe using a single-bladed paddle stroke. **1967** *Nat. Geographic* Sept. 318/1 Halting for lunch, the kayakers beach their craft..then join youngsters splashing in the mole-protected harbor. **1969** *Amer. Dial. Soc.* [illegible] The sport of whitewater kayaking was spawned in the United States... Kayak slalom, the most complex form of kayaking, is a competition in which the kayakist or paddler must pass through a series of gates in the rapids. **1972** *New Earth Catalog* 17/1 A kayak outfitter with kits and..instruction on the art of kayaking down some of Colorado's rivers.

kayak[2] (kai·æk). *Canada.* Also kia(c)k, kyack. [Prob. f. Algonquian.] = Ale-wife[1].

1849 A. Gesner *Industr. Resources Nova Scotia* 121 Sometimes a hundred men, among whom is a sprinkling of Indians, are engaged in taking the 'kiacks' from the stream. **1878** C. Hallock *Sportsman's Gazetteer* (ed. 4) 271 Kyack—*Pomolobus pseudoharengus.* **1965** *Canad. Geogr. Jrnl.* June 209/1 'Alewives' is the common name used in Britain and New Zealand, while the MicMac Indians called the fish 'kayaks', and in Latin it's *Pomolobus pseudoharengus.* **1965** E. Richardson *Living Island* 109 To feast upon the sweet spring salmon and the running kyack-cooks (we still call alewives 'kyacks').

Kayan (kai·ăn). Also Kyan. [Native name.] The name of a people of Sarawak and Borneo, of a member of this people, and of their language. Also *attrib.* or as *adj.*

1846 C. D. Bethune in *Jrnl. R. Geogr. Soc.* XVI. 297 The indigenous population [of Borneo] is included under the names of Dayaks, Kadáyan, Milanau, Káyan, Murût, Dusûn. *Ibid.* 299 The Kayan is the most numerous tribe... Their dialect is different from the Dayak. *Ibid.*, Hill tribes..much oppressed by the Káyans. **1848** H. Low *Sarawak* iv. 97 The Dyaks, the Kyans, and other aboriginal tribes. *Ibid.* x. 331 The swords of the Kyan tribes are of very peculiar construction. **1858** S. St. John *Jrnl. Forests Far East* (1862) II. iii. 45, I then made a speech in the Kayan and also in the Murut languages. **1873** *Proc. R. Geogr. Soc.* XVII. 133 Their [*sc.* the Pakattans'] language is quite different to Malay, Dyak, or Kyan. **1886** in F. Hatton *N. Borneo* (ed. 2) vii. 322 No European blade is more finely tempered than these Kayan weapons. *Ibid.* 329 This Kayan instrument gives forth a soft and soothing kind of music. **1914** *Sat. Rev.* 31 Jan. 147/2 The Ranee carries her love for her people to the extent of even finding excuses for the head-hunting propensities of the Dyaks and Kayans. **1958** C. H. Hartley in T. Harrisson *Borneo Jungle* 149 The orators spoke in Kayan, but subsequently translated their remarks into Malay. **1960** K. F. Wong *Pagan Innocence* vii. 74 (caption) Her baby has been given a good start in life, from the Kayan point of view, by the acquisition of ear-rings to elongate the lobes. **1960** *Guardian* 9 Nov. 10/3 The population [of Sarawak] includes.. Malays, Melanaus, Kayans..and others. **1964** T. Harrisson in Wang Gungwu *Malaysia* iii. xi. 170 A Kayan put down among Kelabits will not understand a word.

kayf (kĕif). Representation of a slang or jocular pronunciation of Café. Cf. *caff.

1962 R. Norman *Goats* i. 8, I had some eggs and bacon in a kayf just around the corner. **1964** *Guardian* 20 Mar. 38/1 This kayf would follows every cool, snide and beat twist of Sixties fashion.

kaylong, var. *kelong.

kayo (kē·i-ō), *a. slang* (orig. *U.S.*). [Reversal of the pronunc. of *O.K. under the influence of next (see *K. 4 b).] = *O.K. *a.*

1923 H. C. Witwer in *Cosmopolitan* Apr. 118/2 Anything you say is kayo with me, kid, unless you tell me good bye! **1928** Wodehouse *Money for Nothing* v. 103 If you think it's kayo, then it's all right by me. **1946** *Amer. Speech* XXI. 138 How these new speech trends [*sc.* abbreviations] in turn may be translated back into the written word is shown by *kayo*, (etc.).

kayo (kē·i-ōu), *v.* and *sb. colloq.* (orig. *U.S.*). [Representation of the pronunc. of K.O. (*s.v.* *K. 4).] **A.** *v. trans.* = *knock *v.* 12 a. **B.** *sb.* = Knock-out *sb.* 2.

1923 H. C. Witwer *Fighting Blood* 324 You never been knocked cold in your life—why go out of your way to get kayoed? **1932** J. T. Farrell *Young Lonigan* iii. 112 He sat down, saying to himself that he was Young Studs

Lonigan..now in training for the bout when he would kayo Jess Willard for the title. **1933** *Amer. Speech* VIII. III. 39/1 The knockout blow or kayo itself is variously called kay..K.O. **1939** Wodehouse *Uncle Fred in Springtime* xx. 303, I still don't see..why he should have slipped kayo drops in. **1968** *Globe & Mail* (Toronto) 3 Feb. 40/4 Quarry with 14 kayos in a 25-1-4 record is known as a stopper. **1972** *Times* 10 Nov. 14/6 There was ..the kayoing of the prop forward by a deliberate punch. **1975** *Cleveland* (Ohio) *Plain Dealer* 3 Mar. 1-C/5 Gardner..now 25, was kayoed by Patterson in the sixth round in 1957, won a gold medal in the 1956 Olympic Games for boxing.

kayser (kai·zəɹ). *Physics.* [Name of J. H. G. *Kayser* (1853–1940), German spectroscopist.] A name proposed for the unit of wave number, cm^{-1}.

1951 W. F. Meggers in *Jrnl. Optical Soc. Amer.* XLI. 1064 Because H. Kayser first founded the correct procedure in determining spectroscopic wave numbers.. I think it appropriate to propose that the name of the unit hitherto called cm^{-1}, be Kayser, with K as the natural abbreviation. **1953** *Ibid.* XLIII. 412 The Joint Commission for Spectroscopy recommends that the unit of wave number hitherto designated as cm^{-1} be named 'kayser', with the symbol *K*... **1968** Hill & Day *Physical Methods Adv. Inorg. Chem.* iv. 157 The same wave numbers are cm^{-1} or kayser (κ) with 1000 kayser = 1 kilokayser (kκ).

Kayser–Fleischer ring (kai·zəɹ,flai·ʃəɹ rɪŋ). *Path.* [f. the names of Bernhard *Kayser* (1869–1954) and Bruno *Fleischer* (b. 1874), German ophthalmologists.] A pigmented ring around the cornea of a variable orange, green, or brown colour, diagnostic of Wilson's disease.

1930 *Amer. Jrnl. Ophthalm.* XIII. 1049/2 We must rely to a great extent upon the presence of a Kayser-Fleischer ring for a differential diagnosis [of Wilson's disease]. **1948** *Amer. Jrnl. Med. Sci.* CCXV. 601/2 Both brothers present an unusually well developed orange-green Kayser-Fleischer ring. **1970** Passmore & Robson *Compan. Med. Stud.* II. xxv. 38 A characteristic brown ring sometimes visible in the cornea, the Kayser-Fleischer ring.

kaza, var. *caza.

1885 *Encycl. Dict.* IV. 468/3 *Kaz-a*,..a district or subdivision of sandjak, marked out for administrative purposes. **1920** *Glasgow Herald* 27 Aug. 9 General Gouraud has arrived at Zahleh, where he announced the reunion with the Lebanon of the four kazas of Baalbek, Bekaa, Ruchaya, and Hasbaya. **1972** D. Dakin *Unification of Greece* xii. 166 A broad band of territory composed of the kazas (districts) of Monastir and Florina in the west; of Gevgeli, Vodena..and Yanitsa in the central region; and Serres and Zihna in the east.

kazachoc (kæzæ·tʃǫ·k). Also **kozatchok** and other forms repr. the acc. or pl. of the Russ. word. [Russ., dim. of *kazák* Cossack.] A Slavic, mainly Ukrainian, dance with a fast and continually quickening tempo. Sometimes used erron. for a step of this dance, properly called the *prisiadka*, in which the male dancer squats on his heels and kicks out each leg alternately to the front.

1928 C. Garnett tr. *Gogol's Mirgorod* 63 No-one could have watched without inner emotion how all danced that most free, most furious dance the world has ever seen, called from its mighty originators the Kozatchok. [illegible] **1962** *Listener* 11 Jan. 71/1 He played Polish and Russian dances of his native land. [illegible] **1966** K. Giles *Provenance of Death* iii. 78 She switched on the player and danced the Kazachok. [illegible]

Kazak (kăzā·k). [ad. *Kazakh*, name of a town in Azerbaijan.] A type of Caucasian wool rug, characterized by large geometric designs in striking colours, and a thick durable quality.

1900 J. K. Mumford *Oriental Rugs* (1901) ix. 107 The regulation rug of the Derbend variety..partakes of the character of the Kazak. **1946** *Penguin New Writing* XXVII. 30 An exquisitely golden yellow and midnight blue thread Kazak rug. *Ibid.* 102, I took a last..look at the Afghans and Kazaks. **1962** C. W. Jacobsen *Oriental Rugs* 233 Kazaks and Karabaghs are thick piled Caucasian rugs. The Kazaks are geometric in design. **1973** *Times* 2 Apr. 6 (Advt.), A fascinating Royal Pakistani Kazak.

Kazakh (kăzā·k). Also K(h)asa(c)k, Kazak, Qazaq. Pl. Kazakhi, Kazakhs. [Russ.] One of a Turkic people of central Asia, forming the basic population of the Kazakh SSR (Kazakhstan); the language spoken by this people. Also *attrib.* or as *adj.*

1832 J. Bell *Syst. Geogr.* IV. 395 Among the names of tribes noticed, are: *the Kuthai Kipchauchs, Kuzzauks,* and *Kara Kalpaks.* **1886** R. N. Cust *Lang. as Illustr. by Bible-Translation* 32 The Türki Branch [of Ural-Altaic], containing Chuvásh, Kazâni, Nogâi..and Kirghiz in the Kazák Dialect. **1907** G. W. Hunter in M. Broomhall *Chinese Empire* 296 There are large numbers of Kirghiz—Khasack tribes—here in the district of Urumchi. They are mostly wandering nomadic tribes, though some have farms and settled abodes. **1918** G. W. Hunter *Examples Turki Dial.* Pref., Whilst doing some Qazaq Turki translation work we found it necessary to read some native

Qazaq books. And we thought it might be useful to keep a record of our studies in Qazaq, and in the various other Turkish dialects... Much still requires to be done in Qazaq, Tartar and the Kirghese dialects. **1945** *Daily Sci. Abstr.* VII. 3 Crossbred cows..gave an average milk yield ..39.2 per cent. more than Kazakh cows... Crossbreeding Kazakh and Schwyz cattle. **1948** D. Diringer *Alphabet* 568 Eastern or Altai-Kirghiz Turkish or Kazakh Turkish, spoken by several hundred thousand nomads in the Altai and T'ien Shan mountains. **1949** N. Jasny *Socialized Agric. USSR* III. xiv. 330 On the prices paid by the Kazakh for collectivization see the end of the preceding chapter. *Ibid.* III. xiii. 323 The Kazakhi, a Mongol pastoral tribe inhabiting mainly Kazakhstan... The Kazakhi should have numbered 4·6 millions in 1939. **1957** *Chambers's World Gazetteer* (rev. ed.) 359/2 Kazakh SSR, or Kazakhstan... Kazakhs.. constitute 60% of the total pop... The written Kazakh language has existed only since the revolution. **1968** Bethell & Burg tr. *Solzhenitsyn's Cancer Ward* I. iv. 49 Next to him Egenberdiev, a middle-aged Kazakh shepherd, was not lying but sitting on his bed, legs crossed as though he was sitting on a rug at home. **1969** *Guardian* 21 June 9/5 The Russians have broadcast propaganda in the Kazakh language from Alma Ata. **1974** *Times* 21 Dec. 7/7 An 800-mile border with China. It crosses mountainous areas which provide pasture land for Kazakh and Chinese hill shepherds.

kazoo. For def. read: (See quot. 1938.) Now also made of plastic or metal and played as a jazz instrument. Also **kazoo** *v.*, to make a sound like that of a kazoo; **kazoo-er, kazoo-ist,** one who plays the kazoo; **kazoo-ing** *vbl. sb.*

1895 *Montgomery Ward Catal.* 245/3 Kazoos, the great musical wonder,..anyone can play it; imitates fowls, animals, bagpipes, etc. **1909** R. A. Wason *Happy Hawkins* (1911) xxvi. 301 The storm that was presently kazooin' along was fierce an' horrible. **1926** *Daily Colonist* (Victoria, B.C.) 14 Jan. 9/1 There will be an abundance of paper hats, balloons, kazoos and other novelties. **1926** Whiteman & McBride *Jazz* ix. 101 [illegible] **1938** *Oxf. Compan. Mus.* 583/2 *Mirliton*, the French name for what English children call (or used to call) 'Tommy Talker', or 'kazoo'. It is a tube with a membrane at each end and two holes in the side, near the two ends, into one of which holes one sings in one's natural voice. [illegible] **1940** *Amer. Speech* XV. 125 George Gershwin's..'Blue Monday' has..*Kazoo Mutes* (to cornets). **1956** J. Latimer *Sinners & Shrouds* ix. 83 'A neglected instrument, the kazoo,' he was saying. [illegible] **1956** M. Stearns *Story of Jazz* (1957) xv. 171 Jack Bland, Dick Slevin, and Red McKenzie played a banjo, a comb wrapped in tissue paper, and a kazoo (a toy horn with tissue paper that vibrates with humming). **1959** *Guardian* 26 Aug. 5/2 The kazooers' parents give tremendous support. *Ibid.*, One hears that there is..a kazoo band in Newcastle. **1966** T. Pynchon *Crying of Lot 49* i. 10 The Fort Wayne Settecento Ensemble's variorum recording of the Vivaldi Kazoo Concerto. **1968** *Blues Unlimited* Nov. 8 Dewey Corley..now plays kazoo and washtub bass. **1970** *Guardian Weekly* 14 Mar. 18/1 Running the gauntlet of a battalion of kazooers. **1970** *Peace News* 8 May 8/4, I think the time has come for us to make our own music. ..To beat on pots and pans, blow kazoos and our combs wrapped in wax paper. **1972** *Guardian* 29 May 5/5 The unusual cigar-shaped instrument, the kazoo—a cheap toy which makes a vibrant noise when blown—can help unmusical children to sing in tune. **1973** *Ibid.* 19 Feb. 8/4 [illegible]

Keating (kī·tiŋ). The name of Thomas *Keating*, a 19th-century chemist, used (usu. *attrib.* or in the possessive) as the proprietary name of an insect powder first manufactured by him.

1876 *Trade Marks Jrnl.* 11 Oct. 576/2 Keating's Persian insect destroying powder. Thomas Keating. **1886** B. Potter *Jrnl.* Dec. (1966) 193 The little room above the saddle room was sprayed with Keatings powder and shut up. [illegible] **1899** *Trade Marks Jrnl.* 14 Mar. 487 Keating's Powder... Insect Destroying Powder. Thomas Keating.. Wholesale Chemists. **1915** A. D. Gillespie *Lett. from Flanders* (1916) 218, I have plenty of Keating's powder left. **1920** D. H. Lawrence *Touch & Go* III. ii. 84 And are you going to comb 'em out, or do you propose to use Keating's? **1926** A. Huxley *Let.* 10 Aug. (1969) 271 The typhoid is said to be very bad..be forearmed and take a lot of Keatings. **1928** F. Stark *Lett. from Syria* (1942) iv. 136 As for fleas, Keating's has succumbed to numbers. **1941** G. Greene in *Spectator* 4 July 8/2 Keating's Powder has taken the place of the military escort.

keatite (kī·tait). *Min.* [f. the name of P. P. *Keat* (born 1923), U.S. chemist who first synthesized it + -ite[1].] An artificial modification of silica produced at high pressures as tetragonal crystals.

1954 [see *coesite]. **1955** *Trans. Brit. Ceramic Soc.* LIV. 665 The polymerized silicic acid now crystallizes.. to the crystalline tetragonal phase, keatite. **1959** *Zeitschr. für Kristallogr.* CXII. 410 The density of keatite (50 /gr·cm^3) is intermediate to that of cristobalite..and quartz. **1962** C. Frondel *Dana's Syst. Min.* (ed. 7) III. 308 Keatite has been synthesized by heating commercial dried silica..in water containing an alkali.

Keatsian (kī·tsiăn), *a.* and *sb.* [f. the name of the English poet John *Keats* (1795–1821) + -ian.] **A.** *adj.* Of, pertaining to, or characteristic of Keats or his poetry. **B.** *sb.* A student or admirer of Keats or his poetry.

1845 T. Wade *Let.* 26 May in H. E. Rollins *Keats Circle* (1948) II. 119 A Keats-ian poem. **1891** *Athenæum* 23 May 667/3 A little manuscript book..of some interest to Keatsians. **1901** H. B. Forman in *Keats Compl. Works* III. 112 He [*sc.* Woodhouse] opens with a Keatsian enough punctuation of the first four lines. **1910** H. Walker *Lit. Victorian Era* II. ii. 299 A Keatsian intensity of sensuousness is his [*sc.* Tennyson's] characteristic. **1934** F. L. & F. Lucas *From Olympus to Styx* vi. 69 In spite of its Keatsian name, Lamia has little romance or even history. **1968** *Guardian* 22 Mar. 9/3 The kind of detail, immensely rewarding to Keatsians, with which Mr Gittings's biography swarms, is well exemplified by his suggestion about the authorship of the essay for 'The Indicator' which Leigh Hunt..was composing extempore..while Keats..lay listening on a sofa. **1970** *Daily Tel.* 19 Sept. 9/6 The day.. beginning as a Keatsian 'season of mellow fruitfulness' lay wrapped in its own swaddling clothes.

So **Keatsi·ana** (see Ana *suff.*, *-iana).

1818 R. Woodhouse *Let.* Nov. in H. E. Rollins *Keats Circle* (1948) I. 66, I shd like to add that to my collection of 'Keatsiana'. **1898** [see *-iana]. **1962** *Daily Tel.* 21 Mar. 17/5 [heading] Keatsiana... I have been commissioned to write a book on Keats.

kebab, keebaub, khubab, kibab, kibaub, qabab, varr. Cabob. (The usual form is *kebab.*) Pieces of meat roasted on a skewer (see also quot. 1970). Cf. *shish kebab.

[illegible] A superb dinner of fifty covers, cooked in the Mogul taste: pilaus, kebabs, curries, and other savoury dishes. [illegible] **1932** *Times Lit. Suppl.* 10 Nov. 836/4 A learned disquisition on 'skewer-cooking', of Eastern origin, leading to various kinds of kebabs. **1954** *Gd. Housek. Cookery Bk.* 450/2 The Muslims who gave the dish its name (which can also appear in such forms as khubab, kibbab and qabab) used their swords to impale the food. [illegible] **1963** R. Carrier *Great Dishes of World* 139 If the weather is too cool for outdoor cookery, kebabs can be grilled indoors with ease. **1970** Simon & Howe *Dict. Gastron.* 231/1 Cubes of vegetables are also called kebabs. [illegible] **1973** *Times* 9 Nov. 14/4, I then bought some kidney kebabs. **1974** *Times* 4 May 11/6 The Open Space in Tottenham Court Road..is within easy reach of..the kebab houses around Charlotte Street. The least frantic at lunchtime seem to be the Cypriana Kebab House..and the Venus Kebab House.

kebaya, var. Kabaya.

The spelling *kebaya* is now usual.

1909 R. O. Winstedt *Papers on Malay Subjects: Life & Customs* II. 49 The long, sleeveless kebaya, [is] now universally worn by women. **1953** A. Keith *Land below Wind* vi. 99 The kebaya, a short Malay blouse, was purchased, and Kuta..fastened it with one insecure brooch. [illegible] My eye was often disappointed by chromatic discords between the ladies' skirts and their kebayas, long-sleeved jackets for which they favoured brightly variegated flower patterns. **1963** *N.Z. Woman's Weekly* 17 June 21/2 [illegible] **1966** D. Forbes *Heart of Malaya* [illegible] **1972** *Guardian* 13 Jan. 5/3 Our beautiful M[alaysia]-S[ingapore] A[irlines] hostesses in their sarong kebayas are dedicated to superb in-flight service.

kebla(h), keblah, varr. Kiblah (in Dict. and Suppl.).

Kechua, var. *Quechua.

keck-handed *a.*, var. *cack-handed *a.*

Ked (ked), *sb.*[3] orig. *U.S.* (See quot. 1967.) Proprietary name of a soft-soled canvas shoe.

1917 *Trade Marks Jrnl.* 14 Nov. 1093 Keds... Rubber, Leather, and Fabric Footwear. United States Rubber Company, New York,..Manufacturers of Rubber Goods. **1961** *Encounter* Apr. 23/2, I am wearing Keds, and feel light on the foam rubber soles. **1966** B. H. Deal *Fancy's Knell* (1967) i. 15 He found the Ked. He picked it up..reluctant to think beyond the small canvas shoe in his hand. **1967** L. J. Healey in R. L. Cohen *Footwear Industry* x. 93 We [*sc.* U.S. Rubber Co.] wanted to call it Peds, but..it came too close to..other brand names. So we batted it around for awhile and decided on the hardest-sounding letter in the alphabet, K, and called it Keds, that was in 1916. **1968** H. C. Rae *Few Small Bones* II. viii. 138 He dug a hole in the junk on the couch, tossing a track-suit and a pair of Keds..on to the floor.

kedgeree. Add: Also *transf.* and *fig.*

1909 in Webster. **1928** R. Campbell *Wayzgoose* ii. 48 English, art, music, vegetables, and song, All to the same consistency you mash. Your life—a Kedgeree! Your mind—a hash! **1938** *Archit. Rev.* LXXXIII. 3/1 Winstanley's Eddystone was a remarkable kedgeree of bits and pieces—its builder had previously been known chiefly for some remarkable waterworks in Hyde Park. **1968** R. West *Sk. Vietnam* ii. 65 Furniture, clothes, shrines.. were heaped on to the lorry in a gigantic kedgeree.

keebaub: see *kebab.

keel, *sb.*[1] Add: **3. b.** A longitudinal member or assembly of members running the length of a rigid or semi-rigid airship at the bottom of the envelope.

In quot. 1877, and perh. also 1888, *keel* has not acquired this specific sense.

[**1877** *Design & Work* 1 Dec. 602/2, I arrived at this principle [of propelling the air boat]:..that though the car must contain the weight of passengers, cargo, and machinery, even to do duty as the weighted keel or plummet, yet it is only in that character it can serve the navigation in aid of propulsion.] **1888** *Peel City Guardian* 22 Sept. 3/3 Connecting the balloon with the arrow-like rod beneath is a keel of the same material as that composing the body of the balloon. **1893** *Eng. Illustr. Mag.* July 746/2 From the outer gallery the [airship] *Attila* looks as if her bottom was gently curved, terminating in the customary orthodox keel... But three feet below the level at which we stand lies a flat projecting bottom. **1910** A. Williams *Engin. Wonders of World* III. 48/2 The distribution of the load over the gas holder in such a way as not to strain any part unduly is, in the case of a Zeppelin airship, simplified by the employment of a girder keel. **1929** E. F. Spanner *About Airships* iii. 28 Throughout the length of the keel there is a more or less uniform lift, varying according to the size of the gasbags. **1955** *Oxford Jun. Encycl.* IV. 20/2 The semi-rigid type, in which a long rigid keel supports the passenger and engine-cars, has been developed mainly by the Italians. *Ibid.*, Keels running through the hull [of a rigid airship] add strength and provide access to various parts of the ship. **1974** J. B. Collier *Airship* 12/1 The distinction between these two types [*sc.* non-rigid and semi-rigid airships] is sometimes hard to draw, but 'semi-rigid' implies..that the airship in question has a rigid keel.

† **c.** In some early aeroplanes and kites, a vertical fin fixed towards the rear of the fuselage and parallel to it, and intended to give lateral stability. *Obs.*

1894 O. Chanute *Progress in Flying Machines* 184 Very good results with central keels have been obtained by M. Boynton with his various forms of 'Fin' kites. *Ibid.* 185 Keels have been frequently proposed for aeroplanes, in which they will produce less resistance to forward motion than obtains with other arrangements. **1907** C. Dienstbach in *Navigating the Air* (Aero Club Amer.) p. xxxix, A multiplicity of 'keels', which might be called 'barbarian' if compared to American moderation. **1910** R. W. A. Brewer *Art of Aviation* xvii. 230 The Antoinette machine has a smaller keel, but some of the monoplanes dispense with this surface altogether. **1911** G. C. Loening *Monoplanes & Biplanes* xii. 255 In the old Voisin type use was made of several vertical keels, partitions, placed not only at the rear, but also between the main surfaces themselves. **1919** H. Shaw *Text-bk. Aeronaut.* vii. 97 The dihedral planes give rise to a greater righting moment, when tilted at a similar angle, than the keel, and so are more efficient.

d. A longitudinal member running along the centre of the bottom of the hull of a flying boat (or the float of a seaplane), or the fuselage of a landplane from one end to the other.

1920 *Flight* 23 Sept. 1019/2 The hull lines are somewhat unusual, the downward sweep of keel and chines in front of the rear step being rather more pronounced than usual. **1930** P. H. Sumner *Marine Aircraft* vi. 164 The type of keel used in the flexible circular flying boat hull is that which is built up as a light girder, comprising a keel proper, keelson and rider piece. The keel proper..is rabbeted on its upper face and receives the vertical keelson. **1933** W. Munro *Marine Aircraft Design* iv. 58 The detail design of frames, bulkheads, stringers, keel, etc., is very definitely affected by the heat treatment and anodic treatment of the material. **1968** *Flight International* 12 Dec. 983/1 Because of the four-leg main undercarriage [of the Boeing 747]..a centre-line keel links the lower part of the forward and rear fuselage. **1969** *Jane's 100 Significant Aircraft* 81/1 Four-engined commercial flying-boat... Structure composed of deep keel, widely spaced transverse frames and heavy stringers.

4. In dogs, the sternum or breast-bone, esp. in the dachshund and other breeds in which it is a prominent feature.

1950 C. L. B. Hubbard *Dachshund Handbk.* iv. 50 Chest oval, well let down between the forelegs, with the deepest point of the keel level with the wrist joints. **1962** R. H. Smythe *Anat. Dog Breeding* i. 19 Dachshunds possess an over-lengthy body and an over-developed sternum, the 'keel'. **1971** F. Hamilton *World Encycl. Dogs* 344 The Standards [for dachshunds] require that the height at the shoulder should be half the length of the body..the lowest point of the keel being on a level with the wrist joint.

6. b. [Norw. *kjøl.*] The spinal ridge of mountains stretching down the centre of Norway.

1856 Ld. Dufferin *Lett. High Latitudes* (1857) xii. 381 The back-bone, or *keel*, as the sea-faring population soon learnt to call the flat snow-capped ridge that runs down the centre of Norway. **1968** G. Jones *Hist. Vikings* II. i. 59 The upturned keel of mountains running south from Finnmark almost to Stavanger and Värmland. *Ibid.* 69 The mountain wildernesses of the Keel.

keel-boat. Add: **b.** *U.S.* (Earlier and later examples.)

1786 in *Mag. Amer. Hist.* (1877) I. 176 Great numbers of Kentucke and keel boats passing every day; some to the Falls, others to Post Vincent—Illinois Country. **1874** E. Eggleston *Circuit Rider* xxvii. 266 A stranger.. reported that he had seen such a man on a keel-boat. **1949** *Indiana Mag. of Hist.* June 147 The first keelboat on the St. Joseph River, the 'Fair Play', arrived at South Bend July 1, 1832.

Hence **keel-boater, -boatman.**

1839 *Knickerbocker* XIII. 344 A..keel-boatman..saw a steam-boat gallantly paddling up against the centre current of that 'Father of Rivers'. **1883** 'Mark Twain' *Life on Mississippi* iii. 41 The keelboatman became a deck hand, or a mate. **1912** I. S. Cobb *Back Home* 296 [He was] the roughest of them all..rougher even than the keel-boaters and the trappers. **1941** F. L. Dorsey *Master of Mississippi* 129 Keelboatmen and 'broadhorn' pushers eyed it with suspicion. **1949** B. A. Botkin *Treas. S. Folklore* II. ii. 191 Besides madmen and devils the Mississippi River also bred giants—the keelboatmen whose acknowledged king was Mike Fink.

keeled, *a.* Add: **c. keeled scale**, in certain reptiles, a scale with a central ridge.

1870 A. R. Wallace *Contrib. Theory Nat. Selection* iii. 99 The large caterpillar..startled him by its resemblance to a small snake... It resembled a poisonous viper, not a harmless species of snake, as was proved by the imitation of keeled scales on the crown produced by the recumbent feet, as the caterpillar threw itself backward! **1907** R. L. Ditmars *Reptile Bk.* xviii. 160 With most of the species [of plated lizard], the scales of the middle portion of the back are strongly keeled. **1970** *New Yorker* 19 Sept. 30/3 It [*sc.* DeKay's snake, *Storeria dekayi*] has keeled scales—a keeled scale is one with a ridge down its middle—which give it a smooth, unshiny appearance.

d. keeled scraper (F. *grattoir caréné*) Archæol., a form of prehistoric flint-tool.

1911 W. J. Sollas *Anc. Hunters* viii. 218 Carefully flaked like the snout of the keeled scraper. **1921** M. C. Burkitt *Prehistory* iv. 75 *Keeled scrapers.* This tool..is very common in Middle Aurignacian times. It has a flat under surface, from which the flakes on the upper surface are struck off in a fan-shaped manner. **1927** Peake & Fleure *Hunters & Artists* iv. 46 Here [*sc.* at La Ferrassie etc.] we find keeled scrapers, of a massive form but carelessly made, and more rarely gravers trimmed obliquely. **1968** *Encycl. Brit.* IX. 448/1 Core scrapers were made on small blocks or were actual cores reutilized as scrapers. Keeled scrapers present a systematic and symmetric removal of tiny blades to form a thick, fluted scraper extremity.

keeler[2]. **1.** (Later examples.)

1854 *Househ. Words* 2 Sept. 54/2 They are pressed into keelers—tubs made of substantial oak, lessening in size to suit the lessening bulk of the cheese as it dries. **1895** *Montgomery Ward Catal.* 578/3 Indurated Wood Fibre Ware... Keelers..Diam. 20 in. **1909** E. I. Denny *Blazing the Way* III. i. 393 A distracted grey-haired lum-e-i, his mother, came to our house to beg for a keeler of water.

keelie. 2. (Later examples.)

1909 *Athenæum* 1 May 528/2 Most people will..appreciate the sterling..character..of the Glasgow 'keelie' of twelve. **1937** *Times Lit. Suppl.* 13 Nov. 870/3 Wondering ..whether the rascally little Glasgow 'keelie'..will succeed in betraying both sides for pay. **1962** 'H. Calvin' *System* ii. 31 A Glasgow keelie who had grown up in a two-roomed slum. **1969** I. & P. Opie *Children's Games* iv. 155 'It has to be played in the dark,' remarks a keelie. **1973** *Times* 18 May (Glasgow Suppl.) p. iv/2 The archetypal Glasgow keelie is a gallus man.

Keemun (kī·mun). Also **Kee-Moon, Kee Mun, Kee-mun.** The name of a district in China used to describe a black tea grown there. Also *attrib.*

1892 J. M. Walsh *Tea* v. 85 Kee-mun..is another of the newest descriptions of China Congou teas... The dried leaf varies considerably in style and appearance, some lots having an evenly-curled and handsome leaf, while others again are brownish and irregular. **1907** *Yesterday's Shopping* (1969) 1 Finest China, Plain (Keemun) —lb. 3/2. **1935** W. Stevens *Let.* 20 Dec. (1967) 303 This morning for breakfast I had some of the best Kee-Moon, and found it to be a delightful tea. **1958** *Catal. County Stores* (Taunton) June 18 A Blend of Pure China Keemun and Finest Formosan Oolong Teas. **1967** V. C. Clinton-Baddeley *Death's Bright Dart* 64 Neither Lapsang, nor Kee Mun, nor Oolong, nor Jasmine..and certainly not Earl Grey. He brewed a pot of Darjeeling.

keen, *a.* Add: **3. b.** Of prices: competitive. Cf. quot. 1862 in Dict., sense 6, and *keenly *adv.* 6.

1964 A. Fiber *Independent Retailer* v. 55 (heading) Mail order has grown rapidly in recent years... As warehouses and offices are situated in low-rent areas and there is no need of a sales staff, overheads are low. Prices, therefore, are often very keen. **1975** *Evening Herald* (Dublin) 8 May 13/3 (Advt.), Dennis Rent a Car. Keenest rates. New Street, Dublin 8.

4. d. Jolly good, very nice, splendid. *colloq.* (orig. *U.S.*).

1914 'High Jinks, Jr.' *Choice Slang* 14 *Keen*, excellent... 'A keen day.' 'A keen time.' **1925** *College Humor* Aug. 76/1 *Keen*, fine; attractive, splendid. **1940** *New Yorker* 16 Nov. 19/3 'My mother's going to buy me four new dresses.' 'That's keen.' **1948** *Hearst's International* Dec. 162/3 'What are you studying at school?' 'Journalism.' 'That sounds keen,' said Sally. **1964** *Punch* 8 July 38/1 It's fab, Henchcliffe, it's gear, moody, groovy, keen and withitly gogo. **1968** N. Fleming *Counter Paradise* vi. 87 He slowed to a standstill beside the second flag. 'Keen,' he said.

6. b. *keen on* (earlier example). Also, sweet on, in love with.

1886 E. Dowson *Let.* 15 May (1967) 78 Is there anything you are keen on? [illegible] **1943** [illegible] 'He's frightfully keen,' she said, 'on you.'

C. a. *keen-bladed* (example), *-eared* (example), *-eyed* (later example), *-lipped* (later example), *-sighted* (example).

1906 *Macm. Mag.* Apr. 457 An escort of sturdy little Japanese armed with service rifles and the keen-bladed *katana*. [illegible] **1952** C. Day Lewis tr. *Virgil's Æneid* iv. 76 Massylian riders galloped behind a keen-nosed pack. **1862** Bagehot in *National Rev.* Jan. 214 If you place the most keen-sighted being in the midst of..an aristocracy, she will sink to the level of those elements.

Keene (kīn). [Name of Richard Wynn *Keene*, who patented the plaster in England in 1838.] *Keene's cement*: a plaster which sets to a very hard white finish and consists of gypsum that contains added alum (or another salt) and has been thoroughly calcined at a high temperature.

1869 [illegible] *Brit. Pat.* 621 2, I would prepare a slab of the required size in cement, by preference Keene's cement. **1927** E. A. Doncaster in C. Martin *Indust. & Manuf. Chem.: Inorg.* II. ix. 117 Keene's cement is now the general name for a number of different plasters prepared for various manufacturers, the original patent having expired. **1947** J. C. Rich *Materials & Methods Sculpture* iv. 60 Dry, powdered mineral colors can be added to Keene's cement, which is occasionally used sculpturally as a very hard plaster.

keener[2] (kī·nəɹ). *U.S.* [f. Keen *a.* + -er[1].] One who drives a hard bargain; also, a person or thing in some way superior.

1839 [see *yes *sb.* 1]. **1860** Bartlett *Dict. Amer.* (ed. 3), *Keener*, a very shrewd person, one sharp at a bargain, what in England would be called 'a keen hand'. Western. **1872** Schele de Vere *Americanisms* 436 *Keener*, a noun made from the adjective, is a Western term for a sharp man. [illegible] **1942** Berrey & Van den Bark *Amer. Thes. Slang* § 542/2 *Keener*, a shrewd person. [illegible] § 542/2 *Bargainer, keener.*

keenly, *adv.* Add: **6.** *Comm.* At a keen price, cheaply.

1928 *Daily Express* 28 Aug. 7 With advantages like this we can quote more keenly.

keeno, var. Keno (in Dict. and Suppl.).

keep, *v.* Add: **16. d.** *to keep wicket*: see Wicket 3. Also *absol.*, to act as wicket-keeper.

1862 *Baily's Monthly Mag.* Aug. 85 The Surrey people ..selecting..a John Walker to keep. **1920** P. F. Warner *Cricket Reminisc.* 161 Lockyer 'kept' for the Players on and off between 1854 and 1866. **1931** *N. & Q.* 14 Feb. 121/2 Alfred Lyttelton..kept wicket for England. **1959** *Times* 29 June 11/4 One of Somerset's clerical keepers of the gloves.. [illegible]

29. c. Colloq. phr. *you* (etc.) *can keep* (something): it arouses no desire, envy, or interest in me; I am not interested in (it).

1956 J. Popplewell in *Plays of Year 1955* XIII. 335 *Robert.* My hobby's writing plays. *Tom.* You can keep it. **1964** M. Drabble *Summer Bird-Cage* i. 8 [illegible] **1971** *Guardian* 11 Dec. 5/1 [illegible] If that is Arctic, they can keep it. **1973** *Ibid.* 12 Apr. 13/3 It makes me a bit sick actually and they can keep their mag as far as I am concerned.

35. (U.S. examples of *to keep school*.)

1847 *Knickerbocker* XXX. 31 A girl whose education does not qualify her for 'keeping school'. **1869** E. Chamberlain *Indiana Gazetteer* (ed. 3) 136 There are in the County..school houses in which schools are kept for a portion of the year, in most of the districts. **1867** 'T. Lackland' *Homespun* II. 264 Mr. John Porringer.. 'kept' this school, and was in the habit of keeping it so long as he lived and liked.

38. b. Of a school: to be held. *U.S.*

1845 *Knickerbocker* XXVI. 277 One afternoon, when 'school didn't keep', [illegible] **1867** 'T. Lackland' *Homespun* I. 123 The District School has not 'kept' since the week began. **1908** M. E. Freeman *Shoulders of Atlas* 68 School ain't going to keep today.

44. b. *to keep oneself to oneself* (examples).

1748 Richardson *Clarissa* IV. 27, I was resolved to keep myself to myself till I saw the issue of it. **1846** *Swell's Night Guide* [illegible] **1848** J. H. Newman *Loss & Gain* [illegible] sure of my convictions? **1905** H. G. Wells *Kipps* I. i. 7

They 'kept themselves to themselves', according to the English ideal. **1960** 'H. Carmichael' *Seeds of Hate* iv. 37 My husband and I prefer to keep ourselves to ourselves. [illegible] She keeps herself to herself so much. **1973** J. Porter *It's Murder with Dover* ii. 17 Miss Marsh has always been one to keep herself to herself.

48. a. *spec.* To retain (food, etc.) in one's stomach, without vomiting.

1955 'A. Gilbert' *Is She Dead Too?* vi. 119 Think you could keep some hot toast down? Well, have a try. **1968** 'S. Woods' *Past Praying For* ix. 72 Nothing but water has been given to Olivier without Dr. Noyes's consent; and even then, he couldn't keep anything down. **1969** [illegible] **1973** 'A. York' [illegible] 'I don't think I could keep it down.'

50. c. *trans.* To avoid or stay away from; not to use; also as *attrib. phr.*; *keep off the grass*: see *grass *sb.*[1] 9.

1949 M. Mead *Male & Female* ii. 42 Tchambuli..placed a keep-off sign on the coconut-palm-trees. **1968** *Listener* [illegible] have been instructed to keep off heavy make-up.

57. f. Esp. in phr. *keep it up*; *spec.* to prolong a party, drinking-spree, etc.; to 'live it up'. (Later examples.)

1752 J. Millward *Let.* in M. M. Verney *Verney Lett.* (1930) II. ii. xxiv. 250 When they [*sc.* the Welsh] get in liquor they are very troublesome and noisy. They kept it up all night. **1788** Grose *Dict. Vulgar T.* (ed. 2), [illegible] **1810** M. Dwight *Journey to Ohio* (1912) [illegible] **1837** [illegible]

j. Esp. in phr. *to keep up* (often *keeping up*) *with the Joneses* (or *Jones's*): to strive not to be outdone by one's neighbours; to emulate one's neighbours. Also *transf.* orig. *U.S.*

1913 A. R. Momand in *Globe* (N.Y.) 1 Apr. 16/3 (*Comic-strip title*) Keeping up with the Joneses—by Pop. [illegible] **1927** Chase & Schlink *Your Money's Worth* i. 7 Certain things we buy.. to keep up with the Joneses, or happily, to surpass the Joneses. **1933** E. Weekley in *Trans. Philol. Soc.* 94 This tendency to personify by the use of a familiar name is due to the same psychology which describes the social ambitions of the suburbs as 'keeping up with the Joneses'. **1953** F. P. Keyes [illegible] He could not be thankful enough that he did not have a nagging wife, one who insisted on making a show, on 'keeping up with the Joneses', as people were beginning to say. [illegible] **1958** [illegible] **1963** [illegible] We like to keep up with the Joneses and are therefore well disposed [illegible] **1971** *Times Lit. Suppl.* [illegible] with Joneses to keep up with, increased in cost, display and ostentation.

l. To continue to maintain a friendship or acquaintance; to keep in touch. (Cf. 57 f.)

[illegible] She did not seem to be able to keep in personal touch with them... She [illegible] **1916** E. V. Lucas *Vermilion Box* [illegible] I heard this morning of the death of one of my oldest friends—[illegible] who was at school with me, and [illegible] whom I knew at Oxford. Both went straight into the army, but kept up. **1947** [illegible] 205 We [illegible] kept up. [illegible] 125 We always kept up [illegible] **1971** 'D. Shannon' [illegible] had known [illegible] and they had 'kept up'.

58. keep-fit *a.*, denoting exercises, etc., designed to keep people fit and healthy; (occas.) a person who does such exercises; also *ellipt.* as *sb.*; **keep-left** *a.*, designating a sign, etc., directing traffic to the left of the road; **keep-out** *a.*, designating a sign that prohibits entry.

1938 M. Carter [illegible] Gardening comes into their day's programme and 'keep fit' exercises. **1939** 'N. Blake' *Smiler with Knife* v. 88 A healthy, bouncing, Keep-Fit sort of girl. **1961** J. Stroud [illegible] A Girls' Keep Fit class was in session. [illegible] including [illegible] and 'Keep Fit'. **1967** O. [illegible] goes to his [illegible] Keep Fit [illegible] **1971** [illegible] The [illegible] *Discovery* [illegible] 'keep left' signs are automatically lit. **1962** C. Watson *Hopjoy was Here* iv. 38 A pair of dogs..coupled on the road's crown and performed a six-legged waltz around a keep-left bollard. **1971** J. McClure *Steam Pig* v. 75 A deserted area surrounded by Keep Out signs. **1974** *Times* 9 May 6/5 To protect your garden a 'keep out' sign is not enough. You also need a tall fence.

keep, *sb.* Add: **6. c.** (Earlier and later examples.) Freq. in phr. *to earn one's keep* (also *fig.*).

1801 Jane Austen *Let.* 3 Jan. (1952) 101 The keep of two will be more than of one. **1815** M. Birkbeck *Notes Journey through France* (ed. 3) 21 M. Tessier hires the whole of the keep of this flock. He pays £62. 10s. sterling to the farmer for the sheep pasture... He buys Lucerne hay for four winter months..making the expense of keep £142. 10s. sterling. **1937** R. Macaulay *I would be Private* I. xv. 137 Now he can just earn his keep digging treasure on those cays. **1963** A. Ross *Australia 63* iv. 99 The four Test stars..whose appearance cost the Tasmanian authorities £A300, again individually and in bulk failed to earn their keep. **1971** *Country Life* 4 Nov. 1237/2 Under favourable growing conditions this will provide late autumn keep, really valuable spring feed, or both. **1972** *Accountant* 19 Oct. 497/3 'All assets must earn their keep,' declared Mr Shaw.

7. For *U.S. colloq.* read *colloq.* (orig. *U.S.*). Also in extended use: in deadly earnest. In *Cricket*, defensively, in order to remain at the wicket. (Add further examples.)

1861 *Ladies' Repository* Oct. 627/1 Pay him! Nothing. He and I played for 'keeps', and I was the best player and won all his. **1871** *Wright County Monitor* (Clarion, Iowa) 29 Nov., Winter has at last come 'for keeps'. **1893** S. R. Crockett *Stickit Minister* 79 She'll even set down the black bag to play for keeps wi' the boys at the bools. **1897** *National Police Gaz.* (U.S.) 26 May 3/1 He is in the business for 'keeps', as they say in America. **1905** *Westm. Gaz.* 19 Sept. 3/2 Any other batsman..would doubtless have played for 'keeps' and taken not the slightest risk. **1923** *Cricketer Ann. 1922–3* 90 To-day, the dominant feature of the game is the individual 'playing for keeps'. **1933** D. L. Sayers *Murder must Advertise* xv. 253 Ten to one 'e'll lose 'im for keeps, now. **1949** D. G. Smith *I capture Castle* III. xii. 214 Maybe when I bring you back we shall find it's gone for keeps. **1970** G. E. Evans *Where Beards wag All* xix. 219 You played *for keeps* sometimes, in other words all the marbles you won became yours. **1972** D. Lees *Zodiac* 107 These bastards are playing for keeps. ..I'm in trouble. **1973** 'H. Howard' *Highway to Murder* ix. 108 Everybody belonged to the rat race where people played for keeps.

keepable, *a.* Add: Hence **keepabi·lity.**

1898 R. H. Wallace *Adulteration of Dairy Prod.* 12 The keepability was also tested, and Swedish butter proved to be on an average more keepable than butter from other countries. **1962** *Times* 3 July (Agric. Suppl.) p. vii/4 The keepability of these pig meat products.

keep-alive (kī·pălaiv), *a. Electronics.* [f. *to keep alive* (Keep *v.* 24).] Applied to devices and phenomena in certain kinds of discharge tube which operate or occur continuously and serve to initiate an intermittent main discharge or facilitate its establishment. Also *ellipt.*, such a device or phenomenon.

1933 J. H. Morecroft *Electron Tubes* viii. 182 In large bulbs a 'keep-alive' circuit is used to maintain the ionization. **1947** A. E. Whitford in L. N. Ridenour *Radar Syst. Engin.* xi. 410 To insure rapid breakdown at the beginning of each pulse, a supply of ions in the gap is maintained by a continuous auxiliary discharge inside one of the cones. This requires an extra electrode, known as the 'keep-alive' electrode. **1947** Crowther & Whiddington *Science at War* i. 47 The final form was a cavity ..with a small pilot discharge or 'keep alive' to help the gas discharge to start. **1959** R. L. Shrader *Electronic Communication* xxix. 842 To make the TR tube more sensitive, a d-c keep-alive voltage is applied across it at all times. This voltage is not quite high enough to support ionization. **1965** Gewartowski & Watson *Princ. Electron Tubes* xv. 554 Tubes in which radioactive material has been placed for the purpose of obtaining short ionization times are said to have a radioactive keep-alive. *Ibid.*, In some tubes a low-current discharge between auxiliary electrodes is operated continuously as a keep-alive mechanism.

keeper, *sb.* Add: **1. f.** *Cricket.* A wicket-keeper.

1744 *Laws* [of Cricket] in *New Dict. Arts & Sci.* (1755) IV. 3459/2 When the ball has been in hand by one of the keepers or stoppers. **1868** J. Lillywhite *Cricketers' Compan.* 49 The best 'keeper' who ever stood behind a wicket. **1926** H. Strudwick *25 Yrs. behind Stumps* 244 There were very few, if any, better keepers than the Notts man. **1927** [see *aunt 4 b]. **1975** *Cricketer* May 9/2 A tall 'keeper's rise from his crouch is less rapid than a smaller man's.

g. *Football.* A goal-keeper.

1957 J. Milburn *Golden Goals* 140 (caption) Milburn rates Ditchburn among the greatest 'keepers he has ever faced. **1974** *Oxf. Mail* 21 Aug. 16 (caption) Bicester's Phillip Pratt (10) heads the ball past Thame keeper Micky Taylor for his second goal.

4. (Later examples.)

c **1810** W. Hickey *Mem.* (1960) xviii. 291 He at that time was the professed keeper of Mrs. Cuyler, a great jack whore, without pretensions to manners. **1846** *Swell's Night Guide* 83 Keepers are the sinews of your trade.

6. d. A simple ring worn in the ears to keep a pierced hole open.

1960 *Woman's Realm* 2 Apr. 69/3 Pure gold keeper rings to keep the ear-piercing open, ready for the first real earrings. **1968** K. O'Hara *Bird-Cage* vii. 55, 1 pair gold keeper ear-rings.

keeper (kī·pəɹ), *v.* [f. the sb. or as a back-formation f. Keepering.] *trans.* To look after as a gamekeeper. So **kee·pered** *ppl. a.*

1921 *Chambers's Jrnl.* Sept. 388 An estate well-preserved and well-keepered. **1958** *Times* 13 Sept. 9/3 The full benefits of hand-rearing can be enjoyed only on ground which is adequately keepered. **1961** R. Jefferies *Evidence of Accused* i. 9 If the estate were keepered.. it would become a rattling good shoot. **1971** *Country Life* 23 Sept. 766/3 In recent years this chalk-stream water [*sc.* River Itchen] has been well keepered and only lightly fished by a maximum of three rods. **1972** *Times* 7 Aug. 22/3 (Advt.), Keepered partridge and pheasant shoot.

keepering. (Earlier and later examples.)

1861 *Baily's Monthly Mag.* Jan. 185 His keepering consisted..in an amiable crusade..against 'them darned rats'. **1963** P. MacTyre *Fish on Hook* iii. 48 Wynrame isn't much of a hand at the keepering. **1971** *Country Life* 12 Aug. 363/3 A shepherd..deplored the increase in the depredations of the fox, since, he said, 'keepering' had gone out of fashion.

keeping, *vbl. sb.* Add: **2. d.** *Cricket.* Wicket-keeping.

1868 in W. A. Bettesworth *Walkers of Southgate* (1900) 291 Stephenson's 'keeping' was also first-rate. **1920** P. F. Warner *Cricket Reminisc.* 156 His [*sc.* Blackham's] keeping to Mr. Spofforth with the 1878 Australian XI. was a revelation.

5. b. Delete † *Obs.* and add later examples.

1853 Mrs. Gaskell *Ruth* III. i. 29 She beguiled a young gentleman, who took her into keeping. **1932** J. M. S. Tompkins *Pop. Novel in England 1770–1800* v. 193 Women of the town flaunt at the races and are taken into keeping. **1964** *Listener* 12 Mar. 444/3 There is no stigma on 'keeping' or 'living', nor on illegitimacy.

keeping, *ppl. a.* Add: Esp. of fruit (cf. Keep *v.* 41).

1816 Jane Austen *Emma* II. ix. 187 There never was such a keeping apple any where as one of his trees. **1842** Hood *Let.* 12 Oct. in F. F. Broderip *Memorials Thomas Hood* (1860) II. iv. 140 Our gardener said they [*sc.* pears] were a *keeping* sort, and would be good at Christmas. **1963** *Times* 11 Feb. 13/5 The majority of pupils ate sandwiches and keeping-apples.

keepsake. b. (Earlier example.)

1839 Thackeray in *Corsair* (N.Y.) 26 Oct. 522/1 A book on Versailles with numerous engravings in the Keepsake fashion.

keertan, var. *kirtan.

keeshond (kē·shǫnd). Also **-hound.** Pl. **-honden.** [Du.] A grey, long-coated dog, a variety of the spitz, of medium size, with a fox-like head, pointed ears, and tail curled over the back; originally a Dutch or German breed.

1926 *Westm. Gaz.* 10 Feb., Kesshonds [*sic*] will find many admirers on account of their showy wolf-coloured coats. Some have been exhibited recently as Dutch barge dogs. **1927** *Daily Express* 15 June 15 The Keeshounds, Dutch dogs,..were shown in this country for the first time. They are about the size of a Chow, grey black in colour, exceedingly strong. **1927** E. C. Ash *Dogs: their Hist. & Devel.* I. 159 The keeshond is the well-known Dutch dog, often seen on the barges. **1950** A. C. Smith *Dogs since 1900* 260 In 1926 the Kennel Club decided that ..all previously registered as Dutch Barge Dogs should be transferred to the Keeshond registers. **1971** F. Hamilton *World Encycl. Dogs* 596 Keeshonden were the dogs of the peasantry—living in villages and on farms as well as barges. *Ibid.* 598 The Keeshond's long coat requires special attention once a week with a stiff brush. **1972** *Country Life* 10 Feb. 328/3 Miss H. M. Loughrey..picked out a Dalmatian, a chow chow,..a keeshound.

keessel, var. *kissel.

keester, var. *keister.

'keet (kīt). *Austral.* Colloquial shortened form of Lorikeet or Parakeet.

1936 A. Russell *Gone Nomad* x. 83 Snow-white cockatoos flashed in flocks across the stream, green 'keets screeched among the gums. **1959** S. H. Courtier *Death in Dream Time* iii. 27 That's where the 'keets come... They're a wonderful sight in the flame-trees..all green and gold.

Keewatin (kīwēi·tin). *Geol.* The name of a district in the Northwest Territories, Canada, used *attrib.* and *absol.* to denote the oldest division of the Archæan in North America and rocks representing it, found in the Canadian Shield region.

1886 A. C. Lawson in *Ann. Rep. Geol. Survey Canada 1885* I. 14cc, The most appropriate name for the series that suggests itself to me is 'Keewatin', the Indian name for the North-west, or the North-west wind, which has been applied to the district within which the rocks occur. *Ibid.* 19cc, The contact of the Laurentian gneiss and the Keewatin schists. **1925** *Bull. Geol. Soc. Amer.* XXXVI. 363 The Couchiching beds pass down under the Keewatin. **1927** *Jrnl. Geol.* XXXV. 143 At present the general use of the term Keewatin in the United States and Canada is as the name of the oldest rock series of the region. *Ibid.*, The view that the Keewatin rocks are the oldest in Canada is moreover inconsistent with much recent geological work. **1970** Dorr & Eschman *Geol*

Michigan iv. 39/1 Nearly 3·5 billion years ago, during Keewatin time, streams from bordering highlands deposited sediments and lavas poured out on the surface of the earth in what is now the Upper Peninsula of Michigan. **1972** L. W. Mintz *Hist. Geol.* xiii. 388 Keewatin rocks, like all other very old sediments, appear to be oceanic in origin and consist chiefly of metamorphosed lavas and turbidites.

kef: see *Kif.

‖ **keftedes** (kefte·ðēz), *sb. pl.* Also **keftedhes, keftethes, keph-.** [Gr. κεφτές, pl. κεφτέδες meat ball, f. Turk. *köfte*.] A Greek dish of small meat balls made with herbs and onions.

1912 E. Craies *Recipes from East & West* 82 Mix..some grated Parmesan cheese, as it makes the keftedes much lighter. **1958** E. Liddell *Morea* ii. vii. 173 The pleasant young woman in charge of the children handed round *keftédhes* and hard-boiled eggs. **1966** *Observer* (Colour Suppl.) 2 Oct. 45/4 Keftethes..minced veal..grated onion..breadcrumbs. **1970** *Times* 29 Apr. 18/4 You will get..keftedes (spiced meat balls).

Keftian (ke·ftiăn), *sb.* and *a.* Also **Keftiu** sing. and *coll.*; pl. **Caphtorim** (Heb. *Kaphtōrīm*). [Cf. *Caphtor*, Heb. *Kaphtōr*, name in O.T. for the place of origin of the Philistines.] **A.** *sb.* Name in Egyptian records of a people of the E. Mediterranean, identified by some authorities with the Cretans; a member of this people.

c **1000** Ælfric *Heptateuch* Gen. x. 14 Mesraim, Cames oper sunu, gestrynde six suna; of þam comon þa Philistei & seo mægð Capturym. **1611** Bible Gen. x. 14 And Mizraim begat..Casluhim (out of whome came Philistim) and Caphtorim. **1894** A. H. Sayce *Races Old Testament* iii. 53 The Caphtorim..were the natives of the coast-land Caphtor. **1903** H. R. Hall in *Ann. Brit. Sch. Athens* 1902–1903 174 The Keftiu..the facial type of the Keftians..is definitely European. *Ibid.* 165 Apparently Müller regarded the Keftiu not as themselves genuine Mycenaeans, but as mere handlers on of Mycenaean objects. **1913** R. A. S. Macalister *Philistines* i. 14 The Egyptians were brought into direct contact with the Keftians. **1931** *Times Lit. Suppl.* 13 Aug. 623/2 G. A. Wainwright in 'Keftiu: Crete or Cilicia?' publishes further evidence in support of his thesis that the Keftiuans of the Egyptian monuments are not Cretans..but Cilicians. **1961** C. F. Pfeiffer *Baker's Bible Atlas* (1962) ii. 41/2 The family of Mizraim had several branches: Ludim, Anamim, Lehabim..and Caphtorim. **1967** H. E. L. Mellersh *Minoan Crete* v. 53 There can hardly be any doubt at all that the Keftiu were the Cretans.

B. *adj.* Belonging to or relating to this people or their language.

1903 H. R. Hall in *Ann. Brit. Sch. Athens* 1902–1903 173 A list of Keftian proper names on a writing-board and a list of words of the Keftian language on papyrus..are the last records of the men of Keftiu. **1913** R. A. S. Macalister *Philistines* i. 13 It has been suggested that it might be a nominative suffix of the Keftian language. *Ibid.* iv. 115 The sign *a*, a man running, shows the simple 'suit-band' which forms the sole body-covering of the Keftian envoys. **1930** J. Garstang *Hittite Empire* ii. 42 It was left to the Keftian traders to carry on commercial intercourse that brought to Egypt the wares of Crete and Mycenae.

keg, *sb.* Add: **1. d.** *spec.* A barrel of beer; beer. *Austral.* and *N.Z. slang.*

1945 E. G. Webber *Johnny Enzed in Middle East* 15/2, I thought it was the keg. **1957** N. Culotta *They're a Weird Mob* (1958) viii. 110 We struggled with the kegs, and got them set up on the bench. **1959** M. Shadbolt *New Zealanders* 108 Wild proletarian parties, in slummy Freeman's Bay, where kegs flowed and fights flared. **1965** J. O'Grady *Aussie English* 16 Containers run from five-ounce glasses to eighteen-gallon kegs.

e. In full *keg ale, beer, bitter.* Ale, etc., from a keg (see quot. 1972).

1964 W. Weeks *Beer & Brewing in Amer.* 25 Keg (draught) beer is kept at low temperatures. **1965** *Brewers' Jrnl.* XCVII. 373/2 Mr. G. Dent drew attention to increased beer sales, and in particular a large increase in Keg beers. **1965** *Times* 12 Feb. 17/1 When demand for keg beer rose, they would seek to expand. **1967** C. Drummond *Death at Furlong Post* xi. 151 The brewery..had rushed out twenty-four gallons of keg beer.. The whole bleedin' world wants keg ale.' **1968** W. Garner *Deep, Deep Freeze* vi. 73 The Duke of York..stocked a good line in keg bitter. **1968** 'D. Rutherford' *Skin for Skin* ii. 18 Crispin ordered two pints of Keg and carried them to the dimmest corner of the Saloon. **1971** *Times* 12 June 14/5 The new drinkers who do like beer are willing to pay extra for keg. **1972** *Which?* Apr. 121/2 One way to think of 'keg' beer is as a 'bottled' beer, but in a five or 11 gallon sealed metal container. Keg is chilled, filtered and (usually) pasteurised beer which has had carbon dioxide added.

kegler (ke·gləɹ, kē·gləɹ). *N. Amer.* Also **keggler.** [G. *kegler* skittle-player.] One who plays tenpin bowling, skittles, ninepins, etc.

1932 *Lincoln* (Neb.) *State Jrnl.* 9 Mar., Floyd Ohlsen began calling the pin topplers, keglers, and the folk in Cleveland began looking at him askance. **1943** Falcaro & Goodman *Bowling for All* vii. 28 The kegler who feels that he has a 300 game..will discover that every delivery gets tougher after that. **1956** *Wall St. Jrnl.* 12 Mar. 1/1 These sportsmen and women, known to each other as 'keglers', will spend about $250 million on their sport, 10 times as much as both major baseball leagues collected in a gate receipts last year. **1958** *Economist* 20 Dec.

1085/1 If the 'keggler' does not knock down all the pins with his two balls he is credited merely with the number of pins he does knock over. **1970** *Daily Colonist* (Victoria, B.C.) 5 Apr. 15/2 The Kokomo, Ind., kegler beat Dick Ritger, of Hartford, Wis., in the championship [bowling] match. **1974** *Plain Dealer* (Cleveland, Ohio) 27 Oct. 12-C/5 In match play, each kegler faces each other one time, with the winner of a contest receiving 30 bonus pins.

‖ **kehilla** (kəhi·lă). Also **kehillah, kille.** Pl. **kehil(l)ot, kehilloth.** [ad. Heb. *qěhillāh* community.] The Jewish community in a town or village.

1882 tr. *L. Kompert's Scenes from Ghetto* 67 She grew up, became beautiful, and in all the *kille* people talked of nothing but Hendel, the daughter of Rebb Paltiel Wolf. **1892** I. Zangwill *Childr. Ghetto* I. 126 Every town which could muster the minimum of ten men for worship boasted its Kehillah. **1950** [see *Judenrat]. **1961** *Observer* 3 Dec. 10/6 The formal communities of Russian Jews, the *kehilot*, were abolished in 1919 and never restored. **1967** N. Cohn *Warrant for Genocide* vii. 163 The official New York, Jewish Community Council bore the same name the Kehilla (Yiddish for 'Kahal') was chiefly concerned with protecting and educating Jewish immigrants. **1971** *Jewish Communities of World* (Inst. Jewish Affairs) (ed. 3) 29 The Buenos Aires Community (*Kehillah*) is the largest. *Ibid.* 30 There are 20 organized local Communities (*Kehilot*) in France. **1973** *Jewish Chron.* 2 Feb. 7/2 We have no intention..of creating a separatist movement or a new *kehilla* or some political group.

kehoeite (kī·hō̆ə·ait, kī·hō̆ɪit). *Min.* [f. the name of Henry Kehoe, who discovered it + -ite¹.] A hydrated hydroxide and phosphate of aluminium, zinc, and calcium, found as white, chalky, amorphous masses in South Dakota.

1893 W. P. Headden in *Amer. Jrnl. Sci.* CXLVI. 24, I propose the name Kehoeite as the name of this mineral, after Mr. Henry Kehoe, to whom I am indebted for the material and who was the first to observe its occurrence. **1964** *Mineral. Mag.* XXXIII. 799 The hardness of kehoeite cannot be determined because of its microcrystalline nature.

keister (kī·stəɹ, kai·stəɹ). *U.S. slang.* Also **keester, keyster.** [Etym. unknown.] **1. a.** A suitcase, satchel; a handbag; a burglar's tool-case; a salesman's sample-case, etc.

[**1882** *National Police Gaz.* (U.S.) 7 Oct. 10/ Prominent among the small army of confidence operators in this city are: 'Grand Central Pete'..'The Guinea Pig'.. 'Keister Bob'.] **1882** G. W. Peck *Peck's Sunshine* 227 The boy took the bugler's keister and went to the elevator. **1910** G. B. McCutcheon *Rose in Ring* 80 Ruby Noakes..who controlled the contraband in his efforts to construct a table out of three 'blue seats' and a couple of property trunks, or 'keesters', as they were called. **1926** *Flynn's* 16 Jan. 637/2 All this chatter 'bout keisters with false bottoms an' muslins is mostly pipe stuff an' hopchin. **1926** Clan Now. 101/2 *Keyster*, handbag or suitcase. **1933** 'P. Cain' *Fast One* vi. 211 Hang on the front..until you see them big pig-skin keesters go in. **1935** Collier & Westrate *Reign Soapy Smith* i. A open sample case of liberal dimensions..the typical 'keister' of the street hawker. **1935** *Jrnl. Abnormal Psychol.* XXX. iii. 363 *Keyster*, a suitcase. **1950** H. E. Goldin *Dict. Amer. Underworld Lingo* 114/2 Ditch that keister. It draws heat (attracts police attention).

b. A strong-box in a safe; a safe (see also quot. 1924).

1913 A. Stringer *Shadow* 36 He made a mental record of dips and yegs and till-tappers and keister-crackers. **1921** P. & T. Casey in *Adventure* (U.S.) 18 July 12/2 They..breaks 'oler the keester o' the bank safes an' gits away with all the stored-up jack. **1924** G. C. Henderson *Keys to Crookdom* 409 *Keister*, burglar's kit; a toilet safe. A handbag that can be strapped and locked. **1931** *Amer. Speech* VII. 110 Can we use can-opener on this keister? **1950** H. E. Goldin *Dict. Amer. Underworld Lingo* 114/2 Easy on the soup (crude nitro-glycerine) with that keister or we'll be 'dead'.

2. The buttocks.

1931 *Amer. Speech* VI. 439 *Keister*, a satchel; also what one sits on. **1951** Wodehouse *Old Reliable* xi. 131 Up in the diner, they'd be leave me flat on my keister and go off and buy candy and orchids for the other girls. **1968** *McLean's Mag.* Dec. 1/1 His job at Christ the King was defined by Father Mooney as 'getting people off their keesters'. **1975** *New Yorker* 10 Mar. 90/2 Just put your keyster in the chair and shut your mouth.

Kekchi (ke·ktʃi), *sb.* and *a.* Also **Quecchi, Quekchi. A.** *sb.* Name of an ancient people belonging to the Maya Empire, the modern descendants of whom now live in Guatemala. Also, a member of this people (ancient or modern); their language. **B.** *adj.* Of or pertaining to the Kekchi or their language.

1883 J. Baily tr. *Juarros's Statistical & Commercial Hist. Guatemala* vi. 198 No one of the kingdoms of the New World has so many different languages as Guatemala, the following 26 being peculiar to it, viz. Quiche.. Quecchi.. **1888** [see] *Brit.* Brit. XXIV. 760/1 Pocomchi Group ..Quekchi..Coban district, Guatemala. **1949** E. A. Nida *Morphol.* (ed. 2) ii. 6 In Kekchi, a Mayan language of Guatemala. **1960** R. C. Ball *Board & Table Games* I. 35 This is another running-fight game played by the Kekchi Indians of Central America. **1964** E. A. Nida *Toward Sci. Transl.* iii. 46 In fact, at one time some Kekchis in Guatemala asked the missionary not to attempt

to explain the 'truths of their faith', for if such matters could be explained and understood, they would then 'cease to be religion'.

Kekulé (ke·kiule). *Chem.* The name of Friedrich August *Kekulé* (1829–1896), German chemist, used *attrib.* and in the possessive with reference to a system of structural formulae devised by him, the principal features of which are the quadrivalency of the carbon atom and the linking of carbon atoms to form chains, and esp. his conception of the benzene molecule as having a closed hexagonal ring of six carbon atoms with alternate double and single bonds between them (later extended to other conjugated cyclic carbon compounds).

1872 *Jrnl. Chem. Soc.* XXV. 824 The author says that he acknowledges the value of Kekulé's ring formula. **1908** A. W. Stewart *Rec. Adv. Org. Chem.* i. 6 But as they actually existed in isomeric forms, the same was to be expected in benzene, if the Kekulé theory were correct. **1938** H. Gilman *Org. Chem.* I. ii. 57 A further, rather serious objection to the Kekulé formula arose from the existence of only three ortho-disubstituted benzenes... The benzene molecule consists of three ordinary double bonds, it is so inferior to the ethylenic hydrocarbons in reactivity. **1951** I. L. Finar *Org. Chem.* I. xx. 408 The difficulty with the Kekulé formula is that it represents benzene with three double bonds, and the oscillation does not account for the difference in behaviour between these and olefinic bonds. *Ibid.* 409 The Kekulé structures contribute far more to resonance (about 80 per cent.) than do the Dewar structures (about 20 per cent.). **1961** L. F. & M. Fieser *Adv. Org. Chem.* iii. 15 In the Kekulé theory of structure performed valuable service but half a century before the nature of the Kekulé bond became understood. **1964** Cram & Hammond *Org. Chem.* (ed. 2) xxiii. 543 Two equivalent Kekulé structures may be drawn for this $C_{14}H_{10}$ hydrocarbon (*sc.* cyclooctadecanonaene?). **1972** E. Clar *Aromatic Sextet* vi. 29 The VB theory assumes nine Kekulé structures for triphenylene which are supposed to contribute equally to the ground state.

kel: see *Kelly, kelly *sb.*² 2.

kelch (keltʃ). *slang.* Also **kelt, -tch, keltz.** [Etym. unknown.] A white person: only in depreciatory use.

1912 F. M. Hueffer *Panel* i. ii, 14 'Do you mean to say that you haven't got a single book of James?'.. 'Never heard the name,' the bookstall boy said. 'But there's plenty by Mrs. Kerr Howe.' 'That kelch!' Major Foster exclaimed. **1938** C. Himes *Black on Black* (1973) 173 Then he met a high-yellah gal, a three-quarter kelts, from Ohio. **1946** Mezzrow & Wolfe *Really Blues* 375 Here to stick my kelt and. **1970** C. Major *Dict. Afro-Amer. Slang* 73 *Kelt, keltch*, white person; Negro passing for white.

‖ **kelebe** (ke·lĕbi). *Gr. Antiq.* [Gr.] (See quot. 1890).

1858 S. Birch *Hist. Anc. Pott.* & *Porc.* I. ii. iv. 162 The Archaic pyxis of Apulian *stamnos*, the *kelebe*, or *crater*, with columnar handles, are later in date. **1890** Cent. *Dict.*, *Kelebe*, in Greek Archæol., a large ovoid, wide-mouthed vase, with a broad flat rim and two handles connecting the rim and the body, and each supported by a column. **1960** R. M. Cook *Greek Painted Pott.* 363 *Kelébe*, obsolete, a conventional name for a column krater.

K-electron: see *K 3 e.

kelep (kē·lep). [Kekchi.] A Central American stinging ant, *Ectatomma tuberculatum*.

1904 W. M. Wheeler in *Science* 30 Sep. 439/2 There is little probability that the kelep can be successfully established in Texas. **1910** — *Ants* i. 9 It (*sc.* United States Department of Agriculture) recently introduced a Guatemalan ant, the 'kelep' (*Ectatomma tuberculatum*) into Texas for the purpose of destroying the very injurious cotton-boll weevil. **1951** C. P. W. Muesebeck et al. *Hymenoptera of Amer. North of Mexico* 785 *Ectatomma*.. *tuberculatum*.. This species, known as the kelep, was introduced into Texas in the early part of this century to combat the cotton boll weevil. It has apparently become extinct.

Kelim, var. *Kilim.

Kellaways (ke·lăwēz). *Geol.* Also formerly **Kellaway, Kelloway(s).** The name of a village near Chippenham in Wiltshire, used *attrib.* to designate a group of clays and calcareous sandstones of Jurassic age lying below the Oxford clay and above the cornbrash, and found in a belt extending from Dorset to Yorkshire.

1813 J. Townsend *Character of Moses* I. vi. 103 Kelloway rock. The next calcareous stratum, first attracted our notice, at Kelloway Bridge. **1888** J. Prestwich *Geol.* II. xiv. 217 On the Continent the Kelloways Rock forms a division of almost equal importance with the Oxford Clay. **1913** *Q. Jrnl. Geol. Soc.* LXIX. 152 (*heading*) The 'Kelloway rock' of Scarborough. [*Note*] The appellation 'Kelloway' is best used for specimens from Leckenby (1859) or in reference to Yorkshire beds, and 'Kellaways' in relation to deposits in Wiltshire and elsewhere. **1933** W. J. Arkell *Jurassic Syst. Gt. Brit.* xii. 346 The thickest development of the Kellaways Rock facies in the West of England was seen in a short cutting on the Midland and South-Western Junction Railway at South Cerney, near Cirencester. **1946** [see

*Callovian *a.*]. **1960** Bennison & Wright *Geol. Hist. Brit. Isles* xii. 307 The sandy limestone with hard flaggy bands of the Upper Cornbrash has a fauna resembling that of the overlying Kellaways Beds.

‖ **keller** (ke·lǝɹ). [G., = cellar.] A beer-cellar in Austria or Germany. Also *attrib.*

1927 *Sat. Rev. Lit.* 3 Dec. 382 Meetings of congenial colleagues in some Keller. **1966** *Guardian* 21 Sept. 10/2 The moonlit sleigh rides, and keller bars. **1969** A. Glyn *Dragon Variation* viii. 229 Drinking beer in a keller, singing German songs. **1969** *Guardian* 21 Oct. 10/5 Austria, with its..boisterous bars and kellers. **1973** *Daily Tel.* 3 Feb. 12/4 Ski-ing, they wear bright, crazy clothes; après-ski-ing, they whoop it up in the *kellerbars*.

Kellgren (ke·lgrĕn). The name of Henrik *Kellgren* (1837–1916), Swedish physician, used *attrib.* to denote a system of massage devised by him. Hence **Kellgrenite** *a.* and *sb.*, (one) that is a practitioner of this system.

1907 *Boston Med. & Surg. Jrnl.* CLVII. 493/1 In the Kellgren method, the fingers of the gymnast remain in contact with the skin, and thus they are enabled to really manipulate the parts beneath. **1928** G. B. Shaw in *Eng. Rev.* Jan. 14 Osteopathy, for instance, is, like Kellgren massage and bone-setting. *Ibid.* 10 The Kellgrenites are not recognized by the Medical Gymnasts. *Ibid.*, Some time ago a Kellgrenite masseur..qualified himself for the British register. **1925** — in *Times* 12 Nov. 10/2, I do not hold any brief for osteopathy, or Kellgren massage, or naturopathy or homoeopathy. **1944** *Everybody's Pol. What's What?* xxiii. 205 A manipulative technique which, like Kellgren massage.., needed about two years training to acquire.

kellin, obs. var. *Khellin.

Kello·vian, *a. Geol. rare.* [ad. F. *kellovien* (A. d'Orbigny *Paléont. française. Terrains Jurassiques* (1842–9) I. 608), f. Kellowa(y)s, former name of *Kellaways.] = *Callovian *a.*

1888 J. Prestwich *Geol.* II. i. 9 Kellovian Limestones and Marls.

Kelloway(s), varr. *Kellaways.

Kelly, kelly (ke·li), *sb.*² [Prob. f. the name *Kelly*, a common Irish surname.] **1.** (With capital initial.) A type of pool (Pool *sb.*² 3) using fifteen balls (see quot. 1934). In full, *Kelly pool.* U.S.

1898 *Handb. Rules of Billiards* (Brunswick-Balke-Collender Co.) (rev. ed.) 78 The game of Kelly pool is played with fifteen numbered balls, and one white ball not numbered. **1913** *Billiards Mag.* Oct. 25 (*caption*) A 'Nine' of billiard notables gathered from different states by Artist Carlson for a pictorial game at 'Kelly'. **1918** *Ibid.* Jan. 55/1 A new idea for the old Kelly pool. **1934** Webster, *Kelly pool*, a variety of fifteen-ball pool in which each player draws a number and, while playing on the object balls in numerical order, aims to pocket the ball of the number corresponding to his own, thereby winning the game. **1948** Menjou & Musselman *It took Nine Tailors* 29 They spent the night in the Kelly pool and rushing in it and I used to indulge in personally were Kelly pool and bowling.

2. Rhyming slang for belly. Also *Derby* (or *Darby*) *kelly* or *kel.*

1906 E. Dyson *Fact'ry 'Ands* viii. 96 Er cold, proud man tumbles on his Darby Kel in the dirt. **1930** R. Sharpe *Chicago May* xxxi. 287/2 Darby Kelly, belly. **1942** T. Rattigan *Flare Path* ii. ii. 138 Just that ride home. Cor, I still feel it down in the old darby kel. **1967** L. Deighton *London Dossier* 20 The only places now making boiled beef available to your darby kellies, or bellies, are the Jewish salt-beef bars. **1970** A. Draper *Swansong for a Rare Bird* vi. 41 My old kelly was rumbling and I fancied a pie and chips.

3. A man's hat; *spec.* a derby hat (cf. sense 2 above). *slang* (chiefly *U.S.*).

1915 *Recruiter's Bulletin* (U.S.) Dec. 33/4 Say, old top, when you go home tonight, find your old brown Kelly down real tight. **1922** *Collier's* 4 Mar. 8/2, I have got to wear a brown jig on my head which weighs at least ten pounds. **1927** E. Hemingway in *Atlantic Monthly* July 11/1 John put his kelly down on the table. It was all wet. His coat was wet, too. **1948** Lait & Mortimer *New York: Confidential* xxxi. 276 Some of the larger clubs run up to $50,000 a year for the privilege of checking your kellys.

4. *slang.* (See quots. 1942 and 1963.)

1934 H. N. Rose *Thesaurus of Slang* viii. 57/1 Oilfield slang.. Hole to set the 'Kelly' in when not in use: *rat hole.* **1939** D. Hager *Fund. Petroleum Industry* ix. 203 In older drilling methods where the kelly was not used, the drill pipe was badly cut by the gripping devices. **1942** Berrey & Van den Bark *Amer. Thes. Slang* § 516/4 Kelly, a square joint on top of the drill stem passing through a square hole in the rotary table. **1946** *Mod. Petroleum Technol.* (Inst. Petroleum) 87 As drilling proceeds the square section of the kelly is lowered through the rotary table until its full length is below the table. The drilling string is then raised, the kelley removed, a fresh length of drill pipe is coupled up and the kelley is replaced. **1963** *Gloss. Mining Terms* (B.S.I.) ii. 10 *Kelly*, the rod attached to the top of the drill column in rotary drilling. It passes through the rotary table and is turned by it, but is free to slide down through it as the borehole deepens. **1974** Scotsman 22 Apr. p. x, Only in oil can you break off kelly and set down on rams while keeping a straight face.

5. In full, *kelly green.* A light green colour. orig. *U.S.*

1936 *Mademoiselle* Sept. 13 (Advt.), A slipon with sleeveless sweater..rust with Churchill green, kelly with brown. *Ibid.* Nov. 49/1 (Advt.), Sweaters..in..yellow, kelly green, gray. *Ibid.* Dec. 3/2 (Advt.), Sweaters with an English pedigree..jockey red, Kelly green or natural. **1958** 'E. McBain' *Killer's Choice* (1960) vi. 69 'What sort of green was it?' 'Almost a Kelly green.' **1966** G. Baxt *Queer Kind of Death* (1967) xvi. 223 Kelly green lanterns at the end of the garden. **1972** 'R. Crawford' *Whip Hand* i. iv. 54 The kelly-green jungle was tangled; it had a gunner.

6. *attrib.* and *Comb.* **kelly bushing, joint** (see quot. 1942); **Kelly's eye** *colloq.* (see quot. 1925).

1935 H. C. George in *Bull. U.S. Bureau of Mines* No. 224. 115 Means of escape from..the kelly board on every drilling and redrilling derrick shall be provided. **1942** Berrey & Van den Bark *Amer. Thes. Slang* § 516/2 *Kelly-board*, a platform at the height of a 'kelly'. *Ibid.* § 516/4 *Kelly joint*, the first joint of pipe attached to the kelly. **1925** Fraser & Gibbons *Soldier & Sailor Words* 134 *Kelly's eye*, no. 1 in the game of 'House'. **1933** L. A. G. Strong *Sea Wall* 256 A game of house was going on, and a voice monotonously droned the numbers: ..'Kelly's eye.' **1945** E. Waugh *Brideshead Revisited* II. i. 230 The voice of the officer in charge of tombola—'Kelly's eye—number one; legs, eleven; and we'll Shake the Bag'. **1957** *Times* 13 June 11/4 Miss L. Brabner ..began calling the numbers over the loudspeakers... 'Kelly's Eye, No. 1; and Legs Eleven' echoed throughout the ship.

Kelmscott (ke·lmzkǫt). The name of Kelmscott House, Hammersmith (named after Kelmscott Manor, Kelmscott, Oxfordshire), the home of William Morris (1834–1896), used in the name of the Kelmscott Press, which was founded there by him in 1891 and worked until 1898; also used *absol.* or *attrib.* to designate the books produced by this press or designed in a similar style.

1891 W. Morris *Story of Glittering Plain* 188 Here endeth the Glittering Plain, printed by William Morris at the Kelmscott Press, Upper Mall Hammersmith, in the County of Middlesex. **1896** G. B. Shaw in *Sat. Rev.* 10 Oct. 387/2 If he [sc. William Morris] had started a Kelmscott Theatre instead of the Kelmscott Press, I am quite confident that..he would have produced work that would within ten years have effected every theatre in Europe. **1920** A. Swinnerton *Dorothy Hill* Feb. 3 The Kelmscott Press sold its hand proofs. **1935** T. B. Reed *Hist. Old English Letter Foundries* 2 The Kelmscott books were bound..in limp vellum. **1947** *Times Lit. Suppl.* 12 Mar. 174/1 The book has somewhat of a Kelmscott air. **1963** *Listener* 21 Mar. 511/1 There is a vigour and vitality in his decorations that make Burne-Jones's Kelmscott borders rather tame. **1966** Berry & Poole *Ann. Printing* 262/1 The beauty of the Kelmscott books was due to harmony of type and decoration, the spacing of words and lines, the positioning of the text on the page, the careful choice of paper and ink, and the excellent presswork... Other private presses followed which served to increase the interest in fine book printing.

kelong (ke·lǫŋ). Also 9 **kaylong.** [Malay.] A large fish trap built with stakes, common along the coasts of the Malay Peninsula. Also *transf.*, a building erected over one.

1878 F. McNair *Perak & Malays* ix. 93 The kaylongs are made of hurdles composed of strips of bamboo, some five feet long, fastened closely together with rattans. **1900** W. W. Skeat *Malay Magic* v. 315 *Kelong* is the name given to one of the kinds of fishing-stakes (something like weirs) common on the coasts of the Peninsula. **1953** R. Graham *War Damage Report* (Malaya War Damage Commission) xiv. 52 As one approaches the coast of Malaya one cannot fail to notice the large number of fish traps, or kelongs, built in some cases far out to sea. **1971** *Daily Tel.* 18 Sept. 11/3 At Pongol Point on the north-east shore you can dine at the Sea Palace restaurant on a kelong or fish trap of stakes driven into the shallow sea bed, where fish are lured by lamps at night into a net. **1974** *Guardian* 6 Sept. 16/4 Fashionable now is a kelong visit. The kelongs are thatched huts over stakes driven into the seabed from which local fishermen cast their nets. Some of these have been 'improved' with bar, restaurant, and changing rooms for tourists to spend the day swimming and sunning and in the evening watch the fishermen bring in their haul.

kelp¹. Add: **4. kelp crab,** a spider crab, *Pugettia producta*, found on the Pacific coast of North America.

1884 *Bull. U.S. Nat. Museum* No. 27. 112 The Kelp Crabs are used by the natives. **1939** Ricketts & Calvin *Between Pacific Tides* i. 80 A dark, olive-green, spider crab, *Pugettia producta*, occurs so frequently on strands of the seaweed *Egregia* and others that it is commonly called the kelp crab. The points on the carapace and the spines on the legs are sharp. **1954** S. F. Light et al. *Intertidal Invertebr. Centr. Calif. Coast* 342 A few organisms (e.g. the kelp crab *Pugettia producta*..) graze directly on the larger attached algae.

† **kelp**² (kelp). *Obs. slang.* Also **kilp.** [Origin unknown.] A hat. Hence **kelp** *v. trans.* (see quot. 1812).

1630 J. Cole in *Ordinary of Newgate, his Account* I. 13/2, I and Thomas Campson..broke open a Hatters.. and robbed it of three Dozen of Kilps. **1753** J. Poulter *Discoveries* (ed. 2) 26 We jostle him up, and one knocks his Kelp off. **1812** J. H. Vaux *Vocab. Flash Lang.* in *Mem.* (1964) 247 *Kelp*, a hat; *to kelp a person*, is to move your hat to him.

Kelper (ke·lpəɹ), *sb.* and *a.* Also **kelper.** [f. Kelp¹ + -er¹.] **A.** *sb.* The name given locally to a native or inhabitant of the Falkland Islands, the shores of which abound in kelp. **B.** *adj.* Of or pertaining to a Kelper.

1960 M. B. R. Cawkell et al. *Falkland Islands* viii. 93 Improved sanitation and street lighting, main drainage and increasing literacy have made this a reasonable sort of..whose business-like nature could readily appreciate. **1960** E. R. Pettingill *Penguin Summer* (1962) i. 7 Some of them were native Falklanders—'Kelpers', they called themselves. **1968** *Economist* 7 Dec. 33/2 More and more [Falkland Islands] young people are emigrating to New Zealand to escape the isolated monotony of kelper life. **1971** *Daily Tel.* 17 Jan. 10/4 Successive British Governments have laid it down firmly that the 'kelpers', all 2,000 of them, must decide their own destiny.

kelpie² (ke·lpi). *Austral.* [f. the name of an early specimen of the breed.] A smooth-coated, prick-eared, Australian sheep-dog, which may be black, black-and-tan, blue, or red; first bred from imported Scottish collies about 1870.

1907 R. Leighton *New Bk. Dog* 472/2 The Kelpie... is not perhaps an example of high, scientific breeding; but he is a useful, presentable dog. **1934** *Bulletin* (Sydney) 14 Feb. 40/2 Lassie was, like all kelpies, highly strung, with brains under her broad, thin-skulled skull. **1936** [see *barb *sb.*² 3]. **1971** F. Hamilton *World Encycl. Dogs* 74 Kelpies were bred back in the early days of settlement from the Collie type dogs which came to Australia with the early farmers. **1972** *Southerly* XXXII. 9, A kelpie and a blue cattle dog had raced to meet them, barking frantically.

kelson, keelson. Add: **1. c.** (Spelt *keelson.*) A structure in the hull of a flying-boat (or the float of a seaplane) analogous to the keelson of a ship's hull.

1920 *Flight* 2 Sept. 948/1 Such members as keelson and chines are of rock elm. **1928** Chatfield & Taylor *Airplane & its Engine* xii. 237 Since the loads on the bottom of the boat in landing are large, additional longitudinal members called keelsons are provided to stiffen it. **1930** [see *keel *sb.*¹ 3 d]. **1942** R. H. Longe in R. A. Beaumont *Aeronaut. Engin.* xviii. 486/1 The keelson..is the backbone of the hull [of a flying-boat], and runs the full length of the hull, from the stem to the stern post. **1966** *Jane's 100 Significant Aircraft* 38/2 Twin-engined flying-boat... Floors notched out for two-thirds of depth to fit over solid keelson... Keelson is continuous member from stem to stern. **1974** *Flight International* 7 Nov. 646/2 The fuselage..is based on a twin-girder keelson running from bow to tail.

kelt¹. Delete from def. 'Now only *Sc.*' and add later examples.

1937 *Evening News* 15 Feb. 8/3 A kelt—that is a fish which has paid its visit to the gravel beds—is a sad wreck and is in no form to put up a fight. **1965** Scotsman 26 Jan. 11/4 If the fish has shed only a few eggs..it is still technically and in law a kelt.

kelt, keltch, varr. *Kelch.

kelter², **kilter.** (Earlier and later examples of *out of kilter.*)

1628 W. Bradford in *Mass. Hist. Soc. Coll.* (1856) 4th Ser. III. 235 We very sight of one (*sc.* a gun) though out of kilter) was a terrour unto them. **1938** 'E. Queen' *Four of Hearts* (1939) xii. 171 Life's been out of kilter around here for some time. **1960** M. Phillips in *Analog Science Fact/Fiction* Nov. 16/1 We've had some reports that some of the government machines are out of kilter, and I'd like you to go over them for me. **1968** J. C. Holmes *Nothing More to Declare* 77 A deeply traditional nature thrown out of kilter, and thus enormously sensitive to anything unprotected. **1973** *Times* 13 Oct. 1/3 There [*sc.* in N. Ireland], an allotment of 12 seats at Westminster is based upon electoral quotas wildly out of kilter with the quotas for England, Scotland, and Wales.

keltz, var. *Kelch.

Kelvin, kelvin (ke·lvin). [The title of Sir William Thomson, Lord *Kelvin* (1824–1907), British physicist and inventor.] † **1.** (With lower-case initial.) A name proposed (but little used) for the kilowatt-hour, the ordinary commercial unit of electric energy. *Obs.*

1892 *Electrician* 6 May 1/1 The President of the Board of Trade having cordially approved and Lord Kelvin having acquiesced; after the word 'unit' in the Provisional Orders of this year, the words 'hereinafter called a kelvin' will be introduced. **1911** *Encycl. Brit.* XXVII. 740/2 The corresponding energy unit being the kilowatt-second, and 3600 kilowatt-seconds or 1 kilowatt-hour (called the Board of Trade unit' or a 'kelvin').

2. (With capital initial.) Used *attrib.* and in the possessive to designate instruments and concepts devised by Lord Kelvin, as **Kelvin balance**, an electrical measuring instrument having a set of horizontal coils arranged in the form of a balance with a sliding weight, which is used to balance the electromagnetic forces produced by a current passed through the coils and so to measure its strength; **Kelvin (double) bridge,** a modification of the Wheatstone bridge used for measuring low

resistances and having two pairs of variable ratio resistors, an adjustable standard resistor, and the unknown one; **Kelvin's law,** the law that the most economical cross-sectional area of a conductor used as a transmission line is that for which the cost of the energy dissipated in any period is equal to the charges during the same period on the capital cost of the line.

1896 *Proc. Physical Soc.* XIV. 166 A convenient method for determining low resistances, such as a Kelvin bridge. **1909** F. A. C. Perrine *Conductors for Electr. Distribution* viii. 160 (*heading*) Kelvin's Law of economy in conductors. **1904** Swenson & Frankenfield *Testing Electro-Magn. Machinery* I. 122 The Kelvin Balance is recognized as a standard instrument. **1906** *Rep. Brit. Assoc. Adv. Sci.* 119 The current leads of standard resistances intended for measurement on the Kelvin double bridge should have a resistance not greater than the standard itself. **1933** E. W. Golding *Electr. Measurements* ii. 14 Kelvin balances are manufactured with ten different ranges up to 2,500 amp. **1957** W. J. John *Med. Electr. Engin.* III. x. 301/2 Kelvin's law is usually expressed in terms of annual costs, though it is equally true of capitalised costs. **1967** H. Cotton *Adv. Electr. Technol.* iii. 92 Calculate the current flowing in the galvanometer of the Kelvin double-bridge circuit.

3. (With capital initial.) Used *attrib.* to designate an absolute scale of temperature (defined thermodynamically in terms of the operation of an ideal heat engine) in which the zero is identified with absolute zero and values are assigned to one or more fixed points so as to make the degrees correspond in size to those of the centigrade (Celsius) scale. So *Kelvin temperature*, a temperature expressed in terms of this scale.

As the scale was orig. defined, the temperature interval between two fixed points (the freezing point and the boiling point of water) was made to be exactly 100 degrees, so that the value of each point was a matter of experiment (which produced values of about 273° and 373° respectively); in 1954 the scale was redefined in terms of a single fixed point, the triple point of water, to which was assigned the value 273·16° exactly (giving a figure of approximately 273·15° for the ice point).

[**1871** J. C. Maxwell *Theory of Heat* viii. 155 (*heading*) Thomson's absolute scale of temperature. **1899** W. Watson *Text-bk. Physics* vi. 334 The temperature..are 283°, 293°..on Lord Kelvin's absolute scale.] **1908** *Amer. Jrnl. Sci.* CLXXVI. 227 The derivation of the Kelvin thermodynamic scale from the expansion of nitrogen. **1932** Glazebrook *Dict. Appl. Physics* I. 270/2 If the same size of degree be adopted for the Kelvin scale, then..the readings, on either scale, will be identical. **1937** M. W. Zemansky *Heat & Thermodynamics* ix. 140 Two temperatures on the Kelvin scale are to each other as the heats absorbed and rejected respectively by a Carnot engine operating between reservoirs at these temperatures. *Ibid.* 144 The Kelvin temperature is therefore numerically equal to the absolute temperature and may be measured with a gas thermometer. **1958** Condon & Odishaw *Handbk. Physics* v. iii. 30/2 This new absolute thermodynamic scale, called the Kelvin scale, does not differ from the thermodynamic 'centigrade' scale by an amount which can be determined experimentally at the present time. **1964** *Units & Standards of Measurement: Temperature* (Nat. Physical Lab.) 3 The thermodynamic Kelvin scale..is recognized as the fundamental scale to which all temperature measurements should ideally be referred.

b. *degree Kelvin* (or *Kelvin degree*): a degree of the Kelvin scale (in size equal to the degree centigrade), symbol °K (see *K 4 f); now formally called a *kelvin* (symbol K) and incorporated into the International System of Units as a basic unit.

1911 *Physical Rev.* XXXIII. 220 The writer has assumed that the degree Kelvin and the degree centigrade are equal. **1930** *Engineering* 9 May 595/3 A perfect gas of one free electron per atom would necessarily contribute three calories per gram-atom per degree Kelvin to the specific heat of the metal. **1941** *Temperature* (Amer. Inst. Physics) 11 When using the Thermodynamic Centigrade Scale one is simply expressing the number of Kelvin degrees that a particular thermal state is above or below the ice point. **1945** G. R. Noakes *Text-bk. Heat* i. 26 The name 'degrees Kelvin' (°K.) is used for the true absolute scale temperature, and 'degrees absolute' ('A., or °abs.) for temperatures measured by other means on any other scale numbered up from absolute zero. **1962** Corson & Lorrain *Introd. Electromagn. Fields* iii. 108, *T* is the absolute temperature in degrees Kelvin. **1968** *Nature* 16 Nov. 651/2 Resolution No. 3 [of the thirteenth General Conference of Weights and Measures] changes the name of the 'degree Kelvin' (symbol °K) simply to 'kelvin', with symbol K... The terms of the revised definition of the unit of thermodynamic temperature are (resolution No. 4): 'the kelvin..is the fraction 1/273·16 of the thermodynamic temperature of the triple point of water'. **1971** *Physics Bull.* Mar. 144/1 The inversion transitions.. indicate that the ammonia molecules are at a temperature of several tens of kelvin. **1973** *Sci. Amer.* Aug. 84/3 Only liquid hydrogen will serve, and it must be stored at an extremely low temperature (22 degrees Kelvin, or 22 degrees Celsius above absolute zero).

Kelvinside (ke·lvinsaid). The name of a residential district of Glasgow, used *attrib.* or *absol.* to designate the supposedly affected and refined accent with which some of its residents speak.

1903 J. H. Millar *Lit. Hist. Scotl.* 317 A mincing and quasi-genteel lingo of their own (the sort of English known in some quarters as 'Princes Street' or 'Kelvinside'). **1928** I. C. Ward *Phonetics of English* x. 76 Some types of Scottish town speech—sometimes called 'Kelvinside' or 'High English'. **1932** N. Gunn *Lost Glen* ii. ii. 17 A thin..woman..fair and full of refinement and a Kelvinside-English accent. **1935** *Trans. Philol. Soc.* 8 The artificial variety of speech which has been called Kelvinside English. **1966** G. Paxton *Proper Names* 143 *Kelvinside*, the north-western area of Glasgow around the University and Kelvingrove Park, where the residents were supposed to speak with a mincing ('Kelvinside') ..accent.

Kemalism (ke·maliz'm). [f. the name of *Kemal* Atatürk (c 1880–1938), Turkish soldier and statesman + -ism.] The political, social, and economic policies advocated by Kemal Atatürk, which aimed to create a modern republican secular Turkish state out of a part of the Ottoman empire. So **Ke·malist**, one who advocates or believes in the theory of Kemalism; an adherent or supporter of Kemalism; also as *adj.*

1920 *N.Y. Times* 5 Dec., Greek troops have dispersed the Kemalists in the district of Nicea. **1921** *Manch. Guardian Weekly* 29 July 65 The Kemalists are evacuating Angora. **1925** *Ibid.* 21 Dec. 4/6 The anti-Kemalist revolt..has as its centre the town of Erzerum, where, curiously enough, Kemalism was born. **1927** *Observer* 18 Sept. 7 In Turkey an utter victory for the Kemalists over the Greeks supplied the opportunity for a comprehensive scrapping of Turkey's Ottoman habits..Quite the most striking of these post-war metamorphoses has been accomplished by Kemalist Turkey, the trader among Oriental nations in the race for 'Westernisation'. **1947** A. Koestler in *Palestine* Nov. 1947 344 Under a de Gaulle régime of the Kemalist type you could after all go on publishing your stuff within limits. **1959** *Encounter* Sept. 36/1 The theoretical justification of Kemalism. **1963** *Times* 23 Apr. 11/3 The pamphlets were said to have been signed 'the Army of Young Kemalists'. **1966** *Tablet* 12 Mar. 299/2 The Kemalist revolution was often brutal.

kempas (ke·mpās). Also 9 **kompas, koompass.** [Malay.] A hardwood timber tree, *Koompassia malaccensis*, native to Malaya, Sumatra, and Borneo; also, the wood of this tree.

1830 T. J. Newbold *Pol. & Statistical Acct. Straits of Malacca* II. viii. 110 The oo and charcoal, (of the Kompas, Kamooi, or other hard woods) are used for. **1883** W. E. Maxwell in *Jrnl. Straits Br. R. Asiatic Soc.* 12 Koompass..Wood yellowish white,..Used for ship-building. **1935** I. H. Burkill *Dict. Econ. Products Malay Peninsula* II. 1286 Kempas sheds all its leaves every year. **1940** E. J. H. Corner *Wayside Trees Malaya* I. 397 The *Kempas* occurs throughout the lowland forest in swampy ground..as well as on the hills and ridges in the forest up to 1,100 ft. **1966** *Handbk. Hardwoods* (Forest Prod. Res. Lab.) 127 Kempas normally seasons fairly well. **1959** 'M. Derby' *Tigress* iii. 48 He waited in the fork of the kempas tree. **1965** C. Shuttleworth *Malayan Safari* i. 16 There are..medium hardwoods such as kapur, keruing, kempas and so on. **1972** T. C. Whitmore *Tree Flora of Malaya* I. 266/2 Kempas is a smaller tree than tualang, it is more often cut for timber.

‖ **Kempeitai** (ke·mpēitai). Also **Kempetai.** [Jap.] The Japanese military secret service in the period 1931–1945.

1947 J. Bertram *Shadow of War* 197 A single communication..to the *kempeitai* revealing my identity would be more than enough. **1953** J. Trench *Docken Dead* iv. 117 Docken had betrayed Richard to the Kempetei. **1961** R. Seth *Anat. Spying* viii. 148 The Kempeitai, like the Nazi Gestapo, was the most powerful and most hated of all Japanese institutions..It derived its power for the most part from the semi-independent position which it held within the army. *Ibid.* 149 Wherever the Kempeitai was, its most important function was counter-espionage. **1962** B. Swett-Escott *Baker St. Irregular* v. 127 If the Gestapo or the Kempetai had kept track of his movements, they would have had a good idea of what was likely to happen. **1964** *This is Japan* 1966 105 A tough *kempeitai* gendarme, with the customary red armband, two holstered revolvers, gold-and-black teeth, and foul *daikon* breath. **1969** J. M. Gullick *Malaysia* iii. 97 The teachers..played a leading part in maintaining the underground resistance in the towns despite torture and other reprisals by the Japanese Security Police (the Kempeitai).

kempite (ke·mpəit). *Min.* [f. the name of J. F. *Kemp* (1859–1926), American geologist + -ite¹.] An emerald-green basic chloride of manganese, $Mn_2(OH)_3Cl$, found in California.

1924 A. F. Rogers in *Amer. Jrnl. Sci.* VIII. 150 The name kempite is given in honor of Professor James Furman Kemp. **1949** *Bull. Geol. Soc. Amer.* LX. 1944 It seems likely that the formula of kempite is $MnCl_2.3MnO.3H_2O$. **1968** I. Kostov *Mineral.* 195 Kempite is orthorhombic-dipyramidal, occurring as a weathering product of manganese ores.

kempt, *ppl. a.* Delete 'arch.' and add later examples. Also *transf.*

1905 *Daily Chron.* 15 June 4/4 The little girls wear clean pinafores; their hair is neatly kempt. **1929** R. Aldington *Death of Hero* iii. xii. 336 The street paving was badly worn, but looked marvellously smooth and kempt to Winterbourne. **1946** S. Spender *European Witness* i. 6 Gardens as well kempt as a short hair-cut. **1951** W. Sansom *Face of Innocence* i. 1 The kempt yellow gravel of

drives. **a 1954** F. B. Young *Wistanslow* (1956) 113 A spacious expanse of greensward, smooth and kempt as the clarest turf of an Oxford college. **1975** *Times* 21 July 10/7 Artificially beautified people looking kempt and highly coloured for the hairdressers' and beauticians' trade fair.

‖ **ken** (ken), *sb.*³ Also 8–9 **kin.** Pl. **ken,** occas. **kens.** [Jap.] A Japanese unit of length equal to six *shaku*; equivalent to approximately 71·5 inches (1·82 metres).

1727 J. G. Scheuchzer tr. *Kæmpfer's Hist. Japan* II. 660 Its Top contains sixty *Kin*, or Mats, according to their way of measuring, or about as many European fathoms. *Ibid.* 407 This bridge is supported, in the middle, by a small Island, and consequently consists of two parts, the first whereof hath 36, the second 87, Kin, or Mats, in length. **1845** *Encycl. Metrop.* XX. 486/2, 1 kin = 2 fathom. **1884** Satow & Hawes *Handbk. for Travellers Cent. & N. Japan* (ed. 2) 17 The *chō* is further subdivided into 60 *ken* and the *ken* again into 6 *shaku*, the shaku being about 11·9 English inches. **1956** K. Tomiki *Judo* i. 22 The floor space for a contest shall be 5 *ken* (30 feet) square of 50 *tatami*.

‖ **ken** (ken), *sb.*⁴ [Jap.] A prefecture; one of the territorial divisions of Japan.

1882 *Encycl. Brit.* XIV. 490/2 His [sc. Sho-tai's] territory was declared first a *han* or feudal dependency, and afterwards a *ken* or prefecture of the Japanese monarchy. **1890** B. H. Chamberlain *Things Japanese* 134 There are two present divisions of the soil of the Empire—one older and more popular one into provinces (*kuni*)..and a recent, purely administrative one into prefectures (*ken*) of which at the present moment..there are forty-three. **1899** Kipling *From Sea to Sea* I. xviii. 378 I was in the western *ken*—districts, as you call them. **1942** W. E. Reischauer *Japan Past & Present* iii. 119 Two years later, in 1871, the fiefs were entirely abolished, and the land was divided into a number of new political divisions called *ken* or 'prefectures'. **1965** E. V. Gatenby *Twelve Doors to Japan* i. 16 Today..they [sc. the *kuni*] have been absorbed into the larger prefectures (to, dō, fu, and ken, of which there are forty-six) and have lost much of their contemporary meaning.

‖ **ken** (ken), *sb.*⁵ [Jap.] A Japanese game of forfeits played with the hands and with gestures.

1890 B. H. Chamberlain *Things Japanese* 125 The Japanese play various games of forfeits, which they call *Ken*, sitting in a little circle and flinging out their fingers, after the manner of the Italian *mora.* **1898** A. Diósy *New Far East* v. 236 Dignified old ladies of the best..of our other rollicking games of forfeits, were frivolities far beneath their notice. **1958** *Japan* (Jap. Nat. Commission for Unesco) xxii. 1030 Ken is a game introduced from ancient China with many variations... The game was held at banquets and feasts in the Edo Period when *ken* meets were also held.

kenaf (kĕnæ·f). Also **kanaff.** [Persian.] = *ambari.

1891 *Kew Bull.* 204 Recently an announcement has been made of the discovery of a new textile plant on the shores of the Caspian. The plant known as Kanaff by the natives is said to yield a soft elastic and silky fibre... It is supposed that Kanaff fibre..will successfully compete with jute and other textile for sacking, ropes, and pack-thread. **1948** *Bot. Gaz.* CVI. 349/1 The identity of the plant [introduced into the U.S.S.R.] and known as *kenaf* (*Hibiscus cannabinus*, L.) which..is capable of producing more than a ton of dry fiber per acre. **1964** A. E. Hamer *Jute Substitute Fibres* i. 2 Kenaf is the name used principally in Russia, America and Cuba..and the word is derived from the local names Kanaff, Kanap or Kanapel, given to the plant where it grows on the Persian shore of the Caspian sea. *Ibid.* ii. 8 No one seems to know why the fibre plant was given the name Kenaf in the Western Hemisphere. **1967** *New Scientist* 14 Dec. 660/1 They point out that kenaf is already a leading substitute for jute fibres. **1974** *Times* 12 Jan. (Ghana Suppl.) p. iv, Achievements [in the production of raw materials] were sparse except in some agricultural sectors like cotton and kenaf.

‖ **kendo** (ke·ndo). [Jap.] The Japanese sport of fencing with bamboo swords.

1921 S. K. Uyenishi *Text-bk. Ju-Jutsu* i. 14 *Kendo* or *Ken-jutsu*, 'the hard way' and 'the hard art'..is the elaboration of the old two-sword play of the samurai or 'two-sworded men'. **1933** *Official Guide to Japan* [Jap. Imperial Govt. Railways] p. cxxxvi, *Kenjutsu* or *Kendō* (also called *Gekken*), the art of handling a sword, corresponds to European fencing and is an ancient art of the country. **1939** R. Kaji *Japan Facing* p. 56 (*caption*) *Kendō*, Japanese fencing, in which the participants use bamboo swords. **1958** *Economist* 1 Nov. 422/1 The government has reintroduced the forbidden *kendo* or military fencing in schools. **1964** R. A. Lidstone *Introd. Kendō* 15 Quite understandably Kendō, with its militant background, lost favour after the last war. **1969** J. Ball *Cool Cottontail* (1967) xv. 37 He had become interested in the basic Oriental martial arts: judo, kendo, aikido, and karate. **1974** *Publishers Weekly* 28 Jan. 100/1 (Advt.), Kendo, or Japanese sword fighting, is an ancient method of training body and mind that now is becoming popular throughout the world.

keneme (kenī·m). *Linguistics.* Also **ceneme.** [f. Gr. κενός empty: see *-eme.] (See quot. 1966.) Cf. *empty word* (*empty *a.* and *sb.* C). Hence **kenema·tics, kene·tics,** the study of kenemes; **kenema·tic, kene·mic, kene·tic** *adjs.*

The forms with initial *c-* predominate over the etymological forms with *k-*.

1939 L. Hjelmslev in *Proc. Third Internat. Congr. Phonetic Sci.* 271 A language is a category of two members, ..One of these planes, the plerematic plane, gives form to the content; the other, the cenematic plane, forms the expression. *Ibid.* 272 The constituents—in plerematics: the pleremes, in cenematics: the cenemes—are usually of two types: central and marginal constituents. **1950** S. Potter *Our Lang.* viii. 86 Beginning with the phoneme, philologists pass on to speak about *morphemes, taxemes* or *tagmemes, sememes* (including *pleremes* and *kenemes*)..as if *-eme* were a brand-new suffix meaning 'linguistic agent'. **1958** C. F. Hockett *Course in Mod. Ling.* lxiv. 575 Phonemes or phonemic shapes..Some phonemes are linguistic *pleremes. Ibid.*, A kenematic unit introduces two new terms for general applicability: *cenematic* and *plerematic*. The cenematic structure of language is phonology; the plerematic structure is grammar plus semology. **1966** M. Pei *Gloss. Ling. Terminol.* 35 *Ceneme*,..1. Linguistically, a phoneme, or anything pertaining to phonology (Hockett). 2. The smallest unit of expression without corresponding content, differing from a phoneme in that it does not necessarily consist of sound, but may include letters and other semantic indicators (Hjelmslev). **1967** [see *glossematics]. **1967** *Word* XXIII. 469 My observations apparently support the structuralist separation of cenemics and plerematics. *Ibid.* 471 In addition to their two cenetic and two plerematic systems, bilingual children naturally have to master two sets of form-to-meaning relationships. **1969** *English Studies* L. 43 The tradition from Saussure and Hjelmslev brought about a phonemic—or cenemic—analysis based on pure relations.

Kenite (kī·nait), *sb.* and *a.* [f. Heb. *kēnī* a gentilic adjective associated with Heb. *ḳayin* a weapon made of metal, Aram. *ḳayn* an ironsmith, maker of iron weapons and tools, Aram. *kēnay*, *kaynāyā* smith, metal-worker, Nabatæan *kynw*, Old South Arabian *'knw*: see -ite¹.] **A.** *sb.* A member of an ancient nomadic people from S. Palestine, frequently mentioned in the Old Testament. **B.** *adj.* Of or pertaining to the Kenites.

From the etymology it has been conjectured that the Kenites were a tribe of metal-workers.

[c **1000** Ælfric *Heptateuch* Gen. xv. 19 Cyneí & Ceneneí; Cetmoneí. **1382** Wyclif *Num.* xxiv. 21 Sothli he saw also Cyney.] **1535** Coverdale *Gen.* xv. 19 The Kenites, the Kenisites, the Kadmonites... *Judg.* iv. 11 Heber the Kenite. a **1879** M. Dods *Handbk. Holy Bible* (1888) I. *Judg.* iv. 11 Which removal is here mentioned, lest any should wonder to find the Kenites in this place. **1916** M. Henry *Exposition Five Bks. Moses Num.* xxiv, The Kenites were now the secrets of the Nations. **1876** *Encycl. Brit.* IV. 569/1 The aboriginal Rephaim and three Arab tribes, the Kenites, Kenizzites, and Kadmonites. **1911** *Ibid.* XV. 739/1 Moses himself married into a Kenite family (Judges i. 16). **1962** G. A. Buttrick *Interpreter's Dict. Bible* III. 7/1 The last mention of the Kenites in connection with the history of Israel is during the time of David before he became king over all Israel (1 Sam. 30:29). **1968** *Encycl. Brit.* XIII. 251/1 Kenites, a clan closely related to the Midianites and Amalekites, frequently mentioned in the biblical narratives about the pre-monarchic history of Israel. The name was derived from Cain, whose descendants they were believed to be (Gen. iv).

† **kennedy** (ke·nĕdi). *Obs. slang.* [Said to be f. the name of a man who was killed by being struck on the head with a poker.] **a.** A poker. **b.** A blow inflicted by a poker, freq. in the phr. *to give* (someone) *kennedy.* Also **kennedy** *v. trans.*, to strike with a poker.

1823 *Morning Herald* in *Spirit of Public Jrnls. for M.DCCC.XXIII* (1825) 77 Mr. Davis bore three irregular tears as long as he could, but at last..he ventured to tell Mykle to put the poker down. **1839** *Ibid.* When saw Mykle do but seize the poker, and threaten to 'Kennedy him' (beat him with a poker). **1859** Hotten *Dict. Slang* 55 *Kennedy*, to strike or kill with a poker. A St Giles' term, so given from a man of that name being killed by a poker. Frequently shortened to *Neddy*. **1864** *Athenæum* 29 Oct. 559 St Giles's perpetuates the memory of a..man..who was killed by a poker by calling that instrument a kennedy.

kennel, *sb.*¹ Add: **1. a.** Also (usu. *pl.*), an establishment where dogs are bred, or where they are cared for in the absence of their owners.

1887 G. Stables *Practical Kennel Guide* (ed. 3) xv. 245 I know some kennels..which are a disgrace to civilised society—dirty and filthy in the extreme. **1896** *Notable Dogs of Yr.* (Advt.), Borzoi kennels and sanatorium for dogs and cats. Patients treated at the kennels. **1925** J. Lucas *Pedigree Dog Breeding* 61 Those intending to board their dogs for more than a week or two should try and visit one or two kennels before making their selection... Many kennels..make a speciality of whelping cases. **1931** N. W. Lewis *Your Dog* iii. 173 Good kennels can be bought..from big kennels that have become overstocked. **1969** [see *boarding kennel*, s.v. *boarding *vbl. sb.* 7]. **1974** R. Rendell *Face of Trespass* ii. 27 I've never left her since she was a puppy... I couldn't put her in kennels. She'd fret.

3. *kennel-maid, -man* (further examples), *-work*; **Kennel Club,** an organization, founded in 1873, which establishes dog breeds, records pedigrees, issues the rules for dog shows and

trials, etc.; also, a branch of this organization; **kennel lameness,** a rheumatic disease in dogs, freq. affecting the forelegs.

1874 C. S. Pearce *Kennel Club Stud Bk.* p. v, The Club shall be called the Kennel Club, It shall endeavour in every way to promote the general improvement of dogs, dog shows, and dog trials. **1925** *Discovery* Oct. 1910/1 The breeds have been taken up by kennel clubs and the dogs are now officially called [Pyrenean Sheepdogs]. **1959** *Listener* 22 Mar. 447/1 The other details were vouched and signed by the clerk of the local kennel club. **1976** F. Hamilton *World Encycl. Dogs* 60 Registration of a pedigree dog in the Kennel Club is a simple matter; it is usually completed by the breeder, the fee of 5s..registers the dog for life. **1841** R. T. Vyner *Notitia Venatica* 43 Kennel-lameness, so called because it is confined to the kennel. **1938** Huntsman's disease, or rheumatism, is common among hounds. **1858** Beaufort & Morris *Hunting* 112 That mortal scourge which among men is known as rheumatism, and among hounds as kennel lameness. **1930** C. Frederick et al. *Fox-Hunting* xvii. 116 'Kennel lameness' is an.. and which, it is feared, continues in some kennels. **1928** *Kinder Waller* *Iona of Treatm.* in Small-Animal Pract. ii. 145 Kennel lameness... This layman's term is meant to indicate a condition brought about by a diet deficient in calcium which affects mainly full-grown animals kept for some weeks or months in kennels and fed chiefly on the ordinary diet. **1907** *Westm. Gaz.* 3 Sept. 1/3 The lady's service was the pioneer, the proper way and the position of kennel-maid. **1959** *Daily Express* 10 Jan. 3/2 The showing of dogs is skilled work for a kennelmaid. **1970** *Daily Tel.* 24 Apr. 17/1 He had bought substantial quantities of veterinary preparations for the manager and had left the treatment to a young kennelmaid. **1954** C. L. B. Hubbard *Compl. Dog Breeders' Manual* x. 103 With reliable staffing the kennel runs smoothly enough, therefore it is essential that the kennel-man be given a clear understanding of his duties. **1973** *Shooting Times & Country Mag.* 24 May 18/3 In our kennels... **1929** *Daily Express* 16 Jan. 3/3 Kennelwork as a career for girls. **1972** Dangerfield & Howell *Internat. Encycl. Dogs* 268 Kennel work cannot be classed as an effeminate career for a male or an overly masculine job for a woman.

kennel, *v.* Add: **1. b.** With *up.* To return to one's kennel (also *fig.*): to keep quiet, to shut up. *colloq.*

1919 'Ian Hay' *Fugitive* iii. i. 63 You've run her to earth; your job's done. Kennel up, hounds! **1929** W. H. Downing *Digger Dial.* 30 *Kennel up*, stop talking. **1932** Galsworthy *Flowering Wild.* xi. 84 Kennel up, Reggie! You've had too much and so have I. **1960** *Ibid.* Apr. 18/3 'My old Tanner..' 'Kennel up, will you?' **1960** *Shooting Times & Country Mag.* 4 Mar. 37/3 After clean straw had been put down in the kennels, the order 'kennel up!' was given, though not instantly obeyed by some.

Kennelly (ke·nĕli). [Name of Arthur Edwin *Kennelly* (1861–1939), U.S. electrical engineer, who in 1902 suggested (as did Heaviside independently) that such a layer existed.] *Kennelly–Heaviside layer* or *region*: = *E-layer* (*E II. 9), *Heaviside layer.

1925 *Science* 22 May 540/1 (*heading*) The Kennelly-Heaviside layer. **1926** *Nature* 24 Oct. 609/2 The hypothesis of an electrically conducting stratum in the upper air was independently enunciated in an article by Prof. A. E. Kennelly..published in the *Electrical World and Engineer* of New York on March 15, 1902. The official date of Heaviside's discussion of his hypothesis is December 19, 1902. ..If names are to be attached to the hypothetical layer it would be usual, in equity, the 'Kennelly-Heaviside' layer, a name which is beginning to be used in America. **1932** *Discovery* Oct. 316/1 The 'Kennelly-Heaviside' layer is responsible for the reflection of long and medium wireless waves. **1938** *Discovery* Feb. 26/2 Professor Appleton..observed that there was yet another layer beyond the Kennelly layer. **1963** G. M. B. Dobson *Exploring Atmosphere* viii. 142 These different ionized layers in the atmosphere have been named and the names which have been placed alongside them in the diagram, but the *E* region is also known as the 'Kennelly-Heaviside' region, and the *F* region as the 'Appleton' region, after their discoverers. **1972** *Science* 5 May 463/2 It was not until 1926 that the existence of this 'Kennelly-Heaviside layer' was demonstrated.

kennetic (kene·tik), *a.* [f. Ken *v.*¹ 11 after Kinetic *a.*] (See quot. 1955.) Usu. in phr. *kennetic inquiry.*

1950 A. F. Bentley in *Science* CXII. 775/1 Kennetic inquiry is a name proposed for organized investigation into the problem of human knowings and knowns, where this is so conducted that the full range of subject matters—all the knowings and all the knowns—form a common field. *Ibid.* 775/2 To form the name 'kennetic', the Scottish 'ken' or 'kenning' has been preferred to any word in the groups centering around 'cognition'. **1955** M. Reifer *Dict. New Words* 116/1 *Kennetic adj.*, pertaining to the study of the acquisition of knowledge viewed as a transaction between the learner and the entire matter to be learned. **1960** H. C. Shands *Thinking & Psychotherapy* ii. 40 It is important for an understanding of the processes of kennetic inquiry to remember always the shifting nature of means and ends. **1970** — *Semiotic Approaches to Psychiatry* 8 Natural philosophy must be primarily interested..than with the understanding of communicative process, and..toward 'kennetic inquiry'.

‖ **kennetjie** (ke·nəkʼi). *S. Afr.* Also **kennetje.** [Afrikaans.] = Tip-cat.

1947 *Cape Times Weekend Mag.* 21 June 2 Whether they were climbing in hedges..or playing *kennetjie* or marbles, the youthful Miss H—— and her companions reaped the knees of healthy children. **1953** *Cape Argus* 18 Feb. 3/6 On these grassy squares we played 'rounders' and on

the paths marbles, tops, hop-scotch, and kennetie. **1959** *Cape Times* 2 Apr. 9/3 South African in England tells me that what they call *kennetjie* in the Boland is sweeping Lancashire, where it is known as tip-cat. **1973** *Standard Encycl. S. Afr.* III. 190/1 In its delightful memoirs P. B. Borcherds mentions the games of 'kennetjie' (tip-cat) in which he took part..at the close of the 18th century... The kennetjie itself was a piece of wood about 5 inches..long and 1½ inch..in diameter, tapered off at both ends.

keno. (This spelling is now usual.) Also 9 **keeno, quino.** For etym. read: [ad. F. *quine* set of five winning numbers in a lottery, f. L. *quini* distributive of *quinque* five.] Add earlier and later examples.

1814 B. F. Palmer *Diary* 30 May (1914) 70, I employ'd in washing & mending my remaining keeno. **1843** J. F. Cooper *Ned Myers* xi, I commenced operations by purchasing shares in a dice-board, a roulette table, and a quino table. **1870** A. S. Evans *Our Sister Republic* xx. 462 A great deal of gambling of some kind..playing keno..in fifteen hundred people, which is devoted exclusively to *quino*, played for money. The cards or tickets are pasted down upon the tables and soon number at least one thousand all told. Each player is provided with a number of grains of corn..of the numbers on his card as the numbers drawn out by the dealer are called..One game..takes about three minutes. **1871** *Figaro* 15 Apr., The police pulled every Keno establishment in the city. **1873** J. H. Beadle *Undevel. West* iv. 93 This game..like faro, the less of the 'cutthroat' about it than the others. **1875** E. King *Southern States N. Amer.* xv. 206 A man..tried to shoot the keno-dealer in the back. **1889** K. Munroe *Golden Days of '49* iv. 40 We'll have a drawin' over in Slim Jim's keno ranch. **1904** W. N. Harben *Georgians* xxiii. 218 They're the ones that they call the 'White Elephant'—a gamblin', keno shebang on a giant scale. **1907** C. E. Mulford *Bar-20* (1914) ix. 108 I'd shore look nice loping around a keno lay-out without my guns on. In' the same town with some cuss jumpin' me, wouldn't I? **1951** E. Kefauver *Crime in Amer.* (1952) xii. 144 Another huge gambling casino..ran dice, keno and card games. **1968** R. F. Adams *Western Words* (ed. 2) 169 Keno..is played with a large globe, called the keno goose.

b. *int.* An exclamation (expressing encouragement or approval).

1868 *Terr. Enterprise* (Virginia, Nevada) 30 Sept. 3/2 When they thus got three beans in a row they were to call out 'Keno!' **1907** C. E. Mulford *Bar-20* (1914) iii. 19 It was Keno. **1907** C. E. Mulford *Bar-20* (1914) xix. 193 He wants to know whether 'th cards are stacked an' why he can't holler 'Keno'. **1920** in J. M. Hunter *Trail Drivers of Texas* I. 205 Shake yer spurs an' make 'em rattle! Keno! Promenade to seats.

kenotron (ke·nŏtrǫn). *Electr.* [f. Gr. κενό-ς empty + *-tron.] A kind of highly evacuated thermionic diode designed for rectification at high voltages.

Kenotron was formerly a proprietary name in the U.S.

1915 S. Dushman in *Gen. Electric Rev.* Mar. 156/1 In the case of a rectifier containing a hot filament as cathode and exhausted to as high a degree of vacuum as possible, there is no conduction except by electrons. In order to distinguish the latter type of hot cathode rectifier from other forms..the designation, kenotron, has been specially coined. **1915** *Electrician* 21 May 242/1 A kenotron has been constructed for rectifying 250 milliamperes at 180,000 volts. **1918** *Official Gaz.* (U.S. Patent Office) 1 Jan. 1/1 General Electric Company... Kenotron. For X-ray apparatus. Claims use since about May, 1916. **1946** W. T. Sproull *X-Rays in Pract.* viii. 120 Kenotrons are widely used in the electric circuits that are provided to supply the high voltage for x-ray tubes. *Ibid.* 121 In an x-ray tube, the space current is said to be 'emission limited', while in a kenotron the current is 'space charge limited', as in most radio vacuum tubes.

kenozooid (ke:nozōu·oid). [f. Gr. κενό-ς empty + Zooid *sb.*] In a colonial bryozoan of the phylum Ectoprocta, an individual consisting of the body wall or Zoœcium alone, without tentacles or an alimentary canal. Also called † **kenozoœ·cium** [f. Gr. κενό-ς empty + Zoœcium].

1909 G. M. R. Levinsen *Cheilostomatous Bryozoa* 46 We can distinguish between four main forms of individuals (Bryozooids)...*Kenozoœcia* (*Kenozooids*), which only have no polypide, but as a rule no aperture and always no operculum. *Ibid.*, The Kenozooecia must be regarded as supporting, fastening and connecting individuals. **1959** L. H. Hyman *Invertebrates* V. 372 Kenozooid is the name applied to heterozooids that lack zooidal differentiation and consist simply of body wall enclosing a cavity.

Kensal Green (kensăl grī·n). The site of a large cemetery in London, used allusively as the type of a cemetery or as a symbol of death and burial. Also *fig.*

1842 Dickens *Let.* 26 Apr. (1974) III. 211 What would I give if the dear girl whose ashes lie in Kensal-Green, had lived. c **1885** A. W. Pinero in M. R. Booth *Eng. Plays of 19th Cent.* (1973) IV. 391 Let me die..unless I am lucky enough to fall in some foreign set-to, I shall be buried some day at Kensal Green. **1903** Kipling *Five Nations* 152 That Kensal-Green of greatness called the files. **1914** G. K. Chesterton *Flying Inn* xxi. 275 Before we go to Paradise by way of Kensal Green. **1967** O. Lancaster *Eye to Future* i. 24 My mother was wholly free from the lachrymose necrophilia of the Victorians; not for her the

KENSINGTON

pious outings to Kensal Green. **1975** P. Somerville-Large *Couch of Earth* iv. 73 The necropolis was huge... We wandered about this devastated Kensal Green, Ismael..kicking at the piles of bones.

Kensington (ke·nziŋtǫn). The name of a borough of London (now part of the Royal Borough of Kensington and Chelsea), used *attrib.* or quasi-*adj.* to designate speech supposedly characteristic of people living in Kensington (cf. next).

1968 J. Lock *Lady Policeman* viii. 62 'Haymarket!' the manner is brusque and the voice exaggerated Kensington. **1971** *Guardian* 22 Oct. 10/3 One spectator exclaimed, 'all the peasants speak Baliganj Bengali!' (read: 'Kensington English').

b. *Kensington (outline) stitch*, a needlework stitch which is formed by putting the needle into the material from the front and returning it some way back whilst splitting the thread.

1881 C. C. Harrison *Woman's Handiwork* I. 33 Feather stitch..is often incorrectly termed 'Kensington' or 'crewel' stitch. **1883** *Century Mag.* Sept. 787/1 They know little of Kensington stitch or of Eastern-woven portières. **1909** *Cent. Dict.* Suppl. s.v. *stitch*, *Kensington stitch*, in *embroidery*, a long and a short outline-stitch, appearing alternately. **1934** M. Thomas *Dict. Embroidery Stitches* 186 Split Stitch... Also known as Kensington Outline Stitch.

Kensingtonian (kenziŋtōu·niăn), *sb.* and *a.* [f. *Kensington* (see prec.) + -IAN.] **A.** *sb.* An inhabitant of Kensington.

1889 G. B. Shaw in *Star* 6 Dec. 2/5 The Kensingtonians are asses to neglect these concerts. **1951** R. Campbell *Light on Dark Horse* 45 Zulus are far more important.. than dogs to Kensingtonians. **1965** G. McInnes *Road to Gundagai* xv. 267 The whole of this great grey sun-baked continent [*sc.* Australia] she regarded much as if it were Hornsey or Tooting Bec, and she a Kensingtonian of high degree. **1968** *Listener* 6 June 733/1 Hot or cool jazz would have been just too much and might have driven out the dreary West Kensingtonians with their dim uniformed escorts. **1974** *Times* 23 May 16/7 A Kensingtonian in her eighties..missed the style and elegance of the old shop.

b. A supposedly refined or affected manner of speech typical of people living in Kensington. *rare.*

1911 A. Bennett *Hilda Lessways* I. x. 91 Hilda..had been deprived of her Five Towns accent at Chetwynd's School, where the purest Kensingtonian was inculcated.

B. *adj.* Of, pertaining to, or characteristic of Kensington; *spec.* denoting refined or affected speech.

1902 A. Bennett *Anna of Five Towns* xi. 290 His broad Five Towns speech contrasting with the Kensingtonian accents of the coroner. **1936** *Times Lit. Suppl.* 27 June 541/1 Superior Margery Seymour, with her Kensingtonian 'mothah and brothah'. **1958** *Listener* 2 Oct. 537/1 A truly Kensingtonian drawing-room. **1960** *Times* 24 Feb. 15/3 Miss Maggie Smith and Miss Moyra Fraser at times vocally suggested mere Kensingtonian refinement. **1961** Wodehouse *Service with Smile* (1962) v. 73 Somehow it seemed worse and more wounding coming from those Kensingtonian lips. **1971** *Listener* 28 Oct. 596/3 Kensingtonian shrieks, with fiddle, indicated Gypsies relaxing.

Kensitite (ke·nzitəit). [f. the surname *Kensit* (see below): see -ITE¹.] A follower of John Kensit (1853–1902), a Low Church extremist who objected to alleged Romanizing aspects of the Anglican Church.

1898 *Tablet* 6 Aug. 207/1 It was disloyalty, with which Mr. Drummond and his brethren were charged by an irate and portly Kensitite with a Hyde Park voice. **1904** *Daily Chron.* 5 Mar. 6/6 The cheering of the 'Kensitites' brought a crowd quickly to the scene. **1907** *Ibid.* 14 May 5/4 Some of the audience endeavoured to expel the Kensitites. **1927** W. E. Collinson *Contemp. Eng.* 50 The Ritualists, strongly opposed by the Kensitites or followers of John Kensit. **1928** *Daily Express* 1 Dec. 11 (*heading*) Kensitite protests at the new Archbishop's election. **1936** A. Huxley *Eyeless in Gaza* iv. 29 The Ritualists and the Kensitites were at it again.

kenspeckle, *a.* (Later examples.)

1916 J. Buchan *Greenmantle* xx. 259 The immediate front of a battle is a bit too public for any one to lie hidden in by day, especially when two or three feet of snow make everything kenspeckle. **1930** H. S. Walpole *Rogue Herries* II. 392 He..wondered what it must be for such a boy to be in charge of so wild and tumultuous and kenspeckle an army. **1971** *Lancet* 6 Nov. 1028/2 He [*sc.* a cockerel] was..a kenspeckle figure in the neighbourhood [*sc.* in Scotland]. **1973** *Perthshire Advertiser* 8 Aug. 13/2 There have been others..who, if not as kenspeckle and dynamic in the public eye, have given of their time, talents and means.

kent, *sb.*² Add: (Earlier examples.) Also *attrib.*

1820 W. Scoresby *Acct. Arctic Regions* II. 296 The fat of the neck, or what corresponds in other animals with the neck, is called the Kent. **1837** R. Hamilton *Nat. Hist. Whales* 106 A band of blubber two or three feet in width, encircling the fish's body at what is the neck in other animals, is called the *kent*, because by means of it the fish is turned over or *kented*. To this band is fixed the lower extremity of a combination of powerful blocks, called the *kent-purchase*, by means of which, the whole circumference of the animal is, section by section, brought to the surface.

Kent (kent), *sb.*³ [Name of a county in England.] In full, *Kent sheep*. (See quot. 1957.)

1809 D. Price *Syst. Sheep-Grazing Romney Marsh* iv. 186 The New Leicester breed..were ripe for the slaughter-house in April, whereas the South Down and the Kents would not be so till the latter end of the summer. *Ibid.* 201 The Kent sheep being 65 lb. lighter than the Leicesters ..makes the gainings of the Kent sheep in the whole 152 lb. of mutton. **1891** R. Wallace *Rural Econ. Austral. & N.Z.* Plate LXIX (*caption*) Romney Marsh or Kent Sheep. **1894** *Country Gentleman's Catal.* 52/1 Sheep—Kent or Romney Marsh. Bred by owner. **1957** *Encycl. Brit.* XX. 476/1 The *Romney* is a long-coarse-wool, white-face, hardy, polled sheep that originated in Kent, Eng. It is sometimes called the Kent or the Romney Marsh. **1960** *Farmer & Stockbreeder* 8 Mar. 15/1 Grassland types [of sheep] include ..Greyface and Kent. **1972** *Country Life* 16 Mar. 607/1 The sheep..can break down organo-phosphates by blood enzymes. Among Dorset Downs 3 per cent could do this efficiently..and among Kent 37 per cent.

kent, *v.*² (Earlier example.)

1820 W. Scoresby *Acct. Arctic Regions* II. 296 By means of it, the fish is turned over or kented.

kentallenite (kentæ·lènəit). *Petrogr.* [f. *Kentallen*, name of a village in Strathclyde (formerly in Argyllshire) + -ITE¹.] An olivine-bearing monzonite.

1900 Hill & Kynaston in *Q. Jrnl. Geol. Soc.* LVI. 532 Taking..the Kentallen rock as our type, we propose that the term kentallenite should be substituted for olivine-monzonite. Kentallenite may be briefly defined as a coarse or medium-grained holocrystalline rock, consisting of olivine and augite, with orthoclase, plagioclase, and biotite in varying proportions. **1960** E. B. Bailey et al. *Geol. Ben Nevis & Glen Coe* (ed. 2) xv. 192 The handsome black kentallenite, once worked as an ornamental stone, is pierced by a few white felspathic segregations. **1969** *Scottish Jrnl. Geol.* V. 11 Kentallenite is a pyroxenic member of the dominantly hornblendic Appinite Suite..of the British Caledonian calc-alkaline igneous province.

kente (ke·ntə). Also **Kente.** [Twi, = cloth.] In full, *kente cloth*. In Ghana, a banded material; also, a long garment made from this material, loosely draped on or worn around the shoulders and waist.

[**1881** J. G. Christaller *Dict. Asante & Fante Lang.* 228 *Kente*, country cloth, a home-made negro-dress, consisting of a number of narrow stripes of cotton-cloth sewed together.] **1957** M. Banton *W. Afr. City* xii. 218 They may forsake western dress for Kente cloth. **1959** A. Abbs *Ashanti Boy* i. 38 The Chief was dressed in a gorgeous silk Kente... He was accompanied by some elders and friends, all wearing colourful Kentes..with quick, characteristic swings of the right arm, each man re-arranged his cloth. *Ibid.* 255 *Kente*, cloth woven on native loom, usually in narrow strips that are sewn together. The designs are geometrical and each one has a distinctive name. **1962** *Times* 23 Nov. 4/2 The Ghanaian girls came past in their Kente dresses of gold, dark blue or deep pink and mauve. **1963** *Economist* 1 June 894/1 The confident swirling of Ghanaian *kente* robes. **1964** *Ibid.* 8 Feb. 488/2 Dr Nkrumah..intends to clutch his people by the lapels (or kente cloths). **1969** *Times* 22 Oct. (Ghana Suppl.) p. viii/4 Not even the..King of the Ashanti, knows much more than six weeks in advance when he is going to hold his durbah..when the Golden Stool of the Ashanti is paraded before a crowd robed in its gaudiest *kente* cloth.

kentia (ke·ntiă). [mod.L. (C. L. Blume 1836, in *Rumphia* II. xvii. 94), f. the name of William *Kent* (d. *c* 1828), botanical collector + -IA¹.] A palm with pinnate leaves of the genus so called, native to Australia and some Pacific islands, or one formerly included in this genus.

1870 B. S. Williams *Choice Stove & Greenhouse Ornamental-leaved Plants* 223 The *Kentias* are handsome robust plants, with pinnate leaves. **1909** *Westm. Gaz.* 8 Jan. 9/2 Over 1,000 Kentia palms from the South Sea Islands. **1937** M. James *Family Garden* i. 46 *Kentias* are sometimes called curly palms. **1951** *Dict. Gardening* (R. Hort. Soc.) II. 1015/2 The Howeas (usually called Kentias in the market) are the most popular of Palms for general decorative work. **1966** E. J. H. Corner *Nat. Hist. Palms* vii. 172 They [*sc.* pigeons] have been thought responsible..for the wide distribution of *Areca* and *Kentia* in the South Pacific.

Kenticism. Restrict *rare* to sense in Dict. and add: **b.** A word, idiom, or expression peculiar to the OE. or ME. Kentish dialect; language characteristic of such dialect.

1933 *Amer. Jrnl. Philol.* LIV. 307 They show traces of the Kentish dialect, which is not surprising as there is Kenticism in the inked glosses in the manuscript. **1965** H. Kökeritz in Bessinger & Creed *Medieval & Ling. Stud.* 294 We can discern in his [*sc.* Wyatt's]..poems the survival of certain Kenticisms.

Kentish, *a.* **3. b. Kentish glory** (examples).

1775 M. Harris *Eng. Lepidoptera* 27 (*heading*) Glory, Kentish. **1869** E. Newman *Illustr. Nat. Hist. Brit. Moths* 47 The Kentish Glory.—Fore wings of the male brown; hind wings orange-colour: all the wings of the female alike, pale smoky-brown. **1899** D. Sharp *Cambr. Nat. Hist.* VI. vi. 406 The 'Kentish glory', *Endromis versicolor*, ..is a large and strong moth, and flies wildly in the day-time in birch-woods. **1971** *Times* 28 Jan. 12/6 The birch which provides the last English home of the Kentish glory moth.

kentrogon (ke·ntrǒgǫn). *Zool.* [ad. Fr. *kentrogone* (Y. Delage 1884, in *Arch. Zool. Expér. & Gén.* 2 Sér. II. 606), f. Gr. κέντρον sharp point + γόν-ος offspring.] A larval form of parasitic barnacles of the order Rhizocephala, in which state the barnacle first attaches itself to the host.

1909 A. Sedgwick *Student's Text-bk. Zool.* III. v. 430 At this stage the young Cirripede is known as the Kentrogon larva. **1909** W. T. Calman in E. R. Lankester *Treat. Zool.* VII. v. 134 (*caption*) *Kentrogon*-stage, after the *Cypris* shell has been cast off and the 'dart' has been formed. **1940** *Chambers's Techn. Dict.* 473/1 Kentrogon (*Zool.*). A stage in the life-history of certain parasitic *Cirripedia* (e.g. *Sacculina*) which succeeds the Cypris stage and precedes the entry of the parasite into the body of the host.

Kentuck (ke·ntɒk), *a.* and *sb.* *U.S.* Also Kaintuck. [Abbrev. of *KENTUCKY.] = next.

1826 T. Flint *Recoll.* 15 A 'Kentuck' is the best man at a pole. **1831** *Constellation* (N.Y.) 143/2 Placing a huge lump of his favorite 'Kentuck' [tobacco] in his mouth. **1834** W. A. Caruthers *Kentuckian in N.Y.* I. 24, I gets a quid of the real Kentuck twist into my mouth. *Ibid.* 25 When we Kentuck boys get at it, it won't all end like a log rollin'. **1842** *Amer. Pioneer* I. 157, I then entered a Kentuck boat and descended the river. **1852** B. Casseday *Hist. Louisville* ii. 69 'And you waded in like a raal Kaintuck,' rejoined Nine-Eyes. **1872** W. J. Flagg *Good Investment* 544/1 You must expect me to defend myself Kaintuck fashion. **1941** L. D. Baldwin *Keelboat Age* 61 The Americans..considered a 'Kentuck' best at the setting poles. **1942** in H. Wentworth *Amer. Dial. Dict.* (1944) 338/2 The Kaintucks were spared a feud with the N.Y.C. police.

Kentuckian (kentɒ·kiăn), *a.* and *sb.* *U.S.* Also 8–9 **Kentuckyan.** [f. next: see -IAN.] **A.** *adj.* Of or pertaining to Kentucky. **B.** *sb.* A native or inhabitant of Kentucky.

1779 G. R. Clark *Campaign in Illinois* (1869) 85 If not deceived by the Kentuckyans, I should still be able to compleat my design. **1784** [see *INDIANIAN]. **1804** C. B. Brown tr. *Volney's View Soil & Climate U.S.A.* 71, I have observed the Kentuckian bank of the river to be formed of similar ridges. **1831** [see *BEAD *sb.* 5 d]. **1886** F. C. Baylor *On Both Sides* 145 A handsome carriage.. drawn by a beautiful pair of Kentuckian thoroughbreds. **1949** B. A. Botkin *Treas. S. Folklore* p. xix, The pioneer saga of planting colonies..and creating names, legends, and ballads in that wild and wonderful country that the first Englishmen found in Virginia, the first Virginians in Kentucky, and the first Kentuckians and Virginians in Texas. **1969** I. Kemp *Brit. G.I. in Vietnam* iii. 43 Staff Sergeant Howell was..a paunchy, cheerful, easy-going Kentuckian.

Kentucky (kentɒ·ki). Chiefly *U.S.* Also **Kentucke.** [From the name of the river; the original meaning of this is uncertain.] One of the south-eastern United States, lying south of the Ohio River and east of the Mississippi; used *attrib.* to designate things originating in, or connected with, this state.

Only a selection of collocations is given here: see D.A.E. and D.A. for fuller lists.

1785 E. Denny *Mil. Jrnl.* (1859) 57 Our fleet now consists of..batteaux, besides two large flats called Kentucky boats. **1785** [see *COFFEE-TREE 2]. **1811** A. Wilson *Amer. Ornith.* III. 85 [The] Kentucky Warbler, *Sylvia Formosa*,..inhabits the country whose name it bears. **1832** in A. Nicoll *Hist. Eng. Drama 1660–1900* (1959) VI. 487 (*title of play*) The Kentucky Rifle; or, the horse and the murderer. **1835** J. H. Ingraham *South-West* II. 175 The young Mississipian..dresses plainly,..often in pantaloons of Kentucky jean. **1849** E. Emmons *Agric. N.Y.* II. 68 An earlier kind of grass than timothy, is the Spear grass, Meadow grass, or Kentucky blue grass. **1872** Schele de Vere *Americanisms* 416 The Coffee-tree (*Gymnocladus canadensis*), often called Kentucky Coffee-tree, or Kentucky Locust, derives its name from the fact that in the days of early settlements the seeds were frequently used as a substitute for coffee. **1875** *Courier-Jrnl.* (Louisville, Ky.) 18 May 4/3 The Kentucky Derby, a dash of 1½ miles for three-year olds. **1901** *Jrnl. Chem. Soc.* LXXIX. I. 984 Several kilograms of Western Kentucky leaf, used mainly as 'fillers'. **1943** J. S. Huxley *TVA* vi. 49 The famous Kentucky blue grass (which won't grow without plenty of phosphorus). **1962** R. B. Fuller *Epic Poem on Industrialization* 40 Lawyer vice-presidents were fast being substituted For the Kentucky colonels As executive heads of industry. **1966** [see *BURGOO 2]. **1968** *Canad. Antiques Collector* Aug. 11/1 Their gunsmiths developed and perfected the 'Kentucky' rifle, the accuracy and superiority of which was proven. **1972** E. Thorpe *Night I caught Santa Fé Chief* xiv. 183, I went into the drug store and ordered Kentucky fried chicken.

kentum, var. *CENTUM.

Kenya (ke·nyă, kī·nyă). The name of an E. African state used *attrib.*, as **Kenya Asian** = **Kenyan Asian*; **Kenya coffee,** a mild coffee grown in Kenya.

1968 *Times* 22 Feb. 1/7 The unrestricted right of entry to Britain of some Kenya Asians must be respected, insisted Mr. Heath. *Ibid.* 1/7 Some Kenya Asians were allowed to opt for British passports. **1971** *Guardian* 27 Feb. 5/1 A Kenya-Asian teacher..said that the Walsall Education Authority refused her a job because of her accent. **1926–7** *Army & Navy Stores Catal.* 3/2 Coffee.-Kenya. Roasted, whole—lb. 2/5. **1937** *Discovery* Oct.

p. xciii/1 Finest Kenya Coffee. **1970** *E. Afr. Standard* 2 Jan. 5/3 In 1965 some 20,000 bags of Kenya coffee were sold locally, consumption per head being around 130 grams per year.

Kenyah (ke·nyă). Also **Kenniah, Kenya.** [Native name.] A member of one of the aboriginal peoples inhabiting parts of Borneo and Sarawak.

1866 C. Brooke *Ten Yrs. Sarawak* I. ii. 73 The branches [of the Malanans] inhabiting the inland and up-rivers.. are Kanowit..Kenniah, Bakatan..and numerous others. **1901** Hose & McDougall in *Jrnl. Anthropol. Inst.* XXXI. 174 The Kenyahs inhabit a district far inland among the head-waters of the Baram river. *Ibid.* 175 Of the many animals that the Kenyahs dare not eat or kill..the common white-headed carrion-hawk..is by far the most important. **1911** J. G. Frazer *Golden Bough: Magic Art* (ed. 3) II. xxi. 385 In the interior of Borneo the Kenyahs generally place before the main entrance of their houses the wooden image of Balli Atap, that is, the Spirit or God (*Balli*) of the Roof. **1957** *Encycl. Brit.* XIII. 346/1 Kenyah economy is based on the cultivation of dry rice in jungle clearings. **1960** *Guardian* 9 Nov. 10/3 The population [of Sarawak] includes..Kelebits, Kenyas, Muruts, and others.

Kenyan (ke·nyăn, kī·nyăn), *sb.* and *a.* [f. *KENY(A + -AN.] **A.** *sb.* A native or inhabitant of Kenya.

1938 T. W. Wallbank in *Jrnl. R. Afr. Soc.* Suppl. Apr. 20 Perhaps a more immediate and vital problem concerns the character of the young, native-born Kenyan. **1960** *Daily Tel.* 21 Jan. 10/3 Moderate Kenyans can regard the claims of the African group only with distrust. **1969** *Listener* 8 May 634/3 This week a number of Asian Kenyans, holders of British passports, are due to arrive in Britain without work permits. **1971** *Sunday Nation* (Nairobi) 11 Apr. 8/5 We must, as Wa Kataka says, try to encourage self-analysis as part of our national political ethic, and the best people, nay the people best qualified to do this, are Kenyans themselves. **1972** *Daily Tel.* 29 Dec. 2/8 The bank clerk, Rajendrakumar Thakrar, a Kenyan, who lived in Clovelly Road, Southampton, was said to be a 'happy man' until a month ago.

B. *adj.* Of, pertaining to, or characteristic of Kenya or its people. *Kenyan Asian*, an Asian, esp. one from India or Pakistan, resident, or formerly resident, in Kenya.

1946 R. Capell *Simiomata* I. 14 Kenyan gossip in the Aegean. **1960** *Guardian* 22 Apr. 6/5 Ex-Mau-Mau members..padded through the Kenyan forests. **1962** *Times* 23 Aug. 11/1 Tony, the epitome of all the New Kenyan morality must destroy. **1968** *Times* 13 Feb. 9/1 That so many Kenyan Asians are now claiming their legal right of entry to Britain is embarrassing. **1968** *Guardian* 5 Sept. 18/8 The settler farmers who opted for Kenyan citizenship. **1970** *Ibid.* 24 Mar. 11/3 Europe's airport lounges are still littered with passportless Kenyan Asians. **1972** *Ibid.* 26 June 6/5 What have Ulster Catholics, Russian Jews and Kenyan Asians in common? They are all members of underprivileged minorities.

Ke:nyaniza·tion. [f. *KENYAN *a.* + -IZATION.] In Kenya, the replacement of settlers and Asians by Kenyan Africans in government posts, the civil service, and other occupations. So **Ke·nyanize** *v. trans.*, to make Kenyan in character, organization, etc.

1963 *Times* 17 May 13/7 Nowhere will Europeans be required more than in administration, since the pace of Africanization—or Kenyanization, to use the locally preferred term—has been until recently far too slow to produce adequate numbers of trained and experienced African civil servants. **1970** *Guardian* 25 Nov. 11/6 The Kenyanisation and Ugandanisation law. **1971** *Inside Kenya Today* Mar. 7/1 The target of Kenyanizing the top management personnel will be reached sooner than later. **1973** *Daily Tel.* 11 Jan. 1/1 Kenya is to order 418 non-citizen Asian traders to shut their businesses and quit the country by June 1 under its Kenyanisation policy.

Kenyapithecus (ke:nyăpi·þikǒs). [mod.L., f. *KENYA + Gr. πίθηκος ape.] A fossil hominid of the genus so called, first discovered in Kenya in 1961 by L. S. B. Leakey; sometimes included in the genus *Ramapithecus*.

[**1961** L. S. B. Leakey in *Ann. & Mag. Nat. Hist.* IV. 690 The following is a diagnosis and preliminary description of this new Kenya Lower Pliocene primate. Super-Family Hominoidea. Family Incertae sedis. Genus *Kenyapithecus* gen. nov... The type of the genus is the new species *Kenyapithecus wickeri* described below. Species *Kenyapithecus wickeri* sp. nov.] **1963** *Times* 1 June 6/1 The pelvis of *Oreopithecus*..is probably farther removed from human ancestry than its East African contemporary, *Kenyapithecus*. **1967** *Observer* 15 Jan. 9/6 Dr. Leakey..endeavours to show..that his *Kenyapithecus* hominids are quite distinct from the ancestors of apes living at the same time. **1971** J. Z. Young *Introd. Study Man* xxxi. 440 The fossils of *Ramapithecus* are less ape-like and more human. They were first named from specimens from the late Miocene of India, but the form known as *Kenyapithecus* from East Africa..is similar.

kenyte (ke·nəit, kī·nəit). *Petrogr.* [f. the name of Mount *Kenya*, Kenya: see -ITE¹.] A light red to pale brown rock consisting of olivine-bearing phonolitic trachyte.

1900 J. W. Gregory in *Q. Jrnl. Geol. Soc.* LVI. 209 It is unadvisable to include the Mount Kenya lavas among the pantellerites, and the name of kenytes is accordingly proposed for them. **1921** H. G. Ponting *Gt. White South* xi. 103 One of the most grotesque of these shapes was a monolith of kenyte... The geologists were much interested in this curiosity, and, as kenyte lava is very brittle, exhorted all not to injure it. **1954** W. C. Smith in *Brit. Antarctic 'Terra Nova' Expedition 1910, Nat. Hist. Rep.: Geol.* II. iii. 41 It is now known that there is an important difference between the rocks of Mount Kenya named kenyte by J. W. Gregory and the lavas of Mount Erebus to which Prior extended the name. The kenytes of Mount Kenya actually contain large insets of nepheline. **1968** *Mineral. Abstr.* XIX. 323/1 A K/Ar date determined from anorthoclase indicates an age of 0·68(±0·14) m.y. for the Antarctic kenyte of the Cape Royds area.

kephalin. Now usu. spelt *cephalin* (see *CEPHALIN²).

Kepler (ke·plər). The name of Johann *Kepler* (1571–1630), German astronomer, used, chiefly in the possessive, to designate things and concepts discovered or investigated by him, as **Kepler's equation,** the equation $\theta = \phi - e \sin \phi$ relating the mean anomaly θ of a planet to the eccentric anomaly ϕ and the eccentricity e of the orbit; **Kepler's laws** (see Law *sb.*¹ 17 c (*a*)); **Kepler's nova** or **star,** a supernova which appeared in 1604 in the constellation Ophiuchus and disappeared in 1606; **Kepler('s) problem,** the problem of solving Kepler's equation for the eccentric anomaly of a planet in a known orbit given the mean anomaly, which is effectively that of finding the position of the planet at any given time.

[**1714** *Phil. Trans. R. Soc.* XXVIII. 1 (*heading*) Problematis Kepleriani, de inveniendo vero motu planetarum, areas tempori proportionales in orbibus ellipticis circa focorum alterum describentium, solutio Newtoniana.] **1721** J. Keill *Introd. Astron.* xxiii. 287 (*heading*) Kepler's problem. **1883** *Encycl. Brit.* XV. 709/1 By far the most important problem is to find the values of θ and r as functions of t, so that the direction and length of a planet's radius-vector may be determined for any given time. This generally goes by the name of Kepler's Problem. **1890** A. M. Clerke *Syst. Stars* vii. 97, 1604, Kepler's star. **1902** F. R. Moulton *Introd. Celestial Mech.* v. 148 (*heading*) Geometrical derivation of Kepler's equation. **1954** C. Payne-Gaposchkin *Introd. Astron.* (1956) xiv. 392 Attempts to identify the remains of Tycho's and Kepler's novae with stars have failed. **1958** Condon & Odishaw *Handbk. Physics* II. iv. 32/2 Kepler's equation defines ϕ as a function of nt.., which functional relation is the subject of a large mathematical literature. **1964** *Yearbook Astron. 1965* 49 The most brilliant 'new star' of which we have an accurate record was Tycho's Star of 1572, which was, of course, a supernova, and which became equal to Venus... Its only subsequent rival has been Kepler's Star of 1604, which also seems to have been a supernova. **1970** G. K. Woodgate *Elem. Atomic Struct.* ii. 28 This is called an accidental degeneracy. It occurs likewise in the equivalent classical problem—the Kepler problem of planetary motion under an inverse square law of force.

Keplerian (keplĭə·riăn), *a.* [f. prec. + -IAN.] Of or pertaining to *KEPLER or his discoveries and investigations; applied *spec.* to (*a*) motion, orbits, and trajectories such as occur when one body moves freely in the gravitational field of another (much more massive) body, viz. an ellipse (in accordance with Kepler's laws) or some other conic section; (*b*) a refracting telescope that has a positive objective and a positive eyepiece and gives an inverted image.

1851 Mill *Logic* (ed. 3) I. III. ii. 313 If the Keplerian operation, as a logical process, be really identical with what takes place in acknowledged induction, the definition of induction ought to be so widened as to take it in. **1909** Webster, Keplerian telescope. **1922** A. D. Udden tr. *Bohr's Theory of Spectra* ii. 37 The orbit of the electron deviates a little from a simple ellipse and is no longer exactly periodic. This deviation from a Keplerian motion is, however, very small compared with the perturbations due to the presence of external forces. **1935** E. A. Milne *Relativity, Gravitation & World-Struct.* xiv. 267 Newton ..determined the nature of the possible motions of the particle and showed that it consisted of Keplerian orbits, or parabolas or hyperbolas with Keplerian properties. *Ibid.*, Newton's solution of the Keplerian problem. **1958** *Listener* 20 Nov. 839/1 The Keplerian universe, which did away with the epicycles, was systematically ignored by Galileo. **1959** K. A. Ehricke in H. S. Seifert *Space Technol.* viii. 92 Figure 8–50 presents a survey of possible Keplerian (in distinction to powered) mission profiles for the inner and outer solar system. **1966** *McGraw-Hill Encycl. Sci. & Technol.* VII. 452/2 If the second lens has a positive power, the telescope is called a..Keplerian telescope and the separation of the two parts is equal to the sum of the focal lengths. **1966** *Daily Tel.* 31 Oct. 9/8 True weightlessness..can best be simulated in an aircraft flying in a so-called Keplerian trajectory. **1968** R. A. Lyttleton *Mysteries Solar Syst.* iv. 124 It is to be remembered that in computing an orbit, the aim is to obtain a Keplerian path about the sun—an ellipse, parabola, or hyperbola—that fits the observation.

kept, *ppl. a.* Add: **1. a.** Also of a man or boy maintained or supported in a homosexual relationship.

1963 *Economist* 27 Apr. 304/1 The complete failure to translate his off-beat characteristics into homosexual or kept-man terms. **1966** 'R. Standish' *Widow Hack* xi. 121 The appalling indignities endured by kept men pass belief. **1969** *Jeremy* I. III. 25/1 At the upper-end of the scene is the kept-boy who has little or nothing in common with the humbler 'rent-boy'.

ker-. Add: Also **che-, co-.** (Further examples.)

1836 *Public Ledger* (Philadelphia) 27 July (Th.), Down I came chewallop..and overset the chair. **1844** 'J. Slick' *High Life N.Y.* II. 88 We drew up co-wallop right afore Jase's house. *Ibid.* 154 Ca-smash went the chair. **1854** M. J. Holmes *Tempest & Sunshine* 2 Then, again, you'll go in co-slush. **1855** *Spirit of Times* 29 Sept. 387/1 And the fust thing you knows he falls and down he comes kerflumix. **1858** S. P. Avery *Harp of Thousand Strings* 44 He fell 'kerslap' upon the hot goose of the pressman! **1884** 'Mark Twain' *Huck. Finn* xxiii. 234 Jes' den, 'long come de wind en slam it to, behine de chile, ker-*blam*! **1903** *Outing* XLIII. 83/1 The sound made by the water when the frog dives, we used to express when we were boys, by the word 'kerplunk'. **1908** *Magnet* I. 1, 'Ker-woosh!' ejaculated the junior, as he sprawled on the floor over Harry Wharton's legs. 'What's that in the way?' **1923** *Public Opinion* 15 June 565/1 With both feet set down kerplunk he closed the interview. **1926** F. M. Ford *Man could stand Up* II. iv. 164 Kerumph—the wagons of coal would fly over until we recalled our planes. **1935** H. G. Wells *Things to Come* xi. 96 Can I go when I grow up? And see the other side of the moon! And plump back *ker-splosh*! into the sea! **1937** *New Masses* 26 Oct. 18/1 Their [*sc.* Hollywood journalists'] vernacular divides the failures into three subdivisions: flop, flopperoo, and ker-plunk. **1939** T. S. Eliot *Old Possum's Pract. Cats* 17 Growltiger to his vast surprise was forced to walk the plank. He who a hundred victims had driven to that drop, At the end of all his crimes was forced to go ker-flip, ker-flop. **1939** J. Cary *Mr. Johnson* 41, I go trow him.. In de river Thames, kersplash. **1942** Z. N. Hurston in A. Dundes *Mother Wit* (1973) 26/2 Ker-blam-er-lam-er-lam! And dat was de last of Brer Engine-driving Monk. **1959** M. Gilbert *Blood & Judgement* vi. 59 The boat hit the surface with a solid ker-splash. **1963** *Punch* 30 Jan. 178/2 The boot..kerplonked to the carpet as straight and true as Newton's apple. **1963** *New Yorker* 29 June 26 That's why I nearly went kerplunk when you walked out of here with him. **1970** *Observer* (Colour Suppl.) 15 Feb. 36/4 They wear..extraordinary bathing costumes with Mighty Mouse zigzags across their chests, so that one half-expects them to rush about the beach shouting 'Pow!' and 'Zap!' and 'Kerrump!'

kera, var. *KRA.

keramat, var. *KRAMAT.

kerasin (ke·răsin). *Biochem.* Also **cer-, -ine.** [irreg. f. Gr. κέρας horn + -IN¹.] A cerebroside, $C_{48}H_{93}NO_8$, which gives lignoceric acid, D-galactose, and sphingosine on hydrolysis and is a white amorphous powder.

1878 Kerasine [see Phrenosin]. **1884** J. L. W. Thudichum *Treat. Chem. Constitution Brain* iv. 178 Kerasin.. is a cerebroside, namely a body which contains the sugar cerebrose, combined with at least two other radicles. **1933** Cameron & Gilmour *Biochem. of Med.* v. 78 In kerasin or cerasin it [*sc.* the fatty acid] is lignoceric acid, $C_{24}H_{48}O_2$. **1951** *Arch. Path.* LI. 338 A lipoprotein fraction containing 62 per cent kerasin has been isolated from two spleens removed surgically from patients with Gaucher's disease.

keratinize, *v.* Add: (Further example.)

1973 *Nature* 9 Feb. 398/1 Ambivalent cells of the vaginal squamous epithelium keratinize under oestrogenic action.

2. *trans.* To make horny, to subject to keratinization.

1909 in Webster. **1924** *Q. Jrnl. Microsc. Sci.* LXIX. 33 The centre [of the tumour] is completely keratinized. **1971** J. Z. Young *Introd. Study Man* xi. 146 The products of the epidermal line are ultimately keratinized.

Hence **ke·ratinized, ke·ratinizing** *ppl. adjs.*

1896 Keratinising [in Dict.]. **1904** *Trans. Ophthalm. Soc.* XXIV. 13 Meibomian secretion organisms, etc., will become adherent to the keratinised plaques. **1912** A. S. Grünbaum *Essent. Morbid Histol.* ii. 11 In epithelial cancer, the result is the formation of somewhat typical cell nests, which are composed of concentric groups of keratinised cells. **1924** *Q. Jrnl. Microsc. Sci.* LXVIII. 105 Keratinized epidermis, e.g. a finger-nail, can be resolved after treatment with sulphuric acid. **1924** *Ibid.* LXIX. 49 In the keratinizing cells the volume of the cytoplasm becomes only slightly greater. **1952** *Brit. Jrnl. Dermatol.* LXIV. 424 The diagnosis of well-differentiated keratinizing squamous carcinoma. **1964** G. H. Haggis et al. *Introd. Molecular Biol.* i. 6 When sufficient vitamin A is present, the cells [in the tissue culture] fail to form a keratinized squamous epithelium as they usually do. **1966** *Lancet* 31 Dec. 1457/2 There is thus gradually built up a picture of increasing epidermal hyperplasia which at length becomes heaped up into a papilloma composed of the keratinising remnants of former follicles.

keratino- (ke·rătino), used as combining form of Keratin, as **kerati·nocyte,** an epidermal cell which produces keratin; **keratino·lysis,** lysis or destruction of keratin; so **ke:ratinoly·tic** *a.*, bringing about keratinolysis; **ke:ratinophi·lic** *a. Bot.*, growing on keratinous material such as hair, feathers, etc.

1956 D. M. Pillsbury et al. *Dermatol.* ii. 7 The epidermis..contains two distinct cell types... The most numerous and important of these are the keratin synthesizing cells (Malpighian cells or keratinocytes) which make up 95 per cent of the epidermis. **1964** *Progress Biol. Sci. Rel. Dermatol.* II. 415 The keratinocytes are derived

from the surface ectoderm. **1972** *Amer. Zoologist* XII., 35/1 Melanin pigmentation of mammalian epidermis results from the interaction of melanocytes and keratinocytes in the synthesis, transfer, transport, and ultimate disposition of melanosomes. **1956** *Experientia* XII. 309/1 Enamel minerals stimulate keratinolysis, and keratinolysis contributes to dissolution of enamel apatite. **1905** *Jrnl. Investigative Dermatol.* XLIV. 308/2 True keratinolysis is within the capability of the dermatophytes. **1952** *Mycologia* XLIV. 177 The keratinolytic power of the fungus is reduced. **1970** *Biol. Abstr.* LI. 9627/1 (*heading*) Keratinolytic and keratinophilic fungi in the immediate surroundings of cattle. **1946** J. S. Karling in *Amer. Jrnl. Bot.* XXXIII. 751/1 Inasmuch as these species appear to be limited in occurrence and growth to tissues which contain keratin they will be described as keratinophilic chytrids. **1971** *Indian Jrnl. Med. Res.* LIX. 1699 Three hundred and thirty-one domestic animals ..were examined for the presence of dermatophytes and other keratinophilic fungi.

kerato-. Add: **keratecta·sia** *Ophthalm.* [Ectasia], protrusion of the cornea; **kera·tic** *a.* *Ophthalm.*, occurring on the cornea; **ke:rato-acantho·ma** (pl. **-omas, -omata**) *Path.* [Acanth(o- + *-oma], a tumour-like overgrowth of the skin, resembling a squamous carcinoma with a keratinized centre, but usually healing spontaneously; molluscum sebaceum; **ke:rato-conjuncti·val** *a.*, of or pertaining to the cornea and the conjunctiva, or keratoconjunctivitis; **ke:ratoconjunctivi·tis** *Path.*, inflammation of the cornea and conjunctiva; any disorder so characterized; **keratode·rma, -ia** *Med.*, a local or general thickening of the horny layer of the epidermis; **keratoge·nic, kerato·genous** *adjs.*, producing, or promoting the production of, keratinous material; **keratohy·alin(e** *Biochem.* [ad. G. *keratohyalin* (W. Waldeyer in *Beiträge zur Anat. und Embryol. als Festgabe Jacob Henle* (1882) 149], the substance which makes up the granules in the granular layer of the epidermis; **kerato·ma** *Path.* [*-oma], a hard patch of thickened epidermis, due either to hypertrophy of the horny layer or to friction or pressure; a callus; **keratomala·cia** *Path.* [Malacia], a disorder in which the cornea becomes soft and opaque, associated with vitamin A deficiency; **kerato·meter** *Ophthalm.*, an instrument for measuring the radii of curvature of the front surface of the cornea by observing images reflected in it; an ophthalmometer; so **keratome·tric** *a.*, obtained by using a keratometer; **kerato·metry,** measurement of the radii of curvature of the cornea; **ke:ratomyco·sis** *Path.*, fungal infection of the cornea; **kerato·pathy** *Ophthalm.* [-pathy], any of various disorders of the cornea; **keratopla·stic** *a.*, promoting keratinization, and hence restoration, of the epidermis; **keratoplasty,** add: [ad. G. *keratoplastik* (F. Reisinger 1824, in *Baiersche Ann. f. Abhandl.*, etc., *aus d. Gebiete d. Chir.*, etc. I. 215)] (examples); **keratoscope,** add: [ad. Sp. *keratoscopio* (Placido 1880, in *Periódico de Oftalm. Prát.* Sept.–Nov. 44)] (examples); **kerato·scopy,** add: [ad. F. *kératoscopie* (Cuignet 1873, in *Rec. d'Ophthalm.* I. 14)] (examples); **kerato·sis** (pl. **-oses**) *Path.* [-osis], (*a*) any of various disorders characterized by circumscribed hyperkeratosis; (*b*) a keratotic lesion; hence **kerato·tic** *a.*, of or pertaining to keratosis.

1887 *Syd. Soc. Lex.*, Keratectasia. **1904** L. W. Fox *Dis. Eye* vi. 172 *Keratectasia.* The term applied to the undue protrusion of an opaque cornea as the result of some inflammatory condition. **1918** J. H. Parsons *Dis. Eye* (ed. 3) xi. 198 As the cicatrix becomes consolidated the bulging may disappear, or it may remain permanently as an ectatic cicatrix (keratectasia from ulcer). **1972** *Biol. Abstr.* LIV. 6213/2 Differential diagnosis is discussed for keratoconus, keratectasia,..and corneal transplant marginal degeneration. **1907** J. H. Parsons *Dis. Eye* xiii. 301 The keratic precipitates..consist of leucocytes which are deposited from the aqueous upon the back of the cornea and stick there. **1955** P. D. Trevor-Roper *Ophthalm.* xxiii. 415 The endothelial cells become distended and tacky, so that any corpuscles that have been exuded into the aqueous from the iris and ciliary vessels are liable to adhere, forming keratic precipitates. **1950** A. Rook in *Proc. R. Soc. Med.* XLIII. 839 MacCormac and Scarff (1936, *Brit. J. Derm.*, **48**, 624), in the first published account of the condition, proposed to name it molluscum sebaceum, but as this term has been employed as synonymous with molluscum contagiosum, we prefer the name kerato-acanthoma which was suggested some years ago by Dr. Freudenthal. **1950** *Q. Cumulative Index Medicus* XLVIII. 1201/2 Kerato-acanthoma (molluscum sebaceum). **1952** *Brit. Jrnl. Dermatol.* LXIV. 425, I wonder if Dr. Ferguson Smith's cases of multiple self-healing epitheliomata..are really very different from these kerato-acanthomata. **1954** *Jrnl. Amer. Med. Assoc.* 5 June 562/2 Keratoacanthoma enlarges rapidly to maximum size (1 to 2 cm.) in four to eight weeks. **1972** *Cancer* XXIX. 1387 Histopathologic study of 108 keratoacanthomas and 14 squamous cell carcinomas failed to reveal any consistent single feature allowing for their distinction. **1941** *Amer. Jrnl. Ophthalm.* XXIV. 900 (*heading*) Keratoconjunctival lesions observed at high altitudes in Bolivia. **1965** *Biol. Abstr.* XLVI. 4694/1 Diagnostic value of the keratoconjunctival test in dysentery. **1887** *Syd. Soc. Lex.*, Keratoconjunctivitis. **1892** *Arch. Ophthalm.* XXI. 445 On kerato-conjunctivitis of rhino-pharyngeal origin. **1954** S. Duke-Elder *Parsons' Dis. Eye* (ed. 12) xxxiii. 554 Kerato-conjunctivitis sicca (Sjögren's syndrome), a general systemic disturbance of unknown origin usually occurring in women after the menopause.., is characterized by deficiency of the lacrimal secretion leading to dryness of the eyes. **1960** *Jrnl. Infectious Dis.* CVI. 162/1 Keratoconjunctivitis in sheep of an infectious character is prevalent in many sheep-breeding districts in Norway. **1972** *Biol. Abstr.* LIV. 325/2 An outbreak suspected to be infectious bovine keratoconjunctivitis in Zebu cattle was reported in Khartoum, Sudan. **1933** *Arch. Dermatol. & Syphilol.* XXVII. 87 In man a diffuse keratoderma is the result of a different mutation than that causing p..ular keratoderma. **1967** H. Montgomery *Dermatopath.* .. v. 68/2 Recently a 34-year-old woman was seen who had had diffuse keratoderma of the palms and soles all of her life. **1972** C. B. S. Schofield *Sexually Transmitted Dis.* xvi. 183 Skin lesions..are found in about 10 per cent of patients with Reiter's disease... These fully developed lesions are known as keratoderma blenorrhagica, and histologically are indistinguishable from pustular psoriasis. **1902** H. W. Stelwagon *Treat. Dis. Skin* iv. 502 Besnier divides the cases into four classes:..(2) the symmetric keratodermia developing in childhood, of an erythematous and irritable character..; (3) symmetric keratodermia, especially of the feet, developing primarily in isolate foci..; (4) accidental keratodermias. **1970** *Dermatologica* CXLI. 321 Localized congenital erythrokeratodermias are a separate entity of keratodermias among erythrodermias and hyperkeratoses. **1923** Keratogenic [see *encephalitogenic *a.*]. **1959** *Science* 26 June 1744/2 The presence of CO_2 at concentrations higher than atmospheric concentrations restricts the ability of the chorion to undergo keratogenic changes. **1971** *Dermatologica* CXLII. 14 The follicular-keratogenic properties of several fatty acids were investigated. **1887** *Syd. Soc. Lex.*, Keratogenous. **1907** *Practitioner* Dec. 849 The keratogenous and analgesic properties of picric acid, as exhibited in the treatment of burns. **1951** *Ann. N.Y. Acad. Sci.* LIII. 474 In the hair, the fibrils..are most pronounced in the keratogenous zone. **1962** *Jrnl. Investigative Dermatol.* XXXVIII. 237/1 Keratohyalin granules occur in abundance in epidermal cells located next to the keratogenous zone. **1887** *Syd. Soc. Lex.*, Keratohyaline. **1889** *Brit. Jrnl. Dermatol.* I. 235 Waldeyer held that they [*sc.* granules] were composed of a solid hyalin-like substance, which he called Keratohyalin. **1937** E. Wolff *Dis. Eye* i. 13 In the cells of the deeper layers are found numerous granules of keratohyaline. **1972** *Biochim. & Biophys. Acta* CCLXI. 416 The three most common amino acid residues in the keratohyalin material ..are glutamic acid, glycine and alanine. **1887** *Syd. Soc. Lex.*, Keratoma. **1902** H. W. Stelwagon *Treat. Dis. Skin* iv. 498 Callositas. Synonyms—Tyloma; Tylosis; Keratoma; Callus. **1931** L. McCarthy *Histopath. Skin Dis.* ix. 438 Keratoma senilis is one of the factors that make up the clinical condition known as old-age atrophy of the skin. **1972** *Biol. Abstr.* LIV. 374/1 (*heading*) A case of senile keratoma on the eyelid. **1876** Dunglison *Dict. Med. Sci.* (rev. ed.) 565/1 Keratomalacia. **1886** C. M. Culver tr. *Landolt's Refraction & Accommodation of Eye* v. 413 Certain affections of the cornea, as keratomalacia, or central corneal ulcers, may bring about a notable flattening of this membrane. **1920** *Biochem. Jrnl.* XIV. 519 The histological and bacteriological evidence shows that keratomalacia among rats consists in a breakdown of the corneal tissue, caused by bacterial invasion. **1969** *New Scientist* 30 Jan. 227/1 Keratomalacia, scurvy and beriberi are also frequent aftermaths of infection in people living on diets deficient in vitamin A, ascorbic acid, and thiamine respectively. **1886** C. M. Culver tr. *Landolt's Refraction & Accommodation of Eye* iv. 330 Javal and Schiötz have adapted a similar disc to their keratometer. **1927** *Amer. Jrnl. Ophthalm.* X. 683/2 In the measurement of irregular astigmatism in the center of the cornea, the keratometer is usually extremely valuable. **1972** Stone & Phillips *Contact Lenses* iv. 105 Since the optic radii of a contact lens are similar to those of the cornea, a keratometer may be used to check them. **1885** *Arch. Ophthalm.* XIV. 175 The hyperbolic lenses..can be manufactured to order to suit each particular case from data furnished by the keratometric measurements. **1927** *Amer. Jrnl. Ophthalm.* X. 678/1 The accuracy of a keratometric record..depends primarily on having a good modern instrument. **1973** W. G. Sampson in *Symposium Contact Lenses* (New Orleans Acad. Ophthalm.) ii. 22 Javal's rule was used in the past to estimate the predicted spectacle cylinder from keratometric measurements. **1891** *Ophthalmic Rev.* X. 250 (*heading*) Contributions to keratometry. **1972** Stone & Phillips *Contact Lenses* iv. 105 The central radius is determined by 'classical' keratometry. **1883** *Ophthalmic Rev.* II. 369 (*heading*) Keratomycosis. **1951** H. L. Birge in A. Sorsby *Syst. Ophthalm.* ix. 305 Keratomycosis (moniliasis) is rarely seen in the cold climates, and usually follows some sort of trauma to the eye with an earthy substance. **1971** *Amer. Jrnl. Ophthalm.* LXXI. 1191/2 A fifth case of keratomycosis due to Allescheria boydii is reported. **1948** D. G. Cogan et al. in *Arch. Ophthalm.* XL. 625 The corneal changes are those generally called band keratitis, but, for obvious reasons, are more properly designated as band keratopathy. **1958** *Circulation* XVIII. 524/2 Lipid keratopathy..consists clinically of a fatty plaque in an area of the cornea that has been previously vascularized. **1972** H. M. Leibowitz in Gasset & Kaufman *Soft Contact Lenses* xxv. 202 Hydrophilic contact lenses have been found to be extremely useful in the therapy of bullous keratopathy. **1887** *Syd. Soc. Lex.*, Keratoplastic. **1907** W. A. Pusey *Princ. & Pract. Dermatol.* i. 126 The keratoplastic action that is noted from the application of various powders upon raw epithelial surfaces may be due partly to their drying action. **1951** A. Grollman *Pharmacol. & Therapeutics* xxv. 522 Chrysarobin is used in skin diseases, especially in psoriasis... In a strength of 10 to 20 per cent it has a keratolytic action while in more dilute form (5 per cent) it exercises a keratoplastic action. **1857** Dunglison *Dict. Med. Sci.* (rev. ed.) 518/1 Keratoplasty. **1888** *Arch. Ophthalm.* XVII. 524 In making his experiments on keratoplasty Wagenmann..first took flaps from the cornea of rabbits which were left in partial communication with it. **1939** E. B. Spaeth *Princ. & Pract. Ophthalmic Surg.* xvi. 489 Circumscribed or partial penetrating keratoplasty has offered up to the present day the best permanent results. **1948** *Sci. News* VIII. 31 This problem has been resolved by keratoplasty, or corneal transplantation. **1972** *Arch. Ophthalm.* LXXXVII. 538/1 Corneas more than 50 hours old and from elderly donors are dependable for use in penetrating keratoplasties. **1886** C. M. Culver tr. *Landolt's Refraction & Accommodation of Eye* iv. 3.. This author [*sc.* Placido] uses, as a 'keratoscope', a disc of card-board, wood or zinc, 23 centimetres in diameter. On one side is drawn a series of concentric circles, alternately black and white. **1910** H. C. Parker *Handbk. Dis. Eye* 81 The cone is observed by inspection..with an instrument known as the keratoscope. **1972** *Jrnl. Optical Soc. Amer.* LXII. 169/1 In a conventional keratoscope, light from a flat or curved target subtending about 15° at the eye is specularly reflected by the anterior surface of the cornea. **1882** C. Macnamara *Man. Dis. Eye* (ed. 4) ii. 40 Keratoscopy may assist us in forming a diagnosis. **1902** E. H. Lendon *Method of Cuignet* 5 The word 'Keratoscopy' has now come..to be equivalent to 'Shadow Test'. **1972** Feldman & Carney in Gasset & Kaufman *Soft Contact Lenses* xxxiii. 269 By using photoelectric keratoscopy and lenses designed by computer to fit the measured eye, we were able to resolve these problems. **1885** *Buck's Handbk. Med. Sci.* I. 419/2 (*heading*) Laminated epithelial plugs (*Keratosis obturans*, Wreden, Burnett). **1888** *Trans. Path. Soc. London* XXXIX. 357 That the keratoses were in their turn due to arsenic no one will, I think, doubt. **1939** *Arch. Dermatol. & Syphilol.* XXXIX. 235 A unique case of tumor-like keratoses developing on the dorsum of the hands after severe sunburn is reported. **1966** Wright & Symmers *Systemic Path.* II. xxiv. 1487/2 The type of keratosis that is caused by arsenic has a microscopical picture much less complex than that of senile and solar keratosis. **1972** *Daily Colonist* (Victoria, B.C.) 26 Jan. 2/1 These brown spots..can be generally called senile keratoses, although they are not all of the same type. **1934** *Brit. Jrnl. Dermatol.* XLVI. 162 Keratotic lesions..were observed with this distribution in nine of the ten cases. **1972** *Arch. Dermatol.* CV. 249/3 A 18-year-old daughter of the patient..began developing firm keratotic papules on her thighs at the age of 1.. **1972** C. B. S. Schofield *Sexually Transmitted Dis.* xvi. 183 The lesions tend.., if dry, to develop keratotic crusts.

keratolysis (kerătǫ·lisis). *Med.* [f. Kerato- + *-lysis.] **1.** (In Dict. s.v. Kerato-.)

2. A condition in which parts of the epidermis (esp. on the palms and soles) peel off periodically.

1895 *Brit. Jrnl. Dermatol.* VII. 37 (*heading*) A case of congenital exfoliation of the skin—(keratolysis exfoliativa?). **1902** J. W. Ballantyne *Man. Antenatal Path. & Hygiene: Foetus* xviii. 320, I place fœtal keratolysis here among the idiopathic maladies..simply to emphasise the fact that sometimes no such pathogenesis is possible. **1939** Lewis & Hopper *Introd. Med. Mycol.* ix. 102 In keratolysis exfoliativa the lesions consist of superficial scaly macules, which may coalesce and are localised to the palms and/or soles.

b. Loosening or removal of the horny layer of the epidermis through the action of chemical or mechanical agents.

1936 C. W. Dodge *Med. Mycol.* xx. 736 The infection ..extends to the thick palmar or plantar surface as a gyrate area of keratolysis. **1956** D. M. Pillsbury et al. *Dermatol.* iv. 22 The chemical disorganization of keratin is called keratolysis. It may be brought about chiefly by two classes of chemical agents: (1) strong alkalies and (2) reducing agents. **1969** *Arch. Dermatol.* C. 10/2 Pitted keratolysis is quite common in our military patients who have spent at least several days in wet conditions.

keratolytic (kerătoli·tik), *a.* and *sb.* *Med.* [f. Kerato- + *-lytic.] **A.** *adj.* Able to destroy keratinous material, esp. the horny layer of the epidermis.

1893 P. H. Pye-Smith *Introd. Study Dis. Skin* xi. 24.. The hardened skin may be covered with salicylic acid... The best method of applying this valuable keratolytic agent is probably..as a plaister. **1917** M. B. Hartzell *Dis. Skin* vi. 47 (*heading*) Keratolytic agents (those which soften the horny cells of the epidermis and cause its exfoliation). **1956** D. M. Pillsbury et al. *Dermatol.* iv. 22 Common examples of this type of keratolytic agent are the thioglycate salts (reducing agents) which are the principal ingredients of home sets for the permanent (cold) waving of hair. **1961** *Sabouraudia* I. 87 Keratolytic enzymes have not yet been isolated from dermatophytes.

B. *sb.* A keratolytic agent.

1932 A. C. Roxburgh *Common Skin Dis.* iii. 25 Keratolytics. For removal of horny thickening. **1940** Becker & Obermayer *Mod. Dermatol. & Syphilol.* iv. 36/2 Resorcin. This drug is somewhat similar in action to salicylic acid, and the two are the only drugs we use as keratolytics. **1971** *Brit. Jrnl. Dermatol.* LXXXIV. 31.. Keratolytics and other topical agents have proved unhelpful.

keratophyre (ke·rătǒfəiᵊɹ). *Petrogr.* [ad. G. *keratophyr* (C. W. Gümbel *Die paläolithischen Eruptivgesteine des Fichtelgebirges* (1874) 44), f. Kerato-: see *-phyre.] (See quot. 1972.)

1889 *Geol. Mag.* Feb. 71 Microscopical examination and chemical analysis show that these rocks consist, in part at least, of soda-felsites or keratophyres. The keratophyres (so named from their resemblance to hornstone) were first described by Gümbel. **1937** A. Johannsen *Descr. Petrogr. Igneous Rocks* III. 49 The keratophyres

are similar in appearance [to normal trachytes] but usually are much weathered and show brown or green tones. **1956** E. W. Heinrich *Microscopic Petrogr.* iii. 51 Trachytes occur as volcanic rocks and minor hypabyssal intrusives. ..Albitized types (keratophyres) occur with spillites. **1972** *Gloss. Geol.* (Amer. Geol. Inst.) 385/2 *Keratophyre*, a name originally applied by Gümbel..to trachytic rocks containing highly sodic feldspars, but now more generally applied to all salic extrusive and hypabyssal rocks characterized by the presence of albite or albite-oligoclase and chlorite, epidote, and calcite, generally of secondary origin. Originally the term was restricted to lavas of pre-Tertiary age but this distinction is not recognized in current usage.

keratosulphate (kerătosɒ·lfēit). *Biochem.* Also (*U.S.*) **-sulfate.** [f. Kerato- + Sulphate *sb.*] A mucopolysaccharide composed of galactose and sulphated acetylglucosamine units which is found in the cornea and in costal cartilage.

1953 K. Meyer et al. in *Jrnl. Biol. Chem.* CCV. 611 A polymer composed of *N*-acetylglucosamine, galactose, and sulfate in equimolar portions.., a novel type of sulfated mucopolysaccharide, we propose to name keratosulfate. **1960** G. K. Smelser in Duke-Elder & Perkins *Transparency of Cornea* 33 Loss of metachromatic staining, following digestion of the sections with testicular hyaluronidase, suggests that keratosulphate appears late in development. **1972** *Jrnl. Pediatrics* LXXXI. 108/2 Keratosulfate was clearly visible on the chromatograms [of the urine] of three patients with Morquio's disease, whereas none of the normal subjects or patients with other mucopolysaccharidoses tested had any visible keratosulfate.

kerb, *sb.* Add: **2. b.** *kerb-edge, -side*; *kerb market, stocks* (see *on the kerb* in 2 a and *curb *sb.* 15); **kerb crawling** *vbl. sb.*, = *gutter crawling* vbl. sb. (*gutter *sb.*¹ 8); also as *ppl. a.*; also **kerb-crawl** *v. intr.* [as a back-formation]; **kerb-crawler; kerb drill,** the exercise of standing on the kerb and looking right, then left, then right again before crossing the road; **kerb service,** see *curb service* (*curb *sb.* 15); **kerb weight** (see quot. 1967).

1971 *Daily Tel.* 13 Jan. 11/1 A man who kerb-crawls is not committing an offence. **1972** *Ibid.* 16 Oct. 3 A campaign against motorists who 'kerb crawl' in search of prostitutes. **1955** 'C. H. Rolph' *Women of Streets* ii. 43 The solicitation of respectable women in the streets by 'kerb-crawlers'..is causing much disquiet. **1969** *Sunday Tel.* (Brisbane) 20 Apr. 3/6 Women who complain of kerb-crawlers who pester them are often hypocritical to sex. **1949** in Partridge *Dict. Slang* 1091/2 Kerb-crawling vbl. sb. and ppl. a. **1971** *Times* 10 Aug. 2/1 The Rev. Peter Hawkins..complaining of kerb-crawling by men looking for girls. *Ibid.* 2/2 Respectable women had complained of being importuned by kerb-crawling motorists. **1948** R. Bailey *Safety Way* iii. 17 In some schools the children practise Kerb Drill with their teachers and learn to judge speeds. **1969** *Highway Code* 5 *Kerb drill.* Before you cross, stop at the kerb, look right, look left, look right again. **1970** R. Rendell *Guilty Thing Surprised* ix. 105 They poured through the gates.., paying no attention to the kerb drill. **1930** D. H. Lawrence *Nettles* 11 Idle we stand at the kerb-edge, Auntie, Dangling our useless hands. **1955** W. Faulkner *Fable* 306 On through the gate into an alley, a blank wall opposite and at the kerb-edge [curb-edge, 1954 U.S. ed.] a big..car. **1905** *Daily Chron.* 28 Apr. 4/4 The kerb market in New York reflects in a general way the tone of the stock market. **1956** P. Highsmith *Blunderer* viii. 69 Walter..bought her chocolate sodas at the kerb-service drugstore. **1905** *Daily Chron.* 11 Nov. 4/7 Fruit from the kerbside barrow is marvellously cheap nowadays. **1923** *Daily Mail* 21 July 3 The growing tendency of motorists to avail themselves of kerb-side pumps. **1959** *Times* 8 Dec. 5/6 The all-day parker has become so accustomed to free kerbside accommodation that the very idea of having to pay a parking charge seems to him an outrage. **1969** *Daily Tel.* 19 Aug. 15/1 A kerbside cafe. **1971** Slaughter & Goodman *Every Man should have One* ii. 18 If they succeed in looking like a model girl..it is meant as a compliment to you, not an expectation of ritzy treatment with champagne flowing and nothing less than an Aston Martin at the kerbside. **1923** *Daily Express* 30 July 2/6 Kerb Stocks. **1958** *Economist* 25 Oct. (Suppl.) 10/1 With a kerb weight of 26 cwt and 240 BHP of power from the 3.7-litre engine, the maximum speed will equal that of any four-seater car in the world, close to 150 mph. **1967** E. Rudinger *Consumer's Car Gloss.* (ed. 2) 64 *Kerb weight*, weight of a car without occupants or luggage but with oil and water and some petrol.

ker-blam: see *ker-.

Kerch (kəɹtʃ). Also **Kertch.** The name of a town in the Crimea used *attrib.* to designate a type of ancient red-figured Greek pottery found there.

1936 G. M. A. Richter *Red-Figured Athenian Vases* I. vii. 213 (*heading*) Kerch style... With the Kerch period we reach the last phase of Athenian red-figured pottery. *Ibid.* 214 Kerch vases have been found at Alexandria. **1960** R. G. Haggar *Conc. Encycl. Cont. Pott. & Porc.* 412/2 In the Greek decadence, popular vases of so-called 'Kertch' ware, from the name of the town where they have been found, were exported to South Russia. **1965** J. V. Noble *Techniques Painted Attic Pott.* (1966) iii. 64 Attic chous in the Kerch style, mid fourth century B.C., in the Metropolitan Museum of Art.

kerel (ke·rəl). *S. Afr.* Also **kêrel.** [Afrikaans, f. Du. = Carl *sb.*¹] A fellow, chap, young man.

1873 *Cape Monthly Mag.* Oct. 215, I have always known you to be a *slimme kerel*. **1896** H. A. Bryden *Tales S. Afr.* 214 *Kerel* (my boy) you have never by chance heard the story of the vrouw there and her Frenchman? **1902** J. H. M. Abbott *Tommy Cornstalk* 81 It is that 'kerel' French who is coming. **1939** S. Cloete *Watch for Dawn* 41, I want no more than justice, kerels. **1974** *Cape Times* 11 Nov. 9/2 Willem Prinsloo..served his lekker peach brandy on the stoep as the kêrels gathered to bid good afternoon to the setting sun.

kereru (ke·rěru). [Maori.] The New Zealand pigeon, *Hemiphaga novæseelandiæ*.

1873 W. L. Buller *Hist. Birds N.Z.* 157 *Carpophaga Novæ Zealandiæ* (New-Zealand Pigeon)..*Native names.* Kuku, Kukupa, and Kereru. **1905, 1949** [see *kuku 1]. **1955** W. R. B. Oliver *N.Z. Birds* (ed. 2) 440/1 New Zealand Pigeon. Kereru..was noticed and recorded by all the early explorers.

Keres (kēⁱ·rēⁱs). Also **Queres.** [American Indian.] **a.** A Pueblo Indian people inhabiting parts of New Mexico. **b.** The language of the Keres, forming the Keresan group. Also *attrib.* or as *adj.*

1893 T. Donaldson *Moqui Pueblo Indians* 91 The Queres group (Keresan stock) are the Pueblos of Santa Ana, San Felipe, Cochiti, San Domingo, Acoma, Zia, and Laguna. **1925** *Amer. Indian Life* 397 The Pueblo Indians..number about 10,000. They are usually classified according to language into four or five stocks, the Hopi of Arizona, the Ashiwi or people of Zuñi, the Keres of Acoma and Laguna to the west and, to the east, of five towns on the Rio Grande, and, also in the east, the Tanoans including the Tewa and the people of Jemez. **1957** *Encycl. Brit.* XVIII. 753/2 At the discovery in 1540 the Hopi, Zuñi and Keres may have been twice.. as numerous as to-day. **1965** *Canad. Jrnl. Linguistics* Spring 99 Hokan-Siouan in Sapir's arrangement includes six major units... The six units are: ..Keres (Powell's Keresan)..and an 'Eastern Group'.

Keresan (ke·rəsăn). [f. prec. + -AN.] A linguistic group consisting of Keres only. Also *attrib.* or as *adj.*

1893 [see *Keres]. **1939** L. H. Gray *Found. Lang.* xii. 408 In North America, twenty-five linguistic families are listed..*Algonkin*... *Iroquois*... *Keresan* with two dialects. **1950** F. Eggan *Social Organiz. Western Pueblos* 2 The more western Keresan-speaking villages. *Ibid.* 223 The Keresan villages of the Rio Grande. **1968** A. M. Josephy *Indian Heritage Amer.* xvi. 149 Keresan-speaking people of Acoma and Laguna in western New Mexico [are] considered western Pueblos. **1969** H. E. Driver *Indians N. Amer.* (ed. 2) xxvii. 517 As one proceeds eastward, to Zuñi, Acoma, Laguna, and other Keresan pueblos, the system [of matriarchy] becomes more bilocal and bilateral.

ker-flip, -flop, -flumix: see Ker- (in Dict. and Suppl.).

kerfuffle (kəɹfʌ·f'l). = Curfuffle *sb.*, *gefuffle.

1959 J. Fleming *Miss Bones* xiv. 150 The kerfuffle over the stolen jewels last week. **1960** E. W. Hildick *Jim Starling & Colonel* viii. 62 Butcher said he didn't know what all the kerfuffle was about. **1965** *New Statesman* 30 Apr. 693/3 After..some abortive backstage kerfuffles at the National Theatre, Wedekind's *Spring Awakening* has scraped past the Lord Chamberlain. **1968** 'B. Mather' *Springers* xii. 130 In the kerfuffle of the last half hour I had forgotten the poor soul's personal needs. **1973** K. Amis *Riverside Villas Murder* ii. 40 A lot of our readers are going to think all this kerfuffle over an old skeleton being snatched is..a bit of a joke.

Kerguelen (kəɹgēⁱ·lən, kɜ̄·ɹgələn). The name of a group of islands in the southern Indian Ocean, used *attrib.* in **Kerguelen cabbage,** a cabbage-like plant, *Pringlea antiscorbutica*, of the family Cruciferæ, which is confined to several islands in this region.

1847 J. D. Hooker *Bot. Antarctic Voy.: Flora Antarctica* II. 240 The illustrious Cook first discovered and drew attention to the 'Kerguelen's Land cabbage' during his first voyage, when accompanied by Mr. Anderson as surgeon and naturalist. The latter gentleman drew up an account of some of the more remarkable plants which he collected there and in other islands, which are preserved in the Banksian library. *Ibid.* 241 The Kerguelen's Land cabbage, cabbage though it be, [is] a cherished object in the recollection of the mariner. **1879** H. N. Moseley *Notes by Naturalist on 'Challenger'* viii. 191 The feature which distinguishes the general appearance of the vegetation of Christmas Harbour..is the presence of the Kerguelen Cabbage in large quantities. The plant grows on the slopes and bases of the cliffs in thick beds. The cabbage is in appearance like a small garden cabbage, but often with a long trailing stalk. **1952** J. A. King in F. A. Simpson *Antarctic Today* xv. 305 The Kerguelen cabbage (*Pringlea antiscorbutica*) is not as abundant [in the Prince Edward Islands] as at Kerguelen. **1960** *Times* 20 Apr. 14/7 There are no trees or shrubs and only some 17 species of other vascular plants, the largest being the so-called Kerguelen cabbage.

Kerman, var. *Kirman.

Kermanji (kəɹmā·ndʒi). [Kurdish.] A language of the Iranian group spoken by the Kurds of Kurdistan.

1882 *Encycl. Brit.* XIV. 157/2 The present Kurdish language which is called Kermánjí..is an old Persian patois, intermixed to the north with Chaldæan words and to the south with a certain Turanian element which may not improbably have come down from Babylonian times. **1914** T. E. Lawrence *Home Lett.* (1954) 295 A grammar of Kermanji (Kurdish). **1959** *Chambers's Encycl.* VIII. 274/1 They [*sc.* the Kurds] speak Kermanji, derived from an old Persian dialect.

Kermanshah, var. *Kirmanshah.

kermesse (kəɹme·s). *Cycling.* [Fr.: see Kermis.] A circuit race.

1963 *Times* 13 June 3/5 They stopped..to put on a.. 26-mile, 10-lap kermesse (circuit race). **1965** *Times* 31 May 4/5 *Kermesse* is a Continental term which British cycling uses to describe a circuit race. It originates from Belgium and is the Flemish for the carnival which usually includes such a race. **1971** D. Armstrong in N. G. Henderson *Cycling Year Bk.* xii. 101 Any rider who has ever won some little local kermesse would label it a classic event if he thought anyone would listen to him.

kern, *sb.*² Restrict *rare* to senses in Dict. and add: **3.** *Meteorol.* [abstracted from G. *kernzähler* kern (nucleus) counter (A. Wigand 1913, in *Meteorol. Zeitschr.* XXX. 13).] A particle which acts as a condensation nucleus; in a **kern counter,** a device in which a sample of air is supersaturated and condensation nuclei made visible and collected for counting: orig. intended to measure the concentration of condensation nuclei in the air, but now known to respond to particles too small to act as such under atmospheric conditions.

1941 *Jrnl. R. Aeronaut. Soc.* XLV. 72 The Aitken nucleus or 'kern' counter determines the number of nuclei in an air sample by subjecting it to a rapid expansion. This causes a considerable degree of supersaturation, and the drops which form around each nucleus fall on to a ruled slide where they may be counted. **1951** H. G. Houghton in T. F. Malone *Compendium Meteorol.* 165/2 The sweeping action of the dust particles on the kerns. **1954** J. C. Johnson *Physical Meteorol.* vii. 206 These dusts, or kerns, as they are called, which make up most of the suspensoids in the atmosphere and are important for their light scattering effects on visibility, contribute little or nothing to the condensation process. *Ibid.* 207 Dust counts made by instruments such as the Aitken kern counter have little correlation with the number of active condensation nuclei in the atmosphere. **1967** R. W. Fairbridge *Encycl. Atmospheric Sci.* 84/2 Such nuclei counters (or kern counters..) have some distinct advantages.

kernel, *sb.*¹ Add: **7. d.** *Chem.* = *core *sb.*¹ 7 f.

1916 G. N. Lewis in *Jrnl. Amer. Chem. Soc.* XXXVIII. 768 In every atom is an essential kernel which remains unaltered in all ordinary chemical changes. *Ibid.* 769 It must not be assumed..that the distinction between kernel and shell is absolutely hard and fast. **1927** E. N. da C. Andrade *Struct. Atom* (ed. 3) xv. 552 The interaction of the outer electrons with one another is large compared to the interaction of the single electrons with the core. We may refer to it as the kernel scheme. **1927** [see *core *sb.*¹ 7 f]. **1957** Sienko & Plane *Chem.* iii. 64 The letters represent the entire core, or kernel, of the atom.

8. b. *Linguistics.* (i) The stem or common basis of a set of inflectional forms, such as *lach-* in the German verb *lachen*; (ii) in full *kernel sentence* (in an early version of transformational grammar), a relatively simple sentence that results from the application of only a few (obligatory) transformations, and to which other sentences may be related by further transformations; a set of such sentences.

1894 O. Jespersen *Progress in Language* vi. 143, I shall have to separate word and case-ending, as far as this is feasible... We want a special term for this distinction; and I propose to call the substantial part of the word, felt as such by the instinct of each generation as something apart from the ending (*eag* in the example chosen [*sc.* O.E. *eage*]), the *kernel* of the word, while *eagan* is the historic 'stem'. **1918** *Amer. Jrnl. Philol.* XXXIX. 91, I think it unfortunate that the author feels it necessary to use the term *kernel* in place of the now generally used name *root*. **1933** L. Bloomfield *Lang.* xiii. 225 This *lach-*, strictly speaking, is a bound form; it is called the kernel or stem of the paradigm. **1957** N. Chomsky *Syntactic Structures* v. 45 We define the kernel of the language (in terms of the grammar G) as the set of sentences that are produced when we apply obligatory transformations to the terminal strings. **1961** R. B. Long *Sentence & its Parts* 494 'Kernels'. This term is applied to stripped-down nucleuses. **1963** J. Lyons *Structural Semantics* ii. 14 It appears that the kernel for English consists of simple, active, declarative sentences. **1969** *Canad. Jrnl. Linguistics* XV. 25 Sentences of the form *she is eager to please* demand that the subject, but no other nominal, of their embedded kernels be deleted. **1969** *Neuphilol. Mitt.* LXX. 204 The oft-repeated claim that all 'well-formed' sentences of a language are derivable from a single syntactic kernel is clearly unfounded.

c. *Math.* [tr. G. *kern* (D. Hilbert 1904, in *Nachrichten von d. Königl. Ges. d. Wissensch. zu Göttingen* (Math.-physik. Kl.) 49).] A function of two or more variables which, multiplied by one or more functions each of just one of the variables, constitutes the integrand of an integral with respect to these latter variables. (Orig. defined for integral equations in which the kernel is known and the other function(s) unknown, but now used in other situations also.)

1909 M. Bôcher *Introd. Study Integral Equations* 13 Comparing Abel's equation..with Liouville's equation.. we see that they come respectively under the following types: $f(x) = \int_a^x K(x,\xi)u(\xi)d\xi$ [and] $u(x) = f(x) + \int_a^x K(x,\xi)u(\xi)d\xi$ in which $f(x)$ and $K(x,\xi)$ are to be regarded as known functions and $u(x)$ is the function to be determined... *K* is called the kernel of these equations. **1924** W. V. Lovitt *Linear Integral Equations* i. 6 An integral equation is also said to be singular if the kernel becomes infinite for one or more points of the interval under discussion. **1962** *Newnes Conc. Encycl. Nucl. Energy* 762/2 The function $P(\mathbf{r}, \mathbf{r}_0, E, E_0)$ is called the slowing-down kernel of the moderator. It cannot, in general, be calculated exactly, but various approximations have been developed. **1971** J. W. Miles *Integral Transforms in Appl. Math.* i. 1 We define $F(p) = \int_a^b K(p,x)f(x)\,dx$ to be an integral transform of the function $f(x)$; $K(p, x)$, a prescribed function of *p* and *x*, is the kernel of the transform.

d. *Math.* The set of all the elements that are mapped by a given homomorphism into the identity element (for the group operation in the case of groups, for addition in the case of rings).

1946 E. Lehmer tr. *Pontrjagin's Topological Groups* i. 11 The set of all the elements of the group *G* which go into the identity of the group *G** under the homomorphism *g* is called the kernel of this homomorphism. **1959** G. & R. C. James *Math. Dict.* 224/2 If a homomorphism maps a ring *R* onto a ring *R**, then the kernel of the homomorphism is the set *I* of elements which map onto the zero element of *R**. **1971** G. Glauberman in Powell & Higman *Finite Simple Groups* i. 9 Since φ maps *G* into an abelian group, *G'* is contained in the kernel of φ.

9. (sense *8 b (ii)) *kernel string, word*; **kernel sentence**: see quot. 1968 and 8 b (ii) above.

1957 N. Chomsky *Syntactic Structures* viii. 89 Both 'the hunters shoot' and 'they shoot the hunters' are kernel sentences. **1968** J. Lyons *Introd. Theoret. Ling.* vi. 256 A kernel sentence..is any sentence which is generated from a single kernel string without the application of any *optional* transformations. **1972** *Science* 23 June 1304/1 The basic units of information in language are the 'kernel sentences'—primitive nondecomposable sentences that can be modified and combined in various ways..to produce a very large number of different messages. **1957** N. Chomsky *Syntactic Structures* viii. 88 Alternative transformational developments from the same kernel strings. **1965** *Language* XLI. 392 Variants occurring under unique operators or kernel-words.

kernicterus (kǝ̄ɹni·ktěrǒs). *Path.* Formerly also kern-, Kernikterus. [ad. G. *kernikterus* (G. Schmorl 1903, in *Verhandl. d. Deut. Path. Ges.* VI. 112), f. *kern* nucleus + *ikterus* Icterus, jaundice.] The staining of nuclei of the brain cells with bilirubin, which sometimes occurs, usu. associated with rhesus incompatibility, in neonatal jaundice, and which causes permanent brain damage; the disease or condition characterized by or associated with such staining.

1912 *Brain* XXXIV. 488 An important analogy may be drawn from the occurrence of 'Kernikterus' in certain cases of familial icterus gravis neonatorum. **1933** *Amer. Jrnl. Dis. Children* XLV. 757 Opinion is almost unanimous on the subject of the pathogenesis of kernikterus, namely, that following some injury, the nerve cells are subsequently stained with the bile pigments carried to them by the blood stream. **1950** *Jrnl. Laryngol. & Otol.* LXIV. 505 Sixteen cases of perceptive deafness associated with kernicterus are reported. In fourteen of these cases the jaundice was due to Rh iso-immunization. **1953** *Lancet* 28 Mar. 613/1 Kernicterus is never found apart from jaundice. **1965** *Amer. Jrnl. Path.* XLVI. 336 Yellow ganglion and glial cells, characteristic features of human kernicterus, were demonstrated in 23 rabbits.

Hence **kerni·cteric** *a.*, afflicted or associated with kernicterus.

1956 *Jrnl. Speech & Hearing Disorders* XXI. 407 (*heading*) Clinical pathologic aspects of kernicteric nuclear 'deafness'. **1967** *Jrnl. Neurochem.* XIV. 192 The kernicteric animals are known to have pathologic lesions in the CNS comparable to those found in brain tissue of infants dying with kernicterus.

Kernig's sign (kǝ̄·ɹnigz sain). *Med.* [f. the name of V. M. *Kernig* (1840–1917), Russian physician.] The inability of a patient to straighten his leg at the knee when lying on his back with the hips fully flexed, an indication of meningitis.

1901 *Brit. Med. Jrnl.* 16 Feb. 396/2 'Kernig's sign' was well marked. **1924** *Ibid.* 20 Dec. 1159/1 The angle at which Kernig's sign begins to be positive is clearly very different for a baby who can suck his own toe and a stocky middle-aged man. **1950** T. R. Harrison *Princ. Internal Med.* viii. 94/2 If the head cannot be so flexed and Kernig's sign is positive, it is likely that the patient has either meningitis or subarachnoid hemorrhage. **1969** I. J. T. Davies *Postgrad. Med.* vi. 201 The cardinal signs of inflammation of the meninges are neck stiffness..and Kernig's sign which is spasm of the hamstring muscles when the knee is extended with the hips fully flexed.

kernite (kǝ̄·ɹnait). *Min.* [f. *Kern*, name of the county in California where it was discovered + -ite¹.] A hydrated form of sodium borate, $Na_2B_4O_7.4H_2O$, that occurs as large transparent crystals and is used as a source of borax.

1927 W. T. Schaller in *Amer. Mineralogist* XII. 24 The new mineral kernite, $Na_2B_4O_7.4H_2O$, was received through Hoyt S. Gale..and is reported to occur in quantity in the southeast corner of Kern County, California... The name kernite is proposed. **1930** *Prof. Papers U.S. Geol. Survey* No. 158. 146/1 The extreme abundance of kernite, its large crystals, its perfect cleavage, and its clearness and transparency render the occurrence one of almost unique character as well as of striking beauty. **1951** C. Palache et al. *Dana's Syst. Min.* (ed. 7) II. xv. 337 The material had earlier been called rasorite..but the name kernite has published priority. **1972** M. H. Battey *Mineral. for Students* 235 Borax and the related mineral kernite, $Na_2B_4O_6(OH)_2.3H_2O$, are the principal sources of borax and of boron.

kernos (kǝ̄·ɹnǫs). *Archæol.* Pl. **kernoi.** [Gr.] An ancient Mediterranean and Near Eastern earthen vessel with small cups around the rim or fixed in a circle to a central stem.

1903 J. E. Harrison *Proleg. Study of Gk. Relig.* iv. 158 *Kernos*, a vessel made of earthenware, having in it many little cups fastened to it, in which are white poppies, wheat, barley..; and he who carries it..tastes of these things. **1955** L. Woolley *Alalakh* xii. 344 Three black impressed sherds, one from a beaker and one from a *kernos*. **1960** T. Burton-Brown *Early Mediterranean Migrations* ii. 35 The kernos shape..is known in Azarbaijan, for there is an unpublished example from there exhibited in the Tehran Museum. **1967** A. D. Lacy *Greek Pott. Bronze Age* v. 260 The kernos or cluster vase..is a composite vessel in which one or more rings of small cups are mounted upon a central stem. Kernoi have been found at Phylakopi in Milo, and in Crete. **1971** *Ashmolean Mus. Rep. Visitors 1970* 16 Purchased:..a Proto-White Painted kernos, Late Cypriot IIIB. **1972** Y. Yadin *Hazor* ii. vii. 101 The cultic aspect of the whole area is further attested by the enormous nests of vessels, which included incense-stands, 'cups and saucers', kernoi, and 'rattles'.

kero (ke·ro). Austral. and N.Z. colloq. abbrev. of Kerosene *sb.*

1930 *Bulletin* (Sydney) 10 Sept. 22/3 Never feed in troughs. With little stalls and cut-down kero.-tins.. calves can be fed quickly enough for anybody. **1938** X. Herbert *Capricornia* ix. 129 Take it and buy some tucker and kero. **1957** 'N. Culotta' *They're a Weird Mob* (1958) vi. 78 Pat threw in ..a 'kero tin' of water. **1961** P. White *Riders in Chariot* xi. 411 The rusted kero stove. **1968** *N.Z. Listener* 15 Mar. 6/2 Just picture us flicking on lights (no candles and kero to mess with). **1970** M. Kelly *Spinifex* xv. 214 The fire's all right. I put out a quart of kero and there's stacks of green stuff.

kerogen (ke·rǒdʒěn). *Petrogr.* [f. Gr. κηρό-ς wax + -gen.] Orig., the carbonaceous material in oil shale that gives rise to crude oil on distillation; in later use extended to denote any organic material in sedimentary rock which, like the oil-yielding kind, is insoluble in the usual organic solvents.

1906 D. R. Steuart in H. M. Cadell et al. *Oil-Shales of Lothians* iii. 142 We are indebted to Professor Crum Brown, F.R.S., for suggesting the term Kerogen to express the carbonaceous matter in shale that gives rise to crude oil in distillation. **1923** P. E. Spielmann *Genesis of Petroleum* ii. 17 In some cases it may have been that kerogen consisting of unicellular algae and of spores like those of the lycopods, had actually been distilled. **1941** *Jrnl. Inst. Petroleum* XXVII. 426 The term 'kerogen'.. is merely a convenient name for the organic matter from which oil is obtained when rocks containing it are heated; hence the kerogens of different deposits may be chemically different. *Ibid.* 427 The insolubility of kerogen in organic solvents is a characteristic property which serves to distinguish true kerogen rocks from such materials as oil- or tar-sands, bituminous limestones, etc. **1958** *Habitat of Oil* (Amer. Assoc. Petroleum Geol.) 758 Kerogens isolated from the marine rocks were obtained as very fine, amorphous, soft powders that varied from dark brown to jet black... When the samples were heated in an open test tube only small amounts of an oily distillate were formed. **1961** *Fuel* XL. 387 Colorado oil-shale kerogen is predominantly a cyclic material, highly saturated with hydrogen, and contains oxygen, nitrogen and sulphur atoms associated with ring structures. **1963** J. P. Forsman in I. A. Breger *Org. Geochem.* v. 148 The term 'kerogen'..may be defined as the insoluble organic matter occurring in sedimentary rocks. *Ibid.*, Since the dividing line between oil shales and other rocks is rather arbitrary, there appears to be no logical reason for restricting the name 'kerogen' to a certain rock type. **1968** *Times* 16 Nov. 9/6 The most ancient kerogen yet known comes from the Fig Tree System of rocks in Swaziland, thought to be 3,100 million years old.

kerosene, *sb.* Add: Also **kerosine** (see note below). Now important as a fuel for some kinds of internal-combustion engines, esp. jet engines.

Kerosene, -ine is now the usual name for paraffin in much of the U.S. (see quot. 1961¹) and in Australia and New Zealand; in Britain its currency is largely restricted to technical contexts. The spelling *kerosine* was adopted in 1925 by the Amer. Soc. for Testing Materials and (in Britain) by the Institute of Petroleum; the *-ene* form remains the usual one in general usage and still occurs in technical contexts.

1864 Kerosine [in Dict.]. **1925** *Proc. Amer. Soc. Testing Materials* XXV. 287 The sub-committee [on nomenclature] wishes to call particular attention to the spelling of the word 'kerosine'. This matter was brought to our attention by the Executive Committee of the Standardization Committee of the Institution of Petroleum Technologists, it being pointed out that the ending 'ene' is very generally applied to pure hydrocarbons. The suffix 'ine' already applied to gasoline is therefore also applied to kerosine. **1933** *Industr. Chemist* IX. 227/2 High sulphur, low quality, off-colour kerosene distillates..may be hydrogenated. **1949** *Thorpe's Dict. Appl. Chem.* (ed. 4) IX. 389/2 The word 'kerosene'..is an alternative name with paraffin oil (British) and coal oil (American). **1950** *Inst. Petroleum Rev.* IV. 9/1 The American Society for Testing Materials and the Institute of Petroleum..desire to reiterate their strong recommendation for international recognition of the spelling *kerosine*. **1951** C. R. Noller *Textbk. Org. Chem.* iv. 53 Previous to 1910, kerosene was the most important product derived from petroleum. *Ibid.*, The demand for the kerosene fraction is increasing again, since it is being used as the fuel for gas turbines and jet engines. **1954** *Chem. & Engin. News* 5 Apr. 1386/3 'Kerosene' is commoner than 'kerosine'. The ASTM [*sc.* American Society for Testing Materials] and ASA [*sc.* American Standards Association] have preferred 'kerosine', probably in order to make it consistent with 'gasoline', and *CA* [sc. *Chemical Abstracts*] has adopted 'kerosine' as the choice of an authoritative group in the petroleum field. **1957** Frazer & Eshelman *Tractors & Crawlers* i. 14 Kerosene and distillate..will burn satisfactorily in engines which are designed for low-grade fuels. **1960** *B.S.I. News* Dec. 25 (*heading*) Kerosine (paraffin) unflued space heaters. **1961** *Amer. Speech* XXXVI. 27 Fuel for lamps: Kurath found *kerosene* in general use through the North. The Atlas field interviewer, on the other hand, encountered *coal oil* in all three Cleveland interviews. Both terms are in widespread use today. 1. The commercial *kerosene*..increases in frequency with the youth and the cultivation of the informant. 2. *Coal oil*.. appears to find its chief strength among the old and the uncultivated. **1961** D. Petrie *Petroleum* xi. 62 The heavier products condense on the lower trays, and the lighter ones like kerosine and petrol near the top. **1966** C. Orr *Particulate Technol.* ix. 439 Organic liquids like benzene, toluene, kerosine. **1966** G. W. Turner *Eng. Lang. Austral. & N.Z.* i. 22 Trade has given currency to such words as *kerosene*: an English lady surprised a New Zealand chemist by asking for four gallons of 'paraffin' and he surprised her by supplying four gallons of 'liquid paraffin'. **1967** W. A. Gruse *Motor Fuels* i. 6 The next higher-boiling fraction, kerosene, will cover the range about 180 to 290°C. **1970** *Sci. Jrnl.* Mar. 44/1 Aviation kerosine has a strong and unpleasant odour.

b. *kerosene box, bucket, case, engine, flare, lamp* (earlier and later examples), *stove, tax, tin*; *kerosene-lit, -soaked* adjs.

1948 V. Palmer *Golconda* x. 76 There was a bookcase made of kerosene-boxes nailed together. **1929** K. S. Prichard *Coonardoo* xxiv. 235 Coonardoo had kerosene buckets of water boiling. **1905** W. B. *Where White Man Treads* 304 He invited me to his whare, and seated me in the seat of honour—the slab bunk—while he made shift with the ubiquitous kerosene case store-all. **1909** *Westm. Gaz.* 3 Feb. 2/2 A gasolene- or kerosene-engine. **1920** D. H. Lawrence *Women in Love* ix. 125 The market-place was hot with kerosene flares. **1869** *Bradshaw's Railway Manual* XXI. p. xii (Advt.), Dietz & Co., petroleum and kerozene lamp manufacturers. **1945** *Coast to Coast 1944* 75 It was a dingy hole all right, with a feeble kerosene-lamp trying to soak up some of the shadows. **1974** 'I. Drummond' *Power of Bug* viii. 113 The smell of gasoline could not have come from..kerosene lamps. **1932** W. Faulkner *Light in August* v. 107 He saw himself enclosed by cabinshapes, vague, kerosenelit. **1908** *Westm. Gaz.* 3 June 12/1 No sooner had the speeches concluded than..a light was set to the stack of kerosene-soaked pipes. **1946** *Coast to Coast 1945* 175 In the shed that had been fitted up with taps and shelves, a sink and an old smoky kerosene stove, Bennie was always happy. **1928** *Manch. Guardian Weekly* 10 Aug. 113/4 Mr. Churchill was deeply chagrined by being compelled to withdraw his proposed kerosene tax. **1891** C. Roberts *Adrift Amer.* xii. 211 There was no difficulty in cooking it as an old kerosine tin furnished a pot, and..I soon had it boiling away. **1908** E. J. Banfield *Confessions of Beachcomber* i. 48 All convenient vessels available, even to the never-to-be despised kerosene tins, were utilised to store the nectar. **1937** *Discovery* June 169/2 Dozens of boys and girls..came along with calabashes, bags and kerosene tins full of locusts. **1969** *Coast to Coast 1967–68* 18 Kate had been boiling clothes in kerosene tins on the kitchen stove and carrying them outside to be rinsed in big tubs on the back veranda.

ker-plonk, -plunk: see Ker- (in Dict. and Suppl.).

Kerr (kēᵊɹ, kāɹ, kǭɹ). The name of John *Kerr* (1824–1907), Scottish physicist, used *attrib.* to designate certain devices, phenomena, and concepts discovered by him or arising out of his work, as **Kerr cell**, a transparent cell containing two plate electrodes in a substance exhibiting a strong Kerr (electro-optical) effect, by means of which an applied voltage may be made to vary the plane of polarization of plane-polarized light and hence (when the cell is placed between crossed nicols) the intensity of a beam; **Kerr constant**, a number expressing the degree to which a substance exhibits the electro-optical Kerr effect, equal to the difference between the extraordinary and ordinary indices of refraction of the substance divided by the wavelength of the monochromatic light used and by the square of the electric field strength; **Kerr effect,** (*a*) the rotation of the plane of

polarization of light when it is reflected from a magnetized surface; (*b*) the production of birefringence in a substance by the application of an electric field.

[**1893** J. J. Thomson *Notes Rec. Res. Electr. & Magn.* v. 501 There is..no reason to expect that the order of the metals with respect to Kerr's effect should be the same as that with respect to Hall's.] **1909** *Chem. Abstr.* III. 2084 The theory of the Kerr effect shows that..the effect of the mirror of ferromagnetic metal placed in the magnetic field is the same as would be the face of a naturally active crystal without the magnetic field. **1910** *Hawkins' Electr. Dict.* 232/2 *Kerr effect*, the effect produced in dielectrics when subjected to electro-static stresses, so that they become double refracting. **1927** *Proc. Nat. Acad. Sci.* XIII. 506 These experiments involved measuring the differences in the lag of the Kerr effect behind the electric field for the liquids in two Kerr cells. **1927** *Phil. Mag.* III. 715 All the substances which possess a large Kerr constant have molecules which are electrically polar. **1937** G. S. Monk *Light* xvi. 318 Gases exhibit a Kerr electric effect which is about 1/1000 that for ordinary liquids. **1953** J. R. Partington *Adv. Treat. Physical Chem.* IV. x. 279 Kerr cells containing this liquid [*sc.* nitrobenzene]..are used in television and cinema apparatus as instantaneous switches or relays. **1953** J. Morgan *Introd. Geom. & Physical Optics* xvi. 361 The Kerr effect depends on the difference in the dipole moments produced by the electric field along different axes of the molecule, and such measurements give information about the molecular structure of the liquid. **1969** McGuire & Flanders in Berkowitz & Kneller *Magnetism & Metall.* I. iv. 145 The Kerr magnetooptical effect refers to polarized light reflected from a magnetized surface. The reflected light is elliptically polarized with the major axis of the ellipse rotated with respect to the original axis of the light. **1972** J. R. Meyer-Arendt *Introd. Classical & Mod. Optics* III. i. 297 Kerr cells are used most often as ultra-fast shutters.

kerria (ke·riă). [mod.L. (A. P. de Candolle 1817, in *Trans. Linn. Soc.* XII. 154), f. the name of William *Ker* or *Kerr* (d. 1814), English botanical collector.] A deciduous shrub of the monotypic genus so called, native to China and Japan, belonging to the family Rosaceæ, and bearing single or double yellow flowers; = Corchorus 2; *white kerria*, a closely related shrub, *Rhodotypos kerrioides*, which bears white flowers.

1823 J. Lindley *Donn's Hortus Cantabrigiensis* (ed. 10) 200 *Kerria japonica*, Kerria. Japan... 1804. **1829** J. C. Loudon *Encycl. Plants* 455 *Kerria*. So named after Mr. William Ker, a botanical collector, who was sent some years since to China, whence he sent many curious plants. The plant named after him is the common Corchorus japonica of the gardens. **1836** *Edward's Bot. Reg.* XXII. 1873 (*heading*) Japan Kerria. **1896** J. H. Veitch *Traveller's Notes* vii. 112 Some lovely flowering bushes of double Kerria are..to be seen. **1900** M. Thorn in W. D. Drury *Bk. Gardening* xi. 407 *Rhodotypos kerrioides* (White Kerria) is a charming and easily-grown shrub. *Ibid.* 465 The double-flowered Kerria (*K. japonica florepleno*) is a favourite wall-shrub. **1939** W. J. Bean *Wall Shrubs & Hardy Climbers* II. 100 The typical, single-flowered kerria is a spreading, twiggy shrub. **1951** *Dict. Gardening* (R. Hort. Soc.) 1103/1 Kerria, White. See *Rhodotypos kerrioides*. **1970** C. Lloyd *Well-Tempered Garden* iv. 356 Japanese quinces, weigelas, kerrias, mahonias..are all stripped by birds at various times.

kerrie. (Examples of *kierie*, now the more usual form.)

1939 A. W. Wells *S. Afr.: Planned Tour* 410/1 Some Afrikaans words are used in English conversation, e.g.: ..*kierie*, walking-stick. **1953** *Cape Times* 5 Jan. 1/7 There were about 1,000 fighting with kieries. **1959** *Ibid.* 10 Nov. 1/1 Two other Europeans had been discharged after being treated for kierie wounds. **1970** *Cape Herald* 15 May 3/3 He found..a kierie, a two-piece pink costume, a night gown... The costume and gown had blood spots on them.

Kerry. Add: **2.** (Earlier and later examples.) Also, *Kerry cow*.

1829 G. Griffin *Collegians* (ed. 2) I. xi. 233 Her little cow was a kind Kerry, and had the best of grazing. **1907** *Macm. Mag.* May 514 We saw a cattle-boat bringing Kerries to Scariff. **1929** E. Bowen *Last September* III. xxiii. 294 Look at those little teeny black cows. Those are Kerry cows. **1975** *Irish Times* 9 May 4/6 Kerries, Ayrshires, Herefords, Charolais, and the Jerseys, the nicest of all.

3. *Comb.* **Kerry blue terrier**, a breed of Irish terrier, of medium size, with a long, silky, blue coat, found in Kerry and neighbouring counties and introduced elsewhere during the nineteen-twenties; also *Kerry, Kerry blue*.

1922 D. Matheson *Terriers* 145 The Kerry (Blue) Terrier. Since the close of the Great War we have had this variety of Irish Terrier introduced to us under its present title. **1930** E. C. Ash *Pract. Dog Bk.* 150 The history of the Kerry Blue is not easy to obtain. **1955** N. Fitzgerald *House is Falling* x. 169 She had firmly attached herself to the big Kerry Blue. **1971** F. Hamilton *World Encycl. Dogs* 455 Kerry pups are born black. The blue color comes later, showing at any time after six or seven months. **1973** *Country Life* 15 Feb. 384/3 The terriers replaced the gundogs..the eventual winner the Kerry Blue terrier.

Kerry² (ke·ri). The name of a town and neighbouring range of hills in the county of Powys, on the Welsh borders, used *attrib.* in **Kerry Hill sheep** to designate a breed of sheep developed there, distinguished by a thick fleece and black markings near the muzzle and feet. Also *absol.*

1908 *Daily Chron.* 1 May 4/6 Many people are under the impression that the famous Kerry Hill sheep come from Ireland. As a matter of fact, they come from Kerry in Montgomeryshire. **1937** F. B. Young *Portrait of Village* v. 102 By the time he had finished stocking it with Wessex Saddleback pigs, Kerry Hill sheep, short-horn cows, runner ducks and Rhode Island Red poultry,..he had made another big hole in his dwindling capital. **1945** J. F. H. Thomas et al. *Sheep* ii. 33 The Kerry Hill (Wales) ..is a polled breed of medium size which carries a dense fleece. **1960** [see *Devon]. **1971** *Farmers Weekly* 19 Mar. 83/1 For many years we kept Kerrys at Downton, now we have commercial Cluns—with some Welsh blood.

kerseymere. 2. (Earlier example.)

1808 Jane Austen *Let.* 30 June (1952) 204 My kerseymere Spencer is quite the comfort of our Even[g] walks.

ker-slap, splash, -splosh: see Ker- (in Dict. and Suppl.).

kerstenite (kǝ̄·ɹstěnait). *Min.* [f. the name of K. M. *Kersten* (1803–1850), German chemist, who first reported it + -ite¹.] A yellow selenite or selenate of lead, of uncertain composition, with a greasy to vitreous lustre.

1868 J. D. Dana *Syst. Min.* (ed. 5) 669 Kerstenite *Dana*. **1930** J. W. Mellor *Comprehensive Treat. Inorg. & Theoret. Chem.* X. lviii. 833 C. M. Kersten reported normal lead selenite... J. D. Dana called the mineral kerstenite—*cf.* cobalt arsenide. The data do not distinguish whether the mineral is a selenite or selenate. **1951** C. Palache et al. *Dana's Syst. Min.* (ed. 7) II. 640 Kerstenite was found as botryoidal masses with a fibrous fracture as an alteration of zorgite at the Friedrichsglück mine near Hildburghausen, Thuringia, Germany. **1962** W. A. Deer et al. *Rock-Forming Min.* V. 188 Several other minerals and compounds have the 'barytes structure'; among them are: $BaSeO_4$, $PbSeO_4$ (kerstenite)..and $(K, Cs)BF_4$.

kertch, var. Curch.

1900 A. Carmichael *Carmina Gadelica* I. p. xxv, On the morning after the marriage the mother of the bride.. placed the 'breid tri chearnach', three-cornered kertch, on the head of the bride before she rose from her bed. *Ibid.*, The feast of the 'bord breid', kertch table, was almost as great as the feast of the marriage table.

Kertch, var. *Kerch.

keruing (ke·ru,iŋ). Also **kruin.** [Malay *kĕruing*.] The light or dark brown hardwood timber of several trees of the genus *Dipterocarpus*, found in Malaysia, Sabah, and Indonesia.

1921 F. W. Foxworthy in *Malayan Sci. Bull.* I. 36 Meranti and Keruing, because of their contained resinous substances, are better firewoods than some of the heavier woods. **1934** A. L. Howard *Man. Timbers of World* (rev. ed.) 152 In the Malay Peninsula there are many species of Dipterocarpaceæ which produce timber known as keruing. **1948** *Ibid.* (ed. 3) 282 Keruing. See Kruin. *Ibid.* 291 Kruin.—This is produced by 15 to 20 species of the genus Dipterocarpus. **1950** C. W. Bond *Colonial Timbers* 54 Keruing is not unpleasant to handle at the bench. **1966** P. F. Burgess *Timbers of Sabah* 102 Keruing is remarkable among tropical timbers for its generally straight grain. **1970** *Timber Trades Jrnl.* 21 Mar. 54/1 There have been few 'spot' offers of keruing circulating, and this timber has been hard to buy in the popular specifications.

kerumph, ker-woosh: see *ker-.

kerygma. Add: Pl. **kerygmata.** (Further examples.)

1936, etc. [see *Didache 2]. **1949** *Scottish Jrnl. Theol.* II. 316 And the miracle of the *kerygma* is just this that through it the once and for all event in Jesus Christ becomes event all over again in the faith of the hearer. **1953** *Ibid.* VI. 314 The essential difference from Christianity, however, lies in the place occupied by such speculations about the kosmos, its beginnings and end, its hierarchies and denizens, in the respective *kerygmata*. **1955** *Ibid.* VIII. 158 Usage in direct speech may also be due to the editor, but the ruling presumption is against that, especially in the *kerygmata*. **1958** *Times Lit. Suppl.* 23 May p. vii/1 And what he writes, without a word of justification, about the unity of the mind of the apostles with that of Christ, and of the apostolic *kerygma* with the activity of Christ, could be used by the Pope himself and welcomed by him as coming from a Protestant source. **1962** *Ibid.* 17 Aug. 628/2 It is also true that historical facts become objects of religious faith only when they belong to salvation-history and are part of the *kerygma*, the proclamation of the gospel, which presents them as sources and types of the redemption offered to mankind. **1970** West & Francis *Scandal in Assembly* xxviii. 151 Without the *Kerygma*, the *didache* reduces itself to a system of ethics as unstable as the customs of men.

Hence **kerygma·tic** *a.*, belonging to or of the nature of preaching; **kerygma·tically** *adv.*

1929 *Church Times* 18 Jan. 82/1 Our ministry..before all else..is 'kerygmatic', the proclamation of a Kingdom and a King. **1949** *Scottish Jrnl. Theol.* II. 88 They hold that the Scriptures of the New Testament witness to Jesus Christ Himself *kerygmatically*, that is in such a way that He Himself, by the power of His Spirit, speaks to men through them. **1955** *Ibid.* VIII. 346 The present Hexateuch represents the elaboration and the heaping up of traditions around the simple kerygmatic theme, the chief premonarchial elements of which can be discerned in old confessions. **1969** E. Simons in *Sacramentum Mundi: Encycl. Theol.* III. 246/1 This dialectic of mediation..which may be termed dialogal, is a kerygmatic event. **1970** J. Donceel tr. *Rahner's Trinity* II. 57 Some new word..may be more precisely and easily understood, hence kerygmatically more useful than the word 'person'.

‖ **kesa-gatame** (ke·sa,gatā·me). *Judo.* [Jap.] The 'scarf' hold, a way of holding the opponent by the edge (the so-called 'scarf') of his jacket in an attempt to immobilize him.

1932 E. J. Harrison *Art of Ju-Jitsu* v. 68 (*heading*) 'Locking in the form of a scarf' (kesagatame; otherwise hongesa, or 'regular scarf'). **1957** Takagaki & Sharp *Techniques Judo* III. viii. 112 *Kesa gatame* is one of the simplest holds to secure. **1968** P. & K. Butler *Judo & Self-Defence for Women & Girls* 141 (*heading*) Kesa-gatame—scarf hold. T applies kesa-gatame, U struggles then submits.

Keshan, var. *Kashan.

keskeedie, keskidee, varr. *kiskadee.

kest (kest), *v. dial.* var. of Cast *v.* esp. in senses 'cast aside, throw away' and (*fig.*) 'do down, outdo'.

1590 Spenser *Faerie Queene* I. xi. 163 That infernall Monster, hauing kest His wearie foe into that liuing well. **1862** C. C. Robinson *Dial. Leeds* 334 A man who has been abroad in bad weather, ill-clad, says,—'Ah sal nivver kest what I've gotten to neet I knawah.' **1892** J. Wright *Gram. Dial. of Windhill, W. Riding* iii. 29 In the following words we have *e* which in many of the examples is no doubt the *i*-umlaut of *a*:..*kest* (ME. *kesten*) to cast. **1893–4** R. O. Heslop *Northumb. Words* II. i. 421 He kest his claes ower syun an' gat caad. **1913** D. H. Lawrence *Sons & Lovers* II. vii. 164 At seven o'clock the family heard him buy threepennyworth of hot-cross buns... He turned away several boys who came with more buns, telling them they had been 'kested' by a little lass. *Ibid.* viii. 200 Just then Wesson entered...'I see you've kested me,' he said, smiling rather vapidly. 'Yes,' replied Barker.

kësterite (ke·stěrait). *Min.* Also **kesterite.** [ad. Russ. *kësterit* (Z. V. Orlova, 1956), f. *Këster*, name of its locality in Yakutia, Siberia: see -ite¹.] A black sulphide of copper, tin, zinc, and iron, $Cu_2(Zn,Fe)SnS_4$.

1958 *Mineral. Mag.* XXXI. 963 *Kësterite*...(Cu,Sn,Zn) S, containing Cu 30·56, Sn 25·25, Zn 11·16, S 23·40%. **1968** I. Kostov *Mineral.* ii. 147 Sakuraiite..is considered an indium analogue of kesterite.

Keswick (ke·zik). The name of a town in Cumbria, used to designate a variety of cooking apple, in full **Keswick codlin(g)**, which has a greenish skin tinged with red and was first introduced by John Sander, who lived in the town.

1814 *Mem. Caledonian Hort. Soc.* I. 374 (*title*) Information regarding the Carlisle and Keswick Codlin Apples. *Ibid.* 376 The Keswick Codlin tree has never failed to bear a crop since it was planted..twenty years ago. **1831** H. Ronalds *Pyrus Malus Brentfordiensis* 5 Keswick Codlin. A favourite sauce apple from the North of England. **1912** Mulford & Clay *Buck Peters* xxii. 198 It's Buck as sure as little apples Kesicks. **1936** H. V. Taylor *Apples Eng.* vii. 74 Those [apples] that became soft and frothy, even on coddling (i.e. parboiling), were known as coddlings—the Keswick Codling, English Codling. **1973** *Countryman* LXXVIII. iv. 42, I could watch the sheets billowing, and the clothes flapping and being confettied by blossom from the crabs, keswicks and white heart cherries.

ket-: see *keto-.

keta (kī·tă). Also † **keth.** [Russ.] = *chum *sb.*⁴

1824 S. Black *Jrnl. Voy. from Rocky Mountain Portage* (1955) 52 In Bears River near the Lake they speared a kind of salmon he names Keth in the Rapids. **1901** A. M. B. Meakin *Ribbon of Iron* xviii. 263 At Nikolaevsk upwards of a hundred and ninety-three thousand pnds of a salmon called 'keta'—*salmo lagocephalus*—are salted annually. **1905** D. S. Jordan *Guide to Study of Fishes* II. iv. 73 In Japan *keta* is by far the most abundant species of salmon. **1933** A. W. Shiels *Seward's Ice Box* 83 Chums or keta are plentiful in all districts. **1962** *Co-op Grocery News Bull.* (Saskatoon) 1 Aug. 1 Formerly known as dog salmon, the chum, with the scientific name Oncorhynchus Keta, has also been called the qualla, keta and calico salmon. It is caught all along the coast of British Columbia.

ketal, ketamine, etc.: see *keto- a.

ketch (ketʃ), *v.*² *dial.* var. (pa. t. **ketched**) of Catch *v.*

1815 D. Humphreys *Yankey in England* I. 21, I guess, he is trying to ketch me—but it won't du. I'm tu old a bird to be ketch'd with chaff. **1865** Dickens *Mut. Fr.* II. iv. xv. 287 Wot is it, lambs, as they ketches in seas, rivers, lakes, and ponds? *a* **1883** [see *knuck 2]. **1911** E. Wharton *Ethan Frome* ii. 60 You'll ketch your death. The fire's out long ago. **1916** W. O. Bradley *Stories & Speeches* 18 You'll never ketch me hollerin' at no Republican gatherin'. **1929** H. W. Odum in A. Dundes *Mother Wit* (1973) 184 If so you gonna ketch hell. **1967** *Atlantic Monthly* Apr. 103/1 You heard about that joke a dollar down and a dollar when you ketch me? **1968** S. Stuckey in A. Chapman *New Black Voices* (1972) 445 Run, nigger, run, de patrollers will ketch you.

ketchak, var. *ketjak.

kete, kête, varr. *kit *sb.*[11]

ketene (kī·tīn). *Chem.* Also -en. [ad. G. *keten* (H. Staudinger 1905, in *Ber. d. Deut. Chem. Ges.* XXXVIII. 1735): see Ketone and -ene.] **a.** Any compound containing an ethylenic double bond adjacent to a carbonyl group, i.e. the structure >C=C=O.

[**1905** *Jrnl. Chem. Soc.* LXXXVIII. 444 Diphenylketen $CPh_2{:}CO$, the first representative of a new type of compound, is prepared by the action of zinc on an ethereal solution of diphenylchloroacetyl chloride.] **1907** *Ibid.* XCI. 1941 There can..be no doubt that the new substance is a keten, having the formula $CH_2{:}CO$. **1942** Fuson & Snyder *Org. Chem.* xxi. 296 Carbonyl compounds which contain the grouping >C=C=O are known as ketenes... Ketenes fall into two classes, aldoketenes and ketoketenes. **1971** *Jrnl. Chem. Soc.* (C) 3645 The addition of ketenes to conjugated dienes generally yields 1,2-adducts.

b. The simplest of the ketenes, $CH_2{:}C{:}O$, a pungent colourless gas which dimerizes on liquefaction and has wide applications in synthesis owing to its high reactivity.

1907 *Jrnl. Chem. Soc.* XCI. 1941 Keten has a peculiar penetrating smell. **1946** [see *acetylation]. **1964** N. G. Clark *Mod. Org. Chem.* xi. 211 On heating acetic acid vapour at 700° to 800° under reduced pressure, it loses the elements of water to form keten. **1972** *Guardian* 16 Oct. 15/2 Although keten and its organic relations have been known and used for generations, no one had realised that they could produce salts.

kethubah, var. *ketubah.

‖ **Kethubim** (keþūvī·m, ket-), *sb. pl.* Also 7 Chetoubim, 20 Ketubim, K'tubhim, -vim. [Heb. *k'thūbīm* writings.] = Hagiographa *sb. pl.*

1690 tr. *J. Le Clerc's Five Lett.* ii. 102 Others..they call Chetoubim, or simply Writings; that is to say, the Psalms [etc.]. **1892** H. E. Ryle *Canon of Old Testament* vi. 127 The *Psalter* is the most important book of the 'Kethubim'. *Ibid.* 132 The Book of Job..had no fitting place in the Canon save in the mixed group of 'the Kethubim'. **1904** *Jewish Encycl.* VI. 151/1 It can not now be ascertained when the name 'Ketubim' and the Greek designation 'Hagiographa' were first given to the collection. **1961** *Jrnl. Biblical Lit.* LXXX. 106 Tanak..stems from the abbreviations of the three divisions, Torah, Nebhi'im, and K'tubhim, of the Hebrew Scriptures. **1967** D. T. Kauffman *Dict. Relig. Terms* 274 *Ketuvim* or *Kethubim*, third section of the Jewish Scriptures: the Sacred Writings beginning with the Psalms and ending with the Chronicles.

‖ **ketjak** (ke·tʃæk). Also **ketchak, 'tjak.** [Balinese, f. *'tjak-a-tjak'*, the sound of the chanted refrain accompanying the dance.] A Balinese dance, with a male chorus. Also *attrib.*

1937 M. Covarrubias *Island of Bali* (1972) viii. 219 *Ketjak*, large groups of men singing in chorus, moving and dancing to the rhythm of the music. **1938** De Zoete & Spies *Dance & Drama in Bali* ii. 85 The *ketjak* dance is primarily a dance of exorcism. **1942** C. Barrett *On Wallaby* ix. 176 The Monkey Dance or Ketchak Ceremonies impressed me more than all other dances and dramas... A Ketchak was held solely for our entertainment. **1954** J. Coast *Dancing out of Bali* viii. 161, I had suggested the gamelan club forming a 'Tjak, or Monkey Dance chorus. **1972** C. Simpson *Bali & Beyond* 26 *Ketjak* is the name of the male chorus in the *sanghyang* and this dance extends the role of that chorus to a *Ramayana*-story presentation in which the number of men is increased to about 150. **1972** *Times* 11 Nov. 13/4 The dancers you see performing the *Barong* or the *Ketchak*..are not professionals.

keto- (kī·to), comb. form of Ketone. **a.** (Before a vowel also **ket-**.) As an inseparable formative element of terms in *Chem.* and *Med.*: **ketal** (kī·tæl) [after Acetal], any compound of the type $R^1R^2C(OR^3)OR^4$, where neither R^1 nor R^2 is a hydrogen atom (see quot. 1926); **ke·tamine**, a crystalline anaesthetic and analgesic substance, $C_{13}H_{16}NOCl$; 2-*o*-chlorophenyl-2-methylaminocyclohexanone; **ke·tazine** [ad. G. *ketazin* (Curtius & Thun 1891, in *Jrnl. f. prakt. Chem.* XLIV. 162)], any compound of the type RR'C:N·N:CR″R‴, made by reacting one molecule of hydrazine with two molecules of (identical or different) ketones; **ke·timine**, any compound containing the grouping >C=NH, formed e.g. by the action of ammonia on a ketone; **ke:toacido·sis**, acidosis due to enhanced production of ketone bodies; hence **ke:toacido·tic** *a.*, of or pertaining to ketoacidosis; **ke:tobe·midone** [perh. f. carbethoxy- + M(ethyl + -id⁴ + -one], an analgesic, $(C_6H_4OH)(CO\cdot C_2H_5)C_5H_8N\cdot CH_3$, with action similar to that of morphine; **ketoge·nesis**, production of ketone bodies; **ketoge·nic**, † **-gene·tic** *adjs.*, producing ketone bodies; applied *spec.* to a diet that is rich in fats and low in carbohydrates and has been used therapeutically to produce ketosis; (*17-*)*ketogenic steroid*, any steroid which yields a (17-)ketosteroid on oxidation with a bismuthate; **α-ketogluta·ric acid**, a dibasic keto-acid, $HOOC\cdot CO\cdot CH_2\cdot CH_2\cdot COOH$, which is formed by oxidation and decarboxylation of isocitric acid in the Krebs cycle; hence **α-ketoglu·tarate**, a salt or ester, or the anion, of α-ketoglutaric acid; **keto·he·xose**, any ketose with six carbon atoms; **α-ke:toi:sovale·ric acid**, a crystalline carboxylic acid, $(CH_3)_2CH\cdot CO\cdot COOH$, from which valine is synthesized by transamination in fungi; hence **α-ke:toi:sova·lerate**, a salt or ester, or the anion, of α-ketoisovaleric acid; **ketoke·ten(e)** [a. G. *ketoketen* (Staudinger & Klever 1908, in *Ber. d. Deut. Chem. Ges.* XLI. 909)], any ketene of the type RR'C:C:O; **keto·lysis**, decomposition of ketone bodies; hence **ketoly·tic** *a.*, causing or pertaining to ketolysis; **ketoly·tically** *adv.*; **ketoste·roid**, any steroid whose molecules contain a ketone group; *17-ketosteroid* (see quot. 1964); **keto·xime**, an oxime of a ketone (i.e. any compound containing the group >C:NOH), formed by the action of hydroxylamine on a ketone; **ke·tyl** [-yl], any salt containing a free-radical anion of the type RR'·Ċ·O⁻, formed by dissolving a metal in a ketone.

1924 *Chem. Abstr.* XVIII. 3518 (*heading*) Velocity of hydrolysis of acetals and ketals. **1926** *Ibid.* XX. 2937 The saponification of ketals (acetals of ketones) was studied in the presence of mineral acid catalyzers. They are liquids of disagreeable odor, insol. in water, sol. in alc., ether, etc., stable towards dil. alkali, rapidly hydrolyzed by dil. acid. **1933** *Jrnl. Amer. Chem. Soc.* LV. 3744 The method described in this paper..furnishes a simple approach to cyclic ketals which are difficult or impossible to make by the usual method. **1965** *Nomencl. Org. Chem.* (I.U.P.A.C.) C. 104 Compounds containing the group >C<(OR¹)(OR²) are termed acetals. [*Note*] The name 'ketal' is abandoned. **1967** I. L. Finar *Org. Chem.* (ed. 5) I. viii. 189 Ketones do not readily form ketals when treated with alcohols in the presence of hydrogen chloride... Ketals may, however, be prepared by treating the ketone with ethyl orthoformate. **1966** *Approved Names* (Brit. Pharmacopœia Comm.) 23 Ketamine. **1967** *Martindale's Extra Pharmacopoeia* (ed. 25) 1530/2 Ketamine hydrochloride is a potent analgesic and anaesthetic with actions similar to those of phencyclidine hydrochloride.., to which it is chemically related. **1968** *Anesthesia & Analgesia Current Res.* XLVII. 775/2 Ketamine has both a stimulatory and a depressive effect on the cardiovascular system. The stimulation predominates with small doses. **1972** *Anesthesiology* XXXVI. 311 With the increased medical and veterinary use of ketamine, it will probably become a popular hallucinogenic street drug. **1894** *Jrnl. Chem. Soc.* LXVI. 348 (*heading*) Transformation of ketazines into pyrazolines. **1911** *Ibid.* C. i. 571 The ketazines from the following ketones have been prepared and examined: benzophenone, fluorenone, and tetramethyldiaminobenzophenone. All the compounds are stable..and when hydrolysed by hot mineral acids yield hydrazine and ketone. **1943** H. Gilman *Org. Chem.* (ed. 2) I. ix. 812 Hydrazones and ketazines have also been hydrogenated by means of platinum..catalysts. **1909** J. F. Thorpe in *Proc. Chem. Soc.* XXV. 309 It is suggested that as the compounds containing the group C:NH are in many respects analogous to the ketones, the general name ketimine should be applied to them, leaving the name imine to be applied to the secondary amines. **1938** Allen & Blatt in H. Gilman *Org. Chem.* I. vi. 568 Monomolecular ketimines..have been prepared; they too are readily hydrolyzed and reduced. **1971** N. L. Allinger et al. *Org. Chem.* xxi. 586 Grignard reagents add to the cyano group of most nitriles to give salts of ketimines. **1958** *Diabetes* VII. 230 (*heading*) The nature and correction of diabetic ketoacidosis. **1961** *Endocrinology* LXVIII. 815 Rather marked insulin insufficiency is necessary for the development of uncontrolled diabetic ketoacidosis. **1969** *Nature* 20 Dec. 1155/1 All the children suffered from vomiting, lethargy and ketoacidosis, and did not grow. **1966** Dunlop & Alstead *Textbk. Med. Treatm.* (ed. 10) 353 The patient in ketoacidotic coma must be sent without delay to hospital. **1972** *Jrnl. Clin. Invest.* LI. 493/2 Intensive therapy for 72–96 hr with parenteral glucose and alkali were necessary before he recovered from his ketoacidotic episodes. **1949** *Jrnl. Pharmacol. & Exper. Therap.* XCVII. 188 Keto-bemidone appears to be one of the most addictive drugs yet discovered. **1958** *A.M.A. Arch. Internal Med.* CI. 745 Comparison of ketobemidone with other common analgesics shows that it is one of the safest drugs to use for analgesia because of the great difference between its analgesic and euphoria-producing doses. **1972** J. Ball *Five Pieces Jade* vi. 73 Have you heard of keto-bedmidone [*sic*]? Or Claradon, that's another name for it... It is extremely addictive. Considerably more than heroin. **1915** *Arch. Internal Med.* XV. 65 Feeding of pancreas may augment ketogenesis by determining increased absorption of fat. **1933** Cameron & Gilmour *Biochem. of Med.* v. 77, 1 gram by weight of glucose can be oxidized in the body along with 4·8 grams of fat, without ketogenesis. **1972** *Diabetes* XXI. 50/2 Ketogenesis from fatty acid is thus tied into the liver cell's energy metabolism. **1915** *Arch. Internal Med.* XV. 63 It seems probable that it falls in the same category as that increased sensitiveness of the diabetic organism to ketogenetic factors after repeated pancreatic opotherapy. **1911** Stedman *Med. Dict.* 452/1 Ketogenic. **1921** *Jrnl. Biol. Chem.* XLIX. 162 A method is described by which the ratio of ketogenic to antiketogenic molecules in the metabolic mixture of a subject may be calculated from the respiratory quotient. **1930** *Sci. Amer.* Nov. 391 This is called a ketogenic diet, because it tends to produce an excessive amount of ketones and their derivatives in the blood. **1952** J. K. Norymberski in *Nature* 20 Dec. 1075/1 The difference between 17-ketosteroids found before and after oxidation of urine affords a measure of '17-ketogenic steroids' which represent an important group of corticosteroids. **1968** R. F. Steiner *Life Chem.* ix. 168 Four amino acids—leucine, isoleucine, phenylanine [*sic*] and tyrosine—can give rise to ketone bodies... The..group of amino acids are termed ketogenic. **1972** *Jrnl. Clin. Endocrinol. & Metabolism* XXXIV. 580/1 Routine urine steroid analysis showed normal excretion of 17-ketosteroids and 17-ketogenic steroids. **1973** *Nature* 2 Mar. 74/1 He was interested in the mechanism of action of ketogenic diets which were then used for the treatment of urinary tract infections. **1911** *Jrnl. Chem. Soc.* C. i. 520 Ethyl α-ketoglutarate, b.p. 114°/13 mm., obtained by esterification of the acid in a closed tube at 120°, is a colourless liquid. **1940** *Jrnl. Biol. Chem.* CXXXVI. 302 Carbon dioxide combines directly with pyruvic acid to yield oxaloacetic acid, the latter then combining with an additional molecule of pyruvate to form α-ketoglutarate. **1971** *Scand. Jrnl. Clin. & Lab. Invest.* XXVIII. 365/1 The known ammonia-detoxifying processes..involve amination of α-ketoglutarate to glutamate and amidation of glutamate to glutamine. **1908** *Jrnl. Chem. Soc.* XCIV. i. 713 When ethyl oxalosuccinate is treated with hydrogen chloride in the cold and the solution boiled, hydrolysis takes place and α-ketoglutaric acid..is formed. **1937** *Biochem. Jrnl.* XXXI. 300, α-Ketoglutaric acid thus appears to arise in the course of pyruvic acid oxidation. **1938** *Ibid.* XXXII. 112 Ten human urines contained between 10 and 40 mg. α-ketoglutaric acid per 24 hr. specimen. **1968** R. F. Steiner *Life Chem.* xii. 219 α-Ketoglutaric acid represents an important junction between the metabolic pathways of carbohydrates and amino acids. **1899** *Jrnl. Chem. Soc.* LXXV. 423 With ketohexoses (lævulose, sorbose), the purple colour appears after a few minutes. **1938** M. L. Wolfram in H. Gilman *Org. Chem.* I. xvi. 1442 The French scientist, Pelouze, described the isolation of a new ketohexose from the juice of the berries of the mountain ash. **1971** N. L. Allinger et al. *Org. Chem.* xxvii. 698 Fructose, another common hexose, has a ketone group at C-2 and is called a ketohexose. **1953** *Jrnl. Biol. Chem.* CCV. 480 Conversely, α-ketoisovalerate accumulation is depressed whenever the supply of valine is sufficient to permit maximal growth rate. **1966** *Biochemistry* V. 409/1 A soil microorganism, *Pseudomonas* P-2, growing on pantothenate as sole carbon source, converts this in part to β-alanine, α-ketoisovalerate, and valine. **1953** *Jrnl. Biol. Chem.* CCV. 475 (*heading*) Isoleucine and valine metabolism in Escherichia coli. V. α-Ketoisovaleric acid accumulation. **1971** *Jrnl. Nutrition* CI. 1165/1 The effect of substituting α-ketoisovaleric acid for L-valine upon the nitrogen balance of a young female was evaluated. **1908** *Jrnl. Chem. Soc.* XCIV. i. 318 The first groups are termed aldo-ketens and the second, keto-ketens. **1937** F. C. Whitmore *Org. Chem.* 281 The ketoketenes are colored, are easily auto-oxidized to give peroxides, and form addition cpds. with cyclic tertiary amines. **1951** I. L. Finar *Org. Chem.* I. xii. 230 If the compound is of the type R·CH:C:O, it is known as an aldoketen; and if $R_2C{:}C{:}O$, then a ketoketen. **1937** *Amer. Jrnl. Physiol.* CXIX. 734 Any reduction in the ketosis consequent to carbohydrate administration must be due to either a decrease in the rate of the former ('antiketogenesis') or to an increase in the rate of the latter ('ketolysis'). **1938** *Jrnl. Biol. Chem.* CXXVI. 106 Liver slices from a well fed monkey have a very low rate of ketolysis as compared with liver slices from well fed rats, rabbits, and guinea pigs. **1921** *Ibid.* XLVII. 435 Glucose thus exhibits, in alkaline solution *in vitro*, a 'ketolytic' action in hastening the oxidation of acetoacetic acid which would appear to be analogous to its 'antiketogenic' action in the body. **1936** *Jrnl. Nutrition* XII. 646 That aspect of the quantitative relationship of ketolytic to ketogenic factors which has interested us especially.., is the relation of carbohydrate combustion to the reduction of ketogenesis. **1938** *Jrnl. Biol. Chem.* CXXVI. 106 The large amount of work..showing the extrahepatic tissues..to be much more active ketolytically than the liver is very convincing. **1939** *Biochem. Jrnl.* XXXIII. 931 (*heading*) The isolation of 17-ketosteroids from the urine of normal women. **1959** *Austral. Jrnl. Exper. Biol. & Med. Sci.* XXXVII. 147 It was decided to employ enzymatic hydrolysis and separation of the individual ketosteroids in an attempt to identify and estimate them individually. **1964** A. White et al. *Princ. Biochem.* (ed. 3) xlviii. 856 Since each of the various urinary metabolites of testosterone has a ketone group at C-17, these substances are referred to as 17-ketosteroids, and their concentration in the urine is a useful index of endogenous production of androgenic hormones. **1888** *Jrnl. Chem. Soc.* LIV. 443 (*heading*) Conversion of ketoximes into pseudonitroles. **1938** C. D. Hurd in H. Gilman *Org. Chem.* vii. 636 The reaction of bromine with ketoximes yields bromonitrosoparaffins. **1971** N. L. Allinger et al. *Org. Chem.* xxii. 605 The oximes of ketones (ketoximes) undergo an overall rearrangement to amides when heated with certain inorganic reagents, followed by treatment with water. **1914** *Chem. Abstr.* VIII. 113 (*heading*) Metal ketyls, a large class of compounds with trivalent carbon. **1934** *Trans. Faraday Soc.* XXX. 23 By using dioxan as solvent high concentrations of the ketyls were obtained and their paramagnetism placed beyond doubt. **1971** N. L. Allinger et al. *Org. Chem.* xix. 493 If benzophenone is converted to the ketyl with sodium, for example, the ketyl concentration is sufficient to give the solution a beautiful blue color.

b. In Combs. in which *keto* may be used *attrib.* (without a hyphen) as an independent

word or joined by a hyphen to the second element, as *keto-acid, -compound, -ester, -form*, etc. (in which *keto-* denotes the presence of a ketone group); also **keto-e·nol** *attrib. phrase*, **-eno·lic** *a.*, applied to the tautomerism between the ketonic and enolic forms of certain compounds.

1911 *Chem. Abstr.* V. 3686 (*heading*) Preparation of amino acids from keto acid phenylhydrazones and aluminium amalgam. **1968** R. O. C. Norman *Princ. Org. Synthesis* vii. 237 A dibasic acid which, as a β-keto-acid, is readily decarboxylated by heat. **1891** *Proc. Chem. Soc.* VII. 91 The resinous matters often formed in large amount on nitrating many phenols are, doubtless, products of the interaction of several molecules of the addition compounds, or of the keto-compounds formed from them in the first instance. **1927** *Jrnl. Amer. Chem. Soc.* XLIX. 849 Acetylacetone may undergo keto-enol tautomerism in the following way.

$$CH_3COCH_2COCH_3 \rightleftharpoons CH_3COCH{:}C(OH)CH_3.$$

1971 N. L. Allinger et al. *Org. Chem.* viii. 172 Keto-enol interconversion is subject to catalysis by acid or base. **1909** *Proc. Chem. Soc.* XXV. 309 It is advisable to apply some general term to this form of isomerism, similar to the phrase keto-enolic isomerism in use with the oxygen derivatives. **1936** *Biochem. Jrnl.* XXX. 745 The present paper contains some observations..on certain substances ..which under similar conditions yield colours with this reagent [*sc.* diazotized sulphanilic acid]. The group has the common character that all its members are capable of keto-enolic tautomerism. **1937** F. C. Whitmore *Org. Chem.* 444 Diacetosuccinic ester, as a beta keto ester gives reactions like those of acetoacetic ester itself. **1958** *Oxf. Univ. Gaz.* 23 Apr. 882 Grignard reactions of keto-esters. **1927** *Jrnl. Amer. Chem. Soc.* XLIX. 856 Acetylacetone in 95% alcohol solution is an equilibrium mixture of the keto and enol forms. **1968** R. O. C. Norman *Princ. Org. Synthesis* i. 18 For ethyl acetoacetate..two factors increase the bonding..of the enol form relative to the keto form.

ketolic (kītǫ·lik), *a. Chem.* [f. Ketol + -ic.] Having the functional groups of a ketol.

1960 *Biochem. Jrnl.* LXXVII. 400/1 A method for the quantitative fractionation of..ring D α-ketolic oestrogens. **1971** *Folia Histochem. & Cytochem.* IX. 391 (*heading*) Oxidative enzymes and α-ketolic steroids in the interrenal cells of the South African clawed toad.

ketonæmia (kītǒnī·miă). *Med.* Also (chiefly *U.S.*) -nemia. [f. Keton(e + Gr. αἷμα blood: see -ia¹.] An abnormally high concentration of ketone bodies in the blood.

1917 *Amer. Jrnl. Med. Sci.* CLIII. 343 Marriott demonstrated ketonemia in such animals. **1941** *Jrnl. Biol. Chem.* CXXXVIII. 128 Ketonemia develops within 39 hours in man, as compared to 2 to 3 days in the dog. **1962** *Lancet* 29 Dec. 1350/2 The diabetic syndrome in this patient was unusual in that he had severe hyperglycæmia, insulin resistance, and no ketonæmia. **1965** Lee & Knowles *Animal Hormones* vii. 112 The effect of removal or destruction of the islet tissue varies in different animals... Ketonaemia does not occur in rabbits and it is slight in calves; in pigs ketonaemia may be severe, yet coma does not occur.

ketone. Add: **2.** Special combs.: **ketone body**, any of the three related compounds acetone, acetoacetic acid, and β-hydroxybutyric acid, which are produced in the body in fatty- and amino-acid metabolism; an 'acetone body'.

1915 *Arch. Internal Med.* XV. 40 There are in the literature only casual references to the influence of organotherapy on the formation and excretion of ketone bodies. **1936** *Jrnl. Nutrition* XII. 647 The term 'rate of ketogenesis'..takes into account both the ketonemia, or level of ketone bodies in the blood, and the ketonuria, or the rate of ketone body excretion in the urine. **1959** A. White et al. *Princ. Biochem.* (ed. 2) xix. 479 The total ketone body concentration in blood, expressed as β-hydroxybutyrate, is normally below 1 mg. per 100 ml., and the average total daily excretion in the urine is approximately 20 mg. **1974** *Nature* 18 Jan. 161/2 Urine samples were tested for 'ketone bodies' with a commercial preparation that contains sodium nitroprusside.

ketonic acid (earlier example); **ketol, ketose** (examples).

1876 *Jrnl. Chem. Soc.* XXIX. 926 The ketonic acids are converted by nascent hydrogen into oxy-acids. **1891** *Ibid.* LX. 1176 Ketoses. **1892** E. F. Smith tr. *V. von Richter's Chem. Carbon Compounds* (ed. 2) 498 Bromine water..oxidises the aldoses to their corresponding monocarboxylic acids..whereas the ketoses (fructose and sorbinose) are not attacked. **1894** G. McGowan tr. *Bernthsen's Text-bk. Org. Chem.* (ed. 2) 237 (*heading*) Ketone-alcohols (ketols). **1938** M. L. Wolfram in H. Gilman *Org. Chem.* I. xvi. 1442 The Lobry de Bruyn dilute alkali interconversion reaction was used..in obtaining the crystalline ketose of lactose (lactulose). **1968** R. O. C. Norman *Princ. Org. Synthesis* xvi. 501 Treatment [of an epoxide] with dimethyl sulphoxide gives an α-ketol. **1972** *New Phytologist* LXXI. 475 The phenomenon of ketose formation in polyol media is much better documented for bacteria where, for example, fructose may be formed from mannitol in yields of up to..95%.

ketonize (kī·tǒnaiz), *v. Chem.* [f. Keton(e + -ize.] *intr.* Of a compound which undergoes keto–enol tautomerism: to change into the keto-form.

1937 F. C. Whitmore *Org. Chem.* 345 The free enediol is unstable and ketonizes to the acyloin at once. **1938** L. F. Fieser in H. Gilman *Org. Chem.* I. ii. 55 The hydroxyl derivatives of benzene show much less tendency to ketonize than do the aliphatic enols. **1962** E. L. Eliel *Stereochem. Carbon Compounds* viii. 242 The enolate..ketonizes in such a way that the proton approaches equatorially.

Hence **ketoniza·tion**, the process of changing into a keto-form.

1931 *Canad. Jrnl. Res.* V. 26 The suggested formation of the lignosulphonic acids by ketonization of a phenolic nucleus is most unlikely. **1937** F. C. Whitmore *Org. Chem.* 280 Ketonization gives acetic acid, esters, acid amides, acid chloride, etc. **1968** R. O. C. Norman *Princ. Org. Synthesis* i. 19 Compared with the ketonization of phenol.., the ketonization of β-naphthol is more favourable by about 11 kcal per mole. **1972** *Jrnl. Chem. Soc. Perkin Trans.* I 987/1 Kinetic control of enolisation and ketonisation is replaced by thermodynamic control at higher temperatures.

ketonuria (kītǒniūə·riă). *Med.* [f. Keton(e + -uria.] The excretion of abnormally large amounts of ketone bodies in the urine.

1913 *Jrnl. Amer. Med. Assoc.* 13 Dec. 2161/1 No one is justified at present in concluding that the liver is the sole organ concerned in the phenomena of ketonuria. **1923** *Glasgow Herald* 25 July 7 Daily injections of insulin ..removed entirely the ketonuria for a period of some days. **1939** *Nature* 29 Apr. 728/1 During rest after work there is a reduction in the glucose tolerance,..associated with a low respiratory quotient and ketonuria. **1972** *Drugs of Choice 1972–73* xxxi. 468/2 One of the more pernicious consequences of hypoglycemia is the hyperglycemia and ketonuria that may follow.

ketosis (kītōᵘ·sis). *Med.* [f. *ket(o- + -osis.] A condition characterized by an increased production of ketone bodies, which is associated with a predominance of fat metabolism and with diabetes.

1917 F. M. Allen in *Amer. Jrnl. Med. Sci.* CLIII. 335 No other name but acidosis exists for the metabolic process which it denotes. Ketonuria and ketonemia have their accurate place but do not cover the ground. Possibly the word ketosis might be suggested and used for special purposes, but the change of established usage would be difficult and seems unnecessary. **1925** *Biochem. Jrnl.* XIX. 948 In man after a relatively short period of starvation..a considerable ketosis appears. **1951** A. Grollman *Pharmacol. & Therapeutics* vii. 151 The paroxysmal cerebral dysrhythmia characteristic of epilepsy may be altered by inducing ketosis by a low carbohydrate, high fat diet. **1960** *Farmer & Stockbreeder* 22 Mar. 136/1 There does appear to be a connection between the heavy feeding of kale, and the incidence of ketosis (acetonæmia), but in rather a roundabout way. **1961** [see *acetone]. **1970** R. W. McGilvery *Biochem.* xxii. 530 However, the meat-fat diet does cause the appearance of an asymptomatic ketosis.

Hence **keto·tic** *a.*, suffering from or associated with ketosis.

1943 *Jrnl. Amer. Med. Assoc.* 2 Jan. 52/2 During episodes of ketotic acidosis it was necessary to give 2,360, 2,500 and 2,795 units of insulin in twenty-four hours before adequate control was accomplished. **1959** *Jrnl. Appl. Physiol.* XIV. 1028/1 The turnover time was increased also by about ⅓ in the ketotic cows. **1961** L. Martin *Clin. Endocrinol.* (ed. 3) ii. 52 Uncontrolled diabetics with high blood sugar levels do not as a rule suffer from anorexia unless they are also ketotic. **1972** *Jrnl. Clin. Invest.* LI. 1440/1 The cause of ketotic hypoglycemia, the commonest form of hypoglycemia in childhood, is not known.

kettle. Add: **1. b.** A bowl- or saucer-shaped vessel in which operations are carried out on low-melting metals, glass, plastics, etc., in the liquid state.

a **1817** [see *potash kettle* (Potash *sb.* 4)]. **1892** P. Benjamin *Mod. Mechanism* 803 A rendering and refining kettle for making..fancy toilet soap. **1895** E. L. Rhead *Metall.* xv. 214 The ore is ground to a pulp in the mill, or arastra, and transferred to kettles with bottoms made of copper. **1929** *Industr. Chemist* V. 487/1 A Pfaudler, all cast-iron, glass-lined, 300-gallon, reaction kettle or chemical still. **1940** H. L. Hind *Brewing* II. xxiv. 576 The vessel[s] in which the mash is boiled..are usually known as kettles in America. **1952** M. R. Mills *Introd. Drying Oil Technol.* iii. 48 Oleo-resinous varnishes are commonly produced in portable kettles of 45–3,000 lb...capacity. **1953** *Archit. Rev.* CXIV. 187/1 The raw materials used in the former process [*sc.* the manufacture of synthetic resins] are highly inflammable, and it was desirable that the 'kettles' in which this process is carried out should be in a separate building. **1955** Kirk & Othmer *Encycl. Chem. Technol.* XIV. 653 Stationary open kettles are used to polymerize large batches of oil. **1967** J. D. Gilchrist *Extraction Metall.* x. 259 Softening—at 750°C, in a wide, open hearth furnace or in the open saucer-shaped lead 'kettle' which presents a very large surface for oxidation, Sn, Sb and As are slowly oxidized out with air or litharge.

2. b. Also, *a different* or *another kettle of fish*: a different state of affairs, a different matter altogether.

1937 *Discovery* Nov. 353/1 H. S. Thompson's 'Garnet in Flight' is another kettle of fish. **1938** R. Warner *Professor* vi. 141 Professor..you're very good at thinking out schemes—brainwork, I mean..if you understand what I mean; but that's only half the battle you know, in fact a different kettle of fish altogether. **1942** E. Waugh *Put out More Flags* iii. 172 Until now the word 'Colonel' for Basil had connoted an elderly rock-gardener on Barbara's G.P.O. list. This formidable man of his own age was another kettle of fish. **1959** J. L. Austin *Sense & Sensibilia* (1962) ii. 14 Looking..at a distant village on a very clear day across a valley, is a very different kettle of fish from seeing a ghost.

4. c. (Earlier and later examples.) Also (now the usual meaning in *Geomorphol.*), a *kettle hole* (see sense *6 b).

1866 *Smithsonian Contrib. Knowl.* No. 197. 3 To form an idea of the appearance of the 'potash kettle' country, we may imagine a region of drift moraines inverted, and.. occupied by cavities of irregular size and depth. *Ibid.* 4 On the north of the Peshattego river..the 'kettles' are very numerous. **1877** T. C. Chamberlin *Geol. Wisconsin: Survey of 1873–79* II. ii. v. 206 The peculiar feature of this range..consists of numerous depressions in the drift variously known as 'Potash Kettles', 'Kettles', 'Potholes', 'Pots and Kettles', 'Sinks', etc. Those which have most arrested popular attention are circular in outline, and symmetrical in form... Large numbers of these depressions are not perfectly circular, but rudely oval, oblong or elliptical, or are extended into trough-like, or even winding hollows. *Ibid.* 214 If masses of the ice became incorporated within the drift, ..their subsequent melting would give rise to a depression constituting one form of the kettles. **1896** T. G. Bonney *Ice-Work* i. i. 34 These 'kettles', when first discovered, were filled with débris, and still contained the large rounded boulders by which they had been mainly excavated. **1926** *Jrnl. Geol.* XXXIV. 315 The ice blocks that formed the kettles in pitted outwash varied in size from a few yards to several miles in diameter. **1942** C. A. Cotton *Climatic Accidents* xxiv. 328 Both bowl-like round pits and elongated trench-like kettles are common. **1970** R. J. Small *Study of Landforms* xi. 384 A bare tract of boulders, gravels and sand separates the two glacier snouts, and is pitted by numerous circular water-filled hollows ('kettles') marking small masses of ice that calved from the glaciers, became trapped in the debris, and subsequently melted.

d. A watch. *slang* (chiefly *Criminals'*).

1889 Barrère & Leland *Dict. Slang* I. 516/2 *Kettle* (thieves), a watch; *red kettle*, gold watch. **1931** [see *groin *sb.*² 3]. **1935** G. Ingram *Cockney Cavalcade* xiv. 234, I pinched his 'kettle' what those two blunderers left behind. **1936** J. Curtis *Gilt Kid* xxv. 244 Next buckshee kettle that comes my way I'll just stick to it. **1960** 'A. Burgess' *Doctor is Sick* xvi. 122 Edwin, student of philology, knew what kettles were, cheap smuggled watches guaranteed to go for a day or two. **1970** *Brewer's Dict. Phrase & Fable* (rev. ed.) 603/2 A tin kettle is a silver watch and a red kettle a gold one.

6. a. *kettle-lid, -scrubber, -stand.* **b. kettle-bail** *U.S.*, a dredge used in taking scallops; **kettle-holder** (earlier example); **kettle hole**, a depression in the ground thought to have been formed by the melting of an ice block trapped in glacial deposits, esp. one that is circular and deep; freq. *attrib.* in *kettle-hole lake* = *kettle lake; **kettle lake**, a lake in a kettle hole; **kettleman**, restrict † to sense in Dict. and add: (*b*) (also **kettle man**) one who attends to a kettle in various industries; **kettle moraine** *Geomorphol.* [orig. applied as a proper name to such a moraine in Wisconsin], moraine characterized by the presence of numerous kettle holes.

1881 E. Ingersoll *Oyster-Industry* 245 *Kettle bail*, a dredge used in catching scallops, which has the blade adjusted to swing in the eyes of the arms, in order to prevent its sinking into the mud of the soft bottom on which it is used. **1887** G. B. Goode *Fisheries U.S.: Hist. & Methods* II. 571 The dredge for a soft bottom differs from the other in having the 'blade' adjusted to swing in the 'eyes' of the arms in order to prevent its sinking into the mud. This is called the 'kettle-bail' style of dredge. **1813** M. Edgeworth *Let.* 1 May (1971) 32 After having admired..a picture of Cromwell and Fanny's kettle-holder we sallied forth. **1882** *Proc. Amer. Assoc. Adv. Sci.* XXXI. 395 The kames of Cherry Valley..are composed of stratified water-worn gravel, ..and, as a series of conical hills and reticulated ridges, enclosing 'kettle holes', form conspicuous objects in the centre of the valley. **1895** J. D. Dana *Man. Geol.* (ed. 4) 970 Kettle-holes are bowl-shaped depressions, usually 30 to 50 feet deep and 100 to 500 feet in larger diameter. Each depression..was the resting-place, and often the burial-place, of a huge mass of ice that became detached during the melting. **1902** Gilbert & Brigham *Introd. Physical Geogr.* vi. 143 Many lakes with steep rims in the midst of much glacial waste are known as Kettle-hole Lakes. **1930** *Q. Jrnl. Geol. Soc.* LXXXVI. 112 Numerous lakes and pools lying in kettle-holes dot the surface of the moraine. **1957** G. E. Hutchinson *Treat. Limnol.* I. i. 90 The exact nature of such basins depends largely on the details of the process of deglaciation..producing an extraordinary number of kettle-hole lakes in North America. **1970** Dorr & Eschman *Geol. Michigan* vii. 151/2 Most of the smaller inland lakes of Michigan occupy kettle-holes. **1914** *Prof. Papers U.S. Geol. Survey* No. 82. 163 Till or similar impervious material appears to be present in considerable amounts, as indicated by the numerous springs, kettle lakes, and similar features. **1968** R. W. Fairbridge *Encycl. Geomorphol.* 587/2 The hollow is frequently water filled, so that it forms a kettle lake, kettle pond or swamp. **1920** 'K. Mansfield' *Bliss* 44 They spent half their time ..dosing him with various awful mixtures concocted by Pip, and kept secretly by him in a broken jug covered with an old kettle lid. **1833** B. Silliman *Man. Sugar Cane* 15 The manner in which the hands are distributed during the cutting season is the following..forty hands with knives..six kettle men. **1960** *Classification of Occupations* (General Register Office) 51/2 Kettleman—gelatin, glue, size mfr... metal mfr... oil seed crushing. **1963** *Lebende Sprachen* VIII. 130/2 Kettleman. [**1883** T. C. Chamberlin *Geol. Wisconsin: Survey of 1873–79* I. i. xv. 275 That portion of the moraine which..was formed by the joint action of the Green Bay and Lake Michigan glaciers, constitutes a succession of irregular hills and ridges, locally known as the Kettle Range, from the peculiar depressions which characterize it... As this moraine will need a specific name to distinguish it from other similar accumulations, the term Kettle Moraine may fittingly be applied to it.] **1889** G. F. Wright *Ice Age N. Amer.* vii. 120 Attention was first directed..by President T. C. Chamberlin to the character and connection of the kettle-moraine in Wisconsin. **1897** W. B. Scott *Introd. Geol.* viii. 155 When such masses melt they form depressions in the mound and give rise to the 'kettle moraines'. **1937** Wooldridge & Morgan *Physical Basis Geogr.* xxii. 387 Mounds and ridges of gravelly drift are referred to in British glacial literature as eskers and kames, or, generally, as kettle-drift or kettle-moraine. **1970** B. B. Luckman in C. A. Lewis *Glaciations Wales* viii. 176 (*heading*) The Kington–Orleton kettle moraine. **1843** Thackeray *Irish Sk.-Bk.* II. 278 Thus it was I drew her Scouring of a kettle... That sweet kettle-scrubber! **1881** C. Schreiber *Jrnl.* 1 Nov. (1911) II. 367 Found a fine old kettle stand..and a few minutes after had the good luck to find the kettle to fit. **1960** H. Hayward *Antique Coll.* 157/1 *Kettle-stand*, a special stand which was introduced with tea-drinking in the later 17th-cent., of two main kinds. (a) A small table ..with a gallery or raised edge round the top... (b) A box-like arrangement set on four legs. **1970** D. Ash *Dict. Eng. Antique Furniture* 92/2 Kettle-stands were lower than contemporary tripod tables so that the kettle, mounted on its lamp-stand, would be at a convenient height.

kettler. Restrict † to sense in Dict. and add: **2.** A colour-mixer's assistant who attends to the boiling of dyestuffs.

1921 in *Dict. Occup. Terms* (1927) § 381. **1960** *Classification of Occupations* (General Register Office) 51/2 Kettler..106 [dyers of textiles].

kettle-stitch. (Earlier and later examples.)

1818 H. Parry *Art of Bookbinding* 2 *Kettle-stitch*, the stitch at head and foot of the book, independent of the bands, to tack or fasten the sheets together. **1846, 1906** [see *catch-stitch* (*catch- 3 a)]. **1973** *Islander* (Victoria, B.C.) 30 Sept. 11/1 Dick shewed me some of the stitches used in book binding. Among them, the kettle stitch, used in hand sewing during the early days of book binding.

kettling, var. Chitling.

1869 *Overland Monthly* III. 130 Then there are the delusive 'kettlings', among the 'low-down' people... I will simply say that it is fried sausages, minus all the unhealthy and absurd meat which most people insist on stuffing into the intestinal integuments.

‖ **ketubah** (ketūvā·). Also **kethubah, ketuba.** [Heb. *kethūbhāh* written statement.] A formal Jewish marriage contract which includes financial provisions for the wife in the event of the husband's death or of divorce.

1841 Borrow *Zincali* I. ii. vii. 344 The walking of the bride..to the house of her betrothed..the reading of the Ketubah. **1891** M. Friedländer *Jewish Relig.* ii. vii. 487 The fulfilment of the conditions agreed upon in the *kethubah*..tended to render divorce a rare event. **1960** *Commentary* June 500/1 The Rabbinical Assembly has modified the *Ketubah* (marriage contract). **1960** [see *get sb.[3]]. **1962** B. Abrahams tr. *Life Glückel of Hameln* iv. 79 When we stood all together under the *chuppah* with the bride and bridegroom, we found that..the Ketuba had not been drawn up! **1974** H. Kemelman *Tuesday Rabbi saw Red* 7, I read the *Ketubah*, that is the marriage contract, which the groom has signed previously.

keurboom (kūə·ɹbum). Also **keur.** [Afrikaans, f. *keur* choice + *boom* tree.] A small South African tree of the genus *Virgilia* (*V. oroboides* or *V. divaricata*) of the family Leguminosæ, having pinnate leaves and racemes of white, pink, mauve, or red, scented flowers.

1731 G. Medley tr. *Kolb's Present State of Cape of Good-Hope* II. 258 The Amaquas Tree. This Tree the Cape-Europeans call *Keur-boom*. It grows so quick, that in Two Years Time it becomes, from a Small Plant, a Tree of Eight or Nine Foot in Height. **1854** L. Pappe *Silva Capensis* 13 *Virgilia Capensis*. Lamk. (*Keurboom*)... Wood rather light and soft. Looks well when polished. **1907** T. R. Sim *Forests & Forest Flora Cape Good Hope* 204 Keur is often cultivated for its ornamental evergreen foliage and sweetly scented flowers. **1925** R. Marloth *Flora S. Afr.* II. i. 72 *Virgilia capensis*. Keurboom... In favourable localities it reaches a height of 40 feet with trunks three feet in diameter. Such trees form a beautiful sight when covered with their pale rose or lilac coloured, sweet-scented flowers. **1955** J. Packer *Valley of Vines* iv. 58 The *keurbome* toss fragrant pale-pink plumes in the wind. **1973** Palmer & Pitman *Trees S. Afr.* II. 903 Botanists, foresters, and gardeners have not yet decided to their mutual satisfaction whether the genus *Virgilia* is composed of one, two, or three different species of keurboom.

kevir, var. *kavir.

kew. Short for Thank you.

1939 G. B. Shaw *Geneva* ii. 30 Sit down. *Begonia* [complying]: Kew. **1961** *Times* 14 Aug. 9/4 Of 20 recipients of sitting space, five said 'Thank you' or 'Thanks'; three said 'Kew'.

Keweenawan (kiwīnǭ·ăn), *a.* Also † **Keweenawian.** [f. the name of *Keweenaw* Peninsula, Michigan + -an, -ian.] Of, pertaining to, or designating the most recent division of the Proterozoic in North America, as represented by rocks in the region of the Great Lakes. Also *absol.*, the Keweenawan period or rocks.

1876 T. B. Brooks in *Amer. Jrnl. Sci.* CXI. 210 We are therefore justified, I think, in regarding the Copper-bearing rocks of Lake Superior as a distinct and independent series, marking a definite geological period which separates the Silurian from the Huronian ages... Since Keweenaw Peninsula forms one of the most striking geographical features in Lake Superior.., I suggest the name Keweenawian for this period. **1883** R. D. Irving *Copper-Bearing Rocks Lake Superior* ii. 25 The Keweenawan rocks form the larger part of Keweenaw Point. **1893** *Jrnl. Geol.* I. 126 The source of the lavas of the Keweenawan. **1930** *Econ. Geol.* XXV. 252 Erosion had peneplained the original folds of the Keewatin and Timiskaming rocks before Keweenawan time. *Ibid.* 262 They are distinctly younger than the Algoman and older than the Keweenawan. **1935** *Bull. Geol. Soc. Amer.* XLVI. 504 The Keweenawan lavas of the Lake Superior district form the inner border of the Lake Superior syncline. They crop out over an area of several thousand square miles, and in one section on Keweenaw Point, Michigan, form a series with a minimum thickness of 15,000 feet. **1970** Dorr & Eschman *Geol. Michigan* iv. 56/1 Keweenawan lavas were derived from molten magmas of the 'basic' or ferromagnesian type.

kewpie (kiū·pi). orig. *U.S.* Also **cupie.** [A dim. form of Cupid.] In full, *kewpie doll.* A chubby doll with a curl or topknot on its head, from a design by R. C. O'Neill (1874–1944). Also *transf.*, of a person.

1909 R. O'Neill in *Ladies Home Jrnl.* Dec. 28/1 The Kewpie wights stay up at nights, All gayly singing rum-te-tum. *Ibid.*, The reason why these funny, roly-poly creatures are called Kewpies..is because they look like little Cupids. **1912** *Ibid.* Oct. 109 The kewpies were invented by Rose O'Neill. They are always doing good, helping Dotty Darling and her Baby Brother to have a good time... So Rose O'Neill has made the Kewpie Kutouts. **1913** *Official Gaz.* (U.S. Patent Office) 13 May 536/1 Rose O'Neill Wilson... Kewpie. Particular description of goods.—Dolls. **1922** A. Haddon *Green Room Gossip* 176 The floor is strewn with at least fifty toys—teddy bears, golliwogs, bunnies, and woolly mascots galore. Kewpies, too, on the dressing-table. **1926** G. S. Chappell *Younger Married Set* iv. 48, I have here..a magnificent kewpie doll. **1929** T. Wolfe *Look Homeward Angel* (1930) xxii. 324 Over the screened hearth, on a low mantel, there was a Kewpie doll, sashed with pink ribbon. **1937** *Official Gaz.* (U.S. Patent Office) 15 June 543/2 Rose O'Neill Wilson. The Kewpies. For Cartoons in Periodical Publications. Claims use since Dec. 21, 1912. **1937** J. Steinbeck *Of Mice & Men* iii. 93 They got a rag rug on the floor and a kewpie doll lamp on the phonograph. **1943** K. Tennant *Ride on Stranger* vii. 60 Outside the turnstiles, vendors vie for custom, waving gilded cupie dolls on canes. **1946** I. L. Idriess *In Crocodile Land* xxvi. 180 With the little black kewpie of a piccaninn aboard her tiny canoe, Wagis paddled downstream towards the river mouth. **1947** *Archit. Rev.* CI. 206/3 There are dolls for sale on the pier, too, Kewpie or cuddly or fashionable. **1952** M. McCarthy *Groves of Academe* (1953) ii. 18 Decorated with the same rag-dolls and teddy-bears, pink kewpies won at shooting-ranges. **1954** *Encounter* Feb. 52/1 Art Young, a white-haired little kewpie, sitting in a corner, was pointed out to me. **1957** M. Millar *Soft Talkers* 62 A kewpie doll. One of those tiny celluloid kewpie dolls you can buy in the dime store. **1960** *Birmingham Mail* 13 June 3/4 Each year Roo..has brought his woman..a cupie-doll. **1960** *Official Gaz.* (U.S. Patent Office) 30 Aug. TM 174/1 Cameo Doll Products Co., Inc., Port Allegany, Pa... Kewpie. For Dolls. **1961** A. Christie *Pale Horse* v. 67 A most terrible Kewpie doll from the hoop-la. **1969** N. Cohn *A Wop Bopa Loo Bop* (1970) viii. 81 She'd be like some kewpie doll, all sheen and varnish and eyes that really roll. **1972** M. J. Bosse *Incident at Naha* iii. 126 Kewpie dolls..staring at you stupidly and insanely from their glass eyes.

b. A sweet made in the form of a kewpie doll.

1916 *Daily Colonist* (Victoria, B.C.) 28 July 12/6 (Advt.), Kewpie Dolls, made of pure barley sugar, a box 10¢. **1918** *T. Eaton & Co. Catal.* Fall & Winter 385/3 Kewpie Kandy Dolls are made from the pure cane sugar. ..Twelve little kewpie dolls in each package.

key, *sb.*[1] Add: **6. c.** *Chess.* (*a*) In full *key move.* The first move in the solution of a set problem, upon which depends the style of the whole solution. Also *attrib.* or as *adj.* (*b*) The whole solution of a set problem.

1827 W. Lewis *Chess Probl.* Pref., I defer for the present publishing the Solutions, that the reader may solve the Problems without being tempted to refer to the Key. **1846** *Chess Player's Chron.* VI. 65 Amateurs wanting either time or perseverance to undertake the solution themselves, ..may obtain the key by addressing a note to the Editor. **1878** S. Loyd *Chess Strategy* 60 Key-moves which threaten an easy mate. *Ibid.* 201 There is always a great deal of chance in solving a problem, such as hitting upon the key by accident. **1923** B. G. Laws *Artistry of Chess Probl.* 5 Key, continuations and principal mates are all of the first order... The key abandoning an important protective Pawn..is little short of being superb. **1938** C. S. Kipping *Chess Probl. Sci.* I. 51 This was followed by six dis. checks..with a check key, and J. L. Millins showed that it could be done with a quiet key... Ua Tane, using White King diagonal battery and Pawn promotion key showed that six were possible. *Ibid.* 56 Petrovic's astounding triple task has not a key which will appeal to everyone. **1945** *Chess in Schools* (Chess Educ. Soc.) 27 *Problems*, composed positions in which a key move is to be found, after which mate is forced in a specified number of moves. **1958** Mansfield & Harley *Mod. Two-Move Chess Probl.* 19 The solver's attention is drawn to it, and Key possibilities are limited. *Ibid.*, Such a Key is thematic and good.

d. The device used to 'key' an advertisement (see *key *v.* 5). orig. *U.S.*

1905 Calkins & Holden *Art of Mod. Advertising* xi. 266 A variation of the 'key' in advertising is the coupon. **1915** H. W. Hess *Productive Advertising* xiii. 199 Accounts may be opened charging up to each key not only (1) number of inquiries, but (2) amount spent on that particular key, [etc.]. **1957** Clark & Gottfried *University Dict. Business & Finance* (1967) 201/1 The key may be a special street or box number, or a department to which the reply is to be addressed, or it may be a number or letter code included in the reply coupon if one is used.

8. b. Phr. *in key* (*with*): in harmony (with); in a style that matches; *out of key* (*with*): out of harmony (with); not matching; unsuitable.

1919 B. Ruck *Disturbing Charm* ii. xiv. 287 Everything in that drawing-room was in key with that mantelpiece. **1920** E. Pound *Hugh Selwyn Mauberley* i. 9 For three years, out of key with his time, He strove to resuscitate the dead art Of poetry. **1931** E. Wilson *Axel's Castle* ii. 34 We shall be thrown fatally out of key with reality. **1934** H. G. Wells *Exper. Autobiogr.* II. vii. 489 That was entirely out of key with James's assumptions.

15*. [Respelling of *ki* in *kilo.*] A kilogramme of a drug. *U.S. slang.*

1968–70 *Current Slang* (Univ. S. Dakota) III–IV. 76 *Key*, kilo of any narcotic; a measurement for marijuana, 215 lbs, $300.00+. **1970** *Time* 13 Apr. 36 A $10 or $20 'key' of Lebanese hash can fetch $1,500 or more in the U.S. **1972** J. Wambaugh *Blue Knight* (1973) iv. 45 On her coffee table she had at least half a key and that's a pound of pot and that's trouble.

16. (sense 7) *key-centre, -change, -system.*

1940 *Scrutiny* Sept. 122 Without establishing a key-centre the fluctuating basses eventually soar..into another homophonic passage. **1931** G. Jacob *Orchestral Technique* iii. 24 The choice..should..rest entirely on simplicity of key—the piece as a whole with all its modulations and sectional key-changes being taken into consideration. **1959** 'F. Newton' *Jazz Scene* 115 The free and continual key changes. **1973** J. Wainwright *Pride of Pigs* 175 He rippled through a key-change bridge passage, then moved into the beat. **1934** C. Lambert *Music Ho!* i. 28 His [*sc.* Debussy's] destruction of the key-system. **1959** D. Cooke *Lang. Mus.* ii. 44 All the modes eventually became major and minor scales..; and so arose our key-system.

b. Passing into *adj.* in the sense of 'dominant', 'controlling', 'chief', 'essential'; esp. designating some person or thing that is of crucial importance to others. See also *key man* (sense 17 below).

1913 E. C. Bentley *Trent's Last Case* xi. 207 When chance or effort puts one in possession of the key-fact in any system of baffling circumstances, one's ideas seem to rush to group themselves anew in relation to that fact. **1916** *Economic Jrnl.* XXVI. 24 We are asked to learn one essential lesson from the war, and that is, not to be caught short of any 'Key' industry. **1926** D. L. Colvin *Prohibition in U.S.* 509 Occupants of key offices such as the Presidency or the Attorney-Generalship. **1927** W. E. Collinson *Contemp. Eng.* 94 Greek was said to occupy a key-position. **1928** J. Boon *Victorians, Edwardians & Georgians* I. 203 There had been considerable difficulty in getting hold of a key witness. **1931** C. A. Lejeune *Cinema* 4 A study of these key-names, unobscured by any commercial considerations of box-office value. **1934** C. Lambert *Music Ho!* ii. 110 It is Stravinsky who is the key-figure of our times. **1941** *Hutchinson's Pict. Hist. War* 19 Mar.–13 May 103 German airmen..were ensconced in key positions. **1945** K. R. Popper *Open Soc.* I. viii. 135 The key-passage of the philosopher king. **1946** *Sun* (Baltimore) 6 Mar. 1/6 The news was spread all over front pages with photographs of the key figures. **1955** *Ann. Reg. 1954* 256 Investments in the key industries were to be gradually reduced. **1959** *Times Rev. Industry* June (London & Cambridge Econ. Bull.) p. i/1 The key factor in an assessment of the economic position and prospects of the country is Mr. Amory's Budget. **1963** *Times* 12 Feb. 7/4 A small number of key-workers among the affected men stayed away from work. **1964** *English Studies* XLV. 243 One wonders..how many of their key-questions will in fact gain in coherence. **1966** *Illustr. London News* 30 July 27/2 Australia and North America are key areas. **1969** *Times* 16 July 5/8 Key abbreviations used by mission control and the astronauts. **1970** J. Ardagh *New France* xi. 529 It was in 1943 that Sartre brought out his key philosophical work. **1970** *Physics Bull.* Nov. 493/1 Two ideas were key in the discovery of the kinoform. **1971** *Black Scholar* Dec. 55/2, I viewed myself as assisting everything that was done, and you must recognise that this is what's key in the liberation of women.

17. key-block, (*a*) a block, usu. of wood, also of metal or, in lithography, stone, used in the printing of chiaroscuro and colour pictures to give the outline, and to provide a guide for the accurate registration of the tint or colour blocks; (*b*) in limestone and marble quarrying, the first block or blocks to be removed from a new layer of stone; **key-chain,** a chain to which a key or keys may be attached; **key-drawing,** (*a*) in lithography and colour printing, an outline drawing which is transferred on to the key-plate and used as a guide to printing the colours; (*b*) *Cinematogr.* (see quot. 1940); **key-holder,** (*a*) 'an electric-lamp holder or socket containing a switch' (Webster 1909); (*b*) a person who keeps the key or keys of a workshop, factory, etc.; **key-**

log *Logging*, a log which is so caught or wedged that a jam is formed and held by it; **key man, key-man,** (*a*) *Logging*, a man who finds and dislodges the key-logs in a jam (*Obs. U.S.*); (*b*) an operator of telegraph keys (*Obs. U.S.*); (*c*) one who plays a leading or important role in a group, an industry, etc.; **key-money** (further examples); **key move** (see sense 6 c above); **key-plate,** (*a*) a key-hole escutcheon; (*b*) in colour-printing from a metal surface, the outline plate answering to a keystone in lithography; **key-ring** (examples); **key-word,** (*a*) a word serving as a key to a cipher or the like; (*b*) a word or thing that is of great importance or significance; *spec.* in information-retrieval systems, any informative word in the title or text of a document, etc., chosen as indicating the main content of the document; so **key-word-in-context,** used *attrib.* of an index or concordance in which key-words are listed alphabetically, preceded and followed by a fixed amount of the immediate context.

c **1870** A. Ashley in *Jrnl. Printing Hist. Soc.* (1969) V. 73 Key blocks. **1910** R. M. Burch *Colour Printing* v. 126 Baxter's process was simply the colouring of an impression from an outline or key block, which could be either a copper, zinc or steel plate, or a litho stone..by successive impressions from colour blocks of wood or metal, one for each tint used. **1934** O. Bowles *Stone Industries* ii. ix. 212 In opening up a new floor the first blocks to be removed are known as 'key blocks'. **1937** *Discovery* Mar. 77/1 In order to get the colours fitting accurately, prints are taken from the first or key-block and pasted on to each successive block. **1966** Berry & Poole *Ann. Printing* 241/1 They vary in quality, many being spoiled by the heavy impression from the wood key block. **1967** *Amer. Speech* XLII. 290 Once this first troublesome block, called the *key block*, is out, further quarrying becomes much easier. **1895** *Montgomery Ward Catal.* 86/2 Polished steel key chains, 18 inches long. **1950** F. P. Walkup *Dressing the Part* (rev. ed.) 368 A ridiculous fad..was the 'zoot suit'..adorned with a keychain that swept the ground. **1973** K. Giles *File on Death* ii. 42 He locked the door... He had a housekeeper-like key chain attached to the top button of his fly. **1937** *Discovery* Oct. 300/1 The use of several colours, flat or stippled, combined with a black or dark-coloured key drawing. **1940** *Chambers's Techn. Dict.* 475/2 *Key drawing* (*Cinema*). In animated cartoon production, *key drawings* indicate situations at special instants, such as at beats in the bar of music, after which the in-between drawings are made to fit with the timing. **1951** D. Bland *Illustration of Bks.* ix. 133 The artist first makes his key drawing in black, regardless of whether it is intended finally to print in black. From this non-actinic prints are prepared, one for each colour, showing the key drawing in a pale blue colour that will not photograph. **1928** *Daily Tel.* 11 May 5/6 Workmen were waiting outside ready to begin work for the day... The key-holder had not arrived. **1970** P. Laurie *Scotland Yard* iii. 66 The Inspector calls the nick to check the keyholder's register and see who can let us in. *Ibid.*, The keyholder is the caretaker of the flats. **1851** J. S. Springer *Forest Life & Forest Trees* 166 It may be thought best to cut off the key-log, or that which appears to be the principal barrier. **1902** S. E. White *Blazed Trail* xxxii. 211 By pulling out or chopping through certain 'key' logs which locked the whole mass. *a* **1951** B. Cronin in *Austral. Short Stories* (1951) 173 Half-way up the kiln a key-log had swung loose. **1851** in C. M. Wilson *Aroostook* (1937) 104 The key man then commences prying while they are pulling. **1907** *Washington Star* 30 Sept. 9 Some of the leading keymen are sounding as their shibboleth the cry of 'government ownership of the telegraph systems'. **1921** *Daily Colonist* (Victoria, B.C.) 15 Oct. 1/1 Following the conference of 'Key' men of the G.W.V.A. last July, arrangements were advanced to have in attendance at the convention the foremost leaders of the navies and armies of the allied nations in the Great War. **1949** I. Deutscher *Stalin* 143 The 'handful' of Bolsheviks consisted of well-organized and disciplined 'key-men'. **1963** *Times* 18 Feb. 8/3 One of the key men in this new method of assessing needs and opportunities is the local prefect. **1906** *Westm. Gaz.* 28 Mar. 5/2 Some house-agents would still extort 'key-money' from tenants. **1963** *Times* 21 Feb. 5/2 A former property manager was alleged at Southend-on-Sea Magistrates' Court today to have exploited the housing shortage to extort key money. **1969** R. Rendell *Best Man to Die* iii. 26 They'd had to find the key money for the flat themselves. **1903** T. A. Strange *Hist. Guide French Interiors* 326 The key plate is formed as a vase-shaped burning lamp. **1909** R. E. Le Blond *Let.* in H. G. Clarke *Baxter Colour Prints* (1919) 103 These prints were first engraved on a steel plate, a key-plate, or, as I should call, a master-plate. **1969** J. Gloag *Short Dict. Furnit.* (rev. ed.) 317 (*caption*) The simplest type of escutcheon, pivoted to cover the key-plate. **1970** *Jrnl. Printing Hist. Soc.* VI. 73 A key plate of the design was printed in black or brown, usually from an..engraved steel or copper plate. **1889** *Cent. Dict.* III. 3280/2 *Key-ring.* 1. A finger-ring from which projects a tongue or blade which is either fixed or movable on a hinge, and serves as the key to a lock. Such key-rings were formerly common, and were often of rich design. 2. A ring used for keeping a number of keys together by being passed through their bows. **1906** E. Nesbit *Story of Amulet* xi. 261 Jane added a key-ring. **1967** *Economist* 11 Feb. 512 An advertising firm..has now been hired to show that gaullism washes whiter. It has begun to do so in American fashion with badges, key-rings, drum-majorettes and all. **1859, 1885** Key-word [in Dict., sense 16]. **1922** J. C. H. Macbeth tr. *Langie's Cryptogr.* iv. 166 Suppose the key-word to be 'bankruptcy'. **1926** *Encycl. Brit.* II. 822/1 As to shop detail, the keyword to mass production is simplicity. **1948** Blunden *Shakespeare to Hardy* (1964) 211 The key-word is the 116th of the Sonnets. **1956** E. W. F. Tomlin in A. Pryce-Jones *New Outl. Mod. Knowl.* 91 Let us take the keyword 'science'. **1967** Cox & Grose *Organiz. Bibliogr. Rec. by Computer* iv. 82 The system can process documents represented by a set of keywords. **1971** *New Scientist* 1 Apr. 32/2 'Keywords' ..are supposed to describe the content of the document as concisely and exactly as possible. **1959** Key-word-in-context [see *KWIC* (*K 4 f)]. **1969** *Computers & Humanities* IV. 23 Current book catalogs of John Wiley and Sons..among many others, already contain all the elements of data proposed for the data bank except a key-word-in-context index. **1971** *Ibid.* VI. 32 Indices and concordances..can be unlemmatized, like the key-word in context concordance to Livy. **1975** *Studies in Eng. Lit.: Eng. Number* (Tokyo) 121 The concordance employs the KWIC (key-word-in-context) principle: Each index word appears, separated by appropriate spacing, in the middle of a line, with its context extending to the left and right.

key, *sb.*[3] Add: Also, **key deer,** a subspecies of the North American white-tailed deer, *Odocoileus virginianus clavium*, found in the Florida keys.

1955 *Sci. News Let.* 29 Oct. 277/2 The Key deer is the smallest of all North American deer, only about 26 to 29 inches tall, 38 inches from nose to tail, and averaging only about 30 pounds. **1964** L. S. Crandall *Managem. Wild Mammals in Captivity* 590 Size seems to decrease from north to south, ranging..down to tiny creatures such as the Florida key deer.

key, *v.* Add: **2. c.** To cause (glued surfaces, pigments, etc.) to adhere.

1922 *Encycl. Brit.* XXX. 34/2 Roughing of the surfaces to be glued was adopted to secure keying. **1963** R. R. A. Higham *Handbk. Papermaking* ix. 226 Adhesives are used in coating to 'key' the pigment to the surface of the paper.

3. a. *to key up* (further examples); also, to render (someone) nervous or tense, freq. as *keyed-up* ppl. adj.; so *to key down*: to lower in pitch or intensity.

1889 *Cent. Dict.* s.v., *Keyed up*, high-strung; excited. **1904** Ade *True Bills* 35 He was all keyed up for Matrimony, and the next thing to do was to choose the Lucky Bride. **1922** R. S. Woodworth *Psychol.* vii. 126 It is the state of *excitement*, or of being 'all keyed up'. **1923** D. H. Lawrence *Phoenix II* (1968) 251 Everything that everybody feels is keyed down, and muted, so as not to impinge on anybody else's feelings. **1926** H. Crane *Let.* 5 Jan. (1965) 231 One really has to keep one's self in such a keyed-up mood for the thing. **1927** *Daily Tel.* 16 Aug. 12/5 He has keyed down his playing to such a pitch that we get no impression at all of his character as a man. **1961** C. McCullers *Clock without Hands* ix. 187 Although he was emotionally keyed up, Sherman yawned.

5. To distinguish (an advertisement) by some device which will identify responses to it. orig. *U.S.*

1905 Calkins & Holden *Art of Mod. Advertising* xi. 264 The advertiser likes to know which particular mediums pull best. To accomplish this the advertising is 'keyed'. Some form of address is used which can be varied in each magazine. **1927** *Daily News* 7 June 5/5 Advertisers who key their advertisements report their best results from the 'Daily News'. **1943** C. S. Forester *Ship* 56 How often had he devised ingenious methods by which to 'key' advertisements to discover which had the greatest pulling power. **1952** *Economist* 20 Sept. 718/1 This 'coupon sales' technique in a limited area..has the virtue in terms of publicity of being elaborately 'keyed'; the sales technicians can follow the success of their advertising and promotion schemes round by round. **1967** *Times Rev. Industry* Oct. 82/2 One keys advertisements and measures returns.

6. *Electronics.* **a.** To switch on or off, or from one state to another, by means of a key or relay, as in telegraphic transmission. **b.** To provide (electronic equipment) with means by which it may be switched abruptly from one state to another. Cf. *keying *vbl. sb.* 1.

1929 K. Henney *Princ. Radio* xvii. 443 (*heading*) Keying a transmitter. **1930** *Proc. IRE* XVIII. 1691 The transmitters are started or stopped and keyed locally or from Burnham. **1933** K. Henney *Radio Engin. Handbk.* xviii. 460 Keying the output is accomplished by means of a magnetic modulator. **1943** F. E. Terman *Radio Engineers' Handbk.* ix. 629 Unless the oscillator is keyed, the low-power portions of the transmitter operate continuously. **1954** E. Molloy *Radio & Television Engineers' Ref. Bk.* ix. 16 Means must be found ..to minimize any 'thumps' in the receiver when the transmitter is keyed. **1961** Gray & Graham *Radio Transmitters* vi. 137 Where 'break-in' is needed it may be necessary to key the transmitter oscillator to avoid generating an interfering signal ..in the receiver. **1966** M. Schwartz et al. *Communication Syst. & Techniques* vii. 280 The carrier is 'keyed' on and off to describe the two telegraph states.

7. To cause (something) to fit *in* with something else or *into* a group, pattern, etc. Also *intr.*

1947 *Sun* (Baltimore) 13 Nov. 18/2 (*heading*) Mr. Reuther's victory keys into a world-wide pattern. *Ibid.*, The Socialist Premier..went so far as to speak of an invisible 'orchestra leader' keying these strikes into a symphonic program whose purpose seemed clear. **1949** *Ibid.* 14 Feb. 8/4 (*heading*) Keying labor law into the facts of life. **1958** *Times Lit. Suppl.* 14 Mar. 133/3 It [*sc.* the Book of Mormon] is much longer than the New Testament. Moreover, it is elaborately 'keyed-in' to the Bible. **1960** E. Bowen *Time in Rome* i. 14 Lean young skyscrapers..key in with Rome's general virtuosity. **1969** *Guardian* 21 July 1/4 A carefully planned schedule which keys into the two-hourly orbit of their mother craft.

8. *trans.* To operate on (esp. to transfer (data) or to set (copy)), or to produce, by manipulating the keys of a keyboard. Also with various advbs.

1963 Gregory & Van Horn *Automatic Data-Processing Syst.* (ed. 2) v. 145 The user makes an inquiry by keying in an address in high-speed or bulk storage. **1964** C. Dent *Quantity Surveying by Computer* vii. 100 Checking can be done using a comparator machine or by reading over the print-out against the dimensions; this last method has the advantage of requiring the data to be keyed only once, instead of twice. **1967** Cox & Grose *Organiz. Bibliogr. Rec. by Computer* iv. 101 Messages requesting search information are printed out directly on the console typewriter. The search criteria are then keyed back through the console and the keyword file is searched. **1972** *Physics Bull.* Sept. 531/3 Instead of manually selecting individual type slugs from a case..the compositor keys in the copy at a specially laid out keyboard.

‖ **keyaki** (ke,a·ki). Also **kiaki.** [Jap.] An important Japanese timber tree, *Zelkova serrata*, or its pale, elm-like wood.

1904 [see *hinoki]. **1907** *Yesterday's Shopping* (1969) 203/1 Japanese Trays..Kiaki..Inlaid Wood..Polished. **1948** A. L. Howard *Man. Timbers of World* (ed. 3) 653 While there is some resemblance to the keyaki of Japan (*Z*[*elkova*] *serrata*), there is an entire absence of that beautiful sheen or lustre which the Japanese wood possesses in a high degree. **1965** J. Ohwi *Flora Japan* 381/1 *Zelkova serrata*... Keyaki. Tall trees with gray-brown smooth bark.

keyboard, *sb.* Add: **1. b.** *pl.* Musical instruments that have keyboards. *colloq.*

1971 *Ink* 12 June 19/2 Toni Brown wrote most of the songs..and plays keyboards. **1971** *Melody Maker* 13 Nov. 40/6 Rod's been playing keyboards since he was six. **1974** *Ibid.* 30 Nov. 46 The other four are the originals: Darryl Way on violin, Francis Monkman on keyboards and guitar, [etc.].

2. Also, a similar set in other kinds of machine. (Further examples.)

1846 H. Highton *Brit. Pat.* 11,070 3 Feb., Each terminus of telegraphic communication..is provided with..one of the key boards in use in single magnetic needle electric telegraphs. **1859** *Abridgments of Specifications relating to Printing* (Patent Office) 594 By depressing any one of the several keys on a key-board a lever is acted on,..thus giving motion to certain guide pulleys and cords which work the pistons by which the type is thrust out on to the guide plate. **1892** A. Powell *Southward's Pract. Printing* (ed. 4) xxxii. 310 *Hattersley's Composing Machine.* The type..is contained in the upper part of an iron framework about 3 feet square and 5 feet high. In the lower part of this framework is the key board. **1893** *Times* 25 Sept. 2/6 A machine..possessing remarkable capabilities for printing and adding up figures has recently reached this country from the United States... The keyboard has a number of buttons or keys, each representing a figure, and these are actuated for pounds, shillings, and pence just in the same way as a typewriter. **1911** *Encycl. Brit.* XXVII. 545/1 Composing machines in which the compositor put together types in the required order..by operating a keyboard which liberated them from magazines and assembled them in the order in which the keys had been struck. **1958** Gotlieb & Hume *High-Speed Data Processing* iii. 57 A keyboard transcriber for preparing magnetic tapes offers the advantage of eliminating records intermediate to the source document and the tape. **1970** O. Dopping *Computers & Data Processing* xix. 321 The programmer sits at an electric typewriter connected to the computer, and keys in his program with the keyboard.

keyboard (kī·bōəɹd), *v.* [f. the sb.] *trans.* = *key *v.* 8. Also *absol.* or *intr.*

1961 H. W. Larken *Compositor's Work in Printing* xii. 166 Concentration on the task of keyboarding the copy to the exclusion of any concern over the performance of the operations of matrix assembly, casting and distribution, gives a remarkably clean proof with a high setting speed. **1965** *Rep. Proc. Computer Typesetting Conf. London Univ. 1964* v. 153 She is now keyboarding at a rate of 120 to 150 words a minute. **1967** Cox & Grose *Organiz. Bibliogr. Rec. by Computer* ii. 43 Let me now digress..to illustrate some keyboards and photocomposing machines, for the more conventional way of generating character codes is to keyboard them. **1967** Karch & Buber *Offset Processes* ii. 40 The RCA-301 computer accepts punched paper tape and produces a new tape, adding justification (otherwise keyboarded by the Teletypesetter operator). **1969** *Sci. Amer.* May 68 After the encyclopedia..has been keyboarded into the computer. **1973** *Ann. N.Y. Acad. Sci.* CCXI. 284 It is now possible..to keyboard matter on an ordinary IBM Selectric typewriter..and feed it into an OCR machine where the characters are read automatically ..and then transcribed onto magnetic tape.

Hence **key·boarded** *ppl. a.*; **key·boarding** *vbl. sb.*, the action or process of keyboarding something; manipulation of the keys of a keyboard.

1965 *Rep. Proc. Computer Typesetting Conf. London Univ. 1964* iv. 143 Due to the rather inferior setting in the first hand keyboarded material counted, we probably opened up the computer parameters a little too widely. **1965** *Practical Printing & Binding* (ed. 3) iii. 41/1 An attachment to speed up the keyboarding of lines which require to be letter-spaced. **1967** V. Strauss *Printing Industry* ii. 96/2 The product of keyboarding is a punched tape which is used to assemble lines of justified type images on composing machines equipped for tape operation. **1970** *Brit. Printer* Mar. 67/3 No operator with less

than five years' keyboarding experience scored higher than 87 per cent in the test.

keyed, *a.* Add: **4.** *keyed-up*: see *key *v.* 3 a. **5.** *Electronics.* **a.** Of electronic equipment or devices: provided with a means by which it may be rapidly switched on or off, or 'keyed' (see *key *v.* 6). **b.** Of a signal: intermittent, abruptly stopped and started, as in telegraphic transmissions.

1942 *Proc. IRE* XXX. 15/1 Among the novel features of the design [of the television camera] are..keyed diodes for black-level setting. **1943** *Gloss. Terms Telecomm.* (*B.S.I.*) 65 *Type A1 waves* (*keyed continuous waves*), continuous waves which are keyed according to a telegraphic code. **1943** F. E. Terman *Radio Engineers' Handbk.* ix. 630 Keying causes the load placed on a power-supply system by the keyed stages to change abruptly. **1961** Gray & Graham *Radio Transmitters* vi. 138 Frequency variation of a keyed oscillator..can be reduced by making the oscillator frequency independent of supply voltages.

keyer (kī·əɹ). *Electronics.* [f. Key *v.* + -er[1].] A device for switching the signal supply to electronic equipment on and off.

1933 K. Henney *Radio Engin. Handbk.* xviii. 466 (*heading*) Tube keyer for transmitter. **1965** L. E. Foster *Telemetry Syst.* iv. 195 The signals are passed through a keyer which converts them from varying-amplitude to constant-amplitude signals of varying pulse width, or pulse duration. **1967** *Electronics* 6 Mar. 346/2 (Advt.), Electronic musical instrument engineer. Should have knowledge of transistor and diode keyers, wave shapers, [etc.].

keyhole, *sb.* Add: **1. b.** (See quots.) *slang.*

1896 Farmer & Henley *Slang* IV. 95/1 *Keyhole*, the female *pudendum*. **1927** *Jrnl. Abnormal & Social Psychol.* XXII. 14 Another term for the female organs is *cabbage*... Other symbols are *keyhole* and *bread*. The former is found infrequently.

c. *Astronautics.* A comparatively narrow area through which a spacecraft must pass to reach its objective. *colloq.*

1962 *Times* 21 Feb. 10/1 He swept towards the so-called 'keyhole in the sky', through which he had to pass if orbit was to be achieved. **1968** *Daily Tel.* 27 Dec. 1/2 It must hit a corridor only 35 miles wide. If it dips below this tiny 'keyhole in space', Apollo 8 and its crew will be burned up.

4. key-hole urchin, a flattened North American sea-urchin, with openings in the test, belonging to the genus *Mellita* or closely related genera.

1897 in Webster, Key-hole urchin. **1904** H. L. Clark in *Bull. U.S. Fish. Comm. 1902* XXXII. 565 *Mellita pentapora* (Gmelin). Key-hole Urchin. **1962** D. Nichols *Echinoderms* v. 76 Among the gnathostomes the clypeasteroid sand-dollars achieve probably the greatest specialization, some, such as the Key-hole Urchin, *Rotula*, becoming remarkably flat and possessing holes through the test.

keyhole, *v.* For *trans.* read *intr.*, and add further examples.

1905 *Kynoch Jrnl.* Oct.–Dec. 172 Some of these weapons ..shot wildly, the bullet invariably keyholing. **1910** Kipling *Land & Sea Tales* (1923) 190 The bullet must have ricochetted short of the butt, and it has key-holed, as we say. **1957** *Amer. Speech* XXXII. 194 *Keyhole*, of a bullet: to enter the target with the side foremost so that a rectangular or oblong hole is cut; of a handgun: to shoot bullets that keyhole.

keying, *vbl. sb.* **1.** (Examples corresponding to *key *v.* 6.)

1918 W. H. Eccles *Wireless Telegr.* (ed. 2) 247 When the high voltage in the repeat side does not exceed 100 or 200 volts it is easy to interrupt that circuit by aid of a morse key. Other suggestions for keying will be found in the section on Wireless Telegraphy. **1931** L. B. Turner *Wireless* v. 131 To start a Poulsen arc it must be struck... In keying, therefore, the oscillation cannot be started and stopped as in a spark transmitter. Instead, the Morse key—or..the set of relay contacts controlled by the key—is made to alter the wavelength slightly. **1966** S. Stein in M. Schwartz et al. *Communication Syst. & Techniques* vii. 280 With rectangular modulation, the states are described by which of a pair of possible frequencies is transmitted. The system is commonly known as frequency-shift keying (FSK). Phase-shift keying (PSK) is the analog of phase modulation.

Keynesian (kēi·nziăn), *a.* and *sb.* [f. the name of J. M. *Keynes* + -ian.] **A.** *adj.* Of or pertaining to the English economist John Maynard Keynes (1883–1946) or his economic theories, esp. regarding State control of the economy through money and taxes. **B.** *sb.* An adherent of these theories. Hence **Key·nesianism.**

1937 *Economic Jrnl.* XLVII. 153 The latest Keynesian analysis does indeed justify..such policies as redistribution of income. **1942** *Fortune* May 61 Mr. Keynes is now a Director of the Bank of England, and he is not the only Keynesian on the Board. **1946** *Ann. Amer. Acad. Pol. & Soc. Sci.* Nov. 284 The distinctive feature of this textbook ..is Keynesianism. **1947** A. P. Lerner in S. E. Harris *New Economics* (1948) IX. xlv. 654 The very simplified form of the Keynesian system..speaks as if there were only one kind of asset. *Ibid.* 655, I say, as a Keynesian, that the rate of interest is determined by the supply and demand for cash. **1951** R. F. Harrod *Life J. M. Keynes* x. 414 Not prepared to accept Keynesian doctrine. **1960** *Commentary* June 463/2 The war left a legacy of wild Keynesianism that continues in a new war economy to sustain prosperity. **1965** *New Statesman* 9 Apr. 560/1 Devious and vulgarised Keynesian calculations of the 'inflationary gap'. **1974** *Times Lit. Suppl.* 20 Sept. 1022/4 The notion of Keynesianism as a system might have been attractive to his vanity but it would have been repellent to his intelligence.

key-note, *sb.* Add: Also **keynote.**

2. b. *attrib.*, as **keynote address** or **speech** orig. *U.S.*, a speech, usu. an opening address, designed to state the main concerns or to set the prevailing tone for a conference or the like; often used at political rallies merely to arouse enthusiasm or promote unity; so **keynote speaker,** one who gives a keynote speech.

1905 *Milwaukee Jrnl.* 28 July 8/1 His address tonight will undoubtedly be a keynote speech. *c* **1908** *Great Issues & National Leaders of 1908* 54 He..began his 'keynote' address. *Ibid.* 133 The 'Keynote' Speech of Democracy. **1948** *Times* 21 June 5 The first two days [of the Republican Convention] will be taken up with..the 'keynote' speech. To heighten suspense and increase the mystery, the 'keynoter' is Governor Green. **1957** *Observer* 13 Oct. 2/5 Mr. Eugene Black, president of the International Bank for Reconstruction and Development, will deliver the 'keynote address' to-morrow. **1967** *Economist* 23 Sept. 1072/2 Mr Thorpe decided to set the tone for the week with a short keynote speech. **1968** *Globe & Mail* (Toronto) 3 Feb. 3/6 A first-rate keynote speaker from the party. **1973** *Observer* 14 Jan. 44/3 Roy Jenkins, MP, will give the keynote speech. **1973** *Black World* Mar. 49 A number of speakers are scheduled for each panel following keynote addresses.

keynote (kī·nōᵘt), *v. colloq.* (orig. *U.S.*). [f. the sb.] *trans.* To express the prevailing tone or idea of (something); to address (a meeting) as a keynote speaker. Hence **key·noting** *vbl. sb.*

1934 in Webster. **1945** *News Review* 10 May 14 Keynoting the surface atmosphere, a Daily Express cartoon showed strap-hanging travellers reading splash-headlined newspapers. **1955** E. Lowry in Sperber & Trittschuh *Amer. Pol. Terms* (1962) 226/1 Keynoting implies the ability to make melodic noises and give the impression of passionately and torrentially moving onward and upward while warily standing still. **1963** *Amer. Mineralogist* XLVIII. 959 Dr. Ira Cram..will keynote the meeting [of the Society of Exploration Geophysicists]. **1965** *Daily Tel.* 9 June 21/6 Keynoting the meeting, Dr. —— highlighted the need for [etc.]. **1974** *State* (Columbia, S. Carolina) 5 Mar. 9-B/2 Judge George N. Leighton, Howard University and Harvard Law School graduate, will keynote the symposium.

keynoter (kī·nōᵘtəɹ). *colloq.* (orig. *U.S.*). [f. Key-note *sb.* + -er[1].] A keynote speaker.

1926 *S.P.E. Tract* xxiv. 123 *Keynoter*, one who outlines the policy of a campaign, sounds the key note. **1948** [see *key-note *sb.* 2 b]. **1962** *Economist* 3 Nov. 462/1 The same evangelistic style he displayed as keynoter of the 1960 Republican convention.

keypunch (kī·pʌntʃ), *sb.* Also **key-punch, key punch.** [f. Key *sb.*[1] + Punch *sb.*[1]] A device for punching holes or notches in cards or paper tape in which one of a set of keys is pressed to produce the hole or holes corresponding to a particular character.

1933 L. J. Comrie *Hollerith & Powers Tabulating Machines* 7 The Powers automatic key punch..has automatic feeding and ejection. **1948** *Math. Tables & Other Aids to Computation* III. 127 Data, transferred by keypunch operators onto teletype tapes. **1970** O. Dopping *Computers & Data Processing* iii. 57 Tape can be punched in keypunches. **1970** *Honey* June 18/2 (Advt.), A keypunch operator..prepares the cards that tell computers what to do.

keypunch (kī·pʌntʃ), *v.* Also **key-punch.** [f. prec. sb.] **1.** *trans.* **a.** To produce holes or notches in (a card or paper tape) by means of a keypunch.

1947 *Jrnl. Chem. Education* XXIV. 62/1 A master table of cards, bearing *x* and *f*(*x*), is key-punched from printed tables. **1959** *Science* 16 Oct. 958/1 A number of other items were then added from cards which had been especially key-punched for the purpose.

b. To put into the form of punched cards or paper tape by means of a keypunch.

1959 *Jrnl. Assoc. Computing Machinery* VI. 13 The completed output definition form is..keypunched directly into cards. **1960** E. Delavenay *Introd. Machine Transl.* 119 The text of the foreword was given in French typescript to Mr Brown...He proceeded to the I.B.M. headquarters where he keypunched it. **1971** *Computers & Humanities* V. 304 The computer phase of the project consists of key-punching an Old Church Slavonic–English composite glossary. **1973** *Ann. N.Y. Acad. Sci.* CCXI. 319 There have been thousands of lines of poetry.. keypunched in order to produce concordances.

2. *intr.* To operate or use a keypunch.

1970 O. Dopping *Computers & Data Processing* iii. 59 In keypunching for automatic data processing, checking methods other than verification punching are normally used.

So **key·punching** *vbl. sb.* Also **key·punchable** *a.*, capable of being represented on punched cards or paper tape.

1954 *Jrnl. Assoc. Computing Machinery* I. 155/1 The punched-card input makes it possible to use keypunching and verifying techniques. **1965** L. E. Foster *Telemetry Syst.* iv. 221 Card-handling equipment provides keypunching, verifying, sorting..and listing services. **1969** P. B. Jordain *Condensed Computer Encycl.* 395 Programming..may end with coding (writing in keypunchable form). **1970** *Computers & Humanities* IV. 210 Gould adapted the *Plaine and Easie* keypunchable notation of 48 characters to a code specifically for the uniquely-written chant.

keypuncher (kī·pʌntʃəɹ). Also **key puncher.** [Partly f. Key *sb.*[1] + Puncher, partly f. *keypunch *v.* + -er[1].] **a.** = *keypunch *sb.*

1965 L. E. Foster *Telemetry Syst.* iv. 224 The semi-automatic area consists of..two oscillograph readers with six associated card punchers, 13 key punchers, four verifiers, three *X-Y* plotters, [etc.].

b. One who operates or uses a keypunch.

1967 Cox & Grose *Organiz. Bibliogr. Rec. by Computer* 144 The keypuncher should transcribe this.

keysender (kī·sendəɹ). *Teleph.* [f. Key *sb.*[1] + Sender.] A device for applying electric impulses representing a telephone number to a circuit by means of a set of keys (numbered 0 to 9), in place of a dial.

1929 *Telegraph & Telephone Jrnl.* XVI. 34/1 The light, airy switchroom, the tap-tap of the keysenders and the humming of the dials, this is my introduction into the subscribers' Paradise—an Automatic Exchange. **1934** *Post Office Electr. Engineers' Jrnl.* XXVII. 32/1 It has recently been decided..to provide mechanical keysenders both for P.B.X. operators and, if desired, for individual telephones. **1948** J. Atkinson *Herbert & Procter's Telephony* (new ed.) I. xi. 216/2 The keysender..is a device introduced as an alternative to the dial on large type private branch exchange switchboards to reduce the time expended by a P.B.X. operator when dialling numbers from the switchboard. By its use, the P.B.X. operator is free to attend to other calls whilst the keysender is still transmitting the digits which have been keyed up. **1970** *New Scientist* 5 Nov. 275/1 Tests at the St Albans exchange showed that the keysender was markedly quieter than the rotary dial and saved an average of 2–3 seconds on each call.

So **key·sending,** the use or operation of a keysender.

1928 C. W. Brown *Automatic Telephony Simplified* viii. 143 Key sending, and not dial sending, would, of course, be provided at manual exchanges when necessary. **1930** *Gloss. Terms Telegraphs & Telephones* (*B.S.I.*) 26 *Key-sending B-position*, a B-position equipped with digit keys for the purpose of making calls direct to an automatic exchange. **1938** Herbert & Procter *Telephony* II. ix. 393 Each manual exchange adopting direct 7-digit keysending utilizes a suitable automatic exchange as a routing centre for the whole of its indirectly routed traffic. **1950** J. Atkinson *Herbert & Procter's Telephony* (new ed.) II. xvii. 558/2 There are three alternative arrangements of the keysending equipment.

keyster, var. *keister.

keystone, *sb.* Add: **1. d.** *ellipt.* = *Keystone State* (in Dict., sense 5). Also *attrib. U.S.*

[**1803** in H. M. Jenkins *Pennsylvania* (1903) II. xii. 316 Pennsylvania is the Keystone of the democratic arch.] **1844** *Congress. Globe* 4 June 662/3 The old Key-stone has never furnished the Union with either President or Vice President. **1948** *Time* 21 June 22/3 The control of keystone Pennsylvania was one of the big question marks of the convention.

5. keystone effect, in *Cinemat.*, the formation of a trapezial projected image as a result of the line of projection not being normal to the screen; a similar distortion of a television picture in which a rectangular object gives a trapezial image; **Keystone State** (examples).

1914 J. B. Rathbun *Motion Picture Making* vi. 135 With the projector installed at one side of the screen, the keystone effect will be horizontal instead of vertical. **1940** D. G. Fink *Princ. Television Engin.* iv. 142 When such a scanned image is reproduced in the receiver, all the picture elements in the lower lines are spread out too far relative to those in the top line. This is the so-called keystone effect. **1967** *Electronics* 6 Mar. 79/1 (Advt.), The electron gun is set at an angle to the phosphor and the deflection system compensates for keystone effects. **1836** *Southern Lit. Messenger* II. 277 The little German farmer..in the Key Stone State. **1948** *Daily Ardmoreite* (Ardmore, Okla.) 27 Apr. 6/2 Republican aspirants matched strength in the politically important keystone state.

b. Used *attrib.* (with capital initial) in allusion to the slapstick comedy films produced by the Keystone film company, formed by 'Mack Sennett' in 1912; esp. of films featuring the 'Keystone Cops', a group of bumbling policemen. Hence *Keystones*, slang for 'police'.

1913 *Writer's Mag.* Nov. 188/2 If you have never written burlesque or vaudeville material..you will do well to watch a dozen of the Keystone comedies. **1918** R. H. Knyvett *Over There* 54 We received at this time the nickname 'Keystone soldiers', some genial ass conceiving that we looked as funny as the Keystone police. **1935** A. J. Pollock *Underworld Speaks* 66/2 *Keystone*, a special, uniformed police officer. **1964** W. Markfield *To Early Grave* x. 168 He was straddling the running board like a Keystone cop. **1967** [see *James Bond]. **1969** J. Wainwright *Big Tickle* 162 Get the keystones in the act. They watch. We perform. **1971** A. Hunter *Gently at Gallop* i. 6 The local Keystones move in demanding

alibis. **1974** N. Freeling *Dressing of Diamond* 107 The extreme infantilism of mentalities unable to distinguish between the Keystone Kops and a shattered child.

keystone (kī·stōᵘn), *v. Television.* [f. the sb.] *trans.* To subject to a keystone effect (see *keystone 5). Hence **key·stoning** *vbl. sb.*

1940 Zworykin & Morton *Television* xv. 473 If the complete sawtooth..is a modulation, the shape will be keystoned. **1940** D. G. Fink *Princ. Television Engin.* iv. 142 (*caption*) Trapezoidal distortion ('keystoning') of scanning pattern in the iconoscope. **1951** S. Deutsch *Theory & Design Television Receivers* xv. 495 This characteristic is counteracted by keystoning the CRT picture in the opposite sense by means of two permanent magnets. **1972** R. G. Middleton *Transistor Television Servicing Guide* (rev. ed.) vii. 86 A thermistor in the vertical-deflection coil circuit will occasionally..develop an abnormal resistance characteristic that will produce keystoning.

keystroke (kī·strōᵘk), *sb.* Also **key-stroke, key stroke.** [f. Key *sb.*[1] + Stroke *sb.*[1]] A depression of a key on a keyboard, esp. as a measure of work.

c **1910** *Comptometer* (Felt & Tarrant Manufacturing Co.) 4 Because a simple key-stroke does it all, the Comptometer saves 60% of time on addition. **1914** E. M. Horsburgh *Mod. Instruments & Methods of Calculation* 101 A beginner..might by a slurred or partial key-stroke make it add a wrong amount. **1921** J. A. V. Turck *Origin Mod. Calculating Machines* 160 Felt was interested in the solution of the problem for detection and correction of the errors in key-strokes. **1966** E. J. & J. A. McCarthy *Integrated Data Processing Syst.* v. 122 Most work is figured on the basis of 100 key strokes a minute. **1972** *Sci. Amer.* May 14/1 (Advt.), Although it's only 3 by 6 inches and weighs 9 ounces..it computes transcendental functions with a single keystroke in less than a second.

Hence **key·stroke** *v. trans.* = *key *v.* 8; **key·stroking** *vbl. sb.* Also **key·stroker,** one who operates the keys of a keyboard, esp. of a keypunch.

1966 *Ann. Rev. Information Sci.* I. 192 One of the impediments to greater use of automation in library and information center operations is the formidable prospect of keystroking an existing corpus of text. **1967** *Library Jrnl.* 1 Mar. 975 *All* publications..have to be key-stroked at some stage in their life-cycle. **1971** J. B. Carroll et al. *Word Frequency Bk.* p. xix. The samples [of text] were keystroked freeform onto IBM cards using IBM 029 keypunch equipment. **1971** *Computers & Humanities* VI. 40 If key-stroking plays a part in data entry, the key-stroker's own deviations will be added to those already in the text.

‖ **kgotla** (kgǫ·tlā). Also 9 **cotla, kotla.** [Tswana.] An assembly of tribal elders among some Bantu peoples; also, the place of such assembly (see also quot. 1857). Cf. *kuta.

1840 B. Shaw *Memorials of S. Afr.* xx. 303 Morokos Kotla had no attractions yesterday; we went and sat down in it, but we could not bear to remain. **1846** H. H. Methuen *Life in Wilderness* viii. 253 We threaded our way between the wattle hedges..and reached the cotla, or place of assembly, set apart in all native tribes for the purpose of holding public meetings. **1857** D. Livingstone *Missionary Trav. & Res. S. Afr.* i. 15 Near the centre of each circle of huts there is a spot called a 'kotla', with a fireplace; here they work, eat, or sit and gossip over the news of the day. **1896** H. A. Bryden *Tales S. Afr.* 192 They reached the large *kotla*, or enclosure, in the centre of the town, where Tapinyani's own residence stood. **1949** *Cape Times* 10 Nov. 2/8 Chief Bathoen, of the Bangwaketse tribe, gave evidence about the June *kgotla*, saying that the Assistant Government Secretary had told the tribesmen that they were gathered to discuss the right of succession of chieftainship. **1966** *New Statesman* 15 Apr. 531/2 The *kgotla*..until recently the arena of decision in Bechuanaland. **1972** *Times* 11 Sept. (Botswana, Lesotho & Swaziland Suppl.) p. iii/4 Until now, the traditional place for dealing with such matters has been the *kgotla*.

khad, var. Khud.

1925 *Dollar Mag.* Mar. 35 Our bungalow is situated on a little knoll with steep khads on three sides down to the river about 150 feet below. **1961** L. D. Stamp *Gloss. Geogr. Terms* 279/2 *Khad*, a torrent in the hills.

‖ **khadar** (kā·dăɹ). *India.* Also **kadir, khádar, khaddar, khâder, khadir,** and with capital K. [Hindi.] **a.** A flood-plain; land susceptible to flooding.

1832 G. C. Mundy *Pen & Pencil Sk.* I. v. 269 The road continued partly along the khâder of the river. **1879** Medlicott & Blanford *Man. Geol. India* I. xvii. 404 The alluvial plain itself..is composed of *bhángar*, or high land..and *khádar*, or low land, the low plain through which each river flows. **1882** W. Theobald *Mason's Burma* I. 4 The delta..of the Ganges, the richest land of Lower Bengal, being composed of *Khadir* land. **1887** J. M. Brown *Shikar Sk.* 247 Pig-sticking in the Kadir, or old bed of the Ganges. **1887** *Encycl. Brit.* XXII. 98/2 Eastward of the khâdar lies the sandy central tableland. **1904** A. Knox *Gloss. Geogr. & Topogr. Terms* 193 *Khadar* (Hind.), low lands fit for rice-growing. **1929** *Blackw. Mag.* Oct. 526/2 There is more game and greater variety of game in a kadir country than in any other kind. **1955** Brown & Dey *India's Mineral Wealth* (ed. 3) xix. 691 In the Brahmaputra valley, and the whole of the Assam plain is one great *khadar* or strath.

b. Recently deposited alluvium.

1919 D. N. Wadia *Geol. India* xxii. 251 The Khadar deposits are, as a rule, confined to the vicinity of the present channels. **1945** H. L. Chhibber *India* I. xiv. 187 The Newer Alluvium is confined to the river channels and their flood plains and is locally termed as *Khaddar*. **1949** M. S. Krishnan *Geol. India & Burma* xx. 519 The Newer Alluvium (called *Khadar* in the Punjab) is light coloured and poor in calcareous matter.

khaddar (kæ·dăɹ). Also **khadder, khadi.** [Hindi.] Indian home-spun cotton cloth.

1921 *Glasgow Herald* 27 Dec. 7 This tent will be made of hand-spun 'khadder'. **1922** *Ibid.* 22 Dec. 9 Delegates are already pouring into Neaya for the session of Congress that opens on December 26, and a huge marquee of khaddar (homespun) material has been erected. **1923** *Daily Mail* 23 Feb. 8 No person wearing clothes other than those made from 'Khaddar'—Mr. Gandhi's own favourite brand of hand-woven cloth—is tolerated in the 'presence'. **1924** C. D. Groth tr. *Rolland's Mahatma Gandhi* 72 Tagore, too, praised this *khaddar*, or *khadi*. **1925** E. S. Jones *Christ of Indian Road* v. 116 The whole city was dressed in white home-spun khaddar, the sign of the Nationalist. **1936** J. Nehru *Autobiogr.* xv. 106 Khadi clothes and third-class railway travelling demand little money. **1960** Koestler *Lotus & Robot* 278 The produce of the cotton mills is cheaper than homespun *khadi*. **1971** *Illustr. Weekly India* 18 Apr. 6 He alighted, this huge khadder-clad mountain of a man, his eyes alert and smiling. **1974** *New Yorker* 3 June 28/2 Gandhi called on all Indians..to raise cotton and to spin and weave it, so that there would be no shortage of *khadi*, or homespun cloth. **1975** *Bangladesh Times* 22 July 5/2 There may be easily manageable industries such as handloom, Khadi, village industries, handicrafts, [etc.].

khaki, *a.* and *sb.* Add: **B.** *sb.* Also (usu. *pl.*), a uniform or garment made from this fabric.

1936 *Amer. Speech* XI. 50 Unless he learns..to restrict the use of *khaki* to cotton uniforms of that shade..he is still a *John* [*viz.* a recruit]. **1956** *Ibid.* XXXI. 192 A marine's uniform wardrobe consists of *greens, blues, khakis*. **1956** H. Gold *Man who was not with It* (1965) xviii. 157 Once I sat wearing nothing but a pair of shiny starched new khakis. **1961** *Harper's Mag.* Oct. 43/2 The only clothing I owned was four pairs of khakis, three sweat shirts, a tweed jacket. **1970** 'T. Coe' *Wax Apple* (1973) i. 8 He was short and wiry, dressed in khakis and T-shirt.

C. *khaki election* (examples); also used of the general elections of 1918 and 1931. So *khaki vote*.

1913 *Everyday Phrases Explained* 164 *The Khaki Election.* This was the General Election of 1900, when the Government appealed successfully to the country for its approval of the South African War. **1917** G. B. Shaw *What I really wrote about War* (1930) x. 273 You may look forward with exultation to a century of triumphant khaki elections. **1918** *Times* 23 Aug. 8/4 Parliament clearly contemplated a khaki election in the best sense, with every soldier, sailor,..and Red Cross worker voting on the sole but splendid qualification of national service. *Ibid.*, Above all, in 1900 there was no khaki vote. That is the great dividing line which the Reform Act has drawn between 1900 and 1918. **1958** *Times* 15 Feb. 8/4 In November, 1918, she was adopted..as a candidate for Parliament. At the 'khaki election' she stood as a Coalition candidate. **1963** *Times* 6 Feb. 15/1 The 'khaki election' of 1900, when the Conservatives swept the country. **1968** A. Marwick *Brit. in Cent. of Total War* iii. 94 The election of 14 December 1918..was a 'Khaki' election in which the dominant element was patriotic hysteria. **1971** D. Ayerst *Guardian* xxx. 472 Sir Warren Fisher.. remained at the Treasury..and Ramsay Macdonald at 10 Downing St. He decided to hold a khaki election.

D. khaki bos, bush *S. Afr.*, (*a*) one of several herbs of the family Compositæ, *Tagetes minuta, Inula graveolens,* or *Schkuhria pinnata*; (*b*) = **khaki weed*; **khaki weed,** *Alternanthera repens*, a member of the family Amaranthaceæ which has spread from South America to South Africa and Australia.

1907 R. W. Thornton in *Agric. Jrnl. Cape of Good Hope* 7 Jan. 76 The Khaki Bush is a species of *Aplopappus*... The plant is an annual shrub, rising on a straight main stem to a height of about from two to three feet, with fine pointed leaves and light yellow flowers, which yield an enormous quantity of fine seed. **1913** C. Pettman *Africanderisms* 258 Khakibush. A species of *Aplopappus*. The name has reference to the dull fawn colour the withered leaves assume... The name is also applied to *Alternanthera Achyrantha*, R. Br., a troublesome weed now spread widely throughout South Africa, the seeds having been introduced from the Argentine Republic with imported fodder. The name was given to this plant because it made its appearance in military camps during the late war in places where it was previously unknown. **1932** R. Marloth *Flora S. Afr.* III. ii. 239 Species [of *Schkuhria*] 12, one as a weed in S.A., *S. bonariensis*, Tr., *khaki bush*. **1932** Watt & Breyer-Brandwijk *Medicinal & Poisonous Plants S. Afr.* 192 *Inula graveolens* Desf., Khaki bush, Khaki weed, Stinkweed (Australia) is an introduced species. *Ibid.* 195 *Tagetes minuta* L., an introduced weed, known as Khaki bush, Mexican marigold, and Kakiebos, is frequently referred to in the daily Press as an excellent parasiticide for cattle. **1947** H. C. Bosman *Mafeking Road* xix. 108 He seemed to have picked out all the useless bits for his pictures—a krantz and a few stones and some clumps of khaki-bos. **1956** A. G. McRae *Hill called Grazing* ii. 15, I had never in all my life seen such a profusion of cockle burr, Scotch thistle and 'khaki bush'. **1907** H. G. Mundy in *Transvaal Agricultural Jrnl.* V. 939 Khaki-weed or Amaranthus weed (*Alternanthera echinata*). **1932** R. Marloth *Flora S. Afr.* III. ii. 239 Species [of *Tagetes*] about 40, Amer., one as a weed in S.A., *T. minuta* (*khaki weed*). **1953** P. Lanham *Blanket Boy's Moon* ii. vi. 131 *Matekoane* is a weed resembling the khaki weed. **1958** N. Gordimer *World of Strangers* xiv. 212 Khaki-weed, the growth of neglect and desolation, standing dead and high. **1965** *Austral. Encycl.* I. 167/2 The South American annual needle burr (*Amaranthus spinosus*) and khaki weed (*Alternanthera repens*) have become serious pests in Queensland.

Hence (often with capital initial) **kha·kied** (kā·kid) *pa. pple.*, dressed in khaki; *fig.* possessed by a militant spirit; **kha·kiism,** militant spirit or policy; **kha·kiite,** an enthusiast for a war policy; **kha·kiness** = **khakiism*. (All *temporary*.)

1900 *Westm. Gaz.* 4 May 2/2 The Portsmouth electors.. did not allow themselves to be persuaded..into khakiness. **1900** *Daily Express* 26 June 5 (Cassell's Suppl.), The departure of khakied troops for the front. **1900** *National Rev.* June 535 There is no reason to suppose that Lord Salisbury has, so far, surrendered to Khakiism. *Ibid.*, The Khakiites are strenuous and determined. **1904** *Westm. Gaz.* 19 Nov. 2/1 The last election, when certainly the confidence-trick was indeed played on a Khakied nation. **1945** *Brit. Jrnl. Psychol.* Jan. 34 These millions of Khakied citizens run the whole gamut of traditional military gradation.

‖ **khal** (kāl). *India.* [Bengali.] (See quot. 1958[2].)

1903 *Jrnl. Trop. Med.* VI. 200/2 This [*sc.* the anahar plant] is steeped in the big khal at Ishapur, and during the fermenting stage mosquitoes are generated very plentifully. **1904** A. Knox *Gloss. Geogr. & Topogr. Terms* 193 *Khal* (Bengali), a creek. **1958** N. Ahmad *Econ. Geogr. E. Pakistan* i. i. 15 The upper part of the plain between the Garai-Madhumati and the Padma,..is full of rivers, streams, and *khals* of various sizes. *Ibid.* 338 *Khāl*, narrow natural channel of water.

khalasi (kălæ·si). Also **calassie, kalashi, -y** (-ʃ-), **kalas(s)i, khalishee, khelasse,** etc. [Hind.] A native servant or labourer, esp. one employed as a seaman. Also *attrib.* in **khalasi watch** (see quot. 1911).

1800 F. Gladwin tr. *Ayeen Akbery* I. ii. 232 The tundeil is the chief of the khelasses, or sailors. **1848** *Alfred in India* 44 Alfred saw the *calassies*, or tent-pitchers, beginning to take down the tent to pack it on the cart. **1848** J. H. Stocqueler *Oriental Interpreter* 115/1 *Kalashy*, an Indian menial. His business is, properly speaking, confined either to what relates to camp equipage, or to the management of the sails and rigging on board a budjrow or river boat. *a* **1865** Smyth *Sailor's Word-bk.* (1867) 421 *Khalishees*, native Indian sailors. **1907** M. Roberts *Flying Cloud* 22 He had sailed with Khalasi crews for ten years. **1911** *Coast Seamen's Jrnl.* 9 Aug. 1/2 There is growing up a system, mostly in steamers, sometimes called Kalashi watches. This means that certain men are kept on the regular watch-and-watch while the other members of the crew are what is called 'day men'. The 'day men' work all day and are supposed to sleep all night. **1917** *Yachting Monthly* XXII. 197/1 Nothing loth the Kalassis [*sc.* bunder boat crews at Karachi] obeyed. **1931** W. H. Parker *Leaves from Unwritten Log-Bk.* xi. 95 All hands were kept on deck (or keeping Kalasi watches) for two days and nights. **1957** D. G. O. Baillie *Sea Affair* 237 The Lascars, known as *khalassies*, belong to the Deck Department. They are Moslems. **1963** P. J. Abraham *Last Hours* 9, I noticed some of the seamen, or Khalassis, in their blue dungaree uniforms. **1973** *Times of India* 16 Oct. 4/5 Those on strike include..winch operators, khallasis, tindals and other workmen.

Khaldian (kæ·ldiăn). [f. *Khaldis* or *Khaldi* the name of the supreme god in Urartu + -an, -ian.] **a.** Also **Khaldæi, Khaldean.** Orig. the divine offspring of Khaldi; more usually, a native or inhabitant of the ancient Armenian kingdom of Urartu. **b.** Also **Khaldic.** The language spoken by this people. Also as *adj.*

1882 A. H. Sayce in *Jrnl. R. Asiatic Soc.* XIV. 412 Khaldis..was also the father of other gods who were called 'the Khaldians' after him. **1898** —— in J. Hastings *Dict. Bible* I. 140/1 The supreme god of Armenia was Khaldis..from whom the inhabitants of the country took the name of 'people of Khaldis'. From this was derived the name of Khaldæi or Khaldeans. **1901** *Westm. Gaz.* 29 July 5/2 The Khaldians..are to be identified with the Chaldaioi of the Greek writers, who figure prominently in Xenophon. **1908** A. H. Sayce in *Encycl. Relig. & Ethics* I. 793/1 The present article deals with Proto-Armenian religion as revealed in the Vannic or 'Khaldian' cuneiform inscriptions. **1925** —— in *Cambr. Anc. Hist.* III. viii. 170 If another title is wanted in place of Vannic, Alarodian would be preferable to Khaldian. **1939** L. H. Gray *Foundations of Lang.* xii. 381 *Khaldic* (also called *Urartaean* or *Vannic*), the language of a kingdom which flourished between 900 and 600 B.C...is known from nearly two hundred inscriptions in Akkadian cuneiform. **1952** O. R. Gurney *Hittites* vi. 124 A direct descendant of Hurrian is the language of the kingdom of Urartu, sometimes called Vannic or Khaldian. **1959** *Chambers's Encycl.* XIV. 243/1 The people were called Khaldis, they worshipped a sun-god and an air-god. **1966** *Ibid.* XIV. 179/1 Urartu was an ancient kingdom in the highlands of Armenia...The people called themselves 'children of Khaldi' (their national god) and so have sometimes been called Khaldians by modern historians.

khalifa. Add: **2.** *S. Afr.* Also **califa, chalifah, kalifa.** A Malay sword-dance which originally had religious significance. Also *attrib.*

1856 *Cape Monitor* 16 Jan. 2/3 (*heading*) Checking the Malay 'Califa'. *Ibid.*, Several Malay priests have been examined..and it would seem from their evidence that the Califa is by no means part of the Mahomedan religion, and..ought only to be played on a certain night in each year. **1861** *Cape Monthly Mag.* Dec. 356 The most characteristic of their customs is the 'Khalifa', a religious ceremonial of the highest solemnity. **1867** M. Kollisch *Mussulman Population Cape Good Hope* 59 The feasts which the 'Khalifa' involved were highly amusing. **1900** *Diamond Fields Advertiser* (Kimberley) 31 May 2/3 Town Hall, Kimberley. Khalifa Representation in aid of the fund for the relief of the sick and wounded in the Transvaal war. **1950** M. Masson *Birds of Passage* xv. 141 But the *pièce de résistance* of the whole affair was a kalifa dance by two hundred Malays. **1953** Du Plessis & Lückhoff *Malay Quarter* iii. 62 During the previous century, there were many bands of Chalifah-players. *Ibid.*, The chief priest..is designated by the term *Chalifah*, which has now, through common usage, come to be applied to the performance itself. **1972** *Standard Encycl. S. Afr.* VI. 147/1 The most spectacular custom of the Cape Malays is the performance of the Chalifah.

Khalka, Khalkha (ka·lkă), *sb.* and *a.* [Native name.] **A.** *sb.* **a.** One of a Mongol people in Outer Mongolia; the people themselves. **b.** The language spoken by them. **B.** *adj.* Of or pertaining to this people, their language, or the territory they inhabit.

1873 *Jrnl. R. Geogr. Soc.* XLIII. 122 The K of Kuren being pronounced by the Kalkas,..would have a word having..a much greater affinity than many Chinese names have to those in use amongst the natives of the soil. **1876** H. H. Howorth *Hist. Mongols* I. viii. 468 Lamaism had spread very greatly among the Mongols, and it was a subject of pride among the Khalkhas to believe that they had..a regenerate Buddha among them. **1926** A. N. J. Whymant *Mongolian Gram.* i. 1 The Khalka (or as it is generally spelt according to Russian orthography, Khalkha) Mongolian possesses seven vowels and twenty consonants. *Ibid.* 4 The Kalmuck presents consonantal difficulties over the Khalkha. *Ibid.* 5 A common word in all dialects of Mongolian is one mostly heard as *chichik* (a flower) among the Khalkhas. **1948** D. Diringer *Alphabet* 318 The three principal [Mongolian] dialects, Khalkha, Kalmuck and Buriat. **1967** D. S. Parlett *Short Dict. Lang.* 87 Modern literature based mainly on Khalkha dialect in an alphabet based on Cyrillic. **1970** H. Kevin *Incredible Mile* xlviii. 132 If Bernard Shaw..had known as much about Mongolian costume as about English dialects..he could have told me..that blue and brown meant a Khalka..that if a woman's sleeves had horizontal pleats they would be of a Khalka.

|| **khalukah** (halu·kă). Also **chalukah, haluka, halukkah, haluqah.** [Talmudic Heb. *ḥ'luqqāh* distribution, f. Heb. *ḥālaq* to distribute.] The distribution of contributions or donations sent by the Jews of the diaspora to support the Jews in Palestine. Now *Hist.*

1880 *Encycl. Brit.* XIII. 686/2 Annual contributions (*haluka*) amounting to about £50,000 a year. **1920** *19th Cent.* Oct. 627 Fear has been expressed that Jewry in Palestine..will degenerate into a new form of 'Chalukah'. **1923** W. F. Livingstone *Galilee Doctor* 48 The Jew who followed his native genius, engaged in commerce, and made a success of it, was independent of the khalukah. **1927** *Observer* 9 Apr. 8/3 This class has not increased owing to the failure of the Haluka by which it was supported. **1941** *Universal Jewish Encycl.* V. 188/1 In..the 1930's and 1940's, the collections dwindled away.. The Halukkah had been replaced by contributions through Zionist organizations. **1944** M. Hadas tr. *Elbogen's Cent. Jewish Life* ix. i. 246 A system of distributing alms to the various national groups (*kolelim*)..was elaborated and given the name *Halukkah*. *Ibid.* 250 He also worked for the education of girls and for freeing the Jewish population from the scourge of the *Halukkah*. *Ibid.* 251 The *Halukkah* used its power to penalize parents who sent their children to the modern schools and to hinder any progress in making them productive. **1947** S. W. Baron *Social & Relig. Hist. Jews* (ed. 2) V. xxiii. 37 The 'Babylonian' congregation..demanded independent control over the distribution of charitable funds for the poor of the Holy Land. It thus adumbrated the nineteenth-century struggles over the distribution of *haluqqah* funds raised in the various countries of the dispersion. **1962** *New Jewish Encycl.* 181/2 Halukkah..begun in olden times and continued until replaced by modern fund-raising. **1970** *New Stand. Jewish Encycl.* 831/1 Halukkah, the distribution of collections made abroad for the support of the poor in Palestine. 'H. Jews' were those living on such contributions.

Khamba (ka·mbă). Also **Kamba, K(h)ampa.** [f. Tibetan *Kham* East Tibet + suffixal element *-ba* or *-pa*.] **a.** A Tibetan people from Kham; one of this people. **b.** The language spoken by this people. Also *attrib.* or as *adj.*

1908 C. Bell *People of Tibet* i. 4 Eastern Tibet.. This area is known as Kam, the people as Kam-pa. *Ibid.* vi. 59 The Lhasa men, they used to say, were fonder of talking than of working; the Kam-pas (eastern Tibetans), irritable and quarrelsome. **1955** G. T. Bull *When Iron Gates Yield* iv. 58 The soldiers..were looking forward to a good scrap, for such is the Tibetan Khamba's temperament. *Ibid.* v. 67 A Kamba Tibetan..would rather his blood be spilt in the snow..than to be found alive and defamed. **1957** R. Ford *Captured in Tibet* i. 22, I had picked up enough words of the Khamba dialect to be able to talk to the local inhabitants. *Ibid.* iii. 47 When Radio Peking switched its Tibetan broadcasts to a more suitable hour, it still had no audience, for the news-reader spoke with a Khamba accent. **1960** 'S. Harvester' *Chinese Hammer* ii. 31 The only fighting in the country is in the east and restricted to Khambas. **1962** H. E. Richardson *Tibet & its Hist.* i. 11 The Khampas..live between the upper Yangtse and the Chinese border. *Ibid.* xii. 201 The Chinese..exploited the traditional mistrust of Lhasa officialdom existing among the Khampa and Amdowa tribes to the east of the upper Yangtse. **1973** *Times* 20 Oct. 3/2 The fourteenth Dalai Lama is..now 38 and has lived in exile in the Indian Himalayas since 1959, when Khamba rebels persuaded him to flee from Lhasa with them after their abortive uprising against the Chinese occupation.

Khamitic, var. Hamitic *a.*

1880 A. H. Sayce *Introd. Sci. of Lang.* II. vi. 181 A number of dialects..are classed together as Ethiopian or Khamitic.

khan². (Later examples.)

1947 *Archit. Rev.* CII. 99 (*caption*) The nearest building, with a row of small domes, is the khan or shopping centre. **1951** A. Christie *They came to Baghdad* i. 8 Captain Crosbie..turned down a small alleyway into a large Khan or Court. **1958** R. Liddell *Morea* vi. 165 The buses going to Kalamata pull up at a khan near the village of Abyochori.

|| **khana** (kā·nă). *India.* [Hind. *khānā* food, dinner.] Food, a meal.

1859 G. F. Atkinson *Curry & Rice* Plate 13 (*caption*) There, now, an invitation to dinner!—to a 'Burra Khanah', literally a grand feed. **1888** Kipling *Phantom 'Rickshaw* 33 The khansamah went to get me food. He did not go through the pretence of calling it 'khana'—man's victuals. **1933** *Discovery* Nov. 349/1 Vital ingredients for the preparation of the *khana* for the Sahib. **1953** E. M. Forster *Hill of Devi* 73 The cooking of our Indian khana. **1955** J. Lehmberger in *Caribbean Anthol. Short Stories* 19 At noon they would be eating their khana, sitting over the open fire.

|| **khanda,** var. Khanjar.

1888 Kipling *From Sea to Sea* (1899) I. vii. 53 'And what do you make in Udaipur?' 'Swords,' said the man..throwing down an armful of..tulwars, and khandas.' **1957** *Encycl. Brit.* XX. 647/2 In that rite, with a two-edged dagger (*khanda*) water is stirred up. **1971** *Daily Tel.* 11 June 3 (*caption*) An alternative to the kirpan is the Khanda, a small dagger.

|| **khanga** (ka·ngă). Also **kanga.** [Swahili.] In East Africa, a fabric printed in various colours and designs with borders, used esp. for women's clothing.

1967 *Sunday Times* 15 Jan. 31/5 *Kangas*..are the most riotous with words and even jokes printed on them. **1969** *Flamingo* (E. Afr.) Dec. 18/2 They often wear the two-piece dress, kanga. **1970** [see *Kitenge]. **1971** *Daily Nation* (Nairobi) 10 Apr. 5/1 (Advt.), Kitenge Khanga Kikoy..at African Drapering. **1975** *Daily Tel.* 7 July 9 Khangas..come from Kenya, and are fine soft cotton rectangles about 60 in by 42 in, printed in literally hundreds of the most fabulous colours and designs... Wear them as a stunning beach dress or sarong type skirt knotted dead front.

|| **khansu** (kanzu·). Also **khanzu.** [Swahili *kanzu* shirt, f. Arab. *kasā* to clothe.] A loose outer garment worn in East Africa.

1966 W. Hastings *Killing in Black & White* i. 3 Steel watched the Africans in the street below. Only the older men..wore loose-fitting khanzus; the young favoured European dress. **1972** D. Creed *Trial of Lobo Icheka* i. ii. 23 Salim, went away, his white khanzu swishing agitatedly about his legs. **1973** *Daily Tel.* (Colour Suppl.) 30 Nov. 63/1 A servant at a club was going to be wearing a *khanzu*, somewhat resembling a nightshirt.

|| **khanum** (kā·nəm). Also **caño, canum, hanim, hanum.** [Pers. *khānum*, Turk. *hanim*: see Khan¹.] In the near East, a lady of rank, the wife of a khan. Also = Mrs., madam, as a title or term of address.

1824 J. Morier *Adventures Hajji Baba* I. xxiv. 256 My situation is that of hand-maid to the khanum, so my mistress is called. **1826** Leyden & Erskine tr. *Mem. Zehir-Ed-Din* 12 The second daughter of [Shir Haji Beg], Kutluk Nigar Khanum, was my mother. **1834** [see *Khatun]. **1848** J. H. Stocqueler *Oriental Interpreter* 145/2 The wives..are called Khanum, 'Lord', and signifies Lady, the wife of a Khan. **1859** C. R. Markham tr. *Gonzalez de Clavijo's Narr. Embassy to Court of Timour* vi. 145 The first wall and tents were for the use of the chief wife of the lord, who was called the Caño. **1900** *Daily News* 9 Jan. 12/6 The Queen of Greece..won the hearts of the Moslem feminine world by returning in person all the visits of the harem ladies. **1929** *Spectator* 21 Aug. 259/1 Closely-veiled figures of Turkish Khanums. **1930** L. Origa *Portrait of Turkish Family* xii. 123 Hasan..peered at my mother..and said uncertainly: 'I went to your house first... I did not know it was like that, hanim efendi.'

khapra (kā·pră). [ad. Hindi *khaprā* destroyer, f. *khapna* to destroy.] In full, *khapra beetle*. A small, brownish-black beetle, *Trogoderma granarium*, of the family Dermestidæ, native to India but widely found elsewhere as a pest of stored grain.

1896 E. C. Cotes in *Indian Museum Notes* III. 119 It [*sc.* a dermestid beetle] is known as *Kapra* in the Delhi bazaar, where it is said sometimes to destroy as much as six or seven per cent. of wheat stored in godowns. **1928** *Bull. Entomol. Res.* XVIII. 251 (*title*) Investigations on the control of the Khapra beetle. **1955** *Hilgardia* XXIV. 1 The khapra beetle, *Trogoderma granarium* Everts, was first identified in the United States in stored wheat and barley..in October, 1953. **1959** *Jrnl. Econ. Entomol.* LII. 313/1 Each food sample was placed in a plastic cell with a single khapra-beetle egg. **1971** *New Scientist* 27 May 503/2 Khapra beetles may be baited by merely suspending large pieces of jute fabric on top of the infested grain.

kharif (kari·f). [(Hind.) Arab. *ḵarīf* gathered, autumn, harvest, autumnal rain.] **1.** In India, the autumn crop, sown at the beginning of the summer rains.

1845 *Encycl. Metrop.* XXIII. 765/1 [Hindustan's] harvests..are equally profitable both in Spring (rabi) and Autumn (kharif). **1882** W. W. Hunter *Indian Empire* 385 The *kharif* or autumn harvest. **1886** A. H. Church *Food-Grains India* 99 Where indigo is grown for the kharif, barley is its usual accompaniment in the rabi. **1911** *Encycl. Brit.* XXVI. 610/2 The *kharif*, or autumn crops, sown in June and reaped in October or November. **1969** *Pioneer* (Lucknow) 13 Aug. 2/3 There has been no kharif sowing in about 2,000 villages in Western Rajasthan due to failure of monsoon this year, according to official sources here. **1975** *Bangladesh Observer* 31 July 5/4 In spite of improved irrigation facilities most areas of India are dependent on rainfall for cultivation. Timely and adequate rainfall is essential for good kharif crops.

2. Also **khareef.** The rainy season in the Sudan.

1920 *Blackw. Mag.* Nov. 668/1 The gazelle here do not drink from the hour the kharif ends. **1951** R. A. Hodgkin *Sudan Geogr.* ii. 10 If we plant a grass seed and a tree seed at the beginning of the *kharif* in a semi-desert they will both start to grow during the months of rain. *Ibid.*, vi. 61 If there is..enough rain during the *kharif*, there will be a good crop. **1961** L. D. Stamp *Gloss. Geogr. Terms* 280/1 *Kharif*, applied to the rainy season in the northern Sudan.

Kharoshthi (kăro·ʃti). Also **Kharoshthi, Kharosthi, Kharosti.** [Skr. *kharoṣṭī*.] The name of one of the two oldest alphabets in India, derived from Aramaic and used for about seven centuries from c 300 B.C. in north-western India. Cf. *Brahmi.

1891 A. Cunningham *Coins Anc. India* 36 A script named *Kharosti*, which was read from right to left, is said to have been one of the forms taught to the youthful Buddha. *Ibid.* 37 As the name was derived from the inventor of the writing, I think it probable that the name of Zoroaster, or *Zardusht* himself, may have been preserved in the term *Kharosti*, as the inventor or introducer of that particular form of writing. The Kharosti alphabet, under this view of its origin, would naturally have been imported into every country under Achæmenian rule. **1902** [see *Brahmi]. **1934** A. Toynbee *Study of Hist.* III. 130 In India, the Kharoshti script is certainly derived from the Aramaic alphabet. **1948** D. Diringer *Alphabet* ii. v. 301 Many Indian scripts..exist to-day used for tongues belonging to various linguistic groups... All these scripts seem to have descended from two prototypes, the *Kharoshthi* and the *Brahmi*. **1958** L. Woolley *Hist. Unearthed* vi. 75 The documents..were in strangely different scripts and languages... Many were in Kharoshthi, a script known from inscriptions in north-west India. *Ibid.* 129 Most of the Kharoshthi documents were written on wooden tablets. **1973** W. B. Lockwood *Panorama Indo-Europ. Lang.* 193 Two alphabetic scripts were in use in ancient India... One of these, the Kharoshthi..is found only from the middle of the third century B.C. to the third century A.D... It seems certain that the Brahmi script, like Kharoshthi, also goes back to a Semitic model.

Khasi (kā·si). Also † **Cossyah, Khas, Khasia(n), Khasiya.** [f. the name *Khasi* (see below).] **a.** A member of a Mongoloid people found in the Khasi and Jaintia Hills in north-eastern India; also, an Indo-Aryan people inhabiting the hills of Kumaon and Garhwal; a member of one of these peoples. **b.** A language of the Mon-Khmer group spoken by them. Also *attrib.* or as *adj.*

1789 in Seton-Karr & Sandeman *Selections from Calcutta Gaz.* (1865) II. 218 We understand the Cossyahs, who inhabit the hills to the north-westward of Sylhet, have committed some very daring acts of violence. **1824** R. Heber *Narr. Journey Upper Provinces India* (1828) I. xvii. 479 The Khasiyah nation pretend to be all Rajpoots of the highest caste. **1838** W. Pryse *Introd. Khasia Lang.* p. iii, The Khasia or Cossyah is the dialect of a small tribe, reckoning..somewhat under 200,000 souls. *Ibid.* p. vi, It is hoped that the majority of the Khasia words..will be found in the Vocabulary. **1882** *Encycl. Brit.* XIV. 295/1 True Khas can be distinguished from the Laos only by the lobe of the ear. **1907** P. R. T. Gurdon *Khasis* i. 1 We can..suppose that the Khasis are an offshoot of the Mon people of Further India. **1911** *Encycl. Brit.* VII. 830/2 In language 27,172,697 of the inhabitants [of Eastern Bengal and Assam] spoke Bengali, 1,349,784 speak Assamese, and the remainder Hindi and various hill dialects, Manipuri, Bodo, Khasi and Garo. **1928** *Trans. Scottish Ecclesiol. Soc.* V. 230 A barbaric tribe of the Khasias burn their dead and raise to their honour menhirs... If a Khasian is in distress of any kind, he prays to some deceased ancestor. **1926** *Other Lands* July 151/2 There are six Khasis, five Nepalis, and two Lepchas from the Darjeeling Hills. **1936** *Discovery* July 205/2 The Khasis and Syntengs are an isolated group of the great Mon-Khmer family. **1937** H. W. Tilman *Ascent Nanda Devi* i. 7 The..Khasiyas, a race of a caste lower than the Brahmans or Rajputs..yet generally allowed to be also immigrants from an Aryan source, adore..the mountain god Siva. **1967** D. S. Parlett *Short Dict. Lang.* 73 *Khasi*,..a dialect apparently transitional between Mon-Khmer and Munda language types.

Khaskura (kāskū·rā). Also **Khaskra.** [Native name.] An Indic dialect spoken in Nepal.

1911 [see *Gurkhali]. **1928** R. L. Turner in Northey & Morris *Gurkhas* iv. 70 The Indo-Aryan dialect,..variously called Khaskurā, Parbatiyā, Gorkhālī, or officially and preferably, Nepālī. **1943** E. Shipton *Upon that Mountain* vi. 112 We had two main occupations on the voyage out. One was to learn Khaskra, a language spoken by the Sherpas. **1950** T. Longstaff *This my Voyage* v. 93 Besides speaking Khaskura, the common tongue of Nepal. **1971** J. Pemble *Invasion of Nepal* i. 10 Nepali is but one dialect among a number of related tongues known generically as *Khas kura* (Khas speech) or *Parbatiya* (relating to the hills). Modern Nepali is often referred to by either of these names.

khassadar (kæ·sădāɹ). Also **kassidar.** [Native name.] In the border region of north-western India and Afghanistan, a local militiaman (see quots. 1930 and 1950).

1901 T. H. Holdich *Indian Borderland* xii. 273 An Afghan force of about 1,000 men, half of whom were Kassidars (irregulars)... The Kassidars were but half-trained troops. **1909** *Daily Chron.* 13 Apr. 3/5 The Khyber Pass will be closed to caravans..due to the aggressive attitude of the Khassadars, or Militia, on the Afghan border. **1923** *Glasgow Herald* 16 July 6 The object of the murder is to discredit the Khassadar movement. **1930** *Aberdeen Press & Jrnl.* 22 Apr. 5/2 The khassadar is an extremely irregular soldier, performing a sort of police duty on caravan routes on the North-West Frontier. **1950** W. K. Fraser-Tytler *Afghanistan* xiii. 260 Khassadars were tribal police responsible for the general safety of the roads running through the areas of their respective tribes. **1965** K. P. S. Menon *Many Worlds* 94 These outposts are manned by Khassadars, a kind of irregular tribal police, formed for the purpose of enforcing the security of the road.

khat, var. Kat.

1858 [see Kat]. **1932** E. Waugh *Black Mischief* iv. 119 Mahmud el Khali bin Sai'ud..sat among his kinsmen, moodily browsing over his lapful of khat. **1960** A. Waugh *Foxglove Saga* vi. 107 The standard British Pig, the very mention of whose name was said to make distant sheikhs and rebellious tribesmen in savage areas of the Empire drop their weapons and chew khat with a renewed vigour. **1969** Verdcourt & Trump *Common Poisonous Plants E. Afr.* 98 Khat is a narcotic drug well-known to Arabs and Somalis. *Ibid.*, If khat-eating is accepted as less harmful than some have supposed, then a very profitable native local industry could be properly controlled.

|| **khatak** (katā·k). Also **khata, khatag.** [Tibetan *k'a-btágs* a scarf of salutation.] In Tibet, a scarf presented to visitors.

[**1863** E. Schlagintweit *Buddhism in Tibet* xiii. 190 The silken scarfs inscribed with sentences which Tibetan politeness requires should be offered by visitors or enclosed in letters..are called in Tibetan Khatak, or Tashi Khatak, 'scarf of benediction'.] **1902** *Daily Chron.* 23 Oct. 3/1 Having presented his *khatag*, and placed in the lap of the Dalai Lama a piece of gold..he took his seat..about ten feet distant from the Grand Lama. **1960** 'S. Harvester' *Chinese Hammer* vii. 70 The traditional ceremony of exchanging white silk *khatas*, scarf tokens of goodwill usually decorated with swastikas.

khatib (katī·b). Also **khateb.** [ad. Arab. *ḵaṭīb*.] A Muslim preacher; one who recites the khutbah.

1625 Purchas *Pilgrimes* II. ix. xix. 1661 These Atollons are subdiuided into many Ilands, in each of which..is a Doctor called Catibe, superior in the Religion of that Ile, who hath vnder him the particular Priests of the Moschees. **1821** J. Leyden tr. *Malay Annals* v. 50 The letter was read by the khateb. **1839** T. J. Newbold *Pol. & Statistical Acct. Straits of Malacca* I. v. 249 The Khatib..recites the khatbeh, an oration or sermon, in praise of God..on Friday, in the mosque. **1875** T. P. Hughes *Notes Muhammadanism* 131 The Khatib or preacher then seats himself on the *Mimbar* (pulpit). **1971** R. Lewis *Everyday Life in Ottoman Turkey* 43 The khatib, who delivered the Friday sermon and led the invocation of God's protection on the Sultan and his family.

Khatti, var. *Kheta.

|| **khatun** (kā·tūn). Also **kadun.** [Pers.] A lady. Also used as a title of courtesy.

1834 J. Morier *Ayesha* I. iv. 80 She once made the sign of the cross..but now she is a kadûn—a khanûm, a head of a harem. **1927** *Blackw. Mag.* Nov. 687/2 My wife and other senior *khatuns*. **1959** P. N. Bazaz *Daughters of Vitasta* vi. 144 The next queen to be noticed by contemporary historians is Gul Khatun. *Ibid.* vii. 153 The last of the Muslim queens to achieve renown was Habba Khatun.

khaya (kəi·yă, kē·yă). [mod.L. (A. de Jussieu 1830, in *Mém. Mus. Hist. Nat.* XIX. 249), ad. Wolof *khaye*.] A tropical African tree of the genus so called, belonging to the family Meliaceæ; the timber of a tree of this kind, better known as African mahogany.

1910 H. Thompson *Gold Coast: Rep. Forests* 38 in *Parl. Papers* (Cd. 4993) LXV. 201 *Khayas* and *Pseudocedrelas* were frequently met with in this direction [*sc.* to the east of N'kwansia]. **1916** C. E. Lane-Poole *List of Trees Sierra Leone* 53 The Khayas yield the African Mahogany so esteemed in Europe. **1936** J. D. Kennedy *Forest Flora S. Nigeria* 157 A characteristic of *Khaya* seedlings is the very long drip-tip on the primary leaves. **1956** *Handbk. Hardwoods* (Forest Prod. Res. Lab.) 138 Mahogany, African... Other names..khaya (United States). **1962** Watt & Breyer-Brandwijk *Medicinal & Poisonous Plants S. & E. Afr.* (ed. 2) 1406/3 *Khaya grandifoliola*... African mahogany, Bigleaf mahogany (Uganda), Khaya mahogany. *Ibid.*, *Khaya nyasica*... Khaya mahogany. **1964** R. W. J. Keay et al. *Nigerian Trees* II. 259 Although known to the trade as 'Benin Mahogany', this species [*sc. Khaya grandifolia*] is not the typical Khaya of the southern Benin forests.

Khazar (kăzā·ɹ). Also **Chazar, Chozar, Khozar.** [Heb.] A member of a people of Turkish origin who from the 8th to the 10th or 11th century occupied a large part of southern Russia. Also *attrib.*

1854 J. H. Newman *Lect. Hist. Turks* II. i. 63 The horde of Chozars, as this Turkish tribe was called,..transported their tents..into Georgia. **1854** R. G. Latham *Native Races Russ. Empire* x. 143 The Khazars poured themselves over eastern Europe in the 7th, 8th, 9th, 10th, and 11th centuries. **1863** [see Turkic *a.*]. **1878** R. G. Latham *Russian & Turk* viii. 210 The Khazars seem to play the same part in the history of Eastern Russia that the Avars played in that of Southern Bavaria... We meet the term *Chazaria* as the land of the *Khazars*. **1882** *Encycl. Brit.* XIV. 60/1 Merchants from every nation found protection..in the Khazar cities. **1903** *Jewish Encycl.* IV. 1/1 The Chazars..established themselves in the territory bounded by the Sea of Azov, the Don and the lower Volga. **1930** C. A. Macartney *Magyars in 9th Cent.* 62 From a short time after the arrival of the Khazars, the Cuban Bulgars disappear from history. **1934** A. G. Chater tr. *Undset's Stages on Road* i. 29 The Khazars of the Crimea..were Tartars who had been converted to Judaism and were ruled by a Jewish king. **1952** E. Hyams *Soil & Civilization* 171 The Jewish religion was adopted by the Chazar princes. **1965** [see *Bulgar *sb.*].

khellin (ke·lin). Also † **kellin.** [Orig. coined as F. *kelline* (I. Mustapha 1879, in *Compt. Rend.* LXXXIX. 442), f. *kell*, given as the Arabic name of *Ammi visnaga*; the *h* originated with Samaan (1931), who gave the Arabic name as *khella*: see -in¹.] A tricyclic crystalline compound, $C_{14}H_{12}O_5$, obtained from the fruit of the North African umbelliferous plant *Ammi visnaga* and formerly used in the treatment of angina pectoris.

The name was orig. given to the glucoside of this compound.

1879 *Jrnl. Chem. Soc.* XXXVI. 1041 The author proposes for this glucoside the name *kellin*, from the Arabic name of the Ammi Visnaga. **1931** K. Samaan in *Q. Jrnl. Pharmacy* IV. 14 Ibrahim Mostapha described a crystalline principle which he named Khellin. **1938** *Chem. Abstr.* XXXII. 2120 Fantl and Salem..obtained from the fruits of *Ammi visnaga* a compd., kellin, $C_{14}H_{12}O_5$. **1947** *Lancet* 26 Apr. 557/2 Khellin causes a conspicuous and long relaxation of all the visceral smooth muscle—the intestines, uterus, bile-ducts, bronchi, and especially ureters. *Ibid.*, In 1945 a fresh interest in khellin arose as the result of the discovery that it acts as an extremely potent coronary vasodilator which, in doses used, has no effect on the general blood-pressure and does not increase the oxygen requirements of the heart. **1957** M. Plotz *Coronary Heart Dis.* xvii. 301/2 Favorable results with khellin therapy in man have been reported by many investigators. *Ibid.*, The most important factor limiting the use of khellin orally has been the high incidence of toxic reactions. **1964** *To-Day's Drugs* (B.M.A.) 132 Khellin had a brief period of popularity about ten years ago, but when the drug was evaluated by controlled clinical trials and by objective methods it was found that it was no better than a placebo.

|| **khet** (kēt). [Hind., Hindi.] In India, a tract of cultivated land.

1878 P. Robinson *In my Indian Garden* 176 In the still air could be heard..from the scattered *khets*, the bark of the prowling fox. **1886** —— *Valley of Teetotum Trees* 63 In all the swampy jheels and crop-grown khets. **1922** *19th Cent.* Oct. 589 The land is divided by one broad distinction into the khet and the jungle—that is to say, into the cultivated and the wild.

Kheta (ke·tă, χe·tă). Also **Khatti, Khita.** [Egyptian name.] Name of an ancient people and kingdom in the Near East: now usually equated with the Hittites. See *Hittite *sb.* and *a.*

1884 W. Wright *Empire of Hittites* ii. 19 The war between Egypt and Kheta was brought to a close by a treaty of peace concluded between Rameses I. and Saplel the Hittite king. **1890** T. Ely *Man. Archæol.* v. 76 On the monuments [of the Hittites] the name *Khita*, *Kheta*, or *Khatti*..frequently occurs. **1948** D. Diringer *Alphabet* 89 Egyptian inscriptions mention the powerful Kheta-empire. **1952** O. R. Gurney *Hittites* 2 Who could doubt that the Kheta-folk of the Egyptian texts and the Hittites of the Old Testament were one and the same? **1957** *Encycl. Brit.* XI. 599/1 Egyptian evidence shows that in the time of the 18th to the 20th dynasties, between the years 1500–1190 B.C., a powerful northern kingdom *Kheta* sought perpetually to obtain political influence over Syria and therefore often fought with Egypt. Already the Pharaoh Thutmosis III. (1501–1447), who had conquered Syria as far as the Upper Euphrates, received presents from the prince of Kheta. **1970** Bray & Trump *Dict. Archaeol.* 104 *Hittites, Hatti* or (to the Egyptians) *Kheta*, a people who infiltrated Anatolia and in smaller numbers the Levant from the north c 2000 B C, but the details of their origin are more than somewhat obscure.

Khilafat (kilā·fat). [ad. Arab. *ḵilāfat* caliphate, office or rule of a caliph.] The spiritual headship of Islam, residing in the person of the Turkish Sultan; used *attrib.* to designate the Muslim anti-British movement in India after the treaty of Sèvres in 1920. Hence **Khila·fatist,** a supporter of this agitation.

1921 [see *extremist]. **1923** *Edin. Rev.* Jan. 182 The *Khilāfat*..is the Vice-gerency of the Prophet. **1925** *Contemp. Rev.* Apr. 430 An influential section of the Moslem community, dissociating itself from the Khilafatist section. **1926** *Encycl. Brit.* Suppl. II. 429/2 The 'Khilāfat' agitation. **1950** R. C. Majumdar et al. *Adv. Hist. India* (ed. 2) 985 The two brothers Muhammad Ali and Shaukat Ali, and Maulana Abul Kalam Azad organized a mass movement of the Muslims known as the Khilafat movement. **1957** M. D. Kennedy *Short Hist. Communism in Asia* vi. 58 The movement thus launched for creating disaffection among Moslems in India was known as the Khilafat movement. **1964** A. Swinson *Six Minutes to Sunset* ix. 163 Nair and many others considered that it had been Gandhi's espousal of the Khilafat cause which had triggered off the rebellion. **1972** P. Hardy *Muslims of Brit. India* 190 The philosophy of the *Khilafat* movement was not that of territorial nationalism, but of community federalism, and of a federalism wherein one party, the Muslim, looked outside the common habitat, India, for the *raison d'être* of the federal relationship.

Khilim, var. *Kilim.

Khirbet Kerak (kō·ɹbĕt ke·răk). *Archæol.* Name of a town on the south-west edge of Lake Tiberias in Syria, used *attrib.* to designate a type of early Bronze Age pottery first found there in the 1940s, which is red and black in colour with highly burnished finish and fluted decorations.

1949 W. F. Albright *Archaeol. of Palestine* iv. 76 The most interesting new pottery of this age is the lustrous red and black burnished 'Khirbet Kerak' ware, which first became known at the site which bore this name, ancient Beth-Yerah, at the south-west corner of the Sea of Galilee. **1952** V. G. Childe *New Light Most Anc. East* xi. 219 The pottery includes wheel-made red-slipped and lattice-burnished vases and—but only in the last phase of a long occupation—hand-made particoloured black and red vases of 'Khirbet Kerak ware'. **1960** K. M. Kenyon *Archaeol. in Holy Land* v. 124 Side by side with the native wares are vessels of Khirbet Kerak ware. ..This type of pottery is also found in northern Syria. **1968** *Encycl. Brit.* II. 610/2 This Khirbet Kerak ware is characteristically Anatolian, but is distinguished by its ornament of ribs and flutings and of raised geometrical patterns, especially spirals.

Khirgese, var. *Kirghiz *sb.* and *a.*

Khita, var. *Kheta.

khiva, var. *kiva.

Khlist (χlist). Also **Chlist, Khlyst.** Pl. **Chlists, Khlisti, Khlysts, Khlysty.** [Russ., lit. a whip.] A member of a sect of ascetic Russian Christians, formed in the 17th century, who believed that Christ could be reincarnated in human beings through their suffering.

1856 R. Farie tr. *A. von Haxthausen's Russ. Empire* I. viii. 254 On Easter night the Skoptzi and Khlisti all assemble for a great solemnity, the worship of the Mother of God. **1874** J. H. Blunt *Dict. Sects* 250/1 *Khlisti*, a name signifying 'Flagellants' given to a Russian sect..formed about 1645 by a deserter from the army named Daniel Philipitch. **1920** A. Paget tr. *Nekludoff's Diplomatic Reminisc.* vii. 71 Rasputin was in fact a *Khlyst*, *i.e.* half 'Shaker', half Flagellant—a strange sect which from time to time rises in Russia from the common depths to the upper classes of society. **1929** A. Huxley *Do what you Will* 152 None but heretics have preached it [*sc.* humility]. The Russian Khlyst, for example. *Ibid.* 153 Rasputin practised what he preached, and sinned—most conspicuously, as was the custom of the Khlysty, in relation to the seventh commandment. **1967** D. T. Kauffman *Dict. Relig. Terms* 274 *Khlysts, Chlists*, or *Klysty flagellants*, Russian ascetics originating in the seventeenth century.

Khmer (k'mēəɹ), *sb.* and *a.* [Native name.] **A.** *sb.* **a.** A native or inhabitant of the ancient kingdom of Khmer in south-east Asia, which reached the peak of its power in the 11th century and was destroyed by Siamese conquests in the 12th and 14th centuries; also (from 1863) such a person in Cambodia; now, a native or inhabitant of the Khmer Republic (established 1970). **b.** The monosyllabic language of this people, belonging to the Mon-Khmer group of the Austro-Asiatic family. **B.** *adj.* Of or pertaining to the Khmers or their language.

1876 *Encycl. Brit.* IV. 723/2 The name given by the people of Camboja to their own race is *Khmêr*. **1881** C. J. F. S. Forbes *Compar. Gram. Lang. Further India* I. iv. 48 The Cambodians acknowledge..the Kouys, as the most ancient stock, and style them the 'Khmerdom', or ancient Khmers. **1921** *Edin. Rev.* July 172 If the Khmêrs were the ancient people of Cambodia, here we have an important land mark in common between them and the Khasis. **1921** E. Sapir *Lang.* iv. 71 The use of prefixed elements to the complete exclusion of suffixes, is far less common. A good example is Khmer (or Cambodgian). **1932** W. L. Graff *Lang.* 417 Mon-Khmer comprises the three civilized languages Mon, Khmer, and Cham. **1943** *Burlington Mag.* Aug. 198/1 Of imitations of Khmer bronzes, on the other hand, a number have passed through my hands. **1959** *Chambers's Encycl.* VII. 499/2 What the Khmer and Indians called this state is not known. **1966** *Listener* 22 Sept. 409/1 The Khmers, as the people of Cambodia are known, have a maturity about their nationalism. **1967** *Economist* 23 Sept. 1088/1 Prince Sihanouk has been at odds with the Chinese since April, when he accused the *Khmers rouges*, his local communists, and their foreign friends of launching a guerrilla campaign in north-western Cambodia. **1970** *Guardian* 5 May 9/5 The men were being lectured, in both French and Khmer, on tactical formations. **1970** M. Pereira *Pigeon's Blood* iv. 48 The closest resemblance..was the medieval Khmer craftsmanship of Cambodia. **1971** *Nat. Geographic* Mar. 316/1 When the Khmer people were docile, the Vietnamese would withdraw all but a token force. **1972** M. Sheppard *Taman Indera* 6 Funan was eventually conquered by a Khmer prince of Kambuja,..in the middle of the sixth century. **1973** *Times* 10 Apr. 6/7 Khmer Rouge and Vietcong guerrillas have launched several rocket attacks.

Khoikhoi (koi·koi). *S. Afr.* Also **Khoi Khoi, Khoi Khoin,** † **Quaiquae,** etc. [Hottentot, lit. 'men of men'.] The Hottentots' name for themselves; also used by others in the sense 'Hottentots'; the language which they speak. Also *attrib.* or as *adj.*

1791 tr. *Le Vaillant's Travels into Interior Parts of Africa* II. 154 A Hottentot man..Khoe-Khoep. **1801** J. Barrow *Acct. Trav. S. Afr.* I. iii. 151 [The name] by which the whole nation was distinguished, and which at this moment they bear among themselves in every part of the country, is *Quaiquae*. **1880** *Encycl. Brit.* XII. 309/2 The common denomination adopted by themselves was Khoi-Khoin (men of men). **1881** T. Hahn *Tsuni-||Goam* i. 1 These Khoikhoi generally go by the name of *Hottentots*. *Ibid.* 5 The Khoikhoi language is entirely void of prefixes. **1897** A. J. Butler tr. *Ratzel's Hist. Mankind* II. 247 The Khoi-Khoi (Bushmen and Hottentot) group of languages. **1910** G. M. Theal *Yellow & Dark-Skinned People of Afr.* iv. 79 Those destitute Hottentots..had more Bushman than pure Khoikhoi blood in their veins. **1930** I. Schapera *Khoisan Peoples of S. Afr.* i. 5 The term [*sc.* Khoisan] is compounded of the names *Khoi-Khoin*, by which the Hottentots call themselves, and *San*, applied by the Hottentots to the Bushmen. **1969** *Oxf. Hist. S. Afr.* I. i. 10 Blood group studies suggest, however, that certain Khoikhoi speakers are closely allied in their blood-group patterns to African negroids. *Ibid.* ii. 43 There were also groups of herders, most of whom spoke Khoikhoi. *Ibid.* 56 The Khoikhoi may have been shepherds before they were cattle-men.

Khoja. Add: **2.** A member of a Muslim sect of converts from Hinduism, found mainly in western India and retaining some Hindu customs. Also *attrib.*

1882 *Encycl. Brit.* XIV. 64/1 Only the military class, the priesthood, and the khodjas are exempt from the payment of taxes. The khodjas consider themselves descendants of the prophet. **1921** C. Eliot *Hinduism & Buddhism* III. lviii. 455 The sects known as Khojas and Bohras owe their conversion to the zeal of Arab and Persian missionaries who preached in the eleventh century. **1931** G. MacMunn *Relig. & Hidden Cults India* 97 A portion of the Assassins escaped to India where they have developed into a wealthy trading fraternity known generally as the Khojas or 'worthy men'. **1937** L. Bromfield *Rains Came* I. vi. 41 The prospect of putting over a sharp deal in Bombay on the Khojas and Parsees. **1970** D. G. Mandelbaum *Soc. in India* II. vii. xxix. 555 There are several Isma'ili jatis, the Bohras and Khojas of Gujarat being the major groups among them... Khoja doctrine held..that the Aga Khan was an incarnation of the 'glorious Tenth Avatar'.

khoker, var. *koker.

Khond (kɒnd). Also **Kandh, Kond, Kondh.** [Native name.] **a.** A Dravidian people inhabiting Orissa in eastern India; one of this people. **b.** The language spoken by the Khonds. Also *attrib.* or as *adj.*

1852 S. C. Macpherson in *Jrnl. R. Asiatic Soc.* XIII. 217 The ancient state of Orissa was formed chiefly from the territories of..the Khonds, the Koles, and the Sourahs. *Ibid.* 221 The Khond religion exists in oral traditions alone. **1867** [see *Gond *sb.* and *a.*]. **1881** *Encycl. Brit.* XII. 778/2 Bishop Caldwell recognizes twelve distinct Dravidian languages:— (1) Tamil,.. (10) Khond. **1909** E. Thurston *Castes & Tribes S. India* III. 398 A Kondh funeral dance in the Ganjam Māliahs. **1937** H. G. Rawlinson *India* viii. 128 Among the Khonds, one of the primitive tribes, a human victim known as the *meriah* was until recently sacrificed in order to secure fertility for the fields. **1946** R. C. Majumdar et al. *Adv. Hist. India* 826 To the first Lord Hardinge's Government belongs the credit of taking steps to stop the human sacrifices practised by the Khonds in Orissa. **1959** *Chambers's Encycl.* VII. 480/2 The most eastern member of the Central group [of Dravidian languages] is the language of the Khonds or Kandhs. *Ibid.* X. 244/2 The most interesting of the aboriginal races are the Konds. **1970** D. G. Mandelbaum *Soc. in India* II. xxii. 422 His trade is carried on with men of the Kond tribe who are rated as a clean jati by the neighbouring Oriyas.

Khorassan (korăsā·n). Also **-asan, Khurasan.** Name of a province in north-east Iran, used *attrib.* and *ellipt.* to designate a carpet or rug made there, usu. with vivid colouring and fine silky texture.

1900 J. K. Mumford *Oriental Rugs* xi. 218 The realism which marks certain carpets of the Feraghan group is fairly outdone in many of the proper Khorassans. **1904** M. B. Langton *How to know Oriental Rugs* ii. 68 The Khorasan is a most satisfactory rug, beautiful in colour, durable, and pliable. **1931** A. U. Dilley *Oriental Rugs & Carpets* Pl. 25 (*caption*) Khorassan Prayer Rug. Characteristic minute floral pattern, Herati, with border of narrow bands. **1931** C. Tattersall *Carpets of Persia* 39 Mashhad. Much like Khorasans, but with shorter and closer pile. **1962** C. W. Jacobsen *Oriental Rugs* 238 No rugs have been imported as Khurasans in the past 20 years. **1971** K. Watson tr. *Hubel's Bk. Carpets* 213 Another peculiarity of the ground-weave of the Khorassan carpet is the weft: the thick straight weft is covered by the thin sinuous weft threads accompanying it.

Khotan (kotā·n). Also **Khoten.** Name of a city and district on the south of the Takla Makan desert in Chinese Turkestan, used *attrib.* and *ellipt.* to designate a carpet or rug made there, usu. with Chinese geometrical patterns or stylized natural designs.

1871 R. B. Shaw *Visits to High Tartary* xii. 259 There is a large covered reception-place with a verandah in front of all. Here an immense Khoten carpet is spread with rugs along the back. **1899** Kipling *From Sea to Sea* I. xvi. 362 The English can only be artistic in spots and by way of the art of other nations—Sicilian tapestries, Persian saddle-bags, Khoten carpets. **1904** M. B. Langton *How to know Oriental Rugs* viii. 228 Bayard Taylor, in his Travels, speaks of the beautiful Khotan carpets which were spread on divans in the 'royal rest' rooms near Yarkand. **1920** E. & P. Sykes *Through Deserts & Oases Cent. Asia* vi. 82 The old Khotan carpets, their colours made from vegetable dyes, were attractive, and the silk carpets are highly prized and very difficult to obtain. **1931** A. U. Dilley *Oriental Rugs & Carpets* x. 224 Khotan rugs of the ancient vintage, with geometric blocked design, numerous borders, half-Chinese and half-Persian,..are the outstanding jewels of the Chinese Turkestan family. **1960** G. & C. W. Digby tr. *Haack's Oriental Rugs* vii. 56 Kashgar and Khotan are the names by which these carpets are more properly known; they show a strong Chinese influence in their design. **1967** *Times* 21 Feb. 21/4 (Advt.), An extremely rare and beautiful 18th century Khotan carpet, 12 ft. 7 in. × 7 ft. 4 in.

Khotanese (kōᵘtănī·z), *sb.* and *a.* [f. *Khotan* (see prec.) + -ese.] **A.** *sb.* The people of Khotan; one of this people; the Middle Iranian language of Khotan. **B.** *adj.* Of or pertaining to Khotan.

1882 *Encycl. Brit.* XIV. 67/2 The Khotanese keep camels, horses..and fowls. **1907** A. Stein *Anc. Khotan* I. 141 It is a point of special interest that dancing..is a pastime freely indulged in by Khotanese of both sexes and of all classes. *Ibid.* 144 In the case of the Khotanese there has been, besides a slight admixture of Turki blood, an admixture also of Tibetan. **1939** C. H. Gray *Foundations of Lang.* 320 Besides Middle Persian proper, we have a fair amount of material in some other Middle Iranian dialects, notably Middle Parthian..and Khotanese or Middle Sakian..in the southern part of East Turkistān. **1948** D. Diringer *Alphabet* II. vi. 350 The material contained in Khotanese manuscripts is of great variety; there are official and business documents, translations of Indian tales, [etc.]. *Ibid.*, There are some indications to show that Khotanese began to be used in writing in the second century A.D. **1957** *Archivum Linguisticum* IX. II. 128 The most likely source would seem to be Khotanese. **1961** H. W. Bailey *Khotanese Texts* IV. 14 The Buddhist literature of which complete or partial translations existed in Khotanese is of large extent. *Ibid.* 17 A celebrated Khotanese painter of Buddhist subjects.

Khowar (kōᵘ·wāɹ). A Dardic language spoken in Chitral in north-west Pakistan. Also *attrib.* or as *adj.*

1882 *Encycl. Brit.* XIV. 9/2 Their [*sc.* inhabitants of Chitral's] language, *Khowār*, is closely allied to the dialects of the Kafir tribes. **1950** T. Longstaff *This my Voyage* x. 199 The singing must be in Khowar, the tongue of Chitral, which sounds like Romany. **1964** *Language* XL. 304 A native Khowar speaker. **1967** D. S. Parlett *Short Dict. Lang.* 73 *Khowar*..., 7 th[ousand] speakers of a dialect in N.W. India related to Kalash and probably (less closely) to Kafir. **1974** 'S. Harvester' *Forgotten Road* i. 11 He had a working acquaintance with dialects of most Kafir valley tribes, even Khowar and the almost extinct Domali language.

Khozar, var. *Khazar.

Khrushchevism (kru·stʃǒfiz'm). [f. the name of Nikita Sergeevich *Khrushchev* (1894–1971), Soviet statesman + -ism.] The practice or principles of Khrushchev, notable for his denunciation of Stalin and his advocacy of peaceful coexistence with the Western powers. So **Khrushche·vian** *a.*, of, pertaining to, or characteristic of Khrushchev or his policies.

1957 *Amer. Speech* XXXII. 294 The party has switched from Stalinism to Khrushchevism. **1961** *Listener* 28 Dec. 1107/2 The world of romance and fairy tale is never far away from the Khrushchevian Utopia. **1962** *Ibid.* 18 Jan. 112/1 There are several kinds of communism in the world—Titoism, Khrushchevism, Maoism in China. **1963** *Economist* 3 Aug. 418/2 Khrushchevism is..to be preferred by the West to Maoism. **1964** *New Statesman* 17 Apr. 592/2 What strikes one in the record of Krushchevism is the great number of vital issues which Khrushchev and his men have posed but not tackled. **1973** *Daily Tel.* 22 Nov. 10/6 We know now that 'Ivan Denisovich' was not the harbinger of better things, but the high point of Khruschevian 'liberalism' from which there has since been a steady retreat.

khubab: see *kebab.

khud. *attrib.* Add: *khud-climbing, -stick.*

1906 *Westm. Gaz.* 13 Aug. 7/1 Killed..in India while khud-climbing. **1925** A. G. Arbuthnot in G. Burrard *Big Game Hunting* 118 Take your telescope, rest it on a rock or on your 'khud stick'. **1928** *Blackw. Mag.* Jan. 25/2 He jabbed his khudstick into the ground.

khurta, var. *kurta.

khyal (ki,ā·l). Also **kheal.** [Skr.] A traditional type of song in northern India, with instrumental accompaniment, usually containing two main themes.

1882 N. A. Willard in R. S. M. Tagore *Hindu Mus.* (ed. 2) I. 102 In the *Kheal* the subject generally is a love tale, and the person supposed to utter it, a female. **1914** A. H. F. Strangways *Mus. Hindostan* vi. 165 Both are *Khyāls*, and by Rabindranath Tagore. **1921** H. A. Popley *Mus. India* vi. 89 The *Khyāl* was introduced later than the Dhrupad, in order to find a place for the graces which are not allowed in the former. *Ibid.*, Khyal singers and Dhrupad singers are usually different. **1971** *Shankar's Weekly* (Delhi) 4 Apr. 24/4 Lines of a khyal are ordinarily made to sprawl over several matras each of a long duration.

Khyber Pass (kəi·bəɹ pɑs). [The chief pass in the Hindu Kush mountains between Afghanistan and north-west Pakistan.] *Rhyming slang* = Arse *sb.* 1. Also *ellipt.* **Khyber.**

1943 M. Harrison *Reported Safe Arrival* 32 Not knowin' wevver they was on their 'eads or their Kybers [*sic*]. **1960** J. Franklyn *Dict. Rhyming Slang* 88 *Khyber Pass*, arse, not the buttocks but the anus. Usually 'He can (you can) (they can) stick it up his (your) (their) Khyber!' It is an expression of disapproval. **1966** *New Statesman* 16 Sept. 408/3 Can it really be..that 'Khyber' is genuine rhyming slang for the posterior? **1968** *Crescendo* Jan. 6/3 If we sit on our Khybers, we will miss out on all the things that make our lives the richer.

kia, var. *kya.

kiaat (kiʸā·t). *S. Afr.* Also **coyatte hout** (*hout* = wood), **kajat, kajatenhout, kiatt, kijaat.** [Afrikaans, f. Du., f. Malay *ki djati*, *kajoe djati* good wood.] The tree *Pterocarpus angolensis*, belonging to the family Leguminosæ, and found in southern Africa; the timber of this tree.

1801 J. Barrow *Acct. Trav. S. Afr.* I. v. 339 Catalogue of Useful Woods... Coyatte hout... Tough. [Uses:] Staves for butter firkins. **1862** L. Pappe *Silva Capensis* (ed. 2) 29 *Atherstonea Decussata* (Cape Teak, or Kajatenhout)... This tree grows 20–25 feet high. **1921** D. E. Hutchins in T. R. Sim *Native Timbers S. Afr.* 108 The stem of Kajatenhout is not wanting in thickness. **1947** *Cape Times* 8 Nov. 12/3 Desks: Large selection in imbuia, kijaat, African mahogany and teak. **1956** *Handbk. Hardwoods* (Forest Prod. Res. Lab.) 159 Muninga—*Pterocarpus angolensis*. Other names..kajat, kajatenhout, kiatt (Union of South Africa). **1957** *Cape Times* 3 Sept. 3/3 There was not a scrap of wool in sight in the theatre-like, kiaat-panelled hall. **1973** Palmer & Pitman *Trees S. Afr.* II. 937 The kiaat is a medium-sized deciduous tree.

kiaki, var. *keyaki.

kian, early form of Cayenne. (Examples.)

1794 A. Thomas *Newfoundland Jrnl.* 13 May (1968) 39 The Cook was holding, at an unseasonable moment, the Saucepan of Mock Turtle to season it higher with Kian. **1845** B. Upton *Let.* 10 Sept. in *Amer. Hist.* (1966) XVII. iv. 87/2 Cayenne pepper (kian) grows wild here on the prairies.

kiang. The more usual spelling of Kyang.

1882, 1885 [see Kyang]. **1906** [see **desert-frequenting* adj.]. **1934** *Times Educ. Suppl.* 22 Sept. p. iv/3 The father is a very old Kiang, a true wild donkey from Tibet. **1970** *Guardian* 26 Mar. 3/1 Tibet's wild donkeys..known as kiangs. **1973** *Times* 20 Feb. (India Suppl.) p. xi/3 The plateau of Ladakh in the high Himalayas harbours forms related to those in Tibet such as the wild yak, the kiang, and the chiru.

|| **kia ora** (ki,ă ǭə·ră). *N.Z.* [Maori.] An exclamation of good will: good health! be well!

1896 in E. E. Morris *Austral Eng.* (1898) 247 You will hear any day at a Melbourne bar the first man say *Keora ta-u*, while the other says *Keora tatu*... These expressions are corruptions of the Maori, *Kia ora taua*, 'Health to us two!' and *Kia ora tatou*, 'Health to all of us!' **1905** W. B. *Where White Man Treads* 273 Kia ora, oh coloured brother! **1914** A. A. Grace *Tale of Timber Town* v. 32 The digger put his pint to his hairy lips, [and] said, '*Kia ora*. Here's fun.' **1933** 'E. Milton' *Waimana* II. iii. 82 Good-byeee, and kia-ora! **1938** R. Finlayson *Brown Man's Burden* 66 They smiled sweetly and said 'Kia ora', as they bought bunches of the bitter cherries. **1966** G. W. Turner *Eng. Lang. Austral. & N.Z.* viii. 169 Some of these [*sc.* Maori terms] seem to have been commoner in the past than they are now, e.g. *Kia ora* 'good health', more often seen in England (as a brand name for soft drinks) than in New Zealand now. **1970** *N.Z. Listener* 12 Oct. 12/5, I heard her soft, sad voice: 'This is the kia ora, Dave.'

|| **kiap** (ki·āp). [Native name.] In New Guinea, a European patrol officer or policeman.

1923 W. R. Humphries *Patrolling in Papua* xv. 162 They had been told by their 'Kiap' to make a road. **1943** S. W. Reed *Making of Mod. New Guinea* v. 174 Fear and respect for the *kiap* seem to be mingled in the native mind: fear of punishment..at the *kiap's* command, and respect..for the authority of the officer. **1969** *New Guinea & Austral.* IV. iii. 9/1 Those who are subservient to Kiap doctrine. *Ibid.*, We cannot be expected to remain within the fence of racial integration whose fencing materials—the persuasiveness of the Kiap, the teaching of the church, and the legislation of a colonial administration—show distinct signs of weathering. *Ibid.* 10/1 His motion was defeated in a council whose views were predictably Kiap-dominated. **1970** *Times* 31 Mar. (Austral. Suppl.) p. ii/4 Australia was content to provide an underpaid administration whose *kiaps* (patrol officers) were largely concerned with exploration of one of the world's wildest countries.

kiawe (ki,a·ve). [Hawaiian.] = Algarroba.

1915 W. A. Bryan *Nat. Hist. Hawaii* xvii. 241 (*caption*) A fine Algaroba tree (Kiawe) (*Prosopis juliflora*). **1917** *Nature* 20 Sept. 57/2 One of the introduced trees of great economic importance is the algaroba tree (*Prosopis juliflora*) or kiawe, as the Hawaiians call it. It is found in a belt on the lowlands along the shores of all the islands, and occupies the soil almost to the exclusion of other plants. The pods are very nutritious, and are eagerly eaten by all kinds of stock. Its flowers furnish an excellent quality of honey. **1937** D. & H. Teilhet *Feather Cloak Murders* v. 102 The feathery boughs of a Kiawe tree. **1965** M. C. Neal *Gardens of Hawaii* 413 The kiawe, a fairly large tree, seldom reaching a height of 60 feet... is the commonest and most valuable tree introduced to Hawaii.

kibab, kibaub: see *kebab.

kibble, *sb.*³ (Further examples.)

1834 *Chambers's Jrnl.* 9 Aug. 223/2 He had been a kibble-boy in the mine. **1843** *Ainsworth's Mag.* IV. 507 Huge quantities of iron, boiler and kibble plates. **1881** *Instructions to Census Clerks* (1885) 84 Copper Miner... Kibble Filler. **1901** G. L. Kerr *Pract. Coal Mining* iii. 35 The centre of the cradle contains an opening which provides space for two buckets or 'kibbles' passing each other. **1924** *Hist. Rev. Coal Mining* (Mining Assoc. Gt. Brit.) iii. 38 The advantages of running two kibbles simultaneously are generally more than overbalanced by the increased complications of guides and shaft fittings required. **1967** *Gloss. Mining Terms* (*B.S.I.*) IX. 10 *Kibble*, a large steel bucket used to remove debris and to transport men and materials in a shaft sinking.

kibbler (ki·bləɹ). [f. Kibbl(e *v.*¹ + -er¹.] A machine which kibbles or grinds coarsely; also, one who operates or tends such a machine. So **ki·bblerman.**

1882 Ogilvie *Imp. Dict.*, *Kibbler*, one who or that which kibbles or cuts, especially a machine for cutting beans and peas for cattle. **1921** *Dict. Occup. Terms* (1927) § 159 *Kibbler, kibblerman*;..attends and feeds machine in which he breaks up oil cake into nodular pieces before grinding into meal. **1922** *Glasgow Herald* 21 Feb. 3 Chaff and root cutters, grinding mills, or kibblers. **1945** J. F. Lockwood *Flour Milling* i. 42 Wheat should always be ground into a coarse meal before testing... A small hand kibbler, such as a coffee grinder, is suitable for this purpose. **1968** *Encycl. Brit.* VI. 804/2 *Crushers or kibblers* are used mainly for oats, corn and linseed, which are bruised, broken or flattened without being ground into a meal.

|| **kibbutz** (kibu·ts). Pl. **kibbutzi·m** (-īm) (occas. **kibbutzes**). [ad. mod.Heb. *ḳibbūṣ* gathering.] A collective settlement in Israel, owned communally by its members, and organized on co-operative principles. Also *attrib.*

1931 tr. *Ann. Rep. Central Tenuvah Federation* (Palestine) *1930* i. 3 'Tenuvah' is a co-operative marketing association... Its members are the kvutzoth (collective farms), moshavim (smallholders' settlements), and kibbutzim (large farm training groups). **1944** H. F. Infield *Co-operative Living in Palestine* 22 The term Kvutza is used alternately with Kibbutz, which has practically the same meaning... The present study uses Kvutza only for the rural settlements and Kibbutz for the superior co-ordinating organization. *Ibid.* viii. 119 (*heading*) Kibbutzim—the roof-organizations. **1957** *Observer* 29 Sept. 13/2 Miss Helga Pilarczyk..danced the Dance of the Seven Veils..with sturdy acrobatics that suggested rather more of the kibbutz than the necrophile. *Ibid.* 1 Dec. 5/2 We saw..people in *kibbutzim*, the voluntary collective farms—some of the serenest people I have ever met. **1958** *Times Lit. Suppl.* 20 June 341/4 It describes how a London-bred girl, with a carping and snobbish mama, ups and joins a *kibbutz* in Israel. **1960** *Times* 15 Oct. 7/6 The kibbutz movement, whose jubilee occurs tomorrow, has been slumping. **1962** L. R. Banks *End to Running* I. iv. 57 Kibbutzes, I remembered from somewhere, were some kind of communistic settlements in Israel, where children were raised away from their parents and nobody owned anything. **1964** M. Argyle *Psychol. & Social Probl.* xvi. 204 Some communities have arranged for children to be brought up in groups... The best-known example of this is the *kibbutz* system in Israel. **1968** C. A. Doxiadis *Between Dystopia & Utopia* ii. 46 Alfonso Reyas, who states that even 'America is a Utopia'..and the Kibbutzim in Israel are the results of utopian theories. **1972** *Sci. Amer.* Dec. 43/2 Children in the kibbutz are cared for from infancy in small, one-age peer groups by trained personnel. They live apart from their parents for most of the day and in most kibbutzes during the night as well. **1973** *Jewish Chron.* 6 July 3/1 (*heading*) Kibbutzim back land appeal by sheikh.

|| **kibbutznik** (kibu·tsnik). [Yiddish, f. prec. + *-nik, Pol. and Russ. noun suffix denoting person connected with (something).] A member of a *kibbutz*.

1949 Koestler *Promise & Fulfilment* II. ii. 221 We..had the treat of an hour's moonlight journey across the Sea of Galilee, with *kibbutzniks* from Ein Geb sprawling all over the deck and singing Hebrew songs. **1950** G. Mikes *Milk and Honey* III. 129 The main characteristics of *Kibbutz*-life are:— (*a*) *Kibbutzniks* have no money at all;.. (*c*) *Kibbutzniks* eat their soup at the end of their meals. **1959** F. M. Wilson *They came as Strangers* III. iii. 203 The kibbutzniks believed that by redeeming the soil of the Holy Land..they were..redeeming..the Jewish people. **1965** *New Statesman* 3 Sept. 314/2 Kibbutzniks played chess in the shade. **1967** *Listener* 13 July 62/1, I wandered along the tranquil shores of Galilee, stepping over trenches and talking to armed *kibbutzniks*. **1973** *Jewish Chron.* 6 July 10/2 (*caption*) Ehud Adiv, kibbutznik and paratrooper.

kibitz (ki·bits), *v. slang* (orig. *U.S.*). Also **kibbitz.** [Yiddish, f. G. *kiebitzen* to look on at cards, f. *kiebitz* lapwing, pewit; interfering onlooker at cards.] *intr.* To look on at cards, or some other activity, esp. in an interfering manner (e.g. by standing close to the shoulders of the players); to offer gratuitous advice to a player; to act as a kibitzer. Also *trans.*, to watch (a game, person, etc.), esp. in an officious or meddling way. Hence **ki·bitzing** *vbl. sb.* and *ppl. a.*

1927 in *Amer. Speech* (1928) IV. 159 The trade journal..devotes an editorial..to the 'kibbitzer'. It defines 'kibbitzing' as a slang expression used to indicate the act of offering gratuitous advice by an outsider. **1951** M. McLuhan *Mech. Bride* 67/1 The children..have retired to a philosophic knoll to do a little uneasy kibitzing. **1955** I. Fleming *Moonraker* iv. 43 Drax..[sorted] his cards..only into reds and blacks, ungraded, making his hand very difficult to kibitz. **1957** R. Stout *If Death ever Slept* (1958) viii. 106 Corey Brigham stood behind them, kibitzing. **1961** A. Smith *East-Enders* vi. 96 Do you remember the café proprietor with whom I kibbitzed on the soarers? **1966** P. J. Nicholson *Rat Race* I. iii. 66 Harry moved about, kibitzing on conversations here and there. **1967** *Sci. Amer.* Sept. 64/2 (Advt.), Reilly—with one shipmate posing and two others kibitzing—is finishing off a watercolor. **1972** *New Yorker* 26 Aug. 51/1 Even when the equipment is operating smoothly, the temptation is all but irresistible to gather around the polygraph and kibitz.

kibitzer (ki·bitsəɹ, kibi·tsəɹ). *slang* (orig. *U.S.*). Also **kibbitzer.** [Yiddish *kibitser*; cf. prec.] An onlooker at cards, etc., esp. one who offers unwanted advice; a busybody, an officious meddler.

1927 Kibbitzer [see *kibitz *v.*]. **1935** A. J. Pollock *Underworld Speaks* 66/2 *Kibitzer*, an onlooker of a card game who seldom plays but frequently criticizes the play of the contestants, and offers advice freely. **1936** Wodehouse *Laughing Gas* xii. 135 The fiend leaped on to the porch and immediately dispelled any notion that might have been lurking in the minds of the checker players that here was a mere kibitzer who had come to breathe down the backs of their necks and offer advice. **1948** *Chess Rev.* Sept. 1, I tried to outline several plans of attack, but, in the excitement, with many kibitzers moving the pieces, I was unable to concentrate. **1950** J. Dempsey *Championship Fighting* x. 52 Keep your eyes closed to the kibitzers or wise guys. **1966** H. Yoxall *Fashion of Life* ix. 85 A member of the club was sitting watching... I went one down and the kibitzer added, 'If I'd known how you played I'd have bet..you wouldn't make it.' **1970** *Globe & Mail* (Toronto) 25 Sept. 30/6 The spectators held their breaths like kibitzers watching a pinball player try to coax the ball into a pocket. **1970** L. M. Feinsilver *Taste of Yiddish* iii. 294 It was not until the growth of radio and films that Yiddishisms like *Kibitzer*, [etc.],..began to achieve wider circulation.

kiblah. Add: Also **kebla, keblah, kebleh, kibla, qibla(h.** Also *transf.* (Further examples.)

1740 W. Stukeley *Stonehenge* v. 24 A *kebla*, or a place towards which we are to address the Deity. **1825** Scott *Talisman* in *Tales Crusaders* III. iii. 91 The Moslem turned towards his *kebla*, the point to which the prayer of each follower of the prophet was to be addressed. **1876** R. D. Osborn *Islam under Arabs* I. iii. 82 There have been few incidents more disastrous in their consequences to the human race than this decree of Muhammad changing the *kibla* from Jerusalem to Mekka. **1883** *Encycl. Brit.* XVI. 553/2 Mohammed..altered the direction of prayer (*ḳibla*)..towards Mecca. **1883** E. O'Donovan *Merv* xi. 106 Other pilgrims were standing on their little carpets with their faces towards the *keblah*..commencing their evening devotions. **1902** M. Roberts *Immortal Youth* i. 1 Ah, yes, to be in London, at the centre of things,..at the kebleh of the universe. **1911** *Encycl. Brit.* XVII. 420/2 In prayer the worshipper faces the *qibla*. **1973** *Times Lit. Suppl.* 28 Dec. 1590/1 A concatenation of aisles on one side of the building indicated the qiblah. *Ibid.* 1590/5 Attested early associations with 'David's Sanctuary' and the first Qiblah.

kiboko (kibōᵘ·ko). *Africa.* [Swahili, = hippopotamus.] A strong, heavy whip made of hippopotamus hide. Cf. Sjambok *sb.*

1921 *Chambers's Jrnl.* 118/1 One of those who had long ruled them with the *kiboko* (sjambok) in times gone by. **1922** C. T. Campion tr. *Schweitzer's On Edge Primeval Forest* iv. 67 A kiboko (or sjambok) of hippopotamus hide. **1947** J. Stevenson-Hamilton *Wild Life S. Afr.* vii. 62 The whips made of the skin of the hippopotamus, variously known in Africa as sjamboks, kibokos, kourbashes, and so on, have considerable value. **1969** *Tanzania Notes & Records* July 11 There were also many references to the 25 or 30 strokes of the *kiboko* used as a punishment in German days.

kibosh, *v.* Add: (Earlier and later examples.) Cf. *kybosh.

1884 'Cruck-a-Leaghan' & 'Slieve Gallion' *Lays & Legends N. of Ireland* 87 The Rector pull'd out an' oul' fourpinny-bit..An' handed the pill that wid kibosh the fun. **1890** *Punch* 16 Aug. 74/3 Wy, they'd queer the best pitches in life, if they kiboshed the Power of the Quid! **1969** *Listener* 27 Feb. 264/2 What a pity that the stipend has not kept pace..with the fall in the value of money (and it even comes to you less PAYE, thus kiboshing manoeuvrability in the field of expenses!).

Kichua, var. *Quechua.

kick, *sb.*¹ **1. a.** *free kick*, substitute for def.: (see quots. 1961.) Add earlier and later examples.

1882 in Charles-Edwards & Richardson *They saw it Happen* (1958) 299 A free kick awarded for a handling of the ball enabled Suter to place it well to the right wing forwards. **1961** F. C. Avis *Sportsman's Gloss.* 26/1 *Free kick*, an uninterrupted kick allowed to a team for an infringement against it, the opposing players having to stand ten yards away. *Ibid.* 268/2 *Free kick*, in Rugby Union football, an optional drop-kick, place-kick, or punt, taken as the result of a fair catch, the player free-kicking from the mark. **1972** *Guardian* 16 Mar. 23/3 From a free kick on the right, Foggo chipped on and the net bulged with relief.

fig. (Further examples.) Esp. in phr. *a kick in the pants*: a grave or humbling set-back; an expression of severe criticism or disapproval; similarly, *a kick in the teeth*.

[**1836** D. Crockett *Exploits & Adv. Texas* i. 14 If a man is only determined to go ahead, the more kicks he receives in his breech the faster he will get on his journey.] **1925** D. H. Lawrence *Refl. Death Porcupine* 105 The novel itself gives Vronsky a kick in the behind. **1933** E. O'Neill *Ah, Wilderness!* (1934) IV. ii. 134 Aw, you deserved a kick in the pants..making such a darned slob of yourself. **1937** Partridge *Dict. Slang* 881/1 It's better than a kick in the pants. **1940** R. A. J. Walling *Why did Trethewy Die?* i. 25 Giving the Methusalahs a kick in the pants. **1963** *Listener* 24 Jan. 152/2 General de Gaulle's statement on the Common Market was described by the East German Deutschlandsender as 'a kick in the pants' for Britain. **1970** *Globe & Mail* (Toronto) 25 Sept. 41/4 As some philosopher once noted, the only difference between a pat on the back and a kick in the pants is about eight inches. **1972** *Guardian* 28 Oct. 12/1 The Liberals' proud victory at Rochdale..has given the two major parties the kick in the teeth that each of them deserves.

d. (Earlier and later U.S. examples.)

1839 *Chemung* (N.Y.) *Democrat* 25 Dec. (Th.), Take the hint without a kick, and shut the open door. **1904** F. Lynde *Grafters* xii. 155 To-day he came around and gave me back my opinion, clause for clause as his own. But I have no kick coming. **1910** C. E. Mulford *Hopalong Cassidy* viii. 57 'We ain't got no kick, have we?' retorted Cavalry. **1948** *Gainesville* (Texas) *Daily Reg.* 3 July 6/2 The admission price will be upped to six-bits, which shouldn't draw any kicks from fans.

2. b. (Further examples.) Hence, a pulse or surge of electricity capable of producing a jerk in a detecting or measuring instrument. *B kick* Telegr. (see quot. 1928¹).

1910 *Hawkins's Electr. Dict.* 233/1 *Kick*. 1. In general, a recoil. 2. Any impulsive movement imparted in telegraphy to delicate instrument parts by a discharge from the line. *Ibid.*, *Kick of coil*, a discharge taking place from an electromagnet coil. **1928** A. E. Stone *Text Bk. Telegr.* xiii. 191 With the non-polarised relays a different method has had to be adopted, in order to eliminate the effects of what is known as the B kick. This term is applied to the break in the continuity of signals received on the non-polarised relays, due to the momentary demagnetisation of their cores when the current in the line is reversed. *Ibid.* 194 The 'kick' observed on the galvanometer. **1930** *Proc. R. Soc.* A. CXXIX. 214 An ambiguity arises in determining the residual range of an α-particle from the magnitude of the 'kick' recorded by the counter. *Ibid.* 216 The third record..shows the kicks smaller and more uniform in size. **1957** [see *kicksorter]. **1959** J. W. Freebody *Telegr.* vii. 200/2 Another difficulty found in the operation of diplex and quadruplex circuits was known as the B-kick which occurred when the B-side relay was operated by a marking current and the current was then reversed by the operation of the A-side key.

c. A strong or sharp stimulant effect, esp. that of liquor or drugs; *spec.* something that makes a drink potent; a thrill, excitement, pleasure; a feeling of marked enjoyment or the cause of such enjoyment; esp. in phr. *to get a kick out of* (something), to be excited or pleased by, to enjoy; *for kicks*, purely for pleasure or excitement, freq. recklessly or irresponsibly.

1844 *Bentley's Misc.* XVI. 597, I then demanded a common cocktail. 'With the kick in it?' said he. 'Oh, by all means,' I replied... It was..somewhat strong; but then that was my fault, for having ordered it 'with the kick in it'. **1899** R. Whiteing *No. 5 John St.* xxi. 216 'My Gawd! won't them chaps from the Collynies 'ave the kick!' he observes, in allusion to their entertainment at the public expense. **1903** *Daily Chron.* 16 Jan. 5/1 With cayenne and mustard (to give their food the missing 'kick' [of alcohol]). **1924** P. Marks *Plastic Age* xi. 101 'Who wrote "La Belle Dame sans Merci"?'..'I think Jawn Keats wrote it. It's one of those bedtime stories with a kick.' **1927** W. E. Collinson *Contemp. Eng.* 81 Home-brew with a kick in it. **1928** *Daily Express* 4 Dec. 10/3, I was told I should get a kick out of that journey—and I certainly did. **1929** *Evening News* 18 Nov. 15/6 A cocktail basis with a real kick (42 deg. proof spirit). **1933** D. L. Sayers *Murder must Advertise* ix. 159 There's a kick in being afraid. **1935** S. Spender *Destructive Element* 82 Strether accepts even the fact that he is living with Madame de Vionnet; in fact, he gets a kick

KICK

out of it. 1941 *Jazz Information* Nov. 22/2 A man who..worked hard and got his kicks and saved a little money. 1942 *R.A.F. Jrnl.* 2 May 35 We get a great kick out of wearing it. 1946 Mezzrow & Wolfe *Really Blues* (1957) 373 *For kicks*, for pleasure's sake. 1951 *Manch. Guardian Weekly* 28 June 2 To seek a heftier 'kick' from real narcotics. 1956 [see *cramp *v.* 5 c]. 1961 Wodehouse *Service with Smile* (1962) x. 155 He added that the beverage had a kick, and Lord Ickenham agreed that its kick was considerable. 1963 *Listener* 17 Jan. 133/1 Anti-social, sexually ruthless, stealing cars for kicks. 1967 M. M. Glatt et al. *Drug Scene* iii. 39 He no longer got a 'kick' or 'flash' from taking drugs. 1974 *Advocate News* (Barbados) 5 Mar. 3/2 The pusher can more easily persuade him to try something with a bit more kick to it.

d. An interest or enthusiasm, esp. one that is temporary; a fashion, fad (cf. sense 4 in Dict.); a subject, line of thought, or manner of behaving; = *bag *sb.* 1 d; esp. in phr. *on the —— kick* = doing, or enthusiastic about, the thing specified by the prefixed word or words. orig. *U.S. slang.*

[1942 Berrey & Van den Bark *Amer. Thes. Slang* § 233/2 The fashion; rage,.. *the kick.*] 1946 *Jazz Record* July 8 The whole jazz world was on a Hawkins kick. 1955 M. McCarthy *Charmed Life* ii. 34 He had been..a magazine editor. He was on that kick, as he called it, when he met Martha. 1955 B. Bailey in Shapiro & Hentoff *Hear Me Talkin' to Ya* xviii. 298 When I was starting up, they used to say the two races couldn't get along playing. They used to say stuff like they were afraid we'd go after their women. All that's been proved false, and everything else on that prejudice kick has been proved false. 1957 M. Millar *Soft Talkers* ii. 20 'He's rather sensitive about being caught by the cops in bed with another man's wife.' 'For Pete's sake, Esther, get off that kick, will you?' 1959 C. MacInnes *Absolute Beginners* 88 Mannie wasn't in on the Angries kick. *Ibid.* 93 They didn't like it when little Emmanuel got on the writing kick? 1963 B. S. Johnson *Travelling People* iv. 66 The star of the production is Maurice Bunde,..fifty-ish but on a tremendous Back-to-Youth kick. 1970 *Globe Mag.* (Toronto) 26 Sept. 8/3 We must get off this kick that every job is a career—it isn't. 1971 *Times Lit. Suppl.* 12 Nov. 1409/1 Somewhere behind the cumulative high, the peace-kick, the good vibes, efficient entrepreneurs..were smiling their mean smiles all the way to the bank.

e. *Athletics.* A sharp burst of speed, esp. towards the end of a middle-distance race.

1955 F. Stampfl *On Running* vii. 108 By making his final burst of 300 yards from home Bannister could hope to draw the sting of Nielsen's powerful kick and late finish. 1966 R. Clarke *Unforgiving Minute* xvi. 143, I was obviously holding his finishing 'kick' with no effort whatever and I certainly ran faster when I did sprint. 1972 *N.Y. Times* 4 June 4/5 Instead, Wottle overtook Bob Wheeler of Duke starting the stretch run and turned back a belated kick by Jerome Howe of Kansas State, who finished second in 3:39.8.

6. b. *pl.* Shoes. orig. *U.S.*

1904 'No. 1500' *Life in Sing Sing* 250/1 *Kicks*, shoes. 1936 K. Mackenzie *Living Rough* xi. 160 My new kicks, every time I took a step, made a sound like the back of a bird store. 1937 Partridge *Dict. Slang* 601/1 *Pair o(f) kicks*, boots, shoes: tramps' c[ant]. 1964 L. Hairston in J. H. Clarke *Harlem* 285 After I brushed my kicks, I looked my wig over in the mirror. 1973 *Black World* Apr. 63 My terrible blue-and-white kicks.

7. (Further examples.)

1938 Wodehouse *Summer Moonshine* i. 18 'She slung your brother Joe out.' 'And with only ten dollars in his kick, mind you.' 1962 R. Cook *Crust on its Uppers* (1964) iii. 27 I'm about to stuff my pony in my kick. 1968 *Sunday Truth* (Brisbane) 22 Sept. 22/8 One of Luke's jobs was to see that the money was banked every week. Luke put it in his own kick.

8. *Comb.*, as **kick-pleat** (or **plait**), a pleat in a narrow skirt to allow freedom of movement; **kick-stand**, 'a device for holding up a bicycle or motorcycle when not in use consisting of a metal bar or rod that is attached by a swivel device to the frame and may be kicked to a vertical position as a prop' (Webster, 1961); **kick-start, -starter**, a device for starting an internal combustion engine, esp. on a motor-cycle, by a downward thrust on a pedal; hence **kick-start** *v. trans.* and *intr.*, to start (an engine) thus; **kick-turn** *Skiing*, a form of standing turn; **kick-wheel**, a potter's wheel worked by a foot pedal.

1934 Webster s.v. *plait*, A kick plait is a variation of the latter [*sc.* box plait] used to give breadth to a narrow skirt. 1960 *Sunday Express* 20 Nov. 14/2 Pencil skirt with back 'kick' pleat. 1947 *Cycling Handbk.* (League of Amer. Wheelmen, Inc.) 23 Heavyweights are generally equipped with coaster brakes,..kickstands, and other accoutrements dear to the hearts of juvenile Americans. 1963 D. Broun *Subject of Harry Egypt* i. 8 There was a metered space..and he eased the cycle into it and over onto the kickstand. 1914 *Motor Cycle* 2 Apr. 138/1, 1912 P. and M., free engine, 2-speed, kick start. 1928 *Manch. Guardian Weekly* 15 June 474/2 The biggest boy is demonstrating..how to kick-start an engine. 1959 I. Jefferies *Thirteen Days* i. 17, I..turned my bike down the slope to save kick-starting. *Ibid.* xi. 188, I kick-started and zoomed up the rock-slope. 1962 'D. Wilson' *Search for Geoffrey Goring* viii. 170 He kickstarted the motor and rode on. 1916 *Motor Cyclists' A.B.C.* 107 A kick starter is fitted to a machine for the purpose of allowing the engine to be started whilst the rider is in the saddle by a downward kick of a pedal. 1919 C. P. Thompson *Cocktails* 235 She mounted on the kickstarter and stamped on it with resolute vigour. 1961 *Engineering* 13 Oct. 486 The clutch, gearbox, kickstarter, transmission and rear stub axle are combined with the engine into a single unit. 1910 W. R. Rickmers *Ski-ing* 27 People with stiff or short legs should take short ski, as otherwise certain necessary movements (*i.e.*, the kick-turn) become difficult or impossible. 1960 *Sunday Express* 18 Dec. 15/4 Kick-turns, in which you stand on one ski on a slope, lift the other and turn it right round, and then bring the second ski round too so that you are facing in the opposite direction. 1893 E. A. Barber *Pott. & Porc. U.S.* xii. 250 Such wares..were produced in large quantities by negro men and boys, who employed the old-fashioned 'kick-wheel' in their manufacture. 1949 K. S. Woods *Rural Crafts Eng.* v. xvi. 233 A kick-wheel is driven by a horizontal movement of one foot on a treadle. 1968 J. Arnold *Shell Bk. Country Crafts* 15 With the employment of an improved kick-wheel for 'throwing', they were able to make tremendous advances. 1972 *Islander* (Victoria, B.C.) 30 July 5/1 The pottery produced is hand built. However, the group does work with a kick-wheel, as well as an electric-wheel.

kick, *v.*[1] Add: **I. 3. b.** Said also of the ball, and of the bowler. Also with *up*. (Further examples.)

1866 'Captain Crawley' *Cricket* 25 You will most likely get a run whether the ball shoots or kicks. 1877 C. Box *Eng. Game Cricket* 453 The ground is said to kick when the ball, after being pitched, rises almost perpendicularly. 1888 Steel & Lyttelton *Cricket* 152 Spofforth was bowling rather more than medium pace, bringing the ball back a foot or more very quickly from the pitch, sometimes kicking to the height of the batsman's head and at others shooting. 1904 P. F. Warner *How we recovered Ashes* xiii. 246 The ball was always turning, and one or two deliveries kicked up rather awkwardly. 1963 A. Ross *Australia 63* i. 33 McKenzie got one to kick in the next over.

c. *Telegr.* Of a relay: to break contact momentarily.

1928 A. E. Stone *Text Bk. Telegr.* xiii. 192 If during this period the tongue of the relay 'kicks', the local circuit is momentarily broken. 1959 J. W. Freebody *Telegr.* vii. 195/2 In a duplex circuit this surge flows through the line coil of the relay and would cause the relay to 'kick' if a similar balancing surge were not also allowed to pass through the relay balance coil.

4. e. *refl.* To reproach or be angry with (oneself); to be annoyed at something one has done or omitted to do.

1891 *Voice* (N.Y.) 29 Jan., In the absence of any of the committee to kick I went home kicking myself. 1892 W. S. Walsh *Handy-bk. Lit. Curiosities* 584 To kick one's self, often used with an infinite variety of adjuncts,—*i.e.*, to kick one's self 'all over the house', 'all over the place', etc.,—means to feel or express violent dissatisfaction with one's self. 1903 *Independent* (N.Y.) 15 Jan. 148/2 He goes away kicking himself. 1907 A. Bennett *Let.* 5 May (1966) I. 90 Those who persuade themselves to act on this assumption from the start will have least cause to kick themselves in the distant future. 1955 L. P. Hartley *Perfect Woman* xxi. 188 All the way to Tilecotes he could have kicked himself for not having made the engagement for next week. 1966 B. Kimenye *Kalasanda Revisited* 48 Mrs. Mulindwa could have kicked herself for making the suggestion in the first place. 1973 *Times* 27 Dec. 13/4 Rangers were rightly kicking themselves afterwards.

f. *trans.* To give up or overcome (a habit, esp. drug-taking). Also *intr. colloq.* (orig. *U.S. slang*).

1936 *Amer. Speech* XI. 123/2 *To kick the habit*, to stop using drugs. 1951 *Nat. Educ. Assoc. U.S. Jrnl.* May 342/2 Later on they find themselves hooked and can't kick the habit unless they receive medical and psychiatric help. 1956 B. Holiday *Lady sings Blues* (1973) xiv. 121 Along about the end of the war I went to Joe Glaser's office and told him I wanted to kick and I'd need help. 1958 *Oxford Mail* 29 July 6/5 Harmony would again be restored if Johnny could 'kick' his craving. 1964 S. Bellow *Herzog* (1965) 334 Between his false teeth (to help him kick the smoking habit, as he had once explained to Herzog) he kept a plastic toothpick. 1971 *Black World* Mar. 56/1 I'll help you, man, cuz I know you want to kick. *Ibid.* Apr. 22/1 Let's kick that habit, let's use soul music. 1972 *Times* 3 Jan. 8/3 In a moment of weakness, I watched an episode of this [television serial] after having kicked the habit for more than 12 months.

5. b. *to be kicked upstairs* (earlier and later examples); also, *to kick (someone) upstairs*.

c 1697 [see Upstairs *adv.* 1 b]. 1750 C'tess of Shaftesbury *Let.* 28 Nov. in Earl of Malmesbury *Lett.* (1870) I. 78 The Bedfordian set will be honourably kicked up or down stairs. 1952 'W. Cooper' *Struggles of Albert Woods* iii. v. 197 The plot was devastatingly simple—Dibdin was to be kicked upstairs and Albert was to take his place. 1962 R. B. Fuller *Epic Poem on Industrialization* 27 Kicking the bosses upstairs—high out of the way. 1967 G. F. Fiennes *I Tried to run a Railway* vii. 78, I got eventually kicked upstairs to Paddington. 1970 *Guardian* 11 Nov. 20/6 Which party has kicked more people upstairs?

II. kick about or **around.**

a. *intr.* To walk or wander about; to go from place to place, esp. aimlessly. *colloq.* (orig. *U.S.*).

1839 C. M. Kirkland *New Home* xxv. 195 We heard that he was better, and would be able to 'kick around' pretty soon. 1846 B. Upton *Let.* 12 Dec. in *Amer. Heritage* (1966) June 93/2, I have been kicking about with scarcely leisure enough to take my meals. 1946 F. Sargeson *That Summer* 56 We're going to have a good time just kicking around.

b. *to be kicking about* or *around*: to lie scattered around, esp. in a casual or untidy fashion; to be available, unused, or unwanted.

1867 'T. Lackland' *Homespun* i. 80 The..doctor, whose instruments..lie kicking about like ordinary household trumpery. 1877 E. Peacock *Gloss. Words Manley & Corringham, Lincolnshire* 148/2 When I went ower to Rotterdam, bacca was that cheap, it was kickin' aboot i' th' streets. 1906 J. F. Kelly *Man with Grip* 99 Now kindly remove that old radium, It's been kicking around for a week. 1955 W. Gaddis *Recognitions* ii. viii. 658 Too much gold, that was their difficulty, gold kicking around all over the place. 1967 'V. Siller' *Biltmore Call* 128 'Is there a sandwich kicking around?' 'Oh, sure, and I just made a fresh pot of coffee.'

c. *trans.* To kick in all directions; also *fig.*, to treat (someone) harshly, unfairly, or contemptuously. Chiefly *U.S.*

1938 C. Porter *Most Gentlemen don't like Love*, Most gentlemen don't like love,—They just like to kick it around. 1939 J. Steinbeck *Grapes of Wrath* ii. 8 A good guy and also he was not one whom any rich bastard could kick around. *a* 1940 F. Scott Fitzgerald *Last Tycoon* (1949) i. 25 You seem to take things so personally... You just ask to be kicked around.

d. To discuss or examine (a subject, idea, etc.); to try out. *colloq.* (orig. *U.S.*).

1939 *Esquire* May 75 Speaking again of Swing: few tunes deserve its name till they've been 'kicked around' by good performers. 1947 F. Wakeman *Saxon Charm* vi. 118 He agreed to write the scene experimentally. 'Maybe I'll get excited about it when I start kicking it around.' 1966 'D. Shannon' *With a Vengeance* (1968) iii. 42 They ..drifted over by the other side of the big room to kick it around a little. 1971 'G. Douglas' *Time to Die* xv. 159 They kicked the details around for a few more minutes and then left them to stew.

kick back.

a. *intr.* (See quot. 1909.)

1909 Webster, *To kick back, Mech.*, to start backwards;—said of an internal-combustion engine in starting with the crank when the spark is advanced and a too early ignition is effected. 1935 T. E. Lawrence *Let.* 13 Feb. (1938) 855 We launched the Dinghy: the quietest and sweetest tick-over of any Dinghy yet! It kicked back, when cold. So we put the ignition back a trifle.

b. *trans.* and *intr.* To return (money, stolen goods, etc.) to the person from whom they were obtained; to pay (money), esp. as a kick-back (see *kick-back). *colloq.* (orig. *U.S.*).

1926 Maines & Grant *Wise-Crack Dict.* 10/2 *Kick-back*, have to return a sucker's money. 1930 *Amer. Mercury* Dec. 456/2 Kick back with that hooch or we give you the works. 1934 *Atlantic Monthly* Aug. 139 The kick-back operates in the following manner. A wage scale is set either by law, as in government contracts, or by agreement between capital and labor. The worker assumes that he is to get so much per day or per hour for his work. At the end of the week, he is required to return or kick-back part of his wages to a designated person, often a foreman or a bookkeeper. 1970 'B. Mather' *Break in Line* v. 59 The luggage coolies..kicked back half of their take to the Pathan hall porter.

kick down. *trans.* and *intr.* To operate a kick-down device (see *kick-down).

1909 *Cent. Dict.* Suppl. s.v., *To kick down*, to bore (a well) by a drill worked as follows: A wooden casing is sunk in the ground or rock for a few feet and the boring-tool works inside of and is guided by this casing... The tool is moved or kicked down by the pressure of the operators' feet. 1959 *Observer* 1 Mar. 21/5 Second can be obtained by kicking down the accelerator. 1963 *Which?* Oct. (Car Suppl.) 116/2 More effort was needed to 'kick-down' on the Zephyr 6.

kick in.

a. *trans.* To break down (a door, etc.) by kicking against the outer side; *spec.* (*U.S. slang*), to break into (a building).

1881 R. L. Stevenson *Treas. Isl.* (1883) v. 39 Then there followed a great to-do.., furniture thrown over, doors kicked in. 1926 J. Black *You can't Win* vii. 78 I'll kick in the first private house that looks good. We'll surely find a coat and maybe a few dollars. 1931 *Detective Fiction Weekly* 17 Jan. 23/1 Harold G. Slater's big jewelry store safe had been 'kicked in' and robbed of twelve thousand dollars.

b. *trans.* and *intr.* To contribute (money, etc.); to pay (one's share). *slang* (orig. *U.S.*).

1908 K. McGaffey *Sorrows of Show-Girl* 45 The lawyer guy kicked in with the balance of the ten thousand. 1908 H. Green *Maison de Shine* 282 If somebody else will get 'em to kick in I'll play the show. 1928 [see *chin *sb.*[1] 1 d]. 1936 Wodehouse *Laughing Gas* xxiv. 254 To encourage the Christmas spirit in whoever was supposed to kick in with my ransom. 1948 *Lawton* (Okla.) *Constitution* 2 July 8/1 The spectators 'kicked in' with a little cash. 1972 *Fortune* Jan. 112/2 Hillard Elkins, producer of *Oh! Calcutta!*, asked him to help back his productions of two Ibsen plays; Lufkin kicked in $10,000.

kick off.

b. Also *fig.*, to start, begin. Freq. const. *with*.

1911 R. Brooke *Let.* 25 Apr. (1968) 300 'Are you ready to kick off?' he said...I *gathered* it merely meant was she ready to go out to San Lorenzo. 1942 F. Sargeson in *N.Z. New Writing* I. 5 To kick off with we'd fool about in the water. 1954 L. Durrell *Let.* 14 Mar. in *Spirit of Place* (1969) 124, I will kick off with Freya Stark and Sir Harry Luke. 1968 *Blues Unlimited* Nov. 17 It kicked off with Bob Hite..ranged through Dave Kelly's bottleneck playing. 1969 G. E. Evans *Farm & Village* xii. 131 The old bo's would come in, and my father and I used to go down to the bar to *kick off with*.

c. To die. *slang* (orig. *U.S.*).

1921 J. Dos Passos *Three Soldiers* ii. i. 61 Another kid's kicked off with that—what d'they call it?—menegitis. 1948 E. Waugh *Loved One* 22 'It belonged to some old Britisher who's just kicked off.' 'I am that Britisher and I have not kicked off.' 1969 C. Burke *God is Beautiful,*

Man (1970) 29 If he don't come back his old man will get sick and kick off too. 1970 R. Lowell *Notebk.* 122 The old bitches Live into their hundreds, while I'll kick off tomorrow.

8. kick out.

d. *Surfing.* (See quots.)

1962 T. Masters *Surfing Made Easy* 64 *Kicking out*, turning up and over the wave to end a ride. 1965 J. Pollard *Surfrider* ii. 20 First let's 'kick out'—shift the weight to the rear of the board and pull it over the top of the wave.

10. kick-and-rush, used *attrib.* to describe football played with more vigour than art; **kick-ball**, for *Sc.* read orig. *Sc.*, and add further examples; **kick-out**, (*b*) *Surfing* (see quots. and cf. sense 8 d above); **kick-the-can** (or **-tin**, etc.), a children's game in which a tin can is kicked (fully described in I. & P. Opie *Children's Games* (1969) 164–6).

1906 *Daily Chron.* 26 Nov. 9/2 It was a kick-and-rush game, played badly. 1930 *Daily Express* 9 Sept. 12/5 The football they played was of the kick-and-rush order. 1971 E. Shorris *Great Spirit* i. 17 You played kickball in the streets. 1972 J. E. Franklin in W. King *Black Short Story Anthol.* 354 During recess the children played kick-ball, tag, and other games. 1967 J. Severson *Great Surfing* Gloss. *s.v.*, A kick-out is a last-ditch effort to keep from losing your board. 1970 *Studies in English* (Univ. Cape Town) I. 32 The *kick-out*..involves stepping on the rear of the surfboard with considerable force and, at the same time, raising the lead foot, lifting the nose of the surfboard out of the water, and making it possible to pivot the board on its tail. 1971 *Ibid.* II. 27 The *kick-out* is an act of desperation. The surfer turns his board violently from the tail and as he leaves the board kicks it—so he hopes—over the top of the wave. 1909 *N. & Q.* 5 June 445/2 Children's games in Orkney... Kick the tinnie. 1959 I. & P. Opie *Lore & Lang. Schoolch.* xviii. 377 Orthodox games like 'Kick the Can' and 'Jacky Shine a Light'. 1959 B. Sutton-Smith *Games N.Z. Children* ii. 58 More popular were those games in which the players helped one another to fight the He, and of these the most widespread was the game known as *Kick the Tin*. 1966 'L. Lane' *ABZ of Scouse* 59 *Kick-ther-can*, a form of street football, using old tin cans. 1971 *Stornoway Gaz.* 10 July 1/8 Children are inventive folk, They make their own best ploys. Smooring, leevo, kick the can, Sufficed when we were boys. 1973 B. Broadfoot *Ten Lost Years* viii. 86 My father called me in from outside, kick the can or one of those games we used to play.

Kickapoo (ki·kăpū). *U.S.* [Amer. Indian.] **a.** One of a tribe of North American Indians of the Algonquian family, now resident in reservations in Kansas, Oklahoma, and Mexico. **b.** The language of this people. Also *attrib.* or as *adj.*

1722 D. Coxe *Descr. Carolana* 50 Nations to the West of this Lake, besides the beforemention'd, are Part of the Outogamis, Mascoutens, and Kikpouz. 1835 C. F. Hoffman *Winter in West* I. 276 The Indians that frequent the neighbourhood of Chicago..are chiefly Pottawattamies and Ottawas, with a few Chippewas.., and a straggling Kickapoo. 1933 L. Bloomfield *Lang.* 72 The languages of..the Great Lakes region (Ojibwa,..Kickapoo,..and so on). 1960 B. Keaton *Wonderful World of Slapstick* (1967) 19 He had a Kickapoo squaw on one side of him, a Kickapoo brave on the other.

kick-back, kickback (ki·kbæk). orig. *U.S. colloq.* [f. phr. *to kick back* (*kick *v.*[1] II).] **a.** A refund, a rebate; the return of money, goods, etc.; a payment (usu. illegal) made to a person who has made possible or facilitated a transaction, appointment, etc. Also *attrib.*

1932 *Editor* 6 Feb. 112/2 *Kick-back*, a return of money. 1934 [see *kick back* b s.v. *kick *v.*[1] II]. 1934 *Sun* (Baltimore) 24 Jan. 1/3 The 'kick-back' system of cutting PWA workers' pay. *Ibid.* 1 Feb. 1/5 These 'kickbacks' were described as levies amounting to from $15 to $25 a week on the musician's salary. 1935 N. Ersine *Underworld & Prison Slang* 49 *Kickback*, loot that must be returned to avoid arrest. 'They took a grand off the hoosiers, but they had to make a kickback when the marks beefed.' 1939 *Ibid.* 13 Feb. 16/5, 150,000 persons and companies throughout nation get 'kickback'... Several hundred Maryland Corporations and individuals received tax refunds during the last fiscal year. 1940 F. Riesenberg *Golden Gate* 308 Longshoremen were finding it tougher than ever to get jobs, even through kick-backs of pay, bottles of liquor, and cigars. 1958 M. Dickens *Man Overboard* xiv. 218 With Mr Pearse and his little kick-backs out of the picture, the food budget was reduced. 1959 *Listener* 3 Dec. 960/1 A number of employers were prepared to offer bribes, pay 'kickbacks'. 1971 *Courier Mail* (Brisbane) 8 Mar. 4/7 The [U.S. official tax] guide says: 'Bribes and kickbacks (a form of bribe) to non-government officials are deductible.' 1972 *Daily Tel.* 19 June 10/5 The promoter claims that another member of the committee approached him demanding a kick-back on the profits and, after he had refused this proposal, the permit was somehow no longer forthcoming.

b. A strong reaction or repercussion; an undesirable result.

1935 M. M. Atwater *Murder in Midsummer* xxii. 210 His bluster was the kick-back of his strained nerves. 1940 *Amer. Speech* XV. 64 This kickback of the idea into the word, wherein..the word is..vested with unusual suggestive power. 1953 Wodehouse *Performing Flea* 177 The feeling that he showed a lack of public spirit in getting away and leaving us to receive the kick-back. 1954 R. Knox *Retreat for Lay People* xiv. 140 Even as a matter of psychology, isn't it probable that all this negative business has a kick-back which is bad for us? 1965 *Listener* 6 May 658/1 We can over-mechanize it [*sc.* education]. One of the kick-backs of this is the University of California situation, over-planning, the over-administering of education.

c. *Railways.* A device whereby the direction of wagons, etc., can be reversed.

1947 *Richmond* (Virginia) *Times Dispatch* 1 Apr. 6/1 The empty [coal] car is then kicked off the dumper by the next loaded car, rolls by gravity to a high 'kickback' at the outshore end of the pier and thence by gravity to the yard for empty cars. 1962 *Times* 26 Oct. (Spencer Steelworks Suppl.) p. xviii/2 The gravity operated kick-back which reverses the wagon's direction.

d. In timber preservation (see quots.).

1947 *N.Z. Timber Jrnl.* Sept. 61/2 *Kick back* (wood preservative), surplus antiseptic released from the wood when pressure is withdrawn after impregnation. 1968 *Gloss. Terms Timber Preservation* (*B.S.I.*) 21 *Kickback*, the amount of preservative forced out of the timber when pressure is released.

kick-down (ki·k,daun). [f. Kick *v.*[1] + Down *adv.*] A device that is operated by the foot; *spec.*, on a motor vehicle, a device whereby one can change to a lower gear, esp. by pressing right down on the accelerator pedal in a vehicle with automatic transmission; also, the act of thus changing to a lower gear. Freq. *attrib.*

1909 *Cent. Dict.* Suppl., *Kick-down*, the apparatus used in kicking down. See *to kick down*. 1954 *New Automotive Encycl.* 35c/1 A valve operated by a solenoid magnet, which is connected to a governor and kickdown switch. 1958 *Times* 15 July 7/6 The unduly light pressure required for the kick-down. *Ibid.*, The ease with which the kick-down change occurred. 1959 *Times* 17 Mar. 14/7 The accelerator kick-down is an over-riding control that can be brought into play for maximum acceleration at any speed up to about 65 m.p.h. 1971 *Daily Tel.* 24 Mar. 11/4 The kick-down switch on the accelerator was unusually light and pleasant to use.

kicker. Add: **5.** *Poker.* A high third card retained in the hand with a pair at the draw.

1892 W. J. Florence *Handbk. Poker* 91 To keep two small cards and an ace is called holding up 'a kicker'. This draw is made by the player, hopeful of getting two pairs, with the additional ace or king. 1895 'Templar' *Poker Manual* 57 Sometimes a player raises on a single pair and a kicker, *i.e.* a high card. 1946 Morehead & Mott-Smith *Penguin Hoyle* 127 To keep an ace or other high card as a 'kicker' seriously decreases the chances of improving.

6. An outboard motor, or a boat driven by one. Also *attrib.* *N. Amer. colloq.*

1928 L. R. Freeman *Nearing North* ii. i. 132 The kicker is hung in a hole cut at a proper height in the long overhang of the stern. 1937 *Times* 13 Oct. 15/6 At Fort Simpson..four of us went upstream to a creek in a 'kicker' (which is a canoe powered by an out-board motor) and swam luxuriously. 1942 L. Rich *We took to Woods* (1944) ii. 33, I want to cut the stern off square, when I get the price of an outboard motor, and make a kicker-boat out of it. 1953 Berrey & Van den Bark *Amer. Thes. Slang* (1954) § 82a/1 *Kicker*, an auxiliary motor on a boat. 1963 R. D. Symons *Many Trails* xvi. 165 Travelling downstream with a good 'kicker' (outboard motor) pushing the canoe at a good speed is easy. 1967 E. B. Nickerson *Kayaks to Arctic* iv. 30 Soon we learned to..differentiate outboard motors, or 'kickers' as they are universally called here.

7. *Printing.* (See quots.)

1930 K. E. Olson *Typogr. & Mechanics of Newspaper* xiii. 421 Every fiftieth or one-hundredth paper is turned slightly askew by an automatic 'kicker' in order to facilitate the making up of bundles of given content. 1967 V. Strauss *Printing Industry* vi. 384/1 Counting of newspapers is made easier by the 'kicker', a metal arm which pushes, say, every 25th or every 50th paper out of line, thereby dividing the flow of papers into smaller batches.

kick-in (ki·kin), *sb.* *Football.* [f. Kick *v.*[1] + In *adv.*] Practice goal-shooting before the start of a match.

1961 *Times* 10 Feb. 19/7 Previously the British method had demanded a general kick-in before the toss-up. 1972 G. Green *Great Moments in Sport: Soccer* ii. 33 The moment the Hungarians..began their 'kick-in' before the start one got the distinct feeling that something unusual was in store. 1973 *Shoot!* 1 Dec. 15/1 The home team's goalkeeper is injured during the pre-match kick-in.

kicking, *vbl. sb.* Add: **b. kicking plate**, a metal plate fixed to the lower part of a door, etc., to prevent damage or wear; **kicking-strap**, (*b*) *Naut.* a rope lanyard fixed to the boom to prevent it from rising.

1940 *Chambers's Techn. Dict.* 476/1 *Kicking plate* (Join.), a plate fixed on the face of the bottom rail of a door, to prevent the damage caused by persons kicking the door to open it. 1959 *Engineering* 16 Jan. 94/3 At the base of each frame is a kicking plate of porcelain-enamelled sheet steel. 1951 G. Pennant *Young Sailor* v. 70 It is to prevent the boom from lifting that a kicking strap is fitted. 1961 R. M. Tetley *Sailing* ii. 47 It is in a gybe that a kicking-strap proves its worth, since it holds down the after end of the boom thereby allowing complete control to be maintained over the sail at all stages of the manœuvre.

kicking, *ppl. a.* Phr. *alive and kicking*, add def.: indubitably alive; very lively and active. (Earlier and later examples.)

c 1831 J. R. Planché *Olympic Devils* in *Extravaganzas* (1879) I. 71 *Plu.* Alive? *Char.* And kicking. 1840 *New Monthly Mag.* LVIII. 497 He is (as the Irishman says), 'alive and kicking'. 1930 R. Lehmann *Note in Music* 130 He seemed to imply with amusement how particularly alive they were. 'Father, mother, two young brothers, and sister Clare, all alive and kicking.' 1966 L. Southworth *Felon in Disguise* xi. 158 I'd feel happier if I knew Donaldson was alive and kicking.

b. *Cricket.* Of the ground, a bowler, or his bowling: causing the ball to 'kick' (see Kick *v.*[1] 3 b in Dict. and Suppl.). Also of a lawn-tennis service.

1885 J. Lillywhite *Cricketers' Compan.* 53 The Gloucestershire batsmen found Palmer unplayable on a 'kicking' wicket. 1888 [in Dict.]. 1891 W. G. Grace *Cricket* iii. 67 My brother was the faster [bowler], and on a rough kicking wicket met with great success. 1924 F. G. Lowe *Lawn Tennis* 12 The only way to take a fast kicking service.

kickininee, var. *kokanee.

kick-off. Add: **b.** *fig.* The start, beginning; an inaugural or opening event.

1875 [in Dict.]. 1919 Wodehouse *My Man Jeeves* 200 The kick-off was scheduled for one o'clock in the morning, when the household might be expected to be..asleep. 1969 *New Yorker* 11 Oct. 43/2 The kickoff starts tonight, with a dinner for the living benefactors of the Museum. 1973 M. Truman *Harry S. Truman* i. 20 The first major crisis came on Labor Day, when we went to Detroit to make the traditional kickoff speech in Cadillac Square.

kicksorter (ki·ksǭɹtəɹ). *colloq.* Also **kick sorter.** [f. Kick *sb.*[1] + Sorter.] An instrument that classifies electrical pulses according to their amplitude and registers the number received in each amplitude range; a pulse-height analyser.

1947 *Rev. Sci. Instruments* XVIII. 90/2 Five instruments, which have been named Pulse Analysers, Pulse Amplitude Analysers, or 'Kicksorters', have been developed in England and Canada. 1957 *Economist* 7 Sept. 767/2 (Advt.), But why 'kick sorter'? Because it sorts out electrical 'kicks' or impulses according to their amplitude —more than 16,000 of them in each of 100 channels and at speeds up to 1,250 pulses per second. 1968 *Brit. Med. Bull.* XXIV. 259/1 The same equipment can also be used to generate a histogram of spike amplitude, comparable to the pulse-height analyser or 'kick sorter' used by the nuclear physicist.

kickster (ki·kstəɹ). [f. *kick *sb.*[1] 2 c + -ster.] One whose behaviour is governed principally or solely by the desire for 'kicks'.

1963 *Guardian* 4 Oct. 20/6 Christine is a kickster... She will go with 15 or 18 men at a time. 1967 J. G. Morgan *Involved* 57 'He looked a real kink, you know, a regular kickster.' 'Educate me,' Frankie said, 'I want to be with you.' 'A nut,' Janet explained, 'a kink who likes going to a scene but won't participate, he gets his kicks out of watching everyone else knocking themselves out.' 1972 *New Society* 9 Nov. 346/3 Apart from the bovver-boy type, there are some girl gangs..and middle class kickster groups... There's a theory it's best to have something to steal, rather than nothing, but this is no help with kicksters.

kick-up. Add: **2. b.** A dance or party. *colloq.* (orig. *U.S.*).

1778 W. Beatty *Jrnl.* 1 Dec. in *Maryland Hist. Mag.* (1908) III. 116 We Collected the Girls in the neighbourhood and had a kick up in the Evening. 1796 Grose *Dict. Vulgar T.* (ed. 3) s.v. *kicks*, A kick up; a disturbance, also a hop or dance. 1899 R. Whiteing *No. 5 John St.* x. 100 There's a little bit of a kick-up to-night with a few of us—sort of sing-song. 1910 'G. B. Lancaster' *Jim of Ranges* vi. 126 'What d'yer do at a kick-up, Jim?' 'Oh, hide-an'-seek..an' kiss-in-the-ring,' explained Jim.

4. (See quots.)

1883 W. S. Gresley *Gloss. Terms Coal Mining* 147 *Kick-up*, see *tipper*. 1893–4 R. O. Heslop *Northumberland Words* II. 423 *Kick-up*, an apparatus at a pit bank, made like an iron cradle, by which a tram is turned upside down and emptied on to the screen. 1909 H. Louis *Dressing of Minerals* 451 In larger mines it is more usual to use cars with fixed sides and to use some form of 'Tippler' or 'Tumbler' for turning the car over and thus emptying out its contents. Tipplers are of two kinds: end tipplers or 'Kick-ups' and side tipplers.

5. = Kick *sb.*[2] 1.

1901 in *Oxf. Eng. Dict.* s.v. *Kick sb.*[2] 1. 1923 H. J. Powell *Glass-Making in Eng.* ii. 22 Feet of goblets, showing hem and kick-up. *Ibid.* v. 74 Stability had been given by pushing upwards and inwards the base of the bulb to form the familiar 'kick-up' of modern wine-bottles.

kicky (ki·ki), *a.* [f. Kick *v.*[1] + -y[1].] **1.** *Sc.* (See quot. 1808.) Also, clever, lively; provoking, teasing, annoying.

1790 A. Shirrefs *Poems* 213 Auld Meg hersel' began the play, Clad in a bran-new hudden gray, And in't, I wat, she look'd fu' gay, And spruce and kicky. 1806 G. S. in J. Cock *Simple Strains* 93 Fu' mony a witty touch, and kicky line, Wad won the praise o' langer heads than mine. 1808 Jamieson, *Kicky.* 1. Showy, gaudy... 2. High-minded, aiming at what is above one's station. 1910 in *Sc. Nat. Dict.* (1960) V. 395/1 Isna' that kicky 'at I canna min' fat comes neest.

2. *Cricket.* Causing the ball to 'kick' (see Kick *v.*[1] 3 b in Dict. and Suppl.).

1888 Steel & Lyttelton *Cricket* iii. 150 It is a slow easy wicket he has to bat on, and not a 'caked', kicky one. 1903 *Windsor Mag.* Sept. 393/2 A very kicky wicket generally averages matters somewhat by supplying one dead shooter.

3. Providing 'kicks', exciting, lively. *N. Amer. colloq.*

1968 *N.Y. Times* 15 Aug. 42 It brought out some kicky styles to preview its new fur boutique. 1969 *New Yorker* 20 Dec. 79 (Advt.), One of our kids said it would be 'kicky' to have one of those 'blow-up' chairs. 1970 J. G. Vermandel *Dine with Devil* i. 5 Yes, all right, let's do the kicky stuff first and then try some high fashion. 1972 M. J. Bosse *Incident at Naha* i. 60 At first impressed, she seemed doubtful after I had unloaded our theories. 'It's kicky and all..but where's it lead to?' 1973 *Time* (Canada ed.) 25 June 8/1 The designs are variously casual, racy, sporty—or kicky, trendy and funky.

kid, *sb.*[1] Add: **5. d.** A young man or woman. *colloq.* (orig. *U.S.*).

1884 *Cheyenne* (Wyoming) *Sun* 3 Nov. 3/1 There were some strange pranks played by the Cheyenne 'Kids' on the occasion of the 'Halloween'. 1896 *Emporia* (Kansas) *Gaz.* 15 Aug. 15 We have discovered a kid without a law practice and have decided to run him for attorney general. 1926 J. Black *You can't Win* iv. 26 I'll tell you what I'll do with you, kid. 1949 *N.Y. Times* 9 Oct. 50/3 A kid [*sc.* a college freshman] from anywhere immediately feels that he belongs to a great family. 1955 J. D. MacDonald *Brass Cupcake* v. 46 Kathy came into my office... I spoke out of the corner of my mouth. 'We can't talk here, kid.' 1974 N. Freeling *Dressing of Diamond* 127 You got to learn. That's a kid's job. Make yourself useful.

6. kid brother orig. *U.S.*, one's younger brother; **kid-brush**, a soft brush used in the process of finishing goatskins; **kid sister** orig. *U.S.*, one's younger sister; **kid** (also **kid's**, **kids'**) **stuff** *colloq.* (orig. *U.S.*), something suitable for children; a very simple or trivial task, etc.

1895 J. L. Williams *Princeton Stories* 143 The evenings would pass pleasantly enough in fighting with Helen, his married sister, across the table, and in guying his kid brother. 1941 *Penguin New Writing* IX. 106 She.. lived with her parents and kid brother in Kennington. 1971 B. Cobb *I fell among Thieves* iii. 39 He was the kid-brother whom I helped as far as I could, seeing that we had no mother. 1885 C. T. Davis *Manuf. Leather* xxxii. 532 The skins..are then wet over with gum-water and brushed with a very soft brush, called a 'kid-brush'. 1920 F. Scott Fitzgerald *This Side of Paradise* (1921) I. i. 36, I let people impose on me..entertain their kid sisters. 1939 'N. Blake' *Smiler with Knife* xi. 159 His manner towards them was affectionate, teasing, whimsical... They might have been his kid sisters just out of the school-room. 1962 'M. Innes' *Connoisseur's Case* xiv. 172, I don't sound a very nice kid sister. But I'm quite fond of him. 1929 F. D. Brooks *Psychol. Adolescence* xviii. 605 The little fellow looked at the book a minute,..and in a very caustic, critical manner sneered, 'Kid stuff.' 1959 J. Braine *Vodi* ii. 39 He only had to say, 'Bloody nonsense' or 'Kid's stuff, Coverack' and close his ears to Tom. 1962 L. Deighton *Ipcress File* xxi. 141 Communists...won't be using kids' stuff like this bomb. 1967 *Spectator* 7 July 9/3 One addiction specialist described it [*sc.* marijuana] to me contemptuously as 'kid-stuff'. 1974 M. Babson *Stalking Lamb* viii. 50 I've taught you the only system that makes real money... Anything else is just kids' stuff.

kid, *sb.*[5] Add: In colloq. phr. *no kid*, no kidding, I am not kidding.

1873 [in Dict.]. 1880 *Punch* Dec., *'Arry.* My gloves was the cheese no kid. 1899 R. Whiteing *No. 5 John St.* xxiii. 234 He do seem to enjoy hisself, no kid! 1916 J. B. Cooper *Coo-oo-ee* i. 14, I tell you, Nelly, she's a woman as will blaze a track right enough, no kid. 1922 Joyce *Ulysses* 418 Got a prime pair of mincepies, no kid. 1964 *Amer. Folk Music Occasional* i. 91 True story, no kid.

kid, *v.*[4] Add: Also, to joke with, tease. Also *intr.* or *absol.*, and const. *along* or *on*; freq. in phr. *no kidding*, I am not kidding; that is the truth.

1839 H. Brandon in W. A. Miles *Poverty, Mendicity & Crime* 163/2 *Kidding on*, to entice one on. 1879 [in Dict.]. 1901 [see *down and out *adj. phr.*]. 1903 G. B. Shaw *Man & Superman* ii. 70 Garn! youre kiddin. 1906 S. Ford *Shorty McCabe* xiii. 273 I'll stand for all the private kidding you can hand out. 1914 E. E. Cummings *Let.* 27 July (1969) 9 There's a dead monkey-fish hard by the boat club... No kidding! 1916 C. J. Dennis *Moods of Ginger Mick* 89, I can see ole Ginger..Grinnin' a bit to kid 'is wound don't pain. 1920 S. Lewis in *Sat. Even. Post* 11 Dec. 11/2 The boss ain't such a bad pill if you know how to kid him along. 1928 D. L. Sayers *Lord Peter views Body* 287 Really? No kidding? 1932 J. T. Farrell *Young Lonigan* iv. 154 'You wouldn't fool us, Gov'nor, would you?' kidded Johnny. Studs thought it wasn't every guy who could kid with his old man, like Johnny could. 1936 J. L. Hodson *Our Two Englands* vi. 103 'No, we don't even get kidded (chaffed) for doin' the housework any more,' a man of thirty told me. 1947 W. Stevens *Let.* 20 Aug. (1967) 565 Next to the passion flower I love fuchsias, and no kidding. 1952 'J. Tey' *Singing Sands* xii. 205 'I'm a policeman.' 'No kidding!' 1959 *Times* 27 June 7/7 If the Australian had not..'kidded himself along',..then his heart might have broken. 1969 *Listener* 9 Jan. 34/3 Mrs O'Hare has, of course, come in for a lot of kidding and wry jokes. 1969 *New Yorker* 30 Jan. 18/3 We asked some reclining youths where the Festival was, and they pointed across a vast valley to some tiny lights... 'You're kidding!'.. We sank to the grass. 1974 *Titbits* 30 May 22/4, I have always known I was impotent but kidded myself that if I could find the right wife everything would miraculously become O.K.

kidang. Now usually **kijang.** Substitute for def.: The Malay name for the Indian muntjac or barking deer, *Muntiacus muntjak*. Add earlier and later examples.

1783 W. Marsden *Hist. Sumatra* 94 Deer: *rooso: keejang*. These are variety [*sic*] of the deer species. 1824 T. Horsfield *Zool. Res. Java* s.v. *Cervus muntjak*, In the Javanese language..the name is Kīdang, which with a slight modification—Kijang—is also employed in the Malayan language. 1839 T. J. Newbold *Pol. & Statistical Acct. Straits of Malacca* I. vii. 436 Of the genus Cervus, are the Kijang or Cervus Muntjac, the Rúsa etc. 1880 *Encycl. Brit.* XIII. 602/2 The kidang or mintjac (*Cervulus muntjac*) and the rusa (*Rusa hippelaphus*) are the chief representatives of the deer kind [in Java]. 1900 W. W. Skeat *Malay Magic* v. 251 The Gold spirit being supposed to take the shape of a *kijang* or roe-deer. 1958 J. Slimming *Temiar Jungle* ii. 28 They'll eat *Rusa*, *Pelandok*, and *Kijang*. 1965 C. Shuttleworth *Malayan Safari* iii. 38 The first visitor to the salt-lick was a *kijang* or barking deer.

kidder[1]. Add: (Examples.) Also, one who jokes or teases.

1888 *Sporting Life* 15 Dec. 3/2 The champion kidder. 1891 J. Newman *Scamping Tricks* xi. 88 [He] was a beautiful kidder and could patter sweet and pretty. 1899 Ade *Fables in Slang* 84 They wanted a..Name.., so the Side-Show-Announcer, who was something of a Kidder.. gave them Zoroaster. 1922 Wodehouse *Clicking of Cuthbert* ix. 223 'Mr Winklethorpe told me I was very good with the wooden clubs,' she said defiantly. 'He's a great kidder,' said Ramsden. 1963 J. N. Harris *Weird World Wes Beattie* (1964) xvii. 196 Mr. Herbert Jackson was known as a real salesman, a man with personality, a great kidder, a hot sport and a number of other things.

kiddie, var. Kiddy *sb.* in Dict. and Suppl.

kiddleywink (ki·dliwiŋk). *dial.* Also **kiddle-a-wink, kiddle-e-wink, kiddleliwink, kiddlewink, kiddley wenk, kiddlywink, kidley-wink.** [Origin unknown; cf. Tiddlywink 1.] An ale-house, esp. in the West Country; a low or unlicensed public house (see also quot. 1859).

1830 *Royal Cornwall Gaz.* 25 Dec. 4/6 One hundred and forty public-houses..opened..are called 'Kidley Winks'. *Ibid.*, A gentleman..suggested to the late Chancellor of the Exchequer the idea of retail-breweries. His name is Kidley Wink—hence the term 'Kidley Wink', as applied to the new beer shops. 1859 Hotten *Dict. Slang* 56 *Kiddleliwink*, a small shop where they retail the commodities of a village store. 1864 'F. Derrick' *Kiddle-a-Wink* ii. 92 The dreary little chamber allotted to him..bad as it was,.. was better than many a Kiddle-a-wink could boast of. 1865 R. Hunt *Pop. Romances W. of Eng.* 2nd Ser. 109 A drunken frolic..at a low beer shop or 'Kiddle-e-Wink'. 1890 *N. & Q.* 18 Jan. 48/2 Can any of your correspondents inform me what is the derivation of the word 'kiddlewink', or 'tiddledy winks'? A friend tells me in the Midland Counties it denotes a house where beer is sold without a licence. 1964 C. Davey *Cornish Holiday* ii. 20 The potency of the drink sold in the innumerable public houses in Downlong, and in the 'kiddleywinks' or ale-houses near the mines. *Ibid.* x. 140 A tin-miner, laced with courage from the 'kiddley-wink', saw lights amongst the rocks.

kiddo (ki·do). *colloq.* [f. Kid *sb.*[1] + *-o.] = Kid *sb.*[1] 5 a, b, and *d; freq. as a familiar form of address.

1896 A. Morrison *Child of Jago* xiii. 135 Josh was up almost before Kiddo Cook reached him. 1905 *Dialect Notes* III. 85 Say, kiddo, what are you going to do this evening? 1916 [see *guess *v.* 6]. 1938 M. Allingham *Fashion in Shrouds* xix. 340 'Have I ever let you down, kiddo?' The pseudo-American accent was slick. 1959 [see *bat *v.*[1] 4]. 1961 *John o' London's* 9 Nov. 517/3 When it comes to choosing between the balance of power and unborn babies, I'm for the kiddos, every time. 1974 N. Freeling *Dressing of Diamond* 128 'How long do I have to stay?'.. 'Just as long as we thinks right, kiddo.'

kiddush (ki·duʃ). Also **K-.** [Heb. *qiddūš*, sanctification.] A ceremony of prayer and blessing over bread and wine, performed by the head of a Jewish household at the meal ushering in the Sabbath or a holy day. Also *attrib.*

1753 *Jewish Ritual* 34 All together with him in Concert, say the *Keedush*, i.e. the Sanctification. 1891 M. Friedländer *Jewish Relig.* 254 On Friday evening, before the meal, we praise God for sanctifying the Sabbath by a prayer called *Kiddush*, 'sanctification'. 1932 A. Z. Idelsohn *Jewish Liturgy* ii. x. 133 The text of the *Kiddush* consists of Gen. 1:31 and 2:1–3. Then there follows the blessing over wine and the closing paragraph... In the home, the benediction over bread follows. 1945 G. Dix *Shape of Liturgy* iv. 88 On festivals there was another common cup blessed and partaken of, besides the cup of blessing, both at a *chabûrah* meeting and at the ordinary family meal of a pious Jewish household. This was the *kiddûsh*-cup. 1960 *Commentary* June 500/1 Perhaps only one out of every two families begin the meal with the *kiddush*. 1973 *Jewish Chron.* 18 May 50/2 A service was conducted..at the Friern Hospital Synagogue and a kiddush was provided for the patients.

kiddushin (kidū·ʃīn). Also **K-.** [Aramaic *qiddūšīn*, pl. of *kiddush.] The section of the Jewish Mishnah treating of betrothal and marriage; also the ceremony of betrothal, and the money or article given by the groom to effect the betrothal.

1883 *Encycl. Brit.* XVI. 505/2 Kiddushin (betrothal and marriage), in four chapters. 1904 *Jewish Encycl.* VII. 485/2 'Kiddushin' is the rabbinical term for betrothal, because the wife becomes thereby the sacrosanct possession of the husband... In the Mishnah, Kiddushin is divided into four chapters and comprises..forty-seven paragraphs. 1936 H. Freedman in I. Epstein *Babylonian Talmud* VIII. p. xi, The Hebrew name Kiddushin ('Consecration') for betrothal, is worthy of note. The Talmud defines it as an act whereby the bride is rendered sacrosanct. *Ibid.* 51 There it was given her as a deposit... But here he gave it to her as *kiddushin*: if she did not want it [as such], she should have thrown it away. *Ibid.* 218 Her father [alone can accept *kiddushin* on her behalf] but not she herself. 1971 *Encycl. Judaica* X. 986 *Kiddushin*.., the last tractate in the order *Nashim*... It deals with matrimonial matters... There is no corresponding word for *kiddushin* in English. It is more than an 'engagement'.., as it can be dissolved only by divorce, and moreover the law of adultery..applies from the moment of *kiddushin*.

kiddy, *sb.* For 'Also 6 **kiddie**' read 'Also **kiddie**'. Add: **4. b.** *Comb.*, as **kiddy brother** = **kid brother*; **kiddy car** orig. *U.S.*, (*a*) a small toy car for a child; (*b*) a perambulator; **kiddy sister** = **kid sister*.

1963 C. Mackenzie *My Life & Times* II. 82 That was my kiddy brother. He's a new boy. 1918 *Sears, Roebuck Catal.* Index, Kiddy car. 1951 I. Shaw *Troubled Air* xx. 331 There wasn't enough left of the cab to make a kiddie car. 1953 Wodehouse *Performing Flea* 58 You'll hear them call for Mr. Warner's bicycle, Mr. Lasky's kiddie car and Mr. Louis B. Mayer's roller-skates. 1959 *Times* 24 Feb. 13/4 A toy making factory..for the manufacture of fibreglass toys, kiddy cars, and rocking horses. 1973 *Sci. Amer.* Jan. 10/2 The invention by her [*sc.* Fräulein Gretel Steiff] of the 'Teddy Bear Doll' will go down in history with the kiddie car and other things that make life happy for children. 1913 C. Mackenzie *Sinister St.* I. i. vii. 103 He also learnt to speak..of..'my people' and 'my kiddy sister'. 1915 A. Bennett *Those Twain* (1916) i. ix. 159 'He's taken a terrific fancy to Maud, my kiddie sister,' said Daisy. 1963 C. Mackenzie *My Life & Times* II. 60 Still laughing uncontrollably..in a fresh recollection of the face of our kiddy sister and the tone of her voice.

kiddy-wink, kiddywink (ki·diwiŋk). *colloq.* Also **kiddiewinkie, kiddywinkle, kiddywinky.** [Familiar extension of Kiddy *sb.*] = Kiddy *sb.* 2 (usu. jocular or affected). Also *attrib.*

1957 P. Wildeblood *Main Chance* 201 Delicious milky-boo for the kiddy-winks. 1959 P. Bull *I know Face* x. 183 My performance..was pretty macabre, and must have frightened the bejesus out of the kiddy-winks. 1962 *Spectator* 22 June 827/2 Morality plays for the kiddie-winkies. 1968 L. Berg *Risinghill* 250 The approach was fine. None of this kiddywinky stuff. They became grown-up emotionally and mentally well in advance of their years. 1970 M. Tripp *Man without Friends* xiii. 142 He's at Bognor with his kiddiewinkie. 1974 *Times* 13 Aug. 8/8 Dad Robinson..puts off the average incompetent father. Still, the kiddywinkles aren't to know.

kidlet (ki·dlėt). [f. Kid *sb.*[1] + -let.] A young child. Also *fig.*

1899 'J. Flynt' *Tramping with Tramps* ii. 31 The other 'kidlets', as they were nicknamed, were as deformed morally as was the adopted girl physically. 1903 J. Dewey *Let.* Mar. in R. B. Perry *Tht. & Char. W. James* (1935) II. lxxxi. 521 We won't attempt to father you with all the weak kidlets which are crying in the volume to be born. 1959 C. MacInnes *Absolute Beginners* 165 A lot of kidlets helping him to do so.

kidley-wink, var. *kiddleywink.

kidnap (ki·dnæp), *sb.* [f. the vb.] The act of kidnapping. Also *attrib.*

1961 Webster *s.v.*, A kidnap plot. The kidnap car. 1973 'I. Drummond' *Jaws of Watchdog* xv. 204 There was no money in killing you, but maybe a lot in a kidnap. 1973 *Observer* 12 Aug. 1/2 (*heading*) Jet kidnap was attempt to capture top guerrilla. 1974 N. Freeling *Dressing of Diamond* 201 A kidnap case—yes..we had it on our telex last night.

kidnapper. Add: Also (*U.S.*) **kidnaper.**

1909 O. Jespersen *Mod. Eng. Gram.* I. 149 Americans write *kidnaper*, which to an Englishman would suggest [kidneipə] or [-napə]. 1969 *Eugene* (Oregon) *Register-Guard* 3 Dec. 3A/2 The two men..freed themselves moments after the kidnaper abducted Miss Birdsong and called police. 1973 *Philadelphia Inquirer* 7 Oct. 19 The kidnapers agreed to call the Millers back several hours later.

kidney. Add: **5. c. kidney dressing-table**, a dressing-table with a kidney-shaped top; **kidney fern** *N.Z.*, a fern, *Cardiomanes reniforme*, with kidney-shaped leaves; **kidney graft**, the operation of transplanting a kidney from one person to another (see Graft *sb.*[1] 3); **kidney machine**, a machine for effecting hæmodialysis; = *artificial kidney* (*artificial *a.* 5), *hæmodialyser; **kidney-pie**, (*a*) a pie containing kidneys; (*b*) *Austral.* and *N.Z. slang*, flattery, humbug, deceit; **kidney punch** (see quot. 1954); **kidney-rotter** *Austral.* and *N.Z. slang* (see quot. 1958); **kidney worm**,

either of two parasitic nematodes, *Stephanurus dentatus*, which attacks pigs, or *Dioctophyma* (or *Dioctophyme*) *renale*, which attacks man, dogs, and other mammals.

1932 *Times Lit. Suppl.* 9 June 420/3 A walnut suite with a kidney dressing-table. 1965 D. Torr *Diplomatic Cover* iii. 46 She leant over her frilly kidney dressing-table. 1867 E. Sauter tr. *F. von Hochstetter's New Zealand* vi. 133 The singular form of the Kidney-fern (*Trichomanes reniforme*). 1926 J. Devanny *Lenore Divine* xx. 185 The kidney-fern was everywhere, sprawling over ground and trees. 1951 J. Frame *Lagoon* 7 My..grandmother.. could find kidney fern. 1966 *Encycl. N.Z.* I. 645/2 The kidney fern, *Cardiomanes reniforme*, with undivided leaves fringed with prominent sori, is common throughout the country. 1962 *Daily Tel.* 26 Nov. 1/3 (*headline*) Kidney graft doctor worse. 1970 Kidney graft [see *graft *sb.*[1] 3]. 1966 *Daily Tel.* 28 Sept. 19/4 Kidney patients would be able to use an artificial kidney machine in their own homes. 1972 *Guardian* 9 Feb. 20/2 Patients who rely on electrical machinery (such as kidney machines) in their homes. 1836–9 Kidney-pie [see Kidney 5]. 1937 Partridge *Dict. Slang* 455/1 *Kidney-pie*, insincere praise. 1896 Ade *Artie* i. 3 Artie..gave him a friendly blow, known to ringside patrons as a 'kidney-punch'. 1954 F. C. Avis *Boxing Reference Dict.* 61 *Kidney punch*, a blow falling at the kidneys—a foul punch liable to result in disqualification. 1964 Kidney punch [see *caper *sb.*[2] 1 c]. 1958 *Tararua* XII. 27 The frameless pack..being more familiarly known by that elegant term, *kidney-rotter*. 1971 *N.Z. Listener* 22 Feb. 51/2 Twenty-odd years ago..we slogged along in hobnailed boots, carried 'kidney-rotter' packs, [etc.]. 1893 W. B. E. Miller et al. *Dis. Live Stock* vi. 410 Various symptoms are popularly attributed to 'kidney worms', especially a weakness or partial palsy of the hinder limbs, inclination to lie down, and awkwardness in the gait. 1905 Moussu & Dollar *Dis. Cattle* iii. 539 This so-called kidney worm of hogs (*Sclerostoma pinguicola*) should not be confounded with the kidney worm (*Dioctophyme viscerale*) of dogs and man. 1934 H. O. Mönnig *Vet. Helminthol. & Entomol.* iii. 170 The 'kidney-worm' of swine occurs in the perirenal fat. 1963 Jubb & Kennedy *Path. Domestic Animals* II. vi. 272/1 *Dioctophyma renale* is the giant kidney worm, the largest of parasitic nematodes... It is usually found in dogs, mink, and other fish-eating mammals, but is recorded in ox and horse.

kid-stakes, kidstakes (ki·dstēɪks). *Austral.* and *N.Z. slang.* Also **kidsteaks.** [Cf. Kid *sb.*[5]] Humbug, pretence.

1916 C. J. Dennis *Songs Sentimental Bloke* 124 *Kid stakes*, pretence. 1919 W. H. Downing *Digger Dial.* 30 *Kid-stakes*, insincere flattery; inveiglement; a wheedling or deceitful speech or action. 1922 A. Wright *Colt from Country* 201 'It was no kid stakes,' declared Bucks. 'The old man had the roll ready.' 1945 G. Casey *Downhill is Easier* 138 All his kidstakes during the afternoon had probably been caused by his jealousy of Peter South. 1949 F. Sargeson *I saw in Dream* ii. xiv. 154 But it looked as if it was all kidsteaks. 1960 A. Kimmins *Lugs O'Leary* ii. 17 This isn't kid-stakes... This is deadly serious.

kief, var. Kef 2. Also **kieff.**

1878 [see Kef 2]. 1907 *Daily Chron.* 7 Dec. 3/5 What opium is in China, Kieff (Indian hemp) is threatening to become in Morocco. 1925 *Blackw. Mag.* Nov. 621/2 The sentry..lays down his rifle to accept the pipe of kieff from this same officer. 1971 *Frendz* 21 May 13 Dope... Turkish... Commonly cut with Honey, earth, kief glue & boot polish to make it black.

Kieffer (kī·fəɹ). The name of Peter *Kieffer* (d. 1890), American gardener, used *attrib.* in **Kieffer pear** to designate a variety of yellow-skinned pear (*Pyrus pyrifolia* var. *culta* × *P. communis*) developed by him. Also *absol.*

[1879 *Amer. Agriculturist* Jan. 21/1 (*heading*) A New Pear—Kieffer's Hybrid. *Ibid.*, Mr. Parry secured the original tree, and has called the new variety 'Kieffer's Hybrid'.] 1880 *Gardener's Monthly* Feb. 49/1 The Kieffer Pear.—A contemporary asks what evidence there is that this is a hybrid between the Chinese Sand Pear, and the ordinary garden variety? 1901 L. H. Bailey *Cycl. Amer. Hort.* III. 1242/2 The Kieffer Pear originated with Peter Kieffer, of Roxborough, Philadelphia, an Alsatian gardener. *Ibid.*, The Kieffer Pear is now very popular in many parts of the country because of its great vigor, healthfulness, productiveness, and the keeping qualities of the fruit. *Ibid.*, 1243/1 Such varieties as Kieffer and Bartlett are usually classed as self-sterile kinds. 1916 —— *Pruning-Manual* v. 116 Young pear trees, particularly of the Kieffer type, make very long and erect growths. 1948 W. Stevens *Let.* 19 Aug. (1967) 610, I..asked him to try to find Kieffer pears for me this autumn. 1961 M. W. Black in Hyams & Jackson *Orchard & Fruit Garden* xxix. 160/1 Each of the other export varieties [of pear] (Beurré Hardy,..Winter Nelis, Kieffer, Josephine de Malines, Doyenné du Comice, etc.) are tested individually.

kie-kie. (Earlier and later examples.)

1847 *N.Z. Jrnl.* No. 191. 106/1 Passed..through a wet wood of supplejack and kiekie. 1905 W. B. *Where White Man Treads* 16 Aye, and kie kie also, with its two kinds of sweet-meats, both on the same vine. 1966 *Encycl. N.Z.* II. 785/1 The only non-berry fruit of significance was that of kiekie (*Freycinetia banksii*), the flavour of which has been likened to that of a pear.

|| **kielbasa** (kīlbæ·să, ki-). Also **kolbasa, -i.** [Pol. *kiełbasa* sausage, Russ. *kolbasa* sausage.] (See quot. 1965.)

1953 S. Bellow *Adventures of Augie March* xx. 435 All these poor punks..with immigrant blood and washday smells and kielbasa and home-brew beer. 1958 'Rysia' *Old Warsaw Cook Bk.* 2 (*heading*) Kielbasa, Polish sausage cut in small pieces and served on tooth-picks with a drink of vodka. *Ibid.* 49 Until the sixteenth century..the preferred meat was..Polish sausage—'kielbasa', or 'bigos'. 1959 *Times* 23 Jan. 12/6 Attaché cases—which must surely contain the innermost secrets of the Kremlin, but which usually turned out to harbour a lump of black bread and a petrified portion of kolbasa. 1965 *House & Garden* Jan. 60 *Kielbasa* or *kolbasi* (Polish sausage), a highly seasoned garlicky sausage. It comes fresh, smoked, uncooked and cooked, but usually must be poached before it is eaten. 1969 R. & D. DeSola *Dict. Cooking* 133/2 *Kielbasa*, red-cased Polish sausage..often served with sauerkraut.

kierie: see *kerrie.

kiering (kiə·riŋ), *vbl. sb.* [f. Kier + -ing[1].] Boiling in a kier or vat.

1922 *Encycl. Brit.* XXX. 590/2 A uniform process of 'kiering' (boiling under pressure with a lye of caustic soda) was introduced. 1954 *Textile Terms & Definitions* (Textile Inst.) 22 *Kier boil* (*Kiering*), the process of prolonged boiling of cotton or flax materials with alkaline liquors in a..Kier. 1963 A. J. Hall *Textile Sci.* iv. 162 Kiering and bleaching. In the first [bleaching stage] the textile material is boiled..within a..kier.

Kierkegaardian (kiəɹkěgā·ɹdiăn, -gǭə·ɹd-), *a.* and *sb.* [f. the name of Sören *Kierkegaard* (1813–55), a Danish philosopher + -ian.] **A.** *adj.* Of or pertaining to Kierkegaard or his philosophy. **B.** *sb.* An adherent or admirer of Kierkegaard's philosophy.

1943 *Horizon* Oct. 260 The great psychological crisis has now been given its actual material content, and one of the most important keys..to the inner Kierkegaardian room, has been found. 1947 *Partisan Rev.* Mar.–Apr. 187 The terminology of the neo-Kierkegaardian, Karl Barth. 1950 *Mind* LIX. 415 What possible appeal..could a Kierkegaardian Christianity have for children? 1963 Auden *Dyer's Hand* 444 In some of Brand's speeches, however, there is an emphasis on the human will which is Nietzschean rather than Kierkegaardian. 1964 *English Studies* XLV. Suppl. 243 A kierkegaardian must be struck by the radical and extreme position taken by Pater. 1971 G. Steiner *In Bluebeard's Castle* iii. 57 The Kierkegaardian concept of 'total possibility', of a fabric of reality open at all points to the rift of absurdity and disaster, has become a commonplace.

Kievan (ki,e·făn, ki,e·văn), *a.* Also **Kievian.** [f. *Kiev*, a city in Russia + -an.] Of or pertaining to the city of Kiev, esp. with reference to the historical period (*c* 900–*c* 1150) when it dominated European Russia.

1927 D. S. Mirsky *Hist. Russia* ii. 8 The golden age of the Kievian political power occurred in the reigns of Vladimir and of his son Yaroslav (1019–1054). 1957 K. A. Wittfogel *Oriental Despotism* x. 418 In Medieval Sweden and Kievan Russia the decisive social relations.. never seem to have matured... We may view them as.. 'marginal' feudal society. 1959 *Listener* 26 Mar. 564/1 The rise of the Kievan State in the tenth and eleventh centuries. 1965 *Language* XLI. 140 A recent study..includes some [foreign] loanwords as borrowed in the..Kievan period.

kieve, var. Keeve, kive 1.

1875 [in Dict.]. 1887 R. Hunt *Brit. Mining* (ed. 2) 910/2 *Kieve*, a vat or large iron-bound tub for washing ores. 1959 *Times* 10 Nov. p. iii/5 Large mashing vessels called kieves from which the wort is drawn off leaving the grains behind. 1967 *New Scientist* 6 July 23/1 The mash is cooked up with water in mash tuns, or kieves.

kiewiet (kī·vit). *S. Afr.* Also **kievietjie, kiewietjie, kiewit, kiewitje, kivit.** [Echoic.] The crowned lapwing, *Vanellus* (or *Stephanibyx*) *coronatus*.

1785 G. Forster tr. *Sparrman's Voy. Cape Good Hope* I. iv. 153 Flocks of *keuvitts*..towards the dusk of the evening, screamed out a disagreeable sound resembling that of the name they bear. 1818 C. I. Latrobe *Jrnl. Visit S. Afr.* vi. 131 Some kivits, or plovers, were the only birds.. we saw, during several hours' ride. 1867 E. L. Layard *Birds S. Afr.* 294 The 'Kiewit' makes known its presence by its loud plaintive call. 1896 H. A. Bryden *Tales S. Afr.* 121 The cry of one or two night birds may be heard—the dikkop and kiewitje plovers. 1936 E. L. Gill *First Guide S. Afr. Birds* 139 Crowned Lapwing, Kiewiet; *Stephanibyx coronatus*. The commonest and most generally distributed of the dry-land plovers in South Africa. 1939 [see *dikkop 1]. 1957 McLachlan & Liversidge *Roberts's Birds S. Afr.* 131 Crowned plover. Kiewietjie... Widespread but local all over South Africa to north-eastern Africa. 1964 P. A. Clancey *Birds Natal & Zululand* 156 *Vanellus coronatus*... Not normally shy, but extremely vociferous, uttering a harsh cry which has given rise to its local name of 'kiewit'. 1974 *Eastern Province Herald* (S. Afr.) 7 Nov. 5 The wide-eyed appearance of this 'kiewietjie' yesterday, nesting on a lawn.., is understandable after it had to sit through the surrounding Guy Fawkes celebrations this week.

kif, var. Kef 2. Cf. *kief.

1880 Bentley & Trimen *Medicinal Plants* IV. § 231 In Algeria the Hemp is cultivated under the name of Kif. 1938 *Mr.* (N.Y.) Dec. 117/1 In one corner two Arabs were sitting quietly smoking, and the smell of *kif* (hashish) was in the air. 1960 *Spectator* 29 July 176 He lived longest in Tangier, in a room in the Medina where he cooked hashish candy..crouched in a Reich orgone box smoking kif. 1962 *Ibid.* 3 Aug. 165 The triumphs of various kif (hashish) smokers over their persecutors. 1963 T. Tullett *Inside Interpol* iv. 45 Hashish..is generally smoked in a 'kif' pipe. 1969 [see *dagga[1]].

kijaat, var. *kiaat.

kijang: see *kidang.

kike (kaik). *slang* (orig. *U.S.*). [Said to be an alteration of *-ki* (or *-ky*), a common ending of the personal names of Eastern European Jews who emigrated to the U.S. at the turn of the 20th c.] A vulgarly offensive name for a Jew. Also *attrib.* or as *adj.*

1904 R. L. McCardell *Show Girl & her Friends* 49 And what do you think? He had the impudence to tell me that Louie Zinsheimer was a kike! 1912 *McClure's Mag.* XXXIX. 230/2 'It's a mascot, be-dad! Jam it, ye kike!' screeched Tracy. 1919 F. Hurst *Humoresque* 211 A little red-haired kike like her! 1919 Mencken *Amer. Lang.* 115 An Englishman..knows nothing of our common terms of disparagement, such as *kike*..and *rube*. 1924 P. Marks *Plastic Age* xviii. 201 You go chasing around with kikes and micks. 1932 J. Dos Passos *1919* 164 The little kike behind the desk had never been to sea. 1940 R. Stout *Over my Dead Body* vi. 84, I don't care if the background is wop or mick or kike..so long as it's American. 1956 D. Karp *All Honorable Men* 74 If you repeat that lie, I'll wring that skinny kike neck of yours with my own hands! 1963 V. Nabokov *Gift* iii. 179 My better half..was for twenty years the wife of a kike and got mixed up with a whole rabble of Jew in-laws. 1963 *Spectator* 21 June 815 He knocks down Stern's wife, calls her a kike. 1972 *National Observer* (U.S.) 27 May 17/3 When kikes are shrewd and dagos or wops are sly and murderous, it is only one step from the epithet to contempt.

|| **kikoi** (kikoi·). [Swahili.] In East Africa, a striped cloth of distinctive design with an end fringe, worn round the waist. Also *attrib.*

1942 *E. Afr. Ann.* 1941–2 27/1 His loin-cloth, 'kikoi', was edged with stripes of red, black and yellow, and showed below his shirt. 1970 *Vogue* Jan. 96/4 Mwembe Tayari is a lively, open free-for-all, with *kikois* (Arab-style sarongs), beaded jackets..and handicrafts. 1971 *Daily Nation* (Nairobi) 10 Apr. 5/1 (Advt.), Kikoy Shirts and Dresses.

Kikuchi (kiku·tʃi). *Physics.* [Name of Seishi *Kikuchi* (b. 1902), Japanese physicist, who first observed the lines.] *Kikuchi line*: each of a series of lines in electron diffraction patterns which are attributed to the elastic scattering of previously inelastically scattered electrons and may be used to determine the orientation of crystalline specimens; so *Kikuchi pattern*.

1934 *Physical Rev.* XLV. 43/1 The [electron scattering] patterns from stibnite consist of spots, Kikuchi lines, bands, circles and parabolas. 1948 *Proc. Physical Soc.* LX. 343 The elementary diffraction theory for single scattering also leads to a Kikuchi-line breadth proportional to the corresponding plane spacing *d/n*. 1966 D. G. Brandon *Mod. Techniques Metallogr.* ii. 113 Kikuchi line patterns can be used to give a more accurate estimate of the specimen orientation than can be determined from the normal spot pattern. 1968 *Mineral. Abstr.* XIX. 83/2 A simplified stereographic projection from a Kikuchi pattern is used to determine the orientation of a crystal. 1970 *New Scientist* 23 July 176/2 Kikuchi patterns, well known to users of the transmission electron microscope.

Kikuyu (kikū·yū). [Native name.] **a.** The name of an agricultural Negroid people, the largest Bantu-speaking group in Kenya; a member of this people; the language they speak. Also *attrib.* or as *adj.*

1894 N. Bell tr. *L. von Höhnel's Discovery Lakes Rudolf & Stefanie* II. vi. 298 Our attention was called to another party of about forty natives approaching us... Whether they were Masai or Kikuyu neither we nor our guide could tell, so the Count went off with one or two of our Kikuyu friends to ascertain. 1904 H. Hinde *Vocab. Kamba & Kikuyu Lang.* p. v, Both the Kamba and Kikuyu languages belong to the Bantu group and their construction is precisely similar to that of Swahili. *Ibid.* p. xviii, It is not unfrequent, in general conversation, both in Kamba and Kikuyu, to ignore the conjugated forms of the verb. 1911 J. G. Frazer *Golden Bough: Magic Art* (ed. 3) I. iii. 76 This curious pretence of being born again regularly formed part of the initiatory rites through which every Kikuyu lad and every Kikuyu girl had to pass before he or she was recognised as a full-grown member of the tribe. 1920 *New Statesman* 10 July 391/1 The Kikuyu, bending over their cultivated plots. 1935 E. Hemingway *Green Hills Afr.* (1936) iii. i. 173 Kamau, the driver, was a Kikuyu. 1936 *Discovery* June 195/1 Brought up to speak Kikuyu from childhood. 1950 D. Jones *Phoneme* 20 In one form of the Kikuyu language of Kenya the sounds ɸ, β and b appear to be..constituting a single phoneme. 1955 *Times* 7 June 6/5 The weakening of the Mau Mau is most noticeable to the east of Mount Kenya, where its Meru and Embu supporters are less determined than the Kikuyus. 1958 *Times* 8 Jan. 6/7 After warnings in Kikuyu had been given over a loudhailer and ignored, the staff opened fire on the rioters' barricade. 1967 M. J. Coe *Ecol. Alpine Zone Mt. Kenya* 1 Kikuyu folk-lore tells how, when the earth was formed, a man named Mogai made a great mountain. 1973 *Nature* 11 May 107/1 At Cambridge he studied Modern French, Medieval French and Kikuyu for Part 1 of the Tripos.

b. Kikuyu grass, a creeping perennial grass, *Pennisetum clandestinum*, native to the highlands of Kenya, and cultivated elsewhere as a lawn and fodder grass.

1913 *Rep. Dept. Agric. Union S. Afr. 1910–11* 241 Kikuyu grass...We received in good condition a rooted plant of the Kikuyu-grass of British East Africa. This has been planted out and has made vigorous growth; it has not yet flowered, and I am therefore unable to name it. **1921** *Kew Bull.* 85 (*heading*) Kikuyu grass. *Ibid.* 93 Kikuyu grass in the presence of water will put on top-growth. **1934** *Bulletin* (Sydney) 31 Oct. 22/2 Now is the time in N.S.W. to plant Kikuyu grass, which is suitable to any class of soil that is not binding. **1951** Edwards & Bogdan *Important Grassland Plants of Kenya* II. 62 In the Highland Forest regions Kikuyu grass occupies the land for a period following clearance of the forest climax. **1970** W. Smith *Gold Mine* xxviii. 67 Rod escorted the Steyners down across the vivid green lawns of Kikuyu grass.

c. A controversy in the Anglican Church, which first arose at the Kikuyu Conference of 1913, regarding the admissibility to Holy Communion of the members of other Christian churches. Also *attrib.*

1914 H. Henson (*title*) The issue of Kikuyu. **1915** C. Kelway *Story of Kikuyu* 8 The mission field is, in fact, full of potential 'Kikuyus'..the vital questions which emerge from the Kikuyu Conference. **1915** L. Pullan *Missionary Princ. & Primate on Kikuyu* 31 This pronouncement on the Kikuyu question is considered official. **1965** S. Neill *Anglicanism* (ed. 3) xiii. 366 The years after the First World War introduced a new period of more dramatic possibilities and intenser strains. The trouble started with the Kikuyu controversy.

‖ **kikyo** (kī·kyo). Also **kikiyo, kikuyo.** [Jap.] A local name for *Platycodon grandiflorum*, a herbaceous perennial of the family Campanulaceæ, native to China and Japan; the Chinese bell-flower.

1884 tr. *J. J. Rein's Japan* I. vii. 145 The splendid blue-flowered Kikiyo (Platycodon grandiflorum DC.).. appear in numbers only at a height of about 1,000 metres. **1899** L. Hearn *In Ghostly Japan* ii. 15 The crest upon the robe was the *kikyō*-flower. **1911** *Encycl. Brit.* XV. 162/1 If some familiar European flowers are absent, they are replaced by others strange to Western eyes..the *kikyo* (*Platycodon grandiflorum*). **1965** J. Ohwi *Flora Japan* 853/1 *Platycodon grandiflorum*... Kikyō... Frequently planted as an ornamental and for medicine.

kilch (kilʃ). [Swiss German *kilch*.] The Swiss name for a small whitefish, *Coregonus pidschian*, found in northern Europe, Asia, and Canada.

1881 K. Semper *Nat. Conditions of Existence* 320 The little fish of the Lake of Constance known as the Kilch. **1931** J. R. Norman *Hist. Fishes* ix. 175 One of the White-fishes (*Coregonus*), an important food-fish of Lake Constance, known locally as the Kilch. **1962** D. W. Tucker tr. *Sterba's Freshwater Fishes of World* 56 The following may be kept successfully...the Blaufelchen or Grosse Schweberenke (*Coregonus wartmanni* Bloch) from the lakes of the Voralpen, and even the very rare Kleine Schweberenke or Kilch (*Coregonus acronius* Rapp).

kilchoanite (kilχōu·ănəit). *Min.* [f. *Kilchoan*, name of the village in the Highlands near which it was first found + -ITE[1].] A colourless, orthorhombic polymorph of a calcium silicate, $Ca_3Si_2O_7$, of which rankinite is another polymorph.

1961 Agrell & Gay in *Nature* 4 Mar. 743/1 A mineral corresponding to phase *Z* has been found in limestones thermally metamorphosed by gabbro at Ardnamurchan, Scotland, and Carlingford, Eire. It has been named kilchoanite after the village in Ardnamurchan near where it was first found. *Ibid.*, Kilchoanite has always been found as a replacement of rankinite and no crystal form has been observed. **1969** *Mineral. Mag.* XXXVII. 517 Two polymorphs of tricalcium disilicate, rankinite and kilchoanite, have been discovered near Tokatoka, New Zealand. **1969** *Trans. Brit. Ceramic Soc.* LXVIII. 225/1 The morphology of synthetic kilchoanite made at ~200°C, as revealed by transmission electron microscopy, is platy with a characteristic lozenge shape.

kiley. *Austral.* [var. Kylie.] = *bat *sb.*[2] 3 e.

1945 [see *bat *sb.*[2] 3 e].

kilhig (ki·lhig). *U.S. Logging.* Also **killig.** [Origin unknown.] A short stout pole used as a lever or brace to direct the fall of a tree.

1905 *Terms Forestry & Logging* (U.S. Dept. Agric. Bureau Forestry) *Kilhig*, a short stout pole used as a lever or brace to direct the fall of a tree (N.W.). **1913** R. C. Bryant *Logging* 83 Kilhig or sampson...It consists of a pole..either sharpened or armed on one end with a spike. **1971** F. C. Ford-Robertson *Terminol. Forest Sci.* 148/2 *Killig* (USA), *Pushpole* (Cw), a stout pole, sometimes notched into the tree stem at one end and braced against the base of a peavey handle at the other, used to push a small tree manually in the desired direction of fall.

kilian (ki·liăn). *Ice-dancing.* Also **killian.** [Origin unknown.] A fast ice-dance executed by a pair of skaters side by side.

1935 *Skating* Dec. 34 The Killian is one of the new English dances and has been described by Mrs. Willie Frick. It is a fast tenstep, done side by side. **1936** *Skating Rev.* III. 13 The lady's mohawk is an ordinary.. mohawk, right foot crossed behind, and not as done in the Kilian, crossed in front. **1938** T. D. Richardson *Ice Rink Skating* vii. 50 The Kilian is a delightful dance, but is strictly more of a pair skating step. **1941** *U.S. Figure Skating Assoc. Rulebk. 1940–41* 76 The Kilian is of Austrian origin. It analyzes as a snappy side by side Man's Fourteenstep—using a choctaw in place of a spread eagle turn. **1962** *Times* 3 Mar. 3/5 A fine tango by the British champions..enabled them to make up ground which they had lost on the kilian. **1968** *Daily Tel.* 29 Feb. 13/8 Skating both the kilian and blues with advanced techniques and deportment too impeccable to criticise, Ford and Miss Towler look well set for a third successive victory in tomorrow's free style.

Kilim (kilī·m). Also **Kelim, Khilim.** [Turk., f. Pers. *kilīm*.] In full *Kilim carpet, rug*, etc. A pileless woven carpet, rug, etc., made in Turkey, Kurdistan, and neighbouring areas.

1881 C. C. Harrison *Woman's Handiwork* III. 139 A Kelim rug..may be bought at a reasonable price..to lay before the fire. **1884** J. R. G. Griffitt *Turkey Carpets & their Manufacture* 12 Carpets without a pile, known as Kilims. *Ibid.*, The Oushak Kilims..differ from the many varieties of Kilims made by the Nomad tribes throughout Asia-Minor. **1895** *Brit. Warehouseman* Feb. 25/2 Arabian 'Kelims' or hand-woven woollen tent hangings. **1900** J. K. Mumford *Oriental Rugs* (1901) Pl. XXIII (*caption*), The real color value of Khilims must become very clear to everyone who sees this plate. **1923** *Daily Mail* 27 June 14 The heavier Kurdish Kelim rugs make the best floor covering. **1926** T. E. Lawrence *Seven Pillars* (1935) II. xix. 124 Feisal would buckle on his ceremonial dagger and walk across to the reception tent, which was floored with two horrible kilims. **1931** A. U. Dilley *Oriental Rugs & Carpets* i. 4 Theoretically, the product of this weaving was flat-surfaced, and would be classified by us as khilim or tapestry. **1960** S. Foot *Emergency Exit* xiv. 113 The front hall, with its gay Kelims. **1967** [see *goat *hair*]. **1972** *Vogue* Jan. 10/2 Kurdish Kelim rugs from £22. **1972** *Country Life* 3 Feb. 258/3 A fine collection of woven goods and felts: carpets, Kilims, tent bands.

Kilkenny (kilke·ni). Name of a county and city in Leinster in the Republic of Ireland, used *attrib.*, as in **Kilkenny cat**, one of a pair of cats fabled to have fought until only their tails remained; *transf.*, of combatants who fight until they annihilate each other; so *Kilkenny fight*. Also **Kilkenny coal**, an Irish name for anthracite; **Kilkenny marble** (see quot. 1959).

1822 *Hive* I. II. 32/2 (*heading*) Kilkenny cats. *Ibid.*, One gentleman stated his opinion, that a *Kilkenny* cat was, of all other animals, the most ferocious;..'..I once,' said he, 'saw two of these animals fighting..and..drove them into a deep saw-pit, and..left them to their amusement. Next morning,..what d'ye think I saw?..there was nothing left in the pit *but the two tails and a bit of flue*.' **1849** *Pict. Guide Birmingham* 162 Whatever may be the ultimate fate of the combatants—and it once seemed likely to be that of the Kilkenny cats. **1861** Mrs. Beeton *Bk. Househ. Managem.* 33 Of *coal* there are various species: as, pit, culm, slate, cannel, Kilkenny, sulphurous, bovey, jet, &c. **1900** J. London *Let.* 2 May (1966) 105 If Russia and England played the Kilkenny cat act, there would be peace in the world. **1901** *Graphic* LXIV. 288/1 The fate of the Kilkenny cats will meanwhile have overtaken the villains. **1910** *Encycl. Brit.* II. 105/2 Other terms [for anthracite]..are ..'blind coal' in Scotland, and 'Kilkenny coal' in Ireland. **1930** F. J. North *Limestones* 145 Black Kilkenny Marble is widely used. **1931** *Sun* (Baltimore) 7 Mar. 8/2 He has been in the center of a good old-fashioned Democratic kilkenny fight. **1931** *Times Lit. Suppl.* 23 July 578/2 All these excitable disputants have disappeared like the Kilkenny cats through the excess of their own zeal. **1959** *Chambers's Encycl.* VIII. 217/1 The Kilkenny marbles are greyish crystalline limestones which become deep blue or black on polishing. **1970** I. Origo *Images & Shadows* ii. 40 The home of my mother's Anglo-Irish parents..was an Italianate house built towards the end of the eighteenth century of the grey local limestone known as 'Kilkenny marble'. **1972** *Guardian* 10 Mar. 1/3 Mr Heath and Mr Wilson..guard the rival despatch boxes with the mutual respect and tolerance of a pair of Kilkenny cats.

kill, *sb.*[1] Add: **2. b.** Phr. *in at the kill*: present at the killing of an animal; also *transf.* and *fig.*

1814 Prince William *Let.* 18 Feb. in P. Ziegler *King William IV* (1971) ix. 115 The game is up with Bonaparte and I shall be in at the kill. **1969** *Amer. Heritage Dict.* s.v., *In at the kill*, present at the moment of triumph.

c. *Lawn Tennis* and *Rackets.* The striking of a ball in such a way that it cannot be returned. Cf. Kill *v.* 7 a.

1903 *Westm. Gaz.* 31 Aug. 8/1 Grant put in some mighty 'kills' from the service line. **1908** *Baily's Mag.* June 483/1 They both of them fairly bombarded the wall, often.. bringing off beautifully low 'kills'. **1920** W. T. Tilden *Lawn Tennis* 87 The server covers and strives for a kill at once. **1969** *New Yorker* 14 June 68/2 Graebner delivers a Wagnerian kill. The ball digs a hole in the turf near Ashe's left foot.

d. The destruction or putting out of action of an enemy aircraft, submarine, etc.; the aircraft, etc., so destroyed. *colloq.*

1944 *Times* 7 Mar. 2/3 The men of this station..can show plenty of evidence of 'kills'. **1951** N. Monsarrat *Cruel Sea* v. vi. 360 But this was to be no swift kill: perhaps, indeed, it was to be no kill at all. **1962** *Daily Tel.* 20 July 1 (*headline*) Atlas rocket 'kill' by anti-missile. **1969** G. Macbeth *War Quartet* 60 We had sailed five weeks Without a kill. **1971** *Daily Tel.* 22 Nov. 7 (*caption*) Mr H. M. Stephen..examining..parts of a Messerschmitt 109 fighter which, as a pilot officer, he shot down on Nov. 30, 1940, while operating from Biggin Hill. It was the war-time base's 600th 'kill'.

e. *Boxing.* (See quots.) *colloq.*

1950 J. Dempsey *Championship Fighting* xxv. 200 His opponent will be after him quickly for 'the kill'—for the knockout. **1954** F. C. Avis *Boxing Reference Dict.* 61 *Kill*, a knock out.

4. kill ratio *U.S.*, the proportion of casualties on each side in a military action.

1968 *N.Y. Times* 11 Aug. 1. 3 Those Nigerians who had escaped the cross-fire had fled northward into the forest, leaving behind 41 dead, the Biafrans said. They put their own losses at three killed and a dozen wounded. The lieutenant was pleased with the kill ratio. **1973** *New Yorker* 17 Feb. 89/1 Our military..can produce sickeningly effective 'kill ratios'.

kill, *v.* Add: **2. b.** *kill out* (later examples).

1966 R. M. Lockley *Grey Seal, Common Seal* x. 147 In New Zealand I saw how the red deer are killing out the young native forest trees in the South Island Alps. **1970** *New Scientist* 31 Dec. 576/1 Broilers are 'killed out' at eight weeks. **1972** *Country Life* 30 Nov. 1504/2 These small birds [*sc.* turkeys]..are killed out at 10–12 weeks of age.

e. Also, *to kill out.*

1950 *N.Z. Jrnl. Agric.* Apr. 364/1 The Southdown has the advantage over the Leicester in that its progeny are quicker maturing and kill out at prime weight and at an earlier age (3 to 4 months). **1971** *Country Life* 30 Dec. 1857/3 Limousin-sired fat cattle killed out at 68 per cent; far above our national average for our native breeds.

3. d. To consume; to eat or drink; *spec.* to empty (a bottle of liquor). *colloq.* (orig. *U.S.*).

1833 *Sk. & Eccentr. D. Crockett* xi. 145, I can kill more lickur..and cool out more men than any man you can find in all Kentucky. **1887** *Lantern* (New Orleans) 20 Aug. 2/2 The lady had killed a dozen [oysters]. **1934** J. T. Farrell *Young Manhood* xviii. 291 'We'll drink to that,' said Fat. They killed the bottle. **1967** N. Fitzgerald *Affairs of Death* vii. 125 We drank with maudlin solemnity to Stella's memory, killing the bottle in the process.

e. In printing or journalism, to cancel or delete (matter) before publication; to discard (type); to suppress or deny (a story, etc.). *colloq.* (orig. *U.S.*).

1865 *Wilkes' Spirit of Times* 16 Dec. 256/1 Two galleys of equal length, one being marked 'Must', the other 'Kill this'. **1887** *Courier-Journal* (Louisville, Ky.) 29 Jan. 5/4 Please kill the deer story sent by Associated Press this morning. **1903** E. L. Shuman *Pract. Journalism* 62 The editor can make room by killing the last paragraphs of the other stories. **1929** [see *copy-desk* s.v. *copy *sb.* C]. **1938** E. Waugh *Scoop* II. i. 133 We're killing this story... Go round to the Press Bureau and have Benito issue an official *dementi*. **1967** Karch & Buber *Offset Processes* ii. 40 'Dead' ads are killed. **1972** *Human World* May 75 This is a dull and confused book. (We killed our review of it as not worth the space.)

f. To turn off or stop (an engine, esp. the motor of a car). *colloq.* (orig. *U.S.*).

1886 *Philadelphia Even. Tel.* 20 Mar., The hose was cut ..and engines killed so that it will take days to bring them to life again. **1907** E. S. Field *Six-Cylinder Courtship* 9, I lost no time in starting. What a blessing that I hadn't killed my engine! **1935** M. M. Atwater *Murder in Midsummer* iv. 41 Jim killed the engine and switched off the lights. **1971** D. MacKenzie *Sleep is for Rich* vi. 196, I moved the hired car into the cobbled courtyard... I killed the motor.

g. *Metallurgy.* To treat (steel when molten) so as to prevent the evolution of oxygen on solidification (now done by adding a reducing agent: cf. *killed *ppl. a.* 2 b); to remove (iron oxides) from the molten metal by this means.

1906 W. Macfarlane *Princ. & Pract. Iron & Steel Manufacture* iv. 46 Higher class steel requires 'killing'—that is, it requires to be kept in the furnace for about half an hour..after it has become fluid and it must be poured at a proper temperature. **1918** A. W. & H. Brearley *Ingots & Ingot Moulds* x. 172 When cast steel was made by the crucible process..it was necessary to continue the heating, and..increase the temperature as far as possible, in order to 'kill' the steel. **1926** *Jrnl. Iron & Steel Inst.* CXIV. 407 On account of the titanium alloy being of only 17 per cent. strength, it is not practicable to kill a heat of steel with titanium only. **1940** Simons & Gregory *Steel Manuf.* vi. 31 There are several ways of killing steel. One ..is the addition in the ingot mould before teeming of about 0·02 per cent of metallic aluminium (Al) to the melt. **1969** R. Stephen *Iron & Steel for Operatives* xii. 56/1 All the iron oxide has been removed, or killed by de-oxidation, and has entered the slag in the ladle.

h. To extinguish or obscure (a light); also, to extinguish (a cigarette). *colloq.*

1934 *Tit-Bits* 31 Mar. 12/1 'Niggers' are not men of colour, but blackboards used to 'kill' unwanted reflections from the powerful lights. **1939** *Evening News* 7 Nov. 4/5 'Kill that baby and put a nigger in its place.' ('Put out that small spotlight and substitute a black screen'). **1940** *Chambers's Techn. Dict.* 476/2 *Kill* (Cinema), colloquialism for extinguish lights. **1942** R. Chandler *High Window* (1943) xix. 135 She killed her cigarette in Morny's copper goldfish bowl, speared the crushed stub absently with the letter opener and dropped it into the

waste-basket. **1959** M. Pugh *Chancer* 153 Could you kill that cigarette..? It's smouldering somewhere. **1967** J. Wainwright *Worms must Wait* lxxvii. 200 The window shattered and the lights were killed almost simultaneously.

4. c. *Theatr. colloq.* (See quot. 1952.)

1933 [see *comedy[1] 2 c]. **1952** Granville *Dict. Theatr. Terms* 106 *Kill a laugh*, to start a fresh line before the laugh evoked by the preceding one has died down.

d. *Athletics.* To put (a rival runner) out of contention in a race by setting a fast pace, or suddenly accelerating. Also with *off.*

1962 B. Hewson *Flying Feet* xi. 132 Derek..slowed the pace to a crawl, obviously hoping to use his finishing kick to kill off Mike and myself. **1968** G. Gretton *Out in Front* v. 74 He set a fast pace which 'killed' Heino, who collapsed and retired.

6. a. Also, to convulse (someone) with laughter; to excite, thrill, delight.

[**1856** C. M. Yonge *Daisy Chain* II. viii. 414 Ethel saw Meta in fits of laughing... 'Ethel! you will kill me!' said Meta, sinking back on the sofa.] **1938** C. Calloway *Hi De Ho* 16 *Kill me*, show me a good time, send me. **1942** Berrey & Van den Bark *Amer. Thes. Slang* § 591/5 Delight the audience,..*kill 'em.* **1951** J. D. Salinger *Catcher in Rye* x. 82, I took her to see this French movie... It killed her. *Ibid.* 83 She killed Allie, too. I mean he liked her, too. **1960** C. Dale *Spring of Love* ix. 176 He kills me sometimes, the things he says. **1971** *Melody Maker* 13 Nov. 31/6 During the Elton John tour in the States, which was a gas, man, we killed them night after night.

b. Delete '(An Irishism.)' and add further examples.

1816 Jane Austen *Emma* III. vi. 106 Nothing killed him like heat—he could bear any degree of cold. **1899** G. W. Peck *Uncle Ike* (1903) xix. 172 'Now wouldn't that kill you,' said the boy... 'That breaks up my scheme to fight the French.' *a* **1953** E. O'Neill *Long Day's Journey* (1956) II. i. 53 No wonder my feet kill me each night. **1962** J. Cannan *All is Discovered* i. 19 My feet are killing me anyway and this dam' strapless bra is rubbing me raw. *Ibid.* vi. 140 The 'middy'-heeled shoes which after the long walk along the hot roads had been 'killing' her. **1965** J. Porter *Dover Two* v. 61 The long cold walk..did nothing to lighten Dover's mood. His feet were killing him.

c. Used in the infinitive form after another verb with adverbial force = 'to a great or impressive degree'; esp. in phr. *dressed* (*got up*, etc.) *to kill*, dressed showily or impressively. *colloq.*

1818 Keats *Let.* 23 Jan. (1958) I. 216 One chap was dressed to kill for the King in Bombastes. **1845** *N.Y. Even. Express* 5 Mar. 2/4 Mrs. Polk..dresses 'to kill'. **1848** Bartlett *Dict. Amer.* 194 To do anything *to kill*, is a common vulgarism, and means to do it to the uttermost; to carry it to the fullest extent; as, 'He drives to kill'; 'she dances to kill'. **1862** J. R. Lowell *Biglow Papers* 2nd Ser. 36 'T was Concord Bridge a-talkin' off to kill With the Stone Spike thet's druv thru Bunker Hill. **1922** Joyce *Ulysses* 280 Got up to kill: on eighteen bob a week. **1957** N. Mitford *Voltaire in Love* xviii. 218 Mme du Châtelet..always took the part of the leading lady, dressed up to kill and covered with diamonds. **1970** G. W. Barrax in S. Henderson *Understanding New Black Poetry* (1973) III. 358 Dress to kill Shoot to kill Love to kill If you will But write to bring back the dead.

7. d. (Earlier and further examples.)

c **1558** *Enterlude of Welth, & Helth* sig. D1[v], With kindnes my her ye do kyll. **1582** G. Whetstone *Heptameron of Civil Discourses* sig. T4[v], You will kill her with kindnesse. **1607** T. Heywood (*title*) A woman kilde with kindnesse. **1761** G. Colman *Jealous Wife* IV. i. 67 You absolutely kill Him with Kindness. **1842** F. A. Kemble *Let.* 31 Mar. in *Rec. Later Life* (1882) II. 189 Lord Morpeth..has a.. mother and sisters, and really should not, on their account, be killed with kindness. **1898** F. P. Dunne *Mr. Dooley in Peace & War* 38 'They'll kill him with kindness if he don't look out,' said Mr. Hennessy. **1925** A. Huxley *Along Road* 61 The country..has not been killed by the deadly kindness of those who, like myself, are nature's townsmen. **1935** I. Brown *Heart of England* viii. 84 Now we purge by persuasion that new Beelzebub, the complex, or kill it by kindness.

f. *to kill two birds with one stone*: see Stone *sb.* 16 b.

g. Ironical phr. *it won't* (etc.) *kill you* (or *him, us*, etc.): that would not be too much to endure.

1858 Trollope *Three Clerks* I. vii. 130 'We are both used to that, I fancy,' said Tudor, 'so it won't kill us.' **1913** J. Vaizey *College Girl* vi. 83 Suppose I ask them? Twopence three farthings each would not kill them! **1945** A. Kober *Parm Me* 123 Even if your father's gonna lay out a few dollars, O.K., so it's not gonna kill him! **1967** 'G. North' *Sgt. Cluff & Day of Reckoning* ii. 16 'You could have stopped in bed...' 'Lie there awake?' 'It wouldn't have killed you.' 'Getting my own breakfast didn't either.'

h. *to kill the goods*: in soap-making, to emulsify the melted fat by a partial saponification.

1885 W. L. Carpenter *Treat. Manuf. Soap* 167 The boiling, and the addition of fat and lye, must be continued until a small sample..has a tolerably firm consistence... Practice alone will enable the operator to judge of the completion of this first operation, called 'pasting'. In English phraseology, it is called 'killing the goods' or raw material. **1888** J. Cameron *Soaps & Candles* 82 Saponification, pasting, or killing the goods. **1894** C. R. A. Wright *Animal & Veg. Fixed Oils* 468 The effect of the action of the hot ley on the melted fatty matter is to 'kill the goods'—*i.e.*, to emulsify the whole, so that no distinct layer of melted fat swims up on taking a sample.

Killarney (kilā·ɹni). The name of a town in Co. Kerry, Ireland, used *attrib.* in **Killarney fern**, the bristle fern, *Trichomanes radicans*, which was formerly abundant in the neighbourhood of Killarney.

[**1844** E. Newman *Hist. Brit. Ferns* 313 It has lately been supposed by many excellent botanists, that there are two Irish species of *Trichomanes*,—the Killarney and the Glouin Caragh plants.] **1863** *Fern Manual* 160 It is not only a native of Ireland, as the popular name (Killarney Fern) would lead us to believe, but is found also in many other parts of the world. **1888** C. T. Druery *Choice Brit. Ferns* vi. 59 We have been very successful in the culture of the Killarney Fern. **1891** L. T. Meade *Sweet Girl Graduate* x. 89 I'll take you into our fern-house... We have got such exquisite maidenhairs, and such a splendid Killarney fern. **1910** C. T. Druery *Brit. Ferns* 257 In Ireland it has been found so frequently as to have acquired the popular name of the Killarney Fern. **1960** P. Taylor *Brit. Ferns & Mosses* 88 The Killarney Fern has a most unusual distribution, being found chiefly in the tropics of the Old and New Worlds, with an extension northwards in Europe to the Pyrenees and Ireland. **1973** *Times* 17 Dec. 2/7 Picking wild plants prohibited in peer's Bill... The 20 flowers are..in danger of extinction... The plants are:..Killarney fern,..Snowdon lily, spiked speedwell, spring gentian, Teesdale sandwort.

kill-crazy (ki·l krēi·zi), *a. colloq.* (orig. *U.S.*). [f. Kill *v.* + Crazy *a.* 4.] Insanely desirous of killing; murderous.

1942 Berrey & Van den Bark *Amer. Thes. Slang* § 184/14 *Kill Crazy Dillinger*.., John Dillinger, Midwest outlaw. **1949** *Sun* (Baltimore) 1 Dec. 3/4 All the prior plans to kill me have been made by the same man—a kill-crazy man. **1959** G. Jenkins *Twist of Sand* xiii. 280 Scientists don't go kill crazy..just for the sake of one lost species. *Ibid.* xiv. 306, I had no plan. I was as kill-crazy as he. **1972** 'J. Ripley' *My Word you should have seen Us* 179, I was pure animal, and kill-crazy.

kill-devil, *sb.* (*a.*). **2.** Substitute for def.: A colloq. name for rum (see also quot. 1846). *Obs.* exc. *Hist.* (Add earlier and later examples.)

1639 J. Josselyn *Jrnl.* 24 Sept. in *Acct. Two Voy.* (1674) 26 Captain Thomas Wannerton..drank to me a pint of kill-devil *alias* Rhum at a draught. **1846** *Swell's Night Guide* 123/2 *Kill-devil*, new rum, from its pernicious quality. **1885** *Century Mag.* Apr. 884/2 Rum, or 'kill-devil', as it was everywhere called, was rendered plentiful by the trade with the West Indies and by the New England stills. **1970** M. Slater *Caribbean Cooking* 7 From sugar came rum, 'kill-devil', the spirit of cane.

killed, *ppl. a.* Add: **1. c.** *Med.* Applied to bacteria and viruses that have been killed or rendered non-infectious, and hence to preparations containing them.

1919 *Med. Ann.* XXXVII. 493 A special series of vaccines have been prepared... They are standardized emulsions in normal saline of killed bacteria. **1925** J. W. Bigger *Handbk. Bacteriol.* xiii. 152 Workers have found it possible to immunize monkeys against experimental pneumonia by the intratracheal inoculation of killed cultures. **1930** *Syst. Bacteriol.* (Med. Res. Council) VII. vii. 117 There remains a considerable doubt whether immunity..can be produced by the use of killed virus. **1964** M. Hynes *Med. Bacteriol.* (ed. 8) xxv. 385 The killed vaccine [of poliomyelitis] does not necessarily prevent subsequent intestinal infection, but it evokes sufficient neutralizing antibodies to prevent the viræmia which precedes central nervous system involvement. **1964** H. Jolly *Dis. Children* xvi. 343 Salk vaccine is a killed preparation containing all three types of virus. **1971** Beeson & McDermot *Cecil-Loeb Text-bk. Med.* (ed. 13) 416/2 Killed poliomyelitis vaccine is now used very infrequently.

2. b. *Metallurgy.* Of steel: treated when molten so as to prevent the evolution of oxygen on solidification and the consequent formation of blow-holes (now done by adding a reducing agent).

1884 W. H. Greenwood *Steel & Iron* xviii. 425 When the metal throws out sparks or teems fiery, it is said not to be 'killed', and it is indicative of the metal not having been sufficiently long in the fire after fusion, and such steel will yield unsound or honey-combed ingots. **1926** *Jrnl. Iron & Steel Inst.* CXIV. 406 In the application of titanium to killed steel the conditions are entirely different, and a complete absence of gas evolution as well as the maximum degree of deoxidation are generally desired. **1935** *Jrnl. R. Aeronaut. Soc.* XXXIX. 1121 Scientific metallurgists would like to see all steels fully killed,..but economic considerations at present compel the manufacture of rimming and partly-killed steel. **1970** A. Cibula in O. Kubaschewski et al. *Gases & Metals* ii. 31 Small amounts of hydrogen and nitrogen are detected, and in killed ingots these gases predominate over carbon dioxide.

killer. Add: **2.** In full, *killer whale.* For '*Orca gladiator*' substitute '*Orcinus orca*'. (Later examples.) Also *fig.*

1884 G. B. Goode *Fisheries U.S.: Nat. Hist. Aquatic Animals* 17 The Killer Whales are known the world over by their destructive and savage habits. **1931** W. G. Carr *By Guess & by God* 89 The British had been developing the mechanical killer-whales since 1904, when, unescorted, a flotilla of A-boats engaged in fleet manœuvres. **1937** Norman & Fraser *Giant Fishes, Whales, & Dolphins* xiii. 290 The colour of the Killer is well marked and is distinctive for the species. The back is black and the belly white. *Ibid.* 291 A feature unusual in cetaceans is the great discrepancy in size between the male and female Killer Whale. **1973** R. Burton *Life & Death Whales* i. 23 Killer whales live in small groups and feed mainly on fish but they also kill seals and other whales.

4. b. A contrivance for killing a large ferocious animal (e.g. a wolf, a shark); also, an explosive implement for the painless killing of cattle, horses, etc. Cf. *humane *a.* 1 d.

1892 *9th Ann. Rep. U.S. Bureau Amer. Ethnol. 1887–88* 259 (*heading*) Whalebone wolf-killers. **1901** *Amer. Anthropologist* Apr.–June 391 Eskimo and Samoan 'Killers'. **1901** *Morning Leader* 18 Dec. 3/7 The deadly instrument known as 'Greener's Killer',..for the painless destruction of old and incapacitated horses, is the invention of the well-known gun manufacturer, Mr. W. W. Greener... The 'killer' consists of a noiseless explosive apparatus resembling a short rifled barrel, which contains a small cartridge with steel-pointed bullet. **1919** T. Hardy *Let.* 7 Nov. in *One Rare Fair Woman* (1972) 187, I am a strange member of the Wessex Pig Society. I accepted the nomination entirely in the hope of helping to popularize the 'killer'. **1968** A. Wilson *Pract. Meat Inspection* i. 1 Cattle are usually stunned by a captive-bolt pistol (Humane Killer). The types usually employed are..3. Schermer mechanical killer.

6. A cow, sheep, etc., that is killed or destined to be killed for food. Usu. *pl. colloq.* (chiefly *Austral.* and *N.Z.*).

1897 I. Scott *How I stole over 10,000 Sheep in Austral. & N.Z.* 9 'You know the killers, don't you?'..i.e. the sheep the boss used for his own mutton at the house. **1907** W. H. Koebel *Return of Joe* 294 We were to be treated to a portion of a valuable stud ram for supper in place of the ordinary 'killer'. **1931** V. Palmer *Separate Lives* 124 He had put it [*sc.* the bullock] among the herd of killers in the home paddock in the hope that the new overseer might use it for beef in mistake. **1937** *Amer. Speech* XII. 104 As killing cattle or killers (cattle ready for killing) they are inferior to corn-fed stock. **1952** *Coast to Coast 1951–52* 210 One afternoon they had brought in some killers to the yards, and Murray went over to the kitchen for the bucket and the knives. **1972** P. Newton *Sheep Thief* iii. 27 The yarding of the killers..was regarded as a tryout for new dogs—with the stench of blood sheep are loth to approach the yard.

7. An impressive, formidable, or excellent person or thing; one who 'kills' people (see Kill *v.* 6 a in Dict. and Suppl.). Cf. *lady-killer* (Lady *sb.* 16). *slang* (orig. *U.S.*).

[**1900** *Dialect Notes* II. 44 *Killer*, 1. One who does things easily. 2. One who recites perfectly.] **1937** *Metronome* Apr. 55 That Zutie drummer-man is really a killer! **1940** *Swing* Jan. 26 *Farewell Blues* is another of those very fast killers. *Ibid.* May 10/3 I'm a killer with my new shepherd plaid suit. **1968** *Blues Unlimited* Sept. 6 'They were going to put out some records, and Luther went to see about it' said Percy, 'his "Dust my broom" was sure a killer.' **1970** *Melody Maker* 19 Sept. 28/6 George Khan has a solo on the up-tempo passage of the same track which is an absolute killer.

8. *attrib.* and *Comb.*, as *killer instinct*; **killer submarine** (see quot. 1955); **killer whale** (see 2 above).

1931 Visct. Knebworth *Boxing* 78 He [*sc.* Dempsey] had more fighting spirit and more of the sheer killer instinct in him than was in all four of them rolled together. **1960** J. Fingleton *Four Chukkas to Austral.* vi. 64 Unless Peter May can instil into his men the killer-instinct they so plainly lack, the next two Tests..will be lost. **1973** *Times* 20 Feb. 12/2 Already he has given Oxford a lift with his drive, enthusiasm, attack and killer instinct. **1955** M. Reifer *Dict. New Words* 116/2 *Killer submarine*, a small submarine designed for hunting down and sinking enemy submarines. **1960** *Economist* 31 Dec. 1378/2 The other kind of vessel that the Navy wants to build in larger numbers is the 'killer' submarine driven by atomic power and equipped with vastly improved sonar devices to seek out and destroy enemy submarines under water.

killer-diller (ki·ləɹ,di·ləɹ). *slang* (orig. *U.S.*). [Rhyming reduplication of *killer 7.] = *killer 7. Also *attrib.* or as *adj.*

1938 *Life* 8 Aug. 56 The hot musician..shudders when he hears Benny Goodman announce his next radio number as a 'killer-diller'. **1939** R. P. Dodge in Ramsey & Smith *Jazzmen* xv. 341 The fast 'killer-diller' arrangement. **1957** W. C. Handy *Father of Blues* x. 147 My old friend Wilbur Sweatman—a killer-diller and jazz pioneer. **1970** *Melody Maker* 3 Oct. 15/6 The inevitable concessions to the gallery made by the band in some of the more 'killer-diller' specialities.

killian, var. *kilian.

killick, killock. Add: **b.** A leading seaman's badge, bearing the symbol of an anchor; hence, a leading seaman. Also *attrib.* or as *adj.*, leading, chief. *colloq.*

1915 'Bartimeus' *Tall Ship* iii. 62 He paid for the misplaced generosity of his well-wisher with his 'Killick'. (*footnote*) Anchor. The distinctive badge of a leading rating. **1920** —— *Unreality* II. iii. 115 Picked up my killick. *Ibid.* iv. 126 The sight-setter raised his brows at the red worsted anchor adorning Bill's sleeve. 'Killick, eh!' he ejaculated. **1925** Fraser & Gibbons *Soldier & Sailor Words* 135 *Killick*.., the lower deck term for the Petty Officers' Anchor arm badge. **1930** G. Wells *Naval Customs* 91 *Killick*,..a slang term for a 'leading seaman'. **1945** 'Tackline' *Holiday Sailor* xiv. 143 Been in barracks for a matter of six months. Killick then, o' course. **1949** Partridge *Dict. Slang* Add. 1092/2 *Killick-scribe*, a Leading Writer.

killing, *vbl. sb.* Add: **b. killing bottle,** a bottle containing a poison for killing captured insects, etc.; **killing-circle,** the area within which, at a certain range, the charge of shot from a gun is sufficiently compact to kill the game; cf. Pattern *sb.* 11.

1877 *Encycl. Brit.* VI. 134/2 Beetles when caught may ..be dropped..into what is known as the 'killing bottle', the bottom of which contains cyanide of potassium covered over with a layer of gypsum. **1960** B. K. Wilson *Lovely Summer* ii. 21 He could not understand how his mild-mannered father could trap the butterflies in his killing-bottle with such enthusiasm. **1974** *Guardian* 20 Sept. 17/8 Insects hitting a panel dangling below the tent will fly up into the dome, where they will be anaesthetised by fumes being given off by a killing bottle. **1886** Walsingham & Payne-Gallwey *Shooting: Field & Covert* vi. 94 The charge of a 20-bore is smaller, lighter, and has a less killing circle than has a 12-bore. **1892** W. W. Greener *Breech-Loader* 148 For ordinary sporting purposes a gun which shall give its largest killing circle at 30 yards with the first barrel, and at 40 with the second, will be found the most convenient of good shots.

2. A large profit; a quick and profitable success in business, etc. *slang* (orig. *U.S.*).

1888 *Texas Siftings* 24 Mar. 13/2 Fred Jarvis..getting $15,000 in The Louisiana State Lottery drawing... Many.. would like to know something relative to the man who was fortunate enough to 'make a killing'. **1912** T. Dreiser *Financier* xxx. 340 Railroad securities..were considered weak under the present circumstances, and a great killing was expected. **1938** 'N. Shute' *Ruined City* i. 11 I'm a banker, of course. I don't take tips, and I don't make any great killings, but in my quiet way I get along all right. **1959** F. Hobson *Death on Back-Bench* x. 136 Instead of making a financial killing which would have made him far wealthier and more powerful than ever before, he was going to be worse off than he had ever been. **1967** M. McCarthy in *Observer* 30 Apr. 12 Their personal..aim..was to make a killing..in Vietnamese real estate. **1972** Wodehouse *Pearls, Girls & Monty Bodkin* iv. 52 If Soapy Molloy made a killing nobody could be more eager to celebrate than Dolly. **1972** *Times Lit. Suppl.* 12 May 544/2 Authors hold on in the hope of making a killing on film rights.

killing, *ppl. a.* **2. e.** (Examples.)

1844 'J. Slick' *High Life N.Y.* I. xvi. 255 Then the tall man in whiskers would begin to look as if he ralv had been a killing critter with the women folk. **1874** L. Troubridge *Life amongst Troubridges* (1966) 83 She was in her dressing-gown and looked too killing, exactly like those fat chinamen flying kites on Amy's Japanese screen. **1889** 'Mark Twain' *Connecticut Yankee* 111 A lecturer.. flooded an audience with the killingest jokes for an hour. **1923** [see *dine *v.* 1 b]. **1960** M. Spark *Bachelors* iii. 47 'That's exactly what I expected you to say,' Marlene said. 'I think you're killing.'

Kilmarnock (kilmā·ɹnǫk). **1.** The name of a town in Ayrshire (now Strathclyde), used (freq. *attrib.*) to designate various articles made there, or practices characteristic of its inhabitants; *spec.* **Kilmarnock bonnet, cowl,** a cap resembling a tam-o'-shanter.

1643 *Edinburgh Testaments* (MS.) LX. 292b (D.O.S.T.), Sextene pair of Kilmarnock stokinges estimat ale to fyftie pundis. **1681** S. Colvil *Mock Poem, or, Whiggs Supplication* 15 He used to take lives With Whingers, and Kilmarnock Knives. **1713** R. Wodrow *Analecta* (1842) II. 262 Ther ar some of your Kilmarnock whittles, that, though they look not soe fair on it as your English knives, yet have a better edge. **1789** D. Davidson *Thoughts on Seasons* 84 An honest wabster..Whase haffet a Kilmarnock hood Kept warm an' snug. **1790** A. Wilson *Poems* 72 His fearsome blue Kilmarnock cowl. **1806** J. Patterson in J. Greenshields *Ann. Lesmahagow* (1864) App. 46 And some..with subtle wrist, Give to their stane, the true 'Kilmarnock twist'. **1822** H. Ainslie *Pilgrimage to Land of Burns* 6 The manner in which the whole man was so properly roofed in with the ancient Kilmarnock bonnet. *Ibid.* 31 Flourishing..their 'Kilmarnocks' manfully round their heads. **1828** *Blackw. Mag.* Jan. 40 Kilmarnock bonnets, pixie caps, and mittens. **1833** Carlyle *Sart. Res.* in *Fraser's Mag.* Dec. 670/1 Even the Kilmarnock nightcap is not forgotten. **1877** J. M. Neilson *Poems* 49 He..cover'd the bald pow o' Willie Shakspeare Wi's big blue Kilmarnock. **1902** *Daily Chron.* 20 Dec. 5/2 The Scottish team of curlers who have departed for Canada wanted to be rigged out with old-style Kilmarnocks. **1946** *Glasgow Herald* 24 Aug., Having recently taken up bowling, I have heard players saying, 'That's a Kilmarnock one' when a shot has shown evidence that the player was determined to run no risks in disturbing the head. **1949** J. Drummond *Behind Dark Shutters* i. 21 They agreed to pipe a water line from Townhead and put in a 'Kilmarnock well' in the Row. **1951** J. Masters *Nightrunners Bengal* ix. 94 The two infantry regiments tried to get the new white Kilmarnock caps to replace their heavy shakos. **1953** J. E. Shaw *Ayrshire* 32 The Kilmarnock bonnet manufacturers strike a specially sentimental note in the hearts of all Scotsmen... It was at one time a national headwear. **1960** H. Hayward *Antique Coll.* 157/2 *Kilmarnock carpets*, double-cloth carpeting was made in Kilmarnock from 1778 and three-ply was perfected in 1824.

2. Kilmarnock willow, a pendulous form of the goat willow, *Salix caprea* forma *pendula*, first discovered by James Smith (*c* 1760–1840), a Scottish botanist, and subsequently distributed by Thomas Lang, a Kilmarnock nurseryman.

1854 *Gardeners' Chronicle* 17 June 392/1 The name Kilmarnock Willow was bestowed upon it to distinguish it from the common Weeping Willow. **1869** *Rep. Comm. Agric. 1868* (U.S. Dept. Agric.) 202 Kilmarnock willow (*Salix caprea* var. *pendula*)..becomes one of the most distinct of the hardy weeping plants which we possess. **1892** A. C. Apgar *Trees Northern U.S.* 166 The Goat-willow is the one generally used for the stock of the artificial umbrella-formed 'Kilmarnock Willow'. **1913** Elwes & Henry *Trees Gt. Brit. & Ireland* VII. 1746 [*Salix caprea*] Var. *pendula*... Pendulous in habit, usually grafted on a stock about 4 ft. high, and forming a weeping shrub, which is known as the Kilmarnock Willow. This was discovered in 1840 on the banks of the river Ayr, and was propagated by Lang, nurseryman at Kilmarnock. **1972** S. C. Warren-Wren *Willows* iii. 61 It [*sc.* a pendulous variety] is *S*[*alix*] *caprea* forma *pendula* Th. Lang, and is commonly referred to as the Kilmarnock willow. Its branches are stiffly pendulous.

Kilner jar (ki·lnəɹ dʒā·ɹ). The proprietary name of a type of preserving jar. Also *Kilner preserving jar.*

1930 *Trade Marks Jrnl.* 19 Nov. 1777 The 'Kilner' jar... Glass fruit preserving jars. Kilner Brothers, Limited,.. London,..manufacturers. **1972** 'S. Woods' *They love not Poison* vi. 82 Jenny came through from the kitchen with a Kilner jar about one third full of cream which she proposed to shake until it separated. **1973** *Listener* 25 Jan. 117/2 The old-fashioned Kilner preserving jar lent itself to both wet and dry methods of bottling. **1974** *Times* 16 Nov. 11/1 Old-fashioned vinegar crocks can be things of beauty... You can otherwise simply use a Kilner jar.

kilo[2] (ki·lo). Abbrev. of Kilometre, -meter.

1888 E. Dowson *Let.* 13 Nov. (1967) 17, I hope..that you are now ..laying in a stock of hygiene many kilos from this foggy, pestilent metrop. **1948** G. H. Johnston *Death takes Small Bites* viii. 195 They drove five kilos in silence. **1973** *Evening Standard* 11 Oct. 1/2 After bogging down in the sand by a wrecked tank, we dug ourselves out and hastily retreated a few kilos until we reached a tank laager.

kilo-. Add: **b.** *kilo-ampere, -bar* [*bar *sb.*[6] 1, 2], *-bit* [*bit *sb.*[4]], *-byte, -curie, -electron volt, -gauss, -hertz, -joule, -parsec.* Also *kilobuck* (joc.) [*buck *sb.*[8]], a thousand dollars.

1901 J. A. Fleming *Handbk. for Electr. Laboratory* I. i. 68 The standard kilo-ampere balance. **1972** *Physics Bull.* Mar. 152/3 A few calculations, taking into account material and engineering limitations show that the voltage produced is low, say up to 10 V, but currents of kiloamperes are readily achieved. **1928** N. Shaw *Man. Meteorol.* II. p. xxxii, *Kilobar*, a name which it is proposed to substitute for the word 'millibar' to express a pressure of 1000 dynes per square centimetre, on the ground that the word 'bar' had been used in Physical Chemistry to denote the C.G.S. unit of pressure (1 dyne per square centimetre) before it was employed in meteorology to express a million dynes per square centimetre. **1971** I. G. Gass et al. *Understanding Earth* iii. 60/2 Olivine Mg_2SiO_4 can recrystallize at pressures of over 150 kilobars to a new polymorphic form called a spinel structure. **1961** Gray & Graham *Radio Transmitters* i. 3 In the case of the transmission of business-machine or telemetered data, it is more usual to express the speed in bits or kilobits (1,000 bits) per second. **1972** *Sci. Amer.* Sept. 126/3 High-speed leased lines provide a data speed of 48 to 240 kilobits per second. **1951** *N.Y. Times Mag.* 22 Apr. 35 'Kilobuck'..is a scientist's idea of a short way to say 'a thousand dollars'. **1955** *Sci. Amer.* June 126/2 This instrument became the prototype of a commercial spectroscope which sells for about a kilobuck. **1970** *Times* 30 June 22/7 Its storage capacity could be 2,048 kilobytes compared with a maximum of 512 kilobytes for the 360/50. **1973** *Physics Bull.* Oct. 626/1 It has a 64 kilobyte store with 4·8 megabytes of exchangeable disc storage. **1946** *Science* 2 Aug. 92/1 We came to the Project thinking in terms of millicuries and found that we had to face problems of curies and sometimes..even kilocuries. **1968** *Listener* 27 June 828/1 This power package contained as its energy source 17 kilocuries of Plutonium 238. **1950** Glasstone *Sourcebk. Atomic Energy* iii. 82/1 One [*sc.* energy unit], equal to a thousand electron volts, called the kilo-electron volt, is represented by Kev. **1971** D. W. Sciama *Mod. Cosmol.* ii. 31 The distinction between X- and γ-rays is a somewhat arbitrary one, but..we may take the dividing line to be an energy of 100 keV (kilo-electron volts). **1895** F. C. Baily in *Rep. Brit. Assoc. Adv. Sci.* 206 Gaussage would be about 50 in small transformers, up to 40,000 in large dynamos. The latter could be conveniently reckoned in kilogausses. **1973** *Sci. Amer.* Oct. 21/1 If the magnetic field at the surface of a continuous-sheet guideway is 20 kilogauss (about the field strength at the pole face of a good magnet), the lift force is 60 pounds per square inch. **1929** *Daily Express* 11 Jan. 3/6 A national common frequency of 1,040 kilohertz (288·5 metres). **1935** Turner & Banner *Electr. Measurements* i. 4 Cycle per second. [*Note*] For radio purposes this is sometimes termed the 'Hertz' and a practical unit called the 'Kilohertz' or 1000 cycles per second used. **1967** *Electronics* 6 Mar. 183/1 Since the power is limited, the bandwidth has to be restricted to 500 kilohertz. **1893** T. O' C. Sloane *Stand. Electr. Dict.* 317 Kilojoule. **1938** *Jrnl. R. Aeronaut. Soc.* XLII. 885 Formation of frost on the wing is accompanied by a rise in temperature which raises the deposit to 0°C. A curve is given showing the energy in kilojoules necessary to remove ice. **1974** *Times* 20 Feb. 12 Each of the colourfully illustrated cards shows a different food, together with its value in carbohydrate, calories and kilo-joules (the measurement that will be used instead of calories when Britain goes metric). **1922** *Encycl. Brit.* XXX. 301/2 The most remote cluster known is distant 67 kiloparsecs or 200,000 light years. **1970** *Sci. Jrnl.* Mar. 71/3 The solar system is located on the mid-plane of the disc 10 kiloparsecs from the galactic centre.

kilocalorie (ki·lŏkæ:lŏri). Also **-calory** (*rare*). [f. Kilo- + Calorie.] = *calorie a.

1894 J. H. Raymond *Human Physiol.* ii. 166 It is estimated that an average man produces daily 2500 kilo-calories, which is about 100 kilo-calories per hour. **1909** *Cent. Dict.* Suppl., *Kilocalory.* **1923** H. Moore *Textbk. Intermediate Physics* xxii. 202 The Continental Engineers' Unit is..the kilogram-calorie or kilocalorie. **1969** *Listener* 16 Jan. 67/3 The food intake was very low, 1,600 kilo-calories per head per day, but the population was healthy and apparently active. **1971** *Nature* 12 Nov. 112/3 Moreover, SI units are nowhere mentioned, the traditional Torr and kilocalories being preferred. **1974** *Sci. Amer.* Sept. 162/3 The 'calorie' of nutritional parlance is actually a kilocalorie.

kilocycle (ki·lŏsəik'l). [f. Kilo- + Cycle *sb.*] **a.** One thousand cycles (of an oscillation or other periodic phenomenon). **b.** *ellipt.* One thousand cycles per second: = **kilohertz.*

1921 *Wireless Board List Radio Telegr. Waves* 3 Frequency is expressed in 'Kilo-cycles' (K.C.). **1926** E. B. Moullin *Theory & Pract. Radio Frequency Measurements* iii. 73 (*heading*) Interpolated frequency in kilocycles per sec. **1927** *Daily Tel.* 14 June 5/3 The decision of the B.B.C. to adopt forthwith the kilocycle method of stating frequencies instead of the wave-length method. **1950** *Engineering* 7 Apr. 379/2 The high-velocity sound-level recorder...is designed for inputs with frequencies between 50 cycles and 15 kilocycles per second. **1959** R. L. Shrader *Electronic Communication* iv. 86 A frequency of 3,500,000 cycles can be expressed as 3,500 kilocycles (3,500 kc), or 3·5 megacycles (3·5 Mc).

kilogramme, -gram. Substitute for def.: A unit of mass (formerly also taken as a unit of weight) that was introduced as a fundamental unit of the metric system and is now one of the base units of the International System of Units, being equivalent to approximately 2·205 lb.; it was intended to be the mass (or weight) of a cubic decimetre of distilled water, but in practice has been defined almost from its inception as the mass of a unique physical standard (orig. the 'Kilogramme des Archives', and since 1889 the International Prototype of the Kilogramme). (Earlier and later examples.)

1797 *Jrnl. Nat. Philos.* I. 197 The kilogramme [is] equal to 2 pounds 5 gros 49 grains. **1946** *Proc. R. Soc.* A. CLXXXVI. 171 The kilogram was originally defined by reference to a 'natural' standard, i.e. the mass of the cubic decimetre of water. The material representation of this standard was the Kilogramme des Archives, a simple cylindrical piece of platinum which was constructed in the latter part of the eighteenth century. **1964** H. S. Hvistendahl *Engin. Units* iii. 20 Study of the archives of the BIPM [*sc.* the Bureau International des Poids et Mesures] has shown that whilst due to the lack of precision in scientific terminology at that time, the word 'poids' (weight) was improperly used as a synonym of 'mass', the kilogramme étalon was definitely intended to represent the unit of what we now always call mass, and not the force by which that mass unit was attracted to the earth. **1969** *Changing to Metric Syst.* (Nat. Physical Lab.) (ed. 3) 5 The spellings of the units of length (metre) and mass (kilogramme) given above are those used by the General Conference of Weights and Measures. In North America the spellings are meter and kilogram. **1969** *Symbols, Signs, & Abbreviations* (R. Soc.) 22 Kilogramme. **1971** *Quantities, Units, & Symbols* (R. Soc.) 22 Kilogram. **1971** *Nature* 8 Jan. 80/1 A thousand kilogrammes of ice. **1972** *N.Z. News* 26 Jan. 1/5 The prices for all classes of fleece wool showed an improvement of from 10c to 16c per kilogram. **1973** *SI Units* (Internat. Stand. ISO 1000) 20 The kilogram is the unit of mass; it is equal to the mass of the international prototype of the kilogram.

b. An object having a mass of one kilogramme and made as a standard of mass or weight.

1831 *Jrnl. R. Inst. Gt. Brit.* II. 65 The rather unsightly appearance of the standard kilogramme of Monsieur Fortin's making. **1856** *Phil. Trans. R. Soc.* CXLVI. 875 Permission was obtained from the French Government to compare the pound with the standard kilogramme of platinum deposited in the Archives on the 22nd of June, 1799, known as the 'kilogramme des Archives'. *Ibid.* 878 The kilogramme, after having been washed with alcohol, was suspended from the right-hand pan of the balance. **1922** *Nature* 19 Oct. 622 During the nineteenth century.. serious doubt arose as to whether the accepted standard kilogramme [*sc.* the 'Kilogramme des Archives' made by Lefèvre-Gineau and Fabbroni in 1799] actually did comply with the original definition of the kilogramme... A proposal to construct a new standard kilogramme brought the whole subject under review in 1872, and the Commission Internationale du Mètre..decided that the international kilogramme should be a copy of the 'Kilogramme des Archives'. **1936** *Rep. Board of Trade Standards Dept. Comparisons Imperial Standards* 3 The opportunity was..taken to undertake..the comparison of the national copy of the kilogram with the principal standard kilograms of the Board of Trade. **1946** *Proc. R. Soc.* A. CLXXXVI. 172 Some forty or more copies of the international kilogram were constructed. The principal copies ..are preserved at the International Bureau of Weights and Measures, Sèvres, Paris... One such copy, No. 18, designated the British Copy of the Kilogram, is kept in this country. **1962** *Brit. Jrnl. Appl. Physics* XIII. 456/1 The international kilogramme, which is a platinum-iridium weight at the head of an extensive hierarchy of similar kilogrammes, was known to be extremely stable.

2. *attrib.* and *Comb.*, as **kilogramme calorie** = *calorie a; **kilogramme-force** (pl. *kilogrammes-force*), a unit of force equal to the

weight of a mass of 1 kilogramme, esp. under standard gravity.

1900 E. Buckingham *Outl. Theoret. Thermodynamics* ii. 12 The small or gram calorie, in distinction from the large or kilogram calorie, which is one thousand times as large. **1930** *Engineering* 21 Nov. 640/1 Curve I shows the maximum liberation of heat in kilogram-calories per cubic metre per hour. **1951** B. L. Goodlet *Basic Electrotechnics* i. 4 Heat energy is measured in kilogram-calories, one of which equals 4186 joules. **1960** *McGraw-Hill Encycl. Sci. & Technol.* VII. 347/1 The kilogram is also sometimes defined as a unit of force, that is, the weight of a 1-kilogram mass, hence the expression kilogram-force. **1972** *Physics Bull.* May 285/1 The addition of small weight pieces by hand to the scalepan system permits forces in newtons and kilograms-force to be obtained readily.

kilometre, -meter. Also with pronunc. (kilǫ·mītəɹ), prob. under the influence of such words as *speedometer, thermometer.*

kilometric, *a.* Add: (Further examples.) *kilometric guarantee*, a guarantee of gross receipts per kilometre of line conceded by the Turkish government in the late 19th and early 20th centuries to (foreign) companies constructing railways within the then Empire.

1902 *Daily Chron.* 24 Jan. 3/3 The revenues to be assigned for the service of the kilometric guarantee have not yet been specified. **1909** *Westm. Gaz.* 16 Dec. 2/1 The Bagdad Railway (with the iniquitous kilometric guarantee). **1909** D. Fraser *Short Cut to India* iii. 39 A heavy kilometric guarantee was provided. *Ibid.* 45 When the kilometric receipts exceed..4500 francs, the whole of the surplus goes to the government. **1943** *Gloss. Terms Telecomm.* (*B.S.I.*) 64 Radio-waves are classified as follows: A. According to wavelength, in metres... Kilometric 10 000 to 1 000. **1973** *Nature* 21 Sept. 144/2 Urey assumes that the..heart-and-soul of a comet is a solid nucleus of kilometric size.

kilo-stone (ki·lŏ-, kī·lo,stōun). Also **kilostone.** [f. *kilo[2] + Stone *sb.*] = *kilometre-stone.*

1925 *Brit. Weekly* 1 Jan. 342/2 Down the road..a woman was crying with maternal, anxious voice, 'Louise! Louise!' The cry was wafted up to the kilostone. **1936** *Discovery* Feb. 63/2 We dispute the author's recollection of the kilo-stones on the Seres Road [in Macedonia].

kiloton (ki·lŏtɒn). [f. Kilo- + Ton[1].] **a.** One thousand tons. Somewhat *rare.* **b.** A unit of explosive power, equal to that of one thousand tons of T.N.T. Freq. *attrib.*

1950 J. O. Hirschfelder et al. *Effects of Atomic Weapons* i. 14 It is necessary for about 3×10^{24} atoms to suffer fission to release the energy equivalent to 20 kilotons of TNT. **1952** *Birmingham* (Alabama) *News* 24 Apr. 14/5 According to informed forecasts, the new bomb will have an explosive power of between 200 and 300 kilotons. **1956** A. H. Compton *Atomic Quest* v. 303 A 20 kiloton explosion (i.e., energy released equivalent to 20,000 tons of conventional explosives). **1959** *Times Lit. Suppl.* 22 May 306/4 The atomic strategy..relies on technology rather than on men and reckons fire-power in terms of kilotons of explosive. **1971** I. G. Gass et al. *Understanding Earth* xxiv. 333/1 The difficulties of discriminating between minor 'background' Earth tremors and the effects of [nuclear] tests of a few kilotons were seen to be formidable. **1973** *Sci. Amer.* Oct. 47/3 One of the aerial bombs, a comparatively low-yield weapon, had an explosive force equal to 'a few hundred kilotons' of TNT.

kilovolt (ki·lŏvōult). [f. Kilo- + Volt *sb.*] **a.** One thousand volts.

In quot. 1861 the sense is 'a milliohm'.

1861 *Electrician* 9 Nov. 3/2, 1 Farad per second = 1 Volt, or unit of resistance. 1/1000 Volt = 1 Kilovolt. **1905** *Nature* 7 Dec. 142/1 The disruptive voltage is V kilovolts. **1927** *Glasgow Herald* 8 Dec. 13 Kilovolt transformers. **1938** R. W. Lawson tr. *Hevesy & Paneth's Man. Radioactivity* (ed. 2) v. 65 The hardest characteristic radiation known, viz. the *K*-radiation of uranium, has a wavelength of 101 X.U. and requires an excitation potential of 115 kilovolts. **1948** *Sci. News* VII. 103 The thickness of metal which can be penetrated is directly related to the kilovoltage of the X-ray set, and a modern portable industrial unit employing 400 kilovolts will penetrate up to 5 inches of steel. **1973** *Sci. Amer.* Nov. 133/1 Bigger sizes are foreseen, but 10 kilovolts at low current is not attractive.

b. *attrib.* and *Comb.*, as **kilovolt-ampere,** a unit of apparent electric power, used when an alternating current is not in phase with the voltage, equal to the product of an r.m.s. voltage of 1000 volts and an r.m.s. current of 1 amp; **kilovoltmeter,** a voltmeter for measuring thousands of volts.

1909 *Cent. Dict.* Suppl., Kilovolt-ampere. **1933** Edgcumbe & Ockenden *Industr. Electr. Measuring Instruments* xxii. 443 These quantities..are directly proportional to the kilo-watts and the reactive kilo-volt-amperes, respectively. **1950** *Engineering* 6 Oct. 281/3 Automatic control is provided to enable either the power factor or the reactive kilovolt-amperes to be maintained constant. **1923** R. Knox *Radiogr. & Radio-Therapeutics* (ed. 4) I. i. 45 The kilovoltmeter for measuring the high-tension current. **1971** Stern & Lewis *X-Rays* xiii. 232 The kilovoltmeter is connected across the primary of the high-voltage transformer.

Hence **ki·lovoltage,** voltage expressed in, or of the order of, thousands of volts.

1933 *Proc. R. Soc.* B. CXII. 366 A kilovoltage of 78 was used..to ensure the production of the characteristic K radiation of tungsten. **1948** [see *kilovolt a above]. **1971** Stern & Lewis *X-Rays* xiii. 233 The quality and quantity of x-ray output for a tube are..not described uniquely in terms of the kilovoltage and milliampere meter readings.

kilowatt. Add: **a.** (Earlier example.)

1884 *Engineering* 17 Oct. 358/1 Professor Koyle thought the term kilowatt sufficiently explicit.

b. *Comb.* **kilowatt-hour,** a unit of energy equal to that produced in one hour by a power of one kilowatt, viz. 3·6 million joules.

1892 [in Dict.]. **1905** *Daily Chron.* 20 July 6/4 If he uses this 300 kilowatts for every minute he is working throughout the quarter, say..720 hours, he will use 300 × 720 kilowatt hours or 'units'. **1951** B. L. Goodlet *Basic Electrotechnics* i. 11 Electrical energy is sold commercially by the kilowatt-hour—1000 watts acting for 1 hour. **1965** Owens & Sanborn *Fund. Electr.* vii. 366 If a lighting circuit uses 2000 watts for a period of one hour, the amount of electric energy consumed is 2 kilowatt-hours.

Hence **ki·lowattage,** power expressed in kilowatts; also *fig.*

1935 *Daily Express* 5 Apr. 8 A human dynamo of enormous kilowattage. **1971** *Sci. Amer.* Sept. 212/1 (Advt.), There was a time when 'the more, the better' was a national attitude—when a surging yearly increase in kilowattage scored prestige points.

kilp, var. *kelp[3].

Kilroy (ki·lroi). The name of a mythical person, popularized by American servicemen in the war of 1939–45, who left such inscriptions as 'Kilroy was here' on walls, etc., all over the world.

The explanation in quot. 1946 is one of numerous unverifiable accounts of the origin of the name.

1945 *Sat. Even. Post* 20 Oct. 6/2 On the crib hung a hand-lettered sign which asserted: 'Kilroy Slept Here'... Wherever he was, Kilroy had been there and left his mark behind: 'Kilroy Was Here', or 'Kilroy Passed Through' or 'You're in the Footsteps of Kilroy'. **1946** *N.Y. Times* 24 Dec. 18/4 As far as the American Transit Association is concerned, the identity of the elusive Kilroy of 'Kilroy Was Here' fame has been established, and in token of this recognition a street car was shipped today to James J. Kilroy of Halifax, Mass. *Ibid.* 18/5 In brief, Mr. Kilroy's claim is based on the following: During the war he was employed at the Bethlehem Steel Company's Quincy shipyard, inspecting tanks, double bottoms and other parts of warships under construction. To satisfy superiors that he was performing his duties, Mr. Kilroy scribbled in yellow crayon 'Kilroy was here' on inspected work. Soon the phrase began to appear in various unrelated places, and Mr. Kilroy believes the 14,000 shipyard workers who entered the armed services were responsible for its subsequent world-wide use. **1951** *Here & Now* (N.Z.) May 10/1 United Nations forces wonder what the jolly rodger they're doing [*sic*] to do when they reach the northernmost point. Mark 'Kilroy' in the snow and come back again perhaps. **1952** P. Atkey *Juniper Rock* xvii. 149 Such English as was possessed by the C.Q.M.S. was of a somewhat dated military character... 'Kilroy was 'eah,' said the C.Q.M.S. helpfully. **1962** *Guardian* 27 Nov. 7/5 For a long time Kilroy was the exclusive possession of the American forces. **1966** *Sunday Times* (Colour Suppl.) 4 Dec. 73/2 GI Jargon '*Kilroy was here*', standard graffito. **1968** M. Allingham *Cargo of Eagles* iii. 48 'A lot of people would like to know where he is just now. I wonder why he left that little pointer behind?' 'A "Kilroy was here" sign..?' **1969** L. Durrell *Spirit of Place* 162 When I am rich I shall have a memorial plaque placed..on Inverness station..reading 'Kilroy was here—but oh so briefly'!

‖ **kilta** (ki·ltă). Also **kilter.** [Origin unknown.] In India, a kind of wicker basket. Also *attrib.* = made of wicker.

1876 C. F. G. Cumming *From Hebrides to Himalayas* II. v. 134 Our provisions were packed..in long native baskets, called *kilters*. **1896** S. J. Stone *In & beyond Himalayas* 39 The provisions and cooking apparatus were carried in kiltas (wicker baskets covered with leather). **1901** Kipling *Kim* xiii. 338 He skipped nimbly from one *kilta* to the next, making pretence to adjust each conical basket. **1927** *Blackw. Mag.* Mar. 312/1 A *kilta* carrying-chair, carried on the back of one man.

kilter (ki·ltəɹ). *Poker.* [prob. var. of dial. Kelter[4].] A hand consisting only of cards of little or no value.

1895 'Templar' *Poker Manual* 55 Suppose you have an utterly valueless hand dealt you, say for example, deuce, four of hearts, six of clubs, seven of spades and nine of diamonds; this sort of hand is termed a 'kilter'. **1904** R. F. Foster *Pract. Poker* 126 The Southern custom of raising the ante on a kilter, and then standing pat. **1948** O. Jacoby *On Poker* xii. 144 There are any number of additional combinations of cards which are given rank in various localities. Some of these are the Dog, the Tiger, the Skeet, the Kilter, [etc.].

kilter, var. *kilta. See also *kelter[2].

kilty (ki·lti). *colloq.* Also **-ie.** [f. Kilt *sb.* + -y[6].] One who wears a kilt; esp. a nickname for a Highland soldier. Also, *attrib.* and as *adj.*

1842 D. Vedder *Poems* 112 In double quick time did the kilties career. **1900** S. R. Crockett *Little Anna Mark* xii. 103 Yon's nae lassie! Yon's a kiltie lad. **1902** J. Milne *Epistles of Atkins* xi. 208 A second Gordon communicates the views of a second Boer also upon the Gordon men. 'The "kilties" are devils to fight.' **1906** *Daily Chron.* 6 Aug. 3/6 A picturesque touch..is given to the town by the presence of the 'kilties' and other regiments of soldiers. **1921** *Daily Colonist* (Victoria, B.C.) 18 Oct. 3/7 (Advt.), St. Margaret Kiltie Dresses for girls 2 to 10 years, in navy, brown, saxe and green, [etc.]. **1927** *Scots Observer* 14 May 16/4 The Kilty piping for money. **1927** H. A. Vachell *Dew of Sea* 261 She..assured him.. that he was the 'kiltiest' boy she had ever met.

kimchi (ki·mtʃi). Also **kimchee, kimch'i.** [Korean.] A raw strongly-flavoured vegetable pickle, the Korean national dish.

1898 I. L. Bishop *Korea & her Neighbours* I. vii. 98 Wine, soup, eggs, and *kimchi*..were produced, and had to be partaken of. **1898** J. S. Gale *Korean Sk.* iii. 100 We found a Korean hut and, to our delight, dined on rice and *kimch'i* (pickle) once more. **1905** J. London *Jacket* (1915) xvi. 210, I know *kimchi*. It is a sort of sauerkraut. **1925** *Blackw. Mag.* Mar. 320/1 Then five different sorts of *Kimchee*, a horrible dish made out of vegetables which have become rotten. **1951** C. Osgood *Koreans & their Culture* v. 85 The women prepare cucumber kimch'i by washing the vegetables, then cutting small pieces off their ends, and slitting the sides in three or four places. **1966** S. McCune *Korea* iv. 33 A unique part of the diet and important for its vitamin content is *kimchi*, an unbelievably 'hot' pickle. **1972** P. M. Bartz *S. Korea* 40/2 Korea emphasizes pickled cabbage (*kimchi*), hot peppers, and other seasonings. **1972** *Korea Times* 19 Nov. 6/3 With the onset of the kimchi-making season, housewives are concerned about preparing jars in which the pickled vegetables are stocked.

Kimeridge, var. Kimmeridge (in Dict. and Suppl.).

Kimmerian, var. Cimmerian *a.* and *sb.* in Dict. and Suppl.

1917 E. Pound in *Poetry* (Chicago) Aug. 251 To the Kimmerian lands and peopled cities..came we. **1950** H. L. Lorimer *Homer & Monuments* ii. 52 The description of the country of the Kimmerians..would suit their settlements on the north coast of the Black Sea. *Ibid.* v. 286 The conflict between Kimmerian and Scyth in South Russia. *Ibid.* 306 The result of Kimmerian and Scythian invasions.

Kimmeridge. Add: Also **Kimeridge.** (Earlier and additional examples.)

Although the village name is now spelt *Kimmeridge*, the form *Kimeridge* is current in geological usage, reflecting the former spelling of the name; cf.:—**1933** W. J. Arkell *Jurassic Syst. Gt. Brit.* xiv. 441 The spelling *Kimeridge* was used by H. B. Woodward.., by Damon.., and by most earlier authorities. The new form *Kimmeridge* was not heard of before Webster and Buckland introduced it in the [early] nineteenth century and seems to have no justification. **1947** —— *Geol. Country around Weymouth* (Mem. Geol. Survey Gt. Brit.) v. 68 Inland towards Kimmeridge village. [*Note*] The name of this village was Cameric in Domesday Book and Kymerich by 1293; by the time of Hutchins' great work on Dorset it had become Kimeridge... The Ordnance map of 1811 (Sheet 16, Old Series) retained one 'm', and so did Damon in his *Geology of Weymouth* (1860 and 1884–8) and H. B. Woodward in '*The Jurassic Rocks of Britain*', (vol. v, 1895), *Mem. Geol. Survey.* In the resurvey of 1892 (Sheet 342, New Series), two 'm's' were adopted and the Geological Survey followed suit in 1898.

1760 *Phil. Trans. R. Soc.* LI. 549 The Kimeridge [*printed* Kimendge] coal..appears in most of the cliffs of the isle of Purbeck. **1774** J. Hutchins *Hist. & Antiquities Dorset* I. 194/1 He says the Kimeridge coal, of which I sent him some pieces, is very much like, but not so large as, the Bovey coal. **1814** T. Webster in *Trans. Geol. Soc.* II. 167 Clay with limestone and bituminous shale, containing the Kimeridge coal. **1816** —— in H. C. Englefield *Descr. Isle of Wight* 187 The whole of the Kimmeridge strata, consisting of a series of argillaceous and calcareous layers, are situated below the Portland oolite. **1818** W. Phillips *Selection of Facts Geol. Eng. & Wales* 61 The shale is so bituminous as to form a coaly matter of considerable extent near Kimmeridge, whence it is termed the Kimmeridge coal. **1830** *Phil. Mag.* VII. 198 The manner in which the fossils are disposed in the Kimmeridge clay at Havre. **1855** *Q. Jrnl. Geol. Soc.* XI. 125 The present example from the Kimmeridge shales appears to be a distinct species. **1860** R. Damon *Handbk. Geol. Weymouth* 54 From the coast of Dorsetshire the Kimeridge clay extends inland, with little interruption, to the coast of Norfolk. **1923** L. D. Stamp *Introd. Stratigr.* xiv. 230 In the upper part is a band of oil-shale ('Kimmeridge Coal'). **1931** Gregory & Barrett *Gen. Stratigr.* x. 155 The Kimeridge Clay varies in thickness from 100 ft. in the Midlands to 1250 ft. in the Wealden bore near Hastings. **1946** L. D. Stamp *Britain's Struct.* xii. 135 Almost everywhere the Kimeridge Clay coincides with low ground. **1967** D. H. Rayner *Stratigr. Brit. Isles* ix. 303 The thickest sequences of Kimeridge Clay come from Wessex and the Weald;..near the type locality of Kimeridge Bay in east Dorset 1,650 feet are known. **1970** *Oxford Times* 12 June 14 This field was mostly Kimmeridge clay, with some sand at the bottom, and not so intractable as the Oxford clay.

Kimmeridgian, *a.* (s.v. Kimmeridge). Add: Also **Kimeridgian.** [ad. F. *kimméridgien* (A. d'Orbigny *Paléont. française. Terrains Jurassiques* (1842–9) I. 610), replacing earlier *kimméridien* (J. Thurmann *Essai sur les Soulèvemens jur. du Porrentruy* (1832) I. 11).] Hence, of or pertaining to this subdivision (which is a stage of the Upper Jurassic next below the Portlandian). Also *absol.* (Earlier and later examples.)

kimzeyite (ki·mzi,ait). *Min.* [f. the name of the *Kimzey* family (see quot. 1958) + -ITE[1].] A dark brown zirconian garnet.

1958 MILTON & BLADE in *Science* 6 June 1343/3 This mineral..is here named 'kimzeyite'. The Kimzey family has been actively associated with mineralogical developments in Magnet Cove for almost a century... Notably William J. Kimzey, his son Joe Kimzey, former state geologist of Arkansas, and Lawton D. and John Kimzey. **1960** *Bull. Geol. Soc. Amer.* LXXI. 1930 Kimzeyite, $Ca_3(Zr, Ti, Mg, Fe'', Nb)_2 (Al, Fe''', Si)_3 O_{12}$, is a new type of garnet occurring as dodecahedrons modified by trapezohedron [*sic*] at Magnet Cove, Arkansas. **1967** *Amer. Mineralogist* LII. 773 The garnet compositions synthesized include the end-compositions..$Ca_3Zr_2Fe_2SiO_{12}$ (kimzeyite).

or relish, on such an occasion. **1905** W. B. *Where White Man Treads* 17 If his kinaki (relish) were fish, his predatory instincts invented the means to catch them. **1921** H. GUTHRIE-SMITH *Tutira* x. 70 This man's body..was eaten as a relish—*kinaki*—with the fern-root. **1949** P. BUCK *Coming of Maori* (1950) II. i. 110 Less difficulty was experienced in providing the meat complement, or *kinaki*, to go with the vegetable foods.

kinase (kəi·nēⁱz, -s). *Biochem.* [f. Gr. κιν-εῖν to move + *-ASE.] **a.** Any substance which converts an inactive precursor (a zymogen) into an active enzyme.

1902 W. H. THOMPSON tr. *Pawlow's Work of Digestive Glands* ix. 160 When a tube was introduced into the fistula, and the succus entericus afterwards collected in separate portions, the amount of kinase in the secretion became steadily less and less. **1903** *Jrnl. Chem. Soc.*

kindergartens are established in factories. **1968** *Harrods Christmas Catal.* 22/1 A kindergarten toy which is lots of fun.

kindergartener, also, a pupil at a kindergarten.

1919 G. A. MILLER *Prowling about Panama* ix. 129 (*caption*) Happy kindergartners, Panama. **1937** D. ALDIS *Time at her Heels* v. 103 The procession would end with the kindergarteners, looking very bustling and important as they hurried towards the little red chairs reserved for them. **1967** *Boston Sunday Herald* 26 Mar. I. 35/7 There will be no school sessions for currently enrolled kindergarteners on that day. **1973** *Jrnl. Genetic Psychol.* CXXII. 255 Eighty children, 40 kindergarteners and 40 third-graders.

2. *transf.* The nickname given to a group of young men with imperialist ideals who were recruited by Alfred, Lord Milner (1854–1925), High Commissioner of South Africa, to aid with reconstruction work after the South African war of 1899–1902. Freq. *Milner's kindergarten.*

1902 *Cape Times* 12 Sept. 6/10 Lord Milner described the nature of the government that was to be set up, which was a Council. He (Mr. Merriman) wondered what sort of rag-tag and bob-tail they would have got to sit on that Council... What did he expect to set up? A sort of kindergarten of young Balliol men—laughter—to govern this great country. **1913** W. B. WORSFOLD *Reconstruction New Colonies* I. x. 259 Mr Curtis was one of the most flagrant examples of Lord Milner's 'kindergarten'. **1958** *Spectator* 22 Aug. 244/3 The ideas of Milner and his dedicated 'kindergarten' about the treatment of the African. **1958** *Listener* 6 Nov. 739/3 Those were the days when the group of Englishmen who helped to bring order out of chaos and to prepare the way for the future came to be nicknamed Lord Milner's 'kindergarten'. **1971** *Oxf. Hist. S. Afr.* II. vii. 331 Locally he appointed young Oxford graduates—'Milner's kindergarten'—to the administrative positions. *Ibid.* 346 To Selborne and the members of the Kindergarten..unification was desirable.

‖ **kinder, kirche, küche** (ki·ndəɹ, ki·ɾχʏə, kü·χʏə). Also in different order. [G.] 'Children, church, kitchen'; a phrase, freq. used ironically, to denote the interests and preoccupations of a housewife.

1935 D. L. SAYERS *Gaudy Night* xxii. 453 The Nazi doctrine that woman's place in the State should be confined to the 'womanly' occupations of *Kinder, Kirche, Kuche.* **1953** E. SIMON *Past Masters* III. iii. 155 Good heavens! *Kinder, Kirche, Küche*, however you pronounce it. Still..at least you haven't given up smoking. **1963** *Economist* 10 Aug. 519/2 Her countrywomen are unhappy housewives trapped in the home by an ideal of *Kinder, Küche, Kirche* which is vigorously pressed by American educators,..and husbands. **1965** R. RENDELL *To fear Painted Devil* x. 124 He was an awfully *kinder, küche, kirche* sort of person, house-proud, passionately neat and tidy. **1968** R. HARRIS *Nice Girl's Story* xv. 112 As a wife, I should expect you to make it [*sc.* the bed]—*kinder, kirche, küche*, you know. **1970** W. GARNER *Puppet-Masters* v. 40 He subscribed wholeheartedly to Hitler's womanly ideal of *Kinder, Küche, Kirche.* **1970** G. GREER *Female Eunuch* 303 The women students..had not yet formed any clear idea of the disabilities which increasingly encumber women as they move..towards *kinder* and *küche.*

kinderspiel (ki·ndəɹʃpīl). [G.] A dramatic piece performed by children.

1902 *Daily Chron.* 19 Dec. 5/2 An opera..and a kinderspiel are being rehearsed. **1930** *Aberdeen Press & Jrnl.* 28 Feb. 3/2 'Blossom Time', a pretty kinderspiel, was part of a delightful entertainment given..by the Sunday School children.

kindhea·rtedly, *adv.* [-LY[2].] In a kind-hearted manner.

1900 H. C. BEECHING in *Monthly Rev.* Nov. 91 The brass lectern, which the good sister..kindheartedly uncovered for him. **1957** B. SPOCK *Common Sense Bk. Baby & Child Care* (new ed.) 148 Whenever her child has a crying spell in the daytime or wakes at night, she kind-heartedly makes another bottle for him. **1972** A. W. READ in *Current Trends in Linguistics* X. 588 Wentworth (1931) complained that *kindheartedly* was missing from all dictionaries, though he knew it from all sources; it was thereupon inserted into the Merriam–Webster of 1934.

kindred, *sb.* and *a.* Add: **B. 2.** Esp. in phr. *kindred spirit* (see quot. 1950).

1849 GEO. ELIOT *Let.* 13 Sept. (1954) I. 307 She says, 'You won't find any kindred spirits at Plongeon, my dear.' **1898** E. HOWARD *To-Morrow* viii. 83 They can see their way to join with a sufficient number of kindred spirits. **1950** PARTRIDGE *Dict. Clichés* (ed. 4) 124 *Kindred spirit, a*, a person like another in character and temperament.

kindy (ki·ndi). Austral. and N.Z. colloq. abbrev. of KINDERGARTEN.

1966 G. W. TURNER *Eng. Lang. Austral. & N.Z.* viii. 173 If a child eats his vegies as he should, he is soon ready for kindy. **1968** *Wanganui* (N.Z.) *Photo News* 8 June 78 (*caption*) Kindy Ball. Guests of honour at the recent Wanganui Free Kindergarten Ball. **1973** *Courier-Mail* (Brisbane) 4 June 15/6 (*heading*) Off to kindy..at home. A new scheme under consideration by the State Education Department will provide children in isolated areas with a pre-school education... A scheme for 'kindy by correspondence' would be available by the start of 1974. **1975** *Telegraph* (Brisbane) 28 Feb. 12 Self-help is basis for special kindy.

kine (kəin), *sb.*[2] *Linguistics.* [Back-formation from *KINE(SICS.] An isolable element of body movement or gesture made in non-vocal communication. Also *attrib.*

1952 R. L. BIRDWHISTELL *Introd. Kinesics* (U.S. Dept. State, Foreign Service Inst.) 15 Most users of kinesic material will not be able to record and analyze every kine played by the actor in any given situation. **1955** *Etc.: Rev. Gen. Semantics* XIII. I. 13 The least isolated particle with discriminational meaning we call a *kine*. **1965** OSGOOD & SEBEOK *Psycholinguistics* iv. 84 A particular motion or posture of a given part of the organism (facial or bodily) is called a *kine* (equivalent to *phone*). *Ibid.* 85 Various 'minimal pairs' of kine patterns (for example, variations in eyebrow position with the rest of the facial pattern constant).

kine-. Most words of Greek etymology beginning with *kine-* are now commonly pronounced with (i).

kine-[2] (ki·ni), var. *CINE (reverting to the Gr. initial κ).

1899 *Daily Chron.* 31 Aug. 3/2 The British Museum authorities have made arrangements for the safe custody of kinenegatives dealing with events of national importance. **1923** *Chambers's Jrnl.* 603/2 The kinegraph registers the short intake of the breath marking his embarrassment. **1927** *Bulletin* 12 Aug. 14/2 An enthusiast for the kine camera. **1928** *Daily Express* 28 Mar. 13 He has turned the music-hall into a home of kine-variety. **1959** *Listener* 3 Sept. 356/2 Kine-recordings are used in large numbers to help distribution.

kinema (ki·nīmă, kəinī·mă). Variant of *CINEMA with initial *k* from the Greek original. Now *rare.*

1914 *Evening News* 29 Sept. 4/5 It was my first step in the path of the kinema actor. **1921** *19th Cent.* Apr. 672 Properly handled, the Kinema could be made to endear the two races to one another. **1928** *Western Morning News* 28 Dec., The new kinema on the site of the old Post Office at Totnes. **1942** *Electronic Engin.* XIV. 708/1 The commercialisation of the sound film in the kinema over twelve years ago. **1954** A. CORNWELL-CLYNE *3-D Kinematogr.* 8 Commercial kinemas had only exhibited coloured short films which had to be viewed through spectacles. **1974** *Listener* 22 Aug. 252/1 A period's grandest buildings..indicate a community's current concern: medieval castles..1920s Kinemas.

b. *attrib.* and *Comb.*, as *kinema-camera, drama, film, producer, projection, theatre*; **Ki·nemacolo(u)r,** proprietary name of a method of producing motion pictures in the natural colours by means of revolving colour screens.

1927 *Manch. Guardian Weekly* (Suppl.) 2 Dec. p. xvi/2 The kinema-camera. **1909** *Trade Marks Jrnl.* 19 May 830 Kinema Color... Kinematographic apparatus and photographic films bearing finished pictures in natural colours for use therewith. Charles Urban,..Wardour Street, London; manufacturer. **1909** *Daily Chron.* 3 June 7/2 'Kinema-color', or animated scenes in nature's actual tints. **1914** *Times* 29 Jan. 4/3 These lectures might perhaps be illustrated by kinemacolour photographs. **1918** H. CROY *How Motion Pictures are Made* 288 By the Kinemacolor process colored motion pictures were made of the Coronation. **1949** R. LOW *Hist. Brit. Film 1906–14* iii. 101 As the sheer novelty of colours wore thin the company perforce became more ambitious and in September 1910 it was announced that the first drama in Kinemacolor was being filmed. **1916** A. BAKSHY *Path Mod. Russ. Stage* 221 The kinema-drama raises some of the most fundamental problems of art. **1915** *Truth* 6 Oct. 567/1 A levy of 1d. per foot on all imported kinema film. **1929** *Melody Maker* Apr. 418/2 A mixed programme of kinema films and variety acts. **1921** *19th Cent.* Apr. 672 The Kinema-producers in California. **1916** *Chambers's Jrnl.* 26 Feb. 207/1 [The lamp's] suitability for kinema projection. **1929** *Melody Maker* Apr. 418/2 In nearly all kinema theatres which have been converted from Music Halls. **1932** B. B. HAMPTON *Hist. Movies* xii. 257 Tally..sold his Kinema Theater and First National franchise to Gore Brothers and Lesser.

kinematic, *a.* Add: **b.** *spec.* in *Mech.* Applied to a set of mechanical elements so disposed in relation to each other that the relative position and motion of each is uniquely determined by the relative position and motion of the other(s).

1876 A. B. W. KENNEDY tr. *Reuleaux's Kinematics of Machinery* i. 43 A machine consists solely of bodies which thus correspond, pair-wise, reciprocally. These form the kinematic or mechanismal elements of the machine. *Ibid.*, If a kinematic pair of elements be given, a definite motion can be obtained by means of them if one of the two be held fast or fixed in position. *Ibid.* 46 The whole now forms a linkage returning upon itself... A combination of pairs of elements in this way we shall call a chain, or more fully a kinematic chain. **1915** R. F. MCKAY *Theory of Machines* viii. 92 When one of the links of the kinematic chain is fixed, the chain is called a mechanism. When the mechanism transmits force..it is called a machine. **1957** R. M. PHELAN *Fund. Mech. Design* i. 1 A chain..in which with one link fixed every point in every other link must move in a definite path is called a constrained or kinematic chain.

2. *kinematic viscosity*: see *VISCOSITY.

kinematically (kinīmæ·tikăli), *adv.* [f. KINEMATIC, -ICAL *adjs.*: see -ICALLY.] From the point of view of kinematics.

1876 A. B. W. KENNEDY tr. *Reuleaux's Kinematics of Machinery* vi. 234 Thus..the force-closed beam-chain became the imperfect but still kinematically far more complete 'parallel motion'. **1915** R. F. MCKAY *Theory of Machines* viii. 93 A direct-acting steam engine and a quick-return motion are kinematically identical, though their general appearance would not suggest such a connection. **1955** J. S. BEGGS *Mechanism* vi. 194 The kinematically equivalent chain thus obtained has all lower pairs, and Eq. (6–5) may be applied. **1971** D. W. SCIAMA *Mod. Cosmol.* xvi. 198 The rotation of the Galaxy can be detected not only dynamically from its flattened shape but also kinematically by observing the motions of stars..far out from the centre.

kinematics. Add: **b.** The kinematic features or properties of something. Const. as *sing.* or *pl.*

1955 J. S. BEGGS *Mechanism* ix. 273 The kinematics of the area wheel is shown in Fig. 9–11. **1973** *Sci. Amer.* Aug. 33/1 The kinematics of the decay required that the mass of the particle be very small, perhaps even zero.

kinematograph. Add: (Further examples.) Hence **kinema·tograph** *v. trans.*, **ki:nemato·grapher**, **ki:nematogra·phical** *a.*, **ki:nematogra·phically** *adv.*, **ki:nemato·graphy.** (Variants of the corresponding *cine-* forms.)

1896 *19th Cent.* July 135 The Kinematograph is already at more than one of them [*sc.* the music-halls], showing a stormy sea, the Thames at Waterloo Bridge, the race for the Derby. **1900** *Nature* 15 Feb. 384/2 Prof. R. W. Wood will exhibit..the Kinematographical Demonstration of the Evolutions of Reflected Wave-fronts. **1907** *Westm. Gaz.* 24 Aug. 6/3 Acting, Sir, is mere kinematography. What we require is something more static, reposeful, and intellectual. **1908** *Daily Chron.* 26 Sept. 7/2 Mr. Charles Urban during the past five months has enjoyed facilities to kinematograph the efforts made during this period to salve the Gladiator. **1911** *Chambers's Jrnl.* 412/1 Here, however, the scientific kinematographer has gone farther. **1925** *Daily Mail* 13 Apr. 6/5 One..important thing..is to get into the hiding-place unobserved by the birds which he happens to be kinematographically on the track of. **1954** A. CORNWELL-CLYNE *3-D Kinematogr.* 7 Colour kinematography and synchronized sound. **1969** T. H. GUBACK *Internat. Film Industry* v. 102 The Kinematograph Renters Society..brought action against the two Maltese films. **1971** (*title*) British kinematography sound and television.

kinematoscope. = *CINEMATOSCOPE. (*Disused.*)

1898 *Windsor Mag.* VIII. 113/1, I knew that conjurors were to be obtained there,..and the kinematoscope. **1926** *Encycl. Brit.* Suppl. II. 960/2 This machine was patented in the United States as the Kinematoscope Feb. 5 1861.

kineme (kəi·nīm). *Linguistics.* [f. Gr. κίνη(σις movement + *-EME.] A meaningful unit of body movement or gesture made in non-vocal communication.

1952 R. L. BIRDWHISTELL *Introd. Kinesics* (U.S. Dept. State, Foreign Service Inst.) 22 Phone: Allophone Phoneme. Kine: Allokine (? Kineme?). **1955** *Etc.: Rev. Gen. Semantics* XIII. I. 13 Each of these *classes* of allokines or variants within a range, we define as *kinemes*. **1965** OSGOOD & SEBEOK *Psycholinguistics* iv. 85 The second step in analysing any gestural 'language'—again, in parallel with linguistics—would be to determine what movements are significant in the code, i.e., what classes of kines constitute *kinemes* (equivalent to *phonemes*) by virtue of having the same significance. **1972** B. G. COOKE in T. Kochman *Rappin' & Stylin' Out* 34 In this paper the gestures of giving and getting skin shall be considered as *kinemes* according to Birdwhistell's classification.

kinemics (kəinī·miks). [f. *KINEM(E + -ICS.] (See quot. 1954.)

1953 J. B. CARROLL *Study of Lang.* 279 Kinemics. **1954** PEI & GAYNOR *Dict. Ling.* 115 *Kinemics*, the study of units of gestural expression.

kinescope (ki·nīskōᵘp). *Television.* [f. Gr. κίνη(σις movement + -SCOPE.] **1.** A cathode-ray tube specially constructed for use in a television set. Chiefly *U.S.*

The registration as a proprietary name was cancelled in 1950.

1932 *Official Gaz.* (U.S. Patent Office) 17 May 568/1 RCA Victor Company, Inc., Camden, N.J. Filed Feb. 6, 1931. *Kinescope* for Cathode Ray Tubes and Thermionic Tubes. **1933** V. K. ZWORYKIN in *Proc. IRE* XXI. 1656 The name 'kinescope' has been applied to the cathode ray tube used in the television receiver to distinguish it from ordinary cathode ray oscilloscopes because it has several important points of difference. **1949** B. GROB *Basic Television* ix. 144 The picture tube, or kinescope, has the funnel-shaped form and internal structure of a conventional cathode-ray tube. **1971** H. E. ENNES *Television Broadcasting* viii. 357 The kinescope upon which the image to be recorded is displayed is a special type of tube employing a flat face, with phosphor and brightness characteristics that allow use of inexpensive film.

2. A film recording made from a television broadcast.

1949 *Richmond* (Va.) *News Leader* 25 Oct. 30/1 Kinescope or television recording, the process of filming a television show off the receiving tube. **1953** *Manch. Guardian Weekly* 2 July 15/1 On June 2 a United States television network transmitted a kinescope version of the B.B.C.'s television of the Abbey Coronation service. **1957** *Economist* 19 Oct. 226/1 It buys a programme or series of programmes from, say, station KETC in St Louis, and

then makes it available to all the other ETV outlets on kinescope. **1957** P. FRANK *Seven Days to Never* ii. 20 He took a course in American television and radio. Every Tuesday afternoon he listened to recordings, and watched kinescopes, of the most popular programs. **1964** L. A. WORTMAN *Closed-Circuit Television Handbk.* v. 93 The major advantage in kinescope recording is that the final film can be shown wherever there is a 16-mm projector.

Hence **ki·nescope** *v. trans.*, to make a kinescope of; **ki·nescoped** *ppl. a.*, reproduced from a kinescope recording; **ki·nescoping** *vbl. sb.*

1949 *Life* 17 Oct. 75 Each show is kinescoped (filmed and sound-tracked) and re-telecast from stations in the rest of the country. **1961** P. LEWIS *Educ. Television Guidebk.* iv. 70 (*heading*) Kinescoping possibilities. **1964** L. A. WORTMAN *Closed-Circuit Television Handbk.* v. 93 One can usually recognize a kinescoped telecast by virtue of its visual quality. **1967** *Telegraph* (Brisbane) 3 Mar. 12/1 The only way then to record a show was by a film technique called 'kinescoping'... At that time the quality was diabolical.

kinesic (kəinī·sik), *a. Linguistics.* [f. Gr. κίνησ(ις movement + -IC.] Of or pertaining to communication effected non-vocally through movements or gestures. Hence **kine·sically** *adv.*

1952 R. L. BIRDWHISTELL *Introd. Kinesics* (U.S. Dept. State, Foreign Service Inst.) 15 It is suggested that any student beginning kinesic recording work on but one part of the body at a time. **1952** *Rep. 3rd Ann. Round Table Meeting Ling. & Lang. Teaching* (Georgetown Univ. Inst. Lang.) 66 A rather thorough preliminary study of *kinesics* has indicated that the kinesic system can be analyzed and described. **1955** *Etc.: Rev. Gen. Semantics* XIII. I. 18 He was kinesically more 'mature' than the other boys. **1959** *College English* XX. IV. 172 Kinesic and paralinguistic phenomena constitute separate patterned systems, which differ in their structure from culture to culture. **1964** CRYSTAL & QUIRK *Prosodic & Paralinguistic Features Eng.* ii. 18 Those kinesic phenomena which are unintended by the individual. **1965** L. PEDERSON in *Lang. Programs for Disadvantaged* (U.S. Nat. Council Teachers of English) 247 Most speech occurs in the form of dialogue with two or more participants actively cooperating at the structural, paralinguistic, kinesic, proxemic, and haptic level. **1967** L. THAYER *Communication* 61 Inspection of the working transcript of the linguistically and kinesically recorded data revealed repetitive and apparently systematic body behaviors. **1968** *Amer. Speech* XLIII. 202 There are many conventionalized kinesic systems, each with its own hierarchy. **1972** J. L. DILLARD *Black English* v. 203 It is quite believable..that gestural and kinesic cues might 'give away' a member of the Negro subculture to people who knew that culture.

kinesics (kəinī·siks). *Linguistics.* [f. Gr. κίνησ-ις movement + -ICS.] The study of those body movements and gestures by which, as well as by speech, communication is made; body movements and gestures which convey meaning non-vocally.

1952 R. L. BIRDWHISTELL (*title*) Introduction to kinesics: an annotation system for analysis of body motion and gesture. **1955** *Etc.: Rev. Gen. Semantics* XIII. I. 12 Kinesics may be defined as the systematic study of the visually sensible aspects of non-verbal interpersonal communication. **1957** *Psychiatry* XX. 74 Kinesics, or gestures and motions, are not instinctive human nature but are learned systems. **1958** A. A. HILL *Introd. Ling. Struct.* xxi. 409 The vocal qualifiers together with kinesics make up the paralinguistic system. **1970** J. FAST *Body Lang.* (1971) i. 11 To understand this unspoken body language, kinesics experts often have to take into consideration cultural differences and environmental differences. **1972** W. M. AUSTIN in A. L. Davis *Culture, Class, & Lang. Variety* viii. 152 Kinesics is the study of body posture, tonus, and movement in man and the other animals.

kinesimeter (kəinisi·mītəɹ). [f. KINESI- + -METER.] An instrument for investigating the properties of different areas of the skin, by which a movable point whose position can be measured may be applied to the surface with a known force.

1885 H. H. DONALDSON in *Mind* X. 402 This machine was devised by Prof. [G.] Stanley Hall, and will be described in a forthcoming paper, under the name of the 'Kinesimeter'. **1901** E. B. TITCHENER *Exper. Psychol.* I. II. 145 (*caption*) Arm-rest, designed for use with kinesimeter.

kinesiology (kəinīsi,ǫ·lǒdʒi). [f. KINESI- + -OLOGY.] The field of study concerned with the mechanics of (human) bodily movement.

1894 N. POSSE *Special Kinesiology* p. v, I have deemed it desirable to change the title into *Special Kinesiology*,—it being a treatise on the mechanics, effects, and classification of special exercises. **1936** *Nature* 14 Mar. 438/2 This book [sc. *Mechanics of Normal and Pathological Locomotion in Man*] should take a prominent place in the literature of kinesiology. **1941** *Res. Q. Amer. Assoc. Health & Physical Educ.* XII. 163 While kinesiology borders on the field of several sciences, its greatest practical interest is in the field of physical education. **1963** M. G. SCOTT (*title*) Analysis of human motion: a textbook in kinesiology.

Hence **kinesiolo·gic, -lo·gical** *adjs.*, **kinesio·logist**, a person who studies kinesiology or kinesics.

1941 *Res. Q. Amer. Assoc. Health & Physical Educ.* XII. 167 He presents a concise kinesiological analysis. *Ibid.* 165 The importance of the part he [*sc.* Aristotle] played as a founder of the study of the mechanics of movement is hardly realized by many contemporary kinesiologists. **1950** K. F. WELLS *Kinesiology* xvi. 350 (*heading*) The kinesiologic analysis of a movement. **1952** R. L. BIRDWHISTELL *Introd. Kinesics* (U.S. Dept. State, Foreign Service Inst.) 3 The term *social* kinesiologist has been selected as the term for one attempting to analyze systematically the data covered by kinesic investigation. *Ibid.* 15 This is particularly difficult for the average American mover when working with the kinesiological systems of other Americans. **1965** OSGOOD & SEBEOK *Psycholinguistics* iv. 85 Kinesiologists require training in objective looking—the untrained observer will be likely to perceive only those movements which are significant in his own 'language'.

kinesis (kəi-, kinī·sis). Pl. **kineses.** [mod.L., f. Gr. κίνησις motion.] † **1.** *Cytology.* Karyokinesis, mitosis. *Obs. rare.* (Cf. *KINETIC *a.* 4.)

1904 *Jrnl. R. Microsc. Soc.* 529 At the first metaphase there is a second division (? longitudinal) which appears preparatory to the second kinesis. **1906** *Ibid.* 282 The two constitutive branches of the definitive chromosomes..separate from one another in each chromosome at the first kinesis.

2. *Biol.* [Adopted (in G.) by W. Rothert 1901, in *Flora* LXXXVIII. 374, after its use as a suffix in *photokinesis*.] An undirected movement of an organism that occurs in response to a particular kind of stimulus.

1905 *Jrnl. Compar. Neurol.* XV. 139 Kinesis is a term which seems to have been first used by Engelmann for the increase or decrease of activity produced by certain agencies. The fact that certain bacteria increase or decrease movement in the light he called photokinesis. Rothert accepted the term kinesis for such changes in the amount of activity produced by chemicals, calling this chemokinesis. **1940** FRAENKEL & GUNN *Orientation of Animals* i. 10 The term *taxis* is to-day used for directed orientation reactions... Undirected locomotory reactions, in which the speed of movement or the frequency of turning depend on the intensity of stimulation, we call kineses. **1955** STORER & USINGER *Elem. Zool.* xxiii. 394 Taxes and kineses..enable insects and many other animals to find and inhabit the small environmental niche or microclimate in which each kind is most successful. **1960** L. PICKEN *Organization of Cells* x. 450 The formation of cell aggregates is undoubtedly favoured by a non-directional movement of the cells, by a kinesis, with thigmotaxis taking over once contact is established.

kinetheodolite (kinīþīǫ·dǒləit). Also (with hyphen) **kine-theodolite.** [f. Gr. κίνη(σις movement + THEODOLITE.] A telescope used to follow the path of a projectile, aircraft, or the like, and mounted so that its elevation and azimuth angles are indicated.

1941 *Illustr. London News* CXCVIII. 380/2 The girls work with two kine-theodolites set far apart and connected at the firing position by a central control post which photographs the shell-bursts. **1946** *Jrnl. R. Aeronaut. Soc.* L. 925/1 Bombs of differing weights were dropped... They were followed visually by kine-theodolites, which gave the flight path. **1958** *New Scientist* 21 Aug. 655/1 The contributions..on the tracking of the Russian satellites..were quite outstanding. Thus the data obtained using kine-theodolites by scientists of the Ministry of Supply were the most accurate optical observations reported. **1961** *Engineering* 26 May 750/3 Woomera has a number of double kinetheodolite sites, one instrument tracking the missile, the other the target aircraft. **1965** K. J. TURNER in M. A. Perry *Flight Test Instrumentation* III. 229 The model trajectories are recorded by..kinetheodolites.

kinetic, *a.* Add: **2.** *kinetic heating*, heat generated by the compression and acceleration of air by a fast-moving body.

1954 *Aircraft Engineering* XXVI. 138/3 Kinetic heating is caused by two related phenomena; first the adiabatic compression of the air as it is brought up to the velocity of the leading edge or nose of the body and second the frictional heating which takes place in the boundary layer as the air adjacent to the surface tends to accelerate the air, through which the plane is passing, up to the flying speed. **1959** L. BROGLIO in *Adv. Aeronaut. Sci.* I. 216 The design of an aircraft or missile is often ruled by the transient-temperatures of its external surface, due to kinetic heating and radiation. **1970** *Progr. Aerospace Sci.* XI. 64 Another effect which occurs in flight..concerns the influence of kinetic heating.

b. *Biol.* Of or pertaining to a kinesis (sense *2).

1905 *Amer. Naturalist* XXXIX. 167 It has been found convenient to distinguish between two factors in the effect of light on Drosophila, a kinetic effect and a directive one. **1914** *Jrnl. Exper. Zool.* XVII. 273 Animals, such as the blowfly larva, which respond phototactically to horizontal light show a simple kinetic response when subjected to uniform illumination from above. **1941** S. O. MAST in Calkins & Summers *Protozoa Biol. Res.* v. 277 Kinetic responses.—If an amoeba is kept for some time in very weak light it becomes inactive; if the light is then increased, the organism gradually becomes active again.

c. *transf.* and *fig.*, esp. active, dynamic, full of energy.

1931 H. G. WELLS *Work, Wealth & Happiness of Mankind* (1932) ix. 382 In every preceding phase where there has been a concentration of wealth it has been far less easily converted into kinetic purchasing power. **1934** —— *Exper. Autobiogr.* II. ix. 658 For the purposes of the

1846 J. G. SAXE *Progress* (1847) 28 How would she [*sc.* the Muse] strive, in fitting verse, to sing The wondrous Progress of the Printing King! 1847 E. D. BANCROFT *Lett. from England* (1904) 113 We both went to a concert at Mr. Hudson's, the great railway 'king', who has just made an immense fortune from railway stocks. 1919 F. HURST *Humoresque* 194 You've never met Mr. Feist, have you, the film king? You two ought to get acquainted—one makes the films and the other makes them famous. 1927 WODEHOUSE *Small Bachelor* ii. 25 She had also been the relict of the late P. Homer Horlick, the Cheese King, and he had left her several million dollars. 1966 'J. HACKSTON' *Father clears Out* 83 He began to tell me what the sheep kings had to put up with. *Ibid.* 123 The young sheep king.

b. *King Willow*, the game of cricket. Cf. quot. 1876 s.v. WILLOW *sb.* 5.

1933 A. G. MACDONELL *England, their England* xvii. 285 The evening papers were already beginning to talk of the Advent of King Willow. 1936 S. R. JONES *Eng. Village Homes* iii. 36 Football thrives lustily, and King Willow reigns on the old turf of the greens in summer. 1972 P. DICKINSON *Lizard in Cup* vi. 92 Loyalty to..the imagined spirit of King Willow.

7. a. *King of Six*, the male of certain polygamous South African birds.

1913 C. PETTMAN *Africanderisms* 260 *King of six*, a King Williamstown name for the Rooibekje...The reference is to the number of females by which the male is generally accompanied during the breeding season. 1931 R. C. BOLSTER *Land & Sea Birds S.-W. Cape* 133 Of the two Bishop Birds, the Black and Yellow one is said to be polygamous, whence the name 'King of Six' in the vicinity of Cape Town.

11. c. (Usu. with capital initial.) The British national anthem, 'God Save the King'.

1932 *Week-end Rev.* 30 Apr. 554/2 Programme to-night as follows:—British Movietone News. Sunshine Susie. Mickey Mouse. The King. 1959 I. JEFFERIES *Thirteen Days* vii. 95 The band played the King and we all stood up. 1967 R. HARRIS *All my Enemies* iii. 34 We applauded, stood for 'The King'.

12. a. *King-Emperor, -Sovereign.*

1902 *Westm. Gaz.* 27 Feb. 11 The King-Emperor is honoured among us [*sc.* Americans] because he stands for the great people whom he rules. 1971 R. RUSSELL tr. *Ahmad's Shore & Wave* xv. 159 'Have you ever attended the King-Emperor's levée?' asked the Diwan Bahadur. 1908 H. H. JOHNSTON *George Grenfell & Congo* I. xx. 448 The Governors-General or heads of departments representing the King-Sovereign in Africa.

13. a. king-carp, a variety of the common carp, *Cyprinus carpio*; **King Country** *N.Z.*, an extensive region in the North Island of New Zealand formerly allotted to the Maoris under a king; so *King Movement, Party*, etc., referring to the followers of this king; **king-hit** *Austral. slang*, (*a*) a knock-out blow; a hard punch; (*b*) a fighter or bully; a leader; hence as *v. trans.*, to punch hard or knock out; **king-list,** a list of the names of kings; **king mackerel,** a game fish of the eastern U.S. coast, *Scomberomorus cavalla*, also called Spanish mackerel or king-fish; **king salmon** *N. Amer.*, the Chinook or quinnat salmon, *Oncorhynchus tshawytscha* (earlier and later examples); **king-side** *a. Chess*, made or done on the king's side of the board; also applied to men situated on that side; **king-size** *a.*, of an extra large size; of larger size than normal; *spec.* designating an extra large cigarette; hence *ellipt.* as *sb.*, a king-size cigarette; also *king-sized* adj.

1908 *Westm. Gaz.* 7 Aug. 10/3 Yesterday a king carp was hooked. 1930 E. PARKER et al. *Fine Angling for Coarse Fish* 162 A very heavy king-carp..weighed 18½ lb. This fish is rarer than the common carp. 1971 B. J. MUUS *Freshwater Fish Brit. & Europe* 136/1 Scaled carps (often known as king carps) are covered by small uniform scales. 1884 J. H. KERRY-NICHOLLS (*title*) The King Country; or, explorations in New Zealand, a narrative of 600 miles of travel through Maoriland. 1910 J. COWAN *Maoris of N.Z.* xxvii. 294 The men..finally faced death..in the famous redoubt at Orakau, on the borders of what afterwards came to be known as the King Country. 1917 G. H. SCHOLEFIELD *New Zealand* iii. 19 A railway ran from Auckland..until it touched..the boundary of the native preserve known as the 'King Country', which stretched like a neutral zone across the island. 1944 A. MULGAN *From Track to Highway* ii. 58 In the end he took refuge in the King country, below the military frontier in the Waikato, where the defeated Kingites had been left unmolested. 1966 *Encycl. N.Z.* II. 223/2 The King Country, or Rohe Potae, was originally a large tract of the western central North Island... Europeans called the area 'the King Country' because it was here that Tawhiao sought refuge following the Maori Wars. 1923 G. COLLINS *Valley of Eyes Unseen* i. 29 Neither blow was a true king hit, however, and neither Chink was anything near knocked out. 1941 BAKER *Dict. Austral. Slang* 41 *King hit*, a knock-out blow. (2) As for 'king dick' [= a leader, boss]. 1944 L. GLASSOP *We were Rats* xiii. 76 'Do this galah over,' he whispered in my ear. 'He's a king-hit merchant.' 1945 BAKER *Austral. Lang.* vi. 120 To bump, comb down,..king hit. 1962 S. GORE *Down Golden Mile* 277 'King-hit me, the bastard,' he muttered. 'With me own gun.' 1970 *Sunday Truth* (Brisbane) 16 Aug. 32/6 You king-hit him with what appears to be savage brutality. 1974 *Sunday Mail* (Brisbane) 29 Sept. 28 D/2 A piece of legislation that has been described as..a king-hit to human rights. 1914 E. A. W. BUDGE *Short Hist. Egyptian People* iii. 27 The famous King-List drawn up for Seti I, and cut upon a wall in a temple built by him at Abydos. 1962 J. GRAY *Archaeol. & Old Testament World* ii. 34 In the king-lists the 'Flood' demarcates between historical dynasties and the early ages. 1939 J. O. LA GORCE *Bk. Fishes* (ed. 2) 338/2 King mackerel serves as Spanish, and bonito may be served as either Spanish or King mackerel. 1953 F. ROBB *Sea Hunters* vii. 104 Old Drum Watts says these aren't barracouta at all—says they're king-mackerel or something. 1965 A. J. MCCLANE *Standard Fishing Encycl.* 449/2 King mackerel reach a much larger size than any other American Spanish mackerel. 1970 M. SLATER *Caribbean Cooking* 11 *Kingfish*, a game fish, sometimes called King Mackerel which can weigh up to 100 lb. 1858 King movement [see *HUI]. 1860 T. BUDDLE *Maori King Movement* 3 This chief..initiated a Maori King movement in the South. 1884 J. H. KERRY-NICHOLLS *King Country* 6 In 1854,.. Te Heuheu ..summoned a native council at Taupo, when the King movement began in earnest. 1959 K. SINCLAIR in J. E. Gorst *Maori King* p. xxiii, The King movement survives today, though its followers are less numerous than a century ago. 1860 T. BUDDLE *Maori King Movement* 72 It becomes..the duty of those entrusted with native interests..to enter promptly into negociations with the King party. 1944 A. MULGAN *From Track to Highway* ii. 55 The King party wanted to keep their king and their flag. 1881 *Amer. Naturalist* XV. 177 These species [in the North Pacific] may be called the quinnat or king salmon. 1959 *Vancouver Sun* 28 Aug. 5/1 Fall is also in the return of the salmon to their rivers—not in the early king salmon runs that come to a few rivers in May. 1941 F. REINFELD *Keres' Best Games of Chess* 86/1 Not only winning a Pawn, but devaluating the remaining Black King-side Pawns. 1954 H. GOLOMBEK *Game of Chess* 12 In the case of King side castling, the King is moved two squares to the right. 1973 *Times* 13 Aug. 12/7 Hartston won quickly with a strong king-side attack against Mestel. [1825 J. CONSTABLE *Let.* 10 Dec. in *Corr.* (1964) II. 419 Sir Thomas has done 4 pictures in Paris—the two of the King & Dolphin are very large King size, & fine—the others, are the Dolphiness and Duchess of Berry head size.] 1942 *Time* 7 Sept. 18/1 (Advt.), Regent Cigarettes, King Size. *Ibid.*, King Size Regent's the cigarette for *moderns* like you. 1949 *Sun* (Baltimore) 1 July 2/1 He..allowed Stryker to read the king-size question summarizing the life of Whittaker Chambers. 1957 H. ROOSENBERG *Walls came tumbling Down* iv. 88 We..had a king-size meal. 1957 P. WILDEBLOOD *Main Chance* 159 'Have you got a cigarette?' 'King-size,' said Mrs. Tull. 'Oh, I can't stand them big ones.' 1966 King size [see *ECONOMY 9]. 1971 *New Scientist* 17 June 707/2 It's a pity that Rudolph de Salis..never had the advantage of his king-size burgers. 1971 WODEHOUSE *Much Obliged, Jeeves* xiii. 133 The snag which had raised its ugly head was one of formidable—you might say king-size—dimensions. 1943 in *Amer. Speech* (1944) XIX. 111/2 So that's it for king-sized beauties. 1953 *Manch. Guardian Weekly* 24 Apr. 7 A kingsized defence programme.

c. King Edward (VII potato), an oval variety of potato with a white skin mottled with red, introduced in 1902 by J. Butler; **king fern,** (*b*) *N.Z.*, a large fern, *Marattia salicina*, with a swollen, starchy rhizome; (*c*) *Todea barbara*, a fern closely related to the royal fern, *Osmunda regalis*, found in Australia, New Zealand, and South Africa.

1926 R. N. SALAMAN *Potato Varieties* xxvi. 274 (*heading*) King Edward VII. *Ibid.* 275 King Edward, whose parentage is unknown, was raised by a gardener in Northumberland... This variety is today the most popular in England. 1949 —— *Hist. & Social Influence Potato* x. 170 The 'King Edward' became a favourite in the kitchen. 1963 *Times* 22 Apr. 2/6 The three most popular potato varieties grown in England and Wales are all relatively old. Arran Pilot..was introduced just over 40 years ago. Majestic..was introduced in 1912 and King Edward VII, with about 25 per cent [of the maincrop acreage], in 1902. 1911 W. R. GUILFOYLE *Austral. Plants* 354 *Todea barbara*. 'King Fern' or 'Swamp Sponge Fern'. 1921 H. B. DOBBIE *N.Z. Ferns* (ed. 2) xxix. 374 *M[arattia] fraxinea* (like an ash leaf). 'Para', 'King Fern', 'Horseshoe Fern'. The largest herbaceous fern in New Zealand; plentiful in the early days, now becoming scarce. 1962 J. H. WILLIS *Handbk. Plants Victoria* I. 10 *T[odea] barbara*...Austral King-Fern (King Fern)..all States except W.A. (but very localized in S.A.), N.Z., S. Afr. 1963 B. PEARSON *Coal Flat* xxii. 379 The three of them huddled under a king fern.

14. king's blue, a shade of blue (see quots.); a substance giving that colour; **King's messenger:** see MESSENGER 3; **king's peg,** a drink consisting of brandy and champagne; **King's (National) Roll,** a roll of employers pledged to employ at least a fixed proportion of disabled ex-service men after the war of 1914–18.

1908 C. MAYER tr. *Zerr & Rübencamp's Treat. Colour Manuf.* II. 200 The blue cobalt compounds known in commerce as smalt, king's blue, cobalt blue, [etc.]. 1951 R. MAYER *Artist's Handbk.* ii. 52 *King's blue*, cobalt blue; formerly smalt. 1970 *Canad. Antiques Collector* Oct. 17/1 Very little, if any, Bristol-blue glass was made between about 1800 and about 1820, when it again came into vogue under the name of king's blue. This was a loyal gesture to George IV. 1890 King's peg [see *HEIDSIECK]. 1899 C. J. C. HYNE *Further Adv. Capt. Kettle* xi. 265 Cranze kept up a steady soak on king's peg—putting in a good three fingers of the liqueur brandy before filling up the tumbler with champagne. 1958 M. PROCTER *Man in Ambush* xiii. 148 This was the shy man who drank champagne laced with brandy, a millionaire's drink... It was called King's Peg. 1919 HAIG in *Times* 11 Nov. 10/5, I..appeal to employers..to give a pledge of their sympathy by enrolling their names on the King's National Roll under the national scheme for the employment of disabled men. 1920 *Times* 16 Feb. 9/4 The King's Roll. First edition, with 9,500 firms, now in the press.

15. King Charles's head [with reference to Mr. Dick in Dickens's *David Copperfield* xiv], an obsession or fixed idea; **King James('s) translation** or **version,** the Authorized Version of the Bible (1611); also *King James*.

1882 W. HOW *Let.* in H. Barnett *Canon Barnett* (1918) I. xxii. 275 Like King Charles's head, there was no keeping you out. 1889 G. B. SHAW *London Music 1888–89* (1937) 124, I am afraid I shall have to drag in the subject of music rather often in this column. I know that it is my King Charles's head. 1929 C. MACKENZIE *Gallipoli Memories* xii. 198 And then, of course, he produced his King Charles's head, which was the landing at Bulair. 1972 F. M. LÓPEZ-MORILLAS in R. Highfield *Spain in 15th Cent.* xiv. 441 His King Charles' head was the role played on the first voyage by Martín Alonso Pinzón. 1973 *Times* 30 June 13/5 My own King Charles's head is the use of 'nerve-wracking' for 'nerve-racking'. 1835 *Penny Cycl.* IV. 374/2 The period of King James's translation. 1931 *Sunday School Times* (Philadelphia) 22 Aug. 458/1 A good English translation should be in good English idiom, and the old King James was that at least. 1932 *Jrnl. R. Anthrop. Inst.* LXII. 283 He made 'a serpent of brass', as the King James version says. 1973 *Sci. Amer.* Aug. 98/3 Add the number on the top of *A* to the number on the bottom of *B*, then find the chapter of Genesis (in a King James Bible) that corresponds to the sum.

king-bird. 1. For '*Paradisea regia*' substitute '*Cicinnurus regius*'. (Later example.)

1958 G. DURRELL *Encounters with Animals* II. 47 Here [*sc.* in a Brazilian zoo]..three king birds of paradise were living... The male is about the size of a blackbird, with a velvety orange head contrasting vividly with a snow-white breast and a brilliant scarlet back.

2. (Later example.)

1926 D. H. LAWRENCE *David* viii. 65 *Jonathan*:..Shall not the leader shine forth? *Saul*: Even so. And the young King-bird shall moult his feathers in the same hour.

3. Substitute for def.: One of several North American tyrant fly-catchers of the genus *Tyrannus*. (Earlier and later examples.)

1778 J. CARVER *Trav. N. Amer.* 475 The King Bird is like a swallow, and seems to be of the same species as the black martin or swift. 1801 *Massachusetts Spy* 25 Nov. 1/2 Just as a parcel of King-birds will pick at a Crow. 1959 VAN TYNE & BERGER *Fund. Ornith.* vii. 205 The Gray Kingbird (*Tyrannus dominicensis*) and the Black-whiskered Vireo (*Vireo altiloquus*)..leave Cuba to winter in South America.

kingdom, *sb.* Add: **4. d.** *to come (in)to one's kingdom*: to acquire authority, power, attractiveness, or the like. Cf. *Luke* xxiii. 42.

1892 KIPLING & BALESTIER *Naulahka* xviii. 211 Now we are come to our Kingdom..Little it profits us. 1930 L. G. MOBERLY *Eternal Dustbin* xiv. 194 That woman has come into her kingdom. 1973 R. RENDELL *Some lie & Some Die* iii. 30 Good luck. Remember me when thou comest into thy kingdom.

king-fish. Add: **2.** A leader, chief, boss; freq. used as a nickname for a particular person, notably for Huey Long (1893–1935), Governor and Senator from Louisiana. *U.S. slang.*

1933 HUEY P. LONG *Every Man a King* xxvii. 277 We from time to time termed various of our political enemies the 'Kingfish', most prominent of which was..a certain corporation lawyer. *Ibid.*, I am participating here anyway, gentlemen. For the present you can just call me the Kingfish. 1934 *Sun* (Baltimore) 21 Aug. 10/2 The Kingfish [*sc.* Huey Long] is ideally equipped for a dictator's rôle. 1939 *Ibid.* 17 July 11/1 King Levinsky, the Kingfish, who earned a fortune..during four spectacular years in the ring. 1946 *Richmond* (Va.) *Times Dispatch* 26 Dec. 1/5 Mr. Brown..is sometimes referred to as the 'kingfish' of City Council. 1968 *Word Study* Dec. 4/1 The term *kingfish* may be applied to an undisputed leader or master. It has been used as a personal appellation, self-applied I believe, to Huey Long, and to a character in the enduring radio show of Amos 'n' Andy.

kingfisher. Add: **3.** In full, *kingfisher blue*. A brilliant blue colour.

1922 *Daily Mail* 11 Dec. 1 (Advt.), All the leading colours including Ivory, Apricot, Jade, Kingfisher, [etc.]. 1956 G. DURRELL *My Family* v. 58 The sea smooth and opalescent, kingfisher-blue. 1970 *Observer* 18 Jan. 1/8 (Advt.), Both dresses in navy, deep oatmeal or kingfisher. 1971 R. RENDELL *One Across* vi. 55 A wool dress of brilliant kingfisher blue.

Kingite (ki·ŋəit). *N.Z.* [f. KING *sb.* + -ITE[1] 1 b.] A follower of the Maori king (see *King Country* s.v. *KING *sb.* 13 a). Also *attrib.* or as *adj.*

1860 T. BUDDLE *Maori King Movement* 60 On this point the kingites carry with them the sympathies of the majority. 1884 J. H. KERRY-NICHOLLS *King Country* 6 In 1857 Kingite meetings were held in Paetai. *Ibid.* 9 Sir George Grey..opened up communication with the chiefs of the Kingites. 1910 J. COWAN *Maoris of N.Z.* xxvii. 286 This Kingite war-song..is still on the lips of the Waikato people. *Ibid.* 289 All might have gone well had the Kingites been able to restrain their more turbulent spirits. 1959 K. SINCLAIR in J. E. Gorst *Maori King* p. xiv, He was driven out by the Kingites on 18 April, 1863. *Ibid.* 263 There were no large Kingite meetings, at which he could have been formally 'installed'.

kingklip (ki·ŋklip). In full, **kingklipfish** (-visch). [Afrikaans, f. Du. *koningklipvisch*: see *KLIPFISH.] One of several South African marine food fishes, esp. *Epinephelus andersoni*, of the family Serranidæ.

[1843 J. C. CHASE *Cape of Good Hope* 169 Koning Klip Fish, King Rock Fish. Scarcer than the preceding, very considerably larger, and less delicate, but in much repute.] 1876 H. BROOKS *Natal* iv. 141 Klipvisch, kingklip fish..are held in very high estimation. 1878 K. JOHNSTON *Africa* xxiii. 393 Not fewer than forty-four varieties of edible fishes have been enumerated, including ..'King Klip'. 1893 H. A. BRYDEN *Gun & Camera S. Afr.* xx. 449 Many of the fish in Cape waters furnish excellent eating; the Roman, kingklipvisch, stomneus, steenbras, and klipvisch being among the choicest. 1923 *Nature* 24 Feb. 271/1 The kingklip (in appearance like a ling). 1930 C. L. BIDEN *Sea-Angling Fishes of Cape* 2 Angling for..kabeljou, stockfish, and king klipfish. 1950 M. MASSON *Birds of Passage* xii. 121 Vendor and buyer alike haggled over the merits of..Kingclipfish. 1971 *Rand Daily Mail* 4 Sept. 1/3 Kingklip, soles and Cape salmon were almost unobtainable.

King Kong (ki·ŋ,kǫ·ŋ). [Name of the ape-like monster featured in the film *King Kong* (1933).] **a.** Used as a nickname for anyone of outstanding size or strength. (In quot. 1966 used ironically.)

1955 E. WAUGH *Officers & Gentlemen* I. vi. 62 He looks like a gorilla. They..sent him here to teach us to climb. We call him King Kong. 1966 'L. LANE' *ABZ of Scouse* 59 *King Kong*, derisory name for a weedy, undersized individual. 1970 M. KELLY *Spinifex* vi. 103 'What about King Kong?'..'He's just an honest murderer.' 1970 K. PLATT *Pushbutton Butterfly* (1971) xi. 123 'He looked like a big ape.'..'Make it a gorilla...That way we can pin it on King Kong.' 1974 *Guardian* 23 Aug. 8/4 Finn MacCool was a legendary Irish giant, a King Kong with a generous heart.

b. Cheap alcohol. *slang.*

1946 MEZZROW & WOLFE *Really Blues* 357/2 *King Kong*, cheap moonshine, corn whisky. 1950 H. E. GOLDIN *Dict. Amer. Underworld Lingo* 117/1 *King Kong*, (South) a potent drink made from the skimmings of boiling sugar cane.

king-maker. Add: Also *transf.* and *fig.*

1899 E. WHARTON *Greater Inclination* 162 John Oberville? I'll tell you what he is—the power behind the throne, the black Pope, the King-maker. 1949 *Sun* (Baltimore) 10 Aug. 1/6 Hunt boasted of responsibility for getting so many Government officials their jobs that he was known 'socially' as 'the kingmaker'. 1959 *Manch. Guardian* 4 Aug. 4/2 Governor Ribicoff, of Connecticut.. has launched his career as a kingmaker by taking command of the Kennedy forces. 1968 'G. BAGBY' *Another Day* iv. 76 This was a young man who was clearly destined to make his mark... The kingmakers had their eye on him. He was going to go places. 1972 *Times* 27 Dec. 5/8 This was a sad example of a politician still trying to play the kingmaker long after his influence had waned.

king-pin. Add: **2.** (Earlier and later examples.) Also, the most important or outstanding person in a party, organization, etc.

1867 *Harper's Weekly* 14 Sept. 590/2 His best position was as a batter. He was a 'King-pin' there. 1914 *Chambers's Jrnl.* Jan. 62/1 The cars are mounted on bogie trucks, the connection being by means of a central or 'king-pin'. 1915 C. J. DENNIS *Songs of Sentimental Bloke* 102 But 'struth! 'E is king-pin! The 'ead serang! 1926 K. S. PRICHARD *Working Bullocks* (1956) 206 'My!' Mary Ann gasped incredulously, 'and you was the king pin last week.' 1957 J. WATEN *Shares in Murder* 99 Then he must be the biggest fence of the lot. The kingpin. The daddy of all fences. 1957 *Economist* 5 Oct. 22/1 The balloting for the seven seats filled by the constituency parties revealed that, with the kingpin removed, the former Bevanite machine is showing signs of disintegration. 1958 *Engineering* 28 Feb. 265/3 Another remarkable feature of the design is a front suspension which uses telescopic dampers as the king pins and steering swivels. 1970 *Daily Tel.* 30 Oct. 2/6 The owner of three shops was the kingpin behind a wholesale shoplifting plot. 1971 M. TAK *Truck Talk* 96 *Kingpin*, the bolt on the underside of the front of a trailer that fits into the tractor's fifth wheel to couple the tractor and the trailer together.

king-post. Add: **2.** On a ship (see quots.).

1927 G. BRADFORD *Gloss. Sea Terms* 95/2 *King post*, a short derrick mast to support the smaller cargo booms. 1948 R. DE KERCHOVE *Internat. Maritime Dict.* 385/1 *King post.* 1. A short heavy mast which serves to support a boom. 2. The centerline pillars in a ship's hold. 1961 F. H. BURGESS *Dict. Sailing* 127 *Kingpost*, a vertical post, sometimes resembling a mast, erected near the hatches, to support and top a derrick boom.

king's man, kingsman. Add: **1. b.** *the King's men*: a name for the dramatic company otherwise known as 'the King's Majesty's Servants' under James I.

1613 R. DABORNE *Let.* 29 Oct. in P. Henslowe *Henslowe Papers* (1907) 76 They shall have the play or noe, they rale upon me I hear bycause the kingsmen hav given out they shall hav it. *Ibid.* 77 Eight pound besyds my rent which J will fully satisfy y[u] eather by them or the kings men as y[u] please. 1886 *Dict. Nat. Biogr.* VII. 286/1 Burbage's position justifies the conjecture..that he had been connected with the lord chamberlain's men, subsequently called the king's men, and originally called Lord Strange's company. 1923 E. K. CHAMBERS *Eliz. Stage* II. 218 The King's men gave eight plays at Court..during the winter of 1614–15. 1951 M. CHUTE *Shakespeare of London* xi. 233 The King's Men gave a Sunday production of *The Merry Wives of Windsor* and then followed it on St. Stephen's night with *Measure for Measure*.

c. *U.S.* One who supported the British cause at the time of the American Revolution. *Obs.* exc. *Hist.*

1809 P. FRENEAU *Poems* II. 11 Whate'er some angry king's-men say, You play a game that must be won. 1857 *Ladies' Repository* XVII. 83/1, I never feed kingsmen if I can help it. 1949 *Sat. Even. Post* 2 Apr. 98/4, I am neither king's man nor rebel.

4. A member of King's College, Cambridge.

1803 C. SMART in *Gradus ad Cantabrigiam* 81 Ev'n gloomiest Kings-men, pleas'd awhile, Grin horribly a ghastly smile. 1852 C. A. BRISTED *Five Yrs. in Eng. Univ.* (ed. 2) 127 He came out the winner, with the Kingsman and one of our three close at his heels. 1968 *Tablet* 17 Aug. 820/1 He was anxious, like many Kingsmen, to be recognised as a Fabian. 1973 *Observer* 1 July 32/3 He was a fellow of King's College, Cambridge..and immediately after his death he received what must have been regarded by most Kingsmen as the ultimate accolade. 1974 *Times Lit. Suppl.* 29 Nov. 1346/3 Basileon. A Magazine of King's College, Cambridge... The founder-editors believed that facility in writing resulted from early practice, and their rather solemn purpose was to give Kingsmen a chance of trying their hand... To a Kingsman of between the wars they still give off that faint but delightful savour.

Kingston[2] (ki·ŋstən). *Naut.* The name of John *Kingston*, 19th-century British dockyard foreman, used *attrib.* and *absol.* († and in the possessive) to designate a kind of conical valve he invented for use in the sides of ships below the water-line which opens outwards with a screwing action.

1846 J. BOURNE *Treat. Steam Engine* xi. 223/1 In modern steam vessels Kingston's valves are..used, which consist of a spindle or plate valve fitted to the exterior of the ship, so that if the internal pipe or cock breaks, the external valve will still be operative. 1859 *Reed's New Guide Bk. Local Marine Board Exam.* 62 (*heading*) Boiler valves and cocks... Kingston valves. 1883 A. E. SEATON *Man. Marine Engin.* xvi. 301 For all large inlets the Kingston valve is preferable, as it acts as a non-return valve in case of the spindle breaking, and can then always be worked by simply forcing it outwards. *Ibid.* 303 A valve may be fitted in lieu of a cock to even the smallest Kingston. *Ibid.*, In the Navy, Kingston valves are fitted to all inlets and blow-off pipes. 1905 E. M. & B. DONKIN tr. *Bauer's Marine Engines & Boilers* 418 Most of the valves in merchant ships consist of ordinary valves opening inwards, but on warships the old so-called 'Kingston' valves are still frequently met with. 1933 'L. LUARD' *All Hands* 124 Drop of air in three main. Open three Kingston. Close air. Open three vent. 1966 P. E. SEGDITSAS *Elsevier's Naut. Dict.* 111 Kingston valve; sea cock.

King Street (ki·ŋ strīt). [Name of the street near Covent Garden, London, in which the headquarters of the Communist Party of Great Britain Executive Committee has been situated since 1920.] Used *transf.* to designate the Communist Party of Great Britain, its members, or its leaders. Also *attrib.*

1958 C. COCKBURN *Crossing Line* iii. 49 The place did good business in those days..because the *Daily Worker* staff and the people from King Street..used it. *Ibid.* v. 83 This was a situation which gave many people at King Street nightmares. 1961 *Guardian* 5 June 8/4 The Labour party..shunning the support of King Street. 1964 C. DRIVER *Disarmers* iii. 72 If the King Street commissars were not so invincibly stupid, they would have insisted that the movement be left severely alone. 1970 *Times Lit. Suppl.* 18 Sept. 1016/3 Even if not a card-carrying member of the Communist Party, Strachey was at this time the grey eminence of King Street. 1972 *Observer* 8 Oct. 29/2 One version of events is that 'King Street' had decided the miners wouldn't end the strike unless they were given 25 per cent. King Street means the headquarters of the Communist Party.

kingy (ki·ŋi). [f. KING *sb.* + -Y[6].] A children's game resembling 'He' but played with a ball; the winner is declared King.

[1916 N. DOUGLAS *London Street Games* 5 There are other ball-games, such as hot rice..and king and missings out.] 1959 B. SUTTON-SMITH *Games N.Z. Children* II. 150 Kingy. In this game the players begin by standing round in a circle with their arms on one another's shoulders. The ball is dropped in the middle of the circle. When it touches someone's foot that person is He. 1969 I. & P. OPIE *Children's Games* ii. 95 'Kingy' is a ball game in which those who are not He have the ball hurled at them, without means of retaliation. 1972 *Where* Apr. 102/2 Transferred to playground or playing field, such games as 'kingy' can lose much of their colour and point, since the existence of natural and human obstacles and hazards are of their essence.

kinin (kəi·nin). [f. Gr. κιν-εῖν to set in motion + -IN[1]; in sense 1 abstracted from *bradykinin* (s.v. *BRADY-).] **1.** *Biochem.* Any of a group of polypeptides of low molecular weight which are formed in tissue (from inactive precursors in the blood) in response to injury and have local effects that typically include pain and the dilatation of blood vessels.

1954 SCHACHTER & THAIN in *Brit. Jrnl. Pharmacol.* IX. 352/1 Since this substance in wasp venom cannot as yet be identified with any substance hitherto described, it is tentatively designated as wasp (or venom) kinin, or simply as kinin. 1962 *New Scientist* 4 Oct. 35/2 If the kinins get into the general circulation they are destroyed before they can produce general effects throughout the body. 1963 *Listener* 7 Nov. 741/2 An important factor in rheumatism may be an accelerated formation and release of certain compounds known as kinins, which are formed from amino-acids. The kinins may cause many of the painful symptoms, such as the swollen joints that are a prominent part of the disease. 1970 PASSMORE & ROBSON *Compan. Med. Stud.* II. xvii. 1/2 In addition to the above kinins, bradykinin homologues..are found in many parts of the animal kingdom, e.g. in wasp and scorpion venoms ..and in the plasma of reptiles and birds.

2. *Plant Physiol.* = *CYTOKININ.

1956 C. O. MILLER et al. in *Jrnl. Amer. Chem. Soc.* LXXVIII. 1375 A substance which markedly promotes cell division in various plant tissue cultures..has been shown..to be 6-furfurylaminopurine. The specific name *kinetin* has been applied to this substance, and the generic term *kinin* is suggested for any substance which similarly stimulates cytokinesis. 1965 *New Scientist* 25 Mar. 788/3 The third and most recently investigated group of plant growth substances is that of the kinins. 1971 F. C. STEWARD *Plant Physiol.* VIA. iii. 356 The photoperiodic responses to red light may also involve kinin effects.

kininogen (kəini·nŏdʒĕn). *Biochem.* [f. *KININ + *-OGEN.] Any biologically inactive precursor of a kinin (sense *1).

1963 *Jrnl. Physiol.* CLXIX. 45P The plasmin preparation..did not form kinin from purified kininogen. 1970 PASSMORE & ROBSON *Compan. Med. Stud.* II. xvii. 2/1 Although plasma and tissues contain readily activated kinin-forming enzymes and kininogens, free active kinin is rarely detected in biological fluids.

kinjal (kindʒā·l). [Native name in the Caucasus (= Russ. *kinzhál*), a. Pers. (Arab., Turk., Urdū) KHANJAR.] (Among Caucasians and Kurds) = HANDJAR, HANJAR, KHANJAR.

1862 *Harper's Mag.* XXVI. 803/1, I laid aside revolver and kindshall. 1889 J. ABERCROMBY *Trip E. Caucasus* 130 At his waist hung a *kinjal* and a long native sabre. 1897 R. D. BLACKMORE *Dariel* viii. 86 His hand was playing with his *kinjal* all the time, for so they call those deadly bits of steel, without which they never think their attire complete. 1924 *Blackw. Mag.* Feb. 149/1 The scar of a Kurdish kinjal.

kink, *sb.*[1] Add: **1.** (Examples of use with reference to hair; also earlier and later *transf.* examples.)

1848 *Yale Lit. Mag.* XIV. 82 (Th.), Come! wake up, and shake the kinks out of your land legs. 1857 MAYNE REID *War Trail* xiii. 67 Yes, there was the same negress with.. the little well-oiled kinks hanging like corkscrews over her temples! 1930 [see *CRONK *a.*]. 1962 *Kenyon Rev.* XXIV. 94 Don't worry about Saturday night. Play around. Work the kinks out. 1970 G. F. NEWMAN *Sir, You Bastard* 259 There existed kinks in the man's career; it was only a question of drawing on the right one.

b. A sudden bend in a line, course, or the like that is otherwise straight or smoothly curved.

1899 S. BARING-GOULD *Furze Bloom* 27 That [wall] on the left makes a kink to respect 'The Brothers' Grave'. 1928 L. S. PALMER *Wireless Princ. & Pract.* v. 132 The curve sometimes exhibits a sudden 'kink' or discontinuity. 1965 G. MCINNES *Road to Gundagai* v. 74 Below the kink the street degenerates rapidly. 1971 *Sunday Express* (Johannesburg) 28 Mar. 7/1 A new grandstand for 2,000 spectators at the kink on the main straight.

2. a. (Earlier U.S. and later examples.) In recent use also = a state of madness; an instance of, the practice of, or suffering resulting from sexual abnormality.

1803 T. JEFFERSON *Let.* 24 Nov. in *Writings* (1897) VIII. 280 Should the judges take a kink in their heads. 1915 F. M. HUEFFER *Good Soldier* IV. ii. 229 By a kink, that I could not at the time understand, Miss Hurlbird insisted that I ought to keep the money all to myself. 1924 [see *ENSNARL *v.*[1]]. 1950 T. S. ELIOT *Cocktail Party* II. 120 And so you suppose you have what you call a 'kink'? 1959 *Encounter* Mar. 22 Hates kissing. Undertakes most kinks..but no buggery. 1959 M. GEE in C. K. Stead *N.Z. Short Stories* (1966) 279 He's got a kink I reckon. He'll end up in the nuthouse. 1965 *Movie* Summer 44/4 The result is the story of the sexual hallucinations of a young girl..played for flat-out kink.

3. *U.S.* A human being in various slang applications. **a.** A Negro. *Obs.*

1865 J. H. BROWNE *Four Yrs. in Secessia* xxxix. 288 'Coming the kink' was to steal a negro from the country, and dispose of him in town. 1944 *Amer. Speech* XIX. 173 *Kink* shows an obvious allusion to the Negro's hair.

b. A criminal.

1914 JACKSON & HELLYER *Vocab. Criminal Slang* 52 *Kink*, a crook; a larcenous criminal. Also used by yeggs to designate a non-criminal tramp, or one who is not initiated into the particular craft of the speaker. 1950 H. E. GOLDIN *Dict. Amer. Underworld Lingo* 117/2 *Kink* (scattered areas of East and near South), a thief, especially an expert in stealing automobiles.

c. A sexually abnormal person; one who practises sexual perversions; loosely, an eccentric, a person wearing noticeably unusual clothes, behaving in a startling manner, etc.

1965 *Harper's Bazaar* Jan. 54/1 His phone is ex-directory because of all the kinks who used to phone at 2 a.m. 1967 [see *KICKSTER]. 1968 B. TURNER *Sex Trap* xv. 149, I believe the psychiatrists have other ideas about what makes a kink kinky. 1972 'J. RIPLEY' *My Word you should have seen Us* 35, I have known queers. I have known kinks.

kinked, *ppl. a.* [s.v. KINK *v.*[2] in Dict.] (Later examples in extended uses.)

1966 *Punch* 5 Oct. 521/1 Others were delighted by the elegance of the language and the sinister kinked logic governing the behaviour of the characters. 1967 A. HUNTER *Gently Continental* viii. 127, I *am* scared. I can't protect Trudi. Frieda is kinked. 1969 D. C. HAGUE *Managerial Econ.* iv. 92 The kinked demand curve is derived from the..curves we have already been using in our analysis of trade association pricing. 1970 D. UHNAK *Ledger* (1971) vii. 97 Stoner Martin massaged the back of his neck. 'This kind of work can sure leave you kinked up.'

kinker (ki·ŋkəɹ). [f. KINK *sb.*[1] + -ER[1].] An acrobat, a contortionist (see also quot. 1948).

1926 *Amer. Speech* I. 282/2 *Kinker*, a performer or acrobat. 1931 *Amer. Mercury* Nov. 353/1 *Kinkers*, circus performers. 1948 MENCKEN *Amer. Lang.* Suppl. II. 684 A contortionist is a *frog, bender* or *Limber Jim*, a freak or snake-charmer is a *geek*, and all performers are *kinkers*.

kinkiness (ki·ŋkinės). [f. KINKY *a.* + -NESS.] The quality or habit of being kinky (in the senses of Dict. and Suppl.); a kinky state.

1924 W. DEEPING *Three Rooms* ii. 15 His black hair was wavy even to kinkiness. 1951 G. C. SHATTUCK *Dis. Tropics* lvii. 694 The hair loses its luster... The natural kinkiness disappears. 1959 C. MACINNES *Absolute Beginners* 22 'What kind of print might you be needing?' I went on, not sure yet what kinkiness I had to cater for. 1966 *Punch* 28 Sept. 483/1 Normally he [*sc.* a cat] prefers to bite men and scratch women, which seems to point to some kinkiness in his make-up. 1968 R. V. BESTE *Repeat Instructions* x. 104 What's wrong with her? Or is a certain kinkiness in spelling now evidence of a security risk? 1971 S. JEPSON *Let. to Dead Girl* iii. 23 If black vinyl thigh boots under a slit midi skirt..wasn't kinky on purpose, he'd be glad someone else would tell him what it was... Trust Fatty to look for kinkiness. 1972 *Daily Tel.* 27 Apr. 9/2 His book..is too smuttily concerned with 'kinkiness' to justify its sub-title.

kinkless (ki·ŋklės), *a.* [f. KINK *sb.*[1] + -LESS.] Without a kink: applied in *Electronics* to a kind of tetrode designed so as to eliminate the irregularity of the current–voltage characteristic of an ordinary tetrode (which shows up as kinks in the characteristic curve).

1943 C. L. BOLTZ *Basic Radio* x. 164 The second method of suppressing the secondary emission is to align the control grid and screen very carefully so that the wires are opposite... Such a valve is a kinkless tetrode. 1953 F. LANGFORD-SMITH *Radio Designer's Handbk.* (ed. 4) I. i. 8 Some 'kinkless' tetrodes are also used as v-f and i-f amplifiers. 1958 W. F. LOVERING *Radio Communication* viii. 170 To avoid the kink in the E_a–I_a characteristic it is necessary to return the secondary electrons to the anode even when the screen is positive to anode... The first solution to this problem was the pentode... A later solution was the kinkless tetrode (or Beam tube).

kinky, *a.* Add: Also **kinkey. 1.** Also *Comb.*, as *kinky-bearded, -haired, -headed, -tailed* adjs.

1844 *Congress. Globe* 6 Jan. App. 42/3 [The Negro's] skull is as thick, his hair is as kinkey, his nose as flat..as they were the day he was first introduced. 1848 W. T. THOMPSON *Major Jones's Sk. Trav.* 146, I happened to call one of the nigger waiters 'boy'. The kinky-headed cuss looked at me sideways, and rolled the whites of his eyes at me. 1861 in *Rebellion Rec.* (1862) I. III. 137 A marked distinction is laid Between the rights of the mistress, And those of the kinky-haired maid. 1865 Kinky-haired [in Dict.]. 1925 W. DEEPING *Sorrell & Son* xxix. 290 The hard-bitten, kinky-haired casualty-sister assisting him with critical and voiceless composure. 1937 *Time* 16 Aug. 58/2 Ill lay kinky-bearded, 64-year-old Thorvald Stauning, Premier of Denmark, after breaking a leg. 1956 C. AUERBACH *Genetics in Atomic Age* 93 When a female mouse is irradiated during the second week of pregnancy, some of her young may be born with kinky tails. Unlike true mutations to kinky tail, these deformities are not inherited. If, later on, visible mutants should turn up in the progeny of such a kinky-tailed mouse, they are not more likely to affect the tail than the eyes or the ears or the coat.

2. For (*U.S. colloq.*) read (*colloq.*). (Later examples.)

1889 *Sportsman* 2 Jan. (Farmer), The kinky ones and the worthy ones who play hole-and-corner with society. 1907 E. M. FORSTER *Longest Journey* viii. 100 This jaundiced young philosopher, with his kinky view of life, was too much for him. 1929 W. J. LOCKE *Ancestor Jorico* xix. 263 A fellow ought to know something about the funny kinky ways of ordinary men and women. 1950 T. S. ELIOT *Cocktail Party* II. 120 But when everything's bad form, or mental kinks, You either become bad form, and cease to care, Or else, if you care, you must be kinky.

b. Lively, spry, energetic. *U.S. local.*

1903 G. S. WASSON *Cap'n Simeon's Store* vi. 107 'He ain't over and above kinky, though, I s'pose likely?' 'No, ..nothin' very antic about him...there must be some buckram left into him, too, the way he keeps a-going.' 1914 *Dialect Notes* IV. 4 *Kinky*, in high spirits. 'You seem to be feeling pretty kinky to-day.'

c. *Criminals' slang.* Of things: dishonestly come by (see also quot. 1954). Cf. CROOKED *a.* 3 b, *BENT *ppl. a.* 5 a, b.

1927 *Collier's* 23 July 15/1 'Why, you can't tell me that you didn't know those five big cars were kinky.' 'Kinky?' ..'Those cars were bent.' 1931 G. IRWIN *Amer. Tramp & Underworld Slang* 117 *Kinky*, criminal; crooked; unlawful. Said of stolen goods, or of an individual known to be without the law. 1942 BERREY & VAN DEN BARK *Amer. Thes. Slang* § 470/1 Booty,..killing, kinky goods, lift. 1954 W. R. & F. K. SIMPSON *Hockshop* 275 Canfield..was never accused..of having 'kinky' gambling paraphernalia. By that I mean dice and cards and roulette wheels that gave the house an unfair advantage.

d. In senses corresponding to *KINK *sb.*[1] 3 c. Of persons: perverted, esp. sexually; *spec.* homosexual. Of things or situations: suggestive of sexual perversion, as of certain items or styles of dress (e.g. *kinky boots*); in weakened sense, bizarre.

1959 C. MACINNES *Absolute Beginners* 16 Suze..meets lots of kinky characters..and acts as agent for me getting orders from them for my pornographic photos. 1960 'A. BURGESS' *Doctor is Sick* xviii. 145 'If you think I'm perverted you're completely mistaken. I'm quite normal.' 'Normal? You? That's a laugh. You're kinky, the same as what I am.' 1960 F. NORMAN *Fings ain't wot they used t'Be* II. i. 52 Fancy anyone being so kinkey about a brown teapot. 1963 *Daily Tel.* 11 Dec. 19/2 The phrase 'kinky advert'..meant that she acted as an advertisement for irregular sexual practices. 1964 *Ann. Reg. 1963* 1 It [*sc.* 1963] was the year..that women adopted the fashionable long 'kinky' boot. 1964 J. BURKE *Hard Day's Night* i. 11 She was dead kinky for sweetbreads. 1964 *Times Lit. Suppl.* 8 Oct. 925/1 Zoo men receive a constant stream of kinky letters. 1966 *Listener* 6 Jan. 23/1 One of the girls—a buxom specimen in kinky patent leathers. 1966 J. PORTER *Sour Cream* ix. 117 On the one hand I had to trust Zinaida, on the other I knew she was perfectly capable of fabricating the whole scheme just to get rid of Katia whom she loathed. It was like something out of Greek tragedy, only kinkier. 1967 A. DIMENT *Dolly Dolly Spy* ii. 19 He produced a pack of Black Russian cigarettes, dead kinky, and tossed me one. 1968 [see *KINK *sb.*[1] 3 c]. 1971 *Daily Tel.* 16 July 11/4 In a moment of excessively kinky passion a husband strangles his mistress. 1972 F. WARNER *Lying Figures* III. 36 Kinky sex makes them feel inadequate.

B. *sb.* **a.** A person with 'kinky' hair. **b.** An object dishonestly obtained. **c.** A sexually abnormal or perverted person. **d.** *pl.* Kinky boots.

1926 J. F. DOBIE *Rainbow in Morning* 4 One considers the negro as a shining apostle of sweetness and light, another as a gentle old darkey, and still another as a 'phallic kinky'. G. THOMAS *Ibid.* 154 The Fayette County and other South Texas 'kinkies' whose songs I have been noting. 1927 *Collier's* 23 July 14/1 *A kinky*, any stolen car. 1941 *Amer. Mercury* Mar. 349/2 The titles of every car Joe sold could be searched clear back to the factory... Yet the cars were strictly kinkies. 1942 BERREY & VAN DEN BARK *Amer. Thes. Slang* § 430/2 *Curly-headed person*, curly, curly-locks, curly-pate, kinks, kinky. 1950 'N. BELL' *I am Legion* vii. 216 O Good Lord... Quite half of them *are* kinkies. 1959 *Encounter* July 83/1 No prostitutes, no queers, no kinkies. 1965 *Punch* 19 May 755/1 Palm memory of the agreeable feel of the two-in-hand will come back, and if snakeboots replace kinkies, the thrill of the double slalom round the sprigs above the eyelet-holes. 1967 A. DIMENT *Dolly Dolly Spy* vi. 82 Porny photos, various drugs and birds for kinkies at Oxford.

kinnikinic. The more usual forms are now **kinnikinnick, kinnikinnik.** Also many other varr. **1.** (Earlier, *attrib.*, and later examples.)

1805 J. ORDWAY in Lewis & Ordway *Jrnls. Western Explor.* (1916) 199 Some Indians had hung up..a Scraper a paint bag.., kinikaneck bags, flints, [etc.]. 1827 T. L. MCKENNEY *Sk. Tour to Lakes* 181 The pipe of an Indian.. and a pouch made of the skin of some animal, in which he carries his *kinnikanic*, a kind of fragrant weed that has a leaf like our box wood. 1839 C. A. MURRAY *Trav. N. Amer.* II. 22 We took out our kinnekinik-bag. 1844 —— *Prairie-bird* II. 179 Volumes of *kinnekenik* smoke. 1860 H. Y. HIND *Narr. Canad. Red River Expedition* I. 315 A sandy ridge..was covered with the bear-berry from which kinnikinnik is made. 1867 'MARK TWAIN' *Amer. Drolleries* (1875) 41 The most popular..smoking tobacco is.. Killikinick. 1889 K. MUNROE *Golden Days* xxvi. 284 Put that in your pipe and smoke it, along with your killikinick. 1890 E. CUSTER *Following Guidon* viii. 101 Kinnikinnic.. is a mixture of willow bark, sumach leaves, sage leaf, and tobacco, and is thoroughly mingled with marrow from buffalo bones. 1920 *Chambers's Jrnl.* 31 Jan. 136/1 The curling wisps of *kinickinick* smoke. 1969 *Islander* (Victoria, B.C.) 23 Nov. 13/2 Since their canoe had been swamped, the two pipe-smoking canoeists had been without tobacco. They used kinnikinik..which the Indians smoke.

2. (Earlier, *attrib.*, and later examples.)

1822 A. EATON *Man. Bot.* (ed. 3) 178 *Arbutus uva-ursi*, bear berry, kinnikinnick... Dry, barren sand plains. 1853 J. W. BOND *Minnesota & its Resources* 303 Some dry Kinne-kin-nick bark is generally carried along, cut very fine for the purpose of smoking. 1910 *Anthropos* V. 420 Nor should we forget to mention the fruit of the *kinnikinik* or bearberry bush (*Arctostaphylos uva-ursi*), which, though insipid enough to a white man, is of such importance in the eyes of some tribes, as the Chilcotins, that it gives its name to one of their minor seasons. 1938 M. THOMPSON *High Trails of Glacier Nat. Park* 86 As we climb into the Hudsonian zone we find extensive carpets of kinnikinnick. 1956 V. FISHER *Pemmican* xxviii. 259 They went another day, and another, eating nothing but rose hips, leaves, kinnikinnik bark, moss, water cress. 1963 *Vancouver Sun* 23 Nov. 21/1 The rolling hills..are park-like with their copses of fir, tamarack, poplar and willow, dotted through open stretches of bitterbrush and kinnikinnik.

kino- (kəi·no), comb. form of Gr. κινεῖν to set in motion, as in **kinoci·lium** (pl. **-cilia**) *Biol.*, a cilium which is capable of moving (in contrast to an immobile cilium, called a stereocilium), *spec.* such a cilium borne singly on each hair cell of the maculæ of the inner ear amid a group of about a hundred stereocilia; **ki·noform** *Physics*, a transparent plate with a contoured surface so made as to introduce phase differences into an incident parallel beam in such a way as to form a single three-dimensional image of the particular object for which the calculation of the surface was made; **ki·noplasm** *Cytology* [ad. G. *kinoplasma* (E. Strasburger *Histol. Beiträge* (1892) IV. 60)], a supposed special kind of cytoplasm which was formerly held to be fibrillar in nature and to give rise to the active parts of a cell (such as the membrane and the mitotic apparatus); hence **kinopla·smic** *a.*

1933 M. FERNÁN-NÚÑEZ tr. *Ramón y Cajal's Histol.* x. 148 These cilia [of ciliated epithelium] are completely free and carry out spontaneous vibratory and whip-like movements, both of flexion and extension in the greater number of cases (kinocilia); in other cases, as in the epididymis, they appear immobile (stereocilia). 1956 *Acta Oto-Laryngol. Suppl.* No. 126. 46 The kinocilium in each cell is closely consistent in structure with the kinocilia that have been observed in the trachea, the fallopian tube, and in a number of unicellular organisms. 1971 *New Scientist* 29 July 283/1 The kinocilium, an apparently active hair among the passive hairs of each hair cell on the basilar membrane of the ear. 1968 P. M. HIRSCH et al. *Digital Holograms & Kinoforms* in *IBM Technical Symposium on Laser Applic., Sept. 1968* (typescript) 244 Our efforts have been in making digital holograms and a new kind of imaging element, which we call a kinoform. 1969 L. B. LESEM et al. in *IBM Jrnl. Res. & Devel.* XIII. 150/1 The kinoform is a new, computer-generated wavefront reconstruction device which, like the hologram, provides the display of a three-dimensional image. In contrast, however, the illuminated kinoform yields a single diffraction order and, ideally, all the incident light is used to reconstruct this one image. *Ibid.* 151/2 Unlike the making of holograms, which can be physical recordings of actual wavefront interference patterns.., it does not appear possible to create a kinoform using completely optical techniques. 1970 *Physics Bull.* Nov. 493/1 Kinoforms, like digital holograms, are also made by calculating the wavefront scattered from the numerically defined object, but instead of adding in a reference beam, the amplitude of the scattered wavefront is assumed constant. A photographic plot of the phase distribution (mod 2π) is made, and translated into a phase plate. *Ibid.* 495/2 Kinoforms can be used as arbitrarily shaped phase plates in lens applications. 1894 *Jrnl. R. Microsc. Soc.* 581 Prof. E. Strasburger concludes that in the cytoplasm two constituents are contrasted in their activity. To one of these, the kinoplasm, the radiations round the centrospheres, the spindle-fibres, and the combining filaments.. owe their origin... The other constituent..is the trophoplasm. 1910 G. M. CALKINS *Protozoöl.* i. 29 Some [observers]..have endeavored to show that archoplasm, or, in a larger sense, kinoplasm, is not only specific, but a kind of 'superior' protoplasm, self-perpetuating and distinct. 1934 L. W. SHARP *Introd. Cytol.* (ed. 3) ii. 43 It appears to be the kinoplasm that is responsible for protoplasmic streaming; it flows through the relatively stationary trophoplasm, carrying with it the plastids and chondriosomes. 1965 K. ESAU *Plant Anat.* (ed. 2) iii. 57 The old concept of the existence of a special kind of active, fibrous cytoplasm, the kinoplasm. 1900 E. B. WILSON *Cell* (ed. 2) vi. 322 Beyond this the two forms of protoplasm show a difference of staining-reaction, the kinoplasmic fibrillæ staining deeply with gentian-violet and iron-hæmatoxylin, while the trophoplasm is but slightly stained. 1905 *Rep. Brit. Assoc. Adv. Sci.* 577 Strasburger considers them [*sc.* blepharoplasts] as kinoplasmic in nature, and thus brings them into relation with his other kinoplasmic structures, the centrosome and spindle. 1927 *Protoplasma* II. 201 The kinoplasmic spheres from which the contractile vacuoles arise in *Spirogyra*.

-kins, *suffix.* Variant of -KIN, as in *babykins, boykins*, etc., and formerly in certain oath-words, as *bodikins, lakens* (see LAKIN[2]), *maskins, pit(t)ikins*.

kinship. Add: **1. b.** *Anthropology.* The recognized ties of relationship, by descent, marriage, or ritual, that form the basis of social organization. So *attrib.* and *Comb.*, as *kinship category, group, structure, term*; **kinship system,** the system of relationships traditionally accepted in a culture and the rights and obligations which they involve.

1866 J. F. M'LENNAN in *Fortn. Rev.* 15 Apr. 580 Kinship through the mother had been in Homer's time undisputed among the Greeks. 1910 J. G. FRAZER *Totemism & Exogamy* I. 20 The Psylli, a Snake clan in Africa, had a similar test of kinship. 1914 W. H. R. RIVERS (*title*) Kinship and social organization. *Ibid.* 1 The aim of these lectures is to demonstrate the close connection which exists between methods of denoting relationship or kinship and forms of social organisation. 1937 R. H. LOWIE *Hist. Ethnol. Theory* x. 171 In 1909 Kroeber, while fruitfully paving the way for work on the linguistic categories embodied in kinship systems, denied any social determinants. 1945 G. & M. WILSON *Analysis of Social Change* vi. 162 Any attempt to bolster up a legal system based on kinship is doomed to failure in an expanding society. 1949 E. E. EVANS-PRITCHARD in M. Fortes *Social Struct.* 101 Nuer themselves..see that it is undesirable to obliterate.. the boundaries between kinship categories. 1949 F. EGGAN in *Ibid.* 121 One of the most significant advances in the study of kinship systems in modern times has been Professor A. R. Radcliffe-Brown's method of structural or sociological analysis. 1951 R. FIRTH *Elem. Social Organiz.* i. 32 Other basic relations..are due to position in a kinship system. 1955 M. GLUCKMAN *Custom & Conflict in Afr.*

iv. 99 Children are desired by a Zulu kinship-group because they strengthen it. 1957 V. W. TURNER *Schism & Continuity in Afr. Soc.* iii. 77 In most primitive societies social control at the local level is associated with position in the kinship structure. 1958 A. R. RADCLIFFE-BROWN *Method in Social Anthropol.* II. iv. 171 The kin of any given person were classified into a limited number of categories, each denoted by one kinship term. 1969 M. FORTES *Kinship & Social Order* (1970) p. vii, My thesis is that the structuralist theory and method of analysis in the study of kinship and social organization..stems directly from Morgan's work. 1970 E. LEACH *Lévi-Strauss* vi. 99 Ties of filiation and..ties of siblingship..provide the basic bricks out of which kinship systems are built up. 1971 *World Archaeol.* III. 217 Ethnographic evidence therefore focuses on metalworking as a kinship or descent group-organized activity.

kinzigite (ki·ntsigəit). *Petrogr.* [ad. G. *kinzigit* (H. Fischer 1860, in *Neues Jahrb. f. Min.* 797), f. the name of the *Kinzig* valley, W. Germany: see -ITE[1].] A metamorphic schistose rock containing garnet, biotite, and varying amounts of quartz, plagioclase, sillimanite, and cordierite. Hence **kinzigi·tic** *a.*, containing kinzigite.

1878 *Jrnl. Chem. Soc.* XXXIV. 208 Garnet-graphite-gneiss was hitherto unknown [in the Black Forest],..being formerly known by the name of Kinzigite. 1947 *Bull. Geol. Soc. Amer.* LVIII. 1024 The kinzigites of the Askainen-Lemu zone are fine- to medium-grained and contain garnet as well as cordierite. 1965 *Mineral. Abstr.* XVII. 227/2 A kinzigitic formation..occurs in the district of Miglierina (Catanzaro). This formation is composed of kinzigites, kinzigitic gneiss, and amphibolic kinzigites.

Kioko (kiō̆u·ko). Pl. **Kioko, -os.** [Native name.] The name of an African people inhabiting Zaïre and Angola, and their language; a member of this people. Also *attrib.* or as *adj.*

1884 *Encycl. Brit.* XVII. 319/1 *Equatorial Group*:.. Wa-Lunda, Kioko, Wa-Shinsh, [etc.]. 1897 A. J. BUTLER tr. *Ratzel's Hist. Mankind* II. 376 It is otherwise with the Songos and Kiokos, who let you deal with them in the usual way. *Ibid.* 404 Here, Lunda will be spoken; there, perhaps only half a mile away, Kioko. 1908 H. H. JOHNSTON *George Grenfell & Congo* I. xi. 194 Round him [*sc.* a person of Hima descent] a community would group itself... Thus..the kingdoms..of the Luba, Lunda, Kioko, and other Bantu countries came into existence. 1948 M. GUTHRIE *Classification Bantu Lang.* Index 86/1 Kioko (A[ngola]). 1966 *Chambers's Encycl.* III. 50/1 Among the African Kioko, however, it is the availability of criminals for food that seems paramount.

kiore (kiō·re). *N.Z.* [Maori.] In full, *kiore rat*. A small vegetarian rat, *Rattus exulans*, native to New Zealand. Also *transf.*

1838 J. S. POLACK *New Zealand* I. ix. 314 The *kiore*, or rat, has been introduced at an early period by European vessels. 1840 —— *Manners & Customs New Zealanders* II. xiii. 125 A thin person [is termed] Kioré or rat. 1843 E. DIEFFENBACH *Trav. N.Z.* II. I. vii. 114 The fat of the native rats (Kiore) killed on such lands should be given to the principal proprietor. 1883 in A. R. Wallace *Australasia* (ed. 3) xxvi. 559 The native rat, called Kiore, has been destroyed by the imported European rat. 1949 P. BUCK *Coming of Maori* (1950) II. i. 102 The Polynesian rat (*kiore, Mus exulans*) arrived in the voyaging canoes... They were probably stowaways. 1959 TINDALE & LINDSAY *Rangatira* 203 The kiore rat, taken to New Zealand by the ancestors of the Maori..was a small vegetarian animal..now extinct..except on a small island off the coast. 1966 *Encycl. N.Z.* III. 50/1 By the early 1920s it was believed that the kiore had become extinct. Today it is known to have survived in a few localities in widely separated areas.

kiosk. Add: (Now usu. with pronunc. kī·ǫsk.) **2.** No longer restricted to France and Belgium.

1933 P. MACDONALD *Mystery of Dead Police* xvii. 186 His quarry was at the change kiosk. 1963 V. NABOKOV *Gift* iii. 156 There was..a triangular island with a kiosk, at which tram conductors regaled themselves with milk. 1964 G. JOHNSTON *My Brother Jack* 36 Somehow we were able to get to the big kiosk-restaurant behind the point. *Ibid.* 37 The kiosk had been very late Victorian, with imitation turrets and spires. 1966 *South Australian Yearbk.* No. 1, 169 There are refreshment kiosk facilities. 1966 *Listener* 18 Aug. 227/2 Breaking into a tobacco kiosk. 1971 *E. Afr. Standard* (Nairobi) 10 Apr. 7/8 Most of the food sold in the kiosks is approved by medical officers.

3. = *telephone kiosk.

1928 *Daily Mail* 25 July 19/4 It is expected that nearly 500,000 new lines will be laid, several thousand new kiosks erected, and several hundred telephones fixed at rural railway stations. 1972 'H. BUCKMASTER' *Walking Trip* 197 'I'd better call Norman..he has a right to know.' 'Here's a kiosk. Have you enough change?' 1974 M. BABSON *Stalking Lamb* xxiv. 179 He broke off the connection, swung open the door and stepped out of the kiosk.

Kiowa (kəi·ŏwă, kəi·ŏwǭ), *sb.* and *a.* Also **Kiawa, Kyaway,** etc. [Native name.] **A.** *sb.* **a.** An Indian people of the south-western U.S.; a member of this people. **b.** The language of this people.

1810 Z. M. PIKE *Acct. Expeditions Sources Mississippi* App. II. 16 The only nations with whom the Pawnees are now at war, are the Tetaus, Utahs, and Kyaways. 1849 G. A. F. RUXTON *Life in Far West* vii. The Kioway loves the pale-face, and gives him warning. 1856 W. W. WHIPPLE in *Rep. Explor. Route to Pacific* (U.S. War Dept.) III. v. 80 Some resemblances are likewise to be observed between the Kioway and the languages of the southern and western tribes of the Sioux or Dakota stock. 1874 [see *COMANCHE *sb.* 2]. 1928 *U.S. Bureau Amer. Ethnol. Bull.* No. 84. 2 Six vowel qualities and twenty-two consonants are found in Kiowa. 1959 E. TUNIS *Indians* 106/1 The fingers of one hand, held up and moved in a shaking circle, was the Kiowa—rattlebrained. 1965 *Canad. Jrnl. Linguistics* Spring 78 Kiowa (Powell's 'Kiowan'), a language of the western Plains, seems now to have been reasonably well established as a part of the Aztecan-Tanoan phylum. 1969 *Observer* (Colour Suppl.) 18 May 25/4 When the son of a wealthy Kiowa achieved an exploit, everyone heard about it.

B. *adj.* Of or pertaining to any of the above; *Kiowa Apache*, an Athapascan people associated with the Kiowa; a member of this people; also, their language.

1821 J. FOWLER *Jrnl.* (1898) 64 The Kiowa cheef with his nation had stoped and intended we shold stop with them. 1865 J. PIKE *Scout & Ranger* iv. 64 Houston's design was to carry the war into the Comanche and Kiowa country. 1885 W. P. CLARK *Indian Sign Lang.* 33 This gesture refers to the Apaches living with the Kiowas at the Wichita Agency, Indian Territory..frequently called Kiowa Apaches. 1928 *U.S. Bureau Amer. Ethnol. Bull.* No. 84. 2 (*heading*) Vocabulary of the Kiowa language. 1937 J. G. MCALLISTER in F. Eggan *Social Anthropol. N. Amer. Tribes* iii. 100 Communication was carried on by means of the sign language, and most of the old Kiowa-Apache knew a little Kiowa, though very few of the Kiowa had any knowledge of the Kiowa Apache tongue. 1963 *Univ. Calif. Publ. Linguistics* XXIX. 102 Western Apache..is one of the Apachean languages. Others are Chiricahua, Mescalero, Jicarilla, Kiowa Apache, Navaho, and Lipan. 1969 *Observer* (Colour Suppl.) 18 May 25/4 A Kiowa warrior was forced by custom to give away some of his wealth.

kip, *sb.*[2] Add: **3.** *Gymnastics.* (See quot. 1972[1].) *U.S.*

1909 in WEBSTER. 1967 [see *KIP *v.*[3]]. 1972 W. VINCENT *Gymnastic Routines for Men* 123 A kip is a vigorous and rapid extension of the hip joint for the purpose of developing momentum to raise the center of gravity of the body. It may be performed on all the events in gymnastics in one form or another. *Ibid.*, Kips may be performed forward (clockwise) or backward (counter-clockwise) with either the legs or the upper body as the moving part and the other as the stabilizing part. 1972 B. TAYLOR et al. *Olympic Gymnastics for Men & Women* v. 108/2 The movement begins with the gymnast jumping to a glide swing on the low bar and continuing into a glide kip position.

kip, *sb.*[3] **1.** Delete † *Obs.* and add later *Comb.* examples.

1922 JOYCE *Ulysses* 541, I saw him, kipkeeper! Pox and gleet vendor! 1965 BROPHY & PARTRIDGE *Long Trail* 140 *Kip-shop*, a brothel.

2. Also (*rare*) **kipp,** and *Comb.* as *kip-house, -shop*. Also, a sleep, the action of sleeping.

1889 BARRÈRE & LELAND *Dict. Slang* I. 521/2 *Kip house*, a tramps' or vagrants' lodging-house. 1893 *Sessions Papers Cent. Criminal Court* 16 Nov. 39 He said, 'I only came here for a *kip*.'..*Kip* means sleep, I believe. 1908 J. M. SULLIVAN *Criminal Slang* 14 *Kipp*, a lodging house. 1925 FRASER & GIBBONS *Soldier & Sailor Words* 136 *Kip*: A sleep. Rest. A bed. A hammock, *e.g.*, 'To do a kip—to have a sleep.' 1932 *Fortn. Rev.* Mar. 325 The jake drinker's..earning capacity is nil, and if he has no lair of his own there is the 'doss house' or 'kip shop'. 1936 J. CURTIS *Gilt Kid* i. 12 He had spent a few nights in kip-shops. 1938 —— *They drive by Night* ix. 103, I got to have a rest. I ain't had no kip. 1943 M. HARRISON *Reported Safe Arrival* 18 Like the Professor, Harry was 'partial to a kip'. 1946 *Penguin New Writing* XXVIII. 123 Conditions under which the transport drivers work, of their cafés and kip-houses. 1962 *Observer* 11 Mar. 34/3 (*caption*) Dossers at a London kip-house. 1971 B. W. ALDISS *Soldier Erect* 78, I had to stay with the captain..while the other lucky sods settled down for a brief kip.

kip (kip), *sb.*[6] *Austral.* and *N.Z.* [Origin unknown.] A small piece of wood from which pennies are spun in the game of two-up.

1898 *Bulletin* (Sydney) 17 Dec. Red Page/2 The *kip* is the piece of wood used in 'two-up', otherwise pitch and toss. 1933 *Ibid.* 5 July 20/1, I see the pennies in the air, The outstretched hand that holds the kip. 1945 [see *BAT *sb.*[2] 3 e]. 1948 V. PALMER *Golconda* xxx. 250 He [was].. becoming more convinced every day that his whole future lay in winning the [Parliamentary] seat. At first the idea had been hardly more than a toss of the kip to him; now it was woven into his daily fantasies. 1964 A. WYKES *Gambling* iii. 62 In this game [*sc.* two-up], two pennies are placed on a flat stick (called the 'kip') and are thrown into the air by the 'spinner'.

kip (kip), *sb.*[7] *Engin.* orig. *U.S.* [Prob. f. KI(LO- + P(OUND *sb.*[1]] A unit of force equal to the weight of 1,000 lb., used in expressing loads.

1915 H. R. THAYER *Struct. Design* II. vi. 87 Shear in kips. [*Note*] 1 Kip = 1000 lbs. *Ibid.* 250 Maximum shear 110 kips. 1949 S. BUTTERWORTH *Struct. Analysis* ii. 30 The actual sway force is 3 kips. 1959 L. C. URQUHART *Civil Engin. Handbk.* (ed. 4) v. 45 The panel load on the upper lateral system is 25 × 150 = 3,750 lb = 3·75 kips. 1962 *Engineering* 8 June 746/3 Each of these pavement designs could be expected to carry a million applications of the 18 kip axle load..before serviceability dropped to 2·5.

kip (kip), *sb.*[8] [Thai.] The basic monetary unit of Laos.

1955 *Britannica Bk. Year* 265/1 The Laotian unit of currency had its name changed from *piastre* to *kip*, without any effect on its purchasing power. 1959 *Economist* 24 Jan. 305/2 The *kip* has been devalued. 1965 *Ibid.* 20 Nov. p. xxxvi/1 An indelible public impression of foreign exchange dealers..inhabiting a rarefied world of eight-ball arbitrage and private jokes about the baht, the kip and the won.

kip (kip), *v.*[2] *slang.* [f. KIP *sb.*[3]] *intr.* To go to bed, sleep. Also, to lie *down*. So **ki·pping** *vbl. sb.*; also *attrib.*, as **kipping-house,** a lodging-house.

1889 BARRÈRE & LELAND *Dict. Slang* I. 522/1 *Kip*, to (popular and thieves), to sleep or lodge. 1899 C. ROOK *Hooligan Nights* i. 10 Next door..that's where me and my muvver kipped when I was a nipper. 1919 *Athenæum* 1 Aug. 695/2 'To kip' is to go to bed—or what serves for a bed. 1925 E. JERVIS *25 Yrs. in Six Prisons* xix. 243, I used to conduct services in the 'kippin'-'aases', or common lodging-houses. 1929 J. B. PRIESTLEY *Good Companions* I. iv. 116 Yes, we'll have to kip down for an hour or two, Annie. 1938 J. CURTIS *They drive by Night* iv. 46 I'm kipping here tonight and all. 1939 *Airman's Gaz.* Dec., This will be very useful if you forced-land and have to kip out in a field. 1961 *New Statesman* 26 May 830/3 Nancy..set her persuasive charms to work to get Billy, Bob and Nick a free sky-sheltered bench to kip on. 1973 *Weekly News* (Glasgow) 11 Aug. 14/4 A driver whose van broke down near Bristol, decided to kip down in the driver's seat.

kip (kip), *v.*[3] *Gymnastics. U.S.* [f. *KIP *sb.*[2] 3.] *intr.* To perform a kip.

1909 in WEBSTER. 1967 LOKEN & WILLOUGHBY *Compl. Bk. Gymnastics* (ed. 2) ii. 12 Go to the bridge from the kip position on back of the shoulders using the kipping action.

Kipchak (kiptʃā·k). Also **Qipchak.** [Russ., ad. Jagatai.] **A.** *sb.* **a.** A member of a Mongolian people of central Asia. **b.** The language of this people, a Turkic dialect. **B.** *adj.* Of or pertaining to this people, or their language.

1865 J. & R. MICHELL tr. *Valikhanof's Russians Cent. Asia* iii. 62 The towns of Almalyk..were chief stations on the high road traversed by the Genoese traders..as well as by the Kipchak ambassadors on their..missions to the great Khan. 1879 *Encycl. Brit.* IX. 85/2 The nomads are mainly Kipchaks and Kara Kirghiz or Buruts. 1898 A. J. BUTLER tr. *Ratzel's Hist. Mankind* III. 319 The Kiptchaks, whose fame for extraordinary valour is known throughout Central Asia. *Ibid.* 348 The Kiptchaks are only a clan of the Kara-Kirghis. 1953 O. CAROE *Soviet Empire* I. iii. 37 The migrations of the Oghuz may have been prompted by Kipchak pressure. *Ibid.* 38 The Kipchak language was of the Turkic family. 1959 *Chambers's Encycl.* VI. 426/2 *Golden Horde*, or West Kipchak Horde, the name given to the western division of the great Mogul empire..after.. 1241... The neighbouring East Kipchak Horde was known as 'White'. 1970 D. M. LANG *Armenia* xi. 273 Other Armenian communities of the diaspora adopted the Mkhitar Code as their own, so that it was translated into Latin, Polish, Georgian, Russian and even Qipchak. 1972 G. CLAUSON *Etym. Dict. pre-13th-Cent. Turkish* p. xix, In XI the Kıpçak were west of the Oğuz in southern Russia.

Kiplingese (kipliŋī·z). [-ESE.] The literary style and characteristics of the writer Rudyard Kipling (1865–1936). Also **Kiplinge·sque** *a.* [see -ESQUE], resembling Kipling in style; so **Kiplinge·squely** *adv.*; **Ki·plingish** *a.*, typical of Kipling or his works; **Ki·plingite** [see -ITE[1] 1 b], an admirer of Kipling; as *adj.*, characteristic of Kipling; **Ki·plingize** *v.* [see -IZE] *trans.*, to make Kipling-like.

1894 'Sunlight' *Year Bk. 1895* 77 A glance at the adaptation is enough to reveal its Kiplingesque roll and emphasis. 1898 *Windsor Mag.* Dec. 131/1 True Kiplingites. 1899 'G. F. MONKSHOOD' *Kipling* 188 Perhaps the most distinctly Kiplingite piece of prose in the whole book. 1899 *Westm. Gaz.* 28 June 3/1 Thorpe..is, merely, the primitive Kiplingesque type of man transferred from the battle-field or the plains of India to the Stock Exchange. 1899 *Daily News* 1 Dec. 8/2 The account of the making of the first axe..is told quite in the heroic style of Kingsley and Morris, flavoured here and there with more than a dash of 'Kiplingese'. 1903 *Times Lit. Suppl.* 2 Oct. 277/3 The whole poem..has another claim upon the attention of the reader as an example of Kiplingized Longfellow. 1905 CHESTERTON *Heretics* iii. 45 The modern army is not a miracle of courage; it has not enough opportunities, owing to the cowardice of everybody else. But it is really a miracle of organization, and that is the truly Kiplingite ideal. *Ibid.* xx. 292 No man has any business to be a Kiplingite without being a politician, and an Imperialist politician. 1909 H. G. WELLS *Tono-Bungay* IV. iii. 492 They served me up to the public in turgid degenerate Kiplingese. 1921 G. B. SHAW *Back to Methuselah* IV. i. 170 You have actually Kiplingized me... He is said to have invented the electric hedge. I consider that in using it on me you have taken a very great liberty. 1928 *Weekly Dispatch* 24 June 15/3 The road to home these days lies across the 49th degree of latitude between Bishops Rock and Nantucket. It is a sad, un-Kiplingish thought. 1931 *Times Lit. Suppl.* 17 Sept. 692/3 It contains also quite a number of amusing and quite Kiplingesque 'Just So' stories. 1966 *Punch* 8 June 826/1 Nor am I sure that the anonymous genius who originally picked 'East of Suez' to describe the sphere of our Asian involvement chose wisely. It's so emotively Kiplingesque, so redolent of imperial splendours and miseries. 1969 *Guardian* 27 Nov. 14/6 No one ran Kiplingesquely amuck; no shot was fired in anger. 1972 J. WAINWRIGHT *Night is Time to Die* 53 *The Green Eye of the Yellow God*... It's Kiplingish... But it was written by J. Milton Hayes. **1973** *Daily Tel.* 24 Nov. 16 Kiplingites will be interested to learn that Kipling Terrace, a Victorian development at Westward Ho!, North Devon, is to be auctioned next month.

Kiplingism (ki·pliŋiz'm). [-ISM.] †**1.** *Cambridge Univ. slang.* A sarcastic term for the errors and solecisms alleged to occur in the edition of the 'Codex Bezæ' (1793) by Thomas Kipling (d. 1822), afterwards Dean of Peterborough.

1803 *Gradus ad Cantabrigiam* 81 A *Kiplingism*; a blunder-but levelled at poor Prisician's head by the learned Dr. Kipling... **1899** *G. Monkhouse Kipling* 15 A 'Kiplingism' was long an expression for a Latin blunder. **1950** M. Marples *University Slang* 110 An anthology of *Kiplingisms*...

2. Views or opinions or style of expression characteristic of Rudyard Kipling (see *KIPLINGESE).

1898 *Daily News* 7 Oct. 6/3 The manner otherwise may degenerate into sheer mannerism, a Kiplingism of Kipling. **1901** *Speaker* 26 Jan. 469/1 Sportsmen may be divided into two classes—those who care more for the chase than the killing and those who merely make 'bags' and break records. But the latter are not sportsmen...and their method is nothing but Kiplingism out of place. **1920** H. G. Wells *Outl. Hist.* 524/1 The crude Darwinism and the Kiplingism of the later Victorian years.

Kipp (kip). *Chem.* The name of Petrus Jacobus *Kipp* (1808–64), German chemist, used in the possessive (and *absol.*, or in *attrib.* use) to denote an apparatus for the generation of gas by the action of a liquid on a solid as and when gas is required.

The apparatus consists essentially of three glass bulbs, of which the upper and lower ones are connected and contain the liquid and the middle one is connected with the lower one and contains the solid; while a tap in the middle bulb is open, liquid rises into the bulb and gas is evolved, whilst closing the tap causes the pressure of the gas to increase until the liquid is forced out of the middle bulb into the lower and upper ones, out of contact with the solid.

1879 *Proc. Cambr. Philos. Soc.* III. 160 A gentle current of hydrogen from the Kipp's apparatus of... **1901** F. G. Benedict *Chem. Lect. Exper.* 5 The Kipp generator, or one of its various modifications, remains today the only portable gas generator for the lecture table... **1912** J. W. Mellor *Mod. Inorg. Chem.* iii. 45 Kipp's apparatus is very convenient when a steady current of hydrogen is needed for some time. **1921** J. R. Partington *Text-bk. Inorg. Chem.* xi. 185 Instead of a flask, a Kipp's apparatus may be used... **1965** D. Abbott *Inorg. Chem.* xi. 135 It [sc. hydrogen sulphide] is most conveniently prepared for laboratory use in a Kipp's apparatus by the action of dilute hydrochloric acid on ferrous sulphide.

kipper, *sb.* and *a.* Add: **A.** *sb.* **3. a.** A person, esp. a young or small person, a child. *slang.*

1905 *Daily Chron.* 30 Mar. 4/7 The expression 'giddy kipper', which Mr. Charles Brookfield has introduced to Mr. Justice Darling's notice. **1907** *Punch* 10 Apr. 254/2 Half-a-dozen dreadfully common young bicyclists were commenting on her discomfiture with delightful inventions of 'Giddy old Kipper', 'Sweet Seventeen', 'Cheero, Maudie—you'll win!' **1923** M. M. Gem *Helstrington's Adventures* 175 If you're enterprising enough to climb one of the trees christened by usage 'The Kipper's Tree'... **1959** I. & P. Opie *Lore & Lang. Schoolch.* ix. 170 A chap who has got duck's disease is most often labelled 'Titch'... Alternatively: ankle biter, ..kipper, microbe, midge, [etc.].

b. An Englishman, an English immigrant in Australia. *Austral. slang.*

1946 R. Rivett *Behind Bamboo* 397 *Kipper*, Englishman. **1946** *Sunday Sun* (Sydney) 8 Aug. Suppl. 15 An able seaman on a kipper warship called the Eagle. **1963** *Times Lit. Suppl.* 24 May 370/2 Quite often they [sc. English immigrants in Australia] are referred to as Kippers. **1967** K. Giles *Death & Mr Prettyman* ii. 57 You kippers—no guts and two faces—see only your under the armpits... What about the east of Suez caper, eh?

4. *Naut. slang.* A torpedo. Cf. *FISH *sb.*[1] 1 h.

1953 A. Mars *Unbroken* iii. 74 As she was only crawling along I aimed my first 'kipper' just a fraction ahead of her bows. **1959** G. Jenkins *Twist of Sand* v. 86, I evaluate its firing power at eighteen torpedoes—I think kipper is a distressing piece of naval slang—in thirty minutes.

C. kipper kite *R.A.F. slang* (see quot. 1943); **kipper tie** (see quot. 1969), a gaudy and very wide neck-tie.

[**1941** L. Walmsley *Fishermen at War* ix. 138 Kipper, I discovered, was airman's slang for a fishing boat. The chief function of this particular station was the escorting of convoys and fishing fleets, and the section which had the latter duty to perform was known as the 'Kipper Patrol'.] **1942** *Gen* 1 Sept. 14/1 A Coastal Command plane is a 'kipper kite'. **1943** Hunt & Pringle *Service Slang* 42 *Kipper-kites*, aircraft engaged on convoy escort duties over the North Sea and usually giving protection to the fishing-vessels. **1966** *Daily Tel.* 20 Jan. 15/6 Neckties are slightly wider and pointed, though not yet as floppy as London's Carnaby Street kipper ties. **1969** *Guardian* 16 Sept. 9/4 Michael Fish [sc. a London designer of men's wear]...can...take credit for popularising the wide tie, named 'kipper' after him. **1973** *Times* 30 May 18/3 He had come from his Suffolk home wearing a kipper tie and black and white patterned shirt, full of energy and ideas.

ki·pper, *sb.*[2] *Austral.* [Native name.] A young Aboriginal who has been initiated and is admitted to the rights of manhood.

1841 C. Eipper *Statement German Mission to Aborigines* 8 With these weapons the natives invest their young men at the age of from fourteen to sixteen years... These young men are then called *kippers*, and for the first time enjoy the privilege of taking an active part in the fight. **1853** H. Jones *Adventures Austral.* 126 Around us sat 'Kippers', i.e. 'hobbledehoy blacks'. **1885** R. C. Praed *Austral. Life* i. 24 A ceremony at which the young men...receive the rank of warriors and are henceforth called *Kippers*. **1966** W. S. Ramson *Austral. Eng.* vi. 129 *Bora*, 'a rite of initiation', *kipper*, used of a youth who has passed through such a rite, and *boyla* and *koradji*, 'an aboriginal medicine-man or witchdoctor', are used only in their original and specific senses.

kipper, *v.* Add: Also *intr.*, *transf.*, and *fig.*

1894 Kipling *Seven Seas* (1896) 56 The Leevin' God, That does not kipper souls for sport or break a life in jest. **1909** R. Beach *Silver Horde* 129 He's an awful spender. I'm half kippered (= drunk) myself. **1924** *Glasgow Herald* 28 Jan. 10 Oily cotton-waste was picked up at the gates of yards and factories, and our hands were duly kippered over smoking lumps of this stuff. **1930** R. Campbell *Adamastor* 20 Hang him up to kipper in the sun. **1973** *Times* 14 May p. iii/3 [Advt.], Central heating designed to prevent the average household from being kippered on one side and frozen on the other. **1969** *Daily Tel.* 30 Dec. 6/5 On the last day of addiction, some twice or thrice as many cigarettes as normal. The next morning you should feel sufficiently kippered as to see the error of your ways.

ki·pperer. [-ER[1].] One who kippers herrings.

1902 *Nature* 4 Sept. 435/2 The 'kipperers' and the 'gutters' have their peculiar troubles. **1920** *Glasgow Herald* 10 July 6 Joint meetings of fishermen, curers, salesmen, buyers, and kipperers were held at both places. **1930** *Aberdeen Press & Jrnl.* 21 Mar. 6/5 A shed standing alongside a kippering kiln... occupied by Mr. David Mackenzie, kipperer. **1955** *Times* 3 May 6/3 Merchants and kipperers refused to pay the new minimum price of 91s. a cran.

kippersol (ki·pəsɒl). *S. Afr.* Also **kiepersol**. [Corrupt. f. *KIEPERSOL.] A small evergreen African tree of the genus *Cussonia*, belonging to the family Araliaceæ.

1838 'R. Iron' *Dream Life* 26 A kippersol tree. *Ibid.* 29 She cut the front off a kippersol, and got out a large piece... and sat down to chew it. Kippersol is like raw quince. **1898** 34 When one has had no food but kippersol juice for two days. **1932** T. R. Sim *Native Timbers S. Afr.* 94 Kippersol [etc]. **1954** R. Lighton *Out of the Strong* 55 Lemon trees, kippersol, and carob... **1961** *Cowrie Bushveld, Bananas & Bounty* v. 76 According to legend the [Cussonia] *umbellifera*, with its broad flat leaves, was first called Kiepersol by the Cape Malays, whose word for monkey is *kie* and for umbrella in the Afrikaans word *persol* and the combination results in the picturesque name of Monkey's Umbrella. **1973** Palmer & Pitman *Trees S. Afr.* III. 1691 Most of the species [of *Cussonia*] are known as 'kiepersol'... The name travelled to the Cape where it was first used for the Cussonia species common around Cape Town, *Cussonia thyrsiflora* Thunb., and is now a general name for all the cussonias with their umbrella-like mops of leaves at the ends of the branches.

kipsie, kipsy (ki·psi). *Austral.* [f. KIP *sb.*[2] + -Y[6], -IE.] A house (see also quot. 1919).

1916 C. J. Dennis *Songs Sentimental Bloke* 124 *Kipsie*, a house, the home. **1919** W. H. Downing *Digger Dial.* 31 *Kippsie*—lean-to; shelter; house; dugout. **1943** *Coast to Coast 1942* 92 He turned and looked our little weather-beaten kipsie over. **1946** *Coast to Coast 1945* 236 Our little kipsy brushed a heavy stillness like that lumber bush that hangs over tree and earth before a storm breaks. **1969** *Courier-Mail* (Brisbane) 29 Nov. 13/1 Our kipsy was an abut-looking little place.

Kipsigis (ki·psigis). [Native name.] The name of a people inhabiting western Kenya, and their Nilotic language; a member of this people. Also *attrib.* or as *adj.*

1931 *Africa* IV. 467 The Kipsigis and the other tribes speaking an almost identical language...are really pastoral in custom and thought, though they have ceased to be truly nomadic, like the Masai. **1939** E. E. Evans-Pritchard in J. G. Peristiany *Social Inst. Kipsigis* p. xx, The Kipsigis are Nandi-speaking. *Ibid.* p. xxxi, The Kipsigis clan (*oret*) is a totemic non-exogamous group of persons, and the neighbouring tribes seems to be responsible for its origin. **1950** J. A. C. Huntingford *Nandi Work & Culture* ii. 7 The Kipsigis (or Lumbwa), are... known as the European as Lumbwa, a misnomer of unknown origin, as none of the neighbouring tribes seems to be responsible for this name. *Ibid.* xii. 131 The Kipsigis text accompanied the translation for the better understanding of Kipsigis thought. **1947** *E. Afr. Ann.* 1946–7 43/1 Nilo-Hamitic tribes such as the Nandi, Suk and Kipsigis. **1964** A. N. Tucker in D. Abercrombie et al. *Daniel Jones* 445 The Kalenjin languages are spoken by two groups of non-Bantu speakers in the Rift Valley Province...the Nandi Group, comprising Nandi, Kipsigis, Keyo ('Elgeyo'), Tugen ('Kamasia'), Nony ('Elgon'), Sabaot ('Sabei'), and others. *Ibid.*..The languages chosen here as representative are Nandi (with occasional reference to Kipsigis) and Western Päkot. **1971** *E. Afr. Standard* (Nairobi) 10 Apr. 7/4 A fight...broke out between Kipsigis and Kisii tribesmen.

kir (kɪr). Also **Kir.** [The name of Canon Felix *Kir* (1876–1968), mayor of Dijon, who is said to have invented the recipe.] A drink made from dry white wine and crème de cassis.

1966 *Times Lit. Suppl.* 19 May 456/5 M. Follain's work should be read with a kir, a benedictine or a calvados. **1967** *Sat. Rev.* (U.S.) 22 Apr. 50/2 In 1967 the drink...is *kir*, which the pros call *blanc Cassis*, or if they are really with-it buying so—Mascat. **1967** L. Deighton *Expensive Place* x. 74 'Waiter,' he called, 'Four kirs.' The white wine and cassis came. **1967** A. Lichine *Encycl. Wines* 172 Around Dijon it [sc. Cassis] is used as a popular aperitif, a little Cassis being put in a glass that is then filled with a fairly neutral, dry white wine... Also called *kir*. **1968** M. Tripp *One is One* xvii. 159 You couldn't get Kir, his customary drink, anywhere. **1974** *Guardian* 22 Aug. 11/6 The summer aperitif 'kir'.

kiradjee, var. *KORADJI.

Kirby (kɜ·ɹbi). The name of Charles *Kirby*, 17th-c. English fish-hook maker, used *attrib.* and *absol.* (†and in the possessive) to denote a design of fish-hook originated by him.

[**1655** Walton *Compleat Angler* (ed. 2) xvii. 313 But if you will buy choice hooks, I will one day walk with you to Charles Kerbyes in Harp Alley in Shoe-lane, who is the most exact and best Hook-maker that the Nation affords.] **1804** T. Best *Conc. Treat. Art of Angling* (ed. 6) ii. 23 Ford and Kirby's hooks are excellent ones. **1823** T. F. Salter *Angler's Guide* (ed. 5) xvi. 243 In choosing fly-hooks, prefer for the single ones whose shank is similar to the Kirby hook, or those which have a loop shank. **1870** H. Cholmondeley-Pennell *Mod. Pract. Angler* i. 9 The round and Kirby bends are very deficient in penetrating power, and disproportionately short in the shank as compared to their breadth of bend, either for appearance or use, more particularly in the matter of flies. **1967** H. Knox *Blacklight* ii. 31 The box held a collection of wickedly barbed Kirby hooks.

kirby-grip (kɜ·ɹbi grip). Also **Kirbigrip** (proprietary name), **kirbigrip**. A type of sprung hair-grip.

1926 *Trade Marks Jrnl.* 6 Jan. 12 Kirbigrip...Hair-pins of ordinary Metal. Kirby, Beard & Co., Limited,... Birmingham; Manufacturers. **1945** 'A. Gilbert' *Don't open Door* xi. 55 Two plain brown combs and a kirbygrip. **1949** 'J. Tey' *Brat Farrar* xxiii. 224 The aged kirby-grip that kept Jane's hair off her face. **1953** Dylan Thomas *Adventures Skin Trade* (1955) i. 18 Mrs. Probert next door, ..butting the air with her kirby-grips. **1955** See *hair-grip* s.v. *HAIR *sb.* 9 b]. **1959** *Woman's Own* 12 Dec. 21/3 Ribbon bows fixed to kirby-grips or combs. **1960** C. Snow *Marianne & Mark* i. 20 Marianne fought a card of kirbygrips. **1973** B. Bainbridge *Dressmaker* i. 20 Nellie, when put out, could appear to be a sprightly, her white hair plastered to her head in waves and a kirby grip to keep it neat.

Kirchhoff (kɪ·əɹtʃɒf, ‖ ki·ɹχhɒf). Also (erron.) Kirchoff. The name of Gustav Robert *Kirchhoff* (1824–87), Ger. physicist, used in *Kirchhoff's law*: **a.** *Electr.* Either of two laws concerning electric networks in which steady currents are flowing: (*a*) (the first law) the algebraic sum of the currents in all the conductors that meet in a point is zero; (*b*) (the second law) the algebraic sum of the products of current and resistance in each part of any closed path in a network is equal to the algebraic sum of the e.m.f.s in that path.

1891 R. Mann *Dict. Ind. Art.* (1896) 269 ... *Kirchhoff's law*... ... Kirchhoff's *law*, *Ibid.* ... **1919** W. H. Kemp *Handbk. Electr. Testing* v. 45 Kirchhoff's laws, though exceedingly simple,...are not so well known as they should be. **1966** W. C. D. Whetham *Theory Exper. Electr.* v. 127 The principles of continuous current-flow which we have now established may conveniently be applied to complex circuits and networks of conductors in the form of two statements known as Kirchhoff's laws. **1950** M. Nelkon *Electr.* v. 130 Kirchhoff's first law is a mathematical statement of the fact that the charges do not accumulate at any junction of an electrical circuit. *Ibid.* 132 We need to apply Kirchhoff's second law to two complete circuits as there are two unknowns.

b. *Physics.* The law that the absorptivity of a body for radiant energy of any particular wavelength is equal to its emissivity at the same temperature for the same wavelength.

1901 G. K. Burgess tr. *Le Chatelier & Boudouard's High-Temperature Measurements* viii. 140 (*heading*) Kirchhoff's law. **1945** F. A. Berry et al. *Handbk. Meteorol.* iv. 288 It is an immediate consequence of Kirchhoff's law that the intensity emitted by a body can never exceed the black-body intensity and can equal it only in the spectral regions where the body is opaque. **1967** R. W. Fairbridge *Encycl. Atmospheric Sci.* 793/1 We assume black-body radiation which is that of a body that is characterized by maximum possible absorption at all incident wavelengths, insuring maximum emissivity according to Kirchhoff's law.

Kirghiz (kiə·ɹgɪ·z), *sb.* and *a.* Also **Khirgese**, **Kirghis**, **Kirgiz**. [ad. Russ. *Kirgiz*.] **A.** *sb.* A widespread Mongolian people of west central Asia, now chiefly inhabiting the Kirghiz Soviet Socialist Republic; a member of this people; their Turkic language. **B.** *adj.* Of or

pertaining to the Kirghiz; *spec.* **Kirghiz pheasant** = *Mongolian pheasant* (*MONGOLIAN *a.* 4). Also **Kirghi-zian** *a.* and *sb.*

1652 P. Heylyn *Cosmographie* iii. 190 These again subdivided into severall Tribes, which they call their Hordes, of which the most considerable are, 1. the Nagaian Tartars, 2. the Zavolhenses, 3. the Thumenenses, 4. the Kirgesi. **1837** De Quincey in *Blackw. Mag.* July 109/2 The murderous attacks of their cruel enemies the Bashkirs and the Kirghizes. **1888** *Encycl. Brit.* XXIII. 661/2 Tatar dialects (Kirghizian, Bashkiri, Nogai). **1898** A. J. Butler tr. *Ratzel's Hist. Mankind* III. 326 The Kirghiz women adorn their plaits with beads, shells, and copper buttons. **1908** T. G. Tucker *Introd. Nat. Hist. Lang.* 134 *Kirghiz*, comprising the speeches of the Black Kirghiz (or *Buruts*) in the part of Turkestan bordering on China, and of the Cossack Kirghiz to the north of the Caspian, the sea of Aral and Lake Balkash. **1921** *19th Cent.* May 871 Kirghizes, Lesghiens, Mingrelians. **1922** C. W. Beebe *Monogr. Pheasants* III. 96 On the south-east the enormous Tian-Shan serve as the boundary between the Kirghiz Pheasant and both *shawi* and *tarimensis*. **1922** *Contemp. Rev.* Sept. 342 The Kirghiz population has retained its nomadic habits. **1924** *Blackw. Mag.* Aug. 203/1 The Russians, who were conscripting young Khirgese men for use on the railway. **1931** A. U. Dilley *Oriental Rugs & Carpets* x. 226 Kirghiz rugs...are country-bred, vigorous and coarse. **1932** R. Jones tr. *Popoff's City of Red Plague* iii. 46 One could not help being struck by the large number of Mongolian types—torn by their Mussovite masters from the remote steppes and forests of Siberia and Asiatic Russia;...Kalmucks, Kirghiz, Yakuts, and similar obscure, semi-barbaric tribes. **1935** Huxley & Haddon *We Europeans* vii. 212 Ethnologists use the word [sc. Turki] loosely and in several different senses:—(b) to designate a certain ethnic group of which the Turks, the Kirghiz and the Tatars are best known. **1946** J. Millar *Horned Pigeon* i. 21 A dashing White Russian officer...spoke frequently of the Kirghizian sheep that drag their fat-containing tails behind them on little sleighs. **1955** V. Cronin *Wise Man from West* xiii. 243 In this country of chasms and precipices roamed the Kirghiz, a predatory tribe. **1961** L. F. Brosnahan *Sounds of Lang.* iv. 119 *Kirl*, is a very large rose tree. **1962** A. Mugg ... **1964** ... *Mongolian pheasant* s.v. *MONGOLIAN *a.* 4]. **1967** D. S. Parlett *Short Dict. Languages* 73 *Kirgiz*,... apparently one of the oldest (or oldest attested) Turkic tribes. **1971** *Whitaker's Almanack 1972* 966 The Kirghiz S.S.R. occupies the north-eastern part of Soviet Central Asia and borders in the south-east on China.

‖ **kiri** (kiə·ri). [Jap.] = PAULOWNIA.

1727 J. G. Scheuchzer tr. *Kæmpfer's Hist. Japan* I. 1. iv. 119 *Kiri*, is a very large but scarce Tree. **1822** F. Shoberl tr. *Titsingh's Illustr. Japan* 255 Sometimes this cane is made of the wood of the kiri-tree. **1877** *Trans. Asiatic Soc. Japan* V. i. 9 The second of the Imperial badges is a representation of the leaf and flower of the *kiri*, or *Paulownia Japonica*. **1893** A. M. Bacon *Jap. Interior* xiv. 177 The blossoms and leaves of the kiri-tree, (paulownia imperialis), which is the sign of the imperial family. **1928** Blunden *Jap. Garland* 12 The broad-leaved *kiri*. **1972** *Nat. Geographic* Sept. 374 Miss Hori still goes into the forests to select the best *kiri*, or Paulownia wood.

‖ **kirin** (ki·rin). Also **Kirin.** [Jap., f. Chinese (see KYLIN).] A fabulous beast of composite form, freq. portrayed in Japanese pottery and art (see quots.); = KYLIN.

1727 J. G. Scheuchzer tr. *Kæmpfer's Hist. Japan* I. 1. x. 125 *Kirin*, according to the description and figure, which the Japanese give of it, is a winged Quadruped, of incredible swiftness, with two soft horns standing before the breast, and bent backwards, with the body of a Horse, and claws of a Deer, and a head which comes nearest to that of a Dragon. **1875–80** Audsley & Bowes *Keramic Art of Japan* I. p. xxxviii, The Japanese have described the Kirin as a supernatural animal, requiring for its creation the concurrence of a certain constellation in the heavens. **1900** F. Litchfield *Pott. & Porc.* vii. 172 Figure subjects are not common in this kind of china, but one finds representations of...the *Kirin*,...a monster with the body and hoofs of a deer, the tail of a bull, and a horn on his forehead. **1908** H. L. Joly *Legend in Jap. Art* 158 The Chinese *Shang Huan Fu*.. female Sennin, shown riding upon a Kirin. **1963** [see *HO-HO]. **1971** L. A. Boger *Dict. World Pott. & Porc.* 170/1 Included among other familiar (porcelain) motifs were...the five fabulous creatures:... Kirin (Japanese unicorn), [etc.].

Kiriwinian (kiriwi·niăn), *sb.* and *a.* [f. *Kiriwina*, the name of the largest of the Trobriand Islands + -IAN.] **A.** *sb.* A native or inhabitant of Kiriwina; the Austronesian language spoken there. **B.** *adj.* Of or pertaining to Kiriwina.

1916 *Jrnl. R. Anthrop. Inst.* XLVI. 356 There is very little of the universally reported native's dread of darkness among the Kiriwinians. *Ibid.* 354 He...acquired sufficient knowledge of the Kiriwinian language to be able to dispense with the services of an interpreter. *Ibid.* 391 Archaic expressions...the natives only partially understand...it is extremely difficult to make them translate the meaning correctly into modern Kiriwinian. **1922** B. Malinowski *Argonauts W. Pacific* xx. 480 The Kiriwinians have to go inland to the industrial districts of Kuboma...to acquire the articles needed. **1964** *Language* XL. 306 Three Kiriwinian dialects.

Kirlian (kɪ·ɹliăn). The name of S. D. and V. K. *Kirlian*, 20th-cent. Russian electricians, used *attrib.* with reference to the process invented by them of directly recording corona discharges from the surfaces of objects on photographic material.

1970 Ostrander & Schroeder *Psi Psychic Discoveries behind Iron Curtain* xvi. 218 In the color Kirlian pictures of plant leaves we saw the same basic colors: blue and reddish-yellow. **1973** *Popular Photogr.* LXXII. 90 Examples of Kirlian photographs—that is, images obtained on film without camera or lens by direct recording of an electric charge transmitted by the given object, to which a high-frequency charge has been applied. **1974** *Science* Jan.–Feb. 19/2 Kirlian photographs may also offer a measure of interpersonal relations.

Kirman (kɪ·ɹmă·n, kɜ·ɹmă·n). Also **Karman, Kerman.** The name of a province and town in south-east Iran, used (freq. *attrib.*) to designate a carpet or rug made there, usu. having soft delicate colouring and naturalistic designs.

1876 O. St. John in F. J. Goldsmid *Eastern Persia* I. vi. 101 Not only flowers and trees, but birds, beasts, landscapes, and even human figures are found on the Karmán carpets. **1900** J. K. Mumford *Oriental Rugs* xi. 187 It has been customary, until very lately, among the rug dealers of the West and Constantinople as well, to attribute the Kirman rugs to Kermanshah. **1931** A. U. Dilley *Oriental Rugs & Carpets* iv. 94 Of all the villages weaving for export, none across a Kirman twice as big as my room at home. **1953** Reed *Oriental Rugs & Carpets* v. 69 By the turn of the century some very good Tabriz and Kirman pieces were being made. **1969** C. W. Jacobsen *Check Points How to buy Oriental Rugs* vii. 132 Take the name Kirman, which is one of the highest priced and best of new rugs being woven today... Many people bought these as Kirmanshah (also spelled Kermanshah). **1970** *Canad. Antiques Collector* Dec. 22/2 On the floor is a fine pile pink Persian Kerman rug [etc.].

Kirmanshah (kɪɹmănʃă·, kɜɹ-). Also **Kermanshah.** The name of a city in west Iran, used confusedly (freq. *attrib.*) to designate a carpet made in Kirman, usu. one with white field and flowered medallion and borders. See prec. word.

The confusion with *Kirman* seems to have arisen because of the similarity of the two names and the great importance of Kirmanshah as a wool-trading town.

1900 J. K. Mumford *Oriental Rugs* xi. 188 In design, the best of the Kermanshahs affect the floral treatment. **1904** M. B. Langton *How to know Oriental Rugs* ii. 78 There are a few antiques in this country; but the modern Kirman or Kermanshah rug, made throughout the district, rivals the old. **1931** A. U. Dilley *Oriental Rugs & Carpets* vi. 90 (*caption*), Kirman, Southeast Persia (mistakenly called Kermanshah). *Ibid.* iv. 94 Some fifty years ago, large flower-strewn medallion patterns with rippled pendants and floriated corners were devised and applied to large carpets by weavers working for Tabriz rug merchants who saw in the fine Kerman workmanship opportunity for commercial enterprise. These rugs and carpets were called Kermanshah, to permanent confusion with the rug-weavings of the city of that name located miles to the west. **1969** C. W. Jacobsen *Check Points How to buy Oriental Rugs* vii. 133 The Kirmanshah was never a current name for these rugs because they were woven in and around Kirman, while the town of Kirmanshah was 1,000 miles to the west.

kirombo (kiro·mbo). [Malagasy.] The cuckoo-roller, *Leptosomus discolor*, a large grey or black and brown bird found only in Madagascar and the Comoro Islands.

1891 *Ibis* Apr. 214 The natives of the north-west of Madagascar give this bird the name of *Kirombo*. It has the curious habit of hovering in the air and uttering a very loud note, striking its wings against the body as it calls. **1953** J. J. Simek *Naturalist in Madagascar* x. 138 The *Vorombeo*, or Kirombo roller. **1964** A. L. Thomson *New Dict. Birds* 172/1 Cuckoo-roller: *Leptosomus discolor*, sole member of the Leptosomatidae (Coraciiformes, suborder Coracii). This monotypic family is restricted to Madagascar and the nearby Comoro Islands; the bird is called 'Courol' in some works, and less frequently 'Kirombo'.

‖ **kirpan** (kɪɹpā·n). [ad. Panjabi and Hindi *kirpān*, f. Skr. *kṛpāṇa* sword.] The sword or dagger worn by Sikhs as a religious symbol.

1894 J. H. Gordon *Sikhs* iv. 41 Every true Sikh must always have five things with him, the names of which all begin with the letter *k*—namely, *..kard* (knife), and *kirpan* (sword). **1923** *Contemp. Rev.* Sept. 293 Guru Govind Singh...prescribed for his Singhs five symbols, of which it is sufficient to note here...the *kirpan*, a weapon which is sometimes a miniature carried in the hair, sometimes a dagger not more than a foot in length, sometimes a sword. **1952** J. Masters *Deceivers* v. 45 He whipped a twelve-inch dagger, the Sikh kirpan, from his belt. **1964** A. Swinson *Six Minutes to Sunset* vi. 120 It is possible that many disciples, in fact, were armed with the *kirpan* (*sword*) ... **1971** H. R. F. Keating *Inspector Ghote plays Joker* vii. 99 At his side there hung a kirpan, traditional knife of the Sikh. **1971** *Daily Tel.* 11 June 3 (*caption*) Sant Mann Singh...a visiting Sikh religious leader who has agreed not to carry his kirpan, a sword, in public.

Kirschner (kɪ·ɹʃnəɹ). [Name of Aage *Kirschner* (fl. 1905), Da. chemist.] *Kirschner value*, a number expressing the proportion of certain fatty acids (esp. butyric acid) in a fat (see quot. 1961).

1911 *Analyst* XXXVI. 337 The Kirschner value is practically a measure of the butyric acid content of the mixture. **1928** E. R. Bolton *Oils, Fats & Fatty Foods* iv. 83 Although in general practice the Kirschner value is a measure of the amount of butter fat present, yet there are a few vegetable oils that give large Kirschner and Reichert-Meissl, and low Polenske values. **1961** *Methods for Chem. Analysis of Butter (B.S.I.)* 8 The Kirschner value is the number of millilitres of 0·1N aqueous alkali solution required to neutralize the water-soluble volatile fatty acids which form water-soluble silver salts distilled from 5g of the fat under the precise conditions specified in the method. **1973** D. Pearson *Laboratory Techniques Food Analysis* vi. 156 Genuine butter fat gives Kirschner values from 20·5 to 26·4.

kirschsteinite (kɪ·ɹʃstəinait). *Min.* [f. the name of Egon *Kirschstein* (see quot. 1957) + -ITE[1].] Iron-monticellite, esp. as a naturally occurring mineral.

1957 Sahama & Hytönen in *Mineral. Mag.* XXXI. 698 For this natural CaFeSiO$_4$ the name kirschsteinite is proposed, in honour of the German geologist, the late Dr. Egon Kirschstein, who died in the events of the World War I in East Africa. *Ibid.*, The analysis corresponds to the following molecular composition: CaFeSiO$_4$ 69·4 mol. %, CaMnSiO$_4$ 4·3, CaMgSiO$_4$ 22·6, excess Fe$_2$SiO$_4$ 3·7. Accordingly, the mineral is to be called magnesian kirschsteinite. **1962** W. A. Deer et al. *Rock-Forming Min.* I. 1 The iron analogue of monticellite, kirschsteinite, CaFeSiO$_4$, is known from slags, but has not been reported from a natural occurrence, and a magnesium-rich kirschsteinite containing 69 per cent. CaFeSiO$_4$ is the most iron-rich mineral of the Fe$_2$SiO$_4$–CaFeSiO$_4$ series yet reported. **1969** *Amer. Mineralogist* LI. 1192 Mineralogical studies on the debris formed during the underground Gnome nuclear explosion in a salt horizon of the Salado formation near Carlsbad, New Mexico have shown that significant quantities of olivine and kirschsteinite are present in the water insoluble fraction.

Kir-Shehr (kiə·ɹʃēɹ). Also **Kirshehir.** The name of a town in central Turkey, used *attrib.* and *ellipt.* to designate the brightly coloured prayer rugs made there.

1900 J. K. Mumford *Oriental Rugs* x. 139 Border medallions...found in Kir-Shehrs of old date. *Ibid.*, Some of the small Kir-Shehr mats have several subdivisions of their field. **1931** A. U. Dilley *Oriental Rugs & Carpets* vi. 165 Kir-Shehr...prayer rugs [are distinguished] by central panels covered by flight of stairs-motif. **1967** *Times* 21 Feb. 21/4 (Advt.), Several fine prayer rugs: Ladik, Mudjur, Kirshehir, [etc.]. **1971** K. Watson tr. *Hubel's Bk. Carpets* 86 Cherry red and grass green are noticeable in the colour scheme of Kirshehirs.

kirtan (kɪ·ɹtan). Also **keertan, kirtana.** [Skr.] (See quots.)

1898 A. H. Fox Strangways *Music of Hindostan* ... ... **1914** *Indian Mus.* (ed. 2) vii. 313 The *Kirtana* or *Hari-kirtana* is a musical performance, sacred and ... a sacred character, the theme of which is always ... a moral. The *Kirtana* is a beautiful combination of *rhetoric*, *rhyme* and *rhythm*. **1960** Koestler *Lotus & Robot* i. i. 65 During Kirtans her body got stiff and benumbed. **1967** Singh & Massey *Indian Dances* xv. 129 The poet Jayadeva, author of the *Gita Govinda* who composed numerous keertans or devotional songs, and whose wife expressed them through dance. **1966** J. & R. Godden *Two under Indian Sun* iii. 82 Indian singing too was usually about the gods, kirtans or sacred poems that told these epics of Krishna, or Rama and his love Sita. **1970** *Listener* 24 Sept. 409/1 The *kirtan*... is the traditional song that seems to be always about spiritual love and pain. **1971** *Illustr. Weekly India* 18 Apr. 10/2 *Tamashas*... consisted in group recitations from mythology and legend, interspersed with *bhajans*, *keertans* and folk songs.

kirving, *vbl. sb.* (s.v. KIRVE *v.* 2). Add: **b.** *pl.* Coal dust or small pieces of coal produced during mining operations.

1956 [see *GUMMING *vbl. sb.*[1] b].

‖ **kisaeng** (kɪ·saŋ, -eŋ). Also **gesang, ki-saing, kisang.** [Korean.] In Korea, a trained female entertainer, the Korean equivalent of the geisha girl.

1895 L. J. Milb *Quaint Korea* vi. 155 The Korean word for the class of women of whom I am writing is ki-sang; but they are generally called geisha. **1904** C. J. D. Taylor *Koreans at Home* ix. 50 The favourite entertainment for those who can afford it is the dancing of 'gesang'. **1905** J. London *Jacket* (1915) 212 Tell us more about the kisaeng and the dancing-girls. **1908** H. N. Allen *Things Korean* viii. 125 The best performances...consist of dances by the class of public dancing girls or gesang. **1931** C. Osgood *Koreans & their Culture* xiv. 259 The kisaeng entertainments seem to have started as a court institution. **1953** D. Portway *Korea* vi. 107 The Korean married woman is much too home-bound to be expected to sit down at a public dinner, and the kisaeng girl takes her place. The kisaeng girl is definitely not a prostitute, though probably the height of her ambition is to become the concubine of some wealthy man. **1973** *Guardian* 31 Jan. 4/6 In South Korea they [sc. Japanese tourists] are herded to mock 'kisaeng' parties designed to camouflage courtesans as trained 'kisaeng' (traditionally skilful, geisha-like entertainers).

kisan (kisā·n). [Hindi *kisān*, f. Skr. *kṛṣāṇa* one who ploughs.] In India, a peasant, an agricultural worker.

1935 *Ann. Reg. 1934* 152 The Communist Party of India...had through a *kisan* (peasant) organisation in the

Punjab sought to make propagandist capital out of economic discontent. **1936** J. Nehru *Autobiog.* ix. 62 The Indian *kisans* have little staying power, little energy to resist for long. *Ibid.* x. 63 Agrarian troubles are frequently taking place in various parts of India and the *kisan* agitation in certain parts of Oudh in 1920 and 1921 was but one of them. **1959** M. Brecher *Nehru* iii. 69 Jawaharlal's discovery of the *kisans* (peasants) spurred him to action on their behalf. *Ibid.* 70 His [sc. Nehru's] reputation as a friend of the *kisan* spread into the interior. **1969** *Hindu* (Madras) 28 July 9/8 Eleven Communist kisans were arrested yesterday by the Kilayur Police when they obstructed the ploughing with tractor of some fields in Kazhakkarai village. **1969** *Pioneer* (Lucknow) 13 Aug. 5/2 The whole legal system should be overhauled to make dispensation of justice speedy and cheap, specially to the poor kisans.

kishke (ki·ʃkə). Also **kishka, kishkeh, kishker.** [Yiddish.] **a.** Beef intestine casing stuffed with a sausage-like savoury filling. **b.** In *sing.* and *pl.* The guts. *slang.*

1936 Mencken *Amer. Lang.* (ed. 4) 217 In New York City the high density of Eastern Jews in the population has made almost every New Yorker familiar with a long list of Yiddish words e.g., ..*kishkes*, *kilt*, [etc.]. **1951** L. W. Leonard *Jewish Cookery* xiv. 190 Stuffed kishke may be roasted with chicken, duck, goose or turkey. **1959** B. Kops *Hamlet of Stepney Green* i. 12 You sweat your kishkers out to give them a good education. **1964** 'Dylan Thomas' ... 120 'Phi! kishka, pishka, Hit him in the kishka. **1967** G. Sims *Last Best Friend* vi. 53 They had...dined at Bloom's in Whitechapel High Street on beetroot borsch and stuffed kishka. **1968** *Guardian* 8 Oct. 2/6 It is not every city where you can go into a café and find yourself offered ..kishke, blintzes, kiskha, [etc.]'. **1968** L. Rosten *Joys of Yiddish* 181, I laughed until my kishkas were sore. **1970** L. M. Feinsilver *Taste of Yiddish* iii. 293 Gentiles are learning to enjoy and pronounce such typical dishes as *kishke* and *kreplach*. **1972** *Listener* 16 Mar. 341/3 Kishkeh vaguely resembles a kosher haggis...stuffed with a mixture of flour meal, grated onion and fat.

Kisii (ki·si,ī). [Name of a district on the east side of Lake Victoria, Kenya.] A Bantu people from Kisii; a member of this people. Also *attrib.* or as *adj.*

1909 C. Eliot *E. Afr. Protectorate* vii. 131 The Kisii people are practically unknown to Europeans. *Ibid.*, Igizzi (the language of the Kisii) would appear from these vocabularies to have some resemblance to the Kikuyu tongue. **1934** N. K. Strange *Kenya To-Day* vi. 82 The ornamental bead belts...and Kisii ware, various shaped vases and bowls made from soapstone...are amongst the attractive curios to be bought throughout Kenya. **1939** J. G. Peristiany *Social Inst. Kipsigis* p. xxi, The traditional enemies of the Kipsigis are the Masai to the south and the Luo and Kisii to the west. **1963** *Times* 31 May 10/2 Mr. Sagini is a Kisii. **1971** *E. Afr. Standard* (Nairobi) 10 Apr. 7/4 Kisii farmer killed in fight after stock theft.

kiskadee (kiskădī·). Also **keskeedee, keskidee.** [Echoic, f. the call of the bird.] A tyrant flycatcher, *Pitangus sulphuratus*, found in Central and South America; also used for related birds of the family Tyrannidæ.

1891 *Timehri* V. 61 One of the most common of birds...is a brown and yellow Tyrant-shrike called the keskeedee (*Pitangus sulphuratus*). *Ibid.* 88 The large kiskadee (*Pitangus sulphuratus*)...whose loud, harsh and fierce cry of kiskis-kadee is to be heard at all times of the day. **1922** *Blackw. Mag.* July 161/1 Glorious clumps of bamboo with kiskadees clinging like yellow blossoms to the bending plumes. **1941** *Penguin New Writing* VI. 75 Some kiskadees were singing in the mango tree. **1955** E. R. Blake *Birds Mexico* 347 While mainly insectivorous like their relatives, Kiskadees also commonly catch small fish. **1958** J. Carew *Wild Coast* i. 7 Hector watched the birds and remembered...a pair of kiskadees. **1964** A. L. Thomson *New Dict. Birds* 316/1 Some larger species with shrike-like or terrestrial habits, such as the Kiskadee Flycatcher (or Great Kiskadee) *Pitangus sulphuratus*...take lizards, frogs, mice, and small birds.

kiskitomas. *U.S.* (Earlier and later examples.)

1809 A. Ritson *Poetical Pict. Amer.* 161 Their nuts, black walnuts, persimins, Kiscatoma nuts, and Chinquapins. [**1810** F. A. Michaux *Histoire des arbres forestiers* I. 20 *Shell bark hickory* [*sic*]...nom le plus en usage dans tous les États-Unis... *Kiskythomas*, par les Hollandais du New-Jersey.] **1832** D. J. Browne *Sylva Americana* 184 The Dutch settlers...near the city of New York, call it Kisky Thomas nut. **1836** W. Dunlap *Mem. Water Drinker* (1837) I. 48 While the rustic jest, or the tale of...wars...mingle with the cracking of the kisskatomasses...and walnuts. **1894** *Jrnl. Amer. Folk-Lore* VII. 98 *Carya alba*, kiskytom, Otsego Co., N.Y.

Kislev (ki·slef, -liū). Also 4–7 **Casleu,** 6–7 **Chisleu,** 8– **Kislew.** [Heb.] The third month of the Jewish civil year and the ninth of the ecclesiastical year, corresponding to parts of November and December.

1382 Wyclif *Zech.* vii. 1 The word of the Lord is maad to Zacharie, in the fourthe day of the nynthe monethe, that is Casleu [*later version* Caslew; **1535** Coverdale Casleu]. **1388** —— *Neh.* i. 1 It was doon in the monethe Casleu [**1535** Coverdale Chisleu], in the twentithe yeer. **1611** *Bible* 1 Macc. i. 54 The fifteenth day of the moneth Casleu. **1838, 1876** [see *HESVAN]. **1880** *Encycl. Brit.* XII. 421/2 Upon the great altar of burnt offering a small altar to Jupiter Capitolinus was erected, on which the first offering was made on 25th Kislev 168. **1904** *Jewish Encycl.* VII. 515 On the twenty-fifth of Kislew the *Hanukkah* festival...commences. **1940** *Universal Jewish Encycl.* II. 632/2 If the day is subtracted, it is taken from the month of Kislev, and the year is termed *haserah*. **1962** D. Bridger *New Jewish Encycl.* 269 *Kislev*, the third month of the Hebrew calendar corresponding to November–December, and consisting of 30, and sometimes 29 days.

kiss, *sb.* Add: **6.** Also *attrib.* **kiss impression** *Printing* (see quot. 1960); **kissproof** *a.*, of lipstick, that will not smudge, come off, etc., if its wearer kisses or is kissed; also *fig.*

1908 R. Dalgin *Advertising Production* 89 If a high-light dot carries little ink, only contact ('kiss impression', we call it) must be used. **1960** G. A. Glaister *Gloss. Bk.* 208 *Kiss impression*, one in which the ink is deposited on the paper by the lightest possible surface contact and is not impressed into it. This technique is required when printing on coated papers. **1962** F. T. Day *Introd. to Paper* 95 The letterpress process employs various machines all of which operate on the same principle, that of bringing inked type surfaces together in a 'kiss impression' with the paper. **1967** Karch & Buber *Offset Processes* ix. 448 Long press runs with a 'kiss' are possible because the offset plate does not touch the paper but contacts the blanket with a very light 'kiss' impression. **1967** V. Strauss *Printing Industry* vii. 448/2 The inking cylinder should be set for a 'kiss impression', a term indicating that the least pressure...[etc.] ... **1936** *Amer. Speech* ... image transfer is to be used. **1934** Dylan Thomas *18 Poems* 26 Happy Cadaver's hunger as you take the kissproof world. **1938** M. Sharp *Nutmeg Tree* iii. 40 She exchanged her Hawaiian Sunset for a new Kiss-proof in flamingo red. **1940** N. Shute *Landfall* ii. 37 'You don't use lipstick.' 'That's all you know. They told me it was kissproof in the shop'. **1950** *Punch* 19 Aug. 39/2 Eight pills of genuine English kissproof lipstick, in the new, passionate tangerine shade. **1962** *New Scientist* 27 Sept. 686/3 Kissproof lipstick was among the most profitable inventions of the present century. **1970** 'J. Canning' *Painted Tent* iii. 194 You can give me a kiss. It's all right now. I don't fret—the stuff's kiss-proof.

b. Phr. *the kiss of death* [f. the association with the kiss of betrayal given to Jesus by Judas in the Garden of Gethsemane (Matthew xxvi. 48–50)], a seemingly kind or well-intentioned action, look, association, etc., which brings disastrous consequences; *the kiss of life*, the mouth-to-mouth method of artificial respiration; also *attrib.* and *fig.*

1948 'N. Shute' *No Highway* iv. 113, I told you that he'd put the kiss of death on it. **1952** H. Waugh *Last Seen Wearing* (1953) 241 I'm starting to take a liking to this guy. The kiss of death. **1960** *Times* 20 July 13/3 Military assistance from Rhodesia would be the kiss of death to Mr. Tshombe. **1960** *Guardian* 10 Dec. 5/1 Let us hope that the critics' approval does not, at the box-office, prove a kiss of death. **1970** *New Scientist* 27 Aug. 403/1 In some countries state participation is essential for a scientific programme, in others it often seems the kiss of death. **1961** *Daily Mail* 22 Sept. 1/7 Mrs. Alice Lowe...used the 'kiss-of-life' to save her 19-month-old nephew Geoffrey Ahmed at Oldham yesterday. **1962** *Guardian* 25 June 4/4 Two children...were given the 'kiss of life' artificial respiration treatment. **1964** *Ibid.* 21 Apr. 18/4 Here was Mr Houghton giving the debate the kiss of life...rather than Mr Boyd-Carpenter responding to treatment. **1969** P. Dickinson *Pride of Heroes* i. 28, I cut the rope...and lowered him to the floor to administer the kiss of life, a technique in which I have taken instruction. **1969** *Private Eye* 5 Dec. 17/2 Finding her six years old goldfish 'Bubbles' on the carpet beside the tank, a Nottinghamshire woman gave it the kiss of life. **1972** *Daily Tel.* 6 Jan. 15/6 Firemen rescued them from their first-floor flat...and tried to revive them on the footpath with the kiss of life and oxygen.

kiss, *v.* Add: **6. c.** *to kiss the hand* (*hands*) (later examples). **h.** *to kiss and be friends*, *to kiss and make up*: to become reconciled; also as a substantival phr. **i.** *to kiss* (a person's) *arse, behind, bum*: to behave obsequiously towards (a person). As *imp.*, esp. in phr. *kiss my arse*: a vulgar rejoinder, stronger than 'go to hell'. **m.** *to kiss and tell*: to recount one's sexual exploits. **n.** *to kiss better* (or *well*): to comfort (a sick or injured person, esp. a child) by kissing him, esp. by kissing the sore or injured part of the body; also *fig.* **o.** *to kiss goodbye*: to bid farewell with a kiss; freq. used *fig.* and ironically. **p.** *to kiss off*, *slang*, (*a*) *trans.* to dismiss, get rid of, kill (see also quot. 1935[2]); (*b*) *intr.* to go away, die.

c. 1955 H. Nicolson *Diary* 6 Apr. (1968) 281 Anthony [Eden] drives to the Palace and kisses hands on his appointment as Prime Minister. **1963** *Times* 31 Jan. 14/2 Mr. F. J. Blakeney was received in audience by The Queen this morning and kissed hands upon his appointment as Her Majesty's Ambassador Extraordinary and Plenipotentiary for the Commonwealth of Australia at Bonn. **1974** *Guardian* 7 Mar. 26/4 Mr Foot...started work to settle the miners [*sic*] dispute even before kissing hands with the Queen.

h. a 1654 J. Selden *Table-Talk* (1689) 36 The People and the Prince kiss and were Friends, and so things were quiet for a while. **1697** W. Dampier *Voy.* I. xv. 414 The Captain and his people kiss'd and were friends. **1894** H. Taylor *Kiss & Make Up* [sc. boring]. **1896** ... **1898** E. Pemberton ... [Bagehot] ... 'kissed hands'... ... was to kiss and make up... **1942** Berrey & Van den Bark *Amer. Thes. Slang* § 334/3 *Become reconciled; make up*...kiss and make up. **1958** *Listener* 2 Oct. 508/1 The party to which I had invited myself was a sort of Kiss-and-make-up. **1969** M. Puah *Last Place Left* xviii. 118 Play the argument bit again...and then play it by kiss-and-make-up.

i. 1705 in *N. & Q.* (1971) Feb. 46/1 You can father it... just as you did another man's philosophical essay upon the wind...when you made bold with several pages from the learned Dr. Bohun, without saying so much as Thank you, for his assistance as kiss my a—se. **1749** Fielding *Tom Jones* I. vi. xi. 288 The Wit...lies in desiring another to kiss your A—for having just before threatened to kick his. **1934** H. Miller *Tropic of Cancer* (1940) 207 If it weren't that I had learned how to kiss the boss's ass, I would have been fired. **1937** 'G. Orwell' *Road to Wigan Pier* v. 196 You 'get on'...by kissing the bums of Labour politicians. **1938** ... **1955** I. MacNeice *Earth Compels* 34 Let us thank God for value in advertisement For those who go their own way, will not kiss The arse of law and order. **1956** R. Hoggart ... **1960** ... **1963** ... **1966** ... to the side. **1965** *Amer. Speech* XXXVIII. 165 To curry favor with a professor: ...*kiss ass*. **1972** *Fairbanks* (Alaska) *Daily News-Miner* 3 Nov. 2/5 McGovern had told an airport antagonist to 'kiss my a...'. The candidate's national political director said the remark had been rather natural for a Democratic nominee. 'After all,' Mankiewicz said, 'he can't kiss his rear goodbye.'

m. 1695 Congreve *Love for Love* II. x. 50 Oh fie Miss, you must not kiss and tell. **1828** Swift's *Night Guide* iii. For those who wish to know her qualifications *as une coucheuse*, try her; for we will not, on all occasions, kiss and tell. **1922** G. B. Shaw *Let.* 30 Dec. in *B. Shaw & Mrs. Campbell* (1952) 235 A gentleman does not kiss and tell.

n. 1868 A. Taylor *Original Poems for Infant Minds* (1814) 72 Who ran to help me when I fell, And would... kiss the place to make it well? My Mother. **1939** E. Bowen *Last September* xvi. 207 She kept...feeling the bump. 'David must 'kiss it better' for her. **1966** *New Society* 23 June 19/1 Mothers...welcome the opportunity of being able to 'kiss their baby better'. They find it easier to have the sick child at home. **1972** *Guardian* 6 May 9/1 I've got this old pain back. 'You must go to the doctor's,' Maggie said, when she'd failed to kiss it well.

o. 1935 C. Day Lewis *Time to Dance* 35 On January 8, 1920, their curveting wheels kissed England goodbye. **1944** 'N. Shute' *Pastoral* xiv. 290 I'll tell Proctor he can kiss his truck goodbye for the rest of the day. **1949** *Listener* 8 Jan. 50/1 Soviet foreign policy will be kiss it goodbye, say that the Chinese are almost ready to kiss the Soviet experts goodbye. **1970** V. Gielgud *Candle-Holders* v. 45 If she chooses one of the Eltham team for a partner, poor George can kiss the trophy goodbye.

p. 1935 *Amer. Speech* X. 22/1 *To throw (someone) down*, 'to kiss off (one's lessons)' to reject... to deceive. **1935** ... *kiss off*, to die. **1942** ... **1945** L. Shelly *Jive Talk Dict.* 28/1 *Kiss off*, death. **1945** R. Chandler *Playback* ...'Kiss off, I just started again.' ... **1961** ... **1968** ... **1970** C. Major *Dict. Afro-Amer. Slang* 73 *Kiss-off*, to die. **1973** ... *Forward* Apr. 1/2...

7. Used in various collocations to denote the comparative ease of an action, etc.; as in (*as easy as*) *kiss my* (or *your*, etc.) *hand*, *finger*, etc.

1891 [see Thumb *sb.* 5 i]. **1909** P. Werlins *Story of Virginia Perfect* xxv. 249 It will be very easy to make respectable friends, and so Miss Malet will find out, though she thinks it is going to be as easy as kiss my hand. **1922** Kipling *Debits & Credits* (1926) 167 The 'ole Semme front washed out as easy as kiss your 'and. **1926** F. M. Ford *Man could stand Up* i. ii. 21 The prospect had seemed as near—as near as kiss your finger! **1949** J. Symons *Bland Beginning* 187 He wanted us to let the book in. It was as easy as kiss your hand. **1958** ... **1968** *Punch* 4 Sept. 330/3 The furs...dropped down like kiss-your-arm into net provided. **1973** V. Canning *Flight of Grey Goose* v. 92 You might be on to it as soft as kiss my hand.

kiss-. Add: **kiss-curl**, a small flat curl worn on the forehead, in front of the ear, or at the nape of the neck.

1856 *Punch* 29 Nov. 219/1 Those pastry-cook's girl's ornamental called kiss-curls. **1867** H. Spicer *Bound to Please* II. 13 Both Jessamy...was nursing a kiss curl, which hung in ringlets than usual in the day. **1920** *Daily Express* 8 Sept. 3/6 Any kind of curls from Nell Gwynn ringlets to kiss curls. **1966** *Punch* 21 Dec. 933/3 'It's the little touches,' he murmurs,...faintly disordering his kisscurl before presenting it to the public. **1968** J. Ironside *Fashion Alphabet* 194 Although a kiss-curl is usually regarded as being on the cheek, like a 'confidante', sometimes it is used as being in the middle of the forehead. **1974** *Guardian* 27 Mar. 12/6 Haley...came on looking much the same as ever, kiss-curl and all.

kissage (ki·sĕdʒ). [f. KISS *v.* + -AGE.] Kissing.

1886 Kipling *Departmental Ditties* (ed. 2) 11 Ere they hewed the Sphinx's visage Favouritism governed kissage, Even as it does in this age. **1898** G. B. Shaw *Let.* 5 Jan. in *Ellen Terry & Shaw* (1931) 287 The best thing for the kne is kissage; for the heart, careful wrapping round by the fireside in that warm tale, with, if possible, a red head... **1960** F. Frankfurter *Felix Frankfurter Reminisces* iii. 28 This system, the objectivity of the marking, and the other considerations—no kissage by favors has always

been the slogan there—creates an atmosphere and habits of objectivity and disinterestedness...and a zest for being very good at this business which is rare.

kissar (ki·sāɹ). [ad. colloq. Arab. *kīsār*.] (See quot. 1964.) Also *attrib.*

1864 C. Engel *Music Most Anc. Nations* ii. 41 A kissar from Abyssinia...is so far different from the common Nubian kissar, that the body is square, without sounding-holes. *Ibid.*, Not only the Abyssinian kissar, but also the common one are called by the Nubians *gbouna*. **1947** J. Pulver tr. *Panum's Stringed Instruments Middle Ages* 12 Engel...points to the striking likeness existing between the type of lyre depicted here and the lyre which, under the name of kissar, is still used by several North African peoples. **1964** S. Marcuse *Mus. Instruments* 291/1 *Kissar*, *kithara*, bowl lyre of E. Africa, a survival of the ancient Gk. lyra, still found in Ethiopia, Sudan, and Uganda. **1967** T. Rees *Ward-Ladle* vi. 118 In the lyres one finds..an open, drum-shaped body, with a stretched skin, a survival of the kissar-type.

‖ **kissel** (ki·səl, kisye·l). Also **keessel.** [ad. Russ. *kisél'*.] A sweet dish made from fruit juice mixed with sugar and water, which is boiled and thickened with potato or cornflour.

1924 A. Gagarine *Russ. Cook Bk.* p. 191 When eaten hot, the kissel should be taken with hot cream, milk or custard. **1943** E. M. Almedingen *Frossia* vii. 266 A hot ...of potato flour. **1963** ... spirit drink, the kissel, frankly dessert, sugared chicken feathers, and cranberry jelly, 'kissel', and pickled cucumbers lying in a wooden bowl. **1951** M. McCarthy *Groves of Academe* (1953) iii. 54 The small presents she was in the habit of bringing them—a dish of Russian kissel with a white napkin over it. **1969** *Guardian* 8 Aug. 7/4 Blackberry kissel...Heat the cornflour with a little cold water, add with the brandy, stir into the blackberry purée. **1971** *Times* 4 Aug. 5/8 Moscow housewives are buying up huge amounts of berries just now—mostly blackcurrants, redcurrants and cranberries—to preserve them as jam or make a thin jelly (*kissel*).

kisser. Add: **2.** The mouth; the face. orig. *Boxing slang.*

1860 *Chambers's Jrnl.* XIII. 348/1 His mouth is his 'potato-trap'...or 'kisser'. **1892** P. H. Emerson *Son of Fens* iv. 43 'Oh,' he say, and dabbed the wet mittens upon my kisser kind of smart. **1927** W. E. Collinson *Contemp. Eng.* 27 In speaking of parts of the body, the Boxer's term darie(-ing or cook, but we did not...use 'I'll hit you on the kisser' (mouth). **1938** D. Runyon *Furthermore* v. 81 He is a tall skinny guy with a long, sad, mean-looking kisser, and a mournful voice. **1953** R. Chandler *Lt.* 15 Mar. in *R. Chandler Speaking* (1966) 48 Chandler...consumed three double gimlets and fell flat on his kisser. **1965** J. Wainwright *High-class Kill* 156 Ope them sweet little, lying little, kisser of yours, and they'll start squashing something that makes sense.

Kissi (ki·si), *sb.* and *a.* [Native name.] **A.** *sb.* An agricultural people inhabiting the regions of Guinea, Sierra Leone, and Liberia near the headwaters of the Niger; one of this people; also, their language. **B.** *adj.* Of or pertaining to the Kissi.

1884 *Encycl. Brit.* XVII. 319/1 *Felup Group*,...Kissi. **1916** [see *KONO[1]]. **1949** E. A. Nida *Morphol.* (ed. 2) v. 107 In Kissi, a language of French Guinea,...the distinction between monosyllabic and polysyllabic verbs requires a different set of affixes. **1967** *Ibid.*, The Army had...recruited a high proportion of its soldiers among the Koranko, Kono, and Kissi tribes in the farthest corners of the Protectorate. *Ibid.* vii. 122 The Kissi, who speak a separate language, have the same religion, the Mende...much influenced by Mandinka, live far inland in the high ground where the French, British, and Liberian frontiers meet. **1970** *Western Folklore* XXIX. 241 Among some of the Liberian Kissi of the frontier regions the musical bow accompanies tribal initiation rites in the forest, being utilized in place of the wooden drum.

kissing, *ppl. a.* Add: **b. kissing bug** *U.S.*, a blood-sucking bug of the family Reduviidæ; **kissing cousin**, a relative or friend with whom one is on close enough terms to greet with a kiss; also *transf.*; **kissing gourami**, a small Malaysian freshwater fish, *Helostoma temminckii*, which kisses with its lips; **kissing trap** *slang*, the mouth.

1899 *Pop. Sci. Monthly* Nov. 33 Several persons suffering from swollen faces visited the Emergency Hospital in Washington and complained that they had been bitten by some insect while asleep... These were the 'kissing bug' scare. **1904** N.Y. *Even. Post* 4 Aug. 1 The doctors were unable to decide whether he had been bitten by a mosquito or a kissing bug. **1932** Metcalf & Flint *Fund. Insect Life* viii. 222 Family Reduviidæ. The Assassin or Kissing Bugs...This is a very large family of mostly flattened, oval bugs... The assassin bugs catch small insects and suck their blood as food. Some species, when handled, inflict painful bites on man. **1973** L. S. Chadwick tr. *Linsenmaier's Insects of World* 120/1 The bite of many reduviids is very painful, even to man; in warm countries certain species even enter houses on occasion at night and suck the blood of people. In North America this may be done by the black 'kissing bug' (*Melanolestes picipes*). Because it is a bloodsucker...**1949**. **1953** ... **1966** ... If you could...**1951** ... **1965** ... Lyttleton's ... He had thirty or forty cousins, 'kissing cousins', **1965** ... close enough to be kissing cousins. **1970** *Guardian* 31 Aug. 7/4 We remember the Dutch more than we remember the people of any other country—we are truly kissing cousins. **1973** *Publishers Weekly* 25 June 32/2 (Advt.), From gems to their kissing cousins, stones. **1958** W. T. Innes *Exotic Aquarium Fishes* 560 *Helostoma temminckii*... Popular name, Kissing Gourami. **1962** R. H. Axelrod *Tropical Fish* iii. 59 The Kissing Gourami [is]...so named for the mutual kissing or pecking of pairs at feeding or scraping algae from the sides of the tank. **1962** D. W. Tucker tr. *Sterba's Freshwater Fishes of World* 794 There is an unexplained peculiarity of this fish: its 'kissing'. **1969** ... **1850** ... *Old Sketch. of Life* ... **1854** C. Bede' *Further Adventures Verdant Green* iv. 35 'Jerry...as he had been his opponent's 'kissing trap'...it is usually called his 'kissing-trap'. Displays an ugly mark! **1942** Berrey & Van den Bark *Amer. Thes. Slang* §121/6 *Mouth*,...kissing trap, loud-speaker, maw, [etc.].

kiss-in-the-ring. (Earlier and later examples.)

1801 J. Strutt *Sports & Pastimes* iv. iv. 285 A boy must touch a girl, and a girl a boy, and when either of them be caught they go into the middle of the ring and salute each other; hence is derived the name of *kiss in the ring*. **1928** W. de la Mare *Broomsticks* 26 There were quantities of things to eat and lots to see, and Kiss-in-the-Ring. **1936** 'R. Crompton' *Sweet William* vii. 171 The games were to be Kiss in the Ring, Postman's Knock, Turn the Trencher, and others of similar kind. **1957** J. Masters *Far, Far the Mountain Peak* i. 58 'Why don't I suggest a game of ring-a-ring-a-roses or kiss-in-the-ring?' **1969** I. & P. Opie *Children's Games* vi. 201 Throughout the nineteenth century 'Kiss in the Ring' was a favourite sport of young people at fairs and holiday gatherings.

kiss-me-quick. Add: **4.** *kiss-me-quick hat*: a hat bearing the words 'kiss me quick' (or some other, usu. jocular, phrase) on the front.

1963 *Guardian* 12 July 9/6 Mature matrons...now...wear kiss-me-quick hats. **1974** J. Wainwright *Evidence I shall Give* xxiv. 120 Whitby...Genuine Olde Worlde charm, wearing a Kiss-Me-Quick hat. A fishing community, taken over by day-trippers.

kist, *sb.*[1] Add: **1. a.** (S. Afr. examples.)

1958 *Personality* 4 Dec. 27/3 Ancient brass-bound *kists* of teak help to furnish the back stoep. **1959** *Star* (Johannesburg) 2 Jan. 7/4 (Advt.), Heavy bowfronted *kists* price cut to £13/15/6. **1971** *Cape Times* 13 Feb. 21/3 (Advt.), Furniture and effects...walnut bedroom suite, easy chairs...several large teak glass fronted cupboards, a carved Zanzibar camphor-wood kist. **1973** *Leader* (Durban) 7 May 15 (Advt.), Imbuia Kist...R29.00.

c. *kist o' whistles*, *whistles*, an organ (ORGAN *sb.*[1] 2). *Now rare.*

1790 ... **1970** *Canad. Antiques Collector* Dec. 10/1 A kissing ball, consisting of evergreen, wrapped round a cluster of apples, provides the 'kissing' element. **1956** R. Crick *Green Willow* vii. 91 In some houses...kissing-boughs hung over doorways. **1969** E. Wilkes *Rose-Garden Game* viii. 201 The old English Christmas globe, called a kissing-bough, which was made up of three interlocking hoops of greenery, hung from the ceiling and lit up with candles. **1971** T. Wright *Dict. Obsolete & Provincial English* II. 614/2 *Kissing-bunch*, a bush of evergreens sometimes substituted for mistletoe at Christmas. **1913** D. H. Lawrence *Sons & Lovers* vi. 116 It [sc. the kitchen] was small and curious. There was a kissing-bunch. **1859** C. W. Wilson *Mapping the Frontier* (1970) i. 77 It will be a mad matter if we cannot get something whereby worth the whole rights of the kissing bush. **1879–81** F. Jackson *Shropshire Word-Bk.* 237 *Kissing-bush*, a bunch of evergreens and mistletoe garnished with ribands and fruit, which is hung from the kitchen ceiling, or hall, at Christmas-tide. **1875** W. Alexander *Sk. Life among my Ain Folk* v. ii. 245 When the leading fiddler...this fine four-to-wow bout, with the usual kissing-trip... this is 'kissing time', and, after an attempt more or less successful on the part of each male dancer to kiss his partner's cheek, at it they go! **1916** F. Norris *Vandover* (1914) v. 58 'Kissing time' first...**1922** Joyce *Ulysses* 354 Edy asked her the time, and Miss Cissy...said it was half past kissing time, time to kiss again. **1952** T. S. Eliot *Murder in Cath.* i. 25 If you will remember me, Lord... **1966** M. 'Malatesta' in ... **1971** [mistletoe] ... from the lid... for all that, 1911 ... in the kissing time...

2. kissing-ball, -bough, -bunch, -bush, a Christmas wreath or ball of evergreens, freq. arranged with fruit and ribbons, which is hung from the ceiling and under which a kiss may be taken; **kissing time,** the time to kiss, freq. used as a joc. reply to children who ask the time.

kissing, *vbl. sb.* Add: **1. b.** Phr. (*when*) *the kissing had* (or *has*) *to stop*: (when) the 'honeymoon' period finished (or finishes); (when) one is forced to recognize harsh realities.

1855 Browning *Toccata of Galuppi's* xiv, in *Men & Women* I. 61 What of soul was left, I wonder, when the kissing had to stop? **1966** *Guardian* 20 Aug. 12/3 In the past 13 years more than 190,000 adult Jamaicans have come to settle in this country... Now the kissing has to stop. **1973** *Times* 26 Oct. 8/3 It left wise principles still exposed to the emergence of extremists of the right... It is a few short steps to the place where the kissing has to stop.

1772 A. Ramsay *New Misc. Scots Sangs* [*Tea-Table Misc.*] 141 The Kist fou o' Whistles, That make sic a Cleiro. **1828** J. Ruddiman *Tales & Sk.* 60 To cram doun our craigs, will we, nill we, their 'kist o' whistles.' [in Dict., sense 1]. **1889** G. D. Sharp *London Mosaic* 1888–89 (1931) 126 M. Gigout, who was performing on the 'Kist o' whistles'. **1891** R. Ford *Thistledown* vi. 106 There was no such thing as an organ, or 'kist o' whistles,' in any Presbyterian kirk in the land. **1936** *Discovery* July 223/2 The formal Highland 'kist of whistles' would spoil the architectural effect. **1967** 'H. MacDiarmid' (title) A kist of whistles. **1969** C. Gesson *Northumberland & Durham Word Bk.* 118 *Kist o' whistles*, an organ in Scotland and Northumberland.

Kiswa (ki·swa). Also **Kiswah.** [Arab.] The black cloth which covers the Kaaba.

1599 Hakluyt *Voyages* II. i. 203 Moreover he delivereth unto him ye *Chiswa Talnabi*, which signifieth in the Arabian tongue, The garment of the Prophet. **a 1817** J. L. Burckhardt *Trav. Arabia* (1829) I. 254 The four sides of the Kaaba are covered with a black silk stuff, hanging down, and leaving the roof bare. This curtain...is called *kesoua*. **1855** R. F. Burton *Pilgrimage* II. xvi. 82 The Kiswa is a black, purple, or green brocade, embroidered with white or with silver letters. **1934** J. A. J. Travers *Mod. Pilgrim in Mecca* viii. 124 In 'the Haram'...remains till the day of the festival. **1957** J. Hopkins ... ... the covering itself, is changed. **1958** R. Ruthen *Holy Cities Arabia* I. xiii. 172 The beautiful silk and wool kiswa with its gold-embroidered band...not having been sent this year, the Kiswa was not changed. **1965** J. Kritzeck ... 190 (*caption*) The Kaaba in Mecca, covered with the Kiswa. **1967** ... ... **1959** *Chambers's Encycl.* IX. 192/2 Each old Kiswa is cut up and sold to pilgrims. **1963** *Ann. Reg. 1962* 275 Saudi refusal to accept the Kiswa, the traditional holy cloth for the Kaaba shrine in Mecca sent annually from Cairo.

Kiswahili (kiswahī·li). [Native word, f. *ki*- prefix designating an abstract or inanimate object + SWAHILI.] A major language of the Bantu family, spoken widely in Kenya, Tanzania, and elsewhere in East Africa, where it serves as a lingua franca.

1864 J. A. Grant *Walk across Afr.* p. xvi, Kiswahili, the dialect of the Wasuahili on the east coast of Africa. **1936** *Discovery* June 196/1 A decree of 1929...recognised a certain Bena customs, written in Kiswahili. **1967** L. Sheperd *Surveying Bush* ii. 21 You should read Swahili... He knew little English, but this was an advantage as it forced me to learn Kiswahili. **1967** C. W. Reckmann *Swahili-Engl. Dict.* p. vi, The Swahili language (Kiswahili) is a Bantu language spoken by people as many as forty million people throughout a great part of East Africa. **1972** *Guardian* 14 June 5/5 The main language of Dar Es Salaam is Kiswahili. **1972** *Daily Nation* (Nairobi) 22 Jan. 6/2 (Advt.), Candidates must be Tanzanian Citizens fluent in both Kiswahili and English. **1973** *Inside Kenya Today* May 39/1 The Sin School...manages to arrange the course for Kiswahili learning, **1974** G. Wigg *George Wigg* v. 110 Facilities for learning Kiswahili were made available in Chinyanja speaking units.

kit, *sb.*[1] Add: **1. b.** (Later examples; also in extended use, and, by metonymy, the contents of a kit, used as a measure of weight.)

1906 *Daily Chron.* 2 Jan. 3/2 A day's work is one hundred kits... 1000 kits of fish. **1934** 'Taffrail' *Seventy North* ix. 185, 20,000 to 25,000 'kits'—...each weighing ten stone might be landed from the trawlers—say 125 to 150 tons. **1936** 'L. Luard' *Conquering Seas* xv. 246 In ten hours of berthing, with a full two thousand kit aboard...over a hundred tons of fish. **1961** *Guardian* 18 Jan. 9/2 He was...pushing a barrow on the fish docks, piled with aluminium 'kits' which, with full, each contain 10 stone of fish.

2. d. A set or outfit of tools, equipment, etc.; *spec.*, a collection of parts sold for the buyer to assemble. Also *fig.* and *attrib.*

1859 G. W. Maskell *Vocabulary* 48 *Kit*...the implements of a burglar. **1907** *Yesterday's Shopping* (1969) 177/2 The Army & Navy Boot Kit contains 3 shoe brushes, ...1 tin of boot polish, complete in leather case. **1935** A. J. Pollock *Underworld Speaks* 68/1 Kit, a safe-cracker's tools. **1951** *Amer. Speech* XXX. 226 Kit, all the accessory parts to assemble one section of the plane. **1963** A. Huxley *Genius & Goddess* i. 3 We've made up a kit of cheap perfume. **1961** Partridge *Dict. Slang Suppl.* 1160/1 *Kit*, paint-kettle and brushes. Housepainters' coll. **1970** *Listener* 9 Apr. 148/3 Beckett forces upon you a do-it-yourself Tantalus-kit. He requires you to seek and not to find. **1970** *House & Garden* Mar. 77/2 'Cena chair'... Comes in kit form (see components below)—from £14 19s. plus post kit. **1970** *Times* 8 Sept. 3/1 Make your own sausages... Kit includes hand-filler, recipes, etc. **1970** 'Carter Brown' *Hang-Up Kid* xv. 154 ...to the coffee table... ...of sausages. **1973** ... D. Marlowe *Echoes of Celandine* 1. 12, I have also a pennant for the Tiger Moth ...**1929** Mercedes Benz SSK three times.

e. An outfit of drums, cymbals, and other percussion devices and accessories used by a drummer in a dance-band, jazz-group, or the like.

1929 *Melody Maker* Mar. 259/3 Lyman plays the drums in the band—at least when he is playing his kit. **1934**, **1965** [see *drum kit* s.v. *DRUM *sb.*[1] 12 a]. **1971** *Melody Maker* 17 Nov. 47/3 It's more important to those play the music than the colour of your kit. A good drummer is going to sound good even if he plays on the table top.

f. A quantity of printed matter on a specified topic for students, etc.

1968 *Globe & Mail* (Toronto) 17 Feb. 36 (Advt.), Savannah's full of colonial names... Why not add your name to an illustrated kit! Write for our free travel kit. **1971** J. B. Carroll et al. *Word Frequency Bk.* p. vi, The materials themselves include text-books, periodicals, encyclopedias, novels, student workbooks, kits, and so on, all of which contain vocabulary to which students are exposed. **1973** *Guardian* 7 June 6/7 The SACK—School and Community Kit—is made up of information and ideas on community projects for schools, together with reference material. **1974** *Catholic Herald* 4 Oct. 4/2 The study kits come in colourful folders and contain pictures, taped songs, fact sheets and quotations from Church leaders.

3. Also, *the whole kit and boiling* (*boodle, caboodle, cargo*). (Cf. *CABOODLE.) *U.S.*

a 1852 F. M. Whitcher *Widow Bedott Papers* (1856) xxiii. 257 The hull kit and cargo on 'em had conspired together. **1859** Bartlett *Dict. Amer.* (ed. 2) 33 *Biling*, a vulgar pronunciation of boiling. The phrase *the whole* (or more commonly *hull*) *kit and bilin* means the whole lot, applied to persons and things. **a 1861** T. Winthrop *John Brent* (1883) xxiii. 272, I motioned we shove the hul kit and boodle of the gamblers ashore on logs. 'Twas kerried. **1888** *Boston Globe* 5 Feb. 1/3 If any 'railroad lobbyist' cast reflections on his character he would wipe out the whole kit and caboodle of them. **1908** *Dialect Notes* III. 327 (East Alabama) *Kit and bilin*, the crowd. Usually in the expression 'the whole kit and bilin'. **1920** S. Lewis *Main St.* 50 The whole kit and bilin' of 'em are nothing in God's world but social[illegible]. **1946** *Newsweek* 16 Sept. 32/2 It gave the farm and the whole kit and boodle to Stanley. **1969** *Listener* 22 May 707/3 The whole kit and caboodle of us were then investigated by the FBI to see how many subversives there were among us.

4. b. kit-car, a motor car sold in parts for assembly by the owner; a build-it-yourself car; **kitset** *N.Z.*, the components and aids for assembling an article (radio, furniture, etc.) or model (aeroplane, etc.).

1953 G. Durrell *Overloaded Ark* xiii. 216, I..set off early one morning in the back of the Schiblers' kit-car. **1964** *Sunday Times* (Colour Suppl.) 29 Nov. 53/3 The modern kit car.. All you need is practical application, a few hand tools. **1970** *Sunday Truth* (Brisbane) 18 Jan. 60/3 Here we loaded the kit-car..and travelled south to Biafra. **1963** *Weekly News* (Auckland) 8 May 56/5 *Transistor radio kitsets.* New low prices. Easy-to-build range from crystal set, £1, to 7T portables, £9/19/6. **1966** *Ibid.* 26 Jan. 6/6 Many other clubs throughout the Auckland province assist the young yachtsman with kitsets and drawings for home building (of yachts).

kit, *sb.*³ Also ⁓ Kitten *sb.* 1 b.

1957 Kitt [see *KITTEN *sb.* 1 b]. **1970** *Times* 8 Sept. 11/3 The kits [sc. young mink] are only born in six months. **1974** A. Dillard *Pilgrim at Tinker Creek* ii. 15 You crouch motionless on a bank..and are rewarded by the sight of a muskrat kit paddling from its den.

kit (kit), *sb.*¹¹ *N.Z.* Also **kete, kête.** [ad. Maori *kete.*] A basket plaited from flax. Hence **kit-ful.**

1814 E. Markham *N.Z. or Recollections of It* (MS.) 44 They make Baskets or Kits as we call them for potatoes. **1841** W. Colenso *Jrnl.* (typescript) I. 120 Opening his kete and taking out his Blanket. **1856** E. B. Fitton *New Zealand* 68 Neatly made baskets, plaited from flax, and known by the name of 'Maori kits'. **1877** W. T. L. Pratt *Colonial Experiences* 31 Potatoes were procurable from the Maories [sic] in flax kits, at from one to five shillings the kit. [illegible] **1884** Lady Martin *Our Maoris* 44 My heart is like an old kête (i.e., a coarsely-woven basket). **1902** W. Satchell *Land of Lost* xviii. 161, I will give you a kitful when you go away. **1936** R. Hyde *Check to your King* xiii. 156 Great flax kits of kumaras..were left outside the door. **1938** R. Finlayson *Brown Man's Burden* 75 So the three men came along with kitfuls of [shellfish] long before Hira came back. **1948** Baker *N.Z. Slang* vi. 35 We have adopted into wide use the term *kit* for a shopping basket. **1958** S. Ashton-Warner *Spinster* 41 When I come to this word 'basket' in my book I never says 'basket'.. I always says 'kit'. **1969** F. Sargeson *Joy of Words* iii. 105 Between the two of them they carried a flax kit of food.

kit, *v.*¹ Add: **2.** To equip (someone or something) with a uniform, an outfit, personal effects, equipment, etc. Freq. with *out, up.* So **kitted** (*out, up*), provided with clothing, accessories, etc.; **kitting up** *vbl. sb.*

1919 W. Lang *Sea-Lawyer's Log* ii. 13 It is pleasant to march down to the kitting-up store and have garments thrown at you, without price. *Ibid.* 16 Now we have been 'kitted up', as the nautical expression has it. **1925** T. E. Lawrence *Let.* 25 Aug. (1938) 482 Sergeant take this man to the Q.M. Stores, kit him at once, and put him into the first train for Cranwell. **1945** *Times* 25 May 2/2 In a day or so these men would be kitted up in smart new uniform and go on leave. **1948** 'N. Shute' *No Highway* ii. 50 I'll get a letter through to Ottawa asking them to kit you up for the trip. **1958** *Technology* May 68/3 Some firms may wish to give students a 'kitting up allowance'. **1960** K. Amis *New Maps of Hell* (1961) iv. 95 There are cases on record of writers having to kit out contemporary narratives with aliens and space-ships in order to make a sale. **1962** *Guardian* 7 Aug. 5/1 A child can have ten days skiing for under £25 and be kitted out by Moss Brothers into the bargain. **1963** *Times* 18 Jan. 3/6 The cars—numbered, polished and kitted with every conceivable gadget to compete against time and the elements—will set course for the Côte d'Azur. *Ibid.* 20 Apr. 3/4 Probably the business locality this weekend will be Bristol, where the England players for the short Antipodean tour will foregather to be kitted and generally vetted. **1970** *New Society* 5 Mar. 384/1 Voluntary labour repaired and kitted out about 20 houses in three months. **1973** *Daily Tel.* 21 Nov. 14/3 The 1500 TC offers good value at £1,295 for a well kitted-out four-door 1½-litre.

Kitab (kitā·b). [Arab. *kitāb*, lit. writing, book.] The Koran; also, a sacred book of certain other revealed religions, e.g. the Bible.

[**1652** J. Notstock tr. *Andrés's Confusion Muhamed's Sect* ii. 53 O Moore,..what thinkest thou of the Scripture which you so much reverence, that..ye..keep it like a God, and call it *Alkitib Alhazim*, (i.e.) a glorious book and *Alcoran alhadin*, (i.e.) the Mighty *Alcoran*.] **1885** T. P. Hughes *Dict. Islam* 280/2 *Al-Kitāb*.., 'the Book', a term used for the Qur'ān, and extended to all inspired books of the Jews and Christians, who are called *Ahlu'l-Kitāb*, or believers in the book. *Ibid.* 484/1 The Muhammadan Scriptures..are usually appealed to and quoted from as.. *al-Kitāb*, 'the Book'. **1912** *Moslem World* II. 168 Mohammed..knew of the existence of the Holy Scriptures of both Testaments, and..fully intended to refer to them... In the first place, *Al Kitāb*, 'the Book', is a clear reference to 'the Bible'. **1927** *Encycl. Islam* II. 1044/2 The 'Book' par excellence is..the Ḳur'ān itself; it is the revelation of God... As the Kitāb is the word of God it has also the meaning in the Ḳur'ān of 'a decree of God' or it becomes the 'impression' which God stamps upon the hearts of man. **1959** S. & N. Ronart *Conc. Encycl. Arabic Civilization* 297 *Al-Kitāb* as a specifically Moslem term denotes the Holy Book, i.e. the Koran, but is also extended to the Scriptures of..Christianity and Judaism, whose adherents are called *Ahl al-Kitab* (People of the Book). *Ibid.* 298 (*heading*) Koran,..the Holy Book of Islam, frequently spoken of by Moslems simply as The Book (*al-kitāb*).

kitchen, *sb.* Add: **1. b.** *hell's kitchen*, an area or place that is regarded as very disreputable or unpleasant; *spec.* a district of New York City once regarded as the haunt of criminals; *thieves' kitchen*, a place inhabited by thieves or other criminals; also *transf.*

1838 N. Hawthorne *Amer. Notebks.* (1932) 15 He.. swore fervently in favor of driving the British 'into Hell's kitchen' by main force. **1868** A. J. Munby *Diary* 29 Jan. in D. Hudson *Munby* (1972) 248 We went to see the Thieves' Kitchen,..a large long antique cellar,..men and lads, perhaps 15 in all, lounging on benches. All thieves. **1894** *Harper's Mag.* July 223/2 Her father had moved into..a ramshackle old barrack just at the edge of Hell's Kitchen. **1894** W. J. Locke *At Gate of Samaria* (1895) xxvii. 319 They went together to East End music halls, bank holiday gatherings, thieves' kitchens, night clubs in the West End. **1900** 'Flynt' & Walton *Powers that Prey* 98 'Think Hell's Kitchen 'ud learn him?' Hell's Kitchen, in the speech of people who do not know what it means to work there, is the foundry. **1909** J. R. Ware *Passing Eng.* 243/2 *Thieves' kitchen* (London Street, 1882), the name satirically given to the then new Law Courts. **1941** C. O. Skinner *Soap behind Ears* 168 She asked me a few routine questions..in the manner of someone questioning a welfare worker concerning life in Hell's Kitchen. **1949** *Sat. Even. Post* 15 Jan. 39/1 It stands between a greasy garage and a tawdry row of brownstone tenements on the edge of Hell's Kitchen, west of Eighth Avenue on 49th Street. **1960** *Observer* 24 Jan. 5/5 Where did I think the biggest thieves' kitchen was to-day? **1973** P. Geddes *Ottawa Allegation* iii. 32 A Whitehall trusty, too, once away from the thieves' kitchen of intelligence and admitted to the counsels of the mandarins. **1974** D. Ramsay *No Cause to Kill* ii. 141 A bar on the East Side, almost the width of Manhattan Island away from Hell's Kitchen. **1974** 'M. Innes' *Mysterious Commission* xviii. 162 The place was certainly no thieves' kitchen. Honeybath..became aware of its respectable opulence.

f. Phr. *to go into* (or *to take tea in*) *the kitchen*: see quots. ?*Obs.*

1889 Barrère & Leland *Dict. Slang* I. 415/2 *To go into the kitchen* (popular), to drink one's tea out of the saucer; an allusion to the vulgar method of drinking very common among servants. **1894** G. F. Northall *Folk-Phrases* 30 To take tea in the kitchen = To pour tea from the cup into the saucer, and drink it from this.

g. A part of a casino at Monte Carlo where gamblers place smaller bets than in the *salles privées*.

1931 W. Holtby *Poor Caroline* i. 25 In the kitchen the whirring of wheels, the jangle of voices..had grown intolerable. **1932** Wodehouse *Louder & Funnier* 255 It may be that your neighbours at the Le Touquet tables have a winsomeness lacking in those who congest the 'kitchen' at Monte Carlo. **1964** A. Wykes *Gambling* xii. 288 The old Winter Casino has 11 rooms. The smallest (i.e., the ones where the stakes are smallest) are known collectively as 'the Kitchen'. This is where the majority of Monte Carlo's 'amateur' gamblers play.

h. The percussion section of an orchestra or band. *slang.*

1931 G. Jacob *Orchestral Technique* vii. 68 We now come to a consideration of the percussion group (commonly known as 'the kitchen'), chief among which stand the *Timpani* (or kettledrums). **1934** S. R. Nelson *All about Jazz* ii. 49 Next in the rhythm section we will have a look at the 'gentlemen of the kitchen'.

5. a. *kitchen-folk, -girl* (later examples), *-man* (later examples), *-mechanic.*

1901 M. Franklin *My Brilliant Career* iii. 13 The dashing snake yarns told by our kitchen-folk at Bruggabrong. **1912** W. Owen *Let.* 1 Feb. (1967) 114, I..must unpack a Crate of *China*... Fit occupation for me, who have far more knowledge of these matters than the kitchen-folk. **1835** J. H. Ingraham *South-West* II. 253 There are some Yankee 'kitchen girls',..who can do more house work..than three or four negro servants. **1957** M. Spark *Comforters* ii. 33 The kitchen girls grumble about the work. **1910** *Granta* 11 June 10 His door was sported, but on a covered dish left outside by a kitchen-man I observed three slices of cold beef. **1930** Kitchen man [see *BED *sb.* 1 f]. **1887** *Lantern* (New Orleans) 23 July 2/2 A..dirty looking kitchen mechanic called Maggie Howard. **1931** 'D. Stiff' *Milk & Honey Route* 215 A hobo camp cook, a kitchen mechanic. **1942** Z. N. Hurston in A. Dundes *Mother Wit* (1973) 224/2 Best you can do is to confidence some kitchen-mechanic out of a dime or two. **1969** L. G. Sorden *Lumberjack Lingo* 67 *Kitchen mechanic*, a dishwasher in a logging camp. Cookee.

b. *kitchen-stair.*

1844 C. M. Yonge *Abbeychurch* ix. 188 Katherine, seeing Elizabeth go towards the kitchen stairs. **1902** *Granta* 3 May 287 It was the Fancy Dress Ball of the season, and the Duchess of Billingsgate was waiting at the head of the kitchen-stair to receive her guests.

c. *kitchen cupboard, -furniture* (example), *-knife, -range* (earlier example), *scissors, stool, unit, -ware* (later examples).

1865 *Trans. Illinois Agric. Soc.* 1862 V. 161 The warm kitchen cupboard. **1916** *Daily Colonist* (Victoria, B.C.) 4 July 12/2 (Advt.), Kitchen cupboard with closed cupboard and drawers in base, separate open shelving on top. **1785** *Daily Universal Reg.* 1 Jan. 2/2 Two waggons, loaded with his Majesty's kitchen furniture. **1969** M. Pugh *Last Place Left* xi. 68 She turned with a kitchen knife in her hand. **1785** *Daily Universal Reg.* 1 Jan. 3/2 Perpetual ovens, in Kitchen Ranges..upon an entirely new construction, heated without the assistance of any flue. **1907** *Yesterday's Shopping* (1969) 214/2 Scissors, kitchen—6 in., 0/11½. **1966** *Olney Amsden & Sons Ltd. Price List* 33 Kitchen Scissors.. 10/6. **1926–7** *Army & Navy Stores Catal.* 140/1 Kitchen stools. Made of Deal.. height 14 in.—3/9. **1948** A. Laski *Keeper* ii. 15 'Now, tell me all about it,' she commanded, perched on the kitchen stool. **1937** *N.Y. Times* 21 Mar. 10/1 The demand for electrical kitchen units is greatest in the Midwest. **1958** *House & Garden* Mar. 120 (Advt.), You can start with an EZEE Sink Unit and gradually build your dream kitchen around it... EZEE are the only kitchen units you can buy in a complete range. **1930** H. Crane *Bridge*, I ran a donkey engine down there on the Canal in Panama..Then Yucatan selling kitchen-ware. **1967** *Times* 11 Nov. 13/8 It was common ground that 'kitchen-ware' meant 'ware' of such a class as to be appropriate to use in a kitchen, but that consumer goods and mere packaging were excluded. **1974** J. Drummond *Boon Companions* xxi. 78 The display windows were still lit. He looked past kitchen-ware and crockery.

d. *kitchen foil, match, paper.*

1958 *Observer* 21 Sept. 8/5 Steam the joint with a few cloves..before wrapping it up generously in kitchen foil. **1961** *Guardian* 24 Mar. 11/6 Wrapping the fish in well buttered kitchen foil. **1955** J. D. Macdonald *Brass Cupcake* iii. 29 [illegible] nibbling on a kitchen match. **1973** R. Thomas *If you can't be Good* (1974) xii. 155 He would stick a cigarette between his lips and light it with a kitchen match. **1846** *Jewish Manual, or Pract. Information Jewish & Mod. Cookery* [illegible] with white kitchen paper, so as to prevent the paste coming off. **1962** F. T. Day *Introd. to Paper* viii. 87 Household rolls, plain or with printed designs, tile and kitchen papers.. are a few of the varieties which are rewound on the winding machine from the larger diameter rolls. **1974** D. Fletcher *Lovable Man* v. 37 A wad of absorbent kitchen paper.

f. *Comb.*, as *kitchen-bed-sittingroom, -diner, -dining-room, -living-room.* A room serving both as a kitchen and as a room of the type designated in the second (or further) element.

1951 Koestler *Age of Longing* ix. 152 The 'study' had at first been a corner of their kitchen-bed-sittingroom, partitioned off by a sheet. **1961** *Times* 11 Dec. 13/7 The kitchen-diner is much favoured and consequently catered for. **1974** *Country Life* 28 Feb. [Suppl.] 30/2 Bathroom, lounge, kitchen/diner, cloakroom. [illegible] *Water* (1964) vi. 66 Nobody had visited the kitchen-dining room [illegible]. **1974** *Times* 29 May 11/3 The kitchen-dining room.. is 29 ft. by 15 ft. **1904** *Westm. Gaz.* 13 Dec. 8/1 Three bedrooms, kitchen-living room, scullery, and wash-house. **1965** D. Chapman *Home & Social Status* iii. 32 [illegible] a kitchen or kitchen-living-room: bedrooms and bathroom. **1963** N. Marsh *Dead Water* (1964) vi. 121 The old kitchen..had been converted into a kitchen-living room.

7. kitchen Dutch [tr. Du. *kombuis-Hollands*], now *rare*, the dialect of Afrikaans spoken by Cape Coloured people in the Western Province of S. Africa; later, used by English speakers as a contemptuous term for Afrikaans; **kitchen evening** *Austral.* and *N.Z.*, a party to which guests bring gifts of kitchenware for a bride-to-be; **kitchen Kaffir,** now *rare*, a *lingua franca* of southern Africa; = *FANAGALO; **kitchen police,** in the U.S. army, enlisted men detailed to help the cook, wash dishes, etc.; the work of these men; **kitchen shower** *U.S.*, = **kitchen evening*; **kitchen tea** *Austral.* and *N.Z.*, = **kitchen evening.*

1894 F. A. Barkly *Boers & Basutos* (ed. 2) vii. 109 By this time they [sc. our two children] could both speak Sesuto and 'Low' or 'Kitchen Dutch' (as it is called in those parts) well. **1899** W. S. Lagman *How to speak Dutch* (ed. 2) Pref., My friend J. F. van Oordt, who has tried to strike the happy medium between 'High Dutch', not often understood by the people, and the 'Kombuis-Hollands' (Kitchen-Dutch) of the uneducated coloured servants. **1959** *Chambers's Encycl.* XII 763/2 By 1875, when the spoken language was firmly established, S. J. du Toit founded a 'Society of True Afrikaners' to propagate the written language; this met at first with violent opposition from the peasant and the politician—both English and Dutch—and Afrikaans was called kitchen Dutch, as the Greek of the Bible was once supposed to be 'bad' Greek. **1964** V. Pohl *Dawn & After* 102 What delighted us most was the originality of Gashep's speech. To us he spoke a kind of kitchen Dutch into which he introduced English and Sesuto words. **1931** *Auckland Star* 22 Mar. 7/2 A kitchen evening was given by Mr. and Mrs. A. R. Gillett at their residence..in honour of Miss Sabina Gardner, whose marriage..to Mr. F. Richardson takes place shortly. **1862** G. H. Mason *Zululand* iv. 58 In adopting [the official dialect]..no doubt, the Bishop has been guided by one of the chief clerks in the native department; who was born and reared amongst the Cape Colony Caffres, and, consequently, prefers it to learning Zulu proper; which, of course, is held in contempt by all officials, and sneeringly called 'Kitchen Kaffir'. **1924** *Cape Argus Mag.* 2 Feb. 5 A wonderful language is Kitchen Kaffir, a weird medley of dialects, interspersed with English words. **1956** [see *BAROTSE]. **1964** 'D. Wilson' *Search for Geoffrey Goring* viii. 144 He speaks a bit of English and some Kitchen Kaffir as well as Swahili. **1971** T. Sharpe *Riotous Assembly* (1973) ii. 15 She had spoken to him in Kitchen Kaffir, the language of command, [illegible]. **1917** Kitchen police [see *K.P.* s.v. *K 4 f]. **1918** *Wells Fargo Messenger* Jan. 87/3 My present position does not require me to perform any of the so-called dreaded duties, such as guard duty, kitchen police, stable orderly. **1928** Farrow *Dict. Mil. Terms* 329 *Kitchen police*, those charged with the scullery work of the kitchen. **1929** F. A. Pottle *Stretchers* (1930) 33 Before first call, six or more unfortunates from each bed were routed out of bed and went on kitchen police. [illegible] **1936** *Amer. Speech* XI. 51 [illegible] **1924** H. Croy *R.F.D. No. 3* 69 It was a 'kitchen shower'. The glittering array was piled high, like a special sale in a racket store—dish-pans, saucepans, pie pans, and so on. **1974** *News & Reporter* (Chester, S. Carolina) 22 Apr. 2-A/5 Mrs. J. J. Key and Miss Mary Smyrn were joint hostesses last Wednesday evening when they honored Miss Marilyn Hicks, bride-elect of the season, with a kitchen shower at the Key home on Columbia Street. **1948** N. Scanlan *Kelly Road* xvii. 195 A 'linen tea' for the bride-elect, and a 'kitchen tea' and a 'China tea' followed. **1965** *Sunday Mail* (Brisbane) 21 Nov. 26 Michelle Bowes and Patricia Donovan gave the bride a kitchen tea on Friday. **1970** G. Greer *Female Eunuch* 116 The more class the families can pre-empt, the more [illegible].

kitchenable (ki·tʃénăb'l), *a.* [f. Kitchen *sb.* or *v.* + -able.] Suitable for cooking and serving at table.

1909 *Chambers's Jrnl.* Feb. 193/2 There is probably no bird upon our game-list which is more eagerly sought after than the wood-cock..for his kitchenable qualities. **1913** G. Bolam *Wild Life Wales* vi. 50 In judging of the probable kitchenable qualities of a bird.

kitchen cabinet. orig. *U.S.* [f. Kitchen *sb.* + Cabinet.] **1.** *Politics.* A group of unofficial advisers (orig. of the President of the U.S.A.), popularly believed to have greater influence than the actual Cabinet (or the elected representatives, etc.). Hence, a private or unofficial group of advisers to the holder of an office.

1832 W. S. Archer *Let.* 8 July in A. C. Cole *Whig Party in South* (1913) 13 [illegible] the Kitchen Cabinet. **1842** *Ainsworth's Mag.* Dec. 554 We will hurl the kitchen cabinet tyrants from their stools. **1860** J. Parton *Life A. Jackson* III. xvi. 183 These were the gentlemen—Lewis, Green, Hill and Kendall—who, at the beginning of the new administration, [illegible] Kitchen Cabinet. [illegible] **1904** *N.Y. Herald* 14 Jan. 5 The kitchen cabinet is a development of the ascendency of Governor Odell in republican affairs. It consists of the body guard of his closest friends and advisers. [illegible] Mr. Brownell is part of the Dewey 'kitchen cabinet'. **1969** *Guardian* 24 Mar. 11/3 Every policy decision is being taken by the Prime Minister and a kind of 'kitchen cabinet' of intimates. **1973** *Times Lit. Suppl.* 13 May 576/4 Every ruler operates on two levels, the personal and the official, that is, he has a kitchen cabinet and a body of ministers.

2. A cabinet for domestic and culinary utensils, etc., in a kitchen.

1895 *Montgomery Ward Catal.* 607/1 Kitchen Cabinets. Very useful about a house where a 'shortage' of closets or cupboards is felt. *Ibid.* 607/1 Kitchen Cabinet Table.. Has a molding board, spice and cutlery drawer, sugar and groceries drawer, flour drawer, and a cupboard fitted for cooking utensils. [illegible] **1933** *Discovery* July 219/2 Most houses and flats are fitted as a matter of course with kitchen cabinets and refrigerators, hanging cupboards in bedrooms and so on. **1972** M. J. Bosse *Incident at Naha* 14 The sugar bowl would have gone back into a kitchen cabinet.

Kitchener² (ki·tʃénəɹ). The name of Herbert Horatio *Kitchener*, first Earl Kitchener of Khartoum and of Broome (1850–1916), British soldier, used *absol.*, *attrib.*, or in the possessive to denote a man of his imposing and taciturn personality, soldiers recruited while he was Secretary of State for War (1914–16), or aspects of appearance or of dress characteristic of him or of these troops. Also allusively. So **Ki·tchenerism,** a quality characteristic of Kitchener.

1903 *Westm. Gaz.* 31 Aug. 1/2 The sober thrift of the 'Kitchener of Russian finance'..is in keeping with his personality. **1916** J. N. Hall *Kitchener's Mob* i. 1 'Kitchener's Mob' they were called in the early days of August, 1914, when London boardings were clamorous with the first calls for volunteers. *Ibid.* iii. 23 'Kitchener's Rag-Time Army I calls it!' growls the veteran of South African fame. [illegible] **1916** H. G. Wells *Mr. Britling* i. iii. 74 He was presented as a monster of energy and self-discipline; as the determined foe of every form of looseness, slackness, and easy-goingness... 'It's Kitchenerism'.. [illegible] **1925** Fraser & Gibbons *Soldier & Sailor Words* 136 *Kitchener's blue*, [illegible]. **1929** *Papers Mich. Acad. Sci. Arts & Lett.* X. 304/1 *Kitcheners*, a name given by the Regulars to soldiers enlisting for the emergency. **1930** C. D. Baker-Carr *From Chauffeur to Brigadier* ix. 127 The new Kitchener Divisions were now beginning to 'come out'. **1937** Partridge *Dict. Slang* 458/1 *Kitchener wants you!* A military c.p. [illegible] **1969** 'A. Cave' [illegible] A large European wearing a pink shirt, baggy grey trousers and a drooping Kitchener moustache. **1971** D. Meiring *Wall of Glass* vii. 69 [illegible] Kitchener [illegible] Sam Browne.

kitchenette (kitʃéne·t). orig. *U.S.* [See -ette.] A small room or alcove in a house, flat, etc., combining kitchen and pantry. Also *attrib.*

1910 *Variety* 7 May 16/4 [illegible] a kitchenette. **1919** *Ladies' Home Jrnl.* Oct. 117 Ovenette and kitchenette cookery. **1921** *Daily Colonist* (Victoria, B.C.) 2 Oct. 38/1 Kitchenettes and bathrooms are also installed. **1922** *Hotel World* 25 Mar. 10/1 Kitchenette suites of 2 rooms. **1924** *Glasgow Herald* 28 Apr. 5 The New York business woman.. wants her kitchenette and her home cooking, be it ever so simple. [illegible] **1934** [see *DRYING-RACK]. **1955** [see *DINETTE]. **1960** *Times* 25 Feb. 1/5 Bed-sitter, with kitchenette, in private house Kensington. **1974** *Country Life* 30 May 1317/1 (Advt.), Granny annexe comprising: living room, kitchenette, bedroom and shower room.

Kitchen rudder. [Named after J. G. A. *Kitchen*, Englishman, who patented the device in 1914.] A steering device for small craft consisting of a pair of curved deflectors either side of the propeller whose position is altered to change the course or speed of the vessel or to cause it to go backwards.

1920 *Engineer* 6 Feb. 140/1 The essential parts of the Kitchen rudders consist of two curved deflectors generally formed as parts of a circular cylinder, partly enclosing the propeller. **1920** *Shipbuilding & Shipping Rec.* 12 Feb. 200/2 (*caption*) Admiralty pinnace with Kitchen rudder. **1923** *Man. Seamanship* (Admiralty) II. 200 Boats are fitted with a clutch and reverse gear, clutch and reversing propellers, or clutch and Kitchen rudder. **1961** F. H. Burgess *Dict. Sailing* 127 *Kitchen rudder*, one comprising two curved blades that may completely enclose the propeller; they are rotated by a wheel on the tiller to open or close for going either ahead or astern; the direction and speed are thus controlled from the tiller, while the engine runs unattended at one speed.

kitchen sink. [f. Kitchen *sb.* + Sink *sb.*¹ 1 c.] **a.** A sink in which dirty dishes, vegetables, etc., are washed. Freq. used as a symbol of women's enslavement to the kitchen.

1873 *Young Englishwoman* May 251/1 [illegible] **1973** *Guardian* 14 Feb. 11/2 Women stayed right where they were: for the most part at the kitchen sink, or in low-paid clerical or light manual work.

b. *fig.*

1888 Kipling *Under Deodars* (1889) 5 All his ideas and powers of conversation..are taken from him by this—this kitchen-sink of a Government.

c. *everything but the kitchen sink* and similar phr.: everything imaginable.

1948 Partridge *Dict. Forces' Slang* 106 *Kitchen sink*, used only in the phrase indicating intense bombardment—'They chucked everything they'd got at us except, or including, the kitchen sink.' 'The kitchen stove' was also used. **1958** *Wall St. Jrnl.* 23 Oct. 4/4 Gen. Trudeau said the military services often slow down development of new weapons 'because we are such perfectionists that we want everything but the kitchen sink in a weapon'. **1965** 'E. McBain' *Doll* x. 128 Brown began searching. 'Everything in here but the kitchen sink,' he said. **1966** —— *Eighty Million Eyes* xi. 189 We'll throw everything but the goddamn kitchen sink at you. **1967** L. White *Crimshaw Memo.* (1968) iii. 61 He goes out and buys himself an XKE Jaguar..it had everything but the kitchen sink on it.

d. Used (with hyphen) *attrib.*, or *absol.*, to designate a group of English realistic painters of the 1950s and later, or their type of art, or a group of English realistic authors (chiefly playwrights) of the same period or their plays or publications.

1954 D. Sylvester in *Encounter* Dec. 61 (*title*) The kitchen sink. *Ibid.* 62/1 The post-war generation takes us back from the studio to the kitchen... The kitchen sink too. *Ibid.* 62/2 It is evident that neither objectivity nor abstraction is the aim of the young painters of the kitchen-sink school. **1956** L. McIntosh *Oxford Folly* 69 On the walls were several drawings and paintings of the 'Kitchen Sink' school of modern English painters, whose sordid subject matter contrasted with their luxurious setting. The one above the mantelpiece depicted a lavatory. **1959** *Listener* 19 Feb. 340/3 'Kitchen sink' painting is not an exclusive English phenomenon: it originated in France, with Rebeyrolle and Minaux. **1960** *Times* 15 Mar. 13/6 Mr. Ronald Duncan is reported as saying that the English Stage Company..presents only left wing 'kitchen sink' drama. **1960** *Guardian* 28 Oct. 9/5 The day of the social-realist, kitchen sink advertisement has dawned. **1963** *Times* 16 Feb. 9/3 If the British new wave were interested only in easy money they would stick to the slag-heap and the kitchen sink. **1965** *Punch* 26 May 762/1 The 'Kitchen-Sink' tag..began as descriptive heading for a post-war school of painters (nudity, violence, squalor, blasphemy, subversiveness and distortion are somehow morally OK qualities in the visual arts) and only later became a pejorative title for a school of playwrights. **1973** *Black World* Apr. 41, I wasn't going to write any more Black 'kitchen sink' dramas.

So **ki:tchen-si·nkery.**

1964 *Listener* 16 Apr. 624/1 We've been attacked for too much pessimism, sordidness, and kitchen-sinkery. **1969** B. Turner *Circle of Squares* i. 7 The longest-ever season of kitchen-sinkery on our stages.

kitchen stove. [f. Kitchen *sb.* + Stove *sb.*¹ 5 a.] **a.** A stove in a kitchen.

1845 *Knickerbocker* XXV. 106 He wished her in the south of France or the kitchen stove, rather than there. **1925** F. Ayscough *Chinese Mirror* 386 Ts'ao Chün, Lord of the Kitchen-Stove. **1959** P. H. Johnson *Humbler Creation* xvi. 108 'Mrs. Fisher is a most awful stick,' Lucy said. 'One might as well say one's lines to the kitchen stove.' **1974** *Times* 21 Sept. 7/2 The kitchen stove, a low stout Rayburn of chipped yellow enamel.

b. *everything but the kitchen stove* and similar phr. = *KITCHEN SINK c.

1927 E. Wallace *Feathered Serpent* i. 9 'Got everything on except the kitchen stove,' said Mr. Crewe pleasantly... 'You're a fool to go out with all that stuff on you.' **1928** [see *BOB *v.*³ 2 b]. **1948** [see *KITCHEN SINK c]. **1959** *Listener* 8 Jan. 63/2 The assumption that, since Christmas is the only time children go to the theatre, they must therefore be given everything, including the kitchen-stove. **1960** M. Cecil *Something in Common* xxi. 235 'I suppose you haven't an ulcer?' Colin said that—apart from the kitchen stove—it was the one item he'd forgotten to bring. **1964** C. Dale *Other People* x. 178 There's Mum, hair tinted, face done up, everything on bar the kitchen stove.

kitcheny, *a.* Delete *rare* and add later examples.

1926 Kipling *Debits & Credits* 328 We eat kitcheny food. **1971** P. Dickinson *Sleep & his Brother* vi. 137 Doll had left the room..and was making kitcheny noises beyond. **1973** *Times* 16 Nov. 15 (*caption*) Most of the things are decidedly kitcheny.

kite, *sb.* Add: **3.** (Examples referring to modifications of the toy kite designed to support a man in the air or to form part of an unpowered flying machine (cf. *AEROPLANE 1).)

1826 Viney & Pocock *Brit. Pat.* 5420, This Patent is obtained for an Invention by which kites are made to act as..sails, for the purpose of navigating or drawing vessels.., for the purpose of raising weights or persons in the air,.. or for the hoisting of flags. The peculiarities of these kites are:—..they have four lines by which their power is controlled or their course directed. **1875** A. M. Clark *Brit. Pat.* 169, This Invention relates to a kind of kite or ærial apparatus to be used for military and other purposes, its chief object being to raise to the desired height in the air, and to support in a sufficiently tranquil position for reconnoitring a scout, look out man, or sentry. **1889** H. S. Maxim *Brit. Pat.* 16,883, My invention is chiefly designed..to provide for the construction of an aëronautic machine which can, while moving forward in the air, be caused to rise or descend at any desired velocity or to travel at any predetermined height above the ground... I provide an adjustable covered framework or kite of very large dimensions... For convenience of description I will hereinafter term this covered framework or kite an 'aëroplane'. **1893** *23rd Rep. Aeronaut. Soc.* 17 What we have to do is to make the wings, whether fixed or flapping, in the requisite form, and add aëroplanes or kites, and find the necessary power to drive them along at their proper angles. **1894** *Proc. Internat. Conf. Aerial Navigation, Chicago, 1893* 253 Among the different free-flying models which I exhibited..in the large hall of the Engineering Society in Vienna there was a model of my gliding aeroplane or kite, which illustrated the support to be obtained from the air.

b. A proposal or suggestion offered or 'thrown out' tentatively in order to 'see how the wind blows'. (Cf. *BALLON D'ESSAI.) See also Fly *v.*¹ 5 a in Dict. and Suppl.

1902 *Nature* 14 Aug. 380/2 A few suggestions have been thrown out by various students which must be regarded more as trial hypotheses than as definite conclusions, indeed they should be looked upon rather as 'Kites'. **1904** *Westm. Gaz.* 5 Aug. 2/2 The new Army scheme..is to be debated on Monday, but whether as a Government proposal or the private kite of the Minister for War remains wholly obscure. **1973** A. MacVicar *Painted Doll Affair* ii. 29 'I'm sorry ye're lumbered wi' me,' he said, sounding anything but sorry. I ignored this blatant kite.

c. An aeroplane. *slang*, esp. *Services'*.

The popular use of *kite* in this sense prob. originated with the 'box-kite' aeroplane (see *BOX *sb.*² 24); but the uses in quots. 1838, 1909 are direct applications of *kite* in sense 3.

[**1838** J. H. Pennington *Aerostation*, (*caption*) Steam-kite, or inclined plane, for navigating the air.] **1909** S. F. Cody in *Aeronaut. Jrnl.* XIII. 15/2, I had to turn my hobby into manufacturing a kite in order to raise money to build a flying machine, or to put 'power' into my kite, consequently the term 'Power Kite'. *Ibid.* 19/1 This is the finished power kite ready to start. The screws are not really propellers. You may call them tractors or propellers, which you like. **1917** in A. J. L. Scott *Hist. Sixty Squadron R.A.F.* (1920) 100 He told me that he had managed to fly his kite back with great difficulty. **1934** T. E. Lawrence *Let.* 19 Mar. (1938) 793 The German kites will be new and formidable. **1942** T. Rattigan *Flare Path* 1. 33 A kite from the Polish squadron. **1952** M. Tripp *Faith is Windsock* xii. 183 The Squadron hasn't lost a single kite in the last three raids. **1969** [see *HARRY *sb.*² 8 b].

d. Phr. *high as a kite*: see *HIGH *a.* 16 b.

4. b. *Criminals' slang.* A communication (esp. one that is illicit or surreptitious); *spec.* a letter or verbal message smuggled into, out of, or within a prison.

1859 G. W. Matsell *Vocabulum* 49 *Kite*, a letter; fancy stocks. **1923** J. F. Fishman *Crucibles of Crime* ix. 203 Sometimes..prisoners manage to plant notes in various parts of the prison which are to be picked up by the intended recipient. This practice of 'shooting' contraband notes is known among the prisoners as 'flying a kite'. **1925** *Flynn's* 3 Jan. 665/2 *Kite*, a message per lip. *Kite*, a letter or note. **1927** [see sense *4 c]. **1953** H. Bryan *Inside* (1954) xvii. 279 Having settled on the girl, one would send her a 'kite', or love letter. **1960** Wentworth & Flexner *Dict. Amer. Slang* 194/2 *Fly a kite.* 1. To write a letter; esp. to smuggle a letter into or out of prison. *Underworld use.* 2. To send an airmail letter, often requesting money or assistance. *Modern use, mainly underworld but gaining some popularity.* **1971** *N.Y. Times* 21 Oct. 52/2 *Kite*, a complaint to the police about an illegal operation, often originating with a disgruntled gambler.

c. *slang.* A cheque (sense 3), *esp.* a blank cheque or a cheque drawn on insufficient funds or forged from a stolen cheque-book.

1927 *Dialect Notes* V. 446 *Fly a kite*, v. (1) To pass a bad cheque. (2) To sell worthless stocks and bonds. (3) To write mournful letters, as of prisoners, to sympathetic old women and charitable institutions. **1928** E. Wallace *Gunner* xxx. 243 He had spent the afternoon searching London for the right 'kites'. There is quite a brisk trade in blank cheque forms. **1936** J. Curtis *Gilt Kid* iv. 45 Used to say that he'd been done for kites, but everyone reckoned it was for poncing. **1962** R. Cook *Crust on its Uppers* i. 21 The real morries..flying dodgy kites with each other at bent spielers till the punter..outs his kiting-book too and scribbles a straight one. **1969** T. Parker *Twisting Lane* 41 He's in for what they call 'kites', dud cheques, you know.

5. b. On a minesweeper, a device attached to a sweep-wire submerging it to the requisite depth when it is towed over a minefield.

1915 *Chambers's Jrnl.* June 386/1 Between the vessels of each pair is the sweep-wire, sunk to the necessary depth in the water by means of towed 'kites', wooden arrangements acting on the same principle as the ordinary air-kites. **1923** *Man. Seamanship* (Admiralty) II. 172 The present form of kite consists of a specially shaped metal plate which has a tendency to dive when towed. This is towed over the stern by a kite wire.

9. a. *kite-like* adj.; **kite bar,** a bar or stripe of an undesirable colour in the plumage of a fancy pigeon. **b.** *kite-line, -maker, -string; kite-faced, -like* adjs.; **Kite mark, Kitemark,** a quality mark, similar in shape to a kite, granted for use on goods approved by the British Standards Institution; also *transf.*; hence **kite-mark** *v. trans.*, to use the Kite mark on; **kite-marked** *ppl. a.*, bearing the Kite mark; **kite-tail** (earlier non-attrib. U.S. example). **c.** in sense 4, **kite-man,** a person who obtains money against bills of exchange or cheques that will not be honoured; *spec.* (see quot. 1967).

1876 R. Fulton *Illustr. Bk. Pigeons* 108 A softer shade of blue, with brown, or what are called by Pouter fanciers 'kite' bars. **1922** Joyce *Ulysses* 537 Alone on deck,.. yellow kitefaced, his hand in his waistcoat opening, declaims. **1901** Kipling *Kim* xiv. 365 From the edge of the sheep-pasture floated a shrill, kite-like trill. **1909** *Daily Graphic* 26 July 10/1 When floating on an up-draught they [sc. the planes] will be expanded as a fan expands, and will present a larger kite-like surface. **1828** *Kaleidoscope* 12 Aug. 48/2 There is no obstacle to interfere with the kite-lines. **1876** 'Mark Twain' *Tom Sawyer* xx. 304 He took a kite-line from his pocket. **1926** M. Leinster *Dew on Leaf* v. 211 The kite-makers were busy making fantastic objects in bamboo and paper. **1928** E. Wallace *Double* v. 64 This was a favourite rendezvous of the swell mob, the 'kite' men, the confidence artists. **1967** J. Phelan *Nine Murderers & Me* 162 Kite-man, one who passes forged cheques. **1952** *B.S.I. Monthly Information Sheet* Oct. 16 (*heading*) The 'Kite' mark on consumer goods. *Ibid.*, In March, 1952, when the Utility schemes for textiles and clothing were brought to an end, most of the industries concerned undertook to co-operate with the B.S.I. in preparing voluntary standards..to ensure that at least the same levels of quality could be maintained. It was also agreed that wherever possible the B.S.I. 'Kite' mark would replace the former Utility mark as a guarantee to consumers that the goods bearing the mark were, in fact, 'up to standard'. **1956** *Observer* 12 Feb. 6/4 It is hoped that eventually the B.S. Kite-Mark will certify the quality of many..consumer goods. **1957** *Times* 23 Sept. 11/4 The first fireguards bearing the kite mark of the British Standards Institution will be in the shops this autumn. **1957** *Economist* 12 Oct. 157/1 'Shoppers' Guide' discusses articles against the measure of the BSI Kite-mark, which is a guarantee only of minimum satisfactory performance. **1958** *B.S.I. News* Nov. 3 Buyers, particularly large-scale purchasers, can do much to safeguard their workpeople from injury by insisting, wherever possible, that the safety equipment which they buy should be Kite-marked. *Ibid.* 17/1 We hear from Kirk and Company (Tubes) Ltd., that they have recently been granted a licence to Kite-mark their malleable cast iron and cast copper alloy pipe fittings under B.S. 1256. **1960** *Ibid.* Apr. 5/2 (*heading*) Kitemarking should be extended. **1966** *New Scientist* 12 May 338/3 The findings will be used by the British Standards Institution when it publishes the 'rules' for 'kite-marked' toothpastes later this year. **1971** *Brit. Standards Yearbk.* p. xvi, The British Standards Mark (known as the Kitemark) is a registered certification trade mark owned by BSI. Manufacturers may apply to BSI to use the mark on their products when their quality control arrangements are considered satisfactory and they have agreed to comply with a Scheme of Supervision and Control involving..inspection, sampling and testing. **1971** *Daily Tel.* 24 Nov. 11/5 The annual Civic Trust Awards,..emphasising architectural and design excellence in a total lived-in environment, have become sought-after kitemarks for planners..and architects. **1972** *Ibid.* 12 June 2/6 It will be illegal for shops to sell crash helmets which do not have the BSI's 'kitemark' seal of approval. **1972** *Which?* June 192 So far, no cabinets with mirrors have been given the Kitemark... If we hear of any Kite-marked cabinets, we will mention them. **1841** Thoreau *Jrnl.* 6 Aug. in *Writings* (1906) VII. 266 Like pasteboards on a kite string. **1971** *N.Z. Listener* 22 Mar. 13/1 Get seen around with a good woman on your kite-string and no-one bothers you regardless. **1869** L. M. Alcott *Little Women* II. x. 150 The others are torn up to..bandage cut fingers, or make kite-tails.

kite, *v.* **1. a.** Add to def.: To move quickly, to rush; to rise quickly. Const. *around, off, up, etc.*

1854 'O. Optic' *In Doors & Out* (1876) 92 You did not use to be fond of 'kiting' round in this manner. **1864** L. N. Boudrye *Jrnl.* 21 Aug. in *Hist. Rec. Fifth N.Y. Cavalry* (1865) 165 A well directed shell..sent them 'kiting' to the woods again. **1870** W. W. Fowler *Ten Yrs. Wall St.* 504 Would seem to be enough to start a panic, or send the market 'kiting' up among the tall figures. **1908** Kipling *Lett. to Family* viii. 72 We have seen a financial panic.. send whole army corps of aliens kiting back to the lands whose allegiance they forswore. **1931** Wodehouse *Big Money* viii. 181 The stock kited sixty points the first day. **1935** C. S. Forester *Afr. Queen* vi. 116 Bet they were surprised to see the old *African Queen* come kiting past. **1965** J. Potts *Only Good Secretary* i. 13 Yes, and her too, kiting off to Long Island when she ought to be here.

3. b. (Earlier U.S. example.) Now usually, to write or cash a dud or temporarily unbacked cheque. Hence **ki·ting** *vbl. sb.*, the raising of money on credit; the passing of forged or unbacked cheques. Cf. *KITE-FLYING *vbl. sb.* 2.

1839 C. F. Briggs *Adventures H. Franco* II. iv. 35 He stuffed half a dozen blank checks into his hat, and said he must go out and kite it to save his credit. **1866** *Congress. Globe* 29 June 3482/1 Every kiting charter like this one—I speak of the National Telegraph Company. **1872** *Ibid.* 3 Apr. 2128/2 (Th.), They may hold the bonds, as has often been done in kiting corporations, and then take the property they have thus swindled the public out of. **1934** H. N. Rose *Thesaurus of Slang* iii. a. 21/1 Issue a check which hasn't sufficient backing: *to kite.* **1942** Berrey & Van den Bark *Amer. Thes. Slang* § 556/1 Kiting, kite flying, extending credit or sustaining a balance in the bank by means of 'kites'. **1950** H. E. Goldin *Dict. Amer. Underworld Lingo* 118/1 *Kite*, v. 1. To issue or pass, as a forged check or bond; (more accurately) to issue or pass a check against insufficient funds... *Kite checks* (New England States). 1. To issue forged checks. 2. To write checks, usually post-dated, against insufficient funds. 3. To raise illegally the face value of otherwise good checks. **1959** N. Mailer *Advts. for Myself* (1961) 438 Her ideas about.. those investors with some credit rating whom an exurban bank, proud of its personal touch, might allow to kite a cheque for twenty-four hours..intrigued the banker. **1960** F. Gibney *Operators* vi. 158 Check kiting..is a different and more complex process than forging... Few up-and-coming businessmen..can claim to have resisted the temptation to write a pressing check just a day or two before some money is due, in the prayerful expectation that their deposit will get into the bank's ledgers before the check they have cashed comes home to roost. **1962** R. Cook *Crust on its Uppers* x. 90 We'll kite 'em at the airport! **1963** J. N. Harris *Weird World Wes Beattie* (1964) xv. 184 He was up to his ears in debt—always kiting checks before payday. **1969** 'E. Lathen' *Murder to Go* (1970) xvii. 173 If it had been a question of..kiting a cheque—well, that wouldn't surprise you at all. Clyde cut corners all his life. **1970** R. Lewis *Wolf by Ears* iii. 137 The technical terms..are 'kiting' and 'lapping'. Money is transferred between two accounts, recording the receipt prior to the balancing date and the payment after the balancing date. That's 'kiting'.

c. To send a communication; *spec.* to smuggle a letter into, out of, or within a prison. (Cf. *KITE *sb.* 4 b.)

1925 *Flynn's* 3 Jan. 665/2 *Kite*, to send a signal; to send a message. **1936** *Detective Fiction Weekly* 4 Jan. 116 A letter which I had 'kited' out of the prison. **1945** L. Shelly *Jive Talk Dict.* 13/2 *Kite*, to air mail or exchange.

kite-flying, *vbl. sb.* [Kite *sb.* 9 b.] **1.** *lit.* The flying of a kite on a string. Also *attrib.*

1804 T. A. Ward *Diary* 5–6 Oct. in *Peeps into Past* (1909) viii. 51/1 He went away..to renew the sport of kite-flying. **1827** [in Dict. s.v. Kite *sb.* 9]. **1849** Dickens *Dav. Copp.* (1850) xvii. 179 He never took an active part in any game but kite-flying. **1926** M. Leinster *Dew on Leaf* v. 199 Towards the end of the kite-flying season, numerous Chinese had made their way to various pieces of rising ground with these toys. **1962** W. O. Mitchell *Kite* ii. 14 Kite-flying would be a lot like taking a drive on Sunday or going on a picnic.

2. *slang.* The raising of money (*a*) by persons collusively exchanging accommodation bills or cheques on different banks, in none of which they possess sufficient funds; (*b*) by one person transferring accounts between banks and creating an illusory balance against which he cashes cheques; (*c*) by a person passing forged, stolen, or unbacked cheques.

1834 [in Dict. s.v. Kite *sb.* 9]. **1848** Bartlett *Dict. Amer.* 195 *Kite flying*,..a combination between two persons, neither of whom has any funds in bank, to exchange each other's checks, which may be deposited in lieu of money, taking good care to make their bank accounts good before their checks are presented for payment. *Kite flying* is also practised by mercantile houses... A house in Boston draws on a house in New York at 60 days or more, and gets its bill discounted. The New York house, in return, meets its acceptance by re-drawing on the Boston house. **1850** Thackeray *Pendennis* II. xxiv. 244 That kite-flying, you know, Mr. M., always takes two or three on 'em to set the paper going. Altamont put the pot on at the Derby, and won a good bit of money. **1951** 'H. Cecil' *Painswick Line* vi. 55 'Kite flying' is a method of borrowing money from a bank, which is not prepared to lend it to you. **1964** Z. Progl *Woman of Underworld* xix. 178 Passing dud cheques (kite-flying) calls for unwavering nerve and audacity. **1970** M. Kenyon *100,000 Welcomes* iv. 23 Forgery? Fraudulent conversion? Kite-flying?

3. Sending up a 'kite' or *ballon d'essai*. Also *attrib.*

1898 'A. Hope' in *Daily News* 4 Apr. 7/1 Principally it [sc. the press interview] was said to be used as a means of what might be called kite-flying. **1899** *Ibid.* 29 June 5/4 The object of Boer diplomacy is to prolong the kite-flying stage indefinitely until commercial interests..urge England to accept anything. **1927** *Daily Tel.* 30 Aug. 8/6 These suggestions are dismissed in British circles as mere 'kite-flying'. **1958** *Economist* 20 Dec. 1058/2 The Belgrade newspaper, *Borba*, which is often the government's mouthpiece, has emphasised..that if other countries would accept immigrants from Jugoslavia these would be allowed to leave... If this was kite-flying, the kite has been allowed to drop to the ground with a thud. **1969** *Courier-Mail* (Brisbane) 2 May 3/8 State D.L.P. secretary (Mr. J. Judge), said: 'This is nothing but kite-flying. The D.L.P. will be making no decision on preferences until the Central Council meets next Tuesday.'

So (as a back-formation) **ki·te-fly** *v. trans.* and *intr.*; **ki·te-flyer.**

1844 *Knickerbocker* XXIV. 258 The most persevering kite-flyers that I know of, are the Reformers. **1860** Bartlett *Dict. Amer.* (ed. 3) 231 *Kite-flier*, a financier who practises the operation of '*kite-flying*'. **1876** *Monthly Packet* June 577 Heavy bets are laid on the best kite-flyers. **1896** Kite-flyer [in Dict. s.v. Kite *sb.* 9]. **1935** 'D. Hume' *Gaol Gates are Open* 8 Higher in the hierarchy of crime come the 'kite flyers', who cash worthless cheques. **1965** *Listener* 22 July 111/2 The irrelevance of his office to foreign affairs..stressed that he was merely kite-flying. **1968** 'L. Egan' *Serious Investigation* (1969) iii. 34 Right now we've got a con man operating on the pensioners, and a kite flyer passing all the rubber checks. **1972** *Daily Colonist* (Victoria, B.C.) 16 Jan. 2/7 Sarasota, Fla... Some 89 contestants jammed Lido Beach for the Association's 4th Annual International Flyoff, held in honor of America's foremost kiteflyer Ben Franklin.

‖ **kitenge** (kite·ŋgi). Pl. **kitenges, vitenge(s.** [Swahili.] In East Africa, a fabric, usu. of cotton and printed in various colours and designs with distinctive borders, used esp. for women's clothing.

1969 *Daily Nation* (Nairobi) 31 Oct. 5/1 Betty was taking a good look at the lovely things on sale, but had almost made up her mind to wear traditional kitenge on the 'Big Day'. **1970** *Sunday News Mag.* (Tanzania) 30 Aug. 8 Take the case of a polygamous peasant whose wives have been turned into working machines to bring wealth to him while all they get in return are pieces of kitenge or khanga. **1971** *Inside Kenya Today* Mar. 42/1 Local fabrics such as Kitenges and Kangas are also used to make clothes at the Institute. **1971** *Standard* (Tanzania) 7 Apr. 1/8 UWT members have been urged to reflect the African culture in their dress and other items that convey our intrinsic values. May Day vitenges will be on sale at 19/20.

kite's-foot, kitefoot. 2. (Earlier and later U.S. examples.)

1788 *Mass. Centinel* 4 June 94/2 Crowley & Clark Have just received a quantity of Kitefoot Tobacco, of a *superior quality* for smoking. **1835** H. C. Todd *Notes Canada & U.S.A.* 50 Maryland produces..the Bright Kite's Foot Tobacco.

kitful: see *KIT *sb.*¹¹

kiting: see *KITE *v.* 3 b.

ki·tless, *a.* [f. Kit *sb.*¹ + -less.] Having no kit (Kit *sb.*¹ 2 b); without (adequate, suitable) clothing.

The definitions in the work cited in quot. 1846 are often unsatisfactory; it nevertheless seems to provide a genuine example of the sense defined above.

1846 *Swell's Night Guide* 123/2 *Kitless*, a bare, naked fellow. **1936** J. Buchan *Island of Sheep* ix. 160 [He] went off with the women..to look after her wardrobe, for she also was kit-less.

‖ **kitsch** (kitʃ). Also **Kitsch.** [G.] Art or *objets d'art* characterized by worthless pretentiousness; the qualities associated with such art or artifacts. Also *attrib.*, *Comb.*, and *transf.*

1926 B. Howard *Let.* in M. J. Lancaster *Brian Howard* (1968) ix. 166 A healthy week..riding, chasing dogs and listening to 'Kitsch' on his wireless. **1939** *Partisan Rev.* VI. 40 Kitsch is mechanical and operates by formulas. Kitsch is vicarious experience and faked sensations. Kitsch changes according to style, but remains always the same. Kitsch is the epitome of all that is spurious in the life of our times. **1941** Auden *New Year Let.* iii. 59 Reason's depravity that takes The useful concepts that she makes As universals, as the kitsch. **1949** Koestler *Insight & Outlook* 410 The more romantic a work of art, or a landscape, the quicker its repetitions are perceived as *kitsch* or 'slush'. **1955** *20th Cent.* June 541 In a time of crassness and stridency perhaps unique in history, a time when an alternative civilization of *kitsch* is not only available to all but clamantly thrust upon them, it is imperative to strive for agreement, order, and coherence in the ranks of the cultivated. **1958** *Observer* 23 Feb. 14/1 What is so extraordinary about some of these *kitsch* masterpieces is the way they can be enjoyed on two planes, both as themselves and as their own parodies. **1958** *Times* 4 July 13/4 Few attempts are made in England to mount productions of plays of the *commedia dell' arte* tradition; and such attempts are in danger of being dismissed as 'art theatre *kitsch*'. **1961** *Times* 11 May 10/4 There are the same highbrows as in England, who consider that the quality of the pure entertainment as such is generally *kitsch* or trash. **1962** *Times* 6 Apr. 17/3 Their attitude to this *kitsch*-culture is highly equivocal. **1965** *Spectator* 22 Jan. 108/1 If leaders of the state choose their job..they must choose to be the victims of the kitsch and whitewash and balderdash. **1967** *Ibid.* 29 Dec. 812/2, I have never seen such kitsch, not even in French provincial towns or Irish church bazaars. **1972** *Listener* 24 Aug. 236/1 A galloping fancy for Victoriana, a sophisticated and uncritical taste for Kitsch and the cute. **1972** *New Yorker* 30 Sept. 24/3 This is one of the liveliest and most popular of their *kitschfests*.

So **kitsch** *v. trans.* (*rare*), to render worthless, to affect with sentimentality and vulgarity; **ki·tschy** *a.*, possessing the characteristics of kitsch.

1951 W. Sansom *Face of Innocence* ii. 16 Situations that have for many become unendurably hackneyed, spoiled by bad artists or kitsched by politics. **1967** *Time* 17 Feb. 104 The kitschy existential slogan: 'Things just happen. No reason, no reason, just a happening.' **1969** R. Petrie *Despatch of Dove* i. 19 Her family owned a furniture factory. 'We make..mostly kitschy bits fit to furnish Grimm's fairy tales.' **1973** *Times* 27 Aug. 5/6 Costumed in a distressingly 'kitschy' manner.

kittel (ki·t'l). [Yiddish (G., overall, smock), ad. MHG. *kitel, kietel* cotton or hempen outer garment, prob. ad. Arab. *quṭn* cotton.] A white cotton or linen robe worn by orthodox Jews on certain holy days; also used as a shroud.

1891 M. Friedländer *Jewish Relig.* 492 The *kittel* or *sargenes* is part of the raiment in which the dead are clothed... In some countries..the bride presents the bridegroom with this article on the wedding-day: and it is worn by the husband on New-Year's Day and on the Day of Atonement and on the *Seder*-evening during the service. **1972** C. Raphael *Feast of Hist.* i. 11/2 A Yemenite Seder... The celebrants are wearing their traditional festive clothes, the men in white '*kittels*'. **1973** *Synagogue Light* Sept. 46/2 Garbed in the white Kittel, head wrapped in a Tallith, he who blows stands near the Rabbi.

kitten, *sb.* Add: **1. b.** (Later examples.)

1957 J. H. F. Stevenson *Mink in Britain* (ed. 2) v. 18 Once the kittens, or kits as they are called, are able to fend for themselves, they do so. **1964** R. M. Lockley *Private Life Rabbit* iv. 54 It was possible to handle and weigh week-old kittens without causing their desertion by the doe. **1972** R. Adams *Watership Down* xlii. 356 Clover's had her litter. All good, healthy kittens. Three bucks and three does.

c. In extended use: a girl-friend; a young woman; often as a form of address.

1870 D. J. Kirwan *Palace & Hovel* xliii. 612 The 'Kitten' is a blonde, with black eyes, a pretty, babyish face,..a profusion of golden hair. **1908** W. De Morgan *Somehow Good* xii. 119 'Kitten,' said Sally's mother to her suddenly, 'I think I shall go away to bed.' **1938** W. G. Hardy *Turn back River* iii. 28 'You'll have to go, kitten,' Clodia whispered hurriedly... For an instant she held her sister close. **1961** 'E. Lathen' *Banking on Death* (1962) vii. 59 He..accepted a sherry from June with a somewhat absent, 'Thanks, Kitten.' **1970** G. Greer *Female Eunuch* 266 There are the cute animal terms like.. *kitten* and *lamb* [to signify a woman].

4. Slang phr. (orig. *U.S.*), *to have kittens*: to lose one's composure; to get into a 'flap'.

1900 *Dialect Notes* II. 44 *Kitten*. In phrases 'get kittens', 'have kittens'. 1. To get angry. 2. To be in great anxiety, or to be afraid. **1937** *Times* 15 Feb. 13/4 Mr. Partridge allows 'jitters'..but not 'having kittens'. **1940** K. M. Knight *Rendezvous with Past* xx. 141 If he knew what I know about you he'd have kittens. **1943** Hunt & Pringle *Service Slang* 38 *Having kittens*, perturbed. 'The Colonel is having kittens'—the Colonel is upset and he is very, very angry. **1950** M. Kennedy *Feast* ii. ix. 42 She's been having kittens all day because Mrs. Siddal says she's got to empty slops. **1952** W. Plomer *Museum Pieces* xxv. 210 My doctor nearly had kittens when I suggested my being dropped to the *maquis* by parachute. **1959** 'A. Gilbert' *Third Crime Lucky* ii. 28 Gertrude was going to have kittens when she discovered that extravagance. **1967** N. Freeling *Strike Out* 76 When one of the horses has something wrong with it—then everybody has kittens.

ki·ttenishness. [-ness.] Kittenish characteristics or behaviour.

1905 *Smart Set* Sept. 15/1 Monsieur de Latour felt, as well as saw, that Madame de Beauregard, for all her kittenishness, was really a very great lady. **1926** *Chambers's Jrnl.* Aug. 610/1 Ages back the American girl abjured all that was in the nature of kittenishness. **1974** F. Nolan *Oshawa Project* xv. 99 'What do you mean..?' she said, every trace of the kittenishness falling away.

Kittitian (kiti·ʃăn), *sb.* and *a.* Also **Kittician.** [f. *St. Kitt(s* + *-itian* as in *Haitian.*] **A.** *sb.* A native or inhabitant of the island of St. Kitts, in the West Indies; also, the form of regional English spoken in St. Kitts. **B.** *adj.* Of or pertaining to St. Kitts or its inhabitants.

1966 *Listener* 26 May 754/2 The factors which drew Kitticians to this particular area. **1970** J. Brown *Un-melting Pot* vii. 104 They also came as islanders: Jamaicans, Barbadians, Kittitians, or Grenadans. **1973** *Observer* (Colour Suppl.) 16 Dec. 25/1 Carol Browne's brother will not speak Kittitian..but he will speak the Jamaican dialect. **1973** *Country Life* 31 May 1544/1 Rodney Lad.. carried off a considerable amount of Kittitian property, leaving the local whites in an unpatriotic mood. **1974** *Advocate-News* (Barbados) 19 Feb. 9 By rescuing sugar the Government was in fact rescuing Kittitians.

kittle cattle (ki·t'l kæt'l). orig. *Sc.* [f. Kittle *a.* + Cattle *sb.*] Used to denote people or animals that are capricious, rash, or erratic in behaviour; also *transf.*, objects, concepts, etc., that are difficult to use, sort out, or comprehend.

Initially in phr. (*kings are*) *kittle cattle to shoe behind*, an elaboration of *kittle to shoe behind* (1600 in Dict. s.v. Kittle *a.*).

1818 Scott *Heart Midl.* in *Tales my Landlord* 2nd Ser. IV. i. 5 Kings are kittle cattle to shoe behind, as we say in the north. **1876** [see Kittle *a.*]. **1881** H. Shanks *Peasant Poets Scott., & Musings* 342 Now, women are (compared wi' men), More contumacious; and when Their 'birse' is up,—my certy! then They're kittle cattle. **1888** *Trans. Highland Soc.* 197 Even as machines are easily deranged so sheep are 'kittle cattle'; no more delicate animal breathes. **1900** E. T. Fowler *Farringdons* i. 15 She knew a great portion of the Methodist hymn-book by heart, and pondered long over the interesting preface to that work, wondering much what 'doggerel' and 'botches' could be—she inclined to the supposition that the former were animals and the latter were diseases; but even her vivid imagination failed to form a satisfactory representation of such queer kittle-cattle as 'feeble expletives'. **1926** Fowler *Mod. Eng. Usage* 717/1 It is well known that *and which* & *but which* are kittle cattle, so well known that the more timid writers avoid the dangers associated with them. **1935** *Times Educ. Suppl.* 28 Dec. p. iv/1 It helps to discount the general impression that, as a race, they [sc. lilies] are kittle-kattle. **1942** F. Smythe *Alpine Ways* 14 The 'he man' mountaineer who scorns tourists as mere kittle kattle, the tourists who jeer at the mountaineer as mad, have yet to gain that fuller understanding which abhors intolerance. **1966** *Sunday Express* 13 Mar. 16/7 Princesses in love..are at times Kittle Cattle.

kitty⁴. Add: **2.** (Earlier examples.) Also, the money (freq. placed in the centre of the table) taken by the winner of a game or round (the usual sense). So *transf.*, earnings, liquid capital, a reserve fund; a sum of money made up of contributions by people involved in a common activity.

1887 J. W. Keller *Game of Draw Poker* 12 *Widow*, or *Kitty*—A percentage taken out of the pool to defray the expenses of the game or the cost of refreshments. **1891** E. Dowson *Let.* 1 July (1967) 206, I shall..refuse to associate myself with penny nap & an unlimited Kittie. **1903** 'J. Flynt' *Rise of Ruderick Clowd* iii. 106 'Forty [cents] out of every dollar you cop out—understand?'—'Who gets the rest?'—'The mob an' the kitty. The kitty is the fall-money reserve. A mob like ours ought to carry a $3000 kitty all the time. It's drawn on when one of us gets arrested an' has to hire lawyers an' get bail. If you get a tumble, for instance, the rest of us'll have to stand by you—see?' **1905** *Daily Chron.* 12 Sept. 4/7 'Kitty wins everything,' is the bookmakers' plaint. **1909** 'O. Henry' *Roads of Destiny* xiii. 213 Your thousand dollars is gone into the kitty of this corrupt country on that last bluff you made. **1924** T. Rohan *Confessions of Dealer* iv. 51 The King of the Knock-Out..counts out banknotes to the tune of £1,000, and places these notes in the bowl or kitty which occupies the centre of the table. The first man to help himself from the kitty is the dealer who bid £100 at the sale. **1929** [see *FULL HOUSE 2]. **1935** A. J. Pollock *Underworld Speaks* 68/1 *Kitty*, money taken from virtually every gambling pot for purpose of profits or expenses, or whangdoodles at the end of a friendly home game. **1969** *Listener* 17 July 77/1 In 1949, the authorities at the hall had enough money in the kitty to install a new aluminium roof. **1971** *Ink* 12 June 14/1 Bernie's salesmen kept bringing in the lolly..until by 1970 they had $2·4 billion in their management kitty.

3. *Bowls.* The jack.

1898 D. Willox *Poems & Sk.* 174 Now, in throwing up the 'Kitty', Oor first player's quite a card, But in playing tae't he seldom Ever gangs within a yard. **1909** *Westm. Gaz.* 6 Sept. 4/1 When the bowl goes near the kitty. **1926** H. Hendry *Poems* 98 Kitty, that licht and lively quean, Wha links athort the bowling-green. **1959** *Chambers's Encycl.* II. 476/1 The small white earthenware ball, known as the jack or 'kitty', at which the bowls are aimed, must not be more than 8 in. in circumference or 10 oz in weight.

kittydid, var. Katydid.

kiva (kī·vă). Also **khiva, kiver.** [Hopi.] A chamber, built wholly or partly underground, used by the male Pueblo Indians for religious rites, etc.; Estufa. Also *attrib.* and *transf.*

1871 in *Utah Hist. Q.* (1939) VII. 54 Found pieces of pottery and arrowheads...Also saw a 'kiver' or underground 'clan room'. **1875** *Scribner's Monthly* Dec. 205/2 This kiva, as it is called in their own tongue, is called '*Estufa*' by the Spaniards, and is spoken of by writers in English as the 'Sweat House'. **1898** *17th Ann. Rep. U.S. Bureau Amer. Ethnol.* 1895–6 611 A pueblo of the size of Awatobi..would no doubt have ceremonial chambers or kivas. **1927** W. Cather *Death comes for Archbishop* iv. ii. 132 It was a smothered fire in a clay oven, and had been burning in one of the kivas ever since the pueblo was founded. **1931** *Discovery* Sept. 279/2 The 'khiva' as they are called by the present Hopi Indians... It seems clear that their purpose was to provide a meeting place for councils and ceremonies for the men of the clan. **1950** F. Eggan *Social Organiz. Western Pueblos* ii. 28 Between the members of the kiva groups there are no special kinship ties. **1958** J. Cleugh tr. *Jungk's Brighter than Thousand Suns* xviii. 295 The mechanism used took the form of a 'critical assembly' under remote control. Processes going on inside the 'Kivas'—the buildings had been called after the sacred ceremonial chambers of the Pueblo Indians, which could only be approached, with the greatest awe, by their priests—were to be observed solely on television screens. **1964** E. A. Nida *Toward Sci. Transl.* viii. 169 For the Zuñis, uttering *melika* in a kiva ceremony would be as out of place as bringing a radio into such a meeting.

Kiwanis (kiwā·nis). [Origin obscure.] In full, *Kiwanis Club.* A society of business and professional men formed in Detroit in 1915 for the maintenance of commercial ethics, and as a social and charitable organization; any similar society formed later elsewhere in the U.S.A. or in Canada. Also **Kiwa·nian,** a member of a Kiwanis Club.

1921 *Daily Colonist* (Victoria, B.C.) 16 Mar. 5/4 An initiation ceremony for the purpose of welcoming into the organization three new members took place at the weekly luncheon of the Kiwanis Club yesterday. *Ibid.* 30 Mar. 7/5 As Kiwanians and citizens, Mr Smith claimed, it was a duty to look after the children. **1922** *Collier's* 29 Apr. 5/2 It had a civic association, and Rotary, Kiwanis, and Lion Clubs. **1926** *Glasgow Herald* 18 Sept. 4 Take Galveston, for example... A newspaper describes how the local Kiwanians and their lady friends..rounded off an evening's delight by indulging in a hand-shaking competition. **1948** *Daily Ardmoreite* (Ardmore, Okla.) 30 Mar. 3/1 A class of older girls recently entertained the local Kiwanis club. **1949** *Amer. Speech* XXIV. 29 Virginians..show little of the exuberant affability of a Midwestern Rotarian or Kiwanian. **1964** Mrs. L. B. Johnson *White House Diary* (1970) 5 Jan. 30 We drove to the little house in Johnson City.., which we are having restored as a Community Center, where, hopefully, 4-H youngsters or the PTA, or Kiwanis, or Lions, or ladies' groups, or whatever can hold their meetings. **1972** *Fairbanks* (Alaska) *Daily News-Miner* 3 Nov. 1/2 Alaska Attorney General John Havelock analyzed and approved the moral basis of Alaska's Oil Legislation in an address before a joint Kiwanis banquet here last night.

kiwi. Add: **2.** (With capital initial.) A New Zealander, esp. a New Zealand soldier; also, a New Zealand sportsman. Also *attrib.*

1918 *N.Z.E.F. Chrons.* 24 May 179/1 In the evening the 'Kiwis' gave a performance to a crowded hall. *Ibid.* 21 June 225/1 The New Zealand boys..will find the 'Kiwi' A.D.S. as reasonably safe as is possible. **1945** Baker *Austral. Lang.* ix. 178 New Zealand football representatives acquired the names *All Blacks, Fernleaves*, and *Kiwis.* **1947** B. Mason in D. M. Davin *N.Z. Short Stories* (1953) 333 We were Kiwis, and a long way from home. **1958** R. France *Race* 12 Laurie was not a real Kiwi, or hard-bitten New Zealander. **1960** B. Crump *Good Keen Man* 58, I suspect she was a real Kiwi mum, with a soft spot for her little Harry. **1961** in J. Reid *Kiwi Laughs* 10 Told in a homely Kiwi idiom. **1963** N. Hilliard *Piece of Land* 35 They fill the place with immigrants... A Kiwi'd never look at the prices some of them want. **1974** *Times* 17 June 12/7 It is hurtful to many Australians to see the Kiwis take all the credit for Antipodean anti-Pom feelings.

3. Also **Kiwi.** A non-flying member of an air force (see also quot. 1938). *slang.*

1918 B. Hall *Diary* 22 Jan. in Hall & Niles *One Man's War* (1929) xxxii. 289 Visited the Avenue Montaigne Headquarters. It is full of non-flying aviators. The American pilots call them Kiwis. **1925** Fraser & Gibbons *Soldier & Sailor Words* 137 *Kiwi*, Air Force slang for a man on ground duty and not qualified for flying service. **1931** *Vanity Fair* (N.Y.) Nov. 78/3 There are terms with which to plaster the green pilot and the non-flyer. *Quirks* or *kiwis* are beginners—sometimes the terms are broadened to include the layman. The origin of *quiik* is somewhat obscure..but *kiwi* is a derogatory reference to the Australian [sic] kiwi-bird, which, having only stub wings, is unable to fly. **1938** *Amer. Speech* XIII. 156/2 *Kiwi*..a person with no practical flying experience; often used as a term of disparagement toward one who speaks with authority concerning flying but whose knowledge is entirely theoretical. **1943** Hunt & Pringle *Service Slang* 43 *Kiwi*, a word brought over by the New Zealand airmen with a new meaning: one who do not belong to air crews. **1960** Wentworth & Flexner *Dict. Amer. Slang* 307/1 *Kiwi*, an air force man, esp. an officer who cannot, does not, or does not fly.

4. = *kiwi berry, fruit*.

1972 *Daily Colonist* (Victoria, B.C.) 2 Aug. 19/1 Have you noticed a small brown fruit called kiwi in local markets lately?.. Sometimes called a Chinese gooseberry. **1973** *Sat. Rev. Soc.* (U.S.) Mar. 53/1 Twenty-six different crops, most of them fruit—almonds, apples..kiwis, nectarines, olives.

5. *attrib.* and *Comb.*, as *kiwi feather, -hunter, -preserve*; **kiwi berry, fruit** = *Chinese gooseberry* (*Chinese *a.* 2).

1968 *N.Z. News* 23 Oct. 2/4 New Zealand exports of Chinese gooseberries will enter the United States—where they are called 'giant Kiwi-berries'—almost duty-free from next season. **1966** *Daily Chron.* 7 July 6/5 Two presents included..a rug of kiwi feathers from New Zealand, and three rare engravings. **1928** R. D. Finlayson *Brown Man's Burden* 47 They had covered his shrunken body with fine kiwi-feather cloaks. **1966** *N.Y. Times* 13 Aug. 12 Chinese gooseberries, also known as kiwi fruit, are in metropolitan markets for the third season in increased quantities. **1970** *N.Z. News* 7 Jan. 4/4 A storage technique first developed for bananas has been tested on Chinese gooseberries, now renamed 'Kiwi fruit'. *Ibid.* 4/5 A 20 lb. pack of Kiwi fruit. **1973** *Massey Ferguson Rev.* (N.Z.) Mar./Apr. 5/4 Chinese gooseberries or 'kiwi fruit', bear extravagantly. **1873** Kiwi-hunter, -preserve [in Dict.].

ki-yi, *sb.* Add: **2.** A dog.

1895 *Harper's Mag.* Nov. 962/1 I'm not really a ki-yi, and while I don't like bicyclists,..I won't bite you. **1904** *Buffalo* (N.Y.) *Express* 20 June 4 A butcher in Brussels made sausage of the carcass of a 200 elephant which had been killed. Doubtless the Brussels kiyis yelped for joy. **1923** J. London *Valley of Moon* I. x. But them skinnin', sap-headed stiffs, with the grit of rabbits and the silk of mangy ki-yi's, a-cheerin' me—me!

Kizil (kizi·l), *a.* and *sb.* Also **Kyzyl.** [ad. Turk. *kızıl* red.] **A.** *adj.* Of or pertaining to a Turkic Tartar people of southern Siberia. **B.** *sb.* A member of this people.

1898 A. J. Butler tr. *Ratzel's Hist. Mankind* III. v. 336 The Kizil Tartars on the Upper Chulym. *Ibid.* 342 The Kizil tribe of the Tomsk Kirghizes. **1909** Webster, Kizil *n.* **1964** tr. *Levin & Potapov's Peoples of Siberia* 358 The Kyzyls belonged to a large Turkic-speaking group living in the Chulym Basin. *Ibid.* 360 The language of the present-day Kyzyls is identical with Kachin.

Kizilbash (ki·zilbaʃ). (Also used as *pl.*) Also 8 -bac, 9- **Kizzil-**, **-bashi.** [ad. Turk. *kızılbaş*, f. *kızıl* red + *baş* head.] **a.** A Persianized Turk of Afghanistan. **b.** A member of any of several cultural or religious minorities in Asian Turkey.

1727 J. G. Scheuchzer tr. *Kæmpfer's Hist. Japan* I. i. vi. 58 The Kizilbacs, or Noblemen, and great Families, in Persia value themselves mightily upon their being of Turcoman extraction. **1815** [see Turk¹ 1]. **1875** *Encycl. Brit.* I. 237/1 The Kizilbashes may be regarded as modern Persians, but more strictly they are Persianized Turks. **1898** A. J. Butler tr. *Ratzel's Hist. Mankind* III. xv. 365 In Persia and Afghanistan the Turks, Kizilbashes, Usbeks, Turcomans, are even more sharply distinguished from the Persians. **1902** *Encycl. Brit.* XXV. 120/1 The Kizilbashes of Kabul. **1920** *Blackw. Mag.* Jan. 121/2 Hosts of Tartar, and Afghan, Persian and Kizilbash. **1960** *Guardian* 28 June 8/3 Mingled with the Pathans of the south [of Afghanistan]..are..Turkomans, Kizilbashes, Kirghiz. **1960** *Spectator* 2 Dec. 889/3 Religious sects living apart like the Kizil Bashis, Shiah, Tartars and Kâra Papachs [in Turkey, *c* 1912].

Kjeldahl (ke·ldāl). *Biochem.* The name of Johann *Kjeldahl* (1849–1900), Danish brewing chemist, used *attrib.* and in the possessive to denote a method of estimation of nitrogen invented by him, in which the organic substance to be analysed is treated with concentrated sulphuric acid and the ammonium sulphate so formed is converted by excess alkali to ammonia, which is then titrated; **Kjeldahl flask**, a glass flask having a round bottom and a long wide neck, used in the Kjeldahl method.

1885 *Jrnl. Chem. Soc.* XLVIII. 688 (*heading*) Nitrogen determination by Kjeldahl's method. **1909** P. B. Hawk *Pract. Physiol. Chem.* (ed. 2) xxii. 382 Place 5 c.c. of urine in a 500 c.c. long-necked Jena glass Kjeldahl flask. **1964** [see *colorimetrically *adv.*]. **1970** R. M. McGilvery *Biochem.* xxviii. 704 The Kjeldahl method has the advantage of being applicable to insoluble materials, such as most foodstuffs, and the disadvantage of not distinguishing proteins from other sources of nitrogen, such as nucleic acids, urea, and the like.

Klá, var. *Kullah.

Klaas (klās). The name of a servant who travelled with the French explorer, François Le Vaillant (1753–1824), used in the possessive in **Klaas's cuckoo**, a bronze and green cuckoo, *Chrysococcyx klaas*, found in the southern part of Africa.

1867 E. L. Layard *Birds S. Afr.* 250 Klaas's cuckoo is not uncommon in most wooded parts of the colony. **1903** A. C. Stark *Birds S. Afr.* III. 188 This cuckoo was first obtained by Levaillant and named by him after his faithful Hottentot servant Klaas... Klaas' Cuckoo frequents both bush and thorn lands. **1936** E. L. Gill *First Guide S. Afr. Birds* 198 Among the birds parasitized by Klaas's Cuckoo are various sunbirds, warblers and kingfishers. **1951** R. Campbell *Light on Dark Horse* xi. 143 The bird known to this day as Klaas's Cuckoo..is the ruby cuckoo. **1964** P. A. Clancey *Birds Natal & Zululand* 222 The young Klaas' cuckoo ejects the nestlings of the foster parent a few days after hatching.

klaberjass, var. *klobbiyos.

Klamath (klæ·məþ). Also †**Clamet.** [ad. Chinook *łámal* Klamath.] **A.** *sb.* A Penutian Indian people of the Oregon–California border; a member of this people; also, their language. **B.** *adj.* Of or pertaining to this people.

1826 J. McLoughlin *Lett.* (1941) I. 33 Mr. Ogden goes..thence towards Lac Sale makes a Circuit West and comes out above the Clamet tribe. **1853** H. R. Schoolcraft *Hist. & Stat. Information Indian Tribes* III. 233 The goods destined for the Klamath River Indians, had been sent to Trinidad. **1881** *Encycl. Brit.* XII. 829/2 The Klamath family..comprises the Lutuami or Klamaths proper. *Ibid.*, *Californian Races*.—This is mainly a geographical grouping, but with these ethnical and linguistic families—the Klamath, Pomo, and Runsien. **1890** A. S. Gatschet *Klamath Indians* I. 209 Triphthongs are not infrequent, since Klamath has a greater tendency to accumulate consonants than vowels. **1935** [see *Dakota *sb.* 2]. **1966** *Canad. Jrnl. Linguistics* Spring 123 The last family represented in Oregon is Klamath-Modoc. Klamath speakers are found in fairly large numbers... Modoc is represented by only about seven or eight speakers.

C. Klamath weed, the local name of a species of St. John's wort, *Hypericum perforatum*.

1922 W. J. Smiley *Weeds Calif.* 54 (*Hypericum perforatum* L.) English names:..Common St. John's-wort, Klamath weed. **1949** *Sunday World-Herald Mag.* (Omaha) 1 May 10/1 Klamath weed probably can be controlled or destroyed chemically. **1972** Debach & Huffaker in C. B. Huffaker *Biol. Control* v. 118 Figure 1... shows the degree of biological control achieved in California of the formerly serious Klamath weed, *Hypericum perforatum* L., by colonization of an imported exotic beetle which feeds upon it.

Klan, short for Ku-Klux-Klan. Also **Klansman**, occas. **Clansman.**

1868 *Century Mag.* XXVIII. 409/1 The Klan now, as in the past, is prohibited from doing such things. **1884** [see Invisible *a.* 1 c]. **1905** T. Dixon (*title*) The Clansman. **1924** J. M. Mecklin *Ku Klux Klan* i. 9 The modern Klan was organized by William J. Simmons in 1915. *Ibid.* 15 Masked men leaped from their cars clad in Klan regalia. *Ibid.* 6 Public sentiment..seems to have supported the Klansmen. **1953** *Manch. Guardian Weekly* 31 Dec. 11 A klansman from Georgia. **1973** *Freedomways* XIII. 30 The..killings and the Attica prison slaughter are reminders of the post-Civil War Klan days. **1973** *Guardian* 13 Mar. 15/3 A former member of the Klan, Senator Byrd has a dismal voting record on social and civil rights issues. **1973** *Black Panther* 1 Sept. 8/5 The Assistant District Attorney tossed out the charges saying that although the law prohibits Klansmen from appearing in public, the faces of the 24 were visible.

‖ **klang** (klaŋ, klæŋ). [G., = sound.] A musical tone composed of fundamental and overtones; = Clang *sb.* 3. Hence **klangfarbe** (-fa·rbə) [G. *farbe*, colour], musical quality of a note, timbre, 'clang-tint' (see Clang *sb.* 4); **klangfarbenmelodie**, melody of timbres.

1867 [see Clang *sb.* 3]. **1890** J. Clauser *Septonate* i. 37 (Funk), in music a tone or a klang is thought, heard, and considered as a unit. **1959** *Listener* 12 Nov. 834/2 It is possible that some of the mannerisms—disjointed rhythms, 'Klangfarbenmelodie' (the playing of notes that have no connection with each other one by one), etc., may fall by the wayside. **1966** *Ibid.* 19 May 736/3 No. 5, is systematically subjected to *klangfarbenmelodie*, emphasis of instrumental timbre.

klapper (klæ·pəɹ). *S. Afr.* [perh. ad. Malay *kĕlapa* coconut or Afrikaans *klapper* rattle.] = *Kaffir orange* (*Kaffir 4); also used for other shrubs or trees with similar fruit.

1863 W. C. Baldwin *Afr. Hunting* vi. 199 We had a capital lunch from some wild fruit about three times the size of an orange, called a clapper. It has a hard shell outside, which one could knock against a tree to crack or break. **1921** T. R. Sim *Native Timbers S. Afr.* 120 *Strychnos pungens*... Klapper (Eng. Jac), Wild Orange. **1932** Watt & Breyer-Brandwijk *Medicinal & Poisonous Plants S. Afr.* 140 The pulp of the fruit of *Strychnos pungens* Solered., Wild orange, Kaffir orange, Klapper..*Strychnos spinosa* Lam. (*Brehmia spinosa* Harv.), Kaffir orange, Klapper..and *Strychnos gerrardi* N.E. Br...is acidulous from the presence of citric acid, and is very refreshing. **1939** tr. *E. N. Marais's My Friends the Baboons* iii. 29 Wild peaches, sour klappers (unknown in the south), medlers, moepels, and various other kinds of fruit made our wilderness a veritable orchard. **1966** E. Palmer *Plains of Camdeboo* x. 177 The spekboom, the wild plum and the klapper, now bright with colour. **1973** Palmer & Pitman *Trees S. Afr.* III. 1857 The monkey orange, or klapper as it is often known,..is an evergreen tree.

‖ **klatsch** (klatʃ). Also **klatch.** [G., tittle-tattle, gossip.] A visit; a coffee-party. Cf. *coffee-klatsch, *kaffeeklat(s)ch.

1953 A. Miller *Crucible* (1956) i. 34 There are accounts of similar klatches in Europe, where the daughters of the towns would assemble at night and..give themselves to love. **1958** R. Buckminster *Naughts in Attic* xvi. 155 So later, when the men returned and set up their camp we had them over for a klatch.

klaxon (klæ·ksən), *sb.* orig. *U.S.* Also **Klaxon.** [Name of the manufacturing company.] An (electric) horn or warning hooter, orig. one on a motor vehicle. Also *klaxon-horn*.

1910 *Sat. Even. Post* 17 Sept. 48 The Klaxon has never taken a life; it has saved thousands. **1911** *N.Y. Times* 16 Oct. 12/7 Speedometer, slip covers, pigskin upholstery and klaxon. **1917** *Contact Airman's Outings* 66 A signal rocket streaked from the first Boche biplane, and the trio dived almost vertically, honking the while on Klaxon horns. **1918** R. H. Knyvett *Over There with Australians* ix. 199 These noises were made chiefly with Klaxon horns. **1920** *Motor Manual* (ed. 23) xv. 150 The electrically-operated Klaxon horn. **1924** H. Gilbert *Biy Market* 243 Enemy Stamp sounded his klaxon. **1949** *Reader's Digest* Apr. 140/2 It was a gray Pierce Arrow, equipped with two bulb horns and an electric Klaxon. **1965** 'J. Le Carré' *Looking-Glass War* 8 He heard the klaxons..moaning out over the drab warehouses airfield like the howl of starving animals. **1973** P. Evans *Bodyguard Man* i. 14 The evening traffic was thick, shrill with sudden braking and klaxon horns.

Hence **kla·xon** *v. intr.*, to sound a klaxon; also *trans.*; **kla·xoning** *vbl. sb.*

1922 E. V. Lucas *Genevra's Money* vi. 38 The almost constant clatter and klaxoning of motor-cars and lorries on the high-roads. **1924** G. Frankau *Gerald Cranston's Lady* iv. 48 Jess, Klaxoning furiously, slackened pace round the dangerous stone-wall turning. **1973** *Daily Tel.* 15 Sept. 11 There are two sides to every situation, once the Press, television and radio have 'klaxoned' the story to the general public. **1973** G. Beare *Snake on Grave* vii. 35 A little white Fiat klaxoning shrilly.

Kleagle (klī·g'l). [f. *Klan + Eagle *sb.*] A title given to an officer of the Ku-Klux-Klan.

1924 J. M. Mecklin *Ku Klux Klan* i. 8 The head of the promotion department is a whole was Imperial Kleagle E. Y. Clarke... The head of the 'realm', or state, was called a King Kleagle, and the house-to-house solicitors, or legwork men, were called Kleagles. **1934** E. T. Bryson *Personal Recoll.* 8 The Kleagle showed considerable irritation..in the conversation which followed. **1949** *Sun* (Baltimore) 31 Jan. 1/2 There was a time when Johnston could have called the roll of the kleagles and merely waited for donations to roll in. **1959** *Time* 13 June 24/2 Samuel Green, the Grand Dragon of the Ku Klux Klan, was frantically exhorting his Kleagles and Cyclops to mass for a hysteria of cross-burning and hate-spieling.

klebsiella (klebzi,e·lă). [mod.L. (V. Trevisan 1885, in *Atti Accad. Fisio-Medico-Statistica Milano* 4th Ser. III. 105), f. the name of *Klebs* (see next) + *-ella* (see -el².)] A coliform bacterium of the ill-defined genus so called, which includes Friedländer's bacillus, *Klebsiella pneumoniæ*, and others associated with respiratory, urinary, and wound infections and occas. with hospital epidemics.

1926 W. W. Gilbern *Elem. Text Bk. Gen. Microbiology* [sic] xxv. 337 (*heading*) Klebsiella infections. **1948** H. E. Dowling *Acute Bacterial Dis.* xx. 364 Cholecystitis and cholangitis are frequent. **1956** *Nature* 1 Sept. 496/2 R. Y. Stanier et al. *Microbial World* xvii. 341 The organism is known as *Klebsiella pneumoniæ*. **1965** *Lancet* 12 May 969/1 In 1 patient with a urinary-tract infection a klebsiella strain was isolated. **1973** A. L. Smith *Princ. Microbiol.* (ed. 7) xxiii. 351 *Klebsiella* are non-motile, gram-negative, aerobic organisms.

Klebs–Löffler (kleb,lö·flaɹ). The names of T. A. E. *Klebs* (1834–1913) and F. A. J. *Löffler* (1852–1915), German bacteriologists, used *attrib.* to designate the bacterium (*Corynebacterium diphtheriæ*) which is the cause of diphtheria in man and similar diseases in other animals.

1895 *Lancet* 22 Dec. 1577/1 The differences between it and the Klebs-Löffler bacillus [were] pointed out. **1897** *Ibid.* 5 June 1533/1 We might now fully speak of conjunctival inflammations as due to gonococci, to trachomacocci, to Weeks bacilli, to streptococci, to Klebs-Löffler bacilli, or to pneumococci. **1897** Muir & Ritchie *Man. Bacteriol.* xv. 329 The organism is..known as the Klebs-Löffler bacillus, or simply as Löffler's bacillus. **1936** *Discovery* XVI. 244/1 An enormous number of cases, mostly drain throats, are rushed off to hospital because the Klebs-Löffler (diphtheria) bacillus has been found. **1972** Pelczar & Reid *Microbiol.* (ed. 3) xxix. 567 Diphtheria is an acute febrile infection caused by *Corynebacterium diphtheriae*, also known as the Klebs–Löffler bacillus.

Kleenex (klī·neks). orig. *U.S.* The proprietary name of an absorbent disposable cleansing paper tissue. Also *attrib.*

1925 *Picture-Play Mag.* Apr. 107/2 (Advt.), This secret of famous stage beauties... is simply the use of Kleenex in removing cold cream and cosmetics... This soft velvety absorbent is made of Cellucotton... Use it once, throw it away. **1925** *Trade Marks Jrnl.* 15 July 1545 Kleenex... Absorbent pads or sheets (not medicated) for surgical or curative purposes or in relation to the health. Cellucotton Products Company,..City of Neenah,.. Winnebago, State of Wisconsin, United States of America; Manufacturers. **1936** N. Coward *Play Parade* (1954) IV. 90 They stuff wads of Kleenex paper in between their toes and under their necks to prevent the make-up soiling their ties. **1943** M. McCarthy *Company She Keeps* (1943) 197 The tears streamed from her eyes... She took a box of Kleenex from a drawer and handed it to her. **1956** E. Ambler *Night-Comers* ix. 84 She had a box of Kleenex... Gypsy Rose Lee' *Gypsy* xxiv. 315 She dabbed at her eyes with a tattered Kleenex, then shoved it back in her purse. **1967** *Listener* 16 Mar. 368/3 The Master himself fulminates in California against some of his more conservative juniors, in an essay not yet reprinted in England (it dismisses Britten's War Requiem as *Kleenex Music*). **1969** G. Greene *Trav. with my Aunt* xvii. 179 She wiped her fingers on the Kleenex and opened the yellow envelope. **1971** *Trade Marks Jrnl.* 12 May 510/1 Kleenex... Babies' disposable napkins, diapers and lining sheets for all such goods... Kimberly-Clark Corporation,..Neenah, State of Wisconsin, United States of America; Manufacturers. **1974** *Listener* 11 July 62/3 The almost unpopulated wilderness of Maine, where TV dinners and Kleenex are in short supply.

kleig, Kleig, varr. *klieg.

Klein bottle (klain-). [Named after Felix *Klein* (1849–1925), Ger. mathematician.] A closed non-orientable surface that can be represented in three dimensions by passing the neck of a bottle through its side and joining its end to a hole in the base.

1941 Courant & Robbins *What is Math.?* v. 262 Another interesting one-sided surface is the 'Klein bottle'. **1950** *Astounding Sci. Fiction* Dec. 75/2 'The Möbius band', Turpelo said, 'has unusual properties because it has one surface... The Klein bottle, with two singularities, manages to be inside of itself.' **1966** H. Eves *Survey of Geom.* II. xv. 337 The surface is homeomorphic to a sphere with two crosscaps, and is called a Klein bottle, after Felix Klein who first described it in 1882. *Ibid.*, Show that a Klein bottle can, by one cut, be converted into a disc and a Möbius strip.

‖ **kleindeutsch** (klai·ndoitʃ), *a.* [G.] Referring to or favouring a United Germany, excluding Austria. Also as *sb.*, a supporter of such a policy. Cf. *grossdeutsch *a.*

1926 A. W. Ward *Germany 1815–90* I. vi. 484 The Austrian members..coalescing..on the question of the exclusion of Austria..under the attractive name *Grossdeutsche*, and designating their opponents who..included E. M. Arndt as *Kleindeutsche*. **1945** F. Darmstaedter *Germany & Europe* ii. 51 In the Frankfurt Assembly the adherents of the 'kleindeutsche' solution (exclusion of Austria) gained the upper hand over those of the 'grossdeutsche' (inclusion of Austria). **1946** [see *grossdeutsch *a.*]. **1968** F. Eyck *Frankfurt Parl. 1848–9* vii. 255 The exclusion of Austria from Germany (which was to become the programme of the *Kleindeutsche*) was proposed the King of Prussia as hereditary German Emperor. *Ibid.* viii. 363 The antithesis of *Kleindeutsch* (Lesser German) and *Grossdeutsch* (Greater German) is not sufficiently subtle to describe the real difference between the parties.

Kleinian (klai·niən), *a.* and *sb.* *Psychol.* **A.** *adj.* Of or pertaining to Melanie *Klein* (1882–1960) or her theories which, though basically Freudian, differed particularly in the field of child psychoanalysis. **B.** *sb.* A follower of the theories of Melanie Klein.

1955 L. Menon *Schools of Psychoanal. Thought* 663/1 Introduction..to Kleinian theory. **1959** *Encounter* Apr. 41/1 A..comprehensive interpretation of various aspects of art in terms of Kleinian theory. **1960** R. Wälder *Basic Theory Psychoanal.* xi. 233 Analysts of the Kleinian school of thought have made different claims. **1961** J. A. C. Brown *Freud & Post-Freudians* iv. 78 There is another group who are disliked by the Kleinians, who are thought by some to represent the main movement in orthodox psychoanalysis. **1963** *Times Lit. Suppl.* 15 Feb. 111/2 It has become possible to write even a lyric poem without being diagnosed as a Freudian, a Kleinian, [etc.]. **1972** *Guardian* 6 July 15/3 The Kleinian examination of infantile development.

kleinite (klai·nait). *Min.* [ad. G. *kleinit* (A. Sachs 1905, in *Sitzungsber. d. preuss. Akad. d. Wissensch.* 1904), f. the name of J. F. C. *Klein* (1842–1907), German geologist: see -ite¹.] A hydrous chloride and sulphate of mercury and ammonium, occurring as transparent or translucent crystals of a yellow to orange colour.

1907 *Chem. Abstr.* I. 2454 Kleinite may be a mixture of a mercury-ammonium chloride..in great preponderance, with an oxychloride and sulphate or oxysulphate of mercury. **1932** *Amer. Mineralogist* XVII. 547 Specimens of kleinite, terlinguaite, montroydite, and mosesite from the Texas locality were then prepared for *x*-ray examination. **1954** *Mineral. Abstr.* XII. 433 Comparison of the data for kleinite..with those of mosesite..suggests that the former has a three-dimensional structure of $[Hg_2N]^{1+}$..with a formula $[Hg_2N](Cl,SO_4)xH_2O$, where *x* is about ½.

klementite (klĕme·ntait). *Min.* [ad. G. *klementit* (G. Tschermak 1891, in *Sitzungsber. d. kaiserl. Akad. d. Wissensch.* (*Math.-nat. Cl.*) Abt. I. C. 40), f. the name of Constantin *Klement* (b. 1856), curator of the Natural History Museum, Brussels: see -ite¹.] A variety of thuringite containing more magnesium than iron.

1892 E. S. Dana *Dana's Syst. Min.* (ed. 6) 656 *Klementite*. In thin scales in quartz veins at Vielsalm in Belgium. Probably monoclinic... Color dark olive-green. **1954** *Mineral. Mag.* XXX. 279 It may be found convenient to retain the name klementite as a variety of thuringite with Mg>(Fe″+Fe‴); the original klementite analysis falls in this field. **1968** *Ibid.* XXXVI. 753 This chlorite has been identified as klementite, a variety of thuringite... In thin flakes the mineral is light green but deeper coloured in aggregates... The composition approximates to $Mg_{3\cdot3}Fe_{1\cdot3}Al_{1\cdot4}Si_{3\cdot3}(O,OH)_{18}$.

klendusity (klendiū·sĭti). *Bot.* [f. Gr. κλ-εῖς bar, bolt + ἔνδυσις entry: see -ity.] The resistance of a plant to disease, through the presence of some characteristic that inhibits infection. So **klendu·sic** *a.*, showing resistance of this kind.

1940 J. I. Wood et al. in *Phytopathology* XXX. 362 The ..continued misuse of such words as *immunity*, *resistance*, *tolerance* and *klendusity* (with *resistance* as a catch-all) tends definitely to confuse readers. *Ibid.* 364 Klendusity: Ability of a susceptible variety to escape infection because of possession of some quality preventing or hindering successful inoculation under conditions conducive to infection in other varieties. **1943** *Phytopathology* XXXIII. 19 A few seedlings in subsequent generations that were derived indirectly from the almost sterile hybrid were about as capable of escaping infection as *L[ycopersicon] chilense* itself, thus showing heritability of klendusity. *Ibid.* 692 This klendusic seedling did not fruit for some time. **1950** *Trans. Brit. Mycol. Soc.* XXXIII. 156 On this side of the Atlantic 'klendusity' and 'suscept' were each thrown out without a hearing, because they are the very opposite of comfortable words. **1958** *Virology* VI. 303 The nature and inheritance of the tendency to escape infection (klendusity) was investigated. **1967** R. K. S. Wood *Physiol. Plant Path.* xii. 400 Before these different forms of resistance are dealt with in detail, it is as well to refer to what some pathologists regard as a form of resistance, and which is usually referred to as 'disease escape'. Less frequently the term 'klendusity' is used for the same thing. **1973** *Guide Use Terms Plant Path.* (Federation Brit. Plant Pathologists) 22 Klendusity: the failure of a susceptible host to become infected, in the presence of the pathogen, because of qualities preventing or hindering the operation of a vector or other inoculating agent... Klendusity may be considered a form of disease escape.

klep (klep), *sb.* Slang abbrev. of Kleptomaniac. Hence as *v. intr.*, to steal.

1889 Barrère & Leland *Dict. Slang* 523/1 *Klep*, a thief; to *klep*, to steal. **1896** Farmer & Henley *Slang* IV. 117/2 *Klep* (popular), a thief. Short for kleptomaniac. **1949** 'N. R. Nash' *Young & Fair* 20 The kleps have started!

klepto (kle·ptō). Slang abbrev. of Kleptomaniac.

1958 *New Yorker* 25 Jan. 29 Some befuddled guest (or klepto, more likely) abstracted a hat. **1962** *Punch* 4 July 24/1 Playwrights of the imminent Klepto school. **1964** 'E. V. Cunningham' *Lydia* (1965) xii. 178 You got it.. right out of Helen Sarbine's purse... What are you—some kind of nut or klepto?

kleptobiosis (kle:ptobai,ōu·sis). *Zool.* In quots. **clepto-.** [f. Gr. κλέπτης thief, κλέπτειν to steal + βίωσις way of life.] Among ants and certain other social insects, an association in which a small species feeds on the refuse of a neighbouring nest inhabited by a larger species, or robs returning workers of the host species of the food they are carrying. Hence **kle:ptobio·tic** *a.*

1901 W. M. Wheeler in *Amer. Naturalist* XXXV. 516, I have therefore adopted the following headings... Cleptobiosis. Wasmann's 'Diebsameisen'; first regular form of compound nest. *Ibid.* 529 All the known cleptobiotic ants are of minute size and of subterranean habits. **1927** H. St. J. K. Donisthorpe *Guests of Brit. Ants* iii. 78 Ants exhibit a variety of associations, symbiotic, mutualistic, and parasitic... Such associations consist of— Cleptobiosis (originally used to denote thievery, now applied to brigandage by Wheeler); Lestobiosis [etc.]. **1971** E. O. Wilson *Insect Societies* xix. 377/2 As a rule, cleptobiotic aculeates prey on only one or a very few species of other aculeates. *Ibid.* 381/2 Members of the melipomine genus *Lestrimelitta* do engage in nest robbing, or 'cleptobiosis'.

kleptocracy. Add: Also, government by thieves; a nation ruled by this kind of government. (Further examples.)

1968 S. L. Andreski *African Predicament* vii. 109 The essence of kleptocracy is that the functioning of the organs of authority is determined by the mechanisms of supply and demand rather than the laws and regulations. **1971** *Guardian Weekly* 10 July 18 The Federal Republic of Cameroon, which he [sc. Leonard Barnes] seems to regard as one of the less wicked kleptocracies. **1975** *Times Lit. Suppl.* 4 July 740/4 The role of corruption within the [Franco] regime...and the growth of a virtual kleptocracy within the administration.

kleptolagnia (kleptolæ·gniă). [mod.L., f. Gr. κλεπτο-, combining form of κλέπτης thief + λαγνεία lust; formed by analogy with *algolagnia.] A morbid desire, associated with fetishism, to achieve sexual gratification through theft.

1917 J. C. Kiernan in *Urologic & Cutaneous Rev.* **1928** H. Ellis *Stud. Psychol. Sex* VII. viii. 491 Kleptolagnia.. is an effort to attain the direct gratification of the sexual impulse by the aid of emotional energy generated by the excitement of the theft. **1960** Hinsie & Campbell *Psychiatric Dict.* (ed. 3) 414/1 *Kleptolagnia*, a morbid desire to steal. **1969** E. M. Brecher *Sex Researchers* (1970) ii. 55 *Psychopathia Sexualis* presents cases of satyriasis and nymphomania,..kleptolagnia (sexual arousal through stealing), [etc.]. **1972** *Cumulated Index Medicus* XIII. 7662/2 A case of kleptolagnia.

‖ **kletterschuh** (kle·təɹʃū). Pl. **-schuhe.** [G.] A cloth- or felt-soled light boot worn esp. for rock-climbing; usu. in *pl.* Sometimes colloquially abbrev. to *klets*. Also *attrib.*

1920 J. P. Farrar in G. W. Young *Mountain Craft* iii. 96 Kletterschuhe are much used in the Dolomites, and they can be used..not only on dry rocks but also in icy climbs... A good kind is the so-called Sexten pattern, the soles of which are built up of layers of cloth. **1950** tr. *Mountaineering Handbk.* (Assoc. Brit. Members Swiss Alpine Club) ii. 23 Kletterschuhe (cloth soled boots) are an advantage for difficult rock climbs because they hold and do not slip. **1951** C. D. Milner *Dolomites* 88 The usual wear in the Dolomites for many years has been felt-soled *kletterschuhe*... Today *kletterschuhe* and rubber soled shoes are both used... An interesting compromise is to have Vibram soles, over a thin leather undersole, on *kletterschuhe* soft uppers. **1963** *Climber* Aug. 11/1 Beginners come to climbing without being aware that there is any kind of mountain footwear other than 'vibs'— barring..the P.A.'s or kletterschuhe to which they eventually hope to aspire. *Ibid.* 12/2 If you're a rock climber rather than an all-round mountaineer..then Vibrams, P.A.'s or 'klets' are the wear for you.

kleywang (klĕi·wæŋ). Also † **calewang; kelewang.** [ad. Malay *kĕlewang*.] A single-edged Indonesian sword.

1783 W. Marsden *Hist. Sumatra* 277 There are weapons of a make between that of a scimitar, and a knife ..as..the calewang. **1839** T. J. Newbold *Pol. & Statistical Acct. Straits of Malacca* I. i. 39 The crew are armed with ..swords (the parang and kleywang). **1900** W. W. Skeat *Malay Magic* 24 The articles of Malay regalia usually consist of a *silasila*..and a few weapons, generally a kris, *kleywang*, or spear. **1936** G. B. Gardner *Keris & Other Malay Weapons* 73 *Golok jambu* or the Kĕlantan *kĕlewang*... This is short and curved. *Ibid.* 75 The *sulu kĕlewang*.. A single edge sword that gets wider and heavier towards the point.

klieg (klīg). *Cinemat.* Also **kleig** and with capital initial. [f. the name of two brothers, A. T. and J. H. *Kliegl*, who invented it in the U.S.] In full, *klieg light*. **a.** Orig., a kind of arc lamp invented for use as a studio light; hence, any powerful electric light used in film-making, or in television.

1925 *Movie Mag.* Nov. 57/1 The scene in a motion picture studio is eternally the same. It doesn't matter what difference there is in the set which raises its skeleton wings towards the Kleigs. **1927** U. Sinclair *Oil!* xii. The kleigs glare upon them, and a dozen moving picture cameras grind. **1931** A. Nadell *Projecting Sound Pict.* xiv. 250 Klieg (arc) lights are passé; they made too much noise; 1,000-watt incandescents are used. **1932** S. Gibbons *Cold Comfort Farm* xvii. 237 The sunshine, cruel as a Kleig light, revealed every wrinkle in his melancholy ..face. **1947** Auden *Age of Anxiety* ii. 39 The polychrome Oval With its kleig lights and crowd engineers. **1951** 'J. Wyndham' *Day of Triffids* viii. 141 At the first burst it was as dazzling as a klieg light. **1957** *New Yorker* 1 June 21/1 My brother and I [sc. Mr. John H. Kliegl] invented the klieg light around 1911—the first practical light for taking motion pictures indoors. It projected a beam by means of carbons, that emitted a light of high actinic power. **1967** *Economist* 7 Jan. 40/2 Inaugurated in a blaze of familiar klieg lights. **1969** *Rolling Stone* 28 June 18/1 The kleig lights shine for a mercifully brief moment upon the aging cosmetic face.

b. klieg eyes, an eye condition caused by exposure to very bright light, characterized by watering and conjunctivitis. Hence *klieg-eyed* adj.

1923 *Sci. Amer.* Oct. 243/1 The burning of the eyeball by the ultra-violet rays... This malady appears so freely among motion-picture actors..that a name, 'Kleig eyes', has been coined for it. **1941** *Amer. Cinematographer* Dec. 589 The ultra-violet glare from those unshielded arcs literally sunburned the actors' eyeballs and created that dread malady, 'kleig eye'. **1973** *Rolling Stone* 30 Aug. 8/3 Most folks got back to San Clemente sated, klieg-eyed and tired.

Kline (klain). *Med.* The name of Benjamin S. *Kline* (b. 1886), U.S. pathologist, used *attrib.* to designate a diagnostic test for syphilis devised by him in which serum, blood, or spinal fluid is mixed on a slide with a lipid antigen and examined under a microscope for precipitation.

1929 *Amer. Jrnl. Syphilis* XIII. 583 (*heading*) The Kline slide precipitation test for syphilis..with a clinical evaluation in syphilitic and non-syphilitic cases. **1944** J. H. Stokes et al. *Mod. Clin. Syphilol.* (ed. 3) x. 471 The Kline finger-puncture blood test procedure avoids the objection of donors to a double venipuncture. **1963** *Amer. Jrnl. Clin. Path.* XL. 551/2 The Kline test with cardiolipin–synthetic lecithin antigen yielded more accurate results in terms of TPI-reactivity than did the tests with cardiolipin–natural lecithin antigens.

Klinefelter (klai·nfeltəɹ). *Med.* The name of Harry Fitch *Klinefelter* (b. 1912), U.S. physician, used in the possessive (less commonly *attrib.*) to designate a syndrome he described (with others) in 1942 which affects males and becomes evident at puberty or after, being characterized by small testes, eunuchoidism, gynæcomastia, and infertility, usu. now restricted to those cases in which the cells have an extra X sex chromosome (most commonly in an XXY constitution, in contrast to the XY of normal males and the XX of normal females).

[**1946** *Med. Jrnl. Austral.* 28 Sept. 446/1 (*heading*) The Klinefelter–Reifenstein–Albright syndrome.] **1950** *Jrnl. Clin. Endocrinol. & Metabolism* X. 630 (*heading*) Dystrophia myotonica, with special reference to endocrine function (Klinefelter's syndrome). *Ibid.* 635 In the 2 male patients the diagnostic criteria for the so-called Klinefelter syndrome were satisfied. **1958** H. J. Roberts *Difficult Diagn.* i. i. 37 The evolution of the belief that micro-orchidism with gynecomastia (the Klinefelter syndrome) probably represents a genetic defect in the sex chromosomes. **1964** L. Martin *Clin. Endocrinol.* (ed. 4) vi. 209 Apart from these two main XXY and XY types, Klinefelter's syndrome has also been described in variants possessing the chromosomal constitutions of XXXY, XXXXY, and XXYY. **1969** *Nature* 16 Aug. 680/2 Züblin and Pasqualini noted a childishness, shyness, lack of drive and a degree of intellectual impairment in patients with Klinefelter's syndrome. **1970** Passmore & Robson *Compan. Med. Stud.* II. xxxi. 15/2 The most important findings were that abnormal females with Turner's syndrome were sex chromatin negative like normal males, and that abnormal males with Klinefelter's syndrome were sex chromatin positive like normal females. **1971** *Daily Tel.* (Colour Suppl.) 10 Dec. 20/1 One of the most common physical intersex conditions affecting a minimum of 2·65 males out of every 1,000 is the Klinefelter syndrome.

b. *ellipt.*

1961 Davidson & Robertson Smith *Proc. Conf. Human Chromosomal Abnormalities* ii. 23 In 5 of the Klinefelter cases our observations were consistent with the belief that the chromosome number was regularly 47. **1967** Bartalos & Baramki *Med. Cytogenetics* x. 155 In Hornstein's experience, only a few of the Klinefelter patients were free from some form of psychopathological traits. **1971** *Daily Tel.* (Colour Suppl.) 10 Dec. 20/1 In Klinefelter there is an extra X chromosome added to the male XY.

Kling (kliŋ). [Malay *Keling* Tamil, ad. *Kalinga* an old name for a strip of coast along the Bay of Bengal.] A disparaging term applied to Indian settlers in Malaysia.

1606 E. Scott *Exact Discourse E. Indians* sig. F4 If it were not for the Sabyndar, the Admirall, and one or twoe more, which are Clyn men borne, there were noe liuing for a Christian amongst them. **1625** Purchas *Pilgrimes* I. iv. ii. 385 The fifteenth of Iune, heere arriued Nockhoda Tingall a Cling-man from Banda, in a Iaua Iuncke. **1839** T. J. Newbold *Pol. & Statistical Acct. Straits of Malacca* I. i. 8 The Chinese, and the natives from India (Chuliahs and Klings,) are by far the most useful class. **1868** C. Collingwood *Rambles of Naturalist on Shores & Waters China Sea* xv. 245 The Klings are, indeed, the only people who can contest the field with the Chinese. **1869** A. R. Wallace *Malay Archipelago* I. ii. 31 The Klings of Western India are a numerous body of Mahometans, and, with many Arabs, are petty merchants and shopkeepers. **1890** Kipling *Barrack-Room Ballads* (1892) 135 The frigate-bird shall carry my word to the Kling and the Orang-Laut. **1968** *Encycl. Brit.* XIII. 400/1 In Malay usage 'Kling' carried associations of disparagement from the start. It..was replaced by 'Tamil' early in the 20th century.

kling-kling (kli·ŋkliŋ). Also **cling-cling, clinkling, klinkling.** [Echoic.] A Jamaican name for a grackle, *Quiscalus niger*.

1847 Gosse & Hill *Birds Jamaica* 219 It is to the first of these notes that the bird before us owes his local names of Tinkling, Tintin, Clinkling. **1949** *Caribbean Q.* I. iv. 42 What to Miss Bottome is a tropical bird, is to Vic Reid a pechary or a klinkling. **1955** R. G. Taylor *Introd. Birds Jamaica* 11 Kling-kling, a bird of medium size with shining black plumage, pointed beak, a pale yellow eye and long boat-shaped tail. **1960** J. Bond *Birds W. Indies* 217 Greater Antillean Grackle *Quiscalus niger*. Local names: Tinkling; Cling-cling; Ting-ting (Jamaica). **1965** I. Fleming *Man with Golden Gun* v. 75 Two large black birds.. whirled in... She said, 'We call them kling-klings but learned folk call them Jamaican grackles.'

klino-. Add: **klinokine·sis** [*kinesis 2], a kinesis in which the movement is one of turning; hence **klinokine·tic** *a.*, **klinokine·tically** *adv.*; similarly **klinota·xis** [Taxis 6], a taxis in which the movement is one of turning.

1937 D. L. Gunn et al. in *Nature* 18 Dec. 1064/2 We propose to divide kineses into (*a*) ortho-kineses..and (*b*) klino-kineses..—variations in angular velocity. **1954** *New Biol.* XVII. 49 A klino-kinesis (a higher rate of turning in dry than in moist air) is also present [in woodlice]. **1970** R. A. & B. M. Maier *Compar. Animal Behavior* v. 81 A second category of kinetic response—klinokinesis—involves changes in the rate of turning associated with shifts in the intensity of stimulation. **1940** Fraenkel & Gunn *Orientation of Animals* v. 45 In some animals there is a klino-kinetic response in which the rate of random turning, or angular velocity, depends on the intensity of stimulation. **1964** *Oceanogr. & Marine Biol.* II. 478 Unoriented klinokinetic movements occur [in two species of Erycinidæ] under uniform overhead light. **1946** *Nature* 13 July 58/2 When behaving photopositively *Hydra* orientates itself klinokinetically. **1940** Fraenkel & Gunn *Orientation of Animals* vi. 59 The kind of reaction in which these regular deviations are a necessary part of the orientation mechanism is here named klino-taxis. *Ibid.* 75 Klino-taxis is an uncommon type of reaction to light. **1958** *New Biol.* XXVII. 72 A fly larva with a single median eye moves away from a single source of light... The movement involves swinging from side to side; as the eye is illuminated from one side so the swing is away from that side... This swinging motion is a klinotaxis. **1970** R. A. & B. M. Maier *Compar. Animal Behavior* v. 81 The first type [of taxis]—klinotaxis—involves a series of successive comparisons of the intensity of stimulation.

klip (klip), *sb.* *S. Afr.* Also dim. **klippie.** [Afrikaans, a. Du. *klip* cliff, rock, stone.] **1.** Also, *blink klip* [Afrikaans *blink* shining]. A diamond; † micaceous iron ore (quot. 1835).

1835 A. Smith *Diary* 22 Jan. (1939) I. 225 [The Bechuana] mix fat with the blink klip and rub it over the body in moderate quantities, but in great quantities upon the head. **1887** J. W. Matthews *Incwadi Yami, or Twenty Yrs. S. Afr.* xiii. 186 The natives had not yet acquired a knowledge of the value of diamonds or *klips* as they were then termed. **1892** J. R. Couper *Mixed Humanity* vi. 48 Flogged to death for stealing a 'klip' (as the Dutch and many of the Kaffirs call a diamond). **1893** T. Reunert *Diamonds & Gold S. Afr.* i. 6 By this time the attention of every one in that neighbourhood was turned to seeking *blink klippe*..and during the following year several diamonds were picked up. **1897** *Pearson's Mag.* July 67/1 Fifteen years on that blathted breakwater, just for being found with a few little klips on you. **1911** L. Cohen *Remin. Kimberley* 35 'I'll show you the klip. Here it is.' With that he pulled out his snuffbox..; inside was a fairly sized beautiful octahedron diamond. **1967** L. G. Green *Like Diamond Blazing* xiii. 144, I met old Hottentots who remembered Luderitz; they said he had two little boxes of *blink klippies*..and that he carried these diamonds with him.

2. A stone, pebble.

1852 C. Barter *Dorp & Veld* vi. 50 Stooping to set large *klips* (stones) behind the wheel, to prevent the wagon from slipping back. **1899** A. Werner *Captain of Locusts* 63 She left me when we were on the trek over into Basutoland..and the boys and I could only cut a cross on the thorn-tree..and put a heap of klippies to mark the spot.

Hence **klip** *v. trans.*, to place a stone behind (a wheel) in order to prevent a vehicle from rolling backwards.

1878 H. A. Roche *On Trek in Transvaal* iv. 91 We crawling into the wagon, the wheels of which were 'klipped', to keep us from running down the hill, trying to nap at intervals.

klipbok (kli·pbǫk). Also **klipbokkie, klipbuck.** [Afrikaans, f. Du. *klip* rock + *bok* buck.] = *klipspringer.

1886 G. A. Farini *Through Kalahari Desert* i. 4 Not even the beasts of the desert, the klip-bok (rock buck), or stein-bok (stone buck)..are to be seen. **1895** J. G. Millais *Breath from Veldt* 92 The most curious thing about the klipbuck is the shape of its feet and the manner in which it uses them in springing up and down its native rocks. **1939** tr. *E. N. Marais's My Friends the Baboons* v. 59 The troop must often have had the chance of catching little klipbuck, dassies, and red hares. **1947** *Cape Argus* (Magazine Section) 23 Oct. 1/9 The dog..brought down a klipbok. **1953** J. R. Ellerman et al. *S. Afr. Mammals* 188 *Oreotragus oreotragus* Zimmerman, 1783. Klipspringer. Klipbokkie.

klipdas (kli·pdas). [Afrikaans, f. Du. *klip* rock + *das* badger.] = *dassie 1.

1853 *Edin. New Philos. Jrnl.* LV. 214 Basking themselves on the sunny side of the krantzes..may generally be seen several of the Klipdas, Cony, Rock Rabbit, or Cape Hyrax (*H. capensis*). **1886** P. Gillmore *Hunter's Arcadia* xxvi. 248 From this descendant of Holland..I bought..the skins of some rock rabbits, the *klip das* of the Dutch. **1953** J. R. Ellerman et al. *S. Afr. Mammals* 157 *Procavia capensis* Pallas, 1766. Dassie; Hyrax. Klipdas. Distribution: one of the commonest mammals in the Union.

klipfish (kli·pfiʃ). Also **clipfish, klepvis, klippfish, -fisch, klipvissie.** [ad. Du. *klipvisch* (f. *klip* rock) and Da. *klipfisk* (f. *klippe* rock).] **1.** *S. Afr.* A viviparous, brightly-coloured, marine fish of the family Clinidæ, living in shallow water or rock pools.

[**1731** G. Medley tr. *Kolb's Present State Cape Good-Hope* I. xx. 256 The Hottentots frequently take Abundance of a Sort of Fish, call'd Rock-fish. These are Fish without Scales.] **1790** E. Helme tr. *Le Vaillant's Trav. Afr.* I. ii. 22 The *klepvis*..is without scales, and taken among the rocks on the sea shore. **1806** J. Barrow *Trav. S. Afr.* (ed. 2) II. i. 38 The *Klip* or rock-fish, the *Blennius viviparus*, makes no bad fry. **1838** J. E. Alexander *Expedition Interior Afr.* I. iv. 88 Abundance of excellent fish are to be procured here; such as the delicious Roman fish, Hottentot, 'Jacob Fever', mullet, stump nose, and clip fish. **1876, 1893** [see *kingklip]. **1902** *Trans. S. Afr. Philos. Soc.* XI. 224 Several names, or parts of names, are derived from the localities in which the fish are found. Thus we have..Klip Visch, Steen Klip Visch (a peculiar redundancy). **1953** U. Krige *Dream & Desert* i. 14 They had..gazed into rockpools full of starfish, slowly waving sea plants and green-and-gold *klipvissies* drifting lazily from crevice to crevice. **1969** *Nature* 8 Nov. 540/1 The shores of South Africa are enlivened by a group of typically littoral viviparous fishes known locally as klipfishes (but as clinids to the academic). **1973** *Farmer's Weekly* (S. Afr.) 18 Apr. 102 The klipfish is fried in butter and carefully filleted on the plate with knife and fork.

2. A codfish split open, boned, salted, and dried.

1835 *Penny Cycl.* IV. 273/2 The klip-fish is cut along the back, and the back-bone taken out, after which it is salted down in the bottom of the vessel. **1881** S. Walpole in *20th Ann. Rep. Salmon Fisheries 1880* 23 in *Parl. Papers* (C 2901) XXIII. 299 Cod are either cured as stock fisch or as klipp fisch... The klipp fish are split and boned before they are salted. **1925** J. T. Jenkins *Fishes Brit. Isles* 136 The bulk of the cod caught in the northern fisheries is split open, washed, and salted in pickle and then dried on rocks (Klippen). This is the so-called Klipp-fish. **1961** H. Angerman et al. in Borgstrom & Heighway *Atlantic Ocean Fisheries* 80/2 The klipfish is subjected to more complicated treatment than the stock-fish.

klipkous (kli·pkōus). *S. Afr.* Also **klipkoes, klipkos.** [Afrikaans, f. Du. *klip* rock + *kous* stocking.] = Sea-ear 1; *abalone.

1731 G. Medley tr. *Kolb's Present State Cape Good-Hope* II. 209 The *Klip-Kousen* are sometimes call'd, by the Virtuosi, *Nabel*-Snails. **1785** G. Forster tr. *Sparrman's Voy. Cape Good Hope* I. ii. 26 A sort of snail or cockle, *klipkous* (*Haliotis*, Linn.) from half a foot to a foot and a half diameter, is usually stewed, but makes in my opinion a very unsavoury dish. **1843** J. C. Chase *Cape of Good Hope* vii. 168 Klip Kous... A shell fish, most delicious, but requiring much trouble in the preparation. **1910** D. Fairbridge *That which hath Been* xxv. 303 The first muscadel grapes and the finest klip-kous from the rocks had invariably found their way across the sands to Meerlust. **1930** C. L. Biden *Sea-Angling Fishes of Cape* xviii. 260 The crushed remains of klipkoes or venus ear—a shellfish, *Haliotis*. **1947** L. G. Green *Tavern of Seas* viii. 66 The perlemoen or klipkous, largest and most beautiful of Cape shellfish. **1950** M. Masson *Birds of Passage* iii. 41 Souvenirs, among which was a Venus ear or klipkos whose brown crust had been scraped away to reveal the iridescent shell. **1966** H. J. Duckitt *Bk. Recipes* 73 The Perlemoen, or Klipkous (stone-stocking) a species of shellfish found on many parts of the South African coast, adhering to the rocks.

klippe (kli·pə). *Geol.* Also **Klippe.** Pl. **klippes,** ‖**-en.** [a. G. *klippe* partly or totally submerged rock.] A part of a nappe which has become detached from its parent mass by sliding or by erosion of intervening parts.

1902 *Encycl. Brit.* XXV. 333/2 These [structures], called Klippen, are abrupt pyramidal masses, the beds in the upper part being not only older than those in the lower, but also 'contorted, fractured, crushed, and mixed up', while the newer are comparatively undisturbed. **1912** *Smithsonian Misc. Coll.* LVI. No. 31. 12 It is well known that some of these isolated masses, those of the Klippes, are 'exotic'; that is to say, no strata of the same facies have ever been found in place. **1942** O. D. von Engeln *Geomorphol.* xv. 332 Klippen are peculiar in that, unlike the outliers which persist beyond the main front of a weathering escarpment, and which have younger beds capping older strata, they have older beds over younger beds. **1954** W. D. Thornbury *Princ. Geomorphol.* x. 273 Chief Mountain, Montana, is a well-known example of a klippe, in which an isolated mass of Pre-Cambrian rock rests upon Cretaceous beds. **1969** M. G. Rutten *Geol. W. Europe* xi. 239 Further southwest, these isolated klippen ..merge into a continuous nappe.

klippie: see *klip *sb.*

klipspringer. Substitute for def.: A small African antelope, *Oreotragus oreotragus*. Add later examples.

1907 P. Fitzpatrick *Jock of Bushveld* 245 The dainty little klipspringers led them many a crazy dance along the crags and ledges of the mountain face, jumping from rock to rock. **1936** R. Campbell *Mithraic Emblems* 68, I always thought to be A klipspringer or chamois. **1947** J. Stevenson-Hamilton *Wild Life S. Afr.* xvi. 115 The klipspringer (*Oreotragus oreotragus*). *Ibid.*, The ubiquitous klipspringer, truly the chamois of Ethiopia. **1960** *Times* 29 Sept. (Nigeria Suppl.) p. xxi/5 But two other local races special to Nigeria, of that curious little pithy-haired antelope, the Klipspringer,..seem doomed. **1972** L. van der Post *Story Like Wind* vi. 180 This man in front of him wore nothing except a loin-cloth of soft yellow klipspringer leather.

Klischograph (kli·ʃograf). *Printing.* [f. G. *klischee* stereotype or electrotype plate: see -graph.] The proprietary name of a type of electronic engraving plate (see quots. 1955 and 1963).

1955 *Trade Marks Jrnl.* 16 Mar. 288/1 Klischograph... Machines for making printers' blocks, printers' clichés and printers' formes; and plain and engraved printers' plates. Dr. Ing. Rudolf Hell..., Kiel-Dietrichsdorf, Germany; Manufacturer. **1963** *Times* 17 July 6/7 In the field of electronic engraving a German firm, Dr. Ing. Rudolf Hell,

are demonstrating their latest type of klischograph. This consists of an electronic scanning instrument, capable of dealing with both black and white and multi-colour reproductions, which operates an engraving stylus to produce a gravure cylinder. Preparation of gravure cylinders can be a lengthy business, and this process, it is claimed, can reduce the time taken from several days to a matter of hours. **1967** Karch & Buber *Offset Processes* ii. 20 These machines [sc. electronic platemakers] include the models of..Klischograph, Photo-lathe and Elgramma.

klister (kli·stəɹ). *Skiing.* [Norw. *klister* paste.] A soft wax for applying to the running surface of skis to facilitate movement, used esp. in warm weather.

1936 B. Lunn tr. *Hallberg & Mückenbrünn's Compl. Bk. Ski-ing* iv. 39 We specially recommend the following waxes: Ostbye (Medium..Klister); Bratlie (Nysnö.. Klister). **1948** —— *Ski-ing Primer* xix. 95 Klister..is generally sold in tubes. There are many brands. **1951** Eugen & Atwater *Ski with Sverre* xi. 94 For wet snow, well above freezing, use a soft klister type wax.

klob: see next.

klobbiyos (klǫ·biyǫs). *Cards.* Also **klaberjass, klobbyosh, klobiosk,** etc. [ad. G. *klaberjass*, f. Du. *klaverjas* a type of piquet.] A type of piquet, esp. popular with eastern European Jews. Also in shortened form **klob.**

1892 I. Zangwill *Childr. Ghetto* I. 124 They played loo, 'klobbiyos', napoleon, vingt-et-un. **1928** *Daily Tel.* 6 Nov. 9/2 Dice, nap, and klobiosk were played... The Magistrate: What is klobiosk? Inspector Dyer: It is similar to the English game known as 'Five hundred'. **1937** D. Runyon *More than Somewhat* 10, I find Little Isadore playing klob with a guy. **1946** Morehead & Mott-Smith *Penguin Hoyle* 94 Klaberjass, a game for two players, is better-known under corrupted names—Clobber, Clob, and the like and particularly under the name Kalabrias, which is a popular Hungarian game. **1946** Mezzrow & Wolfe *Really Blues* ii. 20 Easygoing guys who spent half their lives playing klabiasch, pinochle, and tarok. **1961** A. Smith *East-Enders* ix. 159 They will create a few private gambling hells... There will be klobby-osh and strip poker.

klockmannite (klǫ·kmænait). *Min.* [ad. G. *klockmannit* (P. Ramdohr 1928, in *Centralbl. f. Mineral.* A. 226), f. the name of F. F. H. *Klockmann* (1858–1937), Ger. mineralogist: see -ite¹.] A slate-grey selenide of copper, CuSe, found impure as granular aggregates and synthesized as crystal plates.

1939 *Mineral. Mag.* XXV. 295 Another copper selenide from the same locality [sc. Sierra de Umango, Argentina] has since been named klockmannite. **1969** *Acta Crystallogr.* XV B.2420 (*heading*) Twinning in the superlattice structure of CuSe, synthetic klockmannite.

‖ **klomp** (klǫmp). *S. Afr.* Also **clompie, clumpjie, klompie, klompje, klumpjie.** [Afrikaans, f. Du. *klomp*: see Clump *sb.*] A group, esp. of animals.

1853 W. R. King *Campaigning in Kaffirland* 215 Even at three quarters of a mile, we were able to disperse small 'clumpjies' of Kaffirs and cattle. **1861** T. Baines *Jrnl.* 25 Nov. in *Explor. S.-W. Afr.* (1864) ix. 241 Snyman also saw nothing except one 'klumpjie' of kameels. **1896** H. A. Bryden *Tales S. Afr.* 70 Rather suddenly we came upon a *klompje* of giraffe. **1920** F. C. Cornell *Glamour of Prospecting* xi. 170 Next day we were off well before sun-up, anxious to shoot something for the pot, but it was not till late in the afternoon that Poulley spotted a *klomp* of springbok. **1937** S. Cloete *Turning Wheels* xxviii. 434 We saw some of them [sc. cattle]..a big clompie of a thousand head or more. **1963** —— *Rags of Glory* v. 44 Great *klompies* of them [sc. stallions] could live together with hardly a serious quarrel till a female came along.

klompie (klǫ·mpi). *S. Afr.* Also **klompje.** [Afrikaans, f. Du. *klomp* a small blue brick + Afrikaans *-ie* a dim. ending.] A type of brick.

1926 S. G. Millin *South Africans* iv. ii. 111 Open hearths outlined in stone or in the small bricks called *klompjes*. **1949** L. G. Green *In Land of Afternoon* xv. 199 When you step on to the klompie brick stoep and into the low cool rooms..you are in a more leisurely world. **1950** *Cape Times* 4 Mar. 11/8 (Advt.), Golden Brown Klompies 9 in. × 4 in. × 1¾ in. and 2,500 face bricks.

Klondike (klǫ·ndaik), *sb.* Also **Klondyke.** [f. the Kutchin name *tron-duik* hammer river, a tributary of the Yukon River.] **1.** The name of a region (and river) in the Yukon, NW. Canada, the scene of a gold-rush in the years following 1896. Hence many *attrib.* uses, as *Klondyke fever*, (*gold*) *rush*, etc., applied to this period and to the life lived during it. **b.** *ellipt.* as *sb.*, a mine or quarry of valuable material.

1897 *Slocan* (B.C.) *Pioneer* 31 July 4/2 The Klondike fever has struck Slocan City in a mild form. **1897** *Athenæum* 9 Oct. 483/3 The rich Klondyke of Malory and Geoffrey of Monmouth had not escaped the eyes of previous prospectors. **1898** *Century Mag.* Mar. 697/2 These men made their way home, as best they could, out of the wreckage of the first Klondike rush. **1898** *T. Eaton & Co. Catal.* Spring & Summer 124 Men's Klondike mining coats... Men's Klondike sleeping bag... Men's Klondike shirts. **1912** H. Footner *New Rivers of North* 192 We guessed that we were upon the spot where our last white predecessors had made camp in the year of the Klondike rush. **1948** *Life* 2 Feb. 49/3, 78-year-old Emil J. N. Ott,.. a veteran of the Klondike Rush, who smelts rough gold into ingots. **1958** *Encycl. Canadiana* VI. 15 (*caption*) Hundreds watch as the *Australia*..leaves Seattle for the North at the height of the Klondike fever. **1961** *Tamarack Rev.* XIX. 4 For a long time, before the Klondike days, there was a feud and a battle between them two logging camps. **1964** *Edmonton* (Alberta) *Jrnl.* 11 July 3/6 The past 60 years have obscured or distorted the extraordinary strength and endurance, and the Klondike wealth of the erstwhile sourdough who became a legendary character during his own lifetime. **1965** *Star Weekly* (Toronto) 2 Jan. 37/2 Mrs. MacCleave was a young girl at the height of the Klondike gold rush. **1973** *Times* 1 Feb. 12/8 He combed the reports of Congressional committees rarely read or reported by others. It was a new Klondike.

2. A card-game played with a single pack of fifty-two, the object being to see how many cards can be built up in sequence and suit on a row of aces. *N. Amer.*

1902 L. McKee *Land of Nome* 163 All the games were going—roulette, vingt-et-un,..Klondike, and craps. **1908** U. Sinclair *Metropolis* vi. 91 The smoking-room, where the stout little Major had gotten a group of young bloods about him to play 'Klondike'. **1910** R. Service *Trail of '98* 183 There were crap-tables,..the Klondike game, Keno, stud poker, roulette and faro outfits. **1946** Morehead & Mott-Smith *Penguin Hoyle* 174 Klondike is probably the most widely known solitaire game. **1953** J. Walker *Pardon my Parka* (1958) 127 We sat around.. and we played a vicious gambling patience called Klondike..which cost me a vast amount of money. **1968** *Encycl. Brit.* XX. 875/2 Many solitaires can be played by two, but Klondike is by far the favourite for this purpose.

3. The name given to a herring fishery off the W. coast of Scotland. (Cf. *Klondike *v.*)

1929 W. Keir *Herring Trade on Continent 1928* 16 In the early part of the season the trawlers fished mostly on the 'Klondyke' grounds off the West Coast of Scotland.

Klo·ndike, *v.* Also **Klondyke,** and with small initial. [See prec., sense 3.] *trans.* To export (fresh herring) (as opp. to pickled herring).

1923 *Glasgow Herald* 25 Oct. 6 A regular fleet of steamers 'Klondyking' or running the fresh fish direct from the various landing ports to Germany. **1930** *Aberdeen Press & Jrnl.* 30 Jan. 8 If a boycott was attempted, they would klondyke their supplies into the Dutch and other markets. **1930** *Morning Post* 2 Aug., Reference is made to the quantities of herrings 'klondyked'—which means despatched fresh to the Continent. **1945** [see *Bismarck 3].

Klondiker (klǫ·ndaikəɹ). Also **Klondyker.** [f. *Klondike *sb.* + -er¹.] **1.** A prospector in the Klondike. Also *transf.*

1897 *Brit. Columbia Mining Jrnl.* (Ashcroft, B.C.) 9 Oct. 1/6 The venturesome Klondiker who may select this valley as his road to the diggings may rest assured that his daily bill of fare will not only be ample but of good variety. **1901** *Daily Colonist* (Victoria, B.C.) 5 Nov. 3/2 Steamer Amur arrived from Skagway and way ports early this morning, with gold, salmon, and many Klondikers, among her 78 passengers. **1904** Burgess & Irwin *Picaroons* 102 The Story of the Returned Klondyker. **1917** *Dialect Notes* IV. 420 Terms from New Orleans. *Klondiker*, an heiress-hunter. **1954** A. M. Bezanson *Sodbusters invade Peace* 4 There's an old trail that some Klondikers tried to take their outfits over.

2. An exporter of or dealer in fresh herring from the Scottish fisheries; a ship used for this.

1926 *Glasgow Herald* 19 Dec. 8 The 'Klondykers' are.. the German boats which buy the herring and transport them for sale in Germany. **1938** L. MacNeice *I crossed Minch* ii. xii. 166 The herring fishing was nothing here now. He remembered when the German Klondikers used to come into Loch Seaforth and buy [herrings] from the natives on the spot. **1953** *Press & Jrnl.* (Aberdeen) 17 June, The klondykers, however, sent only about one-third of the 1951 quantity of fresh herring to Western Germany. **1971** *Stornoway Gaz.* 7 Aug. 3/3 The first Klondyker of the season, from Norway, arrived at Stornoway during the week-end for a cargo of herring.

Klondiking (klǫ·ndaikiŋ), *vbl. sb.* Also **Klondyking,** and with small initial. [f. *Klondike *v.* + -ing¹.] **1.** Prospecting in the Klondike during the gold-rush period.

1900 J. London *Let.* 31 Jan. (1966) 87, I spoke at length in previous letter concerning my tramping and Klondiking.

2. Dealing in or exporting fresh herring from the Scottish fisheries.

1927 J. T. Jenkins *Herring* 146 While trawled herring are unsuitable for pickling, they are well adapted for 'Klondyking'—a method of preparing herring for export practically fresh by sprinkling them with salt and ice. **1930** P. F. Anson *Fishing Boats & Fisher Folk E. Coast Scotl.* 20 On an average about 12 per cent [of the herring catch] is exported fresh to Germany. This freshing export trade is known as 'klondyking'. **1973** *Stornoway Gaz.* 27 Jan. 1/1 The herring was disposed of as follows: Freshing and kippering—27 crans; klondyking—3,565 crans. **1973** *Courier & Advertiser* (Dundee) 21 Feb. 9/1 Ullapool.— Herring 3420 crans—1430 crans for home market, £5·50 to £6·70; 220 for freezing, £4·60 to £5; 120 for canning, £4·50 to £5; 1650 for klondyking.

klong (klǫŋ). Also **khlong.** [Thai.] In Thailand, a canal. Also *attrib.*

1898 E. Young *Kingdom Yellow Robe* ii. 26 When agricultural enterprise led to the formation of inland settlements, no roads were made..but canals or 'khlongs' were cut instead. **1928** *Daily Express* 13 Mar. 12/4 A broad flat field..beribboned with klongs (canals). *Ibid.*, Millions of happy frogs sing in high shrill voices, perched on the banks of the klongs. **1967** *Nat. Geographic* July 81 Where are the klongs of yesteryear, all those colorful canals, crisscrossing the city, that had made travel agents and writers burble about Bangkok as the Venice of the East? **1970** M. Pereira *Pigeon's Blood* i. 10 A wooden jetty on the banks of one of the numerous klongs, or canals, which intersected the city. **1972** *Nation* (Bangkok) 22 Nov. 4/4 He had made a survey of the plant and found that the water released from the ponds into the nearby klongs and later to the Chao Phya river needed to be medically treated.

‖ **klonkie** (klǫ·ŋki). *S. Afr.* [Afrikaans, a blending of *klein* small and *jong* small boy + dim. suff. *-kie*.] A coloured boy; occas., an African boy; a coloured man.

[**1913** Pettman *Africanderisms* 268 *Klong*..is in common use in various parts of South Africa, and is applied to coloured males without reference to age, much as the word 'boy' is among the English colonists.] **1953** A. Paton *Too Late the Phalarope* viii. 58 The small klonkie from the black people's location..liked to hang around the store. **1955** D. Jacobson *Trap* i. 32 Strained and shy, the boy's voice came: 'Good night, baas. Thank you, baas.' 'Good night, *klonkie*,' Van Schoor replied. **1960** D. Lytton *Goddam White Man* iv. 102 But don't let me the coloured klonkie living in those huts on the farm feels shame at his failure to provide better for his kids.

klooch (klūtʃ). *NW. Amer.* Also **klootch.** [Chinook Jargon (from Nootka) *klootchman* woman.] An Indian woman. Also (variously spelt) **kloo·chman.**

Quot. 1861 illustrates the erroneous form *klookwuman*.

1837 H. Beaver *Let.* 10 Mar. (1959) 38 'Klout-chemin'..is the term used to express the whole female sex, in whatever degree of relationship, whether rational or irrational. **1860** *Brit. Colonist* (Victoria, B.C.) 24 Mar. 2/3 About 30 Cape Flattery Indians arrived in canoes yesterday..on a visit to the Songish tribe, for the purpose of buying a clootchman for their chief. **1861** in C. Maiden *Lighted Journey: Story of B.C. Electric* (1948) 1, I perceived a clootch-vimmen a' standin' outside of a 'ouse, and she was a-laughing at me. **1865** G. Stuart *Montana as it is* 81 Oregon is the place to hear the 'Chinnook' in all its purity; it has 'played' the English language 'square out' in that land of..'cloockmans' and camus. **1897** M. H. E. Hayne *Pioneers of Klondyke* 25 The klūtch (short for klūtchman, the local name for squaws) dress exactly like the men. **1901** *Daily Colonist* (Victoria, B.C.) 8 Oct. 3/1 Old klootchmen were to be seen with their arms filled with all kinds of goods. **1907** R. Dunn *Shameless Diary of Explorer* 60 Starved dogs, half-naked children, shawled klooches, bucks in prospectors' old clothes, all gathered, stared, shook hands, clucked questions. **1945** R. W. Service *Ploughman of Moon* 176 In the old days he had taken up with a klootchman, and had written home, saying he was married to an Indian Princess. *Ibid.*, On one side of me I had a klootch with a papoose tied to her back. **1956** *Beaver* Summer 44 The Indians had respected the promises of the Great White Queen—or as she was known— 'King George's Klootchman'—and they relied upon the word of the Hudson's Bay Company. **1966** H. Marryat *Cariboo Cowboy* vi. 58 An Indian 'klootch' with three kids playing around the tent. **1969** *Islander* (Victoria, B.C.) 3 Aug. 10/2 Jenny, a Klootchman as Indian women were called, washed, ironed, sewed on buttons and mended socks.

klops (klǫps). [Ger.] A type of meat-ball or meat-loaf.

1936 I. S. Rombauer *Joy of Cooking* (ed. 2) 89/2 German meat balls. (Koenigsberger Klops). **1966** L. Davidson *Long Way to Shiloh* ix. 132 The proprietor was eating a plate of klops at the next table. **1972** N. Froud *Some of our Best Recipes are Jewish* 75 Klops is an East European improved version of a meat loaf.

Klucker, var. *Kluxer.

† **klumene** (klū·mīn). *Chem. Obs.* [f. mod.L. *kalium* potassium (see quot. 1900) + -ene.] = Acetylene.

1853 H. Watts tr. *Gmelin's Hand-bk. Chem.* VIII. 150 A carbide of potassium..gives off, when immersed in water, a peculiar combustible gas, which is klumene. **1900** V. B. Lewes *Acetylene* iii. 63 Edmund Davy, in 1836, named the newly-discovered gas bicarburet of hydrogen..; whilst later the name 'klumene' was bestowed upon it, because it had been derived from a kalium compound—potassium carbide. **1901** *Oxf. Univ. Gaz.* 6 Dec. 204 Olefine and Klumene Compounds. **1902** *Encycl. Brit.* XXV. 35/1 *Acetylene*, klumene or ethine, is one of the gaseous compounds of hydrogen and carbon.

klunk (klʊŋk). *U.S. slang.* Also **clunk.** [Of unknown origin.] A derogatory designation for a person.

1942 Berrey & Van den Bark *Amer. Thes. Slang* § 396 Klunk [in list of terms of disparagement for a person]. **1959** N. Mailer *Advts. for Myself* (1961) 399 What was unique about Jones was that he had come out of nowhere, self-taught, a clunk in his lacks. **1964** S. Bellow *Herzog* (1965) 78 He sat there, in her own words, like a clunk, bored, resentful. **1964** *N.Y. Herald-Tribune* 2 Jan. 8/1 Mr. Wagner has been a remarkably good mayor, and the klunks who don't realize this, they add, underestimate neither the Mayor himself nor the nature of his responsibilities.

klutz (klʊts). *U.S. slang.* Also **klotz, kluhtz.** [Yiddish, f. G. *klotz*, lit. = wooden block. Cf. Clot *sb.*] A clumsy, awkward person, esp. one considered socially inept; a fool. Also as *v.* So **klu·tzy** *a.*, awkward, foolish.

1965 *Sat. Rev.* (U.S.) 28 Aug. 51/2 The dancers look good and the artists look a little *klutzy*. **1968** F. Mullally *Munich Involvement* i. 10 Look, I feel a bit of a klutz in this crumpled day suit. **1968** L. Rosten *Joys of Yiddish* 185 Two *klutzes* were discussing their wives. **1970** L. M. Feinsilver *Taste of Yiddish* 276 Recent bilingual jokes about Christmas, like 'Santa *Kluhtz*'. *Ibid.* 303 Choreographer Kenneth MacMillan has attempted a compromise between dancing and acting that too often leaves Nureyev..with nothing to do but klutz around the stage. **1970** *New Yorker* 17 Jan. 72 The sad, klutzy ballerinas of the Music Hall pollute children's first live experience of dance. **1970** *Time* 2 Nov. 83 Basically I'm the klutz who makes a terrific entrance to the party and then trips and falls and walks around with food in her hair. **1973** E.-J. Bahr *Nice Neighbourhood* ix. 99 Janet is an utter klotz.

Kluxer (klʊ·ksəɹ). *U.S.* Also **Klucker.** [f. (Ku-)Klux (Klan) + -er[1].] A member of the Ku-Klux Klan. Also *attrib.*
1879 A. W. Tourgée *Fool's Errand* xxvii. 141 Ef dere's any mo' Kluckers raidin' roun' Burke's Corners, dar'll be some funerals tu. **1923** *Nation* (N.Y.) 21 Nov. 570 We are not much impressed with the desireability of organizations specially formed to fight the 'Kluxers'. **1929** *Sun* (Baltimore) 31 Jan. 1/2 Johnston is disowned by the kluxers or whatever is left of them in Oklahoma. **1944** J. S. Pennell *Hist. Rome Hanks* 176 Are you a Klucker, Mr. Ocamb? **1948** *Daily Ardmoreite* (Ardmore, Okla.) 25 July 20/4 Some 20 more noted Kluxers..and similar ragtag and bobtail were on hand. **1963** D. B. Hughes *Expendable Man* (1964) v. 146 There's not going to be any color business in this case. I'm not going to have it messed up with Kluxers or with bleeding hearts. **1965** C. Colter in A. Chapman *New Black Voices* (1972) 72 Jus' ask her t' pull up that Kluxer sheet and show you her arm.
Hence **Klu·xery**, conduct or behaviour characteristic of members of the Ku-Klux Klan; **Klu·xism**, the principles and practice of the Ku-Klux Klan or of similar organizations.
1929 *Sun* (Baltimore) 31 Jan. 2/2 When Henry Johnston announced for Governor, carrying the flambeau of kluxism, Mrs. Hammonds became his Joan of Arc. **1949** *Richmond* (Virginia) *Times-Dispatch* 23 Jan. iv. 2-D/2 There have been too many episodes in the Old Dominion of late which reflect the spirit of kluxery.

klydonograph (kləidōᵘ·nograf). *Electr. Engin.* [f. Gr. κλύδων wave, billow + -o + -graph.] An instrument for making a photographic record from which the voltage and polarity of a surge can be inferred, consisting of a point electrode resting (in darkness) on a stationary or moving film behind which is a plate electrode.
1924 J. F. Peters in *Electr. World* 19 Apr. 769 (*heading*) The Klydonograph. An instrument for accurately measuring and recording voltage surges. *Ibid.* 769/1 The word 'klydonograph' was coined by Dr. Roscoe M. Ihrig of the Carnegie Institute of Technology. **1940** *Nature* 22 June 982/1 Three klydonographs were coupled to the line by means of concentric-cylinder type capacitance 'potential-dividers'. **1963** D. J. Malan *Physics of Lightning* xi. 92 Since the Klydonograph indicates voltage, it is necessary to know the surge impedance of the system of conductors in order to calculate the peak current.

klystron (klai·strǫn). *Electronics.* Also **Klystron** (now *rare*). [f. Gr. κλύζειν (stem κλυσ-) to wash or break over + *-tron.] An electron tube for amplifying or generating microwave signals in which a beam of electrons from a thermionic cathode is passed through a gap in a cavity resonator across which is applied a high-frequency voltage, so that the electrons collect into bunches and on reaching a second gap induce a (larger) high-frequency voltage across it. Freq. *attrib.*, as *klystron oscillator, tube.*
1939 R. H. & S. F. Varian in *Jrnl. Appl. Physics* X. 324/1 Such an apparatus we call a 'klystron', from the Greek verb 'klyzo', expressing the breaking of waves on a beach. **1945** H. D. Smyth *Gen. Acct. Devel. Atomic Energy Mil. Purposes* xi. 118 The varying electric field..introduces small, periodic variations in ion velocity, and has the effect of causing the ions to 'bunch' at a certain distance down the tube. (This same principle is used in the klystron high-frequency oscillator...) **1959** G. Troup *Masers* i. 1 In conventional microwave amplifiers and oscillators such as the Klystron, an alternating electromagnetic field interacts with elementary particles, electrons, by virtue of their charge. **1959** *New Scientist* 19 Nov. 982/3 The Jodrell radar pulses originated in a big ultra-high-frequency valve of the type known as a klystron. **1964** *Listener* 16 Apr. 626 (Advt.), EEV 25 kW klystrons are being supplied to the British Broadcasting Corporation for use in the new UHF television transmitters. **1973** *Sci. Amer.* Sept. 74/3 The microwaves are produced by a pair of 30-kilowatt klystron tubes powered by a diesel generator.

K(-)meson. Also (*rare*) **k meson.** *Nuclear Physics.* [f. *K* (in *K(-)particle) + *meson[2].] = *kaon.
1954 *Phil. Mag.* XLV. 1219 The mass of the K-meson, measured by ionization-range is 970±100 m_e. **1954** *Physical Rev.* XCIV. 1794/2 All the *K* mesons observed lived at least 2×10^{-8} sec before coming to rest in the emulsion. **1964** *Listener* 30 Apr. 711/1 Usually the beams consist of particles called pi-mesons which have 270 times the mass of the electron, or the more exotic K-mesons with about 1,000 electron masses. K-meson beams have been particularly valuable recently in the production of further new particles. **1968** M. S. Livingston *Particle Physics* iv. 81 The new particles were of two basic types: *k* mesons with mass values less than nucleon mass, now called kaons; and hyperons, with masses greater than the nucleon mass. **1973** L. J. Tassie *Physics Elem. Particles* ix. 84 There are two kinds of neutral *K*-mesons, the K^0 and the $\bar{K}^0$.
Hence **K-me·sic** *a.* = *kaonic *a.*
1958 *Phil. Mag.* III. 33 Negative K-mesons captured by the heavy nuclei in nuclear emulsion form K-mesic atoms. **1967** *Comments Nucl. & Particle Physics* I. 112 *K*-mesic x-rays from Li, Be, B, and C.

knacker, *sb.*[1] Add: **3.** *pl.* The testicles. *slang.*
1866 T. Edmondston *Etym. Gloss. Shetland & Orkney Dial.* 76 *Nackers*, testes, S. **1877** in E. Peacock *Gloss. Words Manley & Corringham, Lincolnshire* 150/2. **1889** Barrère & Leland *Dict. Slang* I. 523/2 *Knackers*.. (Butchers, &c.), the testicles, also 'knuckers'. **1922** Joyce *Ulysses* 576 Eh, Harry, give him a kick in the knackers. **1940** J. Cary *Charley is my Darling* xlv. 261 I'll murder the bastards.... I'll take the knackers offen them. **1951** *Landfall* Sept. 177 Sling your hook out of this dump before it gets you by the knackers. **1969** G. Greene *Travels with my Aunt* i. v. 42, I may regret him for a while tonight. His knackers were superb.

knacker, *sb.*[3] Add: **2.** *knacker's yard*: Also *transf.* and *fig.*
1961 F. H. Burgess *Dict. Sailing* 128 *Knacker's yard*, the shipbreaker's yard. **1966** 'L. Lane' *ABZ of Scouse* 59 *Knacker's yard*: said of a place that looks a complete mess. **1967** T. Gunn *Touch* 42 The graveyard is the sea... They have all come who sought distinction hard To this universal knacker's yard.

knacker (næ·kəɹ), *v. slang.* [f. Knacker[3] 2 or *knacker *sb.*[1] 3.] *trans.* To kill; to castrate; usu. in weakened sense, to exhaust, to wear out. So as an imprecation. Freq. as *pa. pple.* or *ppl. a.*
1886 H. Baumann *Londinismen* 90/2 *Knacker*, umbringen; he's knackered, er ist abgemurkst worden. **1936** B. Penton *Inheritors* ix. 72 Coons is cheap. They'd knacker us white bushmen if they got the chance and let them Chows and Jimmy Tannas breed like rabbits. **1946** *Penguin New Writing* XXVII. 79 His eyes narrowed but he knew I had him knackered. **1959** M. Pugh *Chancer* vi. 86 'Wasn't it Major Fleming with the Bren-gun?' 'Major Fleming be knackered. It was Ramsay.' **1963** *New Society* 22 Aug. 5/1 Other adoptions are 'get knotted' and 'knackered' which have come to mean innocently enough, 'go to hell', and '*kaput*'. **1971** B. W. Aldiss *Soldier Erect* 258 Gor-Blimey came up, panting like a dog. 'I'm knackered,' he said. Blood was streaming down his face from a cut on his temple. **1971** *Times* 21 May 8/7, I kept thinking I should whip up the pace and then I'd think 'I'm knackered, I'll leave it for another lap.' **1973** C. Bonington *Next Horizon* xxi. 283 We've been above Base Camp for twenty-eight days. If we had to go back to carrying now we'd have to go all the way back down for a rest. We're just too knackered to carry. **1975** *Sunday Times* (Colour Suppl.) 23 Feb. 25/2 Oot a' mornin' daein' thae miracles. I'm *knackered*! Gie's a glass o' that wine. Nae kiddin' son, I'm knackered.

knackwurst (na·kvūɹst). Also **knockwurst.** [Ger.] A type of German sausage.
1939 M. Brinig *Anne Minton's Life* (1940) 241 This very second, with us full of beer and knackwurst and sauerkraut.., there's nothing to worry about. **1965** *House & Garden* Jan. 60 *Knockwurst, knackwurst*, a sausage similar to but larger than the frankfurter, and more highly seasoned. **1968** *Washington Post* 5 July A 19/3 (Advt.), Foremost knockwurst, kosher style. **1968** A. Whitney *Every Man has his Price* xvii. 147 The waiter came, and they ordered—Knackwurst for Deb, and a dish improbably called Kalbshaxe for Robin. **1970** 'E. Lathen' *Pick up Sticks* (1971) x. 86 Beer and knockwurst. **1971** *Sunday Times* (Colour Suppl.) 27 June 50/2 *Knackwurst*: not unlike a thicker Frankfurter lightly flavoured with garlic, and also best served hot.

knaidel (knē̆i·dəl). Also **knaydl.** Usu. in pl. **knaidlach** (knē̆i·dlaχ), **kneidlach.** [f. Yiddish *kneydel*, ad. MHG. (and mod.G.) *knödel* *knödel.] A type of dumpling eaten esp. in Jewish households during Passover. Also *transf.*
1951 L. W. Leonard *Jewish Cookery* i. 6 The *Seder* is a home festival, and every part of it, from *Haggada* to *Knaidlach*, has been enriched with traditions that are passed along to the children from generation to generation. **1955** —— *Jewish Holiday Cook Bk.* 87 Drop..into rapidly boiling clear soup..and cook..till Knaidlach rise to the top. **1960** S. H. Rivkin *Mama's Meichulim: Trad. Jewish Cooking* 150 Looking down at our faces after the long prayers, Papa would say perceptively, 'Children! They are not so hungry for religion as they are for knaidlach.' **1968** L. Rosten *Joys of Yiddish* 185 *Knaydl* is used affectionately for a child..or to describe a round, fat, chubby woman. **1973** *Times* 3 Feb. 13/4 You can try the kneidlach soup (with matzo-meal dumplings).

knall-gas (knæ·l,gæs). *Chem.* Also **knallgas** and with capital initial. [a. G. *knallgas*, f. *knall* bang, detonation + *gas* gas.] Any explosive mixture of gases, esp. one of two volumes of hydrogen with one of oxygen.
1899 R. E. Baynes tr. *Meyer's Kinetic Theory of Gases* iii. 86 (*table*) Knallgas. [*Note*] This is the mixture of hydrogen and oxygen produced by electrically decomposing water. **1923** *Jrnl. Chem. Soc.* CXXIII. 1026 Berthelot measured the *initial* rates of the flame in hydrogen and in carbon monoxide Knall-gas..; he also made a similar series with hydrogen and with ethylene Knall-gas. **1924** *Nature* 8 Mar. 373/2 The explosion of an undiluted $2CO+O_2$ 'knall-gas'. **1927** Bone & Townend *Flame & Combustion in Gases* xxxiv. 432 In cases where excess of either hydrogen or oxygen was present in the system, the observed rate was still..proportional to the partial pressure of the '*knall-gas*' present. **1969** Ottaway & Irvine tr. *Netter's Theoret. Biochem.* iii. 72 The reverse reaction (the Knall-gas reaction) $2H_2+O_2=2H_2O$ releases the same energy [as is absorbed in the forward reaction].

knapsack. Add: **b. knapsack pump** = *knapsack sprayer*; **knapsack sprayer** (or **spray**), a sprayer consisting of a hand-held nozzle supplied from a pressurized reservoir that is carried on the back like a knapsack.
1894 *Country Gentlemen's Catal.* 302/2 Vermorel's knapsack pumps. **1944** *Living off Land* vii. 153 Firefighting weapons are many and varied... The knapsack pump—a 4-gallon container with shoulder straps, brass hand pump and nozzle—is most effective. **1897** *Sears Roebuck Catal.* 162/3 The Celebrated Myers Knapsack Spray Pump. **1909** *Cent. Dict.* Suppl. s.v. *sprayer, Knapsack sprayer.* **1950** *N.Z. Jrnl. Agric.* Feb. 177/2 Obtaining complete coverage of tall [orchard] trees with a knapsack sprayer is most difficult. **1971** *Community* (E. Afr. Community) Apr. 8/1 The new method could well replace the present widespread use of knapsack sprayers which was considered inefficient in insecticide application.

knapsacked (næ·psækt), *a.* [f. Knapsack + -ed[2].] Equipped with a knapsack.
1905 *Westm. Gaz.* 26 Aug. 3/1 The knapsacked mountaineers come and go. **1908** Hardy *Dynasts* III. ii. ii. 76 The sunset slants an ochreous shine Upon the English knapsacked line. **1926** R. Macaulay *Crewe Train* i. i. 5 Knapsacked British walkers.

knarr, var. *knorr.

knaydl, var. *knaidel.

knead, *v.* Add: **3. b.** *trans.* and *intr.* To manipulate or paw repetitively with or as with the action of (the claws of) a cat.
1954 G. Durrell *Bafut Beagles* iii. 69 The cloud seemed to move,..padding and kneading the mountain crests like a cat on the arm of a gigantic chair. **1967** 'T. Wells' *Dead by Light of Moon* (1968) ii. 27 He..began to purr and knead at the blanket. **1968** V. Canning *Melting Man* vi. 144 The cat woke me by kneading determinedly on my chest. **1968** R. Sawkins *Snow along Border* xii. 102 It began kneading dough, claws exposed.

knee, *sb.* Add: **1. b.** A damaged condition of the knee. Cf. *housemaid's knee* (Housemaid c in Dict. and Suppl.), *tennis-knee* (Tennis *sb.* 3 b).
1921 J. C. Jenkins in E. H. D. Sewell *Rugby Football* ix. 195 Unfortunately developed a 'knee' and had to retire in his prime. **1922** Joyce *Ulysses* 127 'Are you hurt? I'm in a hurry.' 'Knee,' Lenehan said. He made a comic face and whined, rubbing his knee. **1971** I. Peebles *Denis Compton* x. 97 Denis, handicapped by his knee, was no longer able to get down the pitch to the slower bowlers.
2. d. *across one's knee*, (of someone, esp. a child) placed face-down on the knee(s) to be spanked.
1866 [see Whipping *vbl. sb.* 1 a]. **1916** 'Taffrail' *Pincher Martin* ix. 154 If yer don't stop it I'll put yer across my knee an' give yer wot for. **1936** 'N. Blake' *Thou Shell of Death* xiii. 230 Manny's the time I've had him across me knee—and Miss Judith, too—and belted them with a slipper. **1959** I. Fleming *Goldfinger* ix. 126 This one has got to go dead or I'll put you across my knee.
11*. An abrupt obtuse or approximately right-angled bend in a graph between parts where the slope varies smoothly.
1880 *Proc. R. Soc.* XXX. 513 An interval of constant stress of even five seconds produces a perceptible 'knee' in the curve. *Ibid.* 514 We get a stepped curve, having a number of 'knees' upon it. **1904** *Physical Rev.* XIX. 114 On the rising curve there is seen to be a more or less well defined 'knee' where the relation of stress to strain undergoes a marked change. This 'knee' might be said to mark the elastic limit. **1926** R. W. Hutchinson *First Course Wireless* viii. 144 Consider now the parts of the curve where the bending is greatest, i.e. the 'knees'. **1957** G. E. Hutchinson *Treat. Limnol.* I. vii. 429 From the upper plane of maximum curvature, termed by Munk and Anderson the knee of the thermocline, to the lower plane of maximum (inverse) curvature. **1967** L. G. Lawrence *Electronics in Oceanogr.* iii. 56 The arrangement makes special use of the knowledge that the saturation 'knee' of a *B-H* loop of a given material can be modified by allowing the magnetism of the earth to contribute to the effective operating point of this knee.
12. *knee-buckle* (earlier Amer. example), *-grip, -pad, -pants, -room, -sock, -trousers* (U.S.); *kneewards* adv.
1754 *South Carolina Gaz.* 1–8 Jan. 4/3 To be sold...shoe and knee buckles, snuffers, gun hammers. **1903** *Westm. Gaz.* 12 Feb. 2/4 The bridle-rein light in the hand, The knee-grip steady and sure. **1925** E. T. Brown *Compl. Motor-Cyclist* 126 The non-essential accessories include a luggage grid, speedometer, leg-shields.., knee-grips, handle-bar gloves. **1858** J. A. Warder *Hedges & Evergreens* 71 As it is easier to work on your knees, you will provide thick knee-pads for them. **1955** E. Pound *Classic Anthol.* i. 71 Saw I white knee-pads decent misery I'd know one man still feels and thinks as I. **1972** P. Driscoll *Wilby Conspiracy* (1973) xi. 145 September, on all fours

with a pair of rubber knee-pads on, was vigorously polishing the slate floor. **1869** *Atlantic Monthly* July 74/2, I made my initial bow before the foot-lights, in my small Canton flannel knee-pants. **1916** *Daily Colonist* (Victoria, B.C.) 1 July 12/4 (Advt.), Boys' Straight Knee Pants, of good quality English tweeds. **1942** *Short Guide Gr. Brit.* (U.S. War Dept.) 20 There are..youngsters in knee pants ..who have lived through more high explosives..than many soldiers saw..in the last war. **1969** Widdowson & Halpert in Halpert & Story *Christmas Mumming in Newfoundland* 162 More modern costumes reported alongside the older disguises include service uniforms, ice-hockey clothing (presumably with padded shoulders and knee pants), [etc.]. **1958** *Times* 19 Aug. 11/6 The head-room is only just sufficient, and the same reservation applies to the kneeroom in the back when the driver's seat is pushed back for a fairly tall driver. **1970** *Times* 16 Apr. 18 More front headroom and rear kneeroom could be devised by reducing the bulk of their cushions and back-rests. **1964** *Punch* 19 Aug. 284/3 Green plaid knee-socks end in sling-back shoes. **1966** T. Pynchon *Crying of Lot 49* ii. 41 A long-leg girdle and a couple pairs of knee socks. **1899** T. Hall *Tales* 162 Since she was a little girl in short dresses and he a boy in knee trousers. **1926** *Brit. Weekly* 24 June 250/5 On the patterned skirt the design grew larger as it reached kneewards. **1968** G. Jones *Hist. Vikings* iii. iv. 255 Ibn Rustah notes the full baggy trousers gathered kneewards vouched for by Scandinavian picture stones.
13. knee-action, (*a*) in a horse, the action or coordination of movement of the knee joint; (*b*) exaggerated raising of the knee by an athlete; (*c*) in motor vehicles, a form of independent front-wheel suspension; **knee-bend**, the action of bending the (human) knee, esp. used of a physical exercise in which the body is raised and lowered without use of the hands; so **knee-bend** *v. intr.*; **knee-board**, (*b*) in a cotton-yarn winding-machine (see quot.); **knee-brace** *Engin.*, a strut fixed diagonally between the lower chord of a truss and one of its supporting columns; hence **knee-braced** *ppl. a.*, **-bracing** *vbl. sb.*; **knee-breech**, sing. of *knee-breeches*; **kneecap** *v. trans.*, to shoot a person in the knee (or leg) as a form of punishment; so **kneecapping** *vbl. sb.*; **knee-chest position**, a position adopted by some women in sexual intercourse (see quot. 1936); **knee-hobbling** *vbl. sb.*, fastening an animal's knees with a hobble; **knee-jerk**, also *attrib.*, and *fig.*, predictable, automatic, stereotyped; **knee-length** *attrib.*, reaching down (or up) to the knee; also *ellipt.*, (a garment of) such a length; **knee-plate**, (*b*) *Shipbuilding*, an angled metal plate used as a knee (sense 7 a); **knee-slapper** *U.S.*, an uproariously funny joke; **knee-sprung** *a. Farriery* (see quot. 1905); **knee-stake** *v. trans.*, in *Leather manuf.*, to soften (a skin) by aid of the knee; **knee-trembler** *slang*, an act of sexual intercourse between persons in a standing position (so **knee-tremble**).
1868 H. W. Woodruff *Trotting Horse* iv. 62 [The colt] continually hit himself in the elbows, by reason of excessive knee-action as it appeared. **1903** A. Adams *Log of Cowboy* xv. 100 They will discuss how to shoe that filly so as to give her certain knee action which she seems to need. **1908** *Westm. Gaz.* 22 July 2/1 My action is low and sweeping, mainly from the hips, but many men..have a lot of what may be called 'knee-action'. **1935** A. C. Baugh *Hist. Eng. Lang.* x. 370 Of late we have heard a good bit about *free-wheeling, safety-glass, knee-action*, while *service stations* and *tourist camps* are everywhere along the road. **1963** R. F. Webb *Motorists' Dict.* 144 *Knee-action suspension*... The front wheels are supported on upper and lower radius arms, the upper of which has an action like the human knee joint where it joins the king pin link support. **1941** *Penguin New Writing* IX. 62, I practise the knee-bend, the stare, and the slow roll. **1961** A. Miller *Misfits* xi. 119 Guido half knee-bends with his rope over his thighs and pulls. **1963** I. Fleming *On H.M. Secret Service* xi. 117 He proceeded to a quarter of an hour of knee-bends and press-ups. **1972** *Village Voice* (N.Y.) 1 June 26/2 La Lanne moves from knee bends and neck stretches to pitching his mattresses and reducing aids without the slightest break in his pace or enthusiasm. **1895** R. Marsden *Cotton Weaving* 257 The board.. generally called the knee-board, an incorrect name if regard be had to its function. This board is usually covered with flannel, and forms a check upon the too easy delivery of the yarn to the draught of the spindle, thereby securing uniformity of tension in the winding. **1912** A. Morley *Theory of Struct.* xv. 423 The kneebraces meeting the stanchions 4·75 feet below the caps. **1959** L. C. Urquhart *Civil Engin. Handbk.* (ed. 4) v. 3 Frequently.. trusses are stiffened in their own vertical planes by inserting knee braces at both ends between the bottom chord and supporting columns. **1915** H. R. Thayer *Struct. Design* II. xii. 448 (*heading*) The knee-braced steel frame. **1940** *Archit. Rev.* Mar. 102/2 The roof unit and ceiling members..in the 27 ft. wide blocks [form] a knee-braced truss. **1950** *Engineering* 31 Mar. 366/1 By using a knee-braced portal structure, broad-flange beams may be employed for spans up to about 70 ft. **1912** H. R. Thayer *Struct. Design* I. iii. 53 It is sometimes necessary to use knee bracing..but it is not as strong as the X bracing.. and it introduces large bending stresses. **1904** *Daily Chron.* 11 Oct. 3/5 Men do not dress now, they merely clothe themselves, and they will not alter this fact by adopting the knee-breech. **1927** *Observer* 22 May 12 (*heading*) A blow to the knee-breech crusade. **1975** *Daily Tel.* 12 Aug. 2/7 Man 'kneecapped' in Carrickfergus. **1975** *Observer* 8 June 4/3 Ulster's gunmen have found they can get hold of Government cash by giving victims a 'knee-capping'—their grim colloquialism for a bullet in the legs. .. Kneecapping..has replaced tarring and feathering as the province's most common form of terrorist punishment... 'This so-called kneecapping is really a misnomer, because the kneecap itself is rarely touched.' **1936** H. M. & A. Stone *Marriage Manual* vii. 249 The woman in the so-called 'knee-chest' position, that is kneeling face downward. **1968** R. Kyle *Love Lab.* ix. 127 The arm can be set for only two positions, the supine and knee-chest. **1908** *Animal Managem.* 150 Grazing should be afforded at every opportunity, and for this purpose knee-hobbling is the best plan to adopt. **1951** J. Holloway *Lang. & Intelligence* v. 79 The knee-jerk reflex may be more or less rapid. **1963** *N.Y. Times* 7 Oct. 30 The place has always been full of liberals... In Washington, we call them crackpots, knee-jerks, do-gooders. **1969** *Time* 30 May 22/3 'What you have here,' he said, 'is the opposite of the knee-jerk liberal—the knee-jerk conservative.' **1970** *Daily Tel.* 2 June 19 In spite of knee-jerk reactions speculating on a Swiss franc revaluation, the Swiss franc never reached its 'ceiling' against the dollar. **1973** *Washington Post* 5 Apr. B 2 There is also some tired business about 'educating abroad' and a knee-jerk Mafia joke because Candoli is Italian. **1895** *Montgomery Ward Catal.* 483/3 Horsehide leggings, either 'knee' or 'thigh' lengths. **1897** *Sears, Roebuck Catal.* 240/1 Ladies'..Union suit..shaped form fitting waist, knee length. **1909** *Daily Chron.* 6 Jan. 7/1 Both sexes wear deerskin breeches and knee-length coats. **1922** Joyce *Ulysses* 517 To lace up crisscrossed to knee-length the dressy kid footwear. **1958** B. Nichols *Sweet & Twenties* xvi. 206 The taste of the twenties was not entirely represented by knee-length frocks and bobbed fringes. **1966** *Guardian* 25 July 6/2 A hovering knee-length is generally expected to be the winter norm. **1967** *Punch* 4 Jan. 1/1 The lengths of female laid bare by minis are apt to be covered anew by costly knee- and thigh-length boots, thick tights and miscellaneous 'warms'. **1858** J. Grantham *Iron Ship-Building* 217 Bulkheads to be five in number;.. to have brackets, or knee plates, riveted horizontally against the ship's side. **1969** *Jrnl. Abstr. Brit. Ship. Res. Assoc.* XXIV. 218 (*heading*) Determination of the effectiveness of a knee plate by plastic theory. **1966** *New Yorker* 5 Nov. 128 'How's the World Treating You', an English comedy at the Music Box, is full of knee-slappers like that one. **1970** W. Burroughs, Jr. *Speed* 84, I needed a phone book which the guard thought a real knee slapper. **1875** *Scribner's Monthly* June 208/1 Particularly when that animal's foundered and knee-sprung. **1905** J. W. Axe *Horse* I. 74 When the knee is displaced forward in advance of the vertical line it is said to be 'bowed', or the horse 'stands over', 'knee sprung'. **1903** L. A. Flemming *Pract. Tanning* 51 When in just the right condition, the skins are knee-staked for the purpose of softening them to get rid of the stretch. **1896** Farmer & Henley *Slang* IV. 119/1 *Knee-trembler*, a standing embrace; a fast-fuck; a perpendicular. **1965** G. Melly *Owning-Up* vi. 67 A member of the band..gave her a knee tremble at the back of the building. **1966** F. Shaw et al. *Lern Yerself Scouse* 62 *We wen up der jigger fera kneetrembler*, we went courting in lovers' lane. **1971** B. W. Aldiss *Soldier Erect* 18 They would be going to the pub for a pint and afterwards Nelson would get her against our back wall for a knee-trembler... He claimed that knee-tremblers were the most exhausting way of having sex.

knee, *v.* Add: **3.** *spec.*, to strike a person (esp. in the groin) deliberately with the knee. Also *fig.*, implying foul play.
1953 *Time* 20 July 13/1 Like most successful rough and tumble fighters, Senator Joe McCarthy always presses in, and is adept at forensic kneeing, gouging and butting. **1955** [see *butt-end *v.* 2]. **1967** K. Giles *Death in Diamonds* ix. 176 He belted the P.C., kneed another in the stomach and tried to bolt. **1968** 'R. Raine' *Night of Hawk* xxxvi. 174, I..knee'd him in the groin. **1972** J. Mosedale *Football* ix. 122 Guyon..spun round and kneed Halas, breaking three of his ribs. **1973** *N.Y. Times* 6 Oct. 4/5 One plainclothesman repeatedly kneeing Mr Ogden in the back.
b. To urge (a horse) on by pressing the knees against its flanks. *U.S.*
1924 C. E. Mulford *Rustlers' Valley* iii. 33 Then he.. turned his own animal southward and kneed it forward. **1926** —— *Cassidy's Protégé* x. 133 The herder,..kneeing his horse, rode swiftly back and forth several times for a hundred feet each way.
7. To renew the knees of (a garment). *U.S.* and *dial.*
1847 H. Howe *Hist. Coll. Ohio* 348 After wearing out their woollen pantaloons, [they] were obliged to have them seated and kneed with buckskin. **1891** R. Kerr *Maggie o' Moss* 36 Corduroys! and them sae clouted, Backside, foreside, knee'd an a'.

kneed, *a.* **1.** Also, *feeble-kneed.*
1904 E. M. Forster in *Independent Rev.* Mar. 280 They are so weak-chested and anæmic and feeble-kneed.

knee-halter, *v.* Add: (Earlier and later examples.) So **knee-haltering** *vbl. sb.*
1835 A. Steedman *Wanderings S. Afr.* I. vii. 215 While the servants knee-haltered and watched the horses, we returned on foot to where the corpse lay. **1908** *Animal Managem.* 126 The practice of grazing may be taken advantage of to accustom horses to knee haltering. **1926** T. E. Lawrence *Seven Pillars* (1935) xi. 85, I was rolled up in my cloak and asleep in a most comfortable little sand-grave before Tafas had done knee-haltering my camel. **1939** S. Cloete *Watch for Dawn* ii. 23 Dismounting, he off-saddled, knee-haltered his horse.

knee-high, *a.* Reaching as high as the knees. Freq. in jocular phrase (orig. U.S.) *knee-high to a grasshopper* (and varr.), i.e. very short.
1743 W. Ellis *Mod. Husbandman* Aug. xvi. 64 By Michaelmas following they were Knee high. **1799** C. B. Brown *Arthur Mervyn* I. xv. 141, I never cried in my life, since I was knee-high. **1814** *Portsmouth* (New Hampsh.) *Oracle* 2 Apr. 3/2 One..who, as farmer Joe would say, is 'about knee high to a toad'. **1824** *Microscope* (Albany, N.Y.) 12 June 55/1 (Th.), He has lived with me ever since he was 'knee high to a musquitoe'. **1833** *Louisville Herald* 20 Mar., It is really the best version of an old story we have heard 'ever since we were knee-high to a frog'. **1833** J. Neal *Down-Easters* I. 78 A bit of a rogue he was too, when he wa'n't more'n knee high to a bumbly-bee. **1841** W. G. Simms *Kinsmen* II. 63 (Th.), Ever since I was knee high to a splinter. **1843** [in Dict. s.v. Knee *sb.* 12]. **1851** *Democratic Rev.* XXVIII. 301 You pretend to be my daddies; some of you who are not knee-high to a grasshopper! **1887** *Harper's Mag.* Oct. 754/2 Their myriads of gray trunks stood knee-high in water. **1892** *Dialect Notes* I. 230 Knee high to a duck. **1925** R. Graves *Welchman's Hose* 25 He gibed at modern poets, 'Show me one Knee-high in stature to a Tennyson.' **1937** *Discovery* June 170/2 The grass grows in thick tufts and is knee high. **1942** *R.A.F. Jrnl.* 18 Apr. 2 Air gunners..envisage him [*sc.* a gremlin] as something fairly big—say, knee-high to an air-gunner. **1957** I. Cross *God Boy* (1958) xxii. 187 Sister Theresa, who is not much more than knee-high to a grasshopper. **1973** H. Carvic *Miss Seeton Sings* (1974) 35 A little Italian cock sparrow about knee-high to a grasshopper.

kneeler. Add: **4.** (Later examples.) Also *attrib.* in *kneeler chair.*
1909 M. B. Saunders *Litany Lane* i. xi, She would follow the Stations of the Cross with a slow dreaminess, and lean longer over her kneeler chair when the services had finished. **1969** E. H. Pinto *Treen* 94 The traditional U-shaped elm kneeling mat or 'kneeler'..had many uses on the farm. It was also used until early in this century in the house. **1972** S. Burnford *One Woman's Arctic* (1973) iii. 70 Some beautiful inset work sewn in to the sealskin kneelers.

kneeling, *vbl. sb.* Add: **3.** *kneeling-desk* (earlier Amer. example), *-mat, -stool* (earlier and later examples).
1647 in *Archives of Maryland* (1887) IV. 321 A kneeling desk, & a picture of Paules. **1907** *Yesterday's Shopping* (1969) 231/4, 2 Housemaid's Kneeling Mats—0 2[s] 6[d]. **1969** Kneeling mat [see *kneeler 4]. **1844** C. M. Yonge *Abbeychurch* v. 64 Do you not remember how much trouble Rupert took to find a pattern for the kneeling-stools? **1954** W. Hannah *Christian by Degrees* xii. 167 To the west of the Sepulchre is a kneeling-stool.

knees up, Mother Brown. A light-hearted popular song beginning thus; a popular dance in which the knees are vigorously raised to the accompaniment of the song. So *ellipt.*, as **knees-up** *sb.*, *spec.* a lively party or gathering. Also occas. in extended uses.
1939 Weston & Lee *Knees up Mother Brown!* 3 Ooh! Knees up Mother Brown! Well! Knees up Mother Brown!..knees up, knees up! Don't get the breeze up Knees up Mother Brown. **1945** *Daily Mirror* 8 May 1/2 We are dancing the Conga and the jig and 'Knees up, Mother Brown'. **1958** *Times* 15 Aug. 9/4 'Knees up, Mother Brown' is an injunction to apprehend nothing but jollity. **1961** A. Wilson *Old Men at Zoo* iv. 213 Matthew ..was involved with a circle of old women out on the spree who were doing 'Knees up Mother Brown'. **1963** P. Willmott *Evolution of Community* vii. 75 We went to another house on the banjo for a 'knees-up'. **1966** K. Martin *Father Figures* iv. 88, I happened to get leave on Armistice Day and crossed to London in time to witness the glorification in Trafalgar Square. We danced round and round all night long, singing 'Knees up, Mother Brown' and other fragments from English folklore. **1966** A. Prior *Operators* xvi. 259 He..turned to see Emmie doing a knees-up, her skirt held high. **1967** L. Deighton *London Dossier* 54 As indigenous to London as a Saturday-night knees-up in the boozer. **1969** *Guardian* 31 Jan. 22/3 An irrepressible bunch of girls chanted 'Knees up Barbara Castle'. **1971** *Ibid.* 18 Sept. 8/5 He and Mr [Harold] Wilson finished up doing Knees Up Mother Brown together. **1974** *New Scientist* 3 Oct. 39/1 Two new video discs were demonstrated last month during the annual video industry knees-up at Cannes. **1975** *Oxford Times* 3 Jan. 16/5 (Advt.), Saturday January 4th. 8 pm. (New Year's Knees-Up).

kneesy, -ie (nī·zi). *colloq.* [Jocular dim. of *knees*: see -y[6].] Amorous play with the knees; the relationship implied by such activity. Also redupl.
1951 'M. Spillane' *Big Kill* v. 102 We got back to the table and played kneesies while we talked. **1954** A. Melville *Simon & Laura* in *Plays of Year* XI. 83 Ramming Le Touquet down *my* throat, as though you'd never heard of kneesy-kneesy under a table at the Café de Paris. **1970** *New Yorker* 26 Sept. 36/3 Ha! No kneesies, no invitations to a midtown matinée?

kneidlach: see *knaidel.

‖ **Kneipe** (knəi·pə). Also *erron.* **Kneip, Knipe.** Pl. **-en, -es.** [G.] A convivial meeting of German university students (and the like) at a tavern or restaurant. So **kneip** *v. intr.* [after G. *kneipen*], to indulge in this conviviality.
1854 I. Aldridge *Diary* 9 June in Marshall & Stock *Ira Aldridge* (1958) 189 Visited a Knipe or a Society of German Students. **1864** H. Mayhew *German Life & Manners* II. 243 Youths whose lives are apparently given up to the mere conviviality of '*Kneiping*' or beer-drinking.

1874 J. M. Hart *German Univ.* ix. 139 In whatever other respects the German student may be irregular, he always *kneips* according to rule. **1880** 'Mark Twain' *Tramp Abroad* iv. 43 Kneips are held, now and then [at the University of Heidelberg], to celebrate great occasions,—like the election of a beer king. **1911** R. Brooke *Let.* 31 Jan. (1968) 275 Last night he took me to a '*Kneipe*'. The students who are working at some special subject band together to form a club... Every Monday evening this Verein has a 'Kneipe', a meeting. **1924** A. Geikie *Long Life's Work* vii. 217 After the meetings during the day, every night a 'Kneipe' at which Zirkel, Lossen, Reusch.. and a host of younger men took part.

Kneipp (nəip). The name of Sebastian *Kneipp* (1821–97), Bavarian priest, used *attrib.* to designate (a system of) hydropathic treatments advocated by him, a special feature of which was walking barefoot through dewy grass. Hence **Knei·ppism.**
[**1891** A. de F. tr. *Kneipp's My Water-Cure* p. xviii, It is evident that in Germany, at least, Pfarrer Kneipp's cure is going to influence the present state of medicine to a considerable extent.] **1895** *N.Y. Med. Jrnl.* LXII. 523/2 Is the Kneipp cure injurious? **1900** Dorland *Med. Dict.* 341/1 Kneippism. **1901** J. H. Kellogg *Rational Hydrotherapy* i. 28 The leading features of the so-called 'Kneippism' are simply a revival of these rude practices of ignorant peasants a century and a half ago. **1911** Stedman *Med. Dict.* 454/2 *Kneipp method*, the treatment of disease by hydrotherapy in various forms—douches, wet packs, full and local baths, compresses, vapor baths, walking barefoot in the dew of early morning, etc. **1933** H. F. Wolf et al. *Textbk. Physical Therapy* xxiii. 268 Such headaches are promptly relieved..by walking barefooted in wet grass (Kneipp cure). **1966** *Punch* 2 Feb. 161/2 Bavaria also offers a range of Kneipp cures. The good Father Kneipp..knew that walking barefoot on wet stone floors would cure constipation.

Knesset (kne·sėt). [Heb., lit. gathering.] The parliament of the State of Israel. Also *attrib.*
1949 *Jewish Chron.* 11 Feb. 1/2 On Monday next Israel's newly elected Constituent Assembly, to be known as the Knesset Gedola, will hold its first meeting in Jerusalem. **1956** *Ann. Reg. 1955* 283 The General Election for the Knesset. **1959** *New Statesman* 14 Nov. 648/1 At the other political pole, the Communists slumped heavily, losing three of their six Knesset seats. **1972** *Guardian* 27 Mar. 4/4 Arab propagandists are wrong in claiming that a map hangs in the Knesset building showing the land from the Nile to the Euphrates.

knick (nik), *sb.*[2] *Geomorphol.* Also **nick.** [a. G. *knick* bend, kink, break.] **a.** = *knickpoint.
1932 *Bull. Geol. Soc. Amer.* XLIII. 416 If one convex nick can not be produced in the profile of a stream that is eroding its valley in an enlarging dome of continuously accelerated upheaval, all the less can a series of nicks be produced. **1941** C. A. Cotton *Landscape* xx. 233 Even where nicks occur at resistant outcrops, this is quite commonly a result merely of retardation at such points of headward erosion due to rejuvenation. The most widely accepted explanation of nicks in valley profiles is that they are the effects of successive lowerings of base-level. **1970** I. Cornwall *Ice Ages* i. 35 At the point where renewed erosion begins there will be a sharp increase in rate of fall, showing a 'knick', or downward break in gradient.
b. The angle formed by a pediment and the adjacent mountain slope.
1936 *Zeitschr. für Geomorphol.* IX. 132 Massive rocks, on weathering, produce steep mountain slopes because of their widely spaced joints. They are also characterized by a relatively sharp 'Knick', whereas in the same area more closely jointed rocks have lower slopes and there is a more gradual transition into the pediment at the base. **1952** *Ann. Assoc. Amer. Geogr.* XLII. 305 Johnson..pointed to the sharp angle (knick) between pediment and inselberg as an indication that lateral corrosion and not weathering-retreat is the cause of destruction of the feature. **1963** D. W. & E. E. Humphries tr. *Termier's Erosion & Sedimentation* ii. 36 Pediments are surfaces cut into hard rocks at the foot of mountains which they continue to erode. At their highest point, they join the mountain side at a break of slope called a 'knick'.

knicker[2]. Add: **1. b.** *attrib.* and *Comb.* *knicker fabric, hose, skirt; knicker-pink* adj.; **knicker yarn**, = *knickerbocker yarn* (*knickerbocker 3 c).
1974 *News of World* 22 Sept. 14/4 Faiman's in Southampton sold us a skirt in nasty 'knicker' fabric. **1899** *Northern Times* (Golspie, Sutherland) 22 June 1/2 (Advt.), In the Gents. Department..Hand-Knitted knicker hose a speciality. **1973** *Sunday Times* 7 Oct. 36/5 Alive and well in a spanking new knicker-pink emporium at 100 Mount Street. **1912** *Woman's Weekly* 27 Jan. p. iii/2 (Advt.), Knicker Skirts, Nightdresses, Pyjamas, &c. **1929** *Encycl. Brit.* XXIII. 879/2 'Knicker yarns' are produced by throwing little bits of highly coloured material into the last cylinders of the card so that instead of being broken up by carding they are carried forward as 'knickers' into the spun thread. **1951** *Good Housek. Home Encycl.* 327/2 'Knicker' yarns may be produced by including bits of coloured material in the final carding.
2. Usu. in pl. A short-legged (orig. knee-length), freq. loose-fitting, pair of pants worn by women and children as an undergarment. In extended use, the shorts worn by boxers, footballers, etc. (the application of the term to men's garments is now almost exclusively *U.S.*). Cf. *Directoire *a.* 1, *knicks *sb. pl.*
1882 *Queen* 7 Oct. 328/3, I recommend..flannel knickers in preference to flannel petticoat. **1895** *Home Chat* 2 Nov. 301/2 Serge knickers..for girls from twelve to sixteen. **1926** *Ibid.* 22 May 507 French Knicker made in Grafton's Voile and Grafton's Chiffonelle. Trimmed with lace. Elastic waistband. **1926** *Vogue* Late Nov. p. xxiii, An Original Directoire Knicker of milanese. **1928** G. B. Shaw *Intelligent Woman's Guide Socialism* i. 2 Laws..are amended and amended and amended like a child's knickers until there is hardly a shred of the first stuff left. **1938** Dylan Thomas *Let.* 16 June (1966) 201, I find a tripper's knicker in the gully. **1951** *Good Housek. Home Encycl.* 156/1 Iron knickers lengthwise, one leg at a time. **1954** F. C. Avis *Boxing Reference Dict.* 62 *Knickers*, boxing shorts. **1966** F. Shaw et al. *Lern Yerself Scouse* 49 *Ee's got both legs in one knicker*, he is not playing [football] well. **1968** 'C. Saint-Laurent' *Hist. Ladies Underwear* x. 145 (*caption*) Little girls wore short knickers before women did. **1974** *Guardian* 19 Feb. 1/5 Women workers in a lingerie factory..waved yellow knickers at Mr Thorpe as a gesture of support.
b. *pl.* as *int.* An expression of exasperation, surprise, contempt, etc.
1971 *TV Times* 23 Sept. 6/1 When things go wrong then I'll say: 'Knickers. I'll have another go.' **1974** *Pacifist* June 20/2 This is where the revolution's happening, man, and knickers to the metropolis! **1974** *Daily Tel.* 5 Oct. 5/3 Asked whether she would tell them more about the man.., she said: 'Knickers. I have told you all I know.'

knickerbocker. Add: **3. a.** (Later examples.)
1962 *Times* 14 Apr. 9/4 Plus-fours, or knickerbockers, as the Americans prefer to call them. **1967** *Daily Tel.* 26 Aug. 7/2 Boots were all around Paris and so were tweed.. knickerbockers as the logical successor to the now-almost-boring trouser suits. **1969** J. Laver *Concise Hist. Costume* ix. 251 The new baggy knickerbockers were known as 'plus-fours'.
b. = *knicker[2] 2.
1872 *Young Englishwoman* Oct. 554/2 Lady's longcloth *knickerbockers*. These drawers fasten behind. **1887** *Lady's World* Oct. 403 It [*sc.* a peasant's blouse] is girdled at the waist by a leather belt, and falls over the short woollen skirt, which just reaches the knees, where it meets the linen knickerbockers. **1895** *Home Chat* 20 Apr. 176 We spoke of satin knickerbockers in connection with this trousseau. **1913** B. L. Blackmore *ABC of Cutting Garments* 140 In girls' knickerbockers, the back band is sometimes buttoned to the front band, instead of the whole garment being attached to the under-bodice or to stays. **1969** R. T. Wilcox *Dict. Costume* (1970) 234 (*caption*) Muslin knickerbockers—girl's 6 to 8—buttoned at sides—pleated cambric frill.
c. *Comb.* **knickerbocker suit** (see sense 3 in Dict.); **knickerbocker yarn**, a yarn flecked with different colours.
1868 C. L. Eastlake *Hints Household Taste* xi. 264 The knickerbocker suit..has been adopted for wear in many country gentlemen's houses. **1879** C. M. Yonge *Burnt Out* iii. 52 A knickerbocker suit, just Charlie's size, had been turned over to his cousin Ada. **1911** *Encycl. Brit.* XXVIII. 906/2 *Flaked Yarn* has a cloudy appearance imparted to it..as in *Knickerbocker Yarn*, by dropping small quantities of dyed fibres into two..rovings at the spinning machine. **1932** E. Midgley *Technical Terms Textile Trade* II. 81 True 'knickerbocker' yarn is obtained by flecking the spotting material during carding. **1950** '*Mercury*' *Dict. Textile Terms* 303 *Knickerbocker tweeds*, rough-faced wool and cotton mixture dress goods made with nub yarns, in mixed colours. The yarn is known as knickerbocker yarn.
d. Knickerbocker Glory, a quantity of ice cream served with other ingredients in a tall glass.
1936 G. Greene *Gun for Sale* i. 11 Have a *parfait*... They do a very good Maiden's Dream. Not to speak of Alpine Glow. Or the Knickerbocker Glory. **1941** M. Treadgold *We couldn't leave Dinah* xvi. 256 'Lyons' Corner House,' capped Caroline, envisaging the increasing possibilities of Knickerbocker Glories. **1963** *Times* 25 Feb. 11/4 Knickerbocker Glory oddly has become the name of a specially luscious mixture of ice-creams and the 'plus fours' deserved a similar grandeur of title. **1973** *Times* 20 Oct. 12/7 At five in the morning it must have had all the charm of a pre-breakfast Knickerbocker Glory.

knickpoint (ni·kpoint). *Geomorphol.* Also **nickpoint** and as two words. [Partial tr. G. *knickpunkt*, f. *knick* (see *knick *sb.*[2]) + *punkt* point.] A break of slope in a river profile, esp. one where a new curve of erosion arising from rejuvenation intersects an earlier curve.
1924 *Bull. Geol. Soc. Amer.* XXXV. 638 Since erosion progresses headward when a region is uplifted by a uniform tilting movement.., the longitudinal profiles of those streams will record a knickpoint at the headward limit of quickened erosion. **1937** Wooldridge & Morgan *Physical Basis Geogr.* xv. 220 The rate of upstream recession of knickpoints will evidently vary with the character of rocks eroded. **1954** W. D. Thornbury *Princ. Geomorphol.* v. 110 Along some streams..nickpoints are found which do not seem to be related to more resistant rock. **1967** J. Challinor *Dict. Geol.* (ed. 3) 142/2 To the present writer it seems that nearly all knick points are due to some commoner and more compelling cause [than rejuvenation], such as the outcropping of a hard bed or glacial erosion. **1968** C. R. Twidale *Geomorphol.* vi. 190 Not all falls and rapids indicate nick points, for hard-rock strata crossing the river's path may well produce a break of slope. **1973** *Nature* 9 Nov. 75/1 Recession of valley nickpoints inland from the coastal margin was followed by valley widening, scarp retreat and pedimentation.

knicks, *sb. pl.* Colloq. shortening of *knickers* or *knickerbockers*.
1895 *Punch* 15 June 285/2 One young piece in grey knicks and cream cloth..took my fancy perdigious, dear boy. **1923** J. Manchon *Le Slang* 173 Knicks..s. pl., un pantalon de femme. **1937**, **1952** [see *cami-]. **1972** J. Wilson *Hide & Seek* vi. 105 Mary, if you'd just stand up I could take your wet knicks off.

knife, *sb.* Add: **1. e.** (Earlier and later examples.) Also *while (one) would say knife.*
1874 M. Clarke *His Natural Life* II. iii. viii. 170 He was over the wall before you could say 'knife'. **1922** Joyce *Ulysses* 158 Toss off a glass of brandy neat while you'd say knife. **1954** A. Macrae *Both Ends Meet* in *Plays of Year* X. 509 With a couple like that you'll be in a lawsuit before you can say 'knife'. **1973** M. Muggeridge *Infernal Grove* i. 71 Like alcoholics after taking the cure—never another drop; well, just a taste perhaps, and then, before you could say knife, back on the meths.
f. *the knife*: used as typical of surgical operations. Also *attrib.*
1880 Tennyson *Ballads* 88 But they said too of him He was happier using the knife than in trying to save the limb. *Ibid.* 95 My sleep was broken besides with dreams of the dreadful knife. **1932** Kipling *Limits & Renewals* 350 And leave you knife-wallahs to kill our patients? **1961** *Woman* 18 Mar. 10/4 Had she seen the new nurse on women's surgical? The knife men always had the luck.
g. *to get* or *have one's knife into* (a person): to exhibit a malicious or vindictive spirit towards; to persecute unrelentingly.
1890 D. C. Murray *John Vale's Guardian* III. xxxvi. 173, I reckon you've got your knife into Mr. Jousserau. **1911** H. S. Walpole *Mr. Perrin & Mr. Traill* vi. 116 This was to be the beginning of persecution. The Reverend Moy-Thompson had got his knife into him. **1930** J. B. Priestley *Angel Pavement* ix. 440 You got your knife into him the first time he came here, and after that of course he had to be blamed for everything. **1963** N. Marsh *Dead Water* (1964) i. 26, I don't know what's got into you. Why've you got your knife into this reporter chap?
h. *night of the long knives*: see *long knife 2.
i. *you* (or *one*) *could cut* (something) *with a knife*: colloq. phr. used to describe an atmosphere (lit. or fig.) so thick that it seems capable of being cut with a knife.
1892 A. W. Pinero *Magistrate* i. 18 There's a fog on the line—you could cut it with a knife. **1954** M. Sharp *Gipsy in Parlour* (1955) xiii. 111 The smell was chiefly cabbage ..and one could have cut it with a knife. **1973** G. Moffat *Deviant Death* v. 68 You could have cut the atmosphere with a knife.
2. a. *knife-back* used *attrib.*; *knife-case* (examples); *knife-thrust* used *fig.*; *knife-tray* (examples); *knife-throwing*; *knife-featured* (example); *knife-happy, -sharp, -skewed* adjs.
1886 *Harper's Mag.* June 119/2 Between these knife-back ledges are plots of sea-green grass. **1966** *Listener* 2 June 789/1 Miniature trains of rubber-tyred, electrically-driven cars (with knife-back seats or flat decks for standing passengers) would run on set routes. **1790** *Pennsylvania Packet* 6 Jan. 1/4 Steel and gilt hat buckles, and A few inlaid mahogany knife cases. *c* **1807** Jane Austen *Watsons* (1954) 344 Nanny..was beginning to bustle into the parlour with the Tray & the Knife-case. **1971** *Canad. Antiques Collector* May 3 (Advt.), Rare and elegant pair of circular inlaid mahogany Knife Cases of the highest quality. English. Circa 1800. **1895** *Century Mag.* Aug. 638/2 A tall, lanky, sharp-boned, knife-featured fellow. **1961** *Amer. Speech* XXXVI. 147 *Knife happy*, overeager to resort to operation, said of a surgeon. **1964** *New Statesman* 21 Feb. 306/3 Sacha Pitoeff scowls away as a canapé-ferrying, knife-happy villain. **1955** E. Pound *Classic Anthol.* i. 73 Down from the spring the knife-sharp waters run. **1973** J. Wainwright *Devil you Don't* 145 The wind came in, knife-sharp, from the North Sea. **1974** H. R. F. Keating *Bats fly Up* ix. 95 Ghote felt a knife-sharp happiness. *a* **1918** W. Owen *Coll. Poems* (1963) 41 Your slender attitude Trembles not exquisite like limbs knife-skewed. **1923** G. Collins *Valley of Eyes Unseen* i. 28 If there's shooting or knife-throwing. **1959** *Times* 24 June 13/1 The knife-thrusts of Ibsen's dialogue. **1967** *Coast to Coast 1965–6* 228 The first knife-thrusts of hunger had developed into a permanent ache of emptiness. **1851** C. Cist *Cincinnati* xiii. 215 Among the principal articles are..knife trays. **1939–40** *Army & Navy Stores Catal.* 168/1 Knife trays. Japanned and filleted.
b. knife-bar (earlier U.S. example); **knife-blade**, (*a*) the blade of a knife; (*b*) something sharp or pointed; (*c*) in *Mountaineering*, a kind of piton (see quot. 1968); also *attrib.* and *Comb.*; **knife-cleaner** (earlier U.S. example); **knife-money** (example); **knife-pleat**, a narrow sharply creased pleat (in a garment, esp. a skirt); so **knife-pleated** *a.* (see also Pleated *ppl. a.*), **knife-pleating** *vbl. sb.*; **knife-rest**, (*b*) *Mil. slang*, a barrier or obstruction composed of barbed wire and timber; **knife switch** *Electr. Engin.*, a switch consisting of a conducting blade or set of blades hinged at one end so that it may be swung out of or into a fixed contact or set of contacts at the other end; **knife-thrower**, one who throws knives (*spec.* as a form of entertainment); also *U.S. slang* (see quot. 1905); **knife-work**, the use of knives as weapons or instruments; also *fig.*
1867 *Trans. Illinois Agric. Soc.* VII. 312 By the arrangement of its parts the knife-bar is placed further forward than in most machines. **1799** Knife-blade [in Dict., sense

2 a]. **1902** *Daily Chron.* 12 Sept. 3/2 The snowy knife-blade *arête*. **1911** J. A. Thomson *Biol. Seasons* i. 44 The knife-blade-like larvæ of the eel. **1950** J. Dempsey *Championship Fighting* x. 49 All fingers, including the thumb, pressing tightly against each other to form a 'knife blade'. **1955** E. Pound *Classic Anthol.* ii. 106 A shallow basin gives the fish no shade, Dive as they will, there's flash of fin's knife-blade. **1968** P. Crew *Encycl. Dict. Mountaineering* 75/2 *Knife-blade*, a long thin piton. The name is mainly applied to chrome-molly pitons of this type. **1971** D. Haston in C. Bonington *Annapurna South Face* xvii. 206 It was a long and tortuous pitch done in one run-out on one of our big ropes. Firstly knee-deep mushy snow, then hard ice to exit, with one miserable knife-blade for protection. **1869** L. M. Alcott *Little Women* II. i. 12 A knife-cleaner that spoilt all the knives. **1901** *Chambers's Jrnl.* Apr. 255/2 Between 1122 and 224 B.C. a very curious knife-money was used in the state of Tsi. This coin was of copper, shaped like a bill-hook, and about seven inches long, with the handle terminating in a ring, doubtless for the purpose of stringing the coins together. **1891** *Cassell's Family Mag.* Nov. 753/2 'Knife-pleats'—as the Americans call them, to distinguish the single from the box-pleat—are turned towards the centre of the back [of the mantle]. **1928** *Daily Mail* 31 July 1/2 Well made with smart knife pleats at sides. **1964** *McCall's Sewing* ii. 30/1 *Knife pleats*, series of pleats that turn in the same direction, are usually equal in width and are pressed straight to the hem. **1905** Knife-pleated [see Pleated *ppl. a.*]. **1937** *Times* 27 Sept. 19/2 A knife-pleated vermilion dinner gown. **1965** *Punch* 12 May p. xvii, Knife-pleated travel skirts. **1895** *Montgomery Ward Catal.* 37/1 Fast Black Sateen Waists... 2 rows knife pleating from shoulder to belt. **1937** *Times* 27 Sept. 19/2 Knife pleating was used in several graceful gowns. **1919** *Athenæum* 15 Aug. 759/1 Knife-rests, *chevaux de frise*. **1921** F. W. Bewsher *Hist. 51st Div.* vii. 114 Stooks of cut strands of wire and overturned knife-rests lay everywhere. **1958** P. Kemp *No Colours or Crest* iv. 54 The entrance to the courtyard was blocked by a heavy 'knife-rest' barbed wire entanglement. **1964** A. H. Farrar-Hockley *Somme* ii. 83 Gaps had been filled with wired knife-rests and concertina rolls pegged down with iron pickets. **1907** H. H. Norris *Introd. Study Electr. Engin.* ix. 225 Open knife switches as described are not commonly used for circuit-breaking purposes above 500 volts and a few hundred amperes. **1962** *Newnes Conc. Encycl. Electr. Engin.* 726/2 Air-break isolators are made in a large variety of forms and range from the simple knife switch to those suitable for the highest transmission voltages. **1905** *Smart Set* Oct. 3/1 'They got a new knife-thrower up to the hotel,' he announced... (A 'knife-thrower', be it known, is parlance for waitress.) **1953** Wodehouse *Performing Flea* 190 He would shoot all round you till you felt like a knife-thrower's assistant, but you were really quite safe. **1973** *Listener* 19 July 80/1 Electrons are fired at an object, and they trace its outline like a knife-thrower at a fair. **1845** W. G. Simms *Wigwam & Cabin* 2nd Ser. 143 But none of your knife-work, le'me tell you. **1931** D. L. Sayers *Five Red Herrings* xxii. 255 Copying a canvas isn't the same thing as painting direct... It's the technique that's a nuisance... I don't feel handy with so much knife-work. **1954** J. R. R. Tolkien *Two Towers* 141 It has been knife-work up there. **1955** J. Morrison in *Austral. Short Stories* (1963) 2nd Ser. 147 Collins the overseer did all the knife-work—castrating, ear-marking, and tailing. **1965** J. Lawlor in J. Gibb *Light on C. S. Lewis* 76 Then proceed to the knife-work of murdering to dissect, in order to sweep the vile body aside to make room for the certified masterpieces.

knife, *v.* Add: **3.** *intr.* To move as with the action of a knife cutting or passing through.
1920 W. Camp *Football without a Coach* 107 If any of these three center men lunges through—'knifes' through, as it is called—he opens the door on either side of him. **1950** J. Dempsey *Championship Fighting* xx. 120 Deflection of the blow by..knifing with the forearm. **1958** *Times* 25 Sept. 3/2 The principal advantage of the American yacht seemed to be her ability to sail closer to the wind and knife more smoothly through the water than Sceptre. **1965** *Harper's Bazaar* June 68/2 If you come across a Salon 1959..knife on to it. **1971** *Flying* (N.Y.) Apr. 30/3 Skirting the coast for awhile before knifing northwest to Bordeaux.

knife and fork. **1. b.** (Further examples.)
1838 *Manch. Guardian* 26 Sept. 3/2 This question of universal suffrage is a knife-and-fork question, a bread-and-cheese question, after all. **1909** J. R. Ware *Passing Eng.* 163/2 *Knife and fork tea* (middle class, 1874). Vulgarisation of high tea. **1929** J. B. Priestley *Good Companions* iii. vi. 624 A sound specimen of their knife-and-fork tea. **1963** Bird & Hutton-Stott *Veteran Motor Car* 150 The first model of Lutzmann was a knife-and-fork copy of the Benz.

knife-board. Add: **1.** (Earlier example.)
1829 in A. Mathews *Mem. Charles Mathews* (1839) IV. ii. 26 *Joe Merriman*,—formerly imp of the ring, slave of the knife-board, and footman to Mr. Jenkins.
2. Also *attrib.* and *transf.*
1931 *Times Lit. Suppl.* 24 Dec. 1043/2 The photograph ..is delightful..of a very primitive 'knifeboard' omnibus with its top-hatted travellers. **1963** *Times* 6 Feb. 11/4 Those devoted to seeing Shakespeare on his home-ground and taking the full treatment will no doubt be thankful that the Stratford Theatre is not of the old and austere style with hard benches in the pit and knife-board seats in the gallery.

knife-edge. Add: **1.** (Further examples.)
a **1930** D. H. Lawrence *Last Poems* (1932) 39 The sudden dripping down of the knife-edge cleavage of the lightning. *a* **1935** T. E. Lawrence *Mint* (1955) ii. xii. 132 He creases everybody's trousers with the knife-edge that Stiffy demands. **1969** *Sci. Jrnl.* Nov. 35/2 Steel wheels or even solid rubber tyres..give a knife-edge response which would render driving at speeds over 80 km/h impossibly dangerous.
2. b. (Further examples.) *Esp.* a sharp crest of rock, ice, sand, or the like. Also *attrib.*, as *knife-edge ridge.*
1907 *Westm. Gaz.* 26 Nov. 3/2 The road thereto lies along a 'knife-edge'. **1925** W. J. Miller *Introd. Physical Geol.* viii. 252 A knife-edge ridge may also develop where glaciers in two parallel valleys erode and steepen the valley sides until only a very sharp divide separates the valleys. **1925** E. F. Norton *Fight for Everest, 1924* i. ii. 31 A ridge which begins in a knife-edge of rock worthy of the Chamounix Aiguilles. **1945** Baker *Austral. Lang.* iii. 58 Knife edges, as certain razor-back sandhills are known. **1963** *Times* 24 Jan. 10/3 Mr. John Brass, chairman of the West Midlands Division of the N.C.B., said yesterday that west Midland coal supplies were poised 'on a knife edge' at the present. **1964** C. Willock *Enormous Zoo* v. 74 The immediate foreground beneath the sandstone cliff..was eroded into a series of gullies and knife-edges like row upon row of yellowed shark's teeth. **1969** C. R. Longwell et al. *Physical Geol.* xii. 266/2 An arête is a jagged, knife-edge ridge created where two groups of cirque glaciers have eaten into the ridge from both sides. **1971** *Country Life* 25 Feb. 408/3 The ridge goes easily at first on firm snow..and finally ends in a snow knife-edge. **1972** *Guardian* 18 Feb. 24/5 The Government's knife-edge victory for the European Communities Bill in the Commons last night. **1974** *Times* 12 Oct. 1/1 A government with a knife-edge majority. *Ibid.* 1/2 Mr Wilson intends to govern on a parliamentary knife-edge for at least two or three years.
3. *Diamond-cutting.* (See quot.)
1909 J. Wodiska *Bk. Precious Stones* 349 *Knife-edge*. The girdle of a brilliant cut to a sharp edge and polished.

knifeless, *a.* (Later example.)
1916 H. G. Wells *Mr. Britling* iii. ii. 412 Knifeless and forkless meals.

kni·fe-man. One who uses a knife as an instrument, a tool, or a weapon; *spec.* † (*a*) in the parlance of N. American Indians, an Englishman *Obs.*; (*b*) *slang*, a surgeon.
1643, **1852** [see Knife *sb.* 2 a]. **1901–2** *Ann. Brit. Sch. Athens* VIII. 294 The few picked 'knife-men' who lay or crouched in the trenches cutting through the compost of bones and pottery inch by inch. **1923** *Dialect Notes* V. 235 *Knifeman*, a man who fights with a knife... Not common. **1956** J. Masters *Bugles & Tiger* i. 19 There was always the sense-sharpening chance of a sudden storm of bullets, a rush of knifemen. **1961** [see *knife *sb.* 1 f]. **1964** *Amer. Speech* XXXIX. 271 *Cutter man, knife man*, a worker who operates the cutter. **1973** 'R. MacLeod' *Burial in Portugal* v. 110 A third man..was coming in at a rush with a knife... Gaunt..felt the jar as his heel took the knife-man low in the stomach.

knifer. Add: **2.** One who carries or uses a knife as a weapon.
1870 Dickens *E. Drood* xxiii. 188 Jacks. And Chayner men. And hother Knifers. **1905** *Times* 11 Dec. 5/2 This is a new move of the hooligans and knifers, who lately committed attacks on enlightened working men.

knifey, knifie (nəi·fi), *sb.* Chiefly *Sc.* [f. Knife *sb.* + -y[6], -ie.] Either of two games played by boys with knives: (*a*) = Mumble-the-peg; (*b*) (see quot. 1969).
1896 E. Turner *Little Larrikin* iv. 40 An angel in little blue knickerbockers playing knifey on a heap of builders' sand. **1901** *Scottish Antiquary* XVI. 49 'Bonnety' and 'Knify'..are the 'Hatty' and 'Knifey' which..the Edinburgh Academy once knew so well. **1934** G. M. Martin *Dundee Worthies* 179 'Knifie'..[was] played with a knife with open blade on any grass plot. **1939** F. L. Combs *Harrowed Toad* 90 Another fieldsman who may have seen the ball but had been indulging spasmodically in 'knifey'. **1951** *Banffshire Advertiser* 16 Aug., I eest tae be the best knifey player in the toonie squeel. **1969** I. & P. Opie *Children's Games* vii. 221 The first boy throws a knife..so that it sticks in the ground not more than twelve inches to the left or right of one of his opponent's feet. The other boy..plucks the knife out of the ground, and moves his nearest foot to the place where the knife went in... Throughout Scotland..it is 'Knifie'. **1973** 'J. Patrick' *Glasgow Gang Observed* xii. 113 A game of 'knifey' began with Rose's front door as target. Bayonets and commando knives..sank into the woodwork.

knifey (nəi·fi), *a.* [f. Knife *sb.* + -y[1].] Resembling the edge of a knife in narrowness or sharpness; also *fig.* (see quot. 1937).
1906 G. A. B. Dewar *Faery Year* 258 Hovering, he presents to the wind but a knifey edge of wing. **1937** Partridge *Dict. Slang* 459/2 *Knifey*, (of a person, esp. a customer) that cuts things painfully fine when dealing in the money-market: stockbrokers'. **1955** H. Smith *Horseman through Six Reigns* xix. 192 Withers that are neither rounded nor knifey.

knight, *sb.* Add: **4. e.** Fig. phr. *knight in shining armour*: in informal or ironic use, a person regarded as a medieval knight in respect of his chivalrous spirit, especially towards women.
1965 V. Canning *Whip Hand* xv. 188 A man..didn't have to be a knight in shining armour. **1967** M. Summerton *Memory of Darkness* x. 124 Most people regard him as a crank. I'm afraid, so far as Dilys is concerned, he makes a very ineffectual knight in shining armour. **1968** A. Diment *Gt. Spy Race* i. 8, I was one of the new knights in shining PVC armour. Come to rescue the lower-middle class maiden from the dragon of boredom. **1973** *Ottawa Jrnl.* 16 Aug. 1/6 'Throughout his whole brief, Mr. Cassidy has attempted to portray himself as a knight in shining armor but on close examination it portrays our whole judicial system as something less than perfect,' said Chief Seguin.
7. a. fig. *knight's move*, an indirect or devious move.
1958 P. Shore in N. Mackenzie *Conviction* 37 The favoured entrant moves, to quote the current jargon, by a series of 'knight's moves' over the management board. **1959** *Listener* 29 Jan. 219/2 Mankind does move forward, even if it is often by the Knight's move. **1963** V. Nabokov *Gift* iv. 228 Any genuinely new trend is a knight's move, a change of shadows, a shift that displaces the mirror. **1967** G. Sims *Last Best Friend* xviii. 168 We took a kind of Knight's move, one step forward and a jump to the side. **1972** W. McGivern *Caprifoil* (1973) ii. 37 You've made a knight's move in thought.
12. c. *knight of the pencil*, a bookmaker; *knight of the road*, (*b*) a commercial traveller; (*c*) a tramp; (*d*) the driver of a lorry, taxicab, etc.
1885 *Punch* 7 Mar. 109 The Knights of the Pencil, Sir, hold that backers, like pike, are more ravenous in keen weather, and consequently easier to land. **1927** *Daily Express* 11 June 7/3 We have mentioned the knights of the pencil. Bookmakers are not permitted to take up positions at will on racecourses under Greyhound Racing Association jurisdiction. **1889** J. Burnley *Romance Mod. Industry* 317 Customers used to come out miles upon summer evenings to meet the 'knights of the road',..and the old travellers on their part would spend two or three days with some of their clients. **1928** *Sunday Express* 12 Feb. 11/4 Secrets of the Commercial Traveller's Bag. By John S. Banks, for twenty years a 'Knight of the Road'. **1928** *Daily Express* 8 Aug. 3/5 If something of this spirit could be instilled into the regular 'knights of the road', if they could be inspired with some notion of the dignity of work and the shame of alms-taking. **1971** R. Reisner *Graffiti* vi. 82 Truck drivers are notorious for their..loudly expressed admiration for women. They are true knights of the road. **1974** L. Deighton *Spy Story* ii. 20 Finally some knight of the road deigned to do a Gloucester Road to Fulham.
d. *Knights of Columbus*, a society of Roman Catholic men founded at New Haven, Connecticut, in 1882.
1901 *N.Y. Tribune* 22 July 3/4 Wednesday the Knights of Columbus and Utah people will unite their forces; on Thursday another double-header. **1929** F. A. Pottle *Stretchers* (1930) iv. 66 Besides the enormous structures of the Y.M.C.A. and..Knights of Columbus,..the general public had provided at Merritt many other agencies of relaxation and amusement quite peculiar to the camp. **1948** *Green Bay* (Wisconsin) *Press-Gaz.* 12 July 16/7 Members of the Green Bay lodge Knights of Columbus were reminded today that the annual Fish Fry will be held at the Shorewood Country club next Tuesday. **1974** *Listener* 21 Nov. 683 The Knights of Columbus are associated with lodge meetings and bingo.

Knipe, var. *Kneipe.

kniphofia (niphōᵘ·fiă). [mod.L. (C. Moench *Methodus Plantas Horti Botanici et Agri Marburgensis* (1794) 631), f. the name of Johann Hieronymus *Kniphof* (1704–63), German physician and botanist + -ia[1].] A perennial herb of the genus so called, native to southern or eastern Africa, belonging to the family Liliaceæ, and bearing spikes of red, yellow, or orange flowers; also called red-hot poker, torch lily, and formerly Tritoma, a synonymous generic name.
1854 *Curtis's Bot. Mag.* LXXX. 4816 Serrulated-leaved Kniphofia... Of late years visitors to the Royal Gardens of Kew have been much struck with the beauty of the flower-spikes of a *Kniphofia* (*Tritoma* of most authors), planted in several of the beds. **1900** W. D. Drury *Bk. Gardening* x. 336 Kniphofias may be accommodated in shrubberies and wide borders. **1935** A. G. L. Hellyer *Pract. Gardening* x. 81 Delphiniums, kniphofias, and ordinary lupins may be planted from 2 feet to 2 feet 6 inches apart. **1971** *Country Life* 2 Sept. 567/2 You will probably put your emphasis..on delphiniums, kniphofias, [etc.].

knish (kniʃ). [Yiddish, f. Russ. *knish*, *knysh* a kind of cake.] A dumpling of flaky dough filled with chopped liver, potato, or cheese, and baked or fried.
1930 A. Gross *Kibitzer's Dict.* 47 Knishes—Dyspepsia. **1932** L. Golding *Magnolia St.* iii. vii. 560 There would be *knishehs* to eat. **1960** *New Yorker* 29 Oct. 36/3 Have a canapé. The knishes are especially delicious. **1965** 'E. Queen' *Fourth Side of Triangle* iv. 164 We take one of these thin little pancakes, or knishes—almost like tortillas, aren't they? **1973** *Times* 12 Apr. 18/7 Arthur Goldberg, the Democratic candidate, was running a gauntlet of knishes, pizza and egg-roll wherever he went. **1973** *Daily Colonist* (Victoria, B.C.) 27 May 2/4 He consumed three meat knishes, two blueberry knishes, four potato knishes and two cream-filled knishes.

knit, *v.* Add: **2. c.** *spec.* To do knitting in plain stitch as opposed to purl.
c **1890** tr. *T. de Dillmont's Encycl. Needlewk.* 196 Piqué pattern..1st and 2nd row—purl 7, knit 1, purl 1, knit 1, [etc.]. **1902** [see Purl *v.*[1] 4]. **1944** A. Thirkell *Headmistress* iii. 61 She was well settled into knit two, purl two. **1972** 'B. Graeme' *Tomorrow's Yesterday* xiv. 142 She would have to undo three rows of knitting. Knit two, purl two, the pattern called for, but she had knitted one and purled three.

knit, *sb.* **1.** Delete †*Obs.* and add later examples. Also, a knitted fabric. **knit stitch,** the plainest stitch in knitting.

1897 *Sears, Roebuck Catal.* 217/3 Turtle Neck Sweater. Extra heavy knit. All wool. **1932** D. C. Minter *Mod. Needlecraft* 68/2 (*caption*) Rows of plain knitting... Smooth knit stitch:..rough purl stitch. **1960** *Guardian* 19 Feb. 8/6 Dress and jacket of corded cotton knit. **1960** *News Chron.* 22 Mar. 11/3 Textured nylon is making news with new chunky knits, and fur fabrics. **1963** *New Yorker* 29 June 44 It looks like sharkskin, but it's really a knit. **1964** *Guardian* 22 Jan. 8/4 Knits are used for dresses, coats, suits.., so that it is possible to dress in knitwear from dawn to dark. **1964** *McCall's Sewing* iv. 53/1 When buying a knit, remember that it is a resilient fabric.

b. The action or process of knitting.

1926 E. K. Middleton *New Knitting* Pref., Left hand knit and left hand purl are simpler and quicker than the old right hand knit.

c. A knitted garment. Freq. in *pl.*

1938 D. Baker *Young Man with Horn* I. iv. 38 His brother Henry..was selling jersey knits. **1965** *Harper's Bazaar* June 66 Sportive, can't-wait-for-winter knits. **1972** *Daily Tel.* 30 Oct. 13/1 Glitter knits are one of the top fashions for winter.

knit, *ppl. a.* Add: **2.** [Or a use of *knit *sb.* 1.] (Further examples.)

1895 *Montgomery Ward Catal.* 283/1 Children's Knit Undershirts. Ladies' Jersey Knit Ribbed Vests. **1897** *Sears, Roebuck Catal.* 203/1 Lumberman's Knit Socks. **1922** E. E. Cummings *Enormous Room* vii. 155 A knit sweater of a strangely ugly red hue. **1962** F. I. Ordway et al. *Basic Astronautics* xiii. 517 The inner liner of the suit is of neoprene-coated fabric with knit stretch sections. **1970** *Catal. L. L. Bean* (Freeport, Maine) Fall 32 Deluxe insulated coverall... Knit cuffs at wrists. **1970** *Women's Wear Daily* 23 Nov. 31/2 So many knit pants in the market are dumb, missy looks.

knitted, *ppl. a.* Add: Also as *sb.*, a knitted garment. Freq. in *pl.*

1958 *Woman* 9 Aug. 31/4 Briony is going to Ardoaghy to buy more knitteds. **1960** *Woman's Own* 19 Mar. 34/2 Two elegant knitteds from one set of instructions. **1963** *Harper's Bazaar* Feb. 41 The knitteds we can't do without.

knitting, *vbl. sb.* Add: **1. c.** A girl or girls. *slang.*

1943 C. H. Ward-Jackson *Piece of Cake* 39 *Knitting*, girl or girls. **1946** J. Irving *Royal Navalese* 104 *Knitting*, girls in the plural. The singular of this is '*A Piece of Knitting*'. **1962** Granville *Dict. Sailors' Slang* 68/2 *Knitting*, girl friend or girls collectively.

3. *knitting bag, bee, book, frolic, machine, pattern, silk* (example), *wire, wool*; **knitting-pin** (earlier example); **knitting sheath** (later U.S. example).

1789 W. Cowper *Let.* 31 Jan. in *Corr.* (1904) III. 347 A basket..[which] contained..a Knitting bag, and a piece of plumcake. **1939** J. Cary *Mister Johnson* 209 The accomplished..traveller..sets up her household, complete with family and knitting bag, even in trains and tram shelters. **1971** A. Christie *Nemesis* ii. 20 Taking her knitting out of its embroidered knitting bag. **1855** *Chicago Times* 19 Mar. 2/6 This girl had been at a knitting bee, at the house of a friend. **1880** *Harper's Mag.* Sept. 508/1 In winter they sometimes had knitting bees. **1843** A. Lambert (*title*) My knitting-book. **1873** *Young Englishwoman* June 311/3 We must refer you to Madame Goubaud's Knitting and Netting Book. **1973** Horne & Bowden *Bk. Knitting & Crochet* i. 10 The contents of the knitting books of the 1840s differed greatly from those of today. **1818** H. B. Fearon *Sk. Amer.* 223 They are invited to the preacher's house, to partake of a supper... This is termed a knitting frolic. **1858** P. L. Simmonds *Dict. Trade Products* 216/2 *Knitting machine*, a machine for weaving and making knitted work. **1927** M. Ostenso *Mad Carews* (1929) iv. 44 Her mother had sent to the city for a knitting machine in the hope that she might be able to supply her neighbours with woollen socks. **1974** *Times* 12 Feb. 11/5 He started..with a couple of hand frame knitting machines. **1885** C. M. Yonge *Nuttie's Father* II. i. 7, I couldn't get a knitting pattern Miss Headworth was to send Lady Ronnisglen. **1961** M. Stewart *Ivy Tree* iii. 46 The colourless voice..might have been discussing a knitting pattern. **1973** 'J. Ashford' *Double Run* xi. 85 Nina, at a critical point in the intricate knitting pattern, ..dropped a stitch. **1857** C. M. Yonge *Dynevor Terrace* I. iv. 49 He had in the other pocket..wools..the long knitting-pins under his arm like a riding-whip. **1867** T. Lackland' *Homespun* I. 23 A great tear trembles on her cheek as she adjusts her needle in the knitting sheath she wears. **1897** *Sears, Roebuck Catal.* 321/2 Knitting and Crochet Silk..is used for knitting mittens, stockings and other articles which require washing. **1850** *Rep. Comm. Patents 1849* (U.S.) 491 The needle itself, and thimble will be exhibited in museums with distaffs, spinning-wheels, knitting-wires, [etc.]. **1860** Geo. Eliot *Mill on Floss* III. vi. vi. 87 Maggie's ball of knitting-wool rolled along the ground. **1965** A. Christie *At Bertram's Hotel* xii. 114 She had a splendid time rounding up knitting patterns, new varieties of knitting wool, and suchlike delights.

knitwear (ni·twēəɹ). [f. Knit *ppl. a.* + Wear *sb.*] Knitted articles of clothing.

1925–6 *Army & Navy Co-op. Soc. Price List* 698 'Braemar' Knitwear for Gentlemen. **1928** *Daily Express* 7 May 4 Two-piece suits, knitwear, leather coats and raincoats. **1957** L. F. R. Williams *State of Israel* 111 There is already some considerable export of cotton goods, as well as of knitwear. **1966** *Illustr. London News* 26 Feb. 5 The knitwear is made of Shetland wool and is fully fashioned with raglan sleeves. **1973** J. Thomson *Death Cap* vi. 91 He was in Bradford, seeing some knitwear manufacturers on business.

knob, *sb.* Add: **1. e.** *with knobs on*: jocular slang phr. = 'that and more' (indicating ironic or emphatic agreement, or in retort to an insult, etc.).

1930 M. Kennedy *Fool of Family* xiii. 129 'I'm waiting for the Marchese Ferdinando Emanuele Maria Bonaventura Donzati.' 'With knobs on,' agreed Gemma airily. 'Who's he?' **1931** J. J. Farjeon *House Opposite* ii. 'You are nothing,' said the Indian. 'And so are you, with knobs on!' barked Ben, and slammed the door. **1938** Wodehouse *Code of Woosters* viii. 196 '"Ha jolly ha!" to you, young Stiffy, with knobs on,' I retorted with quiet dignity. **1941** M. Treadgold *We couldn't leave Dinah* xv. 232 '*Schweinhund*,' screamed Nannerl. '*Schweinhund* yourself with knobs on,' returned Petit-Jean tartly. **1959** I. & P. Opie *Lore & Lang. Schoolch.* iii. 45 Same to you with knobs on. **1969** *Guardian* 22 Feb. 9/6 They shouted something... 'The same to you with knobs on,' Jim shouted back. **1970** A. Price *Labyrinth Makers* xiv. 179 If the A.S. 12 was the answer to Egypt's Russian missile boats, the A.S. 15 was the answer with knobs on.

2. (Later U.S. example.)

1895 *Century Mag.* Aug. 621/2 One of the many knobs from which Daniel Boone is said to have looked first over the Blue Grass land.

4. b. *Austral.* and *N.Z. slang.* A double-headed penny. Also *nob.*

1928 J. Devanny *Dawn Beloved* xv. 163 Sometimes two pennies are filed down to half their thickness and then joined together. A double-tailer is called a 'grey'; a double-header a 'nob'. Of course, anyone caught cheating is liable to get bashed about a bit. **1941** Baker *Dict. Austral. Slang* 41 *Knob*, a double-headed penny.

8. *knob-twiddling, -twister, -twisting*; **knob-cone pine,** *Pinus attenuata*, a species native to California; **knob-nose** *S. Afr.*, name applied to a member or to a tribe of 'Kaffirs' having this distinguishing feature; also *attrib.* (or as *adj.*); **knob-nosed** *a.*, having a knob-shaped nose; *spec.* = **knob-nose.*

[**1882** A. Kellogg *Forest Trees of California* in *Rep. State Mineralogist California* Appendix II. 51 Knobby-cone Pine... The Knobby Pine is a lofty tree of much beauty.] **1884** C. S. Sargent *Rep. Forests N. Amer.* 196 *Pinus tuberculata*... Knobcone pine. **1905** —— *Man. Trees N. Amer.* 22 *Pinus attenuata*, Lemm. Knob-cone pine... Fruit..becoming light chestnut-brown, with thin flat scales rounded at the apex, those on the outer side being enlarged into prominent transversely flattened knobs. **1932** W. Dallimore in *Conifers in Cultivation* (R. Hort. Soc.) 28 *Pinus.. attenuata* Lemm. (*P. tuberculata* Gord.)—Knob-cone Pine—Oregon to California. **1967** N. T. Mirov *Genus Pinus* iii. 170 *Pinus attenuata* is known as 'knobcone pine'. Its area is chiefly in California, although..it extends northward to the mountains of southwestern Oregon. **1900** A. H. Keane *Boer States* vii. 99 Hence the extraordinary differences that are observed between..the degraded Magwamba ('demons' or 'devils'), called 'Knobnoses' by the Transvaal Boers, and the Basutos. **1943** D. Reitz *No Outspan* iv. 59 We went up along the Sami river to Sibasa's country and then to the chief of the knob-nose kaffirs. **1839** W. C. Harris *Wild Sports S. Afr.* xxxix. 350 A friendly tribe of natives, whom, from a peculiarity in the nasal prominence, they dignified with the appellation of 'knob-nosed Kafirs'. **1864** J. A. Grant *Walk across Afr.* 93 A knob-nosed duck. **1887** Knob-nosed [in Dict.]. **1905** *Westm. Gaz.* 2 Oct. 8/2 The knob-nosed lizard (*Lyriocephalus scutatus*) from Ceylon. **1968** *Times* 29 Nov. 1/1 Some are hi-fi maniacs, knob-twiddling perfectionists. **1973** *Times* 7 June 18/6 He thinks commercial competition could benefit the BBC..by encouraging them to do more knob-twiddling, instead of staying tuned to one programme all day. **1940** *Chambers's Techn. Dict.* 478/2 *Knob-twister*. A casual reference to monitor man or recordist in motion-picture production. **1950** *People* (Austral.) 7 June 50/2 He believes that 90 per cent of the 'knob-twisters', as they are called, favor the abolition of the [betting] boards. **1929** *Radio Times* 8 Nov. 450/2 'Earth's End' stations can be received..if you..find your fun in knob twisting. **1958** *Listener* 13 Nov. 799/2 Audiences for a number of these [programmes] have dropped considerably since they were transferred from their old places in the Light Programme. This might be blamed on mass inertia, as applied to knob-twisting.

knobbler. 1. (Later example.)

1971 *Country Life* 19 Aug. 431/2 They were all dead, six of them: five hinds and calves and a small knobbler.

knobbly, *a.* Add: Esp. of knees; *knobbly-knees competition*, a competition in which a prize is awarded to the competitor with the 'knobbliest' knees.

1958 *Times* 15 Aug. 9/4 These knees are..knobbly and ..unsightly. **1968** *Listener* 12 Sept. 349/3 When I witness a theatrical 'happening' and 'audience participation' is urged upon me, I generally feel that I would sooner enter for a knobbly-knees competition at Butlin's.

knobby, *a.* Add: **3.** Full of rounded knolls or hills; hilly. *U.S.*

1869 'Mark Twain' *Innoc. Abr.* liii. 558 It is as knobby with countless little domes as a prison door is with bolt-heads.

knobby (nǫ·bi), *sb.* *Austral.* [f. the adj.] An opal.

1921 K. S. Prichard *Black Opal* I. i. 8 'Look at this.. and this!' he cried eagerly, going over the two or three small knobbies in his hand... 'I'm going in now,' he said, thrusting the opals into the bag. *Ibid.* iii. 23 Paul was holding up a good-looking knobby so that red, green, and gold lights glittered through its shining potch as he moved it. **1971** J. S. Gunn *Opal Terminol.* 24 *Knobby*, opal which is found as a solid lump..rather than in a seam formation.

knobkerrie. Add: (Earlier and later examples.) Also **knob-kerri, knobkier(r)ie, knobkirrie, knob-kurrie, knopkierie.**

1844 *United Service Mag.* July 337 With the precious book..in one hand, and his knob-kurrie in the other, away he trudged. **1926** T. E. Lawrence *Seven Pillars* (1935) vi. lxix. 384 Spattering the brains of a cornered mob of Germans one by one with his African knob-kerri. **1940** Baumann & Bright *Lost Republic* 220 When a Basuto beats his wife, he does it with a knopkierie. **1949** *Cape Argus* 10 Nov. 1/6 Motibeli was struck on the head with a knobkierie. **1961** *Reader's Digest* Feb. 142/1 On the river-bed lay the weapon—an African knopkierie. **1973** *Times* 8 Feb. 1/5 The chanting Africans, waving knobkerries, were cordoned off by police. **1974** *Eastern Province Herald* 2 Sept. 1 Miners attacked each other with knobkerries.

knock, *v.* Add: **2.** Also, *to knock a hole, gap,* etc.; *to knock daylight into* (cf. Daylight 1 c).

1881 *Punch* 17 Sept. 124/1 Ready at the call of duty to frame a new programme or knock daylight into an old one. **1890** A. Conan Doyle *Sign of Four* iv. 68 He knocked a hole..in the lath and plaster ceiling. **1906** Somerville & 'Ross' *Some Irish Yesterdays* 85 You may see him skilfully 'knocking a gap' (*i.e.* unbuilding a wall).

c. (Later examples.)

1898 A. Bennett *Man from North* xi. 95 Two guineas the suit, my boy! Won't I knock 'em in the Wal-worth Road! **1910** Wodehouse *Psmith in City* xix. 167 He told him that he had knocked them at the Bedford the week before. **1947** K. Tennant *Lost Haven* vii. 97 The skirt was flared with cunning little tucks at the waist, so that it fitted her like a glove... 'That ought to knock them,' Mark's granddaughter said aloud. **1954** 'N. Blake' *Whisper in Gloom* I. vi. 83 Wasn't she in pantomime?.. Bet she knocked them.

d. To copulate with; also, to make pregnant. So in phr. *to knock a child* (or *an apple*) *out* (*of*).

1598 Florio *Worlde of Wordes* 94/1 *Cunnuta*, a woman nocked. **1604** Marston *Malcontent* III. iii. sig. E2v Haue beate my Shoomaker, knockt my Sempstres, cuckold my Pottecary, and vndone my Taylor. **1785** Grose *Dict. Vulgar T.*, *Knock*, to knock a woman, to have carnal knowledge of her. **1818** Keats *Let.* 5 Jan. (1931) I. 80 They call good Wine a pretty tipple, and call getting a Child knocking out an apple. **1922** Joyce *Ulysses* 401, I cannot but extol the virile potency of the old bucko that could still knock another child out of her. **1936** J. Steinbeck *In Dubious Battle* iv. 45 Sooner or later some girl'd get knocked higher than a kite. **1963** T. Parker *Unknown Citizen* v. 120 You give your missus so much money a week, you knock a few kids out of her, and that's about it, really. **1967** D. Pinner *Ritual* ix. 96 I've knocked some girls in my time but I've never had such a rabbiter as you. The cruder it is, the more you like it.

e. To rob (esp. a safe or till). *Underworld slang.*

1767 *Sessions Papers* IV. 151/2, I heard him say he got twelve shillings once by *knocking the lobb*... What is that?.. That is breaking open a place. **1924** G. C. Henderson *Keys to Crookdom* App. B. 397 Blowing a peter. Blowing a safe open with explosives. Also called knocking a peter, blowing a pete, getting a box. **1963** *Times* 25 May 12/2 The appellant had been asked if he had told someone in the 'Norfolk' that he got the money by safe breaking. The appellant had replied: 'Aye but you will never prove that I got it by knocking a safe.'

f. To speak ill or slightingly of, disparage, find fault with, criticize captiously. Also *intr.* and *absol. colloq.* (orig. *U.S.*).

1892 'J. Miller' *Workingman's Paradise* 85 Admit it's a business concern and that everybody growls at it, it's the only paper that dares knock things. **1896** Ade *Artie* xii. 106 There's a lot o' people in the ward that's got their hammers out and they're knockin' him all they can. *Ibid.* 110 He's got to make good with 'em to keep 'em from knockin. **1901** 'H. McHugh' *John Henry* 54 I'm not knocking, remember; I'm only saying what I think. I hate a knocker. **1904** *Sun* (N.Y.) 4 Aug. 5 'Of course there'll be plenty of cranks to knock this scheme,' said he. **1906** *Daily Colonist* (Victoria, B.C.) 16 Jan. 4/3, I refer to the practice of allowing any kicker in the city to avail himself of newspaper space to knock some public man or some public institution. **1919** 'Ian Hay' *Last Million* iii. 36 A certain licence is permitted to professional grouchers; but 'knocking' the Cause is the one thing that the New Crusaders will not permit. **1926** *Spectator* 3 Apr. 635/2 A reputation for 'knocking' is enough to ensure being blackballed from some of the best clubs. **1930** Wodehouse *Very Good, Jeeves!* ii. 44 Where does a valet get off, censoring vases? Does it fall within his province to knock the young master's chinaware? **1958** K. Amis *I like it Here* xvi. 205, I shouldn't like you to get the idea I'm trying to knock Portugal and the Portuguese. **1958** *Spectator* 12 Dec. 865/1 On the last page he protests about 'the growing tendency in some newspapers today to write only "knocking" stories about stars as big as Tommy'. But..almost any publicity is good publicity: you can knock around the clock and the moon-faced masses will only hear applause. **1970** *New Scientist* 5 Mar. 478/2 They're knocking Concorde again, the cads. **1974** *Observer* 22 Sept. 14/5 It's fashionable nowadays to 'knock' England for its shortcomings.

5. b. (Further examples.) Also, (i) of an internal-combustion engine, to suffer from knock caused by faulty combustion (see *knock *sb.*[1] 1 c). (ii) Of fuel for an internal-combustion engine: to give rise to knock when burnt in an engine.

1905 *Daily Chron.* 5 May 3/5 You advance your spark.. to the point at which your engine does not knock. **1909** *Motor Cycling* 22 Nov. 32/1 If the engine begins to 'knock', a few vigorous thrusts at the pedals should be given immediately. **1916** Eighinger & Hutton *Steam Traction Engineering* v. 156 The main shaft boxes..will often knock. **1925** R. J. B. Sellar *Sporting Yarns* 186, I shall have to pull up, old chap. Cylinder's knocking! **1927** *Industr. & Engin. Chem.* Jan. 145/1 There is at present no satisfactory method of expressing the tendency of a fuel to 'knock', or detonate. **1937** Wodehouse *Lord Emsworth & Others* ix. 299, I became aware that the engine was not humming so smoothly. It had begun to knock. **1960** V. B. Guthrie *Petroleum Products Handbk.* iv. 21 Engine designers..have done a great deal to minimize the tendency of engines to knock. **1966** *McGraw-Hill Encycl. Sci. & Technol.* III. 309/2 Hydrocarbon fuels with compact molecular structures are less likely to knock.

6. b. *to knock the end in* or *off*: to spoil the whole affair. *slang.* ? *Obs.*

1919 *Athenæum* 8 Aug. 727/2 To 'knock the end in' is to spoil the whole show. **1925** Fraser & Gibbons *Soldier & Sailor Words* 138 *Knock the end off, to*, to spoil anything.

e. Also *to knock all of a* *heap, *down with a* *feather, *for a* *loop, *for* *six; *to knock* silly, *cold, *endways, *rotten, *sideways; *to knock the nonsense*, etc., *out of.*

1856 C. M. Yonge *Daisy Chain* I. v. 48 The girlishness and timidity will be knocked out of him by the boys. **1892** 'Mark Twain' *Amer. Claimant* v. 63 'When I came to breakfast Miss Gwendolen—well, she knocked everything out of me, you know—.' 'Wonderful girl, wonderful.' **1931** *Times Lit. Suppl.* 15 Oct. 787/1 The boy returned to Turin, where his royal relatives did their best to get his revolutionary notions knocked out of him. **1935** L. & A. Maude tr. *Tolstoy's Iván Ilých & Hadji Murád* 273 They'd have knocked the nonsense out of you in the army, and be was worth five of such as you at home!

7. knock about. a. (Earlier and later examples.)

a **1817** Jane Austen *Persuasion* (1818) III. iii. 44 They [*sc.* sailors] are all knocked about, and exposed to every climate, and every weather, till they are not fit to be seen. **1926** T. E. Lawrence *Seven Pillars* (1935) lxxvi. 424 Young Mustafa refused to cook rice; Farraj and Daud knocked him about until he cried. **1969** *Listener* 24 July 103/2 After being knocked about to an appalling extent in the first week of the war, the Poles were rallying until the Russians came in on the other side.

b. (Earlier and later examples.)

1833 *Sk. & Eccentr. D. Crockett* i. 31 David, collecting his clothes,..began to knock about. **1834** W. G. Simms *Guy Rivers* II. viii. 98 I've been a matter of some fifteen or twenty years knocking about..in one way or another. **1929** C. Mackenzie *Gallipoli Memories* iii. 21 He had knocked about all over the Pacific and would have been a splendid companion. **1937** M. Sharp *Nutmeg Tree* xx. 265 You're older, and you've knocked about a bit.

c. *to knock the balls about*: to strike a (billiard, croquet, etc.) ball idly; to play (such a ball game) in a casual fashion.

1864 C. M. Yonge *Trial* II. vi. 123 Tom..had seen the Andersons knocking about the balls in the new gardens.. and proposed to..try to get up a match. **1872** Trollope *Golden Lion* xviii. 305 He knocked the balls about with his cue. **1907** F. E. E. Bell *At Works* vi. 130, I have seen a club with two free tables, where men..have been happily knocking the balls about from 9 a.m. onwards. **1916** A. Bennett *These Twain* II. xiv. 271 'Shall we knock the balls about a bit?' They began a mild game of croquet.

d. To lie around, to be available or in the vicinity; to impend.

1866 F. Hunt *25 Yrs. Experience N.Z.* ix. 45 My carpet bag I left knocking about amongst them with utmost carelessness. **1870** R. P. Whitworth *Martin's Bay Settlement* 23/1, I have commenced to make a dingy..out of old boards that are knocking about. **1889** G. B. Shaw *London Music 1888–89* (1937) 116 There is plenty of musical talent knocking about misused or misdirected. **1897** P. A. Philips *Memories of Past* 21 When we did have them they were pretty severe [fires], such as..W. S. Grahame's in Fort Street (any amount of drink knocking about), the fire in High Street. **1902** Conrad *Typhoon* ii. 22 Observing the steady fall of the barometer, Captain MacWhirr thought, 'There's some dirty weather knocking about.' **1908** W. H. Koebel *Anchorage* ii. 45 Do you know of a billet knockin' about anywhere that'ud suit him, boss? **1916** A. Bennett *These Twain* I. v. 60 'There are one or two ordinaries knocking about the place,' said Edwin, 'but we haven't got a proper bicycle-house.' **1939** [see *eye *sb.*[1] 3 b]. **1948** R. Finlayson *Tidal Creek* i. 17 I'll just see about a box that ought to be knocking about.

e. *to knock about with*: to be a habitual companion of.

1915 T. Burke *Nights in Town* 323 We talked of Love, Wines, Dinners, Music-halls, of the men we had knocked about with, the girls we had loved. **1924** M. Kennedy *Constant Nymph* viii. 122 Look at the sort of people the poor child has knocked about with.

7*. knock around, round = *knock about* (sense 7 b). Also *fig.*

1848 W. T. Thompson *Major Jones's Sk. Travel* 8 I'm gwine..to New York,..and Boston and all about thar, and spend the summer until pickin time, nockin round in them big cities, mong them peeple what's so monstrous smart. **1856** C. E. DeLong in *Calif. Hist. Soc. Q.* (1930) IX. 65 Got up late knocked around. **1874** V. Pyke *Adventures G. W. Pratt* (1890) I. vi. 27 Seems there's a joke knocking around somewhere. **1884** [in Dict., sense 5 d]. **1924** C. Mackenzie *Heavenly Ladder* iii. 55 Not that I'm against your style of services myself. But most of the people round here haven't knocked around like I have. **1938** G. Greene *19 Stories* (1947) 76 He's knocking around somewhere. **1959** P. McCutchan *Storm South* xiii. 198 Mrs. van Neyland's been a married woman, and she's knocked around.

7. knock back. a.** *trans.* To refuse, to rebuff. *Austral.* and *N.Z. colloq.*

1930 V. Palmer in *Bulletin* (Sydney) 19 Feb. 51/1 Not the sort of man we want... I knocked him back. **1939** K. Tennant *Foveaux* IV. ii. 368 Why, she knocks back the boss where she works, if he gets gay. **1944** L. Glassop *We were Rats* xviii. 104 Still goin' to keep knockin' back the sheilas? **1948** V. Palmer *Golconda* xi. 85 Most of them knocked his appeals back lightly and watched the proceedings with tolerant amusement. **1952** J. Cleary *Sundowners* ii. 108 He wouldn't knock it [*sc.* money] back if you offered it to him. **1957** 'N. Culotta' *They're a Weird Mob* (1958) vi. 86 Never knock back O.P's [*sc.* other people's smokes]. **1969** *Private Eye* 12 Sept. 14 (*caption*) Knocking back a free night at the *flea-pit* too. **1973** *Nation Rev.* (Melbourne) 24–30 Aug. 1398/6 Never knock back a dollar, I guess.

b. *trans.* To drink (esp. intoxicants) or eat heartily or heavily; to swallow a drink at a gulp. Also in phr. *to knock it back. colloq.*

1931 Brophy & Partridge *Songs & Slang 1914–18* (ed. 3) 326 *Knock it back.*—To eat; sometimes, to drink. **1939** *Eastbourne Herald* 6 May 13/2 Thirsty Eastbourne really does 'knock back' thousands of gallons in the course of a year. **1947** 'A. P. Gaskell' *Big Game* 8 'Boy oh boy oh boy,' he chanted, 'won't I knock back those handles tonight.' **1951** J. B. Priestley *Festival at Farbridge* III. ii. 476 That's why he knocked a few back—and he doesn't as a rule. **1951** 'J. Wyndham' *Day of Triffids* i. 26, I knocked back the last of my brandy, and went out. **1953** X. Fielding *Stronghold* III. iv. 221, I hear you knock it back a bit. Well, so do I. **1957** C. MacInnes *City of Spades* I. ix. 69 My two friends knocked back their gins. **1957** 'N. Culotta' *They're a Weird Mob* (1958) iii. 42 'What is this pin one on, Joe?' 'Knock one back. Gunna 'ave a drink?' **1961** M. Kelly *Spoilt Kill* ii. 71 When you were fourteen you knocked back a whole jar..at one sitting. **1968** M. Richler in R. Weaver *Canad. Short Stories* 2nd Ser. 194 Hod was knocking back large snifters of brandy.

c. *trans.* To retard, to check. *Austral.* and *N.Z.*

1945 J. Pascoe in *N.Z. Geographer* I. 27 An early winter will knock his flock back. **1946** F. Sargeson *That Summer* 85 The two sprees had knocked me back considerably [financially].

8. knock down. a. Also, to bring down by a shot, or by artillery, etc., fire.

1733 W. Byrd *Journey Land of Eden* in *Writings* (1901) 311 We pursued our Journey thro' uneven and perplexed Woods, and in the thickest of them had the Fortune to knock down a young Buffalo, 2 Years old. **1809** M. L. Weems *Life F. Marion* xi. 98 Many a family goes without dinner unless the father can knock down a squirrel in the woods. **1940** *War Illustr.* 12 Apr. 366/3 But before the Germans had managed to recover from their surprise that only three British 'planes dared to attack them, the young flight leader had knocked down two of them.

f. (Further examples. Cf. Knock-down *a.* 3 (in Dict. and Suppl.), *knocked-down* s.v. *knocked *ppl. a.* 2.)

1945 B. Macdonald *Egg & I* (1946) xix. 194 The six-hundred-gallon water tank arrived, knocked down and looking disappointingly like a bundle of faggots. **1958** *Times Rev. Industry* Dec. 65/3 Motor-car body shells.. knocked down for export. **1973** *Amer. Speech 1969* XLIV. 206 *Knock down*, disassemble freight or merchandise.

h. (Earlier and later examples.) Also *N.Z. slang.* Hence **knocking down** *vbl. sb.*

1852 in *Occasional Papers Univ. Sydney Austral. Lang. Res. Centre* (1966) No. 9. 15 They then go 'upon the burst' as they call it, and drink until all their earnings are 'knocked down'. **1861** H. W. Harper *Lett. from N.Z.* (1914) 65 [Station hands] proceed to 'knock down their cheque', giving it to the landlord and bidding him treat all comers as long as it lasts. **1866** *Bk. Canterbury Rhymes* 19, I knock my earnings down [at the Royal Hotel]. **1874** A. Bathgate *Colonial Experiences* xi. 142 He would get amongst a bad lot and knock down every penny of our hard-earned cash. **1879** J. Grey *His Island Home* iii. 32/1 They were 'knocking down' their cheques and living at the rate of ten thousand a year. *Ibid.*, They appeared to derive intense satisfaction from the knocking down process until their resources were exhausted. **1904** M. Cradock *Sport in N.Z.* I. i. 10 Their rabbit cheques generally find their way to the nearest public house, to be 'knocked down' as soon as received. **1965** J. S. Gunn *Terminol. Shearing Industry* I *Knock down*, to spend a cheque, usually in one quick celebration.

i. (Earlier examples.) Also, to embezzle generally.

a **1854** J. F. Kelley *Humors of Falconbridge* (1856) 86 No knocking down, sir! **1864** T. L. Nichols *40 Yrs. Amer. Life* I. 89 The omnibus-drivers were expected to 'knock down' a certain proportion of the receipts. **1872** J. D. McCabe *Lights & Shadows N.Y. Life* xi. 214 In order to make up the deficiency between their actual wages and their necessities, the conductors and drivers have fallen into the habit of appropriating a part of the money received from passengers to their own use... This practice of 'knocking down', or appropriating money, begins with the conductor, as he alone receives the money paid for fares. **1949** 'J. Evans' *Halo in Brass* xx. 172 Some..clerk who was knocking down on the till.

k. *pass.* Of a ship (see quots. 1891 and 1948). *U.S.*

1873 G. H. Procter *Fisherman's Memorial* 128 A severe gale, which knocked the vessel down and nearly swamped her. **1891** H. Patterson *Illustr. Naut. Dict.* 104 *Knocked down*, said of a vessel when, by the force of the wind acting upon her sails and spars, she is careened to such an extent that she does not recover herself. **1948** R. de Kerchove *Internat. Maritime Dict.* 387/1 *Knocked-down*, the situation of a vessel listed over by wind to such an extent that it does not recover.

l. To earn, get paid. *U.S.*

1929 M. Lief *Hangover* vi. 100 She and Humphrey rented a cottage in Westport..where authors who knocked down $3,000 for knocking out a short story of 5,000 words, built such magnificent houses. **1949** *New Yorker* 5 Nov. 76/2 You wanna know hommuch that animal knocks down a week?

10. knock off. b. (Later example.) Also, to discharge or dismiss from employment, to 'lay off'.

1881 A. Bathgate *Waitaruna* xii. 172 [The boss] would growl at the offending shearer and make use of some vague threat of 'knocking him off'. **1896** 'M. Rutherford' *Clara Hopgood* xxvi. 256 As reg'lar as winter comes Longwood is knocked off—no work. **1955** *Times* 9 June 8/3 The Cunard company put the main restaurant at his service and the staff captain 'knocked off all the men from their duties'.

c. (Earlier and later examples in sense 'to desist from one's work or occupation'.)

1842 *Spirit of Times* 4 June 158/2 My tackle being very light I had to humor him, and 'twas full half an hour before I killed him. Knocked off, and set Joe to work to boil rock [i.e. fish]. **1851** H. Melville *Moby Dick* II. xii. 95 Do you want to sink the ship, by knocking off at a time like this? **1916** 'Boyd Cable' *Doing their Bit* iii. 49 The factory was knocking off for dinner as we came away. **1969** M. Crouch *Essex* ii. 28 One who has just knocked off for his tea-break.

d. (Earlier Amer. example.)

1767 'A. Barton' *Disappointment* I. i. 10 As for McSnip, he intends to knock off business, home to England and purchase a title.

e. *spec.* To write, paint, etc., in a hurried and perfunctory fashion.

1886 in *Amer. Speech* (1950) XXV. 35/1 When he knocked off a few stanzas of poetry. **1925** R. Fry *Let.* 11 Nov. (1972) II. 584 Derain..lets 'em [*sc.* dealers] have any old thing, or rather, what's much worse, any new thing which he's knocked off. **1970** W. Garner *Puppet-Masters* xv. 124 Look, you could knock off a few hundred words on Baxx without so much as scratching the surface of your *magnum opus.*

f. (Earlier and later examples.)

1811 Jane Austen *Let.* 6 June (1952) 288 As *you* knock off a week from the end of her visit, & *Martha* rather more from the beginning, the thing is out of the question. *a* **1817** —— *Persuasion* (1818) III. ii. 26 Every comfort of life knocked off! Journeys, London, servants, horses, table. **1869** *Bradshaw's Railway Manual* XXI. 384 A great deal is knocked off from our claim against the Grand Trunk. **1926** J. Buchan *Dancing Floor* I. i. 11 First string of the 'Varsity mile. Believed..to be going to knock five seconds off his last year's time. **1966** *Melody Maker* 15 Oct. 19 (Advt.), Quality instruments at knocked-off prices. **1972** *Daily Tel.* 30 Mar. 19/2 The gloomy assessment..knocked 12p off ICI's share price in London.

g. *Cricket.* Of batsmen, to score the runs requisite for victory, or to oblige (a bowler) to be taken off by scoring heavily from his bowling.

1851 J. Pycroft *Cricket Field* ii. 30 If in the field..and trying hard to prevent these few runs being knocked off by the last wickets, I know of no excitement so intense. **1860** *Baily's Monthly Mag.* Mar. 34 These two gentlemen scored 123 runs between them, knocking off Caffyn, Jackson, Parr, [etc.]. **1963** A. Ross *Australia 63* 18 Pullar and Cowdrey knocked off the 49 required to win without actually being separated.

h. [*imp.* use of 10 c.] *knock it off!*: leave off! stop it!

1902 *N.Z. Illustr. Mag.* V. 488 Knock it off, boys. **1945** D. Dempsey *It ain't Brooklyn* in *Best One-Act Plays 1944* 28 Will you knock it off, please? **1961** J. Heller *Catch-22* (1962) xxvii. 294 'Hey, knock it off down there,' a voice rang out from the far end of the ward. 'Can't you see we're trying to nap?'

i. *slang.* To steal, to rob. Also *transf.*

1919 *Athenæum* 8 Aug. 729/1 A curious term used by a Tommy, in 'explaining' his deficiencies of kit, is 'Someone knocked it off' for 'Someone pinched (or made away with) it'. **1925** E. Wallace *Mind of Mr. J. G. Reeder* vi. 224 A big-shouldered man whose speciality was the 'knocking-off' of unattended motor-cars. **1938** F. D. Sharpe *Sharpe of Flying Squad* i. 14 They learn to 'knock things off'. **1956** C. Willock *Death at Flight* iv. 42 Mr. Goss had shown himself willing to knock off a pheasant himself. **1959** *New Statesman* 26 Sept. 404/2 After quietly knocking off a couple of retail shoe chains at the end of 1958, he entered the public takeover lists and won control of a Connoisseur's gobbet—Temperance Billiard Halls. **1960** *Observer* 24 Jan. 5/2 The boys either knocked off a hut where they knew gelly was kept or straightened a quarry man. **1963** J. Prescot *Case for Hearing* i. 16 Always dropping in on me..with search-warrants..and turning over that place of mine as if they expected to find some knocked-off gear there. **1969** *Sunday Truth* (Brisbane) 20 July 30/5 Only a few weeks after he finished up at St. Laurence's Christian Brothers College, Luzzcek knocked the place off. **1973** A. Hunter *Gently French* iii. 24 Just met a bloke..in the nick... Him what was in there for knocking-off cars.

j. *slang* (orig. *U.S.*). To kill; to murder.

1919 E. Streeter *Same Old Bill* 28 Im goin to rite just as much as I can. Thats partly sos you wont worry an partly so that if I get knocked off you will have something to amuse you in case you go into a convent. **1929** *Papers Mich. Acad. Sci., Arts & Lett.* X. 304 *Knocked off*, killed. **1942** E. Paul *Narrow St.* xxiv. 217 Hitler..ordered the blood purge which knocked off Roehm, Von Schleicher, and others among his former pals. **1943** P. Cheyney *You can always Duck* iv. 75 A United States Army officer was knocked off in a joint of his off Mount Street. **1948** Partridge *Dict. Forces' Slang* 107 *Knock off*, to kill. **1959** H. Hobson *Mission House Murder* xxii. 145 One of my boys..got knocked off—an' nobody does a damn' thing about who knocked him off. **1973** C. Mullard *Black Britain* I. ii. 24 In one village a white launched a murder campaign because 'he liked knocking off blacks'.

k. *Underworld slang.* To arrest (a person); to raid (an establishment).

1926 F. D. Wilkinson in *Flynn's* 6 Feb. 58/1 'Willie of Detroit is here and is knocking everybody off.' (Meaning, arresting them.) **1930** *Amer. Mercury* Dec. 456/2 *Knock off*, to raid; to arrest. 'The feds knock off the scatter.' **1930** G. Smithson *Raffles in Real Life* xix. 256 About ten days or so after being 'knocked off'..the Chief Warder came to my cell. **1939** 'D. Hume' *Heads you Live* ii. 24 You..acted as a so-called hostess at the Angel Club in Dean Street for a year before it was knocked off. **1960** J. Stroud *Shorn Lamb* iii. 33 There was two other boys wiv Egg when you knocked him off, why ain't they 'ere? **1969** R. V. Beste *Next Time I'll Pay* xi. 157 You're the sort who'd knock off his mother because she hadn't got a lamp on her bike five minutes after lighting up time.

l. *slang.* To copulate with, to seduce (a woman).

1952 S. Kauffmann *Philanderer* (1953) viii. 134 Hell, she isn't much,..but she's all there is around here. And if you don't want her, I don't mind knocking her off. **1965** A. Prior *Interrogators* v. 69 Do you think that young twit Wilkinson is knocking her off? **1970** G. Greer *Female Eunuch* 265 The vocabulary of impersonal sex is peculiarly desolating. Who wants to..'knock off a bit? of belly? of crumpet?' **1974** *Times Lit. Suppl.* 11 Oct. 1109/4 Knocking off his best friend's busty wife during boozy sprees on leave in Soho.

12. knock out. a. Also, to stun or kill by a blow.

1903 *Sun* (N.Y.) 2 Dec. 1 Scott's reputation is excellent, and the managers fear that he has been knocked out and robbed. *a* **1918** W. Owen *Coll. Poems* (1963) 7 Ye get knocked out; else wounded—bad or cushy; Scuppered; or nowt except yer feelin' mushy.

d. *to knock out of time*: also in extended use.

1874 Trollope *Phineas Redux* II. xxviii. 228 You'll come all right after a few weeks. You've been knocked out of time;—that's the truth of it. **1891** *Young Man* Apr. 140/2 A [bicycle] ride of ten miles within the hour may mean comfort and the capability of doing another twenty easily; a ride of eleven miles in the hour may just mean knocking a man out of time. **1970** *Brewer's Dict. Phr. & Fable* (ed. 12) 613/2 *To knock out of time*, to settle one's hash, to double him up.

h. *trans.* To earn. *Austral.*, *N.Z.*, and *U.S. slang.*

[**1871** C. L. Money *Knocking about in N.Z.* ii. 18 They knocked out in this day as much gold as sufficed to make them afterwards two rings.] **1873** V. Pyke *Story Wild Will Enderby* (ed. 4) I. xiv. 62 Two industrious young men who worked very hard for a bare living—'just knocking out tucker', as the phrase went. **1874** —— *Adventures G. W. Pratt* 12/2, I can knock out tucker enough for the pair of us. **1920** *Sat. Even. Post* 27 Mar. 3/2 At that I was knocking out about eighteen hundred dollars per annum selling cigars out of South Bend. **1959** Baker *Drum* (1960) II. 123 *Knock out*, to earn (a sum of money).

i. *trans.* To eliminate, remove forcibly, get rid of, destroy. orig. *U.S.*

1883 'Mark Twain' *Life on Mississippi* 465 The religious feature has been pretty well knocked out of it [i.e., Mardi-Gras at New Orleans]. **1889** *Kansas City* (Missouri) *Times & Star* 17 May, By a vote of 11 to 9 the Missouri senate knocked out the legislative reduction of tolls here by the Bell Telephone Company. **1904** *Sun* (N.Y.) 5 Aug. 4 In power, the Democrats wouldn't knock out protection if they could. **1927** J. N. McIlwraith *Kinsmen at War* xvii. 170, I will have to knock that idea out of Lucy's head too, straightway. **1933** F. Baldwin *Innocent Bystander* (1935) ii. 30, I got a good deal of it knocked out of me. **1944** *Return to Attack* (Army Board, N.Z.) 15/1 In the Bir el Gubi area the 22nd Armoured Brigade..knocked out forty-five enemy tanks. **1955** *Times* 28 June 4/4 It is now believed that even if all the major ports of the United Kingdom were knocked out by atomic attack sufficient food for the population could still be passed through minor ports. **1971** *Daily Tel.* 17 Dec. 1 India claimed to have knocked out forty tanks in a major battle on the Kashmir front.

j. *trans.* (*Founding.*) To separate (a flask) from a casting contained inside it, or (a casting) from a flask containing it.

1906 *Jrnl. Iron & Steel Inst.* LXX. 174 The castings were all made in green sand, and were allowed to cool before being 'knocked out', *i.e.* taken from the sand. **1942** *Engineering* 6 Mar. 195/2 One difficulty was to get cool sand after the castings had been knocked out. **1955** H. E. Crivan in W. C. Newell *Casting of Steel* vi. 227 Heavier, dry sand work can be knocked out over a grid using hammers to loosen the sand.

k. *to knock oneself out*: to make a considerable effort, to apply oneself energetically (to the point of exhaustion).

1936 *Mademoiselle* Mar. 43/2 All the fancier lassies.. are practically knocking themselves out in an effort to get to Hollywood. **1951** Green & Laurie *Show Biz* p. xxi, They like 'knocking themselves out' for *Variety*.

l. To give (a person) enjoyment, to excite. Often *refl.* and in *pass. slang* (orig. *U.S.*).

1942 *Amer. Mercury* July 95 *Knock yourself out*: have a good time. **1944** *New Yorker* 8 July 27/1 There are times when Duke laughs naturally and exuberantly; for example, when the boys..are competing to see who can whistle the lowest note. 'I knock myself out,' he says. **1947** *Band Leaders & Record Rev.* Feb. 20 'When I heard it,' Ella Mae says, 'it knocked me out.' **1953** D. Wallop *Night Light* xix. 236 It's pretty hard to be knocked out with a baby when you know its old man is bored with the whole idea. **1956** B. Holiday *Lady sings Blues* (1973) ii. 26, I used to make them crazy dishes... This used to knock him out. When my time was running out, he made me an offer to stay on and cook for him. **1957** J. Kerouac *On Road* (1958) III. iv. 202 A man who knocked himself out every evening and let the others put the quietus to him in the night. **1966** *Melody Maker* 7 May, I only heard half an hour of Ornette but I wasn't knocked out at all.

13. knock over. a. (Further examples.)

1852 Dickens *Bleak Ho.* (1853) xxix. 285 Such a resemblance..that it completely knocked me over. **1882** W. D. Hay *Brighter Britain!* I. vii. 184 If a single bushman could not have knocked that tree over before dinner time, he would not have been worth wages.

c. *trans.* In warp knitting: to cause (a stitch) to pass *over* the head of the needle on which it was held.

a **1877** Knight *Dict. Mech.* II. 1238/1 *Knocking-over bar*, the bar against which the loops and fabric are drawn as the needles retreat, so that the loops shall be thrown or knocked over the heads of the needles. **1885** W. T. Rowlett tr. *Willkomm's Technol. Framework Knitting* II. iii. 145 These sinkers..must move up and down, and backwards and forwards, so as to sink the thread into loops, bring them forward under the needle beards, land and knock over the old stitches, and..lock in the new stitches, and take them to the back of the needles. **1952** D. F. Paling *Warp Knitting Technol.* i. 6 The presser is now withdrawn and the needle bar continues its downward motion, thus causing the fabric loops to pass further up the needle beards until finally they are knocked-over the needle heads as the latter pass below the level of the sinkers. **1964** H. Wignall *Knitting* ii. 28 The old loop is now cast-off or knocked over.

d. *trans.* (*Underworld slang.*) To rob (a person), to burgle (a building); to steal (from).

1928 *Detective Fiction Weekly* 7 July 52/2, I just got knocked over for that wad we jest lifted... My pocket was picked... I was tapped, touched, if that's any plainer. **1932** *Ibid.* 6 Feb. 129/2, I ain't knocked nothin' over for some little time now. **1937** C. R. Cooper *Here's to Crime* iv. 89 There's the real fun of bank-robbing—running the roads. Old Harve used to love it. I've seen him run roads when he had no intention of ever knocking over a can. **1940** *Illustr. London News* 26 Oct. 548/2 The job looks easy enough—a big hotel at Tropico Springs that any fool could 'knock over'. **1941** K. Tennant *Battlers* i. 9 Life 'on the track' was not so bad, with good places to camp and 'cockies' sheep to knock over.

16. knock up. a. *Bootmaking.* To cut or flatten the edges of the upper after its attachment to the insole.

1905 *Westm. Gaz.* 30 Oct. 7/3 A mechanical device for trimming off the surplus material from the lasted boot before it is 'knocked up', and a machine for the 'knocking-up' process itself, the latter guaranteed to 'knock up' between 400 and 500 pairs of boots per week.

d. (Later examples.) Also, to prepare (food) quickly. *U.S.*

1869 L. M. Alcott *Little Women* II. v. 61 Don't cry, dear, but just exert yourself a bit, and knock us up something to eat. **1890** *Harper's Mag.* May 894/2, I jest killed a chicken, and knocked up a few biscuit. **1931** H. Nicolson *Diary* 14 Aug. (1966) 87 He has got out several tenders for printing... He and Joseph have..knocked up a dummy lay-out. **1967** *Official Jrnl. Patents, Trade Marks & Designs* (Austral.) XXXVII. 1538/2 Plumbing means can be purchased ready made by factories whereas they once might have to be 'knocked up'..from basic materials. **1972** *Shooting Times & Country Mag.* 4 Mar. 21/2 They will knock you up a meal to hold you through the coldest day's fishing or wildfowling.

e. (Further examples.)

1860 *Baily's Monthly Mag.* Oct. 41 Tinley in a trice knocked up 8. **1955** *Publ. Amer. Dial. Soc.* xxiv. 37 This adds up to over $1500 per week which must be *knocked up* —just to meet operating expenses.

f. (This sense is not current in the U.S.)

1973 *National Observer* (U.S.) 3 Feb., Fielding's guidebook considerately explains that a male host may quite casually tell a female American house guest that he will 'knock you up at 7:30 tomorrow morning'. The term, of course, conveys nothing more than a rapping at the door until one is awakened.

h. (Later examples.)

1897 A. Beardsley *Let.* 25 Feb. (1971) 259, I am aghast at the amount of travelling she [*sc.* his sister Mabel] has to get through before the tour comes to an end. I do hope she won't 'knock up' while she is over there. **1941** I. L. Idriess *Great Boomerang* x. 78 They travelled fast then, taking the chance. But their horses knocked up.

j. To make (a woman) pregnant; (*less commonly*) to have sexual intercourse with (a woman). *slang* (orig. *U.S.*).

1813 C. Earle *Diary* 12 Apr. in J. McPhee *Pine Barrens* (1971) ii. 33 William Mick's widow arrived here in pursuit of J. Mick, who she says has knocked her up. **1836** D. Crockett *Exploits & Adv. Texas* vii. 97 Nigger women are knocked down by the auctioneer, and knocked up by the purchaser. **1860** Hotten *Dict. Slang* (ed. 2) 166 *Knocked up*. In the United States, amongst females, the phrase is equivalent to being *enceinte*, so that Englishmen often unconsciously commit themselves when amongst our Yankee cousins. **1925** E. Hemingway *In Our Time* (1926) 165 Hell, no girls get married around here till they're knocked up. **1934** H. Miller *Tropic of Cancer* 241 Nearly all the co-eds had been knocked up some time or other. **1952** B. Malamud *Natural* 133 You haven't knocked up a dame maybe? **1971** H. C. Rae *Marksman* I. vi. 51 He screwed her, knocked her up first go and.. married her..before she could even contemplate abortion. **1973** E. Bullins *Theme is Blackness* 170 The girls all got knocked-up and set up homes, got married, went on the block or on welfare or turned into booze hounds.

knock, *sb.*[1] Add: **1.** (Earlier and later examples of *double knock.*)

1819 Keats *Let.* 13 Mar. (1958) II. 46 The variations of single and double knocks. **1883** C. J. Mathews *Patter versus Clatter* ii. 10 (*Double knock*, L.H.) There, someone come to call. Polly, go and see who it is; stop, child, take off your apron, it's a double knock.

b. A misfortune, a rebuff, a blow; adverse criticism. Freq. in phr. *to take the knock*: to sustain a severe financial or emotional blow, to suffer a setback.

The *fig.* examples in Dict. belong here.

1890 *Globe* 21 Apr. 6/1 A broken backer of horses who has taken, what is known in the language of the turf, as the knock. **1900** E. Wells *Chestnuts* xxiii. 226 When a prominent backer takes the knock racing, he sometimes has the greatest difficulty to avoid his creditors. **1905** 'H. McHugh' *You can search Me* iii. 50 There are only four people in New York city who can write criticisms—the rest of the bunch are slush-dealers, and a knock from any one of them is a boost. **1906** Galsworthy *Man of Property* III. iv. 322 Here's a poor devil whose mistress has just been telling him a pretty little story of her husband... He's taken the knock, you see. **1929** D. Runyon in *Hearst's International* Nov. 73/1 It will be a knock to his reputation. **1930** V. Palmer in *Bulletin* (Sydney) 30 Apr. 38/3 [McCurdie] lay there... 'He's taken the knock,' said a cattle-buyer... In a moment a change came into the atmosphere around the sleeping man. **1930** D. Runyon in *Collier's* 13 Sept. 7/1 They are always doing something which is considered a knock to the community, such as robbing people. **1936** A. Huxley *Eyeless in Gaza* vi. 54 'One's had a pretty bad knock,' he added self-consciously, in that queer jargon which he imagined to be colloquial English... That 'bad knock' was a metaphor drawn from the boxing contests he had never witnessed. **1948** V. Palmer *Golconda* xiv. 111 He saw himself..ready to stand up and take the knock if they got into trouble with the john. **1955** *Times* 19 Aug. 2/5 In a dress suit much too large for him, he is on top of the world by submitting with such cheerful readiness to its knocks. **1959** *Encounter* Aug. 7/1 Like other institutions of the Establishment, it has taken a knock or two in recent years. **1962** *B.S.I. News* June 9/1 Advertising has had some hard knocks from its critics recently. **1973** A. Behrend *Samarai Affair* ii. 24 The pilot..in the event of an accident will.. [be] summoned to appear before the Pilotage Committee to explain his actions and take the knock if held to blame.

c. A knocking noise, or knocking noises, in an engine; *spec.* in a reciprocating internal-combustion engine, noise caused by a very abrupt rise in pressure in the cylinder as a result of too rapid combustion (in spark-ignition engines, the sudden spontaneous ignition of all the unburnt portion of the mixture before the flame from the sparking plug reaches it); faulty combustion of this character. Cf. *knock *v.* 5 b.

1899 J. Perry *Steam Engine* v. 115 In double-acting engines we can often utilise the inertia forces to alter the point in the crank pin path at which the knock occurs, so that it shall not produce such serious effects. **1903** M. P. Bale *Gas & Oil Engine Managem.* iv. 61 Knocking in the Cylinder.—This often arises from premature firing of the charge before the end of the compression stroke is reached, thus throwing a greater pressure than usual on the piston before it commences the power stroke, and causing a jar or knock as the crank turns the dead centre. **1903** *Cassell's Cycl. Mech.* 3rd Ser. 264/1 Locating 'Knock' in Steam Engine. **1908** *Motor Cycle* 15 Jan. 46/1, I have had a number of letters lately referring to the existence of 'knock' in engines that have run a year or two in private hands. **1920** *Cornh. Mag.* Sept. 314 The carbon knock, the ignition knock, and the bearing knock are fairly simple propositions. **1927** W. Deeping *Doomsday* xxv. 265 Half-way up the long hill..'Cherry's' engine developed a sudden and rather fearsome 'knock'. **1933** [see **compression-ignition*]. **1939** Croft & Tangerman *Steam-Engine Princ.* (ed. 2) xiii. 410 By far the commonest causes of knocks are water in the cylinder and loose bearings. **1956** Molloy & Lanchester *Automobile Engineer's Ref. Bk.* v. 5 The well-known effect of ignition timing on knock is due to the fact that the relative timing of the piston and the spark-ignited flame controls the pressure in the end-gas. **1963** C. Campbell *Sports Car Engine* ix. 181 During knock more heat is transferred to the cylinder walls. **1973** A. Parrish *Mech. Engineer's Ref. Bk.* II. 17 Correct choice of mixture strength, ignition timing, fuel (octane number) and good combustion chamber design will allow smooth combustion without knock which occurs if the end gas reaches the condition where self-ignition causes an explosion of all the mixture remaining in the chamber.

3. *Cricket.* An innings; a spell at batting (in a match or at practice).

1889 J. Lillywhite *Cricketers' Ann.* 72 Surrey were fortunate to get first 'knock', and..were able to just reach the second hundred. **1898** G. Giffen *With Bat & Ball* i. 2 At last..I would..bowl for a little while; and then they began to give me an occasional knock. **1900** *Captain* III. 200/1 'You play cricket yourself, then?' 'Oh, I have an occasional knock.' *Ibid.* 210/1 'W.G.' advises every batsman to have a knock..before going in. An over or two at the nets loosens your muscles. **1909** *Pearson's Mag.* Aug. 180/1 Crofton's had won the toss and taken first 'knock'. **1927** *Observer* 27 Nov. 28 His knock..included eight boundaries. **1958** 'N. Blake' *Penknife in my Heart* iii. 50 I'm taking first knock. I've got to be sure you'll go in when it's your turn. **1970** *Times* 26 Aug. 11/8 A fine knock by Mushtaq, who batted for two hours and hit ten fours.

4. *knock for knock*: applied to an agreement between insurers that each will pay his own policy-holders without regard to the question of liability.

1906 *Daily Chron.* 26 July 6/6 Mr. Fairbank said that the 'knock for knock' agreement had never paid with the horse vehicles. **1927** B. C. Hoskins *Insurance Lexicon* 127 *Knock for knock agreement.*—An arrangement made between Companies..for dealing automatically with collisions between vehicles owned by their respective insureds; each Company undertakes to pay for the damage to its own insured's vehicle irrespective of the question of liability as between the parties in collision. **1958** *Manch. Guardian* 11 June 9/6 The knock-for-knock agreement is an arrangement whereby when two insured vehicles have been in collision each insurance company pays for the damage to the car it has insured,..without regard to the degree of blame, if any, of the driver. **1972** *Mod. Law Rev.* XXXV. I. 18 Some types of cases which are handled by small claims courts in other jurisdictions are dealt with in England in ways which obviate the necessity for a claim. Perhaps the most significant example is knock-for-knock agreements among motor-vehicle insurers.

5. a. (An act of) copulation; so *on the knock*, engaged in prostitution. **b.** *Austral.* Phr. *to do a knock with*: (see quot. 1941).

1933 N. Lindsay *Saturdee* 138 Supposin' I was to do a knock with girls, what 'ud I say to them? **1937** Partridge *Dict. Slang* 460/2 *Knock*, a copulation. **1941** Baker *Dict. Austral. Slang* 23 *Do a knock* (*line*) *with*: to take an amorous interest in a member of the opposite sex. **1969** D. Bagley *Spoilers* i. 11 Maybe she was on the knock.

6. Special Comb.: **knockmeter,** an instrument for measuring the intensity of knock in the cylinder of an internal-combustion engine; **knock rating,** (the determination of) the insusceptibility of a fuel to knock.

1934 *Jrnl. R. Aeronaut. Soc.* XXXVIII. 353 Knock intensity is measured by a bouncing pin, in conjunction with either a knockmeter or a gas-evolution burette. **1960** V. B. Guthrie *Petroleum Products Handbk.* iv. 17 The test engine is equipped with a pressure-sensitive pickup mounted in the cylinder head in direct contact with the combustion chamber. A knockmeter is used in conjunction with this pressure-sensitive element to indicate on a scale the intensity of the engine knock... A fuel that is to be tested is brought up to a standard knock intensity, as indicated on the knockmeter, by adjustment of the engine compression ratio. **1932** *Engineering* 8 July 45/3 The marked effect of cylinder temperature upon the relative knock ratings of fuels was observed by Heron in 1928. **1933** *Aircraft Engineering* Aug. 177/1 To have an agreed scale of knock-rating for aviation fuels is no less important. **1959** *B.S.I. News* Aug. 14 Two draft ISO recommendations covering the motor and research methods of determining knock rating.

knock-. Add: **knock-knock** *sb.*, *v.*, and *int.* in various senses (see quots.); **knock-toe,** a galley-punt; **knock-u·pable** *a.* (*nonce-wd.*), likely to be 'knocked up' or weak; **knocku·pedness** (*nonce-wd.*), the state of being 'knocked up' or weak.

1904 *Daily Chron.* 2 July 8/1 The knock-knocking at the door sending a thrill through the pulse. **1936** *Variety* 19 Aug. 25/5 Manager Russell Bovim, of Loew's Broad, Columbus, cashed in handsomely on the 'Knock Knock' craze now sweeping the country. **1941** C. Graves *Life Line* 179 Certain trawlers have the job of sweeping for magnetic mines (known as 'Maggies') and the latest acoustic mines (known as 'knock-knocks'). **1957** O. Nash *You can't get there from Here* 151 Who, rapped Mr. Webster, escapes an escapee? That, knock-knocked Mr. Merriam, is what puzzles me. **1959** I. & P. Opie *Lore & Lang. Schoolch.* v. 82 A craze for Wellerisms is apt to develop in a school in the same way that there are still sometimes crazes for limericks, Little Audrey jokes, Knock-knocks, and Shaggy-dog stories. **1961** Partridge *Dict. Slang Suppl.* 1161/1 *Knock! knock!* A c.p., dating from the middle of Nov. 1936... Orig. ex U.S. It is used, esp. among busmen, by a person about to tell a dirty story or, esp., to make a pun, gen. in doubtful taste. **1974** *Radio Times* 19–25 Oct. 59 'Knock, knock.' 'Who's there?' 'Richard Milhous.' 'Richard Milhous who?' 'Ah...how quickly people forget.' **1903** W. C. Russell *Overdue* vi. 104 It is the Deal galley-punt too, called in the parts she belongs to 'knocktoe'. **1929** F. C. Bowen *Sea Slang* 80 Knock-toe, an old name for the Deal lugger-rigged galley punt, in which there was little room for the feet. **1857** Geo. Eliot *Let.* 5 Apr. (1954) II. 314 For some time I have been unusually weak and knock-upable. **1855** D. G. Rossetti *Let.* 19 Sept. (1965) I. 271, I am very sorry indeed to hear of your knockupedness but I warned you about that window.

knock-about, knockabout, *a.* (*sb.*) Add: **A.** *adj.* **1. b.** Also *transf.*

1914 A. Huxley *Let.* Feb. (1969) 57 The whole city is permeated with the Mission to Undergraduates—the Bp of Oxford being heard twice nightly in a knock-about sermon at St. Mary's. **1924** J. Buchan *Three Hostages* xiii. 183 He liked plays with shooting in them, and knockabout farce. **1969** *New Yorker* 31 May 110/2 His strong colloquial flavour and knock-about verbiage cannot be conveyed in French. **1974** *Daily Tel.* 5 Oct. 16/1 One would have expected the argument in the Election campaign to have been conducted with a high level of rigour and seriousness, which would not be incompatible with some of the traditional knockabout fun.

3. Designating a class of sloop-rigged sailing-yacht (see quots. 1894, 1897); also, designating a sailing-yacht without a bowsprit. Also as *sb. N. Amer.*

1894 in *Forest & Stream* (1895) 12 Jan. 35/2 A knock-about boat is a seaworthy keel boat (not to include fin-keels) decked or half-decked, of fair accommodations, rigged simply, without bowsprit, and with only mainsail and one head sail. **1897** *Outing* (U.S.) Dec. 235/1 The knockabout class, which originated in Boston a few seasons ago, has much to recommend it... It is free from all freakiness. It has no fin-keel... With a moderate sail area it is under control at all times... This class is limited to five hundred square feet of sail. All are keel-boats, and all must be under twenty-one feet on the load water-line. **1919** *Canad. Fisherman* 695/1 The fine new knockabout schooner, General Haig, the first of her kind ever built here. **1927** G. Bradford *Gloss. Sea Terms* 96/1 *Knock-about*, a sloop or schooner without a bowsprit and whose jib sets from a stay at the stern. **1970** *Amer. Neptune* XXX. 196 The term knockabout soon was restricted to the type lacking a bowsprit and was further used attributively to designate any rig in which a long fore overhang took the place of a bowsprit. *Ibid.* 197 The single knock-about yawl was *Arapahoe*, built in 1893.

b. More generally, descriptive of small yachts or dinghies. Also as *sb.*

1904 *N.Y. Even. Post* 21 May 6 There are numerous knockabouts and other small yachts in the Pawcatuck River. **1921** *Yachting Monthly* XXXII. 105/2 The usual lug and mizen clinker built 'knockabout' boats, common to seaside watering places. **1970** *Motor Boat & Yachting* 16 Oct. 39/2 It [*sc.* the balanced lug] is a powerful sail and very well suited to knockabout dinghies.

B. *sb.* **1.** (Examples of sense 'knockabout performance'; also *transf.*)

1899 Ade *Fables in Slang* (1900) 83 These two Troupers began their Professional Career with a Road Circus.. doing a Refined Knockabout in the Grand Concert or Afterpiece. **1930** *Observer* 2 Mar. 15/3 The actors attack both the recitative and knockabout delightfully. **1955** *Times* 9 Aug. 2/5 One called The Cold War, in which Mr. Wisdom, suffering from a fever, tries out different miraculous cures evincing immediately their different after-effects, is the most intricate piece of knockabout seen on the London stage for some time. **1959** *Ann. Reg. 1958* 50 The proceedings were wound up by a forcible speech from the leader of the party, Mr Grimond, in a style of party-political knockabout in which he demonstrated that it was easier to abuse the other parties than to explain the purposes of one's own. **1970** *New Society* 5 Mar. 406/3 For while there is some knockabout..most of the pieces here are technically of a high order.

3. A small motor vehicle for casual use.

1956 W. H. Whyte *Organization Man* (1957) vi. 71 A little knock-about for the wife to run down to the station in.

knock-back, knockback, *sb. dial.*, *Austral.* and *N.Z. colloq.* [Phr. *to knock back* (see *knock *v.* 7** a) used as sb.] A refusal, a rebuff.

1898 *Evesham Jrnl.* 21 May 7/1 He..objected to the powers of the guardians being relegated to the officers. It was a knock-back. **1902** *Eng. Dial. Dict.* III. 478/2 It was a nasty reply—a complete knock-back. **1919** W. H. Downing *Digger Dial.* 31 *Knock-back*, a refusal. **1933** *Bulletin* (Sydney) 18 Jan. 17, I sought casual work... Result, six flat knockbacks. **1941** K. Tennant *Battlers* xvii. 182 She could take a 'knock-back' as though it didn't matter. **1943** F. Sargeson in *Penguin New Writing* XVII. 56 And I said to myself, well, a knock-back from one of yous isn't going to make me lose any sleep. **1960** N. Hilliard *Maori Girl* II. xiv. 161 She had mentioned the knock-back from the Tallahassee Milk Bar. **1962** [see *daddy 2]. **1972** *Sunday Mail* (Brisbane) 9 Jan. 5/2 When he lead[s] the conversation..on to Jesus and Christianity we don't get any knockbacks.

knock-down, *a.* and *sb.* Add: **A.** *adj.* **1.** (Earlier and later lit. examples.)

1794 *Sporting Mag.* IV. 78/2 After the first knock-down blow, Johnson attempted to shift. **1971** *Nature* 15 Oct. 441/1 They are used against houseflies, mosquitoes and cockroaches..; their great advantage is that pyrethrin I has a high kill rate and pyrethrin II a high knock down rate.

2. Also, *knock-down book, fee. U.S.*

1888 *Harper's Mag.* Nov. 934/2 Bills for knock-down fees are presented for payment to auctioneers every month. *Ibid.* 937/2 The knock-down book records the price, buyer, and all particulars of every sale in the Auction-room.

3. Add to def.: sold as, or in the form of, a number of separate parts that require to be assembled. (Earlier and later examples.)

1795 W. Winterbotham *Hist. View Amer. U.S.* III. 305 Articles of exportation [were]..231,776 Barrels of dried and pickled fish..48,860 Shook or knock-down casks. **1952** *Archit. Rev.* CXI. 241/1 Swedish 'knock-down' furniture. *Ibid.*, The production of 'packaged' furniture (an adjective which is preferable to either knock-down or demountable, since it not only sounds better, but is more definitive). **1969** *Sunday Mail* (Brisbane) 2 Feb. 26/6 Cars imported in the complete knockdown form. **1971** *Rand Daily Mail* 4 Sept. 10/3 The boat is imported in a knock-down form and is assembled here. **1972** 'E. Lathen' *Murder without Icing* (1973) vii. 65 Millions of Americans..had stayed up late Christmas Eve to struggle with knockdown toys.

B. *sb.* **2.** Also *knock-down* (*and*) *drag-out*: a free-for-all, a rough-and-tumble fight; also *transf.* and *attrib. U.S. colloq.*

1827 J. F. Cooper *Prairie* I. iv. 93 Making it a real knock-down and drag-out! **1834** *Amer. Railroad Jrnl.* III. 304/1 He was one of our careless unconcerned knock down and drag out looking sort of fellows. **1932** *Tulsa* (Okla.) *World* 13 Mar. v. 2/3 A knock-down-and-drag-out battle between Roosevelt and Smith forces. **1941** Auden *New Year Let.* II. 34 The hard self-conscious particles Collide, divide like numerals In knock-down drag-out laissez-faire. **1949** B. A. Botkin *Treas. S. Folklore* II. iii. 251 When it comes to knock-down-and-drag-out..political bickering..well, as Donald Davidson's typical Middle Georgia country gentleman, 'Cousin Roderick', puts it, 'Politics is for lawyers.' **1952** B. Ulanov *Hist. Jazz in Amer.* (1958) viii. 89 Whole choruses..and ensemble sections..are delightful..in knockdown and dragout choruses. **1968** *Punch* 1 May 624/1 An elderly bum who's just sold him the Lincoln Centre at the knock-down drag-out price of fourteen bucks.

4. The heeling of a ship by the force of the wind.

1888 *Scribner's Mag.* May 526/1 Every bit of that water came in through the hatch at the time of the knock-down. **1926** H. Howard *Yacht 'Alice'* 13 Raised deck amidships which gives excellent room below, increases the structural strength and adds greatly to her stability and righting moment in case of a knockdown. **1951** H. I. Chapelle *Amer. Small Sailing Craft* 242 Such a maneuver takes room and also might lead to a knockdown when the boa pays off, as she has no way on. **1973** J. R. L. Anderson *Death on Rocks* xi. 201 Anthea had seen our knockdown... They stood by us until it was clear that we were not going to sink.

5. An introduction. *U.S.*, *Austral.*, and *N.Z. slang.*

1865 'D. Rattlehead' *Adventures of Fudge Fumble* v. 61, I asked the young man if he would go down some night and give me a 'knock down' to the family, and Miss Kate, more especially. **1885** 'Phudge Phumble' *Adventures Greenhorn in Gotham* 39 He loved Lucy at first sight and couldn't help it. So he managed to get a knock down to her. **1896** Ade *Artie* iii. 24 Take me over and gi' me a knock down to the queen in the corner. **1916** C. J. Dennis *Songs Sentimental Bloke* 125 *Knock-down*, a formal introduction. **1930** *National Education* (N.Z.) May 197/2, I heard one young fellow ask another to 'give him a knockdown to that tart in the green skirt'. I gathered that he was asking for an introduction to a young lady!! **1937** J. Weidman *I can get it for you Wholesale* xvi. 156 'Meet the stable,' he said, waving his hand to take in the girls... 'You want a knockdown to something?' **1945** *Chicago Daily News* 7 Sept. 31/3, I know her well! I'll sell you a knockdown to her for two-bits. **1946** F. Sargeson *That Summer* 66 He called me over and gave me a knock-down, and she was certainly the goods. **1974** *Telegraph* (Brisbane) 28 May 24/1, I think you should meet Fred. And maybe Bert, too. So here's for the knock-down.

knocked, *ppl. a.* Add: **2. knocked-down.** In the form of a number of separate parts that require to be assembled.

1776 [in Dict.]. **1908** *Sears, Roebuck Catal.* 371/1 The patent knocked down construction..give [*sic*] this splendid rocker additional strength and rigidity and permit its shipment in a package 33 inches long. **1950** *Engineering* 3 Feb. 139/2 For carriage in merchant vessels all vehicles..should possess a 'knocked-down' height of less than 7 ft. 9 in. **1960** *Times Rev. Industry* July 74/1 Cars..imported..in completely knocked down packs. **1971** *Timber Trades Jrnl.* 14 Aug. 58/1 Filters in a new 'knocked down' kit form.

knocker. Add: **1. e.** A fault-finder, one who is addicted to captious criticism. (Cf. *knock *v.* 2 f.) *colloq.* (orig. *U.S.*).

1898 in Wentworth & Flexner *Dict. Amer. Slang* (1960) 308/2 That pack of knockers and snapping curs. **1901** [see *knock *v.* 2 f]. **1911** *Daily Colonist* (Victoria, B.C.) 22 Apr. 4/2 The Cranbrook Herald says that the 'pestilential knocker' has been doing his best to injure the Southeast Kootenay by misrepresenting the condition of a party of English settlers. **1923** J. Moses *Beyond City Gates* 154 The 'knocker' of his home town is, on this line of deduction, a 'knocker' of his Empire, a destroyer of thought, labour, and enterprise. **1928** *Sunday Express* 18 Mar. 5/2 All the knockers were there,..yearning to find fault. **1956** W. H. Whyte *Organization Man* (1957) ix. 124 This system virtually ensures that the over-zealous or the 'knocker' type of man will not get ahead. **1958** *Times Lit. Suppl.* 24 Jan. 37/3 Today it would be difficult to get together such a team of 'knockers' as Harold Stearns did for his *Civilization in the United States*. Cheerfulness has been creeping in among the intelligentsia. **1962** M. Harris in P. Coleman *Austral. Civilization* 57 It is said that Australians are 'knockers'; that is, they gain pleasure from seeing superiority in talent, intellect or energy reduced to the scale of average mediocrity. **1969** *Telegraph* (Brisbane) 5 Dec. 3/2 Knockers are people who identify the ordinary Australian bloke as an easy-going, irresponsible oaf who spends more time drinking and arguing with his mates than working. **1972** *Shooting Times & Country Mag.* 1 July 26/3 Today the 'knockers' seem to delight in slamming anything British.

2. d. *Austral.* and *N.Z.* (See quots.)

1933 L. G. D. Acland in *Press* (Christchurch, N.Z.) 4 Nov. 15/7 *Knocker*, a small leather pad fixed near the heel of shears to keep the blades from closing too far. **1938** R. M. Burdon *High Country* viii. 84 A piece of rawhide known as a knocker is now used to prevent the shears clashing when closed, but before this was introduced the clack and snap of steel meeting steel was a noise inseparable from any busy shearing shed. **1941** Baker *Dict. Austral. Slang* 42 *Knocker*, a leather pad fixed near the heel of a pair of hand shears to prevent the blades closing too deeply. **1959** H. P. Tritton *Time means Tucker* 31/1 Shears do not click. The gullets of the blades are filled with soft wood, or sometimes with cork. These are called 'knockers', and they stop the heels of the blade from meeting. **1965** J. S. Gunn *Terminol. Shearing Industry* I. 34 *Knockers*, small pads, usually of leather or softwood, inset near the heel of hand shears. These stopped jarring and prevented the blades from closing too far and cutting the shearer's hand... It is likely that the name developed because they knocked together, but it could be a misspelling of 'nock'.

e. One who buys from, or sells to, persons at their residences; a door-to-door salesman; also, the action of selling (etc.) from door to door. Phr. *on the knocker* (and varr.), (engaged in

buying from, selling to, or canvassing) from door to door; also, (obtained) on credit.
1934 P. ALLINGHAM *Cheapjack* xiii. 166 A 'knocker-worker' is one who sells things at people's front doors. *Ibid.* xv. 186 'The knocker's the only game in the winter' said London Joe. 1936 *Evening News* 11 Dec. 11/1 A valued and regular lady customer drives up..and..orders petrol..finds she has left her handbag at home... The hand..yells out: 'Oi, there's a lidy 'ere wants some juice on the knocker!' 1959 *Listener* 7 May 802/2 That record of progress in Blackpool shows what can be done if we work, in the first place, as our canvassers say, on the knocker. 1959 G. SAVAGE *Antique Collector's Handbk.* 156 'Knockers' are jewellery and antique dealers who operate by calling from door to door in search of something to buy, and their purchases are sold to larger dealers. 1960 A. PRIOR in *Pick of Today's Short Stories* XI. 185 If I kept getting as much jewellery for him on the knocker then perhaps he wouldn't have to sell. 1963 J. F. STRAKER *Final Witness* viii. 81 Once she got a whole pile of stuff on the knocker, and then the firm came and took it back. 1967 *Sun* 17 July 7/2 The 'knocker boys'..trick old ladies into parting with family heirlooms for a fraction of their value. 1968 M. ALLINGHAM *Cargo of Eagles* ix. 116 I've worked the knocker if you know what that means—the door-to-door selling racket. 1970 *Sunday Times* 18 Jan. 37 A knocker was a specially trained salesman working, not under the authority and generally not in the pay of a district sales agent, but for the company itself, out of the Dayton executive offices.
f. Pl. The female breasts. *vulgar.*
1941 J. SMILEY *Hash House Lingo* 25 Fix the knockers—look at the nice breasts on that woman. 1948 N. MAILER *Naked & Dead* (1949) III. ii. 484 Look at the knockers on her, Murray says. 1967 J. KENNAWAY *Some Gorgeous Accident* I. 15 She was slight..but with great little knockers—breasts being for mothers. 1970 *Private Eye* 11 Sept. 16 Hello, luv! Phew, look at them knockers!! 1972 M. J. BOSSE *Incident at Naha* 24 I'm jealous. She has those big knockers, and I'm afraid you like them.
5. *knocker-off*, (*b*) *Underworld slang*, a thief.
1926 E. WALLACE *Door with Seven Locks* iii. 28 Tommy Cawler had been a notorious 'knocker-off' of motor-cars. 1952 'J. HENRY' *Who lie in Gaol* iv. 61 They are mostly house-breakers and petty thieves, or 'knockers-off' in prison parlance.

knockered (nǫ·kəɹd), *a.* [See -ED².] Of a house door: fitted with a knocker.
1921 *United Free Ch. Miss. Rec.* May 137/1 We entered by those massive brass-studded and knockered doors. 1928 *Daily Tel.* 24 Apr. 12/7 Knockered front doors and curtained windows.

knocking, *vbl. sb.* Add: **1.** (Examples, some corresponding to *KNOCK *v.* 5 b.)
1899 J. PERRY *Steam Engine* ii. 30 Knocking or Backlash.—It will be noticed that however good may be the fit of a brass to a pin, when the forces between them are suddenly reversed, there is a blow. 1903 M. P. BALE *Gas & Oil Engine Managem.* iv. 61 Knocking may also arise from the key of the flywheel becoming loose. 1928 [see *DECOKE *v.*]. 1946 *Mod. Petroleum Technol.* (Inst. Petroleum) 245 All reciprocating petrol engines, if run on unsuitable fuel, will produce a characteristic noise known as knocking, the quality of the sound varying from a sharp pink to a low thud according to the design of the engine and the composition of the fuel. 1971 *Guardian* 13 Dec. 1/4 Steam injection could eliminate 'knocking' on unleaded fuel even in high compression engines.
b. (Earlier and later examples with *about*; also with *up*.)
1840 *N.Z. Jrnl.* 1 Aug. 183/1 Any one coming out must expect to have a good knocking about. 1897 *Organ Voicing & Tuning* 9 A knocking-up cup, similar in form to the cone... It is of great substance, and, therefore, heavy, that it may the more readily effect its mission, namely, that of 'knocking up' or reducing the wind-hole of the pipe. 1905 *Westm. Gaz.* 19 Aug. 15/2 It is always the brim that matters in one's knocking-about hat. 1922 A. E. CRAWLEY *Lawn Tennis Do's & Don'ts* 42 As in many other games, it is not normal to get into your stroke at once; hence the need of a few minutes' knocking-up before a game. 1924 F. G. LOWE *Lawn Tennis* 10 When a new stroke has been learnt..it is an excellent idea to practise it against a wall until it becomes perfect. This 'knocking up' will also materially improve footwork and quicken up the player. 1970 *Country Life* 20 Aug. 469/3 (*caption*) 'Knocking-up' slate in front of a miner's house.. Co. Durham. 1973 *Oxf. Mag.* 4 May 9/2 *Knocking-up* is notoriously not done in America, even by tennis players trying to warm up their partners before a match.
c. **knocking-off**: (i) = *KNOCK-OFF *sb.* 2; also (in full *knocking-off time*), the time laid down for the end of a spell of work.
1886 in *Amer. Speech* (1950) XXV. 35/1 The entire mill is kept on operation long after knocking off..to make up for time lost during the day. 1887 in *Ibid.*, Saturday evening last, just about knocking off time. 1894 in *Leeds Mercury Weekly Suppl.* 11 Aug. (E.D.D.). 1922 C. E. MONTAGUE *Disenchantment* ix. 123 He..knocked off work for the day. There was no knocking off for the army. 1944 *R.A.F. Jrnl.* Aug. 292 At knocking-off time..the hooter sounds and everyone climbs thankfully out of their overalls. 1958 *Times* 14 Apr. 6/1 It is not all *that* difficult to control tea-breaks, knocking-off times or shift changes without upsetting your employees. 1974 N. FREELING *Dressing of Diamond* 45 Another cubic metre to shift before knocking-off time.
(ii) *Spinning.* Automatic stopping of the bobbin and flyer frames when a sufficient length of yarn has been wound on to the bobbin.
1883 H. E. WALMSLEY *Cotton Spinning* 15 See that the saddles and weights are accurate, and that both front and back knocking-off motions are in perfect working order. 1901 —— *Cotton Spinning* II. iii. 67 It is still more difficult to invent a motion that will prevent tenters from doffing bobbins before knocking-off does take place. 1908 H. PRIESTMAN *Princ. Woollen Spinning* x. 282 If a mule were empty, and were run with the knocking-off gear out of action,..the rollers would revolve at all kinds of varying speeds. 1927 T. THORNLEY *Cotton Spinning* (ed. 4) iii. 61 This completes the knocking-off motion and the lap may now be doffed.
(iii) *Weaving.* Automatic stopping of the loom when the shuttle fails to reach the box.
1912 T. ROBERTS *Tappet & Dobby Looms* ix. 198 In order to reduce the vibration and strain on the various parts of the loom when knocking-off takes place, strong springs..are employed which serve as cushions for the frogs. 1935 J. W. HUTCHINSON *Mod. Looms* xxix. 265 The sudden knocking-off of the loom may crack the cast iron brushes on the sword pin.
d. **knocking over.** *Machine Knitting.* = *KNOCK-OVER.
a 1877 KNIGHT *Dict. Mech.* II. 1237/1 The stripping or knocking-over wheel..then throws the old loops entirely over the tops of the needles. 1964 H. WIGNALL *Knitting* ii. 28 The needle now moves to its lowest position drawing the new loop through the fabric loop which is now cast off. On the original Lee knitting frame this called for physical effort and this action was called knocking over.
e. Phr. *the last knockings*, the last earnings; so *to be on the last knockings*, to approach the end of one's employment or earnings.
1939 H. HODGE *Cab, Sir?* iii. 31, I..have left it late, and come in 'at the last knockings'. 1958 F. NORMAN *Bang to Rights* III. 137 When I was on the last knockings I tried to get my bird out earning again.

knocking, *ppl. a.* Add: **b.** *knocking copy*, advertising which claims that the product of another manufacturer is inferior to one's own. (Cf. *KNOCK *v.* 2 f.)
1958 *Times* 1 July 11/5 Certain types of announcement covering such things as..politically and religiously controversial statements, 'knocking' copy, and so on. 1960 *Guardian* 19 Dec. 3/2 In advertising, the pamphlet advocates the ending of the 'no knocking copy rule', which restricts competition by preventing one advertiser from 'fairly disparaging' another's product. 1966 G. N. LEECH *Eng. in Advertising* iv. 37 'Knocking copy', designed to discredit competing products, offends against the principle of 'positiveness' mentioned earlier, and is besides considered bad form. 1969 *Listener* 23 Jan. 127/3 Knocking copy means advertisements which point out clearly that one brand is better than another brand.

kno·cking-shop. *slang.* [Cf. *KNOCK *v.* 2 d; Eng. Dial. Dict. s.v. *knocking-house*.] A brothel.
1860 HOTTEN *Dict. Slang* (ed. 2) 166 *Knocking-shop*, a brothel, or disreputable house frequented by prostitutes. 1938 'J. SPENSER' *Crime against Soc.* xiii. 126 She might wonder if you hadn't been gettin' scragged in a knockin' shop for not payin' yer dues. 1945 E. WAUGH *Brideshead Revisited* 300 This eye-sore..always reminds me of one of the costlier knocking-shops. 1959 'O. MILLS' *Stairway to Murder* xiv. 149 I'm a clean-living man... I wouldn't go to any back-street knocking shop. 1969 L. KENNEDY *Very Lovely People* ii. 111 Yes, it seems that some of the girls are running a knocking-shop on the side. 1971 B. W. ALDISS *Soldier Erect* 115 Somewhere there must be a woman whose longings corresponded to mine. Perhaps she could be found, that unknown She, even within the confines of a knocking shop.

knock-kneed, *a.* Add: **a.** (Earlier Amer. example.)
1774 in *Maryland Hist. Mag.* (1911) VI. 41 Charles Blundell, an Englishman,..a very slender made fellow much knock-kneed, with light brown hair very short.

knock-knock: see *KNOCK-.

kno·ckless, *a.* [See -LESS.] **a.** *Nonce-wd.* That enters without knocking on a door. **b.** Of an engine: free from knock (*KNOCK *sb.*¹ 1 c).
1907 W. DE MORGAN *Alice-for-Short* xlvii. 500 Alice was interrupted by the advent of the doctor, knockless but with musical boots. 1928 *Daily Chron.* 9 Aug. 11 (Advt.), A 'knockless', livelier engine; reduced vibration and less wear and tear of engine parts.

knock-me-down, *a.* and *sb.* **A.** *adj.* (Later *fig.* examples.)
1922 JOYCE *Ulysses* 300 And Bloom, of course, with his knockmedown cigar putting on swank with his lardy face. 1944 DYLAN THOMAS *Let.* 31 Dec. (1966) 270, I haven't read more than a scattered few of the poems..stopping with delight..at many knock-me-down lines. 1958 *Times Lit. Suppl.* 26 Dec. 757/1 Its current number deals with the very subject..and it does so in a brisk, knock-me-down way which is extremely refreshing. 1967 *Listener* 16 Feb. 235/1 It was not yet fashionable for poets to present themselves as knock-me-down sales representatives of verse.

knock-off, *sb.* and *a.* Add: **A.** *sb.* **2.** The act of leaving off one's work or occupation; the signal for doing this. Also *attrib.*, as *knock-off signal, time, whistle*.
1899 'Knock-off' time [in Dict., sense B]. 1902 *Daily Chron.* 13 June 6/3 Just at that moment the officer in charge gave what is technically known as the 'knock-off', or the signal to discontinue the play of water on the building. *Ibid.*, After the 'knock-off' signal had been given. 1947 K. TENNANT *Lost Haven* xvi. 249 The knock-off whistle blew at the mill. 1948 D. BALLANTYNE *Cunninghams* (1963) i. 13 From early morning to morning smoko, to lunch, to afternoon smoko, then to knockoff.
3. *slang.* A robbery; *concr.* (see quot. 1963); phr. *on the knock-off*, engaged in stealing.
1936 J. CURTIS *Gilt Kid* x. 100 They [*sc.* gloves]..gave away the fact that he was still on the knock-off. 1963 *Austral. T.V. Times* 18 Apr. 10/2 *Knock-off*, loot or illegally found goods. 1969 J. GARDNER *Complete State of Death* vi. 94 The really profitable knock-offs, like the Train Robbery.
4. A copy or reproduction of a design, e.g. of a textile, china, etc. *U.S.*
1966 *N.Y. Times* 25 Jan. 44 Copying designs to sell for less has a name in the industry. It is called the 'knockoff'. 1970 *Washington Post* 30 Sept. B. 14/1 People who appreciate genuine pate de foie gras..might like to serve it on a decently designed plate, and not on a knock off..of 18th Century English china. 1971 *Time* 25 Jan. 38 [Coco] Chanel had long since refused to join the cabal of Paris designers who tried to prevent style piracy... Private customers paid $700 for the original; buyers, intent on knockoffs, paid close to $1,500.
B. *adj.* **2. a.** Machine Knitting. *knock-off lap* (see quot. 1957).
1884 W. T. ROWLETT tr. *Willkomm's Technol. Framework Knitting* I. ii. 102 In the second lap..the loops are simply taken back into the throats of the sinkers with the old stitches. This second lap is called a 'knock off lap', because it does not form stitches. *Ibid.*, The purpose of such knock off laps is manifestly to bring as much thread as possible into the fabric, which thus becomes thick and soft, and suitable for underclothing or linings for shoes, &c. 1926 J. CHAMBERLAIN *Hosiery, Yarns & Fabrics* vii. 173 As a lap is formed without pressing and clearing the knock-off lap it is in every sense a parallel to the tuck-stitch of ordinary knitting. 1952 D. F. PALING *Warp Knitting Technol.* viii. 106 Knock-off laps can be used in conjunction with striped warps to produce purely vertical stripes on the face of the fabric without any overlap between adjacent colours. 1957 *Textile Terms & Definitions* (Textile Inst.) (ed. 3) 55 *Knock-off lap (warp knitting)*, a length (or lengths) of yarn received by a needle and not pulled through the loop of the previous course.
b. Designating a mechanical part (e.g. of a vehicle) that may be removed or disengaged by knocking.
1896 W. E. HIPKINS *Wire Rope* 47 They are connected and disconnected..by means of suitable couplings. When the rope is under considerable tension 'knock-off' hooks have to be used. 1905 C. HURST *Valves & Valve-Gearing* (ed. 4) vii. 130 In high-class Corliss engines it is often necessary to fit an automatic knock-off gear in connection with the governer, so that should the speed of the engine exceed or fall below certain limits the trips will be thrown into such a position that will prevent engagement of the catches,..the engine thus [being] brought to rest. 1958 *Sunday Times* 26 Jan. 27/1, I would recommend that knock-off hubs are fitted in future; they can be changed in thirty seconds. 1963 A. F. W. COULSON et al. *Man. Cotton Spinning* II. ii. v. 122 The doffing cycle commences when the desired length of lap has been made; the measuring plate..on the measuring motion releases the spring-loaded knock-off lever, which..disengages the brake on the rack-shaft. 1967 *Guardian* 12 Sept. 3/4 Knock-off wheels with proper splined hub and graduated spokes.
c. *Spinning.* Associated with or bringing about knocking off (*KNOCKING *vbl. sb.* 1 c (ii)).
1927 T. THORNLEY *Cotton Spinning* (ed. 4) iii. 56 It is customary to apply a full lap automatic stop motion..to automatically stop the feed and delivery parts every time a lap is sufficiently full... The favourite..knock-off motion is worked from one of the calenders on the Hunter cog principle.

knock-on, *sb.* and *a.* **A.** *sb.* In *Rugby Football* (see KNOCK-).
1845, 1881 [see PUNT *sb.*³ 1]. 1888 [in Dict. s.v. KNOCK-]. 1905 *Daily Chron.* 29 Dec. 9/5 The alleged incident of the referee in the New Zealanders–Newport match sounding the whistle when a knock-on by an All Black proved to the advantage of the Welsh team was..contrary to law. 1960 *Times* 11 Jan. 17/1 A body of dedicated Scots thought that France's last [try] followed a knock-on. 1974 *Sunday Tel.* 23 June 35/7 The referee did not see the knock-on.
B. *adj.* **1.** *Physics.* Ejected, produced, or caused as a result of the collision of an atomic or sub-atomic particle with an atom.
1940 *Nature* 13 July 65/2 A few particles of energy greater than 10^8 ev...were therefore considered to be mesons responsible for the production of 'knock on' showers. 1953 *Jrnl. Brit. Interplanetary Soc.* XII. 203 These [*sc.* nuclear fragments] may be accompanied by knock-on particles (or shower particles) which also move with velocities comparable to the original particle. 1967 A. H. COTTRELL *Introd. Metall.* ii. 14 There are three main types of radiation damage,.. (3) Knock-on damage in which atoms inside the material are knocked from their sites by the impacts of nuclear particles. 1971 *Nature* 6 Aug. 422/2 Knock-on protons produced by 3MeV neutrons would not..produce visible flashes.
2. Designating a mechanical part (e.g. of a vehicle) that may be attached or fastened by knocking. Also *absol.*
1952 *Motor Manual* (ed. 34) vii. 150 There are two principal types of wheel fastening. The less common, known as the knock-on, has at its centre a large nut with two lugs or projections. *Ibid.* p. x, Knock on wheel fastening. 1959 *Times* 12 Sept. 3/3 Moss's car has knock-on wheels. 1965 *Punch* 20 Oct. 567/2 The imitation knock-on hubs had been knocked off by genuine thieves. 1967

I. HAMILTON *Man with Brown Paper Face* ii. 24 The strengthened knock-on wire wheels were original M.G.

knock-out, *a.* and *sb.* Add: **A.** *adj.* **b.** Also *fig.*
1938 *Ann. Reg. 1937* 90 The task of the Government..was to make impossible the greatest danger to civilisation—the knock-out blow. 1955 *Times* 9 May 19/1 A new fashion has, however, arisen—the cry for knock-out competition in private enterprise.
c. Designating (a system used in) a competition or tournament in which the defeated competitors in each round are eliminated.
1896 W. BROADFOOT et al. *Billiards* i. 40 Scarcely a dozen really important handicaps on the old 'knock-out' principle have been played in the last twenty years. 1897 K. S. RANJITSINHJI *Jubilee Bk. Cricket* 281 The first elevens meet in a series of matches, played on the 'knock-out' system. 1920 *Motor Cycle* 22 July 114/1 Competitors in the motor cycle events were run off in pairs on the knock-out principle. 1921 E. B. TURNER in E. H. D. Sewell *Rugby Football* xiv. 244 The competition was run on 'knock-out' lines as it is at the present time, the teams which entered being drawn in ties, and those left in after each round being again paired by lot until only two were left in the final. 1928 *Daily Mail* 25 July 16/4 The singles championship held by the Ayton Tennis Club..was played on the knock-out principle. 1953 E. SMITH *Guide Eng. Traditions* 94 Besides the League there is also a 'knock-out' competition for the English Cup. 1955 *Times* 18 Aug. 4/3 Leeds, Huddersfield, Halifax, and Hull will be expected to go a long way in the various knock-out competitions again. 1966 *Listener* 30 June 936/1 The British Isles are to be invaded by football fans from all over the world bent on seeing the knock-out international competition known as the World Cup. 1974 *Country Life* 2 May 1070/2 By mid-summer we are left with 32 [cricket] clubs who fight it out on a knockout basis.
d. *knock-out drops* (also occas. *sing.*), a liquid drug of which drops are put into liquor to render a person unconscious or stupefied (e.g. in order to rob him). Also *fig. colloq.* (orig. *U.S.*).
1895 J. S. WOOD *Yale Yarns* 152 Our dandy team played a logy, tired sort of game, as if each man had been given knock-out drops, and we all felt blue! 1904 EX-INSPECTOR ELLIOTT *Tracking Glasgow Criminals* 23 The use of drugs..or what is more familiarly known in criminal circles as 'knock-out' drops is common enough in most cities. What is known as 'knock-out' drops is chloral hydrate, and from fifteen to thirty grains of it produces a sleep that lasts three hours. 1926 J. BLACK *You can't Win* xii. 152 Here I learned to beware the crafty shanghaier with his knockout drops. 1955 'N. SHUTE' *Requiem for Wren* viii. 242 About midnight..I took one of Aunt Ellen's things. It was a knock-out drop all right because I didn't wake up till half past nine. 1958 *Sunday Times* 10 Aug. 8/3 The first [film] was a typical knock-out drop—a study of 'inter-specific interdependence'.
e. *Mech.* Designating or pertaining to a knock-out (see *B. 5 a, b).
1907 *Installation News* Nov. 6/2 The 'knock-out' principle as applied to junction boxes. 1925 HODKIN & COUSEN *Textbk. Glass Technol.* xxxii. 439 The knock-out arm moves outwards, so that the bottle falls off the mould base into the trough. 1946 DU BOIS & PRIBBLE *Plastics Mold Engin.* iii. 104 As the press continues to open, the pin attached to the knockout bar rises and pushes the wedge, with molded piece attached, up out of the cavity. 1955 *Die Design Handbk.* (Amer. Soc. Tool Engineers) v. 23 Knockout slugs should be readily removable from the outside of the box; so care should be exercised to see that the material is pierced from the proper side. *Ibid.*, Knockout dies. 1963 H. R. CLAUSER *Encycl. Engin. Materials* 104/1 If necessary knockout pins are cast into the mold.
f. *Founding.* Used in or pertaining to the knocking out of castings and flasks (see *KNOCK *v.* 12 j).
1942 *Engineering* 6 Mar. 195/2 With continuous casting with a mould conveyor, there was great difficulty in adjusting the whole conveyor for uniform speed, so that, when the castings came to the knock-out grid, they would be cool enough for knocking out. 1958 *Ann. Rep. Chief Inspector of Factories on Industr. Health 1957* 15 in *Parl. Papers 1958–9* (Cmnd. 558) XIII. 183 Various types of exhaust systems used to control dust and fumes at central knock-out positions..have been examined. 1973 *Steel Castings Abstr.* XXII. 24 (*heading*) Improving the knock-out properties of silicate bonded sands.
g. *colloq.* Of a person or thing: of overwhelming or surpassing quality. Cf. *B. 4.
1966 *Crescendo* Aug. 41 (Advt.), Chasing a real knockout sound? You'll find the most rewarding instrument..is the Hammond organ. 1968 *Listener* 5 Sept. 307/2 The wit and repartee of the DJ... 'Hi there—it's great to be with you and welcome to another knock-out show.'
B. *sb.* **2.** (Earlier and *transf.* examples.)
1887 in *Amer. Speech* (1950) XXV. 35/1 A knock-out was no more possible with these youngsters. 1891 *Sporting Life* 25 Mar. 7/3 The Barrier man was nearly helpless, and Choynski tried frantically to pull himself together for one good knock out. 1895 G. B. SHAW *Let.* 17 Sept. (1965) 560 Got up [after cycling smash] within the prescribed ten seconds, but had subsequently to admit knock-out. *a* 1918 W. OWEN *Coll. Poems* (1963) 71 One of us got the knock-out, blown to chops. 1944 *Return to Attack* (Army Board, N.Z.) 23/2 The tanks, which had suffered three knockouts from 88-millimetre guns.
4. *colloq.* A person or thing of overwhelming or surpassing quality. Cf. *A. g.
1892 *Idler* June 549 'E's a knockout! 1898 J. D. BRAYSHAW *Slum Silhouettes* 28 Got a rippin' good voice, ain't he? It's a knock-out. 1908 *London Mag.* June 473/2 The tent is a knock-out. 1918 A. QUILLER-COUCH *Foe-Farrell* xi. 163 The view from the top is a knock-out. 1920 WODEHOUSE *Jill the Reckless* (1922) xiv. 213 He had a respect for Wally's opinion, for Wally had written 'Follow the Girl' and look what a knock-out that had been. 1935 A. HUXLEY *Let.* c May (1969) 394 The greatest knockout is the 'Assumption' at S. Vincente, where there are also some small pictures of absolutely staggering beauty. 1953 R. LEHMANN *Echoing Grove* 190 A *whizzing* beauty! Really but really a knock-out. 1958 *Daily Mail* 1 Sept. 10/8 The clever, wicked face of Emlyn Williams expressing the words of Dylan Thomas was once again a knockout. 1970 A. CAMERON et al. *Computers & O.E. Concordances* 24 I've got a version of *Paradise Lost* that is a knock-out. 1971 *Daily Colonist* (Victoria, B.C.) 11 June 2/1 I'm sick to death of women's liberation. I don't see any real knockouts running around with those placards, just a bunch of unhappy uggas. Nobody would want to spend an evening—much less a lifetime—with them.
5. a. *Mech.* A device for 'knocking out' or ejecting something, esp. from a mould or die.
1893 *Funk's Stand. Dict.*, *Knock-out*, a device for throwing out finished work from a punching- or stamping-machine. 1896 O. SMITH *Press-Working of Metals* xii. 243 There is a distinct class of ejecting press- or die-attachments known by the general name of 'knockouts', or sometimes and if limited to the lower die, as 'knockups'. 1915 F. D. JONES *Diemaking & Die Design* vii. 282 The mechanically operated knockout..applied to a punch press..operates more satisfactorily than a rubber bumper. 1919 F. A. STANLEY *Punches & Dies* vii. 152 The die is made..with a knock-out which ejects the work after the trimming operation. 1946 DU BOIS & PRIBBLE *Plastics Mold Engin.* ix. 371 The knockout pins may be made as sleeve knockouts working over a core pin. 1960 EARY & REED *Techniques Pressworking Sheet Metal* xviii. 377 Basically, there are two main types of knockouts: the spring knockout and the solid knockout.
b. A part of a box or other article designed to be forced out to form a hole.
[1907: implied in A. e above.] 1939 H. P. RICHTER *Pract. Electr. Wiring* x. 124 Around the sides [of the outlet box] and in the bottom are found 'knockouts'—sections of metal that can be easily knocked out to form openings for wire to enter. The metal is completely severed around these sections except at one small point. 1955 *Die Design Handbk.* (Amer. Soc. Tool Engineers) v. 23 Typical single and double dies for producing knockouts in conduit boxes, enclosing boxes, and similar products. 1962 *Gloss. Terms Glass Industry (B.S.I.)* 36 *Button* (cap, knockout), a portion of a piece of pressed ware so designed that it can be knocked out or off to make a hole.
c. *Founding.* The process of separating a casting from the flask and sand in which it was made (cf. *KNOCK *v.* 12 j); the place where, or equipment with which, this is carried out.
1942 *Engineering* 6 Mar. 195/3 The method of dealing with mould-making, closing and knock-out must depend upon the nature of the product. *Ibid.*, The cooling of the sand on its way from the knock-out to the reconditioning mill presented some difficulty. 1955 H. E. CRIVAN in W. C. Newell *Casting of Steel* vi. 227 Mechanical jolt knock-outs are in operation. 1972 P. R. BEELEY *Foundry Technol.* viii. 417 The interval before knockout is important from the points of view of moulding box utilisation and of the temperature of the sand in the system. 1973 *Steel Castings Abstr.* XXII. 11 The redesign of castings..and alterations to the plant layout, including the knock-out, have reduced the labour required.
6. A 'knock-out' competition: see *A. c.
1928 *Observer* 4 Mar. 22 Pembroke..have won the finals..of both the football 'knock-outs'. 1959 A. WESKER *Roots* III. 66 The fireman's whist drive. Won seven'n six in the knockout.

kno·ck-over. *Machine Knitting.* [f. vbl. phr. *to knock over* (*KNOCK *v.* 13 c).] The act or process of causing a stitch to pass over the head of the needle on which it was held. Also *attrib.*
1952 D. F. PALING *Warp Knitting Technol.* i. 6 A forward movement of the sinker bar combined with a further downward movement of the needle bar ensures a gradual knock-over. 1964 H. WIGNALL *Knitting* ii. 29 The frame needle goes down to its first knock-over. *Ibid.* 39 When the needle and point are locked together below the top of the knock-over bit they rise and the needle passes through the transferred loop.

kno·ck-up, *sb.* and *a.* **A.** *sb.* A practice or casual game at lawn tennis, squash rackets, etc.
1884 E. W. HAMILTON *Diary* 4 May (1972) II. 609 In the afternoon had a little 'knock up' at lawn tennis. 1922 F. HAMILTON *P.J.: Secret Service Boy* iii. 89 Shall we have a little knock-up against the wall of the stables? It doesn't make half a bad squash court. 1930 W. S. MAUGHAM *Breadwinner* II. 73 Why don't you and Dinah go and have a knock-up? 1946 *Penguin New Writing* XXVIII. 29 Expecting to find the three of them having a knock-up while they waited for him. 1973 G. MITCHELL *Murder of Busy Lizzie* xiv. 159 'I think I *will* go to bed.'..'All right, unless you'd like a knock-up at table tennis first.'
B. *adj.* Designating a knock-up: see above.
1928 *Weekly Dispatch* 24 June 21/7 Many of the world-famous players engaged in final 'knock-up' games at Wimbledon yesterday.

knockwurst, var. *KNACKWURST.

knödel (knö·dəl). Also **knoedel.** Pl. **-s,** ‖ **-n.** [G.] In Germany, a type of dumpling. Cf. *KNAIDEL.
1827 M. WILMOT *Jrnl.* 23 July in *More Lett.* (1935) 276 [H]oltzknecht Knödel—Woodcutter's dumpling, a sort of rice moistened with butter..fried..in little round balls. 1873 *German Nat. Cookery for Eng. Kitchens* x. 176 'Klösse' or 'Knödeln'... The bread used for them must be light, and without crust, either grated, crumbled, or soaked in cold milk or water. 1948 R. SYSONBY *Cook Bk.* (ed. 2) 144 Mariller Knoedel... Wrap some peeled fresh apricots each in a square of the paste... Drop the dumplings into..boiling water. 1968 L. DEIGHTON *Continental Dossier* 345 Baden-Württemberg... 'Knödel' ('dumplings') are famous. 1971 A. R. DANIEL *Baker's Dict.* (ed. 2) 108/2 *Knödel*, Bavarian name for a special kind of small dumpling.

Knoevenagel (knö·vĕnāgĕl). *Chem.* The name of Emil *Knoevenagel* (1865–1921), German chemist, used *attrib.* and occas. in the possessive to designate various reactions in organic chemistry, *esp.* the reaction between an aldehyde or ketone and malonic acid or a related compound containing active hydrogen, catalysed by ammonia or an amine, to yield an acid with the group —CH:CH·COOH.
1907 J. B. COHEN *Org. Chem. Adv. Students* I. vii. 285 (*heading*) Knoevenagel's reaction. 1931 *Jrnl. Chem. Soc.* 745 Catalytic action..effected by the salts of bases with mineral acids in the malonic acid condensation and the Knoevenagel reaction generally. 1938 G. H. RICHTER *Textbk. Org. Chem.* xvi. 289 The Knoevenagel reaction is the condensation of the carbonyl group of aldehydes and ketones with *active methylene* hydrogens; the reaction takes place in the presence of primary and secondary amines, especially diethylamine or piperidine. Knoevenagel was the first to point out the great catalytic power of ammonia and amines for this type of condensation. 1967 *Organic Reactions* XV. 204 (*heading*) The Knoevenagel condensation. 1969 R. C. DENNEY *Named Org. Reactions* 52 An interesting development has been the use of the Knoevenagel condensation to increase the chain length of sugars.

Knole sofa (nōᵘl sōᵘ·fă). [f. the name of the prototype at *Knole* Park, Kent, *c* 1605–20.] A sofa designed in the style of an early 17th-century model, having adjustable sides that may be lowered to make it into a day-bed. So *Knole couch, settee.*
[1868 C. L. EASTLAKE *Hints Household Taste* vi. 142 The sofa at Knole..is an example of thoroughly good design... The sides can be raised or lowered..thus enabling the sofa to be used as a couch or a settee, at pleasure.] 1942 R. KING *Design in Evil* xxi. 207 A Knole sofa sheened with lemon-colored damask. 1943 D. WELCH *Maiden Voy.* xv. 122 Two fringed and tasselled Knole settees. 1945 *Burlington Mag.* LXXXVI. 114/1 The famous Knole couch..is the first comfortable type of couch that was made in England. 1951 A. CHRISTIE *They came to Baghdad* ii. 17 That Knole settee..in electric blue satin.

knoop¹ (nūp), *dial.* var. KNOP *sb.*¹ 2.
1871 G. M. HOPKINS *Jrnls. & Papers* (1959) 209 Then the knot or 'knoop' of buds some shut, some just gaping.

Knoop² (nūp, knūp). The name of Frederick *Knoop* (1878–1943), U.S. instrument-maker, used *attrib.* with reference to an indentation test devised by him, in which hardness is measured by the size of the indentation produced in a substance by a pyramidal diamond indenter of specified shape under a known load.
The pronunc. (knūp) is used by members of the Knoop family.
1940 *Metals & Alloys* Sept. 292/1 The article is a description of the Knoop Indenter, a new instrument for determining the micro-hardness of thin layers of metals. 1945 *Amer. Mineralogist* XXX. 595 The Knoop hardness of gypsum is approximately 32 to 45 or more, depending on orientation. 1961 E. CAMERON *Ore Microsc.* iv. 75 The Knoop hardness (*KV*) is given by the formula KV (in kg/mm²) $= 14230 \times P/l^2$ where *P* is the test load in grams, and *l* is the long diagonal of the indentation in 0·001 mm. 1971 *Nature* 14 Nov. 661/1 A study..on anisotropy in the hardness of crystals using the Knoop indenter. 1971 *Brit. Jrnl. Nutrition* XXVI. 234 The Knoop test of hardness..was applied to human nails.

knop, *sb.*¹ Add: **1. b.** A loop or tuft (often of different colour) formed in a strand of yarn for ornament. Also *attrib.* in *knop yarn*. (See quots.)
1904 GOODCHILD & TWENEY *Technol. & Sci. Dict.*, Knop yarn. 1914 BARKER & MIDGLEY *Analysis Woven Fabrics* 272 *Knop yarn.*—A yarn upon which knops or lumps of yarn of one or more colours appear at intervals. 1929 *Encycl. Brit.* XXIII. 879/2 The knop yarn—in which knops are formed at any required intervals on an otherwise level thread by holding one thread tightly and allowing the second thread to run in slackly to form knops of the required size. 1964 *Which?* Sept. 284/2 *Knop or nub*, a compound yarn with lumps or balls of yarn at regular or irregular intervals. 1968 E. GALE *From Fibres to Fabrics* iv. 44 *Knop yarn.* Two threads are twisted together, with one at regular intervals being given in very rapidly so that it is wound round and round the first thread in the form of a hard knop or lump.

knopite (nǫ·pəit, knǫ·pəit). *Min.* [ad. Sw. *knopit* (P. J. Holmquist 1894, in *Geol. För.*

Förh. XVI. 73), f. the name of Adolf *Knop* (1828–93), Ger. mineralogist: see -ITE¹.] A variety of perofskite occurring in lead-grey crystals, in which calcium is partly replaced by cerium.
1896 *Mineral. Mag.* XI. 158 Minerals at first thought to be perofskite and dysanalyte are here described under the new name knopite. 1927 *Mineral. Abstr.* III. 412 Knopite, a mineral new to Canada, occurs sparingly in small bunches in a basic pegmatite. 1947 *Ibid.* X. 138 Knopite is here a rock-forming mineral of primary magmatic origin, in contrast to its other occurrences as a contact mineral or in pegmatite. 1962 W. A. DEER et al. *Rock-forming Min.* V. 50 A variety [of perovskite] rich in rare earths, chiefly cerium, has been called knopite... In addition to rare earths of the cerium group, the lanthanum group is usually present.

knopkierie: see *KNOBKERRIE.

knopped, *a.* Restrict ? *Obs.* to sense a in Dict. and add: **b.** (Further example); *spec.* of the stem of a glass.
1869 G. M. HOPKINS *Jrnls. & Papers* (1959) 193 A fine sunset..; along the earth-line a train of dark clouds of knopped or clustery make. 1960 H. HAYWARD *Antique Coll.* 158/2 Knopped stem, a type of stem found on 18th cent. drinking glasses composed of a varying number of knops. 1963 *Times* 30 Apr. 14/4 C. Davis bought..a cylinder-knopped wine glass for £155. 1973 *Times Lit. Suppl.* 6 Apr. 374/5 There are knopped and wrythen wine glasses.
c. *knopped yarn*, yarn ornamented with knops or tufts. See *KNOP *sb.*¹ 1 b.
1911 *Encycl. Brit.* XXVIII. 906/2 *Knopped Yarn* is formed by twisting together several strands, one of which is at intervals delivered in greater lengths than the others, in order to allow a loop to be made.

knopper (nǫ·pəɹ). Pl. **knoppern, knoppers.** [G., = gall-nut.] A kind of oak-gall caused by an insect of the genus *Cynips*, formerly used in tanning and dyeing.
1879 *Encycl. Brit.* X. 44/1 The 'knoppern' galls of *Cynips polycera*, Gir., are cones having the broad, slightly convex, upper surface surrounded with a toothed ridge. 1903 H. R. PROCTER *Princ. Leather Manuf.* xviii. 262 *Knoppern* are galls produced on the immature acorns of various species of oaks, principally *Q[uercus] Cerris* in Hungary... They..have been largely replaced by valonia. 1908 E. T. CONNOLD *Brit. Oak Galls* 143 *Cynips calicis*... 'The Knopper Gall'... It occurs principally on *Quercus pedunculata*, but also on *Q. sessiliflora*. 1953 F. N. HOWES *Vegetable Tanning Materials* xxxvii. 260 The best known European galls are those obtained from eastern Europe (the Balkans and adjoining regions), which are commonly known as 'knoppers' or 'knoppern' or 'acorn galls'... There are various kinds of knoppern recognized, such as those formerly obtained from Hungary, Bohemia, Dalmatia and Serbia.

knopple (nǫ·p'l), *v. rare*⁻¹. [?f. KNOBBLE *sb.* or KNOP *sb.*¹] *trans.* = KNOB *v.* 1.
1870 G. M. HOPKINS *Jrnls. & Papers* (1959) 201 Herds of towering pillow clouds, one great stack..was knoppled all over in fine snowy tufts and pencilled with bloom-shadow.

knorr (nǫɹ). Also **knarr, knörr.** [ad. ON. *knörr* ship, merchant ship.] A mediæval type of ship of Northern Europe, having a single sail (see quots.).
1889 P. B. DU CHAILLU *Viking Age* II. xiii. 212 We find them [*sc.* trading ships] mentioned under their different names—viz., Knörr, Kugg, Byrding (ship of burden), Vistabyrding, [etc.]. 1932 C. M. SMITH *Northmen of Adventure* xiv. 320 The round ships went under a variety of names. The largest class was the *knorr*. 1967 H. HARRISON *Technicolor Time Machine* xi. 114 Where the dragon-prowed Viking ship was long and narrow, this *knorr* was wide and stood high out of the water—and was at least a hundred feet long. 1968 G. JONES *Hist. Vikings* III. ii. 188 Captain Folgar..in 1932 took a replica of a 60-foot knörr across the Atlantic. 1971 S. E. MORISON *European Discovery Amer., Northern Voys.* iii. 35 The Norse discoverers of Greenland and Vinland did not use a long Viking ship... There is ample evidence that they used the *knarr*, a beamy type propelled principally by one big square sail made of a coarse woolen cloth called 'wadmal', rigged with an additional sprit to set well close-hauled. 1973 *Country Life* 17 May 1373/2 A Knarr, the cargo ship in which the great [Viking] voyages to Greenland and North America were made.

Knossian (knǫ·siăn, knōᵘ-), *a.* Also **Cnossian.** [f. Gr. Κνωσσός Knossos or Cnossos + -IAN.] Of or pertaining to Knossos, a city in ancient Crete, where, according to classical tradition, King Minos ruled and kept the Minotaur in a labyrinth; historically, the centre of the Minoan civilization as revealed by the ruins of a vast labyrinthine palace (18th–14th centuries B.C.). Also as *sb.*
1894 A. J. EVANS in *Jrnl. Hellenic Stud.* XIV. 283 The incised marks on the slabs of the Knôsian building do not..stand alone. 1895 H. S. JONES tr. *Sel. Passages Anc. Writers* I. i. 4 The Knossians also possess the dance of Ariadne..a relief in white marble. 1900 A. J. EVANS in *Ann. Brit. Sch. Athens* VI. 16 An artistic advance which..was not reached till the fifth century before our era, some eight or nine centuries later than the date of this Knossian fresco. *Ibid.* 59 Out of about 25 distinct signs..near parallels to about 6 occur in the Knossian linear series. 1909 —— *Scripta Minoa* I. I. iii. 21 A perfect clay bar of the same general class as some of those from the Knossian deposit..had been acquired... The first stage represented by the existing West Wing of the Knossian palace goes far back into the Second Middle Minoan Age. 1939 J. D. S. PENDLEBURY *Archaeol. Crete* iv. 180 L.M. II vases of Knossian fabric in L.M. I deposits at Pseira. 1950 G. E. DANIEL *100 Yrs. Archaeol.* vi. 192 To the great Bronze Age civilisation which he discovered in Crete, Sir Arthur Evans gave the name Minoan... Other names were suggested, such as Knossian, Cretan, or Aegean. 1962 *Times* 21 Apr. 9/6 Deposits are 'telescoped', material of several Knossian periods being represented in a single layer of rubbish... It cannot be assigned..to any particular subdivision of the Knossian system. 1963 *Listener* 21 Mar. 495/1 The Cnossian ruler's contemporaries in Egypt.

knot, *sb.*¹ Add: **3. c.** *at the rate of knots*, very fast, quickly. *colloq.*
1892 R. WARDON *Macpherson's Gully* vi. 40 When she's [*sc.* the Teremakau river has] got her back up, travellin' in a hurry, like—tearin' along at the rate o' knots like she is to-day—..she's got to be treated with all doo respeck. 1921 'T. COLLINS' *Rigby's Romance* xxxii. 222, I went for it at the rate of knots, with the fire lathering along behind me roaring like fury. 1932 KIPLING *Limits & Renewals* 80 A natty little grey and black self-driven coupé came from Brighton way at the rate of knots. 1941 BAKER *N.Z. Slang* vi. 53 To travel at the rate of knots.
10. a. *to tie* (a person) (*up*) *in*(*to*) *knots* (or *a knot*): to confuse or nonplus (someone).
1860 *Baily's Mag.* Aug. 368 Never before..were bowlers or fielders so 'tied up in a knot'. 1888 A. G. STEEL in Steel & Lyttelton *Cricket* iii. 167 The team was beginning to get tied up into a knot. 1957 D. ROBINS *Noble One* (1960) xx. 191 He is tied up in knots. He's fighting himself as well as me. 1974 I. MURDOCH *Sacred & Profane Love Machine* 154, I could tie you into such knots, but I won't bother... You won't tell me the truth even now.
19. **knot-catcher** (see quot.); **knot-gall,** a species of oak-gall produced by the cynipid *Andricus noduli*; **knot-head** *N. Amer.*, a stupid person (see also quot. 1940); a stupid horse; **knot-hole,** (*b*) (example); (*c*) a hole formed by the excavation of clay; **knot-horn** = *knot-horn moth*; **knot-stitch** (examples); **knot-writing,** a mnemonic aid consisting of strings in which a number of knots are made.
1927 T. WOODHOUSE *Artificial Silk* 100 The threads or yarn from the cone cheeses are first led up through coils in wires termed knot catchers. 1894 C. R. STRATON tr. *Adler's Alternating Generations* 34 The knot gall is found in June on *Q[uercus] pedunculata, Q. sessiliflora*, and *Q. pubescens*. 1908 E. T. CONNOLD *Brit. Oak Galls* 65 The Knot Gall. 1940 *Amer. Speech* XV. 447/2 *Knot head*, low intelligence. 1961 WEBSTER, *Knothead*, a dull-witted blunderer. 1961 R. P. HOBSON *Rancher takes Wife* i. 21 Harold called in a loud voice to the horse. 'Step up there, you old knothead.' 1962 A. FRY *Ranch on Cariboo* xv. 160 I'd the repertoire of a mule skinner, developed behind a wide variety of knothead horses. 1972 J. AIKEN *Butterfly Picnic* ix. 163 Why hadn't he *said* he was going to, the silly knothead? 1903 *Westm. Gaz.* 31 Dec. 3/2 The little coons..climbed up to the knot-hole, and scrambled down inside. 1964 E. HUXLEY *Back Street New Worlds* x. 98 There are craters..called knot-holes, and from them clay has been scooped and loaded into tub-like steel wagons to proceed..to the kilns. 1967 M. CHANDLER *Ceramics in Mod. World* i. 29 The underlying Lower Oxford clay..is taken from the claypit or 'knot-hole' by a mechanical excavator. 1899 D. SHARP in *Cambr. Nat. Hist.* VI. 424 The males frequently have the basal-joint of the antennæ swollen; hence the term 'Knot-horns' applied by collectors to these moths. 1881 C. C. HARRISON *Woman's Handiwork* I. 84 Beginning with the hemstitch of our grandmothers, we may add..lace stitches, herring-bone, buttonhole..darning and knot stitch. 1964 *McCall's Sewing* ii. 30/1 *Knot-stitch*, stitch used to secure thread at beginning and end of stitching. 1896 A. J. BUTLER tr. *Ratzel's Hist. Mankind* I. 344 In West Australia,..a network of reed serves for a messenger's credentials,—a reminiscence of the once more widely-developed knot-writing.

knotted, *a.* Add: **1. c.** Colloq. phr. *to get knotted*, to 'go to hell'. Usu. in *imp.*, stop annoying me!
1963 *New Society* 22 Aug. 5/1 Other adoptions are 'get knotted' and 'knackered' which have come to mean innocently enough, 'go to hell', and '*kaput*'. 1964 B. W. ALDISS *Dark Light Years* iii. 39 Get knotted, Duffield, you ruddy trouble-maker. 1965 M. FORSTER *Bogeyman* viii. 144 'You are to behave properly.' 'Get knotted,' said Natalie, deliberately. 1968 'H. CALVIN' *Miranda must Die* ii. 19, I don't know why the hell you didn't tell him to get knotted and be done with it. 1969 O. BLAKESTON *For crying out Shroud* v. 46 'That boy in Rome said we looked like a hero.'..'Get knotted. We might as well be dead.' 1972 G. LYALL *Blame the Dead* xii. 80 'I'll lend you a good book about security.' 'Get knotted, Major.'

knottedness (nǫ·tĕdnĕs). [f. KNOTTED *a.* + -NESS.] The character or manner of being formed into a knot.
1909 in WEBSTER. 1962 C. S. OGILVY *Tomorrow's Math* vi. 109 Does the system of lines of magnetic force in space surrounding the knot reflect topologically the knottedness of the curve?

knotter. Add: **3.** With prefixed numeral: a boat or ship that makes (so many) knots an hour.
1908 *Pall Mall Gaz.* 20 Apr. 6/1 Not many of the so-called '30-knotters' could steam at this speed. 1929 'SEAMARK' *Down River* i, Essex noted the stolid little ten knotter ahead.

know, *v.* Add: **1. b.** Phrases: *not to know one's arse from one's elbow* (and similar phrases): a coarse expression suggestive of complete ignorance or innocence; (*not*) *to know from nothing* (*U.S.*): to be totally ignorant (about something).
1930 R. BLAKER *Medal without Bar* xiii. 69 'But nor 'an 'un' (this phrase was his masterpiece of thoughtful emphasis), 'nor 'an 'un of us knows 'is ears from 'is elbow when it comes to learning—learning like you officers have got up your sleeves.' 1936 *Mademoiselle* Mar. 43/1, I find I belong to the wrong gender to take part in such confabulations, and know from nothing. 1942 BERREY & VAN DEN BARK *Amer. Thes. Slang* §150/3 *Be ignorant*, know from nothing. 1944 'N. SHUTE' *Pastoral* iv. 75, I wish I'd had a crowd like that for my first crew. We none of us knew arse from elbow when they pushed me off. 1945 'F. FEIKEMA' *Boy Almighty* (1950) xvii. 162 Them San dietitians, they don't know from nuthin'. 1945 T. SHOR in Mencken *Amer. Lang.* (1948) Suppl. II. 695 A *square* don't know from nothin' and a *creep* is worse'n a jerk. 1966 'L. LANE' *ABZ of Scouse* 29 Don't know Thairsday from brekfuss-time. *Ibid.*, Don't know 'is arse from 'is elbow. 1968 *Encounter* Sept. 22/1 He knows from nothin'.
5. *to know like a book* (see *LIKE *adv.* 1 c).
b. (Earlier and later examples of *know thyself*.)
c 1527 tr. *Erasmus's Dicta Sapientium* sig. A3ᵛ *Nosce te ipsum*, know thy selfe. 1849 LYTTON *Caxtons* III. XVI. x. 183 'Know thyself,' said the old philosophy. 'Improve thyself,' saith the new. 1905 A. MACLAREN *Gospel St. Matthew* I. 43 The proud old saying of the Greeks, 'Know thyself'..would result in this profound abnegation of all claims, in this poverty of spirit. 1929 A. HUXLEY *Let.* 7 Jan. (1969) 306 'Know thyself' was probably one of the stupidest pieces of advice ever given. *a* 1930 D. H. LAWRENCE *Last Poems* (1932) 266 When at last we escape the barbed wire enclosure Of *Know Thyself*, knowing we can never know. 1941 *N. & Q.* Feb. 138 The folly of that impossible precept 'Know thyself'.
9. c. *to know one's* ——: to be well acquainted with something, to be well up in something. E.g. *to know one's business, onions* (see *ONION *sb.*), *stuff* (see *STUFF *sb.*¹).
11. b. *not to know what hit one*: see *HIT *v.* 8 e. Also, *you know*: a phrase used with aposiopesis (the implication to be imagined) or const. *what, whom*, etc. (as a means of avoiding naming the person, etc., referred to).
1867 TROLLOPE *Phineas Finn* (1869) I. x. 84 She told me once..it would lead to my being everlastingly—you know what. She isn't so squeamish as I am, and said it out. 1875 GEO. ELIOT *Let.* 13 Jan. (1956) VI. 116, I had a letter from 'you know whom' last night. 1911 D. H. LAWRENCE *White Peacock* II. i. 219 It's the way she swings her body—an' the curves as she stands. It's when you look at her—you feel—you know. 1925 *New Yorker* 7 Mar. 19/1 Of course there's no use me asking you if you took in all the revues where the girls come out—you know. 1937 C. DAY LEWIS *Starting Point* iii. 44 Never mind, kick him in the you know where—he's used to it. 1948 D. BALLANTYNE *Cunninghams* ii. 12 She is you know [*sc.* in the family way] to a Maori. 1949 D. M. DAVIN *Roads from Home* 99 Too much you know what last night, eh? 1970 *Harrap's French-Eng. Dict. Slang* 201 *Qui-vous-savez*, (said of person one does not wish to name) you know who.
c. *you know*: now freq. as a mere conversational filler. Also, *don't you know?*, a variant of *you know* (cf. *DONCHER).
1880 'MARK TWAIN' *Tramp Abroad* App. D. 611 Nothing gives such an air of grace and elegance and unconstraint to a German or an English conversation as to scatter it full of 'Also's' or 'You-knows'. 1885 A. EDWARDES *Girton Girl* II. iii. 40 Attack me? Why that was only a foolish joke, don't you know? 1896 F. C. PHILIPS *Undeserving Woman* 104 'When?' said George, 'I'd like to put the thing right at once, don't you know.' 1924 D. H. LAWRENCE *Phoenix II* (1968) 304 Little smart man of the shabby world, very much on the spot, don't you know. 1926 G. HUNTING *Vicarion* iv. 63 This represents some years of study, you know, this little exhibition I have given you. 1930 'SAPPER' *Finger of Fate* 225 My wife is such a nervous woman, don't you know. 1947 [see *BOUDIN]. 1965 *Listener* 2 Dec. 914/1 A. They're supposed to be, you know, sexy. B. That's all right, but all men are the same, after one thing, but sometimes, you know, it can be wonderful. 1968 *Ibid.* 16 May 626/2 Too often one hears people on the wireless beginning an elaborate sentence—they flounder about for a bit and then break off with: 'you know'. 1969 WIDDOWSON & HALPERT in Halpert & Story *Christmas Mumming in Newfoundland* 151 You could buy them in St. John's, you know, the false faces. 1974 *Sunday Times* (Colour Suppl.) 3 Feb. 66/4 People get the wrong idea, thinking we might be, you know, glamorous or brilliant or something.
f. *not if I know it* (earlier example).
1865 TROLLOPE *Miss Mackenzie* I. ii. 33 'Tom,' said I, when he asked me to go down to Drunder Street, 'not if I know it.'
Also (the phrases are arranged in the chronological order of their first recorded use in English as far as this is determinable):
to know little (or *nothing*) *and care less*: to be unconcerned *about*; to be studiously ignorant of.
1814 JANE AUSTEN *Mansf. Park* II. xi. 251 'I know nothing of the Miss Owens,' said Fanny calmly. 'You know nothing and you care less, as people, say Never did

tone express indifference plainer.' 1853 LYTTON *My Novel* II. VIII. iv. 322 'Ah!' said Egerton, who, as it has been before said, knew little, and cared less, about the Hazeldean pedigree, 'I..had forgotten it.' 1893 R. L. STEVENSON *Catriona* xxii. 267, I tell ye I ken naething and care less either for him or his breed. 1924 R. H. MOTTRAM *Spanish Farm* I. 71 Madeleine knew little and cared less as to what this might mean, except as it affected the work of the farm. 1925 F. HARRIS *My Life & Loves* III. xii. 183 The great London doctors knew nothing about leprosy and cared less. 1925 O. W. HOLMES in *Holmes–Laski Lett.* (1953) I. 741, I think he generally was kind in his judgment of me, except when Roosevelt was so angry at my dissent in the *Northern Securities* case (about which you probably know little and care less). 1931 F. L. ALLEN *Only Yesterday* v. 88 The shock troops of the rebellion were not alien agitators, but the sons and daughters of well-to-do American families, who knew little about Bolshevism and cared distinctly less. 1937 N. COWARD *Present Indicative* VIII. v. 321 Even at the time we realised in our hearts that the bulk of the public knew nothing about *Sirocco* and cared less.
to know the reason why: to demand (and get) an explanation. Cf. REASON *sb.*¹ 5.
1825 R. S. HAWKER *Cornish Ballads* (1869) 1 And shall Trelawney die? Here's twenty thousand Cornish men Will know the reason why! 1894 SOMERVILLE & 'ROSS' *Real Charlotte* III. xxxix. 87 She had laid out a good deal of money on the house and farm, but she was going to get a good return for it, or know the reason why. 1934 G. B. SHAW *On Rocks* II. 68 My Union Jack men would keep order, or theyd know the reason why. 1941 *Punch* 20 Aug. 155/2 Two months ago Herr Hitler said his armies would sweep through Russia or he would know the reason why. 1942 *Ibid.* 11 Feb. 113/2, I caught him in the wash-house for an explanation or I'd know the reason why, and it appeared I'd ruined his life.
and knows (or *knew*, etc.) *it*: is clearly aware of (what has been stated).
1848 MRS. GASKELL *Mary Barton* I. vi. 103 The son was strikingly handsome and knew it. 1898 G. B. SHAW *Mrs. Warren's Profession* III. 208 I'm not a young man, and I know it. 1930 J. B. PRIESTLEY *Angel Pavement* v. 248 Well, she's pretty enough, and knows it, the little monkey. 1932 E. V. LUCAS *Reading, Writing & Remembering* xi. 182 Meredith was very handsome, and he knew it.
to know what one likes: a phrase used to imply that the speaker knows which works of art, poems, etc., he likes without necessarily having an informed opinion to support his view.
1873 H. JAMES *Compl. Tales* (1962) III. 72, I went with Harold a great deal to the Louvre, where he was a very profitable companion. He had the history of the schools at his fingers' ends, and, as the phrase is, he knew what he liked. 1881 —— *Portrait of Lady* II. v. 67, I don't care anything about reasons, but I know what I like. 1959 *Listener* 9 July 75/3 In reality, she was just a wealthy collector. She knew what she liked. 1974 R. HILL *Very Good Hater* xi. 93 'Are you interested in art?' asked Mrs Housman politely. 'I know what I like,' he answered.
don't I know it: I am well aware of it, you need not tell me.
1874 M. CLARKE *His Natural Life* (1875) II. iii. 192 The old trick. Ha! ha! don't I know it! 1899 KIPLING *Stalky & Co.* 151 'We didn't always knock him about, though!' 'You did when you could catch him... Don't I know it!' 1936 'R. WEST' *Thinking Reed* xii. 419 'I hate it,' she said. 'I hate it.'..'Don't I know it,' said Alan. 1964 J. CREASEY *Look Three Ways* x. 96 'He's in a mess that poor devil is.' 'And don't I know it?' 1970 B. COBB *Catch Me* i. 13 'They've only been married a few months. She's still starry-eyed.' 'Don't I know it!'
before you know where you are (and similar phrases): very soon, very quickly.
1916 A. HUXLEY *Let.* 30 June (1969) 104 Steps must quickly be taken, or we shall find the place full of effigies and all the money spent before we know where we are. 1930 W. S. MAUGHAM *Bread-Winner* ii. 102 Almost before you know where you are, they're young men and women with characters of their own. 1936 WODEHOUSE *Laughing Gas* i. 9 And little by little and bit by bit, before you know where you are—why, there you are, don't you know. 1956 A. WILSON *Anglo-Saxon Att.* II. ii. 341 Gerald said at the end of her story, 'Yes, that's certainly jolly sad,' and, before he knew where he was, he had given her a cheque for the dispensing of charity. 1970 C. WHITMAN *Death out of Focus* xii. 183 You're a clever devil... You'll be an Inspector before you know where you are.
not to know whether one is coming or going (see *COME *v.* 26 e).
to know too much: used in a context of murder, or of a threat to kill, because the victim knows too much to be allowed to live.
1922 CHESTERTON (*title*) The man who knew too much. 1953 A. CHRISTIE *After Funeral* xxi. 163 'And why should anyone want to kill you, beautiful Rosamund?'..'Because I know too much, of course.' 1966 'S. WOODS' *Enter Certain Murderers* xii. 191 At the risk of being melodramatic..you know too much.
to know where one stands (or *is*) *with* (someone): to know how one is regarded by (someone); to know a person's views (on an issue).
1950 J. CANNAN *Murder Included* ii. 33 'Those blunt, downright people are never irritating—you know where you are with them.' 'They're irritating to some people.' 1951 E. PAUL *Springtime in Paris* iv. 90 An honest whore knew where she stood. 1954 L. P. HARTLEY *White Wand* 37 One never quite knew where one was with her. 1966 *Oxf. Univ. Gaz.* 23 Dec. 433/2 If the majority now rise and say they support Council's paragraph (*a*), we shall know where we stand—we shall all be standing! 1972 F. WARNER *Lying Figures* II. 9, I wasn't a pushover. All I wanted was to know where I stood.
(*do*) *you know something?*: shall I tell you this surprising fact?, I am going to tell you something.
1965 I. FLEMING *Man with Golden Gun* viii. 113 Mr. Paradise..said softly 'You know something?' 1971 J. BRUNNER *Honky in Woodpile* v. 37 'You know something?' We looked expectant. 1972 P. DICKINSON *Lizard in Cup* x. 159 You know something? She was reared in a home. 1972 J. WILSON *Hide & Seek* ii. 29 Do you know something, Mary? Mr Harris is the nicest man I know, except for my father.
g. Misc. phrases in which *know* is used *intr.* or *absol.* (usually with something implied and sometimes with specific idiomatic force):
I want to know: well, well! *U.S. colloq.*
1833 J. NEAL *Down-Easters* I. 45, I *want* to know! exclaimed the other down-easter. Well, you *do* know, replied the southerner. 1840 *Knickerbocker* XVI. 20 'I want to know!' said the lady; 'precious soul!' 1888 *Harper's Mag.* Sept. 530/1 'Why, Jered Hopkins!' she said, looking up at him; 'I want to know!' 1904 J. C. LINCOLN *Cap'n Eri* iii. 39 'I want to know!' exclaimed Captain Perez. 'You don't tell me!' said Captain Jerry. 1911 —— *Cap'n Warren's Wards* x. 154 'She said she would be delighted!' 'I want to know!' 1923 R. D. PAINE *Comrades of Rolling Ocean* 169 And you come from North Dakoty! I want to know.
that's all you know: you do not know the facts, you do not understand (used censoriously of the person to whom the phrase is addressed). Also, *that's all you know about it.*
1876 TROLLOPE *Prime Minister* III. xi. 183 'They may do foolish things, dear; and yet..not interfere with politics.' 'That's all you know about it, Plantagenet.' 1879 C. M. YONGE *Magnum Bonum* III. xxxiv. 723 'She thought you a catch in the old days.' 'That's all you know about it!' 1930 E. H. YOUNG *Miss Mole* ii. 20 'And breakfast in bed is not what you want, Hannah.' 'That's all you know about it,' Hannah said. 1961 I. FLEMING *Thunderball* ii. 19 'I wouldn't have thought these people would be interested.'..The young man snorted, 'That's all you know.' 1973 'S. WOODS' *Enter the Corpse* 165 'He hasn't been near them,' said Boney Nelson confidently... 'That's all you know,' Meg retorted.
what do you know?: used as an expression of mild surprise = 'Isn't that amazing?' 'Well I never!' 'Just fancy!' Also, *What do you know about that?*
1914 [see *GET *v.* 21 d]. 1916 'B. M. BOWER' *Phantom Herd* ii. 33 Now what do you know about that, Mig? 1933 E. E. CUMMINGS *eimi* 245 What do you know—out of every 50 chances to make a mistake, those greedy tovariches took advantage of 4 (versus 1 mistake out of 10,000 chances in America). 1943 K. TENNANT *Ride on Stranger* vii. 72 Why, the louse!..He's glad to get rid of us. What do you know about that? 1947 'N. SHUTE' *Chequer Board* iii. 63 Say, what do you know? They ain't got no sewer here. 1952 'C. BRAND' *London Particular* xvi. 216 Well, what do you know, boys?—let's call it a day. 1957 J. KEROUAC *On Road* (1958) xii. 80 And that thousand dollars was..right there on top of the safe, what do you know about that? 1959 'M. NEVILLE' *Sweet Night for Murder* xxi. 200 'Yeah... That's right... I'll say! What do you know!' Which crescendo of surprise was a clear..statement of agreement. 1968 'A. GILBERT' *Night Encounter* v. 80 'Well,' marvelled Frankie, 'what do you know?' 1971 R. DENTRY *Encounter at Kharmel* (1973) v. 89 Well! What do you know? So the Company has been getting off its well-padded bum at last.
wouldn't you (or *he*, etc.) *like to know?*: I have no intention of telling you.
1923 G. ATHERTON *Black Oxen* xx. 105 'Look here!' he said. 'How far do you go?' 'Wouldn't you like to know?' 'I should. Not for personal reasons, for girls..bore me.' 1941 I. BAIRD *He rides Sky* 123 The old crumpet fires off a lot of bilge like..'What do you do in your spare time?' (wouldn't *he* like to know?)... And so on and on. 1942 BERREY & VAN DEN BARK *Amer. Thes. Slang* § 205/6 *I won't tell you*, don't you wish you knew?..wouldn't you like to know?, you'd like to know? 1963 M. BORRELLI *Street Lamp & Stars* xiv. 127 'And what did *you* do, Naso Stuorto?' 'Wouldn't you like to know.' 'I can guess.'
you never know or *one never knows*: something unexpected or surprising may occur.
1924 G. B. SHAW *St. Joan* vi. 94 A flaw in the procedure may be useful later on: one never knows. 1926 F. W. CROFTS *Inspector French & Cheyne Mystery* viii. 103 'I don't see that we should gain much by looking at the outside of the house.' 'You never know... If we see nothing no harm is done.' 1948 'J. TEY' *Franchise Affair* xiv. 147 It would be too great luck that he should be staying at the Midland, but one never knows. 1972 E. BERCKMAN *Fourth Man on Rope* iii. 38 Among the most unpromising debris there might lurk..some jewel as yet undiscovered. *You never know*, faithfully she invoked the formula that spurred the weariest..*you never know*. 1974 J. MANN *Sticking Place* viii. 129 'I'll come with you,' Edward said...He added in a low, ominous voice..'You never know.'
for all I know (or *he knows*, etc.): as far as I am aware, since I know nothing to the contrary.
1930 E. WAUGH *Vile Bodies* viii. 143 But these young people have got hold of another end of the stick, and for all we know it may be the right one. 1934 F. W. CROFTS *12.30 from Croydon* xxi. 292 They stood to gain by Mr. Andrew Crowther's death, and though they didn't stand to gain so much as the other two, for all we know to the contrary any one of them may have been in greater need. 1937 D. RUNYON *More Than Somewhat* v. 104 Leaving the wop yelling very loud, and maybe cussing us in wop for all I know. 1954 W. S. MAUGHAM *Ten Novels* i. 3 Everybody skips, but to skip without loss is not easy. It may be, for all I know, a gift of nature, or it may be something that has to be acquired by experience. 1954 E. CALDWELL *Love & Money* (1955) xiii. 171 How do I know you're telling the truth? For all I know, this might be some more scheming between you and Tess. 1955 D. GARNETT *Aspects of Love* iv. 119 Well, if you believe in mermaids I might be one, for all you know.
I wouldn't know: I cannot be expected to know, that is outside the range of my knowledge. Also, *I wouldn't know about that.*
1939 W. M. RAINE *River Bend Feud* x. 72 Faint wrinkles creased the forehead of the engineer. 'Has he fixed up an alliance with the outside ranchmen?' he asked. 'I wouldn't know about that,' Raleigh answered. 'But if he hasn't, he will.' 1950 J. CANNAN *Murder Included* iii. 44 'The bedroom..was only locked by the deceased during her ablutions.' 'As you say nowadays—I wouldn't know,' said Sir Charles. 1952 M. R. RINEHART *Swimming Pool* xii. 110, I wouldn't know. I've never had one. 1960 L. P. HARTLEY *Facial Justice* xvi. 133 'Every man has his type, of course—.' 'I wouldn't know about that.' 1961 J. B. PRIESTLEY *Saturn Over Water* v. 64, I wouldn't know... I'm just a painter. 1968 B. FOSTER *Changing Eng. Lang.* i. 42 As an avowal of ignorance, British English has long used 'I couldn't say', but this is often replaced now by the *I wouldn't know*... In Britain it started making headway in the 'thirties, and in a British serial film (*Pimpernel Smith*) of 1940 the late Leslie Howard remarked 'In the deplorable argot of the modern generation, "I wouldn't know".' 1969 M. PUGH *Last Place Left* vii. 45 'That's why married people get so complicated in bed, isn't it?' 'Do they? I wouldn't know.'
wouldn't you (*just*) *know?*: 'just fancy!' 'imagine that!'; as one might have foreseen. orig. *U.S.*
1946 H. P. M. BROWN *Sound of Hunting* I. 52 Wouldn't you know? Of all the days to get stuck out there, he has to pick this one. 1966 *Listener* 3 Mar. 325/2 George Scott is an English professor (wouldn't you know?) who's engaged in the *bellum sexuale* with his wife. 1973 *Washington Post* 13 Jan. B.8/7 Wouldn't you just know. Lorne Greene, also known as Ben Cartwright, has gone right out and gotten himself another steady job; this time with the ABC network.
I don't (or *he*, etc., *doesn't*) *want to know*: I am not interested. Occas. const. with person as object.
1948 'N. SHUTE' *No Highway* iii. 79, I was trying to tell her what to do if things look bad. But if she doesn't want to know, I can't do more. 1967 *Listener* 14 Sept. 326/1 After doing a hard week's work I had nothing in my pocket..nothing at all and that went on for four years... After that I said: 'Well, that's it. I don't want to know. I can get a living a lot easier than going to work.' 1969 *Focus* Feb. 16/2 But if you are paying it all in on a Friday, and taking it all out again on Saturday, do not be surprised if the building society does not want to know you. 1973 *Observer* 14 Jan. 7/3 It remains to add that all this, and much more, was well enough known at the time. But the fellow-travellers didn't want to know. 1973 *Times* 19 Sept. 13/4 (Advt.), Graduates you have a problem. If you wanted the summer following graduation free, you missed out on the 'milk round'. Many employers don't want to know by the autumn.
15. *to know a thing or two* (later examples); *to know it all*: not to be aware of one's deficiencies, to be deaf to advice or instruction; also *to think one* (or *he, she*, etc.) *knows it all.* Cf. *know-all, know-it-all* s.v. KNOW- in Dict. and Suppl.
to know the ropes (see Dict. and ROPE *sb.*¹ 4 c); *to know all the answers* (see *ANSWER *sb.* 6 b); *not to know beans* (see *BEAN *sb.* 6 e).
1870 E. G. WHITE *Testimonies for Church* No. 19. 73 You have so long thought, with the peculiar class I have mentioned, that you knew it all, that you will not see your deficiencies when they are presented before you. 1929 J. B. PRIESTLEY *Good Companions* I. iii. 24 Ted..admitted that he knocked about a bit and knew a thing or two. 1944 E. CALDWELL *Tragic Ground* (1947) iii. 31 Jim Howard Vance is a pretty smart fellow. He was talking in there just a while ago like he knows a thing or two. 1972 G. DURRELL *Catch me a Colobus* vi. 111 As I had warned Long John, there comes a time on every collecting trip when you begin to think that you know it all. This is a moment of great danger, for you *never* know it all, however hard you try. *Ibid.*, I made a mistake once by thinking I knew it all, and got bitten by a snake. 1973 WODEHOUSE *Bachelor Anonymous* iv. 33 The serfs and vassals now know a thing or two and prefer to make their living elsewhere. 1973 *Black World* Sept. 97/1 To my once respected student who has taken over the pompous entitlement as chief white critic of inferior Black literature, let me say..: 'Stop knowing it all.'

know-. Add: **know-all, know-it-all:** also *attrib.* or as *adj.*, full of knowledge; esp., deaf to advice or instruction.
1881 TENNYSON *Tiresias* (1885) 49 *We* have knelt in your know-all chapel. 1906 *Daily Chron.* 30 Aug. 3/4 He..maintains in his know-all manner that the two counties of Wigtown and Kirkcudbright..were integral parts always of the kingdom of Scotland. 1923 D. H. LAWRENCE *Birds, Beasts & Flowers* 185 Fool, in spite of your pretty ways, and quaint, know-all, wrinkled old aunty's face. 1935 WODEHOUSE *Blandings Castle* ix. 225 These know-it-all directors make me tired. 1956 H. GOLD *Man who was not with It* (1965) ii. 16 He looked over the apprehensive afternoon crowd with its know-it-all faces. 1959 *Encounter* XIII. II. 57 All big smart know-it-all Marxists. 1974 *Times* 23 Jan. 1/8 We didn't realize, until it was too late, how our know-it-all attitude was undermining the self-assurance of parents.

knowed. Widespread dial. pa. t. of KNOW *v.*
Numerous examples in *E.D.D.* (q.v.). The examples that follow are *U.S.*
1848 J. R. LOWELL *Poet. Wks.* (1873) 252/1 My! when he made Ole Hunderd ring, She *knowed* the Lord was

nigher. **1872** W. Coleman in *Rep. 42nd U.S. Congress 2 Sess. Joint Select. Comm. Condition of Affairs Late Insurrectionary States* XI. 484 Of course I knowed him. **1929** *Amer. Mercury* Sept. 50/1 Got in trouble one time.. Knowed officers couldn't 'rest me. **1942** *Ibid.* July 87, I knowed you'd back up. **1949** in B. A. Botkin *Treas. S. Folklore* III. i. 434, I knowed dad-blamed well they wa'n't no fox in that sourwood.

know-how (nōu-hau). orig. *U.S.* [f. vbl. phr. *to know how* (Know *v.* 12).] Knowledge of how to do some particular thing; technical expertness, practical knowledge.

1838 *New Yorker* 14 July 260/2, I promise, 1st. To do the duties of the office to the best of my know-how, and have a stouter man than myself to help me. **1857** *Spirit of Times* 26 Dec. 270/3 'No, no, Massa,' replied the gentleman from Africa, 'charge fifty cents for killing, and fifty for the know how.' **1899** Kipling *From Sea to Sea* II. 95 He has the money. We have the know-how. He comes in winter to play poker... When he's lost his money we make him drunk and let him go. **1936** J. Dos Passos *Big Money* 282 Charley Anderson, the boy with the knowhow. **1944** H. A. Wallace *Century of Common Man* 52 We have the 'know how' to help many of the poverty-stricken peoples to set their feet on the path of education, manual dexterity, and economic literacy. **1946** *Times* 10 Dec. 2/6 With regard to American investments in this country the Government welcomed such applications if they brought to this country real industrial knowledge, the 'know how', which otherwise we should be without. **1947** Auden *Age of Anxiety* (1948) i. 26 A modern product Of nerve and know-how with a new thrill. **1949** *Listener* 15 Sept. 451/3 A manager may have knowledge of a process in his charge, but not the know-how possessed by the foreman who controls it. **1952** R. M. Hare *Lang. Morals* x. 159 Everything that we are taught how to do must..be reducible to principles, though these may be 'know-hows' hard to formulate in language and more easily taught by example. **1953** *Encounter* Oct. 68/2 There were some who were persuaded that technical knowledge—that intangible but wonderful thing called American 'know-how'—could somehow be made a substitute for capital in the poor countries. **1967** L. B. Archer in Wills & Yearsley *Handbk. Managem. Technol.* 129 It takes special talent and know-how on the part of a properly trained designer to create a successful house style. **1972** P. M. Hubbard *Whisper in Glen* ix. 91 She hasn't got the mere social know-how to carry it off.

knowing, *ppl. a.* Add: **3.** Comb. *knowing-looking* ppl. a.

a **1817** Jane Austen *Northanger Abbey* (1818) I. vii. 81 A gig, driven along on bad pavement by a most knowing-looking coachman. **1927** H. V. Morton *In Search of England* i. 9 [Newbury, in Berkshire] is, like all towns which have any traffic with race-horses, a knowing-looking, bandy-legged town.

5. Restrict ?*Obs.* to 'Const. *of, in*' and add later U.S. examples of *knowing to*.

1905 *Springfield* (Mass.) *Weekly Republ.* 29 Dec. 16 Some of the neighbors were knowing to the event. **1906** *Dialect Notes* III. 144 I'm knowing to that; you're wrong. **1913** *Ibid.* IV. 2 You are knowing to that. **1913** H. Kephart *Our Southern Highlanders* xiii. 297 Reckon Pete was knowin' to the sarcumstance?

knowledge, *sb.* Add: **8. c.** Philos. *knowledge about, knowledge by description*: knowledge of a person, thing, or perception gained through information of facts about it rather than by direct experience (opp. *knowledge by* (or *of*) *acquaintance*, see *acquaintance 1 b).

1885, etc. [see *acquaintance 1 b]. **1945** E. Mayo *Social Probl. Industr. Civilization* (1949) I. i. 15 The student is required to relate his logical *knowledge-about* to his own direct acquaintance with the facts. **1952** B. Mayo *Logic of Personality* iii. 30 Knowledge *about* something is called knowledge by *description*. **1954** [see *acquaintance 1 b]. **1967** *Encycl. Philos.* IV. 350/1 Parallel to this on the side of knowledge of things is the distinction between knowledge by acquaintance and knowledge by description. **1968** A. J. Ayer *Origins Pragmatism* II. iii. 293 The mind has 'knowledge about' an object not immediately there.

16. knowledge factory, term applied pejoratively to a university or college, etc., which places undue emphasis on vocational training; **knowledge industry,** term applied fancifully or pejoratively to the development and use of knowledge, *spec.* in universities, polytechnics, etc.

1928 *World's Work* May 55 Next day we visited the knowledge factory, and..the head teacher asked if I had ever been sent to school. **1968** *Listener* 4 July 6/2 Some students who rioted on British campuses (like some in France and Italy) have been protesting at having found themselves in a knowledge factory when they thought they were headed for something else. They find themselves being trained for the managerial and technocratic élites, whereas what they demand is the right to question the structure of society which makes such élites necessary. **1969** C. Davidson in Cockburn & Blackburn *Student Power* 341 The production of an increase in socially useful and necessary labour power is the new historic function of our educational institutions that enables us to name them, quite accurately, knowledge factories. **1962** F. Machlup *Production & Distribution of Knowledge in U.S.* iii. 45 If the phrase 'knowledge industry' were to be given an unambiguous meaning, would it be a collection of industries producing knowledge or rather a collection of occupations producing knowledge in whatever industries they are employed. **1963** C. Kerr *Uses of University* iii. 87 Basic to this transformation is the growth of the 'knowledge industry', which is coming to permeate government and business. **1968** *Economist* 28 Feb. 51/3 This is a book for the serious investor who..wants to learn something about the operations of the New York Stock Exchange and the 'knowledge industry', with its analysts, theorists..and numerous other 'ists'. **1970** *Globe & Mail* (Toronto) 25 Sept. B2/2 The report notes the emergence of the knowledge industry, growing emphasis on people values.

knowledgeable, *a.* Add: Also **knowledgable. 2.** Delete *colloq.* and add earlier and later examples. Also, cognisant *of*.

1829 G. Griffin *Collegians* (ed. 2) I. xi. 233 She went.. to Shaun Lauther, the knowledgeable man, and put a half-a-crown into his hand, and asked his advice. **1901** *Academy* 21 Sept. 240/2 The review..is not only able and knowledgable; it is also..fair. **1903** *Daily Chron.* 4 Aug. 3/2 His manner is so knowledgable and convincing that they will question nothing of his theories. **1905** *Westm. Gaz.* 2 Feb. 2/1 If any official English politician has a knowledgable opinion of how these Powers are likely to combine or to clash..he should be sought. **1908** *Daily Chron.* 13 Feb. 5/7 All 'knowledgable opinion'..is against the Bill. **1945** Mencken *Amer. Lang.* Suppl. I. 423 The English have many counter-words that fail to make the Atlantic journey, *e.g., knowledgeable.* **1955** *Sci. Amer.* June 102/2 It will be a great day for mankind when we become equally knowledgeable about the lives of microbes. **1973** *Times* 31 July 6/7 Mr Dean drew the erroneous conclusion that the President was fully knowledgeable of the cover-up at the time of the March 13 meeting.

Hence **knowledgeabi·lity.**

1946 *Time* 19 Aug. 98 His portrait shows Caesar to be a man as far beyond mere knowledgeability as a Hitler or a Stalin. **1957** N. Frye *Anat. Criticism* 263 This has a truth that the myopia of knowledgeability is more apt to overlook. **1965** F. Sargeson *Mem. Peon* iv. 62, I had impressed my host by my..knowledgeability.

know·ledgefully, *adv. rare.* [f. Knowledge *sb.* + -ful + -ly².Cf. *knowledge-full* adj. s.v. Knowledge *sb.* 16.] In the manner of one who is fully informed.

1906 *Harper's Mag.* Feb. 474/2 He has written of this very knowledgefully, of course, and very justly.

known, *ppl. a.* Add: **1. c.** *known to the police*: applied to a person with a criminal record. Also ellipt. *known* (itself occas. used as *sb.*).

1909 Galsworthy *Silver Box* III. 75 Is she known here? ..No, your Worship, they're neither of them known, we've nothing against them at all. **1924** A. Christie *Poirot Investigates* ix. 255 Billy Kellett?.. He's known to the police! **1938** F. D. Sharpe *Sharpe of Flying Squad* ix. 112 A long communication telling us that Mrs. Cousins was not 'known to the police'. **1971** E. McGirr *No Better Fiend* 69 The late Mantel had been 'known' since 1928. It was a dismal dirty story. **1973** K. Giles *File on Death* vi. 149 A lot of Irish boys in Granchester... I spotted a couple of 'knowns'.

know-nothing, *sb.* and *a.* Add: **A.** *sb.* **1.** (Earlier example.)

1827 J. F. Cooper *Red Rover* I. ii. 42 The fellow is a know-nothing!

2. (Earlier example.)

1854 *Harper's Mag.* Aug. 400/1 A secret combination designated as 'Know-Nothings'..have operated with much success in local elections in many of the larger places.

B. *attrib.* or *adj.* **1.** (Later examples.)

1959 *Encounter* XII. ii. 32 He..sounds most like a know-nothing native writer. **1972** *Computers & Humanities* VII. 13 Robert Wachal has offered a humorous account of the problems encountered by the know-nothing humanist who wishes to learn about computing.

2. (Earlier example.)

1854 *Southern Lit. Messenger* XX. 540/1 This Know Nothing movement will prove to be..a giant evil.

know-nothingism. 2. (Earlier example.)

1854 W. G. Simms *Southward Ho!* 252 Know-Nothingism had not then become a fixed fact in the political atmosphere.

Knoxian (nǫ·ksiăn), *sb.* and *a.* [f. the proper name *Knox* (see below) + -ian.] **A.** *sb.* An adherent or follower of John *Knox* (*c* 1505–72), the Scottish Reformer who was mainly responsible for establishing the Presbyterian Church. **B.** *adj.* Of or pertaining to John Knox.

1714 J. Collier *Eccl. Hist. Gt. Brit.* II. v. 394/2 In this Abstract, the Knoxians make a scandalous Representation of some Part of the Litany. **1905** *Westm. Gaz.* 4 Mar. 12/3 His Knoxian project was dropped in favour of his works on Cromwell. **1906** *Daily Chron.* 22 May 3/3 Buchanan, in fact, never was a Reformer in the Knoxian sense. **1933** *Times Lit. Suppl.* 30 Nov. 848/2 In Scotland —after the Knoxian supremacy was established—'not one single adherent of the Roman Communion was martyred because of his faith'. **1937** E. Percy *John Knox* IV. xv. 253 A strong anti-clerical party..who would have nothing to do either with Knoxians or with sectaries. **1961** C. H. & K. George *Protestant Mind of Eng. Reformation* II. v. 177 Knoxian precepts from Geneva had altered the entire course of Scottish history.

knubbly, *a.* (Later examples.)

1910 W. de la Mare *Three Mulla-Mulgars* iv. 55 Three big knubbly cudgels. **1917** W. J. Locke *Red Planet* xxi. 277 He held out his hand, a dirty, knubbly, ragged-nailed hand. **1939** 'A. Bridge' *Four-Part Setting* vii. 79 Two loose knubbly little cushions.

knuck (nɒk). Also occas. **nuck.** [Shortening of Knuckle *sb.*, Knuckler.] **1.** *slang.* A thief, a pickpocket. Cf. *knuckle *sb.* 2 c, Knuckler 1. ? *Obs.*

1812 J. H. Vaux *Vocab. Flash Lang.* in *Mem.* (1819) II. 184 *Knuck, knuckler,* or *knuckling-cove*, a pickpocket, or person professed in the knuckling art. **1848** 'N. Buntline' *Mysteries & Miseries N.Y.* I. 33 There is a house in Cherry street..[that] has been known to the 'crossmen' and 'knucks' of the town as 'Jack Circle's Watering place' and 'fence'. **1903** A. H. Lewis *Boss* 168 But knucks, dips, sneaks,..an' strong-arm men have got to quit. **1904** 'No. 1500' *Life in Sing Sing* 251/1 *Nuck*, a thief. **1935** A. J. Pollock *Underworld Speaks* 68/2 *Knuck*, a thief.

2. *pl.* 'A game of marbles in which the winner shoots at his adversary's knuckles' (Clapin). *U.S.*

1840 *Southern Lit. Messenger* VI. 385 To the game of marbles he devotes much of his leisure time, and is counted a proficient particularly in knucks and five in the ring. *a* **1883** G. W. Bagby *Sel. Misc. Writings* (1885) II. 20 He tries to keep somebody's country store, but will close the doors whenever the weather is fine to 'ketch chub' or play knucks. **1886** *Harper's Mag.* Dec 41/2 They were playing 'knucks' together. **1935** *Amer. Speech* X. 159/1 More on Marble Names and Games... Knucks Down. A variant of *knuckle down.*

3. Shortening of Knuckle-duster.

1897 *Sears, Roebuck Catal.* 593/1 Knucks, heavy nickel plated and polished. $0.30 per pair. **1918** C. Sandburg *Cornhuskers* 88, I slipped my fingers into a set of knucks. **1966** Wodehouse *Plum Pie* i. 48 To reason successfully with that king of the twisters one would need brass knucks and a stocking full of sand. **1973** D. Westheimer *Going Public* ix. 130 He produced a two foot length of stout nylon cord with brass knucks at either end. 'You can hit or strangle with it..' he explained.

knuckle, *sb.* Add: **2. b.** *near the knuckle*: near the permitted limit (esp. in regard to decency); *to go the knuckle* (*Austral. slang*), to punch, to fight.

1895 W. P. Ridge *Minor Dialogues* vii. 72, I can stand a joke as well as anyone, but whispering's a bit *too* near the knuckle. If you've got anything to say, say it. **1909** *Westm. Gaz.* 4 May 2/2 A series of articles entitled 'Crimes of Passion', full of abominable details 'as near the knuckle' as the police would allow. *Ibid.* 6 Sept. 1/3 If a play shows that its author has..a sincere respect for his art, it must be stopped if it goes at all 'near the knuckle'. **1930** W. S. Maugham *Cakes & Ale* 147 What I like about 'er is that she gives you a good laugh. She goes pretty near the knuckle sometimes, but she never jumps over the fence. **1944** J. Devanny *By Tropic Sea & Jungle* xviii. 160, I always got on well with the blacks, because I never went the knuckle on them, and never interfered with their women. **1945** L. A. G. Strong *Othello's Occupation* iv. 89 Did you notice how she stiffened, when I slipped in that bit about remembering what she was looking at—what was on the table? I got a bit near the knuckle there. **1962** S. Gore *Down Golden Mile* i. 26 Then he said: 'Want to watch out for them quiet snoozers. Sometimes they can go the knuckle a bit themselves!' **1973** D. Jordan *Nile Green* xiii. 58 He ho-ho'd jovially to show he was joking. It was a little close to the knuckle for my taste.

†**c.** = Pickpocket *sb.* 1. Cf. *knuck 1. *Obs.*

1781 G. Parker *View of Soc.* II. i. 73 *Knuckle*, in the flash language, signifies those who hang about the Lobbies of both Houses of Parliament, the Opera-House and both Play-Houses, and in general wherever a great crowd assemble. They steal watches, snuff-boxes, &c. **1785** Grose *Dict. Vulgar T., Knuckles*, pickpockets who attend the avenues to public places, to steal pocket books, watches, &c., a superior kind of pickpockets. **1846** *Swell's Night Guide* 124/1 *Knuckles*, pickpockets.

6. *knuckle-length, -rapper, -rapping* vbl. sb. and ppl. adj.; **knuckle ball, knuckleball** *Baseball* (see quots.); **knuckle-head,** a slow-witted or stupid person; **knuckle sandwich** *slang,* a punch in the mouth.

1927 *Secrets of Baseball* iii. 37 If you're worrying about when to use a 'knuckle ball'. **1929** *Encycl. Brit.* III. 163/1 There is what is known as the knuckle ball, in which the knuckles of the pitcher's hand play a prominent part in giving the ball erratic motion. **1970** *New Yorker* 24 Oct. 39/2 The knuckleball is thrown not with the knuckles but with the fingertips. **1972** *N.Y. Times* 4 June v. 2/5 Phil Niekro baffled the Mets with his celebrated knuckleball, allowing them only three hits. **1975** *Cleveland* (Ohio) *Plain Dealer* 6 Apr. 13–C/3 Oakland, aiming for a fourth straight world championship, opens at home with Vida Blue pitching against Chicago's veteran knuckleball specialist, Wilbur Wood. **1944** in Wentworth & Flexner *Dict. Amer. Slang* (1960) 310/2 You knuckle-heads. **1948** *Amer. Speech* XXIII. 249/1 Knucklehead, a slow or stupid person. **1971** R. Parkes *Line of Fire* xvii. 158 What I'm trying to get across to you knuckleheads is that it was *not* murder! **1973** J. Cleary *Ransom* i. 25 He's a knuckle-head, he knows nothing and doesn't want to know. **1906** *Daily Chron.* 29 Mar. 6/4 A man..proceeded to measure it with the knuckle-length of his closed fist. **1938** *Times* 9 Mar. 19/4 The knuckle-length coat is in stripes of various sizes. **1910** H. G. Wells *Hist. Mr. Polly* iv. 56 The aunt..was..a knuckle-rapper and sharp silencer: no friend for a slovenly little boy. **1944** R. Lehmann *Ballad & Source* 243 'No,' he said shortly, but not in the knuckle-rapping way I had half feared. **1968** *Globe & Mail Mag.* (Toronto) 13 Jan. 3/1 Anti-Dow demonstrations at U of T drew just polite knuckle-rapping. **1968** *Listener* 18 July 75/1 Did you ever suggest to him..how a field commander might feel when he got one of these knuckle-rappings? **1974** *Farm & Country* 26 Mar. 4/2 Your knuckle-rapping will teach him to think before he is carried away on a flow of words. **1973** A. Buzo *Norm & Ahmed* 12 He tried to hang one on me at Leichhardt Oval

once, so I administered a knuckle sandwich to him. **1973** *Ottawa Jrnl.* 17 July 23/3 Give the guy a knuckle sandwich and let the teeth fall where they may.

knuckle, *v.* Add: **2.** (Later examples.)

1955 *Times* 19 May 15/4 He replied that there was no power on earth to make a local party accept a candidate. He was rather sorry they knuckled under to Transport House in this division. **1964** *Ann. Reg. 1963* 10 Britain, he said, had 'knuckled under' to threats of African violence, but there was little he could constitutionally do about it. **1973** *Nation Rev.* (Melbourne) III. 31 Aug. 1444/1 Now the last group of any size..has knuckled under following a series of splits and coups.

3. b. To dig (one's hand) into a specified position, knuckles first.

1890 Hall Caine *Bondman* I. i. 10 Thrusting his head beneath his chin, he knuckled his left hand under the islander's rib.

6. *Farriery.* With *over*, of the knee or fetlock: ?to project through weakness of the ligaments. Also with *forwards*. (Cf. *knuckle-kneed* adj. s.v. Knuckle *sb.* 6.)

1877 A. Sewell *Black Beauty* (*c* 1878, ed. 5) xl. 198 The knees knuckled over, and the forelegs were very unsteady. **1877** M. H. Hayes *Vet. Notes for Horse Owners* i. 30 'Knuckling over', as a result of hard work, appears to be due to relaxation of the capsular and lateral ligaments [of the fetlock joint]. **1906** J. W. Axe *Horse* IV. 53 The animal knuckles over at the joints. **1907** *Ibid.* V. 298 The movements of the hind-limbs are for a time weak, and the fetlocks knuckle over now and again during progression. **1907** *Ibid.* VI. 347 In young horses it is common to meet with a knuckling forwards of the hind fetlocks.

7. *Golf.* To bend (the knee) inward. Also *absol.*

1909 *Times* 23 Apr. 16/2 An elaborate knuckling of the right knee in putting. *Ibid.*, The 'knuckling' habit in putting has long been exposed as fallacious. **1909** *Westm. Gaz.* 11 May 12/3 You may 'knuckle' if knuckling conduces to comfort, provided that..you get the process over before beginning the actual stroke.

knu·ckle-dust, *v.* [Back-formation from Knuckle-duster.] *trans.* To strike with a knuckle-duster. So **knu·ckle-dusting** *vbl. sb.*

1909 *Daily Chron.* 29 Dec. 1/7 One boy..got the robber's head under his arm and gave it a knuckle-dusting. **1962** V. Nabokov *Pale Fire* 151 The brief affray during which two of the attackers were knuckledusted and knocked out by the brave Lorrainer.

knuckler. Add: **3.** *Baseball.* A knuckle ball (see *knuckle *sb.* 6).

1928 G. H. Ruth *Babe Ruth's Own Bk. Baseball* vi. 79 Eddie used to toss 'knucklers' until he had the hitters blue in the face. **1972** *N.Y. Times* 4 June v. 2/5 The Mets, meanwhile, were subdued by Niekro's knuckler.

knucklesome (nɒ·k'lsəm), *a. rare.* [See -some¹.] Having prominent knuckles. Also *transf.*

1919 W. De Morgan *Old Madhouse* xx. 306 That young woman was bony and knucklesome. **1922** C. E. Montague *Disenchantment* xii. 170 The twisty valley and knucklesome banks of the Somme.

knu·ckle-walker. [f. Knuckle *sb.* + Walker *sb.*¹] Any primate, such as the gorilla or chimpanzee, which has a quadrupedal gait involving the backs of the knuckles (rather than the tips of the fingers or flat of the palm) making contact with the ground. So **knu·ckle-walk** *v.*, **knu·ckle-walking** *vbl. sb.*

1859 R. Owen *On Classification & Geographical Distribution of Mammalia* 75 Of the broad-breasted quadrumana, are the knuckle-walkers or the brachiators, i.e. the long-armed gibbons, most nearly and essentially related to the human subject? **1967** R. H. Tuttle in *Amer. Jrnl. Physical Anthropol.* XXVI. 171 (*title*) Knuckle-walking and the evolution of Hominoid hands. *Ibid.*, Most orangutans assume one of a variety of flexed hand postures, but they cannot knuckle-walk. *Ibid.* 171/2 The knuckle-walking posture of chimpanzees and gorillas is unique among primates. **1967** S. L. Washburn in *Proc. R. Anthrop. Inst.* 23/1 Both chimpanzees and gorillas are knuckle walkers. *Ibid.*, It appears that our ancestors were arboreal apes for many millions of years, that they then shared a common knuckle-walking stage with the ancestors of the chimpanzee and gorilla, and that only later they became bipeds. **1969** A. H. Schultz *Life of Primates* v. 56 In the orang-utan the mode of using the hands in walking..is more in the form of a clenched fist, rather awkwardly twisted sideways, and rarely like real knuckle-walking. **1972** O. J. Lewis in R. H. Tuttle *Functional & Evolutionary Biol. Primates* ix. 212 Thick palmar radiocarpal ligaments are not the exclusive property of knuckle-walkers... It thus seems that knuckle-walking requires no especially striking modifications of the wrist joint. **1973** *Nature* 10 Aug. 373/2 The first component separates quadrupedal cercopithecoids from both knuckle-walkers (*Pan* and *Gorilla*) and the quadrupedal arm-swinger.. *Ateles.*

Knudsen (knu·dsən). The name of Martin H. C. *Knudsen* (1871–1949), Danish physicist, used *attrib.* (or occas. in the possessive) to designate apparatus, phenomena, and concepts connected with his work.

a. *Physics.* **Knudsen cell,** a vessel in which a substance is heated in equilibrium with its vapour, which is allowed to diffuse out of a small orifice; **Knudsen effect, flow,** the effusion from an orifice (also called *Knudsen effusion*) or the flow through a tube of a gas with a high Knudsen number, so that the resistance to flow arises principally from collisions of the molecules with the walls rather than with one another; **Knudsen gas,** any gas in a state characterized by a Knudsen number much greater than one; **Knudsen gauge** or **manometer,** an instrument for measuring the absolute pressure of a rarefied gas by means of the transfer of momentum by the gas molecules between two fixed plates at different temperatures and a suspended vane, which undergoes a rotation dependent on the pressure; **Knudsen number,** the ratio of the mean free path of the molecules of a gas to a length derived from the dimensions of the apparatus in which or past which it is flowing.

1954 *Jrnl. Chem. Physics* XXII. 1414/2 The vapor pressure was determined by collecting a known fraction of the vapor effusing from a Knudsen cell. **1960** *Ibid.* XXXIII. 530/1 The equilibrium vapor from a Knudsen cell was collimated into a molecular beam and allowed to pass through the ionizing region of a mass spectrometer. **1953** *Physical Rev.* LXXXIX. 796/1 The Knudsen effect occurs when two portions of a gas are separated by a very fine capillary or a porous plate with openings so small that collisions between the gas molecules in the capillary or the pores are infrequent compared with the collisions of the gas molecules with the walls. **1954** *Jrnl. Chem. Physics* XXII. 1414/2 The vapor pressure has been redetermined over a temperature range from 1630 to 1970°K. As in the previous measurements the Knudsen effusion method was used. **1937** *Bull. Chem. Soc. Japan* XII. 199 If the mean free path is large in comparison with the diameter the flowing quantity is independent of the viscosity but inversely proportional to the square root of the molecular weight of the gas, and such flow is called Knudsen's or the molecular flow. **1958** R. D. Present *Kinetic Theory of Gases* iv. 61 Effusion through a circular orifice can be considered as a special case of Knudsen flow in which the length of tube is small compared to its diameter. **1958** *Encycl. Physics* XII. 212 Another example which is more intuitive is afforded by a Knudsen gas. This is a gas which is sufficiently rarefied so that intermolecular forces can be completely ignored. **1972** *Chem. Abstr.* LXXVI. 87851 A collision-free Knudsen gas was considered between 2 parallel plates with a time-dependent temp. gradient. **1918** *Physical Rev.* XII. 452 Delicate equilibria at low pressures in this way make it possible to measure exceedingly small quantities of emission products even with the use of a less sensitive form of Knudsen gauge. **1925** F. H. Newman *Production & Measurement Low Pressure* viii. 155 These investigators have also constructed Knudsen gauges capable of measuring pressures as high as 10^{-2} mm. **1959** *Chambers's Encycl.* XI. 188/1 Knudsen gauges can be constructed to cover ranges from 10^{-3} to 10^{-7} mm Hg. **1961** J. Thewlis et al. *Encycl. Dict. Physics* IV. 185/2 The functions of the Knudsen manometer essentially depend on the variation of the thermal conductivity of a gas with pressure, provided the pressure is low enough, i.e. the mean free path is sufficiently great. **1956** G. N. Patterson *Molecular Flow of Gases* v. 159 In both high-vacuum systems and high-altitude flight, the Knudsen number becomes large. **1957** Liepmann & Roshko *Elem. Gas Dynamics* xiv. 353 For flow similar to Couette flow, i.e., flow that is confined between walls, one can easily define these two limiting cases by the ratio of mean free path Λ to channel diameter d. Λ/d is often called the Knudsen number; if $\Lambda/d \ll 1$, intermolecular collisions dominate; if $\Lambda/d \gg 1$, collisions with the boundaries dominate.

b. *Oceanogr.* **Knudsen burette, pipette,** special types of burette and pipette for use in Knudsen titrations; **Knudsen method, titration,** a method for determining the chlorinity and hence salinity of sea-water by titration against silver nitrate solution and reference to a set of tables, *Knudsen's tables*, first published by Knudsen in 1901.

1959 H. Barnes *Apparatus & Methods Oceanogr.* I. iv. 86 *The Knudsen burette.* The silver nitrate solution contained in a reservoir is delivered from a special burette, with an automatic zero. **1966** B. B. Baker et al. *Gloss. Oceanogr. Terms* (ed. 2) 32/1 By using normal water as a comparison standard, Knudsen burettes and pipettes for the analysis, and Knudsen's Tables to compute the results, determinations as accurate as those of a time-consuming gravimetric analysis can be made with a rapid titration of the sea water against silver nitrate solution, employing potassium chromate or other suitable indicator for the end-point. **1954** *Jrnl. Marine Res.* XIII. 246 The keystone of the Knudsen method is the adjustment of the silver nitrate concentration so that at the end of the titration the number representing the burette reading is approximately equal to the chlorinity in per mille. [**1923** Glazebrook *Dict. Appl. Physics* III. 677/2 The pipette.. generally used is of the Knudsen pattern, with a three-way tap instead of a mark.] **1951** *Jrnl. du Conseil Internat. Explor. de la Mer* XVII. 223 By means of a Knudsen pipette 15 ml. were taken out of a number of sea-water samples. **1966** Knudsen pipette [see *Knudsen burette* above]. **1923** Glazebrook *Dict. Appl. Physics* III. 677/2 The salinity and density can, of course, be determined from the 'chlorine' content by Knudsen's Tables. **1966** Knudsen's Tables [see *Knudsen burette* above]. **1962** *Nature* 10 Feb. 520/1 The precision of measuring conductivity on a good salinometer is at least five times better than that of the standard Knudsen titration.

knur. 3. (Later examples.)

1967 *Antique Finder* Aug. 11/3 Knur and Spell. This Georgian tavern game must really be a forerunner of darts. **1972** *Daily Tel.* (Colour Suppl.) 14 Jan. 22/3 In fact Knur and Spell is simply a formalisation of something that every male person in the world must have done at some time—throwing a ball or stone in the air and giving it a tremendous clout with a stick. How maddening when you miss! What a marvellous eye you feel you have when you do it fair and square! The Spell is the device that throws the ball (the Knur) into the air. It is an iron contraption that lies on the ground looking a bit like a rat trap; the framework holds a flat horizontal spring, held down by a catch. On the releasable end of the spring is a little cup containing the knur, a tiny white ball (porcelain, of all things), one inch across and weighing half an ounce. You stand about four feet away from the spell, armed with a 'stick', rather like a billiard cue with a hammer head (called the 'pommel'). You tap the catch on the spell, the knur jumps up about four feet; the object is to hit it farther, over a fixed number of goes, than anyone else.

knut, jocular variant, often pronounced (kənɒ·t), of *nut *sb.*¹ 8* (a fashionable or showy young man). Hence **knu·tty** *a.*

1911 *Granta* 25 Nov. 136/2 He, Timothy Gray, bhoy, lad, knut. **1913** L. A. Harker *Ffolliots of Redmarley* v. 57 He was..a 'knut' of the nuttiest flavour. **1914** *Scotsman* 5 Oct. 8/1 It is clear that he has once been a 'knut' in spite of his oil-stained khaki service jacket and trousers. **1915** A. Wimperis *Gilbert the Filbert* (*song*), I'm Gilbert, the Filbert, The Colonel of the Knuts. **1916** E. V. Lucas *Vermilion Box* 52 Among the people staying here is a knut. He must be almost the last of the tribe; but here he is, just as knutty as though the Algies and Berties were still ruling the roast, and not Mars at all. **1919** C. Orr *Glorious Thing* xvii. 212 He was trying to be knutty, he said. **1929** G. Stowell *Hist. Button Hill* 183 The Knut was an urban and suburban phenomenon of the years 1912 to 1914 inclusive. **1973** *Listener* 6 Sept. 320 The 'silly asses', the 'knuts' who were wiped out on the Somme.

‖ **ko¹** (kōu). *N.Z.* [Maori.] A digging-stick.

1843 E. Dieffenbach *Trav. N.Z.* II. III. 367 *Ko*—a tool with which the natives plant their sweet potatoes. **1868** W. Colenso in *Trans. N.Z. Inst.* I. Essay 15 A *ko*, a rude kind of narrow and pointed spade with a very long handle, to which, at about 18 inches or more from the point, they [*sc.* Maoris] fitted a small crooked bit of carved wood, as a rest for the foot. **1905** W. B. *Where White Man Treads* 2 Plantations of kumara..which the Maori with his primitive ko (native wooden spade) had brought under tillage with much labour. **1941** Alley & Hall *Farmer in N.Z.* i. 3 His [*sc.* the Maori's] instant casting aside of his poor crude *ko* and *timo* (or grubbing stick) in favour of European spades and hoes. **1955** W. J. Phillipps *Maori Carving Illustr.* 16 Designs carved on the upper ends of ko or digging sticks.

‖ **Ko²** (kōu). *Ceramics.* [Chinese, = elder brother. See quot. 1954.] In full, *Ko iú* or *yao* [Chinese, = ware], *Ko ware.* A crackled Sung ware closely related to Southern Kuan; also, a name for other crackled porcelains.

1882 *China Rev.* X. v. 310/1 Speaking of porcelain ware.., it is absolutely necessary to distinguish the kinds called Ch'ái porcelain.., Ū porcelain.., Government porcelain.., Ko porcelain. **1882** *Ibid.* XI. III. 176/2 Ko-iú (Ko porcelain) has, on its ground, hidden lines like spawn. **1904** E. Dillon *Porcelain* v. 63 The word Ko yao is used as a general name for many kinds of [Sung] crackle ware... In a more restricted sense it includes only the early pieces with a greyish white glaze and well-marked crackles. **1954** G. Savage *Porcelain* I. i. 67 *Ko yao* was, by tradition, made by the elder of two brothers Chang... The type has a very dark stoneware body. **1972** *Collector's Guide* June 95/3 Traditionally there is Northern Kuan, Southern Kuan, Altar Kuan and the related Ko ware.

koa. (Later examples.)

1954 J. Sheridan in J. Macdonald *Lethal Sex* (1962) 181 Four men carried the pigs on huge *koa* platters carved with supporting feet. **1965** *N.Z. Listener* 17 Dec. 4/2 The commoners [in Hawaii] were permitted only the heavier and less buoyant Koa boards.

koaftah, var. *kofta.

koala (ko͵ā·lă). Now the usual spelling of Koolah. In full, *koala bear.* (Examples.)

1808, 1859 [in Dict. s.v. Koolah]. **1902** *Daily Chron.* 3 July 3/4 The koala of Australia, has also a very big cæcum. **1937** C. Kearton *I visit Antipodes* x. 132 The Koala Bear was first seen by a young explorer who journeyed to the Blue Mountains in 1798. **1944** A. Russell *Bush Ways* xix. 95 There was..scarce a gum tree but that sheltered an opossum or a koala. **1966** G. Durrell *Two in Bush* v. 173 Fortunately, before it became too late, the Government stepped in and passed laws strictly protecting the Koalas, and slowly over the years their numbers have built up again.

‖ **koan** (kōu·ăn). [Jap., f. *kō* public + *an* matter, material for thought.] In Zen Buddhism, a paradox put to a student to stimulate his mind.

1946 R. Benedict *Chrysanthemum & Sword* (1947) xi. 246 The significance of the koan does not lie in the truths these seekers after truth discover. **1957** *Time* 4 Feb. 66/2 A less physical shock technique is the *koan*, a problem designed to shock the mind beyond mere thinking. **1958** A. Huxley *Let.* 11 Jan. (1969) 844 They might act as Zen koans and cause sudden openings into hitherto unglimpsed regions. **1960** Koestler *Lotus & Robot* II. x. 236 The koan—the logically insoluble riddle which

the pupil must try to solve. *Ibid.* 237 There are said to exist some one thousand seven hundred koans, divided into various categories... The oldest-known koans are the 'Three Barriers of Hung-Lun', an eleventh-century Zen master. **1972** *Times Lit. Suppl.* 28 Jan. 85/1 What he comes up with—his runes and enigmas and impromptu koans—builds gradually into a supplementary creation.

koatuku, obs. var. *kotuku.

kob¹. Add: Esp. the species *Kobus kob.* In full, *kob antelope.* (Later examples.)

1897 Sclater & Thomas *Bk. Antelopes* II. vii. 139 The first specimen of the Kob Antelope that reached Europe alive, so far as we know, was that presented to the Zoological Society of London by Mr. John Foster in 1836. *Ibid.* 140 From Senegal and the Gambia the Kob extends through the interior of West Africa to Togoland. **1964** C. Willock *Enormous Zoo* iv. 40 The kob... To look at, this lovely creature with the back-raked horns is a little like a larger version of the impala... In fact its nearest relative is the large, shaggy, big-horned waterbuck. **1964** G. B. Schaller *Year of Gorilla* (1965) iii. 75 Rwindi, near the southern end of Lake Edward,..is the central location from which to see the herds of buffalo, elephant, hippopotamus, waterbuck, kob antelope. **1967** *Listener* 6 Apr. 459/2 Among the Ugandan kob males may defend small territories... By tradition a certain number of kob are associated with each breeding ground. **1974** *Nature* 29 Nov. 345/3 In the Uganda kob antelope there exists a system of social behaviour designed to prevent clandestine matings.

kob² (kǫb). *S. Afr.* = *kabeljou.

1906 *East London* (Cape Province) *Dispatch* 26 June 3/7 Our well-known and very common kabeljaauw, called for briefness 'cob' or 'kob'. **1913** W. W. Thompson *Sea Fisheries Cape Colony* 155 Kabeljaauw..; Cob or Kob (East London). **1930** C. L. Biden *Sea-Angling Fishes of Cape* v. 108 Kabeljou; Kob (abbreviated name). **1950** *Cape Times* 17 Nov. 13/8 Kob are well worth a try from the usual spots around the sandy sweep of the bay. **1959** *Ibid.* 16 Nov. 2/5 Visitor Lands Big Kob. **1974** *South-Western Herald* (S. Afr.) 2 July 5 'I am now after kob,' said Mr. Thomson.

2. kob water, disturbed, discoloured water in which the kob is often found.

1930 C. L. Biden *Sea-Angling Fishes of Cape* v. 113 On the south-east coast particularly they keep a watchful eye on what is known as 'kob-water'—a discoloration of the sea, either milky, dirty yellow, or what one would liken to pea soup. **1957** S. Schoeman *Strike!* iii. 71 Kob-water usually results from a disturbance of the seabed. **1974** *Argus* (Cape Town) 31 Dec. 4/2 The familiar, ginger-beer coloured water known to anglers as 'kob water' is moving in along parts of the Strandfontein coast-line.

kobeite (kōu·bi͵əit). *Min.* [Partial tr. Jap. *kobeishi* (J. Takubo et al. 1950, in *Jrnl. Geol. Soc. Japan* LVI. 512), f. *Kobe*, name of a locality in Kyoto prefecture, Japan + *ishi* mineral (formative suffix of names of minerals): see -ite¹.] A black, prismatic, hydrated multiple oxide of formula close to $AB_2(O,OH)_6$ (where A represents mainly yttrium, iron, and uranium and B represents mainly titanium, zirconium, hafnium, niobium, and tantalum), but with much less $(Nb,Ta)_2O_5$ than minerals of the euxenite-polycrase series and more TiO_2.

1950 J. Takubo et al. in *Jrnl. Geol. Soc. Japan* LVI. 513 [Eng. abstr. of article in Jap.] The content of TiO_2 is.. exceedingly high compared with that of polycrase or blomstrandite. So the writers propose here to call this mineral 'Kobeite' after the name of the locality. **1957** *Amer. Mineralogist* XLII. 342 The kobeite-bearing cobble ..was collected while panning sands and gravels in the Paringa River, South Westland, New Zealand. **1961** *Mineral. Jrnl.* III. 146 The writers are inclined to recognize kobeite as a distinct mineral species belonging to multiple oxides rich in titanium, zirconium and yttrium group rare earths.

kobo (kōu·bo). Pl. **kobo.** [See quot. 1972¹.] A unit of currency in the Federal Republic of Nigeria, equal to $\frac{1}{100}$ naira.

1972 *N.Y. Times* 9 Aug. 14 Kobo..is a corruption of the word 'copper' and the popular term here [*sc.* Nigeria] for a penny, a copper coin... The kobo, symbolized by a lower-case k, will be produced in coins of one-half kobo, 1, 5, 10 and 25 kobo. **1972** *Times* 9 Oct. (Nigeria Suppl.) p. viii/4 The new currency being introduced by the Central Bank of Nigeria consists of naira and kobo... Kobo was the popular name for the old penny pieces... The new kobo is the same size as the old Nigerian shilling and carries a depiction of cocoa seeds. **1973** *Whitaker's Almanack 1974* 985 Nigeria (Federal Republic of)... Naira = 100 Kobo.

Koch (kǫχ, kǫk). The name of Robert *Koch* (1843–1910), Ger. bacteriologist, used in the possessive (less commonly *attrib.*) to designate certain things related to his work on tuberculosis, as **Koch's bacillus,** *Mycobacterium tuberculosis*, which causes tuberculosis in man, and was first isolated by Koch; **Koch's laws** = **Koch('s) postulates*; **Koch('s) phenomenon,** the altered reaction to an inoculation of (living or dead) tubercle bacilli of an animal already infected with tuberculosis from that of a healthy animal, the infected animal showing a reaction which is quicker and locally more severe but which is not followed by a general infection; (now regarded as a classic example of delayed hypersensitivity); **Koch('s) postulates,** a set of four criteria which should be satisfied before a given disease is attributed with certainty to any particular micro-organism, viz. (*a*) the organism concerned is present in each case of the disease; (*b*) it is possible to isolate it from the diseased animal in a pure culture; (*c*) the introduction of such a culture into a suitable healthy animal produces the disease in it; and (*d*) the organism is recoverable from the animal so infected; **Koch's tuberculin,** either of two kinds of tuberculin (*Koch's old* and *new tuberculin*) orig. devised by Koch; formerly also called *Koch's fluid, liquid, lymph*, and now usually just *tuberculin.*

1885 *Jrnl. R. Microsc. Soc.* 557 (*heading*) Staining of Koch's bacillus. **1890** *Lancet* 22 Nov. 1119/2 He drew attention to the advantage of being able, by the injection of Koch's fluid, to diagnose whether a serious laryngeal affection was carcinomatous..or tubercular. *Ibid.* 1120/1 Experiments with Koch's lymph. *Ibid.* 1121/1 He was inoculated with Koch's liquid for distinct facial lupus, and at the same time for pulmonary tuberculosis. **1891** Dr. Koch's lymph [see Tuberculin]. **1897** Muir & Ritchie *Man. Bacteriol.* ix. 229 (*heading*) Koch's tuberculin. **1898** R. T. Hewlett *Man. Bacteriol.* ii. 31 With regard to the pathogenic organisms..Koch has laid down the following conditions, which have been termed Koch's Postulates, which must be complied with before the relation of an organism to a disease process can be said to be completely demonstrated. **1899** Koch's lymph [see Tuberculin]. **1910** Muir & Ritchie *Man. Bacteriol.* (ed. 5) x. 284 (*heading*) Koch's old tuberculin. *Ibid.* 288 Another preparation has..been introduced, known as 'Koch's new tuberculin'. **1911** *Encycl. Brit.* XX. 783/1 Koch's tuberculin has been of inestimable value in the early diagnosis of tuberculosis, especially in animals. **1929** Topley & Wilson *Princ. Bacteriol. & Immunity* II. xlix. 729 There are two classical examples, which provide striking illustrations of this double aspect of the allergic reaction. One of these is the well-known Koch's phenomenon (Koch 1891). **1939** K. L. Burdon *Med. Microbiol.* ii. 20 These principles have become known as 'Koch's laws' or 'Koch's postulates'. **1944** L. E. H. Whitby *Med. Bacteriol.* (ed. 4) i. 2 Very few organisms pathogenic to man fulfil Koch's postulates rigidly. **1955** *Sci. Amer.* June 103/1 Koch found the bacillus in body tissues, in sputum and in urine. He was able to grow the microbe in an artificial culture, to reproduce tuberculosis by injecting it into new animals and to recover it again from the infected tissue—a procedure which has become standard for connecting a given disease with a germ and is known by the name Koch's Laws. **1963** Humphrey & White *Immunol. for Students of Med.* x. 322 The injection into a normal guinea-pig of as much as 2 ml of Koch's old tuberculin has little effect. But the injection of 0·1 [*printed* 0·0] ml into a guinea-pig, in the eighth or tenth week of tuberculous infection, may kill it within a few hours. **1964** Wheeler & Volk *Basic Microbiol.* xxiii. 263/2 *Mycobacterium tuberculosis* (more commonly called the tubercle bacillus and sometimes Koch's bacillus) was shown by Robert Koch to be the causative agent of tuberculosis. **1970** Passmore & Robson *Compan. Med. Stud.* II. xxi. 17/1 This Koch phenomenon is a specific example of cellular immune response. **1973** *Sci. Amer.* Oct. 28/1 Thus the famous 'Koch postulates' have been satisfied, and the tumor is firmly placed among those transmitted by an infectious agent.

kochubeite (kǫtʃū·bi͵əit). *Min.* Also **kotschubeite.** [ad. G. *kotschubeit* (N. von Kokscharow 1863, in *Bull. de l'Acad. Imp. des Sci. de St.-Pétersbourg* V. 369), f. the name of Count P. A. *Kochubei*, 19th-century Russian mineralogist: see -ite¹.] A mineral of the chlorite group that is a chromiferous variety of clinochlore and occurs as rose-red rhombohedral crystals.

1868 J. D. Dana *Syst. Min.* (ed. 5) 500 *Kotschubeite*.. in the district of Ufaleisk, Southern Ural. **1910** *Encycl. Brit.* VI. 256 Alumina may also be partly replaced by chromic oxide, as in the rose-red varieties [of chlorite] kämmererite and kotschubeite. **1954** *Mineral. Mag.* XXX. 280 The chromiferous chlorites include three varieties: kämmererite..; chrome-clinochlore, with less than 4% Cr_2O_3; and kochubeite, a variety of clinochlore with more than 4% Cr_2O_3. **1958** [see *kämmererite].

Koch–Weeks bacillus. *Med.* [f. the names of Robert *Koch* (see *Koch) and J. E. *Weeks* (1853–1949), American ophthalmologist: see Bacillus.] The bacterium, *Hæmophilus ægyptius*, which is a common cause of infectious conjunctivitis.

[**1897** *Arch. Ophthalmol.* XXVI. 102 Peters and Fuchs claimed never to have been able to find the Koch–Weeks's bacillus.] **1898** R. T. Hewlett *Man. Bacteriol.* xx. 359 Conjunctivitis is usually of three kinds—viz. the acute, caused by the Koch–Weeks bacillus; the gonorrhœal; and the chronic. **1974** Passmore & Robson *Compan. Med. Studies* III. ii. xxxiii. 8/1 Conjunctivitis may occur in epidemic form in schools, offices, etc., the classical epidemic pink-eye, being caused by *H. aegyptius* (the Koch–Weeks bacillus).

Kodachrome (kōu·dăkrōum). [f. Koda(k *sb.* + -*chrome* (Gr. χρῶμα colour).] The registered trade name of a method of colour photography used by Kodak Ltd. Also, a colour film manufactured by this method; a photograph or slide produced from a Kodachrome film. Also *attrib.*

The spellings in quots. 1915 and 1966 are *erron.*

1915 *Chambers's Jrnl.* 25 Sept. 687/2 The kodachchrome process offers a means of enabling the former [*sc.* the amateur] to realise his ambition. **1926** *Trade Marks Jrnl.* 22 Sept. 2158 Kodachrome... All Goods included in Class 8 [*sc.* philosophical instruments, scientific instruments, apparatus for useful purposes, etc.]. Kodak, Limited,..London, W.C.2; Manufacturers and Dealers. **1930** C. W. Ackerman *George Eastman* viii. 285 Ever since the advent of the Kodak system, Eastman had dreamed of photographs in natural colors and the research workers..had just succeeded in a color process for portraiture, which was called Kodachrome. **1935** *Discovery* July 190/2 Recently Kodak have introduced a very interesting three-colour subtractive film for sub-standard cinematography which is to be known as 'Kodachrome'. **1936** *Times* 10 Jan. 7/5 Among other short films shown was one which illustrated the results obtained by the new Kodachrome process. **1951** L. Z. Hobson *Celebrity* (1953) ii. 22 Each grandchild, whether in the flesh or on Kodachrome, was Beauty and Goodness personified. **1960** A. Coren in *Introduction: Stories by New Writers* 102 Brash teased-up Kodachrome lupins and a postcard sky. **1961** G. Smith *Business of Loving* xv. 271 The air had a tint of Kodachrome blue in it. **1966** *Guardian* 28 Dec. 4/6 Pope Pius, in Kodakchrome, smiled at stacks of institutional china. **1969** *Focal Encycl. Photogr.* 815/2 *Kodachrome*, pioneer subtractive process of colour photography worked out by L. D. Mannes and L. Godorosky and introduced by Kodak... It makes use of an integral tripack in which the emulsions contain no colour formers but are subjected to individual dye development. **1970** *Daily Tel.* (Colour Suppl.) 7 Aug. 7 A vivid Kodachrome of the coach roaring down a hillside became the background for the new cheques. **1971** R. Dentry *Encounter at Kharmel* (1973) ix. 148 Bloody marvellous country... I've got a box of Kodachromes to prove it.

Kodak, *sb.* For def. read: The proprietary name of a range of cameras produced by Kodak Ltd. Add earlier and later (including *attrib.*) examples.

1888 *Official Gaz.* (U.S. Patent Off.) XLIV. 1072/1 Photographic Cameras and Sensitized Plates and Film Therefor.—The Eastman Dry Plate and Film Company... 'The word "Kodak".' **1899** Merwin & Webster *Short Line War* v. 62 Near the box was a kodak picture of Miss Porter. **1907** W. James *Pragmatism* viii. 290 We want a Kodak-picture and we press a button. **1913** F. A. Talbot *Pract. Cinematogr.* 22 Contrary to general belief, taking the 'movies' is quite as simple as snapshot photography with a Kodak. **1933** R. L. Sutton *Arctic Safari* 43 We had to exercise considerable self-restraint in the matter of kodak portraiture... I think that we did not use up more than three packs of film. **1966** J. Betjeman *High & Low* 47 These are the walls adorned with portraits, Camera studies and Kodak snaps. **1974** *Times* 26 Jan. 8/3 Americans walk the length of the street market in the Portobello Road snapping it with Kodaks.

b. (Later examples.)

1898 *N.Y. Observer* 3 Mar. 258/1 Some of the rest took kodaks of us. **1901** E. Hornby *Jrnl.* 22 Mar. in *Sinai & Petra* (1907) 180 M. at once took two kodaks of it. **1930** B. Willis *Living Afr.* vii. 98 After I had taken a couple of kodaks.

Kodak, *v.* Add: (Later examples of senses a and b.) Now *rare.* The derivatives *Kodaker, Kodakist*, and *Kodakry* seem to be obsolete.

1924 *New Republic* 24 Dec. 120 Our Main Street is the happy-hunting ground of the ill-willed camera. Picture ahead, Kodak as you go. **1928** *Ibid.* 12 Dec. 90 The young cook who had been one of the last to leave the ship, Kodaking as he went. **1934** J. Collier *Defy Foul Fiend* 318 Willoughby's eye had kodaked the attitude of a beaten child, sulking in tears. **1936** *Time* 14 Dec. 21 A French actress who recognizes Mrs. Simpson and tries to Kodak her gets a blow from the British bodyguard knocking her camera from her hand. **1948** H. M. Gloster *Negro Voices Amer. Fiction* 165 McKay's second novel, *Banjo* (1929), an impressionistic kodaking of life among the colored boys of the Marseilles breakwater. **1954** *Life* 26 Apr. 155 Kodaked by friend as he himself aimed a Kodak, Eastman was photographed on a ship in 1890 by early model which took round pictures.

Kodiak (kǫ·dyæk). Also **Kadiak** (kæ·dyæk). The name of an island off Alaska, used *attrib.* to designate the large brown bear, *Ursus arctos middendorffi*, found there, as well as in Alaska itself and on other islands off the coast. Also *absol.*

1899 R. Ward *Rec. Big Game* (ed. 3) 474 Even more gigantic is the Kadiak bear..of Kadiak Island, Alaska. **1904** C. R. E. Radclyffe *Big Game Shooting in Alaska* 268 The Kodiak brown bear (*Ursus middendorffi*). **1930** *Sat. Even. Post* 13 Dec. 11/2 A Kodiak bear looks as big as an elephant as he ambles..through vegetation that comes only to his stomach. **1955** *Arctic Terms* 47/1 Kodiak bear. The world's largest carnivore, *Ursus middendorfi*, occasionally weighs 1,500 pounds, yellowish to dark brown in color, ranging from the Alaska Peninsula southeast on coasts and adjacent islands to British Columbia. Also called the 'big brown bear', 'brownie', 'Alaskan brown bear', 'Kadiak bear'. **1966** R. Perry *World of Polar Bear* ix. 111 In both size and weight polar bears approximate to the giant Alaskan and Kodiak bears. **1974** 'R. B. Dominic' *Epitaph for Lobbyist* xx. 173 He was going to

walk..into the Zoo... Kodiaks and Himalayas, the tourist booklet said.

koechlinite (kǒ·χylinəit). *Min.* [f. the name of Rudolf *Koechlin* (1862–1939), curator of the mineral collection, Hof-Museum, Vienna + -ite¹.] A molybdate of bismuth, Bi_2MoO_6, found as minute, greenish-yellow plates, and in soft white to yellow masses.

1914 W. T. Schaller in *Jrnl. Washington Acad. Sci.* IV. 354 Koechlinite (bismuth molybdate), a new mineral from Schneeberg, Saxony. **1943** *Amer. Mineralogist* XXVIII. 537 Koechlinite also was found as soft white to yellow masses associated with bismoclite on specimens from Bygoo, New South Wales. Here, too, the mineral has formed by the alteration of a pre-existing bismuth mineral. **1966** *Jrnl. Inorg. & Nucl. Chem.* XXVIII. 1125 The lattice parameters were equal to those of natural koechlinite within the experimental error.

koedoe: see *kudu.

koek(oe)makranka, varr. *kukumakranka.

‖ **koeksister** (kŭ·sistəɹ). *S. Afr.* Also **koeksuster, koesijster, koesister.** [Afrikaans *koe(k)-sister.*] A kind of sugared doughnut popular in South Africa.

1904 *Hilda's Where is it? of Recipes* (Pettman) 128 *Koesisters* (Batavian or old Dutch sweetmeat recipe). **1913** C. Pettman *Africanderisms* 272 *Koesijsters*,—a confection or sweetmeat which has been boiled in fat and dipped in powdered sugar. **1947** L. G. Green *Tavern of Seas* viii. 72 There was a koesister (doughnut) specialist. **1953** *Cape Times* 24 Mar. 16/1 Tea and coffee, *boerebeskuit, koeksusters, melktart*, and scores of other platteland delicacies. **1959** H. Gerber *Trad. Cookery Cape Malays* 9 Some of them [*sc.* Malay housewives] turn out lighter pastries, spongier *koesisters* than others. **1967** E. M. Slatter *My Leaves are Green* 9 The coffee was hot and strong and the flaky koeksisters melted in my mouth. **1974** *Sunday Times* (Johannesburg) (Colour Suppl.) 3 Mar. 7 Those delicious onions with a sour sauce, melk snysels, koeksisters.

koelie: see *coolie, cooly 2 b.

koelreuteria (kǒlroitiə·riă). [mod.L. (E. Laxmann 1772, in *Nov. Comm. Acad. Sci. Imp. Petrop.* XVI. 561), f. the name of Joseph G. *Koelreuter* (1733–1806), German naturalist + -ia¹.] A deciduous tree of the east Asian genus so called, belonging to the family Sapindaceæ, esp. *K. paniculata*, from northern China, which has large panicles of yellow flowers and pinnate leaves which turn bright yellow in autumn.

1789 W. Aiton *Hortus Kewensis* II. 7 Panicled Kœlreuteria. Nat[ive] of China. **1818** *Bot. Reg.* IV. 330 (*heading*) Panicled Koelreuteria. **1914** W. J. Bean *Trees & Shrubs Hardy in Brit. Isles* II. 687 Mr Wilson introduced from China, in 1900, a rather distinct Koelreuteria with very large and often quite bipinnate leaves. **1920** A. D. Webster *London Trees* 71 The Koelreuteria is readily propagated by cuttings of either root or branch. **1946** L. J. F. Brimble *Trees in Brit.* xxvii. 209 *Koelreuteria* is a native of northern China, but it is sometimes cultivated in Britain for ornamental purposes since its yellow flowers and autumnal foliage are so attractive.

koembang, var. *kumbang.

koenenite (kǒ-, kǒ·nĕnəit). *Min.* [ad. G. *koenenit* (F. Rinne 1902, in *Centralblatt f. Mineral.* 493), f. the name of Adolph von *Koenen* (1837–1915), German geologist who first found it: see -ite¹.] A hydroxide and chloride of magnesium, aluminium, and sodium (formerly thought to be an impurity), which forms pale yellow scales when pure, but is normally red owing to enclosed hæmatite.

1902 *Jrnl. Chem. Soc.* LXXXII. II. 611 Koenenite. This new mineral was found..in crevices in the clay of the salt deposits at Volpriehausen, in the Sollinger Wald, Hanover. It is red in colour. **1952** *Mineral. Abstr.* XI. 459 Koenenite is widely distributed in the German salt deposits, occurring in salt-clay with blue halite, in anhydrite 'Hartsalz', and carnallite. **1968** *Zeitschr. für Kristallogr.* CXXVI. 7 Koenenite, $4NaCl.4(Mg,Ca)Cl_2.5Mg(OH)_2.4Al(OH)_3$, is built up of two trigonal substructures: $[Na_4(Ca,Mg)_2Cl_{12}]^{4-}$ and $[Mg_7Al_4(OH)_{22}]^{4+}$. **1968** I. Kostov *Mineral.* II. iii. 195 Koenenite is trigonal with perfect {0001} cleavage.

koesijster, koesister, varr. *koeksister.

‖ **kofta** (kǫ·ftă). Also **koaftah, kooftah.** [Hind. *kofta* pounded meat.] A rissole, made of meat or fish, popular in the East. Also *attrib.*

1888 W. H. Dawe *Wife's Help to Indian Cookery* 71 *Kofta kabâb.* [Recipe given.] **1932** M. R. Anand *Curries* 49 In Kofta Curry..all the tomatoes should be put in at once to form the gravy. **1936** E. P. Veeraswamy *Indian Cookery* 51 Kooftahs..are usually balls of minced mutton or beef, blended with onions, garlic, ginger and certain spices, and cooked in a curry sauce. **1955** H. Day *Curries of India* iii. 27 Koaftah curry is quite simple to make and is a universal favourite all over India and Pakistan. **1971** *Femina* (Bombay) 16 Apr. 57/2 Put *koftas* in it [*sc.* the water] and cook till soft enough.

‖ **kogai** (kōu·gai). [Jap.] Environmental pollution in Japan.

1970 *New Yorker* 23 May 94 Although *kogai* is one of the most controversial and widely covered topics in the Tokyo press, the daily seminars [on pollution] were closed to Japanese reporters. **1971** *Peace News* 17 Sept. 1 (*heading*) Basic theory of Kogai. *Ibid.*, 'Kogai' is the Japanese word which is used to identify the pollution problem. It cannot be literally translated, for it is also used to refer to environmental problems above and beyond simple pollution: factory noise, vibration, obstruction of sunlight, traffic congestion, water shortage, etc.

kohekohe (kǒ·ikǒ·i). *N.Z.* [Maori.] A deciduous tree, *Dysoxylum spectabile*, of the family Meliaceæ, which has pinnate leaves and panicles of fragrant white flowers.

1835 W. Yate *Acct. N.Z.* (ed. 2) ii. 48 Kohekohe (*Laurus kohekohe*).—A fine handsome tree, with a trunk free of branches to a height of forty feet. **1855** H. R. Richmond *Let.* 11 Aug. in *Richmond–Atkinson Papers* (1960) I. iv. 173 There is a beautiful view of the splendid kohekohe bush on your ridge. **1910** L. Cockayne *N.Z. Plants* iii. 29 In some few cases the flowers of a tree are produced on the thick branches, as in the kohekohe (*Dysoxylum spectabile*), and not, as usual, from amongst the leaves. **1950** *N.Z. Jrnl. Agric.* Sept. 215/3 The mainrange forest, consisting of tawa, beech, rimu..kohekohe. **1966** *Encycl. N.Z.* II. 234/2 Kohekohe is a medium-sized tree usually 30–40 ft in height, with a trunk 1–3 ft in diameter.

kohl¹. Add: Hence **kohl** *v. trans.*, to darken with kohl. So **kohled** *ppl. a.*

1947 *Penguin New Writing* XXIX. 10 Altogether the face..of an actor roughed and kohled. **1964** *Punch* 26 Feb. 298/2 She had appeared..in her most glamorous *sari*..eyelids kohl-ed. **1971** R. Dentry *Encounter at Kharmel* (1973) x. 183 Ten-year-old boy sopranos with kohl-ed eyes.

Kohlrausch (kōu·lrauʃ). *Physical Chem.* [The name of Friedrich Wilhelm *Kohlrausch* (1840–1910), German physicist.] *Kohlrausch's law*: that the equivalent electrical conductivity of an electrolyte at infinite dilution may be represented as the sum of two constants, viz. the ionic mobilities of the cation and the anion respectively.

1888 *Jrnl. Chem. Soc.* LIV. 331 Kohlrausch's law that the conductivity of a neutral salt may be represented as a sum of two constants, one of which depends on the nature of the acid and the other on that of the base. **1924** H. J. Creighton *Princ. & Applications Electrochem.* v. 91 This additive relationship..is called Kohlrausch's law or 'the law of the independent migration of ions'. **1964** G. I. Brown *Introd. Physical Chem.* xxxiii. 369 The immediate usefulness of Kohlrausch's law is that it provides a method for finding the Λ_∞ [*sc.* the equivalent conductivity at infinite dilution] value for weak electrolytes from Λ_∞ measurements on strong electrolytes. **1969** R. A. Horne *Marine Chem.* iii. 117 Kohlrausch's law..of the additivity of conductivities..is valid only in the limiting case. Solutions of finite concentration exhibit an appreciable departure from the law.

Kohs block (kōuz blǫk). *Psychol.* Also **Kohs' block.** [f. the name of Samuel *Kohs* (b. 1890), U.S. psychologist.] One of a set of coloured cubes used in psychiatric testing with which the subject is required to reproduce patterns presented to him.

1930 *Child Development* I. 341/2 The Kohs block design test was given to 29 children in the Child Institute of Johns Hopkins University. **1941** *Jrnl. Appl. Psychol.* XXV. 420 The Kohs Block Designs test..has found extensive application in clinical psychological work. **1954** A. Anastasi *Psychol. Testing* x. 245 In the Kohs Block Design.., the subject is presented with a set of identical one-inch cubes, whose six sides are painted red, blue, yellow, white, yellow-and-blue, and red-and-white, respectively. Colored designs are presented on each of 17 test cards, the subject being required to reproduce each design by assembling the proper blocks. **1964** M. Critchley *Developmental Dyslexia* ix. 57 Constructional tasks.. include the assembling of jig-saw puzzles, a game which may not be easy for some of these dyslexics. This difficulty is readily assessed by the test of Kohs' blocks, where some dyslexics fare badly.

‖ **kohua** (kōu·͵hu͵ă). *N.Z.* [Maori *kōhua.*] **a.** A Maori oven. **b.** A three-legged iron pot or kettle. Cf. *go-ashore* (c) (Go *v.* VIII).

1843 E. Dieffenbach *Trav. N.Z.* II. iv. 43 The native oven, *hangi* or *kohua*, made in the well-known manner with heated stones. **1901** A. A. Grace in D. M. Davin *N.Z. Short Stories* (1953) 53 Soon the *kohua* was sizzling over a bright fire. **1905** W. B. *Where White Man Treads* 72 The last day of all he [*sc.* Captain Cook] gave Toia another pot, and..with much pointing at it, said: 'Now go ashore.' So we took that to be its name; for do we not call it at this distant day a 'kohua'? (corruption of 'go ashore'). **1949** P. Buck *Coming of Maori* (1950) II. i. 112 The early trade goods included three-legged iron pots which from their function were also termed *kohua*.

koi (koi). [Jap.] A local name in Japan for the common carp, *Cyprinus carpio.*

1727 J. G. Scheuchzer tr. *Kæmpfer's Hist. Japan* I. I. 136 *Koi* is another sort of it [*sc.* Steenbrass], which also resembles a Carp. **1875** H. W. Bates *Illustr. Trav.* VI. 140/2 Some *koi*, a coarse-tasted fish of the carp species. **1884** tr. *J. J. Rein's Japan* I. viii. 197 The most conspicuous [of the Cyprinodontidæ] in size and importance are the Carp or Koi (Cyprinus carpio L.) and the Japanese Crucian or Funa. *Ibid.* II. ii. 440 In every house in which during the previous year a boy has been born, a flag waves on a long bamboo staff, consisting of a large painted koi (carp) of paper. **1896** L. Hearn *Kokoro* vi. 89 The real koi, the great Japanese carp, ascends swift rivers against the stream. **1971** S. Eliovson *Gardening Jap. Way* 86 The carp (*koi*) is a symbol of strength and perseverance.

‖ **koi-cha** (koi·tʃă). Also **koi cha, koicha.** [Jap.] In Japan, powdered tea mixed to a thick brew and drunk ceremonially.

1727 J. G. Scheuchzer tr. *Kæmpfer's Hist. Japan* II. App. I. 15 This powder is mix'd with hot water into a thin pulp, which is afterwards sip'd. This Tea is call'd *Koitsjaa*, that is, thick Tea, by way of distinction from the thinner Tea, made only by infusion, and it is that which all the rich people and great men in Japan daily drink. **1890** B. H. Chamberlain *Things Japanese* 338 The resulting beverage resembles pea-soup in colour and consistency. There is a thicker kind called *koi-cha*, and a thinner kind called *usu-cha*. **1960** B. Leach *Potter in Japan* iii. 66 That was the first time I have had 'Koi Cha'. It is a variant of the Tea Ceremony in which the powdered tea is mixed to a thick brew and the bowl is passed from hand to hand. **1965** W. Swaan *Jap. Lantern* xvi. 182 It is *koi-cha*, or thick tea, made from leaves which have been ground to a powder. **1970** J. Kirkup *Japan behind Fan* 206 The host is ready to serve his guests with the *koi-cha*, or thick tea.

‖ **koi hai,** var. Qui-hy.

1960 M. Malgonkar *Distant Drum* II. xxiii. 162 The very senior officers, right close to the top and steeped in the tradition of the old *koi-hais*. **1962** *Listener* 8 Nov. 776/1 There was dry knowing laughter among the *Koi Hais* in the shuttered clubs. **1967** *Ibid.* 11 May 610/1 It is..almost a generation since the last of the *koi hais* packed up..and the ghosts of Poona are now faint indeed. **1971** *Illustr. Weekly India* 4 Apr. 22/3 On the boat, coming out to rejoin Maurice, I had learnt quite a lot of Hindustani, unfortunately from a very charming British army officer who had an atrociously *koi hai* accent which I never quite managed to shed. **1973** *Times* 19 Feb. (India Suppl.) p. ix/2 The *Koi Hais* and 'bottoms up, old boy', are confined, in Britain, to the personal column of *The Times*, and in India, to a brittle crust of individuals who seem to prefer the caricature to the real thing.

koilonychia (koiloni·kiă). *Path.* [a. G. *koilonychia* (J. Heller 1897, in *Dermatol. Zeitschr.* IV. 490), f. Gr. κοῖλ-ος hollow + ὄνυξ, -υχος nail + -ia¹.] A condition of the finger-nails in which the outer surfaces are concave instead of convex; spoon-nail.

1902 *Trans. Med. Assoc. Missouri* XLV. 157 The disease..is known technically as koilonychia and ordinarily as 'spoon' nails... Koilonychia may be observed in one or more of the nails of the fingers of one or both hands. **1934** J. M. Vaughan *Anaemias* iii. 43 Koilonychia is not found in 'splenic anaemia', while some degree is extremely common in idiopathic hypochromic anaemia. **1970** *Jrnl. Pediatrics* LXXVII. 1057/2 There is a significant correlation between koilonychia and iron deficiency in infants 9 to 13 months of age.

koine (koi·ni). [Gr. κοινή, fem. sing. of κοινός common, ordinary.] **a.** Originally the common literary dialect of the Greeks (ἡ κοινὴ διάλεκτος) from the close of classical Attic to the Byzantine era. Now extended to include any language or dialect in regular use over a wide area in which different languages or dialects are, or were, in use locally.

[**1886** *Encycl. Brit.* XXI. 653/1 As might be expected, this κοινή, like the κοινή of the Greeks, has a comparatively limited vocabulary.] **1913** D. B. Durham *Vocab. Menander* 8 The year 600 A.D. is a convenient date at which to divide the Koine from the Greek of the middle ages. **1926** *Germanic Rev.* I. iv. 297 Assuming that all our dialects had given way to a High German *koinė*, we should still recognize the characteristic distinctions of the former dialects. **1927** A. H. McNeile *Introd. New Testament* 278 Such [constructions] as were rapidly making their way into the *Koine* Greek. **1933** *Amer. Speech* VIII. Oct. 5/1 The American *koinė* in eliminating the extreme variations in English dialects nevertheless absorbed enough of their peculiarities to make it a highly varied unity. **1956** A. Toynbee *Historian's Approach to Relig.* xix. 270 In the first century of the Christian Era the dissemination of the books of the New Testament in the Attic *koinė*—the 'standard Greek' of the day—ensured their finding readers as far afield..as Britain..and India. **1958** D. Whitelock *Changing Currents Anglo-Saxon Stud.* 6 The general use of the West Saxon literary *koine* in the tenth and eleventh centuries. **1964** S. M. Ervin-Tripp in J. A. Fishman *Readings Sociol. of Lang.* (1968) 197 Superposed varieties [of speech] include many types, from occupational argots to koines used for trade and regional communication, such as Melanesian Pidgin and Swahili. **1965** *Times Lit. Suppl.* 22 Apr. 317/2 A 'disc-jockey' found himself saying 'take time *out*', in accordance with the conventional mid-Atlantic koine these people have to use. **1966** K. H. Albrow in C. E. Bazell *In Memory of J. R. Firth* 2 F. Roberts is in the habit of speaking standard Welsh with a Northern accent as his Welsh koine except at home.

b. A set of cultural or other attributes common to various groups. Also *attrib.*

1924 A. J. B. Wace in *Cambr. Anc. Hist.* II. xvi. 466 During the last two centuries, at least, of the Mycenaean dominion in Greece and the Aegean, there was a cultural *koinė*, and it is at least likely that there was a linguistic *koinė* as well. **1939** J. D. S. Pendlebury *Archaeol. Crete*

vi. 358 Crete had entered the Hellenistic *koine* and its individuality is nearly lost. **1962** *Economist* 28 Apr. 340/3 The Mauretanian and Numidian kingdoms..were centres of an Afro-European 'koiné'.

‖ **koinonia** (koinō͞u·nia). *Theol.* [Gr. κοινωνία communion, fellowship.] Christian fellowship or communion, either with God or, more commonly, with fellow Christians.

1907 W. P. Du Bose *Gospel according to St. Paul* xvii. 243 As the first two truths of our faith in Christ might be called those of the Father and the Son, so the third may be designated that of the Spirit. Or, to put it in the other way, as the first two may be called those of the divine love and the divine grace, so the third may be named that of the divine *koinonia*. **1920** 'W. S. Palmer' *Christianity & Christ* 177 Thinking of the Church I am reminded of the 'Koinonia', the fellowship of early Christians which came of the Pentecostal inflowing of the Spirit of God. **1938** *Theology* XXXVI. 211 The Church's tradition of social and economic justice; the primitive koinonia, the medieval just price and condemnation of usury. **1949** *Scottish Jrnl. Theol.* II. 67 We exist in the Image of the Living God who Himself confronts Himself to become one God in the *koinonia* of the Holy Spirit. **1967** J. Macquarrie *Dict. Christian Ethics* 73/1 The point of departure for Christian thinking about ethics is the concrete reality in the world of a community, a *koinonia*, called into being and action by Jesus of Nazareth.

‖ **koji** (kō͞u·dʒi). [Jap.] An enzyme preparation derived from various moulds, esp. *Aspergillus oryzæ* and closely related species, and used to bring about the fermentation involved in the production of saké, soy sauce, etc.

1878 R. W. Atkinson in *Nature* 12 Sept. 522/2 The rice-grains are found to be covered with large quantities of fine hair-like threads, the mycelium of the fungus added. In this state it is called 'kōji'. **1926** Thom & Church *Aspergilli* vi. 64 Koji in its various forms is used in several fermentations. **1953** J. Ramsbottom *Mushrooms & Toadstools* xxiii. 275 The Koji for the enormous fermentation industries of Japan is *Aspergillus Oryzae*, or closely allied species. **1960** A. E. Bender *Dict. Nutrition* 122/1 Takadiastase. Or Koji, an enzyme preparation produced by growing the fungus, *Aspergillus oryzae*, on bran, leaching the culture mass with water and precipitating with alcohol. **1965** Raper & Fennell *Genus Aspergillus* xviii. 358 *A. flavus*..had been encountered frequently among cultures received from Japanese workers as isolates from commercial inoculum, or 'koji', for fermentation industries. *Ibid.* 391 The manufacture of saké is dependent upon the use of *A. oryzae* for the preparation of the koji used to digest rice starch and protein.

kojic (kō͞u·dʒik), *a. Chem.* [f. *Koji: see -ic.] *kojic acid*: 5-hydroxy-2-hydroxymethyl-γ-pyrone, $HO{\cdot}C_5H_2O_2{\cdot}CH_2OH$, a crystalline pyrone derivative produced from dextrose by some fungi of the genus *Aspergillus* and having mild antibacterial properties.

[**1912** T. Yabuta in *Orig. Communications 8th Internat. Congr. Appl. Chem.* XXV. 455, I have given the name 'Koji acid' to this substance.] **1913** *Jrnl. Chem. Soc.* CIV. I. 180 Kojic acid..obtained from finely powdered *Aspergillus oryzae* forms colourless needles or prisms. **1947** *New Biol.* II. 88 Kojic acid is a mild disinfectant of the same order of activity as phenol. **1971** *Jrnl. Trop. Med. & Hygiene* LXXIV. 164/1 Kojic acid production which is supposed to be a constant property of *A[spergillus] flavus*.

kokama, var. *Kukama.

kokanee (ko·kani). Also **kickininee.** [ad. Interior Salish *kikinee*.] A landlocked dwarf subspecies, *Oncorhynchus nerka kennerlyi*, of the sockeye salmon.

1875 in *Okanagan Hist. Soc. Rep.* (1953) 17 There we would fill our basket with the shining kik-e-ninnies. **1937** *Kootenay & City of Nelson, B.C.* 62 The Kokanee or 'Silver Trout', which is in reality a landlocked Sockeye salmon abounds in the larger lakes of the district. **1940** *Nature* 3 Aug. 172/1 Attempts to explain the origin of kokanee (a variety of sockeye salmon) were made. **1963** *New Scientist* 31 Jan. 228/2 The kokanee, *Oncorhynchus nerka kennerlyi*, might increase overall production when cultured with trout. **1963** *Globe & Mail* (Toronto) 2 Mar. 8/6 The kokanee (or kickininee) is a sockeye salmon that does not migrate to the ocean. Its life span is similar to that of the sockeye in that it returns to the nursery stream after three or four years, at which time it weighs about a pound. **1965** A. J. McClane *Standard Fishing Encycl.* 457/2 The kokanee was originally found in Oregon, Idaho, Washington, British Columbia, and northward into Alaska... In Japan it is found in Lake Akan in northern Hokkaido... Morphologically the kokanee and sockeye are identical. **1970** D. Waterfield *Continental Waterboy* i. 2 Enabling the silver trout or kokanee..to reach the formerly inaccessible river.

‖ **koker** (kō͞u·kəɹ). *Guyana.* Also **khoker.** [Du.] A sluice-gate, a lock-gate; the narrow stretch of water between such gates.

1851 J. A. Tinne *Rep. Outfall Drainage Brit. Guiana* 11 By means of a koker or sluice near the sea, the drains are emptied at low tide and the sluices are shut when the tide rises higher than the water in the drains. **1893** J. Rodway *Hand-bk. Brit. Guiana* 9 Through openings in the front dams, closed at high water by marine gates called kokers, the canals empty themselves into the sea. **1944** *Drainage & Irrigation Schemes* (Brit. Guiana Bureau of Publicity) 3 Investigations have to be made concerning..Capacity of sluices and kokers. *Ibid.* 5 The surplus water being discharged through..kokers when the tide is sufficiently low. **1951** E. A. Mittelholzer *Shadows move among Them* II. ix. 240 The ditch was wide and..where it entered the stream stood what seemed to be a sluice-gate. 'That's the *koker*... It's a Dutch word... The *koker* is left open so that the tide can rise and fill the ditch... Then at high tide the *koker* is shut.' **1958** J. Carew *Black Midas* iv. 61 The procession moved out of sight down the path by the khoker and towards the river. **1965** 'Lauchmonen' *Old Thom's Harvest* iii. 34 The dereck of the first koker from the sea toppled over.

kokerboom (kŭ·kəɹbŭm). *S. Afr.* [Afrikaans, f. Du. *koker* quiver + *boom* tree.] A large aloe, *Aloe dichotoma*, the size of a small tree, whose branches were formerly used to make quivers for arrows; = *quiver-tree* (Quiver *sb.*[1] 2).

1774 F. Masson *Jrnl.* 2 Nov. in *Phil. Trans. R. Soc.* (1776) LXVI. 309 We found a new species of aloe here, called by the Dutch Koker Boom, of which the Hottentots make quivers to hold their arrows. **1812** W. J. Burchell *Jrnl.* 14 May in *Trav. S. Afr.* (1824) II. vii. 199 The natives more towards the western coast, frequently use the branches of the *Aloë dichotoma*, which is therefore called by the Hottentots and Colonists, *kokerboom* or quiver tree. **1920** F. C. Cornell *Glamour of Prospecting* 116 We outspanned about sunset on an open plateau covered with vegetation and studded with many of the queer looking aloes known as *koker boomen*, or 'quiver trees'. **1950** *Cape Argus* 5 Aug. 7/5 As you drive northwards through the mountains to Springbok you may see that weird tree-aloe, the kokerboom, flowering beside the road. **1959** J. D. Clark *Prehist. S. Afr.* ix. 226 The arrows were usually kept in a quiver made from leather..or bark, in particular the bark of the 'Kokerboom' tree which is a species of aloe. **1974** *Eastern Province Herald* (S. Afr.) 28 Jan. 11 Those who lived in houses built of Kokerboom trunks stood back and watched as..the flames leapt and crackled through the wooden structures.

‖ **kokeshi** (kō͞u·keʃi). [Jap.] A kind of wooden Japanese doll.

1959 R. Kirkbride *Tamiko* xxii. 172 He bought Tamiko a kokeshi doll. **1970** J. Kirkup *Japan behind Fan* 63 Some stalls were selling celluloid masks and *kokeshi*. **1973** R. Littell *Defection of A. J. Lewinter* iv. 19 Sarah's 'things'—sea shells..Japanese *kokeshi* dolls, paperweights.

kokila. (Later example.)

1931 L. H. Myers *Prince Jali* iii. 37 A kokila-bird was sounding its single, inexpressive note.

kokkewiet (kǫkĕvī·t). *S. Afr.* Also **cock-o-veet, kook-a-vic.** [Afrikaans, echoic.] = *Bokmakierie.

1896 E. Clairmonte *Africander* vii. 126 The kook-a-vic was piping his shrill note in a bush hard by. 'Kook-a-vic, kook-a-vic, kook-a-vic.' **1926** O. Schreiner *From Man to Man* 49 A cock-o-veet came flying up to her. *Ibid.*, Kokkewiet: The Bush-shrike, a very handsome bird with resonant call notes of great beauty. **1936** E. L. Gill *First Guide S. Afr. Birds* 49 Bokmakierie, Bacbakiri, Kokkewiet... The calls are duets by the inseparable pair. **1970** *Standard Encycl. S. Afr.* II. 401/1 Bokmakierie..is sometimes locally referred to as bakbakiri, kokkewiet, janpierewiet or bush-shrike.

kokko, var. *Koko[2].

koklas(s (kō͞u·klăs). Also **cocklass, kuklass.** [Nepalese, var. of *pokras, pukras*.] The pheasant, *Pucrasia macrolopha*, several subspecies of which are found in the Himalayas and China and have been introduced elsewhere.

1864 T. C. Jerdon *Birds India* III. 525 The Cocklass is of a rather retired and solitary disposition. *Ibid.* 524 *Koklas* or *Kokla*, in various hill dialects. **1898** *Ibis* Jan. 39 The Koklass was not uncommon in the forest above Gund. **1922** *Blackw. Mag.* Mar. 323/2 In the woods below a kuklass crowed hoarsely. **1952** Bates & Lowther *Breeding Birds Kashmir* 278 The Koklas is the commonest pheasant in Kashmir. **1965** P. Wayre *Wind in Reeds* xv. 219, I recognised them at once as Koklass Pheasants. *Ibid.*, Koklass are high mountain birds and come from the western Himalayas and North Eastern Tibet across to Eastern and Northern China. **1974** *Country Life* 25 July 244/3 From a dealer's shop in London, came a pair of Koklass pheasants..[which] the Trust succeeded in breeding..in captivity.

koko[1]. Add: = *Cocoyam. (Later example.)

1938 *Jrnl. R. Anthrop. Inst.* LXVIII. 125 These [*sc.* cultivated crops] are..koko yams (*Colocasia antiquorum*), maize, [etc.].

koko[2] (kō͞u·ko). Also **kokoh, kokko.** [Burmese name for the tree.] The brown hardwood obtained from *Albizia lebbeck*, a tropical, deciduous tree of the family Leguminosæ, or the tree itself.

1862 E. Balfour *Timber Trees of India* (ed. 2) 30/2 In the Prome district a special tax was levied on the felling of 'Kokoh' and 'Padouk' under the Burmese rule. **1881** J. S. Gamble *Man. Indian Timbers* 157 *A[lbizzia] Lebbek*... The Siris Tree. Vern[acular]... Kokoh, Burm. **1911** J. H. Holland *Useful Plants Nigeria* II. 299 *Albizzia Lebbek*... East Indian Walnut, Kokoh or Kokka (Rangoon). Wild in Tropical Asia..; distributed to Tropical Africa. **1930** *Observer* 29 June 18/2 In respect of woods the building itself is an exhibition: every piece of joinery..has come from India... Most of the floors are of a pleasant brown wood called koko. **1937** J. M. Dalziel *Useful Plants W. Trop. Afr.* 211 *Albizzia Lebbek*... The timber is sometimes called East Indian Walnut, and is marketed as Kokko or Koko. **1947** J. C. Rich *Materials & Methods Sculpture* x. 290 *Koko* or East Indian Walnut is a hard, dense, close-grained tropical wood imported from Burma. It is dark brown and is usually available in log form. **1956** *Handbk. Hardwoods* (Forest Prod. Res. Lab.) 128 Kokko varies considerably in size according to locality. *Ibid.* 129 Kokko is said to be used for sliced veneers and in the furniture industry.

kokopan, var. *Cocopan.

‖ **kokopu** (kǫ·kŏpu). *N.Z.* [Maori.] A small freshwater fish, *Galaxias fasciatus*.

1886 R. A. Sherrin *Handbk. Fishes N.Z.* 138 'Kokopu', Dr Hector says, is the general Maori name for several very common fishes in the New Zealand streams and lakes, belonging to the family of *Galaxiidæ*. **1929** W. Martin *N.Z. Nature Bk.* I. xix. 175 The Kokopu or 'Native Trout' (*Galaxias fasciatus*) is known from all parts of New Zealand south of the Bay of Islands. **1949** P. Buck *Coming of Maori* (1950) II. viii. 236 The *kokopu* do not run until March. **1962** *Post-Primary School Bull.* (Wellington, N.Z.) XV. I. 27 Kokopu, *Galaxias fasciatus*, live in most of our streams and smaller rivers. **1966** *Encycl. N.Z.* I. 675/2 *Galaxias fasciatus*, the kokopu or banded galaxias, is found throughout the country.

‖ **kokowai** (kō͞u·kowai). *N.Z.* [Maori.] Red ochre, burnt red clay (see quot. 1949).

1836 J. A. Wilson *Jrnl.* July in *Missionary Life & Work N.Z.* (1889) III. 43 Two large totara posts..daubed with *kokowai*. **1840** J. S. Polack *Manners & Customs New Zealanders* I. xix. 210 The *powaka* is kept neatly painted red with *kokowai*. **1845** E. J. Wakefield *Adventure N.Z.* I. 124 A carved post which was painted with *kokowai*, or red ochre. *Ibid.* II. 87 The *kokowai*-painted monuments which I have mentioned. **1878** *Trans. N.Z. Inst.* XI. 75 *Kokowai* is a kind of pigment, burnt, dried, and mixed with shark-liver oil. **1905** W. B. *Where White Man Treads* 7 The rafters painted with kokowai (iron-ore rust), for a ground colour in red, and adorned with intricate volutes in pipe-clay for white. **1949** P. Buck *Coming of Maori* (1950) II. xiii. 319 The decorative painting of woodwork did not advance very far in Polynesia. In New Zealand, red ochre, or haematite, was termed *karamea* and after it was burnt and powdered it became *kokowai* or *horu*. The *kokowai* was mixed with shark oil to form a red paint. **1963** T. Barrow *Life & Work Maori Carver* 28 The red clay or haematite used by the Maori for paint was first burnt, then powdered, when it became *kokowai*.

kok-saghyz (kǫksagi·z). [ad. Russ. *kok-sagyz*, of Turkic origin.] A kind of dandelion, *Taraxacum koksaghyz*, whose roots contain a latex used for making rubber.

1932 *Bull. Rubber Growers' Assoc.* Sept. 534 The Kak-Saugyiz, gave a material with even more resilience, but it has a lower content of caoutchouc. **1945** K. E. Knorr *World Rubber* x. 182 Experiments with *kok-saghyz*, the Russian dandelion, and *Cryptostegia* were soon abandoned. **1954** H. J. Stern *Rubber* i. 18 Kok saghyz (*Taraxacum kok saghyz*, Rodin). This plant was discovered in 1931... The roots contain about 90 per cent. of the total rubber in the plant. *Ibid.*, The amount of information [about Krim Saghyz] available is less than in the case of Kok Saghyz. **1959** J. C. T. Uphof *Dict. Econ. Plants* 354/2 *Taraxacum kok-saghyz* Rodin., Kok-saghyz. (Compositae).—Herbaceous perennial. Turkestan. Roots are a source of rubber. Cultivated in some parts of Russia. **1971** Roff & Scott *Fibres, Films, Plastics & Rubbers* 681/2 (*index*) Kok-saghyz.

koktaite (kǫ·ktă,əit). *Min.* [ad. Czech *koktait* (J. Sekanina 1948, in *Acta Acad. Sci. Nat. Moravo-Silesiacae* XX. I. 1), f. the name of Jaroslav *Kokta*, Czech mineralogist (see quot. 1948): see -ite[1].] Hydrated calcium ammonium sulphate, $(NH_4)_2Ca(SO_4)_2.H_2O$, occurring in acicular monoclinic crystals and identical with artificial ammonium syngenite.

1948 *Mineral. Abstr.* X. 352 Artificial ammonium-syngenite..has the composition $(NH_4)_2Ca(SO_4)_2.H_2O$ (analysis by J. Kokta..), and agrees in the optical data with the mineral, named koktaite, from Žeravice. **1968** I. Kostov *Mineral.* II. ix. 504 Syngenite and koktaite are isotypic.

‖ **koku** (kō͞u·ku). [Jap.] **a.** A Japanese unit of capacity equal to ten *to*, used for liquids and solids (esp. of rice as a monetary measure); equivalent to approximately 39·7 gallons (180 litres) or 4·96 bushels. **b.** A Japanese unit of capacity equal to ten cubic *shaku*, used for vessels; equivalent to approximately 9·8 cubic feet (0·278 cubic metres).

1727 J. G. Scheuchzer tr. *Kæmpfer's Hist. Japan* I. II. v. 199 The Emperor..order'd, that three *koku's* of rice should be given, or lent to any family, that stood in need of it. **1871** A. B. Mitford *Tales Old Japan* I. 96 His revenue of eight million kokus reverted to the Government. *Ibid.*, The koku of rice, in which all revenue is calculated, is of varying value. **1892** Kipling *Lett. of Travel* (1920) 42 Five Japanese dollars (fifteen shillings) per *koku* of 330 lbs. **1896** L. Hearn *Kokoro* x. 170 The seat of a daimyō of three hundred thousand koku. **1904** *Daily Chron.* 30 Mar. 4/5 Jeyas..reduced the civil list to 9,000 kokus, or 44,500 bushels of rice, which was the way then that revenue was paid. **1911** *Encycl. Brit.* XV. 193/1 Any vessel having a capacity of more than 500 *koku* (150 tons). **1931** G. B. Sansom *Japan* II. viii. 168 Any who could produce from the new fields one thousand *koku* of rice were promised lifelong immunity to tax.

1938 D. T. Suzuki *Zen Buddhism & its Influence on Jap. Culture* I. vii. 153 Hideyoshi gave him three thousand *koku* of rice for his service to him as tea-master.

Kol (kō͞ul). *India.* Also **Col(e), Kole.** [Of disputed origin.] Mundu-speaking tribes of Chota Nagpur and Bengal in India (see also quot. 1896); a member of any of these tribes. Also *attrib.* or as *adj.*, of or relating to any of these tribes or their languages.

1795 J. T. Blunt Jrnl. 2 Feb. in *Asiatick Res.* (1803) III. 61 Not wishing to injure the *Coles* by encamping on the little spots, which, with much care and toil, they had cleared..we took up our abode..in the jungle. **1827** R. Jenkins *Rep. Territories Rajah of Nagpore* ii. 30 The Koorkoo dialect is found to resemble that spoken by the Lurka Koles, on the frontier of Singbhoom. **1847** B. H. Hodgson *On Aborigines India* p. ii, The Kól or Dhánger race. *Ibid.* iii. 149 The Kóls are, indeed, as enterprising as industrious. *Ibid.* iii. 150 Kól is an old and classical name, and the best I think for the great mass of aborigines intervening between the Bhils, the Gonds, and the Ganges—at least till we know them better. **1866** *Jrnl. Asiatic Soc. Bengal* XXXV. II. 154 The present population..are of the race best known to us by the name of 'Kol'. **1871** [see *Dravidian *adj.*]. **1872** E. T. Dalton *Descr. Ethnol. Bengal* v. i. 125 The Kols rejecting all change adhered to their impurity of life. **1896** W. Crooke *Tribes & Castes N.W. Provinces & Oudh* III. 294 *Kol*, a Dravidian tribe found in considerable numbers along the Vindhya Kaimûr plateau. **1903** Risley & Gait *Rep. Census India 1901* I. 282 The Kol language has..two main dialects, Mundari and Ho. **1931** E. A. H. Blunt *Caste System N. India* xiv. 287 The Kol is a tribe of aboriginal jungle folk, akin to the Bengal Mundas. **1957** C. B. Mamoria *Tribal Demogr. India* iv. 62 In the iron-ore industry..the labour force..consists of largely Santhals and Kols.

kola[1]. Now the usual spelling of Cola, used to refer to two trees of the genus *Cola*, *C. acuminata* or *C. nitida*, or their seeds, also called kola nuts.

1830 [see Cola]. **1868** *Curtis's Bot. Mag.* XCIV. 5699 (*heading*) Kola-nut tree. *Ibid.*, The Kola has been introduced into the Royal Gardens, Kew. **1890** *Kew Bull.* 255 Kola nuts contain some constituent analogous to caffeine. **1932** J. Cary *Aissa Saved* xv. 82 A family of kola-nut traders, father, mother, sister, and two children, hurrying to Kolu to sell at good prices. **1937** J. M. Dalziel *Useful Plants W. Trop. Afr.* 101 Elephants eat the fruits of kola and damage the tree. *Ibid.* 103 Kola is generally supposed to contain about equal quantities of caffeine and theobromine. *Ibid.* 104 Amongst some peoples a kola tree is planted to commemorate a joyful event. **1957** M. Banton *W. Afr. City* ix. 170 That afternoon he will 'pin flour'—a small ceremony for which he will invite an *alfa* and some respected local Muslims to pray for the deceased; after praying they will eat kola and balls of rice flour mixed with sugar and water. **1964** E. Huxley *Back Street New Worlds* xiv. 141 Had I been a Nigerian, he'd have offered me a kola nut, symbol of hospitality among his people.

Kola[2] (kō͞u·lā). *India.* (See quot. 1873.)

1873 E. Balfour *Cycl. India* (ed. 2) III. 255/1 *Kola*, Beng. A class of hindoos whose principal avocations are basket and mat-making. **1916** [see *Kolam]. **1937** L. Bromfield *Rains Came* III. xxvii. 412 There were Kathis and Kolas.

kolach (kǫ·lātʃ). Also **kolache, kolachi, kolachy.** Pl. **kolache** (kǫ·lātʃi), **kolaches.** [ad. Czech *koláč*, f. *kolo* wheel, circle.] A small tart or pie popular in Czechoslovakia, topped or filled with a sweet mixture, preserve, etc.

1918 W. Cather *My Antonia* (1926) v. i. 381 'Show him the spiced plums, mother. Americans don't have those,' said one of the older boys. 'Mother uses them to make *kolaches*,' he added. **1947** M. Given *Mod. Encycl. Cooking* I. 528 (*heading*) Fillings for kolachy. **1953** A. Heath *Internat. Cookery Bk.* 187 Cover the kolach with it [*sc.* pastry]..and bake in a hot oven. **1961** H. Watney tr. *Břízová's Cooking Czech Way* 143 Kolache (flat fruit buns) and filled rolls made from yeast dough are typically Czech. **1967** Mrs. L. B. Johnson *White House Diary* 7 July (1970) 545 For dessert a typical specialty of the area—'kolaches', a rich pastry that has a center of dried apricots or prunes. **1969** O. Hesky *Sequin Syndicate* xviii. 169 Mamma's sent you some of her *kolachis*. I don't know if they'll let you eat them?

Kolam (kō͞u·lām). Also **Kolamb.** [Origin unknown.] In India, a group of people (whose name for themselves is *Kōlavar*) speaking a Dravidian language similar to that of the Parji of Bastar.

a **1863** S. Hislop *Papers Aboriginal Tribes Cent. Provinces* (1866) I. 10 The Kolâms and the common Gonds do not intermarry... Their dress is similar; but the Kolâm women wear fewer ornaments. **1885** E. Balfour *Cycl. India* (ed. 3) II. 593 *Kolam* or *Kolamb*, a Gond tribe, along the Kandi Konda or Pindi Hills, on the south of the Wardha, and along the table-land stretching east and north of Manikgarh, and thence south to Dantanpilly, running parallel to the right bank of the Pranhita... The Kolam race are found also in the Amraoti, Wun, and Maiker districts as a wild race. **1916** Russell & Lal *Tribes & Castes Cent. Provinces India* III. 521 Mr. Hira Lāl suggests that the Kolāms may be connected with the Kolas, ..who regard the Kolamallai hills as their original home. He further notes that the name of the era by which the calendar is reckoned on the Malabar coast is Kolamba.

Hence **Ko·lami**, the language spoken by these people.

a **1863** R. Temple in S. Hislop *Papers Aboriginal Tribes Cent. Provinces* (1866) II. p. i, The English words having been..classified, the design was to ascertain and record ..the equivalents in..Gondi, Gayeti,..Kolami. **1916** Russell & Lal *Tribes & Castes Cent. Provinces India* III. 520 The Kolāms..have a language of their own, called after them Kolāmi. **1968** *Encycl. Brit.* VII. 655/1 Five of the Central Dravidian languages..form a closely related group—Kolami, Naiki, Parji, Ollari and Poya Gadaba.

Kolarian (kolē͞ə·riăn). [f. *Kolar* (see quot. 1869) + -ian.] A non-Aryan linguistic stock of India, the Munda group. Also as *adj.*, of or pertaining to this stock.

1866 G. Campbell in *Jrnl. Asiatic Soc. Bengal* XXXV. II. 28, I propose then to call the northern tribes Kolarian or Coolee Aborigines. **1869** —— in *Jrnl. Ethnol. Soc. London* I. 130, I designate these tribes 'Kolarian' (as distinguished from the Dravidians), from the name Kol, Kolee or Coolee, applied to many of them, and the old name 'Kolar' by which India was known in very ancient times. **1915** Risley & Crooke *People of India* (ed. 2) i. 48 The hypothesis of the north-eastern origin of the Kolarians depends on the fancied recognition of Mongolian characteristics among the people of Chutia Nāgpur. **1923** A. L. Kroeber *Anthropol.* iii. 46 Typical representatives [of Indo-Australian people] are the Vedda of Ceylon; the Irula and some of the Kolarian tribes of India. **1928** V. G. Childe *Most Anc. East* iii. 52 Early Indian races, particularly the Dravidians, the Kolarians, and even the Veddahs of Ceylon. **1957** G. S. Ghurye *Mahadev Kolis* i. 6 Kolis of Bombay..do not show the slightest trace of Kolarian tongue in their languages. **1972** W. B. Lockwood *Panorama Indo-Europ. Lang.* 226 Munda or Kolarian are names given to a group of languages spoken today in the central part of India.

kolbasa, kolbasi, varr. *Kielbasa.

Kolbe (kǫ·lbə). *Chem.* The name of A. W. H. *Kolbe* (1818–84), German chemist, used *attrib.* and in the possessive to designate two syntheses devised by him: (*a*) the electrolysis of a salt of a carboxylic acid, R·COOH, to yield a (substituted) paraffin R_2; and (*b*) the reaction between sodium phenoxide and carbon dioxide to yield sodium salicylate.

1885 *Jrnl. Chem. Soc.* XLVIII. 982 The successive reactions in Kolbe's process are then (1) the formation of sodium phenyl carbonate..; (2) the conversion of this substance into sodium salicylate..; and (3) the formation of disodium salicylate and phenol. **1915** *Chem. Abstr.* IX. 294 The K and NH_4 salts, on electrolysis, give first an alk. and later an acid reaction, the electrolysis probably proceeding in accordance with Kolbe's..reaction. **1926** J. Read *Text-bk. Org. Chem.* xxvi. 570 Kolbe's reaction is also given by other phenols. **1942** Fuson & Snyder *Org. Chem.* xi. 133 Still higher members [of the dibasic acid series] may be made by the Kolbe electrolysis of the potassium salt of an acid ester. **1951** I. L. Finar *Org. Chem.* I. xxviii. 557 The original industrial method of preparing salicylic acid was Kolbe's synthesis (1859)... The method now used is a modification of the Kolbe synthesis..known as the Kolbe-Schmidt method. It is carried out by heating sodium phenoxide with carbon dioxide at 120–140° under pressure. **1969** R. C. Denney *Named Org. Reactions* xxiii. 76 Woolford has used the Kolbe reaction to prepare long-chain ω,ω'-dibromohydrocarbons which can only be obtained with considerable difficulties by other methods.

kolbeckite (kǫ·lbekəit). *Min.* [ad. G. *kolbeckit* (F. Edelmann 1926, in *Jahrb. f. d. Berg- u. Hüttenwesen in Sachsen* C. A74), f. the name of Friedrich *Kolbeck* (1860–1943), German mineralogist: see -ite[1].] A blue hydrated silicate-phosphate of beryllium, calcium, and scandium (and possibly aluminium).

1928 *Mineral. Mag.* XXI. 568 *Kolbeckite*. Beryllium phosphate or silicophosphate as cyan-blue monoclinic crystals from Saxony. **1959** *Bull. Geol. Soc. Amer.* LXX. 1649 Sterrettite and kolbeckite are the first known scandium-bearing phosphates. **1966** Z. Lerman tr. *Vlasov's Geochem. & Mineral. Rare Elements* II. vi. 219 Kolbeckite is distinguished by its marked pleochroism. It is a very rare mineral.

kolea (kolē·ā). [Hawaiian.] A local name for the Pacific golden plover, *Pluvialis dominica fulva*, which is native to the northern parts of North America and eastern Siberia, but winters in Hawaii and other parts of the Pacific.

1887 *Proc. U.S. Nat. Mus.* X. 80 *Charadrius dominicus fulvus* (Gmel.) Pacific Golden Plover, Kolea. **1933** E. H. Bryan *Hawaiian Nature Notes* 256 The golden plover, called kolea by the Hawaiians, is a little larger than the mynah bird but longer in the leg. **1944** G. C. Munro *Birds Hawaii* 56 The kolea is a valuable bird to Hawaii as a destroyer of insects... Yet it has so long been considered a game bird and table delicacy that efforts to obtain the protection it deserves have not been very successful.

Koli: see *Kori[2].

kolinsky (kǫli·nski). Also **kolinski.** [f. Russ. *Kola*, name of a port in north-west Russia.] A name for the fur of a *Mustela sibirica*, the Japanese mink. Also *attrib.*

1851 *Illustr. Catal. Gt. Exhib.* III. 803 Skins and Furs... Kolinski. **1892** H. Poland *Fur-Bearing Animals in Commerce* p. li, Kolinski, undressed. **1919** *Queen* 5 July 21 (Advt.), Kolinsky coat in the fashionable Sable Colour. **1923** J. C. Sachs *Furs & Fur Trade* 62 The 'sable' brushes, which are in such request by artists, are generally made of kolinski tails. **1928** *Strand Mag.* Aug. 183/2 A Persian lamb coat with a collar of kolinsky. **1968** J. Ironside *Fashion Alphabet* 155 The Siberian China mink is known as Kolinsky.

‖ **kolkhoz** (kǫ·lkǫz). Also **kolhoz, kolkhos,** etc. Pl. **kolkhoz, kolkhozes, kolkhozy.** [Russ., f. *kol(lektivnoe khoz(yáĭstvo* collective farm.] A collective farm in the U.S.S.R. Also *transf.*

1921 *Russian Economist* I. II. 389 The 'Kolkhoses' are the means of a guerilla war with the peasants... The measure in which..support is given depends on the degree to which a particular kind of 'Kolkhose' approximates to the ideals of Communism. **1931** *Ann. Reg. 1930* 190 There was a real rush of the peasants to enter the *Kolchosy* (the abbreviated Russian denomination of the collectivised farms). *Ibid.* 191 A sensational article by Stalin entitled, 'Intoxicated by Success'. It dealt with the Kolchos movement, and in it the dictator strongly condemned the forcible methods of collectivisation hitherto used. **1931** H. G. Wells *Work, Wealth & Happiness of Mankind* (1932) iv. 184 The Kolkhoz seems to be the old Tsarist Mir in a state of emotion. *Ibid.* x. 508 The great fields of the Kolkhozy. **1943** E. M. Almedingen *Frossia* viii. 285 He knows Russia... He has seen Sovhozes and Kolhozes. **1946** [see *Ejido]. **1949** F. Maclean *Eastern Approaches* I. viii. 133 The eating-house..seemed to be monopolized by the higher *kolkhoz* officials whom I found engaged in preparations for the forthcoming elections to the Supreme Council of the Kazakh S.S.R. **1951, 1952** [see *Agrogorod]. **1957** *Ann. Reg. 1956* 256 The villages [in Hungary] seem quietly to have dissolved most of their local *kolkhoz* and then returned to work. **1958** *New Statesman* 5 Apr. 423/3 The *kolkhozes*..will now be in a position to introduce certain measures hitherto found only on state farms, including a guaranteed minimum wage. **1968** C. A. Doxiadis *Between Dystopia & Utopia* 46 The Kolkhozes in Russia.. are the results of utopian theories. **1972** *Guardian* 4 Aug. 4/4 A kolkhoz, or collective farm, is a symbol of the crude rustic for city dwellers.

Hence **ko·lkhoznik** (pl. **-niki**), a member of a *kolkhoz*.

1955 H. Hodgkinson *Doubletalk* 27 Each worker, or *kolkhoznik*, has to give between 100 and 150 work days on the common land. **1964** *Economist* 18 July 255/1 With their dependants the *kolkhozniki* amounted to about 56 million persons.

kolladie, var. *Korari.

kollergang (kǫ·ləɹgæŋ). [G., = crushing action.] A crushing machine used in milling paper-pulp.

1890 A. Watt *Art of Paper-Making* 82 For the purpose of crushing the knots of the straw, and other hard particles ..a machine termed the 'kollergang' or 'edge-runner' is sometimes employed. **1907** G. Clapperton *Pract. Paper-Making* (ed. 2) iv. 38 The ability of the Scandinavian makers to allow a considerable time for milling the pulp in kollergangs or beaters. **1963** [see *edge-runner* s.v. *Edge *sb.* 12 b].

kolm (kǫlm). *Petrogr.* [Sw.] (See quot. 1954.)

1930 *Jrnl. Amer. Chem. Soc.* LII. 4848 The ash of the kolm constitutes from 20 to 40% of the original, and contains a few hundredths of a per cent. of lead. **1954** S. I. Tomkeieff *Coals & Bitumens* 61/1 *Kolm*, variety of cannel coal occurring locally as lenticles in Swedish alum shales, and containing 30% of ash rich in rare metals, including uranium and radium. **1969** M. G. Rutten *Geol. W. Europe* iii. 44 Within these black shales lenses may occur of a sort of coal, the 'kolm'. Both shales and kolm have in recent times acquired importance for their uranium content.

Kolmer (kǫ·lməɹ). *Med.* The name of John *Kolmer* (1886–1962), U.S. pathologist, used *attrib.* and *absol.* to denote a modification proposed by him of the Wassermann complement fixation test for syphilis using beef heart cardiolipin as the antigen. Also **Kolmer–Wassermann.**

1921 *Med. Clinics N. Amer.* V. 670 We have studied the results of the Kolmer method of performing the Wassermann test in over 2000 tests. **1925** *Jrnl. Laboratory & Clin. Med.* X. 315 (*heading*) Red corpuscle suspension for the Kolmer–Wassermann reaction. **1932** Schamberg & Wright *Treatm. Syphilis* xxxiii. 591 The Kolmer–Wassermann test is not usually positive until the primary lesion has been present for eight days. **1971** S. M. Brooks *V.D. Story* iv. 60 A positive Kolmer or VDRL..practically always indicates nervous system involvement. **1971** *Amer. Jrnl. Clin. Path.* LV. 735 The automated Kolmer test appeared less satisfactory than the manual Kolmer test.

Kol Nidre (kǫl ni·dre). Also **Kol Nidry.** [Aramaic *kol nidhrē*, all the vows.] A mournful Aramaic prayer, opening with the words *Kol nidhrē*, sung by Jews at the beginning of the service on the eve of Yom Kippur: it asks for annulment of vows made in the previous year and for forgiveness of sins. Also, the service itself or the melody to which the prayer is sung.

1881 M. Keizer in I. L. Mombach *Sacred Mus. Compositions* p. viii, (*heading*) New Year & Atonement... Ki Vayoum Hazeh..Kol Nidry..Mechalkyl. **1891** M. Friedländer *Jewish Relig.* ii. 408 Transgressors..desirous to pray in the Synagogue on the Day of Atonement, were admitted... Such was the original object of *Kol-nidre*. **1893** I. Zangwill *Ghetto Tragedies* 47 It was *Kol Nidré* night, the commencement of..the Day of Atonement. **1908** *Daily Chron.* 5 Oct. 5/2 All this time services are held wherever a few of the children of the Ghetto are gathered together. The first, held at 6 o'clock on Sunday evening, at the opening of the fast, is called 'Kol-Nidré'. **1922** Joyce *Ulysses* 470 A..band is heard..playing the Kol Nidre. **1932** C. Roth *Hist. Marranos* 379 The famous ceremony of the Annulment of Vows on the eve of the Day of Atonement, the Kol Nidre service. **1966** H. Kemelman *Saturday the Rabbi went Hungry* (1967) i. 7 The chanting of Kol Nidre ushered in the Holy Day. **1973** *Synagogue Light* Sept. 12/1 On Tish B'Av..we take off our shoes. On Yom Kippur also, commencing before Kol Nidre.

kolo (kō͞u·lo). [Serbo-Croatian, = wheel.] A Yugoslav dance performed in a circle.

1911 E. L. Urlin *Dancing Anc. & Mod.* iii. 45 The Kollo Dance (or Kolo) is seen depicted on very ancient tombs of the Bogomiles in Dalmatia, archaic in type; it must therefore have been of ancient religious origin. **1941** 'R. West' *Black Lamb* I. 398 They had our men and women brought in to dance the kolo to them. **1952** L. & D. Janković *Dances Yugoslavia* 21 The kolo usually has many phases and very often works up from a gentle swaying motion to more animated movement, which dies down to a moderate level according to the directions of the musician and the leader. **1963** *Daily Mail* 24 Aug. 2/2 A frolicking Mr. Kruschev... With his holiday host, President Tito, he joined the girls in the kolo, a popular dance performed in a circle to the accompaniment of singing. **1969** *Daily Tel.* 5 Nov. 13/6 The dancers launched themselves on an old Bosnian dance, a silent kolo from Glamotch.

kolong, var. *Kulang.

kolovratite (kǫlŏvrā·təit). *Min.* [ad. Russ. *kolovratit* (V. I. Vernadsky 1922, in *Compt. Rend. de l'Acad. d. Sci. de Russie* A. 37), f. the name of L. S. *Kolovrat*-Chervinsky (1884–1921), Russian radiochemist: see -ite[1].] A greenish-yellow amorphous or finely crystalline mineral that is probably a hydrous vanadate of nickel and zinc.

1925 *Mineral. Mag.* XX. 290 A new mineral, named kolovratite by V. I. Vernadsky, which..appears to be widely distributed in Fergana... The mineral is a vanadate of nickel. **1927** *Mineral. Abstr.* III. 234 Kolovratite.. from Kara-Chagyr occurs as a fine powder disseminated in carbonaceous quartz-schist. **1962** *Canad. Mineralogist* VII. 314 Our evidence suggests, however, that kolovratite is a hydrous zinc-nickel vanadate, or possibly a silico-vanadate, rather than a nickel vanadate as inferred in the original description.

komatik (kǫ·mătik). Also **kamotik, kamootik,** etc. [Eskimo.] A dog-sledge used by the people of Labrador. Also as *v. intr.*, to travel on a komatik.

1824 W. E. Parry *Jrnl. Second Voy.* 567/1 *Sledge, a*, ka-moo-tik. **1853** *Trans. Lit. & Hist. Soc. Quebec* IV. 337 They..are so hardy that six or eight of them tackled to a heavily laden sledge or 'commettek' will travel as much as twenty leagues in a day. **1905** N. Duncan *Dr. Grenfell's Parish* xi. 133 The sick and starving are sought out by dog-team and komatik. **1919** W. T. Grenfell *Labrador Doctor* (1920) xi. 199 Sails can sometimes be used with advantage on the komatik as an adjunct. **1921** *Beaver* Jan. 16/1 We had the ordinary length of sled or 'komatik', which is about twelve feet long and weighs about a hundred pounds. We carried a load of about six hundred pounds, which..is considered light. **1934** G. M. Sutton *Eskimo Year* xxxvi. 233 Jack and I decided we would *komatik* to the head of the bay to set traps and locate some fox-dens. **1936** *Discovery* Sept. 274/2 An Eskimo *komatik* (sledge) complete with driver and several dogs neatly harnessed. **1940** R. Finnie *Lure of North* 117 In the dead of winter, however, the native komatik reigns supreme, gliding easily where a basket sled would drag. **1965** *Globe & Mail* (Toronto) 6 Jan. 1/3 The dog sled, or komatik, has replaced the bicycle for northern Cubs. **1973** C. Bonington *Next Horizon* xvii. 235 The komatic or sledge... Now I was sitting on the back of a komatic, near the head of Cumberland Sound in Baffin Island.

kombaars (kǫ·mbaɪs). *S. Afr.* Also **kombers(e.** [Du. *kombaars* coverlet, rug.] A blanket, coverlet, or rug.

1812 W. J. Burchell *Jrnl.* 1 May in *Trav. S. Afr.* (1824) II. 175 This *kombaars*, or coverlet, is a genuine South-African manufacture, being nothing more than a Hottentot *karóss* of large dimensions. **1840** B. Shaw *Memorials S. Afr.* i. 28 His *kombaars* tied with leathern strings. **1850** R. G. Cumming *Five Yrs. Hunter's Life S. Afr.* I. ix. 186 In the evening I took my pillow and 'komberse', or skin blanket, to the margin of a neighbouring vley. **1913** C. Pettman *Africanderisms* 274 *Komberse*, a rug, blanket; sometimes a kaross is so styled.

‖ **kombé** (kǫ·mbi). [Mang'anja.] The juice obtained from the seeds of *Strophanthus kombe*, a Central African climbing plant of the family Apocynaceæ, which is one of the sources of the drug strophanthin.

1859 J. Kirk *Zambesi Jrnl. & Lett.* (1965) I. 171, I take a walk in the neighbourhood, find the plant used to poison the arrows. It is an arborescent twiner, getting quite to the top of the trees. The juice of the fruit is used... They call it *Kombe*. **1888** *Buck's Handbk. Med. Sci.* VI. 663/1 Strophanthus (Kombé, Arrow Poison). **1934** G. E. Trease et al. *Text-bk. Pharmacognosy* xx. 512 One of these [species of *Strophanthus*], known to the natives of the Shiré River as *kombi*, was noted by Livingstone in 1861... In 1885 Fraser isolated strophanthin and recommended the use of the seeds in medicine. **1952** *Ibid.* (ed. 6) xviii. 479 The glycosides of *kombé* contain cymarose. **1967** *Martindale's Extra Pharmacopoeia* (ed. 25) 541/2 Strophanthin-K (B.P.C. 1954). Strophanthin; Kombé Strophanthin.

kombu (kǫ·mbu). [Jap.] A brown seaweed, one of the local species of *Laminaria*, used as food, esp. as a base for stock.

1884 Satow & Hawes *Handbk. for Travellers Cent. & N. Japan* (ed. 2) 519 *Kombu*, a broad, thick, and very long species of seaweed, most of which is exported to China. **1905** *Nat. Geogr. Mag.* May 218 Kombu is one of the staple foods of the country [*sc.* Japan], entering into the dietary of almost every family. **1949** *New Biol.* VII. 96 The various preparations of 'kombu' are said to have quite distinctive flavours. **1958** G. Mikes *East is East* 67 *Kombu* (a kind of seaweed) is not less delicious than boiled cabbage. **1969** G. W. Prescott *Algae* x. 350 In Japan, laminarian food (from stipes) is called *kombu*; food from *Alaria* is called *sarumen*. Stipes of some of the larger brown algae are cut, washed in fresh water, boiled and seasoned with preserving fluids.

komfoor (kǫmfō͞ə·ɪ). *S. Afr.* Also **komvoor, konfoor.** [Afrikaans, f. Picard form of OFr. *chaufoire* kettle for hot water.] A small brazier; a chafing-dish; a foot-stove.

1841 *Cape of Good Hope Almanac* (Advt.), Tools of all descriptions... Brushes, kettles and komfoors. **1844** J. Backhouse *Narr. Visit Mauritius & S. Afr.* v. 84 To preserve warmth, the Dutch women use an apparatus to set their feet upon, called a Komfoor. **1940** 'B. Knight' *Walking the Whirlwind* vii. 144 She signalled to the slave near the door to hand round charcoal from the komfoor to relight some of the pipes. **1951** *Cape Times* 6 Sept. 16/1 Samovars are almost as common in Cape Town as komvoors and spittoons used to be in the salesrooms. **1965** A. Gordon-Brown *S. Afr. Heritage* II. 21 Brass Coffee Pot under which is the original Konfoor to contain burning charcoal for keeping the coffee hot. **1972** A. A. Telford *Yesterday's Dress* 81 She sits drinking coffee with her bare feet upon a 'komfoor' which as its name implies was a wooden 'chaufette' containing a pan of charcoal, a comfortable footwarmer.

Komi (kō͞u·mi). [Native name (see quots. 1800 and 1888).] **a.** A people of northern central U.S.S.R.; a member of this people. **b.** The language of the Komi. Also *attrib.*

[**1800** W. Tooke *Hist. Russia* I. i. 13 *Komanes*. They were neighbours of the Madshares or Ugres, and migrated in conjunction with them at the close of the eighth century to Pannonia. They dwelt upon the river Kuma, from which they also had their name. On the other side of the Terek is still a people named Kumuiks; perhaps remains of the old Kumanians. *Ibid.* 42 One part of these were called Kumani or Komani, from the river Kuma.] **1888** *Encycl. Brit.* XXIV. 283/1 The Zyrians..constitute the bulk of the population on the Ural slopes. They formerly inhabited the Kama and Vyatha basins, and call themselves Komi-yurt, or Komi-yas. **1911** *Ibid.* XXVI. 317/2 They [*sc.* the Syrenians] call themselves Komi. **1933** H. Kohn *Nationalism in Soviet Union* 62 These Finnish peoples, the Komi, the Mari, and the Mordvins, were steadily pushed back by the Russians into less fruitful areas with severe climates. **1942** K. W. Deutsch in J. A. Fishman *Readings Sociol. of Lang.* (1968) 600 Syryen (Komi). **1959** *Chambers's Encycl.* XIV. 121/1 (table) Komi. **1964** *Language* XL. 906 Komi is today represented by two literary languages, Komi(-Zyrryan) and Komi-Permyak. *Ibid.* 98 The semantic functions of the Komi and Udmurt forms.

komita(d)ji, var. *Comitadji.

‖ **Kommandatura** (kǫma:ndātū͞ə·rā). Also **-antur, -atur.** [ad. G. *Kommandantur* commandant's headquarters, command post.] The centre of operation of a military government.

1937 Blunden *Elegy* II. 31 We met the walls the Roman host Used for their *Kommandantur* post. **1949** *Ann. Reg. 1948* 238 The machinery of four-Power rule set up under the Potsdam Agreement, the Allied Control Council and the Berlin *Kommandatura*, had become more and more an instrument for political propaganda. **1957** H. Roosenberg *Walls came Tumbling Down* iii. 65 We saw several people lugging shot-guns and rifles to the new *Kommandatura*. **1958** P. Kemp *No Colours or Crest* x. 220 The miserable Eles..had been a frequent visitor to the German *Kommandatur* in Peć. **1969** *Listener* 9 Jan. 35/3 The Soviet Kommandatura in the centre of Prague. **1975** *New Yorker* 22 Sept. 78/2 The site of the Russian Kommandatura in Leipzig had been shown on the map of the city.

‖ **Kommers** (kǫmḗ·rs). Also **commerce, -se, -s.** [G., ad. Fr. *commerce*, L. *commercium* Commerce *sb.*] = Commers.

1839 Longfellow *Hyperion* I. II. iv. 120 Come in; come in. You shall see some sport. A Fox-Commerce is on foot, and a regular Beer-Scandal. **1841** W. Howitt *Student-Life Germany* xvi. 315 For these Commerses, the rule is to drink beer, and this is called a Commers in beer. **1908** C. Sidgwick *Home Life Germany* vi. 56 A *Kommers* is a students' festival in which the professors and other senior members of a university take part. **1960** *Times* 17 Sept. 9/7 One may be fortunate enough to be invited to a *Kommers*, a meeting in which the students settle round tables for the evening, quaffing beer..and singing.

Komodo (komō͞u·do). The name of an Indonesian island, used *attrib.* in **Komodo dragon, monitor** to designate a large monitor lizard, *Varanus komodoensis*, native to this island and neighbouring ones.

1927 *Daily Express* 15 June 15 The curator of the reptiles..introduced me to her latest novelty, a Komodo dragon, seven feet long and as ugly as a bad dream. **1927** *Proc. Zool. Soc.* 256 Lord Rothschild exhibited a mounted specimen of *Varanus komodensis* [*sic*], the so-called 'Dragon' of Komodo Island... The interesting feature of the Komodo Monitor, apart from its bulk, lies in its relationship. **1928** *Ibid.* 1017 (*title*) On a living Komodo Dragon. **1941** J. S. Huxley *Uniqueness of Man* viii. 184 The gorilla, the orang-utang, the Komodo dragon, and other creatures are on the margin of this category [of creatures kept from extinction only by rigorous protection]. **1969** A. Bellairs *Life of Reptiles* II. xi. 468 The Komodo dragon (*Varanus komodoensis*) has been generally regarded as the largest existing lizard and reaches a length of at least 3 metres (10 ft.) and a weight of 163 kg. (360 lb.). It is found only on the small island of Komodo to the east of Java and on one or two neighbouring islands. **1973** *Nature* 7 Sept. 52/1 There is evidence that the Komodo monitor can take prey as heavy as itself.

kompas, var. *Kempas.

Komsomol (kǫ·msŏmǫl). Also **Comsomol.** [Russ. *komsomól*, short f. *Kommunisticheskii Soyúz Molodëzhi* Communist Union of Youth.] An organization of Communist youth in Russia; a member of this organization. Also *attrib.*

1934 T. S. Eliot *Rock* i. 39 That's what they do in Russia. Ever 'ear o' the Comsomol? We'll turn all the churches into workers' clubs. **1944** G. B. Shaw *Everybody's Pol. What's What?* x. 82 Comsomols, Ballillas, Leagues of Youth and cognate bodies and movements which are sprouting everywhere. **1948** J. Towster *Political Power in U.S.S.R.* 137 The closest auxiliary of the Party is the Komsomol or Young Communist League. **1949** *Ann. Reg. 1948* 211 Communists and Komsomols had helped to build mosques. **1958** *Listener* 5 June 934/1 Why should Chekhov's characters proclaim prophetic and optimistic truths about a better future which sounds [*sic*] as if they were gleaned from a Komsomol textbook? **1959** *Ibid.* 3 Dec. 964/2 A Komsomol branch secretary. **1973** *Nat. Geographic* May 606/2 The Komsomol (Communist youth organization) in her district had advertised openings at Togliatti stating that a modern low-rent apartment was part of the employment contract.

komvoor, var. *Komfoor.

kona (kō͞u·nā). [Hawaiian; orig. the name of the leeward areas of the Hawaiian Islands.] A stormy south-west wind in the Hawaiian Islands.

1864 R. Anderson *Hawaiian Islands* i. 28 Occasionally a prolonged gale comes from the south, called a Souther, or 'Kona'. **1866** 'Mark Twain' *Lett. from Hawaii* (1967) 23 In the stormy season—in the season of the terrible kona. **1892** Stevenson & Osbourne *Wrecker* 9 It was blowing a kona, hard. **1934** M. D. Frear *Lowell & Abigail* p. xiv, But by night..how the Kona wind moaned and drove the rain up from the sea. **1967** J. Severson *Great Surfing* Gloss., *Kona wind*, a south wind in the Hawaiian Islands, generally onshore at most of the southern exposure beaches. Occasionally offshore or glassy on the north shore.

konak (konā·k). [ad. Turk. *qonaq* halting-place, inn.] A large house, palace, or official residence, in Turkey, or in the (former) Ottoman Empire.

[**1675** J. Covel *Diary* (1893) 175 We..had a man who alwayes went before to every *Conack*, or stage, and brought in muttons, beafes, veales, and the like.] **1852** C. T. Newton *Let.* 30 Sept. in *Trav. & Discov. Levant* (1865) I. vii. 87 We asked for the konak, or official residence of the Aga. **1878** S. L. Poole *People of Turkey* I. ix. 221 When in the interior I had the opportunity of visiting some Konaks worthy of note. **1897** E. A. Bartlett *Battlefields Thessaly* iv. 88 The Konak was deserted by all except the Mushir's valet. **1908** tr. *P. Draganof's Macedonia & Reforms* ii. 23 The latter drove them from the konak. **1926** *Spectator* 17 July 88/2 The delicate 'konak' or palace of Prince Milosh Obrenovitch. **1957** L. Durrell *Bitter Lemons* 163 There is also a wretched Konak and prison.

‖ **konaki** (kǫ·naki). *N.Z.* Also **koneka.** [Corruption of Maori *kōneke* sledge.] A horse-drawn wooden sledge.

1914 M. Hall *Woman in Antipodes* I. 86, I also got some insight into up-country life..noting..the primitive Maori 'koneka' or sledge. **1953** J. W. Brimblecombe in J. Reid *Kiwi Laughs* (1961) 179 He had to go and haul lime tomorrow with the konaki. **1959** M. Shadbolt *New Zealanders* 15 The trunks were lashed to a konaki, which Mother discovered was the Maori word for a horse-drawn wooden sled. **1961** B. Crump *Hang on Minute* 109 Uncle Wally lifted the [cream-]can..out to the old konaki.

Kond, Kondh, varr. *Khond.

‖ **konditorei** (kǫ·ndītǒrəi·). Pl. **konditorei, konditoreien.** [Ger., f. *konditor* confectioner.] Confectionery; a confectioner's shop, a shop where pastries are sold.

1935 J. Buchan *House of Four Winds* iii. 69 He was in the habit of sampling..whatever Unnutz produced in the way of café and konditorei. **1963** N. Freeling *Because of Cats* ii. 35 When she went to a concert it would be to satisfy a need, like women who went to konditoreien to stuff themselves with wonderful cream cakes. **1973** *Times* 30 May (Austria Suppl.) p. iv/6 *Konditorei* abound, and you will find it hard to resist those delectable Viennese pastries bulging with cream.

‖ **konfyt** (kǫnfĕ͞i·t). *S. Afr.* Also † **comfaat, conf(e)yt.** [Afrikaans, = Du. *konfijt* Comfit *sb.*] Fruit preserved in sugar; preserve.

1862 Lady Duff-Gordon *Let.* 15 Apr. (1927) 144, I have bought some Cape 'confeyt'; apricots, salted and then sugared, called 'mebos'—delicious! **1871** *Cape Monthly Mag.* II. 25 Coffee and cakes, tea and '*comfaat*'. **1929** *Daily Tel.* 15 Jan. 9/4 Now that Cape fruits are being sent over to this country in ever-increasing quantities, the word 'konfyt' has come into our vocabulary. 'Konfyts' are delicious preserves made from these exotic fruits. **1939** S. Cloete *Watch for Dawn* 39 He had konfyt of melon and little oranges. **1947** L. G. Green *Tavern of Seas* ix. 87 Malay families sometimes sell these fine eighteenth century saucepans, kettles and konfyt pots. **1950** *Cape Argus* 8 May 4/4 She became particularly famous for her konfyts and jams. **1955** L. G. Green *Karoo* xi. 140 White housewives made *konfyt* of it. **1974** D. Rooke *Margaretha de la Porte* 229 'Konfyt and meerbos, what a typical Dutch spread,' Fanny commented with a patronizing smile; but she was quick enough to help herself to the glazed watermelon and sugary apricots.

kongkang: see *Kukang.

kongoni (kǫŋgō͞u·ni). [Swahili.] An alternative name for Coke's hartebeest, *Alcelaphus buselaphus cokei*, found in the plains of East Africa.

[**1889** J. C. Willoughby *E. Afr. & its Big Game* 54 These hartebeest (*swahili Kongoni*) are of the species known as *Cokii*, and..are very plentiful here. **1905** H. A. Bryden in H. G. Hutchinson *Big Game Shooting* II. vi. 114 Coke's hartebeest (*Bubalis cokei*) (*Kongoni* of the Swahilis) is the commonest..of this group in East Africa.] **1908** W. S. Churchill *My Afr. Journey* i. 9 Herds of red kongoni—the hartebeeste of South Africa. **1921** *Outward Bound* June 40 Within the noonday shade Of mushroom-headed thorn trees, the great 'Kongoni' wade. **1942** *E. Afr. Ann. 1941–2* 127/2 Zebra, kongoni, steinbuck, Thomson's gazelle, and ostrich are numerous on the plains. **1971** *Nature* 24 Dec. 483/2 Eland..and boran ..cattle were equal in water demands, while kongoni (*Alcelaphus buselaphus*) and wildebeest..are comparable with sheep.

‖ **kongsi** (kǫ·ŋsi). [ad. Chin. *kung-ssu* (as pronounced in Hokkien) company, corporation.] In Malaysia, an association or partnership, esp. an association of Chinese people.

1839 T. J. Newbold *Pol. & Statistical Acct. Straits of Malacca* I. i. 14 Their interior affairs..are arranged by the heads of their respective kongsis or fraternities. **1900** W. W. Skeat *Malay Magic* v. 255 The men of a rival *kongsi* had brought limes. **1964** D. K. Bassett in Wang Gungwu *Malaysia* II. vii. 122 A *kongsi* is an association of Chinese based on regional, occupational or dialect ties. **1966** D. Forbes *Heart of Malaya* x. 116 Three or four hundred jammed together in a junk chartered by a merchant combine, called a *kongsi*. **1969** J. M. Gullick *Malaysia* ii. 48 In its primary sense 'kongsi' denotes a Chinese association or secret society. **1972** *Straits Times* (Malaysian ed.) 23 Nov. 11/3 The tin mine kongsi house in which I found her was occupied by the accused's brother.

koniak, koniaku (kǫ·nyak, kǫnya·ku). Also **konjak, konnyaku.** [ad. Jap. *ko-n-nya-ku, ko-nya-ku.*] A local name for *Amorphophallus rivieri*, a large herb of the family Araceæ, cultivated in Japan for the flour obtained from its roots.

1884 tr. *A. de Candolle's Orig. Cultivated Plants* II. i. 76 The konjak is a tuberous plant of the family Araceæ, extensively cultivated by the Japanese. **1954** J. M. Morris *Wise Bamboo* iv. 54 Anything edible can be put into *sukiyaki*. That night we had leeks, *tofu* (beancurd), *konnyaku* (a gelatine-like vegetable which I could not identify), onions and bamboo sprouts. **1965** *This is Japan 1966* 80/3 All the balloons were fashioned of Japanese rice paper and *konnyaku* (devil's-tongue root) which is a chief ingredient in the popular Japanese dish *sukiyaki*. **1970** J. Kirkup *Japan behind Fan* 4 Lumps of *tofu* (white bean-curd) and *konyaku* (devil's tongue jelly). **1972** Y. Lovelock *Vegetable Bk.* II. 269 The related giant arum or devil's tongue..known to the Japanese as *koniaku*, is often cultivated for food... Their [*sc.* the tubers'] taste and smell is [*sic*] strong and disagreeable, but this is lost after they have been soaked in whitewash (i.e. milk of lime), crushed and cooked. The resulting flour is used for making a kind of pasta and other dishes.

konimeter (kǫni·mĭtəɹ). [f. Gr. κόνις dust: see -meter.] A kind of impinger in which a known volume of air is directed on to a slide coated with a substance to which any dust particles will adhere.

1918 *Jrnl. Chem., Metall. & Mining Soc. S. Afr.* XVIII. 200/1 We are indebted to the perseverance and ingenuity of Mr. R. N. Kotzé, the Union Government Mining Engineer, for a simple solution by means of an instrument he has devised and called a Konimeter. **1937** Drinker & Hatch *Industr. Dust* vi. 99 The dust-collecting efficiency of the konimeter is admittedly low and is selective with respect to size. **1958** *Ann. Rep. Chief Inspector of Factories on Industr. Health 1957* 17 in *Parl. Papers 1958–9* (Cmnd. 558) XIII. 183 A team..travels to particular factories and takes samples, using..a Kotzé konimeter... Dust samples..are analysed..and the particle size and particle counts recorded. **1965** *Guide to Prevention & Suppression of Dust in Mining* (Internat. Labour Office) xvi. 306 Side-by-side sampling with a thermal precipitator and a konimeter was performed on the dust created by a series of different underground operations.

konini (kō͞u·nini). *N.Z.* [Maori.] A native shrub or small tree, *Fuchsia excorticata*, or its berries.

1867 J. D. Hooker *Handbk. N.Z. Flora* II. 765/2 Konine. *Fuchsia excorticata*. **1869** T. H. Potts in *Trans. N.Z. Inst.* II. 48 Beneath, Bell-birds..seek the konini, clinging to its brittle sprays. **1889** T. Kirk *Forest Flora N.Z.* 53 The settlers sometimes term it Kotukutuku or Konini, but more generally fuchsia. **1949** P. Buck *Coming of Maori* (1950) II. i. 88 The berries of the..konini..are transient and were not important as food. **1960** B. Crump *Good Keen Man* 12 When a pigeon shirred up out of a konini I almost dropped my rifle. **1968** M. Johnson *N.Z. Flowering Plants* 57 One seldom sees now a rich display of the purple black berries, konini, that delicacy of the Maori palate of former times.

koniscope (kǫ·niskō͞up). [f. Gr. κόνις dust + -scope.] An instrument for estimating the concentration of dust in the air by observing the depth of colour in a sample of the air when it is expanded into a tube with glass ends and containing moist paper to saturate it with water vapour.

1892 J. Aitken in *Proc. R. Soc.* LI. 435 This new instrument we intend to call a Koniscope. In its present form this instrument consists of an air-pump and a metal tube with glass ends. **1905** *Westm. Gaz.* 26 Aug. 13/1 Dr. Aitken has invented a very simple instrument for this purpose, which he calls the koniscope, or dust detective. **1925** Shaw & Owens *Smoke Probl. Gt. Cities* viii. 145 Condensation of water on the dust particles is utilised in Aitken's..koniscope; a fog is produced and compared with a standard.

konjak, var. *Koniak, koniaku.

konk, var. Conk, the nose.

1846 *United Services Mag.* May 13 Indignant at the liberty thus taken with his konk.

konk, var. *Conk *v.*[2]

1917 E. Middleton *Glorious Exploits of Air* 20 The latter [*sc.* the engine] 'konked'. Down went the aeroplane. **1919** 'Rafbird' *Zooms & Spins* 19 I'm flapping from Puddlemarsh..—came down there with a konking engine. **1919** *Air Pie* 93 What would you do if your engine 'konked' out at 20,000 feet? **1925** Fraser & Gibbons *Soldier & Sailor Words* 138 *Konked* (*konked out*), an Air Force term used of an aeroplane engine stopped working, failed, or broken down. Also a general expression, meaning 'knocked out', dead. **1942** C. S. Lewis *Broadcast Talks* II. iii. 49 Some fatal flaw always brings the selfish and cruel people to the top and it all slides back in misery and ruin. In fact, the machine konks. It seems to start up all right and runs a few yards, and then it breaks down.

Konkani (kǫ·ŋkāni). [Marathi *Konkaṇī*, f. *Konkaṇ* Konkan, a coastal region of western India.] **a.** A native or inhabitant of the Konkan. **b.** The Indic language of the Konkani. Also *attrib.* or as *adj.*

1873 *Cycl. India* (ed. 2) III. s.v. *Konkan*, The mixed Konkani tongue, appears to be only Marathi with a large infusion of Tulu and Canarese words... Mr. H. Mogling however mentions that the Konkani-speaking brahmins of Mangalore, consider it quite distinct from, though cognate with, Marathi. **1885** G. C. Whitworth *Anglo-Indian Dict.* 169/2 The southern Konkanis..are much more in the habit of seeking their living abroad than other natives. **1905** G. A. Grierson *Ling. Survey India* VII. 1 The dialectic differences within the Marāṭhī area are comparatively small, and there is only one real dialect, *viz.*, Kōnkaṇī. *Ibid.* 17 A Kōnkaṇī translation of the Bible appeared at Serampore in 1819. **1948** D. Diringer *Alphabet* 372 Konkani, one of the most important Marathi forms of speech. **1959** *Chambers's Encycl.* VIII. 354/2 There is a western group composed of Kashmiri, Lahnda..Sindhi..Rajasthani,..and Konkani. **1969** *Eve's Weekly* (Bombay) 20 Dec. 55/2 Boy-carollers.. singing Christmas carols in Konkani. **1971** *Hindustan Times Weekly Rev.* (New Delhi) 4 Apr. p. iv/5 Konkani has no script, no grammar, no rules.

Konno(h), varr. *Kono[2].

konnyaku, var. *Koniak, koniaku.

kono[1] (kō͞u·no). Also **ko-no, Ko-No.** [Korean.] A Korean board game.

1895 S. Culin *Korean Games* 100 The games played on diagrams, like our game of Merrells, receive the name of *Ko-no* in Korea, a term my informant could not further define. *Ibid.* 101 In all games of *Kono*, as in Pa-tok, the black men move first. **1960** R. C. Bell *Board & Table Games* I. iii. 98 Five Field Kono... The object of the game is to move the pieces across to the other side of the board to occupy the places vacated by the opponent, and the first player to do so wins the game. **1970** *Nature* 19 Sept. 1206/1 Ko-No (Fig. 1) is probably the simplest fully determined board game known.

Kono[2] (kō͞u·no). Also **Konno(h).** [Native name.] **a.** A Mandingo-speaking people of Sierra Leone; a member of this people. **b.** The language of the Kono people, having affinities with Vai. Also *attrib.* or as *adj.*

1909 R. H. K. Willans in *Jrnl. R. Afr. Soc.* XXX. 130 The history of the Konnoh people is of great interest... The Konnohs at the present day have degenerated into an almost insignificant tribe. *Ibid.* 145 *Soa*..in Mendi and Konnoh means three. **1916** N. W. Thomas *Specimens Lang. Sierra Leone* Introd., Kono, Susu and Kisi were recorded from informants with a very moderate knowledge of English. **1925** T. N. Goddard *Handbk. Sierra Leone* 55 The *Konnos* and *Korankos* are closely connected with the Mandingos. **1925** H. C. Luke *Bibliogr. Sierra Leone* (ed. 2) 148 Kono Hymns. **1926** F. W. Butt-Thompson *Sierra Leone* vi. 44 The present border of the Kono lands. **1951** K. L. Little *Mende of Sierra Leone* iii. 70 The Kono men, who are mostly labourers in the town, wear European shorts and shirt outside the trousers. *Ibid.* vii. 147 The Mende tend to look down on certain peoples, such as the Kono. **1964** C. Fyfe *Sierra Leone Inheritance* 250 During the march to Weeima we were attacked on all sides by the Konnos and several of our people were wounded.

konze (kǫ·nzi). Also **konzi, nkonze.** [Swahili.] A local name for Lichtenstein's hartebeest, *Alcelaphus lichtensteini*, an antelope found in the plains of central and southern Africa.

1877 F. C. Selous *Jrnl.* 20 Dec. in *Hunter's Wanderings Afr.* (1881) 303 In the afternoon I went out with my rifle, and shot a konze antelope... The black mark down the front of the face of the hartebeest is..wanting in the konze. **1908** R. Lydekker *Game Animals Afr.* 111 The konzi, or Lichtenstein's hartebeest. **1964** E. P. Walker et al. *Mammals of World* II. 1446/1 Hartebeests; Ngondo, Nkonze, Kondikondi, Sig (native names).

kooftah, var. *Kofta.

kook (kŭk). *slang.* [prob. abbrev. of *Cuckoo *a.* or Cuckoo *sb.* 3.] **1.** A cranky, crazy, or eccentric person. Freq. *attrib.* or as *adj.*

1960 *Daily Mail* 22 Aug. 4/5 A kook, Daddy-O, is a screwball who is 'gone' farther than most. **1963** *Time* 4 Oct. 37 'Don't think that just because he talked about those way-out rockets he's a kook,' cautioned a fellow officer. **1964** *Economist* 28 Nov. 969/2 Thousands of 'beatniks, kooks and crackpots'. **1965** J. Potts *Only Good Secretary* (1966) ii. 26 Max is kind of a kook. He paints these kooky pictures. **1968** Mrs. L. B. Johnson *White House Diary* 18 Jan. (1970) 623 Mrs. Hughes..said ..'I think that anybody who takes pot because there is a war on is a kook.' **1968** *N.Y. Times* 26 Mar. 32 'Has it ever occurred to you that the kook market has grown?' said a United States auto executive when asked to explain the growing sales of foreign cars. **1970** E. R. Johnson *God Keepers* (1971) xv. 166 It's a kook clique all right. It's ..a happy place. That's kooks to you cops. **1971** *Black World* June 67/1 These marchers were all probably a bunch of kooks like Harry always said. **1973** *Publishers Weekly* 25 June 68/1 A bona fide kook who is never quite able to get in gear till he finally dies paddling his canoe across the Atlantic.

2. orig. *U.S.* A novice, or one who is inexpert, in surf-riding. Also *attrib.*

1961 in *Amer. Speech* (1962) XXXVII. 150. **1966** *Surfer* VII. 9 This letter is to protest about dumb kook girls out in the water. *Ibid.* 17 All most of [these surfers] are is a bunch of loud-mouthed kooks who come down here and clutter up the beach. *Ibid.* 39 Malibu..was also the birthplace of the 'kook box', that monstrosity known as the poor man's paddle board. **1971** *Studies in English* (Univ. of Cape Town) II. 25 The reason for this reticence is that surfers wish to differentiate themselves from kooks, who surf badly.

kooka (ku·kă). *Austral. slang.* = Kookaburra.

1933 *Bulletin* (Sydney) 13 Sept. 21/4 An article on upsetting the balance of nature in which one Shepstone quotes the sad case of the kooka. **1949** *Geogr. Mag.* Feb. 374/1 Hence such names for the Kookaburra as.. *breakfast bird, kooka* and *ha ha pigeon*. **1964** P. Carlon *Price of Orphan* iv. 44 He was going to be grabbed like a kooka sighting a snake.

kookaburra. Substitute for def.: A large, arboreal, brown kingfisher, *Dacelo novæguineæ*, distinguished by its peculiar laughing cry; formerly called the Goburra and also called the laughing jackass. (Later examples.)

1906 *Westm. Gaz.* 13 Oct. 16/3 Just before the hour of sunrise the echoes of the Australian bush are awakened by the extraordinary cackling laughter of the great brown kingfisher, or kookaburra, as the aboriginals call it. **1934** *Bulletin* (Sydney) 3 Jan. 31/3 An assembly of kookaburras in the trees close at hand burst into a chorus of derisive laughter. **1936** F. Clune *Roaming round Darling* ii. 13 A kookaburra laughed sardonically as we passed the Assembly Hall. **1959** *Guardian* 28 Nov. 5/2 The chortled comments of the kookaburra bird. **1966** G. Durrell *Two in Bush* iv. 132 The first of these were three fat young Kookaburras, or Laughing Jackasses as these giant kingfishers are called in Australia.

kook-a-vic, var. *Kokkewiet.

kooky (kŭ·ki), *a. slang.* Also **kookie.** [f. *Kook + -y[1].] Cranky, crazy, eccentric.

1959 *Motion Pictures* Aug. 32/1 Get set for some far-out talk on teen-age romance by the kookiest cat in town—Edd Byrnes. **1961** *Spectator* 15 Sept. 360 I've got this kooky Aunt who reads novels. **1961** *John o' London's* 30 Nov. 615/4 The dialogue is conscientiously kookie. **1962** *Sunday Express* 21 Jan. 17/1 A Hollywood comedienne noted for her 'kookie' performances. **1963** E. L. Wallant *Tenants of Moonbloom* (1964) iv. 48, I feel like dropping the Muse and this kooky life and marrying a nice tired businessman. **1965** [see *Kook 1]. **1971** *Daily Tel.* 21 Aug. 16/1 A 'kooky' young American woman: enthusiastic, energetic, enterprising and 'a bit nuts'. **1973** *Nation Rev.* (Melbourne) 31 Aug. 1452/3 (Advt.), 'No Sex Please, We're British!' The funniest, kookiest night of your life.

So **koo·kily** *adv.*, in a kooky manner; **koo·kiness**, the state of being kooky.
1962 *Sunday Express* 21 Jan. 17/3 Kookiness doesn't go with a kimono. 1968 *Punch* 19 June 899/1 This study of a kooky girl is also..kookily narrated. 1970 *Sudbury* (Ontario) *Daily Star* 26 Feb. 18/4 There's nothing you can do, so accept your mother's kookiness gracefully. Her antics in no way diminish you in the eyes of your friends. 1974 *Observer* 21 Apr. 37/1 'Isadora' kookily takes off on a Freudian odyssey round Europe.

koolack, obs. var. *KULAK.

Kooleen, var. *KULIN.

kooliman, var. *COOLAMON.

koolookamba (kūlūkæ·mbă). Also **kulu-kamba** and shortened form **kulu.** [Native name in Gabon.] A West African variety of the chimpanzee, *Pan troglodytes.*
1860 P. B. Du Chaillu in *Proc. Boston Soc. Nat. Hist.* VII. 360 The cry of the Kooloo-Kamba is very different from that of the *Troglodytes calvus* and chimpanzee. 1896 R. L. Garner *Gorillas & Chimpanzees* 176 The kulu-kamba is..by far the finest representative of his genus. *Ibid.*, A young female kulu. 1929 R. M. & A. W. Yerkes *Great Apes* xxiv. 302 This observer [*sc.* Du Chaillu] means to imply that the three types of chimpanzee called by him Kooloo-Kamba, *Troglodytes calvus*, and chimpanzee, have distinctive calls or cries.

koompass, var. *KEMPAS.

Koonbee, var. *KUNBI.

Kooranko, var. *KORANKO.

koorn kriek, var. *KORINGKRIEK.

kootchar (kū·tʃaɪ). *Austral.* Also **koochee.** [Aboriginal name.] A small, stingless bee, *Trigona australis.*
1884 H. J. Hockings in *Trans. Entomol. Soc. London* 149 The second species [of stingless bee] ('Kootchar') is also black in colour but has a fine yellow streak across the upper part of the thorax. *Ibid.* 154 'Kootchar' are only to be found where a sandy soil is present. 1932 Koochee [see *KARBI]. 1961 *Amer. Mus. Novitates* No. 2026. 5 It was perhaps by error that Hockings attributes this species (under the aboriginal name of *kootchar*) to the coast.

Kootenai, var. *KUTENAI, KUTENAY.

kootie, var. *COOTIE *sb.*²

‖ **kop** (kǫp). [Afrikaans, f. Du. *kop* head, COP *sb.*²] **1.** *S. Afr.* A hill. Cf. KOPJE.
1835 C. L. Stretch *Jrnl.* 1 Apr., The troops..advanced in the direction of T'Slambies Kop, a high point visible from the heights near Graham's Town. 1878 H. A. Roche *On Trek in Transvaal* xiv. 303 One fine Kop or Kopje we passed upon which grazed an immense herd of fine oxen and heifers. 1900 A. H. Keane *Boer States* p. xvii, *Kop*, a crest, an eminence. 1901 L. James in J. Ralph *War's Brighter Side* 347 The three field batteries then came into action against a high tableland *kop* which formed the right of the held position. 1932 C. Fuller *Louis Trigardt's Trek* ii. 26 It is east by north of Taikundo Kop. 1939 tr. *E. N. Marais's My Friends the Baboons* i. 74 On one side the kloof was bordered by a krans, two to three hundred feet high, and on the other by a kop so steep that it could almost be called a krans too. 1971 *Rand Daily Mail* (Home Owner Suppl.) 26 June 5 Two of Johannesburg's most famous 'kops'—Langermann's Kop..and Pullinger Kop.
2. *Assoc. Football.* (With capital initial.) In full, but now less usu., **Spion Kop** (spəi·ŏn kǫp). [f. *Spion Kop*, Afrikaans name of a hill near Ladysmith in S. Afr., scene of a battle in the Boer War (1899–1902).] A high bank of terracing for standing spectators, orig. and especially the one at Anfield, home ground of Liverpool Football Club, but now of more general U.K. application (see quot. 1974²). Also, the spectators themselves, massed on such terracing, and *attrib.*
1926 *Liverpool Daily Post* 29 May 9/3 At last night's meeting of the Liverpool Football Club an important step was made known... The club had decided to..concentrate upon improving Spion Kop at the back of the Oakfield-road goal. 1960 B. Liddell *My Soccer Story* vi. 43 All they [*sc.* the spectators] wanted was the final whistle, so that they could come swarming over the ground from the Kop..and carry us off the field. *Ibid.* viii. 54 Kicking into our favourite goal, at the Spion Kop end. 1966 P. Moloney *Plea for Mersey* 56 There is in Liverpool a school of indigenous verse—known as the Kop choir, which produces, as it were spontaneously, verses to suit every situation that might occur within the game of football. 1966 *Liverpool Echo* (Football ed.) 30 Apr. 1/3 Liverpool went into the lead again in 69 minutes... This set the Kop off again and they gave us pretty well their whole repertoire [of songs and chants]. 1973 B. Shankly in *News of the World Football Ann. 1973–74* 4 For a time it seemed as if the Kop would never stop singing. 1973 *Liverpool Echo* (Football ed.) 22 Dec. 1/5 Liverpool immediately resumed the first half pattern of constant attack, this time towards the Kop goal. 1974 *Times* 18 Mar. 10/8 A match of electric energy..ended with the heaving Kop awash with red banners and scarves. 1974 *Sunday Times* 14 Apr. 26/2 What Anfield sings today, other Kops will be singing tomorrow.
So **Ko·pite, Ko·ppite,** a spectator who frequents the Kop terracing.
1960 B. Liddell *My Soccer Story* viii. 53 He got a tremendous ovation from the generous Koppites at the finish. 1966 P. Moloney *Plea for Mersey* 58 'You'll never walk alone,' the Koppites sing. 1974 *Liverpool Echo* (Football ed.) 6 Apr. 16/9 The president of the Cambridge University Boat Club who comes from Birkenhead is also a loyal Kopite. 1974 *Sunday Times* 14 Apr. 26/2 It makes Liverpool Kopites smile..to hear other fans singing their songs.

kopasetic, var. *COPACETIC *a.*

‖ **kopdoek** (kǫ·pduk). *S. Afr.* [Afrikaans, f. *kop* head, COP *sb.*² + *DOEK cloth.] A head-cloth.
1911 *State* (Cape Town) Dec. 642/2 He deposited his shapeless hat on the floor, tapped his red kopdoek with a clawlike forefinger, and waited for an inspiration. 1957 *Cape Times Week-end Mag.* 6 Apr. 3/5 The Swazis barter the bones for food and clothing..; half a monkey means a new kopdoek. 1974 *S. Afr. Panorama* Feb. 10 In their bright blue-and-pink 'kopdoeks' (headscarves), they lend a colourful note to an already colourful scene.

‖ **kopi** (kōᵘ·pi). [Austral. Aboriginal word.] Gypsum- or selenite-bearing rock or mud.
1898 D. W. Carnegie *Spinifex & Sand* II. ii. 42 We came on a small tract of 'kopi country' (powdered gypsum). *Ibid.* 43 This kopi is peculiar soil to walk over. *Ibid.* III. iii. 91 A sort of powdery gypsum, called 'Kopi' by the natives. 1936 A. Russell *Gone Nomad* ix. 71 The drying bed of the lake was composed of a clinging kopi mud that would have enmeshed a duck. 1957 D. Niland *Call me when Cross turns Over* i. 8 The lie of all that barren land he knew, the shelves, the pits, the mullock dumps of kopi, every mound and every rise and fall. 1971 J. S. Gunn *Opal Terminol.* 24 *Kopi*, gypsum which sometimes carries good opal. In L. Ridge the kopi is different, being flattish crystals of silenite, rather than the gypsum clay of S. Aust. and Qld.

kopje: see *KOPPIE.

Koplik (kǫ·plik). *Med.* [The name of Henry *Koplik* (1858–1927), U.S. pædiatrician.] *Koplik('s) spot*: a small greyish-yellow spot, usually with a red halo, occurring on the buccal mucosa (or sometimes on the intestinal mucosa) in the early stages of measles.
1899 *Med. News* (N.Y.) LXXIV. 734/2 Koplik's spots were seen on the inner surface of the cheeks and under lip. 1939 Duval & Schattenberg *Textbk. Path.* ix. 284 Microscopically the macule or Koplik spot is a dense central collection of lymphocytes surrounded by tissue that is edematous and hyperemic. 1948 H. Pinkerton in W. A. D. Anderson *Path.* xv. 359 Koplik spots have been described in fatal cases in the intestinal mucosa. 1970 Hornstein & Gorlin in Gorlin & Goldman *Thoma's Oral Path.* (ed. 6) II. xvii. 756/2 Koplik's spots result from superficial necrosis of the mucosa and disappear after two to six days.

koppa (kǫ·pă). A letter (ϙ) standing between π and ρ in the early Greek alphabet (= Heb. *koph* ק, Lat. *q*). It was later displaced by κ, but survived as a numeral = 90.
1870 W. W. Goodwin *Elem. Greek Gram.* I. 2 Two obsolete letters—*Vau* or *Digamma*..equivalent to F or W, and *Koppa* (ϙ), equivalent to Q..are used as numerals. 1883 I. Taylor *Alphabet* II. 138 The letters *vau* and *koppa*, which were discarded in the Eastern alphabets, except as numerals, were retained in Latin. 1888 King & Cookson *Princ. Sounds & Inflexion Gr. & Latin* iii. 52 The Phoenician ϙ, *Koppa*, fell into disuse; it survived longest in the alphabet of Chalcis, whence it passed into the Roman alphabet as Q. 1933 C. D. Buck *Compar. Gram. Gr. & Latin* 73 The *wau*, *koppa*, and *san*, which disappeared from the alphabet, were maintained as numeral signs. 1970 *Oxf. Class. Dict.* (ed. 2) 48/1 K was used originally before *a*, C before *e* and *i*, ϙ (Greek Koppa) before *o* and *u*.

koppel (kǫ·pĕl). Also **capel, coppel.** [Yiddish.] A skull-cap worn by male Jews.
1892 I. Zangwill *Childr. Ghetto* I. 118 Old Hyams..had been sitting quiet with brow corrugated under his black velvet *Koppel*. 1972 *Jewish Chron.* 1 Sept. 22/1 (Advt.), For the high festivals: Machsorim—Tallisim—Kittles—Coppels. Unrivalled selection of New Year cards. 1973 *Ibid.* 9 Feb. 22/5 Some pious Jews..make a point of wearing a hat and not a 'capel' for prayer and some Chasidim wear both a hat and 'capel' underneath it.

koppie (kǫ·pi). Now a common spelling of KOPJE.
1848 R. Gray *Jrnl.* 14 Nov. (1849) 76 Large dreary plains interrupted by rocky koppies, abounding with the springbok and the gnu. 1850 N. J. Merriman *Cape Jrnls.* (1957) 136 This will account for our retiring behind a koppie to pitch our tent. 1853 *Edin. New Philos. Jrnl.* LV. 79 The bush..covers over all inequalities of ground..and makes almost every kloof and koppie exactly resemble each other. 1881 [see KOPJE]. 1884 E. P. Mathers *Trip to Moodie's* 143 He was prospecting about and tried the quartz at the bottom of the koppie... The koppie dips into a gully on each side of it. 1926 O. Schreiner *From Man to Man* viii. 290 You will creep on hands and knees over rough koppies. 1957 *Times* 19 Nov. 12/7 In the course of my afternoon walk over the koppies one day I came upon two of my pupils engaged in a strange game. 1966 E. Palmer *Plains of Camdeboo* v. 74 The low ridge of koppies overlooking the gap. 1971 *Rand Daily Mail* (Home Owner) 27 Mar. 9/2 It is a hilly area and there are the Melville koppies to add something extra to the area. 1971 *World Archaeol.* III. 182 With few exceptions, however, smelting sites are not found on the koppies themselves.
Comb. **koppie** (also **kopje**) **walloper,** a diamond-buyer. *Hist.*
1886 G. A. Farini *Through Kalahari Desert* ii. 21 The wily Jew was a 'partner' in a 'company' of ten 'Koppje wallopers'. 1897 H. Raymond *B. I. Barnato* i. 14 The slang camp term..for this [*sc.* diamond-buyer] was 'kopje walloper', derived from the circumstance that in the earliest days the diamonds were obtained from a number of kopjes or small hills in the neighbourhood of the camp, and the dealers travelled on foot from one to the other purchasing the finds as they were turned out at the sorting tables. 1947 *Cape Argus* 20 Dec. 2/5 A 'koppie walloper' was a diamond buyer who went from claim to claim buying stones. The name was used in Kimberley in the early days. 1949 K. L. Simms *Sun-Drenched Veld* viii. 70 Profits dwindled quickly, especially as native workmen stole many diamonds, which they sold to unscrupulous 'kopje wallopers' who made enormous profits without having done any hard work. 1955 E. Rosenthal in Saron & Hotz *Jews in S. Afr.* vi. 114 A 'koppie-walloper', that is one who went from claim to claim buying diamonds as the diggers produced them. 1967 E. M. Slatter *My Leaves are Green* 38 Now don't sell it to any of these kopje-wallopers.

kora¹ (kōᵊ·ră). Also **cora, korro.** [Native name.] A West African stringed instrument resembling a harp.
1799 M. Park *Trav. Afr.* xxi. 278, I have now to add a list of their musical instruments, the principal of which are,—the *koonting*, a sort of guitar..; the *korro*, a large harp, with eighteen strings; [etc.]. 1874 C. Engel *Descr. Catal. Musical Instruments S. Kensington Museum* (ed. 2) 151 Mungo Park enumerates, among the popular instruments which he saw in Senegambia, the *e-korro*. 1935 G. Gorer *Africa Dances* iv. 305 The drums are sometimes supplemented and occasionally supplanted by..the cora, a six-stringed instrument somewhere between a guitar and a harp. 1965 *Economist* 16 Jan. 229/3 The Senegal national anthem..begins..'Strum your koras, strike the balafous'. 1970 P. Oliver *Savannah Syncopators* 47 The great harp-lute, called the *kora*.

Kora² (kōᵊ·ră). Also **Coranna, Koran(n)a,** occas. **Koraqua.** [Of disputed origin.] **a.** Any of a group of Hottentot peoples in southern Africa; also, a member of any of these groups. **b.** The language spoken by them. Also *attrib.* or as *adj.*
1801 J. Barrow *Acct. Trav. S. Afr.* I. vi. 403 The country to the eastward of the Roggeveld, is inhabited by different hordes of Bosjesmans. One of these, called the *Koranas*, dwelling on the right bank of the Orange river, directly east from the Roggeveld, is represented as a very formidable tribe of people. 1806 —— *Voy. Cochinchina* 373 The native inhabitants which are settled on the banks of the Orange river..are a variety of the Hottentot race..called the *Koras*. *Ibid.*, What the Gonaquas were on the eastern coast the *Koras* seem to be to the northward, a mixed breed between the Hottentot and the Kaffer. 1824 W. J. Burchell *Trav. S. Afr.* II. 251 The following specimen of the Kora, or Koraqua, dialect, was obtained. 1831 *Graham's Town Jrnl.* 30 Dec. 3 At a Koranna kraal.., the first cases of Small Pox presented themselves. 1871 J. Mackenzie *Ten Yrs. North of Orange River* 493 A certain word in Koranna, if pronounced in a loud key, means *handkerchief*. *Ibid.* 501 The insecure villages of the pastoral Korannas. 1881 *Encycl. Brit.* XII. 312/2 The Kora dialect, spoken by the Korannas, or Koraquas, dwelling about the middle and upper part of the Orange, Vaal, and Modder Rivers. 1936 J. A. Engelbrecht *Korana* I. 2 The history of the origin of the Korana can only be approximately arrived at. *Ibid.* II. 83 Kora tribes..left the Cape to seek new pastures. *Ibid.* III. 197 A complete linguistic survey of all the areas in which Kora..is still spoken at the present day could not be undertaken. 1955 J. H. Wellington *S. Afr.* II. xvi. 234 The tribes occupying the Cape Peninsula and adjacent areas at the time of Van Riebeeck's arrival were the Goringhaiqua and the Kora (later known as the Koranna). 1961 *Encycl. S. Afr.* 278/1 Today pure Koranas are almost extinct. 1968 *Encycl. Brit.* XI. 751/1 Hottentot is the European name for the Nama, Kora and other languages comprising 14 or 15 subdivisions of the main Hottentot speech. 1974 [see *HOTTENTOT 1].

kora, var. *KOURA.

koraddy, koradi, varr. *KORARI.

‖ **koradji** (kǫ·rădʒi). *Austral.* Also 8 **carradygan, carrahdy,** 9 **coradge(e, karaji, kiradjee, korradgee.** [Austral. Aboriginal word.] Amongst the Australian Aborigines, a medicine-man.
1793 J. Hunter *Hist. Jrnl.* xxi. 523 Having taken leave of their new friends the *Car-ra-dy-gans* (doctors), our party set off. 1798 D. Collins *Acct. Eng. Colony New South Wales* I. 594, I think I may term the car-rah-dy their high priest of superstition. 1845 J. O. Balfour *Sk. New South Wales* 14 The coradgees, who are their wise men, have, they suppose, the power of healing and fore-telling. Each tribe possesses one of these learned pundits, and if their wisdom were in proportion to their age, they would indeed be Solons. 1865 W. Howitt *Hist. Discovery in Austral.* I. 287 One who seemed a coradge, or priest, went through a strange ceremony of singing, and touching his eyebrows, nose and breast, crossing himself, and pointing to the sky like an old Druid. 1867 S. Bennett *Hist. Austral. Discovery* 250 *Kiradjee*, a doctor. 1885 R. M. Praed *Austral. Life* 23 The korradgees, or medicine men, are the chief repositories (of the secrets of their religion). 1892 J. Fraser *Aborigines New South Wales* 63 For some diseases, the karáji, or native doctor when he is called in, makes passes with his hand over the sick man, much in the same way as a mesmerist will do. 1966 [see *BOYLA].

Koranko (kŏræ·ŋko). Also **Kooranko.** [Native name.] **a.** Name of a West African people; also, a member of this people. **b.** The language of this people.
1825 A. G. Laing *Trav. W. Afr.* iv. 200 The manner of courtship among the Korankos is exactly the same as among the Timannees. *Ibid.* 208 Dancing is a prominent feature among the amusements of the Koorankos. 1883 R. N. Cust *Sk. Mod. Lang. Afr.* I. xi. 184 There appear to be Dialects [of Temne], the Quiah,..and the Koranko. 1957 M. Banton *W. Afr. City* vii. 127 Koranko is a closely related dialect [to Mandinka]... The Koranko and Kono inhabit the high lands to the north-east and east. 1962 *Times* 3 Mar. 10/6 The man is a Koranko of west Africa.

Koran(n)a, Koraqua: see *KORA².

korari (kōᵘ·rari). *N.Z.* Also **coorraddie, kauradi, kolladie, koraddy, koradi, kraddy.** [Maori.] The flower stalks of the New Zealand flax, *Phormium tenax*; also occasionally used for the plant itself.
1832 G. Bennett in *London Med. Gaz.* 18 Feb. 750/2 *Phormium Tenax*, or flax plant. *Koradi* of the natives of New Zealand... This valuable plant is indigenous to New Zealand. 1834 E. Markham *N.Z. or Recollections of It* (MS.) 44 Koraddy or Flax undrest split green and worked up. 1840 J. S. Polack *Manners & Customs New Zealanders* I. xvi. 175 The flax (*Korari*)..is of two or three kinds. 1843 *N.Z. Jrnl.* No. 92 177/2 With a piece of kauradi or flax. 1845 E. J. Wakefield *Adventure N.Z.* I. iii. 63 The plant is called *phormium tenax* by naturalists. The general native name for the plant..was *korari*. 1873 V. Pyke *Story Wild Will Enderby* (ed. 4) III. iii. 115 A 'mohiki' is constructed of 'koradies'—*Anglice*, the flowering stalks of the flax. 1879 J. Barr *Old Identities* 53 A 'kolladie' (the flower-stalk of the flax, about 7 ft long) was carried by each as a balancing pole or staff. 1888 A. H. Duncan *Wakatipians* ii. 12 We collected all the dried *coorraddies* (flower stalks of the flax-bush). 1933 *Press* (Christchurch, N.Z.) 19 Oct. 15/7 The individual plants are called *flax bushes* and the dried reed stalks (which Maoris and children tied into bundles to make rafts) are called *flax-sticks*, *koradi sticks*, or *kraddy sticks*. 1949 P. Buck *Coming of Maori* (1950) II. ix. 238 Children armed with a flax flower stalk (*korari*) were taught to spar by their parents.

Kore (kōᵊ·re). Also 9 **Cora, Core.** [Gr. κόρη, = maiden.] **1.** In Greek mythology, the daughter of Zeus and Demeter, known as Persephone when she married Hades.
1844 L. Schmitz in W. Smith *Dict. Gr. & Roman Biogr. & Mythol.* I. 852/1 Core..the maiden, a name by which Persephone is often called. 1849 *Ibid.* III. 204/2 In the mysteries of Eleusis, the return of Cora from the lower world was regarded as the symbol of immortality. 1885 *Encycl. Brit.* XIX. 824/1 Proserpine herself was commonly known as the daughter (Core). 1916 E. Pound *Lustra* 18 Korè is seen in the North Skirting the blue-gray sea. 1930 —— *XXX Cantos* xvii. 79 Korè through the bright meadow, with green-gray dust in the grass. 1968 *Encycl. Brit.* XVII. 644/2 One of the variants of the name Persephone is..a form with a pre-Hellenic suffix that suggests that she was a pre-Hellenic deity..a goddess of the dead. If this is so, her connection with Demeter came later, when she was identified with Kore (the Greek word for 'maiden')..daughter of Demeter; Demeter and Kore were the leading goddesses in the Eleusinian mysteries.
2. *Gr. Sculpture.* (Often with lower-case initial.) A statue of a draped maiden.
1920 *Q. Rev.* July 28 There are, in our Archaic Room, good examples both of the *Kore* and her male companion the *Kouros*. 1934 *Burlington Mag.* Jan. 46/1 The archaic bronzes include..an Ionian kore..and a warrior. 1950 H. L. Lorimer *Homer & Monuments* vi. 356 This arrangement can be plainly seen on a Kore from the Acropolis. 1971 *Ashmolean Mus. Rep. of Visitors 1970* 18 A terracotta scent vase in the form of a standing kore, Rhodian, late 6th century B.C.

Korean (kŏrī·ăn), *a.* and *sb.* Also 7–9 **Corean.** [f. *Korea* country in Eastern Asia + -AN.] **A.** *adj.* Of or pertaining to Korea, since 1954 divided into the *Republic of Korea* (capital, Seoul) in South Korea and the (Communist) *Korean People's Republic* (capital, Pyongyang) in North Korea (respectively S. and N. of the 38th parallel). **B.** *sb.* **a.** A native or inhabitant of Korea. **b.** The agglutinative language of Korea, which is related to Japanese.
1614 R. Cocks *Let.* 25 Nov. in *Diary* (1883) II. 270 He was prevented by a Corean Noble-man. 1727 J. G. Scheuchzer tr. *Kæmpfer's Hist. Japan* I. 63 The Coreans had been subdued. *Ibid.* 76 Encompass'd by the Corean sea. 1813 *Q. Rev.* Oct. 256 Classes and Families of Languages... Tartarian... Corean. 1822 F. Shoberl tr. *Titsingh's Illustr. Japan* 323 A Corean Fisherman and his Wife. 1885 E. W. Hamilton *Diary* 11 Apr. (1972) II. 834 The idea was broached in the Cabinet of endeavouring to make arrangements with the Chinese and Japanese Governments for our occupying Port Hamilton in Korean territory in the event of war with Russia. 1899 A. H. Keane *Man Past & Present* viii. 307 An adaptation of the Chinese symbols to the phonetic expression of the Korean syllables. 1921 [see *ANNAMITE *a.* and *sb.*]. 1966 *Listener* 24 Mar. 423/2 The Korean people..are genuine Tartars. *Ibid.*, Korea proves that it is possible for a poor Asian country to hope for both bread and freedom. The Koreans value that freedom. 1967 D. S. Parlett *Short Dict. Lang.* 70 Agglutinative structure, like Korean, unlike Chinese. 1972 *Korea Times* 19 Nov. 1/6 Written tests in English, Korean and general knowledge will be held.
C. Korean chrysanthemum, a late-flowering hybrid chrysanthemum first developed from *Chrysanthemum coreanum* by A. Cumming, American nurseryman, about 1930; also *ellipt.*; **Korean pine,** a slow-growing pine with dark green leaves, *Pinus koraiensis.*
1931 *Horticulture* 15 Sept. 392/1 No list of Fall flowering perennials would be complete if the Korean chrysanthemum were omitted. When the flowers of this splendid novelty open, they are a pure white color with a chrysanthemum gold center, but as the flowers mature they assume a claret pink shade. 1936 K. Luxford *Culture of Chrysanthemum* 72 The hardy hybrid Korean chrysanthemum is one of the most notable acquisitions to the border of recent years. 1938 A. E. Wright *Outdoor Chrysanthemums* ix. 87 The Koreans have made a tremendous impression in America. 1961 *Amat. Gardening* 30 Sept. 9/2 Some of the later flowering pompon chrysanthemums and Koreans I am lifting and putting into pots. 1866 'Senilis' *Pinaceæ* 115 Pinus Koraiensis: The Corean Pine. 1914 W. J. Bean *Trees & Shrubs Hardy in Brit. Isles* II. 182 *P[inus] koraiensis*, Siebold. Corean Pine..introduced by J. G. Veitch in 1861. 1969 T. H. Everett *Living Trees of World* 55/1 The Korean pine (*P. koraiensis*) occurs both in Korea and Japan, and under favorable circumstances it may grow to 150 feet high.
Hence **Kore·anize** *v. trans.*, to give a Korean character to.
1930 W. F. Sands *Undiplomatic Memories* (1931) 70 Emily, Koreanized as Lady Om, was now reigning sweetly in Seoul. 1972 *Korea Times* 16 Nov. 1/7 A new turning-point in the realization of a 'Koreanized democracy'.

Koreish (kŏrəi·ʃ). Also 7–9 **Coreis(h),** 9– **Koreisch, -eysh, Ḳoraish, -aysh, Ḳur-, Quraysh,** etc. [Arab. *ḳuraiš* Koreish, *ḳuraši* Koreishite.] An Arabic tribe living around Mecca, to which Muhammad belonged; also, a member of this tribe. Also *attrib.* or as *adj.*
1649 A. Ross tr. *Alcoran* Introd., He declaimeth against such as worship Idols, particularly against the Inhabitants of the City of Mecca, and against the Coreis, who were enemies to his designe. 1734 G. Sale tr. *Koran* Prelim. Disc. i. 25 There were several dialects of it [*sc.* the Arabic language], very different from each other: the most remarkable were that spoken by the tribe of Hamyar and the other genuine Arabs, and that of the Koreish... The dialect of the Koreish is usually termed the pure Arabic. 1856 R. F. Burton *Personal Narr. Pilgrimage to El-Medinah* III. xxvi. 157 It was closed by the Kuraysh when they rebuilt the house in Mohammed's day. 1858 W. Muir *Life of Mahomet* I. I. iii. 197 The likeliest is the meaning '*noble*'; but it is also possible that the Coreish..may have conferred upon the word that meaning... Again, it is derived from a metaphorical resemblance to *Coreish*, the name of a fish which eats up all others; or to *cursh*, a high-bred camel. Others refer it to a root which signifies *to trade*. 1861 J. M. Rodwell tr. *Koran* p. viii, Othman..entrusted the redaction to..Zaid.., with whom he associated as colleagues, three, according to others twelve of the Koreisch, in order to secure the purity of that Meccan idiom in which Muhammad had spoken. 1871 C. M. Yonge *Pioneers & Founders* iv. 81 Sabat was an Arab of the tribe of Koreish, the same which gave birth to Mahomet himself. 1880 E. H. Palmer in *Sacred Bks. East* VI. p. lix, Zâid..with three men of the Qurâis (Mohammed's own tribe),..to fix the reading definitely according to the pure Qurâis idiom. 1907 D. S. Margoliouth tr. *J. Zaydan's Umayyads & Abbâsids* 11 The poet Ibn Harmah..was an adoptive member of the clan Khalj which was itself adoptive in the tribe Kuraish. *Ibid.* 14 The Kuraish dealt in slaves as they did in other merchandise. 1916 J. Buchan *Greenmantle* ii. 23 To capture all Islam..the man must be of the Koreish, the tribe of the Prophet himself... There are families..that claim Koreish blood. 1957 *Encycl. Brit.* XV. 646/2 A member of the tribe Koreish.., Muhammad is said to have been a posthumous child. 1959 T. B. Irving in Kritzeck & Winder *World of Islam* II. 189 The Quraysh, centring around Makkah, used the purest Arabic of all; the ratio of any tribe's distance from the Quraysh was a good index for ascertaining the purity or corruption of any Arab dialect.
So **Korei·shite, Qurayshite,** etc., a member of the Koreish tribe.
1708 S. Ockley *Conquest Syria, Persia & Ægypt by Saracens* I. sign. DD5 verso, Koreishæ, or Korashites, a noble Tribe among the Arabs, of which Mahomet was. 1734 G. Sale tr. *Koran* Prelim. Disc. vi. 145 The spoil..he bestowed..on the Meccans only.., highly distinguishing the principal Korashites, that he might ingratiate himself with them. 1871 C. M. Yonge *Pioneers & Founders* iv. 81 He formed a close friendship with his colleague, Abdallah, likewise a Koreishite Arab. 1903 D. B. Macdonald *Devel. Muslim Theol.* App. i. 305 The unlettered Qurayshite prophet, Muhammad. 1968 *New Larousse Encycl. Mythol.* (new ed.) 323/1 The goddess El-'Ozza was also held in high honour among the Koreishites.

koreke (kore·ke). *N.Z.* [Maori.] An extinct native quail, *Coturnix novæzealandiæ.*
1871 F. W. Hutton *Catal. Birds N.Z.* 22 Native Quail. Koreke. Black, streaked with white, and varied with reddish brown on the back. 1882 W. D. Hay *Brighter Britain!* II. 223 The Koreke..the native quail. 1930 W. R. B. Oliver *N.Z. Birds* 374 New Zealand Quail. Koreke. 1966 R. A. Falla et al. *Field Guide Birds N.Z.* 100 Galliformes. The only indigenous member of this order, *Coturnix novaezealandiae*, Koreke of the Maoris, became extinct about 1870.

‖ **korero** (kōᵘ·reᵊro). *N.Z.* Also 9 **corero,** etc. [Maori.] Talk, conversation, discussion; a conference.
1807 J. Savage *Some Acct. N.Z.* xi. 75 Corero..Speaking. 1834 E. Markham *N.Z. or Recollections of It* (MS.) 4 A grand Corrirow or talk. 1845 E. J. Wakefield *Adventure N.Z.* I. 78 There were about sixty men assembled, and they proceeded to hold a *korero* or 'talk' on the all-important subject. 1855 C. W. Richmond *Jrnl.* 20 Aug. in *Richmond–Atkinson Papers* (1960) I. 176 They had a long korero about New Zealand and the Bay, where the Doctor has been quartered for 5 years. 1863 T. Moser *Mahoe Leaves* 30 He had to pass several pahs on the road, at all of which there would be 'koreros'. 1883 J. P. Ward *Wanderings with Maori Prophets* ii. 13 He was brim full of talk, or *korero*. 1930 B. Guthrie *N.Z. Memories* 54 After some *korero* and polite speeches the Maoris stood aside. 1936 'R. Hyde' *Check to your King* xx. 236 The old and conspicuous means of the *korero* as a call to arms. 1938 R. D. Finlayson *Brown Man's Burden* 13 There was to be a korero on politics and the revival of Maori art. 1960 *Weekly News* (Auckland) 22 June 58/3 After New Zealand's first test defeat in South Africa in 1960 by 13–0 all members of the touring team had a korero the next morning. 1971 *N.Z. Listener* 22 Mar. 7/2 The *korero*..would let the Maoris tell the pakehas what they couldn't so easily tell them in a setting shaped by European conventions.

Korewah, var. *KORWA.

korfball (kǫ·ɹfbǫl). Also **korfbal.** [Du. *korfbal*, f. *korf* basket + *bal* ball.] A game of Dutch origin, resembling basketball and netball. Also *attrib.*
1915 J. Feith *Sport in Netherlands* 27 '*Korfbal*', a game not known in England... The ball, which is as big as that used for football, is thrown.., and the place of the goal is taken by a bottomless basket fastened to a pole. 1960 *Sunday Times* 5 June 40/6 Holland beat England 10–8 in a korfball international at Mitcham, Surrey, yesterday, after leading 7–3 at half-time. 1965 J. B. Pick *Phoenix Dict. Games* (rev. ed.) 89 *Korfball* (24 players)..is a net-ball-basketball game of Dutch origin..usually played by teams of six ladies and six gentlemen. Starting humbly in 1902..in the Netherlands.., the game was introduced into Britain..in 1946. *Ibid.* 90 There are now ten British korfball clubs. 1969 D. C. Joynson *Guide for Games* (ed. 2) vi. 157 Korfball Netball. The court is marked as for Netball. Team—5 players. *Ibid.* viii. 211 *Indoor Korfball.* Two teams of 6 players. The pitch may be (a) a basket-ball court; (b) a netball court; (c) a rectangular court 60 to 80 ft. by 30 to 40 ft. 1973 *Observer* 7 Oct. 26/1 There is only one mixed team sport in the world: korfball, and the first of its summarised rules is that you can only hinder someone of the same sex.

korhaan (korhā·n, kŏrā·n, kɹrā·n). *S. Afr.* Also **knorhaan, knoraan, koerhaan, koorhaan, korhaen,** etc. = KORAN². (For 'genus *Eupodotes*, esp. *E. afra*' in definition, substitute 'the family Otididæ'.)
1731 G. Medley tr. *Kolb's Present State Cape of Good-Hope* II. 139 The Knorhan. Among the Wild Fowls at the Cape, there is a Sort of Birds, a Male of which the Europeans there call Knor-Cock: a Female they call Knor-Hen. 1775 [see KORAN²]. 1785 G. Forster tr. *Sparrman's Voy. Cape Good Hope* I. iv. 153 *Knorrhane* is the name of a kind of *Otis*. 1786 *Ibid.* (ed. 2) I. iv. 153 *Korr-haen* is the name of a kind of *Otis*. 1801 J. Barrow *Acct. Trav. S. Afr.* I. iv. 264 A new species of Korhaen or bustard was seen here. 1819 [see KORAN²]. 1867 E. L. Layard *Birds S. Afr.* 287 The Knorhaan (lit. Scolding Cock) is abundant throughout the whole colony, frequenting the open country. 1876 H. Brooks *Natal* 137 The koraan or knoraan of the Dutch settlers (*Eupodotis scolopacea*). 1896 H. A. Bryden *Tales S. Afr.* 250 The bush *koorhaan*..are playing their strange aerial pranks. 1906 W. L. Sclater *Birds S. Afr.* IV. 291 The Red-crested Knorhaan is found singly or in pairs. 1936 *Blackw. Mag.* Mar. 296/2 The country was still rolling grass, with little game except koerhaan, which were plentiful everywhere. 1936 E. L. Gill *First Guide S. Afr. Birds* 150 The 'Korhaans' are the smaller-sized Bustards. 1964 P. A. Clancey *Birds Natal & Zululand* 143 The blue korhaan occurs sparingly in the interior of Natal. 1972 L. G. Green *When Journey's Over* (1973) xii. 143 Casserole of guinea fowl, stewed *korhaan* and pigeon and Namaqua partridges were as familiar at Klipfontein as grouse in the Savoy Grill.

kori¹ (kōᵊ·ri). *S. Afr.* [Sechuana.] In full, *kori bustard* = *GOMPAAUW.
1811 W. J. Burchell *Jrnl.* 27 Oct. in *Trav. S. Afr.* (1822) I. xvi. 393 We shot a large bird of the bustard kind... The present species, which is called *Kori* in the Sichuana language, measured, in extent of wing, not less than seven feet. *Ibid.*, A representation of the head of the Kori Bustard..is given at the end of this chapter. 1872 L. Lloyd tr. *Andersson's Notes Birds of Damara Land* 258 Kori Bustard... This splendid bird is found throughout the year in Damara. 1889 H. A. Bryden *Kloof & Karroo* xvii. 306 Among the bustards stands pre-eminent the great Kori bustard—the gom paauw of the Dutch colonists. 1962 [see *GOMPAAUW]. 1966 E. Palmer *Plains of Camdeboo* xii. 209 The Kori bustard, a dweller of the desert or semi-desert, a gigantic creature weighing up to fifty pounds or more, heavier than the great bustard of Eurasia which is usually held to be the heaviest bird of the air. It has a wing span of up to eight feet, and it stands nearly five feet high.

Kori² (kōᵊ·ri). Also **Koiri, Koli, Koree.** Name of a tribe of low-caste Hindu weavers of northern India; also, a member of this tribe.
1839 J. Tod *Travels in W. India* vii. 143 The chief fragment of this once superb monument is enclosed and half-hidden by the huts of Koli weavers. *Ibid.* xvii. 361 The rest is made up of the agricultural and artizan classes, as Aheers, Kolis, &c. &c. 1868 *Rep. Ethnol. Comm. Specimens Aborig. Tribes Jubbulpore Exhib. 1866–67* I. 11 The Mahars, Korees, and other weaving clans. 1873 *Cycl. India* (ed. 2) III. s.v. *Kori*, All the weaver caste throughout Hindoostan are stated by Colonel Tod to be Koli. They call themselves Julai, but are sometimes styled Kori. 1885 J. C. Nesfield *Brief View Caste Syst. N.-W. Prov. & Oudh* xi. 106 The weaver caste is better known to this day by the tribal name of Kori than by the functional names of Bunkar or Joria. 1896 J. N. Bhattacharya *Hindu Castes & Sects* XI. ii. 233 The Kori and Koli of Northern India are weavers professing the Hindu faith; but they are very low castes. 1896 W. Crooke *Tribes & Castes N.-W. Prov. & Oudh* III. 316 Kori, the Hindu weaver caste. *Ibid.* 318 The Koris are all Hindus... The status of the Kori is very low. 1916 R. V. Russell *Tribes & Castes Cent. Prov. India* IV. II. 133 Bodies of the Kori and Katia weaving castes of northern India have been amalgamated with the Mahārs. 1957 G. S. Ghurye *Mahadev Kolis* i. 6 Koris or Koiris are known over a large part of Northern India. I do not know that any responsible student of Indian ethnography has suggested an ethnic connection between Kori and Koli.

Koriac(k), Koriak, varr. *KORYAK.

korimako (kǫrimā·ko). *N.Z.* [Maori.] A New Zealand honey-eater, the bell-bird, *Anthornis melanura.*
1855 R. Taylor *Te Ika a Maui* 75 In the first oven a korimako was cooked. 1863 A. S. Atkinson *Jrnl.* 29 Sept. in *Richmond–Atkinson Papers* (1960) II. 63, I lay on my back..listening dreamily to the birds, just waked up. The Korimakos in full chorus in the bush and the larks and matatas in the open land about us. 1873 W. Buller *Hist. Birds N.Z.* 94 Certain forest-ranges were famed as Korimako preserves. 1888 *Trans. N.Z. Inst.* XXI. 213 In fine weather the bush along the south shores of Lake Brunner re-echoes with the rich notes of the tui and korimako. 1930 W. R. B. Oliver *N.Z. Birds* 487 Bell Bird. Korimako. *Anthornis melanura.* 1966 *Encycl. N.Z.* I. 193/2 Common Maori names for the species [*sc.* the bell-bird] are korimako or makomako.

korin¹ (kōᵊ·rin). [Local name.] A small West African gazelle, *Gazella rufifrons.*
[1846 J. E. Gray in *Ann. & Mag. Nat. Hist.* XVIII. 214 Senegal Gazelle. *Gazella rufifrons*... The Corinne, F. Cuv[ier]..not of Buffon.] 1852 *Catal. Mammalia Brit. Mus.* III. 60 *Gazella rufifrons.* The Korin. 1893 R. Lydekker *Horns & Hoofs* v. 232 The korin (*G. rufifrons*), from Senegal, is a species of fully 24 in. in height, distinguished by the uniformly sandy yellow colour of the central streak on the face, and the absence of any tufts of hair on the knees. *Ibid.* 407/2 (index) Korin gazelle. 1962 M. Burton *Syst. Dict. Mammals of World* 255 Common Gazelles of E. Africa include..Korin or Red-fronted Gazelle (*G. rufifrons*), Senegal to Sudan.

Kōrin² (kōᵊ·rin). The name of the Japanese artist Ogata *Kōrin* (1658–1716), used *attrib.* in *Kōrin school, style*, to denote a school of Japanese painting, founded in the Edo period and associated chiefly with Kyoto, of which Kōrin was the greatest exponent.
[1884 Satow & Hawes *Handbk. for Travellers Cent. & N. Japan* (ed. 2) 96 The only new school that appeared in the seventeenth century was that of Kōrin, a famous lacquer painter, who appears to have been originally a pupil of the Tosa school.] 1898 M. Tomkinson *Jap. Collection* II. 113 The Kōrin style was a late offshoot of the Yamato-Tosa school..stamped by a bold flowing line and vigorous composition, and usually by a supreme contempt for naturalistic rules. 1909 L. Binyon *Jap. Art* v. 34 Here we have the first attempts at a fusion of the two styles, Chinese and Japanese, which reached its final development in the Korin school. 1912 E. F. Fenollosa *Epochs Chinese & Jap. Art* II. xiv. 129 We can call the chief masters of this Korin school the greatest painters of tree and flower forms that the world has ever seen. 1970 *Oxf. Compan. Art* 632/2 The Kōrin style represented a reversion to classical Japanese tradition.

koringkriek (kōᵊ·riŋkrīk). *S. Afr.* Also **koorn kriek.** [Afrikaans, f. Du. *koren* corn + *kriek(en* to chirp.] One of several long-horned grasshoppers of the family Tettigoniidæ.
1913 C. Pettman *Africanderisms* 277 *Koorn kriek*,..an insect belonging to the *Locustidæ*; it is very destructive to pumpkins, mealie cobs, etc., and does at times great damage to crops. 1954 S. H. Skaife *Afr. Insect Life* ii. 30 Many people in South Africa fear the koringkrieks because they have the reputation of being poisonous. 1955 D. Jacobson *Trap* iii. 73 The only sound was the shrill call of the koringkrieks, rasping frenziedly in one dry scream. 1966 E. Palmer *Plains of Camdeboo* i. 13 The koringkrieks lurching on immense and crooked legs.

Korku (kǭ·ɹku). Also **Korkoo, Kurku. a.** Name of a Kolarian tribe of the Central Provinces of India; also, a member of this tribe. **b.** Their language. Also *attrib.* or as *adj.*
a 1863 S. Hislop *Papers Aborig. Tribes Cent. Provinces* (1866) App. p. xi, All Kurkus are of one caste. They eat from the hands of Hindus, but not from Gonds or Mahars. 1868 *Rep. Ethnol. Comm. Specimens Aborig. Tribes Jubbulpore Exhib. 1866–67* III. 17 Vocabulary of Korkoo words. 1874 Watson & Kaye *People of India* VII. 404 Korkoos are more migratory than Gonds. 1906 G. A. Grierson *Ling. Survey India* IV. 167 There is only one sub-dialect of Kūrkū, the so-called Muwāsī... It does not differ much from ordinary Kūrkū. 1908 H. H. Risley *People of India* ii. 100 Totems found among sixteen castes and tribes, including..the primitive Gonds, Korkus, and Orāons. 1950 N. S. Saigal in A. V. Thakkar *Tribes India* 74 Among Korkus, 3% are Christian converts and 1·5% are Muslim converts. 1961 *Amer. Speech* XXXVI. 223 Korku phonology and morphophonemics... (Analysis of Korku with particular emphasis on low tone-*cum*-aspiration and stress..). 1962 J. J. Gumperz in J. A. Fishman *Readings Sociol. of Lang.* (1968) 468 Similarly in India the North Dravidian tribal languages and Munda languages such as Korku are found deep in the Indo-Aryan territory.

kornelite (kǭ·ɹnĕləit). *Min.* [ad. Hung. *kornelit* (J. Krenner 1888, in *Magyar tudományos akad. értes.* XXII. 131), f. the name of *Kornel* Hlavacsek, who found the original specimen: see -ITE¹.] A violet to pale pink hydrated ferric sulphate, $Fe_2(SO_4)_3.7H_2O$.
1892 E. S. Dana *Dana's Syst. Min.* (ed. 6) 957 Kornelite... Stated to be a hydrous ferric sulphate. 1910 *Mineral. Mag.* XV. 429 Rhomboclase..occurring together with szomolnokite..and other iron sulphates (kornelite, copiapite, coquimbite, &c.) at Szomolnok, Hungary. 1937 *Amer. Mineralogist* XXII. 569 The mineral designated as kornelite is a lower hydrate of normal ferric sulphate than coquimbite. 1965 *Mineral. Abstr.* XVII. 180/2 The minerals kornelite..from Tintic Standard mine, Utah.., coquimbite.., and quenstedite.., have been re-examined.

‖ **koro** (kōᵊ·ro). Also **kora.** [Jap. *kōro* incense-pot, censer.] An elaborate Japanese vase, usu. of bronze, jade, or porcelain, in which incense is burned.
1822 F. Shoberl tr. *Titsingh's Illustr. Japan* II. 234 *Koro*, a fire-terrine. *Ibid.* 228 A..chafing-dish, with its *koro*, or stand, for burning..incense. 1889 M. B. Huish *Japan & its Art* vii. 89 No notice of the contents of a Japanese house would be complete without some reference to the incense-burners (*ko-ro*) which find a place there, and also in the Buddhist temples. These afforded employment for a large number of artists in bronze. 1891 Chamberlain & Mason *Handbk. Travellers Japan* (ed. 3) 13 The *Kōro*, or incense-burner, generally in bronze or porcelain. 1959 *Times* 28 Apr. 20/6 A spinach green jade koro and cover. 1974 *Daily Tel.* 4 June 16/6 The koro is carved in archaic style, has large horned dragon handles and is surmounted by a Buddhistic lion and a cub.

koromiko. For '*Veronica salicifolia*' substitute '*Hebe salicifolia*'. Add later examples.
1921 H. Guthrie-Smith *Tutira* iv. 26 Flourished green tutu..and Koromiko (*Veronica salicifolia*). 1933 *Bulletin* (Sydney) 9 Aug. 21/2 The shoots of the koromiko were chewed and swallowed for dysentery. 1963 S. Ashton-Warner *Teacher* 76 The idea has its duplicate often enough outdoors—in the right-angled leafing arrangements and the pairs of leaves set exactly opposite each other as in the Koromiko.

koroplast, var. *COROPLAST.
1888 *Amer. Jrnl. Archæol.* IV. 417 The independence and capriciousness which characterize the work of Greek koroplasts. 1950 H. L. Lorimer *Homer & Monuments* v. 163 Greek soldiers of fortune would now and again drift to Cyprus and furnish a subject for local koroplasts. 1952 *Antiquity* XXVI. 104 The Tarsus koroplasts were more closely allied..with those of Alexandria than with the more sophisticated craftsmen of Myrina and Tanagra.

korora (ko·rora). *N.Z.* [Maori.] The southern blue penguin, *Eudyptula minor.*
1871 F. W. Hutton *Catal. Birds N.Z.* 53 *Eudyptula minor*... Blue Penguin. Korora. 1905 W. B. *Where White Man Treads* 62 A smaller canoe and men..crowding into it..like 'korora' (penguins) awaiting the tide. 1966 R. A. Falla et al. *Field Guide Birds N.Z.* 22 Southern Blue Penguin. Maori name: Korora.

‖ **korowai** (kǫ·rowai). *N.Z.* [Maori.] A cloak or mat made of flax, ornamented with black twisted thrums. Also *attrib.*
1820 *Gram. & Vocab. Lang. N.Z.* (Church Missionary Soc.) 168 *Kóro ai*, a certain garment. 1845 E. J. Wakefield *Adventure N.Z.* I. 244 The *korowai*..is woven of *muka*, or scraped flax, and ornamented with bunches of twisted tags of the same, dyed black. *Ibid.* 245 A great many varieties of the *korowai* are made. 1900 A. Hamilton *Maori Art* (1901) v. 285 The ground-work of a feather cloak is the same as that for a *korowai* mat. *Ibid.* 326 A plain *korowai* mat with the strings arranged in bands. 1921 H. Guthrie-Smith *Tutira* xi. 92 The foe can be seen cloaked in their *korowai* mats. 1938 R. Finlayson *Brown Man's Burden* 10 In the clear space by the flagpole old Tamarua is delivering a speech of welcome, mere in hand, korowai cloak swishing. 1955 W. J. Phillipps *Carved Maori Houses* p. xii, *Korowai*, cloak ornamented with twisted black thrums.

korradgee, var. *KORADJI.

korrigan (kǫ·rigăn). Also **corrigan,** and with capital initial. [f. Breton (Vannes dial.), fem. of *korrig* gnome, dim. of *korr* dwarf. Cf. Cornish *coryk*.] The name of a fairy or witch in Breton folklore, noted esp. for stealing children.
1855 C. M. Yonge *Hist. Sir Thomas Thumb* 106 She has lost her dear little Louis; the Korrigan has taken him. 1865 T. Taylor tr. *Hersart de la Villemarqué's Ballads & Songs Brittany* p. xiv, Vannes is the home of the legends of gnomes and spirits, of dwarfs and fairies that haunt rocks and woods, streams and fountains, of the *dus* and *mary-morgan*, the *poulpican*, and the *korrigan*. 1883 J. S. Stallybrass tr. *J. Grimm's Teutonic Mythol.* II. xvii. 447 In Bretagne the *korr*, pl. *korred* answers to our elf, the *korrigan* to our elfin. *Ibid.* 469 A Breton story of the *korrigan* changing a child is in Villemarqué. 1949 *Funk's Stand. Dict. Folklore* I. 253/2 *Corrigan*, in the folklore of Brittany, a female fairy: said to have been one of the ancient druidesses, and therefore malicious towards Christian priests. She is fond of pretty human children, and usually gets the blame for all changeling substitutions. 1969 V. Rowe *Loire* i. 43 The sea-cave of the korrigans, a reminder that this is still very much a part of Brittany. The korrigans, in its [*sc.* Brittany's] Celtic folklore, are the pigwidgeons or leprechauns, or hobgoblins, with which most Celtic lands seem always to have been infested.

korrigum (kǫ·rigŏm). [f. Kanuri *kargum.*] A West African antelope, *Damaliscus korrigum*; also called the topi.
1826 Denham & Clapperton *Narr. Trav. N. & Cent. Afr.* 192 Senegal Antelope..was found on the plains of central Africa. The natives call this species *Korrigum*. 1836 *Proc. Zool. Soc.* IV. 103 The Korrigum of Denham and Clapperton's Travels..was a very distinct animal from the Koba. 1895 Sclater & Thomas *Bk. Antelopes* I. II. 61 About the year 1840 Whitfield, a collector employed by Lord Derby..obtained specimens of the korrigum from the vicinity of Macarthy's Island on the River Gambia. 1964 L. S. Crandall *Managem. Wild Mammals in Captivity* 669 The korrigum or Senegal hartebeest (*Damaliscus k. korrigum*), found from Lake Chad to Senegal, was described from an animal living in the Zoological Gardens of London.

korro, var. *KORA¹.

Korsakoff (kǭ·ɹsăkǫf). Also **Korsakow, -ov.** The name of S. S. *Korsakoff* (1854–1900), Russian physician, used *attrib.* or in the possessive to denote a type of psychosis, namely a syndrome, often the result of chronic alcoholism, which is characterized by disorientation, memory loss for recent events, and consequent confabulation.
1900 Dorland *Med. Dict.* 546/1 *Korsakoff's psychosis*, delirium or insanity associated with polyneuritis. 1903 *Jrnl. Mental Sci.* XLIX. 673 (*heading*) Twelve cases of 'Korsakow's disease' in women. 1924 [see *CONFABULATION 2]. 1938 *Arch. Neurol. & Psychiatry* (Chicago) XXXIX. 483 We decided..to study the forces of organization in perception and memory in cases of the Korsakoff syndrome. 1941 *Brit. Jrnl. Psychol.* XXXI. 230 The object of this paper is to describe a variety of paramnesia recently observed in..Korsakow's psychosis. 1967 *Brit. Jrnl. Psychiatry* CXIII. 619 (*heading*) A case of craniopharyngioma presenting as Korsakov's syndrome. 1968 G. A. Talland *Disorders of Memory & Learning* iii. 63 A Korsakoff patient who, when asked to draw a woman..accurately reproduced the style..current ten years earlier. 1970 R. M. Suinn *Fund. Behavior Path.* xi. 288/1 The characteristic signs of Korsakoff's psychosis are disorientation of time and place, anterograde amnesia.., and a marked tendency to fabricate answers to fill in the past.

‖ **koru** (kǫ·ru). *N.Z.* [Maori.] A common motif in Maori carving and tattooing, consisting of a spiral pattern terminating in a bulb.
1938 W. J. Phillipps in *Art in N.Z.* X. 205 The *koru*, or *pikopiko*, which today rarely appears in carving, but is used in the construction of composite patterns for rafters. 1946 —— in *Dominion Mus.* (N.Z.) *Rec. Ethnol.* I. 16 In tattoo it was also customary to incise a koru type of design on the centre of the forehead. *Ibid.* 21 Two pairs of koru, large and small. 1964 T. Barrow *Decorative Arts N.Z. Maori* iv. 64 Painted patterns appear on the underside of rafters in ceremonial meeting-houses and superior dwellings. Their elaborate curvilinear designs are usually based on a small bulb-like motif (*koru*) shaped like the looped top of an uncurling fern frond. 1970 *Dominion Mus.* (N.Z.) *Rec. Ethnol.* II. 31 An example of a carved *koru* design in its simplest form. *Ibid.* 34 The *koru* motifs (the curved stalk-like forms that terminate in a bulb).

koruna (kǫ·rună). Also *erron.* **korona.** Pl. **korunas, koruny.** [Czech, lit. crown.] The basic monetary unit of Czechoslovakia, introduced as the Czech crown after the 1914–18 war (abbrev. *Kč*), and replaced and revalued after the 1939–45 war as the crown of the Czech and Slovak State (abbrev. *Kčs*); 1 Kčs = 100 hellers (*haléř*, pl. *haléře*). Also, a coin corresponding to this unit. Also *attrib.*
The forms *KC* and *KCS* (for Czech *Kč* and *Kčs*) in quot. 1947 are erroneous.
1920 *Czecho-Slovak Trade Jrnl.* Apr. 9/1 Prices..between 1100 and 1500 Kč per cwt... Already 3000 to 3200 Kč are being asked for small quantities. 1930 G. Druce *Wanderings Czechoslovakia* i. 29 It may..be taken that 1 crown (*koruna*, abbreviated to Kč, which signifies the Czechoslovak crown) is worth 1½d... These coins are in denominations of 1 and 5 crowns, and also 20 and 50 hellers (halíř, plural haléřů; 100 h. make 1 kč.). 1947 *Whitaker's Almanack* 891/2 Pre-war revenue (1938) in the currency then in use Czechoslovak *Koruna* (Crown) of 100 *heller* was KC10,120,000... In October, 1945, currency reform was introduced and a new coinage *KCS* (*Koruna* of Czechs and Slovaks) adopted. 1962 R. A. G. Carson *Coins* 387 The new coinage system took as its unit the korona struck in cupro-nickel. 1967 *Economist* 19 Aug. p. xxvi/3 An annual turnover in the region of 6,000 million koruna (just under £900 million).

‖ **korupe** (ko·rupe). *N.Z.* Also **korurupe.** [Maori *kōrupe, kororupe.*] The outer facing of the lintel of a door, often richly carved.
[1844 W. Williams *Dict. N.Z. Lang.* 36/2 *Karupe*,..lintel of a door.] 1897 A. Hamilton *Maori Art* (1901) II. 84 The front of the doorway was finished off by a carved slab, the *korupe*, or *korurupe*, which rested on the carved edges of the *whakawai*. The *korupe* was not put in its place until the spaces in the walls had been filled in with *raupo*. 1927 T. E. Donne *Maori, Past & Present* xix. 160 The *paré* or *korupe*, which was placed over the outside of the doorway to a house. 1949 P. Buck *Coming of Maori* (1950) II. ii. 128 A carved lintel (*korupe, pare*) was rested on the upper ends of the forward flanges of the jambs and so completed a doorway that was unique in Polynesian art.

‖ **koruru** (ko·ruru). *N.Z.* [Maori.] **1.** A wooden carving of a man's head, esp. one placed on the gable of a house.
[1871 W. Williams *Dict. N.Z. Lang.* (ed. 3) 64/2 *Kōrūrū*,..figure placed on the gable of a house.] 1897 A. Hamilton *Maori Art* (1901) II. 85 The junction of the barge-boards was covered by a carved flat face, the *koruru*, which was adorned with feathers, and sometimes surmounted by a full-length figure, the *tekoteko*. The *koruru* was kept in place by a boss at the back. 1916 E. Best *Maori Storehouses* i. 3 The carvings thereof were of the *koruru*..type. *Ibid.* 25 Such minor adornments as a carved head (*koruru*) on the gable. 1949 P. Buck *Coming of Maori* (1950) II. ii. 129 Sometimes a complete human figure (*tekoteko*) was used instead of the *koruru* type and sometimes a combination took place in which the *tekoteko* stood on the head of the *koruru*. 1966 W. J. Phillipps *Maori Life & Custom* xvii. 123 A group of carved heads, koruru, illustrating a variety of presentation.
2. A Maori version of knuckle-bones (see also quot. 1897).
1897 A. Hamilton *Maori Art* (1901) II. 106 *Koruru*,.. 2. A toy with two strings, which when played with makes a whizzing or roaring noise. 1924 E. Best *Maori* II. xi. 92 The game of jackstones, or knucklebones..is known as *ruru, koruru*..and *tutukai*.

korvort, repr. an ignorant pronunc. of CAVORT *v.*
1909 H. G. Wells *Tono-Bungay* III. iii. 356 She described the knights of the age of chivalry as 'korvorting about on the off-chance of a dragon'.

Korwa (kǭ·ɹwă). Also **Korewah. a.** A Kolarian tribe of the Chota Nagpur area of India; also, a member of this tribe. **b.** Their language. Also *attrib.* or as *adj.*
1865 E. T. Dalton in *Jrnl. Asiatic Soc. Bengal* XXXIV. II. 29 The Khairwars.., Korewahs and Coles number from 5000 to 7000 each. *Ibid.* 18 The Korewah iron..is greatly prized. 1872 —— *Descr. Ethnol. Bengal* v. 125 Not one of them would acknowledge that he could speak a word of Korwa. 1875 Watson & Kaye *People of India* VIII. 424 Korwas..are basket makers by profession... Korwas are evidently descended from an aboriginal stock, and are hardly recognised as Hindoos... Korwa dacoits do not intermarry with Korwa burglars. 1906 G. A. Grierson *Ling. Survey India* IV. 148 The most idiomatic Korwā is spoken in Jashpur and Sarguja, in the south of Palamau, and in Mirzapur. *Ibid.* 149 Remarks on Korwā grammar which follow. 1908 H. H. Risley *People of India* i. 44 Among the large groups..the Korwa (74·4) are just included in the long-headed division. 1923 R. B. Dixon *Racial Hist. Man* III. i. 259 Groups of 'casteless' aboriginal peoples, such as the Munda, Korwa, Malé. 1950 R. K. Deshpande in A. V. Thakkar *Tribes India* 112 The Korwa Language has much affinity with that of the Korkus of the Mahadoo hills. 1957 G. S. Ghurye *Mahadev Kolis* i. 6 The Korwas and the Korkus..have a better claim to relationship with the Koris. But the Korwa,..and the Korku are..assigned to the Kol or the Kolarian ethnic stock. 1972 W. B. Lockwood *Panorama Indo-Europ. Lang.* 227 The languages in question are..Koda (35,000) and Korwa (25,000).

Koryak (kǫ·ryak). Also **Korak, Koriac(k), Koriak.** [Russ. *Koryáki* (pl.), the Koryak people.] **a.** A people inhabiting the northern part of the Kamchatka peninsula; also, a member of this people. **b.** The Palæo-Asiatic language of this people. Also *attrib.* or as *adj.*
1780 W. Coxe *Acct. Russ. Discoveries* I. i. 3 The first expedition..was made in 1696, by sixteen Cossacs, under the command of Lucas Semænoff Morosko, who was sent against the Koriacks of the river Opooka. *Ibid.* 5 The peninsula of Kamtchatka..is bounded..on the North by the country of the Koriacs. *Ibid.* iv. 49 In the autumn of 1754 they were joined by a Kamtchadal, and a Koriac. 1790 tr. *J. B. B. de Lesseps's Trav. in Kamtschatka* II. 105 The idiom of the Koriacs has no affinity to that of the Kamtschadales. *Ibid.* 117, I recognized in his features a Koriac prince... I have long owed the reader a description of a Koriac sledge. 1832 J. Bell *Syst. Geogr.* IV. vi. 74 The Koriaks inhabit the country between the Anadyr and the peninsula of Kamtschatka. 1893 *Funk's Stand. Dict.* I. 989/1 *Koriak*,..a member of a race inhabiting northeast Siberia; also, their language, related to Kamchatkan. *Koryak.* 1898 J. Y. Simpson *Side-Lights on Siberia* i. 12 Of the sub-Arctic races that inhabit the north-east of Siberia, it is sufficient to recall the names of the..Tchuktchi,..the wild Koryaks..and Yukaghirs, all differing in speech and appearance, but still related. 1907 W. Jochelson in *Internat. Congr. Americanists* XV. 121 The subterranean Koryak house is still in use among the maritime Koryak, who are not Russianised. 1910 G. Kennan *Tent Life in Siberia* xx. 203 The Korak numerals are:—Innín, One... Meen-ye-geet-kᵉhin, Ten. *Ibid.* 204 It would be a hard day's work for a boy to explain in Korak one of the..problems in Ray's Higher Arithmetic. 1917 W. Bogoras *Koryak Texts* 2 The Koryak dialects may be divided into two large groups,—the western branch, which includes the Maritime Koryak of Penshina Bay and also the Reindeer Koryak; and the eastern branch, which includes the Maritime Koryak of Kamchatka. *Ibid.* 3 The Koryak language, in contrast to the Chukchee,..is furthermore divided into several local dialects. 1928 W. Jochelson *Peoples of Asiatic Russ.* ii. 49 The northern Kamchadal dialect..contains a considerable number of Koryak words. 1953 J. Ramsbottom *Mushrooms & Toadstools* v. 45 The Fly Agaric is among the objects believed by the Koryak to be endowed with particular power. 1954 Pei & Gaynor *Dict. Ling.* 116 *Koryak*, a language, spoken by about 1,000 persons in north-eastern Asia; a member of the Chukchi-Kamchadal family of languages, classified in the *Hyperborean* or *Palaeo-Asiatic* group.

kosh, var. *COSH *sb.*³

Koshare (kōᵘʃā·ri). [ad. a Keresan word.] A member of a Pueblo Indian clown society representing ancestral spirits in rain and fertility ceremonies.
1890 A. F. Bandelier *Delight Makers* 8 Shyuote, what have you heard about the Koshare? 1924 D. H. Lawrence *Mornings in Mexico* (1927) 123 The intermittent black-and-white fantasy of the hopping Koshare, the jesters, the Delight-Makers. a 1956 F. Lawrence *Mem. & Corr.* (1961) 4 These Koshares wore big Spanish hats with plumes. 1959 E. Tunis *Indians* 129/1 Clowns, called Koshare, were provided to lighten the solemnity of the ceremonials.

kosher, *a.* (*sb.*) Delete ‖ and add further examples.
1934 J. Brophy *Waterfront* i. 23 To cook food in the prescribed kosher manner. 1959 *Hotel Managem. & Restaurant Trade Jrnl.* Feb. 28/2 Only kosher meals are served on El Al aircraft. 1971 *Guardian* 14 Oct. 11/2 The artificial meat business has a very good-going trade already in Kosher bacon made from soya beans. 1973 *Jewish Chron.* 19 Jan. 1/5 The Kissingers will also be given kosher food.
b. Also *kosher butcher.*
1932 Kipling *Limits & Renewals* 359, I had a whole tin of salmon once from a kosher butcher. 1973 *Guardian* 16 June 3/6 The last kosher butcher left for Israel.
c. Correct, genuine, legitimate. *colloq.*
1896 Farmer & Henley *Slang* IV. 135/1 *Kosh* (or *Kosher*)... Adj. (common).—Fair; square. 1924 *Cosmopolitan* Nov. 104/2 It don't sound kosher to *me!* 1930 D. Runyon in *Sat. Even. Post* 13 Sept. 7/3 'Everything is very Kosher,' Harry the Horse says. 'You need not be afraid of anything whatever. We have a business proposition.' 1930 *Amer. Mercury* Dec. 456/2 Listen shamus, you got me wrong. I'm strictly kosher. 1953 H. Miller *Plexus* iv. 164, I made little or no effort to keep up with the others, it being no concern of mine what went on in this realm of make believe. All I felt called upon to do was to keep a straight face and pretend that everything was Kosher. 1959 C. MacInnes *Absolute Beginners* 80 It's so as to play down the queer thing in our country, and hide it behind the kosher game. *Ibid.* 157 It's all very well sneering at universities..but really and truly, it would be wonderful to have a bit of kosher education. 1961 L. Gribble *Wantons die Hard* iv. 48 'No financial irregularities?' 'Strictly kosher... It's so good it stinks.' 1966 T. Pynchon *Crying of Lot 49* iii. 60 They got the contracts. All drawn up in most kosher fashion, Manfred. 1973 *Jewish Chron.* 8 June 22/5 As for the children of a couple married in a register office, these are quite kosher.
B. Also, *to keep kosher*, to observe the Jewish law regarding diet.
1963 'R. L. Pike' *Mute Witness* vii. 119, I wouldn't go on the stand and state that my mother kept kosher without a chance to check.
Also **ko·sherness.**
1949 Koestler *Promise & Fulfilment* III. iii. 317 Kosherness..is only one of the relatively minor rabbinical plagues in the young state. 1950 G. Mikes *Milk & Honey* 117 The Kosherness of a meal depends just as much on the way of cooking as on the raw material bought in the shop.

kosher, *v.* Add: (Earlier and later examples.) Hence **ko·shering** *vbl. sb.*
1871 E. Levy *Jewish Cookery Bk.* 40 Take a fresh tongue, porge and..cosher it. 1960 *Commentary* June 499/2 *Kashrut* has survived in a mangled form... Scarcely anyone *kashered* the meat. 1974 *Times Lit. Suppl.* 31 May 592/3 What Mrs Wahlhaus has tried to do is to bring a new and more interesting approach to many traditional recipes, but without enough style or detail; she does not even explain how to kosher. 1974 W. Foley *Child in Forest* II. 218 Meat was meat (until all that Koshering business of soaking it in water and salt had emasculated it).

kosin (kōᵘ·sin). *Chem.* Also † **koussin.** [ad. It. *koussino* (C. Pavesi 1858, in *Giornale di Farm., di Chim.* VII. 49), f. *kouss-o* KOUSSO + *-ino* -IN¹.] Either of two related bicyclic phenols (not orig. differentiated), $C_{25}H_{32}O_8$, which are responsible for the anthelmintic property of kousso and are isolated as pale yellow crystals.
1875 *Pharmaceutical Jrnl. & Trans.* V. 562/2 From the aqueous residue of the alcoholic extract the koussin of Bendall, existing as a calcium compound, is precipitated as a more or less crystalline whitish powder. 1875 *Jrnl. Chem. Soc.* XXVIII. 468 The koso flowers yield about 3 per cent. of kosin. 1911 *Encycl. Brit.* XV. 921/1 *Kousso*... The active principle is koussin or kosin,..which is soluble in alcohol and alkalis, and may be given in doses of thirty grains. 1937 *Jrnl. Chem. Soc.* 563 The pale yellow product was identical with Merck's kosin... By fractional crystallisation we were able to separate this material into two distinct compounds, α-kosin, m.p. 158° (yellow needles) and β-kosin, m.p. 120° (yellow prisms). 1952 *Ibid.* 3103 Extracts of the flowers [*sc. Hagenia abyssinica*] contained large amounts of amorphous material, some of which could be converted into kosins by the action of alkali.

kos-kos, var. *KAS-KAS.

‖ **koss, kos.** = COSS, cos *sb.*²
1826 Leyden & Erskine tr. *Mem. Zehir-Ed-Din* 393, I directed Chikmâk Beg..to measure the distance from Agra to Kâbul; that at every nine kos he should raise a minâr, or turret, twelve gez in height. 1884 [see COSS, cos *sb.*²]. 1893 Kipling *Many Inventions* 193 He may have gone to the next hut... It is only four *koss*. 1901 —— *Kim* ii. 50 Think how far thou art on the road—an hundred *kos* from Lahore already. 1912 —— *Divers. Creatures* (1917) 227, I stood in the line..one *koss*, two *koss* distant.

Kossæan: see *KASSITE *sb.* and *a.*

Kossak, var. COSSACK (see *COSSACK 2 d). Also, *Canad.*, a sealskin or deerskin jacket.
1845 W. D. Cooley tr. *Parrot's World Surveyed* I. i. 10 This portion of the empire is traversed by a line of Kossak posts. 1919 W. T. Grenfell *Labrador Doctor* (1920) vii. 155 He wore it over a deerskin kossak, which is not the custom of cavalrymen.

kosso (kǫ·so). = KOUSSO.
1936 E. Waugh *Waugh in Abyssinia* i. 13 His mother lived by selling *kosso*, a specific against tapeworm, in the streets of Gondar.

koswite (kǫ·zwəit). *Petrogr.* Also **kosvite.** [a. F. *koswite* (Duparc & Pearce 1901, in *Compt. Rend.* CXXXII. 892), f. the name of the *Kosw-inski* Mountains, Pawda, in the Middle Urals: see -ITE¹.] A coarse-grained peridotite consisting mainly of diopside, olivine, and magnetite. Hence **koswi·tic** (**kosvitic**) *a.*
1901 *Jrnl. Chem. Soc.* LXXX. II. 398 Associated with olivine-gabbros in the Solimask district, near the source of the Kosswa river, is a new type of basic eruptive rock to which the name koswite is given... As the magnetite decreases in amount and becomes idiomorphic, there is a passage from koswite to ordinary pyroxenite. 1922 *Mineral. Abstr.* I. 327 The mining district around Nikolae-Pavdinsk on the eastern slopes of the northern Urals..consists largely of basic igneous rocks (dunite.., &c., and the local types koswite, tilaite, [etc.]). 1964 *Doklady Earth Sci.* CLIV. 149/2 The exceptionally high contents of volatile components in kosvites (mainly apatite-ore pyroxenites) increase the acidity of the melt. *Ibid.*, In the case of a kosvitic melt, which itself contains a considerable quantity of volatile components.

kotare (kōᵘ·tare). *N.Z.* [Maori *kōtare.*] The native kingfisher, *Halcyon sancta.*
1873 W. L. Buller *Hist. Birds N.Z.* 69 *Halcyon vagans.* (New-Zealand Kingfisher.)..Native names. Kotare and Kotaretare; 'Kingfisher' of the colonists. 1882 W. D. Hay *Brighter Britain!* II. 221 The Kotare..is a kingfisher. 1930 W. R. B. Oliver *N.Z. Birds* 430 Sacred Kingfisher. Kotare. *Halcyon sanctus.* 1968 *Landfall* XXII. 249 Like worm in kotare's beak my old maori lies limp.

‖ **kotatsu** (kota·tsu). [Jap.] A wooden frame which is placed over the hearth in Japanese houses and covered with a thick quilt to give an enclosed area within which people can warm their hands and feet. Also applied to the hearth and the cover together. Also *attrib.*
1876 W. E. Griffis *Mikado's Empire* II. vii. 416, I got up, entered the best room in the house, and curled up under a kotatsū. 1880 I. L. Bird *Unbeaten Tracks Japan* II. lv. 254 The *kotatsu*..consists of a square, wooden frame, standing over a basin of lighted charcoal, and supporting a large wadded quilt or *futon*, under which you creep... The invitation to creep under the *kotatsu* is as welcome as the 'sit in' of the Scotch Highlands or the 'put your feet in the stove' of Colorado. 1889 M. B. Huish *Japan & its Art* vii. 86 The only other articles of furniture will be the kotatsu, a square wooden frame, which in winter is placed over the hibachi or stove, and is covered with a large wadded quilt or *futon* (under this the whole family huddle for warmth). 1970 J. Kirkup *Japan behind Fan* 126 A limited life of kitchen and *kotatsu* gossip in some narrow-minded village.

kotla, var. *KGOTLA.

‖ **koto.** Add: (Earlier and later examples.) Also † **kotto.**
1795 tr. *C. P. Thunberg's Trav. Europe, Afr. & Asia* IV. 58 The *koto* bears a strong resemblance to our dulcimers, having a number of strings, which are struck with sticks. 1822 F. Shoberl tr. *Titsingh's Illustr. Japan* 234 *Kotto*, a kind of harp. 1891 A. M. Bacon *Jap. Girls & Women* ii. 42 The *koto* is an embryo piano, a horizontal sounding board, some six feet long, upon which are stretched strings supported by ivory bridges. 1893 F. T. Piggott *Mus. Japan* III. 135 The Koto is the chief of modern Japanese instruments. 1932 F. L. Wright

Autobiogr. II. 210 A fine lady plays the Koto, its graceful length laid upon the expanse of matting. **1961** A. Baines *Mus. Instrum.* 208 The *koto*, the national instrument of pre-Westernized Japan. Its six-foot long soundbox is placed on the floor... Along it run thirteen waxed silk strings of equal length and played open, each tuned by its own movable bridge to contribute to a pentatonic scale of the 'major third' type..over two and a half octaves. **1965** W. Swaan *Jap. Lantern* i. 10 The chief item is a concerto for *koto*, the traditional Japanese harp. **1973** E. T. Sithole in T. Kochman *Rappin' & Stylin' Out* 69 You would have expected me to write about the music of Bach..and not about Koto music (Japan).

kotschubeite, var. *KOCHUBEÏTE.

kotuku (kōᵘ·tuku). *N.Z.* Also † **koatuku.** [Maori.] The white heron, *Egretta alba*.

1846 C. Heaphy *Jrnl.* 23 Apr. in N. M. Taylor *Early Travellers N.Z.* (1959) 218 Shot a very fine *koatuku*, or white heron. **1882** W. L. Buller *Man. Birds N.Z.* 52 The White Heron occurs so sparingly in most parts of New Zealand that 'rare as the Kotuku' has passed into a proverb among the Maoris. **1949** P. Buck *Coming of Maori* (1950) II. xii. 284 Feathers of the..heron (*kotuku*) were also valued. **1963** *Evening Post* (Wellington, N.Z.) 26 Oct., In the..action shots..are the..kaka and spectacular kotuku or white heron.

kotwalee. Add: Also **kotwali.**

1964 A. Swinson *Six Minutes to Sunset* ii. 34 Major MacDonald marched a company of troops down Hall Gate as far as the Kotwali. **1969** *Pioneer* (Lucknow) 13 Aug. 6/7 The kotwali police claim to have arrested last night three of the inter-district gang of burglars operating for several years in Pratapgarh, Sultanpur and Allahabad districts.

kotyle. = Cotyle 1.

1926 C. Day Lewis in *Oxf. Poetry* 18 (*title*) Naked woman with kotyle. **1948** A. Lane *Greek Pott.* iv. 26 Kotyle, painted in brown-black on greenish buff clay.

kou (kōᵘ). [Hawaiian.] A Hawaiian tree, *Cordia subcordata*, of the family Boraginaceæ, or its dark brown wood.

1825 W. Ellis *Jrnl. Tour Hawaii* i. 27 The houses, which are neat, are generally built on the sea-shore, shaded with cocoanut and *kou* trees. **1866** 'Mark Twain' *Lett. from Hawaii* (1967) 128 It [*sc.* the coffin] is made of those two superb species of native wood, kou and koa. The former is nearly as dark as ebony. *Ibid.* 207 Among the varied and handsome foliage of the kou, koa, kukui.. its [*sc.* the orange's] dark, rich green cone was sure to arrest the eye. **1913** J. F. Rock *Indigenous Trees Hawaiian Islands* 415 The Kou, which is indigenous in the Hawaiian Islands,..can only be found along the sea-shore here and there... The wood of the Kou was much sought for. **1969** T. H. Everett *Living Trees of World* 292/2 A native of sandy shores of the Indian and the western Pacific oceans, the kou or sea trumpet (*C[ordia] subcordata*) is an evergreen tree that grows up to 50 feet in height and has a dense, spreading crown.

koukou, var. *KUKU 1.

Koula(h), varr. *Kulah[1].

koulak, var. *KULAK.

kouprey (kū̆·pre). [Cambodian native name.] A large wild ox, *Novibos* (or *Bos*) *sauveli*, first discovered in Cambodia in 1937.

1940 *Mem. Mus. Compar. Zool. Harvard* LIV. 421 On comparing the measurements of the skull and skeleton of the kouprey with those of a gaur and a bantin, the kouprey revealed important differences. **1955** F. Bourlière *Mammals of World* iii. 131 Mention should also be made here of the curious Kouprey or Cambodian Forest Ox, whose taxonomic status is still rather uncertain. **1970** *New Scientist* 23 July 177/1 The kouprey (*Novibos sauveli*) was..a large blackish-brown animal, with prominent white legs and a chestnut facial pattern, the bull stands over 6 feet tall. **1971** *Ibid.* 11 Feb. 342/1 The American invasion of Cambodia last year probably put paid to the kouprey *Bos sauveli*, a primitive wild cow.

‖ **koura** (ku·ră). *N.Z.* Also **gorau, kora.** [Maori.] A small freshwater crayfish, *Paranephrops planifrons*.

1847 J. Johnson *Jrnl.* 5 Jan. in N. M. Taylor *Early Travellers N.Z.* (1959) 152 A small settlement, occupied in the summer for the purpose of catching *gorau*, a species of cray-fish, or rather fresh water lobster. **1867** E. Sauter tr. *F. von Hochstetter's New Zealand* viii. 171 Of the Macrura I may mention especially *Paranephrops tenuicomis* Dana... The natives call it Koura. **1873** J. H. H. St. John *Pakeha Rambles through Maori Lands* viii. 151 Rotorua has its *kora*, a large prawn, or a diminutive freshwater lobster. **1945** F. Sargeson *When Wind Blows* 92 In the creek they saw the crayfish and Mr Jones said he called them crawlers, but the Maoris called them koura. **1974** *N.Z. Listener* 20 July 13/1 Here we were fed koura and kouras are a real luxury in the North Island.

‖ **kouros** (kū̆·rǫs). *Gr. Antiq.* Pl. **kouroi.** [Gr. (Ionic form of κόρος boy).] A sculptured representation of a youth.

1920 [see *Kore 2]. **1932** Beazley & Ashmole *Greek Sculpture & Painting* vi. 27 For all the beauty and variety of the late archaic kouroi and korai, it is not in these that late archaic sculpture finds its most perfect expression, but in the action-figures—men not being but doing. **1939** *Ann. Brit. Sch. Athens* XXXVI. 2 The Anarysos Kouros is of Parian marble. It has a reddish tint; but in many places..the surface has perished to some extent. **1963** *Listener* 21 Mar. 511/1 The great *Kouroi* at the National Museum in Athens. **1965** D. E. Strong *Classical World* 58 In figure-sculpture the ideal schemes of the nude *kouros* and the draped female (*kore*) hold the field until the early years of the 5th century.

koutekite (kū̆·tėkəit). *Min.* [f. the name of J. *Koutek* (see quot. 1958) + -ite[1].] An arsenide of copper, Cu_5As_2, found as bluish-grey microscopic grains with a metallic lustre.

1958 Z. Johan in *Nature* 31 May 1554/1 While studying specimens from Černý Důl in Krkonôse (Giant Mountains), Bohemia, a new mineral of composition Cu_3As has been found in the bicarbonate gangue... The new mineral has been named koutekite in honour of Academician J. Koutek, professor of economic geology, Charles University, Prague. **1967** *Mineral. Abstr.* XVIII. 281/2 Koutekite, Cu_5As_2.., has been recognized at Daluis in the upper Var Valley, where it occurs with domeykite.

kovsh (kǫvʃ). Pl. **kovshi.** [Russ.] A ladle or container for drink.

1884 A. Maskell *Russ. Art* vi. 134 The bowls or ladles termed kovsh. **1935** *Burlington Mag.* June 298/1 The large kovsh, or ladle..follows an interesting evolution. **1949** H. C. Bainbridge *Peter Carl Fabergé* vii. 133 Objects in..painted enamels in the way of ikons, kovshi, tea sets, [etc.]. **1960** *Harper's Bazaar* Aug. 11/1 A silver-gilt kovsh: used in Russia for drinking vodka. **1966** *Daily Tel.* 18 Oct. 16/5 At Sothebys... A Russian silver and enamel kovsh, a squat wine jug, went to G. Lawrence for £210.

kowdie, obs. var. Kauri.

1899 T. Kirk *Forest Flora N.Z.* 143 When the timber was first introduced into Britain it was termed 'cowrie' or 'kowdie-pine'.

kowhai. Substitute for def.: An evergreen shrub or small tree, *Sophora tetraptera*, of the family Leguminosæ, native to New Zealand and bearing racemes of yellow flowers. Add earlier and later examples.

1831 G. Bennett in *London Med. Gaz.* 12 Nov. 182/1 *Sophora tetraptera*... This tree is the Kowhy, or Kongia, of the natives, and attains the height of from forty to fifty feet. **1897** D. McK. Wright *Station Ballads* 123 There are dreams in the gold of the kowhai. **1926** *Trans. N.Z. Inst.* LVI. 670 'Kowhai' went through many stages —'goa', 'gohi', etc., before settling to the two forms 'kowhai' in the North and 'gowhai' or 'gowai' in the South. **1947** O. Ruhen in *Coast to Coast 1946* 1 Old gold blossom freckled the untidy lattice of brown boughs that the kowhais threw up on the top of the bank. **1952** 'J. Guthrie' *Paradise Bay* i. 12 In a kowhai bush, among the yellow beak-shaped blossom. **1966** *New Statesman* 16 Dec. 915/3 The beach itself is sheltered by..bush patched with yellow and gold kowhai bloom and white clematis.

kow-tow. Now the usual form of Kotow *sb.* and *v.* (with pronunc. kautau·). So **kow-tow·er.**

1863 *Fraser's Mag.* Dec. 709/2 With one dash of the brush a clever artist at once hits off all the characteristics of his subject, whether it be a bat or moth in the gloaming, or a humble cricket *kow-towing* to a lordly mantis perched on a blade of grass. **1864** [see Kotow *v.*] **1874** A. C. Maclay *Let.* 1 May (1886) 47 Then followed a tempest of *kow-towing* that beggared description. **1883** [see Kotow *v.* b]. **1898** W. G. Gulland *Chinese Porc.* I. p. xxiv, Lord Amherst..would not perform the *kow-tow* (kneeling) before the emperor. **1905** [see *hewgag]. **1907** *Westm. Gaz.* 12 Nov. 14/1 The degrading custom of high native Ministers of the Crown kowtowing. **1920** *Blackw. Mag.* Aug. 225/1 The conventional bowing or kow-tow position. **1961** *Spectator* 8 Sept. 313 They regard the Russians as..kow-towers to the West. **1966** D. Forbes *Heart of Malaya* xi. 129 Miss Khan kowtowed to the laughing monk. **1966** *Listener* 29 Sept. 443/2 Not even the emissaries of the Pope could escape the Great Kow-tow—the ceremony involving the three kneelings and nine prostrations before the throne of the Chinese Emperor. *Ibid.* 444/1 An envoy of the Tsar who arrived in Peking in the sixteen-seventies to discuss Russo-Chinese differences refused to kow-tow to the Emperor. **1972** *Times* 21 Oct. (Hongkong Suppl.) p. i/6 Peking has referred officially to Hongkong's shameful colonial status only once since President Nixon's dignified *kowtow* and the belated entry of the people's republic into the United Nations.

‖ **koyan** (koya·n). Also 9 **coyan.** [Mal. *koyan*.] In Malaysia, a unit of weight equal to 40 *piculs*, equivalent to approximately 5,330 lb. (2·42 tonnes).

1783 W. Marsden *Hist. Sumatra* 157 The *coolah* or bamboo, containing very nearly a gallon, is the general standard of measure among the Rejangs: of these eight hundred make a *coyan*. **1820** J. Crawfurd *Hist. Indian Archipelago* I. III. i. 273 Thirty pikuls [make] one *koyan*. **1839** T. J. Newbold *Pol. & Statistical Acct. Straits of Malacca* I. i. 26 Salt, and..rice, etc. are sold by measure, generally by that of the coyan. **1900** W. W. Skeat *Malay Magic* v. 228 A field which..produces a *koyan* of rice a year will fetch a rent of about two hundred gallons more or less. **1947** R. O. Winstedt *Malays* 125 The produce of rice formerly exceeded the internal consumption by about the annual average of 2,500 *koyan*. **1957** *Federation of Malaya Ann. Rep.* II. iv. 519 The *koyan* (40 *piculs*).. —5,333⅓ lbs.

K(-)particle. *Nuclear Physics.* Also † **k-particle.** = *kaon.

1949 *Nature* 15 Jan. 85/2 The agreement between the sets of values for mesons..gives strong support for the assumption of a spontaneous decay of the *k*-particle. **1954** *Proc. R. Soc.* A. CCXXI. 293 Those charged particles more massive than the π-meson which, at the end of their range in the emulsion, transmute with the emission of a singly-charged particle, are referred to as *K*-particles... The charged particles are in some cases μ-mesons, and others π-mesons. **1955** *Sci. News Let.* 19 Feb. 117/2 K particles are mesons of mass intermediate between electrons and protons. **1966** D. L. Livesey *Atomic & Nucl. Physics* x. 468 The neutral *K* particle must therefore possess a rest mass greater than 550 m_e, the measured energy release being such that the calculated mass is about 1000 m_e.

kra (krā). Also **kera.** [Mal. *kera*.] The long-tailed or crab-eating macaque, *Macaca fascicularis* (= *M. irus*), native to southern and south-eastern Asia.

1821 T. S. Raffles in *Trans. Linnean Soc.* XIII. 247 The Malay name has frequently a close resemblance to the cry of the animal it designates; and this is remarkably the case in the present instance... The Kra is not easily domesticated. **1839** T. J. Newbold *Pol. & Statistical Acct. Straits of Malacca* I. vii. 432 Of the genus Semnopithecus are..the Lotong,..the Kra, or Simia fascicularis. **1911** *Encycl. Brit.* XVII. 472/2 The *lotong*, *kra*, and at least twenty other kinds of monkey. **1932** S. Zuckerman *Social Life Monkeys* ix. 143 No periodicity has been noted in the rarely observed sexual response of bonnet monkeys and adult female kra monkeys (common macaques). **1969** Ld. Medway *Wild Mammals Malaya* 50/2 Long-tailed or Crab-eating Macaque. Kera. *Macaca fascicularis*.

‖ **kraak porselein, kraakporselein** (krāk pōᵊ·ɪsĭlēⁱn). Also (with partial tr.) **kraak porcelain.** [Du.; see quot. 1954.] Blue-and-white Chinese porcelain of the Wan-li period (1573–1619) or later in the seventeenth century, or a European imitation of this.

1954 H. Garner *Oriental Blue & White* v. 37 This type of porcelain is known as *Kraak porselein*, from the Dutch term for the type of Portuguese ship from which the porcelain was first captured in 1603. The ship, the carrack *Catherina*..was..taken to Amsterdam. *Ibid.*, Kraak porcelain had a great influence on the development of European pottery, being extensively copied at Delft. **1956** J. A. Pope *Chinese Porc. Ardebil Shrine* II. 137 In terms of general shape the most noticeable difference between these *kraakporseleins* and other sixteenth-century wares is to be found in the dishes. **1961** M. Sullivan *Introd. Chinese Art* ix. 185 Soon after 1600 a particular type of thin, brittle Wan-li export blue-and-white began to reach Europe. This ware, called *kraak* porcelain.. caused a sensation when it appeared. **1969** *Times* 25 Mar. 15/5 (Advt.), Kraak porselain [*sic*] dishes.

kraal, *sb.* Add: **2. b.** In Sri Lanka, an enclosure into which wild elephants are driven; also, the process of capturing elephants in this way. Also *attrib*. So **kraal-town,** a town formed to accommodate the company assembled to view a kraaling of elephants.

1891 *Outing* (U.S.) Dec. 171/1 An elephant kraal is no simple matter, the drive taking possibly a couple of months to accomplish. *Ibid.*, Gradually, slowly but surely, the herds..are driven toward the kraal. *Ibid.* 174/1 In less than a week's time a town springs into existence. 'Kraaltown!' with its clubs, hotels, saloons, cafés, and 'chummeries', to say nothing of suburban villas, etc. **1933** D. E. Blunt *Elephant* ii. 103 Great difficulty was experienced in finding the best way of capturing the young African elephants, the *keddah* or *kraal* method used in Ceylon proving impossible in the Congo. **1956** R. Pieris *Sinhalese Social Organization* v. ii. 185 The *pannikalē* assisted in driving the elephants into the *kraal* or enclosure.

c. An enclosure in water for holding live sponges or turtles.

1939 *Nature* 13 May (Suppl.) 807/2 Inshore waters in close proximity to sponge kraals. **1961** *Encounter* Apr. 18/2 The turtles swim in dense kraals.

kraal, *v.* Add: Also of elephants.

1891 *Outing* (U.S.) Dec. 174/2 Hurrying them to the kraal we lowered the huge bars and kraaled our first elephant.

krab (kræb), colloq. abbrev. *karabiner.

1963 *Oxf. Mountaineering* 8 To add to my worries I found that I was down to my last krab. **1970** *Sunday Mail* (Brisbane) 3 May 7/6 On the way up, the leader puts pitons in the rock, and runs his rope through krabs behind him so he cannot fall far.

kraddy, var. *korari.

Kraft (krāft). Also **kraft.** [Sw., = strength, in *kraftpapper* kraft paper.] A strong smooth brown paper made from unbleached soda pulp. Also *attrib*. In full, *kraft paper, kraft brown*.

1907 G. Clapperton *Pract. Paper-Making* (ed. 2) 37 'Kraft' brown papers. *Ibid.*, No Kraft yet produced in this country combines the crispness and elasticity which form so distinctive a feature of the Kraft papers produced by the best Scandinavian mills. **1914** E. A. Dawe *Paper* 56 Kraft browns may be described as glazed browns, as they are sometimes finished with a glazed surface both sides. A special kind of pulp is used for krafts. **1920** *Glasgow Herald* 12 June 5 In the case of sulphite and kraft processes the pulp is prepared by cooking chips of

wood under pressure with different liquor solutions. **1930** *Economist* 8 Feb. 303/1 The market for kraft paper must be regarded as declining, but the Swedish newsprint mills are well supplied with orders. **1959** *Gloss. Packaging Terms* (*B.S.I.*) 67 *Kraft liner*, a kraft paper used as the outer or inner facing in the manufacture of solid and corrugated fibreboard, and in the manufacture of fibreboard drums and tubes. **1968** *Economist* 3 Feb. 63/2 Swedish pulps are mainly for kraft, including sack paper, and newsprint. **1969** T. C. Thorstensen *Pract. Leather Technol.* ix. 143 In the manufacture of paper, lignin of the logs of the wood chips is released by cooking the chips with bisulfite in accordance with the Kraft process.

‖ **kragdadig** (kraχdā·diχ), *a.* *S. Afr.* Also **kragdadige.** [Afrikaans, = Du. *krachtdadig*.] Resolute, firm, vigorous. Hence **kragdadigheid** (-hēⁱt), resoluteness, spirit of determination.

1949 *Cape Times* 21 Sept. 8 Where any evidence can be found in this rigmarole of tentativeness for the *kragdadigheid* which is supposed to distinguish Nationalist Ministers we fail to see. **1952** *Ibid.* 21 May 5/3 The Government wanted to make scapegoats of himself and Mr. Carneson so that it could claim *kragdadige* (strong) steps against communism. **1957** *Cape Argus* 11 Feb. 8/8 Signs of that kragdadigheid of which Cabinet spokesmen claim a monopoly. **1958** *Cape Times* 18 Feb. 8/5 Mr. Sauer's answer was that the inquiry was merely United Party propaganda and he was not going to answer it... That is at least a nice *kragdadige* answer. **1963** *Time* 10 May 30/2 Helen Suzman warned that black nationalism as well as white nationalism feeds 'on this type of *kragdadigheid* (toughness)'. **1973** *Sunday Times* (Johannesburg) 15 Apr. 17 Even the Minister of Sport..could not restrain himself from issuing one of those kragdadige statements that get no one anywhere. **1974** *Daily Despatch* (East London) 4 Feb. 8 The new Parliamentary session has barely started and already there are distressingly clear pointers that it will be the season for kragdadigheid.

Krag–Jørgensen (kræg,yō·ɪgənsən). The names of O. H. *Krag* (1837–1912) and E. *Jørgensen*, Norwegian firearm designers, used to designate a type of rifle (and carbine) introduced in Denmark and Norway in the late nineteenth century and adopted in U.S.A. Abbrev. (colloq.) **Krag.**

1899 *Scribner's Mag.* XXV. 20/1 Our arms were the regular cavalry carbine, the 'Krag', a splendid weapon, and the revolver. **1901** *Kynoch Jrnl.* Aug.–Sept. 133/2 He has also the latest pattern..Ross straight pull Krag–Jorgensen. **1902** *Encycl. Brit.* XXV. 658/1 The cavalry is armed with the Krag–Jorgensen carbine. *Ibid.*, The infantry and coast artillery have the Krag–Jorgensen rifle. **1910** *Harper's Weekly* 5 Mar. 16/3 Krag–Jorgensen rifles were cracking merrily in their users' pursuit of the wily amigo... Damn, damn, damn the Filipinos, cross-eyed kakiak ladrones. Underneath our starry flag civilize 'em with a Krag, And return us to our own beloved homes. **1964** H. L. Peterson *Encycl. Firearms* 186/1 The bolt-action Krag–Jørgensen..was developed during the 1880's. *Ibid.* 186/2 The loading gate on the Danish Krag pivots forward, while those of the Norwegian and American Krags pivot horizontally.

krai, var. *kray.

Krakowiak (krăkōᵘ·vi,æk). Also **-wyak.** [Polish, f. *Kraków* (Eng. Cracow), a city and region in southern Poland.] = Cracovienne.

1888 F. Niecks *Frederick Chopin* II. xxx. 233 Chopin has only once been inspired by the krakowiak—namely, in his Op. 14, entitled *Krakowiak, Grand Rondeau de Concert*. **1958** P. Kemp *No Colours or Crest* xii. 278 Waltzes and foxtrots were abandoned for the whirling, stamping folk dances of Poland—the Krakowiak, Oberek, and Kujawiak. **1966** *New Statesman* 1 Apr. 465/2 A yellowing piece of paper testifying that in 1952 its bearer danced the Krakowyak satisfactorily before an audience of experts.

Krama, var. *Kromo.

kramat (kramā·t, krā·măt). Also 8 **crammat**; 9 **grammat, kramet; karamat, keramat.** [ad. Mal. *keramat* adj., numinous, sacred, holy, unusual, having supernatural qualities; sb., holy place, holy person; ad. Arab. *karāmāt*, pl. of *karāma* miracle worked by a saint other than a prophet.] A Muslim holy place or place of pilgrimage (see also quot. 1833). Also as *adj.*, sacred.

1783 W. Marsden *Hist. Sumatra* 203 The place of greatest solemnity for administering an oath, is the *crammat* or burying ground of their ancestors. **1833** G. Greig *S. Afr. Almanac & Directory* 156 A tomb of a celebrated Malay Priest, near the farm Zandvliet, is frequently visited by the Mahomedans, and where they perform Divine Service, or what is called Grammat. **1839** T. J. Newbold *Pol. & Statistical Acct. Straits of Malacca* I. v. 252 In every part of the country are found tombs of men famed for piety... They call such tombs Kramets. **1900** W. W. Skeat *Malay Magic* 61 There is usually in every small district a holy place known as the *kramat*. **1910** D. Fairbridge *That which hath Been* xxiii. 283 Near the sand hills and the sea lies the kopje on the summit of which is the white mosque, the kramat, which marks to this day the resting place of Sheik Joseph. **1944** I. D. Du Plessis *Cape Malays* iii. 31 From the tomb of Sjech Yussuf a series of *Karamats* stretches in a rough circle round the Peninsula. **1947** L. G. Green *Tavern of Seas* xiv. 136 Followers of the Prophet listening to the reading of the Koran within the 'kramat'. **1947** R. O. Winstedt *Malays* 24 Many sacred (*Keramat*) places are the graves of by-gone shamans. **1964** R. Perry *World of Tiger* xiv. 221 The *kramat* tiger of Malaya..is protected by a guardian spirit and can be driven away from a cattle-fold by a child. **1965** R. McKie *Company of Animals* v. 84 He was *kramat* which is as near as any animal can come to being regarded as sacred. **1970** *Straits Times Ann.* 68 Here are three Malay Tombs. They are *keramat*, a sacred or respected place. **1971** *Drum* July 55 As the street fills, there'll be Coons and a band and a jam session, the scene fading out into a shot on Signal Hill, where the Karamat singers will chant the sacred Islamic music.

kran (krān). [ad. Pers. *qrān*.] A Persian coin and monetary unit.

1882 E. O'Donovan *Merv Oasis* I. 249 The Turcomans ..will accept only the old-fashioned *kran* and *toman*. **1902** *Encycl. Brit.* XXXIII. 513/2 The Indian rupee and the Persian kran are widely circulated through Mesopotamia. **1920** *Brit. Mus. Return* 75 in *Parl. Papers* XXXVI. 673 A silver kran of Muzaffar al-Din, Shah of Persia. **1922** *Blackw. Mag.* Mar. 393/1 For two krans an Arab will swear a false oath. **1933** V. Sackville-West *Coll. Poems* 178, I bought these beads in Isfahan; I bought a handful for a kran,—That's sixpence—at the motley stall. **1934** F. Stark *Valleys of Assassins* ii. 130 He allowed me to give him two krans with which he wandered off to buy our horses' dinner. **1958** F. Maclean *Person from England* v. 246 O'Donovan had with him a silver casket,..for which he had paid about six hundred *krans*, or twenty-five pounds sterling.

krans(e. *S. Afr.* = Krantz, kranz.

1785 G. Forster tr. *Sparrman's Voy. Cape Good Hope* II. xi. 48 He looked out for a *klipkrans* (so they generally call a rocky place level and plain at top, and having a perpendicular precipice on one side of it). **1852** C. Barter *Dorp & Veld* 93 We had been directed to look out for a white krans in the mountain. **1927** W. Plomer *I Speak of Afr.* i. 37 Eddies of sound reaching the nervous leaf-like ears of the krans-coloured sheep in the stones overhead cause them to lift their heads. **1939** tr. *E. N. Marais's My Friends the Baboons* iii. 32 The leopard was still ahead of us in the kloof and unless he had fled up the kranses, it was probable that we would meet him again. **1961** L. van der Post *Heart of Hunter* xiv. 189 All day long she [*sc.* the rock-rabbit] darts in and out of the shadows and clefts of our Kranses and rocky hill-tops.

krantz, kranz. Add: Pl. **krantze, krantzes.** (Later examples.)

1903 Kipling *Five Nations* 196 But 'e wasn't takin' chances in them 'igh an' 'ostile kranzes. **1916** J. Buchan *Greenmantle* xxi. 283 A little hill split the valley, and on its top was a *kranz* of rocks. **1924** R. Campbell *Flaming Terrapin* iv. 72 Her pitchy crows..cling with gnarly toes To their steep krantzes. **1952** *Cape Times Mag.* 19 July 6/5 Inquisitive baboons often watch the bathers from the krantze above. **1959** G. Jenkins *Twist of Sand* xiv. 301 Peaks and valleys, fretted with razor-like kranzes and unscaleable cliffs. **1966** E. Palmer *Plains of Camdeboo* v. 89 Koeltas, our guide, led us round the side of a krantz by a pathway as narrow as a bit of string.

‖ **krapfen** (kra·pfĕn). [G.] In Germany and other German-speaking areas: a doughnut (see also quot. 1845).

1845 E. Acton *Mod. Cookery* (ed. 4) App. 581 *Appel krapfen*... Boil..apples..mix them with..almonds, beaten to a paste,..raisins..cinnamon..roll out some butter-crust..cut into four-inch squares,..fill them.. lay another square on each,..bake them. **1877** E. S. Dallas *Kettner's Bk. of Table* 156 Beignets à la Dauphine. .. Throughout Germany they are known as Berliner Pfannkuchen; throughout Austria as Wiener Krapfen. **1966** P. V. Price *France: Food & Wine Guide* 151 *Krapfen* (rather like little light doughnuts).

krater, var. Crater 1.

1857 [see Crater 1]. **1935** *Antiquity* IX. 414 The low stems..are more difficult to place, but they resemble the stems of the kraters in the contemporary group from Lakkéthra in Cephallenia. **1950** H. L. Lorimer *Homer & Monuments* ii. 73 Argive Geometric krater of early type from Amathus. **1969** R. Tashkent *Ambiguous Man* viii. 80 The great swords, the daggers, pins, cups, kraters, spears. **1974** *Times* 8 Mar. 9/3 Mr Hecht bought fragments from Mr Sarrafian as a cover for selling fragments of a krater unearthed in Italy and smuggled out for sale to the Metropolitan.

kratogen (kræ·tŏdʒen). *Geol.* [a. G. *kratogen* (L. Kober *Der Bau der Erde* (1921) i. 21), f. Gr. κράτο-ς strength: see -gen.] An area of a continent that has resisted deformation over a (geologically) long period of time. Hence **kratoge·nic** *a.*

1923 *Bull. Geol. Soc. Amer.* XXXIV. 210 The 'Kratogens', once the area of the most ancient geosynclines, may, after they are peneplained, be widely flooded by epeiric seas. **1934** *Geogr. Jrnl.* LXXXIII. 515 After a few preliminary chapters dealing with the evolution of orogenic and kratogenic (continental) areas, he sets forth his views on..mountain-building. **1939** R. Ruedemann in Ruedemann & Balk *Geol. N. Amer.* I. 48 The geosynclines appear to be the compensating areas of subcrustal flowage between the subsiding oceans and the unmoved or horst-like neutral areas or 'kratogens' of the positive elements. **1969** Bennison & Wright *Geol. Hist. Brit. Isles* v. 91 Certainly, the Midlands of England formed a kratogen which persisted throughout the Ordovician and Silurian. *Ibid.* vi. 129 The marginal facies of south Shropshire and the English Midlands, which were similarly laid down on the margins of the (Midlands) kratogenic block and by shallow seas spreading across it.

kraurosis (krǭrōᵘ·sis). *Path.* [mod.L. (A. Breisky 1885, in *Zeitschr. f. Heilkunde* VI. 75), f. Gr. κραῦρ-ος brittle, dry + -osis.] Atrophy of the skin of the vulva, by some regarded as a distinct disease.

1888 *Trans. Amer. Dermatol. Assoc.* 64 Under the name 'Kraurosis' Professor Breisky, in his book on gynaecology, describes a disease of the vulva, which he considers a hitherto unknown form... The same disease has been described, fourteen years ago, by Robert F. Weir, of New York, as 'Ichthyosis of the Vulva'. **1901** C. H. Roberts *Outl. Gynæcol. Path.* i. 14 In Kraurosis, the tissues of the vulva are smooth and pale, with irritable tender red patches along the margins of the hymen and in the vestibule. **1948** W. Schiller in W. A. D. Anderson *Path.* xl. 1167 Kraurosis rarely follows the artificial menopause produced by operative removal of the ovaries. **1973** A. S. Woodcock in Fox & Langley *Postgrad. Obstetr. & Gynaecol. Path.* iv. 57 Kraurosis and senile atrophy are generally accepted as synonymous.

Hence **krauro·tic** *a.*, affected with kraurosis.

1894 *Brit. Med. Jrnl.* 7 Apr. Suppl. 63/1 Carcinomatous nodules were detected in the kraurotic tissue.

Krause (krau·zə). The name of Wilhelm *Krause* (1833–1909), Ger. anatomist and histologist, used in the possessive and with *of*-adjunct to designate structures in the body which he investigated, as: **a.** A kind of encapsulated plexus of sensory nerve endings found in mucous membranes, the dermis, the conjunctiva, and elsewhere. **b.** (A transverse membrane appearing as) a thin dark line separating adjacent sarcomeres in a striated muscle fibril. **c.** Each of the numerous small accessory lacrimal glands situated under the conjunctiva close to where it joins the eyeball.

a. 1872 H. Power tr. *Stricker's Man. Human & Compar. Histol.* II. xxix. 317 The nerves of the glans [penis] are very numerous and..end within the papillae in Krause's terminal bulbs. **1875** A. Gamgee tr. *Hermann's Elem. Human Physiol.* x. 457 Terminal nerve bodies (Krause's corpuscles) are oval, or more or less rounded vesicles of 0·03–0·06 mm, consisting of an areolar sheath, with nuclei and soft homogeneous contents, in which the nerve-fibre runs, terminating in a point. **1888** A. Flint *Text-bk. Human Physiol.* (ed. 4) xvi. 516 In certain membranes the nerves terminate in end-bulbs, or corpuscles of Krause. **1953** C. E. Osgood *Method & Theory Exper. Psychol.* i. 7 Pain is attributed to free nerve endings; cold to Krause's end bulbs. **1960** R. A. Weale *Eye & its Function* ii. 30 Their shape is reminiscent of one type of pressure receptor, namely Krause's corpuscle. **1972** M. L. Barr *Human Nervous Syst.* iii. 35/2 It was thought at one time that end-bulbs of Krause were receptors for coolness and end-bulbs of Ruffini for warmth, but the evidence that these are special temperature sensors is no longer regarded as conclusive.

b. 1873 *Phil. Trans. R. Soc.* CLXIII. 442 The transverse membrane of Krause and the narrow disk of intermediate substance..of Heppner would seem to be referable to the heads of two contiguous series of muscle-rods, which..often meet in the middle of the clear stripe. **1887** *Buck's Handbk. Med. Sci.* V. 65/2 Krause's membrane or intermediate disk. **1939** W. E. Le Gros Clark *Tissues of Body* v. 99 Cutting through the isotropic disc is a thin dark line which is believed by some histologists to mark the position of a membrane (Krause's membrane) separating the length of the fibril into a series of partially isolated segments. **1970** Fallis & Ashworth *Textbk. Human Histol.* viii. 58 The major cross bands of the myofibrils are the A and I bands... In the center of the I band is a thin dark line known as the Z (Zwischenscheibe) band, or intermediate line of Krause; the Z band attaches to the sarcolemma at the periphery of the cell.

c. 1905 *Trans. Ophthalm. Soc.* XXV. 1 (*heading*) Cyst of Krause's gland. **1933** E. Wolff *Anat. Eye & Orbit* iii. 113 The glands of Krause are accessory lacrimal glands occurring under the conjunctiva from the fornix to the convex border of the tarsus. **1960** R. A. Weale *Eye & its Function* xii. 185 Accessory lacrimal apparatus is provided by Krause's glands.

krausite (krau·səit). *Min.* [f. the name of Edward Henry *Kraus* (1875–1973), U.S. mineralogist + -ite[1].] A pale yellow hydrated sulphate of potassium and iron, $KFe(SO_4)_2 \cdot H_2O$.

1931 W. F. Foshag in *Amer. Mineralogist* XVI. 352 In the Calico Hills, San Bernardino County, California, a small deposit of sulfates of iron [was found]. This deposit was found to contain a fair abundance of a new sulfate of iron and potash... For this new mineral species the writer proposes the name *krausite*. *Ibid.*, Krausite occurs in several different forms. That found in the alunite forms comparatively large but rough crystals... The krausite of the transition zone is in lemon yellow crystals, often clear and with brilliant luster. **1935** J. W. Mellor *Comprehensive Treat. Inorg. & Theoret. Chem.* XIV. 340 Krausite is insoluble in water; and is slowly hydrolyzed when left in contact with water. **1965** *Amer. Mineralogist* L. 1929 This spatial arrangement of the coordination polyhedra accounts for the perfect (001) and good (100) cleavages in krausite.

kraut (kraut). [a. G. *kraut* herb, vegetable, cabbage.] **1.** = Sauerkraut, sourcrout. Also *attrib.* and *Comb.*

[**1845** *Punch* IX. 94/1 Happy midst his native *kraut* [*sc.* cabbage] My princely Albert wanders.] **1855** Geo. Eliot in *Fraser's Mag.* June 311/1 *Kraut* and *wurst* may be

called the solid prose of Thuringian diet. **1895** *Montgomery Ward Catal.* 574/1 Kraut Forks..Combined Clothes Pounder and Kraut Stamper. *Ibid.* 576/2 Kraut Cutters, 8 × 26 inches. 3 cast steel knives. **1937** D. Runyon in *Collier's* 16 Jan. 47/2 Nicely-Nicely now observes that the very choicest spareribs are on Jake's plate, and also the most kraut. **1950** *Amer. Speech* XXV. 315/1 The recipe for krautfurters..under the title 'Kraut and Frankfurters Team in Delightful Casserole Dishes'. **1961** *John o' London's* 13 July 88/3 Such words as..turkeyfurter and krautfurter: the last-named being made from *sauerkraut*.

2. (Often with capital initial.) A German, esp. a German soldier. Also *attrib.* and *Comb.* *slang* (orig. *U.S.*).

1918 G. E. Griffin *Ballads of Regiment* 34 But he always loved a soldier, be he..'Krout' or 'Mick'. **1919** C. B. Hoyt *Heroes of Argonne* 41 The Frogs and Krauts got it fixed up between 'em. **1926** *Sat. Even. Post* 12 June 148/3 'Wait a while, fellers,' he said. 'The krauts are sockin' the crossroads.' **1929** W. T. Scanlon *God have Mercy on Us!* 4 What gives me a pain in the neck is all the time we wasted up at Verdun in the old trenches when we might have been killing Krautheads. **1936** *Our Army* Feb. 14 We and the kraut-eaters were mixing it up to make the world safe for bigger and better wars. **1938** Johnson & Pratt *Lost Battalion* 17 There were 'kraut' machine guns barking at them in the dark. **1945** *Daily Herald* 8 May 4/4 The men just said things like, 'Well, the Krauts are done for.' **1946** J. M. Scott *Other Side of Moon* ii. 32 It is the only building left standing in the village. The Krauts blew up the rest. *Ibid.* v. 86 The Kraut will soon have to fall back. **1954** W. Faulkner *Fable* 376 He's got to be killed from in front, by a Kraut bullet—see? **1957** M. K. Joseph *I'll soldier no More* (1958) xiii. 291 I'll fix you, you kraut bastard. **1962** R. Cook *Crust on its Uppers* i. 24 We were doing some biz near Munich.. back to this day in krautland. **1966** T. Pynchon *Crying of Lot 49* i. 15 Maybe..he should have been in a war, Japs in trees, Krauts in Tiger tanks. **1971** J. Osborne *West of Suez* i. 27 An odd Kraut or two, bellowing at their Fraus. **1974** *New Society* 21 Feb. 435/2 The Hitler stuff, together with the usual illustrations of inflated multi-million mark notes..looks more like what is inelegantly called 'Kraut-bashing'. **1974** L. Deighton *Spy Story* xviii. 195 'That goddamn Kraut sub,' said the Conning Officer.

kray (krai). Also **krai.** [Russ.] In the U.S.S.R., a second-order administrative division, a region, a territory.

1938 N. de Basily *Russ. under Soviet Rule* iv. 153 Practically each republic is administratively divided into provinces or regions (*oblast* or *krai*). These provinces and regions are in turn divided into districts (*rayon*), towns, and villages. **1951** T. Shabad *Geogr. U.S.S.R.* ii. 46 An oblast or kray is usually organized in such a manner as to include..a well-coordinated economic region centering on an important industrial and commercial center and specializing in the production of some particular commodity, while striving for regional self-sufficiency in the greatest possible degree. *Ibid.* 47 Krays, which occur only in the Russian SFSR, are identical to the oblasts, but are usually larger in area—the term 'kray' is given to those divisions which contain autonomous oblasts. **1967** J. P. Cole *Geogr. U.S.S.R.* vi. 101 In Siberia some krays and oblasts are enormous, and extend from the Trans-Siberian Railway as far as the Arctic Coast.

Krebs (krebz). *Biochem.* The name of Sir Hans Adolf *Krebs* (b. 1900), German-born British biochemist, used *attrib.* (or occas. in the possessive) to designate a circular sequence of enzyme-catalysed reactions occurring in mitochondria as part of cell respiration in aerobic organisms, in which an acetyl group (bound to a coenzyme and produced by glycolysis or other catabolic processes) is combined with oxaloacetic acid and then oxidized by a succession of reactions which produces carbon dioxide, serves to convert adenosine diphosphate to the energy-rich triphosphate (by means of the cytochrome system), and regenerates oxaloacetic acid.

1941 *Jrnl. Biol. Chem.* CXXXIX. 483 Oxidation of pyruvate by the Krebs cycle involves its union with oxalacetate. **1950** *Chem. Abstr.* XLIV. 1552 (*heading*) Final stage in biological oxidation processes: Krebs' tricarboxylic acid cycle. **1955** *New Biol.* XIX. 85 The Krebs cycle is a mechanism through which the various synthetic activities of the cell mutually affect each other. **1969** *New Scientist* 9 Oct. 64/3 Because some of the intermediate compounds in the Krebs cycle need to be bled off for other purposes, the bacillus has a special mechanism for stoking up its level from glucose. **1971** *Sci. Amer.* May 128/2 Next are the tens or hundreds of mitochondria, the enclosures where the complicated biochemical machinery of the Krebs cycle is mounted, producing most of the cell's ATP fuel by the oxidation of small organic molecules. **1971** Yudkin & Offord *Harrison's Guidebk. Biochem.* (new ed.) xi. 101 The Krebs cycle is responsible for the oxidation of the acetyl group of acetyl coenzyme A.

kredemnon. (Later examples.)

1865 E. Rimmel *Bk. Perfumes* v. 94 There were many other modes of wearing the hair, such as the *strophos*, the *nimbo*, the *kredemnon*. **1940** *Times* 19 Mar. 6/3 The captains of two East Coast colliers exchanging views..on the merits of their kredemnons. **1959** E. Pound *Thrones* xcvi. 3 Κρήδεμνον..κρήδεμνον..and the wave concealed her, dark mass of water.

kreef (krīf, krēf). *S. Afr.* [Afrikaans, f. Du. *kreeft* lobster.] = *crawfish *sb.* 1 b.

1863 *Queenstown Free Press* 30 June (Pettman), A new theological schism has sprung up amongst the Malays touching the important question whether *Kreef* or crawfish is to be considered ceremonially unclean or not. **1902** [see *Cape lobster* (*Cape *sb.*[3] 4)]. **1936** *Nature* 11 Jan. 74/1 (*heading*) The natural history and utilisation of the Cape crawfish, kreef or spiny lobster, *Jasus* (*Palinurus*) *lalandii*. **1959** *Cape Argus* 7 Nov. 11/7 The three month open season for kreef fishing..opened last Sunday. **1970** *Cape Times* 28 Oct. (S.A. Fishing Rev.) 5/4 Previously the Government had allowed producers to export 6·9 m. tailweight pounds of kreef. **1970** G. Croudace *Scarlet Bikini* vi. 73 'And no *kreef*,' Tony said, referring to the rock lobster.

KREEP (krīp). [f. *K*, chem. symbol for potassium + *REE*, abbrev. for *r*are-*e*arth *e*lement + *P*, chem. symbol for phosphorus (in allusion to its unusual composition).] A substance found on the moon as glassy fragments and as a constituent of fines and breccias, characterized by a high content of potassium, phosphorus, and rare-earth elements and unusually little iron.

1971 *Sci. News Let.* 23 Jan. 62 Another Apollo 12 find ..was that of an exotic component called KREEP by some—for high content of potassium, rare earth elements and phosphorus—found in rock 13. **1971** N. J. Hubbard et al. in *Earth & Planetary Sci. Lett.* X. 343/1 We will refer to this distinctive glass as KREEP glass (from its distinctive potassium, rare-earth element, and phosphorus content). **1971** *Geochim. & Cosmochim. Acta Suppl.* 2 I. 393 Hubbard et al. (1971a) designated this class of material as the KREEP component. *Ibid.* 398 Breccias with glass matrices..are the most abundant form of KREEP in the soil samples. **1973** *McGraw-Hill Yearbk. Sci. & Technol.* 284/2 The activity fell off smoothly in the adjacent regions to the east and west, including the other near-side maria. ..The localization of the high U-Th regions indicates a restricted distribution of KREEP (a potassium, rare-earth, phosphate-bearing phase), with which the high U-Th component is associated in returned lunar samples.

Kremlin. Add: *The Kremlin* (in Moscow): (used for) the government of the U.S.S.R. Also *transf.* (in trivial use).

1933 H. G. Wells *Shape of Things to Come* II. § 9.211 The Kremlin was content to consolidate the kindred Slav Soviets. **1943** *Sun* (Baltimore) 28 Sept. 12/2 Moscow itself has taken pains to assure us that Mr. Browder has no direct pipeline into the secret councils of the Kremlin. **1961** *Evening Bull.* (Philadelphia) 5 Mar., The Kremlin is the only world capital powerfully arrayed against crossing the new frontier. **1966** [see *Elysée]. **1966** H. Sheppard *Dict. Railway Slang* (ed. 2) 7 *Kremlin*, British Rail Headquarters. **1973** *Times* 19 Feb. 18/7 People who referred to their head offices as 'the Kremlin'..were somehow lacking in motivation. **1974** *Times* 10 Apr. 18/2 Kremlin-watchers study the pages of *Pravda* for the slight shift in nuance.

So **Kremlino·logy,** the study and analysis of the Soviet Government and its policies; **Kremlinolo·gical** *a.*; **Kremlino·logist,** such an analyst; also *transf.*

1958 *Oxf. Mag.* 13 Feb. 289/1 For all his interest in Kremlinology..the author is not really very good at it. **1960** *Daily Tel.* 7 Dec. 12/2 Kremlinologists, versed in the mysteries of Marx and Lenin, Mao Tse-tung and Mr Khruschev, tell us that the signs are that Mr Khruschev has won yet another battle, at least on points. **1961** *Spectator* 2 June 806 The approach commonly nicknamed 'Kremlinological'. *Ibid.*, Exaggerated claims on behalf of this method by some of its practitioners..who..equate Kremlinology with Soviet studies as a whole. **1968** *Guardian* 2 Apr. 11/5 The 'Kremlinological' expertise which enabled the White House to play the Kremlin power game. **1970** *Ibid.* 1 Oct. 15/4 The question now teasing Labour Kremlinologists is how long Denis [Healey] will be satisfied to remain a mere member of the NEC. **1971** *Times* 22 Jan. 8/8 Kremlinology gone wild. **1972** A. Ulam *Fall of Amer. University* i. 35 The budding Kremlinologists were put in their place, which often and quite properly turned out to be the C.I.A.

Kremnitz, var. *Cremnitz.

Krems (kremz). [f. *Krems*, the name of a town of northern Austria.] Used *attrib.* to designate a white lead pigment used as a paint base; the same as Cremnitz white.

1854 F. W. Fairholt *Dict. Terms Art* 256/2 Krems White. A carbonate of lead;...it takes its name from the city where it is manufactured, Krems or Crems, in Austria, and is sometimes termed Vienna white. **1940** *Chambers's Techn. Dict.* 479/2 Krems white.

‖ **kreplach** (kre·plaχ), *sb. pl.* Also **creplach, -lich.** [Yiddish *kreplech*, pl. of *krepel*, ad. dial. G. *kräppel* fritter, cogn. w. G. *krapfen.] Triangular noodles filled with chopped meat or cheese and served with soup.

1892 I. Zangwill *Childr. Ghetto* I. 114 *Creplich*, which are triangular meat pasties. **1932** L. Golding *Magnolia St.* III. vii. 555 She will also make *knishehs*, *varennikas*, *creplach* and *blintsies* for the old-fashioned Jewish people. **1954** *Amer. Speech* XXIX. 104 Though partial to gefilte fish, They scorn the kreplach and the knish, They list pastrami, but evince A total apathy to blintz. **1957** L. Stern *Midas Touch* viii. 66 The thought of *kreplach* and *matzoth* dumplings made his mouth water. **1965** *New Statesman* 20 Aug. 250/1 Kreplach (sing. krepl) are a kind of Jewish dumpling served in hot chicken soup on festive occasions... Chinese serve remarkably good kreplach in.. soup... Other close cousins of kreplach are ravioli and especially tortellini. **1972** F. B. Maynard *Raisins & Almonds* 89 My mother was making *kreplach*. She placed a dab of cheese on the dough and pinched it in carefully... The *kreplach* dropped into the boiling pot with a sharp plopping noise.

‖ **kretek** (kre·tek). [Indonesian *kerétèk*.] In Java, a cigarette containing cloves.

1958 H. Forster *Flowering Lotus* i. 6 These were the famous *kretek* cigarettes, in which the tobacco is mixed with cloves; their crackling gives the cigarettes their name. **1966** *Economist* 16 Apr. 235/2 A packet of *kretek*, the clove cigarette smoked by most people, costs almost as much. **1973** *Times* 27 Dec. 10 It also turns out that although Kawung smoking is in general associated with low socio-economic status, and that low status itself affects development of the disease, the relatively small numbers of patients with higher socio-economic status who smoke these cigarettes are also adversely affected, compared with patients of the same status who smoke Kretek or Western cigarettes.

kriegie (krī·gi). *slang.* [Abbrev. of G. *kriegsgefangener* prisoner of war: see -ie.] An Allied prisoner of war in Germany during the war of 1939–45 (see also quot. 1948).

1944 *World's Press News* 31 Aug. 17/1 The *Yorkshire Post* reports receipt of a remarkable publication from Yorkshire prisoners of war in Germany. This takes the form of a special volume called the 'Kriegie Edition' of the *Yorkshire Post*. *Ibid.* 17/2 Kriegie..is an abbreviation of the German for prisoner of war. **1946** Brickhill & Norton *Escape to Danger* 11 The worn track..which kriegies 'pounded' or 'bashed' (walked) for hours at a time. **1948** *Amer. Speech* XXIII. 217 (*heading*) Kriegie talk. *Ibid.* 218 Ex-prisoners from the E.T.O. speak of Jap kriegies, or, 'If there is another war I'll probably be a Russian kriegie,' and my personal usage would apply it to any prisoner of war from now on. **1956** D. M. Davin *Sullen Bell* II. vii. 153 But there I was, a bloody kriegie for the rest of the war.

kriegspiel. Add: **2.** A form of chess invented about 1900 by M. H. Temple. Two players at separate boards play without seeing or being told each other's moves, though they may ask some strictly limited questions of an umpire who conducts the game at a third board.

1903 *Brit. Chess Mag.* Sept. 370 Kriegsspiel..may have merits, but it is not a war game. **1906** H. Cayley in *Chess Amat.* Nov. 46 (*heading*) Kriegspiel or the Chess War Game. **1922** *Brit. Chess Mag.* Oct. 375 As a chess-player he..preferred Kriegspiel, at which he was always the life and soul of the table. **1961** *Times Lit. Suppl.* 1 Dec. 870/5 Kriegspiel itself together with further variations. **1969** A. Glyn *Dragon Variation* viii. 237 There'd be simuls, rapid-transit, five-minute chess, kriegspiels, lectures in the evenings.

So **krie·gspieler, krie·gsspieler,** one who plays kriegspiel (in the senses of Dict. and Suppl.).

1891 *19th Cent.* Feb. 299 Keen Volunteers..are enthusiastic Kriegsspielers. **1916** *Brit. Chess Mag. Chess Ann. 1915* 58 The chartered libertine among P.E.'s, kriegspielers, and chess-players in general.

Krilium (kri·liŏm). [f. *kril-*, altered form of *-cryl-* of *polyacrylonitrile* + *-ium*.] A proprietary name of various mixtures of polyacrylate salts and other carboxylated polymers manufactured as soil conditioners for improving the texture of soil and its ability to resist erosion.

1952 *Sci. News Let.* 5 Jan. 8/2 The soil improvement chemical will come on the market soon under the name of Krilium. **1952** *Official Gaz.* (U.S. Patent Off.) 8 July 226/2 *Krilium* for synthetic resin materials in the form of powders, granules, emulsions, dispersions, and solutions. **1958** Teakle & Boyle *Fertilizers* iv. 116 Krilium is the name given to several compounds selected from over seven hundred chemical synthetics tested by Monsanto. Two of these, VAMA 6—Krilium (6 vinylacetate: maleic acid complex) and HPAN 9—Krilium (9 polyacrylonitrile) have earned good reports. **1967** G. W. Cooke *Control of Soil Fertility* xxix. 458 'Krilium' produced an improvement in the percentage of soil aggregates which were stable in water.

krill (kril). Also **kril.** [ad. Norw. *kril* very small fry of fish.] A small, shrimp-like crustacean of the order Euphausiacea, or a large group of these animals, forming food for fishes and whales.

1907 J. G. Millais *Newfoundland* viii. 164 In June.. whales suddenly become extremely scarce, owing..to the trend seawards of the stream of 'kril' or red shrimp, on which the great *Balænoptera* subsist. **1912** *Rep. Brit. Assoc. Adv. Sci.* 178 A small red crustacean...forms the 'krill' of the whalers. **1928** Russell & Yonge *Seas* v. 117 Euphausiids, or 'krill', as they are called by the Norwegians..are about an inch and a half in length, but are so abundant that they form a large part of the food of many of the northern fishes, and are the chief food of nearly all of the whalebone whales. **1931** *Discovery* XII. 317/1 The catches of plankton show that the shrimp-like crustaceans, or 'kril', on which the whales feed are not.. to be found only in local concentrations, but are spread over immense areas in the open ocean. **1959** A. Hardy *Fish & Fisheries* i. 10 The largest of these whales, the rorquals,..specialise in feeding upon the krill (euphausiacean shrimps) which, though large for plankton animals, are but an inch or so in length. **1970** *Sci. Amer.* Dec. 20/3 Perhaps a crop of suitably large zooplankton such

as krill—the shrimplike animals that are the principal food of the baleen whales in the Antarctic—could be raised in a fertilized lagoon.

krimmer (kri·məɪ). Also **crimmel, crimmer, krimma.** [G., f. *Krim* (Russ. *Krym*) Crimea (see *Crimean *a.*).] A grey or black fur made from the wool of young lambs in or near the Crimea; an imitation of this. Cf. *astrakhan b, *karakul b.

1834 *Penny Cycl.* II. 519/2 The lamb yields a fine and beautiful fleece, which the dealers call a 'crimmel', the bulk of them being imported from the Crimea. **1892–3** *T. Eaton & Co. Catal.* Fall & Winter 10/2 The same styles of garments trimmed in a hundred different ways with krimmer, imitation lamb, persian-lamb, beaver, otter, nutria, sable, [etc.]. **1904** *Westm. Gaz.* 28 Jan. 4/2 Chinchilla or krimmer. **1906** *Ibid.* 3 Nov. 13/1 Grey krimmer. **1923** *Ibid.* 26 Jan. 11/1 A definition of 'crimmer lamb', as a commodity in the fur trade, was agreed upon at Marlborough-street yesterday. **1929** [see *karakul]. **1930** *Economist* 1 Nov. (Russian Suppl.) 37/2 Hare-Peschanik, Persian Lamb, Persian Lamb-Broadtail, Crimmer Lamb. **1949** *Amer. Speech* XXIV. 96 Another common type of Persian lamb is the krimmer... It is characterized by heavier fur and looser curl. Other strains of lamb can be dyed to simulate krimmer.

Krio (krī·,o). [Native name.] An English-based Creole language in Sierra Leone. Also *attrib.* or as *adj.*

1955 P. Strevens *Papers in Lang.* (1965) ix. 116 There is the Freetown Krio, an English-based language, containing many borrowings from Yoruba. **1957** M. Banton *W. Afr. City* i. 6 The Creoles..developed a distinctive dialect version of the English language, known as Krio, which incorporates Portuguese, African, and other loan words, has an African rather than a European syntax, and is incomprehensible to the untrained English ear. Krio is to be distinguished from the native pidgin. *Ibid.* ix. 166 *Geda* is the Krio form of the English 'together'. **1961** *Guardian* 16 Feb. 6/4 Krio, the special Creole patois spoken in Sierra Leone. **1963** *Ann. Reg. 1962* 452 The threat of a collapse into a dialect deriving from English but not easily understood, a dialect less accessible than the base but genuine *lingua franca* of 'Pidgin': Krio, the language spoken at Freetown, was an example. **1972** W. B. Lockwood *Panorama Indo-Europ. Lang.* 119 Creole English has also developed in West Africa. Krio 'Creole', centred on Freetown, Sierra Leone, is of complex origin.

kris, kriss. Now the usual forms of Creese *sb.*

1883 [see Creese *sb.*]. **1895** Conrad *Almayer's Folly* xi. 220 With a shout and a leap he would be in the midst of them, kriss in hand, killing, killing, killing. **1927** R. J. H. Sidney *In Brit. Malaya To-Day* 62 She will give Raja Besi a *kris* by means of which he will be able to kill the Jin. **1935** Wodehouse *Blandings Castle* iv. 107 Malays, when pushed past this point, take down the old *kris* from its hook and go out and start carving up the neighbours. **1948** W. S. Maugham *Here & There* 315 He awakened just as he thought a kriss was being drawn across his throat. **1953** *News Chron.* 2 June 7/1 In the first carriage is the Sultan of Kelantan... He carries a kris (a dagger) made from an elephant tusk. **1964** R. Perry *World of Tiger* vii. 106 The bull succumbed to stabs from poisoned *krises* tied to poles. **1965** R. McKie *Company of Animals* i. 16 An ancient Malay kris with ivory scabbard and garuda hilt. **1972** M. Sheppard *Taman Indera* 127 A short beak and a pair of folded arms can be recognized on many Malay kris-hilts to this day.

Krishna (kri·ʃnă). The name of a Hindu deity or hero (see Krishnaism), used *attrib.* to designate Krishnaism or followers of this cult. Cf. *Hare Krishna.

1875 Monier Williams *Indian Wisdom* xii. 332 Kṛishṇa-worship is comparatively modern. **1895** E. W. Hopkins *Religions of India* xv. 405 One may represent the attitude of a Krishna-worshipper in the epic somewhat in this way. *Ibid.* 411 The regular Vishnuite laudation affected by the Krishna sect. *Ibid.* xvi. 469 The parallels between the latest Krishna cult and the Biblical narrative are found..in the Puranas. **1958** Gerth & Martindale tr. *Weber's Religion of India* (1960) iv. 138 Vishnu..is honored..as patron of the dance drama and erotic orgies of the Krishna-cult. *Ibid.* v. 188 The ardent love of the redeemer of the later Krishna religion. **1970** J. Needleman *New Religions* (1972) x. 212 Who has walked the streets of any American..cities without encountering the young followers of the Krishna Consciousness movement? **1971** [see *Hare Krishna]. **1973** R. Thomas *If You can't be Good* (1974) vi. 44 A fairly old town house..a couple of doors or so down from the Krishna kids.

Kriss Kringle (kri:s kri·ŋg'l). *U.S.* ?*Obs.* Also **Christ-kinkle, Kriskringel, Krisking'l, Kris Kringle.** [See quot. 1919.] = Santa Claus.

1830 J. F. Watson *Annals of Philadelphia* 242 Every father in his turn remembers the excitements of his youth in Belsh-nichel and Christ-kinkle nights. **1849** J. Rees *Myst. City Life* 93 Do you think Kris Kringle will come down the chimney to-night? **1864** *Sacramento Union* 7 Jan. 5/2, I do not know whether the good Saint Nicholas.. answers in those regions to the musical title of Kriss Kringle, as in the Queen City. **1919** H. L. Mencken *Amer. Lang.* iii. 89 Another example of debased German is offered by the American *Kriss Kringle*. It is from *Christkindlein*, or *Christkind'l*, and properly designates, of course, not the patron saint of Christmas, but the child in the manger. A German friend tells me that the form *Kriss Kringle*, which is that given in the Standard Dictionary, and the form *Krisking'l*, which is that most commonly used in the United States, are both quite unknown in Germany. **1928** G. Ade *Let.* 9 Oct. (1973) 136 We really believed that Kris Kringle could..go down chimneys which were not large enough to take care of a nest of barn swallows in the summer time. **1947** *Chicago Tribune* 20 July IV. 3/1 It is the story of an old man whose name was Kris Kringle and who believed he really was Santa Claus.

kriti (kri·ti). Also **krithi.** [Skr. *kṛiti* the act of composing.] In the music of southern India, a song, often devotional in character, which is deliberately composed and not an improvisation on a set theme.

1914 A. H. F. Strangways *Mus. Hindostan* iii. 84 In Mudaliar Chinnaswami's *Oriental Music* sixty of his [*sc.* Tyagarāja's] songs (*Kritis*) are printed in staff notation. **1957** O. Gosvami *Story Indian Mus.* xix. 211 The *Kriti* is the most developed type of musical composition in the South. **1969** *Indian Express* (Bombay) 28 July 3/2 The krithi 'Naradamuni' in this raga was extremely pleasing. **1972** P. Holroyde *Indian Mus.* iii. 107 Kritis are sung in all tempos but without the long monosyllabic passages of the alamkaras or ornamentations.

Kromayer lamp (krōᵘ·məi,əɪ læmp). *Med.* [f. the name of Ernst *Kromayer* (1862–1933), German dermatologist.] A water-cooled mercury-vapour lamp used therapeutically for local ultra-violet irradiation.

1911 *Allbutt's Syst. Med.* (ed. 2) IX. 480 The comparative value of the Finsen-Reyn and Kromayer lamps has been studied by Maar. **1927** *Observer* 18 Dec. 9/2 We [*sc.* Bermondsey Council] have eight large mercury vapour lamps, two carbon arcs, and one water-cooled Kromayer lamp. **1949** E. B. Clayton *Electrotherapy & Actinotherapy* xx. 342 The Kromayer lamp is designed for local irradiation only. It has the advantages that, being water-cooled, it can be held in contact with the skin [etc.].

kromesky, -eski (krome·ski, krǫ·meski). Also **crom-, -esque, -esqui.** [ad. Polish *kroméczka*, little slice.] A croquette made of meat or fish minced, rolled in bacon or calf's udder and fried.

1846 C. E. Francatelli *Mod. Cook* 309 Croquettes of Fowl and Mushrooms. The mince for these is prepared in the same way as for *Kromeskys*. **1861** —— *Cook's Guide* 120 Kromeskys are made with all kinds of croquet preparations, whether of meat, fish, or shell-fish. **1884** *Girl's Own Paper* May 428/1 A dozen oysters will make a moderate sized dish of kromeskies. **1892** *Encycl. Pract. Cookery* I. 478/2 Some authorities spell it Cromesquis, some Kromesquis, and others Kromeskies. **1920** E. Silvester *Sensible Cookery* 54 Kromesques of veal. **1928** *Evening News* 28 Dec. 4/5 Kromeskies of Turkey. **1951** *Good Housek. Home Encycl.* 348/2 A coating batter is used for making fritters, kromeskies, etc.

kromnek disease (krǫ·mnek). *S. Afr.* [Afrikaans, f. Du. *krom* crooked + *nek* neck.] A local name for the spotted wilt virus disease of tomato; also, a similar disease of tobacco.

1932 *Grocott's Mail* (S. Afr.) 2 Apr. 4 'Kromnek' or Kat River Wilt of tobacco is a disease which is not known anywhere in the world outside the Cape Province. **1933** E. S. Moore in *Sci. Bull. Dept. Agric. Union S. Afr.* No. 123. 1 The disease known locally as 'Kromnek' or 'Kat River Wilt' is by far the most serious of the diseases of tobacco in the Stockviström division. *Ibid.* 15 The evidence indicated not only that the tomato disease is of virus nature but that probably it is caused by the same virus which is responsible for the tobacco kromnek. **1941** *Nature* 19 Apr. 480/2 Control of the Kromnek (Spotted Wilt) Disease of Tomatoes. **1957** K. M. Smith *Textbk. Plant Virus Dis.* (ed. 2) 572 Tomato Spotted Wilt Virus... Synonyms: T.S.W. Virus; Kromnek or Kat River Disease Virus.

Kromo (krōᵘ·mo). *Indonesia.* Also **Krama.** [ad. Javanese *krama*, Indonesian *keromo*.] The polite form of Javanese, used by those of lower status when addressing social superiors.

1817 T. S. Raffles *Hist. Java* I. viii. 366 Nearly one half of the words in the vernacular language have their corresponding term in the *Bása Kráma* or polite language, without a knowledge of which no one dare address a superior. **1893** W. B. Worsfold *Visit to Java* xiii. 228 Two forms of modern Javanese are employed in everyday speech. First, the language of ceremony, called Krama; and secondly, the common speech, or Ngoko (meaning literally the thou-ing speech). **1925** P. Radin tr. *Vendryès's Lang.* IV. ii. 257 Among the natives of Java, a superior speaks to his inferior in Ngoko, but the inferior answers in Kromo. **1932** W. L. Graff *Lang.* xi. 423 Official Javanese is called *kromo*. **1948** D. Diringer *Alphabet* II. vii. 424 Modern Javanese gradually breaks up into.. Krama Inggil, a form of speech used in addressing gods and the aristocracy, [etc.].

kromogram (krōᵘ·mŏgræm). *Photogr. Obs.* exc. *Hist.* Also **Kromogram.** [Altered form of *chromogram.] = *chromogram.

1897 *Brit. Jrnl. Photogr.* 8 Jan. 18/1 We understand that kromskops and sets of kromograms will shortly be obtainable from the Photo-chromoscope Syndicate Limited. **1898** [see *kromskop]. **1969** H. & A. Gernsheim *Hist. Photogr.* (ed. 2) xliii. 523 In his Photochromoscope camera (1893) three separation negatives were taken in succession on one plate by means of a repeating back containing red, green and blue-violet filters. From these, diapositives were made by contact printing. When cut into the three separations and laid on the Kromskop viewing instrument (1892) containing filters of the same colour and mirrors, the optically superimposed Kromograms appear in perfect colour.

kromskop (krōᵘ·mskōᵘp). *Photogr. Obs.* exc. *Hist.* Also **Kromskop.** [f. *krom-*, Ives's altered form of Chromo- + *-skop*, altered form of -scope.] A viewer for the three positives of a chromogram, enabling them to be visually combined and seen as a single coloured picture. Freq. *attrib.* (in Ives's use).

1897 *Brit. Jrnl. Photogr.* 8 Jan. 17/2 The Kromskop is the name finally chosen for Mr. Fred. E. Ives's perfected stereo-photo-chromoscope. **1898** F. E. Ives *Krōmskōp Color Photogr.* i. 6 The Kromogram must be placed in the Krōmskōp in order to visually reproduce the object photographed. *Ibid.* xi. 36 It is absolutely necessary to use in any of the Krōmskōp cameras the kind of orthochromatic sensitive plates for which that particular camera has been adjusted. **1928** O. Wheeler *Colour Photogr.* iii. 14 Almost equally obsolete, though still of considerable interest and comparatively practical from the amateur standpoint, is the additive arrangement adopted in an instrument..variously designated a photochromoscope, chromoscope or kromskop. **1968** E. de Maré *Colour Photogr.* ii. 43 Ives also produced the Lantern Photochromoscope which projected the three pictures first through red, green and blue filters and then through three lenses directed to form a single image on a screen. He improved his camera and a viewer, which in a new form he called the Kromscope [*sic*]. It revealed stereoscopic pictures in full colour and this he demonstrated in 1896.

króna (krōᵘ·nă). Also **krona.** Pl. **krónur.** [Icel.; cf. Krone.] The basic monetary unit of Iceland; also a coin representing one króna.

1886 R. Senf *Illustr. Postage Stamp Album* 47 Iceland ..Coinage:..100 öre (= aur) = 1 krona. **1922** P. A. Ólafsson *Iceland* 24/2 The coinage is the same as elsewhere in Scandinavia, i.e. 1 króna (pl. krónur) = 100 aurar. **1938** *Encycl. Brit. Bk. of Year* 321/2 Currency unit: *króna* (exchange 22·15 krónur = £1). **1946** K. Tryggvason in T. Thorsteinsson *Iceland* (ed. 4) 164 Up to the end of the [1914–18] war there was little or no difficulty in keeping the Icelandic króna at par with the Danish Krone. **1958** *Spectator* 6 June 726/3 The Communists in the Icelandic Government are playing the fishing dispute for all they're worth, and trying to evade collective responsibility for economic measures which will drastically devalue the Kronur. **1970** D. Bagley *Running Blind* i. 28 A 100-kronur banknote.

Kronecker delta (krōᵘ·nekəɪ de·ltă). *Math.* [f. the name of Leopold *Kronecker* (1823–91), Ger. mathematician + Delta.] A function of two integers defined as equal to one if the integers have the same value and zero otherwise; symbol δ_{ij} or δ_i^j. Also (the *generalized Kronecker delta*), a function of 2*k* integers that takes the values 0 or ±1 (see quot. 1927[2]) (*rare*).

1927 O. Veblen *Invariants Quadratic Differential Forms* i. 3 The theory of determinants and allied expressions is essentially a theory of alternating sets of quantities, and can be made to depend on certain fundamental alternating sets of quantities which have only the values 0 and +1 and −1. These sets of quantities are known as generalized Kronecker deltas because of their analogy with the Kronecker delta which is already well known. *Ibid.*, The generalized Kronecker delta has *k* superscripts and *k* subscripts, each running from 1 to *n*... If the superscripts are distinct from each other and the subscripts are the same set of numbers as the superscripts, the value of the symbol is +1 or −1 according as an even or an odd permutation is required to arrange the superscripts in the same order as the subscripts; in all other cases its value is 0. **1937** A. A. Albert *Mod. Higher Algebra* (1938) x. 228 Our rule for multiplying matrices implies that $E_{ij}E_{kl} = \delta_{jk}E_{il}$ ($i, j, k, l = 1, \ldots, s$) where δ_{jk} is the Kronecker delta. **1961** P. E. Pfeiffer *Linear Systems Analysis* ii. 30 The value Δ of a determinant whose elements are δ_{ik} is unity, where δ_{ik} is the Kronecker delta.

Kru: now the more usu. spelling of Kroo.

1884, 1897 [see Kroo]. **1957** M. Banton *W. Afr. City* i. 5 Another important group was that of the Kru. They are members of a sea-faring tribe inhabiting part of what is now the Liberian coast. **1970** P. Oliver *Savannah Syncopators* 41 The Kru speaking peoples of Liberia. **1973** *Times* 17 Apr. (Liberia Suppl.) p. vi/5 Elizabeth Tonkin has recently been from Birmingham on a study of Kru linguistics.

krug[1] (krug). [Ger.] A beer-mug or tankard.

1866 C. M. Yonge *Dove in Eagle's Nest* II. xii. 235 He has..excused himself from aiding his two gentlemen-squires in consuming their krug of beer. **1963** T. Pynchon *V.* viii. 203 They seemed to seek some Hofbrauhaus of the spirit like a grail, hold a krug of Munich beer like a chalice.

Krug[2] (krug). The proprietary name of a champagne made by the firm of Krug et Cie. of Reims.

1876 *Trade Marks Jrnl.* 6 Sept. 455/1 Krug & Co. Reims. .. Paul Krug, of and on behalf of the firm of Krug and Co., Reims, France; champagne wine merchants. **1891** in C. Ray *Compleat Imbiber* (1967) IX. 122 All Brands of Champagne in stock..Pol Roger, Krug, Moët et Chandon. **1920** G. Saintsbury *Notes on Cellar-Bk.* v. 71 Taking well-known brands..I do not know that I was more faithful to any than to Krug. 1967 G. Smith in L. Deighton *London Dossier* 125 The best game pies..ideal for demolishing with a bottle of Krug in the car park before the Oxford and Cambridge game at Twickenham. 1974 D. MacKenzie *Zaleski's Percentage* xv. 222 'Champagne for my friend, Inspector-Detective.' It was vintage Krug and perfectly chilled.

Krugerism (krū·gəriz'm). *Hist.* [f. the name of Stephanus Johannes Paulus *Kruger* (1825–1904), president of the Transvaal 1883–1901.] The nationalist (pro-Boer) policy of President Kruger. So **Krugerite** (krū·gərait) *sb.* and *a.*, an adherent of, adhering to President Kruger or his policy.
1896 *Westm. Gaz.* 3 Dec. 5/1 Those who have effusively championed Mr. Chamberlain for what they imagined was his agreement with their Krugerite sympathies. 1897 *Daily News* 25 Jan. 5/6 Krugerites we know, and Rhodesites, but the Schreinerites (politically) all seem to live in London. 1897 *Times* 4 Feb. 3 Pure and unadulterated Krugerism. 1897 *Daily News* 24 Mar. 7/1 The conflict between the two ideals—the Rhodesian or British, and the Krugerite or non-British. 1900 *Pall Mall Gaz.* 29 Mar. 8/1 There are those who suggest that, perhaps, if the scrutineers had not been Krugerites, Joubert would have been found at the head. *Ibid.* 11 June 2/3 In the spring of last year he denounced the corruption of the Krugerite gang. 1902 Kipling *Traffics & Discov.* (1904) 33 Van Zyl wasn't any Krugerite. 1923 B. Ronan *Forty S. Afr. Yrs.* 183 Rhodes..was recognised as the only leader capable of checking the spread of Krugerism in South Africa. 1972 *Sunday Times* (Johannesburg) Colour Suppl. 11 June 9 My father was not a Krugerite, he was a follower of Joubert, who would be called a Progressive today, I suppose.

Kruger rand (krū·gəɹ rænd, rant). Also **Kruger Rand, Krugerrand.** [f. *Kruger* (see prec.) + *Rand *sb.*] A South African gold coin bearing a portrait of President Kruger.
1967 *S. Afr. Digest* 14 July 3/3 The first gold Krugerrand coin was struck at the South African Mint in Pretoria last week by the Minister of Finance, Dr. N. Diederichs. *Ibid.*, The Krugerrand is to be minted in limited numbers and is intended for overseas issue. 1971 *Standard Encycl. S. Afr.* III. 313/2 The Krugerrand, a gold coin of 32·7 mm diameter containing 1 troy oz. of fine gold, was first struck in 1967. The obverse shows the bust of Paul Kruger. 1974 *Harpers & Queen* Sept. 33/2 Keep some Kruger rands under your mattress. 1974 *Daily Tel.* 14 Dec. 18/3 The South Africans are quickly taking advantage of the price rise by minting Krugerrands (made legal tender to avoid capital gains tax) and Britons have been flocking to buy the coins.

Krukenberg (kru·kĕnbɜɹg). *Med.* The name of Friedrich Ernst *Krukenberg* (1871–1946), Ger. scientist, used *attrib.* to designate a kind of metastatic ovarian carcinoma that is usually secondary to a carcinoma of the stomach or colon (described by Krukenberg in 1896).
1911 *Amer. Jrnl. Obstetr.* LXIV. 930 Among a series of metastatic ovarian carcinomata two..showed the picture of a Krukenberg tumor. 1934 R. A. Willis *Spread of Tumours* xxiii. 312 Gastric carcinomas which yield Krukenberg ovarian growths seldom yield metastases in other tissues. 1961 R. D. Baker *Essent. Path.* xvii. 464 The most characteristic of the metastatic tumors [of the ovary] comes from the stomach, produces mucous, signet-ring cells, and is called Krukenberg tumor.

krummholz (krʊ·mhǫlts). [G. *krummholz* crooked wood, the popular name of a dwarf pine, *Pinus mugo* var. *pumilio*, adopted as the name for a particular type of vegetation by A. Grisebach in *Vegetation der Erde* (1872) II. xxiii. 488.] = *elfin-wood* (*Elfin *sb.* 4).
[1903 [see *elfin *sb.* 4].] 1908 *Bot. Gaz.* XLV. 334 The *Krummholz* is composed of two trees only... In its *Krummholz* form it [sc. *Pinus flexilis*] assumes the most fantastic shapes. 1942 R. Peattie *Friendly Mountains* 162 The upper part of the spruce slope dwindles to a 'scrub' forest, or 'krummholz'. 1964 Gleason & Cronquist *Nat. Geogr. Plants* ix. 101 (*caption*) Dense, stunted growth of the sort here shown is known as Krummholz. 1967 *Jrnl. Glaciol.* VI. 820 Shading by krummholz spruce and fir, blueberry and willow.

krummhorn. Add: Also *crumhorn, **krumhorn, krum horn.** a. Delete 'obsolete' and add later examples.
1864 Krumhorn [in Dict.]. 1883 J. W. Mollett *Illustr. Dict. Art & Archæol.* 186/2 *Krumhorn*, an old musical instrument of the cornet kind. 1955 Auden *Shield of Achilles* iii. 76 There I stand in Eden again, welcomed back By the krumhorns, doppions, sordumes of jolly miners. 1969 *Daily Colonist* (Victoria, B.C.) 6 July 23/1 That's if you call 14th, 15th and 16th century harpsichords, clavichords, recorders, krum horns, citterns, lutes and oboes up to date.

Krupp (krʊp, ‖krup). [Name of Alfred *Krupp* (1812–87), German metallurgist, founder of steel and armament works at Essen in Germany.] A gun made at a Krupp factory.
1883 *Whitaker's Almanack* 445/1 She is a casemate ship ..armed with four 10-in. steel Krupps and one 12-in. Krupp. 1887 *Times* (Weekly ed.) 26 Aug. 8/1 The Krupps ..are mounted on Vavasseur carriages. 1900 *Daily News* 23 July 5/4 The Bogue Forts are being re-armed by the Chinese with quick-firing Krupps. 1916 'Boyd Cable' *Action Front* 264 One solitary Krupp dropping in here, and we'd have a pretty-looking mess. 1926 T. E. Lawrence *Seven Pillars* (1935) I. xiii. 95 The Arabs rejoiced when they came, and believed they were now equals of the Turk; but the four guns were twenty-year-old Krupps, with a range of only three thousand yards.
Hence **Krupped** (krʊpt), **Kru·ppized** *ppl. adjs.*, made or carried out in a manner originated by Krupps.
1899 *Army & Navy Register* (U.S.) 3 June 361/3 The great severity of the ballistic tests..necessitates the employment of a Kruppized process. 1902 *Encycl. Brit.* XXXI. 355/2 An A.P. shot should perforate two calibres of wrought iron, one calibre of Harveyed steel, or ⅞ calibre of Krupped armour.

Kruschen (krū·ʃən). *Kruschen salts*, a proprietary aperient; also *ellipt.* As an advertising catch phr. *that Kruschen feeling*, a feeling of vigorous health.
1925 R. W. G. Hingston in E. F. Norton *Fight for Everest, 1924* 350 Kruschen salts, 2 bottles. 1925 R. Macaulay *Casual Commentary* 131 The happy spring when..we are full of that Kruschen feeling. 1928 L. C. Dunsterville *Stalky's Reminisc.* xv. 226 He was very liverish in the early morning and had none of that 'Kruschen feeling' about him. 1936 C. Day Lewis *Friendly Tree* xii. 174 That Crane girl acts like a dose of Kruschen on the staff.

kryo-. Delete the def. of **kryokonite** (see *kryokonite below).

kryokonite (kraiˌǫkōᵘ·nait). Now usu. **cryoconite.** [f. Kryo- + Gr. κόν-ις dust + -ite[1].] A grey powder found in layers at the bottom of holes in glaciers, at one time thought to be meteoric in origin but now thought to consist of dust blown by wind from areas beyond the ice margin. Also *attrib.* in *kryokonite hole.*
1872 A. E. Nordenskiöld in *Geol. Mag.* IX. 356 In the bottom of them [sc. holes in the ice filled with water] we found everywhere..a layer..of grey powder, often conglomerated... The substance is not a clay, but a sandy trachytic mineral... I propose for this substance the name Kryokonite. 1889, 1891 Kryokonite [in Dict. s.v. Kryo-]. 1925 N. E. Odell in E. F. Norton *Fight for Everest, 1924* 311 On the East Rongbuk Glacier were some rather beautiful examples of the so-called 'cryoconite holes' or 'dust holes', in which small particles of morainic material had melted their way down into the surface of the ice, as is so often to be seen on arctic glaciers especially. 1957 *Gloss. Geol.* (Amer. Geol. Inst.) 69/2 Absorption of radiation by the cryoconite causes ablation and formation of cryoconite holes or Dust wells. 1963 J. L. Dyson *World of Ice* xii. 138 In the bottom of every pit is a fine-grained gelatinous material called cryoconite, consisting partly of dust blown by the wind from areas beyond the ice margin. But cryoconite contains a considerable amount of organic material in the form of several kinds of blue-green algae and fungi. 1967 Hamelin & Cook *Illustr. Gloss. Periglacial Phenom.* iii. 87 (*caption*) Cryoconite holes.

kryzhanovskite (kriʒānǫ·vskait). *Min.* [ad. Russ. *kryzhanovskit* (A. I. Ginzburg 1950, in *Doklady Akad. nauk SSSR* LXXII. 763), f. the name of V. I. *Kryzhanovskiĭ* (1881–1947), Russ. mineralogist: see -ite[1].] A greenish-brown hydrated basic phosphate of manganese and ferric iron, $MnFe_2(PO_4)_2(OH)_2.H_2O$, found in Kazakhstan, U.S.S.R.
1951 *Mineral. Abstr.* XI. 189 Kryzhanovskite, a new mineral in the group of phosphates... Rough crystals (2–3 cm.) of prismatic habit occur..in the oxidation zone of pegmatite. 1971 *Amer. Mineralogist* LVI. 5 Since kryzhanovskite is predominantly the ferric equivalent of the phosphoferrite group, the species has valid status and the name is to apply to all members of the phosphoferrite group containing an excess of 50 mol percent Fe^{3+} in the octahedral sites.

K-shell: see *K 3 e.

ktypeite (ti·pēˌait). *Min.* Also -īte. [ad. F. *ktypéite* (A. Lacroix 1898, in *Compt. Rend.* CXXVI. 605), f. Gr. κτυπέ-ω crash, resound: see -ite[1].] A form of aragonite, occurring as pisolites under strain, which decrepitates.
1898 *Jrnl. Chem. Soc.* LXXIV. 604 The violent decrepitation, on account of which the name ktypeite is given. 1902 *Ibid.* LXXXII. ii. 89 Conchite..is identical with aragonite, and the same is probably also true of ktypeite. 1951 C. Palache et al. *Dana's Syst. Min.* (ed. 7) II. 191 Ktypeite..is a name given to the substance of certain fibrous pisolites in the belief that it represented a new polymorph but which is very probably identical with aragonite. 1962 W. A. Deer et al. *Rock-Forming Min.* V. 312 'Ktypeite', from hot spring pisolites, is almost certainly aragonite.

kuaka (kwa·kā). *N.Z.* [Maori.] The bar-tailed godwit, *Limosa lapponica.*
1873 W. L. Buller *Hist. Birds N.Z.* 198 *Limosa baueri.* (Barred-rumped godwit.) ... Native name.—Kuaka. 1882 W. D. Hay *Brighter Britain!* II. 222 The Kuaka..is the bird spoken of as 'curlew' and 'grey snipe' by colonists. 1905 W. B. *Where White Man Treads* 252 These thoughts flow through my brain like a covey of kuaka (snipe). 1966 *Encycl. N.Z.* I. 819/1 In New Zealand the great majority of our migrant birds are waders, and the best known and most abundant of these is the eastern race of the bar-tailed godwit, the kuaka of the Maori (*Limosa lapponica*).

Kuan, Kwan (kuān). [f. Chinese *guān, kuān* official.] Used to denote imperial patronage or official usage in China, as:
1. *Kuan Hua* (hwā) [language, speech] = Mandarin[1] 2.
1814 J. Marshman *Clavis Sinica* ii. 559 The most correct and extensive colloquial dialect is termed..Kwan-hwà. 1845 *Encycl. Metrop.* XVI. 583/2 The learned language of the present day (*Kwan-hwa*) or dialect of the Mandarins. 1848 [see *mandarin dialect* (Mandarin[1] 4)]. 1889 L. C. Hopkins (*title*) Guide to Kuan Hua. 1932 W. L. Graff *Lang.* xi. 421 The modern Chinese of Pekin, the Mandarin, Kuan Hua, or Guoryu..has become the language of officialdom. 1968 *Encycl. Brit.* V. 634/1 Mandarin... Formerly called variously *kuan-hua* 'official speech', whence the term 'Mandarin', or *p'u-t'ung-hua* 'general (*v.* local) speech'.
2. *Kuan Yin* (yin) [Lord of Mercy], a goddess of Chinese Buddhism, to whom intercession for aid or protection is made; a representation in sculpture of this deity.
1845 *Encycl. Metrop.* XX. 490/2 The Kwan-yin, or merciful Goddess of the Chinese. 1871 S. Beal *Catena of Buddhist Scriptures* 121 The work known as the Po-kien ..says:..The spirit..gazing on Kwan-Yin, a covenant saviour (a sworn friend). 1906 S. W. Bushell *Chinese Art* II. viii. 7 Large images of Kuan-Yin enamelled with turquoise blue and other soft colours. 1922 E. T. C. Werner *Myths & Legends China* x. 251 As Mary is the guiding spirit of Rome, so is Kuan Yin of the Buddhist faith. 1930 O. Sirén *Hist. Early Chinese Art* III. iv. 47 In the tall eleven-headed Kuan-yins..a suggestion of movement may be observed. 1943 *Burlington Mag.* Dec. 311/2 Kuan yin figures of the Ming and Ch'ing period are represented in great numbers and show the eagerness of the artist to excel in decorative variations and in manual skilfulness. 1963 P. C. Swann *Art of China, Korea & Japan* vi. 142 The best known of Sung Buddhist sculptures are the indolent *Kuan-yin* 'Goddess of Compassion' (the Sanskrit Avalokitesvara) figures with their full, fleshy bodies seated in the *maharaja-lila* or position of royal ease. 1969 R. Quest *Cerberus Murders* v. 35 Some benign deity —Kuan Yin, perhaps.
3. In full, *Kuan ware, yao* (yɑu) [jade]. A type of thickly glazed celadon made in predominantly greyish colours at Hangchow during the Sung dynasty; similar pottery (as the *Kuan jar*) produced elsewhere in China in later centuries.
1888 F. Hirth *Anc. Porc.* 19 Kuan-yao, or Mandarin Porcelain, is the produce of certain Government factories. 1915 R. L. Hobson *Chinese Pott. & Porc.* I. iv. 48 One of the puzzling features in the study of the Sung wares is the interrelation of the various makes, such as the Ju, Kuan, [etc.]. *Ibid.* v. 59 A new pottery..copied the forms of the older Kuan ware. 1938 *Burlington Mag.* July 37/1 A *Kuan yao* saucer of a delicate bluish grey glaze. 1944 W. E. Cox *Bk. Pott. & Porc.* I. xviii. 426 Neither of the Kuan yao factories so far as we know survived the Mongolian conquest. *Ibid.* II. xx. 592 (*caption*) Bottle vase with glaze of Kuan type of pale blue-green. 1960 *Times* 21 June 22/6 A..blue and white Kuan or wine jar. 1961 M. Sullivan *Introd. Chinese Art* viii. 158 As soon as Southern Sung had established themselves at Hangchow they naturally sought for factories which could produce a ware fine enough to be classed as *kuan.* 1971 P. David *Chinese Connoisseurship* 139 *Kuan Ware.* This was made on the orders of the Palace Works Department.

Kuba (kūbā·). [The name of a town in north-east Azerbaijan, U.S.S.R.] = *Kabistan.
1900 J. K. Mumford *Oriental Rugs* viii. 100 *Caucasian.* ..'Kabistan' or 'Kuba'. 1931 A. U. Dilley *Oriental Rugs & Carpets* vii. 178 The name Kuba is applied both to the old weavings, allied to dragon rugs, and to semi-antique rugs of a Kabistan character. 1963 *Times* 23 Feb. 4/7 A small Kuba carpet, woven in a pattern of dragons and flowers, fetched £2,000. 1972 P. L. Phillips tr. *Formenton's Oriental Rugs & Carpets* 242 *Kuba or Kabistan.* Caucasian carpets of the Shirvan family. The most common decoration is made up of lines of rectangles or squares one above the other along the central part of the field.

kubong (kū·bǫŋ). Also **kubung.** [Malay.] The flying lemur, *Cynocephalus variegatus*, a small, south-east Asian mammal of the order Dermoptera, which glides by means of stretching the membranes linking its limbs and tail; also called the colugo.
1821 T. S. Raffles in *Trans. Linn. Soc.* XIII. 248 *Lemur volans* Linn. Kubung of the Malays. 1929 S. S. Flower *List Vertebrated Animals in Gardens Zool. Soc. Lond. 1828–1927* I. 68 The 'Flying Lemurs', 'Kubong' or 'Colugo' of Malaya and the Philippines..not yet alive in the Gardens, 31 Dec. 1927. 1961 *Listener* 2 Nov. 740/1 The flying Kubong as seen in 'Wings in the Malayan Forest'. 1965 C. Shuttleworth *Malayan Safari* vi. 84, I was anxious to trap kubong (flying lemur). 1969 Ld. Medway *Wild Mammals Malaya* 6/1 Malayan Flying-lemur, Kubong... *Cynocephalus variegatus.*

Kuchaean, Kuchean (kutʃī·ăn). [ad. F. *koutchéen* (S. Lévi 1913, in *Jrnl. Asiatique* II. 315), f. *Kucha*, the name of a town in Sinkiang,

China: see -an.] The western dialect of *Tocharian, Tocharian B. Also *attrib.* or as *adj.*
1939 *Cambr. Anc. Hist.* XII. iii. 97 A..language formerly known as Tocharish but now more correctly called the Kuchean or Turfanese language. 1939 L. H. Gray *Foundations of Lang.* 101 Kuchaean *ñakte* 'god'. *Ibid.* 322 Relatively little of Kuchaean is yet accessible. 1948 D. Diringer *Alphabet* vi. 348 There is..a general agreement to call 'Dialect B' [of Tocharian] *Kuchean.* 1965 *Language* XLI. 108 The Kuchean 3 pl. preterite of *käm* 'come' is *kamem.* 1972 W. B. Lockwood *Panorama Indo-Europ. Lang.* 254 It was subsequently established that Tocharian B was the language proper to Kucha and may therefore be called Kuchean.

‖ **Kuchen** (ku·χĕn). Also **kuchen.** Pl. **Kuchen, Küchen.** [G., lit. 'cake'.] (In Germany or among people speaking German or Yiddish) a cake; now freq. a cake taken with coffee.
1854 Geo. Eliot *Let.* 12 Nov. (1954) II. 185 The Germans eat their Bratwurst and Kuchen from house to house in gladness of heart. 1855 —— in *Fraser's Mag.* June 706/1 Kuchen (generally a heavy kind of fruit tart). 1858 Mrs. Gaskell *Let.* 1 Oct. (1966) 894 M. Mohl treated us to coffee & kuchen. 1861 —— *Grey Woman* i. in *All Year Round* 5 Jan. 300/1 We had nearly finished our coffee, and our '*kuchen*' [sic], and our cinnamon cake. 1894 G. du Maurier *Trilby* I. ii. 164 They will..bring him tea and gin and *küchen* [in *Harper's Mag.* Feb. 348/2 printed kuchen] and *marrons glacés.* 1907 I. Zangwill *Ghetto Comedies* 316 Home-made Kuchen and other dainties. 1972 L. P. Bachmann *Ultimate Act* xxiii. 208 We sat at the kitchen table having coffee and *Kuchen.*

kudos. Add: (Further examples.) Now general *slang* and *colloq.* ¶ Sometimes *erron.* treated as *pl.* (kiū·dōᵘz); so **ku·do** (back-formation) *sing.*, honourable mention, praise for an achievement (see also quot. 1941).
1941 J. Smiley *Hash House Lingo* 34 *Kudo*, good standing with the management. 1950 F. Allen in G. Marx *Groucho Lett.* (1967) 73 A man sitting on a toilet bowl swung open the men's room door and added his kudo to the acclaim. 1961 *Wall Street Jrnl.* (Eastern ed.) 18 Oct. 12/2 This did not win Mr. Eisenhower many kudos in the press. 1963 *Life* 19 Apr. 29/2 A kudo to *Life* for a fine story on baseball's spring training. 1970 G. F. Newman *Sir, You Bastard* vii. 196 News services buzzed, but George Doodie sought no kudos; his name was mentioned only once. 1972 *Sunday Mirror* 17 Sept. 47/1 This below-strength Chelsea side captured the few kudos that were going. 1972 J. Creasey *Splinter of Glass* vii. 55 He wanted Roger to take the kicks if this failed but was prepared to give him the kudos if the use of the newspapers succeeded. 1972 *Homes & Gardens* Nov. 60 It seems almost a kudos to have a lady pilot. 1972 *Bankers' Mag.* Winter 23/2 Kudos are expressed to Messrs. Gene Jackson, Joel Anderson, and John Tolford for their aid.

kudu. Now the most common spelling of Koodoo, although *koedoe* is also used in S. Afr. Substitute for first part of def.: Either of two African antelopes, *Tragelephas strepsiceros*, the greater kudu, or *T. imberbis*, the lesser kudu, which is confined to East Africa. (Later examples.) Also *attrib.*
1901 *Knowledge* July 150/2 The horns take the form of upwardly directed corkscrews, mimicking in fact to a certain degree those of the beautiful African kudu antelope. 1903 J. Y. F. Blake *West Pointer with Boers* xxvii. 367 Where the duiker, spring-bok and koedoe roam. 1931 *Discovery* XII. 61/2, I doubt if the black rhinoceros will long survive, or the sable or the eland or the kudu. 1947 L. Hastings *Dragons are Extra* viii. 176 A Kudu bull with his great spiral horns a good sixty inches long. 1957 *Cape Times* 6 Apr. 5/7 A variety of rhinos, nyalas, koedoe, blou wildebees and other game. 1961 *Guardian* 9 June 9/3 He roused the boys..with a kudu horn he had captured in Matabeleland. 1964 C. Willock *Enormous Zoo* i. 20 Greater kudu..are antelope of scrub-covered mountains. 1971 *Inside Kenya Today* Mar. 52/1 Other craters in Marsabit Forest are frequented by Greater Kudu. 1973 *Nature* 12 Jan. 106/1 The most obvious change has been a decline in woodland-associated species such as lesser kudu, baboon, vervet-monkey, leopard and impala.

kudzu (ku·dzu). [Jap. *kuzu.*] In full, *kudzu vine.* A perennial climbing plant, *Pueraria thunbergiana* (or *P. lobata*), of the family Leguminosæ, native to China and Japan, and cultivated elsewhere as a fodder plant, an ornamental, or an aid in the prevention of soil erosion.
1893 *Garden & Forest* VI. 504/2 In Japan the Kudzu.. has some economic value. 1901 L. H. Bailey *Cycl. Amer. Hort.* III. 1465/2 Kudzu Vine. Perennial with large tuberous starchy roots..fl[ower]s pea-shaped, purple, in axillary spikes late in the season, not showy: pod large and flat. 1948 *Atlantic Monthly* Nov. 60/1 Kudzu, a coarse, rapidly growing legume of incredible efficiency in checking gullies, restoring drainage, and storing nitrogen, came from Japan. 1951 *Dict. Gardening* (Roy. Hort. Soc.) IV. 1713/2 P[ueraria] lobata. Kudzu Vine. Climber with twining, scarcely woody stems growing many feet long in a season; roots thick, fleshy, starchy. 1973 *Daily Colonist* (Victoria, B.C.) 1 Apr. 24/5 The Kudzu vine..is probably the fastest growing of all the perennial vines, making as much as ten feet the first year from seed, and capable of making annual growth of 30 feet or more. 1974 A. Dillard *Pilgrim at Tinker Creek* xii. 212 In summer that path is wrapped past finding in saplings, bushes, kudzu, and poison oak.

‖ **kuei** (kūˌēⁱ). [Chin.] A Chinese bronze food-vessel.
1935 A. J. Koop in L. Ashton *Chinese Art* 77 The *fu* is an oblong tray with straight, steeply sloping sides, a dragon-head loop handle at each end, and a spreading hollow foot cut away in the centre of each side... The *kuei* is somewhat similar but has a more rounded form. 1954 H. Munsterberg *Short Hist. Chinese Art* iii. 60 New shapes ..appear, while others like the kuei and the chung, or bell, become very common. 1963 *Times* 23 Jan. 12/7 Aberdeen University paid £580 for a massive bronze *kuei* (food vessel), decorated with vertical ribs between borders of dragon and monster masks. 1963 P. C. Swann *Art of China, Korea & Japan* i. 26 (*caption*) Chinese ritual bronze food vessel (*kuei*) of about 1000 BC, with dragon-headed handles. 1973 *Oxf. Univ. Gaz.* CIII. Suppl. 5. 55 Two-handled bowl of *Kuei* shape, moulded decoration derived from bronze.

Kufic. Now the more usual form of Cufic *a.*
1792 R. Heron tr. *Niebuhr's Trav. Arabia* I. viii. vi. 270, I copied here [sc. at Beit el Fakih] an ancient *Kusic* [sic] inscription. 1906 *Harmsworth Encycl.* VI. 4223/3 The Kufic script was in use for coins from the end of the 7th to the 13th century. 1913 H. J. R. Murray *Hist. Chess* 171 The titles being in the Kufic character upon a blue ground. 1931 A. U. Dilley *Oriental Rugs & Carpets* Pl. 40 (*caption*) Turkish Rug of Arabesque Design and Kufic Border. 1968 G. Jones *Hist. Vikings* iii. i. 157 Arabic, German, and Anglo-Saxon coins on Gotland; kufic silver, Arabic and Rhenish glassware..at Birka. 1971 R. Russell tr. *Ahmad's Shore & Wave* i. 10 He had the name of the house..written in Kufic script.

kufuffle, var. Curfuffle *sb.*
1960 A. Wykes *Snake Man* iii. 38 After this kufuffle was over and we were on our way again.

‖ **Kuge** (kū·ge). Also † **Cangue; kuge, Kugé.** [Jap.] In feudal Japan, the name of the nobility attached to the Imperial Court at Kyoto; a court noble.
1577 R. Willes in Eden & Willes tr. *Hist. Trauayle W. & E. Indies* f. 255 The heads and beards of his ministers are shauen, they haue name Cangues. 1727 J. G. Scheuchzer tr. *Kæmpfer's Hist. Japan* I. ii. ii. 152 The whole Ecclesiastical Court in general assumes the title of *Kuge*, which signifies as much as Ecclesiastical Lords, and this they do by way of distinction from the *Gege.* 1871 A. B. Mitford *Tales Old Japan* I. 71 The cap and robes worn by the Kugé, or nobles of the Mikado's court. 1880 F. V. Dickins tr. *Chiushingura* (new ed.) 159 They were noble ladies, daughters of *Kugé*, who were peers of the Mikado's creation. 1904 L. Hearn *Japan: Attempt at Interpretation* xii. 265 Next to him stood the kuge, or ancient nobility,—descendants of emperors and of gods. 1957 *Times Lit. Suppl.* 11 Oct. 607/2 The court nobles, almost as useless and cut off from real life by the habits of the Court, of '*ce pays-ci*', as Mme. de Grignan called it, as so many Kuge nobles of old Japan. 1970 [see *hinin].

kugel (kū·gĕl). Also **coogle, kuggol.** [Yiddish, lit. ball, f. MHG. *kugel, kugele* ball, globe.] In Jewish cookery, a kind of pudding served as a main course or as a side-dish.
1846 *Jewish Manual, or Pract. Information Jewish & Mod. Cookery* iv. 55 *Kugel and commean.* Soak..Spanish peas and..Spanish beans..take..fine gravy-beef [etc.]. 1871 E. Levy *Jewish Cookery Bk.* 58 Coogle, or pudding, and peas and beans. 1892 I. Zangwill *Childr. Ghetto* I. 114 Other delicious things there are in Jewish cookery... *Kuggol*, to which pudding has a far-away resemblance. 1914 N. Newnham-Davis *Gourmet's Guide to London* 149 In the great earthenware jar which holds the soup is cooked the 'kugel', a kind of pease-pudding. 1958 J. Grossinger *Art Jewish Cooking* 115 Vegetables are comparatively unimportant in most Jewish homes. In their place, *kugels* and *charlottes*, resembling puddings or pudding-soufflés, were substituted. They may be served as separate courses, as accompaniments to meat or poultry, or even as dessert if they are sweet. 1972 *Listener* 16 Mar. 341/3 It [sc. chollant] was a mélange of meat, potatoes, butter-beans, onions, kishkeh, kugel and hope... The kugel (German for bullet) looked like an anaemic cannon-ball... The basic formula included fat, flour, grebenes or fried onion.

kugelhupf: see *gugelhupf. Also **kugelhopf.**
1886 [see *centre *v.* 4]. 1906 *Mrs. Beeton's Bk. Househ. Managem.* lxii. 1656 *Kugelhopf*, a German cake; a kind of rich dough cake. 1961 J. Heller *Catch-22* (1962) xxiv. 249 Napoleons and *petits fours* from Paris, Reims and Grenoble, *Kugelhopf*, pumpernickel and *Pfefferkuchen* from Berlin. 1966 P. V. Price *France: Food & Wine Guide* 148 The traditional *kugelhopf* (a light yeast cake, usually with a hole in the centre).

kuhn-kan, var. *coon-can.
1936 G. Greene *Journey without Maps* i. iii. 42 They.. played Kuhn-Kan for very small stakes. 1951 —— *Lost Childhood* 14 The mild nursery game of Kuhn Kan.

kuka (kū·kā). [Native name.] A name used in Ghana for *Khaya senegalensis*; also *kuka-tree*; see *khaya.
1882 *Encycl. Brit.* XIV. 153/2 Kuka [a Central African town]..received its name from a kuka or monkey bread tree (*Adansonia digitata*), which attracted the attention of the settlers as a rare thing in the district. 1924 *Blackw. Mag.* Sept. 351/1 A kuka-tree is a stumpy smooth, grey-barked thing of enormous girth. 1932 J. Cary *Aissa Saved* xliii. 230 We should go there by the kuka tree. 1961 F. R. Irvine *Woody Plants of Ghana* 523 *Khaya senegalensis*... Kuka (local name).

kukama (kukā·mā). Also **kokama, kookaam.** [Sechuana.] The southern African antelope, *Oryx gazella*; see Oryx b.
1852 *Catal. Mammalia Brit. Mus.* III. 105 *Oryx gazella.* The Kookaam or Gemsboc. Horns straight, shelving backwards. 1857 D. Livingstone *Missionary Trav. & Res. S. Afr.* iii. 56 The gemsbuck or kukama. 1888 J. S. Kingsley *Riverside Nat. Hist.* V. 326 The *Oryx capensis* of South Africa, or Gemsbok of the Dutch colonists, Kokama of the Bechuanas, is even more striking in its coloring. 1971 R. H. N. Smithers *Mammals Botswana* 238/2 *Oryx gazella.* Kukama. Gemsbok.

kukang. Add: Also **kongkang.** Substitute for def.: The slow loris, *Nycticebus coucang*, found in south-east Asia. (Earlier and later examples.)
1821 T. S. Raffles in *Trans. Linn. Soc.* XIII. 247 *Lemur tardigradus* Linn. Kukang (Malay)... Of this the natives distinguish a large and a small variety. 1900 *Proc. Zool. Soc.* 321 The Slow Loris... 'Kúkang', 'Kongkang', and 'Kongka' of the Malays. *Ibid.*, At one time I used to sleep in a hammock slung in a veranda close to a cage of Kongkangs. 1969 Ld. Medway *Wild Mammals Malaya* 47/1 Slow Loris. Kongkang, Kera duku. *Nycticebus coucang.*

Kuki[1] (ku·ki). [Native name.] Any one of several peoples inhabiting the hills of Manipur and Mizoram, on the Indo-Burmese border; a member of one of these peoples; also, their language. Also *attrib.* and *Comb.*, as *Kuki-Chin* [*Chin *sb.*[2]] (see quot. 1954).
1799 *Asiatick Researches* II. 188 If a *Cúci* assail the house of an enemy..he acquires honour and celebrity in his tribe. *Ibid.* 193 A party of *Cúcis* visited the late Charles Croftes..at Jáfarabád, in the spring of 1776, and entertained him with a dance. 1803 *Ibid.* VII. 186 The *Kookies* choose the steepest and most inaccessible hills to build their villages..which..are called..in the *Kookie* language, *K'hooah.* 1871 E. B. Tylor *Primitive Culture* I. xi. 424 The Kukis of Assam think that the ghost of every animal a Kuki kills in the chase or for the feast will belong to him in the next life. 1872 E. T. Dalton *Descr. Ethnol. Bengal* 75 English..one..Kuki khut. 1885 E. Balfour *Cycl. India* (ed. 3) II. 618/2 The new Kuki clans are presided over by rajas and muntris. 1927 *Blackw. Mag.* June 816/1 The Kuki alone of all these hill tribes understands the value of combination. *Ibid.* 817/1 Roaring and retaliation keep alive a warlike and truculent spirit among the Kukis. 1939 L. H. Gray *Found. Lang.* 389 Arakan-Burmese, comprising Kuki-Chin.., Old Kuki (Rānkhōl, Shō or Khyang, Khami, etc.), and Burmese. 1948 D. Diringer *Alphabet* vi. 367 Manipuri or Meithei, a Kuki-chin speech. 1954 Pei & Gaynor *Dict. Ling.* 116 *Kuki-Chin*, a group of dialects (Lai, Lushei, Meithei, Tashon, etc.), constituting a subdivision of the Arakan-Burmese branch of the Tibeto-Burmese sub-family of the Sino-Tibetan family of languages. 1971 B. W. Aldiss *Soldier Erect* 219 But the Assam Battalion kept coming across the road, Kukis, Karsis, and all the other tribesmen. 1972 *Language* XLVIII. 476 The crucially important and ramified group of Tibeto-Burman languages known variously as Kukish, Kuki-Chin, or Kuki-Chin-Naga, spoken in Assam and Western Burma.

‖ **kuki**[2] (ku·ki). *N.Z. Hist.* Also **kooky.** [Maori, ad. Cook *sb.*] A slave of a Maori chieftain.
1832 A. Earle *Narr. Residence N.Z.* (1966) 60 A chief had set one of his kookies (or slaves) to watch a piece of ground planted with the koomera, or sweet potato. 1845 E. J. Wakefield *Adventure N.Z.* I. 230 The alarm had been caused by some *kuki*, or slaves. 1882 W. D. Hay *Brighter Britain!* I. x. 283 Her father and mother were only kukis.

kuklass, var. *koklas(s.

Ku-Klux. Add: **1.** (Earlier and later examples.) This society was revived in 1915 and spread outside the Southern States, terrorizing various ethnic and religious minorities, and acting violently against white Protestants whom they judged to be opposed to their cause. Later the society fragmented into several State organizations. The Ku-Klux Klan regained strength in the Southern States of the U.S. in the 1950s in opposition to the Civil Rights movement of American Blacks. Cf. *Klan, *Kluxer.
1867 *Citizen* (Pulaski, Tenn.) 29 Mar. 3/1 The Kuklux Klan will assemble at their usual place of rendezvous.., exactly at the hour of midnight, in costume and bearing the arms of the Klan. 1868 *N.Y. Herald* 1 July 6/4 If the Democratic Convention can only be induced..at the dictation of the Knights of the Golden Circle and the Ku Klux Klan, to place upon their ticket some copperhead opponent of the war, the radicals will have a clear track next November. 1915 *Atlanta* (Ga.) *Jrnl.* 6 Dec. 2/4 (*heading*) Charter is granted for the Ku Klux Klan. *Ibid.*, Secretary of State Philip Cook Monday issued a charter to the Knights of the Ku Klux Klan, a fraternal insurance order which was organized on Stone Mountain a few nights ago with weird and mystic ceremony. 1920 *N.Y. Times* 11 Oct. 1/2 The old Ku Klux Klan has been reorganized and is regularly chartered under the laws of Georgia. 1924 *Imperial Night-Hawk* 10 Sept. 6 The district meeting of the Ku Klux Klan held recently at Twin Lakes, Realm of Iowa, was a huge success. 1944 *Atlanta* (Ga.) *Jrnl.* 4 June 1/1 The Knights of the Ku Klux Klan, Inc. has officially ceased to exist. 1945 *N.Y. Times*

21 Oct. 33/3 The Ku Klux Klan, claiming a membership of more than 20,000 in Georgia, is burning its fiery cross again. 1958 *News & Observer* (Raleigh, N. Carolina) 19 Jan. 1/5 A shouting horde of Robeson County Indians tonight routed the Ku Klux Klan here. 1962 A. S. Rice *Ku Klux Klan in Amer. Politics* ix. 114 [In 1949] the secret order splintered into many rival groups, each considering itself..the direct spiritual heir of the Invisible Empire, Knights of the Ku Klux Klan founded in 1915. 1970 in J. F. Kirkham et al. *Assassination & Political Violence* iv. A. § 3. 216 The first Ku Klux Klan, which lasted from 1865 to 1876, was a principal means of administering this violence in the South. *Ibid.* § 4. 218 White Capping seems to have been an important link between the first and second Ku Klux Klans. White Cap methods of punishment and costume seem to have been influenced by the first Klan. *Ibid.* App. D. § 1. 364 It was not until 1956 when the efforts..failed to stem the trend toward integration in the South that the Ku Klux Klans revived... In San Antonio, Tex., a cross was burned..to 'let the niggers, Jews, and Catholics know we're back in operation'.
b. In extended use, of other vigilante groups. Also *transf.* and *fig.*
1930 W. & E. Muir tr. *Feuchtwanger's Success* iv. xvi. 539 The Munich Ku Klux Klan..evidently consisted of young men who did not understand that a match, once it is decided, can't be fought all over again. 1944 A. Huxley *Let.* 28 July (1969) 511 You might pass on to your agent these simple talking points, with which to allay the studios' fears of the medical Ku Klux Klan. 1966 A. Sachs *Jail Diary* xxiv. 215 Then he asks me what I know about the local Ku Klux Klan.
2. (Earlier examples.)
1868 in T. D. Clark *Pills, Petticoats & Plows* (1944) 62 We are inclined to think he is somewhat disloyal, and may be in sympathy with the Ku Kluxes. 1877 J. M. Beard *K.K.K. Sk.* 40 The Ku-Klux themselves were about as intangible examples of ghostliness as were ever wrapped in loose-fitting bombazine.
Hence **Ku-Klux Kla·n(n)er, Kla·nism, Kla·nsman.**
1868 in S. F. Horn *Invisible Empire* (1939) 335 Let every Ku Klux Klansman heed The General Order of General Meade. 1923 *Nation* (N.Y.) 11 July 35 He will help his fellow Ku Klux Klaners. 1924 H. Crane *Let.* 5 Mar. (1965) 177 O'Neill's new play..in which a white woman marries a Negro... He..receives terrible threats and insults through the mail from the *Ku Klux* Klanners. 1924 J. M. Mecklin *Ku Klux Klan* 98, I have yet to come in contact with the first trace of Ku Klux Klanism. 1933 H. G. Wells *Shape of Things to Come* ii. § 1. 141 They became Ku Klux Klansmen, Nationalists, Nazis. 1948 *Time* 15 Mar. 29/2 Last week Georgia's Grand Dragon Samuel Green carefully explained that Ku Klux Klansmen wore masks to protect themselves against the prejudice of Jews, Catholics and foreigners.

kuku (kū·kū). *N.Z.* [Maori.] **1.** Also **koukou, kukupa.** = *kereru.
1835 W. Yate *Acct. N.Z.* iii. 53 *Kou-kou*—The bird so-called is a small owl, a native of New Zealand. 1873 [see *kereru]. 1881 J. L. Campbell *Poenamo* vii. 115 The *kukupa*..was just the bird created expressly for the true cockney sportsman. 1905 W. B. *Where White Man Treads* 19 For food..in his [sc. the Maori's] catalogue, for flavour and bulk, the plentiful kereru, or kuku (pigeon) headed the list. 1936 'R. Hyde' *Check to your King* xiii. 157 Your wild pigeon (*kuku*)..beds itself on the fern. 1949 P. Buck *Coming of Maori* (1950) ii. i. 93 The principal forest birds sought for food were the wood pigeon (*kereru, kukupa, kuku*), [etc.].
2. Either of two common mussels, *Perna canaliculus* or *Mytilus edulis aoteanus.*
1905 W. B. *Where White Man Treads* 2 White seashore sandhills, representing the accumulated deaths of ages of pipi, pupu, kuku, and other molluscs. 1949 P. Buck *Coming of Maori* (1950) ii. i. 106 Shellfish such as..sea mussels (*kuku*)..were cooked..and dried also as reserve food. 1960 N. Hilliard *Maori Girl* 153 Shellfish—*paua, kuku, kina.*

kukui (kuku·ˌi). [Hawaiian.] An evergreen tree, *Aleurites moluccana*, of the family Euphorbiaceæ, native to the Moluccas and south Pacific islands; its large seeds yield an oil used for lighting and other purposes, and the tree is also called the candlenut or candleberry tree.
1825 W. Ellis *Jrnl. Tour Hawaii* vii. 167 Along the narrow and verdant border of the lake at the bottom, the bread-fruit, the *kukui*, and the *ohia* trees, appeared. 1866 'Mark Twain' *Lett. from Hawaii* (1967) 98 These trees were principally of two kinds—the koa and the kukui—the one with a very light green leaf and the other with a dark green. 1890 *Ibis* II. 175 The delicately-indented foliage of the kukui has..a lovely silvery appearance in certain lights. 1913 J. F. Rock *Indigenous Trees Hawaiian Islands* 257 The Kukui is one of the most common of Hawaiian forest trees... The nuts contain 50 per cent of oil, which is known as Kekuna in India and Ceylon, and Kukui in Hawaii. 1937 D. & H. Teilhet *Feather Cloak Murders* viii. 143 Silvery Kukui trees draped their clustered pink flowers over old walls. 1967 *Economist* 9 Sept. 892/1 The luau lights outside nearly every [Hawaii] hotel and restaurant may now consist of a concealed gas jet instead of kukui nut-meat burning gently in a shell.

kukumakranka (kukumakra·ŋkā). *S. Afr.* Also **koekmakranka, koekoemakranka.** [Afrikaans, prob. f. Hottentot name for the plant.] A small, bulbous, perennial plant of the genus *Gethyllis*, belonging to the family Amaryllidaceæ, and bearing fragrant white flowers and an underground fruit.
1793 tr. *C. P. Thunberg's Trav. Europe, Afr. & Asia* I. 116 Kukumakranka (*gethyllis*) is the name given to the legumen or pod of a plant, that grew at this time among the sand-hills near the town, without either leaves or flowers. This pod was of the length of one's finger, somewhat wider at top than at bottom, had a pleasant smell, and was held in great esteem by the ladies. The smell of it resembled in some measure that of strawberries, and filled the whole room. 1811 W. J. Burchell *Jrnl.* 31 Jan. in *Trav. S. Afr.* (1822) I. ii. 55 On Green Point on the Flats in the neighbourhood of Cape Town, grows a celebrated little plant, which still preserves its original Hottentot name, being known by no other than that of *Kukumakranki.* 1857 L. Pappe *Floræ Capensis Medicæ Prodromus* (ed. 2) 39 The elongated, club-shaped, orange-coloured fruit of this plant has a peculiar fragrance, and still preserves its old Hottentot name of *Kukumakranka.* 1932 Watt & Breyer-Brandwijk *Medicinal & Poisonous Plants S. Afr.* 28 An alcoholic infusion of the fruit of *Gethyllis spiralis* L.f. (*Gytheilia spiralis*), Koekoemakranka (Bramakranka), Koekmakranka, was taken by the early Cape colonists for the relief of colic and flatulence. 1950 *Cape Times* 15 May 14 Before I found a palate for mushrooms, it was the kukumakranka we used to go seeking in the veld. *Ibid.* 18 May 12/3 Two explanations of how the koekmakranka got its name have reached me. 1973 Y. Burgess *Life to Live* 179 When the drug did not help she asked the old women to gather 'kukumakranka' leaves for her, and she asked Magriet for brandy to steep them in.

kukupa: see *kuku 1.

‖ **kula** (kū·lā). Also **Kula.** [Melanesian.] In some Pacific communities, esp. in the Trobriand Islands, an inter-island system of ceremonial exchange of items as a prelude to or concomitant of regular trading.
1920 B. Malinowski in *Man* XX. 97 (*heading*) Kula; the circulating exchange of valuables in the archipelagoes of eastern New Guinea. *Ibid.* 101 Glancing at the map we see a number of circles, each of which represents a certain sociological unit which we shall call a *Kula* community. 1927 —— *Sex & Repression Savage Soc.* ii. iii. 93 Those in charge of the overseas expeditions called *Kula* are often supposed to have dreams about the success of their ceremonial trading. 1951 R. Firth *Elem. Social Organiz.* i. 3 The Trobriand islanders of New Guinea exchange in the kula. 1970 *Nature* 12 Dec. 1064/2 The best known exchange mechanism of this kind of recent times is the kula trade of the Trobriand islanders of the western Pacific.
Hence **kula** *v.*, to exchange ceremonial gifts in this manner.
1922 B. Malinowski *Argonauts W. Pacific* iii. 101 Several villages do not kula. *Ibid.* xi. 281 Bokuyoba.. gave the pair..to Kadamwasila..her son..*kula'd* it on to some of his southern partners.

Kula, var. *Kullah. See also next word.

Kulah[1] (kū·lā). Also **Koula(h), Kula.** The name of a town in W. Turkey used *attrib.* or *ellipt.* to denote a rug using the Ghiordes knot made there.
c 1882 *Cardinal & Harford's Price List Oriental Carpets & Rugs* 13 *Koula rugs and mats.* Various qualities, bright colours, Red usually predominant. 1899 *Northern Times* (Golspie, Sutherland) 22 June 1/5 (Advt.), About 600 to 800 Oriental and other Rugs, from 4s 9d upwards, including..Koula. 1900 J. K. Mumford *Oriental Rugs* x. 150 The narrow stripe with undulating pattern, referred to as a characteristic of all Ghiordes, antique and modern, is rarely found in pure Kulahs. *Ibid.*, In design the modern Kulahs have nothing characteristic. 1904 M. B. Langton *How to Know Oriental Rugs* v. 165 The antique Koulah prayer-rug differs from the Ghiordes in various ways. 1931 A. U. Dilley *Oriental Rugs & Carpets* vi. 164 Ghiordes and Kula prayer rugs were made in adjacent cities. 1962 C. W. Jacobsen *Oriental Rugs* 248 With their close neighbours, Ghiordes, Kulahs share the position of being one of the rarest and most sought after of all Oriental Rugs. 1972 P. L. Phillips tr. *Formenton's Oriental Rugs & Carpets* 94 Kula rugs, like most Anatolian rugs..have no special distinguishing motif and so are difficult to identify.

kulah[2] (kū·lā). Also **kulla, kullah.** [Pers. *kulah* a cap.] (See quot. 1969.)
1920 *Chambers's Jrnl.* 29 May 408/1 Dark-blue *pugris*, with red and white silk stripes, tied round a *kulla* or conical skull-cap of scarlet. 1952 J. Masters *Deceivers* xi. 125 The prisoner second from the left wore a Mohammedan kulla under his turban. 1969 R. T. Wilcox *Dict. Costume* (1970) 189/1 *Kulah*, a conical cap of felt or lambskin, the headgear of Middle Eastern monks and dervishes. 1971 R. Dentry *Encounter at Kharmel* iii. 45 The men wore round kullah caps, ragged chogas over long grey tunics, and scarves.

kulak (kū·læk). Also † **koolack, koulak.** [Russ. *kulák* fist, tight-fisted person, pl. *kulaki*, f. Turki *ḳul* hand.] In pre-Revolution Russia, a well-to-do farmer or trader; in the Soviet Union, a peasant-proprietor working for his own profit. Also *transf.*
1877 D. M. Wallace *Russia* (ed. 2) I. vii. 159 Not a few industrial villages have thus fallen under the power of the *Kulaki*—literally Fists—as these monopolists are called. 1886 *Encycl. Brit.* XXI. 84/1 The enrichment of a few 'kulaks'. 1921 *Contemp. Rev.* Jan. 26 'Kulaks'..a nickname for the close-fisted village traders, usurers, and rich peasants. 1925 P. Gibbs *Unchanging Quest* xiv. 109 These peasants think the Duma will..kill all the Koulaks, or Jewish moneylenders. 1929 [see *collective A. 2 f]. 1931 M. Hindus *Red Bread* iv. 66 Legally, a *koolack* is a man who indulges in some form of exploitation, employs hired help or derives an income from rent or interest or the operation of an agricultural or industrial machine. Actually, however, a *koolack* is a successful farmer as success is measured in Russia. 1934 G. B. Shaw *On Rocks* 164 They [sc. the Soviet government] also proscribed the kulak, the able, hardheaded, hardfisted farmer who was richer than his neighbors. 1951 G. Mikes *Down with Everybody* 48 He was a kulak, a spy and an enemy agent, but now he had realised his mistake—namely that it *was* a mistake to be a kulak, a spy and an enemy agent. 1952 R. Campbell *Lorca* 7 Lorca was by birth a landowning 'kulak'. 1957 *Observer* 10 Nov. 5/8 The peasants [in China] have been 'voluntarily' collectivised..but there has been no Russian-style campaign for the 'elimination of the *kulak* as a class'. 1970 *New Scientist* 1 Jan. 15/1 The improved grain husbandry..may favour the rise of 'kulaks' or 'improving landlord' groups.

kulang (kū·læŋ). Also **kolong, kulan, kullung.** = *coolung.
1838 [see *coolung]. 1887 [see *dhandh]. 1910 *Blackw. Mag.* Feb. 285/2 The kulan, too, or demoiselle cranes, are lovely birds..and are found in great numbers all up the river Jumna. 1915 [see *coolung]. 1949 N. B. Kinnear *Whistler's Pop. Handbk. Indian Birds* (ed. 4) 444 The Common Crane and the Demoiselle Crane are not usually distinguished from each other in India and are well known collectively under the names of Kunj and Kulung.

kulchur. ¶ With distortion of spelling to indicate an affected or vulgar pronunciation of Culture *sb.* Cf. *culture *sb.* after sense 5 a.
1940 E. Pound *Cantos* liv. 43 The 10th a charter of labour And the last on keepin' up kulchur. 1959 *Listener* 4 June 997/3 The famous six essential books, Pound's guide to Kulchur. 1971 *Frenzy* 21 May 3/4 There is a spirit of freedom, solidarity and struggle; an intensity of life worth all the ephemeral fantasies to be offered so far by the Rock/alternative Kulchur.

Kulin (kūlī·n). Also **Kooleen, Kulina.** [ad. Skr. *kulīna*, f. *kulin* well-born.] In Bengal, a Brahman of the highest class. Also *attrib.* or as *adj.*
1866 *Atlantic Monthly* Dec. 733/1 The privilege of maintaining a plurality of wives is restricted to a very few.. except in the case of *Kooleen* Brahmins, that superlative aristocracy of caste. 1873 E. Balfour *Cycl. India* (ed. 2) III. 312/1 *Kulin*, a class of brahmans in Bengal, who are deemed by other brahmans to be of very pure descent and in consequence many are anxious to wed their daughters to them. 1911 *Encycl. Brit.* XIII. 511/2 Only an extreme section—the so-called *Kaulas* or *Kulinas*..persist in carrying on the mystic and licentious rites taught in many of the Tantras. 1911 G. B. Shaw *Getting Married* 139 Kulin polygyny, though unlimited, is not really a popular institution. 1932 [see *Kulinism* below]. 1970 N. B. Bonarjee *Under Two Masters* 3 The Kulins became an integral and important part of Bengal's life and culture.
Hence **Ku·linism**, the polygamous system of the Kulins.
1890 in *Cent. Dict.* 1891 H. H. Risley *Tribes & Castes of Bengal: Ethnogr. Gloss.* I. 146 The Bansajas are those Kulins who lost their distinction on account of..their want of charity, discipline, and due observance of marriage law, three qualities which in later times constituted Kulinism. 1932 L. S. S. O'Malley *Indian Caste Customs* i. 10 A Kulin Brahman who had an embarrassing number of female relatives is known to have had eighteen of them..married in a batch to a boy ten years old—.. Kulinism, as the practice is called, has nearly died out. 1968 B. Walker *Hindu World* II. 229 In Bengal.. polygamous relationships..were common among certain brāhmin sub-castes, notably the *Kulin*, 'noble',..and the custom is called kulīnism.

Kullah (kūlā·). Also **Kalá, Klá, Kula.** [Pegu *Goḷa* Indian Buddhist immigrant, f. Skr. *Gauḍa* ancient name of N. Bengal (Yule and Burnell).] (See quot. 1886.)
1800 M. Symes *Acct. Embassy to Kingdom of Ava* xii. 290 On being informed that I was 'a Colar', or stranger,.. they were reconciled. 1858 H. Yule *Narr. Mission to Court of Ava* i. 5 His private dwelling was a small place on one side of the court, from which the women peeped out at the *Kalás.* 1886 Yule & Burnell *Hobson-Jobson* 378/1 *Kulá* or *Klá*, n.p. Burmese name of a native of Continental India; and hence misapplied also to the English and other Westerns who have come to Burma from India; in fact used generally for a Western foreigner. 1890 Kipling *Departmental Ditties* (ed. 4) 82 For the Burmans said That a kullah's head Must be paid for with heads five score. 1929 F. T. Jesse *Lacquer Lady* i. iii. 21 All foreigners are kalās to the Burmans.

kulla(h), varr. *kulah[2].

kullum, var. *coolung.

kullung, var. *coolung, *kulang.

kultur (kultū·r). Also **Kultur.** [G., ad. L. *cultūra*, or F. *culture* Culture *sb.*] Civilization as conceived by the Germans; esp. used in a derogatory sense during the 1914–18 and 1939–45 wars, as involving notions of racial and cultural arrogance, militarism, and imperialism. Also *attrib.* and *transf.*
1914 *Punch* 16 Sept. 239/1 (*heading*) The Imperial Prussian College of Culture. *Telegrams*: 'Kultur, Berlin'. 1914 *Spectator* 31 Oct. 589/1 The idea that the extension of the *Kultur* of a nation can be effected by the extension

by arms of its Empire. 1915 *Times* 30 Mar. 6/4 *Kultur*, in fact, has become the exact opposite of 'culture'. 1915 A. Huxley *Let.* Oct. (1969) 84 We have founded a club, chiefly for the purpose of self protection against Queen's and for the propagation of Kultur. 1916 J. B. Cooper *Coo-oo-ee* xii. 170 People have no time for Germans after their kultur demonstration in Belgium. 1917 A. G. Empey *Over Top* 305 A British rat resembles a bull-dog, while a German one, through a course of Kultur, resembles a dachshund. 1918 Kipling *Kipling's Message*, It is the peculiar essence of German Kultur, which is the German religion, that it is Germany's moral duty to break every tie, every restriction that binds man to fellow-man, if she thinks it will pay. 1926 C. H. Herford *Mind of Post-War Germany* v. 22 The stabilizing forces which post-war Germany derived from her inherited Kultur. 1939 tr. *C. Leiser's Nazi Nuggets* 82 Since the Nazis and the Japanese have been getting cozy and have signed a pact to foster their Kultur. 1973 L. Snelling *Heresy* i. i. 4 How ignorant I am of contemporary Kultur.
Also (with varying degrees of naturalization) **ku·lturbild** [G. *bild* picture, image], a description of the culture (of a period, etc.); **ku·lturgeschi·chte** [G. *geschichte* history], the history of the cultural development (of a country, etc.); history of civilization; **ku·lturgut** [G. *gut* possession], a cultural asset; **ku·lturhund** [G. *hund* dog], **kultur-hound** = **culture vulture*; **ku·lturkampf** [G. *kampf* conflict], the conflict between the German government and the Papacy for the control of schools and church appointments (1872–87); also *transf.*; **ku·lturkreis** [G. *kreis* circle], a cultural group; a cultural complex (the term is associated esp. with the German anthropologists F. Graebner and W. Schmidt); **ku·lturstaat** [G. *staat* state], a civilized country; **ku·lturträger** [G. *träger* carrier], an upholder or defender of civilization.
All usu. with capital initial in Eng. as in German.
1961 *Times* 23 Nov. 16/4 This book had to be a *Kulturbild* rather than a biography. 1964 *English Studies* XLV. 91 Professor Schirmer in his *Kulturbild* attempts to relate John Lydgate to his age by means of his poetry. 1876 *Mind* I. 447 The novel facts and attractive generalisations of *Culturgeschichte* are insensibly casting discredit upon the thoughtful introspection of one's own adult experience. 1938 *Year's Work Eng. Stud. 1936* XVII. 29 Brandenstein's little monograph on the first Indo-European migration is a return to the study of comparative vocabulary as a means for re-imagining some aspects of *Kulturgeschichte.* 1968 *Listener* 4 Apr. 448/1 English music historians have, on the whole, concentrated on the chronicling of technical matters and have avoided *Kulturgeschichte*, as it is practised elsewhere. 1952 *Man* June 83 A member of a lower caste tends to imitate the '*kulturgut*' of higher-caste people... At the same time he is attached to his own *kulturgut.* 1966 *Amer. N. & Q.* June 158/1 No effort was made to segregate the demonstrably regional *Kulturgut* from that which is literary, and even worldwide. 1969 *Language* XLV. 235 The later association of the term with the very poor may represent a sort of..Kulturgut. 1946 Mezzrow & Wolfe *Really Blues* (1957) xi. 196 That Kultur-hound didn't know..that underneath the phony label was a genuine Victor one. 1963 *Listener* 17 Jan. 138/3 So our provincial *Kulturhunde* had thirty minutes of that Vassar-educated Mona Lisa, Miss Mary McCarthy. 1879 *Dublin Rev.* Oct. 350 History of the Prussian 'Kulturkampf'. 1896 W. Miller *Balkans* ii. v. 205 A regular *Culturkampf* raged for nearly twenty years, in which the Turkish officials were far less adverse than the Greek clergy to the Bulgarian demands. 1902 *Encycl. Brit.* XXXII. 271/2 In Germany, when the Pontificate of Leo XIII. began, a disastrous conflict between the Imperial Government and the Church was in progress. It was called the Kulturkampf, as professing to be undertaken on behalf of civilization and culture. 1926 C. H. Herford *Mind of Post-War Germany* i. 6 The Rhinelands, fervently Catholic, and still acutely mindful of Bismarck's *Kulturkampf.* 1936 H. G. Wells *Anat. Frustration* xiii. 150 A vast Kultur-Kampf lies between mankind and peace. 1966 *New Statesman* 18 Feb. 218/3 The *kulturkampf* between Flemish and Walloon French has now reached the Catholic University of Louvain. [1897 L. Frobenius in *Petermann's Mitteilungen* XLIII. 225 (*title*) Der westafrikanische Kulturkreis. *Ibid.* 225/2 Es mag deshalb die Bezeichnung 'Westafrikanischer Kulturkreis' zunächst beibehalten werden.] 1948 A. L. Kroeber *Anthropol.* (rev. ed.) xvii. 770 Part of a wider theory advanced by the *Kulturkreis* (culture-sphere) movement or school of ethnology in continental Europe is that the Indonesian-Melanesian cultures are also characterized by the same block of culture traits that includes those enumerated for West Africa. 1971 *English Studies* LII. 256 It will have important implications..for the whole theoretical question of genre study in literature and the whole historical one of Anglo-Irish relations in the early middle ages, indeed of all of the early Christian *Kulturkreis*, which included the British Isles, Iceland, Scandinavia, and parts of Germanic Europe. 1925 *Manch. Guardian Weekly* 16 Oct. 311 There is no 'Kulturstaat' (civilised State) that would not punish political crimes. 1936 *Mind* XLV. 295 A state must at least preserve its existence, as a condition of becoming a *Kulturstaat.* 1948 J. Towster *Political Power in U.S.S.R.* i. 6 Such conceptions of the state as *Rechtstaat* or *Kulturstaat.* 1920 D. H. Lawrence *Women in Love* i. 13 She was a *Kulturträger*, a medium for the culture of ideas. 1962 *N. & Q.* May 190/2 Two types of borrowing situation are envisaged, the bilingual community with oral/aural mediation, and the unilingual community with *Kulturträger* and written mediation.

‖ **kulturny** (kultu·rni), *a.* [Russ. *kul'túrnyĭ* civilized.] In the Soviet Union: cultured, civilized.
1955 H. Hodgkinson *Doubletalk* 39 Cultured, or *kulturny* behaviour, is highly esteemed in the Soviet Union. 1959 *New Statesman* 23 May 711/3 Aesthetic considerations never played a part in the previous drives for a more *kulturny* mode of life, which were more concerned with manners than with the cultivation of good taste. 1973 J. Shub *Moscow by Nightmare* ix. 97 She let the porter take her one small suitcase—it wouldn't be '*kulturny*' to carry it herself.

kulu, kulukamba: see *koolookamba.

kumara. Now the usual spelling of Kumera. For '*Ipomœa edulis*' in def., substitute '*Ipomœa batatas*'. Add further examples.
1851 H. R. Richmond *Jrnl.* 23 Feb. in *Richmond-Atkinson Papers* (1960) I. ii. 88 There was nothing else to help ourselves to except kumara or sweet potato, in appearance like long thin potatoes, in taste like boiled chesnuts. 1905 W. B. *Where White Man Treads* 2 Parklands, blurred and dotted with plantations of kumara. 1934 *Bulletin* (Sydney) 26 Sept. 20/1 The supplies of pig, eel, shark and kumara were unlimited. 1936 [see *kit *sb.*[1]1]. 1965 *N.Z. Jrnl. Agric.* July 55/1 Most home gardeners purchase their kumara plants from seedsmen. 1971 *N.Z. Listener* 16 Aug. 54/4 Be careful with the kumara... Its energy value is twice that of the potato.

kumbang (ku·mbæŋ). Also **koembang.** [local Javanese *koembang.*] A föhn wind which blows in Java.
1931 A. A. Miller *Climatol.* v. 83 These winds have a foehn-like nature, their dry heat doing considerable damage to the more sensitive crops, especially tobacco. Such a wind is the 'Koembang' of Java. 1954 E. D. Laborde tr. *Robequain's Malaya, Indonesia, Borneo & Philippines* ix. 187 At that time,..the lowlands between Cheribon and Tegal fairly often experience the drying effort of the south-east wind which, after crossing the Pembarisan Hills, becomes a foehn known as the *kumbang.* 1969 M. Sukanto in H. Arakawa *Climates N. & E. Asia* iv. 222 Tjirebon and Tegal (Central Djawa) are also exposed to such a wind during the dry season. In this case, the wind flows down from the western part of Mount Slamet and blows in a northern and northeastern direction. This wind is called the Kumbang.

‖ **kumbuk** (kʊ·mbʊk, ku·mbuk). [Sinhala.] A name used in Sri Lanka (Ceylon) for the arjun, *Terminalia arjuna*, an Indian evergreen tree of the family Combretaceæ.
1869 R. H. Beddome *Flora Sylvatica* I. 17 It [sc. *Terminalia tomentosa*] is called..Koombook in Ceylon. 1894 H. Trimen *Hand-bk. Flora Ceylon* II. 160 T[*erminalia*] *glabra*... Kumbuk. 1923 *Blackw. Mag.* Dec. 860/2 A lovely clear purling stream with the most beautiful Kumbuk trees fringing it on either side. 1971 *Ceylon Observer* (Mag. ed.) 19 Sept. 3/3 She had heard the beast sigh loudly behind the great big *kumbuk* tree.

‖ **kumkum** (ku·mkum). Also **kum-kum, kunku.** [Hindi *kuṅkum* saffron.] A red powder used ceremonially, and by Hindu women to make a small distinctive spot on the forehead; the spot so made. Also *attrib.*
1938 K. van A. Gates tr. *Ramabai Ranade's Himself* vii. 57 A pleasant custom..allows a hostess at a ladies' party to apply a dot of red powder (*kunku*) to the forehead of each guest. *Ibid.* xv. 133 Don't you have the custom of wearing..*kunku* marks, and such things? 1943 Moraes & Stimson *Introd. India* (ed. 3) 113 Hindu women wear on the forehead a red dot (called *kumkum*, which is also the name of the powder from which it is made). An orthodox Hindu widow does not wear a *kumkum.* 1958 *Times* 15 Oct. 24/5 Mrs. Usha Rajwade, who earlier..rubbed on Kum-Kum powder. 1969 *Commerce* (Bombay) 26 July 149/3 Sometime back there was understandable furore in the city of Bombay when some missionary school asked the girls not to apply *kumkum* which is a sign of good fortune for an Indian girl. 1973 *Guardian* 14 Apr. 11/3 God needs offerings—garlands, coconuts, kum-kum powder, incense.

kümmel. (Earlier examples.)
1864 P. Marié *Tribute to Fair* p. x, Our friend here sipping Kümmel. *c* 1870 in H. W. Allen *Number Three St. James Street* (1950) 186/1 Liqueurs. Bottle... Kummel..9/-. 1877 E. S. Dallas *Kettner's Bk. of Table* 269 Caraways in palpable form have now disappeared from our tables, but only to return in the spirit—in Russian bottles labelled Kümmel.

Kümmell's disease (kü·mĕlz dizī·z). *Path.* [f. the name of Hermann *Kümmell* (1852–1937), German surgeon.] Delayed collapse of a vertebra after an injury.
1903 *Index-Catal. Library Surg.-General's Office U.S. Army* 2nd Ser. VIII. 874/2 (*heading*) Kümmel's [sic] disease. 1923 *Jrnl. Amer. Med. Assoc.* 1 Dec. 1866/1 Kümmell's disease is a distinct clinical entity, having as its basic origin a traumatism to the spine, which may be mild or severe. 1951 *Amer. Jrnl. Surg.* LXXXI. 166/2 The diagnosis of Kümmell's disease is not justified without negative lateral roentgenograms taken in the early postinjury period, with later films showing positive findings of collapse.

‖ **kumpit** (ku·mpit). [Native name.] A trading-vessel of the Philippine Islands.
1951 *People* (Austral.) 17 Jan. 48/3 They were finally got away by a Filipino..who paddled over from the Philippines in a 'kumpit' and took them back under a small cargo of rice. 1968 *Economist* 12 Oct. 81/3 There's many a fast kumpit no doubt leaving Sabah bound for the Sulu shores. 1969 R. Thomas *Singapore Wink* xiii. 138 '..the *Wilfreda Maria.*' 'What's that?' 'A *kumpit.*' 'And a *kumpit* is a what?' 'It's an eight-ton ship. I bought it from a Moro pirate. I'm a smuggler.' 1972 *Sunday Times* (Singapore) 24 Sept. 7/3 Strict watch is maintained on all kumpits (Filipino boats) arriving in Sabah.

kumquat (kʊ·mˌkwǫt). [See Cumquat.] **1.** A small, orange-like citrus fruit from a tree of the genus *Fortunella*, native to southern China and Malaysia; now the usual spelling of Cumquat.
1870 *Jrnl. R. Hort. Soc.* II. 46 (*title*) Observations on the kumquat. 1882 [see Cumquat]. 1913 E. H. Wilson *Naturalist in W. China* II. iii. 25 The Kumquat (*C. Japonica*) is sparingly cultivated for its fruits, which, preserved with sugar, are an esteemed delicacy. 1968 *Punch* 13 Mar. 402/2 Mr. Adams is..occasionally affected (a face like 'a hybrid kumquat'). But he's written a master thriller.
2. *Austral.* A very small native citrus fruit, *Eremocitrus glauca*, or the tree producing it.
1889 J. H. Maiden *Useful Native Plants of Austral.* 8 *Atalantia glauca*.. 'Native Kumquat', 'Desert Lemon'... New South Wales and Queensland. 1955 P. White *Tree of Man* (1956) 400 Boys should eat kumquats, the syrup running from the corners of their mouths. 1965 *Austral. Encycl.* IV. 227/2 *Eremocitrus glauca*, native kumquat or wild lime (Rutaceæ): small spiny tree of western New South Wales and Queensland; the globular ½ inch fruits are pleasantly acidic and suitable for preserves, also cool drinks.

‖ **kumri** (ku·mri). [Native term.] A system of shifting cultivation practised in Karnataka, western India (see quot. 1938). Freq. *attrib.*
1904 W. Schlich *Man. Forestry* (ed. 3) II. i. iv. 106 The system is still extensively practised in India under a variety of names, as jhooming, dhya, kumri, taungya cultivation, etc. 1938 *Jrnl. Madras Geogr. Assoc.* XIII. 272 These people started what is known as 'Kumri' or 'Podu' cultivation. This is simply clearing the virgin forest, burning the felled trees and sowing rice, ragi, dhall etc., after the fire has died out, without any previous preparation of the soil. As the quantity of the ashes available was considerable, agricultural crop flourished and all that the 'Kumridar' had to do was a little weeding and protection of the crop from wild animals. He then left that area and selected another virgin forest close by. 1954 O. H. K. Spate *India & Pakistan* xxii. 624 The jungle is largely occupied by tribal remnants practising shifting *kumri* cultivation.

Kumyk (kū·mik). Also 8 **Coumyk;** 20 **Kumik, Kumuk.** [f. *Kumyk* name of a plateau in the Caucasus.] **a.** A Turkish people from the Kumyk plateau; also, a member of this people. **b.** Their language.
1788 G. Ellis *Mem. Map Countries between Black Sea & Caucasus* 15 The Tartars are of three tribes, viz. 1. Terekemens, Turcomans, or Trukhmenians,.. 2. Coumyks. These live to the northward of the former, about the lower parts of the rivers Sundsha, Koisu, and Axai. 1902 *Encycl. Brit.* XXX. 81/2 Having long been more civilized than the surrounding Caucasian mountaineers, the Kumyks have always enjoyed some respect among them. 1948 D. Diringer *Alphabet* 568 Kumuk Turkish. 1970 *Encycl. Brit.* V. 102/2 Feudal relations developed..in the northern Caucasus among the Kabardians and Kumyk. *Ibid.* XXII. 399/2 Kumyk is also spoken in the Dagestan A.S.S.R.

kunai (ku·nai). [Native name in New Guinea.] A large coarse grass, *Imperata cylindrica*, found in tropical Asia, Australia, and the Pacific region; also called *blady grass* (*blady *a.*) and *lalang. Also *attrib.*
1933 *Bulletin* (Sydney) 6 Sept. 39/4 Soon the weeds will have overrun the scar by the river bank, and a field of lush *kunai* will grow. 1945 *Coast to Coast 1944* 102 A gale ..set the tall kunai grass swaying. 1958 *Times* 27 Nov. 12/6 On the waterfront, the buildings were..of native design, thatched with kunai grass. 1962 *Coast to Coast 1961–62* 53 Beyond him was the fantasy of New Guinea green:..in front the kunai plain which reached for five miles to the Finisterre Ranges. 1965 *Austral. Encycl.* IV. 367/1 *Imperata cylindrica* var. *major* (blady grass) is widely spread over eastern and northern Australia, chiefly in coastal districts, and extends through Polynesia and Malaysia to tropical Asia where it is popularly known by other names (kunai, lalang, and kogon).

Kunbi (kū·nbi). Also 6 **Corumbijn,** 9 **Coombie, Koonbee, Kunbee.** [ad. Hindi *kurmī.*] A member of an Indian agricultural caste.
1598 W. Phillip tr. *J. H. van Linschoten's Discours Voy. E. & W. Indies* i. xxxix. 73/2 The Canarijns & Corumbijns [*sic*] are the Countrimen. 1808 R. Drummond *Illustr. Gram. Parts Guzerattee, Mahratta & Eng. Lang.*, Kunbee, Koonbee or Koolumbee, (Guz. and Mah.) *Koonbee* is a term, which is given, whatever be his profession or occupation, to the pure *Sooder* or first of the fourth class, who are, in Guzerat for the most part, Cultivators of the Land. This title, in the Deccan distinguishes the Cultivator, from him, who wears arms, and prefers to be called a Mahratta. 1826 W. B. Hockley *Pandurang Hari* I. x. 230, I begged the coombie, or cultivator, to give me some bread and rice. 1845 *Encycl. Metrop.* XX. 33/1 The Kunbí is a pure Súdra. 1875 *Encycl. Brit.* I. 423/2 The bulk of the population [of

Ahmadnagar] consists of Marhattás and Kunbís, the latter being the agriculturists. **1923** *United Free Ch. Miss. Rec.* June 250/2 Patils of villages, the kunbi, the cartman, the man on the road with whom I passed the time of day, ..all of them were kindly polite and courteous.

Kung-fu, kung-fu (kuŋ,fū·, kʊŋ-). Also without hyphen. [Chinese.] The Chinese form of *KARATE. Also *attrib.*

1966 *Punch* 14 Sept. 388/3 Kung-fu is here. **1968** *Clarendonian* XXII. 270 Chinese Kung-fu is still taught today—but only as a Martial Art to a very select, carefully chosen few. **1970** K. Platt *Pushbutton Butterfly* i. 9 The Chinese now call their form of karate *Kung Fu*... It's mostly leg-fighting. **1971** 'A. Hall' *Warsaw Document* xvii. 213 It was probably *kaminari*, a bastardized form of *kung fu*. **1974** *Listener* 17 Jan. 93/3 The plot.. hinges on Lee wiping out an ex-monk, Kung-fu (martial arts) vice chief. **1974** *Bookseller* 13 Apr. 1961/1 There has been a great upsurge of popular interest recently in kung fu, the ancient Chinese art of self-defence, encouraged by the films of Bruce Lee..and the television series Kung Fu.

kungu (kʊ·ŋgu). Also **kungo.** [Nyanja *nkungu*.] A small East African gnat, *Chaoborus edulis*. Hence **kungu cake**, the bodies of large numbers of these gnats, compressed to form a cake.

1865 D. & C. Livingstone *Narr. Expedition Zambesi* 373 A kungo cake, an inch thick..was offered to us. **1897** H. H. Johnston *Brit. Cent. Afr.* 436 The 'Kungu' fly of Lake Nyasa. **1899** D. Sharp in *Cambr. Nat. Hist.* VI. vii. 467 The kungu cake mentioned by Livingstone as used on Lake Nyassa is made from an Insect which occurs in profusion there, and is compressed into biscuit form. **1902** H. H. Johnston *Uganda Protectorate* I. 413 The kungu fly has a soft little body, scarcely as large as that of a flea, with gauzy wings. **1964** H. Oldroyd *Nat. Hist. Flies* pl. 30 (a) (*caption*) A piece of 'kungu-cake', made up entirely from the bodies of small midges, *Chaoborus edulis*, from an East African lake.

kunku, var. *KUMKUM.

‖ **Kunst** (kunst). [G., = art.] The G. word in Comb. (which are in varying degrees naturalized in Eng. scholarly writings), as **Kunstforscher** (ku·nstfǫrʃər), practitioner of *Kunstforschung*, art historian; also **Ku·nstforschung**, (scientific) study of fine art, art history; **Kunstgeschichte** (ku·nstgəːʃi·χytə), the history of art, art history; **Kunsthistoriker** (ku·nsthistǫːrikər), an art-historian; **Kunstlied** (ku·nstlīt) = *art song* (see *ART *sb.* 18 b); **Kunstprosa** (ku·nstprōᵘzā), literary prose, stylized or highly wrought prose.

1899 R. Fry *Let.* 15 Nov. (1972) I. 175 We are in the thick of a *Kunstforscher* fight. **1923** A. Huxley *Antic Hay* vii. 104 As a connoisseur and *kunstforscher*, Mr. Clew was much esteemed. **1933** *Burlington Mag.* Oct. 145/1 It would almost seem as though the Government had never recognized the existence of such a being as the 'Kunstforscher', perhaps because we have no better name for him in English than the clumsy and not altogether accurate 'art historian'. **1959** *Times Lit. Suppl.* 17 July 420/3 The editor, an industrious and knowledgeable *Kunstforscher*, supplies interesting scraps of information concerning the artist and his pictures, his models, his friends, and his methods. **1966** *Punch* 2 Nov. 683/2 There being no formal *Kunstforschung* in his student days. **1892** W. James *Let.* 7 Oct. (1920) I. 328, I have mapped out a profitable course of winter reading, *Naturphilosophie* and *Kunstgeschichte*. **1936** *Mind* XLV. 515 The range of questions that it opens out is exemplified by the two essays in *Kunstgeschichte*. **1937** H. Nicolson *Diary* 31 Dec. (1966) 315 Ben has been..in Florence preparing himself as a *Kunsthistoriker*. **1971** 'M. Innes' *Awkward Lie* iii. 64 All the professors and judges, the *Kunsthistoriker*, the Ministers of the Crown. **1880** Grove *Dict. Mus.* II. 133/1 The Volkslied has gradually disappeared, giving place to the Kunstlied, of which the accompaniment is an important feature. **1936** C. S. Lewis *Allegory of Love* 223 Usk is trying to write prose which shall have wings like verse—coloured and tunable prose—*Kunstprosa* in a word. **1960** K. J. Dover *Greek Word Order* i. 11, I cannot swear that the decrees of the Ozolian Locrians do not betray the hand of a mute inglorious Gorgias, but I may be allowed to doubt that and to believe that in early documents from the Peloponnese, the North-West and Crete the influence of *Kunstprosa* is minimal. **1965** *Listener* 19 Aug. 281/1 That gigantic waste-paper basket, the Short Story... Straight tales, character-sketches,.. experimental *Kunstprosa*, automatic writing, [etc.].

‖ **Künstlerroman** (kü·nstlər,roma:n). [G.] A *BILDUNGSROMAN about an artist.

1941 H. Levin *James Joyce* 41 The novel of development..becomes a novel of the artist, a *Künstlerroman*. **1957** N. Frye *Anat. Criticism* 307 The confession flows into the novel, and the mixture produces the fictional autobiography, the *Künstler-roman*, and kindred types. **1969** J. Gross *Rise & Fall Man of Lett.* i. 20 *Pendennis* can be seen as a first faltering step towards the *Künstlerroman*, the novel about the making of a novelist.

kunzite (kʊ·ntsəit, kʊ·nz-). *Min.* [f. the name of George F. *Kunz* (1856–1932), U.S. gemmologist; see -ITE¹.] A lilac-coloured variety of spodumene which is valued as a gem and becomes phosphorescent or changes colour when irradiated (see quot. 1962); a gemstone of this mineral.

1903 C. Baskerville in *Science* 4 Sept. 304/2 On account of this unusual and characteristic phosphorescence, as well as the other properties, I propose the name *Kunzite*. **1905** *Westm. Gaz.* 18 Jan. 12/1 To those who love some new thing the latest discovery in precious stones, Kunzite, will..be extremely welcome. **1949** *Rocks & Minerals* Mar.–Apr. 177 Kunzite. A perfect pear shaped stone of good color, 4·70 carats, $50.00. **1962** R. Webster *Gems* I. vii. 129 Kunzite shows a golden-pink or orange glow under long-wave ultra-violet light, and a similar but much weaker effect is seen under the short-wave ultraviolet lamp. Under an x-ray beam kunzite shows a very strong orange fluorescence with a strong and persistent afterglow. When the phosphorescence has died away the stone is found to have changed its colour to a bluish-green; this remains stable provided that the stone is kept away from a strong light. **1973** *Fortnum & Mason Ltd. Christmas Catal.* 23/1 Openwork ring, in 18 ct. gold set with diamonds and one freeform kunzite, £265·00.

Kuomintang (kuomintæ·ŋ, -tā·ŋ). [Chin., lit. 'national people's party'.] A nationalist radical party founded in China under Sun Yat-Sen in 1912, and led, after his death in 1925, by Chiang Kai-Shek, constituting the government before the Communist Party took power in October 1949, and subsequently forming the central administration of Taiwan.

1912 J. O. P. Bland *Recent Events & Present Policies China* iv. 107 In the beginning of September, an arrangement was effected, by the leaders of the T'ung-Meng-hui, to amalgamate with five minor political groups 'for the sake of harmony' under a new name, the Kuo-Min-tang, or Nationalist party. **1913** W. H. Hosking *Great Squeeze* iii. 36 The *Kuomintang* is a coalition of parties and is the only one that counts just now. **1928** T. F. Millard *China* 39, I remember the assassination of one Sung, a Kuomintang leader. **1941** E. Hemingway *Let.* 30 July in *Morgenthau Diary (China)* (U.S. Senate Committee on Judiciary) (1965) I. 458 The bitterness between the Communists and most of the Kuomintang leaders I talked to including the Generalissimo, can hardly be exaggerated. **1948** J. K. Fairbank *U.S. & China* ix. 193 The National Government of China at Nanking in the decade from 1927 to 1937 was the most modern and effective that China had known. It was led by Chiang Kai-shek and controlled by the Kuomintang on the basis of party dictatorship. **1952** C. P. Fitzgerald *Revolution in China* ix. 229 It will be argued that the Kuomintang remnants in Formosa are 'White Chinese', and much Right Wing American misconception of the Chinese Revolution is due to this belief. But..the Nationalists are failed revolutionaires. **1957** *Times Lit. Suppl.* 27 Dec. 782/3 This bitter struggle between Kuomintang and Communists. **1971** K. Hopkins *Hong Kong* 216 For the present the fifty-year-old struggle between the Communists and the Kuomintang continues, muted, with the British administration holding the ring. **1972** S. L. Appleton in *Asian Survey* Jan. 35 Kuomintang and government leaders, summoned to an emergency meeting following the U.N. vote, themselves cited the need for political reforms.

‖ **Kuo-yü** (kuo,yü). Also **Guoryu, Kuoyu.** [Chin., lit. 'national language'.] The name given to the Chinese 'national tongue', a form of Mandarin adopted for official use.

1932 Guoryu [see *KUAN, KWAN 1]. **1934** in Webster. **1954** Pei & Gaynor *Dict. Ling.* 116 *Kuo-yü*, the new 'national tongue' of China, based on the Peiping dialect of North Mandarin; now estimated to be used by about 300,000,000 persons. **1957** *B.B.C. Handbook* 132 There were large audiences to the relays by Radio Hong Kong of BBC programmes in Cantonese, English, and Kuoyü. **1964** *New Statesman* 10 Apr. 581/4 (Advt.), Applicants must be native Kuoyu speakers. **1968** D. Torr *Treason Line* 45 He remembered just enough of the *Kuo-yü* he had learnt in Chungking. **1969** J. M. Gullick *Malaysia* i. 29 The universal teaching in Chinese schools of the national Kuo-Yu dialect should in time provide a *lingua franca*.

Kupferschiefer (ku·pfəɹʃīfəɹ). *Petrogr.* Also 9 **Kupfer-schiefer**; and with small **k.** [a. G. *kupferschiefer*, f. *kupfer* COPPER *sb.*¹ + *schiefer* shale, slate.] A bituminous brown or black shale of the Permian series, which is worked in Germany for copper; = *copper schist, -slate* (COPPER *sb.*¹ 11).

1830 H. T. De La Beche *Sections & Views* 47 Professor Sedgwick..considers that he can trace the equivalents of the copper slate (kupfer-schiefer)..of the Thuringerwald, in the magnesian limestone of England. **1831** *Phil. Mag.* X. 46 Those sandstones and conglomerates which surmount the carboniferous series, and separate it from the *kupfer-schiefer* and magnesian limestone. **1879** *Encycl. Brit.* X. 352/1 The Kupfer-schiefer contains numerous fish.. and remains of plants. **1886** J. Geikie *Outl. Geol.* xxi. 298 The Kupferschiefer (2 feet thick) has long been famous for its ores of copper and other metals. **1921** A. W. Grabau *Textbk. Geol.* II. xxxviii. 506 At the base of this limestone series lies the important black copper-bearing shale, the Kupferschiefer, which rests upon the red sandstone. **1967** D. H. Rayner *Stratigr. Brit. Isles* viii. 253 The Hilton Plant Bed..consists of dolomitic sands and silts whose flora includes some species common to the Marl Slate and Kupferschiefer.

Kupffer (ku·pfəɹ). *Histology.* The name of Karl Wilhelm von *Kupffer* (1829–1902), Bavarian anatomist and embryologist, used in **Kupffer('s) cell** (also *cell of Kupffer*), a phagocytic cell that occurs in the lining of the sinusoids of the liver and has long radiating processes of cytoplasm (described by Kupffer in 1876).

1901 tr. *H. Dürck's Atlas & Epitome Special Path. Histol.* II. 18 The so-called stellate cells, or Kupffer's cells, ..are faintly visible. **1924** H. E. Roaf *Text-bk. Physiol.* xx. 262 Kupffer cells have been seen containing remnants of red blood corpuscles and they may be instrumental in collecting the hæmoglobin for the formation of bile pigment. **1956** *Nature* 24 Mar. 575/2 Blocking of Kupffer cells interferes with the storage of vitamin A ester in the liver of the rat. **1966** C. R. & T. S. Leeson *Histol.* xiv. 314/1 It is probable..that the stellate cells of Kupffer are increased in number in time of need by differentiation of the more primitive, endothelial cells.

‖ **Kur** (kūr). [G.] A cure, a taking of the waters (in Germany or another German-speaking country, as at a KURSAAL); a spa.

1885 Geo. Eliot *Let.* 25 June in J. W. Cross *George Eliot's Life* (1885) II. xiii. 369 It would have marred the Kur for me if I had every day to undergo a *table d'hôte*. **1892** C. M. Yonge *That Stick* I. xx. 238 The end of Constance's holidays was in view, the limit that had been intended for the Kur at Ratzes. **1915** F. M. Hueffer *Good Soldier* i. 10 The music of the Kur orchestra. **1974** *Country Life* 24 Oct. 1212/1 The Kur park and its monuments are still as the Edwardians knew them.

‖ **kura** (ku·rā). [Jap.] In Japan, a fire-proof store-house.

1880 I. L. Bird *Unbeaten Tracks Japan* I. x. 106 There is a *kura*, or fire proof storehouse, with a tiled roof on the right of the house. **1906** R. A. Cram *Impressions Jap. Archit.* iii. 63 Every house of any pretension possesses its 'kura', or storehouse, built of wood and bamboo, but covered two feet thick with clay... After a big fire in a Japanese city, nothing is left but fine ashes and the scorched but reliable kura. **1936** K. Nohara *True Face of Japan* iii. 44 Between the houses of wood, paper and thin plaster, tower the white-washed massive buildings of the kura. **1965** W. Swaan *Jap. Lantern* v. 58 The works of art may be brought in still packed in the special boxes in which they are stored in the *kura*.

kurakkan (ku·rǎk,kān). Also 7 **coracan.** The Sinhala name for a type of cereal grass, *Eleusine coracana* (Indian *raggee*), which is extensively grown in chenas in Sri Lanka (Ceylon) where flour from its grain forms a staple food of the poorer villagers.

1681 R. Knox *Hist. Relation Ceylon* I. iii. 11 There are divers other sorts of Corn, which serve the People for food in the absence of Rice... There is *Coracan*, which is a small seed like Mustard-seed. This they grind to meal or beat in a Mortar, and so make cakes of it. **1824** A. Moon *Catal. Indigenous & Exotic Plants Ceylon* 9 Eleusine... 1 coracana, common,..Kurakkan,..Ceylon, *cult.* **1864** G. H. K. Thwaites *Enumeratio Plantarum Zeylaniæ* v. 371 *E. Coracana*, Gærtn, (nom. vulg. 'Koorakkan'), is extensively cultivated by the Cinghalese as a food grain. **1900** J. D. Hooker in H. Trimen *Hand-bk. Flora Ceylon* V. 277 The Sinhalese 'Kurrakan'..is a very stout prolific form of this, with the spikelets crowded in many series, and a globose rugose seed. It is extensively cultivated for its grain in Ceylon. **1913** L. Woolf *Village in Jungle* i. 10 When the rains fall in November the ground is sown broadcast with millet or kurakkan. **1971** *Ceylon Daily News* 18 Sept. 4/6 In the villages around here the traditional food has been kurakkan and curd.

kurchatovium (kʊ̄ɹtʃätōᵘ·viəm). *Chem.* [ad. Russ. *kurchátovii* (Flerov & Kuznetsov 1967, in *Priroda* Nov. 35), f. the name of Igor *Kurchatov* (1903–60), Russ. nuclear physicist: see -IUM.] (A name proposed for) an artificially produced transuranic element, atomic number 104. Symbol Ku. Cf. *RUTHERFORDIUM.

1967 I. Zvára et al. *Joint Inst. Nucl. Res.* (Dubna, U.S.S.R.) *Preprint D6–3281* (*title*) Experiments on chemistry of element 104—kurchatovium. **1968** *New Scientist* 11 Jan. 85/1 Scientists working..at Dubna..were last year awarded the Lenin Prize for their work on the synthesis of transuranium elements, in particular element 104. This latter element was named Kurchatovium by the Russians. **1970** *Soviet Radiochem.* XII. 536 It was confirmed..that kurchatovium forms a chloride KuCl4, with properties close to those of the chloride HfCl4, and, consequently, is an analog of hafnium and zirconium, i.e., a member of sub-group IVb. **1971** *Inorg. & Nucl. Chem. Lett.* VII. 1115 The present work shows once again that the doubts expressed by the Berkeley group..concerning the chemical identification of kurchatovium are completely unfounded. *Ibid.* 1119 [*Reply of the 'Berkeley group'*.] We believe that these comments raise some valid questions as to whether or not 'element 104 (kurchatovium—Ku) was chemically isolated and identified'.

Kurd (kʊ̄ɹd). Also 7 **Coord**, 8–9 **Curd.** [Native name.] One of a pastoral and agricultural people of Aryan stock, found in northern Iran and Iraq and eastern Turkey, with the adjacent regions of the U.S.S.R. (the area being collectively known as *Kurdistan*). Also *attrib.* or as *adj.* So **Ku·rdish** *a.*, of or pertaining to the Kurds or their language, a dialect belonging to the Iranian group; as *sb.*, the language itself; **Kurdista·n** *a.* and *sb.*, (of) a Kurdish rug.

1616 T. Roe *Let.* 30 Oct. in *Embassy to Court of Gt. Mogul* (1899) II. 310 The King..tooke occasion to take in by force a reuolted Nation to the East of Babilon. The People are Called Coords. **1776** Gibbon *Decl. & F.* I. xiii. 381 The arrows of the Carduchians... Their posterity, the Curds,..acknowledge the nominal sovereignty of the Turkish sultan. **1813** *Q. Rev.* Oct. 257 Languages and Dialects... Median, Zendish, Pehlvish, Persian, Kurdish. *Ibid.* 267 The Kurds speak a corrupt Persian. **1823** Byron *Don Juan* vi. lxxxvi. 29 Asia, where Kaff looks down upon the Kurds. **1836** T. Skinner *Adventures Journey Overland to India* II. v. 69 The wife of the Kurdish traveller. **1854** J. H. Newman *Lect. Hist. Turks* iv. iii. 265 Saladin was a Curd. **1868** W. D. Whitney *Lang. & Study of Lang.* (ed. 2) v. 192 The Persian..with its outliers on the north-west and on the east—as the Armenian, the Kurdish, the Ossetic, and the Afghan. **1882** E. O'Donovan *Merv Oasis* I. 325 A Kurd encampment. **1899** Mrs. L. M. Elton tr. *Nazarbek's Through the Storm* 204 About thirty Kurdish brigands rushed out of the forest. **1920** *Glasgow Herald* 12 May 9 Turkey accepts..a scheme of local autonomy for the predominantly Kurdish areas east of the Euphrates, south of the southern frontier of Armenia. *Ibid.*, The Kurds inhabiting that part of Kurdistan which has hitherto been included in the Mosul vilayet are to be allowed..to adhere to the independent Kurdish State. **1924** *Blackw. Mag.* Nov. 583/2 Pursued by a volley of oaths in English, Scots, and Kurdish. **1926** T. E. Lawrence *Seven Pillars* (1935) ii. xix. 124 It was an ordinary bell tent, furnished with..a fairly good Kurd rug, a poor Shirazi, and the delightful old Baluch prayer-carpet on which he prayed. **1931** A. U. Dilley *Oriental Rugs & Carpets* Pl. 22 (*caption*) Kurdistan, Mina Khani Design. **1955** *Times* 21 May 9/6 The Kurds of Iraq live along the frontiers of Turkey and Persia, among the mountains that form Iraq's natural boundaries. They are a hill people, and recognizably related to the Alpine Swiss, the Sherpas, and other mountaineers. **1963** *Times* 12 June 11/1 Kurdish would be the first language in elementary education in Sulaymaniya. **1970** D. M. Lang *Armenia* i. 39 Many of the neighbours of the Armenians in antiquity have vanished from the map, like the Hittites..; relapsed into barbarism, like the Kurds, descendants of the proud Medes; [etc.]. **1970** L. Sanders *Anderson Tapes* lxv. 174 The brothers had taken a very nice Kurdistan [rug] down to the truck. **1971** *Guardian* 3 Dec. 4/4 An unfortunate Pesh Merga (Kurdish soldier). **1973** *Times* 9 Aug. 5/1 Mr Baluk told the court that he was a member of the Kurdish minority in Turkey. **1974** *Evening Standard* 12 Feb. 48/5 (Advt.), Superb oriental carpets & rugs. Including:..outstanding Kurdistan rugs in fascinating geometric designs.

kurdaitcha (kuɹdəi·tʃā). *Austral.* Also 9 **kooditcha.** [Austral. Aboriginal.] (See quot. 1909.) Also, a malignant supernatural being. Also *attrib.*

1886 E. M. Curr *Austral. Race* I. v. 148 It was discovered in 1882..that the Blacks..wear a sort of shoe when they attack their enemies by stealth at night. Some of the tribes call these shoes Kooditcha, their name for an invisible spirit... The soles were made of the feathers of the emu, stuck together with a little human blood... The uppers were nets made of human hair. **1896** *Proc. R. Soc. Victoria* VIII. 66 The wearing of the Urtathurta and going Kūrdaitcha lūma appears to have been the medium for a form of vendetta. **1909** *Cent. Dict.* Suppl. I. 693 *Kurdaitcha.* 1. Among the tribes of central Australia, a man chosen to avenge the death of one who had died, every death being supposed to be due to the magic influence of some enemy. 2. A kind of shoe, made of emu-feathers matted together with human blood, worn by the kurdaitcha when on his errand. **1932** *Times Lit. Suppl.* 28 Jan. 52/2 The *Kurdaitcha* among the tribes of Central Australia goes forth against his enemy with a medicine-man skilled in this sinister art. **1940** A. Upfield *Bushranger of Skies* (1963) xvi. 156 You ought to make us Kurdaitcha shoes. *Ibid.* xix. 183 A Kurdaitcha man is an evil spirit—always wandering about the poor blackfellows' camp at night. **1953** —— *Murder must Wait* xxii. 146 All good aborigines should be fast asleep..safe from the dread Kurdaitcha. **1959** S. H. Courtier *Death in Dream Time* ii. 19 The bewildering display of aboriginal weapons and implements... death-bones and kurdaitcha shoes. **1962** A. Upfield *Will of Tribe* vi. 57 The fear of the Kurdaitcha Man and the Great Snake.

kurfuffle, var. CURFUFFLE *sb.*

1955 C. S. Lewis *Surprised by Joy* vii. 114, I could put up with any amount of monotony far more patiently than even the smallest disturbance, bother, bustle, or what the Scotch call *kurfuffle*.

‖ **Kurhaus** (kū·r,haus). [G.] A building at a German health resort where the medicinal water is dispensed for drinking and external use; a pump-room; hence, sometimes, a similar building at a watering-place outside Germany.

1855 Geo. Eliot in *Fraser's Mag.* July 61/2 The white *Kurhaus* glittering on a grassy slope. **1857** C. Kingsley *Two Yrs. Ago* III. ix. 269 He drives up to the handsome old Kurhaus. **1935** J. Buchan *House of Four Winds* 12 Mr McCunn..accepted the consultant's prescription, and rooms were taken for him at the Rosensee *kurhaus*. **1962** N. Freeling *Love in Amsterdam* 1. 38 Tuesday night he was at the Kurhaus in Scheveningen.

‖ **kuri** (ku·ri). *N.Z.* [Maori *kurī*.] A Maori dog, long extinct: now usu. = *GOORIE, GOORY.

1838 J. S. Polack *New Zealand* I. ix. 308 The karárāhé, or dog (Canis Australis), which, when young, is known as *kuri*, has been an inhabitant some two or three centuries. **1843** E. Dieffenbach *Trav. N.Z.* II. iv. 46 The New Zealand dog is different from the Australian dingo;..the native name is kuri. **1900** J. Scott *Tales Colonial Turf* 180 It was a sort of poodle..no one would think of asking who owned the kuri. **1930** J. Cowan in J. Reid *Kiwi Laughs* (1961) 97, I wish you'd give it to them hot and strong about the blasted 'kuris' worrying my sheep. **1949** P. Buck *Coming of Maori* (1950) I. iv. 64 The dog (*kuri*) figures a good deal in tradition. **1959** Tindale & Lindsay *Rangatira* 195 *Kuri dog.* The Maori dog was small... It was kept as a pet, it gave some aid in hunting, and it provided..fresh meat... It appears to be a domesticated strain of the feral dog of India..now extinct in New Zealand.

Kurile (kiū·rīl), *sb.* and *a.* [f. the *Kurile islands*, a chain of small islands stretching northwards from Japan to Kamchatka, since 1945 held by the U.S.S.R.] **A.** *sb.* A native or inhabitant of the Kurile islands. **B.** *adj.* Of or pertaining to these islands or this people. Hence **Kuri·lian** *a.* and *sb.*

1764 J. Grieve tr. *Krasheninnikov's Hist. Kamtschatka* I. i. 19 When the ice is carried thither with the beavers on it, then the Kuriles, who follow the ice along the shore, assemble here in great multitudes. **1819, 1843** [see *AINU]. **1845** *Encycl. Metrop.* XXV. 866/2 A small portion of the continent..opposite to the Isle of Sakhalyin..is occupied by the Aïnos or Kurīls, whose Isles extend from Japan to the Southern extremity of Kamchatka. **1875** *Encycl. Brit.* I. 426/2 The Ainos..are distinguished by an exuberance of hair on the head and body, a circumstance which has given rise to their name of 'Hairy Kuriles'.

Kurku, var. *KORKU.

kurl-a-mo, kurl-the-mo: see *CURL *v.*¹ 1 c.

Kurnai (ku·ɹnəi). [Austral. Aboriginal.] **A.** *adj.* Of or pertaining to an Aboriginal tribe of south-eastern Australia. **B.** *sb.* This tribe; a member of this tribe; their language.

1911 J. G. Frazer *Golden Bough: Magic Art* (ed. 3) I. v. 324 The Kurnai tribe of Gippsland in Victoria. **1965** *Austral. Encycl.* I. 72/2 The Kurnai of Gippsland used to cut one or both hands from a corpse. The hand was wrapped in grass and dried, and a possum-fur string attached so that it could be worn about the neck of a near relative. By pushing or pinching the wearer, it was said to give warning of an approaching enemy. **1965** W. E. H. Stanner in R. M. & C. H. Berndt *Aboriginal Man in Austral.* viii. 230 Much mischief has resulted from the supposition that all Aborigines in all important respects resemble the Aranda, Kamilaroi, Kurnai, and Murngin. **1972** *Talanya* I. 29 In Kurnai..he gives the following phrases.

kurnakovite (kʊ̄ɹnākǫ·vəit). *Min.* [f. the name of N. S. *Kurnakov* (1860–1941), Russian mineralogist: see -ITE¹.] A hydrated borate of magnesium, $Mg_2B_6O_{11}·13H_2O$, found as colourless granular aggregates with a vitreous lustre at Inder, Kazakhstan, U.S.S.R.

1940 M. N. Godlevsky in *Compt. Rend. (Doklady) de l'Acad. des Sci. de l'URSS* XXVIII. 638 During my visit to the Inder deposit in 1938, I found in the dump of one of the pits..a solid colourless mineral which proved to be a borate, of the composition $2MgO.3B_2O_3.13H_2O$. The new mineral was named kurnakovite. **1962** *Amer. Mineralogist* XLVII. 402 In the monoclinic and triclinic dimorphs..inderite and kurnakovite respectively, the asymmetric unit has been found to contain three chemically different boron sites. **1970** *Soviet Physics: Doklady* XIV. 1141/2 In kurnakovite an isolated H_2O particle is bonded to four boroxy rings by a single Mg octahedron.

‖ **Kurort** (kū·r,ōᵊrt). Also 9 **curort; kurort.** [G.] In Germany or other German-speaking countries, a health-resort, a watering-place; also *fig.*

1868 Geo. Eliot *Let.* 27 June (1955) IV. 454 We..daily rejoice that we have found such a Cur-ort, suiting both mind and body. **1926** *Spectator* 21 Aug. 276/1 A company is being formed..to erect a modern thermal *Kurort*, and build an up-to-date hotel. **1930** *Discovery* June 171/2, I had no troubles, except a longing for the mountains.. and even the attractions of Prague, the Riesengebirge, and several magnetic *kurorts* failed to hold me back. **1941** V. Nabokov *Real Life S. Knight* xv. 146 She loved inventing some rare illness and going to some famous kurort. **1949** Koestler *Promise & Fulfilment* III. iv. 325 Others, like Naharia, are German Kurorts grafted on by plastic landscape surgery.

Kuroshiwo. Add: Also Kuroshio, Kurosiwo, and as two words (with or without a hyphen). (Further examples.)

1928 Russell & Yonge *Seas* x. 231 In the Pacific.. there is a system of oceanic currents much after the manner of that in the Atlantic, the Japan Current or 'Kuro Shiwo' corresponding to our Gulf Stream. **1967** *Oceanogr. & Marine Biol.* V. 57 Often compared with the Gulf Stream and Kuroshio, the East Australian Current is essentially different because it does not, as do the other two, make up an eddy system extending towards the east across the ocean. **1969** *Sci. Jrnl.* Jan. 16/1 The Kuroshio ..is a swift current that runs along the western edge of the Pacific just as the Gulf Stream does in the North Atlantic.

kurper (kʊ̄·rpəɹ). *S. Afr.* Also **karper, kerper, kurpur.** [Afrikaans, f. Du. *karper* carp.] A name used for several fresh-water fishes resembling carp, esp. *Sandelia capensis*.

1831 *S. Afr. Q. Jrnl.* Oct. 19 Kurper... Inhabits most of the rivers towards the Southern extremity of Africa. **1902** *Trans. S. Afr. Philos. Soc.* XI. 213 In an expedition to the interior, led by Surgeon Pieter van Meerhoff, we are told by him that in the space of 1½ hours they caught 'beautiful Carp (Karper)'... This is undoubtedly the fish (*Spirobranchus capensis*) still called Karper in the Colony by the Dutch. **1913** C. Pettman *Africanderisms* 285 Kurper or Kerper... *Spirobranchus capensis*, a well-known freshwater fish. See Karper. **1947** K. H. Barnard *Pict. Guide S. Afr. Fishes* 80 The Cape Kurper (*Sandelia capensis*) is a very well-known fish in the rivers and vleis of the southwest Cape... In the Eastern Province a closely allied species known as the Rockey or Bain's Kurper (*Sandelia bainsii*) occurs. **1952** *Cape Times* 31 Jan. 9/8 The casualties include yellow fish..mudfish, carp and kurpur. **1971** *Sunday Times* (Johannesburg) *News Mag.* 28 Mar. 8/5 The big kurper are found much deeper than usual.

Kurrichane (kʊritʃa·ne). The name of a place in the western Transvaal, South Africa, used *attrib.* in **Kurrichane thrush** to designate a thrush, several races of which are found in Africa south of the Sahara.

1924 *Ibis* 770 *Turdus libonianus*. Kurrichane Thrush. This thrush is common in Nyasaland. **1936** D. A. Bannerman *Birds Trop. W. Afr.* IV. 314 The Kurrichane Thrush is not unlike a female European Blackbird. **1960** G. Durrell *Zoo in my Luggage* i. 32 A kurrichane thrush treated us to a waterfall of sweet song.

‖ **kurta** (kʊ̄·ɹtā). *India.* Also **khurta, kurtha.** [Hind.] A loose shirt or tunic worn by men and women.

1913 W. G. Lawrence in T. E. Lawrence *Home Lett.* (1954) 485 Me in a dhoti and khurta, white Indian clothes. **1920** *Chambers's Jrnl.* 29 May 408/1 A guard of honour in scarlet *kurtas* (blouse-tunics). **1966** *New Statesman* 2 Sept. 316/3 A painfully thin youth—very Hindu, whose kurtha was always clean. **1968** *Observer* 25 Feb. 3/4 Beatle George Harrison..dresses in the flowing pyjamas and *kurta* worn by people in North India. **1969** *Femina* (Bombay) 26 Dec. 8/4 Allambana of Delhi, run by Mohini Tandon, held an exhibition of embroidered saris and kurtas at Calcutta's Park Hotel. **1972** *Vogue* May 137 Atlantic and Othello Khanh..in Indian cotton kurtas and tie-dyed turbans.

kurtosis (kʊ̄ɹtōᵘ·sis). *Statistics.* [mod.L., f. Gr. κύρτωσις a bulging, convexity, f. κυρτός bulging, convex.] A shape-characteristic of a frequency distribution that reflects the sharpness of the peak (for a unimodal distribution) and the shortness of the tails, and is generally measured by the quantity μ_4/μ_2^2 or its excess over 3 (μ_4 and μ_2 being the fourth and the second moments about the mean of the distribution).

1905 K. Pearson in *Biometrika* IV. 181, I have already called $\beta_2 - 3 = \eta$ the degree of kurtosis. **1931** L. H. C. Tippett *Methods of Statistics* ii. 28 There are several curves having the same standard deviation but varying kurtosis. **1952** W. L. Gore *Statistical Methods for Chem. Exper.* ii. 16 The kurtosis is useful in determining if a frequency distribution differs from the normal error curve. The kurtosis of a normal distribution is equal to 3; smaller values than 3 indicate a flatter distribution than the normal (a platykurtic distribution), while values above 3 indicate a more sharply peaked distribution than the normal (a leptokurtic distribution). **1968** P. A. P. Moran *Introd. Probability Theory* vii. 317 Tables of..the coefficient of kurtosis $\mu_4\mu_2^{-2} - 3$. **1972** P. Laslett *Household & Family in Past Time* iv. 129 These variables were then summed and averaged: measures of skewness and kurtosis were computed.

kuru (ku·ru). [Native name.] A progressive and fatal degenerative disease of the brain which is endemic in an area of the Eastern Highlands of New Guinea and is characterized by ataxia and tremor.

1957 *New Eng. Jrnl. Med.* 14 Nov. 974/1 The current report of our preliminary findings is based on the careful study of 114 cases of this new disease, which the local populace know by the name of 'kuru'. **1958** *Times* 9 Jan. 10/1 It is estimated that about 1 per cent. of the population is affected with *kuru*. **1965** *Sunday Mail* (Brisbane) 10 Oct. 2 Outside the Fore region only the adjacent areas where inter-marriage has certainly occurred show cases of kuru. **1967** *New Scientist* 26 Jan. 190/2 A brain disease known as *kuru*..appeared about 45 years ago in a tribe of cannibals in New Guinea and is now the tribe's most common cause of death. **1971** *Nature* 30 Apr. 589/1 Kuru ..has so far been transmitted from eleven different human patients to eighteen chimpanzees with incubation periods of 14–39 months after intracerebral inoculation. **1973** *Sci. Amer.* Jan. 126/3 The prototype of the slow virus [disease] is probably kuru, that remarkable affliction of the people of the South Fore River of New Guinea, apparently spread by ritual cannibalism of the brain tissue.

‖ **kuruma** (kurū·mā). [Jap.] A rickshaw. So **kuruma·ya**, one who pulls a rickshaw.

1727 J. G. Scheuchzer tr. *Kæmpfer's Hist. Japan* I. 11. iv. 180 Sai Sin first obtain'd leave of the Emperor to be carried about in a *khuruma*, or cover'd Chariot, drawn by two Oxen. **1880** I. L. Bird *Unbeaten Tracks* I. i. 18 From *kuruma* naturally comes *kurumaya* for the *kuruma* runner. **1889** E. Arnold *Seas & Lands* (1891) xiv. 188 The Kurumas are wheeled sharply round and brought up with a general shout of arrival in front of a Japanese inn. **1892** —— *Japonica* 44 The *Kuruma*-men can trot in safety round every corner. *Ibid.* 62 The Tokio citizens call their little cab *kuruma*, which means 'a wheel', and the coolie who pulls it is termed *kurumaya*. **1894** L. Hearn *Glimpses Unfamiliar Japan* I. i. 2 The..charm of Japan.. began for me with my first kuruma-ride... The jinrikisha,

or kuruma, is the most cozy little vehicle imaginable. **1898** I. L. Bishop *Korea & her Neighbours* II. xxiv. 79 Warm winter clothing, a Japanese *kurumaya's* hat..and Korean string shoes completed my outfit. **1904** R. J. Farrer *Garden of Asia* ii. 12 At dangerous corners the kurumaya howls dolefully to make the people avoid the path. *Ibid.* xxiii. 234 Mr. Desire,..taking us up in the kuruma, proceeds to whirl us home to our friends. **1909** *Daily Chron.* 21 Oct. 7/2 A couple of stalwart kuruma-ya who do their eight miles an hour with ease.

Kurume (kuru·me). The name of a Japanese town on the island of Kyushu, used *absol.*, or *attrib.* in **Kurume azalea,** to designate one of a group of small, evergreen azaleas developed there from a variety of *Rhododendron obtusum* early in the nineteenth century and introduced to America and Europe by Ernest Henry Wilson (1876–1930) in 1919.

1920 E. H. Wilson in *Garden Mag.* Mar. 38/1 It was during the Arnold Arboretum expedition to Japan in 1914 that I first became acquainted with these Kurume Azaleas. **1924** E. H. M. Cox *Rhododendrons for Amat.* v. 98 Kurume azaleas should be consistently fed to ensure good flowers. **1949** *Jrnl. R. Hort. Soc.* LXXIV. 145 When first introduced the Kurumes were given an undeserved reputation for tenderness. **1964** J. Berrisford *Rhododendrons & Azaleas* iii. 41 In the eighteen-twenties a cult arose among the feudal gentlemen of Japan and the dwarf evergreen azaleas were bred privately... Thus arose the two-hundred-and-fifty-odd varieties of Kurume azaleas, so called from the town of Kurume where they were later discovered. **1965** 'M. Neville' *Ladies in Dark* x. 101 I'd taken up two Kurumes that she'd ordered. **1970** S. B. Sutton *Charles Sprague Sargent & Arnold Arboretum* x. 258 The pilgrimage to the Kurume Azaleas came at the latter part of an expedition which was, as one expected from Wilson, a success both botanically and horticulturally.

‖ **kurus** (kŭrū·ʃ). Also formerly **ghrush, ghurush, grouch, grush, gurush.** [Turkish *kuruş*.] A Turkish piastre, $\frac{1}{100}$ of the value of a lira; a coin of this value.

1882 *Numismatic Chron.* 3rd Ser. II. 175 Suleyman II. issued his own large silver pieces in 1099, and gave them the name of *ghrûsh*, which recalls the grossi, groschen, and groat of the Western States... Without entering deeply into the question of the exchange value of this Turkish ghrûsh, or piastre, as it was called by travellers—not, however, to be confounded with the small modern piastre —it is interesting to notice that the ghrûsh and the akcheh, which was its lowest 'divisionnaire', were constantly altering their relations. **1906** *N.E.D.* s.v. *Piastre* 2, The English (French, German, etc.) name..of a small Turkish coin, called in Turkish *ghūrūsh*, $\frac{1}{100}$ of a Turkish pound, having in Turkey, in 1900, a circulating value of about 2*d.* **1917** A. R. Frey *Dict. Numismatic Names* 93/1 *Ghrush*,..The name of the coin is variously written Grush, Gurush,..etc. **1927** *Weekly Dispatch* 6 Nov. 18 Turkish stamps have a perpetual 'grouch'... There are eleven stamps, values 1, 2, 2½, 3, 5, 6, 10, 15, 25, 50 and 100 *grouch*, which is the current coinage. **1959** E. Pound *Thrones* xcvii. 24 The olde double-ducat, The olde turkish grouch. **1960** M. Case tr. *Boulanger's Turkey* (Hachette World Guides) p. lxxvi, The monetary unit of Turkey is the Turkish pound (lira), which is made up of 100 kuruş. There are metal coins for 1 K.; 2 1/2 K.; 5 K.; 10 K.; 25 K.; 50 K.; 1 T.L. **1971** *Whitaker's Almanack 1972* 954/1 The Turkish *Lira*..is divided into 100 *Kurus*.

kurvey (kʊɹvēⁱ·), *v. S. Afr.* Also **karwey.** [f. Du. *karwei* hard work, big job (ad. F. *corvée*): see KURVEYOR.] *intr.* To carry goods in an ox wagon. Hence **kurvey-ing** *vbl. sb.*

1873 *Queenstown Free Press* 8 Aug. (Pettman), For various reasons not a farmer kurveys between either Concordia or Springbok and Port Nolloth. **1876** T. Stubbs *Reminiscences* I. 49, I tryed a trip at Kerveying, I took a load to Fort Wiltshire. **1884** M. A. Carey-Hobson *At Home in Transvaal* I. iii. 29 'There will be an end to those visits one of those days,' said the merchant, 'and then good-bye to your karweying, Walters.' **1902** *Encycl. Brit.* XXXI. 81/2 'Kurveying' (the conducting of transport by bullock-waggon) in itself constituted a great industry.

Kushan (ku·ʃān), *sb.* and *a.* Also **Kushana.** **A.** *sb.* A people originating in central Asia who invaded India and established a powerful dynasty (1st–3rd centuries A.D.) in the North-West; also, a member of this people. **B.** *adj.* Of or pertaining to this people, esp. to the dynasty.

1872 *Numismatic Chron.* New Ser. XII. 182 Some time before the Christian era, the chief of the *Kuei-shwang* tribe of the great *Yuchi*..subjected the other four tribes of the nation, and assuming the title of King of the *Kuei-shwang*, or *Kushân*, conquered..Ophiana, Kophene, and Parthia. *Ibid.* 183 The *Asiani* are evidently the *Kushân* tribe. **1935** [see *fire-altar* (*FIRE *sb.* B. 5)]. **1935** [see *HAN¹]. **1948** D. Diringer *Alphabet* vi. 343 More important in the development of Indian writing were the inscriptions of the Kushana kings. **1966** F. Stark *Rome on Euphrates* xi. 261 The Kushans had entered history in the first half of the first century A.D. with their king, Kujula Kadphises. *Ibid.* 262 This expansion of the young Kushan kingdom occurred while the vigorous policy of Nero and his advisers was making itself felt. **1971** *Fashion Panorama* (Ceylon) July–Sept. 25 On the coins of the Kushanas are depicted the portraits of monarchs like Kaniksha, Huvishka, Kadphises etc. The Kushan rulers are depicted in these portraits as wearing trousers and a coat. **1972** *Times* 25 Oct. 21/7 The shrine emerged as a Mazdaean fire-temple dedicated to the memory of Kanishka, legendary emperor of the Kushan dynasty.

kus-kus, var. *KAS-KAS.

kuta (ku·tā). = *KGOTLA.

1943 M. Gluckman *Admin. Organization Barotse Native Authorities* 7 *Kuta* (*khotla*),..council, court, Native Authority. *Ibid.* 12, I think the kuta would be made more efficient if its numbers were reduced; but this..can only be discussed after the kuta organisation has been made regular, instead of muddled. **1955** —— *Judicial Process among Barotse* i. 9 Since the council is not only a court, I use the native term *kuta*... Usually the ruler does not attend the hearings of cases, though the kuta's judgment is referred to him for confirmation. Even if the ruler chooses to sit in the kuta while a case is being tried, it proceeds as if he were not there. **1959** G. D. Mitchell *Sociol.* 83 The council, known as the *kuta*, is both a political and a judicial body.

Kutani (kutā·ni). [f. the name of the village of *Kutani-mura* in the former province of Kaga, Japan.] Used esp. *attrib.*, as *Kutani ware*, a kind of gold and dark red Japanese porcelain.

[**1875–80** Audsley & Bowes *Keramic Art Japan* I. 43 Almost all the good and important pieces of Kaga ware which we have seen are marked with the two characters signifying Kutani.] **1880** A. W. Franks *Jap. Pott.* 80 The amateur prefers the original Kutani ware of dark-red and greyish-white colour. **1890** B. H. Chamberlain *Things Japanese* 283 There were two principal varieties of the ware: *Ao-Kutani*, so called because of a green (*ao*) enamel of great brilliancy and beauty..used in its decoration, and Kutani with painted and enamelled *pâte* varying from hard porcelain to pottery. **1960** B. Leach *Potter in Japan* vii. 170 Contemporary and old examples of Kutani wares. **1967** *N.Y. Times* 22 Oct. 1. 10 The dramatic bronze castings were discovered in Hong Kong, the hand-decorated Kutani porcelains in Japan.

Kutenai, Kutenay (kū·tĕnēⁱ, -ni). Also **Kootenai** and many other variants. [Native name, *Kútonâqa*.] **A.** *sb.* **a.** An Indian people of the Rocky Mountains; also, a member of this people. **b.** Their language. **B.** *adj.* Of or pertaining to this people or their language.

1801 A. Mackenzie *Voy. from Montreal* (map following pref.) Cattanhowes. **1809** D. Thompson *Jrnl.* 9 Sept. in *Washington Hist. Q.* (1920) XI. 99 They all smoked, say 54 Flat Heads, 23 Pointed Hearts & 4 Kootenaes, in all about 80 men. **1831** R. Cox *Adventures Columbia River* II. vii. 152 The Cootonais are the remnants of a once brave and powerful tribe. **1838** S. Parker *Jrnl. Exploring Tour beyond Rocky Mts.* xxiii. 304 The Cootanies inhabit a section of the country to the north of the Ponderas along M'Gillivary's River... They speak a language distinct from all tribes about them, open and sonorous, and free from gutturals. **1846** H. Hale in *U.S. Exploring Expedition 1838–42* VI. 204 (*heading*) Kitunaha, or Coutanies, or Flat Bows. **1877** A. S. Gatschet in *Mag. Amer. Hist.* I. III. 170 The Kootenai, Kitunaha, or Flatbow language. **1891** *7th Ann. Rep. U.S. Bureau Amer. Ethnol. 1885–86* 85 (*heading*) Kitunahan Koluschan Families. **1893** *Rep. Brit. Assoc. Adv. Sci.* 575 The Kootenays believe that they came from the East, and one of their myths ascribes to them an origin from a hole in the ground east of the Rocky Mountains. **1894** *Amer. Anthropologist* Jan. 69 The tomtit, the owl, the robin, and a few other birds are believed [by the Kootenays] to speak Kootenay. **1929** *Amer. Speech* V. 116 Indian tribal names..were usually transcribed by persons whose ears were unaccustomed to any but European languages... We inherited..corruptions..*Kutenai* from *Kutonaqua*. **1932** D. Jenness *Indians of Canada* ii. 20 Kootenayan, Siouan, Iroquoian, and Algonkian, are spoken also in the United States. *Ibid.* xxii. 360 In the firm conviction that the dead would one day return to life at lake Pend-d'Oreille, all the Kootenay bands assembled at that lake in certain winters to hold a religious festival. **1955** P. E. Baker *Forgotten Kutenai* i. 8 The word Kutenai is spelled..Kootenai, Kootenay, Kootenae, Cootanie, and Cootenai as well as Kutenai. **1959** E. Tunis *Indians* viii. 112/2 One tribe, the Kutenai, made bark canoes with exaggerated back-slanting bows, like the ram bows of 1898 battleships. **1965** *Canad. Jrnl. Linguistics* Spring 78 Kutenai (Powell's 'Kitunahan'), spoken in the eastern Plateau area bordering on the north-western Plains, stands in lonely isolation among a variety of languages of sure affiliation. **1969** O. W. Johnson *Flathead & Kootenay* 16 The Kutenais sowed and harvested sacred plants for ceremonial smoking before they ever heard of White men.

kutnahorite (kutnāhōᵊ·rəit). *Min.* Also **kutno-.** [ad. G. *kutnohorit* (A. Bukowsky, 1901 (see *Neues Jahrbuch f. Min., Geol. u. Pal.* (1903) II. 338)), f. *Kutná Hora*, name of the town in Czechoslovakia where it was first found: see -ITE¹.] A reddish-white rhombohedral carbonate of calcium, manganese, magnesium, and sometimes iron, belonging to the dolomite group of minerals.

1907 *Mineral. Mag.* XIV. 402 *Kutnohorite*... A rhombohedral carbonate with the atomic ratios Ca:Mn:Fe:Mg = 7:5:1:2 occurring as reddish-white cleavage-masses. **1955** *Amer. Mineralogist* XL. 751 Kutnahorite occurs at Franklin as anhedral masses with curved cleavage surfaces up to three centimeters in size in a small veinlet cutting the normal franklinite ore. It is translucent, with a pale pink color. **1962** W. A. Deer et al. *Rock-Forming Min.* V. 267 Kutnohorites from Franklin containing approximately 10 per cent. excess $CaCO_3$ show some degree of order in their structure. **1967** *Amer. Mineralogist* LII. 1751 Occurrences of kutnahorite have been recorded from Franklin, New Jersey.., Kutná Hora, Czechoslovakia.., Chvaletice, Czechoslovakia..and Providencia, Mexico.

kuvasz (kū·vaʃ, kū·vas). Pl. **kuvaszok.** [Hungarian, fr. Turkish *kavas* guard.] A large white long-coated Hungarian breed of dog, used as a guard dog in its native country.

1935 *Working Dogs* (Amer. Kennel Club) 103 Being a working dog of the larger size, the Kuvasz should be sturdily built and impress the eye with its strength and activity. **1947** C. L. B. Hubbard *Working Dogs of World* II. 100 The Kuvasz is the best-known Hungarian breed of dog... Kuvaszok on 'sentry-go' are most unpleasant if met by strangers. **1971** F. Hamilton *World Encycl. Dogs* 75 The Kuvasz is spirited, intelligent and courageous.

Kuwaiti (kuwēⁱ·ti). [Arab. *kuwaytī*, f. *Kuwayt* Kuwait.] A native or inhabitant of Kuwait, a principality on the Persian Gulf; the dialect of Arabic spoken there. Also *attrib.* or as *adj.*, of or pertaining to this country or its inhabitants.

1928 A. Rihani *Ibn Sa'oud of Arabia* xxviii. 352 The Ikhwan marched on Jahrah, slaughtered five hundred of the *mushrekin*, the infidel Kuwaitis. **1930** —— *Around Coasts Arabia* III. i. 242 The sail—in its shadow and from its bounties, the Kuwaitis live. **1947** *R. Cent. Asian Jrnl.* 271 The Kuwaitis..had invariably used the Turkish flag. **1950** *Middle East Jrnl.* IV. 18 All know the majesty of the Kuwaiti *boom* entering their harbors to exchange cargoes. **1961** *Daily Tel.* 8 July 16/6 The field commander of the Kuwaiti army. **1967** T. M. Johnstone *Eastern Arabian Dial. Studies* 29 The consonant system in Kuwaiti is the same as for the dialect group as a whole. **1970** *Times* 3 Apr. (Arab League Suppl.) p. iii/3 In the field of social affairs and labour, the Public Assistance Law has been issued with special regard to the stipulations of our religion about social welfare as well as the traditions of the Kuwaiti society and in accordance with the constitution. **1971** *Guardian* 30 Aug. 3/2 Kuwaitis can afford a summer-long holiday. Kuwait is the richest country on earth in terms of income per capita. **1972** H. Osborne *Pay-Day* 1. i. 11 A young Kuwaiti called Rifai. **1975** *Times* 20 Jan. 61/2 Even after the recent Arab purchase by Kuwaiti interests of 14% of Daimler-Benz AG, there seems to be little chance of a change in the laws.

‖ **kuzushi** (kuzu·ʃi). [Jap.] In Judo, a method of unbalancing one's opponent.

1950 E. J. Harrison *Judo* 103 *Kuzushi*, breaking opponent's posture or balance. **1957** Takagaki & Sharp *Techniques Judo* 1. ii. 18 *Kuzushi*, methods of unbalancing the opponent. **1968** P. & K. Butler *Judo & Self-Defence for Women & Girls* ii. 36 Do, please, remember that you can break your opponent's balance to *any* angle you wish... *Tsukuri* is the action you take to break your opponent's posture, *kuzushi* is the effect that it has on her.

kvell (kvel), *v. U.S. slang.* [ad. Yiddish *kveln*, ad. G. *quellen* to gush, well up.] *intr.* To boast; to feel proud or happy; to gloat.

1967 *Listener* 28 Dec. 849/3 *The New York Spy* is a useful and terribly bright guide to New York, conscientiously *kvelling* through 'the city's pleasures', charmed alike by brutal manners, as chronicled by Tom Wolfe, and the Jewish takeover (London swings but Jewish New York *kvells*). **1968** L. Rosten *Joys of Yiddish* 199 Only from your children can anyone *shep* (derive) such *naches* (prideful pleasure) as makes you *kvell*. **1970** L. M. Feinsilver *Taste of Yiddish* 364 'You've got reason to kvell'; 'is he kvelling!'

kvetch (kvetʃ). *U.S. slang.* Also **kvetsch.** [Yiddish *kvetsh*, ad. G. *quetsche* crusher, presser.] A term of personal abuse: *spec.* a person who complains a great deal, a fault-finder. Also **kve·tcher.**

1964 S. Bellow *Herzog* (1965) 61 She's got a disgusting father and a *kvetsch* of a mother. **1964** W. Markfield *To Early Grave* (1965) xi. 187 There was Ozzie Waldman, Ozzie the *kvetch*. For his favor you could die. He gave away nothing. **1966** *New Society* 12 May 9/2 The idiom of the New Yorker—Gentile or Jew—..has a lot of Yiddish words, like schlepp..shiksa and kvetsch. **1968** L. Rosten *Joys of Yiddish* 200 What a congenital kvetcher! *Ibid.*, It will take forever, he's such a kvetch. **1970** S. Ellin *Man from Nowhere* xix. 94 A bagger [*i.e.* investigator] should not dress so conspicuous that even these old *kvetchers* around here turn and look.

Hence as *v. intr.* [ad. Yiddish *kvetshn*], to complain, to whine; so **kve·tching** *vbl. sb.*

1965 *Holiday* July 98 The Beatles..came along in the middle of a wave of *kvetching*—songs constantly stressing the negative. **1968** *Atlantic Monthly* Oct. 70 He is an amiable one, not given to angry kvetching. **1971** *Harper's Mag.* Feb. 111 After listening to Kashouk *kvetch* for a couple of hours, Sol Hurok..put the question direct. 'Tell me, Kashouk,' Hurok wanted to know. 'If you always lose so much money, why do you stay in business?'

kvutza (kvu·tsā). Also **kvutzah, kwuza.** Pl. **kvutzot, -oth.** [mod.Heb. *qĕbhūṣāh*, f. Heb., group.] In Israel, a communal and cooperative settlement, which, with others, may form a kibbutz.

1921 H. M. Kallen *Zionism & World Pol.* xvii. 255 *Kwuzoth* or coöperative workmen's colonies were outfitted. **1929** J. H. Holmes *Palestine To-Day & To-Morrow* iv. 188 (*heading*) The 'Kvutzoth', or communal colonies. *Ibid.* 195 Most of the settlers in Nahalal, for example, had originally been members of a *Kvutzah*. **1934** *Cook's Traveller's Handbk. Palestine, Syria & Iraq* (ed. 6) 48 The post-war settlements of the Zionist Organisation are all based on co-operative principles, and belong to two classes:— (a) The *Moshav*... (b) The *Kvutzah*, which is more or less

communal in character. The entire assets and produce of the village are owned in common... Several of these *kvutzoth* are organised into groups (*kibbutzim*) based on the principle of mutual assistance and exchange of man power. **1944** H. F. Infield *Co-operative Living in Palestine* (1946) 6 The Kvutza is well established. The first of these settlements was founded in 1908... A recent census.. shows seventy-six Kvutzot. **1967** *Encycl. Brit.* VI. 456 The 'Kibbutz' and 'Kvutza', where all property is collectively owned. **1973** *Times Lit. Suppl.* 23 Feb. 212/5 The burning of the Degania B kvutza by the Arabs.

Kwa (kwā). Also **Kwo, Qua.** [Native name.] A branch of a Niger-Congo language family, including Akan, Ewe, Ibo, and Yoruba. Also applied to a native speaker of one of the languages in this family. Also *attrib.* or as *adj.*

1857 H. Goldie *Princ. Efik Gram.* p. viii, If we take Creek Town..as a centre, and describe a circle of one hundred miles radius, we shall either include or touch upon the tribes of Usahadet..Efut..Aqua (Qua), [etc.]. **1883** R. N. Cust *Sk. Mod. Lang. Afr.* xi. 238 Kwa or Qua. Is different from the Efik and is spoken to the South of that Language... It is stated that the Efik are immigrants from the Interior, and that the Qua are the indigenous inhabitants. **1919** [see *IBIBIO *sb.* and *a.*]. **1930** A. Werner *Struct. & Relationship Afr. Lang.* ii. 33 Along the coast of the Gulf of Guinea—say from Cape Mount to the Niger mouth—we have what Westermann calls the Kwa group. **1955** [see *FANTI *sb.* and *a.*]. **1970** P. Oliver *Savannah Syncopators* 39 The Kwa group [of tribes] which stretches from Liberia east to Ibo territory in Nigeria. *Ibid.* 42 The Kwa-speaking peoples of the coastal rain forest. **1971** A. Kirk-Greene in J. Spencer *Eng. Lang. W. Afr.* 128 The Kwa group of languages, of which Yoruba is one.

kwacha (kwa·tʃā). [Chibemba *kwacha* dawn.] **a.** Used as a Zambian nationalist slogan. **b.** The basic currency unit in Zambia. Also, a banknote of this value.

1962 K. Kaunda *Zambia shall be Free* xvii. 160 For a long time I have led my people in their shouts of *Kwacha* (the dawn). We have been shouting it in the darkness; now there is the grey light of dawn on the horizon and I know that Zambia will be free. **1966** *Times* 10 Mar. 8/7 Mr. Arthur Wina, Finance Minister, told Parliament today that in 1968 Zambia will have its own decimal currency. The new unit will be the 'Kwacha', worth 10s. and meaning 'Dawn of Freedom'. **1967** D. C. Mulford *Zambia* v. 198 Speaking to the Conference's 4,000 delegates amid shouts of 'Action Now' and 'Kwacha', Kaunda launched into an impassioned attack. **1971** *Whitaker's Almanack 1972* 775/1 Zambia adopted decimal currency on Jan. 16, 1968, the unit being the *Kwacha*, equivalent to 10s. of the former currency. The *kwacha* = 58 p. sterling. **1972** *Daily Tel.* 18 Sept. 18/8 The company has a fully paid up capital of 2·7 million kwacha, which is to be purchased by Indeco from Unilever of Britain over seven years. **1973** *Guardian* 23 Mar. 14/5 The exchange rate is given as ·7143 kwachas per US dollar.

‖ **kwai-lo** (kwai,lo). Also **kwai-tze** (kwai,tse). [Chinese (Cantonese dial.).] 'Foreign devil', a name given by the Chinese to foreigners.

[**1878** H. A. Giles *Gloss. Far East* 76 *Kwei-tsze* or *Kuei-tzŭ*, devils. A Chinese term for foreigners. **1910** J. S. Thomson *Chinese* i. 70 Now and then an urchin spits at a foreigner's chair and shouts, *Fan kwei lai* (See, here's a foreign devil).] **1944** H. B. Rattenbury *China, my China* i. 9 On the streets they cursed me for..a 'yang kuei-tze'—'a foreign devil'.] **1969** *Times* 9 Dec. (Taiwan Suppl.) p. viii/6 To the *kwai-lo's* innocent suggestion that the Foochow or coastal school might be classified under the generic label 'shanghai', Master Wei responded with a sour oath. **1972** *South China Morning Post* (Hong Kong) 4 Dec., *Kwai Lo*, a foreigner (or literally devil man). **1972** *Times* 21 Oct. (Hongkong Suppl.) p. i/4 A discreet variety of Mao-style padded-coat is admirable wear for young and elderly *kwai-lo* (foreign devils) in the Hongkong winter.

Kwakiutl (kwā·kiut'l). Also 9 **Kwahkewlth,** etc. [Native place-name, *Kwa'gul*.] **a.** An Indian people of the north-west coast of N. America; also, a member of this people. **b.** Their language. Also *attrib.* or as *adj.*

1848 *Jrnl. Ethnol. Soc. London* I. 233 (*in list*) 7. Quagheuil. Inhabiting Broughton's Archipelago. **1874** *Rep. Indian Branch, Canada Dept. Minister of Interior 1873* 33 Quackewelths, with sub-tribes 2,000. **1897** F. Boas in *Rep. U.S. Nat. Mus. 1895* 311 (*title*) Social organisation and secret societies of the Kwakiutl Indians. *Ibid.* 321 Kwakiutl, Salishan, and Chemakum..show certain similarities in form. **1921** E. Sapir *Lang.* iv. 81 Causative duplications..as in Kwakiutl *metmat* 'to eat clams' (radical element *met-* 'clam'). **1933** L. Bloomfield *Lang.* xxvi. 470 Thus, Quilleute, Kwakiutl, and Tsimshian all have different articles for common nouns and for names, and distinguish between visibility and invisibility in demonstrative pronouns. **1937** R. H. Lowie *Hist. Ethnol. Theory* ix. 133 A blue-blood in the caste-ridden Kwakiutl society. **1944** [see *DEVIANT *ppl. a.* 1]. **1951** R. Firth *Elem. Social Organiz.* i. 3 The Kwakiutl Indians of British Columbia compete in the potlach. **1955** W. Gaddis *Recognitions* I. i. 23 With the loss of Camilla he returned to the times before he had known her, among the Zuñi and Mojave, the Plains Indians and the Kwakiutl. **1959** E. Tunis *Indians* 136/1 The Kwakiutl and Nootka languages were similar to one another but unlike any other known tongue. **1965** W. P. Alston in M. Black *Philos. in Amer.* 24 What in our culture would be a look of contempt would be a look of affection among the Kwakiutl. **1969** *Times* 22 Sept. 14/3 A Kwakiutl door post from Cape Commerell is sculptured in the form of a gigantic bear—a forest spirit—its lips rounded in a mysterious hooting cry.

‖ **kwanga** (kwa·ŋgā). [Native name.] In Zaïre (formerly the Congo), a kind of bread made of manioc.

1907 *Daily Chron.* 28 Oct. 7/3 With the exception of a few people..who supply the State with 'kwanga' (native bread)..all the people I saw are taxed with rubber. **1908** H. H. Johnston *George Grenfell & Congo* II. xxviii. 796 In those happy days ten cakes of kwanga..could be bought for one brass rod. **1971** *Guardian* 31 July 10/6 Kwanga..looks..like..candlewax.

kwashiorkor (kwǫʃi,ǭ·ɹkǭɹ). [Native name in Ghana (formerly the Gold Coast).] A wasting disease that is caused by an insufficient intake of protein by the body and chiefly affects young children in tropical countries, producing apathy, oedema of the extremities, desquamation, and partial loss of pigmentation (and is generally associated with diarrhoea and stunted growth), and leading in severe cases to death.

1935 C. D. Williams in *Lancet* 16 Nov. 1151/1 The name 'kwashiorkor' indicates the disease the deposed baby gets when the next one is born, and is the local name in the Gold Coast for a nutritional disease of children, associated with a maize diet. **1951** G. C. Shattuck *Dis. Tropics* lvii. 695 The early stage of kwashiorkor is difficult to recognize. Probably there are many mild cases which never develop the typical syndrome. **1954** *New Biol.* XVII. 20 One of the commonest diseases, called Kwashiorkor, is associated with protein deficiency in the diet when young and especially at the stage when breast-feeding ceases and no very suitable foods to replace it are available. **1959** *Times* 2 Dec. 10/5 Another hospital doctor described the incidence of kwashiorkor as 'needless slaughter' which could easily be prevented by funds for milk and other proper nourishment. **1968** *Observer* (Colour Suppl.) 29 Dec. 18/3 We saw a fretful baby with the unmistakable signs of early *kwashiorkor*, the disease of protein starvation that ravages Africa. **1970** D. B. Jelliffe *Dis. Children Subtropics & Tropics* (ed. 2) vii. 172 The 'sugar baby' type of child..which conserves his subcutaneous fat, may even be obese and recovers very rapidly on a high-protein diet, is the typical example of kwashiorkor without previous malnutrition. **1971** *Progress* (Cape Town) May 9/1 One of the unit's major achievements was to prove that the cure of kwashiorkor.. could be initiated by the administering of a synthetic skimmed milk.

‖ **kwedini** (kwidi·ni). *S. Afr.* Also **khwedini, khwidini, kweding, kweedini.** [ad. Kaffir *kwenkwendini* vocative form of *kwenkwe* boy (Pettman).] A native boy.

1912 *Queenstown Representative* 27 Jan. 5/1 This '*kweendine*' was walking behind the pole driving the bullocks on. **1946** *Spotlight* (Johannesburg) 23 Aug. 6/1 A twelve-year-old *kwedini* asleep across some sacks. **1949** *Cape Argus Mag.* 12 Nov. 7/5 Uneducated, 'raw' Native kwedinis, aged 15 and upwards. **1955** J. B. Shephard *Land of Tikoloshe* v. 37 A Khwidini is 'not old enough for a man, nor young enough for a boy'. *Ibid.* viii. 59 Abakhweta.. do not mix with the younger Khwedinis. **1970** *Daily News* (Durban) 18 Dec. 13 One White trader complained that kwedings (boys) had destroyed more than 1000 white telephone cups.

kweek (kwīk). *S. Afr.* [Afrikaans, f. Du. *kweek* couch grass.] In full, *kweek grass*. A local name for several creeping grasses, esp. *Cynodon dactylon*.

1904 *Transvaal Agric. Jrnl.* Oct. 185 The Transvaal kweekgras is shorter and more of a surface grass than the Bermuda grass. **1929** J. W. Bews *World's Grasses* v. 184 It [sc. *Cynodon dactylon*] is commonly known as 'Bermuda grass' and in S. Africa as 'Kweek grass', though that name is applied to other species of creeping grasses as well. **1937** S. Cloete *Turning Wheels* xviii. 294 The dark-green rings of kweek grass marking the site of his kraals. **1947** *Cape Times* 2 May 9 Only 342 acres were planted to marram grass and kweek. **1954** C. E. Hubbard *Grasses* 335 Like other well-known grasses it [sc. *Cynodon dactylon*] has numerous common names, being known as 'Kweek' in S. Africa, 'Doob' in India, 'Couch' in Australia, 'Bermuda Grass' in the United States, and in the British Isles sometimes as 'Creeping Dog's-tooth-grass' or 'Creeping Finger-grass'. **1970** A. Fulton *I swear to Apollo* 4 The redgrass gave way to kweekgrass and steekgrass that grew not in a soft thick carpet but in sparse little tufts dotted here and there amongst the boulders.

kwela (kwēⁱ·lā). [Afrikaans, ? ad. Zulu *khwela* climb, mount.] A popular dance, or its accompanying jazz-like music, of mid- and southern Africa.

1958 *Time* 16 June 37/1 The haunting sound of penny-whistle jazz has become the favorite music of South Africa's slum-caged blacks—and of a great many white hipsters. In the dusty streets, urchins rock to the penny-whistle's fast *kwela* beat; in shabby speakeasies, women shuffle to its slower *marabi* rhythm. **1958** *Gramophone* Dec. 328/2 Those addicted to the shrill squawking of the Kwela flute will have to hear..*Something New From Africa*. **1960** *Guardian* 1 Apr. 10/7 When night falls, she can dance the kwela, mambo, or high-life with any or all of them [*sc.* Nigerians]. **1961** *Sunday Times* 12 Mar. 15/7 One of the tin whistle bands that make kwela music at street corners in Johannesburg. **1969** J. Brunner *Plague on Both your Causes* xiii. 96 The music—all local stuff halfway between *kwela* and the long Arab-influenced melodic lines of the Sudan—rose to a pitch of frenzy. **1974** *Sunday Times* (Johannesburg) 13 Oct. 15, I would..listen to kwela music and dance with gouty feet.

kya (kəi·ā). Also **kaia, kia.** [ad. Zulu *-khaya* place of abode.] In South Africa and Rhodesia, an African's hut; also, quarters of an African servant.

1909 K. Fairbridge *Veld Verse* 85 Where the high-veld breaks to valley..Stands a kaia looking Northward through the mountains to the plain. **1910** J. Buchan *Prester John* xvi. 257 Inanda's Kraal was a cluster of kyas and rondavels. **1911** *East London* (Cape Province) *Dispatch* 24 Nov. (Pettman), A native living in a kraal at Lydenberg quarrelled with another native, whom he accused of having fired his *kya*. **1935** L. G. Green *Great Afr. Mysteries* (1937) xvi. 192 Each house has a separate *kya* in the back garden for the servant. **1950** *Cape Argus* 4 Mar. 9/5 Fowl-pens, temporary tin Native kias and big heaps of rubble. **1956** N. Gordimer *Six Feet of Country* iii. 38 These two white-washed servants' rooms (some white people called them kyas,..wanting to keep in their minds the now vanished mud huts which the word indicated). **1971** *Guardian* 29 Sept. 19/3 The houseboys' Kias (usually one small room) at the bottom of the garden.

kyack (kəi·æk). *U.S.* [Orig. unknown.] 'A form of packsack consisting of two hollow containers swung on either side of a pack-saddle' (*Dict. Americanisms*).

1901 *Sunset* Mar. 138/1 Our camp now lay in perfect chaos—blankets, kyacks, saddle-bags, and cooking utensils in a jumble. **1904** *Outing* (U.S.) Oct. 98/2 Exactly the same bitter partisanship obtains in the choice of saddle, in the choice of alforjas or kyacks. **1944** R. F. Adams *Western Words* (1945) 88/2 Kyacks might be described as hollow containers, one on each side of the horse, each of sufficient capacity to hold the equal of two five-gallon oil cans placed side by side. **1948** *Sierra Club Bull.* (San Francisco) Dec. 2/2 Stock that..don't mind getting wet up to the kyacks at a stream crossing.

kyack, var *KAYAK².

Kyan, var. *KAYAN.

kyang. (Earlier and later examples.)

1869 A. A. A. Kinloch *Large Game Shooting* I. iv. 13 Kyang are found all over the elevated plateaux and valleys of Thibet. **1960** 'S. Harvester' *Chinese Hammer* xii. 118 The risk of being reincarnated as a female *kyang*, the wild asses who roamed the *tangs* and high plateaux.

‖ **kyat** (kī,ā·t). [Burmese.] The basic monetary unit of Burma since 1952.

1952 *Times* 17 June 10/3 It has been announced in Rangoon that the Union Bank of Burma Act, 1952, will become effective on July 1 next, when the new standard unit of monetary value, the *kyat*, comes into force. The kyat will be the exact equivalent of the present Burma rupee and will be divided into 100 units, each of which will be called a *pya*. **1955** E. Mannin *Land of Crested Lion* xi. 155 We..buy baskets of strawberries from Ghurkas..at half a *kyat* a basket. **1971** *Nat. Geographic* Mar. 361/1 Working seven days a week for nine months of every year, the couple extract an average of 120 pounds of sap from each tree, earning about 7½ *kyats* ($1.56) a day. **1972** *Guardian* 15 Aug. 3/4 Burma's foreign reserves are currently down to..£12 millions... The local currency, the kyat..is worth less than a third its official rate of 14 to the pound sterling.

kyathos. = CYATHUS 1 a.

1889 in *Cent. Dict.* **1935** Richter & Milne *Shapes & Names Athenian Vases* 30 *Kyathos*... Ladle in the form of a cup with foot and long upward curving handle. **1948** A. Lane *Greek Pott.* ii. 9 The *kyathos* or ladle for dipping the mixture off into jugs. **1960** R. G. Haggar *Conc. Encycl. Cont. Pott. & Porc.* 211/1 *Kyathos*—ladle;.. *Oenochoe*—jug for wine; *Kylix*—shallow stem cup.

kybosh. Add: **1.** = KIBOSH 1.

1846 *Swell's Night Guide* 124 Kybosh on, *to put the*, to turn the tables on any person, to put out of countenance. **1896** H. G. Wells *Wheels of Chance* xli, 'I put the kybosh on *his* little game,' he remarks. **1924** *Chambers's Jrnl.* May 296/2 Standofer's fairly put the kybosh on us this time. **1952** J. Cleary *Sundowners* iii. 122 Well, that puts the kybosh on it. **1956** H. G. de Lisser *Cup & Lip* xxii. 246 Good for you... You have put the kybosh on them. **1971** *Times Lit. Suppl.* 7 May 531/2 Not only did the First World War liquidate the Edwardian douceur de vivre. It also put the kybosh on the rationalist's faith in progressive social evolution. **1975** *Sunday Post* (Glasgow) 10 Aug. 7/3 She'd been looking forward to some salmon fishing, but the heatwave's put the kybosh on that.

2. (See quots.)

1845 G. W. M. Reynolds *Mysteries of London* I. xxiii. 60/1 The Thieves' Alphabet... K was a kye-bosh [*f.n.* 1s. 6d.], that paid for his treat. **1968** *Gloss. Brit. Argot*, *Kybosh*, one and a half shillings.

As *vb.* (var. KIBOSH *v.*).

1933 J. Cary *Amer. Visitor* iv. 41 The question is, can I kybosh the whole scheme at the same time?

kye (kəi). *Naut. slang.* [Origin unknown; but cf. E.D.D. *kyish* dirty.] **1.** (See quots.)

1929 F. C. Bowen *Sea Slang* 80 A kye. A rating who is mean with his money. **1946** J. Irving *Royal Navalese* 105 *Kye*, a mean, unworthy sort of fellow.

2. Cocoa or chocolate.

1943 Hunt & Pringle *Service Slang* 42 *Kie*, seaman's slang for cocoa. **1943** Baker *Dict. Austral. Slang* (ed. 3) 46 *Kye*, chocolate. (R.A.N. slang.) **1962** Granville *Dict. Sailors' Slang* 68/2 *Kye*, ship's cocoa which used to be issued in slabs, already sweetened, cut as needed and dissolved in boiling water to the desired consistency. The origin of the word is dialectal, from the adjective *kyish*,

KYLINDRITE

L

L. Add: **I. 2.** (Earlier examples. See also *ELL².)

1843 'R. Carlton' *New Purchase* I. xi. 80 On the first floor were two rooms, and connected with a Lilliputian half-story kitchen forming an L—as near as possible. **1873** T. B. Aldrich *Marj. Daw* etc. 167 Mr. Jaffrey's bedroom was in an L of the building. **1874** *Rep. Vermont Board Agric.* II. 510 To save expense, it is apt to be the case that no cellar is put under the L part of the house.

3. L-head, -headed *adjs.*, applied to (a reciprocating internal-combustion engine having) L-shaped combustion chambers, in which the valves are situated in a side arm.

1916 L. Mantell *Man. Motor Mech.* iii. 17 One of the most frequent errors made by designers is in attempting to obtain high compression in an L-headed engine with a short stroke. **1920** *Sci. Amer.* 3 Jan. 6/3 The intake manifold of several power plants, both on overhead-valve and L-head types of engine, is cast entirely within the detachable cylinder head. **1922** *Encycl. Brit.* XXX. 37/2 The..Vee Renault of 1912,..the..Vee RAF of 1913–14, and the..Vee RAF 4a, all of which had cast-iron L-headed cylinders. **1946** R. F. Kuns in Kuns & Plumridge *Automobile Engines* iv. 60 L-head engines are quiet in operation and are long lived. *Ibid.*, The L-head or, as the English call it, the side-valve engine. **1963** Bird & Hutton-Stott *Veteran Motor Car* 98 Their cylinders were L-headed and cast in pairs.

II. 6*. Other symbolic uses in science.

a. In *Physics L* is used to designate the series of X-ray emission lines of longer wavelength than the *K*-series obtained by exciting the atoms of any particular element (cf. *K 3 e); these arise from electron transitions to the atomic orbit of second-lowest energy, with principal quantum number 2, which is thus termed the *L-shell*, and electrons in this shell *L-electrons*. **L-capture**, the capture by an atomic nucleus of one of the *L*-electrons.

1911, 1923 [see *K 3 e]. **1930** Pauling & Goudsmit *Struct. Line Spectra* x. 172 There are three absorption edges corresponding to the removal of an electron from the L shell. **1930** *Phil. Mag.* IX. 205 The K electron distribution in carbon will be determined mainly by the central nucleus, and the influence of the L electrons will be comparatively small. **1934** H. E. White *Introd. Atomic Spectra* xvi. 326 When a *K* electron is missing,..the binding energy of the *L* electron is approximately that for the corresponding electron in the element with the next higher atomic number. **1956** *Nucl. Sci. Abstr.* X. 1123/2 (*heading*) Effect of the correlations existing between the electron positions on the ratio *ρ* of the probability of L capture to that of K capture. **1968** *Physical Rev.* CLXVI. 945/1 The exchange correction..in the case of Be⁷ increases the *L*-capture probability by a factor of almost 4. **1970** E. P. Bertin *Princ. & Pract. X-Ray Spectrometric Analysis* vi. 182 Elements having atomic number 57 (lanthanum) or higher..are usually determined by measurement of their *L* lines with gas-flow proportional counters. **1972** R. Bolton *Org. Mechanisms* i. 14 The K-shell is now filled... A third electron must..be placed in the higher-energy L-shell.

b. In *Physics l* and *L* denote the quantum numbers of the orbital angular momentum of one electron or a group of electrons, respectively (superseding the *k* (= *l*+1) of the old quantum theory).

The use of *l* as a quantum number, and the values assigned to it, varied until shortly after the publication (in 1926) of Schrödinger's theory of the atom.

LS-coupling, an approximation used in the quantum theory of the atom when the spin–orbit interaction of individual electrons is small compared with the remaining electrostatic interaction between one electron and another, so that the orbital angular momenta of the electrons may be coupled to give a resultant L, their spins coupled to give a resultant S, and these resultants coupled in turn to give the total angular momentum J of the electrons. Also called *Russell–Saunders coupling*. Cf. *jj-coupling* (*J II. 6 c).

[**1925** Russell & Saunders in *Astrophysical Jrnl.* LXI. 61 Their remaining properties may be explained on the assumption that the two displaced electrons have fixed orbital momenta, L_1, L_2, of the amount indicated by Landé, but that the inclination of their planes is quantized, so that the resultant angular momentum *K* may have any geometrically permissible value in the series 1/2, 3/2, [etc.]. **1926** *Proc. R. Soc.* A. CXI. 84 The spectroscopic nature of each term..is specified by a quantum number *l* which relates to the whole set of electrons not in complete groups. It..is taken to be ½, 3/2, 5/2...for S, P, D terms, so that $l = k_1$ [$\equiv k - \frac{1}{2}$] when there is only one electron in an incomplete group. It may perhaps be thought of as the resultant angular momentum of the incomplete group. **1926** *Ibid.* CXII. 81 Spectral terms are to be designated in the usual way as follows:—S, P, D, F, G,...corresponding to the values 1, 2, 3, 4, 5,...for a spectral term quantum number denoted by 'l'.] **1928** H. S. Allen *Quantum* iv. 66 The quantum number j_a is now denoted by *l*, which is called the 'group quantum number'... Its value in this case [*sc.* of a single electron] is $l = k - 1$. *Ibid.*, The inner quantum number *j* is compounded vectorially from *s* and *l*. **1934** H. E. White *Introd. Atomic Spectra* xii. 190 These cases are known as *LS*-, or Russell–Saunders, coupling, on the one hand, and *jj*-coupling, on the other. **1934** [see *K 3 f]. **1970** G. K. Woodgate *Elem. Atomic Struct.* vii. 140 Along the sequence *np*(*n*+1)*s* from light to heavy elements, for example C(2*p*3*s*), Si(3*p*4*s*), Ge(4*p*5*s*), Sn(5*p*6*s*), there is a progression from *LS* to *j-j* coupling. Ge(4*p*5*s*) is an example of intermediate coupling for which neither *L*, *S* nor j_1, j_2 are even approximately good quantum numbers.

c. *Bacteriol.* [*L* said to be f. the *L*ister Institute, where Klieneberger worked.] The designation (now usu. as *L-form*) of an atypical form of certain bacteria which arises from and usually reverts to the normal form but is sometimes stable, and which lacks a cell wall, exhibits a very variable shape, and somewhat resembles a mycoplasma.

1935 E. Klieneberger in *Jrnl. Path. & Bacteriol.* XL. 93 These swollen elements [among the bacillary chains of *Streptobacillus moniliformis*], with an associated fine mycelial system resembling that of pleuropneumonia and agalactia, constitute an independent colonial system containing all strains of *S. moniliformis* so far examined. A delicate streptococcus has also been recovered..from the nasopharynx of healthy guinea-pigs harbouring a similar pleuropneumonia-like symbiont. These two pleuropneumonia-like organisms in association respectively with *S. moniliformis* and a streptococcus..will be referred to..simply as L1 and L2 pending the coining of appropriate generic names. **1950** C. E. Clifton *Introd. Bacteria* vi. 132 *Streptobacillus moniliformis*..resembles the actinomyces in many respects but gives rise after several days' incubation to highly pleomorphic forms characteristic of the pleuropneumonia group... Present evidence indicates that there is but one organism involved and that the actinomyces-like form and the L_1 form represent different stages of growth of one organism characterized by a complex reproductive cycle. **1968** *Zinsser Microbiol.* (ed. 14) lii. 794/2 Some L phase mutants revert back to normal size organisms as soon as the penicillin is removed from the medium. Others are stable in the L phase on solid media but revert back when subcultured in broth. **1973** J. Levy et al. *Introd. Microbiol.* ii. 39 Because they did not have a wall, they were not affected by penicillin or other antibiotics that interfere with cell wall synthesis. The mutation to an L-form is therefore troublesome if the bacterium is a pathogen.

d. *L-band*: a frequency band of electromagnetic waves used for radar, extending from 390 to 1550 megahertz.

1947 J. S. Hall *Radar Aids to Navigation* vii. 233 A cross-band airborne interrogator-responser recently developed consists of an *L*-band (about 25 cm) transmitter operating on a number of preselected frequency channels and a 10-cm receiver. **1967** *Electronics* 6 Mar. 52/1 Tradex, an adaptation of the radar developed for the ballistic missile early warning system, operates at uhf and L band.

7. Abbreviations.

L, learner; Liberal (in politics); low (on the selector mechanism in a car with automatic transmission).

L.A., local authority; LA, L.A., Los Angeles; L.A.F.T.A., Latin-American Free Trade Association; L.C. (A., I., M., T., etc.), landing craft (assault, infantry, mechanized, tank, etc.); *l.c.* (*Printing*) (examples); L.C.C., London County Council; LD, lethal dose: used with following numeral, as LD50, LD_{50}, indicating the percentage of a large group of similar animals that is killed by such a dose; LD or L-D (process), in steel-making, the Linz-Düsenverfahren (process) or Linz-Donawitz-Verfahren (process); L.D.C., less developed country; L.D.V., Local Defence Volunteers; L.E., LE (*Med.*), lupus erythematosus; usu. *attrib.*; L.E.A., Local Education Authority; L.E.M., lunar excursion module; see also *Lem; L.F., low frequency; LH (*Biochem.*), luteinizing hormone; L.M., lunar module; LMF, lack of moral fibre; L.M.S., London, Midland, and Scottish (Railway); L.N.E.R. (earlier L.N.E.), London and North-Eastern Railway; LNG, liquefied natural gas; LOI, lunar orbit insertion; LOS, loss of signal; L.P., long-playing (record); LPG, liquefied petroleum gas; LRL, Lunar Receiving Laboratory (building where astronauts and lunar samples are quarantined for a period after returning from the moon); LRV, lunar roving vehicle; L.s., letter (not autograph) signed; cf. *A.L.(S.)* s.v. *A III; L.S. (*Cinemat.*), long shot; L.S.E. (occas. L.S. of E.), London School of Economics; LSI, large-scale integration (of electronic microcircuits); L.S.T., landing ship, tank(s); LTH (*Biochem.*), luteotrop(h)ic hormone; L.V., luncheon voucher.

See also *LOX *sb.*¹, *LSD² (as main entries).

1936 *Motor Manual* (ed. 29) xiii. 193 'L' plates must be carried at the front and rear of the car. **1936** *Punch* 26 Feb. 248/1 Ermyntrude, inspired by blind jealousy (and aided by some rather L driving by Rachel), emerged from the garage. **1959** *Manch. Guardian* 27 Aug. 6/3 There are still an unknown number of drivers..who may have used L-plates for years without even applying for a test. **1963** P. Roberts *Know the Law Handbk.* vi. 154 (*caption*) An L driver involved in an accident. **1970** G. S. Wilkinson *Road Traffic Offences* (ed. 6) v. 472 On and from 1st January, 1961, he may drive only as a learner-driver, i.e. with L plates. **1869** *Whitaker's Almanack* 83/1 Andover—Hon. D. F. Fortescue, *L.* **1908** *Daily Chron.* 16 Dec. 1/6 Mr. Mackarness (L., Newbury) asked whether [etc.]. **1974** *Times* 11 Oct. 4/1 Accrington... Total vote 42,259 (83·8%)—Lab 20,050 (47·4%), C 15,018 (35·5%), L 7,191 (17·0%). **1951** Toboldt et al. *Automatic Transmissions* i. 15 The selector may be set in any one of five positions, namely, parking (P), neutral (N), low (L), drive (D), and reverse (R). **1932** J. L. P. W. Hewison *Local Expenditure: Address E. Sussex Ratepayers' Assoc.* 4 Let every L.A. be rationed as to the percentage of its income which it may spend on loan charges. **1967** *Punch* 1 Mar. 292/2 The LAs spend between them on nursery schools 0·281 per cent of their total educational expenditure. **1949** H. G. Alsberg et al. *American Guide* 1199 Los Angeles... Airports: Union Air Terminal, & L. A. Mun. Airport. **1953** *Amer. Speech* XXVIII. 54 If you're confused—it's still L.A... (Native votes on pronunciation of 'Los Angeles'.) **1969** *Daily Tel.* 12 Nov. 20/3 The centre of California is LA, a concrete and glass *mélange* embracing 102 incorporated cities and spread over 4,851 square miles. **1972** B. Rodgers *Queens' Vernacular* 109, I can't walk around LA without gettin' the horns—there's so many pretties. **1973** *Black World* Apr. 96 On a junket to L.A. and New York. **1960** *Times Rev. Industry* July 73/1 In February this year Argentina, Brazil, Mexico, Paraguay, Peru, and Uruguay signed the Montevideo Treaty, setting up the Latin American Free Trade Association... The L.A.F.T.A. aims at removing customs duties to fellow members in 12 years. **1966** *Economist* 19 Nov. 826/1 As a member of LAFTA, Venezuela hopes not only to increase exports but also to reap economies of scale, especially in the heavier industries. **1972** *Buenos Aires Herald* 6 Feb. 3/3 A large share of the disequilibrium stemmed from trade with LAFTA (Latin American Free Trade Association) nations. **1833** *Penny Mag.* Monthly Suppl. Oct.–Nov. 468 *l.c.*,..to have words or letters printed in 'lower case', or small letters. **1892** A. Powell *Southward's Pract. Printing* (ed. 4) xvii. 129 *l.c.*, set the word in lower case letters. **1911** A. E. Housman *Let.* 28 Aug. (1971) 119 The type-written text contains the letters: J (cap.) j (l.c.). **1973** *Collins's Authors & Printers Dict.* (ed. 11) 244/2 *l.c.*,..(typ.) lower case, that is *not* caps. **1961** B. Fergusson *Watery Maze* i. 42 The mahogany boat..became the standard Landing Craft Assault or LCA. **1967** *Jane's Surface Skimmer Systems 1967–68* 97/2 Current amphibious vehicle programmes being conducted for the Bureau of Ships are: Landing Craft Assault (LCA), [etc.]. **1898** L.C.C. [see *BETTER-TO-DO *adj.* phr.]. **1907** *Daily Chron.* 3 Sept. 4/7 This is one of the little matters that the L.C.C...might well look into. **1970** Ruck & Rhodes *Govt. Greater London* iii. 45 The Labour Party Opposition were convinced that a scheme of reform was unnecessary and particularly a scheme which involved the abolition of the L.C.C. **1943** *Time* 22 Nov. 24/3 The broad wake of a PT, plus the outline of the LCI, must have looked like bigger game. **1944** *Hutchinson's Pict. Hist. War* 12 Apr.–26 Sept. 344/1 One company of Pioneers, some of whom had spent two hours in the sea after their L.C.I. had been torpedoed. **1943** *Time* 4 Oct. 63/2 As early as 1936 the Navy experimented with tank lighters, and from these tests emerged the LCM (Landing Craft, Mechanized), a 50-footer which carries a crew of four and a medium tank. **1943** *Newsweek* 27 Sept. 23/2 The row of LCT's on the beach belching vehicles looks like a long line of stranded, gasping whales. **1955** 'N. Shute' *Requiem for Wren* iii. 79 This was the L.C.T. Mark 4, the standard tank landing craft, British built and the most common of the lot. **1927** J. W. Trevan in *Proc. R. Soc.* B. CI. 483 Toxicity should be stated primarily in terms of the 'median lethal dose', that is the dose which kills 50 per cent. of a large group of animals. As a convenient abbreviation I would suggest the symbol LD50... For doses which kill other proportions of large groups of animals it is convenient to use the analogous symbols LD75, LD25, for doses which kill 75 per cent., and 25 per cent., and similarly for doses killing other proportions. **1950** *Proc. Soc. Exper. Biol. & Med.* LXXIII. 497/2 Only 400r of X-irradiation was used. This dose corresponds to LD[illegible] for swine. **1958** Thomson & Straube in W. D. Claus *Radiation Biol. & Med.* iv. 101 Two-thirds of an LD[illegible] dose will probably kill less than 1 percent of the animals exposed. **1968** *Observer* 16 June 9/1 Sarin is 30 times more toxic than phosgene, and the amount necessary to achieve what the experts know as 'LD50'..is 40 drops on the skin. **1954** H. A. Trenkler in *One Year LD-Oxygen Refining Process* 11/1 Our method of steel refining with pure oxygen by blowing downwards into a bath of metal..[is] called 'LD-process' (Linzer Düsenverfahren). **1965** *New Statesman* 7 May 709/3 The building of modern plants—LD converters, bigger blast-furnaces, wider sinter plants, automated soaking pits and rolling mills. **1973** *Times* 31 May (Austria Suppl.) p. iii/7 The Japanese have the highest production of LD steel, and by 1972 51 per cent of world steel production was based on the LD process. **1967** *Times* 25 Sept. p. xiv, There is no doubt that the speedy economic advancement of the less developed countries (L.D.C.s) would benefit the whole world economy. **1972** *Advocate-News* (Barbados) 19 Feb. 9/1 The assembled CARIFTA leadership prepared a package for progress, a labilization. **1956** *Radiation Res.* V. 263 The increased susceptibility of X-irradiated DNP to the action of trypsin is not inconsistent with the concept of a labilization of the DNA-to-protein salt-like secondary linkages. **1965** *Dissertation Abstr.* XXV. 4355/2 It is concluded that the labilization of the carboxyl carbon of glycine is a discrete reaction which can be measured independently of reactions related to further metabolism of the α-carbon. **1967** Pike & Brown *Nutrition* vii. 145 Other labilizers of the lysosomal membrane are ultraviolet light and ionizing radiation. **1974** *Nature* 13 Dec. 579/1 The labilisation of the ligand *trans* to an oxo group is a well known effect.

labio-. Add: (*a*) **labiovelar** adj. (later example); also as *sb.*; hence **labiove·larize** *v.*, to pronounce with labiovelar articulation; **la:biovelariza·tion.** (*b*) **labio-lingual** *a.*, pertaining to the lips and the tongue; existing or occurring along a line from the lips to the tongue; hence **la:bioli·ngually** *adv.*, in the labio-lingual direction.

1908 G. V. Black *Work on Operative Dentistry* II. 15 The axio-bucco-lingual plane, or the bucco-lingual plane, ..passes through the tooth bucco-lingually parallel with its long axis. In the incisors and cuspids this is the labio-lingual plane. **1940** O. A. Oliver et al. *Labio-Lingual Technic* 10 Used jointly, labial and lingual appliances represent the labio-lingual technic. **1972** *Nature* 24 Nov. 236/2 The labiolingual compression of the tooth in hominids. **1949** V. H. Sears *Princ. & Technics for Compl. Dent. Constr.* xxiii. 284 The upper incising occlusal unit should be narrow labio-lingually in order to cut through the food with application of little force. **1963** C. R. Cowell et al. *Inlays, Crowns & Bridges* iv. 30 Some anterior teeth are exceptionally thin labiolingually and gold on the lingual aspect may cause the crown..to lose its natural translucence. **1971** *Nature* 23 Apr. 514 The right lateral incisor is represented only by the broken root which measures 4·5 mm mediodistally and 6·0 mm labiolingually. **1895** W. M. Lindsay *Short Hist. Latin Gram.* x. 156 We must distinguish..Labiovelars,.. which become Labials in some languages. **1939** E. Prokosch *Compar. Germanic Gram.* 72 The treatment of the labiovelars in Germanic is similar to that in Latin. **1952** A. Cohen *Phonemes of English* 31 *bail*-[beil] v. *wail*-[weil] dist[inguished] by labial v. labiovelar art[iculation]; plosion v. glide. **1968** Chomsky & Halle *Sound Pattern Eng.* 311 An interesting pattern arises with regard to the labiovelars. We may ask whether these are labials with extreme velarization or velars with extreme rounding. **1937** J. R. Firth *Papers in Linguistics* (1957) vii. 80 The tonal diacritica and possibly also what we have called yotization and labio-velarization may be considered as syllabic features. **1953** K. Jackson *Lang. & Hist. Early Brit.* 440 Some degree of labiovelarisation of the 3 caused by the *u*. **1933** L. Bloomfield *Lang.* 118 These two modifications appear together in labiovelarized consonants. **1964** E. Palmer tr. *Martinet's Elem. Gen. Ling.* ii. 51 Consonants..which possess the timbre of [u] are called labiovelarized.

labium. 1. b. Add to def.: Now usu. called in full *labia majora*, and formerly † *labia externa* (or *external labia*). Also (in full *labia minora*, † *smaller labia*, † *labia interna*, *inner labia*), the two smaller folds of skin situated within the labia majora and extending downwards and backwards from the clitoris: the nymphæ. (Further examples.)

The sing. forms *labium majus, minus* occas. occur.

1634 A. Read *Man. Anat.* I. vi. 85 In the externall part ..first appeare labia, the lippes which are parted by the magna rima or fossa, the large chinke. **1826** J. Lizars *Syst. Anat. Plates* xi. 82 The mucous surface of the labia externa is the seat of syphilitic ulcers. *Ibid.* 84 The nymphæ, or labia interna. [*Note*] Syn. Labia pudendi minores: Alæ minores. **1838** R. Hunter *Text Bk. Human Anat.* (ed. 2) ix. 208 Female organs of generation... 1st, the mons veneris—2d, the two labia majora, or labia pudendi —3d, the clitoris, with its prepuce—4th, the two labia minora, or nymphæ—[etc.]. **1845** *Encycl. Metrop.* VII. 493/1 The external or proper labia are thick folds of integument which bound the vulva on either side; and within these are the nymphæ or smaller labia. **1906** H. Ellis *Stud. Psychol. Sex* V. 134 The inner lips, the nymphæ or labia minora, running parallel with the greater lips which enclose them, embrace the clitoris anteriorly. **1907** W. N. Parker tr. *R. Wiedersheim's Compar. Anat. Vertebr.* (ed. 3) 483 'Labia majora' also occur in certain other Primates, but in most Monkeys 'labia minora' are alone present bounding the vulva, and these belong morphologically to the clitoris and not to the scrotal folds. **1936** H. M. & A. Stone *Marriage Manual* iii. 89, I have often seen labia barely a quarter of an inch wide and I have examined women whose inner labia measured two and a half inches in width. **1953** A. C. Kinsey et al. *Sexual Behav. Human Female* xiv. 577 As sources of erotic arousal, the labia minora seem to be fully as important as the clitoris. *Ibid.* 578 Sometimes the labia [minora] are rhythmically pulled in masturbation. *Ibid.*, We do not yet have evidence that the labia majora contribute in any important way to the erotic responses of the female. **1962** *Gray's Anat.* (ed. 33) 1547 Anteriorly, each labium minus divides into two portions. **1965** Masters & Johnson in J. Money *Sex Research* iv. 67 During the excitement phase of the human sexual response cycle, the labia minora (sex-skin) turn bright pink in color and..engorge to approximately twice their previously normal size.

laboratory. Add: Also with pronunc. (lăbǫ·rătəri). **4. laboratory animal,** any animal (e.g. rat, monkey, mouse) commonly used for experiments in a laboratory; **laboratory frame (of reference)** *Nuclear Physics*, the frame of reference in which a laboratory is stationary, and with respect to which measurements of particle energy, velocity, etc., are generally made (see quot. 1958); **laboratory system** *Nuclear Physics* = ***laboratory frame**.

1899 *Allbutt's Syst. Med.* VI. 517 The so-called 'irritation contracture' observable in the monkey (but not in other laboratory animals). **1937** *Nature* 24 July 155/1 Among those using this fish as a 'laboratory animal'. **1958** O. R. Frisch *Nucl. Handbk.* xix. 22 From a theoretical standpoint it is most convenient to calculate in a frame of reference in which the total linear momentum is zero (centre of mass frame, or..centre of momentum frame..). Experimentally the target particle is usually at rest in the laboratory frame of reference. **1971** *Sci. Amer.* June 76/3 The theoretical analysis of events in the rapidly moving frame can be made with some degree of confidence and transformed back to the laboratory frame. In this way theory can be compared with experiment. **1951** D. Bohm *Quantum Theory* xxi. 525 Collisions usually involve firing particles at other particles that are at rest in the laboratory system.

labour, labor, *sb.* Add: **2. b.** (Earlier and later examples.)

1839 J. F. Bray (*title*) Labour's wrongs and labour's remedy; or, The age of might and the age of right. **1848** *Punch* XV. 261 Thither [*sc.* to Australia] should Labour repair to seek Demand. **1916** A. Richardson *Man-Power of Nation* 55 The time is..opportune for trade unions to recognise their responsibility for the encouragement of the flow of capital for the benefit of industry... This subject of the relationship of labour to economy of output may be said to be hackneyed. **1940** W. Temple *Hope of New World* 61 If there is to be tension at all, let it be between the financial interests of Shareholders and the productive interests of Management and Labour in co-operation. **1970** *Encycl. Brit.* XXII. 652/2 Until after the turn of the century organized labour seldom gained any measure of public sympathy.

c. (With capital initial.) Short for 'the Labour Party'. Also *attrib.* (see sense *8). Quasi-*adv.* in phr. *to vote Labour*.

1906 *Times* 19 Jan. 4/3 (*heading*) The Liberals and the Labour men. *Ibid.* 10/1 Just before going to press the news arrived that Lord Stanley..had been defeated..by Mr. W. T. Wilson (Labour). **1918** A. Huxley *Let.* 25 Nov. (1969) 171 Tell Brett also to remember to vote, and to vote Labour, our only hope. **1920** *Manch. Guardian* 5 Jan. 6/2 Could any conceivable Labour Government have made blunders so gross? **1924** *Ibid.* 2 May 9/1 The Labour party and Labour leaders have always been divided upon the subject of P.R. [Proportional Representation]. **1932** J. Buchan *Gap in Curtain* iii. 149 The younger Tories as a whole were enthusiastic, and, what is more significant, the Left Wing of Labour blessed it cordially. *Ibid.*, Collinson, a young Labour member from the Midlands, declared that Geraldine was the best Socialist of them all. **1945** *Let us face the Future* (Labour Party) x. 10 Labour led the fight against the mean and shabby treatment which was the lot of millions while Conservative Governments were in power. **1949** Lewis & Maude *Eng. Middle Classes* I. iv. 81 Both Conservatives and Labour competed for the middle-class vote. *Ibid.* 82 The new Labour formula was nicely expressed by Philip Snowden. **1956** C. Cockburn *In Time of Trouble* xix. 244 The Labour people, the 'progressive intellectuals'. **1966** M. Edelman *The 'Mirror'* viii. 151 Its brilliance was that at no time did the *Mirror* specifically urge its voters to vote Labour. **1971** B. Hindess *Decline Working-Class Politics* viii. 173 The teenagers of the 1960s..missed the political experience of their parents, the long identification with and support for Labour.

d. Short for *Labour Exchange 2.

1935 M. Harrison *Spring in Tartarus* I. 105 You see, mister, I can't go on the Labour, cause I 'aven't been stood off. I'm on'y ill. **1963** T. Parker *Unknown Citizen* iii. 88 I'll ring you up Monday to tell you how I went on at the Labour. **1971** R. Rendell *One Across* iv. 37 Work's not easy to come by when you've no qualifications... Can't they find you anything down at the Labour? **1972** L. Henderson *Cage until Tame* vi. 45 I'm going for a job the Labour picked out for me.

8. *labour bank* (earlier example); also (in senses 2, 2 b, and *2 c), *labour-bill, bureau, candidate, colony, content, cost, government, law, leader* (later examples), *master, movement, permit, power, song, union*. **b.** instrumental, as *labour-dominated* adj. Also **labour camp,** a penal settlement where the prisoners are obliged to undertake labouring work; **Labour Day** *U.S.*, a legal holiday observed on the first Monday of September; a similar holiday observed in Australia, New Zealand, and elsewhere; **labour force,** (*a*) = **labour power*; (*b*) [cf. *force *sb.*[1] 4 d] a body of workmen; workers, as opposed to employers, considered as a single body; **labour-intensive** *a.* (see *intensive *a.* (*sb.*) 5 b); **Labour-Liberal** *a.* and *sb.*, (a Member of Parliament) combining Labour and Liberal ideas (in early use a Labour M.P. who accepted the Liberal whip); **labour market** (earlier and later examples); **labour note,** a note indicating value in terms of work; **labour-only** *a.*, denoting a sub-contractor who, or sub-contracting which, supplies only the labour for a particular piece of work; **Labour Party,** a political party specially supporting the interests of labour; in the United Kingdom, the organized party formed in 1906 by a federation of trade unions and advanced political bodies to secure the representation of labour in Parliament; **labour relations,** the relations between management and labour; **labour-saving** *a.*, designed to ease or eliminate work (later examples); so **labour-saver; labour ward** (sense 6 in Dict.), a room in a hospital set aside for childbirth.

1832 *Crisis* 28 Apr. 16/1 In Poland-street they had established a Labour Bank. **1898** *Engineering Mag.* XVI. 26 Every improvement in labour-saving machinery diminishes the proportion which the labour-bill bears to the cost of the product. **1832** *Crisis* 11 Aug. 90/3 Perhaps the best preliminary mode..will be by the establishment of Equitable Exchange Labour Bureaus. **1893** *Rep. Agencies & Methods Unemployed* 6 in *Parl. Papers 1893–4* (C. 7182) LXXXII. 377 A detailed account of..labour bureaux and of various organisations dealing with distress. **1908** *New Encycl. Social Reform* 998/2 The recent establishment of a system of public employment bureaux called labor bureaux. **1900** *Jrnl. Soc. Arts* 11 May 510/1 Prisoners.. might serve their time in..quarries, which would be turned into labour camps. **1931** J. S. Huxley *What dare I Think?* iii. 88 Infringement of this order could probably be met by a short period of segregation, say in a labour camp. **1958** *Spectator* 6 June 723/3 Recsk, one of the most abominable labour camps in the world. **1974** *Times* 18 Feb. 14/7 Perhaps the conference helped to save Mr Solzhenitsyn from a labour camp. **1893** H. F. McLelland *Jack & Beanstalk* 16 You'd make a good Labour Candidate. **1921** F. W. P. Lawrence *Labour Party* 3 For nearly every seat there is a Labour candidate. **1948** M. Phillips in H. Tracey *Brit. Labour Party* I. 9 Labour candidates are selected by the Constituency Parties in co-operation with the National Executive Council. **1963** J. Blondel *Voters, Parties, & Leaders* v. 136 Labour candidates are mainly drawn from the middle and lower middle classes. **1888,** etc. Labour colony [see *colony *sb.* 5 c]. **1948** *Spectator* 9 Jan. 38/2 Fine worsteds..have a high 'labour content', and raw material is a low item in the cost of their production. **1896** J. A. Hobson *Probl. Unemployed* 69 The labour cost of distributing a given quantity of goods. **1903** *Westm. Gaz.* 9 July 2/1 The imposition of such duties as will equalise our labour-costs with the labour-costs of our foreign competitors. **1914** [see *assembly line* s.v. *assembly 1 c *attrib.*]. **1966** A. Gilpin *Dict. Econ. Terms* 117 *Labour-costs per unit of output*, the cost of the labour in real terms involved in making each unit of output from a factory. **1886** *N.Y. Times* 7 Sept. 8/1 (*heading*) How Labor Day was observed by all classes of workmen. **1887** *Westm. Gaz.* 6 Sept. 7/1 An Act passed last winter by the State Legislature, making the first Monday in September a legal holiday, to be called 'Labour Day'. **1910** *World Almanac* (N.Y.) 30 An act [of 1893–4] making Labor Day a public holiday in the District of Columbia. **1931** *Daily Express* 2 Sept. 1/5 The governing committee of the New York Stock Exchange, in response to requests of members, decided to close on Saturday, and also to close on Monday, on account of Labour Day, and to resume on Tuesday. **1963** *Times* 26 Feb. 9/1 The city's [*sc.* Melbourne's] big retail stores invented this affair, which has taken over the traditional public holiday called Labour Day, now in any case a quaint anachronism in a country of trade union strength. **1974** *Anderson* (S. Carolina) *Independent* 23 Apr. 1B/1 This is a drastic change from present and past school calendars which start several days prior to Labor Day and finish earlier in the spring. **1959** *Daily Tel.* 19 Aug. 9/2 Nottingham Conservative leaders last night denounced as 'full of inaccuracies and false inferences' a report prepared on behalf of the local Labour party... [The] Conservative vice-Chairman of the Labour-dominated Watch Committee, said: 'The entire report is..full of inaccuracies and false inferences.' **1974** *Listener* 23 May 650/3 The finished film was shown to the Labour-dominated committee. **1885** J. L. Joynes tr. *Marx's Wage-Labour & Capital* (1886) 8 Capital necessarily pre-supposes the existence of a class which possesses nothing but labour-force. **1909** B. Webb et al. *Socialism & National Minimum* I. 38 The parasitic trades, where the employers are able to exact from their workers more labour-force than they replace. **1911** F. T. Carlton *Hist. & Probl. Organized Labor* xvi. 431 In times of depression there are unemployed land and capital as well as an unemployed labor force. **1940** *Economist* 11 May 849/2 Men are being tempted away and the labour force is being shifted round without any consideration of the relative urgency or importance or even skill of different jobs. **1969** *Times* 12 Feb. 9/2 They have run a computer model of an underdeveloped country's economy which relates the growth of g.n.p. to the labour force, capital stock and other factors. **1926** *Encycl. Brit.* II. 653/2 In Jan. 1924.. the first Labour Govt. in this country was formed. **1945** *Let us face the Future* (Labour Party) x. 10 A Labour Government will press on rapidly with legislation extending social insurance..to all. **1971** *New Statesman* 9 July 34/2 Aneurin Bevan had a very simple rule about the role of a Foreign Secretary under a Labour Government. **1897** J. Bryce *Impressions S. Afr.* xxi. 447 There are also certain 'labour laws', applying to natives only, and particularly to those on agricultural locations. **1902** *Ann. Amer. Acad. Pol. & Social Sci.* XX. 240 When a state legislature passes a new labor law, or revises an old one. *Ibid.* 241 Labor laws, however good, cannot enforce themselves. **1967** A. Hepple *Verwoerd* ix. 122 With the intensification of race and labour laws, the Non-whites began to resist. **1920** S. Lewis *Main St.* xvi. 202 You socialists make me sick! I'm an individualist. I ain't going to be nagged by no bureaus and take orders off labor-leaders. **1972** *N.Y. Times* 3 Nov. 21/8 He is scheduled to meet..labor leaders. **1973** *Times* 16 Nov. 20/8 Labour leaders have been appointed to the boards of nationalized industries..in a personal capacity and not in their own industries. **1902** *Westm. Gaz.* 26 Feb. 6/3 If he could do that when he was returned as a Labour Liberal member. **1904** *Daily Chron.* 7 Jan. 5/4 Two English Liberal Members (one Liberal and one Labour-Liberal). **1834** Mill in *Monthly Repos.* New Ser. VIII. 320 We have to lessen the pressure on the

labour-market. **1946** *R.A.F. Jrnl.* May 160 There must be many a man..awaiting his release with some trepidation on account of uncertainty about his future in the labour market. **1965** Seldon & Pennance *Everyman's Dict. Econ.* 152 It [*sc.* the employment exchange] thus helps to make the labour market work by acting as a clearing-house of information about employment. **1974** *Guardian* 24 Jan. 13/3 Parents are not going to stay at home, and the labour market needs them. **1901** *Daily News* 10 Jan. 9/3 The labour master..certified him able to do the work. **1921** *Dict. Occup. Terms* (1927) § 731 *Labour master*, engages and discharges casual labour..employed by dock or harbour authority [etc.]. **1870** *Scribner's Monthly* I. 71 The preacher..beats about..in a dissertation on..the 'labor movement'. **1893** L. T. Hobhouse (*title*) The Labour Movement. **1944** M. Laski *Love on Supertax* ix. 88 Her ignorance of Party matters, Labour Movements, working-class life. **1969** A. Plater *Close Coalhouse Door* III. 66 The Labour movement that was made by the miners out of blood, sweat and tears. **1832** *Crisis* 28 Apr. 12/1 Money was not necessary. Labour-notes were sufficient. **1894** B. Jones *Co-operative Production* I. 89 These labour notes were to supersede the use of metallic coins and ordinary bank notes, and were to become a superior kind of money. **1967** *Times Rev. Industry* Apr. 64/2 It is partly the search for regular and reasonably high earnings that makes men hire themselves to labour-only sub-contractors. **1969** M. Gagg in R. Fraser *Work* II. 132 A third system is labour-only sub-contracting in which a bricklayer or a group of bricklayers will undertake to do the brickwork on a job for so much per yard super (98 bricks). *Ibid.* 133 For the past seven years I have been a labour only sub-contractor. **1974** *Shelter News* Easter 3/2 From the trade union point of view, the labour-only sub-contracting practice gives rise to many problems. **1886** Labour party [in Dict.]. **1892** Roydhouse & Taperell (*title*) The Labour Party in New South Wales. **1896** *Labour Annual* 39 This [of 1895] was the first General Election in which an organized Labour party, independent of either Liberal or Tory, and opposing either or both, has taken part in the United Kingdom. **1905** J. R. Macdonald in W. T. Stead *Coming Men* 222 The Labour Party..will represent trades; it will represent the working class; it will represent a coherent body of fundamental Labour opinion. **1922** *Encycl. Brit.* XXXII. 507/1 The Labour party..included the Independent Labour party and the Fabian Society and one or two smaller Socialist bodies. Locally it was organized in several hundred Local Labour parties. *Ibid.* 884/2 For many years there was a Labor or Socialist Labor national party, which regularly nominated a candidate for the [U.S.] presidency. **1926** *Daily Chron.* 13 May 2/3 What are we to say of the Trade Union and Labour Party leaders who on Saturday, May 1, agreed to the sending out of the strike instructions? **1945** W. K. Richmond *Educ. in England* vi. 116 The generous programme for free secondary education issued..by the Labour Party. **1971** *New Statesman* 9 July 37/1 You could lift almost any quote from the Labour Party's guide for women's sections to illustrate the patronising view they take of us. **1971** *Sunday Australian* 8 Aug. 3/4 The Federal President of the Australian Labor Party..yesterday called on his party to dissociate itself from extremist students. **1927** *Melody Maker* Aug. 777/1 Al Payne should have been leader, but the necessary labour permits could not be obtained, and the band remains in America. **1943** E. M. Almedingen *Frossia* ii. 107, I have no right to employ anyone..without a proper labour permit. **1866** *Leisure Hour* 17 Mar. 171/1 The competition of labourers with each other for employment, which, in a country like ours, where there is always a vast reserve of labour-power, must far more than counterbalance any good to the labourer arising from the competition of the masters for his services. **1896** J. A. Hobson *Probl. Unemployed* 2 Off-time..implies waste of labour-power. **1959** B. Wootton *Social Sci. & Social Path.* ix. 280 He matters in himself, and not merely as a unit of cannon-fodder, labour-power or population. **1943** J. S. Huxley *TVA* 116 The TVA's work in making a comparative survey of rates and conditions in all the fertilizer plants of the region, which helped materially in promoting better labour relations. **1973** *Guardian* 12 Mar. 11/7 Leyland's personnel manager declined to comment. ..'I personally am anti-press. I have had too much trouble with you lot buggering up labour relations.' **1902** *Chambers's Jrnl.* Dec. 830/2 The machine appears to be a real labour and time saver, and is moderate in price. **1929** A. Huxley *Do what you Will* 86 The machine..is..a labour-saver. **1904** *Sci. Amer.* 21 May 404/2 The present enormous industry..was rendered possible only by the introduction of labor-saving machinery. **1932** *Discovery* Jan. 10 Many remarkable labour-saving machines have been evolved in recent years. **1957** *Observer* 11 Aug. 8/4 The old owners have dwindled, departed, and found labour-saving bungalows in quieter spots. **1964** M. McLuhan *Understanding Media* xvi. 161 The American farmer, confronted with new tasks and opportunities, and at the same time with a great shortage of human assistance, was goaded into a frenzy of creation of labor-saving devices. **1971** *Engineering* Apr. 92/2 With its time and labour-saving features. **1888** L. A. Smith *Music of Waters* 275 Isaac D'Israeli, in his 'Curiosities of Literature', mentions the numerous labour-songs used by the ancient Greeks. **1921** Labour song [see *blues]. **1974** *Times* 25 Sept. 14/8 She..sang a labour song about joining the union. **1866** in *Documentary Hist. Amer. Industr. Society* (1910) IX. 133 Each member belonging to the National Labor Union. **1884** J. Hay *Bread-Winners* xi. 183 The labor unions have ordered a general strike. **1944** H. A. Wallace *Century of Common Man* 25 July 84 The people of America know that the second step towards Nazism is the destruction of labour unions. **1973** *Times* 16 Nov. 20/7 Nationalism has had an important effect on labour union attitudes in Canada. **1933** A. W. Bourne et al. *Queen Charlotte's Text-bk. Obstetr.* (ed. 3) xiv. 266 No person is allowed in the hospital labour ward without a mask, which covers both the mouth and nose. **1953** E. Simon *Past Masters* IV. vi. 260 That first time the baby had been born..at home... Now..I was taken straight to the Labour Ward. **1968** S. Bender *Obstetr. for Pupil Midwives* xi. 147 The woman admitted to the labour ward and her accompanying relative or friend are greeted cheerfully.

labour, *v.* **9.** For *Obs.* read *Obs.* exc. *poet.*, and add later example.

1872 Tennyson *Gareth & Lynette* 31 But Kay the seneschal who loved him not Would hustle and harry him, and labour him Beyond his comrade of the hearth.

laboured, *ppl. a.* Add: **3.** Also *laboured-at*.

1876 G. M. Hopkins *Poems* (1918) 9 And lily-coloured clothes provided Your spouse not laboured-at nor spun. Hence **la·bouredly** *adv.*, **la·bouredness.**

1882 *Daily Tel.* 24 Feb. (Cass.), He spoke labouredly and with hesitation. **1930** J. W. Mackail *Largeness in Lit.* 6 Largeness is..the opposite..of thinness, of tightness, of labouredness.

Labour Exchange. Also with lower-case initials. [Labour *sb.* 2 + Exchange *sb.*] **1.** An establishment for the exchange of the products of labour without the use of money. Also *attrib.* Now only *Hist.*

1832 *Crisis* 28 Apr. 16/2 Mr. B. Warden..stated that they had erected a new school, called a Labour Exchange School. **1832** *Crisis* 25 Aug. 99/1 To investigate the Principles upon which the proposed Equitable Labour Exchange was to be founded. *Ibid.* 6 Oct. 122/1 Labour Exchange notes. ..Labour Exchange banks. **1894** B. Jones *Co-operative Production* I. 90 The Labour Exchange was opened on September 3, 1832, under the title of 'The Equitable Labour Exchange'. **1906** G. J. Holyoake *Hist. Co-operation in Eng.* (rev. ed.) I. 265 The Labour Exchange was not Mr. Owen's idea, but he adopted it.

2. An office serving as a means of connection between workers and employers, esp. one forming part of an organization to assist in the finding of employment. Also *attrib.*

1869 C. L. Brace *New West* v. 53 One of the remarkable instances of the intelligence and humanity of this new community was the organization, in 1868, of the 'Labor Exchange'. **1893** *Rep. Agencies & Methods Unemployed* 55 in *Parl. Papers 1893–4* (C. 7182) LXXXII. 377 Registry offices..for those labourers having the title of the 'British Labour Exchange'. **1895** J. A. Hobson *Probl. Unemployed* 130 [illegible] Labour Exchange. **1958** *New Statesman* 22 Feb. 224/1 [illegible] Labour Exchange. **1973** [see *job centre* s.v. *job *sb.*[2] 7].

3. The finding of employment for workers.

1896 J. A. Hobson *Probl. Unemployed* 128 No system of mere labour exchange, however well-conducted, would increase the total amount of employment over a long area of time.

Labourism (lē̆i·bəriz'm). [f. Labour *sb.* 2 b + -ism.] The principles or tenets of the Labour party in politics; the holding or advocacy of these principles.

1903 *Rep. 3rd Ann. Conf. Labour Representative Comm.* 30/1 in *Labour Party Foundation Conf. & Ann. Conf. Rep. 1900–05* (1967) 109 Let them leave [illegible] Liberalism and Toryism and every other 'ism' that was not Labourism. **1909** *Westm. Gaz.* 13 May 3/2 [illegible] plied him with Imperialism, Mr. Keir Hardie with Labourism. **1908** *Ibid.* 4 May 9/2 Two years ago many Liberals coquetted with Labourism, and the result gave them a fright. **1926** J. R. Macdonald in *Public Opinion* 14 Mar. 248/3 Their Toryisms, Liberalisms and Labourisms. **1977** *Time Out* [illegible] of Labourism over the last [illegible] years.

Labourist (lē̆i·bərist). [f. Labour *sb.* 2 b + -ist.] A supporter of the interests of Labour in politics; an advocate of Labourism.

1903 *Handy Notes for Unionist Workers* Aug. 3 The Labourists in Parliament..number over a dozen. **1910** *Daily Chron.* 1 Feb. 1/7 Liberals, Labourists and Nationalists are solid against the veto of the Lords and against Food-Taxes. **1927** *Observer* 5 June 12/3 Six months ago the five seats were held by two Liberals, three Conservatives, two Labourists, and one Liberal. **1969** *Guardian* 23 Jan. 9/2 An ideological struggle [illegible] between the Labourists, who favour [illegible] and the new Leftists.

Labourite (lē̆i·bərait). [f. Labour *sb.* 2 b + -ite[1].] = *Labourist, often used of members of Parliament representing the Labour Party in Britain, Australia, or other countries.

1903 *Daily Chron.* 19 Dec. 5/5 Free Traders and Free Trade Labourites—40. **1909** T. Hodgkin *Let.* 15 Nov. in L. Creighton *Life & Let. T. Hodgkin* (1917) xiii. 326 It is this abominable selfishness of men whether Capitalists or Labourites which seems to wreck all forms of government. **1920** *Blackw. Mag.* June 830/1 The moderate Labourites have reason to remember this. **1923** *National Rev.* Jan. 646 The Labourites were led by Mr. Ramsay Macdonald before the war. **1927** [see *Anglo-Saxon IV. A. *adj.*]. **1955** [see *de Gaullist *a.* and *sb.*]. **1960** *20th Cent.* May 452 He [*sc.* Michael Young] was the first of those Labourites who renounced the tired phrases of inherited socialist doctrine.

Labrador. Add: **Labrador (dog, retriever),** a medium-sized, black or yellow, short-coated retriever belonging to a breed originally developed in Newfoundland and Labrador; **Labrador pine,** the grey or jack pine, *Pinus banksiana*, native to the northern parts of North America.

[**1829** G. Head *Forest Scenes* 41 The dog was of the Labrador breed, extremely powerful, and of enormous stature.] **1842** R. H. Bonnycastle *Newfoundland in 1842* II. ix. 24 They [*sc.* water-dogs] are of two kinds; of the short, wiry-haired Labrador dog, and the long, curly-haired Newfoundland species. **1910** *Kennel Encycl.* III. 1099 Labrador Retrievers..were originally imported from Labrador. **1921** *Blackw. Mag.* Dec. 794/2 All the guns had dogs—beautifully trained Labradors. **1963** A. Smith *Throw out Two Hands* xix. 195 John arrived..with a remarkable collection of people, plus a black labrador who loathed nettles, and a rifle. **1971** C. Fick *Danger Transcript* (1973) 155 Finding out..whether he liked Brahms, silk pyjamas, Labrador retrievers and such. **1973** *Country Life* 8 Feb. 325/1, I have just chosen a new Labrador puppy. **1803** A. B. Lambert *Descr. Genus Pinus* I. 7 Labrador Pine... Habitat in America Septentrionali. **1921** H. Kephart *Camping & Woodcraft* (rev. ed.) I. 239 The gray (Labrador) pine or jack pine is considered good fuel in the far North.

labradorescence (læ:brădǫre·sĕns). *Min.* [f. Labrador(ite + *-escence.] The brilliant play of colours exhibited by some specimens of feldspars, esp. labradorite.

1911 *Encycl. Brit.* XVI. 30/2 This optical effect, known sometimes as 'labradorescence', seems due in some cases to the presence of minute laminae of certain minerals.. arranged parallel to the surface which reflects the colour. **1962** W. A. Deer et al. *Rock-Forming Min.* V. 143 The phenomenon of labradorescence, however, has not been conclusively explained. **1966** J. Sinkankas *Mineral.* viii. 235 Vivid colors also arise from within labradorite feldspar, the effect being known as labradorescence.

Labradorian (læbrădō·riăn), *sb.* and *a.* [f. Labrador + -ian.] **A.** *sb.* A native or inhabitant of Labrador. **B.** *adj.* Of or pertaining to Labrador.

In quot. 1947 the Labradoreans are Eskimos.

1863 H. Y. Hind *Explor. Labrador Peninsula* II. 135 The residents on the coast, or Labradorians as they may well be termed, frequent the St. Augustine in the winter [illegible]. [illegible] Labradorian [illegible] **1909** P. W. Browne *Where Fishers Go* 146 Off Cape Charles the coast again becomes broken and rugged—'hummocky', as it is termed by Labradorians. **1910** C. W. Townsend *Labrador Spring* 68 Dr. Grenfell's hardest work is to teach the Labradoreans the value of fresh air inside their houses. **1947** T. Tanner *Outl. Geogr., Life & Customs Newfoundland-Labrador* II. [illegible] Labradoreans [illegible] the third glacial age, the Labradorean area of radiation was preponderant and the Keewatin played a minor role. **1973** [illegible] Labradorian [illegible]

labretifery (læ:brĕti·fĕri). [f. Labret + L. *fer-*, carrying + -y[3].] The practice of wearing labrets.

1884 *Science* 3 Oct. 345/1 Dr. W. H. Dall then read a paper on the use of labrets, its title being 'The geographical distribution of labretifery'. **1905** C. Davenport *Jewellery* v. [illegible] labretifery [illegible].

‖ **labrys** (læ·bris). [a. Gr. λάβρυς, double-headed axe.] The double-headed axe of ancient Crete.

1901 *Jrnl. Hellenic Stud.* XXI. 108 It seems natural to interpret names of Carian sanctuaries like Labranda in the most literal sense as the place of the sacred axe (*labrys* was the Lydian (or Carian) name for the Greek πέλεκυς, or double-edged axe. *Ibid.* 109 The Cretan coins [illegible] the labrys, set up on its long pillar-like handle [illegible]. **1928** C. Dawson *Age of Gods* viii. 188 Especially remarkable is the word for the sacred Double Axe—Labrys—which forms the root, not only of the Cretan Labyrinthos..but also appears in the title of the god of the Double Axe, Zeus Labrandeus, at Mylasa on the mainland of Asia Minor. **1957** A. MacNab *Bulls of Iberia* i. 6 The emblems which appear on French coins of the Pétain régime.. is precisely the *labrys*, or sacred double-headed axe of Minoan Crete.

labyrinthitis (læbiriŋþai·tis). *Med.* [f. Labyrinth + -itis.] Inflammation of a labyrinth of the internal ear.

1912 Adams & McCrae *Text-bk. Path.* viii. 519 Extension of the disease [*sc.* inflammation of the middle ear].. is most likely to involve the bone in the direction of least resistance, which may lead to extradural inflammation, labyrinthitis, or infection of the lateral sinus. **1934** *Times Lit. Suppl.* 25 Jan. 56/2 Although..the labyrinthitis from which Swift suffered..can create an anxiety-complex, it is almost certainly true that any recurrent disability.. will induce a similar neurosis. **1939** *Anatomical Rec.* LXXIV. 221 Sporadic cases of labyrinthitis appear at times in strains of rats in which no form of ear disease is commonly found. **1972** *Audiology* XI. 322 Variations occur in..viral labyrinthitis.

laburnum. Add to etym. [Adopted as a generic name by P. K. Fabricius (*Enum. Meth. Plant. Hort. Med. Helmstadiensis* (1759) 228).] Substitute for def.: A small tree of the genus so

called, esp. *L. anagyroides* or *L. alpinum* and their hybrids, belonging to the family Leguminosæ and bearing long pendulous racemes of bright yellow flowers followed by pods of poisonous seeds; also, the dark wood of a tree of this kind. (Further examples.)

1812 E. Sang *Nicol's Planter's Kalendar* v. 91 The Laburnum timber which brought so high a price was of the variety called the Tree Laburnum. **1838** J. C. Loudon *Arboretum et Fruticetum Britannicum* II. xli. 592 Hop-poles..are said, when formed of laburnum, to be more durable than those of almost any other kind of wood. **1914** W. J. Bean *Trees & Shrubs Hardy in Brit. Is.* II. 1 Few trees..are so beautiful as the two common laburnums. **1947** A. L. Howard *Trees in Britain* 134 In earlier times ..British-grown laburnum was greatly prized for inlay, turnery and cabinet work... Cut in cross section in the round, it was known by the term 'oyster work'. **1974** W. Condry *Woodlands* xiv. 154 There are long stretches of hedge that are almost purely laburnum which make a marvellous show in June when they are hung with golden chains of blossom. **1975** *Times* 15 Jan. 17/6 An early eighteenth-century oyster walnut and laburnum cabinet.

lac[2]. See also *lakh.

lac[3] (læk). *Biol.* An abbrev. of Lac(tose used (usu. *attrib.*) as a symbol or name (and printed in italics): orig. used to denote the ability (of normal individuals) or inability (of mutants) of the bacterium *Escherichia coli* to metabolize lactose (see quot. 1947), and later to designate (the parts of) the genetic system involved in this ability; as **lac operon,** a group of adjacent genes in the chromosome of *E. coli* which, in the presence of lactose, cause the bacterium to synthesize the enzymes that enable it to metabolize lactose. Also as *sb.*, the (in)ability to metabolize lactose; a lac^+ bacterium; the *lac* operon.

1947 J. Lederberg in *Genetics* XXXII. 505 Particular attention was paid to the isolation of 'lactose-negative' or 'Lac^-' mutants [of *E. coli*]. *Ibid.* 506 (Table), Symbols used for various loci... 'Sugar' fermentation. The ability to ferment is designated '+'; the inability '−'. *Lac* lactose. *Gly* glycerol. **1952** *Nature* 24 May 882/2 The number of papillæ forming on surface colonies of lac^- bacteria, grown on agar containing lactose and another carbon source, is a reflexion of the number of mutations to the lac^+ condition. **1961** Sager & Ryan *Cell Heredity* v. 136 If a cell [of *E. coli*] heterozygous for lac^+ and lac^- is isolated, it produces some progeny still heterozygous and others in which the lac^- has segregated from the lac^+. **1961** Jacob & Monod in *Cold Spring Harbor Symp. Quant. Biol.* XXVI. 197 (*caption*) Genetic map of the *Lac* region of *E. coli*... The lower line represents an enlargement of the *Lac* region, with the two structural genes *z* and *y* and the regulator gene *i*. **1961** S. E. Luria in *Ibid.* 210/1 An altered sensitivity of phage-carried *lac* genes to the specific repressor of the *lac* operon. **1970** J. R. Beckwith in Beckwith & Zipser *Lactose Operon* ii. 5 (*heading*) *Lac*: the genetic system. *Ibid.*, Strains in which the *lac* genes are fused to other bacterial operons, such as the *trp* operon and *purE* operon... Since, in such strains, the *lac* genes are now part of the other operon, nearly all the methodology used for analysis of *lac* can be used for that operon. *Ibid.*, Most of the major concepts of operon structure, expression and regulation have come out of the work on the *lac* operon of *E. coli*. **1971** *Times Lit. Suppl.* 13 Aug. 958/4 Dr Beckwith..[is] the centre of a political storm-in-a-scientific-teacup following his isolation of the lac operon gene.

laccase (læ·kē̆ⁱz, -s). *Biochem.* [a. F. *laccase* (G. Bertrand 1894, in *Compt. Rend.* CXVIII. 1217). f. mod.L. *lacc-a* Lac[1]: see *-ase.] A copper-containing enzyme which effects the oxidation of hydroquinones to quinones, involved in the setting of lac.

1895 *Jrnl. R. Microsc. Soc.* 649 Laccase resembles the diastatic ferments in its properties, except that, instead of hydrolysing, it incites direct oxidation. **1899** J. R. Green *Soluble Ferments* xix. 296 The fungus which yields laccase most readily is *Russula fœtens* Pers., one of the Basidiomycetes, which is fairly common in woods during the summer. **1961** *Jrnl. Biochem.* (Tokyo) L. 264/1 In the present investigation laccases of Japanese and Indo-Chinese lacquer trees have been purified..and their properties compared. **1966** L. I. Ingraham in Florkin & Stotz *Comprehensive Biochem.* XIV. 440 Laccase catalyzes the aerobic oxidation of hydroquinones to *p*-quinone. It contains four copper atoms per mole which react with one mole of oxygen. **1971** *European Jrnl. Biochem.* XXIII. 487/2 Whether the unequal load of laccase II with carbohydrate is associated with a fixation of this enzyme to the cell walls, or whether it plays any role in the postulated aggregation of the low molecular weight laccases II and III to the high molecular weight laccase I..must be elucidated by further experiments.

laccolite. For 'mass' read 'concordant mass'. Add further examples.

1937 Wooldridge & Morgan *Physical Basis Geogr.* viii. 110 Laccolites are closely related in manner of origin to the stratiform intrusions of igneous rock..termed 'sills'. **1946** L. D. Stamp *Britain's Struct.* ix. 80 A special type of sill is one where the lava swells out to form a lens-shaped mass—which according to its particular form is known as a laccolite (with a flat base) or a phacolite (with a curved base).

laccolith. Add: Now commoner in use than *laccolite*. (Further examples.)

1944 A. Holmes *Princ. Physical Geol.* vi. 86 There are few good examples of laccoliths in Britain, though many stocks have been wrongly called laccoliths. Stocks are discordant intrusions, whereas laccoliths, like sills, are concordant. **1960** B. W. Sparks *Geomorphol.* vii. 151 Laccoliths, which are closely related to sills but which were formed from a magma too viscous to spread far, may form local dome-like features when exposed by erosion.

Hence **laccoli·thic** *a.*, of, pertaining to, or characteristic of a laccolith.

1896 *Jrnl. Geol.* IV. 741 The hypothesis..that the granites are batholithic, not laccolithic. **1898** *Ibid.* VI. 705 When vertical displacement with faulting is one of the chief characteristics of the intrusion, a distinction from normal laccolithic intrusion should be recognized. **1933** [see *bysmalithic *a.*]. **1968** R. W. Fairbridge *Encycl. Geomorphol.* 281/2 The Devil's Tower of Wyoming is sometimes claimed to be an eroded laccolithic dome.

lace, *sb.* Add: **8. a.** *lace-box, -curtain* (examples; also *fig.* and *attrib.*, middle- or upper-class, 'respectable', having social pretensions), *-stitch*. **c.** *lace-trimmed* adj. (earlier examples). Also *laceless* adj.

1904 Lace box [see **Bible-box*]. **1969** E. H. Pinto *Treen* 370 Lace boxes, to stand on chests of drawers..enjoyed their greatest popularity during the second half of the 17th century and during Queen Anne's reign. **1895** *Montgomery Ward Catal.* 347/3 Lace curtains and lambrequins. **1934** J. T. Farrell *Young Manhood* xviii. 282 They were all trying to put on the dog, show that they were lace-curtain Irish, and lived in steam-heat. **1949** *Sat. Rev. Lit.* (U.S.) 25 June 33/1 Mrs. Ruskay's folks were lace-curtain Jews; they had a piano and a Polish maid. **1960** *Guardian* 8 July 8/4 The Kennedy millions..were.. wrested by a lace-curtain Irishman from..the Boston Brahmins. **1964** *Publ. Amer. Dial. Soc.* XLII. 35 The most common-place of these is the distinction between *shanty-* and *lace-curtain Irish*, i.e., those who remain in the lower-class communities near the center of the city..and those who move into lowermiddle-class communities and work hard to approximate the ideals of vulgar respectability. **1965** 'E. Queen' *Fourth Side of Triangle* iii. 122 Also, I have the misfortune to be Irish. And not lace-curtain Irish, either! **1970** *Guardian* 5 June 10/2 Britain has a long tradition of what might be called lace-curtain racialism. **1974** J. Stubbs *Painted Face* iii. 56 Every house in the square was veiled in lace curtains. **1901** *Daily News* 4 June 2/6 The shoes, low and laceless, slip on easily. **1968** R. Clapperton *No News on Monday* xi. 129 His feet were thrust into a pair of laceless tennis shoes. **1872** *Young Englishwoman* Dec. 658/2 These medallions are worked in appliqué of muslin or net, in satin stitch, knotted-stitch etc. Lace stitches fill the centre of the flower. **1961** A. Liley *Craft of Embroidery* i. 28 (*heading*) Lace stitch filling... This is also a buttonhole stitch variation giving the laciest effect of all the buttonhole fillings. **1835** Dickens *Sk. Boz* (1836) 1st Ser. I. i. 4 Her previous admiration of 'the board'..fades into nothing before her respect for her lace-trimmed conductor [*i.e.* the beadle]. **1861** Geo. Eliot *Silas Marner* xii. 217 A lace-trimmed cradle.

9. lace-bark, (*b*) substitute for def.: any of several small New Zealand trees of the family Malvaceæ, including *Plagianthus betulinus* and several species of *Hoheria* (cf. *houhere), with toothed leaves and clusters of white flowers; also *attrib.*; add examples; **lace-bug,** an insect of the family Tingidæ, including many species of bugs that feed on plants and sometimes become pests; **lacecap,** a hydrangea whose corymbs are made up of small fertile flowers or a mixture of these with larger sterile ones, giving the effect of lace; also used *attrib.* or as *adj.* to describe flower-heads of this type belonging to hydrangeas or other plants; **lace-wood,** (*a*) *Austral.* = **lace-bark* (*b*); (*b*) the wood of a plane tree, *Platanus occidentalis* or *P. acerifolia*.

1868 *Trans. N.Z. Inst.* Essays 33 Ribbon Wood, or Lace-bark tree (*Plagianthus Lyallii*). A very ornamental shrub tree, with large leaves and flowers. **1906,** etc. [see *houhere]. **1957** *Landfall* XI. 234, I lay in a patch, Of bush—giant manuka, some lace-bark. **1958** S. Ashton-Warner *Spinster* 189 Flaring out above the long grass and from beneath the lace-bark tree. **1966** *Weekly News* (Auckland) 6 Apr. 40 The lacebark which is also known as mountain ribbonwood (Hoheria glabrata), grows on the western side of the Southern Alps. **1895** J. H. & A. Comstock *Man. Study Insects* xiv. 139 The Lace-bugs. Dainty as fairy brides are these tiny, lace-draped insects. One glance at the fine white meshes that cover the wings and spined thorax is sufficient to distinguish them from all other insects, for these are the only ones that are clothed from head to foot in fine white Brussels net. **1923** E. A. Butler *Biol. Brit. Hemiptera-Heteroptera* 196 In the *Tingidae*, or Lace Bugs, we are not rich, for we have but twenty-four in our fauna. **1932** Metcalf & Flint *Fund. Insect Life* viii. 225 Lace bugs look as though they were cut out of fine gauze. **1967** K. M. Smith *Insect Virol.* xi. 225 The vector in both cases is somewhat unusual, it is not a leafhopper but a 'lace bug'. **1972** Swan & Papp *Common Insects N. Amer.* xii. 123 The Tingidæ, or lace bugs, are..small, oval or rectangular in outline, and lacelike in appearance owing to reticulated pattern on the head, thorax, and wings. **1950** W. E. Shewell-Cooper *Compl. Gardener* IV. 339 Besides the hortensia type there is another type of *H[ydrangea] macrophylla*, the Lacecaps, which have a flat head with small fertile flowers in the centre and a ring round the outside formed by the larger sterile flowers. **1966** J. Berrisford *Wild Garden* ii. 29 Hydrangeas will grow extremely well in woodland... Grow the lacecaps too. Their flower-pattern is exquisite... The nearly related *Schizophragma hydrangeoides* is a climbing version of the hydrangea family... Its lacecap flower-heads have a solitary large bract-like sepal. **1967** *Sunday Times* 21 May 14/5 The lovely lacecap hydrangea 'Blue Wave'. **1971** *Guardian* 17 Apr. 7/7 The lacecap-flowered Viburnum mariesii. **1898** Morris *Austral Eng.* 258/1 Lace-bark, Lacey-bark, or Lacewood. **1902** G. S. Boulger *Wood* v. 101 Very choice ornamental woods are employed mainly as veneers. Such are..those of Walnut; and the beautiful Lacewood or Honeysuckle wood of North America (*Platanus occidentalis*). **1930** *Morning Post* 2 Aug. 12/2 The anomaly that Queensland silky oak should be purchased by the United States..and then shipped to this country where it is sold as lace-wood. **1962** J. C. S. Brough *Timbers for Woodwork* (rev. ed.) xvi. 168 Cut upon the quarter, plane-wood shows an exceedingly handsome figure, and American grown wood converted in this fashion was sent over in fairly large quantities some years ago under the name of Lace-wood.

lace, *v.* Add: **4. b.** *spec.* in Bookbinding, to attach the boards to a volume sewn on cords by passing the slips through holes pierced in them. Also with *in*.

1818 H. Parry *Art of Bookbinding* 15 Put the paste-boards on each side of the book..and mark on them, with a bodkin, the places where the bands are to be drawn or laced in. **1835** J. Hannett *Bibliopegia* 30 One board is then placed on each side of the volume, even at the head, and marked with a bodkin opposite to the slips intended to be laced in. **1871** *Amer. Encycl. Printing* 74 When the boards are affixed to the volume by means of the bands being passed through holes made in the boards, they are said to be laced in. **1946** A. J. Vaughan in H. Whetton *Pract. Printing & Binding* xxxi. 382/1 The boards are held to the book by three or more cords. These are the cords upon which the sections have been sewn and the boards are said to be 'laced on'. Should these cords not have been laced through the boards the binding is what is known as a cased book, a cheaper style. **1961** [see *lacing *vbl. sb.* 3 e].

e. *intr.* Of structures that resemble or suggest lacing: to pass *across* a gap or *about* an object. Also *fig.*, to become entwined.

a **1889** G. M. Hopkins *Poems* (1918) 76 Her dearness.. more and more times laces round and round my heart. **1899** H. G. Wells *When Sleeper Wakes* x. 103 The cables and bridges that laced across the aisles were empty. *Ibid.* xxii. 288 A flimsy seeming scaffolding that laced about the great mass of the Council House.

f. *trans.* To pass (film or tape) between the guides and other parts of a projector, tape recorder, or the like so that it occupies the path taken from one spool to the other when the machine is running. Usu. with *up*.

1948 C. A. Hill *Cine-Film Projection* v. 54 Unless the film is your own, you must put its care before everything else, even if you have to stop the show, but this should never be necessary if you always lace the film correctly. **1966** G. Sinstadt *Whisper in Lonely Place* vi. 106 A facia panel opened to reveal a tape deck. He removed a spool from the metal container and laced up the tape. **1968** C. N. G. Matthews *Tape Recording* ix. 80 Press the stop button almost at once or you will come to the end of the tape and have the trouble of lacing it up again. **1974** *Some Technical Terms & Slang* (Granada Television), *Lace up*, to thread film through a machine for projection or transmission.

6. (Later examples.)

1923 H. G. Wells *Men like Gods* I. vi. 89 'Tell me', that engaging phrase, laced his conversation. **1971** *Nature* 2 July 70/2 That work led to two by-products: a Beilby award in 1948 and a fund of sea stories with which to lace his general conversation.

laced, *ppl. a.*[1] Add: **8*.** *laced valley* (Building): a valley between the slopes of two adjoining roofs in which the end tile of each row abuts against a tile-and-a-half tile laid diagonally on the valley board.

1931 C. G. Dobson *Roof Tiling* iii. 39 No lead is required in a laced valley. **1947** R. Greenhalgh *Mod. Building Construction* II. 582/2 Other methods..give swept and laced valleys.

‖ **lacet**[2] (lase). [Fr., 'lace, hairpin bend'.] A hairpin bend in a road.

1895 S. Weyman *Red Cockade* vi. 84 The road there descends not in *lacets*, but straight. **1922** *Glasgow Herald* 18 July 9 A series of lacets brings one to the bleak crest of the Col de Cavolles. **1932** Kipling *Limits & Renewals* 342 The Massif in Spring, the multiplied lacets Hampered by slips or drifts. **1970** 'S. Troy' *Blind Man's Garden* ix. 116 The Route Napoléon with its frequent *corniche* and *lacet*.

lace-up (lē̆ⁱ·sʌp), *a.* (and *sb.*). [Lace *v.* 10.] Of boots or shoes: that are fastened with laces. Also *ellipt.* as *sb.*, a lace-up boot or shoe.

1836, etc. [see Lace *v.* 10]. **1889** J. K. Jerome *Three Men in Boat* xi. 170 He would have his lace-up boots. **1951** [see *button-up* s.v. *button *v.* 4]. **1964** *Observer* 15 Nov. 14/4 Girls' shoes are much worse for the feet than the usual boys' lace-up. **1972** R. Quilty *Tenth Session* 20 Her funny old-fashioned lace-up shoes. **1973** *Times* 7 Nov. 18/3 A..quantity surveyor was clumping along in tan three-inch lace-ups.

Lach, L'ach, varr. *Lech, Lekh *sb.*[5] and *a.*

lachenalia (laʃənē̆ⁱ·liă). [mod.L. (N. J. Jacquin 1787, in *Nova Acta Helvetica* I. 39),

f. the name of Werner de *la Chenal* (1736–1800), Swiss botanist + -ia[1].] A small, bulbous plant of the genus so called, belonging to the family Liliaceæ, native to South Africa, and bearing thick, often spotted leaves, and spikes or racemes of tubular or bell-shaped flowers; also called *Cape cowslip*.

1789 *Curtis's Bot. Mag.* III. 82 Like most of the Cape plants, the *Lachenalia* requires to be sheltered in the winter. **1900** L. H. Bailey *Cycl. Amer. Hort.* II. 866/1 Lachenalias are Cape bulbs that are easily flowered in a cool greenhouse in early spring or even in winter. **1923** *Chambers's Jrnl.* Dec. 786/2 A host of delightful, and to us strange, things—..sombre-hued babianas, and the lachenalias with their weird-spotted foliage. **1961** P. White *Riders in Chariot* viii. 236 A plaster pixie..out on the front lawn, beside the golden cypresses, amongst the lachenalia. **1970** *Guardian* 9 May 9/2 It is more than 200 years since the Cape cowslip was first introduced to England... Lachenalias flower in midwinter.

Lachmann (la·χman). [The name of Karl *Lachmann* (1793–1851), German philologist.] *Lachmann's law*: the rule that in Latin, a short root-vowel in the present-tense stem of a verb is lengthened in the past participle if the present-tense stem ends in a voiced plosive.

[**1850** C. Lachmann *In T. Lucretii Cari De Rerum Natura Libros Commentarius* 54 Participia passiva ea quorum in praesenti consonans est aut liquida aut *s* semivocalis, quantitatem praesentis secuntur..contra ubi in praesenti media est, participia producuntur.] **1913** *Classical Rev.* XXVII. 122 It is true that 'Lachmann's Law'.., admittedly impossible as originally stated, was revived in a much modified form by Pedersen. **1928** *Language* IV. 181 (*title*) Lachmann's law of vowel lengthening. **1965** W. S. Allen *Vox Latina* 70 The same evidence indicates that Lachmann's Law also applies before *s* in the subjunctive *adāxim*..as against *effēxim*. **1970** *Harvard Stud. Classical Philol.* LXXIV. 58 Kurylowicz therefore saw correctly..that a morphophonemic solution was indicated; that Lachmann's law was not a phonetic rule. **1973** W. S. Allen *Accent & Rhythm* 18 Thus, referring to the Latin phenomenon of 'Lachmann's Law' (whereby e.g. *făcio* forms a past participle *fāctus*, but *ăgo* forms *āctus*), Kiparsky..suggests that it can be accounted for by introducing the presumed Latin vowel-lengthening rule V→long/—*g* before, rather than after, the Indo-European consonant-assimilating rule C→voiceless/—*t*.

lachrymal, *a.* and *sb.* Add: The etymologically correct form **lacrimal** is now usual in scientific use. (Further examples of sense 2 of the adj.)

1913 *Gray's Anat.* (ed. 18) 960 The lacrimal [*ed. 17* (1909): lachrymal] ducts or canals..commence at minute orifices, termed *puncta lacrimalia*. **1913** *Cunningham's Text-bk. Anat.* (ed. 4) 824 The lacrimal gland..is supplied by the sympathetic and lacrimal nerves. [*Ed. 3* (1909): lachrymal.] **1919** *Lancet* 10 May 792/2 During the stage of surgical anæsthesia the lacrymal glands cease to secrete. **1919** *Physiol. Abstr.* IV. 183 [*From abstract of article containing the prec. quot.*] The lacrimal gland in anæsthesia. **1950** D. B. Kirby *Surg. Cataract* xii. 276/1 Care must be exercised to prevent the solution from going through the lacrimal duct to the nose. **1954** S. Duke-Elder *Parsons' Dis. Eye* (ed. 12) xxxiii. 543 The lacrimal apparatus consists of the lacrimal glands and the lacrimal passages. [*Edd. 1* (1907)–*11* (1948): lacrymal.] **1968** Passmore & Robson *Compan. Med. Stud.* I. xxiv. 54/2 The cornea is continually bathed by the secretion of the lacrimal glands.

B. *sb.* **2.** *pl.* (Later example.)

1931 M. Summers *Supernatural Omnibus* 31 What an event [in Dickens] was a funeral from a house! The way to all these sadly sentimental lachrymals had been paved before by the lugubrious cortèges of the time of Anne.

lachrymate, *v.* Delete *Obs.*[−0] and add pronunc. (læ·krimē̆ⁱt). Now current chiefly in scientific use (freq. as **lacrimate**: cf. *lachrymal *a.* and *sb.*), with the sense: to discharge moisture from the eyes. (Examples.) Hence **la·chrymating** *vbl. sb.* and *ppl. a.*

1922 *Encycl. Brit.* XXXII. 110/2 If a sufficient number of lachrymating grenades could be thrown. **1944** *Brit. Jrnl. Ophthalm.* XXVIII. 330 The patient lacrimates when he salivates. **1962** W. K. McEwen in H. Davson *Eye* III. x. 272 In man there is the added ability to weep, or lacrimate, which is an excessive outpouring of the lacrimal gland. **1964** *Amer. Jrnl. Ophthalm.* LVIII. 1056/1 Tension in lacrimating patients is more easily measured with the Schiøtz tonometer.

lachrymator (læ·krimē̆ⁱtǫɹ). Also **lacrimator.** [f. *lachrymat-, lacrimat-* (in Lachrymatory *a.* and *sb.*, etc.) + -or.] Any substance which causes irritation and copious watering of the eyes when it comes into contact with them (in the form of a gas, spray, dust, or the like).

1918 *Jrnl. Amer. Med. Assoc.* 30 Nov. 1823/1 Lacrimators (acetone, xylene or benzene bromid). **1922** *Encycl. Brit.* XXXII. 111/2 Lachrymators, on account of the extreme sensitiveness of the eye, can produce an effect in extraordinarily weak concentrations. **1948** *Biochem. Jrnl.* XLII. p. xxvi, The '—SH' theory provides an explanation for the existence of two main groups of lachrymators: (a) substances containing a halogen atom in some such group as $—CH_2Cl$ together with a neighbouring keto or other group which makes the halogen 'positive', (b) substances containing a —CH=CH— group, also with a neighbouring keto or other group which polarizes the double bond. **1963** *Times* 25 Apr. 6/4 Chloracetophenone, a lacrimator in adequate concentrations, had toxic effects.

lachrymatory, *a.* and *sb.* **A.** *adj.* (Further examples.)

1916 *Yorks. Post* 21 July 5/5 A violent artillery preparation with asphyxiating and lachrymatory shells. **1935** *Brit. Med. Jrnl.* 20 July 133/1 Lachrymatory gases were here and off again unless there was a continual rain of tear-gas shells. **1971** *Agric. & Biol. Chem.* (Tokyo) XXXV. 1831 The lachrymatory character and the pungent flavor [of onions] had been decreased by γ-irradiation.

lachrymogenic (læ:krimodʒe·nik), *a.* [f. Lachrym(ation + -o + *-genic.] Giving rise to lachrymation.

1921 *Chem. Abstr.* XV. 398 (*heading*) Estimation of the lachrymogenic power of irradiating substances by the threshold method. **1971** *Jrnl. Agric. & Food Chem.* XIX. 269/2 Thioethanol, thioacetone, thiobutanal, and thiohexanal *S*-oxides were also synthesized and their lachrymogenic properties were evaluated.

‖ **lachsschinken** (la·χsʃiŋkĕn). Also erron. lachschinken. [G., f. *lachs* salmon + *schinken* ham.] Cured and smoked loin of pork.

1923 A. Ward *Encycl. Food* 463 *Lachsschinken*, two trimmed, boneless pork-loins, mild sweet-cured, faced and pushed into large beef casings, sixteen inches or so in length, pressed, tied with strings, dried, and lightly smoked. **1937** *Atlantic Monthly* Mar. 267, I burst into enthusiasm on the subject of *Lachs-schinken* (which, as the name indicates, is a sort of ham that both looks and tastes like salmon). **1958** *Catal. County Stores, Taunton* June 5 Continental sausages... Lachsschinken, eaten cold, finely sliced—4 ozs. 2/7. **1965** *New Statesman* 11 June 914/2 'Sausages of all kinds not in airtight containers'—which include the mortadella and the salami, the krakauer and lachsschinken. **1971** *Sunday Times* (Colour Suppl.) 27 June 50/3 *Lachschinken*: pork fillet, cured, smoked and rolled in a thin casing of pork back fat. Serve with fresh fruit as an alternative to *prosciutto*.

laciness (lē̆ⁱ·sinės). [f. Lacy *a.* + -ness.] Lace-like quality or effect.

1903 *Westm. Gaz.* 9 July 4/2 Flowered silk, or embroideries, or chiffon, and laciness.

lacing, *vbl. sb.* Add: **3. e.** *Bookbinding* (see quot. and *Lace *v.* 4 b).

a **1877** Knight *Dict. Mech.* II. 1244/1 *Lacing*,..securing the book to the sides by carrying the bands or slips through perforations in the boards. **1961** T. Landau *Encycl. Librarianship* (ed. 2) 189/1 *Lacing-in*, in hand-binding, in the 'extra' style, the method of attaching boards to the sewing cords, which are laced through holes in the boards.

4. lacing course *Building*, a special course built into an arch or wall in order to bond different parts together and give added strength.

1886 H. C. Seddon *Builder's Work* i. 66 It is better..to build the arch in half-brick rings, with a few bonding or lacing courses built in at intervals, to tie the separate rings together. **1899** *Notes on Building Construction* (rev. ed.) I. iii. 46 Walls such as those built with flints, or other small stones,..are frequently strengthened by building in with them lacing courses, consisting of horizontal bands either of ashlar, coursed rubble, or brickwork. **1947** R. Greenhalgh *Mod. Building Construction* I. 229/1 In arches of very wide span, the different rings are sometimes bonded to each other by inserting courses of stretchers in the depth of the arch at intervals. These courses are termed lacing courses.

lacis (la·si). [Fr.] A kind of lace made by darning patterns on net.

1865 F. B. Palliser *Hist. Lace* ii. 14 The volume..is that of the Venetian Vinciolo..dating from 1587... The work is in two books. The first of Point coupé... The second of Lacis, or subjects in squares, Fig. 3, with counted stitches, like the patterns for worsted-work of the present day. **1875** *Encycl. Brit.* XIV. 183/2 The productions of this art, which has some analogy to weaving, in the early part of the 16th century came to be known as.. 'lacis' in France... With the development of the renaissance of art, free flowing patterns and figure subjects were introduced and worked in lacis. **1953** [see *filet 1]. **1963** *Times* 1 June 11/6 One development was the introduction of all-white lacis, such as was recorded in the bed furnishings of Mary Queen of Scots with ornament built up in darning stitches on a basis of open-mesh net.

lack, *v.*[1] Add: **4.** Also with *for*.

1892 'Mark Twain' *Amer. Claimant* 40 Here's hoping he'll never lack for friends. **1898** Skeel & Brearley *King Washington* (1899) 170 He was one of the many who had lacked for partners. **1906** E. Phillpotts *Portreeve* II. iv. 154 The outward signs that she had marked upon him did not lack for inner causes.

lackadaisy, *int.* (Earlier examples.)

1748 Smollett *R. Random* I. viii. 56 She exclaimed, 'Good lack-a-daisy! the rogue is fled!' **1796** M. Edgeworth *Parent's Assistant* (ed. 2) I. 164 The carpenter.. said '*lack-a-daisy!*' when he saw that the old theatre was pulled down.

lackey, *sb.* Add: **1. c.** *spec.* As a term of political abuse: a servile follower.

1939 G. E. R. Gedye *Fallen Bastions* ii. 37 The Communists did not hesitate to condemn them [*sc.* the Austrian Socialist leaders], as 'Social Fascists' who did 'lackey service' to capitalist reaction. **1941** *Amer. Mercury* Apr. 417/2 American bankers..have already stepped into the role of lackeys of British Imperialism. **1957** C. Hunt *Guide to Communist Jargon* p. xiv, The Soviet Union stands for peace, and the imperialists for war, in which they are once again supported by their socialist 'lackeys'. **1969** *Listener* 27 Apr. 429/3 Whatever else went under the name of socialism was either 'wilful deception by lackeys of the bourgeoisie' or the self-deception of those who hesitated 'between life-and-death struggle and the role of assistants to the expiring bourgeoisie'. **1972** *Sat. Rev.* (U.S.) 24 June 30/2 As any good Maoist..can tell you, Hussein is nothing more than a Western lackey. **1973** *Black Panther* 21 July 10/2 The U.S. government or its lackeys. **1974** A. Ross *Bradford Business* 129 Bloody fascist lackeys!

lack-wit, lackwit (læ·kwit). [Lack *v.*[1] 7.] A witless or stupid person.

1667, 1809 [see Lack *v.*[1] 7]. **1911** H. S. Harrison *Queed* iv. 42 West..abused himself for a shiftless lackwit who was slated for an unwept grave. **1936** F. Clune *Roaming round Darling* xxi. 210, I was embarrassed exceedingly by an inarticulate lack-wit, while I filled the petrol tank.

Laconian, *a.* and *sb.* Add: **B.** *sb.* **2.** The dialect of ancient Greek spoken in Laconia (Sparta).

1830 Tufnell & Lewis tr. *Müller's Hist. & Antiq. Doric Race* II. App. viii. 502 We have considered the Doric dialect in general, as spoken by the whole race, only marking out the Laconian as its purest variety. **1875** *Encycl. Brit.* XI. 133/2 Three changes characteristic of Laconian came in at a comparatively late date. **1954** Pei & Gaynor *Dict. Ling.* 118 *Laconian*, one of the Doric dialects of ancient Greek.

‖ **lacrimæ rerum** (læ·krimai reᵊrʊm). [L.] With reference to Virgil, *Aeneid* I. 462: the sadness of life; tears shed for the sorrows of men.

[**1841** Carlyle *Let.* 19 July in R. K. Webb *H. Martineau* (1960) vii. 196 Poor Harriet! She was..almost sublime to me there. *Sunt lachrymae rerum.*] **1929** A. Huxley *Do what you Will* 195 Hellenic lovers..may have wept the *lacrimae rerum*. **1951** Auden *Nones* (1952) 71 Only the young and the rich Have the nerve or the figure to strike The lacrimae rerum note. **1959** *Times* 19 Mar. 4/1 He [*sc.* Mahler] told in his music truths of humanity, of the *lacrimae rerum*, and of alleviating beauty. **1969** G. Smith in A. Huxley *Lett.* (1969) 1 Though often tinged with a profound instinct to weep for life, it—and by no means the sadness of *lacrimæ rerum*—is Huxley's hallmark.

lacrimal, lacrimate, -or, see *lachrymal *a.* and *sb.*, etc.

lacsamana, var. *laksamana.

lactalbumin (læktæ·lbiū̆min). *Biochem.* [ad. F. *lactalbumine* (A. Commaille 1866, in *Rec. de Mém. de Méd.* XVII. 155), f. *lact(o-* Lacto- + *albumine* Albumin.] **a.** The fraction of milk proteins which is obtained after the removal of casein and which is soluble in a salt solution such as saturated magnesium sulphate or ammonium sulphate.

1885 [see *lactoglobulin 1]. **1936** *Biochem. Jrnl.* XXX. 956 Probably the so-called lactalbumin is a mixture of the two molecules α and β and the molecule usually referred to as lactoglobulin is γ. **1970** [see *b].

b. (Usu. *α-lactalbumin*.) A protein or mixture of closely similar proteins occurring in the lactalbumin fraction of milk and having a molecular weight of about 17,400.

1937 *Nature* 19 Jun. (Suppl.) 1058/2 In cow's milk there are two albumins, the α-lactalbumin, of molecular weight 17,600 and sedimentation constant $1{\cdot}9 \times 10^{-13}$ and β-lactalbumin (also called Palmer's lactoglobulin) with $M = 39{,}000$ and $s = 3{\cdot}12 \times 10^{-13}$. **1948** [see *lactoglobulin 2]. **1953** *Jrnl. Amer. Chem. Soc.* LXXV. 329/1 It is proposed, therefore, that the protein isolated by the Sørensens also be called α-lactalbumin even though the isoelectric protein is only slightly soluble in water. **1962** *Biochem. Jrnl.* LXXXIII. 271/1 The lactalbumin was prepared from human-milk whey. **1970** R. Jenness in H. A. McKenzie *Milk Proteins* I. ii. 22 When later the proteins β-lactoglobulin and α-lactalbumin were crystallised from the 'lactalbumin' fraction, the Greek letters used by Pedersen were incorporated into the names... The β-protein was named as a globulin because of its insolubility near the isoelectric point in the absence of salt and the α-protein as an albumin because of its source in the 'lactalbumin' fraction of whey... Two α-lactalbumins, A and B, have been found.

lactam (læ·ktæm). *Chem.* [a. G. *lactam* (Baeyer & Oekonomides 1882, in *Ber. d. Deut. Chem. Ges.* XV. 2102), f. *lact-on* Lactone + *am-id* Amide.] Any of the class of cyclic amides analogous to the lactones, characterized by the group —NH·CO— as part of a ring (*lactam ring*) and formed by the elimination of a molecule of water from an amino and a carboxyl group of an acid; freq. with prefixed Gr. letter (see *lactone 2).

LACTAMASE

1883 *Jrnl. Chem. Soc.* XLIV. 202 The authors propose the name *lactam* for bodies formed like acetylisatin, and *lactim* for those formed like isatin. **1910** *Encycl. Brit.* VI. 59/1 The internal anhydrides of aminocarboxylic acids (lactams, betaines). **1936** L. J. Desha *Org. Chem.* xxvi. 522 Lactam-lactim tautomerism involves the shift of a hydrogen atom from nitrogen to oxygen and vice versa. **1949** E. P. Abraham et al. in H. W. Florey et al. *Antibiotics* II. xv. 669 The four-membered lactam ring which was present in this structure [of penicillin] had not previously been found in natural compounds, and the ease with which it could be broken was responsible for the instability of penicillin under a variety of conditions. **1964** Roberts & Caserio *Basic Princ. Org. Chem.* xx. 712 Formation of α- and β-lactams is expected to generate considerable ring strain, and other more favorable reactions usually intervene. **1966** *McGraw-Hill Encycl. Sci. & Technol.* VII. 377/1 Although γ- and δ-amino acids are nontoxic themselves, the corresponding lactams show pronounced strychnine-type toxicity.

lactamase (læ·ktămēɪz, -s). *Biochem.* [f. *Lactam + *-ase.] *β-lactamase*: any of the enzymes (produced by certain bacteria) which cause the breaking of the carbon–nitrogen bond in the lactam ring of penicillins and cephalosporins (so rendering them ineffective as antibiotics).

1964 *Jrnl. Gen. Microbiol.* XXXVI. 206 Amidase-forming coliform organisms did not attack the 6-APA nucleus, whereas β-lactamase-forming coliform organisms opened the lactam ring. **1971** N. Citri in P. D. Boyer et al. *Enzymes* (ed. 3) IV. ii. 24 The role of penicillinase and other β-lactamases in conferring resistance to penicillins and the closely related cephalosporins has been amply demonstrated.

lactase (læ·ktēɪz, -s). *Biochem.* [a. F. *lactase* (M. W. Beijerinck 1889, in *Arch. néerl. d. Sci. exactes et nat.* XXIII. 434), f. *lact-ose* Lactose: see *-ase.] Any enzyme which catalyses the hydrolysis of lactose to glucose and galactose.

1891 *Jrnl. R. Microsc. Soc.* 374 The fermentation [of milk sugar] is effected by a diastase distinct from invertin, which he calls lactase. **1906** *Jrnl. Physiol.* XXXV. 28 The conclusion is from these experiments that lactase is distributed fairly equally through the whole of the mucous membrane of the intestine. **1959** Jenness & Patton *Princ. Dairy Chem.* iii. 80 There are at least three significant origins of lactase enzymes which may be used for the hydrolysis of lactose. These are: (a) certain species of yeasts..; (b) the intestinal mucosa of mammals..; and (c) β-galactosidase from almonds.

lactate (læktēɪ·t), *v.* [f. L. *lactāt-*, ppl. stem of *lactāre* to suckle.] *intr.* To secrete or discharge milk. Chiefly as **lacta·ting** *ppl. a.*

1889 in *Cent. Dict.* **1908** *Arch. Middlesex Hosp.* XIII. 59 In the lactating breast the number of acini is very large. **1913** *17th Internat. Congr. Med.* III. ii. 173 The pregnant and lactating animals survived the extirpation of the adrenal glands much longer than normal, non-pregnant, or male animals. **1948** *New Biol.* IV. 127 Purified preparations of prolactin were found inferior to more crude pituitary preparations..in their power to increase the milk yield of cows already lactating. **1953** E. Gellhorn *Physiol. Found. Neurol. & Psychiatry* xiii. 305 There is still another function of oxytocin..: the ejection of milk from the lactating mammillary gland. **1956** J. S. Folley *Physiol. & Biochem. Lactation* iii. 28 Mammary tissue from lactating rats..exhibits much greater respiratory activity *in vitro*. **1956** *Jrnl. Pediatrics* XLIX. 550/2 She had suckled all her daughter's children and lactated freely each time. **1971** J. Z. Young *Introd. Study Man* xxiv. 319 To estimate reliably the number of children born to a group of women under the conditions that their husbands are freely available and that they are not using contraceptives or lactating is nearly impossible.

lacta·ted, *a.* [f. as prec. f. L. *lact-*, *lac* milk.] Combined with a milk-product.

1889 *Buck's Handbk. Med. Sci.* VIII. 100/2 In Horlick's and Mellin's there is said to be no unconverted starch. Hawley's contains eleven per cent.,..as does also the Lactated Food. **1896** *Rep. Vermont Board Agric.* XV. 25 This is used in making..lactated food for infants and invalids. **1907** *Yesterday's Shopping* (1969) 500/2 Lactated Pepsin Tablets.

lactational (læktēɪ·ʃənăl), *a.* [f. Lactation + -al.] Of or pertaining to lactation.

1903 *Med. Rec.* (N.Y.) 28 Feb. 337/2 Both suicidal and infanticidal promptings are more common in lactational than puerperal cases—that is, in cases in which insanity commenced more than six weeks after confinement. **1970** *Nature* 19 Dec. 1222/2 It has been found that cows excrete less DDT in lactational fluids than human females.

lactim (læ·ktim). *Chem.* [a. G. *lactim* (Baeyer & Oekonomides 1882, in *Ber. d. Deut. Chem. Ges.* XV. 2102), f. *lact-on* Lactone + *im-id* Imide.] Any of the class of cyclic imines which are isomers of the lactams and characterized by the group —N:C(OH)— as part of a ring.

1883, 1936 [see *lactam]. **1951** I. L. Finar *Org. Chem.* I. xxx. 619 Isatin exists in two forms, the term ψ-isatin being applied to the lactam form I, and isatin to the lactim form II. **1971** *Canad. Jrnl. Chem.* XLIX. 2612 Investigation showed the free lactim ether to be a probable intermediate in this reaction.

lacto-. Add: **la·ctochrome** *Biochem.*, a yellow-orange pigment orig. extracted from milk and now identified with riboflavin; **lactofla·vin** *Biochem.* [a. G. *lactoflavin* (Ellinger & Koschara 1933, in *Ber. d. Deut. Chem. Ges.* LXVI. b. 808)] = *riboflavin; **la·ctogen** *Physiol.*, any lactogenic hormone; *spec.* = *prolactin; **lactoge·nesis** *Physiol.*, the initiation of milk secretion; **lactoge·nic** *Physiol.* [*-genic], pertaining to or having the ability to initiate the secretion of milk; hence **lactoge·nically** *adv.*; **lacto-vegeta·rian** *a.*, consisting of milk and vegetables; so **lacto-vegeta·rianism**.

1879 A. W. Blyth in *Jrnl. Chem. Soc.* XXXV. 532 After the liquid from which the galactin has been removed had been freed from the excess of lead by hydrogen sulphide, an alkaloïdal colouring matter, for which I propose the name of 'lactochrome', may be separated by the addition of nitrate of mercury solution. **1914** *Jrnl. Biol. Chem.* XVII. 261 The facts brought out by this investigation point very clearly to a very close relationship existing between the yellow lactochrome of milk whey and the urochrome of urine. **1936** W. L. Davies *Chem. Milk* xi. 218 Milk from other species of mammals also contains lactochrome. It is present in human milk. **1953** Fruton & Simmonds *Gen. Biochem.* xiii. 319 In 1879 Blyth described the isolation from milk of a yellow pigment (named lactochrome) which showed a striking green fluorescence... By 1936 the chemical nature of the yellow pigment of egg yolk and of milk had been established..; it was shown that this pigment, named riboflavin, is identical with vitamin B_2. **1933** *Brit. Chem. Abstr.* A. 847/2 (*heading*) Lactoflavin, the pigment of milk. **1938** *Encycl. Brit. Bk. of Year* 651/1 Recently it has been found that vitamin B_2 consists of three components, lactoflavin, vitamin B_6..and pellagrous preventing or p.p. factor. **1943** Sumner & Somers *Chem. & Methods of Enzymes* xiii. 244 Various flavins were described, *e.g.*, lactoflavin of milk, hepatoflavin of liver... The flavin of these workers was the same substance which is called today 'riboflavin', or sometimes, 'lactoflavin'. **1946** J. F. Fulton *Howell's Textbk. Physiol.* (ed. 15) liv. 1209 Both the onset and maintenance of lactation require lactogen. **1952** S. J. Folley in A. S. Parkes *Marshall's Physiol. Reproduction* (ed. 3) II. xx. 558 When first discovered it [sc. an anterior-pituitary protein hormone] was variously named prolactin (Riddle), galactin (Turner) and mammotropin (Lyons). Recent American practice tends to favour the terms lactogen or lactogenic hormone; the name prolactin is adopted in this chapter in accordance with English usage. **1962** *Endocrinology* LXXI. 218/2 The Raben preparation.., although quite potent as a lactogen in the pseudopregnant rabbit,.. has little activity in the local intradermal pigeon crop assay. **1967** *Proc. Nat. Acad. Sci.* LVIII. 2307 The recent identification of human placental lactogen (HPL), a polypeptide hormone which shares both biological and immunological properties with pituitary growth hormone (HGH). **1939** Riddle & Bates in E. Allen *Sex & Internal Secretions* xx. 1089 Lactogenesis is a *response* to this hormone which excites also—in both sexes—additional responses more ancient phylogenetically and perhaps more significant generally. **1948** Lactogenesis [see *galactopoiesis* (*galacto-)]. **1969** S. R. Wellings in Reynolds & Folley *Lactogenesis* 5 Lactogenesis may be defined as the process by which full lactation is initiated in an already prepared mammary gland. **1933** *Proc. Soc. Exper. Biol. & Med.* XXI. 300 We have never observed milk secretion in normal or ovariectomized virgin guinea pigs uninjected with the lactogenic hormone. **1946** J. F. Fulton *Howell's Textbk. Physiol.* (ed. 15) liv. 1209 The lactogenic activity of pituitary extracts was first shown by Stricker and Grüter in 1928. **1952** Lactogenic [see *lactogen* above]. **1969** A. T. Cowie in Reynolds & Folley *Lactogenesis* 159, I now turn to the rabbit, the species in which the lactogenic role of the anterior pituitary was first discovered. **1969** R. Denamur in *Ibid.* 60 Thus, lactogenically prolactin modifies the polyribosomes by increasing their number in the cell. **1907** *Practitioner* June 845 The lacto-vegetarian diet..lessens auto-intoxication. **1929** *Encycl. Brit.* VII. 359/1 A lacto-vegetarian diet which permits the free use of milk and eggs. **1940** *Nature* 7 Dec. 726/2 It..looks as if the Briton's dietary will gradually shift, at least during the war years, towards lactovegetarianism. **1951** *News Chron.* 13 Dec. 3/2 Man was not designed for a purely vegetable diet. There was no objection whatever to lacto-vegetarian diet (this includes milk and eggs).

2. Used as comb. form of Lactic *acid* or Lactose: as in **lactobio·nic** *a.*, in *lactobionic acid* [tr. G. *lactobionsäure* (Fischer & Meyer 1889, in *Ber. d. Deut. Chem. Ges.* XXII. 362)], 4-(β-D-galactosido)-D-gluconic acid, $C_{12}H_{22}O_{11}$: a syrup produced by oxidation of lactose; hence **lactobi·onate**, a salt of lactobionic acid; **lactoni·trile**, a yellow liquid, $CH_3\cdot CH(OH)\cdot CN$, that is the nitrile of lactic acid and is used in a method of manufacturing acrylonitrile; acetaldehyde cyanohydrin; **lactophe·nol**, a mixture of approximately equal weights of phenol and lactic acid dissolved in glycerol and distilled water, used for mounting biological specimens.

1927 *Jrnl. Chem. Soc.* 546 Barium lactobionate was methylated in a manner similar to that already mentioned. **1964** *Chem. Abstr.* LXI. 16697 Ca lactobionate.. may be used under the Federal Food, Drug, and Cosmetic Act as a firming agent in dry pudding mixes. **1889** *Jrnl. Chem. Soc.* LVI. 485 Lactobionic acid, $C_{12}H_{22}O_{12}$, is obtained when milk-sugar (1 part) dissolved in water (7 parts) is treated with bromine (1 part). *Ibid.* 486 Lactobionic acid is decomposed into galactose and gluconic acid when warmed with dilute mineral acids. **1967** Kirk & Othmer *Encycl. Chem. Technol.* (ed. 2) XIII. 571 The sequestrant and emulsifying properties of lactobionic acid suggest a commercial potential, especially in the food industry, for this product. **1898** *Jrnl. Chem. Soc.* LXXIV. ii. 509 When silicon tetrachloride is heated with mandelonitrile or lactonitrile, silicic acid and complex tarry products are formed. **1935** *Chem. Abstr.* XXIX. 814 For prepg. an aliphatic cyanohydrin such as lactonitrile, reaction is effected between HCN and an aliphatic aldehyde or ketone such as acetaldehyde. **1961** *Ibid.* LV. 8268 The alkyl and aralkyl α-hydroxy nitriles, i.e. lactonitrile, [etc.].., are effective reagents for the extn. of Au and Ag by cyanidation... Crude lactonitrile which is a by-product from the manuf. of acrylonitrile is inexpensive and very efficient. **1896** *Jrnl. R. Microsc. Soc.* 481 M. J. Amann recommends the following fluids for preserving and imbedding mosses..(1) Lactophenol. **1929** W. R. Taylor in C. E. McClung *Handbk. Microsc. Technique* iv. 139 A solution (Lactophenol) composed of lactic acid [etc.]..is very serviceable and may be used for mounting various materials, softening dried material (especially algæ) or decalcifying specimens. **1970** *Watsonia* VIII. 140 Pollen grains stained with cotton blue in lactophenol.

lactobacillus (læ:ktobăsi·lɒs). *Biol.* Pl. **-bacilli.** [mod.L. (M. W. Beijerinck 1901, in *Arch. néerl. d. Sci. exactes et nat.* VI. 213), f. Lacto- + Bacillus.] Any bacterium of the genus *Lactobacillus* (family Lactobacillaceæ), which includes microaerophilic or anaerobic, non-motile, Gram-positive rods which convert glucose and related carbohydrates to lactic acid and are found in the intestinal tract and in fermenting plant and animal (esp. dairy) products.

1924 *Jrnl. Bacteriol.* IX. 375 A study of lactobacilli from the intestines. **1928** B. W. Hammer *Dairy Bacteriol.* xv. 410 The lactobacilli constitute another type of organism bringing about changes that are important in the ripening of cheddar cheese. **1958** W. C. Frazier *Food Microbiol.* iii. 51 Characteristics that make the lactobacilli important in foods are (1) their ability to ferment sugars with the production of considerable amounts of lactic acid, enabling their use in the production of fermented plant and dairy products, or the manufacture of industrial lactic acid, but resulting in the deterioration of some products, e.g. wine or beer; (2) production of gas and other volatile products..; (3) their inability to synthesize most of the vitamins they require,..making them useful in assays for the vitamin content of foods; and (4) the heat resistance..of most of the high-temperature lactobacilli. **1961** R. D. Baker *Essent. Path.* ix. 190 Although many microorganisms are observed in association with caries,..most investigations have centered around the parasitic lactobacilli. These organisms may have a role in the initiation of the process.

lactoglobulin (læktoglǫ·biŭlin). *Biochem.* [a. Da. *lactoglobulin* (J. Sebelien 1885, in *Oversigt o. d. K. Danske Viedensk. Selskabs Forhand.* 4), f. *globulin* Globulin: see Lacto-.] † **1.** The fraction of milk proteins which is obtained after the removal of casein and is precipitated by a salt such as magnesium sulphate or ammonium sulphate. *Obs.*

1885 *Jrnl. Chem. Soc.* XLVIII. 1000 The author has isolated from milk two albuminoïd substances distinct from casein. Lactoglobulin, the first of these, is present in minute quantity only... Lactalbumin, the second albuminoïd, is precipitated by acetic acid in the filtrate from the preceding compound. **1936** [see *lactalbumin a].

2. (Usu. *β-lactoglobulin.*) A protein or mixture of closely similar proteins occurring in the lactalbumin fraction of milk, insoluble near the isoelectric point in the absence of salt and having a molecular weight of about 40,000.

1936 *Biochem. Jrnl.* XXX. 968 This investigation shows that the lactoglobulin prepared according to the method of Palmer from the lactalbumin fraction of cow's milk is a monodisperse protein. **1945** *Jrnl. Amer. Chem. Soc.* LXVII. 1531/2 β-Lactoglobulin consists of 370 amino acid residues with 366 peptide bonds, arranged in 4 sub-units (polypeptide chains). **1948** *New Biol.* IV. 133 The typical milk proteins, casein, lactalbumin and lactoglobulin, are believed to be mainly products of the serum globulin of the blood. **1955** *Nature* 30 July 218/2 Individual cows produce either a mixture of two electrophoretically distinct β-lactoglobulins or only one or the other of these. *Ibid.* 219/2 β_1-Lactoglobulin forms rectangular plates indistinguishable from the well-known crystals hitherto prepared from mixed milk..whereas β_2-lactoglobulin forms diamond-shaped plates. **1961** *Biochim. & Biophys. Acta* XLIX. 591 In cow's milk β-lactoglobulin constitutes approx. half of the whey proteins and 7–12% of the total proteins.

lactol (læ·ktǫl). *Chem.* [a. G. *lactol* (Helferich & Fries 1925, in *Ber. d. Deut. Chem. Ges.* LVIII B. 1246), f. *lact-on* *Lactone 2 + *-ol* -ol.] Any cyclic compound that is formed by the oxygen atom in a —C(OH)— group (esp. one in a sugar) becoming linked to the carbon atom of a carbonyl group in the same molecule (with the transfer of the hydrogen atom of the former group to the oxygen atom of the latter).

1925 *Chem. Abstr.* XIX. 2931 The name lactol (aldolactol, ketolactol) is proposed for the cyclo-form of hydroxyaldehydes and ketones, the length of the O-bridge being indicated by Greek letters (as for the lactones) or by nos. in brackets (as for the sugars). **1967** [see *glucoside].

lactone. Add: **2.** *Chem.* [ad. G. *lacton* (R. Fittig 1880, in *Ann. d. Chem.* CC. 62).] Any of the class of cyclic esters formed (in theory) by the elimination of a molecule of water from a hydroxyl and a carboxyl group of an organic acid, and characterized by the group —O·CO— as part of a ring; freq. with preceding Gr. letter corresponding to the size of the ring (an *α*-lactone having a three-membered ring, a *β*-lactone a four-membered one, etc.).

1880 *Jrnl. Chem. Soc.* XXXVIII. 378 Pyroterebic acid ..is the internal anhydride of hydroxyisocaproic acid... It is the first representative of its class in the lactic series. For this class of anhydrides the author proposes the name 'lactones'. **1929** *Jrnl. Pharmacol. & Exper. Therap.* XXXVI. 355 The pharmacological study of a series of lactones with a view to their possible usefulness as anthelmintics pointed towards beta angelica lactone and the dilactone of the diacetone diacetic acid as the most promising. **1936** L. J. Desha *Org. Chem.* xxi. 414 All hydroxy acids lose water when heated..: α-Hydroxy acids yield lactides; β-hydroxy acids are converted into unsaturated acids; γ- and δ-hydroxy acids form lactones. **1971** G. A. Taylor *Org. Chem. for Students Biol. & Med.* xiii. 188 α- and β-hydroxy-acids cannot be converted directly into the corresponding α- and β-lactones (three- and four-membered rings).

lactonic (læktǫ·nik), *a.* *Chem.* [f. Lacto- (in chem. names: see below).] **1.** *lactonic acid*: [tr. G. *lactonsäure* (Hlasiwetz & Habermann 1870, in *Ann. d. Chem. u. Pharm.* CLV. 139), f. *lact-ose* Lactose + *-on* -one: see -ic.] **a.** = *galactonic acid.

1871 *Jrnl. Chem. Soc.* XXIV. 547 Lactonic acid, which, according to Fittig's views, must also be monobasic..is found to be really a bibasic acid. **1886** E. F. Smith tr. *V. von Richter's Chem. Carbon Compounds* 378 Lactonic acid, $C_6H_{10}O_6$, is produced from milk sugar and galactose by the action of bromine water and silver oxide. It is a deliquescent, crystalline mass, melting at 100°.

b. = *lactobionic acid* s.v. *lacto- 2.

1957 *Chem. Abstr.* LI. Subject Index 1358/1 Lactonic acid, calcium salt, calcium gluconate solubilization with. **1964** N. G. Clark *Mod. Org. Chem.* xvi. 328 Hydrolysis of lactonic acid yields galactose and gluconic acid.

2. [f. Lacton(e + -ic.] Containing the characteristic ring structure of lactones (sense *2).

1885 *Jrnl. Chem. Soc.* XLVIII. 963 (*heading*) Conversion of lactonic acids into lactones. **1888** *Ibid.* LIV. 251 When the lactonic acids, $CHX\langle^{CH(COOH)}_{O\cdot CO\cdot CH_2}\rangle$, obtained by the union of aldehydes with succinic acid, are boiled..the greater portion decomposes yielding..the monobasic unsaturated acids..together with the lactones. **1924** *Ibid.* CXXVI. i. 45 When it [*sc.* isocampholactone] is oxidised by means of nitric acid, there are formed nitroisocampholactone and a lactonic acid. **1968** *Jrnl. Agric. & Food Chem.* XVI. 252 (*heading*) Lactonic compounds of apricot. **1970** *Acta Chem. Scand.* XXIV. 3428/1 (*heading*) Investigation of lactonic acids in the latex of *Euphorbium canariensis* L. **1971** *Agric. & Biol. Chem.* (Tokyo) XXXV. 27 (*heading*) Studies on the new seventeen member lactonic antibiotics.

lactonization (læ:ktǒnəizēɪ·ʃən). *Chem.* [f. Lacton(e + -ization.] Conversion into a lactone.

1909 *Jrnl. Chem. Soc.* XCVI. i. 551 (*heading*) Lactonisation of acid alcohols. **1939** *Jrnl. Amer. Chem. Soc.* LXI. 3198/1 Of the three possible points of lactonization two, C–9 and C–10, should yield secondary hydroxyl groups when the lactone is opened. **1968** R. O. C. Norman *Princ. Org. Synthesis* i. 24 Consider the lactonization of ω-hydroxybutyric acid in comparison with the esterification of acetic acid by methanol.

lactonize (læ·ktǒnəiz), *v.* *Chem.* [f. Lacton(e + -ize.] *trans.* and *intr.* To change into a lactone. Hence **la·ctonized, -izing** *ppl. adjs.*

1912 *Chem. Abstr.* VI. 1133 The HCl seems to lactonize the Et pyruvate, while on pyruvic acid it acts in 2 ways, either esterifying and then lactonizing, or *vice versa*. **1939** *Jrnl. Amer. Chem. Soc.* LXI. 3198/1 The strong tendency of tetrahydrohydroxyabietic acid..to lactonize further complicated the oxidation experiments. **1939** *Ibid.* 3199/2 A rapid method for hydrolysis of lactonized dihydroabietic acid by caustic fusion has been developed. **1967** *Jrnl. Bacteriol.* XCIV. 1975/1 The mutation affects the activity of the lactonizing enzyme in one of two possible ways.

lad, *sb.*[1] **2. b.** Delete † and add later examples; used esp. in phrs. *a bit of a lad, quite a lad.* Also (*colloq.*), a spirited girl.

1913 [see *bit *sb.*[2] 4 c]. **1926** T. E. Lawrence *Seven Pillars* (1935) v. lxi. 346 He, Rahail, was quite a lad: a free-built, sturdy fellow, too fleshy for the life we were to lead. **1935** G. Ingram *Cockney Cavalcade* iv. 54 'That matcher of ours is a lad, Mum!'..'Oh, what's she been up to now?' **1935** *Punch* 13 Mar. 294/1 Women..Describe him freely to my face As quite a lad. **1960** 'H. Carmichael' *Seeds of Hate* xiii. 117 Bit of a lad is Mr. Alan Clark..running round fancy-free for years. **1969** A. Cade *Turn up Stone* iii. 76 Oh dear! The late Clive Neilson was certainly a bit of a lad, wasn't he?

c. A stable-groom of any age; also, a female one.

1848 *Sporting Life* 8 Jan. 242/1 The more important a groom is, the more mysterious, conceited, pedantic he is... The first thing a lad does now-a-days is to set up a watch, after which, if his mind incline towards horses, he buys what he calls a 'printed book' about them. **1862** *Once a Week* 1 Nov. 512/1 Judging..from the quarter in which these betting operations were carried on, they were led to think that the lad who attended to the horse Gosport,..conveyed intelligence to his master. **1894** *Strand Mag.* May 554/1 He was a good lad, tinged with the archaic stable-slang of Thessaly. **1968** D. Francis *Forfeit* iv. 49, I..called on the trainer, whom I saw almost every time I went racing... 'Did you find Sandy Willes?.. She's one of my best lads.' **1971** *Daily Tel.* 5 Apr. 11 (*caption*) A celebration snack for Specify, winner of the Grand National... Celebrating with him are Mr John E. Sutcliffe (left), his trainer, Mr Richard Bullen, his 'lad', and Mrs Sutcliffe. **1972** D. Francis *Bonecrack* 17 The elderly lad who looked after him was standing at the door. *Ibid.* 19 There had been quite a stir in Newmarket when my father had promoted her to head lad. **1972** *Guardian* 6 May 11/3 A famous owner..was watching his horse on the gallops... 'Who's that lad on 'im?' he said to the trainer. 'Oh,' came the reply, 'that's Cynthia.' **1973** *Daily Tel.* 30 Oct. 9/2 Lads and girls serving their five-year apprenticeship get pocket money..with clothing and keep paid for by the trainer.

Ladakhi (ladā·ki). Also **Ladaki**. [Native name.] **a.** A native or inhabitant of Ladakh, a district of eastern Kashmir. **b.** The language spoken in Ladakh, a dialect of Tibetan. Also *attrib.* or as *adj.*

1893 E. F. Knight *Where Three Empires Meet* viii. 123 A good-natured Ladaki, with a stolid face like that of a Chinese idol, puckered up into an inscrutable and perpetual smile. **1896** *Geogr. Jrnl.* VII. 476 Our Ladakis informed us that this man's predecessor..was promptly put in jail by the Wazir for having defrauded some Ladakis who had been trading in Tibet. **1899** A. H. Francke *Ladakhi Songs* 1st Ser. 2 The Ladakhi rhyme is, as many examples prove, a rhyme of sentence. *Ibid.* 4 The orthography of the Ladakhi and Purig dialects has always kept as near to that of the book-language as possible. *Ibid.* 5 Silent prefixed letters..are written with the Ladakhi verb. **1911** *Encycl. Brit.* XVI. 59/1 It [*sc.* Ladakh] was, however, conquered and annexed in 1834–1841 by Gulab Singh of Jammu—the unwarlike Ladakhis, even with nature fighting on their side, and against indifferent generalship, being no match for the Dogra troops. **1939** L. Petech *Study Chron. Ladakh* 6 There is urgent need of collecting and publishing the most important Ladakhi inscriptions. **1959** *News Chron.* 28 Nov. 4/2 He's a Ladakhi, really. An Indian national, like us. He became a Christian.

‖ **ladang** (ladā·ŋ). Also 8 **laddang**. [Malay.] A piece of land under dry cultivation, often a jungle clearing. Also *attrib.*

1783 W. Marsden *Hist. Sumatra* 60 *Paddee*, on Sumatra and the Malay islands, is distinguished into two sorts, *Laddang* or up-land paddee, and *Sawoor* or low-land. **1839** T. J. Newbold *Pol. & Statistical Acct. Straits of Malacca* I. iv. 119 A small quantity only is grown on the ladangs, or dry-land plantations. *Ibid.* v. 263 The ladang rice, however, is affirmed by some to be sweeter and whiter, and to keep better than the produce of the sawah. **1906** Skeat & Blagden *Pagan Races Malay Peninsula* I. i. 119 A blackberry..grows amongst the underwood ('blukar') on the old Sakai clearings ('ladang'). **1935** *Discovery* Sept. 263/1 Nomads of long habit, they [*sc.* the Sakai and the Semang] roam through their territory but no longer from one 'ladang' to another but rather from the borders of kampong to kampong as they lose their..fear of the Malay. **1954** E. D. Laborde tr. *Robequain's Malaya, Indonesia, Borneo & Philippines* vi. 94 In Indonesia most of the cultivation is done on the *ladang* system. **1958** *Times Lit. Suppl.* 19 Sept. 531/3 Nomads of the deep jungle, they clear and cultivate patches of a few acres, known as *ladangs*..; but in a year or two they move on, clearing new *ladangs* in other parts of the jungle. **1965** C. Shuttleworth *Malayan Safari* ii. 32 In a number of ladangs..they had planted hill rice.

ladder, *sb.* Add: **1. c.** (Further examples.)

1847 De Smet *Oregon Missions* 31 It was on this occasion he conceived the idea of the Catholic ladder—'a form of instruction which represents on paper the various truths and mysteries of religion in their chronological order'. **1910** *Daily Chron.* 24 Jan. 8/3 Some kind of ladder of subjects..would be a great gain. **1942** *R.A.F. Jrnl.* 18 Apr. 15 Knowledge that would serve as a ladder to further research. **1951** R. Firth *Elem. Social Organiz.* i. 29 Special attention was..given to such important matters as the breadth of the educational ladder. **1974** 'W. Haggard' *Kinsmen* x. 98 When he'd made a great fortune Duncan Gregg had gone up the ladder a little. But not very much, he was still in trade.

3. b. In knitted garments or stockings: a longitudinal strip of unravelled fabric, so called from the appearance of the threads.

1838 A. Mathews *Mem. Charles Mathews* II. xi. 246 He had been diverted by observing a fracture (or what a sempstress would term a *ladder*) in the back part of His Majesty's black-silk stockings. **1875** [in Dict., sense 3]. **1908** *Daily Chron.* 31 Dec. 4/6 Silk tights are fragile things, sadly given to 'ladders' on the least provocation. **1919** 'C. Dane' *Legend* 128 Someone ought to see that his socks were mended properly, for there was a great ladder down one ankle. **1957** M. Spark *Comforters* iv. 76 There was a ladder in her stocking. **1973** J. Cleary *Ransom* vii. 158 Sylvia looked up from examining the ladders in her stockings.

c. *Naval gunnery.* A series of range-finding shots up to or back from the target.

1922 *Grand Fleet Gunnery & Torpedo Memoranda on Naval Actions 1914–18* (Admiralty) vii. 57 The procedure generally found best by the control officers when the shots should have been straddling but nothing could be seen was to ladder down with a 200 ladder till shorts were clearly seen, and then ladder up till shorts were not seen, when the process was repeated... Although this blind ladder is extravagant in ammunition, it appears that no other course is open under similar conditions of visibility.

6. ladder-back (chair), a chair in which the back is formed of horizontal pieces of wood, suggestive of a ladder; **ladder-back(ed) woodpecker** *U.S.*, one of several North American species of woodpecker with black and white barred markings, esp. *Dendrocopos scalaris*; **ladder fern**, a fern of the genus *Nephrolepis*, which spreads by creeping rhizomes, producing new crowns; **ladder network** *Electr.*, a network having two pairs of terminals and consisting of impedances that are alternately in series and in parallel, so that the circuit diagram has the form of a ladder; **ladder polymer**, a polymer in which pairs of long straight-chain molecules are joined by recurring cross-links; **ladder-proof** *a.*, of fabrics: not liable to ladder; **ladder-stop**, at the top and toe of a stocking, a band of open-work designed to prevent a ladder; **ladder woodpecker** = **ladder-back(ed) woodpecker*.

1908 *Daily Report* 24 Aug. 8/3 Three ladder-back chairs, with cherubs and a crown, brought £46. **1923** *Daily Mail* 11 Jan. 11 Let the table be of the gate-leg variety and the chairs of the style known as ladder-backs. **1966** A. W. Lewis *Gloss. Woodworking Terms* 51 *Ladder back*, chairs made at the end of the seventeenth century with horizontal slats across the back like a ladder. **1973** J. Burrows *Like an Evening Gone* i. 14 Greta was sitting on a ladderback chair by the kitchen table. **1884** E. Coues *Key to N. Amer. Birds* (ed. 2) 485 *Picoides americanus*.. Ladder-backed Three-toed Woodpecker. **1917** T. G. Pearson *Birds Amer.* II. 149/1 The Ladder-back Woodpeckers are divisible into three regional varieties, the American, the Alaska.., and the Alpine. **1964** A. Wetmore et al. *Song & Garden Birds N. Amer.* 92/1 Ladder-backed woodpecker. *Dendrocopos scalaris*. Vast stretches of hot, treeless desert seem a curious habitat for a woodpecker. Yet to the ladderback such country is home. **1884** W. Miller *Dict. Eng. Names Plants* 222/2 *Nephrolepis cordifolia*, Ladder Fern, of New Zealand. **1893** G. Schneider *Bk. Choice Ferns* II. xliii. 583 Nephrolepis... Ladder Ferns... This genus..belts the world in the Tropics, passing a little beyond them both north and south. **1951** *Dict. Gardening* (R. Hort. Soc.) III. 1365/1 *Nephrolepis*,.. Ladder fern. A genus of about 35 species of handsome ferns, widely dispersed over the tropics. **1969** *Coast to Coast 1967–68* 49 Beyond that was the swamp—tea-trees, paperbarks, huge ladder ferns. **1930** A. C. Bartlett *Theory Electr. Artificial Lines* iii. 41 The theory of a general ladder network, in which all the elements may have arbitrary values, will first be considered and..from it will be derived a class of symmetrical ladder artificial lines, of which the T and Π section lines are but simple cases. **1966** H. J. Reich et al. *Theory & Applications Active Devices* xviii. 543 Ladder-network oscillators ..consist of a voltage-inverting amplifier and a ladder type resistance–capacitance feedback network that usually has three or more similar sections. **1971** *New Scientist* 24 June 761/2 The use of conventional straight chain polymers seems to be restricted by an upper temperature limit of about 550°C, but the ladder polymers (so-called because of their integral cross-linked structure) offer more exciting possibilities. **1974** *Sci. Amer.* Mar. 66/3 Ladder polymers, or double chains, are found in amphibole minerals, such as one form of asbestos. **1927** *Observer* 3 Apr. 25 Celanese cami-bockers... In Ladder-proof Self Stripe. **1962** *Economist* 2 June 897/1 The new answers to feminine prayer [*sc.* a new type of seamless stockings] are said..to be ladderproof, although not hole-proof. **1931** *Daily Express* 15 Oct. 12/7 (*caption*) Pair of Lady's Artificial Silk Hose with ladder-stop tops. **1962** *Which?* Apr. 114/1 Most of the leading firms sell a style with a band of open-work knitting at the welt, known as a ladder-stop. **1870** *Amer. Naturalist* III. 474 The resident species not found westward [of the Colorado Valley] were the Ladder Woodpecker (*Picus scalaris*), the White-bellied Wren, [etc.].

ladder, *v.* Delete ? *Obs.* and add later examples. Also, to furnish with a fish ladder.

1901 J. Black *Illustr. Carpenter & Builder Ser.: Scaffolding* 67 The stack was laddered from the bottom to the top with a series of ladders. **1901** *Chambers's Jrnl.* Sept. 585/2 When Mr. Grant..laddered the Moriston falls.. the Crown claimed and gained the new fishings. **1923** *Daily Mail* 22 June 5 Having just laddered the spire of Truro Cathedral, he found every crevice crammed with jackdaws' nests.

2. *intr.* Of garments, esp. stockings: to develop ladders as the result of the breaking of a thread or threads. Also *trans.* Hence **la·ddering** *vbl. sb.*

1922 *Daily Mail* 14 Nov. 12 (Advt.), Your stockings cannot ladder. Laddering and damage to stockings..are entirely obviated. **1927** W. Deeping *Doomsday* viii. 78 At the last moment a stocking had 'laddered'. **1963** A. J. Hall *Textile Sci.* iii. 152 The demand for excessive sheerness and transparency in ladies' stockings is the root cause of the tendency to ladder. **1973** 'S. Woods' *Enter Corpse*

171 Her dress [was] crumpled, and both her stockings were laddered.

3. *Naval gunnery.* To fire shots in a ladder (sense *3 c).

1922 [see *ladder *sb.* 3 c]. **1959** *Chambers's Encycl.* VI. 662/2 In the case of surface vessels, finding the accurate gun range, which is done by 'laddering' i.e. increasing or decreasing the range of successive salvoes until the target is crossed, is not difficult if the enemy maintains his course and speed.

laddic (læ·dik). *Electronics.* [f. *ladd(erlog)ic*.] A ladder-like device consisting of a rectangular block of a magnetic ferrite containing a line of rectangular apertures, the cross pieces and side pieces being of the same cross-sectional area and having wires passing round them in such a way that the device may be used as a logic element.

1959 Gianola & Crowley in *Bell Syst. Techn. Jrnl.* XXXVIII. 45 The Laddic is a ladder-like structure cut out of a rectangular hysteresis-loop ferrite. *Ibid.* 46 Any Boolean function can be realized as the output of a single Laddic of suitable length. *Ibid.* 50 The operation of the Laddic as an AND gate. **1963** *Engineering* 25 Jan. 170/3 Logical circuit applications of both toroidal cores and multi-apertured elements—the transfluxor and laddic—are discussed.

laddie. (Examples in wider use.)

1919 Wodehouse *Damsel in Distress* xv. 171 'I've got a headache.' 'I thought you would have, laddie, when I saw you getting away with the liquid last night.' **1962** *Coast to Coast 1961–62* 66 'Now get yourself a beer,' he said. 'Pour a couple of beers, laddie.' **1973** J. Porter *It's Murder with Dover* iii. 22 I've been ready for the last bleeding half-hour, laddie!

la·ddish, *a.* [f. Lad *sb.*[1] + -ish[1].] Of or pertaining to a lad or lads; like a lad. Also **la·ddishness.**

1841 S. Bamford *Passages in Life of Radical* I. xiv. 91 A young officer..very laddish, and with limbs long enough for windmill arms. **1886** *Wesleyan-Methodist Mag.* 63 Want of sympathy with..the laddishness of lads. **1907** *Daily Chron.* 24 Oct. 8/3 Missing the laddish laugh, the boisterous gaiety, which they had known aforetime.

laddo (læ·do). *colloq.* (orig. *Ir.*). Also **lado.** [f. Lad *sb.*[1]] Lad, boy.

1870 [see *boyo]. **1939** Joyce *Finnegans Wake* 404 Sure, he's lightseyes, the laddo! **1968** J. Wainwright *Edge of Extinction* 57 She's..out of range. The laddoes at the Kremlin'll know that. **1970** R. Hill *Clubbable Woman* v. 140 Your intruders'll all turn out to be like that laddo last night. **1973** —— *Ruling Passion* ii. vi. 125 You're not seriously suggesting that Lewis wasn't killed by laddo, but by someone else.

la-di-da, *v.* Also **lah-de-dah.** [Cf. the sb.] *intr.* To use affected manners or speech.

1901 *N. & Q.* 6 July 20/2, I like to la-di-da with the ladies. **1930** *John o' London's* 15 Mar. 907/3 There is perhaps too much 'lah-de-dahing about' when royalty is concerned.

Ladik (ladī·k). The name of a village in Turkey, formerly Laodicea, used *attrib.* to describe a type of prayer rug made in the district. Also *ellipt.*

1900 J. K. Mumford *Oriental Rugs* x. 155 Ladik rugs resemble..those of Kulah... The Kulah small stripes, however, are not often found in the Ladiks. **1931** A. U. Dilley *Oriental Rugs & Carpets* vi. 165 Among later weavings Ladik prayer rugs display a versatile art that ranges in architectural pattern from impressive columns to dwarf apexes. **1972** *Country Life* 23 Nov. 1412/2 Antique Turkish Ghiordes and Ladik prayer rugs.

Ladin (lădī·n). [L. *Latīn-us, -um.*] The Rhæto-Romanic dialect spoken in the Engadine in Switzerland, closely related to Romansh.

1877 A. H. Keane tr. *Hovelacque's Sci. of Lang.* v. 238 Known also as the language of the Grisons, the Rheto-Romance, the Rumonsh, and Rumansh. But it seems best to call it simply *Ladin*, with Ascoli, who has recently devoted an important work to its elucidation. **1879** *Encycl. Brit.* VIII. 213/2 The language [in the Engadine] is a dialect known as 'Ladin'. **1880** *Ibid.* XI. 205/1 The remainder [of the inhabitants of the Grisons] use the Romansch or the Ladin dialect. **1969** *Daily Tel.* 29 Jan. 16/5 The first book to be translated into English from Ladin, one of the Romansh group of languages which is spoken in the canton of Grisons, Switzerland, will be published..tomorrow.

ladino[1] (ladī·no). [Sp.] **1.** A vicious or unmanageable horse, steer, etc.; a stray animal. Also as *adj.*, wild, vicious, cunning.

1863 H. W. Bates *Naturalist on River Amazons* II. iv. 265 The old Indians told us..the turtles were 'ladino' (cunning), and would take no notice of the beating a second day. **1891** *Dialect Notes* I. 191 *Ladino*, in Spanish, learned, knowing Latin; then crafty, cunning. In Texas as a noun, a vicious, unmanageable horse, full of cunning and tricks. **1929** J. F. Dobie *Vaquero of Brush Country* ii. 14 They were all outlaws, *ladinos*, wild as bucks, cunning and ready to fight anything that got in front of them... Among them were wrinkle-necked maverick cows and bulls that had never had a loop tossed over their heads.

1942 Berrey & Van den Bark *Amer. Thes. Slang* §916/2 *Ladinos, outlaw cattle*, unbranded strays.

2. In Central America, a mestizo or a white person. Also *attrib.* or as *adj.*

1877 *Encycl. Brit.* VI. 682/2 The inhabitants of Cuba are divided into four classes..; those under servitude, constituting the fourth class, divided into the *bozales*, those recently brought from Africa,—the *ladinos*, those imported before the law of 1821 prohibiting the slave trade,—and the *criollos*, those born on the island. **1902** *Ibid.* XXXII. 395/1 The number of Ladinos (whites and persons of mixed blood) [in Salvador in 1887] was returned at 772,200, and of Indians at 234,648. **1934** A. Huxley *Beyond Mexique Bay* 69 Whites and *ladinos* were conspicuously absent from the ranks. *Ibid.*, These Quichés and Cakchiquels from the hills are as foreign in the white and *ladino* capital as Nepalese in the Punjab. *Ibid.* 70 *Ladino* housewives stood bargaining at the stalls. **1959** *Times* 17 Jan. 7/7 The basic division in Chiapas is that between Indian and *ladino*, the *ladino* being a white or a *mestizo*, who follows the customs of the *patrón* and of western civilization. **1973** 'F. Clifford' *Amigo, Amigo* v. 46 She was dressed *ladino* style, blouse and skirt, and was lighter skinned than pure Indian.

3. (Usu. with capital initial.) A language based on Old Spanish and written in modified Hebrew characters, used by some Sephardic Jews, esp. in Mediterranean countries. Also *attrib.* and *Comb.*

1889 in *Cent. Dict.* **1932** C. Roth *Hist. Marranos* xiii. 324 A few works..were printed..in Ladino (or Spanish in Hebrew characters) in the Levant, at Smyrna or at Salonica. **1948** [see *Judæo-]. **1949** *Spectator* 4 Nov. 595/2 A dialect of Hebrew-cum-Spanish called Ladino is still spoken by many Sephardic Jews everywhere; it is common at the Holland Park synagogue in London.. whose members mostly come from the Middle East. **1959** *Israel Digest* 23 Jan. 6/3 Another important project..is the preparation of a Ladino dictionary. **1965** *Listener* 30 Sept. 490/2 He records a meeting with three Ladino-speaking Jews in the railway station in Sofia. **1972** O. Sela *Bearer Plot* xvi. 101 Ashraf and I are both Sephardic. We both speak Ladino which is very close to Spanish.

ladino[2] (lădī·no). [It.] In full, *ladino clover*. A large fast-growing variety of white clover (*Trifolium repens*), native to northern Italy and cultivated elsewhere, esp. in the U.S.A., as a fodder crop.

1924 A. G. Erith *White Clover* xii. 139 In Italy white clover is named Trifoglio ladino or Ladino clover. **1931** *N.Z. Jrnl. Agric.* XLII. 83 In many ways the Italian Ladino type seems to be related to the large-leaved New Zealand Wild White No. 1... The New Zealand form is.. altogether the more valuable pasture type. **1937** Cox & Jackson *Crop Managem. & Soil Conservation* xxiv. 385 *Ladino* is a large-growing type of white clover that provides much more pasture per acre and is of value for hay. Ladino is a native of northern Italy and is established in importance in Idaho, Oregon, and Washington and is gaining in importance in New England. **1950** Carroll & Krider *Swine Production* xviii. 367 Ladino clover hay contains 19 per cent protein. **1956** Gillespie & Hathaway *Textbk. Gen. Agric.* viii. 151 *Ladino* or *Giant White Clover*, also called Lodi, is Italian in origin and has all the appearance of a giant form of the other white clovers... Its demerit is that it is not frost-hardy and so cannot stand British winters. **1969** J. Janick et al. *Plant Sci.* xix. 381/1 White clover..includes the robust ladino variety from Italy.

ladle, *sb.* Add: **1. b.** In Scottish churches: a similar instrument consisting of a wooden box at the end of a long wooden handle used for taking up the collection and communion tokens.

1813 W. Leslie *Gen. View Agric. Nairn & Moray* 412 The elders make these collections by going round to each with a ladle or small box with a handle to it, when the public worship is concluded. **1830** *2nd Rep. Evidence Sel. Comm. State of Poor in Ireland* § 3369 in *Parl. Papers* VII. 459 The elders carrying about what they call a ladle. **1871** W. Alexander *Johnny Gibb* xi. 81 The elders seized the ladles..and perambulated the kirk. **1929** *Life & Work* Oct. 232/2 A ladle preserved in the parish church of Foulis Easter. **1960** *Press & Jrnl.* (Aberdeen) 26 Apr., Long may they continue to use the ladle and metal communion tokens.

ladle, *v.* Add: *Fig.* examples with *out*. Also const. *up*.

1851 H. Melville *Moby Dick* III. xlviii. 287 Stubb was lustily singing out for some one to ladle him up. **1872** W. H. G. Kingston *On Banks Amazon* iii. 93 Wooden spoons were served to enable us to ladle up the soup. **1913** R. Brooke *Let.* 8 Sept. (1968) 508 But it's absurd to ladle out indiscriminate praise, as most people do. **1969** *Listener* 27 Feb. 278/2, I..was concerned at the way the present system..ladles out routine scientific and technological qualifications which one in five graduates later find are not useful to them.

ladro·nism. *Hist.* [f. Sp. *ladrón* (see Ladrone), a hostile Filipino, an insurgent + -ism.] In the Philippine Islands, organized resistance to law or authority among the native population.

1902 *Outlook* (N.Y.) LXXII. 298/1 A local police and an insular constabulary system have been created, and ladronism, or organized robbery and brigandage,..has almost disappeared. **1903** *Daily Chron.* 26 Dec. 5/6 Ladronism had also been successfully treated by the Courts.

lady, *sb.* **4. b.** (*a*) Delete 'now confined to poetic and rhetorical use', and add further examples.

1914 G. B. Shaw *Pygmalion* (1916) i. 107 *The Flower Girl.* Thank you kindly, lady. **1924** I. Gershwin (*song title*) Lady, be good. **1953** *Manch. Guardian Weekly* 10 Sept. 7 Why, lady, take route 128. **1972** P. Ruell *Red Christmas* xiv. 148 'Lady,' he said, 'you talk sense. Just remember, it's guns that count.' **1974** M. Babson *Stalking Lamb* xxiii. 176, I *know* it can, lady. It won't be the first time.

e. *Lady Bountiful* (see Bountiful *a.* 1); *a lady in the case*, indicating that the key to the problem is a lady (cf. *cherchez la femme); *the Old Lady* (*in* or) *of Threadneedle Street*, the Bank of England; *the lady of the house*, the mistress of a household; a housewife.

1727 J. Gay *Fables* I. l. 172 And when a lady's in the case, You know, all other things give place. *c* **1793** Jane Austen *Volume Second* in *Minor Works* (1954) 136 We had scarcely paid our Compliments to the Lady of the House. **1797** J. Gillray *Caricature* 22 May, Political Ravishment, or The Old Lady of Threadneedle-Street in danger! **1816** Jane Austen *Emma* III. xiv. 254 It was with difficulty that she could summon enough of her usual self to be the attentive lady of the house, or even the attentive daughter. **1820** *Black Dwarf* IV. 36 Van went to wheedle—the street of Threadneedle, To get him, poor dog, a loan;.. He ask'd the old lady to cash him a bill. **1821** Byron *Don Juan* v. xix. 243 'Ay,' quoth his friend, 'I thought it would appear That there had been a lady in the case.' **1850** *Househ. Words* 6 July 337 (*heading*) The Old Lady of Threadneedle Street. **1861** Mrs. Beeton *Bk. Househ. Managem.* 9 The more usual plan is for the lady of the house to have the joint brought to her table, and afterwards carried to the nursery. **1863** A. Trollope *Rachel Ray* I. xiii. 260 Luke, is there no young lady in the case? **1884** *Peel City Guardian* No. 26. 2/1 The rest of the 'Old Lady in Threadneedle-street' remained unbroken. **1909** W. S. Gilbert *Fallen Fairies* ii. 37 In all the woes that curse our race There is a lady in the case. **1958** R. Genders *Pansies, Violas & Violets* x. 100 Those who have retired will be able to give the plants their full attention, whilst those who have to go out to work each day may have to entrust the care of the plants to the lady of the house. **1971** *Guardian* 19 Aug. 9/3 Door-to-door sales people were asked why the opening question is always, 'Is the lady of the house home?' **1974** G. Vaizey *Tangled Web* ii. 26 He..is highly respected by the Old Lady of Threadneedle Street.

f. *pl.* Designation of a public convenience for females. Freq. *ladies'*, and with capital initial.

1918 'K. Mansfield' *Jrnl.* (1954) 140 Also, when she goes to the 'Ladies', for some obscure reason she wears a little shawl. **1936** R. Campbell *Mithraic Emblems* 120 No 'Ladies' here or 'Gentlemens' are seen For most of you to hesitate between. **1938** G. Greene *Brighton Rock* iii. i. 98 The white steps down to the ladies'. **1939** C. Morley *Kitty Foyle* 36 How to get undressed in a Pullman berth, how to find the Ladies. **1944** T. Rattigan *While Sun Shines* i. 190, I lost the plans of the Station Defence... We found them again all right. I'd only left them in the Ladies. **1965** G. Melly *Owning-Up* vi. 64 'They're no good,' he'd tell us as two of them swayed past on their way to the ladies. **1974** D. Meiring *President Plan* vi. 42 Comunicado Number Two..was found, as anonymously advised, in the Ladies' of a San Agustín restaurant.

6. b. *Lady Luck* = Fortune *sb.* i.

1932 M. Short (*title*) Lady Luck in 1941. **1936** C. Sandburg *People, Yes* 165 Yes, get Lady Luck with you and you're made. **1961** T. Henrot *Belgium* 119 A thousand ways of flirting with Lady Luck.

d. *lady wife.* (*Arch.* or *genteel.*)

1840 Dickens *Lett.* (1969) II. 7, I wish I could send you some autographs..but I find..that my lady wife has been bestowing them upon her friends. **1895** C. M. Yonge *Long Vacation* xxviii. 292 Mr. White, in his joy at possessing his graceful lady wife, had spared no expense. **1969** *Listener* 27 Mar. 417/3 We don't think you've laid enough emphasis on the colonel's foresight, courage, and heroic lady wife. **1971** 'A. Gilbert' *Tenant for Tomb* viii. 142, I don't know how far your lady wife's in your confidence.

e. *Lady Macbeth*, with allusion to the character in Shakespeare's play *Macbeth*: a remorseless or melodramatic woman, usu. leading or assisting a weak man.

1876 Trollope *Prime Minister* i. xi. 169, I feel myself to be a Lady Macbeth, prepared for the murder of any Duncan or any Daubeny who may stand in my lord's way. **1919** Kipling *Years Between* 92 A boy drowning kittens Winced at the business; whereupon his sister (Lady Macbeth aged seven) thrust 'em under. **1969** M. Pugh *Last Place Left* iv. 26 'I know you're up to something,' Nell repeated. 'You're taking all this far too calmly.' 'All right, Lady Macbeth.' **1974** J. Mann *Sticking Place* x. 153 Hasn't there been enough killing? I am no Lady Macbeth.

13*. *N. Amer.* A female harlequin duck, *Histrionicus histrionicus.* See *lord and lady* (*duck*) (*lord *sb.* 16).

1792 G. Cartwright *Jrnl.* I. p. xii, *Lady*, a water-fowl of the duck genus, and the hen of the lord.

14. a. *lady doctor* (earlier and later examples), *friend* (later examples), *novelist.*

1858 *English Woman's Jrnl.* I. 90 As she went through the streets,..rude cries of 'Come on, Bill! let's have a good look at the lady-doctor!' would meet her ears. **1873** C. M. Yonge *Pillars of House* IV. xlii. 219 To be a lady-doctor was surely her vocation! **1879** Geo. Eliot *Let.* 18 Mar. (1956) VII. 117 This week for the first time I am going to see a lady friend. **1895** Hardy *Jude* v. iii. 343, I couldn't very well tell it to your lady friend. **1912** A. Brazil *New Girl at St. Chad's* iv. 69 'We have a lady doc-

tor, you see,' said Ruth, 'and she's so jolly.' **1923** Lady-novelist [see *Apolline *a.*]. **1928** R. Campbell *Wayzgoose* ii. 35 And still new-comers to the Wayzgoose throng And lady-novelists a thousand strong. **1931** *Weekend Rev.* 17 Oct. 496/1 The night-watchman, after taunting Larry with his inexperience in affairs of the heart, is obliged to stand by and see the youngster making rapid headway in the affections of his own lady-friend. **1961** *Listener* 30 Mar. 574/3 If 'women novelists' are to become 'lady novelists' as a matter of course, I give notice..that in future, when reviewing the novels of male writers of fiction, I shall make a point of referring to these writers as 'gentlemen novelists'. **1975** C. Aird *Slight Mourning* vi. 58 The lady doctor had arrived.

15. c. *lady-laden* (earlier example).

1870 Tennyson *Holy Grail* 54 Where the long Rich galleries, lady-laden, weigh'd the necks Of dragons clinging to the crazy walls.

16. b. lady-of-the-night, an evergreen shrub, *Brunfelsia americana*, native to the West Indies and bearing white and yellow flowers which are particularly fragrant at night; **lady orchid, orchis**, *Orchis purpurea*, a European and Western Asian orchid with white and reddish-purple flowers.

1924 L. H. Bailey *Man. Cultivated Plants* 667 *B[runfelsia] americana*, L. Lady-of-the-Night. **1959** M. M. Kaye *House of Shade* xiii. 167 Lash..had taken her down into the garden—ostensibly to look at the nocturnal flowering Lady-of-the-Night which grew in profusion in a bed some distance from the house. **1960** *Harper's Bazaar* July 80/2 The scent of frangipane and lady of the night. **1855** A. Pratt *Flowering Plants & Ferns Gt. Brit.* V. 208 Kentish country people call it [*sc. O. fusca*] the Lady Orchis; and..though its form is not very suggestive of its name, yet..there exists some slight similarity in each blossom to a lady attired in wide-spread gown and close bonnet. **1933** M. J. Godfery *Monogr. & Iconogr. Native Brit. Orchidaceæ* 171 *Orchis purpurea* Huds. Brown-winged Orchid, Maids of Kent, Lady Orchid. **1951** V. S. Summerhayes *Wild Orchids Brit.* xiii. 251 (*heading*) Brown-winged or lady orchid (*Orchis purpurea*). *Ibid.* 252 The lady orchid belongs to the Southern Eurasian Element of the British orchid flora.

17. a. Ladies' Aid (Society), † (*a*) *U.S.* (*Obs.*) during the American Civil War, a women's organization devoted to sending garments, bandages, etc., to the soldiers; (*b*) *N. Amer.*, an organization of women who support the work of a church by fund-raising, arranging social activities, etc.; **ladies' cabin, car, carriage**, on public transport, a compartment, etc., reserved for ladies; **ladies' cloakroom**, a cloakroom or lavatory for ladies; **lady's horse**, a horse trained to carry a lady riding side-saddle; similarly **lady's hunter**; **ladies' night**, a function at a men's club, etc., to which ladies are invited; **ladies' room** = **ladies' cloakroom*; **lady's waist** *Austral. colloq.*, a small gracefully-shaped glass; a drink served in such a glass.

1866 F. Moore *Women of War* 214 Mrs. Wittenmeyer, as president of the Ladies' Aid Society of Iowa. **1873** *Sentinel* (Woodstock, Ont.) 5 Dec. 3/1 The Ladies' Aid Society in connection with the Baptist Church in this Town will hold a social this evening. **1893** 'O. Thanet' *Stories Western Town* 185 The furnishing of the church.. is in charge of the Ladies' Aid Society. **1895** *Times* (Niagara-on-the-Lake, Ont.) 4 Apr. 1/2 The Social [was] under the auspices of the Ladies Aid of the Methodist Church. **1908** L. M. Montgomery *Anne of Green Gables* xiv. 143 She had taken it off..when returning from the Ladies' Aid. **1913** E. H. Porter *Pollyanna* iii. 21 Part of the Ladies' Aid wanted to buy me a black dress and hat, but the other part thought the money ought to go toward the red carpet they're trying to get,—for the church, you know. **1964** *Calgary Herald Mag.* 21 Mar. 8/9 The 'Apron Social' and tea given in the basement of the Knox Church last evening under the auspices of the Ladies' Aid Society of the congregation. **1832** E. Grosvenor *Diary* July in G. Huxley *Lady Elizabeth & Grosvenors* (1965) vi. 124 There were 20 fellow-passengers, so that the Lady's cabin was utterly untenable. **1925** E. H. Young *William* iv. 42 She sat down on a velvet-covered couch in the ladies' cabin. **1842** Dickens *Amer. Notes* I. iv. 145 There are no first and second class carriages..but there is a gentlemen's car and a ladies' car. **1847** F. A. Kemble *Let.* 29 May in *Rec. Later Life* (1882) III. 183 From Liverpool to Crewe I had companions in the ladies' carriage in which I was. **1860** E. Hall *Diary* 30 July in O. A. Sherrard *Two Victorian Girls* (1966) 263, I am thankful today that 'Ladies' carriages have been given up in *our* country! **1922** E. H. Young *Bridge Dividing* iii. ix. 289 'I have to catch a train.'..'Be careful to get into a ladies' carriage, Henrietta.' **1918** A. Bennett *Pretty Lady* xxiii. 157 She hurried..to the ladies' cloakroom, got her wraps. **1814** Jane Austen *Mansf. Park* I. iv. 71 Fanny should have a regular lady's horse of her own. **1894** Kipling *Day's Work* (1898) 46 An absolutely steady lady's horse—proof against steam-rollers, grade-crossings, and street processions. **1938** D. A. Houblon *Side-Saddle* vii. 64 In Victorian days and even later a lady's horse to be perfect had always to canter with the off fore leading. **1948** *Horseman's Year* 172 (*heading*) Royal Welsh Agricultural Society show... Hunters... Ladies' Hunters. To be ridden side-saddle. **1955** *Horse & Pony Ann. Illustr.* 1954–55 iii. 101 The leading ladies' hunters were Cufflink, Earmark,..Mighty Grand. **1889** G. B. Shaw *London Music 1888–89* (1937) 266 An invitation from the Grosvenor Club to their 'ladies' night' at the Grosvenor Gallery. **1970** K. Giles *Death in Church* i. 18 The atmosphere of a Masonic ladies' night. **1970** G. Greer *Female Eunuch* 142 On ladies' nights..men embrace and fool about. **1880** 'E. Leathes' *Actor Abroad* xviii. 226 Many of them retire to the ladies' room, and changing their costume for evening dress reappear in the ball-room. **1927** D. L. Sayers *Unnatural Death* xi. 120 The attendant in the Ladies' Room. **1948** G. Vidal *City & Pillar* (1949) iii. 64 'I think,' said Emily, when she came back from the ladies' room, 'I think that we should go over to that room on the left and get a drink.' **1971** D. Eden *Afternoon Walk* ix. 124 I'll go to the ladies' room while you do. **1934** *Bulletin* (Sydney) 4 Apr. 20/1 But a daintier goblet I never fingered than the hourglass shape of a lady's waist. **1963** A. Lubbock *Austral. Roundabout* 59 A pony is drunk out of a small glass called a lady's waist.

b. lady's delight (earlier example); **lady's ear-drop**, the common fuchsia.

1843 L. M. Child *Lett. from N.Y.* i. 2, I am like the Lady's Delight, ever prone to take root. **1829** A. H. Lincoln *Familiar Lect. Bot.* xxv. 145 The Ladies'-ear-drop (*Fuschia*,) is a beautiful exotic. It has a funnel-form calyx of a brilliant red colour... This plant is a native of Mexico, except one species brought from the island of New Zealand. Ten species are said by horticulturists to be cultivated. **1882** H. Friend *Gloss. Devon. Plant Names* 33 Lady's Eardrops. The common garden Fuchsia. Still employed by the older people, but not so commonly as of yore. **1887** M. E. Wilkins *Humble Romance* 195 He cut lavishly sprays of dioletra, or lady's ear-drop, snowballs, daffodils. **1908** L. M. Montgomery *Anne of Green Gables* i. 1 A little hollow, fringed with alders and ladies' eardrops. **1967** D. G. Hessayon *Be your own House Plant Expert* (ed. 2) 18/2 Fuchsia (Lady's Eardrops). Spring and summer flowering plant with pendant blooms.

lady-bird. Add: **3.** The pintail duck, *Anas acuta.*

1885 C. Swainson *Provincial Names & Folklore Brit. Birds* 155 Pintail (*Dafila acuta*)... Also called Cracker, Winter duck, Lady bird (Dublin Bay). From its grace of form. **1917** T. G. Pearson *Birds Amer.* I. 128 Pintail. *Dafila acuta*... Winter Duck; Lady-bird; Long-necked Cracker. **1968** C. F. Jackson *Brit. Names of Birds* 61 Pintail..lady bird Ire[land].

lady-bug. *dial.* and *U.S.* = Lady-bird.

1699 *Phil. Trans. R. Soc.* XXI. 50, I have happened upon..three or four sorts of Lady-Bugs. **1787** [see Lady *sb.* 16]. **1844** 'J. Slick' *High Life N.Y.* II. 30 Like lady bugs round a full blown rose. **1886** *Harper's Mag.* June 45/2 We may discover lady-bugs—small red or yellow and black beetles—among our vines. **1889** M. E. Bamford *Up & Down Brooks* 49 Very frequently one will find a lady-bug with the spider. **1910** *N.Y. Even. Post* 4 Apr. (Th.), Los Angeles, April 1.—Millions of ladybugs are receiving free transportation..to the melon fields of the Imperial valley. **1972** Swan & Papp *Common Insects N. Amer.* xx. 403 Lady Beetles (Coccinellidae). The common British name for these beetles, 'ladybird', is still used to some extent in America; 'ladybug' is probably the most familiar name.

lady's finger. **2. a.** Delete ? *Obs.* and add later examples.

1864 'Mark Twain' in Harte & 'Twain' *Sk. Sixties* (1926) 138 'Lady-fingers'..suggestive of..soft dalliance with pastry, ices, and sparkling Moselle. **1906** *Westm. Gaz.* 30 Apr. 10/1 Lady-fingers and ice-cream. *a* **1938** T. Wolfe *Web & Rock* (1939) i. ii. 22 Was..President Taft the easy prey of lady fingers? **1942** C. Barrett *On Wallaby* iii. 39 'Nana' nibbled at a sponge 'lady's finger'. **1974** H. McCloy *Sleepwalker* vi. 108 We adjourned for tea and ladyfingers at quarter of six.

c. (Example of *c.*)

1876 J. Burroughs *Winter Sunshine* vii. 154 Others are indeed lady apples..like the egg-drop and lady-finger.

d. = *okra.

1905 W. R. Beattie in *U.S. Dept. Agric. Farmers' Bull.* no. 232. 12 There are three general types of okra, viz., tall green, dwarf green, and lady finger. **1935** M. Morphy *Recipes of all Nations* 771 Chicken with 'Ladysfingers'... Put in..a few *bamies*, or 'ladysfingers'. This vegetable.. is a variety of okra, or *gombo* [*sic*], but smaller than the okra of the Southern United States. **1972** Y. Lovelock *Vegetable Bk.* 147 Indians call it [*sc.* okra] *bhindi* and, in their restaurants, *ladies' fingers*, although this is a name applied to many other plants, including grapes, bananas and a kidney vetch.

lady's-maid, *v.* [See Lady *sb.* 17.] *trans.* To wait on (one) as a lady's maid. Also **lady's-maiding** *vbl. sb.*

1914 W. De Morgan *When Ghost meets Ghost* i. xxxii. 392 Maggie goes with her, to lady's-maid her. **1923** U. L. Silberrad *Lett. J. Armiter* v. 115 It prevents her suffering under his lady's maiding.

lady's slipper. **2.** (Earlier examples.)

1836 A. H. Lincoln *Familiar Lect. Bot.* (ed. 5) xv. 101 The Impatiens of the garden is sometimes called Ladies'-slipper, sometimes Balsamine. **1874** B. F. Taylor *World on Wheels* ii. iv. 220 The lady-slippers dance upon the air, while wild Sweet Williams stand admiring by.

Lætare (līt̄ēə·rī, ləitā·rī). [L., imper. sing. of *lætārī* to rejoice: see quot. 1921.] *attrib.* (with *Sunday*) or *ellipt.* Mid-Lent Sunday.

1870 Brewer *Dict. Phr. & Fable* 491/1 *Lætare Sunday.* The fourth Sunday in Lent is so called from the first word of the Introit. **1886** E. L. Dorsey *Midshipman Bob* 6 We would notify the Catholic neighbours the day before 'Lætare-Sunday' (as we called our Mass-day). **1921** *Spectator* 19 Feb. 243/2 The Fourth Sunday of Lent or Mothering Sunday, called 'Laetare' from the opening word of the Introit. **1951** R. Knox *Stimuli* iii. vi. 100 The Church gives us Laetare Sunday in the middle of Lent. **1974** *Oxf. Dict. Chr. Ch.* (ed. 2) 792/2 Laetare Sunday..is also known as Mothering Sunday and Refreshment Sunday.

lævodopa, var. *levodopa.

lævulosan, levulosan (lī·viŭlǒsæn). *Chem.* Also **-ane** (see next). [ad. F. *lévulosane* (A. Gélis 1860, in *Compt. Rend.* LI. 333), f. *lévulose* Lævulose, levulose.] **a.** An anhydride, $C_6H_{10}O_5$, of lævulose. ? *Obs.*

1862 H. Watts tr. *Gmelin's Hand-bk. Chem.* XV. 338 Cane-sugar quickly heated to 160°, and kept in the melted state at that temperature for a moderate time, is converted into..lævulosan and dextro-glucose. **1887** *Encycl. Brit.* XXII. 624/1 It [*sc.* lævulose] fuses at 95° C.; at 170° it passes into lævulosan, $C_6H_{10}O_5$, analogous to glucosan.

b. Any polysaccharide composed chiefly of lævulose residues. Cf. *levan.

1913 *Chem. Abstr.* VII. 3511 (*heading*) Hydrolysis of levulosans and application to vegetable analysis. **1936** [see *fructosan]. **1966** Nowakowski & Clarke tr. *Kretovich's Princ. Plant Biochem.* ix. 295 Higher plants also contain enzymes which catalyse the conversion of sucrose into a variety of polyglucosides and polyfructosides. In some plants (members of the Gramineae and Liliaceae) levulosans..function as transport carbohydrates.

lævulosane (in Dict. s.v. Lævulose, levulose). For '(see quot.)' substitute: obs. var. *lævulosan, levulosan a.

lævulose, levulose. For 'fruit-sugar' substitute: The naturally occurring (lævorotatory) form of fructose, D(−)-fructose. Add further examples.

Lævulose is not a form of glucose as the latter word is now used.

1902 *Encycl. Brit.* XXII. 721/1 Glucose and fructose (lævulose)—the two isomeric hexases of the formula $C_6H_{12}O_6$ which are formed on hydrolysing cane sugar. **1948** W. Pigman *Chem. Carbohydrates* xv. 605 Hydrolysis of inulin by enzymes leads to a practically quantitative yield of levulose (D-fructose). **1974** *Nature* 10 May 194/3 Although it is true that some bacteriologists are extremely conservative in the names they use for carbohydrates, surely nobody now uses 'levulose'..in preference to 'fructose' these days.

Lafite (lafī·t). Also (now considered erron.) **Lafitte.** [Fr., place-name.] Used as the designation of the claret produced and bottled at Château Lafite, in the Médoc district of the Gironde, France.

1707 [see Claret *sb.*[2] (*a.*) 1 b]. **1792** [see *Haut-Brion]. *a* **1845** Barham *Ingol. Leg.* (1847) 3rd Ser. 186 Chambertin, Chateau Margaux, La Rose, and Lafitte. **1871** M. Collins *Marquis & Merchant* II. viii. 237 We'll have some Lafitte. **1888** *Athenæum* 21 Apr. 499/1 Your noble magnum of *Lafitte* E'en Rothschild would have deem'd a treat. **1920** G. Saintsbury *Notes on Cellar-Bk.* iv. 48 A '71 Lafite which hailed from Pall Mall. **1931** S. Jameson *Richer Dust* iv. 77 His glass about to be filled for the third time with an excellent Lafitte. **1963** A. L. Simon *Guide Good Food & Wines* 732/1 The spelling of *Lafite* with two *f*'s and two *t*'s instead of one occurs in old records and even on old labels, but all forms are now obsolete other than *Lafite*. There are other Châteaux in the Gironde, however, bearing the same name with different spellings. **1973** *Country Life* 19 Apr. 1052/2 Similar bottles of '66 of Lafite and Latour brought £360 and £300 respectively.

La France (la frañs). [Fr.] An early type of hybrid tea rose, introduced in 1867, and bearing large, pink, scented flowers. Also *attrib.*

1868 *Floral Mag.* VII. 399 La France has evidently a mixture of Tea and Bourbon blood in it. **1906** *Westm. Gaz.* 11 Aug. 2/3 Roses red, roses yellow, roses white, roses of all known shades and varieties—fragrant La Frances, flaunting Jacqueminots, stately American Beauties. **1928** [see *Gloire de Dijon]. **1937** F. B. Young *Portrait of Village* vi. 119 Its flowers are all old-fashioned and mostly sweet-scented:..moss-roses and silvery-pink *La France*. **1952** 'M. Cost' *Hour Awaits* 78 Fanchon, lovely as a La France rose, but English as Mayfair, was also leaning from her window. **1969** V. C. Clinton-Baddeley *Only Matter of Time* 48 He wished so many of the old roses had not disappeared. When did you last see La France? **1973** [see *hybrid B. 1 b].

lag, *sb.*[1] Add: **4. b.** (Further examples.) More widely in general use: a period of time separating any phenomenon or event from an earlier one to which it is related (causally or in some other way); = *time-lag* (Time *sb.* 52); *angle of lag* (Electr.), the fraction of a complete cycle, multiplied by 360° or 2π radians, by which a sinusoidal current lags behind the associated sinusoidal voltage. See also *jet lag* (*jet *sb.*[3] 11).

1886 S. P. Thompson *Dynamo-Electr. Machinery* (ed. 2) xviii. 330, φ is called the retardation or angle of lag. *Ibid.* 331 The retardation will increase with increased speed... There will be less lag therefore if the machine is so designed that it can be driven at a slow speed. **1902** *Encycl. Brit.* XXVIII. 42/1 [His] method consisted in measuring the interval which elapses between the application of a potential difference..and the passage of the spark. This lag of the spark, as we may call it, is a very important quantity. **1909** *Jrnl. Hygiene* IX. 240 He found that there is an initial period after inoculation during which growth is almost absent: the length of time of this 'lag' varies with the age of the culture used for inoculation and with the species of the bacillus. **1923** *Glasgow Herald*

2 Nov. 12/4 The operation of the 'lag' of two months between the period of ascertainment and the months when the wages based on such ascertainment are paid. **1934** L. T. Agger *Alternating Currents* iii. 38 The current goes through all the events in its cycle one-quarter of a period, or 90°, later than the P.D. For this reason it is said to lag behind the P.D. by 90°; or, expressed in another way, the angle of lag of the current is 90°. **1934** Cultural lag [see *CULTURAL *a.* 3]. **1940** *Economist* 7 Dec. 707/1 It must not be forgotten that a very considerable lag must occur between the dates when insured losses are incurred and compensation is paid. In the case of certain shipping losses,..this lag may extend for the duration of the war. **1962** J. Thewlis et al. *Encycl. Dict. Physics* VII. 190/1 Lag in a control system may be defined briefly as delayed response of the output to changes of input. **1966** *McGraw-Hill Encycl. Sci. & Technol.* VII. 153/2 No instrument responds instantaneously to a change in the measurand; the lag is dependent on the natural frequency of the instrument system and its degree of damping.

c. *Comb.* **lag fault** *Geol.*, a type of overthrust formed when the uppermost of a series of rocks moves more slowly than the lower ones; **lag phase** *Biol.*, the period elapsing between the introduction of an inoculum of bacteria into a culture medium (or other new environment) and the commencement of its exponential growth; **lag time**, the period of time elapsing between one event and a later, related, event, esp. between a cause and its effect; (the extent of) a lag.

1900 J. E. Marr in *Proc. Geologists' Assoc.* XVI. 461 These fissures..would have an outcrop similar to those of thrust-planes or over-faults which approached the horizontal; but they would differ from these, inasmuch as no inversion on a large scale would accompany them. We shall speak of them here as 'lag' faults. **1947** *Q. Jrnl. Geol. Soc.* CIII. 100 There are several lag-faults in the district, but..only in the case of the Tirbach lag-faults is the evidence considered to be conclusive. **1963** E. S. Hills *Elem. Struct. Geol.* vii. 191 Lag faults.—These are low angle faults with normal fault displacement, that originate from the upward movement of the footwall block in a region of general thrusting. The hanging-wall block appears to have lagged behind in the regional movements. **1914** *Jrnl. Hygiene* XIV. 260 A seeding taken during the lag-phase grows with diminished lag. **1944** L. E. H. Whitby *Med. Bacteriol.* (ed. 4) i. 6 Multiplication [of bacteria] passes through four phases: (1) Lag phase—lasting from half an hour to eight hours, during which time there is no increase in numbers..; during this time the organism adjusts itself to its new environment. **1972** *Biochim. & Biophys. Acta* CCLXX. 41 When *E. coli* cells are exposed to low temperatures, they enter a prolonged lag phase. **1956** *Nature* 24 Mar. 579/1 Fragments of chorioallantoic membrane..support the growth of hæmagglutinating particles, but there is a lag-time of about ten hours. **1962** F. I. Ordway et al. *Basic Astronautics* xiii. 530 (*table*) Equipment lagtime before response. **1972** *Times* 26 June 12/4 The typical lag times for technological and cultural change. **1973** *Nature* 7 Dec. 327/1 The lag time of four years has been reduced to two at the behest of the governing council.

lag, *v.*[2] Add: **4.** *trans.* To lag behind.

1930 M. G. Malti *Electr. Circuit Analysis* iii. 26 A curve lags the origin if its zero value..occurs after the point $x=0$. **1966** L. A. Manning *Electr. Circuits* iv. 56 The current function lags the voltage by 90 degrees; that is, the current rises to a maximum value a quarter of a cycle later than does the voltage. **1973** *Nature* 21/28 Dec. 444/1 After the time step, Atomic Time will lag UT by 0·7 s.

lager (lā·gəɹ), *v.* [ad. G. *lagern* to store.] To store (beer) (see quots.). Hence **la·gering** *vbl. sb.*

1946 A. Simon *Conc. Encycl. Gastron.* VIII. 98/2 *Lager beer*, a light type of beer which is brewed from malt, with hops and water, fermented, and then stored (*lagered*). **1962** *Economist* 10 Feb. 543/1 A continuous fermentation process can eliminate the 'lagering' time. **1965** O. A. Mendelsohn *Dict. Drink* 193 *Lagering*, the almost obsolete process of aging beer by holding it in large tanks, where it clarifies naturally.

lagetta. (Later example.)

1871 C. Kingsley *At Last* II. xiii. 196 A bit of veritable natural lace, similar to..the famous lace-bark of the Lagetta-tree.

‖ **lagg** (læg). [a. Sw. dial. *lagg* edge of a bog or marshland, bank of a stream or river (see Lag *sb.*[2]).] A natural ditch along the edge of a raised bog.

1939 A. G. Tansley *Brit. Islands & their Vegetation* xxxiv. 675 The marginal watercourses from the *lagg* (a Swedish term) of raised bog. **1968** R. F. Daubenmire *Plant Communities* iii. 149 Raised bog..is a type confined to the area of a wet basin... The surface becomes convex so that the feeble drainage channels (*lagg*) that run across the basin are pushed to one or both sides of the peat dome.

lagniappe (lanyæ·p). *U.S.* Also **lagnappe, lanyap, -yappe.** [Louisiana Fr., ad. Sp. *la ñapa*, in the same sense.] Something given over and above what is purchased, earned, etc., to make good measure or by way of gratuity.

1849 *Knickerbocker* XXXIV. 407/1 Ime sum pumpkins in that line; but he's a huckleberry above my persimmon, and right smart lanyope too, as them creole darkies say. **1883** 'Mark Twain' *Life on Mississippi* xliv. 402 We picked up one excellent word—a word worth travelling to New Orleans to get; a nice limber, expressive, handy word—'lagniappe'. They pronounce it lanny-*yap*. It is Spanish—so they said. **1884** G. W. Cable *Creoles of Louisiana* xvi. (1885) 114 The pleasant institution of *ñapa*—the petty gratuity added, by the retailer, to anything bought—grew the pleasanter, drawn out into Gallicized *lagnappe*. **1936** W. Faulkner *Absalom, Absalom!* viii. 338 As lagniappe to the revenge as it were. **1947** S. J. Perelman *Westward Ha!* (1949) vii. 84 Since the ship was calling there anyway, the trip would be pure lagniappe, an extra dash of stardust unforeseen in our program. **1958** M. Mayer *Madison Ave.* xiv. 217 Finally, as lagniappe, Nielsen told the company the extent of 'dealer push'. **1966** *New Yorker* 1 Oct. 186 This amusing architectural lagniappe. **1971** *N.Y. Times Bk. Rev.* 7/4 And, as lagniappe, they threw in a list of 'spurious words' the scholars had come upon in dictionaries. **1972** *New Yorker* 7 Oct. 15/2 (Advt.), 64 pieces of exquisite Limoges porcelain sculpture..with superb porcelain tiles to play on as lagniappe.

lagoon[1]. Add: **1.** After 'sand-banks' insert 'or a similar barrier'. (Further examples.)

1877 A. Geikie *Elem. Lessons Physical Geogr.* iv. 271 Lagoons along the sea-margin are for the most part shallow and narrow, running parallel with the coast, from which they are separated by a strip of low land formed of sand, gravel, or other loose material. **1939** W. H. Twenhofel *Princ. Sedimentation* xii. 455 The barrier separating a lagoon from its parent body may result from many causes, but under most conditions it is thrown up by the waves. **1952** W. Shepherd *Living Landscape Brit.* iii. 50 Two bays have been cut off from the sea by shingle. These now form a salt-water lagoon, and may slowly silt up to form a marsh. *Ibid.* 53 A line of rocks across the bay encloses a 'lagoon'. **1968** R. W. Fairbridge *Encycl. Geomorphol.* 590/2 The entrance of a lagoon is restricted by the narrow tidal inlets through the barrier islands and the complex of sand bars which form on both the lagoonal and seaward side of the inlet.

2. (Further examples.) Also, the stretch of water inside a barrier reef.

1848 M. Somerville *Physical Geogr.* I. xiv. 215 Encircling reefs differ in no respect from atoll reefs except that they have one or more islands in their lagoon. **1863** J. B. Jukes *School Man. Geol.* vi. 67 There are..many islands in tropical seas in front of which coral reefs are found at a distance of many miles from the beach of the dry land, their outer edge being nearly dry at low tide, but plunging steeply down into fathomless water, while a broad navigable channel or lagoon..extends between this outer edge and the shore. These are called Barrier reefs. **1928** W. M. Davis *Coral Reef Probl.* xi. 271, I made a circuit of the island on trading steamers, following the lagoon for the greater part of the way. **1959** *Chambers's Encycl.* IV. 121/1 Barrier reefs surround islands or lie off the mainland, with an intervening navigable channel or lagoon, the width of which may be many miles.

2*. *Austral.* and *N.Z.* (See quots. 1849 and 1933.)

Adopted from Amer. Eng.: 1766– examples in *D.A.E.* in sense 'a shallow, fresh-water pond or lake, sometimes artificially formed and usually located near or connected with a lake, river, etc.'

1838 W. C. Symonds in *Jrnl. R. Geogr. Soc.* VIII. 422 On the S.E. coast at this island [*sc.* South Island] are several immense lagoons, into which flow rivers. **1844** *Nelson* (N.Z.) *Examiner* 7 Sept. 108/1 A lagoon..quite a lake in fact. **1849** F. Wakefield *Colonial Surveying* II. 59 Lagoons differ from lakes in being generally formed by surface water gathering in low grounds during the winter, from which there is no outlet. **1927** M. M. Bennett *Christison* iii. 37 At the back was a shallow lagoon. **1933** *Press* (Christchurch, N.Z.) 4 Nov. 15/7 *Lagoon*, any tarn, pond, or open water too small to be called a lake.

2.** An artificial shallow pool used in the treatment and concentration of sewage and slurry.

1909 E. C. S. Moore *Sanitary Engin.* (ed. 3) II. xviii. 691 Drying in lagoons is the system which is often adopted at works where sufficient land is available, and it is without doubt the least satisfactory method. **1926** G. M. Flood *Sewage Treatment & Disposal* x. 106 The sludge produced by any method of treatment may be pumped into lagoons in almost every case. **1975** *Daily Tel.* 9 Jan. 16/6 Sometimes 7,000 or 8,000 wading birds came..to the ash lagoons of West Thurrock power station.

lagoon (lăgū·n), *v.* [f. the sb.] *trans.* To treat (by oxidation) in lagoons. So **lagoo·ning** *vbl. sb.*

1911 G. B. Kershaw *Mod. Methods Sewage Purification* xi. 155 (*heading*) Lagooning. **1922** H. E. Babbitt *Sewerage & Sewage Treatment* xx. 495 The results of lagooning at Philadelphia are given in Table 103. **1935** Metcalf & Eddy *Amer. Sewerage Pract.* (ed. 3) III. ii. 23 Odors from sludge lagooned at Houston were so objectionable that another method of sludge disposal was required. **1969** J. G. Brennan et al. *Food Engineering Operations* xvii. 380 Lagooning is extensively used for the treatment of cannery wastes. **1972** J. Skitt *Disposal of Refuse & Other Wastes* iii. 36 Do not lagoon the top layer.

lagoonal (lăgū·năl), *a.* [f. Lagoon[1] + -al.] Of or characteristic of a lagoon or lagoons.

1910 *Encycl. Brit.* IX. 663/2 Of the Gasteropod genera *Cerithium* with its estuarine and lagoonal forms *Potamides*, *Potamidopsis*, &c., is very characteristic. **1950** *New Biol.* VIII. 92 The earliest known birds were quite small, about the size of a crow, and are known from only two specimens..; even these were preserved in an unique lagoonal environment. **1956** *Nature* 31 Mar. 607/2 The coral reef was not bottomed, the strata penetrated consisting entirely of fine detrital limestones..originally deposited in lagoonal waters not less than 30 fathoms in depth. **1963** D. W. & E. E. Humphries tr. *Termier's Erosion & Sedimentation* ix. 192 The marine and lagoonal faunas of the Cretaceous in the Gulf of Guinea. **1971** I. G. Gass et al. *Understanding Earth* xii. 158/2 The reefs and lagoonal sediments which flourished during the Mesozoic and Tertiary periods. *Ibid.* 159/1 For the past 50 million years reefs have been very efficient structures leading to the formation of vast lagoonal areas.

Lagrange (lăgrā·ŋʒ). The name of Joseph Louis *Lagrange* (1736–1813), Italian-born mathematician who worked in Prussia and France, used *attrib.* and in the possessive to designate various concepts introduced by him or arising out of his work, as **Lagrange('s) equation**, each of a set of equations of motion in classical dynamics relating the total kinetic energy T of a system to a set of generalized co-ordinates q_r and forces Q_r, and to the time t, and having the form $d(\partial T/\partial\dot{q}_r)/dt - \partial T/\partial q_r = Q_r$. (In many contexts interchangeable with *Lagrangian*.)

1858 *Rep. Brit. Assoc. Adv. Sci. 1857* I. 12 The force function U is independent of the differential coefficients η',..and, consequently, of the variables ω,.., hence, writing $H = T - U$, the equations take the form $d\eta/dt = dH/d\omega$, $d\omega/dt = -dH/d\eta$,..which correspond to the condensed form obtained by writing $T - V = R$ in Lagrange's equations. **1902** *Encycl. Brit.* XXVII. 568/1 Hence the typical Lagrange's equation may be now written in the form $d(\partial T/\partial\dot{q}_r)/dt - \partial T/\partial q_r = -\partial V/\partial q_r$, or, again, $\dot{p}_r = -\partial(V-T)/\partial q_r$. *Ibid.*, A classical example of the application of Lagrange's equations is to the motion of a top. **1942** Synge & Griffith *Princ. Mech.* xv. 453 Two features of Lagrange's equations should be emphasized. First, there is no unique set of generalized coordinates; however we choose them, the equations of motion always have the form (15.215). Secondly, since only working forces contribute to δW, reactions of constraint are automatically eliminated. [*Note*] Except where forces of friction do work. **1958** Condon & Odishaw *Handbk. Physics* v. ii. 18/1 If the total number of systems is N, $\Sigma n_j = N$ $\Sigma n_j E_j = E$... Using the method of Lagrange multipliers, introduce multipliers β and λ and find the set of n_j's which make $\delta[\log P - \lambda(N - \Sigma n_j) + \beta(E - \Sigma n_j E_j)] = 0$. **1962** J. Riordan *Stochastic Service Syst.* iv. 66, $g(y)e^{-g(y)} = y$. The solution of this, obtained by Lagrange expansion, is $g(y) = \sum_{n=1} (n^{n-1}y^n)/n!$ **1967** M. G. Smith *Introd. Theory Partial Differential Equations* i. 3 Comparing (1.2.8) and (1.2.9) we have the Lagrange equations $\dot{p}_k = \partial L/\partial q_k$ and $p_k = \partial L/\partial\dot{q}_k$.

Lagrangian (lăgrā·ŋʒiăn), *a.* and *sb. Math.* Also **lagrangian, Lagrangean.** [f. prec. + -ian, -an.] **A.** *adj.* Of or pertaining to the work of J. L. Lagrange (see prec.); of the kind introduced by Lagrange or associated with his work; *spec.* applied to (*a*) **Lagrange('s) equation*; (*b*) the difference between the kinetic energy and the potential energy of a system expressed as a function of generalized co-ordinates, their time derivatives, and time. (In many contexts interchangeable with *Lagrange('s)*.)

1858 A. Cayley in *Rep. Brit. Assoc. Adv. Sci. 1857* I. 2 The above-mentioned form is *par excellence* the Lagrangian form of the equations of motion, and the one which has given rise to almost all the ulterior developments of the theory. *Ibid.* 15 When there is no force function..the forms corresponding to the untransformed forms in T and U are as follows, viz. the Lagrangian form is $dq/dt = q'$, $d(dT/dq')/dt - dT/dq = Q$, and the Hamiltonian form is $dq/dt = dT/dp$, $dp/dt = -dT/dq + Q$. **1870** *Rep. Brit. Assoc. Adv. Sci. 1869* II. 10 Equations of motion in the Lagrangean form. **1882** Lagrangian method [see Eulerian *a.*]. **1904** E. T. Whittaker *Treat. Analytical Dynamics* ii. 38 If we introduce a new function L of the variables $q_1, q_2, \ldots q_n, \dot{q}_1, \ldots \dot{q}_n, t$, defined by the equation $L = T - V$, then Lagrange's equations can be written $d(\partial L/\partial\dot{q}_r)/dt - \partial L/\partial q_r = 0$ ($r = 1, 2, \ldots n$). The function L is called the Kinetic Potential, or Lagrangian function; this single function completely specifies, so far as dynamical investigations are concerned, a holonomic system for which the forces are conservative. **1908** J. H. Jeans *Math. Theory Electr. & Magn.* xvi. 484 The Lagrangian equation corresponding to the coordinate r is found to be..$d(\partial T/\partial\dot{r})/dt - \partial(T - w)/\partial r = R$. *Ibid.* 486 Let Q be the charge on the positive plate at any instant, and let this be taken as a Lagrangian coordinate. **1949** *Math. Tables & Other Aids to Computation* III. 466 Tables of 4-point and 6-point Lagrangean interpolation coefficients. **1954** D. ter Haar *Elem. Statistical Mech.* 445 The values of the x_i for which f is [an] extremum while equations (MA4.01) are satisfied can be determined from the equations

$$\frac{\partial f}{\partial x_i} + \sum_{j=1}^{p} \lambda_j \frac{\partial \varphi_j}{\partial x_i} = 0, \qquad i = 1, \ldots, n.$$

The x_i are now functions of the λ_j, but these quantities, which are called the undetermined multipliers or the Lagrangian multipliers, can be eliminated by substituting for the x_i into equations (MA4.01) and solving for the λ_j. **1957** L. Fox *Numerical Solution Two-Point Boundary Probl.* ii. 31 Lagrangian [interpolation] formulae..may be defined as formulae in which all significant differences have been replaced by pivotal values. **1964** *Oceanogr. & Marine Biol.* II. 14 The velocity along a particular trajectory varied more slowly than one would expect from the variability between one trajectory and another, suggesting that the Lagrangian scale of motion was noticeably larger than the Eulerian scale.

B. *sb.* The Lagrangian function (see (*b*) above).

1938 R. C. Tolman *Princ. Statistical Mech.* ii. 22 For more complicated mechanical systems, or even non-mechanical ones, suitable choices of the Lagrangian may be possible which will make Hamilton's principle still applicable. **1961** J. Thewlis et al. *Encycl. Dict. Physics* II. 99/2 This function, called the Lagrangian of the given mechanical system, contains everything for the unique determination of the motion of the system, provided that we know its initial position and velocity. **1964** E. A. Power *Introd. Quantum Electrodynamics* ix. 129 Then for the interaction with the electromagnetic field the substitution $\mathbf{p} \rightarrow \mathbf{p} - (e/c)\mathbf{A}$, $i\hbar\partial/\partial t \rightarrow i\hbar\partial/\partial t - e\phi$ is made in the Lagrangian of the free electron according to the principle of minimal electromagnetic coupling. **1973** *Physics Bull.* Mar. 183/3 Nonlinear lagrangians.

lagting (la·gtiŋ). Also **lagthing.** [Norw. Cf. Lawting.] **1.** A functional division of the Norwegian Parliament, operating primarily for law-making purposes.

1836 S. Laing *Jrnl. Residence Norway* xi. 456 The Storthing then proceeds to elect what is equivalent to our House of Peers, the Lagthing, or division in which the deliberative functions of the legislative body are invested. *Ibid.*, The functions of the Lagthing are not exactly the same as those of our House of Lords... It can only receive bills from the other house, the Odelsthing; deliberate..and approve or reject. **1927** *Glasgow Herald* 21 Mar. 13 The Bill is now going up to the Lagting. **1957** *Encycl. Brit.* XVI. 556/1 After the opening, parliament divides itself into two sections, the *lagting* consisting of 38 members and the *odelsting* of the remainder. **1974** *Whitaker's Almanack 1975* 918/1 The *Storting* (Parliament) itself elects one-quarter of its members to constitute the *Lagting* (Upper Chamber).

2. Also **Løgting.** [Faroese *Løgting.*] The Provincial Parliament of the Faeröe Isles.

1948 K. Williamson *Atlantic Islands* iv. 104 The Løgting imposes a small money tax on every whale.. caught. **1961** *Denmark* (Danish Ministry for Foreign Affairs) 136 Under the Home Rule Act of March 27, 1948, the publicly elected Faroese assembly, the Lagting or Løgting, has legislative powers in various fields. **1972** J. F. West *Faroe* x. 199 The members of the government may address the Løgting, but they may not vote unless they are elected members.

lahar (lā·hāɹ). [Javanese.] A mud-flow of volcanic ash mixed with water.

1929 *Geol. Mag.* LXVI. 433 (*heading*) The mudstreams ('lahars') of Gunong Keloet in Java. **1944** C. A. Cotton *Volcanoes* xiii. 247 Lahars follow mainly channels already existing, filling them temporarily to the brim with rushing torrents but leaving them empty again and eventually depositing their loads of debris on low ground many miles beyond. **1954** W. D. Thornbury *Princ. Geomorphol.* xix. 497 A lahar that accompanied an eruption of the volcano Galunggung, in Java, in 1822, spread over 114 villages. **1972** *Science* 9 June 1119/1 The site of the eastern spillway is now covered by lahar deposits.

Hence **laharic** (lăhā·rik) *a.*, of or pertaining to a lahar.

1968 *Mem. Geol. Soc. Amer.* CXVI. 472 Approximately 300 cubic miles of laharic debris came from the source areas.

Lahnda (lā·ndă). Also **Lahndi.** [Punjabi, 'western' (see quot. 1907).] An Indo-Aryan language spoken in the western Punjab. Also *attrib.* or as *adj.*

1903 Risley & Gait *Rep. Census India 1901* I. 311 Lahndā is a language the existence of which has long been recognized, but under many names. In the last Census Report it was called Jatkī... I therefore think it best to give it the name which is indicated by the natives of the Punjab themselves, i.e. Lahndā or the Language of the West (Panjabi *Lahndē-di Bōlī*). **1907** G. A. Grierson in *Imperial Gazetteer India* I. vii. 371 Lahndā or Western Panjābī is a language which appears under many names, such as Pothwārī, Chibhālī, Jatkī, Mūltānī, or Hindko... Lahndā, i.e. 'Western', has been lately suggested, and has been tentatively adopted, although it..is far from satisfactory. **1911** *Encycl. Brit.* XVI. 80/2 *Lahnda* (properly *Lahndā* or *Lahindā*, western, or *Lahndē-di bōli*, the language of the West), an Indo-Aryan language spoken in the western Punjab... Lahnda is also known as Western Panjabi and as Jatki, or the language of the Jats, who form the bulk of the population whose mother-tongue it is. **1948** D. Diringer *Alphabet* vi. 376 Kashtawari, which is a dialect of Kashmiri.., but is much influenced by the Pahari and Lahnda languages, spoken by its southern and south-eastern neighbours. **1967** D. S. Parlett *Short Dict. Lang.* 76 *Lahnda*... Closely related to Punjabi proper, with some features associated with the contiguous Dardic dialects.

lai[1] (lēi). [OFr. (see Lay *sb.*[4]).] **a.** One of a number of short narrative poems written either in French or in English in England between the twelfth and the fifteenth centuries, of a Celtic type and concerned with love, magic, and music. Often called *Breton lais.* **b.** A medieval French lyric associated with the *trouvères* of Northern France.

1774 T. Warton *Hist. Eng. Poetry* I. Diss. I. sig. a2 At the conclusion of most of the tales it is said that these Lais were made by the poets of Bretaigne. **1824** R. Price in T. Warton *Hist. Eng. Poetry* II. 430 But Marie's was not the only Collection of British Lais, in French. **1838** E. Guest *Hist. Eng. Rhythms* II. 103 The 'short measures' of Skelton..may perhaps be looked upon as the *direct* descendants of the Anglo-Saxon rhythms, though it must be confessed they much resemble, in their flow, the *lais* and *virelais* of the fifteenth century. **1855** H. H. Milman *Hist. Latin Christianity* VI. vii. 547 Slighter pieces which may call to mind the Lais and Serventes of the South. **1865** T. Taylor tr. *Hersart de la Villemarqué's Ballads & Songs of Brittany* 125 A *lai* by a northern *trouvère* on the same subject was discovered by M. de Fréminville, in the Bibliothèque du Roi. **1883** H. Kennedy tr. *B. ten Brink's Hist. Eng. Lit.* I. II. 179 The Breton *lais* retain most fully their native fragrance. They are usually romantic even when the topic is comic... The poem is often pervaded by a tone of elegiac longing. **1905** E. Rickert in *Mod. Philol.* II. 376 The fresh literary impulse that came with the Normans found little to do with the old Saxon heroes. A few tales were transformed into the *lai* or *chanson de geste* or *roman d'aventure*—*Havelock*, *Horn*, *Guy of Warwick*, *Bevis of Hampton*. **1906** [see *conte b]. **1907** St. J. Lucas in *Oxf. Bk. French Verse* p. vii, Various other kinds of lyric poetry begin to appear at the end of the twelfth century, *motets*,..*rondeaux*, *lais*, *ballettes* and *virelais*. **1923** J. Vising *Anglo-Norman Lang. & Lit.* 47 Breton lais in England are mentioned in *Roman de Renard*. **1925** A. Bell *Le Lai d'Haveloc* 26 There is, however, not only a general connection with the 'lais' but also a special one with those of Marie. **1929** [see *conte b]. **1932** *Oxf. Hist. Music* (ed. 2) II. II. v. 285 The initial stanzas of two trouvères *lais* are written on the melody of 'Ave gloriosa virginum regina'. *Ibid.* 286 On the other side the *lai* is related to the *chanson de geste* and to the other popular songs of the time, to which many of the anonymous *lais* show some resemblance. **1940** *Grove's Dict. Mus.* (ed. 4) III. 269/1 The *lais* (long lyrical poems written in twelve pairs of stanzas, each pair having a different metrical form and a different melody from the rest, except the last pair, which repeats metre and melody of the first) are all set to one musical part only. **1954** [see *conte b]. **1965** R. S. Loomis in Bessinger & Creed *Medieval & Ling. Stud.* 237 We possess three Breton *lais*: *Doon*, *Desiré*, and *Gurun*..which evince some knowledge of Scottish geography.

Lai[2] (ləi). [Local name.] A Mongoloid people living in the Chin hills of Burma; a member of this people; also, the language spoken by this people. Also *attrib.* or as *adj.*

1896 Carey & Tuck *Chin Hills* I. I. i. 3 The Northern Chins call themselves Yo, the Tashons, Haka, and more southern tribes Lai. *Ibid.* iii. 23 The Hakas call themselves Lai, and Yo is the general name by which the Chins call their race. *Ibid.* xiv. 152 The clans which claim the title of Lais are the Hakas, Klang-klangs, Yokwas... The first two are universally acknowledged as Lais, and refuse to admit that the others belong to their race. **1897** A. G. E. Newland *Pract. Hand-bk. Lang. Lais* 1 The *Lais* are the great tribe and its offshoots that occupy the Chin Hills... The language of these people is the *Lai* language, called by the Burmese *Baungshè*, by which term we have hitherto known it. Dialects of it are spoken by all the surrounding tribes, but nearly all understand the *Lai* tongue. *Ibid.* 3 Unlike the Southern Chin language, it will be found that the consonants 'f' and 'r' are both used in the *Lai* speech. **1906** J. G. Scott *Burma* I. 106 The Tashôn tribe is..the most numerous, and next to them come the Hakas, also called the Lai... Lai is said to be likely to become the *lingua franca* of the Chin Hills. **1924** C. M. Enriquez *Races Burma* iii. 15 The Haka (a group of about ten villages including the parent village) claim to be the only true Lai. *Ibid.* vii. 43 The Lai tribes are controlled..by Chiefs. **1963** F. K. Lehman *Struct. Chin Soc.* i. 30 The Haka villagers call themselves *lai*, thinking of themselves as better than their cultural near relations to the South.

laicism (s.v. Laicize *v.*). (Later examples.)

1931 *Economist* 4 July 13/1 The political antagonisms which the Spanish Revolution has let loose: the issues of Republicanism versus Monarchism, Syndicalism versus Capitalism, and Laicism versus Clericalism. **1966** *Ibid.* 11 June 1178/2 Their priests, many of whom equate laicism with socialism, have been outraged by its [*sc.* the government's] plans to modify the present confessional structure of education.

laicity (lēii·sīti). [f. Laic *a.* and *sb.* + -ity.] The principles of the laity; the rule or influence of the laity; the fact of being lay; also *attrib.*

1909 Webster, *Laicity*, laicality. **1925** *Brit. Weekly* 19 Mar. 587/3 The text of the document, which is directed against the 'laicity laws', lies before us. **1928** *Daily Tel.* 30 Oct. 12/3 Accused by his own hotheads of betraying the cause of 'laicity' by acquiescing in these details of the Budget. **1939** A. Toynbee *Study of Hist.* VI. 20 This laicity of the Gods was taken so much in earnest that in a Mahayanian sutra a chapter..is formally addressed to the Gods, as a hint that it is an *œuvre de vulgarisation*.

laid, *ppl. a.* Add: **a.** (Further examples.)

1634 J. Taylor *Needles Excellency* (ed. 10) sig. A2 For Tent-worke, Raisd-worke, Laid-worke, Frost-worke. **1860** Geo. Eliot *Let.* 27–28 Aug. (1954) III. 337 It is piteous to see the laid corn, and then hear the rain pouring in the still night! **1865** F. B. Palliser *Hist. Lace* xxiii. 275 Then there is..true stitch, laid-work,..and cut-work. **1960** [see *brick-stitch* (*brick *sb.*[1] 10)]. **1971** *Country Life* 10 June 1440/2 Reaping a small field of badly laid corn with sickles and reaping hooks.

c. Also, *laid-back, -by* (cf. *Lay *v.*[1] 50 f), *-off* (cf. Lay *v.*[1] 54 f), *-on.*

1868 Laid-off [see Lay *v.*[1] 54 f]. *a* **1877** Knight *Dict. Mech.* II. 1246/1 Laid-on (Joinery), a term applied to moldings which are got out in strips and nailed on to the surface of the object. **1906** *Westm. Gaz.* 9 June 16/3 The Valenciennes running around the laid-on tucks surrounding the skirt. **1908** *Ibid.* 23 Dec. 4/1 To get in under the ball you must have a shallow head..or else a very much laid-back face. **1909** *Ibid.* 30 Apr. 4/2 You can take a laid-back club and loft right over it. **1932** W. Faulkner *Light in August* xviii. 394 Any man could look at him and perhaps recognise him: Byron Bunch, that weeded another man's laidby crop, without any halvers. **1934** *Archit. Rev.* LXXVI. 13 Central heating, lighting, cooking, vacuum cleaning, laid-on water and drainage. **1943** Hunt & Pringle *Service Slang* 43 *Laid on*, confirming that transport and supplies are available, that men are on the spot, and, in short, that everything is ready for action. **1955** *Times* 31 May 5/3 The union proposed that the guaranteed level of pay for 'laid off', meaning temporarily unemployed, workers, should be 80 per cent. of gross pay rather than 100 per cent. of take-home pay as originally proposed. **1960** *Farmer & Stockbreeder* 15 Mar. 75/3 The appeal of zero-grazing is that it requires no fencing, no laid-on water, but maintains fertility. **1967** *Guardian* 26 July 10/7 A group of laid-on Turkish peasants in costume, dancing a jig. **1970** *Daily Tel.* 11 Sept. 6/2 Laid-off workers at Coventry are receiving..£5 unemployment benefit. **1973** *Melody Maker* 28 July 36/2 This hit-writer of the early 60s came out with a highly contemporary style which fitted the fashionable term 'laid-back', and her 'Tapestry' album zoomed up to become one of the three biggest-selling LPs of all time. **1974** *New Society* 7 Mar. 589/3 It's all cheerfully grotty and relaxed in the usual laid-back Montreal style.

laika (lai·kă). Pl. **laiki.** [Russ. *laika*, f. *lai* bark.] A dog belonging to a group of Asiatic breeds of the spitz type, characterized by a pointed muzzle, pricked ears, a stocky body with a thick, rough, grey, fawn, white, or black coat and a tail curled over the back.

1905 H. de Bylandt *Dogs All Nations* I. 576 Samoyed dog. (Laika or Siberian dog)... Dog of medium size, well built and cobby, covered with a thick fur. **1928** *Daily Sketch* 7 Aug. 4/3 The Elkhound is of similar type to the Russian Laika (barking) dog, so-called because, unlike others of the lupine group, its tongue is a bark and not a howl. **1948** C. L. B. Hubbard in B. Vesey-Fitzgerald *Bk. Dog* II. 517 The Spitz of the Asiatic sphere shew considerable variation, and are consequently divided into many well-defined breeds. These are collectively known as Laiki, and each Laika is called (in most cases) after its province or its people. **1971** F. Hamilton *World Encycl. Dogs* 599 Laiki are seen all over Northern Russia... When a Laika, out hunting, sees a bird in a tree or bush, it barks ceaselessly.

Laingian (læ·ŋiăn), *a.* [-ian.] Of or pertaining to the theories of the British psychologist R. D. *Laing* (1927–), esp. that a disintegrative mental illness such as schizophrenia is due to 'normal' social or family pressures which are intolerable to the self, and that re-integrative therapy is therefore possible only when such conventionally accepted pressures are removed. Hence as *sb.*, one who adheres to Laing's theories or practises his method of therapy.

1971 *New Statesman* 16 Apr. 535/1 A Laingian stereotype has blossomed. **1972** *Listener* 20 Jan. 95/3 Her psychiatrist..a 'Laingian' who wants to treat the underlying cause..and not the symptoms. **1973** E. Z. Friedenberg *Laing* (1974) i. 27 Edward Chamberlayne, an archetypical victim..of a Laingian entrapment. **1973** *Listener* 10 May 623/2 The now-familiar Laingian argument that language is the first and strongest of the prison-houses with which our civilisation enslaves the free self.

lair (lēᵊɹ), *sb.*[4] *Austral. slang.* Also **lare.** [Back-formation from *lairy *a.*[2]] A flashily dressed man, one who 'shows off'. Also (*rare*) **lai·rize** *v. intr.*, to act like a lair, to show off.

1935 K. Tennant *Tiburon* ix. 106 He was also considered something of a lare among the girls. **1941** Baker *Dict. Austral. Slang* 42 *Lair*, a flashily-dressed man. **1941** K. Tennant *Battlers* iii. 29 But a brainy young lare called 'the mob' together on the pavement outside the shop. 'This can be worked, can't it?' he asked, displaying his slip. **1953** —— *Joyful Condemned* iii. 22 You came lairizing round at our place like you owned it. **1955** H. Drake-Brockman *Men without Wives* 83 A flash young man. What they call on the goldfields 'a regular lare'. Hair much slicked, double-breasted coat, patent leather shoes. **1956** J. Wright in *Coast to Coast 1955–56* 168 But he was what they called a bit of a lair; he couldn't keep a job and had run through three already. **1956** K. Tennant *Honey Flow* xvi. 188 When they dressed in their best, they looked cheap lares, the type you see leaning against the hotel or the general store. **1973** A. Broinowski *Take One Ambassador* iii. 31 Two young lairs from the surf club carried their boards down..and tossed them onto the sea. With insolent grace they hopped on.

lair (lēᵊɹ), *v. Austral. slang.* Also **lare.** [f. *lair *sb.*[4] or *lairy *a.*[2]] To dress flashily, to dress *up*; to act in a lairy manner. Freq. in pa. pple. (*all*) *laired up.*

1941 Baker *Dict. Austral. Slang* 42 *Lair up*, to dress, esp. to don one's best clothes for a festive occasion. **1945** —— *Austral. Lang.* vi. 119 *All laired up* and its synonym *all mockered up* may also be noted. **1955** H. Drake-Brockman *Men without Wives* 83 It's that Rienzi. He's a trimmer. Always laring round. No good to girls. **1962** S. Gore *Down Golden Mile* 64 He climbs out of the cockpit, all laired up in this red rig-out.

lairy (lēᵊ·ri), *a.*[2] Also **lary.** [ad. Leery *a.*[2]] **1.** *Cockney slang.* Knowing, 'fly', conceited.

1846 *Swell's Night Guide* 78 Lairy and cautious to the green ones, never too fast. **1933** J. Masefield *Conway* 211 *Lairy*, slow, slack; also cunning. **1945** B. Naughton in C. Madge *Pilot Papers* 99 We'll have to keep an eye on him. Spivs are lary perishers. Anything goes wrong they'll never risk their own skin. *Ibid.* 108 They appear to be mentally quicker than most young men: ('Lary' is the word they use for it). **1958** *News Chron.* 23 May 4/7 If someone..is conceited he's lairy. **1967** *Spectator* 4 Aug. 130/3 What I was getting at—before that lairy loon butted in about his thumb—was this. Down the East End we're overcrowded an' we're bleedin' poor.

2. Also **leary, leery.** Flashily dressed; vulgar. *Austral. slang.*

In quot. 1906 used as quasi-*adv.*

1906 E. Dyson *Fact'ry 'Ands* xii. 160 Found drownded with a bloke what done-up 'is 'air dead leary. **1916** C. J. Dennis *Songs Sentimental Bloke* 125 *Leery*, vulgar; low. **1936** F. Clune *Roaming round Darling* xxiv. 258 Then climbed..Jack the Ripper on to Tugboat Annie, his hat at a leery angle, and the pommel leerier still. **1941** Baker *Dict. Austral. Slang* 42 *Lairy*, vulgar, flashily or showily dressed. **1966** B. Beaver *You can't come Back* (1968) 146, I just stood still under the big lairy neon, chuckling a bit and trying to roll a smoke.

‖ **laisse** (lĕs). [Fr.] In Old French verse, = Tirade *sb.* 2.

1872 [see *chanson 2]. **1879** [see Tirade *sb.* 2]. **1929** *Encycl. Brit.* II. 567/2 All the lines in a *laisse* or stanza close with the same vowel-sound. **1955** J. T. Shipley *Dict. World Lit. Terms* 246/1 *Laisse*; tirade. Fr. Pros. Running lines (of 8 or 10 syllables) assonanced; later, rhymed; the verse form of the *chanson de geste*.

laissez-aller. (Later *attrib.* examples.)

1871 [see **fatigue man*]. **1923** [see *dozy *a.*[1] 2].

laissez-faire. Add: Hence **laissez-fai·r(e)ist,** one who believes in a doctrine of *laissez-faire*.

1932 G. B. Shaw *Platform & Pulpit* (1962) 252 A Cabinet of talkers and Laisser-fairists. **1944** A. Jones *Right & Left* 16 The Conservative is neither a planner nor a laisser-faire-ist. **1966** *Guardian* 1 Dec. 8/6 Professor Peacock..isn't too keen on being cast as a 'relentless laisser-fairist'.

‖ **laissez-passer** (lɛse pase). Also **laisser-passer.** [Fr., lit. 'allow to pass'.] A permit, a pass.

1914 T. A. Baggs *Back from Front* xx. 94 You must first pass grim Charon and his watchdogs at the entrance, where your passports, laisser-passers, sauf-conduits, are inspected. **1928** *Sunday Express* 1 July 5 The Ballet was given a laissez-passer and were allowed to come to England through Paris. **1936** E. Waugh *Waugh in Abyssinia* 77 Many writers have left accounts of the intricate system of tolls and hospitality by which the traveller was passed on from one chief to another and of the indifference with which the Emperor's *laissez-passer* was treated within a few miles of the capital. **1951** J. B. Priestley *Festival at Farbridge* I. i. 34 He handed over an Order to View as if it were a *laissez-passer* for the captain of the Swiss Guard at Versailles. **1955** *Times* 28 July 8/4 He has been granted by the Greek Foreign Ministry a laisser-passer to the Greek military zone of the Greek-Bulgarian frontier. **1970** R. G. Feltham *Diplomatic Handbk.* 178 *Laissez-passer*, a permit to travel or to enter a particular area.

laitakarite (ləitākā·rəit). *Min.* [ad. Finn. *laitakariittii* (A. Vorma 1959, in *Geologi* (Helsinki) III. XI. 11), f. the name of Aarne *Laitakari* (b. 1890), director of the Geological Survey of Finland (see quot. 1959): see -ite[1].] A white rhombohedral selenide and sulphide of bismuth, Bi_4Se_2S.

1959 *Mineral. Abstr.* XIV. 139/2 The new mineral was named, in the honour of the discoverer of the material from which it was disclosed, laitakarite, and it is supposed to be isomorphous with joseite. **1963** *Canad. Mineralogist* VII. 678 The data given in Table 1 show the probable identity of laitakarite and selenjoseite. **1968** I. Kostov *Mineral.* II. ii. 164 Additional minerals are laitakarite (Bi_4Se_2S) and paraguanajuatite ($Bi_2(Se,S)_3$), which are isostructural with the minerals of the tetradymite group.

laitance (lēⁱ·tăns). Also (*erron.*) **laitence.** [Fr.] A milky scum appearing on the surface of freshly laid cement.

1909 Taylor & Thompson *Treat. Concrete* (ed. 2) xv. 303 The milky laitance which appears on concrete laid under water represents an actual loss of cement. **1930** *Engineering* 23 May 677/3 The value of carefully providing for the adequate junction between successive layers and the elimination of laitance. **1939** *Archit. Rev.* LXXXV. 267 First is the way in which the tile is fixed to its backing, due to insufficient soaking of both the tile and the rendering, robbing the cement of the water necessary for proper hydration; 'killing' the cement by re-mixing new and old compo that has already started to set; 'laitence', a white scum of water and cement which forms on the surface of the bed of floor tiles, if it is pressed up and down too much while the tiles are being levelled. **1968** D. C. Tibbetts in E. G. Swenson *Performance of Concrete* (1969) x. 168 Laitance was removed and necessary holes cut in the sides of the cylinders for the setting of the low water ends of the concrete braces.

lak (læk). Repr. *U.S. dial.* (esp. *Black English*) pronunc. of Like *a.*, *adv.*, and *v.*[1]

1881 J. C. Harris *Nights with Uncle Remus* (1884) xvi. 80 He'd skuze hisse'f, he would, en gallop down de big road a piece, en paw up de san' same lak dat ar ball-faced steer w'at tuck'n tuck off yo' pa' coat-tail las' Feberwary. **1901** F. L. Stanton (*song-title*) Mighty lak' a rose. **1936** M. Mitchell *Gone with Wind* v. 78 Ain' nobody got a wais' lak mah lamb. **1949** *Crisis* (N.Y.) Nov. 303/1 Would you lak t'go t' the sto'..lak a darlin' li'l boy. **1973** B. Gates in S. Henderson *Understanding New Black Poetry* 310 Now, you gon' stop Yo ackin lak a fool!

Lak (læk). The name of a Caucasian language (see quot. 1954).

1954 Pei & Gaynor *Dict. Ling.* 118 *Lak*, a language.. spoken in the Caucasus; a member of the Eastern Caucasian group of the North Caucasian family of languages. **1971** *Language* XLVII. 233 The West Caucasian languages, where..Lak has forty cases and Tabassaran forty-eight. **1972** *Ibid.* XLVIII. 845 Hockett exemplifies 3a with 'Arunta..Lak and Wishram'.

‖ **lakatoi** (læ·kătoi). [Papuan.] In New Guinea, a native dug-out canoe, with two or more hulls.

1885 W. W. Gill in Chalmers & Gill *Work & Adventures New Guinea* II. i. 258 We were fortunate in seeing two *lakatoi* or Gulf-going crafts; the larger one consisted of fourteen immense canoes lashed firmly together and decked. *Ibid.* 259 Each *lakatoi* starting for the Gulf is filled with earthenware pots. **1911** *Encycl. Brit.* XX. 742/2 The Papuans build excellent canoes and other boats... The most remarkable of their vessels is the 'lakatoi', composed of several capacious dug-outs, each nearly 50 ft. long, which are strongly lashed together. **1926** *Mariner's Mirror* XII. 216/2 Fig. 2 is a *lakatoi* of Port Moresby, S.E. New Guinea. **1964** *Sunday Truth* (Brisbane) 9 Aug. 20/5 Some of them have purchased £400 outboard motors which they fit to lakatois—generally two hollowed out logs with a connecting wooden decking. **1964** *Punch* 26 Aug. 319/2 How will the *lakatoi* run before the wind of change? **1968** *Mariner's Mirror* LIV. 348 Today the sails used on lakatoi are rectangular, with the peak held by a sprit.

lake, *sb.*[4] Add: **1. b.** (Further *fig.* examples.)

1974 *Daily Tel.* 30 July 17/1 The Common Market has a 'wine lake' estimated at 8 million litres..—and yesterday a Labour MP called for some of it to be brought to Britain. **1975** *Times* 9 Apr. 15/3 Butter mountains and wine lakes are part of the price which Europe pays for a common agricultural policy.

c. *the Great Lakes* (examples).

In earlier use freq. without the adjective.

c **1665** P. E. Radisson *Voyages* (1885) 187 Those great lakes had not soe soone comed to our knowledge if it had not ben for those brutish people. **1748** H. Ellis *Voyage to Hudson's-Bay* 151 A Communication with the great Lakes behind Canada. **1759** P. Collinson in W. Darlington *Memorials J. Bartram & H. Marshall* (1849) 217, I don't remember ever reading of any [goats] in the country about the lakes. **1803** W. B. Grove *Let.* 25 Feb. in J. Steele *Papers* (1924) I. 367 The Ocean, the Mexican Gulf, the Mississippi & the Lakes must be our boundaries. **1813** *Niles' Reg.* V. 65/1 The position of the great lakes is..well known to the people of the United States. **1840** J. F. Cooper *Pathfinder* I. p. v, Incidents that might be supposed characteristic of the Great Lakes. **1902** *Encycl. Brit.* XXXII. 551/1 Plan of Great Lake steamer. **1904** N. S. Shaler *Citizen* 77 Where the territory borders on the sea or the Great Lakes, the authorities have charge of such harbours as are not in the control of the federal authority. **1966** *Canadian Geogr. Jrnl.* Apr. 113/2 The abnormally low water levels on the Great Lakes.

d. *to jump* (or *go* (*and*) *jump*) *in the lake*: see *jump *v.* I d.

5. a. *lake-bass, -bed, -front, -herring* (examples), *-island, -isle, -shore* (earlier and later examples), *-steamer, -water.*

1795 J. Scott *U.S. Gazetteer* s.v. *Vermont*, A species of fish called lake bass. **1884** G. B. Goode *Fisheries U.S.: Nat. Hist. Aquatic Animals* 424 The White Bass or Striped Lake Bass, *Roccus chrysops*. **1973** R. Lockridge *Not I, said the Sparrow* (1974) vi. 87 There was only one right way to cook lake bass. **1906** *Yorks. N. & Q.* July 100 Their position on the edge of the old lake-bed. **1937** *Discovery* Jan. 24/1 The bones [of the shovel-tusked Mastodon] lay embedded in the hardened mud deposit of an ancient lake-bed in Mongolia. **1880** 'Mark Twain' *Tramp Abroad* 245 The lake-front is walled with masonry like a pier. **1968** *Economist* 13 July 38/3 A lakefront site that would be better as a park. **1842** J. E. DeKay *Zool. N.Y.* IV. 267 The Lake Moon-eye, *Hyodon clodalis*,..is common in Lake Erie. At Buffalo and Barcelona, it is called Moon-eye, Shiner, and Lake Herring. **1875** *Amer. Naturalist* IX. 135, I received..a collection of deep water 'Siscoes'... Compared with Coregonus most of the species have a more slender form; hence their popular name of 'lake herrings', although their resemblance to the sea herring is quite superficial. **1955** *Arctic Terms* 48/1 *Lake herring*, any of various whitefish of the genus *Leucichthys*, caught in great numbers in circumpolar fresh waters. Also called 'cisco'. **1893** W. B. Yeats in *Bookman* May 43/1 It is said that an enchanted tree once grew on the little lake-island of Innisfree. **1890** —— *Countess Kathleen* (1892) 121 (*title*) The Lake Isle of Innisfree. **1917** E. Pound *Lustra* 61 (*title*) The lake isle. **1798** I. Allen *Nat. & Pol. Hist. Vermont* 61 The two Frenchmen were landed..with instructions to follow the lake shore. **1813** *Niles' Reg.* IV. 159/1 Previous to this period, a great deal of prejudice existed against the lake *shore*, as unhealthy. **1849** *Ex. Doc. 31st U.S. Congress 1 Sess. House* No. 5. II. 731 The sandstone on the lake-shore is..covered by fifteen.. feet of sand and clay. **1851** C. Cist *Sk. Cincinnati in 1851* 319 Hence [arise] their efforts to reach Chicago, by way of the Erie lake shore. **1973** *Tucson* (Arizona) *Daily Citizen* 22 Aug. 1/1 The 15½-foot-deep lake gives Tucson the appearance of being a major lakeshore metropolis. **1847** *Knickerbocker* XXX. 456 He has been inspired by looking down through the iron foot-grating of a great lake-steamer. **1888** C. D. Ferguson *Experiences Forty-Niner* i. 11 It was in the month of September, 1849, when..I embarked on the lake-steamer, A. D. Patchen for Chicago. **1890** W. B. Yeats *Countess Kathleen* (1892) 121, I hear lake water lapping with low sounds. **1906** *Westm. Gaz.* Oct. 6 6/2 And far below the blue lake-waters shine. **1920** Joyce *Let.* 5 June (1966) II. 469 It should be read in the evening when the lakewater is lapping.

b. *lake-girt* adj.

1878 H. M. Stanley *Through Dark Continent* I. x. 222 From the summit of this lake-girt isle. **1908** *Daily Chron.* 4 Aug. 3/1 The *noche triste* when the Spaniards found themselves surrounded in the lake-girt capital of the Aztecs.

6. a. lake-basin (further examples); also, the area drained by all the streams entering a lake; **lake country** (earlier example); **Lake District** = Lake-land; **Lakehead** *Canad.*, (*a*) *Hist.*, the western end of Lake Ontario (quot. 1827); (*b*) the city of Thunder Bay, Ontario, and the surrounding region on the north-west shore of Lake Superior; **lake rampart, ridge** = *ice-rampart* (*ice *sb.* 8).

1865 D. Page *Handbk. Geol. Terms* (ed. 2) 272 *Lake-basin*, in geography, the depressed area which contains the waters of a lake; also the entire area drained by the streams that fall into a lake. In geology, the concavity.. in which the waters of a lake rest. **1882** *Proc. Boston Soc. Nat. Hist.* XXI. 326 In the Himalaya, the valleys of Nepal and Kashmir are old orographic lake basins. **1965** W. D. Thornbury *Regional Geomorphol. U.S.* xxiv. 494/2 Fish can be carried by birds from one lake basin into another. **1967** Jennings & Mabbutt *Landform Stud. Austral. & New Guinea* vi. 111 Such an argument would explain why well formed alluvial fans survive outside the catchment of Lake Torrens and why they are absent or present in only a degraded form within the lake basin. **1842** *Amer. Pioneer* I. 211 No where was the pressure or want of money more sensibly felt than in the lake country. **1835** Wordsworth *Yarrow Revisited* 241 *Force* is the word used in the Lake District for Water-fall. **1851** *Art Jrnl.* 1 May 132/2 The scale upon which the scenery of the English Lake district is laid out. **1886** J. Prestwich *Geol.* I. 267 In the Lake District the planes of cleavage also usually strike about E.N.E. **1936** *Discovery* May 150/2 Lovers of the Lake District..feel that the peculiar wild beauty of the innermost fells will be destroyed by the introduction of large acreages of larch and spruce planted in small rows on the hillsides. **1957** G. E. Hutchinson *Treat. Limnol.* I. i. 1 Lakes therefore tend to be grouped together in *lake districts*. *Ibid.*, The whole group of lakes of a given lake district may be compared with another group. **1827** *Gore Gaz.* (Ancaster, Ont.) 25 May 50/4 It appeared, that a person at the Lake Head, had furnished the York Garrison with 800 bbls. of Flour last year. **1955** *Beaver* Summer 37 From the deck of the loaded freighter, bound for the Sault and Welland Canals, the grain strongholds of the lakehead stand like castles against the sunset. **1968** *Globe & Mail* (Toronto) 13 Feb. B7/2 One of the world's largest multiple-line insurance companies requires a sales oriented management man to establish a sales force in the lakehead. **1860** C. H. Hitchcock in *Proc. Amer. Assoc. Adv. Sci.* XIII. 335 We have discovered similar walls of stone in Vermont, and venture to describe this form of drift under the name of Lake Ramparts. **1870** *Amer. Naturalist* IV. 199 Above all these Drift deposits..are the 'lake ridges'—embankments of sand, gravel, sticks, leaves, etc., which run imperfectly parallel with the present outlines of the lake margins.

b. Lake poets (earlier examples), **school** (earlier example).

1816 *Edin. Rev.* XXVII. 278 His [*sc.* Byron's] views fell more in with those of the Lake poets, than of any other party in the poetical commonwealth. **1824** Mill in *Westm. Rev.* I. 516 Mr. Southey..and the other Lake poets..commenced writing with higher objects. **1816** *Edin. Rev.* XXVII. 66 Other productions of the Lake School.

lake, *v.*[3] Add to def.: *spec.* by causing the hæmoglobin in red blood cells to pass out into the plasma. Hence **laked** *ppl. a.* (Further examples.)

1903 *Science* 6 Mar. 369 For the preparation of hæmoglobin the blood was collected in ammonium oxalate, washed, laked with distilled water [etc.]. **1912** Gulland & Goodall *Blood* vi. 48 Dilution of the plasma causes the corpuscles to swell up and become rounded, and if the dilution be carried too far the corpuscle ruptures and the hæmoglobin passes into solution. The blood is then said to be 'laked'. **1925** C. H. Browning *Bacteriol.* vi. 122 If now tetanus toxin is added the suspension soon becomes transparent, *i.e.* it is laked or lysed, owing to the hæmoglobin diffusing out of the red cells. **1946** *Nature* 28 Dec. 953/1 This is..far from reaching the refractive index level of the red cells (which would have resulted in producing 'laked blood' without hæmolysis).

lake-land, lakeland. Add: **2.** Lakeland terrier, a rough-coated, red or black and tan terrier with a stocky body and a broad muzzle, belonging to a breed developed in Lakeland; also *ellipt.*

1928 *Kennel Gaz.* Dec. 1294/1 (*heading*) Any other breed or variety of British, Colonial or foreign dogs not classified... Lakeland terriers. **1931** *Times Lit. Suppl.* 13 Aug. 620/4 Others [*sc.* terriers] in favour are the Border, the Lakeland, the Fox, the Sealyham. **1960** *Guardian* 18 Nov. 12/6 Two Lakeland Terriers were lost for four days in a disused mine. **1971** F. Hamilton *World Encycl. Dogs* 457 It is doubtful if anyone knows exactly how the Lakeland Terrier was developed but it is generally thought the Border Terrier, the Bedlington and the Fox Terrier were used in its formation. *Ibid.* 458 The trimming of the Lakeland for the show ring has reached a high peak of perfection.

laker[1]. Add: **2.** (Earlier example.)

1814 *Edin. Rev.* XXIV. 1 Imitations of Cowper, and even of Milton.., engrafted on the natural drawl of the Lakers.

3. (Earlier example.)

1823 J. F. Cooper *Pioneers* II. xxiv. 261, I see a laker there, that has run out of the school. It's seldom one finds such a creater in the shallow waters.

4. (Later examples.)

1945 *Seafarers' Log* 27 Apr. 6/3 She is a small laker but generally has more beefs than would the SS Queen Elizabeth. **1961** *Guardian* 15 June 1/2 Until the Seaway opened in April, 1959, the 'freshwater' trade was carried in specially-designed lakers and smaller canallers... New 20,000-ton lakers..are coming into service. **1970** *Daily Colonist* (Victoria, B.C.) 23 Sept. 26/1 The 5,300-ton laker Orefax ran aground..off Battery Island in the St. Lawrence River. **1974** *Globe & Mail* (Toronto) 11 Sept. 8/9 The Canadian pilots will continue to serve ocean vessels downbound through the canal and take U.S. lakers westbound.

5. One accustomed to sailing on a lake.

1838 J. F. Cooper *Home as Found* II. 75 After fishing a few hours, the old laker [*sc.* Captain Truck] pulled the skiff up to the Point. **1910** *Blackw. Mag.* Aug. 173/1 He was an experienced 'Laker', but the scene..had completely unmanned him. **1936** K. Mackenzie *Living Rough* 274 When the deep-water sailor goes on the lakes, he has a tendency to..refer to the lakers as farmers, niggerhead sailors, and other salt-water jokes.

lakh. Now the more usual spelling of Lac[2].

a. Of things or persons in general.

1964 E. Huxley *Back Street New Worlds* x. 100 Plenty of Pakistanis are here already—the High Commissioner's estimate is one *lakh*, or 100,000. **1969** *National Herald* (New Delhi) 29 July 6/5 The labour acts which relate to factories, mines, plantation, transport, shops and establishments, wages, safety and welfare, industrial relations and protection of children are expected to benefit lakhs of workers and wage-earners in the state. **1969** *Hindu* (Madras) 3 Aug. 6/4 The area worst hit by the recent floods in the Brahmaputra Valley is the Sibsagar district where about two lakhs of people have been affected, with thousands rendered homeless.

b. *spec.* Of coins or sums of money.

1859 [see Lac[2] b]. **1955** *Times* 3 Aug. 2/6 Detailed prospecting, which has so far cost 19·80 lakhs of rupees, led to the location of iron ore in Kalabagh and its suburbs in the Punjab. **1971** *Weekend* (Ceylon) 12 Sept. 3/2 The project would cost four lakhs of rupees. **1972** *Times of India* 28 Nov. 5/3 The parcel actually contained 900 Japanese-made wrist watches worth Rs. 2 lakhs. **1975** *Bangladesh Times* 18 July 1/3 The Finance Minister said that the Government had already increased the ceiling of private investment for setting up industries from Tk 25 lakh to Taka three crore.

la·kish, *a. rare.* [f. Lake *sb.*[4] + -ish[1].] **a.** Like a lake. **b.** Characteristic of the Lake poets.

1872 G. M. Hopkins *Jrnls. & Papers* (1959) 222 The broad smooth fall of a lakish apron of water. **1946** *Mod. Lang. Q.* Dec. 497 Bitterly as he attacked Southey for his faults, Jeffrey did not accuse him of the 'Lakish' fault of 'mysticism'.

Lakoda (lakōᵘ·dă). [Name of an area in the Pribilof Islands in the Bering Sea.] Used *attrib.* or *absol.*: a type of sealskin used as a material for coats.

1969 *New Yorker* 8 Nov. 179 A midicoat of Lakoda in its natural caramel uses black beaver for the collar and cuffs. **1975** *Ibid.* 7 Apr. 118/2 On the Atlanta trip, he wore a Lakoda-sealskin coat.

lala, var. *ilala.

‖ **laksamana** (la:ksămā·nă). Also 7 **laxaman,** 9 **lacsamana.** [Malay, ad. Skr. *lakshmana* 'having fortunate tokens' (the name of a mythical hero, Rama's half-brother).] The title formerly given to a high dignitary or admiral in Malaya.

1615 in Danvers & Foster *Lett. received by E. India Co.* (1900) IV. 6 On the morrow I went to take my leave of Laxaman, to whom all strangers' business are resigned. **1821** J. Leyden tr. *Malay Annals* xi. 96 Whoever bears the title of sangcuan may succeed to the rank of lacsamana. **1839** T. J. Newbold *Pol. & Statistical Acct. Straits of Malacca* II. iii. 46 Paul de Gama attacked Johore, but was defeated and slain by the Lacsamana. **1969** J. M. Gullick *Malaysia* ii. 40 Command of the navy rested with the Laksamana.

la-la (lā·lā·), *v.* [Redupl. La *int.*] *intr.* To sing or say the syllable *la* repeatedly, esp. in place of the words or notes of a tune. Also *trans.*, to sing (a song) in this way.

1906 *Daily Chron.* 19 Sept. 9/6 Miss Neale..'la la'-ed a simple tune. **1908** *Ibid.* 18 Aug. 7/1 They search out the secret places of past grandeur, la-la-ing as they issue from court and passage. **1974** *Listener* 17 Jan. 84/1 Those boys who couldn't sing didn't just 'la-la': they said the words, and they were called 'Talking Josephs'. The boys who really could sing were called 'Canaries'.

lalang (lā·lăŋ). Also **lallang.** [Malay.] A local name for *Imperata cylindrica*, a large coarse grass, widespread in tropical countries.

1779 T. Forrest *Voy. New Guinea* II. iii. 192 The country hereabouts is now covered with long grass, called lalang. **1887** H. W. Daly *Digging, Squatting, & Pioneering Life S. Austral.* 158 A shake-down..made up of dried lallang—the thick strong grass of the country. **1887** *Encycl. Brit.* XXII. 93/1 There are waste spaces..covered with coarse lalang grass. **1912** *Chambers's Jrnl.* Feb. 98/1 As I stood in the lalang patch..there passed over my head..all the turmoil of a great city. **1918** [see *blady *a.*]. **1925** *Chambers's Jrnl.* June 408/2 The manager..contracted with a Chinese to eradicate some four hundred acres of lalang grass on his estate. **1933** L. Ainsworth *Confessions Planter in Malaya* 190 It was decided to tackle all 'lallang' land (areas that in the past had been cleared of virgin jungle by the Chinese, but which subsequently had been..allowed to become overgrown with lallang weed). **1969** J. M. Gullick *Malaysia* ii. 67 Lallang is the Malay name for *Imperata cylindrica*, a coarse grass which grows freely on untended land in Malaya. **1972** *Malay Mail* (Kuala Lumpur) 25 May 6/3 Poisons to kill weeds and lallang..could have been absorbed by the jering tree.

-lalia, a terminal element repr. Gr. λαλιά speech, chatter, used in forming words denoting various disorders or unusual faculties of speech; as in *dyslalia* (s.v. Dys-), *echolalia, Glossolalia, *idiolalia* (s.v. *idio-).

Lalique (lalī·k). [f. the name of René *Lalique* (1860–1945), French designer of jewels and glassware.] Used *attrib.* and *ellipt.* to designate jewellery and decorative glassware by or after the manner of Lalique.

1902 *Daily Chron.* 31 May 2/3 Any jewel of uncommon design is modish, but prominently a Lalique, strung upon an imperceptible platinum chain. **1927** *Daily Express* 28 Oct. 5/3 Both tray and stopper are decorated in the Lalique manner. **1936** O. Lancaster *Progress at Pelvis Bay* 67 Its modernistic sofas, lalique panels and cleverly concealed lighting. **1939** A. Keith *Land below Wind* xiii. 224 Like figures on a Lalique vase. **1970** G. Savage *Dict. Antiques* 232/2 A good deal of Lalique glass deserves to be classified as 'antiques of the future'. **1972** *Vogue* 15 Mar. (*inside front cover*) Pair of 'Lalique' glass doves—£65·00.

Lallan. Add: **B.** *sb.* (Now usu. *Lallans.*) Esp., in modern use, a revived and modified form of the spoken dialect as a literary language.

1946 M. Lindsay *Mod. Scottish Poetry* 18 It is largely under MacDiarmid's influence..that the younger poets, especially those who use Gaelic and Scots (or Lallans, as I prefer to call their Scots, drawing more, as it does, upon middle Scots) have developed. **1947** D. Young *Plastic Scots* 3 As it is convenient to have some term of distinction for that part of Scottish literature which is written in Braid Scots or Anglic, to refer to it separately from Scots literature written in Gaelic, English, Latin, or any other tongue, I suggest '*Lallans*', adopting the term of Robert Burns. **1959** *Glasgow Herald* 29 May 8 Lallans is an artificial plaything of frustrated xenophobes. **1974** *Encycl. Brit. Macropædia* XVI. 410/1 Hugh MacDiarmid,..the most prominent exponent of Lallans, achieved an international reputation, but the Lallans revival has faded.

lallang, var. *lalang.

lallapaloosa (læ:lăpălū·să, -ză). *U.S. slang.* Also **lala-, lolla-, -palooser, -paloozer,** etc. [Fanciful formation.] Something outstandingly good of its kind.

1904 'H. McHugh' *I'm from Missouri* vi. 89 Saturday night we had our final parade with the fireworks finish, and it was a lallapaloosa! **1909** F. B. Calhoun *Miss Minerva* xxvi. 204 You sho' is genoowine corn-fed, sterlin' silver, all-wool-an'-a-yard-wide, pure-leaf, Green-River Lollapaloosas. **1911** *Dialect Notes* III. 545 A second word-list from Nebraska... *Lallapaloosa*, something fine or grand; a term of approbation. 'You have a lallapaloosa of a hat', 'That's a lallapaloosa.' **1926** Wodehouse *Heart of Goof* i. 40 To-day he had been so preoccupied with his broken heart that he had made his shots absently, almost carelessly, with the result that at least one in every three had been a lallapaloosa. **1933** A. Merritt *Burn Witch Burn!* (1934) x. 135 She thinks this doll woman a lallapaloozer. Yeah, a lallapaloozer, a corker! **1947** W. Stevens *Let.* 28 Feb. (1967) 547, I think that the book, as a book, is a lollapalooza. **1951** *New Yorker* 20 Oct. 28/3 Though I had long ago forgotten the background, characters, and plot, I distinctly remembered it as a lollapalooza. **1966** *Listener* 2 June 811/1 What is the tone of lalapalooza, since Hector was a pup, bumbershoot, hornswoggle, Milwaukee tumour, benzine buggy—or lalapalooza's equivalent, époustouflante? **1970** S. J. Perelman *Baby, it's Cold Inside* 172 All agreed that Luba Pneumatiç was a lollapaloosa, the Eighth Wonder of the World.

lallygag (læ·ligæg), *v. U.S. slang.* Also **lolly-gag.** [Origin unknown.] *intr.* To fool around; to 'neck'; to dawdle, to dally. Also as *sb.*, fooling around. Hence **la·llygagging** *vbl. sb.* and *ppl. a.*

1862 *Harper's Mag.* Aug. 324/1 Mr. Biggs paused and turned the flesh of the succulent lobster over with his finger. The gentleman inside addressed him: '..Try er lobstaw, bossy?' 'Ain't got no money,' said Mr. Biggs, still fingering the morsels. 'Oh, come now, none o' that ere lallygag,' responded the gentleman. **1868** *Northern Vindicator* (Estherville, Iowa) 30 Dec., The lascivious lolly-gagging lumps of licentiousness who disgrace the common decencies of life by their love-sick fawnings at our public dances. **1869** *Tidal Wave* (Silver City, Idaho) 15 Jan. 3/2 They are too pious to encourage dicing, and the feature of their entertainments may be what the boys call 'lally-gagging'. **1870** *Northern Vindicator* (Estherville, Iowa) 19 Feb., The weather once more is 'salubrious' and balmy, and indicates that winter will not lollygag in the lap of spring. **1880** E. L. Wheeler *Boss Bob, King of Bootblacks* vii. 9/1, I kin get lots o' jobs, if I'd take my pay in friendship an' all sech lollygag. **1910** *Sat. Even. Post* 30 July 19/1 Frank lally-gagged through his first term and came back for the second. **1927** D. Runyon *Trials & Other Tribulations* (1947) 112 When your correspondent was a 'necker' of no mean standing back in the dim and misty past, they called it 'lally-gagging'. **1949** *Jrnl. Amer. Folk-Lore* Jan.–Mar. 63 'Lally-gaggin' was Grandmother's word for love-making. **1965** 'E. Queen' *Fourth Side of Triangle* i. 2 Lallygagging around under the awning away from the gassy streets. **1969** S. Greenlee *Spook who sat by Door* xi. 95 We lolly-gag, maybe turn on, or cook up some soul food. **1971** D. Bagley *Freedom Trap* iii. 20, I said there was to be no lally-gagging around with the staff, Rearden; you just stick to doing your job. **1973** *Springfield* (Mass.) *Union* 25 Sept. 14/1 The Dow Jones average of 30 industrials, which lollygagged most of the day, gained strongly in afternoon trading.

lam (læm), *sb.*[3] *U.S. slang.* [f. *lam *v.* 3.] Escape, flight. Esp. in phr. *on the lam*, on the run; *take it on the* (or *a*) *lam* (see quot. 1935).

1897 *Appleton's Pop. Sci. Monthly* Apr. 832 *To do a lam*, meaning to run. **1904** 'No. 1500' *Life in Sing Sing* xiii. 263 He plugged the main guy for keeps and I took it on a lam for mine. **1931** [see *area-way* s.v. *area 2 b]. **1935** A. J. Pollock *Underworld Speaks* 118/2 *Take it on the lam*, to run away; escape. **1953** W. R. Burnett *Vanity Row* xiii. 94 The dolly was on the lam. **1959** *John o' London's* 10 Dec. 322/3 A young man stops the car, points a gun at them, and orders them to drive him to the border. He is a juvenile delinquent, 'on the lam' after a robbery. **1968** *Washington Post* 5 July A20/1 What useful public purpose is served by making it easy for convicts on the lam from a state penitentiary to acquire an arsenal? **1972** G. Baxt *Burning Sappho* ix. 158 Were you stalling for time while your Brunhilde takes it on the lam?

lam, *v.* Add: **3.** *intr.* To run off, to escape, to 'beat it'. *U.S. slang.*

1886 A. Pinkerton *Thirty Yrs. a Detective* 41 After he [*sc.* a pickpocket] has secured the wallet he will..utter the word 'lam!' This means to let the man go, and to get out of the way as soon as possible. **1901** *Smart Set* Oct. 3/2 Well, when he [*sc.* Uncle Remus] was just driven to desperation he 'lammed aloose', and so shall I. **1932** *Evening Sun* (Baltimore) 9 Dec. 31/5 *Lam*, run away from the police. **1935** R. E. Sherwood *Petrified Forest* II. 124 Say, boss—we better lam out of here. *Ibid.* 158, I hear a car coming, boss. We better lam. *Ibid.* 162 When they get around there, we'll lam. **1946** 'P. Quentin' *Puzzle for Fiends* (1947) xvi. 111 When I get my share, I'll lam out of this place so fast you won't see me for dust. **1959** P. Townend *Died o' Wednesday* iv. 61 What was it they always did in any self-respecting cowboy film?—lammed out..and took to the hills. **1973** M. Mackintosh *King & Two Queens* xii. 171 The time of death..[was] four days before Fisher lammed out.

Hence **la·mster, la·mmister,** a fugitive, a person on the run.

1904 'No. 1500' *Life in Sing Sing* 250 *Lamaster*, fugitive from justice; one who forfeits bail-bonds. **1926** *Clues* Nov. 161/2 *Lamster*, fugitive. Also a member of a pickpocket gang that leaves with the loot. **1948** E. L. Irey *Tax Dodgers* (1949) 47 'My line,' he said, 'is keeping quiet.' With that he had told Kelly that he was a lammister, in other words fleeing justice. **1953** W. Burroughs *Junkie* (1972) ix. 81 Gamblers, perverts, drifters, and lamsters from every state in the Union. **1962** K. Orvis *Damned & Destroyed* xii. 81 Smuggling American lamsters into Canada.

Lamarckian. A. *adj.* (Later examples.) **B.** *sb.* (Examples.)

1928 G. H. Carpenter *Biol. Insects* xii. 365 To sum up this brief discussion on the Lamarckian factor in evolution, it must be admitted that belief in it is encouraged on account of the simple manner in which it explains—if it be a true cause—many observed facts of life. **1953** E. Mayr et al. *Methods & Princ. Syst. Zool.* i. 11 Most of them [*sc.* late nineteenth-century taxonomists]..were Lamarckians. **1972** *Science* 12 May 623/1 The Lamarckian postulate that characters acquired by parents during their own lives can be passed on to their offspring. **1972** *National Observer* (U.S.) 27 May 21/3 The controversy between Darwinians and Lamarckians has raged for nearly a century.

Lamarque (lamā·ɹk). [Prob. a. the name of Comte Maximilien *Lamarque* (1770–1832), French general and politician.] A variety of noisette rose first introduced in 1830, bearing large, fragrant, white flowers with a yellow centre. Also *attrib.*

1837 T. Rivers *Rose Amateur's Guide* II. 82 Lamarque is another hybrid Noisette, approaching to the tea-scented rose, in the size and fragrance of its flowers. **1869** S. R. Hole *Bk. about Roses* viii. 122 Lamarque, the parent of Cloth-of-gold, well deserves a place on some sunny wall.. with its refined and graceful flowers. These are large and full, the outer petals of a soft pure white, the inner of a pale straw colour. **1885** C. M. Yonge *Nuttie's Father* II. viii. 101 She came in leading her little son..carrying a little bouquet for the guest of one La Marque rosebud and three lilies of the valley. **1965** G. S. Thomas *Climbing Roses* vi. 99 'Lamarque'..is a plant only for the warmer west outdoors.

lamb, *sb.* Add: **6.** *lamb-chop* [Chop *sb.*[1] 2 b] (also *fig.*).

c **1838** C. Mathews in M. R. Booth *Eng. Plays of 19th Cent.* (1973) IV. 136 He ate three pounds and a half of lamb chops. **1865** Mrs. Stowe *House & Home Papers* 248 All the edible matters..would form those delicate dishes of lamb-chop. **1962** E. Lucia *Klondike Kate* ii. 40 Mrs Bettis was persistent and her daughter was quite a lamb chop, so he finally agreed. **1963** R. Carrier *Great Dishes of World* 145 Place lamb chops in a flat dish just large enough to hold them and pour marinade mixture over them. **1974** 'E. Lathen' *Sweet & Low* xvii. 165 Deep in a choice between lamb chops and pork chops.

LAMBA

7. lamb's fry (in U.S. also **lamb fries**) [cf. Fry *sb.*² 2 b] (further examples); in the U.K., lamb's offal, esp. testicles; in Austral. and N.Z., lamb's liver.

1822 W. Kitchiner *Cook's Oracle* (ed. 4) 492 *Lamb's fry.* Fry it plain..garnish with crisp parsley. **1861** Mrs. Beeton *Bk. Househ. Managem.* 353, 1 lb. of lamb's fry. **1894** [see Fry *sb.*² 2 b]. **1936** S. E. Nash *Cooking Craft* (ed. 3) xii. 106 Lamb's fry consists of the liver, sweetbread, heart, and some of the inside fat. **1944** H. Wentworth *Amer. Dial. Dict.* 345 *Lamb fries*, lamb's testicles. **1951** *Good Housek. Home Encycl.* 530 *Lamb's fry*, sliced lamb's offal, cooked as a fry. **1963** Rombauer & Becker *Joy of Cooking* (ed. 4) 449/1 Skin, cut into quarters: a medium lamb fries. **1966** Baker *Austral. Lang.* (ed. 2) iv. 83 We could pause to consider *lamb's fry*, as a euphemism for testicles from *marked* or castrated lambs... Our later use of *lamb's fry* for lamb liver is one of our most 'refined' additions. **1969** R. & D. De Sola *Dict. Cooking* 138 *Lamb fries*, lamb testicles.

Lamba (læ·mbə), *sb.* and *a.* Also **Ilamba.** [African name.] **A.** *sb.* **a.** An African of a Bantu people in Northern Zambia and Zaïre; also used as collect. sing. = this people. **b.** The language of this people. **B.** *adj.* Of or pertaining to this people or their language.

1908 A. C. Madan *Lala-Lamba Handbk.* p. iii, The Lala and Lamba dialects are so nearly identical, and both so closely allied to the Wisa, that knowledge of either is a sufficient introduction to the other. **1919** H. H. Johnston *Compar. Study Bantu & Semi-Bantu Lang.* I. iii. 207 Lala-Lamba is spoken.. east and north of the Kafue watershed. **1937** I. Schapera *Bantu-Speaking Tribes S. Afr.* xiv. 314 In Lamba (Central Bantu) the word is *umusi*. **1948** M. Guthrie *Classification Bantu Lang.* 78 Njamba; Ilamba (Ilamba). **1949** E. A. Nida *Morphol.* (ed. 2) 51 Ilamba, a language of Tanganyika. **1950** Radcliffe-Brown & Forde *Afr. Syst. Kinship & Marriage* 221 The Lamba on the Kafue river seem to have a very similar family system. **1956** W. V. Brelsford *Tribes N. Rhodesia* vi. 46 Lamba were as a whole, because of its proximity to the copperbelt, is the most mixed area in the territory. **1965** P. Lotz in Saporta & Bastian *Psycholinguistics* (1961) 9/2 In certain languages the whole sentence is construed in a single syntactic key..as in Ilamba of Northern Rhodesia. **1967** W. M. Hailey *Afr. Survey* (rev. ed.) iii. 113 Studies made.. of the Zulu, Shona, and Lamba languages. **1957** V. W. Turner *Schism & Continuity in Afr. Soc.* viii. 255 Among the Lamba of the Ndola District of Northern Rhodesia..traceable cross-cousin marriages formed an extremely low proportion of the total marriages recorded.

Lambadar, Lambardar, varr. Lumberdar.

1883 [see Lumberdar]. **1908** *New Reformer* (Madras) II. 68 Securing the co-operation of the literate among them and the Lambardars to bring about sanitary reforms, etc. **1920** *Glasgow Herald* 12 Jan. 9/6 At one village the Lambardar was obstructive, would give no information, and refused to accompany him. **1960** J. Masters *Venus of Konpara* xxiii. 177 Our lambardar thinks she [sc. a tigress] has not started to eat yet. **1971** R. Dentry *Encounter at Kharmel* iii. 46, I am not the Wazir.. I am only the Lambadar, Raza Khan. *Ibid.* 48 Lambadar is headman. I am Mayor, you see?

lambaste, *v.* Delete *slang* and *dial.* Add: Also **lambast. b.** *fig.* To scold, castigate.

1886 *Harper's Mag.* July 321/2 With an avalanche of facts, sarcasm and ridicule..a more complete lambasting and more vigorous and thorough roasting than Wise gave Bootlea was never known. **1892** Kipling *Light that Failed* viii. 151, I only gave him his riding-orders to—to lambaste you on general principles for not producing work that will last. **1930** *Times* 13 Jan. 14/4 Mr. Maxton was heard to say, 'Mr. Chairman.. It has been said that you were going to "lambast" me at this conference.' **1938** J. Rice *Somers Inheritance* i. iv. 29 His sermons got down to the bed-rock..even if they..failed to lambaste wickedness with quite the fury it deserved. **1947** *People* 22 June 4/2 So he has castigated America for daring to interfere in Palestine, and now he lambasts Russia for meddling in the affairs of the little countries. **1951** *Oxf. Dict. Nursery Rhymes* 28 Halliwell struck a saner note (1842) but greedily copied down as facts any theories related to him, and though he lambasted Ker, he was not above speculation himself. **1956** *Jrnl. Educ.* July 304 Having myself been lambasted more than once by the Italicists because I dared to write on this subject. **1958** *Times* 16 Oct. 8/5 To this lambasting Dr. Rowse was all smiles and soothing words. **1967** *Boston Globe* 20 May 2/2 He lambasted teaching techniques, saying they have 'become a disciplinary practice'. **1969** N. Hare in A. Chapman *New Black Voices* (1972) 428 They lambasted the ultraderivation of many black intellectuals to jazz music. **1972** *Newsweek* 10 Jan. 19/1 If the economy should stay sour, the alternative script called for Mr. Nixon to lambaste the labour bosses.

lambda. Add: **4.** *Physics.* **lambda point,** the temperature (approximately 2·18 K) below which liquid helium in equilibrium with its vapour exhibits superfluidity, and at which there is a sharp maximum and apparent discontinuity in its specific heat; *transf.*, any temperature at which the specific heat of a substance exhibits similar behaviour, increasing at an increasing rate as the temperature is raised to this value and then dropping abruptly; hence *lambda curve, line* (on a phase diagram), *transition*. Freq. written as λ *point*, etc.

1932 W. H. & A. P. Keesom in *Proc. Sect. Sci. Kon. Akad. Wetensch. Amsterdam* XXXV. 742 The specific heat of liquid helium at about 2·19° K falls from a value of 3·0 to a value of about 1·1 certainly within 0·02 degree.. For convenience sake it is desirable to introduce a name for the point at which this jump occurs. According to a suggestion made by Prof. Ehrenfest we propose to call that point, considering the resemblance of the specific heat curve with the Greek letter λ, the lambda-point. **1933** W. H. Keesom in *Ibid.* XXXVI. 149 It is in this sense that..we speak of the lambda-point, and of the lambda-curve. **1940** *Physical Rev.* LVII. 437 Measurements of the temperature variation of the adiabatic and isothermal Young's and rigidity moduli and of the coefficient of thermal expansion of pressed specimens of ammonium chloride in the neighbourhood of the lambda transition at 242·8° K are reported. **1952** J. F. Allen in F. E. Simon et al. *Low Temperature Physics* iii. 73 The phase diagram of liquid helium..is crossed by a line which has been called the λ-line. **1958** Condon & Odishaw *Handbk. Physics* v. xi. 159/2 The λ transition involves no detectable change in spatial structure. **1964** *Physical Rev.* CXXXV. A1696/1 One would expect that the elastic properties of β brass near its lambda point should strongly resemble those of other solids which undergo cooperative order-disorder transitions. **1966** K. Mendelssohn *Quest for Absolute Zero* x. 234 The characteristic feature of superflow..was transport completely free of friction, taking place at a 'critical velocity' which only depended on temperature and vanished at the lambda-point.

5. *Chem.* A millionth of a litre; usu. denoted by λ.

1934 P. L. Kirk in *Mikrochemie* XIV. 13 It seems logical to use the designations mm³=μl=λ. Such a procedure could simplify discussion considerably if the letter lambda were used with this significance, and we shall in future adhere to this usage. *Ibid.*, Since 1 λ of normal solution contains 1 equivalent..1 λ contains 1 micro-equivalent. **1939** *Mikrochemie* XXVI. 32 All the drops were found to have the same volume, 1λ. **1939** E. J. Conway *Micro-diffusion Anal.* i. 4 For the actual designation of the minute quantities or volumes..we have the milligramme (mg.), and the gamma (γ or 0·001 mg.), the micro-microgramme (μγ) and the lambda (λ or 0·001 ml.)—introduced by Kirk. **1961** A. Steyermark *Quantitative Org. Microanalysis* (ed. 2) i. 2 The terms gamma (γ), and lambda (λ) are to be substituted with microgram (μg.) and microliter (μl.) respectively. **1974** *Nature* 15 Nov. p. xi (Advt.), Corning disposable micro-sampling pipettes are made from 'Pyrex' brand borosilicate glass.. The accuracy of the graduated 5 Lambda (λ) is ±1%.

b. *Nuclear Physics.* Used, usu. *attrib.*, to denote a neutral hyperon (and its anti-particle) which has a mass 2183 times that of the electron, a spin of ½, and zero isospin, and on decaying usually produces a nucleon and a pion; ? orig. applied to other hyperons also. Freq. written as Λ.

1954 *Physical Rev.* XCIII. 861/1 We have reported two examples of Λ° particles produced in hydrogen by negative pions (π⁻) of 1·5-Bev kinetic energy. *Ibid.*, We are using here the nomenclature suggested for V events at the International Congress on Cosmic Radiation, Bagnères-de-Bigorre, France. Accordingly Λ°→π⁻+ nucleon+pion +Q. *Ibid.* XCVI. 543/1 The known hyperons, Λ°, Σ⁺, have masses equivalent to 1200 and 1300 Mev, respectively. **1963** *Sci. Amer.* Jan. 40/2 When a K⁻ meson struck a proton (p), a small fraction of the collisions produced a neutral lambda particle (Λ°) and a negative and a positive pi meson. **1963** K. W. Ford *World of Elem. Particles* vi. 179 The sigma particle lives too short a time to move a measurable distance..; decaying almost at once into a lambda and a photon (Σ°→Λ°+γ). **1968** M. S. Livingston *Particle Physics* iv. 85 The clue to this was the observation of V tracks consisting of two charged-particle tracks coming from a common origin, of which one was identified as a proton and the other as a negative pion.. The neutral particle which decayed to give these tracks, however, now called the lambda-zero, (Λ°), must have had a mass greater than the sum of proton and pion masses.

Lambeg (læ·mbeg). The name of a village near Belfast, N. Ireland, used *attrib.* of the large drums traditionally beaten there on ceremonial occasions; also *absol.* Hence **La·mbegger,** one who beats such a drum.

1925 [see Dalmoney]. **18** Nov. 3/6 The booming 'lambegs' [sic], huge goatskin drums which are an important part of the Orange celebrations in Ulster. **1926** R. Flewart *In Praise of Ulster* 23 Of all the bands the Lambeggers alone are of the real vintage. *Ibid.* 24 Nothing in the world is quite like a Lambeg Band.. The Lambeg drum is a local development of four or six gigantic drums. **1949** H. Shearman *Ulster* xxiii. 299 Not far from Lisburn, on the Belfast side, is Lambeg, a place traditionally famous for drums, for it used to be a great centre for..the drumming parties which used to be so characteristic of the Orange organization. The largest type of Orange drum used to be referred to as a Lambeg drum, and when one beat a tattoo on it one was said to beat a Lambeg. **1952** D. O'D. Hanna *Face of Ulster* x. 112 Slowly and inexorably the drumming parties creep past, for the Lambeggars do not march. **1966** S. Heaney in *Listener* 29 Sept. 475/3 Orange drums, Tyrone 1966. The Lambeg balloons at his belly, weighs Him back on his haunches. **1970** *Guardian* 8 Aug. 1/5 The Lambeg drums, the noisiest and most fervent symbol of Protestant supremacy in Northern Ireland, will not be beaten in Londonderry next Wednesday.

Lambert (læ·mbɔɹt). [The name of Johann Heinrich Lambert (1728–77), German mathematician.] **1. a.** In *Cartography* used *attrib.* and in the possessive to designate certain map projections devised by Lambert, *spec.* a conical conformal projection having two standard parallels along which the scale is true. Also *ellipt.* as *Lambert*.

1879 *Encycl. Brit.* X. 207/2 A translation of this essay [of Gauss's] is to be found in the Philosophical Magazine for 1828.., where Lambert's projection comes out as a particular solution of the general problem. **1912** A. R. Hinks *Map Projections* ii. 18 When the conical orthomorphic projection is used, it is always used with two standard parallels, which is Lambert's second, or Gauss'. **1953** A. H. Robinson *Elem. Cartogr.* iii. 41 The Lambert conic projection..has concentric parallels and equally spaced straight meridians that meet the parallels at right angles.. It has two standard parallels, but the spacing of the other parallels on the Lambert increases away from the standard parallels. *Ibid.* iv. 74 The Lambert azimuthal equal-area projection is most useful when centered in the area of interest. **1971** I. G. Gass et al. *Understanding Earth* xv. 224 (*caption*) A possible geometrical fit of the southern continents at the 500-fathom..contour.. Lambert equal area projection.

b. In *Physics* used in the possessive to designate two laws enunciated by Lambert: (*a*) the intensity of the light emitted by an element of area of a perfectly diffusing surface is proportional to the cosine of the angle between the direction of emission and the normal to the surface; (*b*) (see quots. 1911, 1966).

1895 *Electrician* 20 Sept. 674/2 The diffused reflection practically follows Lambert's law. **1911** R. W. Wood *Physical Optics* (rev. ed.) xv. 437 Lambert's law states that each layer of equal thickness absorbs an equal fraction of the light which traverses it. **1952** R. W. Ditchburn *Light* xv. 441 Experimental observations on the transmission in a homogeneous medium which absorbs, but does not scatter, the light are summarized in Lambert's law, which may be written $I/I_0 = e^{-kx}$... The constant 2a is called the absorption coefficient. **1966** D. G. Brandon *Mod. Techniques Metallogr.* ii. 65 The fraction of the incident intensity transmitted through a thin slice, of thickness *t*, can be calculated from Lambert's law: $I/I_0 = \exp(-\mu t)$, where μ is the absorption coefficient.

2. (Written **lambert.**) A unit of luminance equal to one lumen per square centimetre (equivalent to approximately 3180 candelas per square metre).

1915 P. G. Nutting in *Electr. World* 6 Feb. 333/1, I prefer to speak of a brightness of so many 'lamberts'. This term is now well understood in our laboratory. Where required it is easily translated into lumens and any desired units of area. **1915** H. E. Ives in *Ibid.* 20 Feb. 460/1 Dr. Nutting..comes forward with an excellent name, 'the lambert'. But he applies it, in my opinion, to the wrong unit.. Let us say that a surface has a brightness of one 'lambert' if it is as bright as a 'Lambert's law' white surface under unit illumination. **1923** L. C. Martin *Colour* 178 The normal radiation corresponds to 3/8 candles per square centimetre when the brightness is 1 lambert. **1953** Amos & Birkinshaw *Television Engin.* I. 280, 1 lambert = 1 lumen per square centimetre = 0·3183 candle per square centimetre = 2·054 candles per square inch = 929 foot-lamberts. **1964** J. T. Oxley et al. *Basic Astronautics* iii. 37 (*table*) Sun.. Surface brightness, lamberts.. 6·25 × 10⁵. **1966** D. G. Brandon *Mod. Techniques Metallogr.* i. 58 Haine gives the minimum practical viewing brightness [of an electron microscope] as 3 × 10⁻⁴ lamberts.

Lambeth (læ·mbeþ). [The name of a South London borough.] **1.** Used allusively (chiefly *attrib.*) to refer to the Archbishop of Canterbury, whose palace is at Lambeth, or to the Church of England; esp. in **Lambeth Conference,** an assembly of the Anglican bishops, usu. held decennially at Lambeth Palace; **Lambeth degree,** a degree *honoris causa* conferred by the Archbishop of Canterbury.

1859 W. F. Hook *Church Dict.* (ed. 8) 429/2 *Lambeth degrees*, the popular designation given to degrees conferred by the Archbishop of Canterbury, who has the power of granting degrees in any of the faculties. **1867** *Times* 10 Dec. 7/3 The Bishops..complain that the necessary work of their dioceses is too much for them, and demand an addition to their numbers. That demand is not likely to be listened to while they can find time and thought for a Lambeth Conference. **1876** *Encycl. Brit.* II. 369/2 The archbishop also continues to grant degrees in the faculties of theology and law, which are known as Lambeth Degrees. *Ibid.* 694/2 In 1595..the Primate, Whitgift, accepted a series of articles proposed by Dr Whitaker of Cambridge. These, generally known as the Lambeth Articles, were strongly Calvinistic in tone. **1902** *Ibid.* XXX. 120/2 The resolutions of the Lambeth Conferences have never been regarded as synodical decrees, but their weight and influence has increased with each successive gathering; and in particular the 'Lambeth Quadrilateral' of 1888 has already had a great effect as a plan of reunion. **1941** W. Temple *Citizen & Churchman* iv. 70 It is the duty of Lambeth to remind Westminster of its responsibility to God; but this does not mean that Westminster is responsible to Lambeth. **1958** E. L. Mascall *Recovery of Unity* vii. 153 The often-quoted Statement of the Lambeth Conference Committee. **1974** J. Melville *Nun's Castle* iii. 60 She kept to a set of religious observances which.. would have caused raised eyebrows in both the Vatican and Lambeth Palace.

2. Used *attrib.* or *absol.* to designate a kind of glazed and painted earthenware manufactured in Lambeth from the 17th to the 19th century.

1863 W. Chaffers *Marks Pott. & Porc.* 131 *Lambeth.* About 1650 some Dutch potters established themselves here, and by degrees the manufacture of earthenware became important.. The ware made here was a sort of Delft, with landscapes and figures painted in blue.] **1884** C. Schreiber *Jrnl.* 23 Sept. (1911) II. 440 One of the Lambeth Wine bottles. *Ibid.* 27 Nov. 456 A cat..which he pronounced to be Lambeth (dated 1676). **1900** [see *Doulton]. **1948** F. H. Garner *Eng. Delftware* iv. 29 A Lambeth brick is illustrated. **1961** L. G. G. Ramsey *Connoisseur New Guide Antique Eng. Pott., Porc. & Glass* 27 'Lambeth' and 'Bristol' polychrome 'delftwares' of the second half of the seventeenth century. **1974** *Times* 26 Oct. 13/3 Immigrant Dutch potters had come to set up there [sc. in Lambeth], bringing with them the techniques for tin-plating earthenware which had been associated with Delft. The English products of these potters and their successors are called English Delftware, and that made in Lambeth is called Lambeth Delftware.

3. Lambeth Walk, the name of a street in Lambeth, used as the title of a Cockney song and dance first performed by Lupino Lane in the revue *Me and my Gal* in 1937.

1937 Furber & Gay (*song-title*) Lambeth Walk. **1939** *Times* 23 Mar. 14/4 The 'Lambeth Walk' and its successors have destroyed the tyranny of the foxtrot which for 20 years has made the ball-room an unsociable place. **1942** J. W. Drawbell *Dorothy Thompson's Eng. Journey* ix. 83 She can hardly pull herself away from that cheerful, excited London throng—from the people who had bought fern while the bombs fell and can still sing and dance the Lambeth Walk. **1962** *Guardian* 31 Dec. 5/1 The Lambeth Walk, invented in the thirties..to publicize 'Me and My Girl'.

la·mbda, a *lamdan*. **1948** M. Samuel *Prince of Ghetto* xvii. 251 In Radziwill there lived a *lamdan*, a scholarly Jew... All day long he studied, while his wife attended to the shop. **1970** *New Stand. Jewish Encycl.* 1174/1 *Lamdan*,..a person steeped in talmudic learning. **1974** *Times* 26 Mar. 18/3 He lives in the ideal Jewish world, where the rabbi is the lamdan.

lame, *a.*³ Restrict †*Obs.* to sense in Dict. and add: **2.** *U.S. slang.* A socially unsophisticated person; one who is not skilled in the behaviour patterns of a particular group.

1959 *Esquire* Nov. 70/1 *A lame*, one who doesn't know what's happening. A square. **1967** *Trans-action: Social Sci. & Community* Apr. 5/2 One either knows 'what's happening' on the street, or he is a 'lame'... Negroes..have contributed much to the street tongue.. Such expressions as 'a lame', 'taking care of righteous business'.. and 'soul' can be traced to Negro street life. **1968** in A. Dundes *Mother Wit* (1973) 333/1 Who's the lame who says he knows the game And where did he learn to play? **1971** *Black Scholar* Sept. 39/2 'You owe me some buns, lame!' teased a tall, lanky, yellow young man. **1973** J. Wambaugh *Blue Knight* (1973) vi. 93 They're a couple of lames trying to groove with the Kids. They're nothing.

lame, *a.* **3.** *lame duck*, substitute for def.: (*a*) (see Duck *sb.*¹ 9); (*b*) *U.S. Politics*, an office-holder who is not, or cannot be, re-elected; *spec.* (before 1933), a defeated member in the short session of Congress after a November election; also *attrib.*; (*c*) a ship that is damaged, esp. one left without a means of propulsion; (*d*) an industry, commercial firm, etc., that cannot survive without financial help, esp. by means of a government subsidy. Hence as *v. trans.* (*rare*), to help (a disabled person); to *lame-duck it*: to travel with difficulty.

1761, etc. [see Duck *sb.*¹ 9]. **1863** *Congress. Globe* 14 Jan. 307/1 In no event..could the..Court of Claims] be justly obnoxious to the charge of being a receptacle of 'lame ducks' or broken-down politicians. **1876** C. Chapman *First Ten Yrs. Sailor's Life at Sea* x. 411 A lame duck on the water, in the form of a vessel that has more or less damaged while crossing the perilous ocean. **1910** *N.Y. Even. Post* 8 Dec. 8 'Lame Duck Alley'..is the name they [sc. reporters] have given to a screened-off corridor in the White House offices, where statesmen who went down in the recent electoral combat may meet. **1922** *N.Y. Times* 6 Dec. 18/2 Senator Norris is all for the plan to have the convening of Congress moved up to avoid 'lame-duck Congresses'. **1923** *Independent* (Boston, Mass.) 21 Feb. 213/1 The proposed Constitutional amendment.. has been usually designated as the 'lame-duck' amendment. **1932** *Times* 9 Dec. 14/2 A 'lame duck' Administration was in power, and a 'lame duck' Congress still in being. **1933** P. A. Hardy *Hull Rope* xiv. 256 Our old 'lame duck' has not done so badly after all. **1943** N. Balchin *Small Back Room* 70 It's so bloody dangerous lame-ducking it home by yourself. **1963** J. Fowles *Collector* ii. 213, I want to be her friend and lame-duck her. **1970** *New Yorker* 14 Nov. 175/3 My father, with his predilection for lame ducks, was the natural person to try to rescue it. **1972** *Economist* 26 Aug. 8 The *Economist* calls lame ducks those industries whose survival is claimed to depend on state subsidy.. in the United States a lame duck is a politician whose current term is his last, owing to defeat in a primary or general election, or other reasons. **1973** *Times* 5 June 22/6 The Government, being at that stage still keen on its lame duck policy, refused to help, and the board went away to have a further think. **1973** *Listener* 29 Nov. 741/1 It is now the Congress..which will be disposing what a lame duck President may propose.

lamé (la·me). [Fr., f. *lame* Lame *sb.*¹] A material consisting of silk or other yarns interwoven with metallic threads.

1922 *Daily Mail* 16 Dec. 15/3 Fur panels trim evening gowns of lamé tissue. **1930** *Times* 13 Mar. 11/6 The collection included some beautiful Court gowns, one in pink marquisette embroidered in silk for a debutante had a silver lamé train. **1950** P. Bottome *Under Skin* xii. 206 The Paris doll, splendid in turquoise-blue taffeta under a golden lamé coat, was poised within reach of Henriette's hand. **1968** J. Ironside *Fashion Alphabet* 235 Synthetic metal yarns are now used in lamés. **1971** *Pegboss* & *Mason Christmas Catal.* 43/1 Gold..light-weight lamé jersey tube-dress.. also available in silver. £27.50.

la·me-brain. *colloq.* [f. Lame *a.* + Brain *sb.*] A dull-witted or stupid person. Also **la·me-brained** *a.*

1929 Wodehouse *Mr. Mulliner Speaking* i. 16 A girl with an aunt who knew all about Shakespeare and Bacon must of necessity live in a mental atmosphere into which a lame-brained bird like himself could scarcely hope to soar. **1945** S. J. Perelman *Crazy like Fox* 82 'Well, Miss "Lame Brain"', he retorted sardonically, 'maybe you had better stop galvanizing around nights and pay attention!' **1948** G. H. Johnston *Death takes Small Bites* vi. 163 But it's the same bunch of lame-brains kiddin' themselves they're master minds of crime. **1962** K. Orvis *Damned & Destroyed* xix. 142 Not like the usual lame-brained addict. **1968** *New Yorker* 14 Sept. 58 None of your lame-brain philosophers ever cut any ice with me. **1972** *Times Lit. Suppl.* 24 Nov. 1426/3 We have finished feeding indulgent towards the disaffected lamebrains who think this kind of stuff out.

lamella, *a.* Add: **2.** *Physics.* = *Laminar *a.* 2 (in the broader sense).

1931 W. Wilson *Theoret. Physics* I. viii. 190 Problems of lamellar flow in an incompressible fluid are mathematically identical with electrostatic problems in regions free from electric charges. **1942** M. P. Billings *Struct. Geol.* xvi. 299 In lamellar flow..the individual particles move in parallel sheets which slide over one another like the cards in a sheared playing pack. In such a viscous substance as magma beneath the surface of the earth, the flow is laminar. **1966** *McGraw-Hill Encycl. Sci. & Technol.* XIII. 175/1 In some instances, streamline flow can best be depicted as formed from thin layers of fluid which slip past each other (lamellar flow).

lamelloid (læme·l-, læ·mĕloid), *a. rare.* [f. Lamell(a + -oid.] Resembling a lamella.

1866 *Mem. Boston Soc. Nat. Hist.* I. 141 These transverse processes..possess regular lamelloid walls, so as to form rather canals than simple foramina for the artery.

lament, *v.* Add: **2.** Also with *over*.

1831 T. L. Peacock *Crotchet Castle* v. 85 He laments bitterly over the invention of gunpowder, steam, and gas. **1853** C. Kingsley *Hypatia* I. xiii. 271 Why should they lament over other things?

lamented, *ppl. a.* Add: Also *absol.* or as *sb.*, esp. in phr. *the late lamented*, someone recently dead.

1864 C. M. Yonge *Trial* I. ix. 172 Depend upon it, the late lamented will remain in the ascendant till there are no breakers ahead. **1908** *Daily Chron.* 28 Sept. 4/7 An alternative in the Greek language was 'blessed', but English can get no farther in the way of euphemism than 'the late lamented'. **1952** A. Armis *Murder, Maestro, Please* xvi. 111 You're the one whose husband identified the late lamented, aren't you? **1972** A. Hunter *Vivienne* ix. 113 He wouldn't happen to be the late lamented's husband?

‖ **lamentoso** (lamentō·so). *Mus.* [It.] A direction indicating that a passage is to be played in a mournful style.

1876 Stainer & Barrett *Dict. Mus. Terms* 251/1 *Lamentoso*, mournfully, plaintively. **1959** D. Cooke *Lang. Mus.* iii. 137 The *lamentoso* opening of the finale of Tchaikovsky's *Pathétique* Symphony. **1967** *Listener* 2 Feb. 173/3 This serene opening is followed by a striking fugue, in five parts, marked *lamentoso*.

laminagraph (læ·minăgraf). *Radiology.* Also **lamino-.** [f. Lamina + -graph.] = *Tomograph. (Originally a particular design of tomograph.)

1938 J. Kieffer in *Amer. Jrnl. Roentgenol.* XXXIX. 497/1 The laminagraph is a device embodying the principle of roentgenographic body sectioning, or planigraphy.. It was built at the instigation of, and in cooperation with, the director, Dr. Sherwood Moore, who named it. **1941** *Surg., Gynecol. & Obstetr.* LXXV. 474/1 The case..demonstrates the increase of bone detail by laminagraph as compared to the usual x-ray examination. **1960** *Jrnl. Speech & Hearing Disorders* XXV. 137/2 With a multilayer laminagraph cassette or film-holder, it is possible to record simultaneously a number of adjacent planes or body sections with a single exposure.

Hence **la:minagra·phic** *a.*, of or pertaining to laminagraphy; **la·minagram,** a radiograph taken using a laminagraph; **lamina·graphy** = *Tomography.

1938 *Amer. Jrnl. Roentgenol.* XXXIX. 507/2 The extent of the layer sharply rendered during laminagraphic motion is limited only by the size of the object roentgenographed and by the size of the film used. *Ibid.* 503/2 Satisfactory chest laminagrams can be made in one second. **1939** *Radiology* XXXIII. 560/1 The latest method has been spoken of under the name 'laminography' by various authors. *Ibid.* 560/2 Laminagraphy is more..method whereby body sections roentgenography is accomplished by motion of the tube and film in planes parallel to the plane of the body and at the level of the desired section. **1942** *Urologic & Cutaneous Rev.* XLVI. 706/1 Laminography permits the roentgenographic delineation of structures at different levels.. Laminagrams are made by synchronously moving the X-ray film and the tube carrier in opposite directions. **1962** *Jrnl. Speech & Hearing Disorders* XXV. 137/2 Laminagraphy is used most advantageously when skeletal structures are not clearly visualized with standard techniques. **1966** *Amer. Speech* XLI. 229 Clear images of vocal fold cross-sections during actual phonation by laminagraphy [in] *Amer. Speech*. **1968** H. O. Anger in Gottschalk & Beck *Fund. Probl. in Scanning* xiv. 195 A scintillation camera..is capable of..lateral and anterior views..tomographically..the subject or the object from a single rectilinear scan. **1969** R. & E. Brecher *Rays* xix. 258 The modern technique, known generally as body-section radiography but also as laminography, stratigraphy, tomography, and planigraphy, operates by imparting a reciprocal motion to the X-ray tube and film. **1971** *Jrnl. Amer. Med. Assoc.* 16 Aug. 927/2 The ability to record both dynamic studies and 'thick' laminagrams.. reduces the time necessary for the physician to view the films. **1971** *Radiology* CI. 617/1 A rounded soft-tissue mass was seen near the canal of the infraorbital nerve at laminagraphy.

laminal, *a.* Add: **b.** *Phonetics.* Produced by the blade of the tongue.

1956 C. F. Hockett in *Internat. Jrnl. Amer. Ling.* XXII. 202/1 In all the Central Algonquian dialects.. there is..an apical or laminal affricate [c].. The contrast between apical and laminal is the contrast between using the blade as opposed to the tip of the tongue in the [laminal] denti-alveolar as opposed to the apical denti-alveolar articulations. **1966** M. Pei *Gloss. Ling. Terminol.* 141

Lambic (laṅbik). Also **Lambick.** [Fr.; cf. *alambic* a still.] A strong beer brewed in Belgium.

1889 *Cent. Dict.*, *Lambick*, a kind of strong beer made in Belgium by the process called the self-fermentation of worts. **1906** *Daily Chron.* 2 Mar. 5/6 On being offered a glass of champagne he refused it, asking for some 'Lambic', a popular beverage in Belgium. **1925** G. Davison in E. Fodor *Benelux in 1952* 69 The high-density drink..is *Gueze* (so called when it is bottled, and *Lambic* when it is on tap). **1956** A. Campbell *Bk. Beer* vi. 95 In Belgium there is a beer known as *Lambic*..left to stand and mature for about two years in casks, fermenting slowly in wooden casks. **1960** [see *faro³]. **1964** L. Brewer *A to Z of Holidays Abroad* 109 A wheat-and-barley based beer is made, draught called *Lambic*, bottled *Gueze*.

lambie. (Later examples.)

1870 C. M. Yonge *Burnt Out* ix. 149 Is he hurt? Is my little lambie hurt? **1935** N. Mitchison *We have been Warned* vi. 357 Oh, Lilias, what have you got, Lambie—oh, a lovely stone.

la-mbkill. *N. Amer.* (See Lamb *sb.* 7 b.)

1790 L. Castiglioni *Viaggio negli Stati Uniti dell' America* II. 271 *K[almia] angustifolia*.. Sheep poison, Ivy, Dwarf Laurel, Lamb-kill. **1814** J. Bigelow *Florula Bostoniensis* 103 *Kalmia angustifolia*..a low shrub with rose coloured flowers, very common in low grounds, and known by the names *sheep poison, lambkill, low laurel*, &c. **1851** [see Lamb *sb.* 7 b]. **1868** C. A. Cozzens *Grey Flowers of Field* 515 Lambkill..is a low shrub.. with narrow, evergreen leaves in whorls of three. **1939** *Nat. Geogr. Mag.* May 255/1 Other poisonous members of the heath family are the kalmias, frequently called 'lambkill' from the effect they have on grazing animals. **1954** C. J. Hylander *Macmillan Wild Flower Bk.* 281 Sheep laurel..is also known as lambkill because of the severely toxic substance in the leaves. **1965** E. Richardson *Living Island* 120 Not yet completely grown up in lambkill, wild-roses and blueberry bushes. **1974** A. Huxley *Plant & Planet* xxv. 283 In North America species of *Kalmia*..are known as Sheep-kill or Lamb-kill.

Lamb shift (læm ʃift). *Physics.* [f. the name of Willis E. *Lamb* (b. 1913), U.S. physicist (who with R. C. Retherford demonstrated the effect in 1947) + Shift *sb.*] A displacement of energy levels in hydrogen and hydrogen-like atoms such that those with the same values of the quantum numbers *n* and *j* but different values of *l* are not coincident, as predicted by Dirac's theory, but separated by a very small amount (the level with the lower value of *l* being the higher).

1948 *Physical Rev.* LXXIV. 1157 The effects treated are the Lamb shift, the correction of the g-factor..and the correction of the Compton scattering cross section. **1950** *Ibid.* LXXVII. 745 (*heading*) Departure of the Lamb shift from the Z^4 law in He⁺. **1958** Condon & Odishaw *Handbk. Physics* vii. iv. 65/1 The Lamb shift has been interpreted..as resulting from changes in the electron self-energy which results from its interactions with the electromagnetic and electron-positron fields. **1964** F. A. Power *Introd. Quantum Electrodynamics* i. 7 The modern versions of quantum electrodynamics have enabled a very accurate comparison to be made between theory and experiment. Well-known examples are the Lamb shift energy splitting of the $2S_{1/2}$ and $2P_{1/2}$ levels in hydrogen and the radiative corrections to the magnetic moment of the electron. **1970** G. K. Woodgate *Elem. Atomic Struct.* iv. 69 Figure 4.5 shows the fine structure of the *n*=2 and *n*=3 levels of hydrogen, modified to take account of the Lamb shift. **1973** *McGraw-Hill Yearbk. Sci. & Technol.* 111/1 Experimental programs to measure heavy-ion Lamb shifts are now in progress.

‖ **lamdan** (læ·mdən). [Heb. *lamdān*, lit. one who has learned, f. *lāmadh* to learn.] A person learned in Jewish law; a Talmudic scholar.

1907 I. Zangwill *Ghetto Comedies* 124, I am enough of a *Lamdan* (pundit) to answer it. **1925** 'R. Leahy' *Kariel the Watchman* 98 Some asserted that Getzel was a learned man, a *lamdan*.

Laminal, pertaining to the blade (upper front surface) of the tongue; sometimes applied to palatal phonemes ([ʃ], [ʒ], [č] or [tʃ], [ǰ] or [dʒ], [j]). **1968** Chomsky & Halle *Sound Pattern Eng.* vii. 313 The difference characterized by distributed versus nondistributed does not correspond precisely to the distinction between laminal and apical.

laminar, *a.* Add: **2.** *Physics.* Of the flow of a fluid: smooth and regular, the direction of motion at any point remaining constant as if the fluid were moving in a series of layers sliding over one another without mixing; *occas.* restricted to the case in which the layers are plane (cf. *Lamellar *a.* 2).

1895 H. Lamb *Hydrodynamics* iii. 34 This analysis may be illustrated by the so-called 'laminar' motion of a liquid. **1940** H. F. P. Purday *Streamline Flow* i. 6 The term *laminar* in the strict use of the term, the fluid moves in a system of parallel planes, the velocity having everywhere the same direction, but the magnitude of the velocity is a function of the distance from some fixed plane of the system. The motion is not necessarily steady; it may, for instance, be periodic. **1965** D. A. Gilbrech *Fluid Mech.* (1966) vi. 165 In general, laminar flow occurs at low velocities, between close boundaries, when the fluid is very viscous, and when the fluid is of low density. **1968** Passmore & Robson *Compan. Med. Stud.* I. xxix. 8/2 Flow of gas through tubes is laminar at slow speeds, but at faster rates of flow molecular collisions set up eddies and the flow is then turbulent.

b. *transf.* Applied to a body whose shape is such as to produce a laminar flow of fluid in the boundary layer round it (at normal speeds).

1955 J. Kestin tr. *Schlichting's Boundary Layer Theory* xiii. 229 The prevention of transition on laminar aerofoils. **1963** *Ann. Reg.* 1962 390 Handley Page continued the development of their laminar wing project, in which hundreds of small slits in an aircraft wing were used to suck air away from the so-called boundary layer around the wing and thus to reduce friction and hence fuel costs.

laminaria (læminē⁰·riă). [mod.L. (J. V. F. Lamouroux *Essai sur les genres de la famille des Thalassiophytes non articulées* (1813) 20), f. L. *lamina* thin plate or leaf.] A thin, flat, brown seaweed of the genus so called; also known as oar-weed or kelp. Also *attrib.*

1848 A. Henfrey tr. *Schleiden's Plant* xiv. 299 For its [sc. the sea's] trees stand the Laminariae, often 30 feet long, waving their broad hands. **1857** Geo. Eliot *Jrnl.* 28 June in *Lett.* (1954) II. 356 A long stretch of fine pale sand covered with brown-tufted rocks and shining laminaria in abundance. **1883** [see Laminarian *a.*]. **1935** J. E. Tilden *Algae* vi. 269 Vendo, a Japanese physiologist, gives a recipe for making *kombu-mati*, or 'Laminaria roll'. **1963** C. I. Dickinson *Brit. Seaweeds* 82 The larger brown weeds are classified to *Macrocystis* and *Lessonia*.

laminarize (læ·minăraiz), *v. Aeronaut.* [f. Laminar *a.* + -ize.] *trans.* To design an aircraft surface) so as to maximize the area over which the flow in the boundary layer is laminar. So **la·minarized** *ppl. a.*

1960 *Times* 2 Sept. 6/3 Handley Page are about to test in flight a complete laminarized wing. **1962** *Aeroplane* CI. 428/1 This has been achieved by 'laminarizing' nearly half the metal skin of the airframe, including the fuselage forward of the cabin, and the leading-edge back to half-chord on the tailplane and fin. These surfaces are free from protruding rivet heads or skin overlap. **1966** D. Stinton *Anat. Aeroplane* 265 The most promising technique is to laminarize a larger part of the surface of an aeroplane—70 per cent or more—to bring about greatly improved lift/drag ratios.

Also **la:minariza·tion,** the design or use of laminarized surfaces.

1960 *Times* 2 Sept. 6/3 The airliner would cruise at high subsonic speed, although laminarization was still effective above sonic speed. **1964** *New Scientist* 6 Feb. 329/3 Beyond the first generation supersonic airliner, which is already half-way towards being an all-lifting surface, the next step might be laminarization..in combination with an all-wing design. **1966** D. Stinton *Anat. Aeroplane* 265 The conclusion is that laminarization leads to huge savings in fuel weight.

laminate, *v.* Add: **5.** *trans.* To unite so as to form a laminated material.

1945 H. Barron *Mod. Plastics* xi. 236 Latterly there has been a trend to use plastic fibres for weaving into fabrics which are then laminated in the usual way. **1949** B. L. Davies *Technol. Plastics* xiii. 238 The technique of laminating wood veneers using synthetic adhesives was developed to the stage when the very high strength bonds were sufficiently good for the manufacture of aircrews. **1955** Kirk & Othmer *Encycl. Chem. Technol.* XIV. 696 These adhesives are much used for laminating metal foils to paper. **1973** *Daily Tel.* 6 Nov. 5/6 It [sc. the glue] is used in laminating and veneering wood.

laminating *vbl. sb.* (examples other than in comb.).

1939 H. R. Simonds *Industr. Plastics* (1940) v. 115 The paper or other laminating material is impregnated with the varnish. **1965** *Guardian* 31 Mar. 16/2 The British laminating trade is respected even as far away as North America for the severity of its standards.

laminate (læ·minĕt), *sb.* [Substantival use of the adj.] **1.** A manufactured laminated structure or material, as: **a.** = **laminated plastic*; **b.** a fabric or a flexible packaging material consisting of two or more layers held together by an adhesive.

1939 H. R. Simonds *Industr. Plastics* (1940) v. 118 Care must be taken to keep the steel platens free from scratches and dents, for even the most minute defect on the platen will repeat itself on the surface of the laminate. **1952** Kirk & Othmer *Encycl. Chem. Technol.* VIII. 189 Paper- and fabric-base laminates have now been generally accepted as an engineering material. **1964** *McCall's Sewing* xv. 372 *Laminate*, a layer of fabric which has been fused with a layer of foam; currently the term is widely used for fabrics fused with foam rubber. **1967** *Times Rev. Industry* May 84/3 A host of combinations of film, foil and paper can provide the correct balance of product resistance, strength and light, vapour and moisture barriers.. combinations, or laminates, are made possible by the availability of modern dry-can adhesives. **1969** W. R. R. Park *Plastics Film Technol.* vi. 148 For purposes of clarity, a laminate is defined as any combination of distinctly different plastic film materials or plastic plus other materials. **1970** *Financial Times* 17 Apr. 13/8 A layer of plastic is sandwiched between thin strips of steel. This has advantages in weight and strength and can compete with other types of laminates in office partitions, gearbox casings and building cladding. **1970** O. Dopping *Computers & Data Processing* x. 138 Another static magnetic memory is the laminated ferrite memory which consists of a laminate of a number of thin sheets. **1973** *Sci. Amer.* July 30/1 Sandwich materials (such as plaster-board), which are in essence laminates with the active element (a thermostat) are constructed entirely of laminae, or layers, which taken together give the composite its form.

2. = *Lamination 3.

1968 W. E. Wills *Timber* v. 100 Laminated timber can be made with the laminates either vertical or horizontal.

laminated, *ppl. a.* Add: (Earlier and later examples.) Now common as a designation of various manufactured materials made by lamination, as *laminated glass*, a material consisting of two outer layers of plate or sheet glass attached to an inner layer of transparent plastic; *laminated plastic*, a more or less rigid material made by bonding together, usu. by means of heat and pressure, layers of cloth, paper, or the like that have been impregnated or coated with a synthetic resin; *laminated wood*, layers of wood bonded together with the grain in adjacent layers parallel (in contrast to plywood); also *laminated spring*, a leaf spring.

1665 Hooke *Microgr.* 209 Each of them consisting of an infinite number of very thin shells or laminated orbicular plates, **1888** *Lockwood's Dict. Mech. Engin.* 202 *Laminated spring*, a pressed spring consisting of thin plates superimposed one over the other, as distinguished from helical and coiled springs. **1912** *Automobile Engineer* Feb. 63/3 The main laminated springs are lighter than usual. **1930** *Engineering* 3 Jan. 26/1 The expensive of the war undoubtedly unexpected weakness in the laminated springs of motor vehicles. **1930** *Canad. Patent Office Rec.* 12 Aug. 2145/1 (*title*) Laminated glass. **1931** *Official Gaz.* (U.S. Patent Office) 1 Dec. 270/2 The process of making a low moisture absorption laminated wood product which comprises.. impregnating the dried wood under pressure with a phenolic resin..covering a plurality of laminations of the impregnated wood with a surface coating of powdered phenolic resin and molding the laminations into a homogeneous solid mass. **1933** *Product Engineering* Dec. 456/2 Laminated plastics, for example, can be given any of the usual machining operations. **1936** Laminated board [see *block-board* (*Block *sb.* 23)]. **1937** R. S. Morrell et al. *Synthetic Resins* iv. 125 Laminated material can also be made in the form of blocks, bars, rods, and tubes, by rolling the paper in special machines to the cross-section required, and then moulding them in the hydraulic press. **1938** *Encycl. Brit. Bk. of Year* 147/2 Translucent laminated plastics..made its appearance for use in instrument dials and lighting fixtures. **1939** *Chem. Abstr.* XXXIII. 3509 A pressed board consisting of 2 layers of asbestos stone bound together under pressure by a layer of resin-impregnated hard paper is a fire-resistant decorative material. **1943** A. W. Blount et al. *Marine Engin.* Jan. 57/2 Laminated shafts of phenol-formaldehyde bonded..are not made in the form of laminated board..

laminate, cont.

developments have proved that almost any two fabrics can be bonded together, and the desirable characteristics of both are retained. **1974** *Daily Tel.* 12 Jan. 5 (Advt.), Non-shrinking laminated curtains with the look of velvet.

lamination. Add: **1. c.** The process of uniting two or more layers of material so as to form a laminated material or object; the manufacture of laminates.

1945 H. Barron *Mod. Plastics* xii. 259 Low pressure lamination is now a very popular technique. *Ibid.* 269 The continuous lamination of veneers into tubing and ducts by the winding method. **1952** J. P. Casey *Pulp & Paper* II. xx. 1212 The lamination of metal foil to sulfite paper in the manufacture of candy and gum wrappers. **1967** *Times Rev. Industry* May 84/3 No plastic film is constantly used for packaging..but lamination of, say, cellulose film and polypropylene provides a combination of qualities. **1968** J. Arnold *Shell Bk. Country Crafts* xvii. 219 As far back as the sixteenth century..a Manchester bowyer, one Kelsall, laid a strip of ash along the belly of his bow, so he became a pioneer of lamination.

2. Delete 'Also *concr.* in pl. laminæ' (see sense *3).

3. Any of the layers of a laminated material or object.

1858 [in Dict., sense 2]. **1905** S. P. Thompson *Dynamo-Electr. Machinery* (ed. 7) II. iii. 173 A laminated ring core built up of laminations of iron. **1930** *Whitaker's Electr. Engineer's Pocket-Bk.* (ed. 4) 44 Many [transformer] makers prefer laminating the conductor, and, of course, insulating each lamination. **1940** 'Plastes' *Plastics in Industry* v. 59 Another form of laminated product, that made up of laminations of wood, should be especially attractive to the engineering world. **1968** J. Arnold *Shell Bk. Country Crafts* xvii. 218 The bows are built up with laminations of various woods and glass-fibre. **1971** I. G. Gass et al. *Understanding Earth* xiii. 171/1 [Sedimentary] Laminations are defined as layers less than 1 cm in thickness.

laminator (læ·minēⁱtəɹ). [f. Laminat(e *v.* + -or.] A person or organization that makes laminates, esp. plastic laminates.

1941 *Modern Plastics* Oct. 31 (Advt.), Complete lists of molders, fabricators, laminators, material manufacturers, press and equipment companies. **1952** Kirk & Othmer *Encycl. Chem. Technol.* VIII. 185 With the advent of World War II, a tremendous amount of development and testing work was instituted and carried on, not only by laminators, but by government agencies, etc. **1965** *Guardian* 31 Mar. 16/1 Some laminators tried to help their wholesale customers in the late 'fifties by farming out the wholesale business..of laminated fabrics with improved stability.

laminboard (læ·minbɔ⁰ɹd). [f. *lamin(ated)* board.] (A) composite board consisting of numerous thin strips of wood glued face to face between two facing sheets of wood (or laminated plastic).

1937 S. H. Wainwright *Mod. Plywood* iii. 15 The newer forms of laminated wood..generally known as laminboard. *Ibid.* 16 24 Laminboards, ⅛" thick or more, may be used.. for ceilings or walls. *Ibid.* viii. 44 There is no risk of the joints opening, as the laminboards cannot shrink. **1938** *Archit. Rev.* LXXXIV. 272 (*caption*) Veneered laminboard doors. **1966** Mallinson & Leigh *Timber Trade Pract.* (ed. 3) xii. 186 In 1961 out of the total import of all these materials, plywood represented 80 per cent, and laminboard and battenboard accounted for 20%, and of this the bulk was blockboard. **1971** *Cabinet Maker & Retail Furnisher* 24 Sept. 531/2 Workmen tops are laminboard, and the cupboard is faced with Warerite.

laminectomy (læmine·ktǒmi). *Surg.* [f. Lamin(a + *-ectomy.] Excision of one or more of the posterior arches of the vertebræ (each arch being formed by the junction of two laminæ), esp. as a method of access to the spinal canal.

1892 *Med. Ann.* 458 Formerly the operation was called 'trephining'.., but the trephine is no longer used, and the name laminectomy has been substituted. **1921** J. S. Horsley *Operative Surg.* xvi. 290 The operation by which tumors or other lesions of the spinal cord are approached is laminectomy. **1962** *Punch* 24 Oct. 598/1 That was the year of my laminectomy—the last ten months! **1971** *Canad. Jrnl. Surg.* XIV. 229 The spinal cord was decompressed by laminectomy from T3 to T6.

lamington (læ·miŋtən). *Austral.* and *N.Z.* [app. f. name of Lord *Lamington*, Governor of Queensland, 1895–1901.] A square of sponge cake dipped in melted chocolate and grated coconut.

1909 *Kookaburra Cookery Bk.* (ed. 2) 242 (*heading*) Lamington cake. **1944** J. K. Ewers in *Coast to Coast 1943* 61 Mrs. Whiskers making scones; even Filthy Kate enlisted to turn out lamingtons by the score, because they were her specialty. **1952** *P.W.M.U. Cookery Bk.* (9th impr.) 132 Lamingtons..spread icing on all sides of cake.. Roll in desiccated coconut. **1969** *Southerly* XXIX. 4 They had the innocence of oatmeal porridge, the sweetness of lamingtons.

la·mino-. Combining form of *Laminal *a.* b used in *Phonetics*, as **lamino-dental** *a.*, produced by pressing the blade of the tongue against the front upper teeth; **lamino-palatal** *a.*, produced by pressing the blade of the tongue against the palate.

1968 P. M. Postal *Aspects Phonol. Theory* iv. 82 Hence there are four languages with lamino-dental consonants. **1966** *Publ. Amer. Dial. Soc.* XLVI. 34 Aspirated, voiceless, lamino-palatal stops. **1968** P. M. Postal *Aspects Phonol. Theory* iv. 82 But apico-dental and lamino-palatal stops are not distinguished.

laminograph, -graphic, etc.: varr. *Laminagraph, etc.

Lammas, *sb.* Add: **4. lammas growth, shoot** *Forestry* [Equivalent of G. *Johannestrieb* St. John's shoot, in allusion to St. John the Baptist's day, 24 June], a shoot produced by a tree in summer, after a pause in growth.

1929 T. Thomson tr. *Busgen's Struct. & Life Forest Trees* i. 10 The part of the second shoot is called in that case..appears as a new growth to which the name of Lammas Shoot has been given in view of the approximate date of its appearance.. The lammas shoots of the oak are very common. **1958** *Trans. Inst. Shetland Forestry* xx. 302 So-called 'lammas growth' is common in some species.. In vigorous young oak trees the lammas shoot formation may be repeated three or four times a season. **1971** T. T. Kozlowski *Growth & Devel. Trees* I. v. 204 Lammas shoots often form in response to abundance of available water. *Ibid.* 204 Lammas growth often causes profuse branching and knotty lumber.

‖ **lammervanger** (la·mərfaŋəɹ). *S. Afr.* Also **laemer-vanger, lamvanger.** [Afrikaans, f. *lam* lamb + *vanger* catcher.] The martial eagle, *Polemaetus bellicosus*, or the African lammergeyer or bearded vulture, *Gypaetus barbatus meridionalis*.

1830 *S. Afr. Q. Jrnl.* Jan./Apr. 105 *Gypaetus Barbatus*, Cuv.—Arend and Lammervanger of the Colonists. **1835** A. Steedman *Wanderings S. Afr.* II. v. 7 Others [sc. jackals] had been destroyed by the *Lamer-vanger*, or the bearded vulture. **1846** J. C. Brown tr. *Arbousset & Daumas's Narr. Tour N.-E. of Cape Good Hope* xi. 127 The lammer-geyer of the Cape call it the *golden eagle*.., and the Dutch farmers, *lamvanger*, or *lamb seizer*, because it is accustomed to seize, and carry off to its aerie, a lamb or kid. **1920** F. C. Cornell *Glamour of Prospecting* xi. 105, I had wanted a shot on a splendid *lammer-vanger*, a fine specimen of the eagle. **1940** V. Pohl *Adventures Boer Family* xiii. 79 The *lammervanger* is one of the vulture tribe. **1962** *Smith & Boy Scouts* 108 Martial Eagle. Lammervanger. **1961** *Jnl.* Scouts 108 Martial Eagle. Lammervanger. Mr. Jonny Wahs..saw an eagle—commonly known as a *lammervanger*—catching one of his young lambs. **1970** *Daily News* (Durban) 4 June 21 A 13-year-old White youth captured a long taloned golden eagle (*lammervanger*) with his bare hands.

lammie, lammy. Add: (Examples.) Also *lammy coat, suit*.

Sometimes written as *lamby*.

1886 Lammy suit [in Dict.]. **1903** G. S. Bowles *Stretch off Land* 268 The 'lammy-suit' known aboard Torpedo-Boats and Destroyers. **1915** Kipling *Fringes of Fleet* 64, I loathe destroyers.. the smell of the wet 'lammies' and damp wardroom cushions. **1916** 'Taffrail' *Pincher Martin* xi. 191 He undid the toggles of his lammy coat, and gave the muffler another turn round his neck. **1920** *Blackw. Mag.* Jan. 72/2 North Sea fishermen, fully accoutred in their thick [lammy] suits. **1937** 'Taffrail' *Endless Story* xxiii. 357, I arrayed myself in a tolerably dry 'lammy coat'. **1948** Parnell *Dict. Forces' Slang* 108 *Lammy*, hooded coats of lamb's wool worn by officers and men in severe weather. Also known as 'duffle coats'.

lammister: see *Lam *v.*

lamnoid. Substitute for def. of **A.** *adj.*: Resembling a mackerel-shark of the genus *Lamna*. (Add examples.)

1898 D. S. Jordan *Descr. Species of Fish from Japan* title-page, The Type of a Distinct Family of Lamnoid Sharks. *Ibid.* 199 A remarkably distinct new genus of lamnoid affinities. **1925** *Fishes* (rev. ed.) xiv. 190 The most active and most ferocious of the shark tribe, as well as the largest and some of the most sluggish, belong to a group of families known collectively as Lamnoid, because of a general resemblance to the mackerel-shark or *Lamna*.

Lamoot, Lamout, varr. *Lamut.

lamp, *sb.*¹ Add: **2. b.** (Later examples.)

1899 C. Rook *Hooligan Nights* iv. 63 Ole ruby boko put 'is lamps over me, wiv no error, an' he says, 'Why you're the youngster as come in 'ere afore.' **1901** 'H. McHugh' *John Henry* 90 The old hen with the languishing lamps was still on my trail. **1928** *Daily Express* 29 Aug. 7/4 Woman in an assault case at Weymouth: I said I would her lamps for her. Clerk: What does that mean? The Woman: Blacken her eyes. **1938** F. D. Sharpe *Sharpe of Flying Squad* 331 He had his lamps on the copper.

4. a. *lamp-bulb, -chimney* (examples), *-flame, -glass* (later examples), *-glow, -house* (earlier example), *-shine, -stand* (examples), *-worm*.

1911 *Chambers's Jrnl.* Jan. 78/1 The dark film is deposited on the lower half of the lamp-bulb. **1847** Rep. *Comm. Patents 1846* (U.S.) 376, I also claim the lamp chimney, formed of glass, with two constrictions. **1879** A. S. Stephens *Married in Haste* xv. 85 She unscrewed the lamp-chimney..and polished off a stain of black smoke. **1906** Joyce *Let.* 6 Nov. (1966) II. 186 A lamp chimney here costs one lira! **1904** *Westm. Gaz.* 13 Aug. 6/1 Not a single lamp-flame stirs or quivers. **1920** J. Masefield *Enslaved* 52 The lamp-flame purred from want of oil. **1914** D. H. Lawrence *Widowing of Mrs. Holroyd* I. i. 7 She has been lighting the lamp and holds the lamp-glass. **1974** G. Jenkins *Bridge of Magpies* vii. 99 The condensation dripped from the lamp-glass. **1922** Joyce *Ulysses* 473 Their tunics bloodbright in a lampglow. **1929** H. D. Read *Shakes of Polees* vi. 63 To her nose..then takes his lamps to the lamp-house to be cleaned and trimmed before they are given out to be used. **1953** C. Mackenzie *Sinister St.* ii. xviii. 449 The uneasy warmth of the overarching trees would slope their tender lamp-shine with hushed endearments took them slowly into lamp-shine. **1938** W. de la Mare *Memory* 90 He shook his head and gently from his lamp-stand. **1889** *Punch's Almanack* 2 Dec. 16, A vinous lamp-stand. **1958** W. Wells *Tomo Ikemoto* ii. 217 a lamp-stand for in our drawing-room. ... he could now..see the glow-worm. **1909** *Westm. Gaz.* 5 July 7/1 I was in the lamp-locking cabin, through which it was hard to get the little glass-worm's tail, and the light..at the bottom of the shaft.

b. *lamp-locking* (see quots.).

1894 *Gloss. Terms Evidence R. Comm. Labour* 51/2 in *Parl. Papers* 1894 (C.7063) XXXVIII. 411 *Lamp-locking station*, the place in a mine where the safety lamps of all the miners are examined and locked by an official. **1905** *Westm. Gaz.* 10 July 7/1 I was in the lamp-locking cabin, through which it was hard to get the little glass-worm's tail, from the lamp-station at the bottom of the shaft.

c. *lamp-like* adj. (later example).

1920 D. H. Lawrence *Love Poems* 17 But the Moon.. unfurled Her white, her lamp-like shape.

5. lamp-cap, the base of an electric light bulb or lamp into which are sealed the terminals and the neck of the glass globe; **lamp-house,** the part of a photographic enlarger or projector which houses the light-source; **lamp-mat,** a mat on which a table-lamp is placed; **lamp-socket,** = *Lamp-holder; **lamp-standard,** a post or other strong support for a lamp; **lamp-work** (earlier example); **lamp-worker** (see quot. 1902).

1960 W. P. Maycock *Electr. Wiring* iii. 34 The lamp caps are fitted with a central plunger contact. **1971** L. E. Vrenken in W. Elenbaas *Fluorescent Lamps* (ed. 2) 40 The lamp cap serves to connect the electrical circuit a number of different lamp caps have been designed. **1912** J. F. Hoccs *Opening & Operating Motion Pict. Theatre* 49 (*caption*) Lamp house. **1916** R. E. Welsh *A-B-C of Motion Pict.* 17 In the first place, there is a 'lamp-house', a small cabinet which contains the light. **1933** *Discovery* Mar. 90/1 The illuminant itself [is] enclosed in a 'lamp house' and the light which it gives is the only light that can be used in the camera. **1971** L. B. Happé *Basic Motion Pict. Technol.* x. 361 Even in arc lamp can be substituted for a carbon arc in an existing lampshade optical system but it is preferable to have a complete lamp-house designed around the new source. **1842** *Spirit of Times* 15 Oct. 389/2 (Weingarten), Also to Miss Waterman..[a diploma] for numerous specimens of fancy lamp mats. **1860** *Trans. Mich. Agric. Soc.* VII. 700 Some beautiful lamp-mats and other fancy work. **1873** *Young Englishwoman* June 302/1 Embroidered border for lamp-mat. **1887** 'Mark Twain' *Life on Mississippi* xxxviii. 400 Lamp..standing on a gridiron, so to speak, made of high-colored yarns, by the young ladies of the house, and called a lamp-mat. **1893** *Scribner's Mag.* 1897 L. J. Vance *Baroque* viii. 50 Nobody even lamped its attractions. **1898** E. Wallace *Again Sanders* x. 279 These niggers have lamped the gats. **1938** G. Greene *Brighton Rock* III. iii. 161 He can do it with his lamp-socket. **1963** *Lighting Equipment News* Jan. 26/1 The two lamp socket of the fitting. **1963** *Punch* 8 Feb. 2/6 It was..sited by..the lamp-wick can be fitted to conventional lamp standards. **1968** *Daily Chron.* 5 Jan. 1/2 Ran in to the existing lamp socket. **1968** *Daily Chron.* 22 Aug. 6/2 There's a lamp standard. **1968** *Sci. Amer.* 1965 [see Lamplighter]. **1968** *Weekly Times* 31 Jan. 8 The supporting which is about 10 feet high. **1838** Kennedy *Frank Hours of Disc. Blat.* 162 The delight of a smaller lampworking..

lamp, *v.*¹ Add: **4.** *slang* (orig. *U.S.*). To see, look at, recognize; watch. Cf. Lamp *sb.*¹ 2 b.

1916 H. L. Wilson *Somewhere in Red Gap* v. 198 Stella was standing on the centre table..so she could lamp herself in the glass over the mantel. **1921** *Adventure* (U.S.) 18 July 42/2 But she lamps me again with this 10 o'clock—or what you calls it—over there. **1928** *Joyce* 1957 L. J. Vance *Baroque* viii. 50 Nobody even lamped its attractions. **1928** E. Wallace *Again Sanders* x. 279 These niggers have lamped the gats. **1938** G. Greene *Brighton Rock* III. iii. 161 He can't do nothing with lamps. Lamp him..Lamp our guy? *Ibid.* v. i. 190 Come an' lamp the bathing belles. **1953** K. Tennant *Joyful Condemned* xi. 96 One of the fellows from Central has only to lamp you coming in here, and we all go up. **1968** R. Cook *Crust on its Uppers* iii. 36 Were you dying to have a butchers and lamp all the new bird. **1969** R. Busby *Robbery Blue* iii. 26 I'd like to know where he's been lamped if he..takes the pills. **1970** J. Gardner *Air Apparent* xii. 100 'He'll be lamped us on the road.

lampbrush (læ·mpbrʌʃ). *Cytology.* Also (with hyphen) **lamp-brush.** [repr. G. *lampencylinderputzer* lit. 'lamp-glass cleaner', to which lamp-brush chromosomes were likened by J. Rückert (in *Anat. Anzeiger* (1892) VII. 115): see Lamp *sb.*¹ and Brush *sb.*²] Used *attrib.* to designate chromosomes having numerous paired lateral projections or loops, which loops are usu. apparent only during diplotene in a few groups of animals and give the whole chromosome the appearance of a bottle-brush.

[1891 *Jrnl. R. Microsc. Soc.* 135 This large nucleolus then breaks up, with the production of 'bottle-brush' and plumose figures in the caryoplasm.] **1921** *Ibid.* 436 The formation of the curious 'lamp-brush' chromosomes. **1925** E. B. Wilson *Cell* (ed. 3) iv. 359 These are the very loose so-called 'lamp-brush' chromosomes.., characteristic of the middle growth-period in large, yolk-bearing eggs. **1940** *Proc. Nat. Acad. Sci.* XXVI. 342 The typical lamp-brush chromosomes of the oocytes of lower vertebrates. **1965** Peacocke & Drysdale *Molecular Basis Heredity* vii. 72 In the newt, *Triturus cristatus*, large diplotene chromosomes may be isolated. **1958** *Nature* 18 Apr. 1036/2 Lampbrush chromosomes are thought to persist in the lampbrush form for several days in buffer. These 'lampbrush chromosomes' are sufficiently large for the analysis of enzymes on the chromosomes to be detected by light microscopy. Evidence obtained from this experiment supports the conclusion that chromosomes contain protein, RNA and DNA.

lamp-holder. Also **lampholder.** [Lamp *sb.*¹ 4 b.] A device for securing an electric lamp, a lamp-socket.

1886 *Electrician* XIV. 416/1 Fig. 1 is the lamp-holder as fitted in the Royal Courts of Justice. **1907** *Installation News* Sept. 24/2 In wiring Electroliers..it is generally found to be impossible to find room for the lamp holder inside bowls. **1935** *Discovery* June 183/2 The lamp-holder is typical of a vast number of manipulative jobs in industry—the assembly of an ordinary electric lamp-holder consisting of eighteen parts. **1963** *Times* 6 May p. viii/3 Similarly, wires in the interests of safety the traditional brass lampholder was replaced by its plastics successor, this was more or less a slavish imitation of the metal one.

lamplighter. Add: **4.** A North American freshwater sunfish of the genus *Pomoxis*, esp. the white crappie, *P. annularis*.

1877 *1st Ann. Rep. Ohio State Fish Comm.* 77 P[omoxis] *hexacanthus*..Strawberry Bass, ..lamp-lighter of Portsmouth. **1892** C. F. Lummis *Tramp across Continent* 37 We caught perch and other fish, including 'lamplighters' or crappies. **1947** E. Hemingway *A Bk. Fine...* Crappies—and the Lake Bass. **1849** B. W. Dalrymple *Panfish* 84 Here, my friend, are the various names by which you would address that little gamester, the Crappie, depending on where you happened to be at the moment: Bachelor,..Lake Bass, Lake Erie Bass, Lamplighter.

lamprey. b. lamprey-eel (earlier and later examples).

1726 S. Pennalow in *Coll. New Hampsh. Hist. Soc.* (1824) I. 35 Next day, they kill'd Lamprey Eels. **1831** *Lamprey-Eel River.* **1831** R. Cox *Adventures Columbia River* I. vii. 149 We got plenty of salmon trout..which were oily and very strong. **1885** *Amer. Naturalist* XIX. 94 They are of the lamprey eel, that is the chestnut lamprey.

lampro-. Add: **lamprophy·llite** *Min.* [ad. G. *lamprophyllit* (V. Hackman 1894, in *Fennia* XI. 119): see Phyllo-], a silicate of sodium, strontium, and titanium, $Na_2SrTi_2Si_2O_9$, found as golden-brown prisms; **lamprophyre,** add: [a. G. *lamprophyr* (C. W. Gümbel *Die paläolith. Eruptivgesteine des Fichtelgebirges* (1874) 36) (examples); **lamprophyric** *a.*, add: (examples).

1899 E. S. Dana *Dana's Syst. Min.* (ed. 6) 161 App. 40 *Lamprophyllite*..Crystals..flattened in an orthorhombic form and cleavage. **1944** *Amer. Mineralogist* XXVII. 416 The Bearpaw and Kola lamprophyllites were almost identical in refringence, habit, and orientation. **1965** *Scientia Sinica* (Peking) XIV. 1839 The lamprophyllite in general includes many others [ed] may..in the..mineral lamprophyllite..presently been known in the composition of the compound. **1923** *Geol. Mag.* LX. 553 The potash series of lamprophyres, in which orthoclase and abundance of biotite form the expression of excess of potash over soda, is not so common as the soda-lamprophyres. **1959** W. W. Moorhouse *Study of Rocks in Thin Section* xvi. 325 Lamprophyres..are for the most part melanocratic to mesocratic rocks.. Nearly all..of the true lamprophyres are characterized by a definitely alkaline aspect, indicated by the presence of abundant biotite, alkali hornblende, soda pyriboles, nepheline, or analcite, combined with a low silica content. **1966** *Jrnl. & Proc. R. Soc. New South Wales* XCIX. 38/1 Many of the common lamprophyres, such as minettes, vogesites, kersantites and spessartites are almost chemically identical. **1892** *Geol. Mag.* IX. 231 An intermixture of the acid and lamprophyric magmas took place during the intrusion of the porphyrite. **1970** *Econ. Geol.* N.Z. xiii. 205/1 Narrow lamprophyric dykes and sills cut granite, Ohika beds and the Breccia.

lamp-shade, lampshade. [Lamp *sb.*¹ 4 a.] A shade placed over a lamp to diffuse or direct the light. Also *attrib.* and *fig.*

1850 Geo. Eliot *Let.* 30 Nov. (1954) I. 337, I have bought the Luciferis and done my duty about the Lamp shade. **1877** Knight *Dict. Mech.* II. 1246/1 *Lamp-shade*, a screen placed above the light to intercept or mellow it. It may have a dark exterior and reflecting interior surface. **1899** A. Werner *Captain of Locusts* 211 He removed the burnt matches, and the lamp-shade repaired. **1908** *Stratford-upon-Avon Herald* 24 July 7/4 Lamp-shade making prize-winners are Miss Green. **1908** *Daily Chron.* 2 Oct. 4/4 There were the young ladies of gay Bohemia in Directoire dresses and lamp-shade hats. **1963** R. Macaulay *Last Lett. to Friend* (1962) 107 It didn't really *destroy* very much, except some pictures (which I mourn) and curtains and covers and a few books and lamp-shades, etc. **1960** J. Brophy *Front Door Key* 179 She had compromised with the present, babyish neo-1920's fashion, refusing to wear, along with bows and tiers and lampshade hems, very short and very tight skirts. **1967** E. Short *Embroidery & Fabric Collage* iii. 83 Care should be taken with lampshades to avoid fancy shapes and over-fussy trimmings. **1971** *Shankar's Weekly* (Delhi) 18 Apr. 18/4 He pointed to two burly figures who came out of the nearby tea-shop, one of them wiping his lampshade moustaches.

lampuki (læ·mpuki). Also **lampuca, lampuka.** [Maltese.] A large marine food fish, *Coryphæna hippurus* or *C. equisetis*; = Dolphin 2.

1925 J. A. Hammerton *Countries of World* IV. 2673/1 The lampuca, a migratory fish which comes to the islands in autumn, is caught in considerable quantities and is eaten by all classes of the people. **1958** G. G. Lanfranco *Compl. Guide Fishes Malta* 28 *Coryphaena hippurus*, Linn. ..Lampuka; easily distinguished..much valued as food. **1964** G. Butler *Coffin in Malta* iii. 80 The smell of *calamai* and *lampuki* mingling beautifully with onion and tomato grew stronger. **1969** *Vogue* Nov. 68/2 Lampuki is a delicious fish. **1975** *Times* 11 Jan. 11/6 The greatest [Maltese] delicacy is *lampuki*, which is a fish similar in size to seabass..usually grilled.

lampyrid (læ·mpĭrid). [f. mod.L. family name Lampyridæ: see Lampyrine *a.* and *sb.*] An insect belonging to the Lampyridæ, a family of Coleoptera which includes the glow-worms and fire-flies.

[**1841** E. Newman *Familiar Introd. Hist. Insects* v. 249 Glow-worms or *Lampyrites*.] **1895** J. H. & A. B. Comstock *Man. Study Insects* xxi. 550 (*heading*) The Firefly Family or Lampyrids. *Ibid.* 551 Another common diurnal Lampyrid is *Calopteron reticulatum*. **1899** D. Sharp in *Cambr. Nat. Hist.* VI. ii. v. 248 The Lampyrides, or glow-worms, are of special interest, as most of their members give off a phosphorescent light when alive. **1916** *Jrnl. Morphol.* XXVIII. 145 (*title*) Photogenic organs and embryology of Lampyrids. **1961** *New Scientist* 16 Nov. 664/1 They [*sc.* the Cantharids] include..the photogenic Lampyrids.

lamsiekte (la·msi:ktə). *S. Afr.* Also **lamziekte.** [Afrikaans, f. *lam* lame, paralysed + *siekte* disease.] A cattle disease, usually fatal, found on land deficient in phosphorus, caused by the bacterium *Clostridium botulinum* and characterized by paralysis or muscular weakness; bovine botulism. Also *attrib.*

1790 E. Helme tr. *Le Vaillant's Trav. Afr.* II. v. 92 The first [disease], called at the Cape *Lam-Sikte*, is a sudden paralitic stroke. **1798** S. H. Wilcocke tr. *Stavorinus's Voy. E. Indies* II. 64 The *lamziekte*, is when the cattle are not able to stand; it comes on gradually, and is slow in its progress. **1896** R. Wallace *Farming Industries Cape Colony* xiv. 286 Stiff-sickness or 'stijf-ziekte' and 'lam-ziekte' or paralysis, would appear to be two forms of the same disease. **1946** *Nature* 17 Aug. 239/1 In practice, 0·5–1 unit of circulating antitoxin adequately protects a bovine against natural botulism (lamsiekte). **1948** *Cape Argus* 6 Nov. 1/9 The Division of Veterinary Services ..will shortly make available a new, improved and concentrated lamsiekte vaccine. **1974** *Eastern Province Herald* (S. Afr.) 15 Nov. 15 A young couple..who lost about 70 per cent. of their income when 34 of their dairy cows died of botulism (lamsiekte), have been given three cows and lent four.

lamster: see *Lam *v.*

Lamut (lamū·t). Also **Lamoot, Lamout. a.** A branch of the Tungus people living on the shores of the Sea of Okhotsk. **b.** The language spoken by this people, belonging to the Tunguso-Manchurian group of the Altaic language family. Also *attrib.* or as *adj.* Also **Lamu·tic** *sb.*

1764 J. Grieve tr. *Krasheninnikov's Hist. Kamtschatka* III. xxi. 223 The people that they [*sc.* the Koreki] border upon are the Kamtschadales, the Tchukotskoi,..and the Tungusi or Lamuti. **1790** tr. *J. B. B. de Lesseps's Trav. Kamtschatka* II. 383 Vocabulary of the Kamtschadale, Koriac, Tchouktchi, and Lamout languages. *Ibid.*, (*heading*) Koriac... Tchouktchi... Lamout. **1830** P. Dobell *Trav. Kamtchatka & Siberia* I. ix. 189 Lamoots are the Wild Reindeer Tonguesees, who seldom inhabit one spot more than a month or two at a time. **1880** A. H. Sayce *Introd. Sci. of Lang.* II. viii. 201 To find them we must look to the ruder Lamutic and Tungusian. **1888** *Encycl. Brit.* XXIII. 608/2 On the Pacific the chief subdivisions of the race are the Lamuts, or 'sea people', grouped in small isolated hunting communities round the west coast of the Sea of Okhotsk. **1927** F. Whyte tr. *Bergman's Through Kamchatka* xii. 196 One of the older men..was..chanting monotonous Lamut melodies... The Lamuts..are nominally Christians. *Ibid.* xiii. 215 So ended the Lamut winter festival. **1957** G. Clark *Archaeol. & Soc.* (ed. 3) iii. 94 The best preserved..of the Siberian mammoths was found at Beresovka by a Lamut tribesman. **1964** tr. *Levin & Potapov's Peoples of Siberia* 670 The Evens, who were formerly known as Lamuts,..border in the north and northeast with the Yukagirs, Koryaks and Chukchi.

lanai (‖ lana·i, lānəi·). Also 9 **ranai.** [Hawaiian.] In Hawaii (and by imitation elsewhere), a porch or veranda; a roofed structure with open sides near a house. Also *attrib.*

1823 C. S. Stewart *Jrnl.* 28 Apr. (1828) v. 97 The chiefs were all under one *ranai*, or rude bower. **1826** W. Ellis *Narr. Tour Hawaii* xiv. 387 At half-past ten, the bell rung for public worship, and about 800 people..assembled under a large *ranai* (a place sheltered from the sun) formed by two large canvass awnings, and a number of platted cocoa-nut leaves, spread over the place from posts fixed in the fence which enclosed the court-yard around the house of the governor's wife. *a* **1869** L. Smith in M. D. Frear *Lowell & Abigail* (1934) 124 We soon found that the school house did not accomodate one half of the congregation; and we built a large *lanai* in the front yard and covered it with rushes for them to sit upon, a la Hawaii. **1897** 'Mark Twain' *Following Equator* iii. 61 Nearly every house [in Honolulu] has what is called a *lanai*. It is a large apartment, roofed, floored, open on three sides, with a door or a draped archway opening into the drawing-room. **1898** M. H. Krout *Hawaii* v. 93 On these verandahs or in the *lanai* the family practically lives. **1937** D. & H. Teilhet *Feather Cloak Murders* x. 174 The wide windows opening upon an immense lanai, or porch, with cool overhanging eaves. **1945** L. Mumford *City Devel.* (1946) 81 Without it, adequate gardens are impossible and the private lanai, even when provided, is stuffy. **1947** M. Lowry *Under Volcano* ix. 272 He sat on the lanai sipping okoolihao and singing plaintive Hawaiian songs. **1963** D. B. Hughes *Expendable Man* (1964) ii. 31 Sliding glass lanai doors opened to a vast expanse of close-cropped green. **1964** D. Teilhet *Big Runaround* ii. 30 She hadn't ever seen a couch this wide, she said; and I explained that it was called a lanai-couch. **1969** *New Yorker* 31 May 95/2 (Advt.), 1000 air-conditioned rooms, tower suites and lanais surrounding the..swimming pool.

Lancashire. Add: *Lancashire cheese*, a white semi-hard cheese made in Lancashire. Also *ellipt.*

1896 J. T. Law *Grocer's Manual* 811/1 The peculiarity and distinctive mark of Lancashire, as distinct from Cheshire and Cheddar, is that it is a softer cheese, a good toaster, mellower, and very palatable when ripe. **1910** *Encycl. Brit.* VII. 749/2 Lancashire cheese, when well made and ripe, is loose in texture and is mellow; it has a piquant flavour. As a rule it ripens early and does not keep long. **1937** 'G. Orwell' *Road to Wigan Pier* i. i. 15 For supper there was the pale flabby Lancashire cheese and biscuits. **1955** J. G. Davis *Dict. Dairying* (ed. 2) 195 *Lancashire cheese.* Although not well known outside the county of its origin, it is in great demand in Lancashire, especially in the industrial areas of the south, and is reckoned an excellent cheese for toasting. **1960** S. Fraser *Cheeses Old Eng.* 29 It was dark when I arrived in Preston in search of Lancashire cheese. **1971** *Sunday Times* (Colour Suppl.) 28 Mar. 36/1 *Lancashire.* When young it is slightly sharp and soft enough to spread, but it mellows with age.

Lancaster[2] (læ·ŋkæstəɹ). [Name of the county town of Lancashire.] Used *attrib.* in *Lancaster cloth* (see quot. 1950). Cf. *American cloth* (*American *a.* 3).

1939 *Archit. Rev.* LXXXVI. 258/1 *Lancaster cloth.* This is a muslin base impregnated with a linseed oil compound. **1950** '*Mercury*' *Dict. Textile Terms* 312/2 *Lancaster cloth*, a light, washable oilcloth used on shelves, tables, round wash basins, etc. Made with a cotton back and with a face dressing of linseed oil, etc. compound. **1951** *Good Housek. Home Encycl.* 166/2 A less expensive covering is provided..by Lancaster (or American) cloth. **1960** *Design* Sept. 71/2 The indicator boards..are faced with Lancaster cloth in red, grey-green and yellow.

lance, *sb.*[1] Add: **6*.** = *lance-corporal* (Lance *sb.*[1] 8). *colloq.*

1888 Kipling *Wee Willie Winkie* (1889) 74 The reg'ment don't go 'ome for another seven years. I'll be a Lance then or near to. **1961** Partridge *Dict. Slang* Suppl. 1164/1 *Lance*, lance-corporal: coll. late C. 19–20.

6.** In full, *oxygen lance.* **a.** A thin metal pipe through which oxygen under pressure may be passed in order to burn away metal, concrete, or the like using heat generated by the burning of either the metal to be cut or the pipe itself.

1925 *Iron Trade Rev.* 24 Sept. 749/1 The oxygen lance is a means of burning a hole quickly through steel, slag or brick. Essentially it is nothing but a stream of pure oxygen flowing through a small iron pipe. If the oxygen strikes hot iron or steel, the metal burns rapidly... If the oxygen strikes non-metallic substances, like firebrick or slag, the lance pipe itself burns, produces the necessary heat and flux to melt the way through. **1926** *Blast Furnace & Steel Plant* XIV. 19/1 If a layer of slag is encountered, the lance pipe itself burns. **1944** *Ibid.* XXXII. 1077/1 The oxygen lance..has been used..for opening tap holes in blast furnaces and open-hearth furnaces, for tapping slag from soaking pits, for cutting up spills and skulls, and for..piercing or severing..heavy masses of iron and steel. **1945** *Machinery* (N.Y.) Nov. 156/1 Since the oxygen lance can sever metal of practically any thickness, it is an effective 'trouble-shooter' for metal-disposal problems.

b. A metal pipe, often water-cooled, through which oxygen under pressure may be injected into molten metal or directed on to its surface.

1948 *Jrnl. Iron & Steel Inst.* CLX. 221/1 Oxygen can be used in the basic electric-arc furnace for decarburization, either by means of the 'oxygen lance' or by direction of a strong blast of the gas through the slag cover. **1950** *Ibid.* CLXV. 411/1 The use of the oxygen lance for refining a high-chromium steel..enables the heat to be worked at a temperature some 200°C higher than normal. **1959** *New Scientist* 30 Apr. 965/2 Oxygen for the refining action is injected into the through metal water-cooled jets or 'lances'. **1971** *Engineering Index 1970* 3426/2 Effect of blowing practices in the LD converter on oxygen content of steel. ..Statistical methods were employed to study this influence, with particular reference to the effect of..the height of the lance on the bath.

8. lance-bombardier, the rank in the Royal Artillery corresponding to lance-corporal in the infantry; **lance-jack** *Army slang*, lance-corporal, lance-bombardier.

1935 A. H. Burne *Royal Artillery Mess, Woolwich* xi. 230 In 1901 Driver Homewood was appointed kennel-huntsman. He has since received well merited, if not exactly rapid, promotion to the rank of Lance-Bombardier. **1943** Hunt & Pringle *Service Slang* 43 *Lancejack.* Army for Lance-Corporal or Lance-Bombardier. **1960** D. A. Campbell *Dress R. Artillery* ix. 47 In 1920 the rank of bombardier was upgraded to replace that of corporal, the latter rank being abolished in the Regiment... In the same year the appointment of acting bombardier was changed to that of lance bombardier, both these appointments wear a single chevron. **1968** *Listener* 22 Aug. 252/3 Tempting to identify with the lance-bombardier in charge of this guard squad. **1912** H. Wyndham *Following the Drum* vii. 80 A junior corporal is a 'lance-jack'. **1937** D. Jones *In Parenthesis* iii. 28 Tin soldiers, toy soldiers, militarymen in rows—you somehow suffer the pain of loss—it's an ungracious way of life—buttocked lance-jacks crawling for the second chevron. **1953** A. Baron *Human Kind* ix. 68 Foller the Salvation Corporal an' 'is Saintly Lance-Jack. **1971** L. Deighton *Declarations of War* 11 Lance-jack at the time, actually. *Ibid.*, You're not looking too good, Colonel, if you don't mind an ex-lance-jack saying so.

lance, *v.* Add: **8.** *trans.* To cut (a hole) or inject (oxygen) by means of an oxygen lance.

1945 *Machinery* (N.Y.) Nov. 156/1 After a hole had been lanced completely through, the cut was continued to the bottom of the casting. **1946** *Steel* 11 Feb. 114/2 It was planned..to drain the salamander..by drilling and lancing a hole below the taphole in the base of the furnace. **1963** *Times* 22 Apr. p. viii/6 Oxygen is lanced into the furnace as it is being tapped. This causes the slag and lead to run.

lanceolated, *a.* (Later example.)

1901 *Chambers's Jrnl.* May 348/2 The under side [of the *phiale*] is occupied by narrow lanceolated leaves.

lancet. Add: **4.** *lancet-pointed* (cf. 4 b).

1888 *Century Mag.* Aug. 585/1 These parts..are all in the Lancet-Pointed (Early English) style. **1956** *Nature* 10 Mar. 484/1 This is then crushed to a very fine paste by means of a lancet-pointed dissecting needle.

lancewood. Add: **2.** = *Horoeka.

1910 L. Cockayne *N.Z. Plants* viii. 120 The lancewood is neither *Pseudopanax crassifolium* nor *P. ferox*—it is *P. chathamica.* **1966** *Encycl. N.Z.* II. 258/2 Lancewood, Horoeka (*Pseudopanax crassifolium*). *Ibid.* 259/1 *P. ferox*, a rather rare and local tree occurring from about latitude 35° southwards, has the same juvenile form as that of lancewood.

lancing, *vbl. sb.* (Examples corresponding to *Lance *v.* 8.)

1945 *Machinery* (N.Y.) Nov. 156/1 It was decided that oxygen lancing was the only feasible means of cutting the huge piece. **1955** *Jrnl. Iron & Steel Inst.* CLXXX. 74/1 A new development is the small-scale intermittent process to convert iron in the ladle directly into steel by oxygen lancing.

land, *sb.*[1] Add: **3. e.** Also, (*for the*) *land's sake, land sakes, my land(s).*

1846 *Knickerbocker* XXVII. 18 (Th.), Jedediah, for the land's sake, does my mouth blaze? **1848** J. F. Cooper *Oak Openings* I. v. 82 Land's sake! I've forgotten all about them barrels! **1854** M. J. Holmes *Tempest & Sunshine* xvi. 223 For land's sake dont tell Tempest. **1863** A. D. Whitney *Faith Gartney's Girlhood* ii. 12 Land sakes, Miss Faith! I don't know what you mean. **1894** 'Mark Twain' in *Century Mag.* XLVII. 337/2 My lan', what de reason 't ain't enough? **1908** L. M. Montgomery *Anne of Green Gables* xiv. 141 'For the land's sake!' gasped Marilla. ..'I believe the child is crazy.' **1913** A. Huxley *Let.* 30 July (1969) 51 The Americans..say Gee, whiz, bully, my lands, my soul, [etc.]. **1916** A. Bennett *Lion's Share* xlv. 350 'My land!' exclaimed Nick. 'If he sees me here he'll think I've come on purpose to talk about him.' **1930** J. Dos Passos *42nd Parallel* 50 Land sakes, it gives me the creeps to think of it. **1952** V. Wilkins *King Reluctant* i. iii. 45 But land's sake, how did he get into dat ole lonesome graveyard? **1974** K. Benton *Craig & Tunisian Tangle* xiii. 180 We've only got another week, for land's sake.

4. d. *S. Afr.* An area of ground under cultivation; = Field *sb.* 4 a. Freq. in *pl.*

1731 G. Medley tr. *Kolb's Present State Cape Good-Hope* I. xxviii. 357 The Value of the Tenth of the Produce of Lands is computed at 14000 Florins yearly. *Ibid.* 358 The Colonies are encreasing daily, and daily taking in new Lands for Tillage. **1806** J. Barrow *Trav. S. Afr.* (ed. 2) I. i. 5 At the feet of the hills..are several pleasant farms, having gardens well stored with vegetables for the table, vineyards, and extensive corn lands. **1896** H. A. Bryden *Tales S. Afr.* 248 She had..some good tobacco 'lands', which yielded no mean profit each year. **1926** O. Schreiner *From Man to Man* 23 They burnt harpuis bushes on the lands. **1939** tr. *E. N. Marais's My Friends the Baboons* ix. 112 If he raids a land..he will..hand over to her a share of the mealies or fruit. **1941** S. Cloete *Hill of Doves* (1942) xxviii. 398 They were riding through a mealie land. **1966** E. Palmer *Plains of Camdeboo* xviii. 297 Dust enveloped the world. Maurice and Sita could not even see where the lands had been.

9. a, b. In wider use, esp. in *Engin.*: an area left between adjacent grooves, holes, or the like in any surface; e.g. that between the flutes of a twist drill or the grooves of a gramophone record, or the top of a tooth on various metal-cutting tools immediately behind the cutting edge. (Further examples.)

1907 J. V. Woodworth *Grinding & Lapping* ii. 62 The flutes of [the reamers] were milled with spaced teeth... *Ibid.* 63 Cutting the reamer sharp with no lands on the teeth. **1933** H. C. Bryson *Gramophone Record* xi. 82 The engineer has a table showing the widths and depths of the grooves and the amount of land for various cuts per inch. **1949** Baker & Kozacka *Carbide Cutting Tools* ii. 213 The land..is the portion of the tooth which is just behind the cutting edge. **1958** *Proc. IRE* XLVI. 1085/2 The diffusion regions in the lands of the grooved surface of [the silicon]..are then removed in a second step of lapping. **1962** A. Nisbett *Technique Sound Studio* 255 The groove normally used for 78 rpm recordings..about 4 mils land between grooves per inch. **1966** G. Crawford *Basic Engin. Processes* ix. 228 The lands run along the leading edge of the flutes and act as a guide in the hole already drilled. **1971** B. Scharf *Engin. & its Lang.* xi. 97 Studs. These are very useful headless fastening devices which are threaded on both ends, with an unthreaded section (land) in the middle.

10. a. *land certificate, claim, classification, deal, distribution, improvement, market, question, reclamation, reform, room, speculation, taxation, title, use, utilization, work.*

1838 in *Indiana Mag. Hist.* (1926) XXII. 451 Gentle had settled that he was to pay in land and made an assignment on a land certificate. **1867** E. Rudiger *Wills & Probate* 97 A week or so later he receives from the registry the land certificate, which is substantially the same as the original deed, but with the very important difference of having had the details of the mortgage removed from it. **1812** J. McDonough *Papers* (1898) 12 They therefore will..your knowledge of their land claims, to save by those claims before Congress. [illegible] **1974** *Guardian* 11 Jan. 1 Mrs Marcia Williams, Mr Wilson's private secretary, said last night she would not resign over the land deals affair. [illegible] **1849** *Hansard Commons* 4 May 1166 An advance of money..under the Land Improvement Act. [illegible] **1955** *Times* 4 July 8/4 His post in the Tokyo Embassy as an expert on land reform. **1871** *Leisure Hour* 8 Apr. 223/1 An aeronaut cannot get far enough from the sea in England, and requires all the landroom of a country. **1959** *Tamarack Rev.* xiv. 6 The rough half-moon of islands on the western periphery of the North Atlantic contains under 8,000 square miles of landroom for three and a half million people. [illegible] **1885** W. D. Howells *Rise S. Lapham* xx. 366 He's been dabbling in..patent-rights, land speculations. **1974** D. Francis *Knock Down* xiv. 172 It was like property development and land speculation. You could make a great deal of money without breaking the law. [illegible] **1909** *Westm. Gaz.* 20 Aug. 3/1 The land-taxation proposals of the Budget would affect directly less than a million people. **1812** J. McDonough *Papers* (1898) 11 The land titles in the present State of Florida are..in a dissatisfied state, arising from this uncertainty in which their land titles are placed. **1936** *Discovery* May 131/2 Land titles have taken nearly 20 years or more to prepare. **1935** *Ibid.* Aug. 223/1 A careful land use survey. **1961** *Listener* 7 Sept. 347/2 Recent scientific advance in land-use policy in Africa. **1971** *New Scientist* 21 Jan. 134/2 We are ahead of most countries in democratic land-use planning. **1935** *Discovery* Aug. 220/1 Land Utilisation is the problem of the moment. **1936** *Archit. Rev.* LXXX. 1 (*title*) The Land Utilization Survey of Britain: the first part of the report. **1945** R. M. Lockley *Islands round Brit.* 47 Much of the landwork is done by hand with rude implements. **1971** *Daily Tel.* 19 Nov. 13/1 By 1942, the NUS ha[illegible] 2,000 students in its summer landwork camps.

b. *land-locator, -seeker, -speculator, -worker; land-planning.*

1816 U. Brown *Jrnl.* in *Maryland Hist. Mag.* (1915) X. 364 Those present Land Locaters Surveys will hold good until the former can be Established. **1971** *Islander* (Victoria, B.C.) 30 May 5/1 The tragedy..occurred..when two land locators..came to grief on the Bear River glacier. **1936** *Discovery* Feb. 49/1 There has been a certain amount of 'land-planning', though not on the scale undertaken in the United States. [illegible] **1845** J. J. Hooper *Some Adventures Simon Suggs* iii. 37 By the time he had ridden half a mile, he overtook the land-seeker. **1946** C. McWilliams *Southern California Country* vi. 126 They sold prospective settlers so-called 'land-seekers' tickets', under an arrangement whereby the fare could later be applied on the purchase of railroad land. **1798** I. Allen *Nat. & Pol. Hist. Vermont* 24 Land speculators had..on Mr. Allen. **1873** 'Mark Twain' & Warner *Gilded Age* I. xvi. 180 It must have been a 'railroad man', or a politician, or a land-speculator. [illegible] **1887** *Andover* (Mass.) *Rev.* VIII. 154 Only the tradesworkers and the landworkers are specially considered. **1960** *Farmer & Stockbreeder* 15 Mar. 71/3 Landworkers in the Thirsk and Easingwold districts of Yorkshire.

11. *land-power.*

1928 *Observer* 1 Apr. 14/3 Sea-power took the place of land-power in the sixteenth century. **1957** [see *geo-strategy* (*Geo-)]. **1962** *Listener* 29 Mar. 543/1 A world which, seen from Moscow, is divided into three or four land-powers, and a number of similar areas which can be dominated by sea-power.

b. land moccasin (see Moccasin 3) (earlier example); **land otter** (examples); **land pike** (*a*) (earlier examples); (*b*) an inferior type of pig.

1836 M. Holley *Texas* v. 99 The land and water moccasin..are the only venomous snakes, besides the rattlers, found in Texas. **1844** Lee & Frost *Ten Yrs. in Oregon* vi. 71 Beaver was valued at two dollars per skin,..land otter at fifty cents. **1947** V. H. Cahalane *Mammals N. Amer.* 205 The river and land otter has the outline of a small seal or a very big weasel. **1687** R. Blome *Present State Isles & Territories in Amer.* 56 A Land-Pike is..[illegible] land-pike more frequently designates a thin, lank, half-wild swine.

12. land army, (*a*) (see sense 11 a in Dict.); (*b*) a corps of women established in 1917 for work on the land in wartime (in full *Women's Land Army*); also *attrib.*; **land-base, -based** *adjs.*, operating from a base on land, as opp. to one on a ship or water; **land-borne** *a.*, carried by land, effected over land; **land-bridge,** (*a*) a connection (usu. prehistoric) between two land masses; (*b*) an overland route linking countries more directly than previously, esp. one used by containerized freight; **land company** (earlier examples); **land-connection** = *land-bridge (*a*); **land cress,** a biennial herb of the family Cruciferæ, *Barbarea verna*; also, occasionally used for *B. vulgaris*; **land district** *U.S.*, one of the districts into which the state or territory is divided for matters connected with land; **land-drain** (see quot. 1967); also as *vb.*; hence **land-draining** *vbl. sb.*; **land-drainage; landfast** *a.*, firmly attached to the shore; **land fever** *N. Amer.*, eager desire for, or excitement about, securing land (cf. *gold-fever*); **landfill** orig. *U.S.*, the disposal of refuse by burying it under layers of earth; the refuse so disposed of; also *fig.*; **land-floe,** a sheet of sea-ice extending from the land; **land-gift** = *Bhoodan; **land girl,** a member of the Women's Land Army (see *land army* (*b*) above); **land grant,** a grant of land; *spec.* attrib. in **land-grant college** *U.S.*, a college set up orig. under the Morrill Land Grant Act of 1862, which donated public lands to certain States for the establishment of colleges of agriculture, etc.; **land-jobbing** (examples); **land legs** (see *sea legs* pl.), used to designate the ability to walk comfortably on land after being at sea, in a train, etc.; **land-looker** (earlier and later examples); also (*obs.*), a person claiming to have appraised the land in a given area; **land-mine,** (*a*) an explosive mine used on land; (*b*) a bomb dropped by parachute from an aircraft; ‖**landnám** [ON. *land-nám* f. *land* land, territory + *nám*, f. *nema* to take] = *land-take; **land-office,** hence **land-office business,** a thriving business, like that done in a land-office in boom times; a 'roaring trade'; **landplane,** an aircraft which can only operate from land (opp. *Seaplane); **land-poor** *a.* (*U.S.*), poor through owning much land and being unable easily to support the burden of taxation; **Land-Rover, Landrover** [trade name], a sturdy, four-wheel-drive motor vehicle designed esp. for work in rough or agricultural country; **land-scrip** (examples); **land-sealing,** hunting seals on land; **land-shark** (*b*) (further examples); † **land-sharking** *vbl. sb.* *N.Z.* (see quot. 1840) *Obs.*; **land-sick** *a.*, (*c*) sick of being on the land; (*d*) sick as a result of being on land again after a long sea voyage; **land-slide,** delete *U.S.* and add further *fig.* examples, esp. with reference to a sweeping electoral victory; **land-speed,** (*a*) speed (of an aircraft) relative to the ground; (*b*) speed on the ground (e.g. of a motor vehicle); **land-take** [ON. *land-taka*], the action of taking land; *spec.* with reference to the Norse colonization of Iceland, the land taken by a chief as his province; **land-taxer,** one who believes in, or advocates, the taxing of land-values; **land-value,** the economic value of land in all respects, especially as a basis for rating or taxation; hence **land-valuation; land wheel,** the wheel of a plough that runs on the unploughed land; **land wire** = Land-line 2 (in Dict. and Suppl.); **land-yacht,** a land vehicle similar to a yacht. Also *Land form, *Land-ship.

1917 *Times* 4 Aug. 3/4 The work of applying for the Women's Land Army will be carried on by the Board of Agriculture. **1918** *Times* 6 Feb. 3/5 The conditions under which the land army women are recruited have recently been changed. **1940** *Punch* 19 June 660/1 As soon as you join the Land Army you will find that you are in the thick of a whole lot of live stock. **1943** K. Tennant *Ride on Stranger* xxv. 275, I could always sack you, George,..and get some of these land army girls. [illegible] **1962** *Listener* 29 Mar. 540/1 Land-base missiles and sea-going missiles. **1933** Land-based [see *Based *pa. pple.*]. **1941** *Air News* Mar. 9 The intrinsic disparity between carrier- and land-based planes. **1960** *Times* 11 Feb. 11/6 Though land-based missiles cannot be 'hardened' they are still vulnerable to a fairly accurate nuclear attack. **1973** *Sci. Amer.* May 42/1, 500 [nuclear warheads] in land-based missiles. **1888** *Pall Mall Gaz.* 30 Oct. 12/1 Another class of coal—best selected brights—which are landborne, fetch at the pit mouth 10s. **1934** J. L. Myres in E. Eyre *European Civilization* I. 156 The prehistoric spread of domesticated plants and land-borne cultures. **1957** *Economist* 5 Oct. 19/4 Few [Arab states] fear Russian imperialism because, unlike Turkey or Iran, they have never felt the dead-weight of landborne pressure. **1897** W. B. Scott *Introd. Geol.* xx. 353 Fossils of land animals may demonstrate the former existence of land bridges between regions which have long been separated by water. **1898** W. Turner in *Nature* 13 Jan. 259/1 A Neolithic land bridge was produced..and a free immigration of Neolithic man with his domestic animals became possible. **1911** J. L. Myres *Dawn of Hist.* vii. 138 Some think..that the Hyksos conquest of Egypt may have been a further adventure along the land-bridge. **1941** *Manch. Guardian Weekly* 26 Sept. 194/4 There is now also a land bridge to Russia through Iran, and the Government is certain to consider whether and when we can give any military aid to Russia by that route. **1950** A. L. Rowse *England of Elizabeth* iii. 95 He cites the opinion of Master Twyne that a land-bridge once existed between Dover and Calais. **1969** *Jane's Freight Containers 1968–69* 28/1 The Port of Vancouver..put into operation the concept of the 'Land-Bridge'. *Ibid.* 30/2 The land bridge concept which foresees Canada being used as a rail link for containers moving between Europe and the Orient. **1970** *Times* 2 June (Container Suppl.) p. ii/2 What is this concept, land bridge? The term refers to the part of a movement from one port to another, which consists of an overland haul between ports. **1973** A. Quinton *Nature of Things* x. 301 It is generally believed that Britain was connected to the continent of Europe by a land-bridge at some time in the fairly remote past. **1805** *Deb. Congress U.S.* 30 Jan. (1852) 1044 Having never thought of purchasing any land from the Georgia land companies. **1852** Mrs. Knickerbocker I. 283 'Look,' said an old man..to the agent of the land company. **1876** A. R. Wallace *Geogr. Distribution Animals* I. ii. xiii. 402 There is no evidence of a former land-connection between the Australian and Neotropical regions. **1924** J. G. A. Skerl tr. *Wegener's Orig. Continents & Oceans* ii. 19 The former existence of broad land connections between continents which are widely separated at the present day can scarcely be doubted. **1957** J. K. Charlesworth *Quaternary Era* II. xxxvi. 1333 Geologic data provide[illegible] a land-connexion, so often suggested, between Australia and South America during Tertiary time. **1856** W. A. Bromfield *Flora Vectensis* 33 Mr R. [illegible] of Newchurch tells me it [*sc. Barbarea verna*] is often substituted by the people of this island [*sc.* the Isle of Wight] for the common Water Cress, being known by the opposite cognomen of Land Cress. **1878** Britten & Holland *Dict. Eng. Plant-Names* 129 Cress, Land. (1) *Barbarea præcox*, Br... (2) *Cardamine hirsuta*, L. **1944** W. J. Stokoe *Caterpillars Brit. Butterflies* 179 Wintercress *Barbarea vulgaris*..is also known as Yellow Rocket and Land Cress, to distinguish it from Watercress, which, in general appearance, it closely resembles. **1946** *Nature* 21 Dec. 920/1 Investigations under the Dairy Research Institute have included landcress

taint in cream and butter. **1969** *Oxf. Bk. Food Plants* 152/2 Winter Cress or Land Cress (*Barbarea verna*), is a useful but rarely-grown salad plant. **1812** *Deb. Congress U.S.* 9 Dec. (1853) 28 The Board of Commissioners for the western land district, in the State of Louisiana. **1831** J. M. Peck *Guide for Emigrants* 257 The State is divided into land districts, which are designated by Congress. **1883** *Rep. Indian Affairs* (U.S.) 187 An Act to create three additional land districts in the territory of Dakota. **1767** A. Young *Farmer's Lett.* 245 When the ditching is done, the next work is to land-drain the whole fields in such a manner that every part of them may be laid dry. *Ibid.* 251 In some fields..it is very difficult to tell exactly where to make the land-drains. **1841** J. F. Burke *On Land-Drainage* 4 Remains have been found of some very ancient land-drains. **1932** Blunden *Fall in, Ghosts* 9 The trickling land-drain under the culvert did not report the imminence of an enemy. **1967** *Gloss. Sanitation Terms* (*B.S.I.*) 6 *Land drain*, a drain, composed of porous or perforated pipes, laid in a trench filled with gravel, broken stone, or the like, for sub-soil drainage. **1841** J. F. Burke (*title*) On land-drainage, subsoil-ploughing and irrigation. **1950** *Engineering* CLXIX. 143/3 The book should be of great value also to designers of..land-drainage, irrigation and water-supply works. **1841** J. F. Burke *On Land-Drainage* 35 Land-draining..should never be undertaken but with a determination to do it effectually. **1926** *Daily Colonist* (Victoria, B.C.) 24 Jan. 6/4 Amundsen's experience in the Arctic has been on shipboard, on land, and on landfast polar ice. **1973** *Nat. Geographic* Mar. 350 Anchoring block and tackle to land-fast ice, all strain together to haul the bowhead out of the water. **1839** *Picayune* (New Orleans) 23 Apr. 2/2 Then came the *land fever*, which swept over the country like a pestilence. **1845** C. M. Kirkland *Western Clearings* 4 In the days of the land-fever. **1900** E. B. Osborn *Greater Canada* 60 Many years passed before the North-West recovered from the commercial lethargy which followed this attack of land-fever. **1946** E. Hodgins *Mr. Blandings builds his Dream House* 16 Then, suddenly the land fever seized them. **1972** J. Minifie *Homesteader* vi. 40 Many of the harvesters were bitten by the land fever, and filed on land for themselves once the harvest was over. **1942** in *Sun* (Baltimore) (1944) 10 Feb. 8/1 The so-called sanitary or land fill [system]. **1953** *Richmond* (Va.) *News Leader* 2 Sept. 21 A bulldozer struck water in the landfill dump area. **1967** *Boston Sunday Globe* 23 Apr. 20/3 By 1970, it is expected that the majority of the area's dumps and landfill sites will be filled to capacity. **1969** *New Yorker* 17 May 131/2 We intend to put a lot of landfill in the Credibility Gap. **1971** *Pollution: a Review* (Greater London Council) 8 The Greater London Council..operates a code of practice for good management of landfill sites under its own control. **1971** *Guardian* 11 Oct. 3/6 Landfill is considered one of the most economical ways to dispose of refuse. **1823** W. Scoresby *Jrnl. Voy. Northern Whale-Fishery* iv. 101 The drift of the ice towards the south-west,..for three weeks preceding our entrance amid the land floes, had averaged seven or eight miles a-day. **1866** C. E. Smith *Diary* 20 July in *Listener* (1969) 17 Apr. 525/2 We are unable to stir, with a tremendous land-floe on one side of us and, on the other side, a body of ice extending as far as we can see from the mast-head. **1939** *Beaver* June 31 By May literally hundreds of thousands [of eider ducks] have arrived to feed in the sea and rest idly on the edge of land-floe and ice-pan. **1953** Land-gift [see *Bhoodan]. **1957** *Listener* 30 May 889/3 The Land Gifts Movement..aims at persuading landowners large and small to surrender voluntarily a sixth of their land for distribution to the poor. **1964** T. Zinkin *India* vi. 125 Vinoba Bhave..managed to create such a response for his 'Bhoodan'—land-gift—that the bitterness on which the communists had thrived in Telengana vanished. **1918** *Times* 20 Mar. 9/4 The land girls [had] little felt hats and smocks and their red badges of service. **1919** 'I. Hay' *Last Million* 81 We have consorted with..Farmers, Hedgers, and Land Girls. **1928** 'R. Crompton' *William—the Good* iv. 103 He found his sister Ethel wearing a neat land girl's costume and weeding a bed. **1940** *Manch. Guardian Weekly* 8 Nov. 325 One German pilot even turned his guns against land girls working in the fields. **1958** *Times Lit. Suppl.* 11 July 399/5 A young Tunisian land-girl and youth leader. **1974** M. Cecil *Heroines in Love* vii. 175 Down on the farm the Land Girl was swept off her feet by the farmer. **1862** *N.Y. Tribune* 21 Mar., Some years since, the movement for a Pacific Railroad, attended by an enormous land-grant, assumed proportions that indicated the probable success of the movement. **1869** *Bradshaw's Railway Manual* XXI. 431 Expended.. Land grant expenses—$7,205. **1889** *Century Mag.* Jan. 404/2 The land-grant colleges graduate men fitted to superintend farms and workshops. **1900** *Daily Chron.* 28 Aug. 5/1 At the present time no land-grants to emigrants are being made by the Natal Government. **1943** J. S. Huxley *TVA* vi. 30 In 1862..Land Grant Colleges were established—so called because in every State lands were granted from the public domain to endow a College for the teaching of 'Agriculture and the Mechanic Arts'. **1944** F. Clune *Red Heart* 5 He was..a hander-out of liberal land-grants to sycophantic favourites. **1962** H. R. Loyn *Anglo-Saxon Eng.* iv. 158 If arable is at the centre of the land-grant, connected rights in meadow, pasture.. and wood were closely associated with it. **1967** Mrs. L. B. Johnson *White House Diary* 14 Mar. (1970) 498 Federal participation in education is not exactly new, going back as far as 1785 in the Land Ordinance, the land grant colleges of the 1860's, [etc.]. **1781** in *Mass. Hist. Soc. Coll.* (1814) 2nd. Ser. I. 186 Toryism, British interest, and Land-jobbing views, combine numbers without and within doors. **1885** *Century Mag.* Apr. 826 When the bill to establish a State park at Niagara was on its passage,.. the great majority of the country members were opposed to it, fearing that it might conceal some land-jobbing scheme. **1871** *City-Road Mag.* I. 242/1 If Mr. Goschen has had to get his sea-legs on, Jack finds it as difficult to put on his land-legs. **1927** *Sunday Times* 6 Mar. 23/4 The tourists will disembark..and proceed to Teignmouth to spend eighteen days recovering their 'land legs' and developing combination. **1938** H. Nicolson *Let.* 17 Apr. (1966) 337 Have you..recovered your landlegs as yet? After three days in the train one feels the room rocking like after three days at sea. **1840** *Knickerbocker* XVI. 206 Another class of operators..became popularly known as 'land-lookers'. These met you at every turn, ready to furnish 'water power', 'pine-lots', 'choice farming tracts' or any thing else, at a moment's notice. **1845** C. M. Kirkland *Western Clearings* 6 These blunders called into action another class of operators, who became popularly known as 'land-lookers'. **1893, 1900** Landlooker [see *Cruiser 3]. **1902** S. E. White *Blazed Trail* xvi. 116 This is the usual method of procedure adopted by land-lookers everywhere. **1890** *Electrician* 16 Mar. 502/1 Land Mines. These mines..are intended to be placed a few inches below the surface of the ground, and are so constructed that they..fire themselves electrically or mechanically when the measure of the weight of a man is brought to bear upon them. **1915** R. W. Campbell *Private Spud Tamson* xix. 288 A terrific explosion of land mines, which burst beneath the feet of the enemy. *a* **1917** E. A. Mackintosh *War, the Liberator* (1918) 134 Two sappers brought up land mines and laid them. **1940** *N. & Q.* 21 Dec. 440/2 Up to September of this year a land-mine.. signified a receptacle filled with explosive and concealed immediately below the surface of the ground... In popular parlance it has come to mean a mine, that is to say a thin metal container holding a large quantity of explosive, dropped from an enemy aeroplane upon the land. **1959** *Chambers's Encycl.* II. 413/1 The effectiveness of the 'land-mines' dropped by the Luftwaffe on Britain. **1968** M. Richler in R. Weaver *Canad. Short Stories* 2nd Ser. 183 Had a little disagreement with a land mine, son. **1973** 'R. MacLeod' *Nest of Vultures* vi. 129 British anti-personnel Claymore land-mines. **1858** G. W. Dasent in *Oxford Ess.* 185 Chief after chief coming out [to Iceland].. settling himself on some great chief's lot or landnám, who allotted him a portion on condition of the acknowledgement of his supremacy. **1877** C. A. V. Conybeare *Place of Iceland in Hist. European Inst.* 28 The Goðorð was no doubt intimately connected with the landnám of the most powerful of the immigrants. **1915** K. Gjerset *Hist. Norwegian People* (1932) xxv. 140 The chieftains.. claimed large tracts of land by right of settlement and occupation..while the freemen..with their consent, settled in their *landnám*. **1839** *Picayune* (New Orleans) 2 Apr. 2/3 A practical printer..could do a land-office business here. **1877** 'Mark Twain' in *Atlantic Monthly* Nov. 590/1 Naturally, the prophets of Baal took all the trade. Isaac..went a-prophesying around, letting on to be doing a land-office business, but 't wa'n't any use. **1882** Land-office business [in Dict.]. **1935** M. M. Atwater *Murder in Midsummer* v. 51 He was doing a land-office business in gas and pop and candy. **1951** E. Paul *Springtime in Paris* xi. 203 American students..used to do a land-office business in contraband cigarettes. **1972** *New York* 12 June 35/2 Allen & Co...was doing a land-office business touting Planet Oil. **1923** *Daily Mail* 23 June 5 Among landplanes there are huge new troop-carriers. **1932** *19th Cent.* Feb. 205 One squadron of flying-boats and one of torpedo-bomber landplanes. **1941** E. C. Shepherd *Military Aeroplane* 27 The Coastal Command..has.. landplane reconnaissance craft which can also carry bombs. **1942** *Tee Emm* (Air Ministry) II. 61 Land planes are not designed for alighting on the sea. **1969** K. Munson *Pioneer Aircraft 1903–14* 152/2 Unlike Fabre's seaplane, however, this was both a landplane and a biplane, with a twin-girder 'fuselage' on which was lightly attached an aluminium nacelle encompassing side-by-side seats for pilot and passenger. **1873** J. H. Beadle *Undevel. West* 781 In the country, the old settlers are 'land-poor'—so rich that they can not pay their taxes. **1914** *Collier's Mag.* 31 Jan. 22/2 The land-poor farmer is a well-known institution in the Middle West. *a* **1953** E. O'Neill *Long Day's Journey* (1956) iv. 125 All I told them was I couldn't afford any millionaire's sanatorium because I was land poor. **1948** *Trade Marks Jrnl.* 29 Sept. 786/2 Land-rover... Land motor-vehicles and parts thereof... The Rover Company Limited. **1948** *Motor* 3 Nov. 381/1 Also exhibited is the Land Rover, as a closed estate car with seven-seat capacity, a go-anywhere, four-wheel-drive model powered by the '60' engine. **1953** *New Statesman* 13 June 696/3 Commuting barristers and stockbrokers in their shooting brakes and land-rovers. **1959** *Times Lit. Suppl.* 24 Apr. 243/4 It is one of the few recent books about the Sahara desert in which there is no mention of a Land-Rover. **1960** *Times* 5 July (Agric. Suppl.) p. iv/1 The Land-Rover is as much part of the farming scene as the cattle or the sheep. **1971** *Country Life* 25 Feb. 436/2 Then came the Landrover, also a multi-purpose, cross-country vehicle. **1834** A. Jackson in *Messages & Papers of Presidents* (1896) III. 52 Mr. St. Clair..had permitted the clerk in his office to be the agent of speculations in land scrip. **1848** *Indiana Gen. Assembly Doc.* (1849) I. 181 Such land Scrip as had been issued on the Wabash and Erie Canal. **1862** *Congress. Globe* 10 June 2628/1 There is no railroad company..that has the right to locate land scrip. **1943** L. V. Hamner *Short Grass* 174 Surveyors.. bought up a lot of land scrip for almost nothing. **1911** *Chambers's Jrnl.* July 475/2 In the land-sealing..thousands of fur-seals are driven and forced onwards. **1829** in *Ohio Archaeol. & Hist. Q.* (1939) XLVIII. 331 The Counsel is sure to be supported by the presiding Judges.. & thus the Property of Society is Confiscated Legally between these *Land Sharks*. **1839** J. D. Lang *N.Z. in 1839* i. 14 A class of persons in that Colony [*sc.* New South Wales] who were known by the name of Land Sharks.. have turned their eyes all at once to New Zealand. **1865** C. F. Hursthouse *Lett. on N.Z. Subjects* 89 '*Land Sharks*', twenty years ago this was a term rife in Australia and New Zealand. *a* **1910** 'O. Henry' *Rolling Stones* (1915) 218 A class of land speculators commonly called land sharks, unscrupulous and greedy. **1935** A. Sullivan *Great Divide* 342 The Metis are being stirred up by the land sharks to demand their scrip, then the sharks will swallow them. **1839** *Colonial Gaz.* 28 Aug. 627/2 Land-sharking means pretending to purchase, but really obtaining somehow, land from the natives. **1840** *N.Z. Company Rep.* I. 31 The practice of land-sharking, or the acquisition of land from the barbarous natives by private persons, without any reserves for the use of the natives, or indeed any sort of regard for their just rights. **1855** C. W. Richmond *Let.* 28 Apr. in *Richmond–Atkinson Papers* (1960) I. 162 Such agreements favor landsharking and tend to produce strife and contention. **1888** L. A. Smith *Music of Waters* 219, I could understand any land-sick lad longing for a sea-life if he once heard this ballast-throwing song. **1908** *Westm. Gaz.* 13 Feb. 2/1 The joy of the land-sick sailors who cried, 'The sea, the sea!' **1908** *Daily Chron.* 10 June 4/4 It was very curious, that first step ashore... I was thoroughly land-sick. **1922** D. H. Lawrence *Let.* 5 Sept. (1962) II. 714 We were twenty-five days at sea and are still landsick—the floor ought to go up and down, the room ought to tremble from the engines, the water ought to swish around but doesn't, so one is landsick. The solid ground almost hurts. **1924** —— & Skinner *Boy in Bush* 19 Jack was a little tired and a little land-sick, after the long voyage. **1888** *N.Y. Times* 4 Nov. 5/1 A veritable landslide in Mr. Hewitt's favor. **1896** *Westm. Gaz.* 6 Nov. 7/1 We were justified in urging our readers yesterday to accept with caution the earlier views of the extent of McKinley's majority. It is not a 'landslide'. **1936** *Punch* 23 Sept. 362/1 The first volume of the long expected biography of *Arthur James Balfour* (which takes us as far as the Conservative landslide in 1906). **1946** *Ann. Reg. 1945* 239 Another undoubted shock was the Labour landslide in the British general elections. **1955** *Times* 9 May 11/4 The electoral landslide which swept into power the hastily organized Labour Front. **1974** *Times* 27 Feb. 18/5 Modern Toryism..wants a land-slide victory. **1910** R. Ferris *How it Flies* xx. 464 *Land-speed*, the speed of aircraft as related to objects on the ground. **1935** Eyston & Lyndon *Motor Racing & Record Breaking* vi. 56 Record breaking can be somewhat grim, as is shown by the land-speed attempts at Daytona. **1963** *Times* 2 May 11/2 Donald Campbell reached 110 miles an hour on the salt flats here today when his turbo-jet car Bluebird made a first trial run for his world landspeed record bid. **1971** *Guinness Bk. Records* (ed. 18) xi. 160 The highest land speed recorded by a woman is 335·070 m.p.h. by Mrs. Lee Ann Breedlove. **1906** *Ann. Rep. Board of Regents Smithsonian Inst.* 287 Until a Parliament for Iceland was established in 930 these chieftains were the rulers of the island, each in his district or land-take (landnám), as it was called. **1908** W. G. Collingwood *Scandinavian Brit.* 193 In each landtake the bóndi fixed his homestead, neither on the exposed hill-top, nor on the marshy flat. **1927** E. V. Gordon *Introd. Old Norse* 235 The method of land-take used by settlers in Iceland; they carried fire through the land they were to occupy, and around its limits. **1905** *Westm. Gaz.* 13 Apr. 4/1 The land taxers have an idea that valuable sites are being held back by grasping ground landlords. **1909** *Daily Chron.* 30 Apr. 1/6 As land-taxers, we are thoroughly satisfied that we have got a complete system of land valuation. **1928** *Daily Express* 6 June 2/4 Colonel Wedgwood, the famous Socialist land-taxer. **1851** *Fraser's Mag.* XLIII. 117 Luckily..for railway companies,..land-valuation is a remarkably elastic art. **1908** *Daily Chron.* 6 Aug. 8/3 The land-valuation proposals of the Government. **1880** H. George *Progress & Poverty* viii. ii. 365 To abolish all taxation save that upon land values. **1900** W. Smart *Taxation of Land Values* 38 Of late years we have heard much of a proposal called the taxation of land values. **1908** *Westm. Gaz.* 20 Feb. 2/2 The rates charged on the land-value basis. **1962** H. R. Loyn *Anglo-Saxon Eng.* viii. 319 Considerable variation in land-values..occurred between 1066 and 1086. **1743** W. Ellis *Mod. Husbandman* Sept. iv. 27 The Land Wheel being obliged to go on the Turf its Share is kept too high. **1960** *Farmer & Stockbreeder* 15 Mar. 102/3 For a one-man unit the spreader should be land-wheel-driven for ease of hitching on and off. **1970** G. E. Evans *Where Beards wag All* ii. 46 A two-horse iron plough with round coulter,..land and furrow wheel. **1972** *Country Life* 10 Feb. 321/3 It [*sc.* a plough] had no land wheel, so that depth had to be kept by bearing on the stilts. **1876** Preece & Sivewright *Telegraphy* v. 128 Between London and Amsterdam there are 130 miles of land wire over the Great Eastern Railway, then a cable 120 miles long, and then 20 more miles of land wire. **1908** *Westm. Gaz.* 24 Feb. 4/1 The..cable from Ascension touches land in Cornwall.., whence a land-wire passed the signals on to Greenwich. **1930** *Aberdeen Press & Jrnl.* 23 Jan. 7/6 A microphone was installed at 10 Downing Street, and the Premier's words were carried by land wire to Chelmsford. **1928** *Daily Express* 26 May 9/3 There was shown at Olympia last year a 'land-yacht' that was palatial in its appointments. **1967** *Times* 23 Jan. 9 A school that has its own land yacht, wind tunnel, go-kart and canoeing clubs.

‖**Land** (lant), *sb.*[2] Pl. **Länder** (le·ndər), **Laender, Lands.** [G.] A semi-autonomous unit of local government in Germany and Austria.

[**1920** G. Young *New Germany* 321 The transformation of this Constitution from a centralised republic..back to a federation has been reviewed already. The word 'lander' is literally translated for this and other reasons.] **1920** H. W. V. Temperley *Hist. Peace Conf. Paris* III. 347 The word *Länder*..has been deliberately used instead of the word *Staaten*... The word *States* for the members of a federal Constitution seems therefore to be misleading as expressly repudiated and 'Lands' is used, a new word coined by Professor Young. *Ibid.* 348 Article 5. Constitutional power is exercised..in matters pertaining to the Lands, by the Constitutional bodies of those Lands within the lines laid down by the constitutions of those Lands. **1950** Theimer & Campbell *Encycl. World Politics* 42 Austria became a federal republic, consisting of eight *Lands.* **1955** *Times* 26 Aug. 7/2 Many wage agreements are settled on a Land and not a federal basis. **1958** *Listener* 9 Oct. 571/1 In the Laender under its sway..it [*sc.* Austria] produced the nearest approach to the Welfare State that existed before its establishment in the United Kingdom. **1966** *Economist* 13 Aug. 633/1 The ambitious and consequently hard-pressed *Länder* are demanding that..Bonn's share of the tax-collectors' booty should be no more than 35 per cent. **1969** *Nature* 15 Nov. 633/2 The German universities are at present the responsibility of the Länder. **1973** *Times* 30 Jan. 4/6 The four *Länder*—Hamburg, Bremen, Lower Saxony and Schleswig-Holstein—have asked the Bonn Government to sanction a boycott.

land, *v.* Add: **1. c.** *pass.* In Canada, to be given the status of a landed immigrant (see *Landed *ppl. a.* 3).

1910 [see *Landed *ppl. a.* 3]. **1962** *Canada Month* Aug. 16/3 They arrived from an Italian refugee camp in three groups around mid-month, were duly 'landed' by immigration officials. **1974** *Globe & Mail* (Toronto) 16 May 3/3 So far 22,905 have actually been 'landed'—given legal status as landed immigrants—and it's just a matter of time before most of the others achieve the same goal.

2. e. (Earlier and later examples.) Also *intr.* with *out.*

1886 H. Baumann *Londinismen* 93/1 He landed him a little one on his left ogle. **1898** J. D. Brayshaw *Slum Silhouettes* 2 That on'y made Bill madder 'n ever, an' 'e lands aht wiv 'is right, but the Gent. jest ketched 'is arm. **1912** *Chambers's Jrnl.* June 395/2 After sparring for five minutes, and frustrating every attempt you made to 'land' on him, he would sit down. **1928** *Manch. Guardian Weekly* 5 Oct. 274/3 Why didn't his man 'land out' at the insulting blighter?

g. *Machine knitting.* To secure (a loop) on the closed beard of a needle.

1885 [see *Knock *v.* 13 c]. **1926** J. Chamberlain *Knitting Math. & Mech.* v. 98 Using different lengths of beards in the same machine may result in certain loops not being landed, and consequently not cast off. **1952** D. F. Paling *Warp Knitting Technol.* i. 6 The old fabric loops on their upward movement pass over the tips of the beards which are embedded in the needle eyes, and the old loops are landed on to the closed beards.

h. To bring (an aircraft) to earth from the air; to place (an aircraft or spacecraft, or its contents) on the ground or some other surface after a flight.

1916 H. Barber *Aeroplane Speaks* 49 I'll guarantee to safely land the fastest machine in a five-acre field. **1926** *Encycl. Brit.* I. 65/2 Attempts were later made to land machines on this forward deck [of the aircraft carrier]. **1931** *Times* 19 Feb. 17/2 There was a difference of opinion as to who should land the flying boat?—Very definitely. **1932** W. E. Johns *Camels are Coming* ii. 35 Agents..are usually taken over by aircraft; sometimes they drop by parachute and sometimes we land them. **1948** Gregory & Allan *Helicopter* xvi. 190 There are a lot of things that we have to do to this machine before you can take off and land it. **1952** K. W. Gatland *Devel. Guided Missile* vi. 103 (*caption*) Instead of landing the entire space-ship, a secondary rocket will descend to the surface. **1962** *Times* 30 Apr. 12/7 Russia's latest earth satellite has been successfully landed in a predetermined area. **1967** J. Rowland *Jet Man* vi. 59 Now Whittle's experience of aerobatics came in useful, for he had to 'land' the machine in the water. **1968** *Ann. Reg. 1967* 178 The two accidents were a severe setback to American plans to land a man on the moon before 1970. **1972** *Nature* 3 Mar. 3/1 It is simply too dangerous to attempt to land a manned spacecraft in the lunar mountains.

3. b. Also, to obtain (employment). Also *absol.*

1926 Whiteman & McBride *Jazz* viii. 167 That is another reason why the outsider fails to land. He doesn't know about these rogues. **1946** E. O'Neill *Iceman Cometh* (1947) iii. 152 I'll bet you tink yuh're goin' out and land a job, too. **1952** Granville *Dict. Theatr. Terms* 108 *Land a spot*, obtain an engagement.

8. a. (Later examples.) Also with *up.*

1927 H. Crane *Let.* 19 Mar. (1965) 291, I had just landed in town after three months with the bossy cows. **1958** *Listener* 30 Oct. 694/3 They [*sc.* migrants] land up, exhausted, on islands and headlands. **1965** *Ibid.* 2 Sept. 351/2 After unspecified work in a map shop he landed up, furnished with a testimonial from Charles Graves, in the publishing house of Novello.

b. (Further examples.) Esp. of an aircraft or spacecraft, or a person in one: to alight upon or reach the ground, or some other surface, after a flight.

1784 V. Lunardi *Acct. First Aërial Voy. in Eng.* 37 My principal care was to avoid a violent concussion at landing, and in this my good fortune was my friend. At twenty minutes past four I descended in a spacious meadow. **1899** H. G. Wells *When Sleeper Wakes* xxiv. 326 On Blackheath no aëroplane had landed. **1908** —— *War in Air* ii. 60 The balloon was bumping as though its occupants were trying to land. **1911** W. Kaempffert *New Art of Flying* xiv. 238 Ely's remarkable feat in landing on the deck of a warship in the harbour of San Francisco. **1917** [see *Flatten *v.* 2 b]. **1917** 'Contact' *Airman's Outings* ii. 45 The machine in question was probably hit, however, for it did not return, and I saw it begin a glide as though the pilot meant to land. **1930** *Times* 11 Nov. 16/4 She [*sc.* a flying boat] circled the station and then landed in comparatively calm water. **1952** *Oxf. Jun. Encycl.* X. 7/2 When landing, the pilot is guided on to the deck by the Deck Control Officer who signals with 'bats'. **1953** Leslie & Adamski (*title*) Flying saucers have landed. **1969** *Times* 21 July 1/1 The first word from man on the moon came from Aldrin: 'Tranquillity base. The Eagle has landed.' **1973** *Sci. Amer.* Dec. 102/1 If the birds are pursued, they take off, but they do not fly far before they land again.

d. With *on.* Of an aircraft: to land on the deck of an aircraft carrier. Hence **landing-on** *vbl. sb.*

1937 *Aeroplane* 9 June 691/1 The ship was headed into wind and permission to land-on was given to the first Nimrod. *Ibid.* 16 June 724/1 The landing-on is organised similarly to the flying-off. **1939** *Nature* CXLIII. 592/2 'Landing on' had proved safer than driving a car on an English road. **1954** P. K. Kemp *Fleet Air Arm* 95 They took off and landed on without difficulty, completely independent of the sea.

landaulet. Add: **b.** In form *landaulette.* A type of motor car with a leather hood above the rear seats. Also *attrib.*

1901 *Autocar* 17 Aug. 153 (*heading*) The Peugeot landaulette. *Ibid.*, Mr. C. Friswell's Peugeot landaulette... The vehicle may be stated to be of the standard Peugeot type, but with only such alterations made as are necessary for the accommodation of the landau body. **1905** *Daily Chron.* 17 Nov. 8/4 Now the 'landaulette' is the popular car of the moment. **1906** *Ibid.* 15 Sept. 6/2 The cabs would be of the landaulette type. **1922** A. Haddon *Green Room Gossip* viii. 172 The other evening I rolled up to the Palladium in a big landaulette that carries seven persons. The chauffeur and I had it to ourselves. **1968** G. N. Georgano *Complete Encycl. Motorcars* 175 (*caption*) 1912 Dennis 24 hp landaulette. **1973** *Country Life* 18 Oct. 1190/3 It was not until 1909 that..I hired a cumbersome Astor landaulette from the Oxford garage. **1974** *Daily Tel.* 19 Oct. 17/2 Rolls-Royce's most expensive current car is the open landaulette, made in very limited numbers for heads of State at a cost of about £36,000.

landdrost. Add to def.: Under British administration, the office was abolished. (Further examples.)

1801, etc. [see *Heemrad]. **1947** L. Hastings *Dragons are Extra* ii. 35 Any old leader or *landrost* of the Free State or Transvaal had just Botha's sort of serenity. **1952** E. H. Burrows *Overberg Outspan* i. 16 Outwardly the old Dutch form persisted until 1827 when the *Collegies* were abolished, and the *landdrosts* replaced by Resident Magistrates and Civil Commissioners.

landed, *ppl. a.* Add: **2.** Caught, stuck, encumbered *with.*

In some of the examples a use of the pa. pple. of *land v.* rather than a ppl. adj.

1866 W. Gregor *Dial. Banffshire* 100 A'm fairly lantit wee the aul' coo. **1900** G. B. Shaw *Press Cuttings* 34 Sometimes..they get an idea of their own; and then of course youre landed. **1910** A. Bennett *Clayhanger* iv. vi. 508 The right sort of women don't get landed as the wives of convicts. **1943** J. B. Priestley *Daylight on Saturday* xxii. 171 One thing leads to another..an' then, before you know where you are, you're landed. **1947** 'G. Orwell' *Let.* 23 Oct. in *Coll. Ess.* (1968) IV. 382 I've been landed with another long article which I can't dodge out of. **1960** 'N. Shute' *Trustee from Toolroom* i. 7 You'll be landed with a cat for the rest of your lives. **1974** K. Royce *Trap Spider* i. 27, I told you that I'd speak to your son and I'm landed with it.

3. *landed immigrant*: an immigrant to Canada, admitted for permanent residence; so *landed* (*immigrant*) *status.*

1910 *Statutes of Canada* c. 27, s. 2 (p), 'Land', 'landed' or 'landing', as applied to passengers or immigrants, means their lawful admission into Canada by an officer under this Act, otherwise than for inspection or treatment or other temporary purpose provided for by this Act. **1963** *Maclean's Mag.* 20 Apr. 18/3 The only black people freely admitted to Canada as landed immigrants are a limited number of women. **1964** *Calgary Herald* 4 May 25/3 If he is to see his child, he will have to..be accepted as a landed immigrant. **1968** *Globe & Mail* (Toronto) 2 May 2/4 Brooks..was granted landed immigrant status but Stonehill was refused. *Ibid.*, Stonehill's bid to obtain landed status had strong political repercussions in Ottawa. **1973** *Ibid.* 27 Dec. 2/2 Most Canadians don't realize it but any visitor or landed immigrant with less than five years residence can be deported, together with his entire family, for any criminal code offence, including impaired driving or shoplifting. **1975** *Canadian Mag.* (Toronto) 8 Mar. 6/4 [Some] East Indians who have obtained citizenship or landed immigrant status have been involved in illegal immigration rackets.

Landenian (lænde·niăn), *a. Geol.* [a. F. *Landenien* (A. Dumont 1839, in *Bull. de l'Acad. R. des Sci.*, etc., *de Bruxelles* VI. ii. 466), f. *Landen*, name of a town near Liège in Belgium: see -ian.] Of, pertaining to, or designating a stratigraphic stage at the top of the Palæocene series (or the bottom of the Eocene), lying above the Montian. Also *absol.*

1852 *Q. Jrnl. Geol. Soc.* VIII. 254 In his visit to this country..he [*sc.* Dumont] pointed out to me the many characters common to these Thanet sands and his 'Landenian System', which occupies the same position in the Belgian series. **1902** A. J. Jukes-Browne *Student's Handbk. Stratigr. Geol.* xvi. 485 In Belgium the Montian limestone..is overlain by glauconitic sands which are known as the Lower Landenian and unquestionably correspond to our Thanet Beds. **1923** L. D. Stamp *Introd. Stratigr.* xvi. 274 Upper or Continental Landenian strata succeed [the lower or Marine Landenian] conformably, and are known as the Woolwich and Reading Beds. **1955** G. G. Woodford tr. *M. Gignoux's Stratigraphic Geol.* ix. 485 The English Tertiary begins with the transgressive Landenian. **1969** Bennison & Wright *Geol. Hist. Brit. Isles* xv. 336 Sedimentation commenced with the inundation by the sea of the eastern part of the London Basin in Landenian times, depositing the Thanet Beds. *Ibid.*, The marine facies may not have persisted throughout the Landenian even in the east of the London Basin.

lander. Add: **1. b.** A spacecraft, or a part of one, which is designed to land on the surface of a planet or of the moon.

1961 *Astronautica Acta* VII. 130 The rotary drill..is designed to penetrate 1·5 ft or more into the lunar surface and bring samples into the lander for chemical analysis. **1962** F. I. Ordway et al. *Basic Astronautics* v. 176 Hard landers contain retrorockets to reduce the terminal velocity to between 100 and 300 m.p.h... Soft landers.. are built to descend gently onto the surface. **1967** *Technology Week* 23 Jan. 61/1 (Advt.), This calls for a varied series of probes, orbiters and hard and soft landers. **1971** *Listener* 7 Oct. 476/3 Each vehicle consists of two main parts: an orbiter and a lander.

landesite (læ·ndizəit). *Min.* [f. the name of K. K. *Landes* (b. 1899), American geologist + -ite[1].] A brown hydrated phosphate of manganese and ferric iron, occurring as an alteration product of reddingite at Berry Quarry, Poland, Maine, U.S.A.

1930 Berman & Gonyer in *Amer. Mineralogist* XV. 385 This mineral seems to represent a new species for which the name landesite is here proposed in honor of Professor Kenneth K. Landes who has done much work on the pegmatites of Maine. **1964** *Ibid.* XLIX. 1123 The landesite formula is better expressed in the following manner...

$$[Mn^{2+}{}_{1-x}(Fe^{3+}OH)_x]_3\,[(3-3x)H_2O]\,(PO_4)_2$$

in which, for type landesite, x is approximately 0·25.

landfall. Add: **1. c.** Arrival at land after a flight over the sea; also, = *Landing *vbl. sb.* 1 d.

1908 H. G. Wells *War in Air* vi. 194 New York had risen out of the blue indistinctness of the landfall. **1909** —— *Tono-Bungay* iv. i. 449, I remember our prolonged dragging landfall. **1928** C. F. S. Gamble *Story N. Sea Air Station* ix. 121 The airship L.3..made her 'landfall' off Ingham. **1942** *R.A.F. Jrnl.* 3 Oct. 31 You get a feeling of warming pride as a good landfall is made. Once round the beacon and down you come. **1954** 'J. Christopher' *22nd Cent.* 86 They check you each landfall. Hans got his final warning at Luna City. **1959** *Listener* 22 Jan. 160/1 The average drift-migrants [*sc.* birds] that make a landfall are not necessarily lost.

d. The place where an undersea pipeline reaches land.

1974 *People's Jrnl.* (Inverness & Northern Counties ed.) 7 Sept. 2/6 Burmah and British Petroleum..have.. approached Zetland County Council about the possibility of a pipeline landfall in Shetland. **1975** *Petroleum Rev.* XXIX. 387/2 It took twelve months..to select Flotta.. as the landfall for the oil pipeline from the Piper field.

la·nd form. Also **land-form, landform.** [Land *sb.* 10 a.] **1. a.** A physical feature of the earth's surface such as a hill, plain, cirque, or alluvial fan.

1893 Land form [see *Geomorphology]. **1910** *Encycl. Brit.* XI. 633/2 Thus new land forms are created—valleys of curious complexity, for example—by the 'capture' and diversion of the water of one river by another. **1938** L. D. Stamp *Physical Geogr. & Geol.* ix. 142 The land forms in deserts developed in sedimentary rocks are similar to those in damper temperate climates except that slopes are usually sharper and steeper. **1970** R. J. Small *Study of Landforms* i. 8 Classification can be attempted not only of individual kinds of landform (such as slopes, cliffs, terraces, beaches, lakes, volcanoes, planation surfaces and so on) or types of process.., but of landform assemblages.

† **b.** A landscape of any particular kind. *Obs.*

1893 W. M. Davis in *Nat. Geogr. Mag.* 10 July 73 Every land-form passes through a systematic series of changes from its youth, when its form is defined chiefly by constructional processes, past its maturity, when the processes of sub-aerial sculpture have carved a variety of mouldings and channellings, to its old age in which..denudation reduces the mass to base-level. **1899** —— in *Geogr. Jrnl.* XIV. 485 Where the forces of uplift or deformation have lately..initiated a cycle of changes, the destructive forces can have accomplished but little work, and the land-form is 'young'.

2. A kind (of living organism) found on land.

1897 Land form [in Dict. s.v. Land *sb.* 11]. **1926** J. S. Huxley *Ess. Pop. Sci.* 93 There are three or four other species of animals, such as Proteus,..which..are not known in a land-form at all.

landing, *vbl. sb.* Add: **1. d.** The (or an) action of approaching and alighting on the ground or some other surface after a flight. *happy landings!*: see *Happy *a.* 3.

1784 [see *Land *v.* 8 b]. **1909** *Flight* 13 Feb. 93/1 (*heading*) Flight 'landings'. **1912** *Aeroplane* 19 Dec. 621/2 Major Cameron and Capt. Salmon with Mr. Barnwell and, later, Mr. Knight up behind, put in large number [*sic*] of straights each making very good flights and landings. **1916** H. Barber *Aeroplane Speaks* 49 You can..imagine what a difference that would make where forced landings are concerned! **1923** H. G. Wells *Men like Gods* i. iii. 37 The aeroplanes made an easy landing. **1927** G. Aston *Navy of To-Day* v. 31 The airman, and the airman's home, the aircraft carrier, must steam head to wind..when the airmen want to accomplish 'landings' on her deck. **1936** *Discovery* Aug. 238/1 The camera is raised during take-offs and landings. **1956** [see *Emergency 5]. **1967** D. P. Davies *Handling Big Jets* iii. 30 For take-off and landing the weight should be known to within 5,000 lb. **1969** *Times* 21 July 1/1 The landing, in the Sea of Tranquillity, was near perfect and the two astronauts on board Eagle reported that it had not tilted too far to prevent take-off. **1974** *Daily Tel.* 21 Feb. 17/7 He [*sc.* a balloonist] has food and water for 10 days and the gondola is equipped with floatation devices to keep it upright if he is forced to make a water landing.

5. (Later U.S. examples.)

1832 S. Cumings *Western Pilot* 49 There is a pretty good landing at the upper end of the town. **1867** J. N. Edwards *Shelby* xx. 366 The next day the brigade moved to the river near Gaines Landing. **1895** M. A. Jackson *Mem. Stonewall Jackson* (ed. 2) xii. 211 Just before reaching the landing I stopped to look back.

8. *landing area, fee, field, ground, -leg, site, -tower, vehicle*; **landing beam** *Aeronaut.*, a radio beam to guide aircraft when landing; **landing card**, a card issued to a passenger on an international flight or voyage, which is surrendered on arrival; **landing craft**, a naval vessel with a shallow draught designed for landing troops, tanks, etc., in an amphibious assault; hence *transf.* in *Astronaut.*, the section of a spacecraft which is used for the final descent to the surface of a planet or moon; **landing flap** *Aeronaut.*, a flap that can be lowered to increase the lift and the drag and so make possible lower speeds for take-off and landing; **landing gear**, (*a*) *Aeronaut.*, the structure underneath an aircraft that is designed to support it on the ground and to absorb the shock of landing (in modern aircraft made to be retracted in flight); (*b*) the retractable support at the front of a semi-trailer that supports it when not attached to the tractor; **landing light**, (*a*) a light on the runway of an aerodrome to guide an aircraft in a night landing; (*b*) a light attached to an aircraft to illuminate the ground for a night landing; **landing pad**, (*a*) a small area of an aerodrome or heliport, used for the landing and taking off of helicopters; (*b*) a cushioned or strengthened foot which supports a hovercraft, spacecraft, or the like when stationary on the ground; **landing ship (tank(s))**, a large landing craft for the transport of tanks and other vehicles; **landing speed**, the speed at which an aircraft lands (see also quot. 1911); **landing strip** = *air-strip* (*AIR *sb.*[1] B. III. 7); **landing ticket** = **landing card*; **landing wire**, *Aeronaut.*, a wire on a biplane or light monoplane that is designed to take the weight of a wing when the aircraft is on the ground.

1910 R. Ferris *How it Flies* xx. 464 *Landing area*, a piece of land specially prepared for the alighting of aeroplanes without risk of injury. 1951 *Gloss. Aeronaut. Terms* (*B.S.I.*) III. 23 *Landing area*, the part of the movement area primarily intended for the take-off and landing of aircraft. 1974 G. Mitchell *Javelin for Jonah* ix. 115 You may go ahead with the new landing-areas for jump and pole. 1929 Landing beam [see *BEAM *sb.*[1] 24 b]. 1933 *Flight* 1 June 524 A pointer on a simple instrument showed him any deviation from the landing beam. 1945 *Aeronautics* Feb. 30 (*heading*) Diagram showing the aircraft..entering the landing beam. 1932 G. Greene *Stamboul Train* I. i. 3 The purser took the last landing card..and watched the passengers cross the grey wet quay. 1950 P. Bottome *Under Skin* ii. 18 Got your landing card ready, and your passport? 1966 'W. Haggard' *Power House* vi. 58 He could be asked for a landing card and as a through-booking he didn't have one. 1973 *Times* 13 Dec. 11/2 He included landing cards among the paraphernalia of controls. 1940 W. S. Churchill *Second World War* (1949) II. 593 Great efforts should be made to produce the landing-craft as soon as possible. 1942 *R.A.F. Jrnl.* 18 Apr. 32 Two landing craft were sent ashore with reconnaissance parties. 1943 Landing craft [see *ASSAULT *sb.* 8]. 1953 *Jrnl. Brit. Interplanetary Soc.* XII. 275 The landing craft (a small supplementary vehicle designed for vertical descent with rocket braking, carried to the destination by the parent spaceship). 1957 P. Worsley *Trumpet shall Sound* vii. 144 Landing-craft of all kinds poured out their cargo upon the beaches. 1966 D. Holbrook *Flesh Wounds* 93 Three thousand landing craft were ready to move out of all the ports all along the coast, from Falmouth to Harwich. 1969 *Times* 21 July 8/2 At 1,500 ft., the astronauts slowed the landing craft and brought it gently down four miles off the scheduled target in the Sea of Tranquillity. 1922 *Flight* XIV. 660/1 No extra landing fee will be charged in respect of test flights before departure. 1972 *Times* 11 Feb. 1/1 Strong opposition has come from the airlines to a new system of landing fees which is to be introduced at Heathrow. 1921 *Aeronautics* 13 Jan. 26/1 The improvement of landing fields and equipment. 1959 *Chambers's Encycl.* I. 97 The emergency landing fields, which were set aside by the Royal Air Force for special purposes, were usually grass covered. 1936 *Technical Rep. Aeronaut. Res. Comm. 1934–35* I. 30 Now that so many aeroplanes are being fitted with landing flaps it is important to permit the flap to extend along the whole span. 1940 *War Illustr.* 19 Jan. 620 With wheels and landing flaps lowered, the pilot makes his approach. 1966 *McGraw-Hill Encycl. Sci. & Technol.* XIV. 517/1 Structurally, the aileron is similar to the landing flap. 1911 *Rep. & Mem. Advisory Comm. Aeronaut.* No. 59. Nov. 103 The efficiency of landing gear on various sorts of ground may be tried. 1931 *Flight* 9 Jan. 30/1 The landing gear is designed to give very smooth landing and taxying characteristics. 1931 J. E. Younger *Airplane Construction & Repair* iii. 48 Some airplanes are designed with landing gears which fold up into the fuselage and hence offer no direct wind resistance. 1951 *Amer. Speech* XXVI. 308/2 *Landing gear*, a strong support that holds up the front end of a semi-trailer when it is not attached to a tractor. 1971 M. Tak *Truck Talk* 97 *Landing gear*, the retractable supports on a trailer that prop up the front end when the trailer is unhitched from the tractor. 1971 *Physics Bull.* Apr. 217/1 Steels with improved fracture properties needed in nuclear submarines and aircraft landing gear are also under development. 1912 *Aeroplane* 12 Dec. 584/1 The great deterrent at present is the lack of proper landing grounds. 1920 Landing ground [see *flying school* (*FLYING *vbl. sb.* 3)]. 1943 T. S. Eliot in Ld. Sempill et al. *Friendship, Progress, Civilisation* 20 To descend from this flight into generalities on the particular landing-ground of the present occasion. 1962 L. van der Post *Heart of Hunter* I. v. 80 The great pan..had a floor so wide, level and firm that..the biggest aircraft could land on it. I myself had used it as a landing-ground many times. 1951 *Jrnl. Brit. Interplanetary Soc.* X. 101 In the case of a Moonflight..this means a vertical descent using reverse rocket braking in conjunction with a radar-altimeter and landing-legs. 1969 *Sun* 21 July 1/2 The Eagle, leaving its spidery landing-legs behind, soared away. 1917 *Flight* 4 Jan. 15/1 A new system, called 'Triplex glass landing lights', proved to be inferior to petrol flares. 1920 *Proc. Air Conf., London* 11 Aerodromes will be equipped..at night flying is practicable. Permanent electric landing lights..are being installed. 1922 *Flight* XIV. 519/1 Lighting Set (including navigation lights, landing lights and illumination of instruments). 1937 *Times* 11 Apr. 9/3 They see no reason why they should confuse coloured Very lights or landing lights in the air. 1942 *R.A.F. Jrnl.* 5 Oct. 7 From beneath him a landing light ground downwards. 1969 I. Kemp *Brit. G.I. in Vietnam* iii. 69 He..switched on his landing light, illuminating three paratroopers standing on the landing zone signalling us in. 1973 *Times* 11 Apr. 3/7 They used landing lights to make three trips and everyone on board was winched to safety. 1958 *World Helicopter* Apr. 63 Our cover picture shows one of Sabena's fleet of 12-passenger Sikorsky S.58's making a landing at the heliport on the strip between the two 80 ft. diameter landing pads. 1961 *New Scientist* 2 Mar. 528/3 The actual landing pad need still be no more than 150 ft square. 1967 *Gloss. Terms Air-Cushion Vehicles* (*B.S.I.*) 6 Landing pads, strong points, protruding below the rigid bottom of an ACV, which support the vehicle when at rest on land. 1969 *Islander* (Victoria, B.C.) 23 Mar. 10/1 The first landing pad for the young company was an area of land way down Shelbourne Street, at that time the outskirts of Victoria. 1969 *Times* 27 May 8/5 Its landing pads are 37 in. across, each of them fitted with a probe which can sense the surface. 1943 *Life* 11 Oct. 24/2 The first is the LST (Landing Ship, Tank), 327 ft. long and displacing 5,500 tons. 1944 *Hutchinson's Pict. Hist. War* 27 Oct. 1943–11 Apr. 1944 166 (*caption*) Landing Ship Tanks. These two landing ships tanks close inshore at Bougainville are unloading supplies for U.S. Marines and army transport. 1945 ... Cherbourg ... (landing ship tanks) ... vi. 51 Cedric and I put off in a motor fishing vessel to find our Tank Landing Ship. 1950 W. S. Churchill *Second World War* V. vi. 26 The 'landing-ship, tank'..had first been conceived and developed in Britain in 1940. 1961 B. Fergusson *Watery Maze* iv. 106 Rear-Admiral Burrough, with the cruiser *Kenya* and four destroyers, was to escort the two landing ships. 1966 D. Holbrook *Flesh Wounds* ... Ship Tank ... 1969 *Times* 4 Feb. 13/4 The eastern end of the planned Apollo landing site. 1972 *Nature* 3 Mar. 3/1 The landing site of Luna 20 was some 120 km north of the region from which Luna 16 recovered specimens. 1911 R. M. Pierce *Dict. Aviation* 144 *Landing speed*..the speed with which a landing or descent to the earth is made. 1937 *New Republic* 19 May 35/1 The modern air liner's landing speed has gone up as designers have boosted its top speed by refining line and form. 1962 F. W. Brooks *Mod. Airliner* iii. 75 Wheel brakes..now became a necessity because of the increased landing speeds and taxying speeds of the more heavily loaded monoplanes. 1930 *Aircraft Engineering* Jan. 16/1 The standard intermediate field in low altitudes provides two landing strips or runways. 1944 *Times* 1 July 4/3 Squadrons flying from landing strips in Normandy are taking advantage of every break in the clouds. 1956 W. Graham *Sleeping Partner* 62 Llanvryan had been an aerodrome—a glorified landing strip—in the first place. 1973 G. Greene *Honorary Consul* iv. iii. 218 Señor Escobar has a landing strip on his estancia. 1925 E. Gelibrand *Travelling Do's & Don'ts* v. 19 While the cool, collected person gets things done without unnecessary waste of energy, the flustered one..not having his landing ticket ready..is hustled by the impatient ones. 1930 A. Bennett *Imperial Palace* ii. 382 The hand of the official at the bottom of the gangway was full of landing tickets. 1912 Kipling *Divers. Creatures* (1917) 3 They began turning out traffic-lights and landing-towers. 1927 ... 97/2 Landing Vehicle Hydrofoil. 1969 *Observer* 20 July 7/1 The astronauts crawl into the landing vehicle..and spend three hours checking it. 1927 'Contact' *Airman's Outings* 26 ... I found that part of a landing-wire was chafing against its metal socket. 1917 ... *Aeronaut. Engin.* ... Landing wires support the wings on the ground, but when the aircraft becomes airborne, the stresses are transferred to the flying wires, as the wings tend to lift upwards. 1952 ... *Flight* vii. 100 ... are obviously supporting the weight of the wings when the machine is on the ground. They are called the 'landing wires'. Those below the wing..are called the 'flying wires'.

landing-place. Add: **1. c.** A place where a bird, insect, aircraft, etc., can or does land.

1776 T. Pennant *Tour in Scotl. & Voy. Hebrides 1772* II. 24 Woodcocks... Their first landing-places are in the eastern counties. 1889 *Leisure Hour* 642/1 Insect 'landing-places' would thus, according to the theory, acquire considerable importance in affecting the structure of the flower. 1899 *Strand Mag.* Aug. 183/1 Captain Spelterini's sharp eye had quickly chosen an advantageous landing-place, and the anchor was thrown [from the balloon]. 1909 *Flying: the Why & Wherefore* v. 33 Another advantage of flying high is that in case of an engine stoppage the aeronaut will have time to look round and choose a landing place. 1935 C. Day Lewis *Time to Dance* 35 The oil ran out and cursing they turned about Losing a hundred miles to find a landing-place. 1962 K. W. Gatland *Astronautics in Sixties* xi. 338 After reconnaissance spacecraft and soft-landing probes had given information concerning a suitable landing place, a Surveyor-type probe would be put down close to the desired landing point.

‖ **ländler** (le·ndlər). Also *erron.* **landler**. [G.] An Austrian peasant dance, similar to a slow waltz; the music for such a dance.

1876 Stainer & Barrett *Dict. Mus. Terms* 251 *Ländler*, the name given to a dance popular among the Styrian peasants. 1934 C. Lambert *Music Ho!* i. 58 The romantic orchestration, the solid hymn-tune harmonies, the *Landler* rhythms, [etc.]. 1961 *Times* 21 Oct. 11/3 The Schubert ländler that forms the second movement. 1964 *Listener* 6 Feb. 250/2 If the trio-section is taken at the same speed, it may sound unnaturally hurried, since its Ländler-like characteristics justify a more leisurely pace.

land-line. Add: Also **landline** and as two words. **2.** (Earlier and later examples.) Also, an overland (or underground) line for telecommunication by other means.

1865 *Phil. Mag.* XXIX. 409 (*heading*) On the retardation of electrical signals on land lines. 1866 *Bradshaw's Railway Manual* XVI. 454 Telegraph to India... A provisional agreement entered into for leasing the land line in Egypt to the proposed British-Indian Submarine. 1927 *Observer* 12 June 20 Last Sunday I had the curiosity to listen to the Eastbourne programme... There were obvious defects in the landline, which made the general effect thin and poor. 1930 *Bell Syst. Techn. Jrnl.* IX. 406 The linking of ships at sea with the land telephone network. 1962 *B.B.C. Handbook* 109 A separate receiving station where broadcasts are intercepted and fed to the monitors by land line. 1970 'J. Earl' *Tuners & Amplifiers* iv. 77 ... the signal so as to rid it of certain shortcomings that could have been picked up during its transmission through the ether (radio waves) and studio landlines.

Hence **la·ndline** *v. trans.*, to transmit over a land-line.

1966 *Times Lit. Suppl.* 24 Mar. 248 This Crusade is going to be landlined to most major cities in the United Kingdom. 1969 J. Bennett *Dragon* iii. 43 That at least means no radio pictures to Peking... They'll probably take an advance tomorrow and landline our pictures to Canton. 1972 J. Drummond *Boom Companions* xxxi. 98 That first story's been land-lined and it's going to hit nine o'clock television all over the country.

landlord, *sb.* Add: **4.** *attrib.* and *Comb.*

1845 *Douglas Jerrold's Shilling Mag.* I. 515 Judge-made law or the bench, but landlord-made law is worse. 1880 'Mark Twain' *Tramp Abroad* 186 The landlord-apprentice serves as call-boy, then as under-waiter. 1966 *Daily Chron.* 26 June 5/7 With an air of detachment, as though we were not addressing a landlord-ridden assembly. 1924 R. Graves *Mock Beggar Hall* 19 Waiting the landlord-absentee's return. 1959 *Good Food Guide* 586 Both landlord and waiter were Spanish. 1963 *Times Lit. Suppl.* 17 May 350/3 The parasitic landlord-usurers had to be destroyed as a class.

land-lubbing, *a.* [Irreg. f. Land-lubber.] Land-lubberly.

1885 *Punch* 29 Aug. 100/2 The Judge, a land-lubbing chap in a wig. 1927 *Daily Express* 4 Oct. 3 We land-lubbing civilians know less about the Navy than our maiden aunts might be expected to know about alimony. 1960 *House & Garden* Aug. 31/3 Judged by sea-going or land-lubbing standards. 1966 *Economist* 21 May 799/1 Many..seamen are obeying their union's orders ... 1974 *Sci. Amer.* June 112/2 The line-of-sight land-lubbing microwave relay networks are too expensive. What can the engineers do in orbit?

landmark, *v.* [f. the sb.] *trans.* To be or act as a landmark to; to provide with a landmark.

1941 J. F. Powys ... is not necessary here to landmark further successes. 1928 *Sunday Dispatch* 9 Dec. 9/1 ... to landmark all the many who landmarked those ten years ...

landolphia (lændɒ·lfiä). [mod.L. (A. M. F. J. Palisot de Beauvois *Flore d'Oware* (1804) I. 55), f. the name of M. *Landolphe* (1765–1825), commander of the expedition on which the genus was first discovered + -IA[1].] A tropical African climbing plant of the genus so called, belonging to the family Apocynaceæ and yielding a latex formerly used as a source of rubber.

1887 *Curtis's Bot. Mag.* CXIII. 6963 The first notice of the *Landolphia* yielding India-rubber..is by Col. now Sir J. A. Grant, in the appendix to Speke's Journal. 1920 *Westm. Gaz.* 20 Apr. 4/1 Landolphias, woody climbers,..yield the African rubbers. 1951 *Dict. Gardening* (R. Hort. Soc.) III. 1125/1 Landolphias are an important source of caoutchouc.

landrace (læ·ndrēs). [Da.] A large white pig of the variety so called, originally developed in Denmark, now used elsewhere to produce bacon. Also *attrib.*

1935 Layley & Malden *Evolution Brit. Pig* 98 (*heading*) Landrace. The modern model. *Ibid.*, The Danish Landrace..is the model set up to which it is the ambition of our state to mould all British pigs. 1937 G. R. H. Bishop *Improvement of Bacon Pigs* ii. 27 The contented Landrace sows were hardy and prolific. 1948 H. R. Davidson *Production & Marketing Pigs* ix. 126 There [*sc.* in Denmark], in 1895, were set up the first breeding centres to maintain and improve the breed type of the native Landrace. 1958 'R. Crompton' *William's Television Show* viii. 253 'A Landrace pedigree, that pig is,' said the farmer. 1971 *Country Life* 2 Dec. 1588/1 Our breeders [of pigs]..readily accept Large White, Landrace, or crosses of these breeds.

Landsborough (læ·ndzbǝrǝ). *Austral.* The name of a small town in Queensland, used *attrib.* in **Landsborough grass**, a pasture grass, *Iseilema membranaceum*, found in the area, and better known as small Flinders grass.

1883 F. M. Bailey *Synopsis Queensland Flora* 646 The Landsborough grass. A weak, very leafy, brittle grass of a reddish color, one of the most valuable for fodder. 1889 [see *Barcoo]. 1891 R. Wallace *Rural Econ. Austral. & N.Z.* xxii. 294 *Anthistiria membranacea*, Lindl.—Barcoo grass of Queensland; also called Landsborough grass. West and South Australia, New South Wales and Queensland. 1927 M. M. Bennett *Christison* v. 55 Landsborough grass..shimmered gold and silver. 1929 J. W. Bews *World's Grasses* vi. 253 'Barcoo', 'Landsborough' or 'Red Gulf grass', covers large tracts of the north and interior of Australia.

landscape, *sb.* Add: **1. c.** = Oblong *a.* 1 c. Also as *adv.*

1932 Sayers & Smart in W. Atkins *Art & Pract. Printing* I. xii. 139 The frontispiece..may be printed either upright (termed portrait) or broad way (termed landscape). If a full-page illustration be printed landscape, the inscription or caption beneath must read from foot to head. 1951 D. Bland *Illustration of Books* ix. 146 The landscape plate is always a problem. It is unfortunate that the tall narrow format which is so suitable for a page of type does not lend itself to the average photograph. 1956 H. Williamson *Methods Bk. Design* iii. 16 The same formats can..be used for landscape or oblong books. 1966 G. Hamilton-Edwards *In Search Ancestry* ii. 24 This can be done by buying the full quarto size exercise book..and asking a printer to cut it in half, wide-ways, or 'landscape'. ..This gives you two booklets of 8″ × 5″.

2. b. A tract of land with its distinguishing characteristics and features, esp. considered as a product of modifying or shaping processes and agents (usually natural).

1886 A. Geikie *Class-Bk. Geol.* i. 2 The surface of a country is not now exactly as it used to be. We notice various changes of its topography going on now,..the accumulated effect of which may ultimately transform altogether the character of landscapes. 1896 *Rep. 6th Internat. Geogr. Congr. 1895* 749 We thus have six ranks of units: (1) The form-element. (2) The fundamental form [*sc.* land form]. (3) The group of forms or landscape. [Etc.]. 1922 L. Mumford in H. E. Stearns *Civilisation in U.S.* 4 West of the Alleghanies, the common, with its church and school, was not destined to dominate the urban landscape. 1925 *Univ. Calif. Geogr.* II. 37 The works of man express themselves in the cultural landscape. There may be a succession of these landscapes with a succession of cultures. They are derived in each case from the natural landscape, man expressing his place in nature as a distinct agent of modification. 1937 Wooldridge & Morgan *Physical Basis Geogr.* p. ix, Geography cannot dispense with geomorphology, for a real understanding of the characters and development of the physical landscape is an indispensable preliminary to the study of the cultural landscape and of regions. 1944 A. Holmes *Princ. Physical Geol.* xi. 191 In the Grampian Highlands an old peneplain, now dissected into a landscape of late youth or early maturity (though modified by glaciation), is easily recognised by the even skyline. 1954 W. D. Thornbury *Princ. Geomorphol.* xiv. 364 Two contrasting ideas have developed regarding the ability of glaciers to modify by erosion the landscapes over which they move. 1971 I. G. Gass et al. *Understanding Earth* vii. 100/1 (*caption*) Two photographs of the lunar crater Tycho.., whose ramparts rise 5 400 m above the level of its floor, though only 1 600 m above the level of the surrounding landscape. 1974 H. F. Garner (*title*) The origin of landscapes: a synthesis of geomorphology.

4. h. Later *transf.* and *fig.* uses.

1952 G. Sarton *Hist. Sci.* I. x. 256 Let us return again to Athens and try to consider the intellectual landscape from the point of view of a well-educated man. 1953 A. Huxley *Let.* 31 Oct. (1969) 687 The jewelled palaces..may..be actual *choses vues*—items in the ordinary landscape of certain kinds of people. 1963 *Listener* 7 Mar. 405/1 The landscape of international politics is now very different from what it was only two or three years ago.

5. *landscape-garden* vb.; *landscape-gardener* (earlier examples); **landscape architect**, a practitioner of landscape architecture; **landscape architecture**, the planning of parks or gardens to form an attractive landscape, often in association with the design of buildings, roads, etc.; **landscape-gardening** (earlier and later examples); **landscape-painter** (earlier and later examples).

1863 *6th Ann. Rep. Board of Commissioners Central Park* (N.Y.) *1862* between pp. 60–61 (Map), Olmsted and Vaux, Landscape Architects. 1879 *Chicago Tribune* 3 May 1/3 (Advt.), H. W. S. Cleveland, Landscape Architect. 1890 C. Eliot *Let.* 3 Dec. in *C. Eliot: Landscape Architect* (1924) xv. 273 Landscape gardening is that part of the landscape architect's labor which is directed to the development of formal or natural beauty by means of removing or setting out plants. 1927 T. H. Mawson *Life & Work Eng. Landscape Architect* xiv. 160 A young and able landscape architect..had heard me lecture in England. 1967 G. Collens in A. E. Weddle *Techniques Landscape Archit.* ii. 33/1 The employer should start by setting out his requirements as a basis for discussion with the landscape architect. 1972 *Times* 15 Sept. 2/1 For decades..landscape architects' services were not sufficiently appreciated. 1840 J. C. Loudon in H. Repton *Landscape Gardening & Landscape Archit. of H. Repton* (new ed.) p. vii, These writings [*sc.* of Gilpin and Price] are full of the most valuable instruction for the gardener, relative to the general composition of landscape scenery, and landscape architecture. 1865 F. L. Olmsted *Let.* 1 Aug. in *F. L. Olmsted: Landscape Architect* (1928) II. vi. 74, I am all the time bothered with the miserable nomenclature of L.A. *Landscape* is not a good word, *Architecture* is not; the combination is not—*Gardening* is worse. 1891 C. Eliot in *C. Eliot: Landscape Architect* (1924) xx. 366 We cannot avoid seeing behind the fair figures of Gardening and Building a third figure of still nobler aspect..the art which, for want of a better name, is sometimes called Landscape Architecture. 1915 S. Parsons *Art of Landscape Archit.* p. vi, The study of nature assisted by the best examples is the proper field for the study of landscape architecture. 1967 A. E. Weddle (*title*) Techniques in landscape architecture. 1836 F. A. Kemble *Let.* 1 Mar. in *Rec. Later Life* (1882) I. 45 Adam and Eve landscape-gardened in Paradise, you know. 1891 W. Morris *News from Nowhere* iii. 17 The other day we heard that the philistines were going to landscape-garden it [*sc.* the place]. 1941 E. Wilson *Wound & Bow* ii. 119 When the transfer [of the land] had been effected, Mrs. Kipling set out to landscape-garden it. 1974 'M. Innes' *Appleby's Other Story* i. 5 You don't care, Tommy, for wild nature tamed and landscape-gardened? *a* 1763 W. Shenstone *Works* (1764) II. 139, I have used the word landskip-gardiners; because in pursuance of our present taste in gardening, every good painter of landskip appears to me the most proper designer. 1788 A. Seward *Let.* 14 Oct. (1811) II. 172, I should suppose nobody has ever been so well qualified as yourself [*sc.* H. Repton] for the profession you purpose to assume, that of landscape gardener. *a* 1763 W. Shenstone *Works* (1764) II. 125 Gardening may be divided into three species—kitchen-gardening—parterre gardening—and landskip, or picturesque-gardening: which latter..consists in pleasing the imagination by scenes of grandeur, beauty, or variety. 1788 H. Repton in D. Stroud *Humphry Repton* (1962) ii. 37, I mean in this place to keep an account of the time employed and expenses incurr'd in this service at the same rate as if employ'd in my profession of Landscape Gardening. 1938 *New Statesman* 8 Jan. 56/2 In Andrew Young there are touches of a lesser, a more landscape-gardening Frost. 1946 R. Macaulay *They went to Portugal* 137 Landscape gardening always showed him [*sc.* William Beckford] at his most likeable. 1975 *Garden History* III. II. 1 Mavis Batey's essay on Goldsmith, 'An Indictment of Landscape Gardening', is the clearest and best exposition we have had of that frequent eighteenth-century occurrence, the destruction of villages and hamlets to further the creation of landscape gardens. *a* 1763 W. Shenstone *Works* (1764) II. 129 The landskip painter is the gardiner's best designer. 1779 T. Blaikie *Diary Scotch Gardener* (1931) 159 Those Gardens are Layd out under the Derections of Mr Robert one of the first Landskape painters in France. 1937 *Discovery* July 211 The greatest of English landscape painters. 1974 B. Massingham *Turn on Fountains* iv. 65 They met a landscape painter..who confessed that he 'could do nothing with Connemara'.

landscape, *v.* (See at end of sb. in Dict.) Add: **2.** To lay out (a garden, etc.) as a landscape; to conceal or embellish (a building, road, etc.) by making it part of a continuous and harmonious landscape. Also *transf.* So **la·ndscaping** *vbl. sb.*

1927 [implied by *landscaped *ppl. a.*]. 1930 *N.Y. Times* 9 Feb. XI. 2/1 Suburban developers and home owners are paying more attention to landscaping today. 1930 *Publishers' Weekly* 15 Feb. 858/2 Landscaping is about to become the topic of smart conversation, with the result that garden books should sell as never before. 1943 Forshaw & Abercrombie *County of London Plan* vii. 103 Landscaping must play an important part in the layout of these open spaces particularly those which provide a setting for the houses and blocks of flats. 1957 *Listener* 13 June 949 The planners intend to plant trees round the perimeter and generally landscape the whole area. 1959 *Motor* 22 Apr. 410/1 New roads..are 'landscaped into the countryside and not stuck on it'. 1962 *Daily Tel.* 23 May 21/1 Some aspects of road landscaping were still not fully accepted in Britain. 1966 Mrs. L. B. Johnson *White House Diary* 11 Jan. (1970) 350 The check would be given to landscape the new automobile entrance of the National Zoo. 1974 *Country Life* 17 Oct. 1095/1 The National Trust has landscaped the island.

landscaped (læ·ndskēɪpt), *ppl. a.* [f. prec.] Laid out as a landscape; embellished by landscaping. Also *transf.*, of an office.

1927 *Brit. Weekly* 15 Dec. 283/2 Even factories..frequently have lovely landscaped grounds. 1957 V. Nabokov *Pnin* 9 An artificial lake in the middle of a landscaped campus. 1959 A. Harrington *Life in 'Crystal Palace'* (1960) viii. 112 Our landscaped grounds..will flower. 1968 *Guardian* 17 June 8/3 The open 'landscaped' office, or *burolandschaft* as it is called by its German inventors. 1968 *Daily Tel.* (Colour Suppl.) 29 Nov. 56/2 The American reaction now is for open-plan 'landscaped' offices where all workers, regardless of status, are visually connected in one low horizontal block. 1970 *Times* 9 Feb. 13/2 It will be an eight-storey block with about 60,000 sq. ft. in a landscaped site. 1974 *Daily Tel.* (Colour Suppl.) 8 Mar. 18/2 The initial impression one gets of a landscaped office is often one of irregularity and informality.

landscapist. Add: **2.** A landscape-gardener; one skilled in landscaping roads, offices, etc. Also **la·ndscaper**.

1936 T. Sharp *Eng. Panorama* iii. 48 The activities of the landscapists became founded on a realization and acceptance of natural informality. 1963 *Times* 4 June 12/5 Whereas seventeenth-century landscapers might have moved a tree across an estate, we now move them 30–50 miles. 1963 *New Society* 20 June 26/1 The British method of inviting a landscape committee (few of them landscapists) to study the landscaping of the road. 1965 I. Fleming *Man with Golden Gun* vii. 101 He soon came to the end of the young shrubs and guinea grass the landscaper had laid on. 1967 *House & Garden* Mar. 68/1 Several famous landscapists, including Bridgeman. 1974 *Daily Tel.* (Colour Suppl.) 8 Mar. 18/1 Open plan, say many landscapers, was..only too often a mere space-saving exercise.

Landseer (læ·ndsiǝɹ). The name of the English painter Sir Edwin *Landseer* (1802–73) used *attrib.* in **Landseer Newfoundland** to designate a black and white Newfoundland dog of a type once painted by him. Also *absol.*

1877 G. Stables *Pract. Kennel Guide* ix. 98 The black-and-white, or Landseer Newfoundland. This is quite a distinct breed, not as yet properly recognised at shows. *Ibid.* x. 113 A white chest in a Newfoundland or Retriever, indicates a cross with the Landseer or Setter. 1927 W. A. Wetwan in C. C. Sanderson *Pedigree Dogs* 281 In 1836 Sir Edwin Landseer painted a white and black Newfoundland ..imbuing the mind of the great British public with the idea that black and white was the only correct Newfoundland wear... 'Landseers', which are true to type, are a very beautiful dog. 1971 F. Hamilton *World Encycl. Dogs* 164 In 1779 a well-known English naturalist described a very fine specimen in Northumberland, England, and this dog was later identified as a Landseer. *Ibid.* 165 (*caption*) Taaran Taru..is a typical example of the Landseer Newfoundland.

land-ship. [Land *sb.* 11.] **a.** A wagon or other vehicle serving the same purpose on land as a ship on the sea; *spec.* = *tank *sb.*[7] **b.** A ship erected and kept on land for training purposes.

[1627 J. Taylor *Armado* sig. B1v, (*heading*) A Navy of Land Ships.] 1837 *Penny Mag.* 22 July 276/1 The ox-carts of the Pampas..are quaintly termed by the natives, *barcos de tierra*, i.e., 'land-ships'. 1869 *Cassell's Mag.* Jan. 156/2 In some of the best schools of France and Belgium, it has long been a custom to erect a dry land-ship in the playground. 1907 L. Osbourne *Adventurer* xiii. 159 The land-ship..was hardly more than an aluminium shell.. requiring weeks of labor, possibly months, to make her habitable and ready. Ready? For what? To sail those vast and billowy plains? 1916 *Daily News* 19 Sept. 1/2 The new land-ships or tanks did invaluable work. 1916 *Daily Mirror* 22 Nov. 1/1 (*caption*) To-day we are able to publish the first photograph of one of his Majesty's land ships which have been making such successful cruises on the sea of mud on the Somme. 1934 W. S. Churchill *Gt War* II. xxxv. 519/1 The next day, the 20th [February 1915], I sent for Mr. Tennyson-d'Eyncourt..and convened a conference... As the result of it the Landships Committee of the Admiralty was formed. 1972 *Times Lit. Suppl.* 4 Feb. 113/4 The original initiative in tank development, the so-called 'landships', was naval rather than military.

‖ **Landsmål** (la·ntsmǫl). Also **Landsmaal**. [Norw., f. *land* country + *mål* language.] A literary form of Norwegian devised by the Norwegian philologist Ivar Aasen (1813–1896) from the country dialects most closely descended from Old Norse, and considered to be a 'purer' form of the Norwegian language than the official Riksmål or Dano-Norwegian.

The *Landsmål* controversy followed the appearance of Aasen's grammar and dictionary (1848 and 1850); in 1885 *Landsmål* was given equal status with Dano-Norwegian.

1886 *Encycl. Brit.* XXI. 374/1 By the study of the Modern Norwegian dialects and the mother language, Old Norwegian, the eminent philologist J. Aasen was led to undertake the bold project of constructing..a Norwegian–Norwegian..language, the so-called 'Landsmal'. 1906 *Westm. Gaz.* 10 Aug. 2/3 The party programmes are said to lack definiteness, but that of the Liberals comprises.. the official recognition of the 'Landsmal', or Norwegian of Norway as against the Dano-Norwegian, which is at present the language of the Government. 1911 [see *Dano-]. 1924 *Glasgow Herald* 14 Nov. 7 Let him perpend the circumstances of the Czech revival or of the Norwegian landsmaal movement. 1927 *Observer* 6 Nov. 12 Now Norway has a linguistic national movement, called the Landsmaal or Real Norwegian Language-movement, which holds that the language generally used in Norway is not Norwegian but a Danish dialect, and their aim is to root out that dialect and make the Landsmaal compulsory. 1933 [see *Dano-]. 1957 T. K. Derry *Short Hist. Norway* ix. 190 For the peasantry, in the west at least, *landsmaal* had become a shibboleth by which to distinguish the true democrat from the adherents of the language of foreign snobbery. 1961 L. F. Brosnahan *Sounds of Lang.* ix. 205 The Landsmål, a somewhat more artificial creation based on the main dialects of the west of the country.

landsman. 1. b. Delete *rare* and add later examples.

1950 B. Malamud in *Partisan Rev.* Sept.–Oct. 664 With, after all, a *landsman*, he would have less to fear than with a complete stranger. 1971 *Islander* (Victoria, B.C.) 3 Jan. 10/3, I found out the mate was a landsman of mine who came from Helsingfors. 1973 *Listener* 20 Sept. 377/1 You put on your Shabbat suit..and descended on a near-by relative or *landsman*.

landswoman. [After Landsman.] A woman accustomed to live mainly or entirely on the land; one skilled in land-work.

1837 *Penny Mag.* 14 Oct. 398/1 The scene is presented exactly as it appeared to the eye and imagination of a landswoman. 1891 [see after Landsman]. 1923 *Weekly Dispatch* 1 Apr. 7 Miss Ford might be described as 'the complete landswoman'. She can milk, do anything with horses, and do field tasks.

lane, *sb.* Add: **2.** (Further examples.) Also, a route prescribed for aircraft.

1911 [see *air lane* s.v. *AIR *sb.*[1] III. 8]. 1929 *Encycl. Brit.* I. 231/1 Neon lighting is particularly suitable for landing in fog owing to its distinctive colour, and to the fact that long 'lanes' of illumination can be provided. 1941 A. O. Pollard *Bombers over Reich* 105 The clouds parted a little, and the approaching raiders found enemy fighters collected in the open 'lanes' like soldiers guarding breaches in a fortification. 1956 J. C. Swayne *Conc. Gloss. Geogr. Terms* 86 *Lane*, a much used ocean or air route. 1971 E. C. B. & K. Lee *Safety & Survival at Sea* i. 8 Safety sea-lanes, consisting of a series of two-way lanes with a safety buffer zone separating the inward and outward bound traffic, are used in the approaches to New York harbour and other seaports. 1971 *Sci. Amer.* July 1/1 [An automatic weather information station] was moored in the middle of the Gulf Stream, off the Florida Coast, in a hurricane lane. 1974 L. Deighton *Spy Story* xv. 146 The pilot..climbed again, now that he was no longer forced down under the lanes.

c. In *Athletics*, a course for a runner marked out by broad chalk-lines (orig. strings). Hence also in *Swimming*, such a course marked out by ropes buoyed up by cork floats.

1909 in *Cent. Dict.* Suppl. 1911 *Encycl. Brit.* XXIII. 853/2 The course for sprinting races..is marked off in lanes for the individual runners by means of cords stretched upon short iron rods. 1927 *Daily Express* 23 Mar. 13/5 Sprint-racing in 'lanes' instead of in strings will be in force at the next Olympic Games... The 'lanes' are marked by chalk lines, and have been used in America for some time. 1955 R. Bannister *First Four Minutes* 21, I moved out into the second lane so that I could..avoid the danger of being boxed in. 1960 J. Grinham *Water Babe* xiv. 158 Suddenly the roar in the pool turned to a gasp—Di swam on to the lane ropes. 1970 McGregor & Still *Bobby McGregor Story* ix. 77, I stepped on to the poolside of the magnificent Olympic swimming stadium. I took up my position behind lane 2, officially the position for the fifth fastest qualifier. 1971 D. Emery *Lillian* vii. 75 The bends were tighter on the inside lane and therefore harder to round at full speed.

d. A part of a road, wide enough for one file of vehicles, which is marked out by painted lines and is used to segregate traffic according to speed, intended direction, etc. Also *attrib.* and *Comb.*

1926 *Amer. City* Apr. 358/1 One of the most recent developments in highway design is the so-called super-highway where eight or more traffic lanes are provided for on the same right of way. 1933 *Evening Standard* 19 Apr. 7/2 Roads..would carry any volume of traffic, divided into slow, medium and fast 'lanes'. 1951 *Economist* 22 Sept. 685/3 Super-highways; with at least four lanes. 1959 *Times* 31 Mar. 15/6 The motorist who elects to park his car on a main road reduces the width of the road for a complete traffic lane for what may be hundreds of yards. 1960 *Guardian* 21 Nov. 2/4 Where there is good lane discipline, traffic should be able to pass on the near side. 1962 *Economist* 27 Jan. 327/1 Mr Barnes is a great believer in lane-painting to increase the capacity of streets. 1966 [see *fast lane* s.v. *FAST *a.* 11]. 1968 *Autocar* 7 Mar. 61/3 The first week's working of the London experiments with bus lanes in Park Lane and on Vauxhall Bridge. 1970 *Guardian* 4 Aug. 15/2 Lane-changing, the constant pressure to keep up speeds. 1971 *Daily Tel.* (Colour Suppl.) 22 Oct. 25/4 In town traffic, lane discipline is more a matter of cunning than of boldness. 1972 *Police Rev.* 8 Dec. 1597/2 Failure to judge distance at speed and bad lane drill accounted for most of the accidents. 1973 D. Westheimer *Going Public* ix. 127 Drivers on the inbound lane slowed to a crawl.

e. In ten-pin bowling, etc.: = Alley 4.

1960 D. Taylor *Secret of Bowling Strikes* 125 Most old-fashioned lanes have a center peg in the center of the alley. 1964 F. Brundle *Tenpin Bowling Tips* 79 In some localities a lane which allows the ball to take a wide hook is termed fast... Some authorities..speak of lanes as either 'holding' or 'running'. 1970 C. Schunk *Bowling* i. 4 When alleys were first built in the Southern United States, three-fourths of the lanes were constructed for duck pins. 1974 *Plain Dealer* (Cleveland, Ohio) 26 Oct. 5-D/5 Likewise, the lanes can also be too slick or too dry.

4. b. (Earlier and later examples.) Also *Chancery Lane, Mincing Lane.*

1831 P. Egan *Show Folks* 29 The *swell* performers.. who proudly observe, 'I am engaged at the Lane.'.. But the 'Lane', alluded to in this instance, is Horsemonger Lane; where a number of engagements are suffered to *expire*. 1865 *Chambers's Jrnl.* 18 Feb. 106/1 The 'Lane' (as Chancery Lane is familiarly called). 1872 B. Jerrold *London* viii. 77 When on a certain Sunday we turned into Petticoat Lane, we had the key to the activity of the clothes market of Lazarus. The Lane clothes thousands at Epsom. 1909 *Westm. Gaz.* 6 Aug. 11/4 'The Lane', as that of Mincing is fondly known among the wholesale grocery crowd. 1926 F. M. Ford *Man could Stand Up* II. iii. 138 He had lately promised [them] tickets for Drury Lane... The Lane was the *locus classicus* of the race. 1959 B. Kops *Hamlet of Stepney Green* I. 24, I also stand down the Lane [*sc.* Petticoat Lane] on Sundays now and again. I'm what you might call a purveyor of bad taste. 1974 M. Birmingham *You can help Me* i. 11 Wentworth Street, down which the stalls of Petticoat Lane market spill... We never say 'Wentworth Street'; it's 'The Lane' to us. Ibid. ii. 29 Friday is the day for buying flowers in the Lane.

5*. *Astr.* A narrow band or strip in the sky that differs markedly from its immediate surroundings (e.g. in containing no observable stars or in emitting strong radio signals).

1899 *Astrophysical Jrnl.* IX. 157 The wonderful nebulous region about Rho Ophiuchi..and..the great vacant lanes near that star. 1917 *Proc. Nat. Acad. Sci.* III. 678 A study of the negatives of spiral nebulae obtained with the Crossley Reflector has shown that the phenomenon of dark lanes caused by occulting or absorbing matter is much more frequent than had..been supposed. 1964 R. H. Baker *Astron.* (ed. 8) xvii. 506 The hydrogen lanes traced by Dutch radio observers..in longitudes available to them are shown in Fig. 17·22... In a direction 80° from the sun we note three hydrogen lanes, which trace three spiral arms. 1970 *Nature* 12 Dec. 1077/1 This is identified with NGC 1579 which is a small, irregular, diffuse nebulosity..with a prominent dark lane. 1971 *Ibid.* 21 May 197/3 The underlying common feature of spiral galaxies is the existence of elongated spiral arms traced out by gaseous material (neutral hydrogen lanes; ionized HII regions and dust).

6. *lane-side* (later examples), *-way* (later examples); **lane-route** (examples).

1895 *Funk's Stand. Dict.* 1000/1 *Lane-route*, or *ocean-l. route*, one of the routes prescribed for transatlantic steamers in Northern waters, being different for eastward- and westward-bound vessels, to avoid collisions. 1950 *Ocean Passages for World* (Admiralty, Hydrographic Dept.) (ed. 2) B. I. ii. 41/1 The large number of steam vessels crossing the Atlantic..has necessitated the adoption of clearly defined separate routes to be followed by outward and homeward bound ships... These are known as the *North Atlantic Lane Routes*. *Ibid.*, Masters of all ships..who do not..make use of the 'lane routes', should make themselves acquainted with them, for their own safety. 1899 H. T. Timmins *Nooks & Corners Shropshire* ix. 167 An old country woman tending her cow by the laneside. 1923 *Daily Mail* 2 Apr. 6 To see the lanesides in this delicate livery of verdure and bloom. 1914 Joyce *Dubliners* 185 A crowd which had followed him down the laneway collected outside the door. 1933 L. A. G. Strong *Sea Wall* 258 He charged like a bull across the open space and disappeared into the human laneway.

Lane, *sb.*[2] The name of John *Lane*, 19th-c. English horticulturalist, used in the possessive in **Lane's Prince Albert** to designate a large, green cooking apple of a variety introduced by him in 1857.

1875 *Florist & Pomologist* 233 Lane's Prince Albert..is remarkable both for its excellent quality as a culinary apple and for its prodigious bearing qualities. 1902 [see *Bramley]. 1933 Hall & Crane *Apple* xii. 200 Lane's Prince Albert..is a mid-Victorian introduction. 1962 *Listener* 27 Sept. 495/1, I am thinking of cooking apples like Bramleys and Lane's Prince Albert.

Lang (læŋ). Also **lang**. The name of John *Lang*, used *attrib.* and in the possessive (esp. in *Lang('s) lay*) to designate a lay (Lay *sb.*[7] 7 b) used for wire ropes and patented by him in 1879, in which the strands forming the rope are twisted in the same direction as the wires forming each strand.

1883 *Engineering* 14 Dec. 537/1 In the Lang method the strands and the rope are laid in the same direction. 1887 J. B. Smith *Treat. Cable or Rope Traction* viii. 156 (*heading*) The 'Lang lay', or construction of wire ropes. 1896 W. E. Hipkins *Wire Rope* 52 Ropes on the lang principle. 1930 *Engineering* 1 Aug. 135/1 The hauling rope is of the Lang-lay type, built up of six three-cornered strands and a hemp core. 1959 *B.S. Handbk. No. 4: Lifting Tackle* (*B.S.I.*) I. 90 Ropes made Lang's lay require careful handling to ensure that the rope's end does not twist, and so allow 'turn' to come out of the rope. 1966 *McGraw-Hill Encycl. Sci. & Technol.* XI. 627/1 In lang lay, wires in the strand and the strands themselves are laid in the same direction. This type rope wears better, is more flexible, and lasts longer.

langbanite. Substitute for entry:

långbanite (lǫ·ŋbǎnəit). *Min.* [ad. G. *långbanit* (G. Flink 1887, in *Zeitschr. f. Kryst. und Min.* XIII. 1), f. *Långban*, the name of its original locality in Wermland, Sweden + *-it* -ITE[1].] A black silicate and oxide of antimony, iron, and manganese, near $(Mn^{II}, Sb)_4(Mn^{IV}, Fe^{III})_3SiO_{12}$, with a brilliant metallic lustre.

1887 *Jrnl. Chem. Soc.* LII. 782 (*heading*) Långbanite, a new Swedish mineral. 1949 *Mineral. Abstr.* X. 542 Tetragonal braunite with a cubic pseudo-cell can be fitted into the hexagonal structure of långbanite. 1968 *Arkiv för Mineral. & Geol.* IV. 456 Since the original discovery of långbanite, so many specimens have been identified that it may be considered a moderately abundant mineral.

langbeinite (læ·ŋbəinəit). *Min.* [ad. G. *langbeinit* (S. Zuckschwerdt 1891, in *Zeitschr. f. angew. Chem.* 356/2), f. the name of A. *Langbein*, 19th-cent. German chemist + *-it* -ITE[1].] A sulphate of potassium and magnesium, $K_2Mg_2(SO_4)_3$, colourless when pure, which is known only in salt deposits of marine origin and as a synthetic product, and is used in the production of fertilizers.

1898 *Jrnl. Chem. Soc.* LXXIV. II. 169 The crystals of langbeinite are optically inactive. 1932 *Bull. U.S. Geol. Survey* No. 833. 41 Most of the langbeinite..has a distinctive pink color. It is the 'hardest looking' of the saline minerals and has a conchoidal or irregular fracture with no cleavage. 1951 *Mineral. Abstr.* XI. 247 Langbeinite.. on exposure to air falls to powder consisting of picromerite..and epsomite. 1960 R. A. MacDonald in V. Sauchelli *Chem. & Technol. Fertilizers* xv. 387 Langbeinite..is mined and processed by International Mineral & Chemical Corporation in the Carlsbad basin... The product is marketed at a purity in excess of 95 per cent langbeinite.

‖ **langeleik** (la·ŋēləik). Also **langleik**. [Norw.] An early Norwegian stringed instrument, resembling the zither.

1907 *Westm. Gaz.* 10/1 Ancient music of the Vikings was played..on the 'langleik', the crude guitar of the Norsemen, by an aged minstrel..who is a direct descendant of King Harald Haarfager. 1938 *Oxf. Compan. Mus.* 839/2 Norway possesses several distinctive instruments. The Langleik.., used until near the end of the eighteenth century, was a sort of developed monochord. 1961 A. Baines *Mus. Instruments* 210 A tuning given by Panum for the slightly more advanced Norwegian langeleik. 1970 *Daily Tel.* (Colour Suppl.) 13 Nov. 47 The fjord-loving fiddle is the offspring of a local zither-like instrument called the *langeleik*, whose strings were struck with a plectrum.

Langerhans (læ·ŋəɹhanz). *Histology.* The name of Paul *Langerhans* (1847–88), German anatomist, used *attrib.*, in the possessive, and with *of*-adjunct to designate a kind of dendritic or stellate cell found in the epidermis and characterized by the presence of cytoplasmic granules (*Langerhans granules*). (See also *islet of Langerhans* s.v. *ISLET 2 b).

1890 Billings *Med. Dict.* II. 32/1 *Langerhans' cells*, stellate cells found in deeper layers of epidermis, apparently related to nerve terminations. 1934 E. V. Cowdry *Textbk. Histol.* xxix. 449 Cells of the third category are called melanoblasts (Langerhans, stellate or dendritic cells) because they are supposed to form melanin. 1953 *Phil. Trans. R. Soc.* B. CCXXXVII. 162 This technique ..led to the erroneous conclusion that the cells of Langerhans occur in the basal layer. *Ibid.*, Langerhans' cells are invariably unpigmented. 1965 *Jrnl. Investigative Dermatol.* XLIV. 202/2 The characteristic granules within the Langerhans cell are shaped like a tennis racket... The number, as well as the size, of the Langerhans granules is ..variable. 1965 *Ibid.* XLV. 403/1 The capability of Langerhans cells to synthesize melanin remains to be proven. 1969 S. Bradbury *Hewer's Textbk. Histol.* (ed. 9) xvii. 237 It was at one time thought that the Langerhans cells were effete melanocytes, but they have been shown to be active in the uptake of tritiated thymidine and their E/M appearance suggests also that they are active cells. Their function is still unknown.

Langhans (la·ŋhanz). *Histology.* The name of Theodor *Langhans* (1839–1915), German pathologist, used *attrib.*, in the possessive, and with *of*-adjunct to designate: (*a*) (a cell of) an inner layer of large cuboidal cells, one cell thick, that covers chorionic villi and lies beneath the syncytial layer; (*b*) a distinctive kind of giant cell which has many nuclei, arranged in a ring around the periphery or clustered together at one end of the cell, and is observed esp. in tuberculosis and related granulomatous conditions.

1886 *Buck's Handbk. Med. Sci.* II. 146/2 (*heading*) Langhans' cellular layer. 1900 Dunglison *Dict. Med. Sci.* (ed. 22) App. 1276/2 *Langhans's giant cell*, giant-cell of tubercular granulation-tumor. 1906 *Practitioner* Nov. 663 A teratoma is defined as a growth that consists of elements represented in the developing ovum, whether at an early (e.g. syncytia, Langhans' cells, large mono-nuclear cells), or, at a later date (e.g. neuro-epithelial cutaneous, or other elements). 1937 E. E. Hewer *Textbk. Histol.* xxxii. 280 (*heading*) Layer of Langhans. *Ibid.*, The villi during early pregnancy are covered by an inner layer of cubical cells (Langhans' layer) and an outer syncytial layer. 1939 G. G. Kayne et al. *Pulmonary Tuberculosis* I. iii. 38 Granulomata consisting of epithelioid and Langhans giant cells may be seen as early as the third to sixth day. 1968 H. Harris *Nucleus & Cytoplasm* v. 97 This peripheral distribution of nuclei in certain types of multinucleate cell was first discussed by Langhans, and when such cells are found in pathological conditions they are commonly referred to as 'Langhans' giant cells. 1972 Ebe & Kobayashi *Fine Struct. Human Cells & Tissues* 226 Cytotrophoblasts or Langhans cells of the chorionic villus are thought to be the site of production of a proteinic hormone, chorionic gonadotrophin.

‖ **langlauf** (la·ŋlauf). [G.] Cross-country skiing; a cross-country skiing race. Hence **la·nglaufer**, a competitor in such a race.

1927 A. Lunn *Hist. Ski-ing* xvii. 223 The *Langlauf* was the logical development of Norwegian ski-ing. 1929 *Times* 6 Dec. 6/4 They [*sc.* the British] could not hope to compete against foreign Langlaufers and jumpers. 1963 I. Fleming *On H.M. Secret Service* xvii. 191 Bond..somehow stayed upright on the two miles of treacherous Langlauf down the gentle slope to Samaden. 1968 *Punch* 28 Feb. 293/2 And not only did he win the jumping, but also the downhill, the slalom, and the langlauf. It was a great day for Britain! 1970 *Country Life* 17–24 Dec. 1214/3 The young and energetic have turned ski-loping into a competitive sport. The 30 and 50 km langlauf races are included in the Winter Olympics.

Lang lay: see *Lang.

langley (læ·ŋli). *Meteorol.* [f. the name of Samuel P. *Langley* (1834–1906), U.S. astronomer. Orig. proposed (in G.) by F. Linke 1942, in *Handbuch d. Geophysik* VIII. 30, as a unit of solar energy flux, equal to one gramme-calorie per sq. cm. per minute.] A unit of solar energy per unit area, equal to one gramme-calorie per square centimetre (approximately 41,900 joules per square metre).

1947 *Nature* 6 Sept. 327/1 It is herewith proposed that the 'langley' be defined as the gm. cal./cm.², where 'gm. cal.' denotes the 15°C. gm. cal. It is also proposed that the written abbreviation of 'langley' be 'ly'. 1954 J. C. Johnson *Physical Meteorol.* iv. 108 $\sigma = ..8{\cdot}22 \times 10^{-11}$ langley minute⁻¹. σ is called the Stefan-Boltzmann constant. 1970 Day & Sternes *Climate & Weather* v. 145 The solar constant..is alternatively expressed as nearly 2·0 langleys per minute. 1974 *Nature* 15 Nov. 217/1 For the point of equilibrium I take 17 langley per day (1 langley = 1 calorie cm⁻²) above the 1950 value of 847 langley per day.

Langobard (læ·ŋgǒbaɹd). Also **Longobard**. [see Lombard *sb.*[1] and *a.*] = Lombard *sb.*[1] 1 a. Also as *adj.* Hence **Langoba·rdian** *a.*

1788 Gibbon *Decl. & F.* IV. xlii. 216 The original name of Langobards is expressive only of the peculiar length and fashion of their beards. 1902 L. Villari tr. *P. Villari's Barbarian Invasions Italy* II. III. i. 279 As usual the Longobard host comprised a motley throng of Bavarians, Bulgarians, Gepidæ,..and more especially Saxons. *Ibid.* 280 The beginning of the Longobard rule may be said to date from this event. 1925 *Contemp. Rev.* Aug. 212 The people are of Langobardian, French, or even Gothic origin. 1952 E. Hyams *Soil & Civilization* iv. 30 The first we hear of the buffalo, as a domestic animal, in Europe, is as a gift to the Langobard Court, in Italy, about A.D. 600. 1974 R. A. Hall *External Hist. of Romance Lang.* v. 86 The Langobards' linguistic influence was exerted chiefly in Italy.

Langobardic, *a.* (Later examples.)

1839 K. H. Digby *Mores Catholici* IX. viii. 260 These fragments of Langobardic inscriptions. 1895 T. Hodgkin *Italy & her Invaders* V. iii. 140 Vast stores of wealth, taken from the Gepid dwellings, enriched the Langobardic homes. 1960 W. D. Elcock *Romance Lang.* iii. 228 Words of Langobardic origin having undergone certain sound-shifts characteristic of High German. *Ibid.* 259 A Langobardic king, Liutprand,..saw the possibility of exploiting the political situation.

‖ **langostino** (laŋgǒstī·no). [Sp.] = *Dublin (Bay) prawn* (*Dublin). Cf. *Langoustine.

1915 E. R. Lankester *Diversions of Naturalist* xii. 100 A very large Mediterranean prawn..is called 'Barcelona prawn' and 'Langostino'. 1967 [see *Camaron]. 1970 'D. Halliday' *Dolly & Cookie Bird* ii. 20 A plate of fat pink *langostinas* [sic].

‖ **langouste** (laṅgu·st). [Fr.] A crawfish, *Palinurus vulgaris*, and related species; = Crayfish B. 3 b.

1832 W. Macgillivray *Trav. & Res. A. von Humboldt* xxi. 306 The sailors had been searching for *langoustes*. [1835 W. Kirby *On Power of God in Creation of Animals* II. xv. 49 There is one of a most ferocious aspect..called in the London market the *Thorny lobster*..: it is also called..by the French, who esteem it highly, the *Langouste*.] 1917 N. Douglas *South Wind* xix. 232 Those succulent *langoustes* for which the coastal waters of the island are renowned. 1924 *Blackw. Mag.* Sept. 409/1 The fishwives..spread out before them all varieties of fish and shell-fish from langoustes to cockles. 1942 'R. West' *Black Lamb* I. 23 In Nice, as I sat eating langouste outside a little restaurant down by the harbour. 1949 A. Wilson *Wrong Set* 171 The champagne, the langouste, the pine-scented air..were making of this dinner one of those ..never to be forgotten hours of happiness. 1951 E. David *French Country Cooking* 53 Put in the cut-up pieces of *langouste* and let the pan simmer. 1960 I. Fleming *For your Eyes Only* 230 The blue and yellow langouste came a few steps out from under the rock.

‖ **langoustine** (laṅgustī·n). [Fr.] = *Dublin (Bay) prawn* (*Dublin). Cf. *Langostino.

1946 G. Millar *Horned Pigeon* xiv. 192 *Langoustines* in the Union Bar at Alexandria. 1949 [see *Dublin]. 1951 E. David *French Country Cooking* 35 *Langoustines* are the small..shell fish about three inches long, delicate pink, with a very thin shell, resembling Dublin Bay Prawns. 1965 M. Wallenstein *Merlin's Forest* vii. 86 He began to plan their dinner; beginning with crab or *langoustine*. 1966 P. V. Price *France: Food & Wine Guide* 41 The *langoustine* is like a giant prawn or little langouste. 1974 J. Stubbs *Painted Face* xi. 151 You must try the *langoustine*. It is very good.

langsat (la·ŋsat). Also **langseh, lansat**, etc. [Malay.] The edible fruit of *Lansium domesticum*, a tree of the family Meliaceæ, native to Indonesia and Malaysia; also, the tree itself.

1783 W. Marsden *Hist. Sumatra* 83 *Lansai*. The tree which bears this fruit is large; the leaves are of a lightish green and somewhat pointed. [1795 tr. *C. P. Thunberg's Trav. Europe, Afr. & Asia* (ed. 2) II. 276 *Boa lansay* is the Malay name for the fruit of a tree..which is yet unknown to botanists.] 1839 T. J. Newbold *Pol. & Statistical Acct. Straits of Malacca* I. ii. 53 In the valley grow various fruit-trees, such as..the langseh. 1869 A. R. Wallace *Malay Archipelago* I. v. 132 The Dyaks brought us daily heaped-up baskets of Mangustans and Lansats, two of the most delicious of the subacid tropical fruits. 1911 *Encycl. Brit.* XVII. 472/2 The principal fruit trees are the.. *langsat, rambai*, jack-fruit, etc. 1920 W. Popenoe *Man. Tropical & Subtropical Fruits* xvi. 427 The langsat has not yet become generally cultivated outside of the Asiatic tropics. 1938 *Nature* 14 May 866/2 Many of the fruits and vegetables..have been but names, if that, to dwellers in the temperate zones so far;..the mangosteen and the langsat are still awaited. 1940 E. J. H. Corner *Wayside Trees Malaya* I. 463 The Langsat has an oblong fruit (about 1½ × 1″) with thin, pale greyish-buff rind containing much white latex. 1963 J. Kirkup *Tropic Temper* 115, I had some chicken..followed by sour-sweet langsat fruit.

langsuir (læ·ŋsiūǝɹ). Also **langsuyar**. [Malay.] A female vampire with a whinnying cry, that preys on newborn children. Cf. *Penanggalan.

1881 *Jrnl. R. Asiatic Soc., Straits Branch* June 28 If a woman dies in child-birth..she is popularly supposed to become a *langsuyar*, a flying demon of the nature of the 'white lady' or '*banshee*'. 1900 W. W. Skeat *Malay Magic* vi. 320 The Langsuir..takes the form of an owl. 1972 *Daily Tel.* (Colour Suppl.) 12 May 58/3 The Malayan vampire family includes..the Langsuir, the spirit of a woman who died after hearing that her child was stillborn, known by her green robe, her tapering nails of amazing length and her long black hair which conceals a hole in the back of her neck through which she sucks the blood of children.

language, *sb.* Add: **1.** *first language*: one's native language. *second language*: a language spoken in addition to one's native language; the first foreign language one learns.

1875 W. D. Whitney *Life & Growth of Language* ii. 25 We realize better in the case of a second or 'foreign', than in that of a first or 'native' language, that the process of acquisition is a never-ending one. 1876 C. M. Yonge *Womankind* vi. 40 The second language has been really and grammatically learnt. 1943 I. A. Richards *Basic Eng. & its Uses* 14 The history of the nationalist movement in India is an instructive instance. Its leaders and its chief supporters are speakers of English and sometimes use it rather as their first than as their second language. 1962 R. Quirk *Use of English* i. 6 Something like 250 million people for whom English is the mother-tongue or 'first language'. 1971 *Guardian* 23 June 7/3 Indians and Pakistanis..using a second language at school and their first language for many home activities.

b. *language of flowers* (examples).

1834 tr. *C. de la Tour's Lang. Flowers* 95 It is more especially by..modifications that the Language of Flowers becomes the interpretation of our thoughts. 1847 Thackeray *Van. Fair* (1848) iv. 31 Perhaps she just looked first into the bouquet, to see whether there was a *billet-doux* hidden... 'Do they talk the language of flowers at Boggley Wollah, Sedley?' asked Osborne, laughing. 1949 *Enquire within upon Everything* (ed. 122) 462 *Language of Flowers.* The symbolism of flowers has always possessed a certain fascination, especially for the young person of either sex.

d. *Computers.* Any of numerous systems of precisely defined symbols and rules for using them that have been devised for writing programs or representing instructions and data.

1949 E. C. Berkeley *Giant Brains* iii. 29 We must translate into machine language, in this case punched holes in the program tape. 1956 *Jrnl. Assoc. Computing Machinery* III. 272 In the development of an automatic coding system, two major problems arise. The first is to develop a coding language which permits a programmer to specify the computation he wants the machine to perform. Once this has been done, there remains the task of coding a compiler for a particular high speed calculator which will translate the language into actual machine instructions... The language described here is the one translated by the PACT I Compiling Routine into instructions for the IBM Type 701. 1959 E. M. Grabbe et al. *Handbk. Automation, Computation, & Control* II. ii. 186 The purpose of these activities has been to..set up a class of languages that will be easily translatable by machine from one to another, and also easily recognizable to the ordinary human user... Such languages form the input to a class of automatic computer programs called translators, which perform a translation..into a second or target language. The latter may be either (1) an assembly language such as SOAP, SAP, or MAGIC.., or (2) a straight machine language, in pure decimal, binary (or in some cases such as the Univac I and II), alphanumeric. 1961 Leeds & Weinberg *Computer Programming Fund.* ii. 46 The best way of writing down operations is to write them in alphabetical format. A format used for writing down these alphabetical instructions is called the programming language or paper language, to distinguish it from the machine language..acceptable to the machine circuitry. 1964 F. L. Westwater *Electronic Computers* ix. 145 As the benefits of these codes were realised, each manufacturer produced different 'languages'. 1966 A. Battersby *Math. in Managem.* viii. 206 If each manufacturer prepares a compiler routine which will translate instructions in some universal 'language' into a program in his own code, then programs written in the universal language can be run on any machine. 1967 A. Hassitt *Computer Programming* i. 1 An efficient way of learning to use a computing machine utilizes one of the problem oriented languages such as Fortran, Algol, or PL/1. 1970 A. Cameron et al. *Computers & Old Eng. Concordances* 27 If we program in so-called higher-languages, like Fortran, conceivably PL1,..I myself will be very surprised if the next generation of machines will not accept Fortran programming and probably Cobol, Algol, and PL1 programming.

3. *bad language* (example); *strong language* (examples).

[1759 Burke *Philos. Enquiry Sublime & Beautiful* (ed. 2) v. vii. 338 We do not sufficiently distinguish, in our observations upon language, between a clear expression, and a strong expression.] *c* 1863 T. Taylor in M. R. Booth *Eng. Plays of 19th Cent.* (1969) II. 109 Come, cheeky! Don't you use bad language. *a* 1910 'Mark Twain' *Autobiogr.* (1924) II. 88 She made a guarded remark which censured strong language. 1934 R. Macaulay *Milton* vi. 100 Milton's familiarity with the tradition [of scurrility] may account for much of his strong language, even when reviling in English.

e. (Earlier and later examples.)

1860 Dickens *Uncomm. Trav.* (1861) v. 65 Mr. Victualler's assurance that he 'never allowed any language, and never suffered any disturbance'. 1865 —— *Dr. Marigold's Prescriptions* i, in *All Year Round* Extra Christmas No., 7 Dec. 4/1 But have a temper in the cart, flinging language and the hardest goods in stock at you, and where are you then? 1929 C. C. Martindale *Risen Sun* 173, I have heard more 'language' in a 'gentleman's' club in ten minutes than in all that evening in the Melbourne Stadium. 1974 'M. Innes' *Mysterious Commission* vii. 75 'You behave like bloody fools.' 'Language, now, Mr Honeybath, language.'

f. Phr. *to speak (talk) someone's language, to speak (talk) the same language*: to have an understanding with someone through similarity of outlook and expression, to get on well with someone; *to speak a different language (from someone)*: to have little in common (with someone).

1893 'S. Grand' *Heavenly Twins* I. II. vi. 256 What could Evadne have in common with these flippant people..? They did not even speak the same language. (To their insidious slang she opposed a smooth current of perfect English.) 1904 H. James *Golden Bowl* I. xvii. 297 They hung together, they passed each other the word, they spoke each other's language, they did each other 'turns'. 1915 Conrad *Victory* IV. xi. 391 You seem to be a morbid, senseless sort of bandit. We don't speak the same language. 1923 H. Crane *Let.* 13 Apr. (1965) 131 The older poets and writers down here..don't talk the same language as we do. 1930 A. Huxley *Brief Candles* viii. 284 You'll perceive that he speaks your language, that he inhabits your world of thought and feeling. 1938 F. Scott Fitzgerald *Let.* 7 July (1964) 33, I want my energies and my earnings for people who talk my language. 1957 J. Osborne *Look Back in Anger* II. ii. 64 As for Jimmy—he just speaks a different language from any of us. 1961 A. Wilson *Old Men at Zoo* i. 25 Bobby..had presumed that since he and I 'spoke the same language', I should naturally dislike the Director as much as he did. 1971 R. Rendell *No More Dying Then* xix. 166 She really didn't understand him at all, his need to be respectable... They didn't speak the same language.

6. a. *language acquisition, change, course, description, engineering, event, -form, -group, -pattern, sign, structure, -study, -system, -use.* **b.** *language-learner, -learning, -maker* (later examples), *-teacher* (later example), *-teaching, -user, -using*; **language area**, (*a*) an area of the cerebral cortex regarded as especially concerned with the use of language; (*b*) a region where a particular language is spoken; **language barrier**, a barrier to communication between people which results from their speaking or writing different languages; **language-contact** *Linguistics* (see quot. 1964); **language-game** *Philos.*, a speech-activity or limited system of communication and action, complete in itself, which may or may not form a part of our existing use of language; **language laboratory** (*colloq.* **language lab**), a classroom, equipped with tape recorders, etc., where foreign languages are learnt by means of repeated oral practice; **language-particular** *a.*, = **language-specific* adj.; **language-specific** *a. Linguistics*, distinctive to a specified language.

1921 H. E. Palmer *Princ. Lang.-Study* 14 In addition to certain *spontaneous* capacities, we possess what we may term 'studial' capacities for language-acquisition. 1965 N. Chomsky *Aspects of Theory of Syntax* i. 52 The innate structure of a language-acquisition device. 1971 D. Crystal *Ling.* 257 Alternative theories of language acquisition are much needed. 1937 *Bull. Los Angeles Neurol. Soc.* II. 36 (*heading*) Case illustrating capacity for use of symbols after destruction of the major (left) language area. 1939 L. H. Gray *Foundations of Lang.* ii. 25 One may frequently say that such-and-such an individual is from such-and-such a district within the language-area. 1961 *Lancet* 12 Aug. 361/2 The insistence that the degree of disability—e.g., in aphasia—was proportional to the amount of 'language area' destroyed. 1933 *Discovery* Sept. 281/2 Science itself..might go forward with greatly increased efficiency if the language barrier were removed by the adoption of Basic for Abstracts and Congresses. 1961 *Guardian* 18 May 8/2 A German girl tries to talk to him, but the language barrier is impenetrable. 1971 *Physics Bull.* Sept. 514/2 Important work in a number of countries may be missed because of the language barrier. 1912 L. Bloomfield in C. F. Hockett *Leonard Bloomfield Anthol.* (1970) 37 A suggestion of 'concerted effort to shape usage' is..hitched on to a discussion of the universal unconscious processes of language-change. 1968 J. Lyons *Introd. Theoret. Ling.* i. 22 To have developed a general theory of language change and linguistic relationship was the most significant achievement of nineteenth-century linguistic scholarship. 1954 U. Weinreich in Saporta & Bastian *Psycholinguistics* (1961) 378/1 A full account of interference in a language-contact situation ..is possible only if the extra-linguistic factors are considered. 1964 M. A. K. Halliday et al. *Ling. Sci.* 77 Situations in which one language community impinges on another have been called 'language contact' situations. 1966 *Amer. Speech* XLI. 39 A name in a language-contact situation is sometimes the only element which survives the impact of another language. 1921 H. E. Palmer *Princ. Lang.-Study* 54 Most language-courses must necessarily be *corrective courses*. 1973 A. Price *October Men* xi. 155 [He] had been sent on a language course at a provincial English university. 1963 J. Lyons *Structural Semantics* ii. 36 The distinction between language-operation and language-description. 1971 D. Crystal *Ling.* 54 It proved necessary to..redefine many of the categories,.. to make them applicable to the task of language description. 1953 J. B. Carroll *Study of Lang.* iv. 113 Linguistics may play a part in the solution of certain social problems. If so, a new kind of applied science—'language engineering' as it has recently been termed—may come into being. **1957** *Economist* 7 Sept. 851/2 An electronic data-processing machine..is breaking new ground in 'language engineering' by providing words—as many as five consecutive ones—which are missing from the Dead Sea Scrolls. **1964** *English Studies* XLV. 21 This admits under the label of 'English' a great range of different kinds of 'language event'. **1965** R. M. W. Dixon *What is Lang.?* 93 The data to be accounted for are observed language events. **1901** H. Oertel *Lect. Study of Lang.* ii. 102 The results of all higher classification beyond these, such as language-forms, are ideal types. **1932** A. H. Gardiner *Theory of Speech & Lang.* iv. 207 Jespersen..points out that particular phrases used in this way have become so stereotyped as to be real language-forms, e.g. *Well, I never! I must say!* Most curious of all is *I say!* with nothing following. **1934** R. Benedict *Patterns of Culture* (1935) iii. 48 When we describe the process [of the evolution of Gothic architecture] historically, we inevitably use animistic forms of expression as if there were choice and purpose in the growth of this great art-form. But this is due to the difficulty in our language-forms. **1971** D. Crystal *Ling.* 71 The philosophical search for laws of thought underlying language forms. **1921** H. E. Palmer *Princ. Lang.-Study* 145 Language-games may not further the student sufficiently in the habit-forming process. **1933–4** Wittgenstein *Blue & Brown Bks.* (1958) 17, I shall in the future again and again draw your attention to what I shall call language games. These are ways of using signs simpler than those in which we use the signs of our highly complicated everyday language. Language games are the forms of language with which a child begins to make use of words. **1970** A. MacIntyre *Marcuse* vii. 80 Wittgenstein tries to construct language games. **1970** *Times Lit. Suppl.* 23 July 787/1 In this country it was a dominant caste of philosophers..who seemed to be most gainfully preoccupied with the verbal manifestations of mind, having been coached at 'language-games' by Wittgenstein. **1927** Peake & Fleure *Peasants & Potters* 121 A group with common speech, that is to say a language-group. **1964** *English Studies* XLV. Suppl. 11 His systematic sub-division of the principal language-groups ..represents an astonishing linguistic perception. **1963** *Guardian* 4 Oct. 4/3 In a 'language lab' each student has his own booth and a tape-recorder which guides his speaking in French, Russian, or in any other language. **1968** A. Diment *Bang Bang Birds* ii. 18 There was my speech training. Usually a couple of hours a day down in the language labs. **1931** R. H. Waltz in *Mod. Lang. Jrnl.* XVI. 217 (*title*) Language laboratory administration. **1946** *French Rev.* XX. 19 A large Language Laboratory was installed... Phonographs and records were available at all times of the day. **1963** *Listener* 14 Nov. 791/1 In 1942.. my ideas were referred to as the 'language laboratory', a name that has stuck..to this day. **1969** *Ibid.* 3 July 8/2 I've done most through the Language Laboratory. I think it's a marvellous idea to start off a language by listening to what people say in the language. **1973** *Jrnl. Genetic Psychol.* CXXIII. 7 The Ss were brought in groups of 20 to 30 students each, to a language laboratory where they were seated at individual carrels. **1921** H. E. Palmer *Princ. Lang.-Study* 14 Most language-learners at the present day are found to make an almost exclusive use of their studial capacities. **1965** N. Chomsky *Aspects of Theory of Syntax* i. 43 Cyclic regularities..are much more difficult for the language-learner to construct. **1697** J. Sergeant *Solid Philos. Asserted* Pref. § 10 Perhaps there is not one Evident Truth in it..but only such a way of Plausible Discourse or Language-Learning, as may serve equally and indifferently to maintain either side of the Contradiction? **1964** *Language* XL. 134 Chomsky's hypothesis is that the child is innately equipped with a language-learning device. **1867** W. D. Whitney *Lang. & Study of Lang.* v. 197 Language-makers in different parts of the earth. **1952** H. Read in B. Hepworth *Carvings & Drawings* p. ix/1 In this situation the artists of a period are the language-makers, inventing visual symbols. **1968** P. M. Postal *Aspects Phonol. Theory* viii. 164 The function of morpheme structure rules was to represent those language-particular predictable constraints on the possible combinations of feature specifications both within a segment and sequentially. **1970** *Language* XLVI. 377 It is a possible language-particular constraint on pronominalization in complex structures that a pronoun and its antecedent must lie within the same 'chain of command'. **1935** G. K. Zipf *Psycho-Biol. of Lang.* (1936) 19 Conditions present in all speech-elements or language-patterns. **1961** J. B. Wilson *Reason & Morals* iii. 178 Accepted language-patterns..act primarily as conservative forces both in the individual and in society. **1946** C. Morris *Signs, Lang. & Behavior* 350 In this book 'language sign' is often used in place of 'lansign'. **1970** Language sign [see *lansign]. **1972** *Language* XLVIII. 431 The post-Saussurean debate on the arbitrary nature of the language sign. **1965** N. Chomsky *Aspects of Theory of Syntax* iv. 166 However, there are also many language-specific redundancies. **1969** *Computers & Humanities* III. 258 Studies of..relative frequencies of language-specific syllabic patterns. **1970** *Language* XLVI. 784 It is non-language-specific in that it is empirically based on studies in English, some ten Meso-American languages, some twenty-four Philippine languages, and a few scattered languages from other areas. **1933** L. Bloomfield *Lang.* i. 18 H. Steinthal..published in 1861 a treatise on the principal types of language structure. **1971** D. Crystal *Ling.* 59 Areas of language structure other than grammar were disregarded in most traditional accounts. **1921** H. E. Palmer (*title*) The principles of language-study. **1933** L. Bloomfield *Lang.* i. 1 Many people have difficulty at the beginning of language study. **1964** C. Barber *Ling. Change Present-Day Eng.* vii. 149 Your own speech..is always the right place to begin language-study. **1940** A. H. Gardiner *Theory of Proper Names* 67 Regardless of the language-system as a whole. **1946** *Mind* LV. 339 This task should be approached by construction of consistent language-systems. **1966** *English Studies* XLVII. 193 An item in a highly personal language-system. **1921** H. E. Palmer *Princ. Lang.-Study* 58 The language-teacher must possess a considerable knowledge of phonetic theory. *Ibid.* 15 The language-teaching forces of nature. **1964** W. R. Lee in D. Abercrombie et al. *Daniel Jones* 291 The clear purpose is to see in what manner aids can subserve language-teaching. **1956** J. Holloway in A. Pryce-Jones *New Outl. Mod. Knowl.* viii. 42 Discoveries about language-use which are in themselves not necessarily connected at all with metaphysics. **1963** J. Lyons *Structural Semantics* i. 7 The known or apparent facts of language-learning and language-use. **1965** N. Chomsky *Aspects of Theory of Syntax* 6 The grammar of a particular language..is to be supplemented by a universal grammar that accommodates the creative aspect of language use. **1953** *Mind* LXII. 332 The sentence..mentions neither linguistic expressions nor language users. **1959** *Brno Studies in English* I. 29 The consciousness in language-users of the existing quasi-ideographic trends of the written norm. **1961** *Encounter* Mar. 60/1 Intentional action is characteristic of human beings ..as language-users. **1971** D. Crystal *Ling.* 85 We must ..start with the study of individual language users. **1921** H. E. Palmer *Princ. Lang.-Study* 96 These then are the chief things to be done once we have decided to enlist on our behalf the universal and natural powers of language-using. **1954** U. Weinreich in Saporta & Bastian *Psycholinguistics* (1961) 376/1 The language-using individuals are thus the locus of the contact.

langue. Add: **3.** *Linguistics.* A language viewed as an abstract system, accepted universally within a speech-community, in contrast to the actual linguistic behaviour of individuals (opp. *parole *sb.*).

1924 L. Bloomfield in *Mod. Lang. Jrnl.* VIII. 318 This rigid system, the subject-matter of 'descriptive linguistics', as we should say, is *la langue*, the language. **1947** *Word* III. 16 Langue, tho described as a repository, is not to be thought of simply as a pile of words. **1953** W. J. Entwistle *Aspects of Lang.* i. 26 One may..treat language (*langue*) as a generalization which becomes concrete and individual in speech (*parole*). **1957** [see *diachronically* adv. s.v. *diachronism]. **1964** *Language* XL. 214 Current theorizing about the acquisition and functioning of the speaker's langue. **1965** N. Chomsky *Aspects of Theory of Syntax* i. 4 The distinction..is related to the *langue-parole* distinction of Saussure; but it is necessary to reject his concept of *langue* as merely a systematic inventory of items. **1968** *Word* XXIV. 56 Accent, viewed dynamically, constitutes the *parole* which manifests the pattern or *langue*.

‖ **langue de chat** (lɑ̃g də ʃa). [Fr., lit. cat's tongue.] A long thin piece of chocolate; a crisp biscuit of the same shape.

1907 *Yesterday's Shopping* (1969) 48/2 Chocolate for dessert... Langues de Chat. **1926–7** *Army & Navy Stores Catal.* 54/1 Chocolate... Langues de Chat—box 2/6. **1931** R. H. Heaton *Perfect Hostess* 111 Hand round any Biscuits you prefer. Langue de Chat are popular. **1945** A. Huxley *Time must have a Stop* i. 2 It was French chocolate... Those delicious *langues de chat*. **1964** J. Fleming *The Chill & the Kill* ii. 26 The Christmas Hamper..included..Elvas plums, Turkish Delight,..Cape gooseberries, and Langues de Chat. **1970** Simon & Howe *Dict. Gastron.* 239/1 *Langue de chat*, a type of biscuit (cookie) which derives its name from its shape; thin, long and flat like a cat's tongue.

Languedoc. Add: **2.** **langue d'oc**, the language spoken in mediæval France in areas roughly south of the Loire, where the use of *oc* [f. L. *hoc*] for *yes* was characteristic of many phonetic variations; opp. **langue d'oïl, d'oui**, the language spoken in areas north of the Loire, where *oïl* [f. L. *hoc ille*] (mod. Fr. *oui*) was used for *yes*, and which has developed into standard modern French.

1703 *Acct. Theatre of War in France, being a Geogr. & Hist. Descr. Languedoc* 3, I more approve of the Etymology of those who observe, that, time out of mind, the French have been distinguish'd into *Langue d'Ouy*, and *Langue d'Oc*, that is, into such as say *Ouy*, and such as say *Oc* for *Yes*; the first living on this, and the other on that side the River Loire. [**1819** Scott *Ivanhoe* II. iii. 42 The knight..asked..whether he would choose a *sirvente* in the language of *oc*, or a *lai* in the language of *oui*... 'A ballad, a ballad,' said the hermit, 'against all the *ocs* and *ouis* of France.'] **1854** C. M. Yonge *Little Duke* i. 8 The Normans ..had taken up what was then called the Languédoul, a language between German and Latin, which was the beginning of French. **1866** —— *Prince & Page* iv. 53 My own children..scarce knew whether they spoke English, Languédoc, or Languédoul. **1885** H. James *Little Tour in France* xx. 134 Meetings at which poems in the fine old *langue d'oc* are declaimed. **1903** G. E. C. Casey *Riviera Nature Notes* (ed. 2) liii. 377 The various *Provençal Dialects* are..remnants of the old 'Langue d'Oc'... The 'Langue d'Oïl', or northern French was spoken by men more warlike and more barbarous. **1931** G. J. Renier *The English* viii. 156 The Kings of France did what they could to extirpate the *Langue d'Oc* as soon as they had conquered the southern half of France. **1934** M. K. Pope *From Latin to Mod. French* ii. 17 In the twelfth century the vernaculars of the south and the north (the *Langue d'Oc* and the *Langue d'Oïl*, as they were called after their particles of affirmation) were held to be distinct languages. **1961** P. Green tr. *Oldenbourg's Massacre at Montségur* i. 8 The great barons of the North, the land of the *langue d'oïl*,..were by no means all loyal to the French King.

Languedocian (tæːŋgədōᵘ·ʃiăn), *a.* and *sb.* [f. *Languedoc* (see Languedoc).] **A.** *adj.* Of or pertaining to Languedoc, its inhabitants, or their language. **B.** *sb.* **a.** An inhabitant of Languedoc. **b.** The language spoken there, a dialect of Provençal.

1736 [see Unstained *ppl. a.* 1]. **1765** Sterne *Tr. Shandy* VII. xliv. 154 That sprightly frankness which at once unpins every plait of a Languedocian's dress. **1771** C. Burney *Present State of Mus. France & Italy* 391 Agreeable Provençale and Languedocian melodies. **1792** A. Young *Trav. France* I. 32 Languedocian bishops are certainly not English ones. **1823** A. Thiers *Pyrenees* v. 66 The old Romance language, which mixed with..the Spanish in Languedoc, forms..the Languedocian. **1908** *Daily Chron.* 25 May 6/4 He has..found time to write poems in the Languedocian language. **1927** A. L. Maycock *Inquisition* 178 The records of the Languedocian tribunals. **1936** A. W. Clapham *Romanesque Archit.* iv. 100 A series of apostles..typical of later Languedocian sculpture. **1960** *20th Cent.* Sept. 209 We have used the words *Languedoc* and *Languedocian* in some places where *land of Oc* and *occitanian* would be the literal translation. **1972** J. Avias in Herak & Stringfield *Karst* v. 131 Epeirogenic movements (on the Languedocian coast for example) together with climatic changes caused variations in the relative basal marine or oceanic level.

langue d'oïl, d'oui: see *Languedoc 2.

languent, *a.* Add: **2.** Languid. *poet. rare*⁻¹.

1862 G. M. Hopkins *Vision of Mermaids* (1929), Some would plash The languent smooth with dimpling drops.

languisher. (Later example.)

1896 *Godey's Mag.* Feb. 193/2 A few silly languishers flutter and simper, 'How nice! how lovely!'

languor, *v.* Delete † *Obs.* and add later examples.

1891 A. Beardsley *Let.* July (1971) 24 'I should like,' he [*sc.* Burne-Jones] says, 'to see your work from time to time... I know you will not fear work, nor let disheartenment languor you.' **1969** *Harper's Mag.* June 37 America languors with an illness of euphoria brought on by our leaders. **1975** *N.Y. Times* 11 May 73/2 It embraces contemporary English aristocracy at upper-crust social functions,..hedonistic Romans languoring in ancient cities, [etc.].

la·nikin, *a. rare*⁻¹. [Cf. Cheshire dial. *lankin* and *lanniky*.] Lanky.

1862 Borrow *Wild Wales* II. xxvi. 295 He was a tall lanikin figure with a pair of..staring eyes.

lank, *a.* Add: **4.** *lank-legged* adj.

1906 E. Dyson *Fact'ry 'Ands* xiii. 172 Levi Goss..a lank-legged, ungainly object. **1921** W. de la Mare *Veil* 56 Like lank-legged grasshoppers in June-tide meadows. **1937** —— *This Year, Next Year*, And out of window gaze At lank-legged Peggy.

la·nkily, *adv.* [f. Lanky *a.*] In a lanky fashion.

1903 Conrad & Hueffer *Romance* i. 37 The second mate was lankily stalking the deck. **1926** A. Bennett *Lord Raingo* I. xlviii. 215 'Yes, Raingo,' said the tall, gaunt old man, striding lankily into the presence [of the minister]. **1937** A. Waugh *Eight Short Stories* viii. 253 He was lankily over-grown, with a sallow complexion and a pimply chin.

lankly, *adv.* (See under Lank *a.* (*sb.*).) Delete † and add later example.

1924 C. Mackenzie *Old Men of Sea* xi. 182 Mrs. Ringshaw used to stand beside him, her grey hair wet with spray and lankly waving.

lanky, *a.* Add: **b.** *lanky-legged, -looking* adjs.

1932 Auden *Orators* iii. 90 Lanky-legged Lloyd, and Morgan from Aberdovey, Peacock and long-skulled Cornish Davy. **1922** Joyce *Ulysses* 108 Now who is that lankylooking galoot over there.

2. Used as *sb.*, as a nickname or form of address for a lanky person.

c **1863** T. Taylor in M. R. Booth *Eng. Plays of 19th Cent.* (1969) II. 109 Just you try it, lanky! Yah! Hit one of your own size—do. **1942** Berrey & Van den Bark *Amer. Thes. Slang* § 184/7 Nicknames for a tall, lanky person... Harry Longlegs, Lanky, Legs, Lengthy. **1948** D. Ballantyne *Cunninghams* 212 Hiya, Lanky! **1959** I. & P. Opie *Lore & Lang. Schoolch.* ix. 169 *Lankies.* Inevitably there is a fusion of terms between those for the thin and lanky lad and those for the overgrown.

lanosterol (læ·nǫstěrǫl). *Biochem.* [f. L. *lān-a* wool + -o + *-sterol.] An unsaturated sterol, $C_{30}H_{50}O$, which occurs in wool fat.

1929 Drummond & Baker in *Jrnl. Soc. Chem. Industry* 9 Aug. 238T/2 We recommend that the misleading name isocholesterol be replaced by the less committal lanosterol. **1955** *Soap, Perfumery & Cosmetics* XXVIII. 1262/2 Lanosterol. This lanolin derivative is the newest to arouse interest in the cosmetic and related fields... Pure lanosterol is a light-coloured, free-flowing amorphous powder. **1964** *New Scientist* 22 Oct. 220/1 No fewer than thirteen distinct enzymic reactions are required to build up the sterol prototype, lanosterol, from acetic acid, and about as many to modify lanosterol into cholesterol, the characteristic animal sterol.

lansign (læ·nsəin). Short for *language sign.

1946 C. Morris *Signs, Lang. & Behavior* 36 We propose therefore to call sign-sets of the kind in question lansign-systems, and the individual members of these systems lansigns. **1970** *Sci. Jrnl.* Jan. 57 In the 1930s C. K. Ogden, I. A. Richards and A. Korzybski, and more recently C. E. Osgood, D. H. Mowrer and others, tried to show how language symbols and signs (lansigns, as they are sometimes called) are associated with their referents in much the same way as conditioned stimulus becomes associated with an unconditioned stimulus, as in the classical conditioning theory of Pavlov.

lansquenet. Add to forms: (sense 1) **landsknecht** now usual; (sense 2) **lambskinnet.** **1.** For '17th and 18th centuries' read '16th and 17th centuries'. Add later examples.

1911 *Encycl. Brit.* XIV. 521/1 The Landsknecht was the prototype of the infantryman of the 16th and 17th centuries. **1936** *Burlington Mag.* June 294/1 Among the daggers is an elaborate landsknecht one in its sheath. **1944** Auden *Sea & Mirror* in *For Time Being* iii. 56 Our motheaten..stock costumes which with only a change of hat and re-arrangement of safety-pins, had to do for the *landsknecht* and the Parisian art-student. **1959** *Chambers's Encycl.* I. 610 (*caption*) Landsknecht sword, first half 16th century.

2. (Later example.)

1917 'H. H. Richardson' *Fortunes Richard Mahony* 9 Even the 'shepherds' beguiled the time with euchre and 'lambskinnet'.

lantana (læntā·nă). Substitute for def.: [mod. L. (Linnæus *Hortus Cliffortianus* (1737) 349), f. an earlier Latin name for *Viburnum*, to which its foliage bears a slight resemblance.] An evergreen herb or shrub of the genus so called, belonging to the family Verbenaceæ, often a native of sub-tropical America, and bearing heads of red, yellow, or white flowers. (Add later examples.)

1917 *Nature* 20 Sept. 57/2 Two introduced shrubs, Guava and Lantana, now occupy extensive areas [of Hawaii], and have become great pests. **1933** *Times Lit. Suppl.* 9 Nov. 776/4 The scene is the Tweed River district of New South Wales, where banana plantations compete with the lantana creeper for a foothold. **1947** K. Tennant *Lost Haven* (1968) Prologue 3 The loveliness of the place is a faint, sweet corruption; old, grey, wooden wharves..heaps of coal overgrown with wild convolvulus and lantana. **1961** *Amat. Gardening* 21 Oct. Suppl. 31/1 Most of the lantanas form small shrub-like plants with roundish heads of small flowers. **1969** *New Scientist* 20 Feb. 385/1 The world's worst weeds..include purple nutsedge, Bermuda grass..cogon grass, and lantana.

lanterloo. Restrict † *Obs.* to sense in Dict. and add: **2.** Used as a meaningless refrain (cf. etym. in Dict.).

1951 Auden & Kallman *Rake's Progress* i. 17 The sun is bright, the grass is green: *Lanterloo, lanterloo.* The King is courting his young Queen. *Lanterloo, my lady.* **1951** Auden *Nones* (1952) 54 Turning his barrel-organ, playing *Lanterloo, my lovely*, my *First-of-May*.

lantern, *sb.* Add: **8.** **a.** *lantern fruitage, lecture, roof* (later examples), *slide* (earlier and later examples). **c.** *lantern-fruited, -lighted* adjs.

1920 A. Huxley *Leda* 7 Moons of many-coloured light That swing their lantern-fruitage in the night. **1912** W. de la Mare *Listeners* 53 She rested her old eyes From the lantern-fruited yew trees. **1912** W. Owen *Let.* 6 Feb. (1967) 114 Miss Lingley, brother, & friend, who are giving a Lantern Lecture on their tour among Korean Missions. **1938** L. MacNeice *I crossed Minch* ii. viii. 119 At the end of the service a lantern lecture was announced, which reminded me pleasantly of my childhood. **1871** M. S. Jeune *My School Days in Paris* vii. 92 At midnight a procession, lantern-lighted, wound slowly through the garden-walks. **1906** *Westm. Gaz.* 14 July 2/3 And to our fog-bound window came A lantern-lighted ancient dame. **1942** *R.A.F. Jrnl.* 13 June 3 In caves and cellars,.. lantern-lighted, a multitude of people endure. **1967** *Gloss. Caravan Terms* (*B.S.I.*) 2 *Lantern roof*, a roof with raised centre portion usually throughout its length, the side walls of which are provided with windows and ventilators. **1969** *Canad. Antiques Collector* May 16/2 The Great Kitchen..has a lantern roof supported on four cast-iron columns. **1871** G. Fox in *English Mechanic* 13 Jan. 405/3 (*heading*) Lantern slides. **1909** W. Owen *Let.* 4 Jan. (1967) 49 There was a Church Army Mission with lantern slides. *a* **1930** D. H. Lawrence *Phoenix II* (1968) 115 Gilbert's lectures..with lantern-slides, thrilled Woodhouse to the marrow.

9. **lantern bug** = *lantern-fly*; also *fig.* (see quot. 1774); **lantern clock**, a 17th-century bracket clock worked by weights and surmounted by a bell in a frame; **lantern test** *Ophthalm.*, a test for colour-blindness in which the subject is asked to name or match colours shown by a lantern.

1774 J. Burgoyne *Maid of Oaks* i. ii. 14, I would have put out Mr. Lanternbug's stars with one dash of my pincil. **1847** G. A. F. Ruxton *Adventures Mexico & Rocky Mts.* xix. 156 Of bugs and beetles there is endless variety— including the cocuyo or lantern-bug, and the tarantula. **1927** Haldane & Huxley *Animal Biol.* xi. 228 Many lantern bugs have this anterior prolongation of the head. **1913** L. V. Lockwood *Furnit. Collectors' Gloss.* 18/1 Clock..Chamber... These clocks are intended to hang high on the wall on brackets. Called also Lantern and Bird Cage clocks. **1960** H. Hayward *Antique Coll.* 161/2 *Lantern clock*: a clock of typically English design evolved in the early part of the 17th cent., and persisting, especially in the provinces, until well into the 18th cent... All original lantern clocks are weight driven. **1970** *Canad. Antiques Collector* Dec. 12/1 Lantern clocks..were designed to hang on the wall, and were weight driven and regulated by a balance wheel. **1890** *Brit. Med. Jrnl.* 11 Jan. 73/2 The Lantern Test is the one which I recommend for the testing of sailors and railway *employés*. **1966** K. Wybar *Ophthalm.* ii. 26 The Ishihara or Stilling Test... The tests are more subtle than the lantern tests and are of value in identifying the anomalous trichromats (of the protanomalous or deuteranomalous types) who are often able to pass the lantern tests successfully.

lanthana (læ·nþănă). *Chem.* [f. Lantha-n(um + -*a*, after *alumina, magnesia, thoria*, etc.] Lanthanum oxide, La_2O_3, a white powder.

1887 *Chem. News* 12 Aug. 62/1 A specimen of white lanthana prepared by myself some years ago..on being strongly calcined, became of a fawn colour. **1917** H. F. V. Little in J. N. Friend *Text-bk. Inorg. Chem.* IV. 406 Lanthanum sesquioxide or lanthana, La_2O_3, is obtained as a white powder by the ignition of the hydroxide, carbonate, nitrate, oxalate, etc. **1961** W. K. Anderson in Spedding & Daane *Rare Earths* xxii. 565 Use of ceria, yttria, and lanthana as diluents for oxide fuel bodies is a promising developmental area.

† **lanthanate** (læ·nþănēⁱt). *Chem. Obs.* [f. Lanthan(um + -ate¹.] = *lanthanide.

1946 *Nature* 27 July 134/2 Actually, however, samarium has an abnormally large atomic volume, a peculiarity which it shares with the other two lanthanates having bivalent properties. **1953** [see *lanthanoid].

lanthanide (læ·nþănəid). *Chem.* [ad. G. *lanthanid* (V. M. Goldschmidt et al. 1925, in *Skrifter Norske Vidensk-Akad.* (*Mat.-nat. Kl.*) v. 6), f. *lanthan* Lanthanum: see *-ide 2.] **1.** Any of the series of elements with an atomic number between 57 (lanthanum) and 71 (lutetium) inclusive, or (following the later definition by Goldschmidt et al., on the suggestion of A. Sommerfeld, in *loc. cit.* vii. 10), between 58 (cerium) and 71; all these elements occupy a single position in group IIIa of the periodic table, are predominantly trivalent electropositive metals with similar chemical properties, and occur together in monazite, gadolinite, and certain other minerals. Cf. **rare earth*.

1926 *Chem. Abstr.* XX. 1969 (*heading*) Synthetic pyromorphites, vanadinites and mimetites in which lead is partially substituted by lanthanides. **1937** *Jrnl. Chem. Soc.* 662 A large rather coherent group is furnished by the rare-earth elements, comprising the lanthanide family (elements of atomic number from 57 to 71) and yttrium. **1946** J. R. Partington *Gen. & Inorg. Chem.* x. 262 The rare-earth elements in this period (sometimes called lanthanides, to distinguish them from the total number of rare-earth elements which includes scandium and yttrium in earlier periods). **1950** N. V. Sidgwick *Chem. Elements* I. 444 Two of the lanthanides, samarium and lutecium, have been found to be radioactive. **1957** *Sci. News* XLV. 95 In the conventional periodic table..the pigeon-hole allotted to the element lanthanum..contains fourteen additional elements, formerly called the 'Rare Earths', but now usually called the 'Lanthanons' or 'Lanthanides'. **1965** [see *F III. 1 j]. **1973** J. J. Lagowski *Mod. Inorg. Chem.* xvi. 616 The trivalent lanthanide cations also exhibit striking colors in their crystalline salts and in aqueous solution. **1973** *Chem. Soc. Rev.* II. 49 The most common practice is to successively add known amounts of the lanthanide shift reagent..to the compound under study ..and record the n.m.r. spectrum after each addition.

2. *Comb.*: **lanthanide contraction** [tr. G. *lanthanidenkontraktion* (V. M. Goldschmidt et al. 1925, in *Skrifter Norske Vidensk.-Akad.* (*Mat.-nat. Kl.*) vii. 13)], the decrease in atomic and ionic radii with increasing atomic number observed in the lanthanide series; **lanthanide series**, the series of elements from lanthanum (or cerium) to lutetium.

1926 *Chem. Abstr.* XX. 131 'Lanthanide contraction' is the term applied to the volume contraction of the atoms in the rare earth series Ce–Cu [*sic*]. This contraction opposes the progressive increase of at. vol. in each vertical column of the periodic table. **1945** A. F. Wells *Structural Inorg. Chem.* iii. 94 As a result of this 'lanthanide contraction', so called because it is observed in the elements following lanthanum, certain pairs of elements in the same Periodic Group have practically identical ionic (and atomic) radii. **1971** *Jrnl. Inorg. & Nucl. Chem.* XXXIII. 385 The lanthanide contraction as reflected in certain properties of the lanthanide compounds is not a smooth function of *Z*. **1945, 1958** Lanthanide series [see *actinide]. **1965** B. G. Wybourne *Spectroscopic Properties Rare Earths* i. 2 As we proceed through the lanthanide series, the nuclear charge, together with the number of 4*f*-electrons, increases by one at each step.

lanthanoid (læ·nþănoid). *Chem.* [f. *Lanthan(ide + -oid.] Any element of the lanthanide series (including lanthanum).

1953 Barnett & Wilson *Inorg. Chem.* xii. 135 In modern nomenclature the term 'lanthanons', with variations such as 'lanthanides', 'lanthanates', 'lanthanoids' and 'lanthans' is replacing 'rare earths'. **1969** H. T. Evans tr. *Hägg's Gen. & Inorg. Chem.* xxviii. 681 The colored lanthanoid ions are used in many ways in the glass industry. **1971** *Nomencl. Inorg. Chem.* (I.U.P.A.C.) (ed. 2) 11 The name lanthanoids for the elements 57–71 (La to Lu inclusive) is recommended.

lanthanon (læ·nþănǫn). *Chem.* [f. *Lanthan(ide + -*on*, prob. to avoid confusion with the systematic use of *-ide* (-ide).] (See quot. 1947.)

1947 J. K. Marsh in *Q. Rev. Chem. Soc.* I. 126 The term 'lanthanon' (Ln) is proposed to denote any element of the group from lanthanum to lutecium inclusive and to replace such objectionable terms as 'lanthanate' or 'lanthanide' which have recently had some currency. **1951** *Nature* 31 Mar. 526/1 An analysis of the lanthanon (rare earth) fraction from davidite..has revealed an unusual variation in the abundance of the lanthanons. **1961** W. K. Anderson in Spedding & Daane *Rare Earths* xxii. 522 Advances in the technology and availability of the lanthanons..have brought about an upsurge of interest among nuclear technologists. **1973** *Jrnl. Chromatogr.* LXXVI. 459 *Sym.*-EDDA was studied..as a complexant for the trivalent lanthanons and yttrium.

Lao (lɑu), *sb.* and *a.* [Native name.] **A.** *sb.* **a.** A branch of the Thai people (see quot. 1949) in South-East Asia; also, a member of this people.

1882 *Encycl. Brit.* XIV. 294/2 The Laos are closely related in physique and speech to the Siamese proper. **1885** T. de Lacouperie in A. R. Colquhoun *Amongst Shans* p. lii, We know more of the original seat of the *Lao* or *Ngai Lao*, than of the others. **1915** W. W. Cochrane *Shans* I. 198 The teaching that God is more powerful than all the hosts of evil spirits..ought to be attractive..particularly to the Laos among whom the demons seem to be uncommonly active. **1949** *Jrnl. Amer. Oriental Soc.* LXIX. 63/1 The use of *Lao* is confusing because, while it is now specifically applied to a specific Tai people—called Laotians by the French—another branch, not especially close to the Laotian, has been called Ngai-lao, or Ai lao; while the name has also been sometimes used as synonymous with *Tai*. **1961** P. Kemp *Alms for Oblivion* ii. 22 The majority of Laos stood by the French, whereas the Annamites detested them. **1969** I. Kemp *Brit. G.I. in Vietnam* iv. 78 The Lao also look down on them [*sc.* montagnards], calling them *kha*, meaning 'slaves'.

b. A group of dialects (see quot. 1954) spoken in Laos and neighbouring areas.

1939 L. H. Gray *Found. Lang.* 390 To the south-eastern division belong Siamese, Lao, Lü, and Khün. **1948** D. Diringer *Alphabet* 414 Lao is nowadays widely spoken in northern Siam. **1954** Pei & Gaynor *Dict. Ling.* 120 *Lao*, a group of vernaculars spoken in Siam and in parts of Burma, classified as Shan dialects. **1961** *Times* 23 Jan. 13/6 One of the Lao-speaking provinces. **1966** *Economist* 6 Aug. 536/1 The Americans are beginning to bring in a flood of school textbooks in Lao.

B. *adj.* Of, concerning, or pertaining to the Lao or their language.

1882 *Encycl. Brit.* XIV. 294/2 The last surviving descendant of the ancient Lao dynasty. **1915** W. W. Cochrane *Shans* I. 155 The Lao alphabet..has 45 consonants. **1970** *Times* 26 Feb. 11/2 Somehow a Laos for the Lao people will have to be conserved.

Laocoön (lēⁱǫ·kǫ,ǫn). Also **Lacoon, Laocoon, Laokoon.** [ad. Gr. Λαοκόων.] The name of a legendary Trojan priest who, with his two sons, was crushed to death by two sea-serpents (Virgil *Aeneid* ii. 40–56, 199–231), used allusively, esp. with reference to statues representing him and his sons in their death-struggle. Freq. *attrib.*

1601 P. Holland tr. *Pliny's Hist. World* xxxvi. v. 569 This may bee seene in the image of Laocoon..a peece of worke to be preferred..before all pictures or cast images. *a* **1666** Evelyn *Diary* an. 1644 (1955) II. 107 Above all that fountaine of the Laocoon..is a most glorious & surprizing object. **1699** M. Lister *Journey to Paris* 143 The Atteliers or Work-houses of Two of the famous Sculptures Tuby; in which was a Lacoon Copied in White Marble. **1811** B. R. Haydon *Jrnl.* 12 Jan. in *Autobiogr.* (1853) I. ix. 150 Went to the Academy in the evening, and saw the Laocoon placed out as it was four years ago. **1843** Dickens *Christmas Carol* v. 153 Scrooge..making a perfect Laocoön of himself with his stockings. **1910** H. G. Wells *Hist. Mr. Polly* i. 11 If Mr. Polly..had been transparent..he might have realized, from the Laocoon struggle he would have glimpsed, that..he was not so much a human being as a civil war. *a* **1930** D. H. Lawrence *Last Poems* (1932) 172 Leave the fearful Laocoön of his fellowman entangled in iron To its fearful fate. *Ibid.* 282 We have become..fatally entangled in the Laocoön coils of our conceit. **1938** L. MacNeice *I crossed Minch* ii. viii. 108 The usual Chirico nightmare of marble Fathers of the City and laocoons in plaster. **1967** I. Marder *Paris Bit* 84 She was draped like a female Laocoön in yards of inky black silk. **1970** *Sunday Times* 18 Jan. 56/4 The old flexibility and flair may become Laocoön to the computers.

Laotian (lɑu·ʃən, la,ōᵘ·ʃən), *a.* and *sb.* Also **Laosian.** [f. *Laos*, name of a country in South-East Asia: see -ian, *Lao.] **A.** *adj.* Of or pertaining to the country of Laos. **B.** *sb.* A native or inhabitant of Laos; also the language of the Laotian people.

a **1861** H. Mouhot *Trav. Indo-China* (1864) II. xviii. 154 The Laotian priests..make a frightful noise, chanting from morning to night. **1890** J. G. Frazer *Golden Bough* I. i. 42 Before beginning to work at the salt-pans in a Laosian village, the workmen offer sacrifice to a local divinity. **1911** *Encycl. Brit.* XVI. 190/2 Laos is inhabited by a mixed population falling into three main groups— the Thais (including the Laotians..); various aboriginal peoples classed as Khas; and the inhabitants of neighbouring countries. **1931** *Times Lit. Suppl.* 12 Nov. 880/4 A Laotian festival. **1949** *Jrnl. Amer. Oriental Soc.* LXIX. 63/2 The Southern group of Tai in Indo-China includes the Siamese and the Laotians. **1954** Pei & Gaynor *Dict. Ling.* 120 Thai Lao or Eastern Laotian..and Tai Yüan or Western Laotian. **1967** D. S. Parlett *Short Dict. Lang.* 122 Siamese and Laotian boast literature from comparatively ancient times. **1968** *Guardian* 1 May 9/1 The American..was having an argument with four Laotian officials... Suddenly the American threw the contents of his glass at one of the Laotians. **1971** *Ibid.* 8 Apr. 2/1 The writer is an American freelance who speaks fluent Laotian. **1972** *Mainichi Daily News* (Japan) 6 Nov. 19/4 (*caption*) Japanese 'peace corps' members..teach Laotian farmers how to operate a tractor.

lap, *sb.*¹ Add: **4.** **c.** A form of loin-cloth worn by Indians in Guyana.

1769 E. Bancroft *Ess. Nat. Hist. Guiana* 273 This is called a lap, and is the ordinary covering of the Negroes also. **1876** C. B. Brown *Canoe & Camp Life Brit. Guiana* 34 There were two Indians,..dressed in nature's garb, barring the 'lap'. **1899** J. Rodway *In Guiana Wilds* 254 A party of Indians in nothing but their laps. **1924** *38th Ann. Rep. U.S. Bureau Amer. Ethnol.* xxi. 439 To this belt or girdle..the apron or lap may be attached. *Ibid.* 443 Among the Wapishana, the length of the bark 'lap' (tururi) was..a guide to the importance of the wearer. **1958** M. Swan *Marches of El Dorado* i. 56 An Indian in these parts would be ashamed to wear the bead apron or the red cotton lap of his parents; he cleans his teeth and brilliantines his hair.

5. **c.** (Later examples.)

1920 'Sapper' *Bull-Dog Drummond* 23 Perhaps a year —perhaps six months... It is in the lap of the gods. **1965** *New Statesman* 30 Apr. 674/3 Almost all power lies in the laps of the different Laender [in Germany]. **1971** *Guardian* 27 Feb. 5/5 Lord Justice Davies said it was in the 'lap of the gods' what would be the effect on the younger children if they were ordered to go to their mother's home.

6. Restrict † to senses in Dict. and add: *to drop, throw*, etc., (something) *in someone's lap*, to shift a burden to (someone). Also (intr.) *to drop into the lap of*.

1962 B. Knox *Little Drops of Blood* ii. 35 'And Sammy Bell's gear?' 'We'll dump that one in the lap of the Scientific boys.' **1964** Mrs. L. B. Johnson *White House Diary* 7 May (1970) 134, I showed Mr. Fosburgh the Winslow Homer painting and I think he was as amazed as I am that it should have so precipitously and happily dropped into our laps. **1970** 'M. Hebden' *Mask of Violence* (1971) xx. 187 I'll throw this into Pinow's lap. It's German and high-level, and I don't want to be mixed up in it. **1972** V. Canning *Rainbird Pattern* ii. 33 Quite simply—and this is for you, Bush, because I'm dropping it in your lap— Trader has got to be scotched. **1973** M. Woodhouse *Blue Bone* ii. 12, I went..to meet some people who had a development problem they wanted to drop in our laps.

7. **lap belt**, a safety belt across the lap; **lap-board** (earlier examples); **lap-iron**, a piece of iron used as a lapstone; **lap-robe**, a rug or cloth to cover the lap of a person seated in a vehicle; **lap strap**, a safety strap across the lap.

1952 *Los Angeles Examiner* 21 Mar., Wider 'lap belts' than those now used. **1959** *Sunday Graphic* 25 Jan. 4/5 The easy-to-fit and unobtrusive 'lap-belts' which give 65 per cent of the protection afforded by the full harness. **1961** *B.S.I. News* Mar. 7/1 Car safety belts,..three types.. lap belt, diagonal strap and full harness. **1962** A. Shepard in *Into Orbit* 114, I took off my lap belt and loosened my helmet. **1973** *Sci. Amer.* Feb. 81/3 In the Utah statistics (from 1969) only 16.5 percent were wearing the seat belts; the estimate at present is that, notwithstanding all the urgings by authorities, only about 25 to 35 percent use the lap belt and only about 5 percent the lap-and-shoulder combination. **1974** *Country Life* 31 Jan. 191/2 The cab.. has a bench seat with diagonal belts for two and a further lap belt for a third occupant. **1840** *Picayune* (New Orleans) 18 Sept. 2/3 Ashamed! why, I feel as flat as my own lapboard. **1867** A. D. Whitney *Summer in L. Goldthwaite's Life* vi. 125 On the lap-board across her knees lies her work. **1962** *Westm. Gaz.* 8 Oct. 6/2 The lapstone and the lap-iron have gone out of existence. **1875** Mrs. Stowe *We & Neighbors* xxxix. 373 He took her to ride in such a stylish carriage, white lynx lap-robe, and all! **1914** G. Atherton *Perch of Devil* i. 121 He smiled..into her.. eyes and tucked the lap-robe about her. **1948** *Chicago Tribune* 15 Jan. 3/2, I loved the sleighrides too—snuggled under great buffalo hide lap robes. **1955** W. Gaddis *Recognitions* iii. iv. 846 Engulfed in the flow of a tartan lap robe..he stared fixedly at an open book. **1974** 'I. Drummond' *Power of Bug* xvi. 220 The thin cotton lap-robe which protected the passenger's legs and feet from the dust. **1960** *Guardian* 22 July 20/2 For rear seat passengers a lap strap is probably sufficient. **1961** *Times* 10 Jan. 6/6 If the ordinary lap strap..is used, an occupant of the car will tend to 'jack knife' forward. **1968** A. Diment *Gt. Spy Race* ix. 165, I did up the lap strap [on a seat in a passenger aircraft] and went straight to sleep.

lap, *sb.*³ Add: **2.** **e.** *Metallurgy.* A kind of defect that results when a projecting part is folded over against the surface of the metal and pressed in (e.g. during rolling or forging), so that a seam is produced on the surface.

1914 W. Rosenhain *Introd. Study Physical Metall.* xiv. 324 'Laps', 'rokes', etc.,..result from the partial welding up of fissures or of portions of metal which have become accidentally overlapped. **1939** E. C. Rollason *Metall. for Engineers* iv. 55 A defect, somewhat similar to a roke, is caused by poor roll design or by rolling at too low a temperature. The metal spreads to an extent greater than the designed pass and forms fins on opposite sides of the bar, which in subsequent passes are lapped over to give the lap illustrated. **1967** E. Bishop tr. *M. van Lancker's Metall. Aluminium Alloys* viii. 238 Working may scratch the metal and result in corrosion damage.., or form laps..tears..and excessive work-hardening.

4. **b.** *Warp Knitting.* A loop of yarn on a needle.

1884 W. T. Rowlett tr. *Willkomm's Technol. Framework Knitting* I. i. 41 Each warp thread is also laid *over* a needle and forms the 'lap' over one. **1884**, etc. [see *knock-off *sb.* and *a.* B. 2 a]. **1926** J. Chamberlain *Hosiery, Yarns & Fabrics* vii. 173 The knock-off stitch is often used to produce pure longitudinal stripes on warp knitted fabrics in which case the pressed lap is always made on the same needle and only the knock-off lap..is traversed to effect a lateral joining. **1952** D. F. Paling *Warp Knitting Technol.* i. 5 Assuming that two fully threaded guide bars are used, then each needle will be provided with two threads across its beard. These laps may be in similar directions or in opposite directions according to the relative directions of the overlaps. **1964** H. Wignall *Knitting* ii. 44 The needles are then raised to move the laps below the beards.

6. (sense 2) *lap-boarded; lap-seam*; (sense 4) *lap-drum, -roller* (example), *-tenter*; (sense 5 b) *lap time*. Also **lap-dissolve** *v.*, = *dissolve *v.* 7 b; **lap-join** *v. trans.*, to join by means of a lap joint; **lap-system** (see quot.); **lap winding** *Electr. Engin.*, a kind of armature winding in which the two ends of each coil are connected to adjacent commutator segments, so that each coil overlaps the next.

1927 *Chambers's Jrnl.* Sept. 597/2 Lap-boarded houses which overhang the sea. **1927** *Observer* 17 Apr. 3 No sooner has it [*sc.* the title] been read than it lap-dissolves into the director's name... It should be lap-dissolved in for a mere flash. [**1934** H. M. Harwood *Old Folks at Home* i. i. 21 Sometimes the next picture's on before the last one's gone... lap... dissolve... isn't it?] **1962** *Sunday Times* 5 Aug. 20/4 *The Stranger*: All right, pardon me for living, it's just you looked so much like this very attractive party I met down here last year. (*Lap dissolve to what may be the following day.*) **1902** T. Thornley *Cotton Combing Machines* 17 The six webs are..drawn by frequent pairs of press rollers to the lap-head, consisting of two pairs of heavily weighted press rollers..and of the lap drums. **1968** J. Arnold *Shell Bk. Country Crafts* vi. 116 The arrangement is to have a 'V' on one side and an inverted 'V' on the other, the apex of which is lap-joined flush with the top rail. **1850** *Rep. Comm. Patents 1850* (U.S.) 160, I also claim the combination of burring apparatus..with the calender and lap rollers. **1905** *Westm. Gaz.* 21 Mar. 5/1 He says the explosion was caused by a crack in the lap-seam [of the boiler]. **1964** H. Hodges *Artifacts* iv. 77 Bronze vessels of (riveted) sheet metal could be made perfectly watertight, even when the edges were joined by a simple lap seam. **1894** *Gloss. Terms Evidence R. Comm. Labour* 51/2 in *Parl. Papers 1893–4* (C. 7063) XXXVIII. 411 *Lap System*, also called 'trip system', is a system (in the carter's industry) of piece-work, *e.g.*, a driver taking loads of coal a given distance for a stated sum, works under the lap system. **1881** *Instructions to Census Clerks* (1885) 68 Lap Tenter. **1921** *Dict. Occup. Terms* (1927) 164/2 Lap tenter (cotton). **1909** *Westm. Gaz.* 7 Dec. 5/1 The net lap times of the Auvergne races of 1905. **1973** *Times* 28 Apr. 7/2 Both he and his team-mate, François Cevert, were later able to equal Regazzoni's lap time. **1892** S. P. Thompson *Dynamo-Electr. Machinery* (ed. 4) xii. 311 When we go on to those cases in which the winding is entirely exterior to the core, as for drum armatures, or to those in which there is no core at all, namely for disk armatures, we find that there are two distinct modes of procedure, which we may respectively denote as lap-winding and wave-winding. **1937** A. S. Langsdorf *Theory Alternating-Current Machinery* v. 295 The end connections of a distributed winding may be arranged in several ways, all electrically identical... The order of grouping and the resultant shape of the coils give rise to the respective designations of spiral, lap, and wave windings. **1966** *McGraw-Hill Encycl. Sci. & Technol.* XIV. 505/2 Lap windings are adapted to high-current machines because they may have more than two parallel paths, whereas the wave windings are adapted to small-capacity machines and high-voltage machines because of the series connection of the coils.

lap, *sb.*⁴ **b.** Delete '*Gun-making*' and add def.: A polishing tool of some relatively soft material (as lead or cast iron) made to a special shape for use in lapping (see Lap *v.*⁴ in Dict. and Suppl.). Add further examples.

1905 W. S. Leonard *Machine-Shop Tools & Methods* (ed. 3) xxxi. 506 The laps described above are of the simplest and cheapest forms, namely, a plain shaft for the internal, and a collar for the external, lap. **1920** Oberg & Jones *Gage Design* vii. 191 Laps for Ring Gages.—Three laps are shown in Fig. 15 for lapping ring gages... They are made of cast iron and are ground to fit the ring gage to be lapped. Grinding the thread on a lap will insure accuracy. **1932** Hardy & Perrin *Princ. Optics* xvi. 338 The exposed surface of the blank is then ground by holding it against another tool, called a lap, which has previously been given the proper radius of curvature. The lap is rotated at a moderate speed on a vertical shaft and is fed with a mixture of coarse emery and water. **1942** A. F. Collins *Greatest Eye in World* ii. 43 A concave iron lap is then placed over the lenses on the head and the spindle is rotated by an electric motor.

lap, *v.*¹ Add: **2.** **b.** *U.S.* Of a bear: to gather and eat fruits or nuts. Hence *lapping-season*.

1868 *Amer. Naturalist* May 122 They climb in order to 'lap', as the hunter says. *Ibid.*, When mast is not plenty, they lap black-gum berries. **1881** *Scribner's Mag.* Oct. 858/2 This is called the lapping season, as he ensconces himself in a tree lap and breaks the limbs to pieces, in gathering nuts and fruits.

c. *to lap up*: (*fig.*) to receive (praise, news, etc.) eagerly.

1890 A. James *Diary* 20 May (1964) 119 Where do you suppose they have discovered Self-Sacrifice now? In the heroic bosom of Stanley! who on his own showing laps up the *agréments* of African travel as I do my afternoon tea. **1922** S. Lewis *Babbitt* xxx. 359, I was simply astonished, the way those women lapped it up! **1930** D. H. Lawrence *Phoenix II* (1968) 493 People wallow in emotion: counterfeit emotion. They lap it up: they live in it and on it. **1931** G. Atherton *Sophisticates* ii. xix. 210 'Polly, of all women, to start such a thing!' muttered Emerey. 'Or Toddles, for that matter. I've found out it was she who fed Polly with the idea of doing something new and strange. Of course she lapped it up.' **1958** *Listener* 20 Nov. 815/1 The Indian Embassy in Bonn will lap up information about Eastern Germany. **1972** *Times* 20 Apr. 25/1 Americans have lapped the book up, already getting through Dell's first order of 100,000.

lap, *v.*² Add: **9.** **b.** (Earlier and later examples.) Also in *Motor Racing*. Also *fig.*

1847 W. T. Porter *Quarter Race Kentucky* 50, I told you the brown horse was a mighty fast one... But soon I lapped him. **1857** *Lawrence* (Kansas) *Republ.* 11 June 3 This..was a killing pace, but Mahen lapped him inside the first quarter. **1961** J. S. Salak *Dict. Amer. Sports* 259 *Lap*, pass another car for the second or third time. **1966** *Publ. Amer. Dial. Soc.* 1964 xlii. 6 'To be lapped', to be passed by a car the race distance for which already exceeds the car being overtaken by the length of a complete lap. **1969** 'D. Rutherford' *Gilt-Edged Cockpit* i. 18 The leading Ferrari..was in fourth place and about to be lapped by the Mascot. **1973** *Times* 9 Feb. 15/5 We are constantly being lapped in the wages race.

c. *trans.* and *intr.* Of persons engaged in a race, or their vehicles: to travel over (a distance) as a lap; also simply, to traverse.

1923 *Daily Mail* 24 May 10 The course, 37¾ miles in length, has to be lapped six times. *Ibid.* 4 June 13 The Leyland expert put up the highest speed of the day when he lapped the 2¾ miles at an average of 117 miles an hour. **1927** *Daily Express* 2 June 12/4 Major Segrave hopes..to lap the course at a fair speed. **1928** *Ibid.* 26 May 9/2 There are many machines entered which could lap all day at sixty-five miles an hour. **1973** P. Evans *Bodyguard Man* xiii. 93 Just lapping the track gently. Nothing too strenuous.

lap, *v.*⁴ Add to def.: To rub or abrade so as to make a surface smooth (and often correctly shaped) to a high degree of precision, usually by the use of a rotating lap of suitable shape coated or impregnated with an abrasive dust, paste, or liquid. Add further examples.

1905 W. S. Leonard *Machine-Shop Tools & Methods* (ed. 3) xxxi. 506 We sometimes lap a machine-shaft which is required to run at an extremely high speed... Other machine details may be lapped when an exceptionally high degree of refinement is required, but the process is more commonly applied to measuring-tools, such as the collar- and plug-gages, etc. **1928** E. Buckingham *Spur Gears* xii. 444 Hardened gears are sometimes run together under load with some form of abrasive introduced with the lubricant..to smooth the surfaces and correct some of the errors. This process, however, does more grinding or crushing of the abrasive than it does to polish or lap the gear-tooth profiles. **1958** *Proc. IRE* XLVI. 1063/1 Wafers, of dimensions 1 × ½ inch, of this material are lapped to a thickness of 10 mils. **1973** *Physics Bull.* July 427/2 The techniques devised for lapping and polishing x ray reflectors have been modified to allow the same basic principles to be employed in lapping and polishing surfaces more complex than the plane, sphere or cylinder.

lapageria (læpădʒiᵊ·riă). [mod.L. (Ruiz & Pavon *Flora Peruviana* (1802) III. 64), f. the name of Joséphine Tascher de *la Pagerie* (1763–1814), Empress of France + -ia¹.] A climbing shrub of the monotypic genus so named, belonging to the family Liliaceæ, native to Chile, and bearing large, bell-shaped, pendulous, red or white flowers.

1849 *Curtis's Bot. Mag.* LXXV. 4447 (*heading*) Rose-coloured Lapageria. **1886** G. Nicholson *Dict. Gardening* II. 234/2 Lapagerias rank amongst the most beautiful greenhouse climbing plants in existence. **1929** *Times* 1 Nov. 19/6 A tiny, pillared stone temple of exquisite proportions bowered in clematis, lapageria, and a climbing yellow rose. **1971** *Country Life* 8 Apr. 820/2 This is one of the finest outdoor specimens of lapageria I have seen in the British Isles.

laparoscope (læ·părǫskōᵘp). *Med.* [f. Laparo- + -scope.] Any instrument used in examining the abdomen; now *spec.* one in the form of a tube for insertion into the peritoneal cavity in laparoscopy, having a source of light at the inserted end and an optical system for forming at the other end an image of the illuminated region.

1855 R. G. Mayne *Expos. Lex. Med. Sci.* (1860) 571/2 *Laparoscopium*, name of an instrument for ascertaining the condition of the abdomen under disease; applicable to the stethoscope and the plessimeter: a laparoscope. **1941** Dorland & Miller *Med. Dict.* (ed. 19) 776/2 *Laparoscope*, a special form of trocar bearing a light by means of which the peritoneal cavity, especially the surface of the liver and the peritoneum, can be inspected. **1967** P. C. Steptoe *Laparoscopy in Gynaecol.* ii. 11 A cold-light projector..is used with a fibre glass cable for transmission of light to a quartz rod incorporated in the laparoscope. *Ibid.* iii. 22 The laparoscope is introduced through the vaginal fornix into the pouch of Douglas. **1970** *Sci. Jrnl.* June 57/1 Human ovaries may be inspected by inserting a tube (laparoscope) through to body wall which both illuminates the internal organs and transmits an image back to the observer.

laparoscopy (læpărǫ·skŏpi). *Med.* [f. Laparo- + *-scopy.] Examination of the loins or abdomen; now *spec.* [ad. G. *laparoskopie* (H. C. Jacobaeus 1910, in *Münchener Med. Wochenschr.* 4 Oct. 2091/1)], visual examination of the interior of the peritoneal cavity by means of a laparoscope inserted into it through the abdominal wall or the vagina.

1855 R. G. Mayne *Expos. Lex. Med. Sci.* (1860) 571/2 *Laparoscopia*, a term for the examination of the loins, by means of the stethoscope, plessimeter, etc.: laparoscopy. **1890** Billings *Med. Dict.* II. 33/1 *Laparoscopy*, examination of the abdomen. **1916** *Jrnl. Amer. Med. Assoc.* 23 Sept. 982/2 Laparoscopy and Thoroscopy.—Johnsson has been applying the cystoscope in investigation of the interior of the abdomen and thorax. **1937** *Surg., Gynecol. & Obstetr.* LXIV. Internat. Abstr. Surg. Suppl. 560/1 By means of laparoscopy almost the same observations may be made as when the anterior wall of the abdominal cavity of a cadaver is removed. **1967** P. C. Steptoe *Laparoscopy in Gynaecol.* viii. 42 An ovarian cyst is readily recognised by laparoscopy. **1969** *Nature* 15 Feb. 635/2 Work using laparoscopy has shown that oocytes can be recovered from ovaries by puncturing ripening follicles *in vivo*. **1970** M. R. Cohen *Laparoscopy, Culdoscopy & Gynecogr.* i. 3 By means of laparoscopy the gynecologist can obtain the same view of the pelvic organs as is seen in laparotomy.

Hence **laparosco·pic** *a.*, pertaining to or obtained by laparoscopy; **laparo·scopist**, one who uses the laparoscope.

1967 P. C. Steptoe *Laparoscopy in Gynaecol.* iii. 25 It is..possible to present laparoscopic views direct to a large audience. *Ibid.* vi. 30 The experienced laparoscopist can offer sound advice about the technical errors. **1969** *Proc. R. Soc. Med.* LXII. 440/1 Interpretation of laparoscopic findings depends on..examination of the site, size and development of the ovary, the appearance of the surface and the presence of follicles or corpora lutea. *Ibid.* 441/2 The proper use of laparoscopic techniques is one of the most valuable advances..in gynæcology in the last twenty years. **1970** M. R. Cohen *Laparoscopy, Culdoscopy & Gynecogr.* xvii. 111 It is possible for the gynecologic laparoscopist to view upper abdominal organs as well.

lap-dog. (Later *attrib.* examples.)

1905 *Dial* (Chicago) 16 Feb. 114/2 Lap-dog poets. **1963** *Times Lit. Suppl.* 22 Feb. 132/4 A lap-dog lover.

lapel. (Later *attrib.* examples.)

1895 *Montgomery Ward Catal.* 180 Lapel Buttons, Enameled. **1940** *Chambers's Techn. Dict.* 486/2 *Lapel microphone*, a small microphone, worn on the lapel; suitable for use when the speaker is addressing an audience, or when he cannot remain in a stable position. **1967** *Observer* 26 Mar. 9 The hippies themselves do not need to pin on the lapel buttons they sell. **1969** *New Scientist* 16 Oct. 109/2 The indignity of wearing a lapel badge displaying their name, rank and work-place to the world at large. **1972** *Times* 7 Feb. 1/3 They were issued with lapel badges depicting a black coffin on a white background. **1973** *Country Life* 31 May 56/1 For the sportsman's wife... Lapel brooches in three colours of gold.

lapidarist. Delete † *Obs. rare* and add later examples. For '= Lapidary B 1 b' read '= Lapidary B. 1 a, b; also *fig.*'

1886 *Sci. Amer.* 7 Aug. 84/2 The stone called sapphire by Pliny is now known to lapidarists as lapis lazuli. **1926** C. L. Warr *Principal Caird* iv. 135 He was a slow-working lapidarist, polishing every literary pebble. **1967** *Sat. Rev.* 13 May 31 Limited editions presses are the lapidarists of the publishing world.

lapidicolous (læpidi·kŏləs), *a.* [f. L. *lapid-, lapis* stone + *-col-us* inhabiting + -ous.] Of beetles: living under stones or similar objects. Hence **lapi·dicole** *sb.*, a beetle living in this kind of habitat.

1899 D. Sharp in *Cambr. Nat. Hist.* VI. ii. v. 205 These blind lapidicolous Carabidæ are of extremely minute size, and of most sluggish habits. **1948** J. R. Dibb *Field Bk. Beetles* p. xiii, Habitat-group 4. Under Stones. Logs, timber, sacking, old metal objects and discarded material which has been thrown down in the open. Lapidicoles. **1959** E. F. Linssen *Beetles Brit. Isles* I. 57 Beetles found under stones are Lapidicoles. **1965** B. E. Freeman tr. *Vandel's Biospeleol.* ii. 20 Thousands of..lapidicoles..have been recorded.

lapiés (læ·pyez, læ·piez), *sb. pl. Geomorphol.* Also **lapiaz, lapies, lapiez.** [a. F. dial. *lapiaz, lapiés* pl. (used in the Jura), f. pop.L. **lapida* f. L. *lapis* stone.] **a.** (Const. as *pl.*) = *karren, Karren *sb. pl.*; also (const. as *sing.*), a *karrenfeld*.

1902, etc. [see *karren, Karren]. **1903** *Geogr. Jrnl.* XXI. 328 The surface formation met with most commonly in limestone districts, which is usually known by the German term *Karren*, or the French *Lapiaz*. **1921** *Geogr. Rev.* XI. 594 The identification of 'karren' and 'lapiez' is taken from Eugène Renevier's 'Monographie des Hautes Alpes Vaudoises' (Matériaux pour la carte géol. Suisse), p. 499. *Ibid.* 598 The name adopted by Professor Cvijić is 'lapiez', which is the term used in the French Jura. **1924** J. Cvijić in *Geogr. Rev.* XIV. 27 Lapiés are found at all altitudes from sea level to lofty mountain summits. They were first observed and described in the limestone alps of Switzerland, where in the cantons of German speech they are called *Karren* or *Schratten* and in districts of French speech *lapiéz* or *lapiaz* or *lapiés*. *Ibid.* 40 Lapiés are formed principally by chemical erosion of limestone surfaces by meteoric water. *Ibid.* 44 Typical lapiés occur chiefly on moderately steep slopes. *Ibid.*, Scattered limestone monoliths are rapidly formed out of the lapiés ridges. **1954** W. D. Thornbury *Princ. Geomorphol.* xiii. 319 Where relief is considerable, limestone surfaces are bare of terra rossa and there is exposed an etched, pitted, grooved, fluted and otherwise rugged surface to which the name *lapiés* is most commonly applied. Cvijić (1924) has described..the amazing diversity of surface and form which lapiés exhibits in the Dalmatian karst region. He maintained that lapiés is found chiefly on outcrops of naked rock. **1968** R. W. Fairbridge *Encycl. Geomorphol.* 645/1 Well developed lapiés with pinnacles standing up to 5 meters high are common in emerged limestone reefs, particularly in the South Pacific. **1972** Bauer & Zötl in Herak & Stringfield *Karst* vii. 236 At these altitudes [in Austria] lapies are dominant karst phenomena, covering wide areas. *Ibid.* 237 Where the surface of limestones is exposed by recent soil erosion, irregularly shaped, more or less rounded rills and lapies of different depth are found.

b. Used in the *sing.* form **lapié.** *rare.*

1968 R. W. Fairbridge *Encycl. Geomorphol.* 644/2 (*caption*) Lapiés formed in limestone covered by residual clay. .. When the lapié ridges are destroyed, a dolina will be formed, filled with clay.

Lapith (læ·piþ). *Gr. Mythol.* Pl. **Lapithæ, Lapiths.** [f. L. *Lapithæ*, ad. Gr. Λαπίθαι.] One of the Lapithæ, a people of Thessaly, celebrated for their wars with the Centaurs.

1607 Topsell *Foure-f. Beasts* 504 The fight betwixt the Lapithæ and the Centaurs. **1611** Coryat *Crudities* sig. C8 Amongst the woers of Penelope themselues, amongst the huge bolles of the Lapithæ. *a* **1846** [see *Hellenian *sb.*]. **1874** *Guide Græco-Roman Sculptures Dept. Greek & Roman Antiquities Brit. Mus.* I. 57 One of the Centaurs.. attacked by Lapiths while carrying off Greek women. **1883** A. S. Murray *Hist. Greek Sculpture* II. 55 The Lapiths are youthful, beardless, slim, but firmly knit. **1886** *Guide Exhib. Galleries Brit. Mus.* 77 Sepulchral urn. On the front a Centaur carrying off a female Lapith. **1949** *Oxf. Classical Dict.* 179/2 The fight [*sc.* of the Centaurs] with the Lapiths occurs on the famous François vase and in sculpture on the pediment of the temple of Zeus in Olympia. **1968** *New Larousse Encycl. Mythol.* (new ed.) 169/2 (*caption*) A Centaur struggles with a Lapith at the wedding feast of King Peirithous.

Laplace (lăpla·s). **a.** The name of Pierre Simon, Marquis de *Laplace* (1749–1827), French astronomer and mathematician, used *attrib.* and in the possessive to designate various concepts and mathematical expressions devised by him or arising out of his work, as † **Laplace's coefficient**, a Legendre polynomial; **Laplace's equation**, the equation $\nabla^2 V = 0$, esp. its representation in Cartesian co-ordinates,

$$\frac{\partial^2 V}{\partial x^2} + \frac{\partial^2 V}{\partial y^2} + \frac{\partial^2 V}{\partial z^2} = 0,$$

where V is a function of x, y, and z; **Laplace('s) operator** = *Laplacian *sb.*; **Laplace transform**, a function $f(x)$ related to a given function $g(t)$ by the equation $f(x) = \int_0^\infty \exp(-xt)\, g(t)\,dt$; so **Laplace transformation**, the transformation by which $f(x)$ is obtained from $g(t)$.

1845 F. Lunn in *Encycl. Metrop.* IV. 144 If f be the distance of the differential particle dm from the attracted particle $V = \int dm/f$. We have now to find the quantity V; this we shall do by expanding it into a series, the coefficients of which have peculiar properties, depending upon a partial differential equation to which they are subject... We shall..distinguish them by that [*sc.* name] of their illustrious inventor, calling the differential equation and the coefficients..Laplace's equation, and Laplace's coefficients. **1873** J. C. Maxwell *Treat. Electr. & Magn.* II. i. ix. 162 The theory of spherical harmonics was first given by Laplace in the third book of his *Mécanique Celeste*. The harmonics themselves are therefore often called Laplace's Coefficients. *Ibid.* 164 It is shewn in treatises on Laplace's Coefficients that Q_i is the coefficient of h^i in the expansion of $(1-2\mu h+h^2)^{-1/2}$. **1812** *Phil. Trans. R. Soc.* CII. 31 It is exclusively confined to that class of spheroids which, while they differ from spheres, likewise have their radii expressed by rational and integral functions of a point in the surface of a sphere: in this hypothesis Laplace's equation has been rigorously demonstrated. **1813** *Phil. Mag.* XLI. 9 In Laplace's equation $(d^2V/dx^2)+(d^2V/dy^2)+(d^2V/dz^2)=0$, ..V is a function of x, y, and z. **1962** Corson & Lorrain *Introd. Electromagn. Fields* ii. 36 If we introduce into Eq. 2–23 the electrostatic potential V,.. $\nabla^2 V = -\rho/\epsilon_0$. This is Poisson's equation. In a region of the field where the charge density ρ is zero, $\nabla^2 V = 0$, which is Laplace's equation. *Ibid.* iv. 154 Certain cases of symmetry are best treated in spherical polar coordinates. Laplace's equation then takes the form

$$\nabla^2 V = \frac{1}{r^2}\frac{\partial}{\partial r}\left(r^2\frac{\partial V}{\partial r}\right) + \frac{1}{r^2\sin\theta}\frac{\partial}{\partial\theta}\left(\sin\theta\frac{\partial V}{\partial\theta}\right) + \frac{1}{r^2\sin\theta}\frac{\partial^2 V}{\partial\phi^2} = 0.$$

1873 J. C. Maxwell *Treat. Electr. & Magn.* I. 29 One of the most remarkable properties of the operator ∇ is that when repeated it becomes $\nabla^2 = -(d^2/dx^2 + d^2/dy^2 + d^2/dz^2)$, an operator occurring in all parts of Physics, which we may refer to as Laplace's Operator. **1935, 1936** Laplace operator [see *Laplacian *sb.*]. **1944** T. H. Turney *Heaviside's Operational Calculus made Easy* vii. 84 (*heading*) The Laplace transform method of circuit analysis. **1962** D. R. Cox *Renewal Theory* i. 3 One of the main mathematical tools used in renewal theory is the Laplace transform. **1949** S. Goldman *Transformation Calculus* iii. 57 The Laplace transformation transforms $f(t)$, a function of t, into $F(s)$, a function of some new variable s. **1956** *Nature* 21 Jan. 106/2 The approach is mathematical and based throughout on the use of the Laplace transformation. The book is..intended..as a supplement to introductory text-books on feedback systems and on the functions of a complex variable.

b. *Philos.* The name of Laplace (see a above), used in the possessive to designate an imaginary intelligence described by him which, given the values at any instant of certain physical quantities for all the particles in the universe, could predict in detail the whole of the future from the laws of physics.

1911 J. Ward *Realm of Ends* i. 17 The omniscience of Laplace's imaginary spirit with its completed world-formula. **1947** H. Reichenbach *Elem. Symbolic Logic* viii. 390 Such verification may be technically impossible, although in principle it should be possible to foretell the results of a throw of a die from the initial conditions, given the position of the die, the physiological status of the person considered, and other factors. Let us say that Laplace's superman could do it. **1965** P. Caws *Philos. of Sci.* xxxix. 300 This intelligence has been called 'Laplace's demon', and it has become the patron saint of determinism.

Laplacian, *a.* Add: Also **Laplacean** (*rare* in the physical sciences). (Earlier and later examples: cf. *Laplace.)

1836 *Rep. Brit. Assoc. Adv. Sci. 1835* 27 M. Poisson, indeed, carries much further than Laplace himself the Laplacian views of molecular action. **1852** *Cambr. & Dublin Math. Jrnl.* VII. 127 The class of partial differential equations to which the Laplacian equation belongs. **1908** *Westm. Gaz.* 21 Feb. 2/1 According to the Laplacian hypothesis no 'month' can be shorter than the corresponding day [of the parent planet]. **1911** J. Ward *Realm of Ends* i. 15 Laplace, brushing aside freewill as a palpable illusion, proclaimed the implicit omniscience of the mechanical theory. [*Note*] It was reserved for Clerk Maxwell to point out clearly the inevitable limitation of the Laplacean data. **1920** S. Alexander *Space, Time, & Deity* II. 328 The famous puzzle of the Laplacean calculator is full of confusions but contains a truth. A person who knows the whole state of the universe at any moment can calculate, so it urges, the whole future. **1929** V. Bush *Operational Circuit Analysis* x. 184

$$f(\omega)/\omega = \int_0^\infty e^{-\omega\lambda} A(\lambda)\,d\lambda.$$

..In an expression of this sort $f(\omega)/\omega$ is called the Laplacian transform of A. **1936** P. M. Morse *Vibration & Sound* v. 136 Writing the wave equation as $\nabla^2\eta = (\partial^2\eta/\partial t^2)/c^2$ where the symbol ∇^2 is called the Laplacian operator, or simply the Laplacian. It stands for the operation of finding the bulginess of the surface at some point. **1962** Simpson & Richards *Physical Princ. Junction Transistors* 468 In these [equations] ∇, $\nabla.$, and ∇^2 are symbols for the gradient, divergence, and Laplacian operators. **1971** I. G. Gass et al. *Understanding Earth* iii. 64/2 The planets have only 0·1% of the mass of the solar system, but 98% of..the energy of angular momentum. From the Laplacian hypothesis [of their origin] the Sun would be expected to have much more angular momentum. **1972** J. Rawls *Theory of Justice* § 28. 171 They must follow what some may have called the Laplacean rule for choice under uncertainty. The possibilities are identified in some natural way and each assigned the same likelihood.

B. *sb.* The Laplacian operator, i.e. the differential operator ∇^2 ('del squared') that occurs in Laplace's equation.

1935 Pauling & Wilson *Introd. Quantum Mech.* iv. 85 This [Schrödinger] equation is often written as

$$\frac{-h^2}{8\pi^2}\sum_{i=1}^{N}\frac{1}{m_i}\nabla_i^2\Psi + V\Psi = -\frac{h}{2\pi i}\frac{\partial\Psi}{\partial t},$$

in which ∇_i^2 is the Laplace operator or Laplacian for the *i*th particle. **1936** P. M. Morse *Vibration & Sound* vii. 232 We can write the wave equation as $\nabla^2 p = (\partial^2 p/\partial t^2)/c^2$... The operator ∇^2 is called the Laplace operator, or the Laplacian: it measures the concentration of a quantity (or, rather, the negative of the concentration). The value of $\nabla^2 p$ at a point is proportional to the difference between the average pressure near a point and the pressure right at the point. **1962** Corson & Lorrain *Introd. Electromagn. Fields* i. 17 The product $\nabla.\nabla f$ is commonly abbreviated to $\nabla^2 f$, and the operator ∇^2 is called the Laplacian.

Lapland. Add: **2.** **Lapland bunting**, a northern species of bunting, *Calcarius lapponicus*.

1862 [see Lark *sb.*¹ 2]. **1912** W. E. Clarke *Stud. Bird Migration* II. xxiv. 268 Lapland bunting, *Calcarius lapponicus*.—Since our discovery of this species, Eilean Mor has been visited annually. **1953** D. A. Bannerman *Birds Brit. Isles* I. 313 The Lapland bunting was added to the British list in 1826, when Selby described an example sent from Cambridgeshire to Leadenhall Market. **1971** *Country Life* 9 Sept. 616/2 Rare visitors [in Yorkshire].. such as..Lapland Bunting.

‖ **laplap, lap-lap** (læ·p,læp), *sb.*² [Local word.] In New Guinea, a loin-cloth.

1930 M. Mead *Growing up in New Guinea* xi. 191 A gorgeous new *laplap* proclaims his special state. **1957** O. Ruhen in B. James *Austral. Short Stories* (1963) 197 They were not bush natives, because all three wore cloth laplaps. **1967** 'E. Lindall' *Time too Soon* i. 3 A couple of policemen, militarily smart in their new uniforms... Earlier they had worn dark blue *laplaps*. **1973** *Sunday Times* (Colour Suppl.) 10 June 51/3 When Michael Somare walked through the bar of a Bougainville motel.. he heard himself described as a 'bush kanaka in a lap-lap'.

‖ **lappa** (læ·pă). *W. Afr.* [Hausa.] A woman's shawl or skirt.

1954 E. Warner *Trial by Sasswood* (1955) iv. 68 Gloriously decked out from the waist down in a bright new

LAPPED

golden wrap-around *lappa*. **1957** M. Banton *W. Afr. City* ix. 173 A woman may dance with two or three trilby hats on her head and a man with a woman's *lappa* or shawl, round his shoulders. **1966** C. Achebe *Man of People* ix. 100 She rubbed her eyes with a corner of her lappa and blew her nose into it.

lapped, *ppl. a.* **c.** (Later examples.)
1894 J. E. Davis *Elem. Mod. Dressmaking* iv. 83 Where the back basque of the bodice is box-pleated, full in any way, or has a lapped centre seam, and is not sewn together much below the waist. **1964** *McCall's Sewing* ix. 130/2 *Single lapped seam.* This seam is used for joining seams in interfacings and interlinings because it gives the least possible bulk.

lappet, *sb.* Add: **5*.** *Weaving.* **a.** A figure produced on cloth during lappet-weaving; also, cloth bearing such figures, lappet-cloth.
1863 J. Watson *Theory & Pract. Art of Weaving* vi. 207 The framing of a power-loom for weaving Lappets is nearly the same as the framing of one for plain cloth. *Ibid.* 227 In working lappets with the jacquard machine, the length of the figure will depend upon the number of cards used. **1884** *Encycl. Brit.* XVII. 109/2 For window-curtains, hangings, &c., there are manufactured harness and book muslins, lenos, sprigs, spots, and lappets. **1920** R. Beaumont *Union Textile Fabrication* ix. 304 Combinations of Lappet and Gauze.—Pattern origination in gauze, lappet, and plain or straight weaving, provides for additional changes in the materials of which the yarns are spun. The lappet (dark sections in Fig. 187) being a surface warp yarn is quite a supplementary element. **1957** *Encycl. Brit.* XXIII. 460/2 Crossed Weaving.—This group includes all fabrics, such as gauzes, in which the warp threads intertwist amongst themselves to give intermediate effects between ordinary weaving and lace. Also those, such as Lappets, in which some warp threads are laid transversely..to imitate embroidery.
b. A mechanism for producing the figures in lappet-weaving.
1894 T. W. Fox *Mechanism of Weaving* ix. 250 Elaborate figures are beyond the range of lappets, still there are many small effects that can be economically woven by them. **1924** W. P. Crankshaw *Weaving* xi. 121 Lappets, Swivels, Smallwares and Warp Piles... Lappet and swivel mechanisms are used to produce effects which resemble those obtained by embroidery. **1927** T. Thornley *Cotton Spinning* (ed. 4) ix. 311 (*heading*) The lappets or thread boards and wires. *Ibid.*, During recent years metal thread lappets have become very largely used, being much less likely to warp, become damaged or to lose concentricity with the spindles although the first cost of metal lappets is greater than wood ones.

lappie (læ·pi). *S. Afr.* Also formerly **lapje.** [Afrikaans *lappie* (formerly *lapje*).] A dish-cloth, a small rag.
1892 J. Widdicombe *Fourteen Yrs. in Basutoland* vi. 106, I kept them rolled up in a *lappie* (old piece of rag). **1900** B. M. Hicks *Cape as I found It* x. 179 The dish-cloth is a great institution in the Boer household. A dirty bit of 'lapje' (rag) it is. **1926** E. Lewis *Mantis* iv. xiv. 208 Pouring out a saucerful of water and using his handkerchief for a 'lappie' as he called it, he cleaned the cup. **1939** S. Cloete *Watch for Dawn* 325 It was a beautiful little dress..and Kaspar never gave back the lappie in which it was wrapped. **1970** *Cape Times* 16 Sept. 7/6 There had been 'dramatic evidence about the finding of the *lappie* and the hair stuck in the middle'.

lapping, *vbl. sb.*[3] Add: (Further examples.) Also *attrib.*, as *lapping machine, plate.*
1907 J. V. Woodworth *Grinding & Lapping* ii. 65 (*caption*) Flat cast iron lapping plate. *Ibid.* 68 For lapping small thread gages a lapping machine was constructed. **1935** H. J. Davies *Precision Workshop Methods* xiii. 242 The general effect produced by lapping is to remove the crests of a ground surface down to the bases of the intervening hollows. **1950** C. R. Hine *Machine Tools for Engineers* xii. 241 Rough lapping may remove as much as 0·003 in., and finish-lapping as little as 0·0001 in... Commercial lapping operations can produce parts to limits of 0·000025 in. **1950** W. Cooper in A. W. Judge *Grinding, Lapping & Polishing* II. vi. 201 The Newall 10U Universal Lapping Machine..is a type employed..for lapping anvils and parts of gauges and the locating rollers of their jig-borers and measuring machines. **1971** B. Scharf *Engin. & its Lang.* x. 94 Lapping is regularly used in order to finish gear teeth, plug gauges,..straight surfaces, etc.
b. **lapping in,** the action of grinding in a valve (see *Grind *v.*[1] 5 b).
1921 *Daily Colonist* (Victoria, B.C.) 12 Oct. 6/1 (Advt.), The quick-seating feature of 'Burd's' Piston Rings enables them to be perfectly and quickly fitted to the engine wall. No slow, laborious 'lapping-in' is necessary. **1950** W. Cooper in A. W. Judge *Grinding, Lapping & Polishing* II. vi. 188 It has long been recognized by bearing and lubrication engineers that the 'lapping-in' method of bearing conditioning, by abrasive means, is the only sure and certain way to obtain best bearing performance and life.

Lapponoid (læ·pŏnoid), *a.* [ad. med.L. *Lap(p)ŏn-em* (see Lapp *sb.* and *a.*) + -oid.] Descriptive of racial, particularly cranial, features associated with the early Lapp peoples.
[**1882** Quatrefages & Hamy *Crania Ethnica* i. iv. 142 Ce type *Laponoïde*, si l'on peut s'exprimer ainsi,..se confond, suivant nous, avec celui qu'Eschricht, Masch et Nilsson ont les premiers fait connaître.] **1939** C. S. Coon *Races of Europe* viii. 288 Czekanowski defines his Lapponoid in such a way as to include the Alpine of Ripley, as well as the Lapps proper. **1948** A. L. Kroeber *Anthropol.* (rev. ed.) iv. § 71. 151 *Yellow Race.* Lapponoid: Enters into several European crossed races. **1957** V. G. Childe *Dawn European Civilization* (ed. 6) xi. 209 Most of the skulls from sites in North and Central Russia are described as Lapponoid.

Lapsang Souchong (la·psaŋ sŭʃǫ·ŋ). [Cf. Souchong.] A variety of Souchong China tea with a smoky flavour. Also *ellipt.* **Lapsang.**
Lapsang is a 'market name'. In quot. 1942 the spelling *Lapseng* is erron.
1883 *Junior Army & Navy Stores* 71 China Lapsang Souchong. **1935** M. Morphy *Recipes of all Nations* 726 Among the most popular for exportation are the different grades of Lapsang Souchong. **1938** S. Beckett *Murphy* v. 68 'I hope you like the aroma,' said Miss Carridge. 'Choicest Lapsang Souchong.' **1942** G. Mitchell *Laurels are Poison* xvii. 184 Jonathan..took the lid off the teapot, sniffed, said: 'Lapseng? All right, I'll have some.' **1946** *Aristotelian Soc. Suppl. Vol.* XX. 167 X may *think*, without much confidence, that it tastes to him like Lapsang. **1947** A. Huxley *Let.* 8 Jan. (1969) 562 We have taken to drinking maté, to which a smoky flavour as of Lapsang Soochong has been imparted. **1962** I. Murdoch *Unofficial Rose* vii. 69 He took a sip of the sweet Lapsang Suchong. **1966** *Punch* 28 Sept. 476/1 Both knees free for the Lapsang and seedcake. **1973** G. Butler *Coffin for Pandora* ii. 71, I drank my favourite Lapsang Souchong tea.

lapsarian (læpsēə·riăn), *sb.* and *a.* [f. L. *laps-us* fall + *-arian, or as back-formation from *infralapsarian*, etc.] **A.** *sb.* (See quot. 1928.) **B.** *adj.* Of or pertaining to the fall of man. Also *transf.*
1928 *Funk's Stand. Dict.*, *Lapsarian*, one who believes in the doctrine of the fall of man from innocence. **1954** Dillenberger & Welch *Protestant Christianity* 91 The holders of lapsarian theories..attempted to safeguard the priority of God's activity by ascribing all events and all happenings to him. **1969** A. Richardson *Dict. Chr. Theol.* 189/2 (*heading*) Lapsarian controversy. **1970** K. Millett *Sexual Politics* (1971) 181 The awesome lapsarian moment when the female discovers her inferiority.

lapse, *sb.* Add: **8.** Special Comb. **lapse rate** *Meteorol.*, the rate of fall of temperature with height; also *transf.*
1918 *Meteorol. Gloss.* (Meteorol. Office) 183 *Lapse*,..a word suggested for use instead of gradient..to denote the loss of temperature or pressure of the atmosphere with height. So that *lapse-rate*, or *lapse-ratio*, for temperature will be the fall of temperature per kilometre of height. **1928** D. Brunt *Meteorol.* vi. 46 The average conditions in the troposphere are specified by a lapse-rate of 3°F. per 1,000 feet. **1957** Haltiner & Martin *Dynamical & Physical Meteorol.* xiii. 210 The local increase in lapse rate was due to a combination of low-level warming and high-level cooling by horizontal advection. **1972** *Biol. Abstr.* LIV. 1081/2 The lapse rate of soil temperature indicates a large value in summer and a small one in winter.

lapse, *v.* Add: **1.** Also with *out.*
1920 D. H. Lawrence *Women in Love* xxiii. 351 She possessed him so utterly and intolerably that she herself lapsed out. **1928** —— *Phoenix II* (1968) 525 If I could dance all day as well, I might keep going. It's this leaving off that does me in.—And she lapsed out.

lap-streak. Add: Also **lapstrake.** (Earlier U.S. and later examples.) Hence **la·pstraked** *a.* = Lapstreaked *a.*
1771 *Boston Gaz.* 11 Mar. (Advt.) (Th.), Whale-boats and all sorts of Lapstreak Boats. **1959** *Times Lit. Suppl.* 9 Jan. 22/4 How to fit clinker (or lapstrake) planking on a hull. **1961** F. H. Burgess *Dict. Sailing* 130 *Lap jointed, lap straked.* Describes the system of planking as used in clinker-built boats. **1971** *Islander* (Victoria, B.C.) 10 Oct. 2/1 His 25-foot lapstrake boat was built by Vancouver shipbuilders.

laquearia (lækwiˌēə·riă). *rare*$^{-1}$. [L., pl. of *laqueāre* a panelled ceiling.] A ceiling, roof. Cf. Laquear, Laqueary.
1922 T. S. Eliot *Waste Land* ii. 18 Odours..ascended In fattening the prolonged candle-flames, Fling their smoke into the laquearia.

larboarder, *sb. rare.* [f. Larboard *sb.*] One who is on the larboard side of a boat.
1846 H. Melville *Typee* vi. 44 The poor larboarders shipped their oars, and commenced pulling us ashore.

larch. Add: **3.** **larch blister, canker,** a disease caused by the fungus *Trichoscyphella willkommii*, which causes cankers on the bark of larch trees; **larch needle cast,** a disease caused by the fungus *Meria laricis*, which attacks and kills the foliage of larch trees.
1895 W. R. Fisher *Schlich's Man. Forestry* IV. 402 The larch-blister or canker..is most prevalent in damp places with moist air and in frosty and cloudy localities. [**1891** *Jrnl. R. Agric. Soc.* II. 300 In the course of last summer's visit of inspection..the Consulting Botanist found cases of canker in larch plantations all over England and Wales.] **1895** Larch canker [see *larch blister*, above]. **1919** W. E. Hiley *Fungal Dis. Common Larch* ii. 19 At present larch canker is prevalent only in Europe. **1968** F. G. Browne *Pests & Dis. Forest Plantation Trees* ii. 990 *Trichoscyphella willkommii* (Hartig) Nannf. Fungi, Ascomycotina Helotiales. Synonym: *Dasyscypha willkommii* (Hartig) Rehm. Larch canker. Europe, including Britain and Northern Ireland, and also in the north-eastern United States of America. **1921** *Q. Jrnl. Forestry* XV. 61 The larch needle-cast appears to be very widespread in Britain. **1968** F. G. Browne *Pests & Dis. Forest Plantation Trees* ii. 877 Larch leaf blight, Larch needle cast. Widely distributed in north western Asia and northern Europe..and also recorded in New Zealand and the United States of America.

lard, *sb.* **2.** Add to def.: Also, in mod. use, any edible pig-fat, and (in commercial use) a fatty preparation containing or resembling lard. (Further examples.)
1881 *Analyst* VI. 233 Watered lard being now used extensively, owing to the high price of the pure quality, we are giving our special attention to its manufacture. **1887** *Buck's Hand-bk. Med. Sci.* IV. 380/2 Commercial lard is so universally impure, either being mixed with water or salt, or having a portion of its liquid oil removed, that it is in general unfit for medicinal use. **1906** L. L. Lamborn *Mod. Soaps* iii. 44 Two grades of neutral lard are made—one from the leaf, the other from the back fat of the hog. **1913** Bolton & Revis *Fatty Foods* iv. 100 Lard is often adulterated with a judicious mixture of beef fat and vegetable oils. **1944** H. G. Kirschenbauer *Fats & Oils* vi. 63 After a Congressional investigation the compounded products which up to then had been sold as 'pure lard', 'refined lard', etc., were required to be labelled 'lard compounds'. **1974** *Guardian* 27 Dec. 9/2 Rub 4 oz butter and 2½ oz lard into 10 oz flour sifted with a pinch of salt.
3. *lard-cake, -pail*; **lard-bladder** *colloq.*, a fat person; **lard compound,** a substitute for lard made from lard stearin, oleostearin, or esp. cottonseed oil; **lard oil** (examples); **lard stearin(e,** the solid residue left after the expression of lard oil from lard, used for stiffening soft lard, as an ingredient of some lard substitutes and margarines, and in the manufacture of some soaps.
1891 Kipling *Life's Handicap* 195 Mulcahy confused the causes of things, and when a very muzzy Maverick smote a sergeant on the nose or called his commanding officer a bald-headed old lard-bladder..he fancied that rebellion and not liquor was at the bottom of the outbreak. **1928** W. Gibson *Between Fairs* 19 Ay, but I'd have you know there is offence, when an old lard-bladder of a circus-clown, the likes of you, tries to teach her own business to Nanny Ragtag. **1858** C. M. Yonge *Christmas Mummers* v. 59 Mrs. Harper was..preparing a lard cake for tea. **1861** Geo. Eliot *Silas Marner* x. 160 Some small lard-cakes, flat paste-like articles. **1904** L. L. Lamborn *Cottonseed Products* ix. 172 The ingredients of lard-compound are summer white cottonseed-oil and oleo-stearin. **1913** Bolton & Revis *Fatty Foods* iv. 103 We have found products described as 'lard compounds' in which no lard was present at all. **1946** *Thorpe's Dict. Appl. Chem.* (ed. 4) VII. 189/1 Under pressure of reformative legislation..the term 'refined lard' was replaced by the expressions 'compound lard' or 'lard compound'..; later the term 'lard compound' was still further restricted to products containing more than 50% of genuine lard. **1843** *Rep. Comm. Patents 1842* (U.S.) 82 The article of lard offered for sale in the market for domestic use, and now about to be so much in demand as material for the manufacture of lard oil and candles, is prepared from the adipose matter of the omentum and mesentery of the hog. **1920** Oberg & Jones *Gage Design* vii. 198 When a very slow cutting abrasive is required and the amount to be removed by lapping is small, rouge and lard oil may be used. **1957** *Encycl. Brit.* XIII. 723/2 Lard oil is the limpid, clear, colourless oil expressed by hydraulic pressure from pure lard after it has been 'grained' by storage at a temperature of 45°F. **1891** *Fur, Fin & Feather* Mar. 195 Two empty lard pails with their covers..will complete the culinary outfit. **1968** R. M. Patterson *Finlay's River* 178 Soon the tea-pail—an old lard-pail, smoked and blackened by hundreds of camp fires—was singing, swaying a little over the flames. **1885** W. L. Carpenter *Treat. Manuf. Soap* ii. 26 The so-called 'lard-stearin' left in the presses is frequently used as a substitute for tallow in the soap-pan, when the price of it is suitable. **1906** L. L. Lamborn *Mod. Soaps* iii. 46 Lard-stearin of non-edible quality is a soap-stock for certain grades of soap. **1944** H. G. Kirschenbauer *Fats & Oils* vii. 109 Lard stearine and lard oil for edible purposes are obtained from lard by graining and pressing.

larder[1]. Add: **1. c.** The collection of prey formed by a butcher-bird or shrike.
1919 H. F. Witherby *Pract. Handbk. Brit. Birds* (1920) I. 277 J. H. Gurnsey also records a shrew impaled in a 'larder' and Oldham a young bank-vole. **1964** A. L. Thomson *New Dict. Birds* 733/1 Many species [of shrike] have the habit of impaling their prey on thorns..or of hanging it from the fork of a branch... This provision of a 'larder' is responsible for the English popular name 'butcher bird'.
3. **larder beetle** (examples); **larder bird** = Butcher-bird.
1895 J. H. & A. Comstock *Man. Study Insects* xxi. 539 The Larder Beetle, *Dermestes lardarius*..is the most common of the larger members of this family. **1942** E. O. Essig *College Entomol.* xxxii. 559 Small convex scaly beetles usually feeding on dead or dry animal matter. (Skin or Larder Beetles.) Dermestidæ. **1974** *Times* 16 Apr. 12/7 The larder beetle has been left on the shelf, but a related species..is piling up its numbers. **1948** *Brit. Birds* XLI. 200 Because of the habit of pinning up spare food on thorns the Red-backed Shrike (*Lanius c. collurio*) was far better known in Essex when I lived there, as the Larder Bird.

la·rder, *v. rare.* [f. Larder[1].] *trans.* To store up as in a larder.
1904 Rider Haggard *Gardener's Year* (1905) July 251 The first wasp which came into being must have paralysed caterpillars and lardered them in key-holes. **1948** *Brit. Birds* XLI. 200 The male bird..is much more given to lardering than the hen.

larding, *vbl. sb.* **c. larding-needle** (earlier and later examples).
1675 S. Fell *Let.* 4 Mar. in *Househ. Acct. Bk. 1673–78* (1920) p. xvii, Two larding needles. **1855** E. Acton *Mod. Cookery* (rev. ed.) ix. 181 Secure one end of the bacon in a slight larding-needle. **1958** *House & Garden* Feb. 85/1 A larding needle... With this,..you can thread strips of bacon fat through the breast of a chicken. **1970** Simon & Howe *Dict. Gastron.* 239/1 *Larding needle*, a long steel needle with a large eye into which narrow strips of pork fat or larding bacon are threaded.

lardy, *a.* *lardy cake* (earlier and later examples).
1879 C. M. Yonge *Magnum Bonum* I. xiv. 261 Hot tea and 'lardy cake' tendered for his refreshment. **1933** W. de la Mare *Lord Fish* 64 She had brought Griselda not only a pitcher of new milk..but some lardy-cakes and a jar of honey. **1970** Simon & Howe *Dict. Gastron.* 239/2 *Lardy cake*, country-style bread-dough cakes which appear in several English counties: Sussex, Wiltshire, Oxfordshire and Cambridgeshire.

lare, var. *Lair *sb.*[4]

lares: see Lar.

larf. ¶. Jocular spelling of Laugh *sb.* and *v.*, esp. representing Cockney speech.
1847 *Punch* XII. 2/1 She is so innocent..a half-larfin, and a half-poutin. **1851** Mrs. Stowe *Uncle Tom's Cabin* (1852) I. iv. 45 'And what did mother say?' said George. 'Say?—why, she kinder larfed in her eyes—dem great handsome eyes o' hern...' **1894** Kipling *Day's Work* (1898) 62 The folks..larfed—why, they all but lay down themselves with larfin'. **1901** M. Franklin *My Brilliant Career* (1966) xxxi. 196 A sorrowful lookin' delicate creetur', that couldn't larf to save her life. **1965** L. Deighton *Horse under Water* iii. 19 'I'll larf, sir, that's what I will do; larf.' The Chief gave no sign of laughing either now or at any future time: I thought for a moment that LARF was some strange nautical verb. **1968** —— (*title*) Only when I larf. **1971** *Guardian* 8 Apr. 10/2 Give us a larf, pass the time.

‖ **larga** (lā·ɹgă). [Sp.] In bull-fighting, a pass using the cape (see quots.).
1932 E. Hemingway *Death in Afternoon* xv. 170 Quites were made..by the use of largas. In these the cape was fully extended and one end offered to the bull who was drawn away following the extended cape and then turned on himself to fix him in place by a movement made by the matador who would swing the cape over his shoulder and walk away. **1957** A. MacNab *Bulls of Iberia* v. 52 It is now the job of the peones to 'run' it, *correr al toro*. That means, to wave capes at it, get it to charge the capes, and give it long-distance passes. These are called *largas*, the big cape being held by one tip and sent flying out at full length. **1967** McCormick & Mascareñas *Compl. Aficionado* ii. 61 He works the toro with *largas*. *Ibid.*, In the larga, the cape is trailed in the sand with one hand as the torero runs (bregar) the toro.

Largactil (lā·ɹgæktil). *Pharm.* Also **largactil.** A proprietary name of chlorpromazine hydrochloride, $C_{17}H_{19}ClN_2S{\cdot}HCl$, the form in which chlorpromazine is usually administered.
1953 *Trade Marks Jrnl.* 20 May 430/1 Largactil... May & Baker Limited, Dagenham, Essex; manufacturing chemists. **1965** J. Pollitt *Depression & its Treatment* iv. 56 In uncomplicated cases, the intramuscular injection of chlorpromazine (Largactil) 100 mg. is helpful. **1966** *New Statesman* 4 Feb. 168/1 Others might use a quarter grain of Largactil to take the edge off life in London or the Midlands. **1970** G. Greer *Female Eunuch* 90 If all else fails largactil..and other forms of 'therapy' will buttress the claim of society.

‖ **largamente** (lāːɹgăme·nte), *adv. Mus.* [It.] (See quots.)
1876 Stainer & Barrett *Dict. Mus. Terms* 252/1 *Largamente* (*It.*), slowly, widely, freely, fully. **1880** Grove *Dict. Mus.* II. 92/1 (s.v. *largo*), The term *Largamente* has recently come into use to denote breadth of style without change of *tempo*. **1958** *Times* 27 Nov. 6/6 He did cause just one raised eyebrow with the very much slower tempo he adopted for the *largamente* second subject tune in the finale.

‖ **largando** (lāɹgæ·ndo), *adv. Mus.* [It.] = *Allargando.
1893 J. S. Shedlock tr. *Riemann's Dict. Mus.* 429/1 *Largando* (*slargando, allargando*) Ital., 'broadening'; as a rule it is united with *crescendo*. **1972** *Harper's Dict. Mus.* 178/1 *Largando*..Italian, another spelling of *allargando*.

large, *a.*, *adv.*, and *sb.* Add: **A.** *adj.* **8. b.** Further examples of use, esp. in names of plants and animals.
The compar. *larger* and superl. *largest* are also used in specific names, as *larger cabinet beetle, larger red-crested woodpecker, largest red oak*.
Large Black (pig), a pig belonging to the variety so called, developed late in the 19th century and formerly called the Devonshire Black; **Large White** (pig), a heavy bacon pig of the variety so called, first introduced in Yorkshire about 1850 and formerly called the Yorkshire pig.
1787 W. Sargent in *Mem. Amer. Acad. Arts & Sci.* (1793) II. 159 Large Laurel. **1810** F. A. Michaux *Hist. Arbres Forestiers de l'Amérique Septentrionale* I. 39 American large aspen..nom donné par moi. **1813** H. Muhlenberg *Catal. Plant.* 92 Large aspen (*Populus trepida* or *grandidentata*). **1832** D. J. Browne *Sylva Amer.* 255 As it surpasses the aspen in height, we have given it the name of Large Aspen. **1832** J. Rennie *Conspectus Butterflies & Moths Brit.* 259/1 Large Blue..Large Copper. **1832** T. Brown *Bk. Butterflies, Sphinxes & Moths* I. 18 (*heading*) The large white cabbage butterfly. **1837** *Southern Lit. Messenger* III. 660 There are for sale hats, boots and shoes, India rubber articles,..large bread,..everything on earth. **1845** C. M. Kirkland *Western Clearings* 154 'You'd ought to begin with large-hand, Joshuay,' said Master Horner to this youth [instructing him in penmanship]. **1857** H. T. Stainton *Man. Brit. Butterflies & Moths* I. 11 The Swallow-tail and the Large Copper are only to be obtained in the fens of Cambridgeshire and Huntingdonshire. **1859** Bagehot *Coll. Works* (1965) II. 190 A large-hand copy of life. **1867** *Jrnl. R. Agric. Soc.* III. 633 Some of the classes, the large white breed and those not qualified for the specified classes were only scantily filled. **1876** H. E. Scudder *Dwellers in Five-Sisters Court* i. 7 There was a large-bread bakery at Skōlas. **1896** W. J. Malden *Pig Keeping for Profit* i. 12 The Large White started with a strong frame. **1906** J. Long *Bk. Pig* (ed. 2) xii. 156 The Large Blacks are regarded as being of gentle disposition... The chief counties in which the Large Black pig is bred are Cornwall and Devon. **1947** J. Stevenson-Hamilton *Wild Life S. Afr.* xxv. 205 The large grey mongoose (*Herpestes caffer*).—Colour, grizzled grey, limbs darker. A black brush at the end of the tail. Length over all, about 45 inches. This animal is spread throughout the Ethiopian region. **1961** J. Fitzhugh *Pig Breeding* xiv. 160 The National Pig Breeders' Association caters for the Large Blacks, Large Whites, [etc.]. **1966** E. Palmer *Plains of Camdeboo* x. 181 The large grey mongoose is nocturnal and rarely seen. **1970** *Times* 19 Aug. 9/7 There are five or six species (e.g. the Large Blue or Glanville Fritillary) which ought to be protected. **1971** *Farmers Weekly* (Extra) 19 Mar. 37/2 The Farnsworths fatten all progeny from a herd of 60 Welsh sows, put to the Large White boar. **1973** T. G. Howarth *South's Brit. Butterflies* 51 The tiny parasitic wasp, *Apanteles glomeratus* Linnaeus..normally acts as a control to the common Large White caterpillar.
i. *law of large numbers* [tr. F. *loi des grands nombres* (S. D. Poisson 1835, in *Compt. Rend.* I. 478)]: a statistical law which states that if a series of independent trials or observations is made, in each of which there is the same probability of a particular outcome, then as the number of trials is made larger the chance that the observed proportion of such outcomes differs from the probability by less than any given number, however small, approaches a certainty (or, in stronger terms, the observed proportion approaches the probability).
[**1921** J. M. Keynes *Treat. Probability* xxviii. 336 The 'Law of Great Numbers' is not at all a good name for the principle which underlies Statistical Induction. The 'Stability of Statistical Frequencies' would be a much better name for it. The former suggests, as perhaps Poisson intended..., what is certainly false, that every class of event shows statistical regularity of occurrence if only one takes a sufficient number of instances of it. It..encourages the method..by which it is thought legitimate to take an observed degree of frequency or association, which is shown in a fairly numerous set of statistics, and to assume..that, because the statistics are numerous, the observed degree of frequency is therefore stable.] **1937** J. V. Uspensky *Introd. Math. Probability* x. 182 A far reaching generalization of Bernoulli's theorem, known under the name of the 'law of large numbers'. **1949** W. Kneale *Probability & Induction* iii. 139 Many people who have heard of it under the name of the law of large numbers..suppose it to be a mysterious law of nature which guarantees that in a sufficiently large number of trials a probability will be 'realized as a frequency'. *Ibid.* 141 As an illustration of the importance of the law of large numbers in practical affairs it will be sufficient to mention the business of insurance... The greater the number of persons insuring with the company, the greater the probability that the company's finances will remain sound. **1960** S. Goldberg *Probability* iv. 227 The law of large numbers can be used to supply a theoretical counterpart to our intuitive feeling that if an event A occurs f times in n identical trials and if n is large, then f/n, the proportion of times A occurs, should be near the probability $P(A)$ of the event A.
j. **larger-than-life** *attrib. phr.* Cf. Life *sb.* 7 a.
1950 *New Yorker* 23 Dec. 42 Inviting Mr. Churchill..as the living, larger-than-life embodiment of the British people's opposition to appeasement. **1967** *Sunday Times* 23 Apr. 49 The larger-than-life political figures thunder their dogmas through the act. **1972** D. Francis *Smokescreen* ii. 27, I had very little in common with the sort of larger-than-life action man I played in film after film. **1972** *Jazz & Blues* Oct. 22/1 For many years Mezzrow was an almost larger than life personality.
15. a. *large-berried, -billed, -featured, -flowered, -framed* (earlier example), *-fruited, -leaved, -mouthed, -scaled* adjs.; *large-angle, -aperture, -denomination, -signal, -size* adjs. **c.** **large calorie** = *Calorie a; **large-lung** *a. Path.* = *large-lunged* adj.; **large-mouth** (bass) (earlier and later examples); also **large-mouthed bass;** **large-scale** *a.*, drawn to a large scale, on a large scale, extensive, widespread, relating to large numbers; so **large-scale integration** *Electronics*, the development or use of integrated circuits that each contain a large number of components.
1956 *Nature* 3 Mar. 413/1 Large-angle scatters of cosmic-ray particles. **1966** D. G. Brandon *Mod. Techniques Metallogr.* iii. 138 Few electrons are backscattered out of the target, and those which do escape do so principally by large-angle Rutherford collisions. **1935** *Discovery* Jan. 25/1 The picture was taken on sensitised paper, probably with a small short-focus camera having a large-aperture lens. **1966** D. G. Brandon *Mod. Techniques Metallogr.* i. 10 Using a large-aperture reflecting surface to give good resolution and a reflecting plate to project the reflected image into the microscope column. **1785** G. Washington *Diary* 2 Mar. (1925) II. 346 Planted..all the large berried thorns. **1835** J. J. Audubon *Ornith. Biogr.* III. 599 The birds observed were Large-billed Puffins. **1908** E. J. Banfield *Confessions of Beachcomber* i. iii. 123 Many of the birds are..named in accordance with their notes... 'Piln-piln' the large-billed shore plover. **1954** Fisher & Lockley *Sea-Birds* 294 *Phaëtusa simplex*, large-billed tern. **1927** Haldane & Huxley *Animal Biol.* iii. 88 The kilocalorie of 1,000 calories is the unit of energy which is most useful in human physiology. It is sometimes called the 'Large calorie'. **1973** P. Evans *Bodyguard Man* ii. 19 A wallet thick with large-denomination banknotes. **1974** J. Cleary *Peter's Pence* viii. 237 The large-denomination notes would be distributed by those banks. **1847** Thoreau *Let.* 29 Dec. in *Corr.* (1958) 200 He is large featured. **1963** J. Fountain in B. James *Austral. Short Stories* 269 His face, large-featured, serious and brown. **1813** H. Muhlenberg *Catal. Plant.* 53 Large-flowered Custard Apple. **1846** D. J. Browne *Trees Amer.* 2 The Large-flowered Magnolia is most remarkable. **1952** A. G. L. Hellyer *Sanders' Encycl. Gardening* (ed. 22) 142 There are many large-flowered hybrid strains [of cyclamen] in cultivation. **1971** J. Raven *Botanist's Garden* iv. 84 Our native large-flowered Geraniums..afford excellent illustrations..of the second type of plant distribution. **1869** *Rep. Comm. Agric. 1868* (U.S. Dept. Agric.) 438 Large-framed, wide and straight-backed, and deep-bodied, short-horn cows. **1813** H. Muhlenberg *Catal. Plant.* 48 Large fruited Hawthorn. *Ibid.* 88 Large-fruited Shellbark hickory. **1952** A. G. L. Hellyer *Sanders' Encycl. Gardening* (ed. 22) 198 Large-fruited varieties [of strawberry] now in cultivation are all hybrids. **1785** H. Marshall *Arbustrum Americanum* 93 Large-leaved Virginian Mulberry Tree. **1832** D. J. Browne *Sylva Amer.* 212 We have given it the specific name of Large-Leaved Umbrella Tree. **1957** M. Hadfield *Brit. Trees* 399 Large-leaved Lime-tree... This species varies a good deal, and has for long been extensively planted as an ornamental tree. **1974** *Country Life* 28 Nov. 1639/3 The large-leaved rhododendrons..will grow only on acid soils. **1961** R. D. Baker *Essent. Path.* xv. 372 Hypertrophic, or 'large lung' emphysema, is seen at autopsy as voluminous lungs which do not collapse when the pleural cavities are opened. **1884** G. B. Goode *Fisheries U.S.: Nat. Hist. Aquatic Animals* 401 The Large-mouth is known in the Great Lake region..as the 'Oswego Bass'. **1973** *Sat. Rev. World* (U.S.) 4 Dec. 47/3 Fresh-water fishermen..can try for..large-mouth bass. **1878** C. Hallock *Sportsman's Gazetteer* 679 Large mouthed bass. **1883** 'Mark Twain' *Life on Mississippi* 264 Every detail of the pilot-house was familiar to me, with one exception—a large-mouthed tube under the breast-board. **1883** *Century Mag.* July 376/2 There are but two well-defined species, the large-mouthed bass and the small-mouthed bass. **1919** E. Pound *Quia Pauper Amavi* 39 Oh august Pierides! Now for a large-mouthed product. **1956** *Nature* 3 Mar. 413/2 The fish is related to the freshwater large-mouthed bass of the eastern United States. **1887** Large scale [see Scale *sb.*[3] 11 a]. **1897** [in Dict., sense 15 a]. **1907** *Daily Chron.* 9 Dec. 3/3 Schumann is a minor poet among musicians. We remember his lesser things..and remain cold to his large-scale pieces. **1920** T. P. Nunn *Education* ix. 114 This large-scale experiment. **1934** *Discovery* Oct. 303/2 We do not all realise that the first large-scale (6-inch) survey of these islands was made in Ireland. *a* **1942** B. Malinowski *Sci. Theory of Culture* (1944) vii. 72 Every army must get along on its stomach and..also most large-scale organizations. **1952** V. A. Demant *Relig. & the Decline of Capitalism* i. 21 A period which preceded the appearance of large-scale manufacturing industry. **1957** L. F. R. Williams *State of Israel* 33 The Security Council's action brought large-scale fighting in Palestine to an end. **1966** *AFIPS Conference Proceedings* XXIX. (1966 Fall Jt. Computer Conf.) 65/1 We are now entering another phase of the expansion of materials technology, in which complete equipment components will be processed on slices of semiconductor... This phase has already been given several names, some of which are 'large-scale integration' (LSI), 'computer on a slice', and 'array technology'. The term 'large-scale integration' is close to being the most descriptive, although at times the syntax is awkward. A somewhat more precise term is 'large-scale integrated electronics'. We will use LSI to abbreviate both 'large-scale integrated electronics' and 'large-scale integration'. **1967** *Proc. IEEE* LV. 1988/2 Large scale integration (LSI) presents an opportunity to exploit many of the concepts of design automation. **1968** *Times* 24 Oct. 7/7 The permissible number of large-scale accidents in nuclear reactors should be about one every 100 million years. **1970** *Sci. Amer.* Feb. 22/1 The technology that produces such high-density electronic circuits is called large-scale integration, or LSI. Although the term has no precise definition, it is usually reserved for integrated circuits that comprise 100 or more 'gates', or individual circuit functions, laid down with a density of 50,000 to 100,000 components per square inch. **1973** A. Behrend *Samarai Affair* iii. 30 The large-scale model which occupied the centre of the big oval table..was made of painted wood and represented the approaches to the Port of Liverpool. **1869** Large-scaled [see *brown-banded snake*]. **1936** J. T. Jenkins *Fishes Brit. Isles* (ed. 2) 157 The Hake is a large-scaled member of the cod family. **1955** Coblenz & Owens *Transistors* xi. 146 No truly large-signal theory for transistors exists today that can be applied directly by the design engineer. **1962** Simpson & Richards *Physical Princ. Junction Transistors* vii. 139 Other applications in which transistors are used to a considerable extent include the following: 1. Large-signal steady-state amplification. [Etc.] **1904** *Westm. Gaz.* 30 Mar. 2/1 A reason for utilising the gas-engine as a large-size power unit for central engine work. **1960** E. Delavenay *Introd. Machine Transl.* 93 A large-size dictionary.

B. *adv.* **3.** Delete † *Obs.* and add later U.S. examples.
1834 S. Smith *Sel. Lett. J. Downing* 149 Other folks may talk larger and bluster more. **1872** in A. W. Tourgée *Fool's Errand* (1880) ii. v. 411 He had just talked large about the Ku-Klux.
C. *sb.* **5. f.** *gentleman-at-large*: see Gentleman 2 c.
k. *verdict at large*: see Verdict *sb.* 1 c.
8. b. *in the large*: also, in general, as a whole.
1943 *Sun* (Baltimore) 24 Aug. 2/6 In the large, there is something else to be said for this recent destruction of more than one hundred of the enemy's fighter planes. **1961** A. J. Deutsch in 'E. Crispin' *Best SF Four* 75 The missing persons did not return. In the large, they were no longer missed. **1968** *Times* 15 Oct. 16/7 Much of the information needed to produce a uniformly precise map therefore will be missing. However, it is only the picture in the large that will suffer.

larghetto (lāɹge·to). *Mus.* [It., dim. of Largo.] A term indicating that a passage is to be played slowly; also, a movement or passage played in this way.
1724 *Short Explication Foreign Words in Musick Bks.* 40 *Largetto*, or *Larghetto* denotes a Movement a little quicker than *Largo*. **1801** Busby *Dict. Mus.*, *Larghetto*... A word specifying a time not quite so slow as that denoted by *Largo*, of which word it is the diminutive. **1877** G. B. Shaw *How to become Mus. Critic.* (1960) 28 The overture was taken too rapidly at the *larghetto*. **1958** *Listener* 4 Dec. 964/3 The beautiful *larghetto* from the E minor *sinfonia*. **1959** *Times* 2 Feb. 12/4 In Haydn's F major sonata..there was finesse in his phrasing of its central larghetto. **1970** *Oxf. Compan. Mus.* (ed. 10) 566/1 *Larghetto*,..slow and dignified.

lark, *sb.*[1] Add: **3.** *lark-note, -pie, pudding; lark-charmed, -crested, -high* adjs.; **lark bunting,** the prairie bobolink, *Calamospiza melanocorys*, a bird found on the plains of central North America.
1869 *Amer. Naturalist* III. 296 That pretty and musical bird of the high plains, the Lark Bunting (*Calamospiza bicolor*), also occurred [along the Upper Missouri River]. **1963** R. D. Symons *Many Trails* iii. 30 Small black ones [*sc.* birds] with white wing patches, which the children at once called white wings, not knowing that they were lark buntings. **1879** G. M. Hopkins *Poems* (1918) 41 Cuckoo-echoing, bell-swarmèd, lark-charmèd. **1848** E. S. Dixon *Ornamental & Domestic Poultry* 319 Lark-crested Fowls are of various colours; pure snow-white, brown with yellow hackles, and black. **1909** *Westm. Gaz.* 29 Dec. 8/3 Sometimes he wings straight up, lark-high, into the blue. **1946** Dylan Thomas *Deaths & Entrances* 22 A stone lies lost and locked in the lark-high hill. **1866** R. Leighton in *Westm. Gaz.* (1909) 6 Mar. 6/3 Deep in my soul the throbbing lark-notes lie. **1906** *Westm. Gaz.* 14 Apr. 6/2 Yet hear the lark-note piercing the grey morn. **1723** J. Nott *Cook's & Confectioner's Dict.* sig. R6 (*heading*) To make a Lark Pye. **1861** Mrs. Beeton *Bk. Househ. Managem.* 479 (*heading*) Lark Pie (an Entree). **1910** W. de la Mare *Three Mulla-Mulgars* xi. 166 What's lark-pie to a hungry sailor? **1963** T. Fitzgibbon *Game Cooking* 96 Lark pie. **1863** G. Meredith *Let.* 1 Feb. (1970) I. 189 A new Receipt:—I try it at Orridge's tonight. '*Lark Pood'n*'. **1877** E. S. Dallas *Kettner's Bk. of Table* 272 *Lark Pudding or Pie.*—For the perfection of a lark pudding, go to the Cheshire Cheese, in Fleet Street. **1934** J. J. Williams *Seasonal Cook. Bk.* 263 Lark pudding... Grease a pudding basin... Clean and bone the larks.

lark, *sb.*[2] Add: **2.** An affair, line of business, etc. *colloq.*
1934 P. Allingham *Cheapjack* xiii. 167 There are many Jews among the grafters, but they usually stick to the chocolate 'lark'—or auction. **1936** [see *con c]. **1961** *New Statesman* 22 Sept. 376/3 Exhibitionists they may be but they mean business. This wet sitting for hours on end is not my lark. **1964** J. Porter *Dover One* i. 11 There's an outbreak of fowl pest..or something and, naturally, that's far more up his street than one of these vanishing-lady larks. **1967** G. F. Fiennes *I tried to run a Railway* iii. 28 Jeremy came in one day while this lark was going on. *Ibid.* vii. 86, I am up to my ears in this bloody diesel lark.

lark, *v.*[2] Etym. note, last line, read: 'which is found a few years earlier (1809)'.

larked (lāɹkt), *a. poet. nonce-wd.* [f. Lark *sb.*[1] + -ed[2].] With larks overhead, noisy with the song of larks.
1952 Dylan Thomas *Coll. Poems* 173. I hear the bouncing hills Grow larked.

larkiness (lā·ɹkinĕs). [f. Larky *a.* + -ness.] The quality of being larky; sportiveness.
1896 *Columbus* (Ohio) *Dispatch* 22 Aug., In reality he [*sc.* a choirboy] is the incarnation of all that is mischievous; and..if he sings at a cathedral or important church, the more 'larkiness' is found in his composition. **1905** Chesterton *Heretics* 90 It is hard to see at first sight why so human a thing as leisure and larkiness should always have a religious origin. **1924** R. Hichens *After Verdict* ii. xx. 303 The ball-boys stood ready, looking alert and full of suppressed larkiness. **1928** *Observer* 26 Feb. 15/3 Miss Helen Gilliland..has great quality..., but she needs the supreme gift for this work, of larkiness. **1973** *Times Lit. Suppl.* 30 Nov. 1466/4 Wastage or mere larkiness is fatal both to its intention and execution.

larkish, *a.* (Earlier and later examples.)
1823 *Spirit of Public Jrnls. M.DCCC.XXIII* (1825) I. 75 She went to see the lamplighter's burying, and the folks were all very merry 'and quite *larkish*, in a manner'. **1926** F. M. Ford *Man could stand Up* i. ii. 32 The larkish freak of a school-girl.

larkspur. Add: **b.** The blue colour characteristic of the larkspur.
1927 *Sunday Express* 27 Feb., Newest Season's colours including..Grey, Cocoa, Larkspur, Fawn. **1927** *Daily Express* 12 Mar. 3/5 Larkspur, a pastel blue slightly inclining to the mauve.

larky, *a.* Add: (Earlier and later examples.)
1841 *Punch* 25 Dec. 278/2 The old girl has her two nieces home for the holidays—devilish handsome, larky girls. **1909** [see *buck *v.*[2] 2 a]. **1911** E. M. Clowes *On Wallaby* ii. 35 The young people..are loud and larky and irreverent. **1912** D. H. Lawrence *Lett.* (1932) 28 Every blessed place was full of men, in the larkiest of spirits. **1958** *Vogue* July 57 Osborne has shocked the stage with a real-life genus of larky lower-middle class humanists and given them the heroic status of being worried. **1967** *Listener* 8 June 747/2 A character in whom humility and submissiveness are combined with a larky humour. **1974** *Ibid.* 21 Nov. 673/1 The only disk jockey who sounds genuinely happy... Larky, insouciant and very funny.
b. *transf.*
1925 *Blackw. Mag.* July 80/2 (*Rugby School*) The 'swells' were allowed to wear 'larky' waistcoats, *i.e.*, waistcoats of various hues often with flowery designs embroidered on them.

Larmor (lā·ɹmǝɹ). *Physics.* The name of Sir Joseph *Larmor* (1857–1942), Irish-born physicist, used *attrib.* and in the possessive with reference to a theorem enunciated by him in 1897, according to which the effect of a uniform magnetic field on a rotating particle is (to a first approximation) to cause the frame of reference in which the particle is rotating to rotate in turn about the direction of the magnetic flux with a frequency $\gamma B/2\pi$ (where γ is the magnetomechanical ratio and B the magnetic induction), which for an orbital electron in a monatomic molecule represents a precessional frequency of $eB/4\pi m$ (e and m being the charge and mass of the electron); so *Larmor frequency, precession*; *Larmor's theorem*; also **Larmor radius,** the radius of the helical path of a free charged particle spiralling about magnetic field lines.
1923 Larmor's theorem [see *Coriolis]. **1926** E. C. Stoner *Magnetism & Atomic Struct.* iv. 68 The Larmor precession is of great importance in the treatment of the Zeeman effect, this and diamagnetism being different aspects of the same phenomenon. **1951** *Physical Rev.* LXXXIII. 1000/2 The Larmor frequency, $\omega_P = 2\mu_P H/\hbar$, of a sample of protons in the magnetic field is measured by the resonance absorption method. **1957** B. I. & B. Bleaney *Electr. & Magn.* viii. 186 As a consequence of the Larmor precession, each electron acquires a component of angular momentum about the direction of [the magnetic induction] **B**. **1962** W. B. Thompson *Introd. Plasma Physics* vii. 151 In a magnetic field a second length becomes important, the Larmor radius, r_L, i.e. the radius of the circle formed by the particle orbit about the magnetic field lines. **1971** D. W. Sciama *Mod. Cosmol.* ii. 36 The galactic magnetic field thus has an important effect on the motion of cosmic rays if their Larmor radius is substantially less than the size of the Galaxy.

larn (lāɹn), *v. colloq.* [f. dial. form of Learn *v.* 4; see *E.D.D.* for further examples.] *trans.* To teach; to give (a person) a lesson; freq. used ironically as a threat of punishment.
1790 T. Wilkinson *Mem.* I. 117 You are unfit for the stage, Muster Whittington, and I won't larn you—you may go, Muster Whittington. **1851** [see *jes, jes']. **1899** *Manch. Guardian* 13 Mar. 10/1 Said Mr. Dooley, '..we'll larn thim a lesson.' **1902** E. Nesbit *Five Children & It* viii. 204 I'll larn you, you young varmint! **1928** 'Brent of Bin Bin' *Up Country* xiii. 228 The taller ruffian put a bullet in the wall above his head just to larn him, and his companions advised him to be still. **1931** W. Holtby *Poor Caroline* v. 180 'I'll larn her,' swore he to himself. **1949** 'J. Tey' *Brat Farrar* xvii. 183 Bee took him to call on the tenants... 'Gates last; just to larn him,' Bee said. **1956** C. Blackstock *Dewey Death* ix. 216 That'll larn you, you so-and-so.

larnax (lā·ɹnæks). *Gr. Antiq.* Pl. **larnakes** (lā·ɹnăkīz). [Gr. λάρναξ chest, urn.] A chest, ossuary, urn, or coffin, usually of terra cotta, frequently ornamented with designs.
1870 *Catal. Greek & Etruscan Vases Brit. Mus.* II. 191 In the centre Aphroditê seated on a chest, *larnax*, and looking back at a youthful male figure. **1901** A. J. Evans *Mycenaean Tree & Pillar Cult* 77 In a chambered tomb at Milato..was a painted clay ossuary chest or larnax of the usual Cretan type. **1901** *Rep. Brit. Assoc. Adv. Sci.* 444 The one [tomb], a square chamber with a dromos, yielded parts of two painted *larnakes*, thoroughly Mycenean in design. **1904** *Westm. Gaz.* 24 Aug. 7/3 A later cemetery, containing larnax burials, yielded bronze implements, beads, and vases like those in the palace magazines. **1910** *Encycl. Brit.* I. 248/2 The Cretan 'larnax' coffins..have no parallels outside the Aegean. **1939** V. G. Childe *Dawn European Civilization* (ed. 3) ii. 24 Individual burial..in clay coffins (larnakes). *Ibid.* 27 Larnax burials.

larnite (lā·ɹnəit). *Min.* [f. *Larne*, the name of a town in Co. Antrim, N. Ireland + -ite[1].] A mineral consisting of a metastable monoclinic phase of dicalcium silicate, Ca_2SiO_4.
1929 C. E. Tilley in *Mineral. Mag.* XXII. 79 Next in importance to spurrite comes a mineral consisting of calcium orthosilicate. As this is the first recorded natural occurrence of this compound it is proposed to designate it as larnite, from Larne, in the vicinity of which these contact minerals occur. **1950** *Ibid.* XXIX. 182 Both the nagelschmidtite and larnite phases have been seen in slags. **1957** *Amer. Mineralogist* XLII. 384 The larnite zone is very prominent, forming a band about thirty inches thick, dark grey to almost black in color, hard, flinty, and very tough. The dark color is..due to..the presence of a small amount of very fine-grained magnetite. Pure larnite is presumably white. **1966** [see *bredigite].

Larose (larō·z). [Name of some vineyards in the Bordeaux area of France.] A type of claret; the vineyard or area itself.
1841 Thackeray in *Fraser's Mag.* June 720/1 It is my firm opinion that a third-rate Burgundy, and a third-rate claret—Beaune and Larose for instance, are *better* than the best. **1863** G. Meredith *Let.* 3 Feb. (1970) I. 190 Vins..Claret—Larose. **1920** G. Saintsbury *Notes on Cellar-Bk.* iv. 66 It is true that some of the very best vineyards (Léoville, Larose, Ducru-Beau-Caillou..) are situated there [*sc.* at Saint-Julien]. **1958** A. L. Simon *Dict. Wines* 99/2 Larose, Château. There are several wine-producing estates in the Gironde bearing this name, mostly hyphenated with the name or names of present or past owners.

larrigan (læ·rigăn). *N. Amer.* Also 9 **larigan, larrigin.** [Of unknown origin.] A long boot made of undressed leather.
1886 *Engineering News* XVI. 99/1 And the ordinary foot-gear is a pair of cow-hide moccasins (called shoe-packs or larrigans). **1889** *Amer. N. & Q.* III. 308 A *larigan*, or *larrigin*, in Maine and New Brunswick, is a kind of boot or moccasin of yellow leather, having a long leg reaching above the knee. It is worn by lumbermen in the deep snows of winter. **1915** *Outing* (U.S.) Oct. 27/2 A 'shoe-pac' or 'larrigan' is a beef-hide moccasin with eight to ten-inch top, and with or without a light, flexible sole. **1922** *Short Stories* (U.S.) Feb. 128/2 Over six feet in his larrigans. **1931** 'Grey Owl' *Men of Last Frontier* 180, I was much hampered by a pair of still hard-soled larrigans which I had donned. **1961** *Saturday Night* (Toronto) 23 Dec. 18/1 After breakfast that day I rode to school..wearing a pair of cowhide larrigans greasy with linseed oil. **1968** E. R. Buckler *Ox Bells & Fireflies* ii. 25 Little Tim was six foot seven..who could yet sew a larrigan together with the waxed end neater than a woman could hemstitch.
Hence **la·rriganed** *a.*, wearing larrigans.
1904 C. G. D. Roberts *Watchers of Trails* 287 Then, turning on his larriganed heels, he strode up the trail. **1922** *Short Stories* Feb. 129/1 [The dogs] clipped fangs at Cherriman's larriganed legs.

larrikin. Add: [Wright, Suppl. to *E.D.D.*, cites *larrikin* 'a mischievous or frolicsome youth' from informants in Warwickshire and Worcestershire; see also quot. 1882. Cf. *E.D.D.*, *Larack* (*larack about*, to 'lark' about), cited from C. C. Robinson's *Dial. of Leeds* (1861).] (Further examples.) Also *transf.* and *attrib.*
1868 H. W. Harper *Lett. from N.Z.* (1914) vii. 123 We are beset with larrikins, who lurk about in the darkness and deliver every sort of attack on the walls and roof with stones and sticks. **1871** *Evening Post* (Wellington, N.Z.) 27 Apr. 2/4 Such rowdyism I never saw before, even in Melbourne and San Francisco, the hotbeds of larrikins. **1882** F. W. P. Jago *Anc. Lang. & Dial. Cornwall* 205 *Larrikins*, mischievous young fellows, larkers. 'Mischievous larrikins who pull the young trees down.' *The Cornishman.* **1901** *Daily Tel.* 8 Mar. 8/7 The larrikins of the Legislature..could not be visited retrospectively with an adequate punishment. **1925** E. Wallace *King by Night* xlix. 224 The desire of the larrikin for closer association with his social superiors. **1943** D. Stewart in *Coast to Coast 1942* 204 For all his larrikin assurance, Les was afraid to put his feeling for Leila into words. **1966** 'J. Hackston' *Father clears Out* 26 It had headed straight for the sturdiest, roughest, most larrikin-looking mob of cattle. **1972** *Southerly* XXXII. 199 He had come around a corner in a corridor to find her confronting a larrikin from 2D.
larrikinism (further examples).
1871 *Evening Post* (Wellington, N.Z.) 27 Apr. 2/4 (*heading*) Larrikinism v. Public Entertainments. **1879** C. L. Innes *Canterbury Sk.* v. 39 We had not then [*sc.* in 1852–3] the pestilential element of 'larrikinism' which is so rife now, and which makes many public meetings so objectionable. **1927** M. M. Bennett *Christison* xxviii. 264 This last piece of larrikinism took Mimi so completely by surprise that she could not think of anything to say.

larrup, *v.* Add: (Later examples.) Also *intr.*
1922 Joyce *Ulysses* 498, I let him larrup it into me for the fun of it. **1939** C. Fry *Boy with Cart* 26 Heard the first spatter of drops, the outriders Larruping on the road. **1953** Dylan Thomas *Under Milk Wood* (1954) 81 From the larrupped waves the lights of the lamps in the windows call back the day. **1970** [see *cruelty 1 b].

Larry (læ·ri), *sb.*[3] [Etym. uncertain.] Phr. *happy as Larry*, extremely happy.
1905 T. Collins in *Barrier Truth* (Broken Hill) 29 Dec. 1 Now that the adventure was drawing to an end, I found a peace of mind that all the old fogies on the river couldn't disturb. I was as happy as Larry. **1915** L. Stone *Betty Wayside* xix. 254 If it hadn't been for that busybody we'd have been as happy as Larry. **1934** T. Wood *Cobbers* iii. 25 He said he was as happy as Larry to see a fresh face. **1938** S. Beckett *Murphy* ix. 180 Kept in peace they would have been as happy as Larry, short for Lazarus. **1946** F. Sargeson *That Summer* 32 Then he'd hop round happy as Larry. **1966** Baker *Austral. Lang.* (ed. 2) xii. 271 *Happy as Larry*, extremely happy. Possibly but not certainly commemorating the noted Australian pugilist Larry Foley (1847–1917). **1966** B. Kennelly *Collection One: Getting Up Early* 26 The Knockanore woman was happy as Larry. **1967** *Daily Mail* 18 Jan. 12/3 The tobacco and pipe trade..are as happy as Larry that Dr. Cameron..has won the award.

larva. Add: **2. c.** *larva-case, -stage.*
1855 J. Phillips *Man. Geol.* 459 Thin tufaceous limestones, sometimes full of the larva-cases of phryganidae. **1893** J. Tuckey tr. *Hatschek's Amphioxus* 159 Those stages which form the transition from the development of the embryo..to the larvae stages which are self-nourishing.

larvikite (lā·ɹvikəit). *Petrogr.* Also **laurvikite,** † **-vigite** (lauə·ɹ-). [ad. G. *laurvikit* (W. C. Brögger 1890, in *Zeitschr. f. Kryst.* XVI. 29), f. *Laurvik* (now *Larvik*), name of a Norwegian seaport: see -ite[1].] A kind of syenite that has a characteristic coarse texture dominated by rhombs of soda or soda-lime feldspar, with augite as the chief mafic mineral, and is used as a decorative stone.
1895 A. Harker *Petrol.* 304 (Index), Laurvikite. **1911** *Encycl. Brit.* XVI. 30/1 The ornamental stone from south Norway, now largely used as a decorative material in architecture, owes its beauty to a felspar with a blue opalescence..which Professor W. C. Brögger has termed cryptoperthite, whilst the rock in which it occurs is an augite-syenite called by him laurvigite, from its chief locality, Laurvik in Norway. **1962** R. Webster *Gems* I. ix. 157 The Norwegian rock known as laurvikite..is a material extensively used for building façades. **1965** *Norsk Geol. Tidsskr.* XLV. 69 The two monsonitic syenites from the Oslo area, larvikite and tönsbergite, differ mainly in the colour of the feldspar minerals. The blue schiller which is so predominant in the larvikite varieties is normally completely masked by the red colour of the tönsbergite varieties. **1971** *Country Life* 16 Sept. 683/1 By the date of the opening of Highgate [Cemetery]..the Victorians had become eclectic in their choice, employing all manner of stones, often of distant source, simply on account of their pleasing colour or texture. So we find the iridescent Larvikite from Norway together with all colour tones of granite-type rock from Finland or Sweden.

laryngal, *a.* Add: (Later examples.) Also *absol.*
1922 D. Jones *Outl. Eng. Phonetics* (ed. 2) vi. 14 Glottal or laryngal sounds, viz. sounds articulated in the glottis. **1933** L. Bloomfield *Lang.* vii. 99 A glottal or laryngal stop is produced by bringing the vocal chords tightly together and then letting them spring apart under the pressure of the breath. **1939** L. H. Gray *Foundations of Lang.* iii. 49 Two other glottal (or laryngal) sounds of importance are represented by the *h*'s of English *how, ahoy*.., unvoiced and voiced respectively. **1958** Priebsch & Collinson *German Lang.* (ed. 4) 6 It [*sc.* Hittite] has 'laryngeal' sounds—not identical with the Semitic 'laryngals'.

laryngeal, *a.* and *sb.* Add: **A.** *adj.* **2.** Of a sound: produced in or modified by the larynx; = prec. Also *absol.*
1921 E. Sapir *Lang.* 249 Articulations, laryngeal. **1927** R. Bridges in *S.P.E. Tract* xxvi. 177 The method is that in singing the mouth is fixed in the position that gives the required vowel resonance..and that the laringeal note is as it were forced through it. **1932** W. L. Graff *Lang.* 28 *Laryngeals or glottals*, produced by a narrowing or closure of the vocal cords.
3. Corresponding to sense B. 2 below.
1952 *Bull. Board Celtic Stud.* XIV. 296, I reconstruct in terms of the so-called 'laryngeal theory', here, however, without committing myself to the number of laryngeals necessarily to be assumed at a given time. **1958** [see *laryngal a.]. **1958** A. S. C. Ross *Etym.* 7 It is certainly quite impossible for anyone to understand Laryngeal Theory without being thoroughly familiar with junggrammatisch Ablaut.
B. *sb.* **2.** *Philol.* A hypothetical phonetic element with a laryngeal quality supposed to have existed, spec. in Proto-Indo-European, and to have left traces in the vocalic features of extant Indo-European languages.
1942 E. H. Sturtevant *Indo-Hittite Laryngeals* 15 In this book the word *laryngeals* designates certain consonants of Proto-Indo-Hittite. The name is historically a translation of German *Laryngale*, which term was borrowed from Semitic grammar by Hermann Möller to designate five phonemes of his Proto-Indo-European-Semitic. **1951** *Trans. Philol. Soc.* 88 (*title*) A reconsideration of the Hittite evidence for the existence of 'Laryngeals' in Primitive Indo-European. **1963** *Language* XXXIX. 252 It is intrinsically implausible that a presumably nonsyllabic laryngeal next to a syllabic resonant should vocalize. **1969** *Ibid.* XLV. 260 Plain stops immediately followed by a laryngeal became aspirated stops in Indo-Iranian. **1971** F. R. Adrados in *Archivum Linguisticum* II. 95, I refer particularly to those stems in which the presence of laryngeals gives rise to diverging interpretations.
So **lary·ngealist,** an adherent of a laryngeal theory; also *attrib.* or as *adj.*; **lary:ngealiza·tion,** the action or fact of being laryngealized; **lary·ngealized** *a.*, of a sound produced in or affected by the larynx.
1943 K. L. Pike *Phonetics* vii. 127 Laryngealization may conveniently be said to be trillization with superimposed voice. *Ibid.*, In English one often hears laryngealized vowels. **1964** O'Connor & Tooley in D. Abercrombie et al. *Daniel Jones* 176 Glottal stop and laryngealization before word-initial vowel were accepted. **1964** *Language* XL. 138 The laryngealist will retort that another..problem..is also solved by applying a laryngeal solution. *Ibid.* 140 A typical phonemic analysis of the PIE vowels in laryngealist terms. **1968** *Ibid.* XLIV. 529 Much is made of the fact that only one laryngealized stop (i.e. glottalized or aspirated) occurs per word in Quechua. **1968** Chomsky & Halle *Sound Pattern Eng.* 315 Several African and Caucasian languages exhibit the so-called laryngealized or 'creaky' voice. **1971** *Canad. Jrnl. Ling.* Fall 70 He rejects phonemic /ə/..thus discomfiting the more elderly laryngealists.

laryngo-. Add: **laryngo-pharynx,** substitute for def.: = *Hypopharynx 2; (earlier and later examples); **lary·ngophone,** a microphone designed to be placed or attached to the throat so as to pick up the voice directly with little intrusion of other sounds; **lary:ngotra:cheobronchi·tis** *Path.*, inflammation of the larynx, trachea, and bronchi; *spec.* an acute febrile disease (a form of croup) that exhibits these symptoms and occurs chiefly in young children, in which excessive secretion of mucus causes obstruction of the larynx and sometimes the bronchi.
1893 A. W. MacCoy in C. H. Burnett *Syst. Dis. Ear, Nose, & Throat* II. 195 The laryngo-pharynx..is chiefly interesting because of its relationship to the epiglottis and the superior margin of the larynx, which is situated in front of it. **1960** Laryngopharynx [see *inlet *sb.* 4*]. **1927** *Observer* 6 Nov. 19/3 We have a special instrument, the laryngophone, by means of which we can speak to each other in flight. **1941** *Jrnl. R. Aeronaut. Soc.* XLV. 402 The picking up of the throat vibrations is done with a laryngophone. **1932** Dorland & Miller *Med. Dict.* (ed. 16) 685/2 Laryngotracheobronchitis. **1956** Hinshaw & Garland *Dis. Chest* xi. 183 Among the most serious and difficult ailments affecting small children are a group of diseases variously called acute laryngotracheobronchitis, bronchiolitis, fibrinous bronchitis or 'croup'. These are characterized by violent cough, often associated with laryngospasm and bronchospasm and appear to be of infectious origin. **1972** *Daily Tel.* 2 Feb. 13/5 Two American pathologists have been claiming that laryngotracheobronchitis is frequently responsible for cot deaths, since they have found signs of inflammation in the larynxes, windpipes and bronchial tubes of babies who died.

lasagne (lasā·nye). In sing. **lasagna.** [It.] An Italian dish: a variety of pasta cut in long wide strips.
[**1760** Baretti *Dict. Eng. & Ital. Lang.* I, *Lasagna*, a kind of thin paste cut into slices and dry'd, boiled in water or broth, [etc.].] **1846** E. Acton *Mod. Cookery* (ed. 5) 579 The ribbon maccaroni (or lasagnes). **1849** Browning *Englishman in Italy* in *Poems* II. 333 We shall feast our grape-gleaners..With lasagne so tempting to swallow In slippery ropes. **1960** *News Chron.* 17 Feb. 6/2 Excellent ravioli and lasagne (soft ribbon noodles)..at 2s. 6d. a pound. **1961** *Listener* 20 Apr. 719/1 To make *lasagne verdi* (green *lasagne*)..you will need [etc.]. **1966** T. Pynchon *Crying of Lot 49* i. 10 The layering of a lasagna, garlicking of a bread. **1972** *Guardian* 18 Aug. 11/3 Lasagne..perhaps the finest of all pasta dishes. **1974** *Times* 7 Feb. 16/7 To assemble the lasagne, spoon a little of the meat sauce into the base of a large shallow baking dish. Cover with a layer of lasagne, spoon over a layer of cottage cheese and sprinkle with Parmesan.

lase (lēiz), *v.* [Back-formation from next, the ending *-er* being treated as the ending -er[1] of agent nouns.] *intr.* Of a substance, or an atom or molecule: to undergo the physical processes (of excitation and stimulated emission) employed in the laser; to function as the working substance of a laser. Of a device: to operate as a laser.
1962 *New Scientist* 1 Feb. 270/3 This is well illustrated by the uses now being made of the entirely novel verb 'to lase'... It is common currency among those..whose lives are given to the search for materials that will emit light in coherent and narrow-pencilled beams. **1963** *Ibid.* 10 Jan. 65/3 Hitherto, only certain ionic crystals (notably ruby), some gas mixtures and..one or two liquids, had been persuaded to 'lase'; that is, to emit coherent infrared or visible light of a single wavelength. **1963** *Ibid.* 7 Feb. 293/2 Scientists at Standard Telecommunications Laboratories..last week persuaded their version of the gallium arsenide laser to 'lase'. **1963** *Monsanto Mag.* Mar. 10/2 Calculations..indicated that gallium arsenide would 'lase', or amplify light, if suitably stimulated. **1967** *Guardian* 11 Feb. 1/1 Can X-ray lasers be made? There is no problem in getting substances to lase at the right frequency. **1969** *Sci. Jrnl.* Apr. 55/2 The term 'dye laser' is generally used. Different dyes lase at different wavelengths and..it has been possible to span without a break the range 700–1000 nm by using some 20 different dyes. **1970** *Daily Tel.* (Colour Suppl.) 28 Aug. 16/2 Not all atoms can be made to 'lase'. **1971** *Sci. Amer.* July 37/2 When the injection current becomes high enough so that the light in the crystal making a round trip along the junction plane is amplified enough to offset losses due to absorption, to scattering, to leakage out of the mirrors and so on, the diode is said to be lasing.
Hence **la·sing** *ppl. a.* and *vbl. sb.*
1963 *Monsanto Mag.* Mar. 9/3 Many variations have been produced since then, employing other solids, as well as gases and liquids, as 'lasing' materials. **1966** Smith & Sorokin *Laser* vii. 369 The characteristics of these lasing diodes..differ considerably from those of other lasers. **1973** *Physics Bull.* Dec. 723/1 Lasing at 145 nm in high pressure krypton has also been achieved..with intense electron beam pumping.

laser[2] (lē^{i}·zəɹ). [f. the initial letters of '*l*ight *a*mplification by the *s*timulated *e*mission of *r*adiation', after the earlier *maser.] **1.** Any device that is capable of emitting a very intense, narrow, parallel beam of highly monochromatic and coherent light (or other electromagnetic radiation), either continuously or in pulses, and operates by using light to stimulate the emission of more light of the same wavelength and phase by atoms or molecules that have been excited by some means.
Orig. treated as the name of a particular kind of maser (*optical maser*) emitting visible light, *laser* is now the general term for all devices of this kind, whatever the wavelength of the emitted radiation.
1960 *N.Y. Times* 8 July 7/6 The Hughes device is an optical maser, or 'laser', (the 'l' standing for 'light'). **1960** *Aviation Week* 18 July 97/2 The optical Maser is also referred to by the term Laser. **1960** *Daily Tel.* 29 Dec. 9/4 The laser, a device for amplifying light which could conceivably be developed to produce a searchlight beam that would reach the moon, is still a paper project as far as British scientists are concerned. **1961** *Jrnl. Appl. Physics* XXXII. 178 [*Paper received 13 June 1960.*] The Fabry-Perot interferometer has been suggested for use as a high-mode LASER (light amplification by stimulated emission of radiation) resonator. **1961** *Observer* 19 Feb. 5/2 The new 'laser', as it is called, uses a mixture of helium and neon gas to produce a continuous beam of infra-red radiation... Previous devices have produced only brief pulses of light. **1962** *Science Survey* III. 27 The principle of the maser has been extended also to solid materials and, in addition, it has been found possible to make a light maser (or 'laser') that produces, not microwaves, but visible light. **1963** *Electronics Weekly* 2 Jan. 1/4 The new high-power laser uses a six-inch by half-inch ruby. **1963** *Monsanto Mag.* Mar. 9/2 Early lasers absorbed energy from a strong burst of ordinary white light, organized it, then expelled a powerful beam of a different kind of light. **1963** *Daily Tel.* 24 Oct. 19/4 Already in metal working the term 'Gillette power' is used as a measure of the laser's metal-vaporising capabilities. It represents the number of stacked razor blades through which a beam can bore its way. **1964, 1966** [see *hologram]. **1969** *Sci. Jrnl.* Apr. 53/1 Lasers have been operated which produce visible radiation, ultraviolet, infrared and even submillimetre radiation. **1970** [see *holograph *v.*]. **1971** *Sci. Amer.* June 21/3 A laser is a device for generating or amplifying a beam of light whose waves are both monochromatic (all the same wavelength) and coherent (all in step). The light beam emitted by a laser can be made almost perfectly parallel, its divergence angle being theoretically limited only by diffraction effects. **1972** *McGraw-Hill Yearbk. Sci. & Technol.* 266/2 The first purely chemical lasers requiring no external source of energy to initiate or sustain laser excitation have been operated successfully.
2. *attrib.* and *Comb.*, as *laser beam, bomb, light, reflector*; *laser-guided, -ignited* ppl. adjs.; *laser-heat* vb.
1963 *Monsanto Mag.* Mar. 10/3 A laser beam can generate intense heat—10,000°F. or higher—in a small area. **1970** *Daily Tel.* (Colour Suppl.) 28 Aug. 17/1 A laser beam, focused through the lens of the eye, can weld a detached retina back into place by creating scar tissue. **1970** *Daily Tel.* 31 Jan. 4/2 Scientists..maintain that the laser bomb..is a theoretical possibility. **1972** *Guardian* 29 June 4/3 The drawback to the laser bomb is that the plane producing the beam must keep it on target until the bomb's impact. **1967** *New Scientist* 11 May 326/2 With the laser-guided bomb, the large bombers might be able to drop their loads over the target area from high altitudes with greater assurance of putting them on target. **1972** *Science* 9 June 1108/3 The laser-guided bombs now being used are mostly in the 2000 to 3000-pound range. **1971** *Sci. Amer.* June 27/1 The second question—regarding the feasibility of laser-heating a small dense plasma to thermonuclear conditions without the necessity of a confining magnetic field—is receiving increased attention. *Ibid.* 29/1 A method for converting the fusion energy from laser-ignited deuterium-tritium pellets into electrical power was evolved..early in 1969. **1966** *Listener* 28 July 129/3 The editor..will brandish a laser-light pen to indicate alterations which a computer will make. **1971** B. de Ferranti *Living with Computer* ix. 83 In laser light the waves are all in the same plane and in phase. **1969** *New Scientist* 9 Oct. 81/1 The first men on the Moon have..already placed one laser-reflector on the lunar surface.

lash, *sb.*[1] Add: **4.** An attempt; esp. in phr. *to have a lash (at)*, to make an attempt, to 'have a go at'. *Austral.* and *N.Z.*
1941 Baker *Dict. Austral. Slang* 42 *Lash at, have a*, to make an attempt at (something). **1945** J. Pascoe *Canterbury High Country* 28 A few may spend their cheque in a glorious lash at the beer. **1948** D. Ballantyne *Cunninghams* (1963) vii. 38 Hoping to get a lash at the Huns. **1949** R. Park *Poor Man's Orange* (1950) 193 The blithe pipings of old men who, safe [from the fight] up on their balconies, leaned over rails and exhorted everyone to 'ave a lash. **1971** *Sunday Sun* (Brisbane) 17 Oct. 14/2, I am a natural sportsman. Only last week I donkey licked the local kindy kids at drop the hankie. So I went out to Surfers Paradise course to have a lash.

5. (sense 3) *lash-tender* adj.; **lash rope** *N. Amer.*, a rope used for lashing a pack or load on a horse or vehicle.

1806 Lewis & Clark *Orig. Jrnls. Lewis & Clark Expedition* (1905) V. 114 Sergt. Gass, McNeal, Whitehouse and Goodrich accompanyed them [*sc.* Indians] with a view to procure some pack or lash ropes. **1822** J. Fowler *Jrnl.* 18 June (1898) 159 We then took the lash Roaps and tyed up the Horses. **1843** *Amer. Pioneer* II. 162 Each horse was provided with..a lash rope to secure the load. **1888** Lees & Clutterbuck *Ramble in Brit. Columbia* 229 The lash rope is from thirty to forty feet long. **1929** *Collier's* 5 Jan. 33/3 'Wait until I get my lash rope' (i.e., the rope with which he bound his load on his sledge). **1963** R. Symons *Many Trails* vii. 77 Lash ropes were tightened till the pack animals grunted. *a* **1889** G. M. Hopkins *Poems* (1918) 74 Whether..furled Fast or they [*sc.* ash-boughs] in clammyish lashtender combs creep Apart wide.

LASH, Lash, lash (læʃ). The initials of *lighter aboard ship*, used, freq. *attrib.*, to denote a ship, or system of shipping, in which loaded barges are placed directly on board the ship.

1965 *Maritime Reporter* 1 Nov. 21 (Advt.), The LASH System combines an ultra-simple, fast, automated ship with a large number of low-cost lighters. **1967** *Economist* 7 Jan. 51/2 This is the LASH shipping system (Lighter Aboard Ship Inc.) which both the Americans and Germans are now building. A LASH ship is designed to pick up and carry 250-ton lighters, which are towed to and from the ship regardless of tides or port labour schedules. **1969** *Jane's Freight Containers 1968–69* 378/2 No container ships, but 5 Lash vessels..are under construction. **1970** *Times* 12 Aug. 18 Shipowners also hastened the introduction of..lash vessels in which laden barges are floated directly into a large hull.

lash, *v.*^{1} Add: **4.** (Later example.)

1922 Joyce *Ulysses* 736 You cant get on in this world without style all going in food and rent when I get it Ill lash it around I tell you in fine style.

5. (Later examples.)

1959 G. Freeman *Jack would be Gent.* i. 10 He'd never had the money to lash out properly. **1973** 'M. Yorke' *Grave Matters* i. vi. 35 He must have paid plenty for the place, besides what they're going to lash out in alterations.

6. *absol.* (Later examples.)

1876 G. M. Hopkins *Wreck of Deutschland* viii, in *Poems* (1967) 54 Oh, We lash with the best or worst Word last! **1877** A. Sewell *Black Beauty* (*c* 1878, ed. 5) xx. 93 The man, fiercely pulling at the head of the forehorse, swore and lashed most brutally. **1892** A. Conan Doyle *Adventures Sherlock Holmes* viii. 205 The sudden glare.. made it impossible for me to tell what it was at which my friend lashed so savagely.

lash, *v.*^{2} Add: **3.** *Comb.* **lash-up**, (*a*) a makeshift or hastily contrived improvisation; also *attrib.*; (*b*) (see quot. 1925). Hence **lashed-up** *a.*, improvised.

1898 W. P. Drury *Tadpole of an Archangel* 86 Such a godforsaken lash-up of a bridge you never clapped eyes on! **1907** J. Masefield *Tarpaulin Muster* viii. 102 And down they all go—ship, and tea, and mate, and bishop, and general, and Jimmy and the whole lash-up. **1920** *Blackw. Mag.* Feb. 154/1 By 'lashed up' means—(that is to say, 'improvised')—and with a makeshift staff of assistants, a tolerable chart was produced. *Ibid.* 158/1 We..had been obliged to make 'lash-up' (*i.e.*, makeshift) arrangements. **1924** P. P. Eckersley *Captain Eckersley Explains* i. 5 A 'lash-up' or experimental station was erected at the Marconi Works. **1925** Fraser & Gibbons *Soldier & Sailor Words* 140 *Lash-up*, a failure. A fiasco. The break-down of anything. **1929** O. Harland *Golden Plough* iv. 97 Until we come to the present Imbroglio, the Glorious Lash-up of this very age. **1936** 'Taffrail' *Mystery at Milford Haven* 281 The boat..was what a blue-jacket would have called a 'lash-up', a thing of bits and pieces. **1958** *Economist* 13 Sept. 869/2 Black Knight is essentially a lash-up on which to test various designs of nose cone for the 2,500 mile ballistic weapon Blue Streak that should be ready for test in the early 1960s. **1962** W. Schirra in *Into Orbit* 46 It [*sc.* the couch] was a simple bit of furniture compared to the lashup of tubing, fans, filters and tanks which was built around it. **1966** M. Woodhouse *Tree Frog* xxv. 182 We didn't have time for an instrument check. It's just a lash-up really. **1974** *Exchange & Mart* (South) 27 June 53M/3 Rebuilt motor, not a lash-up.

lashing, *vbl. sb.*^{1} **1. b.** Delete (*Anglo-Irish*): this sense is now in general use. Add later examples.

1901 E. W. Hornung *Black Mask* v. 74 There were lashings of sound wine for one and all. **1927** D. L. Sayers *Unnatural Death* xxiii. 278 Nice little dinner—lashings of champagne. **1942** *R.A.F. Jrnl.* 3 Oct. 30 We fought through lashings of rain and mud to our billets. **1962** J. Wain *Strike Father Dead* 164 Real comfort. And plenty of money. Lashings! She earned a good solid packet at this job. **1966** *Lancet* 2 Apr. 765/1 The crusty wholemeal bread..eaten with lashings of butter. **1975** *Country Life* 6 Feb. 336/3 Chicory..requires lashings of water.

lashkar. b. Substitute for def.: A body of Afridi soldiers. (Add further examples.)

1897 *Times* 6 Oct. 3/1 The lashkar is prepared to offer terms on behalf of the Afridi, Mamund, and Malakand tribesmen. **1908** *Daily Chron.* 27 Apr. 1/7 General Willcock's columns yesterday searched out the enemy's lashkars. **1924** *Glasgow Her.* 14 May 8 The rebel lashkars are melting away. **1955** *Times* 31 May 7/6 Inside this frame, lashkars, or tribal columns, are still raised occasionally. **1973** *Times* 22 Mar. (Pakistan Suppl.) p. ii/3 Baluchistan..is led largely by tribes amenable only to the jurisdiction of their own *sardars*..who often command sizeable *lashkars*, or private levies.

lash-up: see *LASH *v.*^{2} 3.

lasiocampid (lẽⁱsIokæ·mpid), *sb.* and *a.* [f. mod.L. family name *Lasiocampidæ*, f. the generic name *Lasiocampa* (N. Contarini *Catal. Uccelli e Insetti Padova* (1843) 37), f. Gk. λάσιος hairy, shaggy + κάμπη caterpillar.] A member of the family Lasiocampidæ, a group of large moths also known as eggars or lappet-moths, including some species whose larvae are called tent-caterpillars; of or pertaining to an insect of this kind.

1895 J. H. & A. B. Comstock *Man. Study Insects* 360 The larvæ of the Lasiocampids feed upon the foliage of trees, and are frequently very destructive. **1912** *Proc. Entomol. Soc. London* p. iv, Prof. Poulton exhibited specimens of the Lasiocampid moth *Mimopacha gerstaeckeri*, Dewitz. **1934** *Discovery* XVII. 98/1 Sometimes the hairs and spines are..exhibited in an ostentatious manner when an enemy appears. This is well illustrated in Lasiocampids. **1964** V. B. Wigglesworth *Life of Insects* vii. 112 The male..may be specially modified for the recognition and location of these scents. That is most evident in the Saturniid, Bombycid, and Lasiocampid moths, in most of which the female is sluggish and sedentary.

lass. Add: **1. d.** A female member of the Salvation Army.

1886 *War Cry* 9 Oct. 9/4 Cadets to be grouped together in Brigades..Lads or Lasses, as the case may be. **1890** W. Booth *In Darkest Eng.* i. vi. 55 Our two lasses go unharmed and loved at all hours. **1907** G. B. Shaw *Major Barbara* Pref. 171 Bill Walker,..having assaulted the Salvation Lass,..finds himself overwhelmed with an intolerable conviction of sin. **1967** W. S. Smith *London Heretics* iii. v. 238 In 1879 two of General Booth's lasses arrived in the Quaker stronghold of Darlington.

Lassa (læ·sā). The name of a village near Mubi in NE. Nigeria, used *attrib.* in **Lassa fever**, an acute febrile virus disease that occurs in tropical Africa with a high mortality rate (first reported at Lassa in 1969); similarly *Lassa virus*.

1970 *Times* 11 Feb. 7/3 Lassa fever infection can involve almost all the body's organs; symptoms may vary. The virus produces a fever as high as 107°F, mouth ulcers, a skin rash with small haemorrhages, pneumonia, infection of the heart leading to cardiac failure, kidney damage, and severe muscle aches. **1970** *Amer. Jrnl. Trop. Med. & Hygiene* XIX. 690/1 There is no epidemiologic evidence to indicate that Lassa virus is an arthropod-borne virus. **1973** *Nature* 12 Jan. 92/1 Man-to-man spread of Lassa virus undoubtedly takes place through contact with blood or infectious secretions, but a reservoir of infection has yet to be identified. **1975** *Ibid.* 15 May 185/1 Lassa fever..seems to be one of the most virulent virus diseases yet discovered.

'lasses, colloq. abbreviation of Molasses. *U.S.*

1775 in O. E. Winslow *Amer. Broadside Verse* (1930) 141/1 The 'lasses they eat every day, Would keep an house a winter. **1807** *Salmagundi* 16 May 205 This manufacture is called by the Bostonians *lasses candy*. **1854** M. J. Holmes *Tempest & Sunshine* xxii. 317 Get along Jack, pokin' your fingers into the 'lasses cup. **1909** F. B. Calhoun *Miss Minerva* 174 Aunt Minerva's in the kitchen right now makin' me a 'lasses custard. **1948** *Chicago Tribune* 14 Mar. vii. 17/2 Now, come spring, it's ''lasses time' right over the border and 'sap's a-bilin''.

lassie. Add: **1. b.** = *LASS 1 d.

1906 'O. Henry' *Four Million* 84 A Salvation lassie shook her contribution receptacle. **1970** *Guardian* 12 May 10/5 Such items as whether Army lassies should wear lipstick in uniform.

lasso, *sb.* Now usu. with pronunc. (lăsū·).

Fowler remarked (*Mod. Eng. Usage*, 1926, p. 315) '*lasso* is pronounced lasoo' by those who use it; but the English pronunciation is lă'sō.' In ed. 2 (1965) Sir E. Gowers changed this to '*lasso* is pronounced lăsoo' by those who use it, and by most English people too'.

1. (Further examples.)

1824 W. Bullock *Six Months' Residence Mexico* 179 It requires the use of a lasso to catch them. This is thrown with great dexterity by every hostler or servant. **1837** W. Irving *Capt. Bonneville* III. vi. 86 The California horsemen seldom ride out without the laso; that is to say, a long coil of cord, with a slip noose, with which they are expert, almost to a miracle. **1940** H. L. Mencken *Happy Days* 284 They lay in wait in dark Greene street with their ..lassos, and knives. **1966** H. Marriott *Cariboo Cowboy* v. 53 Al was sure a good man with a lasso rope.

fig. **1922** Joyce *Ulysses* 50 In long lassoes..the water flowed full. **1924** R. Campbell *Flaming Terrapin* i. 15 He..hurled Lassoes of dismal smoke around the world.

3. *lasso-throw, -throwing.*

1841 G. Catlin *Lett. on N. Amer. Indians* II. 152 A line, with a sort of 'laso throw', came from an awkward hand on the deck. *a* **1861** T. Winthrop *John Brent* (1883) ii. 11 Man to them was power, and nothing else,—a lasso-throwing machine.

lasso, *v.* Add: Now usu. with pronunc. (lăsū·). Also *fig.*

1965 V. Bonham Carter *Winston Churchill* xviii. 253 Fisher happened to be at Naples and it seemed a heaven-sent opportunity for lassoing him there and roping him in again.

lassoer, lassoing. Add: (Earlier examples.) Also lasso·ist.

1838 'Texian' *Mexico v. Texas* 48 The men were collecting the mules, and when these were driven together, the lassoing began. **1883** Sweet & Knox *On Mexican Mustang through Texas* xli. 584 Juan Gonzales..is said to be the champion lassoer in the world. **1884** W. Shepherd *Prairie Experiences* 40 The lassoer picks out the unbranded calves, and drags them off to the fire. **1906** *Daily Chron.* 16 May 5/7 There have been lassoists before, but never, perhaps, such a master of the art as Will Rogers.

lassu (lā·ʃū). [Hung. *lassú*.] The slow part of a Hungarian csardas (opp. *friss*).

1880 Grove *Dict. Mus.* II. 198/2 Every Csárdás consists of two movements,—a 'Lassu', or slow movement, andante maestoso, and a 'Friss', or 'quickstep', allegro vivace. **1944** [see *CSARDAS]. **1961** *Times* 13 May 5/1 A Haydnish *lassu*.

last, *a.*, *adv.*, and *sb.*^{6} Add: **A.** *adj.* **1.** *spec.* in Cricket, (*the*) *last man* (*in*): the batsman who is not out at the end of an innings; the man who goes in to bat last. Hence *the last pair, wicket.*

1773 *Kentish Gaz.* 24 July, Surry. Yaldin, Last man in, 17. Kent. Mr. Hussey, Last man in, 0. **1833** J. Nyren *Young Cricketer's Tutor* 113 Small went in the last man for fourteen runs, and fetched them. **1870** *Times* 20 July 10/3 Southerton appeared as 'last man'. **1897** H. Newbolt *Vitai Lampada* in *Admirals All* 21 An hour to play and the last man in. **1953** R. Webber *Australians in Eng.* 155 Oldfield (123) and Mailey (46 not out) added 124 for the last wicket in only 40 minutes. **1957** R. Campbell *Coll. Poems* II. 101 No last-man-in has ever batted With a more desperate intent. **1963** A. Ross *Australia 63* iii. 88 The last pair, perched not uncomfortably for forty-nine minutes, had put on 20 runs.

b. Also preceded by an ordinal number, to denote how many places from the end of a series an object, name of a person, etc., occurs.

1880 W. F. Skene *Celtic Scotl.* III. 122 Dathi the second last of the pagan monarchs of Ireland. **1938** I. Goldberg *Wonder of Words* ix. 186 There are technical names for words having accented last, second-last, and third-last syllables. **1963** J. Lusby in B. James *Austral. Short Stories* 235 Before breakfast on our second-last day the Eccentric stood facing Mooney on the sand.

d. Delete † *Obs.* and add later examples in N. Amer. use. Also, the final portion of a period of time, esp. in *the last of pea-time*; also *fig.*

1834 W. A. Caruthers *Kentuckian in N.Y.* I. 190 Our parson whines it out like an old woman in the last of pea-time. **1883** *Century Mag.* Oct. 921/2 The snipe usually makes its appearance in New Jersey and New York about the last of March or the first of April. **1904** E. Robins *Magnetic North* I. 63 Things looked pretty much like the last of pea time. **1908** L. M. Montgomery *Anne of Green Gables* xxxvi. 405 There is a distinguished oculist coming to the Island the last of June. **1931** A. E. Martin *Hist. U.S.* II. ii. 24 By the last of May he had formulated his plan and on the twenty-ninth he issued two proclamations.

f. *last across* (*the road*): a children's game in which each tries to be the last to cross a road (or railway) safely in front of an approaching vehicle (or train). Also *fig.*

1904 A. B. F. Young *Compl. Motorist* ix. 230 If it seems good to them [*sc.* children] to play at 'last across', you had better go very gingerly in their neighbourhood. **1914** 'I. Hay' *Knight on Wheels* xi. 108 A frisky calf, encountered by the way, almost wrecked its own prospects of ever becoming veal by an untimely indulgence in the game of 'Come to Mother, or Last Across the Road'. **1928** *Sunday Dispatch* 15 July 11/3 Socialist back-benchers are playing a dangerous game of 'last across' with Mr. Speaker—just seeing how far they can go at question time without being 'named'. **1957** *Times* 12 Mar. 4/6 Engine drivers are threatening to refuse to drive trains over a section of track ..because children are using it for a game of 'last across'. **1958** R. Liddell *Morea* ii. iv. 103 Chickens seemed to be playing 'last across the road';..dogs tried to meet us head on. **1969** 'A. Hall' *Striker Portfolio* v. 45 We finished up playing 'Last Across' and he cut it too fine.

2. *the four last things* (further examples); also with omission of *four*; *last words*, a person's dying words.

1522 More *De Quat. Noviss.* in *Works* (1557) 76/2 The busi minding of thy .iiii. last things, and the depe consideracion therof, is the thyng that shal kepe thee fro synne. **1606** (*title*) Foure-fould Meditation, of the foure last things: viz...of the Houre of Death. Day of Iudgement. Paines of Hell. Ioyes of Heauen. **1734** J. Trapp (*title*) Thoughts upon the four last things... A poem in four parts. **1808** Scott *Marmion* vi. xxxiii. 366 'Charge, Chester, charge! On, Stanley, on!'..Were the last words of Marmion. **1845** G. Bush *Anastasis* p.v, The great scheme of Scriptural Eschatology, or the doctrine of the last things. **1864** C. M. Yonge *Trial* II. xi. 189 Come, come,..there must be some last words. **1897** J. A. Beet *Last Things* xix. 231 This vision of glory..will be the closing scene of our study of the Last Things. *a* **1916** H. James *Middle Years* (1917) iv. 49 Those 'last words' of the *raffiné* that were chanted and crooned in the damask-hung temple of the Grosvenor Gallery. **1945** E. Waugh *Brideshead Revisited* i. vii. 173, I can't remember all he told me..plenary indulgences, four last things. **1948**, etc. [see *FAMOUS *a.* 1 c]. **1975** J. Aiken *Voices in Empty House* viii. 190 The four last things are death, judgement,

heaven and hell. **1975** R. Player *Let's talk of graves* i. 31 Languishing in the Condemned Cell, contemplating the Last Things.

d. Phr. *if it's the last thing I* (etc.) *do*: used to indicate a very strong desire to do something.

1921 E. O'Neill *Emperor Jones* 180, I kills you, you white debil, if it's de last thing I evah does! **1938** M. Allingham *Fashion in Shrouds* ix. 146 He's got to go on that plane. If it's the last thing he does he's got to go back today. **1971** 'D. Cory' *Sunburst* iii. 44, I want to kill him if it's the last thing I do. **1972** P. Newton *Sheep Thief* xv. 119 I'll get that bloke if it's the last thing I do.

5. Also *last lap* [Lap *sb.*^{3} 5 b]: the final circuit of a track, course, etc.; also *fig.*

1885 *Daily News* 1 Sept. 2/5 At half-distance the positions remained unaltered, and, as they began the last lap, it appeared to be any one's race. **1908** *Daily Chron.* 7 Mar. 7/5 Such an event [*sc.* a cycling race]..is declared to have been won on the last-lap sprint. **1922** Joyce *Ulysses* 234 Bang of the lastlap bell spurred the halfmile wheelmen to their sprint. **1926** O. Barfield *Hist. in Eng. Words* 56 There is a..tendency to transmute.. [modern sports] terms into lively idiom. In this way we can use..the *last lap*. **1932** *Discovery* Dec. 393/1 We learned that weather conditions there had improved and that, for the last lap, we might expect better flying conditions.

b. (Further examples.)

1751 Fielding *Amelia* IV. x. vii. 74 This was indeed almost the last man in the World, whose company he wished for. **1840** H. Reeve tr. *A. de Tocqueville's Democracy in Amer.* IV. iv. iii. 269, I am the last man to contend that these propensities are unconquerable. **1967** *Listener* 28 Sept. 395/3 Degas was the last man to believe in untutored brilliance.

6. *last word* (later examples).

1933 *Punch* 17 May 543/3 Although to my mind the Inverness alterations are absolutely the last word [in modernity], opinions are bound to differ. **1936** *Discovery* Apr. 130/1 The book cannot be the last word on M. Coué. **1966** *Listener* 6 Jan. 12/2 The Trombay establishment is the last word in nuclear sophistication. **1973** *Archivum Linguisticum* IV. 12 Perhaps it is most reasonable to leave the last word to Pidgin.

7. Also *last cry* [tr. Fr. *le dernier cri*]: something in the newest fashion. Cf. *DERNIER *a.* c.

1916 W. J. Locke *Wonderful Year* x. 133 A morning coat (last cry of Bond Street).

9. c. Delete † *Obs.* and add later examples.

1918 Galsworthy *Five Tales* 308 The last of daylight from without mingled with faint intrusion from the lamp within. **1943** K. O'Brien (*title*) The last of summer.

h. (Further examples.)

1816 Scott *Old Mortality* in *Tales my Landlord* 1st Ser. II. x. 241 'Ye hae seen the last o' me, and o' this bonny dye too,' said Jenny, holding between her finger and thumb a silver dollar. **1862** Mrs. H. Wood *Mrs. Halliburton's Troubles* III. xvii. 218 Dick little thought the manufactory had seen the last of him [*sc.* Cyril]. **1889** 'Mark Twain' *Connecticut Yankee* 280 That is the last you are going to see of him till he emerges on the other side. **1910** 'M. Rutherford' *More Pages from Jrnl.* 22, I shall be thankful to see the last of you! **1924** G. B. Shaw *St. Joan* vi. 97 Her heart would not burn, my lord; but everything that was left is at the bottom of the river. You have heard the last of her.

10. b. Delete 'Now *rare*' and add later examples. Now always without *the*.

1923 W. de la Mare *Private View* (1953) 244 And at long last the acquisition of a technical mastery in any art ..is by no means nothing but a gain. **1926** *Manch. Guardian Weekly* Feb. 104/1 The Government followed up the references to agriculture..by launching at long last its land policy. **1936** King Edward VIII in *Times* 12 Dec. 14/4 (*abdication speech*) At long last I am able to say a few words of my own. *a* **1936** Kipling *Something of Myself* (1937) vi. 159 At long last we were left apologising to a deeply-indignant people. **1971** J. Aiken *Nightly Deadshade* ii. 21 Someone answers the phone at long last, making me jump.

C. *Comb.* **2.** *last-chance, -gasp, -resort, -war, -wicket*; **last-minute**, (given, done, made, etc.) at the latest possible time.

1962 *Aeroplane* CIV. 18/1 Carbon deposits had clogged the 'last chance' filter, causing the oil starvation of No. 2 bearing in the JT3D-1 turbofan. **1963** J. Lusby in B. James *Austral. Short Stories* 234 He was giving the Eccentric a last-chance test in individual combat. **1974** G. Jenkins *Bridge of Magpies* xvi. 239 It was a desperate last-chance throw. **1921** D. H. Lawrence *Let.* 2 Mar. (1962) II. 643 K.—who is doing the last-gasp touch. **1971** R. Thomas *Backup Men* xxii. 191 All last-gasp businesses with no need for much of a front. **1972** *Sat. Rev.* (U.S.) 6 May 34/3 Our strategists have spoken too often of 'last-gasp' offensives for us to believe that this one will be any different. **1920** *Ladies' Home Jrnl.* May 140/2 Last-minute sketches from Paris. **1929** J. B. Priestley *Good Companions* ii. iii. 329 Jimmy had a last-minute inspiration. **1931** E. Linklater *Juan in Amer.* iii. i. 204 I've been doing some last-minute Christmas shopping. **1948** 'J. Tey' *Franchise Affair* xix. 227 A last-minute reprieve with the rope round the hero's neck? **1974** M. Yorke *Mortal Remains* ii. i. 39 People were trying to buy last-minute tickets for the day's excursions. **1950** J. Dempsey *Championship Fighting* 170 The pull-away should only be used as a last-resort defence. **1965** H. Kahn *On Escalation* i. 10 The threat of a strike or a lockout is ever present as a last-resort pressure for compromise. **1942** S. Spender *Ruins & Visions* 42 The ghastly last-war voices. **1959** *Encounter* July 29/2 The plan..envisaged the defence of Western Europe almost completely in last-war terms. **1909** *Westm. Gaz.* 1 June 12/2 A great last-wicket stand might once in a way occur. **1975** *Times* 6 Jan. 7/7 A last-wicket partnership of 37 between Mallett and Thomson.

last, *v.*^{1} Add: **3.** Also with *out*.

1881 Mrs. J. H. Riddell *Senior Partner* III. 56 What would hinder him lasting out to ninety [years] or a hundred even? **1921** G. B. Shaw *Back to Methuselah* iv. ii. 194 You people lived on the assumption that you were going to last out for ever and ever and ever.

last-ditch, *a.* [See Ditch *sb.*^{1} 5.] Of opposition, resistance, etc.: maintained to the end. Of an effort, etc.: made at the last minute in an attempt to avert disaster. Also **last-di·tcher**, one who fights to the last ditch; **last-di·tchery.**

1909 *Westm. Gaz.* 30 Jan. 2/1 The only part he is likely to take in the Social Revolution is to be what may be called a last-ditcher in the attempt to resist it. **1927** *Daily Express* 19 Nov. 3/1 A constituency which is to be congratulated on a true last ditcher. **1928** *Daily Tel.* 17 July 18/3 There are few performers who have decided not to broadcast. Almost the 'last-ditcher' is perhaps Harry Tate. **1932** *Mind* XLI. 53 Formal Logic dies hard. It still commands the services of numbers of 'last ditchers'. **1936** 'J. Tey' *Shilling for Candles* i. 6 'Might have walked into the water till she drowned,' said Bill, who was a last-ditcher by nature. *Ibid.*, 'Might have died of an overdose of bulls-eyes,' said Potticary, who approved of last ditchery in Arabia but found it boring to live with. **1951** Koestler *Age of Longing* ii. i. 205, I would rather have been one of the last-ditchers at Thermopylae. **1951** M. McLuhan *Mech. Bride* 44/1 A last-ditch stand of denuded minds. **1955** *Bull. Atomic Sci.* Feb. 54/2 If the French in their last ditch stand at Dien Bien Phu had appealed for U.S. nuclear aid, there would have been similar reactions. **1961** W. Vaughan-Thomas *Anzio* viii. 171 Some got as far as the Loyals manning their last-ditch line on the Lateral Road. **1961** M. Beadle *These Ruins are Inhabited* (1963) vi. 76 All that defeated them was a last-ditch appeal. **1971** *Times* 27 Apr. 9/5 Charlton himself surely was offside before McNab made his last ditch effort to recover the situation. **1973** *Times* 27 Aug. 3/1 Colonel Gaddafi, the Libyan leader, today began a last-ditch attempt to forge a full union with Egypt after his unexpected arrival in Cairo.

Lastex (læ·steks). Also **lastex.** The proprietary name of an elastic yarn formed from a combination of rubber (see also quot. 1968) with silk, cotton, or rayon, used in the manufacture of corsetry, etc.

1934 *Trade Marks Jrnl.* 17 Jan. 70/1 *Lastex.* Yarns and threads composed of a mixture of india-rubber and silk, the silk predominating. Dunlop-Revere Thread Co., Ltd., ..Birmingham. **1935** *Times* 12 June 15/6 Bathing suits are of woollen fabrics woven with lastex yarn. **1946** 'G. Orwell' in *Tribune* 8 Nov. 12/2 Someone has just sent me a copy of an American fashion magazine... Here are a few sample sentences..'Gentle discipline for curves in lacy lastex pantie-girdle'. **1951** J. D. Salinger *Catcher in Rye* xviii. 161 He wore those white lastex kind of swimming trunks. **1955** *Punch* 16 Mar. 348/3 Beautifully less! The ideal is now achieved with a caress of nylon, lastex, and lace. **1968** J. Ironside *Fashion Alphabet* 236 Lastex yarns... The rubber is now being replaced by synthetic elastomeric fibres, which are easier to wash and quicker to dry.

lasting, *vbl. sb.*^{3} (Further examples, not in *attrib.* use.)

1880 *Encycl. Brit.* XXI. 830/2 Lasting is a crucial operation, for, unless the upper is drawn smoothly and equally over the last, leaving neither crease nor wrinkle, the form of the boot will be bad. **1907** *Westm. Gaz.* 4 Nov. 8/4 This method of 'lasting' is new. **1968** J. Ironside *Fashion Alphabet* 126 The lasting, i.e. the making of the upper on the last.

lastness. (Further examples.)

a **1665** J. Goodwin *Filled with Spirit* (1867) 92 Lastness or worstness in estate or condition. **1927** 'E. Bramah' *Max Carrados Myst.* 72 Your account..is entirely based on the fact that you were the last... There stands the man we want,..only you and your lastness get between.

lasya (lā·sya). [Skr. *lāsya*.] A graceful style of female dancing in India.

1937 Subrahmanya Śāstri & Śrinivāsa Ayyaṅgār in Śaṅkara Āchārya *Saundarya-Lahari* 160 The Lāsya or female-dance and the Tāndava or male-dance, both being types of the same Nṛtya. **1967** Singha & Massey *Indian Dances* 23 Lasya is that element of the dance which is graceful and delicate and expresses emotions on a gentle level... Krishna's dance with the *gopees* (milkmaids) is in lasya. **1973** *Daily Tel.* 28 Mar. 15/1 What was most striking was her combination of the tandava (masculine) and lasya (feminine) in a way unique to her.

lat^2. [f. the first syllable of *Latvija* Latvia.] A unit of gold currency established by the state of Latvia in August 1922, with a par value of about 25 to the pound sterling, and discontinued in 1941.

1923 *Glasgow Herald* 23 June 10 The last Budget year was closed with a surplus of over 18,000,000 lats gold. **1928** *Daily Express* 29 Aug. 2/5 The Latvian Ministry of Finance has decided to place an order for five-lat silver pieces..with the Royal Mint, London. **1942** *Statesman's Year-Bk.* 1261 Latvia... The U.S.S.R. authorities abolished the lat currency on April 10, 1941.

lat^3. (Usu. in pl. *lats.*) Slang abbrev. of Latrine.

1927 W. E. Collinson *Contemp. Eng.* 92 At Salisbury Plain and Camberley in 1909/10 I learnt a number of camping expressions like..lats (*latrines*). **1940** M. Marples *Public School Slang* 112 Other synonyms [for lavatories] are rears, lats..and dubs. **1957** J. I. M. Stewart *Use of Riches* i. ii. 25 Turk says that conscientious objectors have to clean out the lats in lunatic asylums.

latania (lătẽⁱ·niă, lāta·niă). **1.** *U.S.* [Amer. Sp.] = *LATANIER 1.

1799 in F. Cuming *Sk. Tour Western Country* (1810) 336 Some..cabins [were] covered over with a shrub like a large fan, called latania. **1819** E. Dana *Geogr. Sk. Western Country* 238 On the..outer margin of the cane, the palmetto, or latania, fill the slope between the cane and the inundated lands.

2. [mod.L. (P. Commerson in A. L. de Jussieu *Genera Plantarum* (1789) 39), f. F. *latanier* the name used in Mauritius.] A fan palm of the genus so called, native to Mauritius and neighbouring islands.

1856 B. Seemann *Pop. Hist. Palms* 229 The Latanias are middle-sized trees. **1900** L. H. Bailey *Cycl. Amer. Hort.* II. 887/1 Latanias are tall, spineless palms. *Ibid.* 887/2 Latanias are essentially warmhouse palms. **1910** [see *GEONOMA]. **1966** E. J. H. Corner *Nat. Hist. Palms* vii. 162 (*caption*) The Borassoid *Latania*; male inflorescence.

latanier (‖latanʸe, lătæ·niəɹ). [Fr.] **1.** *U.S.* [See quot. 1939.] One of several fan palms found in the southern United States and central America, esp. the cabbage palmetto, *Sabal palmetto*.

[**1719** tr. *H. Joutel's Jrnl. Voy. Mexico* 14, I could see from the Ships..[an] Abundance of that Sort of Palm-Trees, in French call'd *Lataniers*.] **1827** *Western Monthly Rev.* I. 315 Palmetto, or latanier, peet, and long moss, add an aspect of novelty to the view. **1868** *Putnam's Mag.* I. 594/1 Here and there..is a 'latanier-hut' with adobe walls and a roof thatched with..palmetto. [**1939** W. A. Read in *Zeitschrift für franz. Sprache* LXIII. 1–11. 46 The usual term in Louisiana-French for this palmetto is *latanier*, m., a derivative of Carib *alattani*, a name recorded by Breton for a West Indian palm with fan-shaped leaves.]

2. The Mauritian name for several fan palms, esp. = *LATANIA 2.

1929 *Encycl. Brit.* XV. 108/1 The coco-nut palm, an importation [to Mauritius], the palmiste (*Palma dactylifera latifolia*), the latanier (*Corypha umbraculifera*) and the date-palm.

latch, *sb.*^{1} Add: **3. b.** *Electronics.* A logic circuit which retains whatever output state results from a momentary input signal until the application of a different signal to the same input point or the same signal to a different point. Also *latch circuit*.

1959 E. M. Grabbe et al. *Handbk. Automation, Computation, & Control* II. xvii. 6 A delay element is provided with external gating which enables it to hold information provided on the 'set' input until a 'reset' input of 1 occurs. This configuration..is sometimes referred to as a latch. **1962** Simpson & Richards *Physical Princ. Junction Transistors* xvi. 403 A variant of it [*sc.* the 'flip-flop'], known as the latch, is an asymmetric bistable multivibrator. **1971** J. H. Smith *Digital Logic* iv. 54 A push button might be pressed and at a certain time in a machine sequence the fact that the button was pressed may be needed to change the sequence. The latch circuit is therefore used to 'remember' that a push button has been pressed. *Ibid.*, A *1* signal applied momentarily to the set input of NOR 1 will make O/P_1 a *0* signal... To reset the latch a *1* signal is fed to the reset input which makes O/P_2 fall to *0*. As NOR 1 no longer has a *1* input, O/P_1 changes to *1*, thereby holding O/P_2 at the *0* level.

5. latch bolt (see quots.).

1909 *Cent. Dict. Suppl.*, *Latch-bolt*, any latch or door-bolt, controlled by a spring and having a beveled head which, when the door is closed, is pressed back by meeting the strike and is thrown out again when the door is shut: the common form of self-locking bolt. **1958** *Encycl. Locks* (J. Parkes & Sons Ltd.) 247 A spring bolt, called also a latch bolt, of a lock or latch is one which having been drawn in shoots out automatically as soon as the handle or key is released.

latch, *v.*^{1} **1.** Delete † *Obs.* and add later examples (senses 1 and 1 b) with *on*.

1937 *Esquire* Jan. 146/2 Dar's uh green sedan up front, uh fo' do' job. Latch on it 'n earn dis dime. **1940** *Sat. Even. Post* 13 Jan. 17/3, I latched onto a shark and killed it. **1946** B. Treadwell *Big. Bk. of Swing* 125/1 *Latch on*, grab on to. **1951** I. Shaw *Troubled Air* viii. 144 They're out for something of their own and they latch on to us. **1954** D. Riesman *Individualism Reconsidered* xiv. 220 [He] has latched on to American consumption know-how at its most garish. **1957** *New Yorker* 29 June 68/2 Mr. Kelly has latched on to a sound (indeed, indestructible) idea for keeping a film in motion. **1959** C. MacInnes *Absolute Beginners* 58 Hundreds of pure pink numbers.. who've latched on to the Welfare thing, but don't belong here. **1962** J. Wain *Strike Father Dead* 107 It was a long time before I could latch on to what was happening. Then I got it. **1968** *Listener* 27 June 837/1 When the doctor said, 'You're going to die, you'd better come back into hospital,' I said: 'Thanks for telling me. I'm going to latch on to life and I'm not coming back to hospital.' **1971** *Engineering* Apr. 41/3 The astute entrepreneurs are latching on to the idea. **1972** C. Drummond *Death at Bar* i. 36 Jarvis soon latched on to two portly dowagers—relatives of his hostess.

latch, *sb.*^{3}, var. Letch *sb.*^{1}

latch-key. Add to def.: Now usu. the key of a spring door-lock. Freq. allusive and *attrib.*, with reference to the use of a latch-key by a

younger member of a household (esp. one who comes home from school when his parents are still at work) or a lodger. (Earlier and later examples.)

1825 C. Mathews *Memorandum-Bk.* (ed. 2) 19 At last he recollected he had a latch key in his pocket. **1836** Dickens *Pickw.* (1837) xx. 199, I couldn't find the place where the latch-key went in. **1902** *Daily Chron.* 22 Aug. 3/6 At the beginning of the latchkey life everything looks delightful. **1905** *Ibid.* 17 Nov. 1/7 The names of 2,596 workmen in Devonport, known as latch-key voters,.. were restored to the occupiers' list. **1944** in *Amer. Speech* (1965) XL. 145 Latchkey children. **1945** Baker *Austral. Lang.* 158 The arrival of American servicemen in Australia produced..*latch-key kids* for children left at home on their own by mothers engaged in war industry. **1946** *Life* 8 Apr. 90/2 His was a latchkey existence. **1950** T. S. Eliot *Cocktail Party* i. iii. 70 Lavinia lets herself in with a latch-key. **1960** *Guardian* 23 Nov. 8/5 What happens to the 'latch-key children' in Germany? **1974** *Courier-Mail* (Brisbane) 14 June 11/5 One in every four school children was a latch-key child.

Also as *v.*, to open with a latch-key; to let oneself into (a house, etc.) by means of a latch-key.

1939 E. S. Gardner *Case of Rolling Bones* (1940) v. 60 Mason latch-keyed the office door to find Gertrude down on hands and knees scrubbing at the charred carpet. **1961** Wodehouse *Ice in Bedroom* xv. 119 As he latchkeyed himself into Peacehaven..there was a song on his lips.

latch-string. (Earlier and later examples.)

1791 in W. R. Jillson *Tales Dark & Bloody Ground* (1930) 109 The doors and the window shutters are..secured by stout bars on the inside with a latch-string of leather hanging out. **1859** *Trans. Illinois Agric. Soc.* III. 342 It is but another proof of the well known characteristics of the people of the west, that they are always to be found with 'their latch strings out'. **1937** V. D. Scudder *On Journey* iii. ii. 298 Especially at Commencement time, when the latch-string hangs out for returning alumnae.

late, *a.*^{1} Add: **2. d.** *late developer*: see *DEVELOPER e.

e. *late cut* (see Cut *sb.*^{2} 10 a and Cut *v.* 31 a), in Cricket, a cut, but with the actual stroke delayed until after the usual moment. Hence also as vb.

1887 F. Gale *Game of Cricket* 263 A splendid bat, back player, and great at a late 'cut'. **1906** [see *square-cut* s.v. Square *a.* 14]. **1912** J. B. Hobbs *Recovering the Ashes* 124 An off ball gave Mr. Trumper a chance to late cut one nicely to the cycle path. **1960** *Times* 3 June 21/2 He will.. late cut another in a way possible to few. **1963** *Times* 7 June 4/5 Padgett, after a delicate late cut for four.., was bowled off his pads. **1974** *Observer* 9 June 24/7 When Underwood came on, Gavaskar danced out to drive him straight, then lay back to late cut through the slips.

f. *late-tackle* v. trans. in Rugby and Association Football, to tackle (an opponent) illegally, when he is no longer in possession of the ball. Also as *sb.* So *late-tackling* vbl. sb.

1957 Late-tackling [see *BLATANT *a.* 2 c]. **1960** T. McLean *Kings of Rugby* xi. 198 Raureti for the second time in the match palpably late-tackled Young. **1962** *Times* 26 June 3/6 People who have seen the film are now satisfied that he was not late-tackled at Pretoria. **1971** J. B. G. Thomas *Roaring Lions* vi. 117 McNaughton swung him round to the ground without malice, but it was a late tackle, and Referee Pring..had no alternative but to award a penalty. **1974** *Times* 22 Nov. 11/1 England got bogged down..against a defence..in which a combination of offside tactics, obstruction and the cynical late tackle played a major role. This is what has come to be expected from visitors to Wembley.

g. Applied to a woman whose menstrual period has failed to occur at the expected time. *colloq.*

1962 J. Ludwig in R. Weaver *Canad. Short Stories* (1968) 2nd Ser. 255 Shirley, maybe you're late this month, eh, dollie? **1969** 'V. Packer' *Don't rely on Gemini* (1970) xiv. 119 Penny was two weeks late... There was a very good possibility that she was carrying Neal Dana's child. **1974** D. Fletcher *Lovable Man* i. 21 Linda realised that she was late... It was impossible to consult her family doctor.

3. Hence *late-afternoon*, *late-night*, used attrib. Also *late-late* (*show*, etc.). Also, further examples referring to persons, esp. in *late bird*. Further examples of fig. phr. *late in the day*.

1816 Jane Austen *Emma* I. xvii. 303 It was rather too late in the day to set about being simple-minded and ignorant. **1861** C. J. Lever *Day's Ride* xlvii. in *All Year Round* 23 Mar. 568/1 Rather late in the day, I take it, to ask who Bob Rogers is! **1885** 'Mark Twain' in *Century Mag.* Dec. 202/2 The damp, earthy, late-night smells. **1888** E. Bellamy *Looking Backward* v. 64 If I was inclined to wakefulness nothing would please him better than to bear me company. 'I am a late bird, myself,' he said. **1898** *Daily News* 30 June 6/3 His friend was what might be called a late man. The Duke [of Wellington], as everybody knew, was quite the reverse. The appointment was for eight o'clock in the morning... 'How can you manage to keep it?' 'Oh,' he replied, 'it's the easiest thing in the world. I shall take it the last thing before going to bed.' **1912** A. Huxley *Let.* 16 June (1969) 44 They are also on the point of putting up a war memorial, though none of the people who were in the war want it and it is now a little late in the day. **1944** D. Edwards in *Austral. Short Stories* (1951) 338 The street that glows with late-afternoon sun. **1956** B. Holiday *Lady sings Blues* (1973) xv. 124 If you're an American citizen and unless you go to bed early these nights, you're liable to see me on the late-late show. **1957** J. Osborne *Entertainer* iii. 31 He plays the piano in one of these late-night drinking places. **1965** *Listener* 16 Dec. 1012/1 Just when the smoke..has cleared ..Mr Roger Pemberton [tells]..why he hates the views of Miss Laski... Surely this is a bit late in the day? **1968** *Ibid.* 5 Sept. 306/2 For thousands of young people Peel has the only late-night show worth turning on. **1969** V. C. Clinton-Baddeley *Only Matter of Time* 9 He had never been a late bird... He was seldom out after twelve. **1971** *Black Scholar* Apr.–May 47/1 Some are sleeping because they're the late late TV show viewers. **1971** P. Purser *Holy Father's Navy* xxii. 109 We got them on to the late-afternoon plane to Zagreb. **1972** P. A. Whitney *Listen for Whisperer* ii. 22. 'I've brought her something from my father.'.. 'It's a bit late in the day for such a message.' **1973** *Amer. Speech 1969* XLIV. 277 An unusual example evolved on a late-night television talk show.

b. *late dinner*: esp. in Victorian society, the main evening meal, held later than the children's dinner.

1838 Mrs. Gaskell *Let.* 2 Dec. (1966) 38 Mr Bradford coming home to late dinner and so agreeable. **1873** L. Troubridge *Life amongst Troubridges* (1966) 11 Mrs Quick is the cook... She makes very good things for late dinner but not for our [*sc.* the children's] dinner. **1885** A. Edwardes *Girton Girl* I. vi. 136 The dinner-hour at Tintajeux was five, the 'late dinner' of Andros Bartrand's youth. **1941** Mrs. Belloc Lowndes *I, too, have lived in Arcadia* xviii. 350 Even now the presence of a child at late dinner would certainly occasion surprise, to almost any guest of that child's parents.

4. spec. *late-Victorian* adj.; **late blight**, a disease of potatoes caused by the fungus *Phytophthora infestans*; = *potato blight* (Potato *sb.* 7); **late wood**, a denser section of the annual ring of a tree, formed late in the growing season.

1905 *28th Ann. Rep. Connecticut Agric. Exper. Station* 379 Spraying will be of greater value, especially if early blight has been injurious or the late blight appears before the end of their season. **1909** B. M. Duggar *Fungous Diseases of Plants* x. 165 The late blight and rot of the potato is so generally known that frequently this malady is simply called the 'potato disease'. **1913** W. J. Locke *Stella Maris* iii. 26 Risca's room was transformed from late-Victorian solidity into early-Georgian elegance. **1918** E. J. Butler *Fungi & Dis. in Plants* viii. 277 The potato disease known as 'blight', 'late blight' or 'Irish blight' first attracted general attention in Europe in 1845. **1920** Beerbohm *And Even Now* 3 A tired suit of Late Victorian pattern. **1929** T. Thomson tr. *Büsgen's Struct. & Life Forest Trees* vi. 178 The difference between early and late wood consists in the former appearing more porous and open than the latter and often differently coloured. **1933** F. D. Heald *Man. Plant Dis.* (ed. 2) xvi. 419 The late-blight attacks and kills the tops of the potato plant and invades the tubers, causing either a *dry* or a *wet rot*. **1955** *Sci. News Let.* 11 June 381/1 In an effort to find clues that might help clear up the mystery surrounding late blight.. plant pathologists..have surveyed the world. **1969** J. Mander *Static Soc.* vi. 175 Teasingly sexless in the late-Victorian manner. **1972** *Sci. Amer.* May 92/1 Effect of smog on a Jeffrey pine that was growing in a forest near Los Angeles may be indicated by the narrowed rings and a reduced amount of latewood in the last nine years of the tree's life. **1975** *Nature* 10 Apr. 507/1 Whatever the cause, there is no indication that the width of the subsequent latewood was significantly altered.

6. *late model*, a recent model of a motor vehicle (usu., with hyphen, *attrib.*); *the late unpleasantness* (U.S.), the Civil War (see *UNPLEASANTNESS).

1917 G. Ade *Let.* 26 Apr. (1973) 63 Our own majestic work of art..has more late-model cars parked around it. **1973** R. Busby *Pattern of Violence* i. 15 They had found a late-model Ford Cortina XL unmistakably tooled up for violent crime.

late, *adv.* Add: **2.** *Proverb.* (Later examples.)

1852 C. M. Yonge *Two Guardians* xviii. 364 She obtained from Agnes some admiration for Caroline's conduct, though in somewhat of the 'better late than never style'. **1876** G. H. Lewes *Let.* 10 Jan. in Geo. Eliot *Lett.* VI. (1956) 211, I think that in the next number at any rate a bill might be inserted with effect—better late than never! **1950** G. Greene *Third Man* ix. 77 Oh, Mr. Dexter, we have been so anxious, but better late than never. **1954** A. Huxley *Let.* 16 Sept. (1969) 711, I am sorry your holiday will have to be postponed so long; but better late than never.

7. *late-acquired, -departing, -discovered, -enfranchised, -gone, -hatched, -houred, -learnt, -maturing* adjs.

1933 *Mind* XLII. 279 The physiological and psychological processes which *cause* it to be given may be as complex and as late-acquired as you please. **1923** Blunden *To Nature* 10 Late-departing yelps the fox. **1932** —— *Halfway House* 72, I have contrived that some most secret treasures Shall lie an age untouched, and late-discovered Shall be the source of hope and peace. **1923** D. H. Lawrence *Birds, Beasts & Flowers* 57 About your feet spontaneous aconite..and purple husband-tyranny Enveloping your late-enfranchised plains. **1937** Blunden *Elegy* 30 Child of late-gone gale. **1906** *Westm. Gaz.* 16 Oct. 10/1 Only a few stragglers—most of them birds with their late-hatched broods—are left. **1908** *Daily Chron.* 10 Jan. 4/6 The pantomime crowd is a very good crowd after all, late-houred and not without failings, perhaps, but generous..to a fault. **1865** G. M. Hopkins *Poems* (1948) 36 That yield That I may win with late-learnt skill uncouth. **1936** Auden *Look, Stranger!* 29 On their behalf guard all the more This late-maturing Northern shore.

lateener. (Earlier and later examples.)

1873 *Young Englishwoman* Oct. 515/2 'Here comes a lateener,' says the captain, as a light vessel rigged in the style familiar to mariners of the Mediterranean,..takes a flight up the Broad. **1953** C. S. Forester *Hornblower & Atropos* 245 The lateener's heading this way.

latency. Add: **b.** *Psycho-analysis.* (See quot. 1934.) Freq. *attrib.*, esp. in *latency period*.

1910 A. A. Brill tr. *Freud's Three Contrib. to Sexual Theory* ii. 38 It is during this period of total or at least partial latency that the psychic forces develop which later act as inhibitions on the sexual life. *Ibid.* ii. 39 Sexual activity remains throughout the whole duration of the latency period until the reinforced breaking through of the sexual impulse in puberty. **1913** E. Jones *Papers on Psycho-Anal.* ii. 26 A period of latency follows, usually from the fifth to the tenth years, when the process of sublimation is at its highest activity. **1934** H. C. Warren *Dict. Psychol.* 150/1 *Latency period*, the period of life between the ages of 4 or 5 and ca. 12 years, which separates the infantile or pregenital sexuality from the beginning of puberty or genital sexuality and in which the sexual manifestations are as a rule less prominent. **1949** J. Strachey tr. *Freud's Three Ess. Theory of Sexuality* ii. 57 It is from Fliess that I have borrowed the term 'period of sexual latency'. **1960** *Encounter* Jan. 80/2 The life of latency groups in Britain. *Ibid.* 81/1 These beliefs and customs are passed on entirely within the latency world without any intervention of adults. **1967** B. Russell *Autobiogr.* I. ii. 38, I remember a very definite change when I reached what in modern child psychology is called the 'latency period'. **1968** E. Erikson *Identity: Youth & Crisis* iv. 156 In postulating a 'latency period' which precedes puberty, psychoanalysis has given recognition to some kind of psychosexual moratorium in human development.

2. a. Delay between a stimulus and a response, esp. in muscle; a latent period.

1882 *Proc. R. Soc.* XXXIII. 463 If, after a muscle has been powerfully extended, and while it is returning, by reason of its elasticity, towards its normal condition, a stimulation be applied, the latency may become as short as the 1/200–1/400 second. **1932** *Jrnl. Physiol.* LXXIV. 17 The general interest in the problem of latency increased when it was found, that while recording the action current of the muscle the latent period was absent, or seemed to be absent. **1951** H. Davson *Textbk. Gen. Physiol.* xvii. 483 The latency, as ordinarily recorded, is thus of the order of 3·5 msec. at 23°C; if, however, we take as a measure of the mechanical latent period the time between the stimulus and the moment when the tension [in the muscle] begins to rise from its minimum value, the period is 3·0 msec. **1963** *Jrnl. Pediatrics* LXII. 724 Cry latency is defined as the time which elapses between the moment of painful stimulation and the onset of crying. **1973** *Jrnl. Genetic Psychol.* CXXII. 177 The child was presented with a standard stimulus along with a number of comparison stimuli and told to point to the one that was the same as the standard... *E* recorded, to the nearest half-second, latency to the first response.

b. *Computers.* More fully *latency time*. The delay before a transfer of data begins following an instruction for its transfer, esp. to or from a rotating storage device.

1954 *First Gloss. Programming Terminol.* (Assoc. Computing Machinery) 11 *Latency*, in a serial storage system, the access time less the word time, e.g. the time spent waiting for the desired location to appear under the drum heads or at the end of an acoustic tank. **1961** P. Siegel *Understanding Digital Computers* xii. 258 The access time consists of the latency time plus the transfer time. The latency represents the amount of time it takes to find the chosen address. **1970** O. Dopping *Computers & Data Processing* x. 145 If the drum makes 3,000 revolutions per minute, each revolution takes 20 milliseconds, and the average latency time becomes 10 milliseconds.

La Tène (latẽ·n). The name of a district at the east end of Lake Neuchâtel, Switzerland, where archæological finds were made, used esp. *attrib.* to denote a culture (fl. *c.* 3rd century B.C.) of the second Iron Age of central and west Europe, and objects found there.

[**1866** J. E. Lee tr. *Keller's Lake Dwellings* 239 A La Tène, Near Marin (Lake of Neuchâtel). This station was known and mentioned in my first report on the lake dwellings to the Society of Antiquaries at Zürich, in 1858. *Note.* The terms Tène (shallow) and Ténevière (submerged hillock) are provincialisms of the fishermen in the lake of Neuchâtel, and Professor Desor in his late work 'Les Palafittes', derives them from the Latin '*tenuis*'.] **1890** R. Munro *Lake-Dwellings of Europe* iv. 278 In making a section through the La Tène elevation there is first encountered a bed of water-worn gravel and sand. **1905** H. B. Walters *Hist. Anc. Pott.* II. xxiii. 502 It is interesting for its close relation to the older La Tène pottery. **1917** [see *BEADED *ppl. a.* 4 b]. **1932** *Antiquity* VI. 198 *La Tène.* A trading post on Lake Neuchâtel gives its name to certain cultures of the *Second Iron Age*. **1935** *Discovery* Apr. 102/1 The three tumuli..proved to be Early Iron Age in date, one of them yielding a small bone plaque with advanced La Tène ornament. **1943** J. & C. Hawkes *Prehist. Brit.* v. 94 Already when the first Hallstatt settlers were arriving in this country their kinsfolk in eastern France and south Germany were building up a new culture, that known as La Tène, which came to represent the height of early Celtic achievement. **1961** *Antiquaries Jrnl.* XLI. 44 Barrows, or burials, within square-ditched inclosures have recently been recognized as characteristic of the La Tène cultures both in the Rhineland and in France. **1962** *Times* 29 Mar. 15/5 The British Celts had evolved..their own superb, non-naturalistic late-La Tène art. **1970** Bray & Trump *Dict. Archaeol.* 130/2 An art style with La Tène elements persisted into the Early Christian period.

latensification (lẽⁱte:nsifikẽⁱ·ʃən). *Photogr.* [f. Laten(t *a.* + Inten)sification.] Intensification of an existing latent image on a photographic film or plate by treatment with a chemical, prolonged exposure to light, or other means. Cf. *HYPERSENSITIZATION b.

1940 *Amer. Cinematographer* XXI. 499/2 With no more equipment than a panchromatic safelight, any photographer can increase the speed of a film from two to four times right in his own darkroom. 'Latensification', the name describing this new process, is an outgrowth of the research being done by Du Pont Film Manufacturing Corporation on high speed 35 mm. films. **1948** *PSA Jrnl.* XIV. 675 (*heading*) Latensification studies with sodium perborate. **1956** [see *HYPERSENSITIZATION b]. **1973** H. W. Cleveland in W. Thomas *SPSE Handbk. Photogr. Sci. & Engin.* vi. 417 Latensification is primarily an increase in the rate of development and therefore more effective with development well below γ_∞.

latent, *a.* Add: **a.** *latent partner*, one whose name does not appear as a member of a firm or company.

1913 *Act 3 & 4 Geo. V* c. 20 § 90 Any latent partner of a company whose estates have been sequestrated.

d. (Further examples in *Ophthalm.* and *Biol.*) *latent virus*, a virus causing no apparent disease in a plant or animal, but capable of producing disease in another to which it is transmitted.

1886 *N.Y. Med. Jrnl.* 4 Dec. 626/2 Heterophoria may, like hypermetropia, be partly or entirely latent. **1931** *Phytopathology* XXI. 593 A virus remaining latent or producing a mosaic in some varieties may cause a well-defined necrotic effect in others. **1937** *Science* 20 Aug. 179 (*title*) A latent virus of lily. **1950** K. M. Smith *Introd. Study of Viruses* ii. 17 We have seen already a good example of the 'lighting up' of a latent virus when discussing swine influenza... The virus is latent in its intermediate host, the swine lungworm, and to induce infection in the pig it must be rendered active by the application of a provocative stimulus. **1951** *Nature* 30 June 1061/1 The presence of latent or 'silent' viruses in plants and other organisms is not, of course, new. **1954** S. Duke-Elder *Parsons' Dis. Eye* (ed. 12) xxix. 483 In all types [of squint] if the fusion mechanism is well-developed and the defect slight, visual alignment may be maintained in normal circumstances by a continued effort of fusion: the squint is then latent and can only be made manifest when fusion is impossible (as by covering up one eye). **1962** W. Carter *Insects in Relation to Plant Dis.* viii. 329 Latent viruses are often important components of virus complexes.

h. Biol. *latent period*: a period between a stimulus and a response, esp. in a muscle or an irradiated individual. (See also sense d and *LATENCY 2 a.)

1877 M. Foster *Text Bk. Physiol.* i. ii. 37 A phase antecedent to any visible alteration in the muscle..during which invisible preparatory changes are taking place in the nerve and muscle, is often called the 'latent period'. **1926** L. Hogben *Compar. Physiol.* viii. 140 The rate of conduction can be determined directly by observing the difference in the latent period of muscular contraction, when a nerve-muscle preparation is stimulated at points along the nerve separated by a measured distance apart. **1933** O. Glasser *Sci. of Radiol.* xviii. 319 Should the criterion be the first reaction which can be observed under the microscope in the living cell, the latent period will be very short. **1947** *Radiology* XLIX. 361/2 The incidence of bone tumors was approximately proportional to the dose administered, and the latent period—in no case less than about 200 days—increased gradually with decreasing dose. **1966** *McGraw-Hill Encycl. Sci. & Technol.* VIII. 638/1 Between the stimulus and the first tension development there is a brief latent period of a few thousandths of the twitch duration.

i. Math. *latent root*: a scalar quantity λ which, when subtracted from each of the elements in the principal diagonal of a square matrix *A*, makes the determinant of the resulting matrix equal to zero (and so is a solution of the equation $Ax = \lambda x$, where x is a column vector).

1883 J. J. Sylvester in *Phil. Mag.* XVI. 267 It will be convenient to introduce here a notion (which plays a conspicuous part in my new theory of multiple algebra), viz. that of the latent roots of a matrix—latent in a somewhat similar sense as vapour may be said to be latent in water or smoke in a tobacco-leaf. If from each term in the diagonal of a given matrix, λ be subtracted, the determinant to the matrix so modified will be a rational integer function of λ; the roots of that function are the latent roots of the matrix; and there results the important theorem that the latent roots of any function of a matrix are respectively the same functions of the latent roots of the matrix itself: *ex. gr.* the latent roots of the square of a matrix are the squares of its latent roots. **1958** R. V. Andree *Sel. Mod. Abstract Algebra* ix. 195 In quantum mechanics and elsewhere, the terms latent roots, proper value, eigenvalue, and eigenwerte are often used in place of characteristic root. **1968** E. T. Copson *Metric Spaces* viii. 117 The condition..is known to imply that all the latent roots of *A* lie in $|z| < 1$. Hence the latent roots of $I - A$ lie in $|z - 1| < 1$. Thus $z = 0$ is not a latent root of $I - A$, so that $I - A$ is non-singular.

j. *latent* (*finger-*)*print* = *LATENT *sb.*

1923 J. A. Larson *Single Fingerprint Syst.* i. 8 Take to your office all removable objects with visible or latent prints. *Ibid.* 9 Various powders, fumes, and solutions are recommended and have been used for the development and preservation of latent prints. **1937** *Discovery* Feb. 56/2 When a latent print is discovered, the area can receive a more liberal fuming to yield the maximum contrast. The amount of time required for fuming a single latent image is variable. **1956** 'E. McBain' *Cop Hater* (1963) xix. 145 The tech crew dusted the latent fingerprints. **1962** M. Proctor *Devil in Moonlight* x. 102 'What's he expert on?' 'Latent fingerprints.' *Ibid.* xiv. 145 What they call latent prints, not visible to the human eye. **1974** G. F. Newman *Price* viii. 242 Latent prints brought out on the non-absorbent surfaces with grey powder for photographing.

k. Psychol. *latent learning*, learning that has taken place without conscious purpose and that is not manifested until there is a goal to be achieved.

1929 H. C. Blodgett in *Univ. Calif. Publ. Psychol.* IV. 122 Do these drops in errors which come after the introduction of reward indicate that something to be called a latent learning developed during the non-reward period—a latent learning which made itself manifest after the reward had been presented? *Ibid.* 133 During the non-reward period, the rats were developing a *latent* learning of the maze which they were able to utilize as soon as reward was introduced. **1938** R. S. Woodworth *Exper. Psychol.* vi. 137 'Latent learning', i.e., learning that is not revealed by the animal's path through the maze, until food has been found in the food box. **1957** *New Biol.* XXIV. 123 If rats, not deprived of any primary bodily needs, are put in a maze, they explore it... If such rats are later trained to run from one part of the maze to another to get food, they learn this task more quickly than similar rats which have not previously experienced the maze. This consequence of exploration is called latent learning. **1956** E. R. Hilgard *Theories of Learning* (ed. 2) vi. 214 This second phase demonstrates genuine latent learning of the true path. **1968** Chaplin & Krawiec *Syst. & Theories Psychol.* (ed. 2) vii. 277 Latent learning is hidden learning which goes on unobserved but which, under certain conditions, can be revealed in performance.

l. Sociol. *latent function*, a function which exists unrecognized within a social attitude or action and which will produce results that have not been foreseen.

1949 R. K. Merton *Social Theory* i. 51 Latent functions..being those which are neither intended nor recognized. **1961** M. Spiro in B. Kaplan *Studying Personality* ii. 108 Latent functions are those consequences which—whether intended or unintended—are not recognized by them [*sc.* members of a society].

m. *Sociol.* and *Statistics.* Applied to certain attributes, structures of relations, and the like (see quots. 1957).

1950 P. F. Lazarsfeld in S. A. Stouffer *Measurement & Prediction* x. 362 (*heading*) The logical and mathematical foundation of latent structure analysis. *Ibid.*, The latent structure approach to the treatment of itemized tests. **1952** Goode & Hatt *Methods in Social Res.* xvii. 286 One alternative..was to abandon the factorial approach and in doing so abandon the concept of a latent-attitude continuum. *Ibid.* 295 Another problem posed by latent-structure analysis..is the fact that its computations are both arduous and complex. **1957** Kendall & Buckland *Dict. Statistical Terms* 158 *Latent structure*,..a structure expressed in terms of variates or variables which are 'latent' in the sense of not being directly observable. Certain econometric relations (e.g. in terms of 'utility') are of this type. *Ibid.*, *Latent variable*, a variable which is unobservable but is supposed to enter into the structure of a system under study, such as demand in economics or the 'general' factor in psychology. Unobservable quantities such as errors are not usually described as latent. **1961** J. Rothenberg *Measurement Social Welfare* iv. xi. 290 The spatial arrangement of alternatives is a uni-dimensional scale of some single 'latent attribute' calibrated by the particular alternatives. **1966** B. S. Phillips *Social Res.* xi. 176 Latent structure analysis may prove to be of particular value in bringing to light a systematic set of assumptions or propositions about the probabilistic relationships between manifest data and latent structures. **1968** Lazarsfeld & Henry *Latent Structure Analysis* ii. 15 It is necessary to make some assumptions about the nature of what we called the 'latent variable'. *Ibid.* iii. 47 Once a latent structure model has been specified, with accounting equations..relating the latent probabilities to the manifest probabilities, we must ask whether those equations can be solved uniquely.

latent (lẽⁱ·tĕnt), *sb.* [f. the adj.] A kind of fingerprint, invisible to the naked eye.

1923 J. A. Larson *Single Fingerprint Syst.* ix. 186 Several latents were left at the house which had been burglarized. **1937** *Discovery* 57/2 This method [of fingerprint detection] does not result in the destruction of the latent. **1939** E. S. Gardner *D.A. draws Circle* (1940) iv. 48 'Any latents, Bob?' the sheriff asked. **1973** A. Hunter *Gently French* iii. 23 The recognizable latents were either Quarles' or off-record, probably innocent.

later, *a.* and *adv.* Add: **B.** *adv.* Also used as a farewell, representing *I'll see you later.* (*U.S. slang.*)

1954 *Time* 8 Nov. 42 *Later*, catchall word for 'I'll be seeing you'; also used at the end of letters. **1955** L. Feather *Encycl. Jazz* x. 346 *Later*, parting phrase, short for 'I'll see you later'. **1972** J. Maryland in T. Kochman *Rappin' & Stylin' Out* 214 The players all started heading for the door, stating, 'Peace,' 'Later,' 'Hat Time,' 'I'm in the wind,' etc.

lateral, *a.* and *sb.* Add: **A.** *adj.* **1. b.** *lateral thinking*: a way of thinking which seeks the solution to intractable problems through unorthodox methods, or elements which would normally be ignored by logical thinking.

1966 *London Life* 22 Oct. 20/3 He [*sc.* Dr. Edward De Bono] divides thinking into two methods. One he calls 'vertical thinking'—that is, using the processes of logic, the traditional-historical method... The other type he calls 'lateral thinking', which involves disrupting an apparent sequence and arriving at the solution from another angle. **1967** E. De Bono *Use of Lateral Thinking* 5 Some people are aware of another sort of thinking which..leads to those simple ideas that are obvious only after they have been thought of. This book is an attempt to look at this sort of thinking... For the sake of convenience, the term 'lateral thinking' has been coined to describe this other sort of thinking; 'vertical thinking' is used to denote the conventional logical process. **1972** *Observer* 5 Nov. 20/6 Don't trade on the argument that women have special qualities (intuition, perception, lateral thinking, or what have you) which are denied to men.

3. a. Also const. *to*. (Further examples.) Also, **lateral line**, in fishes and certain amphibia, a system of organs of sensory perception, arranged in a row along the sides of the body; also *attrib.*; **lateral plate**, in the early stages of vertebrate embryos, the ventral part of the mesoderm, from which certain internal organs develop.

1870 Lateral line [in Dict.]. **1874** Foster & Balfour *Elem. Embryol.* iii. 54 A transparent longitudinal line makes its appearance on either side of the notochord along the line of junction of the lateral with the vertical plate. **1913** *Gray's Anat.* (ed. 18) 242 Lateral to the foramen ovale is the foramen spinosum. **1914** W. E. Kellicott *Outl. Chordate Devel.* iv. 278 The lateral plate [of the chick] is separated into somatic and splanchnic layers by the extra-embryonic cœlom. **1926** J. S. Huxley *Ess. Pop. Sci.* xvii. 191 The special sense-organs for perceiving low-frequency vibrations in water which, like a herring or any other fish, it [*sc.* the tadpole] carries on a 'lateral line' along its flank. **1959** A. Hardy *Fish & Fisheries* ii. 24 We may not yet know all the functions of the lateral line system. **1964** H. W. Manner *Elem. Compar. Vertebr. Embryol.* ix. 150 The mesoderm..was in three distinct, potentially different, portions: a medially located somite, an intermediate mesoderm, and a lateral plate mesoderm. **1968** C. Osborne tr. *Stenuit's Dolphin* (1969) v. 91 Fish perceive the variations in water pressure through a biological system known as the lateral line. **1972** *Nature* 31 Mar. 233/1 Electrodes..were placed at bilaterally symmetrical points over the left and right hemispheres,.. 2 mm lateral to the sagittal suture and 1 mm posterior to bregma.

i. *Phonetics.* Of a consonant: formed by partial closure of the air-passage by the tongue, which is so placed as to allow the breath to escape at one or both sides of the point of contact (e.g. English *l*). Also as *sb.*

1891 L. Soames *Introd. Phonetics* 34 Some persons..let the breath escape on one side only, so it seems better to call it [*l*] a *lateral* or *side* consonant. **1899** W. Rippmann *Elem. Phonetics* § 91. 72 For the *l* sounds the narrowing is between the side rim or rims of the tongue and the side teeth (*lateral* formation). **1903** Scholle & Smith *Elem. Phonetics* 75. § 148 In the articulation of 'l' both sides (or only one side) of the tongue form a narrow passage with the molars and side gums, hence the term *lateral* articulation. **1933** L. Bloomfield *Lang.* vi. 97 In some of these the tongue actually touches the roof of the mouth, but leaves enough room at one or both sides for the breath to escape without serious friction-noise; such sounds are *laterals*, of the type of our [l]... In unvoiced laterals, which occur in Welsh and in many American languages, the friction-noise of the breath-stream is more audible than in unvoiced nasals. **1957** *Essays in Crit.* VII. 128 'Woolly' is composed of two vowels, a semi-vowel and a voiced lateral consonant.

j. Applied (orig. in *lateral cut*) to (the cutting of) gramophone records in which the undulations are cut in the plane of the record by the side-to-side movement of the recording stylus, and hence to equipment and techniques involved in this. Opp. *hill and dale* (*HILL *sb.* 1 b).

1917 *Sci. Amer.* 27 Oct. 307 While many of these patents are more in the nature of slight refinements, particularly means of twisting the mounting of the reproducer so that it can be used for hill-and-dale and for lateral cut records at will, a few of them represent a genuine effort to improve the tonal qualities of the conventional disk phonograph. **1934** *Amer. Speech* IX. 312/2 The records will be.. both of the ordinary lateral-cut type and of the new and superior hill-and-dale, long-playing, unbreakable kind. **1935** H. C. Bryson *Gramophone Record* i. 16 From 1900 onwards there existed three types of record: (*a*) Discs with hill and dale cut... (*b*) Discs with a lateral cut... (*c*) Wax cylinders with hill and dale cut. **1942** *Proc. IRE* XXX. 356/2 (*heading*) Frequency characteristics for lateral recording. **1966** *McGraw-Hill Encycl. Sci. & Technol.* IV. 240/1 Modern lateral pickups are of the crystal, ceramic, magnetic, or dynamic type. *Ibid.* 241/2 A lateral disk record. **1968** *Times* 29 Nov. 11/5 A decade later Emil Berliner saw the advantages of a flat disc and also developed the technique of the lateral cut disc. **1975** *Hi-Fi Answers* Feb. 58/2 Test records used..were those by EMI..and Decca's own lateral cut record, LXT 5346. *Ibid.* 59/2 Modulation levels..which upset it at 8 kHz (for both vertical and lateral cuts) and 18 kHz (for lateral cuts only).

B. *sb.* **3.** *N. Amer. Football.* (See quot. 1971.) Also as vb., to make such a pass.

1934 Webster, *Lateral*, a lateral pass. **1949** *Richmond* (Va.) *Times-Despatch* 10 Oct. 13/2 The last carried 27 yards as Quarterback Ben Raimondi threw a lateral to his left to End Cotton Howell. **1961** Webster, *Lateral*, to throw a lateral pass. **1970** *Globe & Mail* (Toronto) 28 Sept. 18/4 He lateralled to Joe Hernandez, who ran the last 45 yards of a 59-yard touchdown play. **1971** L. Koppett *N.Y. Times Guide Spectator Sports* ii. 69 *Lateral*, a pass to a teammate that does not travel forward. **1972** J. Mosedale *Football* iii. 40 The Bronk passed to Bill Hewitt who lateraled to Bill Karr for a 36-yard score. **1974** *Los Angeles Times* 13 Oct. III. 7/1 Greene ran 11 yards to the Badgers' 5-yard line and lateraled to Baschnagel, who sped into the end zone.

laterality. **a.** Delete † *Obs.* and add later examples of the sense: (right- or left-)sidedness; *spec.* the dominance of the right- or the left-hand member of a pair of bodily organs as regards a particular activity or function (such as the hands in writing, or the cerebral hemispheres in controlling speech).

1926 *Brit. Jrnl. Exper. Biol.* III. 317 The symmetry of right or left limbs will depend on the direction of growth.., and from a proximal cut surface an appendage of reversed laterality may thus originate. **1927** *Biometrika* XIX. 181 There is no evidence whatever of even a correlation between ocular and manual lateralities to say nothing of a master eye determining which is the master hand. **1937** S. T. Orton *Reading, Writing & Speech Probl. in Children* i. 48 We have no guide..as to which is the dominant hemisphere except the 'laterality' of the individual, that is, his handedness, eyedness and footedness. **1950** *Brain* LXXIII. 168 In a systematic study of 100 cases of brain injury, these authors report varying degrees of deformation of the [visual] coordinates in patients with unilateral lesions of either hemisphere. They state that the direction and extent of deformation varied simply with the laterality of the lesion; deviations in right-sided cases were in an anti-clockwise direction..; in cases with left-sided lesions, they were in the opposite direction. **1964** *Dissertation Abstr.* XXIV. 3423/2 Predictions of laterality of cerebral hemisphere lesions can be made with a fairly high degree of confidence for those patients with well-lateralized, rapidly expanding lesions. **1964** M. Critchley *Developmental Dyslexia* viii. 51 Mixed laterality was then imagined to be a factor of special importance in dyslexics who might, for example, prove to be left-eyed, right-handed and left-footed. **1967** *Biol. Abstr.* XLVIII. 8522/2 (*heading*) Laterality in the use of the forepaws in cat. **1971** *Nature* 23 Apr. 524/1 Although usually used when speech dominance is questionable, amobarbital testing is also indicated in persons with apparently well established laterality when they are to undergo commissural section.

c. *Phonetics.* Lateral articulation (cf. *lateral *a.* 3 i).

1953 C. E. Bazell *Ling. Form* 41 A voiced (or voiceless) character of *l* is not regarded as relevant even in languages in which voice is functionally discrete, on the grounds that it is not of functional relevance in combination with laterality.

lateralization (læ:tĕrăləizēi·ʃən). [f. *lateraliz(e *v.* + -ation (or Lateral *a.* and *sb.* + -ization).] Laterality, esp. of cerebral activity; the property of being lateralized.

1950 Penfield & Rasmussen *Cerebral Cortex of Man* ii. 42 (*heading*) Table showing lateralization of sensory responses..in the face area. **1954** *Brain* LXXVII. 526 It is hardly possible to trace familial tendencies in the cerebral lateralization of language function. **1960** O. L. Zangwill *Cerebral Dominance* ii. 5 Conrad..regards left-handedness as in itself bound up with incomplete lateralization of higher cerebral function. **1964** *Dissertation Abstr.* XXIV. 3423/1 (*heading*) Lateralization of lesions of the cerebral hemispheres. **1971** *Nature* 23 Apr. 524/1 When brain surgery which may affect language mechanisms is being contemplated, lateralization of speech is sometimes ascertained before operation. **1973** *Sci. Amer.* Apr. 76/3 Although the relation between speech lateralization and hand preference is not perfect, the high incidence of both left-hemisphere control of speech and right-hand preference is probably not coincidental.

lateralize (læ·tĕrăləiz), *v.* [f. Lateral *a.* + -ize.] **1.** *trans.* To move or displace to the side; to render lateral. *rare*$^{-1}$.

1903 *Therapeutic Gaz.* 15 Feb. 74/2 The woven coudé catheter or soft-rubber catheter should be chosen to measure the urethral length,..since the bend..enables it more readily to surmount or pass projections on the floor,..providing the elbow tip is lateralized.

2. *pass.* To be largely under the control of the left- or the right-hand side of the brain.

1954 *Brain* LXXVII. 533 Milner reports visual-spatial defects consistently more severe in right temporal lobe excisions than left. She suggests that this is a function which is usually lateralized on the right. **1970** M. S. Gazzaniga *Bisected Brain* vi. 116 The neural organization required for spoken language is usually lateralized to one cerebral hemisphere. *Ibid.* 85 Position sense is..crisply lateralized.

lateralized, *ppl. a.* Add: **b.** Of consonants (cf. *lateral *a.* 3 i).

1968 P. M. Postal *Aspects Phonol. Theory* iv. 82 There are no languages with only Lateralized consonants, although languages with no Lateralized consonants are not uncommon. **1969** *Word* XXV. 19 It would seem to be a simple solution to interpret these Yakur sounds as lateralized plosives.

lateralward(s (læ·tĕrălwəɹd, -wəɹdz), *adv.* *Anat.* [f. Lateral *a.* + -ward, -wards.] Laterally; to or from the mesial plane of the body.

1913 *Gray's Anat.* (ed. 18) 236 Extending lateralwards ..on either side is the superior nuchal line. **1967** Lateralward [see *geniculum].

laterite. For 'rock' read 'clayey substance' and add to def.: and in other tropical and sub-tropical regions, which is soft when first dug but hardens irreversibly to the consistency of rock when exposed to the air. **b.** Applied *loosely* to various reddish or iron-rich surface materials in the tropics and sub-tropics. **c.** *Soil Science.* Any soil or soil horizon characterized by a high proportion of sesquioxides, esp. of aluminium and iron, and an unusually low proportion of alkali metals, alkaline earths, and combined silica, such as occurs as a product of chemical weathering in hot climates with periods of abundant rain; *esp.* one which hardens on exposure to air or is derived from one already exposed. (Add additional examples.)

The use and proper application of the word have been the subject of much debate.

1807 F. Buchanan *Journey from Madras* II. xii. 441 What I have called indurated clay is not the mineral so called by Mr. Kirwan. It..is one of the most valuable materials for building. It is diffused in immense masses, without any appearance of stratification, and is placed over the granite that forms the basis of Malayala... As it is usually cut into the form of bricks.., in several of the native dialects, it is called the brick-stone (*Itica cullu*). Where, however, by the washing away of the soil, part of it has been exposed to the air, and has hardened into a rock, its colour becomes black, and its pores and inequalities give it a kind of resemblance to the skin of a person affected with cutaneous disorders; hence in the Tamul language it is called *Shuri cull*, or itch-stone. The most proper English name would be *Laterite.* **1893** R. D. Oldham *Man. Geol. India* (ed. 2) xv. 385 The origin of laterite is still wrapt in obscurity. *Ibid.*, According to some geologists this laterite [of Buchanan] is in reality a soil and formed by the direct decomposition *in situ* of the underlying rock. **1898** *Agric. Ledger* (Calcutta) V. ii. 34 If..it is difficult for the Geologist to decide what is 'laterite', it becomes practically impossible for the agriculturist to say what is a 'laterite soil'. Those 'laterite soils', that is, soils lying on or adjacent to what had every appearance of being laterite rock, which I have seen, had all a bright red appearance when dry; but as will be seen when discussing the analyses of the samples.., some at least of these are probably not true laterite. **1906** *Daily Chron.* 24 Aug. 6/5 The soil and the climate of Seychelles are evidently favourable to the growth of Para rubber, which thrives even in laterite soils where no other plants are at present growing. **1909** *Geol. Mag.* Decade V. VI. 431 The term 'laterite' has been used, in the Malay Peninsula at least, for many years by a large body of engineers for what are essentially masses of iron oxide replacing portions of weathered rock and filling fissures in such rocks near the surface. This (Malayan) laterite..is largely used for public works. **1910** *Ibid.* VII. 444 The foregoing representatives of the class of more or less ferruginous and aluminous deposits which in the Guianas..have been termed 'laterite' do not possess, except in the case of the concretionary ironstones, the property..of 'setting' or hardening on exposure to the atmosphere. Parts of them agree to some extent with what has been laid down as the modern scientific qualification for a rock to be termed 'laterite'—the fact that they are 'essentially characterized by the presence of free hydrate of alumina'. **1911** *Ibid.* VIII. 565 Laterite (or rather some varieties of it) is formed by a process..by which certain rocks undergo superficial decomposition, with the removal in solution of combined silica, lime, magnesia, soda, and potash, and with the residual accumulation..of a hydrated mixture of oxides of iron, aluminium, and titanium, with, more rarely, manganese. *Ibid.*, The property of hardening on exposure to the air is characteristic of many varieties of laterite, but it is not essential property. **1927** *Jrnl. Agric. Sci.* XVII. 546 It is..suggested that where the silica/alumina ratio in the clay fraction [of a soil] falls below 2·0 the soil should be described as 'lateritic', and where this ratio falls below 1·33 the soil should be described as laterite. **1932** G. W. Robinson *Soils* xiii. 279 Many of the descriptions of the supposed laterites merely relate to red soils... It is now generally agreed that the terms laterite and lateritic should be restricted to materials characterised by excess of sesquioxides. **1949** *Publ. Inst. Nat. pour l'Étude Agron. du Congo Belge* Sér. Sci. No. 46. 7 The word 'laterite' is used by us for the sesquioxide-rich, highly weathered clayey materials that change irreversibly to concretions, hardpans, or crusts, when dehydrated, and for the hardened relicts of such materials, more or less mixed with entrapped quartz and other materials. **1966** D. Forbes *Heart of Malaya* iii. 35 It [*sc.* a bungalow] was set on a hillside..at the end of a three-mile estate road of red laterite. **1967** *Nat. Geographic* July 110/2 [Thailand.] A foreman said this was laterite, a low-grade iron ore, good for making roads and filling land. **1970** E. M. Bridges *World Soils* ix. 72/1 Laterite..is formed by an accumulation in the soil of sesquioxidic material.

lateritic, *a.* Add: (Further examples.) Also, in *Soil Science*, applied to a soil that is not regarded as laterite but approaches it in composition.

1927, 1932 [see *laterite]. **1938** *U.S. Dept. Agric. Yearbk.* 974 Soils in which the laterization process has become markedly evident but has not reached completion are known as lateritic soils... They are very important in the West Indies... These soils are characterized by a colloidal fraction whose molecular ratio of silica to alumina is approximately 2. **1946** Lutz & Chandler *Forest Soils* xi. 403 The laterization process, in conjunction with podzolization, has resulted in the development of six zonal soils in the forested warm-temperature and tropical regions: (1) laterites, (2) reddish-brown lateritic soils, (3) yellowish-brown lateritic soils, (4) red podzolic soils, (5) yellow podzolic soils, and (6) terra rossa. **1952** L. M. Thompson *Soils & Soil Fertility* xvii. 285 The clay soils which he studied..were lateritic in nature (high percentage of kaolinite or high in hydrated oxides of iron and aluminum), and it is now recognized that lateritic or kaolinitic materials are resistant to erosion. **1970** *Toronto Daily Star* 24 Sept. 14/7 He pointed to the lateritic, or oxide, ores of the tropics and subtropics as an area where future major expansion in the industry will occur.

lateritization (læ:tĕritəizēi·ʃən). [f. Laterit(e + -ization, rendering G. *lateritisirung* (now *-ierung*) (M. Bauer 1898, in *Neues Jahrb. f. Min., Geol. u. Paläont.* II. 203).] = *laterization.

1903 *Geol. Mag.* Decade IV. X. 62 That the formation of laterite is a mere question of average temperature seems unlikely, for lateritization is very prominent at elevations of 6,000–7,000 feet..in South India, where the temperature varies very little above or below 60° F. **1920** *Geol. Mag.* LVII. 212 The depth of lateritization was so great on the right bank of the Congo at a height of about 500 O.D. that [etc.]. **1934** *Discovery* July 202/1 Lateritization (silica leaching) may be considered as the reverse of podsolization (iron and aluminium leaching). **1966** *McGraw-Hill Encycl. Sci. & Technol.* VII. 406/2 The resulting concentrations of iron and aluminum oxides sharply differentiate lateritization from temperate-climate weathering in which the end product is largely clay minerals (hydrous aluminum silicates).

Hence (as a back-formation, or after G. *lateritisi(e)ren*) **la·teritize** *v. trans.*, to convert into laterite; **la·teritized** *ppl. a.*, **-itizing** *vbl. sb.*

1911 *Geol. Mag.* Decade V. VIII. 509 These five rocks.. were probably formed in situ under conditions in which the lateritizing processes were not pushed to a finish. **1920** *Geol. Mag.* LVII. 214 Certain inclined bands in the norite have been thoroughly lateritized for many yards along their outcrop, and seams of magnetite in them..run practically unaltered from norite into laterite. *Ibid.* 212 The lateritized portion was more readily removed by erosive agencies, and the masses of unaltered rock were left standing in relief. **1970** *Nature* 23 May 693/2 The southern half of Western Australia is an old, slightly dissected, lateritized land surface dating from the middle or late Tertiary. *Ibid.*, The Giles Complex of mafic and ultramafic igneous rocks..has been lateritized.

laterization (læ:tĕrəizēi·ʃən). [f. Later(ite + -ization, rendering G. *lateritisirung* (see prec.).] The alteration of rock to laterite; the kind of weathering or soil-forming process that results in laterite and lateritic soils.

1903 A. Geikie *Text-bk. Geol.* (ed. 4) I. 169 Laterite... The peculiar kind of alteration exemplified by this rock and by Bauxite has been termed 'Laterisation' [by Bauer]. **1917** *Mining Mag.* XVII. 74/2 A very definite line must be drawn..between live laterite or that in process of formation, and dead laterite or that in which laterization is no longer actively operative. **1936** *U.S. Dept. Agric. Misc. Publ.* No. 229.22 The laterization process is, perhaps, more strictly a geological process than one of soil building. **1938** *U.S. Dept. Agric. Yearbk.* 973 The soil-forming process called laterization is essentially the progressive hydrolysis of rock minerals, and its full development results in their conversion to silicic acid, aluminum hydroxide, and iron hydroxide or their more or less complete dehydration products—the Laterites. **1952** P. W. Richards *Trop. Rain Forest* ix. 209 This preferential leaching of silica from a soil is called laterization. **1967** M. J. Coe *Ecol. Alpine Zone Mt. Kenya* 76 The clays show a certain number of darker stratifications but, due to the low temperatures, there are no signs of laterisation.

Hence (as back-formations) **la·terized** *ppl. a.*, converted into laterite; **la·terizing** *vbl. sb.* and *ppl. a.*

1917 *Mining Mag.* XVII. 67/2 Water from the laterizing zone when exposed to the air deposits ferric and aluminium hydrates with hydrous silicate of alumina. *Ibid.* 74/2 The upper portion of the laterite..is detrital, the lower part laterized and partly laterized schist merging gradually into altered schist. **1932** *Technical Communications Imperial Bureau Soil Sci.* No. 24. 19 Soils formed under laterising conditions. **1964** *Sci. Amer.* Nov. 97/2 Laterized soils occur most commonly in the tropical belt between the latitudes of 30 degrees North and 30 degrees South.

lates (lē^i·tīz). [mod.L. (Cuvier & Valenciennes *Histoire naturelle des Poissons* (1828) II. 88), f. Gr. λάτος Nile perch.] A large fish of the genus so called, esp. the *Nile perch* (*Nile), *Lates niloticus*; = Latus[2].

1920 *Blackw. Mag.* May 655/1 Playing a giant lates is no joke. **1921** W. Radcliffe *Fishing from Earliest Times* xxv. 326 A picture of a bronze mummy-case containing remains of a small *Lates*.

latest, *a.*[1] Add: **2.** Also *ellipt.* as *sb.* in *the latest*: the most recent story, piece of news, fashion, etc.

1886 H. Baumann *Londinismen* 94/1 What's the latest: was gibt's Neues? **1889** *Kansas City* (Missouri) *Times & Star* 25 June, The latest the dear girls hereabouts are singing..is, Will he love you as today? **1911** A. Bennett *Card* v. 128 This was Denry's 'latest', and it employed the conversation of the borough for I don't know how long. **1916** G. B. Shaw *Matter with Ireland* (1962) 95 If you want to dine in evening dress confronted with a bediamonded wife and flanked by daughters in the very latest,..you will be unhappy in Ireland. **1922** Joyce *Ulysses* 319 Well, says the citizen, what's the latest from the scene of action? **1940** *War Illustr.* 19 Jan. 623 The newsvendor who still stands shivering at his wintry pitch, lustily shouting 'the latest' when you can only discern him dimly by the glow of his cigarette. **1961** Wodehouse *Service with Smile* (1962) iv. 58 'I say,' he said..'have you heard the latest?'

latex. Add: Pl. (see sense *3) **la·texes, latices** (lē^i·tisīz). **2.** Also, *spec.* that of *Hevea brasiliensis* or other plants used to produce rubber. (Add further *attrib.* examples.)

1922 *Glasgow Herald* 4 Dec. 11 Rubber latex is a limpid liquid which is mixable with water. **1937** *Archit. Rev.* LXXXII. 57/3 The hearth-scene surround is constructed of re-inforced concrete, rendered in a mix of latex-rubber and *ciment fondu*. **1951** M. Abercrombie et al. *Dict. Biol.* 125 Latex of a number of species is collected and used in manufacture of several commercial products, most important being rubber. **1955** *Times* 14 June 11/1 Almost the entire production of the estates is in the form of latex..as distinct from sheet rubber. **1960** G. Lewis *Handbk. Crafts* 237 Latex Foam Cushion. In many of the modern suites of furniture latex foam or some other kind of rubber stuffing is used. **1966** L. Cohen *Beautiful Losers* (1970) II. 177 Several comfortable Latex cups assumed exciting holds here and there.

3. Any dispersion in water of particles of a polymer (originally synthetic rubber) that is formed in a polymerization process, such as is used as a binder in paints or for coating paper and leather.

1937 W. J. S. Naunton *Synthetic Rubber* viii. 140 'Rubber dispersions' should be divided into two groups: (1) Latices produced *in situ* by emulsion polymerisation... (2) Dispersed rubbers produced by the dispersion of preformed solid rubbers. *Ibid.* 150 This synthetic latex forms a good cement for assembling the numerous pieces of natural and artificial leather which constitute the modern mass-produced shoe. **1952** J. P. Casey *Pulp & Paper* II. xxi. 1247 Pulp treated with small amounts of elastomer latices has improved strength properties and increased wet strength. **1954** H. F. Payne *Organic Coating Technol.* I. ix. 372 During the past few years stabilized latexes of a variety of polymeric film-formers have been made available to the paint industry. **1969** T. C. Thorstensen *Pract. Leather Technol.* xiv. 226 In a latex system the binder is emulsified in water. When the latex is applied the water evaporates, or sinks into the leather, and eventually a phase inversion takes place.

4. Special Comb. **latex paint**, a paint having a latex as its binding medium.

1954 H. F. Payne *Organic Coating Technol.* I. ix. 372 Another advantage of latex paints is their very fast drying property. **1965** D. H. Parker *Princ. Surface Coating Technol.* xliv. 724 Exterior latex paints should not be applied directly to unpainted wood, for poor adhesion may result.

lath, *sb.* Add: **1. b.** lath and plaster *Rhyming slang*, master.

1857 'Ducange Anglicus' *Vulgar Tongue* 11 Lath-and-plaster, master.

2. b. (Later example.)

1922 A. Bennett *Lilian* II. iii, The entire office, thanks to that lath, Millicent, was disorganised.

c. *Min.* and *Petrol.* A mineral crystal that is thin, narrow, and elongated.

[**1908** L. V. Pirsson *Rocks & Rock Minerals* iv. 36 In some rocks, such..as the syenites, which are mainly composed of feldspar..they have more or less perfectly the shape of flat tables or rude laths.] **1916** A. Johannsen tr. *Weinschenk's Fund. Princ. Petrol.* x. 199 In this [intersertal] texture the interstices between the feldspar laths are filled with glass. **1941** *Proc. Prehist. Soc.* VII. 65 The rock is a strongly ophitic dolerite with..plates of fresh augite and laths of plagioclase. **1959** W. W. Moorhouse *Study of Rocks in Thin Section* v. 160 Intersertal includes diabasic and ophitic textures, in which the feldspar laths are enclosed with large grains of pyroxene.

4. a. *lath-hammer* (later examples). **b.** *lath-sword* (later example). **c.** *lath-cutting.* **d.** *lath-shaped* adj.

1827 *Western Monthly Rev.* I. 80 A lath-cutting machine..cuts them with great rapidity. **1847** *Rep. Comm. Patents 1846* (U.S.) 91 One patent has been granted for improvements in lath-cutting machines. **1901** J. Black *Illustr. Carpenter & Builder Ser.: Home Handicrafts* 35 The laths are nailed to each stud, or joist... For this purpose the best tool to employ is the..lath-hammer. **1964** J. S. Scott *Dict. Building* 188 *Lath hammer*.., a plasterer's hammer for nailing laths. **1888** J. J. H. Teall *Brit. Petrogr.* 435 This [interstitial] substance occurs in irregular masses wedged in between the lath-shaped felspathic constituent. **1973** *Nature* 9 Feb. 374/1 Lath-shaped crystals also occur and their size is about 0·15 × 0·05 mm^3. **1940** Blunden *Poems 1930–40* 250 While with a half-triumphant mind you crost Lath-swords of words on some uncertain matter.

lather, *sb.* Add: Also with pronunc. (lā·ðəɹ).

1926 Fowler *Mod. Eng. Usage* 315/2 Though lah'dher is often heard, *lather* apparently does not belong to the class of words in which ah & ă are merely southern & northern variants (*pass* &c.). **1968** *New Society* 22 Aug. 266/1 *Lather*: non-U to rhyme with 'father' (invariable in television advertisements) / U to rhyme with 'gather'.

1. *fig.* (Later example.)

1940 L. MacNeice *Last Ditch* 10 The sky is a lather of stars.

c. *transf.* A state of agitation, anxiety, irritation, or the like, such as induces sweat. (Cf. quot. 1660 in sense b.)

1839 F. Trollope *Fragment* in *Dom. Manners Amer.* (ed. 5) 271 Don't be in a lather, father, before you are shaved. I'll do your job, I expect, if you won't be in such a tarnation fuss. **1892** Kipling *Lett. of Travel* (1920) 99 Forced inaction frets the man to a lather. **1931** V. Woolf *Waves* 273, I arrived all in a lather at her house..but did not marry her, being..unripe for that intensity. **1948** 'J. Tey' *Franchise Affair* xvi. 181, I suppose Christina is in the usual lather of sentiment? **1945** E. S. Gardner *Case of Gold-Digger's Purse* (1949) v. 44 You're standing there in a lather of indecision. **1970** *Daily Tel.* 24 Sept. 4/8, I can't work myself up into a middle aged lather over long hair.

lather, *v.* Add: **1.** Also *fig.*

1917 P. Gibbs *Battles of Somme* 171 The enemy was 'lathering' the field of observation with every kind of 'crump' and shell.

lathering, *vbl. sb.* (Earlier and later examples in the sense of 'beating'.)

1835 J. P. Kennedy *Horse-Shoe Robinson* I. ii. 25 He shut that up..by giving Huger a most tremenjious lathering. **1843** 'R. Carlton' *New Purchase* I. xix. 169 Vain all pelting with clods and stones—all latherings with long bean poles! **1954** Weingarten *Amer. Dict. Slang* 222/2 *Lathering*, a scolding; a beating, a thrashing.

lathi. Now the more usual form of Lathee. Also *attrib.*

1850, 1895 [see Lathee]. **1920** *Glasgow Herald* 31 Dec. 7 Some disturbance..in which lathis were used. **1924** R. Graves *Mock Beggar Hall* 64 Then the new power, foreseeing grave events Calls out the lathi-wallahs to line the streets. **1930** *Daily Express* 6 Nov. 3/6 The police made a number of lathi charges to disperse the crowd. **1936** J. Nehru *Autobiogr.* 177 My body felt the baton and *lathi* blows of the police. **1972** *Times of India* 28 Nov. 11/2 A judicial inquiry was demanded into the lathi-charge at Gulbarga.

lati-. Add: **latise·ptate** *a.* = *latisept* adj.

1959 A. R. Clapham et al. *Excursion Flora Brit. Isles* 555 Latiseptate. Of a fr[uit] with the septum across the widest diameter.

latifundia. Add: (Later examples.) Hence **latifu·ndiarist** *a.* = Latifundian *a.* Also, the Italian form **latifo·ndi** (sing. *latifondo*). Also, later examples referring to large plantations in Latin America; so **latifundi·sta** [Sp.], the owner of a *latifundium* (Sp. *latifundio*) in Spain or Latin America; also in anglicized form **latifu·ndist**.

1902 *Encycl. Brit.* XXIX. 612/2 Special contracts..are applied to the *latifondi* or huge estates [in Italy]. **1930** C. F. Jones *S. Amer.* xxi. 310 The establishment of the *latifundia* first began on a large scale during Rosas' first campaign. *Ibid.* 447 The *latifundia* system. **1937** F. Borkenau *Spanish Cockpit* i. 48 Abolition of *de facto* serfdom, splitting up of the *latifundia* in the South and the Centre. **1954** Koestler *Invis. Writing* xxiv. 263 The ruin of farmers and the growth of large latifundia. **1961** *Listener* 24 Aug. 266/2 The actual state farms [in Cuba] have been set up on what were almost uncultivated latifundia. **1962** *Economist* 27 Jan. 333/2 There is no *latifundista* class [in the Dominican Republic] to prevent effective land reform. **1963** *Ibid.* 2 Nov. 461/2 The *latifundistas*, local dignitaries and conservative Catholics [in Spain]. **1964** Gould & Kolb *Dict. Social Sci.* 268/1 A polity dominated by the owners of large estates. As this type of society has been called latifundiarist, there is no point in wasting the word *feudal* on it. **1964** *Punch* 26 Aug. 291/1 The tired and near-forsaken *latifundia* in the main [W. Indies] sugar-producing islands. **1967** C. Seton-Watson *Italy from Liberalism to Fascism* viii. 312 Most of the proprietors of the *latifondi* were absentees. **1970** *Time* 2 Nov. 20 He is determined to expropriate the wealth of the big capitalists, the latifundists and the imperialists. **1974** *Times Lit. Suppl.* 8 Feb. 124/2 The 25,000 or so medium farmers [in Chile] whose holdings were below the minimum legally classified as *latifundia*.

‖**latigo** (la·tigo). *U.S.* [Sp.] A strap for tightening a cinch. Also *attrib.*, as *latigo strap*.

1873 A. S. Evans *À la California* 331 The wide band of woven horsehair, known as the cinch, is drawn up by the powerful purchase on the *latigo* strap until it deeply imbeds itself in the animal's belly. **1894** *Dialect Notes* I. vii. 325 Látigo: a thong... The two ends of the cinch terminate in long, narrow strips of leather—*látigos*—which connect the cinch with the saddle and are run through an iron ring called *larigo*. **1952** J. Steinbeck *East of Eden* 149 He was lacing the latigo through the cinch rings. **1962** W. Stegner *Wolf Willow* III. ii. 155 He picked at the latigo with one freezing unmittened hand. **1968** R. M. Patterson *Finlay's River* 237 The latigo laces are long, finely cut leather thongs, taken from a big hide.

latimeria (lætimiə·riă). [mod.L. (J. L. B. Smith 1939, in *Nature* 18 Mar. 456/2), f. name of Marjorie E. D. Courtenay-*Latimer* (b. 1907), director of the East London Museum at the time of the discovery + -ia[1].] A large marine fish of the genus so called, the only living representative of the order of crossopterygian fishes Actinistia, which was discovered in deep water off the south-east coast of Africa in 1938; = *cœlacanth.

1940 *Nature* 13 July 53/1 The exceptionally oily nature of Latimeria..is interesting geologically. **1953** J. S. Huxley *Evolution in Action* v. 128 Latimeria is the name of a rather primitive kind of fish, more nearly related to lung-fish than to modern Teleosts, one single specimen of which was recently brought up alive by a fishing vessel off the coast of South Africa. **1956** J. L. B. Smith *Old Fourlegs: Story of Coelacanth* 236 All [cœlacanths] have been large fishes..with fins like *Latimeria*. **1968** A. S. Romer *Procession of Life* viii. 166 (*caption*) *Latimeria*, about five feet long, is the last surviving coelacanth.

Latin, *a.* and *sb.* Add: **A.** *adj.* **2.** *Latin letter*, a letter of the Latin alphabet.

1535 [in Dict.]. **1953** K. Jackson *Lang. & Hist. Early Brit.* 179 Latin-letter inscriptions. **1965** *Language* XLI. 238 All Serbo-Croatian examples..are cited in conventional Latin-letter orthography.

4. b. *Latin-American* (adj.), of or belonging to those countries in Central and South America in which Spanish or Portuguese is the dominant language (and which are often referred to collectively as *Latin America*); also (sb.), an inhabitant of one of these countries. Also (ellipt.) *Latin*.

1893 *Funk's Stand. Dict.*, Latin American. **1903** *Westm. Gaz.* 22 June 11/1 Mexico..the richest district in the richest of the Latin-American countries. **1906** *Ibid.* 17 Apr. 9/1 Colombia..is taking her place with those Latin-American countries [etc.]. **1911** *Q. Rev.* Oct. 456 Serious competition for British merchants doing trade with the Latin-American States. **1912** *Chambers's Jrnl.* June 358/2 The amount of British capital invested in the countries of Latin-America is very great. *Ibid.* Nov. 720/2 An Englishman..soon wishes himself well rid of the..Latin-American. **1936** *Discovery* Dec. 365/1 An issue [of *Discovery*] devoted to Latin America. **1955** L. Feather *Encycl. Jazz* i. 30 The wedding of jazz with Latin-American rhythms. **1962** K. Orvis *Damned & Destroyed* iv. 30 The pianos segued smoothly into Latin rhythms. **1962** S. de Madariaga (*title*) Latin-America between the Eagle and the Bear. **1965** *Crescendo* Dec. 14/3 The arrangements are all in the Latin idiom and all of well-known tunes, getting off to a really swinging start with a L-A 'Peter Gunn' you *must* hear. **1966** *Ibid.* Nov. 6/1 All the side one tracks have this straight eight-to-the-bar or Latin feel about them. **1973** 'D. Jordan' *Nile Green* xxxi. 145 It's oil sheiks and Latin American generals and Lebanese rentiers who are going to buy your bonds. **1973** A. Mann *Tiara* i. 4 In the Philippines, some crazy Latin American got near enough to Paul VI to attack him with a knife. **1973** D. Robinson *Rotten with Honour* 8 He stood for a moment in the sunshine, snapping his fingers to a Latin beat. **1974** *Radio Times* 14 Sept. 26/3 Let's Go Latin..a fiesta of Latin-American music.

6. *Latin cross* (examples); *Latin square* [Named (as F. *quarré* (now *carré*) *latin*) by Euler 1782, in *Verh. uitgegeven door het Zeeuwsch Genootschap d. Wetensch. te Vlissingen* IX. 90, from the fact that letters of the Latin alphabet were used in forming it.] add: used as the basis of experimental procedures in which it is desired to control or allow for two sources of variability while investigating a third; hence used *attrib.* (also *absol.*) to designate such a procedure; (further examples).

1797 Latin cross [see Cross *sb.* 18]. **1936** A. W. Clapham *Romanesque Archit.* ii. 25 Although occasionally..the transept is of the T-form of the earlier ages, more generally the arrangement takes the Latin-cross form distinctive of the full Romanesque style. **1966** *Listener* 9 June 835/2 It is a Latin-cross church. *Ibid.*, A Latin cross is a more obviously Christian symbol than a regular geometric figure. **1925** R. A. Fisher *Statistical Methods Res. Workers* viii. 229 (*heading*) The Latin square. **1926** —— in *Jrnl. Ministry of Agric.* XXXIII. 510 For the purpose of variety trials, and of those simple types of manurial trial in which every possible comparison is of equal importance, the problem of designing economical and effective field experiments, reduces to two main principles..[of which the second is] the use of arrangements which eliminate a maximum fraction of the soil heterogeneity, and yet provide a valid estimate of the residual errors. Of these arrangements, by far the most efficient.. is that which the writer has named the Latin Square. *Ibid.*, The term Latin Square should only be applied to a process of randomization by which one is selected at random out of the total number of Latin Squares possible. **1935** —— *Design of Exper.* v. 80 The object of arranging plots in a Latin square is to eliminate from the experimental comparisons possible differences in fertility which may exist between whole rows of plots, and between whole columns of plots, as they stand in the field. **1960** D. J. Finney *Introd. Theory Exper. Design* iii. 30 Four different doses of insulin..were tested on rabbits and compared in terms of the subsequent sugar contents in the rabbits' blood... There is..a strong case for using rabbits as blocks and testing each dose, on different occasions on every rabbit. In addition, however, a block constraint based upon day of injection, so that on each day every dose is tested, is a useful precaution against the possibility that laboratory conditions on a particular day may tend to affect all animals in the same direction. A 4 × 4 Latin square with columns corresponding to different rabbits and rows corresponding to different days, enables both constraints to be incorporated. **1971** *Nature* 13 Aug. 499/1 On drug weeks each of six rats received one of six doses, each in a different order (latin-square design). *Ibid.*, An additional 6·0 mg/kg dose was administered to all subjects during the week after the completion of the latin-square.

B. *sb.* **3. c.** A member of any of the various communities in Europe (France, Italy, Spain, etc.) and Latin America whose language is derived from Latin.

1876 R. Brown *Races of Mankind* IV. xvii. 292 The Aryans of Europe are the Skipitar, Celts, Greeks, Latins, Germans of all branches, Lithuanians, or Letts and Slavs. **1908** Beerbohm *Lett. to R. Turner* (1964) 180 And then, of course, there is the pendant-fact that the Latins are born actors. **1936** J. Curtis *Gilt Kid* iii. 35 A kind of wooden..expression had come over her as it does over all Latins when they're scared of having to give something for nothing. **1949** H. van Zeller *We live with our Eyes*

Open 65 A Latin loves differently from a Saxon for instance. **1955** *Publ. Amer. Dial. Soc.* xxiv. 44 Most of these Latins [*sc.* immigrants from Cuba etc.] congregate on the East Coast. **1963** *Times* 2 Mar. 4/5 The Latins are said to be less susceptible to these emotions than we are.

5. *Latin-based*, *-derived* adjs.; **Latin–Greek**, of or pertaining to both Latin and Greek; **Latin Quarter** (F. *Quartier latin*; cf. Quarter *sb.* 14), the district of Paris on the left or south bank of the Seine, where Latin was spoken in the Middle Ages, and where students and artists live and the principal university buildings are situated; also *transf.*; **Latin school** (also **Latin grammar school**) *U.S.*, a school offering Latin (and sometimes Greek) as part of the syllabus; cf. G. *Lateinschule*, Da. *Latinskole*, Du. *Latijnsche school*.

1964 M. A. K. Halliday et al. *Ling. Sci.* i. 121 Old-fashioned Latin-based grammars. **1964** *Language* XL. 93 The inherited tradition of Latin- and Romance-based usage. **1946** H. Jacob *On Choice of Common Lang.* 38 A Latin-derived constructed language. **1965** W. S. Allen *Vox Latina* 109 As early as the fourteenth century one finds spellings with *ngn* for Latin-derived words. **1942** Partridge *Usage & Abusage* (1947) 290/2 Slang tends to be 'Saxon' rather than 'Latin–Greek'. **1960** *Amer. Speech* XXXV. 233 Unvoicing originated mainly in Latin–Greek bilingualism. **1869** 'Mark Twain' *Innocents Abroad* xv. 150 The *grisettes!*..so devoted to their poverty-stricken students of the Latin Quarter. **1878** R. L. Stevenson *New Arabian Nights* (1882) I. 55 He had chosen to study the attractions of Paris from..a furnished hotel, in the Latin Quarter. **1904** J. T. Grein *Dramatic Crit.* IV. 175 It was a generous mixture of the Latin Quarter and the various queer streets where London minor poetry flourishes. **1904** *Daily Chron.* 12 Dec. 4/4 They are good English garden-party hats, but they don't do for midi on an autumn day in the Latin Quarter. **1922** Joyce *Ulysses* 18 And there's your Latin quarter hat, he said. **1930** E. B. Chancellor (*title*) London's old Latin Quarter, being an account of Tottenham Court Road and its immediate surroundings. **1961** M. Beadle *These Ruins are Inhabited* (1963) iv. 54 There is good reason now for wags to call the university 'the Latin Quarter of Oxford'. **1968** *Listener* 4 July 5/2, I left my friends in the Latin Quarter three weeks ago in a mood of exhausted elation. **1651** *Mass. Bay Rec.* (1854) III. 242 Whosoeuer shall..cause Schollers belonginge to the Colledge or any other Latine Schoole..to spend any of theire time [etc.]. **1680** in C. W. Manwaring *Digest of Early Connecticut Probate Rec.* (1904) I. 355, I give to the lattin Schoole in Hartford £50. **1685** *New Plymouth Laws* (1836) 300 That every County Town shall have and maintain a Latine School. **1781** S. Peters *Gen. Hist. Connecticut* 185 Elms.. surround the center square, wherein are..the jail, and Latin school. **1856** B. H. Hall *Collection of College Words & Customs* (rev. ed.) 124 [A young man from the country] shall be examined and 'conditioned' in everything, and yet he shall come out far ahead of his city Latin-school classmate. **1959** C. V. Good *Dict. Educ.* (ed. 2) 311/2 *Latin grammar school*, a secondary school, emphasizing Latin and usually Greek, the purpose of which was to prepare youths for the universities. **1966** *Oxf. Compan. Amer. Hist.* 462/2 *Latin grammar schools*, the earliest type of college preparatory schools in the colonies, were established on the English model. The first, the Boston Latin School (1635), is still one of the principal schools in that city. *Ibid.* 463/1 By mid 18th century Latin schools were supplanted by academies.

latinate (læ·tinĕt), *a.* Also **Latinate.** [f. Latin *a.* and *sb.* + -ate[2] 2.] Of, pertaining to, or derived from Latin; having a Latin character. Also, occas., resembling an inhabitant of a Latin country.

1904 *Atlantic Monthly* Nov. 690/2 Cranmer transferred to the English..the rich sound and rhythm of the mediæval Latin; and that without the use of Latinate words. **1952** D. Davie *Purity of Diction in Eng. Verse* iv. 67 With intent Of being officious, grow impertinent... 'Officious' (in its Latinate sense, as in Johnson's 'Elegy on Robert Levett') defines and is defined by 'impertinent'. **1956** *Essays in Crit.* VI. 260 An anxious, questioning, excited passage, more latinate in diction. **1960** *Times* 16 Mar. 16/7 Miss Miranda, flamboyant and Latinate in temperament, is given..the part. **1962** W. Nowottny *Lang. Poets Use* i. 23 Latinate syntax is important to Milton because it provides him with more ways..of devising contrasts. **1971** D. Crystal *Ling.* 143 The distortions which..Latinate descriptions could impose.

Latinesque (lætine·sk), *a.* [f. Latin *a.* and *sb.* + -esque.] Resembling Latin; having a Latin character.

1887 E. C. Stedman *Victorian Poets* (ed. 13) 448 Its atmosphere, landscape, and notes of sympathy..are so unEnglish that one must possess the author's latinesque training to feel them adequately. **1903** *Westm. Gaz.* 8 Apr. 2/3 A new language, or a Latinesque language. **1960** *Times* 13 Jan. 6/4 Here the dances are lighter and predominantly Latinesque.

Latinical (lăti·nicăl), *a.* [f. Latinic *a.* + -al.] = Latinic *a.*

1892 *Forum* (N.Y.) July 585 He [*sc.* Hardy] is, in point of diction, the most Latinical writer we have had since Dryden and Milton. **1919** [see *aureate *a.* 2].

Latining, *vbl. sb.* (Later example.)

1893 F. J. Furnivall in J. Capgrave *Life St. Katherine* p. xxiv, I don't think *Prata* above can be a latining of *Akker*, acre, field.

Latinish, *a.* (Later example.)

1920 H. G. Wells *Outl. Hist.* 340/2 Neustria, the nucleus of France, speaking a Latinish speech.

Latinism. Add: The influence or authority of the Latin Church. (See Latin *a.* 3.)

1920 *Contemp. Rev.* Oct. 495 The Spanish Court ladies were sheltered..under the vaulted roof of Latinism. **1970** H. Braun *Parish Churches* iii. 32 In Rome itself, Ravenna, and other towns of Latinism the basilican halls continued to be built out of the ruins of paganism.

Latinist. 2. (Later examples.)

1964 *Catholic Herald* 4 Dec. 13/2 (*heading*) The conversion of a convinced Latinist. **1965** *Ibid.* 29 Jan. 1/8 (*heading*) Latinists ask for prayers.

Latinity. Add: **3.** Latin character.

1915 M. C. Fraser *More Italian Yesterdays* x. 191 True to their Latinity, they gave their victim no chance of testing it [*sc.* their mistrust]. **1934** G. B. Shaw *Prefaces* 740/1 By the end of the nineteenth century the press and the theatre had lost all their Latinity.

Latinless, *a.* (Later examples.)

1906 *Athenæum* 21 July 71/3 The Latinless enthusiast who is curious to explore Propertius. **1958** *Duckett's Reg.* Apr. 44/2 The most Latinless lout cannot fail to understand what the monks are doing at Solesmes.

Latino (lăti·no). *U.S.* [Amer. Sp., f. *Latin-American + Spanish ending *-o*.] A Latin-American inhabitant of the United States. Also *attrib.* or as *adj.*

1946 G. Peyton *San Antonio* xxi. 232 The first program on the University's list is an exchange of students with Latin America. That in itself would be a fresh intellectual experience for Texas, where Latinos are usually looked on as sinister specimens of an inferior race. **1966** Mrs L. B. Johnson *White House Diary* 2 Apr. (1970) 377 Six young girls, all Latinos, had encased themselves in cardboard boxes. **1972** *Listener* 9 Mar. 310/1 America..is meant to be a great melting-pot... Its racial components—Blacks, Latinos, Chinese, Japanese, [etc.]. **1973** *Black Panther* 17 Mar. 5/3 A program was drawn up..by an..action group composed of Blacks, Latinos, and Whites. **1974** *Ibid.* 19 Jan. 5/1 Mr. Rhodes' home was broken into..by a man who appeared to be of Latino origin.

Latino- (læ·tino), used as combining form of Latin *a.* 2 and 4 b, as in *Latino-Faliscan*, *-Jazz*, *-Sabellian*. Also (with *Latino* = abl. of L. *Latinus*) *Latino sine flexione*, the basis for the international language *Interlingua*.

1939 L. H. Gray *Found. Lang.* 332 The Italic dialects fall into three groups: Latino-Faliscan, Osco-Umbrian, and Sabellian. **1954** Pei & Gaynor *Dict. Ling.* 121 *Latino-Faliscan*, a branch of the Italic group of the Indo-European family of languages, consisting of the extinct languages Latin, Faliscan, Hernician and Praenestinian. **1958** P. Gammond *Decca Bk. Jazz* xxi. 265 It is worth remembering, when the history of the Latino-Jazz movement is written, that Kenton and the West Coast boys were years ahead of the boppers and the East Coast 'cool' men in hitching their wagon to the Latin star. **1880** A. H. Sayce *Introd. Sci. of Lang.* II. vii. 110 We find in Italy two great stocks, the Iapygian and the Latino-Sabellian. **1928** O. Jespersen *Internat. Lang.* i. 45 In 1903 the famous Italian mathematician G. Peano started his *Latino sine flexione*... The idea is to take the ablative of each Latin noun and one simple form of each verb to be used practically everywhere. **1939** L. H. Gray *Found. Lang.* 35 Esperanto, Ido, Latino sine flexione. **1946** H. Jacob *On Choice of Common Lang.* 16 *Latino sine flexione*, or Interlingua, as Peano called his system.

latitude. Add: **2. d.** *Photogr.* The range of exposures for which an emulsion, printing paper, etc., will give acceptable contrast; *spec.* the ratio (or its logarithm) of the exposures between which the characteristic curve is straight.

1889 E. J. Wall *Dict. Photogr.* 98 The extreme latitude of exposure which most plates possess. **1907** Sheppard & Mees *Investigations Theory Photogr. Process* III. i. 289 The latitude may be defined as the ratio of the exposure at which over-exposure commences to that at which under-exposure commences, and these two points must be arbitrarily defined. **1939** W. Clark *Photogr. by Infrared* iv. 60 The range or latitude of printing papers is thus always less than that of negative materials. **1962** W. G. Hyzer *Engin. & Scientific High-Speed Photogr.* v. 200 Actually the toe region of the curve is usable in recording shadow detail, which somewhat increases the effective latitude of the emulsion.

latitudinous, *a.* Add: **2.** = Latitudinal *a.*

1906 *Westm. Gaz.* 5 July 4/2 The race is not straight up to the limit of the earth's atmosphere and back again, but latitudinous.

lative (lē^i·tiv), *a.* *Gram.* [f. L. *lāt-* ppl. stem of *ferre* to bring + -ive.] Denoting the case used in some languages, e.g. of the Finno-Ugrian group, to express motion up to or as far as. Also *absol.* Cf. *allative *a.*, *elative *a.*

1939 L. H. Gray *Found. Lang.* vii. 194 The termination finds further analogues in..the Uralic lative and illative, the former indicating motion up to, and the latter motion to the interior of. *Ibid.* 195 The dative occasionally has, in Indo-Iranian, Latin, Teutonic, and Slavic, a lative force denoting the place toward which, in contrast to the illative force of the accusative. **1960** B. Collinder *Compar. Gram. Uralic Lang.* 239 The lative ending denoted that something is moving to the locality (or thing) expressed by the word stem. **1964** *Language* XL. 98 The same general discussion concerning the temporal function of the assumed Finno-Ugric lative case is presented twice.

‖**latke** (lɒ·tkə). Also **lutka, lutke.** [Yiddish, a. Russ. *látka* a pastry.] In Jewish cookery, a pancake, esp. one made with grated potato.

1927 *Amer. Mercury* Feb. 206 Luscious potato *latkes*—pancakes made of grated, raw potatoes, [etc.]. **1958** J. Grossinger *Art Jewish Cookery* p. ix, A Jewish cookbook can be almost considered a history book... Just one instance—the *latke* (pancake), which the wives of the soldiers of..Judah Maccabee hurriedly cooked for their men. **1964** W. Markfield *To Early Grave* (1965) iii. 50, I make a few *lathes*, I paint the kitchen chairs. **1967** A. Bailey in L. Deighton *London Dossier* 55 If you hunger after gefillte fish or latkes when in Soho, try Grahame's Sea Fare restaurant. **1971** M. Masson *Jewish Cookery* 44 Fry the *latkes* until a golden brown on both sides. **1974** *Times* 15 Oct. 13/8 He really does need a few more of my potato lutkas.

latosol (læ·tǒsǫl). *Soil Science.* [f. Lat(erite + -o + *-sol.] (See quot. 1949.)

1949 C. E. Kellogg in *Technical Communications Commonwealth Bureau Soil Sci.* No. 46. 79 We should like to suggest that some new term be adopted to comprehend all the zonal soils in tropical and equatorial regions having their dominant characteristics associated with low silica-sesquioxide ratios of clay fractions, low base-exchange capacities, low activities of the clay, low content of most primary minerals, low content of soluble constituents, a high degree of aggregate stability, and (perhaps) some red colour. The word 'Latosol' has been proposed as the name for this group at the categorical level of suborder... It is a collective term for those zonal soils previously called 'lateritic soils' where the characteristics just mentioned were dominant. **1955** K. Lawton in F. E. Bear *Chem. of Soil* ii. 68 In the group of soils considered to be laterites or latosols, aluminum and iron may make up a large proportion of the soil mass... Some ferruginous latosols contain as much as 60 per cent Fe_2O_3 and thereby could be classified as low grade iron ore. **1965** B. T. Bunting *Geogr. Soil* xvii. 199 In S. America, dark latosols occur in humid areas or on base-rich rocks; red latosols on acid materials and brown latosols on ash or basaltic terrain.

Latour (latū·ɹ). [Fr., ellipt. for *Château Latour*, the vineyard where it is produced.] A red Bordeaux wine from the Haut-Médoc district of France.

1833 [see *château b]. **1920** G. Saintsbury *Notes on Cellar-Bk.* iv. 48, I think the best Latour rather better. **1931** S. Jameson *Richer Dust* iv. 79 Nicholas watched his scout pouring claret into Hugh's glass... A Latour of '93. **1935** *Punch* 28 Aug. 238/2 Ah, the Old Latour and the Old Lafite, And the Old Yquem which was not too sweet. **1958** J. W. Lambert in C. Ray *Compleat Imbiber* II. 200 How, despite one's best endeavours, they echo in the mind, those great names—Lafite, Margaux, Latour, Mouton-Rothschild. **1967** A. Lichine *Encycl. Wines* 321 Full-bodied and hard when young, Latour develops into something firm, rich, and noble. **1973** [see *Lafite].

latrine. Add: **2.** *attrib.* and *Comb.* **latrine rumour** *Services' slang*, a baseless rumour believed to originate in gossip in the latrines; also *absol.*

1918 in *Amer. Speech 1972* (1975) XLVII. 73 (*title of unofficial newspaper*) La Trine Rumor. **1925** in Fraser & Gibbons *Soldier & Sailor Words* 140. **1929** F. A. Pottle *Stretchers* (1930) i. 15 A 'latrine', we learned, was not only a building, but also the name for any particularly exciting but quite unfounded rumour emanating therefrom. **1929** A. W. Wheen tr. *Remarque's All Quiet on W. Front* i. 15 Not for nothing was the word 'latrine rumour' invented; these places are the regimental gossip-shops and common-rooms. **1931** S. Southwold in *Martial Medley* 105 This short essay..confines itself mainly to the rumours current among the fighting forces, and generally referred to as latrine-rumours and dump-rumours. **1950** Partridge *Here, There & Everywhere* 76 Late in the [First] War, tersely (*eine*) *Latrine*—the English term being *latrine-rumour*.

latrinogram (lătrī·nogræm). *Services' slang.* [f. Latrin(e + -o + -gram.] = *latrine rumour* (*latrine 2).

1944 [see *dinkum B. *adj.*]. **1946** [see *bush telegraph* s.v. *bush *sb.*[1] 11]. **1947** D. M. Davin *Gorse blooms Pale* 203 According to current latrino-gram we were going to be given a rest. **1966** *Sunday Times* (Colour Suppl.) 4 Dec. 73/2 GI Jargon. *Latrinogram*, latrine rumour.

lats: see *lat[3].

lattee, var. Lathee.

1864 G. O. Trevelyan *Competition Wallah* 170 Placing a lattee, which is the name for the quarter-staff carried by all Indian peasants, under the defaulter's knee.

latter, *a.* (*adv.*) Add: **A.** *adj.* **5. a.** and **b.** (Further examples.) Also used to denote the last of a group of more than two persons or things, or a person or thing that has been mentioned at or near the end of a preceding clause or sentence (see quots.).

1841 [see Former *a.* 2b]. **1853** [see *fill-up *sb.*]. **1903** G. B. Shaw *Man & Superman* 230 When a man teaches something he does not know to somebody else who has no aptitude for it, and gives him a certificate of proficiency, the latter has completed the education of a gentleman. **1922** Joyce *Ulysses* 616 'Eaten alive?' a third asked the sailor. 'Ay, ay,' sighed again the latter personage. *Ibid.* 680 The former returned to the latter..a sum of money.. advanced by the latter to the former. **1926** H. W. Fowler *Mod. Eng. Usage* 316/2 *The latter* should not be used when more than a pair are in question, as in: The difficult problems involved in the early association of Thomas Girtin, Rooker, Dayes, & Turner are well illustrated..; & what was undoubtedly the best period of the latter artist is splendidly demonstrated... Neither should it be used when less than two are in question; the public and its shillings cannot be reasonably regarded as a pair of things on the same footing in: The mass of the picture-loving public, however, may be assured of good value for the shillings—whatever be the ultimate destination of the latter. **1928** [see *lenate *v.*]. **1938** *Amer. Speech* XIII. 28 Faith, hope and charity are virtues, but few possess them, particularly the latter. **1957** B. & C. Evans *Dict. Contemp. Amer. Usage* 267/2 *Latter* is the older of the two comparative forms... Its chief function is as a contrast to *former*. The contrast implies that some group has been separated into two parts, but more than two elements may be involved. We may say *the three latter events*. **1970** Lofts & Adley *Saint* ii. 15 Another story..was written under the *nom de plume* of 'Leslie C. Bowyer'—the latter being his mother's maiden name. **1971** *Guardian* 24 Dec. 17/5 The Berlin Wall stands unbreached, passes are needed to get into Bethlehem and Father Christmas has been arrested in Oxford Street. It's the latter item that fascinates me. **1974** A. Huxley *Plant & Planet* xvi. 177 Light is also essential to the germination of many seeds such as tobacco, foxglove, many primulas and some lettuces. The dark-induced dormancy of the latter can be broken by quite low illumination.

lattice, *sb.* Add: **2. b.** *Her.* A charge representing lattice-work.

1828 W. Berry *Encycl. Heraldica* I, *Lattice*, or *Lettice*, ..is formed of perpendicular and horizontal bars,..and the lattice may be either interlaced, or not. **1889** C. N. Elvin *Dict. Heraldry* 82/2 *Lattice*, *Tirlace*, or *Treilée*, consists of bars crossing one another at right angles, which do not interlace, but are nailed together at the crossings. **1969** Franklyn & Tanner *Encycl. Dict. Heraldry* 199 *Lattise*,..alt. for 'trellis'.

c. In textile manufacture, a lattice-work apron or conveyer used to carry material into or out of a machine.

1884 W. S. B. McLaren *Spinning* x. 213 The wool is taken from it by a roller which combs it off, and passes it on to an endless lattice, marked 'upper lattice'. **1890** J. Nasmith *Mod. Cotton Spinning Machinery* iii. 19 In each case it is customary to attach lattices to the machine, by which the cotton is thoroughly broken up. **1967** Shaw & Eckersley *Cotton* xi. 77 The machine employs a lattice and rollers to feed the thread waste to a revolving cylinder covered with steel spikes.

d. *Electr.* = *lattice network* (see *4).

1934 A. T. Starr *Electr. Circuits & Wave Filters* vi. 198 The lattice needs twice as many components as the bridged-T network. **1950** W. C. Johnson *Transmission Lines & Networks* xiv. 303 The filters most generally used are made up of T or π sections and L 'half sections' connected on an image basis to form a ladder network... A more general structure, called the lattice, is shown in Fig. 14.3. Not only can the performance of any T or π be duplicated at all frequencies by a lattice, but a lattice can be designed to provide characteristics unobtainable with the T or π. **1960** M. E. van Valkenburg *Introd. Mod. Network Synthesis* xii. 339 If there is any symmetrical network realization for a set of specification functions at all, then there is a symmetrical lattice realization.

3*. a. Any regular arrangement of points or point-like entities that fills a space, area, or line; *spec.* a crystal lattice or a space lattice; **Bravais lattice** (bræ·ve) [named after Auguste *Bravais* (1811–63), French physicist], any lattice in which every point has exactly the same environment (as regards the distances and directions of other points of the lattice); *spec.* any of the fourteen different lattices of this kind in three dimensions (cf. **space lattice*); **crystal lattice**, the space lattice underlying the arrangement of atoms or molecules in a crystal; also, the arrangement of points occupied by the atoms or molecules or of the atoms or molecules themselves.

1895 W. J. Pope tr. *Fock's Introd. Chem. Crystallogr.* ii. 12 Frankenheim..found that fifteen different space-lattices are possible, and then, having deduced from the cleavage and general habit of the crystals that fifteen fundamental forms of crystals are possible, he showed that these latter in many respects correspond with the lattices. Frankenheim's views are not in all respects correct. **1917** *Physical Rev.* X. 441 Manganblende, MnS, is a simple cubic lattice like rock salt. **1926** [see *crystal lattice* s.v. *crystal B. 2 c]. **1927** T. Verschoyle tr. *Haas's Atomic Theory* iii. 58 From the one-dimensional line-lattice, let us now pass to the two-dimensional plane-lattice, which will be formed by points with the coordinates $x = k_1 a$, $y = k_2 b$ constructed in a plane coordinate system, where *a* and *b* are two lattice constants, and k_1 and k_2 can assume every integral value. **1934** *Nature* 16 June 916/1 The electron extracted from the atom may only move through the periodic field of the lattice with certain discrete energies. **1935** *Discovery* May 132/2 The patterns in which the atoms or charged atoms are arranged are often called lattices. There are 'face-centred' lattices, and 'body-centred' lattices, hexagonal close-packed lattices, and so on. **1936** *Mineral. Abstr.* VI. 323 Recently structures have been suggested, e.g. for α-AgI, in which equivalent points are not completely filled or are occupied by different sorts of particles. These structures are said to have 'defect lattices'. **1938** W. A. Wooster *Text-bk. Crystal Physics* 280 A Bravais lattice is one of the fourteen possible arrangements of the points in space which have crystallographic symmetry. **1955** *Mineral. Mag.* XXX. 625 There are in all twenty types of lattices, as defined by their symmetry element[s]: one one-dimensional, five two-dimensional, and fourteen three-dimensional. *Ibid.* 626 All lattices are formed by translations, and all are characterized by inversion. **1958** W. K. Mansfield *Elem. Nucl. Physics* v. 39 These displacements distort solid lattices, producing effects similar to cold working, and decrease the electrical and thermal conductivity. **1966** C. R. Tottle *Sci. Engin. Materials* iii. 67 The crystal lattice of an ionic compound depends on the size of the ions and on their valency. **1969** A. P. Cracknell *Crystals* ii. 56 The mathematical condition to be satisfied in the definition of a lattice is quite stringent... In fact there are only five two-dimensional Bravais lattices. **1970** A. J. C. Wilson *Elem. X-Ray Crystallogr.* iv. 52 Thirteen of the fourteen Bravais lattices (all except the triclinic lattice) possess at least one reflexion plane passing through each point of the lattice. **1971** I. G. Gass et al. *Understanding Earth* i. 11/2 The halite lattice is built on a simple pattern in which sodium particles and chlorine particles occupy alternate corners of a continuously repeated set of cubes.

b. *Nuclear Engin.* An array of fuel and moderator in the core of a nuclear reactor.

1945 H. D. Smyth *Gen. Acct. Devel. Atomic Energy Mil. Purposes* ii. 21 The steady production of atomic power requires a slow-neutron-induced fission chain reaction occurring in a mixture or lattice of uranium and moderator. **1959** H. Jacobowitz *Fund. Nucl. Energy & Power Reactors* iii. 50 The Bulk Shielding Reactor comprises little more than an assembly of enriched uranium fuel elements immersed in water. The height of the active lattice is 24 inches. **1960** S. E. Liverhant *Elem. Introd. Nucl. Reactor Physics* vii. 178 If the lattice or matrix arrangement (i.e., a heterogeneous system) is employed, a chain reaction becomes possible with natural uranium and graphite as moderator. **1973** *Jrnl. Nucl. Energy* XXVII. 458 For water-beryllium lattices, the band structure and Nelkin's expansion..is [*sic*] computed for various directions.

3.** *Math.* A partially ordered set in which every pair of elements has an infimum and a supremum.

1933 G. Birkhoff in *Proc. Cambr. Philos. Soc.* XXIX. 442 If we define a lattice to be any set of elements satisfying [axioms] Ī–VĪ, we can express our results as Theorem 3·1: The subalgebras of any algebra constitute a lattice. *Ibid.*, [F.] Klein calls a finite lattice a 'Verband'. **1951** N. Jacobson *Lect. Abstr. Algebra* I. vii. 208 Boolean algebras were the first lattices to be studied. They were introduced by Boole in order to formalize the calculus of propositions. **1964** H. G. Flegg *Boolean Algebra* iii. 18 The algebra of classes is frequently referred to as Boolean algebra. A rigorous treatment of the algebra has been made by Garrett Birkhoff and Saunders MacLane and of its generalizations, lattice theory, by Birkhoff alone. **1965** S. Warner *Mod. Algebra* I. iii. 105 An ordered structure $(E, \le)$ is a lattice and $\le$ is a lattice ordering if for all $x, y \in E$, the subset $\{x, y\}$ of E admits a supremum and an infimum. **1966** *McGraw-Hill Encycl. Sci. & Technol.* VII. 409/2 Lattice theory deals with properties of order and inclusion, much as group theory treats symmetry. *Ibid.*, The real numbers form a lattice, if $x \le y$ is given its usual meaning... Again, the set *J* of positive integers forms a lattice, if one lets $m \le n$ mean '*m* divides *n*'.

4. *lattice-box*, *-fence*, *-floor*, *-mast*, *-ornament*, *steel*; **lattice conductivity** *Physics*, the contribution to the thermal conductivity of a crystalline substance arising from transfer of energy between the vibrating atomic nuclei in the crystal lattice; so *lattice conduction*; **lattice constant** *Cryst.*, the length of a side, or the size of an angle, of the unit cell of a lattice; *spec.* the length of each of the sides of the unit cell of a cubic lattice; **lattice defect** *Cryst.*, an irregularity in a crystal lattice such as a missing atom or an interstitial one; **lattice energy** *Physics*, the energy required to separate the ions of a crystal to an infinite distance from one another; **lattice filter** *Electr.*, a filter consisting of components connected so as to form a lattice network; **lattice network** *Electr.*, a network having four impedances and two pairs of terminals, each terminal of one pair being connected by an impedance to each of the other pair; **lattice plane** *Physics*, any plane containing lattice points; a layer of atoms or molecules in a crystal; **lattice point**, for '(see quot.)' read: (*a*) a point on a graph or in space having integral coordinates (further examples); (*b*) any of the points of which a lattice, esp. a crystal lattice, is composed; **lattice-truss** (example); **lattice vibration** *Physics*, an oscillation of an atom or molecule about its equilibrium position in a crystal lattice; also, a lattice wave; **lattice wave** *Physics*, a displacement of atoms or molecules from their equilibrium position in a crystal which travels as a wave through the crystal.

1865 'Mark Twain' *Celebr. Jumping Frog* (1867) 16 Smiley kept the beast in a little lattice box. **1938** *Proc. Cambr. Philos. Soc.* XXXIV. 475 The lattice conduction is shown to be important in poor conductors. **1971** I. G. Gass et al. *Understanding Earth* v. 86/2 At moderate temperatures..heat transfer in rocks is almost entirely by 'lattice conduction'. **1938** *Proc. Cambr. Philos. Soc.* XXXIV. 474 In good conductors the lattice conductivity is unimportant. **1962** *Physical Rev.* CXXVII. 1888/2 The lattice conductivity was increased by high-temperature annealing. **1971** I. G. Gass et al. *Understanding Earth* v. 86/2 The lattice conductivity of the various rocks tends to converge with increasing temperatures to something like 2 to 3 W m^{-1} $°K^{-1}$. **1923** H. L. Brose tr. *Sommerfeld's Atomic Struct. & Spectral Lines* iii. 154 The crystal of smallest known lattice constant, namely, diamond. **1927** Lattice constant [see sense 3* above]. **1944** *Ann. Reg. 1943* 363 Siegbahn's determination of the lattice-constant of calcite at 18°C. **1969** Lattice constant [see *lengenbachite]. **1973** *Physical Rev.* B. VII. 674 Energy bands, Fermi surfaces, and densities of states of calcium as a function of lattice constant have been calculated. **1938** *Proc. Cambr. Philos. Soc.* XXXIV. 486 We define a free path L_1 for scattering by impurities and lattice defects on an atomic scale. **1959** *Phil. Mag.* IV. 468 When a metal is strained, the lattice defects introduced cause scattering of the phonons and electrons. **1924** *Physical Rev.* XXIII. 497 (*caption*) Lattice energies, from compressibility data, in kg-cal/g-mol. **1942** C. E. K. Mees *Theory Photogr. Process* iv. 183 In the fifth column are given the differences between silver and sodium salts for the electrostatic lattice energy as another measure of deformation energy. **1965** *Geochem. Internat.* II. 416/1 The methods of calculations of lattice energy are based on the assumption that the crystal is ideally ionic, but such crystals do not exist, and results are always approximate. **1861** *Trans. Illinois Agric. Soc.* IV. 259 An octagon, for exhibition of fancy articles..with a good lattice fence to keep people from the exhibition tables. **1964** H. O. Perkins *Espaliers & Vines* vi. 90 The lattice or Belgian fence types. **1935** E. A. Guillemin *Communication Networks* II. x. 409 (*caption*) Behavior of the reactances (891 *a*) versus the frequency variable $x = \omega/\omega_1$ for the low-pass lattice filter whose index and characteristic impedance functions are given by eqs. (889 *a*) and (890 *a*). **1970** J. Earl *Tuners & Amplifiers* ii. 43 Another arrangement employs a pair of quartz elements within a transformer, the idea then being more representative of the ordinary quartz crystal filter. Such a filter, called a crystal lattice-filter, is shown in Fig. 2.18. **1916** Blunden *Harbingers* 24 When the dryer in his oast Had loaded up his lattice-floors, He called a binman at the doors. **1924** *Harmsworth's Wireless Encycl.* II. 1274 *Lattice mast*, term used to describe a tall, composite structure for the support of a lofty aerial. This type of construction is carried out in both wood and metal... Such a mast is triangular in section, and comprises essentially three upright members held together by tie bars of metal and braced by diagonal bracing of stout timber. **1928** A. Williams *Telegr. & Telephony* xxi. 282 An aerial is carried by five 28-ton lattice masts, 287 feet high, each resting on four legs bolted to 20-ton concrete blocks. **1948** R. de Kerchove *Internat. Maritime Dict.* 399/1 *Lattice mast*, steel mast constructed of riveted structural steel shapes or lattice work. **1931** H. W. Bode in *U.S. Pat. 1,828,454* 20 Oct. 1/1 An important general property of a symmetrical lattice network is that its propagation constant and its characteristic impedance are mutually independent. **1934** A. T. Starr *Electr. Circuits & Wave Filters* vi. 196 Such a lattice network has image impedance *R* at all frequencies, no attenuation at any frequency, and a phase shift which depends upon the reactance characteristic of Z_1. **1956** Amos & Birkinshaw *Television Engin.* II. ix. 126 In general a lattice network has two series and two shunt elements as shown in Fig. 68; this particular network is a symmetrical one in which both series elements are equal to Z_1 and both shunt elements are equal to Z_2... For purposes of calculation it is often more convenient to redraw the network in the form of a bridge circuit. **1923** R. G. Collingwood *Roman Brit.* 75 Coarse ware with incised lattice-ornament. **1923** G. Barr tr. *Graetz's Recent Devel. Atomic Theory* iv. 96 We will call each such series of similar parallel planes a system of lattice planes. **1937** *Amer. Mineralogist* XXII. 449 Consider any two successive lattice planes perpendicular to a screw axis. **1973** K. W. Andrews *Physical Metall.* II. ii. 90 If the grating is actually composed of lattice planes, i.e. layers of atoms or molecules properly located in relation to the electron beam, then an image of these layers could be formed. **1926** *Encycl. Brit.* II. 832/2 A lattice point (*Gitterpunkt*) in space of any number of dimensions is a point with integral co-ordinates. **1936** A. H. Wilson *Theory of Metals* ii. 48 We assume as zero approximation that the electron is in the neighbourhood of one particular lattice point. **1955** *Sci. News* XV. 143 The gas is usually regarded as being accommodated on the lattice of the metal by occupying positions in its interstices, rather than replacing metal atoms at some of the lattice points. **1966** Ogilvy & Anderson *Excursions in Number Theory* x. 120 The point (20, 47), having both its coordinates integers, is called a lattice point of the plane. It is a point of intersection of a horizontal and a vertical line of the coordinate grid, or lattice. **1967** A. H. Cottrell *Introd. Metall.* xvii. 261 This array of lattice points is the space lattice of the crystal. It is important to notice that a lattice point is not an atomic site. In certain simple crystal structures..the pattern of atomic sites happens also to form a space lattice, but in many other structures..there is more than one atom in the motif. **1951** *Archit. Rev.* CIX. 389/2 The lattice-steel roof-trusses are supported on the inner leaf of the reinforced concrete walls and act as permanent shuttering for the concrete roof. **1942** R. H. Newton *Town & Davis: Architects* ii. 42 The few sources I have consulted say nothing significant..about Ithiel Town as the inventor of the lattice truss. **1936** A. H. Wilson *Theory of Metals* vi. 200 The coupling between the electrons and the lattice vibrations is due mainly to changes in the density of the solid. **1959** *Phil. Mag.* IV. 468 At low temperatures the dominant lattice vibrations are those of long wavelength which are not scattered by the impurity atoms. **1969** J. S. Blakemore *Solid State Physics* ii. 114 The particle—or phonon—aspect of lattice vibrations is particularly appropriate when we are concerned with energy transformation. **1936** A. H. Wilson *Theory of Metals* vi. 201 We should then be compelled to take into account the dispersion of the lattice waves. **1955** H. B. G. Casimir in W. Pauli *Niels Bohr* 119 At temperatures well below the so-called Debye temperature Θ only lattice waves with a wave-length of many atomic distances are excited. **1971** DONOVAN & ANGRESS *Lattice Vibrations* iv. 83 In the harmonic approximation the lattice waves travel independently, without hindrance, so that the mean free path is infinite and the thermal resistance is zero.

lattice, *v.* Add: **2.** *trans.* To form into a lattice, arrange as a lattice.
1950 *Amer. Speech* XXV. 24 'Homogeneous' piles and 'heterogeneous' piles, depending on whether the fissionable material is latticed with the moderating material.

‖ **latticinio** (latitʃī·nio), **latticino** (-ī·no). [It., f. L. *lacticinium* milk food.] An opaque white glass used in threads for decorative purposes in Venetian glass. Hence *attrib.*
1855 F. B. PALLISER tr. *Labarte's Handbk. Arts Middle Ages & Renaissance* ix. 348 The opaque white glass, the *latticinio* most usually employed in the filagree Venetian glasses, is only a glass coloured milk-white by oxide of tin or arsenic. **1881** C. C. HARRISON *Woman's Handiwork* III. 229 There are the *millefiori*,.. The *latticino*, with graceful milk-white spirals. The *avventurino*, with the lustre of pure gold. **1937** *Burlington Mag.* Nov. 218/2 Venetian *latticino* glass of the sixteenth and seventeenth centuries. **1969** *Canad. Antiques Collector* Sept. 26/2 Latticinio glass is produced by pouring clear glass around fine 'canes' or rods of white..glass to form a thick rod with thin white rods embedded in it. **1972** *Sunday Tel.* 21 May 10/7 Collectors..will be burying their visual senses in millefiori, butterflies, latticinio.

Latvian (læ·tviăn), *a.* and *sb.* [f. *Latvia*, Lett. and Lith. *Latvija*.] **A.** *adj.* Of or belonging to Latvia, since 1940 a constituent republic of the U.S.S.R., lying on the east coast of the Baltic Sea. **B.** *sb.* **a.** A native or inhabitant of Latvia. **b.** The language of Latvia; Lettish. Cf. LETT, LETTISH *a.* (*sb.*).
1920 *Contemp. Rev.* Aug. 283 Troops under German command on Latvian territory. **1924** J. M. MURRY *Voyage* ii. 28 All these new languages. Lithuanian, Latvian, Esthonian, Transcaucasian. **1926** *Spectator* 31 July 176/1 Latvian is certainly not so difficult to learn as Chinese. **1941** J. H. JACKSON *Estonia* ii. 38 A branch of the Estonian race, the Livs, waged endless warfare with the Letts or Latvians. **1955** *Times* 16 Aug. 9/6 The Lithuanians and Latvians, ethnologically Indo-European, are survivors of ancient peoples. **1964** G. BENNETT *Cowan's War* i. 22 The defences of Riga crumbled, which prevented the workers of all three Latvian ports from declaring for Bolshevism in November. **1973** *Listener* 15 Nov. 656/3 Those who have an ethnic or religious fellow-feeling, whether they be.. Latvians or Georgians, Kurds or Nagas.

‖ **lau** (lɑu). [Native word.] An African water monster supposed to live in the swamps of the Nile valley.
1923 H. C. JACKSON in *Sudan Notes & Rec.* VI. 187 There is also a third kind [of python] of which rumours have come to my ears... This serpent is called Lau by the Nuer and Dinka... It is reported to be of gigantic proportions. **1925** *Blackw. Mag.* Sept. 303/2 The *lau* is a composite beast; it is reputed to have a bit of the bird, snake and lizard in it, like the wyvern in coats-of-arms. **1937** *Discovery* Dec. 369/1 The *lau* and the *lukwata*, monstrous beasts whose hideous calls are heard booming through the grey night-mists of the lakes.

lauan (lawā·n). [Tagalog *lawaan*.] The Philippine name for the light hardwood timber produced by trees of the genus *Shorea* or closely related genera.
1894 H. M. WARD *Laslett's Timber & Timber Trees* (ed. 2) xxii. 226 The outside planks of the old Manilla and Acapulco galleons were of Lauan wood. **1936** *Nature* 26 Dec. 1090/2 The top and side clasps [of the tennis racket] are of Malayan lauan. **1971** N. E. HICKIN *Wood Preservation* 88 Shorea. A very extensive genus of hardwood trees found chiefly in Malaysia, Indonesia, Borneo and the Philippines, the principal timbers of which are generally known as meranti, seraya and lauan, suitably qualified.

laubmannite (lɑu·bmænəit). *Min.* [f. the name of Heinrich *Laubmann*, 20th-century German mineralogist + -ITE¹.] A basic phosphate of ferrous and ferric iron, $Fe^{II}_3Fe^{III}_6(PO_4)_4(OH)_{12}$, of yellow- to grey-green colour.
1949 C. FRONDEL in *Amer. Mineralogist* XXXIV. 536 The name laubmannite is proposed for the species. **1970** *Ibid.* LV. 138 A new locality was discovered, the laubmannite occurring as bright yellow-green aggregates and affording a powder pattern virtually identical with the Arkansas material. The location is Leveäniemi in the Svappavaara mining district, Norrbotten Province, Sweden.

‖ **laud** (la͵ū·d). [Sp.: see LUTE *sb.*¹] A Spanish lute.
1876 STAINER & BARRETT *Dict. Mus. Terms* 276 The word [*sc.* lute]..is most probably from the Arabic *el'ood*, as the instruments came into Europe from the Moors through the Spaniards, who still call it *laud*. **1923** *Blackw. Mag.* July 38/1 The Spanish laud or lute Jo had bought in Murcia during the previous year. **1954** *Grove's Dict. Mus.* (ed. 5) I. 400/1 The instrument [*sc.* the bandurria] is in common use in the south of Spain, generally in conjunction with the *laud* and the *guitarra*.

laudanine. Add: [ad. G. *laudanin* (O. Hesse 1870, in *Ann. d. Chem. und Pharm.* CLIII. 49).] (Examples.)
1871 *Jrnl. Chem. Soc.* XXIV. 1064 White crystals are thereby obtained, from which the laudanine is separated by the action of hydriodic acid, with which it forms a difficultly soluble compound. Laudanine has the composition $C_{20}H_{25}NO_4$, instead of the formula $C_{20}H_{25}NO_3$, as formerly given. **1954** A. BURGER in Manske & Holmes *Alkaloids* IV. 57 Laudanine is optically inactive in spite of the presence of an asymmetric carbon atom in its formula.

laudanosine (lǭdæ·nosīn). *Chem.* [ad. G. *laudanosin* (O. Hesse 1871, in *Ber. d. Deut. Chem. Ges.* IV. 696): see LAUDANINE and -OSE².)] Laudanine methyl ether, $C_{21}H_{27}NO_4$, the dextrorotatory form of which occurs in opium and is a strong tetanic poison.
1871 *Jrnl. Chem. Soc.* XXIV. 1065 Laudanosine..dissolves sparingly in cold, but easily in hot benzol; it forms colourless prisms which melt at 89°. **1900** *Chemist & Druggist* LVII. 846/1 It is therefore probable that laudanosine is *d-N*-methyl-tetra-hydro-papaverine. **1951** [see *GNOSCOPINE]. **1972** *Phytochemistry* XI. 461 The major alkaloids of the above-ground parts of *Argemone grandiflora*..were found to be berberine, α-allocryptopine and protopine; (+)-laudanosine..and chelerythine were identified as minor alkaloids.

‖ **laudator temporis acti** (lǭdē̆i·təɪ te·mpŏris æ·ktəi). [See LAUDATOR.] A Latin phrase, from Horace's *laudator temporis acti se puero* 'a praiser of time past when he himself was a boy' (*Ars Poetica* 173), used of one who looks back to the past as a better time.
1736 SWIFT *Let.* 2 Dec. in Pope *Works* (1757) IX. 209 Have you got a supply of new friends to make up for those who are gone?.. I am afraid it is with friends as with times; and that the *laudator temporis acti se puero*, is equally applicable to both. **1753** CHESTERFIELD in *World* 6 Dec. 293, I am neither sour nor silly enough yet, to be a snarling *laudator temporis acti*. **1814** *Edin. Rev.* XXIII. 316 The suspected praises of any of the *laudatores temporis acti*. **1870** A. C. EWALD *Guide Indian Civil Service* 13 Some old Indians—*laudatores temporis acti*—are wont to make the most doleful prophecies regarding the results of abolishing Haileybury and introducing the Civil Service Commissioners in its place. **1923** BEERBOHM *Peep into Past* 10 Not that he [*sc.* Wilde] is a mere *laudator temporis acti*. **1931** BLUNDEN *Votive Tablets* 360 We can hardly avoid being *laudatores temporis acti* when we think even of Oxford..as it has been, and as it now is. **1965** O. BARFIELD in J. Gibb *Light on C. S. Lewis* p. xv, He..ended his days..as a kind of..guilt-oppressed *laudator temporis acti*.

Laue² (lɑu·ə). *Cryst.* The name of Max von *Laue* (1879–1960), German physicist, used *attrib.* with reference to a method of X-ray diffraction developed by him in which a narrow parallel beam of polychromatic X-radiation is directed at a thin crystal and the resulting diffraction pattern is recorded on a photographic film placed either in front of the crystal (with a hole for the passage of the incident beam) or beyond it; as *Laue method, pattern, photograph*; also **Laue condition,** each of the three equations (one for each linear parameter of the unit cell) which must be satisfied for a diffracted beam to occur in a given direction for particular orientations of the crystal and the incident beam; **Laue spot,** a spot on a Laue photograph corresponding to a diffracted ray.
1915 W. H. & W. L. BRAGG *X Rays & Crystal Struct.* xii. 208 Each spot in a Laue photograph represents the reflection of the X-rays by a certain plane (*hkl*) of the crystal structure. **1935** *Jrnl. Chem. Physics* III. 421/1 Oscillation and Laue photographs were prepared with crystals of lepidocrocite from Eiserfeld, Westerfald, Germany. **1940** *Physical Rev.* LVII. 448/1 Crystals that are naturally in a strained condition..show 'asterism' of Laue spots. **1940** *Nature* 7 Sept. 332/2 The diffuse spot pattern changes much more slowly than the Laue pattern on successive photographs. **1955** E. S. GOULD *Inorg. Reactions & Struct.* xx. 310 The Laue method, employing polychromatic radiation, is no longer important as a tool for structure determinations. Laue photographs are used chiefly for lining up crystals preliminary to examination by filtered radiation, for partial indications of symmetry, and for indicating imperfections and deformations in crystals. **1966** D. G. BRANDON *Mod. Techniques Metallogr.* 69 (*caption*) Derivation of Laue conditions for diffraction. **1973** P. WILKES *Solid State Theory in Metall.* viii. 197 For the three-dimensional case we require that for constructive interference all three Laue conditions..be satisfied simultaneously. **1973** *Soviet Physics: Crystallogr.* XVIII. 320/1 The symmetry of the Laue pattern allowed the mineral to belong to the hexagonal system.

Laufen (lɑu·fĕn). The name of a place in W. Germany near Salzburg, adopted by A. Penck (in Penck & Brückner *Die Alpen im Eiszeitalter* (1909) I. ii. 157, 248) and used *attrib.* to designate a minor retreat and advance of glaciation which he believed followed the last major (Würm) glaciation in the Alps.
1927 PEAKE & FLEURE *Hunters & Artists* 4 After the maximum of the Würm glaciation came a shrinkage called the Laufen retreat, but only of relatively short duration. **1939** AUDEN & ISHERWOOD *Journey to War* 292 When we emerged from holes And blinked in the warm sunshine of the Laufen ice retreat. **1957** J. K. CHARLESWORTH *Quaternary Era* II. xlii. 1174 The Laufen oscillation is probably to be equated with the Aurignacian oscillation.

laugh, *sb.* Add: **3.** *to have the last laugh* (and similar phrs.): to be successful in the end; *to have*, or *get, the laugh on*, or *over*, (*someone*): to have (someone) at a disadvantage; so *the laugh is on* (someone).
1909 J. LONDON *Let.* 1 July (1966) 280 The laugh is on me. I confess to having been fooled by Mr. Harris's canard. **1925** *Times* 21 Mar. 12/4 *The Last Laugh*, the German film which was shown at the Capitol in the Haymarket for the first time on Thursday, is another example of the new school of film production, the basis of which is the recognition of imagination in the spectator. **1937** G. & I. GERSHWIN (song) *They All Laughed* 4 They laughed at us and how! But Ho, Ho, Ho! Who's got the last laugh now? **1942** E. PAUL *Narrow St.* v. 40 Guy was hauled up, put on the carpet, and when he learned that the uncle was willing to make a rather generous cash settlement, considered that he had the last laugh on his fellow workers. **1944** G. B. SHAW *Everybody's Pol. What's What?* xxx. 259 The laugh was on the mob, not on Fouquier. **1949** W. S. MAUGHAM *Writer's Notebook* 329 Sometimes we die sitting quietly in an armchair over a whisky and soda... Then, I suppose, we have the laugh over those who..never rested till the end. **1954** A. MARX *Groucho* xxiii. 200 If she happened to make an error..he would say, 'Well, who's got the last laaaaff now?' **1966** 'J. HACKSTON' *Father clears Out* 18 She's got the laugh on me this time, all right. **1968** D. GODFREY in R. Weaver *Canad. Short Stories* 2nd ser. 306 The Yankee came back about the end of August and we had to give him the last laugh. **1975** J. AIKEN *Voices in Empty House* iv. 121 The dead really have the last laugh on the living.
4. b. A cause of laughter; a joke. Freq. ironic, as in phr. *that's a laugh*, etc.
1895 G. B. SHAW *Our Theatres in Nineties* (1932) I. 51 The piece contained three or four 'laughs' which could not possibly have been explained or described at a dinner party. **1921** *Motion Picture Mag.* Oct. 21/2 There is unlimited room for the screen comedy of manners and for comedy that depends for its laughs upon the sheer power of clever situations. **1930** W. R. BURNETT *Iron Man* I. 3 Ain't that a laugh!.. That guy's been sleeping for the last half-hour, and he says we're a lot of company. **1960** J. WAIN *Nuncle* 165 'Your friends paid for it.' That was a laugh. My friends..were a one-way valve for drinks, cigarettes and loans. **1961** A. WILSON *Old Men at Zoo* i. 51 That's a laugh. When Leacock was head of the Aquarium, he did absolutely nothing. **1966** 'H. CALVIN' *Italian Gadget* ix. 149 Embellished or not, the story would be taken as a wild laugh. **1972** D. DEVINE *Three Green Bottles* III. i. 106 She fell for Dr Kendall and he chucked her too. It's a laugh when you think of it.
5. laugh-line, (*a*) in theatrical use, a comic line received with laughter; (*b*) = *laughter-line* (see *LAUGHTER *sb.* 3); **laugh track,** a recording of audience laughter added to a sound track.
1927 M. SULLIVAN *Our Times* II. 106 'Uncle Herbert's Speaker' gave text and minute directions for what modern comedians would call 'putting over the "laugh line"'. **1960** *20th Cent.* Nov. 470 The humour does not consist of laugh lines, but of moods and contrasts. **1967** 'T. WELLS' *Dead by Light of Moon* (1968) vi. 60 Bright green eyes and laugh lines around the mouth. **1969** W. GARNER *Us or Them War* i. 19 'It scared the living daylights out of me. Except it was dark,' he finished, but nobody played up to the laugh line. **1971** P. O'DONNELL *Impossible Virgin* xii. 252 The little laugh-lines at the corners of her eyes..had gone. **1962** *Variety* 22 Aug. 14 Universal is on the laugh-track for extended play. And the logic is unarguable, as witness the current 'Lover Come Back'. **1966** *N.Y. Times* 20 Nov. 19 Perhaps symptomatic of the quiet corrosion around 'The Jean Arthur Show' is the 'laugh track' or 'canned laughter' on the show. **1969** *Punch* 5 Feb. 193/3 The absence of a laugh track which would only foul up the pace. **1970** *Toronto Daily Star* 24 Sept. 30/3 (*caption*) Program's tired laugh-track..and tired non-plot.

laugh, *v.* Add: **1. b.** *don't make me laugh*: expostulatory phr. freq. used ironically; *to make a cat laugh*: see *CAT *sb.*¹ 13 j; *laugh! I thought I'd die*: exclamatory phr. to indicate excessive laughter; *to laugh like a drain*: see *DRAIN *sb.* 1 f; *to laugh on the other, wrong, side* (*of one's face*) (examples); *to laugh out of court*: see *COURT *sb.*¹ 12 c.
1837 CARLYLE *Diamond Necklace* iii, in *Fraser's Mag.* Jan. 7/1 By-and-by thou wilt laugh on the wrong side of thy face mainly. **1894** A. CHEVALIER *Humorous Songs*, Laugh! I thought I should 'ave died, Knock'd 'em in the Old Kent Road! **1898** J. D. BRAYSHAW *Slum Silhouettes* 246 'E does a bunk dahn the street, lookin' fer all the world like a hunder-done pancake. Laugh—I thought I should ha' died. *Ibid.*, An' as fer you, my lady, wait till I've got yer, I'll make yer laugh the uvver side o' yer face. **1925** D. H. LAWRENCE *Mornings in Mexico* (1927) 21 The monkey..mocks at you and gibes at you and imitates you. ..It's funny, and you laugh just a bit on the wrong side of your face. **1951** 'F. O'CONNOR' *Traveller's Samples* 43 'Who are ye laughing at?' I shouted, clenching my fists at them. 'I'll make ye laugh at the other side of yeer faces if ye don't let me pass.' **1958** *Spectator* 22 Aug. 241/3 The fact that a resolute Government (don't make me laugh) could stuff its fingers in its ears and carry on regardless makes no difference; nothing will in fact be done. **1966** *Observer* 20 Mar. 27/3 The jokes..tended to make one laugh on the other side of one's face. **1967** J. B. PRIESTLEY *It's an Old Country* xiii. 142 'Mind you, I'll never believe there was anything between him and Mum——' 'Don't make me laugh,' Vic said, giving Tom a wink. **1975** S.

JOHNSON *Urbane Guerilla* II. 70 Stanton will soon laugh on the other side of his face.
c. *to be laughing*: to be in a fortunate or successful position (see quot. 1930). *colloq.*
1930 BROPHY & PARTRIDGE *Songs & Slang Brit. Soldier* 136 *Laughing*, comfortable, safe, fortunate, especially in contrast with others or with normal circumstances. E.g. 'He's got a job at Brigade Head Quarters, so he's laughing'; 'Once I get to the C.C.S. I'm laughing'. **1968** *Listener* 19 Dec. 812/3 Oh, Ron, he's got a job—£30 a week he can get now, you know. Skilled motor mechanic, and not put-on like it used to be. Runs his own racket as well. Old Ron's laughing. **1975** M. STANIER *Singing Time* 255 So long as you're a jump ahead you're laughing.
7. *to laugh that off*: phrase used ironically (freq. in imperative) as an invitation to dismiss or get rid of (some accomplished fact) with a laugh; *to laugh* (someone) *out of it*: to persuade (someone) out of a depressed or serious, etc., mood with laughter.
1918 F. B. YOUNG *Crescent Moon* v. 96 Eva tried to laugh him out of it, to make him ashamed of being afraid. **1926** MAINES & GRANT *Wise-Crack Dict.* 11/1 *Laugh that off*, controversial triumph. **1936** 'N. BLAKE' *Thou Shell of Death* xii. 224 Why should he want them [*sc.* footprints] preserved if it wasn't he who originally made them?.. Laugh that one off! **1944** L. MACNEICE *Christopher Columbus* I. 57 They have given me all that I asked—Let Talavera laugh that off if he can. **1974** *Times* 15 Jan. 14/6, I claim to have a complete answer to the charge, so laugh *that* off, Sir Peter.

laugh-in (lɑ·fin). [*-IN³.] A demonstration, event, or situation marked by laughter, often staged for this purpose; *spec.* as the name of an American television comedy programme.
1968 *N.Y. Times* 23 Jan. 79/2 The increasing liberality and topicality of Hollywood variety comedy was further evidenced last night in the hour of Dan Rowan and Dick Martin, whose 'Laugh In' had a preseason tryout and now has deservedly won a niche as a regular series at 8 P.M. Mondays... Their hour is an extraordinary quick succession of sight laughs and sketches, many with a deft and good-natured satirical edge to give the show a contemporary pertinency. **1968** *Manch. Guardian Weekly* 21 Mar. 6 As part of their demonstration against the Defence Minister, Mr Healey...students at Cambridge proposed to organise a 'laugh-in'. **1968** *Listener* 26 Dec. 854/2 There's a kind of cathartic quality about Danny la Rue that is a tremendous relief after weeks of trying to admire the Rowan and Martin Laugh-In. **1969** *Time* 6 June 56 At an airport, Fielding's baggage check-in is a laugh-in. **1969** *Guardian* 10 Feb. 8/3 'Rowan and Martin's Laugh-in' is a proudly esoteric American comedy series shown late on Sunday night on BBC-2. **1974** HAWKEY & BINGHAM *Wild Card* ii. 26 It had not been Wallcroft's scene at all, and he'd had to eat a lot of dirt to stay in the [television] business through its *Laugh-in* phase.

laughing, *vbl. sb.* Add: **b. laughing death** = *KURU.
1958 *Times* 9 Jan. 10/1 The newly discovered illness in New Guinea..has become known as the 'laughing death'. ..The malady is comparable in some respects to paralysis agitans. **1967** *Acta Tropica* XXIV. 193 (*heading*) Kuru—the laughing death.

laughing, *ppl. a.* Add: **b. laughing dove,** the African dove, *Stigmatopelia senegalensis*; **laughing gull,** a North American gull, *Larus atricilla*; **laughing jackass** = JACKASS *sb.* 3 (q.v. for examples), *KOOKABURRA.
1881 E. E. FREWER tr. *Holub's Seven Yrs. S. Afr.* I. ii. 47 The most common birds in the Riet River valley are doves, and those almost exclusively of two sorts, the South African blue-grey turtle-dove, and the laughing dove. **1966** E. PALMER *Plains of Camdeboo* xii. 196 The thorn trees were full of laughing dove. **1789** J. MORSE *Amer. Geogr.* 59 American Birds [include].. Laughing Gull, Goose, Canada Goose [etc.]. **1884** *Bull. U.S. Nat. Museum* No. 27. 169 Laughing Gull... Atlantic coast, from Maine (casually) to mouth of the Amazon. **1968** *Times* 10 Oct. 8/8 The laughing gull could be a useful animal for studying colour perception.
c. *laughing-eyed* adj.
1851 H. MELVILLE *Moby Dick* cxxxii. 597 So have I seen little Miriam and Martha, laughing-eyed elves, heedlessly gambol around their old sire. **1896** H. BELLOC *Verses & Sonnets* 57 This is the laughing-eyed amongst them all: My lady's month **1909** *Westm. Gaz.* 7 Aug. 9/1 Pale-faced women were hugging to their hearts their rosy-cheeked, laughing-eyed children.

laughter, *sb.* Add: **3. laughter-line,** one of the small wrinkles at the corners of the eyes or mouth supposedly formed by years of intermittent laughter.
1938 M. ALLINGHAM *Fashion in Shrouds* xii. 180 His light grey eyes were entirely without humour in spite of the laughter-lines beside them. **1950** *Vogue Beauty Bk.* Autumn 26 You should watch for wrinkles—expression lines that run from nose to mouth, laughter lines round the eyes and frown lines on the forehead. **1971** R. FALKIRK *Chill Factor* iii. 35 Laughter-lines still cobwebbed the corners of his eyes.

laughy, *a.* (Later examples.)
1906 B. VON HUTTEN *What became of Pam* II. ix. 172, I suppose you felt teary, but now you must feel laughy. **1913** G. STRATTON-PORTER *Laddie* vii. 201 Then father, all laughy and criey, said: 'Thank God!' **1950** *Sunday Jrnl. & Star* (Lincoln, Nebraska) 29 Oct. (*headline*) Mister 880 is a charming laughy movie.

‖ **lauhala** (lɑuhā·lă). [Hawaiian *lau* leaf + *hala* pandanus.] A Polynesian screw pine, *Pandanus tectorius*; the dried leaves of the tree, or the material plaited from them.
1826 M. GRAHAM *Voy. H.M.S. Blonde to Sandwich Is.* II. 108 A church..has lately been erected here: its walls are of reeds, lined with the woven leaves of the lauhala. **1866** 'MARK TWAIN' *Lett. from Hawaii* (1967) 99 Shady groves of forest trees..the breadfruit, the lau hala, the orange, lime. **1875** I. L. BIRD *Hawaiian Archipelago* viii. 111 Then we come upon a whole cluster of grass houses under *lauhalas* and bananas. **1898** J. A. OWEN *Story Hawaii* iii. 87 Around the room are hung native mats woven from the long leaf of the 'lauhala' tree. **1933** E. H. BRYAN in *Ancient Hawaiian Civilization* 127 Much of the plaited work, such as baskets, mats, pillows, and fans, was made from *lauhala*, the leaf of the pandanus. **1954** J. SHERIDAN in J. Macdonald *Lethal Sex* (1962) 174 With *lauhala* mats and some reed furniture..it will be quite charming. **1957** P. H. BUCK *Arts & Crafts Hawaii* 250 Sandals.. were quickly woven of any tough fiber at hand: *lauhala*, ti, banana leaves.

‖ **laulau** (lɑu·lɑu). [Hawaiian, reduplicated form of *lau* leaf.] A portion of a Hawaiian dish of meat and fish wrapped in leaves and steamed or baked (see quots.). Also, this cover of leaves.
1940 K. BAZORE, *Hawaiian & Pacific Foods* i. 41 *Laulau*, an individual portion of fresh belly pork and of salted butterfish or salmon, wrapped in several taro leaves, then in ti leaves, and steamed or cooked in an imu. **1954** J. SHERIDAN in J. Macdonald *Lethal Sex* (1962) 162 We ate with the Hawaiian family..steamed *laulaus* (salt salmon, butterfish, and pork wrapped in *ti* leaves), *poi* and coconut pudding. **1957** P. H. BUCK *Arts & Crafts Hawaii* 18 Like fish, they might be wrapped in the ti-leaf cover called *laulau*. **1961** E. FODOR *Hawaii* 15 The Manoa campus is transformed..into a kind of miniature Pacific worlds fair where you can sample kim chee, laulaus, sukiyaki and other Pacific foods.

launch, *sb.*¹ Add: The usual pronunc. is now (lǭntʃ) (similarly with the vb. and its derivs.).
2. (Later example.)
1897 W. C. RUSSELL *Last Entry* 241 The schooner.. swept in long floating launches down upon the boat.
4. a. (Further *fig.* and *transf.* examples.)
1749 J. CLELAND *Mem. Woman Pleasure* I. 18, I soon came to a resolution of making this launch into the wide world, by repairing to London. **1857** *Trans. Mich. Agric. Soc.* VIII. 193, I have seen the commencement of railways, and witnessed the 'launch' of the first locomotive. **1969** J. ARGENTI *Managem. Techniques* v. 25 Anxiety that the launch date [of a product] will be missed. **1969** *Punch* 15 Jan. 96/2 The Ford Capri, a sort of shrunken Mustang, is being built in Britain and Germany and will be launched later this month. But, of course, 'the launch', as the trade calls it, is not as simple as that. **1971** *Sunday Express* (Johannesburg) 28 Mar. 5/1 Mr. Uys..vetoed the display of the same model's nipple when the launch advertisement was submitted.
c. The launching of a missile, spacecraft, glider, or the like. (See also sense 7 below.)
1935 C. H. LATIMER-NEEDHAM *Gliding & Soaring* x. 170 The wind velocity should be ascertained and allowed for in any method of mechanical launching or too vigorous a launch may be given unwittingly. **1952** F. GEEN *A.B.C. of Gliding* 95 The easiest launch is a full-height nose launch. **1963** *Ann. Reg. 1962* 397 The closely matched orbits of the two astronauts also required precise timing of their launches. **1966** *Economist* 18 June 1307/1 The first of three such launches which are to put a chain of 24 defence communications satellites in synchronous orbit round the earth. **1969** *Observer* 20 July 9/8 The astronauts ..sleep or doze for nearly five hours before preparing for the launch.
7. *attrib.* and *Comb.* (sense *4 c) *launch crew, date, site, vehicle*; **launch pad** = *launching pad*; **launch window,** a period outside which the planned launch of a spacecraft cannot take place if the journey is to be completed, owing to the changing positions of the planets.
1962 J. GLENN in *Into Orbit* 6 The most junior member of the launch crew. **1969** *Daily Mail* 15 Jan. 5/3 Then suddenly it was now—launch date just around the corner. **1960** *News Chron.* 29 Sept. 9/6 The 100-foot rocket sat immobile on its launch-pad. **1968** *Times* 23 Dec. 6/3 Captain Lovell and Major Anders..climbed into a van which took them to the launch pad. **1969** *Times* 3 June Suppl. p. iii/1 The world's largest tracked vehicle, 'the crawler',.. carried..Saturn 5..from its lofty assembly building to launchpad 39A. **1969** *Listener* 20 Feb. 233/2 When you fly over the Soviet Union, can you see their launch sites? **1965** *New Scientist* 18 Mar. 701/1 The Gemini spacecraft, launch vehicle and target vehicle are all derived from hardware and technology already in existence. **1966** *Sci. Amer.* Jan. 54 Because of various failures in the launch-vehicle guidance system..a lunar landing was not accomplished. **1965** *Newsweek* 29 Nov. 40/3 It is thought they may even try a third shot before the launch window closes in December. **1966** *Sci. News Let.* 3 Sept. 165 The 20-day period centered around the launch date allowing travel between planets on an orbit requiring the least amount of energy. This is the so-called 'launch window' used to hurl space vehicles from earth to the moon. **1968** *Radio Times* 19 Dec. 41/4 As to timing, they must choose a launch window several days long when the Moon is in the right position relative to the Earth, when the sun is in the right position relative to the lunar landing sites, and when Apollo 8 can return to a suitable landing on Earth.

launch, *sb.*² Add: **2.** *Comb.* **launchman,** a man who operates a launch.
1924 J. MASEFIELD *Sard Harker* 146 Everybody was very still, except for the launchman munching his onion. **1928** *Daily Mail* 13 Aug. 13/4 The complement consists of captain, first and second mate, two cooks, two stewards, boatswain, launchman, and able seamen. **1963** M. SHADBOLT in C. K. Stead *N.Z. Short Stories* (1966) 322 So my father produced the launchman and people from the township as witnesses.
3. Special Comb.: **launch-engine** (see quot. *a* 1877).
a **1877** KNIGHT *Dict. Mech.* II. 1266/2 Launch-engines generally consist of a boiler with engines attached thereto, and are used for propelling the launches of large ocean steamers in shallow harbors, etc. **1889** P. N. HASLUCK *Model Engineer's Handybk.* vi. 69 A double cylinder launch engine fitted with reversing motion. **1909** *Westm. Gaz.* 23 Mar. 4/3 A very fine launch-engine, fitted with.. reversing gear.

launch, *v.* Add: **2.** *spec.* To send off (a rocket, spacecraft, or the like, or an astronaut) on its (or his) course. (Cf. 4 b in Dict. and Suppl., from which this use may equally derive.)
1873 MERCIER & KING tr. *Verne's From Earth to Moon* 145 The gun destined to launch the projectile had to be fixed in a country situated between the 0 and 28th degrees of north or south latitude. *Ibid.*, Launched on the 1st of December,..it ought to reach the moon four days after its departure. **1922** *Encycl. Brit.* XXX. 50/1 A fore-castle deck large enough to enable a seaplane to be launched therefrom on a light subsidiary carriage. **1952** *Oxf. Jun. Encycl.* X. 17/2 The German guided missiles.. launched against London from the French coast were driven by their own power and were automatically controlled. **1957** *Britannica Bk. of Year* 443 The first artificial earth satellite..was launched from a site in the U.S.S.R... on Oct. 4, 1957. **1960** J. N. BELL *Seven into Space* i. 15 He knows an excitement so intense that it seems he can no longer contain it. The first American has been launched into space. **1972** A. C. KERMODE *Mech. of Flight* (ed. 8) xii. 390 As with the X15 these [*sc.* lifting bodies] are launched from a mother craft. **1974** *Daily Tel.* 14 Feb. 1/3 Two more spaceships, Mars-6, and Mars-7, which were launched last August, were due to approach the planet next month.
4. b. *spec.* To release (a balloon or its contents) into the air at the beginning of a flight. (Cf. *2.)
1824 *Encycl. Brit.* Suppl. I. 83/1 It was soon found, that a balloon, launched into the atmosphere, is abandoned, without guidance or command, to the mercy of the winds. **1959** *Chambers's Encycl.* I. 103/2 On 19 Sept. 1783..they launched a sheep, a cock and a duck into the air, enclosed in a basket suspended beneath the balloon.
d. To publish (a book); to put (a product, etc.) on the market.
1870 'MARK TWAIN' *Lett. to Publishers* (1967) 45 We'll have someone standing ready to launch a book right on our big tidal wave and swim it into a success. **1919** J. QUINN *Let.* 3 Oct. in T. S. Eliot *Waste Land Drafts* (1971) p. xvii, My part in connection with launching your book is finished. **1926** H. CRANE *Let.* 5 Dec. (1965) 278 Once this first book is really launched and off my mind. **1966** *Listener* 17 Nov. 716/3 The complicated process of launching a new American car. **1969** J. ARGENTI *Managem. Techniques* v. 25 To launch a product is a complex project.
5. (Later example.)
1906 *Westm. Gaz.* 26 Nov. 6/2 The payment was refused on the ground that the Deal lifeboat launched to the same wreck.

launcher. Add: **2.** A device or structure that launches something or is used for launching; *spec.* (*a*) a structure that holds a rocket or missile during launching; (*b*) a rocket from which a satellite is released into orbit.
1911 T. O'B. HUBBARD et al. *Aeroplane* v. 61 There have been many..mechanical launchers invented, but none have met with any success..except the rail-and-falling-weight method originally used by the Wright biplanes. **1944** C. P. LENT *Rocket Research* 4 The photograph lower left is a German rocket launcher. **1945** *Aeroplane Spotter* 18 Oct. 250/2 Zero-type rocket-projectile launchers eliminated the drag of the earlier tube type to such an extent that the Lockheed P-38J and P-38L Lightning is able to carry seven under each wing. **1957** *Oxford Mail* 20 Aug. 1/4, I saw the 25 ft. long Bloodhound on its launcher, together with its special loading trolley, exactly as it would be on one of the coastal sites. **1958** *Times* 12 Aug. 7/3 Now that Britain can provide the launcher, she should certainly explore the field for partners in the actual launching of a satellite. **1959**, **1969** [see *grenade launcher* s.v. *GRENADE *sb.*¹ 3]. **1970** *Daily Tel.* 28 Apr. 1/4 America was first to use a single launcher to put eight satellites into orbit, in June, 1966. **1971** *Nature* 2 Apr. 282/2 An extensive study of the propagation of Earth surface waves, including the design of launchers for such waves, is described. **1973** *Sci. Amer.* Aug. 105/2, I devised an apparatus for launching a continuous stream of water globules indoors... The launcher consists of a small nozzle that directs a jet of water upward at an angle of approximately 45 degrees. **1974** *Daily Tel.* 14 Feb. 4/8 The ship has been fitted with four missile launchers.

launching, *vbl. sb.* Add: (Later examples.)
1967 *Listener* 23 Feb. 263/3 Admittedly, the launchings will be carried out by American rockets from an American site, but the satellites themselves are purely British-built. **1971** *Nature* 6 Aug. 357/2 It is not safe to base a rocket development project on a single launching once a year.

b. *launching platform, site, station*; **launching pad,** the area on which a rocket stands for launching; also *fig.* and *transf.*
1951 COOKE & CAIDIN *Jets, Rockets & Guided Missiles* 138 Under a blazing afternoon sun, at 3.14 p.m., a modified V-2 rocket carrying a WAC-Corporal in its nose rose slowly from its concrete launching pad. **1958** *Daily Mail* 16 Aug. 1/4 The 88 ft. rocket stands poised on its concrete launching pad here tonight looking like a giant silver propelling pencil. **1959** *Encounter* Dec. 74/2 All this is by way of a launching-pad for the idea of the Non-Nuclear Club. **1963** A. HUXLEY *Let.* 17 Feb. (1969) 948 Julian tells me that your book is now definitely on the launching pad. **1973** *Guardian* 31 Jan. 13/7 The NUS sees the rent-strike movement as a launching pad for its main campaign. **1922** *Encycl. Brit.* XXX. 50/1 Ordinary aeroplanes were carried in fighting-ships with a launching-platform. **1957** *Jane's Fighting Ships 1957–58* 413 The missile is using jet-assisted rocket bottles to launch it from its zero-length launching platform. **1944** *Aeronautics* Aug. 27/1 The counter attack, by bombing the launching sites in the Pas de Calais, was intensified. **1958** *Listener* 13 Nov. 766/2 Israel had also agreed to launching-sites on her territory for United States atomic rockets and guided missiles. **1897** *Strand Mag.* June 712/1 We had better not make the launching station a place like the bank of the river, where it can go only one way. **1944** A. HUXLEY *Let.* 9 July (1969) 507 Five thousand launching stations, firing off twenty robots [*sc.* rockets] apiece—and that would be the end of any metropolis. **1958** C. C. ADAMS et al. *Space Flight* p. xi, A space station would serve as a 'launching station' for space ships to the moon, saving fabulous amounts of precious fuel.

launder, *sb.* For '4–9 lander' read '4– lander' and add: **2. d.** *Metallurgy.* A channel for conveying molten metal from a furnace or container to a ladle or mould.
1900 *Kynoch Jrnl.* Oct.–Nov. 20/1 The tapping hole is now cut through the bottom of the furnace, and a wrought iron channel—technically called a lander—fastened round it. **1906** W. MACFARLANE *Princ. & Pract. Iron & Steel Manuf.* x. 110 The Shoot or Launder, along which the steel and slag are conveyed from the taphole to the ladle, is a half-round gutter made of steel plates. **1929** W. LISTER *Pract. Steelmaking* x. 78 With all fixed furnaces the lander is necessarily 10 ft. to 15 ft. long, and sometimes up to 20 ft. long. **1967** P. MCGEOWN *Heat the Furnace* x. 97 We had her running down the lander as the twelve o'clock hooter sounded. **1971** W. K. V. GALE *Iron & Steel Industry: Dict. Terms* 195 The tapping spout of an open-hearth furnace is usually called a launder.

launder, *v.* Add: **1. b.** To transfer funds of dubious or illegal origin, usu. to a foreign country, and then later to recover them from what seem to be 'clean' (i.e. legitimate) sources. Also *transf.*
The use arose from the Watergate inquiry in the United States in 1973–4.
1973 *Guardian* 19 Apr. 14/2 Suitcases stuffed with 200,000 dollars of Republican campaign funds; money being 'laundered' in Mexico. **1973** *Publishers Weekly* 17 Sept. 54/2 A New York lawyer carrying $200,000 in his camera case to be 'laundered' in Switzerland. **1973** J. M. WHITE *Garden Game* 128 Phoenix is a city where the Mafia is well entrenched; its booming real-estate, building and service industries are ready-made havens for 'laundering' the extortion and gambling money from Nevada and California. **1974** *Globe & Mail* (Toronto) 3 Apr. 1 (*headline*) Kerr concedes U.S. criminals 'launder' money in Ontario.
3. *intr.* Of a fabric: to admit of being laundered; to bear laundering without damage to its texture, colour, etc. Used with adverbs.
1908 *Sears, Roebuck Catal.* 916 It will launder as well as a piece of linen. **1909** *Daily Chron.* 22 July 7/5 A single initial..done in satin stitch..is showy, quickly worked and launders well. **1923** *Daily Mail* 19 Feb. 1 (Advt.), This hard wearing fabric, which launders perfectly, can be obtained. **1951** *Good Housek. Home Encycl.* 252/1 Most silks launder well.

launderette (lǭndəre·t). Also **laundrette.** [f. LAUNDER *v.* + -ETTE.] An establishment providing automatic washing machines for the use of customers.
1949 *Vogue* Oct. 102/2 A new and interesting development in housekeeping—the advent of the self-service launderette. **1952** [see *BAGWASH]. **1955** PRIESTLEY & HAWKES *Journey down Rainbow* 117 Who does not understand the sudden success of our communal laundrettes? **1967** *Which?* July 216 Launderettes are shop-like premises, usually equipped with between 8 and 20 large automatic washing machines, supplied with hot water from a central source, and with about one tumbler drier for each three washing machines. **1968** *Listener* 7 Mar. 316/1 To me the world Strindberg created is like some enclosed launderette of the spirit—the underwear goes round and round but the water has been turned off. **1970** G. GREER *Female Eunuch* 227 Perhaps the failure of such community living could be avoided by including a pub and a laundrette in each block. **1973** E. BERCKMAN *Victorian Album* 113 The enemy lumbered into sight carrying two fat plastic bags: obviously bound for the launderette in the High Street.

lau·ndering, *vbl. sb.* [f. LAUNDER *v.*] The process or action of washing, drying, and ironing linen, etc. Also *fig.*
1894 *To-Day* 17 Mar. 182/2 French cambrics..are not to be starched in the laundering, but left soft. **1908** K. GRAHAME *Wind in Willows* x. 221 I'm in the washing and laundering line, you must know, ma'am. **1949** 'G. ORWELL' *Nineteen Eighty-Four* II. x. 221 Her life had been laundering, scrubbing, darning, cooking, sweeping, polishing, mending, scrubbing, laundering, first for children, then for grand-children. **1970** *Which?* Aug. 247/2 Many people were uninformed about certain aspects of laundering. **1974** *Globe & Mail* (Toronto) 3 Apr. 1/8 Mr. Kerr, who rose in the Legislature to answer Dr. Shulman's question last Friday about laundering of money in Ontario, began by saying, 'The washing or laundering of funds does occur in Toronto.' *Ibid.*, The laundering transaction itself is not illegal.

Launder-Ometer (lǭndərǫ·mitəɹ). orig. *U.S.* Also **Launderometer, launderometer.** [f. LAUNDER *v.* + -OMETER.] The proprietary name of a machine for making standard laundry tests, consisting of a number of jars clamped to a rotating shaft in which the washing is carried out.
1928 *Amer. Dyestuff Reporter* XVII. 731/2 We will now ..introduce to you this machine, which we have baptised the 'Launder-Ometer'... It was originally invented or devised by one of our fellow members, Hugh Christison. **1929** *Official Gaz.* (U.S. Patent Office) 21 May 591/1 Launder-Ometer. For laundry washing machines. Claims use since April 27, 1928. **1938** *Year Bk. Amer. Assoc. Textile Chemists & Colorists* XV. 138 The Launder-Ometer..consists of a heavily constructed copper tank.., supported upon a rigid angle-iron frame. Within this tank there is a brass and aluminum rotor that carries the twenty standard pint jars in which the tests are made. **1958** *Observer* 30 Mar. 11/5 Cotton and other coloured materials are tested in a launderometer, a machine which reproduces as many domestic washes as are considered necessary. **1961** K. DURHAM *Surface Activity & Detergency* ix. 229 The Launderometer..is merely a tank containing a simple mechanism for holding and rotating jars, each containing steel balls, into which the fabric and solution are placed; the jars are rotated and the steel balls provide agitation. The rate of rotation and the weight of the balls are both variable.

laundress, *v.* Add: **c.** To serve (a person) as a laundress.
1850 DICKENS *Dav. Copp.* xxvi. 281 'Sir,' said Mrs. Crupp, in a tone approaching to severity, 'I've laundressed other young gentlemen besides yourself.'

laundrette: see *LAUNDERETTE.

Laundromat (lǭ·ndromæt). orig. *U.S.* Also **laundromat.** [See *-MAT.] The proprietary name of a brand of automatic washing machines; also, by extension, a launderette.
1943 *Trade Marks Jrnl.* 14 July 300/1 Laundromat. Domestic electric washing (laundering) machines. Westinghouse Electric & Manufacturing Company..Pennsylvania. **1951** *Amer. Speech* XXVI. 166 The Westinghouse Company has a 'Laundramat', and there are also 'Laundromats'—often called 'Laundermats' and 'Laundrymats'—open for public patronage. **1955** M. MCCARTHY *Charmed Life* (1956) i. 20 The village mind was still churning up the past, tossing the old dirty linen back and forth impersonally, like one of the washing machines in the new laundromat. **1956** A. HUXLEY *Adonis & Alphabet* 148 Junior colleges, jet-plane factories, Laundromats, six-lane highways. **1957** J. KEROUAC *On Road* (1958) III. ii. 187, I..found nothing but laundromats, cleaners, soda fountains. **1963** *Punch* 2 Jan. p. vi/2 Harrods' sale..includes..laundromats and dryers. **1964** *Economist* 30 May 1024/3 Coin operated Laundromats and dry cleaning machines. **1966** *New Scientist* 16 June 694/3 The 'local' is doomed, one day, to operate like a licensed laundromat. **1966** T. PYNCHON *Crying of Lot 49* v. 121 She found the symbol tacked to the bulletin board of a laundromat, among other scraps of paper offering cheap ironing and baby sitters. **1971** B. MALAMUD *Tenants* 219 They drag out of a laundromat, shoe store, [etc.], every Zionist they can find.

laundry. Add: **2. b.** Articles (linen, etc.) that need to be, or that have been, laundered.
1916 W. J. LOCKE *Wonderful Year* iii. 50 The proletariat hung laundry to dry over royal salamanders and proud escutcheons. *Ibid.* v. 67 Women below at the water's edge beat their laundry with lusty arms. **1965** *Which?* Mar. 80/1 What we have done is to see how much it would cost to wash three different amounts of laundry each week, by each of these methods. **1970** *Laundry & Cleaning Internat.* June/July 1/3 The traditional attitude—that the laundry is paid for from the housewife's weekly budget.
4. *laundry bag, list* (also *fig.*), *mark, room, soap, van, -work, -worker.*
1895 *Montgomery Ward Catal.* 23/2 Laundry Bag; size 14 × 25 inches; made from heavy figured drapery sateen, with white cotton drawing cord and tassels. **1971** P. PURSER *Holy Father's Navy* xi. 60, I shovelled my suit into a laundry bag to send to the cleaners. **1958** *Spectator* 4 July 24/2 Mr. Wardle makes a point of dissociating himself from the laundry-list species of biography-making. **1968** *Time* 10 May 22 The at-large ballot is a bewildering laundry list of 75 names. **1972** *Fortune* Jan. 3/1 As the archetype of U.S. corporations, General Motors is charged by its critics with primary responsibility for a laundry list of social ills, including air pollution, congestion in the cities, ugliness in the countryside, [etc.]. **1972** *Times Lit. Suppl.* 24 Mar. 333/1 A huge panorama of names and dates, as dull as a laundry-list. **1975** *Radio Times* 16 Jan. 5/4 Everybody knows..that there are exceptions to this laundry-list of woes, and that the author can be brilliantly served. **1924** G. S. DOUGHERTY *Criminal as a Human Being* 278 An important factor in this work is a collection of laundry marks. **1962** 'J. BELL' *Crime in our Time* II. i. 22 The..police..overlooked a conspicuous laundry mark, 599, on some of the clothing. **1972** L. LAMB *Picture Frame* xviii. 155 Laundry mark on shirt and maker's name on hand-made shoes. **1967** J. REDGATE *Killing Season* (1968) II. vi. 102 There's a new laundry-room just behind the kitchen. **1971** *Country Life* 23 Dec. Suppl. 11/1 Unique single-story residence..kitchen, laundry room. **1937** *Discovery* Feb. 49/1 The Great Geyser had practically ceased activity, and could not be stimulated to life, even with a hundred-weight of laundry soap—the usual stimulant. **1958** M. DICKENS *Man Overboard* xii. 190 A homely smell compounded of many things, like roasting meat and leather and dogs and laundry soap. **1952** M. ALLINGHAM *Tiger in Smoke* xv. 217 He and two other boys stole a laundry van. **1972** *Guardian* 5 Oct. 28/2 The laundry van in which Sapper Stewart died. **1838** H. MARTINEAU *Retrospect of Western Travel* II. 185 The 10th was Sunday... There was no laundry-work going on. **1891** *Daily News* 15 Dec. 6/2 There was possibility of a good deal of family laundry work arising. **1930** *Times Educ. Suppl.* 16 Aug. 360/2 The teacher must hold full diploma in Cookery, Laundrywork and Housewifery. **1963** F. F. LAIDLER *Gloss. Terms Home Econ. Educ.* 52 *Laundrywork*, the washing and finishing of soiled fabrics. **1894** E. BANKS *Campaigns of Curiosity* 196 She did not consider me up to the mark for a laundry-worker. **1906** *Westm. Gaz.* 21 June 8/1 The attention of laundry-workers is drawn to this. **1975** *Times* 15 Feb. 3/3 Greater London Council..suggested that the post of..'laundry woman' should become 'laundry worker'.

laundry (lǭ·ndri), *v.* [f. the sb.] = LAUNDER *v.* Hence **lau·ndrying** *vbl. sb.*
1880 *American Mail Order Fashions* (1961) 14 The wires can be taken out..so that the entire bustle can be laundried with the greatest care. **1892** *Daily News* 15 Sept. 5/5 The great Chinese national industry is laundrying. **1901** D. SLADEN *In Sicily* I. 152 The ditch in which they did their laundrying. **1919** W. DEEPING *Second Youth* xxvi. 224 No, he can't ask you to laundry his man's clothes. **1957** B. & C. EVANS *Dict. Contemp. Amer. Usage* 268/2 The verb is *to launder*, not *to laundry*. Clothes are *laundered*, not *laundried*.

Laurasia (lǭrē̆i·ʃiă). *Geol.* [mod.L. (R. Staub *Der Bewegungsmech. der Erde* (1928) ii. 121), f. *Laur*(*entia*, name given to the ancient forerunner of N. America (from the Laurentian strata of the Canadian Shield by which it is represented today) + *Eur*)*asia* (see EURASIAN *a.* 1).] A vast continental area or supercontinent thought to have once existed in the northern hemisphere and to have broken up in Mesozoic or late Palæozoic times forming North America, Greenland, Europe, and most of Asia north of the Himalayas. Also, these land masses collectively as they exist today.
1931 *Trans. Geol. Soc. Glasgow* XVIII. 578 The long history of the Tethys girdling the continental shelf of the earth between Gondwanaland and Laurasia is a clear indication of the operation of some opposing force tending to pull the continents apart. **1937** A. L. DU TOIT *Our Wandering Continents* ii. 24 Through..alternating 'polar flight' and 'polar drift' Laurasia and Gondwana have successively impinged upon or else parted from one another. **1944** A. HOLMES *Princ. Physical Geol.* xvii. 367 The [earthquake] belts form two rings..: one enclosing North America and most of Asia and Europe (known collectively as Laurasia), and the other enclosing South America, Africa and Arabia, India, Australia, and Antarctica (known collectively as Gondwanaland). **1971** I. G. GASS et al. *Understanding Earth* xv. 229 Gondwanaland..probably formed about 500 million years ago, and Laurasia, west of the Urals, about 370 million years ago. **1974** *Daily Tel.* (Colour Suppl.) 10 Apr. 20/1 Whether Laurasia and Gondwanaland were themselves joined together into a supercontinent is a question still to be resolved.
Hence **Laura·sian** *a.*
1962 L. C. KING *Morphol. Earth* xii. 399 At the beginning of Cretaceous time virtually the whole of the vast continental interior had been reduced to a landscape of low relief—the Laurasian surface. **1973** *Nature* 1 June 278/2 The Atlantic and Indian Oceans originated from the break up of the Gondwanan and Laurasian continents.

laureate, *a.* and *sb.* Add: **B.** *sb.* **1. c.** *transf.*
1816 BYRON *Bards & Rev.* (ed. 5) 21 Laureat of the long-ear'd kind! *a* **1849** H. COLERIDGE *Ess. & Marginalia* (1851) II. 9 Herrick was the laureate of flowers and perfumes. **1930** R. CAMPBELL *Poems* 12 He..demonstrates, this laureate of the pubs, That 'all good poets have belonged to clubs'. **1941** *Scrutiny* IX. 384 According to their view he [*sc.* Proust] is the laureate of a dying society. **1954** G. W. KNIGHT (*title*) Laureate of peace. On the genius of Alexander Pope.
d. *Nobel laureate*, one who has been awarded a Nobel prize.
1947 CROWTHER & WHIDDINGTON *Science at War* 144 Professor W. N. Haworth of Birmingham, the famous organic chemist and Nobel laureate. **1965** *Listener* 2 Sept. 329/2 Three great physiologists, all Nobel laureates. **1975** *Sci. Amer.* May 53/2 (Advt.), Written by a Nobel Laureate in medicine, this is the first comprehensive treatment of DNA synthesis emphasizing its biochemical aspects and recent developments.

laurel, *sb.*¹ Add: **2. e.** As the name of a colour = *laurel-green.
1923 *Daily Mail* 8 Oct. 5/1 (Advt.), Navy, Nigger,.. Amethyst, Laurel, Wine.
3. c. Versailles laurel (see quots.); **wood laurel,** spurge laurel, *Daphne laureola.*
1728 R. BRADLEY *Dictionarium Botanicum* II. s.v. *Laureola*, Laureola..in English, Spurge-Laurel and Bastard-Laurel, or Wood-Laurel, is a small Evergreen, frequent enough with us, blossoming about Christmas.

1873 W. B. HEMSLEY *Handbk. Hardy Trees* 394 *Daphne Laureola*, Wood Laurel. **1882** *Garden* 25 Feb. 134/3 The Versailles Laurel (*latifolia*) is a large, robust, and bold foliaged form. **1951** *Dict. Gardening* (R. Hort. Soc.) III. 1697/2 *P*[*runus*] *Laurocerasus.* Common, Versailles, or Cherry Laurel. Quick-growing shrub up to 15 or 20 ft.
5. *laurel-brake, -green, -leaf* (attrib.), *-leaved* (later example), *-thicket.*
1853 J. P. KENNEDY *Blackwater Chron.* vi. 73 A man could walk about for a week,..particularly if he got into a big laurel-brake. **1893** *Outing* (U.S.) Oct. 61/2 Only in the wilds of the backwoods,..or in the mountains where tracts of laurel brakes give refuge against men and dogs, do the Virginia deer hold their own. **1938** R. GRAVES *Coll. Poems* 92 Grass-green and aspen-green, Laurel-green and sea-green. **1927** PEAKE & FLEURE *Hunters & Artists* 49 The rude Proto-Solutrean examples of the 'laurel-leaf' blades. **1973** *Times* 26 July 18/3 More than 150 unfinished and broken axes lay on the surface, with hammerstones, anvils, laurel-leaf blades, and many thousands of waste flakes. **1855** A. B. GARROD *Essent. Materia Medica* 122 The bark of Canella alba or Laurel-leaved Canella..; growing in the West Indies. **1750** T. WALKER *Jrnl.* in J. S. Johnston *First Explor. Kentucky* (1898) 49 Just at the foot of the Hill is a Laurel Thicket. **1945** *Mass. Audubon Soc. Bull.* Jan. 274 It was June 25 when I sat on a log in a laurel thicket.
6. *U.S.* **laurel magnolia,** either of two species of *Magnolia*, the evergreen *M. grandiflora* or the sweet bay, *M. virginiana*; **laurel oak,** either of two species of oak, *Quercus laurifolia* or *Q. imbricata.*
1806 P. WAKEFIELD *Excursions N. Amer.* xiv. 93 The laurel magnolia reaches to the height of an hundred feet. **1831** J. M. PECK *Guide for Emigrants* II. 52 From the Walnut Hills to Baton Rouge..you begin to discover the ever verdant laurel magnolia, with its beautiful foliage, of the thickness and feeling of leather. **1850** S. F. COOPER *Rural Hours* 476 The small Laurel Magnolia, or Sweet Bay, is found as far north as New York, in swampy grounds. **1893** W. ROBINSON *Eng. Flower Garden* (ed. 3) 520/1 *M*[*agnolia*] *grandiflora*, the great Laurel Magnolia of the southern United States, is—in England—best treated as a wall-plant. **1903** *Flora & Sylva* I. 19/1 The Laurel Magnolia or Sweet Bay..is certainly a very handsome shrub. **1911** *Encycl. Brit.* XVII. 392/1 The most beautiful species of North America is *M. grandiflora*, the 'laurel magnolia'..introduced into England in 1734. **1810** F. A. MICHAUX *Hist. Arbres Forestiers de l'Amérique Septentrionale* I. 23 *Laurel oak*,..dénomination secondaire dans les Etats à l'ouest des monts Alléghanys. **1832** D. J. BROWNE *Sylva Amer.* 271 East of the Alleghanies this species..is called Jack Oak, Black Oak, and sometimes from the form of the leaves, Laurel Oak. **1901** C. T. MOHR *Plant Life Alabama* 131 Between Bon Secour and Perdido Bay low, sandy hills..support a high forest..of laurel oak and Cuban and long-leaf pine. **1947** COLLINGWOOD & BRUSH *Knowing your Trees* 201/1 Laurel oak has been widely used,e specially in the South, as an ornamental, particularly as a shade or street tree.

Laurentian (lǫre·nʃiăn), *a.*² [f. the name of Lorenzo (Laurentius) de' Medici, who founded a library in Florence in the 15th c.] Of or pertaining to the Laurentian Library in Florence or to manuscripts preserved there. Cf. MEDICEAN *a.*
1860 GEO. ELIOT *Jrnl.* in J. W. Gross *George Eliot's Life* (1885) II. x. 216 That unique Laurentian library, designed by Michael Angelo. **1875** *Encycl. Brit.* II. 438/2 His [*sc.* Michelangelo's] principal works are..the Laurentian Library at Florence. **1879** *Ibid.* IX. 332/1 There are three large and valuable libraries in the city [*sc.* Florence]... The Laurentian, founded by Lorenzo de' Medici, and attached to the convent of San Lorenzo. **1883** R. C. JEBB in *Sophocles, Plays & Fragments* I. p. liii, The manuscripts..used..are the following: In the Biblioteca Mediceo-Lorenziana, Florence, L, cod. 32.9, commonly known as the Laurentian MS. **1936** GREENHOOD & GENTRY *Chronol. Bks. & Printing* (rev. ed.) 14, 1437. A public library is founded at Florence on a bequest by Niccoli; becomes known later as Laurentian Library, and today is oldest existing library in Europe. **1960** G. A. GLAISTER *Gloss. Bk.* 214/2 *Laurentian Codex*, an 11th-century codex of the works of Sophocles, Apollonius Rhodius, and Aeschylus... It is now in the Biblioteca Mediceo-Laurenziana, Florence.

Laurentian: see *LAWRENTIAN *a.*

laurvigite, laurvikite, varr. *LARVIKITE.

lauryl (lǫ·r-, lǭ·ril). *Chem.* [f. LAUR(IC *a.* + -YL.] = *DODECYL; **lauryl alcohol,** $CH_3(CH_2)_{10}CH_2OH$, a crystalline, low-melting alcohol which is obtained by reduction of coconut oil and whose sulphate esters are used in detergents.
1915 P. E. SPIELMANN tr. *V. von Richter's Org. Chem.* I. iii. 227 Lauryl ketoxine, $(C_{11}H_{23})_2C{:}N.OH$, m.p. 39°. **1922** *Jrnl. Amer. Chem. Soc.* XLIV. 2649 By reducing 100 g. of ethyl laurate with 85 g. of sodium, 200 cc. of toluene and 600 cc. of absolute alcohol, there is obtained 53 to 57 g. (65–70%) of lauryl alcohol. **1960** E. L. DELMAR-MORGAN *Cruising Yacht Equipment & Navigation* xxiv. 231 Lauryl pentachlor phenol is..very effective [for rot-proofing canvas]. **1968** J. A. MONICK *Alcohols* iii. 174 A high-quality lauryl alcohol is currently produced by the same process used for *n*-decyl alcohol, namely, the catalytic reduction of coconut oil, coconut oil fatty acids, or their esters, under high pressure. *Ibid.* 175 Lauryl alcohol is also found in oils of lime and the flowers of *Furcraea gigantea.* **1972** *Sci. Amer.* Jan. 88/2 Experiments with lauryl sulfate, a common component of commercial detergents, showed that this substance would indeed break the mucosal barrier.

lausenite (lɑu·sĕnəit). *Min.* [f. the name of Carl *Lausen*, 20th-c. U.S. geologist + -ITE¹.] A hydrated ferrous sulphate, $Fe_2(SO_4)_3.6H_2O$, first found as aggregates of minute colourless fibres after a fire in a mine at Jerome, Arizona, U.S.A.
1928 *Amer. Mineralogist* XIII. 594 The name rogersite is already in use..so a new name will have to be assigned to the present mineral. Dr. G. M. Butler has suggested the name lausenite, after Carl Lausen the discoverer of the new mineral. **1968** I. KOSTOV *Mineral.* II. ix. 499 Lausenite and kornelite are monoclinic, found respectively in white and pale violet globular aggregates with radial-fibrous texture.

lautarite (lɑutā·rəit). *Min.* [ad. G. *lautarit* (A. Dietze 1891, in *Zeitschr. f. Kryst. und Min.* XIX. 447), f. Oficina *Lautar*(*o*, the name of the owners of the pampa where it was first found: see -ITE¹.] Calcium iodate, $Ca(IO_3)_2$, found as colourless or yellowish monoclinic crystals in Antofagasta Province, Chile.
1892 E. S. DANA *Dana's Syst. Min.* (ed. 6) 1040 (*heading*) Lautarite. **1943** R. D. GEORGE *Minerals & Rocks* vii. 198 Lautarite, calcium iodate,..is a colorless to pale yellow, easily fusible mineral having a specific gravity of 4·6 and a hardness of 4. **1968** K. A. JONES tr. *Kirsch's Appl. Mineral.* vi. 114 Lautarite $Ca(IO_3)_2$... Monoclinic, H: 3·0–4·0, D: 4·6.

lautite (lɑu·təit). *Min.* [ad. G. *lautit* (A. Frenzel 1881, in *Min. und Petr. Mittheil.* III. 516), f. *Laut-a*, the name of its original locality near Marienberg, E. Germany: see -ITE¹.] An orthorhombic, grey or black sulphide of copper and arsenic, CuAsS (possibly with silver replacing some copper), having a metallic lustre.
1883 *Encycl. Brit.* XVI. 392/1 Lautite (CuAg)As.S. **1940** *Mineral. Abstr.* VII. 489 Lautite associated with arsenic, tetrahedrite, [etc.].., is found in the Gabe Gottes mine at Sainte-Marie-aux-Mines (Haut-Rhin). It forms bunches of crystals up to 6 cm. × 6–7 mm., showing a brilliant (001) cleavage, enclosed in native arsenic. **1967** *Ibid.* XVIII. 97/1 Lautite was obtained by heating a mixture of realgar and electrolytic copper under 1000 kg/cm² to 315° C.

lav. (læv). A colloq. shortening of LAVATORY *sb.* 4.
1913 C. MACKENZIE *Sinister Street* I. vii. 99 Tell the army to line up behind the lav. at four o'clock. **1933** D. L. SAYERS *Murder must Advertise* i. 8 You've only got to warn him not to use the directors' lav., and not to tumble down the iron staircase. **1960** J. R. ACKERLEY *We think the World of You* 51, I asked instead if I could use the lav. **1961** [see *ELSAN]. **1973** J. THOMSON *Death Cap* iv. 64 Gilbert Leacock went out to the lav... I heard the chain being pulled.

lava. Add: **4. a.** *lava bed* (also *fig.*), *block, boulder, field* (also *fig.*), *-lake*; also **lava flow,** a mass of flowing or solidified lava.
1891 *Century Mag.* Mar. 645 The general direction [of march] was towards the lava beds of northern California. **1905** *Westm. Gaz.* 12 Aug. 13/1 She lived over a 'lava-bed of raw primeval passions'. **1937** *Discovery* Mar. 83/1 This arid lava bed. **1866** 'MARK TWAIN' *Lett. from Hawaii* (1967) 58 We climbed a hill a hundred and fifty feet high.. as full of rough lava blocks as it could stick. *Ibid.* 223 It had lava boulders piled around its base. **1902** *Athenæum* 30 Aug. 287/1 Where bed on bed of mountain-pinks Above the lava-boulders blow. **1866** 'MARK TWAIN' *Lett. from Hawaii* (1967) 222 We clambered over the surrounding lava field, through masses of weeds. **1899** *Geogr. Jrnl.* May 50 The most extensive lava-field in the island. **1906** *Daily Chron.* 21 May 7/3 The smoking lava fields of discussion. **1957** G. E. HUTCHINSON *Treat. Limnol.* I. i. 16 The lakes of the Modoc lava field are frequently regarded as lying on tilted block faults. **1866** 'MARK TWAIN' *Lett. from Hawaii* (1967) 222 Gathering up a fresh lot of specimens, having..discarded those he dug out of the old lava flows. **1888** Lava flow [in Dict.]. **1957** G. E. HUTCHINSON *Treat. Limnol.* I. i. 38 Not infrequently the surface of a newly formed lava flow may cool and produce a crust. **1974** *Country Life* 17 Jan. 66/1 The lava-flows on either side of the runway. **1895** Lava-lake [see *blowing-cone]. **1902** *Nature* 4 Sept. 441/1 (*heading*) The lava-lake of Kilauea. *Ibid.* 441/2, I carefully observed the then existing lava-lake during six successive days.
c. (Later example.)
1878 SWINBURNE *Poems & Ballads* 2nd Ser. 182 All its lava-black Cones.

lavabo. Add: Now also (senses 2 b, *c) with pronunc. (læ·vābo).
2. b. (Example.)
1911 WEBSTER *Lavabo*, a wash basin with its necessary fittings, esp. one set in place and supplied with running water and a waste pipe.
c. = LAVATORY *sb.* 4.
1930 D. L. SAYERS *Strong Poison* xiv. 177 The little lavabo in the passage. **1934** H. MILLER *Tropic of Cancer* 83, I find a ticket in the *lavabo* for a concert. **1971** C. JOHNSTON *Mo* 58 The subject of each was unmistakeably the same—Belchamber's lavabos.

laval (lā·văl), *a.* [f. LAVA + -AL.] Of or resembling lava. Also *fig.*
1891 *11th Ann. Rep. U.S. Geol. Survey* II. 199 These great springs are the outlet of a system of under-laval channels and lakes. **1931** V. WOOLF *Waves* 84 It is the speed, the hot molten effect, the laval flow of sentence into sentence that I need. **1932** W. FRANK *America Hispana* I. v. 230 Volcanic earth has made these churches and these statues. Sometimes it is stone, black and yellow as in the north; sometimes, it is adobe. And sometimes, it is tezontli, the laval rock colour of clotted blood. **1963** A. SMITH *Throw out Two Hands* (1966) xxi. 205 We were 500 feet above that exceptionally laval and unpleasant-looking region.

lava-lava (lā·vă͵lā·vă). [Samoan.] In Samoa and some other Pacific islands, a sort of skirt.
1891 R. L. STEVENSON *Vailima Lett.* (1895) xiii. 115 The weird figure of Faauma..in a black lavalava (kilt). **1900** *Fortn. Rev.* Jan. 49 New Zealanders, Chinese, East and West Indians, and half-castes who are more at home in lava-lavas than bifurcated garments. **1944** *Living off Land* viii. 159 He must produce..equipment consisting of a blanket, bowl, spoon, lava-lava. **1949** M. MEAD *Male & Female* 407 Clothes consisted of a short sarong—called a lavalava. **1957** M. B. PICKEN *Fashion Dict.* 207/2 *Lava-lava*, loincloth or waistcloth of printed calico worn by natives of Samoa and other islands in the Pacific. **1971** *Listener* 9 Sept. 334/3, I saw the famous lava-lava—the short, brightly-coloured skirt which they [*sc.* Samoans] wore below the naked torso. **1973** *Observer* (Colour Suppl.) 28 Oct. 53/1 *Lavalavas* (skirts made from a single length of cloth, originally introduced by the missionaries for reasons of modesty, often wrapped round the body from bosom to knee by Samoan women).

Lavallière (lavalyē̆r'). Also **Lavalière,** with lower-case initial, and without accent; **lavalier.** Name of Louise de *la Vallière*, French courtesan (1644–1710), used *absol.* or *attrib.* to designate certain styles in clothing and jewellery. Also *transf.*, a small microphone.
1873 *Young Englishwoman* July 350/1 (*heading*) White chip Lavalliere Hat. The crown is moderately high, with a rather broad brim, turned up in front and down at the back. **1916** *Daily Colonist* (Victoria, B.C.) 16 July 2/1 (Advt.), Our stock of moderately priced Necklets, Pendants and La Valieres is most attractive. **1942** *Horizon* Oct. 250 His collar and ready-made tie (a lavallière). **1951** E. PAUL *Springtime in Paris* vii. 136 Noel, with broad-brimmed felt hat, seersucker jacket and flowing *lavallière*, personified the artistic type. **1955** *Publ. Amer. Dial. Soc.* XXIV. 122 Any minor bauble such as a lavaliere..is a *dangler*. **1959** *Times* 28 Apr. 20/6 A diamond lavalliere. **1969** D. M. DISNEY *Fatal Choice* (1970) x. 80 Marcy was wearing John's farewell gift, a lavaliere set with an aquamarine..and seed pearls. **1970** *New Yorker* 12 Sept. 32/2 Mr. Siegel, whose clothes for public appearances are striking, running to vests and lavalieres, had a big grin on his face. **1972** R. HENDRICKSON *Human Words* 179 Today the small television microphone that hangs on a cord from the neck is also called a *lavaliere*, taking its name from the pendant necklace. **1974** *Amer. Speech 1971* XLVI. 55 A low impedance, unidirectional microphone is, I feel, better than a lavalier microphone.

‖ **lavandera** (lavănde·ră). [Sp.] In Spanish-speaking countries, a washerwoman; = LAVENDER *sb.*¹
1841 BORROW *Zincali* I. II. vi. 316 The lavanderas engaged in purifying the linen of the capital. **1918** W. H. HUDSON *Far Away & Long Ago* vii. 97 One of the most attractive spots to me was the congregating place of the *lavanderas*, south of my street.

lavatera (lavatīə·ră, lavā·təră). [mod.L. (J. Pitton de Tournefort (1706), in *Hist. Acad. Roy. Sci. Mém.* 86), f. the name of the brothers *Lavater*, 17th- and 18th-c. Swiss physicians and naturalists.] A herb or shrub of the genus so called, belonging to the family Malvaceæ, and bearing pink, white, or purple flowers.
1731 P. MILLER *Gardeners Dict.* s.v. *Lavatera*, African Lavatera, with a most beautiful flower. **1790** *Curtis's Bot. Mag.* IV. 109 Annual Lavatera..flowers from July to September. **1916** G. JEKYLL *Annuals & Biennials* i. 4 Where a temporary filling is desired of plants of important aspect, there are the Tobacco plants,..Foxgloves, Solanums, and Lavatera. **1962** *Listener* 19 Apr. 708/3 Sow hollyhocks, lavatera, sweet peas.

lavatorial, *a.* Add: **2.** Of or pertaining to a style of architecture or decoration alleged to resemble that used for public lavatories.
1936 E. EATON *Summer Dust* I. vii. 42 The newer movement, which demands that every rising young lawyer should furnish an *appartement* in the neo-lavatorial style. **1957** *Sunday Times* 17 Nov. 3/3 The Examination Schools, that lavatorial building of awful omen. **1958** B. HAMILTON *Too Much of Water* ii. 42, I had to pull down an absolutely shattering piece of Victorian lavatorial Gothic. **1969** *Observer* 12 Jan. 25/3 A barber's, a lavatorial café, and nothing much else but grime and desolation. **1974** A. ROSS *Bradford Business* 63 Endless lavatorial town hall corridors.
3. Of or pertaining to lavatories; *spec.* of conversation, humour, etc.: making undue reference to lavatories and their use.
1967 *Daily Tel.* 15 Apr. 9/3 The average gent is of an aggressive disposition and a lavatorial sense of humour. **1969** *Daily Tel.* 17 Nov. 9/1 The words..are perfectly familiar, partly lavatorial, partly sexy, ranging in length from four to seven letters. **1969** C. DERRICK *Reader's Report* vii. 120 A great many poets, when talking about their art, drop instinctively and sub-consciously into the

use of obstetrical and even lavatorial imagery. **1974** *Times* 9 Nov. 10/4 The awful lavatorial embarrassment.

lavatory, *sb.* Add: **4.** In the 20th c. one of the more usual words for a W.C. (now giving way to more recent euphemisms: *lav., loo, toilet*, etc.).

In some examples ellipt. for the appliance itself.

1924 F. M. Ford *Some do Not* I. ii. 37 The hotel having been a former Grand Ducal hunting-box, freshened to suit the taste of the day with varnished pitch-pine, bathrooms, verandahs, and excessively modern but noisy lavatory arrangements. **1963** J. T. Story *Something for Nothing* iii. 106 Albert closed the door and sat down on the lavatory. **1965** *Listener* 2 Sept. 351/2 Flush Conscience down the lavatory. **1967** *Ibid.* 3 Aug. 144/3 Certain kinds of help are..needed... Lavatories a disabled person can get in and onto. *Ibid.* 21 Dec. 802/1 'Loo' is holding its own fairly well and most of 'Toilet's' gains have been at the expense of 'Lavatory'. **1973** A. S. Neill *Neill! Neill! Orange Peel!* (rev. ed.) II. 189 What did people use in the lavatory before the invention of paper?

8. *lavatory attendant, basin, brush, chain, cleanser, pan, paper, seat*; also **lavatory humour**, unsavoury or unwholesome humour making undue reference to lavatories (cf. *Lavatorial *a.* 3); so *lavatory joke*; **lavatory period, style**, a period or style of architecture with 'lavatorial' characteristics (cf. *Lavatorial *a.* 2).

1930 I. Low *His Master's Voice* xv. 217 A lavatory attendant might make a similar boast thought Nikulin. **1939** C. Isherwood *Goodbye to Berlin* 29 The little old lavatory attendant in his white jacket. **1964** R. Petrie *Murder by Precedent* i. 20 Lavatory attendants..aren't as likely to insure their lives as professional men. **1926–7** *Army & Navy Stores Catal.* 321 'Vitrella' should be used for cleaning all porcelain baths, lavatory basins, sinks. **1931** *Times* 16 Mar. 22/1 (Advt.), Lavatory basin in best bed room. **1963** *B.S.I. News* Apr. 33 Ceramic lavatory basins. **1939–40** *Army & Navy Stores Catal.* 167/2 Lavatory brush holder. **1962** *House & Garden* Dec. 51/1 Lavatory brush and container, 12 s. **1969** A. E. Lindop *Sight Unseen* xxix. 245 A bottle of bourbon..stood in the stand which usually held the lavatory brush. *c* **1932** Dylan Thomas *Lett.* (1966) 5 Give me..a book by Paul de Kock, and thou, thou old lavatory chain. **1964** V. S. Naipaul *Area of Darkness* iv. 94 At four..we heard him rise and get ready for his walk: lavatory chain, gargling, clattering, doors. **1926–7** *Army & Navy Stores Catal.* 469/3 Harpic Lavatory Cleanser. **1963** V. Nabokov *Gift* ii. 83 The lavatory humour and crude laughter. **1970** *Sunday Times* (Colour Suppl.) 15 Mar. 25/1 He was real blue, if you like. Some of his jokes. Lavatory humour, my Dan used to call it. **1954** N. Tomalin in *Granta* 6 Nov. 23/2 All in all the most amazing thing about the show is the preponderance of lavatory jokes. **1966** A. E. Lindop *I start Counting* xx. 256, I clutched my head over the lavatory pan. **1974** *Listener* 24 Jan. 100/1 After an atomic explosion, the last things to disintegrate would be articles made of vitreous porcelain—for example, wash-basins and lavatory pans. **1926–7** *Army & Navy Stores Catal.* 121/2 Lavatory Paper Holder. **1956** A. S. C. Ross in M. Black *Importance of Lang.* (1962) 103 Non-U *toilet-paper* / U *lavatory-paper*. **1974** *Times* 23 Jan. 14/4 The Americans..had their lavatory paper panic first. **1952** A. Christie *They do it with Mirrors* iii. 24 'It's pretty ghastly, really,' said Gina cheerfully. 'A sort of Gothic monstrosity. What Steve calls Best Victorian Lavatory period.' **1939** G. Household *Rogue Male* 255, I made him sit on the lavatory seat and read me the shipping news. **1973** *Country Life* 20 Sept. 784/2 Three industries of Derbyshire..[are] the manufacture of artificial flowers, lavatory seats and mint sauce. **1937** H. G. Wells *Camford Visitation* i. 5 That University affair they are building in London..is all made, they tell me, of a sort of mineral nougat, lavatory style. **1944** J. Agate *Red Letter Nights* 141 Liliom commits suicide, and is projected into a celestial police court, the architecture of which is copied from the Early Lavatory style of the National Liberal Club.

lave, *a.* **b.** *Comb.* **lave-eared** *a.* (later examples).

1926 *Blackw. Mag.* Sept. 431/1 If a poet..was lave-eared; if he had the eyes of a fawn, then you might be sure that he was a poet, and fear the worst. **1932** Auden *Orators* II. 44 The nasty lave-eared pop-eyed bitch.

lavender, *sb.*² and *a.* Add: **A.** *sb.* **1. c.** Phr. *lavender and old lace*: the title of a novel and play used to describe a gentle and 'old-fashioned' style.

The novel by Myrtle Reed was published in 1902 and the dramatized version by Rose Warner in 1938.

1966 *Guardian* 25 Nov. 14/7 Arthur Pollard..is largely concerned to dispel the notion that Mrs Gaskell is a writer of 'lavender and old lace'. **1966** M. Steen *Looking Glass* v. 88 E. V. Lucas..never wrote twaddle: the lavender and old lace of his titles masked erudition. **1968** Hocking & Healey *Murder cries Out* iv. 54 This astounded gentleman..had received a description of Miss Willoughby as all 'lavender and old lace'.

3*. *Cinemat.* Positive stock, or a positive print, used for producing duplicate negatives; also (quot. 1936), a print made from such a negative.

1936 C. B. DeMille in *Words* Oct. 6/1 A 'lavender' is something often spoken of in the industry... It is a print made from a negative on lavender stock, which is a weak print from a weak negative, because lavender negatives are only copies of the film originally exposed in the camera and are therefore not as sharp. **1959** W. S. Sharps *Dict. Cinematogr.* 106/1 *Lavender*, the name given to an obsolescent type of master positive stock with a lavender tinted base. The name remains in use to describe a master positive. **1973** D. A. Spencer *Focal Dict. Photogr. Technol.* 340 *Lavender*, fine grain motion picture film used for making duplicate black and white negatives..and coated on lavender tinted base to minimise halation and prevent confusion with ordinary positive stocks. Modern duplicating stock is on a grey tinted base and differs sufficiently in appearance from ordinary positive that the lavender tint is not necessary.

4. *lavender-blue* (later example), *-grey* (later example); *lavender-coloured, -hued* adjs.; *lavender-scented* (earlier and later examples); also **lavender bag**, a bag containing dried lavender; **lavender cream**, lavender-scented cream or furniture-polish; also *lavender furniture cream*; **lavender drawer**, a drawer containing or scented with lavender; **lavender drop**, a drop (sense 4) medicated with lavender; **lavender polish** = *lavender cream; also *lavender floor polish*; **lavender sachet** = *lavender bag; **lavender soap**, soap perfumed with lavender; also *lavender toilet soap*; **lavender wax** = *lavender cream.

1865 Geo. Eliot *Let.* 6 Feb. in J. W. Cross *George Eliot's Life* (1885) II. xii. 396, I want to send my love, lest all the old messages shall have lost their scent, like old lavender bags. **1923** A. Huxley *Antic Hay* iv. 55 Give me..a lavender bag under every pillow. **1965** M. Sharp *Sun in Scorpio* III. xxv. 131 Elspet was peddling lavender-bags. **1936** *Burlington Mag.* Jan. 9/1 Vase with lavender-blue glaze splashed or suffused with purple. **1901** *Westm. Gaz.* 7 Sept. 1/3 Our lavender-coloured view of life. **1936** J. C. Powys *Maiden Castle* (1937) 40 A vision of lavender-coloured tights. **1926–7** *Army & Navy Stores Catal.* 1128/2 Hair and toilet preparations... Lavender Cream. *c* **1938** *Fortnum & Mason Price List* 36/1 Furniture polish..Lavender Cream..per jar 1/6. **1863** Dickens *Mrs. Lirriper's Lodgings* i, in *All Year Round* Extra Christmas No., 3 Dec. 9/1 An advertisement..which I mean always carefully to keep in my lavender drawer. **1811** Jane Austen *Sense & Sens.* II. vii. 126 Some lavender drops..which she was at length persuaded to take, were of use. **1969** A. E. Lindop *Sight Unseen* [illegible] 158 A clean smell of lavender floor polish. **1926–7** *Army & Navy Stores Catal.* 297/2 Lavender furniture cream. **1936** *Burlington Mag.* Jan. 4/1 Buff ware with a crackled lavender-grey glaze. **1901** *Westm. Gaz.* 7 Sept. 1/3, I speak from experience, having lately reached the lavender-hued period. **1961** J. Stroud *Touch & Go* v. 49 The aggressive lavender-polish aroma of a house-proud matriarch. **1966** L. Deighton *Billion-Dollar Brain* v. 51 There was a sweet smell of lavender as we walked through a couple of rooms. **1938** R. Field *All this, & Heaven Too* (1939) iv. 65 The [illegible] in her bureau drawers. **1973** 'S. Harvester' *Corner of Playground* i. 49 Who can make tea smelling like a lavender sachet? **1855** Mrs. Gaskell *North & South* II. xxi. 283 Smoothing down the [illegible] Jenny for an armful of lavender[illegible]. **1938** O. Sitwell et al. *Trio* 124, I care for [illegible] less when he is in a melancholy mood, [illegible] have a tendency to become a little too [illegible]. **1974** J. Wainwright *Evidence I shall Give* [illegible] lavender-scented handkerchiefs. **1875** E. Spon *Workshop Receipts* 385 Lavender Soap.—The basis [illegible] scented with oil of lavender. **1949** D. [illegible] *Castle* II. ix. 133 Rose was..varnishing her [illegible] had been her special treat... I had [illegible]. **1961** A. Wilson *Old Men at Zoo* i. 35 The [illegible] with me, all redolent with lavender [illegible] lady wife, as he was apt to call her. **1890–91** *T. Eaton & Co. Catal.* Fall & Winter 42/1 Sweet [illegible] soap. **1970** M. Kelly *Spinifex* vi. 97 A [illegible] scented with years of lavender wax.

B. *adj.* †**2.** *Photogr.* *lavender rays*, ultra-violet radiation. *Obs.*

1840 Sir J. F. W. Herschel in *Phil. Trans. R. Soc.* CXXX. 20 As orange, indigo, and violet, vegetable tints, are used for [illegible] prismatic hues, I may be allowed to express by the term lavender the rays which produce the [illegible], rather for the purpose of abbreviating the [illegible] of ultra-violet..than for that of laying any stress on the observed fact. **1842** *Ibid.* CXXXII. [illegible] If the action of the spectrum be prolonged, [illegible] whitening becomes sensible in the red, and [illegible] beyond the violet into the 'lavender' [illegible]. **1858** Sutton & Worden *Dict. Photogr.* 248 The [illegible] rays beyond the violet end of the spectrum [illegible] called 'lavender rays'. **1911** *Cassell's Cycl. Photogr.* [illegible] *Lavender rays*, a term (now practically obsolete) applied to the commencement of the ultra-violet beyond the visible violet. **1922** A. E. H. Tutton *Crystallogr.* II. li. 1139 This ultra-violet lamp is visible to the eye at close quarters owing to fluorescence of the retina of the eye itself; and the field of vision appears filled with a haze known as 'lavender fog', owing to fluorescence of the crystalline lens of the eye.]

lavendery (læ·vĕndəri), *a.* [f. Lavender *sb.*² + -y¹.] Perfumed with lavender; fragrant. Also *fig.*

1896 G. B. Shaw *Our Theatres in Nineties* (1932) II. 10 Mr Rose has often written pleasantly about these and other more remote and lavendery antiquities. **1968** P. Dickinson *Skin Deep* v. 107 Go now, the smile said, and I'll retain fond lavendery memories of you; stay and I'll bite. **1974** A. Price *Other Paths to Glory* I. v. 49 The lavendery smell of the strange room.

lavenite. Add: Also † **laavenite, låvenite.** (Further examples.)

1887 *Mineral. Mag.* VII. 234 The laavenite occurs as small honey-yellow crystalline grains. **1927** N. H. & A. N. Winchell *Elem. Optical Mineral.* (ed. 2) II. x. 245 Laavenite is a monoclinic zircono-silicate of Ca, Mn, Na with some Ti, Ta, F, OH. **1967** *Mineral. Abstr.* XVIII. 48/2 A new variety of låvenite occurs in a zone of veinlet nephelinization in massive quartz syenite in the alkalic massif of Burpala in northern Baikal [Siberia].

laventine. (Examples.)

1893 *Funk's Stand. Dict.*, *Laventine*, a thin silk: much used for sleeve-linings. **1921** *Daily Colonist* (Victoria, B.C.) 18 Oct. 19/7 (Advt.), Umbrellas at All Prices... Strong durable covers in mercerized cotton, gloria and laventine.

laver, *sb.*¹ Add: **2.** Also with pronunc. (lā·vəɹ). **laver bread** (also **lava bread**), a name in Wales for a food made from the fronds of *Porphyra umbilicalis*, which are boiled, dipped in oatmeal, and fried.

1701 W. Kennet *Cowell's Interpreter* (rev. ed.), *Laver-bread*, in Glamorganshire and some other parts of Wales, they make a sort of Food of a Sea plant, which seems to be the Oyster-green or Sea-Liver-wort. This they call Laver-bread. **1949** *New Biol.* VII. 94 In the days when butter was more plentiful, laver bread was heated with butter, lemon juice and pepper, and served with roast mutton. **1953** Dylan Thomas *Under Milk Wood* (1954) 5 Is there rum and laverbread? **1962** *Listener* 26 July 140/2 Lava bread..is the only truly Welsh food. It is made from seaweed. **1969** N. W. Pirie *Food Resources* v. 127 Welsh devotion to laverbread is an important factor which restrains the managers of nuclear installations from fouling the sea more than they do now.

Laverack (læ·vəræk). The name of Edward *Laverack* (d. 1877), English dog breeder, used *attrib.* in **Laverack setter** to designate the type of English setter (see Setter *sb.*¹ 11 a) developed by him, a large hunting dog having long white fur flecked with other colours. Also *absol.*

1878 C. Hallock *Sportsman's Gazetteer* (ed. 4) 689/2 Belton.—the ticked or spotted Laverack setter. **1904** *Daily Chron.* 15 Aug. 3/3 In America 'Laveracks' and 'Llewellins' are household words among sportsmen. **1971** F. Hamilton *World Encycl. Dogs* 245 These differing methods of breeding led in later years to the main division in the breed as we know it today, between show dogs and working dogs; although it must be remembered that both types originated with the Laveracks. *Ibid.*, (*caption*) An engraving of Mr. Purcell-Llewellyn's Laverack Setter, Countess.

Laves phase (lā·vəs fēiz). *Metallurgy.* [ad. G. *Laves-phase* (G. E. R. Schulze 1939, in *Zeitschr. f. Elektrochem.* XLV. 850/1), f. the name of Fritz-Henning *Laves* (b. 1906), German crystallographer + Phase *sb.*] Any of a group of intermetallic compounds of composition approximately AB_2 in which the relative sizes of the A and B atoms are such as to allow a stable packing arrangement with unusually high co-ordination numbers.

1940 *Chem. Abstr.* XXXIV. 1895 Laves phases are AB_2 intermetallic compds. whose structure and combination mechanism are detd. by the crystal lattice. **1956** Hume-Rothery & Raynor *Struct. Metals & Alloys* (ed. 3) v. 228 The Laves phases crystallize in one of three closely related structures, and are isomorphous with the compounds $MgCu_2$, $MgZn_2$, or $MgNi_2$. **1967** A. H. Cottrell *Introd. Metall.* xiv. 195 In a Laves phase AB_2, each A atom has 16 neighbours (4A and 12B) and each B atom has 12 neighbours.

lavvy (læ·vi). = *Lav.

1961 Partridge *Dict. Slang* II. 1165/1 *Lavvy*, a Scottish, esp. children's, easing of lavatory. **1962** *John o' London's* 27 Dec. 588/1 An outside loo or 'lavvy'. **1971** *Guardian* 20 Jan. 8/2 A house where the lavvy is behind an arras.

law, *sb.*¹ Add: **1. b.** (Later example.) Also *colloq.* (orig. *U.S.*), a policeman, the police; a sheriff.

1929 M. A. Gill *Underworld Slang*, *Law*, police. **1935** A. J. Pollock *Underworld Speaks* 121/2 *The law*, a police officer. **1944** B. A. Botkin *Treas. Amer. Folklore* I. 131 There was plenty of precedent for Roy Bean in the usual Western 'Law' or sheriff. **1953** W. Burroughs *Junkie* (1972) x. 104 We were in the third precinct about three hours and then the laws put us in the wagon and took us to Parish Prison. *Ibid.* xii. 121 Whenever a law needs money for a quick beer, he goes over by Lupita and waits for someone to walk out on the chance he may be holding a paper [containing narcotic]. **1955** *Publ. Amer. Dial. Soc.* xxiv. 106 Some mobs have a strong prejudice against robbing any *law*, whatever he may be. **1958** F. Norman *Bang to Rights* 152 Two law..came up to me and grabbed hold of me. **1962** 'J. Bell' *Crime in our Time* I. 13 He had only one idea. To get rid of 'the law', clinging to his car. He drove from side to side of the road in an effort to force Meehan off. **1972** D. Lees *Zodiac* 5 Soon, car-loads of Maigret-type law would come screaming up the drive. **1972** *Times* 6 June 18/6, I inquired of the Law where I might cash a cheque, and was directed to the nearest travel agency. **1973** M. Woodhouse *Blue Bone* vi. 56 The Oxford law would know about this, I take it?

3. a. (Earlier and later examples of *law and order*.)

1598 Florio *Worlde of Wordes* 201/2 Legitimo..*according to law and order*. **1796** *Deb. Congress U.S.* 20 Dec. (1849) 1689 A military diploma, expressive of his patriotism and attachment to law and order. **1846** *Nat. Intelligencer* (Washington) 24 Mar. 3/4 The 'Law and Order'

party has clearly fulfilled..its public mission. **1893** [see *comfortably *adv.* 5 a]. **1932** G. F.-H. Berkeley *Italy in Making* I. iii. 42 This repression continued until..even his allies finally began to perceive that law-and-order can be bought at too dear a price. **1952** *N.Y. Times* 11 Mar. 11/1 (*heading*) Ex-sergeant rode to power on cry of 'Law and Order', seizing reins in 1933. **1962** S. Wynter *Hills of Hebron* ix. 118 The Commissioner..had come to use words like 'duty', 'law and order' to cover up a lack of imagination. **1967** P. Henderson *William Morris* II. xii. 281 The SFD and the League..started a campaign for free speech, which suddenly brought Morris into conflict with the forces of Law and Order. **1968** *Economist* 5 Oct. 41/3 Mr Nixon and..Mr Humphrey are both making concessions to this overriding concern about 'law and order'. **1970** *New Yorker* 26 Sept. 137/1 If it happens that rightists are successful in capturing most of the law-and-order vote, then, of course, the country will move to the right politically. **1972** *Daily Tel.* 24 May 15 The published motions cover a wider and more immediate political field than usual—from the economy and unemployment to the media, Ulster and law and order. **1973** *Black World* Dec. 19/1 A sense of determinism that is diametrically opposed to the ruler-class 'law-and-order' and individualism.

8. b. (Examples of *to take the law into one's own hands* (or, obs., *fists*).)

1847 E. Brontë *Wuthering Heights* I. vii. 129 Next time, Master Edgar, take the law into your own fists. **1875** B. Jowett tr. *Plato's Dialogues* (ed. 2) III. 63 Young men will take the law into their own hands. **1877** C. M. Yonge *Cameos from Eng. Hist.* 3rd ser. vii. 63 Cade took the law into his own hands. **1902** A. Bennett *Grand Babylon Hotel* xxvii. 300, I have a few questions to put to you, and it will depend on how you answer them whether I give you up to the police or take the law into my own hands. **1942** A. Bryant *Yrs. of Endurance* xiv. 333 The industrial workers and the starving peasants, deprived of their patrimony by enclosures, took the law into their own hands.

16. a. *a law unto* (or *to*) *himself* (or *themselves*, etc.) (later examples).

1878 R. H. Davis (*title*) A law unto herself. **1930** J. S. Huxley *Bird-Watching* iv. 75 Every male [ruff] is a law unto himself. Some grow black ruffs, others ruffs that are white, sandy, brown, grey, pepper-and-salt, and half a dozen other shades. **1942** Partridge *Usage and Abusage* (1947) 129/1 Certain idiosyncratic, law-unto-themselves writers fall into vagaries when..they depart from those rules. **1962** A. Nisbett *Technique Sound Studio* xii. 221 Electronic music is a law to itself... It really can be fundamentally different from conventional music. **1965** M. Spark *Mandelbaum Gate* ii. 23 'What was your father's Law?'.. 'I'm afraid he was a law unto himself.' **1968** *Listener* 27 June 850/1 Hogarth is very much a law to himself.

c. Phr. *law of the jungle*, the code of survival in jungle life, now usu. with ref. to the superiority of brute force or self-interest in the struggle for survival.

1894 Kipling *Jungle Bk.* 31 Baloo was teaching him the Law of the Jungle... Young wolves will only learn as much of the Law of the Jungle as applies to their own pack and tribe. *Ibid.* 34 A man's cub..must learn *all* the Law of the Jungle. **1895** —— *Second Jungle Bk.* 23 Now this is the Law of the Jungle—as old and as true as the sky; ..the strength of the Pack is the Wolf, and the strength of the Wolf is the Pack. **1927** C. A. & M. R. Beard *Rise Amer. Civilization* I. ix. 403 So the law of the jungle prevailed; and in the frightful contest that followed, the rights of neutrals were as chaff before a hurricane. **1950** A. Bryant *Age of Elegance* x. 329 In the manufacturing districts..the old framework of society..broke down completely. Here only the law of the jungle held. **1951** M. McLuhan *Mech. Bride* 107/2 Mr. Queeny derives his 'law of the jungle' versus 'crusading idealist' from this later nineteenth-century phase of the older split. **1969** *New Yorker* 6 Sept. 33 (*caption*) May I remind you that here law and order means the law of the jungle. **1974** *Times* 4 Mar. 7/4 The Duke said that a purely materialistic society inevitably succumbed to the law of the jungle and political dictatorship.

17. b. *law of large numbers*: see *Large *a.* 8 i.

c. (*c*) (Earlier examples of *Grimm's law*.)

1838 W. B. Winning *Man. Compar. Philol.* I. iii. 36 *Grimm's Law*.—I now proceed with the consideration of Grimm's important law..concerning the regular interchange of certain letters in different languages. *Ibid.* 47 On the principle of Grimm's law, we exclude the Perso-Grecians, High Germans, and Goths, from among the earliest colonists of Italy.

(*e*) *Meteorol.* **Buys-Ballot's law** (bīzbæ·lo) [enunciated by C. H. D. *Buys-Ballot* (1817–90), Dutch meteorologist, in 1857] (see quots.).

1875 *Encycl. Brit.* III. 29/1 Buys-Ballot's 'Law of the Winds'..may be thus expressed:—The wind neither blows round the space of lowest pressure in circles returning on themselves, nor does it blow directly towards that space; but it takes a direction intermediate, approaching, however, more nearly to the direction and course of circular curves than of radii to a centre. **1970** J. Hulbert *All about Weather* vi. 73 Buys Ballot's law..says that if a man stands with his back to the wind, the lower pressure will be to his left in the northern hemisphere, and to his right in the southern.

21. a. *law department, firm, language* (example), *reports* (examples), *school* (examples), *student* (earlier and later examples), *studies*.

1849 E. Chamberlain *Indiana Gazetteer* (ed. 3) 45 In the winter of 1838, the institution was chartered as an University, and in 1842, a law department was established. *c* **1876** 'Mark Twain' *Lett. to Publishers* (1967) 95, I suppose *our* law firm are [*sic*] above average. **1945** G. L. Williams *Learning the Law* xii. 130 There are law firms in the East, located in the great seaports, where young solicitors may often find good places. **1965** Mrs. L. B. Johnson *White House Diary* 21 July (1970) 304 His greatest hope was to stabilize the law firm. **1973** *N.Y. Law Jrnl.* 31 Aug. 1/3 A New York attorney and a law firm have been found by Federal Judge Lloyd F. MacMahon of the Southern District of New York to have violated..the Code of Professional Responsibility. **1797** *Encycl. Brit.* IX. 725/1 (*heading*) Law-Language. *c* **1840** Lady Wilton *Art of Needlework* xviii. 298 No. 50 [of *The English Mercurie*], dated July 23, 1588, is the first now in existence... In it are no advertisements—no fashions—no law reports—no court circular. **1902** *Encycl. Brit.* XXXI. 181/2 The *Law Reports* (begun in 1884) are conducted by the large staff of *Times* law reporters, all of them barristers of at least five years' standing. **1972** *Mod. Law Rev.* XXXV. I. 22 The law reports, commonly regarded as our primary literature, tell us regularly and systematically how large claims are being determined. **1818** *North Amer. Rev.* Mar. 428 A Law School is established at the University. **1863** S. Warren *Pop. & Pract. Introd. Law Stud.* (ed. 3) I. i. 89 In Ireland, there is a 'Law School' in the University of Dublin. **1893** W. K. Post *Harvard Stories* 128 'You couldn't do that if you were a biographee,' reasoned Dane Austin, the law-school man. **1966** G. Wilson in K. Boehm *University Choice* 242 It would be a rare law student who could do anything as straightforward as transfer his own house as a result of what he had learnt at law school. **1835** S. Warren *Pop. & Pract. Introd. Law Stud.* iii. 102 Could the eye of the young law-student be brought to..see how heavily his bodily and mental energies will be taxed..how he would husband them! **1945** G. L. Williams *Learning the Law* iv. 36 A teacher must consider..the amount of time actually available to a law student for his studies. **1966** Law student [see *law school* above]. **1845** C. M. Kirkland *Western Clearings* 42 George Burnet had just come home after finishing what he called his 'law studies'.

c. *law-honest* (earlier example); also *law-honesty*.

1838 J. F. Cooper *Homeward Bound* III. xi. 333 Mr. Dodge belonged to a tolerably numerous class, that is quaintly described as being 'law honest', that is to say, he neither committed murder nor petty larceny. **1905** *Daily Chron.* 6 Dec. 7/7 What may be called law-honesty, the kind of honesty necessary in order to avoid falling into the clutches of the law.

22. b. *law-governed* adj.

1938 *Burlington Mag.* Mar. 148/1 The 'law-governed' development of the language of form. **1960** H. Edwards *Spirit Healing* iii. 25 Every change within our comprehension is the result of law-governed forces applied to the subject.

23. law enforcement, enforcement of the law; freq. *attrib.*; so *law-enforcer*; **law-lord**, (*a*) (later examples); **law-office** (earlier and later examples); **law station** *slang*, a police station; **law-writer**, (*c*) (example).

1936 L. Hellman *Days to Come* II. iii. 70 Dowel was found knifed, dead... That gives us a little job of law enforcement to do. **1955** D. W. Maurer in *Publ. Amer. Dialect Soc.* xxiv. 6 A problem which urgently demands the attention of the legislators, law-enforcement specialists, and the judiciary. **1956** 'E. McBain' *Cop Hater* (1963) xvii. 129 If they can do away with law enforcement, the rest will be easy... First the police, then the National Guard. **1960** *Times* 3 Oct. 13/6 Their functions as law-enforcement officers. **1972** A. Roudybush *Sybaritic Death* (1974) xiii. 119 His proposal..to help the overwhelmed police in their law-enforcement task. **1938** *Tablet* 1 Jan. 1/1 The world was pictured as consisting of the law-makers and law-enforcers. **1975** *Listener* 16 Jan. 67/1 The law-enforcers themselves were bent. **1958** *Times* 24 July 8/7 The sons and daughters of law lords and life peers..shall be treated for their style, rank, dignity, and precedence in the same way as the wives..of hereditary barons. **1972** *Mod. Law Rev.* XXXV. I. 63 Two Law Lords with first instance experience in the Divorce Division. **1873** 'Mark Twain' & Warner *Gilded Age* xii. 117 In the anteroom of the law-office where he was writing. **1973** *N.Y. Law Jrnl.* 2 Aug. 16/6 (Advt.), Political Science undergraduate seeks 4 years law study and clerkship in law office to gain credit to take Bar. **1958** F. Norman *Bang to Rights* 50 After a while we came to a law station. **1959** Anon. *Streetwalker* x. 180, I was in the law station. They got me early. **1852** Dickens *Bleak Ho.* (1853) x. 94 Our law-writers, who live by job-work, are a queer lot.

Law: see *Lawd.

law-abiding, *a.* (Earlier and later examples.)

1839 *Congress. Globe* Dec., App. 14/2 Being a law-loving and law-abiding man, he had voted to preserve the laws. **1855** *Ibid.* 26 Jan. 416/2 The people of Oregon are a law-abiding, honest and gallant people. **1973** C. Mullard *Black Brit.* II. vi. 68 A racist, however intelligent, will never accept that blacks are just like other average human beings, honest, industrious, clean, law-abiding, intelligent, etc.

Lawd (lǭd). Chiefly *U.S.* Also **Law, Lawdy.** Local (esp. Black English) variants of *Lord*, *Lordy*, usually as interjections or in humorous contexts.

1846 C. M. S. Kirkland *Bee-Tree* in *Amer. Short Stories* (1904) 204 Law sakes alive! **1851** Mrs. Stowe *Uncle Tom's Cabin* (1852) II. xx. 49 Law, Missis, you must whip me; my old Missis allers whipped me. **1881** J. C. Harris *Nights with Uncle Remus* (1884) xii. 65 'Is dey anybody home?'.. 'Law, no, honey, folks all gone.' *Ibid.* xxxix. 183 Lawdy mussy, Brer Rabbit! Whar my vittles? **1898** 'J. Kerr' *Cheery Bk.* 121 But, Lawdy! How dat coon can kiss! **1901** G. B. Shaw *Captain Brassbound's Conversion* I. 226 *Drinkwater.* Lawd,..wot jagginses them jurymen was! *Ibid.* II. 242 Git the plice ready for the British herristorcracy, Lawd Ellan and Lidy Wineflete? **1926** W. C. Handy *Blues* 64 Feel so low down an' sad Lawd Lost ev'rything I ever had. *Ibid.* 91 Why did I go away? I used to be so gay There'll come a day! Oh Lawdy Lawd how happy I'll be Down home in Florida. **1928** R. Bradford in B. A. Botkin *Treas. S. Folklore* (1949) III. ii. 485 So about dat time de Lawd comed up on old man Noah. **1945** Mencken *Amer. Lang.* Suppl. I. 664 Euphemisms... For *Lord*: *land, law, lawks, lawdy, lawsy*. **1970** R. D. Abrahams *Positively Black* vi. 142 Lawdy, lawdy, lawdy, lawd, I used to be your reg'lar, now I got to be your dog. **1971** *Black Scholar* Sept. 44/2 'Lawd!' he blurts out and begins to giggle. **1973** *Black World* Sept. 29 Hughes sometimes created short, vignette-type poems with the blues feeling..as he did in 'Bad Morning': Here I sit with my shoes mismated. Lawdy-mercy! I's frustrated!

lawful, *a.* **3.** (Later example.)

1890 M. C. Fraser *Let.* Nov. in *Diplomatist's Wife Japan* (1899) II. xxviii. 143 The Japanese are a profoundly lawful people (if I may use the word in its old sense).

6. Describable or governed by laws of nature.

1939 *Nature* 14 Jan. 64/1 Newton and others have found confirmation even for their religious beliefs in the lawful character of physical phenomena. **1958** M. Argyle *Relig. Behaviour* i. 2 It is commonly assumed that human behaviour is lawful and that it can be predicted by means of psychological laws and explained in terms of psychological processes. **1959** M. Bunge *Causality* i. 22 The principle of universal lawfulness..may be taken to read thus: Every single event is lawful, i.e., is determined in accordance with a set of objective laws—whether we know the laws or not. **1975** *Nature* 3 Apr. 416/2 The results show that the tendency..can be brought under lawful control in such a way as to discriminate against the above hypotheses.

lawfully, *adv.* Add: **1. b.** In accordance with laws of nature.

1959 M. Bunge *Causality* i. 22 The principle of universal lawfulness does not assert that facts are determined by laws, but in accordance with laws, or simply lawfully.

lawfulness. Add: (Later example of sense 'respect for law'.)

1924 B. Williams in *History* Jan. 273 The adventures of..the N.W. Mounted Police, in bringing half a continent to lawfulness and peace.

2. The quality of being describable by laws of nature, or of happening or behaving in accordance with certain general principles that always hold good.

1956 E. H. Hutten *Lang. Mod. Physics* vi. 210 To say that causality is the belief in the lawfulness of nature..is therefore quite meaningless. **1959** [see *lawful *a.* 6]. **1972** *Science* 16 June 1208/3 Experiences..that do not show consistent patterns..will be distinguished from those phenomena which do show general lawfulness.

lawkamercy (lǭ:kămǝ·ɹsi), *int.* Var. of *lawk-a-mussy* (Lawk, Lawks *int.*). (In quot. 1909 as quasi-*adj.*)

1909 J. Masefield *Tragedy of Nan* I. 8 Idle lawkamercy girl. **1927** B. L. K. Henderson *Chats about our Mother Tongue* i. 54 Lawkamercy, lad, what's that?

lawless, *a.* **1.** (Later example.)

1956 E. H. Hutten *Lang. Mod. Physics* vi. 215 This confusion has prompted the view that chance events are lawless.

lawlessly, *adv.* (Later example.)

1972 *Daily Tel.* 4 Aug. 2/8 The council is insistent that it is not behaving lawlessly.

law-like, *a.* Add: **c.** *Philos.* Of a statement, explanation, etc.: resembling scientific laws in saying that some consequence would occur in any situation of a certain sort, though differing in containing reference to individuals; also, such as to be a law of nature if established as true.

1949 G. Ryle *Concept of Mind* iv. 89 How does the law-like general hypothetical proposition work? It says, roughly, that the glass, *if* sharply struck or twisted, etc. *would*..fly into fragments. **1961** E. Nagel *Struct. of Sci.* ii. 21 The premises contain at least one 'lawlike' assumption. **1968** M. Black in R. Klibansky *Contemp. Philos.* II. 59 Scientific or 'lawlike' generalizations require, in Peirce's phrase, reference to a 'would-be'. **1973** *Nature* 27 July 241/1 Archaeological explanations should be very general, preferably of lawlike form.

lawman. Add: **4.** A law enforcement officer. *colloq.*

1959 P. Cook in *Granta* 6 June 33/1 Had he actually seen the rough law-men bundle the startled Widow into the Black Maria? **1962** R. Barker *Clue for Murder* ix. 57 Some lawmen took a delight in seeing the criminal squirm. **1972** *Radio Times* 30 Mar. 14/4 A retired lawman, still sporting a tin star, demonstrated how he could kill with either hand. **1973** J. Wainwright *Devil you Don't* 65 He surely hated Davis... He also hated this goddam lawman.

lawn, *sb.*² Add: **2*.** *Bacteriology.* A layer of bacteria uniformly distributed over the surface of a culture medium.

1951 Whitby & Hynes *Med. Bacteriol.* (ed. 5) xxiv. 433 Phage activity is readily observed on solid culture media. A plate is first thickly inoculated with susceptible bacteria to form a 'lawn' of growth. **1970** Passmore & Robson *Compan. Med. Stud.* II. xviii. 102/1 The routine test dilution (RTD) is determined by placing drops of tenfold dilutions of the phage suspension on a lawn of sensitive bacteria.

3. lawn billiards = Troco; **lawn-cutter** = *lawn-mower*; **lawn-party** (later examples); **lawn sand**, a top-dressing of ammonium sulphate and iron sulphate mixed with sand, used as a fertilizer and weed-killer for lawns; **lawn-sprayer**, a sprayer for diffusing a fine spray of water over a lawn.

1873 *Young Englishwoman* Nov. 572/2 Jean would feel obliged if the Editor would tell her..if lawn billiards can be played on a croquet lawn?.. Is there a book of rules on lawn billiards? **1879** Trollope *John Caldigate* I. xvi. 213 Hunting, shooting, fishing,..lawn-billiards. **1882** [see Troco]. **1910** *Encycl. Brit.* III. 934/2 The game [billiards] was at one time played on a lawn, like modern croquet... A later form of 'lawn billiards' again enjoyed a brief popularity during the latter half of the 19th century. **1897** S. Hale *Let.* 24 Mar. (1919) 315 Such a delicious drive,..and the lawn-cutters making hay smells. **1937** R. S. Morton *Woman Surgeon* xxxi. 346 Many interesting people gather in our frequent outings, lawn parties and other expressions of comradeship. **1955** R. Blesh *Shining Trumpets* (ed. 3) viii. 181 Parades, picnics, funerals, Mardi Gras, lawn parties, dances—he and his band were in demand everywhere. **1973** *Lebende Sprachen* XVIII. 38/1 US lawn party—BE/US garden party—Gartenfest. **1907** *Yesterday's Shopping* (1969) p. xlii/4 Lawn sand. **1909** T. W. Sanders *Lawns & Greens* vii. 68 Methods of exterminating daisies are to put a pinch of salt on the crown of each plant, or to sprinkle 'Watson's Lawn Sand' over the infested parts. **1939** R. B. Dawson *Pract. Lawn Craft* xxi. 152 It may be found more convenient for owners of quite small areas of turf to buy a ready compounded lawn sand. **1968** *Punch* 20 Aug. 304/2 Add 1 oz of iron sulphate to 15 oz of dry sludge and you have a moss killing 'lawnsand'. **1943** Wyndham Lewis *Let.* 15 Aug. (1963) 362 Watching the blue jays..having a shower-bath in a lawn-sprayer.

lawrencium (lǫre·nsiǝm). *Chem.* [mod.L., f. the name of Ernest O. *Lawrence* (1901–58), U.S. physicist + -ium.] An artificially produced transuranic element that concludes the actinide series, the longest-lived isotope of which has a half-life of a few minutes. Atomic number 103; symbol Lw.

1961 *Times* 15 Apr. 7/1 The isotope of element 103 was 'created' on February 14 by four nuclear scientists—Albert Ghiorso, Torbjorn Sikkeland, Almon Larsh, and Robert Latimer... The scientists intend to name the new element Lawrencium in honour of..the founder of the laboratory. **1972** K. M. & R. A. Mackay *Introd. Mod. Inorg. Chem.* (ed. 2) xi. 151/1 Lawrencium behaves as expected for the $f^{14}d^1s^2$ configuration by forming only Lw(III)..and resisting oxidation or reduction. **1972** *Sunday Mail Mag.* (Brisbane) 23 Apr. 11/2 Until lawrencium was discovered, californium..was regarded as the world's most valuable metal.

Lawrentian (lǫre·nʃiǎn), *a.* **1.** Also **Laure·ntian.** Of or pertaining to the military leader and author T. E. *Lawrence* ('Lawrence of Arabia') (1888–1935), or his deeds and writings.

1928 T. E. Lawrence *Let.* 20 Jan. (1938) 568 Laurentian, that sudden insult. I thought I'd sloughed off those manners with the names. **1949** Koestler *Promise & Fulfilment* i. 10 It would have necessitated a team of men with Byronic idealism and Lawrentian imagination. **1966** *Punch* 20 July 123/3 The great merit of Mr. Mousa's study is that while his mind is clearly weighted against the Lawrentian myth, he appears to be scrupulously fair to Lawrence.

2. Also **Laurentian, Lawrencian.** Of or pertaining to the English author D. H. *Lawrence* (1885–1930), or his work or style of writing. Hence **Lawrencia·na, Lawrentiana** [*-iana], objects belonging to, literature about or characteristic of, D. H. Lawrence.

1930 L. P. Hartley in *Sat. Rev.* 15 Feb. 203/1 Morgan's description of the sisters has many Lawrentian echoes, especially in the use of the word 'dark'. **1931** R. Aldington *Colonel's Daughter* v. 270 The good old Lawrentian dark abdomen. **1936** A. Huxley *Olive Tree* 211 A characteristically Lawrencian expression. **1938** *Times* 28 Jan. 367/3 It is..possible also to put her down as a mixture of the living and the merely Lawrentian. **1944** H. Treece *Herbert Read* 92 Most of Read's 'Eclogues'..are filled with a dark passionate emotion, a Lawrentian impulse centred in the body. **1944** *Scrutiny* XII. 256 Levels at which the characteristic Laurentian contribution may well appear the reverse of helpful. **1948** H. T. Moore in D. H. Lawrence *Lett. to Bertrand Russell* p. vi, Meanwhile, here is another contribution to Lawrence literature, one of the last important collections of Lawrenciana. **1959** *Listener* 16 Apr. 683/1 Dr. Leavis successfully established a Lawrentian enclave in Cambridge. **1959** *Times Lit. Suppl.* 30 Oct. 630/1 Mr. de Vries has nicely satirized..Lawrencian and Bohemian 'free livers'. **1959** *N. & Q.* Mar. 120/1 There is an immense literature of 'Lawrentiana'. **1972** *Observer* 19 Nov. 4/2 There has been a 'tremendous revival' in Lawrentiana from a purely touristic point of view. **1973** *Daily Tel.* 7 Nov. 15/2 It has a Lawrentian turn of situation, too. There is a knight, Sir Tom, whose wife Margaret is ill-served—he comes home only for weekends and is too gentle anyway for her.

Lawson (lǭ·sən). The name of Peter *Lawson* (d. 1820) and his son Charles (1794–1873), partners in the firm of Lawson and Son, Edinburgh nurserymen, used *attrib.* or in the possessive in **Lawson('s) cypress** to designate *Chamæcyparis lawsoniana*, a conifer from Oregon first introduced to cultivation by them after seeds had been collected in 1854 by Andrew Murray (1812–78), Scottish botanist, who named the tree after the Lawsons.

1858 G. Gordon *Pinetum* 62 *Cupressus Lawsoniana*, Murray. Messrs Lawson's Cypress... A large graceful tree, growing 100 feet high, and two feet in diameter.. with the branches at first curved upwards..and towards the ends hanging down like an ostrich feather. **1866** *Curtis's Bot. Mag.* XCII. 5581 The Lawson Cypress..has for the last few years been a great favourite in our gardens and shrubberies. **1914** L. H. Bailey *Stand. Cycl. Hort.* II. 730/2 Lawson's Cypress..is one of the most beautiful conifers and very variable, about 80 garden forms being cultivated in European nurseries. **1923** Dallimore & Jackson *Handbk. Coniferæ* 200 The extraordinary variability of the Lawson cypress under cultivation has resulted in a large number of forms being given varietal names. **1957** M. Hadfield *Brit. Trees* 120 The Lawson Cypress has a very limited natural distribution..restricted to a narrow belt on the Pacific coast of the United States. **1970** C. Lloyd *Well-Tempered Garden* iii. 215 One or other form of Lawson's cypress will give you a dense hedge. **1974** *Country Life* 12 Dec. 1855/2 A golden Lawson cypress..is a narrow column of bright gold.

lawsoniana (lǭsǝ͡uniˌā·nă). [a. the specific epithet of *Chamæcyparis lawsoniana*.] = **Lawson('s) cypress* (see prec.).

1959 R. E. Harrison *Handbk. Trees & Shrubs S. Hemisphere* 89/1 The well-known golden Lawsoniana with stiff erect pyramidal habit, and lemon-golden foliage. **1965** F. Sargeson *Memoirs of Peon* ix. 270 We sat in the shade of a lawsoniana hedge.

lawsy, var. of *laws* (Law *int.*).

1880 'Mark Twain' *Lett.* (1917) I. 383 A kind-hearted, well-meaning corpse was the Boston young man, but lawsy bless me, horribly dull company. **1884** —— *Huck. Finn* xvi. 142 But lawsy, how you did fool 'em, Huck. **1914** G. Atherton *Perch of Devil* i. 75 'Your room's pretty!'..'mine's pink—but lawsy!' **1945** [see *Lawd].

lawyer. Add: **6.** *lawyer-ridden* adj.; **lawyer cane** (examples).

1908 E. J. Banfield *Confessions of Beachcomber* i. vi. 209 The lawyer cane or vine (*Calamus*)..is a vegetable of tortuous ambitions. **1936** *Geogr. Jrnl.* LXXXVII. 229 They wore a band..made out of lawyer cane, around their middles as a protection against arrows. **1965** *Austral. Encycl.* V. 266/2 Lawyer Cane or Lawyer Vine, a name popularly if impolitely given to species of the rattan genus *Calamus*..and to *Flagellaria indica*. **1824** Mill in *Westm. Rev.* II. 376 Our lawyers, and lawyer-ridden legislators. **1907** *Daily Chron.* 29 Apr. 6/6 Land reform had been too long delayed, because they had been too frightened and lawyer-ridden.

law·yerish, *a.* [f. Lawyer.] Befitting a lawyer; like that of a lawyer.

1918 Galsworthy *Five Tales* 133 His lawyerish mind habitually put two and two together.

law·yerism. [f. Lawyer.] The influence, or principles, of lawyers.

1915 F. S. Oliver *Ordeal by Battle* 221 To fall back on lawyerism was perhaps inevitable in the circumstances; but to think that it was possible to substitute lawyerism for leadership was absurd.

lax (læks), *sb.*³ Colloq. abbrev. of Lacrosse.

1951 E. Taylor *Game of Hide-and-Seek* II. i. 128 One late afternoon after lax-practice. **1966** J. Gardner *Amber Nine* xii. 203 A far cry from the hockey and lax sticks of Roedean or Vassar. **1968** 'P. Hobson' *Titty's Dead* viii. 86 Thank goodness Mummy doesn't know anything about LaX.

lax, *a.* Add: **5. c.** *Phonetics.* Of a speech-sound, esp. a vowel: produced with the speech organs relaxed.

1909 D. Jones *Pronunc. of Eng.* I. iii. 12 The difference in quality between a tense vowel and the corresponding lax vowel..is sometimes very considerable, especially in the case of closed vowels. **1933** Westermann & Ward *Pract. Phonetics for Students Afr. Lang.* vi. 36 These two sounds occur in Bari as the 'lax' forms of *i* and *u*. **1949** R.-M. S. Heffner *Gen. Phonetics* v. 96 Later scholars have substituted the terms tense and lax for narrow and wide. **1964** Jakobson & Halle in D. Abercrombie et al. *Daniel Jones* 97 A peculiar interplay of the lax-tense and compact-diffuse features underlies the vowel harmony. **1973** *Amer. Speech* 1969 XLIV. 199 The diphthongizing of lax vowels..can be analyzed.

laxaman, var. *Laksamana.

laxness. Add: **b.** (Later examples.)

1887 *Pall Mall Budget* 21 Apr. 2/2 This criminal laxness, so alarmingly on the increase, should be nipped in the bud. **1969** *Daily Tel.* 27 May 1/2 Mission Control has now been exonerated of any laxness in this respect, for it turns out that they were not required to remind the lunar module crew to operate this particular switch.

c. *Phonetics.* Of a speech-sound, esp. a vowel: the state of being lax (*Lax *a.* 5 c).

1909 D. Jones *Pronunc. of Eng.* I. iii. 13 The tenseness or laxness of a vowel can often be observed mechanically by placing the finger on the throat between the larynx and the chin. **1956** Jakobson & Halle *Fund. of Lang.* I. iv. 43 In the opposition of tense and lax consonants, the laxness is frequently accompanied by voicing and the tenseness by voicelessness. **1967** D. Steible *Conc. Handbk. Ling.* 71 *Laxness*, a distinctive acoustic feature of English characterized by relative relaxation of the tongue and jaw while in the act of articulation.

Laxton (læ·kstǫn). The name of *Laxton* Brothers, a firm of English nurserymen, used in the possessive to designate several varieties of fruit bred and introduced by them, esp. the apple **Laxton's Superb**, a popular, late-ripening variety of red-skinned eating apple.

1920 *Jrnl. R. Hort. Soc.* XLV. p. xxxvii, Exhibit. Messrs. Laxton, Bedford: Apple 'Laxton's Superb'. **1926** A. J. Macself *Fruit Garden* vii. 98 Laxton's Superb.—One of the best of early Pears, ripening in August. **1933** Hall & Crane *Apple* xii. 195 Laxton's Superb.—A comparatively new seedling introduced by Messrs. Laxton in 1921, which is now beginning to be generally planted. *Ibid.* 203 Laxton's Exquisite follows closely on James Grieve. **1937** A. H. Hoare *Commercial Apple Growing* ii. 50 Laxton's Epicure..received an Award of Merit from the Royal Horticultural Society in 1931. *Ibid.* 51 Laxton's Fortune, also a result of a Cox's Orange Pippin and Wealthy cross, is claiming attention as an early bright-coloured and attractive apple. **1966** C. R. Thompson *Pruning Apple Trees* II. 194 Other varieties..often crop profusely and yield small apples, e.g...Laxton's Fortune and Laxton's Superb. **1974** *Countryman* Summer 107/2 The Fire Blight Order of 1958..banished this bacterial disease of pears by prohibiting Laxton's Superb, the most susceptible variety.

lay, *sb.*⁷ **2.** Delete 'Now *rare*' and add: An oyster- or mussel-bed; = Laying *vbl. sb.* 2 c, Layer *sb.* 4 b.

1902 *Westm. Gaz.* 12 June 10/1 The oyster and mussel lays off the foreshore have hitherto been worked on the large scale. **1905** *Country Life* 25 Mar. 400/2 More than 200 fresh oyster 'lays' have now been staked out on the north side of the Witham.

7. c. *Printing.* Substitute for def.: The arrangement of type in the case from which a compositor takes it; in full, *lay of the case*; also = **lay gauge*. (Further examples.)

[**1683–4** Moxon *Mech. Exerc., Printing* (1962) 194 The manner how the several sorts of Letters are disposed in the several Boxes, is called, Laying of the Case.] **1884** J. Gould *Letter-Press Printer* (ed. 3) 29, I give the following illustration of the upper-case as it is most commonly laid. In some offices, however, the 'lay' is quite different. **1915** *Southward's Mod. Printing* (ed. 3) I. xxvi. 150 A printed plan of the case..will also be useful to experienced compositors, for there are many variations of the lay to be found in printing offices. *Ibid.* II. iv. 45 The Feed or Laying-on Board is, in the Wharfedale machine, at the base of the cylinder... On the front of the board are the gauges, or 'lays', to which the sheet of paper is laid. **1946** A. Monkman in H. Whetton *Pract. Printing & Binding* ii. 20/1 There is no standardized lay in this country although the variations are in the main only concerned with such characters as the ligatures.., figures, and lower-case k and q. **1946** V. S. Ganderton in *Ibid.* xi. 142/1 Lays, sheet-bands, grippers, and wheels shall be in identical positions on the sheet... If..the register is out along the grip edge, the fault may be due to the front lays lifting too early or too late. **1969** *Studies in Bibliogr.* XXII. 125 (*title*) The lay of the case. *Ibid.*, The single lay, as used for instance in Germany and Switzerland, employs one large case for a fount of type. **1970** E. A. D. Hutchings *Survey of Printing Processes* 199 *Lays, machine*, the points against which a sheet is positioned on the machine prior to impression taking place. **1972** P. Gaskell *New Introd. Bibliogr.* 36 Lays for exotic founts were usually adaptations of those used for the Latin alphabet.

d. A woman who is readily available for sexual intercourse; an act of sexual intercourse. *slang* (orig. *U.S.*). Cf. *Lay *v.*¹ 2 b.

1932 J. T. Farrell in *Story* Mar.–Apr. 46 A foursome passed homeward; two of the group were girls whom Jack and George agreed were swell lays. **1934** J. O'Hara *Appointment in Samarra* (1935) vi. 159 If there was ever an easy lay she was it. **1936** J. Dos Passos *Big Money* 254 There never been a girl got a spoken word by givin' that fourflusher a lay. **1955** W. Gaddis *Recognitions* II. i. 317 She's the girl you used to go around with in college? She's a good lay. **1955** G. Greene *Quiet American* II. iii. 173 You'll just keep her as a comfortable lay until you leave. **1958** E. Dundy *Dud Avocado* III. vi. 266 Roving photographer..blows into town on the lookout for a quick lay. **1962** *Listener* 9 Aug. 223/3 His characters are without perspective: engrossed completely in their own lives, hardly seeing..beyond the next drink, the next lay, the next five pounds. **1971** B. Malamud *Tenants* 16 Tonight an unexpected party, possibly a lay with a little luck.

11. lay-edge *Printing*, the edge of a sheet of paper which is used to determine the correct position of the sheet in a press; **lay gauge**, an attachment on a printing press that keeps the paper in the correct position.

1892 A. Powell *Southward's Pract. Printing* (ed. 4) I. 444 Turn over the sheet..and place it upon the feeding or laying-on board, with the same lay-edge towards the grippers as before. **1946** V. S. Ganderton in H. Whetton *Pract. Printing & Binding* xi. 143/2 An untrimmed lay edge is an unknown quantity, and no two sheets stand up to the lays in the same way if they have a feather edge. For exact work, paper should always be trimmed. **1892** A. Powell *Southward's Pract. Printing* (ed. 4) xlix. 434 Now set the lay gauges on the machine, so that the paper when fed to these will..occupy the right position on the cylinder to receive the impression where it is intended to be. **1961** T. Landau *Encycl. Librarianship* (ed. 2) 190/1 Lay edges. The edges of a sheet of paper which are laid against the front and side lay gauges of a printing or folding machine.

lay, *a.* Add: **4. lay analysis,** psychoanalysis undertaken by an analyst who has not been medically trained; so **lay analyst, psychiatrist, psychoanalyst**, one who practises psychoanalysis without medical training; **lay preacher**, an unordained preacher, esp. among Methodists; **lay reader**, (*b*) a reader of a book, etc., on a subject of which he has no professional or specialist knowledge.

1927 *Internat. Jrnl. Psycho-Anal.* VIII. 174 The Central Executive of the International Psycho-Analytical Association informs us it is their intention to bring forward the question of 'Lay Analysis' at the next Congress, so that opinions may be heard and, so far as possible, decisions arrived at in the matter. **1928** A. P. Maerker-Branden tr. *Freud's Probl. Lay-Analyses* i. 25 Let me, therefore, state that the problem of Lay-Analyses expresses itself most succinctly in the question of whether medically untrained laymen should be permitted to practise psychoanalysis. *Ibid.* viii. 171 As soon as the physician has ascertained this, he may safely leave the treatment to the lay-analyst. **1955** M. McCarthy *Charmed Life* (1956) ii. 36 He had been..a lay analyst. **1747** Lay-preacher [in Dict., sense 1]. **1790** J. Wesley *Works* (1872) IV. 493 Joseph Humphrys: the first Lay Preacher that assisted me in England, in the year 1738. **1823** A. Clarke *Mem. Wesley Family* 34 From this conversation we learn..that he was a lay-preacher. **1906** 'Mark Twain' *What is Man?* (1917) iii. 31 In the Adirondack woods is a wage-earner and lay preacher in the lumber camps. **1962** H. Davies *Worship & Theol. in England* IV. ix. 258 John Nelson..one of Wesley's most trusted lay preachers. **1975** R. Lewis *Double Take* ii. 58 He..was a Methodist lay preacher and a supporter of good causes. **1958** 'J. Bell' *Seeing Eye* xiv. 147 He has been more successful as a lay psychiatrist than he has as a general practitioner. **1933** *Harper's Mag.* Jan. 186/1 A lay psycho-analyst finds that the political philosophy of Thomas Jefferson was a product of Jefferson's infantile revolt against his father. **1885** W. James *Coll. Ess. & Rev.* (1920) 282 To the lay-reader, this absolute Idealism doubtless seems insubstantial and unreal enough. **1907** —— *Pragmatism* ii. 74 Farther than that the ordinary lay-reader in philosophy..does not venture to sharpen his conceptions. **1912** *Motor* 17 Dec. 980/1 The subject matter is..written in a manner easily understood by the lay reader. **1947** *Mind* LVI. 156 This is done in so compressed and allusive a manner that..the lay reader could scarcely be expected to grasp it adequately.

5. *lay-minded* adj.

1898 S. Evans *Holy Graal* 134 We Englishmen of today, a lay-minded folk much misguided of philosophic historians.

lay, *v.*¹ Add: **2. b.** To have sexual intercourse with (a woman). Occas. *intr.*, const. *for*: (of a woman) to have sexual intercourse with (a man). Also *intr.*: (of a woman) to be willing to have (extramarital) sexual intercourse. *slang* (orig. *U.S.*). Cf. *Lay *sb.*⁷ 7 d.

1934 J. O'Hara *Appointment in Samarra* ii. 38 I'm going to take Teddy out and get him laid tonight. *Ibid.* vii. 212 'You're wrong about one thing,' said Julian. ..'I didn't lay that girl.' **1936** J. Dos Passos *Big Money* 305 'Gosh,' he was saying at the back of his head, 'maybe I could lay Elsie Finnegan.' **1938** G. Greene *Brighton Rock* v. v. 214 I'm marrying her for your sake, but I'm laying her for my own. **1950** A. Wilson *Such Darling Dodos* 123 As soon as he laid a new wench..there was always a shift round of staff. **1955** 'H. Robbins' *Stone for Danny Fisher* I. vii. 55 'Does she lay, Danny?'.. His face was flushed as his eyes followed the girl on to the porch. **1956** H. Gold *Man who was not with It* (1965) xviii. 164 Whore! Baby-whore! She been laying for you. **1960** J. Updike *Rabbit, Run* (1961) 184 You've laid for Harrison, haven't you? **1966** Auden *About House* 15 A great-great-grandmother who got laid By a sacred beast. **1969** P. Roth *Portnoy's Complaint* 182 All I know is I got laid, *twice*. **1973** B. Broadfoot *Ten Lost Years* viii. 83 The guy who knew her was one of our gang and he was laying her.

7. b. To place documents containing information on the table (see Table *sb.* 5 b) in order to present the information to the members.

[**1813** *Hansard Commons* 17 Mar. 142 Mr. Whitbread then moved, that the Petition be laid upon the table; which was ordered accordingly.] **1923** *Westm. Gaz.* 3 Aug., The Premier promised to lay all the correspondence, if M. Poincaré consents. **1924** *Hansard Commons* 10 Mar. 1931 His Majesty's Government have been willing to lay the complete records, but objections have been raised. **1964** *Erskine May's Law of Parl.* (ed. 17) xiii. 274 A similar order was made..in cases where a paper was laid under an Act that prescribed a period during which objection to it could be taken.

9. *to lay an egg*: fig. phr. used in various colloq. senses, *spec.*: (*a*) (of an aircraft) to drop a bomb; (*b*) orig. *U.S.* (of a performer or performance) to flop.

1918 [see *egg *sb.* 3 d]. **1927** *Daily Express* 2 June 11/2 'Laying an egg' in Air Force slang means dropping a bomb. **1929** *Variety* 30 Oct. 1 (*headline*) Wall Street lays an egg... The most dramatic event in the financial history of America is the collapse of the New York Stock Market. **1940** J. O'Hara *Pal Joey* 38 You would just as well come wearing a shell if you ever took a job [singing] in a spot like this, that is how big an egg you would lay. **1947** [see *egg *sb.* 3 d]. **1949** L. Feather *Inside Be-Bop* iii. 30 The singer had been laying eggs at the Zanzibar..and Shaw was undecided what to do with him. **1958** *Spectator* 6 June 730/2 The second gambit, when a joke is so drearily bad.. that even a studio audience can't laugh at it, is to admit, quite shamelessly, that one has, as they say, laid an egg. **1964** *People* (Austral.) 16 Dec. 45/1 A Stuka caught us in the town of Lamia. The plane duly laid an egg. I was crouched alongside a wall. The bomb landed on the other side of the wall.

12. c. To bet on (a horse).

1877 *Porcupine* 10 Mar. 790/1 Whether it is as immoral to 'bear the market' as to 'lay the favourite';..all these are irrelevant issues. **1887** W. B. Gilpin *Set of Four Hunting & Racing Stories* vi. 68 They refused to lay him except at odds on. *Ibid.* x. 97 His..plans..'to lay the horse all he could without exciting too much suspicion'. **1891** N. Gould *Double Event* 6 The heaviest layers of odds ..had laid Caloola..for considerable amounts. **1901** *Daily Chron.* 24 July 3/2 For the Derby or other important races Davis would lay a horse to the extent of £100,000 in one bet.

43. *to lay low* (see Low *a.* 18 c).

An occas. use erroneously developed from *to lie low*.

a **1861** T. Winthrop *John Brent* (1883) viii. 70 They may..let their chances slide at cards, but my notion is they're layin' low for bigger hauls. **1894** W. T. Stead *If Christ came to Chicago!* 225 The Democrats laid low and said nothing, for reasons of their own. **1907** M. C. Harris *Tents of Wickedness* IV. iii. 359 He..laid low for the first passer-by, and slugged him.

49. lay away. c. *Tanning.* To place (hides) flat in a vat to steep in strong tan liquor for a long period, as the final stage in the process of tanning. Also *intr.* of the hides. Cf. *Lay-away 1.

1885 C. T. Davis *Manuf. Leather* xix. 368 In tanning heavy upper leather the practice..is to first handle the sides on sticks for ten or twelve days, and then lay them away twice in bark. **1901** F. T. Addyman tr. *Villon's Pract. Treat. Leather Industry* 139 Time required for Laying Away.—The hides are removed from the pit and put back three times so that the tan may be renewed. **1922** A. Rogers *Pract. Tanning* x. 302 The hides are sometimes rocked throughout the early stages up to the time when they are laid away. **1966** G. H. W. Humphreys *Manuf. Sole & Other Heavy Leathers* vii. 120 The goods may be laid away, but rarely these days with bark or other ground material as was the practice in former days when 'layers' were in general use.

50. lay by. f. To work (a crop or field) for the last time, before leaving it to grow without further husbandry. *U.S.*

1759 J. Gordon *Jrnl.* 12 July in *William & Mary College Q.* (1902) 1st Ser. XI. 106 Mowing oats & laying by corn. **1784** J. F. D. Smyth *Tour U.S.A.* II. 127, I was also accustomed to sow a quantity of faulty wheat..in my tobacco grounds, when I gave them the last ploughing, or *laid them by*. **1835** J. H. Ingraham *South-West* II. 285 The ploughing generally ceases and the crop is 'laid by' about the last of July. **1868** *Rep. Iowa Agric. Soc. 1867* 158 The ground should be thoroughly rolled;..then lay by with barshear plow. **1947** *Democrat* 25 Dec. 3/4 This year when the corn was 'laid-by' the crotalaria came up voluntarily.

51. lay down. q. Also of a paving material. Hence, to cover (a surface) *with* something.

1893 A. Conan Doyle *Mem. Sherlock Holmes* (1894) 225 The corridor..was laid down with a kind of creamy linoleum.

s. *intr.* To give up or submit; to break down or cease to act; to fail; to retire or withdraw. *U.S.*

1898 *Scribner's Mag.* XXIII. 453/2, I swear I hate to lay down to such a nincompoop. **1901** Merwin & Webster *Calumet 'K'* 64 You've never had to lay down yet, and you don't now. **1911** H. S. Harrison *Queed* viii. 87 Your body's got to carry your mind around, and if it lays down on you [etc.]. **1923** R. D. Paine *Comrades of Rolling Ocean* x. 193 'Any water leaking in?' 'A trickle under the floor, but the bilge pump will take care of it unless she lays down on me.' *Ibid.* xvii. 293 You stand by me and I won't lay down on you. **1927** *Cleveland Press* 4 Feb., Offered him a bribe to 'lay down' on the prosecution of George J. McKay, alleged arch-swindler.

t. To set up or establish (a certain beat). *Jazz slang.*

1950 Blesh & Janis *They all played Ragtime* viii. 149 The backwoods pianists 'laid down the beat' and 'stacked the blues'. *Ibid.* x. 194 He laid down a terrific stomp. **1959** 'F. Newton' *Jazz Scene* vi. 104 The 'rhythm section' laid down a rock-firm beat. **1968** *Melody Maker* 6 Apr. 8/4 The soloist can play anything he chooses to play on the time that I lay down for him. **1968** *Blues Unlimited* Sept. 23 Preston..takes a few vocals, and lays down some swinging rhythm guitar.

54. lay off. a. (Later U.S. example.)

1919 H. L. Wilson *Ma Pettengill* ii. 46 She took me up to her little bedroom to lay my things off and then down to the parlour.

f. For '*dial.* and *U.S.*' read: orig. *dial.* and *U.S.* (Further examples.)

1863 W. Whitman *Specimen Days* (1882–3) 41 Some of the men are cleaning their sabres.., some brushing boots; some laying off, reading, writing. **1886** H. James *Bostonians* I. iii. 26 She would expect him to be strenuous in return; but he couldn't—in private life, he couldn't; privacy for Basil Ransom consisted entirely in what he called 'laying off'. **1955** *Times* 6 June 7/2 But in the course of this week stocks in some factories will begin to run out. Workers will have to be laid off. **1970** G. Greer *Female Eunuch* 242 The lowest paid employees can be and are laid off. **1972** *Daily Tel.* 1 Feb. 2/7 A pay strike by 500 clerical workers..has caused the company to lay off 2,500 car assembly workers.

g. (See quot. 1901.)

1901 J. Black *Illustr. Carpenter & Builder Ser.: Home Handicrafts* 43 What painters term 'laying off', that is to say, going over the work with the brush uncharged with paint and with strokes all in one direction. **1945** C. H. Eaton in *Practical Painter & Decorator* iii. 90 The laying off should be vertical, that is, from ceiling to floor..on walls, and parallel with the main source of light on ceilings. **1951** *Good Housek. Home Encycl.* 65/1 Do not attempt to 'lay off' or brush out the distemper as with paint or varnish. **1963** W. Tee *Painting & Decorating* x. 77 Finally, you lay-off, which means brushing in the direction of the grain if you are painting wood, or in the longest direction if you are painting a metal gutter or pipe.

h. To desist from (doing something); to abstain from or stop using (something); to stop bothering or pestering (a person). Also *intr.*, freq. as *imp.*: cut it out! stop it!

1908 Kipling *Lett. to Family* ii. 17 The railways..had to find room somewhere..before Nature cried: 'Lay off!' **1919** *Amer. Mag.* May 42/2 If you guys don't lay off of me I'll bounce the two of you. **1919** *Saucy Stories* Aug. 107/2 She..resolved to 'lay off the bright lights in the future'. **1930** D. Hammett *Maltese Falcon* xviii. 221 Make him lay off me then. I'm going to fog him if he keeps it up. **1931** E. Linklater *Juan in America* III. vii. 259 Lay off that ritzy laugh or I'll sock you. **1934** J. Agate *More First Nights* (1937) 59 You would think, wouldn't you, that Josephine, having done enough in the way of arousing suspicion, would lay off a little. **1934** Wodehouse *Right Ho, Jeeves* vi. 60 Lay off the sausages. Avoid the ham. **1936** —— *Laughing Gas* iv. 49 That's all she's after—the title. For heaven's sake, Reggie, lay off while there's still time. **1946** K. Tennant *Lost Haven* (1947) xix. 316 'For God's sake, shut up!..Lay off, Alec, lay off.' Alec laid off. **1947** 'N. Shute' *Chequer Board* 62 Aw, lay off, Jim.—You're not in the South now. **1953** J. Trench *Docken Dead* ii. 21 How does one set about telling one's senior officer to lay off one's friend's wife? **1968** M. Richler in R. Weaver *Canad. Short Stories* 2nd ser. 160 'Oh, lay off,' my father said. 'Give the man air.' **1974** D. Gray *Dead Give Away* vi. 65 I'd lay off stirring up trouble for a bit if I were you.

i. *Naut.* and *Aeronaut.* To indicate (on a chart, etc.), to work out (a course). Cf. Course *sb.* 12.

1942 *Tee Emm* (Air Ministry) II. 83 Always lay off Q.D.M.'s and Q.D.Y.'s as true bearings on your chart. **1943** 'T. Dudley-Gordon' *Coastal Command* 17 Drawing pencil lines which lay off courses of ships and aircraft, and indicate areas under patrol or to be searched. **1961** F. H. Burgess *Dict. Sailing* 131 *Lay off a course*, work out a proposed course on a chart.

j. Of a bookmaker: to insure against a substantial loss resulting from (a large bet) by placing a similar bet with another bookmaker.

1951 E. Kefauver *Crime in Amer.* (1952) xvi. 184 The Nevada bookies also protect themselves by laying off their biggest bets with out-of-state operators. **1974** *New Yorker* 25 Feb. 72 An outside man..runs along the line of bookies and keeps an eye on the odds and lays off some of the money.

55. lay on. i. *transf.* Delete 'in jocular use' and add further examples in sense: to provide, arrange.

Also influenced by sense 55 k in Dict.

1861 *Temple Bar* III. 23 Fifteen shillings an hour, to say nothing of refreshments laid on gratis and supplied at discretion. **1940** C. Gardner *A.A.S.F.* 84 Squadron Leader Dodds..said that he'd got my programme..laid on. 'Laid on' was the Army term for everything—and I found myself using it. **1944** N. Coward *Middle East Diary* 103, I was unable to give a concert as the piano..had not been 'laid on'. **1959** *Punch* 13 May 636/2 Universities do not exist to lay on degree courses to follow the idiosyncratic requirements of a particular employer. **1959** 'J. Welcome' *Stop at Nothing* ix. 139 As usual he had everything laid on and a car was waiting. **1964** E. O'Brien *Girls in Married Bliss* vi. 51 He'd have some hatchet-voiced secretary laid on to tell Kate some boring and familiar lie, like that he was in conference. **1971** B. W. Aldiss *Soldier Erect* 185 Pack your night things in a small pack and get weaving, while I lay on transport. **1973** E. Page *Fortnight by Sea* xvi. 177 Try and lay some coffee on. Plenty of it. Good and strong.

j. Also, to place (thread) on a material before couching it down with a separate thread.

1880 L. Higgin *Handbk. Embroidery* i. 8 '*Japanese gold thread*'..must..be laid on, and stitched down with a fine yellow silk. [**1906** A. G. I. Christie *Embroidery & Tapestry Weaving* viii. 166 A bunch of threads may be laid upon the material, and an open chain, buttonhole, or feather stitch worked over in order to fix it in place.] **1959** *Chambers's Encycl.* V. 155/2 Couching or laid work is a form in which the threads are 'laid' on material and couched down with matching or contrasting colour.

k. (Earlier and later examples.) Also, to provide (a telephone line).

1845 *Punch* 1 Mar. 100/1 Announcing that the water was going to be laid on when it wasn't. **1869** *St. Andrews Gaz.* 7 Aug., The special wires which the Scotch papers have 'laid on' between London and Edinburgh. **1870** Dickens *E. Drood* xxii. 170 There is two bedrooms..with gas laid on. **1885** *List of Subscribers, Classified* (United Telephone Co.) (ed. 6) 17 At 'The Clarendon' in Brighton ..they have a telephone laid on. **1909** *Chambers's Jrnl.* July 477/2 A large supply of hydrogen prepared by a new process is laid on for inflation.

l. To give (something) to (a person). *U.S. slang.*

1942 *Amer. Mercury* July 86 Lay de skin on me [shake hands], pal! **1952** G. Mandel *Flee Angry Strangers* 244 He lays some on his buddies 'n they get to like it; right, Buster? **1960** *Time & Tide* 24 Dec. 1599/3 I've fixed up a real wild basket of ribs and a bottle of juice, and I'd like you to fall by her joint and lay it on her. **1968** *New Yorker* 18 May 45/2 He..took out a copy of his newest album. He wrote something on the back of it and picked up one of the hotel bills. 'Let me just lay this album on the man downstairs. Maybe it'll keep him quiet for two or three days.' 1970 *It* 9–24 Apr. 8/4 Of course you can't lay advice on someone.

56. lay out. b. (Further example.) Also, to knock (a person) unconscious; to kill.

1829 [see *COLD *a.* 1 c]. **1890** in Barrère & Leland *Dict. Slang* II. 9 Galletly was saying, 'I've laid one out' to the other prisoners... Witness also saw the knife, and there was blood on it. **1894** *Daily News* 26 May 8/6 If you strike me I will lay you out. **1896** *Wells Jrnl.* 3 Dec. 7/5 A disposition to 'injure, maim, and lay out an opponent, especially if he be a valuable element in the opposing team'. **1916** 'TAFFRAIL' *Pincher Martin* xviii. 337, I gits laid up ten days ago,.. death vacancy. Poor ole Byles was laid out, yer remember. **1929** J. B. PRIESTLEY *Good Companions* III. v. 589 'But do you mean to say he was laid out?' he demanded... 'On the jaw, I think you said?' **1973** *Scotsman* 21 Feb. 17/6 When they hit you with the word, cancer, it scares you to death. Boom! You're laid out. But I've learned a lot about cancer since then.

60. lay up. m. To assemble or stack (plies or layers) in the arrangement required for the manufacture of plywood or other laminated material (usu. prior to bonding into a single structure).

1927 KNIGHT & WULPI *Veneers & Plywood* xxvi. 286 Stock trucks..with suitable guides against which to jog layers of stock as the freshly glued plywood is laid up. **1942** WOOD & LINN *Plywoods* vii. 74 When working on thin 3-ply boards two panels are frequently 'laid up' between each caul. **1949** B. L. DAVIES *Technol. Plastics* xiii. 233 The dried, impregnated or coated material is cut to size..and the sheets are laid up, i.e. piled one upon the other to a predetermined number. **1962** *Newnes Conc. Encycl. Electr. Engin.* 115/2 Normally not less than three layers of tissue are laid up between the electrodes [of an industrial capacitor] for the lower voltages. **1965** *Plastics Tooling & Manuf. Handbk.* (Amer. Soc. Tool & Manuf. Engineers) vi. 114 Successive plies are laid up until the desired thickness is achieved; then the part is allowed to cure.

61. lay-down, (*b*) applied to a hand or contract at cards (esp. Bridge) which is such that success is possible against any defence, so that no harm would be done by exposing the player's cards on the table; also *ellipt.* as *sb.*, such a hand; also *fig.*

1906 *Westm. Gaz.* 8 Sept. 16/3 Enormous cards are held and we have a lay-down great or small slam. **1934** *Amer. Speech* IX. 10/2 A *cold* game is a sure game, and a *cold* contract is a *lay-down*. **1955** I. FLEMING *Moonraker* vii. 75 It was a laydown Grand Slam for Bond against any defence. **1959** *Listener* 12 Mar. 489/1 Seven Clubs, it will be seen, is a lay-down. **1961** *Times* 6 Dec. 8/3 A lay-down slam in Clubs. **1966** 'W. HAGGARD' *Power House* ix. 92 The interview had diverted him. He'd learnt a lot about Harry Fletcher; he'd held a crushing hand and had played it as a laydown. **1974** *Country Life* 17 Oct. 1139/3 The slam is a lay down.

layabout (lē^{i}·,ăbaut). [f. LAY *v.*^{1} 43 + ABOUT *adv.* 8.] An habitual loafer, idler, or tramp. Also *attrib.* or as *adj.*

1932 G. S. MONCRIEFF *Café Bar* viii. 78 These layabouts were rotters. **1932** S. PEARSON *To Streets & Back* xxiv. 234 The 'down and outs' in Hyde Park are permanent 'layabouts'. **1959** *Punch* 19 Aug. 57/1 He simply uses any old-fashioned plot about layabout art-lecturers getting mixed up with funny spies. **1959** H. PINTER *Birthday Party* (1960) 81 Keep an eye open for low-lives [*sic*], for schnorrers and for layabouts. He didn't mention names. **1961** *John o' London's* 21 Sept. 327/1 A colourful tour of layabout London. **1968** *New Scientist* 2 May 217/2 Those of us gifted by nature with inertia but maligned by society as layabouts. **1972** D. HASTON *In High Places* ii. 35 There was another strong twosome..in the hut, otherwise only student layabouts left.

lay-away (lē^{i}-ăwē^{i}). Also **layaway**. [LAY *v.*^{1} 49.] **1.** *Tanning.* A vat or pit in which hides are 'laid away'; = LAYER *sb.* 4 e. Freq. *attrib.*

1885 A. WATT *Art Leather Manuf.* xii. 145 The Layers.—In these pits, which are termed lay-aways by the Americans, the butts are stratified with ground oak-bark. **1885** C. T. DAVIS *Manuf. Leather* xix. 367 When heat is used on the head leaches the liquor sometimes enters the lay-away yard in a hot condition. **1901** F. T. ADDYMAN tr. *Villon's Pract. Treat. Leather Industry* 137 Tan-pits or lay-away pits are large vats, sometimes round, and made of oak, bound with iron. **1922** A. ROGERS *Pract. Tanning* x. 303 At this stage..the hides go to the layaways. These are vats large enough to permit the hide being spread out flat. *Ibid.* 304 All the layaway liquors are not so worked down and out, some being returned to the leach-house to be strengthened and freshened. **1957** *Encycl. Brit.* XIII. 848/1 The goods are moved from the strongest suspender vats to the layaway vats or floaters.

2. *N. Amer.* = *LAY-BY *sb.* 2 b.

1961 in WEBSTER. **1967** *Boston Sunday Herald* 26 Mar. I. 22/3 Her dress for Easter is on layaway... You can get it out. It only costs $5, but I don't own that much now. **1970** *Globe & Mail* (Toronto) 25 Sept. 32/7 (Advt.), No Dealers—All Sales Final—Christmas Lay-A-Ways... Sale at Toronto Warehouse. **1973** *Houston* (Texas) *Chron.* 21 Oct. 2/1 (Advt.), Use our layaway plan for Christmas!

lay-back. [f. vbl. phr. *to lay back.*] **1.** The receding position of the nose of certain breeds of dog, esp. the bulldog.

1894 R. B. LEE *Hist. & Descr. Mod. Dogs Gt. Brit. & Ireland* (*Non-Sporting Division*) 233 The bones of the lower jaw in specimens [of the bulldog] which have the desired appearance, known as 'upturn' and 'lay back' are found to have the contour of a segment of a circle. **1905** H. ST. J. COOPER *Bull-Dogs & Bull-Dog Breeding* 98 This well-known dog..has a grandly shaped head, with small well-carried ears, large under-jaw, turn-up and lay-back. **1909** *Ladies' Field* 28 Aug. 511/1 Bulldogs:..a brindle, good layback and under-jaw. **1968** H. HARMAR *Chihuahua Guide* 239 *Layback*, the receding nose found in some of the short-faced breeds.

2. a. *Mountaineering.* A method of climbing cracks in rocks (see quot. 1968).

1925 *Jrnl. Fell & Rock Climbing Club* VII. 17 The crack ..is then climbed, utilising the lefthand edge of the crack for a 'lay back'. **1957** CLARK & PYATT *Mountaineering in Brit.* xvi. 237 A layback is almost as tiring in descent as in ascent. If the leader's arms give out he cannot retreat, and has to fall off. **1968** P. CREW *Encycl. Dict. Mountaineering* 77 *Layback*, a method of climbing cracks and flakes by gripping the edge with the hands, leaning back and placing the feet flat on the rock at the side of the crack and slightly below the hands. As the climber pulls on the edge of the crack and presses his feet against the rock, the opposing pressures exerted can be sufficient to support the body. **1971** *N.Z. Listener* 19 Apr. 56/5 Pete had a dekko up a chimney. But there was a lay-back and too much exposure at the top. **1973** C. BONINGTON *Next Horizon* x. 145 Only the crack in its [*sc.* the rock's] back provided a mixture of hand-jamming and lay-back holds.

b. In various sports, the movement or position of leaning backwards or lying on one's back.

1948 R. F. HERRICK *Red Top: Reminisc. Harvard Rowing* 173 The chief differences between Washington and eastern rowing at that time were Washington's lack of layback, fast hands on the release and their tremendous emphasis on the catch. **1962** *Times* 6 Feb. 4/5 They [*sc.* the Cambridge crew] have a longer swing and a longer lay-back than recent crews. **1962** *Austral. Women's Weekly* Suppl. 24 Oct. 3/3 [Surfing] *Layback*, a supreme test of skill in trick riding. The rider lies flat on his back, with feet facing the way board is going. **1968** *Daily Tel.* 6 Dec. 15/6 As always, Miss Waghorn used her long legs to full advantage in the split jump, a majestic spreadeagle and well-timed lay-back and grab-parallel spins.

lay-by. Add: Also **lay-bye.**

A. *sb.* **1.** (Earlier example.)

1826 J. KAY *Let.* 7 July in *N. & Q.* (1960) Apr. 148/1, I have given permission for a Laybye to be formed in the bank of the Canal near Thornhill Bridge.

b. A railway siding.

1906 *Westm. Gaz.* 28 Sept. 7/1 A heavy goods train had left the up-line..and run into a short lay-by. **1955** L. T. C. ROLT *Red for Danger* x. 206 He therefore signalled the L.M.S. goods out of the lay-by, but the driver stopped in Charfield station for water.

c. An area adjoining a road where vehicles may park without interfering with the traffic.

1939 [see *draw-in* (*DRAW *sb.* 11 a)]. **1950** *Engineering* 17 Nov. 387/2 Stopping places off the carriageway in the form of lay-bys. **1959** *Manch. Guardian* 1 July 5/4 The emergence of a new type of picnic—the lay-by high tea... On any arterial road you can see the family saloons.. heave into the lay-by. **1971** *Islander* (Victoria, B.C.) 8 Aug. 4/3 We enjoyed a Sunday picnic in a forest layby and then returned to Holland. **1972** *Daily Tel.* 13 Mar. 3/2 Caravanners who park in lay-bys causing litter and hygiene problems will face prosecution. **1973** *Times* 30 Apr. 14/1 My correspondent at the front owns a weekend cottage in Norfolk and was in the habit of collecting his empty wine bottles and baked bean cans and dumping them in the lay-by bins on the way home on Sundays. **1973** *People's Jrnl.* (Inverness & Northern Counties ed.) 4 Aug. 8/2 The holidaymakers had stopped in a lay-by at Oban and Miss Coldrick thinks she left the camera on the car boot.

2. b. A system of payment whereby a purchaser puts down a deposit on an article, which is then kept on one side for him until he has paid the full price. Also *transf.*, and as *vb.* Chiefly *Austral.* and *N.Z.*

1930 *Sydney Morning Herald* 16 Oct. 4 (Advt.), Avail yourself of our lay-by service. **1943** *Amer. Speech* XVIII. 95 [New Zealand] A few trade names have caught the public fancy, and become generalized .. A system of hire-purchase called the 'Lay-by' has resulted in the verb 'layby', pronounced and written as one word. **1944** W. E. HARNEY *Taboo* (ed. 3) 154 He did not complain, for it was to him a tribal law and custom—a lay-by system to protect him when he was old. **1957** *Rhodesia Herald* 16 Mar. (Advt.), Lay-Byes Accepted Now. **1960** *Times* 25 June 9/4 When in Australia..I was bewildered to find this expression 'Lay by' used widely in large shops, until I discovered that it meant that the management would put aside articles for customers. **1969** *Sydney Morning Herald* 24 May 26/8 (Advt.), Goldfish, tropical. Full range plants, access... Lay-by or terms. Get your discount card now.

layer, *sb.* Add: **I. 1. e.** (Further examples of *layer-out.*)

1895 HARDY *Jude* IV. ii. 248 In the afternoon, when everything was done, and the layers-out had finished their beer, and gone, he sat down in the silent place. **1928** *Observer* 10 June 7/4 Poetry, it is generally known, is dead... Our critics are a generation of layers-out. **1953** R. CAMPBELL tr. *E. de Queiroz's Cousin Bazilio* xiii. 264 The professional layer-out was a woman with a pocked face. **1958** L. DURRELL *Mountolive* xvi. 318 The dead man's clothes are the perquisites of the layer-out. **1974** W. FOLEY *Child in Forest* I. 51 His widowed mother.. acted midwife and washer-woman, or layer-out of the dead.

f. = *gun-layer* (*GUN *sb.* 15).

1896 *Daily News* 6 Aug. 7/2 Two gun detachments, including layers. **1898** G. S. ROBERTSON *Chitrál* xviii. 167 All the gunners, even the 'layers', wore bandages over their eyes. **1911** H. A. BETHELL *Mod. Artillery in Field* x. 147 If any officer or layer fails to locate the target correctly, the result is likely to be considerable waste of time and ammunition. **1971** D. A. LAMB *View from Bridge* ii. 16 On the twelfth of February three of the new artillerymen became first class layers.

II. 2. b. A formation of aircraft flying at the same height.

1940 N. MONKS *Squadrons Up!* iv. 113 One Hurricane, trying to get above the enemy aircraft, observed five layers of ten M.E. 110's each, between 10,000 and 15,000 feet. *Ibid.*, Another Hurricane spent about a quarter of an hour alternately dodging and flying in and out among the enemy layers. **1952** *Oxf. Jun. Encycl.* X. 52 The German fighters usually flew in formation well above the bomber formations... The R.A.F. replied by sending up layer formations, the upper layer to engage the fighters while a lower layer dealt with the bombers. **1959** R. COLLIER *City that wouldn't Die* v. 64 Since to-night was a 'fighter night' ..the guns could engage targets only at 12,000 feet and below—2000 feet below the bottom layer of fighters. *Ibid.* x. 167 As the youngest he had inevitably drawn the highest 'layer' on the Southend–Romford patrol line.

4. b. (Earlier example.)

1667 T. SPRAT *Hist. R. Soc.* II. 308 This Brood and other Oysters they carry to Creeks of the Sea..and there throw them into the Channel, which they call their Beds or Layers, where they grow and fatten.

g. *Cartography.* An area on a map depicted in a particular colour or tint chosen to represent all land between two specified heights. Cf. *layer system* in *5.

1918 BRYANT & HUGHES *Map Work* v. 84 Tints, or layers, of colour are used to denote all the land lying between any two named contours. **1932** J. W. CAMERON *Maps & Map-Work* iv. 45 Hill features are represented on the map by: 1. Contours. 2. Hachures. 3. Hill-shading. 4. Colour layers or layer-colouring. **1969** C. B. M. LOCK *Mod. Maps & Atlases* iii. 134 When the fifth (Relief) 'One inch' series was issued, relief was shown by contours in brown, hachures in orange and hill-shading in grey, with layers in buff tints.

5. layer-cake, a cake consisting of layers of sponge held together by a sweet filling, and usually iced; also *fig.*; **layer cloud** *Meteorol.*, a sheet-like cloud, having little vertical development but pronounced horizontal development; **layer colour** *Cartography*, a colour used in the layer system of showing relief on a map; so *layer colouring, -coloured* adj.; **layer lattice** *Cryst.* [tr. G. *schichtengitter* (F. Hund 1925, in *Zeitschr. f. Physik* XXXIV. 849)], a crystal lattice in which the atoms are arranged in layers a few atoms thick that are separated by a distance greater than the interatomic distance within the layers, so that the interlayer forces are relatively weak; **layer pit** or **vat** *Tanning* = *LAY-AWAY 1, LAYER *sb.* 4 e; **layer-pudding,** a steamed pudding, consisting of layers of suet crust pastry with a sweet filling; **layer shading** *Cartography*, the use of layer tints to show relief on a map; **layer system** *Cartography*, on a map, the representation of land between different heights or contours by different colours or tints that are graded so as to show relief at a glance; **layer tint** *Cartography*, a layer colour, or a tint of such a colour; so *layer tinting, -tinted* adj.

1881 F. OWENS *Cook Bk.* 265 *Lemon butter*, good to eat as sauce, or for layer cakes. **1895** *Montgomery Ward Catal.* 431/2 Tins for pies and layer cakes. **1902** *Daily Chron.* 3 May 8/4 Layer cakes can be made in great varieties according to the filling used. **1905** *N.Y. Even. Post* 16 Dec., In the mixing of this literary layer cake most of the humor rose to the top. **1933** 'R. CROMPTON' *William—the Rebel* iv. 94 He..began to eat the last piece of cream-layer cake. **1962** L. DEIGHTON *Ipcress File* xxi. 140 The sunset was a layer cake of mauve and gold. **1965** R. CARRIER *Cookbk.* xxii. 473 Mocha layer cake, a subtle blend of chocolate and coffee spiked with rum. [**1920** G. A. CLARKE *Clouds* iv. 73 The beautifully waved structure seen in nearly all of the layer-type of clouds from cirrus downward to stratocumulus is caused by the propagation upwards or downwards of the wave-motion that is produced by the flowing of air-currents of different velocities and directions over each other.] **1951** *Rep. Progress Physics* XIV. 192 Thick layer clouds often found over the oceans. **1956** *Nature* 18 Feb. 321/1 The great layer-cloud systems which are associated with cyclones and fronts. **1963** G. M. B. DOBSON *Exploring Atmosphere* iv. 78 The persistence of thick layer-clouds over a city in winter gives rise to a very dark, gloomy day. **1922** *Encycl. Brit.* XXX. 417/2 He extended and popularized the use of 'layer' colours exhibiting relief in land. **1969** C. B. M. LOCK *Mod. Maps & Atlases* iii. 135 A single-colour shadow tone in blue was printed in half-tone over the layer colours and the layer tints themselves were carefully chosen so as to reduce the 'step' effect of the layers. **1932** J. W. CAMERON *Maps & Map-Work* i. 12 There are two editions..: (*a*) In outline,..without contours. (*b*) Contoured and layer-coloured. **1924** *Catal. Maps Ordnance Survey* 4 Relief is indicated by..layer colouring in shades of brown and green. **1969** C. B. M. LOCK *Mod. Maps & Atlases* i. 40 For the best results of all, hill-shading is blended with layer-colouring. **1929** *Trans. Faraday Soc.* XXV. 265 Besides cadmium iodide there are known a number of other types of layer lattices. **1966** A. CAMERON *Princ. Lubrication* xxi. 21 The low friction of graphite may not..be directly related to its layer-lattice structure. **1901** F. T. ADDYMAN tr. *Villon's Pract. Treat. Leather Industry* 137 The object of handling is to give body to the plumped skin, so that it may be able

to support the weight which will press upon it in the layer-pit. **1949** D. WOODROFFE *Stand. Handbk. Industr. Leathers* iii. 47 The butts or other leather, already completely penetrated in the previous stages of the tanning process, are placed singly in the layer pits, a layer of ground tanning material..is sprinkled over each butt, and the process continued with more pieces of leather until the pit is full. Finally, a very strong tan liquor..is run into the pit. **1909** *Daily Sketch* 14 Oct. 14/3 Layer pudding. **1951** *Good Housek. Home Encycl.* 489/2 Syrup is used..as a filling for tarts and layer puddings. **1952** MONKHOUSE & WILKINSON *Maps & Diagrams* ii. 61 (*heading*) Layer-shading and tinting. **1971** G. R. P. LAWRENCE *Cartogr. Methods* ii. 25 Information relating to the shapes of the floors of lakes and oceans can be presented in much the same way as relief information but in most cases only submarine contour lines are used, layer shading being found in some atlases. **1903** *Man. Field Sketching & Reconnaissance* (H.M.S.O.) vi. 28 There is also the layer system of showing hills. **1953** A. H. ROBINSON *Elem. Cartogr.* x. 215/2 The larger the scale, assuming a reasonable degree of contour simplification, the more useful the layer system. **1969** C. B. M. LOCK *Mod. Maps & Atlases* iii. 135 The layer system was not used, but, instead, two printings in purple-grey tones on the shadow side of the hills and one printing in yellow on the illuminated side were added to the standard base map. **1918** BRYANT & HUGHES *Map Work* v. 84 In high country the layer tints become so dark as to obscure all detail. **1969** *Geography* LIV. 198 Relief is shown by contour and layer tint. **1934** J. BYGOTT *Introd. Mapwork & Pract. Geogr.* iv. 21 Certain layer-tinted Ordnance maps, especially the layered quarter-inch and half-inch maps. **1966** *McGraw-Hill Encycl. Sci. & Technol.* II. 534/2 On small-scale maps.. the intervals are frequently layer-tinted from green to brown. **1952** H. C. BROOKFIELD in G. H. Dury *Map Interpretation* xvi. 184 The O.S. 1/63,360 Fifth (Relief) Edition, first published in 1929, employed contours at 50-ft. intervals... There was also layer tinting in buff, the tint changing at each 500 ft. **1971** G. R. P. LAWRENCE *Cartogr. Methods* ii. 24 Colour has been used in the depiction of relief for a number of years in the method known as 'layer tinting'. **1969** T. C. THORSTENSEN *Pract. Leather Technol.* v. 70 The bends lie in the layer vats for one or two weeks in a warm vegetable tan liquor, and again the liquor may be strengthened and heated to gain better penetration and fixation of the strong tanning liquors.

layer, *v.* Add: **3.** *trans.* To place or insert as a layer.

1906 *Times Lit. Suppl.* 12 Jan. 14/1 Mr. Lee has succeeded in neatly layering fallacies of argument. **1974** *Nature* 8 Mar. 110/2 Sample of 55 ml of a 110 ml linear density gradient..was layered into a 110 ml jacketed isoelectric focusing column. *Ibid.* 5 Apr. 519/2 Each incubation was then layered onto a 10–30% sucrose gradient in TKM.

layered, *a.* Add: *spec.* in *Cartography*, having relief shown by the layer system.

1922 *Encycl. Brit.* XXXI. 842/2 These remarks apply with special force to the 'layered' maps; changes in the tones of the layers will greatly alter their character. **1969** C. B. M. LOCK *Mod. Maps & Atlases* i. 40 For reasons of economy, hachures were omitted from the fourth or 'popular' edition published after the 1914–18 war... Publication of layered or shaded maps of selected tourist areas followed.

layering (lē^{i}·əriŋ). *Cartography.* [f. LAYER *sb.* + -ING^1.] = *layer colouring, layer shading* (*LAYER *sb.* 5).

1922 *Encycl. Brit.* XXXII. 1174 (Index), Layering of maps. **1937** *Geogr. Jrnl.* LXXXIX. 52 It is possible that the ultimate solution will be a change in the ground tint of the map over areas of rock, though experiment may show that this gives a false impression of layering.

layette. 1. (Earlier and later examples.)

1839 F. A. KEMBLE *Jrnl. Residence Georgian Plantation* (1863) 158, I have worked my fingers nearly off with making..innumerable rolls of coarse little baby-clothes, layettes for the use of small new-born slaves. **1863** G. DU MAURIER *Let.* in *Young G. du Maurier* (1951) 212 She is now..making the layette and things for the son and heir, whom I am beginning to look forward to with a certain amount of curious expectation. **1939** [see *BAJU]. **1974** *Selfridge Christmas Catal.* 47 Baby doll complete with layette.

laying, *vbl. sb.* Add: **1. b.** (Further examples.)

1900 *Daily News* 20 Sept. 6/2 The stoppage of coal traffic, and the consequent laying off of railway coal train crews. **1968** *Listener* 4 July 24/1 Nowadays in the US redundancy can mean displacement for good; then it was, at worst, a long laying-off. **1970** *Nature* 21 Nov. 709/2 Layings-off are, of course, nothing unusual in the volatile aerospace industry.

2. c. (Later example.)

1960 C. M. YONGE *Oysters* ix. 154 The Colchester and other natural oyster beds and the layings along the Essex coast flourished exceedingly during the eighteenth and much of the nineteenth centuries.

3. laying house (*b*), a building in which laying hens are kept; **laying mash, meal,** a special food for laying hens.

1913 H. R. LEWIS *Productive Poultry Husbandry* vii. 128 (*heading*) Plans and specifications of laying houses. **1962** L. E. CARD *Lippincott's Poultry Production* (ed. 9) vii. 179 (*caption*) Interior of..laying house, showing tiered roofs. **1926** *Daily Colonist* (Victoria, B.C.) 12 Jan. 2/1 (Advt.), Feed Prices..Laying Mash, sack $2.75. **1972** A. A. MCARDLE *Poultry Managem. & Production* (rev. ed.) xiv. 308 (*heading*) Quantity of laying mash and grain needed. **1908** *Illustr. Poultry Rec.* Oct. p. ix (Advt.), Laying Meal 12/6 Cwt. **1935** *Poultry Rec.* Jan. (Advt. inside cover), Alfalfa, Laying Meal, Grit, Shells etc.

lay-light. [f. LAY *v.*^{1} + LIGHT *sb.* 5 b.] A window or light made of glazed panels and set into a ceiling to provide natural or artificial light.

1932 H. ROBERTSON *Mod. Archit. Design* vi. 174 The architect is well advised to make a critical study..of lighting through ceiling lay-lights, before determining his layout. **1934** *Archit. Rev.* LXXV. 31 The lay-light of the same type of glass is hung from the ceiling to provide day and artificial lighting. **1948** R. O. ACKERLEY *Introd. Sci. Artificial Lighting* III. iv. 99 Laylights designed to provide alternatively natural and artificial light through the same glazing, like most dual-purpose devices, are liable to perform both functions badly unless great care is taken in the design. **1951** W. R. STEVENS *Princ. Lighting* viii. 186 We find it difficult to provide high illumination..in interiors without discomfort glare unless we use very large sources, such as laylights or indirectly lighted ceilings. **1966** M. M. PEGLER *Dict. Interior Design* (1967) 260 *Laylight*, a glass or translucent panel set flush into a ceiling to admit natural or artificial light.

lay-off. [LAY *v.*^{1} 54.] A rest, respite, spell of relaxation; a period during which a workman is temporarily dismissed or allowed to leave his work; a part or season of the year during which activity in a particular business or game is partly or completely suspended.

1889 *Gallup* (New Mexico) *Gleaner* 27 Mar. 1/3 Fred Diamond is taking a lay-off. **1904** *Minneapolis Daily Times* 8 June 8 The men who have been on for a year get a vacation of ten days. Those who have been working less than a year have to get along with only a five-day layoff. **1908** 'O. HENRY' *Gentle Grafter* vii. 285 Me and my partner..tried to take a layoff from our professional and business duties; but..our work followed us wherever we went. **1909** R. A. WASON *Happy Hawkins* 148 Now take a lay-off if you want to,..then come back here. **1919** T. K. HOLMES *Man from Tall Timber* vi. 58 At the lay-off,..he had given each man enough money on account to make their vacation..a very wet spell indeed. **1923** *Daily Mail* 10 Sept. 8/5 As a consequence of the 'lay-off' during the summer months it often happens that the muscles of the young player are not sufficiently supple for him to face the rigours of the game. **1926** J. BLACK *You can't Win* xix. 297, I decided to take a lay-off [from burgling]. **1927** *Sunday Express* 8 May 10/3 Although his salary on the speaking stage had fluctuated between £20 and £40 a week, there were..many lay-offs. **1928** *Sunday Dispatch* 8 July 22/2 His opponent will be a French boxer, Pierre Callior. As nothing is known of the Frenchman here we shall be.. in the dark..as to what extent..Teddy's long lay-off has affected his form. **1952** B. ULANOV *Hist. Jazz in Amer.* (1958) xxii. 299 The Herman band took a layoff in Detroit in June 1945. **1956** B. NANKEVILLE *Miracle of Mile* xii. 90 Chris Chataway..had only had ten days' training following a lay-off. **1969** *Daily Tel.* 12 Apr. 1/5 Workers protested at proposed factory lay-offs. **1972** *Guardian* 17 Feb. 6/1 Jaguar is one of the British Leyland groups where extensive lay-offs occurred. **1973** *Times* 17 Oct. 25/1 (*heading*) Lay offs threaten 28,000 workers at Vauxhall as car unrest grows.

lay-out. Add: **1.** Also, the plan or disposition of a house, factory, garden, etc. Also *fig.* and *transf.*

1852 *San Diego Herald* 10 Jan. 2/1 The new 'lay out', at the Wholesale Commission Warehouse and store..on California street, proves quite attractive. **1903** KIPLING *Five Nations* 212 The day's lay-out—the mornin' sun Beneath your 'at-brim as you sight; The dinner-'ush from noon till one, And the full roar that lasts till night. **1905** W. ROBINSON *Eng. Flower Garden* (ed. 9) I. ii. 15 In many books on garden design the authors misuse words... One ..writes 'lay-out' for 'plan'. **1923** *Radio Times* 28 Sept. 28 (Advt.), The lay-out [of a radio set] is neat and compact; the cabinet work distinctive. **1930** *Oxford Times* 14 Mar. 13 A new 'lay-out' plan from the builders was before the meeting. **1931** H. G. WELLS *Work, Wealth & Happiness of Mankind* (1932) 2 Almost all our political and administrative boundaries, the 'layout' of the human population, have become..misfits. **1937** *Discovery* Feb. 47/2 In the planning of new factories.., the engineer and the industrial psychologist should..ensure that, both in the general lay-out and in the smallest details, due regard is paid to the view-point of the men who will actually be doing the work. **1937** D. RUNYON *More than Somewhat* i. 7, I go to see Mr. Tuesday at a Fifth Avenue hotel where he makes his home, and where he has a very swell layout of rooms. **1939** [see *FERRY *sb.*^{1} 3 e]. **1941** W. S. CHURCHILL *Secret Session Speeches* (1946) 37, I am satisfied that up to the present a good lay-out of our available forces has been made. **1943** J. B. PRIESTLEY *Daylight on Saturday* xxv. 195 Both of 'em are up against a pretty tough problem on floor space and general lay-out. **1946** *R.A.F. Jrnl.* May 179 A layout was begun for a new medium bomber to replace the Hampden. **1960** *Observer* 24 Jan. 5/2 How they broke in in the first place depended on the layout. It could be through a lavatory window or by tunnelling laboriously from next door. **1962** *Which? Car Suppl.* Oct. 129/1 The pedal control layouts on the *Peugeot* 403B..were generally liked. **1964** *McCall's Sewing* ii. 30/1 *Layout*, the way the pattern pieces are placed on the fabric for cutting. **1974** J. WAINWRIGHT *Hard Hit* 131 The house—this 'Diasc Farm' place—tell me about it. The layout. The approach. The grounds.

b. Typographical specifications or rough designs for a piece of printing; in extended use, the design details of a cartoon film. Also *attrib.*

1910 *Brit. Printer* Feb.–Mar. p. lxv (Advt.), The Printer will be enabled to submit to his customer both original and artistic lay-outs. **1910** E. G. GRESS *Art & Pract. Typogr.* 35/1 Every printshop should have a 'layout' man. **1913** *Technical World* XIX. 464 And the strange part of it is that from two to five good news picture layouts are made daily and enlarged in this small room. **1933** *Planning* I. vii. 2 A few years ago individualism in advertising would have rejected without hesitation the discipline of lay-out and design which has been accepted with such good effect this time. **1946** A. MONKMAN in H. Whetton *Pract. Printing & Binding* vii. 81/1 It is necessary also for the layout man to be expert in copying the numerous type faces. This is not so very important if the layout is for the compositor, because all he requires for his work is the general plan, indicating type sizes and faces and allocations of white space; but when the layout is for the client, markings showing type sizes and other particulars are not necessary. The sketch should be a clear indication of the appearance of the finished job. **1948** H. MISSINGHAM *Student's Guide Commercial Art* II. 157 To see a first-class 'visualizer' romp through one with a flat-leaded layout pencil is a sight to stir the blood. **1955** M. REIFER *Dict. New Words* 119/2 *Layout*,..2. *Cartoon Animation.* The combined elements of the animation procedure which prescribe the relationship of characters and backgrounds to define a scene and its properties. **1958** *Clarendonian* XII. 198 Throughout his service at the Press he has shown great interest in the Layout Department and has been responsible for the design of many Oxford books. **1959** HALAS & MANVELL *Technique Film Animation* III. 215 He [*sc.* the director] works closely with the lay-out artist, who in most studios is also the designer. **1967** KARCH & BUBER *Offset Processes* iii. 46 The layout man usually receives the reading matter in typewritten form, along with glossy photographs, line drawings or other artwork. **1968** J. R. BIGGS *Basic Typogr.* 122/2 The layout which goes to the printer with the typographic instructions 'marked up'..should be quite accurate with nothing left to chance.

2. *spec.* The equipment used for smoking opium. *U.S. slang.*

1882 H. H. KANE *Opium-Smoking* iii. 32 (*heading*) Description of the 'lay-out', pipe, etc. **1887** in *Amer. Speech* (1948) XXIII. 247 A person that had a private layout in his room. **1891** H. CAMPBELL *Darkness & Daylight* (1895) xxviii. 565 A small room at the rear of the temple contained an opium 'lay-out' for two persons. *Ibid.*, In the center of the platform was a tray which contained the smoker's 'lay-out'. **1946** MEZZROW & WOLFE *Really Blues* (1957) 376 *Layout*, set of instruments used to smoke opium.

b. A scheme, plan, or arrangement; a course of action. orig. and chiefly *U.S.*

1867 in *Amer. Speech* (1942) XVII. 71/1 A 'lay-out' is any proposed enterprise, from organizing a State to digging out a prairie-dog. **1901** S. E. WHITE *Westerners* xxxi. 292 'I'm sorry that I have this to do, Billy,' said Lafond. 'I don't want to. It's none of my lay-out.' **1904** W. H. SMITH *Promoters* ii. 53 There isn't a single move in this whole lay-out that we can't justify by history. **1928** *Sat. Even. Post* 4 Feb. 81/3 Here's the layout. The bonds bear 6 per cent. **1945** BAKER *Austral. Lang.* vii. 142 *Layout, setup*, a trickster's plan of action.

c. A number of persons associated in some way; a set, party, gang (of persons); a family. (Often in a depreciatory sense.) *U.S. colloq.* or *dial.*

1869 *Overland Monthly* III. 128 Several persons in our 'lay-out' (*i.e.*, our company) in New Mexico 'swapped' good American horses for mustangs. **1884** 'C. E. CRADDOCK' *In Tennessee Mts.* III. 143 All them Peels, the whole lay-out, war gone down to the Settlement. **1903** A. ADAMS *Log of Cowboy* vii. 47 Surround this layout, lads, and let's examine them more closely. **1904** W. N. HARBEN *Georgians* 203 I'm a-goin' to close in on that Clegg lay-out to-night, an' locate the'r still. **1927** W. ROGERS in C. M. Russell *Trails plowed Under* p. xiii, I tell you they was a pretty sad lookin outfit. They sho was a lonesome layout.

lay-over. [LAY *v.*^{1} 57.] **1.** An additional cloth laid over a table-cloth.

1777 *Monthly Rev.* LV. 108 Two servants appeared with a small table.., and laid a cloth and a lay-over upon it, in our English fashion, of the finest damask.

2. A stop or stay in a place, esp. overnight; a halt, rest, delay. *N. Amer.*

1873 J. H. BEADLE *Undevel. West* xxxv. 756 Two invalids and myself..applied for a 'lay over', unable to go further. **1903** A. ADAMS *Log of Cowboy* viii. 53 Their cattle having grown restless during their enforced lay-over. **1911** *Daily Colonist* (Victoria, B.C.) 18 Apr. (Mag.) 7/4 The object of this [*sc.* an extra train]..is to allow those going through to Alberni to reach their destination the same day. This will do away with the present lay-over en route. **1968** *Listener* 27 June 841/2 Meanwhile, some principal themes of Beat mythology are being laid down:.. the pastoral theme of travelling on along the American highways with layovers for pot, mescalin. **1969** R. STARK *Blackbird* (1970) iii. 23 We have an airline ticket for this evening..with a change at New York. A four-hour layover there, I'm afraid. **1972** *National Observer* (U.S.) 27 May 13/5 Mrs. Bartels parks at Gateway for the three-hour layover.

layshaft (lē^{i}·ʃaft). Also **lay shaft, lay-shaft.** [Prob. f. LAY *v.*^{1} 43.] A short secondary or intermediate shaft driven by gearing from the main shaft of an engine; *spec.* one inside a gear-box that transmits the drive from the input shaft to the output shaft.

1888 *Lockwood's Dict. Mech. Engin.* 205 *Lay shaft*, a small secondary shaft, which is placed beside, or at the end of a horizontal engine, for the purpose of actuating the valves. It is driven from the crank shaft by means of bevel or spur-wheels. **1908** *Westm. Gaz.* 7 May 4/2 The whole of the valve mechanism being contained in a neat, hinged lay-shaft on top of the cylinder heads. **1911** G. W. HAYTER *Motor-Car Mech. for Beginners* (ed. 4) 45 For the next speed the gear wheel, B, is slid into mesh with the wheel marked C on the third or lay shaft, E. **1958** *Times Rev. Industry* Feb. 84/2 The engine, clutch, and layshaft are mounted on a light chassis and protected by a hinged bonnet which allows easy access to the power unit. **1959** *'Motor' Manual* (ed. 36) iv. 73 Spaced along the layshaft are other gears, the total number being equal to the number of ratios which the gearbox can provide.

lay-up. [LAY *v.*^{1} 60.] **1.** A period during which a person or thing is (temporarily) out of employment or use, as a ship in winter.

1927 *Daily Mail* 7 Apr. 3/6 During the winter lay-up of these vessels their passenger accommodation has been thoroughly overhauled. **1929** *Amer. Speech* V. 72 A compulsory stopping on the '[cattle-]drive' is a 'lay up'. **1955** *Times* 10 May 17/4 The winter demand brought tankers out of lay-up. **1959** C. OGBURN *Marauders* (1960) iii. 87 There were frequent lay-ups, and during one of these the men in a car down the train evidently spotted a couple of ducks beside the tracks. **1967** *Coast to Coast 1965–66* 160 The big hall where the Jap pearlfishers dossed during lay-up.

2. a. The operation of laying up in the manufacture of laminated material (see *LAY *v.*^{1} 60 m). **b.** The assembly of layers ready for bonding so produced.

1942 T. D. PERRY *Mod. Plywood* vi. 169 The lay-up.. consists in assembling the layers of veneer and/or lumber with the adhesive. **1950** WEBSTER Add., *Lay-up*, an assembly of layers of veneers or cores for pressing. **1965** *Plastics Tooling & Manuf. Handbk.* (Amer. Soc. Tool & Manuf. Engineers) iii. 35 In wet layup, the workman saturates a piece of fabric in resin, and then may wring out the excess resin and drape the cloth on the laminate. **1972** *Physics Bull.* Nov. 665/2 *Layup*, a laminate that has been assembled, but not cured.

3. *Basketball.* In full, *lay-up shot.* (See quot. 1961.)

1948 A. F. RUPP *Championship Basketball* xv. 130 The drill gives the boys stamina and endurance and helps teach the lay-up shot when going in with the greatest speed. **1958** A. L. COLBECK *Mod. Basketball* iii. 59 In good-class basketball most plays are designed to get a player free for a lay-up shot. **1959** P. ROTH *Goodbye, Columbus* ii. 35, I took my set shot and, of course, missed. With the Lord's blessing and a soft breeze, I made the lay-up. **1961** J. S. SALAK *Dict. Amer. Sports* 260 *Layup shot*,..a shot taken from underneath or very close to the basket. On this type shot, the ball usually is banked off the backboard, but on occasion the player, on a straight run toward the basket, will 'lay' the ball up to the basket without using the backboard. **1967** *Boston Herald* 1 Apr. 16/4 There wasn't anyone in the Boston contingent who could recall him ever blowing three layups in a game before. **1969** Z. HOLLANDER *Mod. Encycl. Basketball* 121 He could also score on jumpers from the corner, driving lay-ups or tip-ins of rebounds. **1974** *Greenville* (S. Carolina) *News* 22 Apr. 13/3 Havlicek sank a layup with 26 seconds left, putting Boston safely ahead.

laywoman. (See after LAYMAN^1.) (Further example.)

1922 *Daily Mail* 11 Nov. 8 The success of this laywoman is a..cheerful omen of good luck for those women who.. will hold the position of practising members of the English Bar.

Laz (lāz). **a.** A group of Caucasian peoples giving its name to Lazistan in north-east Turkey. **b.** Usu. **Laze** (lā·zə) or **Lazi** (lā·zi). A member of any of these peoples. **c.** The south Caucasian language of the Laz people. Also (**Laz**) *attrib.* or as *adj.* Also **La·zic** *a.*

1836 *Jrnl. R. Geogr. Soc.* VI. 191, I embarked at Trebizond..in a galley, and kept along the shore to the Russian frontier..passing in succession the districts of Yomurah, Surmenah, O'f, Rizah, and Lázistán. All these..are known under the general name of Lázistán, and the people are called Láz. **1847** A. KERR tr. *L. von Ranke's Hist. Servia* xi. 219 The first step taken by Selim amongst the Lazes..excited open rebellion against him in his capital. **1897** *Daily News* 26 Feb. 5/4 These Lazes played an active part in the Armenian massacres. **1923** *Daily Mail* 21 Feb. 9 Travelling with Kemal is a bodyguard of 'Lazis'. Tall, robust dare-devils from the Black Sea coast..each carries an abundant supply of daggers and revolvers. **1934** *Geogr. Rev.* LXXXIV. 472 The ports and mouths of rivers have three names—Greek, Turkish, and Lazic. **1939** L. H. GRAY *Found. Lang.* 375 South Caucasian..consists of four languages: Georgian.., Mingrelian, Laz, and Svanian. **1948** D. DIRINGER *Alphabet* v. 322 According to Dr. O. N. Kazara, who is of Laz extraction, the physical type throughout Caucasia is remarkably uniform. *Ibid.*, There are various dialects, the principal of which are Kartlian, Mingrelian with Laz, and Svanian. **1950** [see *ERGATIVE *a.*]. **1954** PEI & GAYNOR *Dict. Ling.* 121 *Laz*, a language spoken in the Caucasus; a member of the South Caucasian family of languages. **1963** *Times* 12 Jan. 9/7 In fact it turned out his father was a Kurd from Elazig, his mother a Laze from Trabzon.

la·ze-off. *rare.* [f. LAZE *v.*] A rest from work.

1924 GALSWORTHY *White Monkey* I. xi. 93 Resenting regular work, enjoying a spurt, and a laze-off.

lazy, *a.* Add: **2. b.** Applied to an eye with poor vision which is consequently little used and tends to deteriorate further; *esp.* the unused eye in squint.

1939 R. B. SIMPKINS *Basic Mech. Human Vision* v. 62 Both exophoric and esophoric conditions are probably the most frequent causes of the development of a lazy eye. **1957** [see *heterophoria* (*HETERO-)]. **1971** E. RUDINGER *Eyes Right* 42 Up to the age of 5 years there is a fair chance of success by encouraging the child to use the lazy eye.

4. *lazy-minded* adj.; **lazy arm,** a type of boom from which a microphone may be slung; **lazy daisy** (**stitch**), a petal-shaped embroidery stitch; **lazy dog** *U.S. Mil. slang*, a type of fragmentation bomb designed to explode in mid-air and scatter steel pellets at high velocity over the target area; **Lazy Susan, lazy susan** orig. *U.S.*, a revolving (wooden) stand on a table to hold condiments, etc.; a muffin stand.

1960 O. SKILBECK *ABC of Film & TV* 76 *Lazy arm*, a small, hand-held microphone Boom. **1962** A. NISBETT *Technique Sound Studio* 257 *Lazy arm*, simple form of boom consisting of an upright and a balanced cross-member from which a microphone may be slung. **1923** *Daily Mail* 10 Mar. 14 The way the 'lazy-daisy' stitch is worked is shown at the side of the sketch. **1948** J. CANNAN *Little I Understood* x. 133 Mildred completed six lazy daisies. **1963** N. MARSH *Dead Water* (1964) viii. 221 A rumpled nightgown embroidered with lazy daisies. **1965** *Times* 23 Mar. 12/7 Asked why something more deadly was not preferred—when such things as napalm and white phosphorus incendiary bombs and 'lazy dog' fragmentation bombs are frequently in use—the spokesmen said the gas was being used in situations where the Vietcong might be holding hostages. **1967** *N.Y. Times* 13 Jan. 8 The Lazy Dog is an advanced antipersonnel weapon introduced last spring. **1968** *Punch* 21 Feb. 258 Tomorrow, we'll get three divisions in here, four, we'll get two hundred B-52s, we'll get ground-to-grounds, and whole batteries of Lazy Dogs. **1879** C. M. YONGE *Burnt Out* xii. 192 George..had been getting more lazy-minded and stupid. **1929** V. WOOLF *Granite & Rainbow* (1958) 105 This lazy-minded man was quite capable..of filling a chapter or two..from a fountain of empty, journalistic phrases. **1917** *Vanity Fair* (N.Y.) Dec. 17 (Advt.), Revolving Server or Lazy Susan. **1966** B. ASKWITH *Step out of Time* ii. 35 The home-made jam on the Lazy Susan in the middle of the table. **1971** *Sunday Australian* 8 Aug. 10/1 The best china is used. Silver pots of steaming tea and coffee spin round with wheels of gateaux on a massive lazy susan.

lazy-tongs. Add: (Later example.) Also *attrib.*

1881 G. M. HOPKINS *Let.* 14 May (1938) 100 My only resource is to ask you if you..can suggest some fetch, some boomerang or lazytongs or round-the-corner means of having at him. **1912** *Proc. Amer. Philos. Soc.* LI. 558 A series of links of the 'lazy-tongs' pattern. **1942** 'M. INNES' *Daffodil Affair* II. 70 A lazy-tongs... It used to be quite a popular toy..something rather like a pair of scissors with a piece of lattice-work pivoted to the blades. **1955** *Archit. Rev.* CXVII. 356/3 The 'lazy-tongs' effect of every door being forced to open simultaneously.

lb. Add: **lbf,** the pound as a unit of force; **lbm,** the pound as a unit of mass.

1961 *B.S.I. News* Oct. 26/1 The B.S.I. committee which deals with units and symbols has accepted a form of notation in which a distinction is made between the two concepts 'mass' and 'force'. In this notation the pound has the abbreviation 'lb'; the unit called a pound-force has the abbreviation 'lbf' and is that force which, when acting on a body of mass one pound, gives it an acceleration equal to that of standard gravity. **1962** S. L. BRAGG *Rocket Engines* ix. 148 A jet velocity of 10^4 ft/s could be obtained, giving a thrust of 3 lbf for a flow of 0·0009 lbm/s. **1966** lbf [see *gravitational system*]. **1967** *Technology Week* 23 Jan. 104/1 (Advt.), Pictured from the top: 300 lbf monopropellant hydrazine rocket for orbit control. **1971** I. H. SHAMES *Introd. Statics* i. 12 (*table*) 1 slug ≡ 32·2 lbm.

l'chaim, l'chay(i)m, varr. *LECHAYIM.

L-dopa (eldō^{u}·pā). *Chem.* and *Biochem.* Also l- (now *rare*), L-, -DOPA. [f. *L 7 c + *DOPA.] = *LEVODOPA.

1939 *Jrnl. Physiol.* XCVI. 51*P* Tyrosine given in the diet is partly excreted as *l*-dopa. **1942** *Ibid.* CI. 345 In tyrosinosis *l*(−)-dopa was found to be excreted in the urine after oral administration of *l*(−)-tyrosine. **1958** *Brit. Jrnl. Pharmacol.* XIII. 92/1 (*heading*) Effect of L-dopa and epinine. **1969** *3rd Symposium Parkinson's Dis.* (R. Coll. Surg. Edin.) 178 L-Dihydroxyphenylalanine (L-dopa) proved far less toxic and more effective than the racemic compound studied earlier. **1970** PASSMORE & ROBSON *Compan. Med. Stud.* II. v. 61/2 [In Parkinsonism] DOPA is given by mouth in doses up to 16 g/day of the racemic mixture or up to 8 g of L-DOPA. **1970** D. B. CALNE *Parkinsonism* x. 94 Disappointing results arise from unwanted effects of L-dopa, such as dyskinesia or psychiatric disturbances. **1970** *Times* 29 Apr. 2/3 L-Dopa, the drug used experimentally in treatment of Parkinson's Disease, has passed the clinical trial stage.

leach, *v.*^{2} Add: **3. b.** Also used with reference to the action of water, esp. rain, on soil; also *absol.* (Earlier and later examples.)

1839 J. BUEL *Farmer's Compan.* ix. 74 The wind and the sun dissipate its virtues, and rains leach it and waste its fertilizing powers. **1917** *Mining Mag.* XVII. 75/2 The rocks..are altered and leached of iron. **1951** W. P. KELLEY *Alkali Soils* vii. 146 It should not be inferred that..all the farmer needs to do in order to reclaim any alkali soil is to drain and leach with water. **1954** W. D. THORNBURY *Princ. Geomorphol.* xvi. 420 Kansan and Nebraskan tills are leached to much greater depths. **1971** D. HILLEL *Soil & Water* v. 124 They suggested that leaching soils at a water content below saturation (e.g., under sprinkling irrigation or rainfall or under intermittent irrigation) could produce more efficient leaching and thereby reduce the amount of water required.

c. (Further examples.) Also, to percolate through and pass *out.*

1931 *Forestry* V. 143 In order to avoid any possible effect of some of the preservative leaching out into the medium, some workers raise the blocks above the surface of the medium. **1961** J. N. ANDERSON *Appl. Dental Materials* (ed. 2) xxiii. 240 Dentures should be kept in water after curing in order to allow as much residual monomer as possible to leach out. **1968** *Listener* 21 Mar. 376/2 Cacodylic acid is alleged to disappear very quickly, to leech down through the soil. **1974** A. HUXLEY *Plant & Planet* xviii. 198 Many [viruses] make use of materials which leach through the [leaf] skin.

4. (Further examples.) Also *fig.*

1964 *Listener* 13 Aug. 225/2 It [*sc.* a modern office block] has neither virtues nor vices; it just sits there like a graceless woman, leeching away a bit more of the city's vitality. **1971** *Nature* 13 Aug. 446/3 Insecticides are leached from soil by water. **1973** *Ibid.* 20 July 165/1 Particulate material was collected on fibreglass filters from which lead was leached with hot 70% nitric acid.

leached *ppl. a.* (earlier and later examples); add def.: (*a*) that has been subjected to the action of percolating liquid; (*b*) (also *leached-out*) that has been removed by percolating liquid.

1837 *Cultivator* Aug. 93/2 (*heading*) Leached ashes as manure. *Ibid.*, Leached or drawn ashes possess a highly beneficial effect, particularly when applied to lands deficient in calcareous matters. **1926** A. LOCKE (*title*) Leached outcrops as guides to copper ore. **1961** *Listener* 12 Oct. 559/1 So white are some leached layers [in a podsol] that they have entered archaeological literature as 'layers of ash'. **1963** D. W. & E. E. HUMPHRIES tr. *Termier's Erosion & Sedimentation* vi. 139 The lowest part is a zone of enrichment to which the leached-out soluble salts from the upper part of the soil are carried. **1972** J. G. CRUIKSHANK *Soil Geogr.* ii. 69 Part of the leached compounds are deposited in the lower zone of the soil, but some will be lost from the system through soil drainage and seepage.

leachable (lī·tʃăb'l), *a.* [f. LEACH *v.*^{2} + -ABLE.] Capable of being leached out.

1944 *Experiment Station Rec.* Apr. 450 Potassium could be changed biologically from a leachable to a nonleachable state. **1955** *Sci. News Let.* 27 Aug. 132/2 Experiments indicate that an average of only about 25% of the thorium and uranium in granite rock is 'leachable'. **1972** *Ann. Rep. Freshwater Biol. Assoc.* XL. 40 This contrasts with dead leaves from which leachable substances are said to disappear in about a week. **1972** J. G. CRUIKSHANK *Soil Geogr.* ii. 46 Maximum mobilisation of leachable substances is related to both the volume and composition of the organic leachate. **1973** *Nature* 13 Apr. 452/2 The concentrations reported include the contribution of acid leachable mercury associated with particulate material.

leachate (lī·tʃē^{i}t). [f. LEACH *v.*^{2} + *-ate*, after *filtrate, precipitate*, etc.] (A quantity of) liquid that has percolated through a solid and leached out some of the constituents.

1952 *Sci. Agric.* XXXII. 606 The pH of leachates from decomposing leaves rises as decomposition progresses. **1971** *Nature* 17 Sept. 211/1 The material was then filtered, washed several times and the leachates were combined. Both the residual material and the leachate were analysed for lead. **1972** [see *LEACHABLE *a.*].

leaching, *vbl. sb.*^{2} Add: Also *leaching out.*

1906 E. W. HILGARD *Soils* ii. 24 A heavy depletion of the land by the leaching-out of this important plant food. **1938** R. W. LAWSON tr. *Hevesy & Paneth's Man. Radioactivity* (ed. 2) xxv. 265 In some minerals the ratio of lead to uranium has been altered by leaching-out processes in the course of the long periods of time involved. **1943** MILLAR & TURK *Fund. Soil Sci.* xv. 377 The leaching out of the alkali tends to leave the soil in an even worse physical condition. **1966** G. H. DURY *Ess. Geomorphol.* 56 It is the simple process of leaching out of contained salt, by groundwater, which rapidly reduces the shearing strength of the deposits.

lead, *sb.*^{1} Add: **1. c.** (Later examples.)

1927 *Amer. Speech* Mar. 278/1 *Shake out the lead*, start action. **1942** BERREY & VAN DEN BARK *Amer. Thes. Slang* §578/26 Get the lead out of your pants, to play allegro. **1948** F. BROWN *Dead Ringer* i. 15 Quit asking.. questions and get the lead out. **1961** *Lebende Sprachen* VI. 101/1 He's as lazy as they come, he's got lead in his pants, shoes. **1964** WODEHOUSE *Frozen Assets* vi. 115 She knows I'm in imminent danger of dying of malnutrition unless she takes the lead out of her pants and gets a move on with that picture. **1967** —— *Company for Henry* xii. 207 Those wedding bells aren't going to ring if you don't take the lead out of your pants and get a move on.

3. (Earlier and later examples.) Phr. *lead in one's pencil*: implying (esp. sexual) vigour in a male.

1816 JANE AUSTEN *Emma* III. iv. 54 When he took out his pencil, there was so little lead that he soon cut it all away. **1922** S. LEWIS *Babbitt* i. 9 A silver pencil (always lacking a supply of new leads). **1941** BAKER *Dict. Austral. Slang* 43 (*This will*) *put some lead in your pencil*, this (esp. a drink of beer or spirits) will make you feel fighting fit. **1946** P. LARKIN *Jill* 190 'Well, ere's more lead in yer pencil.' He finished off his half-pint. **1969** [see *CURL *v.*^{1} 1 c]. **1970** *Kay's Catal.* (Worcester) Autumn–Winter 947/3 Pencil both propels and retracts, contains twelve 3 inch leads. **1970** A. DRAPER *Swansong for Rare Bird* vii. 59 She came over with two glasses. 'If that doesn't put some lead in your pencil, Auk, I don't know what will.' **1972** D. LEES *Zodiac* 107 The couscous is supposed to put lead in your pencil but with Daria I needed neither a talking point nor an aphrodisiac.

4. (*cold*) *lead*, bullets.

1809 T. G. FESSENDEN *Pills Poetical* 32 Thus our sporting democrats,.. When they can't reason with a Fed, In logick substitute cold lead. **1837** W. H. WHARTON *Let.* in *Ann. Rep. Amer. Hist. Assoc. 1907* (1908) II. 190 We

would give Mexico nothing but lead. **1891** M. E. RYAN *Told in Hills* 332 [The message] belongs to the command, and I may get a dose of cold lead before I could deliver it. **1918** C. SANDBURG *Cornhuskers* 50 Three riders emptied lead into him. **1964** F. O'ROURKE *Mule for Marquesa* 146 Get 'em up or we'll pump you full of lead!

6. b. Phr. *to swing the lead*: to idle, to shirk; to malinger. *slang.* Hence in similar phrs. and in *Comb.*, as **lead-swing** *sb.* and *v. intr.*, **-swinger, -swinging** *vbl. sb.* and *ppl. a.*

1917 *To-Day* 6 Jan. 243/3 It is evident that he had 'swung the lead' (using Army phrase) until he got his discharge. **1918** B. K. ADAMS *Let.* 25 Jan. in *Amer. Spirit* 71 Lead-swingers are those that stall along, doing as little as they possibly can, hoping the war will be over before they finish. **1922** C. E. MONTAGUE *Disenchantment* iv. 56 Then grey hairs should be a lot of use to you..when you want to get swinging the lead. **1927** A. BROSNAN *At Number 15* I. 30 'If they wanted a three-man job done they had to put forty on to it to make sure it was done.' 'And so they did. That's organisation, that is. Of course, there was some lead-swingers.' **1927** *Daily Express* 2 Mar. 3/4 He said he ..had been 'swinging the lead' for the purpose of getting a permanent pension. **1930** S. BECKETT *Whoroscope* 1 The vile old Copernican lead-swinging son of a sutler! **1939** R. CAMPBELL *Flowering Rifle* II. 60 It was not we who lead-swung to the Pities, When half the loveliest of our ancient cities Were in the clouds rebuilt. **1940** J. B. PRIESTLEY *Postscripts* 70 A wary..old soldier, a lead-swinger, a dodger of the column. **1952** M. ALLINGHAM *Tiger in Smoke* iv. 77 He went sick... It was so hopeless, so damned silly and forlorn as a lead-swing that in the end he got away with it. **1957** A. GRIMBLE *Return to Islands* ii. 32 Their number was not without its natural quota of cheerful leadswingers. **1968** *Manch. Guardian Weekly* 12 Sept. 9 Mr. Crossman..insisted that 'lead swinging' among the unemployed was confined to a very small minority. **1969** *Daily Tel.* 8 Jan. 26/1 Overall absenteeism in the coalfields is running slightly higher than last year... Out of this total, 4·66 per cent. is classified as voluntary absenteeism ('lead-swinging'). **1972** *Daily Colonist* (Victoria, B.C.) 12 Feb. 4/1 The mayor of Victoria accuses the four Greater Victoria members of the legislature of lead-swinging. **1973** *Daily Tel.* 29 Aug. 6/3 'It would soon put a stop to lead-swingers who take a few days off to paint the house or watch cricket,' the doctor added.

11. b. Also obj. genitive, as *lead-free* adj.

1946 [see *lead glass*, sense *12]. **1960** *Farmer & Stock-breeder* 16 Feb. Suppl. 36/2 Sow Feeder..painted [with] one coat lead-free paint. **1970** *Guardian* 13 Apr. 13/4 Lead-free petrol. **1973** *Country Life* 29 Mar. 854/1 Modifications were also made to the engine to enable it to run on lead-free fuels.

c. *lead-covered, -sheathed* (later example).

1891 KIPLING *Light that Failed* xiii. 253 A hall at the foot of some lead-covered stairs. **1908** *Westm. Gaz.* 22 Apr. 8/3 Over twenty miles of lead-covered cables have been laid in the grounds. **1948** G. V. GALWEY *Lift & Drop* vi. 137 The leads to the switchgear were buried. They were lead-sheathed.

12. lead accumulator, a lead-acid cell or battery; **lead-acid** *a.*, applied to a secondary cell or battery in which the anode is a plate or grid of lead (or lead alloy) coated with lead dioxide, the cathode is a similar plate coated with spongy lead, and both are immersed in dilute sulphuric acid; **lead balloon,** a failure, an unsuccessful venture; **lead bronze,** bronze containing lead, which is used in bearings; **lead bullion,** a mixture of lead and other heavy metals obtained as an intermediate product in the extraction of lead; **lead burning,** the welding of lead; so **lead-burn** *v. trans.*, to weld (pieces of lead); **lead cell,** a lead-acid cell; **lead chamber,** a large reaction vessel made of welded sheet lead which is used in the manufacture of sulphuric acid from sulphur dioxide, air, and steam using oxides of nitrogen as catalysts; so *lead chamber process*; **lead crystal** [CRYSTAL *sb.* 5] = *lead glass* below; **lead-flat** (see quots.); **lead glass,** glass containing a substantial proportion of lead oxide; **lead-glaze** *Pottery*, a glaze containing lead oxide; so *lead-glazed* adj., *lead-glazier, lead-glazing* vbl. sb.; **lead-line,** (*d*) the narrow strip of lead between two pieces of stained glass; a came; so **lead-line** *v. trans.*, to put the lead-lines in (stained glass work); **lead-paper** (later examples); hence **lead-papered** *a.*, covered with or containing lead-paper; **lead-plant** (examples); **lead-poisoning** (earlier and later examples); **lead ratio,** the ratio, in a sample of rock, of the quantity of lead (or a lead isotope) to the quantity of its radioactive parents uranium and thorium (or an appropriate isotope of one of these elements), from which the age of the sample may be determined; **lead-tin** *a.*, containing lead and tin; also *ellipt.*, a lead-tin alloy; **lead wool,** lead in a fibrous state, used for caulking pipe joints.

1903 *Chem. News* 17 July 34/2 Dr. Lehfeldt's paper on 'The Total and Free Energy of the Lead Accumulator' was taken as read. **1928** CRENNELL & LEA *Alkaline Accumulators* i. 5 The lead accumulator suffers from certain inherent defects of which the most important are a rather large weight for a given capacity, [etc.]. **1971** G. F. LIPTROT *Mod. Inorg. Chem.* xviii. 242 The voltage supplied by the lead accumulator is just in excess of 2 volts. **1926** W. S. IBBETSON *Accumulator Charging* iii. 26 Fig. 7 illustrates the actions and results of charging and discharging a simple lead acid cell. **1936** *Motor Manual* (ed. 29) iv. 78 The lead-acid type [of battery] is that most general as its cost is much lower. **1972** *Dry Cells, Batteries & Accumulators* iii. 36 Lead-acid accumulators have a good life in terms of charge/discharge cycles. **1974** *Railway Mag.* Apr. 176/2 The locomotive interior is taken up by no less than 160 lead-acid battery cells, giving a 300V supply. **1960** WENTWORTH & FLEXNER *Dict. Amer. Slang* 314/2 *Lead balloon*, a failure; a plan, joke, action or the like that elicits no favorable response; a flop; anything that lays an egg. **1962** L. DEIGHTON *Ipcress File* xxv. 158 With this boy it went over like a lead balloon. **1970** *Sunday Times* 19 Apr. 31/3 *What the Dickens?* was a lead balloon literary quiz wherein the experts showed only how little they knew. **1937** H. N. BASSETT *Bearing Metals & Alloys* viii. 296 Under the general title of lead bronzes are included.. the copper-tin-lead alloys..and the so-called 'tin-free' bronzes. **1951** *Engineering* 6 July 1/3 The main and big-end bearings are all fitted with steel shells, lined with lead-bronze. **1967** *Jane's Surface Skimmer Systems 1967–68* 123/2 Crankshaft... Lead-bronze bearings with steel cups. **1905** A. H. LOW *Technical Methods Ore Analysis* viii. 56 The determination of bismuth in impure lead or lead bullion may be carried out on the same lines as described for refined lead. **1954** W. H. DENNIS *Metall. Non-Ferrous Metals* iv. 242 The crude lead bullion may contain up to 4 per cent of these reduced metallics. **1963** *Times* 22 Apr. p. iv/4 This plant is producing about 40,000 tons of good ordinary brand zinc annually together with lead bullion and by-product cadmium and sulphuric acid. **1886** D. SALOMONS *Managem. Accumulators* 14 It is frequently necessary to perform the operation of soldering or lead burning. **1937** *Archit. Rev.* XXXI. 272/2 Leadburning is a variety of welding. As a process it has been known for centuries, but only since the invention of the gas welding flame have its possibilities been fully exploited. *Ibid.* 270 (*caption*) After casting the flat sheets [of lead] are bent round and the joint lead-burned to form the point. **1963** H. R. CLAUSER *Encycl. Engin. Materials* 368/2 Lead welding, commonly called lead burning, produces a true weld by fusing the parts together without the addition of any different material. **1897** *Physical Rev.* IV. 353 We owe the discovery of the lead cell to Planté. **1928** CRENNELL & LEA *Alkaline Accumulators* ix. 121 The energy, or watt-hour, efficiency of alkaline cells is about 50–55 per cent., as compared with 75 per cent. for lead cells. **1867** *Chem. News* 5 July 12/1 (*heading*) Lead-chamber process. *Ibid.*, This explains the loss of nitric acid in the manufacture of sulphuric acid, which always takes place when the sulphuric acid in the lead-chamber is below the normal strength. **1909** L. KAHLENBERG *Outl. Chem.* xiii. 198 There are commonly three lead chambers, so connected that the gases enter the top of each and pass out at the bottom. **1946** J. R. PARTINGTON *Gen. & Inorg. Chem.* xxiv. 710 The lead chamber plant..consists of (i) pyrites (or sulphur) burners, (ii) a dust separator.., (iii) a nitre oven.., (iv) a Glover tower, (v) a series of lead chambers with arrangements for supplying steam or water spray, and (vi) a Gay-Lussac tower. **1969** H. T. EVANS tr. *Hägg's Gen. & Inorg. Chem.* xxi. 529 The reaction takes place in reaction chambers, formerly lead chambers, that is, large lead-lined rooms, but now most often of other types. **1973** THOMAS & FARAGO *Industr. Chem.* viii. 133 The lead-chamber process is by no means obsolete.., and is likely to remain in operation for the production of acid not exceeding 78 per cent in concentration..and where high purity is not essential. **1902** J. D. & A. EVERETT tr. *Hovestadt's Jena Glass* x. 364 Foerster recalls the fact that the resisting power of lead crystal glass to acids is increased by long-continued exposure to acids. Were it otherwise, the use of this material for wine glasses would long ago have been given up. **1968** *Canad. Antiques Collector* Dec. 19/2 When lead crystal came into fashion about 1800, it was possible to cut the glass in glittering facets. **1969** R. F. LANG tr. *Henglein's Chem. Technol.* 835 Lead crystal contains lead (instead of Ca) and potassium and has high light refraction; it is much used in colored glasses. *a* **1877** KNIGHT *Dict. Mech.* II. 1270/1 *Lead-flat*, a level roof consisting of sheet-lead laid on boarding and joists. **1907** W. DE MORGAN *Alice-for-Short* xxv. 259 Charles remembers the lead-flat sunk in the roof. **1940** *Chambers's Techn. Dict.* 491/2 *Lead-flat*, a flat roof formed of sheet-lead laid on boarding and joists. [**1830** *Phil. Trans. R. Soc.* CXX. 43 The tri-borate of lead glass is almost as colourless as good flint glass.] **1856** W. A. MILLER *Elem. Chem.* II. xi. 764 Lead glass has..the inconvenience of being readily scratched. **1946** *Nature* 26 Oct. 582/1 Colouring oxides such as iron, copper, etc., all produce more intense colours in heavy lead glasses than in ordinary lead-free glasses. **1965** PHILLIPS & WILLIAMS *Inorg. Chem.* I. xiv. 546 Special glasses are made by adding other oxides: for example, lead glasses have a high refractive index and are used in crystal and flint glass. **1899** *Westm. Gaz.* 27 Mar. 6/2 There seems no reason..why.. the operatives should still continue to be exposed to the evils which the use of lead-glaze entails. **1969** *Canad. Antiques Collector* Jan. 28/3 Lead and lustre glazes came early from the Near East. **1901** *Daily News* 3 Dec. 3/7 He states that there is no difference now in price between the lead glazed and leadless glazed ware. **1968** J. ARNOLD *Shell Bk. Country Crafts* 231 Medieval pottery was mainly in the form of lead-glazed earthenware..and was known as faience or majolica. **1908** *Westm. Gaz.* 23 Nov. 9/3 The deceased came under his notice twelve years ago, when he was a lead-glazier. **1962** H. R. LOYN *Anglo-Saxon Eng.* iii. 110 The so-called Stamford ware, utilizing a type of lead-glazing that may have originated in the Netherlands, appears to have spread from East Anglia. **1907** W. DE MORGAN *Alice-for-Short* xxvii. 283 I'll lend you a hand over the lead-lines. *Ibid.* xii. 136 It was Pope's man, Buttivant, who lead-lines up all the windows. **1973** HARRISON & WATERS *Burne-Jones* iv. 50 All the designers had to supply were the cartoons, which were quite often bold drawings without indication of lead lines. **1922** JOYCE *Ulysses* 659 A crinkled leadpaper bag. *Ibid.* 70 He..read the legends of leadpapered packets. **1952** M. ALLINGHAM *Tiger in Smoke* vi. 108 The final covering was a piece of lead paper off a tobacco package. **1833** A. EATON *Man. Bot.* (ed. 6) 15 *Amorpha canescens*, lead plant... Somewhat woody... Galena. **1848** W. H. EMORY *Notes Mil. Reconn.* 399 The lead plant, or tea plant..is in some places so abundant as to displace almost every other herb. **1939** *Nat. Geogr. Mag.* Aug. 220/1 Chief among the peas is a group of close relatives: lead plant,..prairie clovers, together with indigo plant. **1841–2** T. D. MITCHELL in *Western & Southern Med. Recorder* (Lexington, Kentucky) I. 145 (*title*) Practical notes on lead poisoning. **1972** *National Observer* (U.S.) 27 May 10/1 The American Smelting and Refining Co. of New Jersey was accused in a civil suit of unduly polluting the air and environment with its huge smelters here and of causing lead poisoning in at least 135 children. **1920** *Discovery* Apr. 111/2 It is of course obvious that, if a mineral is altered, it has suffered chemical changes whereby the normal lead ratio is upset, for either introduction or elimination of lead may have taken place. *Ibid.* 112/1 In some cases the lead-ratio can be used..for determining the geological position of rocks which yield their age to no other method of investigation. **1946** F. E. ZEUNER *Dating the Past* x. 325 In practice, the analyst measures the total amount of lead present, and the expression Pbtotal/(U + 0·36 Th), accounting for the presence of both uranium and thorium, is the one which has to be determined in every case. It is called the 'lead-ratio'. **1889** *Jrnl. Chem. Soc.* LV. 677 The first alloys experimented on were the lead-tin alloys. **1890** *Ibid.* LVIII. 336 The two alloys always correspond with two cognate points on the solubility curves of zinc in lead-tin, and of lead in zinc-tin. **1928** H. H. COWLEY *Mod. Electr. Wiring* iv. 54 Either copper or lead-tin alloy is generally employed for ordinary wire fuses. **1931** G. O. RUSSELL *Speech & Voice* viii. 67 The author has a lead-tin, round-walled open organ pipe. **1956** *Monogr. & Rep. Ser. Inst. Metals* No. 18. 73 In the lead-tin alloys, as in many other alloy systems, precipitation is accompanied by recrystallization. **1908** *Chambers's Jrnl.* Jan. 120/1 What is called 'lead wool', consisting of pure lead cut into fine strips by machinery. **1930** *Engineering* 10 Oct. 451/1 The end bracket structure on the ends of each tube formed the lateral forms for the joint concrete. The actual face joint was made with lead wool caulked.

b. lead tetraethyl = *tetraethyl lead.*

1924 *Jrnl. Inst. Petroleum Technologists* X. 337 Lead tetraethyl. **1926** *Encycl. Brit.* II. 127/2 The tendency to knocking is suppressed by adding to the motor spirit substances such as lead tetra-ethyl which, it is assumed, act by being adsorbed by the ferriferous carbon in the cylinder. **1971** *Daily Tel.* (Colour Suppl.) 28 May 16/2 The amounts of lead in the environment have increased dramatically since the introduction of lead tetraethyl as a petrol additive in the Twenties... By controlling the rate at which fuel burns, lead tetraethyl promotes smoother ignition.

lead, *sb.*^{2} Add: **1. e.** (Further examples.) Also, a clue (to the solution of something).

1910 J. LONDON *Let.* 19 Nov. (1966) 323 Again and again I have opened up leads of true life and found that it was wholly misunderstood by my reading public. **1959** *Times* 18 Feb. 8/3 The enquiry arose from a complaint.. that he had been given 'definite leads' to the questions in advance. **1971** *Daily Tel.* 17 Dec. 1/5 Three leads are being followed by detectives investigating the attempted assassination of the Jordanian Ambassador. They are a sub machine-gun.., an hotel bill..and fingerprints. **1973** *Times* 5 May 1/2 The French police have decided to shift their inquiry into the axe murder of Mr John Cartland, a Brighton schoolmaster, to Britain next week in search of new leads.

f. *Journalism.* A summary or outline of a newspaper story; a guide to a story that needs further development or exploration; the first (often the most important) item in an issue, bulletin, etc. Cf. *lead story*, etc., under sense 11 b below.

Quot. 1947 refers to a radio news broadcast.

1927 *Amer. Speech* II. 241 'Lead'..is used as a noun to refer to the initial summary of the story, or as a verb to instruct the printer what to put first. **1947** *Hansard, Commons* 19 Dec. 2113 There is what one calls the 'lead', which is..the first item. **1950** D. HYDE *I Believed* xvi. 189, I had several hundred accredited Worker Correspondents sending in regular reports and receiving regular 'leads' and directives from me. **1952** *Manch. Guardian Weekly* 20 Mar. 3 This discovery destroyed many a newsman's first confident 'lead'. **1961** 'B. WELLS' *Day Earth caught Fire* viii. 119 Stenning's brought in a lead to something that could be big. **1973** A. BROINOWSKI *Take one Ambassador* ix. 128 He's onto some lead about a mob of fanatical ratbags.

2. c. *Austral.* and *N.Z.* (See quot. 1933^1.)

1933 L. G. D. ACLAND in *Press* (Christchurch, N.Z.) 4 Nov. 15/7 *Lead*, the front part of a mob of sheep. *Ibid.* 2 Dec. 15/7 An injudicious turn with a dog in an abrupt gully may stop the lead and cause some sheep to be knocked over. **1946** F. D. DAVISON *Dusty* ix. 90 Tom.. sent [the sheepdog] Sapper to the flank [of the mob] and to turn the lead.

d. Finance. *leads and lags* (also attrib. phr. *lead-and-lag*): see quot. 1965^1. Also *transf.*

1958 *Spectator* 31 Jan. 129/1 The 'leads and lags' are being replaced by a more natural pattern of commercial payments. **1959** *Economist* 14 Feb. 619/2 The customary 'leads and lags' are at work, postponing commercial demands for sterling and accelerating sales of sterling. **1962** S. E. FINER *Man on Horseback* xii. 220 Sometimes the demand for popular sovereignty has preceded nationalism, sometimes it has been the other way about; but the leads and lags were never very lengthy. **1964** A. BATTERSBY *Network Analysis* iii. 37 The lead-and-lag (or ladder) system has the merit of simplicity, and it draws attention to the importance of planning the sequence of individual jobs within a departmental activity. **1965** J. L. HANSON *Dict. Econ.* 253/2 *Leads and Lags*, with reference to international payments and their effect on the balance of payments this term is used on the hastening or delaying of

LEAD

payment, the former by residents and the latter to residents in order to take advantage of expectations of changes in the rate of exchange. 1965 *Listener* 13 May 692/2 Some foreigners, in the habit of acquiring sterling in advance of their commitments, refrained from doing so; that would be a mug's game, they thought, when sterling might be devalued before they had to pay. These are known as the 'leads and lags' in trade payments.

3. c. *N.Z.* (See quot.)

1878 E. S. Elwell *Boy Colonists* 214 They made a 'lead' in the stockyard for branding the cattle. This was something like a 'race' for drafting sheep, with a swing gate... It had a wide entrance gradually getting narrower till it became a lane only just wide enough for one beast at a time to squeeze through.

5. c. *Boxing.* The first punch thrown (of two or more) (see also quot. 1954).

1906 [see *cross *sb.* 22 d]. 1950 J. Dempsey *Championship Fighting* x. 50 The first punch thrown (by either) is a lead. 1954 F. C. Avis *Boxing Reference Dict.*, *Lead*, a forward blow made at a fair distance from the opponent. 1970 *Times* 28 Sept. 13/5 Those sneak right leads I hit him with helped as well. 1971 *Black Scholar* Jan. 43/2 Man, this would make these fighters so mad they would forget about boxing and come out swinging wild. And that was all old Jack wanted. He'd step inside their leads and counter punch them to death!

6. a. Also *fig.*

1869 S. Bowles *Our New West* vii. 136 A quaint old miner of the valley, who, 'prospecting' for society that day, had struck a 'lead' in us.

7. a. and **b.** (Earlier and later examples.)

1831 J. Boaden *Life Mrs. Jordan* I. xi. 264 It gave him the lead in a successful play. 1865 *Punch* 7 Jan. 5/1 As a general rule an actor who plays the 'lead' ought to aim at becoming a general manager. 1937 *Daily Tel.* 14 Aug. 9/1 Many leading men and women (and some who are merely minor leads). 1939 [see *character part*]. 1953 [see *big stuff* (*big *a.* B. 2)]. 1973 *Listener* 21 June 844/2 The lead, Martin Thurley, must surely have studied the slovenly dialect of the area.

8. c. *Mus.* The most prominent part in a piece played by an orchestra, esp. a jazz band; the player or instrument that plays this; the leader of a section of an orchestra; also, the start of a passage played by a particular instrument. Freq. *attrib.* orig. *U.S.*

Further *attrib.* examples are given under sense 11 b below.

1934 S. R. Nelson *All about Jazz* v. 99 He evolved what he called a 'harmony chorus', the instruments all playing harmony, with a solo lead. 1937 *Amer. Speech* XII. 47 The lead melody is carried lower than the clarinet. 1952 B. Ulanov *Hist. Jazz in Amer.* (1958) xvii. 203 Hymie Schertzer's rich lead alto sounds. 1967 [see *attack *v.* 7]. 1968 *Blues Unlimited* Sept. 8 They played mostly Italian music and polkas, with Charlie McCoy on lead mandolin.

11. lead-bars (earlier U.S. example); **lead-net** = Leader[1] 15 b; **lead-rope,** a rope used as a lead for a horse or ox; also *fig.*; **lead sheet** *U.S. slang* (see quot. 1942); also *transf.*, an overcoat; **lead-time** orig. *U.S.*, the time taken to produce some manufactured article (see also quot. 1968); also *transf.*

1840 *Congress. Globe* 5 Mar. App. 227/2 The horse broke loose from the coach, taking with him a part of what are now called 'Lead bars'. 1910 *Chambers's Jrnl.* Mar. 192/2 The lead-net is about fifteen hundred feet long. The salmon strike this. 1846 R. B. Sage *Scenes Rocky Mts.* iii. 24 Holding in one hand the lead-rope of his horse. 1901 Kipling *Kim* vii. 169 'We be all on one lead-rope, then,' said Kim at last, 'the Colonel, Mahbub Ali, and I.' 1958 L. van der Post *Lost World of Kalahari* i. 15 Lifting the lead rope from the horns of the two guide-oxen. 1942 Berrey & Van den Bark *Amer. Thes. Slang* § 578/9 *Lead sheet*, a sheet of music containing the melodic line and lyric only. 1945 L. Shelly *Jive Talk Dict.* 28/2 *Lead sheet*, an overcoat. 1961 R. Russell *Sound* iii. 38 You never got around to writing out a lead sheet! 1945 *Birmingham* (Alabama) *News* 19 May 8/1 The 'lead-time' normally required to bring out new models. 1957 *Manch. Guardian* 4 May, The problem is..difficult, on account of the complex character of the equipment in question and the long lead-time involved. 1964 A. Battersby *Network Analysis* iii. 36 The chain-dotted arrows..represent *lead times* when they connect start events. 1968 J. F. Magee *Industr. Logistics* i. 19 'Lead time' is the response time lag of the system, the time that must be allowed at a stock point to replenish stock, including the time needed to process records, transmit information, and process and ship material. 1971 *Inside Kenya Today* Mar. 28/1 Because of the lead-time in switching the emphasis in the secondary schools, the University is under pressure to increase its Arts intake very rapidly. 1973 *Nature* 28 Sept. 179/1 The long lead time required for such a rendezvous or flyby mission makes it impossible to achieve a fruitful interception with Kohoutek.

b. Used in the sense of 'leading'.

1846 R. B. Sage *Scenes Rocky Mts.* xxxiii. 289 Bidding them adieu, with my lead pack-animal returned to the mountains. 1857 in *Ann. Wyoming* (1939) XI. 83 The carriage sustained no injury, but one of our lead Mules became detached from the wagon. 1869 *Overland Monthly* III. 127 With the Texan driver all oxen are 'steers', and he has his 'wheel-steers', his 'swing-steers', and his 'lead-steers'. 1888 Kipling *Barrack-Room Ballads* (1892) 117 Then the lead-cart stuck, though the coolies slaved, and the cartmen flogged. 1890 *Ibid.* 18 The rattle an' stamp o' the lead-mules. 1910 W. M. Raine *Bucky O'Connor* 189 It was as the man in charge circled round to head the lead cows in that a faint voice carried to him. 1929 *Randolph Enterprise* (Elkins, W. Virginia) 28 Mar. 1/2 Dick Collette played the lead violin and Bryan Gainer, second. 1942 Berrey & Van den Bark *Amer. Thes. Slang* § 523/3 Leader, lead story, a leading news item. 1959 J. Osborne *World of Paul Slickey* I. vi. 55 Congratulations..on today's lead story. 1962 *Amer. Speech* XXXVII. 87 A lead article satirizing American temperance groups. 1963 Mrs. L. B. Johnson *White House Diary* 22 Nov. (1970) 3 In the lead car were President and Mrs. Kennedy. 1967 *Time* 25 Aug. 38 The Group Image, one of the new, first-name-only hippie groups, of which Nancy is the den mother..and Artie the lead guitar. The tribe has about 25 musicians, artists and psychedelic experts in it. 1967 W. Soyinka *Kongi's Harvest* 3 Superintendent... Seizes the lead drummer by the wrist. 1973 'F. Clifford' *Amigo, Amigo* xxi. 175 Ahead, the lead horse whinnied. 1973 *Listener* 6 Sept. 312/3 Carl Perkins..now playing lead guitar behind Johnny Cash. 1975 *Guardian* 7 Jan. 6/7 A mob of Hell's Angels set on members of the Troggs pop group in their dressing-room and during a fight the group's lead guitarist was stabbed five times in the back.

lead, *v.*[1] Add: **2. f.** To aim in advance of.

1892 W. W. Greener *Breech-Loader* 267 Theoretically it is correct to lead a quartering pigeon from five to seven feet. 1968 D. Hamilton *Menacers* xxii. 176, I led him by roughly two feet and pressed the trigger of the Luger.

g. Coll. phr. *lead me to* (something previously mentioned), expressing the ability to perform or a desire to comply, or merely expressing eager assent.

1929 W. E. Miller *To you I tell It* 107 'How wood you like to urn a piece of jack?' 'Leed me to it,' says Figgars. 'What's the propozishion?' 1934 D. L. Sayers *Nine Tailors* iv. 307 'Can you ride a motor-bike?' 'Lead me to it, guv'nor!' 1938 D. Smith *Dear Octopus* II. iii. 90 Lead me to that whiskey.

h. *to lead with one's chin* (Boxing slang), to 'stick one's neck out', to leave oneself unprotected; *fig.*, to behave or speak incautiously.

1949 E. S. Gardner in *Argosy* Apr. 110/3 Let him lead with *his* chin. We'll work undercover. 1954 F. C. Avis *Boxing Reference Dict.*, *Lead with chin*, to have a very bad stance or guard. 1968 *Listener* 18 Jan. 78/2, I thought it was a good idea to say that I was prejudiced to begin with, to lead with my chin. 1973 A. MacVicar *Painted Doll Affair* i. 19 Don't go leading with your chin, Bruce.

19. a. (Earlier examples.) **b.** Also with *to.*

1817 Jane Austen *Sanditon* vi in *Minor Works* (1954) 389 Sir Edw: Denham & Miss Denham, whose names might be said to lead off the Season. 1847 *Punch* 27 Mar. 126/2 To lead off a list of Expiring Acts with one that is to live till the National Debt is paid off..is a delusion. 1911 *Chambers's Jrnl.* July 463/2 From these [wagons] rubber tubes protected by encircling wire lead off to each of the streets.

21*. lead through. *Mountaineering.* Said of two climbers: to act alternately as leaders (see quots.). Hence **leading through** *vbl. sb.*

1945 G. W. Young *Mountain Craft* (ed. 4) v. 184 Nowadays, two such experts..make a practice of 'leading through': that is..the second man on reaching his leader climbs straight on past him and leads the next section. 1955 M. E. B. Banks *Commando Climber* ix. 177 We were leading through, that is to say, one of us would climb a pitch and belay himself to the rock, whereupon the other would climb up to him and then continue beyond to lead the next pitch. 1970 A. Blackshaw *Mountaineering* (rev. ed.) v. 143 For experienced climbers two is the best number since this is quick and allows them to 'lead through'. *Ibid.* xvii. 420 Leading through may not save as much time on alpine rock as it does on British rock.

lead, *v.*[2] **3.** (Later example.)

1842 C. J. Lever *Jack Hinton* (1843) xxv. 172, I..seated myself in the scale..and my saddle being leaded to the required weight, the operation took not a minute.

leadbeater[1] (le·dbītəɹ). The name of Benjamin *Leadbeater*, 19th-c. English naturalist, used *absol.* and in the possessive to designate a pink Australian cockatoo, *Kakatoe leadbeateri*, named after him in 1831 by N. A. Vigors (*Phil. Mag.* X. 55). In full, *Leadbeater's cockatoo.*

1848 J. Gould *Birds Austral.* V. 2 (*heading*) Leadbeater's Cockatoo. 1890 'Lyth' *Golden South* xiv. 127 The birds are very beautiful—the Blue Mountain and Lowrie parrots,..lead-beater, and snow-white cockatoos. 1900 *Daily News* 10 Feb. 7/2 The foreign birds are many, and include Amazon and grey parrots, lead-beaters, rose cockatoos, [etc.]. 1973 A. H. Lendon *Cayley's Austral. Parrots* 89 Pink Cockatoo... Synonyms. Major Mitchell Cockatoo, Leadbeater's Cockatoo, [etc.].

Leadbeater[2]. The name of Mr. *Leadbeater*, taxidermist at the National Museum of Victoria, Melbourne, in 1867, used in the possessive to designate **Leadbeater's possum,** *Gymnobelideus leadbeateri*, a very rare Australian opossum named after him in 1867 by F. M'Coy (*Ann. Mag. Nat. Hist.*, 3rd Ser., XX. 287).

1937 *Discovery* XVIII. 364/1 The smaller gliding 'possums..whose ancestry is probably derived from the same stock as the extinct Leadbeater's 'Possum, feed on insects, nectar, fruit and sap. 1942 C. Barrett *On Wallaby* iii. 36 The rarest of all Australian animals—Leadbeater's possum. 1966 G. Durrell *Two in Bush* iv. 142 To the astonishment of incredulous naturalists, a tiny pocket of Leadbeater's Possum was discovered [in 1961] in the eucalyptus forest not far from Melbourne. 1968 *Times* 23 Jan. (Austral. Suppl.) p. xiii/3 Eric Wilkinson..saw something that made him stare in disbelief—clinging to the trunk of a wattle tree beside the road was a small, dainty animal which much resembled the Leadbeaters possum, presumed extinct.

leaded, *ppl. a.* Add: **d.** Affected by lead-poisoning.

1878 J. H. Beadle *Western Wilds* xxxv. 581 Great care must be taken by the workmen not to get 'leaded', that is, not to inhale the fumes from the melted lead, which are very poisonous. 1906 *Daily Chron.* 28 June 6/4 The children of 'leaded' mothers usually die, or if they live inherit the effects of the poison. 1914 *Dialect Notes* IV. 163 *Leaded*, among miners, ill from lead poisoning.

e. Containing added lead.

1936 *Blackw. Mag.* Mar. 359/2 It was said that Archie had obtained a special supply of leaded fuel, which would allow him to bring in the Kestrel supercharger near the ground. 1939 Carpenter & Robertson *Metals* II. xv. 1317 Alloys in the fifth group are those to which large amounts of lead are added to improve their suitability for certain types of bearings... These alloys are known as the 'plastic' or 'leaded' bronzes. 1963 H. R. Clauser *Encycl. Engin. Materials* 370/1 Sheet lead and sheets of leaded plastics are also being used to control noise. 1968 E. R. Petty *Physical Metall. Engin. Materials* xiii. 266 Operators must be shielded from this penetrating radiation and for this purpose..a foot of concrete or several feet of water (usually in a leaded-glass jacket) are necessary. 1972 *Lancet* 1 July 12/2 All the subjects had been simultaneously exposed to tetraethyl lead during the process of scaling a tank which had contained leaded petrol.

leaden, *a.* Add: **4.** *leaden-hearted* (later example), *-lidded, -locked* adjs.

1938 C. Day Lewis *Overtures to Death* 14 Infirm and grey This leaden-hearted day Drags its lank hours. 1946 W. de la Mare *Traveller* 18 His leaden-lidded eyes. 1963 *Listener* 7 Mar. 429/1 A man..Whiskered and leaden-locked.

5. leaden fly-catcher, a small grey-green Australian bird, *Myiagra rubecula*, of the family Muscicapidæ (see Fly-catcher 2).

1908 E. J. Banfield *Confessions of Beachcomber* I. iii. 95 Leaden Fly-catcher, *Myiagra rubecula* (*plumbea*). 1911 J. A. Leach *Austral. Bird Bk.* 125 Leaden Flycatcher... Upper wings, tail, breast leaden-gray glossed with green. 1965 *Austral. Encycl.* IV. 121/1 The best-known [tropical flycatchers] are the leaden flycatcher (*Myiagra rubecula*) which migrates south to Tasmania, and the black-faced flycatcher.

Leadenhall (le·dənhǭl). The name of an area in London, used *attrib.* in *Leadenhall Market*, a poultry market in London; *Leadenhall Street*, a street in London which from 1648 to 1861 contained the headquarters of the East India Company, hence designating the Company itself.

1587 J. Stow *Summarie Chron. Eng.* 407 The Northwest corner of Leaden Hall (the highest grounde of the Citie of London). 1720 Stow & Strype *Survey Cities London & Westminster* I. (*map facing p.* 1) Leaden Hall St. 1825 T. Moore *Mem. Life R. B. Sheridan* I. viii. 367 The people, by the unanimous outcry with which they rose, in defence of the monopoly of Leadenhall Street..proved how little of the '*vox Dei*' there may..be in such clamour. 1831 J. Boaden *Life Mrs. Jordan* II. i. 28 It showed, how the elegant mothers of Leadenhall Street, might, with the greatest gentleness, strain their young ones to bosoms equally *soft*, while they themselves were nourished by the *blood* and sweat of the unhappy peasant of Bengal. 1882 *Encycl. Brit.* XIV. 828/2 The principal markets..are Smithfield (central meat market and poultry market), Leadenhall (poultry and game), Billingsgate (fish), Covent Garden (fruit and vegetables). 1932 P. Spear *Nabobs* 28 Wellesley's remark about 'the cheesemongers of Leadenhall Street' would have horrified them. 1952 M. Bellasis *Honourable Company* p. viii, The factors and merchants, who first went out from Leadenhall Street to 'shake the pagoda tree', were transformed in the course of two or three generations into..self-effacing public servants. 1961 *Wonderful London* (Evening News) 56 Leadenhall market ..was rebuilt in 1881.

leader[1]. Add: **3.** *Leader of the House of Commons* (examples) (see also quot. 1964); freq. *ellipt.* as *Leader of the House*; so *Leader of the House of Lords* (or *of the Upper House*).

1835 *Ann. Reg.* 1834 335/2 It was requisite to find a new chancellor of the Exchequer, and a new leader of the House of Commons. 1852 Disraeli *Lord George Bentinck* xx. 397 The government abandoned this..project.. scarcely with decency, for the leader of the house of lords was eulogizing its virtues..at the moment it was cast away by the chancellor of the exchequer. 1852 Ld. Palmerston *Let.* 24 Dec. in J. Russell *Later Corr.* (1925) II. xx. 119 If the extensive duties of Leader of the House of Commons can be performed without salary why should any public officer have any? 1855 —— *Let.* 7 Feb. in Queen Victoria *Lett.* (1907) III. xxiv. 131 Proposed cabinet... Organ of the Government or Leader of the House of Lords. Marquis of Lansdowne. 1868 C. D. Yonge *Life and Earl of Liverpool* I. iv. 145 According to the usage of that day, when the Prime Minister was a Commoner, the Home Secretary, if a peer, was the leader of the Upper House. 1869 A. Todd *On Parl. Govt. in Eng.* II. iv. 323 The leader of the House of Commons is at liberty to arrange the order of business appointed for government nights as he thinks fit. 1908 A. E. Steinthal tr. *Redlich's Procedure House of Commons* I. 120 The name and function of the chief member of the Government in the House of Commons, the Leader of the House. 1964 Abraham & Hawtrey *Parl. Dict.* (ed. 2) 111 The term 'Leader of the House' was originally applied to the chief spokesman for the Government in the House

of Commons when the Prime Minister was a member of the House of Lords. *Ibid.*, The Leader of the House receives no salary as such... His chief responsibility is for planning and supervising the Government's legislative programme, and in particular for the arrangement..of the business of the House. *Ibid.* 112 The Leader of the House of Lords is the chief spokesman for the Government in that House. 1974 *Guardian* 30 Apr. 1/4 A statement from Mr Short, Leader of the House, on the registration of interests is promised later this week, but there are deep differences between the parties over whether the register should be compulsory or voluntary. *Ibid.*, A promised personal statement by Mr Short, Deputy Leader of the Labour Party and Leader of the House of Commons, was delayed by several hours last night.

e. *spec.* as a rendering of G. *Führer*, It. *Duce*, or Sp. *Caudillo*: the head of an authoritarian state. Usu. with capital initial. Also *transf.* (in quot. 1934 applied to the leader of the British Fascists, Sir Oswald Mosley).

1918 [see *caudillo]. 1934 H. G. Wells *Exper. Autobiogr.* II. ix. 783 Quite a quantity of pleasant boys and nice young men..were acting as ushers, selling idiotic songs about their glorious Leader. 1937 A. Huxley *Ends & Means* i. 2 The twentieth [century] has already witnessed..the emergence of the sheep-like social man and the god-like Leader. 1939 S. Spender tr. *Toller's Pastor Hall* I. 48 I've never spoken a word against the Leader. 1952 A. Bullock *Hitler* iii. 123 There was persistent.. grumbling at the amount of money the Leader and his friends took out of Party funds for their own expenses. 1960 H. Seton-Watson *Neither War nor Peace* viii. 226 The head of the government was the Chancellor, Adolf Hitler, who was also Leader (*Führer*) of the party, and on the death of President Hindenburg in 1934 replaced him as Head of State with the title of Leader of the German Nation.

8. c. For (?U.S.) read (orig. U.S.); cf. *loss leader* s.v. *loss *sb.*[1] 10. (Earlier and later examples.)

1851 C. Cist *Sk. Cincinnati in 1851* xv. 319 These articles [*sc.* sugar, molasses, coffee, etc.] are the leaders, as they are called, in commercial transactions, with the west. 1888 *Chicago Tribune* 29 Apr. 4/7 Goods advertised and sold below cost are technically known as 'leaders'. 1963 'R. Findlater' *What are Writers Worth?* 14 Most [paperback] firms produce about a dozen titles every month..at the summit the 'leader'—the smash-hit novel on which the selling machine is focused. 1967 *Times Rev. Industry* Feb. 31/1 Establishing new products is both costly and hazardous..while old leaders tend to decline over the years. 1972 *Lebende Sprachen* XVII. 34/1 US leader—BE/US loss leader, BE leading article.

12. (Earlier examples.)

1837 *Southern Lit. Messenger* III. 418/2 The Editor thus commenceth his leader. 1838 Dickens *Let.* 23 Dec. (1965) I. 475, I was very much obliged indeed to you for the paper. I..was greatly amused with the 'leader'.

19*. *Cinemat.* and *Tape Recording.* A short length of blank or uncoated film or tape attached at the beginning or end of a reel for purposes of threading or identification.

1917 C. N. Bennett *Guide to Kinematogr.* xi. 185 Refrain from..cutting or punching holes in the film leaders. 1960 J. M. Lloyd *All-in-One Tape Recorder Bk.* (ed. 4) v. 63 The inside and outside leaders are usually of different colours. 1969 J. Elliot *Duel* III. ii. 233 She went to.. learn the mysteries of..opticals and leaders and parallel and printing sync. 1969 D. N. Wood *On Tape* vii. 82 This brings me to the other main use of the leader tape—to act as a title... It is possible to use a chinagraph pencil on the tape itself, but it is much better to use leaders for this purpose.

19.** *Meteorol.* In full, *leader stroke.* A preliminary stroke of lightning that ionizes the path taken by the much brighter return stroke that follows.

1934 Schonland & Collens in *Proc. R. Soc.* A. CXLIII. 657 These preliminary downward strokes will be referred to as leader strokes and the upward strokes which follow them will be called main strokes. *Ibid.*, Sometimes the leader is so faint that a portion only of the track can be seen. 1937 *Jrnl. Inst. Electr. Engin.* LXXXI. 6/2 Immediately the stepped leader stroke reaches the earth the ..return stroke begins to travel..from earth to cloud. *Ibid.*, The leaders to the second and subsequent strokes of a flash usually travel from cloud to ground in a single flight. 1963 *Meteorol. Gloss.* (Meteorol. Office) (ed. 4) 154 Leader strokes directed upwards from ground to cloud may predominate in the case of very high structures. 1966 *McGraw-Hill Encycl. Sci. & Technol.* VII. 510/1 Cloud-to-cloud strokes also involve a step leader and main return stroke.

20. *attrib.* and *Comb.*, as (sense *3 e) *leader-principle, -worship*; (sense 12) *leader-page, -writer* (earlier and later examples); **leader stroke** (see sense 19**); **leader tape,** uncoated tape intended for use as a leader on a reel of magnetic tape; a length of tape so used.

1932 J. Buchan *Gap in Curtain* i. 54 Each of us must concentrate on one particular part to which his special interest was pledged—Tavanger on the first City page, for example, Mayot on the leader page, [etc.]. 1938 *Observer* 9 Jan. 5/1 (Advt.), Eugene Lyons..Assignment in Utopia ..'A moving and truthful account..'—Malcolm Muggeridge (*D. Telegraph*, leader-page article). 1940 'G. Orwell' *Crit. Ess.* (1951) 80 The absence of the leader-principle. There is no central dominating character. 1960 J. M. Lloyd *All-in-One Tape Recorder Bk.* (ed. 4) v. 64 Leader tape is transparent and is shiny on both sides. 1962 A. Nisbett *Technique Sound Studio* vi. 107 A leader tape (giving summarized details of the contents) and a trailer (several feet of coloured tape to give a visual indication of the end) may be cut on to the recording. 1971 *Hi-Fi Sound* Feb. 42 (Advt.), We also carry a full range of Accessories, Leader Tape, Empty Spools, Splicing Tape etc. 1940 'G. Orwell' *Crit. Ess.* (1951) 83 More bloodshed, more leader-worship. 1882 C. Peabody *Eng. Journalism* xix. 144 It is as a leader-writer and special correspondent that he will be best remembered. 1940 *Manch. Guardian Weekly* 22 Mar. 228 But now it is stated in Berlin that Mr. Kuusinen 'has been promoted to be a leader-writer on an obscure provincial paper'.

lea·dered, *pa. pple.* [f. Leader[1] 12.] Treated in a leading article; made the subject of a leader.

1884 *Pall Mall Gaz.* 29 Nov. 3/2 If it [*sc.* an interview] had been a speech it would have been 'leadered' all round. 1897 *Westm. Gaz.* 25 Sept. 5/2 Seeing that the subject is 'leader-ed' in both papers.

lea·derly, *a.* [f. Leader[1] 3 + -ly[1].] Having the character of a leader.

1918 H. G. Wells *In Fourth Year* ii. 23 Very rarely has it [*sc.* the United States] failed to set up very leaderly and distinguished men [as Presidents]. 1922 —— *Short Hist. World* xix. 104 They distinguished certain families as leaderly and noble. 1973 *Daily Tel.* 24 Nov. 16 The engineering community..is entitled to a more leaderly and statesmanlike response.

leadership. (Earlier and later examples.) Add to def.: The position of a group of people leading or influencing others within a given context; the group itself; the action or influence necessary for the direction or organization of effort in a group undertaking. Also *attrib.*, as *leadership behaviour, school, skill.*

1821 C. W. Wynn *Let.* 11 Mar. in *Corr.* (1920) 268 Charles writes that Tierney has regularly resigned the Leadership of the Opposition. 1915 E. & C. Paul tr. *Michels's Pol. Parties* iv. ii. 261 (*heading*) Analysis of the bourgeois elements in the socialist leadership. 1930 O. Oeser tr. *Bühler's Mental Devel. Child* vii. 166 From the schoolgoing age onwards we find that some have the talent for leadership. 1933 M. S. Viteles *Industr. Psychol.* xxvii. 626 The substitution of morale for discipline and of integration for domination calls for a change in the quality of leadership in industry. 1939 J. D. Brown in C. I. Barnard *Dilemmas of Leadership* 3 To treat the difficult problem of executive leadership. 1939 C. I. Barnard *Ibid.* 24 If a system once accepted..destroys leadership or divides followers—then disorganization, schism, rebellion ..ensues. 1947 Sherif & Cantril *Psychol. of Ego-Involvements* vii. 182 Leadership, then, was seen to be a function of the group and its activities. 1962 K. Orvis *Damned & Destroyed* xiv. 95 He went to a communist leadership school. 1963 J. E. Gerald *Social Responsibility of Press* v. 100 Few of the editors of mass-circulation newspapers since 1830 have risked their careers to exert strong leadership in the community. 1964 Gould & Kolb *Dict. Social Sci.* 380/2 The manifestation of leadership behaviour can be observed only in relation to other persons who act in response to the leader and who are collectively referred to as the *following*. 1964 Mrs. L. B. Johnson *White House Diary* 16 Jan. (1970) 51 We had an early dinner this evening..for the heads of the Senate Committees..and the Leadership on both sides and their wives. 1964 *English Studies* XLV. 50 Administrative and leadership skill. 1972 *Jrnl. Social Psychol.* LXXXVI. 29 Investigations of the relationship between personality traits and leadership behavior have failed to reveal any consistent patterns. 1973 M. Truman *Harry S. Truman* xv. 306 Dad once defined leadership as the art of persuading people to do what they should have done in the first place.

lea·d-in. [f. vbl. phr. *to lead in* (Lead *v.*[1] 2); cf. *leading-in* adj. (*leading *ppl. a.* 1 c).] **1. a.** A wire that leads in from outside, *esp.* one connecting an outdoor aerial with an indoor receiver or transmitter. Freq. *attrib.*

1913 *Wireless World* Apr. p. xxxvii/2 The lead was taken from the mast in the garden down to the instruments, which were now moved to the ground floor. This gave me a lead-in wire of 65 ft. 1913 *Work* 14 June 217/3 Lead in, about 40 ft. insulated. 1924 *Wireless World* 10 Sept. 679/2 (*caption*) By fixing your lead-in in this way opening and closing the window is not interfered with. 1934 *Practical Wireless* V. 62/1 (*heading*) A weather-proof lead-in. *Ibid.*, About 6 in. from the end of the lead-in wire, bind round with a 3 in. length of..copper wire. 1950 *Jrnl. Sci. Instrum.* XXVII. 231 (*heading*) Insulated power lead-in for vacuum systems.

b. A wire in an electric lamp that carries the current between the cap and the filament or electrode. Freq. *attrib.*

1929 *Encycl. Brit.* VIII. 291/1 The lead-in wires which carry the current to the filament have to be sealed through the glass. 1962 N. H. Codling in G. A. T. Burdett *Automatic Control Handbk.* viii. 6 Nickel-steel of 42 per cent composition, when copper-clad, is used for the lead-ins of lamps. 1970 A. Byers *Home Lighting* ii. 45 (*heading*) Lead-in wires.

2. *transf.* and *fig.* An introduction, opening, etc.

1928 *Melody Maker* Feb. 188/2 It is electrifying to hear the solo instrumentalists rip in on some unexpected lead-in. 1952 W. R. Burnett *Vanity Row* xiv. 118 Like a radio announcer with an embarrassingly far-fetched lead-in to the commercial. 1958 [see *lead-out 1]. 1958 *Economist* 15 Nov. 579/2 The fantasy life portrayed [*i.e.*, in certain children's comics] is simply a lead-in to the more elaborate and still more depressing dreamworld of the women's magazines. 1962 W. Nowottny *Lang. Poets Use* iv. 90 That opening is seen to be not an embarrassed and forced lead-in to a technically necessary comparison but rather a first and major step in the development of the whole. 1963 P. Moyes *Murder à la Mode* i. 20 Helen Pankhurst finished her lead-in blurb to the Collections feature. 1963 D. Ogilvy *Confessions Advertising Man* (1964) viii. 131 Don't mess about with irrelevant lead-ins. Start selling in your first frame. 1971 *Daily Tel.* 11 Feb. 30/6 Workers who agree in writing to operate incentive bonus schemes should get a 'lead-in' payment of £1 a week. 1972 D. Haston *In High Places* xii. 155, I heard the full story about Harsh's death and began to get some lead-in to the political infighting that had been going on. 1973 *Listener* 30 Aug. 295/1 Keep the lead-in short; some [news]papers enforce a 14-word limit on opening sentences.

leading, *vbl. sb.*[1] Add: **5.** Also in the usage of other religious bodies, and in philosophy.

1821 *Congregational Mag.* Nov. 579 What is Christian experience, but this working in us, this leading of the spirit of God? 1859 Geo. Eliot *Adam Bede* I. iii. 59, I thought it might be a leading of Providence for me to change my way of life. *Ibid.* 60 The strong love God hath given me towards you was a leading for us both. 1969 *Listener* 23 Jan. 117/3 Hence their [*sc.* Peirce and James'] characteristic teaching that all thought exists in signs or in 'leadings' from one area of experience to another.

leading, *vbl. sb.*[2] Add: **d.** More widely, deposition of lead on a surface.

1946 *Happy Landings* (Air Ministry) July 3/2 Pilots can prevent leading of plugs by clearing engines..or by using higher r.p.m. when flying in cold conditions.

leading, *ppl. a.* Add: **1.** (Further example.)

1817 Coleridge *Biog. Lit.* II. xxi. 126 Suppose too all this done without a single leading principle established.

b. leading dog *Austral.* and *N.Z.* (see quot. 1933).

1897 I. Scott *How I stole over 10,000 Sheep in Australia & N.Z.* ii. 9 We had no 'leading' dog. 1933 L. G. D. Acland in *Press* (Christchurch, N.Z.) 4 Nov. 15/7 *Leading-dog*, a dog trained to run ahead of a mob of sheep to keep them steady. 1934 *Bulletin* (Sydney) 16 May 38/2 Rock, the kelpie leading-dog..had never possessed any aspirations towards leadership.

c. *leading-in* adj.: applied to a lead-in wire (of either kind: see *lead-in 1 a, b).

1876 Preece & Sivewright *Telegraphy* vii. 224 The leading-in wire from the terminal pole, consists of a copper conductor insulated with gutta-percha, and well protected by a coating of tarred tape. 1885 *Phil. Mag.* XX. 141 The envelope may have deposited upon it a metallic film, derived from the leading-in wires to which the carbon filament is clamped. 1891 F. C. Allsop *Telephones* viii. 131 The leading-in wire is joined to the line-wire close to the last shackle or insulator. 1914 S. C. Batstone *Electr.-Light Fitting* vii. 138 G is a long glass stem through which the leading-in wires pass for connexion to the filament F. 1924 *Wireless World* 13 Aug. 543/1 (*heading*) Doing away with the leading-in wire. 1936 Orr & Forrest *Introd. Neon Lighting* i. 2 A lighting tube consists of a length of glass tubing bent to the shape required and closed at both ends. Into each end is inserted an electrode, usually in the form of a hollow cylinder of metal, to which are attached leading-in wires which are carried to the outside of the tube through a vacuum-tight seal.

2. *leading aircraftman, hand; leading lady* (in a film), *man* (earlier example; also, in a film).

1827 L. Rede *Road to Stage* 16 The salary is generally first-rate—at all events next to that of the leading man. 1900 *Daily News* 20 Jan. 6/4 'The leading hand in the team trade', as Mr. Kipling, père, calls the elephant. 1913 Leading lady [see *double *v.* 1 e]. 1921 Leading aircraftman [see *aircraftman]. 1939 I. Baird *Waste Heritage* vii. 88 He hated the way Bette's leading man looked, all slicked-up and Hollywood. 1955 T. H. Pear *Eng. Social Differences* vi. 159 Leading-hand, charge-hand.

leading article. 2. a. (Later example.)

1877 W. S. Gilbert *Sorcerer* I. 15 Sir, it is our leading article.

leading edge. [Leading *ppl. a.*] **1.** The forward edge of a moving body; also *transf.*; *spec.* (*a*) that of a blade of a screw-propeller; (*b*) that of a wing, tailplane, or other part of an aircraft; (*c*) that of one of the plates of the earth's crust.

1877 W. H. White *Man. Naval Archit.* xiv. 579 When the plane is moved obliquely, its leading edge, corresponding to the forward edge of a rudder, may be regarded as continually entering water which was comparatively little disturbed by the previous motion. 1888 *Lockwood's Dict. Mech. Engin.* 205 *Leading edge*, that edge of the blade of a screw propeller which cuts the water, as distinguished from the following edge. 1912 *Aeroplane* 12 Dec. 592/1 Looking over the leading edge of the wings from a constant position the ground disappeared regularly. 1922 Glazebrook *Dict. Appl. Physics* I. 364/1 That part [of the surface]..over which the particles of fluid are being gradually retarded—i.e. the part in the neighbourhood of the leading edge of the surface such as, for example, the immersed surface of a ship. 1939 *Archit. Rev.* LXXXVI. 63/2 At the 'leading edge' of each winglike roof, the wooden slats with which it is faced are slightly separated to allow for..ventilation. 1946 Taylor & Allward *Spitfire* 99/2 Special wings were fitted, the leading edge portion of each being constructed as a fuel tank. 1959 H. Barnes *Oceanogr. & Marine Biol.* i. 49 On sandy.. where there is considerable resistance to the leading edge of either the mud bucket or naturalist's oval dredge, only a small sample is usually brought up and deeper burrowing animals frequently avoid capture. 1967 M. Chandler *Ceramics in Mod. World* vi. 177 For very high speed aircraft the sharp leading edges of engines and wings will also probably have to be made of ceramic materials.

1971 I. G. Gass et al. *Understanding Earth* xx. 289/2 A plate whose leading edge is of continental material will gradually increase in size; for new crust will be added where oceanic crust is generated at its trailing edge but little or no continental crust is being consumed at the leading edge. 1972 *Sci. Amer.* Mar. 33/3 The drifting of the continents is another theme; every continent must have a leading edge and a trailing edge.

2. *Electronics.* The part of a pulse in which the amplitude increases.

1945 *Nature* 15 Sept. 319/2 The beginning or 'leading edge' of the pulse marks a packet of energy which can be re-identified after the vicissitudes of travel, thus permitting accurate measurement of time of travel. 1962 Simpson & Richards *Physical Princ. Junction Transistors* vii. 139 In the amplification of small pulses with sharp leading and trailing edges the frequency range may be very broad. 1972 *Radio Times* 6 Jan. 5/3 Listeners may have noticed a change in the Greenwich Time Signals broadcast since January 1... The exact time is signalled by the beginning or 'leading edge' of the long pip.

leadish, *a.* (Later example.)

1784 *Maryland Jrnl.* 27 July (Th.), There are two great-coats missing, one of which is a leadish-coloured country cloth.

lea·d-off. [f. vbl. phr. *to lead off* (Lead *v.*[1] 19).] (See Lead *sb.*[2] 11.) Also *attrib.*

1886 H. Baumann *Londinismen* 94/2 *Lead-off*, Journalisten-Slang: erste(r) (gew. von einem bekannten Schriftsteller herrührender) Artikel. 1892 [in Dict. s.v. Lead *sb.*[2] 11]. 1922 *Ardmore* (Okla.) *Daily Press* 6 May 3/3 His ability to judge close ones..make[s] him an ideal leadoff man. 1938 D. Baker *Young Man with Horn* (1939) iv. 264 For 'Sam, the Old Accordian Man', it was to be a lead-off by Jeff. 1963 Mrs. L. B. Johnson *White House Diary* 27 Dec. (1970) 22 Our foreman, Dale Malechek, took the lead-off bus... I took the second bus and Lynda..the third. 1970 *Toronto Daily Star* 24 Sept. 17/1 Morton.. was greeted by a Willie Stargell leadoff single.

lea·d-out. [f. vbl. phr. *to lead out.*] **1.** A leading out (in various senses).

1906 *Dialect Notes* III. 158 *Stag lead-out*, a dance-number in which only men who have not brought women dance with the women present. 1958 *Spectator* 3 Jan. 13/2 In stark contrast to the Zilliacus broadcasts these received an implied disavowal in the lead-in and lead-out.

2. *attrib.* or as *adj. Electronics.* Applied to a conductor by which current may enter or leave an electronic device.

1939 *Amat. Radio Handbk.* iii. 45/2 These valves..are ordinary valve types mounted in a metal bulb, welded or brazed together, having the lead-out wires passed through eyelets..mounted in the metal bulb. 1962 F. E. Duffield in G. A. T. Burdett *Automatic Control Handbk.* ix. 17 Fig. 18 illustrates a single inductance type of pressure transducer... The lead-out wires are connected to a pair of sealed terminals. 1967 F. Langford-Smith *Radio-Designer's Handbk.* (ed. 7) 1484 The lead-out groove is reduced in length due to the smaller ending diameters. 1970 J. Earl *Tuners & Amplifiers* ii. 29 There are other types [of IC] which represent a component 'block'..with leadout wires or tags. Transistors usually have three lead-out wires..with a possible fourth connecting to a screen or shield.

lead-pipe. [See Lead *sb.*[1] 10.] Used *attrib.* with *cinch* to denote a complete certainty. *U.S. colloq.*

1898 'J. Kerr' *Cheery Bk.* 71, I never had a 'lead pipe cinch'; I never had a 'pull'; I never had a 'straight' that was not beaten by a 'full'. 1911 H. Quick *Yellowstone Nights* xi. 288 Oh its a cinch, a timelock, leadpipe cinch! 1926 *Punch* 7 July 17/1 The Office of Works does not borrow money even to back what Americans call a lead-pipe cinch. 1949 N. Algren *Man with Golden Arm* 23 Not early enough to move no tables, that's a lead-pipe cinch. 1973 *N.Y. Times* 25 Feb. iv. 2/6 To be sure, speculation in gold is not a lead-pipe cinch; its price can go down as well as up.

lea·d-up. [f. vbl. phr. *to lead up* (cf. Lead *v.*[1] 22).] Something that leads up to something else.

1953 M. T. Monro *Thinking about Genesis* I. i. 26 The lead-up is the ordinary one by which we establish, on rational grounds alone, the existence of God, His attributes, [etc.]. 1959 D. Cooke *Lang. Mus.* iii. 145 The Beethoven is the short, breathless lead-up to the final jubilant outburst of the finale of the Choral Symphony. 1959 D. D. C. P. Mould *Peter's Boat* vi. 78, I had occasion to go into..that country's great mediaeval cathedrals now in Protestant hands... Here was a setting, a magnificent lead-up in stone..pointing to one thing and one thing only, the Mass and the Blessed Sacrament. 1972 *Lebende Sprachen* XVII. 73/1 During the vital lead-up to first flights, acceleration and deceleration tests were made during taxying trials.

leadwork[2] (lī·dwəɹk). Also **lead work.** [Origin unknown.] (See quot. 1900.)

Not connected with *lead-work* s.v. Lead *sb.*[1] 12.

1900 E. Jackson *Hist. Hand-made Lace* 213 *Lead works* or *lerd works*, terms used to indicate Modes or Fillings. Fancy stitches employed to fill in enclosed spaces in needle-point and bobbin laces. 1919 T. Wright *Romance of Lace Pillow* ix. 70 A Lille ground..sprinkled with *dots* (*plaits, leadworks* or *points d'esprit* as they are called). 1953 M. Powys *Lace & Lace-Making* iv. 26 Maltese Lace. This specimen has the leaf or lead work used as an ornamental filling.

leaf, *sb.*[1] Add: **4. a.** Also, the leaves of other plants used for smoking.

1972 *Guardian* 29 Jan. 9/2 Mr Williams had three previous convictions for possession of cannabis... 'A man ..let me have some leaf for five shillings.'

9. Delete 'Now only *dial.*' and add further examples. Also *U.S.*

1886 *Harper's Mag.* July 206/2 Lard, 'made from hog round, say head, gut, leaf, and trimming', is..in demand. 1904 L. L. Lamborn *Cottonseed Products* 166 Neutral lard is composed of the fat derived from the leaf of the slaughtered animal. 1911 *Encycl. Brit.* XVI. 214/2 The finest quality [of lard], used for making oleomargarine, is got from the leaf. 1934 F. Allen in *Meat Trade* II. iv. 100 The following parts [of a pig] are removed: the back bone, the blade bone,..and the flair or leaf. *Ibid.* 113 The leaf, or flair, of the pig is generally regarded as producing the best lard. 1955 W. G. R. Francillon *Good Cookery* iii. 53 The leaf or caul (a lining of fat taken from the inside of the animal)..should be placed over the joint before baking.

10. b. Also, a thin piece of soap or other detergent (larger than a 'flake').

1925 G. Martin *Mod. Soap & Detergent Industry* II. I. ii. 35 Soap Leaves are prepared by passing continuous paper sheets over rollers through a hot solution of soap, the excess of soap attached to the surface being scraped off. The paper is then passed over drying cylinders and from thence to a cutting machine. 1959 *Which?* Nov. 152/2 There were differences between these shampoos and some of the powder or leaf varieties. *Ibid.* 154 Packet of 6 leaves.

c. One of the metal strips of a leaf spring.

1905 R. T. Sloss *Bk. Automobile* vi. 124 The friction of the leaves decreases with the tension of the spring. 1936 F. Clune *Roaming round Darling* ix. 78 We left the car to have a couple of extra leaves inserted in the springs. 1971 B. Scharf *Engin. & its Lang.* xii. 147 The individual leaves are free to slide along each other and adequate grease lubrication must be provided to minimise friction.

15*. The external portion of the ear of a mammal or the nasal appendage of a leaf-nosed bat.

1851 H. Melville *Moby Dick* II. xxxii. 225 The ear [of a whale] has no external leaf whatever; and into the hole itself you can hardly insert a quill. 1955 *Times* 16 July 12/1 (*caption*) A forest-living bat, with a very large nose leaf..caught by members of the expedition to British Guiana.

16. a. *leaf-base, -cell, -point, -rib, -shape; leaf-stalk* (later examples) = Petiole 1.

1865 P. H. Gosse *Land & Sea* 26 This plant [*sc.* a grass] grows in large stools or tussocks formed of the densely-matted leaf-bases of successive seasons. 1894 *Pop. Sci. Monthly* XLIV. 488 The huge leaf-bases [of the banana tree]..tightly inclose each other. 1965 Bell & Coombe tr. *Strasburger's Textbk. Bot.* 158 In many leaves..the leaf base is not specially developed. 1875 Huxley & Martin *Course Elem. Biol.* 49 The terminal leaf-cell soon attaining its full size and not dividing. 1974 A. Huxley *Plant & Planet* vii. 57 In which vein the xylem and phloem fit together, so that the sugars from the leaf cells can be passed into the remainder of the plant. 1871 C. Kingsley *At Last* II. x. 71 The curving leaf-points toss in the breeze. 1895 Kipling *Second Jungle Bk.* 141 The lighting shows each littlest leaf-rib clear. 1909 Groom & Balfour tr. *Warming's Oecol. Plants.* III. xxviii. 99 The properties of water bring forth leaf-shapes entirely different from those of land-plants. 1946 *Nature* 13 July 64/1 When a leaf-shape is transferred from a late to an early flowering species the action of the gene is accelerated. 1946 F. E. Zeuner *Dating Past* xii. 381 If one compares this example of an aromorph with the evolution, for instance, of a highly specialized protective character, such as the leaf-shape of a leaf-insect.., one realizes the difference between an aromorph and an ordinary adaptational character. 1895 *Daily News* 27 Dec. 7/1 Both evergreens and deciduous plants are subject to this process of separation at the bottom of the leaf stalk. 1970 Robertson & Gooding *Bot. for Caribbean* (ed. 2) iii. 27 The petiole or leaf-stalk varies somewhat in length and shape.

b. *leaf-boring* adj. See also *leaf-miner, -mining* (Leaf *sb.*[1] 17); Leaf-cutter; Leaf-cutting *ppl. a.*

a 1887 R. Jefferies *Field & Hedgerow* (1889) 115 The coils and turns upon this leaf..are the work of a leaf-boring larva.

c. *leaf-crowned, -encumbered, -hid, -hung, -lined, -roofed* (later examples), *-shadowed, -whelmed* adjs.

1891 W. B. Yeats *Countess Kathleen* (1892) 125 And no one any leaf-crowned dancer miss. 1925 V. Woolf *Mrs. Dalloway* 21 That leaf-encumbered forest, the soul. 1869 J. R. Lowell *Under Willows* 52 Our leaf-hid Sybaris. 1895 W. B. Yeats *Poems* 16 Down in a leaf-hid, hollow place. 1919 V. Woolf *Night & Day* xi. 145 They swept together among the leaf-hung trees of an unknown world. 1921 W. de la Mare *The Veil* 7 The listening, leaf-hung creek. 1895 *Outing* XXVI. 394/2, I filled one of our leaf-lined pails with berries. 1844 J. Tomlin *Missionary Jrnls.* v. 120 The capital of Siam is a large, but not very magnificent city..consisting mainly of leaf-roofed wooden cottages. 1906 *Westm. Gaz.* 10 Sept. 2/3 Where the much-loved birds in their leaf-roofed halls Will herald my morning in. 1845 E. Cook *Poems* 2nd Ser. 187 The leaf-shadow'd thicket. 1868 J. R. Lowell *Under Willows* (1869) 22 So they in their leaf-shadowed microcosm Image the larger world. *a* 1889 G. M. Hopkins *Poems* (1918) 89 We are leafwhelmed somewhere with the hood Of some branchy bunchy bushybowered wood.

d. *leaf-dark, -dry, -eyed, -light, -shaped* (of an arrowhead) adjs.

1936 E. Sitwell *Victoria of Eng.* xix. 227 Their leaf-dark hair smoothed into the Chinese style. 1946 W. de la Mare *Traveller* 20 He caught but leaf-dry whisper of what they said. 1949 S. Spender *Edge of Being* 20 Behind the hedge of leaf-eyed lovers. 1971 B. Patten *Irrelevant Song* 51 Into myth she faded, Leaf-eyed. *c* 1879 G. M. Hopkins *Poems* (1918) 44 Low-latched in leaf-light housel his too huge godhead. 1921 V. Woolf *Monday or Tuesday* 37 Flaunted, leaf-light, drifting at corners, blown across the wheels. 1872 J. Evans *Anc. Stone Implements Gt. Brit.* xvi. 333 Of leaf-shaped arrow-heads..there are several minor varieties. 1923 C. Fox *Archaeol. Cambr. Region* i. 4 Both the leaf-shaped and the tanged types [of arrowhead] commonly occur in the district. 1940 C. F. C. Hawkes *Prehist. Found. Europe* iii. 78 Hollow-based and leaf-shaped arrowheads..appear in the flint industry.

17. leaf-arrowhead, an arrowhead shaped like a leaf, usu. of the Neolithic period and made of flint (cf. *leaf-shaped* adj. s.v. sense 16 d above); **leaf-bed,** a layer of leaves sometimes found in the upper stratum of the earth's surface; **leaf blight,** one of several plant diseases causing the death of foliage; **leaf blister,** (*a*) a disease of certain fruit trees caused by a parasitic mite; (*b*) a plant disease caused by a fungus of the genus *Taphrina*; **leaf blotch,** one of several plant diseases indicated by discoloured patches on foliage, esp. = *black spot 1; **leaf brown,** the colour of (dead) leaves; **leaf-bud** (later examples); **leaf cast** = *larch needle cast* (*larch 3); **leaf-cup,** (*c*) a leaf folded and used as a cup; **leaf curl,** one of several plant diseases characterized by curling leaves, esp. (*a*) = *leaf-roll; (*b*) a disease of peach, almond, and nectarine trees caused by the fungus *Taphrina deformans*; (*c*) a virus disease of cotton; **leaf-drift,** a place where fallen leaves have been blown together by the wind; **leaf-fall,** (*b*) *Bot.*, the shedding of leaves by a plant; **leaf-fat** (later examples); **leaf gelatine,** gelatine manufactured in sheet form for cooking purposes; **leaf-house, -hut,** a house or hut made of entwined leaves; **leaf-lard** (earlier and later examples); **leaf-mass,** a thick growth of leaves; **leafmeal** [-meal] *adv.* (*nonce-wd.*), with leaves fallen one by one; **leaf-monkey,** a monkey found in south or south-east Asia belonging to one of several species of the genus *Presbytis*; = Langur; **leaf-mould,** (*b*) a disease of tomatoes caused by the fungus *Cladosporium fulvum*; **leaf-nosed** (earlier and later examples); **leaf-plate, -platter,** a leaf or leaves used as a plate or dish for food; **leaf protein,** protein, or a protein, present in leaves, esp. when extracted for use as a possible dietary supplement; **leaf-roll,** a virus disease of potatoes shown by curled-up leaves; **leaf scald,** (*a*) = *leaf scorch*; (*b*) a disease of sugar-cane caused by the bacterial pathogen *Xanthomonas albilineans*; **leaf-scar** (later examples); **leaf scorch,** a plant disease caused by a deficiency of potassium, causing leaves to shrivel and turn brown; also, a virus disease causing similar effects; **leaf-shedding,** (*a*) a disease of pine trees caused by the fungus *Hendersonia acicola*; pine needle cast; (*b*) = *leaf-fall* (*b*); **leaf shelter,** a shelter made of leaves; **leaf-skin,** (*a*) the membrane enclosing the leaf-fat; (*b*) the epidermis of a leaf; **leaf-spot,** one of a large number of plant diseases caused by various fungi which mark the foliage; also *attrib.*; **leaf spring,** a spring consisting of a number of strips of metal curved slightly upwards and clamped together one above the other, each strip being longer than the one beneath; so *leaf springing, -sprung* adj.; **leaf warbler,** a small green or yellow bird of the genus *Phylloscopus*, living in bushes or trees; **leaf-work** (later examples).

1954 S. Piggott *Neolithic Cultures* iii. 99 Leaf-arrowheads are a common feature in Belgium. 1963 L. F. Chitty in Foster & Alcock *Culture & Environment* vii. 188 Flints found range from a microlith and a leaf-arrowhead to a gun-flint. 1873 *Archæologia* XLIV. 278 The upper surface of the leaf-bed was well marked and level, as was also..the upper surface of the moss. 1894 *Nature* 26 July 295/1 If we could only meet with some fairly representative leaf-beds, such as abound in newer formations, the Wealden would yield a flora, both varied, and of enormous interest. 1954 S. Piggott *Neolithic Cultures* x. 295 'Leaf bed' with no large vegetable remains, 3–4 ft. thick. 1850 *Rep. Comm. Patents 1849* (U.S.) 440 During the last summer our seedling pears were for the first time badly affected with 'leaf-blight'. 1920 P. J. Fryer *Insect Pests & Fungus Dis. Fruit & Hops* 709 Cherry and Plum Leaf Blight..appears to be slightly on the increase. 1926 [see *frog eye]. 1960 C. Westcott *Plant Dis. Handbk.* (ed. 2) 125 *Mystrosporium adustum.* Leaf Blight, Ink Spot of bulbous iris. 1914 F. C. Sears *Productive Orcharding* xi. 159 Leaf Blister Mite.—Another pest which is frequently troublesome on both pears and apples is the blister mite. 1960 C. Westcott *Plant Dis. Handbk.* (ed. 2) 194 A single genus, Taphrina,

is responsible for most of the hyperplastic (over-growth) deformities known as leaf blister, leaf curl, or, occasionally, as pockets. 1906 M. C. Cooke *Fungoid Pests Cultivated Plants* 75 Iris Leaf-blotch. 1925 *Gardeners' Chron.* 31 Oct. 353/3 (*heading*) A leaf blotch of the Shasta daisy. 1928 *Daily Express* 7 July 4/2 See that none of your favourites [*sc.* roses] is attacked by leaf blotch. 1971 *Country Life* 18 Feb. 389/2 The diseases of mildew, rust and leaf blotch are prevalent throughout Britain. 1923 *Daily Mail* 19 Feb. 1 (Advt.), French Model Jumper made of..Crepe de Chine,..Jade..Leaf Brown, Navy and Black. 1932 W. Faulkner *Light in August* xx. 444 Patches of Confederate grey weathered leafbrown now. 1906 *Westm. Gaz.* 14 Apr. 8/1 The lilac and elder-bushes.. are beginning to unfold their leaf-buds. 1971 *Country Life* 10 June 1440/2 The ash..never makes the mistake of opening its leaf buds before the last night frost. 1933 *Oxford Forestry Mem.* XV. 7 *Meria laricis* Vuillemin, the leaf cast disease of larch, which was first described by Mer in 1895..is probably the most important fungal disease of European larch in nurseries in this country. 1952 E. Ramsden tr. *Gram & Weber's Plant Dis.* iv. 482/1 Leaf cast is the worst disease of young larch trees. 1901 Kipling *Kim* x. 263 He bought sweetmeats in a leaf-cup from a Hindu trader. 1899 G. Massee *Text-bk. Plant Dis.* 82 The disease [of peaches], which is very widespread, is popularly known as 'leaf curl', or simply as 'curl', owing to the fact that the diseased leaves become much curled, distorted, and thickened. *Ibid.* 323 The well-known disease of the foliage of potatoes known as 'leaf curl' attacks the stem..and gradually creeps up. 1926 W. H. Johnson *Cotton* viii. 259 Upland cotton appeared to be less affected by a peculiar leaf-curl disease. 1951 *Dict. Gardening* (R. Hort. Soc.) III. 1654/2 The most common and probably the most serious [virus disease of potatoes] is Leaf Curl (Leaf Roll). 1965 Ripper & George *Cotton Pests Sudan* i. 7 Leafcurl is transmitted from ratoon cotton and wild host plants to cotton by whitefly. 1967 *Punch* 18 Jan. 96/3 It [*sc.* Burgundy mixture] is a good fungicide to use on leafless trees and bushes, particularly against leafcurl in the peach family. 1905 E. Phillpotts *Secret Woman* I. i. 6 While death, not unlovely, appeared in leaf-drift and touch-wood, in acorn cups..and hollow hazel-nuts. 1958 C. Tomlinson *Seeing is Believing* (1960) 60 And it continues Falling flaking into the leaf-drift. 1914 M. Drummond tr. *Haberlandt's Physiol. Plant Anat.* iii. 143 The arrangements [for peeling bark scales] resemble those which occur in leaf-bases in connection with the autumnal leaf-fall. 1947 G. F. Wilson *Detection & Control Garden Pests* vi. 107 Premature leaf-fall is associated with several factors other than pest attack. 1971 *Homes & Gardens* Sept. 128/2 Most of the cornus family colour richly before leaf-fall. 1845 J. J. Hooper *Some Adventures Simon Suggs* v. 65 They've knocked the leaf fat outen him tonight, in wads as big as mattock handles. 1904 L. L. Lamborn *Cottonseed Products* 166 In the packing plants the leaf fat is taken from the animal immediately after killing. 1956 C. Spry *Cookery Bk.* xxx. 957 The incorporation of gelatine with various liquids is much easier now that powdered or very fine leaf gelatine is sold. 1957 E. Craig *Collins Family Cookery* 606 Ten perfect sheets of French leaf gelatine equals 1 oz. 1953 A. Moorehead *Rum Jungle* vii. 107 Green-ants that stitch their leaf-houses together by holding their babies in their arms and drawing out of the babies' mouths a sticky thread. 1958 *Listener* 14 Aug. 237/1 Johnny and Silas were my two servants and lived in a leaf house near mine. 1910 W. de la Mare *Three Mulla-Mulgars* i. 8 He taught them..to build leaf-huts and huddles against heat or rain. 1949 M. Mead *Male & Female* x. 220 A leaky leaf-hut on the side of a mountain. 1848 *Rep. Comm. Patents 1847* (U.S.) 538 The articles thus referred to are put up in these establishments, from the hams..leaf lard [etc.]. 1885 W. L. Carpenter *Treat. Manuf. Soap* ii. 25 The fat immediately surrounding the kidneys yields the best and purest lard. This, and that which is obtained in flaky layers between the flesh and the skin.., is known as 'leaf' lard. 1888 W. T. Brannt *Pract. Treat. Animal & Veg. Fats & Oils* ix. 344 The leaf lard is.. kept separate from the rest. 1908 G. Jekyll *Colour in Flower Garden* vii. 60 We gradually return to the grey-blues, whites and pale yellows, with..the splendid leaf-mass of a wide and high plant of *Euphorbia Wulfenii*, which..rises to a height far above my head. 1958 C. Tomlinson *Seeing is Believing* (1960) 20 Light, swept perpendicular Into the leaf-mass Flickers out. *c* 1880 G. M. Hopkins *Poems* (1918) 51 Though worlds of wanwood leafmeal lie. 1888 W. T. Blanford *Fauna Brit. India: Mammalia* I. 41 Phayre's Leaf-Monkey is found in dense high forests. 1928 *Jrnl. Bombay Nat. Hist. Soc.* XXXII. 472 (*title*) The langurs or leaf-monkeys of British India. 1966 R. & D. Morris *Men & Apes* viii. 236 Various species of leaf monkeys..frequent salt licks and saline mineral springs in Borneo. 1913 M. T. Cook *Dis. Trop. Plants* vii. 217 Leaf Mould..spreads rapidly. 1931 *Times Lit. Suppl.* 24 Sept. 734/4 Leaf-mould (*Cladosporium*) is a source of much loss to growers of the tomato. 1971 T. F. Preece in J. H. Western *Dis. Crop Plants* ii. 12 *Cladosporium fulvum.* Tomato leaf mould. 1843 *List Mammalia Brit. Mus.* 21 Redman's Leaf-nosed Bat. 1850 A. White *Pop. Hist. Mammalia* 47 The bats are arranged by Mr. Gray in two great divisions—the Leaf-nosed bats and the Simple-nosed bats. 1960 G. Durrell *Zoo in Luggage* ii. 61 A handful of leaf-nosed bats with extraordinary gargoyle-like faces. 1965 R. & D. Morris *Men & Snakes* viii. 178 (*caption*) The leaf-nosed snake, showing an unusual form of serpentine camouflage. *a* 1843 Southey *Commonplace Bk.* (1849) 2nd Ser. 422/1 *Leaf-plates.* 'Their plates and dishes are generally formed from the leaf of the plaintain tree or the nymphæa lotos... These are never used a second time.' 1962 B. Harrisson *Orang-Utan* ii. 57 You bend slightly over and down for mouth and fingers to meet above your leaf-plate. 1901 Kipling *Kim* xi. 281 'And we,' said Kim, turning his back and heaping a leaf-platter for the lama, 'are beyond all castes.' 1937 *Rep. Brit. Assoc. Adv. Sci. 1937* 459 Man also eats leaves to some extent,..and these leaves are quite high in protein. In this case there has been no selection of the leaf proteins by animal or plant, and it is probable that this leaf protein is intermediate in value to man between animal and other vegetable protein. 1953 *Jrnl. Agric. Sci.* XLIII. 136/1 Work on laboratory-extracted leaf protein suggests that such material might provide valuable protein feed as it contains many of the amino-acids essential for poultry nutrition. 1971 N. W. Pirie *Leaf Protein* xvi. 157 People habituated to leaf protein accept its flavour so that a larger proportion can be added to a food. *Ibid.*, Freshly made slabs of leaf protein disperse in water to give a smooth paste but slabs stored in deep-freeze gradually become gritty and have to be passed through a mill. 1972 Goodwin & Mercer *Introd. Plant Biochem.* ix. 236 The major part of the leaf protein is in the chloroplast. 1926 *Sci. Proc. R. Dublin Soc.* XVIII. 177 Twenty-nine halves or thirds of tubers.. were infested with aphides from diseased sprouts... Only two..became infected (with leaf-roll, the aphides being *Myzus pseudosolani*). 1946 *Nature* 14 Dec. 885/2 Bismark is resistant to leafroll. 1960 *Times* 29 July 12/6 The telltale curl..shows a potato plant has leafroll. 1899 G. Massee *Textbk. Plant Dis.* 276 (*heading*) Leaf scald. 1924 *Phytopathology* XIV. 587 (*title*) Java gum disease of sugar cane identical to leaf scald of Australia. 1965 G. C. Stevenson *Genetics & Breeding Sugar Cane* v. 147 Resistance of sugar cane species and hybrids to leaf scald disease..is very much complicated by the presence, in various countries, of several different strains of the parasite. 1897 J. C. Willis *Man. Flowering Plants* I. iii. 167 This [*sc.* the absciss layer] splits down the middle and leaves one half upon the stem, where it forms the *leaf scar* covering the wound. 1965 Bell & Coombe tr. *Strasburger's Textbk. Bot.* 167 In almost all woody plants the leaves..are sooner or later shed, leaving leaf scars on the stem showing their former positions. 1921 *Ann. Rep. Agric. & Hort. Res. Station Univ. Bristol* 121 The correlation which has been noted between the amount of potash supply, root growth and the degree of leaf scorch, points very definitely towards the liberal application of potash manures as a remedial measure. 1929 *Misc. Publ. Ministry Agric. & Fisheries* no. 70. 27 Leaf Scorch..was common on Mangolds in Devon and Cornwall in 1927, and on Sugar Beet in the East Midland and Eastern Provinces in the same year. 1933 *Jrnl. R. Hort. Soc.* LVIII. 253 Leaf scorch of apples is a deficiency disease. 1943 *Bull. Ministry Agric. & Fisheries* no. 126. 30 Leaf Scorch..was formerly attributed to *Sporidesmium putrefaciens* Fuckel but is now regarded..as a later and secondary symptom of Yellows. 1952 E. Ramsden tr. *Gram & Weber's Plant Dis.* iii. 342/2 Leaf Scorch (Potassium Deficiency). Potatoes that lack potassium produce a low-growing, rather dark, open haulm. 1961 *Amat. Gardening* 21 Oct. Suppl. 31/2 Leaf scorch. A common disorder of grapevines under glass, in which the leaves take on a shrivelled appearance. 1891 W. Schlich *Man. Forestry* II. 302 In many cases a fungus (*Hysterium pinastre*) is present, and may occasion the disease, which is called 'leaf-shedding'. 1895 *Daily News* 27 Dec. 7/1 With most evergreens the process of leaf-shedding is exactly the same as in the case of deciduous trees. 1937 *Discovery* Sept. 274/2 Two dilapidated leaf shelters. 1816 'A. Singleton' *Lett.* (1824) 75 (Th.), Being born smokers, [the Negroes] make pouches of the inner leafskin of a swine, peeled thin, which is soft, transparent, and tough. 1974 A. Huxley *Plant & Planet* xxv. 281 Alpine rhododendrons..have very thick leaf-skins reinforced with silica. 1901 H. M. Ward *Dis. in Plants* xii. 114 If the fungus becomes epidemic and myriads of leaf-spots are formed, the destruction of foliar tissue..may end in rapid defoliation. 1908 *Jrnl. South-Eastern Agric. College, Wye* XVII. 316 (*title*) Leaf-spot diseases of the apple. 1933 *Jrnl. R. Hort. Soc.* LVIII. 280 Infected seed as a source of the celery leaf-spot is discussed. 1951 *New Biol.* XI. 78 The leaf-spot disease of bananas..was not recorded in the western tropics until 1934. 1972 *Arable Farmer* Feb. 55/1 Latest list of approved products..includes: Benlate (systemic fungicide from Du Pont for control of Botrytis in green beans and leaf spot in celery). 1893 *Funk's Stand. Dict.*, Leaf-spring. 1896 R. Grimshaw *Shop Kinks* 123 In finishing leaf-springs by grinding care should be taken that the grinding-marks run lengthwise. 1905 R. T. Sloss *Bk. Automobile* vi. 123 Leaf-springs seem to give the best results in automobile construction. 1935 *Times* 22 Oct. 9/2 In applying independent front wheel suspension some designers use coil springs and others leaf springs. 1967 *Autocar* 5 Oct. 73/3 He built the car without dampers, having read somewhere that inter-leaf friction in leaf springs might be sufficient damping. It wasn't. 1958 *Times* 26 Sept. 6/4 Air suspension is confined mostly to passenger-carrying vehicles, for the normal leaf springing is considered satisfactory for goods vehicles. 1973 *Times* 4 Oct. 43/3 All the cars have a leaf sprung back axle. 1926 T. A. Coward *Birds Brit. Isles* 122 (*heading*) The Leaf-Warblers. 1929 W. E. Glegg *Hist. Birds Essex* 71 The Willow-Warbler is a very common summer resident, increasing, and the most numerous of the Leaf-Warblers. 1953 B. Campbell *Finding Nests* vii. 96 The off-nest call [of the wood-warbler]..is easier to pick up than similar notes of the other leaf-warblers. 1974 *Lady* 2 May 622/3 Linnaeus did not distinguish all three common leaf-warblers. 1880 J. L. Warren *Guide to Study of Book-Plates* ii. 10 Outside the inscription is some rather fine leaf-work. 1937 *Burlington Mag.* Aug. 69/1 Decorated with the famous leafwork.

leaf (līf), *sb.*[2] *Services' slang.* Also **leef.** [Var. Leave *sb.* 1 e.] Leave of absence, furlough; = Leave *sb.* 1 e.

1846 *Punch* 3 Jan. 10/2 The shabby Capting (who seames to git leaf from his ridgmint whenhever he likes). 1904 Kipling in *Windsor Mag.* Dec. 4/1 What a lot of 'ard work one misses on leaf! 1916 'Taffrail' *Pincher Martin* viii. 124 Wot's the good o' seven days' leaf ter a bloke wot ain't got no money? 1919 *Athenæum* 8 Aug. 729/1 The soldier going on short leave speaks usually of 'going on pass'; sometimes, however, of 'going on leaf'. Why in this phrase (nowhere else) the voiceless *f* is substituted for voiced *v* in 'leave' is a mystery to me. 1929 *Papers Mich. Acad. Sci., Arts & Lett.* X. 306 *Leef*, leave, pass, furlough. 1946 J. Irving *Royal Navalese* 107 *Leaf*, a corruption of Leave—leave of absence... A sailor goes 'on leaf' and *never* on furlough.

leaf, *v.* Add: **3. a.** Also used *intr.* and *fig.* and with *through*, to go through (a book or papers) by turning the leaves, usu. in a casual manner.

1929 *Publishers' Weekly* 19 Oct. 1928/2 There are plenty of people who..like to leaf through a book before buying. 1936 J. G. Cozzens *Men & Brethren* II. 175 Ernest..leafed over the remaining letters. 1936 L. C. Douglas *White Banners* xi. 245 He found the book, opened it on the table and leafed to the pictures he had found most amusing. 1953 *Encounter* Nov. 34/1 So it is possible to leaf through the Essays, reading a few pages and turning away at pleasure, as Montaigne himself read. 1960 'R. East' *Kingston Black* x. 98 She went on leafing through the transcript. 1960 'S. Harvester' *Chinese Hammer* i. 16 [He] leafed through an old issue of the *New Yorker*. 1973 W. M. Duncan *Big Timer* xxi. 141 He picked up a paper and leafed through it idly.

leafit. (Later *poet.* example.)

1916 Blunden *Harbingers* 60 The lopped tree, be it but stub or stock, Thrives, and begems its leafits in a year.

leaflet, *sb.* Add: **2.** *spec.* the thin flap of a valve in the heart or a blood vessel. (Later examples.)

1936 G. R. Herrmann *Synopsis Dis. Heart & Arteries* xvii. 254 The degrees to which valvular lesions develop depend upon the extent of the allergic or inflammatory processes and the trauma to which these inflamed leaflets are subjected at the time and during the healing processes. 1961 R. D. Baker *Essent. Path.* xiv. 333 Uncomplicated insufficiency occurs when the valve leaflets are held against the wall of the heart or of the great vessels or when the ring of the valve is dilated.

3. b. *attrib.* and *Comb.*, as *leaflet literature, party, writer*; **leaflet raid,** a raid in which leaflets are dropped from an aircraft; also *transf.*; so *leaflet drop.*

1903 *Westm. Gaz.* 13 Aug. 2/3 Mr. C. A. Vince, M.A., chief leaflet-writer to Mr. Chamberlain. 1904 *Ibid.* 21 Jan. 2/1 Not even the profuse distribution of Birmingham leaflet literature can alter this fact. 1940 *Flight* 11 Apr. 337/1 This same officer commanded the Whitley which made the first leaflet raid over Berlin. 1940 Harrisson & Madge *War begins at Home* vii. 148 The first leaflet raid—in which the R.A.F. dropped 6,000,000 leaflets over Germany. 1943 Koestler *Arrival & Departure* III. 94 That is why I took part in those leaflet-parties, though it was not my job. 1961 *Guardian* 11 Nov. 1 (*caption*) Leaflet raid on Lisbon. 1969 *Listener* 31 July 145/2 We would send our lecturers out into the fringes of the jungle to soften people up, we would have special leaflet drops before the troops went in. 1974 *Times* 13 Nov. 2/6 Miners from the Nottinghamshire coalfield..made leaflet raids to pits in South Yorkshire.

leaflet (lī·flėt), *v.* [f. Leaflet 3.] *trans.* and *intr.* To distribute leaflets to (people or places). Hence **lea·fleting** *vbl. sb.*

1962 *Spectator* 24 Aug. 268/3 Thousands of campaigners will be putting this case..by leafletting and pamphleteering. 1968 *Peace News* 25 Oct. 5/3 Civilians who originally had tried to leaflet on bases were quickly kicked off. 1969 C. Davidson in Cockburn & Blackburn *Student Power* 361 We should make our presence felt everywhere—in the campus news media, leafletting and poster displays. 1969 *Oxf. Univ. Gaz.* XCIX. Suppl. vii. 156 The Proctors announced the withdrawal of the regulation prohibiting indiscriminate leafleting. 1972 *Listener* 27 Jan. 119/2 Tariq Ali..refers to the question whether factories should be leafleted. 1973 *Daily Tel.* 9 June 2/4 Our friends in Germany and elsewhere will be encouraged to leaflet British soldiers stationed there. 1973 C. Mullard *Black Brit.* I. iii. 33 The new group were people willing to leaflet, organize and demonstrate.

leafleteer (līflėtiə·ɹ). [f. Leaflet 3 + -eer.] A writer of leaflets; the author of a leaflet. (Often contemptuous.)

1892 *Sat. Rev.* 16 July 70/2 It..is written in clear, plain, simple English, the only 'leaf' we could wish our leafleteers to take from this example. 1903 *Westm. Gaz.* 6 Oct. 2/2 We do not in the least mind Professors becoming leafleteers if so they must. 1970 *Sunday Tel.* 14 June 8/5 The leafleteers have fairly polluted the streets this time.

league, *sb.*[1] Add: **b.** *league-wide* adj. (*poet.*).

1843 J. R. Lowell *Prometheus* in *Poems* (1844) 83 The vast Sarmatian plain, league-wide. 1848 —— *Columbus* in *Poems* 2nd Ser. 11 Some league-wide river. 1951 W. de la Mare *Winged Chariot* 56 Life's league-wide cornfields. 1957 R. Campbell *Coll. Poems* II. 254 Across a league-wide valley, white with sprays.

league, *sb.*[2] Add: **1. b.** *the League* = *League of Nations.

1917 H. N. Brailsford *League of Nations* 324 Without the firm resolve to make the League itself an article, and the first article, in the settlement, our need of security will drive us inevitably to other expedients. The settlement, unless the idea of the League penetrates it and inspires it, must draw its principle from the older statecraft of anarchy and force. 1919 J. M. O'Sullivan in *Studies* Dec. 577 Had not the basal idea of the League been thus early repudiated. 1936 A. Huxley *Let.* 2 Mar. (1969) 401 The..atmosphere wd be cleared and a chance given for the reconstruction of the League on a more satisfactory basis. 1944 J. S. Huxley *On Living in Revolution* iii. 32 The failure of the League merely served to underline the urgent need for *some* international political organization. 1950 Theimer & Campbell *Encycl. World Politics* 260/1 The Assembly did not meet again until April 1946, when it decided to dissolve the League, already replaced by the United Nations. 1952 *Oxf. Junior Encycl.* X. 212/2 When the United States Congress repudiated President Wilson's proposals and failed to join the League, its hope of real success was small. 1971

W. H. McNeill in A. Bullock *20th Cent.* 47/1 Should a government defy the League..all the League members would be obliged to..check aggression by imposing sanctions.

c. *Football League* (earlier and later examples). Similarly, *Baseball, Cricket League*. Also ellipt. *League*.

1879 *Chicago Tribune* 17 May 7/5 A misunderstanding has arisen as to the condition of the Cleveland Club, and its inability to play, which will end in an appeal to the League. **1891** *Amer. Cricket Annual* 10 The organisation of the Metropolitan District Cricket League was certainly a move in the right direction. **1892** *Athletic News Cricket Ann.* 51 Lancashire Cricket League... This organisation..has done for cricket what the League has done for football. **1892** J. A. Leighton (*title*) Leighton's North-Western Rugby Football League card. Season 1892–93. **1910** *Encycl. Brit.* IX. 622/1 In 1888 the Football League, a combination of professional clubs of the north and midlands of England, was formed. **1921** A. Huxley *Crome Yellow* ii. 13 All the players in all the teams of the League. **1930** J. Williamson *Amer. Hotel* 293 The Broadway Central has been the scene of several noteworthy episodes. It was there that the National League was organized in 1876. **1935** *Encycl. Sports* 187/2 A number of [cricket] clubs form themselves into a league; each plays all the others in turn, and the championship of the league falls to the one which wins the most matches. *Ibid.* 292/2 The first league, the Football League, was then formed [in 1888]. **1951** *Football Record* (Melbourne) 8 Sept. 12 Approximately 300 visiting schoolboys were recently entertained by the League at the Melbourne ground. **1957** *Encycl. Brit.* III. 159/2 The professional [baseball] clubs usually compete as members of leagues. **1960** B. Liddell *My Soccer Story* vii. 48 For years the Football Association and the Football League have been trying to help players to prepare for the new life ahead when they finish with football. **1969** *Listener* 20 Mar. 384/2, I cannot believe he would still maintain that Rugby League backs 'usually run across instead of straight'. **1973** *News of the World Football Ann. 1973–74* 78 (*heading*) Football League—Division One. *Ibid.* 100 Re-elected to Division [One] when League was extended after the war. **1974** *Daily Record* (Glasgow) 15 Apr. 27/2 Yesterday Stein admitted, 'The League is almost won. We know that!' **1974** *Guardian* 1 Aug. 22/8 It is not possible to be wrapped up in all the competitions and win them as the League exists at present.

e. *transf.* and *fig.* Cf. *big league* s.v. *BIG *a.* B. 2.

1935 J. T. Farrell *Judgment Day* viii. 185 You better go back and play in a grammar-school league. **1959** N. Mailer *Advts. for Myself* (1961) 389 You want to keep a girl who was born to travel in a big league. **1961** *Listener* 12 Oct. 547/1 At the Riga brewery..I saw a notice board with the photographs of the twelve workers who were topping the production league. **1965** *Listener* 23 Sept. 446/1 The English-speaking peoples are excellent at breakfasts, but after that they would scarcely claim to stand high in the gastronomic league. **1966** J. Chamier *Cannonball* xii. 115 She's out of your league, me lad, and you'll take a most almighty toss. **1970** *Washington Post* 30 Sept. B. 4/3 In such a league Paul Mellon has impeccable collections. **1971** *Austral. Seacraft* June 17/2 To join the big league [in speedboating]. **1971** *Where* Oct. 293/2 Neill has a lightness of touch, and a flair for comedy that were in the Wodehouse league. **1972** *New Society* 27 Jan. 187/1 Rory Gallagher, a minor league superstar blues guitarist. **1972** 'M. Yorke' *Silent Witness* v. 121 She was bored because he obviously wasn't in her league. **1973** *Times* 22 Feb. 5/3 The latest incident is not in the same league as the apparently endless series of espionage scandals in and around Bonn in 1968 and 1969.

3*. *Basket-making.* (See quots.)

1903 T. Okey in R. M. Jacot *Useful Cane Work* I. p. ix, When a single continuous cane is used as a combined bottom stick and stake it is termed a 'League'. **1910** *Encycl. Brit.* III. 482/2 When the 'bottom-stick' and 'stake' are formed of one and the same continuous rod, it is termed a 'league'.

4. (sense 1 c) *league championship, club, cricket, football* (hence *league footballer*), *match, player, star, system*; **league table**, a list of the members of a league in ranking order; also *transf.*, a systematic comparison of performance in any field of competitive activity.

1901 *Dundee Advertiser* 4 Jan. 6 That [*sc.* Guiseley] Club winning the League championship. **1969** *Official Baseball Rules* 16 The League is a group of clubs whose teams play each other in a pre-arranged schedule under these rules for the league championship. **1972** G. Green *Great Moments in Sport: Soccer* iv. 58 The previous season Chelsea had won the League Championship of the First Division. **1938** C. E. Sutcliffe et al. *Story of Football League* 14 A meeting of the League clubs was held on 8th February, 1909. **1973** *News of the World Football Ann. 1973–74* 163 (*heading*) Oldest League Clubs. **1961** F. C. Avis *Sportsman's Gloss.* 125/2 *League Cricket*, that organized, outside the county championship, etc., in competitive league groups, e.g. the Central Lancashire League. **1910** T. Charnley *Let.* 13 Jan. in C. E. Sutcliffe et al. *Story of Football League* (1938) 15 Reports are continually being received that the many unfair and unscrupulous tactics indulged in by some of the players engaged in League football are allowed to pass unpunished by the referees. **1959** I. & P. Opie *Lore & Lang. Schoolch.* xvi. 350 The mid-century schoolchild's sporting enthusiasms are more taken up with league football, [etc.]. **1951** *Football Record* (Melbourne) 8 Sept. 18 Congratulations to Ron Clegg, who won the..award for the best League footballer of 1951. **1909** A. Bennett *Matador* (1912) 30 Knype had yet five League matches to play. **1973** *Irish Times* 2 Mar. 3/2 Cup ties are very different to league matches. **1886** H. Chadwick *Art of Pitching & Fielding* 132 The following are the best fielding averages of the Eastern League players. **1938** C. E. Sutcliffe et al. *Story of Football League* 12 The forces outside the League were..ready to take away League players without paying anything for them. **1967** *Australian* 26 Apr. 12 Injuries to many League stars. **1902** *Encycl. Brit.* XXVIII. 425/1 An elaboration of this competition is the 'League system' of the Association game. This..has not been popular with Rugby players. Still it is prevalent in many districts... In the League system a certain number of clubs form a league to play one another twice each season; two points are counted for a win, and one for a draw. The club which at the end of the season comes out with most points wins the competition. **1912** *Football Chart* (G. F. Stirling, Liverpool), Note position of Club each week in League Table and mark the ups or downs. **1930** *Daily Express* 6 Oct. 16 (*heading*) Saturday's League results and tables. **1959** *Times* 19 Mar. 12/2 He also recited with telling effect a 'league table' of unemployment percentages in western countries, ending with Great Britain as the lowest of all. **1967** Coulthard & Smith in Wills & Yearsley *Handbk. Managem. Technol.* 205 Large and expensive personnel departments, which maintain extensive records, card indexes, files, annual appraisal systems, charts, league tables, and so on. **1970** F. C. Avis *Soccer Dict.* (ed. 3) 57 *League table*, the statement of teams in relation to each other during the season, [etc.]. **1972** *Human World* May 3 In 1971 they were half way up the 'league table' of wages instead of near the top. **1972** *Times* 11 July 2/7 A league table of tar and nicotine in most brands of cigarettes seems certain to be produced. **1973** C. Bonington *Next Horizon* ix. 131 Already a healthy element of competition was springing up between the big league climbers of the Alpine countries... This was a little like a League Table, which we all examined with care as we decided what to do next.

League of Nations. An association of self-governing states, dominions, and colonies created by a covenant forming part I of the Peace Treaty of 1919 'in order to promote international co-operation and to achieve international peace and security'. *League of Nations Society* (*later Union*): a society formed to promote the principles of the League of Nations.

1917 H. N. Brailsford (*title*) A League of Nations. *Ibid.* ii. 37 The programme of the British 'League of Nations Society' is as follows. **1917** A. Huxley *Let.* 30 Sept. (1969) 133, I have spent the morning in correcting.. essays on the possibility..of a League of Nations. **1919** *League of Nations Jrnl.* Jan. 1 The Union..Resulting.. from the amalgamation of the League of Nations Society and of the League of Free Nations Association,..includes members of a society which has been working since May, 1915, for the establishment of a League of Nations, and of a new and vigorous association which was inaugurated in the summer of 1918. **1919** *Treaty of Peace* (H.M.S.O.) xii. art. 376 Disputes which may arise..shall be settled as provided by the League of Nations. **1922** *Encycl. Brit.* XXII. 647/2 Perhaps the most important event which happened in Switzerland in 1920 was the first meeting of the League of Nations in Geneva. *a* **1930** D. H. Lawrence *Phoenix II* (1968) 442 He was the scourge of God: not the scourge of the League of Nations, hired and paid in cash. **1934** H. G. Wells *Exper. Autobiogr.* II. ix. 694 The term 'League of Nations' is of English origin and it seems to have been first used by a small group of people meeting in the house of Mr. Walter Rea... (E. M. Forster in his life of Lowes Dickinson (1934) gives reasons for ascribing the term to that writer, who may have used it for the two possible 'leagues' he sketched in the first fortnight of the war.) These people founded a League of Nations Society, with Lord Shaw as president, early in 1915. **1957** *Encycl. Brit.* XIII. 832/2 The League of Nations was legally inaugurated on Jan. 10, 1920. *Ibid.*, Pres. Woodrow Wilson espoused the cause of a league of nations in May 1916. **1971** W. H. McNeill in A. Bullock *20th Cent.* 47/2 Most of these lands were designated League of Nations 'mandates'.

leaguer, *sb.*[2] (Later examples.)

1943 M. Ward *G. K. Chesterton* (1944) xxvi. 435 Many leaguers..felt..that the spirit of criticism of others was too fully developed. **1949** M. L. Darling *At Freedom's Door* I. ii. 52 In this year's election he stood as a Unionist, and like most Unionists was defeated by a Leaguer. **1970** *Cape Times* 28 Oct. 26/2 There are few American major leaguers earning less than $30 000 a year.

leaguer, *sb.*[3] **b.** (Later S. Afr. examples.)

1881 F. R. Statham *Blacks, Boers, & British* iv. 61 You want to see what can be done with South African wine?.. Visit a great airy shed not far from the Cape Town docks, ..the rough and ready wine has become—what? Look at it and see it as it is drawn from the huge casks—leaguers they call them here. **1959** *Cape Times* 14 Mar. 2/6 Two lorries, one carrying a 5-leaguer tank of wine (some 800 gallons) collided here yesterday. **1970** *Ibid.* 28 Oct. 20/3 (Advt.), A wine quota of 320 leaguers.

leaguite (lī·gəit). Also **leagueite.** [f. League *sb.*[2] + -ite[1].] = Leagueist.

1841 *Times* 4 Feb. 5/4 The leagueites polled 9 dead men. **1892** E. Dowson *Let.* 22 Nov. (1967) 253 It seeming to be very much confined to the actual Leaguites themselves.

leak, *sb.* Add: **1. c.** *Electr.* A path or component of relatively high resistance through which a small current flows.

1896 T. E. Herbert *Electricity in Application to Telegr.* xvii. 81 B is connected to earth as is the end of our 40 ohm leak. **1919** [see *grid leak* (*grid 5 b)]. **1940** *Amat. Radio Handbk.* (ed. 2) ii. 33/2 The grid will take up a potential such that the current from grid to filament equals the current through the leak in either the positive or negative half cycle. **1966** [see *grid leak* (*grid 5 b)].

d. An improper or deliberate disclosure of information (e.g. for political purposes).

1950 H. D. Lasswell *National Security* ii. 34 Americans are accustomed to 'government by leak'. **1957** *Economist* 28 Sept. 1004/2 The allegation of a 'leak' about last Thursday's increase in Bank rate has brought forth understandable indignation from those City dealers whose fingers were burned, and an equally understandable demand by the Labour Party for a full inquiry. **1960** L. Cooper *Accomplices* I. ii. 17 Confidential stuff about a security leak from one of our research stations. **1960** *News Chron.* 30 Apr. 4/2 No agenda, no communiqué, no inspired leaks. **1965** H. Kahn *On Escalation* iii. 56 In.. 1964, the United States and the Chinese engaged in a series of such semiformal leaks and announcements about the war in Vietnam. **1967** *Punch* 4 Oct. 509/3 Long among the most skilled practitioners of leak journalism. **1973** *Guardian* 10 Apr. 15/3 The EEC Commission spent an hour and a half..discussing leaks and how to plug them (or so it is reliably leaked to Miscellany).

2. b. *Electr.* Leakage of electric charge or current (see *leakage 2 b).

1863 R. S. Culley *Handbk. Pract. Telegr.* iv. 65 Suppose..a fault to occur connecting the wire to the earth, and offering a resistance equal to that of 20 miles of the line. This 'leak' will lessen the total resistance of the circuit..as if a wire 20 miles long..were fixed to the line at the fault. **1893** [see *leakance]. **1895** [in Dict., sense 2]. **1906** *Phil. Mag.* XII. 403 With very thin paper..no discharge could be observed, whilst in the case of aluminium leaf 0·0005 mm. in thickness a difference in the rate of leak was observed. **1939** *Post Office Electr. Engineers' Jrnl.* XXXII. 138 (*heading*) The localization of small leaks in the underground transmission line system at Cooling Radio Station.

c. *slang.* An act of urination. Freq. in phr. *to take a leak*, to urinate. Cf. Leak *v.* 2 c.

1934 H. Miller *Tropic of Cancer* 182, I stood there taking a leak. **1968** K. Weatherly *Roo Shooter* 111, I saw Sam get out of the Rover,.. I thought he'd got out for a leak. **1969** G. Greene *Trav. with my Aunt* II. vi. 282 All these hours of standing without taking a leak. **1972** F. Raphael *April, June & Nov.* 283 'The guest toilets at the Palace aren't really all that marvellous.' '..Thanks for the tip, I'll remember to take a leak before I go next time.'

3. leak detector, any device for detecting leaks of fluid; **leak-proof** *a.*, not subject to leaks.

1921 *Chambers's Jrnl.* July 454/2 Each bag..is inflated with air and examined all over its surface with leak-detectors. **1968** *Non-Destructive Testing* I. 215/1 Shell have developed a portable hydrocarbon leak detector to facilitate the overhaul of [gas] mains. **1926** *Kitchen Kook* (Amer. Gas Mach. Co. Inc.) 3 The fuel..is contained in an electrically welded, leakproof, steel tank. **1929** *Daily Express* 8 Jan. 8/5 Waste heat in leak-proof pipes to towns near the coalfields. **1960** *Farmer & Stockbreeder* 15 Mar. 44/2 (*caption*) It's leakproof. **1971** *Engineering* Apr. 92/2 (Advt.), Instant..safe..leakproof joints... A pipeline which is flexible while remaining absolutely leak-proof.

leak, *v.* Add: **1.** (Examples relating to electric charge: cf. *leak *sb.* 2 b, *leakage 2 b.)

1863 R. S. Culley *Handbk. Pract. Telegr.* vii. 106 The dampness of the insulators enables part of the electricity to leak or escape from one wire to another, and to the earth. **1917** G. D. Shepardson *Telephone Apparatus* xiv. 224 Little talking current 'leaked' through the signaling equipment. **1959** *Which?* Winter 37/1 If there is a fault in an electrical appliance and current leaks to the exposed parts.

b. (Earlier U.S. examples.)

1832 Webster, *To leak out*,..to escape privately from confinement or secresy; as a fact or report. **1834** S. Smith *Sel. Lett. J. Downing* 58 If it should leak out that I was going.

2. c. Delete *Obs.* and add later examples.

1957 J. Kerouac *On Road* (1958) 90 The prowl car came by and the cop got out to leak. **1971** D. E. Westlake *I gave at the Office* (1972) 173, I kept thinking he'd come back from the john—how long can one man leak?

4. c. To allow the disclosure of (secret or confidential information). (Cf. sense 1 b and *leak *sb.* 1 d.) Also *intr.* Hence **leaked** *ppl. a.*

1859 G. W. Matsell *Vocabulum* 50 *Leak*, to impart a secret. **1916** W. Owen *Let.* Aug. (1967) 402 Here I am beginning to 'Leak information', (when I have to read daily a solemn W.O. Letter, saying that no talk of the War is ever to be indulged in, even in private letters.) **1954** *Encounter* June 11/1 In practice [the dial number] was a secret in name only, since supervisors were instructed to 'leak' the number 'confidentially' to various employees. **1958** *Punch* 3 Nov. 10/3 It seemed pretty clear from what the F.O. had leaked to us that Bonaparte had crossed the Niemen. **1958** *Ann. Reg. 1957* 195 In a miscalculated effort to prepare the public and Congress for the new doctrine, it had been deliberately 'leaked' well beforehand. **1959** *John o' London's* 26 Nov. 265/3 A.. U.S. Air Force sergeant..promptly scares off the circling sharks by leaking information about her non-existent husband. **1962** *Listener* 25 Oct. 647/2 The Council Fathers are supposed to maintain complete discretion, though almost all of them 'leak' to the press. **1971** *Daily Tel.* 14 July 3/4 It was not sufficient for the tribunal merely to establish by whose hand information..was improperly leaked. **1972** *Times* 30 Sept. 3/1 Legislation covering 'leaked' information is proposed by the Franks Committee on Official Secrets. **1973** [see *leak *sb.* 1 d].

leakage. Add: **2.** (Later examples.) Add to def.: improper or deliberate disclosure of information from an office, etc.; unexplained continuous disappearance of something.

1859 'T. Titcomb' *Titcomb's Lett.* (ed. 12) ii. 185 It is entirely rational and right that your wife should understand the basis of all your requirements of her; and when she does this, the chances are that she will not only be economical herself, but will point out leakages in your prosperity for which you are responsible. **1880** E. W. Hamilton *Diary* 30 Nov. (1972) I. 83 There have of late curiously been some leakages. My own belief is that men like Chamberlain and Forster..are most unguarded in their language outside. **1890** *Daily News* 17 Oct. 7/2 It was discovered that there was a 'leakage' in the stamp transfer forms. **1894** *Westm. Gaz.* 19 Sept. 3/3 That leakage from the faith which is taking place among the poor Catholics. **1898** *Daily News* 17 Feb. 2/7 The prizes.. for three best essays on the cause of the leakage in the membership of the Methodist Church,..have been awarded. **1898** *Ibid.* 20 Oct. 6/5 Some extracts..have found their way into the pages of the 'New York Critic'... I am unable..to account for this leakage. **1900** *Edin. Rev.* July 81 The tide of emigration has been stayed..the leakage is diminishing. **1904** A. B. F. Young *Compl. Motorist* 347/2 Leakage of small moneys, during travel. **1908** H. G. Wells *War in Air* i. 25 Mr. K. Butteridge.. intended to keep his secret safe from any further risk of leakage. **1945** E. Waugh *Brideshead Revisited* 10 Our new commanding officer was making an unusual display of 'security'... 'If I find any of these female camp followers waiting for us the other end, I'll know there's been a leakage.' **1972** *Times* 30 Sept. 1/1 The [Franks] committee proposes an Official Information Act to cover leakage of information which would seriously injure the national interest. **1973** A. Christie *Postern of Fate* III. xvii. 248 There were leakages—as always there are leakages in time of war.

b. *Electr.* A gradual escape of charge or current, esp. as a result of imperfect insulation. Also, in *Magnetism*, an escape of flux from a magnetic circuit or device; flux which does not pass through the secondary of a transformer or induction coil, or through the armature of a motor or generator.

1863 R. S. Culley *Handbk. Pract. Telegr.* iv. 59 On a long line the leakage from wire to wire through damp air cannot be altogether without effect. **1902** *Encycl. Brit.* XXVII. 586/2 Since no substance is impermeable to the passage of magnetic flux, the only form of magnetic circuit free from leakage is one uniformly wound..over its whole length. **1922** Glazebrook *Dict. Appl. Physics* II. 190/1 Allowance must be made for flux which leaks across the intervening space between the poles and does not actually enter the armature, and a magnetic circuit has to be designed so as to keep this leakage as small as possible. **1962** *Newnes Conc. Encycl. Electr. Engin.* 229/1 Danger from leakage may be prevented on metal pole lines by a continuous earth wire connected to the poles. **1962** D. F. Shaw *Introd. Electronics* i. 7 If the coil is long enough for the solenoid formula to apply there will be a considerable leakage of flux between the turns.

5. *attrib.*, as (sense *2 b) *leakage current, flux, path*; also **leakage conductance** *Electr.* = *leakance; **leakage detector,** (*a*) = *leak detector* (s.v. *leak *sb.* 3); (*b*) = **leakage indicator*; **leakage indicator** *Electr. Engin.*, any device for indicating or measuring leakage currents flowing to earth.

1887 *Electrician* 3 June 80/2 The attenuation factor is now $\epsilon^{-R/2Lv}.\epsilon^{-K/2Sv}$, if *K* be the leakage conductance, and *S* the permittance per unit length. **1880** *Jrnl. Soc. Telegr. Engin.* IX. 456 The additional term $(v/i)dx$ is the leakage current of *dx*, viz., the potential of *dx* divided by its insulation resistance. **1962** J. Bell in G. A. T. Burdett *Automatic Control Handbk.* iv. 3 The insulation to earth must be maintained in the megohms region otherwise the inaccuracy due to leakage currents will be significant. **1880** *English Mechanic* Oct. 107/3 Cowan's meters, and Mr. Young's..leakage-detector are prominent exhibits [at the Exhibition of Gas and Electrical Apparatus]. **1901** *Catal. Mech. Engin. Collection Sci. Div. V. & A. Mus.* (ed. 3) I. 220 Leakage detector... This is an instrument for rendering audible the slight sound made by water flowing in a pipe, so that if the noise continues, after certain valves are closed, the existence of a leak is indicated. **1923** Meares & Neale *Electr. Engin. Pract.* (ed. 4) I. xv. 499 In the Howard leakage detector a current transformer is connected in the earthing wire of, say, a switchboard frame, and the secondary of the current transformer is connected to a tripping relay. **1971** *Instruments & Exper. Techniques* XIV. 830 The leakage detector operates in stable fashion when the pickup is situated in a high vacuum. **1896** F. Bedell *Princ. Transformer* xv. 302 The leakage flux varies inversely as the reluctance of the leakage path. **1962** Corson & Lorrain *Introd. Electromagn. Fields* vii. 291 Let us consider a toroid of magnetic material with a localized winding... The leakage flux, that is, the flux of **B** which leaves the core, produces poles on the surface..and these poles contribute an intensity **H** within the iron. **1920** *Whittaker's Electr. Engineer's Pocket-Bk.* (ed. 4) 276 The function of a leakage indicator is to provide information as to the insulation resistance of the whole of the electrical system ..to which it is connected. **1958** J. L. Watts *Electr. Maintenance & Repairs* iii. 43 Where the supply is obtained from a three-wire d.c. system with earthed mid-point, a suitable differential ammeter connected across shunts in the positive and negative mains will serve as a leakage indicator. **1896** Leakage path [see *leakage flux* above]. **1909** *Installation News* III. 64/1 The leakage path to the conduit or earth is now very greatly reduced. **1962** *Newnes Conc. Encycl. Electr. Engin.* 437/1 It is usual to assume that those parts of a leakage path lying in ferromagnetic material will require a negligible proportion of the coil m.m.f.

leakance (lī·kăns). *Electr.* [f. Leak *v.* + -ance as a shortening of *leakage conductance*.] Conductance attributable to leakage or imperfect insulation.

1893 O. Heaviside *Electromagn. Theory* I. 453 A process..of representing a large number of separate leaks by

uniform leakance. *Ibid.*, Distribute the inductive leakance uniformly, in imagination, of course. **1928** Bradfield & John *Telephone & Power Transmission* ii. 18 Since the insulation of the circuit..can never be perfect, there must be a certain leakance from wire to wire,..stated in 'mhos', or sometimes in 'micro-mhos' per mile of loop. *Ibid.* iv. 55 The leakance of aerial circuits varies irregularly between wide limits, owing to weather and other conditions. **1962** *Newnes Conc. Encycl. Electr. Engin.* 845/1 Leakance and capacitance effects are large [in telecommunication lines], so that the attenuation is much higher than could be tolerated in power transmission.

leaking, *vbl. sb.* (Later examples.)

1973 *Time* 16 Apr. 53/2 The leaking and publication of classified information has always been a murky area in criminal law. **1973** *Listener* 15 Nov. 658/1 There was.. some leaking to the Arabs from perhaps two EEC capitals, which had disclosed to them the line the Dutch had been taking in confidential Community discussions.

leaky, *a.* Add: **d.** *Electr.* Retaining electric charge only with gradual loss; connected to or having a high resistance that acts as a 'leak'; *leaky-grid detection*, detection in which the signal is applied to the grid of a valve through a series capacitor and a resistor (the latter being connected as a grid leak or in parallel with the capacitor).

1904 A. Russell *Treat. Theory Alternating Currents* I. xvii. 384 (*heading*) Inductive coil and leaky condenser. **1922** J. Scott-Taggart *Wireless Vacuum Tubes* (ed. 4) iv. 89 Leaky grid condenser rectification. *Ibid.* vii. 132 If we employ a leaky grid condenser we can obtain a suitable negative grid potential without..a battery. *Ibid.* viii. 167 A leaky grid condenser may be connected in the grid circuit. **1934** *Jrnl. Inst. Electr. Engin.* LXXV. 298/2 Leaky-grid detection is used in this receiver. **1962** D. F. Shaw *Introd. Electronics* ii. 35 A leaky capacitor..in which the leakage resistance is represented by a shunt. **1968** *Radio Communication Handbk.* (ed. 4) ii. 17/1 Any d.c. voltage developed across the grid leak by the rectification of a modulated or an unmodulated signal will thus constitute a negative bias for the grid and the anode current in the triode will fall... An excessively strong signal will tend to bias the valve beyond the cut-off point, and therefore a leaky-grid detector ceases to function satisfactorily when the input voltage is too great. **1969** R. G. Middleton *Transistor Television Servicing Guide* ix. 105/1 A leaky transistor, such as 22..increases in temperature.

e. Of persons: lachrymose. Also as *sb.*

1905 H. A. Vachell *Hill* vii. 151 'I ain't the leaky sort,' she added fiercely, still gasping. **1959** I. & P. Opie *Lore & Lang. Schoolch.* x. 187 Croydon boys have twenty names for a cry-baby:..leaky, [etc.].

2. *Genetics.* Of a mutant: producing the protein specified by the mutated gene in a form with reduced activity compared with that produced by the wild type. Of a protein so produced: having reduced activity. Cf. *hypomorphic* adj. (s.v. *hypo- II).

1955 *Proc. Nat. Acad. Sci.* XLI. 347 Under given conditions, however, the coefficient can be used as a comparative index of degree of phenotypic effect, a 'leaky' mutant having a high coefficient. **1959** *Ibid.* XLV. 204 This strain is a leaky derivative of strain 21863 [of *Neurospora crassa*]. **1961** *Nature* 30 Dec. 1227/2 Mutants produced by acridines are seldom 'leaky'; they are almost always completely lacking in the function of the gene. **1966** E. A. Carlson *Gene* xiii. 112 The microbial geneticist today uses the term 'leaky mutant' for hypomorph. **1968** R. C. King *Dict. Genetics* 141 *Leaky protein*, a mutant protein that has a subnormal degree of biological activity.

lean, *a.* and *sb.*[2] Add: **A.** *adj.* **4.** Delete 'Now somewhat *rare*' and substitute: Now *rare* except in various techn. senses, as: **a.** Of mortar or concrete: containing little of the binding material. **b.** Of clay: not very plastic. **c.** Of coal: of poor quality; *spec.* deficient in volatile material. **d.** Of ore: of low grade; containing little valuable mineral. **e.** Of fuel gas: of low calorific value. **f.** Of the mixture in an internal-combustion engine: containing a low proportion of fuel. **g.** Of an emulsion, painted surface, etc.: containing little oil.

a. 1726 J. Leoni tr. *Alberti's Archit.* I. 49/1 For small Stones, a thick lean Mortar is best; to a dry exhausted Stone, we should use a fat sort; tho' the Ancients were of the opinion that in all parts of the Walls the fattish sort is more tenacious than the lean. **1936** *Times Lit. Suppl.* 18 Apr. 325/4 Very rich concrete, one part cement with two-and-a-half sand and gravel (concrete so rich is seldom used), is hardly affected at all by sea water... But 'lean' concrete, one part cement to about ten of sand, gravel, or even pozzolani, disintegrates in a year. **1965** *Economist* 13 Nov. 745/3 Outside mining subsidence areas, one of the lower layers [in construction of 'black-top' roads] is often 'lean' (with little cement) concrete.

b. 1754 *New & Compl. Dict. Arts & Sci.* III. 2128/1 Mortar for furnaces, &c. is made with red clay wrought in water in which horse-dung and chimney-soot has been steeped..; this clay ought not to be too fat, lest it should be subject to crack; nor too lean and sandy, lest it should not bind enough. **1885** *Encycl. Brit.* XIX. 600/2 'Lean' clays—those that have a large proportion of free silica—shrink but little, and keep their form unaltered under the heat of the kiln. **1964** H. Hodges *Artifacts* i. 20 Such clays are sticky or greasy..and shrink seriously on drying... Equally a clay may be too aplastic to work, the material being crumbly, also known as short, mealy, lean, or open.

c. 1883 W. S. Gresley *Gloss. Terms Coal Mining* 154 *Lean*, thin, poor; of inferior quality. **1960** *Gloss. Coal Terms* (*B.S.I.*) 9 *Lean coal*, term used in several European countries for coal with a low volatile matter.

d. 1901 *Daily Colonist* (Victoria, B.C.) 20 Oct. 10/3 There are..rumors..that lean ore has been struck in the lower workings. **1965** G. J. Williams *Econ. Geol. N.Z.* v. 57/1 Yields [of gold] ranging up to 0·75 oz. were reported but most of the quartz is very lean.

e. 1924 *Jrnl. Inst. Petroleum Technologists* X. 804 In handling lean [natural] gases of this type large throughputs are necessary if the operation of extraction is to be profitable. **1960** *Economist* 15 Oct. 271/3 A national high-pressure grid supplying industry direct and local systems with lean gas for enrichment to town gas.

f. 1932 F. J. Camm *Bk. Motors* xxxii. 253 Misfiring may be due to incorrect petrol supply, too 'lean' or weak a mixture or an occasional short circuit. **1949** Frazee & Bedell *Automotive Fundamentals* iii. 174 Too low a float level results in a slightly leaner mixture as too little fuel will leave the jets. **1973** *Physics Bull.* Apr. 241/2 The cvcc engine.. was designed with pollution control in mind. It operates on extremely lean air-fuel mixture from the carburettor which is varied according to the operating conditions.

g. 1934 H. Hiler *Notes Technique Painting* iii. 171 The emulsions made from yolk of egg, some gum or resin, linseed oil and sometimes a little wax, are intimate mechanical but not chemical compounds... When egg-yolk is used..a somewhat yellowish tinted 'fat emulsion' results. If gum arabic is used, the result is a whitish 'lean emulsion'. **1961** M. Levy *Studio Dict. Art Terms* 66 *Lean Surface*, the matt surface of a layer of pigment containing a minimum of oil. It is essential that an underpainting which is to be glazed should possess a lean surface. **1967** J. N. Barron *Lang. of Painting* 75 For obtaining better permanence in paintings..the overlying or upper layers of paint are to be increasingly more 'fatty' and contain more oil than the layers they cover, or the 'leaner' ones.

7. *lean-faced* (later examples), *-ribbed* (earlier and later examples).

1892 W. B. Yeats *Countess Kathleen* ii. 34 A crowd of ugly lean-faced rogues. **1953** R. S. Thomas *Minister* 13 By a lean-faced people in black clothes. **1602** Marston *Antonio's Revenge* IV. i. sig. G1[v], Whilst pale cheekt wisdome, and leane ribd arte Are kept in distance at the halberts point. **1925** E., O., & S. Sitwell *Poor Young People* 7 Neptune beat his lean-ribbed ass The braying sea uphill.

lean, *v.*[1] Add: **6. d.** *to lean on* (someone): to put pressure on (a person) in order to extract something from him or force him to do something against his will (see also quot. 1960).

1960 Wentworth & Flexner *Dict. Amer. Slang* 315/1 *Lean against, lean on*... 2. To beat up someone; to threaten to beat up someone or a member of one's family in order to get information, to persuade someone to suppress information, or to extort money; to act or be tough with someone; to coerce. **1965** J. Porter *Dover Three* xv. 168 If you start leaning on her and you don't make the poison-pen business stick good and proper, she'll crucify you! **1967** K. Giles *Death in Diamonds* vii. 126 I'm going to lean on him until I get to know that contact. **1967** J. Morgan *Involved* 51 You were too much tonight..the way you leaned on Tuttles, that was really something. **1972** J. Brown *Chancer* vii. 101 Sandy Crump had been naughty, not telling me about Shag... I'd have to lean on him harder. **1975** *N.Y. Times* 3 Feb. 6/2 'An Attorney General would resign too if he thought he was being leaned on by the Prime Minister or senior ministers on a pending prosecution,' a former Attorney General said.

e. *to lean over backwards*: see *backwards *adv.* A.

leangle, var. Leeangle.

1867 [see Leeangle]. **1945** Baker *Austral. Lang.* xiii. 224 Those aboriginal words we have incorporated in our language... For example:..wurley, leangle, mulga, [etc.]. **1966** W. S. Ramson *Austral. Eng.* vi. 132 Leangle, 'a club', and wirri, 'a throwing stick', are Victorian, both coming from the Gippsland area.

lean-over (lī·nōᵘvəɹ). [f. Lean *v.*[1] + Over *adv.*] An inclination down or forward; *concr.*, something over which one can lean.

a **1885** G. M. Hopkins *Poems* (1918) 79 So long to this sweet spot, this leafy lean-over. **1936** E. Sitwell *Victoria of Eng.* xiii. 163 For others, again, there is the twopenny lean-over. **1969** E. H. Pinto *Treen* 334 Early pipes had a very forward tilt, or 'lean over', on the bowl.

leap, *sb.*[1] Add: **1. c.** *a leap in the dark* (later examples); *by leaps and bounds* (earlier and later examples); *leap forward*: an advance of a marked or notable character.

1720 Pope tr. *Homer's Iliad* V. xxi. 1587 High o'er the surging Tide, by Leaps and Bounds, He wades, and mounts; the parted Wave resounds. **1891** Gladstone in *Star* 11 Dec. 2/5, I shall proceed by skips and jumps; or, as it is the fashion to say now, by leaps and bounds. **1915** Mrs. Belloc Lowndes *Let.* 10 Mar. (1971) 57 Everything is going up, in price, by leaps and bounds. **1940** By leaps and bounds [see *double *v.* 4 c]. **1947** By leaps and bounds [see *Afrikanerizing *vbl. sb.* and *ppl. a.*]. **1954** T. S. Eliot *Confid. Clerk* II. 63, I make decisions on the spur of the moment, But you'd never take a leap in the dark. **1961** *Ann. Reg. 1960* 170 The convention..would amount to a great 'leap forward' towards a virtual European federation. **1966** *Performing Right* Oct. 4 The leap forward of nearly 20 per cent in this revenue is a triumph for the skill and hard work of our administrative staff. **1973** *Times* 21 Mar. (China Trade Suppl.) p. xi/2 The backyard steel furnaces that sprang up during the Great Leap Forward (1958–60).

2. b. (Later example.)

1780 A. Young *Tour in Ireland* I. 126 All the fisheries are his to the leap at Colraine.

8. leap second [after *leap day*], a second which on a particular occasion is inserted into (or omitted from) a scale of reckoning time in order to bring it into correspondence with another scale.

1971 *Nature* 11 June 345/1 An adjustment will be made to all GMT time signal emissions on January 1, 1972, so that a GMT time of 0 h 0 m 0 s will correspond exactly to an IAT time of 0 h 0 m 10 s; thereafter 'leap seconds' will be inserted or omitted as necessary at the end of a particular GMT month... From then on, GMT will always be exactly 10 s slow compared with IAT. A so-called positive leap second will begin at 23 h 59 m 60 s on the last day of the month selected and end at 0 h 0 m 0 s on the first day of the next month; by contrast, if the leap second is negative, 23 h 59 m 58 s will be followed one second later by 0 h 0 m 0 s. **1972** *Daily Tel.* 28 Dec. 10/2 Shortly before 11 p.m. on Sunday (4 a.m. New Year's Day in Britain) technicians at the United States National Bureau of Standards in Boulder, Colorado, will add one leap second to America's atomic clock to correct it to match the Earth's rotation. **1973** *Nature* 21/28 Dec. 444/1 Currently, the Earth loses about 3 ms a day on Atomic Time, and the leap seconds are added where necessary..to keep UT and Atomic Time in close correspondence.

leap, *v.* Add: **6. c.** *Mus.* To pass from one note to another by an interval greater than a degree of the scale. Also *trans.* (Cf. Leap *sb.*[1] 7.)

1879 G. A. Macfarren *Counterpoint* iv. 10 After several consecutive 2nds, in melody, it is bad to leap, in the same direction, upward or downward to an accented note. **1889** E. Prout *Harmony* (ed. 3) xiii. 143 The third of the chord exceptionally leaping, instead of moving as usual by step. **1927** C. H. Kitson *Counterpoint for Beginners* 17 A part may not leap any interval greater than an octave.

leapable (lī·păb'l), *a.* [f. Leap *v.* + -able.] That can be leaped.

1925 A. S. Alexander *Tramps across Watersheds* 128 Some parts of the precipitous sides approach within leapable distance.

leap-frog, *sb.* Add: **3.** *Mil.* (See quot.)

1918 E. S. Farrow *Dict. Mil. Terms* 340 *Leapfrog*, a method of maintaining constant communication with a moving command by using two or more instruments with a single unit, keeping one in operation while another is moving past it to a position in front.

4. *transf.* Competing for higher wages by 'leap-frogging'. Cf. *leap-frog *v.* c.

1958 *Spectator* 31 Jan. 123/2 Nobody has much sympathy with the wage demands of busworkers, town or country; if you use dubious methods of wage bargaining, like the leap-frog, you must expect few tears to be shed if a leap lands you into a ditch. **1961** *Daily Tel.* 14 Oct. 16/6 'Leap-frog' in pay may be checked. **1974** *Times* 25 May 13/1 The wage 'leap frog'..is the cause of a large part of our present tensions.

5. *attrib.* (in various *fig.* senses).

1904 *Daily Chron.* 13 July 6/5 Mr. Morley exposed what may be called the 'leap-frog' logic of the Protectionists. **1917** *Q. Rev.* July 190 The 'leap-frog' game of fleeting Ministries. **1952** L. Ross *Picture* i. 41 The 'leapfrog' director..whose job it would be to arrange things so that Huston would not have to wait between scenes. **1962** *Gloss. Terms Automatic Data Processing* (*B.S.I.*) 50 *Leap-frog test*, a test program stored in locations which are progressively changed by the program itself in order to test the store. **1972** *Times* 19 Dec. 14/1 An attempt to invoke the 'leap-frog' procedure under section 12 of the Administration of Justice Act, 1969, and go direct to the House of Lords from a decision of a judge of the High Court failed.

leap-frog, *v.* Add: **b.** *Mil.* Of detachments or units, esp. in an attack: to go in advance of each other by turns (see also quot. 1942).

1920 *National Rev.* Nov. 355 Behind them marched other divisions who, on the first momentum of the offensive slackening, were to 'leap-frog' over their comrades and continue the drive. **1922** C. E. Montague *Disenchantment* ix. 133 Leap-frogging waves of assault. **1927** *Daily Tel.* 30 Aug. 8/7 Two pairs of mobile picket groups, moving by long bounds and one pair 'leapfrogging' the other. **1942** *R.A.F. Jrnl.* 16 May 32 The Air Force followed on their heels..leap-frogging over huge stretches of desert... As the armies retreated, they leap-frogged back again. **1966** A. J. Barker *Eritrea* iv. 85 Due to the lack of transport it was possible only to lift two companies forward at any one time, the rest had to march. The two rear companies were picked up in turn and leap-frogged to the head of the main column.

c. *transf.* In wage negotiations: to pursue a policy of demanding higher wages every time a group or groups of comparable wage-earners have succeeded in pulling level or ahead. Chiefly as **leap-frogging** *vbl. sb.* and *ppl. a.*

1955 *Times* 6 June 7/2 And if the British Transport Commission and the Government were to give in now it could never again be fought with certainty, no matter how long the leap-frogging between the two unions went on. **1958** *Times* 30 Jan. 4/3 Sir Robert Grimston..said that there was much concern among the fare-paying public at the continual leap frogging in wages between London and the provinces. **1958** *Times Rev. Industry* June (London & Cambridge Bull.) p. x, Engineers..could not be well granted less than was granted to workers in prosperous industries. This seems to produce a threat of leap-frogging wages. **1959** *Listener* 2 July 6/1 The long-term contract relieves the strain of annual efforts to surpass the previous year's gains, or to leap-frog the advances won in another industry. **1967** *Times* 18 Jan. 16 There is leap-frogging in newspaper offices, such that when one department negotiates a rise the others follow regardless of justification. **1970** *Daily Tel.* 15 June 2/5 For the first time collective negotiations on new claims by all unions will replace individual 'leapfrogging' demands. **1973** *Times* 21 Dec. 1/7 To breach Phase Three..would lead to leap-frogging claims which would erode the miners' position in the league table.

d. Other *fig.* uses.

1935 J. C. Squire *Reflections & Memories* 6 It is a time before the jolly vulgarity of Earl's Court had leap-frogged westward to the White City, and then to Wembley. **1949** I. Deutscher *Stalin* xiii. 498 Only in 1943 did the newly built factories and those that had been 'leap-frogged' from the west to the Urals and beyond begin to pour out great quantities of tanks, planes, and guns. **1961** *Times* 28 Mar. 4/5 They [*sc.* Oxford] were accompanied by Imperial College, with whom they paddled in the familiar leap-frogging pattern to Chiswick Eyot. **1962** *Punch* 5 Sept. 330/2 The leap-frogged zones beyond [the Green Belts]. **1964** T. W. McRae *Impact of Computers on Accounting* vi. 175 In fact, they [*sc.* auditors] 'leapfrog' over the entire EDP system. **1971** P. Gresswell *Environment* 122 Development leap-frogs green belts. **1971** J. Wainwright *Last Buccaneer* III. 313 When a man leap-frogs me in the promotion stakes I'm human enough to feel narked. **1972** *Times* 23 Feb. 27/6 So soon as a case at first instance arose involving the ratio decidendi of *Rookes v Barnard* the parties concerned might use the 'leap-frogging' procedure now available. **1973** *Listener* 17 May 653/1 Haldeman..was put in charge of the advance men, leap-frogging ahead of the candidate and arranging for crowds.

learn, *v.* Add: **4.** (Later examples.) Now regarded as a vulgarism.

1831 J. J. Strang *Diary* 31 Dec. in M. M. Quaife *Kingdom of St. James* (1930) 198, I have succeeded in regulating them and learning them what to do without punishing a single scholar. **1914** *Sat. Even. Post* 4 Apr. 10/3, I learned him that, yuh see. *a* **1935** T. E. Lawrence *Mint* (1955) II. iii. 107 We hear rumours that he..wants to make the Depot all drill, and will not permit other officers to learn the men, or men the officers. **1935** Wodehouse *Luck of Bodkins* xv. 181 The English public school system..isn't at all what an educational system should be... If you ask me, they don't learn the little perishers nothing. **1966** F. Shaw et al. (*title*) Lern yerself Scouse. **1974** *Times* 16 Dec. 12/8 We asked whether he had learned the instrument at school... 'No. He learned it himself and now he's learning me.'

¶ e. Phr. *I'll learn you*: used as a warning of impending punishment. *Non-standard.*

1822 J. Galt *Sir A. Wylie* III. xxxiii. 279 I'll learn you to fill yoursel fu'. **1873** C. D. Warner in 'Mark Twain' & Warner *Gilded Age* xxix. 266 The conductor..reached the bell rope, 'Damn you, I'll learn you,' stepped to the door. **1974** P. Wright *Lang. Brit. Industry* iv. 41 The common *I'll learn you*..when used ironically, has the unstandard meaning of 'I'll teach you never to do that again'.

learnability (lǝ̄:ɹnabi·lĭti). [f. Learnable *a.* + -ity.] The quality or fact of being learnable.

1959 *Brno Studies in English* I. 16 The easiness or the difficulty with which it affects the person trying to acquire it (at the risk of coining another barbarous neologism one might term it 'learnability'). **1966** *Philos. Rev.* LXXV. 435 There are others of great importance: brevity, learnability, etc.

learned, *ppl. a.* Add: **2.** *learned society* (later examples).

1798 *Phil. Mag.* June 95 (*title*) Intelligence. Learned Societies. **1863** Hawthorne *Our Old Home* iv. 136 This bewildered enthusiast had recognized a depth in the man whom she decried, which scholars, critics, and learned societies, devoted to the elucidation of his unrivaled scenes, had never imagined to exist there. **1897** W. James *Will to Believe* 306 All our learned societies have begun in some such modest way. **1958** *Observer* 13 July 3/7 The battle of the learned societies with the Inland Revenue. **1973** *LSA Bull.* Mar. 28 He defined a learned society as one which publishes a journal and holds an annual meeting; a professional society as one which is involved in matters of concern to its members.

¶ e. (Earlier and later examples.)

1784 A. Seward *Let.* 29 Oct. in H. Pearson *Swan of Lichfield* (1936) 71 That amusing part of this conversation, which alluded to the learned Pig, and his demi-rational exhibitions, I shall transmit to you hereafter. **1785** W. Cowper *Let.* 22 Apr. in *Corr.* (1904) II. 314, I have a competitor for fame..in the Learned Pig. **1919** Conrad *Let.* 25 Jan. in G. Jean-Aubry *J. Conrad: Life & Lett.* (1927) II. 216 If the Alliances had been differently combined the Western Powers would have delivered Poland to the German learned pig.

3. d. Of publications: devoted to (esp. some branch of) scholarship.

1883 E. B. Bax tr. *Kant's Prolegomena & Metaphysical Found. Nat. Sci.* 128 There is a good deal to be done before a learned journal..can maintain its otherwise well-merited reputation, in the field of metaphysics as elsewhere. **1942** *Amer. Speech* XVII. 3 Since he [*sc.* the writer of detective stories] introduces characters from all walks of life, and since he usually avoids the more formal style, he is a better informant than..the writers for the learned journals. **1951** Auden *Nones* (1952) 61 Lone scholars, sniping from the walls Of learned periodicals, Our fact defend. **1954** E. E. Evans-Pritchard *Inst. Primitive Soc.* p. v, Monographs about primitive peoples and innumerable papers devoted to them in learned journals. **1961** A. Wilson *Old Men at Zoo* i. 50 Subscriptions to learned periodicals. **1969** M. Pugh *Last Place Left* xxii. 167 So you've just met him. And the rest you know..from the learned journals.

learner. Add: **3.** One who is learning to be competent but who does not yet have formal authorization as a driver of a motor vehicle, cycle, etc. Also *attrib.*, as **learner-driver.** (The abbrev. L is shown on the *learner plates* of the vehicle.)

1930 'A. Armstrong' *Taxi* viii. 103 Conversational freedom between..taximen and private 'learner drivers'. **1934** R. F. Broad et al. *Motor Driving made Easy* (ed. 5) ix. 140 A provisional licence will be issued to enable learners to receive instruction qualifying them for the official test. **1935** *Daily Tel.* 7 Mar. 9/4 New drivers.. must start with a provisional or learner's licence. **1938** E. Waugh *Scoop* III. ii. 276 Bonnet and back bore battered learner plates. **1961** [see *book *v.* 2 c]. **1970** D. Marlowe *Echoes of Celandine* i. 15 A learner-car circling the block. **1973** *Times* 28 June 31/1 The learner driver holding up the traffic as he or she falters down the High Street is still part of the British motoring scene. *Ibid.* 21 Sept. 23/5 Although with its 1,000 cars and its 160 branches it is easy to get the impression that there are British School of Motoring learners everywhere..it still only has something like 2½ per cent of the learner-driver population training with it.

4. *Austral.* (See quot.)

1965 J. S. Gunn *Terminol. Shearing Industry* I. 35 A learner is not a shedhand or barrower, but a budding shearer who has not yet shorn 5,000 sheep (10,000 in Queensland).

learning, *vbl. sb.* Add: **1. a.** spec. *Psychol.* A process which leads to the modification of behaviour or the acquisition of new abilities or responses, and which is additional to natural development by growth or maturation. Freq. opp. *insight*.

1897 Bryan & Harter in *Psychol. Rev.* IV. 29 While there are many exceptional cases of quickness and slowness in learning, it requires from two to two and a half years to become an expert operator. **1901** E. L. Thorndike in *Ibid.* VIII. 442 With the monkeys, however, the association is both more rapid and more permanent, and the approach to suddenness and definiteness in their learning simulates that of human beings. **1901** —— *Human Nature Club* iii. 38 This method of learning may be called the method of trial and error..or..the animal method of learning. **1922** R. S. Woodworth *Psychol.* xiii. 311 To compare human and animal learning..cannot but throw light on the whole problem of the process of learning. **1924** R. M. Ogden tr. *Koffka's Growth of Mind* ii. 41 Certain stages of development are attained only after learning has been added to growth and maturation. **1940** W. Köhler *Dynamics in Psychol.* (1942) iii. 114 If this is the case, retroactive inhibition..must also be a disturbance of the product of learning. **1948** E. R. Hilgard *Theories of Learning* xii. 353 It can be stated with reasonable confidence that there are changes in the nervous system accompanying learning. **1968** Gelernter & Rochester in Evans & Robertson *Cybernetics* 70 This is the learning involved when the machine uses results on one problem to improve its guesses about similar problems. **1970** M. H. Marx *Learning: Theories* p. v, It is no longer possible for one psychologist to be fully expert in all the areas of so broad and diversified a field as learning.

4. (sense *1 a) *learning process, programme, score, situation*; **learning curve,** a graph showing progress in learning; **learning machine,** a machine of the electronic computer type that can 'learn' by recording the results of attempts to solve a problem and giving preference to those which are successful; **learning resources** (also *attrib.*), collective materials for learning, e.g. microfilms, audio-visual aids, made accessible in a library, school, etc.; also *learning-resource* attrib. phr.; **learning theory,** theory attempting to account for the process of learning.

1922 R. S. Woodworth *Psychol.* xiii. 307 Learning curve for the rat in the maze. **1924** R. M. Ogden tr. *Koffka's Growth of Mind* iv. 168 All these facts..would naturally operate to shorten the learning-curve. **1967** M. Argyle *Psychol. Interpersonal Behaviour* x. 183 In fact some manual operatives also learn by doing, and learning curves can be plotted which show their rate of progress. **1968** Johannsen & Robertson *Managem. Gloss.* 74 Learning curves indicate how the rate of learning changes with increased practice and are used to predict labour productivity. **1950** A. M. Turing in *Mind* LIX. 458 The idea of a learning machine may appear paradoxical to some readers. *Ibid.* 459 It is probably wise to include a random element in a learning machine. **1954** *Oxf. Univ. Gaz.* 15 June 1035/2 Work..on the insightful learning machine was also continued. **1963** A. M. Andrew *Brains & Computers* 61 Future machines of this kind will certainly also be learning machines. **1967** R. Whitehead in Wills & Yearsley *Handbk. Management Technol.* 57 Brains, self-organizing systems, economic systems, learning machines, computers, and automated factories are among the many subjects examined by the cybernetician. **1922** R. S. Woodworth *Psychol.* xiii. 302 It makes the learning process easier to follow. **1947** A. W. Melton in *Harvard Educ. Rev.* XXIX. 96 Educators..must know how to manage the learning process. **1949** Shurr & Yocom *Mod. Dance* i. 13 Nothing can substitute for the physical activity which is necessary to the learning process. **1962** R. M. Gagné in *Psychol. Rev.* LXIX. 355/1 Autoinstructional devices and their component learning programs. **1969** *Library Jrnl.* 1 Apr. 1536/3 (Advt.), Student body of 5000 on two campuses; 300 teachers; new learning resources center. **1970** *Ibid.* 15 Feb. 800/3 (Advt.), Curriculum adviser to work..in planning..for new learning resource facility incorporating uses of all

learning materials. **1970** *Globe & Mail* (Toronto) 25 Sept. 39/1 (Advt.), The successful applicant..will be responsible for assisting the principal in organizing, equipping and stocking a 3-level Learning Resource Centre in an innovational secondary school. **1970** *Jrnl. Gen. Psychol.* LXXXIII. 46 The analysis of variance on learning scores. **1948** E. R. Hilgard *Theories of Learning* xii. 335 Many learning situations require the selection of one or another possible mode of action. **1947** *Harvard Educ. Rev.* XXIX. 84 (*heading*) The relation of learning theory to the technology of education. **1962** *Listener* 15 Nov. 793/2 The psycho-analysts..believe that learning theory..is doomed to give only an incomplete and sometimes misleading account of personality development. **1967** M. Argyle *Psychol. Interpersonal Behaviour* viii. 148 Various training techniques derived from learning theory have been developed for the removal of symptoms [of mental disorders]. **1968** E. Lovejoy *Attention in Discrimination Learning* iii. 44 This selectivity is likely to be an important part of a learning theory.

lease (līz), *v.*[4] [f. *leas*, pl. of Lea *sb.*[4]] *trans.* To divide (yarn or thread) into leas.

1884 W. S. B. McLaren *Spinning* 242 The length varies from one to twelve yards, and the forms of making up, leasing, and tying are endless. **1927** T. Woodhouse *Artificial Silk* 67 It is quite possible that all the remaining hanks have already been leased.

lea·se-back. [f. Lease *sb.*[3] or *v.*[3] + Back *adv.*] In full, *sale and lease-back*. The sale of a property, etc., to a purchaser on the understanding that the vendor may take out a lease on the property. Also *attrib.*

1947 J. W. Kearns in *Amer. Bar Assoc. Proc., Section of Corporation, Banking & Mercantile Law* 46/1 The origin of the 'lease-back deal' is probably traced back to high taxes, high taxes coupled with the desire of almost any corporate manager to see a liquid balance sheet with very few fixed assets on it. The fewer the better. The lease-back deal is basically a transaction under which a corporation, which in the course of its business will use a considerable amount of real estate, sells that real estate to another entity which may be a university; it may be a charitable foundation; it may be any other type of organization which is tax free under the Internal Revenue Code. The real estate is usually sold at depreciated book value and simultaneously or concurrently a lease-back is granted to the industrial or merchandising concern that has sold the property. **1949** *Business Week* 22 Oct. 31/1 The device under attack is the sale-and-lease-back of real estate. **1964** *Financial Times* 12 Mar. 1/8 Excess cash produced by sale and lease-back arrangements. **1967** *Times Rev. Industry* July 33/2 The illusions of liquidity which can be created by..leaseback. **1970** *Daily Tel.* 30 Dec. 16 Factory-owners who sell their premises and take a lease-back of their accommodation. **1971** *Ibid.* 27 Apr. 21 (*heading*) Leaseback raises £5m for Grand Met. *Ibid.* 17 June 19 A £6 million plus sale and lease-back agreement. **1972** *Accountant* 26 Oct. 516/1 There is an increased use of real estate sale and lease-back and the non-cancellable leasing of capital equipment for a term approximating its economic life. **1974** *Country Life* 7 Mar. 506/1 It was arranged that immediately upon the transfer of the estate, the son would lease back the property to his father. The leaseback was for a term of five years.

leased, *ppl. a.* (Earlier example.)

1869 *Bradshaw's Railway Manual* XXI. 73 The gross earnings of the leased undertakings.

lease-lend. (Level stress.) Also **lease and lend** and *lend-lease. [f. Lease *v.*[3] + Lend *v.*[2]] At first (in 1941) applied to an arrangement whereby sites in British overseas possessions were leased to the United States as bases in exchange for the loan of U.S. destroyers; later in extended uses. Also *attrib.* and as *vb.*

1941 *Economist* 1 Feb. 139/2 The great reduction in American payments under the 'Lease and Lend' Bill. **1941** *Hutchinson's Pict. Hist. of War* 9 July–30 Sept. 48 Mr Harry Hopkins..in charge of the administration of the Lease-Lend Act, arrives in England again. **1942** *Ann. Reg. 1941* 19 The signing of the Lease and Lend Bill in the United States on March 11..caused great rejoicing in England. **1942** *R.A.F. Jrnl.* 3 Oct. 24 The Canadians argued that by lease-lend the Old Man would get one [a Jeep]. **1943** *Daily Tel.* 23 Oct. 4 Britain's new aircraft-carrier Victorious was 'Lease-lent' to the United States. **1944** G. B. Shaw *Everybody's Pol. What's What?* xv. 120 Under the American Lease and Lend arrangements England and Russia are borrowing their war stores from the United States. **1945** *Reader's Digest* July 47/1 Hatch has been particularly successful in 'breeding up' the poultry and livestock of the country by lease-lending his purebred bulls. **1964** *Daily Tel.* 19 May 28/6 To-night 40 Eastbourne policemen were sent to Brighton. They formed part of a 'lease-lend' arrangement entered into by Brighton, Eastbourne and Hastings police forces to help each other out in case of riots.

lea·sing, *vbl. sb.*[3] [f. Lease *sb.*[4] + -ing[1].] *Attrib.* in **leasing reed,** in weaving, a reed through which the warp threads pass as they come off the bobbins.

1927 T. Woodhouse *Artificial Silk* 108 The ends of the bobbins are threaded through a leasing reed. **1960** *Textile Terms & Definitions* (ed. 4) 90 (*caption*) Leasing.

least, *a.* (*sb.*) and *adv.* Add: **A.** *adj.* **1. c.** (Later examples.)

1870 *Amer. Naturalist* III. 234 The least Tern,..and the Roseate Tern, still breed on our coast. **1915** A. R. Horwood *Story Plant Life Brit. Isles* III. i. 105 Hooker recognises three divisions [of Gamopetalæ]: Chironieæ, including Yellow Wort, Least Yellow Gentian, [etc.]. **1946** T. M. Stanwell-Fletcher *Driftwood Valley* 42 Sometimes he and the least chipmunks..play hide-and-seek round boxes and trees. **1955** E. B. Ford *Moths* xi. 166 This [*sc.* the Burren] is also the Irish locality for the Least Minor. **1960** M. Burton *Wild Animals of Brit. Isles* 114 The Least weasel..has not so far been found in this country. **1975** *Country Life* 16 Jan. 131/1 Least tern nesting area.

e. *law* (or *principle*) *of least action* (Physics): the principle that an actual trajectory of a physical system is always such that, in comparison with any slightly different motion between the same end-points, the integral over the trajectory of the momentum with respect to distance (or more generally, of the sum of the generalized momenta with respect to generalized co-ordinates) has a minimum (or a maximum) value.

[**1748** Moreau de Maupertuis in *Hist. de l'Acad. R. des Sci. 1744* 423 Le chemin qu'elle [*sc.* la lumière] tient est celui par lequel la quantité d'action est le moindre. ..La quantité d'action..est proportionnelle à la somme des espaces multipliez chacun par la vitesse avec laquelle le corps les parcourt. **1748** —— in *Hist. de l'Acad. R. des Sci. et des Belles Lettres de Berlin 1746* 286 J'ai découvert le principe universel, sur lequel toutes ces loix sont fondées. ..C'est le principe de la moindre quantité d'action.] **1814** J. Toplis tr. *Laplace's Treat. Analytical Mech.* ii. 47 Maupertuis..asserted, that in all the changes which take place in the situation of a body, the product of the mass of the body by its velocity and the space which it has passed over is a minimum. This he called the principle of the least action, and it was applied by him to the discovery of the laws of the refraction and the reflection of light,..the laws of equilibrium, &c. **1834** W. R. Hamilton in *Phil. Trans. R. Soc.* CXXIV. 252 Although Lagrange and others, in treating of the motion of a system, have shown that the variation of this definite integral vanishes when the extreme coordinates and the constant H are given, they appear to have deduced from this result only the well known law of least action. **1920** A. S. Eddington *Space, Time & Gravitation* ix. 149 The law of gravitation, the laws of mechanics, and the laws of the electromagnetic field have all been summed up in a single Principle of Least Action. For the most part this unification was accomplished before the advent of the relativity theory, and it is only the addition of gravitation to the scheme which is novel. **1966** J. L. Meriam *Dynamics* viii. 349 Hamilton's principle and the principle of least action have found important but limited applications in engineering problems. Their use will undoubtedly grow with time as the complexity and generality of design situations increase. **1973** *Nature* 28 Sept. 223/1 The mechanics section begins with Hamilton's principle (here called the principle of least action) and gives a concise and elegant account of the relation between invariance and conservation laws.

4. (Earlier and later examples of *to say the least* (*of it*).)

1809 *Deb. Congress U.S.* 20 Feb. (1853) 422 To say the least of it, the people will perceive..an uncommon coincidence. **1811** *Ibid.* 17 Jan. 603 To say the least of such a measure, is to term it an experiment. **1928** R. Campbell *Wayzgoose* i. 26 Muses Nine, Those strapping girls whose love, to say the least, Would make a rabid Mormon of a priest. **1974** D. Scannell *Mother knew Best* vi. 59 Mother said vanity was a besetting sin which Amy resented, to say the least of it.

leather, *sb.* Add: **2.** Also, a stirrup-leather.

1907 *Yesterday's Shopping* (1969) 300/2 Hunting saddle, ..complete with stirrup irons and leathers. **1928** D. Byrne *Destiny Bay* vii. § 2. 314 The shorter your leathers, the less you know about your mount. **1936** J. Cary *Afr. Witch* vii. 137 'You ride too long... Take up your leathers.' Fisk obediently took up his stirrups a hole. **1952** M. Allingham *Tiger in Smoke* xiii. 197 Off you go! Shorten your leathers.

b. Also, a leather jacket or coat. Also *sing.*

1962 *John o' London's* 4 Jan. 20/1 A Banquo little more than an Oberon in his 'leather'. *Ibid.* 31 May 535/4 Two youths in leathers and crash-helmets. **1970** *Daily Tel.* 2 Mar. 14 Ankle-length, shiny, wet-look coats, suèdes and leathers were often trimmed with fur. **1972** Ellis & Newman in T. Kochman *Rappin' & Stylin' Out* 378 Wear 'black leathers'. **1973** P. Dickinson *Gift* ix. 142 Ian got into his leathers, Davy put on two extra layers of clothing, the bike started first kick.

d. As the name of a colour.

1872 *Queen* 15 June 431/3 Costume cloth in all the new colours, including pink,..leather,..and all leading colours. **1923** *Daily Mail* 16 Jan. 1 (Advt.), Coat frock.. Grey, Mole, Leather,..New Brown. *Ibid.* 31 July 1/3 (Advt.), Grey, Smoke, Leather and Navy.

e. *slang.* Various articles made of, or clad in, leather, such as (*a*) a wallet or purse; (*b*) a leather-shod foot; hence a kick; (*c*) a boxing-glove; hence a punch or boxing.

(*a*) **1883** 'Mark Twain' *Life on Mississippi* lii. 511, I pulled off an old woman's leather; (*robbed her of her pocket-book*). **1899** 'J. Flynt' *Tramping with Tramps* 395 'To reef a leather' means that the pickpocket pulls out the lining of a pocket containing the 'leather'. **1914** Jackson & Hellyer *Vocab. Criminal Slang* 54 *Leather*,..Some general currency, but used chiefly by pickpockets. A pocketbook; a wallet; a billbook. **1938** F. D. Sharpe *Sharpe of Flying Squad* 331 *Leathers*, wallets. (An inveterate pickpocket is sometimes called 'A Leather Merchant'.) **1955** *Publ. Amer. Dial. Soc.* xxiv. 114 The ordinary billfold which men normally carry, folded double, in the hip pocket, is called a *leather*.

(*b*) **1931** D. Runyon *Guys & Dolls* (1932) vi. 118 Dave walks over and starts to give Waldo Winchester the leather. **1936** J. Curtis *Gilt Kid* vi. 61 Old boys never could stand the leather.

(*c*) **1936** 'R. Hyde' *Passport to Hell* v. 86 It started off as a pretty little bout, though neither knew much about the leather. **1950** J. Dempsey *Championship Fighting* ii. 12 Meehan..threw so much leather and was so rugged that he and I broke even.

3. *spec.* The skin on the ear-flap of a dog.

1952 C. L. B. Hubbard *Pembrokeshire Corgi Handbk.* 112 *Leather*, the skin on the ear flap. **1960** *Times* 2 Jan. 9/2 The ear leather of the workers [*sc.* spaniels] is shorter than in show specimens. **1968** H. Harmar *Chihuahua Guide* 240 *Leather*, the skin of the earflap.

4. b. *leather goods.*

1902 *Westm. Gaz.* 14 June 8/3 A fire broke out in a leather goods manufactory. **1946** J. W. Waterer *Leather* xiii. 222 An..up-to-date manual of leather goods manufacture. **1971** D. Mackenzie *Sleep is for the Rich* vi. 196 A leathergoods store downtown stayed open during the lunch hour.

5. a. *leather-work* (earlier and later examples); also *leather-hard* adj.

1960 H. Powell *Beginner's Bk. Pott.* II. 64 Leather-hard, the condition of clay when it may be cut. Soap condition. **1967** M. Chandler *Ceramics in Mod. World* iv. 122 Each such blank, after partial drying to render it leather-hard, is turned on a semiautomatic lathe. **1971** *Islander* (Victoria, B.C.) 12 Dec. 7/1 The pot and the slip have reached a stage which is known as leather-hard. This means the pot may be handled safely without risk of damage, but is still capable of receiving the indented design. **1856** C. M. Yonge *Daisy Chain* I. xv. 143 Meta has been making a drawing for her papa, and is framing it in leather work. **1906** Sanford & Phillips *Art Crafts for Beginners* (rev. ed.) vi. 137 The great popularity of leather-work among amateurs is due..to the fact that a small and inexpensive equipment is all that is required. **1971** H. Pluckrose *Bk. of Crafts* 53/2 In the past amateur leatherwork meant punching and thonging.

b. *leather-sealer*; *leather-dressing* (example).

1850 *Rep. Comm. Patents 1849* (U.S.) 357, I claim the adjustable scraper..for the purposes and uses of leather dressing. **1662** *Public Rec. Colony of Connecticut* (1850) I. 377 The leath[r] sealers..shal haue allowed vnto them for each Dicker of Leather they seale, 18d. **1798** I. Allen *Nat. & Pol. Hist. Vermont* 272 Weights and measures, leather sealers &c. are regulated according to law.

c. *leather-bottomed, -coated, -faced, -jacketed, -lined* (also *fig.*), *-topped, -upholstered* adjs.

1783 in E. Parkman *Diary* (1899) 298, 9 black chairs.. five leather bottomed Do. **1854** J. E. Cooke *Virginia Comedians* I. xxii. 127 A rude oaken table and some leather-bottomed chairs. **1903** *To-Day* 4 Mar. 191/2 The implements consist of small leather-coated balls and wooden hockey sticks about seven inches long, which are held as one would hold a pencil. **1906** *Westm. Gaz.* 20 Nov. 4/2 Metal-to-metal clutches are..extending in favour at the expense of the old leather-faced bone type. **1908** *Ibid.* 29 Dec. 4/1 The three-speed gear-box.. to which the power is transmitted through the medium of a leatherfaced clutch. **1916** Joyce *Portrait of Artist* ii. 81 There stood the stout leatherjacketed vaulting horse. **1846** W. H. Emory in Frémont & Emory *Notes Trav. Calif.* (1849) 22/2 The first mouthful brought the tears trickling down my cheeks, very much to the amusement of the spectators with their leather-lined throats. It was red pepper, stuffed with minced meat. **1903** *Work* 18 July 382/1 The clutch..pulls the band (which.., is steel, leather-lined). **1913** W. Owen *Let.* 13 Nov. (1967) 211, 2 shirts (leather-lined extra). **1911** O. Onions *Widdershins* 281 The large leather-topped table. **1936** E. E. Evans-Pritchard in *Ess. Social Anthropol.* (1962) viii. 179 They were preceded in this romp through gardens and cultivations by a small boy beating on a leather-topped drum. **1965** G. McInnes *Road to Gundagai* vii. 119 Soft..armchairs and sofas, a big leather-topped desk. **1923** F. L. Packard *Four Stragglers* II. v. 183 Polly Wickes rose hastily from the..big leather-upholstered Chesterfield.

d. *leather-coated, -faced* (also *-face*), *-jacketed* (cf. *leather-jacket 5) adjs.

1902 W. B. Yeats *In Seven Woods* (1903) 12 And leather-coated men, with slings. **1884** 'Mark Twain' *Huck. Finn* xxviii. 287 You ain't one of these leather-face people. I don't want no better book than what your face is. **1919** W. Deeping *Second Youth* xv. 128 She let this leather-faced old rascal flirt with her quite harmlessly. **1934** T. Wilder *Heaven's my Destination* 3 Brush.. chose a seat beside a tall leather-faced man. **1960** *Economist* 8 Oct 149/1 Among the cartoonists Herblock has drawn Mr Khrushchev as a leather-jacketed gang-leader. **1961** *Encounter* XVII. II. 17/2 The leather-jacketed 'Teddy Boy' gangs of Western Germany. **1969** *Daily Tel.* 2 Sept. 1/3 South and East coast resorts were invaded yesterday by hundreds of leather-jacketed Rockers and teenagers in jeans and steel-tipped boots. **1973** J. Wainwright *Pride of Pigs* 114 The leather-jacketed, stocking-feeted Hell's Angel.

6. leather-back, substitute for def.: a large marine soft-shelled turtle, *Dermochelys coriacea*; (later examples); **leather belting,** machine belting made of leather; also *attrib.*; **leather breeches (beans)** *U.S. dial.*, dried beans or dried bean-pods; beans that have been dried and then cooked in their shells; **leather-cloth** (also **leathercloth**), also, a synthetic product simulating leather; **leather-hunting,** esp. a *colloq.* term for fielding when the batsman is hitting out as freely as he likes; hence *leather-hunter*; **leather-leaf:** for '*Cassandra calyculata*' substitute '*Chamædaphne calyculata*'; (earlier examples); **leather medal** orig. *U.S.*, a medal made of leather instead of metal, sarcastically suggested as an award; **Leather-Stocking**, a North American frontiersman [from a character portrayed by J. F. Cooper]; also *attrib.*

1965 R. McKie *Company of Animals* xii. 168 On the beaches of Trengganu..the leatherback turtles lay their eggs. **1969** A. Bellairs *Life of Reptiles* I. ii. 41 The tendency nowadays is to regard the leatherback as a specialised descendant of turtles of more 'ordinary' type. *a* **1877** Knight *Dict. Mech.* I. 273/1 Leather belting is ordinarily prepared in the following manner. **1877** *Design & Work* 9 June 23 (Advt.), Charles Churchill and Co., importers of American machinery and tools,..lathes, vices, planes,.. American leather belting. **1909** *Westm. Gaz.* 6 Apr. 2/1 The exposure of the graft..behind the duty on hides was made by tanners and shoe and leather-belting manufacturers. **1946** J. W. Waterer *Leather* 304 Included in its members are all the principal manufacturers of leather belting. *Ibid.*, Persons engaged in or intending to engage in the leather belting trade. **1913** H. Kephart *Our Southern Highlanders* 292 Beans dried in the pod, then boiled 'hull and all' are called leather-breeches (this is not slang, but the regular name). **1941** J. Smiley *Hash House Lingo* 35 *Leather breeches*, dried kidney beans. **1943** R. Chase in B. A. Botkin *Treas. S. Folklore* (1949) 470 Such communal tasks as stringing beans for canning, or threading them up to make the dried pods known as 'leather britches'. **1972** E. Wigginton *Foxfire Bk.* 15 He..dried leather britches beans. *Ibid.* 167 *Leather breeches beans*... Take a string of dried green beans down, remove the thread, and drop them in a pot of boiling water. **1929** *Publishers' Circular* 18 May 621/3 A revolution has taken place in the world of leather by the introduction of the synthetic product leathercloth. **1937** *Archit. Rev.* LXXXI. 291/1 The manufacturers of cheaper cars began to use leathercloth as a finish over the normal rigid type of body construction and the snob-appeal of purpose-made bodies was lost. **1961** *Times* 30 May (I.C.I. Suppl.) p. viii/1 The original amalgamation brought together..dyestuffs, leathercloth, paints and non-ferrous metals. **1973** *Daily Tel.* 21 Nov. 14/4 This test car was upholstered in a ventilated leathercloth. **1944** Blunden *Cricket Country* i. 19 The laugh at the unfortunate 'leather-hunter' on a hot chase. **1865** J. Pycroft *Cricketana* xiii. 224, I like science more than swiping, and enjoy 'fielding', but not leather hunting. **1905** H. A. Vachell *Hill* xii. 254 And then, when his 'eye' is in, he will give the Etonians such leather-hunting as they never had before. **1934** W. J. Lewis *Lang. Cricket* 143 *Leather-hunting*, a jocular term for the exertions of the fieldsmen when the ball is hit freely to all parts of the field. **1970** *Sunday Tel.* 20 Dec. 21/7 The voracious Richards was in action once more and M.C.C. are assured of more leather-hunting today. **1818** A. Eaton *Man. Bot.* (ed. 2) 173 *Andromeda calyculata*, leather leaf. **1870** *Amer. Naturalist* IV. 217 The Leather Leaf (*Cassandra calyculata*), and *Andromeda polifolia*, are both worthy of attention. **1831** J. Motte in A. H. Cole *Charleston goes to Harvard* (1940) 89 He must be a cute chap, and deserves to have a leather medal. **1837** *Harvardiana* III. 147 (Th.), A leather medal his reward should be, A leather medal and an LL.D. **1860** *Richmond* (Virginia) *Enquirer* 20 Apr. 2/5 (Th.), The individual who conceived the leather medal idea [for identifying dogs] deserves a leather medal himself. **1889** *Kansas City* (Missouri) *Times & Star* 5 Dec., A leather medal..awaits the first misguided person this season writing it 'Xmas'. **1922** Joyce *Ulysses* 750 He ought to get a leather medal with a putty rim for all the plans he invents. [**1823** J. F. Cooper *Pioneers* I. i. 18 His limbs were guarded with long leggings of the same material as the moccasins, which gartering over the knees of his tarnished buck-skin breeches, had obtained for him, among the settlers, the nick name of Leather-stocking.] *Ibid.* 11 The Leather-stocking has put his hounds into the hills this clear day. **1831** M. Holley *Texas* (1833) v. 43 The character of Leather Stocking, is not uncommon in Texas... The dress of these hunters is usually of deerskin. Hence the appropriate name *Leather Stocking*. Their generic name..is *Frontiers-men*. **1909** *Daily Chron.* 1 July 7/3 With most birds, you must make your approach with all the art of a leatherstocking. **1965** *English Studies* XLVI. 313 In this book Cooper draws repeated parallels between the Leather-stocking hero and Moses.

leathered, *ppl. a.* Add: **b.** Made into, or like, leather.

1797 in G. B. Goode *Fisheries U.S.: Hist. & Methods* (1887) II. 435 By walking it [seal-skin] becomes leathered and soft to the foot. **1869** S. Bowles *Our New West* 444 If you bring a liver not entirely leathered and lungs not over half consumed. **1970** L. Jeffers *My Blackness is Beauty of this Land* 9 His face was leathered, lean, and strong, Gashed with struggle scars.

leather-jacket. Add: **5.** A person, freq. a member of a gang or a delinquent group, dressed in a leather jacket.

1959 *New Statesman* 15 Aug. 180/1 All France has learnt about the bands of 'leather-jackets'... Why are these young rowdies called 'leather-jacket'?.. Certainly leather jackets seem to be the uniform for young American delinquents. **1960** *Britannica Bk. of Year* 557/2 The characteristic dress of juvenile delinquents in several countries produced the term *leather jackets*, meaning delinquents. **1963** V. Nabokov *Gift* v. 289 Not long before his deportation from Russia, when some revolvered leatherjackets had come to arrest him.

leathern, *a.* **2.** *Comb.* (Later example.)

1875 J. G. Holland *Sevenoaks* xii. 158 Blue-jays were screaming among leathern-leaved oaks.

lea·ther-neck, lea·therneck. *slang.* [f. Leather *sb.* + Neck *sb.*[1]] **1.** (See Leather *sb.* 6.)

1890 [see Leather *sb.* 6]. **1916** 'Taffrail' *Carry On!* 27 A Royal marine is a 'bullock', 'turkey', or 'Joey', while a soldier is a 'grabby' or 'leather-neck'.

b. A marine. *U.S.*

1914 *Dialect Notes* IV. 150 *Leatherneck*, a marine. **1919** *A Company, Eleventh Frapper* (U.S. Marines) 17 Apr. 1/2 We learn that between 700 and 800 warworn Leathernecks from the famous 5th and 6th Marines.. arrived at Camp Covington. **1926** *Amer. Speech* I. 354/2 'Leatherneck' for a Marine..is derived from the old custom of facing the stiff neck-band of the marine uniform with leather. **1931** *Punch* 3 June 606/1 I'd just passed the remark to the leather-neck on sentry that we was 'avin' a nice peaceful forenoon when the Admiral's buzzer goes, and I 'ops in to see what 'e wanted. **1955** W. Foster-Harris *Look of Old West* i. 11 Under this collar, the troopers were supposed to wear an atrocity of a stock, of black leather. This is where the name 'leather-neck' came from, since the Marines also had to wear these dog-collar affairs. **1968** R. West *Sk. Vietnam* ii. 37 The U.S. Marine Corps. These legendary troops, nick-named 'leathernecks'.

2. = Rouseabout 2. *Austral.*

1898 *Bulletin* (Sydney) 1 Oct. 14/3 In a shearing shed: The boss is the 'finger', the shearers the 'brutes', the rouseabouts 'leathernecks'. **1899** W. T. Goodge *Hits! Skits! & Jingles!* 155 And he 'pinked' him like a leather-neck when squatters paid a pound! **1945** Baker *Austral. Lang.* xvi. 286 *A leatherneck* is a marine in the U.S.; in Australia he is a station handyman.

leathery, *a.* Add: **b. leathery turtle** = *leather-back* (*Leather *sb.* 6).

1875 *Encycl. Brit.* III. 112/1 The 'leathery turtle',..is herbivorous, and yields abundance of oil. **1901** [see Luth]. **1963** J. Kirkup *Tropic Temper* 270 The 'leathery turtles' are among the world's largest and in Malaya they haunt the beaches of Trengganu. **1966** *Festival Malaysia 1966: Calendar of Events* 8 (*caption*) A giant leathery turtle of the East Coast of the Malay peninsula. **1969** A. Bellairs *Life of Reptiles* I. ii. 41 Some workers have believed that the huge leathery turtle (*Dermochelys*)..is more primitive than the rest.

leavable (lī·văb'l), *a.* [f. Leave *v.*[1] + -able.] Able to be left.

1923 H. G. Wells *Men like Gods* I. i. 10 The affairs of the *Liberal* were just then in a particularly leavable state. **1946** *N.Y. Herald Tribune* 2 June (Books) 5 Her rather rubbishy mother..had finally gone off with her artist, bringing despair to her 'leavable' husband, who could not help alienating the people he loved most.

leave, *sb.*[1] Add: **1. e.** Also *transf.*

1963 *Times* 28 Sept. 9/4 While not personally subscribing to the use of the term 'on leave' by office workers to describe their annual break(s), I can contribute reasons for their doing so. **1973** *Times* 17 Apr. 12/8 She will be on a six month leave-of-absence from the [National Theatre] company. **1974** P. De Vries *Glory of Hummingbird* xiv. 206 It'll only be a leave of absence... But..if there's a blowup, the firm will be able to say you were let go.

2. *to take leave*: used as a formula to draw attention, with a somewhat ponderous affectation of presumption, to a truth or state of affairs. Cf. Beg *v.* 3.

1611 [see Take *v.* 21]. **1814** T. S. Raffles *Substance of Minute on Java* 100, I take leave to observe, that the state of landed tenure here is very different from what it is reported to be in other parts of Java. **1820** [see Take *v.* 21]. **1834** M. Edgeworth *Helen* III. v. 93 We must take leave to pause one moment to remark..that the first little fib in which Lady Cecilia..indulged herself..occasioned her..a good deal of..trouble. **1928** *Sat. Rev.* 28 July 127/1 Stephen has many excellent qualities both of heart and head, though whether her sufferings would have cradled her into a first-rate novelist we take leave to doubt. **1938** 'M. Innes' *Lament for Maker* I. viii. 53 We may take leave to think the silly body stood there in the sleet and cursed the lure of the wanderer roundly.

b. (Later examples.)

1916 A. Bennett *Lion's Share* v. 40 'Mother!' cried Audrey. 'Have you taken leave of your senses?' **1942** 'M. Innes' *Daffodil Affair* II. iv. 58, I think you've taken leave of your senses. **1968** L. Goodman *Sun Signs* (1970) 325 You'll think I've taken leave of my senses, if you've just met that particular Pluto person. **1972** 'M. Innes' *Open House* II. xiii. 129 Nothing of the kind... You must have taken leave of your senses.

5. *leave camp, centre, list, period, permit, rota, -way*; **leave-boat**, a boat carrying troops on leave; **leave draft**, a detachment of troops on leave; **leave-out**, at certain schools, permission to go beyond the school precincts; cf. *leave-day, leave-out day* in Dict.; **leave party** *Mil.*, a group of servicemen on leave; **leave-taker**, one making his farewell; **leave-train**, a train carrying troops on leave.

1917 'Contact' *Airman's Outings* v. 118 Passengers on a Channel leave-boat are quieter than might be expected. **1922** Blunden *Bonadventure* iii. 21 To the Plate and back again, in a cargo ship!.. The voyage, no doubt, would be more arduous than that in the leave-boat from Boulogne to Folkestone. **1945** W. S. Churchill *Victory* (1946) 109 Eight new leave camps are under construction. **1961** *Reader's Digest* Feb. 24/1 Last March..the government was asked why African airmen in Her Majesty's service had been excluded from a Kenya leave centre. **1966** *New Statesman* 14 Oct. 537/3 The lavish structure of permanent installations, family homes, schools, leave centres and the like is enormously costly. **1920** *Blackw. Mag.* May 608/2 The sallow complexions..and leanness of a leave draft from the Palestine front. **1917** 'Contact' *Airman's Outings* v. 108 Only during the intervals of attack is the leave-list unpigeonholed. **1940** M. Marples *Public School Slang* 164 I've got leave-out. **1955** *Times* 18 Aug. 10/6 When I was in College at Winchester, one of our favourite pastimes used to be hitch-hiking on leave-out days. These were free days during term when we could do as we liked. **1916** W. Owen *Let.* 3 July (1967) 398, I had the Leave Party to conduct to the Station the other day. **1954** W. Faulkner *Fable* 128 During three of these two-week leave-periods..the entire squad had vanished from France.., and reappeared one morning two weeks later. **1906** *Daily Chron.* 14 May 5/2 The report..recommends.. that traffic in leave-permits be made illegal. **1940** 'Gun Buster' *Return via Dunkirk* I. x. 76 I've been looking at the leave rota, and see you're down for January 10. **1891** Kipling *Light that Failed* xv. 310 Bess found Dick his cabin in the wild turmoil of a ship full of leavetakers and weeping relatives. **1922** Joyce *Ulysses* 190 The quaker librarian came from the leavetakers. **1917** 'Contact' *Airman's Outings* v. 113 The train, true to the custom of leave trains, was very late. **1918** A. Bennett *Pretty Lady* xx. 132 Then I can't catch my train at Victoria..the leave-train. My leave is up to-night. **1913** T. E. Lawrence *Let.* 29 Sept. (1938) 157 They half suggested a royalty of a pound a head a day, as leave-way to dig.

leave (līv), *sb.*[2] [f. Leave *v.*[1] 3.] In Billiards, etc., the position in which the balls are left for the next player or stroke.

1896 W. Broadfoot et al. *Billiards* x. 319 Every leave was the result of accident rather than of design. **1903** W. Mitchell *Cue Tips* 6 The most interesting and not the least useful way to practise billiards..is..to place the balls in certain favourable positions upon the table and attempt to make as large a break as possible from the 'leave'. **1914** Ld. Tollemache *Croquet* xiv. 74 A well thought-out and finished Leave is one of the hall-marks of a first-class player. **1929** J. Davis *Billiards Up-to-Date* viii. 89 An exception..is seen when, at the commencement of a break, an opponent presents you with a leave which necessitates the use of side. **1936** —— *Improve Your Snooker* xii. 69, I cannot guarantee you will bring off this shot if you are presented with a similar leave. **1968** *Croquet* July 2/2 It is when we consider how to make leaves against good shots that the fun really begins.

leave, *v.*[1] Add: **3. e.** To allow, permit, let. *colloq.* (chiefly *U.S.*).

Cf. *to leave..be* s.v. sense 13 in Dict.

1840 *Southern Lit. Messenger* VI. 508/1 If you ha'n't a mind to go, you can leave it be, it's all one to me. **1863** T. D. Price *Diary* 14 Apr. (MS.), I left him have colt. **1910** J. Hart *Vigilante Girl* iv. 55 It's all right so long as you don't leave her get loose. **1916** 'Boyd Cable' *Action Front* 235 Prickles, me lad, it's deep enough we've dug to lave us get out to our German Gineral. **1935** Z. N. Hurston *Mules & Men* (1970) I. vi. 127 Leave the weeds go. Somebody 'll come chop 'em some day. **1940** J. O'Hara *Pal Joey* 103 But I said to him how can I pay you if you dont leave me wear it and I lose my job.

5. c. *to leave* (*something, much*, etc.) *to be desired* (*to wish*, etc.) (examples).

[**1769** F. Brooke *Hist. Emily Montague* IV. 189 Every anxiety is removed from my Emily's dear bosom: a father's sanction leaves her nothing to desire.] **1780** F. Burney *Diary & Lett.* (1842) I. viii. 335 Etty plays as if inspired, and in taste, expression, delicacy and feeling, leaves nothing to wish. **1835** *Athenæum* 16 May 371/1 Her style, too, leaves little to be desired. **1852** *Harper's Mag.* Aug. 422/2 This edition leaves nothing to be desired by the most fastidious book-fancier. **1876** F. Pollock *Pollock–Holmes Lett.* (1942) I. 6 Kent is a considerable advance, but leaves much to be desired. **1895** F. Espinasse *Life E. Renan* x. 185 Dean Stanley's French accent left much to be desired, but his volubility was indisputable. **1939** M. Allingham *Mr. Campion & Others* I. iii. 65 The staff still left much to be desired and the food..was certainly not cooked by a master. **1953** G. Durrell *Overloaded Ark* ix. 166 Apart from his face, which left much to be desired, his feet were swollen to twice normal size with elephantiasis. **1967** A. Bailey in L. Deighton *London Dossier* 52 The vegetables leave much to be desired, but the Stilton is worth having. **1974** *Times* 22 Jan. 2 When the clubs first applied for licences it was decided not to oppose them. 'Now that we know how they operate, we feel they leave a lot to be desired.'

7. d. (Earlier and later examples.)

1884 E. W. Nye *Baled Hay* 56 That is where we get left. **1908** *Daily Chron.* 16 Nov. 5/2 'Oh, never mind those,' says the admiral; 'what has the Navy got?' 'Got left, as usual,' replies the lieutenant. **1928** D. H. Lawrence *Lady Chatterley* vi. 73 It was no good being really good and getting left with it.

e. *Cricket.* Of the ball, to move away from (the batsman); used of a leg break delivery which turns away from the batsman.

1952 A. Bedser *Bowling* II. vii. 64 A type of leg-spin which, of course, makes the ball leave or go away from the batsman. **1956** R. Alston *Test Commentary* iii. 19 Both batsmen seemed especially fallible to the ball that left them.

13. *to leave... be* (later examples); *to leave* (a person) *cold*: see *Cold *a.* 7 e; *to leave it at that*: to proceed no further with a matter; to refrain from pressing a point; *to leave to* (*himself*, etc.): to let (a person, etc.) alone or without help or interference from another or others; *to leave* (a person) *to it*: to leave (someone) alone, esp. to allow him to proceed with a task in hand.

1833 *Chambers's Edin. Jrnl.* II. 145/1 The individual who writes the present paper was once 'so far left to himself' as to spend several months amidst the heartless frivolities which characterise a winter of fashionable life in the Scottish..capitals. **1902** *Captain* VII. 542/1 We'll leave it at that, then. **1910** 'Saki' *Reginald in Russia* 8 Left to themselves, Egbert and Lady Anne would unfailingly have called him Fluff. **1918** C. Mackenzie *Early Life Sylvia Scarlett* II. ii. 283 The petulant way in which she shook herself free from the embrace at last brought Sylvia up to the point of leaving Lily to herself. **1928** Galsworthy *Swan Song* II. ii. 114 He had looked at her, and left it at that. **1943** J. B. Priestley *Daylight on Saturday* xvi. 114 'You never told her what she ought to do..,' said Freda to Jock. And then she left them to it. **1946** E. O'Neill *Iceman Cometh* (1947) II. 102 Leave Hugo be!.. He's earned his dream! **1948** C. Day Lewis *Otterbury Incident* iv. 45 Ted and I left him to it. **1949** V. Grove *Language Bar* viii. 114 If understanding and sense were not sought after, the ignorant would merely corrupt the 'meaningless' word, and leave it at that. **1958** L. A. G. Strong *Light above Lake* xxi. 148 Toby..left him to it. **1966** *Listener* 15 Dec. 443/1 If the House is content to leave it at that for the present,..then I would ask if we might withdraw the resolution and leave it at that for today. **1967** Singha & Massey *Indian Dances* i. 34 South India had been more or less left to itself. **1970** D. Storey *Contractor* i. 40 I'll leave you to it then. **1972** M. Waste *Summer of Red Wolf* 9 Leave me be for a moment, please. **1974** R. Adams *Watership Down* xxxiii. 258 They'll know which way we've gone and they won't leave us alone. **1974** W. T. Burley *Death in Stanley St.* viii. 127 Wycliffe stood up. 'Good. I'll leave you to it.'

b. [Earlier example of *leave go*.]

1776 in *Essex Inst. Hist. Coll.* (1907) XLIII. 218 Tis said we left go pieces of heavy cannon owing to the cowardice of a body of Connecticut troops.

14. e. Freq. in pa. pple. *left over*, remaining, not used up.

1892 'Mark Twain' *Amer. Claimant* xii. 107 Irish stew made of the potatoes and meat left over from a procession of previous meals. **1899** G. B. Burgin *Bread of Tears* ii. 138 The undigested fragments which were left over after the making of the world. **1907** *Smart Set* Mar. 72/1 You can go to the boss for your time—if there's anything left over. **1934** J. O'Hara *Pal Joey* 114 Choice meats like steak & chops etc. that was left over from the nite before. **1955** M. Patten *Learning to Cook* ii. 59 (*heading*) Foods that have been left over.

leave (līv), *adv.* Chiefly *U.S.* var. of *lieve* (Lief *adv.*).

1840 *Southern Lit. Messenger* VI. 508/1 Never mind.. I'd as leave be here as anywhere else. **1902** A. Deland *Old Chester Tales* 80, I would just as leave. **1904** A. D. McFaul *Ike Glidden* xviii. 144, I would as leave go first. **1922** D. H. Lawrence *Sea & Sardinia* 122 They would fetch you a bang over the head as leave as look at you. **1935** G. Ingram *'Stir' Train* i. 16 'I've got a little instrument here,' Margot showed him a thin scalpel..., 'and.. I would as leave cut into anyone's belly as any surgeon.'

leaver. Add: *spec.* a boy or girl who has just left or is about to leave school. See *school-leaver* (*School *sb.*[1] 9).

1910 *Westm. Gaz.* 17 Jan. 5/1 Of the entrants and leavers examined, approximately 3 per cent. of the children suffered from..eye disease. **1959** *Times Educ. Suppl.* 28 June 289/3 One teacher..wearied to despair by the listlessness and lack of interest of his class of 'leavers', persuaded the boys to.. **1972** *Guardian* 3 Aug. 6/6 Employers..were most favourably impressed with leavers' basic art and design skills.

Leavers: see *Levers.

leaves (līvz), var. *lieves.

1771 J. S. Copley *Let.* 17 Aug. in *Mass. Hist. Soc. Coll.* (1914) LXXI. 147 If Mr. Joy would as leaves undertake the..Room as plaister,.. I should prefer it. *Ibid.* 20 Sept. 160, I had as leaves Miller should paper as any one, provided he does it as Cheap.

leaving, *vbl. sb.* Add: **3.** *leaving certificate* (later examples; and additional sense: see quot. 1923); *leaving scholarship*; **leaving-age**, the age at which a pupil is legally entitled to leave school; **leaving-off time**, the time of ceasing work.

1943 J. Graves *Policy & Progress Secondary Educ.* xix. 125 The curriculum would vary according to the normal leaving age and the different interests and abilities of the children. **1914** I. Hay *Lighter Side School Life* i. 24 Oxford and Cambridge Locals..or, in Scotland, the Leaving Certificate. **1923** J. D. Hackett in *Management Engineering* May, *Leaving Certificate*, a card given to laid-off employees, entitling them to consideration when work is resumed. **1963** J. Fountain in B. James *Austral. Short Stories* 275 Brilliant passes in the Intermediate and Leaving Certificate examinations. **1971** *Guardian* 2 July 7/4 With an examination reform which provides a leaving certificate..raising the school-leaving age to 16 would be a failure. **1907** *Westm. Gaz.* 20 Aug. 10/2 It is the usual practice at leaving-off time on Saturdays for the workmen..to cease work at once. **1889** *Nation* (N.Y.) 7 June 464/1 This sum includes the 'leaving' scholarship given by the Clothworkers' Company.

Leavisian (lī·visiăn), *sb.* and *a.* [f. the name of the English literary critic, Frank Raymond *Leavis* (b. 1895) + -ian.] **A.** *sb.* An admirer or follower of F. R. Leavis. **B.** *adj.* Of, pertaining to, or characteristic of F. R. Leavis or his writings.

1959 *Times Lit. Suppl.* 1 May 256/2 There are Leavisians, there are Empsonians, but, as an embattled band, preaching and practising the Master's doctrine, there are no Ricardians. **1963** *Ibid.* 17 May 357/1 The Arnoldian and Leavisian concern with 'high seriousness'. **1964** *Punch* 29 Apr. 624/2 The phrase 'quality of life'..is something of a Leavisian stock response. **1969** *Listener* 16 Oct. 508/2 Many speakers brought a Leavisian concern for..quality of culture. They kept, in other words, to a view of art as a humanising study. **1972** *Times Lit. Suppl.* 19 May 577/2 The Leavisian conceptual articulation is compact, powerful and cogent.

Leavisite (lī·visəit), *a.* and *sb.* = *Leavisian *sb.* and *a.*

1958 *Times Lit. Suppl.* 17 Jan. 30/4 Mr Wain is still involved with Leavisite criticism. **1962** *Listener* 6 Sept. 364/2 The pages about which the Leavisite, or the general reader, may feel himself below G.C.E. 'O'-level on the Snow line are, happily, few. **1963** A. Hartley *State of England* ii. 50 Dr. Leavis seems to have been one of the rare contemporary examples of a teacher conveying a view of life to his pupils... The word 'Leavisite' is not an empty one. **1969** *Sunday Times* (Colour Suppl.) 21 Dec. 25/4 Arnoldians, Leavisites, Marxists, Fabians, Buberites were all there for us to sort it out. **1970** *Guardian* 16 Sept. 10/3 You can't go on writing Leavisite criticism when you've rediscovered that literature relates to the great tradition.

Lebanese (lebănī·z), *sb.* and *a.* [f. Leban-on + -ese.] **A.** *sb.* A native or inhabitant of Lebanon; also *collect.* **B.** *adj.* Of or pertaining to Lebanon or its inhabitants.

1920 *Glasgow Herald* 5 Apr. 6 The Lebanese..have..dissociated themselves entirely from the action of the Syrian Congress. **1926** *Contemp. Rev.* Feb. 194 A distinguished Lebanese Prince. **1927** *Times* (Weekly ed.) 25 Aug. 208/1 Many Lebanese residing in Egypt became French subjects. **1957** M. Banton *W. Afr. City* v. 79 The Lebanese and Indians are not numerous. **1972** M. J. Bosse *Incident at Naha* i. 39, I..accepted some Lebanese-bought dates. **1973** *Times* 17 June 9/1 Pressures within Lebanon to get them [sc. Palestine guerrilla forces] to leave Lebanese territory. **1973** *Guardian* 21 May 3/7 If the fighting had kept going, they would have been obliged to join in: Lebanese against Lebanese.

lebbek (le·bek). Also **labakh, lebba(c)k, lebbakh, lebbeck, lebek.** [ad. Arab. *labak*.] A large deciduous tree, *Albizia lebbeck*, of the family Leguminosæ, native to the tropics of North Africa and Asia and bearing heads of yellowish-white flowers; = Siris a.

1766 tr. *Hasselquist's Voyages & Travels in Levant* 249 Acacia of Upper Egypt... The Arabs call it Lebbeck. **1803** W. Wittman *Travels in Turkey* xiv. 346 In the vicinity of Cairo..a species of the cassia fistula grows to a considerable height, and affords a very spacious shade... By the Arabs this tree is called lebbak. **1920** E. H. Jones *Road to En-Dor* (ed. 2) iv. 39 Along the long, straight road near Cairo..there was an avenue of lebbek trees. **1921** *Blackw. Mag.* Feb. 155/1 They drove out through the eight-mile tunnel of lebbek-trees. **1929** Broun & Massey *Flora Sudan* 174 *Albizzia Lebbek* Benth. Labakh, Lebbek. **1941** C. Barrett *On Wallaby* vi. 134 The lebek trees were in bloom and delicate foliage rippled in the west wind. **1965** Zand & Videan tr. *'Abd al-Latif al-Baghdadi's Eastern Key* 33 While the fruit of the labakh is green it has a styptic savour like a green date, but when it is ripe it becomes sweet and agreeable and takes on a viscous quality.

|| **Lebensform** (lē·bənsfɔ̄rm). Pl. **Lebensformen.** [G., 'form of life'. Used notably by L. Wittgenstein in the German text of his *Philos. Investigations*.] Any type of human activity that involves values, e.g. the artistic or political or religious life; gen., a style or aspect of life.

1937 G. W. Allport *Personality* viii. 231 These Lebensformen are at best only categories of value. **1959** *Times Lit. Suppl.* 11 Sept. 513/2 Mr. Stuart Hampshire in *Thought and Action* is aware of the scale of his undertaking: 'It is necessary first,' he says, 'to view the using of language as a particular form of human behaviour' (Wittgenstein called it Lebensform).

|| **Lebenslust** (lē·bənzlust). [G., = joy of living.] = *joie de vivre.

1890 W. James *Let.* 27 Aug. in R. B. Perry *Tht. & Char. W. James* (1935) I. 414 Your last two letters..breathed a spirit of youth, a sort of Lebenslust. **1958** P. De Vries *Mackerel Plaza* 84 Security he could give her, yes, but not, I'm afraid, something else demanded by her Lebenslust. **1963** *Economist* 19 Oct. 228/1 ..continues to drive him regularly into Bonn.

|| **lebensraum** (lē·bənzraum). Also **L-.** [G., f. genit. of *leben* life + *raum* space.] Territory which the Germans believed was needed for their natural development (now *Hist.*). Also *transf.*

1905 *Mind* XIV. 266 A universal activity..forms an all-comprehending *Lebensraum* in which the manifold may meet and enter into relation. **1935** [see *life-space* s.v. *Life *sb.* 16 a]. **1939** O. Lancaster *Homes Sweet Homes* 46 These treasures were joined on..overcrowded ledges by a new wave of invaders..and..the problem of *lebensraum* had become acute. **1939** A. Salter *Dual Policy* 25 *Lebensraum*, or a place in the sun, is the historic claim and ambition of Germany, as 'encirclement' is her historic anxiety. **1939** *War Illustr.* 9 Dec. 393/1 Moravia and Bohemia had been overrun by the Nazi armies and declared German Protectorates—part of the German people's 'lebensraum'. **1940** [see *Appeasement 4]. **1951** S. van Valkenburg in G. Taylor *Geogr. in 20th Cent.* iv. 109 Kurt Vohwinkel (..1939)..distinguishes three kinds of German *Lebensraum*. The first kind is the real area occupied solidly by Germans; the second the area where besides Germans there are other people but the German cultural influence prevails; and the third is the one in which Germans are outnumbered by others but still because of their racial and cultural superiority have a right to dominate. **1957** *Encycl. Brit.* VIII. 881/2 Hitler was convinced that..Germany..needed Russian territory for *Lebensraum*. **1959** *Listener* 25 June 1119/1 People, both white and black, either intent on gain or simply seeking *lebensraum* and devoid of real roots. **1960** *Guardian* 11 Mar. 6/6 Music, manuscripts, and her little daughter..complete amicably for lebensraum. **1963** *Times* 30 May 14/3 Lebensraum for the Japanese. **1972** W. A. Pantin *Oxf. Life* iv. 49 The problem of *Lebensraum* for the developing natural sciences..was already beginning to appear in the early-nineteenth-century minutes of the Hebdomadal Council.

|| **lebensspur** (lē·bənʃpūr). Also with capital initial. Usu. as *pl.* **lebensspuren.** [a. G. *lebensspur* (O. Abel *Grundzüge der Palaeobiol. der Wirbeltiere* (1912) 65), f. *leben* life + *spur* trace, track, remains (cogn. w. Spoor *sb.*[1]).] A small track, burrow, cast, or the like left in sediment by a living organism; *esp.* one preserved in fossil form in sedimentary rock.

1960 *Gloss. Geol.* (Amer. Geol. Inst.) Suppl. 37/2 Lebensspur. **1962** R. C. Moore *Treat. Invertebr. Paleont.* W. 178/2 Lebensspuren are very numerous in sedimentary rocks.., and in general they have little chance of being preserved as fossils. *Ibid.* 179/2 Many lebensspuren have now been identified as Lebensspuren, were considered to be remains of marine algae. **1964** Pettijohn & Potter *Atlas & Gloss. Primary Sedimentary Struct.* Plate 70A (*heading*) Lebensspuren on underside of sandstone. **1973** *Nature* 30 Mar. 323/2 In a recently completed series of laboratory studies, lebensspuren were produced by individual macrobenthic organisms on a variety of marine sand and mud substrates (freshly collected from..the Bristol Channel) from which all other macrofauna had been removed by passing it through a 1 mm mesh. But the resultant tracks, trails and burrows gradually disappeared when the aquarium tanks..were left undisturbed.

|| **Lebenswelt** (lē·bənsvelt). [Ger.] = *life-world* (see *Life *sb.* 17).

1962 A. W. Levi *Lit., Philos. & Imagination* 138 What Husserl has called the *Lebenswelt*—the ongoing continuity of 'lived' experience. **1964** *Amer. Philos. Q.* I. 127/1 According to Wild the *Lebenswelt* is the world of directly lived experience. **1966** *Philos. Rev.* LXXV. 394 Husserl's transcendental philosophy and..his concept of the *Lebenswelt*.

Leber (lē·bər). *Ophthalm.* The name of Theodor *Leber* (1840–1917), German ophthalmologist, used in the possessive esp. in **Leber's disease**, hereditary optic atrophy, a rare hereditary disease in which partial blindness in both eyes sets in rapidly, typically affecting young men; also called *Leber's* (*hereditary optic*) *atrophy*.

1890 Billings *Med. Dict.* II. 413/1 *Leber's disease*, hereditary optic atrophy. **1902** *Encycl. Medica* X. 358 Possibly the family cases known as Leber's atrophy may come under this heading. **1933** *Times Lit. Suppl.* 25 May 334/4 The..analysis is applied to statistics of albinism, colour blindness, haemophilia and Leber's disease. **1952** C. P. Blacker *Eugenics* iv. 248 Among these genes [located on the human sex chromosomes] are those believed to determine:..Leber's optic atrophy, a progressive form of blindness. **1971** Duke-Elder & Scott in S. Duke-Elder *Syst. Ophthalm.* XII. iii. 86 Leber's hereditary optic atrophy is a relatively rare condition of unknown aetiology. *Ibid.*, In the main Leber's disease is hereditary.

|| **leberwurst** (lē·bəɹwʊɹst). [G.] = *liver sausage.

1855 Geo. Eliot in *Fraser's Mag.* June 706/1 Goethe..is enthusiastic about the delights of dining on blaukraut and leberwurst (blue cabbage and liver sausage). **1969** [see *bratwurst]. **1971** R. Petrie *Thorne in Flesh* iii. 40 A packet of case spaghetti and a jar of leberwurst.

|| **lebes** (le·bīz). *Gr. Antiq.* [Gr. λέβης.] A deep round-bottomed bowl, usually set on a stand, for holding wine; often used as a wedding-gift (*lebes gamikos*).

1851 *Catal. Greek & Etruscan Vases Brit. Mus.* I. 34 Lebes. Clay..; varnish black.. **1885** *Encycl. Brit.* XIX. 614/1 (*caption*) On the left is a..pyxis.., in the centre are the two tall vases (lebes) on a plinth. **1935** Richter & Milne *Shapes & Names Athenian Vases* 9 Lebes (Greek λέβης), deep bowl with round bottom, made to be set on a stand. *Ibid.* 11 Lebes gamikos (Greek λέβης γαμικός), 'marriage bowl'. High foot, double handles on the shoulder, the bowl in one piece with the foot. **1937** *Antiquity* XI. 246 The large nuptial lebes. **1974** Savage & Newman *Illustr. Dict. Ceramics* 177 Lebes (Greek), a type of bowl of Greek pottery used for mixing wine and water. It is ovoid in form and has a high shoulder, a low neck, and two vertical handles; it usually has a rounded bottom and rests on a stand.

Leblanc[1] (ləbla·ṅ). Also **LeBlanc.** The name of Nicolas *Leblanc* (1742–1806), French chemist, used *attrib.* to designate a (now obsolete) process for the manufacture of sodium carbonate in which sodium chloride is treated with hot concentrated sulphuric acid to form

the sulphate ('salt-cake'), which is then heated with limestone and coal and the resulting carbonate dissolved out with water.

[**1864** *Chem. News* 5 Mar. 111/1 (*heading*) Theoretical researches on the preparation of soda by Leblanc's process.] **1880** G. Lunge *Theoret. & Pract. Treat. Manuf. Sulphuric Acid & Alkali* II. iv. 361 (*heading*) The manufacture of soda by the Leblanc process. **1930** J. A. Timm *Introd. Chem.* xxxiii. 438 During the 75 years which followed, the LeBlanc process grew to be a great industry, spreading to Germany, Austria, and England. **1965** D. Abbott *Inorg. Chem.* iv. 157 Sodium sulphate is manufactured by the first stage of the Leblanc process.

Leblanc[2] (ləbla·ṅ). *Electr. Engin.* The name of Maurice *Leblanc* (1857–1923), French electrical engineer, used *attrib.* to designate apparatus invented by him, as **Leblanc connection**, a method of connecting three single-phase transformer windings to convert three-phase current to two-phase; **Leblanc exciter** or **phase advancer**, a device for advancing the phase of the rotor current of an induction motor, consisting of a direct-current armature and commutator, having three sets of brushes per pair of poles connected to the slip rings of the main motor, and driven somewhat faster than the main motor.

1924 M. Walker *Control of Speed & Power Factor of Induction Motors* vii. 117 The Leblanc exciter, consisting of an armature built like a continuous current armature and excited by the rotor currents following either the armature itself or in a field magnet surrounding the armature. **1948** M. G. Say *Performance & Design Alternating Current Machines* (ed. 2) v. 68 The Le Blanc connection has the advantage of using a standard three-phase transformer core. **1965** J. Hindmarsh *Electr. Machines* ix. 492 (*heading*) Leblanc phase advancer. **1966** Brosan & Hayden *Adv. Electr. Power & Machines* vi. 243 In the Leblanc connexion there are three magnetic cores, and the primary windings may be connected in either star or delta. *Ibid.* x. 476 Another method of power factor control utilizes the Leblanc exciter. **1968** A. R. Daniels *Performance Electr. Machines* v. 97 An alternative method of 3/2 phase conversion is the Leblanc connection.

leccer (le·kəɹ). Also **lecker, lekker.** [*-er[6].] Slang or colloquial alteration of Lecture *sb.* (See also quot. 1900.)

1899 *Daily Tel.* 14 Aug., *Leccers*, lectures. **1900** Farmer *Public School Word-Bk.* 124 *Lecker*, 1. (Oxford). A lecture. 2. (Harrow). The electric light. **1904** [see *-er *suffix*[6]]. **1907** 'B. Burke' *Barbara goes to Oxf.* (1915) 115 I'm awfully sorry that I had to cut your leccer, my mother came up quite unexpectedly. **1911** W. Elmhirst *Freshman's Diary* (1969) 10 Had 1st leccer from the Dean this morning. *Ibid.* 11 Had a caller, who..said good night saying we should meet in another sphere what one I don't know as he said he didn't row & he certainly doesn't come to P.Mods leccers. **1914** C. Mackenzie *Sinister St.* II. III. viii. 455 And you won't come out..to watch people buying copies on their way to leckers? **1928** *Daily Express* 29 June 5/3 A..dilapidated basket filled with gay-coloured 'lekker' notebooks.

lech (letʃ), *sb.*[4] Also **letch.** [Now regarded as a back-formation from Lecher *sb.*, but cf. Letch *sb.*[2]] **a.** A strong desire or longing, esp. sexual. **b.** = Lecher *sb.*

1796 [see Letch *sb.*[2]]. *c* **1830** *Venus School Mistress* Pref. in 'Pisanus Fraxi' *Index Librorum Prohibitorum* (1877) 399 It [*sc.* flagellation] is, however, a *lech*, which has existed from time immemorial. **1868** *Index expurgatorius of Martial* 39 There are various rumours as to the nature of your letch. *c* **1888–94** *My Secret Life* III. 147 Did they fuck with me for fun, for letch, or for money? **1934** G. Greene *It's a Battlefield* 204 This is when a girl gets a baby; when she's got a lech like this. **1938** S. Beckett *Murphy* vii. 126 A man could no more work a woman out of her position on her own ground of sentimental lech than he cd outsmell a dog. **1940** S. Lewis *Bethel Merriday* xxxiii. 387 Your letch for power over everyone around you. **1941** 'R. West' *Black Lamb* II. 204 Those who had a lech for violence could gratify it. **1943** H. A. Smith *Life in Putty Knife Factory* x. 157 If anybody noticed what I was doing, they'd think I was an old letch. **1956** E. Pound tr. *Sophocles' Women of Trachis* 17 All started when he had a letch for the girl. **1958** *Spectator* 10 Oct. 482/1 A post-war working-class family..—the grey letch of a father, his jolly rolypoly wife and their prissy daughter. **1959** J. Braine *Vodi* xix. 220, I don't mind admitting I always had a lech for her. **1960** *Times Lit. Suppl.* 27 May 333/4 Graves is 'a lech', whose current mistress is Purling's wife, Jo. **1964** [see *drool *v.*]. **1970** *Guardian* 13 Feb. 9/6 A rich man can have a beautiful young wife even if he is a gropy old letch! **1971** *Petticoat* 17 July 7/1 Out of ten girls who are invited back to men's flats for coffee, at least eight expect a cup of coffee and are quite shocked and horrified when they find themselves pinned to the bed, five seconds after walking into the lech's lair. **1972** *Sunday Times* 12 Nov. 40/3 Many so-called platonic friendships..are merely one-way leches.

Lech, Lekh (leχ), *sb.*[5] and *a.* Also **Lach, L'ach** (lyaχ). [ad. G. *Lech*, O.Russ. *lyakh*; f. O.Pol. **lęch*.] **A.** *sb.* A member of an early Slavonic people once inhabiting the region around the upper Oder and Vistula, whose descendants are the Poles; also, the name of a legendary ancestor of this people. **B.** *adj.* Of or pertaining to the Lechs or their language. Cf. *Lechish *sb.* and *a.*, *Lechitic *sb.* and *a.*

1893 W. R. Morfill *Poland* 23 In the sixth or seventh centuries some people settled on that river [*sc.* the Vistula] are called Lekhs, a word which has never been satisfactorily explained. The older form probably had a nasal: hence we get in the Latin chroniclers *Lenchitæ*, in Lithuanian, *Lenkas*, and in Magyar, *Lengyel*. **1911** *Encycl. Brit.* XXV. 236/2 In the north Polish is closely connected with Kašube, and this with Polab, making the group of L'ach dialects in which the nasals survived... The two Sorb dialects link the L'achs on to the Čechs and Slovaks, the whole making the N.W. group with its preference for *c, z, s* as against *č, ž, š*. **1929** *Ibid.* XVIII. 161/2 The nearest relative of Polish is Polabian, with which it forms the Lech group. **1939** G. Slocombe *Hist. Poland* (new ed.) 12 The Western Slavs had become divided into three distinct sections: the Serbs..; the Czech group..; and the Lech group, in which were included the Obodrites, the Wiltzi, the Pomeranians.., and other tribes who were in the course of the succeeding centuries to form the Polish nation. **1950** A. P. Goudy in *Cambr. Hist. Poland to 1696* i. 10 Besides the name Polanie, there existed another collective name—Lachy (Lechs). This term is used in the Chronicle of Nestor to indicate the Poles and came into frequent use by the old chroniclers.

lech (letʃ), *v.* Also **letch.** [Back-formation from Lecher *sb.*] *intr.* To behave lustfully, to feel or to be lecherous. Occas., to have a (non-sexual) desire.

1911 J. Masefield *Everlasting Mercy* 68 And drunk and leched from day till morrow. **1940** E. Pound *Let.* 18 Jan. (1971) 334, I have now the text of Erigena, and if I could get hold of the recent publications about him, I could write quite a chunk. Not that I am letching to. **1948** Partridge *Dict. Forces' Slang* 110 *Letch*, to look at women, not necessarily in a lecherous way, in spite of its derivation. **1957** C. Day Lewis *Pegasus* 13 Unblest, Unchecked—what a serpent flame letched at her marrow! **1963** 'M. Corrigan' *Why do Women —?* xiii. 89, I..letch around looking for sex thrills. **1972** M. Farhi *Pleasure of your Death* vii. 173 He was still watching the..shapely ankles when Chastity pulled him to task. 'Don't lech!' **1973** *Guardian* 27 Feb. 10/1 A fortyish factory worker..lives with..an obsessively nubile sister whom he obviously leches after.

lechaim, var. *lechayim.

Le Chatelier (ləʃɑ̂te·lye). [The name of Henry *Le Chatelier* (1850–1936), French chemist.] **a.** Used *attrib.* with reference to a test for the soundness (freedom from expansion) of cement using a small hollow brass cylinder split longitudinally and having pointers close to the split which indicate the extent of any expansion that occurs when the cylinder is filled with cement.

1904 *Specification for Portland Cement* (*B.S.I.*) 7 The cement shall be tested by the Le Chatelier method, and shall in no case show a greater expansion than 12 millimetres after 24 hours aeration and 6 millimetres after seven days aeration. The apparatus for conducting the Le Chatelier test..consists of a small split cylinder of brass. **1930** *Engineering* 18 July 62/1 Dry clay when adsorbing water increased in bulk and exerted..sufficient pressure to spring open the Le Chatelier gauge about 1 lb. per square inch. **1963** A. M. Neville *Properties of Concrete* i. 50 The Le Chatelier test detects unsoundness due to free lime only.

b. Used *attrib.* and in the possessive († and in conjunction with the name of K. Ferdinand *Braun* (1850–1918), German physicist) to designate a principle enunciated by Le Chatelier, which states that if a constraint (such as a change in pressure or temperature) be applied to any system in equilibrium, the equilibrium will shift in such a way as to tend to counteract the effect of the constraint.

1910 *Chem. Abstr.* IV. 1600 (*heading*) New isomerization of benzopinacolins and Le Chatelier's law. **1911** *Ibid.* V. 3361 The author gives the following as the usual statement of the Le Chatelier–Braun principle. **1922** *Proc. Amer. Acad. Arts & Sci.* LVII. 25 The condition which must be satisfied in order that the Le Chatelier Principle may hold with regard to the effect of a change in the initial mass of *one* component, is that the addition of such component shall accelerate or retard the transformation (at equilibrium). **1943** *Thorpe's Dict. Appl. Chem.* (ed. 4) VI. 229/1 It would be expected from the Le Chatelier–Braun principle of mobile equilibrium that increase of pressure would exert a retarding effect upon the rate of a unimolecular reaction. **1954** A. R. Bailey *Text-bk. Metall.* viii. 232 A reaction involving a gaseous reactant is favoured by increase in operating pressure, and one involving a gaseous product by decrease of pressure or by sweeping it away in a gas stream; these points also follow from Le Chatelier's Principle. **1961** A. Holderness *Inorg. & Physical Chem.* xviii. 258 If the temperature of an equilibrium system..is lowered, Le Chatelier's Principle requires the equilibrium to shift so as to tend to raise the temperature again; that is, to evolve heat. **1973** *Jrnl. Chem. Education* L. 124 Since stretching a rubber band is an exothermic process, Le Chatelier's principle predicts that heat applied to a stretched rubber band will contract it.

lechatelierite (ləʃɑ̂te·liərəit, ləʃɑ̂telīə·rəit). *Min.* Also † **-iérite.** [a. F. *lechateliérite* (A. Lacroix 1915, in *Bull. de la Soc. franç. de Min.* XXXVIII. 185), f. *Le Chatelier + -ite[1].] Naturally occurring vitreous silica, SiO_2, formed when siliceous material is intensely heated (as by lightning).

1916 *Mineral. Mag.* XVII. 353 *Lechateliérite*, naturally occurring fused (amorphous) silica. **1928** *Amer. Mineralogist* XIII. 77 Lechateliérite is unique in that it is the only naturally occurring glass that is definite enough to be considered a mineral. **1931** *Jrnl. R. Soc. W. Austral.* XVII. 146 Digging revealed a vertical core of lechatelierite in the soil. This core was hollow and very brittle, and extended downwards for about a metre. **1963** W. A. Deer et al. *Rock-Forming Min.* IV. 180 Silica glass (vitreous silica; lechatelierite): can exist at room temperatures and up to 1000°C... It is an unstable glass at all temperatures below 1713°C. **1964** *New Scientist* 16 Jan. 160/1 The invariable signs of flow and common presence of lechatelierite (pure silica glass) indicate that tektites were formed by rapid fusion at very high temperatures.

|| **lechayim** (ləχəi·im). Also **l'chaim, l'cha-y(i)m, lechaim, lehayim.** [Heb., 'to life'.] A drinking toast: to life!

1932 L. Golding *Magnolia St.* III. ix. 582 'Here's mud in your eye!' says the Chicagoan. '*Lechayim!* To Life!' says Mr. Emmanuel. **1963** *Encounter* Apr. 35/1 The guests stood and raising their glasses honoured me, in Hebrew, with that most beautiful of toasts: *Lechaim!* To life! **1968** M. Richler in R. Weaver *Canad. Short Stories* 2nd Ser. 193 'I been here seven years ago and what we done since, it's remarkable. *L'chaym.*' '*L'chaym.*' **1968** P. Durst *Badge of Infamy* iii. 23 Chaim raised his glass. 'Good health.' '*L'chaim*,' Michael returned. **1968** L. Rosten *Joys of Yiddish* 205 *L'chayim*, pronounced *l-khy-im*, with a resounding German *kh*, to rhyme with 'to fry 'em'. Hebrew: 'To life.' The toast offered, with raised glass, before sipping wine or liquor: 'To your health.' **1973** *Jewish Chron.* 2 Feb. 16/4 Miss Kitt raised her wine-glass. 'L'chayim,' she said.

Lecher[2] (le·χəɹ, le·tʃəɹ). *Physics.* Also **lecher.** The name of Ernst *Lecher* (1856–1926), Austrian physicist, used *attrib.* (esp. in *Lecher wires*) and † in the possessive to designate a pair of parallel wires in which the frequency of a high-frequency electric oscillation may be measured by means of a sliding detector or conductor placed so as to bridge the wires, positions of maximum response or absorption being separated by a distance equal to half the wavelength of the oscillation.

1897 *Phil. Mag.* XLIV. 202 In Lecher's arrangement.. the wires are of equal diameter. **1902** *Encycl. Brit.* XXVIII. 59/2 Many problems of electric waves along wires can readily be investigated by a method due to Lecher, and known as Lecher's bridge. **1929** J. A. Ratcliffe *Physical Princ. Wireless* iii. 35 A pair of parallel wires is often used to guide the waves, instead of the single wire... This arrangement is known as the Lecher wire system. **1947** *Jrnl. Inst. Electr. Engin.* XCIV. 953/2 The tuned circuit consisted of a pair of lecher rods, the output being fed through a resonant line to a fixed end-fed half-wave vertical aerial. **1962** W. B. Thompson *Introd. Plasma Physics* ii. 12 In arc discharges the electron density is 10^{11}–10^{12} cm^{-3} and the plasma frequency ~ 100 Mc/s, so the high-frequency signals were picked up on resonant Lecher wires, rectified by a crystal and detected by a galvanometer. **1968** *Radio Communication Handbk.* (ed. 4) xix. 12/2 Lecher lines..comprise a pair of taut parallel wires, spaced an inch or so apart to form an open wire transmission line, and a bridge to short circuit the wires which can be moved along the line as required.

lecherously, *adv.* (Later example.)

1972 *Daily Tel.* 12 May 12/8 Amorous delusions concerning..a lecherously attentive neighbour and her kindly but pre-occupied husband.

Lechish (le·χiʃ), *sb.* and *a.* [ad. G. *lechisch*; cf. *Lech *sb.*[5] and *a.*] = next.

1888 J. Wright tr. *Brugmann's Elem. Compar. Gram. Indo-Germanic Lang.* I. 12 The Slavonic languages fall into a South-Eastern and a Western group... To the latter [belong] Czech.., Sorabian or Wendish..and Lechish (Polish and Polabian or Elbe-Slavonian). **1908** T. G. Tucker *Introd. Natural Hist. Lang.* 224 The classification of the Slavonic tongues which appears to find most favour with students in that branch is as follows:—..(ii) West Group: e. Tzech. f. Sorbian. g. Lechish [Polish, Polabish (Elbe-Slavonic)]. **1936–7** *Slavonic & East European Rev.* XV. 477 The relationship of the present and past Baltic dialects of Slavonic to the other Lechish languages.

Lechitic (leχi·tik), *sb.* and *a.* Also **Lechite, Lekhite, Lekhitic.** [ad. G. *lechitisch*; cf. med.L. *Le(n)chitae* and *Lech *sb.*[5] and *a.*] Name given by some linguists to certain West Slavonic languages (Polish, Kashubian, Slovincian, the extinct Polabian) showing characteristic features in common and sometimes held to have once formed a single sub-dialect within the Slavonic group. Also as *adj.*, of or pertaining to the Lechs or their language. Also comb. form **Lechito-, Lekhito-.** Cf. prec.

1934 G. C. Engerrand *So-called Wends of Germany* (Univ. of Texas Bull. No. 3417) 35 Its Western subgroup ..is composed of the Polish, Kashub-Slovince, former Polab, Wendish.., Czech, and Slovak languages. If we eliminate the three latter ones from that subgroup, we have a family, to which philologists give the name of Lechitic (L'Ach, Lekh), etc., that is characterized..by the persistency of many old Slavic nasals. **1935** *Times Lit.*

Suppl. 15 Aug. 506/2 Like most of the Polish Slav philologists of the last two generations, he [*sc.* T. Lehr-Spławiński] holds the theory of a Lechitic linguistic community, that is to say, that all the Slav languages of the Baltic region from Polish in the east to Dravanian on the west bank of the Elbe constituted an unbroken chain of mutually related groups. **1939–40** *Slavonic & East European Rev.* XIX. 273 Schleicher states that Kashubian stands as a bridge between the West Lechite (Polabian) and East Lechite (Polish). **1946–7** *Ibid.* XXV. 493 Shakhmatov sees in this signs of an intermingling of certain Lechitic tribes with the Russians in North Russia. **1949–50** *Ibid.* XXVIII. 286 The Poles, as the most conservative and least mobile of the Western (Lechitic) tribes, stayed behind in the original habitat. **1950** A. P. Goudy in *Cambr. Hist. Poland to 1696* i. 9 From the linguistic point of view Slovinzish and Kashubish belong to the Polish group and it is usual to class these languages (or dialects) along with Polish and Polabian under the title Lechitic (grupa lechicka). **1964** M. Samilov *The Phoneme jat' in Slavic* 144 In Lekhitic the nasals have generally preserved their nasality. **1966** H. Birnbaum *Ancient Indo-European Dialects* 194 The dissociation of.. the West Slavs into a Lekhito-Sorbian group..and a Czechoslovak group. **1972** G. Stone *Smallest Slavonic Nation* 96 I. Taszycki's view that West Slavonic had first divided into two sub-groups—Lechito-Sorbian and Czecho-Slovak—was subsequently supported by Zdzisław Stieber.

lechwe (lī·tʃwī). Now the usual spelling of Leche. Also **lechwre, leshwe, letchwe, letshewe**; also *attrib.* (Examples.)

1881 E. E. Frewer tr. *Holub's Seven Yrs. S. Afr.* II. vi. 128 The letshewes were larger and the pukus smaller than blessbocks, and both, like all water bucks, had shaggy, light brown hair, and horns bent forward. **1907** *Westm. Gaz.* 6 Sept. 3/1 The haunts of the jacana, of the waders in general, and of the Lechwe and Situtunga. **1915** *Chambers's Jrnl.* Nov. 701/2 The lechwre is remarkable for its waterloving characteristics. **1920** *United Free Ch. Miss. Rec.* Aug. 138/1 The letchwe is an antelope much addicted to knee-deep water. **1936** P. M. Clark *Autobiogr. Old Drifter* x. 131 My first buck..was a large lechwe with beautiful curved horns. **1946** *Cape Times* 7 Aug., Herds of elephant up to 80 strong were seen, as well as.. the rare letchwe and puku buck. **1949** *Ibid.* 30 July 5/5 An expedition..has found many specimens in the Caprivi Strip, including..three genets, seven leshwe, one putu, [etc.]. **1969** *Times* 24 Oct. (Zambian Suppl.) p. xi/2 One of the rarest species of game in the world, the black lechwe antelope, is faced with extinction. **1972** *Nature* 7 Apr. 265/1 He further suggested that..the Okavango and Caprivi be set aside for species such as lechwe, puku and sitatungu.

-lecithal (le·siþăl), *suffix* [f. Gr. λέκιθ-ος yolk + -al], used to form adjs. describing egg cells with yolks of specified kinds, as *alecithal *a.*, *homolecithal* adj. (s.v. *homo-).

lecithin. Add to def.: In mod. use, any of a group of phospholipids found in plants and animals which are esters of a phosphatidic acid with choline and on hydrolysis yield choline, phosphoric acid, glycerol, and two fatty acids; also used as a generic name for these compounds. **b.** A commercial mixture of lecithin with other phosphatides and often other lipids obtained from natural products and used industrially, esp. that from soya beans. (Further examples.) [First formed as F. *lécithine* (N. T. Gobley 1850, in *Jrnl. de Pharm. et de Chim.* XVII. 411).]

1923 [see *lysolecithin* s.v. *lyso-]. **1926** G. D. Elsdon *Chem. & Exam. Edible Oils & Fats* iii. 15 Lecithin is a complex compound..and may be looked on as a triglyceride in which one of the fatty acid radicles has been replaced by a complex organic base containing phosphoric acid. *Ibid.* xii. 205 Maize oil contains 1·1 to 1·5 per cent. of lecithin. **1951** M. B. Jacobs *Chem. & Technol. Food & Food Products* (ed. 2) III. xlii. 2155 The addition of lecithin to chocolate results in a saving of cacao butter, counteracts moisture, and stabilizes the chocolate. **1951** K. S. Markley *Soybeans* II. xvi. 600 It has become customary in industrial circles to call the phosphatide residue obtained in the commercial manufacture of soybean oil soybean lecithin or simply lecithin. The commercial product contains roughly two-thirds phosphatides and one-third soybean oil. *Ibid.* 601 The term commercial lecithin or lecithin is applied almost exclusively to soybean lecithin, whereas 20 years ago it would have referred to egg lecithin. **1954** *Thorpe's Dict. Appl. Chem.* (ed. 4) XI. 47/2 Lecithins and cephalins, about 2%, are present in the oil [from soya beans] and have been widely used as emulsifiers in the food, textile, cosmetic, soap, and other industries. **1961** H. F. Payne *Organic Coating Technol.* II. xxiii. 970 Lecithin..is a balanced polar–non-polar compound and will concentrate at the interface between polar pigments and less polar oils and resins to reduce the interfacial tension and facilitate wetting. **1967** *Martindale's Extra Pharmacopoeia* (ed. 25) 293 Lecithins occur in all animal and vegetable cells and vary in composition according to the source from which they are obtained. The two chief commercial varieties are egg lecithin (ovolecithin).., and vegetable lecithin..from various vegetable sources, particularly leguminous seeds. **1970** Ambrose & Easty *Cell Biol.* viii. 272 The main lipid constituents of plasma membranes are phospholipids (in particular phosphatidylcholine or lecithin), and cholesterol, a steroid. **1973** *Sci. Amer.* Apr. 85/1 The alveolar fluid contains trace amounts of various large molecules; the most abundant are the two principal lipids—lecithin and sphingomyelin—that represent the bulk of the pulmonary surfactant.

lecithinase (le·siþinēⁱz, -ēⁱs). *Biochem.* [f. Lecithin + *-ase.] = *phospholipase.

1910 *Chem. Abstr.* IV. 2680 Lecithinase is itself almost or entirely free from toxicity, showing its independence from the neurotoxin of cobra venom. **1947** *Jrnl. Biol. Chem.* CLXIX. 704 Four types of lecithinases, each acting on a separate ester linkage of the lecithin molecule, were postulated by Contardi and Ercoli in 1932.., but the existence of only three of these has hitherto been demonstrated. These are (*a*) the enzyme found in cobra serum.. which splits off a single unsaturated fatty acid from the lecithin molecule; (*b*) the enzyme found in rice hulls..and *Aspergillus oryzae*..,which splits off both fatty acids; (*c*) the enzyme found in *Clostridium welchii*..,which separates the phospholipide molecule at its ester linkage between the glycerol and the phosphoric acid. **1959** *Biochem. Jrnl.* LXXI. 619/1 It may be of significance for the biological function of intestinal lecithinase that the optimum conditions for its activity are those which can be expected to prevail in the mucosal cells during absorption of fat. **1970** D. R. Dilley in A. C. Hulme *Biochem. Fruits* I. viii. 181 Proteases and lecithinase increase lysosomal permeability.

lecithotrophic (lesiþotrǫ·fik, lekiþo-), *a.* [f. Gk. λέκιθος yolk of egg + *-trophic.] Of the larvæ of certain marine invertebrates, feeding on the yolk of the egg from which they have emerged.

1950 G. Thorson in *Biol. Rev.* XXV. 10 The lecithotrophic pelagic larvae, mainly developing from fairly large yolky eggs, are of a clumsy shape, rather unfit for locomotion. **1962** D. Nichols *Echinoderms* x. 119 *Heliocidaris* has a lecithotrophic larva (feeding on stored yolk). **1967** *Oceanogr. & Marine Biol.* V. 360 Experiments refer to short-lived lecithotrophic larvae.

leck (lek), *v.* *rare* exc. *dial.* [Cf. E.D.D. *leck* *v.*] = Leak *v.* 2 c.

1922 Joyce *Ulysses* 749 Shes [a cat] as bad as a woman always licking and lecking.

lecker, var. *leccer.

Leclanché (ləklā·nʃe). The name of Georges *Leclanché* (1839–82), French chemist, used *attrib.* and *absol.* to denote a primary cell invented by him that has a zinc cathode in contact with zinc chloride, ammonium chloride (in solution or as a paste) as the electrolyte, and a carbon anode in contact with a mixture of manganese dioxide and carbon powder.

1871 *Chem. News* 6 Oct. 166/2, I find that what I said about the Leclanché battery..has led to a false impression. .. It is used only on circuits of considerable resistance, and not much work, for which it is found very suitable. **1878** *Encycl. Brit.* VIII. 93/1 Good instances of this kind of action are furnished by the bichromate battery of Bunsen and the Léchanché [*sic*] cell, which occupy a sort of middle position between one and two fluid batteries. **1891** E. M. Caillard *Electr.* iv. 264 A form of Daniell's cell has been chiefly adopted in England, but the Leclanché is also excellent for telegraphic purposes. **1907** M. K. Kassabian *Rontgen Rays & Electro-Therapeutics* iii. 64 The Leclanché cell consists of a porous cup and a carbon plate. **1946** J. R. Partington *Gen. & Inorg. Chem.* xxix. 827 Manganese dioxide is used..as a depolariser in the Leclanché cell. **1971** L. T. Agger *Introd. Electr.* xi. 171 The dry form of the Leclanché cell, which was introduced towards the end of the last century, has generally replaced the wet Leclanché and is now by far the most commonly used primary cell.

-lect, terminal element, f. L. *lect-*, *legere* to read, used to designate a regional or social variety within a language as in Dialect, *idiolect; also used in forming a number of technical terms in linguistics, as *acrolect*, *basilect*, *isolect*, *sociolect*, etc. (see quots.). Hence (without hyphen) as *sb.*, a social variety of a language or dialect.

1965 W. A. Stewart in R. W. Shuy *Social Dial. & Lang. Learning* 15, I will refer to this topmost dialect in the local sociolinguistic hierarchy as *acrolect* (from *acro-* 'apex' plus *-lect* as in *dialect*). In most cases what is meant by 'Standard' English is either acrolect or something close to it. At the other extreme is a kind of speech which I refer to hereafter as *basilect* (from *basi-* 'bottom'). **1969** *Florida FL Reporter* VII. i. 48 Although acrolect differs also in sounds and words from basilect, grammatical differences between them create the real blocks to communication. **1971** C.-J. N. Bailey in *Working Papers in Ling.* (Univ. of Hawaii) III. v. 39 In this case, the creole becomes a satellite (*satellect* or *acolutholect*) to the established language (*matrilect*). *Ibid.*, The matrilect serves as the *acrolect* in the continuum at one end, while the *basilect* ..will be separated from the acrolect by a graded (systematic) series of *mesolects*. *Ibid.*, An *isolect* has been defined by me elsewhere as a form of speech different from its isolectal correlate. *Ibid.* 41, I have suggested *paralect* to denote folk creations from the related systems (e.g. middle Arabic, Punti, Slavish). **1972** J. L. Dillard *Black English* iii. 107 Higginson recorded many clause and question forms which are much as they still are in Black English basilect today. *Ibid.* 300 Dialect refers to a set of features delimited geographically; sociolect to a socially distributed set. **1974** J. Nist *Handicapped Eng.* iii. 72 Any departure from that code..marks the speaker as a 'folk-speaker' of his basilect... *Ibid.*, Geographical dialects in the present-day British English automatically become class lects. **1975** *College Composition & Communication* XXVI. i. 104/1 She is being primed to be the ideal teacher of basilect students in spite of her messy lect.

lectin (le·ktin). *Immunol.* [See quot. 1954 and -in[1].] A substance, usu. a protein of plant origin, which has the properties of an antibody but is not produced in response to an antigen.

1954 W. C. Boyd in Neurath & Bailey *Proteins* IIb. xxii. 789 It would appear to be a matter of semantics as to whether a substance not produced in response to an antigen should be called an antibody, even though it is a protein and combines specifically with certain antigens alone. It might be better to have a different word for these substances, and the present writer would like to propose the word *lectin*, from the Latin *lectus*, the past participle of *legere*, meaning to pick, choose, or select. **1971** *New Scientist* 8 Apr. 82/2 Over a dozen different agglutinins or lectins have now been isolated from a variety of plant materials, but most work has centred on just two of them—wheat germ agglutinin and concanavalin A. Both preferentially agglutinate several sorts of transformed cells, including those transformed by polyoma virus. **1971** *Nature* 30 July 299/2 During the past eighteen months plant agglutinins, or lectins.., have become a major topic of conversation in many cancer research laboratories.

lectio difficilior (le·ktio difiki·liǫɹ). *Textual Criticism.* Also **difficilior lectio.** [L., 'harder reading', from the maxim *difficilior lectio potior*.] Of two alternative manuscript readings, the one that is less obvious, and therefore less likely to be a copyist's error; also, the practice of giving preference to such a reading.

1901 F. G. Kenyon *Handbk. Textual Crit. New Testament* i. 13 One proposition is so often stated as a leading principle in textual criticism... It is..formulated by Bengel in the words, *Proclivi scriptioni praestat ardua*, or ..*Difficilior lectio potior*; the harder reading is to be preferred to the easier... The 'difficilior lectio' is preferable.. because a hard reading is likely to be altered into an easy one. **1901** W. Edie tr. *Nestle's Introd. Textual Crit. Greek New Testament* iii. 157 The principle laid down in the maxim, *lectio difficilior placet*..is perfectly sound. *a* **1955** B. Flower tr. *Maas's Textual Crit.* (1958) 13 It is right to prefer as a rule the '*lectio difficilior*'. **1962** E. J. Dobson in Davis & Wrenn *Eng. & Medieval Stud.* 130 A's is the *difficilior lectio* and somewhat better in sense. **1966** *English Studies* XLVII. 284 The 'lectio difficilior' can be the better or the worse reading as the case may be. **1968** Reynolds & Wilson *Scribes & Scholars* v. 150 Many references to the principle of *difficilior lectio* will be found in commentaries, and there is no doubt of its value. **1969** R. Renehan *Greek Textual Crit.* 27 As the rarer verb, it is a *lectio difficilior* and should be received into the text here.

lection, *sb.* **2.** Add: Also *attrib.*

1927 A. H. McNeile *Introd. New Testament* 383 It [*sc.* the Codex Bezae] contains certain lection marks which Brightman holds to be Byzantine.

lection (le·kʃən), *v.* *rare*[−1]. [f. the sb.] To read a lesson from.

1922 Hardy *Late Lyrics* 165, I went where my friend had lectioned The prophets in high declaim.

le·ctorship. [f. Lector + -ship.] The office or post of lector.

1605 H. Wotton *Let.* 18 Aug. in L. P. Smith *Life & Lett. Sir H. Wotton* (1907) I. xii. 331 He hath since been stayed with a Lectorship in Genua. **1906** *Westm. Gaz.* 24 Mar. 3/2 It is hereby expressly stated that the Lectorship cannot be held for life. **1911** A. Brennan *Life St. Lawrence of Brindisi* 42 The Lectorship was but the first step in his ascent to the highest dignities of the Order.

lectotype (le·ktotəip). *Taxonomy.* [f. Gk. λεκτός chosen + *type *sb.*[1]] A specimen from the original material serving as the basis for a description of a new species, selected as the type in the absence of a holotype.

1905 Schuchert & Buckman in *Ann. & Mag. Nat. Hist.* 7th Ser. XVI. 103 *Lectotype*, a syntype chosen, subsequently to the original description, to take the place which in other cases a holotype occupies. **1951** G. H. M. Lawrence *Taxon. Vascular Plants* ix. 204 A lectotype is a specimen or other element selected from the original material to serve as the nomenclatural type, when the holotype was not designated at the time of publication, or when the holotype is missing. **1953** E. Mayr et al. *Methods & Princ. Syst. Zool.* xii. 242 A selection of lectotypes should be undertaken only when it leads to the clarification of a taxonomic problem. **1963** Davis & Heywood *Princ. Angiosperm Taxon.* viii. 280 The lectotype should be chosen, if possible, from among the specimens actually seen by the author when he described the species. **1967** R. E. Blackwelder *Taxonomy* iv. 293 There can be no question that a type (at least holotype, lectotype, or neotype) belongs to the species it typifies. **1970** *Watsonia* VIII. 43 The lectotype of *E[uphrasia] rostkoviana* f. *borealis* was chosen by Mr. Sell and me [*sc.* P. F. Yeo] in 1968. **1975** *Trans. R. Entomol. Soc.* CXXVI. 615 This specimen..is here designated Lectotype and has been labelled accordingly.

lectrice. (Later example.)

1899 G. B. Shaw *Shaw on Theatre* (1958) 73 An attempt to force the Salvation Army to have their hymns licensed by the Archbishop of Canterbury, or the daily papers to have their political leaders licensed by the Queen's Lectrice, would produce an overwhelming agitation at once.

lecturable (le·ktʃŭrăb'l), *a. rare.* [f. LECTURE *v.* + -ABLE.] That can be the subject of a lecture.

1828 DISRAELI *Voy. Capt. Popanilla* v. 48 The voices of boys lecturing upon every lecturable topic.

lecture, *sb.* Add: **1.** (Later examples.)

1904 CONRAD *Nostromo* I. vi. 47 In about a year he had evolved from the lecture of the letters a definite conviction. **1922** JOYCE *Ulysses* 708 What fractions of phrases did the lecture of those five whole words evoke? **1929** R. BRIDGES *Testament of Beauty* I. 24 If we read but of Europe since the birth of Christ, 'tis still incompetent disorder, all a lecture of irredeemable shame.

7. *lecture agency, agent, audience, circuit, course, -goer, -hall, list, note, -room* (earlier and later examples), *-theatre* (earlier and later examples), *-tour* (also as vb.); **lecture-day** (later U.S. examples); **lecture-recital,** a lecture illustrated by music.

1925 A. HUXLEY *Let.* 25 Jan. (1969) 240 You suggest lectures for lucre in the U.S.A.:—I have had several offers from various lecture agencies... The fatigue and the boredom of a lecture tour frighten me. **1949** DYLAN THOMAS *Let.* 1 Dec. (1966) 340 He said that the Lecture Agencies..have nowhere near his own acquaintanceship with the institutions. **1966** N. NICOLSON in H. Nicolson *Diaries & Lett.* (1966) 131 Colston Leigh Inc. was the lecture-agency. **1873** 'MARK TWAIN' & WARNER *Gilded Age* lviii. 527, I am a business man. I am a lecture-agent. **1949** DYLAN THOMAS *Let.* 1 Dec. (1966) 341 Surely a letter from Brinnin, acting as my secretary & Lecture-Agent,..would mean something to the Treasury. **1943** WYNDHAM LEWIS *Let.* 5 Dec. (1963) 372 Seeing the gas-shortage whittles down all lecture-audiences, I had quite a lot of people. **1974** M. FIDO *R. Kipling* 64/2 'Here's poetry at last!' he [*sc.* Professor Masson] burst out to his lecture audience on the day 'Danny Deever' appeared. **1965** *Times Lit. Suppl.* 25 Nov. 1057/3 Well-financed readings on large lecture-circuits..are staple. **1967** O. WYND *Walk Softly, Men Praying* v. 62 He sounded like the agent for a lecture circuit telling me that I was standing on the threshold of great things. **1890** H. FREDERIC *Lawton Girl* 150 It may take the form of.. a lecture course. **1956** *Nature* 10 Mar. 455/2 The American graduate student is usually forced to complete a relatively large number of lecture-courses. **1753** in *Essex Inst. Hist. Coll.* (1884) XXI. 153 The meeting adjourned to the next Lecture Day. **1779** E. PARKMAN *Diary* 94 Mr. Badcock has been with me to speak about ye Singing..on proposed Lecture day. **1897** Lecture-goer [see *class-attender* (*CLASS *sb.* 10)]. **1961** M. BEADLE *These Ruins are Inhabited* (1963) xii. 163 Oxford undergraduates aren't the inveterate lecture-goers and note-takers that American college students are. **1865** *Atlantic Monthly* XV. 369 The platform of the lecture-hall has been common ground for ..all our social..organizations. **1870** 'FANNY FERN' *Ginger-Snaps* 179, I get a comfortable seat in church,..or lecture-hall. **1961** NEW ENG. BIBLE *Acts* xix. 9 He..continued to hold discussions daily in the lecture-hall of Tyrannus. **1967** J. HAWGOOD in Cox & Grose *Organiz. Bibliogr. Rec. by Computer* III. 70 The number of minutes that..it takes him to walk there from college or lecture-hall. **1965** *Listener* 4 Nov. 700/2 It was the first time that either of these names had appeared on the Oxford lecture list. **1892** W. WALLACE tr. *Hegel's Logic* (ed. 2) 426 Cf. *Werke*, vii. I. 314 (lecture-note). **1920** G. SAINTSBURY *Notes on Cellar-Bk.* i. 2 An ordinary 'exercise book'.. devoted to base purposes of lecture-notes. **1944** *Mind* LIII. 269 Sometimes one gets the impression of a collection of lecture-notes. **1973** E. TAYLOR *Serpent under It* (1974) iv. 60 Could *you* continue to teach in a place where ..your students knew you had cribbed your lecture notes? **1961** *Observer* 26 Nov. 28/1 (Advt.), Lecture-Recitals..at Royal Academy of Music. **1817** COLERIDGE *Biog. Lit.* I. x. 219 Numerous and respectable audiences,..honored my lecture-rooms with their attendance. **1936** *Discovery* Oct. 301/2 The various buildings which housed the sectional lecture-rooms. **1849** W. ALLINGHAM *Diary* 30 June (1907) iii. 48 We..passed into the lecture-theatre. **1969** *Listener* 1 May 594/2 The ordinary university lecturer is no more exciting on film than he is in the lecture theatre. **1973** *Nature* 28 Sept. 225/1 Above the blackboards in the main physics lecture theatre of a Scottish university where I once worked there used to be written in large letters: 'Truth will in the end always flow in the direction of the greatest speculative reflection.' **1913** R. BROOKE *Let.* 24 July (1968) 486 The most unpopular person in Canada is Winston. Ever since his lecture-tour. **1921** R. FRY *Let.* 19 Dec. (1972) II. 519, I have just got back to London after my lecture tour in the north of England. **1952** 'J. TEY' *Singing Sands* ix. 138, I hope Mr. Brown doesn't go lecture-touring in the States. **1958** *Times Lit. Suppl.* 2 May 237/2 An actress whom he meets while on a lecture-tour in South America. **1973** R. LEWIS *Of Singular Purpose* i. 5 This lecture tour in America..is the first of many recognitions, I'm sure of it.

lecturee·. *rare.* [f. LECTURE *v.* + -EE[1].] One who attends lectures.

1900 J. H. WYLIE *Council of Constance* 191 To make lecturees independent of lecturers. **1939** W. ALLEN *Blind Man's Ditch* 15 There were the born lecturees; like Miss Wiggin, who had been attending classes for twenty years. **1972** *Listener* 9 Mar. 316/3 Poor American lecturees.

lecturing, *vbl. sb. attrib.* Add:

1897 'MARK TWAIN' *Following Equator* i. 25 The starting point of this lecturing-trip around the world was Paris. **1899** M. BEERBOHM *More* 140 His lecturing-tour through the States.

lecythid (le·siþid), *sb.* and *a.* [f. mod.L. family name *Lecythidaceæ*, f. the generic name *Lecythis* (P. Loefling *Iter Hispanicum* (1758) 189), f. Gk. λήκυθος a flask: see LECYTH.] **A.** *sb.* A tropical American tree of the order Lecythidaceæ. **B.** *adj.* Of or pertaining to a tree of this kind.

1871 C. KINGSLEY *At Last* II. xi. 113 The ground was strewn with large white flowers, whose peculiar shape told us at once of some other Lecythid tree high overhead. *Ibid.* 118 Some other Lecythids..go by the name of monkey-pots.

led, *ppl. a.* Add: **2.** *led lamp.*

1887 P. M'NEILL *Blawearie* 84 Will Hood had a 'led' lamp; it soon was kindled.

ledeburite (lē·dĕbiurəit). *Metallurgy.* [ad. G. *ledeburit* (F. Wüst 1909, in *Metallurgie* VI. 523), f. the name of Adolf *Ledebur* (1837–1906), German metallurgist + -ITE[1].] The eutectic of the iron/iron carbide system which is composed of austenite and cementite, contains about 4·3 per cent carbon, and occurs in cast iron.

1912 W. H. HATFIELD *Cast Iron* i. 16 This well-known structure, presented by the solidified eutectic, Wüst proposes to christen 'Ledeburite', after his distinguished compatriot. **1943** *Jrnl. R. Aeronaut. Soc.* XLVII. 218 A high carbon chromium steel in the cast condition..will have good sliding properties (ledeburite structure) and is widely employed in high pressure pump mechanism. **1972** G. A. CHADWICK *Metallogr. of Phase Transformations* iv. 140 The white iron eutectic, consisting of austenite and cementite, is often referred to as 'ledeburite'.

‖ **lederhosen** (lē·dəɹhōu:zən). [G.] Leather shorts, as worn in Alpine regions.

1937 *Night & Day* 8 July 8/2 The men, with any luck, will refuse to discard their grey flannels for lederhosen. **1953** *Time* 17 Aug. 27/2 The powerful Social Democratic Party is presided over by mild Erich Ollenhauer, a sort of chubby Clement Attlee in *Lederhosen*. **1956** WALLIS & BLAIR *Thunder Above* (1959) xv. 154 A boy in a checked sports shirt and *lederhosen*. **1973** *Listener* 25 Jan. 103/2 (*caption*) Hitler..in lederhosen.

ledge, *sb.* Add: **1. b.** *ledge(d) and brace(d) door* (see quots.).

1901 J. BLACK *Illustr. Carpenter & Builder Ser.: Home Handicrafts* ii. 19 (*caption*) Elevation and vertical section of what is termed a ledge and brace door. **1904** GOODCHILD & TWENEY *Technol. & Sci. Dict.* 352/1 *Ledged and braced door*, the same as a ledged door, with the addition of braces or pieces of wood running diagonally across between the opposite ends of two successive ledges. **1957** *N.Z. Timber Jrnl.* Oct. 73/1 *Ledged-and-braced door*, a door similar to a batten door, but framed diagonally with braces across the back, between the battens.

3. d. *Meteorol.* A layer in the ionosphere corresponding to a point of inflexion in a graph of ionization density against height, i.e. a layer in which the ionization increases less rapidly with height than in the regions immediately above and below it.

1949 *Gloss. Terms Radio Propagation* (*B.S.I.*) 5 Distributions in which the vertical gradient [of ionization] falls to a minimum value greater than zero are sometimes referred to as 'ledges'. **1960** RATCLIFFE & WEEKES in J. A. Ratcliffe *Physics Upper Atmosphere* ix. 437 The complicated loss process..stimulates recombination so that an F1 ledge is produced. **1967** *Proc. IEEE* LV. 17/1 Within the F region the main features of the vertical distribution of electrons are the F1 'ledge' at about 160 to 200 km.. and the F2 'peak' which generally lies between 250 and 400 km.

6. ledge-handle, a handle of distinctive shape found on Bronze Age ware.

1891 W. M. F. PETRIE *Tell el Hesy* vii. 42 The ledge-handles are very striking and quite unknown elsewhere. They belonged to large vessels with upright sides... The ledge is of various degrees... Sometimes it is very deeply and sharply waved..or else slightly curved,..or merely nicked,..or lastly a plain ledge.., without ornament or hollow. **1949** W. F. ALBRIGHT *Archæol. of Palestine* iv. 78 The envelope ledge-handle. This name, given it by P. L. O. Guy, is derived from the fact that the laps of the pushed-up ledge-handle,..are now folded over and fastened down as neatly as though each lap were the flap of an envelope. **1952** V. G. CHILDE *New Light Most Anc. East* (ed. 4) xi. 230 Four occupational layers are superimposed at Ghassul, and some rather suspicious ledge-handles are figured from the site. **1972** Y. YADIN *Hazor* III. x. 121 Large and deep bowls with ledge-handles.

ledger. Add: **A.** *sb.* **2. b.** In Thatching, a wooden rod laid across the thatch to hold it in place. Cf. *LEGGET.

1916 C. F. INNOCENT *Devel. Eng. Building Construction* xiii. 196 After the 'yelms' are laid, a ledger, that is, a pointed stick, is thrust into the straw, the length of it being carried across three or four 'yelms' and tied to the rafters at the opposite end. *Ibid.* 198 This method of securing thatch by rods laid across it is..that most generally used in England. The rods, or 'ledgers,' may be either tied or 'sewn' to the rafters, or they may be held down by 'broaches'. **1949** H. L. EDLIN *Woodland Crafts in Brit.* xi. 67 In most parts of Britain thatching materials are secured to the roofs of thatched houses or stacks by narrow pegs of wood, usually hazel. One common name for these is *spars*, but they have many others... *Withynecks, ledgers* and *roovers* have all been recorded. **1959** G. HOGG *Country Crafts* 123 The 'diamond' pattern which a thatcher produces by laying strips of cleft hazel or other thin wood, which he refers to as 'ledgers', criss-cross along the roof a little below the ridge on each side.

8. *ledger-account* (later examples), *-keeper, -scroll, -work*; **ledger-pole** = sense 2.

1902 G. H. LORIMER *Lett. Merchant* vi. 77 Some one who keeps separate ledger accounts for work and for fun. **1903** *Daily Chron.* 5 Jan. 5/5 It would be a bad day for loyalty when people considered loyalty as an item in the ledger account. **1906** *Ibid.* 18 Sept. 3/5 A female ledger-keeper and accountant in one office worked for 6s. a week. **1901** J. BLACK *Illustr. Carpenter & Builder Ser.: Scaffolding* 86 A combination of chains, clips, and screw bolts, used for securing a ledger-pole to standard. **1949** M. L. DARLING *At Freedom's Door* I. v. 116 Till two or three years ago..Hindu Bhats from Rajputana would come every year with their long ledger-scrolls to record in them any additions to the family. **1908** *Westm. Gaz.* 24 Mar. 6/3 He came to Paris, learnt ledger-work, and obtained a situation in a banking-house.

ledra (le·dra). *rare.* Also **ledrah.** [Cornish *ledr, ledra.*] A cliff, steep hill.

1942 A. L. ROWSE *Cornish Childhood* vii. 197 We picnicked all day on the ledrah. **1966** —— in *Listener* 9 June 845/3 When Devon was purple Cornwall was brown, With harvesting bracken On ledra and down.

lee, *sb.*[1] Add: **4.** *lee-rail* (later examples).

1913 J. LONDON *Let.* 20 Nov. (1966) 410 Sailing with lee-rail continually buried. **1961** F. H. BURGESS *Dict. Sailing* 132 *Lee rail awash, with*, heeled well over.

5. lee-helm, the helm when 'down' (cf. *down with the helm* s.v. HELM *sb.*[2] 1 c); **lee ho!, lee o!** (see quots.).

1883 *Man. Seamanship for Boys' Training Ships R. Navy* (Admiralty) (1886) 78 If carrying too much weather-helm, shift the weights further aft; if lee-helm, further forward. **1948** R. DE KERCHOVE *Internat. Maritime Dict.* 407/1 A sailing craft is said to carry lee helm when the helm has to be kept alee to counteract slackness and keep it on its course. **1962** A. G. COURSE *Dict. Naut. Terms* 120 *Lee helm*, a term used in sailing ships to indicate that the tiller is to leeward and the rudder and wheel to windward. **1927** G. BRADFORD *Gloss. Sea Terms* 99/2 *Lee ho!*, a command given by English yachtsmen preparatory to bringing a boat about; same as *hard a lee*. **1961** F. H. BURGESS *Dict. Sailing* 133 *Lee-o*, the helmsman's warning to a crew before going about.

lee·-boarded, *a.* [f. LEE-BOARD[2].] Fitted with a lee-board.

1897 KIPLING *Five Nations* (1903) 40 Do you know the shallow Baltic..Where the bluff, lee-boarded fishing-luggers ride?

leech, *sb.*[2] Add: **2.** *leech-like* adj.

1905 *Westm. Gaz.* 8 Jan. 3/2 He is prepared to stick to it with almost leech-like tenacity. **1908** *Ibid.* 6 Oct. 10/2 Parasitical and leech-like characteristics. **1963** R. P. DALES *Annelids* ix. 176 The parasitic leech-like branchiobdellids also belong to the Prosopora.

leech, *sb.*[3] For def. read: Either vertical edge of a square sail; the aft edge of a fore-and-aft sail. (Add later example.)

1948 R. DE KERCHOVE *Internat. Maritime Dict.* 407/1 *Leech*, the side of a square sail, or the afteredge of a fore-and-aft sail. Also called skirt when referring to square sails.

b. leech-lining (see quot.).

1883 *Man. Seamanship for Boys' Training Ships R. Navy* (Admiralty) (1886) 53 *Q.* What is a goring cloth? *A.* A side cloth of a topsail,..or lining of a topsail, called by sailmakers the leech lining.

Leeds (līdz). [Name of a city in West Yorkshire.] Used *attrib.* or *absol.* as the designation of a cream-ware type of pottery made at Leeds.

[**1783** *Hartley, Greens Trade Catal.* in *Art Jrnl.* (1911) Jan. 25/1 Designs of sundry Articles of Queen's or Cream colour'd Earthen-Ware, manufactured by Hartley Greens & Co., at Leeds Pottery: with A Great Variety of other Articles. The same Enamel'd, Printed or Ornamented with Gold to any Pattern; also with Coats of Arms, Cyphers, Landscapes, etc., etc. Leeds 1783.] **1863** W. CHAFFERS *Marks Pott. & Porc.* 133 *Leeds pottery*, earthenware, manufactured by Hartley, Greens, and Co. Leeds, 1786. This ware has much perforated or basket work. **1872** C. SCHREIBER *Jrnl.* (1911) I. 139 A very pretty Leeds sucrier and cover. **1876** *Ibid.* 485 We..arranged to come and look at his Leeds ware next week. **1903** MRS. H. WARD *Lady Rose's Daughter* xi. 169 The Leeds and Wedgwood dessert dishes that Cousin Mary Leicester had used for half a century. **1968** *Canad. Antiques Collector* June 17/2 What is Mocha Ware? Sometimes referred to as 'Leeds Ware' or 'banded creamware' it is a creamware decorated with seaweed or tree silhouettes. This was made from 1787 up to 1903.

Lee-Enfield (lī,e·nfīld). The names of J. P. *Lee* (1831–1904), Amer. designer of the bolt action, and *Enfield*, a town in Greater London, site of the British Royal Small Arms Factory, designers of the rifling form, used to designate a type of rifle used by the British Army in the S. African War and, modified, in the wars of 1914–18 and 1939–45. Also *Lee-Enfield bullet.*

1902 *Encycl. Brit.* XXXII. 241/1 The Lee-Metford Mark II. rifle has been further improved in its rifling to resist the wear of smokeless powder,..and is now known as the Lee-Enfield rifle. **1910** *Ibid.* I. 874/1 A cone-shaped sharp-pointed bullet, named the Spitzer bullet, has been

tried in the United States..in a Springfield rifle, which is practically identical with the British..Lee-Enfield. This bullet is lighter than the Lee-Enfield bullet. **1917** A. G. EMPEY *Over Top* 297 *Lee Enfield*, name of the rifle used by the British Army. Its caliber is ·303 and the magazine holds ten rounds. When dirty it has a nasty habit of getting Tommy's name on the crime sheet. **1959** [see *ENFIELD]. **1966** *Guardian* 8 July 1/2 The Royal Navy laid the No. 4 short magazine Lee-Enfield ·303 service rifle to rest... The Army and the Royal Marines said goodbye to the rifle a few years ago. **1970** F. WILKINSON *Guns* 135 In 1895 the Lee-Enfield rifle was introduced and was to remain the standard arm until 1902 when a shorter version was approved. This, the Short Magazine Lee-Enfield, was to continue in service..through two world wars.

leegte, var. *LAAGTE.

lee-lone, var. LEE-LANE.

1893 R. L. STEVENSON *Catriona* II. xxii. 265 What would become of you here, and you your lee-lone in a strange place? **1920** N. MUNRO in *Northern Numbers* 35, I heard a maiden sing, All in the lee-lone Sabbath morn. **1928** D. BYRNE *Destiny Bay* i. § 19. 128 Will you.. leave your cousin Jenico and Miss Ann-Dolly by their lee lone? **1966** T. H. RADDALL *Hangman's Beach* IV. xxvi. 399, I thought 'twas only love letters a gel went off tae read by her lee-lone.

Lee-Metford (lī,me·tfǒɹd). The names of J. P. *Lee* (see *LEE-ENFIELD) and W. E. *Metford*, used *attrib.* to designate a type of rifle in use before the Lee-Enfield rifle. Also *Lee-Metford bullet.*

1897 G. B. SHAW *Our Theatres in Nineties* (1932) III. 257 If he does not actually regard it [*sc.* the Bible] as an amulet, and believe that if a soldier carries it into battle it will magically attract and stop the Lee-Metford bullet. **1898** [see *DUM-DUM]. **1902** [see *LEE-ENFIELD]. **1967** *Everyman's Encycl.* VI. 202/2 The British War Office adopted..the Lee-Metford Mark I in 1888. In 1891 the Lee-Metford Mark II was adopted... This rifle was subsequently..improved, and became known as the Lee-Enfield rifle.

leep (līp), *v. Anglo-Indian.* [ad. Urdu (Hindi) *līpna.*] *trans.* To wash with cow-dung and water.

1895 KIPLING *Second Jungle Bk.* 80 The big wicker-chest, leeped with cow-dung. **1920** *Blackw. Mag.* Oct. 464/1 As you smell the fresh leeped earth of the picquet floor.

leer, *sb.*[3] Add: Also **lehr** (līəɹ, lēəɹ). (Further examples.)

1908 W. ROSENHAIN *Glass Manuf.* x. 165 The split cylinders are taken to a special kiln, generally known as a 'lear', or 'lehr', where they are..raised to a dull red-heat. **1918** P. MARSON *Glass* x. 72 These tunnels, or lehrs, are about 40 ft. long. **1937** *Nature* 18 Dec. 1072/1 There has been a corresponding improvement..in lehrs for annealing the finished product. **1943** *Amer. Speech* XVIII. 309/1 Among the latter were boys who carried hot glassware from the molds to the leer and toward the end of a shift they began a chant, 'Ten more trips to the layer O,' 'Nine more trips to the layer O,' and so on... They said 'layer' distinctly in two syllables. **1949** *Jrnl. Soc. Glass Technol.* XXXIII. 287 The term 'lehr' to denote an apparatus or plant for the continuous annealing of glass first appeared in factory usage in the U.S.A. between 1890 and 1900. The word arose most probably by corruption of the original form 'leer', but whether by accident or design is obscure. **1958** *Times* 22 Dec. 1/4 (Advt.), Practical experience of design and construction of glass furnace lehrs also essential. **1965** E. TUNIS *Colonial Craftsmen* vi. 139/1 A boy carried the new bottle to the leer where he snapped the punty off its bottom. **1971** *Atom* May 118/1 A ribbon of glass up to 11 feet wide leaves the float tank and enters the annealing lehr at temperatures in the region of 600°C.

b. leer man, lehr man, one who works at a leer.

1849 A. PELLATT *Curiosities of Glass Making* 67 The instruction to the lear-man, or fireman, rather to run the risk of melting goods by excess of heat than subject them to fly by insufficient. **1912** G. SOWERBY *Rutherford & Son* 27 The new lear man's shaping all right then. **1965** E. TUNIS *Colonial Craftsmen* vi. 139/1 A leer man stood the bottle on a hot iron tray in the leer.

leerfish (lē·rfiʃ). *S. Afr.* [Partial tr. of Afrikaans *LEERVIS.] A large game fish, *Hypacanthus amia*, of the family Carangidæ, found off the Atlantic coast of southern Africa.

1843 J. C. CHASE *Cape Good Hope* II. 169 Leer Fish—A species of Pike, affording considerable sport to the angler. **1902** *Trans. S. Afr. Philos. Soc.* XI. 217 Probably the Cape Leer-fish was so named by the early Dutch sailors, who brought the name from the East Indies. **1930** C. L. BIDEN *Sea-Angling Fishes of Cape* ii. 54 For the past two or three generations the word has been commonly written and accepted by the English-speaking people as leerfish. **1957** S. SCHOEMAN *Strike!* iii. 106 The average weight of leerfish is 20–35 lbs. **1973** *Eastern Province Herald* 28 Nov. 37 Out in the darkness came the unmistakeable sound of a leerfish hurling itself in frenzy after its prey.

‖ **leervis** (lē·rfəs). *S. Afr.* Also **leervisch.** [Afrikaans, f. *leer* leather + *vis* fish.] = prec.

1853 L. PAPPE *Synopsis Edible Fishes Cape Good Hope* 24 *Lichia Amia*, Cuv. & Val. (Leervisch.).. Taken occasionally in Table Bay, but not in great repute, its flesh being deemed dry and rather insipid. **1913** W. W. THOMPSON *Sea Fisheries Cape Colony* 156 *Lichia amia* L... Leer-visch; Leather-fish; Garrick (Natal). **1945** *Cape Argus* 27 Jan., Many leervis, kob and yellowtail have fallen for spoons, spinners and wobblers of many kinds. **1951** *Cape Times* 13 Nov. 2/3 Anglers know Swartvlei as a place where the sporting *leervis* abounds. **1953** J. L. B. SMITH *Sea Fishes S. Afr.* (rev. ed.) 222 The Leervis always seizes a fish across the middle and then works it about in the jaws until head-on for easy swallowing. **1956** *Cape Times* 2 Mar. 2/5 He caught a 32-lb. *leervis* near the lighthouse at Cape Agulhas. **1974** *Eastern Province Herald* 1 Aug. 21 In mid-December..leervis were caught in large numbers off the breakwater.

leery, *a.*[2] Add: (Earlier and later examples.) Also (esp. *U.S. slang*), doubtful, suspicious. Cf. *LAIRY *a.*[2]

1718 C. HITCHING *Regulator* 20 The Cull is leery, *alias* the Man is shy. **1846** *Swell's Night Guide* 46 The president ..who is generally the most cheeky, *leary*, downy cove they can tumble to. **1896** ADE *Artie* iii. 29 The old lady's a little leary of me, but I can win her all right. *Ibid.* xii. 105 I'm leary of it. **1905** *Dialect Notes* III. 63 He is *leery* of book agents. **1909** R. A. WASON *Happy Hawkins* 122, I was rather leery about Jabez. **1923** L. J. VANCE *Baroque* vi. 56 If you hadn't..made me leary that maybe you'd tip your friends off. **1942** E. PAUL *Narrow St.* xx. 165 The Surrealists discovered that they were not, in fact, Communists and that Moscow was leery of their antics. **1956** B. HOLIDAY *Lady sings Blues* (1973) xii. 108, I was leery of any man who could throw those things back at me in a quarrel. **1960** B. CRUMP *Good Keen Man* 113 Harry didn't entirely agree with my suggestion that [my dog] Flynn was probably a bit leery of boars since the one that 'killed' him at Maran. **1965** *Listener* 9 Sept. 391/3, I am..leary of theatrical revivalism. **1966** *Economist* 5 Feb. 489/1 At first, centre voters may be fed up with the government in power, but they are leery of voting for the other side, so they stop halfway and vote Liberal. **1969** *Southerly* XXIX. 9 Leery though I am of Greeks, Sophocles sums up my preoccupations effectively. **1970** *New Yorker* 10 Oct. 174/2 Many tennis authorities have been a little leery about placing her on a level with Lenglen. **1971** [see *GEHEIMRAT]. **1973** *Tucson* (Arizona) *Daily Citizen* 22 Aug. 28/1 The Braunlichs will also tell you that, sad as it is, middle America is leery of things it gets for free.

2. *U.S. slang.* Careful.

1911 H. QUICK *Yellowstone Nights* xii. 300 But be leery that we don't get stuck for non-performance.

lee side. Add: **b.** *Geol.* The 'down-stream' side of a mound of rock which has undergone erosion by a glacier.

1886 [see *drag-line* s.v. *DRAG *sb.* 9]. **1920** A. W. GRABAU *Textbk. Geol.* I. xiv. 374 The eroded bottom of such a valley often shows hummocky surfaces, sloping and smooth on the side from which the glacier moved (stoss-side) and with striated surfaces, but rough and clifted on the side away from the movement (lee side). **1969** J. L. DAVIES *Landforms Cold Climates* ix. 171 The characteristic roche moutonnée presents a streamlined appearance with smoothed, more gently sloping, upstream end and sides and a steeper lee side sometimes smooth but characteristically plucked.

left, *a., adv.,* and *sb.* Add: **A.** *adj.* **1.** Phr. *two left feet*: signifying a clumsy person.

1915 WODEHOUSE *Psmith Journalist* xviii. 132 Mr. Dawson..gave it as his opinion that one of the lady dancers had two left feet. **1959** E. L. MASCALL *Pi in the High* 7 Or dance with two left feet the *valse triste*. **1975** D. RAMSAY *Descent into Dark* iii. 92 Clumsy..you've got two left feet.

c. *left jabber*, a boxer whose characteristic punch is the left jab.

1950 [see *CROSS *sb.* 22 d].

3. *left bank*: applied *spec.* to a part of Paris lying south of the Seine noted for its 'advanced' intellectual views; = *Latin Quarter* (*LATIN *sb.* 5); also *attrib.*; hence *left-bankish* adj.

1893 H. S. EDWARDS *Old & New Paris* I. iii. 10/1 On the 'left bank',..stand the Institute, the Pantheon, [etc.]. **1911** W. J. LOCKE *Glory of Clementina Wing* ix. 129 Paris of the Left Bank, of the studios, of struggle and toil. **1929** E. WILSON *I thought of Daisy* iv. 243 She was staying ..in a little Left Bank hotel. **1932** 'F. ILES' *Before the Fact* ii. 42 Some unpretentious little restaurant on the left bank. **1943** D. GASCOYNE *Poems 1937–42* 45 In a Left-bank café. **1949** *Oxf. Jun. Encycl.* III. 335/1 The streets.. on the left bank are noted for the numerous book-stalls. **1952** A. WILSON *Hemlock & After* II. iii. 168 She had carefully preserved her Left Bank student get-up for this bourgeois gathering. **1958** *Times Lit. Suppl.* 14 Feb. 85/1 The intense form of *Angst* that one associates..with French intellectualism is not confined to the Left Bank. **1958** *Manch. Guardian* 27 Feb. 7/6 The highly complicated 'left-bankish' and somehow heartless fairy story which it tells in its three acts grows no more likeable with renewed acquaintance. **1964** *Economist* 2 May 478/1 Mr Gomulka strongly criticised their 'left-bank' attitudes. **1974** 'S. HARVESTER' *Forgotten Road* vi. 72 Their meal took on a sort of Chelsea or Greenwich Village or Left Bank atmosphere.

b. (Further examples of *left* in political contexts.) Also **left-leaning** *a.*, sympathetic towards the left in politics.

1919 T. E. LAWRENCE *Let.* 27 Sept. (1938) 293 So long as we are the more liberal ('left' in the Parliamentary sense) we call the tune... Our remedy and safeguard will be to trend continually 'left'. **1953** M. LOWRY *Sel. Lett.* (1967) 330, I am even 'left' of de Voto on the subject. **1957** J. OSBORNE *Entertainer* VIII. 62 A chap at my school ..managed to get himself in to the Labour Government, and they always said he was left of centre. **1961** *Times* 23 Jan. 13/6 The left-leaning Captain Kong Lae was, moreover, partly educated in Siam. **1962** J. BRAINE *Life at Top* x. 135, I asked the Warden who Graffham was. 'He's very Left,' the Warden said. **1962** *Listener* 19 July 87/1 A left-of-centre party not unlike the British Labour Party. **1964** GOULD & KOLB *Dict. Social Sci.* 383/2 The word *left* was used in England from the 1920s onwards,..sometimes covering Communists and Socialists and sometimes Liberals as well. **1966** T. PYNCHON *Crying of Lot 49* iii. 50 Peter Pinguid was really our first casualty. Not the fanatic our more left-leaning friends..chose to martyrize. **1972** *Times* 4 Aug. 13/4 The constitution of..the Donovan Commission, was on any showing a fair way left of centre.

c. Further special collocations: (in sport) *left arm, -armer, back, half (-back).* Also *left field* (Baseball): the part of the outfield to the left of the batter as he faces the pitcher; also, a fielder in this position; also *fig.*, a position away from the centre of activity or interest; *left fielder*: a fielder in the left field; also *fig.*

1955 *Times* 9 May 15/1 Goddard opened the bowling with him, left arm over the wicket at a gentle medium pace. *Ibid.* 10 June 4/2 Splendid bowling by Hampshire's young, left-arm bowler Sainsbury, whose seven wickets for 25 runs was the best performance of his career. **1974** *Times* 11 Nov. 8/2 Titmus chopped Bright's orthodox left-arm spin into his stumps. **1960** E. W. SWANTON *W. Indies Revisited* 277 New bowling talent will have to be unearthed, for instance..a slow left-armer. **1897** *Encycl. Sport* I. 419/2 [Assoc. Football] The left back and half-back deal with the opposing right wing. **1955** *Times* 9 May 14/3 Eckersley, the Blackburn Rovers left-back, who has not played for England since November, 1953, has been selected to accompany the Football Association party. **1960** B. LIDDELL *My Soccer Story* xvi. 98 The tragedy of Munich robbed England of one of her finest post-war left-backs in Roger Byrne. **1974** *Guardian* 18 May 19/2 Willie Ormond, the Scotland manager, has.. kept Danny McGrain..at left-back. **1857** *Spirit of Times* 29 Aug. 404/3 Enterprise Club. Maxfield, catcher;.. Webber, left field. **1867** H. CHADWICK *Base Ball Player's Bk. Reference* 51 A ball similarly hit to the right or left fields. **1896** KNOWLES & MORTON *Baseball* 77 Harry Athol..played left field for the Thespians in all their games. **1949** *Minot* (N. Dakota) *Daily News* 22 July 8/8 Marinari spoiled Lettau's chance for a no-hitter, lining a solid single to left field in the fifth frame. **1961** *Amer. Speech* XXXVI. 147 *Out in left field*, disoriented, out of contact with reality. **1970** *Time* 9 Mar. 19 An increasing number of candidates are emerging from leftfield to give voters surprising options. **1974** *Publishers Weekly* 11 Mar. 48/3 Novak's use of religious metaphor may put him in left field (Reinhold Niebuhr was there before him). **1867** H. CHADWICK *Base Ball Player's Bk. Reference* 70 Suppose the left-fielder should be the third striker on the list. **1957** *Encycl. Brit.* III. 160/2 The outfielders are called right fielder, centre fielder and left fielder with relation to a man standing on home plate and facing out across the diamond. **1973** *Publishers Weekly* 29 Jan. 259/1 Sober, necktie-wearing citizens will get a bang out of the book, left-fielders not at all. **1897** *Encycl. Sport* I. 418/2 [Assoc. Football] Three half-backs are played nowadays... They are called..the right, centre and left half-back. **1914** *London Opinion* 19 Sept. 445/1 A 'rising young left-half' for Sludberry Rangers. **1960** B. LIDDELL *My Soccer Story* vi. 40 Bobby Paisley was at left-half.

4. *left-eyedness*; **left-brained,** having the left-hand side of the brain as the dominant half; **left-footer** *slang*, a Roman Catholic.

1890 W. JAMES *Princ. Psychol.* I. ii. 39 Most people.. are left-brained, that is, all their delicate and specialized movements are handed over to the charge of the left hemisphere. **1902** *Daily Chron.* 22 May 3/4 Each half [of the cerebrum] governs the opposite side of the body, and ..as we are right-handed, so we may be called left-brained. **1937** S. T. ORTON *Reading, Writing & Speech Probl. in Children* i. 52 A boy found..difficulty in using the rifle because of his right-handedness and left-eyedness. **1964** M. CRITCHLEY *Developmental Dyslexia* viii. 51 Others attached less importance to the role of left-handedness than to left-eyedness. **1944** J. H. FULLARTON *Troop Target* 26 'What about the R.C.s?' 'Oh, yes. Leave the left-footers behind as gun-picquets.' **1959** I. & P. OPIE *Lore & Lang. Schoolch.* xvi. 344 In Lancashire Roman Catholics are known as 'Micks', and in Dundee as 'Left-footers'.

B. *adv.* (Later example.)

1918 *Daily Chron.* 2 Dec., In Kiel, where the revolution started, matters appear to be going 'left' with a vengeance.

C. *sb.* **2. a.** (Further examples in *Boxing.*) Also, a blow dealt with the left hand.

1897 *Encycl. Sport* I. 136 (*heading*) Stop for lead-off at body with the left. **1912** *Chambers's Jrnl.* 394/2 Out went Reid's murderous 'left' to our unutterable surprise, and down went the man on the platform. **1914** J. H. & A. LAMBERT *Boxing* (ed. 2) 41 If possible send in a straight left to the head. **1930** *Daily Express* 8 Sept. 1/6 Siki fell to a left on the body and was counted out. **1948** 'P. WOODRUFF' *Whatever Dies* 126 An odious person who could be neatly floored by one skilful left to the chin. **1967** G. F. FIENNES *I tried to run a Railway* i. 5 Out shot a telescopic left, and I had the shiner of all time for weeks.

c. (Further examples.) Now usu. applied to a political group holding radical or socialist views.

1939 *John o' London's Weekly* 2 June 321/1 A defiant glare at the Left..with an equally defiant glare at the Right. **1940** W. TEMPLE *Thoughts in War-Time* iii. 24 The Right tends to have a fuller sense of historical continuity than the Left. **1950** THEIMER & CAMPBELL *Encycl. World Politics* 260/2 The communists are referred to as the 'extreme' or 'far' left. **1971** W. LAQUEUR *Dict. Politics* 310 Popularly the Left has favoured rapid social change... During the thirties, Left was associated with opposition to fascism. **1974** 'W. HAGGARD' *Kinsmen* ix. 93 The tiresomely modern bishop..was..very far to the Left.

4. A shot fired at game with the left barrel

of a double-barrelled shotgun; a bird or beast hit by such a shot.

1893 H. A. MACPHERSON et al. *Partridge* II. iii. 131 Now thoroughly awake, you kill three neatly, quickly followed by a smart right and left—one in front and one behind—at a brace that come straight at you. **1908** R. H. BENSON *Conventionalists* I. iii. 82 On Saturday he had killed three rights and lefts, and had not missed more than one single bird flying alone. **1910** *Blackw. Mag.* Jan. 140/1, I got a right and left with the big gun. **1958** M. BRANDER *Rough-shooter's Sport* xx. 217 When..a covey of grouse was flushed.., I only managed to drop one bird. The others, however, performed more than adequately, each bringing down a right and left. **1974** *Field* 5 Dec. 1311/1 Congratulate anyone on a good piece of dog work..as one would if he achieved a right and left.

5. *Surfing.* The (use of the) left foot. Cf. *GOOFY *a.* 2.

1968 W. WARWICK *Surfriding in N.Z.* 17/3 Calculate where the waves are shaping up the best for your style of riding. Obviously..goofy footers will favour lefts. **1970** *Surf '70* (N.Z.) 17/2 Wayne Charlton was one of the best goofy footers to ever surf the left at Fitzroy.

left, *ppl. a.* Add: **1.** *left luggage* (earlier and later examples).

1861 H. RHYS *Theatr. Trip Canada & U.S.* xi. 96 Arrived at the depôt, I discovered in the doubtful light the 'left luggage' room. **1945** G. B. GRUNDY *55 Yrs. at Oxf.* 167 He left it in the left-luggage office. **1963** [see *baggage-room*]. **1971** 'E. CANDY' *Words for Murder Perhaps* xiii. 158 He..came straight back, leaving his case and holdall in the left luggage.

2. a, b. (Further examples.)

1873 'S. COOLIDGE' *What Katy Did* i. 7 In almost every large family, there is one of those..left-out children. **1908** *Westm. Gaz.* 18 Apr. 3/1 (*title*) The little brothers. Or, the land of the left-behind. **1909** *Daily Chron.* 14 Jan. 1/5, I believe the left-out millions are more miserable. **1941** E. BOWEN *Look at Roses* 242 Emma's left-behind silver things. **1965** *Economist* 21 Aug. 674/2 They [*sc.* Negroes who rioted, esp. in Los Angeles] believe—some of them rightly—that they could have risen out of the ghetto of the left-behind but for their colour. **1965** B. SWEET-ESCOTT *Baker St. Irreg.* i. 38 The section was for a few weeks engaged in organising 'left behind' parties all over the British Isles. **1974** W. FOLEY *Child in Forest* II. 220 Getting my swollen inflamed feet back into Leah's left-offs.

left-branching, *a. Linguistics.* [LEFT *adv.*] (Of grammatical constructions) having the majority of its constituents on the left of its tree diagram. Also **left-branching** *vbl. sb.*

1961 N. CHOMSKY in *Proc. Symposia Appl. Math.* XII. 14 Left-branching should offer no problem... A hearer will tend to group left-branching units of a complex sentence (as, e.g., in 'many more than half of the rather obviously much too easily solved problems') as units quite readily. **1965** —— *Aspects of Theory of Syntax* i. 13 A left-branching structure is of the form [[[···]···]···]—for example, in English, such indefinitely iterable structures as [[[[*John*]'s *brother*]'s *father*]'s *uncle*]... There are no clear examples of unacceptability involving only left-branching or only right-branching.

left hand. Add: **3. left-hand drive,** a (motor vehicle) steering system with the steering wheel and other controls fitted on the left side; also, such a vehicle. Hence **left-hand driving.**

1913 A. L. CLOUGH *Dict. Automobile Terms* 187 Left-hand Drive. **1931** *N.Y. Times* 19 July IX. 8/8 It was not..until 1909 that left-hand drive and centre control were introduced, reputedly by Henry Ford. **1933** P. MACDONALD *Mystery of Dead Police* vii. 51 It's a left-hand drive. **1956** *Collier's Year-Bk.* 670/1 Sweden is the only Scandinavian country with left-hand driving, and the desirability of changing to right-hand driving has been discussed off and on for many years. **1966** J. WEATHERHEAD *Force of Innocence* iv. 28, I was finding my left-hand drive difficult in London. **1975** *Guardian* 20 Jan. 7/3 All this year's production will be left-hand drive.

left-handed, *a.* Add: **3.** (Later examples.)

1914 'HIGH JINKS, JR.' *Choice Slang* 14 *Left handed compliment*, one that may be taken either as a compliment or in the opposite way. **1953** *Time* 3 Aug. 36/1 An enthusiastic patter of applause came from the British press, including a left-handed compliment from the *Manchester Guardian* that he was not at all like the movie-type American. **1972** *Ulster Folklife* XVIII. 94 In the dialect of Donegal..*left-handed* betokens 'malicious, underhand'; *a left-handed blessing* is a euphemism for a malediction or curse, and *a left-handed friend* is 'an enemy'. **1974** A. DOUGLAS *Noah's Ark Murders* vi. 54 'I'm not trying to date you.' 'Well, that's a left-handed compliment,' she complained.

5. (Later examples.)

1925 T. DREISER *Amer. Trag.* (1926) II. xxii. 308 The pleasures of this left-handed honeymoon were at full tide. **1935** A. J. POLLOCK *Underworld Speaks* 70/2 *Left handed wife*, a kept woman.

left-handed, *adv.* [f. LEFT HAND.] Towards the left; with the left hand.

1848 *Sporting Life* 1 Jan. 241/2 He also bats left-handed. **1851** *Illustr. London News* XVIII. 133/2 This nut is cut..left-handed. **1909** *Chambers's Jrnl.* Oct. 651/2 The great stag..swinging left-handed..passed Culworth. **1909** E. H. MILES *Lessons Lawn Tennis* (ed. 3) xv. 79, I do not know why ladies should not beat right-handed men players if the latter were compelled to play left-handed. **1928** *Observer* 19 Feb. 24/4 You leave the Oundle road and turn left-handed for Uppingham. **1929** *Morning Post* 30 Dec. 13/1 Hounds..swinging left-handed past Edgecote House. **1974** *Country Life* 7 Mar. 477/1 We rode left-handed beyond the Letham woods as our fox set his mask for Canty hall.

left-hander. Add: **a.** (Further examples.) Also in other games.

1937 [see *CHINAMAN 4]. **1937** S. T. ORTON *Reading, Writing & Speech Probl. in Children* i. 49 Prejudice..is so strong as to amount to the belief that the left-hander is abnormal. *Ibid.* 52 Parson..went so far as to hold that all left-eyed and right-handed individuals were native left-handers who had been shifted by training. **1940** G. MARX *Let.* 5 Sept. (1967) 25 A tennis player with the weirdest assortment of strokes... He's a left-hander. **1961** RUSSELL & ESPIR *Traumatic Aphasia* iv. 29 The left hemisphere is usually dominant..for left-handers. **1970** *Daily Tel.* 29 Dec. 10 Not all the evidence..supports his inference that left-handers have exceptional ability, even if they do range from Leonardo da Vinci to Sir Compton Mackenzie. *Ibid.*, A nice assortment of..40 offensive slang words for left-handers, from kack to cuddy-wifter. **1974** *Times* 6 Nov. 13 He continues to make runs for Western Australia..and has the advantage of being a left-hander.

d. A left-handed compliment.

1959 *Times* 28 Apr. 11/4 'Not bad' might appear a good enough specimen of the simplest type of left-hander.

leftie, var. *LEFTY.

leftish (le·ftiʃ), *a.* [LEFT *sb.* 2 c + -ISH[1].] Inclined to the political views of 'the left'. Hence **le·ftishness.**

1934 H. G. WELLS *Exper. Autobiogr.* II. ix. 809 The violent persecution of Jewish and leftish writers in Germany. **1934** WYNDHAM LEWIS *Let.* 29 Nov. (1963) 226 The strong *Leftish* political colouration of so much of the newest poetry. **1959** *Listener* 6 Aug. 195/1 There were leftish magazines on the tables. **1966** *Economist* 15 Oct. 254/2 This probably has little to do with the [Syrian] regime's 'leftishness'. **1972** *Observer* 6 Aug. 19/3 The leftish Left, the revolutionary Left.

Leftism (le·ftiz'm). Also **leftism.** [f. LEFT *sb.* 2 c + -ISM.] The political views or principles of 'the left'.

1920 *Oxf. Mag.* 19 Nov. 94/1 Mr. Clutton-Brock has consented to read a paper on 'Left-ism'. **1921** N. ANGELL *Fruits of Victory* v. 165 No sooner does the Left of some party break off and found a new party than it is immediately confronted by its own Leftism. **1945** 'G. ORWELL' in *Contemp. Jewish Record* VIII. 169 During the past few years there has been what amounts to a counter-attack against the rather shallow Leftism which was fashionable in the previous decade. **1960** *Guardian* 13 June 9/3 The 40th anniversary of the publication of Lenin's book on Leftism. **1967** C. SETON-WATSON *Italy from Liberalism to Fascism* iv. 160 Labriola was the first Italian to present socialism not as the natural offspring of the leftism of the Risorgimento but as a philosophical system. **1971** *Guardian* 4 Aug. 10/4 There is still a lot of old fashioned and sentimental Leftism (hanging over from much bad verse written in the late 1930s). **1973** *Listener* 19 July 91/2 The infantile Leftism of the fellow-traveller.

Le·ftist. Also **leftist.** [f. as prec.] An adherent of 'the left' in politics. Also *attrib.* or as *adj.*

1924 *Contemp. Rev.* July 20, I would support either a violent reactionary, or extreme Leftist. **1937** E. SNOW *Red Star over China* II. iii. 67 The Leftist Kuoming tang general. **1951** E. PAUL *Springtime in Paris* xi. 206 The anti-Communist Leftists, Existentialists, Trotskyists, Titoists and Anarchists published plans for a rival meeting. **1960** *Guardian* 12 Apr. 8/3 Most of the leaders of the Labour party were probably Leftist rebels at the age of twenty. **1960** *Economist* 8 Oct. 134/2 Many of the speeches were vaguely leftist. **1962** *Listener* 4 Jan. 36/2 It refutes the leftist legend that Dollfuss was simply a Fascist. **1964** L. NKOSI *Rhythm of Violence* 29 She's a bit of a Leftist, but I thought you wouldn't mind. **1967** H. V. DICKS *Marital Tensions* 61 This vivacious, carefree girl with her disdain for tradition was matched by this progressive, leftist scholar. **1974** *Times* 12 Oct. 5/4 It is difficult to find Spanish politicians who do not say they are leftists.

leftness (in Dict. s.v. LEFT *a., adv.,* and *sb.*). (Further examples.)

1884 [see *BILATERALITY]. **1890** [see *DOWNNESS].

left-over, *a.* and *sb.* [LEFT *ppl. a.* 2. Cf. *LEAVE *v.*[1] 14 e.] **A.** *adj.* Remaining over; not used up or disposed of.

1897 R. M. STUART *In Simpkinsville* 65 A bundle of left-over flowers. **1905** *Westm. Gaz.* 28 Dec. 2/1 If..they find themselves with a left-over stock of life-force. **1907** *Smart Set* Feb. 13/1 She tacitly avoided him, and his left-over moments had..been spent philandering. **1967** N. FREELING *Strike Out* 103 The rice had left-over ham and chicken in it. **1968** *Listener* 4 Apr. 438 (*caption*) Is that leftover macaroni cheese..still in the fridge? **1972** *Guardian* 30 Dec. 13/3 Many senior EEC officials..make regular bookings out of Brussels, and the casual visitor has to battle for the left-over seats.

B. *sb.* Something remaining over; *esp.* a portion of some article of food left over from a meal. Freq. *pl.* Also *transf.*

1891 *Cassell's Family Mag.* May 374/1 They all like change of diet, so I provide all sorts of things, with the result that the 'left-overs', as I call them, are appalling. **1897** R. M. STUART *In Simpkinsville* 64, I try to keep the Potter's field a-bloomin' with my left-overs. **1906** *Daily Chron.* 19 Sept. 4/4 We are almost yawning at the 'left-overs' of the scandal banquet. **1950** H. J. MASSINGHAM *Curious Traveller* iv. 71 Now only the shoddy left-overs from the export trade can be bought at inflated prices. **1964** *Punch* 8 Jan. 73/2 Adlibbing madly on Mrs. Hannah Glasse's 'Domestic Cookery Made Easy' (1747) he will combine assorted garden leftovers, crab-apples, damsons, blackberries, radish-pods. **1974** *Sunday Express* 21 Apr. 23/1 (Advt.), As for the babies' left-overs—well, I really scoffed those.

b. A survival.

1902 KIPLING *Traffics & Discov.* (1904) 169 'E's a left-over from Majuba—one of the worst kind. **1911** L. ABBOTT *Amer. in Making* 94 The dread of this Executive power is a curious left-over from Colonial days. **1927** H. E. FOSDICK *Pilgrimage to Palestine* 252 In this ancient monastery these left-overs of a bygone age guard their relics. **1971** I. G. GASS et al. *Understanding Earth* vii. 102/1 Meteoritic debris (probably representing left-overs from the time of formation of the solar system).

leftward, *adv.* and *a.* Add: Freq. in political contexts: (tending) towards 'the left'.

adv. **1957** *Economist* 28 Dec. 1119/1 The Singapore city council elections last Saturday may be taken as an accurate indication of the political trend in the island colony. That trend is clearly leftward. **1973** *Listener* 15 Nov. 668/1, I was rather Conservative as a young man. I've moved gently leftward.

adj. **1936** M. SCHACHTMAN in J. G. Wright tr. *Trotsky's Third International after Lenin* p. xxii, In the message to the Sixth Congress entitled 'What Now?' Trotsky touches upon this Leftward evolution in the European working class. **1939** H. G. WELLS *Holy Terror* III. i. 220 The Group turned its attention to the existing leftward papers. **1949** I. DEUTSCHER *Stalin* 403 Stalin's leftward switch in Russia was not only an earnest affair; it had the grandeur of national drama. **1957** *Times* 11 May 7/2 It is no surprise that in the borough elections the leftward movement seen in national by-elections has been repeated—though not, it seems, carried any further. **1973** *Guardian* 10 Mar. 1/5 Mr Roy Jenkins..is calling on those who share his views..to dig in their feet against what is seen as a dangerous Leftward drift.

le·ftwardly, *adv.* = LEFTWARDS *adv.*

1908 HARDY *Dynasts* III. I. iii. 335 With that in eye he has bundled leftwardly Thomière's division.

le·ftwardness. *rare.* [f. LEFTWARD *a.* + -NESS.] The quality of being leftward in politics.

1944 *Politics* Sept. 247/2 What does *Politics* offer them? A center for leftwardness? **1966** *New Statesman* 5 Aug. 203/2 His leftwardness is smilingly excused, but we are not reminded that Eluard was a communist.

le·ftwards, *adv.* (Later example.)

1971 *Guardian* 3 July 11/8 When the Chinese civil war began in 1946 Liu wobbled to the Right... Then he lurched Leftwards.

left wing. [f. LEFT *a.* + WING *sb.*] **1.** In football and similar games: the position of a player on the left side of the centre(s); a player occupying this position; the part of the field in which a left wing normally plays. Cf. WING *sb.* 7 b.

1882 in Charles-Edwards & Richardson *They saw it Happen* (1958) 300 He was instantly robbed by Strachan, who passed it [*sc.* the football] to the left wing. **1889** [see WING *sb.* 7 b]. **1921** in B. James *England v Scotland* (1969) vi. 125 Remember he was against probably the finest left wing in the three countries; certainly the cleverest outside-left, Alan Morton. **1974** *Liverpool Echo* (Football ed.) 4 May 1/3 He raced down the left wing..to cross the ball into the goalmouth.

2. In Politics. (See LEFT *a.* 3 b.) Freq. *attrib.* Also *transf.*

1884 W. JAMES *Will to Believe* (1897) 171 In theology, subjectivism develops as its 'left wing' antinomianism. *Ibid.*, If the Hegelian gnosticism, which has begun to show itself here and in Great Britain, were to become popular philosophy, as it once was in Germany, it would certainly develop its left wing here as there, and produce a reaction of disgust. **1898** [see LEFT *a.* 3 b]. **1905** W. JAMES *Meaning of Truth* (1909) v. 124 If the formula ever became canonical, it would certainly develop both right-wing and left-wing interpreters. **1921** H. CRANE *Let.* 25 Dec. (1965) 74 You have met about all the personalities in the younger left-wing at all worth while. **1923** G. D. H. COLE *Trade Unionism & Munitions* p. i, One of the principal contentions of the 'left-wing' elements in the Trade Union and Socialist movements. **1940** W. TEMPLE *Thoughts in War-Time* iii. 23 The Left Wing tends to identify the Government and the community. **1957** *Times Lit. Suppl.* 1 Nov. 653/1 Mr Humphreys has been very ambitious here, in an attempt to analyse the relationships between a rich magazine publisher with Left-wing political ambitions and his family. **1972** *Times* 5 Sept. 2/3 The left-wing challenge over Europe is expected to unseat at least one member of the Labour Party National Executive Committee.

Also **left-wi·nger, -wi·ngery, -wi·ngism; left-wi·ngish** *a.*

Quots. 1891, 1896, 1967 are sense 1, the remainder sense 2.

1891 *Peel City Guardian* IX. 7/3 A beautiful bit of passing by the Peel left wingers. **1896** Left winger [see WINGER 2]. **1923** G. D. H. COLE *Workshop Organiz.* 17 The rise of 'left-wingism' inside the Trade Union movement. **1924** *Glasgow Herald* 5 Apr. 9 The unscrupulous, untiring representative of the leftest of left-wingers. **1951** R. CAMPBELL *Light on Dark Horse* 249, I have never been Left-Wingish. **1955** *Times* 2 May 8/5 Mr. Zilliacus is a left winger who has often been a thorn in the side of the party leadership. **1963** *Guardian* 10 May 22/7 Vague Left-Wingism. **1967** J. POTTER *Foul Play* (1968) viii. 90 Good left wingers are in short supply. **1968** *Economist* 7 Sept. 25 Voted on to the general council were Mr Cousins's new ally from the engineers, Mr Hugh Scanlon, and the draughtsmen's militant leader, Mr George Doughty. This gives the sensation of more imminent left-wingery. **1972** *Listener* 2 Nov. 615/3 A period atmosphere of Thirties left-wingery. **1975** *Daily Tel.* 22 Feb. 10/4 Perry Worsthorne was allowed about 30 seconds to question how far Jenkins' left-wingery was an opportunist gimmick, to 'shock and provoke' his opponents.

lefty (le·fti). Also **leftie.** [-Y[6].] **1.** A left-handed person. Also *attrib.* or as *adj.*

1886 *Sporting Life* 7 Apr. 2/4 In last Wednesday's [baseball] game Nashville presented her left-handed battery,..to offset our 'lefty' battery. **1927** *Glasgow Herald* 7 Apr. 12 He was a patriotic Roman youth who allowed his right hand to be burned off..and was henceforward designated by a term which..would be rendered by his comrades of to-day as 'Leftie'. **1969** *New Scientist* 6 Nov. 277/2 Such illustrious lefties as Leonardo da Vinci, Michelangelo..and Paul McCartney.

2. A left-winger in politics. Also *attrib.* or as *adj.*

1935 C. ODETS (*title*) Waiting for Lefty. **1937** in Partridge *Dict. Slang* (1951) 1097/2 (*caption*) Counter-blast to lefties. **1939** R. CAMPBELL *Flowering Rifle* II. 40 As I who've lived beneath the two regimes And have not dreamed the Leftie Teacher's dreams. **1967** *Listener* 10 Aug. 164/1 The lefties are almost completely in control of the nation's communication. **1970** K. AMIS *What became of Jane Austen?* 204, I mean the kind of person who.. buys unexamined the abortion-divorce-homosexuality-censorship-racialism-marijuana package; in a word, the Lefty. **1972** *Times* 6 Oct. 14/7 A leader of the left who is no fair-weather Lefty but the genuine article. **1972** *Observer* 22 Oct. 29/1 These groups pump out quantities of magazines and news-sheets, frequently repetitive, full of Lefty names and too-ready inferences. **1974** *Oxford Times* 8 Mar. 11/3 This word, victimisation, has become a substitute, in leftie jargon, for just punishment.

leg, *sb.* Add: **2. a.** *on the leg*, also of a horse (example); *to give a person a leg up* (further examples); hence *leg-up* sb., a help, support, boost; *to hang a leg* (see HANG *v.* 4 c); *to have a leg*: to be physically attractive, to have a fine appearance (*Obs.*); *to show a leg*: to get out of bed; to make one's appearance; *to be tied by the leg*: to be prevented from doing something by some circumstance.

a **1817** JANE AUSTEN *Persuasion* (1818) II. vi. 116 She, poor soul, is tied by the leg. She has a blister on one of her heels. **1831** B. HALL *Fragments Voy. & Trav.* I. 247, I say, Master Doughy, do you mean to relieve the deck tonight? Here it's almost two bells, and you have hardly shewn a leg yet. **1832** F. TROLLOPE *Dom. Manners Amer.* I. xviii. 281 We should be obliged to pass the whole of Monday there, as the coach..would not arrive..till Tuesday morning. Thus..we were to be tied by the leg for four-and-twenty hours. **1854** 'C. BEDE' *Further Adventures Verdant Green* vii. 61 He used to sing out, 'You must show a leg, sir!' and..kept on hammering at the door till I *did*. **1865** MILTON & CHEADLE *N.W. Passage by Land* i. 13 The dogs kept tumbling off..until hauled back again with the help of a 'leg up' from the people inside [a stage coach]. **1879** G. MEREDITH *Egoist* I. ii. 16 And, says Mrs Mountstuart, while grand phrases were mouthing round about him '*You see he had a leg*'. **1901** *Chambers's Jrnl.* 27 July 554/2 He had..strong introductions to a great financier in Park Lane, who seemed to have good reasons for obliging him in such matters as club nominations and social 'legs-up' generally. **1901** J. N. MCILWRAITH *Curious Career R. Campbell* iv. 45 He might not have managed to mount had not Gib been at hand to give him 'a leg up'. **1908** *Westm. Gaz.* 28 Mar. 2/2 He first wore breeches at the Coronation of Queen Victoria, and there was a curious anticipation of a phrase immortal in literature in his statement that his first Court suit revealed to him 'that he had a leg'. **1916** 'TAFFRAIL' *Pincher Martin* ii. 19 All hands! turn out, turn out, turn out! show a leg, show a leg, show a leg! **1936** C. S. LEWIS *Allegory of Love* ii. 72 We can all but hear the voices shouting 'Show a leg—show a leg'. **1937** B. DE HOLTHOIR tr. *Duhamel's Pasquier Chron.* I. 73 Never mind, if you think it will give him a leg up. **1946** J. IRVING *Royal Navalese* 156 *Show a leg!*.. The boatswain's mates' early morning shout..is a direct link with pre-Nelsonic days when certain women were permitted to live on warships in harbour. **1950** A. L. ROWSE *England of Elizabeth* vi. 233 The family owed its leg-up in the world to Robert's grandfather **1957** *New Yorker* 12 Jan. 25/1 For Nora, who came from a poor and an ugly lower-middle-class home, political action was a leg up. **1965** *Observer* (Colour Suppl.) 30 May 34 *On the leg*. A horse whose legs look too long for his body—he has a lot of daylight underneath him. **1969** 'P. ALDING' *Murder among Thieves* xii. 12 Kerr awoke to find someone was rocking his shoulders.. 'Come on, me sleeping beauty, rise and shine, show a leg,' said P.C. Mottram, with indecent cheerfulness. **1969** *Listener* 9 Jan. 43/2 The boys are here..because local parents think it will give them a social leg-up. **1973** *Weekly News* (Glasgow) 11 Aug. 26/2 Then he got the leg-up on a horse called Native Copper. **1974** *Times* 5 Feb. 24/5 (Advt.), Want a leg up? There's more than one way with N.O.P. Secretary.

c. *on one's hind legs*, in a standing attitude (further examples); *not a leg to stand on* (earlier and later examples).

1594 NASHE *Unfort. Trav.* sig. B4 Faine he would have patcht out a polt-foot tale, but (God knowes) it had not one true leg to stand on. **1825** J. NEAL *Bro. Jonathan* I. 8 As if the Yankee man were determined to leave the.. brigadier without a leg to stand upon, as a lawyer would say. **1856** DICKENS *Dorrit* (1857) II. viii. 393 He had better confess, for he had not a leg to stand on. **1910** H. BELLOC *Pongo* xix. 289 The Pongo was to get on to his very short hind legs and talk of the gravity of the situation and all his party was to listen in awed silence. **1925** V. WOOLF *Mrs. Dalloway* 114 Solemnly Richard Dalloway got on his hind legs and said that no decent man ought to read Shakespeare's sonnets. **1944** E. S. GARDNER *Case of Black-Eyed Blond* (1948) xvi. 158 Mildred had gone to a lawyer, and the lawyer had advised her that she didn't have a legal leg to stand on. **1960** M. SPARK *Bachelors* ii. 21 She hasn't a leg to stand on in the case. He's divorcing her, she's not divorcing him. **1964** J. MASTERS *Trial at Monomoy* i. 26 That's why I'm on my hind legs now, asking you folks to keep calm. **1973** J. WAINWRIGHT *Pride of Pigs* 179 You haven't a leg to stand on... You don't even out-rank me.

14. d. *Lace-making.* A strand of the network which connects the patterns in lace. Usu. *pl.*

1865 F. B. PALLISER *Hist. Lace* xxii. 263 Early guipure of Venice or darned network, in which the raised flowers were strung together by legs or brides. **1900** E. JACKSON *Hist. Handmade Lace* 214 Legs,..the connecting threads thrown across spaces in needlepoint and bobbin laces. **1922** MRS. R. E. HEAD *Lace & Embroidery Collector* 232 *Brides*. Fr. Syns.: bars, legs (Eng.).

e. *U.S. Broadcasting.* A branch or supplementary network attached to the main network and providing coverage for a particular region (see also quot. 1937).

1937 *Printers' Ink Monthly* May 39/1 *Leg*, a regional chain, i.e., one link of stations in a network. **1951** E. E. WILLIS *Foundations in Broadcasting* iii. 47 Supplementary stations are added to the basic network in order to expand the coverage of a particular program. Often these supplementary stations are organised into groups or legs, which provide coverage of an entire section. The networks all have West Coast legs, for example. **1966** *McGraw-Hill Encycl. Sci. & Technol.* XI. 252/2 An appended supplementary network circuit, feeding more than one station from an intermediate point along a reversible or a round-robin system, is called the leg of a network. Network legs are usually..one-way circuits from the AT&T office to the leg office and the stations they feed.

15. c. A part of, or stage in, a journey, race, competition, etc.

1920 *Blackw. Mag.* Feb. 166/1 On each new 'leg' of our zigzags, our eyes were straining over ever-new horizons. **1927** *Nat. Geogr. Mag.* Aug. 185/2 (*heading*) First nonstop leg of the journey was 1,400 miles. **1938** W. L. HUGHES *Bk. Major Sports* xxx. 345 Each man on a relay team is said to run one leg of the race. **1953** R. CHISHOLM *Cover of Darkness* xiv. 151 We began a square search, flying five-minute legs. **1955** *N.Y. Times* 23 Jan. 3/6 Wiggins swam his leg of the relay in 56 seconds flat. **1958** *Times* 8 Sept. 6/3 Where an alien's visit to Britain was split into two parts by a trip to another country, each 'leg' of the United Kingdom visit had, by law, to be dealt with separately by the immigration authorities. **1972** *Nature* 31 Mar. 196/2 The first leg of this route traverses Arctic tundra regions of Alaska and Canada's Northwest Territories. **1973** C. BONINGTON *Next Horizon* xiii. 196 We squeezed out of the snow cave for the last leg down to Scheidegg.

17. leg art *slang* (orig. *U.S.*) = *CHEESE-CAKE 2; **leg drive,** in rowing, drive imparted by movement of the rower's legs; **leg-guard** (further examples); **leg man, woman** orig. *U.S.*, an assistant who does leg work, *spec.* a journalist who goes from place to place gathering information; **leg piece,** (*b*) (examples); **leg-pull** [f. the phr. *to pull one's leg*: see LEG *sb.* 2 a], the act of deceiving a person in a playful way, a humorous deception (so *leg-puller, -pulling* sbs.); **leg-rest** (earlier and later examples); **leg-ring,** an aluminium strip wrapped round a bird's leg to mark it; hence **leg-ringing** *vbl. sb.*; **leg-room,** space for the legs, *spec.* in a car; **leg-rope** *v.*, also *N.Z.*; **leg-rope** *sb.* (*Austral.* and *N.Z.*), a noosed rope for securing an animal by one hind leg; also **leg-roping** *vbl. sb.*; **legs eleven,** a jocular catch-phrase in the game of bingo (or housey-housey), etc., for 'eleven'; also *ellipt.* as *legs*; **leg shop** *colloq.*, a theatre in which 'leg-shows' are produced (*Obs.*); **leg-show** *colloq.* (orig. *U.S.*), a theatrical production in which dancing girls display their legs; **leg-stretcher,** (*a*) a walker; (*b*) a walk (see *to stretch one's legs* s.v. STRETCH *v.* 3 c); **leg woman** (see *leg man* above); **leg work,** work which involves running errands, going from place to place in search of information, etc.

1940 *Amer. Speech* XV. 359/1 *Leg art*, exploitation of sex appeal in pictures. **1958** *Spectator* 10 Oct. 481/1 The Cameo Royal, the leg-art cinema by London's Leicester Square. **1928** *Observer* 1 July 30/3 They are lacking in leg-drive, and their boat does not run evenly between the strokes. **1968** *Encycl. Brit.* XIX. 668/2 Fairbairn..emphasized leg drive and arm pull and considered smooth bladework more important than what he called the 'showy style' of body work. **1844** *Bell's Life* 12 May 1/3 (Advt.), Robert Dark, the Inventor and sole Manufacturer of..the improved leg guards, begs respectfully to inform the lovers of the Game of Cricket that they can be supplied at the shortest notice. **1890** [see *face-screen* (*FACE *sb.* 26)]. **1952** C. DAY LEWIS tr. *Virgil's Aeneid* VII. 160 Working polished leg-guards from malleable silver. **1923** *Nation* (N.Y.) 24 Oct. 454/2 Newsboys and 'legmen' and a foreign news service keep the streets of Mecca aware of all that goes on. **1951** E. PAUL *Springtime in Paris* xi. 195 The Paris police, leg men and cameramen from the Paris newspapers began tailing Nordmann and Tixier-Vignancour day and night. **1960** *Woman's Own* 5 Mar. 9/1 Jeannie supposed he'd have to get another secretary. He had two already, one more or less a leg-man, another who came in by the day. **1967** *Economist* 1 July 26/1 He was Mr Macmillan's leg man during the break-up of the Central African Federation. **1918** G. B. SHAW *Let.* Sept. in W. Loraine *Robert Loraine* (1938) xiii. 247 So long as you have a mouth left and one lung to keep it going, you will still be better than the next best: my pieces are not leg pieces. **1923** J. MANCHON *Le Slang* 179 *Leg-piece*,..ballet. **1915** *Truth* Nov. 848/1 What you describe as a swindle was only a brain-wave of mine ending in a leg-pull. **1938** *Times Lit. Suppl.* 12 Feb. 112/3 In point of fact they appear to have been no more than a not very subtle leg-pull. **1950** T. S. ELIOT *Cocktail Party* III. 148 You always did enjoy a leg-pull, Julia. **1965** M. SPARK *Mandelbaum Gate* iv. 106 Sometimes, Abdul, I wonder if you're just treating me to a big leg-pull. **1970** J. ARDAGH *New France* xi. 549 His whole operation might be partly a leg-pull at the expense of serious literature. **1923** *Motor Cycling* 7 Nov. 2/1 You are a confirmed leg-puller. Just fancy kidding me about the speed of the bus I bought from you. **1969** *Mind* LXXVIII. 31 Most samples are fair samples (God is not a leg-puller). **1908** *Westm. Gaz.* 30 June 2/1, I, too, have lived in Australia, where leg-pulling is one of the chief joys of life. **1926** T. E. LAWRENCE *Seven Pillars* IX. civ. 553, I could hardly tell my own self where the leg-pulling began or ended. **1946** *R.A.F. Jrnl.* May 178 'Liz' and 'Sally'..will take any amount of leg pulling—and give it. **1963** *Times* 30 May 17/2 Mr. Rovere is patently civilized, thoughtful and well-informed in his leg-pulling. **1833** J. C. LOUDON *Encycl. Archit.* III. vi. 1050 A Leg Rest..is sometimes used in dining-rooms by old gentlemen after the ladies are gone. **1854** MRS. GASKELL *North & South* (1855) II. v. 68 He was busy.. contriving a leg-rest for Dixon, who was beginning to feel the fatigues of watching. **1970** *Stoke Mandeville Dict. Managem. Paraplegic Patients* 23 *Leg-Rest*, attachment to wheelchairs used to keep a paralysed leg elevated. **1973** *Green Shield Stamps Catal.* No. 12. 133 (*caption*) Garden Chair with Leg Rest. Adjustable to 8 positions. [**1909** A. L. THOMSON in *Brit. Birds* II. 362 Various investigators..are endeavouring to obtain fuller and more accurate data with regard to migration, by liberating birds marked with metal foot-rings.] **1938** *Brit. Birds* XXXI. 242 Each bird was marked with a light, numbered, metal leg-ring of the British Birds Marking Scheme. **1959** *New Biol.* XXIX. 111 Leg-rings in different colours have been distributed to a large number of I.G.Y. bases in the hope that some of the mysteries associated with this bird's [*sc.* the Antarctic skua's] activities can be solved. **1963** *Times* 5 June 14/4 During the time he has been in Cyprus, Mr. Nicholson has helped in many major ornithological achievements, notably in the hazardous but successful leg-ringing of young Eleanora's falcons. **1928** *Punch* 18 Apr. p. xxii (Advt.), Carries four full-sized people. Ample leg-room... A wonderful performer for such a small horse-powered car. **1958** *Times* 19 Aug. 11/6 They [*sc.* Ford designers] succeeded in giving the passengers sufficient legroom. **1972** *Drive* Spring 147/3 The height, legroom and squab level of the driver's seat can be adjusted. **1878** E. S. ELWELL *Boy Colonists* 235 She kicked out at Ernest, who was trying to get the leg-rope on. **1911** H. FOSTON *In Bell Bird's Lair* 36 Each cow was to be carefully 'leg-roped'. *Ibid.*, Ted was shown how to 'leg-rope' a cow and bail up. **1950** *N.Z. Jrnl. Agric.* Feb. 169/2 Milk stools, door handles, leg ropes, posts and walls must be kept clean. *Ibid.* Nov. 465/3 The tail of the cow should be prevented from waving about and, if necessary, the animal comfortably leg-roped. **1963** *Landfall* Mar. 13, I..let the cow out into the race where, taking the leg-rope with her, she squittered off wild in the eyes. **1912** B. E. BAUGHAN in D. M. Davin *N.Z. Short Stories* (1953) 189 Much to my surprise, there was no leg-roping, and hardly any bailing-up [of the cows being milked]. **1919** W. H. DOWNING *Digger Dial.* 32 *Legs eleven*,..the number eleven in the game of 'house'. **1933** L. A. G. STRONG *Sea Wall* 256 A game of 'house' was in progress and a voice monotonously droned the numbers: '..legs... Kelly's eye.' **1945** E. WAUGH *Brideshead Revisited* II. i. 230 Kelly's eye—number one; legs, eleven; and we'll Shake the Bag. **1965** BROPHY & PARTRIDGE *Long Trail* 144 *Legs eleven*,..in the game of House, eleven. **1871** 'MARK TWAIN' *Screamers* xxviii. 144 They're playing 'Undine' at the Opery House, and some folks call it the leg shop. **1882** J. J. JENNINGS *Theatr. & Circus Life* 238 Burlesque with its blonde attributes kept the country in a rage..and the minor musical attractions of the *quasi* legitimate stage have usurped its principal feature—the leg show. **1900** *Amer. Jrnl. Sociol.* VI. 447 Next follows a cinematograph reproduction of a prize-fight, and then, in striking contrast with the first, a 'leg show' of the most shameless character. **1930** BROPHY & PARTRIDGE *Songs & Slang 1914–18* 137 At a leg-show of these days you saw far less of the female form than is now exhibited in the streets. **1930** J. DOS PASSOS *42nd Parallel* v. 420 Doc wanted to go to see a legshow. **1952** 'J. GUTHRIE' *Paradise Bay* vi. 57 He took me to exciting spectacles which I now know were leg shows. **1969** *Listener* 20 Mar. 399/1 We often use the cliché of the tired business-man to define the low response..that sustains leg-shows. **1616** T. CORYAT *Traveller for English Wits* 42 Your..most obliged Countreyman..the..Legge-stretcher of Odcomb in Somerset, Thomas Coryate. **1942** 'M. HOME' *House of Shade* iii. 57 Marigny was suggesting a leg-stretcher... The two made their way out to the glare of the sun. **1970** *Daily Tel.* 23 May 9 Now lonely, neglected, and often overgrown, they provide delightful leg-stretchers, on the springy turf, for the motorist with an hour or so to spare. **1960** M. G. EBERHART *Jury of One* (1961) i. 9 An old school friend, a fashion writer, had needed an assistant, a leg woman. **1891** *Dialect Notes* I. 207 Reporters characterize a task in which there is more running than writing by the expression *leg-work*. **1942** E. PAUL *Narrow St.* xxxi. 285 The Greek madonna did the leg work faithfully while La Absalom cackled orders through a rift in the portières. **1959** G. COBDEN *Murder for his Money* iv. 51 He wouldn't come himself for Patey was no man of action, but he might send..a man we used a great deal for leg work. **1972** *Daily Tel.* 21 June 13/8, 1,700 men..do the surveying leg-work needed for keeping local maps up to date. **1973**

L. Heren *Growing up Poor in London* vii. 179, I would earn a few bob working on the edge of big stories... The reporters who came down from Fleet Street were nearly always willing to pay for leg work.

b. leg break (examples); hence *leg-breaker*, a leg-break bowler; **leg-cutter** (see quot. 1966); **leg glance, glide,** a shot in which the ball is glanced fine on the leg side; **leg play** (see quot. 1934); **leg side** = Leg *sb.* 6 b; **leg slip,** (a fielder in) a position corresponding to that of the slips (see Slip *sb.*³ 14 a), but on the leg side; **leg spin,** a type of spin which causes the ball to turn from leg side to off (so *leg-spinner*); **leg-stroke** (examples); **leg stump** (later examples); **leg sweep,** a sweeping stroke which sends the ball to leg; **leg theory,** the technique of bowling to leg with a concentration of fielders on the leg side; **leg trap,** fielders stationed for catches close to the wicket on the leg side.

1888 A. G. Steel in Steel & Lyttelton *Cricket* iii. 114 The 'leg break' ball is usually bowled from round the wicket. **1927** W. E. Collinson *Contemp. English* 19 Only gradually did the mysteries of yorkers, full tosses and leg-breaks penetrate to us. **1955** *Times* 9 May 15/1 Mansell.. bowled 17 steady overs of leg-breaks. **1969** M. Parkinson *Cricket Mad* x. 57 We had our suspicions confirmed in the Indian's first over which contained five leg breaks. **1905** *Strand Mag.* June 703/2 Armstrong is a leg-breaker. **1956** R. Alston *Test Commentary* i. 13 Benaud, potentially a most dangerous leg-breaker. **1956** N. Cardus *Close of Play* 37 What is action break? In what way is it a term with a meaning more demonstrable..than 'seamer' and 'leg-cutter'? **1963** *Times* 25 Apr. 4/5 He went to Lord's last year to play his first match for Warwickshire and within a couple of overs had bowled everything from a leg-cutter to an inswinger. **1966** B. Johnston *Armchair Cricket* 102 *Leg-cutter*, is really a fast leg-break which is bowled by 'cutting' across the seam of the ball. **1883** *Cricket* 19 Apr. 39/1 Horan came in after lunch, and soon commenced to score in his own peculiar style, leg glances being his favourite stroke. **1966** B. Close *Close on Cricket* iii. 35 *Leg glance*, this is a refined stroke played against a ball pitched on or outside the leg stump but not far enough up to drive efficiently. **1920** D. J. Knight in P. F. Warner *Cricket* 33 The first of the leg strokes, the leg glide, is a glorious one to watch, and is exceedingly paying. **1955** *Times* 12 July 12/4 Just before luncheon he had lost Brookes, who, after a beautiful leg glide, missed one from Smith to be out leg before. **1888** R. H. Lyttelton in Steel & Lyttelton *Cricket* xvi. 411 It is the bowlers who have most cause to grumble at the modern leg play. **1898** K. S. Ranjitsinhji *With Stoddart's Team* (ed. 4) v. 96 His [*sc.* S. P. Jones's] cutting and his leg play being particularly well-timed. **1928** *Daily Express* 28 June 3/4 If leg-play were more severely penalised and wickets less like polished concrete, the balance between bat and ball would be more redressed and the game restored to its former attractiveness. **1934** W. J. Lewis *Lang. Cricket* 147 *Leg play*, 1. Stopping a breaking or curling ball with the leg instead of with the bat... 2. The playing of balls on the leg side. **1816** W. Lambert *Cricketer's Guide* (ed. 6) ii. 32 If the Ball should come 4 or 5 inches on the leg side, the Striker should move his right foot back at the moment of hitting, playing the Ball between his left leg and the wicket. **1956** N. Cardus *Close of Play* 26 No opposing captain dreamed of setting a close leg-side field for him. **1969** P. Pocock *Bowling* i. 19 The majority of seam bowlers swung the ball away from the batsman, using six fielders on the offside and three on the leg side. **1956** R. Alston *Test Commentary* xvi. 143 Miller swept him through the hastily retreating leg-slips. **1963** A. Ross *Australia 63* vii. 134 Benaud gave Davidson his most aggressive field for some time; neither long leg nor third man, three slips, gully, leg slip and backward short leg. **1888** A. G. Steel in Steel & Lyttelton *Cricket* iii. 116 The ball, coming from a great distance round the wicket and with a considerable amount of leg spin, would be gradually working away to the batsman's off side. **1966** B. Close *Close on Cricket* v. 55 We have a quantity of good finger-spin bowlers but few leg-spin bowlers in first-class cricket. **1927** *Observer* 29 May 28/4 It was a clear case for old Brown's leg-spinners. **1965** P. Walker *Winning Cricket* iv. 57 Nowadays you have to bowl leg spinners with the accuracy of an orthodox right hander to achieve even moderate success. **1906** *Westm. Gaz.* 12 July 4/1 This leg-stroke off a straight ball has two great merits—it scores runs and it puts the bowler off. **1955** *Cricket—How to Play* (M.C.C.) 47 (*heading*) Practising leg-strokes. **1937, 1956** Leg stump [see *Chinese *cut*]. **1846** W. Denison *Cricket Sk. Players* 17 His [*sc.* G. Brockwell's] 'leg sweeps' are very powerful, and generally speaking they are along the ground—not lifted. **1955** A. Ross *Australia 55* xii. 168 His leg-sweep was comfortably finished before the ball had got anywhere near him. **1898** G. Giffen *With Bat & Ball* x. 153 Cooper bowled the leg-theory almost as remarkably as the off-theory is practised nowadays. **1923** *Daily Mail* 11 Aug. 7/4 Newman, following the fashion of the match, bowled the leg theory with a crescent of fieldsmen close in on the leg side. **1956** N. Cardus *Close of Play* 26 He would have discovered a way of retaliation against leg-traps and leg-theories. **1923** *Wisden's Cricketers' Almanack* 329 His stock ball is the in-swinger and here again he often hits the wicket and has not got to rely on his leg-traps. **1924** N. Cardus *Days in Sun* 50 The good balls were pushed for singles through the leg-trap—when they did not get wickets. **1953** R. Warner *Escapade* ii. iii. 93 It's a leg-trap... You see, my dear, the batsman is absolutely forced to play every ball to the leg. **1963** *Times* 5 June 4/2 He took a pace to Titmus only to turn him gently into the leg trap.

leg, *v.* Add: **5. b.** To seize or hold by the leg.

1876 *Coursing Calendar* 149 Birkdale..came round on the outside and legged the hare, which Stolen Moments killed. **1951** L. G. D. Acland *Early Canterbury Runs* 385 To *leg* a sheep is to haul him from the pen to the board by his hind leg, a practice much objected to, especially by owners of heavy sheep.

7. *Cricket.* To send to leg.

1902 *Westm. Gaz.* 11 July 5/2 The newcomer at once started scoring... Nicholl followed him by legging Hopley to the covered stands. **1903** *Star* 8 July 3/4 His first ball was legged by Ranji for what would really have been 3 with a couple of smart sprinters.

legal, *a.* Add: **1. e.** *legal cap*: ruled writing paper used chiefly for legal documents. *U.S.*

a **1877** Knight *Dict. Mech.* I. 455/2 Foolscap and legal cap are of various sizes, from 7½ × 12 to the size of a flat cap-sheet folded 8½ × 14. **1902** W. N. Harben *Abner Daniel* i. 3 Old man Bishop..was carefully reading a long document written on legal-cap paper. **1937** E. J. Labarre *Dict. Paper* 142 *Legal cap*, a size of paper 14″ × 8½″.

f. *legal beagle, eagle,* rhyming collocations designating a lawyer, *spec.* one who is keen and astute.

1949 *Law Library Jrnl.* XLII. 187/1 Legal Eagles and Stuffed Owls in Detroit. **1953** B. Glemser *Dove on his Shoulder* xii. 230 Since these letters are evidence they ought to be handed over to our legal beagles. **1961** I. T. Ross *Requiem for Schoolgirl* vii. 118 He's got some sort of legal beagle who protects him. **1963** N. Freeling *Because of Cats* vii. 116 According to the legal eagles you'll never get away with prosecuting them. **1967** Wodehouse *Company for Henry* vii. 117 You allowed your mind to wander when the legal eagle was doing his stuff. **1968** M. Allingham *Cargo of Eagles* iv. 57 That's the only good thing I've ever heard about your infernal legal eagle. **1974** *Economist* 30 Nov. 75 Legal eagles agree... The meeting, only the second of the justice council in EEC history.., was civilised.

g. *legal positivism*: see *positivism.

2. a. *legal aid*: official assistance allowed under certain conditions towards the expense of litigation (cf. Aid *sb.* 2 and *2 b); *legal capacity*: the authority under law of a person to engage in a particular undertaking, or maintain a particular status; *legal fiction*: see Fiction 5 a; *legal memory* (see quots.).

1890 (*title*) Constitution and bye-laws of the Deutsche Rechts-Schutz Verein (German Legal Aid Society). **1903** *Act* 3 *Edw. VII* c. xxxviii. § 1 Where it appears.. desirable in the interests of justice that [any poor prisoner] should have legal aid in the preparation and conduct of his defence, and that his means are insufficient to enable him to obtain such aid..the committing justices..may certify that the prisoner ought to have such legal aid. **1928** [see *interstate *a.*]. **1959** *Daily Tel.* 18 Mar. 19/6 In eight years legal aid has been given to more than 238,000 litigants. **1967** E. Rudinger *Wills & Probate* 91 The other [leaflet] explained how to get legal advice under the legal aid scheme. **1971** 'M. Underwood' *Trout in Milk* ii. 27 Thanks to 'legal aid' it was much easier to get started at the criminal than the civil Bar. **1972** *Times* 28 Jan. 16/4 All concerned—the parties' advisers, the legal aid committees,..—should put children's cases at the top of the list.

1890 Lely & Geary *Chitty's Pract. Treat. Law of Contracts* (ed. 12) vii. 194 The age of twenty-one years..has been fixed, as the period when an absolute and unlimited legal capacity to contract shall commence. **1968** *Lebende Sprachen* XIII. 82/2 The customer will bear any loss incurred by the bank resulting from the fact that the bank.. does not obtain knowledge of any restrictions of the customer's..legal capacity.

1861, 1876 [see Fiction 5 a]. **1937** R. H. Lowie *Hist. Ethnol. Theory* v. 51 The same holds for the principle of legal fiction, which Maine also treated at length. **1966** Black & Brown *Outl. Eng. Law* v. 43 Sometimes such a presumption takes the form of a *Legal Fiction.* For instance if two persons die in such a way that it is uncertain which of them died first the law presumes..that the younger outlived the older.

1766 Blackstone *Comm.* II. ii. iii. 31 It seems unaccountable, that the date of legal prescription or memory should still continue to be reckoned from an aera so very antiquated. **1882** *Encycl. Brit.* XIV. 650/1 By the Statute of Westminster the First,..the beginning of the reign of Richard I. was fixed as the date of limitation for such actions. This is the well known 'period of legal memory'. **1882** C. Sweet *Dict. Eng. Law* 525 When a person alleges in legal proceedings, that a custom or prescription has existed from time whereof the memory of man runneth not to the contrary..this is..called time of living memory, as opposed to time of legal memory, which runs from the commencement of the reign of Richard I. **1969** J. B. Saunders *Words & Phrases legally Defined* (ed. 2) IV. 171/1 By the ancient rule of the common law, enjoyment of an easement has to be proved from time 'whereof the memory of man runneth not to the contrary', that is to say, during legal memory or since the commencement of the reign of Richard I.

5. b. The exact fare without any tip; a passenger who pays such a fare. Chiefly *taxi-drivers' slang.*

1923 J. Manchon *Le Slang* 179 The legal = the legal fare. **1939** H. Hodge *Cab, Sir?* vii. 84 The last, and probably most common, cause of the inadvertent 'legal' is the super-sensitive meter. *Ibid.* 85 Some 'legals' are simply mean, and give excuses instead of a tip. **1963** M. Levinson *Taxi!* vii. 88 If his next passenger gives him another 'legal' (the exact fare) he will naturally take a very dim view.

legalese (līgălī·z). *colloq.* [f. Legal *a.* + -ese.] The complicated technical language of legal documents.

1914 C. J. C. Hyne *Firemen Hot* 189 He signed his name at the foot of a bald formal agreement, written in the most incomprehensible legalese. **1966** A. Sachs *Jail Diary* xvii. 155 After all these weeks I am not used to reading legalese any more. **1967** 'J. H. Roberts' *February Plan* i. iii. 81, I won't go into the legalese, Mr. Corman, but he claims he had a contract with you. **1973** *Black Panther* 5 May 13/3 Is it any measure of excellence to assemble glossy paragraphs of smart untried legalese? **1973** *N.Y. Times* 11 Aug., Gordon Strachan spoke openly of his resort to legalese.

legation. 3. b. (Earlier examples.)

1832–3 *Reg. Deb. Congress U.S.* 22nd Congress 2 Sess. App. 90/2 The proceedings..are not recorded in the legation. *Ibid.*, The instructions are not in the legation [in London].

legend, *sb.* Add: **7. c.** The written explanatory matter accompanying an illustration, map, etc. Also *attrib.*, as *legend-line.*

1903 *Westm. Gaz.* 8 Jan. 2/1 The sort [of satire] I should employ if—if I were writing legend-lines for a halfpenny comic paper. **1951** D. Bland *Illustration of Bks.* ix. 142 After the blocks have been made and the proofs approved, a paste-up should be prepared for the printer, to include legends or captions. **1963** *Which?* July 200/1 The symbols used on a map should be..explained clearly in the legend (or key). **1970** *Watsonia* VIII. 31 Figure 4. For legend see above. **1974** *Times Lit. Suppl.* 23 Aug. 910/5 The information provided in caption and legend is clear and exactly what is needed.

8. *legend-king, -lust; legend-haunted* adj.

1905 *Westm. Gaz.* 25 Mar. 5/3 Its famous, legend-haunted Jews' quarter. **1908** *Daily Chron.* 5 Oct. 7/1 As a poor and lonely boy he wove his day-dreams by the legend-haunted cliffs of Tintagel. **1930** Blunden *Poems* 319 And thence, before to-morrow's dawn, it springs That they are one with elves and legend-kings. **1911** E. Pound *Canzoni* 42 Nay, on my breast thou must Forget and rest and dream there For thine old legend-lust.

b. Applied to the estimated or planned displacement, speed, etc., of a ship before construction or testing. Also *absol.*

1908 *Westm. Gaz.* 31 July 1/3 A ship.. in the Bay would exceed her legend speed by a knot or two. **1921** *Glasgow Herald* 17 Dec. 12 The British Government may construct two new ships, not to exceed 35,000 legend tons each. **1936** W. S. Churchill in *Second World War* (1949) I. i. ix. 144 If you ask your people [the Admiralty] to give you a legend for a 16-inch-gun ship, I am persuaded they would show you decidedly better proportions than could be achieved at 14-inch.

Legendre (ləʒɑ̃·ndr'). *Math.* The name of A. M. *Legendre* (see Legendrian *a.*), used *attrib.* and in the possessive to designate certain expressions investigated by him, esp. (*a*) the differential equation $(1-x^2)d^2y/dx^2 - 2xdy/dx + n(n+1)y = 0$, (*b*) its solutions $y = P_n(x)$, where $P_n(x)$ is a polynomial that is the coefficient of h^n in the expansion of $(1-2xh+h^2)^{-1/2}$, and (*c*) the associated functions $P_n^m(x)$, equal to $(1-x^2)^{m/2}\frac{d^m}{dx^m}P_n(x)$.

1875 I. Todhunter *Elem. Treat. Laplace's Functions* i. 1 The coefficient of a^n will be a function of *x* which we shall denote by $P_n(x)$, and shall call Legendre's Coefficient of the nth order. **1880** *Encycl. Brit.* XIII. 21/2 Legendre's function. **1885** A. R. Forsyth *Treat. Differential Equations* v. 152 We have now obtained the complete integral of Legendre's equation in all cases when *n* is a real constant, by deducing two integrals which are linearly independent..of one another. **1902** *Encycl. Brit.* XXXII. 798/1 (*heading*) Legendre associated functions. **1930** *Engineering* 26 Dec. 812/3 There are also tables of Legendre functions. **1938** S. Dushman *Elem. Quantum Mech.* vi. 153 The Legendre polynomials form an orthogonal system, since $\int_{-1}^{1} P_k(x)P_n(x)dx = 0$ (for $k \neq n$) [or] $= 2/(2k+1)$ (for $k = n$). *Ibid.* 158 Since the value of any function $P_k(\cos\theta)$ exhibits $k-1$ loops, there are $2(k-1)$ circles parallel to the nodal circles at which the function has the same absolute value. It is for this reason that the Legendre coefficients of zero order are known as zonal harmonics. **1953** A. D. & K. H. V. Booth *Automatic Digital Calculators* xvi. 189 The Legendre polynomials.. are such that, if any function is expanded in a series of the polynomials, the first *n* terms of this series will give the best *n*th degree polynomial approximation to the given function, in the least squares sense. **1962** Corson & Lorrain *Introd. Electromagn. Fields* iv. 172 The solutions of Legendre's equation are called Legendre polynomials, which we denote by $P_n(\cos\theta)$, there being a different polynomial for each value of the index *n*. **1970** G. K. Woodgate *Elem. Atomic Struct.* ii. 17 The solutions are proportional to the associated Legendre functions.

Leger (le·dʒəɪ), *sb.*² Shortened f. *St. Leger* (see Saint *a.* 4 c).

1871 [see Egg *sb.* 4 b]. **1880** Trollope *Duke's Children* I. xvii. 206 Who would like to bet me fifteen to one in hundreds against the two events,—the Derby and the Leger? **1924** Galsworthy *Forest* ii. i. 41 I'd give all mine [*sc.* my ancestors] to know what's won the Leger. **1961** F. C. Avis *Sportsman's Gloss.* 230/2 *Leger Day*, the day upon which the St. Leger is actually run.

legger² (le·gəɪ). [f. Leg *sb.* + -er¹.] **1.** In a slaughter-house, a butcher or packer who works on the legs of the carcasses.

1905 F. W. Wilder *Mod. Packing House* v. 116 The number of men necessary in [different sized] beef killing groups..are given in the following table... 2 front leggers ..3 hind leggers. **1923** R. A. Clemen *Amer. Livestock & Meat Industry* xv. 332 A string of butchers now follow each other in rapid succession. The 'leggers' remove the hind legs at the hoof and the forelegs at the knee. **1949** K. Stronach in A. E. Woodhouse *N.Z. Farm & Station Verse* (1950) 188 The legger slits as a surgeon does And the puncher strips the pelt. **1966** *Mate* (Auckland) Aug. 42 Legger, the man who cuts out the legs from the newly killed lamb in the freezing works.

2. (See quot.)

1927 T. Woodhouse *Artificial Silk* 95 In the manufacture of stockings on such frames two machines are used... One of these machines, termed the 'legger', knits the upper and longer part of the stocking, whereas the other machine, termed the 'footer', knits the remainder of the stocking.

legger³ (le·gəɪ). *U.S. colloq.* Shortened f. *boot-legger. Also (with preceding hyphen) as the second element of *Combs.*, an illegal seller (of something indicated in the first element).

1926 [see *hip *a.*]. **1929** *Variety* 5 June 58/2 Leggers claim the tonic is as potent as a fifth of gin selling for three times the price. **1934** *Time* 29 Jan. 49/1 Hundreds of U.S. citizens have smuggled copies through the customs or bought them from book-leggers. **1937** *Time* 4 Jan. 11/2 Unlike Prohibition's liquor 'leggers', they are not growing rich. **1945** *Chicago Daily News* 12 July 12/2 Most of the counterfeits have been used by steakleggers down East. **1945** Mencken *Amer. Lang.* Suppl. I. 366 But of more interest..are the words showing recent vogue affixes, *e.g.*..*-legger*, as in *bootlegger* and *meatlegger.* **1973** *Times Lit. Suppl.* 20 Apr. 451/5 A sinister group of organ-leggers who go beyond the resurrectionists and try to keep the hospitals supplied with adequate numbers of fresh organs for transplants.

legget (le·gət). Also **leg(g)at(t), leg(g)et(t).** [Etym. unknown.] In Thatching, a tool made from a flat board attached to a handle and studded with nails, used for dressing and driving the reeds into place. Cf. *ledger *sb.* 2 b.

c **1555** in *Norfolk Antiquarian Misc.* (1883) II. 9 A Cardynall legat & a trowell wᵗ suche other tooles. **1787** W. Marshall *Rural Econ. Norfolk* II. xxxii. 61 The eaves being thus completely set, they are adjusted and formed; .. nor are they formed by cutting; but by 'driving' them with a 'legget'. **1927** *Observer* 24 July 5/3 The tools and appliances used in Devon are more or less similar to those used in Norfolk. The 'leggett', however, is called a 'driff'. **1941** [see *biddle]. **1949** K. S. Woods *Rural Crafts Eng.* iv. xiii. 204 The beating tool is called a 'legatt'; it is a square of thick wood studded with horseshoe nails set diagonally, nail and space, in the rows. **1961** *Guardian* 21 Sept. 5/2 Ramming home one corner with a Chaucerian tool known as a leggett. **1969** E. H. Pinto *Treen* xxv. 406 The leggat, legget or reed bat..is used for patting or beating reeds into position. **1971** *Country Life* 18 Nov. 1403/3 The Norfolk thatching reed is then beaten up tight under the hazel rods using an instrument called a leggett. **1972** *Daily Tel.* 28 Oct. 15/2 Each 'yelm' is pegged into place and beaten with a flat piece of wood or 'spud' (reed thatchers use a corrugated surface, called a 'leggat').

‖ **leggiero** (ledʒē°·ro), *a.* *Mus.* [It.] Of musical movement: light and nimble. Also used as *adv.*

1880 Grove *Dict. Mus.* II. 113/2 Leggiero passages are usually, though not invariably, *piano*, and they may be either legato or staccato. **1939** *Internat. Cycl. Mus.* 988/2 *Leggiero* (Italian), lightly, usually applied to a rapid passage for the pianoforte. **1960** *Times* 4 June 9/5 Katchen's *leggiero* playing is predictably delightful.

legging, *sb.* Add: **b.** *Cricket* = Pad *sb.*³ 3 c. Now *rare* or *Obs.*

1858 in *Cricket Q.* (1963) I. 21. **1875** *Baily's Monthly Mag.* May 11 Beldham also, without either leggings or gloves, scored 72 against Brown..when fifty-two years old. **1934** W. J. Lewis *Lang. Cricket* 146 *Leggings*, leg-guards of a primitive type.

legging, *vbl. sb.* Add: **2.** Propelling a boat through a canal-tunnel by human labour (see quot. 1949). Cf. Leg *v.* 3. Also *attrib.*

1861 S. Smiles *Lives Engineers* II. viii. x. 421 It [*sc.* the tunnel] was little larger than a sewer, and admitted the passage of only one narrow boat, seven feet wide, at a time, involving very heavy labour on the part of the men who worked it through. This was performed by what was called *legging.* **1949** *Archit. Rev.* CVI. 13/2 In the early tunnels towpaths were never constructed and boats were propelled through them either by shafting or legging... Legging was carried out by two men, one on each side of the boat lying on their backs and pushing against the tunnel sides with their feet. **1963** *Times* 4 May 6/4 They will stare in amazement at the legging boards used in the navigation of tunnels in the days when horse power meant what it said. **1975** *Times* 13 Mar. 4/7 The police have concentrated on the eighteenth-century 'legging' tunnel, built by Grindley, so called because boatmen had to propel their craft through by thrusting their legs against the roof.

3. *Austral.* and *N.Z.* Of dogs, the biting of an animal on the leg. Cf. *Leg *v.* 5 b.

1933 L. G. D. Acland in *Press* (Christchurch, N.Z.) 4 Nov. 15/7 Some dogs get a bad habit of biting sheep on the leg in yards; others of laming other dogs while fighting. Both these vices are called *legging.* **1960** [see *force *sb.*¹ 7 e].

leggism (le·giz'm). *colloq.* [f. Leg *sb.* 5 + -ism.] The practice, or an act, of 'black-legging' (Cf. Black-leg, -legs 2).

1847 *Sporting Life* 28 Aug. 16/2 The system of adopting leggism as a legitimate part and parcel of horse-racing is one that must soon explode. **1896** Farmer & Henley *Slang* IV. 178/2 *Leggism*,.. the character, practices, or manners of a leg. **1937** Partridge *Dict. Slang* 477/1 *Leggism*, the art or the character of a 'leg'.

leggo (legōᵘ·), a representation of a colloq. or vulgar pronunciation of *let go!*

1884 'Mark Twain' *Huck. Finn* xxx. 310 Leggo the boy, you old idiot! **1889** J. K. Jerome *Three Men in Boat* ii. 24 What are *you* up to?.. leggo, can't you? **1906** Kipling *Actions & Reactions* (1909) 207 Leggo my collar! **1932** L. Golding *Magnolia St.* ii. xiii. 450 Leggo of my 'ead! **1961** 'F. Richards' *Bunter the Ventriloquist* xxv. 145 'Ow! Leggo!' Bunter, wriggling, blinked round in alarm at the captain of the Remove. **1973** 'D. Shannon' *Spring of Violence* (1974) x. 168 You're crazy, man. Leggo of me.

leggy, *a.* Add: Also *transf.*, long-stemmed.

1860 O. W. Holmes *Prof. at Breakfast-Table* x. 310 The white meeting-house, and the row of youthful and leggy trees before it. **1932** *Times Educ. Suppl.* 9 July 267/4 If plants are crowded under glass they will grow 'leggy'. **1965** H. G. W. Fogg *Small Greenhouse* v. 39 Put the boxes, etc., on a shelf close up under the glass of the greenhouse, to prevent the seedlings from becoming drawn and 'leggy'.

legh (lī). Also **leigh.** [Etym. unknown.] = *Irish deer, elk* (*Irish *a.* 2 b).

1774 T. West *Antiquities Furness* p. xlvi, That the legh was a native of Furness, is evident from the heads of those animals frequently found in Furness. **1795** A. Radcliffe *Journey* 483 A remarkably large breed of deer, called Leghs, the heads of which have frequently been found buried at a considerable depth in the soil. **1835** Wordsworth *Guide through District of Lakes* (ed. 5) ii. 38 The leigh, a gigantic species of deer which has long been extinct. **1974** J. W. Smyser in Wordsworth *Prose Works* III. 403 Wordsworth drew on West for his information about the 'leigh'.

Leghorn. 1. (Earlier and later examples.)

1740 *Pennsylvania Gaz.* 22 May 7/2 Leghorn hats. **1742** *Boston News-Let.* 24 June 2/2 Just imported..from London..Leghorn Hats for women. **1969** Gish & Pinchot *L. Gish* iii. 33 Dorothy and I were cool in our full-skirted summer frocks and wide-brimmed leghorn hats. **1975** R. Player *Let's talk of Graves* ii. 46 Her big grey saucer eyes, shaded by the big Leghorn hat.

legion. Add: **1. b.** (Examples of *foreign legion*).

1897 E. A. Bartlett *Battlefields Thessaly* vii. 144 There was a small foreign legion of about five hundred men, made up chiefly of Italians and English. **1924** M. Magnus (*title*) Memoirs of the foreign legion. **1957** P. Kemp *Mine were of Trouble* ii. 19 The Foreign Legion, or Tercio, was founded in the early 1920s by General Millán Astray. **1968** *Encycl. Brit.* XIII. 905/2 The term 'foreign legion' is often used for irregular volunteer corps of foreign sympathizers raised by states at war.

4. b. *Legion of the lost* (*ones*): people who are destitute or abandoned; *spec.* (see quot. 1961).

1870 D. J. Kirwan *Palace & Hovel* xlii. 587 Those fair and frail members of the Legion of the Lost. **1892** Kipling *Barrack-Room Ballads* 63 To the legion of the lost ones, to the cohort of the damned. **1899** —— *From Sea to Sea* I. ix. 299 Raising a Legion of the Lost for colonial service—of men who would do their work in one place for ever and look for nothing beyond it. **1961** Partridge *Dict. Slang* Suppl. 1167/1 *Legion of the lost, the*, those elderly or mentally infirm persons in homes or institutions who have been abandoned by relations and friends and who receive neither visits nor letters.

c. *American Legion*, a national association of ex-servicemen instituted in 1919 in the U.S.A.; *British Legion*, a similar association founded in 1921 and incorporated by Royal Charter in 1925 (since 1971 called the Royal British Legion); also *ellipt.*, as *Legion.*

1919 G. S. Wheat *Story Amer. Legion* 8 At that dinner [in Paris, 16 Feb. 1919] the American Legion was born. *Ibid.* 32 That was the crux of the initial success of the Legion. **1921** *Times* 16 May 6/4 The arrangements for uniting various ex-Service men's societies into one big organization were completed..yesterday. The new organization will be known as the British Legion. **1953** *New Statesman* 13 June 696/2 A thousand business dinners, Legion reunions and family gatherings. **1968** *Encycl. Brit.* I. 764/1 Nonpolitical and nonsectarian, the American Legion's membership requirement is honourable service and an honourable discharge. **1970** *British Legion Jrnl.* Jan. 11/3 'Don' as he was affectionately known by all was a true member of the Legion, always to the fore in all branch and club activities. *Ibid.* 21 Are you going to Jersey for your holiday this year? The Appeals Department of the British Legion will send you..an interesting brochure. **1974** A. Price *Other Paths to Glory* vi. 72 Secretary of the Elthingham branch of the British Legion. *Ibid.* 73 He invariably sought information first from the local Legion secretary. **1974** T. Kenrick *Two for Price of One* xii. 101, I love my country... Been in the [American] Legion twenty years.

legionnaire (līdʒənē°·ɪ). Also **legionaire.** [ad. F. *légionnaire*, f. *légion* Legion 1 b.] A member of the American, British, Foreign, or other Legion.

1818 Shelley *Let.* 20 Nov. (1964) II. 55 The marks of the chisels of the legionaires of the Roman Consul are yet evident. **1927** *Daily Express* 26 Apr. 1/5 Each legionnaire automatically becomes a member of the branch of the Canadian Legion nearest to the settlement on which he lives. **1927** *Daily Tel.* 14 June 7/2 The American Legion ..assembles in Paris in September... The legionnaires will be over two weeks on the Atlantic. **1932** Kipling *Limits & Renewals* 322 His speech..ran From pure Parisian to gross peasant, With interludes North African If any Legionnaire were present. **1965** C. D. Eby *Siege of Alcázar* (1966) xi. 222 Captain Tiede and his Legionnaires filed into the Alcázar through the swimming pool door. **1974** *Northern Times* (Golspie, Sutherland) 23 Aug. 3/2 The Legion piper, Mr. Donnie McKenzie, played Mr. Bain's favourite pipe tune, while Legionaires paid their last respects.

legist. Add: *spec.* One of a group of legal philosophers in the early Han dynasty in China. Also *attrib.* or as *adj.*

1956 A. Toynbee *Historian's Approach to Relig.* ii. 22 In China the uncompromisingly rationalistic Legist school of philosophy was eventually driven off the field by a Confucian school which tempered its Rationalism with a conservative respect for a pre-rationalist tradition. **1957** *Chinese Culture* (Taipei) I. i. 77 As we know, School of the Legists prevailed during the earlier Han dynasty. **1965** *New Statesman* 24 Dec. 1004/1 Eventually the Romans did adopt that model but not, unfortunately, 'deliberately planned and executed in advance by an act of far-sighted and well-calculated statesmanship' (as Princes Hien and Hiao had done in China with the help of a 'sophist of the Legist school, Shang Yang').

legit., legit (līdʒi·t), colloq. abbrev. of Legitimate *a.*, *sb.*, esp. of sense 2 b of the adj. Also in phr. *on the legit*, within the law.

1897 *National Police Gaz.* (U.S.) 26 May 6/1 Bob is envious of Corbett's success as a 'legit'. It pained him to see Jim strutting through four acts of a real play. *Ibid.*, Bob now wants to go into the 'legit'. **1904** *Daily Chron.* 22 Oct. 7/4 At the Hippodrome, four more 'legits' make their first appearance in 'variety'. **1908** 'One of the Old Brigade' *London in Sixties* xiv. 177 Scene shifters, stage carpenters, actors, everything and everybody strictly 'legit' should have the preference of guzzling and swilling to the memory of the immortal poet. **1923** H. Ruby *Let.* 16 Aug. in G. Marx *Groucho Lett.* (1967) 184 He clicked as a legit actor on Broadway. **1931** G. Irwin *Amer. Tramp & Underworld Slang* 138 *On the legit*, honest. **1936** N. Coward *Tonight at 8.30* I. 93 When she stabs herself—she takes such a time about it—that's legits all over. **1946** Mezzrow & Wolfe *Really Blues* ii. 21 Once Sid got..a hundred cases of booze on the legit..but..would sooner have his throat cut than push them at legit prices at the drugstores. **1952** W. R. Burnett *Vanity Row* (1953) v. 41 In the early 'twenties he'd served time..for bootlegging. .. Since then.., he'd been..on the legit. **1955** E. Bowen *World of Love* xi. 206 Left no children—anyway, no legits. **1957** J. Osborne *Entertainer* v. 38 I'd gone legit. for a while..and I'd been in 'The Tale of Two Cities'. **1961** *Times* 29 Nov. 11/5 It provided funds for penetration into 'legit' business for the gang lords. **1965** J. B. Priestley *Lost Empires* i. ii. 16 She used to be legit and pretends to be very haigh-classy,.. though she's only playing feed to a comedian. **1967** J. Horton in T. Kochman *Rappin' & Stylin' Out* (1972) 22 Identified as white, a lame, and square, I had to build up an image of being at least 'legit' (not working for the police). **1969** B. Malcolm in A. Chapman *New Black Voices* (1972) 384 First time I was sweet sixteen Marriage license, zircon ring—all legit. **1970** C. Wood *Terrible Hard* viii. 112 I've never been 'legit' as you might say.., just an old-fashioned song and dance girl. **1972** R. Hill *Fairly Dangerous Thing* ii. vii. 185 A business acquaintance. Runs an escort service. Legit. High class. **1972** *New Society* 7 Dec. 559/1 The age of legit drinking will come down from 18 to 17. **1973** 'H. Howard' *Highway to Murder* x. 127 This dough isn't strictly legit.

legitimacy. 2. (Earlier example.)

1812 *Niles' Reg.* I. 404/1, I never hear an American citizen speak of the '*legitimacy*' of princes without indignation or pity.

legitimate, *a.* Add: **2. b.** *legitimate drama* and absol. (Earlier and later examples.) Also in other collocations. So as *sb.*, an actor of legitimate drama.

1799 *Sporting Mag.* XV. 135/2 A lady to whom the public are so much indebted for the support which the legitimate drama has received from her exertions, and who..has disdained the pantomime and spectacle to which the German muse so often stoops. **1812** *Theatrical Inquisitor* Oct. I. 72 Mr. E treads closely upon the heels of the legitimate stage. **1838** Dickens *Let.* 16 Jan. (1965) I. 355 Let the Legitimate Drama put this, and Joan of Arc..into her pipe. **1909** P. G. Williams in *Sat. Even. Post* 5 June 17/2 The vaudeville actor is much more thrifty than his colleague in the legitimate. **1933** P. Godfrey *Back-Stage* xvi. 207 The principal comedian of *Have a Nibble*..scandalizes the 'legitimates' by discarding the jacket of his sprightly plus-four suit. **1947** N. Marsh *Final Curtain* v. 84, I haven't got the wind for dancing.. and the 'legitimate' gives me a pain in the neck. **1952** *N.Y. Herald Tribune* 28 Aug. 16/7 A revision of New York City's building code to spur the construction of new legitimate theaters. **1968** *Globe & Mail* (Toronto) 17 Feb. 24 Nor is it [*sc.* the city] avoiding the inevitable responsibility of building a smaller legitimate house. **1972** *N.Y. Times* 3 Nov. 1/1 The new hotel would include..a legitimate theater. **1975** *Scottish Field* Jan. 9/1 With his feet now firmly planted in both acting spheres—the so-called legitimate theatre and the pantomime lark—this young-looking veteran [*sc.* Rikki Fulton]..feels fit to accept any professional challenge.

c. (Earlier example.)

1812 *Niles' Reg.* I. 404/2 The 'legitimate' sovereigns of Russia, Austria and Prussia.

e. In *Jazz colloq.*, designating 'serious' music as distinct from jazz or popular music.

1927 *Melody Maker* Apr. 359/2 The number lends itself exceptionally well to the symphonic treatment it has been given, the orchestration is very fine and the modulated passages and general arrangement make it, although a little too 'legitimate' for dancing, perfect from a concert point of view. **1933** *Fortune* Aug. 94 Other jazz heroes such as the Dorseys..have become more or less legitimate musicians for radio purposes. **1946** Mezzrow & Wolfe *Really Blues* (1957) xvii. 341 The New Orleans drum patterns..were closest to 'legitimate' music. **1969** *New Yorker* 20 Dec. 52/3 It would have been interesting if he had made similar measurements during a performance by a 'commercial'—that is, a jazz or dance-band—player..to compare with those of a 'legitimate', or symphonic, player.

le·glessness. [f. Legless *a.* + -ness.] The condition of being legless.

1902 *19th Cent.* Feb. 254 Sir Richard Calmady's leglessness is never for an instant forgotten. **1911** G. B. Shaw *Doctor's Dilemma* Pref. p. xiv, The leg may mortify—it is always safer to operate—..evolution is towards motors and leglessness.

‖ **legong** (legǫ·ŋ). [Indonesian.] A stylized Balinese dance performed by young girls. Also *attrib.* Also, one of the performers of such a dance.

1926 E. Mordaunt *Further Venture Bk.* xxiii. 291 For three days preparations have been going on for a..Legong ..given by a man who has won a law-suit. **1930** H. Powell *Last Paradise* xii. 109 Runis and Madé Rêi were nine years old... They danced the sacred traditional measures of the *legong.* **1937** M. Covarrubias *Island of Bali* (1972) viii. 224 As the archetype of the delicate and feminine, the *legong* is the finest of Balinese dances... The *legong* is performed at feasts, generally in the late afternoon when the heat of the day has subsided. *Ibid.* 228 A very popular dance that seems related to the *legong* is the *djogéd*, performed by a girl in a variation of the *legong* costume and in the traditional *legong* steps. **1971** *Walkabout* (Austral.) Nov. 73/1, I was sitting at this beautiful bar, listening to the distant flurry of the Legong dance.

legouane, var. Iguana. See also Leguan.

1790 E. Helme tr. *Le Vaillant's Trav. Afr.* I. 391 The Hottentot who fired at the Hippopotamus came home late..and was obliged to meet the sarcasms of my Hottentot wits, who tried to persuade him he had fired at a *Legouane* (a kind of large lizard, common in the rivers of Africa).

legrandite (ləgrǫ·ndəit). *Min.* [f. *Legrand*, the name of a 20th-c. Belgian mine manager who collected the first specimen + -ite¹.] A basic hydrated zinc arsenate, $Zn_2AsO_4OH.H_2O$ occurring as colourless to yellow transparent monoclinic crystals at Lampazos, Mexico.

1932 Drugman & Hey in *Mineral. Mag.* XXIII. 175 A chemical analysis..showed it to be indeed a new mineral, a basic zinc arsenate, and the name 'legrandite' is proposed for it in recognition of the collector. **1971** *Soviet Physics: Doklady* XVI. 421/2 Analyses of legrandite from various deposits reveal up to 2% Mn isomorphously replacing Zn.

legume. 2. Delete *Obs.* and add later examples.

1870 J. Yeats *Nat. Hist. Commerce* ii. 137 The legumes of temperate climates are familiar plants. **1969** *Oxf. Bk. Food Plants* 34/2 There are a number of other tropical legumes which are only of very minor or local importance as sources of food. *Ibid.* 44/1 Cowpea..is an annual legume, originating in Africa. **1974** A. Huxley *Plant & Planet* xxx. 365 Various legume seeds have proved the best sources [of vegetable protein] so far.

lehiite (lī·hai,əit). *Min.* [f. *Lehi*, the name of the city in Utah near which it occurs + -ite¹.] A basic hydrated phosphate of calcium, potassium, sodium, and aluminium, of a white to grey colour.

1930 Larsen & Shannon in *Amer. Mineralogist* XV. 329 (*heading*) Lehiite, a new mineral. **1942** *Ibid.* XXVII. 294 Material identified by this writer as lehiite differs somewhat from that described by Larsen and Shannon. It forms dense, light gray layers on the outer shells of the nodules, and is made up of fine to moderately coarse fibers generally in subparallel bands. **1955** M. H. Hey *Index Min. Species* (ed. 2) 236 *Lehiite*, $(K,Na)_2Ca_5Al_8(PO_4)_8(OH)_{12}.6H_2O$ (?).

lehr, var. Leer *sb.*³ (in Dict. and Suppl.).

‖ **Lehrjahre** (lē·ryārə), *sb. pl.* [G. *lehr*(*en* to teach + *jahre* years; cf. G. *lehrling* apprentice.] Apprenticeship, usu. *fig.*

1865 J. A. Symonds *Let.* 15 May (1967) I. 539 The retrospective view you take there of your last two years is the just one. They have been Lehrjahre in a high sense. **1891** E. B. Bax tr. *Schopenhauer's Sel. Ess.* p. x, They settled at Hamburg... It was here that Arthur Schopenhauer spent his *lehrjahre.* *a* **1892** G. C. Robertson *Elem. Gen. Philos.* (1896) xx. 201 The Socratic stage (407–399)—his *Lehrjahre* as they have been called—when he was the pupil of Socrates. **1973** *Times* 14 June 16/2 Julian Fane has written an updated, nineteenth-century *Lehrjahre* book.

lehua (lehu·ă). [Hawaiian.] An evergreen tree, *Metrosideros collina*, of the family Myrtaceæ, native to the Polynesian and Melanesian islands of the Pacific Ocean and bearing panicles of scarlet flowers; also called *ohia* or *ohia lehua.*

1888 W. Hillebrand *Flora Hawaiian Islands* 125 *M*[*etrosideros*] *polymorpha*... The most generally prevailing tree on all Islands between 1500 and 6000 ft., usually gregarious. Nat[ive] name: 'Ohia lehua', or simply 'lehua'. .. The wood is very hard, furnishes the best fuel, and is also used for building houses. **1917** *Nature* 20 Sept. 57/2 Lehua, resembles, in the appearance of the trunk, our white oak, but bears beautiful clusters of scarlet flowers with long, protruding stamens. **1937** D. & H. Teilhet *Feather Cloak Murders* ix. 152 From the lehua trees the fragrance hung. **1965** M. C. Neal *In Gardens of Hawaii* 637 The lehua, a favorite native Hawaiian tree, is the commonest kind..in some forests.

lei (lēⁱ). [Hawaiian.] A Polynesian garland made of flowers, feathers, shells, etc., often given as a symbol of affection.

1843 J. Jarves *Hist. Hawaiian or Sandwich Islands* iii. 65 Garlands of flowers, necklaces of shells, and *leis*, beautiful wreaths fabricated from red or yellow feathers, encircled the limbs of the females. **1883** W. H. D. Adams *Mountains & Mountain-Climbing* 300 These crimson tassels, deftly strung on thread or fibres, are much used by the natives for their *leis*, or garlands. **1905** A. R. H. Moncrieff *World of To-Day* IV. 208 Bedecked by *leis*, garlands of bright flowers or feathers. **1956** J. Masters *Bugles & Tiger* xxii. 292 The garlands round my neck [at Honolulu] were called leis but they smelled much the same as the ones in India. **1966** Mrs. L. B. Johnson *White House Diary* 25 Oct. (1970) 434 There were more 'Blue Ladies'—Mrs. Marcos' hostess committee—with fragrant leis of sampaguita for each of the First Ladies. **1970** N. Armstrong et al. *First on Moon* xiv. 364 Jan Armstrong was wearing a carnation lei sent by friends in Hawaii. **1975** *Times* 2 May 7 (*caption*) Governor George Ariyoshi welcoming the Queen to Hawaii with a lei of orchids.

lei: see *leu.

Leibnitz (ləi·bnits). *Philos.* Also **Leibniz.** [Name of the German philosopher and mathematician: see Leibnitzian *a.* and *sb.*] *Leibnitz'(s) law*: the principle of the identity of indiscernibles (see Indiscernible *sb.* 2).

Leibniz is now the more usual spelling.

1941 O. Helmer tr. *Tarski's Introd. Logic* § 17. 55, *x* = *y* if, and only if, *x* has every property which *y* has, and *y* has every property which *x* has... [This] was first stated by Leibniz..and hence may be called Leibniz's law. **1965** *Philos. Rev.* LXXIV. 341 Physicalism violates Leibniz' law, which requires that if two things are identical they have all their non-intensional and nonmodal properties in common. **1968** *Aristotelian Soc. Suppl. Vol.* XLII. 99 Let us consider Leibniz's law firstly as a principle about the identity of individuals.

Leica (ləi·ka). [f. *Leitz* (see below) + Ca(mera.)] The proprietary name of cameras made by the German firm of Ernst Leitz Wetzlar Gesellschaft.

1925 *Brit. Jrnl. Photogr.* 26 June 387/2 New apparatus. .. Leica cine film camera... Quite an innovation in pocket cameras is one just issued by the well-known firm of Leitz, designed to take about 5 feet of ordinary standard perforated cinematograph film. **1930** *Trade Marks Jrnl.* 23 July 1135/1 Leica... Photographic cameras. Ernst & Leitz Gesellschaft mit beschränkter Haftung, (a Company organised under the laws of Germany),.. London,.. and Wetzlar, Germany; Manufacturers. **1933** W. Alexander *Mod. Photogr. with Mod. Miniature Cameras* ii. 29 *Leica* and *Contax* have now become words.. familiar in the mouths of men throughout..the earth. **1934** H. Nicolson *Let.* 9 Oct. (1966) 184 Lindbergh.. helped me to unload my Leica camera. **1936** P. Fleming *News from Tartary* v. 36 The Leicas turned out very satisfactorily. **1948** 'N. Shute' *No Highway* v. 136 The print was an enlargement from a Leica frame. **1953** C. Day Lewis *Italian Visit* v. 53 Armed with good taste, a Leica and a guide. **1958** G. Greene *Our Man in Havana* I. iii. 25 The victim's Leica had been smashed as well. **1959** P. H. Johnson *Unspeakable Skipton* vii. 57 An American..was photographing the swans with his Leica. **1973** R. Thomas *If you can't be Good* (1974) iii. 28 Using.. an old Leica with some fast film, Sarah turned out striking, informal portraits.

Leicester. Add: Also **Leicestershire.** (Earlier and later examples of variety of sheep.) In *Austral.* and *N.Z.* freq. as *English Leicester.* Cf. **Border Leicester.*

1798 J. Middleton *View Agric. Middlesex* xiii. 348 The Lincoln and Leicester sheep come to Smithfield in perfection. **1798** Jane Austen *Let.* 17 Nov. (1952) 29 One of his Leicestershire sheep, sold to the butcher last week, weighed 27 lb. and ¼ per quarter. **1804** A. Young *Gen. View Agric. Hertfordshire* 189 The cross is with the new Leicester: he sells no lambs. **1809** [see *Kent *sb.*³]. **1874** A. Bathgate *Colonial Experiences* xv. 210 In cultivated lands the Leicester is the favorite. **1923** W. Perry et al. *Sheep Farming in N.Z.* iv. 32 Thus Romney Marsh, English Leicesters, and Lincolns, have been improved or adapted to suit New Zealand conditions. **1950** H. G. Belschner *Sheep Managem. & Dis.* ii. 21 The Leicester crosses well with the big-framed Merino, but it is not so much in favour for this purpose in Australia... The following description of the English Leicester is taken from the flock book of the Australian Society of Breeders of British Sheep. **1956** G. Bowen *Wool Away!* (ed. 2) xii. 142 The English Leicester is another of the British breeds which has played a notable part in developing New Zealand's sheep farming, but is now more or less on the way out.

b. Leicester (occas. **Leicestershire**) **cheese,** a firm-textured full milk cheese originally made in Leicestershire.

1880 J. P. Sheldon *Dairy Farming* 241/1 The finest qualities of..Leicester cheese are generally admitted to be, with the single exception of genuine Stilton, the best cheese produced in these islands. **1902** *Encycl. Brit.* XXVII. 355/2 Derby cheese in its best forms is much like Leicester, being 'clean' in flavour and mellow. **1950** J. G. Davis *Dict. Dairying* 124 Leicester cheese is made from evening's and morning's milk. **1965** T. Fitzgibbon *Art Brit. Cooking* 133 Leicestershire cheese is a hard-pressed, flaky textured cheese..usually coloured a pale orange with annatto. **1970** *Guardian* 6 June 13/2 Leicester, an excellent dessert cheese with a mild flavour.

Leichhardt (ləi·kāɹt). The name of the German explorer of Australia, Friedrich Wilhelm Ludwig *Leichhardt* (1813–48), used *attrib.* in **Leichhardt-tree, -pine** to designate a tree native to Australia and India, *Nauclea orientalis*, of the family Rubiaceæ, which bears heads of yellow flowers; also *absol.*; **Leichhardt's bean** (see *bean *sb.* 4).

1860 F. von Mueller *Essay on Plants collected by Mr. Eugene Fitzalan* 12 The opportunity is an apt one for offering here some remarks on the 'Leichhardt-tree' of the settlers of Rockhampton. **1874** M. K. Beveridge *Lost Life* 40 Groaning beneath the friendly shade That by a Leichhardt-tree was made. **1885** H. Finch-Hatton *Advance Australia!* 258 The Leichhardt is a very symmetrical tree that grows to a height of about sixty feet, and has leaves rather like a big laurel. **1888** F. M. Bailey *Queensland Woods* 76 S[*arcocephalus*] *cordatus*... Leichhardt-tree or Canary-wood... A large tree with a thick, soft, corky bark of a more or less yellowish colour. **1907** Mrs. A. Gunn *We of Never-Never* 64 The camp had been fixed up in the silent depths of a dark Leichhardt-pine forest. **1908** E. J. Banfield *Confessions of Beachcomber* ii. i. 252 'Koo-badg-aroo' (Leichhardt-tree, *Sarcocephalus cordatus*), resembling a strawberry in shape, but brown, spicy and hot. **1944** W. E. Harney *Taboo* (ed. 3) 27 Those huge Leichhardt trees..would then be fashioned into canoes.

leifite (lī--, lē̆i·fəit). *Min.* [ad. Da. *leifit* (O. B. Bøggild 1915, in *Meddelelser om Grønland* LI. 427), f. the name of *Leif* Ericson (fl. 1000), Norse voyager: see -ite¹.] An acidic aluminosilicate and fluoride of sodium occurring as colourless hexagonal prisms.

1917 *Jrnl. Chem. Soc.* CXII. ii. 147 (*heading*) Leifite, a new mineral from Narsarsuk, Greenland. **1968** I. Kostov *Mineral.* ii. v. 406 Leifite, $Na_2AlSi_4O_{10}F.H_2O_{<1}$, found as acicular hexagonal crystals in pegmatites is similar to cancrinite.

leigh, var. *legh.

leightonite (lēⁱ·tŏnəit). *Min.* [f. the name of Tomas *Leighton* (b. 1894), Chilean mineralogist + -ite¹.] A hydrated sulphate of potassium, calcium, and copper found as transparent, pale blue to greenish blue, triclinic crystals at Chuquicamata, Chile.

1938 C. Palache in *Amer. Mineralogist* XXIII. 34 Leightonite, as the new species will be named, is triclinic, as shown by optical examination. **1962** *Canad. Mineralogist* VII. 276 It is interesting to note that leightonite has a chemical homologue—polyhalite, $K_2Ca_2Mg(SO_4)_4.2H_2O$. **1963** *Acta Crystallogr.* XVI. a10/1 Leightonite, $K_2Ca_2Cu(SO_4)_4.2H_2O$, is triclinic, but it is pseudo-orthorhombic (face-centred) to the extent that no departure from 90° has been found in the angles of the unit cell as determined by the usual X-ray diffraction techniques.

leio-. Add: **leiotrichous** *a.* (examples); hence **leio·trichy,** the condition of having straight lank hair.

1855 R. G. Mayne *Expos. Lex. Med. Sci.* (1860) 581/2 *Leiotrichus*, having smooth hair: leiotrichous. **1909** Leiotrichous [see *cymotrichous *a.*]. **1924** A. C. Haddon *Races of Man* (ed. 2) 5 For practical purposes these varieties of hair-form may be grouped as follows: (1) Leiotrichy.., or straight hair, [etc.]. **1935** Huxley & Haddon *We Europeans* iv. 114 Leiotrichy: the condition of straight lank hair, hanging straight down, as among the Chinese and certain other yellow-skinned peoples of Asia, and among the Eskimo. **1936** Leiotrichy [see *cymotrichous *a.*].

Leishman (lī·ʃmăn, ləi·ʃ-). *Med.* The name of W. B. *Leishman* (1865–1926), British pathologist, used *attrib.* and in the possessive with reference to his work in pathology, as **Leishman('s) body** = *Leishman–Donovan body; **Leishman('s) stain,** a mixture of eosin and methylene blue used to stain blood smears.

1903 *Brit. Med. Jrnl.* 28 Nov. 1401/1 (*heading*) Further notes on Leishman's bodies. *Ibid.*, There is never any contour line suggestive of a cell-wall, as with the Leishman bodies themselves. **1904** *Ibid.* 28 May 1250/1 A film prepared from the peripheral blood was examined.., Romanowsky and Leishman's stains being used. *Ibid.* 1252/1 These bodies stain purple or violet with the Leishman stain. **1961** *Lancet* 5 Aug. 315/2 In staining blood-films we obtained just as good results as with Leishman's stain prepared by the standard method. **1972** W. C. Johnson in J. H. Graham et al. *Dermal Path.* xix. 455/2 A definite diagnosis [of leishmaniasis] depends upon the demonstration of the Leishman bodies.

Leishman–Donovan body (-dǫ·nŏvăn-). *Med.* [f. prec. + the name of C. *Donovan* (1863–1951), Irish physician.] One of the numerous ovoid structures consisting of a single non-flagellated leishmania found in the macrophages of sufferers from leishmaniasis.

1904 *Brit. Med. Jrnl.* 28 May 1249/2 (*heading*) Note on the occurrence of Leishman–Donovan bodies in 'cachexial fevers' including kala-azar. **1966** Wright & Symmers *Systemic Path.* II. xxxix. 1601/2 (*caption*) Large numbers of Leishman–Donovan bodies are seen in the cytoplasm of many of the macrophages in this field.

leishmania (līʃmæ·niă, ləiʃ-). *Zool.* and *Med.* Pl. **-ia, -iæ, -ias.** [mod.L., f. *Leishman + -ia¹.] **a.** Any protozoon of the genus *Leishmania* (family Trypanosomidæ), comprising three species which are parasitic in man (and occas. other mammals), occurring as non-flagellated Leishman–Donovan bodies, and which are transmitted by sandflies of the genus *Phlebotomus*, wherein they occur as flagellated individuals in the alimentary canal. **b.** Any flagellate of the family Trypanosomidæ when existing in a leishmanial form.

[**1903** R. Ross in *Brit. Med. Jrnl.* 28 Nov. 1401/2 Laveran has given the name *Piroplasma donovani* to these organisms; and the specific name must therefore be permanently adopted. But if, as I suppose, they are found to belong to a new genus, it would be only fair to give the name *Leishmania* to that genus. In that event the full name would be *Leishmania donovani*, Laveran.] **1914** *Trop. Dis. Bull.* III. 141 Large numbers of leishmania were found in the lesions. **1926**, etc. [see *leptomonas]. **1952** M. E. Florey *Clin. Applic. Antibiotics* I. viii. 247 The lesion had again broken down and leishmania were found in smears. **1961** [see *leptomonad]. **1962** J. D. Smyth *Introd. Animal Parasitol.* v. 63 Leishmanias are unusual in living entirely within the cells of the reticulo-endothelial system. **1968** E. J. L. Soulsby *Helminths, Arthropods & Protozoa Domesticated Animals* 567 Endothelial and macrophage cells contain masses of leishmaniae.

Hence **leishma·nial, leishma·nian** *adjs.*, caused by leishmaniæ; typical of a leishmania as it occurs in man and other mammals (i.e. as a non-flagellated Leishman–Donovan body). Also **leishmani·asis** (pl. **-ases**) [-asis], **-manio·sis** (pl. **-oses**) [-osis], any of several diseases, principally kala-azar (visceral leishmaniasis), oriental sore (cutaneous leishmaniasis) and espundia (muco-cutaneous or American cutaneous leishmaniasis), which are caused by species of *Leishmania*; (**dermal**) **lei·shmanoid** [-oid, after Varioloid *a.* and *sb.*], a condition occurring as a sequel to kala-azar and characterized by an eruption of whitish patches on the skin.

1911 Stedman *Med. Dict.* 470/1 *Leishmaniosis*, infection with a species of *Leishmania.* **1912** *Trop. Dis. Bull.* I. 363 (*heading*) Fourth series of haematological researches on leishmanial anaemia. **1912** *Brit. Med. Jrnl.* 2 Nov. 1194/2 (*heading*) Papers dealing with leishmaniasis. **1914** *Chem. Abstr.* VIII. 1466 (*heading*) Possibility of the excitation of leucopoiesis in Leishmanian infection in childhood. **1916** *Jrnl. Amer. Med. Assoc.* 22 Nov. 1635/2 (*heading*) Tartrate of antimony and potassium in treatment of superficial leishmaniosis. **1920** W. E. Masters *Essent. Trop. Med.* i. 55 (*heading*) The leishmaniases. **1922** U. N. Brahmachari in *Indian Med. Gaz.* LVII. 127/1 In view of the fact that the eruptions are due to leishmania infection whose virus has been modified by antimonial treatment, I propose to call this form of cutaneous leishmaniasis *dermal leishmanoid* just as small-pox modified by vaccination is called varioloid. **1942** [see *leptomonad]. **1966** Wright & Symmers *Systemic Path.* II. xxxix. 1596/2 Cutaneous leishmaniosis (oriental or tropical sore) is caused by the protozoon, *Leishmania tropica.* **1967** A. C. Allen *Skin* (ed. 2) xiv. 541/1 The so-called post-kala-azar dermal leishmanoid is a familiar sequel of visceral leishmaniasis... The leishmanoid begins as patches of erythematous macules on the face and body. **1967** *New Scientist* 17 Aug. 349/2 Small rodents which are carriers of the disease leishmaniasis which causes suppurating sores in humans. **1968** E. J. L. Soulsby *Helminths, Arthropods & Protozoa Domesticated Animals* 565 Developmental stages of the genus [sc. *Leishmania*] occur in the leishmanial form in vertebrates and in the leptomonad form in the insect vector and in culture. **1968** Weinman & Ristic *Infectious Blood Dis. Man & Animals* I. viii. 160 These lack the physiological character possessed by the leishmanial parasites of mammals of being able to grow at temperatures of 34°–38°C.

Leisler (ləi·zləɪ). The name of the early 19th-c. German zoologist, T. P. *Leisler*, used in the possessive to designate **Leisler's bat,** a small black bat, *Nyctalus leisleri*, named after him in 1817 by H. Kuhl (*Deutsch. Fledermäuse* 38), and formerly called the hairy-armed bat.

1904 J. G. Millais *Mammals Great Brit. & Ireland* I. 76 Leisler's Bat is considerably smaller than the noctule. **1910** G. E. H. Barrett-Hamilton *Hist. Brit. Mammals* I. 83 Leisler's Bat, or species closely resembling it, is found in the wooded districts of boreal and transitional Europe and Asia. **1941** H. Cory *Mammals Brit. Isles* 255 Leisler's Bat..is rare in England, but occurs in greater numbers in eastern Ireland. **1960** *Times* 14 June 14/7 Of the Leisler's bat..I recall no obvious distinguishing character.

leisure, *sb.* Add: **5. e.** *lady* (or *woman*) *of leisure*, a woman who has no regular employment or whose time is free from obligations to others.

1948 R. M. Ayres *Missing the Tide* v. 164 She was no longer a lady of leisure in her own house but a paid companion with irksome duties to perform. **1951** M. McLuhan *Mech. Bride* (1967) 40/1 The woman of leisure might wear long skirts, but the working woman was put into adolescent short skirts. **1955** L. P. Hartley *Perfect Woman* xxii. 190 Or as the lady of leisure, reading a book? **1975** D. Ramsay *Descent into Dark* i. 26 She had a whole afternoon to play lady of leisure.

6. a. and **c.** (Further examples).

1816 Jane Austen *Emma* I. ii. 25 He had still a small house in Highbury, where most of his leisure days were spent. **1899** T. Veblen (*title*) The theory of the leisure class. **1907** F. H. Burnett *Shuttle* xxxiv. 347 In another generation there will be a male leisure class [in America]. **1907** Belloc *Cautionary Tales* 31 Learn To pass your Leisure Time In Cleanly Merriment. **1912** J. H. Moore *Ethics & Educ.* vi. 22 The school in its origin was a leisure-class institution. **1941** E. Wilson *Wound & Bow* i. 61 They [*sc.* Estella and Pip] are left with their leisure-class habits and no incomes to keep them up. **1947** O. Barfield in *Essays presented to Charles Williams* 121 It has no particular significance if poetry is to be regarded *only* as..a pleasurable way of diverting our leisure hours. **1951** M. McLuhan *Mech. Bride* (1967) 40/1 Competitive drives and ambitious impulses will be transferred increasingly to leisure and home occupations. **1954** *Encounter* Mar. 78/2 A master-race recruited solely from the leisure-class and endemic to English shores: 'Of such was The Breed.' **1961** D. Jenkins *Equality & Excellence* vii. 143 Problems of adjustment..connected with housing and leisure-time activities. **1963** *Punch* 15 May 710/3 Fashion shows of men's outerwear, underwear, leisure-wear, rainwear. **1964** A. Wykes *Gambling* iv. 80 Watching other people play games has been one of man's favorite leisure-time occupations for a long while. **1965** *New Society* 26 Aug. 5/3 Buxton is trying to..latch on to the leisure revolution, without becoming a coach tours nightmare. **1966** *Guardian* 16 Feb. 6/3 A case for subregional leisure centres—which would help people in small towns to feel less dependent on cities—was made out yesterday by Professor Arthur Ling. **1968** *Economist* 4 May 38/2 One should not conclude that Frenchmen have reached the stage of the fabled leisure civilisation. **1968** *Daily Tel.* 28 Dec. 21/1 The ever-increasing attraction of boating as a leisure-activity and a sport. **1969** *Times* 7 Nov. 14/2 With the increasing demand for leisure wear, formal wear people have softened their lines. **1972** *Accountant* 17 Aug. (Suppl.) 11/3 Expanding construction company specialising in growth market of the leisure industry. **1974** *Times* 12 Feb. 11/7 Silhouette's swimwear and leisurewear sales manager.

leisure (le·ʒiŭɹ, U.S. lī·ʒiŭɹ), *v.* *rare.* [f. the sb.] **a.** *intr.* To have or enjoy leisure. **b.** *trans.* To make leisurely.

1928 Blunden *Undertones of War* 304 There to tarry in careless ways,.. Leisuring after fiery days. **1929** Bridges *Testament of Beauty* i. 32 Science comforting man's animal poverty and leisuring his toil. **1970** G. F. Newman *Sir, You Bastard* vi. 173 Sneed rose late..and leisured with the papers in the coffee-house opposite his flat.

leisured, *a.* Add: **2.** (*the*) *leisured class*(*es* (later examples).

1923 W. S. Maugham *Our Betters* II. 85 American wealth has reached a pitch where it was bound to give rise to a leisured class. **1926** B. Webb *My Apprenticeship* i. 9 The rulers of the country..ought in the main to be drawn from a leisured class. **1929** D. H. Lawrence *Pansies* 43 Obviously he's not one of the leisured classes. **1949** A. Wilson *Wrong Set* 111 The many leisured-class hypotheses by which Mr Cockshott obviously lived. **1960** M. Bradbury *Phogey!* iii. 105 The leisured classes demonstrated their leisure by indulgence in useless pursuits (personal relationships, scholarship, etc.).

leitmotiv. Add: (Later examples.) Also in extended use.

1896 H. Ellis in *Savoy* i. 70 Zola..introduced this sort of *leit-motiv* into literature. **1898** G. Meredith *Let.* 6 July (1970) III. 1303, I long to hear from him of [the] Leit-motif—though indeed he has taken the world more or less into his confidence. **1899** Kipling *Stalky & Co.* 84 A tune whose *leit-motif* was the word 'stinker'. **1912** Wodehouse *Prince & Betty* iv. 61 The name Scobell had been recurring like a *leit motif* in Mr Crump's conversation. **1937** Koestler *Spanish Testament* iv. 94 It provides the *leitmotif* of German foreign policy in Spain. **1955** *Times* 28 May 8/4 But the method remains, the orchestral tapestry of leitmotifs is more resplendent than ever, the drama is even more closely knit into the texture of sound. **1970** G. Greer *Female Eunuch* 151 Self-sacrifice is the leit-motif of most of the marital games played by women. **1972** *Composer & Conductor* Aug. 1/1 Ninety-nine music graduates out of a hundred..will say that the Leitmotiv (or Leitmotif, or leading motive)..was invented by Wagner. Wrong... The correct answer is: Friedrich Wilhelm Jähns, and even he applied it not to Wagner but to Weber. **1974** *Times Lit. Suppl.* 15 Feb. 162/4 There are plenty of *leitmotivs* which recur time and time again.

lek, *sb.*¹ Substitute for def.: A patch of ground used by groups of birds of certain species, esp. blackcock, during the breeding season, as a setting for the males' display and their meeting with the females; the display itself or the season during which it takes place. Also **lek** *v. intr.*, to take part in a pattern of behaviour centred upon a lek; **le·kking** *vbl. sb.*

1942 E. A. Armstrong *Bird Display* xv. 215 The lek [of blackcock] is larger than the ruffs' assembly ground and the individual territories are not so clearly defined. **1964** A. L. Thomson *New Dict. Birds* 432/1 The lek is usually maintained at the arena for considerable periods of time and many species revive the performance at the same display-ground year after year. *Ibid.* 432/2 During lek displays the birds tend to stimulate one another. **1970** *Country Life* 26 Feb. 490 (*caption*) A pair of blackcocks lekking. **1971** *Ibid.* 16 Sept. 693/2 Then came January, when instinct bade her [*sc.* a partridge] to that lekking field that had served as a partridge lekking area since time immemorial. **1972** C. Willock *Death in Covert* (new ed.) i. 6 A randy old blackcock displaying triumphantly on a Highland lek. **1974** *Country Life* 26 Sept. 865/3 The strongest birds are nearest the centre of the [black grouse] lek because it is there that the females first land.

‖ **lek** (lek), *sb.*² Pl. **lekë, leks.** [Albanian.] A unit of currency in Albania.

1927 *Times* 18 Nov. 24/4 The monetary unit chosen is the gold franc (5 lek).., with a parity of approximately 25 to the £. **1937** M. Comencini *Coins Mod. World* 2 Albania Currency unit: the Franka Ari or Gold Franc of 5 Lek or 200 Qindar. One Lek corresponds to 8 5-Qindar Leku.. bronze coins. **1962** R. A. G. Carson *Coins* 408 The republic of Albania, established in 1925, issued a coinage with the franka as the unit, divided into 5 lek, divided in turn into 40 qindar. **1966** *New Statesman* 8 Apr. 500/3 The plans are already drawn, 120 million lekë (nearly £500,000) promised by the state. **1967** *Spectator* 22 Dec. 775/3 Whitaker's [Almanack]..is no sure guide to the rate at which the British traveller can hope to exchange his pittance for Albanian *leks.* **1974** *Albania Today* (Tirana) Jan.–Feb. 16/3 The rate was fixed at 83.72 leks per ruble, as against 100 leks per ruble previously.

‖ **lekach** (le·kaχ). [Yiddish.] A traditional Jewish cake made with honey.

1932 L. Golding *Magnolia St.* iii. viii. 573, I will fill large bags for them with *ingber* and *strudel* and *lekkach.* **1955** L. W. Leonard *Jewish Holiday Cook Bk.* 20 A good standby for holiday entertaining is the traditional Lekach, or Honey Cake. **1960** S. Becker tr. *Schwarz-Bart's Last of Just* (1961) v. 239 That's a *lekach*! A honey cake. **1973** Carr & Oberman *Gourmet's Guide to Jewish Cooking* 124 Foods associated with Rosh Hashanah are honey and honey cake (Lekach).

‖ **lekane** (leka·ni). *Gr. Antiq.* [ad. Gr. λεκάνη a bowl or dish.] A small shallow bowl, usually with handles and a cover. Also dim. **lekanis** (pl. **-ides**).

[**1905** H. B. Walters *Hist. Anc. Pott.* I. iv. 164 The word λεκάνη, however, seems to indicate a large bowl rather than a covered jar, and no satisfactory name has yet been found.] *Ibid.* 177 A method of divination sometimes practised..consisted in placing waxen images in a lekane full of water. *Ibid.* xi. 469 A new form is that known as the *lekane*, a jar for holding sweetmeats; it has vertical handles and a cover of elaborate form, often surmounted by a small vase. **1918** J. D. Beazley *Attic Red-Figured Vases* xix. 188 Milchhöfer mentions four lekanides in the style of the Meidias painter. *Ibid.*, The toilet-vase called lekanis is rarely decorated with pictures before the free period. **1935** Richter & Milne *Shapes & Names Athenian Vases* 23 The word λεκάνη..was widely used to signify flattish bowls employed for different purposes... The word lekanis, on the other hand, apparently signifies a bowl used for a special purpose. **1946** G. M. A. Richter *Attic Red-Figured Vases* 13 The lekanis [was] a covered dish often used as a wedding present. **1950** H. L. Lorimer *Homer & Monuments* vii. 448 The well-known Boeotian lekane in the British Museum. **1967** R. S. Folsom *Handbk. Gk. Pottery* 132 Shapes of pots: Kantharos, tripod pyxis, lekane, skyphos, and cup.

Lekhite, Lekhitic: see *Lechitic *sb.* and *a.*

‖ **lekker** (le·kəɪ), *a.* *S. Afr. colloq.* [Afrikaans, f. Du. *lekker* (cf. G. *lecker*), rel. to Du. *likken* Lick *v.*] Pleasant, sweet, nice.

1900 A. Carter *Let.* 8 Feb. (MS.), On Monday these left and in moving round the mountain was 'verneuked' as Hannes said 'lekker'. **1926** E. Lewis *Mantis* ii. viii. 123 To Mr Dan Hugo nothing tasted so *lekker* as a good cup of coffee at that hour. **1953** F. Robb *Sea Hunters* viii. 137 'Fish soup and baked fish to follow. Lekker!' Olley drooled. **1961** *Personality* 16 May 27 It's a lekker language. **1963** A. Smith *Throw out Two Hands* vii. 82 Charl Pauw..had the thickest of South African accents... 'But the place is lekker, I tell you, its lekker.' **1970** *Rand Daily Mail* 28 Feb. 7/4 Some South African English colloquialisms, again mainly of Afrikaans origin, he regards as downright 'barbarisms', such as..lekker.

lekker, var. *leccer.

‖ **lekythos** (lī·kiþǫs, le·-). *Gr. Antiq.* Pl. **lekythoi** (-oi). Var. Lecythus.

1851 *Catal. Greek & Etruscan Vases Brit. Mus.* I. 27 Lékythos... Clay ash-coloured; round the body brown and crimson bands. **1899** R. Glazier *Man. Hist. Ornament* 77 A vase produced specially for funeral purposes was the Athenian Lekythos, the body of which was covered with white slip, then painted in polychrome with subjects of singular appropriateness. **1931** *Times Lit. Suppl.* 31 Dec. 1054/1 Athenian white lekythoi. **1948** A. Lane *Greek Pott.* iv. 54 The white slip was easily chipped or abraded, and the technique was mainly reserved for..lekythoi that were buried with the dead. **1960** R. G. Haggar *Conc. Encycl. Cont. Pott. & Porc.* 246/2 *Lekythos*,..a narrow-necked flask or vase well adapted (from its long narrow neck) for the slow pouring of oil, chiefly used in funeral rites. **1972** *Oxf. Univ. Gaz.* CII. Suppl. No. 3. 20 White ground lekythos by the Inscription Painter, Attic, mid-5th century B.C.

Lem (lem). [f. the initials.] A lunar excursion module (*lunar *a.* 1 c). See also *L.E.M.* s.v. *L 7.

1962 *Listener* 26 July 150/2 We had a discussion of Project Apollo, the American scheme for getting a man on the Moon. The secondary space-craft for this formidable task has been dubbed a 'lunar excursion module', or Lem for short. **1967** *Economist* 11 Nov. 627/1 The lunar excursion module—the Lem or bug—..will make the actual touch-down on the moon when the great day comes.

‖ **lembing** (lĕmbi·ŋ). Also 9 **limbing, lambing.** [Mal. *lembing.*] A Malay spear characterized by a ridged blade.

1839 T. J. Newbold *Pol. & Statistical Acct. Straits of Malacca* II. xii. 211 The arms of the Orang Laut..are the limbing, or lance; the tampuling, a large hook, [etc.]. **1894** N. B. Dennys *Descr. Dict. Brit. Malaya* 370 For the javelin, or half-pike, the Malays have the name *lambing.* **1936** G. B. Gardner *Keris & Other Malay Weapons* iv. 85 Plate 80 shows an iron *lèmbing.* **1947** R. O. Winstedt *Malays* ix. 165 The origin of the lance (*tombak*) and spear (*lembing*)..await study.

lemma². Add: **2.** *Bot.* In grasses, the lower bract of a floret.

1906 C. V. Piper in *Contrib. U.S. Nat. Herbarium* X. 8 We have taken the liberty to introduce the word lemma to apply to the ..'flowering glume' of authors. **1934** A. Arber *Gramineae* vii. 110 The idea that the grass flower is unique, and requires a special vocabulary..has led to..a series of names [for bracts], such as *gluma florifera, palea inferior, flowering glume* and *lemma*, of which the two latter are the more generally familiar in England and America. *Ibid.* viii. 141 In Ichnanthus the lemma may show remarkable winglike appendages at the base. **1968** F. W. Gould *Grass Systematics* ii. 51 Glumes, lemmas, and paleas are floral bracts. *Ibid.* 53 Lemma characters of taxonomic importance are shape, texture, size in respect to the glumes, nervation, awn development, and surface features.

lemmatization (le:mătəizēⁱ·ʃən). [f. next + -ation.] The action or process of lemmatizing; an instance of this.

1967 *Computers & Humanities* II. 75 Method:... 3. Alphabetic sorting into word forms with context. 4. Lemmatisation. **1971** A. J. Aitken in R. A. Wisbey *Computer in Lit. & Ling. Res.* 14 The methods of lemmatization..so far mentioned necessitate informing the computer explicitly of the destination in terms of headword of every single instance of each word which it has to treat. **1972** *Computers & Humanities* VI. 212 Not all lemmas could, of course, be made to come out correctly from the computer... In fact, the accomplished wrong lemmatizations are more notable than the missing correct ones.

lemmatize (le·mătəiz), *v.* [f. Gr. λημματ-, λῆμμα Lemma¹ + -ize.] *trans.* To sort (words as they occur in a text) so as to group together those that are inflected or variant forms of the same word.

1967 *Computers & Humanities* II. 78 We have..tested programs for concordances, for lemmatizing with computer dictionary, and for transcribing from historical to phonologic alphabet. **1971** J. B. Carroll et al. *Word Frequency Bk.* p. xiii, The AHI Corpus is coded for capitalization. It is not parsed or lemmatized. **1971** A. J. Aitken in R. A. Wisbey *Computer in Lit. & Ling. Res.* 13 From a text prepared in this way the computer could deliver an output resembling a fully sorted collection for a traditional dictionary (already ordered and lemmatized) without further human attention. **1973** *Computers & Humanities* VII. 132 The vocabulary lists were next lemmatized by hand. *Ibid.*, The computer program made no attempt to lemmatize words or to distinguish homographs, but simply counted the number of occurrences of each distinct word-type.

Hence **le·mmatized** *ppl. a.*

1969 *Computers & Humanities* IV. 134 Method: Punching frequency lists and lemmatized texts; transferring to tapes; [etc.].

lemme (le·mi). Colloq. contraction of *let me* (see Let *v.* 12, 14). Cf. *gimme.

1876 'Mark Twain' *Tom Sawyer* ii. 19 Come now; lemme just try... Now lemme try. **1894** Kipling *Day's Work* (1898) 64 Lemme hide back o' you peoples, so's they won't see what I'm at. **1905** H. G. Wells *Kipps* I. i. 27 Ann—lemme kiss you. **1910** C. E. Montague *Hind let Loose* iv. 58 Lemme alone. I'm an old man. Gimme a drink. Lemme alone. **1923** 'R. Crompton' *William Again* iv. 64 'Lemme help!' he pleaded. **1930** E. Pound *XXX Cantos* xix. 86 And in came the street 'Lemme-at-'em' Like a bull-dog in a mackintosh. **1946** K. Tennant *Lost Haven* (1947) ii. 34 Lemme go... Oo-h, you're breaking my arm! Auntie, make 'im lemme go. **1972** C. Weston *Poor, Poor Ophelia* (1973) xxii. 138 Okay, man, lemme think.

lemming. Add: **1. b.** Used *fig.* to denote a person bent on a headlong rush, often towards disaster. Also *attrib.* or quasi-*adj.*; *lemming-like* adj.

[**1959** M. Gilbert *Blood & Judgement* iii. 35 Home-going office workers..potent in mass as a lemming migration.] **1968** M. Bragg *Without City Wall* I. x. 116 To opt out..in a way, you could say that was just as lemming-like as what you're doing. **1969** D. F. Horrobin *Sci. is God* i. 9 This lemming unconcern may have dangerous consequences. **1969** *New Yorker* 12 Apr. 61/2 In Dr. Langseth's view, going to the moon is an impulse ingrained in the national character, as though Americans were astronautical lemmings. **1970** *Islander* (Victoria, B.C.) 15 Feb. 12/1 No one had the slightest idea of what was happening, yet all had joined in the mad lemming-like scramble for the waterfront. **1970** P. Moyes *Who saw her Die?* xx. 256 It was Saturday, the lemming rush was in full spate, the suburbs pouring their millions in bus, tube, train and car into the central sea. **1972** 'J. Bell' *Death of Poison-Tongue* viii. 80 Lemmings..was only the present vogue word..to describe a collection of mindless people moved by a common purpose. **1972** *Guardian* 11 Dec. 12/6 The only way to stop multiple motorway crashes is by educating us all in roadcraft so that our individual intelligence becomes more powerful than our lemming instincts. **1975** *Sunday Times* 16 Feb. 51/1 Last week there were ample signs that the lemming-like rush to pile in at any price was wearing itself out.

lemna. Add to etym.: (Linnæus *Genera Plantarum* (1737) 417). (Earlier and later examples.)

1789 G. White *Selborne* II. liv. 269 Gold and silver fishes..will also feed on the water-plant called *lemna* (ducks' meat). **1801** M. Edgeworth *Belinda* II. xxi. 298 'This,' replied Belinda, 'is what "Th'unlearned, duckweed; learned, lemma [*sic*], call," and it is to be found in any ditch or standing pool.' **1967** C. D. Sculthorpe *Biol. Aquatic Vasc. Plants* vii. 204 As a result of its small size and the ease with which it may be cultured..*Lemna* is eminently suitable for laboratory experiments.

lemon, *sb.*¹ Add: **1. b.** A person with a tart or snappy disposition (quot. 1863). More usually (*slang*), a simpleton, a loser; a person easily deluded or taken advantage of (see also quot. 1950).

1863 P. S. Davis *Young Parson* xxvii. 222 Mrs. Trimble ..had a great deal to say, and no little acrimony in her way of saying it. Indeed, she was what the knowing ones denominated 'a lemon'. **1908** J. M. Sullivan *Criminal Slang* 21 Sucker or *lemon*, a victim of criminals and tramps. **1916** J. B. Cooper *Coo-oo-ee* xiv. 208 There was always a danger of offending a man who has been runner-up in a boxing championship if you make him appear 'like a lemon'. **1931** Wodehouse *Big Money* i. 27, I don't know why it is, rich men's sons are always the worst lemons in creation. **1950** Partridge *Slang To-day & Yesterday* (ed. 3) iii. 313 If she is unpopular, she is *a pill, a pickle, a lemon.* **1966** J. Porter *Sour Cream* x. 137 Criminal carelessness, that's what it was! Leaving me standing here like a lemon. **1973** 'A. Hall' *Tango Briefing* i. 17 They'd sent me down to show me something and they knew I couldn't see it and I felt a bit of a lemon.

c. *slang* (orig. *U.S.*). Something which is bad or undesirable or which fails to meet one's expectations.

Phr. *the answer is a lemon*: used to denote that a reply is unsatisfactory or non-existent.

1909 *Sat. Even. Post* 20 Feb. 38/2 The wheel goes around; wherever the little indicator at the point of the pin stops, there is your prize—or your lemon. **1912** C. Mathewson *Pitching in a Pinch* x. 220 The papers were mentioning him as the '$11,000 lemon'. **1914** 'High Jinks, Jr.' *Choice Slang* 14 *Lemon*, a disappointment. **1922** M. Arlen *Piracy* I. v. 59 'What would happen if *we* went on strike?'.. No one among them..dreamed of answering. The answer was a lemon. **1927** *Daily Express* 13 Dec. 17/1 Middlesbrough seem to have 'picked a lemon', for the draw gives them South Shields as opponents. **1930** P. MacDonald *Link* iv. 75 The answer at first seems to be a lemon, but they're at least the sort of questions that make one think. **1931** *Morning Post* 19 June 6 'I sold five lemons for £210,' said a witness... 'Lemon' was a term used in the trade for second-hand cars of little value. **1959** M. T. Williams *Art of Jazz* (1960) ix. 85 This great record would have been a lemon commercially in 1925. **1961** C. Mabee *Seaway Story* vii. 70 He first politely wished success to New York's lemon, the new twelve-foot Erie Barge Canal. **1963** *Guardian* 21 Jan. 16/6 The French nuclear deterrent..is a military lemon of the first order. **1969** N. Freeling *Tsing-Boum* x. 68 One makes requests through official channels and the answer is a lemon. **1972** *Sat. Rev.* (U.S.) 17 June 7/3 Mechanics are less than delighted to see lines of lemons converging on their service department. **1972** *Sydney Morning Herald* 26 Aug. 1/2 The effect of this on consumers is too many lemons or part lemons coupled with near impossibility of obtaining redress from the manufacturer.

d. Phr. *to hand* (someone) *a lemon*: to pass off a sub-standard article as good; to swindle (a person), to do (someone) down.

1906 H. Green *At Actors' Boarding House* 36 Him gettin' handed a lemon in that English act, puts us up. [**1922** Wodehouse *Clicking of Cuthbert* x. 233 'It did indeed begin to appear as though our beloved monarch.. had been handed the bitter fruit of the citron.' The quaint old idiom is almost untranslateable, but one sees what he means.] **1939** E. S. Gardner *D.A. draws Circle* (1940) vi. 87 The way things are now, I co-operate with them. If they hand me a lemon, I can walk up and down the streets cussing them out for letting politics interfere with the administration of justice. **1970** *New Yorker* 12 Dec. 131/1 These senators felt that the President had handed them two lemons, had gone to the mat for his choices when he didn't have to.

e. *slang.* The head.

1923 Wodehouse *Inimitable Jeeves* i. 13 'What might you have missed?' I asked, the old lemon being slightly clouded. **1952** *Coast to Coast* 195 If you had any brains in that big lemon you'd wipe me. You'd get away.

f. *U.S. slang.* An informer, one who turns State's evidence (see also quot. 1931).

1931 *Amer. Speech* VI. 439 *Lemon*, one who testifies for the prosecution. **1935** G. Ingram *'Stir' Train* ii. 30 'You think you got the low-down on me: well, see me put it on you!' 'You talk like a "lemon"!' **1935** A. J. Pollock *Underworld Speaks* 70/2 *Lemon*, one who turns state's evidence.

5. a. *lemon cheesecake* (earlier example), *cordial, pie, sauce, tea* (later examples).

1728 E. Smith *Compleat Housewife* (ed. 2) 120 To make Lemon Cheese-cakes. **1836** *Mag. Domestic Econ.* I. 182 Lemon cordial. **1909** A. Arnold *Century Cook Bk.* Suppl. 584 Lemon pie. 2 lemons..sugar..butter..4 eggs..cornstarch. **1911** C. Harris *Eve's Second Husband* 154 Then you ate lemon pie, pound-cake and boiled custard. **1972** J. Potts *Trouble-Maker* (1973) ii. 10 Their first square meal in three days. Corn and chicken. Homemade relishes. Lemon pie. **1747** H. Glasse *Art of Cookery* ii. 36 To make Lemon-Sauce for boiled Fowls. **1861** Mrs. Beeton *Bk. Househ. Managem.* 220 (*heading*) Lemon sauce for boiled fowls. **1948** *Good Housek. Cookery Bk.* i. 15 Something piquant should be served with a dish that is very bland, as..lemon sauce with steamed sponge pudding. **1932** L. Golding *Magnolia St.* II. xi. 425 Reb Feivel sat sucking lemon-tea through a cube of sugar. *a* **1963** S. Plath *Crossing Water* (1971) 62 It'll be *lemon*-tea for me. **1975** *Times* 8 Feb. 7/4 There is no licence, but the lemon tea is fresh and good.

7. lemon cheese (curd), lemon curd, a confection made with lemons, butter, eggs, and sugar, and used as a spread or filling; **lemon cling** *U.S.*, a variety of clingstone peach; **lemon-drop** (examples); **lemon-game** *U.S. slang*, a type of confidence trick (see quots.); also *ellipt.* **lemon; lemon-house,** a building where lemons are stored; **lemon meringue (pie),** an open pie consisting of a pastry case with a lemon filling and a topping of meringue; **lemon oil,** an essential oil obtained from lemons; **lemon platt,** a flat sugar-stick, flavoured with lemon; **lemon-squeezer** (earlier and later examples); also *fig.*; also, *Austral.* and *N.Z. colloq.*, a hat with a peaked crown and broad flat brim worn by New Zealand troops; **lemon-thyme** (earlier and later examples); **lemon-verbena** (examples); also **lemon-scented verbena; lemon-wood** (later example); (*b*) a name for several tropical American trees or their light-coloured wood, esp. the Cuban *Calycophyllum candidissimum.*

1853 G. W. Francis *Dict. Pract. Receipts* (ed. 3) 211/2 Lemon cheese curd. **1891** R. Wells *Mod. Flour Confectioner* 101 Lemon cheese. **1909** *Daily Chron.* 17 Aug. 6/4 Boiling lemon cheese over a gas cooking apparatus. **1848** *Rep. Comm. Patents 1847* (U.S.) 196 Fifteen specimens.. of the lemon cling..measured over a foot in circumference. **1895** *Army & Navy Co-op. Soc. Price List* 17/2 Lemon curd, for making Cheesecakes. **1948** J. Betjeman *Sel. Poems* 35 Lemon curd and Christmas cake. **1968** V. S. Pritchett *Cab at Door* iii. 36 On Thursday, she made her second baking, concentrating..on..her Eccles cakes, her puffs, her lemon-curd. **1807** M. E. Rundell *New Syst. Domestic Cookery* 203 (*caption*) Lemon drop. **1854** C. M. Yonge *Heartsease* II. xiv. 316 Here were some lemon-drops for papa. **1938** D. Runyon *Furthermore* x. 187 A young guy by the name of The Lemon Drop Kid, who is called The Lemon Drop Kid because he always has a little sack of lemon drops in the side pocket of his coat, and is always munching at same. **1908** J. M. Sullivan *Criminal Slang* 15 *Lemon game*, defrauding a sucker at a pool game. **1914** Jackson & Hellyer *Vocab. Criminal Slang* 55 *Lemon*,.. a confidence game in which skill at pool is the bait, though its successful negotiation is based upon the dishonesty or avarice of the victim. **1937** E. H. Sutherland *Professional Thief* iii. 68 The lemon is an agreement between the inside man, an expert pool player, and a prospect, by which the prospect will win bets on the pool games played by the expert. Through a supposed fluke the expert wins the game which the prospect had bet he would lose, and the prospect thereby loses his money. **1901** *Chambers's Jrnl.* Nov. 719/2 On the very day of the picking they must be carried to the lemon-house, and great care must be taken that the fruit is not exposed to the sun or bruised in any way. **1916** D. H. Lawrence *Twilight in Italy* 85 We passed through, and stood at the foot of the lemon-house. **1914** S. Lewis *Our Mr. Wrenn* i. 13 Hey, Drübel, got any lemon merang? Bring me a hunk, will yuh? **1922** *Hotel World* 15 Apr. 15/1 Lemon meringue pie. **1959** N. Mailer *Advts. for Myself* (1961) ii. 126 There was roast chicken with stuffing, lemon meringue pie and chocolate cake. **1973** J. Wilson *Truth or Dare* vi. 75 It was lemon meringue pie for dinner. **1896** J. T. Law *Grocer's Manual* 408/2 The essence of lemon coming into commerce..is greatly made up of..the ethereal oil which is present in lemon oil. **1957** *Encycl. Brit.* XIII. 908/1 Among the important by-products resulting from the processing of lemons, after removal of the juice, are citric acid..lemon oil and pectin. **1916** Joyce *Portrait of Artist* (1969) i. 7 The moocow came down the road where Betty Byrne lived: she sold lemon platt. **1965** *Amer. N. & Q.* III. 117/2 'Lemon Platt', commonly sold as 'Yellow Man' at fairs in the North of Ireland,..derives its name..from its flavor. **1900** M. Thorn in W. D. Drury *Bk. Gardening* xi. 469 Lemon-scented Verbena should be represented in gardens where shrubs with fragrant leaves are cherished. **1969** D. Goold-Adams *Cool Greenhouse Today* xvii. 198 Lippia (Lemon-scented Verbena). Half-hardy deciduous shrub from Chile with insignificant flowers but grown in the greenhouse for the glorious scent of its crushed leaves. **1781** *Salem Gaz.* 3 July, Isaac Greenwood..makes Flutes.. Back-Gammon Boxes Men and Dies, Chess-Men, Billiard-Balls, Maces, Lemon Squeezers. **1856** 'Ockside' & 'Doesticks' *Hist. & Rec. Elephant Club* 118 One..had been hit over the head with the lemon-squeezer. **1887** *Century Mag.* Aug. 489/1 The 'Chunkers' were frequently of the 'lemon-squeezer' pattern. **1949** *Nat. Geogr. Mag.* Aug. 235 Knapsack's a Nuisance in the 'Lemon Squeezer' [*sc.* a narrow defile]. **1953** Baker *Australia Speaks* vii. 177 A few other words of wartime vintage..*lemon squeezer*, the peaked hat worn by New Zealand troops (apparently originated by the troops themselves). **1957** T. S. Eliot *On Poetry & Poets* 113 It might be called the lemon-squeezer school of criticism. **1959** B. Kops *Hamlet of Stepney Green* I. 10 Julius Caesar, such a silly geezer, caught his head in a lemon squeezer. **1964** *N.Z. News* 24 Nov. 2/1 The 'lemon squeezer' was no longer suitable headgear for ceremonial

rifle exercises and would never be worn by the New Zealand Army again, said the Chief of General Staff. **1629** J. Parkinson *Parad.* cxxxi. 454 Lemon Tyme. The wilde Tyme that smelleth like unto a Pomecitron or lemon, hath many weake branches trayling on the ground. **1657** R. Verney in M. M. Verney *Mem.* (1894) III. xi. 409 Sweet Marjoram & Lemon Time. **1861** Mrs. Beeton *Bk. Househ. Managem.* 220 *Lemon thyme.* Two or three tufts of this species of thyme, *Thymus citriodorus*, usually find a place in the herb compartment of the kitchen garden. **1971** *Country Life* 20 May 1252/2 Lemon-thyme has a lovely little golden cultivar which should be in all gardens. **1974** Page & Stearn *Culinary Herbs* 44 Those who find the flavour of garden thyme too dominating may prefer the milder and fruity flavour of lemon thyme. **1869** C. L. Brace *New West* iii. 37 Lemonverbenas..are small trees. **1952** J. & L. Bush-Brown *America's Garden Bk.* (ed. 2) xxi. 723 Plants suitable for pot culture... Lemon Verbena. **1971** *Country Life* 20 May 1207/1 A huge lemon verbena (*Lippia citriodora*) is said to be pre-1903. **1924** Record & Mell *Timbers Trop. Amer.* 513 *Aspidosperma tomentosum* Mart... Lemon wood... Color mostly bright, clear canary-yellow. **1934** A. L. Howard *Man. Timbers of World* (rev. ed.) 148 Degame wood. *Calycophyllum candissimum*... Lemon-wood. **1947** J. C. Rich *Materials & Methods Sculpture* x. 290 Lemonwood, also referred to as Degame, is a yellowish or creamy-white hardwood that is sometimes used for carving. Cuba is the major source of this wood. **1969** T. H. Everett *Living Trees of World* 162/2 The lemonwood of New Zealand..has masses of honey-scented yellowish green flowers and leaves that emit a lemon-like odor when bruised. **1972** *Handbk. Hardwoods* (Building Res. Establishment) (ed. 2) 66 Degame. *Calycophyllum candidissimum.* Other name: lemonwood (United States).

b. Abbrev. of Lemonade, *lemon-juice*; also *bitter lemon*, a mineral drink.

1885 *List of Subscribers* (United Telephone Co.) p. xv. Kindly send us..one gross of seltzer, one gross of soda, one gross of lemon, and half that quantity of splits. **1898** J. D. Brayshaw *Slum Silhouettes* 228 'Oh! a lemon an' dash'll do me,' she says... So I calls fer two lemons, wiv a dash of bitter. **1956** R. Postgate in C. Ray *Complete Imbiber* I. 182 'Port 'n lemon', which was an evidence of feminine folly some years ago, was I suppose a sort of proletarian equivalent of pink champagne. **1962** *Guardian* 27 Aug. 3/1 His sister..drinks nothing but orange juice or bitter lemon. **1965** I. Fleming *Man with Golden Gun* viii. 110 Mr Hendriks..nursed a Schweppes Bitter Lemon.

lemonade. Add: **b.** *attrib.* and *Comb.*, as *lemonade bottle, crystal, powder, syrup.*

1972 *Country Life* 30 Nov. 1481/3 The screw-topped or marble-stoppered lemonade bottles of long ago. **1902** J. T. Law *Grocer's Manual* (ed. 2) 528/2 *Lemonade powders or crystals*, these usually consist of a compound of.. bicarbonate of soda,..tartaric acid,..icing sugar,.. essence of lemon with..essence of pineapple. *c* **1938** *Fortnum & Mason Price List* 43/1 Lemonade crystals.. per bot. 9½d. **1896** J. T. Law *Grocer's Manual* 413/1 *Lemonade powders*,..usually consist of a compound of.. bicarbonate of soda,..tartaric acid,..icing sugar. **1938** L. MacNeice *I Crossed Minch* II. x. 140 A plump schoolgirl..had a tin..containing lemonade powder. **1822** M. Edgeworth *Let.* 20 Feb. (1971) 355 You are welcome to the lemonade-Syrop. I have sent my aunt another bottle.

lemony, *a.* Add: **b.** *Austral.* and *N.Z. slang.* Irritated, angry, esp. in phr. *to go lemony at*: to become angry with (someone).

1941 in Baker *Dict. Austral. Slang* 31. **1946** D. Stivens *Courtship of Uncle Henry* 75 He's as lemony as hell when he opens the door and doesn't say a word to me. **1945** Baker *Austral. Lang.* vi. 121 A man in a temper is said.. *to go lemony*..at a person. **1952** *Coast to Coast* 97 Ironbark got lemony. He bellowed like a thousand bulls.

Lemuria (lĭmiūə·riă). [f. Lemur + -ia¹.] A hypothetical continent stretching from Africa to south-east Asia, formerly supposed to have existed in the Jurassic period.

1864 P. L. Sclater in *Q. Jrnl. Sci.* I. 219 In Madagascar and the Mascarene Islands we have existing relics of this great continent, for which as the original focus of the '*Stirps Lemurum*', I should propose the name Lemuria! **1876** A. R. Wallace *Geogr. Distribution Animals* I. iv. 76 Lemuria..is undoubtedly a legitimate and highly probable supposition, and it is an example of the way in which a study of the geographical distribution of animals may enable us to reconstruct the geography of a bygone age. **1880** —— *Island Life* II. xix. 398 Atlantis is now rarely introduced seriously... But 'Lemuria' still keeps its place —a good example of the survival of a provisional hypothesis which offers what seems an easy solution of a difficult problem..long after it has been proved to be untenable. **1944** 'Palinurus' *Unquiet Grave* iii. 85 To have set foot in Lemuria is to have been close to the mysterious sources of existence... Wild ghost faces from a lost continent who will soon be extinct. **1957** P. J. Darlington *Zoogeogr.* x. 590 Some persons have claimed a broad connection or union of Africa, Madagascar, and India even in the Tertiary and have named it Lemuria, but there is decisive evidence against it.

lemurian (lĭmiūə·riăn), *a.* [f. Lemur + -ian; but sense 1 is derived from *Lemuria.] **1.** (With capital initial.) Of or pertaining to *Lemuria.

1871 *Nature* 30 Mar. 429/1 The Máldive and the Láccadive coral-islands belong strictly to the Lemurian region.. and I am not sure that the latter does not reach the mainland of India. **1893** A. Newton *Dict. Birds* 354 The hypothesis of a Lemurian continent was..unnecessary.

2. Of or pertaining to lemurs; characteristic of lemurs.

1891 *Ann. Rep. Board of Regents Smithsonian Inst. 1889–90* 621 An unsymmetrical face, the nasal overture of a pheleiform type, and lemurian attachment of the under jaw. **1893** A. Newton *Dict. Birds* 355 Lemurian remains have been found fossil in France.

lemurid (lĭ·miūrid, le·m-). [f. mod.L. family name *Lemuridæ*, f. *Lemur* (Linnæus *Systema Naturæ* (ed. 10, 1758) I. 29), ad. L. *lemures* ghosts.] A member of the family Lemuridæ.

1884 *American* VIII. 218 True monkeys are scarce, but galagos and certain other lemurids are common. **1972** T. A. Vaughan *Mammalogy* vii. 115/2 The fossil record of lemurids is from Pleistocene and sub-Recent deposits in Madagascar.

lemu·riform, *a.* [f. Lemur + *i* + -form.] Resembling the lemurs. Also as *sb.*

1887 A. Heilprin *Geogr. & Geol. Distribution Animals* 174 Lemurs or lemuriform insectivores (Adapis, Necrolemur). **1972** *Nature* 24 Mar. 180/1 *Archaeolemur* and *Hadropithecus* are cited as the few lemuriforms with symphysal fusion. **1973** *Ibid.* 30 Mar. 353/1 The author has his first sub-order Prosimii embrace Tupaiidae and Tarsiidae as families of equal rank to five lorisiform and lemuriform families.

Lenape (lĕnā·pe). Also **Lenne-** or **Lenni-Lenape.** [See quot. 1819.] **a.** An Algonquian Indian people, also called Delaware Indians, formerly inhabiting the north-eastern United States; a member of this people. **b.** The language of this people.

1728 P. Gordon *Let.* 2 Sept. in S. Hazard *Pennsylvania Arch.* (1852) I. 230 Our Lenappys or Delaware Indians know nothing of it. **1785** T. Jefferson *Notes State Virginia* (1801) 198 Delawares, or Linnelinopies. **1819** J. Heckewelder *Hist. Indian Nations* (1876) p. xl, *Lenni Lenape* being the national and proper name of the people we call 'Delawares', I have retained this name, or for brevity's sake, called them simply *Lenape*, as they do themselves in most instances. Their name signifies 'original people', a race of human beings who are the same that they were in the beginning, unchanged and unmixed. *Ibid.* iii. 76 'It was we,' say the Lenape, Mohicans, and their kindred tribes, 'who so kindly received them on their first arrival into our country.' **1826** J. F. Cooper *Last of Mohicans* II. vii. 191 The Delaware, or Lenape, claimed to be the progenitors of that numerous people, who once were masters of most of the eastern and northern states of America. **1849** E. G. Squier in W. W. Beach *Indian Misc.* (1877) 9 (*heading*) A translation of the Walum-Olum, or bark record of the Lenni Lenape. **1885** D. G. Brinton *Lenâpé & their Legends* iii. 35 *Lenape*, therefore, does not mean 'a common adult male', but rather 'a male of our kind', or 'our men'. **1888** Brinton & Anthony (*title*) A Lenâpé-English dictionary. **1913** *Handbk. Indians of Canada* (Geogr. Board of Canada) 125/1 The early history of the Lenàpe is contained in their national legend, the Walum Olum. **1934** F. W. Hodge *McKenny & Hall's Indian Tribes of N. Amer.* III. 32 The Delawares were situated principally upon tide-water in New Jersey, Pennsylvania, and Delaware. Their own appellation of *Lenne Lenape*, or original people, has been almost forgotten by themselves, and is never used by the other tribes. **1959** E. Tunis *Indians* 21/1 It was taken from the Lenape, an Algonquian language.

lenate (lĭ·nēⁱ·t), *v. Phonology.* [f. L. *lēn(is* soft + -ate³.] = *Lenite *v.* Hence **lena·ted** *ppl. a.* Also **lena·tion** *sb.* = *Lenition.

1909 J. Strachan *Introd. Early Welsh* 12 When an adjective in the positive degree precedes, the noun is lenated. *Ibid.*, After proper nouns there is lenation of a following noun or adjective. *Ibid.* 13 In poetry, when the genitive precedes the noun, it may lenate. **1928** E. Ekwall *Eng. River-Names* p. lxxii, Quite different is the state of things in regard to lenated *t* (*d*). *Ibid.* p. lxxiii, British **b, d, g** were lenated to *v, ð, ʒ*, which latter often disappears.

lend, *sb.*² Add: (Further examples.) Also *Austral.* and *N.Z. colloq.*

1749 J. Steuart *Let.* 29 Dec. in *Publ. Scottish Hist. Soc.* (1915) 2nd Ser. IX. 464, I..sent him inclosed a letter..in which I desire the lend of 20£ sterlin for 18 months. **1946** F. Sargeson *That Summer* 77 Could you give me the lend of a bob? **1965** *Listener* 2 Sept. 339/1 Thanks for the lend of your earhole, mate.

lend, *v.*² Add: **2. e.** (Later examples.)

1940 *Times* 11 Dec. 5/4 In war-time a good many people take to what is vaguely called 'lending a hand' in the domestic circle. **1951** E. Paul *Springtime in Paris* iv. 69 The local heroes all were known, except two passing strangers who had lent a hand at the barricade and died anonymously. **1961** *New Eng. Bible Luke* x. 40 Tell her to come and lend a hand.

h. *to lend colour* (*to*): see *colour *sb.*¹ 12 e.

lend-lease. (Level stress.) = *lease-lend. Also *attrib.* and in extended uses. Also as *vb.* So **lend-leased** *ppl. a.*

1941 *Economist* 15 Feb. 214/1 Future disposition of the armaments now being produced is before Congress in the 'Lend-Lease' Bill. **1942** *Times* (Weekly ed.) 9 Sept. 9/2 Thousands of barrage balloons were lend-leased to the United States soon after Pearl Harbour. *Ibid.*, Lend-leased British anti-aircraft guns help to defend American cities. **1942** *R.A.F. Jrnl.* 3 Oct. (recto rear cover), The contribution of experienced pilots and planes in the fight to clear our side of the Atlantic is an element of the Lend-Lease programme in reverse. **1945** W. S. Churchill *Victory* (1946) 178 Your friendship and great help as Lend-Lease Administrator and Secretary of State will always be remembered with gratitude. **1949** I. Deutscher *Stalin* 512 More than 400,000 lorries were supplied to Russia under Lend-Lease. **1951** Koestler *Age of Longing* i. 18 Your hand, my child, is on lend-lease to a vicious old man. **1957** *Times Lit. Suppl.* 18 Oct. 625/1 Great Britain has undertaken to lend-lease to the United States nothing less than the Victorian Age in its entirety. **1962** *Listener* 15 Feb. 307/3 The abrupt ending of Lend-Lease. **1972** *National Observer* (U.S.) 27 May 3/1 Talks in the State Department are aimed at ending a U.S.-Soviet dispute over lend-lease that goes back to World War II. From 1942 to 1945, the United States supplied Russia with some $10.8 billion in military and civilian equipment under the lend-lease program.

lengenbachite (leŋĕnba·χəit). *Min.* [f. *Lengenbach*, the name of the quarry in Valais, Switzerland, where it was found + -ite¹.] A sulphide of silver, copper, lead, and arsenic, $(Ag,Cu)_2Pb_6As_4S_{13}$, occurring as steel-grey blade-shaped crystals.

1904 *Nature* 1 Dec. 118/2 Mr. R. H. Solly exhibited and described various minerals from the Lengenbach quarry, Binnenthal. Three of these were new, viz. marrite and bowmanite..and lengenbachite. **1944** *Trans. R. Soc. Canada* XXXVIII. iv. 59 An *x*-ray study of lengenbachite shows that this mineral is monoclinic. **1969** *Mineral. Abstr.* XX. 227/2 Lengenbachite gave space group *P*1 or $P\bar{1}$, lattice constants of a subcell a' 35·10 ± 0·03, b' 5·75 ± 0·01, c' 36·92 ± 0·03Å, $\alpha' \sim 90°$, β' 92°35′, $\gamma' \sim 90°$.

length, *sb.* Add: **2. c.** *Bridge.* Four or more cards of the same suit held in a Bridge hand.

1927 M. C. Work *Contract Bridge* iii. 43 The game-goer may be bid with a blank suit or a worthless singleton if the trump length be satisfactory. **1930** E. Culbertson *Contract Bridge Blue Bk.* xxii. 285 To build up, if possible, a great minor suit length in the strong hand. **1948** —— *Contract Bridge for Everyone* (1949) 77 When your principal length or strength is in the suit your opponent has bid, do not overcall. **1958** *Listener* 2 Oct. 541/1 West, from the bidding, is probably aware of his [*sc.* North's] great Club length. **1973** *Sunday Times* (Colour Suppl.) 20 May 90/2 It is easy to enter for East holding length in diamonds by playing the Ace and the Queen.

4. d. *Swimming.* The length of the swimming-bath taken as a measure of distance swum. Also *attrib.*

1912 F. Sachs *Compl. Swimmer* 237 They..arrange their races to suit the baths, and their handicaps..are measured by its length, *i.e.* '3 lengths (90 yards) handicap'. **1931** G. H. Corsan *Diving & Swimming Bk.* viii. 74 Have the fastest swimmers swim a three lengths race. *Ibid.*, Finish with relay races of two lengths. **1972** B. Turner *Solden's Women* xvii. 154 I'm not such a good swimmer as Patricia was. Three lengths at the baths is about my limit.

8. b. length-mark, a phonetic symbol used to indicate the relative length of a vowel sound.

1926 Armstrong & Ward *Handbk. Eng. Intonation* p. vii, Length marks (: long and half-long) are used to indicate length only and not difference in vowel quality. **1932** D. Jones *Outl. Eng. Phonetics* (ed. 3) 65 The letter i without the length-mark stands for the members of the English *i*-phoneme used when the sound is relatively short. **1965** *English Studies* XLVI. 359 No allophonic length-marks are used.

10. (Later examples.) Hence *length bowler.*

1910 *Blackw. Mag.* Jan. 91/1 Only at the last gasp was any serious effort made to knock him off his length. **1937** *Daily Herald* 5 Jan. 14/1 [Verity] The best length bowler in England. **1956** N. Cardus *Close of Play* 176 The old-fashioned 'length' bowlers, ball after ball on the same spot. **1958** D. Bradman *Art of Cricket* 97/1, I prefer to think in terms of a 'good length ball' and to define it thus—'The type of delivery which has the striker in two minds as to whether he should play forward or back.' **1969** *Listener* 1 May 622/3 At first Powell hit the ball all around the field and, just as it looked as if Miller might be finding his length, the item ended.

b. In racket games: the quality of making shots which pitch well back in the court and deny the opponent an easy return; the placing of a shot in this way; the 'form' required to make such shots consistently.

1924 G. W. Hillyard *40 Yrs. Lawn Tennis* viii. 136 He..went on hitting..until he did get his 'length', and then it was..a case of woe betide the other man. **1930** *Morning Post* 19 July 14/6 The Italian's fine mixture of pace and length was pitted against Lott's youth, power, and cunning. **1948** S. Noel *More about Squash Rackets* i. 24 Angles, drop-shots..and reverse angles are all the stock-in-trade of the professional, in addition to a sound length game. **1961** J. H. Giles *Squash Rackets* viii. 41 It [*sc.* the lob shot] can also be used as an attacking shot, providing as it does a complete change of pace and flight from the orthodox drive and length shots. **1964** R. Laver *How to play Winning Tennis* vii. 57 You can get good length with the topspin I use. **1966** *Observer* 8 May 19/5 She was quick to switch from her steady baseline game into a counter attack whenever Miss Niessen lost her length.

11. c. *slang.* A penis; sexual intercourse; so *to slip* (someone) *a length*: (of a man) to have sexual intercourse with.

1949 Partridge *Dict. Slang* Add. 1173/2 *Slip (her) a length*, to coit with (a woman). **1952** C. MacInnes *June in Spring* vi. 156 'Is it hard to get a job on board a ship without experience?' 'Not if you work for nix and don't mind the stokers slipping you a length.' **1968** H. C. Rae *Few Small Bones* III. viii. 216 Beefy, randy-arsed wives crying out for a length. **1970** C. Wood *Terrible Hard* v. 58 Come

on, Suggy, you're 'is batman, 'e's never slipped you a crafty length 'as 'e?

18. lengthman, a man appointed to maintain a certain stretch of road or railway.

The form *lengthsman* in quot. 1902 is an isolated use.

1902 *Times* 22 Sept. 2/5 Every lengthsman or fettler on the Government railway gets 8s. a day for eight hours' work. **1921** *Dict. Occup. Terms* (1927) § 577 *Lengthman*,.. an underman in a gang engaged on maintenance of a specific section..of permanent way. **1959** *New Scientist* 16 Apr. 852/1 The mixed plant community was largely maintained by..the regular cutting with scythe and sickle by ..the County Council 'lengthmen'. **1968** *Telegraph* (Brisbane) 3 June 18/1 Our legislators should modernise transport for railway lengthmen. **1970** *E. Anglian Daily Times* 31 Aug. 4/5 In days of cheaper labour many county council roadmen known as 'lengthmen' were each responsible for the maintenance of a limited number of miles of road in which they took great pride and knew all the peculiarities. **1971** *Times* 8 Apr. 15/3 An old man who lived at Spelbrook... His home was..the lengthman's cottage. **1972** L. Lamb *Picture Frame* xviii. 157 A road (or 'length') man, with broom and shovel strapped to his bicycle cross-bar.

Lengua (le·ŋgwa). [f. Sp. *lengua* tongue (see quot. 1904¹).] **a.** A member of a tribe of South American Indians inhabiting the Paraguayan Gran Chaco area; also *attrib.* or as *adj.* **b.** The language of this tribe.

1822 S. Coleridge tr. *Dobrizhoffer's Acct. Abipones* I. 125 The equestrian nations remaining in Chaco, and still formidable to the Spaniards, are the Abipones,..and Oekakalots, Guaycurus, or Lenguas. **1904** W. B. Grubb *Among Indians Paraguayan Chaco* vii. 57 The *labret* is an extension of the lower lip, which has the appearance of a protruding tongue. Hence the Spanish term *Lengua* was applied indiscriminately by the early colonists to any tribe who adopted this custom. *Ibid.* x. 94 Unless the circumstances are known, some expressions in Lengua are quite meaningless. **1908** *Westm. Gaz.* 11 Sept. 8/2 During the past year sections [of the Bible] have been printed in Lengua, a language spoken by the Indians of the Paraguayan Chaco. **1911** J. G. Frazer *Golden Bough: Taboo* (ed. 3) ii. 38 The Lengua Indians of the Gran Chaco hold that the vagrant spirits of the dead may come to life again. **1950** J. G. Kerr *Naturalist in Gran Chaco* ix. 175 The main work of the Mission was..among a set of Lengua (i.e. Mushcui) Indians known as the Paisiapto or black-food people. **1973** B. J. Susnik in J. R. Gorham *Paraguay: Ecological Ess.* 121 (*table*) Lengua... Since 1850, contacts (bartering) with Spanish Americans.

Leninism (le·niniz'm). [f. *Lenin*, the assumed name of Vladímir Il'ích Ulyánov (1870–1924), the founder and leader of the Bolsheviks and of the Soviet State + -ism.] The political and economic doctrines of Marx as interpreted and applied by Lenin to the governing of the Soviet Union, to the theory of the international proletarian revolution, and to the dictatorship of the working class. So **Leninism–Stalinism,** Lenin's doctrines as interpreted and applied by Stalin.

1918 *Times* 19 Jan. 5/1 (*caption*) From Tsardom to Leninism. **1928** E. & C. Paul tr. *Stalin's Leninism* I. vi. 53 This second formulation was directed against some critics of Leninism, against the Trotskyists. *Ibid.* II. iii. 94 The endeavour of 'practical' persons to have no truck with 'theories' runs counter to the whole spirit of Leninism and is a great danger to our cause. **1935** *Economist* 12 Jan. 73/2 'Leninism' is a series of brilliant footnotes to the Marxist philosophy made by an experimenter. **1948** J. Towster *Political Power in U.S.S.R.* 3 The teachings of this theory are called Marxism, Leninism,..Marxism–Leninism, or Leninism–Stalinism. **1959** *Times Lit. Suppl.* 21 Aug. 479/3 The remainder of the book follows more familiar lines, Leninism being opposed to Stalinism. **1964** E. H. Carr *Socialism in One Country* III. I. xxxv. 500 In Bolshevik doctrine Leninism meant the adaptation of Marxism to the conditions not of a particular country, but of a particular historical period. **1966** P. Heath tr. *Wetter's Soviet Ideology Today* 328 If history has declined to develop in the manner prescribed by Marx, the endeavour must be made to adapt her to this plan. Hence the explicitly voluntaristic element in Leninism. **1966** L. Lempert tr. *Vinogradov's Socialist Nationalisation of Industry* 19 Leninism maintains, and historical experience confirms, that the ruling classes do not yield power of their own free will. **1971** *Times Lit. Suppl.* 21 May 589/1 Scholastic disputes about orthodoxy are no longer a feature of studies of Marxism and Leninism.

Leninist (le·ninist), *a.* and *sb.* [f. *Lenin* (see prec.) + -ist.] **A.** *adj.* Of, pertaining to, or characteristic of Lenin, his followers or his doctrine. Hence *Leninist–Marxist* (cf. *Marxist–Leninist* adj. s.v. *Marxist *sb.*¹ and *a.*¹), *Leninist–Stalinist.* **B.** *sb.* A follower or supporter of Lenin or his doctrine.

1917 *Times* 10 Nov. 6/4 General Korniloff has been placed under the same ban as M. Kerensky, and renewed instructions for the arrest of both have been issued by the Leninist committee. *Ibid.* 23 Nov. 7/2 Trotsky, one of the Leninist chiefs, has just declared that violence done by his supporters is a right, but violence in resisting them is immoral. **1920** *Q. Rev.* Apr. 474 The Socialists and the Leninists. **1928** E. W. Dickes tr. *Marcu's Lenin* 187 The Leninists, as the closer adherents of Ulianov now called themselves. **1934** H. G. Wells *Exper. Autobiogr.* II. ix. 806 His [*sc.* Stalin's] was not a free impulsive brain nor a scientifically organized brain; it was a trained Leninist-Marxist brain. **1949** H. Read *Existentialism, Marxism & Anarchism* 16 Humanism is a term which..even an intransigent Marxist like Lukacs does not disdain—he calls the Leninist theory of knowledge a militant humanism. **1950** tr. *M. Djilas's On New Roads of Socialism* 29 The Soviet Government and the subordinate governments have..organized against her [*sc.* Yugoslavia] an economic blockade and violent pressure..by which all Leninist principles on relations amongst Socialist countries have been trampled underfoot. **1953** *Mind* LXII. 68 The Leninist..is able to demonstrate the inexorable nature of the 'withering away of the state'..because his definition of the state requires that it disappears when classes have been abolished. **1964** D. Caute *Communism & French Intellectuals* I. iii. 54 A demand that henceforth the intellectuals cultivate the spirit of the Party in the Leninist-Stalinist sense of the term. **1966** P. Heath tr. *Wetter's Soviet Ideology Today* ii. 32 If the Leninist concept of matter seeks to constitute a definition, it ought to explain what the nature of matter is. **1971** *Times Lit. Suppl.* 21 May 589/1 A critique of specific points of Leninist doctrine. **1973** E. Hyams *Final Agenda* ii. 20 He was a literal, not a nominal Leninist, and he believed..that wisdom.. lay in Lenin's profound distrust of the bureaucracy.

Leninite (le·ninəit), *a.* and *sb.* [f. *Lenin* (see above) + -ite¹.] = *Leninist *a.* and *sb.*

1917 *Times* 7 Dec. 7/3 Trotsky, on behalf of the Leninite 'Government', has telegraphed to all the representatives of Russia abroad. **1918** *Times* 1 Jan. 5/1 The Leninites.. earmarked for their own disposal both the food and the credits that had been set aside for the soldiers' sustenance. **1920** E. E. Cummings *Let.* 22 June (1969) 72, I do *not* need money..being a good (if innocuous) Leninnite [*sic*] or Trotskyite. **1920** *Glasgow Herald* 3 July 6 The Constitution of 1919 which recognised the existence of soviets, though not in the Leninite sense.

lenis (lĭ·nis), *a.* and *sb. Phonology.* [a. L. *lēnis* soft.] **A.** *adj.* Of one of two or more homorganic consonants: articulated with less energy. Opp. *fortis B. *adj.*

1929 G. K. Zipf in *Harvard Stud. Classical Philol.* XL. 63 Then, if our theory be true, the German *p* must be *fortis*, the German *b lenis. Ibid.* 64, I hazard the guess.. that the same condition of *lenis–fortis* obtains there also. **1962, 1964** [see *fortis B. *adj.*]. **1969** *Word* XXV. 20, l is sometimes realized as a very lenis unreleased t and ŋ as a nasalized vowel. **1971** F. W. Householder *Ling. Speculations* xi. 206 In Iranian Azerbaijani..the back velar stop is normally lenis only, unopposed by any fortis aspirated [*k*].

B. *sb.* (pl. **-es**). A lenis consonant.

1932 W. L. Graff *Lang.* vii. 274 In Alsace people..of French tongue often pronounce voiceless sounds or lenes as in Southern German. **1933** L. Bloomfield *Lang.* 99 Pressure and action are gentle in lenes, vigorous in fortes. .. In English the unvoiced stops are aspirated fortes, but other types occur as non-distinctive variants, notably the unaspirated lenis type after [s]. **1935** G. K. Zipf *Psycho-Biol. of Lang.* iii. 63 In the *voiceless fortes* and the *voiceless lenes* we have stops which differ appreciably in magnitude. **1965** *Amer. Speech* XL. 7 Four groupings within the lenes.

lenite (lĭ·nəit), *v. Phonology.* [Back-formation from *lenition.] **a.** *trans.* To make lenis in articulation. **b.** *intr.* (Of consonants), to become lenis. Hence **le·nitable** *a.*; **le·nited** *ppl. a.*

1912 F. W. O'Connell *Gram. Old Irish* 5 A true lenited *f* occurs in Modern Irish and is pronounced *h. Ibid.* 61 The absolute forms of the copula lenite the following anlaut. **1953** K. Jackson *Lang. & Hist. Early Brit.* 550 The Bretons lenited the consonants. *Ibid.* 556 British *c*, lenited to *g. Ibid.* 474 The consonants ordinarily regarded in Breton as lenitable. **1967** —— *Hist. Phonol. Breton* 309 The geminates, which were not lenitable, constitute a special case. **1971** *Canad. Jrnl. Ling.* Fall 20 Affrication of the yod element would create a consonant cluster which would not lenite. **1972** H. Kurath *Stud. Area Ling.* ix. 153 It should be further noted that 'lenited' /t, k/ appear as the voiced plosives /d, g/ only in the British branch of Insular Celtic.

lenitic (leni·tik), *a. Ecol.* [f. L. *lenitas* mildness + -ic.] Of fresh-water organisms or habitats: situated in still water. Cf. *lotic *a.*

1916 Needham & Lloyd *Life Inland Waters* vi. 315 Organisms..may be roughly divided into two primary groups for which are suggested the following names: I. *Lenitic* or still-water societies. II. *Lotic* or rapid-water societies. **1931** R. N. Chapman *Animal Ecol.* xiv. 285 Quantitative methods for the study of benthonic organisms have made important contributions to our knowledge of the quantity of life in lenitic environments.

lenition (lĭni·ʃən). *Phonology.* [f. L. *lēnis* soft + -ition, after G. *lenierung.*] In Celtic languages, the process or result of making or becoming lenis; softening of articulation; (see quots.). Also *attrib.*

1912 F. W. O'Connell *Gram. Old Irish* 5 In Old Irish a single consonant between two vowels was more loosely articulated than in absolute anlaut, and this phonetic change has been termed both *aspiration* and *lenition.* **1913** J. Morris-Jones *Welsh Gram.* § 103. 162 Continental scholars use 'Lenition' as a term embracing the Welsh 'soft mutation' and the corresponding Irish 'aspiration'. **1935** *Mod. Lang. Notes* L. 518 The term *lenition* might to advantage have been recorded; in recent Celtic grammars this term has taken the place of the older *aspiration.* **1953** K. Jackson *Lang. & Hist. Early Brit.* 424 IE. and Latin *d*..in lenition position initially and internally they became Late Brit. *d.* **1954** Pei & Gaynor *Dict. Ling.* 121 *Lenition*, in Celtic languages, the phonetic change which consonants undergo when occurring between vowels, as well as the change of the initial consonant of a word under the influence of the final sound of the immediately preceding word. **1963** J. P. Hughes *Sci. of Lang.* xiv. 251 A tendency arises..to shift the single medial consonant to a spirant... This process is prominent in the Celtic languages, and is known as *lenition.* **1971** *Canad. Jrnl. Ling.* Fall 17 There is some advantage to (1a), which treats lenition in terms of point of articulation classes.

leno. Add: (Earlier and later examples.) Hence, the type of weave used for this fabric. Also *attrib.* and *Comb.*, as *leno brocade, weave*; **leno loom,** a loom which produces leno weave.

1821 M. Browne *Diary* 11 Aug. (1905) 173 We at last got a leno cap and an under cap to wear with it. *c* **1828** J. R. Planché *Green-Eyed Monster* 8 Leno slip, over white satin, ornamented with leno puffs of white and pink. **1940** *Chambers's Techn. Dict.* 494/2 *Leno*, a fabric with an openwork or an embroidered effect, produced by cross-weaving; fabrics of this character that are of regular texture are usually termed *gauze. Ibid.*, *Leno brocade*, a brocade cotton, or cotton and rayon cloth, produced by a combination of ordinary and cross-weaving. **1964** H. Hodges *Artifacts* x. 141 Gauze or leno..is produced by crossing adjacent warps before passing the weft, and recrossing the warps again before passing the next weft. **1964** *McCall's Sewing* iv. 52/2 (*caption*) Leno weave; gauze weave. *Ibid.*, Some major fabrics woven on leno looms are marquisette, netting, mesh shirting. **1968** J. Ironside *Fashion Alphabet* 239 *Leno*,..a type of weave—an open-work fabric with warp yarns twisted before weaving.

lens, *sb.* Add: **1.** (Examples of wider applications of the word.)

1931 [see *electron lens* s.v. *electron² 2 b]. **1945** *Jrnl. Sci. Instrum.* XXII. 239/1 Another material useful for ultrasonic lenses, especially when the liquid is incompatible with plastics, is lithium. **1951** V. E. Cosslett *Pract. Electron Microsc.* ii. 35 Use is made of a surrounding shield of iron to concentrate the field into a small region near the middle of the lens. **1972** *Science* 16 June 1236/1 The spherical concave lens focused the sound at a nominal 3 cm from the transducer and provided a field 1 mm wide, extending from 2 to 4 cm in range.

2*. *Geol.* A body of ore or rock similar in shape to a biconvex lens.

1903 *Bull. U.S. Geol. Survey* No. 213. 113 The principal mines..have revealed valuable ore bodies of two great types, those which occur as lenses, roughly parallel to the bedding, and those which occur in fracture or fissure zones. **1935** *Economist* 21 Dec. 1283/3 Further lenses of valuable ore would be discovered in that section. **1939** *Proc. Prehist. Soc.* V. 40 Towards the top of the ferruginous gravels appears a lens of non-ferruginous, grey, clayey sand. **1969** Bennison & Wright *Geol. Hist. Brit. Isles* vi. 128 These Lower Palaeozoic rocks occur as discontinuous outcrops or lenses in what has been termed the Meneage Crush Zone... Included lenses may be up to 1 mile in length.

3. (sense 1, 1 b) *lens aperture, barrel, -board, -holder, mount, -work*; *lens-tube* (earlier and later examples); **lens cap,** a cap that fits over the end of a lens tube, used to protect the lens and, in early cameras without shutters, for regulating exposures; **lens coating,** a thin transparent coating applied to a lens to reduce reflection of light at its surface; **lens hood,** a tube, usually circular in cross-section and with outwardly sloping sides, fitted in front of a lens to shield it from light coming from outside the field of view; **lens louse** *slang* (see quots.); **lensman** = *camera-man* (*camera 3 d); **lens paper,** a kind of soft, thin, absorbent paper suitable for wiping lenses; **lens tissue** = **lens paper*; **lens turret,** a mounting fitted to the front of a camera and carrying several lenses, any of which can be brought into use by rotating the mounting.

1916 *Brit. Jrnl. Photogr.* LXIII. 166/2 (*heading*) Some matters concerning lens apertures. **1958** *Oxford Mail* 19 May 7/4 The length of exposure and the size of the lens aperture are linked to ensure that the right amount of light reaches the film at every shutter speed. **1971** L. B. Happé *Basic Motion Pict. Technol.* ii. 62 The brightness of the image formed by the lens is determined not only by the diameter of the lens aperture but also the size of the image. **1940** *Chambers's Techn. Dict.* 495/1 *Lens barrel*, the metal tube in which one or more lenses are mounted. **1958** *Newnes Compl. Amat. Photogr.* iv. 60 Camera body—the choice is between folding bellows or extending lens-barrel. **1967** Karch & Buber *Offset Processes* v. 143 The lens barrel contains a slot..used to insert filters for color work. **1892** *Photogr. Ann.* II. 289 The most important feature is the novel and convenient mode of attaching the front lens board to the baseboard. **1941** R. M. Allen *Photomicrogr.* ii. 67 Most manufacturers provide some type of fixture, preferably with focusing means incorporated in it, for carrying the lenses on the lens board. **1967** Karch & Buber *Offset Processes* v. 143 The lensboard, located directly in front of the copyboard, houses the lens. **1882** *Photogr. at Home: its Appliances & Apparatus for Amateurs* 11 A little shield covered with black velvet.. occupies the place of the ordinary lens cap. **1897** *Sears, Roebuck Catal.* 474/2 The front of the camera can be removed if desired and exposure made with a lens cap. **1965** Mrs. L. B. Johnson *White House Diary* 17 June (1970) 290, I wanted to be darned sure I didn't lose the lens cap on the camera. **1966** LaCour & Lathrop *Photo Technol.* iv. 48/2 It should seldom be necessary to clean a lens which has been protected from dust and fingerprints with a lens cap. **1952** C. B. Neblette *Photogr.* (ed. 5) ii. 48/1 Lens coatings are a remedial measure and do not entirely

remove reflections. **1966** LaCour & Lathrop *Photo Technol.* iv. 48/2 Finger prints on the lens..are detrimental to the lens coating. **1876** tr. *G. Tissandier's Hist. & Handbk. Photogr.* 223 The ordinary lens-holder being removed from the front of the camera. **1894** S. H. Gage *Microscope* (ed. 5) i. 4 (*heading*) Adjustable lens holder with universal joint. **1891** W. E. Woodbury *Encycl. Photogr.* 405 (*heading*) Lens screen or hood. **1908** *Brit. Jrnl. Photogr.* LV. 245/1 The lens-hood..has recently revived, owing to the necessity of shading the lens from direct light in the case of anastigmats which possess large aperture. **1955** E. Hillary *High Adventure* xii. 210, I clipped on [to my camera] the lenshood and ultra-violet filter. **1968** L. A. Mannheim tr. *Brandt's Photogr. Lens* xv. 166 Lens hoods not only have to shield the lens against stray light, but also protect it against accidental finger marks and rain or snow. **1928** *Amer. Speech* III. 368 Actors who strive for the most advantageous positions are also called 'lens lice'. **1950** J. Hall in *Daily Mail* 24 May, Bane of the news-reel cameraman is what he calls a 'lens louse'. They..never miss a chance of getting in front of a news-reel camera. **1951** *N.Y. Herald Tribune* 26 Aug. IV. 4/4 It was common to see a Leatherneck lensman wield a 45-automatic pistol in one hand and a 16-mm. camera in the other, firing both simultaneously at the enemy only a few hundred yards away. **1964** *Punch* 5 Aug. 183/1 He's one of the best lensmen in the business. **1972** I. Hamilton *Thrill Machine* vi. 27, I held back with the pen and ink men while the lensmen pushed forward to the press barricades. **1892** *Photogr. Ann.* II. 43 Unscrew the back combination and use the front alone *in situ*, thus gaining the length of the lens mount. **1938** H. Windisch *New Photo-School* vi. 166 A focusing screen is applied to the lens mount. **1972** Horne & Markham in A. M. Glauert *Pract. Methods Electron Microsc.* I. II. iii. 354 Optical bench manufacturers make a large number of lens mounts and carriers. **1925** A. F. Collins *Amat. Photographer's Handbk.* iv. 61 Dirt that forms on the surfaces [of a lens] in an almost imperceptible film can usually be wiped off with a dry lens paper, which is a very soft Japanese tissue paper especially made for this purpose. **1973** *Nature* 27 July 233/1 Each portion [of ovary] was placed on defatted lens paper on a stainless steel mesh grid in a vitreosil dish. **1941** A. Sussman *Collins's Amat. Photographer's Handbk.* (rev. ed.) iii. 75 Use some dry lens tissue, or an old linen handkerchief that has been freshly laundered. **1955** S. C. Gilmour *Paper* 305 *Lens tissue*, British-made paper of very thin substance and transparent texture, resembling Japanese tissue. Long fibres, great strength for the substance, and extreme absorbency are characteristics. Used for cleaning optical and microscope lenses. **1858** Sutton & Worden *Dict. Photogr.* 260 The diaphragms within the lens-tube entirely prevent the reflection of light from the inside of the tube. **1918** Lens-tube [see *hood *sb.* 5 m]. **1971** L. B. Happé *Basic Motion Pict. Technol.* iv. 142 On 8 mm amateur cameras..the lens hood is usually limited to a deep flange extension to the lens tube itself. **1951** R. Spottiswoode *Film & its Techniques* iii. 64 Camera noise readily seeps through a lens turret and tends to interfere with dialogue recording. **1963** *Movie* July/Aug. 26/3 Brault's insistence on leaving in blank frames as he shifts the lens turret in mid-reel occasionally gets in the way. **1971** L. B. Happé *Basic Motion Pict. Technol.* x. 311 A lens turret mounting is often preferred for unit lenses and is more convenient for automated operations. **1888** G. M. Hopkins *Let.* 1 May (1938) 144 Photography proper now is mere scaffolding..a poor bastard art succeeds the lens-work and disguises what that gives.

lens (lenz), *v. Geol.* [f. the sb.] *to lens out* (intr.): of a body of rock: to become gradually thinner (along a particular direction) to the point of extinction.

1921 G. H. Cox et al. *Field Methods Petroleum Geol.* 11 The effects of irregularities in sands may be considered to be of three types; those in which the sand lenses out entirely, those in which it loses its porosity, and those in which the porous sand continues but is of changing thickness. **1965** G. J. Williams *Econ. Geol. N.Z.* viii. 108/2 Mining went down to the 500-ft level below which the calcite bodies themselves lens out—as proved by angled diamond drill-holes.

lensoid (le·nzoid), *a.* [f. Lens + -oid.] = Lentoid *a.*

1965 G. J. Williams *Econ. Geol. N.Z.* iv. 35/2 The quartz bodies are lensoid, seldom more than 5 ft in thickness, discontinuous longitudinally and overlapping in places. **1973** *Nature* 27 July 215/1 The cherts occur chiefly as intercalations, sometimes of wide lateral extent and sometimes lensoid, between the extrusives.

Lent, *sb.*¹ Add: **Lent-term** (examples).

1861 T. Hughes *Tom Brown at Oxf.* I. iv. 68 We're only half through Lent term. **1950** *Cambridge 1950* (Varsity) 8 The Cardinal's Ball..seems to have taken over the position of chief social event of the Lent term from the now banned Granta Ball. **1974** *Univ. Exeter Calendar 1974–5* 4 Wed. 8 [January] Lent term begins.

Lenten. Add: **B. 3. Lenten rose,** a variety of *Helleborus orientalis*, blooming in late winter and early spring.

1884 J. Wood *Hardy Perennials* 137 (*heading*) Helleborus Orientalis... Sometimes also called the Lenten Rose, as it may often be seen in flower during Lent. [**1897** S. Hibberd *Familiar Garden Flowers* I. 35 As the trumpet daffodils are called 'Lent lilies', so the spring flowering hellebores are called 'Lent roses'.] **1900** W. D. Drury *Bk. Gardening* x. 330 Equally deserving of praise are the Lenten Roses (*H. orientalis*), whose flowers embrace all the shades of rose and purple, as well as white and cream. **1948** P. M. Synge *Flowers in Winter* 57 Lenten Roses.. always look well in a mixed winter bowl of flowers. **1970** C. Lloyd *Well-Tempered Garden* v. 378 The main flush of blossom from Lenten roses is borne from February till April.

lentic (le·ntik), *a. Ecol.* [f. L. *lentus* slow, calm + -ic.] = *Lenitic *a.*

1935 P. S. Welch *Limnology* ii. 13 The lentic environments, sometimes known as the standing-water series, include all forms of inland water (lakes, ponds, swamps, and their various intergrades) in which the water motion is not that of a continuous flow in a definite direction. **1940** L. H. Hyman *Invertebrates* I. iii. 80 Fresh-water habitats classify as lentic or standing-water bodies: lake-pond-swamp series, and lotic, or running-water formations. .. Among the smaller lentic environments, such as pools, ditches, ponds, and swamps, the animals are practically all littoral and benthonic. **1960** N. Polunin *Introd. Plant Geogr.* xv. 498 The stalks of inhabitants of swift currents tend to be much shorter than those of their lentic relatives.

lenticle (le·ntik'l). *Geol.* [ad. L. *lenticula* (see Lenticular *a.* and *sb.*).] A lentil (sense 4*) or a lenticular piece of rock.

1898 J. E. Marr *Princ. Stratigr. Geol.* iv. 35 The lenticles will be wider in a direction at right angles to that of the strike. **1902** *Encycl. Brit.* XXVIII. 654/2 Lenticles or eyes of uncrushed diorite may be traced. **1930** [see *hornblendite]. **1931** Cissarz & Jones *German-Eng. Geol. Terminol.* 179 Gash veins..are usually of small lateral and vertical extent, in the form of small lenticles.

lenticular, *a.* and *sb.* Add: **A.** *adj.* **1. b. lenticular stereoscope** (earlier example).

1852 *Phil. Mag.* III. 17 (*heading*) The lenticular stereoscope.

2. a. Delete *rare* and add further examples. Also, employing a lens or lenses.

1903 *Sci. Amer.* 7 Feb. 98/2 The lenses revolve at a given speed..proportioned to the diameter of the illuminant and the lenticular apparatus. **1961** *Listener* 26 Oct. 670/1 The first lenticular light-house.

2*. *Photogr.* **a.** Embossed with minute lenses, as *lenticular film*, a film having the non-emulsion surface formed into minute lenses (usually cylindrical lenses, giving a corrugated pattern), so that two or more images (as of different primary colours to make up a colour photograph) can be interspersed on the same area of film.

1934 *Photogr. Jrnl.* LXXIV. 206/1 The list of patents relating to the duplicating of lenticular films grows. **1950** A. W. Judge *Stereoscopic Photogr.* (ed. 3) xvii. 297 A more recent method of making Lenticular stereograms.. employs a lenticular screen, made up of contiguous cylindrical-type lenses, of very small width, placed in front of the sensitive emulsion. **1962** W. G. Hyzer *Engin. & Scientific High-Speed Photogr.* i. 42 A lenticular plate, comprised of an array of spherical lenslets, is employed to produce a corresponding array of spots on the photo-sensitive film. A primary lens having an aperture of *f*/6·3 is used to project the image of the event onto the front surface of the lenticular plate, whereupon each individual lenslet converges the image rays intercepted by its surface onto a tiny spot at the focal plane. Sequential images may be obtained by producing relative motion between the lenticular plate and the photographic emulsion. **1966** R. J. Ross *Television Film Engin.* xi. 445 A method of recording on 35 mm lenticular film was at one time employed by the National Broadcasting Company. **1967** *Electronics* 6 Mar. 79/1 (Advt.), The tube's light output is 30,000 foot lamberts, which results in a light level of 15-foot lamberts on a 3′ × 4′ lenticular screen.

b. Applied to a method of colour photography using a film with cylindrical lenticulations and filters with bands of the primary colours parallel to the lenticulations during exposure and projection.

1936 R. M. Fanstone *Colour Photogr.* xvii. 157 (*heading*) Lenticular colour photography. **1942** C. B. Neblette *Photogr.* (ed. 4) xxxii. 797 The Lenticular process is essentially a screen method in which the screen is formed optically on the emulsion during exposure. **1964** E. S. Bomback *Man. Colour Photogr.* iii. 56 The first commercial film based on the lenticular process was marketed by Kodak in 1928 as 16 mm. cine film..called Kodacolor.

lenticularity (lentikiŭlæ·riti). [f. Lenticular *a.* + -ity.] Lenticular form or quality.

1912 E. H. C. Craig *Oil-Finding* viii. 141 Many of the discrepancies between prediction and results nowadays are attributed to lenticularity of the oil-bearing strata. **1925** A. Beeby-Thompson *Oil-Field Explor. & Devel.* I. iv. 111 No cautious operators fail to appreciate the importance of lenticularity and lateral variation of sands. **1928** *Bull. Amer. Assoc. Petroleum Geologists* XII. 248 Lenticularity is a characteristic feature of the Lower Pico beds. **1965** G. J. Williams *Econ. Geol. N.Z.* xix. 345/1 Lenticularity of the sands may have prevented up-dip migration of the oil, although a slight arching of a 4° south-westerly dip may have assisted.

lenticulated (lenti·kiŭlēⁱtĕd), *a. Photogr.* [f. Lenticul(e + -at(e³ + -ed¹.] = *Lenticular *a.* 2* a.

1925 *Brit. Jrnl. Photogr.* LXXII. 65/2 (*heading*) Lenticulated films for colour cinematography. **1950** A. W. Judge *Stereoscopic Photogr.* (ed. 3) xvii. 296 The positive (print or transparency) is viewed through a similar lenticulated screen to that used for the negative. **1970** R. J. Ross *Color Film for Color Television* xv. 147 (*heading*) Lenticulated film systems using black-and-white emulsions on a specially processed base material.

lenticulation (lentikiŭlēⁱ·ʃən). *Photogr.* [f. Lenticul(e + -ation.] **a.** The condition of being lenticulated. **b.** Each of the minute lenses of a lenticular film.

1916 *Brit. Jrnl. Photogr.* LXIII. 117/2 It is known to obtain the lenticulation of such films by rolling them at a suitable temperature between a smooth cylinder and a cylinder carrying in intaglio the engraving of the embossing to be formed on the film. **1932** *Discovery* Dec. 383/1 If the embossed lenticular film is copied on another embossed film, the lenticulations, being so minute, give rise to interference or 'watering', which causes grave defects in the quality of the copy. **1975** *Movie Maker* Feb. 87/2 Many of these materials have a very bright surface with a whiter-than-white look.., and have some type of surface texture in the form of diamond shaped lenticulations or similar. *Ibid.* 90 (Advt.), 'Hi-Flect' Blankana-White with the unique hexagonal lenticulation

lenticule. Add: **b.** *Photogr.* A minute lens of a lenticular film; = *Lenticulation b.

1942 H. C. Colton in C. B. Neblette *Photogr.* (ed. 4) xxxii. 799 Kodacolor film contained about 600 lenticules per inch. **1966** R. J. Ross *Television Film Engin.* xi. 445 A banded color filter acted on the light passing through the camera lens, with the bands parallel to the lenticules, producing three color separation records on a single frame of film.

lentigerous, *a.* (Example.)

1883 E. R. Lankester in *Encycl. Brit.* XVI. 680/2 The two lines of development of the Molluscan eye..the punctigerous and the lentigerous.

lentil. Add: **4*.** *Geol.* A mass of rock distinct in character and having the shape of a biconvex lens; *spec.* one regarded as a subdivision of a formation.

1895 J. W. Powell in A. Keith *Descr. Knoxville Sheet* (U.S. Geol. Survey Atlas), The kinds of rocks are indicated..by appropriate symbols... The following are generally used..Limestones..Lentils in strata. **1910** *Ann. N.Y. Acad. Sci.* XIX. 177 The gray sandstone of the Grès Noirs more than 20 feet thick, containing a thin irregular lentil of coal. **1953** *Bull. Amer. Assoc. Petroleum Geologists* XXXVII. 2410 Formations may be subdivided into members, lentils, 'tongues', beds, *et cetera.* **1970** *Earth-Sci. Rev.* VI. 275 Examples of informal rock units are:..(*b*) beds (e.g., quarry layer, coal beds, oil sands, tongues, lentils, etc.).

lento. Add: **b.** *Philol.* Applied to a word or phrase pronounced more slowly than in normal speech. Cf. *allegro B. 2.

1939 [see *allegro B. 2]. **1964** J. Vachek in D. Abercrombie et al. *Daniel Jones* 205 Yet the pronunciation with [əl], characteristic of a *lento* style of speech,.. can hardly be credited with exercising a decisive influence on the much more frequent *allegro* form with the syllabic [l̩]. **1968** *Language* XLIV. 87 Another main type of downdrift is in steps, with deliberate lento articulation. **1973** *Word 1970* XXVI. 39 The last two words in lento speech.

Lentz (lents). The name of Hugo *Lentz*, 20th-c. German engineer, used *attrib.* to denote a type of poppet valve invented by him for use in steam engines, and a locomotive valve gear employing such valves operated by a camshaft.

1925 *Marine Engineer* XLVIII. 19/2 In developing the Lentz poppet valve for marine engines, those first fitted were applied to engines of the triple and..quadruple-expansion types. **1930** *Engineer* 31 Jan. 132/1 L.N.E.R. locomotive with Lentz valve gear. **1949** C. J. Allen *Locomotive Pract. & Performance 20th Cent.* iv. 40 The most extensive application of poppet-valves to locomotives in Great Britain has been that of Lentz valves to the 'Hunt' class 4-4-0s of the late L.N.E.R. **1966** O. S. Nock *Brit. Steam Railway Locomotive 1925–65* viii. 99 (*caption*) L.M.S.R. Horwich 2-6-0 with Lentz R.C. poppet valve gear.

Lenz's law (le·ntsiz, le·nziz). *Electr.* [Named after H. F. E. *Lenz* (1804–65), German physicist, who first enunciated it (in *Ann. d. Physik u. Chem.* (1834) XXXI. 483).] The law that the direction of an induced current is always such as to oppose the change in the circuit or the magnetic field that produces it.

1866 E. Atkinson tr. *Ganot's Elem. Treat. Physics* (ed. 2) x. vi. 696 On the induction produced between a closed circuit and a current in activity when their relative distance varies, Lenz has based the following law, which is known as Lenz's law. **1931** L. B. Loeb *Fund. Electr. & Magn.* i. 35 Unless Lenz's law holds we could by induction effects get energy out of nothing. **1962** Corson & Lorrain *Introd. Electromagn. Fields* vi. 222 Lenz's law is a particular case of Le Chatelier's principle. **1973** *Physics Bull.* Dec. 715/2 Any change in the external field induces an appropriate supercurrent which (by Lenz's law) is just sufficient to counter the effect of the field variation.

Leonardesque (lĭ:ŏnāɹde·sk), *a.* [f. the name of *Leonardo* da Vinci (1452–1519) + -esque.] Resembling in subject or style, or in the manner of, the works of Leonardo da Vinci.

1864 Crowe & Cavalcaselle *New Hist. Painting Italy* II. xxii. 547 Nothing can exceed the Leonardesque precision of the drawing or the softness and fusion of the impasto. **1904** E. McCurdy *Leonardo da Vinci* 100 How eminently Leonardesque it was to make the angel point at S. John. **1939** *English* II. 276, I had early waited for a Leonardesque sweetness and subtlety which visited his

LEONBERG

features. **1960** *Times* 24 Feb. 15/1 Sir Kenneth [Clark].. had for some time discerned a Leonardesque presence in the painting. **1971** A. SMART *Renaissance & Mannerism in Italy* xvi. 135 There is nothing here of Leonardo's mystery, but rather a calm objectivity and a cold grace that are the reverse of Leonardesque 'romanticism'.

Leonberg (lī·ǒnbǝɹg). The name of a town in south-western Germany used *attrib.* or *absol.* to designate a large dog, a cross between a St. Bernard and a Newfoundland, often golden in colour, of a breed first developed there about 1855. Also **Le·onberger.**

1907 K. LEIGHTON *New Bk. Dog* XVII. 518/1 The Leonberg dog..is supposed also to be a worker among flocks and herds. **1945** C. L. B. HUBBARD *Observer's Bk. Dogs* 179 The Leonberg is now regarded on the Continent as a distinct race. **1954** M. K. WILSON tr. *Lorenz's Man meets Dog* vii. 75 A great, strong Leonberger,.. a member of one of the largest breeds of dog, adopted as mistress the youngest sister. **1962** J. M. BERNSTEIN tr. *Levi's Two-Fold Night* x. 86 Two enormous dogs..of the rare Leonberg breed. **1971** F. HAMILTON *World Encycl. Dogs* 158 The popularity of the Leonberger increased and by 1872 other breeders were competing with similar crosses to obtain large, handsome, utility dogs.

leone (li,ōᵘ·n). [f. the name of Sierra *Leone.*] The principal unit of currency in Sierra Leone; a banknote of the value of one leone.

1964 *Times* 4 Aug. 6/5 The new basic unit is the leone with a value equal to 10s. **1972** *Whitaker's Almanack 1973* 987 Sierra Leone..Leone of 100 Cents.

Leonese (līǒnī·z), *a.* and *sb.* [f. Sp. *León*, the name of a town and region in Spain + -ESE.] **A.** *adj.* Of or belonging to León, an ancient kingdom of Spain and now a province, or to the town of León in this region. **B.** *sb.* **a.** A native or inhabitant of León; also *collect.* **b.** The language of León, a dialect of Spanish with Portuguese affinities.

1845 R. FORD *Hand-bk. for Travellers Spain* II. viii. 558 The minor traits of Leonese character are influenced by local differences. *Ibid.* 559 The houses of the humble Leonese, like their hearts, are always open to an Englishman. **1865** H. O'SHEA *Guide to Spain* 236/2 The present Jesuits..with their usual refinement, tact, and educational talents, will soon..ungothicise the good Leonese. *Ibid.* 245/2 The Leoneses differ considerably in character, according to the nature of the different regions which they inhabit. **1887** *Encycl. Brit.* XXII. 351/2 *Leonese.* Proceedings on inadequate indications, the existence of a Leonese dialect has been imprudently admitted in some quarters. **1893** H. E. WATTS *Spain* ii. 52 Almanzor marched into the Christian kingdom,.. scattering Castilians and Leonese as the Goths had been scattered three hundreds years before. **1932** W. L. GRAFF *Lang.* x. 377 Spanish group, with the Castilian, Andalusian, Aragonese, and Leonese dialects. **1936** W. J. ENTWISTLE *Spanish Lang.* v. 39 Mozarabic co-operation in many important settlements brought the use of Arabic terminology to a maximum in Leonese. **1964** *Archivum Linguisticum* 2 At that stage he [*sc.* Diez] did not yet identify the dialect at issue as Leonese.

leonine, *a.*¹ Add: **1. c.** Designating that form of leprosy called leontiasis, and the lion-like facies characteristic of it.

The allusion to the resemblance to the lion's face can be traced back to the ancient Arab physicians.

[**1749** J. BARROW *Dictionarium Medicum Universale*, *Leontiasis, Leontion,* or *Leonina lepra*, a name for Elephantiasis, or leprosy.] **1813** T. BATEMAN *Pract. Synopsis Cutaneous Dis.* 295 Haly Abbas says the countenance was called leonine, because the white of the eyes becomes livid, and the eyes of a round figure; and Avicenna observes that the epithet was applied to the disease, because it renders the countenance terrible to look at, and somewhat of the form of the lion's visage. **1867** *Rep. Leprosy* (R. Coll. Physicians) 242 The prominent blotches on the forehead gave a sombre character to his countenance; not as yet approaching the leonine expression of tubercular elephantiasis. **1899** T. L. STEDMAN *20th Cent. Pract.* XVIII. 623 The lower part of the frontal skin is drawn downwards and conceals the eyes, as in mad persons and lions. This is why the affection is also called leonine. **1959** R. G. COCHRANE *Leprosy in Theory & Pract.* 367 The 'leonine' appearance in Hansen's disease is..attributable to the nodular leprosy. **1970** G. J. HILL *Leprosy in Five Young Men* 65 Patient 5 was a large dark-skinned man with moderately severe leonine facies.

leonite (lēi·-, lī·ǒnəit). *Min.* [ad. G. *leonit* (C. A. Tenne 1896, in *Zeitschr. d. deut. geol. Ges.* XLVIII. 637), f. the name of *Leo* Strippelmann, 19th-c. German salt-works director: see -ITE¹.] A hydrated sulphate of potassium and magnesium, $K_2Mg(SO_4)_2.4H_2O$, found as transparent, colourless, or yellowish prismatic crystals.

1897 *Jrnl. Chem. Soc.* LXXII. II. 269 There is no crystallographic relation between this mineral and blödite.., so that the older but unpublished name, leonite, is used in preference to kaliblödite. **1932** *Bull. U.S. Geol. Survey* No. 833. 44 In the Joe Mitchell well at a depth of 1,368 feet pale-yellow leonite with a waxy luster is intimately mixed with kainite... In places leonite occurs in larger blebs. **1970** *Mineral. & Petrogr. Acta* XVI. 14/2 This appears to be the first occurrence of leonite for Vesuvian fumaroles. Leonite is known to be associated in salt deposits of oceanic origin, at Stassfurt with halite, at Leopoldshall with kainite, and at Ascherleben.

leopard. Add: **2.** (Later examples.)

1920 *New Statesman* Apr. 20/1 For the moment the public is not likely to get a thorough grounding in economics, nor does the Press leopard show any signs of changing his spots. **1930** D. JERROLD *Lie about War* 35 As for the leopard who failed to change his spots, why blame the war? **1955** W. GADDIS *Recognitions* II. v. 487 You wanted to marry a Christian, you wanted to marry a good Catholic. Well leopards can't change their spots. **1972** G. OAKLEY *Church Mouse* 20/2 The schoolmouse..said that..Sampson was a leopard in sheep's clothing and that a wolf couldn't change its spots. **1973** *Times* 21 Nov. 19/8 There is no evidence to show that the Communist Party leopard has changed its spots.

4. Delete †*Obs.* and add later examples. Also, the skin of the leopard; a coat made from this.

1924 *Vogue* early Sept. 42 (*caption*) Even smarter..is a suède coat lined and trimmed with leopard. **1930** M. BACHRACH *Fur* xv. 197 All Leopards are open-handled and..there is very little natural grease on the skin. **1938** —— *Selling Furs Successfully* ix. 91 It is preferable when manufacturing Leopards into garments that as few seams as possible show after the garments are finished. **1951** R. T. WILCOX *Mode in Furs* vii. 157 Such peltries as bear, lynx, fox, wolf and goat were popular though lamb, civet cat and leopard are noted too [in the early 20th century]. *Ibid.* 208 (*caption*) Hooded circular cape of Somali leopard. **1973** E. MCBAIN *Let's hear It* iii. 44 'My good jewelry.. [has] gone.' 'Anything else?' 'Two furs. A leopard and an otter.'

b. Delete † and substitute for def.: *attrib.* or quasi-*adj.* Made of leopard skin or material resembling leopard skin.

1938 M. BACHRACH *Selling Furs Successfully* ix. 100 'This Leopard coat is rather heavy' is sometimes remarked by customers. **1951** R. T. WILCOX *Mode in Furs* vii. 199 (*caption*) Leopard jacket belted with dark blue antelope—leopard gloves with antelope palms. **1958** *Listener* 28 Aug. 316/3 Scowling Continental 'helps' in leopard slacks. **1974** *Times* 11 Nov. 28/7, 1 sable, skins worked down; 1 absolutely beautiful dark leopard coat. Both made by top furriers.

6. a. *leopard spot*; *leopard-coloured* (examples), *-spotted* adj.; **leopard-man,** a member of a leopard society (see below); **leopard-skin** *attrib.*, made of leopard skin; resembling a leopard skin in appearance; mottled; **leopard-skin chief, priest,** among the Nuer people of East Africa, a mediator or arbitrator who settles disputes (so called from the leopard skin which by custom he wears); **leopard society,** in West Africa, a native secret society whose members dress as leopards and attack their victims in the manner of leopards.

1847 EMERSON *Poems* 73 Gayest pictures rose to win me, Leopard-coloured rills. **1889** W. B. YEATS *Wanderings of Oisin* 78 Or in autumnal solitudes Arise the leopard-coloured trees. **1929** F. W. BUTT-THOMPSON *W. Afr. Secret Soc.* xiv. 283 *Tongo-players*, the Sierra Leonean society..said to have been started about the Eighties..as an organisation of leopard-men hunters. **1936** G. GRIFFIN tr. *Schebesta's My Pygmy & Negro Hosts* iv. 67, I think that I have been the first to obtain any detailed information about these 'Anyoto'—the dreadful 'leopard-men'. **1973** G. GALE in Johnson & Gale *Highland Jaunt* II. iv. 143 He now was happy..telling the bar about the Leopard Men in West Africa. **1895** F. B. & W. H. WORKMAN *Algerian Memories* x. 93 Besides the oasis of Biskra..a number of others were visible, the dark colour of which, contrasting with the lighter hues of the plain, gave the leopard-skin appearance. **1929** E. SITWELL *Gold Coast Customs* 8 Courie shells..outline The leopardskin musty Leaves. **1975** *Times* 25 Feb. 6/7 Bagpipers of the Royal [Nepalese] Army in leopard-skin gaiters. **1940** E. E. EVANS-PRITCHARD *Nuer* iv. 190 There is no central administration, the leopard-skin Chief being a ritual agent whose functions are to be interpreted in terms of the structural mechanism of the feud. **1956** —— *Nuer Relig.* iv. 110 In this particular ceremony several groups were opposed to each other, and the leopard-skin priest was acting in his priestly capacity as mediator between them. **1959** G. D. MITCHELL *Sociol.* v. 89 If one man kills another he will go immediately to a person known as a leopard-skin chief... He is in no sense a chief but rather a person who performs certain ritual acts. **1915** K. J. BEATTY *Human Leopards* i. 6 To deal with this extraordinary class of crime the Government of the Colony of Sierra Leone decided that drastic and exceptional legislation was necessary, and a Bill entitled the Human Leopard Society Ordinance, 1895, was introduced and passed. **1929** F. W. BUTT-THOMPSON *W. Afr. Secret Soc.* i. 20 Most of the criminal associations are 'animal' societies... They include Alligator, Baboon, Boa, Leopard, Panther societies. **1968** *Encycl. Brit.* XIII. 975/2 There were many leopard societies, of which the most renowned was the *anyota* society of the Bali tribe, eastern Congo. **1939** T. S. ELIOT *Old Possum's Pract. Cats* 13 Her coat is of the tabby kind, with tiger stripes and leopard spots. **1972** *Times* 23 Nov. 8/2 The presence of communist cadres within Government-held areas could produce more 'leopard spots', to use the accepted phrase, than the map [of S. Vietnam] suggests. **1931** V. WOOLF *Waves* 239 Different lights fall, making the ordinary leopard-spotted and strange.

b. leopard frog *U.S.*, a green frog with black markings, *Rana pipiens*; **leopard lily** orig. *U.S.*, a name used for several spotted lilies, esp. *Lilium pardalinum* (cf. *panther-lily* (PANTHER 5)); **leopard-spotted goby,** a small brown goby with orange spots, *Gobius forsteri*, found close to the shore in parts of the western coast of Britain and France; **leopard-tree** *Austral.*, a name for either of two species of *Flindersia*, *F. maculosa* or *F. collina*; also used for the South American tree *Cæsalpinia ferrea*; **leopard-wood,** (*b*) *Austral.* = *leopard-tree.*

1839 D. H. STORER in Storer & Peabody *Rep. Fishes, Reptiles & Birds Mass.* 237 *Rana halecina*..[is] better known in this state as the leopard frog from its ocellated appearance. **1840** THOREAU *Jrnl.* 16 June in *Writings* (1906) VII. 141 Twelve hours of genial and familiar converse with the leopard frog. **1948** *Sierra Club Bull.* (San Francisco) Mar. 140 Migration is a part of the story of the American merganser, hibernation of the leopard frog. **1973** *Sci. Amer.* Oct. 26/3 The leopard frog (*Rana pipiens*) is particularly susceptible to a kidney carcinoma. **1902** *Out West* Sept. 349 The leopard-lily lights the heather dun. **1938** J. H. MCFARLAND et al. *Garden Bulbs* 136 Lilium pardalinum. Sometimes called the Western Tiger Lily, this highly esteemed California native also has the common names of Leopard Lily and Panther Lily. **1949** H. MOLDENKE *Amer. Wild Flowers* 323 A great favorite of the Southeast is the leopard lily or pine lily, *L. catesbaei*, found in pinelands and acid swamps on the coastal plain from North Carolina to Florida and Louisiana. **1969** HAY & SYNGE *Dict. Garden Plants* 318/2 [*Lilium*] *pardalinum* Leopard Lily. Summer. Fl[ower] turkscap, orange flushed and spotted with red or maroon, pendulous. **1959** A. HARDY *Fish & Fisheries* x. 212 Mr. P. G. Corbin..is naming it after its discoverer, *Gobius forsteri*; it will also be known by the English name of leopard-spotted goby. **1971** *Nature* 30 Apr. 581/1 Closer examination should reveal the presence of the leopard-spotted goby along the Scottish west coast. **1927** *Austral. Encycl.* I. 474/2 *F[lindersia] maculosa* (Leopard Tree, so called from its spotted trunk) is a small tree (20–30 feet), found in the dry interior. **1933** *Bulletin* (Sydney) 20 Sept. 20/2 The leopard tree starts as a straggly, spiny bush, from the centre of which the stem shoots up. **1965** *Austral. Encycl.* V. 288/2 Leopard-tree, a name used for two species of *Flindersia*—the graceful inland *F. maculosa*, which has spotted bark, and the tall rain-forest species *F. collina* (broad-leaved leopard tree or leopard ash). The South American tree *Caesalpinia ferrea*, much grown as an ornamental flowering and shade tree in coastal Queensland, is also called leopard-tree and leopard-wood. **1888** F. M. BAILEY *Queensland Woods* 76 *F[lindersia] maculosa*... Spotted tree or leopard-wood... Wood bright yellow, nicely marked. **1911** C. E. W. BEAN *'Dreadnought' of Darling* xv. 140 It seems a wonder that Australians on the coast do not make a much bigger use of these delicate Western trees for their gardens, especially the leopard-wood. **1936** F. CLUNE *Roaming round Darling* xviii. 177 Spotted a splendid leopard-wood, reputed to attract lightning more than any other tree.

Leopardian (lī,ǒpā·ɹdiăn), *a.* [f. the name of Count Giacomo *Leopardi* (1798–1837) + -IAN.] Of, pertaining to, or characteristic of the Italian poet and scholar Leopardi, or his works.

1881 *Fraser's Mag.* XXIV. 571 In England we have had as yet no notice of the flood of Leopardian recollections, memoirs, and posthumous correspondence that has recently appeared in Italy. **1934** *Times Lit. Suppl.* 21 June p. xi/1 This return to the Leopardian tradition in the more recent poets has been one of the most striking developments. **1947** *Horizon* Apr. 195 Articles and books have been written on 'Leopardian optimism'. **1970** I. ORIGO *Images & Shadows* viii. 181, I remember telling the distinguished Leopardian scholar and critic, Giuseppe de Robertis..that..I was just beginning a second life of the poet. He began to laugh. 'I see that you have caught it, too,' he said, '*il vizio leopardiano*.'

leotard (lī·ǒtāɹd). [The name of Jules *Léotard* (1830–70), French trapeze artist.] A close-fitting one-piece garment worn by acrobats and dancers; a similar fashion garment. So **le·otarded** *a.*

1920 J. W. MANSFIELD *Let.* Jan. (MS. in G. & C. Merriam Co. files), Leotards..are used by acrobats and aerial performers. **1930** *Theatre Arts Monthly* Jan. p. viii/3 (Advt.), The improved Nat Lewis leotards. Lovely, yet sturdily constructed for hard usage. **1939** ADELER & WEST *Remember Fred Karno* ii. 39 The gymnasts' costume worn by Westcott consisted of the classic leotard, a sort of vest specially designed to leave the arms free, spangled neckpiece and trunks. **1953** *Ballet Ann.* VII. 66/1 The simplest of costumes—white *tutus* for the girls, black *leotards* for boys. **1957** *N. Y. Times Mag.* 3 Mar. 42/1 (Advt.), Low and behold, the *leotard*..the shape they said could never be built into a corselette. **1957** *Life* 12 Aug. 91/2 (*caption*) Short skirt worn over striped leotards. **1957** *Vogue* 15 Aug. 42/1 (Advt.), Worsted knit leotard pants. **1958** *Daily Express* 8 Aug. 2/7 Leotards will be the rage with teenage girls this autumn—ballet tights made of stretch nylon. **1966** T. PYNCHON *Crying of Lot 49* iii. 63 One of the girls, a long-waisted, brown-haired lovely in a black knit leotard. **1969** *Sears Catal.* Spring/Summer 21 Swimsuit. Knit of stretch nylon. Popular one-piece styling takes added fashion interest with its smart leotard look. Suit can also be worn as a leotard. **1972** *Listener* 20 Jan. 93/3 Leotarded, limbs akimbo. **1972** *Village Voice* (N.Y.) 1 June 40/4 Her dance, more leotarded than veiled, was attitude rather than movement.

Léoville (leovil). [Fr.] A red wine from any of three vineyards in the commune of Saint-Julien, district of Haut-Médoc, department of Gironde, France.

[**1833** C. REDDING *Hist. Mod. Wines* v. 149 St. Julien de Reignac..is the eighteenth commune of the Medoc vine country... The inferior growths of La Rose and Léoville are the produce of this commune.] **1875** TROLLOPE *Prime Minister* (1876) I. x. 155 'Oh yes, I remember the wine. You call it '57, don't you?' 'And it is '57;—'57, Léoville.' **1903** H. JAMES *Ambassadors* III. 73 Another degustation of the Léoville, another wipe of his mustache. **1966** H. YOXALL *Fashion of Life* xxv. 238 The Léovilles, it seems to me, are the characteristic wines of St Julien.

Lepcha (le·ptʃă), *sb.* and *a.* Also **Lapcha.** [Native name.] **A.** *sb.* A member of a Mongoloid people, native to Sikkim; the Tibeto-Burman language of this people. **B.** *adj.* Of or pertaining to this people or its language.

1819 F. HAMILTON *Acct. Kingdom Nepal* II. i. 118 The most eastern principality, in the present dominions of Gorkha, is that of the Lapchas, called Sikim. *Ibid.* 125 At this custom-house or mart is a Lapcha collector. **1839** *Jrnl. Asiatic Soc. of Bengal* VIII. 624 These neighbours of the hills are the Limboos, Kerantis, Lepchas, Murmis, and Bhotias. **1840** *Ibid.* IX. I. 393 Hill tribes,.. whose language, exhibiting a mere dialectic difference from the Lepcha, may be expressed in symbols not dissimilar. **1848** J. D. HOOKER in L. Huxley *Life J. D. Hooker* (1918) I. xiii. 256 The Lepchas or mountaineers of Sikkim I like extremely. **1862** H. DE SCHLAGINTWEIT et al. *Results Sci. Mission India & High Asia* II. 268, I had with me natives from a great variety of tribes, Górkhas, Kerántis, and Neváris from Nepál, and Límbus, Lépchas, and Bhútias from Síkkim. **1877** E. L. BRANDRETH in *Jrnl. R. Asiatic Soc.* X. I. 10 These determinatives are generally affixed in the languages of Nepál and in the Dhimal language; prefixed in the Lepcha language, [etc.]. *Ibid.* 15 In Lepcha, also, not only the adjective, but the demonstrative pronoun, as in Tibetan, follows the substantive. **1912** A. GORDON *Life A. H. Charteris* xiv. 339 The aboriginal Lepchas, a gentle race, are devil worshippers. **1940** F. S. CHAPMAN *Helvellyn to Himalaya* iv. 70 We wandered downhill to Dikchu...Most of the people here were sallow-faced Lepchas. **1948** D. DIRINGER *Alphabet* vi. 356 The Lepcha character seems to have been invented or revised by the Sikkim raja. **1965** *Evening Standard* 17 Sept. 6/2 The Sikkimese are Buddhists, and ethnically consist of Bhutias from neighbouring Bhutan; of Tibetans and of the neighbouring Lepchas. **1973** *Times* 12 Apr. 8/6 The Bhutias asserted themselves over the Lepchas, the aboriginal inhabitants, with whom they are now more or less integrated.

leper, *sb.*² and *a.* Add: Now often avoided in medical use because of its connotations. (Further examples of the *sb.*)

1948 R. G. COCHRANE in *Leprosy Rev.* XIX. 39, I feel that it is necessary for me to launch a protest at the constant use of the word 'leper' in medical literature. *Ibid.*, The Conference of the Leonard Wood Memorial held in Manila in 1931 recommended that the word 'leper' should not be used, but I fear this recommendation is completely ignored even by those leprologists who attended the conference. **1964** *Observer* 8 Nov. 33/5 To use the word 'leper' as a synonym for 'untouchable' is to perpetuate the ignorance and prejudice of former days. **1970** *Daily Tel.* 11 May 12/7, I regret to see that your columns have again been defaced twice by the word 'leper'... Its use has been banned by such..bodies as the World Health Organisation and the International Leprosy Association. **1970** *Ibid.* 21 May 18 Those most entitled to say whether they suffer more from the disease or from the stigma attached to the word 'leper' are the patients themselves. The campaign against the word was started by patients in the famous American leprosarium in Carville. **1974** *Ibid.* 30 Jan. 16 Like some friendly leper, Mr Mick McGahey, the Communist vice-president of the National Union of Mineworkers, moves among Labour politicians spreading terrified unease with every jovial slap on the back. **1974** *Times Lit. Suppl.* 15 Mar. 263/4 In 1941 the reappearance of leprosy led to the prohibition of the movement of Aborigines below the 20th parallel—the 'leper line'. **1975** *Daily Tel.* 11 Apr. 3/1 He decided to dedicate his life to the Indian people, and for many years was in charge of 13 hospitals, two for lepers.

lepido-. lepidocrocite, substitute for def.: a red to reddish-brown hydroxide of iron, FeO(OH), which is found as scaly or fibrous orthorhombic dipyramidal crystals, often in association with goethite, $HFeO_2$ (with which it was formerly identified); (add further examples); **lepidotri·chium** [Gk. θρίξ, τριχ- hair], (usu. in pl. **lepidotri·chia**) in most teleost fishes, the bony rays supporting the outer part of the fins.

1919 *Amer. Jrnl. Sci.* XLVII. 322 Optical studies of the reddish, scaly crystals called rubinglimmer and lepidocrocite, led Lacroix to propose that they be classed together as lepidocrocite and separated from goethite, with which they had formerly been identified. **1944** C. PALACHE et al. *Dana's Syst. Min.* (ed. 7) I. 644 The name goethite properly belongs to the species here described, since the name was originally given by Lenz to the material from Eiserfeld now known to be lepidocrocite. Goethite, however, is in general use to designate the compound $HFeO_2$. **1951** *Amer. Mineralogist* XXXVI. 31 Lepidocrocite has the hydrogen atom in a discrete OH group... Hence the decomposition of lepidocrocite occurs at a lower temperature than goethite. **1967** *New Scientist* 13 July 92/3 Professor Lowenstam has discovered that some chitons have a second iron mineral in their teeth—lepidocrocite. **1971** R. J. GETTENS *Two Early Chinese Bronze Weapons* iv. 23 The iron blade which served as the cutting edge has been drastically altered by corrosion... X-ray powder diffraction analysis of samples of rust showed that it consists of two hydrated iron oxides, goethite..and lepidocrocite. **1904** E. S. GOODRICH in *Q. Jrnl. Microsc. Sci.* XLVII. 472 In the majority of Teleostean fish the median and paired fins are covered with a smooth scaleless skin, below which lie the dermal fin-rays. For reasons which will appear later, I propose to call these rays the lepidotrichia. **1963** P. H. GREENWOOD *Norman's Hist. Fishes* (ed. 2) iii. 29 The outer part of the fin [of sturgeons] is supported not by horny rays but by bony fin rays, actually modified scales, called lepidotrichia.

lepidopteran, *a.* Add: (Later examples.) Also as *sb.*

1923 J. S. HUXLEY *Ess. Biologist* ii. 96 It is not only the burnt child who dreads the fire (although a study of moths and candles will convince us that 'Lepidopteran' cannot be substituted as subject of the proverb). **1971** *Nature* 13 Aug. 484/1 Similar responses of hymenopteran parasites to the mandibular gland secretions of their lepidopteran hosts may occur in other species. **1973** *Ibid.* 3 Aug. 253/1 One abundant lepidopteran is the southern armyworm, *Prodenia eridania*, a polyphagous species. **1973** PROCTOR & YEO *Pollination of Flowers* iv. 96 The lepidopteran proboscis is very differently constructed from that of the Diptera.

lepidopterist. (Later examples.)

1971 *Daily Tel.* (Colour Suppl.) 8 Jan. 21/2 If she is an amateur lepidopterist..the candids should obviously show her chasing butterflies. **1975** *Sci. Amer.* Jan. 92/3 The first person in the area who is known to have captured a dark-colored peppered moth was an active lepidopterist, R. S. Edleston, who obtained a specimen in 1849.

lepidopterology (le:pidǫptĕrǫ·lŏdʒi). [f. LEPIDOPTERA *sb. pl.* + -OLOGY.] The branch of entomology which deals with the study of Lepidoptera. Hence **le:pidoptero·logist**; **le:pidopterolo·gical** *a.*

1899 *Proc. 4th Internat. Congress Zool. 1898* 232 An important phenomenon that my studies in Lepidopterology have revealed. *Ibid.*, England and English-speaking America possess the greatest number of Lepidopterologists. *Ibid.* 337 Matters other than Lepidopterological must be settled by a general committee of Zoologists. **1921** (*title*) Bulletin of the Hill Museum: a magazine of lepidopterology. **1967** V. NABOKOV *Speak, Memory* (ed. 2) vi. 123 Since the middle of the century, Continental lepidopterology had been..a simple and stable affair. *Ibid.* 129 Among the very few lepidopterological images in English poetry, my favorite is Browning's.

leporicide. Add: **2.** [-CIDE 2.] The killing of hares. *nonce-wd.*

1914 W. DE MORGAN *When Ghost meets Ghost* I. xviii. 200, I..went..dreading that I should find Achilles [a dog] awaiting applause for an achievement in—in leporicide, I suppose.

lepospondyl (lepospǫ·ndil), *sb.* and *a.* [f. mod.L. name of suborder *Lepospondyli* (K. A. Zittel *Handbuch der Palæontologie* (1887–90) I Abth. III. 348), f. Gr. λέπος husk + σπονδύλος vertebra.] An extinct amphibian belonging to the suborder Lepospondyli, distinguished by vertebrae shaped like hourglasses; of or pertaining to an amphibian of this type. Also **lepospo·ndylous** *a.*

1901 H. GADOW in *Cambr. Nat. Hist.* VIII. iv. 79 The vertebræ [of Stegocephali] exhibit three types... I. Lepospondylous and pseudocentrous.—The vertebra consists of a thin shell of bone surrounding the chorda dorsalis. **1902** C. R. EASTMAN tr. *Zittel's Text-bk. Palaeontol.* II. 118 (*caption*) Lepospondylous vertebrae of *Hylonomus.* **1933** A. S. ROMER *Vertebr. Paleontol.* v. 112 In the lepospondyls the vertebral centrum was a single hollow spool-shaped ossification. *Ibid.*, The divergence of the various lepospondyl groups must have taken place at an extremely early date. **1958** C. K. WEICHERT *Anat. Chordates* (ed. 2) xix. 850 Stegocephalia... Fossil forms: labyrinthodonts, lepospondyls, and phyllospondyls, varying primarily in structure of vertebral column. **1971** E. C. OLSON *Vertebr. Paleozool.* vii. 296 (*caption*) A lepospondyl (microsaur) skull. *Ibid.* 297 Like the other lepospondyls they [*sc. Lysorophus* and its relatives] are highly specialized.

lepper (le·pəɹ). A local variant of LEAPER, freq. of horses in hunting parlance.

1907 J. M. SYNGE *Playboy of Western World* III. 66 There you are! Good jumper! Grand lepper. **1920** *Baily's Mag.* Jan. 37/2 There are plenty of 'leppers' at Newmarket this season. **1931** A. J. CRONIN *Hatter's Castle* I. ii. 41 We'll kick off on the leppers. All aboard for the Donegal Hunt. **1937** R. WESTERBY *Wide Boys Never Work* 173 'I put a couple of nicker on Tenderloin. He's a good lepper at that.'..'Got all the jargon, haven't you? "Good lepper"!—strewth!' **1948** MENCKEN *Amer. Lang.* Suppl. II. 362 Among American horse fanciers a jumping horse is called a *lepper.*

leprolin (le·prolin). *Med.* [f. LEPRO(SY + *-lin*, after TUBERCULIN.] = *LEPROMIN.

1904 E. R. ROST in *Indian Med. Gaz.* XXXIX. 168/2 The next thing that was attempted was the manufacture of a leprolin on the lines of the manufacture of tuberculin. **1934** *Brit. Med. Jrnl.* 21 Apr. 703/2 The typical reaction to leprolin..when applied to healthy persons in an area free from endemic leprosy, differs from the intracutaneous tuberculin reaction in remaining for some days negative or doubtful. **1947** R. G. COCHRANE *Pract. Textbk. Leprosy* vii. 64 Neural (N) Type [of Leprosy]... These cases..are of relatively good prognosis as regards life.. and usually react positively to leprolin. [*Note*] Now termed lepromin. **1971** *Internat. Jrnl. Leprosy* XXXIX. 719/2 We deemed it worthwhile to determine the capacity of lymphocytes from leprosy patients to produce lymphotoxin in the presence of PHA and leprolin.

leprologist. Delete *rare* and add later examples.

1948 [see *LEPER *sb.*²]. **1950** G. W. MCCOY in R. L. Pullen *Communicable Dis.* xlii. 631 Promin..has been used in recent years by a group of leprologists working..at Carville, Louisiana. **1961** G. GREENE *Burnt-Out Case* v. i. 156 There is an old Danish doctor..who became a leprologist. **1970** *Daily Tel.* 21 May 18 Dr Browne is one of the leading leprologists in the world.

leproma. Add: Pl. **lepromas, lepromata.** (Earlier and later examples.)

1895 N. WALKER tr. *Hansen & Looft's Leprosy* ii. 5 The leprous nodes [of nodular Leprosy] or nodular Lepromata are of different size and colour; their consistence is at first firm and hard... Their form is usually semi-spherical, but they are often oblong. **1947** R. G. COCHRANE *Pract. Textbk. Leprosy* viii. 81 We are convinced..that the majority of lepromas commence from pre-lepromatous macules..; some, however, develop from the simple macular lesions. **1970** G. J. HILL *Leprosy in Five Young Men* iv. 103 The mixture [*sc.* lepromin] is essentially a sterile emulsion of lepromas—that is, nodules excised from patients with lepromatous leprosy.

lepromatous *a.* (further examples); also, characterized by or exhibiting lepromas: used *spec.* to designate one of the two principal forms of leprosy (see quot. 1938).

1938 *Internat. Jrnl. Leprosy* VI. 390 [Report of the Subcommittee on Classification of the First International Congress on Leprosy.] Objections have repeatedly been raised to both of the current names of the two types [of leprosy] (i.e., 'neural' and 'cutaneous')... It is the opinion of the committee... (*b*) That because 'cutaneous' has proved particularly confusing its use should be discontinued, and replaced by the term 'lepromatous'. *Ibid.*, Lepromatous (L) type.—All cases of the 'malignant' form of leprosy, relatively nonresistant and of poor prognosis, usually negative to leprolin, exhibiting lepromatous lesions of the skin and of other organs, especially the nerve trunks. **1962** *Lancet* 26 May 1116/2 If the altered tissue response could be maintained by repeated B.C.G. vaccination, this might prove a beneficial adjunct to chemotherapy in lepromatous leprosy. **1971** *Nature* 7 May 48/1 Leprosy, caused by *Mycobacterium leprae*, has two clinicopathological forms: lepromatous, associated with impaired delayed hypersensitivity, and tuberculoid, with intact cutaneous reactivity.

lepromin (le·promin). *Med.* Also **-ine.** [a. G. *lepromin* (P. Bargehr 1927, in *Zeitschr. f. Immunitätsf. und exper. Ther.* XLIX. 347): see LEPROMA and -IN¹.] A boiled saline extract of lepromatous tissue. So **lepromin test,** a test involving intradermal injection of lepromin and examination for a nodule at the site (see quot. 1951).

1932 *Monthly Bull. Philippine Health Service* XII. 300 The leprolin (lepromin) used by Mitsuda, Bargehr, de Langen, de Vogel, Mariani, Muir and Hayashi were prepared in different ways. **1940** ROGERS & MUIR *Leprosy* (ed. 2) 248 The lepromin test is thus of use in measuring the natural resistance of the patient to leprous infection. **1951** WHITBY & HYNES *Med. Bacteriol.* (ed. 5) xiv. 267 The lepromin test consists of the intradermal injection of a small quantity of the extract. The test is positive in milder types of leprosy and in many healthy contacts; it is negative in the more severe types of the disease which have a bad outlook. **1959** G. GREENE *Congo Jrnl.* in *In Search of a Character* (1961) 89 Lepromine used to determine the resistance of an undetermined patient. **1973** BRYCESON & PFALTZGRAFF *Leprosy for Students Med.* vi. 62 Infection with *M[ycobacterium] tuberculosis*, immunization with BCG or previous skin testing with lepromin may, but does not necessarily, induce lepromin positivity in a normal person.

leprophil (le·profil). [f. LEPRO(SY + -PHIL.] One who is attracted to sufferers from leprosy. So **leprophi·lia** [Gr. φιλία affection], such an attraction.

1959 G. GREENE *Congo Jrnl.* in *In Search of a Character* (1961) 24 Should one class Father Damien among the leprophils? **1961** —— *Burnt-Out Case* I. ii. 19 You know very well that leprophils exist... Schweitzer seems to attract them. They would rather wash the feet with their hair like the woman in the gospel than clean them with something more antiseptic. **1963** P. WEST *Mod. Novel* II. i. 97 Leprophilia is the extremist answer. **1964** P. FEENY *Fight against Leprosy* ix. 90 The most serious charge..is that he was a 'leprophil', that is, that he fell in love with leprosy... This disease does hold an inverted glamour for a handful of people. *Ibid.* 91 He [*sc.* Damien] rushed into his work on Molokai like a back-row forward rushing into a scrum. He saw suffering and ran to alleviate it... Leprophilia does not enter into it.

leprophobia (leprofōᵘ·biă). Also **lepraphobia** (leprăfōᵘ·biă). [f. LEPRO(SY + -PHOBIA; cf. LEPRA (*b*).] A morbid or insane fear of leprosy; *spec.* such a fear showing itself in the conviction of a person actually healthy that he is suffering from leprosy.

1894 GOULD *Dict. Med.* 670/1 *Leprophobia*, morbid or insane dread of leprosy. **1911** STEDMAN *Med. Dict.*, Lepraphobia. **1948** *Leprosy Rev.* XIX. 40 Euphemisms will not eradicate leprophobia. **1948** E. MUIR *Man. Leprosy* xiv. 98 Leprophobia may centre round any well-known symptom of leprosy. **1973** BRYCESON & PFALTZGRAFF *Leprosy for Students Med.* iv. 40 Do not treat for leprosy unless the diagnosis is established. Nothing is harder to cure than leprophobia.

leprosarium (leprosɛ̄ᵊ·riɒ̆m). Pl. **leprosaria.** [f. L. *lepros(us* leprous + -ARIUM.] A hospital for sufferers from leprosy.

1846 DUNGLISON *Dict. Med. Sci.* (ed. 6) 430/1 An hospital for the reception of the leprous, *Leprosarium*. **1927** *Lancet* 23 July 212/2 (*heading*) The leprosarium at Makogai. **1935** *Nature* 4 May 757/2 A leprosarium being provided at Darwin. **1938** *Internat. Jrnl. Leprosy* VI. 399 The services of an ophthalmologist, a rhino-laryngologist and a dentist should..be made available in all leprosaria. **1966** *New Statesman* 15 Apr. 525/3 (Advt.), We will channel your gifts... *Against disease.* To hospitals, clinics, leprosaria. **1970** [see *LEPER *sb.*² and *a.*].

leprosery, leproserie (leprǫ·səri). [ad. F. *léproserie* (also used) or Sp. *leproseria.*] A leper-house or -colony.

1884 *N.Y. Med. Jrnl.* 6 Sept. 275/2 In many parts of the country [*sc.* Brazil] *léproseries* have been established outside the city walls, to which are consigned all lepers excepting those of the very wealthiest families. **1891** J. L. ALLEN in *Century Mag.* Feb. 592 Mother Marianne would herself have written, but she was called away to the leprosery. **1897** *Dict. Nat. Biogr.* XLIX. 218/1 He founded the leprosery of St. Thomas the Martyr. **1961** G. GREENE *Burnt-Out Case* I. ii. 13 There was a rule that the leproserie should take contagious cases only.

leptazol (le·ptăzǫl). *Pharm.* [f. ANA)LEPT(IC *a.* and *sb.* + AZ(O- + -OL.] A white, crystalline, bicyclic compound, $C_6H_{10}N_4$, which stimulates the respiratory and motor centres, is used as an analeptic, especially after poisoning by narcotics, and was formerly employed in convulsive psychotherapy.

1946 *Analyst* LXXI. 308 Leptazol (pentamethylenetetrazole) is commonly encountered in the form of a 10% w/v solution containing 0·25% of sodium phosphate. **1953** *Brit. Jrnl. Psychol.* XLIV. 58 After an intravenous injection of Leptazol, given during the course of electro-encephalographic studies, he had a generalized convulsion. **1968** W. C. BOWMAN et al. *Textbk. Pharmacol.* xxii. 603 In animals it [*sc.* meprobamate] protects against convulsions which occur after strychnine, leptazol and electric-shock.

leptocaul (le·ptokǭl), *sb.* and *a.* *Bot.* [f. LEPTO- + Gr. καυλός stem, stalk.] A tree having a relatively thin primary stem and branches; also *attrib.* or as *adj.* Hence **lepto·cau·lous** *a.*; **le·ptocauly** *sb.*, development of this type. Cf. *PACHYCAUL.

1949 E. J. H. CORNER in *Ann. Bot.* XIII. 392 Leptocauly. I use this name to indicate the modern tree with relative[ly] slender primary axis and branches in contrast with the pachycaulous cycad. *Ibid.* 393 The leptocaul, or modern tree, thus comes to dominate in height and spread and distribution.., forming the modern forests. **1964** E. J. H. CORNER *Life of Plants* ix. 154 'Leptocaul' (with thin primary stem) denotes the slender willow construction. *Ibid.* 155 Leptocaul plants predominate in temperate and subtropical climates. *Ibid.* (*caption*) Difference between pachycauly and leptocauly as shown by sections of the young twigs of figs. *Ibid.* xv. 275 Some of the herbaceous forms relate, like the banana, directly to the pachycaulous, others to the leptocaulous. **1973** F. EHRENDORFER in V. H. Heywood *Taxonomy & Ecology* xvi. 319 There is evidence for repeated changes from little-branched monopodial and pachycaulous types..to strongly branched, sympodial, leptocaulous, growth-forms. **1974** *New Phytologist* LXXIII. 977 Leptocauls do not become pachycaul on islands.

leptocentric *a.*: see *LEPTOME.

leptocephalus (leptose·făləs). [mod.L. (L. T. Gronovius *Zoophylacium Gronovianum* (1763) I. 135), f. LEPTO- + Gr. κεφαλή head.] The transparent leaf-shaped larva of a fish of the order Anguilliformes, or eels, or one belonging to the genus *Elops* or *Albula*. The larva was first described as a distinct genus; see *leptocephalan, leptocephalid* (LEPTO-), MORRIS *sb.*³

1769 T. PENNANT *Brit. Zool.* III. 125 We communicated it [*sc.* the fish] to that accurate Ichthyologist Doctor Laurence Theodore Gronovius, of Leyden, who has described it in his *Zoophylacium*, under the title of *Leptocephalus*, or small head. **1880** A. C. L. G. GÜNTHER *Introd. Study Fishes* xiii. 179 No instance is more remarkable than that of the so-called *Leptocephali*, which for a long time have been regarded either as a distinct group of Fishes, or as the larval stages of various genera of fishes. **1931** J. R. NORMAN *Hist. Fishes* xvi. 336 The first British *Leptocephalus* was discovered in 1763 by one William Morris near Holyhead. **1971** *Nature* 2 Apr. 278/3 In January 1930, the Danish Dana Expedition captured a leptocephalus on the Agulhas Bank, south of Africa, which was 184 cm long.

leptokurtic (leptokə̄·ɹtik), *a.* *Statistics.* [f. LEPTO- + Gr. κυρτ-ός bulging + -IC.] Of a frequency distribution or its graphical representation: having greater kurtosis than the normal distribution.

1905 K. PEARSON in *Biometrika* IV. 173 Given two frequency distributions which have the same variability as measured by the standard deviation, they may be relatively more or less flat-topped than the normal curve. If more flat-topped I term them platykurtic, if less flat-topped leptokurtic, and if equally flat-topped mesokurtic. **1954** *Brit. Jrnl. Psychol.* XLV. 96 These curves were clearly leptokurtic as well as skewed. **1966** *New Scientist* 28 July 213/1 The leptokurtic or 'peaked' distributions that geodesists often meet with.

Hence **le:ptokurto·sis** [*KURTOSIS], the property of being leptokurtic.

1907 *Phil. Mag.* XIII. 372 There is..sensible skewness and sensible leptokurtosis. **1937** YULE & KENDALL *Introd. Theory Statistics* (ed. 11) ix. 165 By a slip leptokurtosis is there [*sc.* in *Biometrika* (1905) IV. 169 ff.] inadvertently applied to distributions for which $\beta_2 < 3$ [instead of $\beta_2 > 3$]. **1949** DARLINGTON & MATHER *Elem. Genetics* 401 *Kurtosis*, the departure of a symmetrical frequency distribution from the normal by excess (platykurtosis) or deficiency (leptokurtosis) in its shoulders as opposed to tails and centre.

leptology. Add: † **2.** Used for *LEPTONOLOGY. *Obs. rare*⁻¹.

1928 *Amer. Naturalist* LXII. 208 The underlying basis of crystal form is now known... In fact, a complete science of the fine structure of matter—leptology—is being built up as a result of modern physical research.

leptome (le·ptōᵘm). *Bot.* Also **leptom.** [ad. G. *leptom* (G. Haberlandt *Physiologische Pflanzenanatomie* (1884) vii. 229), f. Gr. λεπτ-ός thin + *-OME.] (See quot. 1965.) So **leptoce·ntric** *a.*, having the leptome surrounded by hadrome.

1898 H. C. PORTER tr. *Strasburger's Text-bk. Bot.* 102 The vascular portion is also termed the xylem or hadrome, and the sieve-tube portion the phloem or leptome. **1902** *Encycl. Brit.* XXV. 409/1 The tissue developed to meet the demands for conduction..is known as leptom. **1914** M. DRUMMOND tr. *Haberlandt's Physiol. Plant Anat.* vii. 347 The protein-conducting elements..form..the delicate leptome portion..of the strand... If the leptome has no fibrous sheath, it of course becomes synonymous with phloem. *Ibid.* 349 If the hadrome is central and the leptome peripheral, the bundle may be termed hadrocentric. .. The opposite or leptocentric..condition is exemplified by the leaf-trace bundles in many monocotyledonous rhizomes. **1940** *Chambers's Techn. Dict.* 495/2 *Leptocentric vascular bundle*, a concentric vascular bundle, in which a central strand of phloem is surrounded by xylem. **1965** K. ESAU *Plant Anat.* (ed. 2) xii. 272 The term *leptom* deserves special mention. It refers..to the soft-walled conducting part of the phloem.

leptomonad (leptǫ·mǭnæ̆d). *Zool.* [f. LEPTO- + MONAD.] **a.** = *LEPTOMONAS a. **b.** Any flagellate of the family Trypanosomidæ when existing in an elongated form with a flagellum emerging from the anterior end and arising near a kinetoplast at this end, which form is assumed only in the invertebrate host (and in culture); freq. *attrib.* or as *adj.*

1909 *Jrnl. R. Microsc. Soc.* 362 (*heading*) New leptomonad in muscids. **1931** R. R. KUDO *Handbk. Protozool.* xi. 145 Genus *Leishmania* Ross... In culture the organism develops into leptomonad forms. **1942** D. L. BELDING *Textbk. Clin. Parasitol.* xi. 143 The species of the genus *Crithidia* occur in the leishmanian, leptomonad and crithidial forms. *Ibid.*, The species of the *Phytomonas* pass through both leishmanian and leptomonad stages. **1961** M. HYNES *Med. Bacteriol.* (ed. 7) xxviii. 433 In man leishmaniæ appear as ovoid organisms with no flagella, but in insects and in culture they turn into flagellated leptomonads. **1962** J. D. SMYTH *Introd. Animal Parasitol.* v. 52 Genus *Leptomonas.* The leptomonads are exclusively parasites of invertebrates. *Ibid.* 64 The morphological changes [of leishmanias] within the sandfly gut are simple. In the mid gut they become leptomonad flagellates which multiply rapidly, spreading forwards to enter the oesophagus and pharynx by the fourth or fifth day. When introduced into the mammalian skin by a bite, the flagellates become rounded and assume the leishmanial form.

leptomonas (leptǫ·mǭnæs). *Zool.* Pl. **leptomonas.** [mod.L., f. LEPTO- + *-MONAS.] **a.** Any flagellate of the genus *Leptomonas* (family Trypanosomidæ), which comprises species parasitic in invertebrates (esp. in the alimentary tract of insects) and existing in both leptomonad and leishmanial forms. **b.** = *LEPTOMONAD b; freq. *attrib.* or as *adj.*

[**1880–1** W. SAVILLE-KENT *Man. Infusoria* I. 243 Genus IV. *Leptomonas*, S.K... Animalcules free-swimming, persistent in shape, elongate fusiform or aciculate, bearing a single long undulating flagellum at the anterior extremity, no distinct oral aperture yet detected. The above generic title combined with the following specific one is here introduced for the reception of the monoflagellate animalcule figured and briefly described..by O. Bütschli.] **1926** C. M. WENYON *Protozool.* I. 312 They are the true trypanosomes typically seen in the blood of vertebrates or their invertebrate hosts: the leptomonas, crithidia, and herpetomonas, which have only an invertebrate host..; the leishmania, which..have both a vertebrate and an invertebrate host..; and the phytomonas, which have both an invertebrate and plant host. *Ibid.* 319 Flagellates of the genus *Leishmania* resemble those of the genus *Leptomonas* in having only the leishmania and leptomonas forms. **1931** BLACKLOCK & SOUTHWELL *Guide Human Parasitol.* ix. 71 In *Phlebotomus argentipes* which has been fed experimentally on infected persons, the leishmania develop in the gut into leptomonas. **1942** J. T. CULBERTSON *Med. Parasitol.* ix. 86 The organisms of the genus *Leishmania*..have in their development only two stages: a leishmania and a leptomonas. **1971** BECK & BARRETT-CONNOR *Med. Parasitol.* iv. 38/2 *Phlebotomus* flies, while feeding, regurgitate leptomonas forms..into the wound.

lepton² (le·ptǫn). *Nuclear Physics.* [f. Gr. λεπτ-ός small, slight, slender + *-ON¹.] Any of the sub-atomic particles that do not participate in the strong interaction and have a mass less than that of a nucleon and a half-integral spin (viz. the electron, the muon, and the neutrinos, and their anti-particles); in recent use extended to include any other (hypothetical) particle, of whatever mass, which does not participate in the strong interaction. (Orig. introduced in a wider sense: see quot. 1948.)

1948 L. ROSENFELD *Nucl. Forces* p. xvii, This can be achieved..by postulating a special kind of interaction between a nucleon and a pair of light particles, or leptons, consisting of an electron and a neutrino. [*Note*] Following a suggestion of Prof. C. Møller, I adopt—as a pendant to 'nucleon'—the denomination 'lepton' (from λεπτός, small, thin, delicate) to denote a particle of small mass, irrespective of its charge. **1959** *Sci. News* LII. 101 A similar conservation law holds for leptons, a group of particles including neutrinos, electrons, and muons. **1964** *Cambr. Rev.* 24 Oct. 53/1 The..equality of the Fermi coupling constant for many different particles, baryons, mesons and leptons. **1967** *New Scientist* 8 June 578/1 These, the leptons, consist of the electron, the muon, two types of the massless, chargeless neutrino..and the anti-particles of these four. **1969** R. E. MARSHAK et al. *Theory Weak Interactions Particle Physics* i. 2 The fact that the weak interaction takes place at approximately the same level of strength between leptons.., between leptons and hadrons.., and between hadrons is a major distinguishing feature of the weak interaction. *Ibid.* 7 The names lepton, meson, and baryon were originally invented to denote light, intermediate mass, and heavy particles respectively, but we now know that the particles in each of the three classes share certain important properties in common; both the leptons and baryons are fermions..whereas the mesons are bosons. However, of greatest importance is the presence of the strong interaction for the meson and baryon classes but not for the lepton class. **1971** *New Scientist* 17 June 669/3 If heavy leptons..exist, Pontecorvo argues that they must have neutrinos also. **1973** *Sci. Amer.* Oct. 111/2 There may well exist a spectrum of leptons of increasing mass. The heavy leptons, if they exist, could be produced, like muons, in pairs in electron-positron annihilation reactions. *Ibid.* Nov. 36/2 The few particles that feel only weak or electromagnetic forces are classed as leptons. Electrons, muons and neutrinos are the only known leptons.

b. lepton number, a quantum number assigned to sub-atomic particles that is ±1 for leptons and 0 for other particles and is conserved in all known interactions.

1958 *Physical Rev.* CX. 1483/2 Such selection rules..consist of giving opposite lepton numbers to μ^- and e^-. **1967** L. M. LEDERMAN in E. H. S. Burhop *High Energy Physics* II. 342 Conservation of lepton number now is a matter of N_μ and N_e being separately conserved. **1973** *Sci. Amer.* Aug. 33/2 The preference of the neutrino for the electron and of the antineutrino for the positron has been included in the theory of weak interactions by assigning a..lepton number to the various weakly interacting particles. The convention is that the electron and the neutrino have a lepton number of +1, whereas their anti-particles, the positron and the antineutrino, have a lepton number of −1.

Hence **lepto·nic** *a.*, of, pertaining to, or involving leptons; *leptonic number* = *lepton number.*

1957 *Ann. Physics* II. 422 Particles labelled by leptonic charge and electrical charge permit a complete identification with the known leptons. **1958** *Proc. 2nd Internat. Conf. Peaceful Uses Atomic Energy* (United Nations) 53 The law of conservation of leptons..states that if a leptonic number is assigned to each particle then the sum of leptonic numbers must be conserved in all reactions. **1969** *Nature* 23 Aug. 780/2 The Λ hyperon has leptonic decays of a very similar kind to those of the Σ.

† **leptono·logy.** *Obs. rare.* [ad. G. *leptonologie* (F. Rinne 1916, in *Neues Jahrb. f. Min., Geol. u. Paläont.* II. 48), f. Gr. λεπτός (neut. λεπτόν) small, slight, slender: see -OLOGY.] (See quot.)

1917 *Jrnl. Chem. Soc.* CXII. II. 166 Alongside the science of stereochemistry is developing a stereophysics, and these, together with the study of crystal structure, form a new branch of science, which the author terms 'Feinbaulehre der Materie' (the study of the ultimate structure of matter), or Leptonology.

leptosomic (leptosōᵘ·mik), *a.* [f. Gr. λεπτό-ς fine, small, thin + σῶμ-α body: see -IC.] In Kretschmer's system, designating a type of physique characterized by leanness and tallness. Also **leptosoma·tic** *a.*, in same sense.

Orig. used by Kretschmer as a synonym of asthenic (*ASTHENIC *a.* b) in his tripartite classification of human physique into pyknic, athletic (*ATHLETIC *a.* 3), and asthenic types. Later he employed a bipartite classification in which only pyknic and leptosomic types were recognized.

1936 E. MILLER in W. J. H. Sprott tr. *Kretschmer's Physique & Character* (ed. 2) 272 Kolle..[discovered] the existence of many cases with leptosomic physique who suffered from cyclothymia in one form or another. **1937** Leptosomatic [see *ATHLETIC *a.* 3]. **1959** *Chambers's Encycl.* XI. 335/1 The two characteristic types of physique

were termed the asthenic, or leptosomatic, and the pyknic respectively. **1960** J. COMAS *Man. Physical Anthropol.* vi. 341 E. Schreider believes that Kretschmer's classification can be reduced to two bipolar types, pyknic and leptosomic, with the inclusion in the latter group of the following varieties: the asthenic, the true leptosomic and the athletic. It should be mentioned that in later editions of his book, Kretschmer abandoned the third (athletic) type, and fell back upon a dichotomy consisting of the pyknic and asthenic (leptosomic).

Hence **le·ptosome** *sb.*, (a person with) a leptosomic physique; also as *adj.*, = *LEPTOSOMIC *a.*

1931 *Times Lit. Suppl.* 10 Dec. 1004/2 The two main classes of white man, which he calls 'linear' and 'lateral' (corresponding to the 'pyknic' and 'leptosome'..of other anthropologists). **1935** *Nature* 9 Feb. 236/1 Of Kretschmer's three types, the leptosome corresponds to the Nordic, the athletic to the Dinaric and the pycnic to the Mediterranean and the Alpine. **1960** J. COMAS *Man. Physical Anthropol.* vi. 341 According to the Italian biotypologists, the average or normal type is located between the leptosomes (slender type) and the pyknics (broad type). **1971** J. Z. YOUNG *Introd. Study Man* xxxix. 573 Kretschmer tried to force everyone into one of three classes, pyknic (round and fat), leptosome (long and thin), and athletic (broad and strong). *Ibid.* 576 Leptosomes were introverted.

leptospira (leptospəi·ră). *Bacteriology.* Pl. **-spiræ.** [mod.L., f. Gr. λεπτό-ς fine, small + σπεῖρα coil.] Any bacterium of the genus *Leptospira* (family Treponemataceæ), structurally similar to the genus *Spirochæta* and consisting of a few species either free-living or parasitic, of which *L. icterohæmorrhagiæ* is parasitic in rats and the cause of Weil's disease in man.

[**1917** H. NOGUCHI in *Jrnl. Exper. Med.* XXV. 759 It calls for a new genus, and on account of its fine and minute windings, the name *Leptospira* is suggested.] **1918** *Ibid.* XXVII. 588 Figs. 1 to 4 are intended to show the appearance of the leptospiræ in an air-dried specimen. **1922** *Lancet* 18 Nov. 1058/1 Uhlenhuth and Zuelzer record a leptospira (*L. pseudoicterogenes*) in salt springs. **1966** WRIGHT & SYMMERS *Systemic Path.* I. xxi. 629 Leptospirae are demonstrable in the blood or urine in only half the fatal cases [of Weil's disease].

Hence **leptospi·ral** *a.*, of, characteristic of, or caused by leptospiræ; *leptospiral jaundice*, infectious or spirochætal jaundice, Weil's disease.

1924 *Brit. Med. Jrnl.* 23 Feb. 314/1 In sections of the liver..leptospiral forms were abundant. **1935** *Ibid.* 24 Aug. 339/1 Men who have been exposed to risk of leptospiral infection. **1937** *Proc. R. Soc. Med.* XXX. 746 Dr. J. Smith has informed us in a letter that he investigated an outbreak of leptospiral jaundice in a fox farm near Aberdeen, and that three foxes died of the disease. **1960** *Guardian* 10 Nov. 9/4 Leptospiral jaundice (Weil's Disease)..is contracted from the urine of rats. **1973** *Times* 31 Oct. 14/2 Some leptospiral serotypes cause serious disease in Man.

leptospire (le·ptospəiᵊɹ). *Bacteriology.* [Anglicized form of *LEPTOSPIRA.] = *LEPTOSPIRA.

1957 R. S. BREED et al. *Bergey's Man. Determinative Bacteriol.* (ed. 7) 907 Pathogenic leptospires were first isolated from human cases of Weil's disease... Since that time [*sc.* 1915] other leptospires..have been recognized as causing disease in man and other animals. **1969** *New Scientist* 20 Feb. 414/2 The leptospiral structures, too, bud out from the surface of the leptospire itself.

leptospirosis (le:ptospəiᵊrōᵘ·sis). *Med.* and *Vet. Sci.* [f. *LEPTOSPIR(A + -OSIS.] Infection with, or a disease caused by, leptospiræ.

1926 STEDMAN *Med. Dict.* (ed. 9) 558/2 *Leptospirosis*, infection with some species of *Leptospira*. **1934** *Brit. Med. Jrnl.* 7 July 10/1 Weil's disease (spirochaetal jaundice, infective jaundice, leptospirosis) has been recognized in many different countries. **1961** *Times* 6 Dec. 14/6 The bulls..have been tested for tuberculosis, brucellosis, and leptospirosis. **1970** *Daily Tel.* 11 May 12/7 In one area the Medical Officer of Health has made notifiable seven tropical diseases—plague, cholera, malaria, yellow fever, leprosy, typhus, and also leptospirosis. **1973** *Massey Ferguson Rev.* (N.Z.) Mar.–Apr. 8/1 Leptospirosis in man and cattle is recognised as a serious rural problem in parts of New Zealand.

leptotene (le·ptotīn). *Cytology.* [ad. F. *leptotène* (H. von Winiwarter 1900, in *Arch. de Biol.* XVII. 55): see LEPTO- and *-TENE.] The first stage of the prophase of the first meiotic division, in which the chromosomes are apparent as fine slender threads. Also *attrib.* or as *adj.*

[**1900** *Jrnl. R. Microsc. Soc.* 654 The reticulum gives rise to a chromatic thread.., which at first fills the nuclear cavity (leptotænic stage).] **1912** *Jrnl. Exper. Zool.* XIII. 360 The pre-synaptic leptotene. *Ibid.* 362 Are the leptotene-threads of this period chromosomes? *Ibid.* 368 Figs. 73 *a* and 73 *b* show two early leptotene-nuclei of this species. **1925** E. B. WILSON *Cell* (ed. 3) vi. 541 In the case of animals..the leptotene-spireme is not continuous but consists of separate segments. **1964** G. H. HAGGIS et al. *Introd. Molecular Biol.* vii. 197 The chromosomes at leptotene must each consist of two chromatids so closely apposed as to give the appearance of a single structure. **1970** AMBROSE & EASTY *Cell Biol.* x. 325 The cell nuclei at this stage contain chromosomes in the form of very fine single threads, hence the name leptotene, meaning 'slender ribbon'.

lerky (lə̄·ɹki). [dial., of unknown origin.] In the Nottinghamshire area, the local name of a children's game (see quot. 1902).

1902 *Eng. Dial. Dict.* III. 576/1 *Lerky*, a noisy game, played with any old tin; this being placed in a ring, while all except one hide themselves, then rush out if unobserved and kick the tin out of the ring. Somewhat similar to hide-and-seek. **1913** D. H. LAWRENCE *Sons & Lovers* I. iv. 75 Paul was towed round at the heels of Annie, sharing her game. She raced wildly at lerky with the other young wild-cats of the Bottoms. **1969** I. & P. OPIE *Children's Games* iv. 167 The game seems to have been well known in city streets before the First World War... Other names: 'Ecky'..'Kick the Bucket'..'Lerky' (Nottinghamshire).

lerp. *Austral.* Substitute for def.: A sweet secretion, or the scales formed from it, produced by larvæ of jumping plant-lice of the family Psyllidæ on the leaves of eucalypts and other plants. Also *attrib.* (Add later examples.)

1907 W. W. FROGGATT *Austral. Insects* 363 Their popular name of 'Lerp Insects' [comes] from the habit of the larvae of many species of forming 'lerp scales', shell-like protective coverings formed from exudations from the insects. **1945** K. C. MCKEOWN *Austral. Insects* 104 The Psyllidae, or Lerp-insects, form an important group in Australia. *Ibid.* 106 In its immature stages the insect lives as a squat little larva or nymph beneath the lerp-scale. **1962** *Proc. Linn. Soc. New South Wales* LXXXVII. 283 The encyrtid parasites described in this paper form one of the lesser groups of parasites of lerp-forming psyllids on eucalypts. **1965** *Austral. Encycl.* V. 290/2 Lerp-insects, a large and common group of jumping plant-lice..which suggest miniature cicadas... Some of them give themselves, through sugary exudations, protective and often picturesque coverings known as lerp scales. *Ibid.* IV. 479/2 The lerp scales secreted by the larvae are often of beautiful design and characteristic of the species. **1970** T. E. WOODWARD et al. in *Insects of Australia* (Commonwealth Sci. & Industr. Res. Organization) xxvi. 418/2 Lerp formation has probably evolved because of the need to protect the nymphs from desiccation. *Ibid.* 419/1 Most species of *Glycaspis* and *Lasiopsylla* are lerp-builders, but some form large bubble-shaped galls with an orifice at the base plugged with the same waxy or sugary material as is used by other species to build lerps.

Les (lez), **Les.,** colloq. abbrev. of *LESBIAN *a.* 2 and *sb.* Also **Lesie, Lessie, Lessy.** Cf. *LEZ, LEZZ.

1929 M. LIEF *Hangover* 235 'Certainly,' responded the Les, 'where is she?' **1959** C. MACINNES *Absolute Beginners* 52 Jill is a Les. and, what is more, you may not believe this, but a Les. ponce. **1965** L. MEYNELL *Double Fault* I. viii. 71 These Lessies are touchy; they just can't stand it when the girl friend leaves them. **1968** L. BERG *Risinghill* 122 What is a 'Lesie'..? **1972** L. P. DAVIES *What did I do Tomorrow?* vi. 74 There's so much homo and lessie stuff knocking about these days. **1972** *New Society* 11 May 301/1, I reckon she's a les you know. **1973** J. JONES *Touch of Danger* xxxii. 188 They're all leses, those extra-girl types.

Lesbian, *a.* Add: **b.** Short for *Lesbian wine.*

1775 E. BARRY *Observations Wines of Ancients* vi. 99 The best Greek Wines, the Chian, Lesbian, Coan, &c. were equally prepared in the same manner. **1824** A. HENDERSON *Hist. Anc. & Mod. Wines* I. viii. 123 The dessert-wines most commonly mentioned as in use among the Greeks are the Thasian and Lesbian. **1846** R. FORD *Gatherings from Spain* xiv. 163 Manzanilla..may be compared to the ancient Lesbian, which Horace quaffed so plentifully in the cool shade, and then described as never doing harm.

2. (Freq. with small initial.) [After the alleged practice of Sappho, the poetess of Lesbos; cf. SAPPHIC *a.* and *sb.*, SAPPHISM.] Of a woman: homosexual, characterized by a sexual interest in other women. Also, of or pertaining to homosexual relations between women. Also as *sb.*, a female homosexual.

1890 BILLINGS *Med. Dict.* II. 47/1 *Lesbian love*, tribadism. **1892** C. G. CHADDOCK tr. *Krafft-Ebing's Psychopathia Sexualis* v. 429 We are indebted to Parent-Duchalet ..for interesting communications concerning Lesbian love. **1925** A. HUXLEY *Let.* 21 Apr. (1969) 246 After a third-rate provincial town, colonized by English sodomites and middle-aged Lesbians, which is, after all, what Florence is, a genuine metropolis will be lively. **1931** R. CAMPBELL *Georgiad* I. 13 No Lesbian governess had got the start of him. **1933** H. S. WALPOLE *Vanessa* IV. 781 She disliked people to take it for granted that unless she was Lesbian she was uninteresting. **1936** C. DAY LEWIS *Friendly Tree* i. 23, I shall never write real poetry. Women never do, unless they're invalids or Lesbians or something. **1940** 'G. ORWELL' *Inside Whale* 132 Gruff-voiced Lesbians in corduroy breeches..could walk along the streets without attracting a glance. **1947** E. TAYLOR *View of Harbour* x. 170 'I think I look like a Lesbian,' Beth said doubtfully. **1972** *Jrnl. Social Psychol.* LXXXVII. 52, 50 adult female respondents who would define themselves ..as being Lesbian or homosexual or both. **1973** [see *HOMOSEXUAL *a.*]. **1974** *Ms* July 118/1 Testimony from a group of bright articulate lesbians covering: lesbian sexuality, problems of lesbian mothers..and the lesbian lifestyle.

Hence **Le·sbianism,** female homosexuality.

1870 A. J. MUNBY *Diary* 2 May in D. Hudson *Munby* (1972) 283 Swinburne..expressed a horror of sodomy.. and an actual admiration of Lesbianism, being unable..to see that that is equally loathsome. **1895** A. DOUGLAS *Let.* in H. M. Hyde *Trials Oscar Wilde* (1948) 360 Thus in England there are no laws against 'Lesbianism' or intercourse of an erotic character between women, and yet there are several women in London whose friendship with other women does carry a taint and a suspicion, simply because these women are obviously 'sapphic' in their loves. **1897** H. ELLIS *Stud. Psychol. Sex* I. iv. 82 Casanova remarked that the women of Provence are especially inclined to Lesbianism. **1965** [see *HOMOSEXUAL *a.*]. **1971** C. WOLFF *Love between Women* iii. 40 No theory has so far been evolved which deals exclusively with lesbianism.

lesbic (le·zbik), *a.* = *LESBIAN *a.* 2.

1892 D. H. TUKE *Dict. Psychol. Med.* II. 865/2 For many years a whole literature of romance and plays has been occupied in the description of Lesbic love, to the great damage of girls and neuropathic women. **1922** JOYCE *Ulysses* 205 Sons with mothers,.. lesbic sisters, loves that dare not speak their name.

lesbo, colloq. abbrev. of *LESBIAN *sb.* Also **lesbie.**

1940 J. O'HARA *Pal Joey* 175, I am all set to be m.c. in a crib where the Lesbos even come and watch the dress rehearsals. **1969** C. HIMES *Blind Man with Pistol* xiii. 145 'One was a man; a good-looking man at that.' 'Man my ass, they were lesbos.' **1970** S. ELLIN *Man from Nowhere* liv. 271, I don't dig Lesbies.

leschenaultia (leʃēnǭᵘ·tiă). Also **lechenaultia.** [mod.L. (R. Brown *Floræ Novæ Hollandiæ* (1810) 581), f. the name of L. T. *Leschenault* de la Tour (1773–1826), French botanist and traveller + -IA¹.] A herb or evergreen shrub of the Australian genus so called, belonging to the family Goodeniaceæ and bearing red, blue, white, or yellow flowers.

1825 *Curtis's Bot. Mag.* LII. 2600 (*heading*) Handsome Lechenaultia. **1916** L. H. BAILEY *Stand. Cycl. Hort.* IV. 1844/2 The leschenaultias require special care in watering. **1955** A. ROSS *Australia 55* 44 Thick clusters of blue and red leschenaultia. **1966** *Times* 11 Nov. (W. Austral. Suppl.) p. iv/2 The exquisite sky-blue leschenaultia, fragile as gossamer, shows its versatility by appearing, in different regions, in scarlet, crimson, blood-red, yellow, orange and green-blue. **1967** A. M. BLOMBERY *Guide Native Austral. Plants* 279 It [*sc.* the genus] is probably best known by the Blue Lechenaultia of W[estern] A[ustralia]. **1972** *Southerly* XXXII. 18 The coffin draped in scarlet with a single bunch of blue leschenaultia, gathered by her friends in the hills.

Lesch-Nyhan (leʃ,nəi·hăn). *Med.* The names of Michael *Lesch* and William L. *Nyhan* (b. 1926), U.S. physicians, used with reference to a rare hereditary syndrome they described in 1964 which affects young boys (usu. causing early death) and is marked by compulsive self-mutilation of the head and hands, esp. the lips, together with mental retardation and muscular movements of choreiform and athetoid character.

1966 REED & FISH in *Arch. Dermatol.* XCIV. 195/2 Indications are that this syndrome is probably sex-linked. .. Since Lesch and Nyhan first described the signs and symptoms as an entity, the condition should be called the Lesch–Nyhan syndrome. **1969** *Sci. News Let.* 11 Oct. 327 Skin cells from patients with a genetic defect known as the Lesch–Nyhan syndrome lack an enzyme essential for incorporation of purine bases into new nucleic acids. **1975** *Amer. Jrnl. Human Genetics* XXVII. 219 The mothers of Lesch–Nyhan cases will be either heterozygous (+/−) or homozygous normal (+/+). In the latter case the Lesch–Nyhan patient would have received a complete mutation from his mother.

Lesghian (le·zgiăn), *sb.* and *a.* Also **Lesghi(e), Lesghien, Lesgian, Lezg(h)ian, Lezgin.** [ad. Russ. *Lezgin.*] **A.** *sb.* A member of a tribe of the north-eastern Caucasus; also (in earlier quots.), one of a mountain people of Daghestan. Also, the language of these people. **B.** *adj.* Of or pertaining to these people. Also **Le·sg(h)ic** *a.*

1854 MAX MÜLLER *Suggestions in Learning Lang. Seat of War in East* 116 *Lesghic Branch.* Lesghistan, or the country of the Lesghi, also called Daghestan..lies between the rivers Koisu, Alazani, and the Caspian Sea. The Lesghi or Leski..may..be the same as the 'Legae' mentioned by Strabo. *Ibid.* 117 The Lesghians are Mahometans. **1875** C. HENEAGE tr. *M. von Thielmann's Journey Caucasus* I. iv. 280 The name Lesghie, used by the Russians in a general sense, and especially applicable to the inhabitants of Southern Daghestan, does not convey to the latter either the idea of a population or even of a single tribe. **1878** [see FINNIC *a.*]. **1879** *Trans. Philol. Soc. 1877–79* 602 Here..*Leghian* must be struck out, since it denotes no particular tribe or people. **1921** *19th Cent.* May 871 The bon-bons of the new faith were cast indiscriminately among Circassians,..Lesghiens..and Negroes. **1939** L. H. GRAY *Foundations of Lang.* xii. 376 This alleged family has four sub-groups, characterised respectively by a sibilant.., a spirant plus a sibilant.., a spirant (e.g. in..Chechen, Avar, Lesghian..) and a spirant plus a sonant. **1959** B. GEIGER et al. *Peoples & Lang. Caucasus* 38 Lezgian..has the status of a literary language in the Dagestan ASSR. **1968** *Encycl. Brit.* XIII. 1012/2 The Lezgian group of languages includes..the Agul, Tabasaran, Rutul, Tsakhur, Budukh and Dzhek languages.

Lesie, var. *LES.

lespedeza (lespedī·ză). [mod.L. (A. Michaux *Flora Boreali-Americana* (1803) II. 70), blunderingly (by a misreading of the surname) f. the name of V. M. de *Céspedez* (fl. 1785), Spanish governor of East Florida.] A herb or shrub of the genus so called, belonging to the family Leguminosæ, native to North America, Asia, or Australia, and bearing clusters of white, pink, or purple flowers; esp. a plant of this kind used in the southern United States as a hay or fodder crop; also called *bush clover*.

1891 *Garden & Forest* 25 Feb. 88/2 A tall, bushy Lespedeza (*L. Prainii*) is a highly ornamental shrub some ten feet high, bearing large panicles of fine purple-blue flowers. **1900** L. H. BAILEY *Cycl. Amer. Hort.* II. 903/1 There are a number of native Lespedezas. **1929** W. FAULKNER *Sartoris* (1932) 5, I expect every spring to find corn or lespedeza coming up in the hyacinth beds. **1943** J. S. HUXLEY *TVA* vi. 45 Lespedeza and other legumes.. bind soil and provide nitrates. **1965** RIPPER & GEORGE *Cotton Pests Sudan* i. 6 Lespedezas, which come from China and Korea, are now widely used in the South East of the U.S.A. to renew worn out soil. **1975** *Country Life* 13 Feb. 373/1 In the crops of shot birds we found..the seeds of beggar weed, wild peas and lespedeza.

less, *a.* (*sb.*), *adv.*, and *conj.* Add: **A.** *adj.* **1. c.** (Later examples.) Now more frequently found but still regarded as incorrect.

1862 M. D. COLT *Went to Kansas* 84, I may see them all doing with still less comforts. **1873** *Nature* 1 May 15/2 The determination of position in the given manifoldness is reduced to a determination of quantity and to a determination of position in a manifoldness of less dimensions. **1874** *Rep. Brit. Assoc. Adv. Sci. 1873* 53 To return to the history of logarithmic tables to a less number of figures. **1904** *Amer. Jrnl. Philol.* XXV. 234 There might have been less barbed wire, less flaring flowers. **1971** *Guardian* 16 Dec. 16/1 The 47-page prospectus..shows that there are less restrictions..than is generally supposed. **1972** 'E. LATHEN' *Murder without Icing* (1973) ii. 21 You've seen less hockey games than my wife.

4. (Examples of 'minus' sense.)

1869 *Bradshaw's Railway Manual* XXI. 304 Dividends were declared at the rate of 5 per cent. per annum in the preference shares, amounting, less income tax, to 1,218 *l.* **1910** *Chambers's Jrnl.* Oct. 661/1 If I borrow £100..I pay my interest, less tax. **1911** *Rep. Labour & Social Conditions Germany* (Tariff Reform League) III. VI–VII. 20 All meat is sold less the bone. **1930** *Times* 25 Mar. 24/2 A full year's dividend on the Preference Shares, less tax, absorbing £16,800. **1972** *Times* 2 Sept. 18/8 Cost of paint ..Less VAT input tax..£500.

7. b. (Later examples.)

1652 T. GATAKER *Antinomianism* 5 In those words of mine nothing les was intended, then this Autor would.. enforce them to speak. **1836** C. FOX *Jrnl.* 23 Sept. (1972) 32 'A gentleman' was announced, who proved to be nothing less than Professor Sedgwick! **1856** DICKENS *Dorrit* (1857) II. i. xxiii. 79 You couldn't do it when your Uncle George was living; much less when he's dead. **1863** J. A. FROUDE *Hist. Eng.* (ed. 3) II. xi. 298 But Elizabeth meant nothing less than to recall Sidney. **1865** *Daily Tel.* 2 Dec. 7/1 We may rest satisfied that the dispute will end in nothing less than a battle royal. *c* **1874** D. BOUCICAULT in M. R. Booth *Eng. Plays of 19th Cent.* (1969) II. 174 You are not mistress in your own house, much less lady of the manor. **1895** *Bookman* Oct. 22/2 His policy became nothing less than a series of gigantic blunders.

C. *conj.* Delete † *Obs.* and read *Obs.* exc. *U.S. dial.* and *colloq.*

When written *'less* it represents a contracted form of *unless*.

1892 KIPLING *Many Inventions* (1893) 41 'Less you want your toes trod off, you'd better get back. **1900** 'FLYNT' & WALTON *Powers that Prey* 62 If any of 'em knows us they'll beef dead sure, 'less we square 'em. **1929** W. FAULKNER *Sound & Fury* 22 They wont know you got wet... Less me and Jason tells. **1973** *Black World* Apr. 61 Don't no broad be puttin' me down, less I go upside her head. **1973** J. DRUMMOND *Bang! Bang! you're Dead!* xxxvii. 129 You gonna be like Jinx Janeiro soon, less you listen to me.

lessen (le·s'n), *conj. U.S. dial.* Also **less'n.** Unless. Cf. LESS *conj.* (in Dict. and Suppl.).

1881 J. C. HARRIS *Nights with Uncle Remus* (1884) xix. 94 But less'n we gits dat Moon out er de pon', dey aint no fish kin be ketch. **1912** MULFORD & CLAY *Buck Peters* (1921) xxiii. 207 'Ain't that yore pinto?' queried Slick. 'Less'n I'm blind,' agreed the cow-punch. **1929** W. FAULKNER *Sound & Fury* 35 And they aint going to be no luck in saying that name, lessen you going to set up with him while he cries. **1938** M. K. RAWLINGS *Yearling* iii. 23 None o' the dogs bayed him, Pa. Lessen I didn't hear, for sleepin'. **1970** H. WAUGH *Finish me Off* (1971) 133 He's not bothering with girls less'n they're going to do him some good.

lesser, *a.* and *adv.* Add: **3.** *attrib.*, as *lesser breed*, applied allusively after Kipling (*Recessional*: see quot. 1897) to persons of inferior status; *lesser light*, applied allusively (after Gen. i. 16 'the greater light to rule the day, and the lesser light to rule the night') to a person of less eminence or importance.

1897 KIPLING *Recessional* in *Times* 17 July 13/6 Such boasting as the Gentiles use Or lesser breeds without the Law. **1955** C. PEARL *Girl with Swansdown Seat* iv. 135 Wilfrid Scawen Blunt..was an English gentleman who shamed his class by championing the lesser breeds in Egypt and India. **1963** BROWN & FOOTE *Early Eng. & Norse Studies* 39 The next stanza, which begins with the hope that the text will not be miswritten nor mismetred by scribes and lesser breeds without the law of final *-e.* **1971** A. PRICE *Alamut Ambush* ix. 113 The authentic supercilious voice of England—the lesser breeds shall not show unfitting qualities of sportsmanship towards each other! **1973** 'H. HOWARD' *Highway to Murder* vii. 82 You and your kind think you can order the lesser breeds around. **1608** SHAKES. *Pericles* II. iii. 41 None that beheld him, but like lesser lights, Did vaile their Crownes to his supremacie. **1873** R. BROWNING *Red Cott. Nt.-Cap* i. 34 Pilgrimage, Concourse, procession with, to head the host, Cardinal Mirecourt, quenching lesser lights. **1893** G. MOORE *Mod. Painting* 45 In the seventeenth century were Poussin and Claude; in the eighteenth Watteau, Boucher, Chardin, and many lesser lights. **1906** GALSWORTHY *Man of Property* III. v. 328 His interest was soon diverted from these lesser lights of justice by the entrance of Waterbuck, Q.C. **1943** K. TENNANT *Ride on Stranger* xiii. 152 One of the lesser lights of the announcers' staff had his face slapped by a young lady. **1964** *Ann. Reg. 1963* 22 At all events the most senior Ministers..turned up on Saturday 27 April, and the lesser lights on the Sunday morning. **1974** *Times* 21 Jan. 14/7 His deepest affection was reserved for the Romantics—Coleridge, Shelley, Keats, and the lesser lights.

Lessie, var. *LES.

lessive. (Earlier example.)

a **1790** B. FRANKLIN *Works* (1836) II. 104 One [way] is, to soak it [*sc.* the grain] all night in a *lessive* or lye.

lessness. (Further examples.)

1868 A. SANDEMAN *Pelicotetics* 188 The Multiple Test of the greaterness and lessness of the ratio. **1961** R. B. LONG *Sentence & its Parts* iv. 99 Sometimes negation is semantically specialized to convey meanings of 'lessness', as in *I don't have a dime.* **1970** S. BECKETT (*title*) Lessness.

lesson, *sb.* Add: **4. b.** *to read* (one) *a lesson*: see READ *v.* 11 b.

Lessy, var. *LES.

leste (le·ste). [a. Pg. *leste* east wind.] (See quot. 1967.)

1864 *Chambers's Encycl.* VI. 248/2 Sometimes a waft of the *lesté*, or east wind, raises it [*sc.* the temperature in Madeira] to 90°. **1911** *Encycl. Brit.* XVI. 499/2 The Leste is commonly accompanied by clouds of fine red sand. **1967** R. W. FAIRBRIDGE *Encycl. Atmospheric Sci.* 1155 *Leste.* This is a hot, dry, easterly wind occasionally encountered in Madeira and the Canary Islands at any season except in summer. Essentially it is an extension of the harmattan that blows across the Sahara and is accordingly hot and dry... It is similar to the sirocco of the Mediterranean and the leveche of Spain.

let, *v.*[1] Add: **12. e.** *let 'em all come*: a catch-phr. denoting cheerful defiance.

1903 *To-Day* 19 Aug. 99/1 'Let 'em all come,' said Billy Frew, cheerfully. **1909** J. R. WARE *Passing Eng.* 167/2 *Let 'em all come*.., cheery defiance. Outcome of the plucky way in which the British, in the first days of the new year, accepted the message of congratulation by the Emperor of Germany to President Kruger on the repulse of the Jameson raid.

14. a. Also (*U.S. colloq.*) in irregular phr. *let's you and me* (or *you and I*, or *us*): let us (do something).

1929 W. FAULKNER *Sartoris* III. 186 Let's you and I take 'em on for a set. **1950** J. D. MACDONALD *Brass Cupcake* vi. 55 Let's you and me duck out of here. **1953** M. DICKENS *No More Meadows* ii. 123 Let's you and me have a drink together first. **1961** R. B. LONG *Sentence & its Parts* i. 23 In informal *let's us go too* strongly stressed *us* is an appositive which actually repeats its principal, the *us* of *let's*. **1964** MRS. L. B. JOHNSON *White House Diary* 24 Mar. (1970) I. 101 Lady Bird, after this is over, let's you and me go out and have a drink.

18. e. (Earlier and later examples.) Also with following *adj.*

1812 M. EDGEWORTH *Absentee* in *Tales Fashionable Life* VI. xiii. 269, I didn't hide, nor wouldn't from any man living, *let alone* any woman. **1961** R. B. LONG *Sentence & its Parts* xi. 264 The use of adjectives as complements of transitive verbs is quite limited... She isn't even pretty, let alone beautiful. **1966** *Listener* 20 Oct. 569/3, I cannot say that I ever felt anything like twice as old (let alone twice as wise) as my Polish friends. **1974** L. DEIGHTON *Spy Story* ix. 100 He'd never be considered for a high security clearance, let alone a job in the Service.

20*. let drive (see DRIVE *v.* 11).

22. let go. c. Phr. *let it go at that* (see *GO *v.* B. 21 e).

e. (Further examples.) *to let oneself* (or *it*) *go*: (*b*) to neglect one's appearance, personal habits, etc.

1923 A. BENNETT *Riceyman Steps* v. i. 239 Her sole concern..was the condition of the shop. Ought she to clean it, or ought she to 'let it go'? **1960** *Woman* 23 Apr. 17/3 The first step towards 'letting yourself go'. **1963** N. STREATFEILD *Vicarage Family* ii. 20 There is a flower garden. It's been let go rather but I saw some nice rose trees. **1970** G. GREER *Female Eunuch* 186 She tries not 'to let herself go', keeps young-looking. **1971** R. RENDELL *One Across* v. 48 She's made a nice job of my hair, hasn't she? I wouldn't want Ethel to think I'd let myself go.

24*. let slide (see SLIDE *v.* 5 b).

29. let down. b. Add to def.: to fail in supporting, aiding, or justifying (a person, etc.); freq. in phr. *to let the side down*. Also *intr.*, to diminish, deteriorate; to relax. Chiefly *U.S.*

1913 E. C. BENTLEY *Trent's Last Case* xv. 307 That's good. I judged you would not let me down. **1925** W. DEEPING *Sorrell & Son* xii. 110, I leave it to you, Stephen. I know you'll not let me down. **1927** *Daily Express* 12 Dec. 2/4 A boy who lets his group down..is made to feel ashamed of himself. **1952** M. STEEN *Phoenix Rising* i. 22 Why..should she present herself to him against this sordid background..letting down her own side? **1958** 'A. GILBERT' *Death against Clock* x. 137 He couldn't guess his Frau was going to let down the side like that. **1969** *Guardian* 8 Sept. 7/2 'House and Garden' let the side down..by advancing very confident pro-reproduction arguments. **1971** J. TYNDALL *Death in Lebanon* xii. 223 George..let the side down by his boat running out of juice. **1973** D. ROBINSON *Rotten with Honour* 20 You have a way of looking at people as if they're about to let the side down. **1974** N. FREELING *Dressing of Diamond* 99 He's my partner... He doesn't let me down, I won't let him down.

1866 'MARK TWAIN' *Lett. from Hawaii* (1967) 250 This Injun don't seem to know anything but 'Owry ikky', and the interest of that begins to let down after it's been said sixteen or seventeen times. **1870** —— *Lett. to Publishers* (1967) 33, I shall watch this Galaxy business pretty closely, and whenever I seem to be 'letting down', I shall withdraw from literature and recuperate. **1926** *Publishers' Weekly* 20 Feb. 563 Sales are increasing instead of letting down. **1964** MRS. L. B. JOHNSON *White House Diary* 14 Jan. (1970) 45 The meeting had broken up a little after midnight, and after a little letting down.. Lyndon had arrived at 2 A.M. for—shall we call it dinner.

d. (*c*) *intr.* Of an aircraft or its pilot: to descend prior to making a landing. Cf. *LET-DOWN *sb.* 2.

1946 *Shell Aviation News* No. 100. 8/3 Another frequently used system was 'Lorenz' blind approach, which assisted aircraft to let-down in adverse visibility. **1947** *Jrnl. R. Aeronaut. Soc.* LI. 391/2 There remain the periods when so many accidents occur, just after taking off, or when the aircraft is letting down to land. **1958** 'N. SHUTE' *Rainbow & Rose* ii. 42 Over Macquarie Harbour I started to let down. **1971** K. WHEELER *Epitaph for Mr. Wynn* (1972) xxxii. 399 I'll be letting down now... On the ground in ten minutes.

g. Also, *to let* (a person) *down eas(il)y.*

1754 RICHARDSON *Grandison* VI. xxii. 120 It will give him consequence in the eye of the world, and be a gentle method of letting his pride down easy. **1863** *Country Gentleman* 2 Apr. 227/3 The object of these ambiguous expressions is to 'let the applicant down easy'. **1866** *Harper's Mag.* Sept. 537/1 How to have the Colonel transferred, or 'let down easy',..was the question. **1907** [see *COTTON *v.*[1] 8]. **1928** G. B. SHAW *Intelligent Woman's Guide Socialism* lvii. 274 The State..must let the loser down easily; and there is no other way of doing this except the way of purchase and compensation.

k. To lengthen (a garment); to lower (a hem) in order to lengthen a garment.

1890 *Monthly Packet* Christmas 182 Mrs. Thorpe was thinking that Babie's pink frock wanted 'letting down' an inch. **1952** E. COXHEAD *Play Toward* v. 125 They both wore cotton [dresses], Sophia's showing where it had been let down, and Madeleine's dating back to the war years. **1953** K. TENNANT *Joyful Condemned* xxxiii. 316 Philippa was..letting down the hem of Margot's green organdie. **1974** N. FREELING *Dressing of Diamond* 121 She liked to sit and sew; a frock whose hem needed letting down.

l. *to let one's hair down*: see *HAIR *sb.* 8 l.

m. To deflate (a tyre).

1968 M. WOODHOUSE *Rock Baby* ix. 91, I..hoped she hadn't had any bright ideas like letting down my tyres or removing the distributor cap. **1973** 'M. YORKE' *Grave Matters* v. i. 81 There was no trace of a hole in it [*sc.* a tyre] when the wheel was brought in for repair. It must have been let down. Deliberately.

31. let in. a. Also *fig.*, to include; to allow (someone) to share (confidential information, privileges, etc.). Freq. const. *on.*

1904 G. S. PORTER *Freckles* xiv. 307, I guess you'll have to let me in on that, too. You mustn't be selfish, you know. **1910** E. A. WALCOTT *Open Door* xiii. 162 Let me in on the game, Tommy. **1923** L. J. VANCE *Baroque* vii. 39 I'll let you in on a secret. **1928** E. WALLACE *Double* xv. 239 He had been 'let in' by acquaintances on the Stock Exchange to several good things. **1942** BERREY & VAN DEN BARK *Amer. Thes. Slang* § 197/5 *Inform; give inside information*,..let *or* leave in on.

c. (Further examples.)

1932 D. C. MINTER *Mod. Needlecraft* 214/1 Joins can be made decorative by letting in a piping cord down the centre. **1968** J. ARNOLD *Shell Bk. Country Crafts* 257 Handles are constructed by letting in a cane as a foundation.

e. (Further examples of *to let* (*oneself* or a person) *in for*.)

1913 GALSWORTHY *Fugitive* II. 48 Mr Malise, I know what I ought to be to you, if I let you in for all this. **1925** D. H. LAWRENCE *Let.* 29 Dec. (1962) II. 873 We sort of let ourselves in for these things. **1938** E. WAUGH *Scoop* I. v. 86 We've been having a row with you lately. Something about a libel action one of our boys let you in for. **1955** *Times* 29 June 12/6 But never once has she let-in her passengers for a major breakdown in foreign parts, or otherwise far from home. **1971** E. LEMARCHAND *Death on Doomsday* i. 16 You're..spelling out the horrors we've let ourselves in for with appalling clarity. **1973** *Listener* 15 Nov. 661/1 Princess Anne, did you explain to Captain Phillips..what he was letting himself in for?

g. *Motoring.* To engage (the clutch) by releasing one's pressure on the clutch pedal.

1933 D. L. SAYERS *Hangman's Holiday* 173 Mr. Egg acknowledged the courtesy with a wave of his smart trilby, and let his clutch in with quiet determination. **1960** I. JEFFERIES *Dignity & Purity* iv. 61, I let in the clutch and zoomed off. **1968** *Listener* 19 Dec. 811/2 The Fiddler chuckled as he let in the clutch. **1973** C. EGLETON *Seven Days to Killing* xx. 213 He slipped the handbrake, raced the engine and then let the clutch in fast.

h. As *ppl. adj.* (See quots.)

1882 J. SOUTHWARD *Pract. Printing* xxvii. 257 Let-in notes are, as the name indicates, let into the text. **1894** *Amer. Dict. Printing & Bookmaking* 336/1 *Let-in notes*, another term for cut-in notes, or those let into the text, as distinct from side-notes. **1973** *Collins's Authors & Printers Dict.* (ed. 11) 247/2 *Let-in notes*,.. those let into the text, as distinct from side-notes.

32. let off. b. *to let off steam*: see *STEAM *sb.*

c. (Earlier examples.)

1814 J. CONSTABLE *Let.* 25 Oct. in *Corr.* (1964) II. 135 Mr. Roberson our curate was so polite as to ask me to dine with him as he had a party the other day, but I begged to be 'let off'. **1816** JANE AUSTEN *Emma* II. xiii. 244 It will be a good thing over..and I shall have been let off easily.

f. As *sb.* (*d*) (Earlier examples.) (*f*) A release or exemption from punishment or obligation. (*g*) *Rifle-shooting.* The pulling of the trigger.

1836 D. CROCKETT *Exploits & Adventures Texas* 52, I was for backing out and fighting shy; but there was no let-off, for the cock of the village..determined not to stay whipped. **1854** *Punch* 23 Sept. 114/2 A light let-off that will be for the murderer of more than half-a-million! **1864** in *Cricket Q.* (1963) I. 21 He had a couple of let offs. **1876** *Baily's Monthly Mag.* July 45 After this let off, Lord Harris hit in his usual free and dashing style. **1894** P. H. HUNTER *James Inwick* i. 10 He was ahint wi' his rent, and no' like to get muckle o' a let-aff frae the laird. **1913** A. G. FULTON *Notes on Rifle Shooting* 7 Position, holding, aim and let-off can be learned to perfection, and these things are the basis of successful Service rifle shooting. **1932** J. A. BARLOW *Elem. Rifle Shooting* ii. 9 Of the three essentials, holding should be placed foremost in order of importance, aiming next, and trigger pressing, trigger pulling, let-off, or whatever you like to call it, last. As far as possible I have purposely refrained from referring to let-off as trigger pressing. **1960** *Pistol Shooting* (Nat. Small-Bore Rifle Assoc.) (ed. 2) 21 The most difficult thing in shooting is to acquire an invariably good trigger let-off.

33. let on. a. For '*dial.* and *U.S.*' read 'orig. *dial.* and *U.S.*' (Further examples.)

1914 C. MACKENZIE *Sinister St.* II. IV. ii. 862 You'd better not let on you know he used to have a shop of his own. **1923** T. E. LAWRENCE *Let.* 5 Feb. (1938) 399 My private opinion is that she's read it, and he hasn't: and can't: but is much afraid to shock her by letting on. **1946** K. TENNANT *Lost Haven* (1947) xi. 173 Maybe Orry didn't like to let on he'd made a mistake in the first place. **1974** M. INGATE *Sound of Weir* viii. 62 'Would you say that he is very feeble?' 'Stronger than he lets on if you ask me. He don't need t' walk like that.'

b. To pretend. orig. *dial.* and *U.S.*

1822 J. GALT *Provost* xlvii. 354 The Provost maun ken nothing about it, or let on that he does na ken. **1828** *Yankee* (Portland, Maine) 23 Apr. 132/3 [In the South] to et on signifies to make believe. **1846** W. CROSS *Disruption* v. 48 She..had the sense to..let on to be just as ill pleased as her mistress. **1876** 'MARK TWAIN' *Old Times Mississippi* 137 If I wanted to..'let on' to prove what had occurred in the remote past..what an opportunity is here! **1929** *Randolph Enterprise* (Elkins, W. Virginia) 28 Mar. 1/1 We..found out that Mr. Van let on to take the proposal seriously. **1961** *John o' London's* 12 Jan. 41/4 In the positive it [*sc.* let on] means 'pretend' or 'simulate' as in 'He let on to be angry'.

34. let out. a. Also *fig.*, to excuse, to release (from some obligation). Also, to release (the clutch of a motor vehicle).

1869 B. HARTE *Luck of Roaring Camp* 41, I ran the whole way, knowing nobody was home but Jim,—and—and—I'm out of breath—and—that lets me out. **1884** 'MARK TWAIN' *Huck. Finn* vi. 43 They said he [*sc.* the Negro] could *vote* when he was at home. Well, that lets me out! **1922** WODEHOUSE *Girl on Boat* v. 101 But this is splendid! This lets you out. **1945** E. WAUGH *Brideshead Revisited* II. i. 222 Oh, that's quite different. It lets you out completely. **1958** C. WATSON *Coffin scarcely Used* vi. 63 The driver..glanced swiftly behind him before letting out the clutch. **1973** J. LEASOR *Host of Extras* viii. 152, I..let out the clutch and we were off. **1974** 'E. LATHEN' *Sweet & Low* xiii. 129 'They've started casting around for motives.' 'That lets all of us out.'

e. Add to def.: to slacken, *spec.* to increase the width of (a garment) by allowing extra material at the seams; to alter (the seams) in order to increase the width of a garment.

1791 F. BURNEY *Jrnl.* Dec. (1972) I. 97 Miss Cambridge said she thought I was grown fat... I assured her I had been obliged to have a Gown let out, that had been made last year. **1863** QUEEN VICTORIA *Let.* 22 Sept. in R. Fulford *Dearest Mama* (1968) 271 She has increased very much in size—her waist being quite broad and her clothes having all to be let out. **1877** A. SEWELL *Black Beauty* (*c* 1878, ed. 5) xxxiii. 158 When the crupper was let out a hole or two, it all fitted well. **1888** KIPLING *Under Deodars* (1889) 69, I shall say that I am going to Phelps's to get it [*sc.* a frock] let out. **1926** S. T. WARNER *Lolly Willowes* I. 17 Nannie would let out another tuck in Laura's ginghams and merinos. **1953** K. TENNANT *Joyful Condemned* xxxvi. 358 You've grown quite plump... I looked out a dress of mine... I'll let out the seams. **1974** R. HARRIS *Double Snare* viii. 54 'The dress is too tight...' 'Perhaps it could be let out.'

i. (Earlier example.) Also, to drive (a motor vehicle) very fast.

1849 F. PARKMAN *Calif. & Oregon Trail* 427 Let out your horse, man; lay on your whip! **1938** H. G. WELLS *Apropos of Dolores* vi. 321, I don't like these minor tracks. I can't let her [*sc.* his car] *out*. **1968** A. MARIN *Clash of Distant Thunder* (1969) xii. 91 There is a nineteen-kilometer stretch of road..that is almost straight, and I let the Alfa out all the way. **1970** 'D. HALLIDAY' *Dolly & Cookie Bird* iv. 50, I..overtook..the lorry..and then let her right out. I did a ton up that road, and probably more.

35. let up. b. (Earlier and later examples.) Also, to relax.

1787 G. WASHINGTON *Diaries* (1925) III. 185 The Plows, after the rain let up, proceeded to finish this part of field No. 5. **1841** J. F. COOPER *Deerslayer* I. vi. 172 Let up, you painted riptyles—let up! **1933** D. L. SAYERS *Murder must Advertise* v. 94 Never let up! Never go to sleep! **1936** H. HAGEDORN *Brookings* xv. 254 He was resourceful, steady, determined, he never let up on a man; and just wore people out. **1943** J. B. PRIESTLEY *Daylight on Saturday* xxx. 243 Bob's a chap that needs to let up, and now he can't let up... So he's angry inside all the time. **1970** C. MAJOR *Dict. Afro-Amer. Slang* 76 *Let up*, command to restrain from verbally abusing someone. **1974** E. LEMARCHAND *Buried in Past* viii. 138 The girl was..on the brink of tears... 'Why not let up a bit?' he suggested.

c. As *sb.* For '*U.S.*' read 'orig. *U.S.*' Add further examples.

1837 *Congress. Globe* 25th Congress 2 Sess. App. 47/3 There was no let up in the matter: the people had so ordered it, and the gentleman ought to be satisfied. **1841** J. F. COOPER *Deerslayer* I. viii. 240 There's no let-up in an Indian's watchfulness when he's on a war-path. **1891** E. ROPER *By Track & Trail* ix. 125 The snow was falling fast, and there was no appearance of a 'let-up'. **1892** *Eng. Illustr. Mag.* Sept. 884 (E.D.D.), There is no let-up, no change of undergarments, no camp. **1912** BELLOC *Green Overcoat* vii. 141 There was a gap in their conversation... A let up. An interval of repose. **1956** D. D. C. P. MOULD *Celtic Saints* xii. 126 The basis of Celtic prayer..is very simple... The background is ascetic, penitential, and without let-up. **1968** *Globe & Mail* (Toronto) 17 Feb. 23/5 The only let-up in policy is that non-profit groups get it [*sc.* a theatre] cheaper than shows booked by impresarios. **1973** *Nature* 21 Sept. 117/2 There is little prospect of a letup in fuel shortages for the next few years.

letch, see *LECH *sb.*[4]

letch *v.*, see *LECH *v.*

letchwe, var. *LECHWE.

let-down (le·t,daun), *sb.* [f. vbl. phr. *to let down* (LET *v.*[1] 29).] **1.** (See LET *v.*[1] 29 j.) Add further examples.

1933 N. COWARD *Design for Living* II. iii. 68 The human race is a let-down, Ernest; a bad, bad let-down! **1934** J. T. FARRELL *Young Manhood* ii. 41 He was lassitudinous in a mood of let-down, already lonesome for yesterday. **1938** J. STEINBECK *Long Valley* 135 Mike knew it was all over. He could feel the let-down in himself. **1946** W. STEVENS *Let.* 19 Feb. (1967) 523 There is not the..let-down between the two that one finds so often. **1960** M. SPARK *Bachelors* x. 182 And now *she*'s in for a let-down, though she won't admit it. **1971** *Daily Tel.* 4 Nov. 8/4 What intrigues Mr Barstow is the inevitable let-down which is the result of getting older, of getting bored, of finding that passion can flicker out.

2. The descent of an aircraft or spacecraft prior to landing. Cf. *LET *v.*[1] 29 d (*c*).

1945 *Jrnl. R. Aeronaut. Soc.* XLIX. 74/1 Extra fuel to allow for errors in navigation, errors in weather forecasts and technical inabilities to make let-downs and landings through conditions of ice, clouds, or bad ground visibility. **1949** *Flight* 30 June 754/2 From its pilot, Col. Gray, we learned that, on the let-down from 35,000 ft, while the B-29 was still in formation, its port inner engine had overspeeded to 4,500 r.p.m. **1960** 'N. SHUTE' *Trustee from Toolroom* v. 97 The note of the engines changed as the let-down began. **1969** *Guardian* 13 Oct. 18/2 The precision of landing depends predominantly on the accuracy with which the spacecraft's orbital position is known at the moment let-down begins.

3. The action of a cow yielding milk. Cf. LET *v.*[1] 29 h.

1960 *Farmer & Stockbreeder* 5 Jan. 69/3 A time-lag between the end of let-down and stripping is too prevalent. **1965** LEE & KNOWLES *Animal Hormones* ii. 32 Thus the ejection of milk at suckling, or the 'let-down' of milk as it is referred to by agricultural workers, is a neurohormonal reflex.

Also *attrib.* or as *adj.*, in the senses of the vbl. phr. *to let down* or of the *sb.* (see above).

1907 M. C. HARRIS *Tents of Wickedness* II. vii. 193 The next was one of these 'let-down' mornings which everybody must remember having awakened to. **1945** E. BOWEN *Demon Lover* 72 The car was a two-seater, with a let-down hood. **1948** [see *fan marker* (*FAN *sb.*[1] 11)]. **1956** *Nature* 24 Mar. 582/1 In studying the function of the milk gland, with the view of increasing milk production, particular attention has been devoted in recent years to the milk-ejection mechanism (the let-down reflex). **1957** R. H. SMYTHE *Conformation of Dog* vi. 88 The exhibition Greyhound has always been noted for length of tibia, low set-on hocks, well let-down stifles. **1964** *Yearbk. Astron. 1965* 135 At a fixed height the lunar let-down engine will fire to reduce the descent rate and, landing legs having been extended, the vehicle will complete a vertical descent on to the lunar surface, hovering for short periods before making the final touchdown. **1973** M. MACKINTOSH *King & Two Queens* ii. 24, I had talked myself out of the let-down feeling and was determined to make the most of the trip. **1974** P. FLOWER *Odd Job* i. 6 Somebody had called it an escritoire... It had a letdown flap you could write on.

lethal, *a.* Add: **A.** *adj.* **1. c.** *lethal chamber*: also for the destruction of human beings, and *fig.*

1928 'M. HOFFE' *Many Waters* III. ix. 106 It [*sc.* the Bankruptcy Court] lacks of all the glitter of splendid sin; it is simply the paltry lethal chamber of the vanquished. **1933** *Punch* 4 Oct. 366/2 To a modern it is clear that age must be abhorrent. The best modern thought advocates a lethal chamber for all over fifty years of age.

d. *Genetics.* Of an allele or chromosomal abnormality (such as a deletion): resulting in the death of an individual possessing it before the normal span or before sexual maturity, or (if recessive) capable of causing such premature death when homozygous.

1917 *Proc. Nat. Acad. Sci.* III. 620 The difficulty which was experienced in getting pure stock was due to the fact that the chief factor for beaded—Bd'—is lethal, killing all flies homozygous for it. *Ibid.* 620 This remarkable genetic situation, wherein both types of homozygotes are prevented from appearing by the action of lethal factors lying in opposite chromosomes, may be termed a condition of 'balanced lethal factors'. **1939** STURTEVANT & BEADLE *Introd. Genetics* (1940) x. 160 It may be noted at once that the lethal genes ordinarily studied are recessive; a gene with a dominant lethal effect is necessarily lost before it can be studied. **1962** I. H. HERSKOWITZ *Genetics* xxviii. 239 Autosomes II, III, and IV of wild-type flies were individually made homozygous to detect the presence of recessive mutants..that are lethal (causing death of all individuals before adulthood), or semilethal (causing more than 90 and less than 100 per cent mortality before adulthood), or subvital (causing significantly less than normal but greater than 10 per cent survival to adulthood)... About 25% of all autosomes tested this way carried a recessive lethal or semilethal mutant. **1973** K. MATHER *Genetical Struct. Populations* ii. 21 Not all genes that affect viability are, however, completely lethal. Of some 3000 chromosomes tested in *Drosophila willistoni*..over 35% carried genes that were lethal or semi-lethal. About half of the remainder carried other genes affecting the viability of flies homozygous for them.

B. *sb. Genetics.* A lethal allele or chromosomal abnormality (see *A. 1 d).

1917 *Proc. Nat. Acad. Sci.* III. 621 We must therefore believe that lethals are very frequent among recessive factors also. **1926** J. S. HUXLEY *Ess. Pop. Sci.* 59 Sometimes the impairment of vitality is so great that the organism pure for these factors cannot exist at all: such factors are styled lethals. **1934** *Mycologia* XXVI. 360 (*heading*) A lethal for ascus abortion in Neurospora. **1956** [see *DEFICIENCY 1 e].

lethal (lī·þăl), *v. rare.* [f. LETHAL *a.*] *trans.* To kill (animals) painlessly; = LETHALIZE *v.* So **le·thalling** *vbl. sb.*

1922 *Daily Mail* 7 Dec. 6 (Advt.), Cat. Dying from internal injuries..: Lethalled. **1925** *Observer* 27 Sept. 13/6 Proper lethalling establishments where cats can be put to sleep free of charge.

lethality. Delete *rare* and add further examples. Also *fig.*

1912 *Phil. Trans. R. Soc.* B. CCII. 2 The lethality of the venom was determined for frogs, rabbits,..and pigeons. **1930** G. BLAKE *Press & Public* 20 Mr. Baldwin..has lately discovered for himself the dread lethality of the powers I have described. **1953** *Sci. News* XXVIII. 18 Myxomatosis in the European rabbits is unique in its very high lethality. **1957** *Times* 26 Apr., Our early missiles will carry high-explosive warheads. Their lethality will be high. **1958** 'P. BRYANT' *Two Hours to Doom* 81 The radioactive cloud..would retain its lethality for hundreds of years. **1973** D. JORDAN *Nile Green* xxviii. 125 Significant strides have been made in the past two decades on the lethality of conventional and nuclear warheads.

lethally, *adv.* (in Dict. s.v. LETHAL *a.*). Delete † and add later example.

1971 *Nature* 10 Dec. 328/3 Lethally irradiated F_1 hybrid mice.

lethargy, *v.* (Later example.)

1893 F. THOMPSON *Poems* 75 It grew lethargied with fierce bliss.

lethed, *a.* Delete † *Obs. rare*$^{-0}$, substitute *rare*, and add example.

1895 F. THOMPSON *Sister Songs* 2 On the dull earth's lethèd ear.

let-in, *ppl. a.*: see *LET *v.*[1] 31 h; **let-off,** *sb.*: see *LET *v.*[1] 32 f.

let-out (le·t,aut), *sb.* [f. vbl. phr. *to let out* (LET *v.*[1] 34).] **1.** (See LET *v.*[1] 34 k.)

2. An excuse, a justification, a method of avoiding (a difficulty), a release from (an embarrassing situation). Also *attrib.*

1935 M. HODGE *Grief goes Over* in *Famous Plays* 331 That's not the trouble. I am glad she is. It's a let-out. **1942** H. C. BAILEY *Dead Man's Shoes* xx. 81 The open verdict let him out, old boy... He may want a good let out over Clavell. **1955** J. L. AUSTIN *How to do Things with Words* (1962) I. 10 Yet he provides Hippolytus with a let-out, the bigamist with an excuse for his 'I do' and the welsher with a defence for his 'I bet'. **1957** *Economist* 5 Oct. 17/1 And there is no easy let-out for the complacent in the notion that landladies are by definition a narrow-minded lot. **1961** *John o' London's* 17 Aug. 195/2 When we have made all possible allowance for apotropaic amulets..and any other let-out you can think up, we are left with a solid residue of crude, unimaginative..obscene paintings. **1971** C. BONINGTON *Annapurna South Face* xi. 138 'It would be best if you could stay at camp tomorrow and dig a platform for the second box before we arrive.' I was so tired that this sounded an easy let-out, and I agreed. **1971** 'D. HALLIDAY' *Dolly & Doctor Bird* xi. 153 Thanks for the let-out. I was just too bloody nosy. **1973** *Fremdsprachen* XVII. 57/1 It is understood that the proposed agreement would provide for a let-out clause. **1973** *Listener* 13 Sept. 346/3 The really essential problem is the increase of population..of the poor world. This is a very nice let-out for the rich countries.

3. *attrib.* or as *adj.* (See quot. 1954.)

1949 *Amer. Speech* XXIV. 92 A peltry is said to be let in if it is cut and resewed in such a manner as to fashion it broader and shorter than the original. Similarly, it is said to be let out if it is rendered longer and narrower than its first shape. **1954** WEBSTER Add., *Let-out*, of furs, subjected to a method of preparation involving cutting the pelt in strips and reassembling them to form a longer, narrower piece in which good features of color and texture are emphasized. **1967** *Boston Sunday Herald* 7 May III. 3/3 (Advt.), The recent weakness in prices in the fur market enabled us to hand pick choice lush let out natural Mink Jackets.

letovicite (letovi·tsait, -vi·sait). *Min.* [ad. G. *letovicit* (J. Sekanina 1932, in *Zeitschr. f. Krist.* LXXXIII. 117), f. *Letovic-e*, the name of its original locality in Moravia, Czechoslovakia + *-it* -ITE[1].] An acid ammonium sulphate, $(NH_4)_3H(SO_4)_2$, found as colourless prismatic crystals in coal-mine waste-heaps.

1932 *Mineral. Abstr.* V. 145 Associated with this are gypsum and a new mineral (letovicite) as minute colourless pseudo-hexagonal plates. **1968** I. KOSTOV *Mineral.* II. ix. 503 Letovicite, lecontite, and mirabilite..are optically negative.

Letraset (le·trăset). The proprietary name of a system of alphabet transfers used for lettering.

1957 *Trade Marks Jrnl.* 27 Feb. 200/1 Letraset... Transfers. Art & Technics Ltd.,..London,..publishers. **1962** HANSELL & OLLERENSHAW *Longmore's Med. Photogr.* (ed. 7) xliv. 519 For both prints and slides it is a simple matter to produce a positive transparency at the same scale as the intermediate, with the lettering correctly disposed... Instant Letraset is well suited to all these requirements. **1964** *Trade Marks Jrnl.* 11 Mar. 397/1 Letraset Instant Lettering... Transfers (decalcomanias) of letters, numerals and punctuation symbols, in sheet form. Letraset Limited,..London,..manufacturers and merchants. **1966** R. ROBERTS *Typogr. Design* iii. 51 The invention of such hand transfer lettering systems as Letraset. **1971** *Physics Bull.* Sept. 527/2 His diagrams are beautifully drawn..with lettering carried out with stencils or even Letraset.

letshewe, var. *LECHWE.

let's pretend (le·ts prite·nd), *sb. phr.* [f. LET *v.*[1] 14 a + -'S 3 + PRETEND *v.* 3 d.] A game of pretence or make-believe. Also *attrib.*, *transf.*, and *fig.*

1948 *United Nations World* Dec. 92/1 The New Look.. was a 'Let's Pretend' fashion. **1963** *Ingenue* Dec. 45 A Short..wrap of let's-pretend beaver (called Norba). **1964** W. GOLDING *Spire* iv. 85 But there comes a point when vision's no more than a child's playing let's pretend. **1967** *Listener* 21 Sept. 371/1 The dividing line between reality and 'let's pretend'. **1969** N. FREELING *Tsing-Boum* xii. 85 Then—everything was queer then. Double-think and let's-pretend. **1970** *Guardian* 23 Dec. 9/6, I told all the children..that Santa Claus wasn't real but a let's-pretend person.

letter, *sb.*[1] Add: **1. c.** *letter-by-letter* (further examples).

1951 [see *COMPARATIST]. **1961** T. LANDAU *Encycl. Librarianship* (ed. 2) 194/2 *Letter by letter filing*, a method of filing..entries in a list,..the basic principle being that each heading, whether consisting of one word or more, is regarded as one unit. **1964** F. BOWERS *Bibliogr. & Textual Crit.* III. vii. 92 Palaeographical 'explanations' have a marked tendency to take it for granted that every letter must be confused with another, as if the scribe were puzzling out the word in detail, letter by letter... In reprints,.. the letter-by-letter misreading posited by palaeographers is impossible. **1967** COX & GROSE *Organiz. Bibliogr. Rec. by Computer* II. 13 The ability to sort word-by-word or letter-by-letter.

e. *colloq.* (freq. in *pl.*). A university degree or other honour (denoted by its initial letters following the name of the holder). Also (esp. *U.S.*), some other mark of distinction, usu. for achievement in sport, e.g. an abbreviation or monogram representing the name of a college or other institution.

1888 KIPLING *Plain Tales from Hills* 54 It was pleasant to be singled out by a Commissioner with letters after his name. **1915** *Chicago Daily Maroon* 10 June 1/4 The Board of Athletic Control will meet today to award letters to this year's members of the track, baseball, tennis and gymnastic teams. **1951** PARTRIDGE *Dict. Slang* (ed. 4) 1098/2 *Letters*, degree-letters after one's name: coll.: mid-C. 19–20. **1964** MRS. L. B. JOHNSON *White House Diary* 10 June (1970) 163 An Illinois boy, totally deaf, who had won awards in science and in English, and a school letter in wrestling, was scheduled..to become a scientist. **1970** *New Yorker* 23 May 54/1 He had earned his high-school letter in four sports. **1974** L. LAMB *Man in Mist* x. 66 The heads of departments knew me... Letters after their names as long as Dick's hatband.

2. a. (Later examples.)

1854, 1884, 1892 [see *LIFT *v.* 3 g].

4. c. letter of credit: see CREDIT *sb.* 2 c and 10 b; **letter of intent**, a letter or similar document containing a declaration of the intentions of the writer.

1961 WEBSTER, *Letter of intent*, a written authorization enabling officers of the federal government in time of imperative need for war materials and supplies to order the making or furnishing of such materials and supplies before the issuance of a formal contract and providing reimbursement for the contractor's expenses if no contract is subsequently issued. **1970** R. JOHNSON *Black Camels* v. 75 The news of his oil strike reached New York with a copy of the letter of intent he had exchanged with Sheikh Rasul.

d. (See quot. 1825.) *dial.* and *U.S.*

1825 JAMIESON *Suppl.* s.v. *Letter*, a spark on the side of the wick of a candle; so denominated by the superstitious, who believe that the person to whom the spark is opposite will soon receive some intelligence by letter. **1854** B. F. TAYLOR *Jan. & June* 220 Two 'letters' are snuffed from the candles. **1902** in *Eng. Dial. Dict.*

e. An article, report, or the like, setting out the social, political, or cultural, etc., tendencies in a specified place.

1782 M. G. J. DE CRÈVECOEUR (*title*) Letters from an American Farmer; describing certain provincial situations, manners, and customs, not generally known; and conveying some idea of the late and present interior circumstances of the British colonies in North America. **1848** *Manch. Guardian* 22 Nov. 5/1 (*heading*) Letters from London. (From a Private Correspondent.) **1874** GEO. ELIOT *Let.* 7 Jan. (1955) VI. 4 The Parisian letter is nicely done... The single paragraph on the pressure of radiation is worth more than the price of the paper. **1913** W. J. LOCKE *Stella Maris* v. 52 He was..the contributor.. of a weekly London Letter to an American syndicate. **1955** *Radio Times* 25 Apr. 16/3 Letter from America, by Alistair Cooke. **1966** K. MARTIN *Father Figures* ix. 173 Dore was an excellent lobby correspondent, who for many years had written paragraphs about Parliament, mainly for the London Letter. **1974** *Spectator* 21 Dec. 787/3 American letter—The President gets smart—[by] Al Capp.

6*. = *French letter* s.v. *FRENCH *a.* 3.

c **1888–94** *My Secret Life* II. 318 My cock and the letter would not agree. **1916** [see *CAP *sb.*[1] 13 d].

7. a. *letter-boy*, *-file* (later examples), *-form*, *-name*, *-post* (later example), *-sequence*, *-shape*, *-string*, *-tray*.

1816 JANE AUSTEN *Emma* II. ix. 176 A stray letter-boy on an obstinate mule. **1895** *Montgomery Ward Catal.* 40/1 The 'Boss' File is the very best letter file ever sold at the price. **1911** O. ONIONS *Widdershins* 294 The shelf on which I kept my letter-files... My files contained..my agent's letters. **1955** E. POUND *Section: Rock-Drill* lxxxvii. 33 Windeler's vision: his letter file the size of 2 lumps of sugar, But the sheet legible. **1908** *Westm. Gaz.* 22 Aug. 14/1 He looks forward to the invention of letter-forms that will be much simpler and much more legible than the traditional symbols. **1937** *Discovery* Dec. 362/2 The letter-form of the calligrapher. **1963** *Times Lit. Suppl.* 4 Jan. 16/3 The changing use of type and letterforms in English. **1889** *Cent. Dict.*, Letter-name. **1934** PRIEBSCH & COLLINSON *German Lang.* v. iii. 261 Rhyme-forms and spelling out names by means of new letter-names applied to the alphabet. **1961** R. B. LONG *Sentence & its Parts* xvii. 388 Examples are *IQ* and *TV*, in which the component letters are given their individual letter-name pronunciations. **1929** D. H. LAWRENCE *Let.* 9 July (1962) II. 1163 If I seal them letter-post they may hold them and make more fuss. **1953** *Language* XXIX. 72 The letter-sequences QL, TSR, SSS..never occur in English spelling. **1964** W. R. LEE in D. Abercrombie et al. *Daniel Jones* 288 There are various letter-shapes to grasp. **1964** *Language* XL. 168 The omission of a given letter-string from a dictionary is no assurance that the combination is not an English word. **1907** *Yesterday's Shopping* (1969) 356/2 Letter Trays. Fitted with spring clip, 4to..0/9.

8. **letter-balance** (examples); **letter-ballot**, a ballot in which the papers are sent by post; **letter-bomb**, an explosive device sent through the post as a weapon of terror; **letter-box**, (*c*) *Mountaineering* (see quot. 1968); (*d*) in espionage (see quot. 1961); **letter-carrier** (later U.S. examples); **letter-man**, (*b*) *U.S.*, a sportsman who has received a letter of distinction (see sense 1 e above); **letter-mark**, a letter used as a contraction or symbol; **letter-perfect** *a.* (earlier example); (*b*) literally correct, verbally exact; *fig.* flawless, unexceptionable; **letter plate**, a plate for fixing to the outside of a door or wall and having a rectangular aperture, covered by a flap, through which letters may be put; **letter-scale**, a scale for weighing letters; **letter-space** *Printing*, a space inserted between the letters of a word; so **letter-spaced** *a.*, **letter-spacing** *vbl. sb.*; **letter-weigher**, a device for weighing letters; **letter-weight** (examples); **letter-word**, a runic symbol or ideogram signifying both the name for something and a specific single letter; a word wholly or partly consisting of a letter or letters which are abbreviations in themselves.

1880 G. N. LAMPHERE *U.S. Govt.* 240/1 Supplying the post-offices entitled thereto with blanks,..twine, letter-balances, and cancelling-stamps. **1901** *Chambers's Jrnl.* Sept. 577/1 A German firm brought out a folding letter-balance, on the pan of which were engraved the British postal rates and the legend, 'Made in Germany'. **1961** *Lebende Sprachen* VI. 70/1 Letter scale (or: balance), die Briefwaage. **1898** *Engineering Mag.* XVI. 126/1 This resolution..submitted to the Society at large in the form of a letter-ballot. **1948** *Times* 4 Aug. 4/5 Each of the letter bombs which he has examined contained sufficient explosive to cause fatal injuries if detonated at close quarters. **1949** KOESTLER *Promise & Fulfilment* xiii. 149 His brother was killed by a letter-bomb sent by the Stern Group. **1972** *Times* 20 Sept. 1/3 One of the unexploded letter bombs. **1973** *Guardian* 28 June 26/5 A letter bomb exploded yesterday in the hands of a Londonderry solicitor..the first casualty of the letter-bomb campaign in Northern Ireland. **1974** *Ibid.* 26 Jan. 26/5 A letter bomb exploded at the head office of Pilkington Brothers, the glass manufacturers. **1951** E. COXHEAD *One Green Bottle* iii. 78 The rib bore them upward, perpetually varied,..now parallel cracks, now a groove, now a letter box... The climb above the crux was even more delightful. **1955** J. THOMAS *No Banners* xvi. 144 A circuit had to be organized into leak-proof compartments with letter-boxes or couriers as the only links between them. **1961** R. SETH *Anat. Spying* iii. 44 The 'letter-box' is a long-standing method of channelling information to the chief. It may be an accommodation address..or it may be a loose brick in a wall. Sometimes one member of a group acts as a 'letter-box' under his own name and at his home address. **1968** P. CREW *Encycl. Dict. Mountaineering* 79/1 *Letter-box*, a rectangular hole in a narrow rock ridge; a hole formed by jammed blocks or flakes of rock anywhere on a rock face. **1969** HURD & OSMOND *Smile on Face of Tiger* vii. 241 The silly girl popped out of the Home Office and they [*sc.* minutes] were in her usual letter-box within the hour. **1971** R. PETRIE *Thorne in Flesh* xv. 187 Aury, apparently, had been a live letterbox for a French security organisation. **1967** *Boston Sunday Globe* 23 Apr. 13/1 Letter carriers in the 25 cities and towns in the Boston Postal District are delivering notices to 565,000 families and business firms, advising them of their own Zip code for use in return addresses. **1973** WODEHOUSE *Bachelors Anonymous* ix. 123 'You're American, aren't you?' said Amelia Bingham. 'I thought so. It was your saying "letter-carrier" instead of postman.' **1973** *Black Panther* 15 Sept. 5/2 Attrition is also used to get rid of workers, especially letter carriers. **1926** *Chicago Tribune* 19 Sept. II. 5/4 The letter men in the line are Wolf and Neff, guards, and Rouse, center. **1974** *Spartanburg* (S. Carolina) *Herald-Jrnl.* 21 Apr. B1/1 Now he is well on the way to becoming Spartanburg High School's first four-sport letterman in at least seven years. **1907** *Congregational Year Bk.* p. xxix, The following letter-marks and signs are adopted:—B. (Baptist); C.H. (Countess of Huntingdon's Connexion). **1845** *Ainsworth's Mag.* VII. 83 'I am letter-perfect,' said another [actor]. **1867** *Harper's Mag.* Aug. 405/1 Where [legal] papers are to be served, and copies must be letter-perfect. **1894** 'MARK TWAIN' in *Century Mag.* Apr. 822 Tom's conduct had remained letter-perfect during two whole months. **1929** F. M. FORD *Let.* 10 July (1965) 186 If you will send me the better copy to look through I will certainly make it as letter-perfect as I can. **1969** *Time* 31 Jan. 1 The performances by such Bergman regulars as Max von Sydow and Gunnar Björnstrand are letter-perfect. **1898** F. W. MACEY *Specifications in Detail* 247 Knockers and letter plates are..made in iron and gun-metal, and vary in price from 3s. to £1 10s. **1923** *Work* 17 Nov. 161/2 A section through the door, showing the relative positions of the letter-plate and the box, is given by Fig. 3. **1971** *Country Life* 1 Apr. 752/3 Letter plates are easily bought. **1895** *Montgomery Ward Catal.* Index, Letter scale. **1900** A. UPWARD *Wonderful Career E. Lobb* 307 Be it enough To move the index of a letter-scale But in the estimation of a hair. **1961** Letter scale [see *letter-balance*]. **1934** WEBSTER, Letter-space. **1967** KARCH & BUBER *Offset Processes* iv. 125 Modification is possible to condense, expand,..letter-space and drop-out shadows in Benday screens. **1959** R. HOSTETTLER et al. *Technical Terms Printing Industry* (ed. 3) 120/2 One point letter-spaced, espacé un point. **1954** *Southward's Mod. Printing* (ed. 8) I. xxix. 178 Letterspacing. Spaces are also used for rendering uniform the thick perpendicular lines of capitals and fancy letters. **1961** T. LANDAU *Encycl. Librarianship* (ed. 2) 195/1 *Letter-spacing*, extra spacing between the letters of a word, especially in a running title or title-page. **1973** *Collins's Authors & Printers Dict.* (ed. 11) 247/2 *Letterspacing*,..shown in MS. by a stroke between letters, and # above (one word). **1862** *Illustr. Catal. Internat. Exhib., Industr. Dept., Brit. Div.* II. No. 5825, Inkstand, pen-tray, blotting-book, book-slide, letter-weigher. **1907** *Yesterday's Shopping* (1969) 366/2 Folding Portable Letter Weigher..to weigh up to 16 oz...each 1/8. **1880** GEO. ELIOT *Let.* 7 Feb. (1956) VII. 249 He left me the beautiful letter-weight. **1923** P. SELVER tr. *Čapek's R.U.R.* I. 5 A large 'knee-hole' writing-table on which stand an electric lamp,..letter-weight, [etc.]. **1927** JOYCE *Let.* 2 Mar. (1957) I. 250 A Chinese student sent me some letterwords I had asked for. The last one is 山. It means 'mountain' and is called 'Chin'. **1934** PRIEBSCH & COLLINSON *Germ. Lang.* v. 253 The emergence of 'letter-words' like *D-zug*.. *L-zug*.. *P.S.*..*G.m.b.H.* **1948** D. DIRINGER *Alphabet* 519 A manuscript containing the poem..of King Alfred's time..which describes in verse each runic letter-word.

letter-card. (Later examples.)

1898 G. B. SHAW *Philanderer* II. 107, I wish you'd ask her not to write on letter-cards. **1939** E. AMBLER *Mask of Dimitrios* x. 194 He bought a *pneumatique* letter-card. **1954** I. MURDOCH *Under Net* viii. 116 We..bought two letter cards. **1971** D. POTTER *Brit. Eliz. Stamps* ix. 94 A popular item in the suite of stationery is the familiar letter-card. They combine the convenience of envelope and paper with the secrecy of an ordinary letter.

lettergram (le·tǝɹgræm). [f. LETTER *sb.*[1] + TELE)GRAM.] A telegram delivered by the postman with the ordinary mail. (Disused.)

1911 *World's Work* XVII. 447/2 'Lettergrams' of fifty words..for an inclusive fee of about two shillings over any distance. **1915** J. WEBSTER *Dear Enemy* 19 When you feel so bursting with talk that only a hundred word telegram will relieve an explosion, at least turn it into a night lettergram. **1921** *Chambers's Jrnl.* Dec. 834/1 An attendant handed him a night lettergram. **1966** *Commonwealth of Australia Post Office Guide* 294 Lettergrams are reduced rate messages which may be lodged only at capital cities and certain other offices having extended hours of service. They are transmitted by telegraph and delivered to addressees by the earliest post on the following working day.

letter-press. 1. Substitute for def.: **a.** The text of a piece of printing, distinguished from illustrations, etc. **b.** Material printed from a relief surface, distinguished from lithographic or intaglio printing. Add further examples.

1892 A. POWELL *Southward's Pract. Printing* (ed. 4) I. 2 *Letterpress printing.* In this the subject is printed from a relief *above* the surface. *Ibid.* 3 Letterpress printing is done with types, blocks of wood, casts in metal, india-rubber, celluloid; electrotypes, &c. **1925** *Southward's Mod. Printing* (ed. 5) II. xxxiii. 274 These methods [of illustration] may be tabulated thus:—For printing by the letterpress method—Wood engravings; Engravings on type metal; Process blocks in line; Process blocks in half-tones... All the above blocks are in relief. **1939** *Guide Exhib. in King's Library* (Brit. Mus.) 9 The Old Testament types are explained in two paragraphs of woodcut letterpress in the upper corners of the design. **1946** W. L. HAYES in H. Whetton *Pract. Printing & Binding* xxv. 287/1 Relief printing, as its name implies, is that in which the printing surface stands in relief, that is, above the surrounding non-printing area; identified in this category is letterpress from type, plate, half-tone and line blocks, wood and lino cuts. **1959** *Penrose Annual* LII. 113 It was ..decided to use a plate etched as shallow as was practicable and to print by 'letterpress-offset'. **1966** *Print* (Wynkyn de Worde Soc.) iv. 39 The basic characteristic of the lithographic process is that the printing image is flat: not raised as in letterpress or recessed as in gravure. **1970** K. LINDLEY *Woodblock Engravers* ii. 29 They [*sc.* Bewick's woodcuts]..enabled the letterpress printer once more to think in terms of books equal to those with copperplate illustrations. **1971** M. MOORMAN in D. Wordsworth *Jrnls.* 107 In 1809–10 W[ordsworth] wrote the letter-press for Wilkinson's drawings of the Lake District.

letterset (le·tǝset). [f. *LETTER(-PRESS + *OFF)SET *sb.*] (See quot. 1963.) Cf. *DRIOGRAPHY.

1963 *Publishers' Weekly* 5 Aug. 87/1 A new name, 'Letterset', for what was formerly called 'dry litho' has been coined. Letterset refers to printing through the use of relief wrap-around plates and an intermediate blanket cylinder to transfer the image to the surface being printed. **1967** V. STRAUSS *Printing Industry* vi. 288/1 In 1962 Miehle and du Pont coined the term 'letterset'..for indirect relief printing. Dry offset of the past and contemporary letterset are based on the same principle but differ widely in actual performance. **1973** S. JENNETT *Making of Bks.* (ed. 5) x. 166 Indirect letterpress or letterset are terms used for letterpress printing on to a blanket cylinder and then on to paper, after the manner of litho offset; the printing surface is a shallow etched metal or plastic plate curved round the plate cylinder.

letting, *vbl. sb.*[1] **1.** (Later examples of *letting down*.)

1917 J. AGATE *Buzz, Buzz!* 9, I hold such statement of the actual and practical scope of current criticism not to be a letting-down of the art we hold dear. **1940** *Chambers's Techn. Dict.* 496/1 *Letting-down*, the process of tempering hardened steel by heating it until the desired colour is reached and then quenching. **1958** *Ibid.* 990/1 *Letting down*, the reduction of altitude from cruising height to that required for the approach to landing. **1966** J. & R. GODDEN *Two under Indian Sun* iii. 68 The letting down of hems or takings in, as dresses were handed down.

Letto- (le·to), combining form repr. mod.L. *Lettōn-*, *Letto*, used with adjs. or sbs. denoting other languages or peoples, signifying 'Lettish and..', as *Letto-Lithuanian*, *-Slavonic*, etc.

1880 A. H. SAYCE *Introd. Sci. of Lang.* II. vii. 107 The old middle or intransitive voice..has been lost in Keltic and Letto-Slavonic. **1913** L. W. LYDE *Continent of Europe* 419 These are the Letto-Lithuanians and Finnic (Esht) peoples of the Baltic Provinces. **1920** *Glasgow Herald* 21 July 8 A Polono-Lithuanian, a Polono-Lettish, and a Letto-Lithuanian question still unsettled. **1935** HUXLEY & HADDON *We Europeans* vii. 221 The Letto-Lithuanians tend to be of medium height, but taller where remote from Slav influence.

‖ **lettre** (letr'). The French word for 'letter' in: **a. lettre de cachet** (letr' dǝ kaʃe), lit. 'letter of seal', a warrant issued in the France of the *ancien régime* for the imprisonment of a person without trial at the pleasure of the monarch. Also *transf.* Cf. CACHET 1.

1718 VANBRUGH *Let.* 30 Aug. in *Athenæum* (1890) 30 Aug. 290/2, I am far from having the least doubt of his intentions to me; I fear only those same letters [*sic*] de cachet, that surprise folks every now and then. **1745** H. WALPOLE *Let.* 1 Aug. in *Corr.* (1941) IX. 23 Before the play itself is suppressed by a *lettre de cachet* to the booksellers. **1799** MALTHUS *Diary* 16 July (1966) 159 [Norway] These men seemed to have been placed by lettres de cachet exactly similar to those in France. **1824** SCOTT *Redgauntlet* III. vi. 163 There are sharp laws in France against refractory pupils—*lettres de cachet* are easily come by. **1849** THACKERAY *Pendennis* I. vii. 70 Why are there no such things as *lettres-de-cachet*—and a Bastille for young fellows of family? **1895** S. WEYMAN *From Mem. Minister of France* 138 M. de Clan..is for shutting him up. Getting a *lettre de cachet*..and away with him. **1916** A. HUXLEY *Let.* 31 Mar. (1969) 96 A lettre de cachet from a Lord Lieutenant..can put one snugly away in the Jail! or Jug without any prospect of a trial for periods quite indefinitely coextensive with the war or even eternity. **1957** C. BROOKFIELD tr. *Durkheim's Professional Ethics & Civic Morals* vii. 87 Louis XIV, clearly, was able to issue his *lettres de cachet* against anyone he wished.

b. Used as the first element in **lettre bâtarde, lettre de forme, lettre de somme,** to designate groups of early type-faces based on manuscript forms then current, now more often described as *BASTARDA, *TEXTURA, and *ROTUNDA or fere-humanistica types.

1887 T. B. REED *Hist. Old Eng. Letter Foundries* i. 53 The Gothic letter employed by the inventors of printing for the *Bible*, *Psalter*, and other sacred works, was an imitation of the formal hand of the German scribes... This letter, as a typographical character, took the name of Lettre de Forme. *Ibid.*, The term [*sc.* *lettre de forme*]..was used by both Tory and Ycair to denote a class of letter which the former denominated *Canon*, or cut according to rule, as opposed to the more fanciful *lettres bâtardes*. *Ibid.* 54 The Lettre de Somme of the Germans..became in the hands of the fifteenth century printers a rival to the Gothic. **1922** D. B. UPDIKE *Printing Types* I. iv. 60 Fifteenth century Gothic type-forms may be roughly subdivided into Pointed, sometimes called *lettre de forme*, Round, sometimes called *lettre de somme*, and a vernacular Cursive black-letter, like the French *lettre batarde*... These three type-forms were the black-letter equivalents of the formal, less formal, and cursive manuscript-hands of the Roman period. *Ibid.* 63 The *lettre de somme* is said (without much authority) to derive its name from the *Summa* of St. Thomas Aquinas, for which..it was early employed. **1927** R. B. MCKERROW *Introd. Bibliogr.* 289 The *lettre de forme*,..derived from the most formal script, ..is in general of a rather narrow and pointed character, tending everywhere to angularity. *Ibid.*, Certain of the smaller sizes used by Wynkyn de Worde and Grafton.. seem very definitely to have that feeling of breadth and openness which is characteristic of the *lettre de somme*. *Ibid.*, The *lettre bâtarde* represents the cursive hand of its time. **1962** N. E. BINNS *Introd. Historical Bibliogr.* (ed. 2) xiv. 182 The rigid and formal types of this group are known as *Textura* types, or..*Lettre de Forme*. *Ibid.*, A broader, rounder, and less formal kind of type known as *Fere-Humanistica*, or *Lettre de Somme*. *Ibid.* 185 The early German and French printers also used various cursive types..known as *Bastarda*, or *Lettre Bâtarde*.

‖ **lettrine** (letrī·n). [Fr.] An initial letter, often decorated, and larger than the size of the text it accompanies.

1932 JOYCE *Let.* 11 Nov. (1957) II. 326 He liked her alphabet and has written to the manager of Burns and Oates about using these *lettrines* for a reprint of Chaucer's A.B.C. poem. **1934** *Ibid.* 25 Apr. (1966) III. 302, I enclose ..two *lettrines* of hers. **1970** *Private Library* III. 174 Wood engravings such as the 'lettrine' (initial letter) that figures at the beginning of this article.

lettrism (le·triz'm). Also in Fr. form **lettrisme.** [ad. F. *lettrisme*, f. LETTER *sb.*[1] + -ISM.] Applied to a movement in French art and literature, characterized by a repudiation of meaning, and the use of letters (sometimes invented) as isolated units. So **le·ttrist, lettriste** *sb.* and *a.*

1946 *Time* 2 Dec. 31/2 Lettrism, founded by Isidore Isou..is a theory of poetry as 'rhythmic architecture'. The rapidly growing hordes of Lettrists..prefer meaningless combinations of letters to dictionary words. **1948** *Spectator* 9 Apr. 432/2, I have been reading this week some poems written in the new mode of '*lettrisme*'. **1949** *Commentary* VIII. 183/2 'The new art,' declared Isou (his real name is Isidore Goldstein), 'accepts as its subject matter the letters reduced to, and become simply, themselves, replacing completely all poetic and musical elements which go beyond the letters in order to shape them into coherent works.' Thus was born Lettrisme, which may be regarded as a resurrection of some old avant-garde theories, or as a postwar symptom comparable to the explosions of Surrealism and Dada after World War I. **1951** *Amer. Mercury* LXXII. 659 In no time at all he [*sc.* Isidore Isou] was writing 'lettrist' poetry and he and his followers were assaulting the ears of everybody in the cafés of the Latin Quarter. **1962** *Times Lit. Suppl.* 13 Apr. 246/4 Hausmann's lettrist interview with some hypothetical French *lettristes*. **1964** *Ibid.* 3 Sept. 796/3 Lettrism is a creative movement..which claims to be able first to revolutionize every aesthetic discipline..from poetry to the theatre, by way of painting, and then to renovate the other cultural domains, whether philosophical or scientific. **1971** J. WILLETT in A. Bullock *20th Cent.* 244/2 The Lettrists in Paris and the Brazilian concrete poets of the 1950s, who were alike in their concentration on the appearance and sound of words or individual letters.

lettuce. Add: **1. b.** *slang* (orig. *U.S.*). Money.

1929 *Amer. Speech* June 357 If you wish to boast of having a great deal of money, you may speak of having.. wads of it, or a wad of lettuce, meaning a big roll of bills. **1932** J. DOS PASSOS *1919* 57 He still had more'n fifty iron men, quite a roll of lettuce for a guy like him. **1967** WODEHOUSE *Company for Henry* v. 84 How are you fixed for lettuce, Hank?.. Dough. Cash. Glue... Money. **1974** J. WAINWRIGHT *Cause for Killing* 216 'They spend money, in Beirut...' 'Phoenicia Street,' murmured Gantley. 'Anything... Any out-of-this-world luxury. Any service. Anything! You have the lettuce... Phoenicia Street can oblige.'

c. = **lettuce green*.

1963 *New Yorker* 1 June 115 *These* shirts..in..cedar, lettuce, navy or red.

3. **lettuce green,** a medium shade of green; also *attrib.*

1897 *Sears, Roebuck Catal.* 222/2 Delicate tintings of.. lemon, lettuce green, scarlet. **1929** E. WILSON *I thought of Daisy* i. 3, I saw lettuce-green cocktail glasses. **1970** *Guardian* 2 June 7/8 Mimosa, Lettuce Green, Lavender Blue. **1971** D. BEATY *Temple Tree* 157 A lettuce-green shirt.

letty (le·ti). *slang.* [ad. It. *letto* bed.] A bed, a lodging.

1846 *Swell's Night Guide* 71 While the old rum cull.. cannot wag from his Letty,..the accumulation of dirt thrives monstrously. **1875** T. FROST *Circus Life* xvi. 279 'Letty' is used both as a noun and a verb, signifying 'lodging' and 'to lodge'. **1933** E. SEAGO *Circus Company* v. 80 All the people of the smaller show dwell in wagons of their own, and do not seek for 'letties' in the town. **1957** J. OSBORNE *Entertainer* III. 77 *Jean*: We can't all spend our time nailing our suitcases to the floor, and shin out of the window. *Archie*: Scarper the letty.

let-up, *sb.*: see *LET *v.*[1] 35 c.

Letzeburgesch (letsǝbu·rgeʃ). Also **Letzeburg, Letzeburgisch, Lezebuurjesh,** etc. [Local name.] The name of the West Moselle Franconian dialect of German spoken by the natives of Luxembourg. Cf. *LUXEMBURGISCH.

1921 R. J. CASEY *Land of Haunted Castles* 152, I asked a young woman of Gosseldange..who..the 'Letzeburgers' were. She replied, proud of her ability to tell me in English: 'They are people who live in the Stadt Luxemburg... They speak a language that is very *difficile* to understand. It is not French. It is not German. It is not Luxemburg. It is Letzeburg.' **1944** *Luxembourg* (Geogr. Handbk. Ser. B.R. 528, Admiralty, Naval Intelligence Div.) 43 The native dialect, *Letzeburgesch*, is a Moselle Franconian dialect belonging to the West Middle German group. **1956** B. MILES *Attic in Luxembourg* xxviii. 215 Letzeburgesch, the Luxembourg dialect..although based upon old Teutonic origins,..borrowed extensively from Celtic, Roman and French. **1961** R. E. KELLER *German Dial.* 10 *Letzeburgisch* might be considered with some hesitation as a *Halbsprache* but *Schwyzertütsch* is for Kloss beyond the pale and he merely concedes that certain circumstances make it difficult to decide whether it might not after all qualify for the status of a *Halbsprache*. **1964** S. H. MULLER *World's Living Lang.* i. 16 The speech of 300,000 citizens of Luxemburg has diverged from standard German so much that it must be considered a distinct language, Luxemburgian (self-designation *lezebuurjesh*). **1965** W. B. LOCKWOOD *Informal Hist. German Lang.* 146 Some 300,000 persons..speak as their native idiom a now fairly uniform German dialect *Luxemburgisch* or, as they call it themselves, *Letzeburgsch*. **1972** *Guardian* 19 Sept. 14/3 The vast majority of Luxembourgers speak Letzeburgesch at home. **1973** *Times* 26 May (Benelux Suppl.) p. v/4 Luxembourg..is a case apart. Its 320,000 inhabitants speak a regional language called *Letzeburgisch* on which the official language, French, is superimposed.

leu (le·,u). Pl. **lei** (le·,i). [Rumanian, = lion.] The basic monetary unit of Rumania.

1879 *Coin Chart Manual* (N.Y.) 2 Value of U.S. Dollar. ..Roumania... 5 Lei 18½ Ben Paras. **1896** *Hist. Banking* III. VII. 333 Up to 1877 Roumania had no paper money. In that year, however, the Government issued non-compulsory notes of 5, 10, 20, 50, 100, and 500 lei. **1902** *Encycl. Brit.* XXXI. 291/1 In Rumania the unit of account is the *leu* of 100 *bani*. **1921** S. GRAHAM *Europe* viii. 107 The twenty, the hundred, the thousand-crown notes are almost identical... Roumanian lei are also much the same in appearance. **1940** G. CROWTHER *Outl. Money* viii. 305 Germany pushes up the value of the Reichsmark relatively to, say, the Roumanian leu; that enables her to quote a very attractive price, in lei, for wheat, without it costing her too much in Reichsmarks. **1967** *Economist* 19 Aug. (Suppl.) p. xlii/3 All families with a joint income of 2,000 lei..are now obliged to buy their flat.

Leucadian (liukēi·diăn), *sb.* and *a.* [f. *Leucadia* (see below) + -IAN.] **A.** *sb.* A native or inhabitant of Leucadia or Leucas, an island in the Ionian Sea. **B.** *adj.* Of or pertaining to Leucadia.

1615 G. SANDYS *Rel. Journey* I. 4 It was a custome amongst the Leucadians in their yearely solemnities, as a propitiatory sacrifice to Apollo, to throw some one from the top, condemned before for his offences. **1890** J. G. FRAZER *Golden Bough* II. iii. 213 From the Lover's Leap, a white bluff at the southern end of their island, the Leucadians used annually to hurl a criminal into the sea as a scapegoat. **1952** R. CAMPBELL tr. *Baudelaire's Poems* 181 Since then I watch on the Leucadian height. **1968** *Encycl. Brit.* XIII. 998/2 It was from the extremity..of this cape that in ancient times the 'Leucadian leap' was made, an ordeal whereby at the feast of Apollo accused persons were tried, those who survived the leap being picked up by boat.

leuchæmia. Add: Also **-emia.** The form **leuchæmia** is obs., and **leukæmia, leukemia** are the usual forms. [First formed as G. *leukämie* (R. Virchow 1848, in *Arch. f. path. Anat.* I. III. 563).] Substitute for def.: A progressive disease of man and other warm-blooded animals characterized by the hyperplastic transformation and greatly increased activity of leucopoietic tissue, leading to abnormal accumulations of leucocytes (freq. of immature or abnormal form) first at the site of leucopoiesis and then (usually) in the blood and elsewhere. (Further examples.)

1938 M. N. Richter in H. Downey *Handbk. Hematol.* IV. xlii. 2889 The type of cell and tissue primarily involved, the extent and distribution of infiltrations, and the presence or absence of immature cells in the peripheral blood make the lesions observed in different cases and different types of leucemia quite diverse, the only feature common to all being the increase in number of white corpuscles and their relative immaturity. 1942 C. L. Heel tr. *Engelbreth-Holm's Spontaneous & Exper. Leukæmia in Animals* i. 3 While leukæmia has not been observed in the lower vertebrates, the condition is known in many kinds of birds. *Ibid.* ii. 29 Detailed information is available about leukæmia in dogs, pigs, cattle, and..rodents. 1951 *New Biol.* XI. 97 In the treatment of leukaemia, a cancer-like disease of the white blood cells, several different classes of chemicals show some value. 1955 *Sci. News Let.* 19 Mar. 182/3 Leukemia, always fatal cancer of the blood, is showing up in survivors of the world's first military atom bombing. 1960 F. G. J. Hayhoe *Leukaemia* ii. 10 Patients with chronic leukaemias nearly always survive more than a year from the time of first symptoms, commonly for 3 to 5 years, and occasionally for very much longer. *Ibid.* 11 Subleukaemic and aleukaemic forms are more often encountered in the acute than in the chronic leukaemias, and they usually become fully leukaemic at a later stage in the progress of the disease. 1964 *Daily Tel.* 3 Jan. 13/3 A second British child suffering from leukemia ..arrived with his mother in Ajaccio, Corsica, to-day for treatment with a new serum which is claimed to cure the disease. 1966 Wright & Symmers *Systemic Path.* I. iv. 181/2 Post-mortem findings in acute leukaemia... Gross enlargement of the liver and spleen, such as is common in the chronic forms of leukaemia, is unusual.

leuchæmic, *a.* (In Dict. s.v. Leuchæmia.) Add: Now usu. **leukæmic, -emic** (see prec.). Add to def.: also, characteristic of or resembling leukæmia; *spec.* marked by an increased number of leucocytes in the blood. (Further examples.)

1922, 1946 [see *leucosis]. 1942 M. M. Wintrobe *Clin. Hematol.* xvi. 616 In many cases peculiarities can be observed in leukemic cells which are like those of neoplastic cells. 1949 [see *aleukæmic *a.*].

B. *sb.* An individual with leukæmia.

1964 *New Scientist* 13 Feb. 402/3 Nineteen of the 24 leukaemics were positive, one was doubtful and four were negative. 1973 *Nature* 12 Jan. 99/1 We have found that the response of lymphocytes from normals or leukaemics to low..doses of PHA [*sc.* phytohaemagglutinin] is often greatly reduced in the presence of serum obtained from patients with untreated AML [*sc.* acute myeloid leukaemia] and ALL [*sc.* acute lymphoblastic leukaemia].

leucin. Now always spelt **leucine**. Add to def.: an amino-acid that is one of the principal constituents of proteins. [First formed as F. *leucine* (H. Braconnot 1820, in *Ann. de Chim. et de Physique* XIII. 119).]

leuco-. Add: Many medical words with first element *leuco-* are also spelt *leuko-*. (*b*) In *Med.* used to represent 'leucocyte' (as in *leucopenia, -poiesis*). (*c*) In *Chem.* [after its use in *Dyeing*: see *b], used to form the names of some colourless compounds that are chemically transformed to coloured ones (as in *leucoanthocyanin*).

leu:coanthocya·nidin *Chem.*, any colourless substance which yields an anthocyanidin on heating with mineral acid; **leu:coanthocy·anin** *Chem.*, a leucoanthocyanidin; *spec.* any that is a glycoside; **leuco-, leukoci·din** († -ine) *Bacteriology* [a. F. *leucocidine* (H. van de Velde 1894, in *La Cellule* X. 434): see -cide 1], any leucotoxin produced by a microorganism; **leucocra·tic** *a. Petrol.* [ad. G. *leukokrat* (W. C. Brögger *Eruptivgesteine des Kristianiagebietes* (1898) III. 264), f. Gr. κρατ-εῖν to rule, prevail], (of a rock) light-coloured; rich in light-coloured minerals; **leu·coderm** (also *Leuco-*) *sb.* and *a.* [Gr. δέρμ-α skin], (of, pertaining to, or being) a person of a white-skinned race; **leucode·rmia** *Path.* = *leucoderma*; **leucodermic** *a.*, (*b*) (naturally) white-skinned; **leucopenia** (examples of the form *leukopenia*); **leucophore** [a. G. *leukophore* (R. Keller 1895, in *Pflügers Archiv Ges. Physiol.* LXI. 147)] = *iridocyte; **leu:copho·sphite** *Min.*, a hydrated basic phosphate of potassium and ferric iron found as white or greenish fine-grained masses; **leucoplacia,** now spelt *leuco-, leukoplakia*; (further examples); hence **leuco-, leukopla·kial** (now *rare*), **-pla·kic** *adjs.*; **leucoplast** (examples); **leuco-, leukopoiesis** (-poi,ī·sis) *Physiol.* [*-poiesis], the production of leucocytes; so **leuco-, leukopoie·tic** *a.*; **leuco·pterin** *Chem.* [*pterin], a white pigment found esp. in certain butterflies; 2-amino-4,6,7-trihydroxypteridine, $H_2NC_6N_4(OH)_3$; **leuco-, leuko·ta·xin(e** *Physiol.* [Gr. τάξις arrangement, order], a nitrogenous material found in injured tissue and inflammatory exudates which on injection causes inflammation, increase in the permeability of capillaries, and the attraction of leucocytes to the site; **leuco-, leukoto·xin** *Med.*, any substance which destroys leucocytes.

1935 G. M. & R. Robinson in *Jrnl. Chem. Soc.* 745 Probably class (b) [of leuco-anthocyanins] consists of relatively simple glycosides or diglycosides, whereas members of class (c) are sugar-free and should be regarded as leuco-anthocyanidins. 1962 J. Clark-Lewis in T. A. Geissman *Chem. Flavonoid Compounds* viii. 218 Choice between the terms leucoanthocyanin and leucoanthocyanidin seems so far to have rested on the preference of individual authors, but there are sound reasons for using leucoanthocyanidin for the sugar-free molecules, and as the general term, and for reserving the term leucoanthocyanin for leucoanthocyanidin glycosides. The terminations thus have the same significance as in anthocyanin and anthocyanidin. All the compounds of this class and known constitution so far discovered in nature are leucoanthocyanidins, i.e. do not contain sugar residues. 1967 J. B. Harborne *Compar. Biochem. Flavonoids* ix. 302 Leucoanthocyanidins (or condensed tannins) can be classified into three groups: (1) low molecular weight substances, which are probably dimers formed by linkage of a flavan-3,4-diol with a catechin.., (2) soluble oligomers, containing 4 to 8 flavan units, and (3) insoluble polymers (flavolans) of 10 or more units. 1967 *New Scientist* 4 May 270/3 The culprits that cause colour change in African mahogany are katechin and leucoanthocyanidin. 1920 O. Rosenheim in *Biochem. Jrnl.* XIV. 185 In the young leaf, however, the pseudo-base does not occur in the free state, but in combination with either a carbohydrate or possibly another complex. For this combination the general name leuco-anthocyanin is proposed. 1960 L. H. Meyer *Food Chem.* vii. 251 Catechins and leucoanthocyanins are present in the tissues of those woody plants studied such as apples, peaches, grapes, almonds, and some pears, while they are absent in herbaceous plants. 1962 T. Swain in T. A. Geissman *Chem. Flavonoid Compounds* xvi. 536 It has been presumed by many workers that the term leucoanthocyanin, like the term catechin, refers to the monomeric C_{15} molecule. 1971 *Ann. de Technol. Agricole* XX. 32 Different methods for the dosage of leucoanthocyanins of white wines have been compared. 1894 *Jrnl. R. Microsc. Soc.* 732 The virulent cocci [of *Staphylococcus*]..secrete a special substance. This, which causes the death of the leucocytes, is termed 'substance leucocide' or leucocidine. 1909 J. G. Adami *Princ. Path.* I. III. viii. 489 These leukotoxins are also known as leukocidins. 1970 Ambrose & Easty *Cell Biol.* xiv. 470 Some bacteria not only resist phagocytosis but produce substances, known as leucocidins, which kill phagocytes. 1909 A. Harker *Nat. Hist. Igneous Rocks* v. 112 The former [*sc.* camptonite, is] a melanocratic type..and the latter [*sc.* mænaite] leucocratic. 1954 H. Williams et al. *Petrogr.* ii. 33 In Johannsen's classification four rock classes are distinguished according to the volume-content of dark minerals, the limits being placed at 5, 50, and 95 percent. Shand also distinguishes four classes, but with different limits, as follows: leucocratic rocks, with less than 30 percent mafic minerals; mesocratic rocks, with 30–60 percent; melanocratic rocks, with 60–90 percent; and hypermelanic rocks, with more than 90 percent mafic minerals. 1965 G. J. Williams *Econ. Geol. N.Z.* xiv. 216/1 The Separation Point granite..is a massive white leucocratic soda-granite. 1924 A. C. Haddon *Races of Man* (ed. 2) 13 Occasionally in leucoderms, sometimes in Negroes, and as a rule in Mongoloid peoples, a fold of skin..covers the inner angle of the eye. *Ibid.* 84 The western steppe lands seem to have been the original home of fair (leucoderm) dolichocephals. 1935 Huxley & Haddon *We Europeans* iv. 115 A broad and convenient classification of skin-colour is as follows: (1) Leucoderms, or white-skinned (Caucasian) peoples; (2) Xanthoderms, or yellow-skinned peoples; (3) Melanoderms, or black-skinned peoples. 1888 *Syd. Soc. Lex.*, *Leucodermia*, see Leukoderma. 1908 *Practitioner* Aug. 349 They [*sc.* freckles] are an example of excess of pigment in the skin, a condition known as hyperchromasia, in contradistinction to achromasia, or leucodermia, in which there is a deficiency of pigment in the skin. 1926 H. H. Wilder *Pedigree of Human Race* vi. 348 Members of the Leucodermic race in Europe or America. 1961 R. D. Baker *Essent. Path.* ii. 18 In some inflammations the total white blood cell count is decreased (leukopenia). 1964 W. G. Smith *Allergy & Tissue Metabolism* ii. 16 Marked reductions in the number of leucocytes (leukopenia) and platelets (thrombocytopenia) circulating in the blood were described. 1924 L. Hogben *Pigmentary Effector Syst.* ii. 24 In the skin of the Chameleon there are present, immediately below the epidermis, cells charged with yellow pigment variously described as guanophores (Schmidt), leukophores or ochrophores (Keller), iridocytes (Pouchet), or interference cells (Brucke). 1963 M. Fingerman *Control of Chromatophores* i. 4 When the guanine consists of fine granules that can migrate the term leucophore is usually employed. 1932 E. S. Simpson in *Jrnl. R. Soc. W. Austral.* XVIII. 71 No previously described mineral approaches this in composition except minervite, a potassium aluminium phosphate, from which it differs in possessing a much greater basicity... It appears therefore to be a new species for which the name Leucophosphite is suggested. 1963 *Prof. Papers U.S. Geol. Survey* No. 475–C. 103/2 Leucophosphite and gypsum represent in large part only a recombination of the elements already present in the phosphatized wood when uplift of the enclosing Moreno Formation exposed it to weathering. 1972 *Amer. Mineralogist* LVII. 397 Leucophosphite, $K_3[Fe^{3+}{}_4(OH)_3(H_2O)_3(PO_4)_4].2H_2O$, possesses an atomic arrangement based on a discrete octahedral tetramer. 1920 W. E. Masters *Essent. Trop. Med.* vi. 477 Leucoplakia may also affect the penis, vulva and vagina. 1962 *Lancet* 1 Dec. 1170/2 Leucoplakia (or lichen sclerosus) diagnosed by the clinician on naked-eye appearances shows variable and non-specific histological features. 1962 *Ibid.* 8 Dec. 1228/2 Here operating is undesirable, except in those few patients in whom leukoplakia develops as well. 1973 *Daily Colonist* (Victoria, B.C.) 7 Sept. 2/1 Leukoplakia is a thickening of the membrane of a mucous surface, commonly on the lip or in the mouth. 1908 *Practitioner* Sept. 354, I believe that if sufficient examinations were made in syphilitic patients..leucoplakial patches would be found in the mucous membrane from time to time. 1923 *Surg. Gynecol. & Obstetr.* XXXVI. 189/1 The leukoplakial conditions of the urinary tract have received relatively little attention. 1907 *Arch. Middlesex Hosp.* IX. (6th Rep. Cancer Res. Lab.) 65 Those leucoplakic conditions of vulva, tongue, and lips which..often precede the development of squamous cell carcinoma. 1917 J. Bland-Sutton *Tumours, Innocent & Malignant* (ed. 6) xxx. 331 In some patients an ulcer appears in a leucoplakic patch. 1962 *Lancet* 1 Dec. 1170/2 If all vulvas described as leucoplakic on clinical grounds are subjected to biopsy, approximately 5% are found to be cancerous at the outset. *Ibid.* 8 Dec. 1228/2 Changes in vulval skin..are very common... Among other features, they are often white, but this does not mean that they are precancerous or leukoplakic. 1886 *Jrnl. R. Microsc. Soc.* 640 In the lower plants..the formation of leucoplasts is a subsequent process, a transformation of the coloured into a colourless chromatophore. 1887, 1902 Leucoplast [see *chloroplast]. 1964 *Oceanogr. & Marine Biol.* II. 199 All species [of the genus *Caulerpa* of green algae] possess amyliferous leucoplasts as well as the ordinary green plastids. 1913 Dorland *Med. Dict.* (ed. 7) 512/2 *Leukopoiesis*, production of leukocytes. 1942 M. M. Wintrobe *Clin. Hematol.* i. 26 Erythropoiesis in the spleen is at first more pronounced than leukopoiesis but it is short-lived. 1973 Woodliff & Herrmann *Conc. Haematol.* viii. 113 Disorders of leucopoiesis are usually reflected by changes in the peripheral blood. 1913 Dorland *Med. Dict.* (ed. 7) 512/2 *Leukopoietic*, forming or producing leukocytes. 1927 A. Piney *Recent Adv. Hæmatol.* iii. 35 Hyperplasias of a character similar to those occurring in the leucopoietic tissue may affect the erythropoietic one. 1973 Woodliff & Herrmann *Conc. Haematol.* viii. 113 A progressive malignant proliferation of the leucopoietic tissues. 1927 *Chem. Abstr.* XXI. 224 (*heading*) Leucopterin, the white wing pigment of cabbage butterflies (*Pieris brassicae* and *P. napi*). 1954 *Sci. News* XXXIV. 91 The purines and pterines contribute a major source of colour to the wings of butterflies... These compounds are only present in small amounts representing in the case of the white leucopterin of Pierid butterflies about 0·18 milligramme per specimen. 1937 V. Menkin in *Proc. Soc. Exper. Biol. & Med.* XXXVI. 167 For the sake of convenience the name leukotaxine is tentatively proposed for this active crystalline nitrogenous substance which is evidently released by injured tissue and is readily recovered in inflammatory exudates. 1947 *New Biol.* II. 135 The most reasonable hypothesis is..that leucotaxine is released from the killed and damaged cells of the injured skin we are considering, and is mainly responsible for the escape of fluid from nearby undamaged capillaries. 1957 *Amer. Jrnl. Physiol.* CLXXXIX. 99 (*caption*) A preliminary intravenous injection of 25 mg of cortisone acetate was performed in the tested rabbit to inactivate the masking effect of any leukotaxine possibly present in the acid exudate to be injected. 1964 W. G. Smith *Allergy & Tissue Metabolism* iii. 39 This material, which Menkin called leucotaxin, can upon injection into the skin induce increased capillary permeability and chemotactic attraction of polymorphs to the injection site. 1908 *Practitioner* Mar. 392 Roentgen-rays appear to lead to the production of veritable 'leucotoxins'. Normal leucocytes, exposed in vitro and in vivo to the action of such leucotoxins present in the serum of animals, which have been exposed to Roentgen-rays, undergo a specific disintegration. 1931 *Biol. Abstr.* V. 488/2 The virulent streptococci do not possess in the same degree the power of leukotoxin production. 1956 *Proc. Soc. Exper. Biol. & Med.* XCIII. 493/2 A leucotoxin develops in the blood of the rabbit in hemorrhagic shock, and..this leucotoxin severely impairs the antibacterial potential of the animal.

b. In Combs. in which *leuco* may be used *attrib.* (without a hyphen) as quasi-*adj.*, or be joined by a hyphen to the second element: chiefly in *Dyeing*, where *leuco* is used to denote the reduced, water-soluble colourless form of a dye which is fixed on the fibre and subsequently oxidized to the dye proper by the air; as *leuco-base* (so *leuco-basic* adj.), *-compound, -cyanide, -dye, -form, -fuchsin*.

1886 E. Knecht tr. *Benedikt's Chem. Coal-Tar Colours* 79 These compounds, called 'leuco-bases', are colourless and yield colourless salts with acids. By oxidation they are transformed..into the colour-bases, which differ from the 'leuco-bases' by containing one atom of oxygen. 1947 L. S. Pratt *Chem. & Physics Org. Pigments* viii. 140 The dyestuff is prepared by condensing *o*-chlorobenzaldehyde with dimethylaniline and then oxidizing the leuco base to the color base. 1958 J. R. Baker *Princ. Biol. Microtechnique* xvii. 309 Schiff's reagent..is often regarded as a leucobase, but this is an error; for a leucobase becomes coloured on oxidation and could not possibly serve in Feulgen's reaction. 1971 E. Gurr *Synthetic Dyes* 108 The leuco bases of triphenylmethane dyes (e.g. crystal violet and malachite green) are extremely light sensitive. 1956 *Nature* 14 Jan. 92/2 Leuco-basic fuchsin is specific for deoxyribonucleic acid. 1970 *Watsonia* VIII. 23 Root tips ..were..stained in leucobasic fuchsin. 1888 *Jrnl. Chem. Soc.* LIV. 493 The anthraquinone-dyes yield leuco-compounds on reduction. 1906 *Notices Proc. R. Inst. Gt. Brit.* XVII. 107 What we are pleased to call leucocompounds, are in the majority of cases by no means colourless. Indigo-white itself is not white but yellow in its alkaline solution which we call a vat. Other vat-dyes have leucocompounds which are even more strongly coloured. 1961 Cockett & Hilton *Dyeing Cellulosic Fibres* viii. 280 All methods used in practice to apply vat dyes to cellulosic fibres involve, at some stage, the conversion of the insoluble vat dye to the soluble sodium salt of the so-called leuco compound of the dye. 1931 *Trans. Faraday Soc.* XXVII. 571 The pure alcoholic leuco cyanide solution is very suitable for a laboratory method [of measuring ultra-violet light]. 1965 J. Kosar *Light-Sensitive Syst.*

viii. 370 Aside from the photographic applications, light-sensitive leucocyanides are useful for detecting, measuring, and recording short wave ultraviolet light. 1954 *Textile Terms & Definitions* (Textile Inst.) 24 *Leuco dye*, a reduced form of dye from which the original dye may be regenerated by an oxidation process. 1973 J. F. Willems in R. J. Cox *Proc. Symposium Photogr. Processing Univ. Sussex* 95 These leuco dyes are strong reducing agents, which in the adsorbed state on the silver halide grain start the development. 1959 *Nature* 15 Aug. 545/1 The production of a coloured dye by transformation of the leuco form. 1945 *Chem. Abstr.* XXXIX. 6288/1 (Index), Leucofuchsin. 1965 E. Gurr *Rational Use Dyes Biol.* 94 Solutions of reduced dyes, such as Schiff's reagent (leuco fuchsin), leuco acid fuchsin and leuco patent blue in distilled water, are oxidized on heating and consequently restored in colour. 1967 *Jrnl. Med. Lab. Technol.* XXIV. 48 (*heading*) Nitric acid leucofuchsin technique for myelinated nerves.

leucocyte. Add: Also **leukocyte** (similarly **leukocytic,** etc.). (Further examples.)

1911 *Jrnl. Amer. Med. Assoc.* 25 Feb. 581/1 Alexin increases in the serum under conditions which favor disintegrating of leukocytes. *Ibid.*, The 'alexin' is synonymous with Metchnikoff's leukocytic ferment. *Ibid.*, Leukocytic extract is essentially the same in its action as alexin. *Ibid.*, The alexin varies with the degree of leukocytosis. 1947 Leukocytosis [see *granulocytosis]. 1951 A. Grollman *Pharmacol. & Therapeutics* xxxi. 732 Mercurial poisoning is sometimes accompanied by a leukocytosis. 1971 J. Song *Path. Sickle Cell Dis.* xviii. 355 Fever, vomiting, nausea, and leukocytosis. 1973 *Nature* 1 June 290/1 We cultured peripheral blood leukocytes for 24 or 48 h with phytohaemagglutinin.

leucosis. Add: Pl. **leucoses. d.** Also **leukosis.** [a. G. *leukosis* (Ellermann & Bang 1908, in *Centralbl. f. Bakteriol. Parasitenkunde u. Infectionskrankh.* (*Erste Abteil., Originale*) XLVI. 609).] = *leukæmia; *esp.* any of various leukæmic diseases of animals. Also *bovine leucosis*, a leukæmic disease of cattle; (*avian* or *fowl*) *leucosis complex*, a group of poorly-differentiated leukæmic diseases of poultry which are typically transmissible.

1922 V. Ellermann *Leucosis of Fowls & Leucemia Probl.* i. 9 Finally, it must be mentioned that Ellermann & Bang have proposed the word 'leucosis' as a common designation for leucemic and aleucemic cases, myeloses as well as lymphadenoses. 1927 A. Piney *Recent Adv. Hæmatol.* iii. 35 The name 'leucosis' is suitable for all forms, either lymphatic or myeloid, and is much to be preferred to leukæmia. 1935 Whitby & Britton *Disorders of Blood* xviii. 357 (*heading*) The leukæmias (leucoses). 1936 *Jrnl. Amer. Vet. Med. Assoc.* LXXXIX. 681 (*heading*) A study of transmissible fowl leukosis. 1941 *Ibid.* XCIX. 214/1 (*heading*) Studies on production of specific antibodies against the agent of the fowl–leucosis complex. 1943 Barger & Card *Dis. & Parasites Poultry* (ed. 3) vii. 177 The cause of the avian leukosis complex or fowl paralysis is a filter-passing agent. 1946 *Physiol. Rev.* XXVI. 48 The term leukemia or leukosis implies neoplasia of a blood cell (hemoblastosis) including both the aleukemic and leukemic types of the disease. 1960 *Proc. 7th Congr. European Soc. Haematol.* II. 291 (*heading*) The haematology of bovine leucosis. 1961 *Brit. Vet. Jrnl.* CXVII. 326 The separation of fowl paralysis granuloma from the leucoses has been suggested by Campbell. 1970 *Times* 13 Apr. 11/4 Only non-vaccinated maiden heifers and young bulls will be allowed in, and the stock will have to be cleared..of foot-and-mouth disease, tuberculosis, brucellosis, leucosis, leptospirosis, and Johnes disease.

leucotic (lⁱukǫ·tik), *a.* Also **leukotic.** [f. Leuc(osis + -otic.] Of, pertaining to, or affected with leucosis (in any sense).

1888 *Syd. Soc. Lex.*, *Leucotic*, of, or belonging to, Leucoma. 1935 Barger & Card *Dis. & Parasites Poultry* vii. 162 The leukotic type of the disease. 1960 *Proc. 7th Congr. European Soc. Haematol.* II. 291 In 'leucotic' herds ..17 animals died. 1961 *Brit. Vet. Jrnl.* CXVII. 316 It has been calculated by Blaxland (1956) that seven to eight million pounds sterling are lost annually in Britain because of the leucotic complex, which in this particular context is taken to include the various types of fowl paralysis.

leucotomy (lⁱukǫ·tǒmi). *Surg.* [ad. F. *leucotomie* (E. Moniz *Tentatives Opératoires dans le Traitement de certaines Psychoses* (1936) viii. 195): see Leuco- (here signifying the white matter of the brain) and -tomy.] = *lobotomy (in *spec.* sense).

1937 E. Moniz in *Amer. Jrnl. Psychiatry* XCIII. 1379 (*heading*) Prefrontal leucotomy in the treatment of mental disorders. 1947 *Times* 13 Feb. 6/4 At intervals in the past 10 years details of the brain operation of leucotomy performed on patients with certain types of mental illness have appeared in the Press. 1950 M. Greenblatt et al. *Stud. in Lobotomy* ii. 10 The English, however, use rather uniformly the more correct term 'leucotomy', in their description of the psychosurgical technic, whereas, in this country [*sc.* the U.S.A.], many still prefer the term 'lobotomy'. 1951 *Lancet* 21 July 91/2 (*heading*) Rostral leucotomy. 1958 *Listener* 5 June 931/2 People make moral objections to pre-frontal leucotomy even as a remedial measure. 1964 M. Argyle *Psychol. & Social Probl.* vi. 87 Other kinds of physical treatment, like insulin treatment and leucotomy, have declined in importance since the new methods have been discovered. 1967 [see *gyrectomy]. 1969 *Times* 22 Mar. 3/2 The symptoms of prefrontal leucotomy..include lack of inhibitions and inability to concentrate.

So **leu·cotome** [a. F. *leucotome* (E. Moniz 1936, in *Bull. de l'Acad. de Méd.* CXV. 390): see *-tome[1]], an instrument used to perform leucotomy; **leuco·tomize** *v. trans.*, to perform leucotomy on; **leuco·tomized** *ppl. a.*

1937 E. Moniz in *Amer. Jrnl. Psychiatry* XCIII. 1380 Sections were made in the subcortical white matter by a leucotome with a steel loop, tending to crush the white matter... At the present time we are using a leucotome with a steel band that cuts rather than compresses. 1951 *Lancet* 21 July 92/1 From the number of patients so leucotomised who have come to me for more extensive operations, the results do not appear very satisfactory. 1959 *Times* 24 July 4/4 Where not so long ago some of these patients might have been leucotomized we now no longer do so. 1964 M. Argyle *Psychol. & Social Probl.* xii. 149 Authoritarians, psychotics and leucotomized patients are all poor at person perception. 1969 *New Scientist* 30 Jan. 229/2 The destruction of localized areas in the brain has also been achieved by..the excision of a segment of tissue with an instrument (a leucotome).

leucovirus, var. *leukovirus.

leucovorin (lⁱukǫ·vǒrin). *Biochem.* [f. mod.L. *Leuco*(*nostoc* (f. Gr. λευκο- Leuco- + Nostoc), the generic name + *citro*)*vor*(*um* (f. Citr-, citro- + L. *vor-āre* to devour + *-um*) the specific epithet, of the bacterium whose growth it was originally found to promote: see -in[1].] = **folinic acid*.

1951 H. P. Broquist et al. in *Federation Proc.* X. 167/1 A crystalline substance with the properties of the 'citrovorum factor' (CF) was prepared from pteroylglutamic acid... The amounts of this crystalline substance, 'leucovorin', required for half-maximum growth per ml. of PGA-deficient medium were 0·00015 μg. for *Leuconostoc citrovorum* 8081. 1954 *Poultry Sci.* XXXIII. 111/1 Leucovorin (synthetic citrovorum factor) appeared to be inferior to folacin in the nutrition of the chick. 1966 [see *folinic *a.*]. 1971 *Cancer* XXVIII. 899/1 The combination of methotrexate with Leucovorin rescue is a well-tolerated therapeutic maneuver.

leukæmia. Now the usual form (with leukemia) of Leuchæmia (q.v. in Dict. and Suppl.).

leukæmogenic (lⁱukī:modʒe·nik), *a. Med.* Also **leukemogenic.** [f. *leukæm(ia + -o + *-genic.] Capable of producing leukæmia; pertaining to the production of leukæmia.

1942 *Jrnl. Nat. Cancer Inst.* (U.S.) III. 231/2 Previous experiments indicated that X-rays are leukemogenic. 1953 *Cancer Res.* XIII. 267/2 Exposure to leukemogenic doses of x-rays. 1971 *New Scientist* 8 July 64/2 If these tumours were then removed.., these could then be shown to have leukaemogenic activity characteristic of leukaemia viruses. 1971 H. J. Woodliff *Leukaemia Cytogenetics* v. 47 Chromosomal abnormalities may be produced by many of the agents discussed above and many of these are also thought to be leukaemogenic, such as viruses, ionising radiation, chemicals and drugs. 1973 *Nature* 9 Feb. 397/1 Thymus dependent lymphatic leukaemia was induced [in mice] by irradiation or by a leukaemogenic virus.

So **leukæ·mogen,** a substance or agent capable of producing leukæmia; **leukæmoge·nesis,** the production or development of leukæmia.

1942 *Jrnl. Nat. Cancer Inst.* (U.S.) III. 239/1 A systematic study of changes in the blood pictures during the early phases of leukemogenesis was beyond the scope of this work. 1944 Dorland & Miller *Med. Dict.* (ed. 20) 804/1 Leukemogen. 1946 *Physiol. Rev.* XXVI. 48 Organisms are subject to accidental exposure to chemical, physical and other agents which may be powerful leukemogens. 1953 *Cancer Res.* 268/1 Genetic factors determine susceptibility to specific leukemogens. 1961 *Lancet* 30 Sept. 748/1 We have been interested in the role of the thymus in leukæmogenesis. 1971 *Brit. Med. Bull.* XXVII. 67/1 The back-lash effects could in fact be mistaken for the advent of a new leukaemogen. 1973 *Nature* 12 Jan. 95/1 Viruses are associated with leukaemogenesis in both laboratory and outbred animals.

leukæmoid (lⁱukī·moid), *a. Med.* Also **leuc-, -emoid.** [f. *leukæm(ia + -oid.] Resembling (that found in) leukæmia but due to some other cause.

1926 *Amer. Jrnl. Med. Sci.* CLXXII. 529 Was it leukemia, simulating Banti's in an aleukemic early stage, or was it Banti's with a terminal leukemoid picture? 1940 *Acta Med. Scand.* CIII. 568 In another case of sepsis with marked hyperglobulinemia there was a leucemoid reaction with crowds of plasma cells. 1946 *Physiol. Rev.* XXVI. 62 The aerobic glycolysis values of preleukemic, leukemic and leukemoid lymph nodes, spleens and livers were 50 per cent to 100 per cent above normal. 1960 F. G. J. Hayhoe *Leukaemia* xvi. 318 The haematological findings in the myeloproliferative and lymphoproliferative syndromes provide a wide borderline between leukaemic and leukaemoid pictures,.. but we shall now be concerned with the unquestionably leukaemoid reactions occasionally observed in association with certain infections, metastasizing tumours, and some non-leukaemic blood diseases.

leuko-, var. Leuco-.

leukocyte, -cytic, etc.: see Leucocyte in Dict. and Suppl.

leukovirus (lⁱū·kovəirǒs). *Virology.* Also **leuco-.** [f. *leuko- (in *leukæmia and *leukosis*, *leucosis) + Virus.] Any of a group of pleomorphic viruses consisting of enveloped single-stranded RNA, different members of which cause leucosis or tumours in mammals and birds.

1968 F. Fenner *Biol. Animal Viruses* I. i. 26 The viruses which are associated with leukosis of chickens and murine leukemia have as their genetic material a molecule of single-stranded RNA of about 12 million daltons atomic weight. The pathogenic potential of both groups of agents is similar, as are the virions. These viruses can therefore be grouped together into a new group, for which we propose the name 'leukovirus'. *Ibid.* 28 Most infections with leukoviruses are latent for prolonged periods, but they may eventually cause fatal disease, usually apparent as a disturbance of the lymphoid or hemopoietic systems. 1970 *Nature* 14 Nov. 622/1 The RNA tumour viruses (leucoviruses) seem to have a different mode of replication from other RNA-containing viruses. 1971 Maramorosch & Kurstak *Compar. Virol.* xvi. 514 The more euphonious term *leukovirus* suggested by Fenner (1968) has recently been approved by the International Commission for the Nomenclature of Viruses and will be used here. 1972 *Sci. Amer.* Jan. 25/3 A group of viruses, variously called the RNA tumor viruses, the leukoviruses or the rousviruses.., replicate by another mode of information transfer.

lev (lev). Also (erron.) **leva.** Pl. **leva, levas, levs.** [Bulg. *lev* (pl. *leva*), lion.] The basic monetary unit of Bulgaria.

1902 *Encycl. Brit.* XXVI. 451/2 Bulgaria..has adopted the metric system... The monetary unit is the *lev*,.. nominally equal to the franc. 1908 *Daily Chron.* 31 Oct. 1/6 The East Roumelian tribute amounts to 2,951,000 leva. 1921 S. Graham *Europe* ii. 33 Bulgarian francs or levas are..worth a bare three-farthings each today. 1928 *Daily Tel.* 16 Oct. 18 The Budget 1927–28 showed a surplus of sixty-five million levs. 1928 *Morning Post* 20 Oct. 11/5 The booty is estimated at many hundreds of thousands of levas (the leva is valued at about 600 to the £). 1959 *Chambers's Encycl.* II. 674/2 In May 1952 a drastic monetary reform was carried out whereby the lev was linked to the Russian rouble at the rate of 1·70 leva = 1 rouble. 1972 D. Dakin *Unification of Greece* xiii. 188 Bulgaria had borrowed from France 245 million *leva* in 1904 and 1907.

‖ **levada** (levā·dă). [Pg.] In Madeira, a canal for irrigation.

1885 J. Y. Johnson *Madeira* vi. 68 Besides the great levadas there are minor ones in every parish. 1920 *National Rev.* Nov. 408 Levadas are narrow canals cut out of the solid rock of volcanic basalt of which the island consists: watercourses of masonry, which intersect Madeira like a network, for the purposes of irrigation. *Ibid.* 409 One of the more hazardous levada walks. 1963 *Times* 28 Mar. 14/6 We..set off from the rest-house on the *levada* at Queimadas. 1975 *Country Life* 13 Feb. 400/1 Every farm is irrigated by *levadas*—a network of channels which carry rainwater down from the mountains.

‖ **levade** (ləvā·d). [Fr., f. *lever* to raise.] (See quot. 1954.)

1944 E. Byng *World of Arabs* 246 The three widely celebrated specialties of Vienna's 'Spanish' Riding School ..are known as the levade, the piaffe, and the capriole. 1953 G. Brooke *Introd. Riding* vii. 69 A system that he believed would enable him to demonstrate the 'levade' and 'croupade' and similar gymnastics. 1954 A. Podhajsky *Spanish Riding School* (ed. 2) 22/1 Training above the ground is usually started either with the Levade or Pesade. In both these exercises which are constructed from the Piaffe, the hind quarters, deeply bent in the haunch, support the entire weight of the body, whilst the fore quarters with the fore feet drawn up under the body, rise more or less high above the ground; the duration of this position depends upon the dexterity and strength of the stallion. If the horse raises the fore quarters so high that his body reaches an angle of 45° with the earth, this is called 'Pesade'; on the other hand, if it raises the fore quarters less high, one speaks of a Levade.

Levallois (ləva·lwa). *Archæol.* [f. the Fr. place-name *Levallois* in north central France, NW. of Paris.] Used *attrib.* as a term for one of the main palæolithic cultures, post-Acheulian and pre-Mousterian. Hence **Levalloisean** (-wa·ziăn), **-ian** *adjs.* Also **Levalloisoid** (-wa·zoid) *a.*, related to, or similar to, this culture. Also in *Comb.*, as **Levalloiso-Mouste·rian** *a.*

1921 R. A. S. Macalister *Text-bk. European Archæol.* I. 239 The transition from Lower to Middle Palaeolithic is marked by the introduction of two important types of implement; the miniature coup-de-poing..and the Levallois scraper. 1932 [see *Clactonian *a.*]. 1934 *Jrnl. R. Anthrop. Inst.* LXIV. 342 Partly Chellean, partly local Acheulean with a hint of Levalloisean. 1937 *Ann. Reg. 1936* 52 Associated with Levalloisian culture. 1937 Garrod & Bate *Stone Age Mt. Carmel* I. I. ii. 8 The majority of these [*sc.* flints] were of Levalloiso-Mousterian type, but a fair number could be referred to the Lower Aurignacian. 1938 *Encycl. Brit. Bk. of Year* 49/1 A Mousterian using the Levallois technique. 1938 *Proc. Prehist. Soc.* IV. 19 A Levalloisian flake industry forms part of the latest Acheulian. 1952 *Ibid.* XVIII. 10 They include Acheulean hand-axes and Levalloisoid flakes. 1959 J. D. Clark *Prehist. S. Afr.* ii. 40 The Middle Stone Age is essentially the time of the specially prepared core and the flake tool derived from it. This is the technique known in Europe as 'Levallois' technique' and in Africa

as 'prepared platform technique'. 1961 L. D. Stamp *Gloss. Geogr. Terms* 532 In East Anglia Clactonian is contemporaneous with Acheulian; Levalloisian with Mousterian.

levan (lī·væn). *Chem.* [f. Lævo-, lev(o- + *-an*, after *dextran*.] A lævulosan (fructan); *esp.* any fructan of the kind produced by certain bacteria, in which the linking of adjacent fructose units is between the second carbon atom of one unit and the sixth of the next.

1902 R. G. Smith in *Proc. Linn. Soc. New South Wales* XXVI. 603 From a review of the lævo-rotatory gummy substances that are hydrolysed to levulose, it appears that this bacterial gum has not hitherto been described. I therefore propose for it the name levan, which was suggested by the polariscopic nature of the gum and derived glucose, and also from the fact that another bacterial gum, which is derived from dextrose, and which yields dextrose on hydrolysis, is known as dextran. 1948 W. Pigman *Chem. Carbohydrates* xv. 604 Fructosans or levans are found widely distributed throughout the plant kingdom..and generally serve as reserve polysaccharides in place of, or in addition to, starch. 1953 Whistler & Smart *Polysaccharide Chem.* xi. 276 D-Fructose polymers occurring in plants..are designated as fructans while those elaborated by microorganisms are called levans... Actually the levans are structurally similar to some of the plant polyfructoses. 1965 T. Akazawa in Bonner & Varner *Plant Biochem.* xii. 287 Although the size of these substances is small, 3–30 hexose residues per molecule, their structure is basically the same as that of the high molecular bacterial levan produced by several microorganisms... The name grass levan has therefore been given to this group of plant compounds. A further interesting point is..that fructosans of inulin type, β-(2→1), and of levan type, β-(2→6), are the main constituents of stem and ear tissues, respectively, of cereal plants.

levantinism (lǐvæ·ntiniz'm). [f. Levantin(e *a.* and *sb.* + -ism.] The spirit or culture of the Levant.

1949 Koestler *Promise & Fulfilment* III. iv. 330 What kind of a civilization will Israel's be? Will it be..the superficial veneer of Levantinism? 1961 *Guardian* 6 Feb. 8/2 Mr Ben-Gurion warned his countrymen against letting 'Levantinism' creep into their national life. 1973 *Observer* 6 May 6/6 Israel's leaders worry about 'levantinism'.

leva·ntinize, *v.* [f. as prec. + -ize.] *trans.* To make Levantine in form or character.

1929 *Times* 5 Nov. 17/4 The Government's policy of 'Europeanizing' a somewhat levantinized Administration. 1930 *Times Lit. Suppl.* 4 Sept. 691/1 The rest of the Turks remained farmers..and thereby escaped the contamination of the Levantinized Ottomanism. 1946 Koestler *Thieves in Night* 112 Our small community will..become levantinised, submerged in the Arab sea.

‖ **levari facias** (livā·ri fē̆ⁱ·ʃiæs). *Law.* [L., = cause to be levied, f. *levari*, to be levied, f. *levāre* to raise + *facias* cause, 2nd pers. sing. pres. subj. of *facĕre* to do, make.] (See quot. 1768.)

a 1625 H. Finch *Law* (1627) IV. xlvi. 471 A *Leuari facias*... *Leuari facias* to leuie execution of the profits of his land and Chattels. 1768 Blackstone *Comm.* III. xxvi. 417 A third species of execution is by writ of *levari facias*; which affects a man's goods and the *profits* of his lands, commanding the sheriff to levy the plaintiff's debt on the lands and goods of the defendant;.. little use is now made of this writ. 1818 [see Fieri-facias]. 1888 *Encycl. Brit.* XXIV. 696/2 *Levari facias* is the means of levying execution for forfeited recognizances. 1959 Jowitt *Dict. Eng. Law* II, *Levari facias*..had been practically superseded before 1883 by the writ of *elegit*; and the Bankruptcy Act, 1883, s. 146 (2), enacted that it should no longer be issued in any civil proceeding.

‖ **leveche** (leve·tʃe). [Sp.] A hot, dry, more or less southerly wind of south-eastern Spain, the local counterpart of the sirocco.

1887 *Encycl. Brit.* XXII. 296/2 The eastern part of this [southern] zone is the part of Spain which is liable to be visited from time to time by the scorching and blasting *leveche*, the name given in Spain to the sirocco. 1927 [see *gibli]. 1962 J. E. Van Riper *Man's Physical World* vii. 222/2 The Mediterranean area also is the home of a hot, searing, dust-laden wind off the Sahara, known in various localities as *sirocco, khamsin, leveche*, or *samiel*.

levee, *sb.*[1] Add: **1. b.** *Geol.* A low broad ridge of water-laid sediment running along the side of a stream channel; also, any of various similar natural embankments, as those formed by mud flows or lava flows, or along a submarine channel.

1870 in L. C. Cramton *Early Hist. Yellowstone Nat. Park* (1932) 129 Passing over a sand levee, grown up with sagebrush, we found ourselves on the open beach of the great Yellowstone Lake. *Ibid.* 137 The shoreline is bordered by a levee of obsidian, lava pebbles, and calcareous fragments, cutting off and inclosing ponds of water behind it. 1910 *Proc. Indiana Acad. Sci. 1909* 260 Deltas occasionally take the form of long, narrow ridges upon one or both sides of a stream, resembling the natural levees in the 'goosefoot' of the Mississippi. 1942 *Jrnl. Geomorphol.* V. 222 (*heading*) Mudflow levees. 1957 G. E. Hutchinson *Treat. Limnol.* I. i. 99 Levees may form along the water courses. 1962 E. A. Vincent tr. *Rittmann's Volcanoes* i. 33 When the supply of lava diminishes and finally comes to an end, the still-fluid lava inside the stream continues to flow out and the mantle of scoriae collapses, leaving a more or less even flow of scoriaceous block lava (clinker lava), flanked on both sides by upstanding block walls, called lava moraines (scoria moraines, lava levées). 1964 *Bull. Amer. Assoc. Petroleum Geol.* XLVIII. 1141/2 Trawl No. 23 was taken from a natural levee [of the Congo Submarine Canyon] and although the water depth was more than 500 fathoms greater than that of trawl No. 22, there was no marked decrease in diversity and abundance of animal life. 1968 R. W. Fairbridge *Encycl. Geomorphol.* 651/2 Alluvial streams flowing on flood plains commonly develop natural levees. Each levee is a low, wide ridge located immediately adjacent to the channel. 1972 G. A. Macdonald *Volcanoes* v. 84 Overflows spread lava a few feet on either side of the river... Repeated overflows gradually build up natural levees.

levee, *v.*[1] (Earlier examples.) Also, to shut or keep *off* by means of a levee.

1832 R. Baird *View of Valley of Mississippi* xxii. 269 Much has been done to levee or embank the Mississippi River. 1837 J. L. Williams *Territory of Florida* 45 Where there is clay enough in the soil, to form good embankments, the waters might be leveed off. 1847 J. Palmer *Jrnl. Trav. Rocky Mts.* 121 Several islands in the river might be *leveed* and successfully cultivated.

So **levee·ing** *vbl. sb.*

1845 *Indiana Senate Jrnl.* 364 An act to authorize the leveeing of Blue river, in Shelby county.

‖ **levée en masse** (ləve an mas). [Fr.] Mass mobilization, orig. in Revolutionary France, in response to invasion; = Levy *sb.*[1] 3. Also *fig.*

1813 F. Burney *Let.* 12 Oct. (1905) VI. 95 Were he not essentially necessary in some department of civil labour and use, he would surely be included in some *levée en masse*. 1832 *Edin. Rev.* Apr. 254 A *levée* [sic] *en masse* was decreed. The zeal of the Polish patriots was unbounded. 1895 T. A. Walker *Man. Public Internat. Law* 135 It is, in fact, clear law that a combatant to be lawful must be formally authorised by a recognised Government, or be a member of a *levée en masse* rising on the approach of an invader. 1940 *Economist* 22 June 1067/1 There has been an outcry..for an immediate *levée en masse*, for the calling up now of every able-bodied man not busy on war work. 1943 J. M. Thompson *French Revolution* xxii. 424 A decree of February 24th, '93, ordered the levying of three hundred thousand men—..less than half the necessary number was procured. It became clear that the Convention must fall back on..a *levée en masse*, or wholesale compulsory enlistment. This plan..finally took shape in Carnot's decree, as it is generally called, of August 23rd. 1949 I. Deutscher *Stalin* v. 143 They were the 'activists' ..behind which there moved into battle a genuine political *levée en masse*. 1972 *Times* 29 Dec. 11/7 A levée en masse is one thing, assassination is quite another.

level, *sb.* Add: **2. b.** *on the level*, (in a) fair, honest, or straightforward (way); reliable, true. Freq. as *adv. phr.* = honestly; truthfully. *colloq.* (orig. *U.S.*).

1872 G. P. Burnham *Mem. U.S. Secret Service* p. vii, *On the level*, meeting a man with honorable intentions. 1896 Ade *Artie* vi. 50, I see barrel-house boys goin' around for hand-outs that was more on the level than you are. 1901 'J. Flynt' *World of Graft* iii. 89 When a man who has been a known thief makes up his mind to quit stealing and live 'on the level', they say in the Under World that he has 'squared it'. 1932 A. J. Worrall *Eng. Idioms* 50 You may be quite sure that the business is quite on the level. 1936 N. Marsh *Death in Ecstasy* xvii. 209 I've had no more'n my fair share. Same goes for Raveenje. He's on the level all right. 1958 R. Graves in *Times Lit. Suppl.* 15 Aug. p. x/3 He also prefers pools to premium-bond gambling—in which a bloke can't choose his own combination of numbers, so how does one know that it's on the level? 1970 G. F. Newman *Sir, You Bastard* ii. 47 If you're on the level, we won't object.

1896 Ade *Artie* v. 42 On the level, I'm surprised you ain't on to that. 1914 Wodehouse *Man Upstairs* 63 'You look good to muh,' he said gallantly. 'The idea!' said Maud, tossing her head. 'On the level,' Mr. Shute assured her. 1923 R. D. Paine *Comrades of Rolling Ocean* iii. 44 This was no fault of mine, on the level. 1931 E. Linklater *Juan in Amer.* III. iii. 231 'You're kidding,' said Buddy. 'On the level!' replied Olympia. 1942 T. Rattigan *Flare Path* II. ii. 137 On the level. I couldn't really.

3. d. A position (on a real or imaginary scale) in respect of amount, intensity, extent, or the like; the relative amount or intensity of any property, attribute, or activity. Freq. preceded by a sb. denoting the property, etc., referred to, as *danger, energy, noise level*.

1897 *Lancet* 5 June 1541/1 The pulse had been rising, and by 8.30 P.M. had reached its normal level (72 in the sitting posture). 1926 *Encycl. Brit.* III. 281/1 Nothing is to be gained by amplifying a signal below the 'noise level' at the location of the receiver. 1931 A. W. Nye in L. Cowan *Recording Sound for Motion Pict.* ii. 31 The sensation level of any sound reaching the ear is the number of decibels it is above the threshold level of audition. *Ibid.*, A change of the level of a sound by 1 db is approximately the smallest that the ear can detect. 1934 G. B. Shaw *On Rocks* I. 208 By the last returns the export of Spanish onions has again reached the 1913 level. 1935 [see *danger level* s.v. *danger *sb.* C]. 1942 W. B. Boast *Illumination Engin.* x. 166 Recommended levels of illumination must provide an adequate safety factor..to maintain visibility well above threshold values for critical tasks. 1948 W. E. Styler in M. Beloff *Hist.* xv. 320/1 Unemployment reached previously unknown levels, and overseas markets collapsed. 1958 H. G. M. Spratt *Magn. Tape Recording* vii. 207 When recording it is essential to provide some means of indicating the level of the signal applied to the tape to ensure that it is neither too high nor too low. *Ibid.* 208 Low signal levels. 1967 [see *danger level* s.v. *danger *sb.* C]. 1968 Miller & Sawers *Technical Devel. Mod. Aviation* vii. 223 The level of general passenger comfort aboard the four-engined jets. 1971 *Times* 17 Mar. 21 (Advt.), I am encouraged by the current general level of orders and I am sure that because of the action which has been taken since the merger, [etc.]. 1973 Harrison & Waters *Burne-Jones* iii. 25 Possessing a high level of natural skill Edward Jones made rapid progress. 1973 *Nature* 23 Nov. 183/1 That could result in dangerous levels of sulphuric acid and sulphates in city air. 1974 *Daily Tel.* 11 Mar. 16 New house starts over the last three months are down to 40 per cent. of their level at this time last year.

e. *Physics.* More fully, *energy level*. An amount of energy associated with an atom or other quantized system and capable of being possessed by one of its constituents, being usu. measured relative to the minimum possible energy of that constituent; also, a discrete state of a quantized system characterized by such energy; *spec.* a state or group of states of an atom characterized by the quantum numbers *n, L, S,* and *J*, as distinguished from a 'term' (a group of levels: see *term *sb.*) and a 'state' so called (a constituent of a level: see *state *sb.*).

1922 A. D. Udden tr. *Bohr's Theory of Spectra* III. iv. 116 The values of the atomic energy corresponding to these [stationary] states are frequently referred to as the 'energy levels' of the X-ray spectra. 1925 Russell & Saunders in *Astrophysical Jrnl.* LXI. 69 When the series limit—or ionization level—can only be found inaccurately..the common convention of measuring terms from this level becomes inconvenient... A desirable alternative might be to set the zero-level at the *lowest* term and measure the others upward from this... In such cases, the numerical values referred to the lowest level might be called 'levels' to distinguish them from 'terms', referred to ionization as zero-point. 1926 R. W. Lawson tr. *Hevesy & Paneth's Man. Radioactivity* vii. 75 This suggests that even in the nucleus of an atom there are different energy levels. 1934 H. E. White *Introd. Atomic Spectra* v. 78 [In sodium] just as in hydrogen these energy levels represent certain possible energy states of an electron, and transitions between them represent spectrum lines. 1935 Condon & Shortley *Theory Atomic Spectra* iv. 122 All except *s* configurations split into two levels, corresponding to $j=l+\frac{1}{2}$ and $j=l-\frac{1}{2}$... The (one or) two levels into which each configuration is split are together said to constitute a doublet term. 1955 R. D. Evans *Atomic Nucleus* iv. 122 After the emission of the β ray, each residual nucleus of Si^{28} is left in an excited level at about 1·78 Mev above its ground level. 1962 D. F. Shaw *Introd. Electronics* ix. 170 There are important modifications to the arrangement of energy levels when, as in a crystalline solid, the atoms are separated by distances of the same order of magnitude (10^{-8} cm) as the atomic diameters themselves... The valence electrons may no longer be associated with a particular atom. They become a group of 'free electrons' in energy levels which belong to the lattice as a whole. 1962 R. E. Dodd *Chem. Spectroscopy* ii. 78 Transitions between rotational levels [of a molecule] without change in vibrational or electronic energy, give lines in the far infra-red and microwave region. 1970 G. K. Woodgate *Elem. Atomic Struct.* i. 5 The spin–orbit interaction is the largest relativistic effect and is responsible for fine structure. Each term splits into levels whose separations are of the order of 1–1,000 cm^{-1}. *Ibid.* 6 The levels are split further into states by the application of a laboratory magnetic field... This is called the Zeeman effect.

f. *Statistics.* In full, *level of confidence* or *significance*. A number chosen as the maximum (or minimum) value of the probability with which any statistical result must be false (or true) for that result to be accepted as having been demonstrated.

1925 R. A. Fisher *Statistical Methods for Research Workers* vi. 157 Taking the four definite levels of significance, represented by P = ·10, ·05, ·02, and ·01, the table shows for each value of *n*, from 1 to 20, and thence by larger intervals to 100, the corresponding values of *r*. 1931 L. H. C. Tippett *Methods of Statistics* iii. 48 Adopting the 0·05 level of significance, a deviation in the mean greater than twice its standard error is statistically significant. 1937 Yule & Kendall *Introd. Theory Statistics* (ed. 11) xxii. 425 There are..two values of *P* [*sc.* a probability] which are widely used to provide a rough line of demarcation between acceptance and rejection of the significance of observed deviations. These values are $P=0{\cdot}05$ and $P=0{\cdot}01$, and are said to define 5 per cent. and 1 per cent. levels of significance. 1950 W. Feller *Introd. Probability Theory* I. vii. 142 However, no sample size can give absolute assurance that $|p'-p|<0{\cdot}005$... Since absolute certainty is unattainable, we settle for an arbitrary confidence level α, say $\alpha=0{\cdot}95$, and require that $|p'-p|<0{\cdot}005$ with probability 0·95 or better. 1972 *Jrnl. Social Psychol.* LXXXVII. 39 The situations variable was significant at the ·01 level for all four dependent variables. *Ibid.* 48 A *t* value of 2·33 or greater indicates that differences between the high and low criterion group means are significant at the ·01 level of confidence.

g. Contextually in *Broadcasting*, etc.: the sound level or signal level as it shows up in the different pieces of equipment. Also *attrib.* in **level test,** a test of signal levels to determine whether changes in control settings, microphone positions, etc., are required.

1940 E. McGill *Radio Directing* ix. 184 If many sound effects are devised it will be found that a great amount of rehearsal time will be consumed in trying to bring to perfection the levels and balances of sounds against orchestra

..under unfavorable acoustic circumstances. 1941 *B.B.C. Broadcasting Terms* 17 Level test. 1962 A. Nisbett *Technique Sound Studio* iv. 79 The purpose of control of levels and lining-up of equipment..is to make the best use of the region between which noise and distortion overtake the recording. 1966 *Listener* 4 Aug. 181/3 Delius's *Requiem*..seemed harassed by eccentric studio management, with levels all over the place. 1966 B. Glemser *Dear Hungarian Friend* xiii. 223 We must do a level test... Just talk naturally..and we will see what we pick up. 1969 J. Elliot *Duel* III. iii. 251 He wants you to say a few words..just for the level.

4. b. A plane or status in respect of rank or authority; position in a hierarchy. Freq. with a qualifying adj.

1933 L. Bloomfield *Language* iii. 49 Provincial colorings of standard English are tied up with differences of social level. 1937 A. Huxley *Ends & Means* x. 148 Examples of non-violence on the governmental level are seldom of a very heroic kind. 1944 *Amer. Speech* XIX. 234/1, I have often been amused at the constant recurrence of certain catchwords and phrases in [Government] memoranda... *level* ('This matter will be handled at the regional *level*'). 1945 *N.Y. Times* 24 June IV. 6/4 One of the reporters asks if he knows of any obstacle to our perfect cooperation with the Russians. Quick as a flash General Eisenhower replied, 'On my level, none.' 1948 Masterson & Phillips *Federal Prose* vi. 30 Until a program for personnel induction at the infant level can be coordinated with the Federal Prose tutorship objectives, this situation will continue to create embarrassments at the administrative and higher levels. 1952 *Economist* 20 Sept., How long it takes to get even a simple low-level decision. 1955 *Times* 10 May 10/1 The western Foreign Ministers have agreed in Paris to invite Russia to a four-Power conference, though the level at which the meeting should be held is not yet decided. 1960 *B.S.I. News* Jan. 9/2 This sort of progress can only be achieved through full consultation from and with users at the national level. 1962 *Sci. Survey* III. 263 There exist equally interesting relationships between odours and animal behaviours on a different and more profound level. 1971 *Guardian* 24 Sept. 22/7 On instructions from director-level, the estimates..had been prepared. 1974 *Daily Tel.* 1 Apr. 6/7 At the next level of responsibility are the 14 regional health authorities. 1974 *Nature* 17 May 210/2 In talking about the recognition of [alien] life at the microscopic level the decision is largely an aesthetic one. 1974 Green & Hooper *C. S. Lewis* x. 253 The stories can be read and enjoyed on at least two levels: by the child who perhaps knows nothing..of any of the authors whose works Lewis knew; and by the reader who knows many.

c. *Linguistics.* (See quots.)

1935 *Trans. Philol. Soc.* 61 Now to illustrate this empirical analysis of meaning at the phonetic, morphological, syntactical, and semantic levels. 1942 C. F. Hockett in *Language* Jan.–Mar. 3 Linguistics is a classificatory science. The starting-point in such a science is to define (1) the universe of discourse and (2) the criteria which are used in making the classification. Selection and preliminary ordering of data determine the *range* of analysis; the choice of criteria fixes the *level* of analysis. In linguistics there are various ranges,.. and two basic levels, *phonological* and *grammatical*. 1958 C. Rabin in *Aspects of Translation* 130 Items which are the same at all levels (e.g. numbers) do not function as level-markers. 1959 M. Halle *Sound Pattern Russ.* i. 25 The rules of translation which make up the grammar can all be subsumed under the formula 'replace x by y under condition z'... A set of rules yielding representations of a particular type is called a *linguistic level*. 1964 E. Bach *Introd. Transformational Gram.* iv. 59 The word 'level' is also used occasionally in another sense to refer to the ordering of rules within a single level and also to the ordering of the PS, transformational, and phonological rules with respect to each other. 1964 M. A. K. Halliday et al. *Ling. Sci.* i. 10 From these three types of patterning are derived the three principal levels: substance, form and context. 1964 R. H. Robins *Gen. Ling.* 12 By extension the term level of language is used to designate those aspects of a language on which at any time the linguist is focusing his attention. 1969 *Pocket Oxf. Dict.* (ed. 5) (Suppl. Austral. & N.Z. Words) 1017 There are 'levels of usage' in Australia and New Zealand as there are elsewhere and the cautionary labels colloq., sl. (= slang), etc., are therefore employed here in the customary manner as a guide to currency. 1973 *Archivum Linguisticum* IV. 17 In the field of English intonation studies, bones of contention ..spring readily to mind: levels versus configurations.

12. level test (see *3 g); **level tube** = *bubble-tube* s.v. *bubble *sb.* 6.

1890 Level tube [see *bubble-tube* s.v. *bubble *sb.* 6]. 1950 J. Clendinning *Princ. & Use of Surveying Instruments* v. 121 The level tube consists of a glass tube, partially filled with liquid, the inner surface of which is carefully ground.

level, *a.* and *adv.* Add: **A.** *adj.* **7. b.** (Earlier example.) So *level head*, a well-balanced person.

1869 'Mark Twain' *Innoc. Abr.* xl. 426 The wanderers were right, and the heads of the same were level. 1906 'O. Henry' *Four Million* 204 James Williams belonged among the level heads.

9. (Earlier U.S. and later U.K. examples.) Also *levelest* in the same sense, and similarly *level worst*, etc.

Of these only *level best* is standard in the U.K.

1851 *An Arkansaw Doctor* 87 (Th.), We put our horses out at their level best. 1884 'Mark Twain' *Huck. Finn* xxviii. 270 The old man..was on hand and looking his level pisonest. 1891 *Harper's Mag.* July 208/2 The pony will not do his level worst again. 1898 H. S. Canfield *Maid of Frontier* 97 She told me..that she was goin' to do her levelest to make our little home comfortable. 1920 Galsworthy *In Chancery* II. vii. 186 Val walked out behind his mother, chin squared, eyelids drooped, doing his level best to despise everybody. 1933 M. Lowry *Ultramarine* 205 You've been doing your level best to make life a misery to me since we left home. 1937 V. Bartlett *This is my Life* xi. 179 Everyone was doing his level best to make me feel nervous. 1953 R. Lehmann *Echoing Grove* II. 89 When the pain nagged he thought about the relation between worry and his acid juices, and did his level best to stop worrying. 1969 *Listener* 24 Apr. 556/1 He did his level best to suppress the views of other members of the embassy.

10. *level-backed, -balanced, -bellied, -browed, -grown, -lidded, -mouthed, -ranked, -tempered* adjs.; **level-compounded** *a. Electr. Engin.*, applied to a compounded generator in which the windings are such as to produce the same voltage on full load as on no load (and usually on intermediate loads also); = *flat-compounded* adj. s.v. *flat *a.* 15; **level luffing,** luffing in which the load is maintained at constant height; freq. *attrib.*, as *level luffing crane*.

1926 Kipling *Debits & Credits* 232 Level-backed and level-bellied watch 'em move. 1917 D. H. Lawrence *Look! we have come Through!* 66 She Put back her fine, level-balanced head. 1938 Belloc *Sonnets & Verse* 199 Level-browed divine Touraine. 1915 W. T. Maccall *Continuous Current Electr. Engin.* viii. 204 The point, B, at which the generator is level-compounded, *i.e.* has the same P.D. as at no load, can be made to occur at any one load. 1957 A. T. Starr *Appl. Electr.* vii. 182 An over-compounded generator really acts as a level-compounded generator plus a booster. 1971 L. T. Agger *Introd. Electr.* xvi. 298 In the level-compounded generator L the p.d. on full load is the same as on no load, and it varies only slightly in between. This is useful where the load changes rapidly, as in traction systems. 1866 G. M. Hopkins *Jrnl.* (1959) 136 Beech branches..with level-grown pieces of pale window-like green. 1926 E. Bowen *Ann Lee's* 146 Miss Phelps' blue, calm, level-lidded eyes. 1922 H. H. Broughton *Electr. Handling of Materials* III. 33 Few level-luffing arrangements have been devised by crane makers on the Continent. 1963 R. Hammond *Mobile & Movable Cranes* iv. 110 Level luffing is achieved by the Babcock 'swan neck' supported by a fixed guy rope. 1971 *Engineering* Apr. 65 (Advt.), Level luffing cranes. 1948 C. L. B. Hubbard *Dogs in Brit.* 301 The muzzle is long, powerful and level-mouthed. 1867 G. M. Hopkins *Jrnls. & Papers* (1959) 153 Very level-ranked sunset. 1939 D. Cecil *Young Melbourne* i. 23 Level-tempered and rational, she found scenes and caprices as tiresome as they did.

level, *v.*[1] Add: **1. e.** *Phonology.* To alter (a sound) so that it falls together with a similar sound. Usu. const. *under*.

1884 H. Sweet *First Middle Eng. Primer* 5 The old diphthongs *ea, ēa, eo, ēo* became monophthongic, *ea* being levelled under O.E. *æ*, written *e* in M.E., and *ēa* under O.E. *ǣ*, so that such a pair as the O.E. *heard* and *þæt* were both pronounced with the same vowel. 1888 —— *Hist. Eng. Sounds* 178 In North. *ā* has been preserved unrounded up to the present day in the Scotch dialects, where it has been levelled under new long *ā*. 1972 *English Studies* LIII. 503 There is, however, another type of development by which ME *ǫr* and ME *ōr* were levelled under the same sound as early as the fifteenth century.

3. Also *refl.*

1907 *Daily Chron.* 25 May 1/7 Another halfpenny may possibly be put on the loaf before prices level themselves again.

4. c. Also with *out*. (Further examples.)

1938 *Times* 24 Jan. 21/4 Later he took up with the heads of the motor industry..the question of their co-operating with the Government to level out the production and sales of cars. 1971 *Fremdsprachen* xv. 227/1 Much can be done to level out these differences by proper use of incentive and social purpose funds derived from the profits of enterprises.

d. To get rid of, put *away*, by levelling. Also *intr.*, with *away*: to become level.

1910 Galsworthy *Sheaf* (1916) 132 All the natural weaknesses and limitations of the dwellers shall be.. levelled away and minimized. 1921 —— *To Let* II. ii. 140 Those two crumpled rose leaves, Fleur's caprice and Monsieur Profond's snout, would level away if he lay on them industriously.

5*. *to level off* (or *out*): to bring an aircraft into horizontal flight (*intr.* and *trans.*); (of an aircraft) to assume horizontal flight, to flatten out.

1928 *Lit. Digest* 12 May 74/2 A 'pancake landing' occurs when the ship is leveled off several feet above the ground. 1928 *New Republic* 15 Aug. 331/2 In a straight dive down, coming out of a 'stall'..which stopped only when I levelled off and began to fly straight. *Ibid.*, When it gets into the diving position, it responds to all the controls and can be gradually levelled out. 1937 D. & H. Teilhet *Feather Cloak Murders* viii. 135 Climbing in the still blue atmosphere to five thousand feet, the ten-passenger Sikorsky amphibian levelled off above Honolulu. 1952 A. Y. Bramble *Air-Plane Flight* xviii. 306 Some air-planes take a considerable time to reach cruising air speed, with cruising power setting, after levelling off. 1963 Level off [see *auto-[1] b].

b. *fig.* To cease increasing or decreasing.

1958 *Listener* 10 July 40/1 If the recent signs of improvement in American business are not followed up, if production levels off again and unemployment rises.. then [etc.]. 1968 *Guardian* 24 Apr. 11/5 The American war effort can begin to level off and eventually to be reduced. 1968 *Times* 16 Dec. 7/1 Yields have been tending to level off, or even to fall. 1972 *Guardian* 21 July 21/5 There are a few signs..that the property market is beginning to level off and prices are steadying somewhat.

8. To be honest or truthful; to tell the truth, speak frankly, behave honestly or deal straightforwardly (*with*). *slang* (orig. *U.S.*).

1920 H. C. Witwer *Leather Pushers* (1921) 174 'Are you levelin' with the Kid in this one?' 'We level in all of 'em!' 1931 *Amer. Mercury* Dec. 416/2 Hymie, the mug, falls in love with her right off, Don't laugh, I'm levellin', honest to God. 1936 R. Chandler in *Black Mask* Mar. 28/1 'I was on the cops, but they bounced me.' I liked his telling me that. 'You must have been levelling,' I said. 1951 I. Shaw *Troubled Air* vii. 107 You're not levelling with me. 1962 K. Orvis *Damned & Destroyed* ix. 64 But see that you level with me about the new pusher. 1966 E. McGirr *Funeral was in Spain* 100 'Think Songbird was levelling?' 'Oh yes... Mr. Songbird wouldn't mislead you.' 1972 'R. Crawford' *Whip Hand* I. viii. 49 I'll level with you. I've been paid to find your brother. 1973 *Tucson* (Arizona) *Daily Citizen* 22 Aug. 55/8 Not often enough will a company truly level with its employes. It won't say, 'These are hard times and here's why.' 1974 L. Deighton *Spy Story* iv. 97 I'd better level with you, son... From now on, control is through me.

le:vel-hea·dedness. [f. *level-headed* adj. (Level *a.* 10) + -ness.] Calmness or balance of mind or judgement.

1876 *Rep. Vermont Board Agric.* III. 156 That same steadiness, or, in horse parlance, level-headedness,.. is quite as essential on the race track..as any where else. 1886 *New Englander* (New Haven, Connecticut) Feb. 179 This unexampled success is due..to the levelheadedness of its clerical guardians. 1896 *Alma Mater* 11 Nov. 43/2 A man whose great characteristic is level-headedness. 1916 *Daily Chron.* 13 Oct. 4/5 The British Air Service is now a great army,.. all endowed with two sterling qualities required by the pilot of the air, courage and levelheadedness. 1927 *Daily Express* 9 Aug. 8/2 A tribute to the level-headedness of the country which in his puny way he tries to wreck. 1937 Koestler *Spanish Testament* iv. 84 English journalists..with their traditional feeling for level-headedness and decency, have often had to complain of this difficulty.

levelling, *vbl. sb.* Add: **2. b.** (Further examples.) Also with *off, out*.

1871 H. Sweet *King Alfred's West-Saxon Version of Gregory's Pastoral Care* p. xxxvii, The change is not phonetic,.. but is due to inflectional levelling, the nom. terminations being made uniform, regardless of gender. 1888 J. Wright *Old High-German Primer* 16 The regular operation of this law was often disturbed by new formations made by levelling. 1903 G. B. Shaw *Revolutionist's Handbk.* vii, in *Man & Superman* 201 To them the limit of progress is, at worst, the completion of all the suggested reforms and the levelling up of all men to the point attained already by the most highly nourished and cultivated in mind and body. 1932 F. R. Leavis in *Scrutiny* I. 137 Mass-production, standardization, levelling-down—these three terms convey succinctly, what has happened. 1953 *Manch. Guardian Weekly* 13 Aug. 7/2 Housewives..cheered the heavy drop in the price of beef and thankfully attributed it to..the 'levelling-off' of inflation. 1955 *Times* 17 Aug. 5/5 Efforts by the employers to make them [*sc.* piece rates] realistic have foundered on the insistence of the men that they shall be changed only by levelling up. 1962 Simpson & Richards *Physical Princ. Junction Transistors* vi. 112 The beginning of this levelling-off process can be seen in the figure. 1964 F. Bowers *Bibliogr. & Textual Crit.* I. ii. 13 This evidence also suggests that authors' papers and not a levelling-out scribal transcript formed the printer's copy. 1971 *Cabinet Maker & Retail Furnisher* 24 Sept. 518/1 Important social changes and a levelling-off of income groups are playing a strong part in the expanding consumer market. 1972 *Guardian* 30 Mar. 14/3 Labour came to office with a strategy of levelling-up, of faster growth to finance greater equality.

level-pegging, *vbl. sb.* (passing into *adv.*). (Level stress.) [Cf. Pegging *vbl. sb.*] On equal terms (competitively), neck-and-neck, neither falling behind nor getting ahead (used of two individuals, groups, etc.). Also (as a back-formation) **level-peg** *v.*

1927 W. E. Collinson *Contemp. Eng.* 30 Cribbage is scored with pegs on a triangular board with two rows of holes on each side and supplied us with the technical expressions: level pegging or neck and neck (*from the racecourse*), [etc.]. 1959 *Motor* 16 Dec. 704/1 Two competitors..[one of which] lost one less mark than [the other] on the fourth section which was the first section on which they were not level pegging. 1962 *Listener* 13 Dec. 1027/2 The sociologist is level-pegging with the psychologist to replace the priest in the cure-of-souls business. 1965 *Observer* (Colour Suppl.) 28 Mar. 16 While air and sea were roughly level-pegging at a million passengers each in 1958, last year's split was 3·5 million to the airlines and 714,000 to the sea. 1973 J. Wainwright *Touch of Malice* 155 They were level-pegging—in rank, in age and in service. 1974 *Listener* 21 Feb. 244/2 ITV first made the running in the coverage of election news... Initiative reverting to the BBC? Let's call it level pegging.

lever, *sb.*[1] Add: **3. f.** (Examples.)

1865 *N. & Q.* 27 Jan. 27/2 (Advt.), The prettiest gift for a lady is one of Jones's gold levers at 11*l* 11*s*. 1895 in *N. & Q.* (1941) 20 Sept. 160/1 It couldn't have been a stop watch. It was a lever.

5. lever frame: delete '*U.S.*' and add examples.

1869 *Bradshaw's Railway Manual* XXI. App. 116 (Advt.), Patent locking lever frames. 1950 *Times Rev. Industry* 21/1 All points and signals are worked from a mechanical or manual lever frame. **1955** *Railway Mag.* May 307/2 At Stockport No. 2 signalbox, the existing mechanical lever frame has been retained. **1963** KICHENSIDE & WILLIAMS *Brit. Railway Signalling* vi. 74 (*caption*) The interior of London Bridge signal box..showing the miniature lever frame.

lever, *v.* Add: **1. b.** To make way by leverage.

1883 S. BARING-GOULD *John Herring* I. i. 9 When he took his weight off,.. the plough levered out of the ground.

2. Also *refl.* with *into.*

1910 *Westm. Gaz.* 24 Mar. 2/3 The Moderates have levered themselves into a position they have no claim to occupy on the Council.

le·verage, *v. U.S.* [f. the sb.] *trans.* and *intr.* To lever; *spec.* to speculate or cause to speculate financially on borrowed capital expecting profits made to be greater than the interest payable. Hence **le·veraged** *ppl. a.*; **le·veraging** *vbl. sb.*

1937 *Harper's Mag.* June 63 Acey leveraged the arm upward. **1957** *Robert R. Young & Alleghany Corp.* 2 Founded in 1929.., Alleghany was a classic example of the highly leveraged holding companies of that period. **1968** *N.Y. Times* 20 Feb. 64 Short-term trading,.. selling short and leveraging through borrowing are all speculative techniques which carry with them greater risk of loss. **1971** *Atlantic Monthly* July 49 He gave her the benefit of his experience, leveraging her up to the ears in convertible bonds. **1972** 'A. SMITH' *Supermoney* IV. i. 209 The corporation discovered that the more it borrowed, the higher the earnings and the higher the stock, so it began to leverage. **1973** *N.Y. Law Jrnl.* 26 July 3/3 Tight credit tends to put some of the marginal builders (that are very highly leveraged and have tiny working capital positions) under additional pressures.

‖ **lever de rideau** (lĕve də rido). [Fr.] = *curtain-raiser.* Also *fig.*

1860 *Players* I. 107 As a lever de rideau it was favourably received. **1891** G. B. SHAW *How to become Mus. Critic* (1960) 193 Signor Lago has produced a few miserably-mounted fragments of worn-out Italian operas by way of *levers de rideau* for Cavalleria. **1906** W. DE MORGAN *Joseph Vance* v. 38 The Man went up into 'the Nursery' to look at the bricks in the chimney... This was a mere *lever-de-rideau*—the principal stage business of the day being an examination of the Drains. **1970** *Brewer's Dict. Phr. & Fable* (ed. 12) 638/2 *Lever de rideau*, a short sketch, etc., performed on the stage before the main play begins.

Leveresque (lĭvĕre·sk), *a.* [f. the name of Charles *Lever* (1806–72), Irish novelist.] Characteristic of the novels of Charles Lever in matter or style. Also **Le·verish** *a.*

1903 *Westm. Gaz.* 18 Mar. 4/2 There are some good stories, old or new, told in a racy and Leverish style. **1905** *Daily Chron.* 15 June 3/1 Of Anglo-Irish lords, of Leveresque landowners, of eighteenth-century spendthrifts. **1922** *Glasgow Herald* 6 June 3 The Leveresque pictures of Irish life.

Levers (lī·vəɹz). Also *erron.* **Leavers.** The name of John *Levers* (1786–1848), who effected improvements in lace-making machines in the early 19th c., used *attrib.*, *absol.*, or in the possessive in the names of the lace-making machinery he developed, and of the lace thus produced.

1828 J. LEVERS *Brit. Pat.* 5741 18 Dec., My improvements in machinery for making lace consist in a certain combination and arrangement of mechanism to be adapted to lace machines constructed upon the principle commonly called or known by the name of Levers' principle. *Ibid.*, The movements of all the working parts of an ordinary Levers' machine are well understood by practical mechanics. **1865** F. B. PALLISER *Hist. Lace* xxxvi. 425 The machines now in use are the Circular, Leaver, Transverse Warp and Pusher. **1867** W. FELKIN *Hist. Machine-Wrought Hosiery* xviii. 281 In February, 1835, T. Allcock.. took out a patent..for a new kind of Levers'. *Ibid.* xix. 294 Goods made upon Levers' Jacquard machines. *Ibid.* xxii. 329 Velvet patterns on circular Levers' bobbin net. **1890** *Chambers's Encycl.* VI. 474/2 The lace-making machine now principally used is known as the Levers machine. **1911** *Encycl. Brit.* XVI. 44/2 The Leavers lace machine does not make either a buttonhole stitch or a plait. **1959** D. E. VARLEY *Hist. Midland Counties Lace Manufacturers' Assoc.* i. 4 John Brown, John Leavers, and Clark and Mart, inventors of the traverse warp, the leavers and the pusher bobbin-net machines respectively, were all Nottingham artisans. **1968** J. IRONSIDE *Fashion Alphabet* 235 *Leavers lace*, any lace made on the machine invented by John Leavers, an Englishman, in 1813. This was the first really satisfactory lace-making machine.

levigator (le·vigĕitəɹ). [f. LEVIGAT(E *v.* + -OR 2 c.] An iron or steel disc, several inches thick and about a foot in diameter, which is rubbed over the surface of a lithographic stone to smooth it.

1914 H. J. RHODES *Art of Lithogr.* ii. 11 The operator.. guides the revolving levigator (disc) over the surface, applying sand and water as required. **1965** ZIGROSSER & GAEHDE *Guide to Collecting Orig. Prints* iv. 39 The metal [for a carbograph] is scratched with carborundum crystals and a levigator, producing a fair approximation of a mezzotint ground with much less labor and time. **1967** E. CHAMBERS *Photolitho-Offset* i. 7 The method of preparing the stone for printing is..a mechanical operation in which a smaller stone, or a flat metallic jigger or levigator..is rubbed..over the surface.

Levi's, Levis (lī·vəiz). orig. *U.S.* Also (in *attrib.* use) **Levi, Levies,** and with small initial. [f. name of the original Amer. manufacturer, *Levi* Strauss.] A type of (orig. blue) denim jeans or bibless overalls, with rivets to reinforce stress-points, patented and produced as working clothes in the 1860s, and adopted as a fashion garment in the 20th century.

The form *Levi's* is a proprietary term in the U.S.

1926 R. SANTEE *Men & Horses* 125 My Levis was brand-new. **1928** *Official Gaz.* (U.S. Patent Office) 18 Sept. 519/1 Levi's..for Overalls. **1934** *Street & Smith's Western Story Mag.* 10 Mar. 132/1 The cowboy's..overalls are called 'Levi's' from the name of Levi Straus, of San Francisco, the pioneer overalls manufacturer of the West. **1935** *N.Y. Herald-Tribune* 28 Apr. XI. 13/2 Old timers advise the prospective dude rancher to..buy in the West a pair of Levi overalls. *Ibid.* 13/3 Levi's, or copper riveted, blue denim riding pants, have been found to be excellent. **1941** *Yankee* Dec. 39/1 Red-flannel underwear, Peavey axes, copper-riveted Levi's, etc. **1944** *Life* 15 May 66/2 Blue jeans ('levis') or corduroys, rolled at the bottom, are worn by almost all boys. **1950** *Time* 27 Feb. 88 When dude ranches became popular in the '30s, Haas introduced 'Levis for Ladies'. **1957** M. B. PICKEN *Fashion Dict.* 212/2 Levies, work pants of extra heavy denim, having pockets attached with rivets. Originally worn by ranchers, lumbermen, and industrial workers... Also overalls, especially the bibless type. **1957** J. KEROUAC *On Road* (1958) III. vi. 217 Dean was wearing washed-out tight levis and a T-shirt. *Ibid.* IV. iii. 266 Stan was wearing a levi outfit, jacket and all. **1961** *Sunday Express* 2 Apr. 14/6 The big rush this summer will be:—*for* parchment-coloured American Levi jeans. **1970** *Guardian* 15 Oct. 11 Whatever social stigma might have been attached to wearing a pair of blue jeans could hardly have been held to apply to 'white Levi's'. **1973** C. BONINGTON *Next Horizon* vii. 100, I can always see her in my mind's eye—bare footed, clad in a pair of old Levis and a simple sweater worn outside her trousers. **1973** E. BULLINS *Theme is Blackness* 59 Her sandy hair is tied by a bandanna and the blue Levis are faded dull.

levitate, *v.* Add: **1.** (Later example.)

1971 *Daily Tel.* 19 Nov. 14/4 The demonstrators linked arms in a great circle..and repeated the invocation in the hope that the entire building would levitate.

2. b. Add to def.: Also in scientific use: To cause (something heavier than the surrounding fluid) to rise or remain suspended without visible means (e.g. using magnetic forces).

1952 *Jrnl. Electrochem. Soc.* XCIX. 206/2 Slugs of any shape of various conductive metals could be levitated in the space between the coils. **1961** *Ann. Reg.* 1960 398 Sceptre 4..was to be rebuilt with an aluminium ring 'levitated' inside its ring-shaped chamber. **1971** *Daily Tel.* 5 Apr. 7/2 Superconducting magnets are now being investigated to levitate fast-moving trains into the air above the rails. **1973** *Nature* 9 Feb. 359/2 It is now feasible both to levitate and to propel a hovertrain using only a linear induction motor.

fig. **1954** C. P. SNOW *New Men* xxxiv. 240 The touch of the metal..levitated me to the forgotten happiness of a joyous summer night.

levitation. Add: **1.** (Later examples: in mod. use chiefly *transf.* from 1 b.)

1902 *Q. Rev.* July 125 Many such victims of levitation [*sc.* deep-sea fishes] have been picked up at sea. **1909** H. G. WELLS *Tono-Bungay* III. iii. 364, I lay in my customary glider position, horizontal and face downward, and the invisibility of all the machinery gave an extraordinary effect of independent levitation. **1966** *New Statesman* 18 Feb. 242/3 (Advt.), Atmospheric levitation. Learn to glide.

b. (Earlier example.)

1874 GEO. ELIOT *Legend of Jubal* 191 On all points he adopts the latest views; Takes for the key of universal Mind The 'levitation' of stout gentlemen.

c. The process of raising or supporting by invisible means something heavier than the surrounding fluid. Cf. *LEVITATE *v.* 2 b.

1939 *Gen. Electric Rev.* XLII. 231 (*caption*) As if by magic the shallow metallic dish rises into the air and appears to obey the gestures of the demonstrator's hand. Actually, the levitation results from a special application of electromagnetic principles. **1952** *Jrnl. Electrochem. Soc.* XCIX. 205 Stable levitation..of various metals in the solid state was obtained between the coils..both in air and in vacuum. **1952** *Jrnl. Appl. Physics* XXIII. 552/1 If electromagnetic levitation can be expanded to a larger scale, the following advantages can be anticipated: 1. The specimen touches no crucible or container during the heating, melting and drainage stages. 2. The heated or molten specimen can be protected by a suitable atmosphere or a vacuum. [Etc.] **1956** *Philips Res. Rep.* XI. 45 The possibilities..of levitation by auxiliary gravitational forces, by reaction forces and by forces in electromagnetic fields are investigated. **1961** *Lancet* 25 Nov. 1181/1 Levitation. It occurred to me that supporting a patient on air might solve some of the problems of nursing patients whose illness necessitates the avoidance of contact with their beds. **1971** *Observer* 28 Nov. 9/4 'Magnetic levitation'..could turn out to be the most important advance in transport technology since the internal combustion engine.

3. *fig.*

1909 *Q. Rev.* Jan. 78 In other words, it [*sc.* Labour] obeys its own law of economic levitation, if we may be permitted to coin a phrase. **1962** W. NOWOTTNY *Lang. Poets Use* vi. 123 The poet..has not yet figured as one who soars into a sphere where diction takes forms so conspicuously unprosaic..that [etc.]... It is my purpose now to begin to enquire at what point levitation into such a sphere may be said to have taken place. **1966** *New Statesman* 28 Jan. 138/3 In his third phase..Tippett concentrates on the spiritual 'levitation' effected by his winging lines and springing rhythms.

So **levita·tional** *a.*, of or pertaining to levitation.

1903 *Edin. Rev.* Apr. 329 These people sought..for a levitational quality akin to the dormitive quality of opium, but never found it. **1912** J. STEPHENS *Crock of Gold* v. xiv. 206 Birds have atmospheric and levitational information which millions of years will not render accessible to us. **1969** *New Scientist* 17 Apr. 117/1 If the curve is anti-spatial (particle extinction) the field is anti-gravitational or levitational.

levitron (le·vitrọn). *Physics.* [f. LEVI(TATE *v.* + *-TRON.] A type of fusion reactor in which stability of the plasma inside a toroidal container is achieved by the combination of a magnetic field parallel to the sides of the torus, produced by an external winding, with a second field everywhere at right angles to the first, produced by a toroidal current-carrying core magnetically levitated inside the tube.

1960 COLGATE & FURTH in *Physics of Fluids* III. 999/2 The toroidal analog of the linear hard-core tube is the 'levitron', a toroidal pinch tube with a central ring core, levitated by magnetic field or held in place transiently by inertia. **1966** *McGraw-Hill Encycl. Sci. & Technol.* X. 234/2 Several Levitrons are built or under construction.

levity[1]. Add: **4.** A saying or expression marked by levity.

1930 BLUNDEN in *Nation* 6 Dec. 327/1 Coleridge, wonderfully well edited by his grandson.., lacks his epigrams and levities.

levodopa (lī:vodōu·pă). *Chem.* and *Biochem.* Also **lævo-** (*rare*). [f. LÆVO-, LEVO- + *DOPA.] The lævorotatory L form of dopa (see *L 7 c and *DOPA).

1970 *Brit. Med. Jrnl.* 7 Feb. 331 (*heading*) Treatment of Parkinsonism with laevo-dopa. **1970** *Jrnl. Amer. Med. Assoc.* 16 Mar. 1857/3 Levodopa increased the severity of depression. **1971** *Lancet* 19 June 1272/2 It seems unlikely that levodopa will have a favourable therapeutic effect on acne vulgaris. **1972** *Approved Names 1970* (Brit. Pharmacopœia Comm.) Suppl. 1, *Levodopa*, (−)-3-(3, 4-Dihydroxyphenyl)-L-alanine. **1974** *Times* 8 Jan. 14/6 The weakness and tremor of Parkinson's disease are thought to be caused by a lack of the chemical dopamine. Treatment with levodopa corrects that deficiency, as the drug is converted to dopamine within the body.

levy, *sb.*[2] Add: (Earlier examples.) Also *local U.K.*, a shilling (*Obs.*).

1829 C. SEALSFIELD *Tokeah* II. ii. 22 'But them fips and levies,' throwing a dirty leather bag with a dozen small silver coins upon the table, 'must first go.' **1832** F. TROLLOPE *Dom. Manners Amer.* I. 171 He drew from thence [*sc.* from his pocket] rather more dollars, half-dollars, levies, and fips, than his dirty little hand could well hold. **1864** HOTTEN *Slang Dict.* 170 *Levy*, a shilling.—*Liverpool.*

levy, *v.* Add: **1. f.** To impose a levy on (a person). Also *refl.*

1902 *Westm. Gaz.* 17 June 9/1 The members will be levied 1s. yearly to support their candidate. **1921** *Ibid.* 24 May 2/4 When the stoppage ceases the miners will levy themselves in order to meet these promissory notes.

levyist (le·vi,ist). [f. LEVY *sb.*[1] + -IST.] One who imposes, or advocates imposing, a levy.

1923 *Glasgow Herald* 5 Mar. 4 Without the investment of money in any form which the levyist could reach.

lew, *sb.*[2] **2.** (Examples.)

1825 J. JENNINGS *Observations Dial. W. Eng.* 52 *Lew*, shelter; defence from storms or wind. **1863** [see LEE *sb.*[1] 1]. **1887** [see LEE *sb.*[1] 1 b]. **1899** W. RAYMOND *Two Men o' Mendip* i. 7 The primroses an' cowslips too be out beautiful in the lew between Black-rocks. **1909** S. REYNOLDS *Poor Man's House* VII. vi. 209 We crouched, all humped up, in the lew of a drifter's bows, whilst the rain water washed around our boots and coat-tails.

Lewis[3] (lū·is). [f. the name of the inventor, Col. Isaac Newton *Lewis* (1858–1931) of the U.S. Army.] In full, *Lewis* (*machine*) *gun.* A light, magazine-fed, gas-operated, and air-cooled machine gun. So **Lewis-gunner** and other *attrib.* and *Comb.* uses.

1913 *Aeroplane* 4 Dec. 606 The Lewis Machine-Gun. **1914** E. A. POWELL *Fighting in Flanders* iii. 72 The Lewis gun..is air-cooled. **1916** *War Illustr.* V. 11/2 Lewis gun-team. **1917** *Ibid.* VI. 466 A 'Lewis' gunner. **1919** *King's Royal Rifle Corps Chron.* 1916 81 Stokes mortars and Lewis gun fire subdued the enemy's resistance. **1923** KIPLING *Irish Guards in Gt. War* I. 134 Strong training at bombing and Lewis-gunnery. **1926** T. E. LAWRENCE *Seven Pillars* (1935) XV. 104 If we strengthened them by light automatic guns of the Lewis type..they might be capable of holding their hills. **1937** *Granta* 3 Feb. 219/1 Memories of John taking command of two Lewis gun

crews in a dispersal under barrage fire. **1946** [see *BREN]. **1964** H. L. PETERSON *Encycl. Firearms* 190/1 The Lewis machine gun..was the first machine gun ever fired from an airplane (1912), and it was adapted for ground and naval use as well. **1974** M. BUTTERWORTH *Man in Sopwith Camel* vi. 58 Flip the cocking handles of the twin Lewises.

Lewis[4] (lū·is). *Chem.* [The name of Gilbert Newton *Lewis* (1875–1946), U.S. chemist, who introduced the concepts.] *Lewis acid*, any compound or ionic species which can accept an electron pair from a donor compound; similarly *Lewis base*, one which can donate an electron pair to an acceptor compound.

1944 I. M. KOLTHOFF in *Jrnl. Physical Chem.* XLVIII. 54 The following terminology is suggested. Acids which satisfy the Lewis definition are called Lewis acids or proto-acids. **1961** G. R. CHOPPIN *Exper. Nuclear Chem.* ix. 148 Ketones, ethers and many other oxygen containing organic solvents may act as Lewis bases. **1962** COTTON & WILKINSON *Adv. Inorg. Chem.* x. 179 Various Lewis bases, such as amines, phosphines, ethers, and sulfides, form 1:1 complexes with BX_3 compounds. *Ibid.* 180 There is good evidence that the relative strengths of the boron halides as Lewis acids are in the order $BBr_3 \geq BCl_3 > BF_3$. **1969** LOWRIE & CAMPBELL-FERGUSON *Inorg. & Physical Chem.* xix. 219/2 All Brønsted acids and bases are also Lewis acids and bases respectively. However, the term Lewis acid can be applied to substances which do not contain protons and are not therefore Brønsted acids. **1973** J. J. LAGOWSKI *Mod. Inorg. Chem.* xiv. 522 Cationic halogen species can be stabilized by Lewis bases.

lewis (lū·is), *v.* [f. LEWIS[1]] *trans.* To fasten by means of, or after the manner of, a lewis.

1837 *Civil Engin. & Arch. Jrnl.* I. 72/1 When the stone is broken..it is separated on the bed by a very large iron crowbar or gavelock, and this is either lewised or chained, and raised by the large crane or 'gin'. **1883** *Proc. Assoc. Municipal Engin.* IX. 88 The only ties are wrought-iron 'lewis' bolts, 'lewised' into the old arch stones and turned down and cemented into the new ones.

lewisia (lū,i·ziă, -i·siă). [mod.L. (F. Pursh *Flora Americæ Septentrionalis* (1814) II. 368), f. the name of Meriwether *Lewis* (1774–1809), American explorer + -IA[1].] A small perennial herb of the genus so called, belonging to the family Portulacaceæ, native to western North America, and bearing solitary or clustered pink or white flowers and leaves arranged in a rosette at ground level.

1863 *Curtis's Bot. Mag.* LXXXIX. 5395 (*heading*) Spatlum, or Reviving Lewisia. **1917** C. F. SAUNDERS *Western Flower Guide* 52 The large root of *Lewisia* is a conspicuous feature. **1961** *Amat. Gardening* 23 Dec. 2 The alpine gardener could feast on the roots of his lewisias. **1963** *Times* 11 May 11/5 Lewisias have come through with flying colours. **1974** J. BERRISFORD *Window Box & Container Gardening* vii. 60 In peaty composts..one might try..beautiful lewisias.

Lewisian (lū,i·siăn), *a. Geol.* [f. *Lewis*, name of the northern section of the largest island of the Outer Hebrides + -IAN.] Of, pertaining to, or characteristic of Lewis: applied to the earlier of the two main groups of Pre-Cambrian rocks in NW. Scotland. Also *absol.*, the Lewisian rocks or strata.

1859 R. I. MURCHISON in *Q. Jrnl. Geol. Soc.* XVI. 240 If this most ancient gneiss required a British name, it might indeed with propriety be termed the 'Lewisian System', seeing that the large island of the Lewis is essentially composed of it..; but the term 'Laurentian' having been already applied to rocks of this age in North America by our distinguished associate Sir W. Logan, I adhere to that name. **1887, 1911** [see *HEBRIDEAN *a.* and *sb.*]. **1938** A. K. WELLS *Outl. Hist. Geol.* vi. 49 The dykes can be followed right up to the junction of the Lewisian with the Torridonian. **1943** *Jrnl. R. Anthrop. Inst.* LXXIII. 75/1 The inconceivably old Lewisian rocks of the north-west Highlands. **1957** G. E. HUTCHINSON *Treat. Limnol.* I. i. 57 Peach and Horne (1910) record many irregular basins formed in this way on the Lewisian gneiss of the western seaboard of Sutherland and Ross, and also in the Outer Hebrides.

lewisite[1] (lū·isəit). *Min.* [f. the name of W. J. *Lewis* (1847–1926), British mineralogist + -ITE[1].] An antimonate and titanate of calcium, iron, and sodium, $(Ca,Fe,Na)_2(Sb,Ti)_2O_7$, which is found as small yellow to yellowish brown octahedral crystals and may be regarded as a titanian romeite.

1895 HUSSAK & PRIOR in *Mineral. Mag.* XI. 83 We have given the name of Lewisite to this new titano-antimonate from Brazil in honour of Prof. W. J. Lewis, during whose tenure of office the study of Mineralogy in the University of Cambridge has been so much encouraged. **1932** *Mineral. Abstr.* V. 185 Lewisite is a cubic mineral, whose composition is probably $(Ca,Fe,Na)_2(Sb,Ti)_2(O,OH)_7$. **1951** C. PALACHE et al. *Dana's Syst. Min.* (ed. 7) II. 1022 Romeite... In Minas Geraes, Brazil, with cinnabar in eluvial sands at Tripuhy near Ouro Preto (lewisite).

Lewisite[2] (lū·isəit). Also **lewisite.** [f. the name of Winford Lee *Lewis* (1878–1943), U.S. chemist + -ITE[1].] A dark oily liquid (colourless when pure) which is a powerful respiratory irritant and causes painful blisters on contact with the skin; 2-chlorovinyldichloroarsine, $ClCH{:}CHAsCl_2$.

1921 FRIES & WEST *Chem. Warfare* ii. 23 One of the most interesting and valuable of the compounds which would have found extensive use had the War continued, is an arsenic compound called Lewisite. **1923** R. F. HORTON *Mystical Quest of Christ* xxi. 223 A new poison gas, called by the barbarous name Lewisite, is of such potency that, released over London by twenty or thirty aeroplanes, it would asphyxiate the whole population in three or four hours. **1937** A. HUXLEY *Ends & Means* xii. 216 In 1937 the 'instrument of God for the protection of the people' was all the armaments existing in 1914..plus arsenic smokes, plus Lewisite. **1938** *Times* 12 Jan. 11/6 There are probably millions of people..who do not realize..that the much discussed lewisite can be dealt with by a scrubbing brush and soap, if treated quickly. **1943** [see *B.A.L.* s.v. *B III]. **1970** *Amer. Rev. Respiratory Dis.* CII. 173 Almost one half of the men who had been exposed repeatedly to mustard gas or lewisite before 1945 had a persistent productive cough.

Lewisman (lū·is,mæ̆n). [f. *Lewis* + MAN *sb.*[1]] A native or inhabitant of Lewis, the northern section of the island of Lewis with Harris in the Outer Hebrides.

1927 [see *FIFER[2]]. **1938** L. MACNEICE *I crossed Minch* i. 4 Drinking beer with some Lewismen while one of them sang a love-song in Gaelic. **1971** *Stornoway Gaz.* 10 July 4/1 The initiative for advancing the prosperity of Lewis must come from the Lewisman himself.

lewistonite (lū·istọ̆nəit). *Min.* [f. *Lewiston*, the name of the city in Utah, U.S.A., near which it was found + -ITE[1].] A basic phosphate of calcium, potassium, and sodium, $(Ca,K,Na)_5(PO_4)_3(OH)$, found as colourless to pale green hexagonal crystals; potassian hydroxyapatite.

1930 LARSEN & SHANNON in *Amer. Mineralogist* XV. 326 Some of the hexagonal prisms which are very much like the dehrnite have less alkali and much more water. Chemically they are so different from dehrnite as to give them species rank and the name lewistonite is proposed after Lewiston, Utah. **1942** *Ibid.* XXVII. 297 Lewistonite ..shows wide variation in its properties. Stout hexagonal crystals associated with oolites of pseudowavellite are divided into six biaxial negative segments. **1968** I. KOSTOV *Mineral.* II. ix. 458 Dehrnite and lewistonite for alkali-bearing apatites.

‖ **lex domicilii** (leks dọmisi·li,əi). *Law.* [L.] The law of the country in which a person is domiciled; the determination of the rights of a person by establishing where, in law, he is domiciled.

1832 BARNEWALL & CRESSWELL *Rep. Cases King's Bench* X. 909 The lex domicilii is to be regarded, yet it is not adverted to where the domicile was when the contract was entered into. **1961** *Times* 26 Oct. 19/2 Capacity and consent were matters that fell to be decided by the *lex domicilii.* **1965** *Mod. Law Rev.* XXVIII. v. 540 Unilateral acts of divorce, for example, the *talak* under Moslem law or the Jewish *gett*, have raised important questions whether an English court would recognise these acts as divorces obtained according to the *lex domicilii* of the parties.

lexeme (le·ksīm). *Linguistics.* [f. LEX(ICON + *-EME. Cf. *MORPHEME.] A word-like grammatical form intermediate between morpheme and utterance, often identical with a word occurrence; a word in the most abstract sense, as a meaningful form without an assigned grammatical role; an item of vocabulary.

1940 B. L. WHORF *Lang., Thought, & Reality* (1956) 160 C. F. Voegelin has accomplished the difficult and signal work of analyzing an immense number of baffling stem compounds of Shawnee into their component lexemes (stems) and other morphemes (formatives). **1946** M. SWADESH in C. Osgood *Ling. Struct. Native Amer.* 319 Lexeme building is accomplished mainly by readaptation of lexemes, paradigmatic forms, or syntactic constructions to new uses without change of form. **1950** *Archivum Linguisticum* II. 10 Discriminating between 'word' and 'lexeme'. **1954** S. NEWMAN in H. Hoijer *Lang. in Culture* 87 It is taken for granted..that no English lexeme is a perfect semantic equivalent of a Zuni lexeme. **1958** C. F. HOCKETT *Course in Mod. Ling.* xix. 171 The lexemes in the two-word sequence *twenty-eighth* are *twenty, eight,* and *-th*... *Red-haired* is two words; but it is a single minimum free form, since the ICs are *red hair* (free) and *-ed* (bound); and it is three lexemes, *red, hair,* and *-ed.* *Ibid.* 174 The definition of lexeme follows unpublished work of Bernard Bloch. **1963** *Amer. Speech* XXXVIII. 50 The meaning of *actor*, a minimum free form, is predictable from its structure, as is the meaning of *wants*, a lexeme. **1963** J. LYONS *Structural Semantics* iii. 40 Certain lexemes and expressions are quickly learnt. **1964** E. PALMER tr. *Martinet's Elem. Gen. Ling.* i. 25 A lexeme like *travaill-* is normally listed in the lexicon in the form *travailler.* **1968** J. LYONS *Introd. Theoret. Ling.* 197 The orthographic word *cut* represents three different inflexional 'forms' (i.e. three different grammatical words) of the lexeme *cut.* **1971** G. ANSRE in J. Spencer *Eng. Lang. W. Afr.* 156 If the phoneme is the minimal phonological unit and the morpheme is the minimal grammatical unit, then [the] lexeme will be the minimal lexical unit. **1971** *Archivum Linguisticum* II. 48 The view has already been expressed that words are names (of lexical items) derived from the combination of roots (or lexemes) and affixes (or morphemes). *Ibid.* 50 The concept and term of 'collocation' has to be seen partly in relation to that of 'root' or 'lexeme'.

lexemic (leksī·mik), *a.* and *sb. Linguistics.* [f. *LEXEM(E + -IC.] **A.** *adj.* Of or relating to lexemes. **B.** *sb. pl.* The branch of linguistics concerned with the study of lexemes.

1954 S. NEWMAN in H. Hoijer *Lang. in Culture* 89 The most difficult problem in this lexemic study is that of establishing valid methods. **1957** *Language* XXXIII. 588, I believe that..*all borrowing is lexemic*, i.e. a transfer of lexemes by imitation from one language into another, using 'lexeme' here to mean a free construction of one or more morphemes. **1962** E. F. HADEN et al. *Resonance-Theory for Ling.* iii. 37 The zero-resonance formula expresses the fact that in morphology *easily* and *gladly* are equivalent. Of course they can be dealt with separately in lexemics. **1963** J. LYONS *Structural Semantics* vi. 122 A definite level of analysis—phonological,..lexemic. **1973** *Archivum Linguisticum* IV. 119 These sememic graphs.. must then be converted into syntactic or 'lexemic' representations.

‖ **lex fori** (leks fōə·rəi). *Law.* [L.] The law of the country in which an action is brought, as determining the nature and modes of the proceedings.

1836 H. WHEATON *Elem. Internat. Law* I. II. ii. 188 The extrinsic evidence by which the existence and terms of the contract are to be proved in a foreign tribunal is regulated by the *lex fori.* **1841** CLARK & FINNELLY *Rep. Cases Lords Appeals & Writs of Error* V. 13 The law on this point is well settled in this country, where this distinction is properly taken, that whatever relates to the remedy to be enforced, must be determined by the *lex fori*, the law of the country to the tribunals of which the appeal is made. **1960** *Times* 5 Mar. 10/2 Further the *lex fori*—English law —did not provide a cause of action and relief appropriate to the enforcement of the foreign right. **1970** *Internat. & Compar. Law Q.* XIX. 173 The Soviet system contains an extremely large number of imperative rules which impose the application of the *lex fori.*

lexic (le·ksik), *a. rare.* [f. Gr. λεξικ-ός: see LEXICON.] = LEXICAL *a.*

1900 *19th Ann. Rep. U.S. Bureau Amer. Ethnol. 1897–8* 832 Primitive languages are essentially structural or morphologic, only incidentally lexic.

lexical, *a.* Add: **1.** *lexical meaning*, the meaning of a base in a paradigm, e.g. of *love* in *loves, loved, loving*, etc.; *lexical change, class, form, item, morpheme, rule, set, unit, word* (see quots.).

1933 L. BLOOMFIELD *Lang.* x. 166 To contrast the purely lexical character of a linguistic form with the habits of arrangement to which it is subject, we shall speak of it as a lexical form. *Ibid.* xvi. 271 Languages with an elaborate part-of-speech system..have parallel forms with the same lexical meaning for use in different syntactic positions. *Ibid.* 277 The relative frequency of the various lexical and grammatical units (morphemes and tagmemes) in a language can be studied. **1942** BLOCH & TRAGER *Outl. Ling. Analysis* iv. 68 The meaning of the base itself..is called lexical meaning. **1951** G. A. MILLER *Lang. & Communication* iv. 89 In the Oxford English Dictionary there are nearly half a million lexical units. **1958** C. F. HOCKETT *Course in Mod. Ling.* 429 Back-formation can..lead to lexical change, in the form of new morphemes. **1962** H. C. CONKLIN in Householder & Saporta *Probl. Lexicogr.* 124 Minimally, a *lexical set* consists of all semantically contrastive lexemes which in a given, culturally relevant context share exclusively at least one defining feature. **1963** BLOOMFIELD & NEWMARK *Ling. Introd. Hist. Eng.* iv. 145 Lexical morphemes are those whose grammatical characteristics can be accounted for by identifying them as members of morphological classes. *Ibid.* vi. 257 All lexical units generated from the same grammatical unit by the same lexical rule are said to belong to a single lexical class. *Ibid.* 282 The lexical word *debtors* has the root *debt* and the affixes *-or* and *-s.* **1964** R. A. HALL *Introd. Ling.* liii. 254 Archaic features may be preserved in different lexical items in different places. **1965** N. CHOMSKY *Aspects of Theory of Syntax* ii. 85 The lexical rule..now allows us to insert *sincerity* for the first complex symbol. **1966** G. N. LEECH *Eng. in Advertising* ii. 21 There are thousands of examples of lexical items composed of a sequence of words: *put out* (= 'extinguish'). **1967** R. A. WALDRON *Sense & Sense Devel.* v. 102 A system of this kind, a limited group of words forming some kind of range or scale of mutually excluding terms is often called a *lexical set.* **1967** *Lingua* XVII. 34 Lexical words imply absence of grammatical meaning and vice versa. *Ibid.* 113 All members of the same paradigm are labeled as the same 'word' (lexical unit, lexeme). **1971** J. B. CARROLL et al. *Word Frequency Bk.* p. l, The basic color terms have often been studied as a lexical set, or semantic field. **1972** P. H. MATTHEWS *Inflectional Morphol.* ii. 11, I shall use orthographic forms in small capitals..to refer to Latin verbs *qua* lexical items.

Hence **le·xicalist** and **le·xicalness** (see quots.).

1967 *Lingua* XVII. 35 There seem to be degrees of both 'lexicalness' and 'grammaticalness' in English. **1970** N. CHOMSKY in Jacobs & Rosenbaum *Readings Eng. Transformational Gram.* 188 We might extend the base rules to accommodate the derived nominal directly (I will refer to this as the 'lexicalist position').

lexicalize (le·ksikăləiz), *v. Linguistics.* [f. LEXICAL *a.* + -IZE.] Usu. in *pass.* **1.** To accept into the lexicon, or vocabulary, of a language.

1937 C. E. BAZELL in *Jrnl. Eng. & Germ. Philol.* Jan. 3 But in a form early lexicalized a shortening previous to

the considerable dialectal differentiation of the dialects would be in harmony with Indo-European tendencies. **1954** PEI & GAYNOR *Dict. Ling.* 122 *Lexicalize*, to incorporate a word, etc., into the lexicon of a language. **1972** A. MAKKAI *Idiom Struct. Eng.* 81 Should the morpheme *flation* become widely accepted.., it would be lexicalized, and Webster's new edition would list it. **1972** *Times Lit. Suppl.* 13 Oct. 1229/3 The uncharted waters of contemporary slang, where what appears ephemeral today may become lexicalized tomorrow, and conversely.

2. To express (a difference that is already expressed in the grammatical structure, or could be) by means of a different lexical item.

1968 J. LYONS *Introd. Theoret. Ling.* viii. 352 Consider the..two sentences:..*Bill died*..*John killed Bill.* In such instances, we may say that the relationship of the transitive to the intransitive is 'lexicalized'.

Hence **le·xicalized** *ppl. a.*; also **le:xicaliza·tion**, the action or process of lexicalizing.

1949 *Archivum Linguisticum* I. 10 The adverb differs from a combination of noun and case by a strong tendency to lexicalisation. *Ibid.* 181 'Immotive'..or 'lexicalised' words. **1961** *Brno Studies in English* III. 28 All this richness was gradually done away with (except for a very small number of adverbial or lexicalized survivals). **1968** J. LYONS *Introd. Theoret. Ling.* viii. 369 The difference between obligatory and optional 'lexicalization' in three-place constructions. *Ibid.*, *Kill* is the 'lexicalized' two-place causative form of *die.* **1971** F. W. HOUSEHOLDER *Ling. Speculations* vii. 103 Lexicalization is here assumed to follow at least some transformations.

lexico- (le·ksiko). [f. Gr. λεξικό-ς.] In some mod. linguistic terms denoting 'lexical and..', as in *lexico-dynamics* (see quot.); *lexico-behavioural, -grammatical* adjs. See also *LEXICOSTATISTIC *a.*

1964 E. A. NIDA *Toward Sci. Transl.* iii. 36 Bloomfield ..sought to define the semantic value of symbols in terms of lexico-behavioral distinctiveness. **1970** LANCASTER & GILLESPIE in *Ann. Rev. Information Sci. & Technol.* V. 38 The structure and uses of a large, dynamic controlled vocabulary for one-line implementation have been discussed by Harley and Lancaster... The authors have coined the term 'lexicodynamics' to express the concept of construction, maintenance, use and change of controlled vocabularies for IR purposes. **1953** Y. R. CHAO in *Language* XXIX. 379 (*title*) Popular Chinese plant words, a descriptive lexico-grammatical study. **1964** E. A. NIDA *Toward Sci. Transl.* xi. 243 In treating such lexico-grammatical features, both form and content must be dealt with, since special forms, e.g. poetry, liturgy, parables, proverbs, epigrams and epistolary formulae, are all important factors in determining meaning. **1971** *Archivum Linguisticum* II. 42 Valéry and also Eliot spoke, I believe, of their experience of composing first the formal rhythmic frame of a poem and letting the lexical and grammatical structures follow, but there are also discernible abstract arrangements of a lexico-grammatical order.

lexicology, lexicological, *a.* Add: (Later examples.) So **lexicolo·gically** *adv.*

1937 J. ORR tr. *Iordan's Introd. Romance Ling.* iv. 287 In some, the arbitrary character of the linguistic signs is more apparent..and these Saussure calls lexicological languages. **1949** *Jrnl. Theol. Stud.* L. 104 The fifth is devoted to doctrine; the sixth and seventh to remarks on syntax and lexicology respectively. **1952** *Archivum Linguisticum* IV. 71 Units which the application of formal and semantic criteria enables the grammarian to detach from lexicologically analysable groupings.

lexicon. Add: **1. b.** Also, the vocabulary or word-stock of a region, a particular speaker, etc.

1954 [see *EUROPEAN *sb.* 2]. **1963** *Amer. Speech* XXXVIII. 143 French-speaking Canadians..are developing a 'standard' form of Canadian French..with the same categories of variation (phonetics and lexicon) from the speech of the mother country as are found in American and Canadian English. **1972** *Archivum Linguisticum* III. 1 They constitute a regular part of his stylistic lexicon. **1973** K. JOHNSON in T. Kochman *Rappin' & Stylin' Out* 142 These racial-identity labels are part of what can be called 'the black lexicon' (words that are used exclusively by black people) formulated to designate concepts derived from the unique experience of black people within their culture. **1973** *Times* 31 July 6/7 He [*sc.* Mr. Ehrlichman] said the term 'deep six'—meaning throw in the river—had not been 'part of my lexicon'.

2. *Linguistics.* The complete set of meaningful units in a language; the words, etc., as in a dictionary, but without the definitions. (Opp. GRAMMAR *sb.*)

1933 L. BLOOMFIELD *Lang.* x. 162 The total stock of morphemes in a language is its lexicon. **1964** R. H. ROBINS *Gen. Ling.* 63 The categories of phonetics, phonology, and grammar are general; the components of the lexicon of a language are particular. **1968** J. LYONS *Introd. Theoret. Ling.* iv. 159 He can afford to make a less exhaustive classification of the lexicon.

3. (With capital initial.) The proprietary name of a game played with cards marked with the letters of the alphabet.

1932 *Trade Marks Jrnl.* 22 June 798 Lexicon... Card games. John Waddington Limited,..Leeds; manufacturers. **1945** D. WHITELAW *Lexicon Murders* i. 15 This card..was one from a pack of Lexicon cards, one bearing the letter V. *Ibid.* iii. 59 A Wop, eh...ever play Lexicon? **1960** *Guardian* 9 Dec. 9/7 Didn't we all learn to spell by playing Lexicon? **1965** R. PETRIE *Running Deep* ii. 28 One of them produced a packet of Lexicon from her bag and spread the letters over the table. **1974** 'J. LE CARRÉ' *Tinker, Tailor* xxii. 187 Smiley appeared to examine the lexicon cards, reading off the words longways and sideways.

lexiconize (le·ksikǫ̆nəiz), *v.* [f. LEXICON + -IZE.] **a.** *intr.* To compile a lexicon. **b.** *trans.* To reduce or make into (the form of) a lexicon.

1892 G. MEREDITH *Let.* 3 Jan. (1970) II. 1056 Your Lexiconizing is clever and I cannot go beyond it. **1908** F. GALTON *Mem. my Life* 254 They admit of being so classified or 'lexiconised', that it would be possible for him to tell..whether a similar set had been already registered. **1952** C. P. BLACKER *Eugenics: Galton & After* 49 Galton spent about eight years on this work... It was only in the last of these, the *Finger-Print Directory*, published in 1895, that the process of 'lexiconizing' or indexing the material was completed.

lexicostatistic (le:ksiko,stăti·stik), *a. Linguistics.* [f. *LEXICO- + STATISTIC *a.*] Of or relating to the statistics of vocabulary. Also **le:xicostati·stics** *sb. pl.* const. as *sing.*, a branch of linguistics closely allied to *GLOTTOCHRONOLOGY.

1952 M. SWADESH in *Proc. Amer. Philos. Soc.* XCVI. iv. 452 (*title*) Lexico-statistic dating of prehistoric ethnic contacts. **1956** S. C. GUDSCHINSKY in *Word* XII. 175 Lexicostatistics is a technique which attempts to provide dates for the earlier stages of languages much as carbon 14 dating provides dates for archaeological finds. **1961** [see *GLOTTOCHRONOLOGY]. **1964** R. H. ROBINS *Gen. Ling.* viii. 318 The attempt to quantify linguistic divergence from a common source [etc.]..is known as lexicostatistics or glottochronology. **1965** *Canad. Jrnl. Ling.* Spring 94 Certain special methods, such as lexicostatistic list comparison, rest on comparative method. **1969** R. A. HALL in *Neuphilol. Mitt.* LXX. 199 Despite the failure of glottochronology, due to basic theoretical faults, a certain amount of interest has been maintained in other, less aprioristic aspects of lexicostatistics.

Also **le:xicostati·stical** *a.*, **le:xicostati·stically** *adv.*

1955 *Internat. Jrnl. Amer. Ling.* XXI. 138 (*heading*) Lexico-statistical skewing from dialect borrowing. **1963** *Language* XXXIX. 60 (*heading*) Lexicostatistically determined borrowing and taboo.

lexis (le·ksis). [ad. Gr. λέξις diction, word, f. λεγ- to speak.] **1.** The diction or wording, in contrast to other elements, of a piece of writing (see also quot. 1950).

1950 *Mod. Philology* XLVII. IV. 243/1, I have already distinguished, in the first part of this essay, between speech as action (*praxis*) and speech as meaningful (*lexis*). **1957** N. FRYE *Sound & Poetry* p. xxiii, Singing and chanting are, in modern times, radically different methods of associating melos and lexis.

2. *Linguistics.* **a.** = *LEXICON 2; items of lexical, as opp. esp. to grammatical, meaning; the total word-stock of a language. **b.** The study of words as lexical items.

1960 E. DELAVENAY *Introd. Machine Transl.* v. 67 During the early days of research the priority of lexis over morphology in preparing the way for machine translation was taken for granted. *Ibid.* 131 *Lexis*, used here, and by M.T. linguists, to designate the words of a language, contained in its dictionary or lexicon, as opposed to the morphology and syntax of that language. **1961** *Language* XXXVII. 325 A distinction..between grammar and lexis seems to be necessary if the patternings are to be economically stated or defined. **1962** R. QUIRK *Use of English* v. 72 The word-stock—also known as the *vocabulary* or *lexis.* **1963** R. M. W. DIXON *Ling. Sci. & Logic* ii. 45 Theories of grammar and lexis are both needed. **1964** *English Studies* XLV. 24 Patterns of vocabulary, or *lexis*, which describe the company words keep. **1966** *Listener* 14 Apr. 534/2 Dr. Steiner's problems with the gaucho's lexis and the Amerindian's grammar..are linguistic and not literary problems. **1971** *E. Afr. Jrnl.* Mar. 35/2 Presentation of lexis is balanced by the introduction of grammatical structures.

‖ **lex loci** (leks lōu·səi). *Law.* [L.] The law of the country in which a legal transaction is performed, a tort is committed, or a property is situated; freq. followed by a defining word or phrase.

1832 BARNEWALL & CRESSWELL *Rep. Cases King's Bench* X. 905 The decisions of both English and Scotch courts shew that the construction of personal contracts depends on the lex loci contractus. **1836** H. WHEATON *Elem. Internat. Law* I. II. ii. 173 A necessary consequence of the rule relating to the application of the *lex loci rei sitæ.* **1848** WHARTON *Law Lexicon* 375/1 All the formalities, proofs, or authentications of them [*sc.* contracts], which are required by the *lex loci*, are indispensable to their validity everywhere else. **1858** J. WESTLAKE *Treat. Private Internat. Law* iv. 64 The fact of status must be referred to the domicile, but the incapacities attendant on the given status to the *lex situs* for immovable property, and to the *lex loci actus aut contractus* for other matters. *Ibid.* xi. 318 Both the form and the legality, the extrinsic and intrinsic validity, depended on the *lex loci celebrationis.* *Ibid.* 335 The *lex loci*..by which the conduct of married persons is to be regulated,..must always be referred, not to the place where the contract was entered into, but where it subsists for the time. **1896** A. V. DICEY *Digest Law Eng.* i. 66 '*Lex loci solutionis*' means the law of the country where a contract is to be performed. *Ibid.* xxvii. 669 It is not necessary that the *lex fori* and the *lex loci delicti* should be identical; it is sufficient if they are similar. **1970** *Internat. & Compar. Law Q.* XIX. 26 This at once gives rise to the problem whether it is the *lex fori* or *lex loci* which gives the plaintiff this cause of action.

ley (lī, lēi). [Var. LEA *sb.*[1]] The supposed line of a prehistoric track in a straight line usually from hilltop to hilltop with identifying points such as ponds, mounds, etc., marking its route (see also quot. 1932).

1922 A. WATKINS *Early Brit. Trackways* 12 The sighting line was called the ley or lay. *Ibid.* 13 Previous writers, treating, say, of Roman or of mediæval roads, not knowing of the existence of the ley, assume that they are speaking of original primary structures, when they are only describing a route evolved from a number of the leys I describe. **1925** —— *Old Straight Track* 220 When you get a good ley on the map, go over it in the field, and fragments and traces of the trackways may be found, always in straight lines. **1932** D. MAXWELL *Detective in Surrey* v. 86 A ley..is an invisible and imaginary line, drawn from one point in the landscape to another, mathematically straight... The key positions..are points where two or more leys cross... The crossing places..would be places of meeting. **1971** *It* 2–16 June 24/4 The leys..interlaced over the whole country. **1974** *Bookseller* 26 Jan. 192 (Advt.), Alfred Watkins' theory about leys which connect ancient sites.

ley, var. LEA *sb.*[2] (Later non-*dial.*) examples.)

The form *lea* is now rarely used.

1932 *Discovery* Feb. 61/1 Some progressive farmers are alternating four years of corn-growing with four years of temporary grass leys, on which bullocks and grass-land sheep are fed. **1948** L. D. STAMP *Land of Britain* iv. 65 The length of time the grass is left down is usually determined by the farmer's own judgment..a common average being seven years. This is the system of 'leys' or 'ley farming'—taking the plough round the farm. **1957** *New Biol.* XXIV. 42 In many areas it was not convenient to change arable land to leys of any considerable duration. **1962** *Listener* 1 Feb. 214/1 The old permanent pastures are being replaced by temporary leys, with the plough 'going all round the farm'. **1972** *Oxford Times* 28 July 8 Don't be in a hurry to plough up and re-seed leys and permanent pastures.

ley, var. LYE *v.*[2] (See quot. 1823.) Hence **leyed** *ppl. a.* (cf. LYED *ppl. a.*).

1823 E. JAMES *Acct. Expedition Rocky Mts.* I. 114 Another very acceptable dish was called leyed corn. *Ibid.* 195 They sometimes prepare this hard corn for eating by the process of leying it, or boiling it in a ley of wood ashes for ..an hour or two. **1825** W. BIGGS *Narr. Captivity among Kickapoo Indians* (1922) 35 Sandy hill cranes boiled in leyed corn.

leycesteria (lestiə·riă). [mod.L. (N. Wallich in W. Carey *Roxburgh's Flora Indica* (1824) II. 181), f. the name of William *Leycester* (fl. 1820), Chief Justice of Bengal + -IA[1].] A shrub of the genus so called, belonging to the family Caprifoliaceæ, native to India, and bearing yellow or purple flowers; also called Himalayan honeysuckle or pheasant-berry.

1838 J. C. LOUDON *Arboretum et Fruticetum Britannicum* II. 1060 The beautiful Leycesteria..is a rambling shrub, with the general appearance of a honeysuckle. **1899** G. JEKYLL *Wood & Garden* ix. 101 We come back to Leycesteria, put rather in a place of honour. **1961** *Amat. Gardening* 23 Dec. 9 Quick growing shrubs which make long wands of growth annually..are..buddleia, leycesteria.

Leydig (ləi·dig). *Anat.* The name of Franz von *Leydig* (1821–1908), German anatomist, used *attrib.*, in the possessive, and with *of-* adjunct to designate various anatomical structures described by or associated with him, *esp.* the interstitial cell of the testis, a large, polyhedral cell occurring in large numbers in the connective tissue around the seminiferous tubules and believed to be the site of androgen production in the testis.

1904 *Amer. Jrnl. Anat.* III. 167 (*heading*) The embryonic development of the interstitial cells of Leydig. *Ibid.*, Boll..believed that Leydig's cells composed the walls of capillaries. **1936** NEAL & RAND *Compar. Anat.* xii. 448 The interstitial Leydig cells of the testis. **1956** *Nature* 21 Jan. 144/1 In all cases the Leydig cell cytoplasm of the interstitium had become sprinkled with lipid droplets. **1962** *Gray's Anat.* (ed. 33) 1522 The interstitial cells of the testis (cells of Leydig) are large polyhedral cells lying in the connective tissue between the seminiferous tubules.

Leyland (lēi·lănd). The name of Christopher John *Leyland* (1849–1926), of Haggerston Castle, used *attrib.* or in the possessive in **Leyland('s) cypress**, to designate a hybrid conifer, × *Cupressocyparis leylandii* (*Chamæcyparis nootkatensis* × *Cupressus macrocarpa*), first raised by him from seedlings collected at Leighton Hall, Welshpool, in 1888.

[**1926** A. B. JACKSON in *Kew Bull.* 114 As this new cypress has already been named *Cupressus Leylandii* by Mr. Leyland, we propose to describe it under that name.] **1933** W. J. BEAN *Trees & Shrubs Hardy in Brit. Is.* III. 124 Leyland's Cypress..is a tree of dense pyramidal habit. **1960** N. J. PROCKTER *Garden Hedges* vi. 107 Leyland's Cypress..is still looked upon as something new, and I regret to say is at present little known. **1970** C. LLOYD *Well-Tempered Garden* iii. 215 The great

favourite now (although it is by no means new) is Leyland's cypress. **1974** A. MITCHELL *Field Guide to Trees of Britain* 68 At least three [hybrid cypresses] exist, but only one, the Leyland Cypress, is commonly grown.

lez, lezz (lez), varr. *LES. Also **le·zzy.**

1956 B. HOLIDAY *Lady sings Blues* (1973) ix. 90 I've known black chicks in show business who were as feminine as me, but before long they got acting like lezzies because it's so easy. **1958** F. NORMAN *Bang to Rights* III. 146, I didn't see why there should be a law against queers and lezes. **1966** I. JEFFERIES *House-Surgeon* iii. 37 'Is she like that with all the house-men?' 'Not the girls; she's lezzy as hell.' **1966** 'L. LANE' *ABZ of Scouse* 62 *Lezzy*, a lesbian woman. **1970** E. BERCKMAN *She asked for It* xi. 135 So you're a couple of Lezzes, you and little Monica?

L-form: see *L 6* c.

Lhasa (lā·sa). Also **lhasa, Lhassa.** The name of the capital of Tibet, used *attrib.* in **Lhasa apso** to designate a small long-coated dog, often gold or grey and white, belonging to a breed originally developed there, and formerly called the **Lhasa terrier**, a name also once used for the *Tibetan terrier* (*TIBETAN). Also *ellipt.*

1904 C. B. *Fry's Mag.* June 364/1 The little toy dog of Tibet, which Kennel Club edicts have declared shall properly be known as the 'Lhassa Terrier', is no novelty in our midst, although his numbers in an unkind climate possibly do not reach a round dozen. **1905** P. LANDON *Lhasa* I. 403 The Lhasa terrier is an entirely distinct breed. **1935** W. HUTCHINSON *Dog Encycl.* II. 1142/2 [The] Lhasa Apso..has only become known in England in very recent years. **1948** C. L. B. HUBBARD *Dogs in Brit.* 303 The Lhasa Apso has had many names given it in the course of its comparatively short British history. **1955** W. GADDIS *Recognitions* I. vi. 206 The lhasa turned to stare at the Coca-Cola machine. **1974** *Radio Times* 14 Feb. 3/2 Polluche is a Lhasa apso—literally translated, it means a Lhasa terrier.

Lhooshai, var. *LUSHAI *a.* and *sb.*

‖ **li**[3] (lī). [Chinese.] (See quots.)

1912 J. J. M. DE GROOT *Relig. of Chinese* iv. 95 The original *li* and *teh*, the only classical rules and ethics which keep man..in perfect harmony with the order of the universe. **1937** D. BODDE tr. *Fung Yu-Lan's Hist. Chinese Philos.* iv. 46 The 'Rites' (*Li*) were used in diplomatic relations. **1942** D. D. RUNES *Dict. Philos.* 168/1 *Li*, reason; law; the rational principle. This is the basic concept of modern Chinese philosophy. To the Neo-Confucians.., Reason is the rational principle of existence whereas the vital force (ch'i) is the material principle. **1953** *Oxf. Jun. Encycl.* V. 101/1 The word *li* used by Confucius, which is often translated by the English word 'religion', really means something more like ceremonial or ritual, the correct observance of which is needed for maintaining friendly relations with the Power or Powers of the unseen world. **1955** A. FANG in E. Pound *Classic Anthol.* p. xvi, Read in that context, Confucius must be understood as trying to integrate music with rites (*li*), just as he tried to integrate poetry with music. **1962** A. F. WRIGHT in Wright & Twitchett *Confucian Personalities* 7 The *li*, spread by fathers, village elders, and government officials, and supplemented by the discipline of ordered family life, would in turn foster social virtues: filial submission, brotherliness, righteousness, good faith, and loyalty.

liable, *a.* Add: **3. c.** *dial.* and *U.S.* Likely.

1886 F. T. ELWORTHY *West Somerset Word-Bk.* 433 Speaking of a wounded hen pheasant a farmer said, 'Tis very liable he's a-croped into one o' these here hovers.' **1890** R. D. BLACKMORE *Kit & Kitty* I. ix. 95 Ould dog be put out at six o'clock riglar, and 'tis liable he'll hurn straight to 'e. **1901** MERWIN & WEBSTER *Calumet 'K'* xi. 198 He's liable to call our men out to-night, ain't he? **1903** *N.Y. Even. Post* 22 Aug., Norman Hunter's new record..is liable to stand unmolested for many years. **1935** H. W. HORWILL *Dict. Mod. Amer. Usage* 189/1 'Boston is liable to be the ultimate place for holding the convention.' 'If the lawmakers get back before the frosts kill the vegetation, many of them are liable to think it a reproach to the nation that grass should be growing in the streets of the national capital.' **1957** B. & C. EVANS *Dict. Contemp. Amer. Usage* 273/1 An American might say *we are liable to be in Chicago next week* without meaning that that would be a calamity.

liaise (li,ĕi·z), *v.* orig. *Services' slang.* Also (erron.) **liase.** [Back-formation from LIAISON.] *intr.* To make liaison *with* or *between.* Hence **liai·sing** *vbl. sb.* and *ppl. a.*

1928 C. F. S. GAMBLE *Story N. Sea Air Station* xiii. 221 [Lord Fisher said in 1916] 'I want a soldier..to keep in touch with the Navy and so "liaise" or exchange inventions which may be suitable.' **1941** *Amer. N. & Q.* Dec. 141/1 The kind of grammatical economy found in a recent (British) Home Guard instruction sheet—in the event of certain circumstances, it stated, two groups were ordered to 'liase' with two others. **1942** *New Statesman* 1 Aug. 75/1 'To liaise'..was at first frowned on by the pundits: its usefulness..soon came to outweigh its objectionableness. **1942** *Tee Emm* (Air Ministry) II. 128 He then hurriedly climbed into a Spit and shot off to 'liaise' with his old Squadron. **1946** A. LEE *German Air Force* 25 Göring never acquired the happy knack of liasing satisfactorily with Germany's senior army generals. **1948** L. MACNEICE *Holes in Sky* 71 The liaising aircraft mounts. **1952** *World Rev.* Sept. 20 The manufacturer, too, has advertising people on his own staff who liaise with the advertizing agency. **1958** L. DURRELL *Mountolive* vi. 139 For convenience it can work to us and liaise with our Service Departments. **1959** *Guardian* 15 Oct. 10/7 He would expect absolute obedience from his subordinates... It remains to be seen whether he could also 'liaise' successfully. **1962** *Times* 28 Mar. 3/1 He will liaise between the dressing room and the press. **1965** *New Statesman* 10 Dec. 919/3 Paris..was..in a state of great confusion... There seemed little liaising to be done. **1970** *Country Life* 1 Oct. 846/3 It would seem advisable for the host to detach well briefed members of the shoot staff to liase with the field trial stewards. **1974** *Times* 18 Feb. 20/8 It would seem that the Government statisticians do not: (a) Liaise with other departments, [etc.].

liaison. 2. a. Delete † and add later examples.

1870 *Putnam's Mag.* May 545/2 The knowledge gained from these new sources..has..given new zest to the alleged *liaisons* of the Republic. **1974** *Country Life* 5 Dec. 1814/2 Florence..remained..a home from home for the British... It is a liaison that seems to have lasted happily down the years.

b. (Earlier example.)

1816 BYRON *Let.* 24 Dec. in *Works* (1900) IV. 29 She is by far the prettiest woman I have seen here... I believe I told you the rise and progress of our *liaison* in my former letter.

3. *Phonetics.* Also in English contexts, and in *Music*, in wider sense.

1905 *Daily Chron.* 7 Feb. 4/7 The nightly false 'liaison' made by a clever actress... 'Take Lady Agatha-r-out,' she says with terrible distinctness. **1917** G. B. SHAW *How to become Mus. Critic* (1960) 292, I will not blame the singer for putting in a little *liaison* of her own at the reprise. **1929** *Amer. Speech* V. 87, I noticed recently a curious instance of consonant *liaison* (if that is the term for a carrying-over that is commoner in French than in English). **1962** *Listener* 6 Sept. 369/3 To our singers the style, the flavour, the true placing of the sounds, the liaisons—every aspect is elusive and deceptive.

4. *Mil.* Close connection and co-operation between two units, branches, allies, etc., esp. during a battle or campaign. Also *transf.*

1816 H. CLARKE *Hist. War* I. xliii. 702/1 Other advantages of a great and important nature arise from the combinations of the various corps of their invading armies maintaining their *liaison* or correspondence, by means of the..staff-establishments attached to every division. **1915** *Oxf. Mag.* 29 Oct. 18/2 The 'overseer' of the Press.. an unrivalled artist in the liaison of departments. **1920** G. H. PERRIS *Battle of Marne* xi. 225 With the I Army pulling north-west, the II Army pulling south-east,..how could anything more than a pretence of liaison be kept up? **1920** *Q. Rev.* July 138 It acted rather as a liaison between the Admiralty and the Press Bureau than as a branch of the latter. **1922** *Encycl. Brit.* XXXII. 967/2 The welfare supervisor..is thus able to refer all matters calling for attention direct to the general manager and may be regarded by him as a *liaison* between him and the various departments dealing with the women employees. **1930** A. W. MYERS *Lawn Tennis* ix. 113 Mind and body must be working in liaison. **1964** B. B. SCHOFIELD *Russ. Convoys* ii. 27 The main trouble during the first years of the war was the shortage of aircraft—otherwise the liaison between the two services was as good as it could be.

5. *attrib.* and *Comb.*, esp. **liaison officer**, an officer in the Services who is concerned with the liaison of units, etc.; also *transf.*

1915 'I. HAY' *First Hundred Thousand* xix. 285 He is one of that most efficient body, the French *liaison* officers, who act as connecting-link between the Allied Forces. **1916** *War Illustr.* 9 Dec. 405 (*caption*) Army liaison dog leaving with a message attached to his collar. **1917** *Times* 5 June 7/1 Members of Parliament have tended less and less in recent times to fulfil their primary duty as *liaison* officers between Parliament and the constituencies. **1918** *Daily Chron.* 19 June 2/2 This position as 'liaison Minister' between the House and the War Cabinet. **1930** *N. & Q.* 5 Apr. 250/2 This [book] is a *liaison* treatise of which ethics is the warp, and economics the very much less important woof, while the whole is coloured with a tinge of metaphysics. **1942** *R.A.F. Jrnl.* 3 Oct. 23 'L' [stands] for 'Liaison pilot'. **1946** *Sun* (Baltimore) 18 Feb. 11/5 The Navy has tabbed entertainment with the high-sounding name liaison unit. **1954** A. HUXLEY *Let.* Apr. (1969) 704 You unquestionably *are* the man to act as liaison officer between pure science and the rest of the world in this matter of the nature of the Mind. **1964** *Amer. Speech* XXXIX. 233 Compares duration of liaison consonants, as in *des airs*, with that of medial consonants, as in *désert.* **1966** *B.B.C. Handbk.* 75 The main duty of the department is to act in a liaison capacity. **1973** *Times* 14 Nov. 8/1 The Israeli liaison officer said he had still not received permission for the journalists to go on to Suez.

liar. Add: **1. a.** *I'm a liar*, (in trivial use) I am mistaken.

1940 *Sunday Express* 31 Mar. 3/5 'That's not my brother Sid you met in here last Thursday. Or was it Friday?' We said we didn't remember... 'I'm a liar. It was Wednesday.' **1972** W. GARNER *Ditto, Brother Rat!* xv. 105 Last winter, was it? No, I'm a liar. The spring. That's right.

c. Delete † *Obs.* and add **liar dice**, a gambling game resembling poker dice, in which the thrower conceals the dice thrown and sometimes declares a false score; also *ellipt.* (in *pl.*).

1946 J. SCARNE *On Dice* (ed. 2) xvii. 386 Liar or Doubting Dice. A popular game on transpacific liners and in the Far East, it is now gaining rapidly in popularity in the United States. **1956** M. MCMINNIES *Flying Fox* I. iv. 55 Everybody was round the bar playing liar dice. **1959** R. KIRKBRIDE *Tamiko* v. 37 'Which do you play, Balin?' 'Which?' 'Liars, Horses, Cameroon—.' 'I don't play dice.' **1966** O. NORTON *School of Liars* i. 2, I spent two months in graduating from the empty lounge to the bar, two more in..reaching the inner group, the liar-dice school. *Ibid.* ii. 23 We sat there playing liars until twenty past two. **1971** C. BONINGTON *Annapurna South Face* ix. 107 After the meal we played liar dice or Scrabble, with our tape-recorder blasting out music in the background.

d. *the liar* (Logic): the name of the paradox involved in a speaker's statement that he is lying or is a (habitual) liar; so *liar paradox, paradox of the liar.*

1871 T. M. LINDSAY tr. *Ueberweg's Syst. Logic* v. §77. 245 This case happens when, and only when, the *truth of the judgment* is itself *the object of the judgment*, or belongs to the object of the judgment. The ancients have empirically discovered this case, without..giving an account of its logical nature. What is called '*The Liar*' represents it. Epimenides, the Cretan, says, all the Cretans are liars. **1906** J. N. KEYNES *Stud. & Exerc. Formal Logic* (ed. 4) App. B. 457 The sophism known as Ψευδόμενος or *The Liar.* **1908** B. RUSSELL in *Amer. Jrnl. Math.* XXX. 240 Hence his statement is false, and yet its falsehood does not imply, as that of 'I am lying' appeared to do, that he is making a true statement. This solves the liar. **1940** —— *Inquiry into Meaning & Truth* iv. 62 The inference from the paradox of the liar is..as follows. **1959** E. W. BETH *Found. Math.* VI. xvii. 485 The natural first reaction to the liar paradox is to ascribe the contradiction to the fact that the statement involved refers to itself. **1967** *Encycl. Philos.* V. 46/1 But one, the Liar,.. is still of great interest to us. **1970** R. L. MARTIN (*title*) The paradox of the liar. **1971** *Philos.* XLVI. 133 (*heading*) Tarski, Frege, and the liar paradox.

liatris (ləi,æ·tris, ləi·ătris). [mod.L. (J. C. D. von Schreber in *Linnæus's Genera Plantarum* (ed. 8, 1791) II. 542), of unknown derivation.] A North American perennial herb of the genus so called, belonging to the family Compositæ and bearing spikes or clusters of purple or white flowers.

1811 *Curtis's Bot. Mag.* XXXIV. 1411 (*heading*) Spiked Liatris. **1870** W. ROBINSON *Wild Garden* xv. 139/2 Plants for very moist rich soils... Liatris, in var. **1931** M. GRIEVE *Mod. Herbal* II. 746/2 Several varieties of *Liatris* are largely used in Southern United States to flavour tobacco. **1961** *Amat. Gardening* 14 Oct. 3/4 Cimicifugas, golden rods, liatris and lobelias.

lib, colloq. abbrev. LIBERATION, freq. preceded by *adj.* (as *gay lib*) or a *sb.* in the possessive (as *men's lib, women's lib*). See the defining words.

1970 *Atlantic Monthly* Mar. 116 The Lib Movement was rich in documentation of the conditioning processes. **1971** *Daily Tel.* 2 Dec. 7/2 Children's lib. notwithstanding, it would be hard to write a children's book without setting up some sort of standard for the child reader to admire. **1973** *Guardian* 3 Feb. 13/1 Lillian Thomas is a member of the Suffrage Fellowship Movement..and is delighted with the Libs. **1973** *Black World* Dec. 12/1 The various 'lib' movements, therefore, are white derivatives of the Black movement. **1974** *Listener* 25 Apr. 520/3 With Scots Lib, as with Women's Lib, it's no good the oppressors expecting the past to be forgotten when convenient.

liba·tionary, *a.* [f. LIBATION + -ARY[1].] = LIBATORY *a.*

1896 W. J. LOCKE *Study in Shadows* vi. 93 Mme. Popea scattered scraps of stuff about her room, in a kind of libationary joy. **1909** *Westm. Gaz.* 16 Feb. 5/2 The new Empress-Dowager had finished performing the libationary sacrifices to the memory of the late Empress-Dowager.

liba·tioner. [f. LIBATION + -ER[1].] One who pours out libations (to a god).

1920 *Brit. Mus. Return* 47 in *Parl. Papers* XXXVI. 673 Black stone squatting figure of Ser, a divine father and libationer of Amen.

li·bber, colloq. abbrev. *LIBERATIONIST. Cf. *LIB, *WOMAN *sb.*

1971 *Tel.* (Brisbane) 19 May 17/1 Women's libbers are preparing to do battle with the police in Baltimore. **1972** *Village Voice* (N.Y.) 1 June 26/4 Now the star-maker has decided to calm the libbers with another token. **1973** *Times* 1 Nov. 12/6 *The Female Woman* sorts out.. the contemporary confusion of ideas about the sexes which the Libbers have..worse confounded. **1973** *Daily Tel.* 24 Nov. 11/8 The..debate set things off by producing a truly appalling female whose anti-male views were so extreme and so crudely expressed that orthodox Libbers in the audience showed dismay.

liberal, *a.* and *sb.* Add: **A.** *adj.* **1.** (Further examples.) Freq. in *liberal arts.*

1753 W. SHIPLEY in D. G. C. Allan *William Shipley* (1968) 229 (*title*) Proposals for raising by subscription a fund to be distributed in premiums for the promoting of improvements in the Liberal Arts and Sciences, Manufactures, etc. **1906** P. ABELSON (*title*) The seven liberal arts, a study in mediæval culture. **1950** E. H. GOMBRICH *Story of Art* xv. 215 The so-called Liberal Arts such as rhetorics, grammar, philosophy and dialectic. **1951** [see *CLINIC *sb.*[2] 3]. **1961** *New Scientist* 16 Mar. 662/1 The better public schools..should be converted to liberal-arts colleges on the American pattern. **1965** *Listener* 11 Mar. 387/2 (Advt.), The major part of the work will be teaching Sociology,..but appropriately qualified candidates will be expected to teach Liberal studies. **1973** *Jrnl. Genetic Psychol.* CXXII. 183 The educational problems of the troubled liberal arts college student.

4. b. (Earlier and further examples.) Also in application to *Judaism.*

1823 (*title*) The liberal Christian. **1828** (*title*) Which society shall you join, liberal or orthodox? **1862** *Dublin Rev.* Nov. 48 Our friends the 'liberal' Catholics may be interested in a note to F. Faber's treatise. **1876** O. B. Frothingham *Transcendentalism New Eng.* vi. 128 It may be inferred that Transcendentalism in New England was a movement within the limits of 'liberal' Christianity or Unitarianism as it was called. **1900** *Jewish Q. Rev.* July 618 (*heading*) Liberal Judaism in England. *Ibid.*, These liberal Jews have no organization. **1920** R. Macaulay *Potterism* vi. v. 253 Modernist liberal-catholic vicars asked him to preach. **1957** *Oxf. Dict. Chr. Ch.* 807/1 The 'Liberal Catholics' who formed a distinguished group in the RC Church in the 19th cent. were for the most part theologically orthodox, but they favoured political democracy and ecclesiastical reform... 'Liberal Protestantism'.. developed into an anti-dogmatic and humanitarian reconstruction of the Christian faith. **1965** *Sunday Times* 5 Feb. 5/3 A plan for a national conference of non-orthodox synagogues, Reform (progressive) and Liberal. **1968** B. M. G. Reardon (*title*) Liberal Protestantism. **1974** *Times Lit. Suppl.* 19 Apr. 424/4 Judaism is divided into Orthodox, Conservative and Reform varieties following the American terminology, and not into the British Orthodox, Reform and Liberal camps.

5. *Liberal-Labour*, of or pertaining to (persons associated with or sympathetic to) both the Liberal and the Labour parties. So *Liberal Labourism*. Cf. *Lib-Lab *a.*

1909 *Daily Chron.* 14 July 1/7 Mr. Hancock, the Liberal-Labour candidate for Mid-Derbyshire. **1929** M. Beer *Hist. Brit. Socialism* (new ed.) II. iv. xvi. 315 In 1898 Gladstone died, and with him one of the main pillars of Liberal Labourism disappeared from British politics.

6. *liberal-anarchic, -bourgeois, -cultural, -democratic, -empiricist, -humanist, -minded* (later examples), *-scientific* adjs.; *liberal-anarchism, -mindedness* (earlier example).

1964 *New Society* 13 Feb. 17/2 The progressive schools have been liberal-anarchic, the product of free enterprise in unorthodox educational ideas. *Ibid.*, Liberal-anarchism will no longer do. **1951** Koestler *Age of Longing* vi. 103 Where did you pick up this idea out of the liberal-bourgeois philosophy of law? **1953** A. K. C. Ottaway *Educ. & Soc.* v. 88 The supporter of the pure liberal-cultural tradition. **1940** Liberal-democratic [see *culture *sb.* 5 a]. **1949** *Mind* LVIII. 254 More than anyone, except perhaps Bertrand Russell, he [*sc.* L. T. Hobhouse] may be regarded as the inheritor of the liberal-empiricist mantle of John Stuart Mill. **1957** N. Frye *Anat. Crit.* 6 It would be easy to compile a long list of such determinisms in criticism, all of them, whether Marxist..liberal-humanist..or existentialist, substituting a critical attitude for criticism. **1925** Beerbohm *Observations* 16 Too proud to fight?.. or too liberal-minded?—or what? **1961** New Eng. Bible *Acts* xvii. 11 The Jews here were more liberal-minded than those at Thessalonika. **1971** 'D. Halliday' *Dolly & Doctor Bird* v. 71 Mini Adult Show for the Liberal Minded. **1783** *Gentl. Mag.* LIII. ii. 938 What the liberal-mindedness of the present age amounts to [etc.]. **1958** *Times Lit. Suppl.* 17 Jan. 26/4 The obvious charge which can be brought against this picture of a suppressed liberal-scientific element is the undeniable fact that it never showed any signs of formulating a practical alternative to current political or ethical machinery.

B. *sb.* **1. c.** In extra-European politics, and in wider application.

1832 *Liberal* (St. Thomas, Ontario) 20 Sept. 3/4 We shall first notice the slanderous imputations cast upon the Liberals, that they are a discontented set of men, ever on the watch to find occasion for complaint and clamour. **1854** *N.Y. Tribune* 22 Apr. 5/5 The 'Liberals' of Maine have called a 'State Democratic Mass Convention' at Portland. **1918** H. V. Evatt *Liberalism Austral.* x. 66 The Sydney press claimed that its own free traders were the Liberals. **1940** *N.Y. Times* 23 Jan. 20/4 Since then [*sc.* the Russian Revolution] Liberal has been a word of confusion. Everybody who was not a Conservative became a Liberal or Radical or Red, whichever came first to the mind. **1955** D. Viklund tr. Tingsten's *Probl. S. Afr.* x. 116 A Communist in South Africa is often, according to the general usage of the word, a liberal. **1957** *New Yorker* 12 Jan. 25/1 Both she and Robbie were campus liberals; they had met at a gathering that had something to do with the Spanish war. *a* **1964** H. Hoover in W. Safire *New Lang. Politics* (1968) 232/2 Fuzzy minded totalitarian liberals who believe that their creeping collectivism can be adopted without destroying personal liberty and representative government. **1969** *New Yorker* 14 June 44/2, I don't think he is a liberal. He's tight with his money, and he wants to see the poor work for their money.

liberalist. (Later examples.)

1958 A. Paton *Hope for S. Afr.* ii. 7 Further, the word 'liberal' has for the white enemies of South African Liberalism another meaning; it shares this further meaning with the words 'liberalist' and 'liberalistic'. This meaning is derogatory and carries the stigma of 'loose', 'careless', 'promiscuous'. **1972** *Times Lit. Suppl.* 27 Oct. 1272/2 Liberal or (in the pejorative corruption of her Afrikaner opponents) liberalist.

liberalistic, *a.* (Later examples.)

1958 [see prec.]. **1958** G. M. Carter *Politics of Inequality* iii. 104 W. A. Maree..brought out well the Nationalist view of the difference between what he called 'the liberalistic approach and the approach of nationalism' to education.

liberalization. Add: *spec.*, the removal by a government of restrictions placed upon the import of goods, the movement of capital, etc.

1940 *Economist* 5 Oct. 431/2 The liberalisation of bank loans to farmers still left unresolved the problem of the bad..credit proposition. **1961** *Ann. Reg. 1960* 496 During the year the liberalization of imports, especially from the dollar area, continued.

liberalize, *v.* Add: **1. d.** To remove restrictions on (the import of goods, outflow of capital, etc.). Also **liberalized** *ppl. adj.* (later example in the above sense).

1940 *Economist* 5 Oct. 431/2 All the joint-stock banks decided..to liberalise their policy of agricultural loans. **1955** *Times* 6 Aug. 6/3 The French Government has agreed to the German request that the 'liberalized' sector of French trade with the O.E.E.C. countries shall be restored.

liberaloid (li·bĕrăloid), *a.* [f. Liberal *a.* + -oid.] Resembling liberal (attitudes, etc.); in a bad sense, exhibiting liberal characteristics, pseudo-liberal.

1951 R. Campbell *Light on Dark Horse* xxiii. 346 The imbeciles of the King's Party in the French Revolution.. tamely handed over the keys of all the forts to the Reds because of the liberaloid mentality that they had acquired from reading the masturbations of that *pisse-froid* Rousseau. **1963** *Spectator* 29 Mar. 400 The mixture of French puritan bourgeois and liberaloid ideologist that emerges.

liberate, *v.* Add: **b.** To free (an occupied territory) of the enemy; also *ironically*, to subject to a new tyranny.

1944 G. B. Shaw *Let.* 4 Dec. in *To a Young Actress* (1960) 181 All your Italian friends must be starving now that we have 'liberated' them. **1955** *Ann. Reg. 1954* 303 Chu Teh..expressed China's intention of 'liberating' Formosa. **1961** *Listener* 28 Dec. 1100/2 President Sukarno's warning to Indonesian troops to be ready to 'liberate' West New Guinea. **1964** A. McKee *Caen* xix. 314 'This place sure has been liberated,' said an American M.P. to an H.C.R. crew, when eventually they reached the waste of brick and stone which had been Vire. **1971** A. Bullock *20th Cent.* iii. 76/2 The West had the great advantage of being liberated by the Americans and British, neither of whom wanted to stay. **1975** *Times* 1 May 1/2 At 11.30 am local time (03.30 GMT), according to Hanoi Radio, Saigon was 'liberated'.

c. To loot (property), to misappropriate. *slang*.

1944 *Daily Express* 7 Oct. 4/3 (*caption*) Excuse me, Canon, but I rather think you've liberated my matches. **1946** E. Linklater *Private Angelo* viii. 86 Those soldiers, who said they had liberated the turkeys and the geese, had taken a most drastic way of giving them their freedom. **1957** [see *careen *v.* 5]. **1965** G. Melly *Owning-Up* vi. 59 He..wore a sombrero liberated, I suspect, from the wardrobe of some Latin American group he had worked with in the past. **1970** *Daily Tel.* (Colour Suppl.) 1 May 9 Shoplifting—'liberating' to hippies—costs the English retail trade the staggering sum of £75 million a year. **1974** S. E. Morison *European Discovery of America: Southern Voyages* viii. 164 Drake's flagship *Golden Hind* carried no bell, but his men 'liberated' one from the church of Guatulco, Mexico, in 1579.

d. To free from social or male-dominated, etc., conventions. Freq. as *ppl. a.* Cf. *emancipated *ppl. a.* 2.

1970 R. Lowell *Notebk.* 191 The liberated girl with a build. **1970** *Globe & Mail* (Toronto) 25 Sept. 12/7 Liberated school. The co-eds from women's colleges in Virginia are going to attend formerly all-male Davidson College. **1970** *New Yorker* 5 Dec. 49/1 It is not only men liberated. It is *women* liberated. **1971** *E. Afr. Jrnl.* Mar. 9/1 Aunt Bimp..goes into the construction business .. She knows nothing about cement and sand but the corruption of the system has served to liberate her. **1973** D. Jordan *Nile Gr-gn* ix. 41 He resents me because I'm a liberated woman who can support herself. **1975** D. Ramsay *Descent into Dark* i. 25 Put '*Ms* Joyce Chandler'. Let them get the idea you're liberating yourself from the Kitchen on principle.

liberation. Also in the senses of *liberate *v.* b, c, and d.

1945 *Sun* (Baltimore) 28 Sept. 11/2 Liberation is only four months old. **1952** *Ann. Reg. 1951* 'Liberation' of the island [*sc.* Formosa] remained one of the primary stated objectives of the régime. **1956** *Ann. Reg. 1955* 314 On 15 August, the tenth anniversary of the 'Liberation' of Korea by the Soviet Russian army. **1966** *Punch* 5 Jan. 14/2 Each member of the crew would apply for his equipment to be 'written off', and for replacements to be issued. And each member of the crew would become the proud owner of the liberation loot. **1966** G. Jackson *Let.* 23 Feb. in *Soledad Brother* (1971) 94 Some..commit unpardonable crimes.. that must in the end bar them from partaking in the benefits of the liberation that is planned for tomorrow. **1967** *Daily Tel.* 14 Mar. 21/1 Negro and Puerto Rican parents who are organising a boycott of a school in Harlem began establishing 'liberation schools' today... They are demanding a say in running the school and in choosing a headmaster. **1970** [see *game-playing* sb. and adj. s.v. *game *sb.* 16 b]. **1971** *Black Scholar* Jan. 58/1 Those in the struggle have to deal with black separatists because they stand today as a potent obstacle to full black liberation. **1973** *Black Panther* 3 Mar. 8/1 The Black Panther Party is..an organization dedicated to the liberation of oppressed peoples in different communities. **1973** *Black World* Oct. 5 Development for us is liberation. It's liberating this person who, until now, has been suffering under colonialism..and under all kinds of superstitious beliefs. **1973** *Listener* 20 Dec. 841/1 The Arabs..are to give financial and diplomatic support to the African liberation movements.

liberationist. Add: Also, an advocate of women's liberation.

1970 G. Greer *Female Eunuch* 13 The organized liberationists are a well-publicized minority. **1971** *Guardian* 18 Jan. 9/2 The whole point..of bra-burning seems to have vanished from some English liberationist minds.

Liberian (ləibiə·riăn), *a.*[1] [f. *Liber(ius* (see below) + -ian.] Of or pertaining to Liberius (Pope, 352–66). So *Liberian basilica*, one of the early churches of Rome, formerly believed to have stood on the site of S. Maria Maggiore; *Liberian calendar*, a calendar attributed to the pontificate of Liberius; *Liberian catalogue*, a list of the Popes until and including Liberius.

a **1773** A. Butler *Lives Saints* (1779) X. 316 The Liberian calendar places him [*sc.* St. Callistus] in the list of martyrs. **1840** E. Cox tr. *J. J. I. von Döllinger's Hist. Church* I. i. 36 The Liberian catalogue would make him to have been bishop during the lifetime of the apostle. **1858** N. Wiseman *Recoll. Last Four Popes* ii. iv. 295 The re-collection..will come back..in images..of solemn entrance into..an open basilica... The Liberian speaks to you of Bethlehem..the Sessorian of Calvary. **1913** E. R. Barker *Rome of Pilgrims & Martyrs* i. iv. 50 Philocalian (Liberian) Calendar of 354.—This Calendar is named *Liberian*, since it was compiled in 354 under Pope Liberius. It is called *Philocalian*, after its author Philocalus. *Ibid.* vi. 85 Sixtus III. (432–440) embellished the Liberian basilica. **1957** *Encycl. Brit.* VII. 4/2 Damasus was nominated.., but the intransigents of the Liberian party ..set up against him another deacon, Ursinus.

Liberian (ləibiə·riăn), *a.*[2] and *sb.* [f. *Liber(ia* (see below) + -ian.] **A.** *adj.* Of or pertaining to Liberia, a West African state founded in 1822, or its people. **B.** *sb.* A native or inhabitant of Liberia; also, a Liberian ship.

1854 A. H. Foote *Afr. & Amer. Flag* xxxiv. 386 The Liberians are freemen. *Ibid.* 388 Captain Cooper will not take exception at the remark, that it is 'the day of small things' with the Liberian navy. **1855** *Wesleyan-Methodist Mag.* I. 307 The Liberians..have acquired lands which no European power could peaceably gain from the natives. **1868** J. A. Horton *W. Afr. Countries* xvii. 270 The entrance to Monrovia, the capital of the Liberian Republic, reminds one of the entrance to a purely native town. **1882** *Encycl. Brit.* XIV. 508/1 The Liberian variety of coffee held in such high esteem. **1906** H. H. Johnston *Liberia* I. i. 7 The governing class..consists of approximately twelve thousand Negroes and Mulattos of American origin, to whom may be added..about thirty thousand 'civilised' Liberians of local origin. **1914** W. H. Page *Let.* 5 July in B. J. Hendrick *Life & Lett. W. H. Page* (1925) III. ix. 120 About half the Liberian Cabinet..have asked for an audience with me this week. **1944** *Amer. Speech* XIX. 164 In Liberia the descendants of returned American slaves who constitute the ruling caste of the country used to call themselves Americo-Liberians... But I am informed by Mr. Ben Hamilton, Jr., formerly of the Liberian consulate in Los Angeles, that..'Liberians consider the term Americo-Liberian opprobrious... Hence they prefer to be called civilized or Monrovian Liberians to distinguish themselves from the natives of the hinterland.' **1952** A. G. L. Hellyer *Sanders' Encycl. Gardening* (ed. 22) 121 *Coffea..liberica*, 'Liberian Coffee', white, fragrant, 15 to 20 ft., Trop. Africa. **1971** B. Callison *Plague of Sailors* iii. 98 Now, when you get aboard the Liberian... **1973** R. Thomas *If you can't be Good* (1974) ii. 13 A Dutch freighter, flying a Liberian flag of convenience. **1973** *Times* 17 Apr. (Liberia Suppl.) p. ii/6 By 1970..this group employed 1,500 Liberians and was the largest of Liberian concerns.

libertarian, *sb.* (*a.*). Add: **2.** (Further examples.) Also as *adj.*

1906 *Westm. Gaz.* 2 Oct. 2/1 No wonder the libertarian woman rebels. **1966** *New Statesman* 22 Apr. 602/2 She is a libertarian and was not happy under the dictatorship of Hassan II. **1969** *Listener* 15 May 666/1 The political activists..belong to what is known as the libertarian Left. **1972** *Science* 12 May 615/3 It gives fair play to the objections of civil libertarians. **1973** *Observer* (Colour Suppl.) 11 Nov. 44/4 Now he's suddenly the darling of the civil libertarians.

libertinous, *a.* Delete † *Obs. rare*⁻¹ and add later examples.

1906 *Daily Chron.* 14 Aug. 3/2 The tale of a bold bad knight, who made libertinous love to a virtuous young woman. **1966** *Punch* 30 Nov. 800/1 We red-eyed supporters of riotous and libertinous Sundays have no other lord before Ted.

li·bertist. *rare.* [f. Libert(y *sb.* + -ist.] An advocate of liberty.

1887 *Voice* (N.Y.) Aug. 11 But not for a moment can the radical personal libertist accept such a heresy.

liberty, *sb.*[1] Add: **5. b.** *to take liberties* (or *a liberty*) (further examples).

1749 J. Cleland *Mem. Woman Pleasure* I. 219, I had seen him taking the last liberties with my servant-wench. **1862** Borrow *Wild Wales* I. xii. 124 The creature [*sc.* a cat] soon began to take liberties, and in less than a week after my arrival at the cottage, generally mounted on my back, when it saw me reading or writing. **1924** A. A. Milne *When we were very Young* 57 Excuse me, Your Majesty, For taking of The liberty, But marmalade is tasty, if It's very Thickly Spread. **1967** *Listener* 23 Feb. 271/1 A scene in which he is wrongfully accused of 'taking a liberty' with one of the female guests.

9. d. *at liberty* (later examples).

1931 *Amer. Mercury* Nov. 351/1 *At liberty*, out of work. **1933** P. Godfrey *Back-Stage* v. 70 It takes many years before the superseded actors and actresses will admit to themselves that the professional terms 'at liberty' and 'disengaged' are no longer applicable to them in a temporary sense.

10. liberty act, a circus act performed by liberty horses; **liberty boat** *Naut.*, a boat carrying liberty men; **liberty bodice,** a close-fitting under-bodice; **liberty bond,** one of the interest-bearing bonds of the 'Liberty' loans issued by the U.S. government in 1917–18; **liberty boy,** (*c*) *U.S.* a supporter of a freedom movement; (*d*), (*e*) (see quots. 1826, 1842); **liberty cabbage** *U.S.*, sauerkraut; **liberty cap** (examples); **liberty horse** (see quot. 1946); **Liberty-loan,** one of the four issues of liberty bonds; **liberty man** (later examples); **liberty** (or **Liberty**) **ship,** a type of merchant vessel built in the United States by rapid mass-production methods during the 1939–45 war; also *ellipt.* **Liberty.**

1933 P. Godfrey *Back-Stage* xvii. 214 The training of a team of spirited thoroughbreds for the 'Liberty' or 'Haute École' acts. **1837** *United Service Jrnl.* Aug. 474 They knew ..that the liberty-boat would be on shore for them at that hour. **1901** *Daily Chron.* 16 Nov. 4/3 The destroyer..ran down a liberty boat..with the loss of three lives. **1956** A. Thorne *Baby & Battleship* i. 33 They..had no intention of coming back until it was nearly time to catch the Liberty Boat. **1916** *Child* May 433/1 The 'Liberty Bodice' Factory, of Market Harborough, have made a speciality of the 'Liberty Bodice'... The bodice is made of durable but soft and elastic material, and is porous and pliable and arranged with well-placed straps carried over the shoulders to take the weight of underclothes and stand the pull of suspenders. **1932** S. Gibbons *Cold Comfort Farm* xiv. 193 Give me my liberty bodice. **1968** J. Ironside *Fashion Alphabet* 70 *Liberty bodice*, bodice to the waist,.. worn by girls. The bodice has built-up shoulders and buttons at front or back. It..has gone out of fashion for all but very young children. **1973** *Radio Times* 18 Jan. 18/1 (Advt.), The wiser you are, the more you appreciate the comfort of a liberty bodice. **1918** in Webster Add., Liberty bond. **1919** E. E. Cummings *Let.* 25 Nov. (1969) 64 Very nice of you all to include me in the liberty bond donation. **1922** *Encycl. Brit.* XXXI. 760/2 The Liberty Bonds and Victory Notes were issued under authority of the Acts of Congress approved April 24 1917, Sept. 24 1917, [etc.]. **1928** Liberty bond [see *drive *sb.* 1 g]. **1774** in C. F. Aspinall-Oglander *Admiral's Widow* (1942) 51 They are distinguished here by the name of Tories, as the Liberty Boys—the tarring feathery gentry—are by the title of Whigs. **1781** S. Peters *Gen. Hist. Connecticut* 393 The liberty boys were..honoured with the presence of ministers, deacons [etc.]. **1826** *New Monthly Mag.* ii. 79 While the paying spectator..applauded, when his feelings prompted, the *liberty boy* [*sc.* free-ticket holder].., if he clapped at all, would clap with gloved hands. **1842** *N.Z. Gaz.* II. 112 People from ships called 'liberty boys' are only allowed to come on shore on Sundays for recreation. **1858** *Texas Almanac 1859* 33 The Liberty boys..joined Austin's Company. **1927** *Haldeman-Julius Q.* July–Sept. 7/2 Here we were..calling sauerkraut 'Liberty Cabbage'. **1967** *Listener* 18 May 642/1 In America it was more than a restaurant owner's life was worth to keep sauerkraut on the menu: it was changed to liberty cabbage. **1803** *Lit. Mag.* (Philadelphia) Dec. 172 A liberty pole..decorated with party coloured flags and liberty caps. **1835** *Mechanics' Mag.* 10 Jan. 256/2 It is wholly at variance with classic authority to place the Pileus or Liberty Cap on the head of the figure representing Liberty. **1843** L. M. Child *Lett. from N.Y.* xl. 274 This age and country, in which liberty-caps abound. **1930** E. Smith *Red Wagon* xxvii. 225 The time came to exhibit his beloved liberty horses, four dapple-grey cobs exactly resembling painted rocking-horses. **1946** M. C. Self *Horseman's Encycl.* 264 Liberty horses are those which perform in the circus without a rider. **1952** N. Streatfeild *Aunt Clara* 103 The comedy horse turn was coming to an end, the liberty horses would follow. **1972** *New Statesman* 7 Jan. 14/2 Yasmin Smart is the only lady ring-master in the world... Since the age of 10 she has been learning to train Liberty horses. **1917** Ade *Let.* 12 June (1973) 64 We find it hard work to induce the farmers and other small investors to take the Liberty Loan bonds. **1921** E. L. Bogart *War Costs* 208 The First Liberty Loan Act of April 24, 1917, authorized a bond issue of $2,000,000,000 and advances to allies of $3,000,000,000. **1922** B. J. Hendrick *Life & Lett. W. H. Page* II. xxii. 273 The American Government finally paid this over-draft out of the proceeds of the first Liberty Loan. **1909** *Daily Chron.* 25 Feb. 1/6 The packet boats which convey the 'liberty' men to Chatham after the day's routine. **1964** R. Braddon *Year Angry Rabbit* xii. 110 A few hundred liberty men on each side.., their flights delayed by bad weather, returned to the firing line too late. **1941** *Marine Digest* 28 June 8/2 The emergency cargo ships, known as the EC-2 type, Liberty ships, will have an overall length of about 425 feet, width 57 feet, approximately 10,000 deadweight tons, and will be oil-burning. **1942** W. S. Churchill *End of Beginning* (1943) 183 The launching of the *Patrick Henry*, the first Liberty ship. **1945** *Seafarers Log* 3/2 The first of the Liberties to be scrapped, the Banvard was delivered into service on April 8, 1943. **1961** W. Vaughan-Thomas *Anzio* iv. 47 DUKWs were already chugging in from the big Liberty ships lying out to sea. **1966** C. R. Tottle *Sci. Engin. Materials* vii. 170 Some of the wartime 'liberty' ships fractured when lying in port, without operational loads.

Liberty (li·bəɹti), *sb.*[2] [The name of a London drapery firm, Messrs. *Liberty* and Co.] Used *attrib.* to designate materials, styles, colours, etc., characteristic of textile fabrics or articles sold by Messrs. Liberty.

1888 Mrs. H. Ward *R. Elsmere* I. ii. vii. 173 Bits of Liberty stuffs with the edges still ragged, or cheap morsels of Syrian embroidery. **1888** *Daily News* 23 Apr. 6/4 Her dress was of two kindred shades of almost indescribable colour, belonging to the class now commonly known..as Liberty tints. **1891** *Ibid.* 19 Jan. 3/1 'Liberty styles' are to be had in every large drapery establishment. **1892** 'F. Anstey' *Voces Populi* 2nd Ser. 112 Putting on a turban and a Liberty sash. **1892** G. & W. Grossmith *Diary of Nobody* ix. 126 Carrie has arranged some Liberty silk bows on the four corners. **1894** W. J. Locke *At Gate of Samaria* (1895) ix. 103 It had long been dismantled of the Liberty curtains, Persian rugs, and cheap Japaneseries. **1900** *Munsey's Mag.* July 517/2 Tying a brown liberty silk veil over my hair. **1901** *Daily News* 7 Mar. 8/4 Dresses and costumes (familiarly known as 'Liberty' gowns and frocks). **1903** A. Bennett *Leonora* viii. 238 She had changed her Liberty dress for the dark severe frock of her studious hours. **1913** R. Brooke *Let.* 22 Nov. (1968) 535 Hindus..in Liberty-coloured garments. **1923** R. Macaulay *Told by Idiot* i. vi. 31 She looked round the Liberty room. **1957** M. B. Picken *Fashion Dict.* 283/1 *Liberty s[atin]*, trade name for soft, closely woven, piece-dyed satin fabric with raw-silk warp and single spun-silk filling. **1972** *Daily Tel.* 6 May 21/3 In spite of its French name, the 'new art' [*sc. art nouveau*] started in England and was originally known as the 'Liberty Style' deriving its name from the famous store which was started by Arthur Lazenby Liberty in 1875. Libertys sold many articles in this new style. **1973** *Country Life* 8 Feb. 365/2 Coat and skirt in a natural-coloured Swiss cloth with a lining and blouse in Liberty lawn.

b. Used *absol.*

1898 *Daily News* 19 Nov. 6/2 Another instance of the vogue enjoyed by English materials on the Continent is the universality of the word 'Liberty'. **1903** *Daily Chron.* 19 Sept. 8/4 Soft satin, called in Paris Liberty, is again being employed as a blouse fabric. **1909** *Westm. Gaz.* 1 Oct. 8/4 With pannier draperies over an under-skirt of Liberty.

‖ **liberum arbitrium** (li·bərǝ̆m āɪbi·triǝ̆m), Lat. phr. (occurring in Livy 4. 43. 5): full power to decide, freedom of action.

1652 N. Culverwel *Act of Oblivion* in *Lt. Nature* [II.] 38 The great Creator of Heaven and Earth, must wait upon mans *liberum arbitrium*. **1880** W. James *Coll. Ess. & Rev.* (1920) 194 Shall I move my index finger or my little finger to show my '*liberum arbitrium indifferentiae*'?

libidinal (libi·dinăl), *a.* *Psychoanalysis.* [f. L. *libidin-*, *libīdo* lust + -al.] Pertaining to or connected with libido.

1922 J. Riviere tr. *Freud's Introd. Lect. Psycho-Anal.* 283 For a son, the task consists in releasing his libidinal desires from his mother, in order to employ them in the quest of an external love-object in reality. **1949** M. Mead *Male & Female* xiii. 278 The pleasure of irresponsibility, untidiness, undirected libidinal behaviour. **1957** *Essays in Crit.* VII. 333 There remains the charge of brutality and its libidinal content. **1970** E. Fromm *Crisis of Psychoanal.* (1971) p. ix, One can speak of a non-neurotic character trait when libidinal impulses are transformed into relatively stable and socially adapted traits.

libido (libī·do, -əi·do). *Psychoanalysis.* [f. L. *libīdo* desire, lust.] Psychic drive or energy, particularly that associated with the sexual instinct, but also that inherent in other instinctive mental desires and drives. Also *transf.* and *attrib.*, as *libido theory*.

1909 A. A. Brill tr. *Freud's Sel. Papers on Hysteria* vi. 147 The anxiety neurosis goes along with the most distinct diminution of the sexual libido or the psychic desire. **1913** C. G. Jung in *17th Internat. Congr. Med.* XII. i. 66 This infantile fixation, which is understood as an unconscious attachment of the sexual libido to certain infantile phantasies and habits. **1922** J. Strachey tr. *Freud's Group Psychol.* 37 Libido is an expression taken from the theory of the emotions. We call by that name the energy..of those instincts which have to do with all that may be comprised under the word 'love'. **1929** *Encycl. Brit.* VIII. 399/2 The tenets of leading psycho-analysts assume a libido, with either one or two fundamental departments, sex, or sex and ego... The physiological basis of the libido and its emotions is hazy, to say the least. **1932** *Brit. Jrnl. Psychol.* Jan. 249 The attempt to find the foundation of group behaviour in instinctive tendencies fails, as does the psycho-analytic 'libido' theory. **1944** J. S. Huxley *On Living in Revolution* xv. 192 *Libido* is the nearest to such a term [*sc.* energy], but its use implies complete acceptance of orthodox psychoanalytic theory and has certain unsatisfactory connotations. **1953** *Encounter* Nov. 25/2 We accept the fact that our sexual libido is thwarted. It is time for us to recognise that our political libido is just as complex-ridden, repressed, and twisted, if not even more. *Ibid.* 26/2 The political libido can be defined as the individual's need to feel himself as part of a community, his urge to belong. **1971** *Sci. Amer.* Sept. 529/3 He shares the libido for new military systems that his entire guild displays. **1972** *Ibid.* Aug. 46/2 It has also been observed that removal of the ovaries does not reduce the libido of human females.

Lib-Lab (li:b,læ·b), *a.* Abbrev. of *Liberal-Labour* (see *liberal *a.* 5); also as *sb.* Hence **Li:b-La·bbery.**

1903 *Review of Reviews* Aug. 113/1 The Progress of the Lib-Labs. The Lib-Lab party is carrying all before it. **1944** G. B. Shaw *Everybody's Pol. What's What?* xxx. 263 Despotic dictators came into fashion as fast as Lib-Lab prime ministers lost face. **1952** H. Nicolson *King George V* vii. 94 Of the 53 Labour members elected in 1906, twenty-nine belonged to the Independent Labour Party and twenty-four were affiliated to the Liberal Party and known as 'Lib-Labs'. **1960** T. I. J. Jones in A. J. Roderick *Wales through Ages* II. 203 There were a few working class members of parliament before that date [*sc.* 1900]..but they counted themselves Liberals or at the most 'Lib-Lab'. *Ibid.* 205 The early years of the twentieth century saw the appearance of a new more militant type of leader ..and his appearance was to bring about the decline of the 'Lib-Lab' ideal. **1960** *Guardian* 18 Feb. 3/6 A famous 'Lib-Lab.' family. **1963** *Ann. Reg. 1962* 14 Mr Gaitskell stiffly shot down the idea of a formal Lib-Lab alliance. **1965** *Economist* 26 June 1496/1 This is the hoary dilemma of Lib-Labbery. **1970** *Guardian* 6 Aug. 10/3 North Cornish Liberals are to urge the Liberal Assembly to fight every parliamentary seat and to have no truck with Lib-Labbery. **1972** *Times* 30 Sept. 15/3 A Lib-Lab party would be well placed in terms of the central argument which is concerned with inflation. **1973** *Daily Tel.* 19 Oct. 16 Mr Wilson shudders at the very thought of a return to the old 'Lib-Lab.' combination.

library[1]. Add: **1. b.** *free library* (examples).

1847 *Howitt's Jrnl.* I. 119/1 A scheme of free libraries. By Dr. Smiles. *Ibid.* 119/2 Samuel Brown, the author of the system of Free Libraries, (or, as he styled them, 'Itinerating Libraries',) was a merchant of the small town of Haddington. **1850** *Manch. Guardian* 28 Dec., A Free Library and Museum for Manchester. **1855** *Act* 18 & 19 *Vict.* c. 70, An Act for further promoting the Establishment of Free Public Libraries and Museums in Municipal Towns. **1902** *Daily Chron.* 4 Mar. 3/2 These things are surely axioms to the free-library reader. **1960** M. Sharp *Something Light* vi. 49 The meeting-place for intellectuals was the Free Library.

d. A theatre-ticket agency.

1827 W. Clarke *Every Night Bk.* 108 It is..necessary ..to procure tickets of admission prior to the opening of the doors: they may be had at the libraries of Ebers, or Andrews, in old Bond Street. **1902** W. H. Chantrey *Theatre Accounts* i. 7 The next source of income is derived from Library Bookings. It is usual for the management of a Theatre to allow a discount of from 5 to 10 per cent. upon seats booked by Libraries. **1959** *Financial Times* 23 June, The word 'library' is used in the trade to designate a central source of tickets—a 'ticket wholesaler' might be a better term. **1973** *Sunday Times* (Colour Suppl.) 28 Oct. 101/1 Ticket agencies..are..organised in the Combined Libraries Association ('library' is theatrical term for ticket agency).

2. b. (Later examples.)

1844 A. W. Kinglake *Eothen* xviii. 282 The books were thorough-bred Scotch... He prided himself upon the 'Edinburgh Cabinet Library'. **1974** *British Bks. in Print* I. p. ccxxvi/1 Home University Library. Oxf. U.P.

c. (*c*) A collection of films, gramophone records, music, etc.

1926–7 *Army & Navy Stores Catal.* 1095 Circulating library for music rolls. Arrangements can be made..for Members to subscribe to a Player Roll Library. **1937** *Amer. Speech* XII. 47/1 *Library*, collection of sheet music. **1957** *B.B.C. Handbk.* 101 Some types of material are recorded specially for the library, as for example, folk-music, dialect, natural history, and sound effects. **1962** H. Orton *Survey Eng. Dial. Introd.* 20 The British Broadcasting Corporation's Permanent Sound-Record Library. **1969** *Listener* 12 June 836/3 Proms are no longer ..regarded as the poor man's record library. **1974** *Guardian* 26 Mar. 32/3 The transcription service..provides a library of BBC programmes on slow-speed recordings. **1974** *Times* 7 Oct. 13/4 Borrowers frequently make tape-recordings..from the material borrowed and can thus acquire..an extensive library of recorded music without paying the composer a penny.

2*. *Computers.* An organized collection of routines, esp. of tested routines suitable for a particular model of computer.

1950 *Proc. R. Soc.* A. CCII. 576 It is convenient to have a 'library' containing sub-routines for performing such standard operations as the evaluation of a sine, or a scalar product. **1951** *Ibid.* CCVI. 539 There are..some eighty sub-routines in the EDSAC library. **1951** *Math. Tables & Other Aids to Computation* V. 46 Routines for solving standard problems will be established on tape and filed in a 'library'. **1958** *Oxf. Mag.* 29 May 469/1 Their 'library' consists of 'routines', 'sub-routines', and even 'quickies', programmes already existing and tested. *Ibid.*, As the library grows in extent, programming for new problems becomes easier. **1964** F. L. Westwater *Electronic Computers* ix. 143 Soon, 'libraries' of standard subroutines were available for each particular computer. **1966** A. Battersby *Math. in Managem.* viii. 195 Standard computing routines are always available for widely-used techniques such as network analyses or linear programming, and the range over which they extend is constantly widening. They are often referred to as 'library programs' or under the general term 'software'. **1967** *Technology Week* 23 Jan. 11/2 (Advt.), Software for Sigma 5 includes..a library of mathematical, business and utility routines.

3. *library book, card, centre, chair, committee, company, desk, house, material, stamp, style, table, ticket*; **library binding,** a special strong binding of books for lending libraries; **library edition,** an edition of good size and print and strongly bound, *spec.* a uniform edition of a writer's works; also (quot. 1917) an edition of a newspaper for depositing in certain libraries; **library frame, glasses, spectacles,** spectacles with heavy frames suitable for use when reading; **library shot** (see quot.); **library steps,** a step-ladder for use in libraries.

1903 J. D. Brown *Man. Libr. Econ.* xxvi. 333 The principal leathers for public library bindings are pig-skin, Persian and Levant moroccos, and roan. **1952** W. McGill in *Library World* Dec. LIV. 90 (*heading*) A note on how some publishers produced reinforced or special library bindings for public libraries nearly 50 years ago. **1863** 'G. Hamilton' *Gala-Days* 146 There was the long service, Sunday school and library books. **1916** A. Bennett *Lion's Share* xxiv. 180 The women play golf all day on that

appalling golf course, and then after tea they go into the town to change their library books. **1973** K. Giles *File on Death* iii. 66 My old girl retires to bed around eight, with her library book. **1966** A. Sachs *Jail Diary* xviii. 164 The station commander..has spoken to his wife and she doesn't mind if I use one of the family's library cards. **1960** *Library Assoc. Rec.* Aug. 261/2 A library centre is a static service-point, whether or not under the control of paid staff, which does not comply with the definitions for full-time or part-time branches. **1883** *Heal & Son Catal.* 181 Dining room and library chairs. **1970** *Country Life* 31 Dec. 16/2 (Advt.), William IV mahogany library chair upholstered in antiqued Havana brown leather. **1831** *Congress. Rec.* 7 Feb. 618 It was referred to the Library Committee. **1965** D. Davinson *Academic & Legal Deposit Libr.* ii. 24 All British universities have a library committee formed from the Senate or similar body. **1745** B. Franklin *Let.* 11 Dec. in *Writings* (1905) II. 296 Our Library Company sends for about twenty pounds sterling worth of books yearly. **1837** W. Jenkins *Ohio Gazetteer* 99 The public institutions are a bank, a library company and a mechanics society. **1895** M. A. Jackson *Mem. Stonewall Jackson* (ed. 2) xi. 197 Between them is a library desk. **1869** Geo. Eliot *Let.* 19 Feb. (1956) V. 16 Ticknor and Field have intimated some intention to bring out a library edition of all my books. **1917** *Times* 2 July 1 (*top right-hand corner*) Library Edition. **1939** A. Huxley *Let.* 19 Feb. (1969) 440 The idea of a library edition makes me feel most horribly posthumous. **1960** S. Unwin *Truth about a Publisher* x. 150 They were..available only in the huge 38-volume Library Edition. **1948** *Optical Practitioner* II. vi. p. vi (Advt.), Frames of unusual design specially made... The 'London' Library frame. **1962** L. S. Sasieni *Princ. & Pract. Optical Dispensing* i. 10 The heavy frame..is often called a 'library' frame. **1971** W. Garner *Andra Fiasco* ix. 56 She stared at him, reaching for a pair of library-frame glasses. **1959** *A.O.P. News* 31 Dec. 9/1 Modern spectacles for men, such as library glasses..can give the wearer a distinguished appearance. **1837** W. Jenkins *Ohio Gazetteer* 281 The library house is a handsome brick edifice. **1962** *Listener* 16 Aug. 259/2 John Elliot's production adroitly wove the tropical library material among the studio scenes. **1953** K. Reisz *Technique Film Editing* 280 *Library shot*, shot used in a film but not recorded specially for it; shot taken from a library or store of shots kept for future use. **1962** *Listener* 22 Feb. 346/1 The other, the one with the library spectacles, also lowered his paper. **1962** *Gloss. Ophthalmic Lenses & Spectacle Frames* (*B.S.I.*) 58 *Library spectacles*, spectacles of heavy weight with broad sides and usually of tortoiseshell or plastics. **1861** *Catal. Cathedral Libr. Calcutta* App. 120 To stamp the title page and other parts of each volume with the Library-stamp. *c* **1762** Ince & Mayhew *Universal Syst. Houshold Furnit.* 3/1 (*heading*) Plate XXII. Two Designs of Library Steps; the First intended for a large Room. **1793** T. Sheraton *Cabinet-Maker & Upholsterer's Drawing-Bk.* App. 9 There are other kinds of library steps which I have seen, made by other persons, but..these must have the decided preference. **1848** Thackeray *Pendennis* (1849) I. iii. 24 He would sit..perched upon the topmost bar of Doctor Portman's library steps with a folio on his knees. **1970** B. Cartland *Secret Fear* i. 5 He would notice the rosewood library steps below the place in the cabinet from which the book had been extracted. **1952** *Vision* VI. iii. 15/2 (*caption*) Heavy semi-rimless library style in dark mottled material. **1741** in *Publ. Colonial Soc. Mass.* (1925) XVI. 712 [The cross table was] compos'd of three Library Tables. **1853** Geo. Eliot *Let.* 29 Oct. (1954) II. 121, I wish to exchange my present one [*sc.* bookcase] for a Library Table, of which I am sorely in need. **1969** *Canad. Antiques Collector* Dec. 9/1 One of the most remarkable series of library tables (bureaux plats) to be found anywhere. **1971** *Country Life* 10 June Suppl. 52 (Advt.), A rare small George III library table with pull-out steps, reading top and two slides, constructed with fine quality solid mahogany. Height 32″. Circa 1790. **1877** M. W. Chapman in H. Martineau *Autobiogr.* III. 63 Heaps of concert tickets, museum tickets, library tickets. **1950** O. Blakeston *Pink Ribbon* vi. 72 Amelia lent me her library tickets. **1965** C. Fremlin *Jealous One* xiv. 110 The usual contents of a hand-bag—comb, purse, powder compact, cheque book, library ticket.

Librium (li·briǝ̆m). *Pharm.* Also librium. The proprietary name of a white crystalline compound, $C_{16}H_{14}N_3OCl.HCl$, used as a tranquillizer.

1960 *Official Gaz.* (U.S. Patent Office) 15 Mar. TM 90/1 Hoffman–La Roche Inc... *Librium* for psychotherapeutic agent. **1968** *New Scientist* 21 Mar. 623/2 Hostile tendencies can often be remarkably controlled by drugs, like Librium and diazepam, which are *not* sedatives, and which do not depress the general level of cerebral activity. **1970** *Times* 22 Sept. 10 In the first cafe he went into someone sold him six librium pills. 'It was my sort of cafe, my sort of people—of course they had gear.' **1972** T. P. McMahon *Issue of Bishop's Blood* (1973) xvi. 228, I decided the pills Dolly had left with Julio were Librium. I use them when the shakes get bad. **1975** N. Mitchison *All change Here* ix. 92, I took a tablet of librium, but had the father and mother of a nightmare.

Libyan, *a.* and *sb.* Add: (Earlier and additional examples.) Also, of or pertaining to (or an inhabitant of) the modern state of Libya.

1592 Nashe *Pierce Penilesse* 34 And such a spirit it was that possest the Libian Sapho, and the Emperorer Dioclesian. **1607** Topsell *Four-footed Beasts* 41 Beares are tamed..for sports among the Roxolani and Libians. *Ibid.* 115 The Lybian Roes..are..inferiour to the Lybian horses. **1725** E. Combe tr. *Huet's Weakness of Hum. Und.* 117 He might be call'd a Libyan, because he stay'd a long while at Cyrene, a City in Libya. **1882** [see *Kabyle]. **1921** *Handbk. Libya* viii. 147 Head-quarters of the Libyan Battalion. **1937** H. Pope *St. Augustine of Hippo* i. 18 In the purple-producing island of Girba.. Libyan was spoken, as Berber indeed is to this day. **1939** L. H. Gray *Found. Lang.* xii. 366 *Graffiti*..of uncertain date and in a script which marks them off from ancient Libyan and from modern Tuareg alike. **1956** H. S. Villard *Libya* xii. 154 In the suppression of native rebellions most of the educated Libyan leaders were exterminated. *Ibid.* 155 The Libyans..continue to place restrictions on the entry of Italian citizens into Cyrenaica. **1971** *Encycl. Judaica* XI. 205 During the Six Day War (1967)..widespread strikes of Libyan oil workers.. brought the flow of oil to a temporary stop. **1975** *Times* 11 Jan. 12/6 The Libyans, still the most active of the governmental supporters of Arab extremists.

Libyo-. Add: *Archæol.* **Libyo-Phœnician, Libyphœnician.** A Phœnician living in Libya, a person of mixed Libyan and Phœnician ancestry, or a Libyan vassal or ally of the Phœnicians.

1876 *Encycl. Brit.* V. 160/2 In Africa her [*sc.* Carthage's] subjects consisted of three classes—(1) Libyo-Phœnicians, [etc.]. **1948** *Antiquity* XXII. 142 Hanno tells us that he set sail..under orders..to proceed outside the Pillars of Heracles and found cities of Libyphoenicians. *Ibid.*, Libyphoenicians can mean either Phoenicians from Africa or African allies or vassals of the Carthaginians. **1962** D. Harden *Phoenicians* 223 Some think that it is the Berber inhabitants of the *territorium* who are the Libyphoenicians so frequently mentioned in ancient texts.

licence, *sb.* Add: **5.** *licence number, plate.*

1937 D. & H. Teilhet *Feather Cloak Murder* ii. 48 'Did you get the licence number?'..the grey car had vanished. **1972** L. Lamb *Picture Frame* xx. 178 We found his van... It was Mallender's licence number. **1926** *Amer. Speech* I. 686/1 American: Number plates. English: License plates. **1962** 'E. McBain' *Like Love* (1964) xiv. 189 You didn't happen to notice the licence plate number, did you? **1974** R. C. Dennis *Conversations with Corpse* xiv. 140, I..landed..a 1968 license plate.

li·cenceless, *a.* [f. Licence *sb.* 2.] Not possessing a licence.

1906 *Daily Chron.* 22 May 7/7 Six licenseless motorists on the Oxford to Birmingham highway were..discovered. **1923** *Glasgow Herald* 10 Oct. 7/1 Licenceless owners will make themselves liable to prosecution if discovered by the Post Office inspectors.

lichee, lichi, varr. Litchi in Dict and Suppl.

lichen, *sb.* Add: **3. b.** With mod.L. adjs., as **lichen planus,** a skin disease characterized by an eruption of wide, flat-topped, shiny, purple-coloured papules; **lichen simplex,** (*a*) a type of eczema characterized by the presence of small red papules; (*b*) (*lichen simplex chronicus*) a disorder characterized by areas of lichenification.

1798 R. Willan *Descr. & Treatm. Cutaneous Dis.* i. ii. 41 The extent of the disease being thus limited, I shall proceed to describe the varieties of it, which have occurred to my observation, under the denominations of Lichen simplex, Lichen agrius, Lichen pilaris, Lichen lividus, and Lichen tropicus. **1842** T. H. Burgess tr. *Cazenave & Schedel's Man. Dis. Skin* 191 Acute lichen simplex requires no other treatment than diluents and tepid baths. **1866** E. Wilson in *Brit. Med. Jrnl.* 13 Oct. 399/1 (*heading*) On lichen planus: the lichen ruber of Hebra. **1910** C. F. Marshall tr. *Gaucher's Dis. Skin* 101 Lichen simplex may occur on all parts of the body, but it chiefly affects the forearms and dorsal surface of the hands, the neck and shoulders, the external and posterior surfaces of the legs, and the internal surface of the thighs. **1934** Dore & Franklin *Dis. Skin* viii. 121 Lichen planus is generally a chronic complaint. **1966** W. D. Stewart et al. *Synopsis Dermatol.* xxvi. 485 Lichen simplex chronicus..is a common pruritic disorder resulting in a localized patch of dermatitis that has a characteristic lichenification. **1971** *Acta Dermato-Venereol.* LII. 216/1 In an epidemiologic house-to-house survey..in Kerala in South India, 7639 individuals were examined for oral lichen planus.

5. lichen-acid, any lichen substance which is an acid; **lichen substance,** any of about 65 compounds, most of which are acids, which are found uniquely in lichens.

1893 *Jrnl. R. Microsc. Soc.* 497 (*heading*) New lichen-acid. **1967** M. E. Hale *Biol. Lichens* viii. 119 Lichen acids and pigments increase the opacity of the upper cortex. **1900** *Jrnl. R. Microsc. Soc.* 235 (*heading*) Lichen-substances. **1954** Asahina & Shibata (*title*) Chemistry of lichen substances. **1967** Lichen substance [see *depside].

lichenification (ləike:ni-, ləi:kĕnifikē̆·ʃən). *Med.* [ad. F. *lichénification* (L. Brocq 1892, in *II. Internat. Dermatol. Congr.* 522), f. F. *lichen* Lichen *sb.*: see -ification.] Hardening and thickening of the skin caused by scratching or other continued irritation; an area of skin so affected.

1892 L. Brocq in *Brit. Jrnl. Dermatol.* IV. 326 At a certain time the skin shows quite a peculiar aspect, with a great exaggeration of its natural lines, representing a sort of network with meshes more or less large and regular, and a decided infiltration of the integuments... Such is the morbid process which I call Lichenification. *Ibid.* 329 In the second group..we only meet with the cutaneous lesions we have described as primary lichenifications. **1934** Dore & Franklin *Dis. Skin* iii. 41 The lichenification itself may be painted with crude coal-tar. **1968** A. J. Rook et al. *Textbk. Dermatol.* I. xii. 222/2 In all forms of lichenification pruritus is the predominant symptom. **1971** *Jrnl. Investigative Dermatol.* LVII. 299/2 Lichenification involves a change in the epidermis, characterized by hyperkeratosis, acanthosis and scattered parakeratosis.

lichenified (ləike·nifəid, ləi·kĕnifəid), *ppl. a.* *Med.* [f. Lichen *sb.* + -if(y + -ed[1] (to parallel *lichenification).] Showing lichenification.

1892 *Brit. Jrnl. Dermatol.* IV. 327 The cases entering into the second group, which we have called Secondary Lichenifications, are not of the nature of real and true lichens, but are simply lichenified dermatoses. **1927** R. C. Low *Common Dis. Skin* xii. 161 The situation of these lichenified areas, usually on the back of the neck, is a rare one in lichen planus. **1967** H. Montgomery *Dermatopath.* I. ix. 190/2 In my experience, one cannot distinguish between chronic lichenified disseminate neurodermatitis and localized forms of neurodermatitis histologically.

lichenized, *ppl. a.* (in Dict. s.v. Lichenize *v.*). Add: **b.** Of a fungus or an alga: living in association with (respectively) an alga or a fungus so as to form a lichen; adapted or evolved to live as a component of a lichen.

1942 *Proc. Sect. Sci. Kon. Akad. Wetensch. Amsterdam* XLV. 276 We believe to have found in the lichenized algal covers of Pleurococcus, Apatococcus and allied species, a better object of study in this respect. **1952** *Symbolae Bot. Upsalienses* XII. 10 It is certainly difficult in many cases to establish whether or not an organism should be regarded as a lichen. One type of questionable or easily misinterpreted cases is provided by such lichen-fungi as are sometimes lichenized and sometimes live without algae. *Ibid.*, A lichenized fungus means in the present paper always a fungus which lives in symbiosis (s. str.) with an alga and which is not a parasymbiont. **1960** *Amer. Jrnl. Bot.* XLVII. 677 (*heading*) Some new and interesting species of Trebouxia, a genus of lichenized algae. **1973** *Nature* 4 May p. xv (Advt.), Taxonomic research in the ascomycetous fungi or their lichenised relatives.

lichenometry (ləikĕnǫ·mĕtri). *Geol.* [f. Lichen *sb.* + -o + -metry.] The dating of moraines or other surfaces recently exposed for lichen colonization by measurement of the size of lichens growing on them.

1957 R. E. Beschel in *Arctic* (Montreal) X. 60/1 This method that I call lichenometry works well for the last 1,000 years under alpine conditions. **1967** *Jrnl. Glaciol.* VI. 818 The basic premise of lichenometry is that the diameter of the largest lichen thallus growing on a moraine, rock glacier or other surface is proportional to the length of time that the surface has been exposed to colonization and growth. **1973** *Oxford Times* 27 July 6/5 There will also be a botanical study of the glaciers and a lichonometry survey to assist the dating of withdrawal of ice.

Hence **lichenome·tric, -me·trical** *adjs.*

1958 *Arctic* (Montreal) XI. 254/1 (*heading*) Lichenometrical studies in West Greenland. **1959** *Biol. Abstr.* XXXIV. 861/2 In several glaciers of the group, a lichenometric study was performed. **1964** *Geogr. Bull.* (Ottawa) Nov. 80 (*heading*) A lichenometrical study of the north-western margin of the Barnes Ice Cap: a geomorphological technique. **1967** *Jrnl. Glaciol.* VI. 819 The wide geographic distribution of the species, together with its long life span and consistent size–age relationship make *R[hizocarpon] geographicum* ideal for lichenometric dating.

lick, *sb.* Add: **1.** (Later examples.) Also (*U.S. colloq.*) *a lick*, somewhat, a bit (usu. in neg. contexts).

1902 W. N. Harben *Abner Daniel* 94 But all day yesterday an' to-day he hain't worked a lick. **1919** H. L. Wilson *Ma Pettengill* vii. 215, I was fool enough to argue with him a bit, trying to see if he didn't have a lick of sense. **1938** C. H. Matschat *Suwannee River* vii. 110, I knocked him loose an' hit him a lick. **1939** Joyce *Finnegans Wake* 415 Seven bolls of sapo, a lick of lime, two spurts of fussfor. **1957** W. C. Handy *Father of Blues* v. 66 We had been complaining violently against an Irishman who couldn't cook a lick. **1971** *Black Scholar* Sept. 37/2 His grandfather was a preacher and he couldn't read a lick. **1973** *Black World* Jan. 63/2 His wife Fanny can't cook a lick. **1973** M. & G. Gordon *Informant* xlix. 188 If you've got a lick of sense, you'll mosey back into the woodwork.

b. (Further examples.) Also, a hasty tidying up, a casual amount of work.

1860 W. White *All round the Wrekin* xx. 207 We only gives the cheap ones a lick and a promise. **1899** E. F. Heddle *Marget at Manse* 43 That lassie gi'es a lick and a promise when I tell her to sweep! **1922** A. Bennett *Lilian* i. vi. 57 The dirty kitchenmaid was giving the stone floor of the porch a lick and a promise. **1934** L. A. G. Strong *Corporal Tune* iii. ii. 230 The room, instead of its usual vigorous cleaning, got what Nelly would have called a lick and a promise. **1942** C. Morley *Thorofare* xl. 355 You ought to be writing the Adventures of a Crustacean. You've only done a lick and a promise. There's six more inches to fill. **1948** M. McCarthy in *Partisan Rev.* May–June 325 The Dublin Gate players..had a slapdash style of acting that suggested an Irish housemaid flailing about with a dust-cloth—they gave their roles a lick and a promise and trusted to the audience's good-nature to take the will for the deed. **1967** V. Lincoln *Private Disgrace* (1968) xi. 91 She had only a basin of water and a rag with which to give the insides of the windows a lick and a promise. **1969** D. Clark *Death after Evensong* vi. 142 A pale sun gave Rooksby a lick and a promise of better things to come. **1972** J. Burmeister *Running Scared* iii. 51 The isolation ward..was given a lick and a promise once a month by an unsupervised maid.

2. (Further examples.) Also *Canad.*

1747 *Virginia Land Patents & Grants* in *Amer. Speech* (1940) XV. 280/2 Crossing the said Run above a Lick.

1750 T. Walker *Jrnl. Explor.* (1888) 51 At the mouth of a Creek..is a Lick, and I believe there was a hundred Buffaloes at it. **1784** J. F. D. Smyth *Tour U.S.A.* I. xviii. 141 Licks are particular places..where the clay or earth is impregnated with saline particles. *a* **1816** B. Hawkins *Sk. Creek Country* (1848) 29 Parallel with this, are some licks in post and red oak saplin flats. **1825** J. Pickering *Jrnl.* 21 Dec. in *Emigration* (1830) v. 49 Deer will go miles to the salt spring, or 'licks' as they are called. **1832** J. McGregor *Brit. Amer.* II. 556 Both buffalo and deer resort to them for the purpose of licking the salt off the shrubs hence the name *lick*. **1957** *Beaver* Summer 37/2 The goat evidently was headed for the same lick from which the sheep were returning.

b. = *lick-log*.

1920 Webster *Lick*, an artificial saline preparation given to sheep and cattle to lick. *c* **1920** W. D. Powdrell *Dairy Farming N.Z.* v. 38 A lick of rock-salt should be provided. **1950** *N.Z. Jrnl. Agric.* July 67/3 By using cobalt either as a topdressing, as a drench, or in licks all classes of stock could be run without any trouble [on this cobalt-deficient country]. **1963** *Times* 4 Feb. 4/7 A large feed block or lick is made available to cattle.

6. (Further examples.) Also *colloq.* and *N.Z.*

The phr. *to go* (or *run*, etc.) *for the lick of one's life* appears to be restricted to Australia and N.Z.

1809 T. Donaldson *Poems* 135 Ere I get a pick, In comes young *Nannie* wi' a lick. **1835** *Gent's. Vade Mecum* (Philadelphia) 14 Feb. 3/4 When you come to put in the scientific licks, I squat. **1847** J. S. Robb *Streaks of Squatter Life* 106 He was puttin' in the biggest kind a licks in the way of courtin'. **1851** —— in T. A. Burke *Polly Peablossom's Wedding* 111, I saw comin' my gray mule, puttin' in her best licks, and a few yards behind her was a grizzly. **1868** *Putnam's Mag.* June 715/1 The father..did an occasional 'lick of work' for some well-to-do neighbor. **1892** *Dialect Notes* I. 230 *To mend one's licks*, to quicken one's steps. 'When the dog got after me, I mended my licks.' **1905** *Dialect Notes* III. 86 You'll have to hit a different lick, if you expect to accomplish anything. **1906** H. D. Pittman *Belle of Bluegrass Country* xv. 224 I'll have to take care of the whole gang, and never get a lick of work out of one of them. **1932** W. Faulkner *Light in August* (1933) i. 7 She's hitting that lick like she's been at it for a right smart while. **1934** J. Masefield *Taking of Gry* 43 'They're [ships] going a good lick, sir,' I said. **1938** *Amer. Speech* XIII. 6/1 *Lick* n., an easy job. 'None of these jobs is a lick.' **1944** J. H. Fullarton *Troop Target* i. v. 46 'Go for the lick of your life down the lane,' commanded Rangi. **1946** F. Sargeson *That Summer* 84 With all of us going for lick of our lives, there'd only be time for a wisecrack now and then. **1948** D. Ballantyne *Cunninghams* (1963) xviii. 203 Clive ran..full lick into the sea. **1949** *Marshfield* (Wisconsin) *News-Herald* 19 July 4/1 The power lobby got in its licks through a subcommittee of the Senate Appropriations Committee passing on the bill for funds for the Department of Interior. **1951** L. MacNeice tr. *Goethe's Faust* i. 33 Lord, these strapping wenches they go a lick! **1966** *Sunday Mail Mag.* (Brisbane) 3 Apr. 6/3 A section of the miners agreed that the happiest solution to the sorry affair would be to lynch Mr. Chapple. The little Cornishman got wind of this thinking and, treating it seriously..went for the lick of his life. **1974** P. Ruell *Death takes Low Road* x. 127 Caroline contrived to be first down the gangway and set off along the quay at a good lick.

7. In jazz, dance-music, etc.: a short solo or phrase, usu. improvised and often interpolated into a piece of written music; = *break *sb.*[1] 9 c; freq. in phr. *hot lick* (cf. *hot *a.* 8 g).

1932 *Melody Maker* June 509 They manage to steal a 'lick' from an American record. **1933** *Metronome* Apr. 29 Please do not..think I want 'hot licks' to memorize in all keys. **1933** *Fortune* Aug. 47/1 His licks (musical phrases) are original. **1935** [see *go *v.* 22 b]. **1935** [see *ad lib. B *adj.*]. **1935** *Vanity Fair* (N.Y.) Nov. 38/3 Hot artists.. add their licks to the exciting music that flourishes there. **1952** B. Ulanov *Hist. Jazz in Amer.* (1958) xix. 237 The panic was on to push vibrato aside, pick up his licks, and produce his sound. **1970** *Globe & Mail* (Toronto) 26 Sept. 27/3 The blues riff is even better, full of Charlie Parker-like bebop licks.

b. Plan, idea. *U.S. colloq.*

1955 S. Allen *Bop Fables* 54 So here's the lick. Take this beat-up bovine to market. **1970** C. Major *Dict. Afro-Amer. Slang* 76 *Lick*, plan, idea, outline of a situation.

lick, *v.* Add: **1. b.** *to lick* (another's) *boots* = *to lick* (another's) *shoe*; *to lick the* (or *one's*, etc.) *chops* (*Jazz slang*), to tune up or warm up before a 'session'.

1890 Kipling *Barrack-Room Ballads* (1892) 23 An' you'll lick the bloomin' boots of 'im that's got it. **1909** G. B. Shaw *Press Cuttings* 9 And now comes this unmannerly young whelp Chubbs-Jenkinson, the only son of what they call a soda-king, and orders a curate to lick his boots. **1930** —— *Apple Cart* i. 34, I had rather be a dog than the Prime Minister of a country where the only things the inhabitants can be serious about are football and refreshments. Lick the king's boots: that is all you are fit for. **1937** [see *cat *sb.*[1] 2 c]. **1937** *Étude* Dec. 835/1 *Licking their chops*, getting warmed up to swing. **1959** J. Braine *Vodi* x. 138 He had to use his..willingness to lick anyone's boots, no matter how dirty, to get the money. **1970** C. Major *Dict. Afro-Amer. Slang* 77 *Licking the chops*, the tuning up musicians do before a jam session. **1974** *Guardian* 19 Dec. 10/2 If Lifestyle (BBC-2) keeps licking boots like this, Cherry Blossom will sprout out of its ears.

7. Read: To run, ride, or move at full speed. orig. *dial.* (widespread outside the U.K.). Also in the U.S., *to lick it*. (Further examples.)

1850 L. H. Garrard *Wah-to-Yah* i. 16 The mad animal ..charged. How they did 'lick it' over the ground! **1856** J. Collie *Poems* 124 Sae aff gaed Death what he cou'd lick. **1886** *Outing* Dec. 198/1 He'd nothin' ter do but ter lick it like blazes, with the little dog a-follerin' along. **1903** J. Lumsden *Toorle* iv. i. 76 Jock! lick awa' in, an' blaw up. **1947** 'A. P. Gaskell' *Big Game* 80 He sped her [*sc.* a car] along. Boy, she can lick. **1953** M. Traynor *Eng. Dial. Donegal* 169/1 *As hard as one can lick*, as fast as one can go. *To lick along*, to go fast. **1966** W. S. Ramson *Austral. Eng.* iv. 65 *To lick*, meaning 'to travel fast' and common in..*as hard as one can lick*.

8. lick-hole *Austral.*, a place where lick-logs are placed for stock to lick; **lick-log** (earlier and later examples); *to stand up to one's lick-logs*, to make a firm stand; **lick-up,** (*c*) used *attrib.* to designate a type of paper-making machine (see quots.).

1928 'Brent of Bin Bin' *Up Country* ix. 143 No horse ..was safe..in the lick-hole country of its myriad spring-heads. Pool found a way with rock-salt to make the lick-holes a trap. **1936** M. Franklin *All that Swagger* xvi. 148 A hint without evidence is a snake in the grass, like that boomer you dispatched to-day at the lickhole. **1834** D. Crockett *Narr. Life* 170, I was determined to stand up to my lick-log, salt or no salt. **1852** G. W. L. Bickley *Hist. Tazewell County* 226 Capt. Moore..was at a lick log ..salting his horses of which he had many. **1948** E. N. Dick *Dixie Frontier* 105 Small troughs were cut in the trunk of a fallen tree and occasionally salt was placed there, making what was known as a 'lick log'. **1929** Clapperton & Henderson *Mod. Paper-Making* xv. 212 The 'lick-up' machine may have either a vat or Fourdrinier wet end, but it does not possess a wet press... On this type of machine the underside of the web sticks to the cylinder and receives the polished surface. **1952** F. H. Norris *Paper & Paper Making* xv. 208 The 'lick up' type of machine..may have either a cylinder mould to pick up the stuff out of a vat and form the paper as on a board machine, or it may have the normal Fourdrinier wet end. On a 'lick up' machine, the wet felt also acts as an over-felt, and as there is no wet press, the web is transferred to the wet felt at the top couch roll.

licker. Add: **a.** Also in sense 6 of the verb.

1894 A. Morrison *Martin Hewitt* ii. 66 'There's no footprint here nor outside.'.. 'That's a licker,' he said. **1895** J. T. Clegg *Works* I. 375 Iv that's ony credit to Walsden it's a licker to me! **1902** *Eng. Dial. Dict.* III. 587/1 Fatther, this sum is a licker; will yo' do it for mha? **1907** *Daily Chron.* 31 July 4/7 The licking his Majesty once suffered..[and] the half-crown the late Queen gave the licker for his pluck. **1908** A. S. M. Hutchinson *Once aboard Lugger* vi. viii. 456 Into a chair Bill collapsed... He gasped 'George, this is a licker, a fair licker.'

b. *licker-in*, the cylinder in a carding-machine which receives the cotton, wool, etc., from the feed-rollers and passes it on to the main cylinder. Also *attrib.*

1850 *Rep. Comm. Patents 1849* (U.S.) 198, I do not claim a licker-in, nor the first main cylinder as such. **1884** [see *burring *vbl. sb.*[1]]. **1884** W. S. B. McLaren *Spinning* 84 To assist the process..the licker-in rollers are sometimes made hollow, and steam is allowed to fill them. **1888** [see *breast *sb.* 9 h]. **1892** [see *Garnett *sb.*]. **1946** A. J. Hall *Stand. Handbk. Textiles* iii. 101 The cotton in lap form from the scutching machine is fed on to one of the small rollers (termed the licker-in).

lickety (li·kĕti), *adv. colloq.* (chiefly *U.S.*). Also 9 **lickitie; lickerty, licketty, -ity, -oty.** [Fanciful; cf. Lick *sb.* 6 (in Dict. and Suppl.), *lick *v.* 7.] Usu. prefixed to another word, as **lickety-split,** at full speed; headlong. Also (nonce-wd.) as *vb.* Also *lickety-cut, -smash, -wallop*, etc.

1817 D. McKillop *Poems* 33, I rattl'd owre the A, B, C, as fast as lickitie An' read like hickitie. **1831** *Boston Even. Transcript* 4 June 2/2 He ran down the street licketty cut, and is probably at home by this time. **1847** J. S. Robb *Streaks of Squatter Life* 116 Away they started, 'lickety-click', and arrived at the winning-post within touching distance of each other. **1848** in *Amer. Speech* (1935) X. 40 *Lickoty liner*, going very fast. **1858** *Harper's Mag.* May 766/2 There they had it, lickety-switch, rough-and-tumble. **1859** Bartlett *Dict. Amer.* (ed. 2) *Lickety Split*, very fast, headlong; synonymous with the equally elegant phrase 'full chisel'. 'He went lickety split down hill.' **1863** L. M. Alcott *Hospital Sk.* iii. 40 When my mate, Eph Sylvester, caved, with a bullet through his head, I got mad, and pitched in, licketty cut. **1869** Mrs. Stowe *Oldtown Folks* xxviii. 358, I tell you if they didn't whip up an' go lickety-split down that 'ere hill. **1886** [see *bump *v.*[1] 2 e]. **1897** [see *band *sb.*[3] 4 b]. **1911** R. D. Saunders *Col. Todhunter* ix. 122 You're worse 'n a old huntin' dog that goes sky-hootin' off lickety-split after a rabbit. **1928** 'Brent of Bin Bin' *Up Country* ix. 141 They rattled the vehicle lickety-smash at a hand-gallop across the flower-strewn plains. **1934** W. Saroyan *Daring Young Man* (1935) 131 And then he was running lickety split across the school grounds. **1949** O. Nash *Versus* 111 Firemen, what is your destination?.. You have lickety-splitted by so often that my thoughts Are utterly split-licketed. **1949** Powys & Bolton tr. *Guitry's Don't Listen Ladies* in *Plays of Year* I. 566, I was coming along the street not thinking of anything, and suddenly you came out and shot past me, lickerty-split. **1949** Wodehouse *Uncle Dynamite* iii. 48 If I was you, I'd hop into that car of yours and drive lickerty-split to London and get another bust. **1955** E. Pound *Classic Anthol.* ii. 87 We took out our cars lickety-clickety at the call. **1960** V. Nabokov *Bend Sinister* ii. 17 The old men overtook him in their turn, clattering lickety-split through the mist. **1961** B. Fergusson *Watery Maze* x. 245 While going lickety-spit..they had come up against the formidable 15th Panzer Grenadier Division.., and they were lucky not to have been chased into the sea. **1972** A. Fowles *Double Feature* xiii. 240 If one of his outriders radios in that Chau Chieu is there he'll come licketty split. **1972** *Last Whole Earth Catalog* (Portola Inst.) 305/3 Just like that. Stopped in here a few minutes, then took off up that creek lickety-split.

lick-spittle. Add: **b.** The practice of toadying. Hence (or as a back-formation from the *vbl. sb.*) as *v. trans.*, to toady to (a person).

1914 A. Harrison *Kaiser's War* 112 A social system of formality, lick-spittle, bullying, and brutality. **1927** *Daily Express* 2 May 12/3 Christ criticised the sins of the Church His mother attended, and got His reward. He did not lickspittle the wealthy.

licuala (likyuwā·lă). [mod.L. (C. P. Thunberg 1782, in *Kungl. Svenska Vetenskapsakad. Handl.* III. 284), f. Makassar *lekowala*.] A small palm tree of the genus so called, belonging to the family Palmaceæ, native to Malaysia, New Guinea, and northern Australia, and having fan-shaped leaves and prickly stalks.

1872 *Gardeners' Chronicle* 14 Dec. 1657/2 The Licualas are Fan Palms of Asiatic origin, requiring hothouse cultivation. **1900** L. H. Bailey *Cycl. Amer. Hort.* II. 911/1 The large fan-shaped leaves of the Licualas are somewhat tender and easily injured. **1930** *Discovery* Nov. 380/1 A few *Licuala* and *Pinanga* palms were seen. **1952** 'W. March' *October Island* x. 125 Zalacca, pinanga, licuala, corypha, And the sealing-wax palm. **1966** E. J. H. Corner *Nat. Hist. Palms* xiii. 310 When the forest is cleared, the thickets of *Licuala* may be left, and they catch the eye with their striking foliage and long sprays of pink, orange, and red berries.

lid, *sb.* Add: **1. b.** (Later example.)

1942 'B. J. Ellan' *Spitfire!* p. x, Shut the *lid*, i.e., close the hood [over the pilot's cockpit].

e. In various slang or colloq. phrases with *down, off, on*, esp. *to put the lid on*, to bring to a close or climax; to conceal or 'clamp down on'.

1915 *Lit. Digest* 4 Sept. 467/1 In fact, excepting the ordinary saloons,..the 'lid' is down, secure and tight. **1964** J. P. Clark *Three Plays* 13, I hope he keeps The lid down on his wife for I fear She is fretting already. **1873** M. F. Mahony *Chron. Fermors* I. xii. 225 What wonder if the lid was constantly getting off her temper. **1904** *Public Ledger* (Philadelphia) 12 Sept. 16 Commissioner of Police McAdoo..has taken frequent occasions to deny that the 'lid' was off, to use the slang definition of a lax police administration. **1910** W. M. Raine *Bucky O'Connor* 96 'Playing with the lid off back there, ain't they?' The sheriff's nod indicated the distant faro-table. *Ibid.* 218 I'll back that opinion with the lid off. **1926** A. Huxley *Let.* 14 Nov. (1969) 276 Then a vast book by the Italian Sociologist Vilfredo Pareto, very good..he really does take the lid off and show you the works. **1927** R. A. Freeman *Certain Dr. Thorndyke* ii. xviii. 272 'My eye,' exclaimed Miller... 'This puts the lid on it—or rather takes the lid off.' **1951** E. Paul *Springtime in Paris* ii. 17 A few jubilant days when the lid was off following Liberation. **1962** *Which?* May 160 (*heading*), 14 cars with the lid off. **1968** *Listener* 19 Dec. 819/2 Are you the Editor of the *Sunday Blast*, the paper that 'rips the lid off'? **1973** *N.Y. Law Jrnl.* 4 Sept. 5/3 Inevitably when the lid blows off and riots and bloodshed and vandalism begin, the courts will be called on to do something effective about it. **1974** 'M. Innes' *Appleby's Other Story* ii. 15 What will happen, I ask myself, when the police take the lid off? What..will be the resulting smell? **1909** *Punch* 30 June 452/2 Your astonishing letter puts the lid on it. **1914** 'High Jinks, Jr.' *Choice Slang* 14 *Lid* (*to put on*), to put the lid on a town means to close the saloons, gambling houses and all other resorts except summer resorts. **1914** G. B. Shaw *Misalliance* 77 *Tarleton*. .. Young man: youre a fool; but youve just put the lid on this job in a masterly manner. **1914** H. A. Vachell *Quinneys'* ii. xx. 288 'Blackmail!' gasped Quinney. 'I prefer to call it a weapon, sir, which you are forcing me, sorely against my will, to use.' 'This puts the lid on.' **1922** C. Sidgwick *Victorian* ix. 69 'That puts the lid on,' said Jane... 'You've done for yourself now.' **1928** T. Gann *Discoveries Cent. Amer.* xii. 168 Then came the earthquake, which must fairly have put the lid on, as far as Uk was concerned. *a* **1930** D. H. Lawrence *Phoenix II* (1968) 236 Inland, in the isolation, the lid is on, and the intense watchful malice of neighbours is infinitely worse than any police system. **1930** J. Buchan *Castle Gay* xiv. 216 You can see for yourself how that would put the lid on it. **1966** Mrs. L. B. Johnson *White House Diary* 6 Aug. (1970) 410 Liz came in with a harried look, wanting to know what she could tell the press. She needed to 'put the lid on' if Luci had departed. **1974** *Times* 6 Feb. 19/4 (*heading*) Putting the lid on distributors' profits.

f. A hat, a cap. *slang.* (Cf. *flip *v.* 9.)

1896 [see *glad *a.* 4 e]. **1916** *Story-Teller* Feb. 828/2 'Dash my wig—where's my lid...' He snatched his cap up off the bunk. **1916** C. J. Dennis *Songs Sentimental Bloke* 21, I dips me lid. *Ibid.* Gloss. 125 *Lid*, the hat. *To dip the lid*, to raise the hat. **1929** Wodehouse *Mr. Mulliner Speaking* ix. 304 You've no idea what a blister you look in that lid. **1946** B. Marshall *George Brown's Schooldays* xlvi. 178 Keep that lid of yours off your bloody ears if you don't want to look like a rotten sheeny. **1956** B. Holiday *Lady sings Blues* (1973) i. 11 All the big-time whores wore big red velvet hats then with bird-of-paradise feathers on them. These lids were the thing. **1960** Wodehouse *Jeeves in Offing* xii. 132 It is almost as foul as Uncle Tom's Sherlock Holmes deerstalker, which has frightened more crows than any other lid in Worcestershire.

g. (See quot. 1971.) *slang.*

1967 *Time* 8 Sept. 18 The high price of 'commercial' marijuana ($10 to $15 for a 'lid' from which some 40 cigarettes can be rolled). **1968** J. D. MacDonald *Pale Grey for Guilt* (1969) xii. 152 We had almost two lids of Acapulco Gold. **1969** *Rolling Stone* 17 May 6/3 We've got this guy rom Sand City we just caught with a lid. **1970** K. PLATT *Pushbutton Butterfly* (1971) iv. 43 He would be selling grass, meth, acid, lids, match boxes,..or mescaline. **1971** E. G. LANDY *Underground Dict.* 120 *Lid*, one ounce of marijuana, a quantity by which it is sold.

lid, *v.* Delete *rare* and add later examples.
1913 *Chambers's Jrnl.* Oct. 729/2 The cans..then move along to be lidded. **1950** *N.Z. Jrnl. Agric.* Nov. 429 (*caption*) A case packed to the correct height is shown in the illustration. Severe damage may occur to fruit on the lidding press unless the pack is crowned correctly. **1959** *Listener* 22 Jan. 191/2 Lid the flan with pastry. **1960** *Encounter* Mar. 21/1 They lidded that box again.

lidar (lai·dāɹ). [f. LI(GHT *sb.* + *RA)DAR.] A system for detecting the presence of objects or ascertaining their position or motion which works on the principle of radar, but uses laser radiation instead of microwaves.
1963 *Bull. Amer. Meteorol. Soc.* XLIV. 568/1 Scattering at 180°, or back-scattering, is the basis for both the microwave radar and the lidar (laser radar). **1963** *New Scientist* 20 June 673/3 The difficulties already encountered in detecting lidar pulses from the Moon will make astronomers wary of attempting to use such methods on the planets. **1968** *McGraw-Hill Yearbk. Sci. & Technol.* 228/1 The purpose of the lidar was to determine where the spray cloud drifted after release by the aircraft, so that the area of forest 'treated' could be accurately determined. **1970** *Daily Tel.* (Colour Suppl.) 28 Aug. 19/1 At Duisberg in the Ruhr a £20,000 laser system—called 'Lidar'..—which was supplied by a British firm, Laser Associates, monitors the pollution coming from industrial chimneys.

liddle (li·d'l), *a.* Representing a foreign or dialectal pronunciation of, or used hypocoristically for, LITTLE *a.* So **li·ddly** *sb.*, a little child.
1906 KIPLING *Puck of Pook's Hill* 224 Come along o' me while I lock up my liddle hen-house. **1929** R. HUGHES *High Wind in Jamaica* i. 8 Rachel, Edward, and Laura, the little ones (or Liddlies, as they came to be known in the family). **1941** M. TREADGOLD *We couldn't leave Dinah* vi. 106 They are nice liddle horses, *nicht wahr*, Karl? **1945** [see *CHUTZPAH]. **1970** C. DRUMMOND *Stab in Back* vii. 165 She takes 'er delicate liddle tray and bird-like appetite to one of the upstairs rooms. **1973** K. GILES *File on Death* v. 144 P'raps we could 'ave a liddle natter.

‖ **lidia** (lī·ðiă). [Sp., lit. 'fight'.] A bull-fight, esp. the earlier stages in which the cuadrilla prepare the bull for the faena; the process whereby the torero obliges the bull to conform to his movements. So **lidiador** (lī·ðiaðōᵊr), a torero considered as controlling his art and the actions of his picadors, and the responses of the bull.
1893 CHAPMAN & BUCK *Wild Spain* v. 57 It was a gay and imposing scene..when the *lidia*, or tournament, took place. *Ibid.* 59 De Bedoya's 'Historia del Toréo'..gives Francisco de Romera as the first professional *lidiador* of the modern epoch. **1932** E. HEMINGWAY *Death in Afternoon* 445 *Lidia*, the fight... *Lidiador*, one who fights bulls. **1952** J. MARKS *To Bullfight* iv. 50 This task consists in calling the toreros to order if they infringe any of the rules that govern the course of *la lidia*, which is the actual conduct of the fight. **1957** A. MACNAB *Bulls of Iberia* i. 11 After a while they start learning to distinguish the cloth from the body. Some breeds..learn quickly, and unless the man knows his stuff properly ('gives correct *lidia*' is the technical expression), he is apt to find himself hanging on a horn. *Ibid.* xv. 230 Antonio is far too good a *lidiador* to..request the President to change the Act. **1967** MCCORMICK & MASCARENAS *Compl. Aficionado* i. 25 As with tragedy, the lidia to the noble bull has about it an aura of inevitability. *Ibid.* viii. 240 He wants the bull to follow the muleta smoothly, as he educated the animal to do throughout his entire lidia.

Lido (lī·do). [Venetian It. *lido*:– L. *litus* shore.] The name of a spit of land, a famous beach resort near Venice, now used *gen.* for: such a spit enclosing a lagoon; a bathing-beach or resort; a public open-air swimming-pool.
[**1611** CORYATE *Crudities* 160 Venice..is distant from the maine Sea about the space of 3 miles. From the which it is deuided by a certaine great banke called *litto maggior*, which is at the least fifty miles in length.] **1673** J. RAY *Observations Journey Low-Countries* 149 These Lagune are..separated from the main Gulf or Adriatic Sea by a bank of earth (il Lito or Lido they call it). **a 1680** EVELYN *Diary* an. 1645 (1955) II. 433 A loud acclamation is Echod by the greate Guns of the Arsenale, and at the *Liddo*. **1860** E. HALL *Diary* 17 Oct. in O. A. Sherrard *Two Victorian Girls* (1966) II. 273 We took the boat to the Lido..and caught our first view of the Adriatic. **1930** *Morning Post* 16 July 5/4 The question of the safety of bathers in the Serpentine 'Lido' was raised at an inquest..yesterday. **1931** *Daily Express* 16 Oct. 8/2 £60,000 lido for England. The bathing pool and sun-bathing beach which the Hastings Corporation has just decided to construct [etc.]. **1934** *Discovery* Aug. 215/1 The broad sandspit or *lido* separating the lagoons from the sea. **1935** 'N. BLAKE' *Question of Proof* xiii. 259 What are you doing with all those deck-chairs..? Going to set up as a Lido proprietor? **1953** B. GOOLDEN *Truth is Fallen* xi. 175 'I haven't got any bathing things here.' 'Why on earth should we want them? It is not a Lido.' **1961** *Guardian* 24 Apr. 7/3 The Lido deck. **1969** V. CANNING *Queen's Pawn* xiii. 231 He went..out on to the One Deck Lido. **1971** *Country Life* 6 May 1106/3 Luino is a clean and pleasant place with several hotels, a lido and camping. **1975** M. KENYON *Mr Big* xxi. 204 All his free time was spent..semi-nude at the Serpentine lido.

lidocaine (li·dokēⁱn). *Pharm.* [f. ACETANI)-LID(E (from which the compound is derived) + -O + *-caine*, after COCAINE.] = *LIGNOCAINE.
1949 *Q. Cumulative Index Medicus* XLV. 105/1 Lidocaine, caudal anesthesia in delivery. **1954** *Jrnl. Pharmacol. & Exper. Therap.* CXII. 432 In spite of the wide application of lidocaine in dentistry and medicine, the physiological disposition of this drug has received only limited attention. **1972** *Sci. Amer.* Aug. 45/3 For heart-attack patients with normal or higher-than-normal heart rates lidocaine (a drug without any influence on the heart rate) is given intravenously almost universally in coronary-care units to suppress ventricular ectopic activity. **1972** *Chest* LXI. 682/1 We discuss and emphasize the danger of administering lidocaine in the presence of atrial tachyarrhythmias with rapid ventricular response.

lie, *sb.*¹ Add: **3. lie-detector** orig. *U.S.*, an instrument intended to indicate when a person is lying by detecting changes in his physiological characteristics.
1909 C. E. WALK *Yellow Circle* iv. 69 It is a lie detector. ..You set some wheels going. **1922** *Rep. 45th Ann. Meeting Amer. Bar Assoc.* 619 (*heading*) The Berkeley Lie Detector and other deception tests. **1933** *PMLA* XLVIII. 609 These views lead to such revolting pseudo-scientific nonsense as the use..of a *lie detector* apparatus in order to convict defendants. **1962** [see *GALVANIC a. a]. **1971** *Daily Tel.* 28 July 4/8 About 30 employees..have been given lie-detector tests in the fight against pilfering. It is believed to be the first use of 'polygraph interviews', as the tests are called, in New York shops. **1974** 'A. GARVE' *File on Lester* ii. 9 When a politician talks of frankness most voters reach for their lie-detectors.

lie, *sb.*² Add: **1.** *the lie of the land* (later examples).
1950 E. H. GOMBRICH *Story of Art* 1 To show the newcomer the lie of the land without confusing him with details. **1956** M. LOWRY *Let.* 13 Nov. (1967) 392 If any-one is to blame it is I, for not giving you the lie of the land before. **1966** D. VARADAY *Gara-Yaka's Domain* xi. 123 The quick powers of grasping a situation with which all game are endowed, showed themselves in the speedy summing-up by the leading boar, as he got the lie of the land.
4. (See also *LYE *sb.*²)
5. A period of resting or lying (esp. in bed). See also *lie-down, -in, -up* below.
1930 L. COOPER *Ship of Truth* i. 30 Sunday was their one chance of a long lie. **1938** D. DU MAURIER *Rebecca* xvii. 271 Have a good long lie tomorrow morning. Don't attempt to get up.
6. lie-about, an idle person, one of no fixed occupation, a disreputable 'character'; = *LAYABOUT; **lie-down** *colloq.*, a rest (on a bed, etc.); a form of protest in which the participants lie on the ground and refuse to move; **lie-in** *colloq.* = sense *5; also, as a form of protest, = prec.; **lie-up,** the fact of lying inactive in a place.
1937 M. ALLINGHAM *Dancers in Mourning* ii. 27 He took out a wallet which would have disgraced a lie-about. **1956** *Daily Mail* 26 Apr. 1/1 They are called champions of the prize ring but on Tuesday they appeared as two fat and horizontal lie-abouts. **1961** *Guardian* 27 Jan. 9/4 This former lie-about has got himself married. **1840** H. MOZLEY *Let.* 13 Oct. in D. Mozley *Newman Family Lett.* (1962) 93, I should be very glad of a lie down but cannot. **1850** C. KINGSLEY *Alton Locke* I. v. 80 You must keep moving all night..or else you goes to a twopenny-rope shop and gets a lie down. **1919** W. S. MAUGHAM *Moon & Sixpence* xlvii. 202 When..we hadn't even got the price of a lie down at the Chink's, he'd be as lively as a cricket. **1928** ST. JOHN ERVINE *Four One-Act Plays* 65 Yes, Aggie, you go an' 'ave a lie-down, see, and you'll be all right. **1936** *Time* 7 Dec., Second Sit-Down, Lie-Down... Twelve women and forty-five men, picketing the Berkshire Knitting Mills in Reading, Pennsylvania, by lying flat on its ice-covered front walk..were arrested. **1970** D. BALSDON *Oxf. Then & Now* III. v. 114 It is..the small body of demonstrators with whom we are here concerned—in particular the sit-down or lie-down to impede the Vice-Chancellor and Proctors in the exercise of their proper duties on November 5th, 1968. **1974** M. BIRMINGHAM *You can help Me* ii. 43, I won't risk our clients to you in your concussed state... Why don't you go and have a little lie-down? **1867** T. WRIGHT *Some Habits Working Classes* III. 206 The luxury of 'a long lie in', is the earliest and most universal of the delights of a working man's Sunday. **1916** 'TAFFRAIL' *Pincher Martin* xvi. 300 Lucky dogs!..You've got a lie in. I envy you. This is a night for poor old Peter to be at sea. **1932** C. L. MORGAN *Fountain* II. iv. 120 He left orders you was both to have a lie-in this morning. **1959** G. FREEMAN *Jack would be Gent.* ix. 192 I'm going to 'ave a bit of a lie in..seeing I'm on 'oliday. **1964** *Tuscaloosa* (Alabama) *News* 20 Apr. 1/8 The reported demonstration plans grew—from an auto stall-in on access roads to the fair to sit-ins, lie-ins and alike on other major highways, bridges and in tunnels throughout the city. **1971** *Time* 27 Dec. 40 Last week pollution protesters staged a lie-in at government offices in Tokyo. **1908** J. W. TYRRELL *Across Sub-Arctics of Canada* (ed. 3) 222 The two hundred mile tramp..had hardened our muscles so much that, with the ten days' 'lie-up' on the bank of the Nelson River,..we were now in first-class walking trim. **1926** *Blackw. Mag.* Dec. 850/2 We settled ourselves down for a happy four months of 'lie-up'.

lie, *v.*¹ Add: **10. d.** (Later examples.)
1876 'MARK TWAIN' *Tom Sawyer* i. 4 But in spite of her, Tom knew where the wind lay, now. **1886** F. T. ELWORTHY *West Somerset Word-Bk.* 434 Which way do the wind lie 'smornin?
e. Of horses, yachts, etc., in a race: to occupy a specified ordinal position. Also *transf.*
1951 E. RICKMAN *Come Racing with Me* iii. 24 What is that with the light blue sleeves lying fourth? **1955** J. CHRISTOPHER *Year of Comet* ii. 49 Who's lying fourth? **1972** D. FRANCIS *Smokescreen* iv. 55 He took the first half mile without apparent effort, lying about sixth. **1974** *Country Life* 24 Oct. 1189/3 Busted is lying third in this year's table of sire's winnings.
12. (Later examples in legal use.)
1958 *Times* 26 Apr. 6/7 If a chief constable is dismissed by a county council an appeal lies to the Home Secretary. **1964** *Mod. Law Rev.* XXVII. III. 322 Nowadays, after the revival of certiorari as a remedy lying for intra-jurisdictional defects, the scope of review on habeas corpus must be defined with more accuracy. **1970** *Internat. & Compar. Law Q.* 4th Ser. XIX. II. 306 The *Erbersatzanspruch* lies against the heirs, and consists of a sum equal to half the value of the portion, to which a legitimate intestate heir would be entitled. **1971** *Mod. Law Rev.* XXXIV. VI. 691 Where X and Y have a regular course of dealing and are likely to make contracts in the future, a *quia timet* injunction will lie to prevent Z, a third party, from inducing breaches of such contracts as may be made in the future.
16*. lie about, to lie here and there; to be left lying carelessly or in disorder.
1852 C. KINGSLEY *Hypatia* (1853) I. xiii. 274 Why, these poor blackguards lying about are very fair specimens of humanity. **1891** R. BUCHANAN *Come live with Me* II. xiii. 168 Ye might leave it [*sc.* poison] lying about, and mischief might happen. **1891** W. MORRIS *News from Nowhere* v. 31 Most children, seeing books lying about, manage to read by the time they are four years old. **1934** G. B. SHAW *Simpleton of Unexpected Isles* i. 4, I hate to see dust lying about. Look! You could write your name in it.
21. lie down. a. Also, to give up; to be remiss or lazy.
1904 W. H. SMITH *Promoters* i. 21 When they finally lie down, we'll just say, 'All right, we'll go ahead alone.' **1916** *Lit. Digest* 8 Jan. 87/1 It is natural enough that the accusation of 'lying down' and quitting has been cast up in turn at each of the participants in the conference. **1918** E. POUND *Let.* 3 Apr. (1971) 134 It is the best that can be done. Hope Kahn won't think I am lying down on the job. **1926** J. BLACK *You can't Win* xiv. 193 An ambitious fighting young lawyer who never 'laid down' on a client.
d. (Further examples.)
1914 G. B. SHAW *Androcles* (1916) I. 17 You know, I should feel ashamed if I let myself be struck like that, and took it lying down. **1931** E. F. BENSON *Mapp & Lucia* viii. 229 She had to swallow her medicine... I had no idea..that she would take it lying down like that. **1974** M. GILBERT *Flash Point* viii. 64, I heard what the beak said to you. I had an idea you weren't going to take it lying down.
23. lie in. d. To remain in bed (after one's usual hour of rising). Cf. *lie-in* (*LIE *sb.*² 6).
1893–4 R. O. HESLOP *Northumb. Words* II. 449 *Lie*, in the combination *lie-in*, to sleep longer than intended. **1911** E. M. CLOWES *On Wallaby* v. 144 On Sundays her husband and son 'lay in', as she called it, till midday, while she gave them their breakfast in bed.
27. lie over. d. *U.S.* To suspend travelling; to stop.
1849 *Ex. Doc. 31st U.S. Congress 1 Sess. Senate* No. 64. 186 But I shall make an early drive and 'lie over' to-morrow at the first water. **1903** A. ADAMS *Log of Cowboy* 181 We overtook a number of wagons loaded with wool, lying over, as it was Sunday.

Lie (lī). The name of Sophus *Lie* (1842–99), Norwegian mathematician, used *attrib.* to denote certain concepts investigated by him, as **Lie algebra,** a vector space extending over a field in which a product operation (×) is defined such that for all x, y, z in the space $x \times y$ is bilinear, $x \times x = 0$, and $(x \times y) \times z + (y \times z) \times x + (z \times x) \times y = 0$; **Lie group,** a topological group in which it is possible to label the group elements by a finite number of coordinates in such a way that the coordinates of the product of two elements are analytic functions of the coordinates of the two elements and the coordinates of the inverse of an element are analytic functions of the coordinates of that element.
1935 *Bull. Amer. Math. Soc.* XLI. 344 A Lie algebra L over a non-modular field F will be called normal simple over F if H is an algebraically closed extension of F and L_H is a simple algebra. **1939** H. WEYL *Classical Groups* vii. 188 The process of averaging over a compact Lie group presupposes our ability to compare volume elements at different points of the group manifold. **1965** H. J. LIPKIN *Lie Groups for Pedestrians* i. 14 The use of the Lie algebra therefore simplifies the solution of the eigenvalue problem for the Hamiltonian by defining a number of integrals of the motion. **1967** G. STEINER *Lang. & Silence* 33 One cannot 'translate' the conventions and notations governing the operations of Lie groups..into any words or grammar outside mathematics. **1969** *Sci. News* 31 May 538 The mathematical name of these patterns is Lie groups or unitary symmetry groups. They have been used to predict the existence of new [subatomic] particles.

‖ **lié** (lie), *a.* [Fr., pa. pple. of *lier* to bind.] Connected with, intimately acquainted with, attached to (a person or group of persons).
1855 E. TWISLETON *Let.* 1 May (1928) xiv. 264 Milnes ..has always been *lié* with Lord Palmerston. **1897** E. DOWSON *Let.* c 14 Nov. (1967) 397, I gather he is rather *lié* with Whibley whom I greatly dislike & do not want to meet. **1906** W. DE MORGAN *Joseph Vance* xi. 86 In case it should strike you that I have said..very little about Nolly, I hereby declare that this is not because I did not love him, for we soon became very *liés*. **1947** E. JENKINS *Young Enthusiasts* 50 Alex and I..each had a young man... Alex was *liée* with a naval officer. **1955** A. L. ROWSE *Expansion Eliz. Eng.* i. 22 Esmé Stuart was a danger: he was a Catholic and *lié* with the Guises.

‖ **Liebchen, liebchen** (lī·pχyen, lī·bχyen). [G.] A person who is very dear to another; a sweetheart, a 'pet', darling. Commonly used as a term of endearing address.
1876 GEO. ELIOT *Dan. Der.* IV. VIII. lxii. 232 'Stay a minute, *Liebchen*,' said Lapidoth. **1941** M. TREADGOLD *We couldn't leave Dinah* xv. 226 How providential that there should be this trustworthy boy to keep the *liebchen* so contented. **1972** G. BAXT *Burning Sappho* iv. 71 You are no fool, *liebchen*. You are the most clever woman in the world. **1972** J. ROSSITER *Rope for General Dietz* xiii. 185 'Thank you, *Liebchen*,' she said softly.

‖ **lieber Gott** (lī·bəɹ gɒt). [G.] Dear God, chiefly as *int.*
1898 M. A. VON ARNIM *Elizabeth & her German Garden* 50 The April baby came..to ask about the *lieber Gott*, it being Sunday. **1912** R. BROOKE *Old Vicarage Grantchester* (1916) 7 And I know How the May fields all golden show..Gild gloriously the bare feet That run to bathe... *Du lieber Gott!* **1929** R. HUGHES *High Wind in Jamaica* v. 120 Lieber Gott! What do you think I am, eh? **1954** M. STEWART *Madam, will you Talk?* xxi. 160 Kramer snarled: '*Lieber Gott*, will you listen to me?' **1969** A. MARIN *Rise with Wind* vi. 75 Weber said grimly, '*Lieber Gott*, what a profession to be in.'

Lieberkühn. Add: Also **-kuehn. 2.** *Anat.* The name of *Lieberkühn* used with *of*-adjunct, or occas. in the possessive, to designate the Lieberkühnian follicles or glands, as *crypts, follicles*, or *glands of Lieberkühn*.
1844 DUNGLISON *Dict. Med. Sci.* (ed. 4) 420/2 Lieberkuehn's glands or follicles. **1859** R. B. TODD *Cycl. Anat. & Physiol.* V. 346/2 The intestinal tubes—or, as they are commonly called, the follicles of Lieberkuehn—are the first to demand our notice. **1866** G. HARLEY *Histol. Demonstr.* 114 The arrangements of the various coats, and also the villi and Lieberkühn's follicles, can be seen under a low power. **1949** ADAMS & EDDY *Compar. Anat.* xi. 278 The glands of Lieberkühn, which supply the succus entericus, have their openings at the bases of the villi. **1970** C. K. WEICHERT *Anat. Chordates* (ed. 4) v. 189/2 The intestinal wall contains myriads of intestinal glands which are of two main types. The first of these are the simple tubular glands, or crypts, of Lieberkühn, found throughout the entire length of the small and large intestines.

Liebermann–Burchard (lī·bəɹmæn,bu·ɹkāɹt). *Biochem.* [The names of Carl *Liebermann* (1842–1914) and H. *Burchard*, German chemists.] *Liebermann–Burchard reaction*, the reaction of unsaturated sterols with acetic anhydride and sulphuric acid in chloroform, which produces various coloured solutions; used esp. as a test for cholesterol, when a blue-green colour is produced; so *Liebermann–Burchard test*.
1904 W. R. ORNDORFF tr. *Salkowski's Lab. Man. Physiol. & Path. Chem.* ix. 92 (*heading*) Liebermann-Burchard reaction. **1915** STEDMAN *Med. Dict.* (ed. 3) 509/2 Liebermann-Burchard test. **1934** *Jrnl. Biol. Chem.* CVI. 746 The very weak color produced by digitonin with the modified Liebermann-Burchard reaction. **1956** E. V. TRUTER *Wool Wax* vii. 185 At present, the only technique for quantitatively determining alcohols of the isocholesterol group is based upon the spectrophotometric measurement of the colour developed in the Liebermann-Burchardt test. **1968** *Indian Jrnl. Med. Res.* LVI. 1776 A method for the estimation of total cholesterol in whole blood, serum or plasma, based on the Liebermann-Burchard reaction is presented.

‖ **Liebestod, liebestod** (lī·bəstōt). [G., lit. 'love's death'.] An aria or a duet proclaiming the suicide of lovers (see also quot. 1964); hence, such a suicide; also *fig.*
1889 G. B. SHAW *London Music 1888–89* (1937) 249 Isolde's Liebestod was a failure. **1928** in D. McCarthy *Drama* (1940) 112 Each pair sinks into the euthanasia of a matter-of-fact liebes-tod. **1947** A. EINSTEIN *Mus. Romantic Era* xvi. 283 Yet this festival opera [*sc.* Aida] ends *pianissimo* and *con sordini* with a *Liebestod*, which is not Romantically philosophical, but purely human. **1959** *Listener* 20 Aug. 280/2 Would the imagination..not die a *Liebestod* at the very moment it attains its goal? **1964** *Conc. Oxf. Dict. Opera* 223/2 *Liebestod*, the title used today for Isolde's death scene in Wagner's *Tristan und Isolde*, but used by Wagner of the mystic love duet in Act 2. **1971** G. STEINER *In Bluebeard's Castle* i. 25 It is permissible to see..in the Wagnerian *Liebestod* surrogates for the lost dangers of revolutionary action.

Liebfraumilch (lī·pfraumilχ, lī·b-). Also **Liebfrauenmilch.** [G., lit. 'milk of Our Lady'.] A white wine orig. produced at Worms; also loosely applied to German white wines.
1833 C. REDDING *Hist. Mod. Wines* vii. 204 The Liebfrauenmilch..is a well-bodied wine, grown at Worms. **1846** TENNYSON *Let.* 12 Nov. in H. Tennyson *Alfred Lord Tennyson* (1897) II. i. 6 Dickens..was very hospitable, and gave us biscuits..and a flask of Liebfraumilch. **1930** W. S. MAUGHAM *Cakes & Ale* ii. 21 We want some of the Liebfraumilch, the '21. **1951** *Good Housek. Home Encycl.* 508/2 The best Hock, which is sold under a number of well-known names, e.g. Johannisberger,.. Liebfraumilch. **1967** A. LICHINE *Encycl. Wines* 323/1 Rheinhessen wines, distinctive in their own right, are so named; the remainder call themselves Liebfraumilch. **1973** *Guardian* 28 June 11/6 Liebfraumilch is an invented name for almost any ordinary German white wine not worthy of its own district label.

Liebig. Add: *Liebig('s) condenser*, a device for condensing vapour, consisting of two concentric tubes, the vapour and condensate passing through the inner one and a cooling liquid through the outer one.
1867 BLOXAM *Chem.* 46, A is a stoppered retort, the neck of which fits into the tube of a Liebig's condenser. **1903** S. YOUNG *Fractional Distillation* i. 6 When a Liebig's condenser is used there is no advantage in having either the inner or the outer tube very wide. **1963** J. W. DAVIS *Adv. Level Pract. Chem.* 158 For preparations in an advanced level course the most suitable water-condensers are short Liebig condensers.

‖ **Liebling, liebling** (lī·pliŋ, lī·bliŋ). [G.] = *LIEBCHEN, LIEBCHEN.
1868 C. M. YONGE *Chaplet of Pearls* I. vii. 79 She is a good little *Liebling*. **1970** J. CLEARY *Helga's Web* iv. 60 'And you're not servile?' 'No, *liebling*.' **1972** J. AIKEN *Butterfly Picnic* vi. 105 Is that you, Liebling?

‖ **Lied, lied** (līt). Pl. Lieder (lī·dəɹ). [G.] A song, esp. one characteristic of the German Romantic period. So **lieder-singer; lieder-singing** *vbl. sb.*
1852 J. C. PATTESON *Let.* in C. M. Yonge *Life J. C. Patteson* (1874) I. iv. 89 He sang some of Medelssohn's [*sic*] Lieder very pleasantly. *Ibid. Let.* 115 As soon as a Lied or Sonata began, away would go my books. **1854** [see SEGUIDILLA]. **1876** STAINER & BARRETT *Dict. Mus. Terms* 274/2 The German lied, the sacred lied or chorale.. was founded upon the ecclesiastical modes and remained unchanged until the days of the Minnesingers. **1924** M. KENNEDY *Constant Nymph* xvi. 222 She listened sadly to German Lieder. **1936** H. READ *Meaning of Art* (ed. 2) 53 Thus we read in our newspaper that Miss X 'is too deficient in variety of *tone-colour* to make a good lieder singer'. **1937** *Sunday Times* 21 Feb. 7/1 In Lieder singing the words, or rather what is at the back of the words, play a large part in determining the appropriate musical style. **1947** A. EINSTEIN *Mus. Romantic Era* xiv. 184 Berlioz, of course, also wrote lieder and various other kinds of songs. **1955** *Times* 16 May 11/5 Programme included operatic arias,..two groups of lieder and English and Italian songs. **1959** *Times* 27 Apr. 5/6 (*heading*) Miss Gerda Lammas: a great Lieder singer. **1960** *Guardian* 22 Apr. 9/2 An intimate tone and style are..essential, to lieder singing. **1963** AUDEN *Dyer's Hand* 505 If one takes, say, a sea-shanty out of its proper context and listens to it on the gramophone as one might to a *lied* by Schubert, one is very soon bored. **1970** *New Yorker* 3 Oct. 36/1 Mother had a nice repertory of the more assailable lieder.

liegeful (lī·dʒfūl), *a. rare.* [f. LIEGE *sb.* + -FUL.] Loyal, faithful.
1872 A. DE VERE *Legends St. Patrick* 72 If ye be liegeful, sirs, decree the day. *Ibid.* 155 Pure of heart, and liegeful unto Christ. **1887** —— *Legends & Rec. Church & Empire* 264 Liegeful I know hath been your wedded life.

lien¹. Add: **2.** *fig.* (Later examples.)
1922 A. BENNETT *Lilian* II. v. 107 She had no lien, no attachment. **1925** *New Statesman* 3 Oct. 687/1 They desire two sets of negotiations to proceed simultaneously, and if they admit they may be pursued separately, there will nevertheless be so many liens between them, that the success of one will be dependent on the success of the other.

lieno-. Add: **lieno-renal** *a.*, pertaining to the spleen and the kidneys: applied *spec.* to a short ligament connecting the spleen and the left kidney.
1887 G. D. THANE *Ellis's Demonstrations Anat.* (ed. 10) viii. 475 The peritoneum may be followed..to the outer part of the left kidney, where it is reflected along the back of the splenic vessels to the spleen, forming one layer of the lieno-renal ligament. **1932** W. WRIGHT in E. P. Stibbe *Pract. Anat.* 301 To the left it [*sc.* the cavity] is closed by the gastro-splenic omentum, the spleen, and a fold passing from the spleen to the left kidney, the lieno-renal ligament. **1967** G. M. WYBURN et al. *Conc. Anat.* i. 32/2 The lieno-renal ligament, which contains the splenic vessels.

lier, obs. var. LEER *sb.*³ (in Dict. and Suppl.).

Liesegang (lī·zəgaŋ). *Physical Chem.* The name of Raphael Eduard *Liesegang* (1869–1947), German chemist, used *attrib.* (esp. in *Liesegang ring*) and in the possessive to designate (the formation of) concentric rings or parallel bands of precipitate following the diffusion, one into the other, of two dissolved substances that react to form a slightly soluble precipitate.
1913 *Chem. Abstr.* VII. 3797 The Liesegang figures produced on gelatin plates. **1917** M. H. FISCHER tr. *Ostwald's Introd. Theoret. & Appl. Colloid Chem.* v. 215 A colloid-chemical method for discovering the addition of agar to fruit jellies and marmalades makes use of the influence which such addition has upon the form and the structure of Liesegang rings when formed in such jellies. **1932** *Jrnl. Physical Chem.* XXXVI. 299 The concentric rings in the 'common gall stones' of inflammatory origin are..a manifestation of the Liesegang phenomenon. **1944** A. VAN HOOK in J. Alexander *Colloid Chem., Theoret. & Appl.* V. 517 Bucher..has devised a new method of blood analysis depending on the sensitiveness of Liesegang Ring formation to very slight variations in blood composition and quality. **1946** *Thorpe's Dict. Appl. Chem.* (ed. 4) VII. 307/2 A large number of stratified deposits occur in nature, and a study of Liesegang's rings has suggested explanations for some of these structures. **1953** *New Biol.* XV. 123 Periodic precipitations of the Liesegang type may occur. **1959** *Science* 15 May 1366/2 The best way to avoid the formation of multiple macroscopical Liesegang bands is to operate with equivalent concentrations of reagents. **1971** *Jrnl. Colloid & Interface Sci.* XXV. 591/2 The membrane-like behavior of the $PbCrO_4$ Liesegang ring system in agar gel has been confirmed..with a wide variety of inorganic ions.

lieutenant. Add: **2. c.** An officer in the Salvation Army.
1884 [see *CAPTAIN *sb.* 5 b]. *c* **1897** A. E. HOUSMAN *Lett.* (1971) 45 Lieutenant Isabella..comes Dealing blows with her umbrella. **1970** *Guardian* 2 May 10/4, I love the Salvation Army through which I found my Saviour... Yours faithfully, Lars Juhlin. Lieutenant.

lieves, var. of *lieve* LIEF *adv.* Cf. *LEAVES.
1863 'G. HAMILTON' *Gala-Days* 241 We'd just as lieves work out of doors..as not.

life, *sb.* Add: **1. a.** *while there is life there is hope* (and similar phrases) (earlier and later examples); *there is life in the old dog yet* (and variants): an assertion of continuing competence, strength, etc., notwithstanding evidence to the contrary.
1539 R. TAVERNER *Erasmus's Proverbes* f. 36ᵛ, The sycke person whyle he hath lyfe, hath hope. **1671** J. CROWNE *Juliana* v. 56 Madam, he breathes, and whilst there's life, there's hope. **1727** J. GAY *Fables* xxvii. 93 While there is life, there's hope, he cry'd. **1808** *Monthly Pantheon* I. 366/1 Whilst there is *life* you know there *are hopes*! **1859** S. ALLEN *Let.* 1 Dec. in D. Ayerst *Guardian* (1971) x. 134 'Are not the advertisements grand?'..'There is life in the old dog yet.' **1908** E. J. BANFIELD *Confessions of Beachcomber* II. ii. 301 While there is life there is hope is evidently Nelly's creed. **1940** *Time* 15 July 49/1 Tallulah Bankhead demonstrated that there's life in Pinero's old girl yet.
c. (*a matter*, etc.) *of life and death*: also, ..*of life or death*.
1837 DICKENS *Let.* c 20 Apr. (1965) I. 249 It is matter of life or death to us, to know whether you have got Ainsworth's MS yet. **1898** W. J. LOCKE *Idols* x. 134 The marriage could be concealed no longer. It was a matter of life or death. **1950** K. WINSOR *Star Money* III. xxix. 249, I never have made any man a matter of life or death to me.
3. c. *for dear life* (examples); *anything for a quiet life; for once in his life*.
1624 T. HEYWOOD *Captives* (1885) III. iii. 169 Anythinge For a quiett lyfe. **1800** M. EDGEWORTH *Parent's Assistant* (ed. 3) VI. 123 Any thing for a quiet life. **1837** DICKENS *Pickw.* xlii. 463 Anythin' for a quiet life, as the man said ven he took the sitivation at the light-house. **1843** W. T. THOMPSON *Major Jones' Chron. Pineville* 93 He..was climbing for dear life. **1846** 'MRS. MARKHAM' *Hist. Eng.* (ed. 12) xxxvi. 402 *George*. I think, mamma, that the fire of London was a happy event for the king, as it made him exert himself, for once in his life, to do some good. **1872** B. JERROLD *London* ii. 23 Hard-visaged men, breathlessly competing for 'dear life'. **1921** H. CRANE *Let.* 17 Oct. (1965) 68 The man who would preserve them [*sc.* feelings] must duck and camouflage for dear life. **1968** 'L. MARSHALL' *Blood on Blotter* v. 40 I'm a born appeaser... Anything for a quiet life.
d. (Further example.) Phr. *not on your life*, not on any account, by no means.
1896 W. C. GORE in *Inlander* Jan. 149 'Say, Jack, are you going to bolt?' 'Not on your life.' **1905** *N.Y. Even. Post* 19 Aug. 2 The congressman was asked if there had been any gambling during the trip. 'Not on your life,' he said. **1913** KIPLING *Divers. Creatures* (1917) 294 'Not on your life!' says Lundie. **1944** *Living off Land* iv. 62 Say that you are lost, properly bushed. You come across a river. Well, that river is not bushed—not on your life it isn't. **1962** F. NORMAN *Guntz* i. 7 My life (I thought) what chance am I going to have if I produce this letter. **1972** H. CARMICHAEL *Naked to Grave* v. 56 'Why not get in touch with your lawyer?' 'Not on your life!.. It would be a tacit admission of my guilt.'
4. *spec.* in *Cricket*, that quality in the pitch which causes the ball to rise abruptly or unevenly after pitching.
1888 A. G. STEEL in Steel & Lyttelton *Cricket* iii. 148 On wet hard wickets..there is still life and pace in the ground; but in the sodden dead state, directly the ball touches the ground it..loses all life and pace. **1906** A. E. KNIGHT *Compl. Cricketer* 348 'Life' from the pitch implies the pitch and sting at or with which the ball leaves the ground.
5. a. (Further examples.)
1720 DEFOE *Capt. Singleton* 73 These indeed were the Life and Soul of all the rest, and it was to their Courage that all the rest ow'd the Resolution they shewd. **1797** R. M. ROCHE *Children of Abbey* I. xvii. 309 They had assembled a number of their neighbours, among whom were a little fat priest, called Father O'Gallaghan, considered the life of every party, and a blind piper. **1814** JANE AUSTEN *Mansf. Park* II. i. 9 Sir Thomas was indeed the life of the party. **1897** M. CORELLI *Ziska* xv. 324 Armand Gervase..was making himself the life and soul of everything at the Mena House Hotel. **1932** L. GOLDING *Magnolia St.* III. ix. 595 He's very much the official life-and-soul-of-the-party. **1939** [see *BORSCH]. **1965** *Melody Maker* 17 July 9 Offstage..Dudley doesn't strike you as being the life and soul of the party. **1970** G. GREER *Female Eunuch* 33 When the life of the party wants to express the idea of a pretty woman in mime, he undulates his two hands.

7. a. *as large as life* (further examples). Hence *larger-than-life*; so *larger-than-lifeness* (nonce). Also *life itself*.
1802 C. WILMOT *Let.* 17 Dec. in *Irish Peer* (1920) 129 A beautiful piece of clockwork representing Apollo with his lyre... It was as large as life. **1822** M. EDGEWORTH *Let.* 9 Mar. (1971) 368 We 6 went together to see Belzonis tomb—the model first and afterwards the tomb as large as life. **1836** T. C. HALIBURTON *Clockmaker* (1837) 1st Ser. 143 As large as life and twice as nateral. *c* **1840** LADY WILTON *Art of Needlework* xxi. 334 Birds..being, in proportion to other figures, certainly *larger* than life, and 'twice as natural'. **1871** 'L. CARROLL' *Through Looking-Glass* vii. 150 It's as large as life, and twice as natural! **1891** G. MOORE *Impressions & Opinions* 89 The illusion is complete; it is just, as the phrase goes, like life itself. **1898** G. B. SHAW *Mrs. Warren's Profession* II. i. 176 This is George Crofts, as large as life and twice as natural. **1926** G. HUNTING *Vicarion* i. 21 What she had seen and heard had been life itself! **1930** J. DOS PASSOS *42nd Parallel* 46 Doc Bingham was sitting as large as life in a rocking chair. **1937** M. ALLINGHAM *Dancers in Mourning* i. 12 A larger-than-life edition of his stage self. **1947** L. MACNEICE *Dark Tower* 70 Larger-than-lifeness need not be part of the recipe. **1953** K. AMIS *Lucky Jim* i. 7 Anyway, there it was in the *Post* as large as life. **1959** *Viewpoint* July 12 Larger-than-life faces on television. **1966** R. A. DOWNIE tr. *O. del Buono's Bond Affair* 18 Allen Dulles insisted on regarding James Bond as a larger-than-life character.
8. a. *of one's life*, denoting the most important event of its kind in one's life. See also TIME *sb.* 6.
1887 A. M. SULLIVAN *Let.* 13 Nov. in H. Keller *Story my Life* (1903) III. iii. 340 We took Helen to the circus, and had 'the time of our lives'! **1936** *Discovery* Jan. 14/2 They got the shock of their lives. **1939** W. SAROYAN (*title*) The time of your life. **1961** L. VAN DER POST *Heart of Hunter* i. 25 The men sat with their heads bowed over arms clasped round their knees like long-distance runners recovering from the race of their lives.
c. (Further examples.) In *Physics* applied *spec.* to the average duration of existence of the members of a population of identical particles or states (equal to the period in which the population decreases by a factor e).
The half-life is equal to the (mean) life multiplied by $\log_e 2$ (about 0·693).
1903 RUTHERFORD & SODDY in *Phil. Mag.* V. 607 In one gram of these elements less than a milligram would change in a million years. In the case of radium, however, the same amount must be changing per gram *per year*. The 'life' of the radium cannot be in consequence more than a few thousand years. **1926** R. W. LAWSON tr. *Hevesy & Paneth's Man. Radioactivity* vii. 64 In this so-called 'normal state' the hydrogen atom can persist permanently, whereas the 'life' of all other stationary states is very short. *Ibid.* xii. 111 Each group [of radioactive substances] is arranged in the order of diminishing half-value period, and begins with the member of longest life. **1926** *Sci. Abstr.* A. XXIX. 170 Using the observation that so long as these lines are absorbed, atoms must be in the s_2 and s_3 states, a determination is made of the mean life of these states. **1942** J. D. STRANATHAN *'Particles' of Mod. Physics* xiii. 535 There is some indication that the mean free path may be longer, and the mean life correspondingly longer, for high energy mesotrons than it is for low energy mesotrons. **1947** *Forum* (Johannesburg) 12 Apr. 15/3 Even with the aid of boreholes, which have yet to be sunk, the 'life' of the dam can be extended only until the end of September. **1958** *Times* 23 July 5/2 Its..turbo-jet engines will be permitted an initial 'life' between overhauls of 1,000 hours. **1968** M. S. LIVINGSTON *Particle Physics* x. 178 The quantity $2\pi\Gamma/h$ is the probability of decay per unit time, or the reciprocal of the mean life τ of the state. Mean life is defined as the time for the population of the state to be reduced to $1/e$ of its initial value... This means that, because of the finite lifetime of an excited state, the energy of the state cannot be sharply defined but is intermediate within the energy spread Γ. **1971** *Gloss. Electrotechnical Power Terms* (*B.S.I.*) IV. i. 26 *Life*, of a lamp. Time during which a lamp has been operated before becoming useless.
d. Imprisonment for life; a life sentence. *slang.*
1903 [see *CELL *v.* b]. **1924** E. WALLACE *Room 13* i. 10 He shot a copper and got life. **1967** [see *BLOW *v.*¹ 27 b]. **1975** *Times* 29 Apr. 4/6 Although the sentence is life, they all want parole.
9. So *a bad life, a first-class life*.
1921 A. HUXLEY *Let.* 23 Mar. (1969) 194 This perpetual lack of perfect physical health is intolerable. This was brought home to me more acutely than usual today by the refusal of the London Life Association to insure me... It is..humiliating to be a Bad Life. **1938** *Times Lit. Suppl.* 24 Sept. 618/2 Elizabeth all her days was reckoned a 'bad life'. **1970** *Times* 5 Dec. 9/3 If one is not accepted as a first class life, the most common procedure is for an insurance company to increase the premium.
11. *Cricket.* (Earlier and later examples.) Similarly in *Baseball*.
1865 *Bell's Life* 24 June 7/1 Mr. Voules (who had 'a life' when he had made but a single) was first to leave. **1868** *Cincinnati Commercial* 24 May 8/2 Meagher had a life given him by Gould not accepting the grounder Meagher hit to him, and Brainard's wild throw to first gave him his second. **1955** *Times* 9 July 4/5 Immediately after luncheon Goddard was given a life when he slashed at Tyson and Evans dropped a fast head-high catch. **1974** *Times* 25 Nov. 10/2 Ali also had a life from Barrett at mid-off.
12. a. *such is life!*: see SUCH *dem. adj.* and *pron.* 2; similarly *that's life, life's like that*; *to live one's (own) life*: to conduct oneself without reference to the opinions of others; *this is the life*: an expression of satisfaction; *it's a great life (if you don't weaken)*: an ironic comment on the difficulties of one's situation; *what a life!*: an expression of discontent; *how's life?*: how are you faring?
1796 W. J. TEMPLE *Diary* 7 Apr. (1929) 167 This interruption is very teasing; but such is Life. **1843** DICKENS *Mart. Chuzz.* (1844) xxix. 347 'Sairey,' says Mrs. Harris, 'sech is life. Vich likeways is the hend of all things!' **1849** N. KINGSLEY *Diary* (1914) 52 For my part [I] could almost wish myself in the same Latitude..but such is life. **1853** C. BRONTË *Villette* I. xiii. 229 Thinking meantime my own thoughts, living my own life, in my own still shadow-world. **1865** [see SUCH *pron.* 2]. **1903** 'T. COLLINS' (*title*) Such is life. **1911** D. H. LAWRENCE *White Peacock* III. iii. 397 At home you cannot live your own life. **1917** *Ladies' Home Jrnl.* Mar. 46 (Advt.), *This is the life.* There are two ways to live nowadays. One way is the life that is daily chock full of healthy activity, wholesome fun and lots of fresh air. **1919** J. BUCHAN *Mr. Standfast* v. 105 'Back to Glasgow to do some work for the cause,' I said lightly. 'Just so,' he said, with a grin. 'It's a great life if you don't weaken.' **1919** WODEHOUSE *My Man Jeeves* 234 She's glued to a chair, with this-is-the-life written all over her, taking it in through the pores. **1924** J. BUCHAN *Three Hostages* xvi. 227 That's life, my dear. We've got to go on to the finish anyhow, trusting that luck will turn. **1926** S. JAMESON *Three Kingdoms* x. 301 After all, she had chosen to stand apart from him and to live her own life, as the moderns have it. **1926** R. MACAULAY *Crewe Train* xi. 213 This was the life. **1930** J. B. PRIESTLEY *Angel Pavement* i. 49 She groaned as she stuck another sheet of paper into the typewriter. 'What a life!' **1933** D. L. SAYERS *Murder must Advertise* i. 10 There goes my thousand quid! Oh, well, that's life. **1935** N. L. MCCLUNG *Clearing in West* xxv. 205 Still Will had his own life to live and must make his own choice. **1935** N. MITCHISON *We have been Warned* v. 511 I've been very busy... How's life? **1943** K. TENNANT *Ride on Stranger* xviii. 202 Oh, it was a great life, if you liked that sort of a life. **1959** M. GILBERT *Blood & Judgement* ix. 102 We weren't sharing rooms... She was living her life, I was living mine. **1968** P. DICKINSON *Skin Deep* vii. 140 No, it's..not the sort of thing that makes the newspapers... Ah well, life's like that. **1970** *New Statesman* 26 June 924/3 Whatever Ned Kelly was really like.. he can scarcely have been like Mr Jagger... The famous last words 'Such is life'—could as well have been 'Pass the salt'. **1972** G. BELL *Villains Galore* viii. 104 'Nothing ventured, nothing lost either,' muttered Boote miserably. 'Gawd! What a life!' **1973** J. MCCLURE *Four & Twenty Virgins* ii. 22 The sports car accelerated away before she reached the end of the path. 'That's life,' said Kegg.
d. (Further examples.)
1851 H. MAYHEW *Mayhew's Characters* (1951) 309, I liked to see 'life', as it was called, and was fond of the company of women. **1918** C. MACKENZIE *Early Life Sylvia Scarlett* II. iv. 332 I've got a fancy..to show you a bit of life. **1937** A. CHRISTIE *Death on Nile* II. i. 41 He's made a good deal of money and he's seeing life, I fancy. **1972** L. MEYNELL *Death by Arrangement* i. 9 The spires of Oxford could go on dreaming..for all he cared; he set about getting himself a degree in the university of life.
e. *the life of the mind*: intellectual or aesthetic pursuits, scholarship; meditation, the realm of the imagination.
1926 E. HEMINGWAY *Men without Women* (1927) 216 Live the full life of the mind, exhilarated by new ideas. **1950** P. BOTTOME *Under Skin* xxiii. 204 If we try to escape into the life of the mind we find you there before us. **1972** G. WIGG *George Wigg* i. 28 He was an inspired teacher.. arousing in us a feeling for literature and poetry and the life of the mind. **1972** *Guardian* 1 Nov. 14/5 Universities exist to promote the life of the mind... They should create and discover knowledge.
13. (Later examples.) So *Life and Times*, a biography combined with a study of the public events of the character's lifetime; *life-and-work(s*, a biography combined with a study of the writings of the subject.
c **1889** W. PATER *Let.* 30 Apr. (1970) 94, I wish I could undertake a life for your admirable Series. **1933** J. THURBER (*title*) My life and hard times. **1951** G. GREENE *End of Affair* v. v. 204 'You seem interested in General Gordon.' 'They want me to do a Life.' **1957** *Times Lit. Suppl.* 25 Oct. 640/2 Mr Wilson's life-and-work summaries are excellent. **1959** *Listener* 9 Apr. 643/2 This is to be a life-and-works, not pure biography. *Ibid.* 3 Dec. 1005/1 It is a 'Life', not a 'Life and Times', that he has written. **1962** *Ibid.* 15 Nov. 804/1 To use a man's letters and all related correspondence to produce a reasonably short life and times. **1975** *Listener* 16 Jan. 93/1 *Cavafy* is..more biographical than critical, but a 'life' was needed and this is the fullest so far.
16. (Further, esp. 20th-c., examples.) **a.** *life-activity, -anger, -body, -centre, -chance, -course, -current, -demand, -drama, -electron, -flame, -flow, -habit, -idea, -instinct, -mate, -meaning, -mystery, -orientation, -path, -pattern, -principle, -quick, -responsibility, -situation, -space, -story, -stream, -stuff, -urge, -wish*.
1914 R. M. JONES *Spiritual Reformers 16th & 17th Cent.* p. xvii, Undivided faith attitudes always liberate within the field of consciousness energy for life-activity. **1937** R. A. WILSON *Birth of Lang.* 83 The modern error..of characterizing life-activity as mechanism. **1924** LAWRENCE & SKINNER *Boy in Bush* 203 It was the anger, the deep, burning life-anger which was the kinship. **1920** S. ALEXANDER *Space, Time & Deity* II. 355 Hunger and thirst..are the affections of its life-body. **1923** D. H. LAWRENCE *Birds, Beasts & Flowers* 54 Fragile-tender, fragile-tender life-body, More fearless than iron all the time. **1902** *Westm. Gaz.* 2 Apr. 10/2 As a life-centre Lake Eyre has long lost its importance. **1942** R. A. KNOX *In Soft Garments* xi. 85 We have got to go back to the life of Jesus of Nazareth, isolating the life-centre from which this vast organism of Christianity has sprung. **1944** *Politics* I. 273/2 However strongly life chances may be differentiated, this fact in itself..by no means gives birth to 'class action'. **1958** W. J. H. SPROTT *Human Groups* 60 The life-chances of children..are almost entirely determined by their position in the kinship scheme. *a* **1930** D. H. LAWRENCE *Phoenix* (1936) v. i. 609 This reversal of the life-course. **1970** R. J. HOLLINGDALE tr. *Schopenhauer's Ess. & Aphorisms* 144 The entire life-course, i.e., the inner and outer history, of each one [*sc.* man] differs.. from that of all the others. **1899** W. JAMES *Talks to Teachers* 257 The occasion and the experience..are nothing. It all depends on the capacity of the soul to be grasped, to have its life-currents absorbed by what is given. **1929** D. H. LAWRENCE *Pansies* 21 A new demand on his intelligence, A new life-demand. **1872** *Porcupine* 12 Oct. 443/2 He wanted to be left to work out his own Life-Drama. **1915** D. H. LAWRENCE *Rainbow* x. 262 On Easter Sunday the life-drama was as good as finished. *a* **1930** —— *Etruscan Places* (1932) 58 So within each man is the quick of him..some spark, some unborn and undying vivid life-electron. **1906** *Macm. Mag.* Apr. 436 Two of these life-flames were burning brightly..in the adjacent theatre. **1960** *Spectator* 14 Oct. 556 English is the true education of the life-flame. **1903** *Ibid.* 11 Apr. 565 The life-flow of justice..ceased to course through her heart. *a* **1930** D. H. LAWRENCE *Last Poems* (1932) 107 People who complain of loneliness must have lost..their life-flow Like a plant whose roots are cut. **1891** C. L. MORGAN *Animal Sk.* 214 To watch his life-habits with sympathetic interest. **1923** D. H. LAWRENCE *Kangaroo* xvi. 338 As some great life-idea cools down and sets upon them. **1908** E. F. BENSON *Blotting Bk.* i. 22 His was the hot blood that could do any deed when the life-instinct commanded it. **1922** Life instinct [see *death-instinct* (*DEATH *sb.* 19)]. **1906** *Westm. Gaz.* 17 Feb. 6/3 Each with the life-mate who should guide his way. **1922** JOYCE *Ulysses* 413 Faithful lifemate. **1923** D. H. LAWRENCE *Kangaroo* xii. 283 It is gruesome, with no life-meaning. **1927** —— *Lovely Lady* (1932) 230 He deemed it [*sc.* sex], as the Chinese do, one of the great life-mysteries. **1936** WIRTH & SHILS tr. *Mannheim's Ideology & Utopia* I. i. 22 The point of view of life-orientation and conduct. **1966** G. E. EVANS *Pattern under Plough* xii. 124 Only tremendous transformations of life-orientation have succeeded in tearing them away from this universal form of religiosity. **1950** *Psychiatry* XIII. 2 He [*sc.* Freud] provides the texts..for contradictory life-paths and social policies. **1955** AUDEN *Shield of Achilles* iii. 76 A fortuitous intersection of life-paths. **1920** T. P. NUNN *Education* i. 6 There is no limit to the number of life-patterns into which good or blameless actions may be woven. **1972** *Sci. Amer.* Jan. 39/1 Although sex-role ideology may be developed in early childhood, it is usually not until adolescence that a girl begins to apply her system of beliefs to her life pattern. **1851** H. MELVILLE *Moby Dick* III. xxi. 146 This same..cunning life-principle in him. **1950** L. S. THORNTON *Revelation & Mod. World* iii. 90 *Totality* and *identity* are two aspects of one life-principle by which the creative Word calls into himself that response which he creates. **1923** D. H. LAWRENCE *Stud. Classic Amer. Lit.* vi. 119 Nowadays society is evil. It finds subtle ways of torture, to destroy the life-quick, to get at the life-quick in a man. **1928** —— *Lady Chatterley* x. 131 What man with a spark of honour would put this ghastly burden of life-responsibility upon a woman. **1936** WIRTH & SHILS tr. *Mannheim's Ideology & Utopia* I. i. 10 Constantly varying social strata and life-situations. **1969** *America* 5 July 17/2 The mass media have formed many of our responses to life situations. **1935** *Psychol. Abstr.* Jan. 3/2 The psychological life-space (*Lebensraum*) is a general hodological space, which shows certain relativities. **1957** R. K. MERTON *Social Theory* (rev. ed.) 384 The social life-space of an individual. **1853** Life-story [in Dict.]. **1930** C. BEATON *Diary* Dec. in *Wandering Yrs.* (1961) 200 Buy her *Life Story* for three dollars. **1960** 'R. SIMONS' *Frame for Murder* xiii. 165 Squere..produced a bundle of papers. 'This is him. His entire life story.' **1879** BROWNING *Dramatic Idyls* 128 'Look unto me and be ye saved!' saith God: 'I strike the rock, outstreats the life-stream at my rod!' **1941** WYNDHAM LEWIS *Let.* 10 Aug. (1963) 295 The character..is so deeply stained with the deposits on the obscure bed of the life-stream. **1880** *Wesleyan-Methodist Mag.* Aug. 621/1 To say that the life-stuff of the lowest fungus and that of the most powerful human brain are *identical*, is absurd. **1956** A. H. COMPTON *Atomic Quest* 160 The life-stuff of intense effort. **1922** D. H. LAWRENCE *Let.* 21 Sept. (1962) II. 717 But I won't mention the life-urge any more. **1926** W. DE LA MARE *Connoisseur* 18, I had become an automaton—little better than a beetle obeying the secret dictates of what I believe they call the Life-Urge. **1944** R. LEHMANN *Ballad & Source* 13 A life-wish so crackling with energy that it could overcome no matter what minatory fate.
b. *life-affirming, -denying, -enhancing, -sapping* adjs.; *life-brightener, -denier, -enhancer*.
1947 A. EINSTEIN *Mus. Romantic Era* xii. 165 He became a priest, the 'Abbé Liszt', who sought in Rome a sort of defense against his overflowing, life-affirming virtuosity. **1966** *Observer* 6 Nov. 27/4 This instinctual, familial, life-affirming note of Tolstoy's. **1906** W. DE MORGAN *Joseph Vance* xxviii. 268 'Come, Joe, some news this time I hope!' I should have liked to be able to say yes, for he looked..as if he sadly wanted a life-brightener. **1922** JOYCE *Ulysses* 497 It's a lifebrightener, sure. **1955** L. P. HARTLEY *Perfect Woman* xxiii. 202 Jeremy, with his insistence on rules and regulations, his instinct for decorum in all things, seemed to her a spoil-sport and a life-denier. **1962** J. B. PRIESTLEY *Margin Released* II. v. 137 He would be twisted..malevolent, life-denying. **1973** *Times Lit. Suppl.* 21 Dec. 1554/1 Dame Rebecca's Augustine is.. introspective and life-denying, disgusted by physical existence... He bequeathed to posterity a complex of life-denying and art-denying ideas. **1955** S. SPENDER *Making of Poem* II. vi. 102 The golden Romantic poet then is more than life-enhancer... He is the magician who.. turns all his experience..into molten imagination. **1964** *Economist* 8 Aug. 530/2 That purely modern life-enhancer, the private car. **1896** B. BERENSON *Florentine Painters* xi. 67 The contemplation of his [*sc.* Leonardo da Vinci's] personality is life-enhancing as that of scarcely any other man. **1960** *Guardian* 18 Nov. 7/6 His passionately serious novel..is a life-enhancing work. **1971** *Ibid.* 4 Jan. 10/2 The search for safer life-enhancing drugs. **1909** *Daily Chron.* 3 Sept. 1/1 The weather improved, but there still remained a light life-sapping wind which drove despair to its lowest recess. **1928** A. HUXLEY in *Sunday Dispatch* 16 Dec. 12/6 No people, it seems to me, has suffered more than the English from that life-sapping malady of too much machinery.

c. *life-clouded, -oriented, -sentenced* adjs.
1921 D. H. LAWRENCE *Tortoises* 18 Life establishing the first eternal mathematical tablet, Not in stone..or bronze, but in life-clouded..tortoise-shell. **1968** *Sun* (Baltimore) 4 July A 16/3 Speakers were using such terms..as life-oriented curriculum..and multi-media and multi-mode curriculum. **1901** *Chambers's Jrnl.* Nov. 744/2 This hapless man had completed seventeen of the twenty years which all life-sentenced prisoners must serve before release on license.
d. *life-blissful, -divine, -empty, -stupid* adjs.
1923 D. H. LAWRENCE *Birds, Beasts & Flowers* 54 Flaked out and come unpromised, The tree being life-divine, Fearing nothing, life-blissful at the core Within iron and earth. **1921** —— *Let.* c 8 May (1962) II. 653 Everybody nice, but rather spent, rather life-empty. **1922** —— *Fantasia of Unconscious* vii. 115 We are really far, far more life-stupid than the dead Greeks.
f. *life-class* (earlier and later examples), *-drawing*.
1891 A. BEARDSLEY *Let.* 13 Oct. (1971) 30, I eventually selected the Impressionist Academy as my school of art.. It will not be so very long before I get into the life class. **1967** 'L. EGAN' *Nameless Ones* iv. 43 He was built like Tarzan, and could have earned a living posing for life classes. **1915** W. OWEN *Let.* 4 Apr. (1967) 329 Great talent in Life-Drawings and Oil Portraits; studied in Paris. **1956** K. CLARK *Nude* iv. 117 A splendid drawing of a nude model, one of the first 'life drawings' of a woman.
17. life-company, a life-insurance company; **life-craft,** a small craft, carried on board a larger one, by which escape may be made in an emergency; **life expectancy,** expectation of life; also *transf.* and *attrib.*; **life-force,** vital energy; so **life-forcer,** a believer in a philosophy of the *élan vital*; **life-gun,** a gun used for sending life-saving apparatus to ships; **life-history** (further examples); also *transf.* with reference to inanimate things; (*b*) life-story, the narrative of the career of a person; **life-index** (see quot. 1915); **life-insurance:** so *life-insurance policy* [POLICY *sb.*²]; also *fig.*; **life-member,** one who has acquired lifelong membership of a library, society, etc.; so **life membership; life net** *U.S.* (see quot. 1969); **life-office** (examples); **life-peer, -peerage** (later examples); so **life-peeress; life-policy** = *life-insurance policy*; **life-raft** (later examples; also in wider use); **life-ring** *N. Amer.*, a life-buoy; **life science,** any of the sciences (such as zoology, bacteriology, or sociology) which deal with living organisms; such sciences collectively; **life-span** [SPAN *sb.*¹ 4], lifetime; period of duration (of an animate or inanimate thing); **life-support** *a.*, applied to equipment designed to make possible the continued normal functioning of the body in hostile or dangerous environments; **life-tenant** (examples); so **life-tenancy; life test,** a test made on a sample of components in specified operating conditions, either for a certain length of time or until failure occurs, to determine the reliability of the components; hence (with hyphen) as *vb. trans.*, to perform a life test on; **life-testing** *vbl. sb.*; **life-token** = *life-index*; **life vest** *U.S.* = *life-jacket*; **life-world** *Philos.* [tr. G. *lebenswelt*], all the immediate experiences, activities, and contacts that make up the world of an individual, or of a corporate, life.
1907 *Westm. Gaz.* 10 Apr. 10/1 That is sufficient justification for the life-company amalgamation. **1970** *Sci. Jrnl.* June 9/1 Plastic lifecrafts carried aboard spaceships much as lifeboats are carried by ocean liners, have been advocated. **1970** *New Scientist* 22 Oct. 178/1 A computer would calculate the position of the lifecraft with an accuracy of one to ten miles. **1935** *Jrnl. Amer. Med. Assoc.* 17 Aug. 514/2 Today, life expectancy at birth is 57 years. **1956** A. H. COMPTON *Atomic Quest* 330 Life expectancy in our country increased by 50 per cent, from 46 years to 69 years. **1962** *Daily Tel.* 30 Apr. 24/3 Details of the Board's plans for closures in Scotland..await completion of a 'life-expectancy' survey. **1972** *Guardian* 15 Aug. 4/5 One of the reasons for higher life expectancy in India today is a better public health system. **1896** W. CALDWELL *Schopenhauer's Syst.* ix. 500 The will is the life-force that pulsates through man's nature. **1903** G. B. SHAW *Man & Superman* III. 109 And these are the creatures in whom you discover what you call a Life Force! *Ibid.* 137 Wagner once drifted into Life Force worship, and invented a Superman called Siegfried. **1920** D. H. LAWRENCE *Lost Girl* xii. 309 Even the will of God is a life-force. **1952** C. DAY LEWIS tr. *Virgil's Aeneid* VI. 137 The life-force of those seeds is fire, their source celestial. **1975** A. FRASER *Whistler's Lane* x. 160 The relentless, uncheckable advent of spring..this all-powerful life force which flowed so strongly. **1931** T. S. ELIOT *Thoughts after Lambeth* 9 These two depressing life-forcers [*sc.* Bertrand Russell and Aldous Huxley]. **1935** AUDEN & ISHERWOOD *Dog beneath Skin* I. (chorus betw. sc. ii & iii) 43 The naughty life-forcer in the norfolk jacket Was the rebels' only uncle. **1910** *Chambers's Jrnl.* Mar. 159/2 The life-gun which is used by the rescuers for shooting lines to the vessel. **1870** D. J. KIRWAN *Palace & Hovel* xxvi. 393 Those street hawkers..will relate their checkered life-histories with great eagerness. **1873** *Monthly Microsc. Jrnl.* X. 53 (*title*) Researches on the life history of a cercomonad. **1909** 'MARK TWAIN' *Is Shakes. Dead?* 141 Philosophers, burglars..surgeons—you can get the life-histories of all of them but one [*sc.* Shakespeare]. **1920** *Discovery* Apr. 111/2 The average value of the uranium present during the life-history of the mineral. **1927** R. FRY *Let.* 31 Aug. (1972) II. 609 The old man..poured out his whole life-history. **1935** B. MALINOWSKI *Coral Gardens* II. VI. 232 The development of speech within the life history of the individual. **1950** K. A. BISSET (*title*) The cytology and life-history of bacteria. **1962** Life history [see *ASTROPHYSICS]. **1884** STEEL & TEMPLE *Wide-Awake Stories* 404 Outside a person's life is an object which faithfully reflects the conditions of his life: this life-index is always very difficult of access. **1915** *Encycl. Relig. & Ethics* VIII. 44/2 'Life-token' or 'life-index' is the technical name given to an object the condition of which is in popular belief bound up with that of some person, which indicates his state of health or safety. **1862** R. H. NEWELL *Orpheus C. Kerr Papers* 1st Ser. 360 He's not an economical man if he don't destroy his life-insurance policy. **1891** E. G. WHITE in *Seventh-Day Adventist Bible Commentary* (1915) VI. 1070/2, I reprieve him from the condemnation of death giving him My life insurance policy—eternal life—because I have taken his place and have suffered for his sins. **1955** *Granta* 26 Nov. 20/2 I'm gonna sell 'em a life insurance policy. **1970** T. HUGHES *Crow* 29 Words came with Life Insurance policies—Crow feigned dead. **1867** *Harper's Mag.* Aug. 349/2 These life-members of my charity. **1907** R. FRY *Let.* 5 Mar. (1972) I. 282 I'm so glad they've made you a life member of the museum. **1926** A. E. HOUSMAN *Let.* 15 Jan. (1971) 233 You may be perplexed by communications from the London Library. I am taking steps to have you made a life member. **1972** H. KEMELMAN *Monday the Rabbi took Off* ii. 22 The by-laws made all past presidents life members of the board. **1859** in H. R. Fletcher *Story R. Hort. Soc.* (1969) xii. 187 It is proposed to raise the money by Donation, by Life Memberships of 40 Guineas and 20 Guineas. **1867** *Harper's Mag.* Aug. 349/1 A most laudable charity—put me down for a life-membership by all means. **1909** *Daily Chron.* 11 Mar. 6/3 New York... Many leapt from the windows and were caught in the life nets. **1947** *Chicago Tribune* 20 July (Comics) 4 Let's see some action—grab that life net! **1969** *Publ. Amer. Dial. Soc.* LII. 33 *Life net*, a net used to catch people who must jump from a building. **1879** Life-office [see *experience table*]. **1972** *Accountant* 28 Sept. 388/2 Rates of interest quoted by life offices at the moment are not so attractive as they were when borrow-all policies were popular. **1948** H. NICOLSON *Diary* 28 May (1968) 141 If they reform the House of Lords, they are certain to make me a life-peer. **1961** *Spectator* 20 Jan. 63 Six new life peers were created. **1973** *Times* 16 May 18/5 The making of life peers rather than hereditary peers (the present Conservative Government has given no hereditary titles) will gradually leave the Crown in increasing isolation as an hereditary institution. **1958** *Times* 24 July 8/7 (*heading*) Life peerages for four women. **1967** *Listener* 20 Apr. 533/1 Her Labour Party allegiance..took her through local government..to a life peerage. **1958** *Times* 24 July 8/7 The wives..of life peers, and the sons and daughters of.. life peers and life peeresses shall be treated..in the same way as the wives..and children of hereditary barons. **1881** *Harper's Mag.* Jan. 274/2 Most of their bargains with the public are made in the shape of life policies. **1907** 'MARK TWAIN' in *North Amer. Rev.* Jan. 14 If I hadn't taken out a life policy on this one the premiums would have bankrupted me long ago. **1942** *Mind* LI. 288 His watchword, in thinking not only of the means but of the ends of a life-policy, is 'here, or nowhere, is my America'. **1972** *Accountant* 28 Sept. 388/1 There are various schemes by which funds can be borrowed to buy shares at the outset, with a life policy being used to repay the loan in due course. **1903, 1922** Life-raft [see *CARLEY]. **1958** [see *air-sea rescue* (*AIR *sb.*¹ III. 1)]. **1962** S. CARPENTER in *Into Orbit* 60, I had a smaller version of it in my liferaft as part of the emergency kit. **1973** *Times* 9 Mar. 26/8 In emergency situations, such as aircrashes, liferafts are automatically inflated. **1912** L. J. VANCE *Destroying Angel* xiv. 189 He managed..to jam the life-ring over her head and under one arm before the next wave bore down upon them. **1972** *Daily Colonist* (Victoria, B.C.) 11 Jan. 7/1 The aircraft sighted..a life-ring bearing the ship's name. **1958** M. A. GRAUBARD (*title*) The foundations of life science. **1959** *Vistas in Astronautics* II. 139 (*heading*) The utilization of a satellite laboratory for life science studies. **1970** C. J. & O. B. GOIN (*title*) Man and the natural world: an introduction to life science. **1973** *Freedom* 2 June 3/4, I regard my own specialism, psychology, as a continuing part of the Darwinian revolution in the Life Sciences. **1918** W. B. YEATS *Per Amica Silentia Lunae* 38 Some.. have foreknown the event and pricked upon the calendar the life-span of a Christ, a Buddha, a Napoleon. **1937** B. H. L. HART *Europe in Arms* xxii. 290 He [*sc.* Napoleon] took short views, since his horizon was his own life-span. **1953** J. S. HUXLEY *Evolution in Action* iv. 103 Their life-span..may extend over several decades. **1957** G. E. HUTCHINSON *Treat. Limnol.* I. ii. 177 The life span of the lake. **1966** C. R. TOTTLE *Sci. Engin. Materials* x. 235 The life-span of the neutron depends on its kinetic energy, and on the material through which it passes. **1974** A. HUXLEY *Plant & Planet* xviii. 189 Plants are too varied to have an average lifespan. *Ibid.*, As soon as we move to multi-cellular plants the lifespans increase. **1959** *Adv. Space Sci. & Technol.* I. 174 (*heading*) Life support system. **1962** F. I. ORDWAY et al. *Basic Astronautics* xiii. 509 A life support system for a manned base on the Moon.. will be exceedingly complex. **1962** D. SLAYTON in *Into Orbit* 20 NASA decided we would adopt the U.S. Navy pressure suit for our spacesuit, so Wally Schirra..started work on the life-support system which we would need to keep the pilot alive and comfortable. **1969** *Guardian* 21 July 1/6 Before take-off the spacecraft's pressurisation system is tested while the astronauts are still in their life-support suits. **1970** *McGraw-Hill Yearbk. Sci. & Technol.* 280/2 All submersibles require life-support systems. **1908** *Westm. Gaz.* 24 Nov. 4/1 The life-tenancy individualism which Mr. Carnegie recommends to us is sharply distinguished from the feudal individualism which obtains in old countries. **1962** H. R. LOYN *Anglo-Saxon Eng.* iv. 178 An abbey or church..could receive estates as a gift, and then yield them back to the donor on a life-tenancy. **1837** Life-tenant [in Dict., sense 16 e]. **1973** *N.Y. Law Jrnl.* 4 Sept. 17/1 Ordinarily a life tenant of real property is entitled to collect all the income but must pay real estate taxes, mortgage interest if any, insurance cost and routine maintenance expenses. **1911** *Chem. Abstr.* V. 1371 (*heading*) Life test of metallic filament lamps. **1929** *Jrnl. Sci. Instrum.* VI. 247 The life test load accommodated by this regulator consists of twenty 2-volt 1-ampere miners' lamps screwed into bus bars. **1958** *Biometrika* XLV. 521 A second difficulty encountered in using conventional experimental methods in industrial life tests is the expense and time involved in waiting for all of the test items to fail. **1959** *Engineering* 20 Feb. 256/2 First on the market, some months ago, was the Hamilton electric watch, which was life tested for several years before the first one was sold, and which has proven itself a most reliable and accurate timekeeper. **1961** *Times* 31 May 18/4 Yet another machine life-tests switches. **1972** R. C. WINANS in D. Baker et al. *Physical Design Electronic Syst.* IV. viii. 368 As a control on time-dependent characteristics, life tests are made on samples from each production lot. These life tests are usually for periods ranging from 100 to 1000 hours, under maximum rated power, temperature, and/or voltage conditions. **1926** J. W. T. WALSH *Photometry* xvi. 449 An important branch of the work of a photometric laboratory is, frequently, the life testing of lamps. **1957** *Ann. Math. Statistics* XXVIII. 432 Life-testing situations where population properties are of greater interest than sample properties usually involve inanimate objects such as automobile tires, light bulbs, etc. **1899** R. C. TEMPLE in *Folk-Lore* X. 403 It now seems to have found a definite place among the recognized technicalities of writers on folklore under the guise of the life-token. **1915** Life-token [see *life-index*]. **1962** S. CARPENTER in *Into Orbit* 155 A tiny life vest which weighed less than a pound and could be folded up into a package not much bigger than a man's hand. **1970** *Washington Post* 30 Sept. D6/1 (Advt.), Fiberglas hardtop,..2 wipers, 6 life vests, 2 fire ext. and bell. **1940** A. SCHUETZ in M. Farber *Philos. Ess. in Memory E. Husserl* 173 Human existence itself is referred to an existent life-world as a realm of practical activity. **1960** D. CAIRNS tr. *Husserl's Cartesian Meditations* § 8. 19 Not just corporeal Nature but the whole concrete surrounding life-world is for me..only a phenomenon of being, instead of something that is. **1964** *Philos. Rev.* LXXIII. 418 Merleau-Ponty analyzes what..Husserl had termed the 'life-world' ('Lebenswelt'). **1969** M. FARBER in R. Klibansky *Contemp. Philos.* III. 167 There are life-worlds for ordinary experience, varying from person to person, from group to group, and from time to time. There are also life-worlds as viewed on the basis of the sciences. **1972** D. FØLLESDAL in Olson & Raul *Contemp. Philos. Scandinavia* 426 We all live in a 'life-world' which is constituted by everyone in community. The term 'life-world' ('*Lebenswelt*') first appeared in an unpublished article on Kant which he [*sc.* Husserl] wrote in 1924, and the life-world became the main theme of his last major work, *The Crisis of the European Sciences* (1936).

life-and-death, *a.* [See LIFE *sb.* 1 c.] Involving life and death; vitally important.
1822 MILL *Let.* 14 Nov. in *Works* (1963) XII. 14 The life-and-death style in which I speak and write about it [*sc.* the Utilitarian Society]. **1824** [see LIFE *sb.* 1 c]. **1834** HOOD *Tylney Hall* I. iii. 24 Joe made shift to explain that he was charged with what he called a life and death letter to Sir Mark. **1857** *Edin. Rev.* CVI. 198 The third, the part he assigned to himself in the life-and-death struggle of his country. **1857** *Chambers's Jrnl.* Nov. 338/1 These are really a life and death matter to our neighbours. **1888** MRS. H. WARD *R. Elsmere* II. III. xxiii. 226, I go about haunted by the *seriousness*, the life-and-death interest people throw into music. **1939** DYLAN THOMAS *Let.* 11 Sept. (1966) 236 If you look at *Tropic of Cancer*..not as a universal life-&-death book,..you must enjoy..it. **1951** S. SPENDER *World within World* ii. 97 At once I was aware of nature as a life-and-death force. **1973** *Times* 21 Dec. 4/2 On Christmas Day and Boxing Day..all inland telegram deliveries..will be suspended. Special arrangements will be made to deliver 'life and death' messages.

life cycle. Also **life-cycle.** [f. LIFE *sb.* + CYCLE *sb.*] **1.** *Biol.* (In Dict. s.v. LIFE *sb.* 17.)
1873 *Monthly Microsc. Jrnl.* X. 57 Thus the entire life cycle of this form is seen. **1894** [in Dict. s.v. LIFE *sb.* 17]. **1967** M. E. HALE *Biol. Lichens* ii. 27 The life cycle of fungi is completed when the vegetative thallus produces fruiting bodies that contain spores.
b. The course of human, cultural, etc., existence from birth or beginning through development and productivity to decay and death or ending.
1938 H. READ *Coll. Ess. Lit. Crit.* I. i. 19 The classical and romantic periods are related to each other in a 'life-cycle' which is the recurring cycle of the growth, maturity, and decay of culture. **1949** M. MEAD *Male & Female* xvi. 339 Here he is, only in middle age, and his life is over—.. no new fields to conquer... So while he is not out of a job ..the very nature of the life-cycle in America is such that he feels like an old man. **1959** G. D. MITCHELL *Sociol.* vi. 103 Religion is mostly important to mankind at times of personal crisis or when new and socially significant stages of the life-cycle are approached: at birth, initiation, marriage, death. **1967** B. S. COHN in P. Bohannan *Law & Warfare* 145 The general rule is that, when a household is celebrating a life-cycle rite, all adult males from the *khandan* and at least one adult male..from every other household in the hamlet is invited. **1969** W. E. MOORE in Lindzey & Aronson *Handbk. Social Psychol.* (ed. 2) IV.

xxxii. 316 A rather different perspective on social organization and social process results from dealing with the entire human life cycle.

2. *transf.*, esp. in *Econ.* and *Comm.*

1965 H. I. ANSOFF *Corporate Strategy* (1968) ii. 31 Objectives..will vary from one type of firm to another depending on the firm's past profitability, its prospects, and its stage in the life cycle. 1969 J. ARGENTI *Managem. Techniques* 70 Many product life-cycles are declining. 1971 *Nature* 12 Feb. 486/2 The photo-oxidation of sulphur dioxide to sulphuric acid in the atmosphere is relevant both to the formation of aerosols in polluted air and to the life cycle of sulphur compounds in the atmosphere. 1971 *Daily Tel.* 21 June 17/6 The four-stage life-cycle of every [manufactured] product—exploration, growth, maturity, decline. *Ibid.* 17/8 Companies usually have a range of products at various stages of the life-cycle.

life-day. (Later examples.)

1893 'MARK TWAIN' *Lett.* (1917) II. 592, I shall tackle Adam once more... I've been thinking out his first life-days today. 1940 AUDEN *Another Time* 61 And we The life-day long shall part no more.

li·fe-form. Also **life form.** [f. LIFE *sb.* + FORM *sb.*] **1.** *Biol.* A habit or vegetative form exhibited by any particular plant or which characterizes a group of plants.

Various life-form classifications have been proposed. That of C. Raunkiær (or a modification of his system) based upon the position of the buds relative to the soil surface during the unfavourable season is the one generally employed.

1899 *Natural Sci.* XIV. 109 Hence groups of similar adaptational form, 'Lebensform' of German authors, need by no means coincide with natural families or groups of species. For example, *Empetrum* and *Erica*, or *Aloe* and *Agave*, possess similar 'life-forms',..yet their floral characters indicate widely separate genetic affinities. 1913 *Jrnl. Ecol.* I. 16 (*heading*) Raunkiær's 'life-forms' and statistical methods. 1926 TANSLEY & CHIPP *Study of Vegetation* ii. 11 Life form is the characteristic vegetative form of a species; in the first place whether it is a tree, shrub, herb, or a member of one of the lower group of the plant kingdom. 1960 [see *growth-form* s.v. *GROWTH[1] 5]. 1964 V. J. CHAPMAN *Coastal Vegetation* i. 9 There is no generally recognized life-form system for the algae. 1971 D. W. SHIMWELL *Descr. & Classification of Vegetation* iii. 74 Raunkiær..recognized fifteen main types of life form.

2. A living creature; any kind of living thing.

1905 *Daily Chron.* 17 Aug. 5/7 In the beginning, before life forms appeared, the sun shone on the ocean. 1908 *Lit. Guide* 1 Aug. 115/2 Proof..of an unbroken chain of psychical continuity between the lowest and highest life-forms. 1971 *Daily Tel.* (Colour Suppl.) 30 Apr. 17/3 If only..it were possible to introduce some terrestrial life-form on to Venus. 1971 I. G. GASS et al. *Understanding Earth* ix. 139/2 Once life had become universally distributed over the face of the globe, it must have prevented the further generation of new life-forms.

life-guard. 4. For *U.S.* read orig. *U.S.*, and add later examples.

1921 *Daily Tel.* 29 Aug. 9/6 They were sustained by this means until the life guards arrived to take the women ashore. 1933 *Boys' Mag.* XLVII. 122/1 When a party bathe, one or two of the best swimmers should be posted as life-guards. 1974 HAWKEY & BINGHAM *Wild Card* xiii. 119 A guy who's drowning..who, if he's not subdued, will take the lifeguard down with him.

life-in-death. A phantom state, a condition of being or seeming to be neither alive nor dead; something having the form or appearance of the supernatural, an apparition, a spectre. Also, = *death-in-life* s.v. *DEATH *sb.* 2.

1817 COLERIDGE *Anc. Mar.* in *Sibylline Leaves* 14 Her skin was as white as leprosy, The Night-Mair Life-in-Death was she, Who thicks man's blood with cold. 1901 *Daily Chron.* 27 Dec. 3/1 They lie in a sort of life-in-death until the touch of a mighty hand grants them their full development. 1904 *Ibid.* 22 Sept. 3/4 In a life-in-death existence she still languishes as an almost forgotten link with the past of forty years ago. 1925 R. W. KETTON-CREMER in *Oxf. Poetry* 22 From something not of earth, nor quite of Death—Some phantom Life-in-Death. 1932 W. B. YEATS *Words for Music* 1, I hail the Superhuman; I call it Death-in-Life and Life-in-Death.

li·fe-line. [f. LIFE *sb.* + LINE *sb.*[2]] **1. a.** (See LIFE *sb.* 17; also used by firemen.) (Earlier and later examples.) **b.** A diver's signalling line.

1700 in *N. & Q.* (1941) 12 July 22/2 Lyfline. 1790 *Gentl. Mag.* LX. in *N. & Q.* (1962) Jan. 17/1 A line..which was called his *Life-line*, as it was found..to have been serviceable in preserving ships and men. 1877 *Encycl. Brit.* VII. 297/2, *e* is the 'life' or 'signal' line, which is attached to the diver's waist, and by which he makes signals and is hauled to the surface. 1896 *Strand Mag.* XII. 351/1 As the strain of the air-pipe was downward, and that of the life-line upward, I concluded that the pipe must be fast below. 1904 *Daily Chron.* 26 Oct. 6/7 Fireman Herbert White lashed a branch hose to his body with a life-line. 1968 *Globe & Mail* (Toronto) 3 Feb. 11/4 Firemen..used a life-line gun to reach the boat and Mr. MacAdam. 1975 *Times* 6 Jan. 4/4 Lifelines used on Mr Heath's ill-fated yacht..were yesterday called sub-standard by the British Safety Council.

2. In *Palmistry*: a mark on the palm of the hand supposed to indicate one's length of life.

1890 L. COTTON *Palmistry* II. 36 If the head line is separated, at its departure, from the life line, it indicates great self-confidence 1894 [see *HEAD-LINE 5]. 1919 BEERBOHM *Seven Men* 154, I had seen in my own hand..a clean break in the life-line. 1971 M. MCCARTHY *Birds of America* 312 He felt a sharp pain in..his palm, the part bounded by his life line.

3. *fig.* The line of life: see LINE *sb.*[2] 1 g. Also, an essential supply route, a line of communication, etc.

1860 HAWTHORNE *Transformation* II. xiii. 209 If there were one of those friends whose life-line was twisted with your own, I am enough of a fatalist to feel assured that [etc.]. 1891 E. S. UFFORD in I. D. Sankey *Gospel Hymns No. 6* 30 Throw out the Life-Line a-cross the dark wave, There is a brother whom some one should save. 1905 *Daily Chron.* 13 Feb. 3/1 Every man who has lived so long..and kept the life-line so straight and true as Mr. Holyoake. 1936 *Lit. Digest* 17 Oct. 13c Cut what Britain calls her 'life line'. 1941 *Times* (Weekly ed.) 15 Oct. 8 The King spent Wednesday at Liverpool seeing what is being done to hold this end of the Atlantic lifeline. 1963 *Times Lit. Suppl.* 1 Mar. 149/1 Liberals who keep a life-line open between the actual world..and the one they would like to see exist. 1970 *Daily Tel.* 7 Oct. 1/1 A £54-million 'lifeline' was thrown by the Government yesterday to the farming industry, which has been faced with having to cut back production because of rising costs. 1975 D. RAMSAY *Descent into Dark* ii. 60 Who the hell thinks about honour when her lifeline's being cut?

lifemanship (ləi·fmænʃip). [f. LIFE *sb.* + *-MANSHIP.] Skill in getting the edge over, or acquiring an advantage over, another person or persons. So (as a back-formation) **li·feman.**

1950 S. POTTER (*title*) Some notes on lifemanship. *Ibid.* 14 Day by day our centres send out young men, yes and women too, to assess the lifemanship approach. *Ibid.*, The Lifeman is never caddish himself. 1952 *Granta* 29 Nov. 8/2 They are the men who thought Lifemanship was something to practise rather than detect. 1953 *Encounter* Oct. 62/2 'Phoney' and 'corney'..are concepts, categories for criticising a whole school of bad 'lifemanship'. 1958 *Spectator* 29 Aug. 285/1 His narrative abounds in bashful lifemanship. 1959 *Times* 9 Apr. 15/3 You mean, don't you, that he was nearly the greatest European lifeman? 1964 *Discovery* Oct. 35/1 Lifemanship consists largely of a surreptitious and diplomatic control of what is ordinarily non-verbal and unconscious communication.

life-or-death, *a.* [See *LIFE *sb.* 1 c.] = *LIFE-AND-DEATH *a.*

1897 G. B. SHAW *Our Theatres in Nineties* (1932) III. 146 We should have had the ablest manager of the day driven by life-or-death necessity to extract from contemporary literature the proper food for the modern side of his talent. 1932 —— *Platform & Pulpit* (1962) 241 They will finally get rid of Parliament because they have a life-or-death pressure of necessity behind them. 1973 J. WAINWRIGHT *Touch of Malice* 134 The impression of life-or-death efficiency.

life-preserver. Add: **4.** *transf.* and *fig.*

1851 *London at Table* i. 8 The 'life preserver', as the half-pint bottle has been termed. 1852 GEO. ELIOT *Let.* 27 Mar. (1954) II. 15 Your cordial assurance..is one of those pleasant things—those life-preservers—which relenting destiny sends me now and then to buoy me up. 1941 J. SMILEY *Hash House Lingo* 35 *Life preservers*, doughnuts. 1953 *Manch. Guardian Weekly* 14 May 11 Chambers admits that..he was busy preparing a 'life preserver' in the form of stolen documents he could use later to silence anyone who might rat on him.

lifer. Add: **3.** One who leads a life of a specified character.

Properly the second element of a compound.

1906 *Daily Chron.* 11 May 6/4 The Gospel did not commend itself to the simple lifers of the country-side, but spread like wildfire among the complex lifers of the Greek cities.

4. A life-peer.

1959 *Economist* 31 Jan. 397/1 An infusion of 'lifers'' half blue blood. 1969 *Sunday Tel.* 30 Mar. 2/8, I will not..turn out for Lifers.

life-rent, *v.* [f. the sb.] *trans.* To assign in liferent; to use and enjoy property during one's life.

1700 *Edin. Gaz.* 2–5 Sept., The Lands of Hiltoun..presently Life-rented by the Lady Rosyth, are to be Set in Tack by way of publick Roup. 1819 SCOTT *Let.* 25 Nov. (1934) VI. 28 My wife's brother has left my children a considerable fortune which is at present life-rented by his lady. 1890 J. RANKINE *Erskine's Princ. Law Scotl.* (ed. 18) 218 And money may be liferented, the interest..being due to the liferenter. 1937 *St. Andrews Citizen* 6 Mar. 2 The residue of his estate be life-rented to his sister.

li·fe-saver. [See LIFE *sb.* 16 b.] **1.** Something that may save one's life.

1883 [see LIFE *sb.* 16 b]. 1944 *Living off Land* v. 112 Failure to take a life-saver [*sc.* quinine or atebrine] on account of the bitter taste sounds childish yet there are such childish persons.

2. A person assigned to watch against accidents to bathers (at a beach resort, etc.). orig. *U.S.* (The customary term in Australia and N.Z.).

1887 *Courier-Jrnl.* (Louisville, Kentucky) 10 Feb. 8/3 The Police and the Life-Savers still prying into the mystery of Smyser's Pond. 1903 *Boston Even. Transcript* 20 Aug., According to a decision of the Election Commissioners a City Point life saver cannot vote. 1931 V. PALMER *Separate Lives* 99 'It's the life-savers!' he shouted. 'They're practising! Come on.' 1934 T. WOOD *Cobbers* 169 You bask in the sun and watch the life-savers, picked volunteers. 1958 *Observer* 16 Nov. 27/2 He is a handsome cove, in the Australian surf life-saver tradition. 1963 V. B. CRANLEY *27,000 Miles through Austral.* xi. 78 A new life-boat donated to the local surf life-savers. Their performance of swimming and rescue work was really outstanding. 1967 C. O. SKINNER *Madame Sarah* ix. 176 Madame Sarah was presented with a Life Saver's certificate. 1968 W. WARWICK *Surfriding in N.Z.* 1 Since life-saving was practised in and around the sea, it was only natural that lifesavers were the first to try surfriding.

3. *fig.* Some quality, characteristic, or circumstance that helps a person to endure adversity.

1909 *N.Y. Herald* in *Daily Chron.* 8 Mar. 5/2 Both have that great life-saver for men who have to endure periods of stress and storm—a sense of humour. 1934 F. SCOTT FITZGERALD *Let.* 18 Dec. (1964) 258 Again thanks for the money. It was a life-saver. 1973 C. EGLETON *Seven Days to Killing* ix. 101 He spotted a Falk street plan of Paris. It cost him six francs but it proved to be a life-saver.

li·fe-saving, *a.* [See LIFE *sb.* 16 b.] Of or pertaining to the saving of life from drowning, shipwreck, etc. Hence *life-saving station*, a coastal or beach building with life-saving equipment and life-savers. Also as *vbl. sb.*, the saving of life (from drowning).

1858 *Statutes at Large U.S.A.* XI. 320 Twenty-eight life-saving stations on the coast of New Jersey. 1877 *Harper's Mag.* Dec. 50/2 The life-saving car is passing from the vessel to the shore with living freight. *Ibid.*, The life-saving station. 1903 *N.Y. Times* 25 Sept. 14 The plight of the sloop had been signalled at the life-saving station at Sandy Hook. 1906 *Daily Colonist* (Victoria, B.C.) 30 Jan. 4/1 There should be somewhere within reach adequate life-saving stations. 1915 W. E. DOMMETT *Submarine Vessels* vii. 80 (*caption*) Bluejacket wearing life-saving helmet. 1931 V. PALMER *Separate Lives* 101 Dot and Peter were playing life-saving in the sand: Dot, lying flat on her back with her frock rucked up over her knees, and Peter restoring circulation the way he had seen the men do. 1932 N. PALMER *Talking it Over* 139 The same short stretch of surf..usually marked off short to give the life-saving patrol a chance of handling the crowd. 1933 *Boys' Mag.* XLVII. 122/1 Learn methods of life-saving as soon as possible. 1967 C. O. SKINNER *Madame Sarah* ix. 175 Then came a delegation from the Life-Saving Society of Havre. 1973 A. MANN *Tiara* ii. 14 McCarthy had won prizes for swimming and life-saving.

Hence (as a back-formation) **li·fe-save** *v. trans.*, to rescue from death by drowning; to act towards as a life-saver. Also *intr.*

1938 'J. BELL' *Port of London Murders* x. 180 You mind what you're about. I'm not going to life-save you a second time. 1968 P. DICKINSON *Weathermonger* i. 11 Geoffrey thought he might possibly be able to swim round to the harbour... But he couldn't do it if he had to lifesave Sally all the way. 1973 'M. HEBDEN' *Dark Side of Island* iii. 32 Can you life-save?.. Are you a good swimmer?

life-size, *a.* Add: **B.** as *sb.* The size of life; a life-size portrait or statue. Also *fig.*

1850 *Art Jrnl.* 1 Mar. 95/2 A figure of life-size. 1864 *Illustr. London News* 16 July 55/1 Certain technical shortcomings are revealed by this first attempt by Mr. Mount in the life-size. 1885 W. M. ROSSETTI in *Encycl. Brit.* XVIII. 681/2 Cato as the emblem of wisdom, and (in life-size) numerous figures of classic worthies, prophets, and sibyls. 1959 *Spectator* 14 Aug. 181/2 This reduces M. Debré from the more-than-lifesize proper to all prime ministers..to more normal dimensions.

life-sized, *a.* [LIFE *sb.* 16 c.] = LIFE-SIZE *a.*

1847 *Art Union* 1 May 149/2 A life-sized half-length figure. *Ibid.* 1 June 203/2 A life-sized cartoon. 1879 CLEMENT & HUTTON *Artists 19th Cent.* I. 26 'The Birds of America'..was completed in 1839.., containing 448 plates, life-sized and colored. 1898 H. A. GUERBER *Story of Greeks* xxix. 69 The temple and grove [at Olympia] were..adorned, with a great many statues representing the other gods and all the prize-winners, for it was customary to place a life-sized statue of each of them in this beautiful place. 1969 *Harper's Mag.* Feb. 106 Stuffed life-sized dolls drop to the floor of the stage.

li·fe-style. [f. LIFE *sb.* + STYLE *sb.* 24.] **a.** A term originally used by Alfred Adler (1870–1937) to denote a person's basic character as established early in childhood which governs his reactions and behaviour. **b.** *gen.* A way or style of living.

1929 A. ADLER *Probl. Neurosis* [i. 7 The style of life is founded in the first four or five years of childhood.] *Ibid.* 8 This fragment of memory records the two typical motives of the main life-style. 1939 H. ORGLER *Alfred Adler* i. 35 We have only really understood an individual when we have revealed the unitary life-style behind this ostensible duplicity. 1946 'G. ORWELL' *Crit. Ess.* 137 True to his life style, Koestler was..promptly arrested and interned by the Daladier Government. 1947 M. MCLUHAN in *Sewanee Rev.* LV. 11. 80 While ostensibly setting about the freeing of the slaves, they became enslaved, and found in the wailing self-pity and crooning of the Negro the substitute for any life-style of their own. 1950 COLE & BRUCE *Educational Psychol.* i. 2 The teacher is confronted with such unique individuals, already shaped into recognizable life styles. 1952 M. MACKENZIE *Contrast Psychol.* p. iii, The Freudian believes that the unconscious conflict, underlying an Anxiety State, springs from frustrated 'sexuality'; the Adlerian from an obstructed 'life style'. 1961 *Guardian* 22 Mar. 10/5 The mass-media..continually tell their audience what life-styles are 'modern' and 'smart'. 1972 *Jrnl. Social Psychol.* LXXXVI. 121 When a man's life style is incongruent with the demands of a task he must perform he will experience stress. 1972 M. MEAD *Blackberry Winter* vii. 82 But in other ways those years in

Bucks County gave me a view of a much earlier life-style, one that corresponded with my grandmother's girlhood. 1973 T. TOBIN in Ade *Lett.* 7 His commercial successes in the theater enabled him to become financially independent, and Ade chose the lifestyle with which he was most comfortable. 1973 *Times* 11 Apr. 18/4 (*heading*) Council of churches want freedom for students to create their own life-styles.

lifetime. Add: Also *attrib.* or as *adj.*, for the duration of a life, during one's life, while one is alive. Phrases: *all in a* (or *one's*) *lifetime*, implying resignation to whatever happens; *of a lifetime*, implying that an event, situation, or thing will never be equalled or repeated.

1849 N. KINGSLEY *Diary* (1914) 44 My consolation is that it is all in my lifetime and thus make myself quite contented. *Ibid.* 52 This must count as 'all in my life-time'. 1898 F. P. DUNNE *Mr. Dooley in Peace & War* 187 Well, tubby sure, 'tis thryin' to be dhrivin' a coal wagon or a sthreet-car; but 'tis all in a lifetime. 1929 WODEHOUSE *Mr. Mulliner Speaking* vii. 222 But let me tell you, my lad, that you're throwing away the laugh of a lifetime. *Ibid.* ix. 314 It must be a cocktail. The cocktail of a life-time. 1961 —— *Ice in Bedroom* xxiii. 188 You take one step in its direction and you're going to get the headache of a lifetime. 1962 *Amer. Speech* XXXVII. 16 New England did contribute the largest number of white lifetime migrants to the state of New York. 1974 *Country Life* 2 May 1065/2 The gifts tax..would be a tax on lifetime gifts.

2. = LIFE *sb.* 8 c (in Dict. and Suppl.).

1858 [in Dict.]. 1920 *Discovery* Apr. 111/1 The helium now found in a mineral can be only a fraction..of the total amount which has been generated within it, and which alone could give a true estimate of its life-time. 1939 *Physical Rev.* LV. 506/1 The lifetime for a mesotron at rest has been estimated..to be of the order $2\text{–}4 \times 10^{-6}$ sec. 1950 *Ibid.* LXXVII. 153/1 (*heading*) Lifetimes of mercury and potassium atoms in excited states. 1965 H. I. ANSOFF *Corporate Strategy* (1968) ii. 24 If it turns out that the life-time of some projects exceeds the budget period, the period is extended for purpose of analysis. 1968 [see *LIFE *sb.* 8 c]. 1968 *Times* 29 Nov. 13/3 The running-down rate of the pulsar implies that it has a lifetime of the same order as the Crab nebula. 1972 *Sci. Amer.* May 50/1 The Viking spacecraft will consist of an orbiter and a lander, each with a lifetime of many months. 1974 *Daily Tel.* 17 Apr. 17/4 There is no possibility of action on the lines of the group's proposals in the lifetime of the present Parliament.

life-timer. [f. LIFETIME.] One serving a life-sentence. (In quot. *fig.*)

1926 J. BLACK *You can't Win* v. 48 Life-timers of society, they were slowly sinking without a straw to grasp at.

li·fe-way. orig. *N. Amer.* [f. LIFE *sb.* + WAY *sb.*[1]] Way or manner of life.

1961 in WEBSTER. 1969 *Beaver* (Winnipeg) Spring 62/2 Wouldn't it be expected to contribute notably to an interpretation of their [*sc.* Eskimos'] life-ways in art? 1973 J. M. ANDERSON *Structural Aspects of Language Change* 203 Especially in regard to human lifeways and institutions, [etc.]. 1975 *Nature* 22 May 280/2 He [*sc.* Professor John Bowker (University of Lancaster)] drew attention to the extent to which religions function as bounded systems of information process, in which resources are made available to human beings for the construction of a life-way from birth to death.

lifey, *a.* Delete 'Now *Sc.*' and add later *fig.* example.

1910 *Chambers's Jrnl.* Nov. 706/2 Those [sapphires] found in Ceylon, which are lighter in colour and 'lifier' than any of the others.

lift, *sb.*[2] Add: **1. b.** (Later examples.) Cf. *LIFT *v.* 11 e.

1929 M. DE LA ROCHE *Whiteoaks* v. 70 'Don't they ever send a car for you?' 'Good Lord, no. Sometimes I get a lift.' 1944 J. S. HUXLEY *On Living in Revolution* ix. 106 We found that a bus recorded on the time-table was in reality non-existent; cadged a lift on a road foreman's car to Denness. 1955 *Times* 26 Aug. 7/4 After giving a 'lift' to a hitch-hiker one will have lost only a tablespoonful or two of petrol, perhaps a teaspoonful of oil, and a saltspoonful or two of rubber off the car's tires. 1974 'J. LE CARRÉ' *Tinker, Tailor* xxxiii. 293 Declining a lift, Smiley said the walk would do him good.

2. b. Delete ?*U.S.* Also, a cheering or encouraging influence or effect, a sense of elation.

1861 T. HUGHES *Tom Brown at Oxf.* I. xiv. 281 He heard Drysdale's view halloa above all the din; it seemed to give him a lift. 1873 A. G. MURDOCH *Lilts on Doric Lyre* 10 Sae jist to gie their hearts a lift..They cannilie put owre a dram. 1887 [see *ASPIRATIONAL *a.*]. 1936 J. DOS PASSOS *Big Money* 498 Dick put down three bourbons in rapid succession but he wasn't getting any lift from them. 1951 E. PAUL *Springtime in Paris* i. 12 Raoul realized that Katya got an enormous lift from secrecy and mystery, and helped her enjoy it. 1957 *Sat. Even. Post* 30 Mar. 102/2 The girl had to rush back to the pusher and complain that it didn't give her a lift. 1975 T. ALLBEURY *Palomino Blonde* xxiii. 135 Hallet had been demented with worry..and the 'lift' that he had got from his talk with Farrow had melted away.

5. d. (Earlier example.)

1829 J. MACAULEY *Nat. & Civil Hist. N.Y.* I. 170 This ..lock has an extent within the gates of one hundred and fourteen feet, with a breadth of thirty—the lift is nine feet.

e. (i) The upward force acting on an aircraft or other body in the air; *spec.* that produced by its motion through the air; the force on an aerofoil that acts at right angles to its direction of motion through a fluid.

1902 *Encycl. Brit.* XXV. 104/1 The sustaining power, or 'lift',.. in horizontal flight must be equal to the weight. *Ibid.*, The present data indicate that, with concave surfaces, angles of 2° to 5° will produce adequate 'lift'. 1919 H. SHAW *Text-bk. Aeronaut.* iii. 43 The upper surface of an aerofoil is considerably more important than the lower surface from the point of view of lift, as the suction over the top surface is numerically much greater than the pressure beneath. 1937 DODGE & THOMPSON *Fluid Mech.* vii. 127 Usually the component opposing the motion is referred to as the drag, while the cross-stream component is called the lift, even though it may not always be acting vertically upward. 1948 *Sci. News* VII. 23 In aerodynamics it is customary to resolve the reaction of the air on a surface into two components, namely lift, which is that part of the force acting upwards..and is thus desirable, and drag, which is the component at right angles to the lift and..resists the forward motion of the surface through the air. 1959 *Chambers's Encycl.* I. 110/2 In straight level flight the lift equals the weight. 1973 *Nature* 28 Sept. 182/1 Most flying insects depend, for their lift and thrust, on conventional aerofoil action which sets up a bound vortex around the moving wing to create a steady-state flow of air. 1974 *Encycl. Brit. Macropædia* I. 371/1 An airship derives lift from two sources: (1) by displacement of air as a balloon (static lift) and (2) from the reaction of airflow over its envelope and control surfaces when it is under way (dynamic lift).

(ii) The (maximum) weight that an aircraft can raise (including or, more commonly, excluding its own weight).

1910 *Blackw. Mag.* July 4/1 The compartments [of the Zeppelin] contained 351,150 cubic feet of hydrogen, giving a lift of eleven tons. 1929 *Nature* 14 Dec. 916/2 Recourse to the Servo-motor gear was not found necessary (if this holds good at full speed the gear may be removed, and then about ½ ton will be added to the useful 'lift'). 1971 *Daily Tel.* 19 Aug. 2/6 It is much lighter and can be lifted by the Puma tactical transport helicopter, which has a total lift of up to 5,500 lb.

f. *Pros.* An element of high intensity in an alliterative measure, marked by stress or tone. (G. *hebung.*) Cf. *DIP *sb.* 1 g.

1894 [see *DIP *sb.* 1 g]. 1927 E. V. GORDON *Introd. Old Norse* 293 The rhythm consists of regular alternation of strong and weak metrical elements, known as lift and sinking respectively. 1953 C. L. WRENN *Beowulf* p. xxxvii, [Alliteration] is never repeated on the last lift. 1961 [see *DIP *sb.* 1 g].

g. *Dance.* A movement in which a dancer lifts his partner in the air.

1921 *Dancing Times* Aug. 867/2 Miss Jules Andre..filled the roll [*sic*] of 'boy' in..many..numbers. Her lifts and adage work were delightful. 1943 K. AMBROSE *Ballet-Lover's Pocket-Bk.* 40 With the invention of each new ballet, new lifts are devised. 1944 'BRAHMS' & 'SIMON' *Titania has Mother* ii. 11 'He's a frightfully bad dancer, mother. His lifts!' She shuddered. 1950 *Ballet Ann.* IV. 69 She is equally at home in the most intricate acrobatic lifts..as [her body] is swung upwards in the air by her attendant cavaliers.

h. Transport by air (cf. *AIR-LIFT 2); also, a number of persons or an amount of supplies so transported. Cf. *LIFT *v.* 11 i.

1942 F. D ROOSEVELT in W. S. Churchill *Second World War* (1951) IV. xxx. 481 The following shipping can be made available by the United States..: Transports, other than combat leaders, with a lift of 52,000 men. 1947 VISCT. MONTGOMERY *Normandy to Baltic* 137 Our resources..made it impossible to fly in the whole of the Airborne Corps in one lift. 1947 M. NEWNHAM *Prelude to Glory* ixiv. 350 The entire force was carried in one lift. 1949 *Flight* 15 Dec. 756/1 We eventually had sixteen crews, consisting of three members each, engaged whole-time on the Lift.

i. The establishment by a sheepdog of control over a flock of sheep. Cf. *LIFT *v.* 11 g.

1942 R. B. KELLEY *Animal Breeding* xi. 115 A little 'eye'..can be associated meritoriously with a steady 'lift' and..restricts the dog from over-running a cast. 1946 F. DAVISON *Dusty* ix. 117 The [sheepdog] trial had four phases; the cast,..the lift, when the dog, having found them [*sc.* his sheep], established control over them; the fetch,..and the carry. 1955 *Galloway Gaz.* 1 Oct. 6 His dog 'Garry' won the Rosebowl for the best outrun and lift. 1964 *Weekly News* (Auckland) 29 Apr. 37/3 Fleet is losing points hand over fist now. He has failed to obtain a good 'lift': in other words he hasn't been able to head them unobserved and then start them moving gently and firmly. 1973 *Country Life* 25 Oct. 1292/1 From its position at 12 o'clock, the dog begins the critical 'lift', with a quiet authority that brooks neither refusal nor panic in the sheep.

10. b. chair-lift, a device for transporting people up a mountain slope, usually consisting of seats suspended from a continuously moving overhead cable; **ski-lift,** a chair-lift, or any of various types of apparatus for hauling skiers uphill. Also absol. *lift.*

1940 F. ELKINS *Compl. Ski Guide* II. 161 New 3500-foot 'T-bar' lift to connect with top of chair lift. 1947 *Penguin New Writing* XXX. 27 Dory found himself going up on the ski-lift with a Frenchwoman. 1953 C. J. ALLEN *Switzerland's Amazing Railways* viii. 93 A simpler application of the *téléphérique* principle is found in the chair-lift, known in French as a *télésiège* and in German as a *Sesselbahn.* 1955 W. PLOMER *Shot in Park* 50 The ski-lift smoothly moves. 1958 *Times* 18 July 11/7 Skiing is also popular..in the Thredbo Valley, where Australia's first chair lift, a mile long, began to work this winter. 1970 *Country Life* 17–24 Dec. 1214/3 Recently, Norway has been developing 'Alpine' resorts where the ski-lifts, the equipment and the ski-schools closely resemble good centres in the Alps. 1972 M. YORKE *Silent Witness* ii. 12 The lifts, and even the cable-car..had stopped, for the snow..had been falling steadily. 1972 D. HASTON *In High Places* vii. 82 Once above the ski-lift level it was still possible to have the whole of a range to one's self on a certain day.

14. *spec.* of paper.

1808 C. STOWER *Printer's Gram.* xvi. 405 Having thus doubled the first lift on the peel, he [*sc.* the warehouseman] raises it, holding it aslant, that the shorter fold of the sheets may open from the peel, in order to convey it over the pole. 1841 W. SAVAGE *Dict. Art of Printing* 444 In the warehouse, each separate portion of printed paper, whether it consists of five or six sheets or more, that is placed upon the poles to dry, is termed a lift. 1861 [in Dict.]. 1888 C. T. JACOBI *Printers' Vocab.* 75 *Lift*, applied to a handful of printed work in the warehouse. 1967 V. STRAUSS *Printing Industry* x. 632 (*caption*) You see the lift of stock to be cut on the bed or table of the cutter. On the left the lift is lined up with one edge, in the rear it is lined up with another edge. 1971 D. POTTER *Brit. Eliz. Stamps* xv. 175 Batches of 1,000 sheets are broken down into 'lifts' of 25.

18. (sense 10) *lift-boy, -button, -cage*; **lift coefficient** *Aerodynamics*, a ratio representing the lift developed by unit area of an aerofoil in relation to the air speed, and defined as the lift divided by the product of the aerofoil area (in plan) and the square of the air speed (and, in mod. use, by half the air density also); **lift-drag** *a. Aerodynamics*, relating to both lift and drag; applied *spec.* to the ratio of the lift to the drag; **lift-fan,** a fan in a hovercraft which provides the air-cushion; **lift-gate,** (*b*) a gate opening on to a lift (sense 10); (*c*) *U.S.* in a motor vehicle, a hinged back panel that opens upwards; **lift-slab** *attrib.*, applied to a labour-saving system of building whereby pre-cast components are raised by jacks to the position desired; **lift truck** = *fork-lift truck*; **lift valve,** a valve which opens by the valve head moving (vertically) out of its (horizontal) seat; **lift-web,** a strip of webbing joining the harness and the rigging lines of a parachute; **lift wire** *Aeronaut.*, a wire on a biplane or light monoplane that extends from the wing to the fuselage and is designed to transmit part of the lift to the latter during flight.

1904 'SAKI' *Reginald* 15 Lift-boys always have aged mothers. 1967 L. MEYNELL *Mauve Front Door* vi. 82 Chauffeurs, waiters, lift-boys..they are the operators. 1955 W. TUCKER *Wild Talent* xiv. 186 The man punched the lift button. 1970 P. GEDDES *November Wind* vi. 64 Havill watched him press the lift button. 1951 R. SENHOUSE tr. *Colette's Last of Chéri* 213 The lift-cage heavily splashed with as much lacquer and gold as a sedan-chair. 1971 R. PETRIE *Thorne in Flesh* xi. 145 A boy lounged on a stool in the silent lift-cage. 1919 H. SHAW *Text-bk. Aeronaut.* iii. 39 As the angle of incidence increases the lift coefficient also increases rapidly, until an angle of about 13° is reached, beyond which the coefficient increases less rapidly, and reaches its maximum value in the neighbourhood of 15°. 1933 *Techn. Rep. U.S. Nat. Advisory Comm. Aeronaut.* No. 463. 18 As speeds above half the velocity of sound are exceeded..the flow breaks down as shown by a drop in the lift coefficient. 1966 *McGraw-Hill Encycl. Sci. & Technol.* I. 85/2 The maximum lift coefficient (the stall value) of the wing is 1·1–1·5. 1919 Lift-drag ratio [see *CEILING *vbl. sb.* 6 b]. 1935 P. W. F. MILLS *Elem. Pract. Flying* i. 6 Variations in incidence..affect lift and drag disproportionately, and thus produce variations in the quantitative relation between lift and drag—that is to say, in what is called the lift-drag ratio. 1960 *Times Rev. Industry* Oct. 58/3 [The] airstream direction detector system..enables an aircraft to be flown on the best lift-drag curve to maintain economic flight conditions. 1962 *Flight Handbk.* (ed. 6) v. 98 The Republic AP-100, in which six J85 engines feed three lift fans. 1967 *Jane's Surface Skimmer Systems 1967–68* 49/2 A drive-shaft runs vertically upward to the 12-blade lift-fan. 1948 G. V. GALWEY *Lift & Drop* i. 14 The crowd gathered at the lift gates. 1951 J. WYNDHAM *Day of Triffids* i. 19, I found a large '5' painted on the wall opposite the lift gate. 1961 WEBSTER *Lift gate*, an upper rear panel (as on a station wagon) that opens upward as a tail gate opens downward. 1963 *Aerospace-Automotive Drawing Standards* (Soc. Automotive Engin.) 1 *Liftgate*, a hinged backwindow. 1970 *Motor Trend World Automotive Yearbk. 1971 Buyer's Guide* 112/3 The rear seat for a Gremlin is an optional extra along with the counter-balanced 'lift-gate' that comes with it. 1974 E. MCGIRR *Murderous Journey* 33 The liftman..was fiddling with the lift gate. 1951 (*title*) Youtz-Slick lift-slab building method (Inst. Inventive Res., San Antonio, Texas). 1960 *Economist* 22 Oct. 378/3 The 'lift slab' principle..was developed in America, the columns are first cast and erected, then pre-cast floor slabs are lifted by synchronised hydraulic jacks. 1962 *Daily Tel.* 30 Nov. 25/4 A 400,000-gallon watertank resting at the base of a tower before being raised 110 ft by the Lift Slab method... The 95 ft-diameter watertank was raised..in about 40 hours. 1963 H. GARNER in R. Weaver *Canad. Short Stories* (1968) 2nd Ser. 56 Even with a lift truck hurrying the parts to the forge we were falling behind. 1971 [see *DOZER[2]]. 1887 *Encycl. Brit.* XXII. 505/1 In many stationary engines lift or disk valves are used, worked by tappets, cams, or eccentrics. 1898 *Engineering Mag.* XVI. 108/1 Compression has been on the increase ever since the adoption of the lift valve. 1971 B. SCHARF *Engin. & its Lang.* xii. 178 Poppet valves. These are spring loaded lift valves which are commonly used, e.g. in internal combustion engines. 1942 *Tee Emm* (Air Ministry) II. 134 Pass the left hand *in between* the left harness lift web and the body and grasp

the right harness lift web. 1947 M. NEWNHAM *Prelude to Glory* viii. 33 To reduce the risk of backward landings men were told..if necessary to turn their bodies by manipulation of the parachute lift-webs. 1958 P. KEMP *No Colours or Crest* iii. 41, I took a frantic pull on my liftwebs to ease the impact. 1915 W. E. DOMMETT *Aeroplanes & Airships* ii. 26 When the machine is in flight, the upward pressure on the wings is taken by 'lift' wires or stays passing to a framework under the fuselage. 1942 C. C. REDMAN in R. A. Beaumont *Aeronaut. Engin.* xvii. 482/1 Wires running..inwards from the tip portions of the upper surfaces to inboard points of the lower surfaces adjacent to the fuselage—are known as 'flying' or 'lift' wires.

lift, *v.* Add: **1.** Occas., to lower after raising from an elevated position. Cf. quot. 1841.

1920 E. O'NEILL *Beyond Horizon* II. i. 73 Lifting Mary to the floor.

b. Also with *down.*

1838 DICKENS *O. Twist* II. xxi. 25 Sikes dismounted..holding Oliver by the hand..and, lifting him down directly, bestowed a furious look on him. 1890 A. CONAN DOYLE *Sign of Four* viii. 138 'He acted according to his lights,' said Holmes, lifting him [*sc.* a dog] down from the barrel. 1898 G. B. SHAW *Candida* i. 106, I cant lift a heavy trunk down from the top of a cab. 1920 E. O'NEILL *Beyond Horizon* II. ii. 95 He lifts her down to the grass. 1940 W. FAULKNER *Hamlet* III. i. 212 He finds the basket by smell and lifts it down from the limb and sets it before her.

f. To help (sick or weak cattle) to stand up. Cf. *LIFTING *vbl. sb.* 1 b.

1899 H. G. GRAHAM *Social Life Scotl. 18th Cent.* I. 155 Cattle..after the long confinement and starving of winter, were mere skeletons, and required to be lifted on their legs when put into the grass.

g. = *face-lift* vb. (*FACE *sb.* 27). So **lifted** *ppl. a.* Also *transf.*

1922 *Ladies' Home Jrnl.* Sept. 28/2 For a skillful surgeon to 'lift' a woman's face—that is, to remove crescent-shaped pieces of skin, near the ears, and at the hair line, thus lifting the cheeks that have begun to sag and so removing the lines of age about the mouth—is actually a simple operation and practically without danger. 1928 G. B. SHAW *Let.* 31 May in *To a Young Actress* (1960) 127 How will you face old age: With a 'lifted' face, with grease paints and an iceball and rouge, with peroxided hair, an old hag..pretending to be a young witch. 1931 *Daily Express* 2 Sept. 3/5 A woman can now have her face lifted one day and appear among her friends the next. 1934 R. MACAULAY *Going Abroad* i. 12 Mrs. Aubrey, bored, felt that they wanted her to have her face lifted, de-wrinkled ..and given a lick of paint. 1951 G. MIKES *Down with Everybody* 71 Modern nationalism is an attempt to see ourselves without the warts; and many historians, writers and poets are the masseurs and cosmeticians of the national beauty parlours, trying to dye our greying hair golden-blonde and trying to lift our faces. 1959 *Cambr. Rev.* 30 May 549/2 Whole courts have had their faces lifted, with stone-work freshly dressed or replaced, stucco renewed. 1974 M. CECIL *Heroines in Love* vi. 149 She..could cling on to her youth..by having her face lifted.

h. (*not*) *to lift a finger*: see *FINGER *sb.* 3 a.

2. d. To raise in price, value, or amount. Also *ellipt.*

1907 *Daily Chron.* 7 Nov. 1/7 Home Rails were lifted all round..several rises being substantial. 1928 *Chambers's Jrnl.* Feb. 99/2 He kept on lifting the betting, merely to increase his plunder. *Ibid.* 115/2 Jackson..opened the pot for a pound. The American..raised it five, and Captain Reginald lifted another five. 1962 A. NISBETT *Technique Sound Studio* 272 To lift programme level 'a stop' is to increase it by turning the fader (potentiometer) from one stud to the next.

3. (i) Of an aircraft: to rise *off* the ground.

1879 *English Mechanic* 4 July 410/3 The small flying model..only just lifted off the pavement. 1899 H. G. WELLS *When Sleeper Wakes* xxiv. 327 The aeropile..was running down its guides to launch. It lifted clean and rose. 1907 *Daily Chron.* 9 Oct. 4/5 She will have to get rid of at least 250 lb. of ballast before she will lift. 1973 J. DRUMMOND *Bang! Bang! You're Dead!* xliv. 151 By the time Sorensen and Pittaway were lifting off the Wapping tarmac, certain constables..were already deploying... They saw the helicopter about the same time as Mariner did.

(ii) Hence in recent use *off* has changed from being a preposition to being an adverb in Astronautical contexts.

1959 W. A. HEFLIN *Aerospace Gloss.* 57/2 *To lift off*, to take off in a vertical ascent. 1961 BURCHETT & PURDY *Cosmonaut Yuri Gagarin* ii. 27 The giant ship lifts off..in a hurricane of white-hot flames. 1971 *Sci. Amer.* Oct. 49/2 On July 21, 1969, *Eagle* lifted off from the moon with its 22-kilogram cargo of lunar rocks and soil.

f. To rise in tone or volume of sound.

1912 GALSWORTHY *Inn of Tranquility* 157 He seemed to enjoy the sounds of conversation lifting round him. 1918 —— *Five Tales* 340 The wayward music lifted up again.

g. *Printing. intr.* Of a forme of type, to stay in one piece when raised from the surface on which it has been assembled; = RISE *v.* 13 c. Also *trans.*, to raise (lines of type), esp. in moving them from a composing stick to a galley, or in preparation for the distribution of used type.

1854 T. FORD *Compositor's Handbk.* 247 *Lift*, this term applies to the raising of a form from the stone. It is said to Lift when no letters drop out. The same term is applied at press when the pressmen are required to Lift a form before it is worked off. 1884 J. GOULD *Letter-Press Printer* (ed. 3) 34 Before lifting the forme off the stone, raise it a little and observe carefully if any letters, &c., are loose and likely to fall out. If the forme 'lifts', take it from the imposing-stone to the proof-press. 1892 A. POWELL *Southward's Pract. Printing* (ed. 4) xxi. 184 Lock up finally, so that the forme will lift. *Ibid.* 185 The next thing to be done is to 'see if it [*sc.* the forme] will lift'; that is, if it can be raised up from the imposing surface without any letters falling out. 1932 SAYERS & SMART in W. Atkins *Art & Pract. Printing* I. iv. 48 If the job contains lines interspersed of the same size and fount (as in display) 'lift' these and place together. 1961 H. W. LARKEN *Compositor's Work in Printing* viii. 95 When type matter is being lifted, it should be handled firmly. *Ibid.* 96 When lifting single lines from a galley or forme, use the side of the galley or the furniture. *Ibid.* 97 Type that is to be distributed should..be lifted in the same manner as that employed for removing it from the stick... The lifted type is allowed to rest on the third finger of the left hand. 1967 KARCH & BUBER *Offset Processes* 544 When each piece of type in a form stays in place after being locked in a chase, it is said to 'lift'.

8. Delete 'In early use' and add later examples.

1905 E. WALLACE *Four Just Men* ix. 165 They was waitin' to cross towards Charing Cross Road when I lifted the clock. 1968 J. LOCK *Lady Policeman* xix. 159 Goods from three or four stores would be found in them. Others would 'lift' a shopping bag first in which to put all the other 'lifted' goods. 1973 J. WAINWRIGHT *Devil you Don't* 107 Lift a bleedin' gun from somewhere.

11. (Later examples of sense 'to drive (cattle) away or to market'.) Also, to remove, discontinue (restrictions, an embargo, etc.).

1840 *Edin. Even. Courant* 19 Sept., We anticipate rather dull sales now, for a week or two, until the St Faith's droves are lifted. 1890 'R. BOLDREWOOD' *Squatter's Dream* iv. 45, I haven't lifted a finer mob this season. 1890 *Argus* (Melbourne) 14 June 4/2 We lifted 7000 sheep. 1906 *Daily Chron.* 12 Sept. 5/7 It was freely said that if we only applied the suspensions would be lifted. 1936 A. RUSSELL *Gone Nomad* I. ii. 11, I hope his droving mission, that of 'lifting' a thousand head of cattle for the markets of the south, was attended with the success it merited. 1941 I. L. IDRIESS *Great Boomerang* x. 75 Red Bill and his gang lifted their cattle. They headed south-west and got safely across to the Paroo. 1974 *Nature* 25 Jan. 171/3 Even if the embargo is suddenly lifted, it will take several weeks for the oil to reach United States ports. 1974 *Daily Tel.* 4 Apr. 17/1 The university authorities yesterday lifted the temporary suspension on deliveries of supplies to the campus.

d. *Hunting.* (See quot. 1968.) Also, to disperse (scent).

1781 P. BECKFORD *Thoughts on Hunting* x. 147 By lifting his hounds too much, he will teach them to shuffle. 1843 *Ainsworth's Mag.* IV. 125, I seldom allow hounds to be lifted, except to a beaten fox. *Ibid.*, To lift, in that case, is proper and justifiable. 1863 C. MORDAUNT *Diary* 6 Mar. in Mordaunt & Verney *Ann. Warwickshire Hunt* (1896) I. 264 [The hounds] had to be lifted several times to holloas. 1919 J. MASEFIELD *Reynard* II. 85 He heard the sounds Of a cantering huntsman, lifting hounds; The ploughman had raised his hat for a sign, And the hounds were lifted and on his line. 1929 *St. Andrews Citizen* 16 Mar. 7 Fife Foxhounds had three poor days last week. Although the weather was good, the bright, warm sunshine 'lifted' scent. 1968 J. GORDON *Beagle Guide* 172 *Lift*, to remove hounds from a lost scent with the idea of trying to hit the line further on.

e. To give a lift to (in a carriage, motor vehicle, etc.). Cf. LIFT *sb.*[2] 1 b.

1884 E. W. HAMILTON *Diary* 17 Aug. (1972) II. 672 A very hot walk. We got 'lifted' back in a carriage; and afterwards played lawn tennis. 1954 M. SHARP *Gipsy in Parlour* xxii. 211 Up she drove, lifted by Mr Simnel the chemist, Taunton-bound. 1959 I. JEFFERIES *Thirteen Days* vii. 87 He'd like to lift me back to Richon fairly soon as the roads were likely to tighten up during the day. 1960 *Sunday Express* 13 Nov. 14/5 A young R.A.F. hitch-hiker I 'lifted' from Shepherd's Bush to High Wycombe. 1965 I. FLEMING *Man with Golden Gun* vi. 90 Get in the back. Lift you down to your car. 1971 M. RUSSELL *Deadline* ii. 22 Can you lift me in your wagon, Wally?

f. *Artillery. trans.* and *intr.* To increase the range of fire from that being used at a given point in an attack.

1916 in A. Farrar-Hockley *Somme* (1964) iii. 94 Avoid a pause at 0000, at minus three in each field battery, where one section will lift on to the support line. 1917 J. MASEFIELD *Old Front Line* ii. 30 The flash of our shells, breaking a little further off as the gunners 'lifted'. 1922 *Encycl. Brit.* XXX. 255/2 The bombardment is 'lifted' from the first line to reinforce that on the second line. 1962 *Ordnance Technical Terminol.* (U.S. Army Ordnance School) (AD 660 112) 176/2 *Lift fire*, to advance the range of fire by elevating the muzzle of a weapon. 1964 A. FARRAR-HOCKLEY *Somme* iii. 96 The 18-pounders lifted on time as they passed the wire. *Ibid.* 98 Some aghast to see the supporting artillery fire already lifting ahead of them.

g. Of a sheepdog: to establish control over a flock of sheep. Cf. *LIFT *sb.*[2] 5 i.

1921 *Kelso Chron.* 12 Aug. 2 This bitch started well... Her haulding, lifting, and penning were good, her bringing and driving very fair. 1942 R. B. KELLEY *Animal Breeding* xiii. 127 When he [*sc.* a pup] has reached this point sit him down and make him lift the flock quietly. 1946 F. D. DAVISON *Dusty* ix. 90 The paddock, what with hills, broken ground and patches of scrub, was not the easiest in the world to lift sheep from. 1949 C. W. G. HARTLEY *Shepherd's Dogs* v. 33 Much will depend upon the manner in which the sheep are 'lifted'.

h. To arrest, take into custody.

1923 G. WATSON *Roxburghshire Word-Bk.* 200 Tam's gruppen an' liftit. 1934 D. ALLAN *Hunger March* III. ii. 208 They've lifted Smith. 1968 'J. ROSS' *Diminished by Death* ii. 27 The youth stood. 'Am I being lifted?' 'Not at the moment. You are helping us with our inquiries.' 1972 *Times* 24 Jan. 2/1 If you have a father who is lifted, he has sons and cousins who will take his place. 1973 'J. PATRICK' *Glasgow Gang Observed* iii. 32 A fund..to raise ten pounds bail money for two of their number who had been 'lifted' the night before for fighting.

i. To evacuate (soldiers) from a beach; to air-lift. Also *transf.* Cf. *LIFT *sb.*[2] 5 h.

1941 J. MASEFIELD *Nine Days Wonder* 19 The first men lifted were not always soldiers. 1963 *Times* 24 Jan. 10/3 An emergency rail freighter service ordered by Lord Robens, chairman of the National Coal Board, is lifting thousands of tons of coal into the worst snowbound areas of south-east and south-west England. 1972 *Daily Tel.* 11 Apr. 17 Medical supplies, tents and food were being lifted in by helicopter last night.

12. a. Further examples of *Hort.* sense referring to potatoes, bulbs, etc. Also occas. *intr.*, in phr. *to lift well*, of the crops or plants concerned: to produce a good yield or be in good condition when lifted.

1888 L. CASTLE *Flower Gardening* 232 November... Lift Gladioli corms, storing them in a dry place; also Dahlia tubers. 1888 HARDY *Wessex Tales* II. 67 The next day went about his swede-lifting and storing. 1891 'H. HALIBURTON' *Ochil Idylls* 106 The dreels [of potatoes] are to lift, An' the neeps are to pu'. 1931 *Morning Post* 19 July 5/1 What to do with the bulbs at this season when, apparently, they are sleeping, has for long been a rather controversial point. Should they be left or lifted? 1959 *Times* 7 Sept. 19/2 Some crops [of potatoes] in Lincolnshire are lifting well, others are below average. 1971 'L. BLACK' *Death has Green Fingers* vii. 83 Suppose whoever it was had lifted the roses already. 1973 *Times* 20 Oct. 14/6 Nurseries..cannot lift and pack all their orders in a month.

14. lift-on, lift-off, used esp. *attrib.*, a method of hoisting containers from one vessel or vehicle to another; also **lift-on** *attrib.*; **lift-out** *attrib.*, made to lift out; **lift-up** *attrib.*, made to lift up.

1956 *Sun* (Baltimore) 16 Oct. 18/3 The relative merits of 'roll-on, roll-off' shipping, where trailers would be rolled aboard, and of 'lift-on, lift-off' service involving only a truck van. 1967 *Freight Management* Jan. 15/1 (Advt.), Last year Southampton handled thousands of containers by lift-on/lift-off. *Ibid.* 46/3 Basically roll-on is more expensive than lift-on. *Ibid.* 47/3 (*caption*) The tanks..can be used on both roll-on or lift-on vessels. 1968 *Economist* 14 Sept. p. xxxiv/1 The North Sea is now the focal point of a fight between two new forms of transport, the roll-on, roll-off ferry services..and lift-on, lift-off container services. 1969 *Jane's Freight Containers 1968–69* 73 (*caption*) Simultaneous roll-on and lift-on of trailers make possible a trip a week to Puerto Rico. *Ibid.* 160/3 Lift-on Lift-off Unitised Loads. 1926–7 *Army & Navy Stores Catal.* 314/2 These boilers are..fitted with a shaking grating and lift-out ashes pan. 1968 *Harrods Christmas Catal.* 3/4 Beauty case with inside pockets and lift-out tray. 1974 *Country Life* 14 Mar. (Suppl.) 41/1 Arm Chair with lift out seat covered in green velvet. 1917 *Installation News* Jan. 5/1 The Cabinets comprise a substantially constructed stained box, fitted with lift-up lid, lock and key. 1950 *N.Z. Jrnl. Agric.* Aug. 132/1 The lift-up gate opens by sliding up between guides fixed to uprights. 1956 *Railway Mag.* Feb. 121/1 There is a separate sheet steel case with lift-up cover containing the engineman's telephone. 1970 *Guardian* 19 Nov. 11/6 Two swing-out drawers, one with a lift-up mirror.

lifter. Add: **2. c.** *Cricket.* A ball, usu. one from a fast bowler, that rises sharply after striking the pitch.

1959 *Times* 28 July 4/7 He was caught off almost the only lifter of the day in his second [over]. 1974 *Daily Tel.* 12 June 34/1 Gavaskar got an awkward lifter from Old and gave a soft catch to the gully.

lifting, *vbl. sb.* Add: **1. b.** The raising of sick or weak cattle to enable them to stand. Cf. *LIFT *v.* 1 f. So *at the lifting*, very weak.

1812 W. SINGER *Agric. County of Dumfries* 220 They become quite lean, almost 'at the lifting', as the farmers say. 1899 H. G. GRAHAM *Social Life Scotl. 18th Cent.* I. 155 This period and this annual operation when all neighbours were summoned to carry and support the poor beasts, were known as the 'Lifting'. 1901 M. FRANKLIN *My Brilliant Career* (1966) v. 18 My mother and father and I spent the day in lifting our cows... This cow-lifting became quite a trade.

c. In competitive walking, the raising of the rear heel before the front foot touches the ground.

1867 *Athlete 1866* 119 Lifting, the usual method of walking unfairly, is done by getting a spring from the toe of one foot on to the heel of the other. 1898 F. A. COHEN in W. A. Morgan *'House' on Sport* 433 What is technically called 'lifting' is, except perhaps in a final burst, seldom of any real advantage.

2. b. lifting beam, a beam, fitted to a crane hook, to which a load may be attached in two or more places; **lifting plate** (see quot. 1888); **lifting screw,** a hook with a threaded shank which can be screwed into an object to facilitate its lifting (see also *LIFTING *ppl. a.*).

1963 R. HAMMOND *Mobile & Movable Cranes* vi. 167 Aluminium-alloy lifting beams are very useful for getting the most out of crane-lifting capacity. 1969 *Jane's Freight Containers 1968–69* 130/1 There are new lifting beams used with existing straddle cranes to handle 24 ft. and 20 ft. containers. 1888 *Lockwood's Dict. Mech. Engin.* 208 *Lifting plates*, plates of wrought or malleable cast iron furnished with holes both for rapping and screwing, and let into or screwed on the faces of patterns; and by which they are lifted from the sand, a lifting screw being inserted into the tapped hole in the plate. 1925 J. G. HORNER *Pattern Making* (ed. 5) iii. 56 Screws..either twisted into

the wood of the pattern or..fitting into corresponding tapped holes in the lifting plates attached to the pattern face, are used. 1885 —— *Pattern Making* xxii. 158 Figs. 206, 207 show two different forms of these [*sc.* rapping] plates, *a* being the plain hole for rapping, *b* the tapped hole for the reception of a lifting-screw. 1925 —— *Pattern Making* (ed. 5) iii. 59 A central hole bored through the boss for the lifting screw. 1944 E. D. HOWARD *Mod. Foundry Practice* 383/1 (Index), Lifting screw.

lifting, *ppl. a.* Add: Also, in *Aeronaut.*, providing lift; **lifting body,** a (wingless) spacecraft with a shape designed to produce lift, so that some aerodynamic control of its flight is possible within the atmosphere; **lifting screw,** a rotor operating in a horizontal plane so as to provide lift for a flying machine (see also *LIFTING *vbl. sb.*).

1902 F. WALKER *Aërial Navigation* v. 79 The lifting screws. 1908 *Jrnl. R. Soc. Arts* LVII. 53/1 The Helicoptère, or lifting-screw flying machine. 1919 H. SHAW *Text-bk. Aeronaut.* ix. 111 In some machines it is arranged that the tail carries a portion of the load, when it is known as a 'lifting' tail. 1923 *Daily Mail* 12 Feb. 7 While aloft the pilot can change the action of his planes so that they cease to act as vertical lifting-screws and function like the surfaces of an aeroplane. 1935 P. W. F. MILLS *Elem. Pract. Flying* i. 1 Aeroplanes..cannot fly backwards owing to their fixed thrust direction and the arrangement of their lifting surfaces. 1964 *Britannica Bk. of Year* 868/1 *Lifting body*, a wingless, somewhat bathtub-shaped vehicle for aerospace travel that combines some of the heat-handling capacity of a capsule with some of the maneuverability of a wingless aircraft. 1966 *McGraw-Hill Encycl. Sci. & Technol.* VI. 388/2 The disk loading [of a helicopter]..expresses the design gross weight as a function of the swept areas of the lifting rotor. 1969 K. MUNSON *Pioneer Aircraft 1903–14* 104/2 The Blériot III was also a floatplane, with annular lifting surfaces fore and aft. 1972 A. C. KERMODE *Mech. of Flight* (ed. 8) xii. 390 These lifting bodies are but a step towards a shuttle service operating to and from a space station orbiting the earth.

li·ftless, *a.* [f. LIFT *sb.*[2] 10 + -LESS.] Not provided with a lift.

1916 W. J. LOCKE *Wonderful Year* xvii. 245 She was living..on the fifth floor of a liftless block of flats in Wandsworth. 1921 *Spectator* 16 Apr. 484/2 In a liftless household.

lift-off, *a.* and *sb.* Also **liftoff, lift off.** [f. vbl. phr. *to lift off* (LIFT *v.* 1 b, *3).] **A.** *adj.* Removable by lifting.

1907 *Yesterday's Shopping* (1969) 385/3 Art cloth box, with lift off lid. 1960 *Farmer & Stockbreeder* 16 Feb. 140/1 (Advt.), Lift-off wide doors..give remarkably easy access. 1970 *Gloss. Industrial Furnace Terms* (*B.S.I.*) 9 *ft-Lioff cover furnace*, a base over which a cover is placed for heating the charge; separate bases are normally available so that the same cover can be used for several charges. 1974 *Country Life* 6 June 1500 George III Decanter Stand... The lift off tray is divided into nine crenellated compartments.

B. *sb.* **1.** *Parachuting.* A method of leaving an aircraft by opening the parachute while standing on a wing, so as to be carried away by the air current.

1930 P. WHITE *How to fly Airplane* xxii. 303 Two men are about to execute what is known as a 'lift-off' from the wings of a bombing plane. 1946 W. F. BURBIDGE *From Balloon to Bomber* iii. 46 There are two main methods of leaving aircraft. One is known as the 'lift-off' and in using this method the airman climbs on to the wing..and, by releasing his parachute, is lifted off. But this method has been superseded by the 'fall free'. 1957 *Encycl. Brit.* XVII. 252/2 In general use, there are two ways of leaving an aeroplane by parachute: (1) the drag-off or lift-off method, and (2) the jump or free-fall method.

2. [after *BLAST-OFF, *TAKE-OFF.] The vertical take-off of a rocket, helicopter, or the like; the moment at which an aircraft begins to leave the ground.

1956 in W. A. HEFLIN *U.S. Air Force Dict.* 299. 1958 *Time* 8 Dec. 15/2 With great restraint, Shotwell and his 40-man launch team quietly waited in their bunker a full seven minutes after the lift-off before they dared shout. 1961 *Aeroplane* CI. 92/1 Lift-off took place at Cape Canaveral at 07.20 hrs. local time. 1962 *New Scientist* 22 Feb. 426/3 A plate in 10-gauge aluminium weighed 6·1 lb complete with nine-inch peg and gave adequate protection for fully 50 lift-offs [of VTOL aircraft] on grass. 1962 *Engineering* 27 July 99/1 The flow characteristic..is designed to facilitate part power operation and easy 'lift off' [of hovercraft] from the surface. 1966 J. A. MORRIS *Bird Watcher* (1968) ii. 27 The launch vehicle exploded soon after lift-off. 1967 D. P. DAVIES *Handling Big Jets* vii. 177 Take care when operating in cross winds. On take-off set in a little into-wind aileron control..; this will stop the down wind roll which will otherwise occur just before lift off. 1967 *New Scientist* 16 Nov. 406/1 From lift-off at Cape Kennedy..to splash-down in the Pacific Ocean northwest of Hawaii..the entire operation appears to have been an almost flawless performance. 1970 N. ARMSTRONG et al. *First on Moon* xi. 245 If a serious malfunction should be detected, either by the men in Eagle or by the men on the ground, an immediate liftoff could be ordered.

b. *fig.* Initiation or commencement of activity; 'getting off the ground' (of a project or scheme).

1967 *Oxford Computer Explained* 7 Prior to lift-off on 1 August it was necessary that the actual stock in the warehouse be counted and the stock figures loaded to the computer. 1970 *Daily Tel.* 25 Apr. 17 (*heading*) Shell has lift-off with its space promotion. *Ibid.*, After weeks of harrowing doubt, the..'biggest, most widespread promotion the world has yet seen' achieves lift-off next week.

lig, *v.* Delete 'obs.' and add: Now freq. *colloq.*, to idle, to lie about.

Early examples of this form are cited in several sections s.v. LIE *v.*[1] A.

1960 *20th Cent.* Feb. 154 The ponce's air of having a function, an occupation..which totally distinguishes him from the mere 'ligging' layabout. 1967 *Melody Maker* 21 Jan. 6 When I was demobbed in 1960 I had no intention of going back to my trade as a fitter. I ligged around and joined Mike Peters. 1967 *Sun* 22 Feb. 6/6 *Lig, loon*, to kick one's heels or lounge about. 1969 *It* 4–17 July 10/2 It's a time for ligging in the streets and doing your thing, man.

ligamentum (li:gămeˑntə̆m). *Anat.* Pl. **-menta.** [L. *ligāmentum* band, tie, bandage.] Used in numerous mod.L. collocations to designate ligaments of the body.

1713 W. CHESELDEN *Anat. Humane Body* I. ix. 30 One large Gland..seated in a Sinus at the bottom of the Acetabulum of the Os Innominatum, which is compress'd by the Ligamentum *Teres.* 1840 G. V. ELLIS *Demonstrations Anat.* 128 The ligamentum nuchæ is a narrow ligamentous structure situated in the cervical region between the trapezius muscle of each side. 1877 W. TURNER *Introd. Human Anat.* II. vi. 335 By its circumference or ciliary border the iris is not only continuous with the ciliary processes, but is connected by fibres, termed *ligamentum pectinatum*, with the posterior elastic lamina of the cornea. 1913 *Cunningham's Text-bk. Anat.* (ed. 4) 308 The laminæ of adjoining vertebræ are bound together by the ligamenta flava (O.T. subflava).., which consist of yellow elastic fibres. 1956 *Nature* 10 Mar. 467/2 Elastic fibres from ox ligamentum nuchæ, when treated with..acetic acid solution.., are apparently devoid of collagen.

ligan, var. LAGAN. (Later examples.)

1906 *Westm. Gaz.* 13 June 4/2 These are, says Mr. Clifford, the 'ligan' of history. 1909 *Daily Chron.* 20 Mar. 5/5 The custody of flotsam, jetsam, and ligan. 1952 *Brewer's Dict. Phr. & Fable* 534/2 *Lagan*, or *Ligan*, goods thrown overboard, but marked by a buoy in order to be found again.

ligand (li·gănd). *Chem.* [f. L. *ligand-us*, gerundive of *ligāre* to bind.] **1.** Each of the atoms or groups attached to the central (usually the metal) atom of a co-ordination complex.

1952 *Jrnl. Chem. Soc.* 4757 Inferences from spectral absorption to thermochemical stability are therefore very speculative, particularly if different types of ligand, say ions and neutral dipoles, are being compared. 1964 J. W. LINNETT *Electronic Struct. Molecules* viii. 138 This implies that..the tendency to form a multiple bond is about the same for the PF_3 ligand as for CO. 1964 *Oceanogr. & Marine Biol.* II. 250 The relative accumulation factors for metals in marine organisms are related, in general, to the stability of the metal ions with ligands. 1971 *Arch. Biochem. & Biophysics* CXLVII. 226/1 At each monomer active site the two iron atoms can be bridged by two small, inorganic ligands.

2. Special Comb.: **ligand exchange,** exchange of ligands between complexes; **ligand field,** the electrostatic field produced by the ligands in the vicinity of the central atom; so *ligand field theory*, the branch of chemical theory which deals with the effect of ligands on the energy levels of the central atom or ion; *spec.* a theory based on an electrostatic model modified by molecular orbital considerations.

1964 *Jrnl. Amer. Chem. Soc.* LXXXVI. 765 (*heading*) Rates of rapid ligand exchange reactions by nuclear magnetic resonance line broadening studies. 1973 *Jrnl. Chromatogr.* LXXXVII. 513 A rapid ligand-exchange chromatographic method for the separation of α-amino acids from peptides is presented. 1956 *Nature* 18 Feb. 305/2 A proper consideration of the effect of the ligand-field explains why the six-coordinate complexes of nickel are equally stable with the cupric complexes. 1960 L. PAULING *Nature Chem. Bond* (ed. 3) v. 174 In some respects the ligand field theory is closely related..to the valence bond theory. 1966 COTTON & WILKINSON *Adv. Inorg. Chem.* (ed. 2) xxvi. 661 This modified CFT [*sc.* crystal field theory] is often called ligand field theory, LFT. However, LFT is sometimes also used as a general name for the whole gradation of theories from the electrostatic CFT to the MO [*sc.* molecular orbital] formulation. 1970 W. L. JOLLY *Synthesis & Characterization Inorg. Compounds* xxii. 324 If we were able gradually to decrease the ligand field strength to zero, we would find that each term would gradually approach an energy corresponding to one of the states of the free ion. 1971 ORCHIN & JAFFÉ *Symmetry, Orbitals & Spectra* vii. 170 Interpretation of the spectra of inorganic complexes is greatly simplified and successfully integrated by the use of ligand field theory.

So **li·gancy** = *co-ordination number* s.v. *CO-ORDINATION 5; **li·ganded** *a.*, bound to a ligand or ligands.

1960 L. PAULING *Nature Chem. Bond* (ed. 3) ii. 63 A sharp distinction is to be made between the number of atoms bonded to a central atom (the ligancy or coordination number of the central atom) and the number of covalent bonds formed by the central atom (its covalence). *Ibid.* xiii. 538 The changes from the standard sodium chloride and rutile arrangements, with ligancy 6, to cesium chloride and fluorite, respectively, with ligancy 8, are nearly the same. 1965 PHILLIPS & WILLIAMS *Inorg. Chem.* I. v. 157 The presence of shared edges and especially of shared faces in a coordinated structure decreases its stability; this effect is large for cations with large valence and small ligancy. 1967 *Jrnl. Biol. Chem.* CCXLII. 3705/2 If the $\alpha\beta$ dimer that reacts with ligand is free in solution, the now liganded dimer, $\alpha^*\beta^*$, can associate with another liganded dimer to produce a liganded tetramer, $\alpha_2{}^*\beta_2{}^*$. 1968 *Inorg. Chem.* VII. 1945/2 There are certain bonding situations common in boron chemistry and uncommon elsewhere that lead to unusual nomenclature problems. These include..excess connectivity or ligancy. 1973 *Jrnl. Molecular Biol.* LXXVI. 238 Hybrids formed in a mixture of HbS and CN met-Hb A would, upon deoxygenation, contain one liganded α chain and one liganded β chain.

ligase (li·gēⁱz, -ēⁱs). *Biochem.* [f. L. *lig-āre* to bind + *-ASE.] (See quot. 1961.)

1961 *Rep. Comm. Enzymes Internat. Union Biochem.* vi. 39 Enzymes catalysing the linking together of two molecules, coupled with the breaking of a pyrophosphate link in ATP, etc., will be known as ligases (pronounced with a short 'i' in English). These enzymes have hitherto been known as synthetases... A new systematic name was necessary. 1965 *Canad. Jrnl. Biochem.* XLIII. 1605 Succinate: CoA ligase (ADP)..catalyzes the formation of succinyl CoA. 1972 *Sci. Amer.* Jan. 31/1 The most unusual of them is an enzyme that is named polynucleotide ligase, which repairs breaks in DNA molecules.

liger (ləi·gəɹ). [f. LI(ON *sb.* + TI)GER *sb.*] The offspring of a lion and a tigress. Cf. *TIGON.

1938 *Times* 28 May 7/7 Two young Whipsnade-bred tigresses have been sent to London with the intention that they shall be paired with lions to produce the so-called ligers. 1948 *Time* 17 May 27 Daisy, a Salt Lake City zoo tigress, gave birth to the first known liger..ever born in the U.S. 1964 *Sunday Mail Mag.* (Brisbane) 8 Nov. 4/6 A 'Liger' is the term applied to the result of a cross between a lion and a tigress. 1975 *Times* 17 Sept. 7/3 The world's only living liger, the cross-breed of a male lion and a female tiger, has died at a Japanese zoo. It was the last of a litter of three born on September 8.

ligger, *sb.* Add: **2. b.** (See quots.) *dial.*

1828 W. CARR *Dial. Craven* (ed. 2) I. 289 *Ligger*.., a branch cut or laid down horizontally in a hedge. 1869 R. B. PEACOCK *Gloss. Lonsdale* 51 *Ligger*.., a branch of thorn or other tree cut half through and laid along the top of a plashed hedge. 1898 B. KIRKBY *Lakeland Words* 92 *Liggers*, long branches which a diker cuts partly through and ligs down to form a dike.

c. = *LEDGER *sb.* 2 b.

1953 J. ARNOLD *Countryman's Workshop* 160 *Liggers*, hazel strips which hold down thatch. 1965 J. G. JENKINS *Trad. Country Craftsmen* ii. 36 The method of making liggers is somewhat similar to that of spar making but the ends are bevelled for neat joining rather than pointed. 1966 *Punch* 10 Aug. facing p. 216 (Advt.), The ridge..is thickly capped with sedge grass..and isgaily embellished with diamond and herring-bone patterns of 'liggers'. 1971 *Country Life* 18 Nov. 1403/3 The finish [of a thatch] that is visible consists of cleft hazel rods called liggers about four feet long neatly secured by spars, which gives a tidy and attractive appearance.

light, *sb.* Add: **1. i.** (Later examples.)

1866 'MARK TWAIN' *Lett. from Hawaii* (1967) 152 The sick Portyghee watched his chance..harnessed the provisions and ate up nearly a quarter of a bar'l of bread before the old man caught him, and he had more than two notions to put his lights out. 1891 *Star* 10 Feb. 3/6 He had been heard to say, 'I should like to put her light out,' and had fired at her bed-room window. 1910 W. M. RAINE *Bucky O'Connor* 25 Mebbe I'd a-put his lights out for good and all. 1935 A. J. POLLOCK *Underworld Speaks* 92/2 *Put his lights out*, to kill. 1955 'A. GILBERT' *Is she Dead Too?* vii. 133 Say she put out the old girl's light, that ain't going to encourage the widower to pay his addresses to her.

2. c. Now also, to reach a full understanding or realization; to be converted (esp. to Christianity).

1812 *Niles' Reg.* III. 195/2 It is indispensably necessary that every man should 'see the light'. 1889 *Kansas City* (Missouri) *Times & Star* 14 Oct., Up to a few weeks ago, he was opposed to a revival of navigation on the Missouri, but now he has seen the light and says he's for it strong. 1903 *N.Y. Even. Post* 10 Sept., It is altogether likely that they, too, will see the light before another week has passed. 1933 H. G. WELLS *Shape of Things to Come* III. iv. 275 Men who saw the light and spoke were only one species of a larger genus of human beings whose minds worked differently from the common man's. 1944 H. JAMES et al. (*song-title*) I'm beginning to see the light. 1966 'L. LANE' *ABZ of Scouse* 94 *See ther light*, to plead guilty or to reform.

5. b. *bright lights*: see *BRIGHT *a.* 10 b; *lights out* (Mil.): the last bugle-call of the day, giving the signal for all lights to be extinguished. Hence in non-military use.

1868 *Queen's Regulations Army* §845 Between tattoo and reveille no trumpet or bugle is to be sounded,.. with the exception of the call 'lights out'. 1905 *Captain* XIII. 42/2 It's off... We aren't allowed to talk after lights-out! 1914 R. BROOKE *Let.* 3 Oct. (1968) 621 Faint lights burning through the ghostly tents, and a distant bugler blowing *Lights Out.* 1922 C. E. MONTAGUE *Disenchantment* iv. 56 They would argue after Lights Out. 1942 *R.A.F. Jrnl.* 13 June 14 There would be no lights-out time, no check-up to ensure every man was in. 1950 A. BARON *There's no Home* ii. 19 The wooden gates..could be closed every evening at Lights Out. 1965 G. JACKSON *Let.* June in *Soledad Brother* (1971) 78 One of those tall ultrabright electrical fixtures used to illuminate the walls and surrounding area at night casts a direct beam of light in my cell at night... Consequently I have enough light, even after the usual twelve o'clock lights-out, to read or study by. 1969 I. & P. OPIE *Children's*

Games viii. 246 The statues have to come to life, and do..the things they think monsters or fairies..would do... The children then commands 'Lights out'..and they have to close their eyes.

f. *out like a light* (with preceding verb or auxiliary): having lost consciousness, having fainted, or gone to sleep, at once.

1934 [see *co v. 85 g]. **1956** B. Holiday *Lady sings Blues* (1973) xix. 155 When it came time to come out for the third curtain call I said, 'Bobby, I just can't make it'... I went out like a light. **1963** A. Hunter *Gently Floating* ii. 29 They came to the bridge, were halted by lights... The lights changed. Gently drove over. **1970** N. Kenyon *100,000 Welcomes* i. 8 I'll drop you at the next lights. **1971** *Daily Tel.* (Colour Suppl.) 22 Oct. 7/2 That's right, you bumbling old fool, slow down as we come to the next lights and we'll miss the green. **1972** *Accountant* 19 Oct. 495/1 Stock markets have been in neutral waiting for the lights to change.

g. Usu. *pl.* Traffic lights. Also *fig.*

1938 E. Bowen *Death of Heart* III. vi. 439 The driver twitched his head once or twice. Then the lights went against him; he pulled up.

6. a. *to throw light upon* (earlier example).

1841 Carlyle *On Heroes* v. 209 When he did speak, it was to throw new light on the matter.

e. (Earlier and later examples.)

1854 Mrs. Gaskell *Company Manners* in *Househ. Words* 20 May 330/1 Why have we not other intellectual games of some kind. Wit, Advice, Bout-rimés, Lights... every one knows these. If they would only not think it beneath them to be called upon..to play at them. **1903** H. G. Wells *Brynhild* vii. 208 Valiant Chevell was generally the director of his set scenes [in a charade], but the direction of the first light was taken out of his hands. **1945** H. Phillips *Word Play* xvi. 84 It is permissible to play tricks of this kind with the Lights—beheading or curtailing the words.

f. The answer to a clue in a crossword puzzle.

1925 'Torquemada' *Cross-Words in Rhyme* Introd. Those who wish a separate entertainment..from each Light in their cross-words. **1965** *Listener* 16 Sept. 425/1 Some of the clues are two lines of verse, each by a different author. The names of the two authors have three or more consecutive letters in common and these letters form the light. **1967** *Sci. Amer.* Sept. 268/2 The horizontal words..are called the cross-lights or simply the lights.

9. a. (Further examples.)

1887 *Lantern* (New Orleans) 7 May 3/1 Some of the leading lights of the National League. **1915** T. Dreiser *Genius* II. xl. 469 What Eugene thought and what White thought of this prospective situation was that the other two would be found wanting, and that the under Colfax would be the shining light. **1944** Berrey & Van den Bark *Amer. Thes. Slang* § 388/4 Principal or most important person..leading card or light. **1945** K. Tennant *Ride on Stranger* xvi. 180 An eminent legal light. **1974** E. Ambler *Dr. Frigo* III. 240 The procession could..move off. I was among the least of the lesser lights and so was among the first out.

12. *fig.* (Later examples.) Usu. opp. to *shade*.

1812 *Dramatic Censor 1811* 182 This may be what our modern playmakers call *light and shade*. **1937** *Printers' Ink Monthly* May 29/1 Light and shade, variations from quietness to tenseness, softness to shouting and which has a tendency to keep a production from a dull sameness. **1952** Granville *Dict. Theatr. Terms* 110 *Light and shade*, the niceties of intonation, inflection, modulation, etc., in the reading of a part.

15. a. *light-effect, -output, -ray, -scatter, -signal, -socket, -song, -source, -switch*; **b.** (objective) *light-absorber, -absorbing, -absorptive, -avoiding, -emitting, -gathering, -loving, -passing, -producing, -reflecting, -reflective, -refracting, -throwing* adjs.; (instrumental, etc.) *light-actuated, -sensitive, -stilled* adjs.

1957 *Technology* Dec. 361/2 Light absorbers for use in products affected by ultra-violet radiations from the sun. **1967** E. Chambers *Photolitho-Offset* vii. 85 The term density refers to the light-absorbing ability of the [silver] layer. **1965** E. R. A. Higham *Handbk. Papermaking* viii. 210 Opacity is dependent on the number of light-absorptive or light-reflective fibre surfaces in a sheet. **1936** *Discovery* Nov. 355/1 Light-actuated apparatus for home use is now on the market. **1924** J. A. Thomson *Sci. Old & New* xxvi. 141 The Fierasfer..is a light-avoiding fish. **1965** B. E. Freeman tr. *Vandel's Biospeleol.* iv. 39 The light-avoiding planarians are simple to keep in captivity. **1902** *Westm. Gaz.* 29 Sept. 3/1 The energy required for producing pendulous movements of atoms and molecules giving light-effects must be very small as compared with the total energy employed. **1962** R. G. Haggar *Dict. Art Terms* 192/1 J. M. W. Turner..carried research into light effects further than any previous artist. **1964** *Oceanogr. & Marine Biol.* II. 351 The decrease in light-emitting capacity of a methanol solution of luciferin. **1869** *Chambers's Jrnl.* 10 Apr. 231/1 Under the high power and vast light-gathering capacity of Sir W. Herschel's four-foot reflector. **1960** *Farmer & Stockbreeder* 8 Mar. 134 The Meopta 12 × 60 has the rare combination of high magnification and brilliant light-gathering power even at night and under bad climatic conditions. **1895** J. H. & A. Comstock *Man. Study Insects* xxi. 562 The Light-loving Anomala, *Anomala lucicola*..also feeds on the leaves of grape. **1974** A. Huxley *Plant & Planet* viii. 99 During the twelve-hour period of the average night..the plant is regarded as 'dark-loving', while in the other twelve-hour period it is 'light-loving'. **1950** *Sci. News* XV. 43 The brightnesses thus catalogued are, however, only apparent... So, in order to compare the light-outputs of the stars, we introduce the idea of Absolute Magnitude.

1958 *Newnes Compl. Amat. Photogr.* 120 Recent developments have been in the direction of maintaining high efficiency and light-output operating at lower voltages. **1961** G. Millerson *Technique Television Production* iii. 38 Construction differences [in lenses]..can vary their respective light-passing abilities, although their stop numbers may be measurably identical. **1845** *Harmony of Comprehensible World* Essay II. xii. 221 Between some bodies there may be no light-producing sympathy, because the internal vibrations of their constituent molecules may not be such as to develop light. **1964** V. B. Wigglesworth *Life of Insects* viii. 126 It is among the insects that some of the most brilliant and certainly the most complete types of light-producing organs are to be found. **1880** 'Mark Twain' *Tramp Abroad* I. xvi. 149 Of Light was the Figure wove. **1950** *Sci. News* XV. 130 If light rays cannot bring about a photo-chemical change unless they are absorbed. **1854** Geo. Eliot tr. *Feuerbach's Essence Christianity* v. 51 Tears are the light-reflecting drops which mirror the nature of the Christian's God. **1951** S. Spender *World within World* iii. 160 Their minds like little caves of calculating darkness which the light-reflecting stone was never penetrated. **1963** Light-reflective [see *light-absorbent* above]. **1889** E. Carpenter *Civilization* 81 It [sc. modern science] takes the emerald, and breaks it up; treats of its color and light-refracting qualities. **1957** Partridge *English gone Wrong* ii. 29 The light-refracting heads of the Communist philosophers and propagandists. **1958** *A.M.A. Arch. Industr. Health* XVIII. 203 A plot of the light-scatter decay was divided into exponential portions by a slope-analysis method. **1962** G. Millerson *Technique Television Production* iii. 45 Lens..spaces, there are several obvious causes for lack of clarity: dirty lenses, light-scatter in the lens. **1936** *Discovery* May 151/1 It was not until the appearance of a new type of light-sensitive cell, known as the rectifier or semi-conducting cell, that photo-electric exposure meters became popular. **1964** *Sun* 18 Sept. 454/2 V. Ivanko also contains strains which have light-sensitive seeds. **1962** *Science Survey* III. 46 The retina is light-sensitive because it contains one or more photosensitive pigments located in its visual receptors. **1920** A. S. Eddington *Space, Time & Gravitation* iii. 50 It would, in fact, be possible for an observer travelling along KP to receive a light-signal..announcing the event O, just as he reached P. **1930** *Morning Post* 19 July 13/6 An extension of the system of light signals for road traffic. **1964** *Amer. Jrnl. Physics* XXXII. 262/2 At an arbitrary instant t..a light signal S is emitted at the origin of the coordinate system. **1960** H. Pinter *Caretaker* ii. 48 There used to be a wall plug..but it doesn't work. I had to fit it in the light socket. **1935** A. H. Haffenden (*title*) Light-song. **1946** L. B. Lyon *Hough Walk Hame* 27 Our anguish has a hand, that gropes For the light-song of the sun. **1963** *Edin. Rev.* July 113 Because a spectrum line changes with change of velocity of light source and other disturbing causes, the value of its record is thereby increased. **1961** G. Millerson *Technique Television Production* iii. 48 Occasionally on static captions, and visible light-sources in the scene, are the worst offenders. **1928** W. de la Mare *Memory* 81 Peace beyond telling share with the light-stilled eye. **1892** F. C. Allsop *Pract. Electr.-Light Fitting* iii. 39 Lamp or branch switches are designated either by the number of lamps they are intended to control, or..by their current-carrying capacity. They are thus called 1, 2, 3 light, or 1, 2, or 3 ampere switches. **1926** G. Hunting *Vicarion* iv. 63 He went back to his light-switch, closed the closet door which stood ajar, and brought his chair toward them again. **1972** H. Carmichael *Naked to Grave* i. 14 He heard the click of a light switch in the bedroom. **1894** 'Mark Twain' in *Century Mag.* Jan. 336 He asked questions that would have brought light-throwing answers. **1902** *Westm. Gaz.* 1 July 2/1 An excellent translation of a light-throwing and thought-provoking book.

16. light-adaptation, self-adjustment of the eye to increased intensity of light by means of a decrease in the sensitivity of the retina; also, in extended use, any reversible change in an organism that occurs in response to increased light; so **light-adapted** *pa. pple.* and *ppl. a.*, in the state that results from light-adaptation; **light barrier**, (*a*) a limit to the resolution possible with an optical microscope arising out of the finite length of light waves (*nonce-use*); (*b*) the speed of light as the limiting speed attainable by any object; **light-box**, (*a*) (earlier example); (*c*) a box-like piece of equipment containing a light and usu. having translucent glass on one side which provides an evenly lighted surface; **light bucket** *Astr.* (*colloq.*), a telescope, regarded as a device for collecting and focusing a large quantity of low-intensity radiation; **light bulb** = *Bulb *sb.* 4; **light-buoy**, a buoy equipped with a warning light which flashes intermittently; **light button**, a knob or disc which, when pressed, turns a light on or off; **light-change** *Astr.*, a change in the amount of light received from a variable star; **light check** *Theatr.* (see quot. 1952); **light cone** *Physics*, a surface in space-time which appears conical when represented in three dimensions and comprises all the world-points from which a light signal would reach a given point (defining the apex) simultaneously (and which therefore appear simultaneous to an observer at the apex); **light cord**, a cord which hangs from a ceiling or lamp stand and operates an electric light when pulled; **light cue**, (*a*) *Broadcasting*, a cue indicated by a light being switched on; (*b*) *Theatr.* (see quot. 1961); **light-cure** *rare* or *Obs.*, a cure effected by sunlight or artificial light; also *attrib.*; **light curve** *Astr.*, a graph showing the variation in the light received over a period of time from a variable star or other heavenly body; **light-demander**, a tree that will not tolerate shade; so **light-demanding** *a.*, of trees or, occas., other plants, needing full light; **light-fastness**, resistance to discoloration by light; so **light-fast** *a.*; **light-filter** *Photogr.* = *colour-filter* (see *colour *sb.* 18); **light-fixture**, the flex, socket, and other equipment which is used with a light bulb; **light fog** *Photogr.* (see quot. 1940 and Fog *sb.*² 4); **light-grasp** *Astr.*, light-gathering power (of a telescope); **light guide**, a cylinder or strip of transparent material, or a bundle of them, along which light can travel with little loss, by means of total internal reflection; **light gun** = *light pen*; **light meter**, an instrument for measuring the intensity of light; *esp.* an exposure meter; **light microscope**, a conventional microscope, in which ordinary light is transmitted to the specimen; **light organ**, in luminescent animals, the structure emitting light; **light pen**, a hand-held, pen-like device that incorporates a lens, photoelectric cell, and amplifier and may be used to feed information by wire to a data-processing system by placing or moving the tip on the screen of a cathode-ray tube or other surface so that electrical impulses are transmitted to the system; **light pipe** = *light guide*; **light-pressure**, pressure exerted on a body by light incident on it; **light quantum** *Physics* = *photon*; **light-scattering**, scattering of light, *spec.* of monochromatic light by a solution as a method of determining the molecular weight of dissolved polymers and investigating their conformation; **light-sensation**, in the study of visual perception, the sensation produced by light; **light show**, a display of changing coloured lights or varied film strips, freq. accompanying popular music; also *attrib.*; **light-stand**, a stand to support a light; **light station**, a group of buildings which includes a lighthouse and associated buildings for housing personnel, supplies, and equipment; **light-tight** *a.* (further examples); **light-time** *Astr.*, the time taken by light to travel from a distant source to the observer; **light trap**, (*a*) *Photogr.*, a device for excluding light from a room or other space without preventing access into it; (*b*) a device for attracting, catching, and sometimes killing night-flying insects; so **light-trapped** *a.*, provided with a light trap; **light value** *Photogr.*, a number representing on an arbitrary scale the intensity of light from a particular direction; *light-value shutter*, a shutter having the aperture and shutter speed settings linked so that they can be altered together in such a way as to keep the amount of light admitted during an exposure constant; **light valve**, a device which regulates the amount of light passing through it, esp. one whose transmission can be varied by an electrical signal; **light-well**, a shaft designed to admit light from above into inner rooms or a staircase of a building; **light-year**, add: it is approximately equal to $9{\cdot}46 \times 10^{12}$ km. ($5{\cdot}87 \times 10^{12}$ miles); also *fig.*; (further examples).

1900 W. H. Rivers in E. A. Schäfer *Textbk. Physiol.* II. 1060 If the eye remained in a condition of light-adaptation, red and blue..became gradually blacker. **1962** H. C. Weston *Sight, Light & Work* (ed. 2) i. 8 Thus, after full light-adaptation, complete dark-adaptation may require an hour. **1964** *Oceanogr. & Marine Biol.* II. 352 Prolonged laboratory culture, temperature, and light- or dark-adaptation had relatively little effect on luminescent activity. **1900** W. H. Rivers in E. A. Schäfer *Textbk. Physiol.* II. 1073 He found that in complete dark-adaptation the recurrent image followed the original immediately and was brighter than to the light-adapted eye. **1935** *Discovery* May 138/1 A source of light which appears only..to the light-adapted eye, that is to one coming in from daylight, is quite obvious to a dark-adapted eye. **1950** *Sci. News* XV. 25 A [spontaneous] change in fixation direction is quite possible, particularly when the eye is not fully light-adapted or where there is too large an object for precise fixation. **1959** *Listener* 31 Dec. 1161/1 When one gets down to sizes round about the wavelength of light..one runs into a barrier, which might be called the light barrier, that no microscope working by means of light can break through. **1964** M. McLuhan *Understanding Media* (1967) I. vi. 68 No further acceleration is possible this side of the light barrier. **1968** A. Diment *Great Spy Race* x. 180 The faster than light spaceships will bring the stars down into our backyard, for once one has broken the so-called 'light-barrier' there is no limit to speed. **1849** Thackeray

Pendennis I. xix. 173 Helen..went for a light-box and his cigar-case. **1940** J. O. Kraehenbuehl *Electr. Illumination* vii. 108/2 The light boxes commonly used may be divided into two classes: those which are covered with some form of transmitting medium which is translucent... **1957** *Screen Printer & Display Producer* July 16/3 Pin-holes were spotted out over a light box. **1962** H. C. Weston *Sight, Light & Work* (ed. 2) vi. 195 These devices consist of a light-box of suitable size, the cover-glass of which allows the emission of light in a regular pattern. **1968** *New Scientist* 31 Oct. 260/1 One piece of equipment is a 34-ft 'light bucket' for seeking out point sources of gamma rays in the universe. **1970** *Nature* 7 Feb. 492/2 Infrared telescopes, more properly called flux collectors—light buckets in the language of astronomy—are cheap compared with similar equipment for the visible spectrum. **1884** Light bulb [see Bulb *sb.* 4]. **1946** E. Hodgins *Mr. Blandings builds his Dream House* (1947) i. v. 78 The cost of your house doesn't get you moved into it with light bulbs in all the sockets. **1973** N. Kenyon *Mr Big* xix. 185 His muscled black tangled limbs trailing flex and popping lightbulbs from the overhead fixtures. **1894** W. Le Queux *Gt. War in Eng. in 1897* xxii. 236 A cruiser..was lying near the Hernelt light-buoy. **1930** W. de la Mare *Desert Islands* 99 Lightship or beacon or winking light-buoy. **1972** The older light buoys exhibit light day and night; but they are gradually being superseded by buoys which automatically light up at sunset and extinguish themselves at dawn. **1929** D. Hammett *Dain Curse* x. 95 [My] hand touched the light button. I had seen enough to light it. Light scorched my eyes. **1902** R. Busby *Frightened* xvii. 172 The time-switch light-button on the wall. **1890** A. M. Clerke *Syst. Stars* ix. 139 The light-change of S. Cancri, the second of the Algol variables, was discovered by Mr. Hind in 1848. **1928** *Sci. Amer.* Sept. 12/2 Algol..has a secondary eclipse of the light-change. **1933** P. Godfrey *Back-Stage* 145 'Light checks' are any alterations to the opening lighting of a scene. **1952** Granville *Dict. Theatr. Terms* 110 *Light check*, a dimming of lights. **1922** E. P. Adams tr. *Einstein's Meaning of Relativity* iii. 91 ...the 'light-cone'. **1964** A. O. Barut *Electrodynamics* i. 8 All time-like vectors lie inside the light cone and the space-like ones are outside. **1964** *Listener* 17 Dec. 976/2 Encouragement also comes from the usual diagrams in physics text books representing 'light cones'. **1929** *Radio Times* 5 Nov. 389/1 They will sit at the [control] panel, flashing 'light cues', fading and cross-fading studios. **1960** J. Hartmann *Theatre Lighting* iii. 37 Light cues are written down during the progress of a rehearsal. **1961** Bowman & Ball *Theatre Lang.* 201 *Light cue*,...the cue for the commencement of some planned change in illumination. **1901** *Chambers's Jrnl.* Dec. 644/2 Hospitals..have already obtained apparatus for the light-cure of lupus. **1904** *Daily Chron.* 17 Apr. 5/3 'Yesterday morning King Edward..paid a lengthy visit to Professor Finsen's light-cure institution. **1890** A. M. Clerke *Syst. Stars* viii. 125 The light curve of [U Geminorum] takes more or less the form of a double peak with a saddle between. **1965** *Astrophysical Jrnl.* CXXIII. 11 The light-curve for the 1952 eclipse, as measured in the Sudan, is much flatter than the curves..from the 1947 observations. **1968** *Project Icarus* (Mass. Inst. Technol.) i. 7 The rate of rotation of an asteroid and the axis of its rotation can be found approximately by careful analysis of the shape and variation of its light curve. **1975** *Sci. Amer.* Mar. 26/3 The X-ray light curve of Centaurus X-3 is the curve of a typical eclipsing binary system. **1892** W. Schlich *Man. Forestry* II. v. 308 As regards light-requirement it [sc. the Weymouth Pine] stands half-way between light-demanders and shade-bearers. **1928** E. S. Troup *Silvicultural Syst.* v. 67 By the gradual opening..is applied to strong light-demanders, larger gaps would be necessary. **1966** *Times* 21 Apr. 16/6 Some trees are such emphatic light-demanders..that they will not thrive if there is any overhead shade. **1889** W. Schlich *Man. Forestry* I. ii. 117 Certain species [which] throughout life..are called 'light demanding'. **1952** H. L. Edlin *Forester's Handbk.* viii. 113 Trees described as light-demanding will only succeed if grown in the open. **1964** *Oceanogr. & Marine Biol.* II. 101 In its [sc. Ernst's] opinion *Ulothrix* is a light-demanding alga. **1957** M. Bohmer-Fashion *Textile Terms* (ed. 1) 231/1 Lightfast. **1961** *Jrnl. Oil & Colour Chemists' Assoc.* XLIV. 847 Bright red paints based on cadmium sulpho-selenide pigments, having relatively high light fastness. **1963** E. P. Pellew *Dyes & Dyeing* III. 48 Light fastness is usually made by partially covering a dyed skein with a piece of wood..and exposing it to direct sunlight. **1959** *B.S.I. News* June 23/2 The colour, colour-strength, transparency and light-fastness of these inks in terms of comparison with master standard inks. **1962** J. T. Marsh *Self-Smoothing Fabrics* vi. 75 Dimethylol dihydroxy ethylene urea appears to be quite outstanding in its effect on the light-fastness of dyed goods in general and those containing reactive dyes in particular. **1971** *Jrnl. Oil & Colour Chemists' Assoc.* LIV. 857 Better light-fastness of pigments, non-yellowing light-stable resins. **1901** *Chambers's Jrnl.* June 367/2 For use either in orthochromatic or colour photography, light-filters..are now commercially produced. **1958** *Newnes Compl. Amat. Photogr.* iv. 83 The smaller increases in exposure needed when light-filters are employed. **1923** *T. Eaton & Co. Catal.* Spring & Summer 357/5 Light fixture, for dining-room or living-room. **1939** D. Parker *Here Lies* 27 He bought..storm-windows, and light-fixtures. **1889** E. J. Wall *Dict. Photogr.* 77 *Light fog* makes its appearance generally all over the plate. **1915** *Photo-Era* XXXV. 170/1 Plate and films must be loaded..with the utmost care to avoid light-fog. **1940** *Chambers's Techn. Dict.* 449/1 *Light-fog* (Photog.), fog in an emulsion, caused by intrusion of extraneous light into a camera or other apparatus which is intended to be light-tight. **1946** *Nature* 6 July

18/1 Wood..used the instrument in his charge for those types of astronomical observation for which it was eminently suitable by virtue of its short focal-length, large field of good definition and powerful light-grasp. **1961** *Listener* 7 Sept. 353/1 For televising relatively faint objects, such as planets, it is necessary to use a powerful telescope with considerable light-grasp. **1951** *Jrnl. Sci. Instrum.* XXVIII. 188/1 (*heading*) A divided light guide for coincidence counting of scintillations due to alpha particles. *Ibid.*, A forked light guide was constructed from..Perspex rod. **1972** *Science* 9 June 1138/1 Luminescence was detected through a fiber-optic light guide. **1970** O. Dopping *Computers & Data Processing* xi. 179 An extension of the CRT terminal is the light pen, or light gun, which can be used for identifying details in the picture displayed by the computer and for making sketches which the computer can record. **1972** *Computers & Humanities* VII. 5 With the use of a light gun the linguist can select from alternative expressions in phrase structure trees. **1922** *Gen. Jrnl.* CLVI. 563/2 Mr. Haydn T. Harrison next interested the members with a description of the Benjamin 'Lightmeter', which is a simple portable apparatus to measure illumination, and enable one to give intelligent and expert advice on factory lighting. **1943** D. Baker *Trio* ii. 92 A light-meter on a cord, some photographic boxes, an oversized black camera. **1960** A. Bronowski *Take one Ambassador* viii. 121 Peering at his light-meter reading on his Asahi Pentax. **1941** Light microscope [see *electron microscope* s.v. *Electron¹ 2 b]. **1962** *Lancet* 5 Aug. 291/1 There are great difficulties in interpreting the images of these small chromosomes because they are almost at the limit of light-microscope resolution. **1896** D. Sharp in *Camb. Nat. Hist.* VI. 279 The structure of the light organs (of *Pyrophorus*) is described by Heinemann. **1965** Russell & Yonge *Seas* 192 Some of these cuttlefish from the deep have over twenty light organs in various parts of the body. **1954** N. B. Marshall *Aspects Deep Sea Biol.* xi. 273 May not some of the light organs which stud the body [of certain fishes] also attract prey? **1969** R. F. Chapman *Insects* In most beetles the light organs are ventral. **1963** *Proc. IRE* XLVI. 1213/1 Hardware semi-permanent photodiodes have been fabricated with intrinsic response times of less than 75 nanoseconds. They have been used with success in many applications among which are..direct..detection or a transistorized 'light pen' for high-speed oscilloscope readout. **1964** *Discovery* Oct. 35/1 (*caption*) Display console of a computer which illustrates actual graphs, characters and drawings stored within the machine in digital form. The operator can make corrections to the display with a 'light pen' which automatically corrects the stored information. **1966** *Sci. Amer.* Sept. 92/2 The stylus-photocell arrangement could be used..to make the cathode-ray-tube display serve for other purposes than making graphs and diagrams. **1973** *Courier & Advertiser* (Dundee) 21 Feb. Light pens will 'read' information. **1966** E. Page *Family & Friends* viii. 124 He pulled at the light cord, and the light went off. **1962** *Times* 6 Apr. 13/5 Light pipes..used to carry light round corners. **1908** *Westm. Gaz.* 23 Oct. 5/3 There is also a small light-pressure effect. **1908** *Encycl. Brit.* Index, Light-pressure. **1931** A. E. Lyttleton *Mysteries Solar Syst.* v. 178 With comminution of cometary particles occurring mainly on the perihelion passages of the comet, the light-pressure will be effective. **1925** R. W. Lawson tr. *Hevesy & Paneth's Man. Radioactivity* (ed. 2) ix. 105 Converted into light quanta. **1948** *Sci. News* VI. 75 In the quantum theory a light signal cannot be subdivided indefinitely. **1939** W. Heitler *Quantum Theory of Radiation* xxxiv. 324 This phenomenon is called light-scattering. **1938** R. W. Lawson tr. *Hevesy & Paneth's Man. Radioactivity* (ed. 2) ix. 105 It follows from the modern 'Quantum Theory' that there are 'smallest possible' amounts of light, which we might call..light-quanta. **1948** *Sci. News* VI. 75 In the quantum theory the electromagnetic field is not a continuous medium but consists of a great number of light quanta, or 'photons'. **1974** *Sci. Amer.* Oct. 68/1 Carbohydrates are the direct result of the photosynthetic activity of green plants... The energy needed to promote this reaction is provided by light quanta from the sun. **1966** H. C. Macpherson *Mod. Astron.* iv. 94 Dr. Wright, photographing Mars,..concluded that the Martian atmosphere was at least 120 miles in depth and possessed appreciable absorbing and light-scattering power. **1935** *Trans. Faraday Soc.* XXXI. 1324 We may therefore conclude that the main factor in the light-scattering of an isotropic protein is the molecular weight of the protein and that its scattering power is a true measure of its molecular dimensions. **1965** Peacocke & Drysdale *Molecular Basis Heredity* iv. 34 The hydrodynamic and light-scattering measurements both indicate that in solution its configuration is that of a stiffened coil, rather than that of a rigid rod or of a completely random coil. **1972** Billingham & Jenkins in A. D. Jenkins *Polymer Sci.* I. ii. 147 Despite the complexity and expense of the technique, light scattering remains one of the most useful techniques for the determination of weight average molecular weights of polymers. **1895** E. B. Titchener in *Amer. Jrnl. Psychol.* VII. 82 *Lichtempfindung*, light sensation. **1914** Williams & Waterlow tr. *Mach's Analysis of Sensations* x. 211 The habit of..giving attention to a large and spatially cohering mass of light-sensations. **1924** R. M. Ogden tr. *Koffka's Growth of Mind* iii. § 13. 134 The most varied light-, dark-, and colour-sensations. **1937** *Discovery* July 216/1 The nature of light-sensation, colour-tone, colour-blindness. **1966** E. Denson in *Berkeley Barb* 1 Apr. 4/1 Led by Tony Martin's light show, which fills the

huge wall behind the bands and their 30 foot row of amplifiers and electronics with red shapes shifting in time to the music, the hall is filled with swaying, writhing people. **1967** *Rampart* 9 Mar. 12/1 The light show atmospheric techniques of projecting slides and wild colors on the walls during rock dances. **1969** *It* 11–24 Apr. 13/1 It is regarded that lightshows began when the 'underground' or 'psychedelic' revolutions began, then it is doubtful that lightshows will ever recover from the damage inflicted during the capitalists' rape of those movements. **1971** E. E. Lamby *Underground Dict.* 120 Light shows are given in auditoriums, coffeehouses, etc. They are put on for the purpose of simulating a hallucinogenic experience. **1836** N. P. Willis *Inklings of Adventure* I. 208 In another moment the light stand was swept from between us, and he struck me down with a blow that nearly felled me a giant. **1867** A. D. Whitney *Summer in L. Goldthwaite's Life* vi. 119 On this little green stand, a round white-draped light-stand with her work-basket and a few books. **1966** *Aids to Navigation Manual* (U.S. Coast Guard) xxix. 3/1 The mission of a light station is to serve..as an aid to navigation..a light on a fixed structure. **1956** *Navigation Dict.* (U.S. Naval Oceanogr. Office) 124/1 *Light station*, a group of buildings including a lighthouse and additional buildings housing personnel, fog signal, radiobeacon, and any other equipment associated with the lighthouse. **1969** *Islander* (Victoria, B.C.) 21 Dec. 16/1 It was December 1934 at Pachena Point, a lonely lightstation on Vancouver Island's stormy west coast. **1971** *Bahamian Rev.* Nov. 5/1 Mrs. Pierre grew up on light stations, and her father was a light-keeper. **1911** T. E. Lawrence *Let.* 31 Mar. (1938) 101 How to render light-tight a dark-room. **1942** *R.A.F. Jrnl.* 2 May 13 One of the..operators had just completed a spool, and my print was ready to leave the light-tight box to get it into the light-tight cane. **1970** *Jrnl. General Psychol.* LXXXII. 208 Behind the opening was a white booklet and a 12 volt d.c. light, both enclosed in a light-tight housing. **1920** A. S. Eddington *Space, Time & Gravitation* iv. 59 Then you must know the speed of light in order to know what light-time to allow. **1959** The light-time being determined by the time taken by the light from the 'moment of..' **1953** *Astrophysical Jrnl.* CXVI. 211 The problem of the determination of the light-time orbit will occur with increasing frequency as the observational data become more accurate and extend over long periods of time. **1868** P. K. Escombe *Methods Hydrodynamics* xi. 185 (*heading*) Light traps. **1969** G. R. C. Bayley *Compl. Photographer* ix. 99 Many otherwise efficient ventilating systems are rendered almost useless by the light trap. **1931** A. D. Imms *Recent Adv. Entomol.* vi. 141 In many countries practical entomologists have made extensive use of light-traps as a means for the quantitative attraction and destruction of noxious species of Lepidoptera. **1939** W. L. & W. M. Miles *Insect Pests Glasshouse Crops* iii. 34 Light-traps also attract the moths and might be used with advantage in cases of persistent infestation. **1965** M. J. Langford *Basic Photogr.* xv. 286 Light traps..are designed to allow entry or exit of staff without introducing light, some form of 'light trap' is essential. **1973** *Entomologist's Rec.* LXXXV. 95 On the night of May 19th I had an unusual, and yet unfortunate, boss of moths in my light trap. **1956** *Focal Encycl. Photogr.* 677/1 Many darkrooms..have light-trapped entrances so that the staff can pass freely in and out without admitting light. **1958** *Newnes Compl. Amat. Photogr.* iv. 75 The leading end of the film projects through a light-trapped slot on the side of the camera. **1956** *Focal Encycl. Photogr.* 680/2 Exposure values, as used on shutters, are also frequently referred to as light values. **1957** T. L. J. Bentley *Man. Miniature Camera* 161 The latest development in the between-lens type of shutter has come the so-called light-value shutter. **1958** *Newnes Compl. Amat. Photogr.* ii. 38 The light-value shutter is a modern device..designed to make speed and aperture setting more easy, and making use of the light value system. **1970** *Which?* June 186/2 You then point the meter at the subject and the needle will move along the light value scale. ..On some meters, instead of transferring the light value from one scale to another, you move a pointer until it overlaps the needle. **1928** *Trans. Soc. Motion Pict. Engin.* XII. 731 (*heading*) The light valve. **1930** *Discovery* July 234/1 There light valves [each a specially designed form of Kerr cell] modulated the beams from the arcs. **1971** E. P. Harris *Basic Motion Pict. Technol.* v. 165 In variable density recording the intensity of illumination passing into the lens system from a lamp and condenser lens is modulated by a light valve consisting of a pair of narrow metal ribbons suspended under tension in a magnetic field at right angles to the direction of the film movement. **1925** V. G. Childe *Dawn European Civilization* v. 82 The palace was really composed of several apartments..grouped round light-wells. **1958** *Listener* 23 Oct. 644/1 The nineteenth-century office block, with the quiet internal lawn shrunk to the scale of the light-well. **1949** A. Huxley *Let.* 26 Feb. (1969) 593 Hubble..showed us the first pictures taken by the 200-inch..light-years distant. **1956** The selected..the nebulae went on with uniform density to a billion lightyears. **1957** T. Aubrey *Nation* 22 Transferred a ship and all it contained from one point in space to another, light-years away. **1962** F. I. Ordway et al. *Basic Astronautics* vi. 289 (*caption*) Known stars within five parsecs (16 light years) of the Sun. **1971** *Guardian* 22 July 11/4 Professor Peter Hungerford..said..abortions should be the decision of the mother alone. This is light years from FPA policy. **1973** A. Holden *Girl on Beach* 143 He really is..a spare-time amateur art critic, light-years removed from a creative artist.

b. *Astr.* Combs. modelled on *light-year*, denoting the distance travelled by light in the time specified; so *light-day, -minute*, etc.

1923 G. D. Birkhoff *Relativity & Mod. Physics* ii. 20 Since it required $(t_2 - t_1)/2$ seconds for the light to travel from the one particle to the other, *B* must have been at a distance of $x = (t_2 - t_1)/2$ 'light-seconds' from *A* at the time *t*. **1925** D. L. Thomson in J. A. Thomson *Sci. & Relig.* 215 A light year is over five million million miles, and the sun is only eight-and-a-half light minutes from the earth. **1963**

Nature 18 May 651/2 If the flashes are real, either the optical source itself is of the order of light-days in size, or ..it must contain substructures of this scale. **1964** *Astrophysical Jrnl.* CXL. 15 Consider a region one light-month. i.e., 7×10^{16} cm, in radius. *Ibid.*, A maximum flash duration of only a few hours is possible for a region a light-month in radius. **1970** *Sci. Amer.* Dec. 25/3 The rapid changes in flux imply that if quasi-stellar objects are as remote as their red shifts indicate, they must have diameters reckoned in light-months, or even less. This means that such objects are on a scale only slightly larger than that of our solar system, which is about one light-day in diameter.

light, *a.*[1] Add: **2. b.** Applied to elements whose specific gravity (or atomic number) is relatively low; *light metal*, a metal of low specific gravity, esp. aluminium or magnesium; so *light alloy*, an alloy based on such a metal.

1912 Rosenhain & Archbutt in *Proc. Inst. Mech. Engin.* Apr. 323 It was decided in the first place to confine the investigations to alloys consisting principally of aluminium, which may be conveniently grouped under the term 'light alloys'. **1924** *Proc. Physical Soc.* XXXVI. 418 The other light elements, hydrogen, helium, lithium, carbon and oxygen gave no detectable effect beyond 7 cm. **1926** *Industr. & Engin. Chem.* Oct. 1016/1 The production of the light metals has only been rendered possible by the comparatively recent work of chemists and chemical engineers. **1936** R. P. Bell tr. *Bjerrum's Inorg. Chem.* 213 The metals fall naturally into two groups: the light metals with densities below four, and the heavy metals with densities above seven. The light metals are the most electropositive, i.e., they have a specially great tendency to form positive ions. *Ibid.*, The light metals react readily with many substances. **1948** 'N. Shute' *No Highway* i. 12, I couldn't find anything about light alloy structures in fatigue prior to the year 1927. **1949** A. J. Field tr. *A. von Zeerleder's Technol. Light Metals* i. 1 There is at the present time no standard value of this property [*sc.* specific gravity] recognized as a qualification for the title 'light metal'. **1959** *Times Rev. Industry* Apr. 55/1 Re-organization within the light-metal industries. **1962** *Appl. Spectrosc.* XVI. 162/1 The data..show the great advantage of using a chromium target tube for light element analysis. *Ibid.* 159/1 Light elements are defined as those elements having an atomic number less than 25. **1969** *Jane's Freight Containers 1968–69* 533/1 Containers: non-standard collapsible light-alloy.

4. *light line* = *light water-line* (see Light *a.*[1] 4).

1894 W. H. White *Man. Naval Archit.* (ed. 3) 47 The displacement of a ship between her light and load lines could be estimated, and would give the true 'dead-weight capability'. **1923** *Man. Seamanship* (Admiralty) II. 270 The portion of the ship's bottom, between the light and loadline, termed the *tapboot*, is difficult to protect from corrosion. **1948** R. de Kerchove *Internat. Maritime Dict.* 416/2 *Light line*, the line of immersion at which a vessel floats when in ballast draft or light trim.

c. *light industry*: an industry making use of relatively light and therefore easily handled materials. (Cf. Heavy *a.*[1] 5 b in Dict. and Suppl.)

1921 *San Francisco Chron.* 20 Sept. 22/1 There may be maintained in a Commercial District..Light Industries, clearly incidental to the operation of an Amusement Park. **1930** *Economist* 1 Nov. (Russ. Suppl.) 8/2 Only 22 per cent. is allotted to the building and equipment of factories devoted to 'light' industries. **1944** J. S. Huxley *On Living in Revolution* 128 Encouragement must be given to light and secondary industries, for only so can a reasonably balanced economy grow up in colonial areas. **1957** [see *heavy *a.*[1] 5 b]. **1961** E. A. Powdrill *Vocab. Land Planning* iv. 66 'Light industry' is any industry which does not commit a nuisance by noise, smell, fumes, soot or grit. **1974** E. Ambler *Dr. Frigo* III. 156 The transformation he envisaged—roads, housing,..hydro-electric schemes, light industry, fertiliser plants.

6. Also *light car*, a small economical car made from light materials.

1908 *Westm. Gaz.* 16 Mar. 5/2 This being essentially a light-car year, more than ordinary interest is manifested in the 8-h.p. two-cylindered polished chassis. **1914** *Light Car Manual* 1 Manufacturers have..solved the difficulty of how to produce a car which shall give all the comfort anyone could desire, and yet..compare favourably with the cost..of a motorcycle and side-car. The whole secret of this solution is summed up in the words 'light car'. **1963** [see *fore-car* (*fore- 5)]. **1970** C. F. Caunter *Light Car* (ed. 2) p. xv, The popularity of the light car, particularly in its minicar form, had the effect in the 1960's of reducing the average size of motor cars in general.

b. Applied to small, relatively light-weight aeroplanes, such as most private (non-commercial) passenger aeroplanes.

1923 *Flight* XV. 168/1 For want of a better term we have referred to the machine as a 'light plane', much as in the automobile world cars below a certain size and weight are termed light cars. **1933** *Meccano Mag.* Mar. 192/1 Light aeroplanes will fly to and from these places from the main routes. **1965** Nayler & Ower *Aviation* v. 56/1 The Piper Aircraft Corporation led the world in the sales of light aircraft in 1959. **1971** P. J. McMahon *Aircraft Propulsion* xi. 331 Occasionally designers of light aircraft look to the possibilities of using automobile type engines as power units. **1971** *Flying* (N.Y.) Apr. 39/2 In fact, it may well be the best all-around lightplane in the world.

18. Also in phr. *light duty*: military service which does not entail full-time work.

1916 A. Huxley *Let.* 13 Feb. (1969) 91 He is still on light duty—so gets plenty of leave from Salisbury Plain. **1953** A. Baron *Human Kind* xiii. 100, I shouldn't be here by rights. I'm a light-duty man.

19. (Further examples.)

1827 L. T. Rede *Road to Stage* 16 In small theatres, the light comedian must play the seconds in tragedy. *Ibid.* 60 In light comedy it is continually requisite to execute music. **1841** Macaulay in *Edin. Rev.* Jan. 524 A great and rapid reform in..our lighter literature was the effect of his [*sc.* Collier's] labours. **1844** J. Cowell *30 Yrs. among Players* 43 The light and low comedy. **1872** D. G. Rossetti *Let.* 20 Sept. (1967) III. 1064 Your idea of George's possibly finding an outlet in light literature does not seem promising to me. **1874** W. Lennox *My Recoll.* I. 186 The highest walks of light tragedy. **1888** G. O. Seilhamer *Hist. Amer. Theatre* I. 23 Comedy parts or light tragedy roles. **1897** *National Police Gaz.* (U.S.) 26 May 6/4 Miss Blanche is, perhaps, the cleverest little lady on the burlesque and light comedy stage. **1929** *Radio Times* 8 Nov. 406/3 *Journey's End*..is not a memorial service; nor, at the other extreme, is it light entertainment. **1958** *Times Lit. Suppl.* 5 Dec. 701/3 The climax, both exciting and comic, just succeeds in lifting this novel out of the light-entertainment class. **1961** *John o' London's* 18 May 567/4 A more profitable career as a light-comedy lead. **1974** P. de Vries *Glory of Hummingbird* iii. 27 Some more pretty good nature lyrics and then a batch of light verse. **1975** *Times* 10 Feb. 3/7 Any programme of cuts..would have to be closely vetted by the IBA to ensure that they maintain a balance between light entertainment programmes and more serious productions.

b. *Light Programme*, one of the regular programme services of the B.B.C., chiefly featuring popular music and light entertainment. Also *ellipt.* as *the Light*. (On 30 Sept. 1967 its name was changed to 'Radio 2'.)

1945 *Radio Times* 27 July 1/1 Alongside these six regionalised Home Services there is to be available a new alternative, the BBC Light Programme... It will be built for the civilian listener. **1956** 'M. Innes' *Old Hall, New Hall* III. iii. 205 No *real* American professor could be *quite* like that—not outside the Light Programme. **1959** S. Gibbons *Pink Front Door* iii. 37 He had missed a particularly good boxing match on the Light. **1962** L. Deighton *Ipcress File* xxviii. 180, I kept the radio tuned to the Light for the 6.30 bulletin. **1966** *B.B.C. Handbk.* 45 The Light Programme seeks to provide a friendly and companionable service for those who are in the mood for entertainment and relaxation. **1968** S. E. Ellacott *Everyday Things in Eng. 1914–68* xi. 162 At the end of the war the Home Service and the Light Programme were established (1945).

23. a. *light-density, -land* (later examples); **b.** *light-boned, -built* adjs.; **c.** Special Comb.: **light bread** *U.S.* (see quot. 1966); **light fantastic** (see Fantastic *a.* and *sb.* A. 6 b), as noun phr., the movements of dancing; **light-heavyweight** (see quot. 1954); also *attrib.*; also *ellipt.* as *light-heavy*; **light oil,** any of various fractions of relatively low specific gravity obtained by the distillation of coal-tar, wood-tar, petroleum, etc.; **light water,** (*a*) water containing the normal (about 0·02%) or less than the normal proportion of deuterium oxide (so *light water reactor*, a nuclear reactor in which the moderator is light water); (*b*) a foam formed by water and a fluorocarbon surfactant which floats on flammable liquids lighter than water and is used in fire-fighting.

1951 Auden *Nones* (1952) 14 Of light-boned children under great green oaks. **1974** J. Stubbs *Painted Face* ii. 48 She was light-boned and well-fleshed. **1821** *Western Carolinian* (Salisbury, N. Carolina) 27 Mar., Crackers and light Bread will always be found in his shop. **1880** 'Mark Twain' *Tramp Abroad* II. xlix. 225 Hot light-bread, Southern style. **1966** *Publ. Amer. Dial. Soc.* 1964 xlii. 20 *Lightbread*, any yeast-raised bread, to distinguish it from biscuit. **1970** C. Major *Dict. Afro-Amer. Slang* 77 *Light bread*, white bread. **1974** *Amer. Speech 1971* XLVI. 62 The notion that *light bread* is a recessive term is especially plausible because baking is rarely done at home and because supermarkets sell *white bread* or simply *bread*. **1953** J. Cary *Except the Lord* v. 15 He was a light-built man, very dark in complexion, with a somewhat hollow face, and a long sharp chin. **1956** E. Muir *One Foot in Eden* 18 The crescent shadow Of the light-built bridge. **1967** *Times* 28 Feb. (Canada Suppl.) 35/3 This method of operation has implications..for light-density branch lines. **1967** *Jane's Surface Skimmer Systems 1967–68* 30/1 The H.M.2 is designed for ferry services on light density routes of short stage lengths. *c* **1843** J. S. Coyne *Binks the Bagman* (1852) i. 10 Then you're fond of sporting on the light fantastic? **1892** A. C. Gunter *Miss Dividends* ix. 128 'You dance very nicely,' she murmurs. 'Yes, for a man who has not tripped the light fantastic for years.' **1913** Galsworthy *Dark Flower* I. vii. 34 When I was your age I twirled the light fantastic with the best. **1953** K. Amis *Lucky Jim* x. 114, I thought you'd all be on the floor by now... I'm not going to permit any more of this skulking about in here. It's the light fantastic for you. **1974** L. Deighton *Spy Story* vi. 57 The inlaid sprung floor would still have supported a light fantastic or two. **1973** R. L. Simon *Big Fix* (1974) xv. 104 It was the guy..who looked like a promising light-heavy. **1975** M. Kenyon *Mr Big* vii. 63 'These the heavyweights?' 'Light-heavy. Watch Hudson, in the blue trunks.' **1903** *National Police Gaz.* (U.S.) 18 July 3/1 And now there is a new champion in a new class—George Gardiner, of Lowell, Mass., the holder of the light-heavyweight title. **1913** J. G. B. Lynch *Compl. Amat. Boxer* (App.) 234 Standard weights. ..Light-heavy weight, 12 stone 7 pounds and under. **1954** F. C. Avis *Boxing Reference Dict.* 77 *Light-Heavyweight*, a standard weight division for professional boxers weighing more than 11 st. 6 lb. but not more than 12 st. 7 lb.; for amateurs 11 st. 11 lb. and 12 st. 10 lb. respectively. **1960** M. Golesworthy *Encycl. Boxing* 210/2 Light-heavyweight—Started in America in 1903 by Lou Houseman, manager of Jack Root, who had outgrown the middleweight division. The limit was set at 12 st. 7 lbs. (175 lbs.) and it remains at that figure today. The division was first recognised in Britain in 1913. **1968** *Encycl. Brit.* IV. 43/1 In 1920 light heavyweight competition was added. **1960** *Farmer & Stockbreeder* 29 Mar. 73/1 The only complaint—a little rain needed on some of the light-land farms. **1974** *Times* 15 Apr. 8/3 Dry weather over the past three weeks, rather too long a period for some light-land farmers, has made possible some catching up on the delays of March. **1867** Bloxam *Chem.* 452 The light oil which first passed over is rectified by a second distillation, and is then sent into commerce under the name of coal naphtha. **1898** F. H. Thorp *Outl. Industr. Chem.* 264 The distillate collected [from wood-tar] below 150°C. is called 'light oil', and is chiefly used as a substitute for oil of turpentine in varnish and paints. **1936** *Economist* 22 Feb. 399/2 Increasing demand for the heavier oils has enabled a higher proportion of refinery production to be marketed in that form, and has reduced the proportion subjected to 'cracking' to obtain light oils, such as motor spirit. **1964** N. G. Clark *Mod. Org. Chem.* xviii. 372 When light oil and crude benzole are distilled to give 'Benzole' for internal-combustion engines—over 70 per cent are treated in this way—a fraction embracing benzene, toluene and the xylenes is collected. **1933** Light water [see *heavy *a.*[1] 2 d]. **1947** Crowther & Whiddington *Science at War* iii. 142 Vast quantities of water have been electrolysed, and separated into 'light water' and 'heavy water', the former containing hydrogen atoms of mass 1 only. **1956** *Nature* 4 Feb. 204/1 Studies have been made of the pressurized light-water reactor and of the sodium-graphite reactor. **1968** *Guardian* 21 Aug. 1/4 The air show's special fire brigade—using helicopters carrying 'light water'. **1971** *Sunday Times* 12 Dec. 45/6 American light water reactors are simple and in some ways cruder. **1972** *Aircraft Engineering* Jan. 28/1 Using Light Water aqueous film forming foam, the team cut a knock-down path to the cockpit within five seconds of reaching the fire. **1973** *Daily Colonist* (Victoria, B.C.) 7 June 42/2 The cost of a solar steam generating plant would now be about double the cost of light-water nuclear plants.

light, *a.*[2] Add: **2. b.** *light red*, (*a*) pale red; (*b*) a pale red or reddish orange pigment produced from iron oxides.

1803 J. C. Ibbetson *Accidence of Painting in Oil* 17 *Light red*, so called, is either calcined green vitriol mixed with a quantity of other substance, and called Venetian red; or calcined yellow oker. **1934** H. Hiler *Notes Technique Painting* ii. 125 *Light red, burnt ochre*... It is quite opaque, and may be defined as a scarlet modified by the addition of a little yellow and grey. **1958** M. L. Wolf *Dict. Painting* 142 The red iron oxides found as natural deposits include Indian red, light red,..and others of lesser importance. **1970** R. D. Harley *Artists' Pigments* ix. 109 Light red came into current use as a colour name during the eighteenth century, when it was generally used to indicate a brownish red prepared by burning yellow ochre.

3. *light-haired* (earlier example); **Light Sussex,** a white variety of hen.

1843 Mill *Logic* II. III. xxiii. 149 The probability..that any given inhabitant of Stockholm is light-haired. **1909** T. W. Sturges *Poultry Manual* xiii. 359 The Light Sussex and the Buff Orpington are both blended in the White Orpington. **1938** L. Pearce-Gervis *Compl. Poultry Keeper & Farmer* vi. 153 The top prices..are still made by the Surrey Chicken... For this market either Pure Light Sussex or a cross in which there is a Sussex strain is necessary, for the white flesh must be maintained. **1965** P. Wayre *Wind in Reeds* xv. 224 A flock of four hundred Silky crossed with Light Sussex bantams.

light, *v.*[1] Add: **5. b.** (Earlier and later examples.) Also *to light in* (or *into*): to attack; to go at. *U.S. colloq.*

1866 'Mark Twain' *Lett. from Hawaii* (1967) 32 And you want to know what made me light out of bed so sudden last night? Only a 'santipede'. **1878** J. H. Beadle *Western Wilds* xii. 187 They double-quicked into town and lit in generally. **1889** K. Munroe *Golden Days* xiv. 156 You've got the levellest head of any man that ever lit into the diggings. **1917** Freeman & Kingsley *Alabaster Box* i. 3 He'll light into those hot doughnuts. **1948** 'J. Tey' *Franchise Affair* xxii. 262 The girl had lit out... She had dressed in a hurry and gone. **1967** *Boston Sunday Herald* 7 May IV. 5/5 Chris did demonstrate he is prepared for a scrap in the coming campaign when he lit into Mrs. Hicks' proposals. **1969** *Listener* 27 Mar. 433/2 Like a latter-day Huck Finn, he lights out for the territory. **1969** *New Yorker* 19 Apr. 81/1 If the astronaut missed mentioning a rock I knew was there, I'd light into him afterward, just like a football coach critiquing a fumble on a film of a game. **1973** *Observer* 15 Apr. 13/2 Inveighing against that new parliamentary building..lighting into..the proposed gymnasium.

light, *v.*[2] **2. c.** (Further examples.)

1943 J. B. Priestley *Daylight on Saturday* ix. 55 Blandford opened..a very fine silver cigarette-box, and both men lit up and were then silent. **1959** C. Williams *Man in Motion* i. 6, I ripped open a packet of the cigarettes, found some matches..and lighted up. **1970** H. E. Roberts *Third Ear* 9/2 *Light up*, to light a marijuana cigarette.

lighter, *sb.*[2] Add: **1.** Also *lighter-up* (see quot. 1921).

1909 *Westm. Gaz.* 21 Apr. 8/1 Robert Brown, lighter-up [at locomotive shed] slight cut on left eyebrow. **1921** *Dict. Occup. Terms* (1927) § 709 Lighter-up, carries live coals from fire hearth in shed to engine fire box.

2. b. = *cigarette lighter* (see *cigarette 2); also any similar mechanical contrivance for

lighting a gas-fire, etc. So *lighter-fluid, -fuel*, the fuel used to work a lighter.

1895 *Montgomery Ward Catal.* Spring & Summer 554/2 There are two lighters, to be used with alcohol. **1907** *Yesterday's Shopping* (1969) 243/3 *The 'Telescopic' Gas Lighter*. The best substitute for matches. **1913** Kipling *Divers. Creatures* (1917) 274 He smelt of rare soaps and cigarettes—such cigarettes as he handed me from a golden box with an automatic lighter. **1930** Sayers & 'Eustace' *Documents in Case* xxxvii. 105, I came in to retrieve a garment or lighter that he had borrowed. **1947** 'N. Blake' *Minute for Murder* ii. 47 A thin cylindrical object..rather like a lighter-fuel container. **1955** W. Gaddis *Recognitions* II. i. 317 Setting fire to his hand dipped in lighter fluid. **1956** E. Ambler *Night-Comers* iv. 85 She had a box of Kleenex and a can of lighter fluid... She began to wipe off the grease. **1959** *New Statesman* 19 Sept. 354/2 The pipe had an aircooled aluminium stem, the lighter was butane-fuelled and had a Cadillac 'V' on the side. **1960** 'H. Carmichael' *Seeds of Hate* xix. 157 There had been a stain on the sleeve..and he had removed it with lighter fuel. **1961** *Esquire* Aug. 59/2 He kept the bottles on a shelf in a clump of lighter fluid and Never-Leak cans. **1974** M. Gilbert *Flash Point* xii. 102 'You haven't got a cigarette by any chance?'..Patrick got out his case. The girl..took one, and got out her own lighter.

lighter-than-air, *attrib. phr. Aeronautics.* Designating a flying machine whose weight is less than the weight of the air which it displaces and which rises as a result of its own buoyancy; also applied to the use of such a machine or machines in flight.

[**1869** J. B. Pettigrew in *Notices Proc. R. Inst. Gt. Brit.* V. 103 A machine lighter than the air must necessarily rise through it. **1887** tr. *J. Verne's Clipper of Clouds* vii. 60 The first inventors did not think of apparatus lighter than air. *Ibid.* viii. 78, I solved the problem of aviation. That is what a balloon will never do, nor will any machine that is lighter than air.] **1903** *Work* 11 Apr. 155/1 The Barton Air Ship..is a combination of the 'lighter than air' and 'heavier than air' system—that is to say, it is a machine in which a system of movable aeroplanes is interposed between the car and the balloon. **1907** *Daily Mail* 19 Mar. 5/6 The Berlin 'Zentralblatt für Bauverwaltung'..protests against the subsidising with State funds of airships, dirigible balloons, and other 'lighter-than-air' vehicles. **1912** S. F. Walker *Aviation* i. 7 'Lighter than Air' apparatus which we call balloons. **1923** Hart & Laidler *Elem. Aeronaut. Sci.* i. 8 The problem of 'lighter-than-air' flight. **1953** S. Spender *Creative Element* i. 28 Tennyson was thinking of a battle of lighter-than-air dirigibles. **1963** *Ann. Reg. 1962* 539 [Auguste Piccard.] He and his twin, Jean Félix, were intensely interested in lighter-than-air flight. **1974** *Times* 13 Feb. 14/3 The role of lighter-than-air craft as cargo carriers.

ellipt. **1887** tr. *J. Verne's Clipper of Clouds* v. 39 To these enthusiasts for 'lighter than air' a no less enthusiast for 'heavier than air' had said things absolutely abhorrent. *Ibid.* x. 94 In spite of all the jealousy of the two enemies of 'lighter than air', they could not help being surprised at the perfection of this engine of aerial locomotion. **1910** *Blackw. Mag.* Feb. 206/1 Neglecting the lighter than air as a military auxiliary.

light face. *Typogr.* [f. Light *a.*[1] + Face *sb.* 22.] A kind of type in which the letters are made up of thin strokes. Also *attrib.* Hence **light-faced** *a.* Cf. *heavy face* (Face *sb.* 22, *heavy *a.*[1] 15).

1871 *Amer. Encycl. Printing* 275 *Light-Faces*, numerous varieties of job type, in which the lines of the letters are unusually light or thin. **1898** J. Southward *Mod. Printing* I. xxii. 140 The first would be called a *light face*, and the second a *heavy face*. *Ibid.* xxiv. 155 In the case of light-faced letters, they are spoiled for good work after the first time of using. **1917** F. S. Henry *Printing for School & Shop* vii. 90 Dainty, light-faced type. **1962** Corson & Lorrain *Introd. Electromagn. Fields* i. 1 Lightface type will indicate either a scalar quantity or the magnitude of a vector quantity. **1963** Kenneison & Spilman *Dict. Printing* 115 *Light face*, descriptive of a type-face of fine appearance as opposed to medium and bold lines. **1970** W. P. Jaspert et al. *Encycl. Type Faces* (ed. 4) p. ix, We speak also of bold-faced and light-faced types, referring to the thickness or thinness of the strokes of the letter.

lighthouse. Add: **b.** *lighthouse keeper* (earlier example), *-man, -tender*.

1738 *Mass. Bay Acts & Resolves* (1904) XII. 513 That the sum of Fifty one Pounds..be paid..for his services as Light House Keeper. **1889** *Athenæum* 23 Feb. 257 (Advt.), The manners and ways of coastguardsmen, lighthousemen, and other amphibious creatures. **1901** *Daily Colonist* (Victoria, B.C.) 1 Nov. 3/2 The lighthouse-tender Manzanita called in yesterday morning for a short stay. **1921** *Ibid.* 16 Oct. 16/1 The lighthouse tender Berens then took up the search and..discovered the small boat sheltering behind Mouatt Reef. **1958** W. Armstrong *True Bk. Lighthouses & Lightships* xiv. 142, I remembered, too, stories of how two lighthouse tenders and a pilot-cutter were sunk by mines in World War I. **1969** *Times* 19 Dec. 2/5 (*heading*) Lighthouse man missing.

lighting, *vbl. sb.*[2] Add: **1.** *attrib.* and *Comb.*: *lighting man, power, rate, socket, wire*; **lighting bridge** *Theatr.*, a narrow platform, suspended over a stage, on which lights are operated; **lighting cameraman** *Cinemat.* and *Television* (see quot. 1960); **lighting plot** *Theatr.* (see quot. 1961); **lighting tower** *Theatr.*, a tall structure on which lights are fixed.

1928 C. H. Ridge *Stage Lighting* vii. 111 The roof and lighting bridge are all built in reinforced concrete. **1933** P. Godfrey *Back-Stage* iv. 43 Above the proscenium arch is the lighting bridge, which will accommodate as many as ten electricians with 'spotting lanterns'. **1967** *Punch* 16 Aug. 242/3 For *Figaro* and Verdi's *Macbeth* at the first [Edinburgh] festival twenty-one years ago, John Christie had to bring in a lighting bridge and sixty floods and perches from Glyndebourne. **1960** O. Skilbeck *ABC of Film & TV* 78 *Lighting Cameraman*, chief cameraman of a unit, who is responsible for ordering the illumination which gives pictorial quality to a shot. **1966** *Listener* 15 Dec. 889/3 The lighting cameraman can earn £1,000 a week on a major picture. **1972** A. Fowles *Double Feature* iv. 67 I'd been Richard Powell, lighting cameraman. **1972** I. Hamilton *Thrill Machine* vii. 30 For difficult interiors he should have had a lighting man, for interviews he should have had an audio man as well. **1931** C. S. Parsons *Amat. Stage Managem. & Production* ii. 18 Lighting plots should always be well rehearsed. **1933** P. Godfrey *Back-Stage* iv. 44 The lighting-plot shows the position of every lamp, the colour of the medium, and the varying intensity of the light required. **1961** Bowman & Ball *Theatre Lang.* 201 *Lighting plot*,..a list, with diagrams, showing the lighting to be used in each scene of a production. **1884** F. Krohn tr. *Glaser de Cew's Magneto- & Dynamo-Electr. Machines* 183 A greater part of its lighting-power is due to the incandescence of the electrodes. **1928** *Chambers's Jrnl.* Jan. 21/1 An automatic regulator wherewith to make..and supply lighting power to the lamps. **1858** Lighting-rate [in Dict.]. **1928** *Chambers's Jrnl.* Jan. 79/1 All that has to be done to start charging is to insert the charger-plug in the nearest lighting socket. **1935** *Discovery* Nov. 326/2 A new power-driven spray-painting outfit which..can be run from an ordinary lighting socket. **1933** P. Godfrey *Back-Stage* iv. 43 In the wings stand a number of lighting towers twelve feet high, each capable of carrying a dozen 1000-watt flood lamps. **1967** R. Courtney *Drama Studio* ix. 76 At least one Lighting Tower is needed on the floor area. **1889** Lighting wire [in Dict.].

3. (Earlier example with *up*.)

1807 Southey *Lett. from Eng.* I. viii. 89 Between eight and nine the lighting-up began.., every window being filled with candles.

b. lighting-up time, the time when lights are switched on, esp. the time when lights on vehicles are required by law to be switched on.

1900 J. K. Jerome *Three Men on Bummel* iii. 55 On sunny afternoons you used to ride about with that lamp shining for all it was worth. When lighting-up time came it was naturally tired, and wanted a rest. **1935** H. G. Wells *Things to Come* iii. 23 It [*sc.* a newspaper] should show the customary insets beside the title of the weather forecast and the lighting-up time. **1947** *Whitaker's Almanack* 159/2 The legal importance of *Sunrise* and *Sunset* is that the Road Transport Lighting Act, 1927 defines Lighting-up Time for vehicles as being from one hour after sunset to one hour before sunrise. **1957** *Times* 11 May 7/1 Lighting-up time, 9.8 p.m. **1963** *Times* 20 May 3/3 M.C.C., going in with three-quarters of an hour left, lost Atkinson before the umpires decided it was lighting-up time.

lightning. Add: **2.** Also, any strong, freq. low-quality, alcoholic spirit. Chiefly *U.S.* Cf. *chain-lightning* (s.v. Chain *sb.* 19 in Dict. and Suppl.).

1858 *Calif. Spirit of Times* (San Francisco) 7 Aug. 1/4 Having in his possession a few kegs of liquid lightning upon which he was avariciously desirous of reaping a speedy profit. **1873** J. Miller *Life amongst Modocs* viii. 94 In one of the saloons where men were wont to..drink lightning. **1945** L. Shelly *Jive Talk Dict.* 28/2 *Lightning and thunder*, whisky and soda. **1958** L. van der Post *Lost World of Kalahari* ii. 53 The fiery Cape brandy known to us children as 'Blitz' or Lightning.

b. *pl.* One of the top grades of white jute.

1929 Woodhouse & Kilgour *Spinning Flax & Jute* i. 2 (*caption*) Jute bale marks... Lightnings. **1929** *Observer* 17 Nov. 4/3 Jute... Lightnings November–December quoted £25. **1964** R. R. Atkinson *Jute* i. 23 White jute is assorted into three main classes... The top class is sub-divided into Firsts, Lightnings, and Hearts.

3. c. *lightning-quick, -rapid* adjs.

1925 V. Woolf *Common Reader* 41 That is the quality that first strikes us in Greek literature, the lightning-quick, sneering, out-of-doors manner. **1947** C. S. Lewis *Miracles* xii. 118 What they painfully reconstruct from a million dots..he really produced with a single lightning-quick turn of the wrist. **1961** *Times* 25 Mar. 4/7 That strangely reluctant take-away of the club and the lightning-quick swing caused plenty of trouble yesterday. **1922** D. H. Lawrence *England, my England* 45 The officer was giving the last lightning-rapid orders to fire.

d. (Further examples.) Also, done, produced, or acting with the speed of lightning; spec. *lightning artist*, an artist who paints or draws pictures very quickly as an entertainment; *lightning-change*, a rapid change of costume made by an actor or performer; also *attrib.*; *lightning strike*, a sudden strike (Strike *sb.* 9) which takes place without any warning; *lightning tournament*, a Chess tournament in which each player must make his move within a prescribed few seconds; also *lightning chess, player*, etc.

1873 W. Mathews *Getting on in World* 242 Now..people travel by 'lightning lines', going from New York to Chicago in twenty-nine hours. **1875** 'Mark Twain' in *Atlantic Monthly* Aug. XXXVI. 192/1 If one of the boats has a 'lightning' pilot, whose 'partner' is a trifle inferior, you can tell which one is on watch by noting whether that boat has gained ground or lost. **1889** G. O. Seilhamer *Hist. Amer. Theatre* II. xxiii. 299 He may..be accorded the distinction of being the first 'lightning change artist' on the American stage. **1908** *Sears, Roebuck Catal.* 342/5 Silver Dip. The new lightning cleaner. **1910** *Year-Bk. Chess* 143 (*heading*) Two special lightning tournaments were held. **1920** *Radiograph* July p. ix/1 The first step was an endeavour to try to stigmatise the..action as 'A Lightning Strike'. **1927** W. E. Collinson *Contemp. Eng.* 70 A few technical expressions of the Halls which are commonly known to English townsfolk, but not likely to find their way into all the dictionaries, e.g...lightning artists (*caricaturists working at great speed*). **1927** E. Glyn *'It'* 240 She had promised to sit for him just for a lightning sketch. **1928** G. B. Shaw *Intelligent Woman's Guide Socialism* lxxxiii. 448 A lightning strike of waitresses in a restaurant. **1940** E. C. Shepherd *Britain's Air Power* 5 In the German 'Lightning War' on Poland the aeroplane was used to the full as an offensive weapon. **1945** *Tee Emm* (Air Ministry) V. 33 Fields, hedges and houses flash past at lightning speed under your wings. **1946** *Happy Landings* (Air Ministry) July 7/2 In addition to making a lightning survey of local air safety measures..a..lecture was given. **1948** C. Day Lewis *Otterbury Incident* iv. 39 Lightning sketches by Miss E. Toppingham. *Ibid.* 48 Toppy's sister, the lightning artist, had done particularly well. **1949** H. Golombek *World Chess Championship 1948* 24 He is..one of the world's best lightning players. **1951** 'Assiac' *Adventure in Chess* III. iii. 99 Most proficient at 'lightning Chess'. **1952** Granville *Dict. Theatr. Terms* 111 *Lightning-change artiste*, the (more usual) quick-change artiste, one who imitates..a number of well-known personalities. **1955** *Times* 3 Aug. 8/7 A shortage of labour caused a lightning strike of about 200 men at the Albert and William Wright docks at Hull to-day. **1971** E. Pritchard *Chess for Pleasure* xii. 148 In the evenings there may be *lightning* tournaments, when games are played at 10 seconds a move.

e. lightning beetle = *lightning-bug*; **lightning box,** a box used in producing stage-lightning; **lightning-bug** (earlier and later examples); **lightning conductor** (earlier and later examples).

1854 B. Jaeger *Life of N. Amer. Insects* 75 Some months since a lady..presented me two of these living Lightning beetles. **1825** P. Egan *Life of Actor* ii. 67 Lightning boxes, sheep hooks, and three harlequin's bats. **1855** 'Q. K. P. Doesticks' *Doesticks, what he Says* xxvi. 237 The prompter was stretched on the top of a canvas volcano, with the bell-rope in his hand, and his hair full of resin, from the 'lightnin-box'. **1928** A. Rose *Stage Effects* 26 Fig. 20 is a simple but useful and convenient form of lightning box... One or more electric lights are to be in the centre of the box. **1778** J. Carver *Trav. N. Amer.* xviii. 491 The Lightning Bug or Fire Fly is about the size of a bee. **1842** *Southern Lit. Messenger* VIII. 199/2 It will never do to tell *us* that there is any humbug in this business, or even that it is a mere *lightning-bug*. **1936** T. S. Eliot *Coll. Poems 1909–1935* 138 The small flare of the firefly or lightning bug. **1947** *Chicago Tribune* 21 June 2/4 He asserted that to 'talk about Henry Wallace intimidating Harry Truman on the veto is like describing a lightning bug as blotting out the rays of the sun'. **1971** *Lebende Sprachen* XVI. 10/1 US lightning bug—BE/US firefly. [**1791** E. Darwin *Bot. Garden* I. Additional Notes 25 The design of these conductors is to permit the electric matter accumulated in the clouds to pass through them into the earth..and it would seem that the finer the point..the better, as it would take off the lightning while it was still at a greater distance.] **1814** W. Bentley *Diary* (1914) IV. 262 The post remained, retained on the side of the steeple by the Lightning conductors. **1967** *Everyman's Encycl.* V. 429/1 He [*sc.* B. Franklin] estab. the identity of lightning with electricity..and suggested the use of lightning conductors on large buildings.

lightning (lǝi·tniŋ), *v.* [f. the sb.] = Lighten *v.*[2] 6. Also *fig.*

1903 *Westm. Gaz.* 16 Nov. 8/2 The two metal balls..thundered and lightninged as they delivered the message. **1926** H. Caine in *Strand Mag.* Jan. 20/1 Mr. Gladstone leapt to his feet, whereupon the air of the House thundered and lightninged for a short ten minutes. **1935** in Z. N. Hurston *Mules & Men* (1970) I. i. 27 You know, when it lightnings, de angels is peepin' in de lookin' glass.

lightningy (lǝi·tniŋi), *a.* [f. Lightning + -y[1].] Suggestive of lightning. (In comb. with *thunder*.)

1906 Galsworthy *Man of Property* II. ii. 141 They had never seen anyone look so thunder and lightningy as that little June!

lights, *pl.* Add: **b.** Colloq. phr. *to scare the (liver and) lights out of* (someone): to scare (someone) greatly.

1884 'Mark Twain' *Huck. Finn* xxix. 307 It most scared the liver and lights out of me. **1956** E. Pound tr. *Sophocles' Women of Trachis* 18 You might start by questioning Likhas, scare the lights out of him, and he might tell you.

lightship. Also *attrib.* and *Comb.*

1887 *Chambers's Jrnl.* 1 Jan. 1 (*title*) The lightshipman. **1904** *Westm. Gaz.* 8 Oct. 9/2 The light-ship men were unable to come on shore. **1905** *Daily Chron.* 10 Oct. 4/5 Such strictures upon the lightship keepers' employment would hold equally with reference to the crews of say the light vessels off the Scilly Islands. **1958** W. Armstrong *True Bk. Lighthouses & Lightships* xiii. 131 'Better be safe than sorry' was always the unspoken motto of lighthouse and lightship men.

light-weight, *sb.* and *a.* Add: Also (without hyphen) **lightweight. A.** *sb.* In *Boxing*, now usu. a competitor weighing between 126 and 135 pounds. Also anything (e.g. a motorcycle) that is relatively light in weight.

1898 *Pearson's Mag.* Feb. 165/1 The torpedo-boat and the destroyer may fairly be called the light-weights of naval warfare. 1908 *Daily Report* 20 July 9/1 Light-weights of 2 h.p. are quite capable of averaging well over 20 m.p.h. 1960 M. Golesworthy *Encycl. Boxing* 211/1 *Weight Divisions*... Lightweight—1868 at 9 st. 7 lbs. (133 lbs.). 1912 raised to present limit of 9 st. 9 lbs. (135 lbs.) by Willie Ritchie.

b. *fig.* A person or thing of little importance, profundity, or intelligence.

1885 A. Edwardes *Girton Girl* III. xi. 205, I am not good at these high passions!..In everything I am a light weight... In life I walk gently. 1902 G. H. Lorimer *Lett. Self-Made Merchant* (1903) ii. 19 That is the spot where a young man has the chance to show that he is not a light-weight. 1951 A. L. Rowse *Eng. Past* 24 Harington was the Queen's godson—clever,.. naughty, a light-weight, perpetrator of puns and practical jokes. 1962 *Listener* 25 Jan. 185/2 Of the other stories in this book, those by the long-established writers..are light-weights. 1965 *English Studies* XLVI. 209 After Holiness or Justice, Courtesy may seem as something of a light-weight.

c. A garment, usu. a suit, made from light-weight material.

1972 J. Quartermain *Rock of Diamond* v. 26 He stood five ten clad in a dark-blue lightweight. 1973 R. Busby *Pattern of Violence* v. 80 He..picked up a man's suit jacket from the floor. It was a grey lightweight.

B. *adj.* (Further examples.) Also of clothes, fabrics, etc.

1892–3 *T. Eaton & Co. Catal.* Fall & Winter 31/2 Light weight overcoats in grey, brown, fawn. 1902 *Daily Chron.* 8 Nov. 8/3 Every woman who walks much clings to a light-weight Russian blouse. 1930 J. T. Hatfield et al. *Curme Vol. Ling. Stud.* 52 The intervening verbs are chiefly light-weight grammatical words. 1940 *Chambers's Techn. Dict.* 499/2 *Light-weight concrete* (Build.), concrete of low unit weight.., made by using aggregates such as pumice, sawdust, and cork, with cellular concrete. 1953 *News Chron.* 2 June 2/2 Squatting to drink it [*sc.* tea] in the light-weight (six pound) proofed cotton nylon tents. 1957 *Times Lit. Suppl.* 13 Dec. 751/2 One cannot reproach him with having produced a rather lightweight book. 1958 *Times* 10 Nov. 14/5 An entertaining, light-weight lecture recital by Mr. Antony Hopkins. 1963 A. J. Hall *Textile Sci.* iii. 158 In these days when light-weight apparel is much favoured the density of a textile fibre is important. 1968 M. Woodhouse *Rock Baby* xi. 109 He'd been.. wearing a lightweight raincoat. 1973 *Nation Rev.* (Melbourne) 31 Aug. 1443/3 A recent column attacking Whitlam for appearing on TV with 'a lightweight variety performer'.

Hence **light-weighted** *a.*, of light weight; carrying a light weight.

1867 J. R. Lowell in *Atlantic Monthly* Nov. 625/2 We ..see the rather light-weighted great man wheeled round the room..to converse with his guests. 1905 *Westm. Gaz.* 21 Mar. 8/2 The London and Brighton Handicap Steeplechase on Friday may be won by the light-weighted Dam.

ligno-. Add: **li:gnosu·lphonate,** any of the salts or esters of the lignosulphonic acids, some of which are used as adhesive binders, as pigment dispersants, in the tanning industry, and in the manufacture of vanillin; **li:gnosulpho·nic acid,** any of various compounds in which sulphonic acid groups are attached to lignin molecules, formed in the sulphite process for producing wood pulp.

1908 *Jrnl. Chem. Soc.* XCIV. i. 717 From the liquid obtained by boiling pine-wood with calcium hydrogen sulphite, the author has prepared barium lignosulphonate, $C_{40}H_{44}O_{17}S_2Ba$. 1952 F. E. Brauns *Chem. Lignin* v. 113 The isolation of a basic calcium lignosulfonate from a commercial spent liquor on a technical scale is carried out in the 'Howard process'. 1963 J. F. Harris et al. in B. L. Browning *Chem. Wood* xi. 578 Lignosulfonates, in particular the carbohydrate-free products, have found a variety of profitable applications. The most notable of these are their uses in oil well-drilling muds and in the production of synthetic rubber. 1908 *Jrnl. Chem. Soc.* XCIV. i. 717 In lignosulphonic acid, part of the sulphurous acid is tightly and part loosely bound. 1931 *Canad. Jrnl. Res.* V. 42 The first lignosulphonic acids studied were obtained by heating benzene-alcohol extracted spruce meal with sulphurous acid. 1967 Kirk & Othmer *Encycl. Chem. Technol.* (ed. 2) XII. 369 The sulfite process for producing pulp can lead to products which consist of lignosulfonic acids, or various lignosulfonates. These are commonly referred to in the trade as lignin sulfonic acids and lignin sulfonates.

lignocaine (li·gnokēⁱn). *Pharm.* [f. Ligno- (as the L. equivalent of Xylo-, the compound having been orig. named *xylocaine* because of its chemical relationship to xylene) + *-caine*, after Cocaine.] A white crystalline aromatic amide, $(CH_3)_2C_6H_3NH{\cdot}CO{\cdot}CH_2{\cdot}N(C_2H_5)_2$, used as a local anæsthetic for the gums and mucous membranes, usually in the form of its hydrochloride and by injection, but also as tablets, sprays, or creams.

1954 *Anaesthesia* IX. 96 The development of a technique utilising the short acting relaxant suxamethonium chloride..and the analgesic lignocaine hydrochloride (Xylocaine). 1959 *Brit. Dental Jrnl.* CVI. 48/2 Extensive clinical experience has shown that procaine and, probably more so, lignocaine are really very satisfactory as local anæsthetic agents in dental practice. 1970 Passmore & Robson *Compan. Med. Stud.* II. viii. 6/1 Lignocaine..is therefore the safest drug in the treatment of acute ventricular arrhythmias even though it is toxic in high dosage.

lignum². (Later examples.)

1903 'T. Collins' *Such is Life* 79 When a certain class of bushman says 'mallee', he means any sort of scrub except lignum. 1933 *Bulletin* (Sydney) 13 Dec. 29/4 Beneath the surface is a mass of lignum roots. 1941 I. L. Idriess *Great Boomerang* xiii. 92 Carpeted with yellowish-green lignum —huge bunches of cane-like shrubs like man-high beehives.

ligroin (li·gro,in). *Chem.* Also ligroïn. [Etym. unknown: in quot. 1881 a. G. *ligroin.*] Any of various naphtha fractions with ranges of boiling points between 90 and 150°C, used as solvents.

1881 *Jrnl. Chem. Soc.* XL. 1181 The author has investigated the causes of explosions resulting from the products of the distillation of mineral naphtha when used for burning purposes. These products are known as 'gasoline', 'benzoline', 'ligroin', and 'lubricating oil'. 1942 H. Barron *Mod. Synthetic Rubbers* vii. 82 Straightforward rectification of petroleum yields: Petrol... Benzine... Ligroin... Kerosene [etc.]. 1965 *Biol. Abstr.* XLVI. 680/2 Experimental solvent extractions of wood rosin from pine stump chips were carried out in ligroin at 0, 20, 50, and 80°C. 1968 B. J. Hazzard tr. *Asinger's Paraffins* iv. 402 Ligroin with a boiling range of 95–150°C (density = 0·745) containing 3·2 per cent of aromatics, 24·2 per cent of naphthenes, and 72·6 per cent of paraffins..can be 60–65 per cent nitrated in a single pass through the reactor.

ligularia (ligiŭlēᵊ·riă). [mod.L. (H. Cassini 1816, in *Bull. Sci. Soc. Philomatique* 198), f. L. *ligula* strap, referring to the shape of the ray-florets.] A herbaceous perennial plant of the genus so called, belonging to the family Compositæ, often native to China or Japan, and bearing yellow flowers.

1839 G. Don *Sweet's Hortus Britannicus* (ed. 3) 382 (*heading*) *Ligularia DC.* Ligularia (*ligula*, a little tongue; ray-flowers ligulate). 1862 *Curtis's Bot. Mag.* LXXXVIII. 5302 (*heading*) Kæmpfer's Ligularia. 1886 G. Nicholson *Illustr. Dict. Gardening* II. 263/1 Ligularias require generally a free, moist soil, and prefer a rather peaty one. 1966 J. Berrisford *Wild Garden* ii. 149 The ligularias are handsome plants for rich moist soil. 1971 *Country Life* 2 Sept. 567/2 My biggest colour splodges are..various yellow composites: heleniums..ligularias, [etc.].

Ligurian, *a.* and *sb.* Add: **b.** *sb.* Also, the Indo-European language of the ancient Ligurians; the Gallo-Italian dialect of this region.

1835 G. C. Lewis *Ess. Romance Lang.* i. 50 The diffusion of the Latin over Italy, in the place of the Etruscan, the Oscan, the Ligurian, and other native dialects, has been already noted. 1888 [see *Iapygian *a.* and *sb.*]. 1909 *Trans. Amer. Philol. Assoc.* XL. 81 The net result of the study is that by far the greater part of what is called Ligurian is strictly Gallic. 1927 C. H. Grandgent *From Latin to Italian* 6 In the northwest of Italy we find the Piedmontese dialect, and, to the south, along the Gulf of Genoa from Monaco to Carrara, the Ligurian. 1933 R. S. Conway et al. *Præ-Italic Dial. Italy* II. ii. 70 Of Ligurian properly so called very little can be said to be known. 1939 L. H. Gray *Foundations of Lang.* xi. 335 In the general area of Lago Maggiore, some seventy-two inscriptions..have been discovered... Their language may well represent the sole known remnants of Ligurian, which would seem to have stood midway between Italic and Celtic. 1968 *Encycl. Brit.* XII. 711/2 The dialects may be divided into six major classes: 1. Gallo-Italian: (*a*) Piedmontese; (*b*) Ligurian; (*c*) Lombard; (*d*) Emilian [etc.]. 1974 R. A. Hall *External Hist. Romance Lang.* iii. 51 Ligurian, used along what is now the Italian and French Riveria [*sic*], and perhaps to the north of this area in present-day Piedmont and Savoy.

ligustrum (ligɒ·strŏm). [L. *ligustrum* privet, adopted by Linnæus (*Hortus Cliffortianus* (1737) 6) and earlier botanists as the name of a genus.] = Privet¹.

1664 Evelyn *Kalendarium Hortense* in *Sylva* 71 July... Flowers in Prime, or yet lasting... Oleanders red and white, Agnus Castus, Arbutus, Yucca, Olive, Ligustrum, Tilia, &c. 1900 M. Thorn in W. D. Drury *Bk. Gardening* xi. 441 Ligustrums (Privets) are represented in many gardens only by *L. ovalifolium* (oval-leaved) and its golden-leaved form. 1957 *New Yorker* 23 Nov. 46/3 I've put four pyramid ligustrums on the sculpture balconies. We always use plants to bring out an exhibition. 1973 W. J. Bean *Trees & Shrubs Hardy in Brit. Is.* (ed. 8) II. 570 L[*igustrum*] *confusum* is..represented at Kew by a plant 10 ft high in the Ligustrum collection.

Lihyanic (liyā·nik), *sb.* Also **Lihyanite, Lihyanian; Lihyani** (liyā·ni). [f. Arab. *liḥyān* + -ic.] The name of an ancient Semitic language known only from north Arabian inscriptions of the 2nd and 1st centuries B.C. Also (all forms), as *adj.*

1911 *Encycl. Brit.* XXIV. 626/1 A more commendable proposal is to call the inscriptions Liḥyānī, since the tribe of Liḥyān is sometimes mentioned in them... Other brief inscriptions..have been discovered... Their writing is a somewhat later form of the Liḥyānī, and the dialect.. seems to be very similar to Liḥyānī. 1932 W. L. Graff *Lang.* xi. 402 From North Arabia we have a certain number of inscriptions dating from the 2d or 1st century B.C. and revealing a language, called Lihyanite, closely related to Arabic proper. 1936 *Encycl. Islam* III. 27/1 The Thamūdaean graffiti..are a development (later or parallel) of the Liḥyānī script. 1937 F. V. Winnett *Study of Lihyanite & Thamudic Inscriptions* 51 The earlier supplanting of Dedanite by Lihyanite points to a Lihyanite conquest of Dedan (al-'Ula) in the early 5th century B.C. 1939 L. H. Gray *Foundations of Lang.* 363 North Arabic is first recorded in Liḥyānian and Thamūdian inscriptions (the former between the second or first century B.C. and the fourth or sixth A.D...) and in Ṣafāitic graffiti. 1948 D. Diringer *Alphabet* ii. ii. 227 The Lihyanite inscriptions can be divided into two groups. 1968 *Encycl. Brit.* I. 663/2 The four South Semitic alphabets.. are known as Sabaean, Lihyanic, Thamudenic and Safahitic.

like, *a.*, *adv.* (*conj.*), and *sb.*² Add: **A.** *adj.* **1. b.** ¶ *like that*: spec. (usu. accompanying the crossing of the speaker's fingers) as an indication that two people described are very friendly or intimate; *like another* [cf. Fr. *comme un autre*]: that is ordinary or unexceptional; that is only one of a number of similar things, possibilities, etc.

1904 H. James *Golden Bowl* I. xiii. 236 Isn't the whole thing..perhaps but a way like another for their gaining time? 1926 F. Scott Fitzgerald *Great Gatsby* ix. 206 'We were so thick like that in everything'—he held up two bulbous fingers—'always together.' 1929 D. Hammett *Red Harvest* xxii. 219 'You're a friend of Whisper's?' 'You bet.' He held up two thin fingers pressed tightly together. 'Just like that, me and him.' 1936 D. Powell *Turn, Magic Wheel* i. 37 Theatrical people..just got here from London—they're like that with Cochrane—they know Dame Sybil Thorndike personally. 1966 I. Murdoch *Time of Angels* xviii. 193, I suppose it's a skill like another. 1971 M. Russell *Deadline* xv. 182 'Of course you had to get on terms with Gregory.' 'Now we're like *that*.'

2. e. Colloq. phr. (*a bit*) *more like* (*it*): nearer what it (etc.) should be or what is desired; better; also, closer to the truth. Cf. More *adv.* 1 h.

1888 Kipling *Phantom 'Rickshaw* (1889) 81 'That's more like,' said Carnehan. 'If you could think us a little more mad we would be more pleased.' 1891 S. Weyman *New Rector* II. xv. 12 'This is better than No. 383, Mrs. Baxter?' 'Well, sir,..it is a bit more like.' 1907 D. H. Lawrence *Phoenix II* (1968) 8 'Ah,' said Beelzebub, 'this is a bit more like it, a bit hotter. The Devils feel at home here.' 1962 D. Mayo *Island of Sin* viii. 63 Paid companion, hell. Whipping-boy was more like it. 1964 A. Wilson *Late Call* iv. 141 Shopping in the Town Centre provided something more like, and she ambled around, taking her time. 1968 P. Durst *Badge of Infamy* vii. 61 'Would you like some coffee?' 'Now that's more like it. Sure why not?'

9. b. (*be*) *like to do, have like to do*, be on the point of doing, almost do, look like doing. Cf. *like *v.*² 2. *U.S. colloq.*

1808 L. Summer in *Southern Hist. Mag.* (1892) I. 52 Strother was 150 votes behind Roberts & like to have lost his election in consequence of his opposing Madison. 1830 in *Jrnl. Illinois State Hist. Soc.* (1930–1) XXIII. 214 The boat went under a tree top and like to took me off. 1836 F. A. Chardon *Jrnl. Fort Clark* (1932) 70 Michael Belhumeres horse fell with him, had like to have broken his neck. 1854 in *Southwestern Hist. Q.* (1931–2) XXXV. 217 The supper—I had like to have said table-spread, for I can call it by no other name—it is a dirty old wagon cover sp[r]ead on the ground. 1855 in *Calif. Hist. Soc. Q.* (1929) VIII. 340 Like to never got back myself. 1884 'Mark Twain' *Huck. Finn* xxxix. 396 We like to got a hornet's nest, but we didn't. 1889 'C. E. Craddock' *Despot of Broomsedge Cove* xviii. 327 That's what like to have happened to me. 1916 'B. M. Bower' *Phantom Herd* vi. 100, I like to died a-laughing. 1930 G. B. Johnson in B. A. Botkin *Treas. S. Folklore* (1949) iv. iii. 608 'I like to have got killed' means 'I almost got killed'. It is surprising how many phrases used by Negroes are exactly the phrases used by English folk. 1973 *Black World* Apr. 63 Damn brim like to covered broad street.

B. *adv.* **1. c.** *to know* or *read* (someone or something) *like a book*: to know very well, understand perfectly; *to speak* or *talk like a book*: (*a*) to talk elegantly; to use literary or pedantic language in conversation; (*b*) to speak knowledgeably and accurately. *colloq.*

1825 J. Neal *Bro. Jonathan* II. xxvi. 444 I can..read you off, like a book. 1829 *Mass. Spy* 28 Jan. (Th.), You talk like a book, Mr. Bond. 1833 J. Neal *Down-Easters* 26 An educated and travelled Yankee..talking like a book, even to his washerwoman. 1839 'H. Franco' *Adventures H. Franco* I. xi. 73 'Know him like a book,' replied Mr. Lummucks. 1843 W. T. Thompson *Major Jones' Chron. Pineville* 74, I knows the Curloos like a book. 1844 T. C. Haliburton *Attaché* 2nd Ser. II. 176 Let a man or woman come and talk to me..and I'll tell you all about 'em right off as easy as big print. I can read 'em like a book. 1853 Lytton *My Novel* II. vii. xxi. 281 'If you can contrive to affect to be angry with him for his extravagance, it will do good.' 'You speak like a book, and I'll try my best.' 1875 [see Speak *v.* 1 d]. 1933 Wodehouse *Mulliner Nights* 101 That terrible old woman saw through my subterfuge last night. She read me like a book. 1940 H. G. Wells *Babes in Darkling Wood* i. i. 25 Don't you talk like a book, Mr Jimmy... Don't you go using long words. 1960 *Sunday Times* 27 Nov. 11 'Speaks Welsh like a book, the professor'—and, what's more, he also writes it like a book though he learnt it late in life.

d. *N. Amer. colloq.* Followed by an adj.: in the manner of one who is ——. Cf. *like crazy* (*crazy *a.* 4 c), *like mad* (Mad *a.* 1 c). Also in less analysable constructions.

1959 *She* May 21/2 Like wow...wonderful. 1961 G. Smith *Business of Loving* xi. 231 Sometimes we get rather soppy about each other..and laugh like young. 1962 [see *blow *v.*¹ 14 e]. 1970 *Time* 31 Aug. 19 Afterward, a girl came up to me and said, 'You kinda look interested in this; did you know there are civil rights for women?' And I thought like wow, this is for me.

6. a. (Later examples.)

1941 *Coast to Coast* 192 She made a challenge of it, like she always does. 1961 *Word Study* Apr. 7/1 Like I said... The Beatnik School of Language Degradation must be responsible for the sudden mushrooming of the above expression. 1966 *New Yorker* 10 Dec. 149 Murray the K tells it like it really is. 1968 *Globe & Mail* (Toronto) 17 Feb. 34 (Advt.), Send for your copy now. Like we said, it's free. 1973 *New Society* 6 Dec. 608/1 His successors can build their modest, unpretentious monuments, like the British have been doing for years. 1974 'E. Lathen' *Sweet & Low* xviii. 174 Like I said, it was one o'clock... Everybody claims they were asleep.

e. Delete † (*obs.*) and add later examples.

1860 in Bartlett *Dict. Amer.* (ed. 3) 244 The old fellow drank of the brandy like he was used to it. 1886 *Harper's Mag.* June 109/2 None of them act like they belonged to the hotel. 1895 J. Prior *Renie* xvii. 191 'E made a noise like 'e were sorry or summat. 1898 H. S. Canfield *Maid of Frontier* 100, I sprung from the chair like a man had shot me through the head. 1932 T. S. Eliot *Sweeney Agonistes* 30 When you're alone in the middle of the bed and you wake like someone hit you on the head. 1940 W. Faulkner *Hamlet* i. i. 7 For a while it looked like I was going to get shut of it. *Ibid.* ii. 52 It seemed like we begun to hear it right away. 1969 *Observer* (Colour Suppl.) 23 Mar. 23/2 They look at me like I'm dirt. 1973 'H. Howard' *Highway to Murder* vii. 76 That sounded like I was being left to hold the baby.

7. (Earlier and later examples.) Also, *colloq.* (orig. *U.S.*), as a meaningless interjection or expletive.

1778 F. Burney *Evelina* II. xxiii. 222 Father grew quite uneasy, like, for fear of his Lordship's taking offence. 1911 A. Bennett *Hilda Lessways* i. vi. 49 He hasn't passed his examinations like... He has that Mr. Karkeek to cover him like. 1929 'H. Green' *Living* vi. 57 'E went to the side like and looked. 1950 *Neurotica* Autumn 45 Like how much can you lay on [*i.e.* give] me? 1961 *New Statesman* 22 Sept. 382/2 'You're a chauvinist,' Danny said. 'Oh, yeah. Is that bad like?' 1966 *Lancet* 17 Sept. 635/2 As we say pragmatically in Huddersfield, 'C'est la vie, like!' 1971 [see *fighting chance* s.v. *fighting *vbl. sb.* 3 b]. 1971 *Black Scholar* Apr.–May 26/1 Man like the dude really flashed his hole card. 1973 *Black Panther* 17 Nov. 9/4 What will be the contradictions that produce further change? Like, it seems to me that it would be virtually impossible to avoid some contradictions.

like, *v.*¹ Add: **6. a.** Colloq. phrases: *I like that!*, an ironical expression of surprise or disgust at someone's impudence, conceit, untruthfulness, etc.; (*to do*, etc., something) *and like it*, (to endure or perform something unpleasant) with a good grace, without complaint; *to like it or lump it*, see Lump *v.*² 2. For *to know what one likes* see *know *v.* 11 f.

c 1869 Taylor & Dubourg in M. R. Booth *Eng. Plays of 19th Cent.* (1973) III. 255 *Lilian.* Bertie, you are a duffer. In these [arithmetic] questions the strength is always presumed to be equal. *Fitz-Urse.* I like that. As if one fellow was ever just as strong as another fellow. 1902 J. Brynildsen *Dict. Eng. & Dano-Norwegian Lang.* s.v. *like*, *I ~ that..!* det var jo rart! 1927 A. B. Cox *Mr. Priestley's Problem* ix. 146 'George, go and send them away.' 'Oh come,' protested George. 'I like that.' *a* 1930 D. H. Lawrence *Phoenix II* (1968) 182 'I feel so bare and brazen without a whiff of powder on my nose.' He gave a shout of laughter. 'I like that!' he said. 1941 H. G. Wells *You can't be too Careful* iii. vii. 138 'You *made* me.' 'I like that.' 1943 Hunt & Pringle *Service Slang* 11 *And like it!* A Naval expression anticipating a grouse and added to any instruction for an awkward or unwanted job. 1955 J. Bingham *Paton Street Case* vi. 97 'I'll have a pint.' 'You won't,' said Stan. 'It's May's birthday. You'll have a double Scotch, Len, and like it.' 1971 'D. Halliday' *Dolly & Doctor Bird* ii. 21 I'll do it again, and you'll stand by and like it. 1974 I. Murdoch *Sacred & Profane Love Machine* 73 'If we didn't stupefy ourselves with drink we wouldn't get so comatose.' 'I like that. You taught me to drink.'

b. *if you like*: if you wish to phrase or consider something in a particular manner; often used as a vaguely intensive expression, = 'indeed', 'perhaps'. *colloq.*

1875 T. E. Bridgett *Our Lady's Dowry* ii. xi. 336 They were placed in churches by simple faith, or credulity if you like, but not by wilful fraud. 1909 W. H. Hudson *Afoot in Eng.* ii. 20 'What!' I exclaimed. 'Lady Y—: that funny old woman!' 'No—middle-aged,' he corrected... 'Very well, middle-aged if you like.' 1955 L. P. Hartley *Perfect Woman* xxxvi. 325 Eighteen years faithful to an unfaithful man: there was a proof of staunchness, if you like! 1968 'A. Gilbert' *Night Encounter* iii. 37 That was a surprise, if you like, you wouldn't have thought Mr. Nicholas had enough humanity in him to give her a child. 1970 *Sunday Times* (Colour Suppl.) 15 Mar. 25/1, I could tell you some stories about Max [Miller], but I won't. Used to make me laugh, though. He was real blue if you like. 1973 *Listener* 15 Nov. 664/1 *Princess Anne*: We're shown the best, if you like. But nonetheless it's life.

like, *v.*² **2.** (U.S. examples.)

1800 A. M. Thornton in *Rec. Columbia Hist. Soc.* (1907) X. 117 Joe..said they had liked to have been lost in Pohick run. 1854 J. E. Cooke *Virginia Comedians* I. xlix. 282 She liked to fainted just now.

-like, *suffix.* Add: **2. a.** (Further examples.)

1910 A. H. Adams *Galahad Jones* 208 I'd be useful-like to keep a look out. 1937 M. Scott *Barbara Prospers* 214 Her bein' shaky-like. 1953 'N. Blake' *Dreadful Hollow* 38 We have a stronger parson now—more active-like.

b. (Further examples.)

1895 A. A. Grace *Maoriland Stories* 105, I suppose you won't care to stop the night with a chap, friendly-like. 1907 W. H. Koebel *Return of Joe* 50 Things seemed panning out so strange-like. 1967 *Observer* 10 Sept. 17/3, I went out with her, but all the coloured girls began to look at me weird-like: I had to pack it up quick.

likelihood. 2. c. For '*Obs.* exc. *Sc.*' read 'Now *rare* exc. *Sc.*' and add further example.

1894 G. Meredith *Ld. Ormont* III. xii. 241 The likelihood is, she'll feel bound in honour to serve him faithfully for the rest of their wedded days.

likely, *a.* and *adv.* **A.** *adj.* **2. b.** Colloq. phr. *not likely*: certainly not; I refuse.

1893 G. B. Shaw *Widowers' Houses* ii. iv. 47 Catch him going down to collect his own rents! Not likely! 1914 —— *Pygmalion* (1916) iii. 153 Walk! Not bloody likely. .. I am going in a taxi. 1922 C. E. Montague *Disenchantment* x. 139 The German sentries said, 'Go back, or we shall have to shoot.' The Englishmen said 'Not likely!' advanced to the German wire, and asked again for an officer. 1934 G. B. Shaw *On Rocks* ii. 240. Take the land with one hand and give back its cash value to the landlords with the other! Not likely. I ask you again, do you take us for fools?

B. *adv.* **2.** Now freq. in N. Amer.

1919 E. O'Neill *Ile* in *Moon of Caribbees, etc.* (1923) 10 I'm afeard there'll be trouble with the hands by the look o' things. They'll likely turn ugly...if you don't put back. 1931 —— *Mourning becomes Electra* i. i. 37 You've likely heard gossip. 1963 *Monsanto Mag.* Mar. 10/2 Beams of different frequencies likely could be obtained by varying the lasing material slightly. 1964 *Amer. N. & Q.* Jan. 76/2 The American public never has understood, and likely never will, the chaotic and complex character of the Indian problem during the 19th century. 1970 *Toronto Daily Star* 24 Sept. 12/8 The heroin addict likely will retrace or repeat many strokes of the pen, showing he is not in full control of his neuromuscular functions. 1971 *Publishers' Weekly* 22 Nov. 14/1 It is possible to predict that within a few years the microfiche likely will move from the library into the study and home.

liker. (Further examples.)

1871 M. Arnold *Let.* 18 Aug. (1895) II. 62, I am one of the true likers of the Continent. 1890 W. C. Gannett *Blessed be Drudgery* 41 Each of us is ringed about by two circles... The outer circle is the circle of our Likers, the inner is the circle of our Lovers.

likkewaan (li·kəvān). Also **lagavaan, likawaan.** [Afrikaans.] = Leguan. Cf. *iguana 2.

1907 P. Fitzpatrick *Jock of Bushveld* 315 There was not even a lagavaan slide, a game path, or a drinking place. 1914 *Farmer's Annual* (S. Afr.) 334 Likavaan Skin. How to cure it. Take a quantity of alum and place it along with a small quantity of water in an old iron pot. 1936 Williams & May *I am Black* xviii. 187 That boy was Shabala. It was he who had captured the likkewan [*in 1949 edition corrected to* likkewaan]. 1949 *Cape Argus Mag.* 14 May 10 The dour old likkewaan with his whiplash of a tail, can still be seen creeping through the shadows in search of birds' eggs to suck. 1971 H. C. Bosman *Bekkersdal Marathon* 123 'What I mean is all right, for instance,' Chris Welman continued, 'is to push a small likkewaan down the back of a visitor's neck, and to pretend to him that it's a mamba.'

likuta (likū·ta). Pl. **makuta.** [Native word; etym. uncertain; perh. f. Nupe *kuta* stone.] A coin and monetary unit introduced in the Democratic Republic of the Congo (now Zaïre) in June 1967, whose value is one hundredth of a zaïre.

1967 in Webster Add. 1971 *N.Y. Times* 27 June 5 The monetary unit [of the Democratic Republic of the Congo] is the Zaïre (Z.) divided into 100 makuta (K.) each worth two American cents. 'Makuta' is the plural form of the word 'likuta'. One Z. equals two U.S. dollars. 1972 *Times* 6 Dec. (Europe & Third World Suppl.) p. iii/2 (Advt.), The Zaïre is divided into 100 Makuta and the Likuta (the singular of Makuta) in turn consists of 100 Sengi.

lil (lil), *a.* Also **li'l.** Colloq. contraction of Little *a.*

1881 J. C. Harris *Nights with Uncle Remus* (1884) lx. 258 One ole Bear..hab one, two lilly Bear... Lil boy Bear, 'e des lahff'.. Lil gal Bear, 'e look skeer. 1912 C. E. Mulford in *Pearson's Mag.* May 627/1 Purty, all right. Brown hair an' I reckon brown eyes. Nice li'l girl. 1912 C. Mackenzie *Carnival* v. 54 Would you like him to give you a lil girl like me? 1932 L. Golding *Magnolia St.* iii. iii. 516 Good ole Mick!.. Plucky lil bastard! 1953 K. Tennant *Joyful Condemned* xxxix. 391 We'll go out and see your Mom. Take her a lil present, maybe. 1967 K. Allsop *Hard Travellin'* xxiii. 274, I can fiddle a li'l bit. 1967 Perry & Aldridge *Penguin Bk. Comics* iii. 99 Much of Li'l Abner has been absorbed into American folk-lore. 1969 'I. Drummond' *Man with Tiny Head* 7 Where's girl? Where's lil girl? 1975 M. Kenyon *Mr Big* xviii. 165 Phoebe said, 'That's a helluva greeting for a li'l ladykin.'

lila (lī·la). *India.* [ad. Skr. *līlā* play, sport, amusement.] In Hindu mythology, the cosmic dance or playful activity of the Supreme Spirit manifested in the Universe (see quots.); also, the name of an Indian dance representing this.

1828 H. H. Wilson in *Asiatick Researches* XVI. 115 His [*sc.* Krishna's] feats, in which his juvenile characters are regarded, are his *Līlā*, or sport. 1861 —— *Ess. & Lect. Relig. Hindus* I. 124 In this description of creation.. the deity [*sc.* Krishna] is still spoken of as a young man... The acts of divinity are his *Līlā*, or sport. 1917 L. L. P. Noble in A. Coomaraswamy *Dance of Śiva* (1924) 63 The Spirit..is all rapture, all bliss, in this play (*līlā*) Free, divine, in this love struggle. 1924 A. Coomaraswamy *Dance of Śiva* 62 The conception of the world process as the Lord's pastime or amusement (*līlā*) is also prominent in the Saiva scriptures. 1926 *Indian Art & Lett.* II. 78 The *Svabhāva*, or nature of Being-Power, is *Lilā*, or Play, a term which means free spontaneous activity. 1962 A. Huxley *Island* x. 167 Shiva-Nataraja dances the dance of endless becoming and passing away. It's his *lila*, his cosmic play. 1967 Singha & Massey *Indian Dances* xi. 109 Their dance is the *lila* or play of cosmic forces and serves as a preface to the drama which is to follow. 1972 N. Hein *Miracle Plays of Mathurā* iv. 70 Not only Vishnu's creation of the cosmos is viewed as līlā, but also his actions within the cosmos when he enters his creation in the form of his various incarnations.

lilac. Add: **2. c.** The scent of lilac, esp. as used in cosmetics, etc.

1895 *Montgomery Ward Catal.* 259/3 Perfumes... White Lilac. 1898 *Illustr. London News* 1 Jan. 34 (Advt.), Lance Perfumes..acacia, violet, lilac..price one shilling. *c* 1938 *Fortnum & Mason Price List* 54/1 Soaps..Lilac. 1951 A. Langenbach *Wines Germany* xxviii. 126 The finest Rheingau growths..have a delicious lilac-like bouquet. 1972 *Guardian* 22 Aug. 9/4 Jackson's have revived the floral perfumes... The current range comprises sweetpea ..purple lilac, and jonquil.

3. *lilac-blossom, -time*; **lilac-breasted roller,** *Coracias caudata*, a bird found in the southern half of Africa.

1890 O. Wilde *Dorian Gray* ii. 15 Lord Henry..found Dorian Gray burying his face in the great cool lilac-blossoms. 1974 A. Goodard *Vienna Pursuit* iii. 115 Some lilac blossom had been blown..across the pavement. 1908 Haagner & Ivy *Sk. S. Afr. Bird-Life* 29 The beautiful Lilac-breasted Roller (*C[oracias] caudatus*) is green, blue and brown, with the breast of a purplish-lilac tinge with white shaft-streaks, and the abdomen blue... This bird is not uncommon in the Transvaal bushveld. 1947 J. Stevenson-Hamilton *Wild Life S. Afr.* xxxiv. 293 The bright plumage..of the lilac-breasted roller. 1971 *Country Life* 28 Oct. 1126/2 [In Ethiopia] I first saw those gorgeously-coloured birds, the lilac-breasted rollers. 1906 W. Robinson *Eng. Flower Garden* (ed. 10) 881/1 To have a good Lilac-time it is essential to have the newer varieties raised in France. 1910 A. Noyes *Barrel Organ* in *Coll. Poems* I. 129 Go down to Kew in lilac-time (it isn't far from London!).

lilacky (lai·lăki), *a.* Also **lilacy.** [f. Lilac 2.] Of a lilac colour.

1863 Queen Victoria *Let.* 23 Dec. in R. Fulford *Dearest Mama* (1968) 281 The sky is a leaden, lilacy blue—with no white clouds. 1910 *Westm. Gaz.* 19 Feb. 2/2 Walls on walls of lilacky limestone.

lilipi (li·lipi). Chiefly *N.Z. Hist.* Also **lilipu, lillipe, lillip(p)ee.** [Origin unknown.] (See quot. 1861.)

In quot. 1830, from a W. African context, *lilipee* may be some other word.

1830 H. Crow *Mem.* ix. 146 For the sick we provided strong soups and middle messes, prepared from mutton, goats'-flesh, fowls, &c. to which were added sago and lilipees. 1860 *Taranaki Punch* I. iv. 8 Something..that, although it is *not* lilipi yet, when pronounced by our old friend.., sounds very like it, namely 'lilly pay'. 1861 J. von Haast *Rep. Topogr. & Geol. Explor. Nelson Province* i. 22 [We confined] ourselves to a small pot of lillipe (or boiled flour) twice a day. 1874 L. J. Kennaway *Crusts* 147 The poor children..lived principally on heavy bread and 'lillippee', (a mixture nothing more nor less than paper-hanging paste). 1880 J. C. Crawford *Recoll. Trav. N.Z. & Austral.* 99 One old lady..presented us with a dish of *lillipee*, which is simply flour and water. 1915 J. Hay *Reminisc. Earliest Canterbury* i. 34 They [*sc.* the Maoris] were very fond..of wheat and sugar boiled together, to which they gave the name of 'Lilipu'.

lilium (li·liŏm). [L. *līlium*: see Lily. Adopted by Linnæus in his *Species Plantarum* (1753) I. 302 as the name of a genus.] = Lily 1.

1902 *Westm. Gaz.* 10 Dec. 10/1 Of all flowers none are affected by frost so much as roses,..violets, and liliums. 1903 *Ibid.* 26 Nov. 12/1 White liliums are fetching 4s. a bunch. 1923 *Chambers's Jrnl.* Dec. 786/2 The finest liliums and choice gladioli grow superbly. 1958 M. West *Second Victory* vi. 91 Hot-house blooms that must have come from a dozen private homes—cyclamen, orchids, liliums and azaleas. 1970 *Lily Year Bk.* XXXIII. 88 [In Australia] at least four leading lilium nurserymen of a decade ago no longer distribute lilium bulbs in appreciable numbers.

Lille (līl). The name of a city in the Nord department of France, used *attrib.* to designate a kind of pillow or bobbin lace.

1865 F. B. Palliser *Hist. Lace* xvi. 211 The semé of little square dots on the ground—one of the characteristics of Lille lace—is still retained. 1900 E. Jackson *Hist. Hand-Made Lace* 176 Lille laces have always been favourites in England, the black especially. 1953 M. Powys *Lace & Lace-Making* iv. 31 The Lille lace edgings generally have a straight edge. 1960 *Connoisseur's Handbk. Antique Collecting* 164/1 *Lille lace*, pillow lace very popular in later 18th cent. England, made sometimes in black, but never as a dress lace.

Lillet (li·le). Also **Kina Lillet.** The proprietary name of a French aperitif; also, a glass of this wine.

1930 H. Craddock *Savoy Cocktail Bk.* 41 *Campden Cocktail*, ½ Dry Gin. ¼ Cointreau. ¼ Kina Lillet. 1951 R. Postgate *Plain Man's Guide to Wine* iii. 59 St. Raphael is slightly sweeter, Byrrh and Lillet slightly sharper. 1960 *New Yorker* 29 Oct. 51/2 Sipping a Lillet-and-gin. 1972 N. Freeling *Long Silence* I. 59 Now—Campari, Lillet, Chambéry? 1973 *Vogue* Jan. 85/2 A twist of orange with Lillet.

lillipe, lillip(p)ee, varr. *lilipi.

lilly-pilly. Also **lilli-pilli.** Substitute for def.: An Australian evergreen tree, *Eugenia* (or *Acmena*) *smithii*, of the family Myrtaceæ, or the timber obtained from it. Also *attrib.* (Later examples.)

1881 *Off. Rec. Sydney Internat. Exhib. 1879* 723 Lilly Pilly. *Eugenia Smithii*... Forms a beautiful shrub when cut back... The wood makes good axe handles. 1936 F. Clune *Roaming round Darling* iii. 26 In 1881 King George V planted a lillipilli-tree. 1944 *Living off Land* ii. 45 Lilly-pilly or Brush Cherry. 1946 *Coast to Coast 1945* 56 A tree-fern, maidenhair in the mossy banks, and a clump of lilli-pillis marked the dell. 1965 *Austral. Encycl.* III. 412/1 *Acmena* is a small genus containing the common lilly-pilly (*A. smithii*), which is widely spread in eastern Australia along rivers and streams and in the rainforests... It has small terminal sprays of insignificant greenish flowers followed by purplish to white fruits, and is much cultivated as a hedge plant. 1965 *Courier-Mail* (Brisbane) 16 Nov. 20 The feathered folk in the lilipillies of the great gully below began to welcome Piccaninny Daylight.

Li-Lo (lai·lōᵘ). Also **Lilo, lilo.** [f. *to lie low.*] The proprietary name of a type of air-bed or inflatable rubber mattress. Also *fig.*

1936 *Trade Marks Jrnl.* 16 Sept. 1150 Li-Lo... Airbeds, air-pillows and air-pillow-bags, all made principally of india-rubber. P. B. Cow & Company Limited,.. London,..manufacturers. 1939 'N. Shute' *What happened to Corbetts* i. 1 Sophie their nurse was lying on a Li-Lo on the oil-stained floor, covered with an eiderdown. 1949 O. Lancaster *Drayneflete Revealed* 63 And Dido on her lilo à sa proie attachée. 1954 W. Noyce *South Col* iv. 58 The really experienced expeditioners, Ed and George Lowe, had their Li-Los inflated. 1960 V. Gielgud *To Bed at Noon* i. iv. 27 If the chairs were no more than adequate there were lilos and cushions in abundance. 1969 *Listener* 24 July 114/3 We've been looking at grand pianos... This is one likely spot where a nudist with an eye to the main chance might well secrete himself. A lilo across the wires. 1972 N. Bentley *Events of that Week* 39, I left Theresa on a Li-lo and put my sandals on. 1973 *Guardian* 13 Oct. 10/1 The social service Lilo has become too easy to loll on.

lilt, *v.* Add: **1. b.** Also with *out.*

1916 A. Bennett *Lion's Share* xxv. 191 Musa lilted out the delicate, gay phrases of Debussy.

3. (Later examples.)

1901 Kipling *Traffics & Discov.* (1904) 79 He lilted a little on his feet when he was pleased. *Ibid.* 80 He went to England, and he became a young man, and back he came, lilting a little in his walk.

lilting, *ppl. a.* Add: Also of one's gait: (sense 3 of vb.).

1903 *Longman's Mag.* Jan. 271 Swinging down the street with an easy lilting stride..marched two Englishmen, soldiers both. 1965 E. Bhavnani *Dance in India* xvi. 208 In a lilting change of movements, boys and girls hold hands or link arms and dance round and round in a circle.

lily. Add: **2. c.** The scent of lily of the valley, esp. as used in cosmetics, etc.

1890–91 *T. Eaton & Co. Catal.* Fall & Winter 42/2 Morse's perfumes..new-mown hay, lily of the valley, 25c. per bottle. 1970 *Guardian* 12 May 9/3 This..sprightly fragrance with..notes of carnation,..lily-of-the-valley and roses.

3. b. Used as a term of abuse, esp. of a man to imply lack of masculinity.

1923 G. Saintsbury *Second Scrap Bk.* v. 39 But in order once more to consider and console that lily, the Educational Expert, let us turn to 'grind'. 1929 Hostetter & Beesley *It's a Racket!* 231 *Lily*, an easy victim, exceptionally gullible person. 1930 D. H. Lawrence *Nettles* 19 And Mr. Mead, that old old lily Said: 'Gross! coarse! hideous!' 1933 S. Spender *Poems* 28 Here the pale lily boys flaunt their bright lips. 1958 J. Raymond *England's on Anvil!* 142 In this he differed from men like William ('Cory') Johnson, Oscar Browning, A. C. Benson and the rest of the Eton-and-King's lilies who were such a lush feature of the period.

4. d. *pl.* The bound feet of Chinese women, in allusion to their Chinese designation *kinleen* 'golden water-lilies'. Also (in sing.) *attrib.* So **lily-footed** *a.*

1841 W. B. Langdon *Descr. Catal. Chinese Collection in Philadelphia* 15 The footstools upon which their 'golden lilies' rest, are covered with embroidered silk. 1886 C. M. Yonge *Chantry House* II. xx. 188 Is he going to wed a fair Chinese with lily feet? 1922 W. S. Maugham *On Chinese Screen* xviii. 72 They rest there for a while on their small feet, their golden lilies, gossiping elegantly. 1933 N. Waln *House of Exile* i. i. 26 We could not walk, as..Mai-da's mother..had 'lily' feet. 1937 E. Snow *Red Star over China* i. ii. 26 Yang Hu-Cheng..was a two-wife man. The first was the lily-footed wife of his youth.

4*. Phr. *to paint* (or *to gild*) *the lily*: to embellish excessively, to add ornament where none is needed.

1595 Shakes. *John* iv. ii. 11 To gilde refined Gold, to paint the Lilly; To throw a perfume on the Violet,.. Is wastefull, and ridiculous excesse. 1919 H. Jenkins *John Dene of Toronto* vii. 113 'Where's Finlay?' asked Colonel Walton. 'He's painting the lily... Seeing how near he can get to this Bergen fellow.' 1928 *Manch. Guardian Weekly* 28 Sept. 243/3 Nature and history have already been so kind to that ancient and charming townlet on the Dart that improvement would be a gilding of the lily. 1935 J. Buchan *House of Four Winds* 22 It's rather like painting the lily, you know. 1953 *Manch. Guardian Weekly* 19 Feb. 13/2 While it may seem to be painting the lily, I should like to add somewhat to Mr Alistair Cooke's excellent article. 1958 J. Raymond *England's on Anvil!* 15 In Englishing the passage, Peter Motteux..contrives at once splendidly to gild the lily and tone down the anti-Protestantism. 1968 *Encycl. Brit.* XII. 842/1 The favourite technique of decoration of Mogul jades is insetting with gold and precious stones..an example of painting the lily that would hardly have commended itself to the Chinese jade carver.

5. *lily-coloured, -green, -scented, -sweet, -yellow* adjs.; **lily-pad** (earlier and later examples); **lily-pond,** a pond in which water-lilies are grown; **lily-trotter,** a water-bird of the family Jacanidæ, esp. *Actophilornis africana*, found in tropical Africa, or *Microparra capensis*, the lesser lily-trotter, found in east Africa; also = Jacana.

c 1866 G. M. Hopkins *Poems* (1918) 9 Lily-coloured clothes provide Your spouse not laboured-at nor spun. 1875 Browning *Inn Album* ii. 72 My big and bony, here, against the bunch Of lily-coloured five with signet-ring. 1739 tr. *Art of Painting in Miniature* (ed. 4) 13 Lilly-Green, Sap-Green, and Gamboge..must be temper'd with fair Water only. 1965 S. Gibbons in J. Gibb *Light on C. S. Lewis* v. 87 Here she was, the right descendant of Grendel, with her lily-green complexion. 1843 *Knickerbocker* XXII. 1 Huge moccasin darting away beneath the dense reeds and lily-pads of the swamp. 1875 J. G. Holland *Sevenoaks* v. 65 A deer, feeding among the lily-pads. 1888 *Nation* (N.Y.) 19 July 57/2 The trout breaking at the edge of the lily-pads. 1946 K. Tennant *Lost Haven* (1947) 2 Shallow blue water from which the great white paperbarks tower shadowing the lily-pads. 1958 G. Durrell *Encounters with Animals* i. 38, I had watched her standing on the lily-pads. 1901 G. Jekyll *Wall & Water Gardens* xx. 161 Such a scene as Mr. Robinson's Lily pond in North Sussex..could scarcely be bettered. 1974 R. Harris *Double Snare* iv. 27 From the direction of the lily pond comes the croak of little frogs. 1796 Coleridge *Poems* 18 Summer's lily-scented plume. 1869 Browning *Ring & Bk.* III. vii. 50 The sword I wear shall pink His lily-scented cassock through and through. 1936 R. Campbell *Mithraic Emblems* 31 Out of a wound that never heals Rills forth the lily-scented blood. 1931 V. Woolf *Waves* 290 Let us commit any blasphemy of laughter and criticism rather than exude this lily-sweet glue. 1920 *Blackw. Mag.* May 649/2 The busy lily trotter, hurrying across the broad flat water-lily leaves. 1951 R. Campbell *Light on Dark Horse* 82 Those strange little birds, the lily-trotters. 1958 G. Durrell *Encounters with Animals* i. 35 It is with the aid of these long toes and the even distribution of weight that they give that the jacana manages to walk across water, using the water-lily leaves and other water-plants as its pathways. It has thus earned its name of lily-trotter. 1971 *Country Life* 30 Sept. 830/1 The lakes [in Tanzania] give you close views of ibises, egrets,..lily trotters. *c* 1865 G. M. Hopkins *Poems* (1948) 123 Lily-yellow is the west.

B. *adj.* **b.** *lily-faced* (later example); **lily-liveredly** *adv.*

1929 D. H. Lawrence *Pansies* 48 It's either you fight or you die, Die, die, lily-liveredly die. 1934 Dylan Thomas *Let.* 14 Jan. (1966) 92 As the black man must have first regarded the features of his lily-faced brother.

li-lying, *vbl. sb. rare*⁻¹. = *lily-work* (Lily A. 5).

1874 G. M. Hopkins *Jrnls. & Papers* (1959) 248 The touching and passionate curves of the lilyings in the iron-work.

lily-white, *a.* Add: **2.** In favour of, committed to, or pertaining to a policy of racial segregation. orig. *U.S.*

1903 *N.Y. Times* 23 Sept., The report that the President was seeking reconciliation with the 'Lilywhite' faction, which eliminated the negro from the last State Convention. 1909 *Westm. Gaz.* 13 Feb. 2/2 That..is what they call the lily-white policy!.. It is the unprincipled white politician who finds anti-negro agitation a popular plank in his platform. 1953 *Manch. Guardian Weekly* 8 Oct. 5/1 Before 1948 groups of landlords had managed to maintain 'lily-white' communities by signing 'restrictive covenants'. 1968 *Morning Star* 10 Aug. 1/1 While the lilywhite Republican convention was nominating Richard Nixon in an atmosphere of ballyhoo and frenzied hysteria, the same city of Miami saw Negroes demonstrating for their rights.

b. Irreproachable, lacking faults or imperfections.

1961 in Webster. 1970 *New Yorker* 9 May 33/3, I think the city should be lily-white on this, so the first thing I'm going to try to do is convert all our city vehicles to low-pollution engines. 1973 *Times* 18 Jan. 2/7 Robert Mark [the commissioner] is determined to have a lily-white police force. He will have a lily-white police force looking pretty in the street.

Lima. Add: **Lima bean** (earlier and later examples).

1819 [see *butter-bean]. 1831 M. Holley *Texas* (1833) xi. 123 He had known winters here so mild, as not to kill the Lima bean. 1969 *Northwest* (*Sunday Oregonian Mag.*) 14 Dec. 19/1 The pulp [of pawpaw] is yellow or orange and contains several brown seeds about the size and shape of a Lima Bean.

b. *ellipt.* = Lima bean. *U.S.*

1856 F. S. Cozzens *Sparrowgrass Papers* vii. 85 Put the Limas to the right..and as for the rest of the seeds sweep them into the refuse basket. 1865 *Trans. Illinois Agric. Soc.* V. 758 Pole Beans—Amongst these the Limas deservedly rank the highest. 1942 E. Paul *Narrow St.* vi. 51 The Épicerie Danton had limas, normal-sized and 'baby', Canterbury, scarlet runners, pintos, [etc.].

limb, *sb.*¹ Add: **2. b.** (Later examples.)

1885 in Farmer *Slang* (1891) II. 18/2 Between you'n me, red stockings ain't becomin' to all—ahem—limbs. 1898 M. Deland *Old Chester Tales* 237 But it was she who informed him that he might stay until his 'limb' permitted him to walk. 1902 H. L. Wilson *Spenders* xxxi. 369 One of my maids who slipped on the avenue yesterday and fractured one of her—er—limbs. 1904 *Courier-Jrnl.* (Louisville, Kentucky) 5 Sept. 1 Her limbs were void of shoes or stockings. 1924 W. M. Raine *Troubled Waters* i. 12 She dexterously arranged the skirt without being able to conceal some inches of slender limb rising from a well-turned ankle.

4. b. Also, the part of a compound core of a transformer, electromagnet, etc., on which a coil is wound.

1902 *Encycl. Brit.* XXVII. 584/1 These [portions] are: (1) the magnet 'cores' or 'limbs', carrying the exciting coils whereby the inert iron is converted into an electromagnet; (2) the yoke, which joins the limbs together and conducts the flux between them; and (3) the pole-pieces. 1934 H. Cotton *Design Electr. Machinery* viii. 162 With core-type transformers the cross-section of the limbs may be rectangular in the case of small transformers, but it is more usual to adopt for all sizes a cross-section which fits as closely as possible into a circumscribing circle. 1943 *Gloss. Terms Electr. Engin.* (*B.S.I.*) 40 Those parts of the [transformer] core surrounded by windings are termed legs or limbs and those not so surrounded are termed yokes.

d. *out on a limb*, in an isolated or stranded position; at a disadvantage. orig. *U.S.*

1897 A. H. Lewis *Wolfville* 59 Seven of us..seein' whatever can we tie down an' brand, when some Mexicans gets us out on a limb. 1934 *Amer. Speech* IX. 11/2 A player is *out on a limb* when he is allowed to play a hand at an overambitious contract. 1939 F. Scott Fitzgerald *Let.* Winter (1964) 50 She might not consider the rearrangement of someone else's words a literary composition, which would leave you out on a limb. 1943 J. B. Priestley *Daylight on Saturday* xxvii. 210 Somebody does something dam' silly, which probably means that some poor devils somewhere are left out on a limb. 1948 J. Steinbeck *Russ. Jrnl.* (1949) iii. 41 No one is willing to go out on any limb. No one is willing to say yes or no to a proposition. He must always go to someone higher. 1959 *Economist* 18 Apr. 214/2 President Nasser is out on a bit of a limb, but in this uncomfortable situation he can take comfort from the thought that there is still no other pan-Arab leader in sight. 1972 *Guardian* 7 Feb. 10/6 Once in the Commission, the British Civil Servant will feel out on a limb, away from the main-stream of his department. 1973 *Times* 23 May 16/5 At the international law of the sea conference Britain could find herself isolated and out on a limb.

5. *limb-dance, -fitter, -fitting*; **limb-bud** *Embryol.*, in an embryo, a small protuberance from which a limb develops; **limb-kinetic** *a.* *Path.*, denoting a form of apraxia (see quot. 1966); **limb-wood** (see quot.).

1906 G. R. Satterlee *Outl. Human Embryol.* v. 55 Outgrowths of mesenchyme occur from the lateral portion of the trunk. These projections are called the limb-buds, and are the *anlages* for the arms and legs. 1926 [see *cultivate *v.* 2 b]. 1965 L. B. Arey *Developmental Anat.* (ed. 7) xii. 210 The limb buds appear late in the fourth week as lateral swellings. *a* 1885 G. M. Hopkins *Poems* (1918) 79 While cripples are, while lepers, dancers in dismal limb-dance. 1967 *Economist* 8 Apr. 121/3 Hangers granted a rise..to its limb-fitters from the start of this year. 1920 *Glasgow Herald* 3 Dec. 8 The limb-fitting centres in the United Kingdom have been increased from 6 to 20. 1959 *Chambers's Encycl.* I. 652/1 Roehampton, the chief limb-fitting centre in England. 1914, 1933 Limb-kinetic [see *ideokinetic* adj. s.v. *ideo-]. 1966 *McGraw-Hill Encycl. Sci. & Technol.* I. 494/1 Several forms of apraxia are usually distinguished. The lowest order apraxia is called limb-kinetic or motor... Limb-kinetic apraxia refers to a loss of coordination usually affecting one upper limb only. Gross movements may be performed fairly well, whereas fine individual movements of the fingers are lost. 1901 J. Black *Illustr. Carpenter & Builder Ser.: Home Handicrafts* 62 [For mosaic work] black is obtained by using ebony or bog oak..green, by.. a species of native green oak, known as 'limb wood'.

limb, *sb.*² Add: **4.** Special Comb.: **limb-darkening** *Astr.*, the apparent darkening of the face of the sun towards its edges.

1931 *Monthly Notices R. Astr. Soc.* XCI. 893, I(*r*) is of the form *a–mx* (limb darkening linearly proportional to distance from the centre). 1938 *Astrophysical Jrnl.* LXXXVII. 45 (*heading*) The effect of an adiabatic layer upon solar limb darkening. 1962 *Science Survey* III. 103 Visual observations..show that the sun's disc is brightest at the centre, becoming gradually dimmer towards the outer edge, or limb. This phenomenon is known as limb darkening, and is due to the fact that we look less and less deeply into the hotter layers of the sun as we view its surface more obliquely.

limb, *v.* Add: **1. b.** To remove branches from (a tree).

1835 H. Evans *Jrnl.* 2 July in *Mississippi Valley Hist. Rev.* (1927) XIV. 202 Weather beaten cotton wood trees limbed and shattered by the storms of the prairies. 1839 E. Holmes *Rep. Explor. Aroostook River* 53 The best mode undoubtedly is to fall the trees and 'limb' them (that is, cut off the limbs,) in June. 1889 *Harper's Mag.* Jan. 231/1 It seemed to be built principally of alder poles well limbed off and placed, roughly speaking, side by side. 1971 *Timber Trades Jrnl.* 3 Apr. 58/2 The chainsaw has long been used for limbing hardwoods.

Limba¹ (li·mbă), *sb.* and *a.* Also **Limbah.** [Native name.] **A.** *sb.* **a.** A member of a West African people inhabiting Sierra Leone. **b.** The language of the Limbas. **B.** *adj.* Of or pertaining to this people or their language.

1902 *Encycl. Brit.* XX. 624/1 Sierra Leone is inhabited by about a dozen distinct African peoples, the most important being the Mende, Temne, Limba [etc.]. 1925 T. N. Goddard *Handbk. Sierra Leone* iii. 55 The *Limbas* may be found in the north of the Temne country. 1925 H. C. Luke *Bibliogr. Sierra Leone* (ed. 2) 149 The Wesleyan Methodist Catechism in Limbah. 1951 K. L. Little *Mende of Sierra Leone* 71 The Limba wear gowns of native cloth and tight shorts of the same material. 1954 R. Lewis *Sierra Leone* xii. 118 When a Limba chief dies, weird moans issue from it [*sc.* a cave]. Limbas can hear the dead man being brought to Kumba the Krifi. 1964 C. Fyfe *Sierra Leone Inheritance* 20 He killed the Limbas and sold them.

limba² (li·mbă). [f. Gabon name *limbo.*] The West African tree *Terminalia superba* or the hardwood obtained from it; = *afara.

[1937 J. M. Dalziel *Useful Plants W. Trop. Afr.* 83 The timber [of *Terminalia superba*] from the Belgian Congo is known as 'limbo (or limba) clair' when at least two-thirds of the diameter is light-coloured.] 1955 *Times* 6 May 12/6 Limba, a light honey-coloured hardwood from Africa. 1960 *House & Garden* Nov. 34/1 The new limba furniture has all the charm and grace one expects. *Ibid.*, An abundance of good designs in limba, tola, lacquer or oak.

limbal (li·mbăl), *a.* *Ophthalm.* [f. Limb(us + -al.] Of or pertaining to the limbus of the cornea.

1947 F. H. Adler *Gifford's Textbk. Ophthalm.* (ed. 4) xi. 221 Fine capillary loops from the anterior ciliary vessels pass into the cornea, forming the limbal arcades. 1958 H. B. Stallard *Eye Surg.* (ed. 3) viii. 570 The limbal incision covered by a conjunctival flap is preferable, for adjacent blood-supply is better and the firm closure of the wound is quicker than in the corneal section. 1972 *Virchows Archiv* A. CCCLV. 277 The limbal region of the bulbar conjunctiva was covered by a stratified squamous epithelium.

limbed, *a.* (Later examples.)

1899 *Echo* 9 Mar. 1/4 Every reader of Dickens remembers the frail ex-prisoner of the Bastille, white-haired and feeble-limbed. 1954 T. Gunn *Fighting Terms* 44 My hate throbs yet but I am feeble-limbed.

limber, *sb.*¹ Add: **3. limber-neck,** a kind of botulism affecting poultry, caused by the toxin produced by a type of the bacterium *Clostridium botulinum.*

1910 C. S. Valentine *How to keep Hens for Profit* 271 The disease called 'limberneck', in which the affected bird is unable to control the head, which droops to the ground, is said by many who have had large experience with it to be the result of stomach irritation brought on by eating maggots. 1927 E. T. Brown *'How to do it' Poultry Bk.* xxiii. 242 Limberneck or false cholera..is due entirely to a form of poisoning. 1931 Dickinson & Lewis *Poultry Enterprises* ix. 216 When flies and maggots infest putrid meat and other spoiled feeds, fowls eating them are likely to show symptoms of limber neck disease. 1964 M. Hynes *Med. Bacteriol.* (ed. 8) xx. 296 These types [of botulism toxin] are also most often responsible for 'limber-neck' of chickens and ducks.

limber, *a.* Add: **4.** *limber-footed* (later example).

1904 *Westm. Gaz.* 8 May 14/2 Here leaps the limber-footed, listening hare.

limber, *v.*¹ Add: Also with *up*, and *intr.* Hence **li·mbering** *vbl. sb.*, **limbering-up** *vbl. sb.*

1901 *Daily Colonist* (Victoria, B.C.) 8 Oct. 3/2 When her machinery is limbered up after it has been used a while she will do much better. 1921 *Ibid.* 5 Apr. 10/2 During the recent fine weather several of the [tennis club] members have been out limbering up after the inactivities of the Winter months. 1921 *Blackw. Mag.* Aug. 262/1 Dempsey had passed the afternoon in a 'limbering-up hike'. 1927 *Dancing Times* Dec. 301/1 Most dance students know that limbering is the basis for every kind of dance work. 1929 Wodehouse *Mr. Mulliner Speaking* ii. 64 He waggled his right leg for a moment to limber it up, backed a pace or two and crept forward. 1957 *Oxf. Pocket Bk. Athletic Training* (ed. 2) 27 An athlete who fails to limber-up properly deliberately handicaps himself. 1962 *Listener* 29 March 566/1 As images and figures in compositions..they limber or stretch or occasionally pose. 1963 H. Garner in R. Weaver *Canad. Short Stories* (1968) 2nd Ser. 23 The limbering up we were getting from our work. 1971 R. Dentry *Encounter at Kharmel* i. 1 He did twenty four half knee-bends to limber up.

limbered (li·mbəɹd), *a.* [f. Limber *sb.*¹] Having a limber.

1920 *Blackw. Mag.* Feb. 279/2, I lent him some men and a limbered waggon. 1942 *R.A.F. Jrnl.* 3 Oct. 18 The officers saw a German gun-team galloping their limbered gun away.

limberly (li·mbəɹli), *adv.* [f. Limber *a.*] In a limber or supple manner.

1891 *Harper's Mag.* Nov. 891/2 His long spare arms swing limberly before a long spare body. 1909 'O. Henry' *Roads of Destiny* xii. 186 They..slouched limberly over to the railroad eating-house. 1964 *N. Y. Times Mag.* 23 Aug. 30 It was sculpted to Chanel's small, athletic figure and moves with it as limberly as a leopard's own leopard skin.

limbic, *a.* Add: Also, of or pertaining to the limbic lobe or limbic system; *limbic system*, a region of the brain comprising the limbic lobe and certain neighbouring areas (see quot. 1967¹).

1952 *Electroencephalogr. & Clin. Neurophysiol.* IV. 407 The limbic system is comprised of the cortex contained in the great limbic lobe of Broca..together with subcortical cell stations. 1957 H. H. Jasper et al. *Reticular Formation Brain* 665 There is no doubt that all the varied elementary somatomotor and vegetative mechanisms so clearly influenced by limbic stimulation remain essentially undisturbed after bilateral limbic lesions. 1967 C. R. Noback *Human Nervous Syst.* viii. 224/1 The difficult-to-define limbic system consists of the limbic lobe..some subcortical nuclei (septal nuclei and amygdala), and neural pathways to other nuclear stations of the brain... Some investigators include the 'other' nuclei (habenular nucleus, hypothalamus, midbrain tegmentum, thalamus, and interpeduncular nucleus) in the limbic system. *Ibid.* 226/2 The midbrain and thalamic reticular nuclei receive inputs from several limbic pathways. 1970 *Nature* 28 Feb. 797/1 Other behaviour patterns..have been elicited by stimulating hypothalamic and limbic regions in many species. 1971 J. Z. Young *Introd. Study Man* x. 136 Sohilder's interpretation of the limbic system is as 'a mechanism for emotion'.

limbo¹. Add: **2. c.** A type of anti-submarine mortar. Also *attrib.* or as *adj.*

1955 *Times* 20 June 4/6 The frigate Grenville fired live projectiles from her Limbo anti-submarine weapon. *Ibid.*, The Limbo..is a multi-barrelled mortar of large calibre, linked automatically with a submarine detector of advanced design. 1956 *Jane's Fighting Ships 1956–57* 240/2 Have some side armour as well as deck protection; limbo type anti-submarine rocket throwers. 1957 *Jane's Fighting Ships 1957–58* 42/1 The two Limbos can each fire a pattern of large depth bombs with great accuracy. 1961 T. D. Manning *Brit. Destroyer* 24 The Squid has been improved on by Limbo which..is not fitted in destroyers but only in frigates.

limbo³ (li·mbo). [Native name.] A dance in which the dancer bends backwards and passes under a horizontal bar raised only a few inches off the ground. Also *attrib.*

1956 *Caribbean Q.* IV. iii. & iv. 204 The firemen also had a characteristic dance similar to the 'limbo', with the body bent sharply backward, knees projecting, and lower legs almost on the pavement. 1958 *Daily Express* 22 Apr. 7/3 The Princess should take home with her memories of gay calypsos and limbos. 1963 *Pix* 2 Mar. 38 Peters said that he first encountered the Limbo on his Pacific cruise.. when it was merely known as 'going under the pole'. 1966 *Observer* 13 Feb. 40/4 Trinidad's champion limbo dancer.. can ease himself blindfold under a bar which is only 8 in. from the floor and is sometimes soaked in spirit and set alight. 1971 *Country Life* 18 Feb. 372 And watching limbo ladies slither and men blow fire. 1972 E. Hargreaves *Fair Green Weed* i. 12 He..found the pair of them sitting on the patio, watching some limbo dancers on the beach. 1973 *Sunday Advocate-News* (Barbados) 21 Jan. 11/3 More variety should be introduced into these shows, including more steelband music, limbo dancing and folk dancing.

limbric (li·mbrik). (See quot. 1960.)

1930 *Economist* 18 Oct. 713/1 There is still some business to be done in other lines, such as mercerised cotton brocades and limbrics. 1940 *Chambers's Techn. Dict.* 500/2 *Limbric*, a plain grey cotton cloth of medium quality; used for curtains, etc. after being piece-dyed or printed. 1940 *Manch. Guardian Weekly* 25 Oct. 307 It seems unlikely..that the restrictions of silk piece-goods and made-up goods will cause a larger trade to be done in cotton poplins, cambrics, limbrics, etc. 1960 *Textile Terms & Definitions* (Textile Inst.) (ed. 4) 91 *Limbric*, a light-to-medium weight, closely woven, plain-weave cotton cloth made from good quality yarns.

Limbu (li·mbū), *sb.* and *a.* Also **Limboo.** [Native name.] **A.** *sb.* **a.** A member of a Mongoloid people of eastern Nepal; this people collectively. **b.** The Tibeto-Burman language of the Limbu people. **B.** *adj.* Of or pertaining to the Limbu people or their language.

1819 F. Hamilton *Acct. Kingdom Nepal* i. 54 Among the Kirats was settled a tribe called Limbu... Their languages are said to be different. 1840 *Jrnl. R. Asiatic Soc. Bengal* IX. 495 The Limboos form a large portion of the inhabitants in the mountainous country lying between the Dood-Koosi and the Kanki rivers, in Nipal. 1854 J. D. Hooker *Himalayan Journals* I. v. 138 The Limboo language is totally different from the Lepcha, with less of the *z* in it, and more labials and palatals, hence more pleasing. 1880 H. A. Oldfield *Sk. from Nipal* I. iii. 53 The country of the Kirantis and Limbus at the present time is divided into fifty-two small subahships. 1909 G. A. Grierson *Ling. Surv. India* III. i. 283 The Limbus are one of the principal tribes of Eastern Nepal... We have no information about the number of speakers of Limbu in Nepal. 1928 Northey & Morris *Gurkhas* xv. 217 The Limbu race is now divided into a large number of tribes. These tribes were formerly centred, in groups, in ten different districts of the Limbu country. 1939 R. Godden *Black Narcissus* v. 54 She was learning to distinguish between the State natives and the hill clans, the Lepchas and Bhotiyas and Limbus. 1970 Rose & Fisher *Politics Nepal* i. 12 The Limbus,.. are making a serious effort to revive their local language as a literary medium after long years of its suppression under Gorkha rule. 1971 K. Kent in C. Bonington *Annapurna South Face* App. H. 310 Later, the Eastern tribes (Rais and Limbus) were also recruited, and, together with some lesser-known but other martial tribes, also became known under the general title of Gurkhas.

Limburger (li·mbɒ̄ɹgəɹ). [a. Du. and G. *Limburger.*] **1.** *attrib.* with *cheese*, or *ellipt.* A soft strong-smelling cheese made in the province of Limburg in Belgium. Also *Limburg cheese.*

1817 C. Campbell *Traveller's Compl. Guide through Belgium & Holland* (ed. 2) iv. 92 The disagreeable smell of the Limburg cheese, called *Herve*, though unpleasant to many, is justly ranked among the delicacies of the rich. 1859 W. H. J. Weale *Belgium* p. xxvii, *Limburg* and *Herve cheese* are also very good, but their odour is intolerable. *c* 1870 *More Yankee Drolleries* ii. 15 Limburger cheese, an abomination guilty of the most powerful odour. 1883 *Authors & Publishers* (G. P. Putnam's Sons) 45 The cheese-monger may not like either Stilton or Roquefort, he may even have his doubts as to the absolute wholesomeness of the more athletic brands of Limburger. 1887 *Harper's Mag.* Mar. 644/1 The obese Teuton delights in 'loud' Limburger and Gruyere. 1905 W. H. Simmonds *Pract. Grocer* III. 88 Limburger is a Belgian cheese of the soft class, which is allowed to ripen before use. 1908 [see *Hamburger 2]. 1966 Marquis & Haskell *Cheese Bk.* iii. 197 Strong cheeses like Limburger and Liederkranz are not for cocktail parties because after an hour or so at warm temperatures they will all but clear the room.

2. a. A native or inhabitant of Limburg. **b.** (Also **Limbu·rgerish.**) The dialect of the Limburgers. Also as *adj.*

1932 *Times Lit. Suppl.* 16 June 446/4 But this Limburgerish would not be a living language if it could be reduced to simple rules. *Ibid.*, Something for which Limburger grammarians have no doubt an appropriate name. 1963 N. Freeling *Gun before Butter* 51 He was no fool. A Limburger... The people from 'below the rivers' are quicker..than the Dutch of Holland.

limbus. 2. (Examples in *Anat.*)

1699 M. Lister *Journey to Paris* 71 The Membrane or Valve on the Left side of the *Foramen Ovale*..extended almost over the hole, without any *Limbus* round its edges. 1877 W. Turner *Introd. Human Anat.* vi. 368 Another membrane..arises from a denticulated spiral crest, the limbus or crista spiralis. 1954 S. Duke-Elder *Parsons' Dis. Eye* (ed. 12) i. 3 The cornea is set into the sclera like a watch glass so that the latter overlaps the cornea all round the periphery; the junction of the two tissues is known as the limbus. 1961 *Lancet* 22 July 166/2 The blade of the knife emerges 0·5–1·0 mm. on the corneal side of the limbus when the section is complete.

lime, *sb.*¹ Add: **3. e.** A vat containing a solution of lime for removing the hair from skins; the solution itself.

1885 C. T. Davis *Manuf. Leather* xxxii. 525 When sufficiently softened the skins are next placed in the 'limes'... The goat-skins remain in the 'limes' about 14 days. 1903 L. A. Flemming *Pract. Tanning* 6 When vat room is scarce, it is good practice to haul the skins out after they have been in the lime a few days. 1946 J. W. Waterer *Leather* ii. ii. 137 The practice of passing the 'packs' of hides..through a series of liquors, commencing with an old or 'mellow' lime. 1969 T. C. Thorstensen *Pract. Leather Technol.* vi. 87 When the hide is introduced into the lime.

5. *lime-burning, -cask, -grout, -hater* (so *-hating* adj.), *-lover* (so *-loving* adj.), *-ooze*; *lime-free* adj.; **lime-rock** (earlier and later examples); **lime-silicate** *a.* *Petrol.*, applied to a rock which was originally an impure limestone or dolomite and has been thermally metamorphosed, with the result that the lime has combined with silica present as impurities to form calcium silicates; **lime-sink** (earlier example); **lime soap,** a mixture of insoluble calcium salts of fatty acids formed as a precipitate when soap is used in hard water and manufactured for various industrial purposes; **lime-soda** *attrib.*, applied to a process for softening water by treatment with lime and sodium carbonate; **lime-sulphur,** an insecticide and fungicide containing calcium polysulphides which is made by boiling lime and sulphur in water; **lime-work,** (*b*) (later example).

1860 J. S. C. Abbott *South & North* 196 [Slaves] employed..in lime-burning or fishing. 1865 Mrs. Stowe *House & Home Papers* 94 Seating himself on a lime-cask which the plasterers had left. 1935 A. G. L. Hellyer *Pract. Gardening* iii. 34 It is wise to make enquiry when ordering rock plants and shrubs as to whether any of those supplied are likely to require lime-free soil. 1974 *Country Life* 28 Nov. 1639/1 V[*iburnum*] *furcatum*..needs a lime-free soil. 1875 R. R. Brash *Eccl. Archit. Ireland* 8 The interior of the walls is filled with small stones and lime-grout. 1907 R. Farrer *My Rock-Garden* x. 148 The plant [*sc. Cenisia excisa*]..is a real lime-hater. 1935 A. G. L. HELLYER *Pract. Gardening* xv. 112 The best plan is to set aside a portion of the rock-garden for lime-haters. *Ibid.* 284/1 (*index*) Lime hating plants. 1971 *Country Life* 18 Feb. 368/1 For some reason, many of the lime-hating shrubs have only white flowers. 1907 R. FARRER *My Rock-Garden* i. 12 At this point I will not embark on the awful question of lime-lovers and peat-lovers. 1971 MRS. D. UNDERWOOD *Grey & Silver Plants* iv. 35 The root fibres of the lime-lovers will wrap themselves round individual [limestone] chippings. 1916 *Nature* 2 Nov. 172/2 It seems probable from the evidence now before us that some of Forrest's newly discovered Chinese rhododendrons..must be reckoned as lime-loving species. 1956 MCCLINTOCK & FITTER *Collins' Pocket Guide Wild Flowers* 306 Limestone Polypody..is the lime-loving counterpart of the Oak Fern. 1974 *Country Life* 28 Nov. 1660/1 Used mushroom compost..is greatly appreciated by lime-loving cherries. 1867 J. N. EDWARDS *Shelby* xxix. 498 He..disappeared for a moment beneath lime ooze, half tanned hides and the smell of a charnel house. 1665 in *Early Rec. Providence* (Rhode Island) (1893) III. 66 Those Lime Rocks about Hackletons lime Killne shal be perpetually Common. 1969 *Islander* (Victoria, B.C.) 22 June 7/1 He had to tow empty scows from the cement works to Tod Inlet and back with full scows of limerock for making cement. 1888 J. J. H. TEALL *Brit. Petrogr.* 464/2 Lime-silicate hornfels. 1902 A. HARKER *Petrol.* (ed. 3) xx. 306 The carbonic acid is completely eliminated, and the whole converted into a lime-silicate-rock (the German 'Kalksilikathornfels' or 'Kalkhornfels'). 1965 G. J. WILLIAMS *Econ. Geol. N.Z.* x. 153/1 Lime-silicate dyke rocks in the ultramafic rocks of southern Westland. 1837 J. L. WILLIAMS *Territory of Florida* 9 Ponds and lime sinks are numerous between the..rivers. 1857 W. A. MILLER *Elem. Chem.* III. vi. 373 The tallow is melted by injecting hot steam into the vat which contains it, and milk of lime is added... An insoluble lime soap is thus formed. 1884 [in Dict. s.v. SOAP *sb.* 2]. 1918 C. M. WHITTAKER *Applic. Coal Tar Dyestuffs* iii. 36 Lime soaps may be removed by treatment with spirits of salts. 1952 KIRK & OTHMER *Encycl. Chem. Technol.* VIII. 524 The use of lime-soap thickened lubricants for the wheels of chariots dates back as far as 1400 B.C... In modern times, however, the manufacture of lubricating greases—also by means of lime soaps—started about 1854. 1961 COHEN & LINTON *Chem. & Textiles for Laundry Industry* iii. 46 Pure lime soap is more or less white when it is formed... If we accept a laundryman's concept of lime soap, on the other hand, we are talking about a dingy, gray, boardy impregnation that builds up in fabrics and defies all efforts to attain good whiteness. 1930 *Engineering* 15 Aug. 219/1 It [*sc.* the base exchange method] gives rise to no precipitate whatever, and this avoids what in the lime-soda process is often a cause of difficulty. 1950 B. E. HARTSUCH *Introd. Textile Chem.* iv. 92 The lime-soda method for softening water is the oldest and is still most used for very large softening plants. 1970 KIRK & OTHMER *Encycl. Chem. Technol.* (ed. 2) XXII. 98 The lime or lime-soda process is based upon precipitation of calcium as calcium carbonate and magnesium as magnesium hydroxide. 1907 *Bull. Bureau Chem., U.S. Dept. Agric.* No. 101. 12 The next set of experiments was to determine the composition of lime-sulphur mixtures boiled the same length of time.., but containing varying quantities of lime and sulphur. 1913 *Jrnl. R. Hort. Soc.* XXXIX. 378 It appears unsafe to spray many varieties of gooseberries with either lime-sulphur or liver of sulphur. 1937 A. M. MASSEE *Pests of Fruits & Hops* xiv. 266 In the post-blossom sprays..lime-sulphur is used as an acaricide as well as a fungicide, and it is then used at a strength of 1 per cent..for the control of the Fruit Tree Red Spider and Apple Scab. 1968 R. HAY *Gardening Year* 472/3 Lime sulphur can also be used against big bud mites on black currants, American gooseberry mildew and peach leaf curl. 1971 *Country Life* 1 Apr. 743/3 We struck east across the A515..to re-cross the road farther north by a limeworks.

lime, *sb.*[2] Add: **1. c.** *ellipt.* for *lime-green sb.* and *adj.* (LIME *sb.*[2] 2); also for LIME-JUICE, as in phr. *gin and lime*.

1923 *Daily Mail* 19 Feb. 1 (Advt.), Smart skirt... Colours: navy,.. gold, lime, cardinal and black. 1937 *Discovery* July 217/2 Dresses are burgundy, pine-blue, lime. 1938 L. MACNEICE *Earth Compels* 23 A gin and lime or a double Scotch. 1972 *Vogue* Jan. 12/2 The colours.. are remarkable—lime and raspberry, lemon, orange, rose.

2. lime-marmalade, marmalade made from limes; **lime-punch** (earlier example); **lime-squash,** a drink made with the juice of the lime (cf. *lemon-squash*).

c 1938 *Fortnum & Mason Price List* 44/1 Marmalade.. Lime—per glass 1/3. 1968 'J. FRASER' *Evergreen Death* x. 80 He did like that lime marmalade they used to get. 1972 *New Statesman* 26 May 709/1 Coffee, bread and lime-marmalade. 1774 P. V. FITHIAN *Jrnl.* (1900) 206 We had after Dinner, Lime Punch and Madaira. 1909 *Daily Chron.* 15 June 4/4 For drinking, lime-squash is superior to lemon squash. 1939–40 *Army & Navy Stores Catal.* 28/2 Lime squash.

lime, *sb.*[3] Add: *lime-walk* (earlier examples), *-wood* (later examples).

c 1662 T. BROWNE *Let.* in *Works* (1931) VI. 307 Uncertain it is whether in any *Tilicetum*, or Lime-walk, abroad it be considerably exceeded. 1816 JANE AUSTEN *Emma* III. vi. 103 Some are gone to the ponds, and some to the lime walk. 1885 F. MILLER *Wood-Carving* iv. 27 Lime-wood was almost exclusively used by Gibbons in his drops and festoons of fruit. 1932 O. EVAN-THOMAS *Domestic Utensils of Wood* 118 Chrism spoon, limewood, entirely carved with sacred objects. 1965 J. ARONSON *Encycl. Furnit.* (1966) 285/1 *Limewood*, light-colored, close-grained wood that cuts as well across as with the grain, rendering it excellent for carving.

lime (ləim), *sb.*[5] Colloq. abbrev. of LIMELIGHT. Freq. in *pl.* Also (in sing.) *attrib.* Phr. *in the lime* (Austral.), in the limelight.

1892 J. NIE *Robinson Crusoe* 6 Here! Where's the lime-light man? Of course, used up all his limes for Crusoe. 1895 B. DALY in Chevalier & Daly *A. Chevalier* II. 248 The footlights are turned low, and the hissing noise behind explains that Sam Pennett, the carpenter, is getting his limes ready for use. 1931 *Daily Express* 22 Sept. 17/1 She ..used to keep in her shop a working model of a theatre, complete to the last 'lime' and 'float'. 1935 RIDGE & ALDRED *Stage Lighting* iii. 18/2 Producers will frequently ask for the 'limes' when they mean front-of-house arc lanterns. 1941 BAKER *Dict. Austral. Slang* 43, *Lime, in the*, in the limelight. 1958 B. NICHOLS *Sweet & Twenties* xiii. 177 There was one figure which the limes should have picked out. 1961 PARTRIDGE *Dict. Slang* Suppl. 1169/2 *Lime, in the*, popular; much publicised: Australian... I.e. 'in the limelight'. 1966 *Guardian* 18 Feb. 10/5 At 16 he was a lime boy, looking after the lights.

lime, *v.*[1] **5. b.** Add to def.: To give (wood) a bleached effect by treating it with lime. Cf. *LIMED *ppl. a.* 2 b.

1966 M. M. PEGLER *Dict. Interior Design* (1967) 266 Woods other than oak can be limed.

d. (Later examples.)

1903 L. A. FLEMMING *Pract. Tanning* 6 The length of time to thoroughly lime the skins depends on the thickness of the skins. *Ibid.* 7 The best results accrue when only skins of like nature and size are limed..together. 1925 J. R. ARNOLD *Hides & Skins* 553 Hides and skins which are prepared for dehairing by sweating or painting..are also limed.

lime, *v.*[4]: see *LIMER[3].

limeade (ləimēɪ·d). [f. LIME *sb.*[2] 1 + -ADE 1 c.] A drink made from lime-juice sweetened with sugar.

1892 F. DAVIES *Temperance Drinks* 81 *Limeade*, plain syrup one gallon, lime-juice a quart, essence of lemon two ounces, essence of lime one ounce. 1936 J. DOS PASSOS *Big Money* 313 Give me a limeade and no sweetnin' in it. 1953 R. CHANDLER *Long Good-Bye* xiii. 79 She had finished her limeade. 1966 *Guardian* 28 Dec. 4/6 The abbot and I..shared a bottle of fizzy limeade.

Limean: see *LIMENIAN.

limed, *ppl. a.* Add: **2. b.** Of wood, esp. oak, that is treated with lime to give it a bleached effect.

1930 *Heal & Son Catal.* 17 Twin Bedsteads in Limed Oak. 1933 *Archit. Rev.* LXXIII. June 230 The Holy Table is in waxed and limed oak. 1952 J. GLOAG *Short Dict. Furnit.* 311 *Limed oak*, oak that has been pickled with a coating of lime... Limed oak surfaces are generally left unpolished. 1957 *N.Z. Timber Jrnl.* Oct. 73/1 *Limed oak*, chloride of lime and water applied to oak furniture to produce a bleached effect. 1973 A. ROOS *Dunfermline Affair* 162 The limed-oak furniture was all built in. Wardrobe, vanity units, dressing-tables.

Limehouse (ləi·mhaus), *v.* [*Limehouse*, a district in the east of London.] *intr.* To make fiery (political) speeches such as Mr. Lloyd George made at Limehouse in 1909. Also as *sb.* and **Li·mehousing** *vbl. sb.*

1913 *Daily Mail* 1 Aug. 5 (*heading*) Mr. Lloyd George himself again... Limehousing at Carnarvon. 1914 *National Rev.* June 543 Mr. Lloyd George went to Ipswich and Limehoused on the eve of the poll. 1920 *Glasgow Herald* 20 Mar. 7 It is exactly what he used to say in the old Limehouse days, though his Limehousing now is of a different kind. 1920 *Punch* 31 Mar. 259/1 Guerrilla tactics in the House, suspension, recognition, pacifism, office, original budgeting, limehousing.., social reform. 1932 *Times Lit. Suppl.* 9 June 426/2 He [*sc.* Bonar Law] introduced..a 'new acerbity' into Front Bench debating, or what his opponents might have called the Conservative counterpart of 'Limehouse'. 1937 PARTRIDGE *Dict. Slang* 484/1 *Limehouse*, 'to use coarse, abusive language in a speech'. 1963 *Punch* 16 Jan. 96/1 Enough of the actor to wallow in invective—'Limehousing' they called it.

limelight. Add to def.: Formerly much used in theatres to light up important actors and scenes, and so direct attention to them. Hence freq. *fig.*

1877 G. B. SMITH *Shelley* I. 45 Transcendent as were his virtues when compared with his faults, the lime-light of a malevolent scrutiny has been turned on the latter. 1881 P. FITZGERALD *World behind Scenes* I. 48 The use of so intense a light as the limelight has favoured the introduction of a new effect in the shape of transparent scenery. 1882 [in Dict.]. 1908 *Daily Chron.* 25 Jan. 3/2 The beauty of his person..helped to throw the limelight upon him. 1922 *Blackw. Mag.* Aug. 150/1 He did not..pose in the limelight to the same extent as his respected chief. 1934 A. HUXLEY *Let.* 1 Oct. (1969) 384 The town hardly gets its full share of the limelight because of the hero. 1952 GRANVILLE *Dict. Theatr. Terms* 111 *Fond of limelight*, greedy for notice. One who claims the centre of the stage. 1955 *Times* 5 May 16/2 German bonds took the limelight in the foreign bond market. *Ibid.* 17 June 9/3 The publicity given to the submission of identical tenders for public authorities' contracts has brought the question into the limelight. 1967 *Guardian* 3 Feb. 7/3 [He] did more than his bit of backing in to the limelight, and his declarations of his own genius aren't to everyone's taste. 1975 R. LEWIS *Double Take* iv. 127 In our business exposure to the limelight of the courts is like the kiss of death.

2. *attrib.*

1874 *Porcupine* 11 Apr. 26/3 And [he], with the willing aid of the stage-carpenter, scenic artist, and limelight man, has made our blood curdle. 1874 *Cassell's Mag.* May 432/1 The lime-light splendour of the tropics. 1876 *Porcupine* 22 Apr. 59/1 There is plenty of bustle and 'business', and lots of pistol-shots and obliging limelight rays. 1892 J. NIE *Robinson Crusoe* 6 Here! Where's the lime-light man? 1897 G. B. SHAW *Let.* 11 June (1965) 774 A ten inch moon, a limelight sky. 1938 W. LORAINE *Robert Loraine* 31 And recount the events of the evening—the mistakes of the limelight man, the hacking cough from the third row.

limelight, *v.* [f. the sb.] *trans.* To illuminate by limelight. Usu. *fig.* Also **li·melighted, -lit** *ppl. a.*; **li·melighting** *vbl. sb.*

1909 *Daily Chron.* 10 Apr. 4/6 The most limelighted person in Europe this morning is Queen Wilhelmina of Holland. 1909 *Westm. Gaz.* 10 Apr. 2/3 We had sympathised with the beautiful lime-lit heroine. 1927 *Daily Express* 21 Feb. 2/4 Unfeminine modern women go lime-lighting their way through the world. 1927 *Observer* 10 Apr. 29 This is not an occasion when the interests of motorists can be served by limelighting. 1940 *Nation* (N.Y.) 28 Sept. 263/1 What are the facts that justify these limelighted conferences in Berlin and Rome? 1964 *Punch* 12 Aug. 213/1 This keenness to pin something on the limelit.

Limenian (limī·niăn). Also **Limean** (limī·ăn). [f. *Lima*, capital of Peru: see below.] A native or inhabitant of Lima. Also *attrib.* Also the Spanish forms **limeño** (lime·nyo) a male, and **limeña** a female, native or inhabitant of Lima.

1824 B. HALL *Extracts Jrnl. Coasts Chili, Peru & Mexico* (ed. 2) I. iii. 90 San Martin's expedition took the Limenians quite by surprise; for they had always held Chili in contempt. 1856 C. R. MARKHAM *Cuzco & Lima* ix. 302 The Limenians, with all their indolence,.. bore themselves towards the Creoles with insolent pride. 1891 E. B. CLARK *Twelve Months Peru* iii. 55 It is considered highly improper for a Limeña, either married or single, to walk the streets alone. 1927 T. WILDER *Bridge San Luis Rey* I. 4 It was rather strange that this event should have so impressed the Limeans. *Ibid.* III. 49 Limean gossips.. declared them to be Castilian. 1962 N. MAXWELL *Witch-Doctor's Apprentice* ii. 13 Any place with no roads..and only one good hotel is not apt to be frequented by Limenos. 1971 *Guardian* 18 Aug. 3/2 [A] cynical Limeño lawyer. 1974 'A. HAIG' *Peruvian Printout* 122 The car was a hired Cadillac... Few Limeños ever hired those vehicles; they were strictly for tourists.

limer[3] (ləi·məɹ). *West Indies.* [Etym. unknown.] A person who hangs about the streets. Hence (as a back-formation) **lime** *v.*[4] *intr.*, to hang about the streets; also **li·ming** *vbl. sb.*

Said to have been used in the 1940s but printed evidence is lacking.

1970 *Express* (Trinidad & Tobago) 6 Jan. 4 'Limers' are a menace on High Street at night. 1972 *Ibid.* 4 Feb. 21 One ride in a route taxi and a little liming in Frederick Street..would uncover whatever the wooding group may have forgotten. 1973 *Sunday Express* (Trinidad & Tobago) 1 Apr. (Suppl.) 13/1 Staying a minute more to lime. *Ibid.* 8 Apr. 13/3 Now I confess that in the past I have often had cause to voice a complaint about his penchant for ole talking and liming. 1974 *Sunday Advocate-News* (Barbados) 3 Mar. 16/6 The limers' attire ranges from the sophisticated to the ridiculous.

Limerick. Add: [The chief town of the county of Limerick in Ireland.] (Earlier examples.)

1896 A. BEARDSLEY *Let.* c 1 May (1971) 128 I have tried to amuse myself by writing limericks on my troubles. *Ibid.* 2 May 129 Your continuation of the limerick is superb.

2. Used *attrib.* to designate: **a.** Gloves of fine leather made originally at Limerick (see quot. 1960).

1804 M. EDGEWORTH *Pop. Tales* I. 245 Are you blind Mr. Hill? Don't you see that they are Limerick gloves? 1842 J. P. LAWSON *Gazetteer Ireland* 607/2 The glove trade ..has now declined, and those articles sold as Limerick gloves are actually manufactured at Cork. 1853 MRS. GASKELL *Ruth* II. vii. 173 She..brought down a pair of Limerick ones, which had been long treasured up in a walnut-shell. 'They say them gloves is made of chicken's-skins,' said Sally. 1960 CUNNINGTON & BEARD *Dict. Eng. Costume*, *Limerick gloves*, 2nd half 18th and 1st half 19th c...made of very fine leather, said to be made from the skins of unborn lambs.

b. A particular bend or pattern of fish-hook or a fish-hook with such a bend (said to have been orig. made in the town of Limerick) in which the wire of the hook is bent abruptly through a large angle behind the barb but thereafter is bent more shallowly to the point at which it continues as the straight shank. Also *ellipt.* as *sb.*, a Limerick hook.

1828 H. DAVY *Salmonia* 141, I have even made a hook, which..I think, I could boast as equal to the Limerick hooks. 1835 T. T. STODDART *Art of Angling in Scotl.* iii. 16 O'Shaughnessy's Limericks..are not always exactly the thing, excepting those used for salmon..which are really excellent. 1856 'STONEHENGE' *Man. Brit. Rural Sports* 235/1 The round-bend hook is that which is most used in England, the Limerick pattern being chiefly in vogue in Ireland. 1928 *Chambers's Jrnl.* Jan. 2/2 He..picked out his lure without hesitation—a Number 5, Limerick-bend, double-hooked 'Blue Charm'. 1956 L. V. BATES *Artificial Flies* iii. 55 The Limerick is probably stronger but the fine elegant proportions of the Dee hook make for better hooking.

c. A type of embroidered lace made originally at Limerick. Also *ellipt.* as *sb.*

1842 J. P. LAWSON *Gazetteer Ireland* 607/2 Limerick lace has now obtained a high celebrity. This beautiful manufacture was introduced into the city in 1829 by Mr Walker, an English gentleman. 1886 B. LINDSEY *Irish Lace* ii. 9 A spirit of enterprise and commercial adventure..originated the Limerick manufacture. 1905 N. H. MOORE *Lace Bk.* 196 Limerick lace is a combination..of cut-work and embroidery, and hardly comes under our definition of lace. 1953 M. POWYS *Lace & Lace-Making* iv. 35 *Limerick lace*, embroidered net, Irish, 19th century... This type of lace is made in all the countries, including America and India. It is light, pretty and easy to produce. Limerick remains the finest of the kind. 1959 *Times* 21 Sept. 12/4 Her old family Limerick lace veil was held in place by a mother-of-pearl coronet. 1967 C. GASKIN *Edge of Glass* iv. 86 [Her] long ancient gown collared in yellowed Limerick lace.

limestone. Add: **c.** *limestone-cliff* (earlier example), *-land* (later example), *-water*; **limestone polypody:** substitute for def.: a fern, *Gymnocarpium robertianum*, restricted to areas of limestone rock; (later examples).

1699 M. LISTER *Journey to Paris* 88 The high Ragstone Mountains and Lime Stone Cliffs. 1811 *Niles' Reg.* 12 Oct. 101/1 Our steepest lime stone lands are very favorable to sheep. 1888 C. T. DRUERY *Choice Brit. Ferns* 118 The young fronds [of the Oak Fern] also, when unfolding, exactly resemble the pawnbroker's sign of three balls, which those of the Limestone Polypody do not. 1908 E. STEP *Wayside & Woodland Ferns* 87 The Limestone Polypody..will be found nowhere except on limestone rocks. 1960 P. TAYLOR *Brit. Ferns & Mosses* 164 The Limestone Polypody occurs in the mountains of Europe, temperate Asia and North America. 1831 J. M. PECK *Guide for Emigrants* 233 Those persons who have been unaccustomed to lime stone water..frequently have eruptions of the skin. 1872 E. EGGLESTON *End of World* ix. 65 Having..quaffed the hard limestone water.

Limey (ləi·mi). *colloq.* and *derogatory.* Also **limey.** [abbrev. LIME-JUICER.] **a.** In the former British colonies (esp. Austral., N.Z., and S. Afr.), an English immigrant. Also *attrib.*

The later examples prob. reflect the U.S. use in b.

1888 D. SLADEN *Austral. Ballads & Rhymes* 31 They'd seen old stagers and limey new chums. 1947 J. BERTRAM *Shadow of War* 251, I can remember scores of fights among the 'Limeys'. 1954 T. S. ELIOT *Elder Statesman* III. 93 Everyone would sneer at the fellow from London, The limey remittance man for whom a job was made. 1962 *Times* 6 Jan. 7/1 The [English] boys [at Sasolburg, Orange Free State] were constantly taunted by schoolmates as 'Pommie', 'Limey', and 'Rooinek'. 1964 O. E. MIDDLETON in C. K. Stead *N.Z. Short Stories* (1966) 198 We'd all be drawing the dole like every other mother-loving beachcomber in this..Limey hole!

b. *U.S.* An English ship; an English (or British) sailor; hence *gen.*, an Englishman, a Briton. Also *attrib.*

1918 A. N. DEPEW *Gunner Depew* 18 So, all over the world, British ships are called 'Lime-juicers' and their sailors 'Limeys'. *Ibid.* 48 Ask any Limey soldier and he will tell you the same. 1918 R. D. PAINE *Fighting Fleets* v. 87 Squads of the American navy patrol began to stroll about..displaying no sympathy..for the shipmate who.. loudly announced that he could whip any three 'Limies' that ever trod a British deck. 1919 *Texas Rev.* IV. 86 In our fleet a British ship is regularly called a 'limey', from the old lime-juicers. The British seaman is likewise a 'limey'. 1924 *Chicago Tribune* 18 Oct. 1/5 (*heading*) Midway Signs Limey Prof. to Dope Yank Talk. 1930 J. DOS PASSOS *42nd Parallel* II. 169, I..shipped out East on a limey, on an English boat. 1921 E. LINKLATER *Juan in Amer.* II. vii. 115 'An Englishman,' he marvelled, 'the first limey I ever saw shot in Chicago.' 1933 'J. SPENSER' *Limey* i. 4 'English, eh?' said the manager. 'I ain't too keen on you Limey's.' 1952 J. STEINBECK *East of Eden* lii. 498 Fights in the bar-rooms with the goddam Limeys. 1954 M. BRUCE *Tramp Royal* II. ix. 121 This ship differed from a Limey vessel in the way the watch bells are struck. 1955 G. GREENE *Quiet American* IV. ii. 241, I don't like Limies. 1968 M. PYKE *Food & Society* iii. 32 The pejorative identification of foreigners with unfamiliar foods. This makes..the English limies. 1969 *Listener* 26 June 881/3 Was it my limey accent that called it forth or did they say it to every customer? 1973 'D. JORDAN' *Nile Green* xi. 49 Guy always plays up the limey accent when he's in the States.

limicole (ləi·mikōᵘl). [f. mod.L. group-name *Limicolæ* (see quot. 1930), f. L. *līmus* mud + *colere* to inhabit.] An oligochæte worm living in mud or water.

[1930 J. STEPHENSON *Oligochaeta* xvi. 607 Oligochaeta may be divided roughly into Terricolae and Limicolae—the earth-dwellers and the mud- and water-dwellers respectively.] 1963 R. P. DALES *Annelids* ix. 180 The common division of the Oligochaeta into the Microdrili or Limicoles (the aquatic families) and the Megadrili or Terricoles (the earthworms) is mainly one of convenience. 1965 B. E. FREEMAN tr. *Vandel's Biospeleol.* vii. 73 The limicoles are more interesting to the biospeleologist.

liminess (ləi·minės). [f. LIMY *a.*] The quality of being limy.

1902 E. A. WOODRUFFE-PEACOCK *Thoroughbreds & their Grass-Land* II. 7 Among plants..a score of species that could be named, suddenly appearing in a turf at once demonstrate a growing liminess in a clay soil.

liming: see *LIMER[3].

limit, *sb.* Add: **2. g.** In various card games, as (*a*) *Poker*, an agreed maximum stake or bet; so *attrib.*, as *limit game*; (*b*) *Bridge*, a call which shows that the strength of the caller's hand does not exceed a certain value; usu. *attrib.*, as *limit bid, raise*.

(*a*) 1892 W. J. FLORENCE *Handbk. Poker* 90 Before a game is commenced it is agreed that so many chips shall be the limit... No game ever should be played without a limit. 1928 *Chambers's Jrnl.* Jan. 116/1 Once again the betting came to Poker Jack, and this was his chance. He coolly raised Rymington the limit, and left his two opponents half-stupefied. 1963 *Esquire's Bk. Gambling* II. iv. 109 It is virtually impossible to bluff in a limit game. With a limit, poker is more like Screeno... Big Poker..as any unlimited-stakes player will be happy to tell you.. requires the most brutality.

(*b*) 1959 REESE & DORMER *Bridge Player's Dict.* 138 A limit-bid is one that describes the strength of a player's hand within fairly narrow limits. *Ibid.*, A raise of partner's suit is generally a limit-raise, expressing the full value of the hand. 1959 *Listener* 24 Dec. 1118/2 Three Spades was a limit bid, which East might have passed. 1964 *Official Encycl. Bridge* 331/1 *Limit*, (1) the highest stake permitted in a bridge club... (2) A bid which shows a maximum as well as a minimum range of values in the bidder's hand. 1974 *Times* 23 Feb. 11/1 A player must take the decision whether..to go straight for a limit bid.

h. *colloq.* The very extreme; the last point or stage; the worst (etc.) imaginable or endurable; the maximum penalty. Phr.: *go the limit*, to behave in an extreme way; to last the stated number of rounds or the full time, as in a boxing match; to allow sexual intercourse; *over the limit*, having exceeded a stated bound or point. orig. *U.S.* (Apparently a fig. use of *2 g.) Cf. *the frozen limit* (*FROZEN *ppl. a.* 1 b).

1904 *Montgomery* (Alabama) *Weekly Advertiser* 26 Aug. 4 We can always depend on Kansas to go the limit in the freak line. 1906 *N.Y. Even. Post* 7 May 1 Desertion is bad enough..but to fire at one's comrades while in the act of turning against them is—well, the limit. 1907 *Westm. Gaz.* 16 Aug. 2/1 They [*sc.* wages] are low everywhere.. but Belfast is what Americans would call 'the limit'. 1908 A. J. DAWSON *Finn* xxiii. 353 I'll be teetotally damned if that ain't the limit! a 1911 D. G. PHILLIPS *Susan Lenox* (1917) I. 389 We've made the plunge. We'll go—the limit. 1914 W. G. LAWRENCE in T. E. Lawrence *Home Lett.* (1954) 502 Bankers and business people of all sorts are the real limit. 1916 'TAFFRAIL' *Pincher Martin* v. 79 Oh, to hell with you and your rotten excuses!.. You're about the frozen limit! 1919 G. B. SHAW *Heartbreak House* I. 18 Really! your father does seem to be about the limit. 1925 E. F. NORTON *Fight for Everest, 1924* 110 If vitality is low in the early hours at Camp III at 21,000 feet, it can be guessed how near the limit 6 a.m. found us at 27,000. 1925 G. MALLORY in *Ibid.* 237 That cutting against time at the end after such a day just about brought me to my limit. 1925 L. J. SMITS *Spring Flight* viii. 89 I'd marry a girl who had gone the limit just as willingly as I would a strict one, perhaps a little sooner. 1927 *Amer. Speech* III. 29 The boxer 'goes the limit' if he succeeds in lasting the specified number of rounds. 1947 'N. SHUTE' *Chequer Board* iii. 68 If you get anything to go before court martial, for example, I'll see they get the limit. 1949 A. CHRISTIE *Crooked House* xvi. 126 This house is the absolute limit!.. I don't see why I should have to be burdened with such peculiar parents. 1966 *Daily Tel.* 11 Aug. 26/6 Attempting to drive while over the 80 mg/100 ml. limit can be punished... Being in charge of a vehicle while 'over the limit' can lead..to up to four months' imprisonment. 1968 N. BENCHLEY *Welcome to Xanadu* iii. 52 She'd heard girls in school talk about going the limit, or all the way.

5. limit dog, one shown in a class limited to dogs having certain required qualifications; **limit gauge** *Engin.*, a gauge used for determining whether a dimension of a manufactured item falls within the specified tolerance; so **limit gauging,** the use of limit gauges to ensure the interchangeability of parts; **limit load** *Aeronaut.*, the maximum load that an aircraft or part of one is expected to bear in particular conditions of operation; so *limit load factor*, the load factor corresponding to this load; **limit point** *Math.*, a point every neighbourhood of which contains a point (usu., a point other than the limit point) belonging to a given set; **limit switch** *Engin.*, a switch that prevents the travel of an object past some predetermined point and is mechanically operated by the motion of the object itself.

1903 *Forest & Stream* 21 Feb. 151/2 Limit dogs was won by St. Elvan. 1909 *Daily Chron.* 11 Feb. 5/6 The first prize for limit dogs over 45 lb. 1905 A. PARR *Machine Tools & Workshop Pract.* i. 10 (*caption*) Limit gauge. 1909 *Westm. Gaz.* 7 Dec. 5/1 When it comes to be measured in a special limit-gauge, the slightest discrepancy is discovered. 1970 W. J. PATTON *Mod. Manuf.* xvii. 454 Instead of measuring actual dimensions, we usually check conformity to tolerance specifications in a production run by fixed limit gauges, often termed 'GO' and 'NOT GO' gauges. 1920 *Proc. Inst. Mech. Engin.* Nov. 1076 Limit gauging may be applied to many kinds of fit. 1964 S. CRAWFORD *Basic Engin. Processes* xiv. 296 Limit-gauging systems have played an essential part in the development of the technique of quantity production. 1950 D. J. PEERY *Aircraft Struct.* iii. 69 The maximum loads which an airplane may be expected to encounter at any time in service are designated as limit loads or applied loads. The load factors associated with these loads are known as limit load factors... For loads which are under the control of the pilot, flight restrictions are used so that the limit load factor is never exceeded. 1967 *Technology Week* 23 Jan. 66/2 As an industry, we have done remarkably well, from a safety point of view, operating large aircraft with limit load factors of only 2·5 in rough air. 1972 T. H. G. MEGSON *Aircraft Struct.* xii. 413 Having decided on an ultimate load then the limit load may be fixed. 1905 *Trans. Amer. Math. Soc.* VI. 90 A geometrically closed set of points is a set that includes all its geometrical limit points. 1926 C. WALMSLEY *Introd. Course Math. Analysis* i. 44 For general sequences we define a limit (or limit point, limiting point, limiting value, limiting number) of any sequence s_1, s_2, s_3..as any number *L*, within an arbitrarily small neighbourhood of which..there lie numbers of the sequence; a number of the sequence itself not being a limit of the sequence unless it is repeated indefinitely often as a term of the sequence or there are other terms of the sequence within the arbitrarily small neighbourhood. 1959 E. M. PATTERSON *Topology* (ed. 2) ii. 29 The points $x = 0$ and $x = 1$ are limit points of the set $0 < x < 1$ on the Euclidean line; and in this case every point of *X* itself is also a limit point. 1930 *Engineering* 9 May 595/2 Automatic control at the end of travel is provided by geared limit switches, and intermediate positions are signalled by a travelling nut on the limit switch. 1956 *Railway Mag.* May 338/1 Limit switches are provided in the hoisting motion, and in the derricking motion, to limit the amount of travel in both directions. 1974 *BP Shield Internat.* Oct. 2 (*caption*) The computerised electronic requirement that operates all valves, limit switches, pressure transducers.

limited, *ppl. a.* Add: **2.** *limited edition*, an edition of a book, or reproduction of an object, limited to some specific number of copies; *limited express* or *train* (U.S.): cf. *limited mail* (in Dict.); *limited mail* (earlier example); *limited monarchy* (earlier examples); *limited war*, one in which the weapons used, the nations or territory involved, or the objectives pursued, are limited or restricted.

1648 R. FILMER (*title*) The anarchy of a limited or mixed monarchy. 1710 in T. B. Howell *State Trials* (1812) XV. 62 The nature of our constitution is that of a limited monarchy. a 1792 BURKE *Address Brit. Colonists N. Amer.* in *Works* (1812) V. 148 England has been great and happy under the present limited Monarchy. 1866 DICKENS *Mugby Junction* in *All Year Round* Extra Christmas No. 10 Dec. 17 Driving..at limited-mail speed. 1879 F. R. STOCKTON *Rudder Grange* ix. 93 Time flew like a limited express train. 1890 *Harper's Mag.* Aug. 409/1 Coming up by the limited train, Miss Lee was not favorably impressed. 1903 *Connoisseur* VI. 252/2 This library has the trail of the 'Limited Edition' serpent over it all. 1904 *Dial* (Chicago) 16 Oct. 238 It is not a book for the limited express. 1920 H. CRANE *Let.* 28 Jan. (1965) 32 A limited edition hastily gathered up would be the only possible method of presentation. 1930 Limited edition [see *HOG-WASH]. 1944 H. TREECE *Herbert Read* 41 The Quixote in him that writes prefaces to Limited editions. 1948 H. J. MORGENTHAU *Politics among Nations* xx. 291 Another variety of a limited war..has been well described. 1955 *Times* 4 July 9/7 Britain, then, has to be prepared for both nuclear and limited war. *Ibid.* 8 July 9/5 The recent war in Korea is an example of a major limited war, and the present operations in Kenya of a minor limited war. 1966 MRS. L. B. JOHNSON *White House Diary* 20 Aug. (1970) 413 It was a good speech [by Lyndon].. 'Perhaps it reflects poorly on our world that men must fight limited wars in order to keep from fighting larger wars.' 1971 *Ideal Home* Apr. 143 This Piccadilly mug is produced by Wedgwood in a limited edition of 4,000. 1972 *Country Life* 16 Nov. (Suppl.) 57/1 Pure silk 36 inch hand-rolled head-scarves, in four designs..and in limited editions for £12.50. 1973 *Daily Tel.* 27 Apr. 18 The National Maritime Museum..is adding two new clocks to its collection of historic timepieces next month. The two, however, are replicas of 18th-century clocks which.. Thwaites and Reed are producing in limited editions.

b. *limited partnership*, inactive partnership, where liability is limited to the value of the capital contribution; so *limited partner*.

1907 *Act 7 Edw. VII* c. 24 § 1 This Act may be cited for all purposes as the Limited Partnerships Act, 1907. *Ibid.* § 4 Limited partners, who shall at the time of entering into such partnership contribute thereto a sum..valued at a stated amount and who shall not be liable for the debts or obligations of the firm beyond the amount contributed. 1931 *Pitman's Business Man's Guide* (ed. 9) 372/1 There is a further division into 'general' and 'limited' partners. The former correspond to the 'active' partners, and the latter to the 'sleeping' or 'nominal' partners. 1951 R. W. JONES *Thomson's Dict. Banking* (ed. 10) 381/1 The idea of the limited partnership has not appealed to the business world and..only some 1,300 such partnerships have been registered. 1970 *New Yorker* 10 Oct. 37/2 Richard Roth, Jr., a limited partner, joined the firm. 1973 *N.Y. Law Jrnl.* 31 Aug. 1/4 It omitted compensation agreements between Empire and its counsel..with whom Mr. Goldberg was then associated as a limited partner.

3. For 'quasi-*sb.*' read 'In *absol.* use'. Add later examples.

1910 KIPLING *Rewards & Fairies* 145 If you're off to Philadelphia..the Limited will take you there. 1913 E. WHARTON *Custom of Country* v. xliii. 556 A bigger and brighter blur ahead, into which they were plunging as the 'Limited' plunged into the sunset. 1938 S. V. BENÉT *Thirteen o'Clock* IV. 295 'Jerry Pye!' she said, 'I don't know what's come over you. You never take me on any limiteds!'

b. = *limited company.*

1905 *Westm. Gaz.* 20 Nov. 8/1 Company floaters have

gone very fast indeed, some limiteds, it is said, not having sufficient capital. 1907 *Daily Chron.* 26 July 3/4 This is my experience in a West-end house..classed with the Limiteds.

limiter. Add: **2. b.** *Electronics.* A device whose output is restricted to a certain range of values irrespective of the size of the input.

1919 R. STANLEY *Text-bk. Wireless Telegr.* (new ed.) II. xiv. 274 To a certain extent the H.F. amplifying valves of the Marconi Co., known as the V.24 type, are limiters, since their characteristic curves are of short range. 1930 H. M. DOWSETT *Handbk. Technical Instruction for Wireless Telegraphists* (ed. 4) xx. 297 Short wave signals vary greatly in intensity due to various causes, and the use of the limiter makes the signal currents more suitable for operating relays or recording apparatus. 1962 A. NISBETT *Technique Sound Studio* 258 Limiters are necessary in AM transmitters, and in the input to disc recorders in situations where overmodulation may occur by accident. 1972 *Sci. Amer.* Sept. 101/3 Since the amplitude of the FM carrier is constant, limiters can be used to reduce impulse noise.

limitrophe, *a.* and *sb.* Add: **A.** *adj.* (Earlier example.)

1763 EARL OF BUCKINGHAMSHIRE *Let.* 6 Dec. in *Despatches & Corr.* (1902) II. 113 The inconveniences which might arise if a country limitroph to Russia was governed by a sovereign allied and connected with the great European Powers.

B. *sb.* (Later example.)

1963 V. NABOKOV *Gift* iii. 155 Thus 'France' corresponded to his warningly raised eyebrows; some kind of 'limitrophes' to the hairs in his nostrils.

‖ **limmu, limu** (li·mu). *Assyriology.* [Assyrian *limmu* period, circuit, administrative year.] The year of office to which the holder gave his name; hence, the office itself. Cf. EPONYM 2.

1862 H. C. RAWLINSON in *Athenæum* 19 July 83/1, I am quite ready to abandon this notion if any more suitable explanation can be found of the office of *Limu*, from which the Assyrians commonly dated; but in the mean time I cannot accept of the unmeaning title of Eponymy, nor can I admit of the etymology which would make the *Limu* merely the 'seer' of the new moon. 1888 Z. A. RAGOZIN *Assyria* (ed. 2) v. 146 The year was then designated as the 'Limmu' of So-and-So. It is thought..that the magistrates themselves, in their capacity of time-keepers, had the special title of Limmu... Every king was *limmu* at least once, generally the second full year of his reign. 1901 R. W. ROGERS *Hist. Babylonia & Assyria* I. xii. 323 Historical inscriptions..often mention the *limmu* or eponym of a certain year, just as they give the name of the king who was reigning. 1956 A. TOYNBEE *Historian's Approach to Relig.* I. v. 57 The Assyrian list of holders of the office of *limmu*. 1963 F. S. LEIGH-BROWNE tr. *Læssøe's People Anc. Assyria* iii. 40 The Assyrians named the year after the official who (for that one year) occupied the office of *limmu*.

limnetic (limne·tik), *a.* [f. Gr. λιμνήτ-ης living in marshes + -IC.] Of, pertaining to, or living in the open part of a freshwater lake or pond, away from the margin or bottom.

1899 G. C. WHIPPLE *Microsc. Drinking-Water* viii. 105 The limnetic or pelagic organisms are those that make their home in the open water. They float or swim freely and are drifted about by every current... Then there are organisms that may be said to be facultative limnetic forms, that is, they are sedentary or free-swimming at will. 1903 *Amer. Naturalist* XXXVII. 503 This work is in the main an extension of Häcker's earlier papers..on the autonomy of the male and female pronuclei and of their derivatives in the development of limnetic Copepods. 1923 *Ecology* IV. 372 (*heading*) Limnetic A[ssociation]. 1955 C. C. DAVIS *Marine & Fresh-Water Plankton* i. 11 Horizontally, the relatively shallow area close to shore, characterized by rooted emergent or floating vegetation, is called the littoral region, while the region of open water is known as the limnetic (or pelagic) region... The limnetic region is much more extensive than the littoral in most lakes. 1957 G. E. HUTCHINSON *Treat. Limnol.* I. i. 19 The region has had a long limnetic history and has contained large tectonic lakes since the middle Tertiary.

limnic (li·mnik), *a. Geol.* [ad. G. *limnisch* (C. F. Naumann *Lehrb. d. Geognosie* (1850) I. II. iii. 814), f. Gr. λίμνη pool of standing water, marshy lake: see -IC.] Formed or laid down in an inland body of standing fresh water such as a lake or a swamp.

[1911 *Proc. Amer. Philos. Soc.* L. 28 Naumann recognized the distinction between deposits formed on the sea border and those in fresh-water lakes... These types he terms paralisch and limnisch. *Ibid.* 29 The causes in paralisch areas are different from those in limnisch basins.] 1940 A. C. NOÉ tr. *Stutzer's Geol. Coal* vi. 159 Coal basins which have originated near the sea are called 'paralic basins'; and coal deposits which were formed inland..'limnic deposits'. This classification goes back to Naumann. Recent peat bogs can also be similarly classified. In the interior of a country—for instance, on the northern edge of the Alps—are found limnic moors, while the paralic moors are situated on the coast, as in Friesland. 1968 M.-TH. MACKOWSKY in Murchison & Westoll *Coal* xiv. 327 In the Upper Carboniferous large paralic coal deposits formed along the northern border of the Variscan mountains. Somewhat later the limnic coals of the Saar, Lorraine and central and south-eastern France developed in troughs between the mountains. 1971 *Nature* 30 Apr. 561/1 Tobien *et al.* have excavated a primitive, forest-adapted *Hipparion* fauna..from tuffaceous and limnic beds on the flank of the Höwenegg volcanic centre in southern Germany.

limno- (li·mno), comb. form of Gr. λίμνη lake, marsh, as in **li:mnobio·logy** *rare*, the biology of lakes and ponds; so **li:mnobiolo·gic, -biolo·gical** *adjs.*

1899 *Ann. Rep. Board of Regents Smithsonian Inst.* 1897–98 505 Switzerland has furnished perhaps the greatest number of investigators and stations for limnobiology. *Ibid.* 510 The Mount Prospect Laboratory,.. placed under the direction of Mr. G. C. Whipple, whose contributions to limnobiologic questions are well known. 1909 *Cent. Dict.* Suppl., Limnobiological. 1957 G. E. HUTCHINSON *Treat. Limnol.* I. p. x, The first volume is intended to cover geographical and physico-chemical limnology, the second will deal with limnobiology and the ecological, typological, and stratigraphic problems of lake development.

limnology. Substitute for defs. of both senses a and b: (The study of) the physical, chemical, geological, and biological aspects of lakes and other bodies of fresh stagnant (sometimes also of fresh flowing) water. (Add earlier and later examples.)

1893 *Geogr. Jrnl.* I. 353 He [*sc.* Friedrich Simony].. became the founder of the special branch of science termed by Forel, Limnology, or the scientific study of lakes. 1929 *Jrnl. Ecol.* XVII. 106 A generation or so ago the general conception of the word limnology was that it covered the subject of the physiography of lakes, and that freshwater biology..was more or less a separate science... As the biology of fresh waters developed this branch of science was fairly often included within the range of the word limnology, but still the term continued to be frequently used for the physiographical aspect of lake investigations only, and the notion of limnology as a mere branch of physical geography separated from the biological side of the question persisted. That definition must now be entirely abandoned... The subject of limnology, then, is the study of everything connected with fresh waters—both stagnant and running—their geography, physical features, chemistry, biology, geology of the substrata and basins, in short the Natural History of fresh waters in the broadest sense of the word. 1938 *Jrnl. Wildlife Managem.* II. 94 (*heading*) Fish biology and limnology of Crater Lake, Oregon. 1946 *Nature* 21 Sept. 421/1 Up-to-date knowledge and theory of limnology—the freshwater equivalent of oceanography—has been combined with practical experience of waterworks. 1971 *New Scientist* 4 Feb. 236/1 The professor of limnology at the University of Lund.

So **limnolo·gical** *a.*, of or pertaining to limnology; **limnolo·gically** *adv.*; **limno·logist,** one who studies lakes.

1896 *Rep. 6th Internat. Geogr. Congr.* 1895 601 Such study [of British lakes] has been for the most part occasional and unsystematic, being only in rare cases worthy of being termed limnological. 1909 WEBSTER, Limnologically. 1910 MURRAY & PULLAR *Bathymetrical Survey Scottish Fresh-Water Lochs* I. 9 Several foreign limnologists have visited the Scottish lochs. 1933 *Geogr. Jrnl.* LXXXI. 380 A German limnological expedition was at work in the Dutch East Indies. 1957 G. E. HUTCHINSON *Treat. Limnol.* I. i. 22 Limnologically, this difference is not of any particular significance. *Ibid.* 27 The Lago di Nemi, famous in antiquity.., is also a lake of considerable limnological interest. 1967 *Economist* 29 July 414/3 The Lake Tahoe Area Council..has spent thousands of dollars on engineering and limnological studies. 1970 *Nature* 25 July 420/2 The English Lake District has been and is still a mecca for limnologists.

limnoplankton (li:mnoplæ·ŋktǫn). [a. G. *limnoplankton* (E. Hæckel 1891, in *Jenaische Zeitschr. Naturwiss.* XXV. 253), f. *LIMNO- + PLANKTON.] Plankton found in fresh water.

1893 G. W. FIELD tr. *Hæckel's Planktonic Stud.* in *Rep. U.S. Comm. Fisheries* 1889–91 580 The totality of the swimming and floating population of the fresh water may be called limnoplankton. 1909 GROOM & BALFOUR tr. *Warming's Oecol. Plants* xxxviii. 161 Limnoplankton appears to be one of the most cosmopolitan of formations. 1955 C. C. DAVIS *Marine & Fresh-Water Plankton* i. 28 The limnoplankton is the plankton of lakes.

limnoria (limnǭᵊ·riă). [mod.L. (W. E. Leach 1815, in *Trans. Linn. Soc. Lond.* XI. 370), ad. Gr. Λιμνώρεια a water-nymph.] A marine isopod crustacean of the genus so called, which includes *L. lignorum*, a borer that attacks timber (= GRIBBLE[2]).

1868 BATE & WESTWOOD *Hist. Brit. Sessile-Eyed Crustacea* II. 353 Many kinds of wood, including old oak, are devoured by the *Limnoria*. 1936 *Times Lit. Suppl.* 18 Apr. 325/3 Against teredo's rival limnoria, luckily less widespread, no poison is effective. 1967 J. A. C. NICOL *Biol. Marine Animals* (ed. 2) v. 231 The gribble *Limnoria* is an isopod which tunnels into wood and feeds on wood particles.

limo (li·mo), colloq. abbrev. of *LIMOUSINE. *U.S.*

1968 *New Yorker* 23 Nov. 96 'You ride in the limo, dear,' he said,..helping her out of the Daimler. 1972 R. K. SMITH *Ransom* I. 18 If there's traffic too close to the limo, we forget it. 1972 *New Yorker* 23 Dec. 36/2 Just at the moment when the heavy groups freaking out in rented limos seemed to have wiped everybody out, this sing-song music..began to flower. 1973 R. MOORE *Fifth Estate* i. 16 The company should be sending a limo for me. I'd be happy to drop you off. 1975 *New Yorker* 24 Mar. 32/3 This place has got a laser beam on the perimeter to keep out marauders, as well as free Rolls-Royce and Mercedes limo service.

Limoges (limōᵘ·ʒ). The name of a city in central France used (freq. *attrib.*) to designate painted enamels, porcelain, etc., made there.

1844 DISRAELI *Coningsby* II. v. viii. 310 He has..a collection of Limoges ware that is the despair of the dilettanti. 1856, 1861 [see *CHAMPLEVÉ *sb.* and *a.*]. 1870 C. SCHREIBER *Jrnl.* (1911) I. 77 Some beautiful Limoges enamels. 1938 [see *CHAMPLEVÉ *sb.* and *a.*]. 1963 L. DEIGHTON *Horse under Water* xvii. 71 Coffee came..in a silver pot attended by Limoges cups and saucers. 1966 MRS. L. B. JOHNSON *White House Diary* 23 Jan. (1970) 356 Exquisite Limoges—white with a chaste gold border—for her formal china. 1966 J. LAVER *Victoriana* 157 Royal Worcester patera of 1867 decorated in the 'Limoges enamel' style..suggested by the painted enamel work on copper which was carried out at Limoges during the Renaissance. 1973 *Country Life* 11 Oct. 1063 French lamp in white Limoges porcelain with sumptuous gold decorations..£14.95.

‖ **limon** (limoṅ). *Geol.* [Fr., = (*a*) silt, alluvium, (*b*) *limon*, loess:– L. *līmus* mud.] A fine sandy soil, probably of similar origin to loess, which is widespread in northern France and Belgium.

1890 *Rep. Brit. Assoc. Adv. Sci.* 1889 560 At the foot of the Alps,..the löss is dark grey; but west of the secondary chain the same deposit is yellowish, and composed almost entirely of silicious materials, with only a very little carbonate of lime. This *limon* or löss..is very generally modified towards the top by the chemical action of rain. 1931 H. ORMSBY *France* v. 168 Where the covering of *limon* was thick, on the lower slopes the farmer has been able to cover the scars of war with fields of grain. 1972 MANSFIELD & POWRIE *France & Benelux* viii. 137 The most typical of these chalklands is Picardy,..remarkably fertile, owing to the thick covering of *limon*.

Limousin (limuzæ̃ṅ). Also 8 **Lemosine, Limosine.** [Name of an old province in central France.] **1.** A native or inhabitant of the former province of Limousin or of the region round Limoges; also, the dialect of this region. Also *attrib.* or as *adj.*, of or pertaining to this region, its inhabitants, or their dialect.

1653 T. URQUHART tr. *Rabelais's Works* (1664) II. vi. 33 When all comes to all, thou art a Limousin, and thou wilt here by the affected speech counterfeit the Parisiens. 1706 J. STEVENS *New Spanish & Eng. Dict.* Pref., In Catalonia ..they have besides a peculiar language of their own... This is a sort of Limosine, or that of the Country of Limoges, being an old barbarous French, formerly brought into this Country out of France. 1716 C. CAMPBELL tr. *Dameto's Hist. Balearick Islands* I. v. 64 That which is in use at this time commonly called the Lemosine Language being originally from Limosin a Province of France, whose Capital is Limoge. 1787 A. YOUNG *Jrnl.* 10 June in *Trav. France* (1792) I. 18 Chestnuts on a calcareous soil, contrary to the Limosin maxim. 1911 S. M. HOGAN *St. Vincent Ferrer* iv. 29 St. Vincent himself tells us that he preached in his mother-tongue—Limousin. 1932 W. L. GRAFF *Lang.* 377 Provençal group, with different varieties: Provençal..Limousin, Gascon. 1964 *Archivum Linguisticum* XVI. 34 In quoting this dialect [*sc.* North Occitan] I give Limousin forms. 1968 F. WHITE *Ways of Aquitaine* iv. 55 The churches of the Limousin abbeys form a school of their own, deeply impressive because of their darkness and their strength. 1973 J. WAINWRIGHT *Devil you Don't* 62 The V.V.S.O.P. Cognac—pale, liquid ecstasy with a bouquet of Limousin oak.

2. An animal belonging to the French breed of beef cattle so called. Also *attrib.*

1970 *Times* 6 Apr. 10/7 The Limousin, a hardy, deep fawn coloured animal, is a beef breed which has impressed many visitors to Paris shows. 1971 *Country Life* 30 Dec. 1857/3 Limousins from the Limoges area..are the most recent invaders of the northern English counties. 1973 *Ibid.* 1 Mar. 548/1 A German Zimmenthal bull, and.. a Limousin bull in France.

limousine (li·muzīn). [Fr., f. prec.] A (luxury) motor car with a compartment for the passengers and a separate compartment for the driver. Also *attrib.*

Orig. the driver's seat was outside though covered with a roof. Since the 1930s the word has been more usual in North America than in the U.K.; recently it has been used, esp. in the U.S., for vehicles conveying passengers to and from large airports.

1902 A. C. HARMSWORTH et al. *Motors* 55 With certain kinds of engines, too, it is difficult to adopt any other form of car than the Tonneau, or for the wet weather the Limousine. 1905 *Westm. Gaz.* 22 Nov. 9/2 A touring car.. fitted with a brougham or limousine body. 1916 A. BENNETT *Lion's Share* xxx. 217 A few days later an automobile—not Audrey's but a large limousine—bumped, with slow and soft dignity, across the railway lines. 1920 *Chambers's Jrnl.* May 279/2 He heard the purring of limousines gliding into Pall Mall. 1922 W. J. LOCKE *Tale of Triona* v. 47 Whom she saw drive away in luxurious limousines. 1960 I. WALLACH *Absence of Cello* (1961) 27 Marian..struggled to the bus [at N.Y. airport] that was known for some reason as a 'limousine'. 1968 *Globe & Mail* (Toronto) 3 Feb. 3/4 Limousine service..to take them to the nearest city bus stop. 1970 *Toronto Daily Star* 24 Sept. 38/1 Jones and most of the borough education directors—who are paid the same salary—are given chauffeur driven limousines. 1972 *Daily Tel.* 16 May 9/3 The open limousine was followed by one in which Prince Philip travelled with Mme Pompidou. 1972 *Times* 27 June (Tokyo Suppl.) p. vi/2 There are regular limousine services

from the airport. 1973 V. CANNING *Finger of Saturn* i. 10 The people from the limousine got out... The limousine driver watched them go. 1973 *Country Life* 13 Sept. 686 If the terms 'limousine' and '1½ litre' appear incompatible, see, sit in, and drive the Triumph 1500... In the generous luxury you get in a well-padded limousine..long-distance travelling is limousine-smooth. All the other limousine touches are there too.

b. limousine liberal: a wealthy liberal. *U.S.*

1969 *Times* 4 Nov. 10/5 The little man truly representing the ordinary people, hitting out strongly at Mr. Lindsay's 'limousine liberal' appeal. 1970 *N.Y. Times* 26 Oct. 36 Canada is most fortunate to have a Premier who is willing to tell the bleeding hearts and limousine liberals what he thinks of them.

limp, *a.* Add: **1. c.** *limp wrist* (see quot. 1960); also *transf.* and (usu. with hyphen) as *attrib. phr.*

1960 WENTWORTH & FLEXNER *Dict. Amer. Slang* 319/2 *Limp wrist* adj., homosexual; said of male homosexuals; effeminate... A homosexual or effeminate man. 1963 *Amer. Speech* XXXVIII. 171 An effeminate young man, a sissy..*limp wrist*. 1969 *Guardian* 18 Mar. 1/3 Washington ..has concluded that if Britain continues to follow a 'limp wrist' policy after the open affront of the shooting affair, the gambling interests would draw obvious conclusions. 1970 C. MAJOR *Dict. Afro-Amer. Slang* 77 *Limp wrist*, having latent homosexual tendencies.

limp, *v.*[2] Add: **1. c.** *spec.* Of a damaged ship, aircraft, etc.: to proceed slowly or with difficulty.

1920 *Conquest* Apr. 291/3 The 'standard patch' has rendered invaluable assistance in helping stricken ships to limp into port. 1935 C. DAY LEWIS *Time to Dance* 37 But he tinkered and coaxed, and they limped Over the Adriatic on into warmer regions. 1971 *E. Afr. Standard* (Nairobi) 10 Apr. 1/1 Mr. Sprinzel, driving car No. 16, a Range Rover, with his co-driver David Benson, limped back to Nairobi yesterday afternoon. 1973 *Daily Tel.* 1 Jan. 1/4 The Fleetwood trawler Wyre Captain, 490 tons, limped into port at Thorshavn, Faroe Islands, yesterday, with a damaged bridge and no navigation instruments.

limpet. Add: **b.** Of officials alleged to be superfluous but clinging to their offices. Also *attrib.*

1905 *Westm. Gaz.* 9 Mar. 2/2 Lord Spencer..had some pertinent criticisms to make of the Limpet Government. 1922 *Daily Mail* 22 Nov. 8 He is rationing the departments and ejecting the 'limpets'. *Ibid.* 23 Nov. 10 Ministries are multiplying their accumulation of limpets and paying them too well. 1927 CARR-SAUNDERS & JONES *Survey Social Struct. Eng. & Wales* 54 After the war, attempts were made to rouse our animosity against 'limpets'.

c. limpet-hammer, a stone tool believed to have been used by prehistoric peoples to knock limpets off rocks.

1885 S. GRIEVE *Great Auk* 57 We were puzzling ourselves as to what could be the use of the numerous oblong stones we met with among the [limpet] shells, and..he.. informed us they were limpet-hammers. *Ibid.*, Subsequent enquiries have only helped to confirm us in the opinion that the large oblong stones found at Caisteal-nan-Gillean are really limpet-hammers.

2. A type of explosive device that is attached magnetically to a ship's hull; so *limpet bomb, charge, mine.*

1942 W. S. CHURCHILL *Secret Session Speeches* (1946) 50 Explosions occurred in the bottoms of the *Valiant* and *Queen Elizabeth*, produced by limpet bombs fixed with extraordinary courage and ingenuity. 1949 F. MACLEAN *Eastern Approaches* II. iii. 202 We fixed a couple of 'limpets' to her stern. These were half-spheres of metal, made to contain a pound or so of high explosive and fitted with a magnetic device to hold them to the side of the ship. 1955 J. THOMAS *No Banners* xix. 175 He wanted permission to use six limpet charges. 1959 *Chambers's Encycl.* IX. 428 Midget submarines placed these limpet mines which, being magnetized, clung to the hull and were fired by a time fuze. 1962 G. WELLER *All about Submarines* (1963) viii. 115 The German officers were sure that limpet mines had been attached to the keel of their battleship. 1971 D. BAGLEY *Freedom Trap* ix. 198 What do we use instead of a limpet?.. I can't invade her in the middle of the Grand Harbour.

limping, *ppl. a.* Add: **c.** *transf.* Of certain relationships that are held to be legal in one country but not in another.

1963 *Listener* 17 Jan. 122/1 There is probably no more embarrassing experience for a couple than to find that they are regarded as validly married in one country, but living in sin in another. Even worse, perhaps, is the position of a man regarded by one country's laws as divorced and therefore single, but by another country as married and liable to be convicted for bigamy if he remarries. This situation, vividly called 'the limping marriage', is deplored by most systems of law including our own. *Ibid.* 123/2 It goes some way towards mitigating the evil of the 'limping marriage'. 1970 *Internat. & Compar. Law Q.* XIX. I. 12 It was scarcely in the child's interests to make her a 'limping infant' by pronouncing an order which would be refused recognition in a country with which she had significant links. *Ibid.* 17 There was the particular danger..that the adoption order would not be recognised in some countries, producing a 'limping' adoption.

limu, var. *LIMMU.

linac (li·næk). *Physics.* [f. *lin*ear *ac*celerator.] A linear (i.e. straight) particle accelerator.

1950 *Amer. Jrnl. Physics* XVIII. 126/1 At Berkeley, L. W. Alvarez was the chief supporter of the 'linac', and he supervised the construction of the present 40-ft. machine. 1966 *New Scientist* 17 Nov. 359/2 Above the underground linac are two Van der Graaff accelerators. 1969 *Sci. Jrnl.* Jan. 55/1 Although many linear radio-frequency accelerators (called linacs) have been built, the more usual construction is circular so that the same field element can be used many times in the complete acceleration cycle.

linage. Add: **c.** Also, the charge made (by a newspaper, etc.) according to the number of lines occupied by an advertisement, etc.

1961 in WEBSTER. 1968 *Listener* 20 June 818/2 (Advt.), Linage 6s. 6d. a line. 1971 *Timber Trades Jrnl.* 14 Aug. 61 Linage. Minimum 20 words. 10p per word.

linalool (linæ·lo,ǫl). *Chem.* [a. G. *linalool* (F. W. Semmler 1891, in *Ber. d. Deut. Chem. Ges.* XXIV. 207), f. G. *linalo*(*ëol* linaloe oil + *-ol* -OL.] An optically active tertiary alcohol, $(CH_3)_2C{:}CH{\cdot}CH_2{\cdot}CH_2{\cdot}C(CH_3)OH{\cdot}CH{:}CH_2$, found in linaloe oil and used in perfumery: the *d*-isomer is *CORIANDROL and the *l*-isomer also occurs in the oils of rose, bergamot, lavender, and thyme.

1891 *Jrnl. Chem. Soc.* LX. I. 540 Linaloe oil appears to be a mixture of several compounds... The principal constituent which is termed linaloöl, boils at about 195–190° [*sic*]. 1922 *Jrnl. Amer. Chem. Soc.* XLIV. 2966 The odorous constituents of the peach consist chiefly of esters of the aliphatic terpene alcohol linalool. 1951 P. Z. BEDOUKIAN *Perfumery Synthetics & Isolates* 292 Commercial linalool is obtained from several sources, principally from oil of bois de rose, linaloe, shiu and coriander... Linalool occurs to the extent of about 80 per cent in the first two oils. 1968 *Times* 5 Nov. 17/2 Linalool and linalyl isobutyrate.. are both fragrant oils found in coriander and lavender.

Hence **li·nalyl,** the radical $C_{10}H_{17}$— present in ethers and esters of linalool.

1900 *Jrnl. Chem. Soc.* LXXVIII. II. 101 Analyses of two samples of essence of bergamot..show that during the process of ripening..the linalyl acetate increases from 33·8 to 37·3 per cent. 1964 *Economist* 26 Dec. 1448/1 Linalyl acetate goes into colognes to provide the lavender notes.

linamarin (linămæ·rin). *Chem.* [a. G. *linamarin* (Jorissen & Hairs 1891, in *Pharmaceut. Post* XXIV. 659), f. L. *līn-um* flax + *amār-us* bitter + -IN[1].] A bitter crystalline compound, $C_{10}H_{17}NO_6$, which occurs in flax; the glucoside of acetone cyanohydrin.

1892 *Jrnl. Chem. Soc.* LXII. I. 502 Linamarin forms groups of colourless needles having a cool and bitter taste. 1934 *Onderstepoort Jrnl. Vet. Sci. & Animal Industry* III. 116 The toxic substance present in the plant *Dimorphotheca cuneata* has been isolated and identified as the cyanogenetic glucoside 'linamarin'. 1964 D. D. DAVIES et al. *Plant Biochem.* x. 385 Feeding experiments have shown that the synthesis of the cyanogenetic glucosides, lotaustralin and linamarin, in *Trifolium repens* is related to the metabolism of isoleucine and valine. 1973 *Daily Colonist* (Victoria, B.C.) 18 July 18/4 Cassava's toxicity takes on a new importance. This toxicity is caused by a substance called linamarin..together with small amounts of a closely related substance, lotaustralin.

linchet: see *LYNCHET.

Lincoln[1]. Add: **1. Lincoln imp,** a grotesque carving in Lincoln cathedral; a door-knocker, ornament, or trinket modelled on this; also *attrib.*; **Lincoln Longwool,** a sheep of the breed so called, characterized by its large size and long fleece; **Lincoln Red,** an animal belonging to a breed of red shorthorn cattle so called, used as producers of both milk and beef; **Lincoln wool,** wool from a Lincoln Longwool.

1926–7 *Army & Navy Stores Catal.* 156/3 Bedroom door knockers..Lincoln Imp, 3½ in., 1/6. 1941 E. BOWEN *Look at Roses* 245 They were mementoes—photos..a Lincoln Imp, a merry-thought pen-wiper. 1967 *Listener* 13 July 48/3 Lincoln Cathedral has its world famous devil—the Lincoln imp. 1894 *CGA Catal.* 46/1 Sheep. Lincoln Long Wool Shearling Rams, bred by owner: 1st and 2nd, £10. 1919 W. C. COFFEY *Productive Sheep Husbandry* xix. 163 Breeders in England organized the Lincoln Long-wool Sheep Breeders' Association in 1892. 1972 *Country Life* 16 Mar. 606/2 Lincoln Longwool and some of the white-faced breeds are used for crossing. 1903 *Farmer & Stockbreeder* 20 July 1195/3 The Lincoln Red is a profitable farmer's beast, growing to great weight. 1966 *Guardian* 13 July 18/1 Cockerington Earl..won the supreme championship of the Lincoln Red cattle. Lincoln Reds are the heftiest of the beef cattle. 1974 *Country Life* 17 Jan. 65/3 The stocky, sturdy, thick-coated Lincoln reds... Lincoln red cattle are..most attractive animals. 1930 L. G. D. ACLAND *Early Canterbury Runs* vii. 177 They cut over twelve hundred bales of Lincoln wool.

Lincoln[2] (li·ŋkə̆n). Name of Abraham *Lincoln* (1809–65), sixteenth President of the U.S., in **Lincoln rocker,** a type of rocking-chair with straight upholstered back and seat and open arms, popular in the mid-19th c.

1952 J. GLOAG *Short Dict. Furnit.* 311 *Lincoln rocker*, a name occasionally used in the United States for a mid-19th century type of rocking chair. 1967 *Boston Sunday Herald* 26 Mar. I. 51/3 (Advt.), Grape carved Lincoln rocker.

Lincolnesque (liŋkə̆ne·sk), *a.* [f. the name of Abraham *Lincoln* (1809–65), 16th president of the U.S. + -ESQUE.] Resembling or having the qualities of Abraham Lincoln. So **Lincolnian** (liŋkōᵘ·niăn), *a.*

1910 H. P. WILLIS *Stephen A. Douglas* xiii. 283 A characteristic Lincolnian anecdote which the speaker applied to the repetition of the 'state fraud' of the Springfield resolutions. 1923 *Public Opinion* 15 June 565/3 That is Lincolnesque in its homeliness. 1962 E. AMES *Daughter of House* (1963) I. i. 30 The two long Lincolnesque furrows of the cheeks. 1963 *Economist* 20 July 257/2 The [Republican] party's Lincolnian heritage. 1968 MRS. L. B. JOHNSON *White House Diary* 9 Apr. (1970) 659 The star of that day was Hector Legge, of Dublin, Ireland, 'Dean' of the traveling journalists, a towering figure.. with a Lincolnesque face.

Lincolniana (li:ŋkə̆niā·nă). [f. the name of Abraham *Lincoln* (see prec.) + *-IANA.] Matter such as books, objects, and writings, relating to or characteristic of Abraham Lincoln.

1921 *Double-Dealer* Feb. 67/1 Another notable essay in Lincolniana appears with the imprint of Walter M. Hill of Chicago—'The Assassination of Lincoln' by E. W. Coggeshall. 1932 *N. & Q.* 19 Nov. 366/2 A considerable amount of hitherto unprinted Lincolniana. 1949 *Jrnl. Illinois State Hist. Soc.* June 137 The volume is one of the few in the vast area of Lincolniana that influenced later research and publications. 1959 *Listener* 19 Feb. 326/1 That vast body of ephemeral Lincolniana.

Lincolnshire (li·ŋkə̆nʃəɹ). The name of a county on the east coast of England, used *attrib.* in **Lincolnshire Curly-coat(ed),** a pig of the extinct breed so called; **Lincolnshire limestone,** a bed of oolitic limestone of Upper Jurassic (Bajocian) age, extensively developed in Lincolnshire and adjoining counties; **Lincolnshire Longwool** = *Lincoln Longwool*; **Lincolnshire Red** = *Lincoln Red*.

[1847 H. D. RICHARDSON *Pigs* iii. 36 The old Lincolnshire breed was light coloured, or even white, with, in most specimens, a curly and woolly coat.] 1917 W. POWELL-OWEN *Pig-Keeping* xi. 130 Lincoln Curly Coated. This Eastern counties breed is..most useful for crossing purposes. 1921 H. A. DAY *Pig-Keeping* 13 The Lincolnshire Curly-Coated. I see no reason why this breed should not be given a trial in districts where it is seldom met. 1972 *Country Life* 16 Mar. 606/2 Oxford Sandy and Black..are only one of five breeds of pigs that have become extinct, the others being the Cumberland, the Dorset Gold Tip, the Lincolnshire Curly Coat and the Yorkshire Blue and White. 1873 *Q. Jrnl. Geol. Soc.* XXIX. 226 The main feature of my Second Part will be the description and consideration of a series of beds grouped by Mr. Judd under the name of the 'Lincolnshire Limestone'. *Ibid.* 284 A few words as to the extent of the area occupied by the Lincolnshire Limestone. It ranges through the whole of the county of Lincoln, stretching into South Yorkshire on the north, and through Rutland into Northamptonshire on the south. 1969 BENNISON & WRIGHT *Geol. Hist. Brit. Isles* xiii. 301 The latter [*sc.* Collyweston Slate] forms a transition to the Lincolnshire Limestone which, appearing in Northamptonshire, thickens northwards to form the conspicuous west-facing escarpment on which Lincoln Cathedral stands. 1874 R. O. PRINGLE *Live-Stock of Farm* vii. 251 The old Lincolnshire long-wools were ungainly animals. 1897 W. HOUSMAN *Cattle* iii. 96 The excellence and purity of the Lincolnshire Red Shorthorn breed.

lincomycin (liŋkomai·sin). *Pharm.* [f. mod.L. *linco*(*lnensis*, specific epithet (see def.) + *-MYCIN.] An antibiotic, $C_{18}H_{34}N_2O_6S$, produced by cultures of the bacterium *Streptomyces lincolnensis* var. *lincolnensis* and given orally or by injection (usu. as the hydrate of the hydrochloride) to combat various Gram-positive bacteria, esp. staphylococci and streptococci.

1963 *Bacteriol. Proc.* 1963 94/1 Lincomycin, developed by The Upjohn Co., is a new antibiotic, chemically distinct from others now in use. 1966 *Practitioner* July 90 Lincomycin hydrochloride should prove useful in general practice against the common gram-positive infections. 1969 *New Scientist* 8 May 277/1 The group A haemolytic streptococci..are becoming increasingly resistant to tetracycline, erythromycin, and lincomycin. 1973 *Brit. Pharmacopœia* 1973 266/2 Lincomycin Hydrochloride is the monohydrate of the hydrochloride of methyl 6,8-dideoxy-6-(1-methyl-4-propyl-2-pyrrolidinecarboxamido)-1-thio-D-*erythro*-D-*galacto*-octapyranoside. 1975 *Daily Colonist* (Victoria, B.C.) 31 Jan. 3/6 Lee Smith said..that United States reports on the drugs clindamycin and lincomycin..were 'wildly exaggerated' in the news media.

lincrusta (linkrʌ·stă). Also **Lincrusta.** [f. L. *līn-um* flax + *crusta* rind, bark: after LINOLEUM.] A special type of thick embossed wall-paper.

1882 *Encycl. Brit.* XIV. 676/2 Mr. Walton, the original patentee of linoleum, has adapted a preparation of oxidized oil and cork or other thickening material embossed with patterns for wall decorations under the name of 'Lincrusta Walton'. 1891 *Jrnl. Soc. Chem. Industry* X. 150/2 Improvements in the Manufacture of Linoleum, Lincrusta, Cere-Cloth, and the like. 1921 *Spectator* 9 Apr. 464/1 'Lincrusta' wears so well that it seemed a sin to take it off. 1923 U. L. SILBERRAD *Lett. J. Armiter* i. 29 Semi-detached house, lincrusta dadoes, basement kitchen—it would suit him. 1939 *Archit. Rev.* LXXXVI. 258 *Lincrusta*. This is a plastic material from which a relief ornament is pressed. 1963 *Punch* 25 Sept. 457/1 He rolled his eyes..towards the lincrusta ceiling. **1973** M. Bence-Jones *Palaces of Raj* viii. 142 Lord Curzon replaced some of the Lincrusta with damask.

lindane (li·ndēⁱn). [f. the name of Teunis van der *Lind*en (b. 1884), Dutch chemist, who investigated the isomers of benzene hexachloride + *-ane* (perh. after *CHLORDANE).] The gamma isomer of benzene hexachloride, $C_6H_6Cl_6$, used as an insecticide; it is a colourless crystalline compound that is toxic to mammals but relatively harmless to plant life, and is used in the form of dusts, sprays, and aerosols.

1949 *Lindane* (Interdepartmental Comm. Pest Control, Bureau Entomol. & Plant Quarantine, U.S. Dept. Agric.) 3 The coined name, 'Lindane', is established for the gamma isomer of the chemical 1, 2, 3, 4, 5, 6-hexachlorocyclohexane of a purity not less than 99 percent. **1956** *Nature* 25 Feb. 367/1 In solid form 'Dieldrin' and 'Lindane' were the most effective toxicants against the larvæ of *Aedes aegypti* and *Culex fatigans*. **1961** *New Scientist* 6 July 26/3 A new liquid seed dressing..is based on lindane (a very pure form of gamma-BHC), which is considered to be the insecticide least hazardous to wild life. **1970** *Nature* 31 Oct. 403/2 Fresh and rain water in Britain contains small but not insignificant amounts of the chief persistent organochlorine pesticides, DDT, lindane and dieldrin.

Linde (li·ndə). The name of Carl P. G. R. von *Linde* (1842–1934), German physicist, used *attrib.* to designate a process for liquefying gases by means of repeated cycles of compression, cooling, and expansion, used in the extraction of nitrogen and oxygen from air by exploiting the difference in their boiling points.

1902 *Encycl. Brit.* XXX. 283/2 The efficiency of the Linde process is small, but it is easily conducted and only requires plenty of cheap power. **1928** J. K. Roberts *Heat & Thermodynamics* v. 111 Much less preliminary cooling is necessary to liquefy a gas by the Linde method than by the Cascade process. **1937** M. W. Zemansky *Heat & Thermodynamics* xiv. 251 In the Linde process for the production of liquid air, the initial temperature, initial pressure and final pressure are chosen so that, after passing through a throttling valve, a drop in temperature is produced. **1968** B. J. Hazzard tr. *Asinger's Paraffins* ii. 105 After the removal of carbon dioxide and benzene by washing, coke-oven gas is liquefied by the Linde process, all the constituents except the hydrogen and some nitrogen condensing.

lindgrenite (li·ndgrĕnəit). *Min.* [f. the name of Waldemar *Lindgren* (1860–1939), U.S. geologist + -ITE¹.] A basic copper molybdate, $Cu_3(MoO_4)_2(OH)_2$, found as transparent, green, monoclinic, platy or tabular crystals.

1935 C. Palache in *Amer. Mineralogist* XX. 491/2 The author takes great pleasure in designating this well-defined mineral lindgrenite in honor of Professor Waldemar Lindgren. His great contributions to the knowledge of the ore deposits and their paragenesis makes it peculiarly fitting that his name should appear in the special literature of mineralogy. **1953** *Amer. Mineralogist* XXXVIII. 905 The physical and optical properties of the Idaho lindgrenite are virtually identical with those of the Chilean material. **1955** *Mineral. Mag.* XXX. 726 This occurrence of lindgrenite at Brandy Gill is perhaps one of the most interesting additions to the remarkable number of rare and unusual minerals..in the Caldbeck district.

lindworm (li·ndwɒɹm). Also **lindorm.** [ad. Da. and Sw. *lindorm*. Cf. *LINGWORM.] A monstrous and evil serpent, common in Scandinavian legend.

1814 H. Weber *Illustrations of Northern Antiquities* 60 The terms worm, drake, dragon, and serpent, are indiscriminately applied to these monsters, as well as lind-drake and lind-worm; probably from their haunt being generally under a linden or lime tree. **1896** W. A. Craigie *Scandinavian Folk-Lore* 439 The lindorm is a favourite monster in Swedish as well as Danish tradition. **1910** F. Bond *Wood Carvings Eng. Churches* I. 63 When a wyvern has no wings as at Limerick, he is, in heraldry, a lindworm. *Ibid.*, No rigid distinction can be made between the dragon, wyvern, and lindworm. **1971** Barber & Riches *Dict. Fabulous Beasts* 97 *Lindorm*, a snake-like creature, which devoured cattle and ate bodies... The lindworm is a heraldic dragon or wyvern without wings, presumably the same as the lindorm.

Lindy Hop (li·ndi hɒp). Also **lindy-hop.** [f. *Lindy*, nickname of C. A. Lindbergh (1902–74), the American pilot who in 1927 was the first to make a solo non-stop transatlantic flight + Hop *sb.*² 2.] A Negro dance originating in Harlem (New York); also *attrib.* Also *ellipt.* **Lindy.** Hence **Lindy** *v. intr.*, to dance the Lindy Hop; **li·ndyhopper**, one who dances the Lindy Hop.

1931 *Zit's Theatrical Newspaper* 2 May 11 The winners of the all-Harlem Lindy Hop contest..drew rounds of applause nightly. **1936** *Life* 14 Dec. 64 The Lindy Hop originated at the Savoy. **1946** Mezzrow & Wolfe *Really Blues* (1957) xvi. 286 We'd get five teams of lindyhoppers from the Savoy Ballroom. **1948** K. Davis *Human Soc.* 79 The late and unlamented dance step called the 'Lindy Hop' was a fad. **1951** [see *JITTERBUG *sb.* 2]. **1959** *Sears, Roebuck Catal.* Spring & Summer 831/2 Betty White's Teen-Age Dance Book... Clear instructions for fox trot, lindy, mambo, cha cha, etc. **1969** *New Yorker* 15 Feb. 31/3 Their feet began to tap, and they tore into the Lindy. **1969** Disch & Sladek *Black Alice* ii. 19 No one could challenge his tango, but when he heard fast music his impulse was to lindy rather than to twist. **1970** G. Greer *Female Eunuch* 108 Here I am, a negro who cannot do the lindy-hop or sing the Blues! **1970** A. Miller in A. Chapman *New Black Voices* (1972) 538 Few can ignore the effects of the cake walk, fox trot, lindy hop, the twist on the movement habits of people in the States. **1973** *Guardian* 7 Aug. 8/4 When I was a kid I would go and dance for the white folks. Do the splits and the Lindy Hop, and they would throw me pennies.

line, *sb.*² Add: **I. 1.** *spec.* as used by climbers (usu. opp. *rope*).

1907 *Yesterday's Shopping* (1969) 700/2 Fine alpine line. **1923** G. D. Abraham *First Steps to Climbing* ii. 35 A light Alpine line is also supplied but that is mostly used by experts on exceptionally difficult courses... For the beginner the ordinary rope is advisable. **1935** D. Pilley *Climbing Days* xi. 224 We set aside the ordinary Alpine rope, and used 120 feet of Alpine line. **1950** *Mountaineering Handbk.* (Assoc. Brit. Members Swiss Alpine Club) ii. 27 Line can be used on ice or rock..or for rappel slings... Doubled, it can be used as a light rope. **1957** Clark & Pyatt *Mountaineering in Brit.* ix. 160 One development in technique was..the increasing use of line in preference to full size rope.

e. (Earlier and additional examples.) Add to def.: A telephonic connection; an individual 'number' or extension. Cf. *hold the line* (*HOLD *v.* 6 h), *hot line* (*HOT *a.* 12 c). Also *fig.*, esp. in phr. *to get the lines crossed*, to become confused.

1847 *Handbk. to Electric Telegraph* 11 So rapid is the transmission of the electric current along the lines of wire, that..to carry the wires eight times round the earth.. would occupy but one second of time. **1900** C. H. Chambers *Tyranny of Tears* I. 2 Miss Woodward. (*Speaking into telephone—very sweetly.*) ..Mr. Parbury's just coming in now—he'll speak to you—keep the line. **1921** *Conquest* Jan. 127/2 The 'busy tone' is sent back to the calling subscriber if the line he wants is busy. **1934** *Punch* 21 Mar. 332/1 The notepaper should carry—(1) The name of the firm. (2) Its address. (3) Fictitious address for creditors. (4) Telephone number (at least ten lines). **1944** H. McCloy *Panic* (1972) 15 Ronnie showed the doctor how to get an outside line and he dialled a number. **1951** *Oxf. Jun. Encycl.* IV. 448/1 The Post Office took over all 'trunk' long-distance lines in 1896, and 6 years later opened the first of several large London exchanges, the 'Central', with 14,000 lines. **1970** B. Knox *Children of Mist* iv. 77 Thane lifted the telephone. When the desk constable answered he asked for a line... Then he began dialling. **1972** J. Wilson *Hide & Seek* iii. 61 What? I can't hear you. It's a terrible line. **1973** *Times* 16 Apr. 14/6 It clearly has the advantage of keeping all the lines from getting crossed and establishing the priorities of policy. **1973** K. Royce *Spider Underground* iii. 50 He told me he couldn't see me then and to get off the line. **1974** *Times* 15 Mar. 8/2 Mr Nixon has admitted that he ordered a cover-up of the plumbers' activities, but suggested that his staff got their lines crossed and took this to be an order to cover up the Watergate affair as well.

(ii) Hence, any wire or cable that serves as a conductor of electric current, for whatever purpose.

1886 G. Kapp *Electr. Transmission of Energy* viii. 205 Overhead lines, whether used for electric lighting or transmission of energy, are exposed to the effects of lightning. **1902** *Encycl. Brit.* XXV. 35/2 Alternate current is used for lighting and continuous current for the tramway line. **1920** *Whittaker's Electr. Engineer's Pocket-Bk.* (ed. 4) 407 Since the induced voltages due to lightning are the same whatever the working voltage of the line, the heavier insulation on extra high voltage lines renders them less subject to lightning trouble. **1930** *Engineering* 25 Apr. 548/2 Minimum expenditure on the transmission and distribution systems from those points, connoting the use of overhead lines. **1957** *Encycl. Brit.* XXI. 887/1 On the teleprinter at the other end of the line, the responses of the armature of a single electromagnet..cause the corresponding character to be printed. **1962** A. Nisbett *Technique Sound Studio* iv. 79 In a building the size of a broadcasting studio centre there is a danger not only of high frequency losses due to capacitance, but also induction of programme signals, hum, etc., from other lines.

II. 7. f. (Further examples.) Also in phr. (taken from American football, but influenced by sense *20 b) *to hold the line*, to maintain, support, a position, viewpoint, etc. In *Cricket*, the line of flight of the ball from the bowler's hand.

1887 [see Dribble *v.* 4 a]. **1956** B. Holiday *Lady sings Blues* (1973) xi. 102 But 52nd Street couldn't hold the line against Negroes forever. **1960** I. Wallach *Absence of Cello* (1961) 48 Her voice had a factious quaver as she dug in and prepared to hold the line on Perry's team. **1961** *Times* 18 Aug. 3/3 At 18 Pullar was bowled by Davidson, playing across the line. **1962** *Listener* 19 Apr. 672/2 'Holding the line'..of costs, prices, and wages is vital to what he believes to be the continuance of American prosperity. **1963** A. Ross *Australia 63* iii. 87 He moved solidly behind the line, early in position for anything that kept low. **1968** W. Safire *New Lang. Politics* 190/2 'Holding the line against inflation' remains a cliché, taken from a football metaphor ('Hold-that-line!') which in turn comes from a military expression. **1969** *Times* 25 Aug. 9/2 Harris, eventually, was leg-before, hitting enthusiastically across the line.

g. *Ballet.* The total effect of the disposition of the dancer's limbs, body, and head in movement or repose. (Cf. sense 7 d.)

1912 J. E. C. Flitch *Mod. Dancing & Dancers* xi. 170 Her purity of line is never broken by..inartistic feats of athletic dexterity. **1922** Beaumont & Idzikowski *Man. Classical Theatr. Dancing* 26 Beauty of line is one of the dancer's greatest assets. **1936** A. Haskell *Prelude to Ballet* xvii. 85 Fluidity and large movements whose line can be extended indefinitely are the essential characteristics of the Russian School. **1948** *Ballet Ann.* II. 91 He has a fine classical technique and excellent line. **1960** *Times* 7 Mar. 3/7 She is already a dancer of great charm..with a particularly striking sense of line that showed to advantage in lifts.

h. *Mus.* Instrumental or vocal melody; a structured sequence of notes or tones.

1923 R. H. Myers *Mod. Mus.* vi. 80 His music has line ..and the enormous merit of condensation. **1955** *Times* 26 Aug. 3/5 In spite of the cello's natural inclination to ruminative melancholy..it has plenty of cantilena... But it is line, always line, not harmony, that is the essence of the matter. **1961** *Listener* 14 Dec. 1046/3 What do singers mean when they talk about 'maintaining the line'? .. It means striking a level in the voice from which all expression is controlled. *Ibid.*, This 'line' of the singer is a physical conception. *Ibid.* 21 Dec. 1089/3 The music takes shape by means of a simple recitative-like vocal line, modal, flexible, limpid, with an orchestral part of matching directness and simplicity. **1962** *Radio Times* 22 Feb. 43/1, I was concerned at the time with the idea of inventing melodic line and harmonic texture directly from the fund of the twelve notes available within the octave. **1967** *Melody Maker* 28 Jan. 7/5, I consider jazz to be a lot of horns and one of those top speed bass lines.

i. Each of the narrow strips into which an image is divided for transmission and reproduction by television, corresponding to a single (usually side-to-side) passage of the scanning spot across the camera tube or picture tube: often with prefixed number, as *625-line(s)*, indicating the number of lines making up a complete picture.

1929 *Proc. IRE* XVII. 1586 He first arrived at a correlation between the number of 'halftone lines per inch' and the corresponding television 'scanning lines'. *Ibid.*, Halftones of letters and photographs were made up, and their appearance compared with the television image on a 48-line system of the same original. **1938** *Encycl. Brit. Bk. of Year* 633/2 The service was continued, using exclusively a standard of 405 lines 50 frames interlaced scanning. **1961** G. Millerson *Technique Television Production* ii. 20 To reduce flicker problems, the beam is made to read the odd lines (odd field) of the image first (i.e. lines 1, 3, 5,..) and then return to scan the even lines between them (i.e. lines 2, 4, 6,..). **1963** *Ann. Reg. 1962* 27 They duly authorized the B.B.C. to start a second television channel by 1964 on U.H.F. and with an improved picture of 625 lines. **1974** *Sci. Amer.* Jan. 115/2 Each 1·25-second signal comprises a 'line' of picture data that is analogous to the line of a television picture. About 850 lines..complete a weather-signal picture.

8. c. Delete the reference to Fraunhofer's lines (cf. *FRAUNHOFER) and add to def.: Hence in extended use, a component of emitted radiation at what is nominally a single discrete wavelength (in practice, over a narrow range of wavelengths containing one at which the intensity is a maximum). (Further examples.)

1932 *Sci. Abstr.* A. XXXV. 1561 (*heading*) Line emission in infra-red. **1962** *Science Survey* III. 67 For a normal lamp, emitting a line in the visible spectrum, the width..of the line would be of the order of 10,000 Mc/s. **1971** D. W. Sciama *Mod. Cosmol.* ii. 21 He calculated that a sensitive radio receiver should be able to detect the 21 cm line as emitted by clouds of hydrogen gas in the Galaxy. **1971** *Nature* 31 Dec. 505/2 The atoms made up of the smaller mass particles would then radiate their characteristic lines at longer wavelengths.

13. c. In *Politics* (orig. *U.S.*), a particular policy or set of policies which a politician may maintain or expect others to follow; = *party-line.* Also *transf.*

1892 *San Francisco Examiner* 9 Nov. 1/7 (*heading*) In the line! California joins the Democratic procession by a decisive majority. **1934** H. M. Chevalier tr. *Malraux's Man's Fate* 149 He knew, too, that Moscow would maintain its line. *Ibid.* 169 'There's a general line that directs us—must follow it.' 'And give up our arms! A line that leads us to fire on the proletariat is necessarily bad.' **1938** *Ken* (Chicago) 7 Apr. 46/2 The Intelligence Service of the Foreign Office is a state within a state, virtually Britain's second, secret Government as far as foreign policy is concerned. It often pursues a line different from the Government's policy. **1943** *San Francisco Chron.* 25 May 14/2 The Nazis have done Senator Happy Chandler of Kentucky the honor of picking up his line... Chandler's line may not get far in this country, but the Nazis are not slow to appreciate it. **1944** M. Laski *Love on Supertax* v. 60, I think the line was made perfectly clear. **1955** *Times* 2 June 6/6 The issue before the court is not so much whether Mr. Lattimore is guilty under the indictment as whether such a nebulous charge as following the Communist 'line' is sufficiently defined to enable him to offer an adequate defence. **1958** *Economist* 29 Nov. 767/2 They think that the liberal line—uncontrolled immigration—can be held for a few more years, but not indefinitely. **1960** *News Chron.* 25 Feb. 2/5 Mr. Barber denied that a 'line' had been agreed on as to the shape of the reports to be sent by..British reporters. **1974** Hawkey & Bingham *Wild Card* xxiii. 188 The official line on what had happened was, at best, grossly understated.

d. *transf.* A marked tendency, a policy or trend (in any activity). In weakened use (*slang*): a glib or superficially attractive mode of address or behaviour, plausible talk. So *to do a line with* (Austral. and N.Z.), to (try to) enter into an amorous relationship with.

Not clearly separable from senses 28 a, b. Cf. also *to shoot a line* (sense 13 g below).

1903 'H. McHugh' *Out for Coin* vi. 83 Are you handing me a line of bogus conversation? **1920** F. Scott Fitzgerald *This Side of Paradise* I. ii. 76 Lordy, Isabelle—this *sounds* like a line, but it isn't. **1923** *Cosmopolitan* Apr. 82/1 'Where have I been all your life, good lookin'?' 'If you think that line will get you anything here, you're crazy!' **1933** J. G. Cozzens *Cure of Flesh* I. 61 He falls in love with Coral and says that some day, when he makes good, he will come back and marry her. Coral thinks it's just a line with him. **1941** *Illustr. London News* CXCVIII. 488/2 The jacket mentions Huckleberry Finn. Mr. Baum is not, of course, on that level; but that's his line. **1941** [see *KNOCK *sb.*¹ 5]. **1942** T. Rattigan *Flare Path* I. 102 They'll think it's a line, sir. **1944** J. H. Fullarton *Troop Target* viii. 63 He was doing a heavy line with the saddler's daughter. **1946** F. Sargeson *That Summer* 91, I could do a line with Maggie. **1946** K. Tennant *Lost Haven* (1947) x. 156 Do you know young Len's doing a line with Gran'pa's little angel? **1953** *Encounter* Oct. 1/2 Appearing at this time, and amidst these problems, *Encounter* seeks to promote no 'line', though its editors have opinions they will not hesitate to express. **1956** A. L. Rowse *Early Churchills* ii. 33 He has a fine line in Churchillian invective. **1958** D. Reeman *Prayer for Ship* viii. 202 He gave me a terrific line about the hold-up. Said it was his partner's fault. But he promises definitely it'll be here tomorrow evening. **1967** *Observer* 6 Aug. 4/6 The sect's most telling line—plugged in all its broadcasts and pamphlets—is that the end of the world is due shortly, probably about 1975. The Arab-Israeli war in June was seen as the first step to Armageddon.

e. *to get a line on*, to acquire information about (a thing), to come to know. So *to give (someone) a line on. colloq.* (orig. *U.S.*).

1903 *Sun* (N.Y.) 18 Nov. 4 'These dressmakers'..cannot get a line on the styles except at the Horse Show. **1920** B. Cronin *Timber Wolves* 138 It ain't over wise to give anyone a line on to what's doing. **1923** R. D. Paine *Comrades of Rolling Ocean* iii. 41 How about these other birds. Give me a line on them. **1928** D. L. Sayers *Unpleasantness at Bellona Club* xiv. 165, I did tumble to it that you'd got a line on me when you sent me down with that detective fellow to Charing Cross. **1935** Wodehouse *Luck of Bodkins* v. 50 If you want to get a line on how she feels, she gave me a letter to give you... Here it is. **1942** *Penguin New Writing* XII. 85 'They got a line on him,' said the R.P. **1947** *Chicago Tribune* 22 July 1/5 If we can find any one who saw her at a dance after 10:30 p.m. we may be able to get a line on whom she was dancing with and whose company she was in when she left.

f. *to lay* (or *put*) *it on the line*: (*a*) to hand over money; (*b*) to state (something) clearly, plainly, or categorically; (*c*) (with direct object) to put (one's career, etc.) at risk. Also with *place*, and the verb *to be*. Chiefly *U.S.*

1929 D. Runyon in *Hearst's International* Aug. 73/1 My rent is away overdue for the shovel and broom..and I have a hard-hearted landlady... She says she will give me the wind if I do not lay something on the line at once. **1940** J. O'Hara *Pal Joey* 100 You fellows always put it on the line for me every pay day. **1950** J. D. MacDonald *Brass Cupcake* i. 13 Lay it on the line. You can't take it with you... Put it on the entertainment account. **1954** J. Symons *Narrowing Circle* xxxvii. 188 'I'll see you're not the loser. You put it on the line with Jake Beverley and he'll put it on the line with you...' 'Let me lay it on the line then, Jake.' **1956** E. Pound tr. *Sophocles' Women of Trachis* 17 Put it on the line, what do you know? Get it out clearly. **1967** 'E. E. Sumner' *Chance Encounter* v. 94 I'll lay it on the line for you, if you like. Are you thinking of asking my girl to marry you? **1968** M. Luther King *Trumpet of Conscience* ii. 40 Our lives must be placed on the line if our nation is to survive its own folly. **1968** *Listener* 22 Feb. 244/3 America must fight in Vietnam..because it has laid its prestige on the line. **1968** *Guardian* 26 July 9/7 Mayor Stokes is putting his career on the line. And the people know it—they won't let him down. **1970** *Ibid.* 9 May 2/4 It was clear to the [American] President that his credibility was on the line with the leaders in Hanoi. **1972** *New Yorker* 26 Aug. 17/2 He had decided to put his artistic reputation as a talented and original director of opera on the line at the outset of his American career with an unorthodox..production of Bizet's 'Carmen'. **1972** J. Quartermain *Rock of Diamond* xxiv. 153 I'll lay it on the line, Raven. You can say yes or no. **1973** *Black Panther* 7 July 8/3 The situation is as bad as before the takeover and it only serves to give the Indian people more reason to put their life on the line. *Ibid.* 6 Oct. 3/2 Egil Krogh..put it squarely on the line: 'Anyone who opposed us we'll destroy.'

g. *to shoot a line* (cf. Shoot *v.* 23 g), to 'put on an act', to talk pretentiously, to boast. So *line-shoot* vb. (*-shooting* ppl. a. and vbl. sb.) and sb., *-shooter*; also *shooter of lines. colloq.*

Cf. sense 13 d above.

1941 N. Coward *Blithe Spirit* I. ii. 50 The whole thing's a put up job—I must say, though, she shoots a more original line than they generally do. **1942** *Penguin New Writing* XIII. 24 Occasionally..it publishes a serious article... But this is regarded as a 'bind',..while its author is invariably dismissed as a 'line shooter', i.e. a conceited person. *Ibid.*, The other day..our C.O. introduced a discussion on tactical evasion by saying: 'I do not want this to develop into a "line-shooting" competition.' **1942** *R.A.F. Jrnl.* 30 May 17 For keeping up the spirits, line-shooting is at least as good as beer-drinking. *Ibid.*, The man who shoots a heavy line about the work he is doing is probably very keen on his job. **1943** Hunt & Pringle *Service Slang* 44 *Lineshoot*, a tall story. **1944** G. Netherwood *Desert Squadron* i. 2 Some of the chaps also came from other well known fighter units. From the 'line-shooting' that ensued, one would think that the squadron which was then in the process of formation could never hope to be as well known as the one they had left—and so on and so forth. **1944** T. H. Wisdom *Triumph over Tunisia* 121 One of the most thorough and decisive of the air operations in the whole campaign was carried out by the Hurri-bombers. And this is no squadron line-shoot. **1946** G. Gibson *Enemy Coast Ahead* 144 These things were happening every night, so there was nothing to shoot a line about. **1951** M. Kennedy *Lucy Carmichael* vii. iv. 377 When Melissa shoots a line..don't protest or argue. Take it up and embroider it. **1952** T. Rattigan *Deep Blue Sea* I. 38 Funny thing about gongs. .. They don't mean a damn thing in war—except as a line-shoot, but in peace time they're quite useful. **1958** *Times Lit. Suppl.* 3 Oct. 564/2 A champion shooter of lines. In a party of outstanding climbers and travellers he could be relied on to cap any story. **1960** *Times* 19 July 18/4 One must bear in mind that what his Lordship had called..'shooting a line' was not necessarily inconsistent with a genuine belief. **1960** V. Gielgud *To Bed at Noon* I. xi. 73 He believed Tom to have been line-shooting as far as his swimming prowess was concerned. **1973** *Listener* 15 Mar. 342/1 [He] was an awful line-shooter. He claimed to have been at Oxford, but..he hadn't been at Oxford. **1973** *Times* 20 Sept. 20/8 He was awarded (his friends thought inadequately) the MBE by the British and by the French the Croix de Guerre. He never shot a line about his escapades but made them into entertaining stories.

14. b. *Fashion.* The outline or dominant features of composition of a dress or suit. Freq. with qualifying term or preceded by a letter of the alphabet (to indicate the outline shape of the garment). Cf. sense 23 a below.

1918 in C. W. Cunnington *Eng. Women's Clothes* (1952) iv. 141 What was called the *'barrel line'* brought out by Callot two seasons ago..certainly is a lovely line. **1930** *Times* 13 Mar. 11/6 The curved line was seen in all the long coats. *Ibid.* 27 Mar. 11/6 There is a distinguished coat in black matasol, which has a slimming line. **1932** *Punch's Almanack* 7 Nov. 8 (*caption*) The line of to day. **1955** *Britannica Bk. of Year* 489/2 Fashion produced a new 'line' in women's clothes, the *H-line.* **1958** *Woman's Own* 24 Dec. 14/3 Which year brought out the following trends: (*a*) the New Look; (*b*) the Trapeze Line; (*c*) the A-line. **1968** J. Ironside *Fashion Alphabet* 92 *Line*, the silhouette of a garment that makes it look fashionable or unfashionable. **1968, 1970** [see *EMPIRE *sb.* 8 b]. **1975** *Vogue* 1 Mar. 84/1 Overall, a clear narrowing of the silhouette, most marked at Saint Laurent, presaging an even sparer line for autumn.

17. b. (Earlier example.) Also ellipt. *the line.*

1779 in W. B. Reed *Life & Corr. J. Reed* (1847) II. 134 Perhaps we would be as well off with Mason and Dixon's line continued. **1845** F. Douglass *Narr. Life F. Douglass* xi. 101 We owe something to the slaves south of the line as well as to those north of it. **1909** 'O. Henry' *Roads of Destiny* xxi. 358 If you had come from below the line, I reckon I would have liked you right smart. **1949** *Sat. Even. Post* 26 Mar. 38/2 The critic thunders, and below 'the line' the shades of Marse Robert and Jeff Davis inevitably are summoned forth to meet the charge.

c. *Bridge.* A line across a score-card. So *above the line*, denoting points scored for game, honours, overtricks, or rubber, or for the failure of opponents to fulfil their contract; *below the line*, denoting points scored for tricks bid and won, and counting towards game.

1905 H. A. Vachell *Hill* vii. 144 My partner..made the Little Slam, and scored nearly six hundred below the line. **1908, 1927** [see *CONTRACT *sb.*¹ 1 g]. **1933** A. G. Macdonell *England, their England* vi. 78 Gone down 650 points above the line whereas he ought to have made two no-trumps. **1967** P. Anderton *Play Bridge* i. 15 They win ten tricks so they score three times the value of Spades below the line, i.e. 90 points plus another 30 points above the line as a bonus for making one more than their contract. **1970** S. Hughes *Art of Coarse Bridge* i. 12 This kind of spectacular finale happens far more often than one might expect, but it takes an awful lot of scoring above the line before anyone actually has..the right cards to do it with.

d. In phrases indicating the boundary between a debit and a credit in one's account, or between ordinary and extraordinary expenditure. (See also quot. 1973.) Also *to pay on the line*, to pay promptly.

1934 J. O'Hara *Appointment in Samarra* i. 20 There were only a few of the Lantenengo crowd who could get a favour out of Ed without paying cash on the line for it. **1938** S. V. Benét *Thirteen o' Clock* 249, I kept on schedule with the work, but I couldn't with the money. Each week, I'd be just a little over the line. **1940** *Economist* 13 Apr. 683/1 The figures 'below the line' in the Exchequer Return show the result of the issue of the 4 per cent. **1948** *Ibid.* 31 Jan. 195/1 Aggregate Government expenditure, including..the 'below line' expenditure. **1959** N. Mailer *Advts. for Myself* (1961) 66, I paid on the line every time. **1966** A. Gilpin *Dict. Econ. Terms* 1 Since 1947 it has been customary for the British Budget to contain a full statement of the estimated expenditure and revenue for the following year, some items being shown as 'above-the-line'..and others as 'below-the-line'. Most current items appear above the line and most capital items below. **1973** *New Society* 28 June 736/2 The growth of petrol promotions has coincided with a growth of giveaways, gimmicks and competitions in the marketing of a wide range of products... Termed 'below-the-line' marketing, it is encroaching on the 'above-the-line' (advertising) share of manufacturers' marketing budgets.

III. 19. *spec.* = Queue *sb.* 3. *U.S.*

1930 M. Sullivan *Our Times* III. xii. 502 People..were herded by policemen into lines stretching away from the marble entrance. **1969** D. C. Hague *Managerial Econ.* xi. 222 The second kind of stock problem is the queueing problem... A queue (what Americans call a 'line')..will form. **1974** *State* (Columbia, S. Carolina) 15 Feb. 1-B/2 At least one employe went as far away as Forest Drive for gasoline and nearby stations, selling gasoline, quickly acquired lines. **1975** *N.Y. Times* 1 Apr. 35/5 It's to stand silently on unemployment lines with other surplus members of America's work force, waiting to sign for your unemployment check.

c. A row of machines or work stations where a product is progressively assembled, or a succession of operations performed on it, as it passes from one end to the other during manufacture or processing. Cf. *assembly line*, *production line*.

1926 *Encycl. Brit.* II. 822/1 All of these lines, with their various machines and operations, are converging on the point where the leaves are assembled into springs. **1937** *Times* 13 Apr. p. xii/2 The raw material is delivered at one end of the machining line with the component passing from machine to machine until it reaches the view table. **1940** *War Illustr.* 16 Feb. 113 In one of Britain's 'shadow' factories bombers on the line will soon be ready to take the air. **1971** *Cabinet Maker & Retail Furnisher* 24 Sept. 531/1 Features of the production facilities at the new factory.. include a fully automated machining line and the longest finishing line in the U.K.

d. In business or management organization, the chain of command or responsibility; the persons responsible for the administration and organization of a business (as opposed to the staff). Hence *line manager, management.*

1960 Nanassy & Selden *Business Dict.* 27 Following are the basic types of internal organization of a business: (1) *line*: The owner gives orders directly to the workers. As the business grows, the owner appoints a few executives, who are responsible to him... (3) *line-and-staff*: Authority flows from top to bottom, with responsibility falling on staff supervisors and special experts. **1964** M. Argyle *Psychol. & Social Probl.* viii. 111 In several British factories it was found that the division between 'line' supervisors and 'staff' technicians tended to disappear—technologists must have supervisory responsibility. **1967** C. Margerison in Wills & Yearsley *Handbk. Managem. Technol.* 25 The accountants considered that they had responsibility for the end-product and sought to control certain actions of line managers. Line managers resented this interference with their authority and started to obstruct the accountants in their ordinary accounting function. **1967** Coulthard & Smith in *Ibid.* 206 A good deal of the failure of these techniques stems from the inability of personnel men to convince line management of their own vital role combined with the assumption by line management that the creation of a specialist department covering personnel policies, training, management development, etc., automatically relieves them of responsibility. **1972** *Accountant* 28 Sept. 391/1 If the internal auditor sees himself as a someone who can review and report upon the functions of line management on matters other than security, then there is one fundamental issue that has to be faced. **1974** *Times* 25 Mar. 17/4 It was a pity that so few line managers were present as it was their present and future competence that was being discussed.

20. b. In the war of 1914–18, the trenches collectively; the front line. So *up the line* (see quots.).

1916 H. W. Fowler *Let.* 5 Mar. in *S.P.E. Tract* XLIII. (1935) 136 What may be going on up the line who knows? **1917** W. Owen *Let.* 4 Feb. (1967) 430, I am now indeed and in truth very far behind the Line; sent down to this old Town [*sc.* Abbeville] for a Course in Transport Duties. **1917** A. G. Empey *Over Top* 313 *'Up the line.'* Term generally used in rest billets when Tommy talks about the fire trench or fighting line. **1919** W. H. Downing *Digger Dial.* 52 *Up the line*, in action. 'Up the line, with the best of luck'—a satirical phrase applied to men who, after being for some time in a safe occupation, were returned to fighting units. **1964** B. Gardner (*title*) Up the line to death.

21. c. *all along the line* (further examples); also, *all (the way) down the line.* Also, *somewhere along the line*, at some point (in time).

1924 R. Fry *Let.* 27 June (1972) II. 553 Both he and Courbet did elaborate portraits of the same patron... Courbet wins all along the line. **1936** A. Huxley *Eyeless in Gaza* xxi. 297 A refugee from Germany... Aryan, but communist—ardently and all along the line. **1962** J. Wain *Sprightly Running* v. 189 There is always the wistful hope ..that these young will not merely benefit from meeting each other, but will, somewhere along the line, actually be *taught* something. **1965** *Listener* 16 Sept. 402/2 It is difficult to estimate its direct effect, because all along the line there are people working hard to try to make sure that those defects do not come back on the patient. **1965** *New Yorker* 20 Nov. 162/3 Somewhere along the line, the surf and wind went out of his playing. **1969** B. Turner *Circle of Squares* xviii. 143 I've helped him all along the line, not always knowing why. **1972** *Guardian* 6 July 2/2 It has been clear that they had had to refer to Moscow for instructions all along the line. **1975** N. Luard *Robespierre Serial* xi. 87 You've lied to me, all the way down the line. *Ibid.* xvi. 146 I'm not going to let that little bastard get away with it. He's screwed us all down the line from Riyadh to Geneva.

22. (Earlier U.S. examples.)

1786 *Mass. Centinel* (Boston) 11 Jan. 3/1 The new arrangement ordered by Congress, for the more safe and regular conveyance of the Mails, by the line of stages. **1818** *Niles' Reg.* XIV. 14/2 A regular line of waggons and packets are established between the city of New-York and Detroit. **1837** W. Jenkins *Ohio Gazetteer* 56 The post office is supplied by daily lines of Coaches from Cincinnati to Dayton.

23. a. *line by line*: delete † and add later examples (freq. used, with hyphens, *attrib.*).

Similarly *line for line* (also, with hyphens, *attrib.*). (For *line-for-line* in Fashion cf. sense *14 b.)

1876 J. Weiss *Wit, Humor, & Shakespeare* iii. 78 There was a worthy old deacon, who, repeating Watts's hymn line for line after his clergyman, said, 'Return, ye rancid sinners!' **1934** T. S. Eliot *Eliz. Ess.* 17 A line-by-line examination of almost any Elizabethan play..would be a fruitful exercise. **1951** L. MacNeice tr. *Goethe's Faust* 9, I aimed at a line-for-line translation. **1958** [see *FASHION *sb.* 9 c]. **1964** E. A. Nida *Toward Sci. Transl.* ii. 17 Dryden felt that there were three basic types of translation: (1) metaphrase, a word-for-word and line-for-line type of rendering; (2) paraphrase..; and (3) imitation. **1969** *Guardian* 29 July 7/3 Line-for-line copies of his [couture] collection. **1971** *Computers & Humanities* VI. 7 Comparisons are made on a line-for-line basis. **1971** *Gloss. Electrotechnical, Power Terms* (B.S.I.) III. iv. 13 *Line by line scanning*, scanning in which the sweep is effected in straight, substantially horizontal strips extending over the entire width of the picture. **1973** *Country Life* 6 Dec. 1970/1 A perfect line-for-line copy of a couture Dior trouser suit.

e. *pl.*, as an imposition (further examples).

1907 *Massacre of Innocents* ii. 13 Vardon, do me five hundred lines. **1914** 'I. Hay' *Lighter Side School Life* vii. 182 Mr. Duckworth..had occasion to set Master Smith fifty lines for inattention. **1959** I. & P. Opie *Lore & Lang. Schoolch.* xv. 325 At my junior school the boys had different doors from the girls and if a boy went through the girls' door he had a 100 lines to write out. **1961** D. Woodward tr. *Simenon's Premier* ii. 36 He took lessons without appearing to see his pupils..and his only reaction was, if one of them grew restless, to give him two hundred lines. **1974** *Age* (Melbourne) 12 Oct. 12/2 *Doing lines*, being kept in to write out good resolutions, such as 'I must not put squashed frogs in girls' sandwiches'.

IV. 24. c. Phr. *line of command.*

1930 *Nautical Mag.* Jan. 41 (*title*) The line of command. *Ibid.* 43 When the machinery fails, then the old line of command is called upon to take its full responsibility. **1962** *Rep. Comm. Broadcasting 1960* 161 in *Parl. Papers 1961–2* (Cmnd. 1753) IX. 259 The planning and operation of a national programme of television can never be simple, even when there is a single objective to be pursued, when effective control resides in a single authority, and when there is a direct line of command.

26. a. *line of communication*: see *COMMUNICATION 6 c.

b. *line clear*, a signal indicating that a line is unoccupied and that a train may therefore proceed; *line of rail* (see quot. 1965; cf. *end of steel*, s.v. *END *sb.* 6 e).

1869 *Cornh. Mag.* Mar. 282 Signalman at 3 tells signalman at 2, 'line clear, send train.' **1907** *Daily Chron.* 16 Oct. 7/4 Martin should have pulled up until he got the line-clear signal. **1936** *Gloss. Terms Railway Signalling* (*B.S.I.*) 9 The block indicator shows 'Line Blocked' or 'Normal', 'Line Clear' and 'Train on Line'. **1963** Kichenside & Williams *Brit. Railway Signalling* v. 46 The signal controlling entry to the block section can only be cleared ..when the block indicator for the section ahead is at 'line clear'. **1965** *Economist* 8 May 655/2 The figure [of unemployed] exceeds 75,000 and..they are concentrated in the few towns along the so-called line-of-rail, the thin strip of urbanisation in which is concentrated..the country's [*sc.* Zambia's] economic activity. **1971** *E. Afr. Jrnl.* Mar. 17/2 A few co-operatives, along the line-of-rail especially, produce poultry products for sale.

e. Chiefly *Canada* and *N.Z.* A settlement road, a bush road.

Such roads often later developed into roads of standard size and quality, and the word *line* appears in many road-names in both countries. The term may be of English dialectal origin.

1828 *Brockville* (Ontario) *Gaz.* 26 Dec. 3/4 A teamster by the name of M'Pherson from the Scotch Line. **1830** W. S. Moorsom *Lett. from Nova Scotia* ix. 344 The greater part of this line is either a rough horse-path, or in the same state as that described under the name of a 'new cut'. **1841** *N.Z. Jrnl.* No. 43. 224/2 Colonel Wakefield is also about to direct a line or bridle road (the basis of the future road) to be cut. **1853** J. M. Richmond in *Richmond–Atkinson Papers* (1960) I. iii. 133 There is what we call a *good* bush road to Rata Nui but beyond it there are two miles of bush walking along what is called a 'line'; a line is made by cutting the supple jacks and small shrubs with a bill-hook. **1863** E. H. Walshe *Cedar Creek* 103 They wished even for the corduroy expedient a little farther on, when the line became encumbered with stumps left from the under-brushing. **1880** W. H. Patterson *Gloss. Words Antrim & Down* 63 *Line*,..(2) a road. The new roads are so called. **1890** E. H. Searle *Angela* i. 2 This track was known to the neighbourhood as 'Mount's Line'. **1933** 'P. Slater' *Yellow Briar* 172 This grain was hauled down the 6th line and stored till the spring in Isaac Chafee's warehouse. **1943** *Amer. Speech* XVIII. 87 In some country districts [in New Zealand] (the Manawatu, for example) the roads are named *lines*—McDonell's Line, Richardson's Line, Union Line—presumably from early boundary or surveyors' lines. **1961** Price & Kennedy *Notes Hist. Renfrew County* [*Ontario*] 110 McNaughton's Plan of 1836 shows Queen's Line as an opened road. **1971** M. Tak *Truck Talk* 99 *Line*, a road, route or highway.

f. A row of traps or of poison bait.

Widely used in English-speaking areas outside the U.K.

1854 Mayne Reid *Young Voyageurs* 190 Moreover, he [*sc.* the wolverine] will follow the tracks of the trapper from one to another, until he has destroyed the whole line. **1871** R. L. Dashwood *Chiploquorgan* viii. 109 We followed an old 'sable line',..a line of traps set for that animal. **1949** *Sat. Even. Post* 22 Jan. 98/2 It is usually a glum day for the trapper when he pays his periodic visit to his line and sees in the snow the tracks of a wolverine joining the tracks that he made himself on his previous swing around. **1960** B. Crump *Good Keen Man* 31 Working from the same hut at first, we laid cyanide lines up every ridge within reach of the camp. The dodge was to work in pairs, one laying blobs of flour flavoured with oil-of-aniseed for bait, the other adding crushed cyanide to each heap of flour. We'd do this for three days, then go back over the lines cutting the ears off the dead possums for tokens. **1968** K. Weatherley *Roo Shooter* 39 Whenever a fox got on the line they lost about a quarter of their morning's catch. It would go round all the traps killing and tearing the rabbits until it was disturbed or caught in an unsprung trap.

g. A pipe or tube (of great or indefinite length in relation to its thickness).

1862 W. J. M. Rankine *Man. Civil Engin.* III. ii. 739 From..reservoir to..town the main pipes may form a double line, so that in the event of a failure of one line, a supply..may be conveyed through the other line. **1895** W. T. Brannt *Petroleum* vii. 237 Beside the lines leading from the oil region to Baku..there are a number of branches which lead from the 21 principal lines to the refineries. **1921** W. F. Durand *Hydraulics of Pipe Lines* v. 231 The buried line cannot be inspected or repaired or repainted on the outside, and these conditions will.. reduce the serviceable life of the line. **1962** F. I. Ordway et al. *Basic Astronautics* x. 411 As the propellants flow through the feed lines to the pump a certain amount of pressure will be lost due to friction. **1966** A. E. C. Vizard in P. Hepple *Natural Gas* 55 By using large diameter lines at relatively high pressures the potential carrying capacity of a single line can be greater. **1974** *Sunday Express* 14 Apr. 1/3 Detectives investigating the death of a diver.. have found that his support line was cut. The line carried oxygen and communication cables to two divers 350 feet down.

h. *Golf.* (See quot. 1910.)

1887 W. G. Simpson *Art of Golf* II. ix. 166 If their advice as to the line and strength be followed, and the putt comes off, it is supposed..that there was no other way of doing it. **1910** *Encycl. Brit.* XII. 223/2 *Line*, the direction in which the hole towards which the player is progressing lies with reference to the present position of his ball. **1971** Trevino & Fraley *I can help your Game* (1972) v. 72 (*caption*) The putt has been stroked but I maintain my immovable body position, concentrating on keeping the blade square to the line.

i. *up the line*: on leave. *Naut. slang.*

1942 *Gen* 1 Sept. 13/2 When a sailor goes on leave he goes 'up the line'.

j. Phr. *the end of the line* (*transf.* and *fig.*). Cf. *the end of the road* (*END *sb.* 3 h).

1948 *Amer. Speech* XXIII. 29 Calcutta commandos.. reached the End of the Line [*sc.* China] by flying..over the Hump. **1955** J. Potts *Death of Stray Cat* (1956) vii. 75 Lillian..turned to face Floyd, as a signal that this was the end of the line for him. **1959** E. Burgess *Divided we Fall* xx. 228 It looks like the end of the line for Roylake. Unless he can think up something—fast! **1967** Wodehouse *Company for Henry* v. 79 'Don't tell me we're there already.'.. 'Yes, this is the end of the line.' **1974** 'J. Graham' *Bloody Passage* x. 133 They have nowhere to go. This is—how do the Americans say it?—the end of the line.

28. b. (Further examples.) Also, *not one's line*, not one's vocation or calling, not among one's pursuits or interests; *to step* (or *get*, etc.) *out of line*, to behave in an unconventional or unexpected manner.

1791 J. Lackington *Mem.* xxv. 191, I cannot help noticing that in one of his [*sc.* Wesley's] publications (stepping out of his line) he betray'd extreme weakness and credulity. **1857** C. Kingsley *Two Yrs. Ago* I. p. xviii, 'He..wanted to call me out.' 'Did you go?'.. 'I told him that wasn't my line.' **1932** D. Runyon *Guys & Dolls* ii. 37 Reasonably safe for anyone who does not get too far out of line. **1937** M. Sharp *Nutmeg Tree* xix. 249 'Wouldn't you like to be Lady Waring?'.. 'No, I wouldn't..it's not my line.' **1938** D. Runyon *Furthermore* iii. 45 He is out of line in giving Frankie the hot foot. **1943** J. B. Priestley *Daylight on Saturday* xiii. 87 The welfare worker act.. wasn't her line at all. **1962** P. Gregory *Like Tigress at Bay* iii. 28 As long as he doesn't get out of line too often, I'll keep him on. **1962** J. Ludwig in R. Weaver *Canad. Short Stories* (1968) 2nd Ser. 244 Women weren't Sidney's line. **1973** N. Graham *Murder in Dark Room* viii. 58 You do it his way or else. I stepped out of line when I checked on Redman.

c. *line of business*: in the 18th- and 19th-century theatre, the kind of parts for which an actor or actress was specifically engaged. Cf. *BUSINESS 20.

1775 F. Abington *Let.* in D. Garrick *Private Corr.* (1832) II. 106 Knowing the impossibility of my attempting that line of business while I am necessarily engaged in so many plays. **1807** A. Holbrook *Mem. Actress* 33 Another shocking custom is, that of giving no distinct line of business; for people, let them possess what talent they may, excel more in certain parts than in others. **1831** P. Egan *Show Folks* 27 Waiting in turn to engage young men for different *'lines of business'* to complete their companies. **1845** *Ainsworth's Mag.* VIII. 150, I have alluded to country actors..acting characters not in their 'line of business'. **1849** *Theatrical Mirror* 17 Sept. 20 We were surprised to see Mrs. W. Daly playing the part of Lady Macbeth, being quite out of her lines of business. **1901** C. Morris *Life on Stage* vii. 40 These were the principal 'lines of business', and in an artistic sense they bound actors both hand and foot.

d. Phr. *one's line of country*, one's pursuit, field of interest, area of study, etc. (Freq. in neg. contexts.) Also, *line of work.*

1861 T. Hughes *Tom Brown at Oxf.* III. viii. 138 This sort of thing isn't my line of country at all. **1926** R. Macaulay *Crewe Train* II. v. 115, I don't advise you to join it [*sc.* the R.C. church]. I don't think it's your line of country, exactly. **1943** N. Balchin *Small Back Room* viii. 94 What? Pinching strange females?.. That's more his line of country than yours. **1951** W. Empson *Struct. Complex Words* 15 A mistake made by Richards..is a great deal more illuminating than the successes of other writers in this line of country. **1957** G. Faber *Jowett* v. 94 Josephine's absorption in her new 'line of work'. **1966** M. Brewer *Man against Fear* i. 15 I'd like to help... But it's not my line of country. Only the police can catch them. **1972** *News & Observer* (Raleigh, N. Carolina) 30 Dec. 4/3 No one lives in the sticks or is asked his line of work very often.

30. (Earlier and later examples.) Also, goods of a particular design.

1834 *Chambers's Edin. Jrnl.* III. 9/3 Even those [travelling salesmen] whose *line* seems the most hopeless and frivolous. **1930** H. Cousins in V. A. Demant *Just Price* v. 102 No business can expect that all its 'lines' will be successful. **1959** *Punch* 16 Sept. 177/1, I can do a nice line in powder compacts. **1971** *Cabinet Maker & Retail Furnisher* 1 Oct. 15/3 Rather than let a slow selling line stand on the shop floor it is reduced immediately.

b. The amount which one underwriter (or one company) accepts as his share of the total value of the subject matter covered by insurance.

1899 Hooper & Graham *Mod. Business Methods* 144 The names and the amounts on the back of a policy.. would appear thus... Each of the above persons is said to 'take a line' in the policy. **1905** [see *front-ranker*]. **1931** *Times* 14 Mar. 12/6 Many of those [*sc.* insurance companies] who have written large lines..are known to have been influenced by a desire [etc.]. **1974** W. L. Catchpole *Business Guide to Insurance* xxiv. 202 If the chosen underwriter..agrees to accept a substantial line at an equitable risk, he becomes the leading underwriter on the slip.

c. *line of credit*: a loan by one country to another, to be utilized by the second for buying goods from the first; credit extended by a bank to a commercial concern to a certain amount; the amount so extended.

1958 *Listener* 18 Sept. 407/1 A line of credit for £8,000,000 from Australia will have helped matters. **1971** *Daily Tel.* 1 Jan. 1/1 A total of 41 million Canadian dollars..was repaid on the Canadian line of credit.

31. *line-end* (later examples), *-length*, *-numbering*; *line-numbered* adj.

1908 *Daily Chron.* 23 Oct. 9/4 Now he types instead of stamping the last words so as to obtain an even line end. **1930** T. Sasaki *On Lang. R. Bridges' Poetry* I. v. 21 The strongest stress..is the one at the line-end. **1961** T. Landau *Encycl. Librarianship* (ed. 2) 226/1 *Line division mark*, a vertical line or double vertical lines used in bibliographical transcription to indicate the place of the ends of lines... Also called line end stroke, dividing stroke. **1905** *PMLA* XX. 814 The uniform background of the recurrent line-lengths. **1929** H. Crane *Let.* 30 Aug. (1965) 344 The line-lengths are longer than in any other section. **1905** *Academy* 14 Oct. 1072/2 We can see him turning over the page, line-numbered. **1959** *N. & Q.* Sept. 313/2 The recent line-numbered edition by W. J. B. Owen. **1953** Amos & Birkinshaw *Television Engin.* I. ii. 33 The system of line numbering must be explained... The lines are numbered according to the positions they occupy in the raster, number 1 being the top line and, in the British system, 405 the bottom line. **1966** *English Studies* XLVII. 296 His marginal references to Folio lines and passages (using the line-numbering of the Globe edition).

32. line angle *Dentistry*, the angle at the junction of two surfaces of a tooth or cavity; **line-at-a-time printer** = **line printer*; **line-backer**, in Amer. and Canad. football (see quot. 1961); **line-ball**: also in *Lawn Tennis*; **line blanking** *Television*, the suppression of signals that would contribute to the picture during fly-back of the scanning spot between the transmission of successive lines; freq. *attrib.*; **line block**, a block bearing a design in relief from which an illustration made up of lines without variations in tone may be printed; an illustration printed in this way; also *attrib.*; **line-book**, (*a*) *Printing* (*Obs.* exc. *Hist.*), a book in which compositors working in companionships (chiefly 19th c.) kept account of the lines of set type credited and debited to them; (*b*) (also **lines-book**) *R.A.F. slang*, a record of boasts (see 13 g. above); **line-bred** *a.*, produced by line-breeding; **line-camp** *N.Amer.*, a camp, esp. a cabin, for ranch hands in an outlying part of a large ranch; **line-casting** *a.*, of a composing machine, casting type a line at a time; so **line-cast** *a.*, **line-caster**; **line density**, also *gen.*, density or concentration of lines; **line drawing** (earlier and later examples); also *fig.*; **line-drawn** *a.*, made by line-drawing; **line-drive** *Baseball*, a ball driven straight and low above the ground; **line drop** *Electr. Engin.*, the voltage drop between two points on a transmission line (as a result of resistance, leakage, or other causes); **line-ending**, (*a*) = *line-filling*; (*b*) the end of a line of poetry; **line-engraved** *ppl. a.*, inscribed with a line engraving; **line-engraver**, one who does line engraving; **line engraving** (earlier example); **line-fence** *N. Amer.*, a boundary

fence between two farms or ranches; **line finder** *Telephony*, a selector which searches for the calling subscriber's line when he lifts his receiver so that the line can be connected to a group of selectors available to any caller; **line-finishing** = *line-filling*; **line frequency** *Television*, the number of scanning lines produced per second; **line gale** *U.S.* = *line-storm*; **line gauge** *Printing*, a ruler showing the size of a type or types; **line graph** = Graph *sb.*¹ 2 (in Dict. and Suppl.) (as distinguished from a *bar graph*, in which vertical rectangles represent the values of the dependent variable); **line haul** *U.S. slang* (see quots.); **line loss** *Electr. Engin.*, loss of electrical energy along a transmission line (as a result of resistance, leakage, or other causes); **line management, manager**: see sense 19 d above; **line officer,** a military or naval officer of the line; **line pipe,** pipe specially manufactured for use in pipelines; **line printer,** a printer that is capable of printing a whole line of characters in each cycle of operation and is usu. operated under the control of a computer; **line-ride** *v. intr. U.S.*, to perform the action of line-riding; **line-rider** *N.Amer.*, one engaged in line-riding; **line-riding** (earlier example); **line scan,** (*a*) the motion of a scanning beam or spot along a line; (*b*) the electrical signal which causes this; (*c*) an apparatus or technique which scans an object or scene line by line; so **line scanning** *vbl. sb.*; **line-sequential** *a. Television*, applied to a system of colour television in which each line of the picture is in one of the three primary colours, the colour changing for each successive line; **line shaft, shafting,** a shaft, or shafting, of relatively great length from which a number of separate machines are driven by countershafts or endless belts; **line-side** *attrib.*, adjacent to a railway line; **line space,** the space provided for a line of typescript; so *line-space lever, mechanism*, etc., the device that turns the platen of a typewriter to a new line of writing; **line-spacing,** the space between successive lines of typescript; *attrib.*, of the device that moves the platen to a new line; **line spectrum,** a spectrum containing lines distributed apparently at random (rather than in groups as in a band spectrum); hence, an emission (of light, sound, or other radiation) composed of a number of discrete frequencies or energies; **line-storm** (earlier and later examples); **line-synchronizing** *a. Television*, applied to a pulse transmitted in a television signal at the end of each line which initiates fly-back of the scanning spot in the receiver, so keeping the scanning process in synchronism with that in the transmitter; also abbreviated to **line-sync**; **line-tub,** a tub in which a whaling line is kept; **line-width** *Physics*, the width of a spectral line as measured by the difference in wavelength, wave number, or frequency between its two sides; **line-work** (later example); (*b*) (see quot. 1968); (*c*) work as a lineman. Also *LINE STANDARD.

1908 G. V. Black *Work on Operative Dentistry* I. 295 Line angles. **1930** W. H. O. McGehee *Text-bk. Operative Dentistry* xi. 338 Flatten the gingival and axial walls, making a definite line angle at their junction. **1963** C. R. Cowell et al. *Inlays, Crowns, & Bridges* iii. 15 Complete the proximal box, using a chisel to plane its vertical walls and to sharpen the line angles. **1955** *Jrnl. Assoc. Computing Machinery* II. 294 Line-at-a-time printer (92 characters per line), operating at a speed of 150 lines per minute. **1963** Gould & Ellis *Digital Computer Technol.* xi. 141 The line-at-a-time printers have, in the main, been adapted from the tabulating machines of punched card practice. **1961** Webster, *Linebacker*, a football player stationed within one to four yards of the line of scrimmage and expected to make quick tackles close to the line of scrimmage on running plays and to protect against short passes. **1968** *Globe & Mail* (Toronto) 13 Feb. 29/1 Darryl Burgess, a 225-pound linebacker from St. Mary's. **1969** *Eugene* (Oregon) *Register-Guard* 3 Dec. 10/2 Oregon linebacker Tom Graham..played well enough to make both UPI and AP All-Coast teams as a rookie. **1970** *Toronto Daily Star* 24 Sept. 18/5 We can always move Corrigall to linebacker. **1973** *Washington Post* 13 Jan. C5/4 The most misguided portion of the show comes during Jones' interviews of Jim Brown, the former football player-turned-actor, and Ray May, a linebacker for the Baltimore Colts. **1891** F. C. Burnand *Miss Decima* 22 Chorus (outside—watching a game of Lawn Tennis)..Ah! 'Line' ball. **1952** Howe & Ducloux tr. *Kerkhof & Werner's Television* iv. 76 The total line blanking of the picture signal is 0·15 L. **1957** Amos & Birkinshaw *Television Engin.* (rev. ed.) I. ii. 31 The synchronising signals are not the only form of intelligence which must be transmitted between lines; an additional signal, known as the line-blanking signal, must also be inserted. **1966** G. H. Hutson *Television Receiver Theory* I. iii. 31 The line blanking period is divided into..the front porch, the line sync. pulse and the back porch. **1896** A. Beardsley *Let.* 29 Sept. (1971) 173 The rest of the drawing has come out so hardly and coldly in the line block. **1924** E. Pound *Let.* 3 Dec. (1971) 191, I think the idea of ten or twelve Blacks of size that cd. go by post, and that cd. be done in line block, might be useful. **1936** *Burlington Mag.* Mar. p. xiv/1 Line-block illustrations from the author's own drawings. **1956** *Nature* 18 Feb. 301/1 The illustrations are well chosen, both the line-blocks and the half-tones. **1972** P. Gaskell *New Introd. Bibliogr.* 272 The detail of all but the very best photographic line blocks tends to be slightly rougher at the edges than that of wood engravings. **1876** J. Gould *Letter-Press Printer* 33 The system adopted in some of the smaller houses is for each compositor to make up and impose his own pages, the making-up being passed from one compositor to the companion who follows him, accompanied by the line book. **1942** *Observer* 4 Oct. 7/2 'There I was, upside down, in cloud, ten-tenths, at 1,500 ft...' But you never get to the end of your story if you were so foolish as to begin like that. 'Lineshoot!' they would cry. 'Line!' And most squadrons have a Line Book in which such statements are written down, to their authors' perpetual shame. **1943** C. H. Ward-Jackson *Piece of Cake* 40 *Lines book*, in which are recorded exaggerated statements made at one time or another by Mess members. **1945** E. Taylor *At Mrs. Lippincote's* xxiii. 194 Quick, the line-book! **1972** P. Gaskell *New Introd. Bibliogr.* 193 He [*sc.* the clicker] kept an account of the number of lines that each man set, both in a line-book and by marking the copy. **1891** R. Wallace *Rural Econ. Austral. & N.Z.* xxxi. 400 The impression that tuberculosis is more prevalent among high line-bred shorthorns than among the ordinary country-bred cattle. **1960** *Times* 19 Sept. 3/4, 20 dams were chosen..these being line-bred. **1971** *Amer. N. & Q.* Apr. 126/2 The quarter horse, developed from cross-breeding Spanish stock imported to America via Florida (Chickasaw horses) and what Nelson Nye calls 'line-bred orientals' from England, was originally a sport animal. **1888** *Century Mag.* Mar. 667/2 But some of the men are out in the line camps, and the ranchman has occasionally to make the round of these. **1949** *10 Story Western* May 12/2 He had been telling them all how he was going to winter here at the Buffalo Crossing line camp. **1963** R. D. Symons *Many Trails* v. 52 Most outfits had what they call 'line camps' strung along the limits of their range, from which 'line riders' operated. **1973** S. Jennett *Making of Bks.* (ed. 5) xv. 286 The italic [of Linotype Baskerville] is a little loose fitting, its width, as in other line-cast type-faces, being governed by the roman. **1972** *Phys. Bull.* Sept. 533/1 These 'second generation' photosetters..are reasonably cheap and are considerably faster than the latest line-casters which are also tape driven. **1913** *Inland Printer* July 486 (Advt.), There are thousands of publishers all over the United States who have been waiting for a line-casting and composing machine so simple and easy to operate that it would prove practical in the small shop. **1916** Legros & Grant *Typogr. Printing-Surfaces* iv. 15 Line-casting machine type-metal undergoes a wastage or depreciation. **1973** S. Jennett *Making of Bks.* (ed. 5) v. 83 The Intertype Fotosetter was also an adaptation, of the Intertype line-casting machine. **1963** *Reshaping of Brit. Railways* (Brit. Railways Board) 65 Line densities are not the only measure of the use made of the railway. **1971** *Fremdsprachen* XV. 276 It can be used to restore old drawings and to improve line density for microfilming. **1891** A. Beardsley *Let.* 25 Dec. (1971) 32, I am anxious to say something somewhere, on the subject of *lines* and line drawing. **1959** *Listener* 9 July 76/3 It [*sc.* an overture] is finer line-drawing than the Gordon Jacob work. **1966** *Ibid.* 6 Jan. 36/3 Over 300 [flowers] are illustrated in close-up colour photographs and 100-odd more in line drawings. **1967** E. Short *Embroidery & Fabric Collage* ii. 51 There is a tendency to produce line drawings which might just as well have been done with a pencil. **1903** *Westm. Gaz.* 17 Oct. 4/2 An order of the King in Council was published with two line-drawn illustrations. **1931** *Randolph Enterprise* (Elkins, W. Virginia) 9 July 5/3 Boyles turned in the star catch of the day by racing..to pull down a line drive with one hand. **1968** *Washington Post* 4 July C1/3 Mantle was safe as Ron Hansen's throw, after snagging a line drive by Andy Kosco, was a trifle tardy. **1894** Line drop [see *line loss* below]. **1962** *Newnes Conc. Encycl. Electr. Engin.* 86/1 D.C. boosters are normally low-voltage d.c. generators employed for adjusting a supply voltage, in line-drop compensation and as an aid in controlling the charging of large accumulator batteries. **1928** E. G. Millar *Eng. Illuminated Manuscripts XIVth & XVth Cent.* i. 9 Many of the line-endings..were added in the fifteenth century. **1962** W. Nowottny *Lang. Poets Use* v. 120 The method he adopts is the eccentric placing of line-endings. **1802** *Monthly Mag.* XIV. 253/1 The best line-engraved prints preserved their superiority. **1881** *Stamp-Collector's Ann.* 5 The fall of the penny stamp and all its line-engraved family. **1936** *Discovery* Dec. 386/1 Practically all [18th-century tradesman's cards] are line- or stipple-engraved. **1965** *Stamp Collecting* ('Know the Game' Series) 44/1 Stamps printed from recess-plates are said to be line-engraved. **1873** *Illustr. London News* 15 Mar. 247/3 This eminent line-engraver. **1965** Doughty & Wahl in D. G. Rossetti *Lett.* I. 9 Charles Warren (1767–1823), line-engraver and a noted illustrator. **1802** *Monthly Mag.* XIV. 253/1 The line engraving is now attaining its deserved preeminence. **1845** J. Comly *Reader & Bk. Knowl.* 96 Always keep good line-fences. **1854** S. H. Hammond *Hills, Lakes & Forest Streams* xxv. 250 Later still, the old line fence was pulled away. **1893** E. R. Young *Stories from Indian Wigwams* 34 One morning I.. went off to help a couple of Indians about their line fences. **1946** *Chicago Daily News* 23 Mar. 1/8 He got into an argument with the boy's parents over the building of a line fence between their properties. **1954** C. Bruce *Channel Shore* 12 From there a person could look east and west along..the northern fields..separated by line fences. **1922** Glazebrook *Dict. Appl. Physics* II. 834/2 The line finder corresponds to the answering plug in a manual exchange. **1950** J. Atkinson *Herbert & Procter's Telephony* (new ed.) II. i. 19/2 If..the volume of traffic and the number of 1st selectors are considerable, then line-finders may become more expensive than subscribers' uniselectors. **1968** E. H. Jolley *Introd. Telephony & Telegr.* viii. 232/2 The subscribers' lines are multiplied over the bank of contacts of the line-finders so that each subscriber's line appears on each line-finder. **1906** E. Johnston *Writing & Illuminating* xii. 205 Line-finishings are used to preserve the evenness of the text when lines of writing fall short. **1936** O. S. Puckle tr. *M. von Ardenne's Television Reception* i. 11 The total number of lines in the complete picture is 240, scanned sequentially and horizontally at 25 picture traversals per second... The line frequency is thus 6,000 impulses per second. **1973** *Newnes Colour Television Servicing Manual* I. i. 26/1 Sawtooth voltage at line frequency is developed across the inductive network. **1836** *Knickerbocker* VII. 17, I must take the oars myself, for that blamed line gale has kept me in bilboes.. a dog's age. **1948** *Words into Type* 544 *Line gauge*, a printer's measuring rule, marked off in nonpareils and picas, sometimes showing other type measurements also. **1967** Karch & Buber *Offset Processes* 544 *Line gauge*, a printer's ruler usually having 6 and 12 point graduations. Sometimes with other point scales, as: agate, 9-point, 10-point, etc. **1956** Line graph [see *DECODE *v.*]. **1972** *Scholarly Publishing* III. 274 Bar graphs, line graphs, pie charts, and other illustrative devices. **1942** Berrey & Van den Bark *Amer. Thes. Slang* § 770 *Line haul*, a scheduled truck route. **1971** M. Tak *Truck Talk* 99 *Line haul*, a scheduled truck run or movement of freight between cities. **1894** A. T. Snell *Electr. Motive Power* iv. 126 The line loss remains constant when the percentage of the line drop is kept the same for variations of supply pressure. **1953** C. F. Hockett in Saporta & Bastian *Psycholinguistics* (1961) 64/1 To supply one hundred-watt light bulb, a generator must transmit one hundred watts of power, plus a bit more to make up for line-loss. **1970** D. Waterfield *Continental Waterboy* iii. 29 And you have line loss, particularly with very long transmission lines. **1850** R. Glisan *Jrnl. Army Life* (1874) i. 2 This rank.. avails its possessor..in everything except commanding troops when a line officer is present. **1909** *Westm. Gaz.* 1 Feb. 2/1 Wives of line-officers, engineers, servants. **1925** R. Graves *Welchman's Hose* 29 They hadn't one Line-officer left, after Arras. **1923** *Amer. Petroleum Inst. Bull.* 31 Dec. 117/2 The work of this committee has been..to the end that a specification might be had that would: (1) Minimize losses arising out of the use of casing line pipe, tubing and drill pipe, in oil field operations. **1930** L. D. Burritt in Walker & Crocker *Piping Handbk.* xiv. 719 Line-pipe threads are of the same form and taper as American Standard threads, but the pipe is threaded with a longer length of thread than is standard pipe. **1967** *Times Rev. Industry* Feb. 45/3 The rising demand for line pipe made sense of a connexion between South Durham and Stewarts and Lloyds, which has been marketing X60 seamless line pipe up to 18 inches in diameter for many years. **1955** *Jrnl. Assoc. Computing Machinery* II. 294 Output: Line printer of a BULL tabulating machine. **1962** *Mod. Lang. Rev.* LVII. 171 The Cambridge Mathematical Laboratory possesses a line-printer which is directly operated by EDSAC 2. **1970** O. Dopping *Computers & Data Processing* iv. 73 A normal speed for a computer line printer is 1,000 lines per minute. *Ibid.* xi. 164 A mechanical line printer has one printing device for each printing position in the line (for each 'column'). The number of printing positions is often between 100 and 160. **1883** *Rep. Productions Agric. 10th Census 1880* (U.S. Census Office) 971 The cattle-raisers were obliged to fence or to 'line-ride' to keep their cattle from trespassing. *Ibid.* 973 The cowboys engaged in this work are called 'line-riders'. *Ibid.*, The cattle of northwest Texas are in a large measure controlled or held on their ranges by a system of 'line-riding'. **1920** Line-rider [see *fence-rider* (*FENCE *sb.* 11)]. **1942** E. E. Dale *Cow Country* 119 This by no means did away with the work of the line rider, though it was made somewhat easier. **1963** Line rider [see *line-camp* above]. **1938** J. H. Reyner *Testing Television Sets* iv. 46 If the time base is operating the appropriate noise will be heard—a rapid ticking on the frame scan and a high squeal on the line scan. **1957** D. G. Fink *Television Engin. Handbk.* x. 12 The harmonic components of the line-scan spectrum may thus be thought of as carrier waves, each with a 60-cps modulation envelope. **1962** *Daily Tel.* 28 Aug. 13/5 Line scan is a system for reconnaissance and mapping at low levels. **1966** D. G. Brandon *Mod. Techniques Metallogr.* 257 The line scans being automatically repeated 50 μm apart. **1971** *Daily Tel.* (Colour Suppl.) 22 Jan. 22/1 False colour photography.. does not record gradations of temperatures exactly, and these can be very important. The instrument which does this is the infra-red linescan, which scans the scene line by line like a television scanner, building up a composite picture from the heat records. **1935** *Television Today* I. 300/2 The line scanning is usually spoken of as the scanning motion. **1971** H. E. Ennes *Television Broadcasting* iii. 125 Picture information is contained in the fundamental and harmonics of the 60-Hz field frequency and the 15,750-Hz line-scanning frequency. **1949** *Electronics* Dec. 68/3 The change of color is introduced between successive lines in the scanning field, that is, the system is in the line-sequential class. **1965** G. du Cloux tr. *Holm's Colour Television Explained* (ed. 2) iii. 55 The R.C.A. appeared to have arrived at the ultimate solution with a line-sequential, or possibly even a dot-sequential system. **1881** *Spon's Dict. Engin.* Suppl. III. 1093 For the bearings of line shafts cast iron is..the best. **1936** Line shaft [see *jack shaft* s.v. *Jack *sb.*¹ 33 a]. **1974** *Encycl. Brit. Macropædia* XI. 253/2 In the days when all machines in a shop were driven by one large..prime mover, it was necessary to have long lineshafts running the length of the shop and supplying power..to shorter countershafts, jackshafts, or headshafts. **1872** J. Richards *Treat. Wood-Working Machines* 95 Pulleys for line-shafting running at high speed should be light and true. **1966** *McGraw-Hill Encycl. Sci. & Technol.* XII. 240/2 The delivery of power to the machines in a shop has generally been converted from line shafting to individual electric motors for each machine. **1961** Webster, *Lineside*, adjacent to a railway line. **1967** *Listener* 26 Jan. 123/1 This can be prodigiously expensive if it involves disturbance of lineside property. **1975** *Daily Tel.* 18 July

2/8 By next year it is expected that there will be fewer faults in the 547 lineside signals and 465 points controlled by the new box. **1951** *Oxf. Jun. Encycl.* IV 472/2 When it reaches the end, a line space lever is pushed to move the paper up to a new line and return the carriage to the right. **1962** *Which?* Dec. 357/1 Some of the models..had only two positions for their line space selector, the others all had three. **1957** *Encycl. Brit.* XXII. 645/1 The machine was soon renamed the Remington. Among its original features which were still standard..in the 1950s are the paper cylinder with its line-spacing and carriage-return mechanism. *c* **1961** *Imperial Type Faces* (Imperial Typewriter Co.), The number of words which can be typed on a quarto page..var[ies] according to the pitch of the letter and line-spacings. **1873** *Phil. Mag.* XLVI. 406 When the gas is near atmospheric pressure, the line-spectrum of nitrogen is brilliant. **1885** [see *BAND *sb.*[2] 13]. **1923** GLAZEBROOK *Dict. Appl. Physics* IV. 780/1 Luminous spectra can be divided into two classes, namely continuous spectra..and discontinuous spectra... Discontinuous spectra may be subdivided into line and band spectra. **1955** MILLER & NICELY in Saporta & Bastian *Psycholinguistics* (1961) 165/2 Acoustically, this means that the voiceless consonants are aperiodic or noisy in character, whereas a periodic or line-spectrum component is superimposed on the noise for voiced consonants. **1957** [see *BAND *sb.*[2] 13]. **1962** H. D. BUSH *Atomic & Nucl. Physics* iv. 96 There are two main features of a β-particle spectrum, a continuous spectrum with energies ranging from zero to a maximum value..and a line spectrum consisting of a number of discrete energies superimposed on the continuous spectrum. **1962** A. NISBETT *Technique Sound Studio* 254 A sound which is composed of individual frequencies (fundamental and harmonics or partials, or a combination of pure tones) has a line spectrum. Bands of noise have a band spectrum. **1850** N. KINGSLEY *Diary* (1914) 115 A fine day with a strong West wind; rather think the line storm is over. **1939** R. FROST *Coll. Poems* 38 The line-storm clouds fly tattered and swift. **1940** W. T. COCKING *Television Receiving Equipment* xix. 281 When a very large amplitude of line sync pulse is applied to the line generator it is tripped at half-line intervals during the frame sync pulses. **1969** C. R. G. REED *Princ. Colour Television Syst.* vi. 71 The line sync pulse duration is 4·7 μs. **1935** *Television Today* I. 300 The time duration of the line synchronising pulse is usually about 10 per cent. of that of each line. **1953** AMOS & BIRKINSHAW *Television Engin.* I. i. 16 A synchronizing signal is sent out every time the scanning beam at the transmitter reaches the end of a line; this signal is termed the line-synchronizing signal (abbreviated to line-sync signal) and has the function of initiating line flyback at the receiver. **1839** *Knickerbocker* XIII. 382 Line-tubs, water-kegs, and waifepoles, were thrown hurriedly into the boats. **1851** H. MELVILLE *Moby Dick* III. xlviii. 287 Reaching out after the revolving line-tubs, oars and other floating furniture. **1946** *Nature* 28 Sept. 450/1 Measures of effective line-width, made..upon the brilliant reversal of the Hα (λ 6563) contour. **1962** *Science Survey* III. 67 The current reports of the line-width of the radiation produced by the helium-neon optical maser show the line-width is approximately one cycle per second. **1971** *New Scientist* 3 June 565/2 Molecular linewidths are of the order of 10^{-2} cm^{-1} at room temperature. **1972** *Physics Bull.* Feb. 83/2 A dye laser tuned to give a sodium linewidth of a hundredth of an ångstrom. **1904** *Brit. Printer* Apr. 86/2 Line work negatives are printed on to zinc. **1911** H. QUICK *Yellowstone Nights* ii. 32 I'm just through with a summer's line-work in the West. **1962** *Times* 10 Jan. 13/4 The pen drawing..is admirable..projecting in its free and open linework all the completeness of an oil composition. **1968** *Gloss. Terms Offset Lithogr. Printing* (*B.S.I.*) 10 *Line work*, copy or reproduction consisting of solid elements only, as distinct from half-tone.

line, *v.*[2] Add: **3. b.** *trans.* and *intr.* To guide or control a boat or canoe from the bank or shore of a stretch of inland water by means of a rope or ropes. *N. Amer.*

1907 J. G. MILLAIS *Newfoundland* 305 Several times they packed everything for a mile or two, but negotiated most of the worst rapids by 'lining' down them. **1912** H. FOOTNER *New Rivers of North* 125 No one has ever descended it alive, but there is a tradition that a party of Iroquois Indians in the 'company's' employ once lined a boat up. **1923** L. R. FREEMAN *Colorado River* 356 The low stage..gave them room to work below instead of lining from a ledge, eighty feet above the water. **1944** T. ONRAET *Down North* ii. 29 The skiff was too heavy for carrying, and to line it down as we had done in the rapids above was impossible. **1969** E. W. MORSE *Fur Trade Canoe Routes* I. i. 5 Provided that the shoreline was reasonably free of snags, the canoe was lined (tracked).

5. *to line out*: also, to delete, obliterate.

1963 S. WEINTRAUB *Private Shaw & Public Shaw* iii. 94 G.B.S...both edited and altered the language of the..contract,..boldly lining out large passages and inserting new ones.

8. a. With *up*: more widely, to align, arrange, deploy, produce, or make ready (someone or something). orig. *U.S.* Also in various slang uses (see quots.). Also, to aim in a direct line *upon* an object.

a **1884** KNIGHT *Dict. Mech.* Suppl. 665/1 *Peep sight*, a form of hind sight for rifles. It has an opening through which the muzzle sight is lined upon the object. **1902** *Westm. Gaz.* 20 Aug. 8/2 (citing a New York newspaper), I..shall not really feel like myself till I get my coat off and line-up a few trust presidents in front of me for general inspection and drill. **1904** 'G. B. LANCASTER' *Sons o' Men* 41 They were fence-making down at the homestead, and there was no man in the district could line up standards in the same day with Muggins. **1906** *Forum* (N.Y.) Oct. 253 The university president must refuse to be lined up by any clique or party. **1910** *Chambers's Jrnl.* May 282/2 After the conflagration, the smaller débris is collected into heaps and reburned, until the ground is sufficiently cleared to admit of being lined up for planting. **1913** G. J. KNEELAND *Commercialized Prostitution N.Y.* 65 She was 'lined up' about a year ago by a gang that 'hangs out' in a cigar store on East 14th Street. Since then she has been a regular prostitute. **1926** J. BLACK *You can't Win* xiii. 181 We located a big poker game in a soft spot and decided to line up the players. **1931** W. G. MCADOO *Crowded Yrs.* x. 142, I did not see how Clark could possibly line up two thirds of the..votes. **1932** E. WALLACE *When Gangs came to London* viii. 58 You can tell the police all about this... But don't tell more'n the truth, or ever try to line me up by my voice. **1934** WODEHOUSE *Right ho, Jeeves* ix. 94, I tell you I have everything nicely lined up. **1939** COFFEE & COWEN *Family Portrait* II. i. 74 But I'd lined up a big job here—(*adds importantly*) with the Romans. **1941** BAKER *Dict. Austral. Slang* 43 *Line up to*, to approach, accost a person. **1953** E. TAYLOR *Sleeping Beauty* x. 175 Don't line up another one [*sc.* drink] for me. **1958** *New Statesman* 6 Sept. 263/1 Mr. Lim soon called for 'a united Socialist front', which would line up his Labour Front party with the right wing against the extreme left. **1962** A. NISBETT *Technique Sound Studio* 270 Sine tones are useful for studying frequency response and for lining up equipment. **1970** *Language* XLVI. 318 All of the sentences have been 'lined up' with respect to the end of phonation. **1973** G. GREENE *Honorary Consul* I. i. 26 It pays to be a consul. .. Permission to import a new car... I suppose he's got a general lined up in the capital to buy it.

c. *Baseball.* To hit a line-drive; to hit (a ball) hard and low. Freq. const. *out*.

1887 *Courier-Jrnl.* (Louisville, Kentucky) 26 May 2/6 He smashed the first ball that came over the plate, and lined out a beautiful hit past second base. **1948** *Daily Ardmoreite* (Ardmore, Okla.) 28 Apr., He..lined out to centerfield and walked twice in five trips to the plate. **1970** *Globe & Mail* (Toronto) 25 Sept. 31/3 Bob Robertson lined a double down the rightfield Line. **1972** *N.Y. Times* 4 June v. 2/5 Willie struck out, lined to Carty in left field, popped to second base and walked.

9. c. *line out* (*intr.* and *trans.*), to transplant (seedling trees) from beds into nursery lines, where they are grown on before being moved to their permanent situation.

1931 *Forestry* V. 17 Care in handling between lifting from seed-beds and lining out is of the utmost importance. **1938** C. P. ACKERS *Pract. Brit. Forestry* v. 180 Seedlings may be left for 1, 2, or 3 years in the seed-beds: they are then lined out and become transplants. **1957** *N.Z. Timber Jrnl.* Oct. 73/1 *Line out*, to transplant seedlings from seed-beds to rows in a nursery. This normally takes place after the first or second year in the seedbed; further lining out may take place again in the same or another nursery. **1970** H. L. EDLIN *Collins Guide Tree Planting & Cultivation* vi. 90 Trees are always transplanted in the nursery along straight lines, and the work is therefore often called lining-out.

‖ **linea** (li·niă). *Anat.* Pl. **lineæ**. [L. *līnea*: see LINE *sb.*[2]] Used in numerous L. or mod.L. collocations to designate lines apparent in or on the body, or structures which form a line.

1611 Linea alba [see LINE *sb.*[2] 1 c]. **1713** W. CHESELDEN *Anat. Humane Body* I. vi. 20 From the lesser Trochanter down the back-part of this Bone 'till within Four Inches of the End, is a Ridge call'd Linea Aspera. **1861** *Guy's Hosp. Rep.* VII. 297 Linear atrophy of the skin..refers to a condition resembling the *lineæ gravidarum*, but affecting both sexes, and probably all parts of the body. **1967** G. M. WYBURN et al. *Conc. Anat.* I. 7/2 A slight furrow formed by an underlying fibrous raphe, the linea alba, extends from the xiphoid process to the symphysis pubis. *Ibid.* vi. 159/1 Pectineus..is inserted into the femur above the linea aspera.

lineage. 2. c. Delete † *Obs.* and add: *Anthrop.* Patri- or matrilineal descent within a social group traced from a single ancestor; also occas. the traditional line of descent for the handing down of skills and knowledge pertaining to a particular craft or profession.

1871 L. H. MORGAN in *Smithsonian Contrib. Knowl.* No. 218. 151 There were but five other nations of the same immediate lineage of whom we have any knowledge. **1877** —— *Anc. Society* ii. 69 The gens came into being upon three principal conceptions, namely; the bond of kin, a pure lineage through descent in the female line, and non-intermarriage in the gens. **1934** R. H. LOWIE *Introd. Cultural Anthropol.* xiv. 254 A clan including only descendants of a single ancestor is a 'lineage'. Commonly it includes members of two or more lineages, but the concept remains the same. **1949** M. FORTES *Social Struct.* 62 Genealogies are cited to show that the founding ancestors of the lineages occupying the townships..came there some ten to twelve generations ago. **1951** R. FIRTH *Elem. Social Organiz.* ii. 53 The Tikopia lineages are patrilineal, membership being traced through the father along the male line to an original male ancestor. **1952** GERTH & MARTINDALE tr. *Weber's Anc. Judaism* I. ii. 28 Cain is the tribal father of the smith and the musician... It may, thus, be assumed that at the time of the establishment of this lineage such artisans, in Palestine as in India, were guest people. **1957** M. BANTON *W. Afr. City* vii. 123 Marriage is an arrangement between two lineages. **1963** *Listener* 7 Feb. 231/1 Arabic documents held by the mosques and the clerical lineages in Northern Nigeria and in Northern Ghana. **1971** *World Archaeol.* III. 217 Each village has a number of smiths of varying degrees of training and competence each assisted by novices who together form a lineage.

attrib.

1949 M. FORTES *Social Struct.* 65 Continuity in the social structure..is maintained by the lineage system. **1951** R. FIRTH *Elem. Social Organiz.* ii. 54 They and their immediate lineage members form a recognized class of 'chiefly houses'. **1957** V. W. TURNER *Schism & Continuity in Afr. Soc.* v. 152 They consolidate the rest of the..lineage membership against them. **1964** GOULD & KOLB *Dict. Social Sci.* 391/2 Relations between the local groups, which may vary in size and locality over time, can none the less be seen as persistent and relatively stable. In this case lineages may compose a total structure or system, the lineage structure or the lineage system.

lineage, var. LINAGE.

lineal, *a.* and *sb.* Add: **A.** *adj.* **2. a.** (Further examples.)

1871 L. H. MORGAN in *Smithsonian Contrib. Knowl.* No. 218. 11 Every system of consanguinity must be able to ascend and descend in the lineal line through several degrees from any given person. *Ibid.* 46 If *Ego* is placed between the father and son the lineal and first collateral lines would become intelligible. **1929** *Encycl. Brit.* XIX. 84/1 The collateral kin of a generation were merged with the lineal.

B. *sb.* **2.** (Later examples.)

1881 E. W. HAMILTON *Diary* 3 Apr. (1972) I. 124 The Probate Duty is to be raised $\frac{1}{4}$% all round, and the 1% (lineals) in the Legacy Duty scale is to be abolished; accordingly lineals will all benefit to the extent of $\frac{1}{4}$%. **1909** *Hansard Commons* 29 Apr. 515/1 In the cases of spouses and lineals..I propose to exempt from the new..duties all legacies.

linear, *a.* and *sb.* Add: **A.** *adj.* **2. b. Linear A,** the earlier of two related forms of writing discovered at Knossos in Crete by Sir A. J. Evans between 1894 and 1901; **Linear B,** the later form, found also on the mainland of Greece, and now shown to be a syllabary imperfectly adapted to the writing of Mycenæan Greek.

[**1902–3** A. J. EVANS in *Ann. Brit. School at Athens* IX. 52 This early system of linear script—which may be conveniently termed Class A as opposed to Class B of the latest Palace Period at Knossos—had a wide extension in the island.] **1907** R. M. BURROWS *Discoveries Crete* vi. 84 The linear writing of class A is now in regular use. *Ibid.* 92 The hoard of clay tablets..shows that its linear writing, called by Mr. Evans Class B, is more advanced. **1909** A. J. EVANS *Scripta Minoa* I. i. v. 31 Documents belonging to the Linear Class A only occur in this particular stratum... In deposits clearly belonging to the remodelled building the inscribed documents all belonged to Class B. *Ibid.* 35 Common to both the linear scripts A and B. *Ibid.* 36 The system of numerals..of the Linear Class B. **1948** A. E. KOBER in *Amer. Jrnl. Archaeol.* LII. 89 Inscriptions of Linear A have been found at several sites. **1950** E. L. BENNETT in *Ibid.* LIV. 81/1 Translations of and commentary upon additional Linear-B tablets from Knossos. *Ibid.* 204/1 The Linear A flourished in the rest of Crete.., and the Linear B at Knossos only. *Ibid.* 218/2 The Linear A ideogram L85 appears to have..the same form. *Ibid.* 219/1 The shapes of four other signs of Linear A..are reflected in the Linear B signs. **1952** J. L. MYRES *Evans' Scripta Minoa* II. 1 In this new 'Linear A' script about one-third of the signs are derived from linearized hieroglyphs. *Ibid.* 2 It was doubtless..local unconformity that provoked the drastic reform of the 'Linear B' script at Knossos. **1953** VENTRIS & CHADWICK in *Jrnl. Hellenic Stud.* LXXIII. 84 Evans believed that Linear *B*..was an administrative revision of Linear *A*, designed to express the same 'Aegean' language. **1966** C. H. GORDON *Evidence for Minoan Lang.* ix. 32 The Linear A and B texts overlap in time. **1972** *Sci. Amer.* Oct. 37/1 The Cretan system of writing, which we call Linear A, was crude but it was adequate for keeping rough accounts.

c. *Mus.* = *HORIZONTAL *a.* 4.

1944 W. APEL *Harvard Dict. Mus.* 409/1 *Linear counterpoint*, a term introduced by E. Kurth..in order to emphasize the 'linear', i.e., horizontal aspect of counterpoint. .. The term is also used..for what the Germans call *rücksichtsloser* (reckless) *Kontrapunkt*, i.e., the modern type of counterpoint which pays little attention to harmonic combination and euphony. **1955** [see *HORIZONTAL *a.* 4]. **1958** A. JACOBS *New Dict. Mus.* 211 *Linear counterpoint*, term—senseless, because all counterpoint is a matter of lines—sometimes used for a type of 20th-century counterpoint (e.g. Stravinsky's) held to be musically valid through the value of the separate lines themselves. **1959** *Listener* 8 Jan. 80/2 In the slow movement the orchestral texture begins in linear style, spare and canonic. **1962** *Ibid.* 18 Jan. 147/1 Linear and rhythmic techniques suggested by medieval music. *Ibid.* 147/2 A conscious employment of linear and metrical 'series' derived from Indian ragas and talas.

3. (Further examples.) Add to def.: capable of being represented by a straight line on a graph (in Cartesian co-ordinates); involving or possessing the property that a change in one quantity is accompanied by or corresponds to a directly proportional change in a related one.

1910 *Encycl. Brit.* IX. 146/2 The limiting tension beyond which the above law of proportionality [between tension and extension] fails to hold is often called the 'limit of linear elasticity'. **1940** *Chambers's Techn. Dict.* 503/1 *Linear amplification*, amplification in which the output current or voltage is strictly proportional to the input voltage. **1941** *Proc. R. Soc.* A. CLXXVII. 382 The disintegration of boron by slow neutrons has been investigated using an ionization chamber filled with boron trichloride in conjunction with a linear amplifier. **1942** *Electronic Engin.* XIV. 711/1 The conversion must be accomplished in a linear manner, i.e., the amplitude change is directly proportional to the frequency change. **1962** D. F. SHAW *Introd. Electronics* vii. 126 The preservation of the shape is a unique property of the sine wave and ..is a feature which it possesses for all linear circuits. *Ibid.* viii. 147 A circuit is linear if the individual components behave in such a manner that the amplitude of the current through each component is directly propor-

tional to the amplitude of the applied voltage and the relationship between the phases of the voltage and current is independent of the current and voltage magnitude. **1973** *Physics Bull.* Oct. 606/1 A linearizer circuit is used to ensure a true linear relationship between conveyor load and indicated reading.

c. *Educ.* Designating or pertaining to programmed learning aimed at step-by-step progress in which the material is broken down into small steps each of which must elicit a correct response before the next one is presented; freq. contrasted with branching methods.

For *linear programming* in a different sense see *7.

1958 B. F. SKINNER in *Science* 24 Oct. 974/2 A first step is to define the field. A second is to collect technical terms, facts, laws, principles, and cases. These must then be arranged in a plausible developmental order—linear if possible, branching if necessary. **1961** *Barron's Nat. Business & Financial Weekly* 30 Oct. 14/2 We are disciples of neither Crowder nor Skinner. Our programs will make use of either branching or linear techniques, depending on which seems best suited to the subject matter. **1962** A. A. LUMSDAINE in J. E. Coulson *Programmed Learning* 135, I believe that linear programs should almost invariably be constructed first, even if branching is later to be introduced. **1964** *Times Rev. Industry* Feb. 100/2 A novel feature is the method it uses—the linear (non-branching) technique of programmed learning. **1969** G. KENT *Blackboard to Computer* viii. 109 The basis of linear programming is that the subject matter to be understood is always presented to the student in small quantities. **1970** W. K. RICHMOND *Concept Educ. Technol.* iii. 103 That the majority of linear programmes are inordinately dull is, of course, a charge which will be strenuously denied by anyone who has laboured to produce one. *Ibid.*, This aseptic dullness is explained by the linear programmer's dependence upon a bird-brained psychology.

4. (Later examples.)

1923 C. R. STOCKARD in *Amer. Jrnl. Anat.* XXXI. 111. 278 The two groups into which almost all ordinary persons fall more or less exactly may..be termed the Linear Type and the Lateral Type. The linear type is the faster growing high metabolizing thin but not necessarily tall group, while the lateral type is slower in maturing and is stocky and rounder in form. **1932** *Field Archæol.* (Ordnance Survey) 30 This term 'Linear Earthwork' is used to describe earthworks like Wansdyke..and the numerous Grim's Dykes... They consist of a bank and ditch and may be of any length from a few yards..to 10 miles. **1959** *Jrnl. Soc. Archit. Historians* XVIII. 40/1 Soria..customarily described his Linear City as a vertebrate animal. **1964** M. ARGYLE *Psychol. & Social Probl.* v. 62 Fewer of the delinquents are of linear (thin and bony) physique. **1966** *Guardian* 5 Apr. 2/6 A new linear city of half a million people..near Inverness..is the ambition of Professor Robert Grieve.

7. Special collocations: *linear accelerator* (see *ACCELERATOR e); *linear motor*, a motor (esp. an induction motor) which produces motion directly in a straight line (as opposed to rotary motion); *linear programming*, a mathematical technique for maximizing or minimizing a linear function (such as output or cost) of several variables (such as resources) when these are required to satisfy a set of linear equations and inequalities; (see also 3 c above).

1957 E. R. LAITHWAITE in *Proc. Inst. Electr. Engin.* CIV. A. 461/1 [The word 'linear' has already been used in connection with particle accelerators, but as there is little likelihood of confusion between these devices and electric motors, there appears to be no objection to the use of the word for the latter.] *Ibid.*, The use of linear motors as liquid-metal pumps is examined. **1966** *Listener* 13 Oct. 535/3 British Rail were the first to support Laithwaite's work on the linear motor, and suggested to him in 1960 its possible application to rail traction. **1973** *Sci. Amer.* Oct. 21/1 The evolution of electromagnetic flight is inextricably linked to the problem of propulsion. Two types of 'linear motor' are being studied for this application. One is called the linear induction motor, the other the linear synchronous motor. **1949** G. B. DANTZIG in *Econometrica* XVII. 203 It is our purpose now to discuss the kinds of restrictions that fit naturally into linear programming. **1953** COOPER & HENDERSON in W. W. Cooper et al. *Introd. Linear Programming* I. 1 Linear programming is concerned with the problem of planning a complex of interdependent activities in the best possible (optimal) fashion. **1966** A. BATTERSBY *Math. in Managem.* iv. 85 Transporting coal to power stations or gasworks, allocating cash to local branches, formulating foods and drawing up a maintenance schedule: these are all areas in which linear programming is at work today. **1967** E. DUCKWORTH in Wills & Yearsley *Handbk. Managem. Technol.* vi. 110 When the optimum order quantities have been decided, problems may occur in scheduling these through factories in the optimum manner... Quite complex methods of the linear programming or queueing theory type may be needed. **1971** *Sci. Amer.* Feb. 84/2 Among the intended tasks for ILLIAC IV is linear programming, a mathematical technique for allocating the use of limited resources to maximize or minimize a specified objective. *Ibid.* 85/2 In order to apply linear programming to an entire economic sector one must incur considerable expense in gathering the data to be used in the model.

linearism (li·nĭăriz'm). [f. LINEAR *a.* + -ISM.] Linearity; emphasis upon line or contour as opposed to colour or mass.

1935 *Archit. Rev.* LXXVII. 61 Far less of a constructor than a decorator, his famous 'linearism' was an echo of the same salient characteristic in Aubrey Beardsley's graphic designs. **1941** *Burlington Mag.* July 17/1 His bold method of brush drawing..is rendered possible by this two-dimensional linearism.

linearistic (linĭări·stik), *a.* [f. LINEAR *a.* + -ISTIC.] Pertaining to or characterized by a linear quality; of a linearized character.

1908 A. J. EVANS in R. R. Marett *Anthropol. & Classics* 41 Many of these signs are linearistic degenerations of animal figures.

linearity. Add: (Later examples.)

1947 A. EINSTEIN *Mus. Romantic Era* xvi. 245 Wagner wrote a Prelude..in the 'linearity' of which he seemed to overstep the stylistic limits of his own time. **1955** HOMANS & SCHNEIDER *Marriage, Authority & Final Causes* 20 Almost all the eastern Asian societies in which he finds mother's brother's daughter marriage preferred are organised in patrilineages. He concerns himself with linearity, [etc.]. **1958** *Times* 5 Dec. 8/6 Ectomorphy describes the linearity of the build. **1964** M. MCLUHAN *Understanding Media* xviii. 172 The linearity precision and uniformity of the arrangement of movable types.

b. *spec.* in *Math.* and *Physics*, the property of being linear in sense 3 of the adj. (in Dict. and Suppl.); proportionality of two related quantities (such as input and output).

1904 [see *EXTRAPOLATE *v.* 2 a]. **1943** *Electronic Engin.* XVI. 55/1 The insertion of a valve amplifier between the time base and the cathode-ray tube often involves some sacrifice in linearity. **1946** *Nature* 7 Sept. 330/2 Linearity is also important in the optical pick-up from the record, in the sense that the illumination of the photo-electric cell must be strictly proportional to the width of the white part of the illuminated area of the record. **1962** SIMPSON & RICHARDS *Physical Princ. Junction Transistors* xii. 269 Because the gain is high and the linearity fair, negative feedback may be used to exchange gain for linearity.

linearize, *v.* Add to def.: To make linear. Hence **li·nearized** *ppl. a.* (Further examples.)

1938 *Proc. Physico-Math. Soc. Japan* XX. 319 A method of linearization of wave equations for the electron. **1956** E. H. HUTTEN *Lang. Mod. Physics* iii. 94 Linearisation is a familiar trick in physics. **1957** L. FOX *Numerical Solution Two-Point Boundary Probl.* iii. 48 The only practical method of general application seems to be to 'linearize' the equations and solve them by an iterative process. **1962** W. B. THOMPSON *Introd. Plasma Physics* ii. 10 Then we obtain the linearized, approximate equations. **1968** CHOMSKY & HALLE *Sound Pattern Eng.* 391 What we have so far defined is simply a linearized version of rules of the form we have been discussing all along. **1969** *Physics Bull.* Nov. 463/1 Analysis is made difficult by the nonlinearity of the governing equations; any attempt to introduce approximations which linearize them seems to result in a loss of the basic features of the flow. **1969** *Canad. Jrnl. Ling.* XV. 25 It was speculated..that the reason that languages had embedding transformations.. was to 'linearize' or spread out in linear form the deeply embedded concoctions which the human mind can produce. **1970** J. EARL *Tuners & Amplifiers* iv. 93 This can.. reduce the distortion and linearise the power response. **1973** *Newnes Colour Television Servicing Manual* I. ii. 57/1 One half of the double triode is concerned essentially with linearising the field scan by means of negative feedback. **1973** *Physics Bull.* Oct. 606/1 Since the output from the digital ratemeter is nonlinear with load, it is necessary to use a short integrating time (about 0·1 s) in the ratemeter to obtain a true load signal after linearization.

Hence **li·nearizer**, that which linearizes, *esp.* a device which linearizes the response of a measuring instrument or other mechanism.

1973 [see *LINEAR *a.* 3]. **1973** *Physics Bull.* Dec. 745/1 The system is precalibrated for particular applications and has a built-in linearizer.

linearly, *adv.* Add: **d.** In a way that involves only terms of one dimension; in a linear or proportional manner.

1851 *Phil. Mag.* I. 295 If a quadratic function (U) be linearly converted into another (V), any minor determinant of any order of V must be a syzygetic function of all the minor determinants of U of the same order. **1859** G. SALMON *Lessons Introd. Mod. Higher Algebra* vi. 34 From the four equations thus formed we can eliminate linearly the four quantities, x^3, x^2y, xy^2, y^3. **1885** [see *LEGENDRE]. **1955** *Bull. Atomic Sci.* June 208/1 The extent of fall-out varies linearly with the fission yield of the bomb. **1963** *Amer. Jrnl. Physics* XXXI. 336/2 A necessary and sufficient condition..is that θ and T be related linearly. **1969** *Listener* 12 June 831/1 Criminals in human society..increase more rapidly than linearly with the total number of victims available. That is to say, if we double the population we more than double the number of criminals.

lineman. Add: **4.** *Amer.* and *Canad. Football.* A forward.

1907 *St. Nicholas* Sept. 1013/2 There was some discussion last year as to whether a line man could run from his position in the line and take the ball from the quarter. **1913** *Collier's* 13 Dec. 27/1 He was an aggressive, hard-fighting, and alert lineman, who did his best work under fire. **1959** *Times* 30 Nov. (Canada Suppl.) p. xx/2 There are five back-fielders and seven linemen. **1970** *Globe & Mail* (Toronto) 26 Sept. 35/5 The linemen and linebackers are all experienced. **1971** *New Yorker* 15 May 54/3 He was a forward, or lineman, in Rugby. **1972** *Nature* 2 June 297/3 Our manuscript read 'linemen', a designation for American football players who crouch in a line with hands on the ground prior to the attack.

linen, *a.* and *sb.* Add: **B.** *sb.* **3. a.** *to wash one's dirty linen in public*, to discuss an essentially private matter, esp. a dispute or scandal, in public.

1867 TROLLOPE *Last Chron. Barset* II. xliv. 2 There is nothing..so bad as washing one's dirty linen in public. **1931** *Times* 3 Aug. 9/1 If the Government had made tactful..representations..to the Holy See,..the whole matter could have been quietly settled without any washing of dirty linen in public. **1935** D. L. SAYERS *Gaudy Night* iv. 80 Even if the poison campaign led to no open disaster..a washing of dirty linen in public was not calculated to do Shrewsbury [College] any good. **1972** *Daily Tel.* 3 May 16, I know it is not done to wash dirty medical linen in public.

c. Abbrev. *LINEN-DRAPER b.

1955 J. PHELAN *Tramping the Toby* 223 *Linen*, a newspaper. **1962** R. COOK *Crust on its Uppers* i. 21 Everything they've ever read in a linen or a clever-clever book.

4. a. *linen-closet, -room, -tape, -thread.*

1885 'M. RUTHERFORD' *M. Rutherford's Deliverance* iii. 41 She cared nothing for the linen-closet, the spotless bed-hangings,..the true household gods of the respectable women of those days. **1955** W. TUCKER *Wild Talent* xiv. 181 The butler..saw Paul's questioning glance at the other two doors. 'The nearest one is a linen closet, sir,' he said. **1900** E. GLYN *Visits of Elizabeth* 33 Aunt Maria.. said it was her day for seeing the linen-room. **1873** *Young Englishwoman* Mar. 150/2 A piece of linen tape..keeps the buttons securely in place. **1880** E. GLAISTER *Needle-work* v. 49 The letters are made in linen tape, unbleached, the yellower the better. **1897** *Sears, Roebuck Catal.* 321/1 Marshall's Linen Thread (100 yd. spools, black only). **1908** *Westm. Gaz.* 28 Dec. 5/2 The whole front is a mass of hand embroidery done in heavy linen-thread. **1975** *Times* 6 Mar. 7/6 The greatest problem with owning old lace is getting it repaired... An added problem is the difficulty of obtaining the hair-fine linen thread.

5. linen basket, a receptacle for dirty clothing; **linen crash** = CRASH *sb.*[2]; **linen cupboard,** a cupboard designed to hold bed-linen and table-linen; also, the contents of such a cupboard; **linen duster,** a duster (see DUSTER 4) made of linen; **linen-horse** = HORSE *sb.* 7 c; **linen-press,** a frame or receptacle for pressing or holding linen; **linen shower** [*SHOWER *sb.*[1]] *N. Amer.*, a party at which a bride-to-be is given presents of household linen, etc.; **linen tea,** a tea arranged in order to provide house-linen for a crèche, day nursery, etc.

1907 *Yesterday's Shopping* (1969) 125/3 Linen baskets. Barrel shape, buff wicker. **1970** G. F. NEWMAN *Sir, You Bastard* viii. 239 He used his shirt to wipe his damp arm-pits, then threw it out on to the linen basket. **1895** *Montgomery Ward Catal.* 24/2 Linen crash or toweling. **1904** *Sci. Amer.* 21 May 409/3 The complainant, a manufacturer of linen crash, had adopted as a trade-mark the words 'Stevens Crash'. **1873** *Young Englishwoman* Mar. 155/1 The plain, economical housewife would do well to follow 'Myra's' plan of her linen-cupboard. **1939** A. THIRKELL *Brandons* ix. 234 She let Miss Morris help Nurse..to go through the linen cupboard and mark some new sheets. **1972** P. RUELL *Red Christmas* v. 48 The linen cupboard was, forecastably, full of linen. It was more of a room than a cupboard. **1867** *Galaxy* III. 635 His cloak..or linen duster..serves as a cover to hide the manipulations of his agile fingers. **1949** *Chicago Daily News* 11 Feb. 21/3 Grandma was a fashion plate in her smart linen duster for Sunday motoring. **1845** R. COBBOLD *Hist. M. Catchpole* II. xxv. 139 The large linen-horses belonging to the gaol stood in the passage. **1906** *Mrs. Beeton's Bk. Househ. Managem.* lxix. 1810 Silks.. should always be dried in the shade, on a linen-horse. **1851** MRS. STOWE *Uncle Tom's Cabin* (1852) I. xviii. 296 The store-room, the linen-presses, the china-closet,.. all went under an awful review. **1970** *Canad. Antiques Collector* June 4 (Advt.), Exceptional Pennsylvania walnut linen-press..Circa 1790. **1904** *N.Y. Tribune* 27 Oct. 7 The managers of the Home for the Friendless invite the public to a linen shower and reception. **1921** *Daily Colonist* (Victoria, B.C.) 3 Apr. 8/3 A delightful linen shower was given in honor of Miss Dorothy Woods, Thursday evening at the home of Mrs. B. E. Lefevre. **1974** *Evening Herald* (Rock Hill, S. Carolina) 18 Apr. 9/4 Hostesses for a linen shower were Mrs. H. D. Long and Mrs. Arthur Snyder. **1916** *Yorks. Post* 19 Apr. 4/7 The first crèche which held such a 'linen tea'... Yesterday's meeting was for the purpose of collecting house linen to furnish the place. **1948** Linen tea [see *kitchen tea* (*KITCHEN *sb.* 7)].

linen-draper. Add: **b.** A newspaper. *Rhyming slang.*

1857 'DUCANGE ANGLICUS' *Vulgar Tongue* 12 *Linen-draper*, paper. **1936** J. CURTIS *Gilt Kid* xxiv. 234 It might be just as well to keep under cover for a little until perhaps the linen-drapers gave him the office that the chase had not been taken up. **1972** *Lebende Sprachen* XVII. 8/3 *Linen draper*, paper.

li·ne-out. Pl. line-outs, lines-out. [f. the vbl. phr. *to line out* (LINE *v.*[2] 8 b).] In Rugby football: (see quot. 1900).

1889 [see TOUCH *sb.* 12]. **1900** A. E. T. WATSON *Young Sportsman* 284 *Line out*,.. the arrangement of forwards opposite to one another when the ball is about to be thrown in from touch. **1906** GALLAHER & STEAD *Compl. Rugby Footballer* v. 80 The line-out work. **1931** *Times* 16 Feb. 5/2 Barrington made a clever mark from a knock-on in a line-out, but failed with the kick at goal. **1955** *Times* 18 July 12/6 The Province won four scrummages and lines-out out of five. **1965** *New Statesman* 19 Mar. 465/1 Just before half-time Wales scored, and thereafter dominated the game, winning the line-outs by intelligent forward play. **1973** *Scotsman* 21 Feb. 18/6 He used to prefer being stationed at No. 7 but agrees that, with the new laws tending so to compress the line-out, it is difficult to win much really usable ball from an opponent of markedly superior height.

liner[1]. Add: **3.** The lining of a garment, esp. one made of an artificial fibre. So *liner suit* (see quot. 1969).

1947 *Horizon* Sept. 203 They took off their helmet liners. **1962** F. I. ORDWAY et al. *Basic Astronautics* xiii. 517 The inner liner of the suit is of neoprene-coated fabric... The outer liner is an aluminized coverall. **1969** *Guardian* 7 Jan. 7/2 There is an undergarment called a liner suit which makes PVCs more comfortable to wear. It is a two-piece affair made from knitted nylon with the inside of cosy brushed cotton. **1970** *Washington Post* 30 Sept. B5/3 (Advt.), Zip-in-or-out orlon liner. **1971** C. BONINGTON *Annapurna South Face* 241, 4-oz. Dunloprufe nylon with open-cell foam liner. *Ibid.* 242, 1 pair overmitts [made of] proofed nylon with Borg fur liner.

4. In full, *liner note*. (See quot. 1953.) orig. *U.S.*

1953 *Britannica Bk. of Year* 638/1 *Liner*, the text accompanying an album of gramophone records. **1955** *Sat. Rev.* (U.S.) Jan. 41 The covers of these new jazz albums..are being covered..with thousands and thousands of words known as 'liner notes'. **1960** D. CERULLI et al. *Jazz Word* (1962) 106 They couldn't come up with any less information than on some liners today. **1968** *Jazz Monthly* Feb. 21/2 The enthusiastic sleeve note by Brian Rust suggests that he may be on the downward path towards acceptance of those degenerate swing bands whom he has damned in nearly all his liner writing! **1969** *Rolling Stone* 17 May 17/3 As Coleman observes in the liner notes, 'Ornette Denardo is hard to keep up with if you don't tell him what to do.'

liner[2]. Add: **7. b.** A cosmetic used for tinting a part of the face; a brush or pencil for applying this; *spec.* = *eye-liner, eyeliner.*

1926 M. SMITH *Bk. Play Production* xi. 182 Liners are smaller sticks of grease paints..used to make lines on the face, such as wrinkles, 'eyebrows', etc. **1958** OSBORNE & CREIGHTON *Epitaph G. Dillon* II. 44, I always touch mine up with a brown liner... The rings under my eyes. **1966** *Harper's Bazaar* Sept. 70/1 Wrapping up lid, indeed eye, in a cocoon of pale grey shadow and liner. **1972** *Daily Tel.* 24 Jan. 11/1 A narrower streak of colour on the lid by the lashes, in place of liner.

8. b. (Earlier examples.)

1829 W. N. GLASCOCK *Sailors & Saints* I. ii. 25 We liners, you know, are not in the habit of leading small craft to their anchorage. **1855** *Chambers's Jrnl.* II. 270/2 Not an hour was lost in expediting the fitting out of our liner, for war was raging.

c. One of the aircraft of a regular line, esp. one for passenger transport; an air-liner; a space-ship.

1905 KIPLING *Actions & Reactions* (1909) 125 A Planet liner, east bound, heaves up in a superb spiral and takes the air of us humming. **1919** H. GOLDING *Wonder Bk. Aircraft* 69 (*caption*) Off! The pilot of the 'liner' is just giving orders to remove the blocks from the wheels of the under-carriage. **1933** *Boys' Mag.* XLVII. 24/1 Mile after mile of seemingly endless country unfurled itself beneath the flying wings of the giant liner. **1951** A. C. CLARKE *Sands of Mars* iii. 24 The observation gallery..completely circled the liner. **1959** *Times Lit. Suppl.* 17 Apr. 230/3 These well-known liners were conceived with the future needs of the air lines in view. **1969** *New Scientist* 2 Oct. 20/2 The Boeing liner will have rather more than twice the capacity of the Concorde.

d. One of a fleet of lorries.

1955 *Times* 29 June 13/3 The commission have offered the trunk service vehicles, which provide regular daily services between certain towns..'liner' services, compared with 'tramps',..in relatively large lots together with their respective terminal depôts.

13. *attrib.*, as **liner train,** a fast through-running freight train made up of detachable containers on permanently coupled wagons.

1962 *Guardian* 30 Oct. 3/2 'Custom built' services..so that customers can..'buy space' on fixed formation trains —'liner trains'—whose wagons can..bear their name and line of business. **1963** *Reshaping of Brit. Railways* (Brit. Railways Board) 142 The description 'Liner Train' is applied to a conception of transport based upon joint use of road and rail for door-to-door transport of containerised merchandise, with special purpose, through-running, scheduled trains providing the trunk haul... The Liner Train..is a train of chassis which will remain continuously coupled... The speed will be a maximum of 75 and an average of 50 miles an hour. **1964** *Observer* 28 June 8/5 If Dr. Beeching's figures are accepted at their face value, the cost of carrying goods by liner trains will be so much less than by heavy lorries..that he should be able to undercut the roads by a comfortable margin. **1970** *Daily Mail* 16 Feb. 1/6 The plant sends rear axles and brake drums by liner trains to other Ford factories.

li·nerboard. [f. LINER[1] + BOARD *sb.*] A paper-board used as a facing on fibre-board.

1961 *Paper Technol.* II. 11. 145 We have produced virtually every type of paper and board known today, for example, ranging from glassine to kraft liner board. **1969** *Jane's Freight Containers 1968–69* 439/2 Rectangular container, two layers corrugated fibreboard covered by linerboard. **1972** *Evening Telegram* (St. John's, Newfoundland) 28 June 4/3 What venture capital went into the Stephenville Linerboard Mill?

linesman. Add: **3. c.** In *N. Amer. Football*, (*a*) an official on the sideline with certain duties governed by the rules of the game (see quot. 1969[2]); (*b*) = *LINEMAN 4. In *Ice Hockey*, an official whose chief duty is to give offside decisions.

1897 *Encycl. Sport* I. 425/2 Two linesmen..mark the distance gained and lost, and are aids to the umpire. **1935** *Encycl. Sports* 528/1 The officials in this game are referee, umpire, linesman and two assistants, and a field judge. **1947** E. A. MCCOURT *Music at Close* 119 The crash of opposing linesmen, the spirited end runs, the long spiralling kicks awoke in him a strong desire to be a participant. **1955** *Globe & Mail* (Toronto) 31 Jan. 19/1 Flaman tried desperately to break from linesman Bill Roberts, who had put the clutch on the rampaging Bruin defenseman. **1969** *Official Rule Bk. Nat. Hockey League 1969–70* 44 The duty of the linesman is to determine any infractions of the rules concerning off-side play at the blue line, or center line, or any violation of the 'Icing the Puck' rule. **1969** *Official Playing Rules Nat. & Amer. Football Leagues* 93 Linesman is to mark with his foot..the yard-line touched by forward point of ball at end of each scrimmage down. *Ibid.*, He and the Umpire are to determine whether ineligible linesmen illegally cross line prior to a pass.

4. One who attends to the upkeep of road-side verges.

1888 *Geysers & Gazers* 7 The road was impeded by a land-slip... As soon as we cleared one slip away, another came into view. I asked the driver why the linesman had not attended to the matter. **1971** *Country Life* 18 Nov. 1372/4, I was privileged to spend many hours in the company of a linesman... He was responsible for the upkeep of the roadside verges of some four to five square miles.

line standard. [LINE *sb.*[2]] **a.** *Metrology.* A standard of length in the form of a metal bar on which are engraved two lines, the distance between which (under specified conditions) is the standard length.

1888 *Encycl. Brit.* XXIV. 478/1 Standards of length are of two types, the defining points being either at a certain part of two parallel lines engraven in one plane (a line-standard), or else points on two parallel surfaces, which can only be observed by contact (an end-standard). **1906** HALLOCK & WADE *Outl. Evolution of Weights & Measures* x. 223 The line standard, of course, can be used with a microscope with cross-hairs, or a micrometer microscope, much more readily than an end standard. **1966** KAYE & LABY *Tables Physical & Chem. Constants* (ed. 13) 7 The yard equal to 0·9144 metre supersedes the definition in terms of the former imperial standard yard, a bronze line-standard constructed in 1845. **1967** A. J. T. SCARR *Metrology & Precision Engin.* v. 67 The calibration of a line standard is determined by comparison with a master scale.

b. *Television.* The number of lines constituting a complete picture.

1959 C. DANIEL in *Rep. Comm. Broadcasting 1960* (1962) 340 in *Parl. Papers 1961–2* (Cmnd. 1753) IX. 259 We understand that most of the rest of Europe are likely to adopt..a line standard of 625 lines. **1973** *Newnes Colour Television Servicing Manual* I. i. 2/2 The degaussing action takes place automatically when there is a change in the line standard during push-button selection.

li·ne-up. orig. *U.S.* [f. LINE *v.*[2] 8 a.] The assembling of a number of persons in a line, e.g. for inspection or identification; an instance of bringing into a line; a list of players in a game, orchestra, etc.; the players on such a list. Also of things. Also *fig.*

1889 *Kansas City* (Missouri) *Times & Star* 11 Mar., The line-up of the Kansas City ball club this season. **1904** *Springfield* (Mass.) *Weekly Republ.* 3 June 1 Thus we have a line-up of corporations against the people. **1907** J. LONDON *Road* 89 Then came the line-up, forty or fifty of us, naked as Kipling's heroes. **1911** H. S. HARRISON *Queed* xviii. 224 He studied his trustee list now more purposefully than he had ever pored over his faculty line up. **1913** G. J. KNEELAND *Commercialized Prostitution N.Y.* 62 Here they come to make deals for their women..or in vice resorts, to plan line ups when a 'young chicken' is about to be broken into the business. *Ibid.*, A 'line up' is the ruin of a girl who flirts with men and accepts their advances and immoral suggestions. Finally she yields to an invitation to visit a furnished room and the word quickly passes among the 'gang'. One by one the boys and men, perhaps only two or three, perhaps more, visit the room. **1914** J. B. BICKERSTETH *Land of Open Doors* 172 Guess I'm in the line up for hell all right—with no return ticket either. **1915** *Policeman's Monthly* Sept. 5 (*caption*) The famous 'line-up' at New York Police Headquarters. **1926** *Clues* Nov. 161/2 *Lineup*, suspects arrested the night before lined up for police inspection. **1928** A. G. HAYS *Let Freedom Ring* 289 The prisoners were brought before witnesses—not in a line-up with others of the same general type but separately. **1931** E. LINKLATER *Juan in America* II. xviii. 197 But she didn't know the line-up... It's Bauer that knows that, because Bauer's going to do the job. **1934** —— *Magnus Merriman* xxvi. 292 He knew the whole line-up from old man Plato to Bergson. **1938** *Amer. Speech* XIII. 71/1 *Line-up*, a record of all expected train movements for the day in the section where the gang is working. **1952** *Manch. Guardian Weekly* 10 Jan. 3 A police line-up. **1955** A. J. MCCARTHY *Jazzbook 1955* 39 The traditional line-up of trumpet, clarinet, trombone, piano, banjo, string bass and drums. **1965** J. POLLARD *Surfrider* ii. 18 The 'lineup' is where the waves line up to break. **1967** *Word Study* Mar. 1/2 To the general public, the English teacher is the witness qualified to identify the felon in a linguistic lineup. **1968** *Radio Times* 28 Nov. 55/4, 11.10. Late Night Line-Up. A last look around the daily scene. **1970** *Language* XLVI. 317 The negative and positive values of time refer to the computer 'line up' point. **1972** *Jazz & Blues* Sept. 11/3 This was the basic line-up which became the Domino band after the success of 'The Fat Man'. **1973** 'J. PATRICK' *Glasgow Gang Observed* v. 52 The phrase 'line-up'..was normally used to describe the queue of boys waiting to have sexual intercourse in one of the 'gang bangs'.

ling, *sb.*[1] Add: **3. ling-cod** *N. Amer.*, a North Pacific species of cod, *Ophiodon elongatus*, also called cultus cod. (Cf. sense 2.)

1955 W. DAWSON *Ahoy There!* 205 Besides salmon, we catch cod..ling-cod of up to (in our case) twelve pounds. **1964** *Canad. Geogr. Jrnl.* Mar. 91/3 This is particularly true in the case of the ling cod, the spear-fisherman's favourite quarry. **1971** *Islander* (Victoria, B.C.) 21 Mar. 2/3, I have taken in recent years, six species of rockfish..also lingcod.

Lingala (liŋgā·lă). Also **Ngala.** [Native name.] A Bantu language spoken by the Bangala people in the Mangala area of Zaïre, widely used in trade and public affairs. Also *attrib.* or as *adj.*

1903 W. H. STAPLETON *Compar. Handbk. Congo Lang.* p. r, Bangala is a name originally given..to a people.. from the settlement of Mangala... The dialect chosen and called Ngala is spoken by the Boloki..and Bokomoi on the South bank. **1922** H. H. JOHNSTON *Compar. Study Bantu & Semi-Bantu Lang.* II. vi. 130 At present we can only surmise that Group KK contains the following sub-groups:..KK₄, the Lingala. **1945** C. M. DOKE *Bantu* 27 Ngala..is today called 'Lingala', though it is based on the speech of the Mangala people. **1965** *B.B.C. Handbk.* 103 Moscow radio started transmissions in..Lingala, Nepalese. **1973** *Black World* May 95/1 The Bolinga, the name for the Black Cultural Resources Center at Wright State University in Dayton, Ohio, is a Lingala word meaning 'love'.

lingberry: see *LINGONBERRY.

Lingby, var. *LYNGBY.

lingenberry: see *LINGONBERRY.

linger, *sb.* Add: **2.** *U.S.* (See quot.)

1895 *Nation* (N.Y.) 9 May 358/3 The enervating influence of the climate, giving rise to that which in the south-western United States is called the 'Texas lingers'.

linger, *v.* Add: **7. c.** *Hort.* To delay the blooming of (flowers) by artificial means.

1906 *Daily Chron.* 12 Sept. 4/4 If you force, you exhaust the [rose-]tree; it languishes. I prefer to 'linger' it.

lingerie. Add: **b.** *attrib.*

1866 MRS. GASKELL *Wives & Daughters* II. xxiv. 248 There was a 'lingerie' shop, kept by a Frenchwoman. **1905** *Daily Chron.* 13 Mar. 8/1 The lingerie blouse made a most emphatic appearance in Paris..this winter. *Ibid.*, It is the lingerie shirt that wins. **1909** *Westm. Gaz.* 22 Feb. 5/2 As to the lingerie gown, its importance in the wardrobe cannot be questioned. **1909** *Public Ledger* (Philadelphia) 24 June 7/7 (Advt.), Fine lingerie waists, trimmed with dainty laces & embroideries. **1909** *Daily Chron.* 10 Sept. 7/2 Embroider..if the sacque is of piqué or lingerie materials. **1964** *McCall's Sewing* ii. 30/1 *Lingerie hem*, a rolled hem that is caught with two overcast stitches at intervals of $\frac{1}{4}$ to $\frac{3}{8}$ inch gathering it in puffs. *Ibid.* 30/2 *Lingerie seam*, supposedly 'rip-proof'; made by pressing both edges of a seam to one side and top-stitching with a zigzag stitch along the edge. **1974** *Country Life* 3–10 Jan. 55/1 Harvey Nichols have.. palazzo suits by Mr Dino. In the lingerie department there are things by Iful, Yves Stillman, Leonara. **1974** *Evening Herald* (Rock Hill, S. Carolina) 18 Apr. 9/4 A lingerie shower was given by Mrs. Preston Ramsey and Mrs. Harold Wolfe.

lingering, *vbl. sb.* Add: **c.** *Hort.* Retarding the time of blooming by artificial means.

1907 *Daily Chron.* 13 Feb. 6/4 Lingering is retardation without frost; it keeps September roses blooming until January.

‖ **lingoa geral** (liŋgōu·ā dʒərā·l). Also **lingua geral.** [Pg., ad. *lingua geral*, lit. 'general language'.] A trade language based on Tupi and used as a lingua franca in Brazil.

1856 [see *JUPATI]. **1860** MAYNE REID *Odd People* 46 We shall use that [name] by which it is known in the 'Lingoa geral', and call it a *malocca*. **1876** *Encycl. Brit.* IV. 235/1 Most of the semi-civilized Indians of Brazil.. speak the Lingoa-Geral, a language adapted by the Jesuit missionaries from the original idiom of the Tupinambaras. **1932** W. L. GRAFF *Lang.* 431 The whole of Brazil that came into contact with the whites eventually became bilingual, knowing their own dialect and the Guarani *lingoa geral*. **1948** C. NIMUENDAJÚ in J. H. Steward *Handbk. S. Amer. Indians* III. 257 After their pacification, the *Mura* began to adopt the Lingua Geral... Later they substituted Portuguese for the Lingua Geral. *Ibid.* 311 *Tapanyuna* is not an *Apiacá* word, but a Lingua Geral term which means 'negro'. **1973** E. BROOKS et al. *Tribes of Amazon Basin in Brazil 1972* ii. 58 Tukano is still spoken widely among Indians of Santa Isabel..and many still speak *lingua geral*, the Jesuit's blend of Tupi and Guarani.

lingonberry (li·ŋɒ̄nberi). *Canad.* Also **lingberry, lingenberry.** [f. Sw. *lingon* mountain cranberry + BERRY *sb.*[1]] A dwarf variety of mountain cranberry, *Vaccinium vitis-idæa* var. *minus*, native to far northern temperate regions; also, another northern species of cranberry, *V. oxycoccus*.

1955 *Arctic Terms* 50 *1 Lingberry*,..a name for the fruit of the mountain cranberry. **1960** J. J. ROWLANDS *Spindrift* 156 In Sweden the cranberry is known as the lingonberry. *Ibid.* 161 This berry [*sc.* mountain cranberry] is probably close to the lingonberry of Scandinavian countries. **1961** *Harper's Bazaar* May 106/2 Sweden prides itself on..pancakes..dressed with lingonberries. **1970** *Beaver* Winter 23 Wild cranberries..are also known as lingenberries. **1971** D. NABOKOV tr. *Nabokov's Glory* (1972) vi. 24 Supper at the station (hazel hen with lingonberry sauce).

lingua. Add: **2. b.** (Examples of pl. form.)

1971 J. SPENCER *Eng. Lang. W. Afr.* 31 A very complex infrastructure of scores of vernacular languages as well as a number of regional lingue franche. **1974** R. A. HALL *External Hist. Romance Lang.* 21 The distribution of the Romance languages is best treated under four heads:.. (3) use as *lingue franche*.

fig. (Later examples.)

1955 *Times* 2 July 5/2 Cold war recrimination became the east-west *lingua franca*. **1958** *Times* 16 Sept. 3/2 Mr. Morrice handles them [*sc.* motifs] by cunningly intermingling realism with the *lingua franca* of ballet.

linguacious, *a.* Add: **1.** Also **linguaceous.** (Later examples.)

1827 J. F. COOPER *Prairie* I. xi. 329 On the summit, Obed fully expected to encounter Esther, of whose linguacious powers, he had too often been furnished with the most sinister proofs. **1950** J. Y. T. GREIG *Thackeray* xiii. 142 She needed it [*sc.* the gift of listening] with a witty and linguacious fellow for a husband. *a* **1953** DYLAN THOMAS *Quite Early One Morning* (1954) 64 And see, too, in that linguaceous stream, the tall monocled men..who lecture to women's clubs.

lingua geral: see *LINGOA GERAL.

lingually, *adv.* Add: **b.** On the lingual side; towards the tongue.

1902 *Dental Digest* VIII. 925, I have seen some cases of compound cavities which started at a point of structural defect perhaps two millimeters distant lingually from the contact. **1910** *Practitioner* Jan. 115 Internally (lingually), the neck of the tooth..is embraced by a thin shallow flap of gum. **1963** J. OSBORNE *Dental Mech.* (ed. 5) viii. 112 During eruption..the facial muscles push the teeth lingually. **1970** *Nature* 25 July 356/2 A crest extends lingually to delineate the extent of the talonid basin. **1971** *Ibid.*, 30 July 311/1 At the incisor sockets the alveolar margin seems to be preserved lingually but is broken labially.

Linguaphone (li·ŋgwăfōun). Also **linguaphone.** [f. LINGUA 2 + *-phone* after GRAMOPHONE.] The proprietary name of a language-teaching system based on the use of gramophone records in conjunction with textbooks (see quots.). Also *attrib.*

1908 *Westm. Gaz.* 29 Feb. 7/1 (Advt.), The Linguaphone. This machine..teaches languages by a marvellous system of phonographic records. **1913** *Chambers's Jrnl.* Nov. 830/2 In the linguaphone system..the records are prepared by distinguished speakers of the various languages. **1925** *Trade Marks Jrnl.* 4 Nov. 2397 Linguaphone..instruments and apparatus for teaching. Jacques Roston,..London,..teacher of languages and translator. **1926** *Glasgow Herald* 28 Feb. 9 The Linguaphone Institute has produced a system by means of which one can, in his own home, from book and gramophone record obtain a working knowledge of languages. **1936** P. FLEMING *News from Tartary* 162, I had left Peking with a minute.. Chinese vocabulary, based on the first half-dozen records of a linguaphone course. **1962** L. DEIGHTON *Ipcress File* xxv. 161 Suppose I buy you a Linguaphone course. **1966** A. E. LINDOP *I start Counting* xv. 174 They had bought some Linguaphone records—so everything was 'quel' this and 'quel' that. **1974** *Country Life* 14 Feb. 304/2 Brito e Cunha breaks off the native-tongue chat..for a caution of 'Hold hard!' or 'I say, sta-idy there!' in linguaphone Leicestershire.

linguate (li·ŋgwĕt), *a.* [f. L. *linguātus* (not recorded in this sense) or f. LINGU(A 1 + -ATE[2].] Tongue-shaped. As *sb.*, a tongue-shaped flint instrument.

1940 *Proc. Prehist. Soc.* VI. 8 Pointed linguates with untrimmed butts. **1963** S. PIGGOTT in Foster & Alcock *Culture & Environment* iv. 84 In Ireland, a fourth distinctive group of linguate daggers can be recognized.

linguatulid (liŋgwæ·tiŭlid), *sb.* and *a.* [f. mod.L. name of former class *Linguatulida*, f. the generic name *Linguatula* (G. F. von Froelich 1789, in *Naturforscher* XXIV. 148), f. L. *linguātus* 'having a tongue', eloquent.] A parasitic worm-like arthropod of the genus *Linguatula*, the adult form of which attacks the nasal passages of certain carnivorous mammals, esp. canids; also, a member of the group Pentastomida, including parasitic arthropods, most of which attack snakes; = *tongue-worm* (*b*) (TONGUE *sb.* 16). Also as *adj.*

1923 *Nature* 17 Mar. 381/2 (*title*) On the linguatulid arachnid, *Raillietiella furcocerca*. **1929** PATTON & EVANS *Insects, Ticks, Mites* I. 666 *Porocephalus* is the name of a Linguatulid genus. *Ibid.*, It [*sc.* a parasite] can hardly be acquired by eating the flesh of snakes, the commonest final host of these Linguatulids. **1942** *Ann. Trop. Med. & Parasitol.* XXXVI. 60 (*title*) Linguatulid infestations of man. **1961** C. H. POPE *Giant Snakes* (1962) 193 Linguatulids, some fifty species of which have been described, are degenerate arachnids closely related to ticks and mites.

linguine (liŋgwī·ne). Also **linguini.** [It., pl. of *linguina*, dim. of *lingua* tongue.] An Italian pasta made of tongue-shaped ribbons.

1948 LO PINTO & MILARADOVICH *Art of Italian Cooking* (1955) i. 24 *Linguini*, narrow, plain noodles cut about $\frac{1}{8}$ inch wide. **1954** G. M. LAPOLLA *Good Food from Italy* 85 When peas are used, rarely is the sauce poured over the linguine or fettuccelle. **1965** M. ECHARD *I met Murder* (1967) vii. 59 Dining on fettuccine or linguini or ravioli. **1968** *N.Y. Times* 26 Jan. 22 The kitchen produces first-rate linguine with clam sauce. **1969** R. LOWELL *Notebk. 1967–68* (1970) 255 The ebb tide flings up wonders: rivers, linguini, Beercans, bloodstreams, eddies. **1972** R. K. SMITH *Ransom* I. 4 Just a little linguine, and a little veal parmigiana.

linguism. Restrict *nonce-wd.* to sense in Dict. and add: **2.** Advocacy of languages on a regional basis.

1967 *Economist* 18 Feb. 596/1 A good deal of the criticism heard..of Indian politics—for instance of 'linguism', 'casteism', and so on—is a roundabout way of wishing that human beings were different from what they are.

linguist. 2. Delete †*Obs.* and add later examples.

1922 O. JESPERSEN *Lang.* 64, I think I am in accordance with a growing number of scholars in England and America if I..apply the word 'linguist' by itself to the scientific student of language (or of languages). **1940** *Amer. Speech* XV. 187 The *Handbook* [of the Linguistic Geography of New England]..seems to suggest that American linguists are naïve. **1964** R. H. ROBINS *Gen. Ling.* 2 The general linguist, in the sense of the specialist or the student concerned with general linguistics. **1966** M. R. D. FOOT *SOE in France* viii. 212 He was a young linguist, a research student at Manchester University. **1966** *New Statesman* 11 Nov. 701/1 One or two of my friends even abandoned literature altogether and became fully fledged philologists—'linguists', as my companion at dinner would have said. **1973** A. P. SORENSEN in D. R. Gross *Peoples & Cultures of Native S. Amer.* 331 Some linguists wondered..whether the comparative method could or even should be applied to American Indian languages at all.

5. *attrib.* or appositive, as *linguist-anthropologist, -philologist, -reader.*

1951 S. F. NADEL *Found. Social Anthropol.* 46 The 'virtuosity' of the linguist-anthropologist..is only the fullest preparation for his task. **1960** *Amer. Speech* XXXV. 217 In a treatment of specific historical changes of morphemes and phonemes the linguist-philologist should attempt to recapture what actually happened. **1964** *Language* XL. 203 A sample passage from a 1956 article by Starkweather is enough to boggle the unprepared linguist-reader.

linguistic, *a.* and *sb.* Add: **A.** *adj.* (Later examples.)

1911 V. WELBY *Significs & Lang.* v. 17 The implicitly false mental image, source of the false linguistic image. **1921** E. SAPIR *Lang.* vi. 156 In a book of this sort it is naturally impossible to give an adequate idea of linguistic structure. **1935** B. MALINOWSKI *Coral Gardens* II. vi. v. 232 Within the linguistic theory of the present book, in which the distinction between 'form' and meaning is in the last instance illusory. **1936** J. R. KANTOR *Objective Psychol. Gram.* xiii. 195 It is undoubtedly necessary to include many other speech parts if we are to cover linguistic phenomena adequately. **1953** J. B. CARROLL *Study of Lang.* i. 5 It was only natural..that the engineer should have perceived the possibilities of developing various sorts of 'linguistic machines', such as a machine for instantly converting human speech into..printed alphabetic symbols. **1957** W. HAAS in *Studies in Ling. Analysis* (Philol. Soc.) 33 'Zero' in Linguistic Description stands for what is acoustically nothing. **1957** G. RYLE in C. A. Mace *Brit. Philos. in Mid-Cent.* 263 Philosophical problems are linguistic problems—only linguistic problems quite unlike any of the problems of philology, grammar, phonetics..etc., since they are..about the logic of the functionings of expressions. **1964** M. A. K. HALLIDAY et al. *Ling. Sci.* i. 18 If language is described according to the version of linguistic theory outlined, the task of the language learner..will be made easier. **1966** *English Studies* XLVII. 270 Instead of adverb transforms we find occasional instances of 'linguistic shortening', which in itself is a means of expressing emotiveness. **1967** R. TEXTOR *Cross-Cultural Summary* 67 The rationale for including linguistic affiliation is..that 'genetic relationships in culture and past historical connections among societies are commonly revealed..among the languages spoken by the peoples in question'. **1968** D. HYMES in *Internat. Encycl. Social Sci.* 366/2 Linguistic description has focused on the form of languages, neglecting the structuring of their use. **1968** CHOMSKY & HALLE *Sound Pattern Eng.* i. 4 The essential properties of natural language are often referred to as 'linguistic universals'. **1972** L. R. PALMER *Descr. & Compar. Ling.* ix. 227 Sound laws do not enable us to *predict* linguistic events as a law of chemistry predicts material change.

b. Special collocations: *linguistic analysis*, (*a*) the analysis of language structures in terms of some theory of language; (*b*) *Philos.*, analysis of language as the medium of thought; so *linguistic analyst*; *linguistic anthropology*, anthropological research based on the study of the language of a selected group; so *linguistic-anthropological* adj.; *linguistic atlas*, a set of tables or maps recording regional or dialectal variations of pronunciation, vocabulary, or inflexional forms; *linguistic form*, any unit or pattern of speech that has meaning; *linguistic geography*, the study of the geographical distribution of languages, dialects, etc.; so *linguistic geographer*, *linguistic-geographical* adj.; *linguistic map*, a map in a linguistic atlas; a map showing the distribution of linguistic features; *linguistic philosophy* = *linguistic analysis (*b*); so *linguistic philosopher*; *linguistic psychology*, the study of human psychology through the data provided by language; cf. *PSYCHOLINGUISTICS; *linguistic science*, the science of language; the systematic study of linguistic phenomena; so *linguistic scientist*; *linguistic stock*, the group to which a set of related languages belongs.

1932 A. F. BENTLEY *Let.* 15 Nov. in Ratner & Altman *J. Dewey & A. F. Bentley* (1964) 51, I have at length found a region of investigation in which some tentative results can be secured, and I am permitting myself to send you a copy of the resulting book, *Linguistic Analysis of Mathematics*. **1943** *Amer. Speech* XXVII. 60/1 Outline of linguistic analysis. **1945** *Mind* LIV. 195 Positivists, as is well known, do not search for answers to the philosophical questions; what they try to bring about, in *all* cases, is the disappearance of the questions by means of what they call linguistic analysis. **1949** *Amer. Speech* XXIV. 55 His charts make it possible to suggest the potentialities of 'slur' as a factor in linguistic analysis. **1957** J. R. FIRTH in *Studies in Ling. Analysis* (Philol. Soc.) p. vi, Palatograms, kymograms..specifically keyed to the linguistic analysis. **1966** J. J. KATZ *Philos. of Lang.* iii. 16 The leading philosophical movements..have concerned themselves with what they call 'linguistic analysis'. **1945** *Aristotelian Soc. Suppl. Vol.* XIX. 7 If anyone was ever a 'linguistic analyst', surely Socrates was. **1957** G. RYLE in C. A. Mace *Brit. Philos. in Mid-Cent.* 263, I gather that at this very moment British philosophy is dominated by some people called 'linguistic analysts'. **1962** *Listener* 17 May 851/1 You might well meet a philosopher described as a linguistic analyst. **1964** E. A. NIDA *Toward Sci. Transl.* iii. 36 The linguistic-anthropological approach to meaning has in many respects paralleled developments in symbolic logic, though the immediate area of study in the two fields is different and the approach seemingly quite divergent. **1968** D. HYMES in *Internat. Encycl. Social Sci.* 354/2 Through Boas the interest became an intrinsic part of American linguistic anthropology. **1923** H. R. LANG in *Romanic Rev.* XIV. 264 It will be clear from this that the study of the charts of this linguistic atlas affords a deep insight into the various phases of the decline of the dialects of Italy. **1930** *Dialect Notes* VI. ii. 67 The Linguistic Atlas of New England will provide an organized collection of the present forms of the spoken language. **1939** *Amer. Speech* XIV. 64/2 Ten sets of 300 phonograph records representing all the present dialects of Germany. Recorded by Telefunken under the auspices of the Linguistic Atlas. **1952** DIETH & ORTON (*title*) A questionnaire for a linguistic atlas of England. **1954** G. BOTTIGLIONI in Martinet & Weinreich *Ling. Today* 261 The way in which the plan of a linguistic atlas is organized and carried out. **1975** *Times* 6 Jan. 4/7 The next project will be the publication of a complete linguistic atlas which will trace on maps not only the use of specific words but of dialect sounds as well. **1921** E. SAPIR *Lang.* iv. 62 Linguistic form may and should be studied as types of patterning, apart from the associated functions. *Ibid.* vi. 127 In dealing with linguistic form, we have been concerned only with single words and with the relations of words in sentences. **1943** *Amer. Speech* XVIII. 228 The flier forced down in Libya.., would have little interest in the linguistic form of the utterance, 'I am an American', in Arabic but he might forfeit his life by not knowing how to say it. **1964** M. A. K. HALLIDAY et al. *Ling. Sci.* i. 20 The least obvious distinction perhaps is that between grammar and lexis, since these are two aspects of linguistic form. *Ibid.* ii. 21 When we describe linguistic *form*..we are describing the meaningful internal patterns of language. **1952** *Word* VIII. iii. 275 Who but Rohlfs combines a background of solid 19th century German scholarship with a thorough training as a linguistic geographer? **1948** *Neophilologus* XXXII. 175 In the absence of English linguistic-geographical data, no more than tentative suggestions regarding the relation between English and its Continental cognates are as yet possible. **1926** *Germanic Rev.* I. iv. 281 Linguistic geography, as geography, is an aspect of human geography. **1930** *Dialect Notes* VI. ii. 74 A course in the methods and the interpretation of the results of linguistic geography. **1933** Linguistic geography [see *HISTORICAL *a.* 2 d]. **1934** H. KURATH in *Proc. Amer. Philos. Soc.* LXXIV. 228 Linguistic geography undertakes to ascertain the distribution of linguistic features (dialectal features). **1939** —— (*title*) Handbook of the linguistic geography of New England. **1954** G. BOTTIGLIONI in Martinet & Weinreich *Ling. Today* 255 Linguistic geography owes its origin to the comparative method. **1968** D. HYMES in *Internat. Encycl. Social Sci.* 359/2 Linguistic geography, or dialectology, and typological comparison, together with general linguistics, often are distinguished as well. **1944** *Amer. Speech* XIX. 135 The book includes sixteen linguistic maps and nineteen illustrations. **1951** *Mind* LX. 104 These words should rejoice the heart of any present-day Linguistic Philosopher. **1963** W. H. WALSH *Metaphysics* i. 16 The brief ascendancy of the Logical Positivists came to an end and their place was taken by the so-called Linguistic Philosophers. **1957** J. L. AUSTIN in *Proc. Aristotelian Soc.* LVII. 9 There are, I know, or are supposed to be, snags in 'linguistic' philosophy, which those not very familiar with it find, sometimes not without glee or relief, daunting. **1962** *Listener* 22 Feb. 353/1 He was reputed to be the high priest of linguistic philosophy. **1966** J. J. KATZ *Philos. of Lang.* iii. 15 The leading philosophical movements in twentieth-century philosophy have been referred to..as 'linguistic philosophy'. **1953** J. B. Carroll *Study of Lang.* iii. 70 The study of verbal behavior..has variously been called the *psychology of language*, *linguistic psychology*, or *psycholinguistics*. **1966** M. Pei *Story of Lang.* (rev. ed.) xii. 286 The number of unsolved problems in the field of linguistic psychology is tremendous. **1922** O. Jespersen *Lang.* 21 Nor did linguistic science advance in the Middle Ages. **1933** L. Bloomfield *Lang.* ii. 21 Linguistic science arose from relatively practical preoccupations, such as the use of writing, the study of literature and especially of older records, and the prescription of elegant speech. **1938** *Year's Work Eng. Stud. 1936* 27 Philology (which scholars tend more and more to call 'linguistic science' or 'linguistics'). **1971** D. Crystal *Ling.* 36 It is also sometimes called linguistic science. **1934** *Amer. Speech* IX. 88/1 Linguistic scientists will find a rich ground for study if they will stop thinking of the written or printed Standard Language as solely a secondary, or derivative, form of speech. **1921** E. Sapir *Lang.* x. 221 What are the most inclusive linguistic groupings, the 'linguistic stocks', and what is the distribution of each. **1953** Beals & Hoijer *Introd. Anthropol.* xvii. 524 As more and more languages are studied and compared intensively with each other, we may expect that the number of linguistic stocks will decrease.

B. *sb.* **b.** (Further examples.)

1847 in Webster. **1902** *PMLA* XVII. 104 Both linguistics and literature are proper university studies. **1908** H. G. Wells *War in Air* iii. § 4 He thought of himself performing feats with the sign language and chance linguistics. **1938** [see *linguistic science* above]. **1953** J. B. Carroll *Study of Lang.* iv. 113 Linguistics thus appears to have a bearing on all the social sciences. **1964** M. A. K. Halliday et al. *Ling. Sci.* i. 9 The term 'linguistic sciences' covers two closely related but distinct subjects: linguistics and phonetics. **1964** R. H. Robins *Gen. Ling.* ii. 66 The linguist..may have to rely on sciences other than linguistics and on unsystematized 'common sense'. **1972** L. R. Palmer *Descr. & Compar. Ling.* xiii. 300 There are few discussions of this subject [*sc.* etymology]..in modern handbooks of linguistics.

c. appositive and *Comb.*

1958 *College English* XX. 12/2 Linguistics-based metrical analysis. *Ibid.* 17/2 A few linguistics-manufactured accessories. **1965** *Canad. Jrnl. Ling.* Fall 40 The long history of the linguistics–literary study opposition.

linguistically, *adv.* (Later examples.)

1921 B. Russell *Analysis of Mind* viii. 141 The subject ..is introduced, not because observation reveals it, but because it is linguistically convenient and apparently demanded by grammar. **1935** B. Malinowski *Coral Gardens* II. vi. iv. 229 The magical word..has got some affinity with the name which linguistically defines the relation of man as speaker to the object addressed. **1942** *Language* XVIII. 7 Non-distinctive elements..are no more significant linguistically than is any other concurrent action of a speaker. **1956** E. H. Hutten *Lang. Mod. Physics* ii. 26 Truth is said to be linguistically neutral: whatever is true is true in any language. **1971** W. P. Robinson in W. H. Whiteley *Lang. Use & Social Change* 78 Linguistically, the code has the possibility of exploiting the full grammatical and lexical potential of the language.

linguistician. Delete *rare*⁻¹ and add earlier and later examples.

1895 E. W. Fay in *Amer. Jrnl. Philol.* XVI. 10 This identification of the earlier 'linguisticians' has been latterly abandoned. **1949** *Studies in Ling.* VII. 59, I intend to use *linguistician* regularly henceforth instead of *linguist* 'worker in linguistics'. **1950** *Ibid.* VIII. 1 To one of these, *linguistician*, I not only cannot subscribe [etc.]... This meaning, exemplified by such words as *mortician* and *beautician*, implies pretentiousness rather than precision. **1954** *English Studies* XXXV. 91 In the absence of any..description by native linguisticians, these observations by an experienced teacher of foreign students ..deserve the attention. **1967** C. L. Wrenn *Word & Symbol* 7 If..texts may be properly explained by allegory and symbolism without any exact knowledge of their language, then the English language..may as well be left to the linguisticians.

li·ngworm. Also lyngorm. [ad. ON. *lyngormr* 'heatherworm'. Cf. *lindworm.] A fabulous serpent.

1870 Magnússon & Morris tr. *Völsunga Saga* xiii. 45 The fashion and the growth of him is even as of other lingworms. **1883** J. S. Stallybrass tr. *J. Grimm's Teutonic Mythol.* II. xxi. 690 The beautiful Thora Borgarhiörtr had a small lÿngorm given her, whom she placed in a casket, with gold under him. **1972** J. Simpson *Icelandic Folktales & Legends* iii. 103 The 'Heath Snake' mentioned here, the *lÿngorm*, was a mythical creature which, like the dragon, had a particular affinity for gold.

linin. Add: **2.** *Cytology* (now chiefly *Hist.*). [a. G. *linin* (F. Schwarz 1887, in *Beitr. z. Biol. d. Pflanzen* V. 9), f. Gr. λίνον (= L. *linum*) thread.] A substance which composes the fine threads seen in interphase nuclei; a thread or network composed of this substance (see quot. 1932).

1887 *Jrnl. R. Microsc. Soc.* 979 As components of the nucleus, Schwartz [*sic*] distinguishes the following substances:—..(3) linin and paralinin, the substance respectively of the nuclear threads, the 'nucleo-hyaloplasm' of Strasburger, and of the intermediate matrix or 'nuclear sap'. **1905** *Rep. Brit. Assoc. Adv. Sci.* 567 The nucleus contains an achromatic network—the linin—in which the chromatin granules are embedded. **1925** E. B. Wilson *Cell* (ed. 3) i. 88 The [nuclear] framework itself appears to consist of two constituents, namely, a continuous 'achromatic' basis, and of more or less discontinuous granules or clumps of 'chromatin' suspended in it... The first of these was found to be oxyphilic and was accordingly designated by Strasburger as nucleohyaloplasm, by Carnoy as the plasmatic network (composed of 'plastin') and later by Schwarz ('87) as linin, a term still in common use. **1932** C. D. Darlington *Recent Adv. Cytol.* 498 *Linin*, a structural component of the nucleus. The term has been applied to the descriptions of various artefacts and has no definite meaning. **1948** W. Andrew tr. *E. D. P. de Robertis's Gen. Cytol.* vii. 137 A fine lightly staining reticulum, the linin. **1969** Brown & Bertke *Textbk. Cytol.* 574/1 *Linin*, achromatic material connecting chromioles in the interphase nucleus, in contrast to only one substance, karyotin, composing the reticulum.

lining, *vbl. sb.*¹ **5.** *lining paper* (later examples).

1938 *Burlington Mag.* July 34/2 Pasted inside [a hanging food cupboard] are the remains of a seventeenth-century lining-paper. **1962** F. T. Day *Introd. to Paper* viii. 87 Rolls of lining papers of all kinds consume a large volume of paper in many grades.

lining, *vbl. sb.*² Add: **1.** Also *lining-up*.

1940 *Chambers's Techn. Dict.* 503/2 *Lining-up*, the operation of arranging the bearings of an engine crankshaft, etc. in perfect alignment. **1959** W. S. Sharps *Dict. Cinematogr.* 107/1 *Lining up*, the setting of camera or other controls, in order to obtain a correctly framed picture. **1967** E. Chambers *Photolitho-Offset* v. 51 Modern layout and lining-up tables are in many respects similar to stripping benches or shiners, with straight edges often in the form of steel rules, and micrometer-adjustable.

4. (Later examples.)

1917 *Encycl. Relig. & Ethics* IX. 27/2 In the ordinary parish churches metrical Psalms only were sung. 'Lining out' by the 'clerk', or precentor, was the order, singing in unison without organ accompaniment the rule. **1968** P. Oliver *Screening Blues* ii. 82 'Lining out' in which a lead singer paces a line and the congregation follows with the same line or a refrain response with a linear reply.

link, *sb.*¹ **c.** Add to def.: In mod. usage sometimes treated as a singular.

1861 H. B. Farnie *Fife Coast* 115 The links lying at the house door, is a very famous one in the annals of golf. **1890** H. G. Hutchinson *Golf* xiii. 311 The links of St. Andrews..holds premier place. *Ibid.* 317 It is a good links. **1904** *Daily Chron.* 20 Aug. 9/5 On a suburban links. **1919** Wodehouse *Damsel in Distress* x. 122 His first act.. had been to ascertain whether there was a links in the neighbourhood. **1933** H. S. Colt in M. A. F. Sutton *Golf Courses* 124 When..the links is stretched..an excellent test of golf is provided. **1972** R. Quirk et al. *Gram. Contemp. Eng.* iv. 181 The following nouns invariably end in -*s*: alms..innings..links (..a golf-links).

link, *sb.*² Add: **1. e.** *to let out the links*, to act with more power, to put more into something.

1839 *Spirit of Times* 6 Apr. 54/2 The horses came to the post... At this time Oscar began to let out a few additional links, and with a desperate rush parted company with Dandy, and won the heat handily. **1868** H. Woodruff *Trotting Horse* xxxiv. 282 Lancet..in the third heat, let out the links in such a manner that he trotted it in 2 m. 25½ s. **1880** P. H. Burnett *Recoll. Old Pioneer* 110 [The buffaloes] let out a few more links, and ran much faster. **1942** *Dict. Amer. Eng.* III. 1429/1 *To let out links*, to make increased exertion or effort.

2. c. (Later examples.) Also *links of love*; (dial.) *link-hide*, *-meat*.

a **1825** R. Forby *Vocab. E. Anglia* (1830) II. 197 *Link*, ..a sausage... We call two together a *latch of links*. In some other counties, a far more correct expression is used, 'a *link* of sausages'. **1869** R. B. Peacock *Gloss. Lonsdale* 51/2 *Links*, black puddings. **1891** 'H. Haliburton' *Ochil Idylls* 133 An' links o' puddin's, black to see, An' yowe-milk kebbuck. **1895** W. Rye *Gloss. E. Anglia* 129 *Link hides*, sausage skins, the intestines of a pig prepared and stuffed... *Link-meat*, mince-meat. **1922** Joyce *Ulysses* 58 Shiny links packed with forcemeat fed his gaze. **1942** *Weekly Telegraph* (Sheffield) 28 Nov. 10/1 A war-time member of the naval service sends the following glossary, ..*links of love*, sausages. **1962** Granville *Dict. Sailors' Slang* 71 *Links of love*, sausages. Cf. *Bags of Mystery* and *Mystery Torpedoes*.

3. a. (Later examples.)

1905 H. W. Nevinson *Bks. & Personalities* 172 Link by link from its small beginning we see the fateful chain of character wrought out. **1928** *New Ventures in Broadcasting* (B.B.C.) iv. 36 A link might be established with local groups. *Ibid.* App. B. 104 There is a very close link between the broadcasting company 'Ravag' and the Vienna Society for Popular Education. **1948** *Internat. Road Federation London Bull.* July–Dec. 9/1 When that is done the first link in the United States of Europe will have been forged. **1968** *Times* 19 Feb. 6/6 The contrast between most backbench speakers and the highly professional commentators who did the links was sharp. **1975** *Sunday Times* 23 Feb. 15/3 The major Press conference announcing the link between the kidnapping and the Dudley shooting produced..more than 700 lines of inquiry.

e. A means of travel or transport established between two particular places.

1869 *Bradshaw's Railway Manual* XXI. 115 The use of the intervening link from Askerne to Knottingley..is also permanently secured. **1928** *Econ. Geogr.* IV. 221 A river link..occurs between Tura and Tavda... These links are characteristic of undeveloped regions. **1934** *Highways & Bridges* 10 July 4/1 The opening of this important section..will prove..a road link of the greatest value. **1950** *Internat. Road Federation London Bull.* Jan. 10/2 A ferry link will cross Cabot Strait. **1961** *Assessment Highway Requirements S. Wales & Monmouthshire* (British Road Federation) 19 It is advised that the link between Haverfordwest..and Milford Haven..should be widened to provide a single 3-lane carriageway. **1975** *Vogue* 1 Mar. 139/2 (Advt.), Air-link services via Barcelona, Genoa, Marseilles.

f. A means of telecommunication established between two particular points.

1911 *World's Work* XVIII. 578/2 Signals had been flashed through the air from Canada to Great Britain and ..the Atlantic was spanned by a new and invisible link. **1926** *Encycl. Brit.* III. 1047/2 The superheterodyne method..is sometimes used for the 'wireless link' between studio and transmitting station in place of the land-line. **1928** *Daily Tel.* 23 Oct. 8/3 President Coolidge, speaking over the radio-link between White House and the workshop of the great inventor, lauded Mr. Edison as the embodiment of the finest traditions of American citizenship. **1957** *B.B.C. Handbk.* 59 The vision signals from remote outside broadcast points are carried back to the main television network by BBC microwave or VHF radio links. **1962** A. Nisbett *Technique Sound Studio* i. 18 The links between the various centres may be landlines or radio links. **1964** J. K. S. Jowett in F. J. D. Taylor *Goonhilly Project* 2 A broadband link to the inland network..is used for demonstration purposes—in particular, transatlantic interchange of television programmes. **1972** *Sci. Amer.* Feb. 15/1 Microwaves do not bend with the curvature of the earth, so that for long links it is necessary to use repeaters that receive, amplify and retransmit the signal.

g. [tr. Russ. *zvenó*.] The name of a small labour unit on a collective farm in the U.S.S.R. Hence *link leader*; **link system,** a system of organizing collective farming into links.

1939 L. E. Hubbard *Econ. Soviet Agric.* xvii. 165 Each brigade was further subdivided into a number of detachments known as *svena* or links, often consisting of relations or members of families living in close proximity. **1950** *Times* 22 Feb. 3/5 Mr. A. A. Andreev..was said to have encouraged during the past 10 years the 'link' system of labour, which..is less effective..than the..'brigade' system. The article blamed the 'link' system for a shortfall in grain and sugar beet deliveries. **1950** *Soviet Studies* I. 261 Piece-work for individuals and small groups was introduced and the work of the link came to be planned... 'Link', the smaller regular working group of collective farm members (averaging about ten people). Several 'links' make a 'brigade'. *Ibid.* 290 Much benefit was derived from..consultations of link leaders. **1958** R. D. Laird *Collective Farming Russ.* iv. ix. 125 As a result of the link system, labor discipline amounted to a major problem. *Ibid.* xi. 154 The brigade leader has a much greater opportunity to effect 'labor-discipline' than did the link leader. **1965** *Economist* 18 Dec. 1283/1 The 'links' are a veiled compromise between the American type of large-scale farming and the Soviet collective method.

h. In Hockey and in Association and Rugby Football = *linkman² b (*c*). Also *attrib.*

In some examples not a clearly distinguishable technical term.

1958 Pelmear & Morpurgo *Rugby Football* viii. 319 Next came the innovation of the stand-off half, thereby making two links. **1962** G. Green in B. Glanville *Footballer's Compan.* ii. 209 Didi, floating about mysteriously in midfield, was always the master link. **1963** *Rugby World* June 24/3 Which is preferable—the fly-half as a link or as a tactical general and spearhead in attack? *Ibid.* 25/2 Neither..is primarily in the 'link' category. **1966** *Hockey Coaching* (Hockey Assoc.) ii. 111 The half-back line is the link between the forwards and backs. **1969** B. James *England v. Scotland* x. 233 The superiority of Baxter and Law, the Scottish midfield link players, over their English counterparts. **1970** *Cape Times* 28 Oct. 26/3 Finch has improved considerably since he was moved to right-back after filling the left-back and link positions. **1971** *Times* 15 Feb. 9/1 Rest were handicapped by Purdy, playing at link, with a hand which became increasingly painful.

7. link buttons, a pair of buttons linked by a thread, etc.; **link road,** a road serving to link two or more major roads or centres; **link rod,** (*a*) a rod which joins the levers on the steered stub axles of a motor vehicle; (*b*) each of the rods which connect pistons to wrist pins on the master rod in a radial internal-combustion engine; **link-verb** = Copula 1; **link-word,** any part of speech performing a linking function.

[**1834** E. W. Brayley *Graphic & Hist. Illustrator* 125 Linked Cloak Buttons.., of silver, and exactly alike.] **1895** *Montgomery Ward Catal.* 170 All our cuff buttons, except link buttons, have patent lever backs. *Ibid.*, (*caption*) Gold filled, engraved link buttons... $1.10. **1964** *McCall's Sewing* ii. 30/2 *Link buttons*, two flat buttons held together with several threads covered with blanket stitches. Used as cuff links. **1934** *Highways & Bridges* 24 July 5/2 New link road from the Bedford–Hitchin road..to the Bedford–Luton road;..a 60 ft. link road from the Bedford–Ampthill road..to the Luton road. **1948** T. Sharp *Oxf. Replanned* 9 The construction of the new Southern By-pass and the important link-road approximately along the line to Roman Way will make it even better for industrial purposes. **1961** *Assessment Highway Requirements S. Wales & Monmouthshire* (British Road Federation) 18 A similar problem exists on the link road between Treharris and Cardiff. **1970** *Milestones* Spring 35/1 A..link road from the M23 to the A23 near Gatwick airport will start from a roundabout over the motorway at Burstow. **1972** *Times* 26 Oct. 3/1 It was the eighth of 32 spans that make up the M4 link road bridge. **1925** W. Deeping *Sorrell & Son* xv. 137 The driver of the lorry..was..repeating the same words over and over again... 'The bloomin' link-rod dropped. I can't think 'ow it came to 'appen. Just when they was passin' me—too. The bloomin' link-rod.' **1928** A. L. Dyke *Aircraft Engine Instructor* ii. 14 The master rod connects to the top or No. 1 piston. The other eight pistons are connected to the eight link rods, the other ends of which bear against bronze bushings on the knuckle pins. **1929** H. T. Rutter *Mod. Motors* II. ix. 333 (*caption*) Front axle of Daimler, showing link rod. **1946** J. W. Vale *Aviation Mechanic's Engine Manual* i. 16 The master rod forms a bearing on the main crankpin and the remaining link rods form a bearing on the knuckle pin arrangement of the master rod assembly. **1970** K. Ball *Fiat 600, 600D Autobook* ix. 108/1 Remove the cotter pins from the nuts securing the head pins of both link rod and nearside track rod. **1892** H. Sweet *New Eng. Gram.* I. 94 We call such verbs *link-verbs*, because they serve to connect the predicate with its subject. *To be* is a pure link-verb, that is, a pure form-word, devoid of independent meaning. **1933** O. Jespersen *Essent. Eng. Gram.* (1939) xiii. 126 It..serves to connect this with the subject as what is technically termed a *copula* or *link-verb*. **1963** F. T. Visser *Hist. Syntax* I. iii. 191 Link-verbs like *to abide*..differ from the link-verb *to be* in that their original meaning is not entirely lost. **1871** Link-word [in Dict.]. **1892** H. Sweet *New Eng. Gram.* I. 95 Other link-words, while having the same grammatical function of connecting subject and predicate, have also definite meanings of their own. **1947** W. S. Allen *Living Eng. Struct.* 235 'Who', 'what', 'which',..etc., are very important as link-words. **1968** *Brit. Med. Bull.* XXIV. 200/2 The computer compiles lists of the words used, and nouns, qualifying words and 'link' words can be sorted out by human intervention before retrieval programs are written.

link, *v.*¹ Add: **1. a.** (Later examples.)

1928 *New Ventures in Broadcasting* (B.B.C.) iv. 31 The aim of the B.B.C. is to link together the various national systems for the benefit of the Empire. **1935** R. C. Woodthorpe *Shadow on Downs* ix. 237 Men began to put up pillars of concrete and link them easily by girders of steel. **1959** *Science* 16 Oct. 954/3 Design effort must be directed toward ensuring that records can be linked in spite of such discrepancies. **1962** K. W. Gatland *Astronautics in Sixties* xi. 344 One proposed method of linking two vehicles in orbit has been outlined.

b. (Later examples.)

1962 *B.B.C. Handbk.* 47 All these studios outside London..can be linked into the network at short notice. **1968** *Globe & Mail* (Toronto) 17 Feb. B4 One of Mercantile's selling points, particularly to the Canadian export–import community, is that we can link into the Citibank international system. **1969** L. Jensen in J. N. Rosenau *Linkage Politics* v. x. 311 How a state links itself with the external environment depends upon what it believes will maximize its power.

2. Also followed by *together*.

1922 Joyce *Ulysses* 77 Those two sluts..linked together in the rain. **1962** K. W. Gatland *Astronautics in Sixties* xi. 341 The ability of two vehicles to match speed and link together in orbit.

3. a. *to link up*, to connect, combine, join *up*.

trans. **1897** *Geogr. Jrnl.* IX. 364 The mouth of the valley ..which I visited for the purpose of linking up the rough survey Garwood and I made. **1927** E. O'Neill *Marco Millions* ii. ii. 45 Their necks, waists, and right ankles linked up by chains. **1942** *R.A.F. Jrnl.* 2 May 11 Efforts are being made to link up other countries. **1942** *Tee Emm* (Air Ministry) II. 68 The fundamental method which links up all the information obtained.

intr. **1915** H. G. Wells *Boon* 211 Every one with ideas ..had to refer to that doctrinal core, had to link up to it. **1925** A. S. M. Hutchinson *One Increasing Purpose* i. xxiv. 147 Did I tell you that or has its connection with what you said only linked up in me since we parted? **1929** *Radio Times* 8 Nov. 393/3 Music lines through Belgium to the whole of Germany..are envisaged for..1930, while it may also be possible to link up to Scandinavia through Hamburg. **1974** 'E. Lathen' *Sweet & Low* xxiii. 217 The Russians and the Americans linked up in space... Every radio in New York was tuned to that docking.

b. *to link up with* (used as in sense 3 a): (*a*) in general contexts.

1899 E. G. White *Testimonies for Church* (1904) VIII. 188 You were willing to link up with them if they would second your propositions. **1903** *Studio* XXVIII. 159/1 To discuss the efforts of the lesser known men who link up the painters of 1830 with those of 1870. **1912** *Q. Rev.* July 231 The limit is entirely a question of alighting—a problem linked up with 'variable speed' aeroplanes. **1915** H. G. Wells *Boon* 174 Here is the sort of thing that I invite the intelligent reader to link up if he can with the very natural phenomenon of [etc.]. **1922** Joyce *Ulysses* 385 Our grandam, which we are linked up with by successive anastomosis of navelcords. **1928** *Sweet Shop* Nov. 6/3 The display man..should link up his shop with the advertisement. **1930** *Times* 15 Mar. 19/4 Our company has always been linked up with the trade to..South America. **1957** E. Bott *Family & Social Network* viii. 217 Many of the individuals and groups to which an urban family is related are not linked up with one another. **1967** E. Short *Embroidery & Fabric Collage* iii. 78 Napkins, tea-cosies, etc., can be designed to link up with the tablecloth or mats.

(*b*) By some means of transport or system of communication.

1907 *Jrnl. Soc. Arts* LV. 374/1 The linking up of railway stations with outlying country districts by means of mechanically propelled road vehicles. **1909** *Chambers's Jrnl.* XII. 658/1 It is freely mooted that Berlin and Munich will also be linked-up with this system [of airships]. **1910** *Ibid.* XIII. 329/1 Switches linked it [*sc.* a monorail system] up with other lengths of line. **1934** *Highways & Bridges* 24 July 4/4 A new road..would be needed..to link up with the main road. **1937** *Discovery* May 163/2 The network of air lines which now links up the United States with Central and South America. **1961** *Assessment Highway Requirements S. Wales & Monmouthshire* (British Road Federation) 3 Wasting money on local improvements which will not in the end link up with an overall system. *Ibid.* 16 It should link up with the by-pass there.

‖ **link** (liŋk), *a.* [Yiddish, f. G. *link* left, left-handed, clumsy.] Not pious, not orthodox (in religion).

1889 *Referee* 3 Feb. 2/3 'Dolly', who was a Jewess, but one who was link rather than froom, was about forty years old. **1892** I. Zangwill *Childr. Ghetto* II. 90 'Suppose,' she said slowly, 'I wanted to marry a Christian?..if I was to marry a very *link* Jew, you'd think it almost as bad.' **1907** —— *Ghetto Comedies* ii. 380 But I am so *link* (irreligious).

linkage. Add: (Further examples.) Also, a link; an association or correlation; the process of linking or connecting (see also quots.). Also *attrib.*

1904 *Brit. & Colonial Printer* 10 Mar. 14/2 A linkage system transmits the movement to the slide bars. **1928** A. S. Eddington *Nature Physical World* xiv. 306 If the two structures were identifiable then the atom would involve a complete causal connection of the two types of phenomena. But apparently no such causal linkage exists. **1940** *Chambers's Techn. Dict.* 503/2 *Linkage* (Elec. Eng.), a measure of the product of the magnetic flux passing through a closed electric circuit and the number of turns in the circuit, the unit being one line passing through a circuit having one turn. **1957** *Educational & Psychol. Measurement* XVII. 207 (*title*) Elementary linkage analysis for isolating orthogonal and oblique types and typal relevancies. **1959** B. Higgins *Econ. Devel.* iv. xvi. 405 Any particular investment project may have both 'forward linkage' (may encourage investment in subsequent stages of production) and 'backward linkage' (may encourage investment in earlier stages of production). The task is to find the projects with the greatest *total* linkage. *Ibid.* xvii. 413 Favoring deliberate unbalancing of the economy to maximize the 'linkage' effects of investment. **1959** *Science* CXXX. 954/1 The term *record linkage* has been used to indicate the bringing together of two or more separately recorded pieces of information. **1962** K. W. Gatland *Astronautics in Sixties* xi. 344 Radar.. may be relied upon to achieve linkage of the spacecraft. **1962** *Which? Car Suppl.* Oct. 143/1 Modified carburettor and linkage to give smoother operation. **1963** F. W. Frey in L. W. Pye *Communications & Political Devel.* xvii. 301 The ratio of the number of existing power linkages..to the number of theoretically possible linkages. **1969** J. N. Rosenau (*title*) Linkage politics. **1970** *Nature* 24 Oct. 387/2 There follows a discussion of the linkages between population growth and food supplies.

b. *Genetics.* (An) association between characters in inheritance, such that if one parent has a pair of characters, there is a probability greater than 50% that any offspring inheriting one of the characters will also inherit the other, which effect is due to the two characters being controlled by alleles located on the same chromosome; formerly called (*gametic*) *coupling* (*coupling *vbl. sb.* 6 e); also, the amount or degree of this association (varying between 50% and 100%). Also *attrib.*

1912 *Biol. Bull.* XXIII. 175 There are no wingless black flies in the F_2 generation, which the Mendelian expectation calls for. Their absence can only be explained by strong linkage of the yellow factor and the factor for wings. *Ibid.* 178 There are actually 1,858 long grey flies to 916 long black, or a ratio of 2 to 1. This is the linkage ratio when two strongly or completely linked factors are concerned. **1915** T. H. Morgan et al. *Mechanism Mendelian Heredity* iii. 58 In the case of yellow and white just given the linkage between the two factors is very strong. **1928** *Hereditas* X. 126 The linkages P_1–V and B–V have been reported by Wellensiek. **1940** *Chambers's Techn. Dict.* 503/2 *Linkage group*, a group of hereditary characteristics which remain associated with one another through a number of generations. *Ibid.* 504/1 *Linkage map*, a diagram showing the position of the genes in a chromosome or group of chromosomes. **1958** *Oxf. Univ. Gaz.* 23 Apr. 892 Genetic investigations on mice with special reference to evolutionary processes and linkage. **1959** *Listener* 3 Dec. 967/1 We shall need to know how many chromosomes there are, and what may be the importance of the phenomenon called 'linkage' in keeping the genes on one chromosome together. **1970** Ambrose & Easty *Cell Biol.* x. 339 The genes [of *Drosophila*] fell into four linkage groups, which corresponded with the haploid number of four chromosomes. *Ibid.* 340 However far apart two genes are,..they will never show less than 50 per cent linkage due to multiple cross-overs.

linked, *ppl. a.* Add: **a.** (Later examples.)

1966 *Punch* 5 Oct. 506/2 Tiny terrace houses are now considered acceptable, at £12,000 to £20,000 a time, because they are called 'Town' or 'Linked' housing. **1970** P. Laurie *Scotland Yard* v. 123 Some traffic-light systems, like the linked set in Oxford Street, are extremely complicated.

c. Of industries: allied to and dependent on one another.

1942 S. Florence in H. B. Newbold *Industry & Rural Life* ii. 43 Certain industries may be linked to other industries..and all the linked industries would have to be dispersed together... An instance of such a complex of linked industries is found in Birmingham. **1961** Estall & Buchanan *Industr. Activity & Econ. Geogr.* v. 108 'Linked' or related industries often require similar types of labour skills... A further advantage is the easy interchange of materials and products between the linked establishments. **1961** E. A. Powdrill *Vocab. Land Planning* iv. 77 They [*sc.* factories] must be..near some other industry providing supplies or markets (i.e., 'linked').

linkedness (li·ŋktnės). [f. Linked *ppl. a.*] Interconnection.

1908 E. V. Lucas *Over Bemerton's* xiv. 137 (*heading*) The linkedness of life is illustrated.

linking, *vbl. sb.* Add: (Later examples.)

1929 C. J. Friedrich *Alfred Weber's Theory of Location of Industries* vi. 206 The connection is based..upon the linking of the market of the one with the other. **1957** E. Bott *Family & Social Network* iv. 99 Marriage becomes a linking of kin groups rather than..a union between individuals. **1960** J. H. Jones in E. Davies *Roads* ii. 26 Linking is the interconnection of two or more traffic signals in such a manner that the beginning of the green period at one signal is related to that of the previous signal.

Also **linking-up.**

1908 *Westm. Gaz.* 19 Nov. 10/4 The 'linking-up' Bill... The process provided for by the [Electricity Supply] bill is known as 'linking-up'. **1909** *Daily Chron.* 19 Feb. 6/5 The increase in the traffic was..the natural result of the linking-up policy adopted. **1923** J. S. Huxley *Ess. Biologist* iv. 160 This first linking-up of sex with mind produced, eventually, a large proportion of the beauty of the organic world. **1934** *Highways & Bridges* 26 June 11/3 A 'Seven-Province Program' for the linking up of 13,670 miles of road. **1958** *Britannica Bk. of Year* 213/1 This linking-up was welcomed by delegates. **1972** *Radio Times* 7 Dec. 6/1 That's the date which has been set for the linking up in space of a Russian Soyuz with an American Apollo.

linking, *ppl. a.* Add: *spec.* **b.** *Gram.* = Copulative *a.* 1. Also *linking verb*.

1935 G. O. Curme *Parts of Speech* iv. 66 *Linking verbs*... The copula..performs..the function of announcing the predicate... Its use spread because there was an absolute need of such a linking word. **1952** A. H. Marckwardt *Introd. Eng. Lang.* ii. 115 Adjectives may also be..separated from the nouns they modify by a linking or copulative verb. **1967** W. N. Francis *Eng. Lang.* ii. 58 The construction is called *linking* or *copulative*... The following are typical linking constructions. **1972** R. Quirk et al. *Gram. Contemp. Eng.* xii. 820 The verb in sentences with subject complement is a 'copula' (or linking verb), which of itself has little meaning but functions as a link between the complement and subject.

c. *spec.* in *Broadcasting* and *Cinemat.*: providing continuity between programmes, scenes, etc., as music, camera shots, commentary.

1941 *B.B.C. Gloss. Broadcasting Terms* 17 *Linking Material*, words or music forming part of a continuity structure. **1960** N. Kneale *Quatermass & Pit* ii. 63 Fade in linking music. **1962** A. Nisbett *Technique Sound Studio* vii. 129 Trimming linking music to length. *Ibid.* viii. 136 Linking narration will..be necessary. **1967** *Listener* 6 Apr. 466/3 All the undeniably pretty pictures in the outdoor location and linking shots cannot breathe life into them.

d. *linking r*: a letter r in word-final position that is normally pronounced before a following vowel but is silent before a following consonant (as in *far, far away*).

1950 J. S. Kenyon *Amer. Pronunc.* (ed. 10) 164 Observe that linking *r* is the use between words of an *r* that is spelt and was formerly pronounced. *Ibid.* 165 Linking *r* is sometimes omitted in Southern British. **1956** D. Jones *Outl. Eng. Phonetics* (ed. 8) xxi. 196 When a word ending with the letter *r* is immediately followed by a word beginning with a vowel, then a *r*-sound..is usually inserted in the pronunciation... *r* inserted in this way is called 'linking *r*'.

li·nkman². Also **link man.** [f. Link *sb.*² + Man *sb.*¹] A person serving as a link between groups of people, etc.

Quot. 1909 prob. represents an extended use of Linkman (in Dict.).

1909 J. R. Ware *Passing Eng.* 168 *Linkman* (*W. London*), general servant about kitchen or yard. **1918** *Linkman* Apr. 1 (*title*) The Linkman—a literary and artistic quarterly review of congenial interests. **1969** *Guardian* 29 July 5/5 He is to..run an advice centre for residents, acting as a day-to-day 'link man' between the people and the available social and welfare services. **1972** *Where* May/June 150/3 We need a scheme which..makes Governors better linkmen between the school and their community. **1972** *Oxford Times* 20 Oct. 24/4 (*heading*) Social services Linkman. *Ibid.*, He will act as linkman between the department and volunteers. **1973** *Times* 20 Oct. 2/3 Mr Heaton, in his closing speech, claimed the prosecution had changed their allegations concerning his role in the burglary from perpetrator of the crime to link man and alibi for Miss Dugdale.

b. *spec.* (*a*) a commissionaire; (*b*) in Broadcasting, a person providing continuity in a radio or television programme consisting of several items; (*c*) in Hockey and Association Football, a player in any of the mid-field positions.

Sense (*a*) is probably an extended use of Linkman (in Dict.).

1939 H. Hodge *Cab, Sir?* xv. 222 A commissionaire is still a linkman to us. **1947** *Gloss. Technical Theatr. Terms* (Strand Electr. & Engin. Co.) 20 *Link men*, staff engaged at the Entrances and Exits of the theatre to pass the public to and from the street. **1960** *Listener* 23 June 1114/1, I must enter a protest against commentators, interviewers, announcers, link-men..and all the glorious company of contemporary communications. **1963** *Times* 25 May 3/6 McLintock is a foil and Gibson the link man. **1965** *Daily Express* 13 Aug. 15/5 Linkmen. 'They have to sort out the initial problems in defence,' said Wade, 'and then offer themselves as the focus for a pass from defence before going forward in supporting roles to the attack.' **1966** *Observer* 16 Oct. 23/5 There will be little change of format in the 26-week run. No chat, no singing, no dancing. No 'linkman' saying 'good evening' and 'goodnight'. **1968** K. Bird *Smash Glass Image* v. 59 One of my qualities as newsreader and linkman was that I remained cool in a crisis. **1968** *Listener* 10 Oct. 469/3 They seemed much more like linkmen waiting for tips outside an expensive hotel than dangerous and purposeful revolutionaries. **1970** *F.A. News* Apr. 340/1 In the days before 'sweepers' and 'link men', Clayton was the ideal old-type 'dual-purpose' wing half. **1974** *Listener* 10 Jan. 58/3 'Well.. can you answer very briefly..is Britain really on the edge of disaster?'..Timings, for linkmen, are of course inexorable.

Link Trainer (li·ŋk trē̇·nəɹ). [f. Edward *Link*, its American inventor.] A flight simulator on which pilots are trained. Also *ellipt.* as **Link.**

1937 *Flight* 28 Oct. 416/2 Practice with a Link Trainer invariably results in a light touch upon the controls of a real aircraft whether flying blind or not. **1939** *War Illustr.* 4 Nov. 243 An ingenious apparatus used in the training of R.A.F. pilots is the Link Trainer. **1940** *Flight* 26 Dec. 548/2 After Link Trainer work, dual instruction in the air in turns, landings and spins, and ground instruction in parachutes, the pupil goes on his first solo. **1942** *R.A.F. Jrnl.* 18 Apr. 3 The lessons you learned on the Link Won't help you evade a Gremlin. **1943** *Ibid.* Aug. 36 Link Trainer Instructor....*Group* I. **1945** *Tee Emm* (Air Ministry) V. 41 Here's a nice little Link Trainer exercise. **1952** *New Biol.* XIII. 51 In some respects this apparatus resembled the Link Trainer, but for a number of reasons it was constructed so that, unlike the Link, it remained stationary. **1960** C. H. Gibbs-Smith *Aeroplane* 299 *Link Trainer*, a synthetic training device, comprising a hooded cockpit, for training in instrument flying, radio aids, etc.

link-up (li·ŋk,ʊp). Also **linkup, link up.** [f. Link *v.*¹] The act or result of linking up; *spec.* (*a*) of troops, or in a military context; (*b*) of spacecraft.

1945 H. Nicolson *Let.* 29 Apr. (1967) 452 You can imagine what an exhilarating week this has been. The surrounding of Berlin; the link-up with the Russian armies. **1945** W. S. Churchill *Victory* (1946) 121 Russian and American troops made a link-up at Torgau. **1952** C. S. Lewis *Let.* 2 Oct. (1966) 244 Much later..came the link-up between his..interest in *Arthuriana* and a new interest in Byzantium. **1958** *Time* (Atlantic ed.) 26 May 20/3 Gaillard had moved to prevent any link-up between the insurgents in Algeria and their sympathizers in France. **1965** *New Scientist* 23 Dec. 852/2 In the technique used in the *Gemini* link-up, *Gemini VI* was first placed in an orbit which was.. elliptical. **1968** K. Bird *Smash Glass Image* v. 64 What.. was the link-up between all these events and what had happened before I came away. **1969** *Daily Tel.* 17 Jan. 22 (*caption*) The link-up between the two Russian manned space-craft took place 150 miles above the earth yesterday. **1970** *Globe & Mail* (Toronto) 26 Sept. 10/3 In the Cambodian fighting, less than a mile had to be spanned for a linkup of Government forces. **1972** *Radio Times* 7 Dec. 6/3 It would be nice to believe that the space link-up is co-operation for co-operation's sake. **1973** D. Francis *Slay-Ride* vii. 78 A world-wide racing investigatory link-up, something along the lines of Interpol.

lin-lan-lone. (Later example.)

1954 J. Betjeman *Few Late Chrysanthemums* 73 The dear old village! *Lin-lan-lone* the bells (Which should be six) ring over hills and dells.

linn². (Later U.S. examples. Cf. *lynn.)

1860 M. Curtis *Woody Plants N. Carolina* 79 Southern Linn. (*T[ilia] pubescens*, Ait.)—This is confined to the Lower Districts of the Southern States. **1884** C. S. Sargent *Rep. Forests N. Amer.* 514 A good deal of black cherry, lin, and locust.

lino² (ləi·no), colloq. abbrev. of Linoleum. Also **linoed, lino'd** *adj.*, covered with linoleum.

1907 C. E. Dawson in *Process Engravers' Monthly* Jan. 15, I at last happened upon some samples of cork lino. **1920** *Glasgow Herald* 10 Apr. 4 The Earl caught the gloves, but the hat fell on the lino. **1933** R. C. Hutchinson *Unforgotten Prisoner* iii. xiv. 420 He went up the lino'd stairs. **1966** M. Patten *Home Making in Colour* 36/2 Lino is easy to lay..and is a good background for rugs and carpets. **1966** J. Chamier *Cannonball* iv. 29 The edges of his coat swept papers skidding onto the linoed floor. **1973** P. Dickinson *Green Gene* ix. 173 The lino-covered staircase. **1973** J. Wainwright *Pride of Pigs* 120 The usual collection of household goods..rugs, carpets and lino.

lino³ (ləi·no), abbrev. of Linotype.

1907 *Daily Chron.* 3 Dec. 4/4 He gave me a sketch of his paper. It was set up by 'linos'.

linocut (ləi·nŏkʊt). [f. Lino(leum + Cut *sb.*²] A design cut in relief on a block of linoleum; a print obtained from this. Hence **lino-cutting** *vbl. sb.*; **lino-cutter,** a person who makes linocuts; also, a tool used in lino-cutting.

1907 C. E. Dawson in *Process Engravers' Monthly* Jan. 14 (*title*) Lino-cuts. A new method in blockmaking for posters and other bold work. *Ibid.* 16 This work, which I call *lino-cutting*, is..so easy that almost any simple design can be cut double crown size in an evening. *Ibid.* Feb. 28 Old chapbooks..are most usefully suggestive to the lino-cutter. **1919** R. Fry *Let.* 19 Nov. (1972) II. 471 Get Pamela to show you the lino-cut she did of a swan. **1927** C. Flight *Lino-Cuts* iii. 19 The chief difficulties in lino-cut colour printing lie..in the arrangement of the design in form and colour. **1948** H. Missingham *Student's Guide Commercial Art* 71 *No. 5 lino cutter*, a sharp steel blade on a pen-nib shank, which can either be used in the normal lino tool handle or in a penholder. **1956** J. Symons *Paper Chase* xvii. 137 Round the walls were a variety of paintings ranging from *collages* to linocuts. **1972** J. Mann *Mrs Knox's Profession* xv. 116 One of those triangular blades screwed into a thick handle—a lino cutter. **1973** *Times* 6 Oct. 14/7 Miss Enid Lawrence..was a versatile artist, painter, lino-cutter... Her name is particularly associated with that of Claude Flight and the development of the lino-cut.

Linofilm (ləi·nŏfilm). *Printing.* [= *line o' film*, after Linotype.] The proprietary name of an electronic photo-composing system. (See quots.) Also *attrib.*

1956 *Trade Marks Jrnl.* 5 Dec. 1212/2 Linofilm... Phototypographic machines and tape perforating machines for use in setting and composing type photographically. Mergenthaler Linotype Company.., Brooklyn, City and State of New York, United States of America; manufacturers. **1961** *Spectator* 14 Apr. 509/3 He referred to..lino-film, and the demarcation problems which must be solved if photo-composing is to continue. **1965** J. Moran *Composition of Reading Matter* vii. 75 The Linofilm Magnetic-to-Punched Tape Converter accepts magnetic tape from standard computer and data processing equipment, and produced 15-hole Linofilm tape. **1973** S. Jennett *Making of Bks.* (ed. 5) v. 86 Most electronic photo-composing machines, however, operate from punched tape only about an inch wide, with six to nine levels or rows of perforations..—the Linofilm, taking tape of fifteen levels, is a notable exception.

Linograph¹ (ləi·nŏgrɑf). *Printing.* [= *line o'* + -graph, after Linotype.] The proprietary name of an American composing machine which casts type a line at a time.

1913 *Inland Printer* May 240 (Advt.), The Linograph is now ready. **1931** *Brit. Printer* XLIII. 229/1 The author passes in review all the different types such as the Linotype, the Typograph, the Monotype,... the American Linograph [etc.]. **1965** J. Moran *Composition of Reading Matter* vi. 59 The Linograph closely resembled the Linotype, but differed in the arrangement of the magazine, which was vertical... The Linograph works were bought by Intertype in 1940.

linography (ləinǫ·grăfi). [f. L. *līn-um* flax (see Line *sb.*¹) + -graphy.] (See quots.) Also **linograph²,** a picture produced by linography.

1888 *Jrnl. Soc. Chem. Industry* VII. 588/1 Linography. This is a name given to photographing on linen or calico, to serve as a basis for painting in oil. **1945** E. Epstean tr. *Eder's Hist. Photogr.* xxxvii. 325 Linography. Photographic reproductions (mostly enlargements) were produced on linen, to be afterwards colored, by a variation of Talbotypy... J. Lüttgens, in Hamburg, states that he used this process in 1856. **1970** *Canad. Antiques Collector* Dec. 26/1 Included in the collection were signed linographs, etchings and posters.

linolenic acid (linŏle·nik, -lī·nik). *Chem.* [tr. G. *linolensäure* (K. Hazura 1887, in *Sitzungsber. d. K. Akad. d. Wissensch.* (*Mat.-Nat. Kl.*) XCV. ii. 1055), f. *linolsäure* Linoleic *acid* with insertion of -*en* -ene.] A liquid unsaturated carboxylic acid, $C_{18}H_{30}O_2$, which is found as a glyceride in linseed and most other drying oils: 9,12,15-octadecatrienoic acid.

1887 *Jrnl. Chem. Soc.* LII. 913 The acids from drying oils contain both linolic acid, $C_{16}H_{28}O_2$, and linolenic acid, $C_{16}H_{26}O_2$. **1921** *Ibid.* CXIX. 1307 The product of the action of zinc on linolenic acid hexabromide was a mixture of α- and β-linolenic acids, although only the α-modification occurs naturally. **1951** R. Mayer *Artist's Handbk.* iii. 111 Poppy oil..owes its property of yellowing less than linseed oil to the smaller percentage of linolenic acid it contains. **1969** J. I. Routh et al. *Essent. Gen., Org. & Biochem.* xxxiii. 643 Unsaturated fatty acids such as linoleic and linolenic are essential components of cellular lipids that must be obtained in the diet, since they cannot be synthesized by the body.

Hence **linole·nate,** a salt or ester of linolenic acid.

1909 *Jrnl. Chem. Soc.* XCVI. i. 357 In the mixture of esters used about 22% was ethyl α-linolenate. **1950** *Jrnl. Nutrition* XLI. 485 When fed with suboptimum doses of linoleic acid, the resultant activity of the additional linolenate equalled that of linoleate.

Linotype. Substitute for def.: The proprietary name of a composing machine invented by Ottmar Mergenthaler (1854–99) that sets type line by line (see quot. 1892). (Further examples.) Also *attrib.* Hence **li·notyped** *ppl. a.*, of type set in this way; **li·notyper** = Linotypist; **li·notyping** *vbl. sb.*, the process of setting type by the use of this machine.

1892 A. Powell *Southward's Pract. Printing* (ed. 4) xxxii. 318 The Linotype..sets up not types, but type-matrices, and then, when a line is complete, passes the matrices on to a foundry which forms part of the apparatus, and a full line is cast... Distribution is avoided. The matter once used..is returned to the metal pot and melted ready for fresh work... The advantages of the machine are its great speed and the economy effected by melting instead of distributing. A disadvantage is the fact that.. there is no mode of correction other than to reset the whole line... Still,..the machine has decided merits. It is in use in many newspaper offices. *Ibid.* (Advt.), The result of a contest between four American Composing Machines..showed that on the Linotype the operator averaged 12,250 Ens an Hour, corrected matter. **1896** *Peterson Mag.* VI. 305/1 Stenographers, typewriters, compositors, and linotypers. **1902** A. W. Bickerton (*title*) The perils of a pioneer: a protest in linotype proof. **1902** *U.S. Census Bull.* No. 242. 73 A new departure in the art of linotyping. **1903** *Stationer, Printer & Fancy Trades' Reg.* 1 Aug. 364/2 At present linotype operators work on piece, that is to say, they are paid according to the work they do. **1908** *Westm. Gaz.* 21 Apr. 7/2 The linotyped calumny of millionaire journalism. **1911** H. S. Harrison *Queed* vii. 83 The little knot of linotypers and helpers..now listened. **1926** *Brit. Gaz.* 12 May 2/2 The *Thanet Advertiser*, Ramsgate, reports that its linotype operators have returned to work. **1946** A. Monkman in H. Whetton *Pract. Printing & Binding* iii. 33/1 As the name implies, the Linotype..is a slug-casting machine. Introduced about 1886 it is now used in practically every newspaper office and also in many general printing offices. **1951** E. Paul *Springtime in Paris* xi. 211 Comrade Rappaport, a linotype operator from *L'Humanité*. **1963** W. Clowes *Guide to Printing* iii. 22 The 'Linotype' machine incorporates both keyboard and caster. **1973** S. Jennett *Making of Bks.* (ed. 5) iv. 68 The product of the Linotype is not a line of separate letters, but a solid metal strip, or 'slug', bearing on one of the long edges the characters that go to make up the whole line. **1974** *Northern Times* (Golspie, Sutherland) 21 June 5/2 His son..was a Linotype Operator on the 'N.T.' [*sc.* Northern Times].

linoxin. Substitute for entry:

linoxyn (linǫ·ksin). *Chem.* [a. Du. *linoxyne* (G. J. Mulder *Scheikundige Verhand. en Onderzoekingen* (1865) IV. i. 120), f. L. *līn-um* flax (see Line *sb.*¹): see Oxy- and -in¹.] Any of various gelatinous or resinous substances obtained by oxidation of linseed oil by air.

1876 J. Harley *Royle's Man. Materia Med.* (ed. 6) 714 By saponification linseed oil is resolved into glycerin and linoleic acid... This, when exposed in thin layers to the air, gradually increases in weight, and is converted into.. oxylinoleic acid, $C_{16}H_{28}O_5$,H_2O. At 212° it loses water, becomes of a blood-red colour, and forms linoxyn, $C_{32}H_{54}O_{11}$. **1925** *Jrnl. Soc. Chem. Industry* 7 Aug. 407T/2 When linseed oil is exposed for five or six days of warm weather in films so thin as to contain less than 0·05 g. of oil per 100 sq. cm. area, a linoxyn can be obtained from the dried (autoxidised) product having the formula $C_{57}H_{94}O_{16}$. **1969** R. F. Lang tr. *Henglein's Chem. Technol.* 774 Linoleum serves as floor covering and contains as basic materials linoxyn (oxidized linseed oil), cork flour, and resins. **1972** *Materials & Technol.* V. xiii. 439 The insoluble linoxyn is highly cross-linked and polymerised.

linsang. Substitute for first part of def.: A civet-like mammal of the genus *Prionodon*, which includes *P. pardicolor* of south-east Asia and *P. linsang*, found in this region and Sumatra, Java, and Borneo. (Add earlier and later examples.)

1821 T. Hardwicke in *Trans. Linn. Soc.* XIII. 236 Viverra? Linsang... The general colour of the animal is a yellowish white. **1969** Ld. Medway *Wild Mammals Malaya* 91/1 Banded linsang... *Prionodon linsang*... In Malaya widespread on the mainland at all elevations but nowhere common.

Linson (li·nsǫ̆n). The proprietary name of a tough fibrous paper fabric used esp. in book-binding.

1948 *Trade Marks Jrnl.* 18 Feb. 127/2 Linson... Book-binding materials. R. & W. Watson Limited,..Renfrewshire; manufacturers. **1952** A. W. Lewis *Basic Bookbinding* ii. 11 Linson is available in numerous colours and surface finishes. **1957** *B.B.C. Handbook* 248 Mrs. Dale's Diary..ordinary edition, 6s. 3d., Linson bound, 3s. 7d. **1963** *Times Lit. Suppl.* 26 Apr. 316/1 Bound in Linson or one of its..equivalents. **1967** J. S. Hewitt-Bates *Bookbinding* (ed. 8) iii. 18 Non-woven materials such as Linson are ideal for school work.

lint¹. Add: **4*.** The material which forms the bulk of the fibres in the cotton boll (cf. *linter¹ b), which is separated from the cotton seeds by ginning and which after processing is the ordinary cotton of commerce.

1877 *Encycl. Brit.* VI. 483/1 When this [*sc.* ginning] is done there remains of the bulk, as gathered from the tree, about one-third of clean cotton fit for manufacturing purposes, and two-thirds of seed. The separation of the seed from the lint is accomplished by different methods. **1883** 'Mark Twain' *Life on Mississippi* xxxiii. 325 In sixteen hundred pounds crude cotton four hundred are lint, worth, say, ten cents a pound. **1967** Shaw & Eckersley *Cotton* iii. 10 The seed cotton..may be ginned three times but only the 'lint' from the first ginning is used in normal cotton spinning.

5. lint-head *U.S. dial.*, a worker in a cotton mill; (in contemptuous use) a person of whom one disapproves.

1933 E. Caldwell *God's Little Acre* vii. 108 I'd rather be a God-forsaken lint-head and live in a yellow company house. **1940** C. McCullers *Heart is Lonely Hunter* (1943) ii. iv. 128, I would have just been a preacher or a linthead or a salesman. **1969** 'J. Morris' *Fever Grass* xvi. 138, I didn't kill that big linthead. *You* did!

linter¹. Add: **b.** *pl.* A product composed of the short downy hairs or 'fuzz' adhering to the cotton seeds (from which it is removed by the linter), which is unsuitable for spinning into yarn and is used as a source of cellulose, etc.

1903 E. A. Posselt *Cotton Manuf.* I. 49 The fibres, short or long, thus obtained, are technically known as 'Linters' and are delivered by the condenser of the linting machine as a sheet or film. **1904** L. L. Lamborn *Cottonseed Products* iv. 50 The purpose of delinting is to remove more completely the short fibres which form the 'linters'... The products of delinting are the linters. **1927** T. Woodhouse *Artificial Silk* iii. 13 The short fibres from cotton seeds—to which the name of cotton linters has been given —are utilized for the cellulose solutions. **1967** Shaw & Eckersley *Cotton* iii. 11 The 'linters' from the second and third ginning are used in the waste trade. **1972** *Sci. Amer.* Dec. 48/2 The early man-made fibers were essentially recast molecules of cellulose originating in wood fibers or cotton linters (very short fibers). **1974** *Ibid.* Apr. 52/1 Paper..has been and is made from rags, straw, cotton linters, bagasse..and flax.

lintling, var. Lintelling *vbl. sb.*

1833 J. C. Loudon *Encycl. Cottage, Farm, & Villa Archit.* 526 The cart-sheds to have a joist..built into the wall at each pillar, and chacked to the lintling beams.

‖ **Linzertorte** (li·ntsəɹtǭɹtə). [G., f. *Linzer* adj., f. *Linz* the name of an Austrian city + *torte* tart.] A kind of tart with a jam filling, decorated on top with strips of pastry in a lattice pattern.

1906 *Mrs. Beeton's Bk. Househ. Managem.* lii. 1542 Linzertorte. (German Gâteau.) **1936** Lucas & Hume *Au Petit Cordon Bleu* 153 Linzertorte... Place a flan ring on to a baking sheet. Line carefully with the paste, fill with raspberry jam... Cover the top criss-cross with thin strips of paste. **1961** J. Heller *Catch-22* (1962) xxiv. 249 Linzer and Dobos Torten from Vienna, *Strudel* from Hungary [etc.]. **1969** H. MacInnes *Salzburg Connection* iv. 47 What chance had she ever had of being taught how to run a house or bake Linzertorte?

lion, *sb.* Add: **10. a.** *lion-king, -limb, -mask, -paw*; **b.** *lion-taming* (examples); **c.** *lion-coloured* adj.; **d.** *lion-faced, -throated* adjs.

1920 E. Pound *Hugh Selwyn Mauberley* 25 The coral isle, the lion-coloured sand. **1964** *Listener* 30 July 163/3 Sun-swept, lion-coloured plains. **1856** C. M. Yonge *Daisy Chain* ii. xix. 551 He leant on his lion-faced boy's arm, and walked down to the Minster. **1919** W. S. Maugham *Moon & Sixpence* lv. 242 A look—how shall I describe it?—the books call it lion-faced. **1946** R. Graves *Poems 1938–45* 35 The Lion-faced Boy at the Fair. **1971** D. Beaty *Temple Tree* 232 Reincarnation. The new Lion King of Ceylon. **1972** *Times* 29 June 16/4 Their epics extol the 'Lion-king' Sundiata who founded the Manding Empire. **1885** G. M. Hopkins *Poems* (1918) 62 Why wouldst thou..lay a lionlimb against me? **1906** *Westm. Gaz.* 9 May 8/2 A large vase decorated with lion-masks. **1933** *Burlington Mag.* July 36/1 The cabriole legs with their goats' heads and lion masks. **1934** *Ibid.* Oct. p. xv/2 The tripod terminates in lion-paw feet. **1972** *Country Life* 15 June (Suppl.) 43/2 Regency Rosewood Bookcase with..brass lion paw feet. **1870** O. Logan *Before Footlights* 354 After stating that lion-taming was a gift of nature with him. **1944** *Mind* LIII. 162 Others found it [*sc.* welfare] wholly comprehended in trout-fishing and lion-taming. **1927** E. Sitwell *Rustic Elegies* 91 Where two lion-throated fountains fell. **1957** R. Campbell *Coll. Poems* II. 56 From lion-throated blooms ablaze.

11. lion comique *Obs.*, a leading comic singer in a music-hall or the like; **lion dance,** a traditional Chinese dance in which the dancers are masked and costumed to resemble lions; so **lion dancer; lion dog,** also, a dog belonging to one of several breeds resembling miniature lions in colour or type of fur, or once used for hunting in country inhabited by lions; **lion-head,** a variety of goldfish, *Carassius auratus*, having an enlarged head; **lion house,** a building in which lions are kept at a zoo; **lion-hunter** (earlier example); **lion-huntress,** a female 'lion-hunter'; **lion marmoset,** a small Brazilian monkey, *Leontideus rosalia*; = *lion-monkey*, Marikina.

1899 Beerbohm *More* 120 The Lion Comique bawled out..some such crude, conventional ditty. **1927** *Observer* 11 Dec. 8 When did the music-hall die?.. Where is the 'lion comique'..? **1937** *N.Y. Times* 10 Feb. 2 The parades, according to a spokesman from the Chinese Benevolent Association, will take the form of lion dances. **1952** W. Eberhard *Chinese Festivals* i. 57 The 'lion dance'..we still see in the streets of old Peking—and sometimes even in San Francisco's Chinatown. **1964** *Catal. National Museum Kuala Lumpur* 5/2 (*caption*) Lion dance tableau, authentically North China, is one of several in one main gallery. **1966** D. Forbes *Heart of Malaya* vi. 77 The Boat People of Hong Kong feast..with roast pig, lion dances and boat races on the twenty-third day of the third moon in our month of April. **1968** *Encycl. Brit.* VII. 33/1 In China itself there may be lotus dances, stilt dances, butterfly dances, lion dances (inspired by Buddhist stories from the parts of India where lions are known). **1927** Bredon & Mitrophanow *Moon Year* xii. 394 Akin to the strolling players are the 'Lion Dancers' who wander from village to village. Each troupe is composed of two or three mountebanks with rude but picturesque properties. **1975** *Times* 17 Feb. 3/3 London's Chinatown in Soho celebrated the Chinese new year yesterday... Lion Dancers..wound and jigged all day around Gerrard Street. **1921** V. W. F. Collier *Dogs China & Japan* xii. 183 Tibetan lion-dogs are bred to resemble lions, and they, like the Chinese, appear to be willing to call any shaggy coated dog a lion-dog. **1938** E. C. Ash *New Bk. Dog* x. 430 Tibetan Lhasa Apso... This being the true Tibetan Lion-dog, golden or lion-like colours are preferred. *Ibid.* xi. 466 (*caption*) A Rhodesian Ridgeback (Lion Dog). **1958** *Bk. Dogs* (Nat. Geographic Soc.) 354/2 Ancient ancestors of the Peke were honored dogs of the imperial palace in Peking... At ceremonies two of these Lion Dogs preceded the emperor, two followed. **1971** Dangerfield & Howell *Internat. Encycl. Dogs* 284/1 Many breeds have been called lion dogs. With the exception of the Rhodesian Ridgeback, these have all been small dogs with a real, or imagined, lion-like aspect. **1928** *Daily Express* 5 July 8 Fancy goldfish can be very expensive... I paid £30 each for a lionhead and an oranda recently. **1972** Y. Matsui *Goldfish Guide* xi. 176 The premier goldfish is the Ronchū or Lionhead. **1895** C. J. Cornish *Life at Zoo* 64 The present Lion House, with its fine outdoor summer palaces, and its indoor winter cages..seems to leave nothing to be desired. **1909** *Westm. Gaz.* 30 Jan. 7/1 We had better get them to the lion-house. **1974** *Times* 29 Apr. 10/8 Work to replace the outdated Lion House could not be started. **1829** R. C. Sands *Writings* (1834) II. 199 During the interval..two lion-hunters..came into the box and introduced themselves. **1828** Scott *Jrnl.* 1 July (1941) 271 A professed lion-huntress, who travels the country to rouse the peaceful beasts out of their lair. **1850** Thackeray in *Punch* 24 Aug. 89 The Lion-Huntress of Belgravia. Being Lady Nimrod's Journal of the Past Season. **1926** A. Huxley *Two or Three Graces* 229 The old familiar stories about that famous lion-huntress were being repeated. **1906** *Westm. Gaz.* 19 Feb. 12/2 Here may be seen the beautiful lion marmoset from Brazil. **1936** E. G. Boulenger *Apes & Monkeys* vii. 195 Of the long-tusked marmosets, two of the most ornate are the 'emperor' and the 'lion'... The lion is one of the most vividly coloured of all mammals, the long silky fur being of a flaming orange hue. **1965** *Amer. Jrnl. Physical Anthropol.* XXIII. 261 (*title*) The skull of the lion marmoset, *Leontideus rosalia* Linnaeus.

lip, *sb.* Add: **1. d.** = Embouchure 3; the condition or strength of a wind instrumentalist's lips.

1889 in *Cent. Dict.* **1933** *Metronome* July 26 He's got the ideas, but his lip's weak yet. **1960** *Jazz Rev.* Nov. 10 My lip went bad after a year in the Earl Hines band. **1972** *Rolling Stone* 9 Nov. 10/2 Having not played for several months, Miles had lost the eternally fragile trumpeters' lip.

2. *a stiff upper lip* (later examples).

1969 C. Booker *Neophiliacs* vi. 134 The tradition of 'stiff upper lip' epics looking back to wartime greatness. **1973** G. Greene *Honorary Consul* i. i. 11 *Machismo*..the Spanish equivalent of *virtus*..had little to do with English courage or a stiff upper lip. **1973** *Guardian* 10 Feb. 3/3 Stiff upper lip all round on Mrs Gandhi's taunt. **1973** *Times* 24 Dec. 14/4 (*heading*) How the Italians are facing up to austerity with a stiff upper lip.

3. d. A lawyer, esp. a criminal lawyer. *U.S. slang.*

1929 *Sat. Even. Post* 13 Apr. 54/3 A lawyer is a mouthpiece or a shyster or a lip. **1930** *Amer. Mercury* Dec. 456/2 Get a lip for a writ an' I'll lam. **1950** H. E. Goldin et al. *Dict. Amer. Underworld Lingo* 127/1 The lip took a hundred skins (dollars) and never showed (appeared) in court.

6. a. (*b*) Freq. in terms of Phonetics, as *lip-action, -closure, -consonant, -position, -protrusion, -rounded* adj., *-rounding, -spreading*.

1933 L. Bloomfield *Lang.* vi. 107 It is relative tenseness, too, which in addition to lip-action, makes the Italian vowels very different from those of English. **1922** O. Jespersen *Lang.* 278 That lip-closure which is an essential part of the ordinary [m]. **1867** Lip consonant [see *back *a.* 1 c]. **1877** H. Sweet *Handbk. Phonetics* 32 Lip. S.G. *w* in '*wie*', '*wo*' is an example of a pure lip consonant. **1929** *Amer. Speech* IV. 414 The lip position of the Gascon sound was identical with that of the Parisian *b*. **1932** D. Jones *Outl. Eng. Phonetics* (ed. 3) xxi. 177 A correct English ʃ may be acquired by..retracting the tip of the tongue and exaggerating the lip-protrusion. **1921** E. Sapir *Lang.* viii. 186 In *foti* 'feet' the long *o* was colored by the following *i* to long *ö*, that is, *o* kept its lip-rounded quality. **1910** *Mod. Lang. Rev.* V. 93 The lip-rounding is less energetic. **1950** D. Jones *Phoneme* xxxi. 224 A.. case..for representing the Japanese u by w on the ground that it has less lip-rounding than the European u-sounds. **1964** R. H. Robins *Gen. Ling.* 97 The English front vowels are mostly accompanied by lip-spreading. **1964** P. Strevens in D. Abercrombie et al. *Daniel Jones* 121 Pupils may be exhorted to..'smile a little as you say that', in order to achieve voicing or lip-spreading.

7. lip-brush, a small brush used to apply lipstick; **lip-click,** a clicking noise made with the lips; **lip gloss,** a glossy cosmetic applied to the lips; **lip-line,** the outline of a person's lips; **lip microphone** (see quot. 1941); **lip pencil** (see Pencil *sb.* 2 c); **lip-print,** the imprint made by a person's lips; **lip-read** *v. trans.* and *intr.*, to apprehend (someone, or what someone says) by observing the movement of the lips; so **lip-reader; lip-rouge,** red cosmetic for the lips; **lip-smacking,** the act of smacking one's lips (see Smack *v.*² 1); also as *ppl. adj.*; **lip-sync(h), -synchronization** (see quots.); so **lip-sync** *v. intr.*, **lip-synchronized** *ppl. a.*, **lip-syncing** *vbl. sb.*

1947 *Glamour* Aug. 96 If you're aiming for makeup perfection..you'll naturally want the finest in lipbrushes. **1958** Lipbrush [see *aerosol 2]. **1960** *News Chron.* 21 June 6/2 Make the most of the mouth one has... This can only be done skilfully with a lip-brush. **1933** E. Sitwell *Eng. Eccentrics* 18 The lip-clicks of the earthworms which are, it may be, amongst the earliest origins of our language. **1939–40** *Army & Navy Stores Catal.* 437/1 Max Factor...Lip gloss—2/6. **1972** *Country Life* 4 May 1127/3, I have now come to..like lip gloss... These shiny lip colours have several good uses. You can use them alone for a pale, glossy look or you can put them over a lipstick to add gloss. **1972** *Vogue* June Special 90 Blueberry Wine Lipstick matching Lip Gloss. **1951** W. Sansom *Face of Innocence* ii. 21 A smell of coffee and cigars blended with the black perfume and the red lipline of Eve's presence. **1960** *Farmer & Stockbreeder* 16 Feb. Suppl. 5/3 It is seldom the young who apply their lipstick badly. Their natural lip-line is still clear. **1941** *B.B.C. Gloss. Broadcasting Terms* 17 *Lip microphone*, type of ribbon microphone designed to be held close to the mouth and to eliminate extraneous sounds reaching it from either side. **1949** *Electronic Engin.* XXI. 354 There will be a commentator's box equipped with a lip microphone. **1904** Lip pencil [see s.v. Pencil *sb.* 2 c]. **1948** *Woman & Beauty* Dec. 57 (*caption*) One of the new Gala lip pencils. **1934** 'J. Rhode' *Poison for One* ii. ii. 92 Has it [*sc.* a drinking-glass] been examined for what I may term lip-prints? **1970** *New Scientist* 3 Sept. 455/2 They have collected lip prints from 280 people, using a technique akin to finger printing. **1892** *Strand Mag.* Mar. 250/2 He..lip read the advocates who examined him. *Ibid.* 251/1 He would like to test the lad's ability to speak, and to lip read. **1906** *Chambers's Jrnl.* 29 Dec. 80/1 An arithmetic class where questions..were not only heard (or, rather, lip-read) but answered. **1946** *Sat. Even. Post* 9 Mar. 10/2 Tele addicts contend they see football better than from any seat in the stadium, and can lip-read the signals. **1962** A. Nisbett *Technique Sound Studio* x. 170 My deaf friend preferred to switch it off and lip-read. **1973** C. Carfax *Sleeping Salamander* vii. 66 Watching, I lipread rather than heard his words. **1912** *Strand Mag.* Jan. 15/1 In those days this youthful lip-reader had no name for her gift of seeing speech. **1941** V. Woolf *Between Acts* 242 There was Dodge, the lip reader,..a seeker like her after hidden faces. **1974** R. C. Dennis *Conversations with Corpse* ii. 10 A pity you aren't a lip reader. **1926** Maines & Grant *Wise-Crack Dict.* 11/1 *Leave a good impression*, use lots of lip rouge. **1930** J. Dos Passos *42nd Parallel* i. 76 She smeared liprouge on his nose. **1947** N. Lindsay *Halfway to Anywhere* vii. 124 With lip smacking over swigs of cooking sherry, it was felt that luxury could go no farther. **1958** [see *capsulize *v.*]. **1966** Auden *About House* 21 Lip-smacking Imps of mawk and hooey. **1961** A. Berkman *Singers' Gloss. Show Business* 55 *Lip sync*, to move the lips in synchronization with a recorded sound; to pantomime with a recording. **1970** M. Tormé *Other Side of Rainbow* (1971) iii. 50 A decision was made to prerecord one of Judy's songs, which she would lip-sync on the show. While lip-syncing is anathema to most singers, it was Judy's particular teacup. **1972** *Cinema Rising* Aug. 2/2 A feature-length Western, and a lip-synch musical. **1959** W. S. Sharps *Dict. Cinematogr.* 107/1 *Lip synchronization.* Abbreviated to Lip sync. The recording of sound, usually at the same time as its associated picture, so that on projection of the completed film, the words uttered synchronize exactly with the performer's lip movements as shown in the picture. **1957** Manvell & Huntley *Technique Film Music* ii. 27 *The Jazz Singer*..introduced in certain sequences lip-synchronized singing by Al Jolson.

lip, *v.*¹ Add: **2. c.** To insult, abuse, be impudent to (someone). *dial.* or *colloq.*

1898 B. Kirkby *Lakeland Words* 93 He lipt mi rarely. **1902** *Eng. Dial. Dict.* III. 618/1 He's lipt mi as Ah was never lipt afoor. **1941** *Penguin New Writing* III. 65 Young Ernie was lippin' me just before you come in. **1972** A. Draper *Death Penalty* ii. 18 If anyone lips you, just swallow it.

lipase (li·pē̆iz, -s). *Biochem.* [a. F. *lipase* (A. A. M. Hanriot 1896, in *Compt. Rend.* CXXIII. 753), f. Gr. λίπ-ος fat: see *-ase.] Any enzyme which catalyses the hydrolysis of fats and oils to fatty acids and alcohols; *esp.* one present in the pancreatic juice.

1897 *Jrnl. Chem. Soc.* LXXII. ii. 150 The active enzyme, for which the name *lipase* is suggested, is also capable of acting..on the natural oils and fats. **1946** *Nature* 14 Sept. 375/1 It thus appears that there are at least two distinctly different enzymes (or enzyme systems) present in these glycerol extracts: (1) a lipase, hydrolysing esters of glycerol..; and (2) an esterase, hydrolysing esters of lower alcohols than glycerol. **1955** *Sci. Amer.* Oct. 70/2 Some enzymes (e.g., amylase and lipase) work in the digestive system, breaking up the crude food material into simpler parts which can be transported by the bloodstream. **1970** R. W. McGilvery *Biochem.* xvi. 342 The globule of stored triglyceride within an adipose tissue cell is not attacked by lipoprotein lipase... Breakdown of the stored material depends on the action of other lipases within the cell.

lipe, var. *lype.

lipid (li·pid). *Biochem.* Also **-ide** (-əid). [a. F. *lipide* (G. Bertrand 1923, in *Bull. de la Soc. de Chim. biol.* V. 102), f. Gr. λίπ-ος fat: see -ide.

For the origin of the now more common form *lipid* see *Chem. & Engin. News* (1952) 5 May 1910.]

1. Any of the large group of fats and fat-like compounds which occur in living organisms and are characteristically soluble in certain organic solvents but only sparingly soluble in water; it is generally taken to include esters of higher aliphatic acids, together with various groups of related and derived compounds, and freq. also steroids and carotenoids.

1925 W. R. Bloor in *Chem. Rev.* II. 244 Three terms have been suggested for the group, namely, 'Lipins' by Gies and Rosenbloom, 'Lipides' by the International Congress of Applied Chemistry, and the old term 'Lipoids' by the author. The term lipins has been used in a different sense by Leathes, and was later adopted by McLean in his monograph as a name for a subgroup containing the cerebrosides and the phosphatides. The term lipoids is understood by many to exclude the fats, although used in the wider sense by many workers on the Continent. For these reasons, and for the sake of uniformity, the author recommends the use of the term Lipides as the general group name. **1927** M. Bodansky *Introd. Physiol. Chem.* iii. 49 The term 'lipides' or 'lipids'..referring to the fats and fat-like substances. **1937** *Nature* 6 Nov. 787 'Lipides' may be confused with 'lipins', and both these words together with 'lipoids' generally signify substances of a fat-like nature yeilding on hydrolysis fatty acids or derivatives of fatty acids, and containing in their molecule either nitrogen or nitrogen and phosphorus... Some confusion may therefore arise since 'lipids' as used by the author signifies not only the above substances but also the simple fats and waxes and even the sterols. **1946** W. R. Fearon *Introd. Biochem.* (ed. 3) x. 172 The third great family of bio-organic compounds, namely, the lipides. **1954** A. White et al. *Princ. Biochem.* xviii. 453 Lipids are the most concentrated source of energy to the organism, yielding per gram over twice as many calories as do carbohydrates and proteins. **1955** J. A. Lovern *Chem. Lipids* ii. 37 Lipids by definition are soluble in the 'fat' solvents, such as ether, alcohol and chloroform. **1958** G. A. Maw *Aids to Org. Chem. for Med. Students* (ed. 5) p. xxxiv, Substances such as paraffin wax, petroleum jelly and mineral oils, although of a fatty or oily nature, are not classified as lipids since they do not occur in living material. **1961** *Chem. Abstr.* LV. 27470 Chemistry of lipides. **1968** Passmore & Robson *Compan. Med. Stud.* I. x. 5/2 The steroids form a large class of lipids, which includes the hormones of the adrenal and sex glands, vitamin D and the bile acids. **1973** *Sci. Amer.* Aug. 89/1 Today there are 10 well-characterized human diseases that are known to be caused by the excessive accumulation of lipids in tissue cells.

2. *attrib.* and *Comb.*, as *lipid storage* (freq. used *attrib.*, designating a disorder otherwise known as a *lipidosis); *lipid-soluble* adj.

1964 G. H. Haggis et al. *Introd. Molecular Biol.* vi. 151 Lipid-soluble substances are substances with a relatively high solubility in oils. **1955** H. J. Deuel *Lipids* II. vi. 623 (*heading*) Lipid storage under abnormal conditions. **1960** *Jrnl. Neurol., Neurosurg. & Psychiatry* XXIII. 211/2 The group of so-called lipid storage disorders..includes.. Hand-Schuller-Christian disease, gargoylism, Niemann-Pick's disease, Gaucher's disease, and amaurotic family idiocy. **1973** *Sci. Amer.* Aug. 88/1 When the enzyme that catalyzes a particular lipid reaction is inactive or absent, excessive amounts of that lipid begin to accumulate in certain tissues. In human beings lipid-storage disorders often result in mental retardation and enlargement of the spleen and the liver. Most of the known lipid-storage disorders are fatal.

lipidosis (lipidōᵘ·sis). *Med.* Pl. **lipidoses.** [f. *lipid + -osis.] Any disorder characterized by an excessive accumulation of a lipid in certain tissues.

1941 S. J. Thannhauser *Lipidoses* i. 2 A summary of the present knowledge..of lipid substances is.. an appropriate introduction to the group of diseases, which are called 'lipoidoses' or 'lipidoses'. **1961** R. D. Baker *Essent. Path.* iv. 47 The lipidoses are conditions in which fatty substances collect in macrophages, sometimes as part of the macrophage (reticulo-endothelial) system, sometimes in ectopic clusters of these cells and rarely in other types of cells. These diseases are also termed lipid-storage diseases. *Ibid.* 49 Niemann-Pick's Disease. This is a truly systemic lipidosis of the macrophage system, occurring typically in infants, with splenomegaly and hepatomegaly. **1966** Wright & Symmers *Systemic Path.* I. v. 245/1 Different lipids are concerned in each of the primary lipidoses, a different enzyme failure presumably being characteristic of each disease.

lipin (li·pin). *Biochem.* Also **-ine.** [f. Gr. λίπ-ος fat + -in[1], -ine[5].] † **a.** Any lipid containing nitrogen. (Originally used more specifically: see quot. 1910.) *Obs.* **b.** = *lipid.

1910 J. B. Leathes *Fats* 3 Lipines [will be used] to denote compounds of fatty acids containing nitrogen but no phosphorus or carbohydrate group. **1911** Rosenbloom & Gies in *Biochem. Bull.* I. 51 We venture to propose the accompanying chemical classification of fats and the many substances related to or resembling them, which collectively can be very appropriately and conveniently called lipins. **1937** [see *lipid 1]. **1946** *Q. Jrnl. Microsc. Sci.* LXXXVII. 444 The lipines used in this investigation were lecithin, cephalin, sphingomyelin, and galactolipine, all obtained from the brains of sheep. **1952** *Chem. & Engin. News* 5 May 1910/1 'Lipide', however, has proved popular, replacing the older 'lipoid' and 'lipin', with a better defined meaning.

Lipiodol (lipəi·ŏdǫl). *Med.* Also **lipiodol.** [f. Gr. λίπ-ος fat + Iod(ine *sb.* + -ol.] A proprietary name of a liquid containing about 40% iodine which is obtained by treating poppy-seed oil with iodine and is used as a contrast medium in radiography.

1923 *Brit. Med. Jrnl.* 4 Aug. 174/1 Between 1 and 2 c.cm. of lipiodol is injected into the spinal theca through a suboccipital puncture. **1925** *Trade Marks Jrnl.* 20 May 1096 Lipiodol... A pharmaceutical preparation for human use. Laurent Lafay,..Paris, France; chemist. **1928** Chandler & Wood *Lipiodol in Diagn. Thoracic Dis.* i. 1 The properties of lipiodol are opacity to the X-rays, a high specific gravity, an absence of any irritating action on mucous membranes, a reputed antiseptic and analgesic action, lubrication, and so firm a combination of the iodine with the oil that none of the ordinary effects of iodine are experienced. **1972** *Acta Chirurg. Scand.* CXXXVIII. 481/1 Oily contrast medium (lipiodol) injections from 10 to 30 ml were given in an attempt to demonstrate tumours..in the liver.

Lipizzan, Lippizan (li·pitsăn), *a.* [See -an.] Of or pertaining to Lipizza or Lippiza, the home of the former Austrian Imperial Stud, esp. designating a strain of horse originally bred there. So **Lipizza·na, Lippiza·na, -a·ner,** a horse of this breed; also *attrib.*

[**1911** M. C. Grimsgaard *Orig. Handbk. Riders* Suppl. 320 (*caption*) 'Muestoso-Moschina', a Lipizza stallion, one of the eight famous stallions at the..Spanish Court Riding-School in Vienna.] **1928** *Observer* 17 June 27/3 Twelve of the famous Lippizaner horses. *Ibid.*, The Lippizana is a perfectly separate and peculiar breed of horse, in appearance much like an Arab, but..more massive. **1946** P. Bottome *Lifeline* x. 100 Each horse was..a *Lippizaner.* **1954** A. Podhajsky *Spanish Riding School* (ed. 2) 4/1 General Patton..fulfilled the request made to him that the Lipizzan stud..be brought back to Austria. **1967** A. Arent *Gravedigger's Funeral* (1968) xii. 189 A Lippizaner, a white stallion from the Spanish Riding Academy in Vienna. **1971** *Islander* (Victoria, B.C.) 30 May 2/4 Those beautiful leaps and dancing steps made by the white Lipizzan stallions of the Spanish Riding School. **1972** *Guardian* 4 Oct. 4/6 The 58 photogenic white Lippizaner stallions of the Spanish Riding School which has just celebrated its four hundredth anniversary.

† **lip-lap** (li·plæp). *Obs.* Also **liplap.** [Native name.] In the Dutch East Indies, a half-caste or Eurasian; a child born in the East Indies.

1798 S. H. Wilcocke tr. *Stavorinus's Voy. E. Indies* I. ii. v. 315 Children born in the Indies, are nicknamed *liplaps* by the Europeans, although both parents may have come from Europe. **1893** *Academy* 11 Feb. 122 The acclimatisation..of liplaps, signos, and mannas, as the Dutch half-castes are variously called.

lipo-. Add: Also used in forming terms in *Biochem.* and other fields.

lipæ·mia (in Dict. as var. *lipohæmia*) (later examples); hence **lipæ·mic** *a.*; **lipoamide** (lipo,ǣ·məid) *Biochem.*, the amide of lipoic acid; **li·poate** *Biochem.*, the anion, or a salt or ester, of lipoic acid; **lipocaic** (-kēi·ik) *Biochem.* [see quot. 1936], a substance extracted from the pancreas which is found to prevent the accumulation of fat in the livers of animals from which the pancreas has been removed; **lipocho·ndrion** (pl. **-cho·ndria**) *Cytology* [ad. G. *lipochondrie* (E. Ries 1935, in *Zeitschr. f. Zellforschung u. mikrosk. Anat.* XXII. 528), f. Gr. χονδρίον, dim. of χόνδρος granule, prob. after G. *mitochondrie* *mitochondrion], a lipoid granule in the cytoplasm, esp. one seen in live preparations and possibly related to the Golgi apparatus; so **lipocho·ndrial** *a.*; **li·pochrome** *Biol.* [ad. G. *lipochrom* (C. F. W. Krukenberg *Vergleichend-physiologische Studien* II. iii. 93), f. Gr. χρῶμ-α colour], any of various mainly yellow or red pigments which are found naturally in both plants and animals and which are soluble in fats or fat solvents (see quot. 1951); **lipofu·scin** [Fuscin], any of various brownish pigments of animals, esp. those characteristically deposited in the cells during old age; (α-)**lipo·ic acid** *Biochem.*, a carboxylic acid, S–S–$CH_2CH_2CH(CH_2)_4COOH$, found in yeast and liver extracts which is a cofactor in the decarboxylation of pyruvate *in vivo*; **lipolytic,** substitute for def.: having the property of decomposing or hydrolysing fats (later examples); hence **lipo·lysis,** the hydrolytic breaking down of fat; **lipoly·tically** *adv.*; **li·pophile, lipophi·lic** *adjs.* [-phil, -phile], having an affinity for lipids; readily dissolving, or soluble in, lipids; **lipopho·bic** *a.* [-phobic], tending to repel lipids; not readily soluble in lipids; **li:popolysa·ccharide** *Biochem.*, any complex containing lipid and polysaccharide moieties; **lipopro·tein** *Biochem.*, any complex containing lipid and protein moieties, *spec.* one which is soluble in water or salt solution (as distinct from a proteolipid); **li:posarco·ma** (pl. **-o·mata**) *Path.* [Sarcoma], a sarcoma of fatty tissue; **lipo·sitol** *Biochem.* [*in)ositol], any phospholipid containing inositol in its molecule, *spec.* the one found in soy-beans; **li·poyl** *Biochem.* [-yl], the radical $C_7H_{13}S_2$·CO— which is derived from lipoic acid.

1915 *Jrnl. Biol. Chem.* XXIII. 317 Alimentary lipemia is due to nothing more than the addition of these glycerides. **1961** *Lancet* 26 Aug. 492/2 After fat ingestion, visible lipæmia normally reaches a maximum in about four hours. **1906** *Bio-Chem. Jrnl.* II. 22 Case XV, also not lipaemic, was allowed a fat-rich diet, but five days later the lipaemic condition was absent, and has remained so. **1961** *Lancet* 26 Aug. 492/2 Sera from 10 patients..were visibly lipæmic before sodium *d*-thyroxine was given. **1960** *Biochim. & Biophys. Acta* XXXVII. 314 The turnover numbers at 25° vary from 1000 with DL-lipoic acid to about 80,000 with DL-lipoamide. **1972** *Zeitschr. für physiol. Chem.* CCCLIII. 875/2 We measured the overall reaction of the multi-enzyme complex.., the decarboxylase and the lipoamide oxidoreductase. **1954** V. H. Cheldelin in Sebrell & Harris *Vitamins* III. xviii. 580 The cyclic disulfide may react to produce an acyl lipoate. **1970** R. W. McGilvery *Biochem.* xi. 215 The oxidizing agent is a coenzyme containing a disulfide bond, lipoate, which is attached to a lysyl residue in the peptide chain of transsuccinylase. **1936** L. R. Dragstedt et al. in *Amer. Jrnl. Physiol.* CXVII. 180 We have chosen the name 'lipocaic' for this substance. It is derived from the Greek words 'λιπος', 'fat' and 'καιω', 'I burn'. A more general term suggesting that the hormone plays a rôle in the utilization of fat was sought but without success. **1955** H. J. Deuel *Lipids* II. vi. 672 A number of facts lead one to question whether or not lipocaic can be classified as a hormone in the usual sense of the word. **1936** *Biol. Abstr.* X. 219 During differentiation the cells..are relatively small..; ergastoplasm is absent and the reserve material consists of yolk globules and lipochondria. **1946** *Jrnl. Exper. Zool.* CI. 361 Apart from finding yolk, pigment granules, and mitochondria, these workers [*sc.* Ries and Fischer] observed large osmiophilic fat granules... These elements were called lipochondria. **1946** [see *liposome 1]. **1950** J. R. Baker in *Proc. Linn. Soc.* CLXII. 71 Since the particular artifact studied by Golgi represents so badly what is actually present in the living cytoplasm, it no longer seems desirable to connect the great neurologist's name with this cellular constituent. A descriptive name is surely preferable. Ries's name 'Lipochondrien' (Ries, 1935) is convenient, but a Greek ending is more suitable for a word that must be used internationally. I therefore suggest *lipochondrion* (plural *lipochondria*). **1968** [see *liposome 1]. **1971** *Acta Embryol. Exper.* 43 (*heading*) The cytoplasmic inclusions of the salamander oocyte. III. Lipochondria. **1946** *Jrnl. Exper. Zool.* CI. 390 Only lipochondrial substances were involved. **1887** *Encycl. Brit.* XXII. 420/2 A red pigment of the lipochrome series. **1928** [see *lipofuscin* below]. **1951** H. J. Deuel *Lipids* I. vi. 511 The term lipochrome was proposed by Krukenberg to cover a number of animal and plant pigments which had been known by such diverse names as luteins, carotin, zooneryth rin, tetronerythrin, chlorophane, xanthophane, and rhodophane. Although this designation was originally limited to pigments with yellow or reddish tints, by implication it obviously should include any fat-soluble pigment such as chlorophyll. **1968** Lipochrome [see *lipofuscin* below]. **1923** *Chem. Abstr.* XVII. 1667 Lipofuscin is not limited to ectodermal cells, although it is found there chiefly. **1928** *Amer. Jrnl. Path.* IV. 293 The pigment present in these last organs..is a yellow to brown granular substance which is frequently tinged with fat stains, and therefore has been called lipochrome in this country, and lipofuscin in Germany. These two names are used to designate the substance in most English and American literature, but they actually represent different pigments. **1964** *Oceanogr. & Marine Biol.* II. 408 Another brown pigment [in the echinoderm *Diadema*] appears to be a lipofuscin. **1968** Passmore & Robson *Compan. Med. Stud.* I. xiii. 16/1 Lipofuscin, one of the commonest cellular pigments, is known by a variety of names (wear and tear pigment, haemofuscin, lipochrome, brown atrophy and age pigment), a selection which demonstrates its complexity as well as ignorance of its function, and indicates that it contains some lipid and some iron. **1951** L. J. Reed et al. in *Science* 27 July 93/2 This work has led to the obtaining of a crystalline compound from processed insoluble liver residues, which is highly active for the growth of *Streptococcus lactis* in the absence of acetate... This compound is being called α-lipoic acid. *Ibid.*, The crystalline compound reported in this paper is designated as α-lipoic acid to indicate that it is the first member to be obtained of a series of chemically related substances which possess acetate-replacing and pyruvate oxidase factor activity. **1962** H. A. Krebs in A. Pirie *Lens Metabolism Rel. Cataract* 351 Cofactors such as..pyridoxal phosphate, or lipoic acid may play a role in controlling reaction rates by virtue of being shared cofactors. **1968** R. F. Steiner *Life Chem.* vi. 100 The reduced form of lipoic acid contains two sulfhydryl groups..and can accept an acetyl group from active acetaldehyde. **1903** Dorland *Med. Dict.* (ed. 3) 380/1 Lipolysis. **1907** *Science* 27 Sept. 413/1 Since the bile salts are known to increase lipolysis, the effects of the sodium salts of cholic, glycocholic and taurocholic acids in *n*/500 solutions were tested on lipolytic hemolysis. **1972** *Jrnl. Lipid Res.* XIII. 651 (*heading*) Effect of cell size on lipolysis and antilipolytic action of insulin in human fat cells. **1912** *Jrnl. Amer. Chem. Soc.* XXXIV. 845 Preparations possessing lipolytic activity. **1955** H. J. Deuel *Lipids* II. ii. 15 No correlation between sex, age, or food intake and lipolytic activity of adipose tissue was observed in rats. **1972** *Jrnl. Lipid Res.* XIII. 325 (*heading*) Hydrolysis of fully esterified alcohols..by the lipolytic enzymes of rat pancreatic juice. **1912** *Jrnl. Amer. Chem. Soc.* XXXIV. 829 Lipolytically inactive substances. **1917** *Jrnl. Biol. Chem.* XXIX. p. xxvi, Experiments..resulted in the production of lipolytically active substances by the action of alkali on castor bean globulin, caesin, and gelatin. **1938** A. D. Whitehead tr. *Jordan's Technol. Solvents* i. 12 The aliphatic..and aromatic..hydrocarbons..are electrically neutral or non-polar since they contain no hydrophile groups. They are therefore hydrophobic or lipophile. **1950** *Chem. & Engin. News* 26 June 2181 (Advt.), The Atlas HLB System..is based on the hydrophile–lipophile balance of each emulsifier. **1965** *Acta Endocrinol.* XLIX. 538 Whether these findings can be attributed to the lipophile properties of the sulphatide facilitating its entrance into the cell cannot be decided. **1946** *Arkiv for Kemi, Mineral. och Geol.* XXIIa. xviii. 29 The lipophilic end should contain an aromatic structure. **1954** Jirgensons & Straumanis *Short Textbk. Colloid Chem.* ii. 16 Substances which, like rubber, polystyrene or polyvinylchloride do not contain hydrophilic groups are insoluble in water. They are composed of lipophilic..groups such as CH_3, —CH_2—, and others, which have some affinity for the molecules of fats, fat solvents and other oils. **1971** *Nature* 21 May 186/2 Morphine has a highly lipophilic molecule. **1946** G. M. Sutheim *Introd. Emulsions* i. 4 Hydrophilic substances..are named oleophobic or lipophobic. **1961** E. O'F. Walsh *Introd. Biochem.* ii. 33 The polar end of the lecithin molecule, here represented as a Zwitterion, is hydrophilic and lipophobic. **1954** *Chem. Abstr.* XLVIII. 9453 Injection of a lipopolysaccharide from *Salmonella abortivoequina* increases the phagocytic activity of the granulocytes. **1958** *Immunology* I. 181 The stimulation of non-specific immunity by lipopolysaccharides could not be correlated with the serum properdin level at the time of challenge. **1970** W. J. Lennarz in S. J. Wakil *Lipid Metabolism* v. 164 Lipo-

polysaccharides, the complex heteropolysaccharides typical of Gram-negative enteric bacteria, are currently under extensive investigation. **1909** *Chem. Abstr.* III. 82 It is probable that in fatty degeneration there is a splitting off of fat from lipoproteins of this character. **1929** *Jrnl. Immunol.* XVI. 448 The constituents in fowl sera responsible for these non-specific precipitations are indicated to be lipo-proteins and neutral fats. **1955** H. J. Deuel *Lipids* II. v. 371 The lipoproteins are widely distributed in living matter, where they occur in cell nuclei, mitochondria, cell membranes, chloroplasts, in egg yolk, in milk, and in blood. **1971** L. W. Burley in Johnson & Davenport *Biochem. Lipids* iv. 86 'Proteolipids'..differ from lipoproteins in being soluble in certain organic solvents but insoluble in aqueous solutions. **1893** Dunglison *Dict. Med. Sci.* (ed. 21) 637/2 Liposarcoma. **1916** E. H. Kettle *Path. Tumours* x. 84 Liposarcomata..are undoubtedly rare. **1979** Passmore & Robson *Compan. Med. Studies* III. xxx. 16/2 Liposarcomata are most common in old men. **1943** D. W. Woolley in *Jrnl. Biol. Chem.* CXLVII. 581 It is proposed to call the new substance soybean lipositol, since it is a lipid which contains inositol. **1949** H. W. Florey et al. *Antibiotics* II. xliv. 1386 The antibacterial activity of 50 units of streptomycin is..completely antagonized by as little as 0·2 mg. of lipositol. **1969** S. R. Williams *Nutrition & Diet Therapy* iii. 29/1 Other important phospholipids are cephalins and lipositols, which are like the lecithins except that they contain other factors in place of choline. **1960** *Biochem. Jrnl.* LXXVII. 347/1 There is a close correlation between the rates of the enzyme-catalysed oxidation of DPNH by the lipoyl derivatives used and the rates of reoxidation of the red intermediate..by the same lipoyl derivatives. **1970** R. W. McGilvery *Biochem.* xi. 215 The reaction is now complete except for the regeneration of the original disulfide bond in the lipoyl group.

lipodystrophy (li:podi·strŏfi). *Path.* Also mod.L. **li:podystro·phia.** [f. Lipo- + Dystrophy.] Any of various disorders of fat metabolism; *intestinal lipodystrophy,* a rare disease (usu. called Whipple's disease) of uncertain ætiology, chiefly of middle-aged men, presenting with joint pains, steatorrhœa, wasting, and lymph node enlargement; *progressive lipodystrophy* [tr. mod.L. *lipodystrophia progressiva* (A. Simons 1911, in *Zeitschr. f. d. ges. Neurol. u. Psychiatrie* (Orig.) V. 36)], a condition in which there is a progressive loss of subcutaneous fat, usu. from the upper half of the body while the lower half retains its fat. So **li:podystro·phic** *a.*

1907 G. H. Whipple in *Bull. Johns Hopkins Hosp.* XVIII. 391/1 In searching for a name to designate this condition great difficulty was experienced. It would seem that no suitable name can be applied to it until the etiological factor is determined. The term Intestinal Lipodystrophy is suggested. **1925** *Q. Jrnl. Med.* XVIII. 224 Progressive lipodystrophy is a rare disease characterized by symmetrical and progressive loss of subcutaneous fat over the face, neck, arms, thorax, and abdomen, with a relative or absolute abundance of subcutaneous fat over the lower limbs. *Ibid.* 231 The active stage having ended, the lipodystrophic process becomes stationary. **1933** *Lancet* 23 Dec. 1417/1 The blood chemistry in lipodystrophia progressiva is normal. **1946** *Ibid.* 18 May 730/2 There is no satisfactory classification of the lipodystrophies. *Ibid.* 731/1, I venture to argue how Ziegler's lipodystrophic case might have a similar disturbance of fat-metabolism without an obvious lipæmia. **1964** *Pediatrics* XXXIII. 609/1 The syndromes of partial and total lipodystrophy have been reviewed... For inclusion in the syndrome of partial lipodystrophy symmetrical loss of fat from the face with or without truncal fat loss, and little or no loss of distal adipose depots was required. In total lipodystrophy fat loss was generalized. **1971** *Daily Colonist* (Victoria, B.C.) 28 Dec. 2/1 Lipodystrophy means a disturbance in the metabolism of fatty tissue.

lipoid, *a.* Add: (Further examples.)

1907 *Biochem. Jrnl.* II. 22 The lipoid material being mainly composed of an ester of cholesterin. **1946** *Nature* 13 July 41/1 Solution of the odorous substance in the lipoid or aqueous phase of these flagellæ is, therefore, the point from which all theories of odour are derived. **1974** *Ibid.* 4 Feb. 293/2 The uterine or milk glands in tsetse flies..release a nutritive liquid of proteinaceous and lipoid nature for the maturing intrauterine larva.

B. *sb.* [a. G. *lipoid* (E. Overton *Studien über die Narkose* (1901) 54).] **a.** Any fat-like substance other than a true fat. **b.** = *lipid.

1906 *Jrnl. Chem. Soc.* XC. ii. 780 (*heading*) The influence of diffusibility and the solubility of lipoids on the permeability of the..cell. **1921** *Physiol. Abstr.* VI. 9 Fats and the substances resembling them ('lipoids'). **1925** [see *lipid 1]. **1933** I. Smedley-MacLean in *Ann. Rev. Biochem.* I. 135 The term lipoid is retained..to denote the ether-soluble constituents of a tissue, without regard to their nature. **1946** W. R. Fearon *Introd. Biochem.* (ed. 3) x. 173 Lipoids..These are biological compounds resembling the lipides in certain physical properties, namely, solubility in fats. **1952** [see *lipin]. **1955** Gaiger & Davies *Vet. Path. & Bacteriol.* (ed. 4) x. All the normal body tissues contain fat in one form or another, the chief forms being neutral fat, fatty acids, soaps and lipoids (lecithin, cholesterol and myelin). **1958** *Times Lit. Suppl.* 17 Jan. 34/1 Living matter consists essentially of carbohydrates (or 'sugars'), lipoids (or 'fats') and proteins. **1971** *Nature* 9 July 138/3 'Lipoid' is used instead of the more usual 'lipid' throughout [the book].

lipoidal (lipoi·dăl), *a.* [f. prec. + -al.] Resembling or containing fat.

1919 *Amer. Jrnl. Anat.* XXV. 251 Lipoidal vacuoles are becoming more abundant in the outer and especially the middle cortical zone. **1920** *Proc. R. Soc. Med.* XIII. (Path. Section) 8 The antitrypsin is lipoidal in nature. **1928** P. Bailey in E. V. Cowdry *Special Cytol.* I. xv. 493 Small globules of lipoidal substances are seen in all the cells. **1954** A. J. Marshall *Bower-Birds* ii. 10 Considerable aggregations of heavily lipoidal Leydig cells can now be seen.

lipoidosis (lipoidōᵘ·sis). *Med.* Pl. **lipoidoses.** [f. as prec. + -osis.] = *lipidosis.

1932 *Amer. Jrnl. Dis. Children* XLIV. 1117 Niemann-Pick's disease..is called phosphatide cell lipoidosis. Christian's disease..is called cholesterol cell lipoidosis. **1936** *Proc. R. Soc. Med.* XXIX. 585 The three principal types of these lipoidoses are Schüller-Christian's disease, the spleno-hepatomegaly of Gaucher and the splenohepatomegaly of Niemann-Pick. **1964** *Lancet* 13 Jan. 64/1 ..when infections, blood dyscrasias, lipoidoses, reticuloses, ..and congenital fibrosis..are excluded there remain some instances in which no obvious disorder can be demonstrated.

liposome (li·posōᵘm). *Biol.* [ad. G. *liposom* (E. Albrecht 1904, in *Verhandl. deutsch. path. Ges.* VII. 64): see Lipo- and *-some[4].] **1.** A natural globule of fat or lipid suspended in the cytoplasm of a cell.

1910 *Anat. Rec.* IV. 211 The protoplasm of renal cells, muscle fibers, etc., shows usually a large number of small more or less refractive droplets (liposomes) when examined in the fresh condition. **1946** *Jrnl. Exper. Zool.* CI. 374 In the following discussion, while being aware of the arbitrariness of the choice, the term 'lipochondria' will be used as an alternative for 'liposomes'. **1968** *Ross Cell Struct.* xxvi. 351 The lipid inclusions of amphibian embryo cells have been studied by Holtfreter and Karasaki. Holtfreter called the larger bodies liposomes, and the small ones lipochondria.

2. A minute artificial globule consisting of one or more layers of phospholipid enclosing an aqueous core, used experimentally as a model for biological membranes.

1968 Sessa & Weissmann in *Jrnl. Lipid Res.* IX. 310 (*heading*) Phospholipid spherules (liposomes) as a model for biological membranes. *Ibid.*, Throughout this review, the artificial structures will..be referred to as 'spherules', ..A. D. Bangham has used the term 'smectic mesophases' and colloquially we have called them 'liposomes' or 'Banghasomes'. As the literature dealing with these structures accumulates, the term 'liposome' is gaining favor, and should win general acceptance. **1970** *New Scientist* 11 June 511/1 In 1965 Alec Bangham..showed that phospholipids in aqueous salt solutions above the phase-transition temperature of the lipid. **1972** *Nature* 19 May..These vesicles are generally termed liposomes, spherulites, smectic mesophases, and sometimes Banghasomes (after Dr. Bangham).

lipotropic (lipotrǫ·pik, -trō·pik), *a. Physiol.* [f. Lipo- + *-tropic.] Tending to prevent or remove an accumulation of excess fat in the liver. So **lipotro·pism,** lipotropic property or phenomena.

1935 C. H. Best et al. in *Nature* 18 May 821/2 The term 'lipotropic' is used to designate substances which decrease the rate of deposition and accelerate the rate of removal of liver fat. **1950** *Jrnl. Biol. Chem.* CLX. 601 (*heading*) Growth and lipotropism. I. The dietary requirements of methionine, cystine, and choline. **1951** A. Grollman *Pharmacol. & Therapeutics* xxvii. 607 Because of their dietary deficiency, choline and, to a lesser extent, methionine and inositol have been used as lipotropic agents in cirrhosis of the liver, hepatitis, etc. **1953** *Canad. Jrnl. Med. Sci.* XXXI. 474 (*heading*) Further studies on lipotropism in the domestic duck. **1968** A. White et al. *Princ. Biochem.* (ed. 4) xxi. 503 Any material capable of contributing methyl groups for choline synthesis has the property of being lipotropic.

Lippes loop (li·pēz lūp). [f. the name of its inventor, the American physician, Jack Lippes.] An intrauterine contraceptive device in the shape of a double S.

[**1962** J. Lippes in C. Tietze *Intra-Uterine Contraceptive Devices* 71, I wanted a device such that the muscular contractions of the uterus would not press on the entire piece of plastic but on one section at a time only... With this in mind, I designed a device which I call a 'loop'.] **1964** *Lancet* 31 Oct. 958/1 The five types available are: the Margulies spiral, the Lippes loop, the Birnberg bow, the Hall-Stone stainless steel ring, and the Zipper nylon ring. **1966** E. Mears in M. Pollock *Family Planning* iv. 50 With..the Lippes loop the inserter is loaded, introduced..beyond the internal os, rotated..and the device expelled by pushing the plunger. **1967** *Time* 7 Apr. 73 [Dr. Jack Lippes] has in Buffalo came a series of double-S designs, now known as the Lippes loop, which has probably enjoyed the widest acceptance both in the U.S. (150,000 users) and overseas (up to 4,000,000). **1971** H. J. Davis *Intrauterine Devices for Contraception* vi. 95 The shield..conforms most closely to the average uterine cavity size. The Lippes loop is the next most anatomically correct, commercially available device.

lipping, *vbl. sb.*[1] Add: **2.** A strip of wood or the like fixed to the edge of a board, door, table-top, etc.; the act of fixing such a strip.

1963 *Gloss. Terms Timber (B.S.I.)* 62 *Lipping,* a strip of wood or other material applied to the edge of a flush door, table top, etc. **1966** A. W. Lewis *Gloss. Woodworking Terms* 53 *Lipping,* fixing a strip of solid wood to the edge of a board, usually by means of a tongue and groove. **1971** *Timber Trades Jrnl.* 14 Aug. 58/2 The pre-glued edging can be either conventional plastics or wood lippings.

Lippizan(a, Lippizaner: see *Lipizzan *a.*

Lippmann (li·pmæn). *Photogr.* The name of Gabriel *Lippmann* (1845–1921), French physicist, used *attrib.* with reference to a method of colour photography invented by him in which colours are produced by interference effects in an emulsion containing very fine silver halide particles.

1902 *Westm. Gaz.* 26 Sept. 4/2 Among all experts in this branch of photography it is agreed that the Lippmann interference process is the only one which gives all the colours of Nature in a direct manner, but these experts are unfortunately also agreed that it is an exceedingly difficult process to work. **1908** *Astrophysical Jrnl.* XXVII. 346 This flexible silver mirror is immediately laid, silver surface down, on a wet Lippmann plate. **1910** R. M. Fanstone *Colour Photogr.* ii. 7 The Lippmann Process presented difficulties, and was somewhat costly to work. **1942** C. E. K. Mees *Theory Photogr. Process* xxi. 872 A Lippmann emulsion..is practically grainless since very small silver halide crystals, 10 to 50 mμ in diameter, and are used for scientific purposes.. They are difficult to use, and extremely slow. **1963** F. W. H. Mueller in *Photographic Theory: Liège Summer School 1962* i. 1, 31 Lippmann emulsions, sometimes called grainless emulsions, contain very small silver halide crystals, 10 to 50 mμ in diameter, and are used for scientific purposes... They are difficult to use, and extremely slow. **1973** T. H. James in *Focal Encycl. Photogr.*..The finest grain is produced by... *Handbk. Photogr. Sci. & Engin.* xvii. 930 The finest grain is produced by the special procedures used to make Lippmann emulsions.

lippy, *a.* Add: **2.** *colloq.* or *dial.* Impertinent, insolent; talkative, verbose.

1875 W. D. Parish *Dict. Sussex Dial.* 70 *Lippy,* impertinent; apt to answer saucily. **1893** W. K. Post *Harvard Stories* 195 Ain't he getting pretty lippy? The lippy dude! **1906** *Punch* 4 Apr. 259/1 'Another of you lip-lappy as Passive Resisters. **1968** V. Canning *Melting Man* v. 128 You're a lippy bastard. **1971** K. Roberts *Classic Slum* 203 Any child who responded to a body blow like he at once disciplined as 'forward' or 'lippy'. **1971** R. Thomas *Backup Men* iv. 32 It might learn them not to be so goddamned lippy.

lipstick (li·pstik), *sb.* Also **lip stick, lip-stick.** [f. Lip *sb.* + Stick *sb.*[1] 8.] A stick of cosmetic for colouring the lips, usu. a shade of pink or red; hence, cosmetic for the lips. Also *attrib.*

1880 E. James *Amat. Negro Minstrel's Guide* 4 Prepared burnt cork, ready for use, 25 and 50 cents per box; lip sticks, 25 cts. *Ibid.* 8 An application of lipstick..around the natural part of the lips will extend that feature to a size quite remarkable. **1919** H. L. Wilson *Ma Pettengill* iii. 93 Metta was even using a lip stick! **1922** A. Bennett *Lilian* iv. vii. 116 She also knew what was the best lipstick. **1926** *Spectator* 18 Sept. 434/2 What is the matter with powder, paint and lipstick? **1928** D. L. Sayers *Unpleasantness at Bellona Club* ii. 21 She applied a lipstick brush tenderly. **1962** F. Kennedy...the tubes used for ashtrays or unwanted lipstick...The clear colours and good texture of the lipsticks. **1974** 'J. Ross' *Burning of Billy Toober* xii. 111 There were two silt lipstick cases.

Hence **li·pstick** *v. trans.* and *intr.*, to apply lipstick to (one's lips); **li·psticked** *ppl. a.*; **li·psticking** *vbl. sb.*; **li·psticky** *a.*, covered or sticky with lipstick; also *fig.*

1926 *Ladies' Home Jrnl.* Apr. 24 She..had recently lipsticked a red mouth into startling contrast to her natural pallor. **1928** *Sunday Express* 15 Apr. 15/4 She may be made of wax, with large, liquid eyes, a lipsticked mouth, and real hair. **1932** *Daily Tel.* 21 May 13/1 They chatted, smoked, 'lipsticked', read, sewed, etc. **1931** *Punch* 28 Oct. 476/1 Her chronicle of minor adventures, if occasionally a shade too lip-sticky to suit the taste, is a roughly vivacious and entertaining. **1933** J. B. Priestley *Wonder Hero* vi. 249 The first thing he did was to see Charlie a large smacking lip-sticky kiss. **1946** A. G. Macdonell *Crew of Anaconda* xxi. 260 Florinda did a little powdering and lipsticking. **1949** *Landfall* June 185 The lipsticky erasers of her daughter into informed adolescence... **1962** *Times* 2 Oct. 15/2 She would open her handbag, lipstick her mouth, and so on..home. **1966** J. B. Priestley *Salt is Leaving* i. 5 Sheila stopped lipsticking. **1969** *Daily Tel.* 3 Dec. 18/5 '..she's gone' 46 My eyes are still wandering over that lip-sticked mouth. **1974** *Listener* 17 June 69/1 Jonathan Miller used to imagine the BBC as an actual aunt, powdered and lipsticked for the camera.

Liptauer, liptauer (li·ptau,ər). Also **Liptai, Liptau, Liptoi.** [G., f. *Liptó* place-name in Czechoslovakia.] A soft cheese originally made in Hungary, usu. coloured and flavoured with paprika and other seasonings. Also *attrib.*

1902 J. T. Law *Grocer's Manual* (ed. 2) 143/2 *Bringen or Liptau Cheese,* a Continental kind made in Hungary. Constituents:—Water, 34 per cent.; fatty matter, 28; Caseine, 23; salts, 5. **1935** M. Morphy *Recipes of All Nations* 348 Liptauer cheese..which is extensively made in Germany and Austria, is of Hungarian origin. It is made from sour milk. **1955** G. Freeman *Liberty Man* i. ii. 32 The fillings came from a delicatessen. Liverwurst and liptauer cheese. **1964** *Punch* 19 Feb. 262/3 Mortadella,

lachscinken, kabanosi or Liptauer cheese. **1964** E. Huxley *Back Street New Worlds* vii. 75 In one [supermarket]..you can buy..Czech braun, gacciatori, or Liptauer cheese. **1967** T. A. Layton *Cheese & Cheese Cookery* 228 Liptoi is a goats' milk cream cheese; Liptoi is sold in Hungary packed in small bladders and owes its name to the county of Lipto (now in Czechoslovakia)... The German name of this cheese, *Liptauer,* is often used in shops and restaurants both in Hungary and abroad.

liquesce, *v.* Add: Also *fig.*, to merge *into.*

1920 *19th Cent.* Dec. 977 The perpetual tendency of privilege, royal as well as any other, to liquesce into the common stream of humanity.

liquescent, *a. transf.* (Later examples.)

1967 *Listener* 31 Aug. 287/3 Huge eyes fluctuating between fear and wonder and roguish amusement—liquescent, all light suddenly sinking out of them. **1969** *Daily Tel.* 10 Dec. 12/4 Debussy's 'Prélude à l'après-midi d'un faune' could scarcely have been bettered for delicate dynamic nuances and shapely liquescent phrasing.

liqueur, *sb.* Add: **2.** (Examples.) Also = *liqueur chocolate.*

1907 *Yesterday's Shopping* (1969) 937 Table glass services... 12 Clarets... 12 Champagnes... 6 Liqueurs... 12 Tumblers. **1925** *Heal & Son Catal.: Glass,* Table glass. ..Champagne..Claret..Sherry..Port..Liqueur..Tumbler. **1965** E. Brown *Big Man* xi. 96 Andy took from his shoulder bag the box of liqueurs. **1967** K. Giles *Death in Diamonds* iii. 59 Elizabeth was..eating chocolates. 'My present to myself,' she said, 'Austrian Liqueurs and none for pigs.' **1968** G. Beard *Mod. Glass* ii. 101 (*caption*) Automatic production six sizes of 'Five Star' 1 oz. liqueur to 12½ oz. goblet.

3. liqueur chocolate, a chocolate with a liqueur filling; **liqueur-glass** (earlier and later examples).

1904 'Saki' *Reginald* 84 Some liqueur chocolates had been turned loose by mistake among the refreshments—really liqueur chocolates, with very little chocolate. **1950** O. Blakeston *Pink Ribbon* vii. 76 Some flying beetle, rather like a small liqueur chocolate, zoomed past my ear. **1969** J. Elliot *Duel* i. iii. 74 A small box of liqueur chocolates..shaped like miniature bottles. **1850** E. Ruskin *Let.* 9 Mar. in M. Lutyens *Effie in Venice* (1965) i. 155 The little Liqueur glasses of Rossolio. **1904** 'Saki' *Reginald* 16 There are liqueur glasses, and crystallized fruits. **1962** E. O'Brien *Lonely Girl* iv. 48 Joanna opened the wine and served it in liqueur glasses to make it go far. **1975** *Sunday Times* (Colour Suppl.) 23 Feb. 11/4, 6 sherry glasses which when half filled double superbly as liqueur glasses.

liquid, *a.* and *sb.* Add: **A.** *adj.* **6.** (Later examples.)

1930 J. M. Keynes *Treat. Money* II. xxv. 67 On the other hand, bills and call loans are more 'liquid' than investments. **1962** C. H. Kreps *Money* i. i. 12 The assets categorized as near money are those that are highly liquid; that is, they are convertible into money quickly, easily, and without loss. **1974** *Times* 12 Nov. 14/4 These small engineering businesses..are now..low on liquid assets.

7. liquid air, air in a liquid state; **liquid compass,** a form of magnetic compass used in ships in which the card and needle are mainly supported by floating in a bowl filled with liquid; **liquid controller** *Electr.* = *liquid rheostat;* **liquid crystal** *Physical Chem.* [tr. G. *flüssiger krystall* (O. Lehmann 1890, in *Ann. d. Physik und Chem.* XL. 404)], a turbid liquid that exhibits double refraction (indicative of internal anisotropy and hence some degree of ordering in its structure, as in an ordinary crystal) and exists as a distinct state of certain pure substances between the melting point and some higher temperature, at which it becomes an ordinary liquid; **liquid extract** (occas. as one word) *Pharm.* = *fluid extract* s.v. *fluid *a.* 1 a; **liquid fire,** any very 'fiery' (in taste) or highly combustible liquid, now *esp.* one that can be sent as a burning jet in warfare; **liquid fuel,** fuel that is a liquid, now esp. as used in rocketry; so **liquid-fuelled** *a.*; **liquid glue,** glue that keeps a liquid form till applied; **liquid lunch** *colloq.*, a midday meal at which drink rather than food is consumed; **liquid manure** *Hort.*, a water extract of manure used as a fertilizer; **liquid oxygen,** oxygen in a liquid state; **liquid paraffin** *Pharm.*, an almost tasteless and odourless oily liquid that consists of hydrocarbons obtained from petroleum and is used as a laxative and in dressings; **liquid petrolatum** *N.Amer.* = *liquid paraffin;* **liquid rheostat** *Electr.*, a rheostat which uses an electrolyte solution as the resistive element; **liquid soap,** soap in liquid form; **liquid starter,** a liquid rheostat used as a starter of an electric motor.

1899 *McClure's Mag.* XII. 397 (*heading*) Liquid air. A new substance that promises to do the work of coal and ice and gunpowder, at next to no cost. *Ibid.* 399/1 A liquid-air engine, if powerful enough, will compress the air and produce the cold in my liquefying machine exactly as well as a steam engine. **1901** *Daily Colonist* (Victoria, B.C.) 6 Oct. 3/4 The most sensational thing in the scientific world today is liquid air. **1925** E. F. Norton *Fight for Everest, 1924* 91 Waiting for us with hot soup in a liquid-air flask. **1946** *Nature* 20 July 105/1 A laboratory liquid-air plant utilizing Freon-12 as a pre-cooling fluid has been designed and is now being constructed. **1865** *Rep. Brit. Assoc. Adv. Sci. 1864* 14 The distinctive peculiarities of the liquid compass are an air-tight metallic case, within which is placed the magnetic needle, and of such size and weight as to be of very nearly the same specific gravity as the liquid in which it is intended to float. **1959** E. C. Goldsworthy *Seamanship & Navigation* iii. 41 This liquid compass is preferred to the dry-card compass where there is much vibration, as the liquid damps the oscillations of the card caused by the movements of the vessel. **1916** C. C. Garrard *Electr. Switch & Controlling Gear* v. 357 The rating of the liquid controller is, therefore, based upon the maximum horse-power dissipated. **1957** W. J. John *Mod. Electr. Engin.* II. iii. 58 (*caption*) A liquid controller of the type shown here provides smooth control from normal down to crawling speed. **1891** *Jrnl. Chem. Soc.* LX. i. 250 Liquid crystals, when heated between cover glasses slightly above the point where they pass into ordinary liquids, retain on cooling the original direction of their optical axes. **1938** *Ann. Reg. 1937* 352 Tobacco mosaic virus is a nucleoprotein of special character, existing as mesomorphic fibres in the cell sap and spontaneously forming liquid crystals of gigantic cell-size when isolated. **1962** *Times* 30 Apr. 7/1 When detergents are mixed with a limited amount of water, several distinct 'mesomorphic' phases may be formed. These phases are also known as 'liquid crystals' because some of their properties are akin to those of a solid while others are characteristic of a liquid. **1972** *Physics Bull.* May 279/1 These fascinating compounds can no longer be regarded as freaks of nature, for out of every two hundred organic compounds at least one may be a liquid crystal. **1864** *Brit. Pharmacopœia* 219 (*heading*) Liquid extract of Bael. **1930** J. W. Cooper *Pharmacy* xiii. 115, 1,000 grm. of couch grass is used to produce 1,000 mil. of liquid extract. **1935** Liquidextract [see *fluid *a.* and *sb.* A. 1 a]. **1968** *Biol. Abstr.* XLIX. 1161/1 This liquid extract [from the bark of *Oroxylon indicum*] lowered the vascular permeability of rats sensitized with egg protein. **1604, 1667** Liquid fire [in Dict., sense A. 1]. **1815** J. Smith *Panorama Sci. & Art* II. 579 It is in this way that the various kinds of cordial waters are prepared... The term liquid-fire has not unaptly been given them. **1838** E. Eden *Let.* 28 Nov. in *Up Country* (1866) I. xxvi. 282 Runjeet produced some of his wine, a sort of liquid fire. **1862** *Temple Bar* July 512 Ordinary phosphorus is readily soluble in bisulphide of carbon: when thus in solution constituting the liquid denominated by Captains Disney and Norton 'liquid fire'. **1915** *Illustr. London News* 13 Mar. 321 The enemy attacking a trench into which they had sent liquid fire. **1889** Groves & Thorp *Chem. Technol.* I. 293 (*caption*) Liquid fuel. **1912** W. S. Churchill *Let.* 11 June in *World Crisis 1911–14* vi. 32 This liquid fuel problem has got to be solved. **1913** *Chem. Abstr.* VII. 3827 One of the principal differences between solid and liquid fuels is in the proportion of C combined with H. **1920** *Conquest* Nov. 47/2 In the near future liquid fuel will have to be 'rationed' again. **1935** *Jrnl. R. Aeronaut. Soc.* XXXIX. 507 Germany consumes two million tons of liquid fuel per year. **1946** *Jrnl. Brit. Interplanetary Soc.* VI. i. 2 At first the research was conducted with powder rockets, but after a few years a change was made to liquid fuel. This was at first tried out on a test bench, and then, in July, 1929, what is claimed to be the first liquid-fuel rocket to take-off was launched near Worcester—but it exploded at 900 feet. **1963** Bird & Hutton-Stott *Veteran Motor Car* 75 Daimler's 'high speed' liquid-fuel engine. **1966** *Electronics* 17 Oct. 35 For the next decade, at least, liquid fuel will propel United States astronauts into space. **1969** *Times* 3 June (Suppl.) p. ii/4 Goddard launched on March 16, 1926, the world's first liquid fuel rocket. **1960** *Times* 18 Oct. 13/6 The Minuteman cannot carry as large a warhead as the liquid-fuelled i.c.b.m.s. **1967** *Technology Week* 20 Feb. 3/2 The hardened, storable liquid-fueled *Titan* carries the largest U.S. missile warhead and can also reach targets beyond the range of the current *Minuteman I.* **1875** E. Spon *Workshop Receipts* 41/1 Liquid glue... Soft water, 1 quart; best pale glue, 2 lbs.; [etc.]. **1927** Knight & Wulpi *Veneers & Plywood* xxv. 276 Probably the general public..thinks of liquid glues, whenever glue is mentioned, but such preparations are in no way typical of manufacturing materials or procedures. **1966** *McGraw-Hill Encycl. Sci. & Technol.* VI. 219/2 Liquid glue is commonly made from fish collagen because this has little tendency to gel, but it can also be made from animal glue by treatment with acid or certain salts to inhibit gelation. **1970** G. F. Newman *Sir, You Bastard* ii. 65 The caretaker, aroused from his post-liquid-lunch slumber, confirmed that the couple had parted. **1972** B. Everitt *Cold Front* vii. 55 He..refused all offers of liquid lunches and bore me off..for a great deal of solid pasta. **1837** C. W. Johnson (*title*) On liquid manures. **1842** J. C. Loudon *Suburban Horticulturist* iii. 59 Arrangements should be made for collecting all the liquid manure into two adjoining tanks. **1869** S. R. Hole *Bk. about Roses* vi. 83 The rich extract, full of carbonate of ammonia..may be used..as liquid manure in the Rosary. **1911** O. Onions *Widdershins* 247 The hares and foxes were down four days ago, and the liquid-manure pumps like a snow man. **1914** J. London *Let.* 21 Sept. (1966) 429, I have a fairly decent brood-barn, with liquid-manure tank attached. **1973** R. Genders *Epicure's Garden* ii. 167 An occasional application of liquid manure will also prove beneficial. **1878** *Jrnl. Chem. Soc.* XXXIV. 10 A jet of liquid oxygen escaping from the tube when the pressure was taken off. **1885** [see Refrigerant *sb.* 3]. **1919** *Chem. Abstr.* 791 Spontaneous explosion of the charcoal in liquid-oxygen containers. **1954** *Economist* 11 Sept. 12/1 Rocket motors do not need atmospheric oxygen, although they frequently carry liquid oxygen, and rocket power is, unlike jet power, effective in outer space. **1956** *Spaceflight* Oct. 5/1 Propellants which are in common use to-day (e.g. liquid oxygen and petrol) yield exhaust velocities of the order of 2½ km./sec. **1884** *Jrnl. Chem. Soc.* XLVI. 1073 The 'liquid paraffin' of the German Pharmacopœia, is an oily liquid consisting of a mixture of hydrocarbons of the methane series. **1943** J. B. Priestley *Daylight on Saturday* ii. 6 She was a dark girl with a long sad nose, and dosed herself with liquid paraffin. **1962** *Which?* Jan. 26/2 Liquid paraffin is the only common lubricant laxative. **1905** *Pharmacopœia U.S.* 336 Liquid Petrolatum, a mixture of hydrocarbons, chiefly of the methane series, obtained by distilling off most of the lighter and more volatile portions from petroleum, and purifying the liquid residue. **1926** *Daily Colonist* (Victoria, B.C.) 20 Jan. 7/3 (Advt.), Liquid petrolatum, heavy, special at 53c. **1951** A. Grollman *Pharmacol. & Therapeutics* xix. 386 Liquid petrolatum is also available in the form of a flavored emulsion. **1905** P. Dawson in M. Maclean *Mod. Electr. Pract.* VI. v. iv. 214 Amongst the advantages claimed for liquid rheostats may be mentioned the gradual variation of the resistance, their compactness and cheap first cost, and the absence of damage from sparking when the current is interrupted. **1957** W. J. John *Mod. Electr. Engin.* II. iii. 59/1 A common method of providing resistance is by means of a liquid rheostat, consisting of a tank of electrolyte containing two electrodes, one fixed and one movable. **1600** Liquid sope [in Dict. s.v. Soap *sb.* 2]. **1907** *Yesterday's Shopping* (1969) 536/3 Moline, a liquid soap, delicately perfumed for the toilet. **1920** A. Keane tr. *Deite's Man. Toilet Soap-Making* (ed. 2) 236 The liquid soaps are mostly solutions of potash soaps in glycerine, sugar, or alkali and strong alcohol solutions. **1966** J. S. Cox *Illustr. Dict. Hairdressing* 90/2 *Liquid soap shampoo.* The base is usually green soft soap and sometimes in addition either cocoanut oil, olive oil or eucalyptus oil. **1907** G. W. O. Howe tr. *Thomälen's Text-bk. Electr. Engin.* 453/1 (Index), Liquid starter. **1916** C. C. Garrard *Electr. Switch & Controlling Gear* v. 353 The liquid starter, in which a solution of caustic soda or washing soda replaces the resistance wire or grids as used in the ordinary form of starter, is the most rugged of all forms of starting gear. **1932** E. Molloy *Pract. Electr. Engin.* IV. 1319/2 Liquid starters are best suited for medium and large-size motors which require to be started infrequently.

B. *sb.* **3.** *liquid-cooled, -filled* adjs.; *liquid-cooling;* **liquid-drop,** used *attrib.* in *Physics* to denote a theoretical model in which an explanation of the properties and behaviour of the atomic nucleus is sought by likening it to a droplet of liquid, as regards the forces between its constituents; **liquid–liquid** *a.*, pertaining to or involving two different liquids, or liquids in two different ways.

1931 *Jrnl. R. Aeronaut. Soc.* XXXV. 180 The developments to be described are..applicable to all water or liquid-cooled engines. **1967** *Jane's Surface Skimmer Systems 1967–68* 25/1 The power plant..comprises a 240 hp Rolls-Royce LV-8 liquid-cooled eight-cylinder engine. **1933** J. D. Frier *Aero Engines* I. vii. 94 A great advantage of liquid cooling is that..the engine temperature is limited to the boiling-point of the liquid employed. **1960** *McGraw-Hill Encycl. Sci. & Technol.* IV. 604/2 In liquid cooling, the engine and radiator may be separated and each placed in the optimum location. [**1939** Meitner & Frisch in *Nature* 11 Feb. 239/1 On account of their close packing and strong energy exchange, the particles in a heavy nucleus would be expected to move in a collective way which has some resemblance to the movement of a liquid drop.] **1939** Bohr & Wheeler in *Physical Rev.* LVI. 426 On the basis of the liquid drop model of atomic nuclei, an account is given of the mechanism of nuclear fission. **1955** R. D. Evans *Atomic Nucleus* xi. 365 The liquid-drop model is the antithesis of the independent-particle models. The interactions between nucleons are assumed to be strong instead of weak. **1970** D. F. Jackson *Nucl. Reactions* ii. 17 The static liquid drop model of the nucleus.. was the first successful nuclear model and was used to describe the bulk properties of nuclei such as nuclear masses and binding energies. **1960** E. L. Delmar-Morgan *Cruising Yacht Equipment & Navigation* ii. 33 The liquid-filled 'dead-beat' instrument [*sc.* compass] has now taken its place. **1967** Karch & Buber *Offset Processes* iv. 128 Modification of copy is effected simply by liquid-filled cylindrical lenses or prisms in front of the copy or film. **1940** Glasstone *Text-bk. Physical Chem.* x. 725 (*heading*) Distribution in liquid–liquid systems. **1951, 1952** [see *gas-liquid* adj. (*gas *sb.*[1] 7)]. **1958** *Chambers's Techn. Dict.* 991/1 *Liquid–liquid extraction,* process whereby two non-mixing liquids are brought together for an exchange of substances dissolved in them. **1968** Coulson & Richardson *Chem. Engin.* (ed. 2) II. xii. 486 The separation of the components of a liquid mixture with a solvent in which one or more of the desired components is preferentially soluble is known as liquid–liquid extraction. This process has been..very extensively applied to the separation of hydrocarbons in the petroleum industry.

liquidate, *v.* Add: **7.** [after Russ. *likvidírovat'* to liquidate, wind up.] To put an end to, abolish; to stamp out, wipe out; to kill.

1924 *Yale Rev.* XIII. 477 In this way the 'Labor Opposition', the 'Workers *Pravda*', and a few other recalcitrant groups were all 'liquidated'. **1926** C. Sheridan *Turkish Kaleidoscope* xvi. 125 The evening paper, *L'Akcham,* came out with large headlines: 'How to Liquidate a Strike'. **1930** *Economist* 1 Nov. (Russ. Suppl.) 2/2 Only in 1929, when the growth of the Socialist section of agriculture was enabling the State to become independent of the supplies of the Kulaks, could the Government begin to 'liquidate' them. **1939** V. A. Demant *Relig. Prospect* iv. 90 The Trotskyists..are 'liquidated' as being insufficiently dialectical to see that the policy of the Russian State at any moment has absolute finality. **1943** C. S. Lewis *Abolition of Man* iii. 37 Once we killed bad men: now we liquidate unsocial elements. **1957** Partridge *English gone Wrong* ii. 31 *Liquidate,* therefore, is an erudite synonym of 'to wind up', hence, in its euphemistic transferred sense, it means 'to eradicate in a thoroughly ruthless manner', 'to destroy, especially by mass murder'. **1970** *Nature* 26 Dec. 1248/2 All existing sources of industrial pollution are to be 'liquidated'. **1971** *Sunday Times* 13 June 12/6 When the army units fanned out in Dacca on the evening of March 25..many of them carried lists of people to be liquidated.

liquidating *vbl. sb.* and *ppl. a.* (Later examples.)

1931 *Economist* 20 June 1331/1 The market capitalisation of the common shares of these concerns was equal to only 74 per cent. of the 'liquidating value' of the assets behind them. **1964** *Ann. Reg. 1963* 103 It provided that.. the permanent heads of the three territorial Treasuries would constitute a liquidating agency to wind up the affairs of the Federation. **1975** [see *liquidator c].

liquidation. Add: **2. b.** *Chess.* The partial clearing of the board, by an exchange of pieces, to obtain an obviously winning position; simplification.

1965 Love & Hodgkins *Further Chess Ideas* xv. 124 Sometimes, too, under pressure of an enemy attack and with good end game prospects if ever one should be reached, wholesale exchanges are most welcome. This is called liquidation. **1965** W. H. Cozens tr. *Euwe & Kramer's Middle Game* II. x. 185 The problem of liquidation is to select the precise moment when pieces, or some particular piece, should be exchanged... Judicious liquidation involves steering a middle course between the one extreme of premature simplification and the other extreme of interminable 'wood shifting'. **1966** *New Statesman* 2 Dec. 854/3 True enough, White is a P up, but the Black heavy artillery is well placed. Yet, hey presto: a miraculous 'liquidation', and a won ending in a few moves.

3. Also, the selling of certain assets in order to achieve greater liquidity. (See quot. 1965.)

1909 *Westm. Gaz.* 2 Mar. 4/2 As many people..think that the word 'liquidation' must necessarily be associated with bankruptcy.., I am asked to state officially that the liquidation of the old company is only one step in the course of reconstruction for the purpose of obtaining fresh capital. **1929** *Observer* 17 Nov. 4/2 The Rhodesian share market was in a depressed condition, owing to the liquidation taking place on American account. **1939** J. A. Schumpeter *Business Cycles* I. iv. 149 Abnormal liquidation destroys many things which could and would have survived without it. **1965** *McGraw-Hill Dict. Mod. Econ.* 299 *Liquidation,* the process of selling assets, such as inventories or securities in order to achieve a better cash position.

b. [f. *liquidate *v.* 7.] The action or process of abolishing or eliminating; the doing away with or killing of unwanted persons.

1925 tr. *L. Trotzky's Whither England?* vi. 145 History is liquidating liberalism and preparing for the liquidation of pseudo-labor pacifism. **1932** *Week-end Rev.* 2 Jan. 24/1 The Russians..took starvation almost as a matter of course, just as they..take as a matter of course the liquidation of unfortunate individuals with contra-revolutionary idealogies. **1949** F. Maclean *Eastern Approaches* i. ii. 24 There was nothing new in the 'liquidation', as it was called, of public figures. For some years past numerous politicians and others had met with this fate, variously branded as 'Trotskists' [*sic*], 'wreckers'..and so on. **1952** *Sat. Rev.* (U.S.) 20 Sept. 37/1 Liquidation..was extended ..to persons in..the Party... The liquidation occurred during the purges (a revolting combination).

4. The action or fact of partaking of an alcoholic drink. *rare.*

1889 F. E. Gretton *Memory's Harkback* 311 As regards *liquidation,* champagne..is now almost as *vin ordinaire.* **1909** 'O. Henry' *Roads of Destiny* vii. 106 His desire for liquidation was expressed so heartily that I went with him to a café..where we had some vile vermouth and bitters.

liquidator. Add: (Later examples.)

1965 Seldon & Pennance *Everyman's Dict. Econ.* 260 The Official Receiver is appointed provisional liquidator, and his appointment may be confirmed or he may be replaced at the first meeting of the creditors. **1971** *Daily Tel.* 9 Feb. 15/5 The creditors' call for a provisional liquidator for Mineral Securities..followed on intense activity ..with leading Australian international financiers. **1971** *Times* 10 Dec. 8/6 The appointment of a provisional liquidator and special manager was essential in order to collect those balances and get the other assets in.

c. A person who implements a policy of liquidation. Cf. *liquidate *v.* 7.

1949 [see *grave-digger 3]. **1963** *Listener* 24 Jan. 171/1 The psychotic actions of totalitarian liquidators. **1975** J. Gardner *Killer for Song* ii. 16 Special Security had employed him as an agent—a liquidating agent, designed, quite literally, to cut down security risks. They had even called him the Liquidator.

liquidity. Add: **1. d.** *Econ.* The interchangeability of assets and money; hence **liquidity preference,** the holding of assets in money or near-money in preference to securities or interest-bearing investments; **liquidity ratio,** that proportion of total assets which is held in liquid or cash form, esp. by a bank.

1923 R. G. Hawtrey *Currency & Credit* (ed. 2) v. 83 The liquidity of the Bank of England is secured by its power of printing notes, and the interchangeability of its deposits with cash is absolute. **1925** G. G. Munn *Bank Credit* i. 9 Gold is the ultimate in liquidity, the ultimate intermediate of exchange, and is *ipso facto* irredeemable. **1936** J. M. Keynes *Gen. Theory Employment* xiii. 168 Liquidity-preference is a potentiality or functional tendency, which fixes the quantity of money which the public will hold when the rate of interest is given. *Ibid.* xvii. 241 Money itself rapidly loses the attribute of 'liquidity' if its future supply is expected to undergo sharp changes. **1940** *Economist* 27 Jan. 152/1 The attainment of a more than adequate liquidity ratio is perhaps the main feature..over the past year. **1940** G. Crowther *Outl. Money* ii. 76 The more highly developed banking systems are more prone to suffer from such a 'liquidity preference' than the less developed countries. **1958** *Economist* 26 July 281/2 The real spark-plug of expansion—improved international liquidity—cannot be discussed in the Commonwealth context alone. **1961** *Ann. Reg. 1960* 488 The 'liquidity ratio' is the proportion of 'liquid assets', i.e. cash, short money and bills, to gross deposits. **1962** C. H. Kreps *Money* i. i. 12 One view of the fundamental nature of interest regards interest as compensation for loss of liquidity. *Ibid.*, Using this liquidity-preference approach, we may say that wealth held in absolutely liquid form—in the form of money, that is—yields its owner no income. **1969** *Times* 5 May (Suppl.) p. iv/3 The market would lose its liquidity... As liquidity declined.., public confidence would lessen considerably. **1972** *Accountant* 23 Mar. 385/1 It was necessary to increase liquidity during the year to finance the continuing substantial capital programme.

liquidize, *v.* Add: Also *intr.*, to become liquid.

1969 *Sunday Times* (Colour Suppl.) 9 Feb. 13/2 The lettuce liquidising in the vegetable compartment.

b. [f. *liquidizer.] To purée, emulsify, or blend in a liquidizer. Hence **li·quidizing** *ppl. a.* and *vbl. sb.*

1959 *Which?* June 49/1 It was satisfactory in its mixing and liquidizing. **1966** *Punch* 2 Nov. 676/3 A very nice Melde dish is made by heating the vivid green..liquidised leaves with cheese to serve on toast as 'Anglo-Saxon Rarebit'. **1972** S. Atterbury *Waste Not—Want Not* ii. 53 After liquidising, pour the soup into a saucepan. *Ibid.* 54 Do not add too much liquid before the pods are liquidised or the subtle flavour might be swamped. **1972** Beale & Johnston *Mixer & Blender Cookery* xxi. 132 Liquidise ½ or 1 banana and add to thick custard. **1972** *Garde ta Foie!: Cambridge Cookery Bk. for Shelter* (Cambridge Shelter Group) 7 Stew gently with lid on until soft. Sieve or liquidise. **1973** *Daily Tel.* 27 Jan. 1/2 After his thirst and hunger strike ended, MacStiofain was given soups, tea and liquidised vegetables.

liquidizer (li·kwidəizəɹ). [f. Liquidiz(e *v.* + -er[1].] A machine used in the preparation of food, to make purées, emulsify, etc. Also *attrib.*, as *liquidizer attachment.*

1950 *Consumers' Res. Bull.* (U.S.) Feb. 9 (*heading*) 'Liquefiers', 'liquidizers', or 'blenders'. **1958** *Observer* 24 Aug. 7/5 The most usual 'extras' are mincers, juice squeezers, and liquidisers—the last consisting of whirling blades inside a glass goblet, to make purées, soups, drinks, etc. **1962** *Which?* May 144/2 All the mixers had liquidiser attachments, and we tested them to see how well they pulped tomatoes, made breadcrumbs, chopped nuts, ground coffee beans, chopped parsley and crushed ice. **1969** H.-P. Pellaprat *Everyday French Cooking* 48 Put through a fine sieve or the liquidiser. **1972** Beale & Johnston *Mixer & Blender Cookery* iv. 20 The liquidizer reduces dried peas, beans, lentils, rice, etc., into powder speedily.

liquidus (li·kwidŭs). [L. *liquidus* Liquid *a.*, adopted in this sense (in G.) by H. W. B. Roozeboom 1899, in *Zeitschr. f. phys. Chem.* XXX. 387.] A line in a phase diagram, or a temperature (corresponding to a point on the line), above which a mixture is entirely liquid and below which it consists of liquid and solid in equilibrium; freq. *attrib.* as *liquidus curve, temperature.* Also (more fully *liquidus surface*), an analogous surface in a three-dimensional phase diagram.

1901 *Proc. R. Soc.* LXVIII. 174 Cooling curves will.. give the approximate moment of complete solidification of an alloy, and enable us to plot in a rough way the 'solidus' curve, as Roozeboom calls it; but the solidus curve thus obtained is not nearly so accurate as the 'liquidus' or freezing-point curve. **1923** Glazebrook *Dict. Appl. Physics* V. 245/2 The liquidus and the solidus meet, in the case under consideration, at the points A and C, which are the melting points of the pure metals, and B, which is known as the eutectic point. **1948** Glasstone *Textbk. Physical Chem.* (ed. 2) x. 762 Separation of solid.. commences at *y* and is complete at *z*'; for this reason the liquidus curve is sometimes called the freezing-point curve, and the solidus curve is called the melting-point curve. **1960** C. J. Phillips *Glass: Industr. Applic.* v. 54 The viscosity at the liquidus temperature is high enough to prevent nucleation. **1967** A. H. Cottrell *Introd. Metall.* xvi. 252 (*caption*) Liquidus surface of a eutectic system. **1972** *Physics Bull.* Nov. 352/3 Glass is a metastable state; it can be brought to the stable state by holding at temperatures near the liquidus.

liquor, *sb.* Add: **7.** *liquor-bar, house, law, licence, question, -saloon* (earlier example), *-seller* (examples), *-selling, -shop* (later examples), *-store* (later examples), *trade, traffic* (earlier and later examples); **liquor prescription** *Canad. Hist.*, a doctor's prescription of alcohol for 'medicinal' purposes, to evade the prohibition regulations.

1813 W. Dunlap *Mem. G. F. Cooke* II. xxx. 278 The fountain of mischief, the liquor-bar, was shut. **1924** W. M. Raine *Troubled Waters* ii. 21 The postmistress handed him a letter and two circulars from liquor houses. **1852** *Boston Bee* 29 July, The Life Boat..takes the Bee to do, for its course in relation to the Liquor law. **1858** A. Lincoln *Coll. Works* (1953) II. 493, I do not believe in the right of Illinois to interfere with..the Liquor Laws of Maine. **1866** G. Meredith *Let.* c 27 Nov. (1970) I. 345 You will become a fanatical Retired Admiral advocating Maine Liquor laws for every natural appetite on earth. **1908** *Daily Chron.* 27 Feb. 4/4 Certain liquor-law restrictions which had existed under the second Empire. **1975** *Listener* 16 Jan. 76/1 The counties would have.. their own educational system, their liquor laws. **1850** *Hunt's Merchant's Mag.* XXII. 87 (*caption*) Statistics of Liquor Licenses in New York City. **1956** B. Holiday *Lady sings Blues* (1973) xix. 157 According to the law.. nobody who has a police record can hold a liquor licence. **1971** *Sunday Express* (Johannesburg) 28 Mar. (Home-finder) 7/2 (Advt.), 1. Dance hall. 2. Restaurant—Liquor licence. **1921** *Daily Colonist* (Victoria, B.C.) 18 Mar. 1/3 The suspension of 19 physicians in Manitoba for unlawfully issuing liquor prescriptions has resulted from an inquiry. **1855** Liquor question [see *liquor-shop* below]. *c* **1918** C. Stelzle *Why Prohibition!* 291 Michigan was about to vote on the liquor question. **1863** *Daily Even. Bulletin* (San Francisco) 29 Sept. 3/2 At 1 o'clock they went into a liquor saloon kept by a woman on Kearny Street. **1855** P. T. Barnum *Life* 359 The liquor seller, the moderate drinker, and the indifferent man. **1877** *Harper's Mag.* Dec. 146/2 A method which practically makes the government the liquor-seller. *Ibid.*, All liquor-selling is not equally dangerous to the community. **1855** 'Q. K. P. Doesticks' *Doesticks, what he Says* xxxi. 276 The great excitement was on the liquor question; it was Noggs and no liquor shops, or Boggs and a few liquor shops. **1877** J. Habberton *Jericho Road* xix. 167 There was not even a streak of light visible under the door of any liquor-shop in the town. **1911** *Encycl. Brit.* XVI. 769/2 The effect has been a very large reduction in the number of liquor shops. **1855** 'Q. K. P. Doesticks' *Doesticks, what he Says* xii. 98 Fire in a liquor-store—hose burst; brandy 'lying round loose'. **1887** *Nation* (N.Y.) 15 Dec. 468/3 To keep a liquor-store in Philadelphia. **1939** F. P. Grove *Two Generations* 39 Take the proceeds to the liquor store. **1964** *Calgary Herald* 24 July 23/2 A liquor store on the site would devalue residential property to the immediate west. **1972** R. Bloch *Night-World* (1974) xi. 72 He passed the lights of the liquor store. **1908** *Daily Chron.* 12 May 4/4 Unfortunately for the Labour party they have got entangled with the liquor trade vote. **1848** J. Marsh (*title*) A discourse on the extent and evils of the Sunday liquor traffic in cities. **1877** *Harper's Mag.* Dec. 146/2 This work is a compilation of evidence on 'the problem of law as applied to the liquor traffic'. **1915** W. J. Bryan *Mem.* 2 Oct. (1925) 434 The brewers and distillers were connecting them with the liquor traffic to their detriment.

liquordom. (Later example.)

1918 T. H. Walker *Principal J. Denney* 119 His hatred of liquordom..sprang from devotion to his Master.

liquorice, licorice. Add: **4.** *liquorice all-sorts* (see *all-sorts* s.v. *all E. 13), *drop, jujube, lozenge, lump, toffee, treasure, -water* (earlier and later examples); **liquorice bootlace** = *boot-lace c; **liquorice-stick,** (*a*) in Dict. (later examples); (*b*) *Jazz slang,* a clarinet.

1928 *Sweet Shop* Nov. p. ii (Advt.), Original Liquorice All Sorts. **1931** [see *all-sorts* s.v. *all E. 13]. **1946** *R.A.F. Jrnl.* May 162 His C.O...with the broad ring and a row of gongs that reminded Joe of liquorice all-sorts. **1952** *Blackw. Mag.* July 30/2 Lollipops, pincushions, liquorice boot-laces, bottles of home-made wine. **1956** 'R. Crompton' *William & Space Animal* v. 135 Doughnuts an' trifle an' liqu'rice boot laces. **1906** 'O. Henry' *Four Million* (1916) 173 Liquorice drops—the kind that make your cheek look like the toothache. **1967** R. Mackay *House & Day* 74 Have you licorice drops? **1891** *Confectioners' Union* IV. 530 (Advt.), Goods packed in 4 lb. boxes.. Bright liquorice jujubes. **1857** J. A. Symonds *Let.* 25 Jan. (1967) I. 86 Will you send me also another box of Liquorice Lozenges. **1893** *Official Catal. Internat. Manufacturing Confectioners Exhib.* 60 (Advt.), Linseed Liquorice and Chlorodyne Lozenges. **1926** 'R. Crompton' *William—the Conqueror* iii. 54 Large paper bags of bullseyes, liquorice lumps, barley sugar and chocolate cigars. **1879** Liquorice-stick [see *goose-yoke*]. **1935** *Vanity Fair* (N.Y.) Nov. 71/3 *Agony pipe, wop stick,* and *licorice stick* for clarinet. **1958** N. D. Hinton in *Publ. Amer. Dial. Soc.* xxx. 39, I have found a common belief that jazzmen refer to a clarinet as a 'licorice stick'... Jazz musicians do in fact use these terms ..in a very peculiar way. **1967** C. Drummond *Death at Furlong Post* iv. 39 Bee dispensed liquorice sticks..and fizzy drinks. **1930** *Confectioners' Union Directory Trade Marks & Trade Names* 18/2 (Advt.), Buttered Liquorice Toffee in Dainty Pieces. **1924** 'R. Crompton' *William—the Fourth* viii. 125 'Have a liquorice treasure?' she said. **1860** Dickens *Gt. Expect.* (1861) I. ii. 29 That intoxicating fluid, Spanish-liquorice-water. **1913** C. Mackenzie *Sinister St.* I. iv. 59 They used to go calling up and down, 'Fine liquorice-water!.. Bring out your pins and have a bottle of liquorice-water.' **1960** *Guardian* 6 Dec. 7/2 The bottles of liquorice-water which provided grog for his young games of pirates.

lira. Add: **2.** A monetary unit in Turkey.

1871 *Murray's Handbk. Turkey in Asia* (new ed.) 27 The coinage consists of copper, silver, and gold, at the rate of 100 piastres to the Turkish pound or *lira.* **1884** F. Boyle *On Borderland betwixt Fact & Fancy* 237 The high wall on our left was that of the pasha's grounds. The one-eyed calender informed me that he could get permission to visit them next day, for a baksheesh of two liras. Thirty-six shillings seemed too much to pay for a stroll through a burnt-up garden. **1904** *Daily Chron.* 17 Feb. 3/6 A thousand Turkish liras weigh about 14 lb. **1912** T. E. Lawrence *Let.* 10 Feb. (1938) 136 The foot seal must be worth a lira at the least. **1975** J. Rathbone *Kill Cure* iii. vii. 128 'How much?'..'Forty dollars or twenty-five sterling. Turkish lira, much more.'

lirate (ləi·rĕt), *a.* [f. L. *lira* ridge, furrow: see -ate[2].] Of a shell: having ridges. Hence **lira·tion,** marking of this kind.

1894 J. W. Taylor *Monogr. Land & Freshwater Mollusca Brit. Isles* I. i. 28 Cingulate, or Lirate, when the whorls are furnished with spiral ribs or ridgings. **1901** *Proc. Zool. Soc.* II. 357 *Actis calotropis...* A very delicate species, vitreous,..delicately spirally lirate. **1904** *Ann. &*

Mag. Nat. Hist. 7th Ser. XIII. 459 This liration bears small tubercles connected by short cross-ridges with the dentations of the keel.

‖ **lirio** (li·rɪo). [Sp. *lirio* iris.] The American Spanish name for the water hyacinth, *Eichhornia crassipes*.

Quot. 1844 refers to a different plant, perhaps the frangipani, *Plumeria rubra*.

[**1844** J. G. F. Wurdemann *Notes on Cuba* v. 140 The quaint lirio's trumpet-shaped flowers painted yellow and red, and bursting in bunches from the blunt extremities of each leafless branch.] **1926** D. H. Lawrence *Plumed Serpent* v. 94 A long canal paved with bright green leaves from which poked the mauve heads of the lirio, the water hyacinth.

Lissajous (li·saʒu). The name of Jules Antoine *Lissajous* (1822–80), French physicist, used *attrib.* and in the possessive to designate the plane figures (mostly crossed loops and simple curves) traced by a point executing two independent simple harmonic motions at right angles to one another and with frequencies in a simple numerical ratio (described by Lissajous in *Compt. Rend.* (1855) XLI. 814).

1877 Rayleigh *Theory of Sound* I. p. vii, Lissajous' Figures. **1902** *Encycl. Brit.* XXV. 50/1 If both forks vibrate, an observer looking through the microscope sees the bright point describing Lissajous figures. **1939** *Brit. Jrnl. Psychol.* Oct. 129 The frequency of the tones employed.. is checked by obtaining a Lissajous's figure against a constant frequency source applied to the other deflectors of the cathode-ray tube. **1943** *Electronic Engin.* XVI. 170 Two measuring techniques for the Lissajous' figures produced on the oscilloscope screen are described. **1973** *Sci. Amer.* Aug. 76/2 The two tones were next matched precisely on a cathode ray oscilloscope by tuning the generated tone until a clear Lissajous figure with a ratio of 1:1 appeared repeatedly on the screen.

lissoir (liswā·r). *Archæol.* [Fr.] A smoothing, polishing tool.

1911 W. J. Sollas *Ancient Hunters* viii. 214 A small object, about the size and shape of the human tongue, possibly the end of a 'lissoir' (smoothing implement). *Ibid.* xii. 368 The ivory 'lissoir' or smoother of the Eskimo..is represented in the Magdalenian industry. **1932** *Jrnl. R. Anthrop. Inst.* LXII. 265 Skin-rubbers were made from antler of *Dama Mesopotamica*, cut obliquely and smoothed, in the manner of the Magdalenian *lissoir*. **1937** Garrod & Bate *Stone Age Mt. Carmel* I. i. ii. 15 Bone points and pierced animal teeth and a *lissoir* of *Dama Mesopotamica* antler. **1964** *New Scientist* 9 Apr. 88/1 A single, well-used bone-tool came to light; it appears to be a sort of 'lissoir' for working leather thongs.

lissom, *a.* Hence li·ssomely *adv.*

1902 W. de la Mare *Songs of Childhood* 54 Though danced she lissomely. **1927** M. Sadleir *Trollope: a Comm.* 322 Trollope worried to find it limping on its way, when usually his stories moved so lissomely.

list, *sb.*³ Add: **2. d.** (Further examples.)

1809 Jane Austen *Let.* 24 Jan. (1952) 257 We..could have staid longer but for the arrival of my List shoes to convey me home. **1856** Dickens *Dorrit* (1857) I. xiii. 106 Mr. Casby rose up in his list shoes. **1908** A. Bennett *Old Wives' Tale* iv. iii. 480 Sophia wore list slippers in the morning. It was a habit which she had formed in the Rue Lord Byron—by accident rather than with an intention to utilize list slippers for the effective supervision of servants.

3. (Later examples.)

1855 P. T. Barnum *Life* 109 Mallet had agreed..to deliver twelve yards of broadcloth 'lists' to Shepard. **1886** F. T. Elworthy *West Somerset Word-Bk.* 442 In flannels and in wool-dyed cloths it is usual to have a list or narrow border on each side of the cloth.

list, *sb.*⁶ Add: **a.** In specific senses. (*a*) The titles of the books (to be) published by a particular publisher. So *autumn list*, *backlist, *spring list*.

1860 G. H. Lewes *Let.* 4 Jan. in Geo. Eliot *Lett.* (1954) III. 243 It will be well now to begin announcing it in lists—if not the title at any rate the fact of a new novel being in the press. **1919** *Publisher's Let.* Aug. in T. S. Eliot *Waste Land Drafts* (1971) p. xvi, Mr. Eliot's work is no doubt brilliant, but it is not exactly the kind of material we care to add to our list. **1922** T. S. Eliot *Let.* 25 June in *Waste Land Drafts* (1971) p. xxii, Knopf said that it was too late for his autumn list this year. **1930** E. Waugh *Vile Bodies* ii. 28, I suppose you could get the book re-written in time for the Spring List? **1938** H. R. Dent in J. M. & H. R. Dent *House of Dent* xxiii. 300 It used to be said..that a publisher kept poetry on his lists more for the look of the thing than anything else. **1951** M. Sharp *Lise Lilywhite* xix. 161 Mr Villiers..published chiefly poetry... He had no list, in the trade sense, nor had he travellers. **1964** R. Church *Voyage Home* viii. 166, I should send the book to the house of Dent, whose list it would suit admirably. **1967** E. Grierson *Crime of One's Own* viii. 60 Christmas operated like a guillotine on the Autumn lists, leaving only a bare four weeks of selling time.

(*b*) An official register of buildings of architectural or historical importance that are statutorily protected from demolition or major alteration. Cf. *list *v.*⁴ 1 e.

1947 *Act* 10 & 11 *Geo. VI* c. 51 §30 With a view to the guidance of local planning authorities..in relation to buildings of special architectural or historic interest, the Minister shall compile lists of such buildings, or approve.. such lists compiled by other persons or bodies of persons. *Ibid.*, So long as any building..is included in any list compiled or approved under this section, no person shall execute..any works for the demolition of the building or for its alteration or extension in any manner which would seriously affect its character. **1968** P. Ward *Conservation & Devel. Historic Towns & Cities* iii. 98 Lansdown Parade..is also a Grade II listed building on the Ministry of Housing and Local Government's list of architecturally or historically important buildings.

(*c*) In the National Health Service, a general practitioner's register of patients.

1949 *Britannica Bk. of Year* 412/2 Doctors starting their careers..had few patients on their lists. *Ibid.* 413/1 The doctor was free to accept or reject anyone applying to go on his list. **1971** *Reader's Digest Family Guide to Law* 242 A doctor..does not have to give reasons for his refusal to accept a patient on his list. **1974** M. Birmingham *You can help Me* iii. 56, I asked him if he did not sometimes hanker after..a few wealthy private patients so that he could afford to keep his list shorter.

d. list-betting, betting on the list of horses displayed in a list shop; **list broker,** a trader in mailing lists; so **list-broking** *vbl. sb.*; **list house** = **list shop*; **listman,** one who works in a list shop; a bookmaker; **list price** (examples); **list shop,** an illegal betting shop where prices on future important races were displayed; **list system** (also *party list system*), a system of voting, common in continental W. Europe, in which voters cast their vote for a list of candidates rather than for an individual candidate; so **list vote, voting.**

1874 *Porcupine* 18 July 248/2 Mr. Chaplin, M.P., with other horse-owners, have..chuckled greatly at the prospect of list-betting no longer interfering with their speculations. **1928** *Daily Express* 24 Mar. 1/1 The..gaming laws..were primarily intended only to abolish notorious gaming houses and list-betting in shops and houses. **1959** *Economist* 7 Feb. 498/1 Publishers now send out circulars to people on mailing lists, bought from a growing class of 'list brokers'. **1967** *Guardian* 27 Dec. 4/2 She is a list broker, which means that she trades in names and addresses. *Ibid.* 4/4 The magnitude of list-broking in the United States. **1970** *Daily Tel.* 12 Oct. 17/3 'List broking' in this country could well develop into the sophisticated service industry it is in America. **1902** 'N. Gubbins' *Dead Certainties* 71 Most of the 'list-houses' (in Long Acre and elsewhere), whose name was legion, had their shutters up on the morning after Lord Zetland's horse had defeated Pitsford. **1922** *Daily Mail* 6 Nov. 11 Most of the listmen got scared to death over particular animals in these final handicaps. **1937** Partridge *Dict. Slang* 486/1 *Listman*, a ready-money bookmaker. **1871** *English Mechanic* 10 Nov. 206/2 The list price for a ½ horse-power engine is £60. *c* **1883** J. Montagu *Let.* in Troubridge & Marshall *John Lord Montagu of Beaulieu* (1930) 30 Now my old machine [*sc.* tandem bicycle] cost £26 list price, and we finally got it for £23:10s. owing to discount for ready money. **1928** *Publishers' Weekly* 30 June 2603 The reprint is usually about one-third of the list price of the earlier edition. **1955** *Radio Times* 22 Apr. 51/1 Hand in an old electric shaver.. and claim £2 allowance off the list price of a Remington 60. **1967** *Autocar* 28 Dec. 38/3 All 'list' prices are taken from *Autocar*'s 'Recommended New Car prices'. **1875** *Encycl. Brit.* III. 619/1 'List shops', where the proprietors kept a bank against all comers, and backers could stake their money in advance on a horse..sprung up..leading to..flagrant dishonesty. [**1901** T. R. & H. P. C. Ashworth *Proportional Representation* vii. 162 The Liste Libre, or Free List system,..applies the proportional principle not to individual candidates but to parties.] **1908** J. King *Electoral Reform* vii. 87 In the Party List System the elector gives his vote for the party list, on which the candidate is enrolled, when he gives a vote to any candidate. **1911** *Encycl. Brit.* XXXIII. 115/2 In the 'list systems'..candidates are grouped in lists. **1926** Hoag & Hallett *Proportional Representation* v. 60 Most of the countries which use list systems..have been successful in securing reasonable accuracy in the assignment of seats to parties. **1971** G. K. Roberts *Dict. Political Analysis* 115 *List system*, a system of election, based on proportional representation of parties or similar groups, each of which presents a list of candidates. The voter then casts his vote for one of these lists. **1911** J. H. Humphreys *Proportional Representation* viii. 180 List votes form a pool from which the candidates of the list draw in succession as many votes as are necessary. **1954** B. & R. North tr. *M. Duverger's Pol. Parties* i. i. 44 The list vote (*scrutin de liste*), operating within the framework of a large constituency, obliges the..local branches of the party to establish amongst themselves a strong system of articulation within the constituency, so that they can agree upon the composition of the lists. *Ibid.* 45 Belgium, where at the end of the nineteenth century party structure was amongst the strongest in Europe: it coincided with list-voting. **1958** W. J. M. Mackenzie *Free Elections* ix. 75 List voting is almost always associated with formulae for distributing seats.

list, *v.*⁴ Add: **1. c.** *U.S.* To place (a property) in the hands of a real-estate agent for sale or rent; to add to the list of properties advertised by a real-estate agent. Cf. *listing *vbl. sb.*² 3.

1906 W. A. Carney *Real Estate Business* v. 20 A real estate broker..should have listed considerable property. *Ibid.* 21 He can sometimes list a real bargain. **1908** *Amer. Real Estate Seller* July 2 Every real estate dealer should have a form contract and use it. He should not list a property that he has not a contract on. **1909** *Ibid.* Aug. 6 The real estate dealers should combine and pass a resolution to list property exclusively. **1911** *National Realty Jrnl.* Mar. 14/2 The land owner, the investor, will also find it to his interest to recognize an active agent and list property with him. **1921** J. B. Spilker *Real Estate Business* v. 25 Only those properties which in the mind of the sales manager are saleable, and only those properties which are secured at a fair price and reasonable to both the buyer and seller, should be listed for sale. **1945** G. H. Beurhaus *Who handles your Real Estate?* (rev. ed.) vi. 19 The broker..proceeds to list property. **1972** J. L. Gale *Listing Real Estate* p. xix, Once we learn the ground rules for listing residences, we can then go on and successfully list property of any kind.

d. To enter (a name and address) in a telephone directory.

1959 R. Stout *Crime & Again* 91 'I'll see if she's listed.' I went to my desk for the Manhattan phone book. **1971** *Post Office Telephone Directory Section 101: London Postal Area* 12/1 A special Greater London Business directory has been introduced, listing certain businesses within about thirty miles of Charing Cross.

e. To protect (a building, etc.) by placing it on a statutory preservation register. Cf. *list *sb.*⁶ a (*b*).

1968 *Act Eliz. II* c. 72 § 52 A building which, immediately before the date of the compulsory purchase order, was listed. *Ibid.* § 54 Matters which may be taken into account by the Minister in listing buildings. **1972** E. Lemarchand *Cyanide with Compliments* vi. 74 The lovely little seventeenth-century timber-framed house... It's recently been listed. *Ibid.* xiii. 170 Some local preservation enthusiasts succeeded in getting the house listed as of architectural and historic interest. **1973** *Daily Tel.* 24 Mar. 14/6 Church House..was listed on Feb. 27 because of its architectural or historic merit. But workmen knocked a hole through the front wall on Tuesday.

listed, *ppl. a.*¹ Add: **2.** Included in a list, directory, or catalogue; *listed building*, one protected from demolition or major alteration by being included in an official list of buildings of architectural or historical importance.

1907 *Installation News* Jan. 11/1 Conduits have now to be manufactured to exact listed diameters. **1965** H. I. Ansoff *Corporate Strategy* vi. 117 Investment funds which trade in listed securities have the additional advantage of knowing the full field of choice. **1968** R. H. McCall in P. Ward *Conservation & Devel. Historic Towns & Cities* iii. 110 Of 439 Listed buildings in the City [*sc.* Winchester], 9 on the Statutory List have been wholly or partly demolished. **1971** P. Gresswell *Environment* 150 There are now about 120,000 'listed' buildings in England. *Ibid.*, An owner can also be threatened with compulsory purchase if he fails to keep a listed building in reasonable repair. **1973** *Country Life* 29 Mar. 866/2 The highest price, about £54,000, was paid for the Wealden farmhouse, a listed building that probably dates from the 17th century.

listed, *ppl. a.*² Add: **b.** *Basket-making.* Having an extra (decorative) skein on a handle.

1912 T. Okey *Introd. Basket-Making* xii. 145, I have carried out a listed handle with skeins of ordinary chair cane. **1953** [see *listing *vbl. sb.*⁴].

listen, *sb.* Add: **2.** (Earlier and later examples.)

1788 'Aspasia' in *Amer. Museum* iv. 565 Every time the door opens, or a foot is on the stairs, you are on the listen. **1935** *World-Radio* 5 July 9/1 People like me, who..are constantly on the listen with half an ear for something. **1968** J. Philips *Hot Summer Killing* (1969) iii. ii. 138 Take a listen while I try to find Jerry. **1968** C. Watson *Charity ends at Home* x. 122 So what I did was to pull off to the side and have a proper listen under the lid. **1970** P. Bair *Tribunal* ii. i. 60 'Did you have a nice talk?' 'I had a long listen.' **1971** *It* 2–16 June 19/3 Give it a listen.

b. listen-in, a period of listening to a broadcast, telephone conversation, etc. Cf. *listen *v.* 2 e.

1922 *Daily Mail* 30 Nov. 7 A listen-in. The Queen.. listened to a recitation sent out from Marconi House. **1946** *Philadelphia Bulletin* 1 Aug. 3 An occasional listen-in on the..[telephone] line later convinced company men.

listen, *v.* Add: **2. c.** Also, *to listen out*, to listen for a sound, e.g. on a radio receiver.

1910 A. Bennett *Clayhanger* iii. v. 362 Don't latch the door. Pull it to. I'll listen out. **1945** *Tee Emm* (Air Ministry) V. 35 Owing to the fact that they were listening out on channel 'A' instead of channel 'B', he failed to make contact. **1946** L. E. O. Charlton *R. Air Force July 1943 to Sept. 1944* 21 (*caption*) This photograph..depicts a scene in the flying control room—'listening out' to bring the Lancasters back to base. **1959** *Listener* 16 July 111/3 Initially I was afraid that the work would founder in an over-poetic fog but Mr. Bradnum was worth listening out for. **1971** J. Wainwright *Last Buccaneer* i. 49 We need receiving equipment to listen-out—to pinpoint every wavelength. **1974** D. Kyle *Raft of Swords* xiii. 140 He searched the air waves... For several days he had 'listened out' to a Russian ship with three operators aboard.

e. *to listen in*, to listen to a broadcast programme, etc.; to listen secretly to a telephone conversation. Also const. *to*, *on*, and *transf.*

1905 *Electrician* 20 Jan. 532/1 At the end of the first section the operator on the ship listened in for a reply. At last he took off the telephone. **1915** A. F. Collins *Bk. Wireless* p. vii, A boy sitting..at home with..a telephone receiver to his ear *listening-in* to the news of the world. **1920** *Wireless World* Jan. 594/2 While 'listening-in', the switch..is placed over to the right. **1926** *Daily Chron.* 13 May 3/1 By the primitive process of passing it from lip to lip the news sped 'like wildfire' amongst the London millions who were not listening in, but were just sitting in their offices or lunching in the restaurants, or walking about the streets. **1928** *Chambers's Jrnl.* Jan. 27/2 None of us could help 'listening in' to the fun that was going on in the kitchen. **1931** *Boys' Mag.* XLV. 99/2 Patients..are able to listen-in to the Radio programmes by means of headphones. **1939** Mrs. Belloc Lowndes *Diary* 5 Oct. (1971) 180, I asked him if he ever listened in, whereupon he said in an explosive tone: 'never'. I observed that one learnt a great deal from listening in. **1973** 'H. Carmichael' *Too Late for Tears* xv. 175 His wife wasn't involved... If she had been she wouldn't have wanted us to listen-in on that phone conversation. **1973** 'M. Innes' *Appleby's Answer* v. 49 They have forgotten about you and resumed their tittle-tattle. Whereupon you listen in. **1973** *Radio Times* 15 Nov. 73/3 For thousands of children..who 'listened in' each Friday afternoon, Romany *was* the countryside.

f. *spec.* To listen to a broadcast programme.

1929 *Radio Times* 8 Nov. 395/3 We sat listening..with a portable set. **1935** *World-Radio* 5 July 19/3 (Advt.), Below 100 Metres, Listen to the World. **1936** *B.B.C. Empire Broadcasting* 2 Dec. 2/3 Your greeting, Big Ben, and then the National Anthem, moved us profoundly—it took quite a time to listen without real emotion. *Ibid.* 9 Dec. 2/1 Whenever there was a sporting commentary a host of people used to come to my bungalow to listen. **1946** *B.B.C. Year Bk.* 11 With the restoration of peace there was a natural tendency for the citizens of other countries to listen, at first, only to their own newly freed broadcasting services. **1970** *B.B.C. Handbk.* 98 There is magic in ensuring that most people in the world can listen in a language they can really understand.

4. *intr.* To sound (in a certain way). Freq. with *to* = to strike (one) as. *U.S.*

1908 K. McGaffey *Sorrows of Show Girl* 78 That listened very well indeed, and we all climbed into a cabbage and vamped over. **1912** C. Mathewson *Pitching in a Pinch* vii. 143 All is fair in love, war, and baseball except stealing signals dishonestly, which listens like another paradox. **1923** R. D. Paine *Comrades of Rolling Ocean* xiv. 250 Here's where I slip it out..to help square the repair bill for my joy-ride. How does it listen to you? **1923** L. J. Vance *Baroque* xxvii. 174 [It] don't listen reasonable to me. **1945** Mencken *Amer. Lang.* Suppl. I. 317 It has been suggested..that *it listens well* may be from *es hört sich gut an*.

listenable (li·s'năb'l), *a.* [f. Listen *v.* + -able.] Easy or pleasant to listen to; willing to listen. Hence **listenabi·lity,** the quality of being listenable.

1920 C. Morley *Haunted Bookshop* vi. 95 He felt very talkative, as most older men do when a young girl looks as delightfully listenable as Titania. **1946** *Newsweek* 18 Feb. 90 In listenability, it is more like the Concerto for Orchestra. **1958** *Oxford Mail* 12 Aug. 6/6 Viewers are invited to..listen to..works carefully selected from the classics for their 'listenable' qualities. **1964** *Listener* 31 Dec. 1066/2 Talks producers might take a little more trouble in making them [*sc.* scripted talks] listenable, if they were given credits in *Radio Times*. **1966** G. N. Leech *Eng. in Advertising* iii. 27 Characteristics of advertising language..readability (or 'listenability'). **1970** *Guardian* 9 June 8/2 Berta was full of listenable opinions on her father. **1971** *Hi-Fi Sound* Feb. 78/2 It can be seen how singularly useless the THD results are, when attempting to assess the 'listenability' of an amplifier.

listener. Add: **3.** One who listens to a broadcast. Also *attrib.* Also *listener-in*. Cf. *listen *v.* 2 e, f.

1922 *Daily Mail* 21 Nov. 7 The limited service has already established itself in high favour with 'listeners-in'. **1923** *Radio Times* 28 Sept. 12/1 It seems to me that the B.B.C. are mainly catering for the 'listeners' who own expensive sets. **1926** *Daily Chron.* 13 May 3/1 By the magic of wireless it was, perhaps, the listeners-in who heard it first. **1929** *Radio Times* 8 Nov. 388/1 The recent broadcasting of *Aïda* has prompted a Forest Hill listener to send in..a very delightful story. **1936** *B.B.C. Ann.* 87/2 The BBC has recently established, at its Head Office, a special unit, with the object of co-ordinating information ..and studying new methods of 'listener research'. **1950, 1951** [see *audience 7 d]. **1970** *B.B.C. Year Bk.* 23 For the great majority of listeners..there will be little evidence of sudden upheaval.

Hence **li·stenership,** the estimated number of listeners to a broadcast programme or to radio (*spec.* as opp. to television).

1943 *Business Week* 30 Jan. 44 Increased emphasis on news broadcasts and commentators boosted listenership particularly between 5 and 7 p.m. **1958** *New Statesman* 2 Aug. 142/2 In America, reports *Time*, sound-radio is enjoying a 'spectacular comeback'; latest figure of 'listenership' show it 'up 8 per cent over last year, 25 per cent over its pre-TV peak in 1947'. **1971** *Daily Tel.* 17 Apr. 19/2 Listenership levels are still an imponderable. It is unlikely that the British public will listen to local radio as much as, say, the Americans.

listening, *vbl. sb.* Add: **b. listening key** *Teleph.* (see quot. 1940); **listening post** *Mil.*, an advanced position used to discover movements or the disposition of the enemy; also *transf.*

1906 J. Poole *Pract. Telephone Handbk.* (ed. 3) x. 159 (*heading*) Kellogg combined listening and ringing key. **1940** *Chambers's Techn. Dict.* 505/1 *Listening key*, the lever key which the operator throws, to put her head-set on to a cord circuit and speak to a subscriber. **1916** *War Illustr.* V. 69/1 At a listening-post. **1928** Blunden *Undertones of War* xv. 167 The men lying at each listening-post were freezing stiff. **1945** *Life* 19 Nov. 119/2 The barbed wire was up everywhere, and the few listening posts that we did have at Vichy, at General Weygand's North African Headquarters—were in perpetual danger of sabotage by the well-meaning but essentially stupid remonstrances of the more emotional Left press. **1961** *Guardian* 29 May 9/5 Vienna is Europe's busiest listening post. **1965** Mrs L. B. Johnson *White House Diary* 3 Nov. (1970) 335 John Gronouski was seated on my left and I enjoyed hearing him talk about Poland and how it serves as a sort of listening post for what is going on in Red China. **1971** J. Tunstall *Journalists at Work* iii. 86 Fairly standard features are centres like Hong Kong and Beirut which are used as 'listening posts' and jumping-off points for covering China and South East Asia, and the Middle East respectively. **1972** *Guardian* 25 Jan. 15/8 Paris uses the Commission mainly as a listening post to find out what the others are up to.

2. (Also *listening-in.*) The action of listening to a radio broadcast, a record-player, etc.; also, the action of listening (esp. secretly) to a telephone conversation. (Cf. *listen *v.* 2 e, f.) Also *attrib.* and *transf.*

1904 *Electr. World & Engin.* 7 May 875/2 The removal of the operator's plug, or her 'listening-in', restores the circuits to their proper condition for subsequent use. **1921** *Wireless World* 10 Dec. 581/2 'Listening in' was indulged in. **1925** A. Huxley *Those Barren Leaves* ii. v. 149 But of what use is leisure, when leisure is occupied with listening-in and going to football matches? **1927** *Sat. Even. Post* 24 Dec. 60/2 These telephones were connected with a listening-in device concealed behind a picture on the wall. **1929** *Radio Times* 8 Nov. 389/3 The sounds heard had emerged from the loud-speaker of the caretaker... The caretaker was temporarily oblivious to everything except his listening. **1939** *War Illustr.* 21 Oct. p. ii/1 Its [*sc.* the B.B.C.'s] dud programmes have led to a great falling-off in listening. **1940** *Manch. Guardian Weekly* 2 Feb. 87/4 From South West Germany it is stated that controllers have been appointed to keep in check the prohibition on listening-in. **1941** *B.B.C. Gloss. Broadcasting Terms* 15 *Listening log*, list, in prescribed form, of the broadcast programmes heard by a member of the listening public. **1953** J. B. Priestley *Festival at Farbridge* ii. ii. 217 Dan Cobbley was another radio personality, although he was on a lower listening figure level. **1957** *Encycl. Brit.* X. 619/2 The new possibilities for continuous listening helped enormously. On the debit side was the tendency especially of solo artists to record their own works on LP disks, repeating endlessly the same established repertoire. **1966** *Listener* 10 Feb. 221/1 The portrayal of..the jostling and jockeying of the foreign ambassadors, made really good listening. **1969** *John Edwards Mem. Foundation Q.* V. 126 The transcriptions of the songs..are as nearly accurate as I can make them. After countless listenings I cannot make out some of the words. **1971** *Gloss. Electrotechnical, Power Terms* (*B.S.I.*) iii. ii. 31 *Listening-in*, listening to a call in progress.

listening, *ppl. a.* Add: **b.** (Also *listening-in.*) That listens to a broadcast, recording, etc.

1926 *Punch* 14 July 39 (*caption*) Husband (to listening-in Wife). What's the matter, dear? Is it bad news of Stravinsky? **1935** *Discovery* Sept. 277/1 They are providing other stations to the ear, to give to the listening public to get more enjoyment from the 'audio' programmes. **1941** *B.B.C. Gloss. Broadcasting Terms* 15 *Listening group*, group of listeners meeting regularly with the twofold object of hearing a particular series of broadcast talks..and engaging in discussion. **1957** *B.B.C. Handbk.* 104 Audience Research set up permanent Listening Panels to report their reactions to the programmes they heard. **1970** *Ibid.* 112 A special listening section keeps track of the activities of foreign radio stations. **1974** *Times* 30 Nov. 10/4 The practitioners [of religious broadcasting]..are impelled by some notion of what the listening public really wants.

lister³. Add: (Examples.) Also *attrib.*

1887 *Sci. Amer.* LVI. 65 When grain is planted by the so-called 'combined lister and drill', the listing forms a ditch or furrow several inches deep, in which the seed is deposited. **1897** *Sears, Roebuck Catal.* 537/2 (*heading*) Subsoil lister with seed boxes complete with runners. **1946** *Harper's Mag.* Oct. 307/2 In my day a lister cost $70 —a great deal of money. **1965** *Austin* (Texas) *Morn. Avalanche* 23 Feb. ii. 6/4 Lister shares for any make tractor $3.50 each.

listerella (listĕre·lă). *Bacteriology.* Pl. **-ella, -ellæ.** [mod.L., f. the name of Joseph (later Lord) Lister (1827–1912), English surgeon + L. *-ella* (see -el²).] = *listeria.

[**1927** J. H. H. Pirie in *S. Afr. Inst. Med. Res.* III. xx. 164 The 'Tiger River Disease' is present among gerbilles... The causative organism of this disease is a small Gram-positive bacillus, for which, from its most striking pathogenic effect, I propose the specific name *hepatolytica*, and the generic name, *Listerella*, dedicating it in honour of Lord Lister, one of the most distinguished of those connected with bacteriology whose name has not been commemorated in bacteriological nomenclature.] **1940** *Vet. Jrnl.* XCVI. 139 Listerella were isolated from the uteri of the sheep slaughtered immediately after aborting. **1940** (*heading*) A new technique for isolating listerellae from the bovine brain.

Hence **li:sterello·sis** [-osis] = *listeriosis.

1939 *Science* 6 Oct. 337/1 More recently Biester and Schwarte reported spontaneous bovine listerellosis in Iowa. **1940** [see *listeria]. **1948** *Jrnl. Bacteriol.* LV. 473 Listerellosis as manifested in the ovine is extremely acute. **1961** R. D. Baker *Essent. Path.* ix. 122 Even with the small gram-positive rod, *Listeria monocytogenes* (listerellosis) is limited largely to meningitis, although it apparently causes abortions and stillbirths.

listeria (listɪə·riă). *Bacteriology.* Pl. **listeria, -ias.** [mod.L., f. *lister(ella + -ia¹ (see quot. 1940).] A bacterium of the genus *Listeria*, formerly called *Listerella*, esp. *L. monocytogenes* which is a widespread pathogen of man and animals.

[**1940** J. H. Pirie in *Science* 19 Apr. 383, I have been informed..that the new name *Listerella* which I proposed for a genus of bacteria in 1927 had already been given to a Mycetozoan by Jahn in 1906 and to one of the Foraminifera by Cushman in 1930. My proposed name, therefore, becomes a homonym, but as the genus has acquired some importance in both human and veterinary pathology and references to 'Listerellosis' are becoming fairly common in literature, I think that a name as near to my original proposal as possible is desirable. I therefore propose *Listeria*, as the name for the genus of bacteria as defined by me in Publication No. XX of the South African Institute for Medical Research.] **1961** *Lancet* 2 Sept. 514/1 The amniotic fluid was examined again..and listeria grew copiously from it.

Hence **liste·rial, liste·ric** *adjs.*, caused by or derived from listerias; **listerio·sis,** infection with or disease caused by listerias.

1941 *North Amer. Vet.* XXII. 345 (*heading*) An outbreak of listeriosis in sheep. **1961** *Jrnl. Clin. Path.* XIV. 193 (*heading*) Human listerial meningitis. **1961** tr. *H. P. R. Seeliger's Listeriosis* 144 A generalized listeric septicæmia. *Ibid.* 145 [Seeliger] established that 'Granulomatosis infantiseptica' was a listeric infection. **1961** *Lancet* 2 Sept. 515/1 Listeriosis should no longer be considered a rare disease in man. **1972** *Amer. Jrnl. Vet. Res.* XXXIII. 591 (*heading*) Effects of listerial hemolysin on rabbit heart. **1974** *Obstetr. & Gynecol.* XL. 91 (*heading*) Listeriosis as a cause of fetal wastage. *Ibid.* 96/1 Penicillin or tetracycline..should be started immediately when the possibility of Listeric infection is entertained.

Listerize, *v.* (Earlier example.)

1888 *Lancet* 2 June 846/1 In this way the patients are 'Listerized', to use a hospital term.

listing, *vbl. sb.*² Add: **3.** *N. Amer.* The placing of a property on the list of a real-estate agent; an estate agent's register of properties that he has for sale; a property so listed. Cf. *list *v.*⁴ 1 c.

1906 W. A. Carney *Real Estate Business* v. 21 Where values are changing it is necessary to confirm and correct the listings every month or two. **1909** *Amer. Real Estate Seller* Aug. 6 It may be well to explain exclusive listing here... When you have a piece of property to sell and you empower one and only one agent to dispose of it for you—your property is being listed exclusively. **1925** P. L. Melberg *Realty Salesman* 138 Well bought is half sold... **1926** Hinkle & Deihl *Real Estate Merchandising* vii. 95 Every salesman's attention..is called to listings that have been renewed... **1963** *Urban Successful Real Estate Practice* 78 Many brokers will take a listing at any price, hoping to get soon after the owner will accept. **1968** *Globe & Mail* (Toronto) 17 Feb. 45 (Advt.), To inspect this new listing please call Viola Roper. **1972** J. L. Gale *Listing Real Estate* p. xix, Listings are the inventory, the stock on the shelves, the merchandise of the real estate broker.

4. An entry in a catalogue, telephone directory, or other list.

1962 K. Orvis *Damned & Destroyed* v. 43, I..reached out for the telephone directory. Helen Ashton had no listing. **1965** J. Clavé (*title*) College textbooks, Supplement I: a classified listing of 9,500 textbooks used in 36 colleges. **1969** *John Edward's Mem. Foundation Q.* V. 125 This complete listing of Cash's recordings was compiled by John Smith. **1971** *Amer. N. & Q.* Sept. 10/1 Every lowe listing [in a bibliography] is found either in the main body or in an appendix. **1971** *Post Office Telephone Directory Section 101: London Postal Area* 6 The alphabetical listings under 'Telephone Service'.

listing, *vbl. sb.*³ Add: The action of List *v.*³ 4.

1805 R. Parkinson *Tour Amer.* 165, I was near two months getting a plough made, therefore I hired for the listing (as they call it). **1887** *Sci. Amer.* LVI. 65 The listing is a ridge in the face that is close to the edges of the furrow on each side, a row of weeds springs up. **1935** *Nature* 17 Aug. 251/1 'Blown-out' lands are those where the drift of loose soil has been to some extent countered. This process, known to farmers as 'listing', consists of specially deep ploughing with the aid of motor-driven tractors... The furrows may be as many as fifteen feet apart.

li-sting, *vbl. sb.*⁴ *Basket-making.* [f. List *v.*³ + -ing¹.] (See quots. 1912, 1953.) Hence **listing-skein.** Cf. *listed *ppl. a.*²

1912 T. Okey *Introd. Basket-Making* 135 *Listing*, an additional skein or skeins wired in with the lapping of skein handles. **1953** A. G. Knock *Willow Basket-Work* (ed. 5) 36 *Listing*, a method of adding a raised form of ornamentation to a lapped handle by the use of one or more additional skeins, called listing-skeins... The listing on any handle can be added after the handle has been lapped. *Ibid.*, Diagram 26 shows a simple listed handle employing three listing-skeins. **1961** L. G. Allbon *Basic Basketry* vii. 53 Another method of decorating a wrapped handle would be with listing. **1972** D. Wright *Baskets & Basketry* ii. 61 Listing is worked over the interwoven No. 3 or 4 cane.

listing, *ppl. a.*² [f. List *v.*⁵ + -ing².] Of a ship: heeling, inclining to one side.

1923 *Public Opinion* 30 Mar. 312/3 Six projectiles struck the listing Iowa.

Lisztian (li·stiăn), *a.* (*sb.*) [f. the name of *Liszt* (see below) + -ian.] Of, pertaining to, or characteristic of the Hungarian pianist and composer Ferencz (Franz) Liszt (1811–86) or his music. Also as *sb.*, an adherent or imitator of Liszt.

1890 G. B. Shaw *London Music 1888–89* (1937) 308 Such Lisztian hero-worshippers as Herr Stavenhagen and the late Walter Bache. **1921** A. Huxley *Crome Yellow* xxi. 247 A brilliant Lisztian tremolo. **1934** C. Lambert *Music Ho!* iii. 165 The Lisztian symphonic poem. **1947** A. Einstein *Mus. Romantic Era* vii. 70 Genuine Lisztians or innovators like Smetana..wrote no more symphonies. **1947** N. Cardus *Autobiogr.* 208 In the 'Sanctus' we have the inspired religious tone of how many Lisztian and other symphonic poems. **1963** *Times* 24 Jan. 14/5 The late Lisztian keyboard style. **1971** *Daily Tel.* 26 Apr. 9/2 [He] produced moments of Lisztian abandon in which the whole orchestral palate crowded in on the keyboard.

lit, *ppl. a.* Add: (Further *fig.* examples.)

1922 M. A. von Arnim *Enchanted April* i. 17 She listened to her impetuous, odd talk and watched her lit-up face. **1936** R. Campbell *Mithraic Emblems* 18 My own lit heart, its rays of fire.

b. *slang.* Drunk (see also quots. 1933 and 1971). Freq. const. *up.*

1914 'High Jinks, Jr.' *Choice Slang* 12 *Lit up*, intoxicated. **1918** M. Gidder *War Birds* (1927) 82 We walked into the vamp's house. We all got lit and had a hell of a time. **1922** *Daily Mail* 16 Dec. 10, I am afraid I was rather tight—certainly lit up. **1933** *Amer. Speech* VIII. ii. 27/1 When one has contracted the habit or is under the immediate influence of the drug, he is *lit up*. **1938** G. Greene *Brighton Rock* iii. i. 109 If I hadn't been a bit lit this wouldn't have happened. **1939** M. Allingham *Mr. Campion & Others* i. ii. 37 Driving round the country with a topper over your eyes and a blanket round your neck at three o'clock in the morning... You must have been lit. **1958** Wodehouse *Spring Fever* xviii. 189 A lit-up Augustus Robb should be considered, provide a spectacle which nobody ought to miss. **1961** R. Hyams *Not in our Stars* xvii. 220 Some of the lads a bit lit, eh?' 'Who's this in the hedge?' **1971** E. E. Landy *Underground Dict.* 121 *Lit up*,..under the influence of a narcotic.

lit., Lit., colloq. abbrev. of (*a*) Literature. Cf. *Eng. Lit. (*b*) Literary *a.* Also used *absol.*, = literary student, literary magazine. Also *lit. crit.*, literary criticism; *lit. ed.*, literary editor; *lit. sup.*, literary supplement.

(*a*) **1890** [see *Eng. Lit.]. **1890** *Geo. Eliot Let.* 11 Feb. (1956) V. 73 The lecture or march ten..apart from both in Greek and Roman lit. **1946** L. Durrell *Let.* 20 Oct. in *Spirit of Place* (1969) 87 In Athens I am going to see Seferis and Katsimbalis and give modern lit a boost in the press. **1950** *Observer* 8 Mar. 22/2 Chadwick's opposition to tacking on 'Lit' to 'Lang', followed by his decision to leave the English Tripos. **1964** W. Markfield *To Early Grave* (1965) xii. 251 Perhaps if I should ever give a regular lit. course. **1977** P. Gedges *Ottawa Allegation* xiii. 173 She..worked in publishing... She was into Canadian Lit. before he could draw breath. **1979** *New Yorker* 21 Apr. 109/3 You don't get much of that in Russian lit.

(*b*) **1895** W. B. Yeats *Let.* 20 Jan. (1954) ii. 245 Not one word was said about the Irish Lit Society and Fred Dowden expressed scorn for the Irish Lit movement. **1895** W. Gore in *Inlander* Nov. 64 *Lit*, literary student. **1900** *Dialect Notes* II. 43 *Lit*, a literary Monthly, Quarterly, etc., a student publication. **1930** *English Jrnl.* XIX. 632 Whatever wit or lightness of heart characterizes the magazines appears in the East; the Western 'lits' are in dead earnest. **1932** H. Nicolson *Diary* 29 Oct. (1966) 122 Kingsley Martin..wants me to become the literary editor... But I..could not expect to make more than £1,000 a year as Lit. Ed. **1932** V. Woolf *Let.* in K. Martin *Editor* (1968) ii. 30, I used to try to write regularly for The *Times Lit. Sup.* **1935** N. Mitchison *We have been Warned* v. 437 He showed her his reviews... The *Lit. Sup.* had been dull, the *New Statesman* annoying. **1936** L. C. Douglas *White Banners* vii. 171 Naturally the 'Lits' began to view Professor Ward with a new respect. **1962** M. Lamb *Summer Bird-Cage* xi. 60 'I could have guessed that from his books.' 'They lack compassion.' 'How beautifully, how lit. critically you put it!' **1963** 'N. Blake' *Deadly Joker* ii. 33 The Americans had..begun to make an industry of lit. crit. **1968** *Listener* 21 Jan. 102/1 the late-agent literit. **1968** E. McGirr *Lead-Lined Coffin* ii. 44 Ecstatic but making profitable divine noises over the Sunday lit. sups. **1970** *Times Lit. Suppl.* 6 Apr. 401/5 The refined worry-beads of lit-crit jargon.

litchi. Add to etym.: [First used as a generic name in P. Sonnerat *Voyage aux Indes Orientales* (1782) III. 255.] Substitute for def.: The fruit of an evergreen tree, *Litchi chinensis*, of the family Sapindaceæ, native to southern China but widely cultivated in tropical countries elsewhere; the fruit is a large berry with a rough, brown skin and sweet, white flesh, which is eaten fresh or preserved. (Later examples.)

1908 *Daily Chron.* 12 Nov. 3/4 Lychees, pine-apples, pears, cranberries, dates, figs, medlars and mangoes swell the number of fruits. **1933** *Punch* 8 Feb. 144/1 We never dreamed that it [*sc.* tinned fruit] would appear..in such dazzling variety..from loquats to li-chees. **1938** *Nature* 14 May 866/2 The litchi has arrived in much larger quantities lately and been much appreciated. **1953** R. Campbell *Mamba's Precipice* 137 Monkeys lived on the beautiful lychees and loquats. **1969** *Listener* 3 July 15/1 You'd never know, looking at me, I had eaten..a lychee in a flower called Reading. **1960** *Oxf. Bk. Food Plants* 104/1 Litchis are most usually eaten fresh, but are also sometimes canned for export or preserved in syrup. **1972** A. F. Simmons *Growing Unusual Fruit* 175 The litchee can..be grown in Britain in a large green-house.

-lite. Add: Also used in forming the names of some rocks, as *ijolite, Phonolite.

literacy. Add: Also *transf.*

1943 *Amer. Mag.* Mar. 104/1 To help many of the priority-stricken people to set their feet on the path of education, manual dexterity, and economic literacy. **1962** *B.B.C. Handbk.* 33 Our skills in the understanding of the medium [*sc.* television] and our own visual literacy are growing all the time. **1969** *Times* 4 Nov. 25/2 A project aimed at raising the level of economic literacy.

literæ humaniores, var. *litteræ humaniores.

literal, *a.* and *sb.* Add: **A.** *adj.* **3. c. literal-minded** *a.*, having a literal mind; characteristic of one who takes a matter-of-fact or unimaginative view of things. Hence *literal-mindedness*.

1869 *Wesleyan-Methodist Mag.* Jan. 28 An old friend, whom we used to call 'Bacon', because he..was a literal-minded man. **1905** J. L. Lowes in *PMLA* XX. 816 A strangely literal-minded age..which..was taking..interpretation of the charming fiction of the Prologue. **1927** *Glasgow Herald* 12 Oct. 10 The gentleman married the lady on the strength of her literal-mindedness. **1941** (see *analyst 6). **1944** *Mind* LIII. 258 One of my aims is..to formulate, in a manner which is free from the literal-mindedness..a literal-mindedness, the epistemological idea and significance of pure semantics. **1957** *Essays & Stud.* X. 18 It delivers us from too much matter-of-fact and from the dreary flats of literal-mindedness. **1974** M. Price *R. Kipling* 80/2 'If you won't retract the lies.' ('I'll blow out your goddam brains!') 'If I don't do certain things, you'll kill me?' Rudyard Kipling's most literal-minded, with infuriating English primness and literal-mindedness.

B. *sb.* **3.** *Computers.* An operand in a program which directly specifies the value of a constant, or defines itself rather than serving as an address or label.

1960 *Nebula: a Programming Lang.* (Ferranti, Ltd.) iii, 1. A literal does not always have a numerical quantity. For example, the item address may have a value which is '216 London Road' and this whole phrase (viz., 216 London Road) is then a literal. **1960** D. N. Chorafas *Programming Syst. for Electronic Computers* ix. 107 Commercial Translator distinguishes among three different types of constants: defined constants, literals and basic constants. *Ibid.*, A literal is a purely numerical constant which is introduced in a procedure statement as the need arises. **1968** S. M. Bernard *System/360 Cobol* i. 52 A literal is a self-defining value; that is, it does not have to be separately defined by the programmer. i. Non-numeric literals..2. Numeric literals. **1968** N. Chapin *360 Programming* ii. 19 The symbolic addresses used by the programmer take four main forms: self-defining values, literals, symbolic names, and relative addresses. **1970** O. Dopping *Computers & Data Processing* xix. 311 In certain languages..a literal must be surrounded by quotation marks. In those systems, the literal may even be a text constant. In that case, we could..write the instruction MOVE 'SUM' TO ..ordering the computer to print the word sum. **1971** L. Coddington *Quick Cobol* ii. 16 Literals are: 377, 'FINAL'..a numeric literal..and alphanumeric literals (YES, NO, SMITH, ZZZZ) which consist of letters. These are not true data-names ('SMITH' is not only the name of a location in store, but the contents of that location. The numeric literals need no explanation.

literally, *adv.* **3. b.** (Further examples of the improper use.)

1902 *Daily Chron.* 10 Dec. 7/2 A contemporary states that Koebill has been 'literally coining money' in England. **1906** *Westm. Gaz.* 15 Nov. 2/1 Mr. Chamberlain literally bulldozed over it with her (?)... **1928** K. Macaulay *Mystery at Geneva* xiv. 72 The things 'they' say! They even say..that 'literally' bears the same meaning as 'metaphorically' (who was literally a mother to him,' they will say). **1960** V. Nabokov *Invitation to Beheading* iii. 31 And with his eyes he literally scoured the corners of the cell. **1973** *Good Food Guide* 170 'Crabs and lobsters are literally so to be found crawling round the floor waiting for an order,' reports an affable visitor.

literarism (li·tĕrăriz'm). [f. Literar(y *a.* + -ism.] = Literaryism.

1942 Partridge *Usage & Abusage* 173/1 *Literarisms* are either the journalese of the cheap literary litterateurs might also be stigmatized as high-brow) or such unusual words as are used only by the literary or learned. **1963** A. Huxley *Lit. & Sci.* i. 5 Snow or Leavis? The bland scientism of *The Two Cultures* or..the violent and ill-mannered, the one-track, moralistic literarism of the Richmond Lecture? **1970** F. R. Leavis in *Times Lit. Suppl.* 23 Apr. 441/4 The term 'literarism' was in fact coined by the late Aldous Huxley for use against me.

literary, *a.* Add: **3. b.** *literary dinner, lunch(eon), party, prize.* Also *literary adviser*: one who gives advice or information on literary matters; *literary agent* (see quot. 1960); also *literary agency*; *literary circle* (see Circle *sb.* 21); *literary criticism* = Criticism 2 (of works of literature); so *literary critic*, *literary-critical* adj.; *literary editor*: (*a*) the editor of the literary section of a newspaper; (*b*) the editor of a book of collected writings; so *literary-editor* vb., *-editorship*; *literary executor* (see Executor 3); *literary world* (see World *sb.* 16 b).

1881 M. Edgeworth *Let.* 6 Jan. (1971) 469 He..criticises so well—not as a mere literary critic vapouring about authorities. **1840** Macaulay in *Edin. Rev.* Jan. 520 In 1698, Collier published his 'Short View'..., a book which threw the public mind into excitement. **1845** H. C. Robinson *Diary* 27 Jan. (1967) 254 Mrs. Jameson.. is now devoted to her literary club... **1898** H. G. Wells 28 June 527 (Advt.), Literary Agency—Mr. F. G. Tomlins..is desirous to announce that a Twenty years' experience with the Press and Literature..enables him to give advice and information to Authors, Publishers and Persons wishing to communicate with the Public. **1853** C. M. Yonge *Heir of Redclyffe* I. xv. 251 She wants to..get a more perfect exclusiveness for that literary circle. **1929** *Ibid.* (of G. H. Lewes *Let.* 12 Feb. in Geo. Eliot *Lett.* (1954) II. 156 I am, as I hope, literary editor and shall probably be a literary circle. **1857** L. G. Robert *...* **1885** *Daily* 13 Feb. I think I shall come his own literary Agent. **1862** ... *Let.* 10 May in [unclear] **1896** *Star* (N.Y.) [?] ... **1898** ...

145 Obstacles..may be a blessing in disguise to half-baked literaries. **1928** *Amer. Speech* IV. 129 In many districts a 'literary' is held every Friday night, when the 'Sandbillers' of this district recite and sing and debate. **1929** E. G. Barnard *Rider Cherokee Strip* 157 We spent a happy winter at this work and visiting our neighbours and going to the literaries.

‖ **litera scripta** (li·tĕrā skri·ptă). [L.] The written word.

1864 J. S. Le Fanu *Uncle Silas* I. xxi. 256 Hence-forward all is circumstantial evidence..except the *litera scripta*, and an odd scrap of paper..must contribute.. what it can. **1910** R. Bridges in *Essays & Stud.* I. 48 The *litera scripta* has an enormous power.

literate, *a.* and *sb.* Add: **B.** *sb.* **4.** (*Lady*) *Literate in Arts*, the title conferred on the holder of a higher certificate for women issued at St. Andrews University. Abbrev. *L.L.A.*

This diploma was discontinued in 1931.

1881 *St. Andrews Univ. Cal.* 205 Any Candidate who passes in four subjects, [etc.]..will receive the title of Literate in Arts (L.L.A.). **1891** R. F. Murray *Scarlet Gown* 122 An L.L.A. or two. **1899** *Chambers's Jrnl.* ... **1922** *Daily Record* 30 July 3 Lady Literate in Arts. **1901** ... *L.L.A. Examination, Diploma, & Title for Women* (Univ. St. Andrews) 8 There is no limit as to age in the L.L.A. Examination.

‖ **littérateur.** An occas. spelling (in English works) of Littérateur.

1806 [see Littérateur]. **1865** E. Dowson *Let.* 21 Dec. (1967) 325, I suppose you have heard of the abortive petition which he started among French littérateurs for the pardon of Oscar Wilde. **1928** W. Stevens *Let.* 14 Oct. (1967) 250 Here I am, the least of littérateurs during the last year or two.

literation. Add: (Examples.)

1928 Hardy in F. E. Hardy *Later Yrs. Hardy* v. 174 His aim in the exact literation of Dorset words is not necessarily to exhibit humour and grotesqueness. **1928** J. Sykes *Mary Anne Disraeli* ii. 85 'D'Israeli and Mrs. D'Izzy'—an unconscious literation of the familiar diminutive.

b. The method or style of making letters.

1926 *Times Lit. Suppl.* 10 June 390/2 To sacrifice..the exquisite literation that in the old hands delights us like a poem.

literature. Add: **3. b.** (Later examples.)

1939 [see *normalizable *a.*]. **1965** [see *recollement 2]. **1971** *Nature* 25 June 490/1 We have searched the literature for reliable radiometric ages for Late Pre-Cambrian glaciogenic rocks, but they seem to be rare. **1973** *Sci. Amer.* June 55/3 A voluminous scientific literature was accumulated on..the spontaneous metal-catalyzed reactions each year on the normal vibrational modes of molecules in liquids and on optical phonons in crystals.

c. (Later examples.)

1938 Wodehouse *Code of Woosters* ii. 8 It is some literature from the Travel Bureau. **1962** *Observer* 4 Mar. 3/2 (Advt.) Please send me literature from the Yugoslav National Tourist Office. **1973** D. Francis *Slay-Ride* viii. 98, I talked my throat dry, gave away sheaves of persuasive literature.

literatus. (Later examples.)

1972 *Times Lit. Suppl.* 19 May 563/3 He had already made literati his own dominion. *Ibid.* 18 July 889/2, I am..a writer of literate to the literati, and I join with a literatus in conversation about the trash we read in childhood.

lites. Delete 'obs.' and add later example.

1959 B. Vesey-Fitzgerald *Cat Owner's Encycl.* 69 Lungs (commonly known as 'lites'), whether of cow, sheep or horse, are very cheap.

‖ **litham** (li·þă·m). Also lisam. [ad. Arab. *lithām* veil.] A veil of cloth wound round the head leaving only the eyes uncovered and worn by the men of the Tuareg people of the central Sahara desert.

1839 E. W. Lane tr. *Thousand & One Nights* I. vi. 467 The Khaleefeh then put on himself the fisherman's jubbeh and turban, and, having drawn a lithám over his face, repaired to the fisherman. **1860** H. B. Tristram *Great Sahara* ... **1874** ... *Bib. Jrnl.* IX. 219/1 The *lithám* or shawl-muffler of the Tuareg, wound round the mouth to keep out the blown sand of the desert. **1903** W. J. H. King *Search for Masked Tawareks* xv. 133 He so stood..slightly raising with his long slender fingers the upper fold of his *litham* or mask. **1966** M. Woodhouse *Tree Frog* xxi. 157 Mohammed Jalil al Mutraq sat still..the veil, the litham of blue..secret in front of him.

lithian (li·þiăn), *a.* *Min.* [f. Lith(ium + -ian 2.] Of a mineral: having a (small) proportion of a constituent element replaced by lithium.

1930 W. T. Schaller in *Amer. Mineralogist* XV. 571 Lithium—lithian. **1953** *Ibid.* XXXVIII. 91 The following abbreviations of lithian muscovite illustrate the close structural similarity to normal muscovite. **1964** *Ibid.* XLIX. 398 (*heading*) Lithian hureaulite from the Black Hills.

lithic, *a.*¹ **2.** (Further examples in *Archæol.*)

1946 F. E. Zeuner *Dating Past* vii. 208 The lithic industries of the Grotte de l'Observatoire were described by Boule. **1971** *Nature* 6 Aug. 383/2 Although other human remains that may be attributed provisionally to *Homo erectus* have been found in Olduvai, the discoveries at [site] WK are the first occasion on which a well represented lithic industry has been directly associated. **1971** *World Archaeol.* III. 144 There is a need for studies of lithic technology.

lithifaction (liþifæ·kʃən). [f. Lithi(fy *v.* + -faction (cf. *petrifaction*).] = Lithification.

1893 *Compte Rendu* 5me *Sess. Congrès Géol. Internat.* 160 The formations of the Coastal plain range in age from Pleistocene to early Cretaceous or late Jurassic.; all, indeed, are commonly unconsolidated and lithifaction is local and exceptional. **1921** I. G. Gass et al. *Understanding Earth* xix. 164/1 Sediments in various degrees of lithifaction. **1972** *Islander* (Victoria, B.C.) 1 Oct. 13/3 According to James T. Fyles, geologist with the department of mines, these concretions are formed during the process of lithifaction of sand.

lithification. (Earlier and later examples.)

1892 *Amer. Jrnl. Sci.* Dec. 468 Even the former moderate temperature..would be sufficient to produce incipient change—at least lithification, if not metamorphism. In fact, lithification of sediments will probably take place under heavy pressure even at ordinary temperature. **1971** I. G. Gass et al. *Understanding Earth* xiii. 185/1 The lithification of the soft sediments after deposition.

lithify, *v.* Add: Chiefly as li·thified *pa. pple.* and *ppl. a.* (Further examples.)

1895 *Geog. Jrnl.* ... **1917** ... the normal beach-rock—lithified sands containing a few boulders or smaller blocks. **1965** J. Dewey *Termier's Erosion & Sedimentation* x. 215 When clays are lithified by compaction and cementation, they become mudstones or limestones with small sand cracks on their surfaces. **1971** *Nature* 2 Apr. 287/1 The rock was thus already consolidated, and lithified, at the beginning of the Pleistocene.

lithiophorite (li:þi,ǫ·fŏrait). *Min.* [ad. G. *lithiophorit* (O. Breithaupt 1870, in *Jrnl. f. prakt. Chem.* CX. 205): see Lithium, -o, and -ite¹.] A mineral consisting of aluminium, lithium, and manganese $(Al,Li)MnO_2(OH)_2$, found as bluish-black monoclinic crystals.

1871 *Jrnl. Chem. Soc.* XXIV. 205 Lithiophorite is amorphous, occurs in compact botryoidal and reniform masses, in flat shell-like forms, and in pseudomorphs after calcspar. **1953** *Amer. Mineralogist* XVII. 142 Material which in the past has been classified as psilomelane may actually be..lithiophorite, previously considered as a variety of psilomelane containing lithium and aluminium, but which in x-ray pattern shows to be a distinct species. **1952** *Acta Crystallogr.* V. 676/2 Lithiophorite has one of the major structure with alternate sheets of MnO_2 and $(Al,Li)(OH)_2$ octahedra placed one on top of the other. **1971** *Mineral. Mag.* XXXVIII. 618 Lithiophorite is..one of the major manganese ore minerals..of nodular manganese in Australian soils.

lithistid, *a.* and *sb.* Add to etym.: [(O. Schmidt *Grundzüge einer Spongien-fauna des Atlantischen Gebietes* (1870) ii. 21).] (Example of sb. and later example of adj.)

1884 J. E. Taylor *Our Common Brit. Fossils* i. 26 Sections of it show it to belong to the *lithistids*. **1972** P. Mellersh *Invertebr. Anat.* (ed. 2) v. 98/2 Desmas are of various forms which become fused together to form a solid meshwork, or lithistid skeleton.

litho (li·þo, now more usually lai·þo). Add: Also, abbrev. of Lithographic, Lithography (in Dict. and Suppl.). (Further examples.)

1903 *Brit. Printer* Jan. 30/2 The *Gazette* is the organ of the litho trade. **1941** Wheeler *Newnes Mod. Printing* (ed. 3) II. xxiii. 28/4 A number of transfers may be laid down on one stone, or with litho plates, even up to..hundreds, or 200 impression on the back. **1946** A. Kirk in H. Whetton *Pract. Printing & Binding* xvi. 190/1 Part of the very extensive multi-colour catalogue business of America..is produced by the web by litho-offset at very high speeds. **1948** *Sci. News* VII. 100 A simple processing technique resulted in litho printing plates equal for use. *Ibid.*, Litho printing, therefore..an ideal medium. **1958** *Times* 13 Jan. 12/1 Photo litho..is worked. **1959** *Jrnl. Inst. Printing*... to litho offset ink is normally stiffer than that used for letterpress. **1973** *Nature* 27 Apr. 561/2 Ordnance Survey maps..should be printed by litho. **1973** *Times Lit. Suppl.* 7 Dec. 1492/2 One of the other original printings of the 1:2500 sheets were quickly by the Ordnance Survey by litho directly from the typescripts. **1973** *Times Lit. Suppl.* 7 Dec. 1492/1 One advantage of which was that the quality of the edge. **1973** *Ibid.* 1518/1 Letterpress, although progressively used by litho, still prints about 30 per cent of bookwork.

Also (abbrev. of Lithograph *v.*) as *vb.*

1934 J. A. Lee *Children of Poor* (1949) vi. 205, I would print or litho circular leaflets and gummed-paper strips.

litho-. Add: **li:thofa·cies** *Geol.*, a facies (Facies 2 b) distinguished by its lithological character (see quot. 1949²); **lithogenesis** (examples); also, the formation of rock; **lithogeny** (b) *Geol.* (rare) = *lithogenesis* (in Dict. and Suppl.); **lithophane** (examples); also **lithophanie**; **li-thophi·le, lithophi·lic** *adjs. Geol.* and *Chem.* [ad. G. *lithophil* (V. M. Goldschmidt 1923, in *Skrifter utgit av Videnskapsselsk. I. Mat.-nat. Kl.* iii. 5); see -phil, -phile], applied to elements which are commonly found as silicates and are supposed to have concentrated in the outermost zone when the earth was molten; **li·thopone** [Gr. πόνος work, anything produced by work], a mixture of zinc sulphide and barium sulphate used as a white pigment in paint, linoleum, and printing ink, and as a filler in paper; **li·thosere** *Ecol.* [*sere], a plant succession having its origin on bare rock; **li·thosol** *Soil Sci.* [*-sol], any azonal soil consisting largely of imperfectly weathered rock fragments; **lithosphere**, add: in recent use, applied to the crust and the upper part of the mantle; formerly also used for the crust together with the whole interior portion of the earth, or the crust together with the entire mantle; (further examples); hence **lithosphe·ric** *a.*; **li:thostrati·graphy** *Geol.*, stratigraphy based on the physical and petrographic characters of rocks, rather than on fossils; so **li:thostratigra·phic, -graphical** *adjs.*; **lithotint** (later examples); so **li:thotinted** *ppl. a.*

1946 M. Kay in *Progr. Rep. Res. Comm. Amer. Assoc. Petroleum Geologists* 23 Jan. 4 The forms present in any population are influenced by age, and by..the lithic factors of lithofacies. **1949** R. C. Moore in *Mem. Geol. Soc. Amer.* XXXIX. 16 It seems clear that 'facies' should not be used in two or more senses..as the field of differentiation, and I suggest the term 'lithofacies'..as appropriate for such meaning. *Ibid.* 32 The rock record of any sedimentary environment, including both physical and organic characters, is designated by the term 'lithofacies'. **1958** *Bull. Amer. Assoc. Petroleum Geologists* XLII. 279 The term 'lithofacies' seems to have been introduced by the Russian geologist Eberzin (1940: *fide* Markevich, 1957). When and by whom it was launched in America is not clear from available references. Soon after 1945 it appeared in publications and was used in the sense given above. **1968** R. W. Fairbridge *Encycl. Geomorphol.* 921/2 (*heading*) Braided stream lithofacies. **1969** *Canad. Jrnl. Sci.*... Lithogenesis, the production or origin of minerals or rocks; lithogeny. **1937** Wooldridge & Morgan *Physical Basis Geogr.* vi. 82 By the period of lithogenesis, during which the rocks later to form the range are accumulated. **1906** G. W. Tyrrell ... **1963** ... *Stoneyside Lithogenesis* ... **1965** D. W. & E. E. Humphries tr. *Termier's Erosion & Sedimentation* 405 The cycle of geological phenomena comprises lithogenesis or petrogenesis, orogenesis, then glyptogenesis. **1888** J. J. H. Teall *Brit. Petrogr.* 437 Lithogeny, that department of petrology which treats of the formation of rocks. **1958** *Contrib. Cushman Found. Foraminiferal Res.* IX. 106/2 The Illinois cyclothem comprises a widely recurrent, repetitious succession of rock types whose lithogenesis reconds the environmental changes of the place and time in considerable detail. Various aspects of the lithogenesis, biology, and ecology of the sedimentary units in the Illinois and related kinds of cyclothems have been described. **1947** Ms. Kinnear ... 10/1 The lithophane process..involved the use of white biscuit plaques of varying thickness, which when held to the light, showed pictures made in various tints... **1948** ... *Jrnl. Soc. Arts* ... **1973** ... Some of the German porcelain factories made effective use of the lithophane for lamp shades and sconces. **1970** G. Savage *Dict. Antiques* 244/2 Lithophanes were first produced in Germany, and the object of the craftsman being to remove sufficient wax to give the desired amount of light transmission, building up his picture from dark and light passages. **1858** W. Chaffers *Marks Pott. & Porc.* (ed. 4) 431 At the Berlin manufactory Lithophanie..was invented. **1964** L. Dillon *Porcelain* xiv. 264 Another application of porcelain was to the 'transparencies' or *lithophanie*, in which the design, as seen by transmitted light, was given by variations in the thickness of the paste. **1923** *Mineral. Abstr.* II. 159 Corresponding with the zones of the earth, the following groups of chemical elements are distinguished: (1) Siderophile...; (2) Chalcophile...; (3) Lithophil elements of silicate fusions (O, Si, Ti, F, Cl, Al, Ca, Na, K, Be, Mg, Ca, V, &c.); (4) Atmophil elements. **1950** Rankama & Sahama *Geochem.* iv. 91 The lithophile metals are those whose ions have 8 electrons in the outermost shell. **1965** Phillips & Williams *Inorg. Chem.* I. xvi. 591 Goldschmidt quoted the following elements: Fe Co Ni [etc.]..as concentrating in the terrestrial sulphides (lithophil) and others in the silicates (lithophil). **1973** *Nature* 28 Sept. 201/1 The entry into or rejection of lithophile ions from silicate lattices is dependent on size and valency. **1881** *Jrnl. Soc. Arts* 16 Aug. 606/1 Elements like Be and Th..are strongly lithophilic. **1884** Knight *Dict. Mech.* Suppl. 551/2 Lithopone. **1902** *Jrnl. Soc. Chem. Industry* 31 Mar. 427/1 'Lithopone' is prepared by mixing together solutions of barium sulphide and zinc sulphate, the precipitate of ZnS, BaSO₄ being then washed, dried, and ignited. **1923** U. R. Evans *Metals & Metallic Compounds* II. 159 Lithopone is a comparatively cheap pigment. **1961** J. P. Casey *Pulp & Paper* (ed. 2) III. xx. 1830 Lithopone has been used in paper. Titanium oxides in cement paper by reacting with the acid in the slurry used to make lithopone. **1916** F. E. Clements *Plant Succession* iv. 182 While the xeroseres of rock and sand dune may be almost equally dry, the differences of hardness and stability result in very dissimilar stages. There may be distinguished as lithoseres and psammoseres. **1966** N. Polunin *Introd. Plant Geogr.* 234 A different characteristic xerosere we will take a lithosere initiated on bare rock. **1939** *U.S. Dept. Agric. Yearbk.* 1938 1171 *Lithosols* (skeletal soils).. An azonal group of soils having no clearly expressed soil morphology and consisting of a freshly and imperfectly weathered mass of rock fragments. **1968**

H. C. T. STACE et al. *Handbk. Austral. Soils* iii. 35 Lithosols are found throughout Australia wherever natural erosion has been active enough to maintain a thin soil cover. **1893** A. GEIKIE *Text-bk. Geol.* (ed. 3) 38 (*heading*) The solid globe or lithosphere. **1910** LAKE & RASTALL *Text-bk. Geol.* i. 8 The Lithosphere or solid part of the earth, so far as it is open to our inspection, consists of rocks. **1950** RANKAMA & SAHAMA *Geochem.* ii. 32 The Sial crust, which is the surface layer of the silicate shell of the Earth (the lithosphere), is composed of three groups of rocks of different origin. **1957** G. E. HUTCHINSON *Treat. Limnol.* I. iv. 222 The water content of the major part of the lithosphere, the great mantle of ultrabasic rock which composes most of the earth, is unknown. **1971** I. G. GASS et al. *Understanding Earth* xvi. 248/1 As a result of seismological studies, it has been realised that the Earth's outermost skin, or lithosphere, which exhibits appreciable strength and rigidity, extends well beneath both continental and oceanic crusts to depths of 50 or even 100 km. **1973** M. W. MCELHINNY *Palaeomagnetism & Plate Tectonics* v. 156 In this 'new global tectonics'.., now generally referred to as Plate Tectonics, it is supposed that a mobile, near-surface layer of strength (the lithosphere) plays a key role.. The lithosphere, which generally includes the crust and uppermost mantle, has significant strength, and is of the order of 100 km thickness. The asthenosphere, which is a layer of effectively no strength on the appropriate time scale, extends from the base of the lithosphere to a depth of several hundred kilometres. **1970** *Nature* 5 Sept. 1016/1 At ridge crests the lithospheric plates are thinned by the elevation of the geotherms as a result of mantle upwelling and emplacement. **1971** I. G. GASS et al. *Understanding Earth* xi. 153 (*caption*) Biospheric, lithospheric and atmospheric evolution on the primitive Earth. **1950** *Bull. Amer. Petroleum Geologists* XXXIV. 2365 Our repeated efforts to treat stage and zone as true time-stratigraphic units have met with failure. Since these are biostratigraphic or lithostratigraphic in character..we have no logical choice but to place them also in a category by themselves. **1970** *Earth-Sci. Rev.* VI. 270 The formation is the fundamental unit in lithostratigraphic classification. **1964** J. CHALLINOR *Dict. Geol.* (ed. 2) 144/2 Lithostratigraphical unit. **1969** *Proc. Geol. Soc.* Aug. 141 For the description of stratified rocks lithostratigraphical procedure is already generally agreed. **1956** *Bull. Amer. Assoc. Petroleum Geologists* XL. 2711 (*heading*) Factors in lithostratigraphy. **1969** *Proc. Geol. Soc.* Aug. 155 The boundaries between stages are based on biostratigraphy where possible, and on lithostratigraphy otherwise. **1938** *Archit. Rev.* LXXXIV. 177/3 Hullmandel's lithotint process, patented in 1840,..used a resin solution..which could be painted on to the polished tint stone so that it printed a modulated, instead of a flat tint. **1969** D. BLAND *Hist. Bk. Illustration* (ed. 2) vii. 250 The forerunner of the chromo-lithograph was the lithotint. *Ibid.* 251 One of the best litho-tinted books is *Original Views of London as it is* (1842).

lithodipyra (li:þodipaiə·rā). [mod.L., f. LITHO- + DI-[2] + Gr. πῦρ fire, as repr. 'stone twice fired'.] The name given to a kind of artificial stone by members of the Coade family when in 1769 they took over the factory in Lambeth where it was made (until *c* 1837), which stone (also called *Coade stone*) was claimed to have greater frost and heat resistance than natural stone and was much used for statues, monuments, and decorative work.

c **1778** (*title*) Coade's lithodipyra or artificial stone manufactory. For all kinds of statues, capitals, vases, tombs, coats of arms, & architectural ornaments &c. &c. **1910** *N. & Q.* 2 July 15/1 A monument to Edward Wortley Montagu, made of Coade's Lithodipyra, is in the west walk of the Cloisters of Westminster Abbey. **1928** *Connoisseur* Oct. 81 (*caption*) Plaque in Lithodipyra, from a design by J. Bacon, R.A. installed on the east front of Hooton Hall, Cheshire 1788. **1954** *Archit. Rev.* CXVI. 296/1 George Coade died in 1770 and can, therefore, have had little to do with the development of 'Coade's Lithodipyra Terra-Cotta or Artificial Stone Manufactory'.

lithograph, *sb.* **1.** (Earlier and later examples.)

1828 DISRAELI *Voy. Capt. Popanilla* ix. 96 It was a sublime lithograph. **1970** *Oxf. Compan. Art* 666/1 Lithographs have..taken on a number of appearances, ranging from simple linear designs made with pen or crayon to colour prints with the most varied effects of transparency and texture.

lithographed, *ppl. a.* (Earlier examples.)

1826 DISRAELI *Viv. Grey* I. II. x. 151 A bundle of Stewart Newton's beauties, languishing, and lithographed. **1829** H. C. ROBINSON *Diary* 13 Aug. (1967) 102 Knebel had shown me a lithographed manuscript.

lithographic, *a.* Add: **1.** (Further examples: cf. *LITHOGRAPHY 3.)

1885 *List of Subscribers, Classified* (United Telephone Co.) (ed. 6) 123 Lithographers and Lithographic Printers. **1892** A. POWELL *Southward's Pract. Printing* (ed. 4) i. 3 Lithographic printing is done with stones, zinc plates, &c. **1915** *Southward's Mod. Printing* (ed. 3) II. xxxiii. 281 Rotary lithographic machines work from zinc or aluminium plates carried on a plate cylinder. **1970** E. A. D. HUTCHINGS *Survey of Printing Processes* v. 75 Lithographic plates are made from both zinc and aluminium. **1972** P. GASKELL *New Introd. Bibliogr.* 269 Nineteenth-century lithography was a separate trade.. Towards the end of the century, however, powered lithographic cylinder machines were developed which had a productivity comparable with that of the letterpress machinery of the period, and they were followed around 1900 by lithographic rotaries which ran at yet higher speeds. These new machines, used in conjunction with photographic transfer methods of plate preparation, pointed the way to the integration of lithographic with general letterpress printing which took place during the first half of the twentieth century. **1973** *Brit. Printer* July 68/3 It was only with the increase in lithographic printing that photosetting began to 'find its feet'.

4. lithographic offset = *OFFSET *sb.* 10 b; **lithographic paper,** paper suitable for lithographic printing; **lithographic varnish,** a preparation of linseed oil used in inks for lithographic printing.

1915 *Southward's Mod. Printing* (ed. 3) II. xxxiii. 282 Lithographic Offset Printing. Offset printing is popularly supposed to have owed its introduction to a 'miss' made in printing on a lithographic cylinder machine. **1946** H. WHETTON *Pract. Printing & Binding* 446/1 Lithographic offset printing. **1960** G. A. GLAISTER *Gloss. Bk.* 239/2 Modern lithographic offset printing is mostly done in rotary presses. **1937** E. J. LABARRE *Dict. Paper* 167/1 Since the introduction of the offset printing process, however, all papers have become possible as 'lithographic' papers. **1963** R. R. A. HIGHAM *Handbk. Papermaking* vii. 202 Lithographic papers..are similar to offset cartridge, but generally cheaper and in a lighter substance. **1903** A. SEYMOUR *Pract. Lithogr.* x. 52 The most useful and the commonest form of reducing medium is a linseed oil product, known in its prepared state as a lithographic varnish. **1951** R. MAYER *Artist's Handbk.* x. 312 Lithographic varnishes, from which many types of inks for other printing purposes are also made, are heat-bodied linseed oils.

lithographically, *adv.* (Further examples.)

1952 in G. H. Bourne *Cytol. & Cell Physiol.* p. iv, (*Imprint*) First edition 1942. Reprinted (with corrections) 1945. Second edition 1951. Reprinted lithographically.. at the University Press, Oxford, 1952 from corrected sheets of the second edition. **1965** *Economist* 6 Mar. 1026/2 No model of Picasso can hide in obscurity and the olive-shaped outline of the head of Françoise Gilot is known, lithographically, to millions. **1967** E. CHAMBERS *Photolitho-Offset* xvii. 264 The artist has complete freedom of expression and application to produce an image from which impressions can be obtained lithographically.

lithography. Add: **3.** Also, a planographic printing process using metal or plastic plates with a sensitized coating on which the matter to be printed is fixed chemically, before the non-printing areas of the plates are damped and the remainder printed with greasy inks on flat-bed or cylinder presses. Cf. *DRIOGRAPHY, *OFFSET *sb.* 10 b, *PHOTOLITHOGRAPHY.

1906 *Brit. Printer* Dec. 296/2 The tendency of the day is undoubtedly towards..the utilization of lithography by typographers doing certain classes of work. **1932** *Jrnl. Soc. Chem. Industry* 9 Sept. T313/1 Lithography is based essentially on the adsorption of fatty acids by the metal. **1946** A. KIRK in H. Whetton *Pract. Printing & Binding* xvi. 190/1 Although the printing of daily newspapers is outside the present scope of lithography, the uninterrupted weekly production of *The Australian*, Melbourne, since 1929 may be instanced as but one outstanding achievement..in this direction. **1973** *Brit. Printer* July 68/3 The radically reduced cost of photosetting machines in recent years, their improved capability, and the continuing growth in the use of lithography have all been instrumental in increasing the demand for photosetting and allied equipment.

Lithol (li·þǫl). *Dye Chem.* Also **lithol.** Any of various azo pigment dyestuffs, many of which are the salts of diazo coupling compounds of β-naphthol and aromatic amino-sulphonic acids. *Lithol red*, any of various salts of the diazo coupling compound of 2-naphthylamine-1-sulphonic acid with β-naphthol, used as red pigments of moderate colour fastness.

Formerly a proprietary name in the U.S.

1903 *Official Gaz.* (U.S. Patent Office) 16 June 1891/1 Dyestuff. Badische Anilin & Soda Fabrik, Ludwigshafen, Germany. Filed April 24, 1903... The word 'Lithol'. Used since June 6, 1901. **1930** A. W. C. HARRISON *Manuf. Lakes & Precipitated Pigments* xii. 146 Lithol red being the most profitable of all the bright pigment reds, as regards such desirable qualities as strength, price and reasonable fastness to light in full shades, it is made in larger quantities than any other pigment red. *Ibid.* xiii. 176 Two cheaper bright yellows which may be used are Pigment Chlorine G.G. and Lithol Fast Yellow G.G., which are formaldehyde condensation products. **1947** L. S. PRATT *Chem. & Physics Org. Pigments* viii. 116 Lithols, as a class, are prepared by coupling the diazonium salts of Tobias acid..with β-naphthol, and then converting the resulting practically insoluble sodium lithols into the corresponding barium, calcium, or strontium products. *Ibid.*, The lithols are the most important single group of organic pigment colors. **1967** KARCH & BUBER *Offset Processes* vii. 269 Lithol reds range in shades from an orange to deep maroon. In between these extremes are some very brilliant and deep shades which are used where extreme permanency to light is not important.

li·thoprint, *v.* [f. LITHO(GRAPHY + PRINT *v.*] *trans.* To print by photolithography, usu. in reference to the production of copies of a typescript by this means. Also as *sb.* = LITHOGRAPH *sb.* Hence **li·thoprinting** *vbl. sb.*

1935 *Amer. Botanist* Jan. 39 The book is lithoprinted from type-written copy. **1947** *Amer. Speech* XXII. 136 It is too bad that this most important of volumes on its subject could not have been put in more solid form than lithoprinting. **1957** TRAGER & SMITH *Outl. Eng. Struct.* 7 The preliminary drafts were reproduced in lithoprinted form. **1969** *Sunday Times* 6 Apr. 30 His enchanting drawings (transformed into big, clear-coloured lithoprints in limited editions of 100 each) are in a gallery run by his mother.

1971 *New Scientist* 28 Jan. 206/2 The several hundred local natural history publications, often duplicated or lithoprinted, offer an easy first outlet for young artists.

lithops (li·pǫps). [mod.L. (N. E. Brown 1922, in *Gardeners' Chron.* 28 Jan. 44/2), f. Gr. λίθος stone + ὤψ face.] A small succulent plant of the genus so called, belonging to the family Aizoaceæ, native to Namaqualand, South Africa, and resembling small stones.

[**1922** *Gardeners' Chron.* 28 Jan. 44/2 Lithops, N. E. Brown. Very dwarf succulent plants, in nature growing buried in the ground with their tops scarcely, or not at all, rising above the level of the surface. *Ibid.*, Dr. Marloth is quite wrong in his identification of the plant, which I have no doubt whatever is a species of Lithops.] **1938** H. A. DAY *Flowers of Desert* iv. 148 It is most difficult to tell which are plants and which are pebbles when the two are mixed, as they are in the habitat of the *Lithops*, for the plants grow in the stony deserts of South Africa. **1966** E. PALMER *Plains of Camdeboo* xvi. 269 We have never found a Lithops on Cranemere, the nearest species that we know of being Burchell's Lithops to the north and another species, *Lithops terricolor*, to the west.

lithospermum. Add to etym.: [Adopted by Linnæus in his *Systema Naturæ* (1735) as the name of a genus.] Substitute for def.: A herb or sub-shrub of the genus so called, belonging to the family Boraginaceæ, native to Europe, northern Asia, or North America, and bearing white, yellow, or blue flowers. (Later examples.)

1900 J. M. ABBOTT in W. D. Drury *Bk. Gardening* ix. 295 Lithospermums are showy rockwork plants. **1937** *Daily Express* 10 May 12/4 The lithospermum's name [*sc.* 'Heavenly Blue'] is no exaggeration of its colour. **1966** J. BERRISFORD *Wild Garden* II. 166 With their brilliant blue flowers and evergreen leaves the lithospermums are most desirable garden plants.

lithothamnion (li:þoþæ·mniǫn). [mod.L. (R. A. Philippi 1837, in *Arch. Naturgesch.* III. 387), f. LITHO- + Gr. θάμνος shrub.] A calcareous, marine, red alga of the genus so called. Also *attrib.* Hence **lithotha·mnic** *a.*

1895 G. MURRAY *Introd. Study Seaweeds* 241 *Lithophyllum* forms thin stony plates of erect habit, while *Lithothamnion* gives rise to massive stony branches. **1935** J. E. TILDEN *Algae* viii. 362 Portions of the *Lithothamnion* crust were kept for many months. **1967** *Oceanogr. & Marine Biol.* V. 551 Both fauna and flora of this lithothamnic bottom are very abundant. **1972** *Sci. Amer.* June 62/1 A hitherto minor group of coralline red algae, the lithothamnions, now began to play an increasingly important role.

Lit. Hum. Abbrev. of *LITTERÆ HUMANIORES.

1939 L. MACNEICE *Autumn Jrnl.* xii. 49 If it were not for Lit. Hum. I might be climbing A ladder with a hod.

litmus. Add: *fig.*, as in *litmus test*.

1957 *Essays in Crit.* VII. 80 Their possession is as good a cultural litmus test as any I can think of. **1971** *Daily Tel.* Nov. 13/2 The litmus test comes with the old issue of whether the NUS should re-join the IUS. **1972** *Times* 21 Nov. 21/7 The litmus test will be the effect of the rise from 4p to 10p in the basic charge for an inpayment.

litoptern (litǫ·ptɜːn). *Palæont.* Also **litopternan.** [f. mod.L. name of order *Litopterna* (F. Ameghino 1889, in *Actas Acad. Nac. Córdoba* VI. 492), f. Gk. λιτός smooth + πτέρνη heel-bone.] An extinct South American ungulate mammal of the order so called. Hence **litopternine** (litǫ·ptɜːnain) *a.*

[**1891** E. D. COPE in *Amer. Naturalist* XXV. 688 The articulation in the Litopterna is of ungulate character.] **1925** C. R. EASTMAN tr. *Zittel's Text-bk. Palaeontol.* III. 123 The Litopternine suborder of ungulates is restricted to South America. **1927** HALDANE & HUXLEY *Animal Biol.* xi. 243 (*caption*) Extinct South American Litopternan. **1933** A. S. ROMER *Vertebr. Paleontol.* xvi. 316 Small litopterns..present in the oldest known Tertiary beds of South America, had primitive bunodont cheek teeth. **1971** *Nature* 15 Jan. 172/1 In some mammals, tapirs, elephants, the extinct astropotheres, pyrotheres, and litopterns, the external nares are posterior and large.

‖ **lit-par-lit** (lìparlì), *a. Geol.* [Fr., = 'bed by bed'.] Designating the intrusion of innumerable narrow, more or less parallel, sheets or tongues of magma into the bedding of rocks. Also as *adv.*

1896 *Q. Jrnl. Geol. Soc.* LII. 635 In..1890..Mr. Horne obtained confirmatory evidence of the 'lit par lit' introduction of granitic materials into the crystalline schists south-west of Strath Halladale. **1902** *Ibid.* LVIII. p. lxxv, M. Michel Lévy recognizes two types of intermixture: the one taking place by superposition, the other by injection lit par lit... In the latter type the compound rock consists of alternating folia of granite and sedimentary material. **1909** F. H. HATCH *Text-bk. Petrol.* (ed. 5) I. 18 When the intrusion is repeated between many planes of stratification, it is known as *lit-par-lit* intrusion. **1925** B. N. ODELL in E. F. Norton *Fight for Everest 1924* 291 It is composed of dark horizontally banded biotite gneiss alternating with bands of light granite, though in the upper part of the cliff the latter is represented by pegmatite, and the whole appears to represent a large-scale example of *lit-par-lit* injection. **1962** A. E. J. & C. G. ENGEL in A. E. J. Engel et al. *Petrologic Stud.* 42 The amphibolites are only locally shredded, injected *lit-par-lit*, or replaced by granite. **1968** K. R. MEHNERT *Migmatites* ix. 278 Banded gneisses and migmatites of the 'lit-par-lit'-type follow.

litre[2]. Add: (Earlier and later examples.)

In 1901 the litre was redefined as the volume of a kilogramme of water under specified conditions (see quot. 1957), which made it equal to approximately 1·000,028 cubic decimetres; this definition was abandoned in 1964 in favour of the original one (see quot 1965).

1797 *Jrnl. Nat. Philos.* Aug. 197 A vessel of a cubical form, having for its side one decimetre, or a cylindrical vessel of the same solid contents, has received the name of litre. It contains about two pounds of water, or twenty-five ounces of wheat. **1902** *Nature* 10 Apr. 538/1 Annexe iv..recapitulates the decisions of the Troisième Conférence Générale held at Paris last October, as to the definitions of the metric units, metre, kilogramme and litre. **1923** GLAZEBROOK *Dict. Appl. Physics* III. 777/2 The definition of the litre has no reference to the units of length, and the original intention that it should be equal to 1 cubic decimetre has been quite abandoned. *Ibid.* 778/2 In his final résumé of the whole observations M. Benoit gives 1 litre = 1000·027 c.c. as the most probable value. **1957** E. R. COHEN et al. *Fund. Constants Physics* ii. 5 The liter is defined as the volume of a kilogram of water, at standard atmospheric pressure and at the temperature of its maximum density, approximately 4°C. On a level of sufficient precision this is an ambiguous definition, since it does not specify the isotopic constitution of the water... We adopt..1 liter = 1000·028 ± 0·004 cm³. **1965** *Nature* 6 Feb. 553/1 The twelfth General Conference of Weights and Measures was held..during October 6–13, 1964... Resolution 6 (the litre) abolished the definition of the litre established in 1901 by the third General Conference, declared that the word 'litre' can be used as a special name given to the cubic decimetre and recommended that the name 'litre' shall not be used to express the results of volume measurements of high precision... It now reverts to its original meaning. The intention is that this litre shall only be used for ordinary transactions in trade and not for scientific purposes. **1969** *Physics Bull.* Feb. 58/2 It was a regrettable international decision in 1964 which reinstated the name litre for a cubic decimetre.

b. Preceded by a number or a word denoting a number so as to form adjs. denoting the capacity (i.e. the inside volume of the cylinders) of a motor vehicle or its engine, and used *ellipt.* for a vehicle having the specified engine capacity.

1927 A. HUXLEY *Let.* 25 Feb. (1969) 284 We have gone and bought a really rather tremendous car—an Itala six cylinder two-litres. **1951** *Engineering* 26 Oct. 533/2 It is a full six-seater, powered by a new 3-litre six cylinder engine developing 90 brake horse-power at 4,100 r.p.m. **1955** *Times* 3 May 10/5 Bruigi, driving a three-litre Ferrari, crashed into a cement road sign near Teramo. **1974** *Country Life* 17 Jan. (Suppl.) 34/1 Put the Triumph 2000 beside any other 2-litre in its class and there's very little in it for mpg.

litter, *sb.* Add: **6.** (sense 4) *litter-bag, -basket, -bin, -box, -bug* (*BUG *sb.*[2] 4 a), *-carrier, -cart, -lout*; (sense 5) *litter-mate, -sister.*

1968 *Punch* 19 June 892/2 Drop it sadly in your heavy white kraft paper car litterbag. **1958** J. CANNAN *And be a Villain* vii. 160 Oh, the mess they leave on our little common..in spite of the litter baskets. **1972** J. BROWN *Chancer* xiv. 191 There's a row of litter baskets each side of the gates. **1947** *Archit. Rev.* CII. 198 (*caption*) On the right a typical litter-bin. **1972** *Guardian* 28 Jan. 5/6 Clerical staff..have been..sweeping up, and emptying litter bins. **1953** *News Chron.* 2 June 4/4 What a mess there will be on Coronation Day unless people use the litter-boxes. **1947** *N.Y. Herald-Tribune* 16 Feb. 2/7 (*heading*) 47,000 subway 'litterbugs' pay $107,000 in fines in 1946 drive. **1959** *Times* 23 July 7/7, I rate such persons in the same category as litter bugs. **1971** *Guardian* 8 Dec. 24/8 He picks up any litter he can find..and he is apt to give litterbugs a severe dressing-down. **1915** J. LONDON *Let.* 26 Jan. (1966) 445 He has no litter-carriers to carry manure. **1967** Litter-cart [see **cloth-capped* adj.]. **1927** *Children's Newspaper* 25 June 8/2 It is time the Litter Lout was taken seriously in hand. **1972** *Guardian* 29 Mar. 24/1 The packaging industry had been made a scapegoat for the actions of the litter lout. **1921** *Genetics* VI. 122 These are assumed to be litter-mates for whom all or nearly all tangible environmental factors may be assumed to be in common. **1946** *Nature* 24 Aug. 258/2 A biological test of a solution in arachis oil of the vitamin A thus obtained showed growth-promoting activity in rats (ten litter-mate growth comparisons) of the order indicated by spectroscopic assay. **1972** *Sci. Amer.* Feb. 22/3 Rodents are small, inexpensive and bear large litters, so that littermates with the same genetic background can be assigned to different conditions. **1960** *Farmer & Stockbreeder* 1 Mar. 79/2 Mr. G. H. Lewis was not to be denied in the junior class, litter-sisters from him..taking first and second places. **1973** *Country Life* 15 Feb. 385/1 Miss S. G. Weall's fawn dog..collected two firsts and his litter sister..two seconds.

litter, *v.* Add: **6. a.** Also with *over*.

1888 MRS. H. WARD *R. Elsmere* I. I. iv. 90 The house was littered over with stanzas from the opening canto of a great poem on Columbus. **1890** A. CONAN DOYLE *Sign of Four* v. 84 The table was littered over with Bunsen burners, test-tubes, and retorts.

‖ **litteræ humaniores** (li·tərai hiumæni,-ǭə·riz). Also **literæ humaniores.** [L., lit. 'more humane letters'.] The humanities, secular learning as opposed to divinity; esp., at the University of Oxford, the study of Greek and Roman classical literature, philosophy, and ancient history; also, = *Greats* (GREAT C. 10).

1747 CHESTERFIELD *Let.* 24 Nov. (1932) III. 1057 Studies of the *Literæ Humaniores*, especially Greek. **1760** STERNE *Tr. Shandy* (ed. 3) II. xii. 61, I would not depreciate what the study of the *Literæ humaniores*, at the university, have done for me. **1883** *Sat. Rev.* 3 Nov. 581/2 We cannot conceive a better accompaniment to the study of *literæ humaniores*. **1907** 'B. BURKE' *Barbara goes to Oxf.* 43 'Greats', you must know, is a nickname for the school of 'Literae Humaniores'. **1911** BEERBOHM *Zuleika D.* iii. 30 He..was reading, a little, for Literae Humaniores. **1926** FOWLER *Mod. Eng. Usage* 240/2 *The Humanities*, or *Litteræ humaniores*, as an old-fashioned name for the study of classical literature. **1962** K. CHORLEY *Arthur Hugh Clough* iv. 72 In Clough's day there were but two schools open to men reading for honours—namely, Mathematics and *Literae Humaniores*. **1965** J. A. W. BENNETT in J. Gibb *Light on C. S. Lewis* 48 But *litterae humaniores* were his foundation, and they did in every sense make him more humane, enlarging his responses not restricting them. **1972** *Univ. Oxf. Examination Decrees* I. 120 The Subjects of the Honour School of Literæ Humaniores shall be (I) Greek and Roman History, (II) Philosophy, (III) Greek and Latin Literature.

li·tterer. [f. LITTER *sb.* 4 + -ER[1].] One who throws or drops litter.

1928 *Sunday Dispatch* 29 July 12/2 Every corner of the finest streets in London is disfigured with the..manifold refuse of the litterer. **1958** *Observer* 10 Aug. 8/7 The Act.. includes parish councils among the authorities empowered to proceed against litterers.

littering, *vbl. sb.* Add: **1. d.** The action of throwing or dropping litter.

1960 *Times* 14 Sept. 12/7 There is something of the threat of doom in the perfunctory notice of New York pavements 'Littering 25 dollars'. **1970** D. E. WESTLAKE *Hot Rock* (1971) I. i. 13 He threw the Kleenex on the sidewalk. Littering. **1972** *Jrnl. Soc. Psychol.* LXXXVII. 324 The interview dealt with littering..'should it be everyone's responsibility to pick up litter?'

little, *a., adv.*, and *sb.* Add: **A.** *adj.* **1. b.** (Further examples.)

1785 J. LATHAM *Gen. Synopsis Birds* III. I. 90 Little Egret... Size of a Fowl: length near a foot: weight one pound. **1802** G. MONTAGU *Ornith. Dict.* I. s.v. *Egret*, Little Egret. **1908** R. LYDEKKER *Sportsman's Brit. Bird Bk.* 248 Of the little egret..sixteen individuals appear to have been recorded from the British Islands during the last century. **1919** Little-smelt [see *GRUNION]. **1953** *Calif. Almanac* 36/2 On certain moonlit nights in the spring great numbers of small fish, termed 'grunion' or 'little smelt' appear along the beaches. **1971** *Country Life* 18 Feb. 356/3 Neither the little egret nor the common heron have been proved to breed in the [Ebro] delta, though both are present.

c. *Little Witham*: in phrases.

1560 J. HEYWOOD *Fourth Hundred Epygrams* sig. A7 Whens come great breeches? from little wittam. **1595** R. WILSON *Pedlers Prophecie* sig. B4v At litle Wytham seuen yeares I went to schoole. *a* **1661** FULLER *Worthies* (1662) Lincs. 153 *He was born at Little Witham*... It is applyed to such people as are not overstock'd with acutenesse. **1787** GROSE *Provincial Gloss.* s.v. *Essex*, He was born at Little Wittham. A punning insinuation that the person spoken of wants understanding. Ray places this proverb in Lincolnshire. **1932** E. WEEKLEY *Words & Names* x. 151 The stupid are said to be 'born at Little Witham'.

2. ¶ **b.** *little language*: also *transf.*

1863 *Fraser's Mag.* Feb. 152/1 She carried on hip a prize baby, a most 'doody' thing, to quote the 'little language'. **1866** MRS. GASKELL *Wives & Daughters* II. xxiv. 244 Some innocent sentences of love,..little sentences in 'little language' that went home to the squire's heart. **1922** O. JESPERSEN *Lang.* viii. 144 It would not do, however, for the child's 'little language' and its dreadful mistakes to become fixed. **1944** H. G. WELLS *'42 to '44* 142 The first thing two lovers set about doing is..to devise a little language of their own.

c. In collocations *little brother, sister*: younger (cf. 2 a). Also *fig.*

1611 BIBLE *Song of Solomon* viii. 8 We haue a little sister, and shee hath no breasts. **1799** JANE AUSTEN *Let.* 21 Jan. (1952) 57 Our own particular little brother got a place in the coach last night. **1859** THACKERAY *Virginians* II. xii. 93 Your brother and mine are gone to see our little brother at his school at the Chartreux. **1876** C. M. YONGE *Womankind* xvi. 126 In no case should they go without a more real chaperon than a maid or a little sister's governess. **1940** G. B. SHAW *Geneva* (1946) IV. 123 Ruritania is, so to speak, our little sister, and..if you laid a finger on her..we should be obliged to knock the stuffing out of you. **1949** R. CHANDLER (*title*) The little sister. **1974** *Country Life* 21 Nov. 1573/3 At present this dehydrated food is only available to caterers and food manufacturers, although its little sister Vegex (dehydrated vegetarian version) is on sale in some health-food shops.

d. *little brother* (see quot. 1928 and cf. *BIG *a.* 3 g); also *attrib.* and in extended use.

1928 *Daily Tel.* 10 July 16/2 Ninety-six youths will leave Tilbury to-day to take up farm work in Australia... Captain R. T. Thornton..will visit many of the 1,400 'Little Brothers' who have gone to the Commonwealth. **1962** *Times* 25 Apr. 11/6 'Little brother' organizations formed mostly of young people.

e. Phr. *to laugh like little Audrey*: to laugh heartily (esp. at a serious situation). Also *attrib.*, of a type of joke, a *CRUELLIE.

1939 C. MORLEY *Kitty Foyle* xiv. 143 She laughs like Little Audrey. **1959** I. & P. OPIE *Lore & Lang. Schoolch.* v. 82 Crazes for limericks, Little Audrey jokes, Knock-knocks, and Shaggy Dog stories. **1972** J. AIKEN *Butterfly Picnic* ix. 165 If I choose to..laugh like little Audrey when I'm all knotted up..who the hell's got the right to forbid me?

13. Little American (cf. *Little Englander*), **Americanism; little black dress** (or *frock*, etc.), a simple black garment suitable for a woman to wear at most kinds of relatively formal social engagements; **little chief hare** *N. Amer.* [tr. Chipewyan *bucka-thrae-ggayaze*], a North American pika, *Ochotona princeps*; **little death** [cf. F. *petite mort*], a weakening or loss of consciousness, spec. in sleep, during an orgasm, etc.; **Little Entente**: see *ENTENTE (quot. 1923); **little green man,** an imaginary inhabitant of outer space; an imaginary person of peculiar appearance (in quot. 1906 an actual person tattooed green); **little house**: delete (now *dial.*) and substitute (now *Austral., N.Z.*, and *dial.*); (further examples); **Little Irelander** (cf. *Little Englander*); **little Joe,** in the game of Craps (see quots.); **little magazine,** a name designating any of various periodicals devoted to serious literary or artistic interests (see also quots.); also *attrib.*; so (as colloq. abbrev.) **little mag; Little Mary** *colloq.*, the stomach; **(poor) little me** († I), used to convey the speaker's mock-depreciation of (and the supposed vulnerability of) himself; **little mother,** a young girl who behaves maternally towards her younger brothers or sisters, or her dolls; **little Neddy** (also **little Ned**) [*NEDDY 3], one of a number of committees under the National Economic Development Council; **little old** (followed by a pronoun or a name): used as an endearing or mock-depreciatory mode of reference; **little Orphan(t) Annie,** the name of an orphan child in a poem by J. W. Riley and a strip-cartoon by Harold Gray, used allusively in various senses (see quots.); **little review** = **little magazine*; **little science,** term used of scientific and technological investigation that does not require large resources; **little season,** a fashionable season in London in the winter; **little theatre,** a small playhouse, esp. one used for dramatic experiment (in quot. 1771 the name of an actual theatre); also *attrib.*; **little Venice,** a name given to various local areas felt to resemble Venice in canal scenery; also *attrib.*; **little Willie,** a term first used of Crown Prince Friedrich Wilhelm Viktor August Ernst of Germany (1882–1951) and applied to persons as a term of disparagement and to weapons (see quots.); (quot. 1907 is prob. an unconnected casual use); **little woman,** (*a*) one's wife, freq. with *the*; (*b*) a private dressmaker or odd-job woman.

1904 *Press* (Philadelphia) 11 Aug. 6 Judge Parker's whole contention is that of the little American. *Ibid.*, His little Americanism invites fuller discussion. **1902** H. JAMES *Wings of Dove* xviii. 427 She might fairly have been dressed tonight in the little black frock..that Milly had laid aside. **1949** D. SMITH *I capture Castle* xi. 192 Perhaps it gives you a glorious, valuable feeling to wear little black suits of fabulous price. **1951** *Woman & Beauty* May 1/2 Invest your all in one good little black dress. **1968** J. IRONSIDE *Fashion Alphabet* 19 *Little black dress.* This highly useful garment was at first almost the trademark of the British designer, Molyneux, who perfected it as an 'after 6' look in the cocktail party era between 1920 and 1939. The ultimate in sophistication then, it is still much in demand. **1973** *Country Life* 13 Dec. 2067/1, I have included a little black dress in my photographs this week, because I think it is the right alternative to glitter for winter '73. **1868** *Proc. Calif. Acad. Sci.* IV. 6 *Lagomys princeps* Richardson—'Little Chief Hare'; Rat-rabbit. **1898** F. RUSSELL *Explor. Far North* p. vi, The timid squeak of the little chief hare was often heard. **1947** V. H. CAHALANE *Mammals N. Amer.* 581 Some imaginative naturalist has given the animal the title of 'little chief hare'. **1960** *Canad. Audubon* Jan.–Feb. 28/3 The industrious little pika has yet another name, the Indian name, 'Little Chief Hare'. **1932** A. HUXLEY *Brave New World* v. 89 The sexophones wailed like melodious cats under the moon, moaned in the alto and tenor registers as though the little death were upon them. **1939** —— *After Many a Summer* II. ii. 198 Like all the other addictions, whether to drugs or books, to power or applause, the addiction to pleasure tends to aggravate the condition it temporarily alleviates. The addict goes down into the valley of the shadow of his own particular little death. **1959** W. GOLDING *Free Fall* v. 108 The little death shared or self-inflicted was neither irrelevant nor sinful. **1959** D. KROOK *Three Traditions Moral Thought* x. 275 That other aspect of the sexual act that Augustine finds so disturbing, that oblivion which the poets called the 'little death', the overwhelming of the will and the reason, need have no terror in it. **1969** G. SIMS *Sand Dollar* x. 126 She attained her climax with a deep shudder... Her features took on a delicate nature in the 'little death'. **1971** R. RENDELL *One Across* III. xxi. 166 A little death would make the unbearable present pass... The pubs wouldn't be open yet. **1973** D. BAGLEY *Tightrope Men* I. 9 That everyday miracle of the reintegration of the psyche after the little death of sleep. [**1906** KIPLING *Puck of Pook's Hill* 185 The little green man orated like a—like Cicero.] **1961** PARTRIDGE *Dict. Slang* Suppl. 1170/2 *Little green men*, mysterious beings alleged to have been seen emerging from flying saucers. **1966** K. GILES *Provenance of Death* iv. 121 Are you saying you are being watched perhaps by little green men? **1967** M. KENYON *Whole Hog* iv. 42 There was a desert-island cartoon..and a little-green-men-from-Mars cartoon. **1969** C. HODDER-WILLIAMS 98·4 iv. 45 I haven't the faintest idea..little green men. **1972** *Daily Tel.* 23 Sept. 7/7 Neither he, nor I, believes that 'little green men' from Mars..are watching us and closely inspecting our planet. **1789** J. PARKER in *New England Hist. & Geneal. Reg.* (1915) LXIX. 300 Charles worked on my Little house. **1902** L. MANN *Mountain Flat* ii. 23 A gate led from the first yard into another in which were the pig sty, the henhouse, the tool-shed and what they called decently 'the little house'. **1939** F. THOMPSON *Lark Rise* i. 10 Later, the place of honour in the 'little house' was occupied by 'Our Political Leaders', two rows of portraits on one print. **1941** BAKER *N.Z. Slang* vi. 53 Other expressions..to make a sale, to vomit; little house, a privy;..poled for a goal. **1959** *Sunday Times* 13 Feb. 5/1 This may not be pleasing to certain little Irelanders who wish to live in complete isolation. **1830** *Dialect Notes* I. 61 *Big Dick*; ten; *little Joe*. **1936** T. S. STRIBLING *Unfinished Cathedral* xii. 216 Tapping a coin..chanting at each shot, 'Come up, Little Joe! Don't deceive yo' pappy!' **1968** *Scottish Daily Mail* 16 July 2/1 If you throw crap dice and a combination of seven is showing on top, what is facing down?..Little Joes? **1962** *Listener* 24 May 900/2 Shaw had no use for the 'little mag' mentality. **1900** *Book-Lover* (San Francisco) Autumn (recto rear cover), In this age..publishers we may tell our friends *Impressions* is a little magazine, simply done to tell the truth about books and other matters. **1913** *Writer's Mag.* Oct. 140/1 *The Black Cat* certainly is the same little magazine that it has been for years. **1926** *Atlantic Monthly* Mar. 391/2 As these little magazines often contain only thirty-two pages, cost..two or three cents a copy to produce, and sell for fifteen; you can afford to dispense with advertisements. **1947** *Partisan Rev.* XIV. 475 This group formally includes members of the faculties of the universities and the few writers in the large cities who do independent critical work pitched beyond the level of commercialism. These find their outlet in the little magazines. **1952** *Times Lit. Suppl.* 29 Aug. Suppl. p. xlvii, The birth of the Little Magazine may also now be seen in retrospect as heralding the decline of the greater. **1958** *Spectator* 18 July 116/3 Little-magazine society. **1971** *Ann. Rep. Curators Bodl. Libr.* 1969–70 19 Acquisitions in contemporary literature of the United States and the Commonwealth, including 'little magazines', continued. **1972** *Guardian* 30 Aug. 8/2 Little magazines have been the pioneers of twentieth-century literature from Wyndham Lewis's Blast in the First World War to F. R. Leavis's Scrutiny in the thirties and forties and on to the present day. **1903** *Punch* 14 Oct. 258/1 And what is the subject of the piece [Barrie's *Little Mary*]? Who is Little Mary? It is nobody: it's simply a nursery name that the child-doctor invents as a kind of polite equivalent to what children ordinarily allude to as their 'tum-tum'. *Ibid.*, Good-natured British audiences have strong Little Maries. **1906** *Daily Chron.* 8 Nov. 6/5 To wear it over their chest, not to speak of Little Mary, as people all now call their other dangerous spot. **1923** U. L. SILBERRAD *Lett. J. Armiter* iv. 82 Then I get a cold in Little Mary, my vulnerable spot. **1933** W. H. HARRISON *Humour in East End* 18 'I've got a little Mary too!' Swift as lightning came the reply, with a shrewd glance at a corpulent waistcoat: 'Not arf yer ain't, guv-nor.' **1781** N. MUNDY *Let.* 21 Oct. in A. E. Newdigate-Newdegate *Cheverels* (1898) iii. 48 How very ill poor little I am used kick'd quite out & not allowed room. **1818** M. EDGEWORTH *Let.* 15 Oct. (1971) 126 Could I four years ago have believed it if I had been prophesied to me that I poor little I should this day have been driving about London with Honora alone? **1895** A. W. PINERO *Second Mrs. Tanqueray* III. 112, I really thought you'd forgotten poor little me. **1895** E. WHITING *No. 5 John St.* xxx. 267 The wonder why the irrepressible..powers could not let 'poor little me' alone. **1933** A. BENNETT *Regent* i. 16 'What about about poor little me?' cried the driver, who was evidently a ribald socialist. **1923** E. BOWEN *Encounters* 9 Nobody takes any notice of little me. **1961** M. BEADLE *These Ruins are Inhabited* (1963) ix. 124 A cold snap prompted the writing of two letters home. **1968** A. DIMENT *Bang Bang Birds* viii. 149 He had put instructions from his bosses to liquidate little me. **1828** M. WILMOT *Let.* 23 Apr. (1935) 316 Blanche..is the little *Mother* of the house upon the occasion. **1897** *Sears, Roebuck Catal.* 333/1 Little Mother's Outfit... Contains a copy of 'Little Mother's Fashion Book', showing designs and directions for making dolls' dresses. **1963** A. WILSON *No Laughing Matter* II. 70 Sukey had better deal with them. She likes being the little mother. **1963** *Daily Tel.* 16 Oct. 31/3 The committee will be among the first 'little Neddies' set up under the auspices of the National Economic Development Council. **1964** *Economist* 5 Dec. 1112/1 The 'little Neds' (the separate councils..for different industries). **1967** *Punch* 24 May 766 The Little Neddy for the hotel and catering industry has just published the result of a survey it commissioned among foreign travel agents. **1969** *Times* 13 Jan. 11/3 Even a small shift in the distribution of domestic resources could meet the needs of the expansion postulated by Little Neddy, provided the profits were already there. **1905** *Little old* [see *MONEY *sb.* 6 a]. **1961** C. COCKBURN *View from West* vii. 71 Iceland..was menacing poor little old England in a truly devilish manner. **1966** J. POTTS *Footsteps on Stairs* (1967) xiv. 177 Why couldn't I have been looking out for the little old me? **1968** J. SANGSTER *Touchfeather* vi. 84 'He wasn't carrying a gun, probably considering that with just little old me to deal with there'd be no need. **1973** A. ROSS *Dunfermline Affair* 80 Askwith and Gibson made polite noises, but little old Abbie refused to dissemble. **1973** *Guardian* 12 Mar. 9/1, I started at 15..and it was three or four years before I got to Little Orphant Annie. **1913** J. W. RILEY *Poems* 9 Little Orphant Annie's come to our house to stay. **1924** H. GRAY in *Daily News* (N.Y.) (Pink ed.) 5 Aug. 26 (*title of cartoon strip*) Little Orphan Annie. **1938** D. SMITH *Dear Octopus* I. 39 You stood there in the doorway..looking exactly like little Orphan Annie. **1952** M. STEEN *Phoenix Rising* vi. 142, I won't have my friends..made uneasy by your bogey tales. You're worse than Little Orphant Annie! **1960** *Woman* 20 Feb. 6/3 She cast herself in the role of Little Orphan Annie. **1966** *Newsweek* 19 July 36/3 Like a Little Orphan Annie dress by Mary Quant or Caroline Charles will do, or for the male, a set of Mod threads. **1966** M. G. EBERHART *Witness at Large* (1967) i. 13 You'll soon be out of a job, Little Orphan Annie. **1967** *Guardian* 2 Aug. 6/7 So now I'm in position of Little Orphan Annie in a letter to the *Times*. **1958** *Times Lit. Suppl.* 24 Jan. 37/3 As Mr. Granville Hicks has pointed out, the little magazine..is able to print what is deemed..adventurous, original, 'careful, critical not creative, moved more by the spirit of Henry James than by that of Baudelaire. **1961** A. M. WEINBERG in *Science* 21 July 161/1 We must make Big Science flourish without..allowing it to trample Little Science—that is, we must nurture small-scale excellence as carefully as we lavish gifts on large-scale spectaculars. **1963** D. J. PRICE (*title*) Little science, big science. **1970** *Sunday Times* (Colour Suppl.) 16 Aug. 22/2 Little Science changed into recognisable Big Science during the Second World War. **1972** *Science* 9 June 1093/1 Those laboratories that were largely scientists' the personal joys of little science are greater than those of big science. **1928** *Daily Tel.* 13 Jan. 15 Lady Chamberlain's Tuesday afternoon At Homes at the Foreign Office..were one of the features of the 'little season'. **1938** *Burlington Mag.* Feb. p. xvi/i Highly successful sales..in the so-called 'little season' which ended just before Christmas. **1959** J. FLEMING *Miss Bones* xv. 159 This..is the Little Season, isn't it? ..There are numerous Embassy parties..first nights in the theatre..and what have you. **1771** SMOLLETT *Humph. Cl.* I. 246 His detestation of the mob..has been heightened..by the consequences of his last appearance in the Little Theatre in the Hay-market. **1813** JANE AUSTEN *Pride & Prej.* III. ix. 166 London was rather thin, but however the Little Theatre was open. **1916** H. B. LEAVITT *ye Vox. Theatr. Managem.* xxxvi. 574 'Little Theatres' have for some time been playing important roles in the dramatic life of Paris, Berlin and London. **1919** *Writer's Mag.* Jan. 127/1 (*caption*) The 'little' theater movement expanding. **1926** *Stage Year Bk.* 45 Another little theatre, the Portmanteau.. opened its doors with a programme of one-act plays. **1929** [see **community theatre*]. **1958** *Listener* 21 Aug. 283/2 The shabby treatment accorded here to our little theatres. **1965** E. O'HARA *Aug.* ii *Wicked Month* ii. 37 She worked for a little theatre magazine. **1971** *Author* LXXXII. 118 A recording author, Mr. X wrote a 'philosophical' conversation piece which was staged at a London little theatre. **1973** *Philadelphia Inquirer* (Today Suppl.) 17 Oct. 36/1 There is only one fully funded..Little Theatre in the Philadelphia area. **1934** M. ALLINGHAM *Death of Ghost* i. 2 Little Venice [on the Regent's Canal] in 1930.. The room..took up the entire first floor of the old house on the canal. **1951** D. NEWTON *London West of the Bars* xxiii. 192 Robert Browning, back from Italy, settled for a time in a tall house overlooking 'Little Venice'. **1960** C. MACINNES *Mr. Love & Justice* 21 Edward was sitting with his girl in the park at Little Venice, up by the Harrow road. **1968** *Guardian* 12 July 20/6 The justices granted a provisional restaurant licence for two barges to be moored..near the lock of the Rochdale Canal... It was hoped this section of the canal..would become a Little Venice patronised by yachtsmen. **1970** *Ibid.* 8 Aug. 7/3 It has taken a solid government survey to bring the Little Venice Boat Show back to London... The anniversary signals the completion in 1820 of the Regent's Canal. **1907** F. H. BURNETT *Shuttle* xxiii. 235 Little Willie' isn't quite as easy as he looks. **1915** D. O. BARNETT *Let.* 27 May (1915) 154 At Ypres..it's been about twenty minutes last night they fired a *Little Willie* on to our trench, **1916** *Daily Mirror* 16 Dec. 7/5 Over there's something ..a thing that's much more powerful than the German one (otherwise known as *Little Willie*). **1925** A. CHRISTIE *Secret of Chimneys* xii. 121 That some one calculated the number on the *Little Willie*. **1935** FRASER & GIBBONS *Soldier & Sailor Words* 304 *Big and Little Willie*, names given the Kaiser and German Crown Prince to a series of cartoons... The names soon..were applied to a variety of objects. For instance, two experimental tanks, which were begun on making, but 1915, '.. 'Little Willie' first 'moved' on September 8th. **1927** 'D. YATES' *Blind Corner* iv. 125 'God give it's Little Willie,' said Ellis, and sucked in his breath. **1945** E. WAUGH *Brideshead Revisited* iv. 85 Little Willie's tail [of Little Willie, **1964** BROPHY & PARTRIDGE *Long Trail* 145 Little Willie. *Big Willie*, the Crown Prince and the Kaiser. Journalese. So used occasionally by the troops, who applied the terms to all manner of things..e.g.—a long-range naval gun operating on the Western Front. **1624** J. CHAMBERLAIN *Let.* 20 Mar. (1939) II. 551, I send Dr. Bargraves sermon and the little woman's verses for my Lady. **1795** W. B. STEVENS *Jrnl.* 11 Aug. (1965) III. 280 The Little Woman's passions swell... She is expecting her Husband to be her Slave. **1801** M. O'CONNELL *Let.* 6 Nov. in D. O'Connell *Corr.* (1972) I. 54 To comply with the earnest request of your little woman whose entire happiness is wrapped up in you. **1852** DICKENS *Bleak Ho.* (1853) xxxi. 314 My little woman will be looking out for me. **1936** 'E. WEST' *Thinking Reed* vii. 162 We went off to a little woman who does maisonettes. **1959** *Sunday Express* 26 July 10/5 I've..found a 'little woman' ..one of those treasures who will dressmake at home. **1970** G. GREER *Female Eunuch* 286 Loving mockery of the little woman. **1973** G. GREENE *Honorary Consul* v. I. 175 The material was quite inexpensive, and I had it run up by a little woman.

14. little-bitty: see *BITTY *a.* 3; **little-boy** *attrib.*, pertaining to, suited to, or resembling a small boy; infantile; so **little-boyish** *a.*; **little-boy-lost**: (used without hyphens) as a title of literary works by William Blake and other writers after him; *attrib.* or as *adj. phr.*, resembling a small boy who has lost his way; also *absol.*; **little boys' room** (cf. **little girls' room*), a genteelism for a gentlemen's lavatory; **little-girl** *attrib.*, resembling or characteristic of a little girl, e.g. a type of collar, a voice; so **little-girlish** *a.*; **little-girlhood, -girlishness; little-girl-lost** (cf. **little-boy-lost*); **little girls' room,** a genteelism for a ladies' lavatory.

1847 THACKERAY *Van. Fair* (1848) v. 39 Out of the little-boy class into the middle-sized form. **1923** D. H. LAWRENCE *Stud. Classic Amer. Lit.* (1924) vi. 143 Old-fashioned Nathaniel, with his little-boy charm. **1929** *Phoenix II* (1968) 537 Men..spend years training up the little-boy-baby-face type, till they've got her perfect. Then the moment they've married her, they want something else. **1951** M. MCLUHAN *Mech. Bride* (1967) 68/2 The heart-throb little-boy-lost has been made almost the male trend to assume the little-boy role. **1955** E. BLISHEN *Roaring Boys* iv. 182 Two faces, well-washed and little-boyish. **1958** H. WAUGH *30 Manhattan East* (1969) 149 That little-boy quality that wasn't little-boyish at all. **1963** W. BLAKE *Songs of Innocence* in *Compl. Writings* (1972) 120 (*title*) The little boy lost... The night was dark... **1941** AUDEN *New Year Let.* III. 31 The hitherto-unconscious creed Of little men who had lost each other. **1963** A. LASKI *Little Boy Lost*... **1965** P. MORTIMER *Pumpkin Eater* xviii... **1972** *Guardian* 9 Mar. ... [remaining quotations for *little-boy*, *little-girl* senses continue].

little, *v.* (Later *poet.* examples.)

1928 HARDY *Winter Words* 194 Can littlest life beneath our life be littled? **1957** A. CLARKE *Later Poems* (1961) 27 Nor, littled by itself, I found one That had never turned from me.

little-go. Add: **3.** *transf.* (various senses).

1852 MRS. GASKELL *Let.* 19 May (1966) 191, I (boldly) little-go last night. **1858** *Leisure Hour* 15 July 448/2 This preliminary spread, or 'little go'. **1909** J. R. WARE *Passing Eng. Vict.* 169/1 *Little go*, first imprisonment, first invented by a fallen university man. **1960** WENTWORTH & FLEXNER *Dict. Amer. Slang* 321/1 *Little go*, an unimportant, unexciting, or incomplete attempt, effort, race, or fight.

little man. Add: **2.** Also, a person working or producing on a small scale; a small craftsman or tradesman; a local man available to do light work.

1825 H. WILSON *Mem.* IV. 103 That little man in St. James's Street, who sells bon-combs. **1890** W. BOOTH *In Darkest Eng.* vi. 214 What is to be done with the little man? **1936** *Economist*..a Poor Man's Bank..doing for the 'little man' what all the banks should do for the 'big man'? **1937** *Ann. Reg. 1936* II. 63 The potential customers being most numerous among the 'little men', *i.e.*, small shopkeepers and owners of one-man businesses. **1952** *Economist* 30 Aug. 514/1 Diversified investment buying by the general public, especially by the 'little man'. **1959** *Motor Manual* (ed. 36) xiii. 269 For cars so built it is seldom possible to have an adequate towing bracket made up by the 'little man round the corner' who happens to have a welding plant and some iron. **1964** *Guardian* 12 Dec. 4/4 What most of us will have to do for quite a while is to make ends meet, with a bit of painting in his spare time. **1966** *Listener* 15 Sept. 382/2 The Bideford 'little man' who leans on rope... *Ibid.*, There will always be a place for the 'little man'.

b. The undistinguished and ordinary 'man in the street'.

1935 E. SUTTON tr. H. Fallada (*title*) Little man, what now? **1935** *New Statesman* 8 June 857/1 The old mob-mongering John Bull has disappeared, and his place has been taken by the 'little man'. **1936** G. ORWELL *Keep Aspidistra Flying* iii. 64 To turn into the typical little bowler-hatted sneak—Strube's 'little man'. **1941** AUDEN *New Year Let.* III. 31 The hitherto-unconscious creed Of little men who had... **1942** *R.A.F. Jrnl.* May 16a Joe represents the typical 'little man' in blue, doomed to a lowly rank. **1952** M. LASKI *Village* xiii. ... **1960** *Times* 28 Sept. 5/4 In this connexion, the little man's world... **1972** ... **1975** *Times* 2 Jan. ...

5. A young male child: see MAN *sb.*[1] 4 f.

Little Neck. *U.S.* The name of a locality in Long Island, used *attrib.* in **little neck clam** to designate small specimens of the quahog, *Mercenaria mercenaria*, or other similar clams. Also *absol.*

1884 *Bull. U.S. Nat. Mus.* No. 27. 234 Another name [for the small round clam] is 'Little Neck', derived from a locality on the north shore of Long Island, known as Little Neck, whose clams had a superior flavour. **1899** J. HATTON *in People's* 17 Dec. ... **1935** J. C. LINCOLN *Cape Cod Yesterdays* 49 Everyone ate quahogs and Little Necks *au naturel*. **1972** *Guardian* 1 Sept. 9/3 Clams from the Maine coast, Quahogs, 'steamers', Littlenecks.

little people. 1. Fairies, gremlins.

1726–31 [in DICT. s.v. LITTLE *a.* 13]. **1807** E. PHILLPOTTS *Lying Prophets* I. xi. 102 The little people take all manner of shapes. **1941** [see *GREMLIN 1]. **1973** *Times* 17 Mar. 14/8 The two larger clovers..were said to afford protection against 'unkind Little People'.

2. Children.

1753 M. W. MONTAGU *Let.* 22 July (1967) III. 15 How often do I fancy to my self the pleasure I should take in seeing you in the midst of your little people! **1876** C. M. YONGE *Three Brides* I. viii. 182 My little people are so anxious to have me with them. **1934** H. G. WELLS *Exper. Autobiogr.* ... **1975** *Sunday Times* 25 Mar. 8/1 (Advt.), Big savings for little people from Boots.

3. The poor; ordinary or undistinguished people.

1827 [see LITTLE *a.* 5 b]. **1901** C. MORRIS *Life on Stage* xxiv. 313 'I won't sing this way,'... **1951** M. MCLUHAN *Mech. Bride* ... **1971** ... the little people, who..have to contend with the frustrations brought about by bureaucracy.

Littler (li·tlǝɹ). The name of William Littler (1724–84), English potter, used in the possessive to denote a rich blue colour applied to porcelain or stoneware.

1957 MANKOWITZ & HAGGAR *Conc. Encycl. Eng. Pott. & Porc.* 130/1 *Littler's blue*, a brilliant 'royal' blue or lapis-lazuli found on porcelain attributed to Longton Hall, and on salt-glazed stoneware believed to have been made by William Littler & Aaron Wedgwood, c. 1750. **1971** *Country Life* 28 Oct. 1162/1 William Littler..was the originator of a rich blue enamel which has become known to collectors' jargon as 'Littler's blue'.

Little's disease. *Med.* [Named after William John Little (1810–94), English physician, who described it.] Cerebral spastic paralysis of infants.

1885 *Boston Med. & Surg. Jrnl.* CXII. 237/1 (*heading*) Cases of congenital muscular rigidity, or Little's disease. *Ibid.*, This congenital muscular rigidity was first described as an affection by Little, of London, in 1853, and Rupprecht..suggested that the name of Little's disease be applied to it. **1887** *Buck's Handbk. Med. Sci.* V. 200/2 Little's disease. Spasmodic tabes of children. **1938** *Times Lit. Suppl.* 22 Jan. 50/2 Recent researches would appear to have established that the trouble from which Byron suffered was not an ordinary club-foot; that he was the victim of some obscure nervous malady—Little's disease, otherwise spastic paraplegia, has been suggested. **1964** S. DUKE-ELDER *Parsons' Dis. Eye* (ed. 14) xxvi. 569 Congenital spastic diplegia (*Little's disease*), a bilateral spastic paralysis present from birth, considered at one time to be due to meningeal haemorrhage as a result of birth injury, is probably a degenerative cerebral process of obscure ætiology.

littly, little (li·tli), *sb.* [f. LITTLE *a.* + -Y[6], -IE.] A small child or person; *pl.*, small children, the younger children of a family, etc.

1893–4 K. O. HESLOP *Northumb. Words* II. 453 *Littlie*, a very little person. **1916** 'BOYD CABLE' *Action Front* 47 *White Sausage* v. 41 How's she? How's the littlies? **1967** B. JEFFERIS *Our Black Summer* vi. 122 Can you see Hilary instructing the littlies at Sunday school?

littly (li·tli), *adv.* [f. LITT(LE *a.* + -LY[2].] In a small, modest, undistinguished way.

1895 *Ibid.* ... **1909** BEERBOHM *Around Theatres* (1924) II. 177 To strut agreeably, littly, as in the average production.

littorina (litǒrai·nă, -ī·nă). [mod.L. (A. d'A. de Férussac *Tableaux Systématiques des Animaux Mollusques* (1822) p. xxxiv), f. L. *litus* shore.] A gastropod mollusc of the genus so called; = PERIWINKLE[2] 1.

[**1820–5** J. & G. B. SOWERBY *Genera Recent & Fossil Shells* ...] **1857** J. REEVE *Conchologia Iconica* X. ... **1874** ... **1963** D. W. & E. E. HUMPHRIES tr. *Termier's Erosion & Sedimentation* ii. 64 The Littorinas of California..have each been able to dislodge 95 gm of grains in twenty-four hours.

2. Littorina (or **Litorina**) sea, the Baltic sea at the end of the Boreal period. Also *Littorina beach, minimum, period, stage, transgression* (see quots.).

1912 R. A. S. MACALISTER *Text-bk. European Archæol.* I. ii. 43 The Baltic Sea became of much the same configuration as at present, but rather larger in extent and more salt; the name given to this stage of its history, once more derived from the name of its mollusc, is *Littorina Sea*. The greatest depression of this phase of land-movement may be called the *Littorina Minimum*. The Littorina period is also known as the *Tapes* period. **1928** C. DAWSON *Age of Gods* iii. 48 The climate of Scandinavia became much warmer, owing perhaps to a change in the currents of the Baltic. **1948** *Austral. Jrnl. Sci. Res.* B. I. 191 The Supralittoral is the region above high-water spring tide levels which has been invaded by species of 'Littorina' [etc.]. **1955** ... *Littorina* ... **1964** ... **1970** ...

littorinid (litǫ·rinid). [f. mod.L. family name *Littorinidæ*, f. generic name *Littorina* (see prec.).] A marine snail of the family Littorinidæ.

1948 *Austral. Jrnl. Sci. Res.* B. I. 191 The Supralittoral is the region above high-water spring-tide levels which has been invaded by species of 'Littorina'... The species are Littorinids, and one might call the Littorinid zone. **1952** *Jrnl. Ecol.* XL. 87 We agree..in defining the upper limit of the 'littoral' as the region where littorinids become dominant. **1964** V. J. CHAPMAN *Coastal Vegetation* ii. 17 Several specimens of the periwinkle..in the form of littorinids. **1974** *Nature* 3 May 11/3 Littorinids are present on the rock shores of all continents.

liturgical, *a.* Add: *Liturgical Movement* (see quot. 1957).

1950 *Tablet* 17 Aug. 197/2 The proficiency of these youngsters has been acquired in our Catholic schools, the grand hope of the Liturgical Movement, and Catholic girls are not lagging behind Catholic boys. **1952** A. G. HEBERT *Liturgy & Society* vi. 126 The so-called Liturgical Movement is concerned with things vastly more important than ritual. **1957** *Oxf. Dict. Chr. Ch.* 815/1 *Liturgical Movement*, a movement in the Christian Church for the restoration of the active participation by the people in the official worship of the Church. **1959** *Times Lit. Suppl.* 29 May 325/3 Since the war the leaders of the Liturgical Movement. ... the Liturgical Movement.

liturgism (li·təɹdʒiz'm). [f. LITURGY + -ISM.] Excessive concentration on liturgy; a disproportionate concern with liturgical detail.

1926 *Q. Register* Feb. 117 With this prevailing liturgism, religion ceased to be an important force in common life. **1963** *Times Lit. Suppl.* ... **1964** E. L. MASCALL *Corpus Christi* iii. 79 It is possible to point to parishes whose priest is an enthusiastic amateur liturgiologist, where the layfolk have to adapt themselves at regular intervals to fresh modifications in the rite and ceremonies of the Mass in agreement with the stage now reached by their pastor in his researches. Liturgism of this kind can rapidly degenerate into rubrical fussiness and a liturgical antiquarianism run riot.

Litvak (li·tvæk). Also **Litvok.** [Yiddish, f. Pol. *Litwak* Lithuanian.] A Jew from Lithuania or its neighbouring regions.

1892 I. ZANGWILL *Childr. Ghetto* I. 38 To a Dutch or Russian Jew, the 'Pollack', or Polish Jew, is a poor creature; and..the 'Pullack' looks down upon the 'Litvok', or Lithuanian. **1970** L. M. FEINSILVER *Taste of Yiddish* 60 Four major dialects developed. Of these, the one spoken by so-called Litvaks, the Jews of Lithuania, White Russia and Northeastern Poland, is essentially standard or literary Yiddish. **1970** I. SIEFF *Mem.* i. 5 'What can you expect? He's a Litvak.' A kind of Scot in Jewry? Whatever the reason, many Litvaks led the way in emigration when the circumstances were most daunting. **1971** B. MALAMUD *Tenants* 209 A middle-aged Litvak, a stocky man in mud-spotted trousers.

Litz (lits). *Electr.* Also **litz.** [f. *LITZ(ENDRAHT.] Used *attrib.* to designate wire composed of many fine strands twisted together and individually insulated, so as to reduce the skin effect and the associated increase in resistance at high frequencies.

1927 *Wireless World* 13 Apr. 447/1 (*heading*) Preparing Litz wire. **1948** *Observer* 17 June 26/3 The most efficient inductance to use in a 'tuned anode' radio-frequency amplifying circuit is one wound with 'Litz' wire. **1959** K. HENNEY *Radio Engin. Handbk.* (ed. 5) iii. 7 Since the smaller the conductor, the less the skin effect, if the conductor is broken down into many small strands, twisted..each strand insulated from the others, the resistance at high frequencies is reduced. Such wire is called Litz wire. **1968** *Radio Communication Handbk.* (ed. 4) iv. 14/2 Tuned circuits of higher Q than can readily be obtained in the conventional type of i.f. transformer (even when wound with Litz multi-stranded wire).

litzendraht (li·tsǝndrāt). *Electr.* [a. G. *litzendraht* Litz wire, f. *litze* braid, cord, lace, strand + *draht* wire.] Litz wire.

1921 J. H. MORECROFT *Princ. Radio Communication* ii. 124 It is..important in getting the stranded wire (sometimes called litzendraht) to see that each strand is of a great number of well-insulated strands..properly interwoven. **1933** K. HENNEY *Radio Engin. Handbk.* vii. 151 The former [*sc.* skin] effect is minimized by the use of conductors with insulated strands—so-called *litzendraht*. **1962** *Newnes Conc. Encycl. Electr. Engin.* 445/2 *Litzendraht*, stranded cable for high-frequency currents.

livability: see *LIVEABILITY.

livant, var. LEVANT *v.*[2]

1912 D. H. LAWRENCE *Let.* 5 Nov. (1962) I. 154 F. had carefully studied *Anna Karenina*, in 1905 I think, and I say to her you read the *Fellow* when *Vronsky*..I am the fellow she livanted with.

live, *a.* Add: **2. b.** (Earlier and later examples.)

1857 *Knickerbocker* L. 456 A neighbouring flash-house, kept by a live Yankee of the name of Martin. **1870** *Scribner's Monthly* I. 71 Quite as likely..the 'advanced' preacher selects a 'live' subject, a theme for the times. **1932** E. V. LUCAS *Reading, Writing & Remembering* 4 A varied, learned and very live and human being. **1960** *Times* 27 Feb. ... **1974** *Times* 16 Apr. 4/8 Bungabah..could be a 'live' Imperial Cup prospect for his...

c. Corresponding to actual facts.

1927 CARR-SAUNDERS & JONES *Survey Social Struct. Eng. & Wales* 152 The Unemployment figures were obtained from the 'live registers' of the employment exchanges in Great Britain. **1931** *Times Educ. Suppl.* 9 May 165/3 The 'live' register has, it is true, dropped from 923 to 757, but these figures are not likely..by the children who left school at the Easter recess.

d. Of a performance, heard or watched at the time of its occurrence, as distinguished from one recorded on film, tape, etc. Also *quasi-adv.*

1934 *B.B.C. Year-Bk.* 248 Listeners have..complained of the fact that recorded material was too liberally used.. but..transmitting hours to the Canadian and Australian zones are inconvenient for broadcasting 'live' material. **1937** M. LOWELL *Listen In* 109 People do not like 'canned' entertainment when they can obtain 'live' entertainment just as easily. **1944** *Ann. Rev.* 1943 348 It was still felt..that attendance at concerts and listening to 'live' performances belonged to a special art of its own. **1947** *Penguin Music Mag.* II. 21 The standard of playing ..has been maintained. **1953** ... there was an unprecedented demand for live performances. **1953** *Sight & Sound* XXIII. 31 ... **1953** E. SMITH *Guide Eng. Traditions* 21 The development of the gramophone and wireless broadcasting..has made many thousands familiar with the great musical composers and anxious to hear 'live' performances of their works. **1955** *Radio Times* 22 Apr. 15/3 At the moment, Northern Ireland has no means of originating 'live' television programmes. **1958** *Listener* 25 Sept. 465/1 Long experience of the effect of gramophone record sales and broadcasts provides pretty convincing evidence that they have..the general interest of the public in attending live performances. **1970** *New Scientist* 2 July 13/1 We now accept full live coverage of soccer games in Mexico as a matter of course. **1974** *Daily Tel.* 2 May 13/5 Live audiences. **1974** *Guardian* ... London Casino..since 1953 the home of Cinerama. **1974** *Times* 24 Nov. 8/8 The hearing was televised live. **1975** *Daily Tel.* 18 Apr. 13/8 When people spoke 'live' or on tape it was their own words.

4. For 'Charged with electricity' read: Connected to a source of electrical potential. In a single-phase supply: being the conductor on which the supply voltage is developed (with respect to the neutral). Also *ellipt.*, a live conductor or terminal. (Further examples.)

1913 D. S. MUNRO *Pract. Electr. Wiring* xx. 281 When the neutral wire is well earthed at earth potential..but may, of course, increased risks, if the live wire be in contact with..some part of the apparatus..with its connections to earth from one single. **1938** J. W. SIMS *Electr. Installations* 148 When the fuse unit is enclosed in an iron case there should be an earth terminal fitted inside the case. **1966** F. W. WHITFIELD *Electr. Install. & Regulations* vi. 125 Two-pin sockets and plugs. Live and neutral only are catered for in these units. **1970** *Which?* Aug. 256/2 Britain objected to black for the live, since in Britain it was being used for the neutral. **1973** G. Burdett *Householder's Electr. Guide* vi. 34 Two-core sheathed flex is now made with the new core colours of brown for the 'live' core and blue for the neutral.

6. Applied *spec.* to an axle.

1875 [in Dict.]. **1903** *Work* XXV. 199/3 The two systems of driving are the live axle and the double sprocket chain. **1904** A. B. F. Young *Compl. Motorist* (ed. 2) iv. 116 The driving of the rear wheels being direct through a powerful chain and live axle. **1929** Newton & Steeds *Motor Vehicle* xxi. 259 A live axle is one that rotates or houses shafts that rotate, while a dead axle is one that does not rotate or house rotating shafts. **1963** Bird & Hutton-Stott *Veteran Motor Car* 8 In modern usage 'live axle transmission' means the combination of propellor shaft and bevel- or worm-geared live axle—but many chain-driven cars also had live axles of a different sort. **1971** *Daily Tel.* 24 Mar. 11/3 It has a water cooled engine driving the rear wheels and a live rear axle.

7*. *Acoustics.* Of a room or enclosure: having a relatively long reverberation time (opp. *dead *a.* A. 14 b).

1931 L. Cowan *Recording Sound for Motion Pict.* xviii. 266 The music is reproduced in the live end, which would correspond to the stage of an auditorium, and the microscope is placed in the comparatively deader end, which would correspond with the audience position. **1962** A. Nisbett *Technique Sound Studio* ii. 33 The answer might at first sight appear to be to do away with reverberation entirely and try to create entirely 'dead' studios... But listening tests indicate that most people prefer a moderately 'live' acoustic. **1974** *Which?* Aug. 243/3 In a very live room—one with a lot of hard surfaces that can reflect sound—upper and middle frequencies will stand out more.

8. live action (see quot. 1960); also *attrib.*; **live fence** orig. *U.S.*, a hedge; also *live-fencing*; **live load** *Engin.*, a temporary or varying load imposed on a structure by its being put to use (cf. *dead load* s.v. *dead *a.* A. 29); **live weight**, the weight of an animal before it is slaughtered and prepared as a carcase; also *attrib.* and as quasi-*adv.*; **live wire** (see sense 4) orig. *U.S.*, *fig.* esp. a person full of energy; also (with hyphen) *attrib.*

1957 Manvell & Huntley *Technique Film Music* iii. 136 A great deal of Hollywood's comedy music is linked with the cartoon world and..a number of techniques associated with the animation studios are increasingly employed during the making of live-action scenes. **1960** O. Skilbeck *ABC of Film & T.V.* 79 *Live action*, normal cinematography as opposed to Animation, Titles, etc. **1964** *Listener* 5 Nov. 735/3 There was a very long-drawn-out rescue scene in which live-action and graphics were mingled in a fashionable but not here very convincing way. **1804** J. Roberts *Pennsylvania Farmer* 84 When the hedge is full grown, then there is a perfect live fence. **1858** J. A. Warder *Hedges & Evergreens* I. i. 13 Live-fences, or—as they are commonly called—Hedges, are a means of enclosure that belongs to an advanced state of civilization. **1866** *Bull's Wellington* (N.Z.) *Almanack* 28 Sow..furze for live fences. **1882** W. D. Hay *Brighter Britain!* 192 We have done something towards making live-fences. **1958** *Chambers's Techn. Dict.* 991/1 *Live fence* (Highways), a hedge. **1829** *Mass. Spy* 25 Mar. (Th.), Messrs. G. Th. and Son have imported 75,000 hawthorns, for 'live fencing'. **1866** Live load [see *factor *sb.* 8]. **1908** M. S. Ketchum *Design of Highway Bridges* ii. 41 The live loads on railway bridges are properly a series of moving concentrated loads. **1918** Cowley & Levy *Aeronautics* viii. 163 When the aeroplane is just smoothing out from a steep dive the angle of attack is suddenly increased and the loading rapidly attains a value which in practice is several times its normal. This [*sic*] in all essentials a live load is allowed for in the safety factor. **1974** *Sci. Amer.* Feb. 93/1 A building must ordinarily sustain three kinds of force. The first is the 'dead load' of the structure itself and its fixed contents; the second is the 'live load' of the building's occupants and of such movable components as elevators. ..The third force is produced by the movement of air. **1852** *Trans. Mich. Agric. Soc.* III. 151 Two hundred lambs..weighing some one hundred pounds..live weight. **1960** *Farmer & Stockbreeder* 12 Jan. 50 More than 3½ lb. of food to make 1 lb. of liveweight gain. **1971** *Farmers Weekly* 19 Mar. 85/1 The cost per pound of gain would exceed the price per pound liveweight received at slaughter. **1972** *Country Life* 30 Nov. 1504/2 These small birds [*sc.* turkeys]..are killed out..when they achieve liveweights of about 8½ lb. **1903** 'H. McHugh' *Back to Woods* 12 Bunch went down to the skating pond one day with $18 and picked four live wires at an average of 8 to 1. **1903** *Everybody's Mag.* IX. 30/1 If you cut 'Browny' you cut a live wire and were socially paralyzed. **1909** *Sat. Even. Post* 13 Mar. 24/1 As a legislator..he was probably known to many people as an aggressive 'comer' of the live-wire kind. *a* **1911** D. G. Phillips *Susan Lenox* (1917) II. v. 110, I sized you up as a live wire the minute I saw you. **1931** A. Christie *Sittaford Mystery* xxiv. 197 He appears to be one of the live wires of this investigation. **1935** Auden & Isherwood *Dog beneath Skin* I. ii. 39 And I'm the live wire of the *Evening Moon*. **1952** J. C. Masterman *To teach Senators Wisdom* i. 17 He was, if anyone was, the live wire of the Senior Common Room. **1971** *Daily Tel.* 21 Aug. 14/4 The live-wire management here will miss few opportunities given the right political and economic background in which to operate. **1973** D. Francis *Slay-Ride* x. 121 Our Lars was a live wire once himself. Did a lot of motor racing.

live, *v.*[1] **2.** Also in phr. *to live off the country* (or *the land*): to obtain sustenance from the produce of the countryside without payment.

1884 *Century Mag.* Feb. 503/1 In his marches he had been obliged to live, to a great extent, off the country. **1913** H. Footner *Jack Chanty* 68 The Indians..live off the land during the summer. **1934** *Discovery* Mar. 63/1 It is possible to 'live off the land' to an extent never dreamed of by earlier explorers. **1949** *Milwaukie* (Oregon) *Rev.* 4 Aug. 1/4 The main cause for the Communist army success was the fact that it was well fed, living off the country as it marched through China. **1966** *Globe & Mail* (Toronto) 5 Jan. 25/3 Eskimos in the area [Boothia Peninsula] live off the land.

3. b. *live and let live* (later examples); also *attrib.*

1928 Blunden *Undertones of War* xv. 161 Our future.. depended on the observance of the 'Live and Let Live' principle. **1936** 'R. West' *Thinking Reed* xii. 418 Servants ..haven't the live-and-let-live of equals. **1957** L. F. R. Williams *State of Israel* 217 The belief that he was on the point of reaching some kind of live-and-let-live understanding with Israel. **1959** *Films & Filming* May 22/3 Father, in spite of the live-and-let-live attitude of his brother, persecutes the elderly Dr. Boris Winkler. **1972** J. Brown *Chancer* iii. 46 What's got into you, love?.. I thought you were always one for live and let live. **1975** J. Aiken *Voices in Empty House* vi. 161 It takes all sorts, as I always say. Live and let live.

4. b. Also, *to live dangerously*: to take risks habitually; to live with little regard for one's safety.

[**1910** T. Common tr. *Nietzsche's Joyful Wisdom* 219 The greatest enjoyment of existence is to *live in danger*!] **1930** A. Henderson *Contemporary Immortals* 78 The numerous attempts at assassination give point to Mussolini's avowed motto, after Nietzsche: 'To live dangerously!' **1938** L. MacNeice *I crossed Minch* vii. 93 The young people lack spirit... If only they would appreciate 'living dangerously'! **1939** F. Scott Fitzgerald *Let.* 7 Oct. (1964) 407, I have 'lived dangerously' and I may quite possibly have to pay for it. **1962** *Listener* 9 Aug. 226/1 The spectators in the Bridgearama theatre saw the British pair..living dangerously. **1969** *Outdoor Life* Mar. 88/2 To fish the stream when the water is running full pipe is to live dangerously.

f. (*c*) *to live it up*: to live gaily and extravagantly.

1951 *San Francisco Examiner* 14 Feb. 12 Lieutenant Thumhill is really livin' it up! **1957** P. Frank *Seven Days to Never* vii. 200 They come to Havanna to live it up. They live about two years in two days. **1959** H. Hobson *Mission House Murder* iii. 24 Off-key characters who live it up like crazy. **1961** C. McCullers *Clock without Hands* xiii. 249 Nobody lives for always, but when I live I like to live it up. **1970** N. Armstrong et al. *First on Moon* ii. 39 Those who lived it up in the cocktail lounges that night were also emotionally moved. **1973** C. Bonington *Next Horizon* xiv. 203 We certainly had little chance of living it up on our *Daily Telegraph* expense accounts.

5. spec. *to live one's own life*: to follow one's own plans or principles; to live independently.

1853 C. Brontë *Villette* I. xiii. 229 Thinking meantime my own thoughts, living my own life in my own still, shadow-world. **1893** O. Wilde *Lady Windermere's Fan* II. 56 There are moments when one has to choose between living one's own life..or dragging out some false..existence that the world in its hypocrisy demands. **1921** R. Macaulay *Dangerous Ages* iii. 52 Now I must Live my own Life, as the Victorians used to put it. **1933** 'E. Cambridge' *Hostages to Fortune* IV. ii. 221 All Jane's set, with their ceaseless chatter about..living their own lives, ..had that macabre, sullen look.

8. b. *where one lives*: at or to the right or vital point. *U.S. slang.*

1860 J. G. Holland *Miss Gilbert's Career* xxii. 386 When that little wife of mine says, 'Tom you're a good fellow, God bless you,' it goes right in where I live. **1878** J. H. Beadle *Western Wilds* xxxvi. 597 The Mormons never got a cent of it. This hurt Brigham—right where he lived. **1886** *Century Mag.* Feb. 511/1 If I could only have reached him where he lives, as our slang says. **1900** 'Flynt' & 'Walton' *Powers that Prey* 122 'Sock it to him!' 'Hit him where he lives!'

9. Also, *to live and learn*: a catch-phr. used, freq. on the acquisition of some new knowledge, to indicate that one learns through experience; *I'll* (or *he'll*, etc.) *live*: there is no need for serious worry or concern (freq. used in a trivial way).

c **1620** in *Roxburghe Ballads* (Ballad Soc.) (1871) I. 60 A man may liue and learne. **1803** M. Wilmot *Let.* 6 Aug. in *Russ. Jrnls.* (1934) I. 32 Humph! thinks I. One must live and learn. **1844** [in Dict.]. **1849** Dickens *Dav. Copp.* (1850) xx. 208 Live and learn. I had my doubts, I confess, but now they're cleared up. **1956** J. Symons *Paper Chase* xi. 88 'Are you hurt?' 'I shall live.' **1957** A. Huxley *Let.* 12 Dec. (1969) 836 Problems which I have been trying to solve for the last four months, without any success; for they are, so far as I can see, insoluble. So there we are. One lives and learns. **1963** N. Freeling *Gun before Butter* III. 145 'Potato salad..doesn't sound much fun.' 'I'll live.' **1967** O. Hesky *Time for Treason* xviii. 146 'Better, darling?' Miller asked Miriam anxiously... 'Yes, she'll live,' Tami said sourly. **1971** M. McCarthy *Birds of America* 111 Well, live and learn! **1972** 'R. Crawford' *Whip Hand* I. vi. 34 The doctor asked him how he felt. 'I'll live.'

12. Delete 'Of shop-assistants' (since *to live in* is now used more generally). Add further examples. Also, spec. *to live together*, to cohabit. Also *transf.*, to have its place.

1813 Jane Austen *Pride & Prej.* III. xv. 266, I..am only concerned that their living together before the marriage took place, should be so generally known. **1875** [in Dict.]. **1879** Trollope *John Caldigate* III. viii. 106 They had not become man and wife... They had lived together. **1890** J. Watson *Nature & Woodcraft* vi. 71 The farm servants of Cumbria 'live in'. **1891** [in Dict.]. *a* **1916** H. James *Sense of Past* (1917) IV. iii. 242 A pot of about the size..of that one..with something or other on the cabinet or wherever, the place where it 'lives', as we say, rather branching out on either side of it. **1919** R. Fry *Let.* 6 Oct. (1972) II. 457, I don't think they're married; they've lived together for twelve years. **1938** E. Bowen *Death of Heart* III. vi. 426, I should like to know how she knew I'd been at her diary. I put it back where it lives. **1953** A. Upfield *Murder must Wait* v. 45 The nurse girl..didn't live in... She came every day. **1958** J. Cannan *And be a Villain* i. 37, I couldn't find any brandy. Do you know where it lives? **1961** N. Streatfeild *Silent Speaker* iv. 61 Mrs. Simpson had led Olivia into the little room where the vases lived. **1963** M. McCarthy *Group* ii. 42 She and Dick had 'lived together' on quite a different basis. **1971** *Woman's Own* 27 Mar. 21/3 One advantage of the permissive society is that it's all right to live together before marriage. **1971** *Daily Tel.* 16 Oct. 2/8 The development is designed..to provide extra accommodation for undergraduates to enable all 400 to 'live in'.

12*. *to live with*: **a.** To live with as if husband and wife; to cohabit with (Cohabit *v.* 2).

1749 J. Cleland *Mem. Woman Pleasure* II. 36. I had now liv'd with Mr. Norbet near a quarter of a year. **1813** Jane Austen *Pride & Prej.* III. viii. 147 She was more alive to the disgrace, which the want of new clothes must reflect on her daughter's nuptials, than to any sense of shame at her eloping and living with Wickham, a fortnight before they took place. **1854** Geo. Eliot *Let.* 23 Oct. (1954) II. 179 If you hear of anything that I have said..in relation to Mr. Lewes beyond the simple fact that I am attached to him and that I am living with him,.. believe that it is false. **1871** [see *allow *v.* 7]. **1879** Trollope *John Caldigate* II. iii. 33 Did she ever live with you?.. As your wife? **1923** D. H. Lawrence *Stud. Classic Amer. Lit.* vii. 123 Do you imagine Adam had never lived with Eve before that apple episode? **1928** E. Wallace *Flying Squad* xvi. 159 People are under the impression that you're living with me. **1963** M. McCarthy *Group* i. 11 The knowledge..of Kay's having 'lived with' Harald filled them with a sudden sense of the unsanctioned. **1972** M. J. Bosse *Incident at Naha* 31 'Virgil asked me to live with him two months ago,' I stated.

b. *fig.* To put up with; to come to terms with. *colloq.*

1937 T. S. Eliot in B. Dobrée *From Anne to Victoria* xliii. 603 Were one a person who liked to have busts about, a bust of Scott would be something one could live with. **1941** F. D. Roosevelt *Let.* 1 July in H. L. Ickes *Secret Diary* (1955) III. 567 Both of these are elements that we have to live with whether we like it or not. **1961** *Listener* 2 Nov. 694/2 We know that, on account of the balance of military power, we have got to live with it [*sc.* Communism]. **1964** 'W. Haggard' *Antagonists* ii. 16 That was awkward, but the experienced Mr Palliser could live with it. **1965** *Ottawa Jrnl.* 29 Apr. 25/3 Canada is so bound by the General Agreement on Tariffs and Trade that in effect it would have to live with its big auto deficit with the U.S. if there were no U.S.–Canada agreement. **1973** J. Porter *It's Murder with Dover* vi. 56 Gary was illegitimate... Not that I ever made any secret about it. It was something Gary had to learn to live with.

c. *to live with oneself*: to retain one's self-respect.

1962 P. Gregory *Like Tigress at Bay* xiv 143 Would he be able to live with himself, later? **1971** 'J. J. Marric' *Gideon's Art* xi. 98, I think he'll find it difficult to live with himself if he's taken off [the job]. **1973** R. Perry *Ticket to Ride* ii. 32 The note of hysteria in her voice stopped me dead. The sensible thing to do would have been to continue on my way but if I did I knew I'd find it awfully difficult to live with myself.

liveability, livability (livăbi·liti). [f. Liv(e)-ab(le *a.* + -ility.] **a.** Survival expectancy, *spec.* that of poultry. **b.** Suitability for habitation.

1914 *Springfield* (Mass.) *Republican* 28 June II. 9/2 You increase the egg yield, the fertility of eggs for hatching and the 'livability' of every chick hatched. **1922** E. W. Nelson in V. Stefansson *Northward Course of Empire* p. xviii, He [*sc.* Stefansson] has developed here and elsewhere the story of the 'livability' of the Far North, and shown that this hitherto dreaded region offers a welcome. **1934** W. A. Lippincott *Poultry Production* (ed. 5) iv. 133 The livability or vigor of a given chick might be looked upon as a part of its individuality. **1945** Nelson & Wright *Tomorrow's House* vi. 75/2 Soft, general illumination which can give this room the same air of livability as the living-room itself. **1950** *N.Z. Jrnl. Agric.* Apr. 331/1 This trial..is..a means of measuring the productive powers, the livability, and the quality of pullets intended as breeding stock. **1960** *Farmer & Stockbreeder* 8 Mar. 158/1 Spinks pullets are uniform, batch after batch, giving high egg production, good size and liveability. **1971** *Farmers Weekly* 19 Mar. 90/2 What about liveability, egg numbers, egg size and feed conversion?

li·ved-in, *a.* [Cf. Live *v.*[1] 12 b.] Occupied, inhabited.

1873 'S. Coolidge' *What Katy did at School* i. 3 The lived-in-look of the best parlor. **1950** C. Edwards in M. Cecil *Heroines in Love* (1974) viii. 203 Your home has that wonderful lived-in feeling. **1960** *House & Garden* July 50/1 Halls..can insulate the more lived-in parts of the house. **1970** *Daily Tel.* 19 May 16/4 It [*sc.* a play] has a lived-in look and all the acting is robustly expressive. **1975** *New Yorker* 21 Apr. 103/1 Mr. Ritman has a knack for making his sets look lived in.

live-in (li·vin), *a.* [f. vbl. phr. *to live in* (Live *v.*[1] 12).] Resident, residing in the establishment (as opp. to living out or at home). So **li·ve-out** *a.*, that lives out.

In quot. 1966 used in the sense 'deliberately living in as a form of protest'. Cf. *-in *suffix*[3].

1955 'C. H. Rolph' *Women of Streets* 148 Worked for four months as live-in domestic. **1966** *Maclean's Mag.* 18 June 1 The 'live-in' principles used by civil rights workers in the southern United States. **1969** *New Yorker* 18 Jan. 41 He has a live-out cook, but he likes to bend over the stove himself. **1970** *Daily Tel.* (Colour Suppl.) 9 Jan. 19/1 In 1920..the House at Maiden Bradley was run with the help of 26 live-in staff. **1971** H. Waugh *Shadow Guest* ii. 10 We had a live-in couple to cook and caretake and a live-out woman to clean.

liveness. Add: (Further examples.) Also *attrib.*

1926 N. M. Gunn *Grey Coast* ii. 21 That liveness of the body with its whirl and suppressed gaiety. **1931** L. Cowan *Recording Sound for Motion Pict.* xviii. 261 (*heading*) Liveness of sets. **1940** *Scrutiny* IX. 94 A translator of Mr. Archer's liveness of interest and taste. **1961** *Jrnl. Acoustical Soc. America* May 604 Liveness effects on the intelligibility of noise-masked speech.

liver, *sb.*[1] Add: **7.** *liver extract, paste, pâté, pudding* (earlier examples); *liver-shaped* adj.; **liver money** (see quots.); **liver pad** (earlier example); **liver rot** (later examples); also, a type of anæmia in sheep, cattle, and, occasionally, other animals, caused by the liver fluke, *Fasciola hepatica*; **liver salt**, a powder with purgative properties which is intended to be taken, in solution, for the relief of dyspepsia or a bad 'liver'; usu. short for the proprietary name *Andrews Liver Salt* and used in *pl.*; **liver sausage** [tr. G. *leberwurst*], a soft sausage filled with cooked liver, or a mixture of liver and pork, with various seasonings; cf. *liverwurst; **liver-spots**, read -**spot** and add: also, one of the small brown spots characteristic of this condition; (further examples); **liver-spotted** *a.*, having liver-coloured spots or liver-spots.

1910 Martindale & Westcott *Extra Pharmacopœia* (ed. 14) 820 Early cases of cirrhosis have been well treated with glycerinated liver extract. **1959** *Brit. Med. Jrnl.* 28 Mar. 833/2 Parenteral liver extracts still enjoy some popularity. **1935** R. M. *Trawler* 132 Firstly, there is the 'liver-money'. The livers of all fish caught..are taken aft and tried down for the oil they produce... The proceeds of sale are divided among the crew. **1962** J. Tunstall *Fishermen* ii. 55 *Liver money*, received for the amount of cod and haddock livers landed. **1879** G. W. Peck *Peck's Fun* 38 A boarder at a Leadville hotel investigated his beef-steak and found that it was a fried liver pad. *c* **1938** *Fortnum & Mason Price List* 51/1 *Potted meats*..Liver Paste—per tin 1/-. **1961** M. Spark *Prime of Miss Jean Brodie* iv. 121 Some sandwiches of liver paste. **1964** S. Bellow *Herzog* (1965) 70 Cheese, liver paste, crackers. **1935** E. Craig *Family Cookery* 50 Calf's *liver pâté*... Put the liver, uncooked bacon and ham twice or three times through a mincer. **1951** E. David *French Country Cooking* 222 *Liver pâté*. The French Amieux brands are always reliable. **1964** 'J. Melville' *Murderers' Houses* iv. 75 Velia was making a liver pâté for Sunday supper. **1716** C. Ludwig *Teutsch-Englisches Lexicon*, *Leberwurst*, a haggass or haggess, a liver-pudding, a pudding made of liver and lights or lungs. **1723** J. Nott *Cook's & Confectioner's Dict.* sig. S2v To make Liver Puddings. Boil a Hog's Liver..; take an equal Quantity of grated Bread, two Pound of Beef-suet. **1937** A. Fraser *Sheep Farming* xvi. 144 Liver-rot has at one time or another caused tremendous losses among sheep. **1972** *Country Life* 2 Mar. 524/2 In the early 19th century men were still prepared to argue that liver-rot was due to poisonous dew. **1896** *Trade Marks Jrnl.* 10 June 517 Andrews Liver Salt.. purifies & strengthens the whole system... A medicinal preparation for human use. The firm trading as Andrews & Co.,..Newcastle-on-Tyne; manufacturers. **1938** L. MacNeice *I crossed Minch* x. 140 De Valera's sham industries—putting liver salts into tins. **1951** J. B. Priestley *Festival at Farbridge* III. iii. 584 The High Street chemist who had sold him shaving soap and liver salts. **1969** M Russell *Hunt to Kill* II. 116 'I'm fine. Bit of a gut-ache. Too much booze.' 'Get home, take liver-salts.' **1855** Geo. Eliot in *Fraser's Mag.* June 706/1 He is enthusiastic about the delights of dining on *blaukraut* and *leberwurst* (blue cabbage and liver sausage). **1868** [see *blood-sausage* (*blood *sb.* 19)]. **1965** *House & Garden* Jan. 60 Liver sausage ranges in seasoning from extremely bland to highly spiced and pungent. **1971** *Sunday Times* (Colour Suppl.) 27 June 50/3 Liver sausages are found in every European country. **1942** S. Spender *Ruins & Visions* III. 51 A tree..rotted By a liver-shaped fungus on the bank. **1954** M. Rickert *Painting in Brit.: Middle Ages* viii. 197 The border..contains some motifs found earlier in East Anglian manuscripts, as the little round liver- and heart-shaped leaves. **1964** 'E. McBain' *Axe* vi. 101 He wore a dark black suit and he kept his hands, brown with liver spots, tented in front of his face. **1971** *Physics Bull.* July 410/3 Seborrheic keratosis, commonly called liver spot—a blemish occurring frequently in the middle aged and aged. **1973** N. Graham *Murder in Dark Room* iv. 24 He had a thin, yellow face dotted with liver spots. **1922** R. Leighton *Compl. Bk. Dog* vi. 86 Prince IV..a liver-spotted specimen. **1971** V. Canning *Firecrest* i. 7 Everything about him was contained, precise and impeccable..the fingernails of his liver-spotted hands immaculate. **1975** D. Bagley *Snow Tiger* xxiv. 205 Critchell placidly continued to fill his pipe with liver-spotted hands.

liverishness. [f. Liverish *a.*] Symptoms attributed to disordered condition of the liver.

1904 *Westm. Gaz.* 5 Oct. 10/1 Ordinary attacks of liverishness or biliousness are swiftly dispersed. **1928** *Daily Express* 14 July 15/7 Yellow, perhaps, suggests liverishness. **1930** R. Lehmann *Note in Music* v. 199 Never, he thought, would he forget..his liverishness.

Liverpool (li·vəɹpūl). [The name of an English city on the River Mersey.] **1.** Applied *attrib.* to the delftware and porcelain manufactured in Liverpool in the eighteenth century. Also *ellipt.*

1863 W. Chaffers *Marks Pott. & Porc.* 130 *Liverpool.* A pottery, called Herculaneum, was established on the Mersey by Richard Abbey, in 1794. **1869** C. Schreiber *Jrnl.* 5 Nov. (1911) I. 60 A Liverpool printed mug of Gen. Wolfe. **1957** Mankowitz & Haggar *Conc. Encycl. Eng. Pott. & Porc.* 131/2 Liverpool is at times difficult to distinguish from Bristol. **1964** M. Drabble *Garrick Year* iv. 56 One of my Liverpool ware teapots was broken. **1972** *Country Life* 6 Jan. 21/3 Liverpool or Delft tiles decorate the fireplace surround. **1972** *Collector's Guide* Aug. 9/1 (Advt.), An extremely rare Liverpool Vase, c. 1760 (William Ball's factory).

2. Special combs.: **Liverpool button** (see quot. 1896); **Liverpool house** *Naut.*, a deck-house; **Liverpool pantile**, a hard ship's biscuit; **Liverpool pennant** (see quot. 1933); **Liverpool sound**, the music, popular in the early 1960s, played by pop singers and groups in Liverpool, chiefly the Beatles; **Liverpool weather** *Naut. colloq.*, windy and 'dirty' weather.

1896 Farmer & Henley *Slang* IV. 212/1 *Liverpool-button*, a kind of toggle used by sailors when they lose a button. **1908** E. Noble *Grain Carriers* III. i. 146 Shanghaied crews are usually persons dressed in a rag and a Liverpool button. **1869** in C. N. Longridge *Cutty Sark* (1933) II. 205 Liverpool house fitted on centre 30 feet long. **1948** R. de Kerchove *Internat. Maritime Dict.* 422 *Liverpool house*, a superstructure extending from side to side and situated amidships in large (steel built) sailing vessels. **1902** B. Lubbock *Round the Horn* viii. 311 He handed me a regular bad-looking Liverpool pantile from the bread-barge. **1925** B. Hayes *Hull Down* iii. 28 The biscuits, commonly called 'Liverpool pantiles', were so hard you could break them only with a hammer. **1933** J. Masefield *Bird of Dawning* 287 *Liverpool pennants*, rope yarns used instead of buttons. **1963** *Daily Mail* 20 Sept. 10/4 It's been boys, boys, boys from Merseyside who have dominated pop music since the 'Liverpool sound' started to carry all before it towards the end of last year. **1963** *Daily Tel.* 10 Dec. 13 A show by the Beatles, the 'Liverpool sound' group, was not typical of 'pop' concerts. **1929** F. C. Bowen *Sea Slang*, Liverpool weather. **1934** G. H. Grant *Consigned to Davy Jones* (1935) xviii. 260 If we was on a windbag..it would be what we calls Liverpool weather.

liverwurst (li·vəɹvŭᵊrst, -wɒ̄ɹst). [Partial tr. of G. *leberwurst.*] = **liver sausage*.

1869 *Atlantic Monthly* Oct. 483/1 Our Dutch neighbors make *liver-wurst* ('woorsht') or meat pudding, omitting the meal, and this compound stuffed into the large intestines, is very popular in Lancaster market. **1929** E. Wilson *I thought of Daisy* iii. 150 The boy-friend's passed out! Too many of those rich liverwurst sandwiches! **1934** H. Miller *Tropic of Cancer* 41 Throw in some sweetbreads,..Throw in some fried liverwurst. **1945** S. Lewis *Cass Timberlane* (1946) vi. 39 Coca-Cola, liverwurst, stuffed olives, and chocolate layer cake. **1954** P. Highsmith *Blunderer* (1956) xxii. 139 Kimmel ordered a liverwurst sandwich. **1965** A. Roudybush *Season for Death* (1966) xix. 111 Spreading two slices of rye bread thickly with liverwurst.

livery, *sb.* Add: **2. c.** An emblem, device, or distinctive colour on a vehicle, product, etc., indicating its owner or manufacturer.

1938 H. A. Vallance *Highland Railway* xiv. 155 He introduced on the Highland Railway the style of painting which was afterwards so well-known on the south coast. Passenger engines were painted yellow... For goods engines a dark green livery was adopted. **1966** J. Thomas *Callander & Oban Railway* x. 169 The Caledonian 2–4–2 tanks..had the plain conical chimney..and the Prussian-blue passenger livery then standard on the line. **1970** *Guardian* 27 July 16/2 The Antonovs [*sc.* planes] are painted battleship grey, with red hammer and sickle emblems, not the normal livery of Aeroflot passenger aircraft. **1972** *Times* 13 Oct. 17/7 London Transport's intention can be simply stated. It is that the livery of the bus fleet will remain red, with a very strictly limited number offered to advertisers for all-over painted designs. **1973** *Times* 1 Dec. 17/7 As the designers responsible for the BEA livery we were invited, along with two other design groups, to make proposals for 'British Airways'. **1974** P. Lovesey *Invitation to Dynamite Party* ii. 22 An enormous express locomotive painted in the brilliant golden ochre and dark olive green livery of the London, Brighton and South Coast Railway Company.

9. (Examples.) Also *attrib.*

1845 F. Douglass *Narr. Life F. Douglass* 16 His stable and carriage-house presented the appearance of some of our large city livery establishments. **1888** C. D. Ferguson *Experiences Forty-Niner* i. 15 We placed our horses in a livery on Third street. **1902** W. N. Harben *Abner Daniel* 29, I could 'a' gone to a livery an' ordered out a team. **1903** A. Adams *Log of Cowboy* xiii. 81 Long before we reached the Mulberry, a livery rig came down the trail to meet us. **1936** E. G. Barnard *Rider Cherokee Strip* 210 They met a traveling man who was driving a good livery team to a buckboard. **1940** W. Faulkner *Hamlet* ii. 32 The village consisted of a livery barn and lot and a contiguous shady though grassless yard.

10. *livery horse* (earlier example).

1838 H. Colman *1st Rep. Agric. Mass.* (Mass. Agric. Survey) 17 The number of stage and livery horses kept in the county cannot fall short of one thousand,..who depend on the purchase of hay.

livery, *a.* **2.** (Examples.)

1937 in Partridge *Dict. Slang* 487/2. **1968** R. Jeffries *Traitor's Crime* iii. 34 You had too much port: port always makes you livery.

lives, *U.S.* var. of *lieve*(*s*), Lief *adv.*

1772 M. Mascarene *Let.* 14 Sept. in *Mass. Hist. Soc. Coll.* (1914) LXXI. 189, I had full as lives have it [*sc.* the portrait] on a larger [plate]. **1856** A. Cary *Married* 22, I would just as lives stand here as not. **1858** J. R. Lowell *Two Gunners* in *Poetical Wks.* II. 126 I'd jest ez lives eat tripe. **1891** *Harper's Mag.* Oct. 820/1, I will get Provided Usher to watch with me. He'd just as lives.

livett, var. Libbet[1].

1908 G. Sanger *70 Yrs. a Showman* xiv. 48 We could see the big sticks—'livetts' they were termed—hurtling towards..the prizes.

liveyere (li·vyəɹ). *Canad.* Also **liveyer, livier, livyer**(**e**. [Prob. f. the phr. *live here* (see below).] (See quot. 1909.) Also a permanent resident (as opp. to visiting fishermen, etc.).

Dict. Canadianisms suggests that the term 'may be < *livier* < OF *livree* (cp. *livery*), formerly in English villages a manorial worker having certain hereditary rights to a cottage and a small piece of land, thus being regarded as a permanent resident'. But cf. Livier in Dict.

1863 J. Moreton *Life & Work in Newfoundland* iii. 35 It is said of any uninhabited place that there are no *liviers* in it. **1881** Earl of Dunraven in *19th Cent.* IX. 93 A 'livier' signifies a person who lives all the year around in a locality, in contradistinction to one who visits it during the fishing season. **1895** W. Grenfell *Vikings of To-day* 57 The summer Labrador settlements are on islands or outside headlands, and here both Newfoundlanders and 'Livyeres' dwell, the latter retiring up the bays and inlets, to be nearer wood and game, when the former return to Newfoundland. **1901** *Chambers's Jrnl.* Jan. 68/2 The residents along this coast are termed 'livyres' (live heres), to distinguish them from the nomadic fisherfolk. **1905** N. Duncan *Dr. Grenfell's Parish* i. 12 The shore fishermen of the remoter Newfoundland coasts, the Labrador 'liveyeres', the Indians of the forbidding interior. **1907** D. Wallace *Long Labrador Trail* xxiii. 273 Even tea and molasses, usually found amongst the 'livyeres' (live-heres) of the coast, were lacking. **1909** *Toilers of Deep* July 176/1 The permanent inhabitants of the Labrador coast, the 'liveyers' are about three thousand in number. **1925** *Dialect Notes* V. 335 *Livyere*, one who lives here (in Newfoundland), usually an outport man. **1955** *Arctic Terms* 50/1 *Liveyere*, a corruption of 'live here', used in south-eastern Labrador to designate a semisettled inhabitant who customarily has a summer dwelling near coastal fishing grounds and a winter dwelling in an inland valley or in a mission settlement. **1966** A. R. Scammell *My Newfoundland* 47 The steamer had just turned the point, whistled hoarsely a couple of times to alert the livyers. **1969** H. Horwood *Newfoundland* xxv. 204 Most of the Livyers today are black-haired and a little darker than most Europeans... But many Livyers with blond hair, blue eyes, and fair skin keep cropping up too.

livid, *a.* Add: **c.** Furiously angry, as if pale with rage. *colloq.*

1912 *Collier's* 9 Mar. 21/1 He sprang to his feet, livid. 'That's a lie,' and he stopped suddenly, startled by his own violence. **1918** C. Mackenzie *Early Life Sylvia Scarlett* II. ii. 292 He was livid with fury. He asked if I thought he was made of money. **1936** M. Kennedy *Together & Apart* II. 151 Betsy is *livid*. She says now she will fight to the last ditch to get complete custody of the children. **1949** R. Chandler *Little Sister* ii. 10 Orrin would be absolutely livid. Mother would be furious too. **1959** J. Verney *Friday's Tunnel* xxiv. 214 Friday's livid because he thinks you've punctured his bike. **1973** 'D. Shannon' *No Holiday for Crime* (1974) x. 162 Mr. MacFarlane would be livid to have it [*sc.* whisky] impounded as evidence.

living, *vbl. sb.* Add: **1. d.** *living-in, -out* (further examples).

1942 *R.A.F. Jrnl.* 13 June 14 The misunderstanding has arisen..over the living-out system on which the Commandos work. The men are neither billeted nor fed by the Army. **1955** M. Laski *Apologies* 59 I'd never have living-in servants again. **1962** L. Deighton *Ipcress File* xi. 71 Murray decided that this was a good time to ask about his living out allowance. **1970** 'J. Bell' *Hydra with Six Heads* vii. 76 There was a living-in cook. **1974** 'P. B. Yuill' *Bornless Keeper* viii. 71 Check with Exeter [Gaol] he isn't on their wonderful living-out rehabilitation scheme.

7. a. *living-dining room*; **b.** *living price* (earlier example); **living area** = **living space* (*c*); **living floor** *Archæol.*, the site of a prehistoric camp indicated by bones, tools, etc., found there; **living space**, (*a*) tr. *lebensraum; (*b*) space for accommodation; (*c*) a habitable area in a room or house; **living standard** [cf. Standard *sb.* 12 a], the level of consumption in terms of food, accommodation, clothing, services, etc., estimated for a person, group, or nation; **living wage** (earlier and later examples).

1962 *Listener* 18 Jan. 135/1 A new house..which I recently visited is centred on a large living area. **1965** in P. Jennings *Living Village* (1968) 103 The garages are built at road level, and the living areas of the houses are built over the garages. **1969** K. Giles *Death cracks Bottle* vii. 74 The inglenook which the architect had fashioned in the living area. **1933** *Archit. Rev.* LXXIV. 20 The exhibit is part of a projected structural unit which includes

an entrance hall and kitchen in addition to the living-dining room. **1945** Nelson & Wright *Tomorrow's House* iv. 41/1 The living-dining-room was a makeshift..but nevertheless an expedient to save space and money. **1968** *Globe & Mail* (Toronto) 17 Feb. 45 (Advt.), Living-dining room with open fireplace. **1946** *Nature* 2 Nov. 637/2 In 1943, further evidence was obtained pointing to the conclusion that on the site, now known as Glorgesailie site 10, there was a series of actual living floors or camp sites of Acheulean men. **1947** L. S. B. Leakey in *E. Afr. Ann.* *1946–7* 69/1 On the ancient land surfaces were uncovered the actual 'living floors' or camp sites of ancient Acheulean hunters. **1965** Hole & Heizer *Introd. Prehist. Archeol.* iii. 35 The amount of bone and stone tools suggests seasonal, or perhaps permanent year-round camps. In Africa these are called 'living floors'. **1834** *Congress. Globe* 3 May 362/2 Mr. Forsyth said that..70 to 76 cents was a very living price for fish oil. **1939** W. S. Churchill *Into Battle* (1941) 127 The Swiss..may rob them of their living-space. **1944** F. Clune *Red Heart* 20 The Australian native..can make a 'living-space' in a country where white people can't. **1959** *Listener* 10 Dec. 1032/1 The 5,000 or so species of pest insects are man's most dangerous competitors for food and living space on this planet. *Ibid.* 17 Dec. 1071/1 The bed-sitting-room was divided by curtains into three living-spaces. **1959** *Chambers's Encycl.* IV. 338/2 Hitler's next objective was to sweep Czechoslovakia out of the road to 'living space' in the east. **1961** L. D. Stamp *Gloss. Geogr. Terms* 296/2 *Living space*, an English translation of the German *Lebensraum* but most authors prefer to use the German word when discussing its geopolitical implications. **1961** *Listener* 28 Dec. 1110/1 Architects will be able to plan, on the ground floor, one, two, or three living spaces—for example, a kitchen partially opening into a dining-room, with a separate living-room. **1966** *Ibid.* 19 May 729/3 The average living space per person in Moscow is officially admitted to be eight square metres. **1972** M. Jones *Life on Dole* i. 15 There might have been a couple of lodgers, unmarried men..whose living space was a cot in the corner. **1957** P. Worsley *Trumpet shall Sound* vi. 121 The aim of the Government was to raise living-standards to those of the Europeans. **1964** Gould & Kolb *Dict. Social Sci.* 690/1 The concept of *living* standards, or levels, would include and perhaps emphasize, 'material' quantitative elements such as the consumption of goods and services. **1968** *Listener* 10 Oct. 469/3 This was the chief reason..why..Western achievements and Western living standards were so consistently lied about. **1974** *Times* 16 Jan. 13/2 Improving personal living standards. **1888** E. Bellamy *Looking Backward* xxviii. 450 The wonder to me is, not that industries conducted as these are do not pay you living wages, but that they are able to pay you any wages at all. **1967** *Listener* 23 Feb. 248/1 Some of them already have to work up to fifty-six hours a week to make up their weekly pay to a living wage of £20. **1974** *Guardian* 31 Jan. 13/8 Another miner..sounds bitter... 'We are broken down old men; we're limping and we are injured. And all we want is a living wage.'

living, *ppl. a.* Add: **2. b.** *living fossil*: a plant or animal that has survived the extinction of others of its group.

1935 C. J. Chamberlain *Gymnosperms* iv. 61 The cycads of today may well be called 'living fossils'. **1953** J. S. Huxley *Evolution in Action* v. 127 There is the persistence of a few survivors from a once-abundant group—so-called 'living fossils', like the duckbill platypus; and that of a whole successful group. **1955** *Sci. Amer.* Apr. 108/2 A living fossil is defined as an organism that has survived beyond its era. A standard example is the tuatara of New Zealand, which looks like a lizard but is in fact the 'sole survivor of an order of reptiles which flourished in the great Age of Reptiles and is now extinct except for this one species'. **1966** C. A. W. Guggisberg *S.O.S. Rhino* ii. 31 The two-horned Sumatran rhinoceros.. has come to us practically unchanged from the Tertiary Age, another 'living fossil'. **1974** *Encycl. Brit. Micropædia* VI. 276/3 Living fossil, an organism long believed extinct that is discovered to be still in existence.

c. *the land of the living* (further examples). See also Land *sb.* 3 c.

1708 Lady Cave *Let.* in M. M. Verney *Verney Lett.* (1930) I. xiv. 266 Sir Thomas is glad to hear Col. Oughton is in the land of the living..having not heard a word from him. **1925** E. Phillpotts *Voice from Dark* ii. 22 And is Mr. Bitton still in the land of the living? **1964** *Roman Breviary* 346 P, You are my refuge, my portion in the land of the living.

f. *living corpse*; *living dead*: as *adj. phr.*, having lost hold of life, not using one's life abundantly; as *sb. phr.*, such people.

1860 J. W. Palmer tr. *Michelet's Love* IV. viii. 243 It may be said that she came out of the asylum a living corpse, and it was not long before she died in reality. **1917** R. Graves *Fairies & Fusiliers* 28 You'll find me buried, living-dead In these verses that you've read. *a* **1930** D. H. Lawrence *Last Poems* (1932) 168, I know the unliving factory-hand, living-dead millions Is unliving me, living-dead me, I, with them, am living dead. **1958** *Listener* 14 Aug. 225/2 What they thought of 'squares', and the living dead in general.

3. *living chess*: a game of chess in which living persons act as the chessmen; *living newspaper* (see quot. 1966); *living theatre*: the theatre, as opposed to the cinema.

1905 *Lasker's Chess Mag.* II. III. 131 So that the reader may visualize the phases through which the game has passed we will show two historical cameos of living chess. **1929** *Brit. Chess Mag.* LXIX. 126 A picturesque display of living chess was given by the schoolboys. **1935** *N.Y. Times* 31 Dec. 10/1 Elmer Rice, regional director, announced that..the Biltmore will open before the end of January with the first topical production of the unit known as the 'Living Newspaper'. The initial show..will offer in graphic terms the background of the Italo-Ethiopian war. **1938** *Times* 28 Dec. 11/3 The 'living' theatre, as it has come to be called in distinction from the cinema, must be in a very healthy condition. **1940** *Ann. Reg. 1939* 40 The Chancellor made the entertainment tax lighter for the 'living theatre'. **1941** J. S. Huxley in *Fortnightly* July 11 The theatre project indeed began to create new types of popular drama like the Living Newspaper, which undoubtedly stimulated social self-consciousness. **1963** *Guardian* 12 Mar. 2/6 It was deplorable that Plymouth had no living theatre. **1966** J. R. Taylor *Penguin Dict. Theatre* 159 *Living newspaper*, a sort of topical documentary revue in a series of short scenes based on current social and political problems, devised in the 1930s in the U.S.A. by the Federal Theatre, and used elsewhere for propaganda during the war. **1970** A. Sunnucks *Encycl. Chess* 291 *Living chess*. During the past 600 years the game played with living pieces has from time to time been presented in different countries. **1971** J. Willett in A. Bullock *20th Cent.* 243/1 The Theater Project..evolved a new form of lecture-cum-sketch in the Living Newspaper.

6. *living picture*, (*a*) = *tableau vivant* (Tableau *sb.* 4); (*b*) a motion-picture.

1875 *N.Y. Herald* 24 Nov. 2/4 Mr. Matt Morgan's Classical Tableaux of Living Pictures, illustrated by a Corps of Beautiful Ladies. **1895** G. B. Shaw in *Sat. Rev.* 6 Apr. 443/2, I sat out the entire list of sixteen 'Living Pictures'. Half a dozen represented naiads, mountain sprites, peris, and Lady Godiva, all practically undraped. **1897** *Knowledge* 1 Sept. 216/2 Last winter saw the 'living pictures' adopted as the craze of the season for music-halls, bazaars, and variety entertainments generally. **1897** *Sears, Roebuck Catal.* 342/2 [The book] includes also elaborate directions for exhibiting Living Pictures and Tableaux. **1899** H. V. Hopwood *Living Pictures* vi. 207 The first requirement in the projection, as in the taking, of Living Pictures is absolute rigidity of the apparatus. **1906** J. A. Manson *Indoor Games* 30 Living Pictures, or Tableaux Vivants, if considered as an indoor game, must be placed on the highest level... It need hardly be said that neither acting nor speech is required in a living picture. **1962** E. Larsen *Film Making* i. 18 Soon nearly every variety show contained 'living pictures' as a programme item.

living-room. Also **living room.** [Living *vbl. sb.* 7.] **1.** A room that is set aside for ordinary social use (as opp. to a bedroom, etc.).

1825 [in Dict. s.v. Living *vbl. sb.* 7]. **1857** C. Vaux *Villas & Cottages* 119 Under the..living-room is a basement-kitchen. **1867** 'T. Lackland' *Homespun* I. 139 The joy with which grand parents welcome us in the great living-room. **1879** F. R. Stockton *Rudder Grange* i. 7 There was a kitchen, a living room, a parlor and bedrooms. **1911** C. Harris *Eve's Second Husband* 310 She occupied one chair in the living room all day. **1911** H. S. Harrison *Queed* xix. 239 Queed..went upstairs to the comfortable living-room. **1917–18** *T. Eaton & Co. Catal.* Fall & Winter 416/4 Massive living-room arm chair. **1931** A. J. Cronin *Hatter's Castle* I. i. 12 The common room, the living room of the house where its inhabitants partook of meals, spent their leisure and congregated in their family life. **1933** *Discovery* July 218/2 The growing popularity of the living-room as the central dominant feature of the modern dwelling..has completely altered the design of the house. **1933** H. Walpole *Vanessa* II. 287 The room at the top of the first stairflight that had once been the drawing-room with the fine gilt chairs,..the rosy cupids on the ceiling, was now the general living-room. **1953** G. Greene (*title*) The living room. **1967** E. Short *Embroidery & Fabric Collage* iii. 78 [The cushion] can transform a divan from a bed into an acceptable piece of living-room furniture. **1971** *Daily Tel.* (Colour Suppl.) 12 Nov. 30/1 In the living-room is the Steinway upright.

2. tr. *lebensraum.

1938 *Times* 18 Nov. 17/6 The Danube valley and the Balkans are..the natural 'living-room' of industrial Germany. **1940** 'G. Orwell' in *New English Weekly* 21 Mar. 321/1 Hitler envisages, a hundred years hence,..a continuous state of 250 million Germans with plenty of 'living room'. **1942** E. Paul *Narrow St.* xxxiv. 305 The leaders had told them Hitler would behave, if appeased, that he did not want their lands and goods but only 'living room'.

Livingstone (li·viŋstŏn). [Origin unknown.] A name used *attrib.* to designate the **Livingstone daisy**, a small, annual, succulent plant, *Dorotheanthus bellidiformis* (*Mesembryanthemum criniflorum*), native to the Cape Province of South Africa, and bearing daisy-like flowers.

1934 B. P. Mansfield in *Gardeners' Chronicle* 19 May 330/1 The Livingstone Daisy. From..seed catalogues.., I gather that *Mesembryanthemum criniflorum* is introduced as a 1934 novelty. **1950** W. E. Shewell-Cooper *Compl. Gardener* III. 243 The 'Livingstone Daisy' is a delightful annual with fairy-tale flowers in many lovely shades of pink, crimson, pale or golden-yellow or apricot. **1960** *House & Garden* Aug. 53/2 South Africans like..the Livingstone Daisy. **1971** *Daily Colonist* (Victoria, B.C.) 16 May 36/5 There is another Mesembryanthemum, *M. criniflorum*, known as the Livingstone Daisy, which is a delightful little spreading carpeter.

Livonian (livōᵘ·niăn), *sb.* and *a.* [f. med.L. *Livonia*, Livland, a former Baltic province of Russia.] **A.** *sb.* **a.** A native or inhabitant of Livonia. **b.** The language of the Livonian people. **B.** *adj.* Of or pertaining to Livonia, or to the Livonian people or their language.

1652 Heylyn *Cosmographie* II. 167 *Derpt* or *Derbren*.. situate on the Beck..in the midst of the Province; and taken at the same time by the Moscovite; who transporting the Livonians into other places, planted these parts with Colonies of his own people. **1677** [see Prussian *sb.*]. **1757** J. Dyer *Fleece* IV. 139 The Livonian gulph Receives her sails. **1824** J. D. Cochrane *Narr. Journey Russia & Siberian Tartary* I. 23 A young Livonian Baron..gave me letters of recommendation to the frontiers of Siberia. **1841** [see Lettish *a.* (*sb.*)]. **1872** [see Lettic *a.* (*sb.*)]. **1882** *Encycl. Brit.* XIV. 723/2 [The] plateaus..of Haanhoff and of the Livonian Aa. *Ibid.* 724/1 The Livonians..have nearly all passed away. **1926** *Spectator* 31 July 176/1 If one of their German servants or retainers wanted to marry a Livonian girl the Lutheran priest would seek to stop him. **1933** L. Bloomfield *Lang.* 68 The other languages of the Baltic branch, Carelian,..Ludian, Vepsian, Livonian, [etc.]. **1936** *Times Lit. Suppl.* 27 June 532/2 Descended, to quote his own words, from 'a middle-class East Prussian, a Westphalian peasant's son, a Livonian *bourgeoise* and a Livonian noblewoman'. **1954** *Trans. Philol. Soc.* 86 In the Livonian dialect of Lettish all gender has vanished, a state of affairs clearly due to the influence of Livonian, a Finno-Ugrian, genderless language. **1972** W. B. Lockwood *Panorama Indo-European Lang.* 140 Languages belonging to the Baltic division of Finnic... The best known of the languages concerned is Livonian which in the thirteenth century still predominated in the area of the Gulf of Riga, i.e., in the historic province of Livonia. It is not yet entirely extinct.

Livornese (livoɹnī·z), *sb.* and *a.* [f. name of the Italian city of *Livorno* (Leghorn) + -ese.] **A.** *sb.* The people of Leghorn. **B.** *adj.* Of or pertaining to Leghorn.

1819 Shelley *Let.* 17 Nov. (1964) II. 157 The astonishment of the Livornese when she returns from her cruise. **1821** *Ibid.* 8 June 296 The Livornese merchants, who sell board & lodging. **1912** N. Douglas *Fountains in Sand* vii. 78 The old distinction between Livornese and Tunisian Jews is slowly becoming effaced.

‖ **livre de chevet** (livr də ʃəvɛ). [Fr., lit. pillow-book.] A bedside book; a favourite book.

1923 A. Huxley *On Margin* 122 For some days I made Dr. Legat's book my *livre de chevet*. **1939** *Times Lit. Suppl.* 13 May 284/1 It may also be recommended as a *livre de chevet*. **1958** L. Durrell *Mountolive* v. 111 It was Darley..who introduced me to the current Alexandrian *livre de chevet* which is a French novel called *Moeurs*. **1972** *Times Lit. Suppl.* 5 May 529/2 To restore the *Heroides* to its rightful place as a *livre de chevet* for the literary connoisseur.

‖ **livre de circonstance** (livr də sirkoṅstaṅs). [Fr.] A book composed or adapted for the occasion.

1945 *Yale Rev.* Winter 356 It is one of those rare *livres de circonstance* with a relevance that may extend far beyond the moment to an entire generation. **1957** C. Vereker *Devel. Political Theory* iii. 93 Begun as a *livre de circonstance*, Hooker's great work raised most of the abiding problems of political authority. **1973** *Times Lit. Suppl.* 1 June 606/5 Mr Neelkant's *livre de circonstance* provides a good deal of ammunition for defenders of the Indo-Soviet Treaty by setting out in detail the many ways in which Russian aid has contributed to India's industrial development.

livyer(e, varr. *liveyere.

‖ **liwa** (lī·wa). [Arab. *liwā'*.] A province or large administrative district in any of several Arabic-speaking countries.

1925 S. H. Longrigg *Four Centuries Mod. Iraq* 308 In 1872 he was Mutasarrif of his own Liwa. **1931** *Special Rep. Progress of 'Iraq, 1920–1931* (Colonial Office) vi. 49 It has been pointed out above that in 1921 the country was divided into 10 liwas. **1957** *Encycl. Brit.* XII. 590C/2 Tobacco is grown in the Sulaimaniya and Ariba *liwas* [of Iraq]. **1959** *Chambers's Encycl.* VII. 705/2 The administration [of Iraq] is centralized, the *mutasarrifs* of the *liwas*..being all appointed from Baghdad. **1962** *Times* 14 Aug. 9/7 The main centre of activity has now shifted northwards to the *liwa* of Mosul. **1974** *Internat. Jrnl. Middle East Stud.* V. 165 Stafford at the time was the Administrative Inspector of the *liwas* of Mosul and Arbil.

lizard. Add: **1. d.** Lizard skin. Also *attrib.*

1895 *Montgomery Ward Catal.* 100/1 Ladies' pocket books... Light American lizard grained leather. **1926–7** *Army & Navy Stores Catal.* 495/3 Manicure cases... Polished lizard with silver gilt fittings. **1957** M. B. Picken *Fashion Dict.* 215/1 *Lizard*, leather made from lizard skins. **1968** J. Ironside *Fashion Alphabet* 238 *Lizard*, scaly skin of the lizard from Java and India. Used mainly for shoes and handbags, it is harsh and not pliable. **1974** A. Laski *Night Music* 134 Her good navy lizard shoes and bag.

2. b. *Austral.* and *N.Z. slang.* A musterer of sheep; a man employed to maintain boundary fences.

1931 'W. Hatfield' *Sheepmates* xv. 121 You'd be better out in the camp with me than crawlin' around a fence like a fly-catcher lizard. **1933** L. G. D. Acland in *Press* (Christchurch, N.Z.) 4 Nov. 15/7 *Lizard*, slang for musterer, I suppose, because they both crawl over the hills. **1937** A. W. Upfield *Winds of Evil* xviii. 175 What bloke wouldn't be depressed at coming down to a fence lizard?.. Come down to fencin' and you want to know why a bloke's depressed. **1945** Baker *Austral. Lang.* II. iii. 63 Shepherds have been known variously as *lizards*, *crawlers*.

c. = **lounge lizard*.

1935 E. Pound *Let.* Feb. (1971) 269 Alas, as you are writing English, you can't call *them there bloody* gallants, 'cake-eaters' or 'lizards', 'dudes', 'gigolos', 'young scum' [etc.].

8. lizard-bird (later example).

1911 Kipling in Fletcher & Kipling *School Hist. Eng.* i. 9, I remember the bat-winged lizard-birds.

lizardite (li·zăɹdəit). *Min.* [f. *Lizard*, the name of a peninsula in Cornwall + -ite[1].] A basic silicate of magnesium, $Mg_3Si_2O_5(OH)_4$, which is a variety of serpentine.

1956 Whittaker & Zussman in *Mineral. Mag.* XXXI. 122 A new name is desirable by means of which specimens [of serpentine] possessing the single-layer ortho-cell can be distinguished from chrysotile and from antigorite. Since a well-known locality for serpentine minerals is the Lizard area of Cornwall, England, and since specimens of the subgroup in question are known to occur there.., the name lizardite would seem appropriate. **1967** *Prof. Papers U.S. Geol. Survey* No. 384B. 101/1 We have further investigated this stability using lizardite from the type locality at The Lizard, Cornwall, England..and natural admixtures of lizardite and clinochrysotile from Snarum, Parish of Buskerud, Norway, from New Almaden mine, Santa Clara County, Calif., and from Forest Hills, Hartford County, Md. **1968** *Zeitschr. für Kristallogr.* CXXVI. 163 The measured *x* and *z* parameters of single-layer lizardite from Raduša chromite Mine, Yugoslavia, indicate a distortion of the basic serpentine layer similar to that found in chrysotiles.

Lizzie (li·zi). *slang.* Also **lizzie.** [Abbrev. of the female Christian name *Elizabeth*.] **1.** A lesbian. Also, an effeminate young man; also *lizzie boy*.

1905 *Dialect Notes* III. 87 *Lizzie* (*boy*), an effeminate young man. **1912** N. L. McClung *Black Creek Stopping-House* xi. 98 She's married to a no-good Englishman, a real lizzie-boy. **1949** A. Wilson *Wrong Set* 98, I wish you wouldn't talk to those Lizzies. **1966** J. B. Priestley *Salt is Leaving* vi. 75 She has a Lizzie crush on me. **1970** J. Symons *Man who lost his Wife* I. x. 70 You'd never have thought I was a lizzie, would you? And butch at that.

2. A motor car, esp. an early model of a 'Ford'. Also *Tin Lizzie*.

1913 'B. L. Standish' *The Desert* (Advt. on rear cover), So, when you get tired of rolling around in your Lady Lizzie..hie yourself to the nearest news dealer. **1915** *Chicago Herald* 2 Nov. 4/2 You're right about the guy who is able to make a tin lizzie out of the cans they tie to him. **1922** *Blackw. Mag.* Jan. 37/2 We then prepared to start for home; but 'Lizzie'..refused..to think of starting. *Ibid.* Feb. 253/1 An extra bad pothole put Lizzie's back axle out of action. **1949** D. M. Davin *Roads from Home* 233 The pace they drove their old tin lizzies. **1956** J. Tickell *Moon Squadron* 39 These special duty 'Lizzies' had to be stripped of all guns, armour and wireless equipment—except radio telephones—in order to allow room for the passengers.

3. (See quot.)

1925 Fraser & Gibbons *Soldier & Sailor Words* 145 *Lizzie*, a big gun: also its shell. A term originating at the Dardanelles and suggested by the firing of the big fifteen-inch guns of H.M.S. *Queen Elizabeth*.

4. Lisbon wine.

1934 M. Ellison *Sparks beneath Ashes* 182 She drinks 'Lizzie' and methylated spirit. **1936** J. Curtis *Gilt Kid* v. 48 A glass of cheap Lisbon red wine, lizzie they called it.

llama. b. (Earlier example.)

1864 'P. Paterson' *Glimpses Real Life* xxv. 244 The hosier's bill is £70—all kinds of stockings, silk,..embroidered, striped, llama,..and otherwise.

llano. Add: Hence **Llane·ro, llane·ro**, an inhabitant of a *llano* (see also quot. 1878).

1878 A. H. Keane in H. W. Bates *Cent. Amer.* 519 (*heading*) American tribes and languages... Llaneros... An Apache tribe. **1908** *Daily Chron.* 27 Oct. 4/6 The llanos,..on which the llaneros live a precarious life. **1910** *Encycl. Brit.* II. 158/2 The chief divisions of the Apaches were the Arivaipa,..Llanero, [etc.]. **1960** H. S. Ferns *Brit. & Argentine* iii. 96 Now José Paez, the *llanero*, began to rally the cowboys to the side of the revolution. **1973** *Country Life* 20 Dec. 2114 Simon Bolivar..was..one of the greatest horsemen who ever lived... His outstanding skill..enabled him to control the wild Llaneros of the coastal plains who..were the backbone of his army.

Lloyd-Georgian (loid,dʒǭ·ɹdʒiăn), *a.* and *sb.* [f. the name of David *Lloyd George* (1863–1945), British politician.] **A.** *adj.* Of, pertaining to, or associated with Lloyd George. **B.** *sb.* A follower or supporter of Lloyd George or his policy.

1909 *Daily Graphic* 12 Oct. 6/2 The self-sacrificing Ministerial millionaires..with the Lloyd-Georgian iron entering into their souls. **1928** [see *Asquithian *a.* and *sb.*]. **1928** *Daily Tel.* 25 Sept. 12/2 The kaleidoscopic contortions of Lloyd-Georgian politics. **1963** [see *dressage].

So **Lloyd-Georgeite** = **Lloyd-Georgian* sb.; **Lloyd-Georgery** [-ery 2], **Lloyd-Georgism**, the political policy of Lloyd George.

1910 *Blackw. Mag.* Dec. 772/2 Lloyd Georgism (not Liberalism) was defeated in the Division last January. **1921** *Spectator* 19 Mar. 352/1 They must now be counted as Lloyd Georgeites rather than as Unionists. **1958** *Ibid.* 11 July 48/1 Most of his time was spent talking..second-rate Lloyd-georgeries about the opinions of posterity.

lloydia (loi·diă). [mod.L. (R. A. Salisbury 1812, in *Trans. Hort. Soc. Lond.* I. 328), f. the name of Edward Lhwyd or *Lloyd* (1660–1709), Welsh antiquary and Keeper of the Ashmolean Museum, who discovered the British species on Snowdon.] A small alpine bulbous plant of the genus so called, belonging to the family Liliaceæ, native to the northern hemisphere, and bearing white or yellow flowers; also called the Snowdon lily or mountain spiderwort.

1850 Hooker & Arnott *Brit. Flora* (ed. 6) 442 Lloydia... *L. serotina* Reich. (Mountain L[loydia])... On the Welsh mountains, rare. **1866** Lindley & Moore *Treas. Bot.* II. 690/2 Lloydia. A liliaceous plant, from five to six inches high. **1956** Walters & Raven *Mountain Flowers* v. 104 The Devil's Kitchen is well named, and I wish anybody luck who tries to find *Lloydia* in flower there. **1966** W. Condry *Snowdonia National Park* vi. 112 Lhuyd found quantities of the tiny lily later called Lloydia in his honour.

Lloyd Morgan's canon (loid mǭ·ɹgănz kæ·nən). *Psychol.* [f. the name of Conwy *Lloyd Morgan* (1852–1936), British psychologist.] (See quot. 1937.)

[**1894** C. Lloyd Morgan *Introd. Compar. Psychol.* iii. 53 In no case may we interpret an action as the outcome of the exercise of a higher psychical faculty, if it can be interpreted as the outcome of the exercise of one which stands lower in the psychological scale.] **1937** *Discovery* May 162/1 No action must be interpreted as a higher psychological process if it can be interpreted as one lower in the scale. This is known as 'the Lloyd Morgan canon'. **1964** A. Koestler in *Listener* 14 May 786/2 A principle which became a kind of eleventh commandment for psychologists, known as 'Lloyd Morgan's canon'. **1968** E. Boring in *Internat. Encycl. Social Sciences* X. 495/2 He [*sc.* Lloyd Morgan] is best known for what has come to be called Lloyd Morgan's canon, which demands parsimony in the inference of an animal's place on the scale of mind from its behavior.

Lloyd's (loidz). [f. the name of Edward *Lloyd* who opened a coffee-house in London in 1688, and supplied shipping information to his clients.] Name of a London association of underwriters and agency for arranging insurance (formerly marine insurance only, but now nearly all kinds); it also issues daily shipping intelligence, as *Lloyd's List* (of which the occasional newspaper *Lloyd's News* was a precursor). *Lloyd's Register* (*of Shipping*), an independent society which surveys ships to ensure compliance with standards of strength and maintenance; its annual classified list of such ships. Also *Lloyd's policy*, an insurance policy underwritten by Lloyd's; so *Lloyd's underwriter*.

1819 M. Edgeworth *Let.* 4 Mar. (1971) 174 He won £30,000 by a bit of gambling insurance on 2 missing.. ships. The ships re-appeared... He never could shew his face at Lloyds afterwards. **1829** G. Griffin *Collegians* (ed. 2) II. xxvi. 248 A more crazy and precarious mode of conveyance could not be found, even among the ships marked with the very last letter in Lloyd's list. **1833** [see Underwriter 2]. **1846** [see Register *sb.*[1] 4 c]. **1876** F. Martin (*title*) History of Lloyd's and of marine insurance in Great Britain. **1882** J. Ashton *Social Life in Reign of Anne* I. xviii. 224 Lloyd's was then in Lombard Street, and indeed to this day, on Lloyd's policies, is stated that this policy should have the same effect as if issued in Lombard Street. **1911** *Encycl. Brit.* XV. 833/2 Lloyd's... Originally a mere gathering of merchants..in a coffee-house kept by..Edward Lloyd in Tower Street, London, the earliest notice of which occurs in the *London Gazette* of the 18th of February 1688. *Ibid.* XIX. 554/2 In 1696 Edward Lloyd..started a thrice-a-week paper, *Lloyd's News*, which..was the precursor of the *Lloyd's List* of the present day. *Ibid.* XXIV. 957/2 *Lloyd's Register*, as at present constituted, has existed since 1834. **1922** *Ibid.* XXXI. 492/1 Lloyd's underwriters have shown a great deal of enterprise in accepting risks of a novel kind. **1959** *Chambers's Encycl.* VIII. 623/2 The incorporation of Lloyd's as a chartered body on 25 May 1871.

Lo, *sb. U.S.* [Humorously from Pope's line '*Lo, the poor Indian*', etc., *Essay on Man* I. 99.] An American Indian. Also *Mr.* (or *Mrs.*) *Lo*.

1871 *Republican Rev.* (Albuquerque, New Mexico) 2 Sept. 1/4 Cowardly Lo prefers to attack none but very small parties of teamsters, farmers, or lone mail riders. **1873** G. W. Perrie *Buckskin Mose* vi. 88 This document set forth that the bearer was a good Lo. *Ibid.*, He appeared again at the head of our train, in the company of thirty or forty other Los. **1874** J. G. McCoy *Hist. Sk. Cattle Trade* 260 Crossing the plains was an undertaking fraught with great danger; especially as Mr. Lo was decidedly fond of horses. **1880** A. A. Hayes *New Colorado* (1881) iii. 40 Colonel Craig..and his men began, not unsuccessfully, the repression and suppression of Mr. Lo. **1885** [see *evolute *v.* 2]. **1904** *Minneapolis Daily Times* 12 June 6 The march of civilization has convinced Lo that fighting is not as profitable as it used to be. **1904** *N.Y. Even. Post* 6 Aug., On Florida's shield stands a placid and buxom Mrs. Lo, with fringed skirt falling to the knee. **1947** R. P. T. Coffin *Yankee Coast* 220 He..went to join Lo, the Poor Indian on the Happy Hunting Grounds.

lo, *int.*[1] Add: **b.** Freq. in phr. *lo and behold* (usu. jocular).

1808 Lady Lyttelton *Let.* June in *Corr.* (1912) i. 20 Hartington..had just told us how hard he had worked all the morning..when, lo and behold! M. Deshayes himself appeared. **1841** Lytton *Night & Morning* II. III. v. 144 The fair bride was skipping down the middle..when, lo and behold! the whiskered gentleman..advanced..and cried—'*La voilà!*' **1849** Dickens *Dav. Copp.* (1850) xxii. 234 What does he do, but, lo, and behold you, he goes into a perfumer's shop. **1930** J. B. Priestley *Angel Pavement* ii. 60 And then—lo and behold—it was there all the time. **1947** T. Williams *Streetcar Named Desire* x. 151 You come in here and sprinkle the place with powder and spray perfume and cover the light-bulb with a paper lantern, and lo and behold the place has turned into Egypt and you are the Queen of the Nile!

lo, 'lo, *int.*[2] Colloq. abbrev. of Hallo, Halloa, Hello, Hullo, Hulloa *ints.*

1921 J. Dos Passos *Three Soldiers* I. i. 13 ''Lo, buddy,' came a voice beside him... 'Goin' to the movies?' **1922** Joyce *Ulysses* 287 Lo, Joe, says I. How are you blowing? **1938** F. D. Sharpe *Sharpe of Flying Squad* ii. 16 ''Lo,' said Moisher. ''Lo,' said Harry. **1968** R. Clapperton *No News on Monday* viii. 99 ''Lo, son. You the detective?' he murmured.

loa[1]**.** Substitute for etym. and def.: [Native name in Angola, used as a specific name in *Filaria loa* (T. S. Cobbold *Entozoa* (1864) xiv. 389) and as a generic name in *Loa loa* (C. W. Stiles 1905, in *Bull. U.S. Dept. Agric. Bur. Anim. Ind.* no. 79. 50).] A filarial worm of the monotypic genus so called, found in tropical Africa and infecting the eyes and subcutaneous tissues in man. Also *attrib.* (Add earlier and later examples.)

1864 T. S. Cobbold *Entozoa* xiv. 389, I had independently arrived at the conviction that the *Loa* was a totally distinct worm from the *Filaria oculi*. *Ibid.*, The parasite in question is rather more than an inch in length, it is.. termed *Loa*. **1958** *Ann. Trop. Med. & Parasitol.* LII. 158 (*title*) The relationship between human and simian Loa in the rain-forest zone of the British Cameroons. **1971** Price & Hopps in R. A. Marcial-Rojas *Path. Protozoal & Helminthic Dis.* lii. 917/1 (*heading*) The eye worm, loa worm. *Ibid.*, He [*sc.* Guyot] described the worm under the native name 'loa'.

loa[2] (lōᵘ·ă). Pl. **loa** or **loas.** [ad. Haitian Creole *lwa*.] A deity in the voodoo cult of Haiti.

1933 J. H. Craige *Black Bagdad* xv. 267 Thanks to the spells I have made in your behalf, the *loas* have held their hands over you, and you are free. **1937** M. J. Herskovits *Life in Haitian Valley* II. iv. 81 The ordinary plates or calabashes on which food is offered the *loa*, or African deities. **1959** H. Charteris tr. *Métraux's Voodoo in Haiti* iii. 120 A *loa* moves into the head of an individual having first driven out..one of the two souls that everyone carries in himself. **1960** *Spectator* 5 Aug. 218/3 The peculiar congregation, led by the Voodoo Priestess, start summoning up their favourite spirits or 'loas', who—dead on cue—take possession of the celebrants one after another. **1966** *Punch* 10 Aug. 235/3 The 'loa', the demon-gods who take possession of the initiates at Voodoo ceremonies and impose on them their own voice, features, and character.

load, *sb.* Add: **2. c.** The material carried along by a stream in suspension, by saltation, or by traction (by some writers material carried in solution is included); the amount of material so carried; hence, by extension, the material carried by various other natural agents of transportation, as glaciers, winds, and ocean currents.

1888 J. W. Powell in *Science* 16 Nov. 229/2 In erosion and corrasion the material which is transported may be called the 'load'. The load is transported by two methods, a portion floats with the water, and another portion is driven along the bottom. **1907** R. D. Salisbury *Physiogr.* iv. 122 The sediment moved by a stream, whether in suspension or at the bottom, is its load. **1950** W. H. Twenhofel *Princ. Sedimentation* (ed. 2) vi. 226 These figures show that the suspended loads of rivers draining dry areas are larger than the dissolved loads. *Ibid.* 227 The loads of standing bodies of water are small in terms of any unit of volume... There is not a great deal of information respecting loads in the open ocean away from shallow water. **1968** R. W. Fairbridge *Encycl. Geomorphol.* 627/2 Load is an additional variable which changes together with flow along a stream wherever it is joined by a tributary. **1970** *Jrnl. Glaciol.* IX. 227 (*heading*) Contrast between the debris loads of polar and temperate glaciers.

3. f. *Electr. Engin.* The electric power that a generating system is delivering or required to deliver at any given moment; **base load**, the minimum value of the load during any period, generally met (in a grid system) by the continuous operation of the most efficient stations, without the intermittent and varying contribution of the less efficient ones.

In the earliest quots. identical with 3 c (quot. 1900 for which belongs here).

1888 *Proc. Inst. Mech. Engin.* Oct. 508 The efficiency of its working was limited to a constant load and a uniform speed, as when the dynamo was supplying a constant current with constant pressure. *Ibid.*, In many electric lighting installations..motors were required that would work economically between wide variations of load. **1891** *Min. Proc. Inst. Civil Engin.* CVI. 15 The cost of labour per unit..would continue to decrease as long as the duration of maximum load increased, up to a certain limit. **1894** [see *booster 2]. **1903** *Electr. World & Engin.* 23 May 866/2 It is necessary at times of fall and winter peak loads to operate the steam plants in the three combination sub-station and subsidiary steam plants which the company was operating three years ago. **1928** *Daily Express* 4 June 15/3 We have, in twenty-six years, built up a huge base-load..with an annual output of over 25,000,000 units. **1956** *Nature* 4 Feb. 204/2 The prospect of competitive nuclear power with low operating costs means that this plant will carry the base load. **1966** *Economist* 14 May 734/1 Running charges have, since 1961, differed by day and by night, since costs for the best, base-load stations at night are so much lower than the average running costs in the mixed bag of stations, from good to awful, used by day. **1974** *Times* 15 Jan. 14/3 A mass switch-off..which would record several million watts being wiped off the national energy load.

g. *Electronics.* An impedance or circuit that receives the output of a transistor or other device, or in which the output is developed.

1918 *Physical Rev.* XII. 180 Variations in potential difference are set up between cathode and grid, and these cause variations in the current in the circuit *FPR*, the power developed in the load *R* being greater than that fed into the input circuit. **1931** *Proc. IRE* XIX. 49 With the pentode..the maximum output was obtained at approximately a 10,000-ohm load. **1943** C. L. Boltz *Basic Radio* xv. 243 In a receiver the load on the output is a loudspeaker or telephone. **1957** B. I. & B. Bleaney *Electr. & Magn.* xiv. 362 In many applications the size of the load is fixed; if, for example, the load is a loud-speaker, its impedance..is generally in the range 5 to 15 ohms. **1962** Simpson & Richards *Physical Princ. Junction Transistors* xi. 251 We require a current in the load of 2 mA. *Ibid.*, Since R_{e1} is assumed to be by-passed by a capacitor, the a.c. load consists of $R_{c1}+R_{e1}$.

h. Colloq. phr. *to take a load off* (*one's feet*): to sit or lie down; to relax.

1945 A. Kober *Parm Me* 35 How's about taking a load off your feet? **1968** J. Hudson *Case of Need* III. i. 175 'Sit down,' she said. 'Take a load off.'

4. Esp. in phr. (*to take*) *a load off one's mind*: (to bring someone) relief from anxiety.

1852 Lytton *My Novel* (1853) III. x. vi. 132 It is a load off one's mind. **1857** Dickens *Perils Eng. Prisoners* iii, in *Househ. Words* Extra Christmas No., 7 Dec. 31/2 It takes a load off my mind to leave her in your charge. **1951** E. Caldwell *Episode in Palmetto* vii. 136 It's a big load off my mind to hear you say that.

b. *slang.* An occurrence of venereal disease; = *dose *sb.* 2 d. Cf. Load *v.* 4 (quots. 1799, 1818).

[**1878** *N. & Q.* 10 Aug. 105/1 *Load*, an eruption, measles, smallpox.] **1937** Partridge *Dict. Slang* 488/1 *Load*, a venereal infection. **1965** F. Sargeson *Mem. Peon* ii. 28 They displayed their rubber goods, and..were doubly protected against finding themselves landed with either biological consequences or a load.

c. An amount of work, teaching, etc., to be done by one person; freq. with defining word prefixed, as *case-load, teaching-load, work-load.*

1946 *Nature* 17 Aug. 216/2 The scientific study of conditions affecting the work-load involved in various processes. **1950**, etc. [see *case-load* s.v. *case *sb.*[1] 14]. **1958** J. C. Herold *Mistress to an Age* (1959) III. xiv. 294 His teaching load amounted to three and a half hours daily for five days a week. **1961** *Lancet* 5 Aug. 303/1 He cemented the relationships by careful inquiry into examinations and study load. **1964** in *Rep. Comm. Inquiry Univ. Oxf.* (1966) II. 450 Please give as accurate estimates as you can for your average weekly load this term. **1966** *Ibid.* 465 A quarter were critical of the heavy teaching load with its consequent adverse effects on research. **1971** *Black Scholar* Jan. 64/2 (Advt.), Normal load is 6 courses per year. **1971** *Sat. Rev.* (U.S.) 18 Dec. 56/2 Teaching loads at white schools often are only a fraction the size of those at black schools.

5. a. Delete 'Now only *dial.* and *U.S.*' and add later examples; now esp. with *on*. Also, a satisfying amount to eat; (*U.S. slang*) a dose of narcotics.

1897 Barrère & Leland *Dict. Slang* II. 22/1 A man who walks unsteadily, owing to intoxication, is said to have a load on. **1922** Joyce *Ulysses* 160 After their feed with a good load of fat soup under their belts. **1929** J. B. Priestley *Good Companions* I. i. 26 You've got a load on and no mistake. **1929** [see *bang *sb.*[3]]. **1934** J. O'Hara *Appointment in Samarra* (1935) vii. 202 What a load you had. Did you get home all right? **1942** Wodehouse *Money in Bank* (1946) xxvi. 229 Drunk!..He's got a load on that would sink an ocean liner. **1948** V. Palmer *Golconda* ix. 65 We're not to blame if men get a load on and begin to fight. **1968** C. Nicole *Self Lovers* ii. 38 I'm sorry about last night. I was carrying a load. Else I'd have recognised you.

6. *a load*: delete † and add later examples; esp. a great quantity of something undesirable or nonsensical.

1943 C. H. Ward-Jackson *Piece of Cake* 41 *Load of guff*, a lot of humbug or nonsense. **1964** [see *crap *sb.*[1] 7 b]. **1965**, etc. [see *cod, abbrev. of *codswallop]. **1967** *Jazz Monthly* Dec. 12 Playing a load of rubbish, while sounding quite competent to the casual listener. **1968** [see *cobbler 1 c]. **1974** A. Morice *Killing with Kindness* iv. 38 No man is an island... That's what Mike used to say. Mind you, I always thought it was a load of rubbish.

7. e. Phr. *to get a load of* (freq. *imp.*): to look at, perceive, make oneself aware of, scrutinize; to listen carefully to. *slang* (orig. *U.S.*).

[**1929** D. Hammett *Dain Curse* (1930) xix. 217 The red-head nurse was getting a load at the keyhole.] **1929** D. Runyon in *Hearst's International* Oct. 64/1, I am not so sure..Blake will care to be anybody's husband, and especially Madame La Gimp's after he gets a load of her. **1941** I. Baird *He rides Sky* 143 What do you think would have happened if Queen Bess had got a load of the Air Force? **1958** E. Dundy *Dud Avocado* I. ix. 157 Come over here... Get a load of this script. **1966** [see *casting-couch* s.v. *casting *vbl. sb.* 4]. **1972** D. Bloodworth *Any Number can Play* xxii. 221 Get a load of that chick over there.

8. *load-bearing, -carrying* adjs.; **load-carrier,** a vehicle with the capacity to accommodate a load; **load cast** *Geol.*, a rounded protrusion on the underside of a stratum (usu. one of sandstone), owing to its having sunk before consolidation into the underlying bed (which is usu. shale); so **load-casted** *a.*, modified or covered by a load cast; **load-casting,** the formation of load casts; **load-cell,** an electronic device for weighing large quantities of material; **load factor,** for 'production' read 'productive capacity' and add: also, in *Aeronaut.*, (*a*) the ratio (or its reciprocal) of the weight of an aircraft to the maximum the wings can support, or that of the force exerted on a part of the structure in ordinary horizontal flight to that exerted in some other condition; (*b*) the ratio of the number of passenger seats occupied to the number available; (earlier and additional examples); **load line,** (*b*) *Electronics*, a straight line that crosses the characteristic curves (of output voltage against output current) of a valve or transistor and has a gradient and position determined by the load, so that it represents the possible operating conditions of the device; **load-shedding,** reduction of the supply of electric current over a specific area, esp. with a view to reducing consumption; also *transf.* and *fig.*; also *load-spreading.*

1925 Hull & Ingberg *Fire Resistance of Concrete Columns* 658 Pittsburgh gravel concrete was used in the load-bearing portion and cinder concrete from bituminous cinders in the outer portion. **1947** *Horizon* Oct. 63 No columns or load-bearing walls intervene. **1961** *Architect & Building News* 21 June 822/1 The building has massive load-bearing walls facing on to the Fellows' Garden and the College Park. **1974** *Times* 18 Feb. 12 Facilities include such items as load-bearing ceiling girders. **1962** *Times* 3 May 19/4 The rear seat can be folded flat to convert the car into an exceptionally roomy load-carrier. **1974** *Country Life* 21 Nov. 1579/3 A very comfortable car and a handy load carrier. **1960** R. W. Marks *Dymaxion World of B. Fuller* 55/1 For this reason the truss has an enormous load-carrying ability. **1953** P. H. Kuenen in *Bull. Amer. Assoc. Petroleum Geologists* XXXVII. 1048 The base is sharply cut and flat or forms pockets in its substratum, 'flow casts' in Shrock's terminology (1948). As this term tends to cause confusion it is here suggested to call them 'load casts'. **1969** Bennison & Wright *Geol. Hist. Brit. Isles* v. 101 Sedimentary structures, including flute casts, load casts, graded bedding, etc., have been described from the arenites. **1957** *Jrnl. Geol.* LXV. 248/1 (*heading*) Load-casted current markings. **1972** F. J. Pettijohn et al. *Sand & Sandstone* iv. 123 Load-casted ripples. **1953** *Bull. Amer. Assoc. Petroleum Geologists* XXXVII. 1051 They were not formed after deposition by load casting. **1972** F. J. Pettijohn et al. *Sand & Sandstone* iv. 124 If one turbidite flow follows on the heels of another, conditions are more favorable for load-casting. **1958** *Engineering* 28 Feb. 39 (Advt.), A standard range of loadcells designed for industrial weighing. **1891** R. E. B. Crompton in *Min. Proc. Inst. Civil Engin.* CVI. 3 What, for want of a better term, is hereafter called the 'load-factor', that is, the relation which the actual output of a plant..bears to what would be its output if worked continuously day and night, at the full load. **1922** *Encycl. Brit.* XXX. 21/2 The 'load factor' is the number of times the weight of the craft which the wings will support; a measure of the strength. **1943** *Jrnl. R. Aeronaut. Soc.* XLVII. 195 Allowing for the 65 per cent. load factor which seems to be about the maximum that can be expected on any commercial service under normal peace-time conditions, that would call for an aircraft providing accommodations for 57 passengers. **1950** Load factor [see *limit load* s.v. *limit *sb.* 5]. **1962** *Times* 16 May 15/3 The passenger load factor (the proportion of passenger capacity used) dropped by 1·7 per cent. to 48·1 per cent. **1970** D. Waterfield *Continental Waterboy* iii. 29 The B.C. Power Commission has a load factor of around 52%. **1931** *Proc. IRE* XIX. 49 The maximum output will be obtained when the slope of the load line equals minus the slope of the plate current curve. **1962** Simpson & Richards *Physical Princ. Junction Transistors* vii. 140 The straight line passing through the battery-voltage point..has a slope ($-1/R_l$) corresponding to the resistance R_l of the load and is called the load line. **1947** *Times* 10 Feb. 2/2 If the saving that was essential was not forthcoming the company would have to resort to load shedding or temporary cuts in supply. **1948** *Ann. Reg. 1947* 8 The cuts in coal and the 'load-shedding'—a term now incorporated into the vocabulary of the citizen—were due to 'the wretched private coal-owners'. **1963** *Guardian* 29 Apr. 8/6 Lord Longford has announced his resignation as chairman... This is part of the load-shedding of some of his extensive social work. **1951** *Engineering* 6 Apr. 402/2 Electricity load spreading..necessary..owing to heavy demand for electricity.

load, *v.* Add: **3. c.** *Electr.* To provide with additional inductance (e.g. by means of a loading coil) in order either to counteract the effect of capacitance and so reduce the distortion and attenuation of signals (in the case of a telephone line or other transmission line), or to reduce the resonant frequency (in the case of an aerial); more widely, to provide with a load (*load *sb.* 3 g) consisting of any kind of impedance.

1901 M. I. Pupin in *Trans. Amer. Inst. Electr. Engin.* XVII. 452 Though a given cord may be properly loaded for some wave-length it will not be properly loaded for shorter wave-lengths. **1922** Glazebrook *Dict. Appl. Physics* II. 852/2 The effect of loading a line in such a way is approximately the same as though inductance were uniformly distributed along the circuit. **1923** E. W. Marchant *Radio Telegr.* iii. 26 The frequency of the oscillation in the aerial can be varied by varying its inductance; that is, coils of copper tube may be inserted which will have the effect of 'loading' it, and so bringing down the frequency of the oscillations. **1962** *Newnes Conc. Encycl. Electr. Engin.* 846/1 Some low-frequency lines are 'loaded' with added inductance to give some approximation to the distortionless condition. **1970** J. Earl *Tuners & Amplifiers* iv. 82 Each source, whether it be radio tuner..or ceramic pickup or tape head, requires to be loaded by a specific value of impedance or within a range of impedance.

4. Also *to load up with* (something).

1880 'Mark Twain' *Tramp Abroad* xxxviii. 435, I loaded them up with paragoric and put them to bed. **1892** —— *Amer. Claimant* iii. 21 He loads up the house with cripples and idiots and stray cats. **1943** K. Tennant *Ride on Stranger* (1968) i. 3 Other men get married without being loaded up with kids, kids, and then more kids.

5. b. To insert a photographic film or plate in (a camera); also with the film as object.

1902 *Year Bk. Photogr.* 13 (Advt.), The 'Roll Film' Automan will be preferred by many on account of the ease with which it may be loaded and unloaded in daylight. **1936** *Discovery* Aug. 237/1 This unique camera..weighs 305 pounds when loaded. **1956** A. L. M. Sowerby *Dict. Photogr.* (ed. 18) 88 A long roll of film..was loaded into the camera at the factory.

c. To fill (a tobacco-pipe).

1927 'F. Lonsdale' *On Approval* II. 66 He crosses to stool down R, and loads his pipe.

9. a. (Earlier example.)

1870 J. K. Medbery *Men & Mysteries Wall St.* 136 To 'load' one's self with stocks is to buy heavily.

11. *Psychol.* To weight (a result or outcome), to contribute to or be correlated with; also *intr.* (const. *on*), to be correlated (with something else).

1931 *Psychol. Rev.* XXXVIII. 408 The ministry is loaded high for interest in people and in language but low for science. **1952** R. B. Cattell *Factor Analysis* xviii. 340 The factor loading (situational index) is not a measure of the mean amount of the contribution of the factor to the situation. For example, the discovery that in a certain collection of books, the factor of weight is loaded 0·6 in thickness and only 0·2 in height simply indicates that for a given weight (overall size) these books vary more in thickness than they do in height—as books on a tidy shelf should. **1970** Liebert & Spiegler *Personality* vi. 132 As part of the factor analysis, the ratings on each of the 50 trait elements, on which the subjects were rated, were correlated with each of the factors which had been found... The elements which loaded (correlated) most highly both in a positive and in a negative direction (recall that the magnitude of a correlation is independent of its sign) are listed in Table 6–3. **1971** *Jrnl. Gen. Psychol.* LXXXIV. 242 Both of these [test variables] were originally predicted to load the insight factor. **1972** *Jrnl. Social Psychol.* LXXXVIII. 190 The items loading on Factor I seem to reflect the parent's interests.

loadability (lōᵘdăbi·liti). [f. Load *v.* + -ability.] The degree of ease with which goods may be loaded or transported.

1945 F. Hamann *Air Words* 35/1 *Loadability*. (1) Cargo volume of commercial aircraft. (2) Ease of access in aircraft. **1955** *Times* 10 June 3/1 The definition of loadability..should include other traffic characteristics such as bulk in comparison to weight, stowage potential, [etc.]. **1960** *Economist* 22 Oct. 392/3 Generally, road hauliers would like to charge more for the smaller loads ('loadability' is as important by road as by rail). **1967** *Freight Management* Jan. 46/2 As far as loadability is concerned, roll-on offers many advantages. **1974** *Country Life* 26 Dec. 2019/3 The Citroen Safari for sheer loadability with comfort.

loadberry (lōᵘ·dbĕri). *Shetland dial.* Also **lodberry.** [Cogn. w. Norw. dial. *ladberg*, ON. *hlaðberg* a lading-rock, a natural quay or pier.] A flat rock forming a natural landing-place; *spec.* a small enclosed landing-place for the unloading of boats.

1764 in W. R. Mackintosh *Glimpses Kirkwall* (1887) 173 The said loadberry or north east part of the Ness of Quanterness. **1871** *Black's Picturesque Tourist of Scotl.* (ed. 19) 596 Lerwick is now before us, a great portion of the houses lining the shore, and standing in the sea, with loadberrys and piers attached. **1950** *Menzies' Guide to Shetland* 10 The most interesting feature of this part of the town [*sc.* Lerwick] is the series of lodberries, or enclosed courtyards with wooden doors surmounting steps down to the water. **1975** *Country Life* 13 Feb. 367/1 Many of these houses had their own slipways with stores or lodberrys.

loaded, *ppl. a.* Add: **1. d.** *fig.* Charged with some hidden implication or underlying suggestion; biased, prejudiced.

1942 *College English* Oct. 16 General Semantics..being metaphysical in a particularly partial and dogmatic sense ..can yield us only a vocabulary of 'loaded' general words, calculated to distort rather than to illuminate the writings of any other school. **1957** *Observer* 29 Dec. 9/2 Is our popular preference for plays of less blatant sexuality a mark of higher civilisation or merely of greater hypocrisy? Is the Dionysiac cult..more childish or simply more honest than the religious practices that have succeeded it in the West?..

These are loaded questions, I admit. **1958** *Times* 7 July 13/2 You cannot solve the riddle of he universe by giving the answer 'Yes' or 'No' to a loaded question. **1961** *Listener* 7 Dec. 991/3 He chose to use emotionally loaded words like 'scare' and 'infection'. **1975** D. Bagley *Snow Tiger* xi. 96 It is improper of Mr Smithers to ask such a loaded question... He is usurping the function of this commission.

2. Of a camera: with a film inserted. Of a film: inserted (in a camera) (cf. *load *v.* 5 b). In *Electr.*, of a telephone line, etc. (cf. *load *v.* 3 c). In insurance: of a life (cf. Load *v.* 10).

1888 *Judge* (U.S.) 20 Oct. 27/1 (Advt.), The Kodak Camera..Loaded for 100 instantaneous views. **1901** *Trans. Amer. Inst. Electr. Engin.* XVII. 475 Propagation of electrical waves over a periodically loaded loop..is compared to that over a uniform loop. **1903** *Phil. Mag.* V. 313 The loaded line discussed in this paper is an electrical circuit of two long parallel conducting wires having self-induction coils inserted at regular intervals. **1922** *Encycl. Brit.* XXXII. 709/2 In 1911 Messrs. Siemens introduced a form of balata dielectric as a substitute for gutta-percha in loaded submarine cables... The effect was to reduce materially the attenuation constant and increase the range of speech in loaded cables. **1928** *Daily Express* 10 May 11/6 The application was refused in the first instance, and only afterwards accepted as a 'loaded' life. **1937** *Discovery* Apr. 112/2 Little..envelopes, in which glass plates and flat films are sold ready loaded. **1968** *Radio Communication Handbk.* (ed. 4) xvi. 33/1 While the centre loaded aerial may be the best radiator, on the lower frequency bands, particularly 1·8 Mc/s and 3·5 Mc/s, mechanical considerations may have to influence the loading coil position.

3. b. Drugged; under the influence of drugs; containing a drug. *U.S. slang.* (Cf. *load *sb.* 5 a.)

1923 J. F. Fishman *Crucibles of Crime* vi. 126 It was discovered that each of them [*sc.* handkerchiefs] has a small ink mark in one of the corners..these handkerchiefs had been dipped in cocaine... The mark in the corner notified the 'snowbird' that it was 'loaded'. **1928** J. Tully *Circus Parade* xviii. 237 When guys are loaded up on heroin it'll give 'em more nerve. **1953** W. Burroughs *Junkie* xii. 120 To get really loaded, you would need four papers. *Ibid.* xv. 147 He was loaded on H and goof-balls. **1962** K. Orvis *Damned & Destroyed* xi. 74 'I'm gorgeous,' she said. 'Loaded and gorgeous.'

4. Rich; extremely wealthy. *slang* (orig. *U.S.*).

[**1910** O. Johnson *Varmint* iv. 60 He's just loaded with the spondulix.] **1948** *Call-Bulletin* (San Francisco) 16 July 10/4 Lonely but loaded, a carpenter newly returned from Guam was taken for a ride by six young women. **1949** A. Hynd *We are Public Enemies* i. 20 The boys were loaded after the safety deposit opening and the money was burning their pockets. **1952** 'J. Tey' *Singing Sands* xi. 172 He slapped his pocket, 'I'm loaded.' **1957** C. MacInnes *City of Spades* I. iv. 23 'Is your Dad rich?'..'He's reasonably loaded.' **1971** D. O'Connor *Eye of Eagle* v. 32 Adriana's a very popular girl and there are guys here who are absolutely loaded.

loader[1]. **3.** Add to def.: Applied similarly to other things, such as agricultural machinery (e.g. *front-end loader* s.v. *front *sb.* (and *a.*) 14), washing machines (e.g. **front-loader*); see also *side-loader* s.v. *side *sb.*[1] 27. (Further examples.)

1968 *Which?* May 149/1 This [washing] machine is a top loader, but has a horizontal stainless steel drum. **1975** *Radio Times* 22–28 Feb. 56/1 Fully automatic top-loader. Takes loads of 4 lbs., 7 lbs., or 10 lbs... 7 programme automatic front loader. Takes load of 9 lbs.

4. Special comb.: **loader gate** *Coal-mining*, a passage along which coal is conveyed away from a long-wall face.

1964 A. Nelson *Dict. Mining* 258 *Loadergate*, a gate road equipped with a gate conveyor or a gate-end loader; the gate to which the face conveyors deliver their coal. **1973** *Times* 3 May 4/1 Then I saw all the loader gate workmen running towards me. Someone shouted 'run, water has broken in.'

loading, *vbl. sb.* Add: **1. f.** *Electr.* Addition of inductance, or the inductance added (see *load *v.* 3 c); any impedance that acts as a load (*load *sb.* 3 g).

1903 *Phil. Mag.* V. 325 Loading..presents the greatest possibilities upon long cable circuits. **1922** Glazebrook *Dict. Appl. Physics* II. 850/2 The increase in voltage resulting from the increased impedance..increases the leakage losses, and these set a limit to the possible improvement in transmission efficiency by loading. **1959** K. Henney *Radio Engin. Handbk.* (ed. 5) xxviii. 29 Uniformly surrounding the conductor with a thin layer of magnetic material of high permeability..is known as continuous loading. This has been used, to some extent, in the construction of some long submarine-cable circuits for telegraphy as well as telephony. However, loading is usually introduced in telephone circuits by connecting loading coils in series with the conductors at intervals. **1968** *Radio Communication Handbk.* (ed. 4) xvi. 34/1 With a centre loaded coil on 3·5 Mc/s it may be possible to vary the transmitter frequency over about 25 kc/s before it becomes essential to re-adjust the loading. **1973** *Physics Bull.* Dec. 716/1 This changes the impedance of the loop and hence it alters the loading of the RF circuit and this can be detected in the RF drive output.

g. The (maximum) current or power that an electrical appliance is designed to take.

1938 E. M. Ackery *Electr. Heating for Public & Commercial Libraries* iv. 42 The coke-boiler was replaced by a 400 gallon thermal storage tank, fitted with immersion heaters with a total loading of 100 kw. **1951** *Good Housek. Home Encycl.* 310/1, 2 or 3 kilowatts is a sufficient loading for the average-sized tank. **1973** *Daily Tel.* 4 Dec. 11/4 Fan heater with loading of 2 kw: 1·8p an hour (2 units).

4*. The weight supported by a wing divided by its area. More fully **wing loading* (cf. **power loading*).

1918 [see *live load* s.v. *live *a.* 8]. **1919** H. Shaw *Text-bk. Aeronaut.* xv. 181 The loading of a machine, which is the weight carried per unit area of surface, varies in different types. **1936** *Discovery* Mar. 73/2 Most birds fly at a loading of $1\frac{1}{2}$ to $2\frac{1}{2}$ lb. per sq. ft. **1973** *Sci. Amer.* Dec. 103/2 The glider can travel much faster than the vulture at a given gliding angle. This is owing partly to..its higher wing loading (the ratio of weight to wing area).

4.** *Psychol.* The extent to which any given factor or variable contributes to or is correlated with some resultant quality or overall situation, usu. represented by a number arrived at by statistical analysis of the results of a series of tests.

1931 *Psychol. Rev.* XXXVIII. 407 Our next problem is to assign a weight or loading of each of the general factors to each of the variables... Engineering, for example, has a high loading of interest in science, a rather low loading of interest in language. **1935** L. L. Thurstone *Vectors of Mind* viii. 201 (*heading*) The elimination of negative factor loadings. **1947** L. E. Tyler *Psychol. Human Differences* xv. 364 The task of the factor analyst then becomes one of determining these weights or loadings. The raw material for the mathematical work in each case is the original table of intercorrelations. **1952** [see *load *v.* 11]. **1971** *Jrnl. Gen. Psychol.* LXXXV. 72 The loadings have been rounded to two figures and the order of both factors and variables rearranged to facilitate inspection. *Ibid.* 212 Because many of the memory tests require recall of the items memorized, and because the operations of divergent and convergent production are so much dependent upon retrieval of information from the memory store, it might be expected that..either production tests would have some memory loadings, or memory tests would have some loadings on production factors.

5. *loading-board, -tool, -tower, -yard*; **loading bay,** a bay (Bay *sb.*[2]) or recess in a building where vehicles, etc., are loaded and unloaded; **loading coil** *Electr.*, an inductance coil used in the loading of telephone lines or aerials (see sense *1 f and *load *v.* 3 c); **loading gauge** *Railways*, (*a*) the maximum height and width allowed for rolling stock to ensure adequate clearance under bridges and in tunnels; (*b*) a device suspended over railway lines for checking the dimensions of rolling stock.

1963 *Listener* 31 Jan. 202/2 The whole street is sometimes used as an open-air loading bay and temporary warehouse. **1971** R. Busby *Deadlock* x. 152 The roller doors of the loading bay were shut. **1910** W. M. Raine *Bucky O'Connor* 36 The loading board was lowered and the horses led from the car. **1901** *Ann. Rep. Amer. Telephone & Telegr. Co.* 4 The efficiency of these lines will be largely increased by the use of 'loading coils'. **1922** Glazebrook *Dict. Appl. Physics* II. 853/1 Loading coils are encased in iron cases and are mounted in manholes—if the circuit..is a cable circuit—or are mounted at the cross-arms of poles if the circuit..is an open-wire line. **1974** *Encycl. Brit. Micropædia* VI. 286/1 Auto radios generally use loading coils because whip antennas are much too short to resonate at broadcast frequencies. **1883** F. S. Williams *Our Iron Roads* (ed. 4) 266 Among the minor appurtenances of a railway station is the wagon loading gauge. **1901** *Young Engineer* I. 53 The fire-box may be extended to the full width of the loading gauge. **1930** *Engineering* 22 Aug. 230/2 The upper part [of a coke wagon] slopes inward to suit the loading gauge. **1874** J. W. Long *Amer. Wild-Fowl Shooting* 20, I usually made a practice of reloading as fast as possible between shots, carrying an ammunition-box and loading-tools with me. **1901** *Chambers's Jrnl.* May 12/1 Steam-cranes and movable loading-towers..lower the coal into the hold of the vessel. **1909** *Westm. Gaz.* 9 June 11/1 Between them is a loading yard 200 ft. by 60 ft.

loadmaster (lōᵘ·d,mɑːstəɹ). *Aeronaut.* [f. Load *sb.* + Master *sb.*[1]] The crew member of an aircraft who is responsible for the load or cargo.

1961 *Newsweek* 6 Mar. 9 The United Nations and C-130 loadmasters did most of the work. **1967** *New Scientist* 17 Aug. 328 The aircraft [*sc.* Douglas C-47s] carry a crew of eight—pilot, co-pilot, navigator, flight mechanic, load master (who also drops the flares), two gun loaders, and a Vietnamese Air Force liaison officer. **1973** *Daily Tel.* 2 Oct. 2 If she passes her 14-week officer training course she will join an aircrew as a loadmaster. **1974** *Ibid.* 22 Feb. 14/4 Flying Officer Howard has logged 3,500 flying hours all over the world as an air loadmaster in VC-10s of No. 10 Squadron based at Brize Norton. The job, formerly known as air quartermaster, was given air crew status in 1962.

loaf, *sb.*[1] Add: **2. e.** Minced or chopped meat moulded into the shape of a loaf and cooked; generally eaten cold, in slices. Usu. with qualifying word, as *beef, ham, meat, veal loaf.*

[**1787** Lady Newdigate *Let.* 21 Oct. in A. E. Newdigate-Newdegate *Cheverels* (1898) v. 71 We made an excellent Dinner upon our Cold Loaf.] **1895** 'M. Ronald' *Century Cook Bk.* 308 Liver loaf, or false pâté de foie gras..is better cold with salad, or used like pâté de foie gras. A loaf of any game may be made in the same way. **1902, 1907** [see *ham loaf* s.v. *ham *sb.*[1] 3]. **1939** Auden & Isherwood *Journey to War* 90 We had fruit-juice, meat-loaf, salad and cake. **1964** J. Masters *Trial at Monomoy* iv. 140 She stared at rows of Spam, corned beef, meat loaf, ham loaf. **1975** *Times* 7 Mar. 5/2 She recommended home-made vegetable soup and meat loaf, followed by apple or rhubarb crumble.

f. *slang.* [Prob. from *loaf of bread*, rhyming slang for 'head'.] The (human) head; hence, the mind, common sense; *esp.* in phr. *to use one's loaf.*

1925 Fraser & Gibbons *Soldier & Sailor Words* 145 *Loaf*, head, e.g., 'Duck your loaf—*i.e.*, keep your head below the parapet'. **1938** J. Curtis *They Drive by Night* xiv. 155 Bloody seconds counted in a job like this. You certainly had to use your loaf. **1943** Hunt & Pringle *Service Slang* 44 *Use your loaf* is the injunction often heard when someone is particularly slow in following orders. But this phrase, in its finer meanings, says: 'Use your common sense. Interpret orders according to the situation as you find it, and don't follow the book of words too literally.' **1949** 'N. Blake' *Head of Traveller* ii. 36 Do try to use your loaf. **1957** P. Frankau *Bridge* 73 He uses his loaf where you and I just muddle along. **1971** B. W. Aldiss *Soldier Erect* 79 You want to use your bloody loaf, Stubbs, or we'll never win this war the way you're carrying on. **1973** *Jewish Chron.* 2 Feb. 12/1 Use your loaf. Didn't Sir Jack Cohen of Tesco..start the same way?

g. *loaf o(f) bread*: rhyming slang for 'dead'.

1930 Brophy & Partridge *Songs & Slang 1914–18* 137 Loaf o' Bread, dead. **1935** Auden & Isherwood *Dog Beneath Skin* III. i. 123 O how I cried when Alice died The day we were to have wed! We never had our Roasted Duck And now she's a Loaf of Bread.

6. *loaf-tin*; **loaf-cake,** a plain cake made in the form of a loaf.

1828 E. Leslie *75 Receipts* 62 Loaf Cake. **1844** *Knickerbocker* XXIV. 423 The biscuit would not rise, her loaf-cake was heavy. **1906** Kipling *Puck of Pook's Hill* 195 Hobden said that the loaf-cake..was almost as good as what his wife used to make. **1941** F. M. Farmer *Boston Cooking-School Cook Bk.* (ed. 7) 624 Loaf and layer cakes. **1883** *Facts, or Experiences Recent Colonist N.Z.* iii. 30 The materials required are:—..one deep tin-pan, six loaf tins, one wooden spoon. **1932** E. Craig *Cooking with E. Craig* 305 Place..in a greased loaf tin. **1972** K. Stewart *Times Cookery Bk.* xviii. 235 Divide the dough... Place carefully in..two small, greased, 1 lb loaf tins.

loafer. Add: (Earlier U.S. examples.)

1830 *Mechanic's Press* (Utica, N.Y.) 10 July 274/1 Nor are they topers at taverns, or *benchers* at groceries, or loafers who 'chase misfortune o'er the towpath'. **1835** *Knickerbocker* VI. 63 The late Ben Smith, Loafer. I present an outline sketch of one of that species of the *genus homo*..which Custom has christened with the expressive appellation of Loafer!

2. [cf. Sp. *lobo* wolf.] A timber wolf, *Canis lupus nubilus*, found in the south-western part of North America; = Lobo. Also *attrib.*

1878 McDaniels & Taylor *Coming Empire* v. v. 314 The Mexicans call these big wolves 'lobos' and the Texans call them 'loafers' which is a corruption of the Mexican word. **1908** *Sat. Even. Post* 4 July 16/3 One of the loafers had run in, leaping out of the black like a streak of gray light. **1948** *Ibid.* 10 July 80/3 With loafers,..mountain lions or bears, he was absolutely ruthless. **1974** *Islander* (Victoria, B.C.) 16 June 13/2 The lobo wolf derived its name from a word in the Blackfoot Indian language meaning 'buffalo wolf' and sounding like 'lobo', the Spanish word for wolf. They were also known as the buffalo grey or loafer wolf as they followed migrations of bison.

3. Usu. *pl.* The proprietary name of a shoe for wearing on informal occasions.

1939 *Trade Marks Jrnl.* 19 Apr. 520/2 *Loafer.* Boots, shoes, slippers, sandals, leggings and gaiters. Fortnum and Mason Ltd.—London. **1948** M. Sturges-Jones *In Wedlock Wake* 11 Pullover sweaters, bobby socks, and leather loafers. **1963** T. Pynchon *V* vi. 136 Profane kicked off his shoes—old black loafers of Geronimo's—and concentrated on dancing in his socks. **1971** 'V. X. Scott' *Surrogate Wife* 161, I saw him standing there in lean slacks..and suède loafers. **1972** *Last Whole Earth Catalog* (Portola Inst.) 406/2 The big thing for guys is jeans or slacks, button-down shirts or T-shirts and brown loafers without socks.

b. A type of jacket for informal wear.

1959 *Trade Marks Jrnl.* 19 Aug. 688/2 *Loafers...* Jackets. Chas. MacIntosh & Company Limited,..London. **1969** P. Roth *Portnoy's Complaint* 125 My clip-on tie and my two-tone 'loafer' jacket.

loafery. Add: **b.** A place where people loaf.

1898 *Daily Tel.* 10 Feb. 7/3 The Whitechapel Guardians have been considering a proposal to call their workhouse by another name... Perhaps 'House of Repose' or 'The Loaferies' would be appropriate. **1903** *Liberty Rev.* July 7 A new trap is set for it—the free loafery at the corner.

loaiasis (lōᵘă,əi·ăsis). *Path.* Also **loasis** (lōᵘ,ē·sis), **loiasis** (lōᵘ,əi·ăsis). [mod.L., f. Loa + *-iasis.] Infection with, or disease caused by, loas.

1913 Castellani & Chalmers *Man. Trop. Med.* (ed. 2) ix. 1442 (*heading*) Loasis. **1919** *Ibid.* (ed. 3) lxxxviii. **1972** Löiasis is a subcutaneous and subconjunctival filariasis caused by *Loa loa*. **1923** E. A. & T. M. Neatby *Man. Trop. Dis. & Hygiene* 382 The species responsible for the filariases and loasis. **1956** *Nature* 25 Feb. 367/1 Studies on loiasis in the Cameroons and Nigeria. **1961** *Lancet* 5 Aug. 323/2 In 1948 Gordon started an investigation of the African filarial infection, loaiasis. **1963** T. J. Brooks *Essent. Med. Parasitol.* III. 259 Loaiasis, or eye worm infection.

loa loa (lōᵘ:ă lōᵘ·ă). [Taxonomic name of the causative organism: see *Loa[1].] = *loaiasis.

1923 E. A. & T. M. Neatby *Man. Trop. Dis. & Hygiene* 396 Loa loa is a filarial infection, confined to West Africa, the worm (*Filaria loa*) resembling *F. bancrofti.* **1963** *Times* 8 Feb. 15/7 It was found that exactly the reverse held true for the form of filarial disease in West Africa, known as loa loa. Here the number of microfilariæ in the bloodstream is maximal at noon and minimal at midnight.

loam, *sb.* Add: **5.** *loam-foot.*

1940 T. S. Eliot *East Coker* i. 8 Lifting heavy feet in clumsy shoes Earth feet, loam feet, lifted in country mirth. **1955** D. Davie *Brides of Reason* 28 Come with me by the self-consuming north (The North is spirit), to the loam-foot west And opulent departures of the south.

loam, *v.* Add: **3.** *Austral. intr.* and *trans.* To search (a region) for gold by washing the loam from a hill's base until the increasing number of gold grains leads to the lode. So **loa·ming** *vbl. sb.*

1916 R. Mackay *Recoll. Early Gippsland Goldfields* vi. 29/2 The science of loaming was either then unknown, or known to very few. *Ibid.* 30/1 The loaming system will tire the strongest and most wiry man that ever swung a pick. **1935** *Bulletin* (Sydney) 13 Feb. 21/1 He'll be loaming up a hill and following a trace, Until he gets above the gold. **1953** *People* (Austral.) 23 Sept. 39/1 Loaming for gold he explains, entails, roughly, taking samples of loam from the topsoil, washing it in a dish, counting the colors and following them in intensity until a likely spot to sink a shaft is found. **1966** 'J. Hackston' *Father clears Out* 50 Old Tom was to make himself useful about the plant, and loam the surrounding country for the reef.

Loamshire (lōᵘ·mʃəɹ). Name given to an imaginary rural county, much used in novels and plays; also (*pl.*) a regiment from this county. Also *attrib.*

1859 Geo. Eliot *Adam Bede* I. v. 108 He was only a captain in the Loamshire Militia. *Ibid.* xii. 230 Jolly housekeeping—finest stud in Loamshire. **1866** —— *Felix Holt* I. i. 18 Transome Court was a large mansion..with a park and grounds as fine as any to be seen in Loamshire. **1912** G. W. E. Russell *Afterthoughts* xvii. 158 In Loamshire 'my foot is on my native heath', and I have been renewing my youth by contact with my early friends. **1920** 'Sapper' *Bull-Dog Drummond* i. 24 Captain Hugh Drummond, D.S.O., M.C., late of His Majesty's Royal Loamshires. **1954** K. Tynan in *Observer* 31 Oct. 6/1 Look about you; survey the peculiar nullity of our drama's prevalent *genre*, the Loamshire play. **1962** *Listener* 6 Dec. 959/1 They [*sc.* English novelists] also have to avoid the pitfall of regionalism and dialect. They have to avoid Loamshire. **1974** Green & Hooper *C. S. Lewis* x. 247 In the version which Green read..Digory..stayed in a farm cottage with an old countryman called Piers and his wife, who spoke with a rather laboured 'Loamshire' accent.

loan, *sb.*[1] **5.** *loan-market*; **loan-blend,** a compound word consisting of both native and foreign elements; a hybrid (see Hybrid *sb.* 2); **loan capital,** the part of the capital of a company or the like that is borrowed for a specified period; **loan-farm** *S. Afr.*, land loaned to a farmer by the government; **loan-form,** a form adopted by one language from another; **loan-place** *S. Afr.* = **loan-farm*; **loan-shark** orig. *U.S.* (see quot. 1928 and Shark *sb.*[1] 2); so **loan-sharking** *sb.*, lending money at exorbitant rates of interest; **loan-shift,** a change in the meaning of a word resulting from the influence of a foreign language; a word so affected; **loan-translation** [= G. *lehnübersetzung*], an expression adopted by one language from another in more or less literally translated form; a *calque.

1950 *Language* XXVI. 215 Loanblends show morphemic substitution as well as importation. **1974** R. A. Hall *External Hist. Romance Lang.* 9 On occasion, a newly formed word may consist, in part, of a native term ..and, in part, of a borrowing. The result is a *loan-blend*, as in Fr[ench] *bar-serveuse* 'bar-maid', with the first part of the loan-compound kept and with the second part replaced by Fr. *serveuse* 'maid, waitress'. **1848** *Bradshaw's Railway Almanack* (ed. 2) 57 The guarantee is extended to the payment of interest on £1,000,000 (the authorized Loan Capital) at whatever rate it may be borrowed. **1964** *Financial Times* 31 Jan. 19/5 No share or loan capital of the Company or any of its subsidiaries has within two years preceding the date hereof been issued or is proposed to be issued for cash or otherwise. **1804** J. Barrow *Acct. Trav. S. Afr.* II. 380 The number of these loan farms registered in the office of the receiver of the land revenue, on closing the books in 1798, were..1832. **1955** J. H. Wellington *S. Afr.* II. III. xiv. 208 To create a stable border farming population in the place of the cattle farmers trekking from loan farm to loan farm, land was offered to settlers on a quit-rent basis. **1966** E. Palmer *Plains of Camdeboo* ii. 21 Probably in the 1770's the land was issued as a loan place and became the temporary property of one farmer. Loan farms were apportioned in the simplest possible way and were held at a nominal rent. **1902** *Amer. Anthropologist* IV. 31 Penobscot *nachigadonkah* is a Passamaquoddy loan-form. **1844** Mill in *Westm. Rev.* XLI. 593 The already existing pressure upon the loan market. **1870** J. K. Medbery *Men & Mysteries Wall St.* 11 Its loan market holds the keys of trade. **1844** J. Backhouse *Narr. Visit Mauritius & S. Afr.* 585 A *loan place* which is a place obtained from the Government, that has not yet been surveyed, is half-an-hour's walk in every direction from the house or centre. **1939** J. S. Marais *Cape Coloured People* iv. 140 During the previous twenty years land had been granted to farmers as 'loan places'. **1905** Taylor & Gibson *Log of Water Wagon* 41 Loan sharks have been following the Lithia all day. **1911** *Collier's* 4 Feb. 8/1 Mr. Ham became interested in the 'World's' lucrative and lengthy list of loan-shark advertisements. **1913** *Munsey's Mag.* Nov. 218/1 In New York the loan-sharks were doing a business of twenty million dollars per annum. *Ibid.* 221/1 At the convention of the Legal Aid Society in Pittsburgh..the loan-shark evil was discussed. **1928** *Daily Tel.* 5 May 9/5 It is hoped by this plan virtually to put out of business the 'loan shark', who exacts usurious rates of interest from the person of small means. **1972** *Sunday Sun* (Brisbane) 27 Aug. 22/1 He was a loan-shark extortionist, and he had a very cute way of making sure customers paid up. **1970** *New Yorker* 15 Aug. 24/1 Other illicit activities engaged in by Cosa Nostra ..included..hijacking, loan-sharking. **1971** *Daily Colonist* (Victoria, B.C.) 10 Sept. 25/5 We are all aware that narcotics, prostitution, gambling and loansharking make up the bankroll of organized and syndicated crime. **1950** *Language* XXVI. 220 Loanshifts in general occur most readily when there is both phonetic and semantic resemblance between foreign and native terms. **1964** *Ibid.* XL. 95 The problems of translators and their role in introducing neologisms and loanshifts. **1974** R. A. Hall *External Hist. Romance Lang.* 9 French *réaliser* 'to bring into existence' has..undergone a shift of meaning to 'become aware', under the influence of Eng[lish] *realise*. Such a process of reinterpretation is known as a *loan-shift*, and its result as a *calque*. **1933** L. Bloomfield *Lang.* xxv. 456 The Slavic languages translate the term [*sc. conscientia*] by 'with' and 'knowledge', as in Russian ['so-vest] 'conscience'. This process, called loan-translation, involves a semantic change: the native terms or the components which are united to create native terms, evidently undergo an extension of meaning. **1958** C. Rabin in *Aspects of Translation* 140 Loan-translation (calque)..is very common [in modern Hebrew]. We find, for example, *gan yeladim* 'garden of children', which (except for the difference in the..order of elements) reproduces G. *Kindergarten* even to the pl. of *Kinder*. **1964** C. Barber *Ling. Change Present-Day Eng.* iv. 101 Another type of loan from a foreign language is the *calque* or *loan-translation*. **1965** *Ulster Dialect Archive Bull.* (Ulster Folk Museum) iv. 11 Fairy lore appears in *fergorta*..occurring also as a loan-translation *hungry-grass*. **1974** *Verbatim* Dec. 2/1 Hebrew..has a little series, mostly loan translations,..on a semantically similar pattern.

loasis, var. *loaiasis.

lob, *sb.*[2] **6. lob-tailing** *vbl. sb.* and *ppl. a.* (earlier and later examples); so **lob-tail** *v. intr.*

1933 B. Willoughby *Alaskans All* 134 Nearby, a third [whale] would pop up and 'lob-tail'—that is stand on its head with its tail out flaying the sea with thundering blows that sent clouds of spray in every direction. **1851** H. Melville *Moby Dick* II. xliv. 298 Five great motions are peculiar to it [*sc.* the tail of a whale]... Fourth, in lobtailing. **1937** *Discovery* Oct. 310/1 In 'lobtailing', it [*sc.* the cachalot] stands on its head with its tail some thirty feet out of the sea.

lob, *sb.*[5] **1.** (Earlier examples.)

1851 J. Pycroft *Cricket Field* ix. 178 Practise high lobs—a most useful variety of ball. **1865** F. Lillywhite *Guide to Cricketers* 59 A good lob-bowler and excellent long-stop. **1871** 'Thomsonby' *Cricketers in Council* 40 The best lob bowlers by a mere turn of the wrist impart an enormous amount of twist to the ball. *Ibid.*, 'Lob' bowling is, we believe, rather undervalued at the present day.

lob, *v.* Add: **3. b.** *to lob* (*in*), to arrive. *Austral. slang.*

1916 C. J. Dennis *Songs Sentimental Bloke* 56 'Twas at a beano where I lobs along To drown them memories o' fancied wrong. *Ibid.* 125 *To lob*, to arrive. **1934** *Bulletin* (Sydney) 12 Dec. 25/2 Scrubby lobs in one sundown while Old Dave is over with the storekeeper. **1950** K. S. Prichard *Winged Seeds* ii. 24 You never knew who'd lob into the camp. **1970** *Sunday Truth* (Brisbane) 5 July 30/5 When they had 15 pines on board, the farmer lobbed on the scene.

4. b. To send (a player) a lobbed ball.

1921 A. W. Myers *20 Yrs. Lawn Tennis* 135 Having discovered the wisdom of lobbing Barrett, Hackett.. allowed McLoughlin to kill anything smashable. **1928** *Daily Tel.* 5 June 17/1 As soon as one is certain of not being lobbed. **1972** D. Delman *Sudden Death* (1973) vi. 170 He lunges for the backhand volley... He is off balance, out of position, and I lob him wickedly.

lobar, *a.* Add: **b.** *spec.* in *Path.* Applied to an acute form of pneumonia lasting about nine days, most commonly caused by pneumococcal infection, and marked by fever, pains in the chest, coughing, and blood-stained sputum, and by inflammation concentrated in one lobe of the lung.

1858 J. Copland *Dict. Pract. Med.* II. 761/2 The French pathologists, and after them some recent English writers, have distinguished the disease [*sc.* pneumonia] into lobar, lobular, and vesicular... Of these the lobar is the most common. **1873** [in Dict.]. **1876** J. S. Bristowe *Treat. Theory & Pract. Med.* iii. 406 Lobar pneumonia commences with hyperæmia of the small vessels which are distributed in the walls of the air-cells and bronchial passages. **1961** R. D. Baker *Essent. Path.* ix. 151 There may be lobar pneumonia in one lung and lobular in another. Lobar pneumonia has this difference, that almost all cases are caused by the pneumococcus, whereas many cases of lobular pneumonia are caused by other bacteria, such as streptococci or staphylococci. **1966** Wright & Symmers *Systemic Path.* I. x. 368/2 In most cases the onset of lobar pneumonia is abrupt.

lobate, *a.* Delete *Nat. Hist.* and add further examples.

1919 D. W. Johnson *Shore Processes* iv. 188 Where the current of a river's distributaries strongly predominate[s] over shore currents and wave attack, the delta shoreline will be of the 'lobate' type, as in the case of the Mississippi delta. **1933** Schuchert & Dunbar *Textbk. Geol.* (ed. 3) xix. 430 When the ice later retreated northward over the Great Lakes region, its front became deeply lobate. **1957** G. E. Hutchinson *Treat. Limnol.* I. i. 85 Genuine morainic dams are probably commoner in regions of moderate preglacial relief that has been covered by continental ice sheets or by large lobate glaciers. **1967** *New Scientist* 2 Feb. 263/2 The hills in the north of the ring define lobate shapes in a manner highly reminiscent of viscous lava flows on Earth.

lobby, *sb.* Add: **3. b.** (Earlier example.)

1808 *Deb. Congress U.S.* 2 Feb. (1852) II. 1536 If we move to Philadelphia we shall have a commanding lobby.

c. In extended use: a sectional interest (see Interest *sb.* 4), a business, cause, or principle supported by a group of people; the group of persons supporting such an interest.

1952 *Economist* 26 July 254/2 American..interests have maintained their effective lobby against the project [*sc.* the St. Lawrence Seaway]. **1954** *Ibid.* 7 Aug. 425/1 M. Mendès-France..has to face powerful colonial lobbies in parliament. **1958** *Listener* 21 Aug. 273/1 The United States Government, sensitive to the Jewish lobby.. backed the Jews. **1959** *Ibid.* 4 June 968/2 They even tackled the vested privileges and subsidies of the powerful alcohol lobby. **1971** *Daily Tel.* 9 Mar. 10/6 The anti-pollution lobby might claim that a spot of exaggeration is justified in such a cause.

4. lobby chest (see quot. 1803); **lobby man,** (*a*) *U.S.* (see quot. 1934); (*b*) a lobbyist.

1803 T. Sheraton *Cabinet Dict.* 261 Lobby chest, is a kind of half chest of drawers, adapted for the use of a small study, lobby, or small lodging room. **1970** *Canadian Antiques Collector* Nov. 15/2 All sorts of compact, changeling furniture..lobby chests and Rudd's tables, [etc.]. **1934** Webster *Lobbyman*, one who works as attendant or porter, or does chores, in a lobby. **1958** *Times Lit. Suppl.* 21 Feb. 93/4 The high-pressure methods of the United States lobby-men, whose contacts and antecedents are open to inspection.

lobby, *v.* For *U.S.* read orig. *U.S.* Add: **1.** (Further examples.) Also *transf.*

1955 *Times* 17 June 4/6 M.P.s were lobbied yesterday by delegates of the Uganda National Congress. **1971** P. Gresswell *Environment* 154 We can always lobby our councillors. **1974** *Times* 18 May 8 [Wilberforce] was lobbying heads of state.

2. (Earlier and later examples.)

1837 *Cleveland* (Ohio) *Herald* (Weekly ed.) 6 Oct. 2/6 Gen. Bronson..spent a considerable portion of the last winter in Columbus, lobbying to procure the establishment of a Bank at Ohio City. **1916** Galsworthy *Sheaf* iii. 55 Animals..cannot lobby in the House of Commons, withdraw votes or commit outrages. **1962** *Listener* 20 Dec. 1041/1 In France the planners, being part of the civil service machine, have always been able to lobby from inside.

lobby-gow (lǫ·bi,gau·). *U.S. slang.* [Etym. unknown.] An errand-boy, messenger; a hanger-on, underling, esp. in a opium den or in the Chinese quarter of a town.

1906 I. Swift *Sketches of Gotham* 41 The lobbygows—the errand men of the Chinese—the whites, who execute commissions for them,..saw and noted this Queen also. *a* **1911** D. G. Phillips *Susan Lenox* (1917) II. x. 248 The lobbygows—men who live by lying in wait in the darkness to seize and rob the lonely, friendless fast woman. **1911** C. B. Chrysler *White Slavery* xi. 80 A 'lobbygow'—a Chinaman who acts as stool pigeon and informer for the police. **1911** G. Bronson-Howard *Enemy to Society* ix. 295, I 'ain't gunna have her think Stevey's tied up with a bunch of lobby-gows. **1930** D. H. Clarke *Louis Beretti* ii. 22 He ran errands for the girls, which made him a lobby-gow in the original meaning of the word. **1956** 'T. Betts' *Across Board* xii. 177 He flung away fortunes in grubstakes to bums, heels, and lobby-gows.

lobbyist. Add to def.: Also, one who promotes a 'lobby' (see *lobby *sb.* 3 c).

1945 *Sun* (Baltimore) 23 Oct. 1/4 Hoffman identified Arundel in a House speech as a Washington 'lobbyist' who, he was informed, picked up the \$75,000 check which paid for the festivities. **1961** *Encounter* Jan. 6/2 Skilful lobbyists with large funds and a powerful influence on the Algiers administration. **1971** *Nature* 4 June 278/2 The food industry lobbyists convincingly argue that the FDA is not responsibly handling the authority it already has. **1971** *Daily Tel.* 9 Nov. 15/6 Legislative provisions, which are promoted by a group of hysterical lobbyists who are spreading the fear of a world catastrophe because of chemical poisoning. **1974** 'R. B. Dominic' *Epitaph for Lobbyist* i. 6, I don't like high-powered lobbyists and their greasy favors.

lobe. Add: **1. i.** *Electr.* A portion of the radiation pattern of an aerial which represents a group of directions of stronger radiation and is bounded on each side by directions in which there is minimum radiation.

1926 *Bell Syst. Techn. Jrnl.* V. 297 It is interesting to observe the variation in the diagrams... A lobe starts as a small bud, it grows in size until it reaches the unit circle,

it then becomes dented. **1947** J. S. Hall *Radar Aids to Navigation* i. 13 All antennas radiate small amounts of power in directions other than the main lobe. **1959** Davies & Palmer *Radio Stud. Universe* iii. 39 In the simplest form of interferometer each narrow lobe of the aerial polar diagram produces its own small drift curve. **1968** *Radio Communication Handbk.* (ed. 4) xiii. 40/2 With increased spacing between the two aerials, more lobes appear... This type of pattern is not very useful, but if the intervening space is filled with aerials spaced λ/2, one pair of lobes grows at the expense of all the others, giving a sharp main beam with a number of relatively small minor lobes.

j. *Calligraphy.* A curved projecting part of a letter.

1957 N. R. Ker *Catal. Manuscripts Containing Anglo-Saxon* p. xxvii, In minuscule of the eighth and ninth centuries **a** is a pointed letter, the back of which projects above the place at which it is joined by the lobe. **1969** M. B. Parkes *Eng. Cursive Bk. Hands 1250–1500* p. xxvi, The letter b comprises a stem or mainstroke which rises above the general level of the other letters and a lobe made with a curved stroke to the right of the stem.

lobectomy (lobe·ktŏmi). *Surg.* [f. Lob(e + *-ectomy.] Excision of a lobe of some organ, esp. of a lung or the brain.

1911 Dorland *Med. Dict.* (ed. 6) 464/1 *Lobectomy*, excision of a lobe of a gland, as the thyroid. **1932** *Arch. Surg.* XXV. 898 (*heading*) Lobectomy and pneumectomy in dogs. **1940** *Brit. Jrnl. Psychol.* XXX. 371 Thirty-two cases, each representing a partial lobectomy. **1949** *Amer. Jrnl. Physiol.* CLVII. 135 Liver lobectomy. **1965** [see *computer 2 b]. **1967** S. Taylor et al. *Short Textbk. Surg.* xvi. 208 Treatment is wide removal of the tumour with surrounding healthy lung... Lobectomy may suffice, but pneumonectomy may be required.

lobelia. b. (Earlier example.)

1849 N. Kingsley *Diary* (1914) 94 Lobelia is the great cure, but some are against it.

loblolly. Add: **1. b.** *U.S. colloq.* A mud-hole.

1865 *Memphis* (Tennessee) *Daily Argus* 19 Nov. 3/1 We noticed a party of two or three men attempting to clean off one of them..but as fast as they cleaned away the loblolly, the lob-lolly rolled back again. **1899** Ade *Doc Horne* i. 6 In those days a mud-hole with this deceptive dry crust on top was called a 'loblolly'. **1903** A. Adams *Log Cowboy* xi. 164 His ineffectual struggles caused him to sink farther to the flanks in the loblolly which the tramping of the cattle had caused. **1944** in *Publ. Amer. Dial. Soc.* II. 58.

3. (Read) = *loblolly pine.*

1819 E. Dana *Geogr. Sk. Western Country* 195 Contiguous to the Florida line, a space, occupying in width from 50 to 60 miles, is timbered with cypress, loblolly, and long and short leafed pine.

lobo. Add: (Further examples.)

1918 C. E. Mulford *Man from Bar-20* ix. 88 What you saw was a bear or a lobo or a cougar come up to see th' fire. *Ibid.* 93 The lobo wolf in the canyon. **1973** R. D. Symons *Where Wagon Led* I. v. 79 But there were still a few buffalo wolves about. We called them lobos.

b. *transf.*

1907 'O. Henry' *Heart of West* 220 I'm not one of them lobo wolves..who are always blaming on women the calamities of life.

lobola. Add: Also **loboler, lobolo.** (Earlier and later examples.) Also, the price or present given for a bride according to this custom. Also *attrib.*

1825 N. Isaacs *Jrnl.* 25 Oct. in *Trav. E. Afr.* (1836) I. iii. 49 Jacob had become enamoured of Enslopee's sister, and had sent three head of cattle to 'Lololer'. **1905** *Westm. Gaz.* 19 Apr. 9/2 The native custom of passing cattle, known variously as 'lobolo', 'ikazi', and 'bohadi', in connexion with marriage. *Ibid.* 8 Sept. 5/1 Many of the natives are able to find the lobola for the luxury of additional wives. **1953** D. Lessing *Five* v. 248 And when our sister marries, he will have her cattle and her lobola money. **1957** G. Muldoon *Trumpeting Herd* ix. 95 According to the Angoni custom, the aunts and not the parents are responsible for fixing the Lobola, or payment for the bride. **1971** *Weekend World* (Johannesburg) 9 May 9/6, I have since paid part of the lobola to her parents. **1972** *Police Rev.* 24 Nov. 1520/3 [Rhodesia] A man may take, under tribal rites, as many wives as he can afford and he pays a bride price (called lobolo) for each.

lobotomy (lobǫ·tŏmi). *Surg.* [f. Lob(e + -o + -tomy.] Incision into a lobe; *spec.* incision into the frontal lobe of the brain, esp. in the treatment of mental illness.

1936 Freeman & Watts in *Med. Ann. District of Columbia* V. 326 (*heading*) Prefrontal lobotomy in agitated depression. **1946** *Lancet* 29 June 953/2 Pain of organic disease relieved by prefrontal lobotomy. **1950** [see *leucotomy]. **1972** *Lancet* 8 July 69/2 Schizophrenia is no longer a standard reason for performing any kind of lobotomy. **1973** *Nature* 23 Mar. 222/3 Unlike the classical lobotomy operation, which reached its height of popularity in the 1950s and was used on up to 50,000 people in the US, psychosurgery now rarely involves actually cutting into the brain.

fig. **1959** N. Mailer *Advts. for Myself* (1961) 91 Success had been a lobotomy to my past, there seemed no power from the past which could help me in the present. **1972** *Village Voice* (N.Y.) 1 June 80/2 Even private screenings are not immune to artistic lobotomy, out-of-focus, and other chronic hazards.

Hence **lobo·tomize** *v. trans.*, to perform lobotomy on; **lobo·tomized** *ppl. a.* (also *fig.*, sluggish, stupefied).

1943 *Jrnl. Amer. Med. Assoc.* 27 Nov. 810/1 The lobotomized individual is friendly, good natured and indifferent to others' opinions. **1952** B. Wolfe *Limbo* (1953) xxii. 368, I can tell you a thing or two about lobotomizing little rapists down to good little pacifists. **1955** *Sci. Amer.* June 34/3 One interesting case was that of a lobotomized patient. When she was given LSD, she reverted to her pre-lobotomy depression. **1963** H. C. Shands in H. I. Schneer *Asthmatic Child* vi. 78 Her brother had developed an intractible schizophrenic illness and had been lobotomized. **1971** 'T. Coe' *Jade in Aries* (1973) iv. 38 Manzoni can have our unfortunate friend committed, castrated, lobotomized and deported.

fig. **1953** W. Burroughs *Junkie* (1972) ix. 84 One night, I got lobotomized drunk in Frank's and went to a queer bar. **1968** *N.Y. Times* 5 Feb. 29 It was the life and times of a tightly clustered and rather faceless group, which ended with a robotlike square dance. Mr. Sheppard's lobotomized shuffle was a joy to watch. **1971** *Atlantic Monthly* Apr. 50 [He] was explaining his reasons for calling the police onto the campus, a speech greeted for the most part with a lobotomized silence.

lobster[1]. Add: **2. b.** A slow-witted, awkward, or gullible person; a fool, dupe; a bore. *U.S. slang.*

1896 Ade *Artie* x. 91 Every time I ever see him he was a lobster. **1900** —— *Fables in Slang* 54 He went to College, where he proved to be a Lobster. **1947** T. H. White *Elephant & Kangaroo* (1948) xix. 157 When she was giving breakfast to Father Byrne, after a Station, she used to lean forward whenever the old lobster spoke. **1965** *English Studies* XLVI. 468 The noun 'lob' 'dupe' became the root of *lobster* 'dupe; victim'.

5. *lobster mayonnaise, patty, soup*; **lobster bisque,** a thick cream soup made of lobster; hence, the colour of this soup; **lobster cocktail:** see *cocktail 4; **lobster Newburg,** lobster cooked in a thick cream sauce containing sherry or brandy; **lobster thermidor,** cooked lobster mixed with a cream sauce, returned to its shell, sprinkled with cheese, and browned in the oven.

1895 'M. Ronald' *Century Cook Bk.* 569 Lobster bisque. **1929** E. Wilson *I thought of Daisy* i. 8 She seemed appetizing in her lobster-bisque dress. **1967** L. Deighton *London Dossier* 49 Bentley's..sells lobster bisque freshly tinned. **1974** *Times* 15 Jan. 14/8 Their amazing lobster bisque did much to console me. **1889** A. B. Marshall *Cookery Bk.* vi. 100 Lobster Mayonnaise à l'Osborne. *Ibid.* 101 Lobster Mayonnaise with Aspic. **1913** J. Vaizey *College Girl* xxvii. 368 Iced soup, lobster mayonnaise, salmon and green peas. **1969** *Queen* 17–30 Sept. 50/3, I would not dispute the quality of the lobster mayonnaise at the Marbella Club. [**1895** 'M. Ronald' *Century Cook Bk.* II. iii. 139 Lobster à la Newburg.] **1914** 'Saki' *Beasts & Super-Beasts* 172 The lobster Newburg and the egg mayonnaise. **1968** H. Franklin *Crash* vii. 82 We..had a dozen oysters, Lobster Newburg and Chablis from the barrel. **1817** I. D'Israeli *Curiosities of Lit.* III. 240 Keep up the fire, and lively play the flame Beneath those lobster-patties. **1845** E. Acton *Mod. Cookery* (ed. 2) xvi. 349 For lobster patties, prepare the fish as for a *vol-au-vent*, but cut it smaller. **1942** Mrs. Belloc Lowndes *Let.* 19 Nov. (1971) 235 There were lobster patties, and queer looking Maid of Honour cakes. **1972** C. Drummond *Death at Bar* i. 36 A large tray of lobster patties. **1723** J. Nott *Cook's & Confectioner's Dict.* sig. S3ᵛ (*heading*) To make Lobster Soop. **1865** R. Riddell *Indian Domestic Econ.* (ed. 6) 37 Lobster soup. [Recipe.] **1960** *Good Housek. Cookery Bk.* (rev. ed.) 78/1 Simple lobster soup. **1973** J. Cleary *Ransom* iii. 74 Lobster soup—why the hell did I buy that? **1933** E. A. Robertson *Ordinary Families* xiii. 291 Lobster thermidor always brings on a sort of gastric aphasia. **1969** R. Airth *Snatch!* ix. 90 She'd made this lobster thermidor.

lobstering, *vbl. sb.* (Later examples.)

1957 *Times Lit. Suppl.* 8 Nov. 676/2 The story opens with lobstering in the Orkneys. **1971** *Country Life* 9 Dec. 1643/1 There is good lobstering and mackerel fishing for the retired.

lobsterish (lǫ·bstəriʃ), *a.* [f. Lobster[1] + -ish[1].] Resembling a lobster; red-faced.

1914 Chesterton *Wisdom of Father Brown* x. 259 [He] thrust his laughing, lobsterish face into the room. **1946** G. Millar *Horned Pigeon* x. 140 Clifton, a bald, lobsterish little man with freckles all over his muscle-rounded back.

lobstick, var. *lop-stick* (*lop *sb.*[3] 4).

local, *sb.*[1] The generally accepted form is now **locale** (see quot. 1926). (Further examples.)

1926 Fowler *Mod. Eng. Usage* 331/1 The 'erroneous form' (OED) *locale* is recommended... If something happens in a locality, the locality becomes that something's locale, or place of happening. **1957** *New Statesman* 2 Nov. 551/3 Madison Avenue is already associated in the public mind (more than subliminally) with tricky manœuvring, having all but replaced Wall Street as a suspect locale. **1970** S. J. Perelman *Baby, it's Cold Inside* 226 My existence would acquire new purpose..if I could visit the actual locale of the movie. **1973** *Nature* 6 July 61/1 These locales were chosen with considerable attention to site safety.

local, *a.* and *sb.*[2] Add: **A.** *adj.* **2. a.** *local time* (later examples).

1968 H. Franklin *Crash* i. 9 Our estimated time of arrival at Cairo is 17.45 local time, 15.45 G.M.T. **1973** 'E. Ferrars' *Small World of Murder* ii. 20 She had not adjusted her watch to local time. **1974** 'A. Haig' *Peruvian Printout* 99 Arrive Lima 0730hrs local time.

c. *local authority*, an administrative body in local government (cf. Authority 3). Also *attrib.*

1861 Mill *Repr. Govt.* xv. 273 Things..which would be best left to local authorities if there were any whose authority extended to the entire metropolis. **1897** *Lancet* 20 Feb. 537/1 [A Bill] which has for its object the superannuation of the officers and servants of local authorities. This latter term has a wide significance, as it includes practically all local bodies having sanitary and parochial functions, outside boards of guardians and other authorities, to which the Poor Law Officers Superannuation Act of last session relates. **1909** *Daily Chron.* 22 July 5/3 There would soon be a growth in the number of local authority training colleges. **1937** *Discovery* Jan. p. viii, The difficulty of persuading local authorities to provide funds. **1956** J. M. Richards in A. Pryce-Jones *New Outl. Mod. Knowl.* 380 The best British local-authority housing. **1972** M. Gilbert *Body of Girl* iv. 40 We put her into a local authority home..and she stayed there until she was fourteen. **1973** *Inverness Courier* 31 July 4/4 It was noted with regret that a teacher from Kingussie High had been offered, but turned down, a local authority house, and it was left with the Clerk to tell the county housing factor.

d. *local call*, a telephone call within a prescribed area around a telephone exchange (opp. a long-distance call); *local cluster* Astr., a cluster of stars (within the Galaxy) to which the sun belongs; also = **local group*; *local exchange* (see quot. 1940); *local group* (also with capital initials) Astr., the cluster of about twenty galaxies to which our own galaxy belongs; † also = **local cluster*; *local line*, a railway line used by local or stopping trains (opp. *main line*); *local paper*, a newspaper distributed only in a certain area and usu. featuring local, as distinct from national, news; *local radio*, radio that serves a local area only; *local room* U.S., the reporters' room in a newspaper office; *local supercluster* or *supergalaxy* Astr., a supercluster to which it is thought the 'local group' belongs; *local talent*, talented people, *spec.* (*colloq.*) the attractive women, in a particular locality.

1927 E. Murray *Post Office* viii. 138 A fixed annual charge for the installation together with a uniform fee for each effective local call. **1975** D. Bagley *Snow Tiger* xv. 124 The exchange has a bank of batteries... We're all right for local calls. **1922** H. S. Jones *Gen. Astron.* xiv. 359 We must therefore conclude that our stellar universe has a longest diameter of at least 300,000 light-years... It seems probable that the Sun is near the centre of a large local cluster situated eccentrically in this larger system. **1938** W. M. Smart *Stellar Dynamics* i. 2 There is some evidence that the stars in the neighbourhood of the sun form a loose cluster—known as the local cluster—with characteristics of distribution somewhat different from those of the galactic system as a whole. **1971** *New Scientist* 29 July 245/1 The Supergalaxy is, in turn, composed of smaller clusters of galaxies, including the local cluster of about a dozen members, our Galaxy being one of them. **1940** *Chambers's Techn. Dict.* 507/2 *Local exchange*, the exchange to which a given subscriber has a direct line. **1918** J. C. Kapteyn in *Astrophysical Jrnl.* XLVII. 106 Within this boundary the B0–B5 stars are about 12 times and the B8–B9 stars about 5·7 times more numerous than in the surrounding regions. This alone proves..that we have to do with a local group which probably does not extend in depth much farther than it does laterally. **1922** H. S. Jones *Gen. Astron.* xiv. 359 On the above hypothesis it must be assumed that the B-type stars belong mainly to the local group, for stars of this type do not increase in number with decreasing apparent magnitude as rapidly as do other types. **1936** E. Hubble *Realm of Nebulæ* vi. 125 The known members of the 'local group' are the galactic system with the Magellanic Clouds as its two companions; M31 with M32 and NGC 205 as its companions; M33, NGC 6822 and IC 1613. **1939** Skilling & Richardson *Astron.* xvii. 543 The two large spirals.., M31 and M33, belong to what Hubble calls the Local Group of nebulae. **1965** *Listener* 2 Dec. 891/2 The Andromeda Spiral and the [Magellanic] Clouds belong to what is termed the 'local group'. **1971** D. W. Sciama *Mod. Cosmol.* iii. 40 They [*sc.* galaxies] show considerable clustering, ranging from pairs of galaxies through clusters with fifteen or twenty members like the local group, up to clusters such as the one in Virgo containing several thousand galaxies. **1869** *Bradshaw's Railway Manual* XXI. 86 The question was accordingly referred to the arbitration of Captain Galton, who decided that the Midland might work the local line with Cheltenham..but that it ought not to work the main line. **1967** G. F. Fiennes *I tried to run a Railway* iv. 43 At Seven Kings we went down the local line. **1837** Dickens *Pickw.* xlix. 532 If it gets into one of the local papers, it will be the making of me. **1883** Local paper [see Paper *sb.* 8]. **1947** G. Greene *19 Stories* 78 No book-shops, just *Film Fun* and the local paper. **1967** R. Rendell *New Lease of Death* ix. 88 Elizabeth Crilling sat..reading the Situations Vacant in last week's local paper. **1966** *Economist* 1 Oct. 22/2 That aim should be the creation of a legal framework within which it would be possible to establish, without subsidy, a large number of low-powered local radio transmitting stations. **1971** *Guardian* 17 Nov. 10/6 The real question..is whether local radio can make a good profit and still be local radio. **1974** *Ibid.* 23 Mar. 10/1 Capital Radio, the general commercial radio in London..[is] in competition with four BBC networks and BBC Local Radio. **1890** *Scribner's Monthly* Aug. 157/2 We were all talking about it one night ..in the local room. **1903** E. L. Shuman *Pract. Journalism* 90 Almost the only open door to the editorial room is through the local room. **1948** *Chicago Tribune* iv. 18 Jan. 2/3 The usual banter that goes on in a local room after presstime. **1958** *Nature* 29 Nov. 1479/2 We assume that the local supercluster is in a state of differential rotation and differential expansion about its centre in the Virgo cluster. 1971 *New Scientist* 29 July 245/2 They analyse the distribution, first of normal bright galaxies known to belong to a local supercluster of galaxies, and then of quasars and some peculiar galaxies. 1953 *Astron. Jrnl.* LVIII. 30 (heading) Evidence for a local supergalaxy. 1974 *Encycl. Brit. Macropædia* VII. 830/1 Evidence found in the early 1950s gave strong support to the concept of a 'local supergalaxy'. 1947 M. Gilbert *Close Quarters* xii. 175 You can play darts and engage the local talent in gossip. 1972 R. Quilty *Tenth Session* 138 He's not the sort who would import local talent just for the hell of it. 1975 *Times* 18 Feb. 13/3 So much 'local' talent, so much unearthed by chance.. Is the crafts revival the illustration of the desire for independence and self-sufficiency?

4. b. *Radio and Television.* **local oscillator**, an oscillator in a receiver that generates oscillations (*local oscillations*) with which an incoming signal is heterodyned.

[1908 R. A. Fessenden in *Electrician* 4 Sept. 787/2 The heterodyne receiver, in which a local field of force actuated by a continuous source of high-frequency oscillations interacts with a field produced by the received oscillations and creates beats of an audible frequency.] 1913 *Proc. IRE* July 10 In the apparatus using a local oscillation generator in combination with a standard rectifier receiver electrical beats are produced and utilized. 1919 R. Stanley *Text-bk. Wireless Telegr.* (new ed.) II. viii. 143 With an independent local oscillator C.W. reception can take place with very loosely coupled circuits. 1924 [see *heterodyne a.]. 1967 Wharton & Howorth *Princ. Telev. Reception* v. 74 The function of the mixer is to multiply together the received and local oscillator signals so as to produce an output at the intermediate frequency. 1972 *Sci. Amer.* Feb. 76/3 Radio telescopes receive signals that are at too high a frequency to be recorded directly on magnetic tape. Independent local oscillators must therefore be used to 'heterodyne' the radio-frequency signal..to a much lower intermediate frequency.

c. local colour, (*c*) something picturesque in itself. Also **local colouring, colourist.**

1854 *Chambers's Jrnl.* 7 Jan. 8/2 Local colouring—*couleur locale*—is a modern expression signifying the accordance..of the adjuncts in a work of art..with the subject. 1904 F. M. Colby *Imaginary Obligations* 7 Stupendous 'local color' work going on at every railway junction, and you feel it is not. 1931 A. T. Slosson (title) A local colorist. 1934 *Amer. Speech* IX. 111/1 Villages with 'local color'. 1949 A. Huxley *Let.* 6 Mar. (1969) 593 About the country in which they lived you might consult, for local colour, a travel book by..Freya Stark. 1959 *Listener* 15 Oct. 616/1 [Henry] James never needed such ironic advice, since he was not a local colourist.

B. *sb.* **2. c.** (Earlier and later examples.) Also, a train which stops at all or most of the stations on a line (opp. an *express train*).

1879 Webster *Suppl.*, Local..an accommodation railway train, which receives and deposits passengers and freight along the line of the road. 1955 Auden *Shield of Achilles* i. 21 Any junction at which you leave the express For a local that swerves off soon into a cutting. 1975 S. Johnson *Urbane Guerrilla* i. 21 The downtown local was already at the platform.

g. A local branch of a trade union. *N. Amer.*

1888 *Nation* (N.Y.) 3 May 356/3 The Knights of Labor have locals of engineers and firemen. 1911 M. W. Ovington *Half a Man* 98 Strong organizations in the South, as the bricklayers, send men North with union membership, who easily transfer to New York locals. 1949 *Hayward* 18 Apr. 29/1 The local announced..miners would refuse to work in the pits with him. 1967 *Boston Herald* 1 Apr. 2/7 Nicholas P. Morrissey, New England regional director of the Teamsters Union, said Boston Local 25 will vote Sunday at 10 a.m. in the Charlestown armory. 1971 D. Ramsay *Little Murder Music* 121 Statement of Detective Anthony Crawley, deputed to question members of Local 6, American Federation of Musicians. 1973 *Evening Telegram* (St. John's, Newfoundland) 24 June 3/1 A trawler..had taken aboard approximately 100,000 pounds of fish, according to Jack Dodd, president of the fishermen's local.

h. (Usu. *the local.*) The public house in the immediate neighbourhood. *colloq.*

1934 *Evening News* 11 Sept. 10/1 After a modest beer or two at the 'local', bedtime calls about nine o'clock. 1937 'T. Shy' in L. Russell *Press Gang!* 178 What about a snort at the local? 1943 *R.A.F. Jrnl.* Aug... Someone..has done him a good turn by..standing him a drink in the 'local'. 1954 L. M. Boston *Children of Green Knowe* 120 The story about it is widespread. I've been told it in much the same form in different 'locals' all over the country. 1957 J. Braine *Room at Top* x. 92 The Siege Gun was our local. 1970 G. Greer *Female Eunuch* 142 Women don't nip down to the local.

localitis (lōᵘkălai·tis). *colloq.* [f. Local *a.* + -itis.] (See quots.)

1943 *Newsweek* 12 Apr. 18 The 'Pacific first' strategists are now reduced to those afflicted with *localitis*, a military disease..common to those, usually in remote spots, who see their local areas as the axial hub of all strategic movements. 1961 Webster *Localitis*, undue concern (as on the part of a military commander) with a particular area or the problems of a particular situation resulting in failure to visualize adequately the whole of which it is a part. 1964 *Listener* 8 Nov. 747/1 He was suffering from the complaint known in the Foreign Service as 'localitis'—such an intense obsession with his own particular field that he could not see beyond it. 1964 *Economist* 25 July 396/1 Among the world's occupational diseases, there is one that afflicts ambassadors. Americans call it 'localitis'..when an envoy is so captivated by the country..to whom he is accredited that he keeps urging his home government to follow a policy which that country would favour.

localizability (lōᵘ:kăləizăbi·lĭti). [f. Localizable *a.* + Ability.] The quality of being localizable.

1957 *Think* May 10 Whitehead's reference to the fallacy of simple location, the reminder that existence need not be tied to localizability, was useful in its day. 1966 *Amer. Philos. Q.* III. 233/2 Spatial localizability..is not so shared. 1969 *Nature* 21 June 1207/1 Since the early days of quantum theory there has been much discussion concerning the concept of the localizability of a particle. 1972 *Science* 12 May 592/3 The ultimate localizability of space-time measurements.

localizer (lōᵘ·kăləizəɹ). [f. Localiz(e *v.* + -er^1.] One who or that which localizes; *spec.* in *Aeronaut.*, a device for transmitting a narrow vertical radio beam along a runway by means of which an incoming aircraft can be brought into line with it and any lateral deviation automatically corrected. (In quot. 1872, 'a reporter of local items'.)

1872 *Newton Kansan* 22 Aug. 3/1 This quiet season..furnishes poor food for localizers. 1889 *Cent. Dict.*, *Localizer*, a small coil of definite resistance placed at each station of an electric fire-alarm system, which is brought into the circuit when the alarm is given, thus enabling the observer at the receiving-station to know the locality from which the alarm is sent. 1892 Mrs. M. Butler *Jrnl.* 26 Jan. in H. Tennyson *Tennyson & his Friends* (1911) 216 He [*sc.* Tennyson]..preferred to believe that Homer's descriptions were entirely imaginary. When I said that I thought that a disappointing view, he called me 'a wretched localizer'. 1942 H. L. Smith *Airways* 364 On the landing field engineers placed a 'localizer', which was a radio device emitting a peculiar beam. 1945 *Aeronautics* Feb. 29/3 The equipment linked up as a whole is then designed to guide the aircraft steadily along a localiser beam to keep it in dead line with the centre of the runway. 1946 *Jrnl. R. Aeronaut. Soc.* Oct. 750/1 The first radio aid to approach and landing, consisting of a Marconi very high frequency track beacon or 'localiser' with associated marker beacons, was installed at Heston aerodrome in 1935. 1965 *Sun* 28 Oct. 8/5 As the plane approaches London Airport it fixes on a 'localiser' beam which brings it into line with the runway.

locally, *adv.* Add: **4.** Also *Comb.*

1896 R. S. S. Baden-Powell *Jrnl.* 5 Dec. in *Matabele Campaign* (1897) xix. 480 The locally-born children are as healthy..as you could wish. 1955 J. Betjeman in R. S. Thomas *Song at Year's Turning* 11 Locally-printed volumes. 1968 *National Fisherman* Aug. 26-A/3 U.S. firms are willing and ready to buy locally caught tuna. 1975 *Country Life* 16 Jan. 143/1 Perhaps Jay was asked ..to revise a locally-made design.

locant (lǫ·kănt). *Chem.* [f. L. *locant-*, *locans*, pres. pple. of *locāre* (see Locate *v.*).] A number or letter in the name or cipher of a compound that indicates the position in its molecular structure of a constituent atom or group.

1946 G. M. Dyson in *R. Inst. Chem. Lect. New Notation Org. Chem.* 10 A number immediately preceding a symbol is referred to as a locant, as, for example, in '2ZN' or '6, 7QC'; in the second example both 6 and 7 are locants. 1952 *Chem. & Engin. News* 2 June 2337/1 He [*sc.* G. M. Dyson] also thinks that a subscript might be used to indicate a number of identical locants; for example 1,1,1,2,2,3,3,4-octachloro- might be written $1_3 2_2 3_2 4$-octachloro-. 1965 *Nomencl. Org. Chem.* (I.U.P.A.C.) C. 29 The starting point and direction of numbering of a compound are chosen so as to give lowest locants to the following structural factors (if present). 1968 R. S. Cahn *Introd. Chem. Nomencl.* (ed. 3) viii. 79 Locants used for the ring system are the usual numerals—the '2' in the examples above refer to the 2-position of the naphthalene nucleus.

Locarno (lǒkā·ɹno). The name of a town in Switzerland used to designate the conference held there and the treaties signed as a result in 1925 between Germany and several other European countries for the preservation of peace and the continuation of existing territorial boundaries. So *transf.*, a similar conference, treaty, or agreement. Hence **Loca·rnist** *sb.*, a supporter or advocate of the policy adopted at the Locarno Conference; also as *adj.*; **Loca·rnize** *v. intr.*, to settle disputes by pacific means.

1926 *Glasgow Herald* 19 Feb. 8 Sir Austen had called Locarno only a beginning... Some business men, having settled a quarrel among themselves, said that they had 'Locarnized'. *Ibid.* 5 Mar. 9 We shall go to Geneva to work there as 'Locarnists' in the Locarno spirit. 1927 *Daily Tel.* 21 June 13/3 After Locarno there was..a genuine desire to get on better terms with the Reich. *Ibid.*, The demand for the evacuation of the Rhineland..has met with opposition, even in strongly Locarnist quarters. *Ibid.* 6 Sept. 9/5 All the talk about Eastern or Danubian Locarnos is very much in the air, when even the Western Locarno is seen to be in difficulties already. 1937 *Ann. Reg.* 1936 [74] The answers of Germany and Italy to the invitation sent on July 24 to a new 'Locarno' Conference were still being awaited. *Ibid.* [189] The Hitler Government could count on the help of Italy in its negotiations for a new Locarno Pact. 1937 Koestler *Spanish Testament* ii. 60 The Europe of the Locarno period..would have reacted..with a storm of indignation. 1971 A. Bullock *20th Century* 72/1 Stresemann had refused to conclude an Eastern Locarno which would mean accepting Germany's postwar frontiers with Poland and Czechoslovakia, and in April 1926 he balanced his acceptance of Locarno by a new treaty of friendship with Soviet Russia.

locatable (lŏkēi·tăb'l), *a.* Also **locateable.** [f. Locat(e *v.* + -able.] Capable of being located.

1936 *Discovery* Apr. 122/1 Thus we find..four parallel alleys, or 'zones', locatable at the 'wavelength' for the stream. 1964 *English Studies* XLV. 250 Any locatable regional dialect. 1968 E. Russenholt *Heart of Continent* III. vii. 101 Officials of 'the Company' and some free traders are locateable.

locater, var. Locator (in Dict. and Suppl.).

location. Add: **5.** (Earlier and later S. Afr. examples.) For 'Also..natives.' read 'Also, in South Africa, the quarters or area set apart for black South Africans; occas. also used of an area in which Coloureds live.'

1833 G. Greig *S. Afr. Almanac & Directory* 191 The population consists of a mixture of Bastards and Hottentots, who are divided into about 60 parties, each of which has a district location allotted to it. 1835 D'Urban in W. M. Macmillan *Bantu, Boer, & Briton* (1929) 128 He may be placed in a location in His Majesty's Colony [*sc.* the Cape]. 1851 J. J. Freeman *Tour S. Afr.* xv. 361 They are located by the Government, and on these locations they cultivate lands and build their native huts. 1926 O. Schreiner *From Man to Man* ii. ix. 316 You..stood looking down at the..little brown huts of the Kaffir Location sleeping at your feet. *Ibid.*, A Kaffir servant might be seen hurrying from the Location to the town. 1945 R. Hargreaves *Enemy at Gate* 241 The 'location' occupied by the half-breed fraternity. 1961 T. Matshikiza *Chocolates for my Wife* v. 45 Several beasts were sacrificed and thousands of location residents partook of the traditional royal roast. 1971 *Rand Daily Mail* 4 Sept. 5/1 Transkeians were against the establishment of locations in their territory.

c. In the production of motion pictures, an exterior place, away from a film-studio, where a scene is filmed; freq. in phr. *on location*. Also *attrib.* orig. *U.S.*

1914 *Scribner's Mag.* Mar. 276/1 It was his duty..to pick out 'locations', as are called the scenes and backgrounds of a moving-picture play. 1918 H. Croy *How Motion Pictures are Made* v. 120 If an exterior is chosen for the first scenes it has been selected in advance by the 'location man' and the director. *Ibid.* iv. 148 Now many actors are..in the studio or on location. 1928 'I. Hay' *Poor Gentleman* iii. 42 They're converting the whole place into what is called a Location, where they can stage dramas of English country life. 1935 *Time* 8 July 32/3 The fault most likely to creep into pictures made on location comes from their producers' natural reluctance to throw away bits of local color even when these impede their story. 1957 M. Summerton *Sunset Hour* iv. 62, I have been working hard... Working out location shots. 1971 *Daily Tel.* (Colour Suppl.) 12 Nov. 51/1 On location in Yugoslavia and at Pinewood Studios I talked to four people deeply involved in the filming of *Fiddler on the Roof*. 1973 J. Leasor *Host of Extras* iii. 41 I'm hiring them out to a film company. Two weeks guaranteed, on location. Corsica.

7. The action of discovering, or the ability to discover or determine, the position of a person or thing.

1900 *Geogr. Jrnl.* Oct. 382 These birds [*sc.* penguins] must have a wonderful power of location. 1962 A. Nisbett *Technique Sound Studio* 259 The script is also marked ..with notes for quick groove location. *Ibid.* 276 These help in the exact location of editing points on tape.

locational (lokēi·ʃənăl), *a.* [f. Location + -al.] Of or pertaining to location.

1909 in *Cent. Dict.* Suppl. 1926 *Cleaning & Dyeing World* Oct. 19 He has one advantage, however, which the extensive advertiser does not have, and that is locational identification. 1957 *Economist* 5 Oct. 15/1 These two locational accidents could have served as the theme of the conference. 1960 Robbins & Terleckyj (title) Money metropolis: a locational study of financial activities in the New York region. 1971 *Nature* 18 June 426/2 In the population sector, locational attraction is a function of existing floorspace and available land in a zone.

locator. Add: Also **locater.**

4. Something which locates; *spec.* a device for indicating the position or direction of something. Also *attrib.*

1902 *Cyclists' Touring Club Gaz.* Aug. 359/1 A spicule of flint..pierced my tube, but kindly remained in evidence as a locater. 1919 *Nature* 30 Oct. 182/1 Sound-locators were also used to board anti-submarine craft. 1951 *Gloss. Aeronaut. Terms* (B.S.I.) III. 29 *Locator beacon*, a non-directional radio-beacon of low power, associated with a recognized instrument landing system. 1971 J. B. Carroll et al. *Word Frequency Bk.* p. xix, The editorial outputs prepared from the tape files included..a locator list that can be used to determine the source of every token in the Corpus. 1973 *Black Panther* 21 July p. B, The automatic car locator system.

loc. cit. (lǫk sit), abbrev. of L. *loco citato* or *locus citatus*, '(in) the place cited', i.e. in the book, etc., that has previously been quoted.

1854 H. H. Milman *Hist. Latin Christianity* I. ii. iii. 149 In the words of the ecclesiastical historian,.. by such a deed a deep stain was fixed on Cyril and the Church of Alexandria. [fn.] Socrat. loc. cit. 1937 M. Leach *Amis & Amiloun* p. xc, Kölbing, *loc. cit.*, and also in his edition of *Amis and Amiloun*. 1969 Y. Kamisar in A. B. Downing *Euthanasia* 132 Chesterton, 'Euthanasia and Murder', loc. cit.

Loch Fyne (lǫχ fəin). The name of a sea loch in West Scotland, used *attrib.* to designate a

type of fishing-boat having a standing lug mainsail.

1906 *Yachting Monthly* II. 15/1 The first boat built on Loch Fyne lines, and approximating in size to the ordinary fishing boat, was the May. *Ibid.* 19 Col. Dunlop's Loch Fyne ketch Marsailidh. 1930 *Ibid.* XLIX. 209/2 More odd rigs have been tried out here than in any other class in existence, including split lugs, Loch Fyne skiff rigs, and sprit sails. 1974 R. Simper *Scottish Sail* 31 The Loch Fyne skiffs were rather lightly constructed and seem not to have lasted very long.

Lochinvar (lǫkinvā·ɹ). The name of the hero of a ballad in Sir Walter Scott's *Marmion*, used allusively for a young male eloper; also *transf.* (see also quot. 1951).

1879 C. M. Yonge *Magnum Bonum* I. xii. 233 His bride..had had a young Lochinvar, and even in her wedding dress, favoured by sympathising servants, had escaped down the back stairs of a London hotel, and been married at the nearest Church. 1890 'R. Boldrewood' *Colonial Reformer* III. xxviii. 129 Much he marvelled at this Australian edition of 'Young Lochinvar'. 1906 'O. Henry' *Four Million* (1916) 125 He..received the hearty thanks of the backyard Lochinvar. 1936 J. Buchan *Island of Sheep* ix. 170 The young Lochinvar business was rather out of my usual line. 1951 E. Hill *Territory* 311 Lochinvars sold the women to the drovers and the stations at £10 a head. *Ibid.* 444 *Lochinvar, the*, old time term for catching lubras to work cattle, etc. 1966 [see *extramural a. b]. 1970 *New Yorker* 3 Oct. 83/1 The majority of young artistic Lochinvars..have turned..to the tools and processes of modern industrial technology. 1972 'J. & E. Bonett' *No Time to Kill* iv. 45 She looked.. expectant, waiting..for the return of her young Lochinvar. But young Lochinvar..had found another bride, and she had married Eldred.

Lochlann (lǫ·χlan). *Hist.* [a. Irish *Lochlann* Scandinavia, *Lochlannach a.* and *sb.* Scandinavian.] A viking, (ancient) Scandinavian, Norseman.

1857 W. Reeves *Adamnan's Life St. Columba* 332/2 About the same time the Fortrenns and Lochlanns fought a battle. 1861 F. O'Curry *Lect. Manuscript Materials Anc. Irish Hist.* x. 225 A book for the saints, and a book for the Fomorians, Lochlanns or Danes. 1880 W. F. Skene *Celtic Scotl.* I. i. vi. 304 Forty-eight of the number of Icolumkill slain by the Lochlanns. 1905 *Westm. Gaz.* 15 Aug. 2/1 The ships of the Lochlanns lie in the river, and never send a man against her. 1922 Joyce *Ulysses* 46 Galleys of the Lochlanns ran here to beach, in quest of prey.

lochlet (lǫ·χ,lĕt). *rare.* [f. Loch^1 + -let.] A little loch.

1925 A. S. Alexander *Tramps across Watersheds* 40 These lochlets with their ancient relics are mostly meadows now.

Loch Ness. [Loch^1.] The name of a loch in Scotland used *attrib.* of a water-monster alleged to exist in its waters. Also *fig.*

1933 *Inverness Courier* 9 June 5/5 The Loch Ness 'monster' was seen near the west end of the Loch. 1934 *Discovery* Jan. 14/1 That elusive creature the sea-serpent is again in the news, this time in the shape of the Loch Ness 'monster'. 1937 *Ibid.* Nov. 334/2 Though the Loch Ness monster itself were laid before him in all its magnitude, he would still surely find an outlet..for his nervous, driving, shrewish disposition. 1959 A. Hardy *Fish & Fisheries* xiv. 364, I am deliberately not discussing the so-called Loch Ness monster. If there is some strange creature there it is clearly not a sea-beast bigger than a seal which might make its way up the shallow River Ness. 1965 *New Society* 9 Sept. 7/3 Britain's brain drain appears, then, to be something of a Loch Ness monster. It surfaces sporadically, then vanishes from view. 1971 *Stornoway Gaz.* 10 July 1/6 Preparations are now complete and they set off in a few days' time to try and capture that elusive denizen of the deep—the Loch Ness Monster.

lock, *sb.*^{2} Add: **1. b.** *lock and key*, delete 'rarely *attrib.*' and add: also *fig.* (freq. *attrib.*) with allusion to the structural complementarity or mutual specificity of a lock and its key.

[1894 E. Fischer in *Ber. d. Deut. Chem. Ges.* XXVII. 2992 Um ein Bild zu gebrauchen, will ich sagen, dass Enzym und Glucosid wie Schloss und Schlüssel zu einander passen müssen, um eine chemische Wirkung auf einander ausüben zu können.] 1901 C. A. Mitchell tr. *Oppenheimer's Ferments* v. 65 Two haptophore groups coinciding with one another ('lock and key') and a subsequently active zymophore group. 1924 K. G. Falk *Chem. Enzyme Actions* (ed. 2) v. 116 E. Fischer's lock-and-key simile for the mutual getting together of substrate and enzyme, each fitting in with the other, gives a mechanical picture of the action. 1950 *Sci. News* XV. 120 So far, blood group antibodies have been described as having the property of reacting by agglutination with red cells which contain the specific antigen, and with no others, on the lock-and-key principle. 1969 *Times* 25 Apr. 13/6 It seems that some of the proteins in the mixture are able to recognize and bind to certain sites on the RNA molecule by a lock-and-key mechanism. 1974 *Physics Bull.* Dec. 581/1 The lock and key interaction between enzyme and substrate does not usually involve strong covalent bonds.

5. *lock, stock, and barrel* (earlier and later examples). Also as *advb. phr.* (See also Stock *sb.*^{1} 28 b.)

1842 W. T. Thompson *Major Jones' Courtship* (1844) 66 All moved, lock, stock, and barrel. 1855 S. A. Hammett *Wonderful Adventures Captain Priest* xii. 76 He sold off his feathered stock, 'lock, stock, and barrel'. 1909 [see *cagey a.]. 1961 B. Ferguson *Watery Maze* xii. 292 One of the ministers would take over lock, stock and barrel the administration. 1974 P. Erdman *Silver Bears* i. 12, I bought us a Swiss bank: lock, stock, and barrel.

10. Add. to def.: also, a similar chamber used between air at atmospheric pressure and either water (e.g. outside a submarine) or a vacuum (e.g. outside a spacecraft).

1914 S. F. Walker *Submarine Engin.* iii. 35 The air lock ..is a chamber with doors at each end, arranged so that only a small quantity of air or water can enter each time the lock is opened. 1959 'Wyndham' & Parker *Outward Urge* ii. 66 The duty-man operated the lock, and presently Troon was outside. ... Taking the place of the Thrill Trunk..came the Escape Chamber method, a permanent chamber or lock entered by way of a watertight door.

11. b. *lock and block (system)*: a system of railway signalling by which a train does not enter a section of line until the preceding train has left it, the signal being locked at 'danger' and only released when the preceding train leaves the section.

1902 *Encycl. Brit.* XXXIII. 146/2 'Lock-and-block' has been used to a limited extent on a good many lines in England and a half-dozen in America. 1905 *Westm. Gaz.* 12 Jan. 7/1 The failure was partly due to faulty line circuits of the lock and block instruments. 1959 *Engineering* 1 Dec. 436/1 Signals..operated mechanically..with the Sykes lock-and-block system. 1966 *Railway Mag.* Nov. 748/1 The Sykes lock-and-block, although odd fashioned, ..has a long record of reliable service in the operation of dense traffic.

c. *Rugby Football.* A player in the second row of the scrummage (see quots.); this position. Also *attrib.*, as *lock-forward*, *-man*.

1906 Gallaher & Stead *Compl. Rugby Footballer* vii. 100 Working the [New Zealand] Scrum... The lock [etc.]. *Ibid.* 104 Immediately behind these hookers..is he whom we call the lock man... His duty is to hold or lock the two hookers in. 1913 L. R. Raphael *Mod. Rugby Football* (1918) xvii. 225 The middle man in the second row, the 'lock', bound the 'hookers' together, not his own row. 1956 V. Jenkins *Lions Rampant* ii. 13 Mr Siggins..was one of the finest lock-forwards of his day. 1969 *Times* 10 Sept. 4/3 It was strange to see the former hefty England wing, Woodward, at lock in the blues' scrummage. 1971 *Sunday Express* (Johannesburg) 28 Mar. 20/6 Springboks fullback Ian McCallum, prop Hannes Marais and lock Frik du Preez are in the opening fixture.

12. a. Delete † *Obs.* and add later examples.

1954 E. Dowty *Teach Yourself Judo* iii. 41 There are only a few basic types of lock, and these can be performed by anyone sufficiently interested. 1974 'J. Le Carré' *Tinker, Tailor* xxii. 189 Guillam selected Tarr's right arm and flung it into a lock against his back, bringing it very near to breaking. ... He would sooner see the whole bank go down the drain..than get beaten by us. Unless we develop an even better lock on him—and that won't be easy.

15. b. The turning of the front wheels of a motor vehicle to change its direction of motion; the full extent of such turning.

1908 *Autocar Handbk.* (ed. 2) xvi. 125 There should be plenty of 'lock' for the wheels, which, with an inconsequence not unusual in our language, means that the wheels shall be quite free to be deflected through a large angle. 1939 L. MacNeice *Autumn Jrnl.* xxiv. 79 The quick lock of a taxi. 1959 *Observer* 1 Mar. 21/5 From lock to lock it takes 3½ turns, ... a heavy throttle foot should provoke tail wag on a slippery surface. Turning circle is 37 ft. 1967 *Autocar* 28 Dec. 45/1 The 35 ft. 1 in. mean turning circle with 4·25 turns lock-to-lock is not excessive. 1974 L. Meynell *Fairly Innocent* xi. 148, I must have got on to the wrong lock... I don't really understand about going backwards.

18. (Later example.)

1922 Joyce *Ulysses* 509 Mary Shortall that was in the lock with the pox.

19. a. (sense 9) *lock-bar, -bridge, -canal, -charge, -cut, -house* (later example), *-pen, -station, -thief, -wall*.

1890 F. L. Packard *Four Stragglers* 312 The lock-bar worked through the side of the pier wall. 1866 Dickens *Mut. Fr.* III. vi. The crowd back by his plank lock-bridge to the towing-path side. 1903 *Westm. Gaz.* 2 Jan. 3/1, I imagine that the Panama waterway is to be a lock canal. 1897 J. Hamilton *Jericho Road* ii. 20 Dont you b'lieve she could run the dam at Mount Zion, and dodge paying lock-charges? 1909 *Westm. Gaz.* 16 Aug. 3/3 Motor boats..probably find their way down lock-cuts made more difficult and less obvious than usual. 1907 *Daily Chron.* 30 Apr. 1/2 An assistant lockkeeper..found the body of a child floating in the lock-cut. 1865 Lock-house [in Dict., sense 9 d]. 1907 *Westm. Gaz.* 20 Aug. 12/1 The lock-pen..opens and shuts now to let through the *Queen Elizabeth* in solitary state. 1855 E. S. Hale *If, Yes & Perhaps* (1868) 15, I would start in the morning by the lock at the station at Brookport on the canal. *Ibid.* 22 At night I walked the deck till one o'clock..to keep guard against the lock-thieves. 1885 ... 66 He ran along the lock-wall to open his gates when he saw us coming.

20. *lock-box U.S.*, a delivery letter-box provided with a lock; **lock-in**, the action or fact of locking in a person or thing (see Lock *v.*^{1} 2 and *3 e); also *attrib.*; **Locknit**, the proprietary name (but see quot. 1935) of a fabric knitted with an interlocking stitch; also **lock-knit**; freq. *attrib.*; **lock-nut**, add: also, a nut specially designed to prevent accidental loosening once it has been tightened; (further examples); **lock-step** (further example); also *fig.*, a rigid or unchanging pattern; also *attrib.*, rigid, unchanging.

1882 E. Crapsey *Nether Side N.Y.* 150 C. H. Chester, M.D., Lock Box 4, Reading, Pa. 1966 M. E. Freeman *By Light of Soul* 34 She saw one letter slanted across the dusty glass of the box. It was not a lock box, and she would not ask the postmaster for the letter. 1955 E. Pound *Section: Rock Drill* lxxxviii. 42 A First Polio (Ske) in his lock-box. 1920 *Contemp. Rev.* Dec. 823 To the lock-out of the masters, the workers replied with the 'lock-in' movement—the temporary capture of the factories and workshops. 1970 *Globe & Mail* (Toronto) 25 Sept. B7/6 There may be some giving in on peripheral items such as reducing the length of lock-in clauses. 1975 *Times* 25 Jan. 13/5 First, there is the 'lock-in' factor. Given additional tax burdens, the first reaction of any jobber is to feel less inclined to sell. 1935 *Trade Marks Jrnl.* 8 May 588/2 Lansil Locknit. Registration of this Trade Mark shall give no right to the exclusive use of the word 'Locknit'. Knitted piece goods composed wholly or mainly of artificial silk. Cellulose Acetate Silk Company, Ltd., Lancaster. 1936 *Times* 14 Feb. 9/5 A three-piece in pale blue chalk stripe locknit and plain jersey. 1951 *Good Housek. Home Encycl.* 431/1 Lock-knit or open weave articles should be dried flat. ... Hams hung from the roof among strings of locknit underwear. 1973 R. Rendell *Some lie & Some Die* ix. 84 Passing her iron across a pair of pink locknit knickers. 1907 *Westm. Gaz.* 21 Nov. 4/2 The steering is..of the worm and sector pattern, having an adjustment that is easily effected by releasing a lock-nut and slightly turning the steering column. 1964 S. Crawford *Basic Engin. Processes* xiv. 304 The feet sections are secured by tightening the locknut with the special adjusting key. 1972 *Practical Motorist* Oct. 165/3 If this resistance is felt before the vertical position..adjustment can be made by loosening the lock-nut 'C'. 1836 T. Power *Impressions Amer.* I. 379 They [*sc.* convicts] were marched from the building in squads, using what is called the 'lock-step', and were regimented into the prison chapel. 1955 *Sci. News Let.* 16 Apr. 235 A 'what will people think' disease is driving us all into lock-step conformity and toward straitjacketed thinking and lock-step living. 1963 *New Society* 7 Nov. 15/1 The prescribed lockstep of school life. 1972 *New Yorker* 30 Oct. 155 Mrs. Handy's lockstep methods (copy the great novelists, read the 'Masters of the Far East', stay away from girls) produced a handful of published novels. 1973 *Business Week* 18 Mar. 22/1 The break could occur if Ireland did not follow Britain into the EEC. For the republic marches in an economic lockstep with Britain. 1975 *Where* Apr. 109/3 Students working..in their own way, and at their own pace, freed from the 'lock-step' of the classroom. 1973 *Times* 25 June 14/3 The 'whole thing' was an attempt..to break what he calls 'the lockstep'—the educational process that leads in a straight line from kindergarten through graduate school, and often onward into the walled-in offices of academia.

lock, *v.*^{1} Add: **3. e.** Const. *in*. To trap or fix firmly or irrevocably; to fix in position.

1953 Berry & Van den Bark *Amer. Thes. Slang* (1954) § 633/2 *Locking in*, adjusting the television image for a clear picture. *Ibid.* § 634/10 *Locked in*, televised image properly synchronized. 1959 *Economist* 16 May 643/1 This year tend to 'lock in' many traders with their present holdings. 1968 *Sunday Times* 3 Mar. 51 The Extra campaign had to be aggressive because people are so locked in to engines. 1968 *Telegraph* (Brisbane) 8 Nov. 14/7 Anything I knew..was too late to help Ford. They already were locked in on their program. 1972 *Sci. Amer.* Nov. 96/3 Lowering its temperature back to normal locks in any deformations due to bondings.

5. *lock out*. **c.** *Electronics and Computers.* Temporarily to prevent the operation or use of. Cf. *Lock-out 2.

1952 R. C. Walker *Relays* viii. 222 Voltage selective systems have been devised in which the value of the applied voltage operates one of a group of thermistors and locks out all the others. 1962 *Gloss. Terms Automatic Data Processing* (B.S.I.) 47 During an autonomous peripheral transfer, the storage locks concerned may be locked out to prevent reference to those blocks until completion of the transfer. 1972 *Computer Jrnl.* XV. 194/2 Another circumstance in which an investigation..is called for is when a record that has been locked out for a particularly long period is being updated relatively frequently and being locked out for an unreasonable time.

6. *to lock up* (Printing): (later examples).

1927 R. B. McKerrow *Introd. Bibliogr.* i. 16 The furniture employed to fill up the chase is 'locked up' by the insertion and driving home of wedges or 'quoins'. 1972 P. Gaskell *New Introd. Bibliogr.* 80 The quoins were driven home with a mallet and 'shooting stick' to lock the forme up tightly.

7. d. (Earlier U.S. examples.)

1839 *Hist. Virgil A. Stewart* 23 (Th.) They are enemies, and let them lock horns. 1855 *Knickerbocker* XLVI. 95 As neither of the trains stop at way-stations, I expect nothing more than to see the two lock horns at the corner of my kitchen.

12. *to lock on to*: **a.** *intr.* Of radar or other equipment: to locate and then to track automatically; to accept as a target or reference object that is thereafter maintained as such (usu. automatically).

1949 *Jrnl. R. Aeronaut. Soc.* May 439/2 The aerial system has been designed to 'lock on' to the responder signal and to 'follow it' during its motion through the atmosphere quite automatically and with great accuracy. 1964 *Discovery* Oct. 7/3 The stabilized instrument platform has been developed..to lock on to the sun (or moon) within two minutes of lift-off. 1966 *New Scientist* 25 Aug. 405

Stars were often mistaken for aircraft lights. In 27 cases pilots chasing a target aircraft had 'locked on' to a star for periods between one and ten minutes and actually tried to fly up to it. 1968 *Times* 10 Dec. 6/7 The satellite was to have used six star trackers which would lock on to reference stars.

b. *trans.* To cause (a piece of equipment) to lock on to some object.

1954 K. W. Gatland *Devel. Guided Missile* (ed. 2) iv. 118 After a short period, radar tracking and aiming devices are 'locked on' to target, and from then on the whole attack is automatic. 1964 *Guardian* 1 Dec. 1/4 Their fourth attempt to 'lock' Mariner-4 on to the star Canopus. 1971 *Nature* 8 Oct. 367/1 The flight took place aboard a Skylark sounding rocket, which was stabilized and 'locked on' to the strong X-ray source Sco X-1 during the four minutes of observing time.

lockage. Add: **2. c.** The passage (of a vessel) through a lock.

1913 J. B. Bishop *Panama Gateway* v. iv. 375 The average number of lockages through the..Canal..was 39 per day.

locked, *a.*^{2} **2.** (Earlier example.)

1810 D. Thomas *Trav. Western Country* 30 The mill-dams on this stream are locked.

locked, *ppl. a.* Add: **d.** *locked-coil*: used *attrib.* to denote a rope or cable which has the outer strands of such a shape as to lock together and form a smooth cylindrical surface.

1886 *Cassell's Family Mag.* Dec. 59/1 A new kind of rope, called the locked-coil rope, has recently been brought out. 1952 T. Bryson *Mining Machinery* (ed. 3) x. 246 The desire to increase further the wearing surface of ropes led to construction of locked-coil ropes.

e. *locked-room*: used *attrib.* to denote a mystery, or a mystery story, involving a locked room.

1942 H. Haycraft *Murder for Pleasure* vi. 104 *The Mystery of the Yellow Room*..remains..the most brilliant of all locked room' novels. 1954 J. Symons *Narrowing Circle* iii. 14, I listened to the dictaphone on my desk, which was a deliberately old-fashioned locked-room style detective story. 1965 'D. Shannon' *Death-Bringers* (1966) vii. 83 He'd never believed there were such things as Locked Room mysteries in real life. 1970 —— *Unexpected Death* (1971) xiii. 194 Lock the door on the outside and shove the key under the door. No locked-room puzzle.

f. *locked groove*: on a gramophone record, a circular groove into which the normal spiral groove runs.

1958 in *Chambers's Techn. Dict.* Add. 1962 A. Nisbett *Technique Sound Studio* xii. 207 On disc, using locked grooves to provide rhythmic repetitions.

g. With *in*: of a surfer, enveloped in and being carried along on a wave.

1965 *S. Afr. Surfer* I. iii. 7 Its breathlessly fast hollow waves afford the lucky surfer an easy 500 yard locked-in ride. *Ibid.* 35 Each situation, from being locked-in to wiped-out, is entirely dependent on how the surfer uses the wave. 1971 *Studies in English* (Univ. of Cape Town) Feb. 27 If the wave is very hollow, as at Gansbaai, then the wave may arch over the surfer and he will get covered up and enjoy a tube ride. This is called being locked in.

locker, *sb.*^{1} **7.** *locker-room* (later examples).

1906 *Westm. Gaz.* 11 July 8/1 Two extra payments are a penny for a bath, including towel and soap, and 6d. deposit for the use of a large locker in the locker room. 1931 *Maclean's Mag.* 1 Aug. 28/4 Mere males are lucky to find sanctuary in locker room, grill and bar. 1954 [see *bathroom]. 1972 *Newsweek* 10 Jan. 50/2 On one side of the crowded Kansas City locker room, veteran quarterback Lenny Dawson dressed hurriedly and disappeared.

lock-on (lǫ·kǫn). [f. vbl. phr. *to lock on* (*to*): see *Lock *v.*^{1} 12.] (The commencement of) automatic tracking.

1960 *Aeroplane* 4 Mar. 278/1 The Bloodhound target radar lock-on technique will be demonstrated and a model of the missile is to be 'fired'. 1967 *Sci. Amer.* May 83 (Advt.) When an air-to-air training missile 'sees' its target, a new microminiaturized signal amplifier..tells the pilot that lock-on has been achieved. 1971 *Times* 25 Feb. 18 As Ankers swooped below that altitude, his radar remained inactive. 'C'mon, radar,' Mitchell implored. 'Get the lock-on.'

2. The establishment of a rigid physical connection.

1967 *Sunday Times* 23 Apr. 8 Back at the surface, a lock-on device enables divers to transfer to a larger chamber on board, releasing the sub to return to work with a fresh diving team. 1968 *New Scientist* 6 June 509/1 As it will need to contend with the underwater currents playing around the submarine, it will need extremely sensitive means of controlling its position just before lock-on. 1969 *Jane's Freight Containers* 1968–69 374/2 The top lift cradle has..power-guided lock-on at each of the four corners. Safe lock-on indicators are provided.

lock-out. Add: Also **lockout.** (Earlier and later examples.)

1854 *Westm. Rev.* V. 120 How far the 'lock-out' has been forced upon the masters by partial 'strikes' of the men,..what proportion of the workpeople have been victims rather than combatants 'locked-out',..these are some among many much disputed points. 1880 [see *close-down]. 1926 *Times* 3 May 2/1 Lord Londonderry, speaking for the mine owners, said the notices which were put up were not lockout notices. 1955 *Times* 17 Aug. 7/2 The employers announced that the strike would be met by a lockout, and there seems to be no prospect that they will change their attitude. 1970 T. Lupton *Managem. & Social Sci.* (ed. 2) iii. 62 The hidden sanction of strike or lockout always underlies bargaining. 1971 *Daily Tel.* 20 Oct. 16 It is somewhat exceptional in industrial relations for employers to resort to a lock-out against organised strikes and overtime bans. 1974 *Socialist Worker* 7 Nov. 16/1 Truck workers at British Leyland's AEC plant marched through Southall last Thursday chanting and determined after the month-long lock-out.

2. *Electronics and Computers.* The automatic temporary prevention of the operation or use of a relay or other device. Usu. *attrib.*

1942 T. Croft *Electr. Machinery & Control Diagrams* viii. 105 With the dual relay, the controller, both the closing and the lockout portions of the switch are operated. 1946 'Electr. Engineer' *Ref. Bk.* VII. 105 When the directional relay closes contacts for fault current fed out of the feeder, the secondary coil operates the associated attracted armature relay which initiates a lock-out operation. The overcurrent relay cannot operate because the secondary coil is not connected to the series coil. 1952 I. Frazee et al. *Automotive Electr. Syst.* vii. 369 Vibrating and lockout circuit breakers consist of a coil winding and a set of contact points..When current in excess of the rated value flows, a plunger..opens the contact points. 1960 *McGraw-Hill Encycl. Sci. & Technol.* XIII. 348/2 The problem in the application of lockout circuits is that of concurrently competing circuits, among which one has to be picked for some action. 1961 N. Chapin *Programming Computers for Business Applic.* viii. 198 The high-speed transfer-of-data phase during which the tape interface copies into or fills from high-speed storage..is sometimes called the lockout phase. 1972 *Computer Jrnl.* XV. 294/2 The information in the record is then read, any necessary checking done, and the update performed. The updated version of the record is then inserted in the file and finally the lockout is cancelled.

lockschen, lockshan, lockshen, varr. *Lokshen.

locksman^2 (lǫ·ksmæn). [f. pl. of Lock *sb.*^{1} + Man *sb.*^{1}] In Kingston, Jamaica, a member of the Ras Tafari cult who wears his hair long and plaited as a mark of his membership.

1960 M. G. Smith et al. *Ras Tafari Movement in Kingston, Jamaica* iv. 23 The Locksmen, whose hair is matted and plaited and never cut, neither their beards. 1966 *Guardian* 3 Feb. 8/1 The long-haired Rastas, the Locksmen, are the ones Jamaicans laugh at in the streets.

lock-up, *sb.* (*a.*). Add: **1. a.** (Earlier and later examples.) Also *attrib.*

1839 T. J. Green *Jrnl. Texian Expedition* xvii. 300 To elude the vigilance of the officer at lock-up time. 1910 A. Herbert *Let.* 5 June (1969) 37 On which we were 5 minutes late for lock-up! 1914 'I. Hay' *Lighter Side School Life* iv. 104 Rules, roll-calls, bounds, lock-ups. 1968 *Eton College Chron.* 22 Mar. 6222/1 Sat. 9 Mar. Lock-up 7.15 p.m.

b. Also *attrib.*

1906 *Daily Report* 26 Aug. 5/4 As a promising speculative lock-up holding, the shares are worth buying at the present prices. 1970 *Observer* 17 Nov. 4/3 The shares may be regarded as a good lock-up investment.

c. *Printing.* The action of preparing plates or formes for printing or placing them in the press; also, a contrivance for holding the plates or formes in a press. Also *attrib.* Cf. Lock *v.*^{1} 6 (in Dict. and Suppl.).

1888 C. T. Jacobi *Printers' Vocab.* 76 *Lock-up chases*, special chases made in order to dispense with large quantities of furniture in filling up spare room in formes or on the press. *Lock-up iron*, the iron stick used for tightening up formes as they stand instead of laying them up. 1925 H. Carter *Let.* 4 May (1965) 203 Lockup & Presswork.. $40.00. 1960 G. A. Glaister *Gloss. Bk.* 240/2 *Lock-up table*, any of several varieties of imposing surface specially equipped for the accurate imposition of formes for colour registration. 1964 *Gloss. Letterpress Rotary Printing Terms* (B.S.I.) 15 Lock-up..1. A mechanical arrangement for holding the printing plates or formes on the press. 2. The action of locking the printing plates or formes on the press. 1967 Karch & Buber *Offset Processes* ii. 23 ... *Ibid.* 24 ... includes lock-up and imposition.

d. (Earlier and later examples.)

1839 *Knickerbocker* XIV. 110 He was seized, and carried to the 'lock-up'. 1972 *Police Rev.* 8 Dec. 1599/1 There would be a place to cater for the Colchester lock-ups. 1973 M. Brian *Pattern of Violence* i. 14 Sam..was at present residing within the central lock-up in..police headquarters, ready to appear before the court.

e. Short for *lock-up garage.

1910 *Bradshaw's Railway Guide* Apr. 1036 Southgate Private Hotel..Lock-up for Bicycles. *Ibid.* 1070 Motor garage, with 12 lock ups. 1973 E. Lemarchand *Let or Hindrance* ix. 103 'Was the car standing out while you were in Cornwall?' 'No. They gave me a lock-up.' 1974 G. Mitchell *Javelin for Jonah* v. 68 We followed Jonah..to the garages..frisked him and pinched the key to his lock-up.

f. *lock-up book, cubicle, garage, prisoner*; **lock-up shop** (further examples).

1854 G. H. Lewes *Diary* 12 Nov. in *Geo. Eliot Lett.* (1955) V. 123 Bought Polly a lock-up book for her Autobiography]. 1910 *Bradshaw's Railway Guide* Apr. 1019 Most up-to-date Motor Garage, with 20 lock-up cubicles. 1935 *Archit. Rev.* LXXVIII. 168/1 A general garage and a few garages..let lock-up garages. 1963 *Times* 21 Feb. 8/7 The rent of all council houses and lock-up garages provided by Maidstone Town Council is to be increased by 12½ per cent. 1846 D. Corcoran *Pickings* 33 To the right of the column we perceived a prisoner whom we at once knew was above and beyond the ordinary class of lock-up prisoners. 1906 *Daily Chron.* 10 Dec. 5/7 Many people patrolled the district in which Fell's warehouse and Beardwood's lock-up shop are situated. 1947 Lock-up shop [see *amusement 7].

loco, *sb.*^{1} Add: **a.** *loco-plant, loco-weed*: (earlier and later examples).

1879 *Special Rep. U.S. Dept. Agric.* No. 12. 211 The losses among cattle, caused by eating the poisonous loco weed, will perhaps not exceed 1 per cent. 1884 *Amer. Naturalist* XVIII. 1148 Experiments..prove that *Crotalaria sagittalis*, the Rattle-box, is a 'loco-plant'. 1904 'O. Henry' in *McClure's Mag.* Apr. 617/1 If you have ever seen a horse that has eaten loco-weed you will understand what I mean when I say that the passengers get locoed. 1948 *Miami* (Okla.) *Daily News-Rec.* 30 June 8/2 Little is heard today of the once troublesome loco weed. 1955 W. Foster-Harris *Look of Old West* ix. 260 The most famous of the lethal stuff is undoubtedly loco, or crazy, weed... Loco grows all over the West, and a locoed horse is easy to spot.

c. *loco weed* = *Marijuana, Marihuana.

1935 A. J. Pollock *Underworld Speaks* 72/2 *Loco weed*, mariahuana; hemp; hashish. 1960 *Time* 25 Jan. 87/2 In U.S. slang marijuana is called..loco weed. 1972 *Sunday Sun* (Brisbane) 2 July 14/3 Detectives from the CIB Drug Squad in Brisbane are becoming quite familiar now with words like..rope and locoweed.

loco, *sb.*^{3} Add: (Earlier and later examples.)

1833 S. Breck *Recoll.* (1877) App. 274 With the *loco*.. he may start from one city in the morning and return again in the evening. 1869 *Bradshaw's Railway Manual* XXI. 14 Supt. of Loco. Dept., C. K. Domville, Belfast. 1955 *Times* 28 May 10/2 The last named company received important contracts including one for 94 diesel electric locos for the Irish State Transport. 1974 A. MacLean *Breakheart Pass* iv. 63 To haul this heavy load with a single loco?.. Thirty hours, I'd say.

b. lo·co-spotting, train-spotting; the action of noting the numbers (and sometimes other details) of locomotives seen; so **loco-spotter,** (as back-formation) **loco-spot** *v.*

1959 *Junior Radio Times* 25 Sept. 1/1 What is the loco-spotter looking for? Chiefly the engine number, which normally is painted on the cab side and also on the smoke-box door; secondly, the name if the engine has one; and thirdly, the code of the shed to which the engine is allocated. *Ibid.*, One of the objects of locospotting is to see—or 'cop'—all the engines in a particular class, marking off the number of each engine as it is observed. 1960 W. E. Hildick *Boy at Window* xvii. 131 It was a train the boy remembered well from his loco-spotting days. 1968 *Listener* 21 Mar. 368/2, I loco-spotted 45076..and 45254. 1971 *Where* Dec. 365/3 They..for many years ran the Loco Spotting Club.

loco (lōᵘ·ko), *a.* orig. *U.S.* [a. Sp. *loco* (see Loco^1). Cf. Locoed *ppl. a.*] Mad, insane, off one's head.

1887 *Outing* X. 7/1 You won't be able to do nuthin' with 'em, sir; they'll go plumb loco. 1904 Conrad *Nostromo* I. vi. 37 He was old, ugly, learned—and a little 'loco'—mad, if not a bit of a sorcerer. 1910 C. E. Mulford *Hopalong Cassidy* iv. 38 Are you loco? Do you mean to let th' rest of th' outfit see that? 1922 *Chambers's Jrnl.* Mar. 167/2 Some of them would be loco over it. 1929 Ade *Let.* 8 Feb. (1973) 139 We have gone a little loco on shopping, because..prices seem low. 1934 R. Macaulay *Going Abroad* i. 13 The young people were, so far as anyone could judge, completely loco. 1965 D. Francis *Odds Against* 124 He'd been quietly going loco and making hopeless decisions. 1973 'A. Hall' *Tango Briefing* iv. 46 You heard of ergot?.. There was a case in France, remember? Half a village went loco.

loco (lōᵘ·ko), *adv.* *Mus.* [It. *al loco*, at the place.] (See quots.)

1801 Busby *Dict. Mus.*, *Loco*,..a word in opposition to *8va Alta*, and signifying that the notes over which it is placed are not to be played an octave higher, but just as they are written. 1970 *Oxf. Compan. Mus.* (ed. 10) 578/2 *Loco*.., 'place', used after some sign indicating performance an octave higher or lower than written and reminding the performer that the effect of that sign now terminates. Often the expression used is *Al loco*, 'at the place'.

loco-descriptive, *a.* (Later example.)

1966 *English Studies* XLVII. 68 Loco-descriptive poems of and guides to the Lakes.

locoman^2. [Loco^3.] A driver of a locomotive; an engine-driver.

1941 *Penguin New Writing* XII. 9 We loco-men carry on whatever the conditions overhead. 1970 *Daily Mail* 3 Mar. 1/2 Locomen at King's Cross plan a one-day stoppage today which will hit services to the North and some commuter trains. 1972 *Guardian* 14 Aug. 1/1 The locomen's union insists that drivers should be paid more money.

locomobile. Add: **b.** *sb.* (Examples.) Also *attrib.*

1895 W. R. Fisher *Schlich's Man. Forestry* V. 748 The elevator and macerating cylinder are driven by a locomobile. 1900 *Sci. Amer.* 27 Jan. 54/1 The steam carriage which is popularly and commercially known as the 'Locomobile'. 1900 *Sun* (N.Y.) 23 May 7/6 A locomobile operator was arraigned for driving a locomobile, which is a steam automobile. 1915 *Lit. Digest* 21 Aug. 387/2 Goodyear Cord Tires... Adopted for the new Locomobile as

standard equipment. *a* 1936 Kipling *Something of Myself* (1937) vii. 177, I bought me a steam-car called a 'Locomobile', whose nature and attributes I faithfully drew in a tale called 'Steam Tactics'. 1962 R. B. Fuller *Epic Poem on Industrialization* 169 No, the ephemeralization Of doing more with less, Took gold along with tonnage And three-ton Locomobiles.

locomote, *v.* (Earlier and later examples.)

1834 *Knickerbocker* IV. 20 Who but our author would represent him [*sc.* a bard], 'locomoting' on a long, dog-trot over the bogs of his neighborhood. 1894 *Proc. R. Soc.* LV. 163 They [*sc.* the leucocytes]..do not locomote over the floor of the counter. 1970 *Amat. Photographer* 22 Apr. 9/3 That foot..pointing daintily downward; except for a prima ballerina I've never yet seen a dame who could locomote like that. 1974 *Nature* 15 Mar. 240/1 Colchicine and vinblastine caused the ruffling activity of fibroblasts locomoting in culture to spread from a restricted area to all parts of the edge of the cells.

locomotive. Add: **B.** *sb.* **1.** (Later examples.)

1959 E. K. Wenlock *Kitchin's Road Transport Law* (ed. 12) 26/1 No one under 21 is allowed to drive a locomotive, motor tractor or heavy motor car on a road. 1971 *Morning Star* 3 July 3/3 More classes of vehicles will be able to use motorways... They include 'locomotives', which are load-carrying vehicles weighing more than 7½ tons.

c. *U.S. slang.* A cheer. Also *attrib.*

1901 *Princeton Alumni Weekly* 131/2 But he saw you trying to join in a locomotive cheer last Saturday. 1907 *Ibid.* 321/2 The boys gave a rousing locomotive and then stood in silence. 1961 Webster, *Locomotive*, a cheer characterized by a slow beginning and a progressive increase in speed and used esp. at school and college sports events.

4. *locomotive works.*

1848 *Mass. Private & Special Statutes* 13 Mar., A corporation, by the name of the Boston Locomotive Works, for the purpose of manufacturing locomotive engines. 1889 G. Findlay *Working & Managem. Eng. Railway* vii. 118 Crewe, which previous to the establishment of the locomotive works was inhabited only by a few farmers and cottagers, has now developed into a flourishing town. 1966 G. F. Allen *Brit. Rail after Beeching* xii. 357 Of the Southern Region's locomotive works, Brighton had already been shut down and Ashford (Kent) had been slated for closure.

loco-move, *v.* (Later example.)

1873 Leland *Egyptian Sk.-Bk.* 88, I only remember one instance when a man who made locomotion his business was unwilling to locomove.

locoum (lōᵘ·kŏm). Also **locum, lokoum, loukoum**(i. [Turk. *lokum*.] Turkish delight.

1887 F. M. Crawford *Paul Patoff* III. xxiii. 239 Two little white saucers filled with pieces of loukoum-rahat, the Turkish national sweetmeat, commonly called by school-boys fig-paste. 1894 [see *delight sb. 4]. 1913 *Chambers's Jrnl.* Apr. 313/2 The elemes and locoums, packed respectively in layers and in cubes. 1921 S. Graham *Europe* ii. 38 Pride intervenes only to stop them begging... But you see the nicest of girls with pinched white faces trying to sell *loukoum*. 1960 *Spectator* 8 Jan. 51/1 In Istanbul,..go for Turkish delight,..a quite different confection from what passes as Turkish delight in Britain. You ask for *lokoum*. 1962 J. Fleming *When I grow Rich* iv. 53 A little *locum* for a friend..a present of a little Turkish delight. 1967 *Vogue* Jan. 4/2 A coral crusted dish of loukoumi. 1970 Simon & Howe *Dict. Gastron.*, Turkish delight, the popular English name of Turkey's sweetmeat *rahat lokum*. 1972 J. Rathbone *Trip Trap* vi. 65 A servant poured Turkish coffee; *lokoum* was offered in tiny silver dishes.

locule. = Loculus 2. (Later examples.)

1953 *New Biol.* XIV. 44 The cotton fruit, called a boll, is divided into three or four locules, each with several seeds. 1967 M. E. Hale *Biol. Lichens* ii. 28 The locules in the stroma are separated by branched pseudoparaphyses.

locum. Add: **b.** Short for Locum-Tenency.

1903 *Lancet* 9 May 101/2 (Advt.), Hospital Locum wanted..for three weeks or less. 1946 *Ibid.* 2 Mar. 322/2 When doing a locum I attended a family of actors.

locum tenens. Add: **b.** The post of a locum tenens; a locum-tenency.

1899 *Lancet* 5 Aug. 86/2 (Advt.), Locum Tenens or good Assistantship by doubly-qualified man. 1908 A. S. M. Hutchinson *Once aboard Lugger* vi. vi. § 2. 437 There's this locum tenens I was going to take up in the North.

locum-tenent. Add: **b.** = Locum tenens.

1899 *Lancet* 19 Aug. 547/1, I met with a serious accident ..in consequence of which I had to engage a locum-tenent. *Ibid.*, This sort of thing should make men very careful as to locum-tenents before engaging them.

locus, *sb.*^{1} Add: **1. b.** *Genetics.* A site or position on a chromosome at which a particular gene is located; *loosely*, a gene.

1913 *Jrnl. Exper. Zool.* XV. 591 White and eosin are allelomorphic to each other, that is, they occupy the same locus in the sex chromosome. 1915 T. H. Morgan et al. *Mechanism Mendelian Heredity* ii. 37 There are three pink eye colors in Drosophila, one whose locus is in the third chromosome (pink). *Ibid.* vii. 155 A mutant factor is located at a definite point in a particular chromosome; its normal allelomorph is supposed to occupy a corresponding position (locus) in the homologous chromosome. 1919 *Anatomical Rec.* XV. 358 In another case of duplication the duplication piece contains only the locus for sable as far as known. 1949 [see *allelomorph]. 1962 *Lancet* 6 Jan. 10/1 The colour-blind locus is thought to be about 10 units of crossing-over from the locus for hæmophilia and about 25 units from the locus for Duchenne's type of muscular dystrophy. 1970 *Nature* 25 July 342/1 Considerable numbers of gene loci are required to code for the primary structures of the immunoglobulin molecules made in any one organism. 1971 *Ibid.* 13 Aug. 498/1 Haemoglobin type in sheep is controlled by a pair of alleles at a single locus.

c. *locus of control* (Psychol.) (see quot. 1972^1).

1966 Mandler & Watson in C. D. Spielberger *Anxiety & Behavior* 286 A locus of control scale has been developed which differentiates individuals according to the degree to which they appraise themselves or the environment to control the occurrence of reinforcement. 1971 *Jrnl. Gen. Psychol.* LXXXV. 98 There is movement from external to internal locus of control from trial to trial. 1972 I. G. Sarason *Personality* (rev. ed.) i. 9 Locus of control refers to the degree to which an individual sees himself in control of his life and the events that influence it. 1972 *Jrnl. Social Psychol.* LXXXVI. 233 The work of Rotter and his associates on perceived locus of control has resulted in a considerable body of evidence.

4. locus classicus (earlier example); **locus communis** (later example); **locus desperatus** (see quot. 1966); **locus standi** (later examples).

1853 Bagehot *Coll. Works* (1965) I. 202 These lines are, as it were, the *locus classicus* of fairy literature. 1843 Mill *Logic* II. v. ii. 339 *Loci communes* of bad arguments on some particular subject. 1922 F. Klaeber *Beowulf* 214 This passage remains..a 'locus desperatus'. 1966 A. J. Bliss *Dict. Foreign Words & Phrases Current Eng.* 231 *Locus desperatus*, a passage in a text transmitted by manuscript whose meaning is so corrupt as to be almost beyond conjecture. 1969 *English Studies* p. lxxv, The result is an editorial 'locus desperatus' which still to some extent defies repeated scrutiny and modern photographic aids. 1969 R. Renehan *Greek Textual Crit.* 2 The textual critic..must decide in each case whether the original reading..has been or can be recovered by modern conjecture or whether the passage is a *locus desperatus*. 1970 *Anglia* LXXXVIII. 367 Faced with such a *locus desperatus*, even a conscientious editor might decline to grapple afresh with the battered folio. 1911 J. Ward *Realm of Ends* x. 212 Death..means that the soul in consequence, so far as it is thus deprived of its *locus standi*, is..in the position of a deserter from the general order. 1974 *Times* 9 Feb. 20 The power of the Department [of Trade and Industry] should, of course, be discretionary but the Panel should be given a *locus standi* with the Department.

locust, *sb.* Add: **5. b.** (Earlier and later examples.)

1863 D. M. Barnes *Draft Riots N.Y.* 82 Go in they did forthwith, and, where moral suasion had failed, the locusts succeeded. 1865 G. A. Sala *My Diary in Amer.* II. 211 The New York policeman wears a handsome uniform. At his side hangs a club or bludgeon... This club is made of 'locust wood'..and by rowdies the policeman is often generically called..a 'locust'. 1904 *N.Y. Tribune* 19 June 4 The policemen did not carry their 'locusts'. 1930 E. H. Lavine *Third Degree* 78 A detective picked out the largest and heaviest locust in the group.

6. locust-bean (later examples); **locust-borer** (examples); **locust stick** = *locust club*; **locust wood,** the wood of a locust-tree; **locust years,** years of poverty and hardship (see also quot. 1962^1).

1958 L. Durrell *Balthazar* ii. 32 He would pick a stick of sugar-cane off a stall as he passed..or a sweet locust-bean. 1972 *Country Life* 30 Nov. 1481/1 Locust beans don't attack the teeth as jube-jubes did. 1839 H. Colman *2nd Rep. Agric. Mass.* (Mass. Agric. Survey) 100 Locust-Borer... [He] washed his locust trees with spirits of turpentine, and in that way compelled the borer to leave them. 1972 Swan & Papp *Common Insects N. Amer.* 448 Locust borer: *Megacyllene robiniae*... The larvae bore in the sapwood of black locust. 1919 Wodehouse *Coming of Bill* (1920) I. i. 15 The policeman..relieved his feelings by dispersing the crowd with well-directed prods of his locust stick. 1742 W. Ellis *Timber-Tree Improved* II. xxxii. 166 Where the Natives can't get Locust-wood, they use this to make their Bows. 1874 *Rep. Vermont Board Agric.* II. 777 *Clytus robiniae*. The larvae feed upon locust wood. [1611 Bible Joel ii. 25 And I will restore to you the yeeres that the locust hath eaten.] 1948 W. S. Churchill *Second World War* I. i. v. 52 (heading) The Locust Years, 1931–1935. 1962 *Listener* 19 July 107/3 Sir Winston Churchill applied the phrase, the locust years, to the middle thirties, when vigorous rearmament should have begun. 1962 W. McElwee (title) Britain's locust years, 1918–1940. 1964 P. Magnus *King Edward VII* xiii. 244 (heading) Locust years. 1970 *Times* 27 May 8 Yet before these locust years of Labour, we had the Conservative years of rising prosperity.

locustal (lokv·stăl), *a.* [f. Locust *sb.* + -al.] Of, pertaining to, or connected with locusts.

1891 *Chambers's Encycl.* VII. 187/1 Temperature may also have something to do with locustal migrations.

locustarian. (Example.)

1895 *Nature* 5 Dec. 108/1 Mr. Scudder..has given much attention to the sounds made by locustarians.

locutionary (lokiū·ʃənări), *a.* *Philos.* [f. Locution + -ary^1.] Of or pertaining to an utterance by a speaker. Cf. *Illocutionary *a.*

1955 J. L. Austin *How to do Things with Words* (1962) viii. 94 The act of 'saying something'..I call, i.e. dub, the performance of a locutionary act. 1962 *Times Lit. Suppl.* 21 Sept. 743/2 A locutionary act is the speaker's act of saying whatever it is he says. 1973 *Ibid.* 5 Oct. 1161/5 The locutionary act was the act of saying something—i e, the act of uttering sounds as constituting a sentence.

lodberry, var. *Loadberry.

Loddon (lǫ·dən). The name of a tributary of the River Thames, used *attrib.*, esp. in **Loddon lily,** to designate the summer snowflake, *Leucojum æstivum*, a small, white-flowered, bulbous plant once common on the banks of this river.

1882 *Dickens's Dict. Thames* 28/3 It [*sc.* the summer snowflake] is very abundant in the meadows by the Loddon, and hence called 'Loddon lilies'. 1938 R. Gathorne-Hardy *Wild Flowers Brit.* viii. 53 The Loddon Lily is to be found principally on the tributaries and main river of the Thames Valley. 1971 *Country Life* 2 Sept. 575/1 On the banks of the River Loddon linger a few 2 ft.-high clumps of the snowy-white Loddon lily, or summer snowflake, *Leucojum aestivum*. 1973 *Times* 15 Dec. 10/7 The Loddon or Summer Snowflake Leucojum aestivum used to grow in great numbers on the banks of the Loddon near Reading. It has now become quite scarce.

lode. Add: **6. lode-light,** a light said to be seen sometimes above a vein of ore.

1883 *Encycl. Brit.* XVI. 443/1 The appearance of the so-called lode-lights may be explained by the production of phosphoretted hydrogen. 1894 C. le N. Foster *Ore & Stone Mining* 107 Appearances of flame above mineral veins..are sufficiently well established to have received a special name 'lode lights' in Cornwall.

loden (lōᵘ·dən). [a. G. *loden* thick woollen cloth.] A heavy waterproof woollen cloth. Used *attrib.* to designate garments made of this material, as *loden cloak, cloth, coat, jacket, mantle, skirt*; also *absol.* Also, a dark green.

1911 Galsworthy *Little Dream* in *Plays* (1929) 204 There enters a lean, well-built, taciturn young man dressed in Loden. 1914 G. Atherton *Perch of Devil* II. 354 She..wrapped herself in a dark lodenmantle, a long cape with a hood that she had worn..in Bavaria. 1916 J. Buchan *Greenmantle* vii. 98 Long shooting capes made of a green stuff they call *loden*. *Ibid.* xv. 196 Blue jeans, loden cloak. 1920 D. H. Lawrence *Women in Love* xxix. 450 The two..daughters of the professor, with their plain-cut, dark blue blouses and loden skirts. 1951 V. Nabokov *Speak, Memory* iv. 61 He wore an ulster unless the weather was very mild, when he would switch to a kind of greenish-brown woollen cloak called a *loden*. 1952 *New Yorker* 13 Dec. 128/2 A rugged coat for the country, made in Austria of greenish or brownish loden cloth and cut something like a hunting coat. 1956 *San Francisco Examiner* 9 Sept. I. 21 (Advt.), The original Alpine Lodencoat. 1957 *Times* 25 Nov. 11/1 This coat is reversible, in loden cloth and water-proofed poplin. 1964 *N.Y. Post* 4 Nov. 11 Russ Togs in black, brown, loden, navy, and menswear grey. 1966 *Listener* 3 Nov. 641/1 People dressed in green Loden jackets. 1969 R. T. Wilcox *Dict. Costume* (1970) 198/1 *Loden*, a waterproof cloth resembling Irish frieze, made by the Tyrolean peasants from the wool of their mountain sheep. It is woven and dyed in several colors but especially a bluish green known as loden green. 1973 *Guardian* 10 Apr. 13/1 The greatest source of inspiration is the traditional Loden coat.

lodge, *sb.* Add: **4. b.** A residence or hotel. (Freq. as the second element of house- or hotel-names.)

1818 Jane Austen *Persuasion* III. ii. 31 As to her young friend's health, by passing all the warm months with her at Kellynch-lodge, every danger would be avoided. *Ibid.* v. 80 Anne walked up..to the Lodge, where she was to spend the first week. 1854 Dickens *Hard Times* I. iii. 12 To his matter of fact home, which was called Stone Lodge, Mr. Gradgrind directed his steps. He had..built Stone Lodge. 1869 *Bradshaw's Railway Manual* XXI. 307 Directors... W. C. Stobart Esq., Etherley Lodge, near Bishop Auckland. 1953 A. Christie *Pocket Full of Rye* iv. 24 Call it a lodge, indeed! Yew Tree Lodge!.. The house was what he..would call a mansion. 1971 *Author* LXXXII. 173 In hotels and auto lodges he listens to many a late-night argument. 1972 *Automobile Assoc. Members' Handbk.* 1972–73 154/1 Linton Lodge, Linton Road [Oxford].

7. (Later examples.)

1969 in Halpert & Story *Christmas Mumming in Newfoundland* 181 The (Protestant) Society of United Fishermen..soon had no fewer than forty-two lodges. 1970 *Britain: Official Handbk.* (H.M.S.O.) xvi. 428 The basic unit of organisation in most British trade unions is the local branch (sometimes called a lodge). 1974 *Socialist Worker* 7 Dec. 8/3 And there were more than 30 other workplace units—such as chapels and lodges—represented.

15. (sense 7) *lodge-meeting, official, room*; (sense 10) *lodge-cover, -covering, -fire, -skin, -trail.*

1878 J. H. Beadle *Western Wilds* ix. 137 The former [*sc.* buffaloes] furnished them with food, clothing, lodge-covers,..and a dozen other conveniences. 1847 F. Parkman in *Knickerbocker* XXX. 234 The squaws of each lazy warrior had made him a shelter..by stretching..the corner of a lodge-covering upon poles. 1837 W. Irving *Capt. Bonneville* I. 111 Knots of gamblers will assemble before one of their lodge fires, early in the evening. 1903 C. T. Brady *Bishop* iii. 47 Most of the Churches have a week-night prayer-meeting, and the other nights are taken up with lodge meetings. 1926 *Scribner's Mag.* Sept. 327/2 A lodge meeting to the average negro is one of the big events of the week. 1933 J. Buchan *Prince of Captivity* II. i. 147 The weekly lodge meetings. 1909 *Daily Chron.* 30 Dec. 1/4 The fifteen lodge officials and delegates prosecuted for offences against the Industrial Disputes Act. 1974 *Times* 6 Dec. 3/1 All lodge officials are completely vindicated. 1911 J. C. Lincoln *Cap'n Warren's Wards* vi. 88 I'm more used to lodge rooms than I am to clubs. *a* 1831 J. Smith *Jrnl.* in M. S. Sullivan *Trav. J. Smith* (1934) 4 They [*sc.* Indian lodges] do not smoke except from a sudden change of wind and then no longer than it takes a squaw to spread a smoke wing of the Lodge skin. **1891** *Century Mag.* Mar. 776 We had already devoured..a small sack made of smoked lodge skin. **1845** J. C. Frémont *Rep. Exploring Expedition* 114 We resumed our journey.. following an extremely good lodge-trail.

lodgement. Add: **1. d.** *Mining* = Lodge *sb.* 13 a.
1883 W. S. Gresley *Gloss. Terms Coal Mining* 159 *Lodgment* (S[cotland]), see *sump* and *lodge*. **1886** J. Barrowman *Gloss. Scottish Mining Terms* 43 *Lodgment*, a reservoir or storage place underground for water for convenience of pumping.
5. lodgement-level (see quot. 1877).
1877 *Encycl. Brit.* VI. 63/2 Driving a gallery..along the course of the coal seam, which is known as a 'dip head level', and a lower parallel one, in which the water collects, known as a 'lodgment level'. **1886** J. Barrowman *Gloss. Scottish Mining Terms* 43 *Lodgment-level*, a room driven level course at a short distance to the dip of a pit and used for storage of water.

lodge-pole, *sb. N. Amer.* **1.** (See Lodge *sb.* 15.)
1805 M. Lewis in Lewis & Clark *Orig. Jrnls. Lewis & Clark Expedition* (1904) II. 88 Found a new indian lodge pole today. **1855, 1865** [in Dict. s.v. Lodge *sb.* 15]. **1903** A. Adams *Log of Cowboy* xxi. 330 He..with The Rebel went back about a mile to a thicket of lodge poles. **1946** G. Foreman *Last Trek of Indians* 242 They pulled down the lodges of three of their villages and sold the lodgepoles and lumber.
2. lodge-pole pine, a pine native to mountainous regions of north-west America, *Pinus contorta* var. *latifolia*.
1859 G. A. Jackson *Diary* 9 Jan. in *Colorado Mag.* (1935) XII. 205 Cut the top off a small lodge pole pine. **1884** C. S. Sargent *Rep. Forests N. Amer.* 564 The forests largely composed of the lodge-pole pine..cover the outlying eastern ranges of the Rocky Mountains. **1905** *N.Y. Even. Post* 29 Apr., The lodgepole pine..bears the common name of 'lodgepole' from the fact that the Indians used its long slender trunks as supports for their wigwams, or lodges. **1949** *Sierra Club Bull.* (San Francisco) June 6 Lodgepole pines, singing birds and scampering chipmunks, ..all familiar elements of the mountain scenes. **1964** *N.Z. News* 24 Nov. 3/1 The lodgepole pine, becoming common on the Waiouru Plains and a cause of concern because of its possible spread in the South Island, is in fact a favoured species for pulping in U.S. **1972** *Islander* (Victoria, B.C.) 23 Jan. 6/1 Amongst the lodge-pole pines and blue spruce. **1972** *Country Life* 7 Dec. 1559/1 Foresters in Scotland can be criticized for planting an excess of sitka and Canadian lodgepole pine.

lo·dge-pole, *v. U.S.* [f. prec.] *trans.* To beat with a lodge-pole. Hence **lodge-poling** *vbl. sb.*
1848 G. F. A. Ruxton in *Blackw. Mag.* LXIV. 139 Nor are they [*sc.* squaws] so schooled to perfect obedience to their lords and masters as to stand a 'lodge poling'. **1850** L. H. Garrard *Wah-to-Yah* (1927) ix. 116 Often..their negligent spouses are lodge-poled (beaten) for such accidents. **1968** *Amer. Speech* XLIII. 218 It was customary to discipline a squaw by *lodge-poling* her, and the term became figurative for any beating.

lodgerdom (lǫ·dʒəɹdəm). [f. Lodger + -dom.] Lodgers collectively; the world of lodgers; a district in which lodgers are common.
1905 *Daily Chron.* 6 Mar. 4/6 Even dingy Lodgerdom would disclaim the place. **1907** *Ibid.* 23 May 3/3 A very pleasant, humorous-pathetic story of lodgerdom. **1927** *Observer* 14 Aug. 6 Discomfort..that goes with the bondage of..lodgerdom.

lodging, *vbl. sb.* Add: **6. lodging-hall** *U.S.*, a lodging-house; **lodging paper,** a handbill advertising lodgings; **lodging turn,** an occasion or period for which a railway employee has to lodge at his place of destination before returning to his place of departure.
1860 J. G. Holland *Miss Gilbert's Career* xii. 208 We left Arthur Blague..sitting on his bed in the lodging-hall. **1817** Jane Austen *Sanditon* vii, in *Minor Works* (1954) 402 No fewer than three Lodging Papers staring me in the face at this very moment. **1952** *Ann. Reg. 1951* 15 Economy measures, some of which (such as more lodging turns) had caused serious strikes. **1955** *Ann. Reg. 1954* 30 About two-thirds of the..footplate men..came out on strike against the introduction of new 'lodging turns', i.e. nights spent, usually in railway hostels, away from home.

lodging-room. (Later U.S. examples.)
1849 *Ex. Doc. 31st U.S. Congress 1 Sess. House* No. 5. II. 1089 One hewed-log lodging-room for hired men. **1906** *Springfield* (Mass.) *Republ.* 7 Feb. 2 Lodging Rooms to Let.

lodicle, var. Lodicule (in Dict. and Suppl.).
1888 *Encycl. Brit.* XXIV. 531/2 Within the pale[a] are two minute, ovate, pointed, white membranous scales called 'lodicles'.

lodicule. Substitute for def.: A green or white scale, the lowest part of a grass flower. (Later example.)
1968 F. W. Gould *Grass Systematics* ii. 55 Lodicules play a role in the opening of the flower at anthesis.

loerie (lu·ri). *S. Afr.* Also **loorie, lourie.** [Afrikaans, f. Du. *lori* Lory.] = Touraco.
1798 A. Barnard *Jrnl.* in A. W. C. Lindsay *Lives of Lindsays* (1849) III. 408, I began to collect my Cape trifles for my friends at home,—some beautiful loories alive—some still more beautiful swallows dead. **1810** W. J. Burchell *Jrnl.* 7 Dec. in *Trav. S. Afr.* (1822) I. ii. 20 In the aviary, I saw the *Touracoo*, called *Loeri* by the colonists. **1812** [see Lory]. **1850** J. S. Christopher *Natal* 33 The beautiful and soft-voiced loerie, the golden cuckoo, the green pigeon..and many others too numerous to particularise. [**1908** *East London* (Cape Province) *Dispatch* 4 Dec. 4 The *vlei-lourie*, perhaps better known hereabouts as the 'rain-bird', the natives regarding it as a weather prophet.] **1932** *Discovery* July 230/2 The Louries, magnificent in green, blue and carmine..nest outside the door of the mission [at Kilimanjaro]. **1950** *Cape Argus Mag.* 18 Mar. 7/7 Loeries with their beautiful crimson-and-green plumage fill the air with their liquid call. **1957** V. W. Turner *Schism & Continuity in Afr. Soc.* x. 293 The woman may dream that her dead relative has appeared to her equipped like a hunter, wearing the red wing-feather of a lourie in her hair. **1971** *Evening Post* (Port Elizabeth, S.A.) *Suppl.* 12 June 5 When we think of birds in general, it is the pleasanter members of that enormous family that come to mind—the brilliantly-plumaged sunbird, the dove, the loerie.

loess. Delete 'found in the valley of the Rhine and of other large rivers' and substitute: which occurs extensively from north-central Europe to eastern China, in the American mid-west, and elsewhere, esp. in the basins of large rivers, and which is usually considered to be composed of material transported by the wind during and after the Glacial Period. (Add further examples.)
1882 F. Richtofen in *Geol. Mag.* IX. 302, I believe I am correct in stating that, among those who have had extensive experience in Loess regions, all who have pronounced an opinion of late years are agreed that subaërial deposition is the only mode of origin by which all its peculiar features can be easily explained. **1906** *Westm. Gaz.* 4 Dec. 2/1 North of the river [*sc.* the Hwang Ho] we come into the land of the loess, a loose light soil of prodigious fertility and the joy of the agriculturist. **1938** C. L. Whittles tr. *Reifenberg's Soils Palestine* ii. 25 Microscopic examination shows that loess consists mainly of extremely small particles of quartz together with calcareous particles which frequently, from their markings, etc., have been derived from fossils. **1972** J. G. Cruickshank *Soil Geogr.* ii. 59 During the deglaciation phases of the Pleistocene glaciations, wind-blown silt was deposited on a spectacular scale in extensive, stoneless loess.

loessial (lö·siăl), *a.* [f. Loess + -ial.] Composed of loess.
1928 *Bull. Amer. Soil Survey Assoc.* IX. 34 These [*sc.* glacial soils] include the till, moraine, drumlin..and other typical forms and a portion of the loessial deposits. **1974** *Nature* 22 Mar. 320/2 Silt, in excess of that which could be derived from weathering of the substrata and generally considered loessial, is found in many British soils.

loessic (lö·sik), *a.* [f. Loess + -ic.] = *loessial *a.*
1909 in *Cent. Dict.* Suppl. **1940** *Nature* 6 July 14/1 In periglacial regions such arid episodes are represented by loessic deposits. **1952** F. E. Zeuner *Dating Past* (ed. 3) vi. 158 The Middle Older Loess of the section is a complex of loessic hillwash material derived from higher up the slope, and of brecciated loess with large molluscan shells, interrupted by a brown soil.

lo-fi (lōu·fəi·). *colloq.* [Repr. *low fidelity*, after *hi-fi.] Sound reproduction less good in quality than 'hi-fi'. Also *attrib.* or as *adj.*
1958 *Observer* 15 June 14/6 For heaven's sake let us have the real Bob Cats, even in lo-fi, and let the record companies make their issues under that kind of policy. **1967** *Sat. Rev.* (U.S.) 29 July 53 Despite Mr. Kolodin's warning of the 'lo-fi', we would urge the purchase of this set as a significant item in Toscanini's recorded legacy. **1968** *Which?* Oct. 312/1 You can buy your hi-fi equipment one piece at a time, and play it through the 'lo-fi' parts of, say, your old radio. **1970** *Daily Tel.* 16 Mar. 24/3 It was because of the cassette's 'lo-fi' that Philips first attacked the bottom end of the market. **1970** J. Earl *Tuners & Amplifiers* v. 100 The medium-frequency a.m. system is.. possibly adequate for 'lo-fi' reception on small transistor sets.

loft, *sb.* Add: **5. d.** A place where sails are manufactured: = *sail-loft*.
1938 T. North *Yacht Sails* xii. 113 When a new sail leaves the loft it should be perfect. **1959** W. R. Bird *These are Maritimes* iii. 92 He learned his trade in his father's loft in West Pubnico... There are only two other 'lofts' in the Maritimes. **1973** *Observer* (Colour Suppl.) 3 June 18/3 His personal sails are no better than his customers', 'They go through the loft as part of the system.'
7. c. *fig.* Elevation, uplift.
1925 *Brit. Weekly* 12 Nov. 159/2 We need more loft in our thinking than our fathers had.
8. loft-bombing (see quot. 1956).
1956 *Time* 24 Sept. 36 Its [*sc.* a low-flying fighter-bomber's] bombing can be made extremely accurate, but if it uses any ordinary bombing system, such as dive-bombing, it is apt to be vaporized by the fireball springing up under its tail. The best way to avoid this misadventure is 'loft-bombing', which uses the speed of the airplane to make the bomb behave like an artillery shell. *Ibid.*, The main advantage of loft-bombing..is not the range of the bomb, but the time that it spends in the air while the airplane is making its get-away. **1960** *Aeroplane* XCIX. 352/2 The first L.A.B.S. manœuvre was completed by an A3D, in which loft bombing had been pioneered by Cmdr. H. F. Lang.

loft, *v.* Add: **2.** (Later example.)
1951 E. Sewell in *Durkett's Reg.* July 105 Blenheims will keep till Christmas, if lofted cool and dry.
3. Also in other games.
1927 [see *infielder 2]. **1950** W. Hammond *Cricketers' School* facing p. 96 W. Hammond hits a 6; position correct for lofting the ball over mid-on. **1963** *Times* 8 June 4/2 Soon afterwards Hunte lofted Allen over mid-off for four, before Allen for the second time in the day, had the last word with a batsman trying to attack him. **1970** *Globe & Mail* (Toronto) 28 Sept. 18/6 Rookie Paul McKay lofted the final Hamilton punt, a high 47-yard spiral. **1972** J. Mosedale *Football* vi. 91 Tittle lofted the ball..over the heads of the other players. **1974** *News & Reporter* (Chester, S. Carolina) 22 Apr. 10-A/6 Guy Meadow lofted a sacrifice fly and Clayton, up for the second time in the inning, singled again.
b. *transf.* and *fig.*
1883 J. Martine *Reminisc. Haddington* 120 He [*sc.* a goat]..thought nothing of pouting and 'lafting' folk. **1902** Barrie *Little White Bird* xxiv. 282 We had lofted him out of the story, and did very well without him. **1948** G. H. Johnston *Death takes Small Bites* v. 110 Her eyes, bright with strain, lofted above the murals to the great range behind the town. **1952** [see *day-beam* s.v. *day *sb.* 23 a, b]. **1960** L. P. Gartner *Jewish Immigrant* ix. 251 The banner of the Hebrew national..renaissance was lofted. **1971** D. Meiring *Wall of Glass* vii. 57 He would..loft his bag up the hotel stairs himself..instead of handing it to the orderly.

lofted, *ppl. a.* Add: **2. c.** Of a ball: hit into the air. Also in other games.
1904 *Daily Chron.* 20 Aug. 9/5, I saw a lofted ball..miss the head of a player in front by not more than six inches. **1955** *Times* 29 June 4/3 He made the winning hit, a lofted straight four, on the stroke of time. **1963** *Times* 27 May 5/3 The same player scored from a 30-yard penalty, while Ramsden replied with a well lofted hit from 40 yards.

lofting, *vbl. sb.* Add: **2*.** *Aeronaut.* The action of a loftsman.
1939 *Jrnl. R. Aeronaut. Soc.* XLIII. 140, I should like to mention two or three operations coming under the control of the engineering department... One of these is the mould loft. The art of lofting as it is now generally followed in American aircraft production was, of course, borrowed from the ship-building industry. **1956** W. A. Heflin *U.S. Air Force Dict.* 304/1 *Lofting*, as applied to airplanes, the act or process of laying out full-size drawings of an airplane that is to be built, of designing and making templates, etc. **1972** *Lebende Sprachen* XVII. 73/2 The production of aircraft part drawings, preparation of magnetic tape for machine-tool control, and many lofting, drawing and checking jobs are to be automated.

lofting (lǫ·ftiŋ), *ppl. a.* [f. Loft *v.* + -ing^2.] Of a stroke in golf: that lofts.
1905 *Westm. Gaz.* 25 Aug. 3/1 Why to go for a low-running shot or for a high lofting shot, respectively.

loftsman (lǫ·ftsmæn). *Shipbuilding* and *Aeronaut.* [Cf. *mould-loft* s.v. Mould *sb.*^{3} 17.] One who reproduces a draughtsman's specifications for a ship or aircraft in full size on the floor of a mould-loft.
1901 T. Walton *Steel Ships* vii. 178 After the plans of a vessel have been prepared in the drawing office, the first man who takes her in hand is the loftsman. His domain is the mould loft, where..he proceeds to reproduce the 'lines' plan upon the loft floor to actual size. *Ibid.* 180 The loftsman's work is to rectify any..irregularities, and to produce perfect harmony and fairness in all lines which make up the form of the hull. **1909** Webster, *Loftsman*, one who lays down the lines of a ship in a shipbuilding loft. **1921** *Dict. Occup. Terms* (1927) § 668 s.v. *Scriever*, *Loftsman*, lays off from plans, on wooden floor,..outlines of ship frames and plates, [etc.]. **1947** *Aircraft Engin.* July 222/1 The loftsman would simply pass a curve pleasing to the eye through a few points determined by the designer. **1955** *Times* 13 July 2/5 Aircraft Loftsmen for the Aircraft Full-scale Layout Section. **1957** *Technology* Apr. 53/2 At a single step the loftsman will become a white collar worker, no longer grubbing about on hands and knees on a dirty floor marking off a life-size scrieve board. **1974** *Sci. Amer.* May 16/3, I worked for eight years..as lumberman, machinist, machine designer and loftsman before retiring to college.

log, *sb.*^{1} Add: **1. b.** *as easy* (or *simple*) *as falling* (or *rolling*) *off a log*.
1839 *Picayune* (New Orleans) 29 Mar. 2/2 He gradually went away from the Lubber, and won the heat, 'just as easy as falling off a log'. *c* **1880** 'Mark Twain' *Speeches* (1923) 97 A man who could have elected himself Major-General Adam or anything else as easy as rolling off a log. **1904** 'A. Dale' *Wanted: a Cook* 207 It was so easy that the inelegant simile of 'rolling off a log' impressed me as being absolutely justifiable. **1913** F. H. Burnett *T. Tembarom* xvii. 223, I dropped into it by accident,..and that made it as easy as falling off a log. **1949** N. Marsh *Swing, Brother, Swing* iv. 67 Don't keep asking if I can understand things that are as simple as falling off a log. **1973** *Times* 10 Feb. 11/3 Acting? said Ernest Borgnine. Why, there was nothing to it, really. 'For me,' he said, 'it's as easy as falling off a log.'
g. *Surfing.* (See quots.)
1967 J. Severson *Great Surfing* Gloss., *Log*, a very heavy surfboard. **1970** *Studies in English* (Univ. of Cape Town) I. 28 His board may be described as a barge or a log, both of which describe a big cumbersome surfboard, one that is difficult to manoeuvre.
4*. A piece of quarried slate before it is split into layers.

1725 D. Eaton *Let.* 13 Feb. (1971) 9 The reason why the slaters could not go on was bycause they could not run their slate out of the log for want of frost. **1939** *Evening News* 2 Feb. 8/6 When the slate is taken to the surface it is called 'log', and is then left in the 'slate-patch' to wait for the frost to break it into layers. **1946** N. Wymer *Eng. Country Crafts* x. 108 Then the props are systematically removed, and the slate is allowed to crash down, breaking up into large slabs which can be levered up and roughly broken by hammer into 'logs'. **1975** *Times* 9 Aug. 12/7 Collyweston slate is unusual in that it is produced by the action of frost on the stone logs.

6. d. Any record in which facts about the progress or performance of something are entered in the order in which they become known; e.g. (*a*) a record of what is found, or how some property varies, at successive depths in drilling a well; a graph or chart displaying this information; (*b*) a record kept by a lorry driver in which details of journeys are noted; (*c*) a record kept of what is broadcast by a radio or television station from moment to moment.
1913 *Jrnl. Geol.* XXI. 671 This company has prepared logs of various..salt wells. **1920** L. S. Panyity *Prospecting for Oil & Gas* xiii. 162 It is the duty of the driller..to keep a record or log of the well. This consists in noting the various formations drilled through, the casing points, and the showings of water, oil or gas. **1924** G. W. Grupp *Econ. Motor Transportation* ix. 187 Nothing is more interesting than..making..a motor-truck performance log. **1925** K. G. Fenelon *Econ. Road Transport* 241 A daily log prepared by the driver of each vehicle, showing the nature of the work performed, the tonnage carried, the time taken, etc. **1937** *Printers' Ink Monthly* May 39/2 *Log*, an account of every minute of broadcasting, all errors being considered. An accurate journal required by law. **1956** *Nature* 21 Jan. 120/2 The study of these continuous velocity logs in conjunction with seismic reflexion records shot from the surface is leading to a better understanding of the origin of reflexions. **1957** M. R. J. Wyllie *Fund. Electr. Log Interpretation* (ed. 2) ii. 105 Even in dirty formations the neutron log can sometimes give a fairly good estimate of porosity. *Ibid.* 110 Logs which make use of the scattering of gamma-rays to determine the density of formations penetrated by boreholes..are rapidly being improved in efficiency. **1960** J. M. Weller *Stratigr. Princ. & Pract.* xvii. 614 Electric logs consist of curves that are continuous records of self-potential and resistivity measured in wells and plotted against depth... In a general way..they indicate differences in lithologic characters of strata and many lithologic changes are shown with great precision. **1963** *Amer. Speech* XXXVIII. 44 *Log*, *log book*, the driver's daily report required by the I.C.C. **1965** W. S. Barry *Airline Managem.* x. 149 Station logs report troubles that have occurred during embarkation or disembarkation. **1968** *Radio Communication Handbk.* (ed. 4) xx. 4/2 Log Keeping. The Post Office requires all amateurs to keep a log book containing full details of all transmissions... Entries must be made at the time of operation, and no gaps should be left in the log. **1974** *Sci. Amer.* May 133/3 These men filled out sleep logs (for pay) and answered many questions.

8. (in sense 'made of logs'): *log barn, barrack, building, causeway, chapel, church, city, college, heap, kitchen, meeting-house, pen, pound, prison, room, shanty* (earlier example), *stable, tavern, tenement, wall, way* (later examples); (in sense 'for use in dealing with logs') *log skid*.
1795 *Pittsburgh Gaz.* 6 June 1/2 To be Sold..two cabins, a log barn. **1845** S. Judd *Margaret* I. iii. 12 On the east side of the road was a log barn. **1948** *Time* 11 Oct. 21/1 A country which still remembered Indians, wild turkeys, log barns, [etc.]. *a* **1861** T. Winthrop *Life in Open Air* (1863) 32 All residents of Damville dwelt in a great log-barrack. **1806** Z. M. Pike *Acct. Expedition Sources Mississippi* I. App. 36 [The fur-trading establishment] at Lower Red Cedar Lake..consists of log buildings. **1828** *Gore Gaz.* (Ancaster, Ontario) 18 Oct. 134/2 The stumps are all taken out—and the log causeways, where these are necessary—are covered with a thick coat of earth. **1831** T. Buttrick in R. G. Thwaites *Early Western Trav.* (1904) VIII. 54 In some places, in low grounds, there would be log-causeways for a considerable distance. **1810** F. Asbury *Jrnl.* (1821) III. 298 Saturday, at William Adams's log-chapel I preached to a small assembly. **1847** F. Parkman in *Knickerbocker* XXIX. 313 We found the log-church ..belonging to the Methodist Shawanoe Mission. **1895** M. A. Jackson *Mem. Stonewall Jackson* (ed. 2) 382 The little log church is..full. **1817** S. R. Brown *Western Gaz.* 106 Vangeville,—A log city..has fifteen or twenty old log houses. **1795** P. Freneau *Poems* 374 On the Demolition of a Log-College. **1850** W. H. Foote *Sk. Virginia* 349 Could we..look into the school of the worthy pastor, then gaining its eminence as 'a log college'. **1818** L. D. Clarke in *Firelands Pioneer* (1920) XXI. 2324, I spread ashes where log-heaps had been burned. **1819** E. Dana *Geogr. Sk. Western Country* 36 The Creoles never having before smelted, except by throwing the ore into log heaps. **1856** A. Cary *Married* 295 Having made a log-heap fire, Martin put the table-cloth about his shoulders. **1933** E. C. Guillet *Early Life Upper Canada* 277 In new settlements during July the whole countryside was illuminated by the burning of log heaps. **1874** E. Eggleston *Circuit Rider* v. 56 The wide old log-kitchen, with its loom in one corner. **1948** *Florida Hist. Q.* July 40 Close to many of the larger houses were log kitchens where cooking and eating took place. **1823** *Baptist Mag.* IV. 74 We have a good log meeting-house on Salt Creek. **1789** M. L. Weems *Let.* in *M. L. Weems: Works & Ways* (1929) III. 148, I lodged in a log-pen. **1832** *Louisville Directory* 102 The ditch was surmounted by a breast work of log pens filled with the earth obtained from the ditch. **1853** 'P. Paxton' *Stray Yankee in Texas* 118 A fish spear is to him [*sc.* the old Texan] a groin,..a house no house, but a log-pen. **1737** in *Coll. New Hampsh. Hist. Soc.* (1863) VII. 358 A log pound 30 ft. square, six feet high, with a good gate, and a lock and key. **1802** G. Barrington *Hist. N.S. Wales* 184 (Morris), The governor resolved on building a large log prison. **1845** C. M. Kirkland *Western Clearings* 212, I went to prison; nothing but a log prison. **1743** D. Brainerd *Let.* 30 Apr. in J. Edwards *Acct. Life D. Brainerd* (1749) 201 It is a Log-Room, without any Floor, that I lodge in. **1903** S. E. White *Conjuror's House.* x. 119 Virginia entered a small log room..and sat down in a musty red armchair. **1847** H. Howe *Hist. Coll. Ohio* 492 They fell to work..erecting bark huts and log shanties. **1923** Log-skid [see *break-down 3]. **1957** *N.Z. Timber Jrnl.* Oct. 73/2 *Log skids*, a platform on which logs are stacked in the forest to assist loading on to trucks. **1834** *Southern Lit. Messenger* I. 120 In the log stable..I saw a number of them. **1810** F. Cuming *Sk. Tour Western Country* 44 We stopped to feed our horses at a small log tavern. **1874** E. Eggleston *Circuit Rider* xvi. 147 Marton was conducted three miles down the river to a log tavern. **1829** J. F. Cooper *Borderers* III. i. 27 The log tenement, the stacks, ..were sending forth clouds of murky smoke. **1841** —— *Deerslayer* (ed. 2) I. ii. 47 The furniture was of the strange mixture that it is not uncommon to find in the remotely situated log-tenements of the interior. **1840** *Knickerbocker* XVI. 247, I looked around on the bare log-walls and ceiling. **1822** Logway [see *ground-hornet* (*ground *sb.* 18 b)]. **1874** B. F. Taylor *World on Wheels* II. vii. 245 Days when, over the old road, ran the yellow mud-stained coach,..pitching along its log-ways. **1973** A. Price in *Winter's Crimes* 5 202 The driver..had driven the cart off the logway.
b. *log-heaving, -raising*; *log-hauler, -lumberer*. **c.** *log-built* ppl. a.
1835 C. F. Hoffman *Winter in West* I. 79 We stopped to breakfast at a low log built shantee. **1937** *Discovery* Nov. 344/2 This sole surviving example of the log-built churches, once common in the forest region of Essex. **1919** W. T. Grenfell *Labrador Doctor* (1920) xiii. 233 The log-hauler would not deliver the goods to the rotary saw. **1962** *Amer. Speech* XXXVII. 134 *Log hauler*, an engineer on a logging train. **1823** W. Faux *Memorable Days Amer.* 180 Log-heaving, that is, rolling trees together for burning, is done by the neighbours in a body, invited for the purpose. **1909** *Westm. Gaz.* 11 Aug. 5/1 The pulp-maker..is not content, like the log-lumberer, to remove the grown trees, but takes the young plants as well. **1864** 'E. Kirke' *Down in Tennessee* iii. 43 In April, 1862, he and his band came upon a party of neighbors collected at a log raising in Fentress County. **1897** E. W. Brodhead *Bound in Shallows* 169 Law, the log-raisin's and corn-huskin's they used to have!

9. log-basket, a basket, or similar receptacle, for holding logs by a fire; **log-canoe, -cock** (earlier examples); **log-deck** (see quot.); **log-drive** (see Drive *sb.* 3); **log-headed** (later example); **log-man,** (*a*) (later example); **log-paddock,** a small field fenced in with logs; **log-rule** (see quot. 1905); **log-runner,** read: *Austral.*, a ground-dwelling bird of the genus *Orthonyx* found in northern New South Wales, Queensland, and New Guinea; add examples; **log-running,** the operation of sending logs down a river; **log-scale** (see quot. 1905); **log sheet,** a log-book in which the driver of a commercial motor vehicle enters particulars of his working and rest hours; **log-slate** (earlier example).
1902 *Westm. Gaz.* 17 Dec. 8/2 A really nice log-basket in wrought iron. **1972** *Country Life* 14 Dec. 1697/2 A split-willow log basket—22 in. long, 18 in. wide and 12 in. high, it costs £4.00. **1752** P. Stevens in N. D. Mereness *Trav. Amer. Colonies* (1916) 315, I..set out..in the morning accompanied by an officer and ten soldiers, who brought us in two log canoes. **1806** M. Lewis in Lewis & Clark *Orig. Jrnls. Lewis & Clark Exped.* (1905) IV. 132 The large woodpecker or log cock. **1853** 'P. Paxton' *Stray Yankee in Texas* 58 (Th.), The log-cock, with his gaudy head-dress. **1905** *Terms Forestry & Logging* (U.S. Dept. Agric. Bureau Forestry) 42 *Log deck*, the platform upon a loading jack. **1904** *N.Y. Even. Post* 3 May 2 The annual log-drives have begun in the upper Hudson watershed. **1926** *Spectator* 24 July 149/1 Anyone..would have been thought log-headed or obstinate. **1845** C. M. Kirkland *Western Clearings* 175 He turned his hand to the plough, and was the 'patient log-man' of a poverty-stricken household. **1900** H. Lawson *On Track* 29 He was putting up a two-rail fence along the old log-paddock. **1895** *Montgomery Ward Catal.* 369/1 Log rules, either Scribner or Doyle scale. **1905** *Terms Forestry & Logging* (U.S. Dept. Agric. Bureau Forestry) 15 *Log rule*, 1. A tabular statement of the amount of lumber which can be sawed from logs of given lengths and diameters. 2. A graduated stick for measuring the densities of logs. The number of board feet in logs of given diameters and lengths is shown upon the stick. **1898** E. E. Morris *Austral Eng.* 272/1 *Log-runner*, an Australian bird, called also a Spine-tail. **1901** A. J. Campbell *Nests & Eggs Austral. Birds* I. 252 A nest I found in the Big Scrub, Richmond River, which I believe belonged to the Orthonyx, or Log Runner, was in a damp situation. **1931, 1934** [see *chowchilla]. **1965** *Austral. Encycl.* V. 359/1 Log-runners construct large domed nests of leaves and moss, with a side-entrance placed usually on the ground or on the top of a low stump. **1878** *Lumberman's Gaz.* 6 Apr., The Green Bay *Advocate* of March 28 says that log-running is commencing all around. **1901** S. E. White *Westerners* xxi. 199 In the log running Michail Lafond was the man always called upon to skim over the bobbing logs. **1877** *Mich. Supreme Court Rep.* XXXVI. 168 The scale of the manufactured lumber exceeded the log scale. **1905** *Terms Forestry & Logging* (U.S. Dept. Agric. Bureau Forestry) 42 *Log scale*, the contents of a log, or of a number of logs considered collectively. **1958** *Listener* 14 Aug. 226/2 The lights come on in the cabs [of the lorries], while the drivers make out their log sheets. **1959** E. K. Wenlock *Kitchin's Road Transport Law* (ed. 12) 78/2 A current record (popularly known as a log sheet) containing the prescribed particulars must be compiled by the driver of every vehicle, [etc.]. **1964** *Times* 11 Feb. 11/6 The practice of keeping duplicate sets of log sheets,..is so common that it is hardly remarked upon. **1834** *Knickerbocker* III. 83 Adding on the log-slate another 'ditto' to the long column of them.

log (lǫg), *sb.*^{3} and *a.* Also † **log.** (with point). Abbrev. of Logarithm, Logarithmic *a.*
See the last paragraph of the note to Logarithm; (*log* is no longer confined to a position before a number).
1631 [see Logarithm]. **1805** J. W. Norie *Epitome Pract. Navigation* Expl. Tables p. xv, Thus the log. of 295 is 2·469822. **1858** I. Todhunter *Algebra for Schools* 308 Given log 2 find log ·0025. **1890** G. F. Matthews *Man. Logarithms* 18 How many positive integers are there whose logs. to the base 3 have 6 for a characteristic? **1960** F. Land *Lang. Math.* ix. 119 Either of the forms $1296 = 6^4$ or $4 = \log_6 1296$ describes the relationship between the number 1296, the base 6 and the index 4. **1971** *Nature* 17 Dec. 419/2 At every stage in dark adaptation, the log threshold for test flash detection..is raised in proportion to the log brightness of the after image.
1785 C. Hutton *Math. Tables* 150 To find the log. sine of 1° [etc.]. *Ibid.*, To find the log. tang. of 2° [etc.]. **1805** J. W. Norie *Epitome Pract. Navigation* Expl. Tables p. xv, The log. sine of 3 points is 9·744739. **1890** G. F. Matthews *Man. Logarithms* 49 The Logarithm of the sine of A is called the logarithmic sine of A and written log sin A. **1967** *Oceanogr. & Marine Biol.* V. 134 In a recent account of headland-bay beaches Yasso (1965) found that their plan geometry, which results from wave movements, closely fits a log-spiral. **1974** *Daily Tel.* 14 May 3 (Advt.), At last there's a pocket calculator which gives you log and trig functions instantly..at a price that makes sense.
2. Special Comb.: **log log,** (*a*) *sb.*, the logarithm of the logarithm (of a number); also *attrib.*, indicating or involving such quantities; (*b*) *adj.* (usu. hyphenated), applied to a graph or to graph paper having a logarithmic scale along both axes; **log-normal** *a.* (*Statistics*), such that the logarithm of the variate is distributed according to a normal distribution; hence **log-normally** *adv.*; **log phase** *Biol.* = *logarithmic phase* s.v. *logarithmic *a.*; **log table,** a table of logarithms; usu. *pl.*
1910 *Encycl. Brit.* IV. 975/1 Dr John Perry added log log scales to the ordinary slide rule in order to facilitate the calculation of a^x..according to the formula $\log \log a^x = \log \log a + \log x$. **1933** S. Dawson *Introd. Computation of Statistics* i. 28 Log.-log. paper, in which both sets of values are represented by lines proportional to their logarithms. **1957** Kendall & Buckland *Dict. Statistical Terms* 169 *Loglog transformation*, the transformation of a probability *P*..according to the formula $Y = \log_e(-\log_e P)$. **1962** *Lancet* 5 May 949/2 An exponential function yields a straight line when plotted on log-linear graph paper, while a power law function gives a straight line when plotted on log-log paper. **1966** D. G. Brandon *Mod. Techniques Metallogr.* 209 The slope of the log–log plot of the current–voltage characteristic near the threshold field ..is of the order of 30. **1945** J. H. Gaddum in *Nature* 20 Oct. 465/1 It is proposed to call the distribution of *x* 'lognormal' when the distribution of log *x* is normal. *Ibid.*, Examples of lognormal distributions have been found in estimates of the numbers of plankton caught in different hauls of the net, and in the amounts of electricity used in medium-class homes in the United States. **1951** *Biometrika* XXXVIII. 434 It is assumed that the population distribution of abundance is log-normal. **1971** J. B. Carroll et al. *Word Frequency Bk.* p. xxi, This model.. is called the lognormal model, because it postulates that the total vocabulary underlying a corpus is distributed according to the familiar 'normal distribution' when the logarithms of the frequencies are used. **1945** J. H. Gaddum in *Nature* 20 Oct. 465/1 The size of the particles of silver in a photographic emulsion were lognormally distributed. **1951** *Biometrika* XXXVIII. 427 (*heading*) The expected frequencies in a sample of an animal population in which the abundances of species are log-normally distributed. **1967** *Proc. Ussher Soc.* I. 277 Testing on a logarithmic scale, however, reveals the existence of two lognormally distributed populations with a discontinuity at about 0·15% Mg. **1938** H. L. Hind *Brewing* I. xv. 367 This method..was termed the Log phase method because it is used to measure acidity during the logarithmic phase of the growth of the bacterium in wort. **1959** F. S. Stewart *Bigger's Handbk. Bacteriol.* (ed. 7) i. 10 The log phase is of relatively short duration, lasting at most for some hours. **1974** *Nature* 4 Jan. 67/1 Stationary and log-phase cultures of *E. coli* B, *E. coli* K 12 Sr..and *B. subtilis* were exposed to 160° C. *c* **1935** J. A. Hammerton *New Popular Educator* 467/1 For practical purposes the indices of 10 have been tabulated in what are called Tables of Logarithms... The student now needs this tool, 'log tables'. **1962** R. B. Fuller *Epic Poem on Industrialization* xx. 143 Napier developed between 1614–1620 His logarithms, his complete log tables. **1969** D. C. Hague *Managerial Economics* vi. 132 Given time, patience and log tables, we could draw up a table like Table 7 for ourselves.

log, *v.*^{1} Add: **1. b.** (Earlier and later examples.) **c.** To remove the logs or trees from (an area). Also const. *off, over, up*. Chiefly *N. Amer.*
1717 in *Mass. House of Representatives Jrnl.* (1919) I. 272 Bridger [is trying]..to compel the Inhabitants..to Pay Him Forty Shillings..for each Team they send to Log and get Timber. **1818** L. D. Clark in *Firelands Pioneer* (1920) XXI. 2322 He and Lines went logging of the land to sow with wheat. **1829** J. MacTaggart *Three Yrs. Canada* II. 206 When the large wood is hewn down and *logged*, that is, cut into lengths and laid round these stacks in a rude pile, the fire can more readily be applied to them. **1829** in E. C. Guillet *Valley of Trent* (1957) 355 After this we logged up and cleared three acres. **1833** *Chambers's Edin. Jrnl.* II. 167/2 He..acquaints his neighbours around him, according to the extent of the land he has to log. **1839** A. Langton *Jrnl.* in *Gentlewoman Upper Canada* (1950) 114 Six or seven acres were logged up during the day. **1902** S. E. White *Blazed Trail* ii. 5 We own, however, five million on the Cass Branch which we would like to log on contract. **1904** —— *Blazed Trail Stories* iii. 46 Suppose you log a knoll which..must grow at least a half million. **1919** B. W. Sinclair *Burned Bridges* 302 As soon as the land is logged off it is open for soldier entry. **1921** H. Kephart *Camping & Woodcraft* (new ed.) I. 113 With this one tool a good axeman can..quickly fell and log-up a tree large enough to keep a hot fire before his lean-to throughout the night. **1948** *Milwaukee* (Wisconsin) *Jrnl.* 18 July 6/5 By 1889 he had built a farm home and 'tourist home' from timber he had cut and logged himself. **1959** A. McLintock *Descr. Atlas N.Z.* 45 Once provisional State forest was logged over for timber it was then released for agricultural development. **1963** E. C. Guillet *Pioneer Farmer* I. 318 Some men were known to log several acres a year entirely alone—without even oxen.
d. *to log up* (see quots. 1889 and 1905). So **logging-up.** *N.Z. colloq.*
1889 *Colonia* I. i. 26 'Logging-up' is generally done in the autumn, when there are strong gales of wind blowing. The bush which has been felled in the winter, is set fire to, and after a day or two when the ground is sufficiently cool for walking on, the still-burning logs are rolled together and piled up with rubbish, so that they may be burnt clean away. **1891** R. Wallace *Rural Econ. Austral. & N.Z.* xv. 232 When the burning is badly done the seed cannot be properly sown; the rubbish lies thick over the ground and the whole has to be gone over again and 'logged-up', else the land is thrown temporarily out of use.. while the owner waits for the remaining rubbish to decay. **1905** J. M. Thomson *Bush Boys N.Z.* ii. 32 These [big unburned trees] are 'logged-up' afterwards, that is rolled together and piled round the stumps, so as to dry thoroughly preparatory to 'firing' them again. **1908** B. E. Baughan *Shingle-Short* 84 [Trees] logged up for burning.
4. (Later example.)
1839 C. F. Briggs *Adventures Harry Franco* I. xix. 194 The captain ordered Mr. Ruffin to log me, and swore he would send me back to the States in irons.
5. (Further examples, other than *Naut.*)
1924 J. Bruce *Power Station Efficiency Control* v. 105 If an analysis is to be made of the boiler-room operating results, the indications from the various instruments must be carefully logged at least every half-hour. **1966** Rubin & Haller *Communication Switching Syst.* viii. 294 Every message which is accepted into the system is logged on a storage device. **1969** Bennison & Wright *Geol. Hist. Brit. Isles* i. 18 One further parameter of particular importance in logging bore-hole strata is the measurement of thermal conductivity. **1974** *Physics Bull.* Jan. 30/2 Up to now data from brake tests have been logged using ultraviolet recorders or human observers.
b. Also, to travel at (a certain speed) as measured by a log; to 'do'. Also of an aircraft or pilot: to attain a cumulative total of (so many hours, miles, etc.) in the air. Also *transf.*, of a machine and the time spent in operation.
1928 *Chambers's Jrnl.* Feb. 116/2 The liner was logging a steady seventeen knots. **1955** *Times* 22 Aug. 8/5 During the past five days..Secretary of State for Air, who has been learning to fly, has logged 13 hours' solo flying, it was stated yesterday by an Air Ministry spokesman. **1956** *IRE Trans. Electronic Computers* V. 138/2 To date 670 hours of operation have been logged on this unit since debugging. **1966** *Listener* 4 Aug. 179/2 The *Graf Zeppelin*.. was the first aircraft to log over a million miles. **1972** *Lebende Sprachen* XVII. 73/2 Over the past two years, our HS 125s..have proved themselves to be increasingly valuable as management tools while logging more than 1,200 trouble-free hours.

loganberry. Substitute for def.: [f. the name of J. H. *Logan* (1841–1928), American lawyer and horticulturist, who first cultivated it.] A fruit produced by crossing a raspberry with a blackberry, or the plant producing it. (Add earlier and later examples.) Also *attrib.*
1893 *Bull. Calif. Agric. Exper. Station* No. 103. 3 The Logan Berry..[has] the shape of a blackberry, the color of a raspberry, and a combination of the flavors of both. **1897** *Gardeners' Chronicle* 24 July 47/1 One of the most interesting of recent contributions from American experiment-stations is Professor L. F. Kinney's bulletin on the Logan-berry. This fruit has been..much talked of in recent years. **1926** *Daily Colonist* (Victoria, B.C.) 11 July 3/2 The Growers' Wine Company, with headquarters in this city, starting out this season with the objective of putting up 100,000 gallons of loganberry wine, has already more than 50 per cent. of that production. **1929** D. H. Lawrence *Phoenix II* (1968) 539 Women..aren't something new on the face of the earth, like the loganberry or artificial silk. **1929** J. Masefield *Hawbucks* 154 Would you care to have some loganberry plants? **1944** E. Carr *House of All Sorts* 151 The seasoned widow brought loganberry wine of her own brew. **1973** R. Genders *Epicure's Garden* I. 67 The loganberry..has been widely grown for bottling and canning.

logarithmic, *a.* Add: *logarithmic amplifier* Electr., an amplifier which produces an output in logarithmic proportion to the input; *logarithmic decrement*: see *decrement 2 b; *logarithmic phase* Biol., the period during which the population of a culture of bacteria increases exponentially with time.
1914 *Jrnl. Hygiene* XIV. 260 This selected strain holds the field during the second or logarithmic phase. **1938** Logarithmic phase [see *log phase* s.v. *log *sb.*^{3} and *a.* 2]. **1954** C. P. Snow *New Men* vi. 101 There was one of those counters whose ticking I had come to expect in any Barford laboratory; there was a logarithmic amplifier, a D.C. amplifier..which would give a measure..of the 'neutron flux'. **1957** C. Wu in C. F. Bonilla *Nucl. Engin.* iv. 128 When a wide range of the neutron flux, as much as six to eight decades, is to be measured, a logarithmic amplifier must be used. **1971** J. S. Hough et al. *Malting & Brewing Sci.* xviii. 480 The next stage is the dividing of the cells at a constant rate, referred to as the 'exponential phase' or 'logarithmic phase'.

logatom (lǫ·gætǫm). *Telephony.* [f. Gr. λόγ-ος word + Atom.] A meaningless syllable formed arbitrarily, usually from initial and final consonants and a vowel, for use in testing telephone systems.
1937 W. H. Grinsted in *Siemens Mag.* (Engin. Suppl.) Jan. 4/2 Measurements are easier and more definite if one goes to the limit of intelligibility or uses nonsense syllables (logatoms). *Ibid.* 8/2 A few approximate contour curves, ..for logatom articulation, have been plotted. **1942** Knight & Prickett *Poole's Telephone Handbk.* (ed. 8) xx. 491 These syllables are composed of consonant, vowel and consonant, based on the esperanto alphabet, which are termed logatoms. **1973** D. L. Richards *Telecommunication by Speech* iii. 165 Table 3.13 shows also a list of English single syllable words, which have been constructed by selecting from the English logatoms just described those that were actual English words.

Logbara, var. *Lugbara *sb.* and *a.*

log-book. Add: **1.** (Further example.)
1813 *Theatrical Inquisitor* II. 362 It [*sc.* the voyage] was divested of all log-book lumber.
b. *Aeronaut.* A book in which particulars of aircraft flights, flying hours, etc., are recorded.
1911 R. M. Pierce *Dict. Aviation* 150 *Log-book*, a book in which the particulars of a balloon-trip or airship-flight are entered or kept. **1917** Grahame-White & Harper *Air Power* iv. viii. 154 Turning to his log-book, he will look up this sign, and identify the place on the coast he is approaching. **1951** O. Berthoud tr. *Clostermann's Big Show* 19 Flying hours..quickly mounted up in my pilot's log-book.
c. (*a*) The registration book of a motor vehicle. (*b*) (See quot. 1971.)
1958 *Listener* 20 Nov. 835/2 The internal combustion engine, or 'I.C.E.' as my log-book calls it. **1963** *Amer. Speech* XXXVIII. 44 *Log book*, the driver's daily report required by the I.C.C. **1965** *Mod. Law Rev.* XXVIII. v. 571 English lawyers will still have to admit that a car's logbook is not admissible evidence of the engine number. **1971** M. Tak *Truck Talk* 100 *Log book*, the daily log in which truckers list their activities. **1973** *Times* 28 Nov. 9 Vehicle Registration Documents (Logbooks) will be stamped by the post office to record each issue of [petrol] coupons.

log cabin. orig. *U.S.* [See Log *sb.*^{1} 9.] A cabin, or small house, built of logs.
1770 in H. R. Shurtleff *Log Cabin Myth* (1939) 25 The court doth appoint..to agree with a workman to build a log cabbin..for a Court House. **1803** F. Asbury *Jrnl.* (1821) III. 119 Kindness will not make a crowded log cabin, twelve feet by ten, agreeable. **1817** S. R. Brown *Western Gaz.* 48 There are six families living in log cabins. **1835** *Southern Lit. Messenger* I. 546 Most of the log cabins have been exchanged for neat white cottages. *Ibid.* II. 53 We behold the low log-cabin of a school-house. **1881** W. M. Thayer (*title*) From log-cabin to White House: the story of President Garfield's life. **1937** A. Huxley *Let.* 3 June (1969) 421 We have a log cabin..on Frieda Lawrence's ranch. **1970** *Globe & Mail* (Toronto) 28 Sept. 26/6 (Advt.), Advertiser..has original log cabin for disposal. **1974** *Guardian* 23 Jan. 11/1 America has always frenetically nurtured its pioneer myths... Log cabins, and old Abe Lincoln.
attrib. **1819** R. L. Mason *Narr. in Pioneer West* (1915) 54 It is very common for a log cabin tavern without a door or window (perhaps a log out to answer both purposes) to sup and lodge twenty persons. **1840** *Nashville* (Tennessee) *Whig* 17 Aug., They are the representatives of a hardy race of honest log cabin pioneers. **1840** *Boston Atlas* 11 Sept., Crow,.. For the Party laid low By the log-cabin boys Of old Tippecanoe. **1841** *Congress. Globe* 22 June 92 Mr. Clark of New York said all this log-cabin slang was quite out of date. **1887** [in Dict. s.v. Log *sb.*^{1} 9]. **1965** A. Colby *Patchwork Quilts* 78 Log cabin quilts were popular in England and America from about the middle of the nineteenth century, and were so called because the square blocks were composed of a square centre patch surrounded by strips of material or 'logs'..overlapped..in much the same fashion as the log cabins were built. **1973** *Sat. Rev. World* (U.S.) 4 Dec. 46/3 Boulder Flat, a log-cabin settlement at 11,000 feet, from which the last inhabitant departed more than one hundred years ago.

loge^2. Add: **1. b.** A concierge's lodge.
1969 *Guardian* 2 Aug. 3/7 The tiny loges, ill-lit and ill-ventilated, in which too many concierges are condemned to live. **1972** R. Mayne *Europeans* v. 72 The views of the man in the street, or the woman in the concierge's *loge*.
2. (Later examples.)
1900 Ade *More Fables* (1902) 188 When he was in a Loge at the Play-House with Exclusive Ethel and her Friends. **1904** A. Bennett *Great Man* xxiv. 260 They occupied a 'loge' in the..Folies-Bergère. **1968** *Globe & Mail* (Toronto) 17 Feb. 24 (Advt.), Loges $5.00. **1974** *Plain Dealer* (Cleveland) 13 Oct. C2/2 The goal..is to keep the building occupied at least 300 nights a year. Naturally, it won't be completely finished for the Sinatra opener. The loges won't be ready.

logged, *ppl. a.* Add: **b.** Also *logged-off*.
1908 *Chambers's Jrnl.* 2 May 352/1 The people who are taking up the 'logged-off' lands are usually accustomed to getting along in a small way. **1911** *U.S. Dept. Agric. Farmer's Bull.* No. 462. 5 The merchantable timber has been stripped from large areas, leaving what is known as 'logged-off' or 'cut-over' land. **1921** *Daily Colonist* (Victoria, B.C.) 25 Mar. 4/3 The report contains a great array of information referring to the sub-division of logged-off lands and expired timber licences for settlement on the coast.

logged, *a. U.S.* [f. Log *sb.*^{1} + -ed^2.] Built of logs.
1784 Washington *Diaries* (1925) II. 294 A Logged dwelling house with a punchion Roof. **1834** *Knickerbocker* III. 32 Immediately on the road, appeared a large rude double logged cabin. **1972** J. Bunting *Lionheads* 155 The bunker, a mud carapace with logged sides, is ten meters inland.

logger, *sb.*^{1} Add: **2.** = *data logger* (*datum 3).
1958 L. E. C. Hughes *Electronic Engineer's Ref. Bk.* 832 Another service..by an automatic logger..is to provide the equivalent of the manuscript log of, say, temperatures in a much more useful form. **1967** *Times Rev. Industry* Apr. 63/3 The basic logger..can be connected to up to 20 data channels to monitor, for instance, the operation of a small process plant.

loggerhead. Add: **2. b.** (See quot.)
Known to be older than 1885.
1909 A. C. Fox-Davies *Compl. Guide Heraldry* 193 The leopard's face... For some unfathomable reason these charges when they occur in the arms of Shrewsbury are usually referred to locally as 'loggerheads'.
3. b. (See quot.)
1904 *Athenæum* 27 Feb. 280/1 The inkstands..include many of the prototypes of the circular heavy inkstand, still used, and known to many under the old name of 'loggerheads'.
6. b. (*b*) (Later examples.)
1870 *Amer. Naturalist* III. 159, I saw a Loggerhead attack a snake. **1906** *N.Y. Even. Post* 8 Aug. 2 Charleston S.C. pet canaries are being killed by a bird that is known as the 'loggerhead'. A loggerhead strikes at the canaries through the bars of the cage. **1939** Forbush & May *Nat. Hist. Birds Amer.* 398 The Loggerhead is an indefatigable destroyer of grasshoppers, for which it seems ever on the watch.
8. *to be at loggerheads* (later examples).
1955 *Bull. Atomic Sci.* Mar. 90/3 Uranium men and oil and gas producers had long been at loggerheads due to the fact these natural substances frequently occur on the same site, though at different horizons. **1955** *Times* 19 May 4/2 The jury would not have much difficulty in getting rid of that suggestion, because those two were obviously at loggerheads. **1975** J. Gardner *Killer for Song* i. 13 'James, it's good to see you.' His expression was at loggerheads with the words.

loggia'd, *a.* Provided with loggias.
1903 *Westm. Gaz.* 9 Dec. 3/1 A great loggia'd palace, gaunt, time-stained, damp-eaten.

logginess. [f. Loggy *a.* 2.] A state of heaviness or sluggishness.
1924 *Scribner's Mag.* July 88/2 He ate sparingly.. rather as insurance against any sensation of logginess. **1969** P. Highsmith *Tremor of Forgery* xxv. 237 He awakened with the now familiar logginess of brain that always took fifteen seconds to clear.

logging, *vbl. sb.* Add: **2*.** The process of taking and recording information about something. (Cf. Log *v.*^{1} 5 in Dict and Suppl., *log *sb.*^{1} 6 d.)
1941 F. H. Lahee *Field Geol.* (ed. 4) xviii. 574 For.. learning more about the lithology and fluid content of rocks in the walls of a bore-hole, and..for more accurately fixing the top and bottom contacts of rocks of varying character..electrical surveying, or electrical logging..has become common practice. **1958** L. E. C. Hughes *Electronic Engineer's Ref. Bk.* 832 With the aid of the automatic logging control..the alarm circuits on each unit controller trigger off printed messages. **1965** G. J. Williams *Econ. Geol. N.Z.* xvi. 255/1 Down-hole resistivity logging.. showed that the kerogen content of the shale is thin-bedded, and that it can be measured rapidly by this means. **1967** *Electronics* 6 Mar. 269/1 Suggested applications include scanning, multiplexing..data logging..and telemetering.
3. *logging-chain, company, establishment, railway, swamp, wheel*.
1825 A. Anderson *Diary* 10 Sept. in G. Sellar *Narr.* (1916) vii. 103 Walked to Toronto... Am no judge of oxen... Besides them had to pay for logging-chain and an ox-sled. **1905** *Terms Forestry & Logging* (U.S. Dept. Agric. Bureau Forestry) 36 *Dump hook*, a levered chain, grab hook attached to the evener to which a team is hitched in loading logs. A movement of the lever releases the hook from the logging chain without stopping the team. **1910** J. Hart *Vigilante Girl* xxvi. 356 He was carrying in his hand a light logging-chain which was attached to his ankles. **1949** *Sat. Even. Post* 15 Jan. 71/2, I rushed around to the toolbox, dragged out one of the heavy logging chains. **1903** A. B. Hart *Actual Govt. Amer. Colonies* 326 Logging companies buy up immense areas of land for timber. **1948** *Time* 9 Feb. 36/3 Logging companies protested it was a poor policy to rob them of 800 loyal, trained workers when there was a shortage of labor. **1851** J. S. Springer *Forest Life & Forest Trees* 67, I have seldom taxed my judgement as severely on any subject as in judiciously locating a logging establishment. **1888** J. Muir *Picturesque Calif.* 460 It is moved from camp to camp by the 'logging' railway. **1926** *Daily Colonist* (Victoria, B.C.) 7 July 1/7 Construction of logging railways and similar facilities will be proceeded with..for the reopening of the logging camps in September. **1969** *Islander* (Victoria, B.C.) 9 Mar. 12/1, I used shank's mare along the logging railway to what was referred to as Headquarters Camp. **1848** Bartlett *Dict. Amer.*, *Logging swamp*, in Maine, the place where pine timber is cut. **1851** J. S. Springer *Forest Life & Forest Trees* 46 We have sometimes heard the voice of prayer even in the logging swamp. **1905** *Terms Forestry & Logging* (U.S. Dept. Agric. Bureau Forestry) 42 *Logging wheels*, a pair of wheels, usually about 10 feet in diameter, for transporting logs.

loggy, *a.* Add: **2.** (Later examples.)
1886 *Outing* VIII. 58/1 They do very well sailing free but on the wind are loggy. **1902** *Westm. Gaz.* 18 Oct. 2/2 They seemed..'loggy' and slow to get going. **1966** H. Marriott *Cariboo Cowboy* vii. 66 A fellow doing quite a bit of riding needs two or three horses at least, because riding one horse day after day makes the horse loggy and leg-weary.
3. Abounding in logs.
1851 A. O. Hall *Manhattaner* 2 The sandy, boggy, loggy, grassy, and snaggy strips of land.

log-house. (See Log *sb.*^{1} 9.) (Earlier and later examples.)
1662 in H. R. Shurtleff *Log Cabin Myth* (1939) 80 As fare Westwardly as the logg house. **1784** J. F. D. Smyth *Tour U.S.A.* II. 9 Constructing temporary habitations (log houses) to reside in. **1879** A. W. Tourgée *Fool's Errand* vii. 34 This log house had..given way to a more pretentious structure of brick. **1965** Mrs L. B. Johnson *White House Diary* 7 Sept. (1970) 318, I arrived at Honeymoon Cabin, a real log house. **1974** 'S. Harvester' *Forgotten Road* v. 54 Clusters of log houses..formed the village.

Logian (lǫ·dʒiăn), *a.* [f. *logi-a* Logion + -an.] Containing the Logia of Jesus.
1909 V. H. Stanton *Gospels as Hist. Documents* II. 48 To call the source we are considering simply 'the Logian document' cannot, I think, be open to the same objection. **1911** J. C. Hawkins in *Stud. Synoptic Probl.* 107 The convenient practice which has grown up of calling it the 'Logian source'. **1921** *Contemp. Rev.* Mar. 263 An expanded form of the original Greek Logian document.

logic, *sb.* Add: **1. a.** Also, since the work of Gottlob Frege (1848–1925), a formal system using symbolic techniques and mathematical methods to establish truth-values in the physical sciences, in language, and in philosophical argument. (Later examples.)
1903 B. Russell *Princ. Math.* I. i. 4 But now Mathematics is able to answer, so far at least as to reduce the whole of its propositions to certain fundamental notions of logic. **1932** Lewis & Langford *Symbolic Logic* v. 118 This logistic method requires that the first branch of logic to be developed should be the calculus of propositions. **1967** A. E. Blumberg in *Encycl. Philos.* V. 13/1 What distinguishes modern from ancient and traditional logic is not only its reliance on symbolic techniques and mathematical methods but also its vastly greater formal power and range of application. **1969** F. Mondadori in R. Klibansky *Contemp. Philos.* III. 352 The phenomenological foundation of logic will make a basic use of Gödel's theorem.
2. b. In phr. *the logic of* ——, indicating the application of logical methods to other subjects of investigation or study; the inferential procedures or structure of some field of inquiry.
1845 Mill in *Westm. Rev.* XLIII. 319 By the logic of a science we understand its method; its particular modes of investigation, and the nature of its evidence. **1882** A. Bain *John Stuart Mill* iii. 87, I was at the meeting, and listened to Herschel's address. One notable feature in it was the allusion to the recent works on the Logic of Science, by Whewell and Mill especially. **1934** *Mind* XLIII. 101 Little puzzles about the logic of classes. **1937** A. Smeaton tr. *Carnap's Logical Syntax of Lang.* iv. § 70. 256 All the foregoing systems of the logic of modalities.. have, it seems, applied the quasi-syntactical method. **1942** R. G. Collingwood *New Leviathan* xxxi. 252 As mathematics is the logic of physics, so law is the logic of politics. **1945** *Mind* LIV. 175, I now regard semantics as the fulfilment of the old search for a logic of meaning, which had not been fulfilled before in any precise and satisfactory way. **1971** Kopnin & Narsky in R. Klibansky *Contemp. Philos.* IV. 321 The elaboration of these problems has led to the necessity of investigating the logic of contemporary scientific knowledge.
3. b. Phr. *the logic of the situation*, the facts which dictate what action is rationally to be taken.
1876 W. James *Coll. Ess. & Rev.* (1920) 34 The very essential logic of the situation demands that we wait not for any outward sign. **1945** K. R. Popper *Open Soc.* II. xiv. 90 The detailed determination of his action by what we may call the logic of the situation. **1946** E. Wilson *Mem. Hecate County* (1951) iv. 117 The logic of the situation impelled me to force her backwards, dropping one hand to her waist. **1960** *Rep. Proc. Conf. Univ. U.K.* 58 For the ambitious young man, the logic of the situation (which fortunately doesn't wholly govern his conduct) is this: time given to teaching is time taken from research; and his future depends not on teaching, but on research. **1961** *Observer* 19 Nov. 11/8 He said that Mr. Gaitskell had taken up the position of outside-right, and that the logic of the situation was for him to reorganise his team so that it would at least be facing in the same direction. **1969** H. Perkin *Key Profession* v. 214 The logic of the situation was that the C.A.T.s should seek to complete their upgrading by seeking recognition as universities.

3*. *Computers* and *Electronics.* The system or principles underlying the representation of logical operations and two-valued variables by electrical or other physical signals and their interactions; the forms and interconnections of logic elements in any particular piece of equipment, in so far as they relate to the interaction of signals and not to the physical nature of the components used; also, the actual components and circuitry; logical operations collectively, as performed by electronic or other devices.
1950 W. W. Stifler *High-Speed Computing Devices* v. 62 For the convenient operation of a general-purpose machine, they [*sc.* Burks, Goldstine, and von Neumann] point out, it is essential that some steps be taken to translate the nonconforming command quoted above to the same stereotype form. This translation of description of all possible operations to prescribed forms has been called the logic of the machine by these authors, and the term is now in general use. **1952** *Math. Tables & Other Aids to Computation* VI. 42 It is possible, with this new approach, to obtain many of the advantages of a digital computer and also the essential advantages of an analog differential analyzer. The result is a different type of digital 'logic' from that used in the general purpose digital computer. **1954** *IRE Trans. Electronic Computers* Mar. 33/2 The authors have presented the more general aspects of the machine in block diagram form and in addition have given pertinent illustrations of the instrumentation of the logic. **1956** *Ibid.* IV. 134 (*caption*) Logic for self-timing full length carry. **1962** Simpson & Richards *Physical Princ. Junction Transistors* xvi. 401 If we now define the most positive value of the output to represent a 'one' and the most negative value a 'zero' in a binary system of arithmetic (the positive logic system) the emitter followers form an 'and' gate, i.e. all *n* inputs must be 'one' if the output is to be 'one'. **1967** *Electronics* 6 Mar. 26 Santa Clara was making milliwatt resistor–transistor logic for the project. **1968** *Proc. Inst. Electr. Engin.* CXV. 1385/2 He separated the ternary circuits into two sets of binary circuits, one based on a positive logic and the other on a negative one. Then he used translating circuits between the two logics and achieved a true ternary output with the aid of a combining circuit. **1970** *New Yorker* 11 Apr. 34 The computer logic is so fast that it has to loaf at several intervals while the input and output devices..are printing information. **1971** *New Scientist* 25 Mar. 692 The technology [of fluidics] was developed basically to provide a system of control logic and power amplification in the adverse environment of space. **1973** *Nature* 20 Apr. 494/2 The transistor has endowed tremendous scope for performing electronic functions (for example switching, which permits binary logic, and amplifying, which makes possible many other forms of signal processing).
b. *attrib.*, as *logic design, designer, diagram, function, module, network, operation, state*; **logic circuit,** a circuit for performing logical operations and consisting of one or more logic elements; **logic element,** a device (usu. electronic) for performing a logical operation, in which the past or present values of one or more inputs determine the values of one or more outputs in accordance with a simple scheme which most commonly involves, in effect, only two possible values for the signals; **logic gate,** a logic circuit that is a gate (*gate *sb.*^{1} 8 g).
1953 *Communications & Electronics* (N.Y.) Nov. 593/1 The search for simple abstract techniques to be applied to the design of switching systems is still..in its early stages. The problem in this area which has been attacked most energetically is that of the synthesis of efficient combinational that is, nonsequential, logic circuits. **1959** K. Henney *Radio Engin. Handbk.* (ed. 5) ix. 7 Very complicated logic circuits involving many thousands of diodes have been assembled for use in electronic computers. **1968** *Times* 18 Oct. 16/5 Computer logic circuits based on the ternary system of arithmetic have been devised by two engineers... They have devised logic circuits that perform the basic arithmetical operations in ternary numbers. **1956** *IRE Trans. Electronic Computers* V. 132/2 The evolution of a set of standard logical circuit blocks allows this design without direct reference to the circuits, thus reducing logic design to the application of a set of rules expressing the input and output capabilities of each of the logical circuit blocks. **1972** D. Zissos *Logic Design Algorithms* p. v, Complex switching circuits are used extensively, not only in the logic design of systems such as digital computers and message-switching networks, where they are the subject of study by specialist logic designers, but also for purposes of industrial control and automation. **1956** Logic designer [see *logical *a.* 1 b]. **1962** *Gloss. Terms Automatic Data Processing* (*B.S.I.*) 59 *Logic diagram*, a graphical representation of the logic design. **1971** J. H. Smith *Digital Logic* iv. 54 There are a number of different circuits which will carry out this function and it is normal to use the symbol of the circuit employed when drawing logic diagrams. **1959** C. V. L. Smith *Electronic Digital Computers* iv. 106 This technique makes it possible to make large arrays of logic elements in very compact form. *Ibid.* (*heading*) Symbols for logic elements. **1969** P. B. Jordain *Condensed Computer Encycl.* 291 A sequential logic element has an output determined by present and past input signals: it has some degree of retention, or memory. **1971** *New Scientist* 25 Mar. 692 The technology of fluidics is divided into..pure fluidic techniques which use a logic element with no moving parts, and the method using moving part logic. **1959** K. Henney *Radio Engin. Handbk.* (ed. 5) ix. 7 Since diodes have two distinctly differing impedance levels, they lend themselves to use in performing predetermined decision or logic functions just as switches or relays may. **1961** *Times* 3 Oct. (Computer Suppl.) p. v, What the technical terms (binary, logic gates, programme and many others) mean. **1967** *Electronics* 6 Mar. 45/3 Each cell will have a working structure, or 'base', of 13 logic gates, plus 13 flip-flops. **1971** J. H. Smith *Digital Logic* iv. 40 It is usually unnecessary to design logical circuitry today, as many manufacturers supply logic modules to carry out the more common forms of logic. **1974** *Physics Bull.* Mar. 113/1 Ortec will be displaying the new fast logic modules in the M300 series. **1961** *Engineering* 17 Feb. 269/1 The first step in designing the computer was the synthesis of the logic network to perform the necessary functions. **1970** *Nature* 12 Sept. 1092/2 The most exciting possibility for these devices lies naturally in their use as shift registers and for logic operations in computers. **1973** *Sci. Amer.* Nov. 14/1 (Advt.), In a digital circuit, the important thing is to know the logic state of a node, whether it's above the threshold voltage and therefore a logic high, or below the threshold voltage and therefore a logic low.
5. *logic-book*; (sense 3) *logic-chopper, -chopping*; **logic-tight** *a.* [after Watertight *a.* 1], impervious to logic or reason.
1685 tr. *Arnauld & Nicole's Logic* 17 We should give a reason for omitting so many questions as are found in the common Logic-Books. **1895** W. James *Coll. Ess. & Rev.* (1920) 394 An hypothesis, we are told in the logic-books, ought to propose a being that has some other constitution and definition than that of barely performing the phenomenon it is evoked to explain. **1906** *Daily Chron.* 13 Mar. 6/2 Mr. Balfour..made his reappearance in his old part of the Logic-chopper. **1924** R. Graves *Mock Beggar Hall* 32 Put it another way, thou logic-chopper. **1956** J. Blish *Earthman, Come Home* i. 33 You have no ties, no faith. You will have to excuse ours. We cannot afford to be logic-choppers. **1904** W. James in *Mind* XIII. 458 This is a kind of intellectual product that never attains a classic form of expression when first promulgated. The critic ought therefore not to be too sharp and logic-chopping in his dealings with it. **1960** Koestler *Lotus & Robot* I. iii. 132 The Schoolmen confined themselves to verbal logic-chopping. **1912** B. Hart *Psychol. Insanity* vi. 82 The delusion is preserved in a logic-tight compartment. **1968** P. McKellar *Experience & Behaviour* x. 269 The widespread tendency..to surround their favourite beliefs with logic-tight compartments.

logical, *a.* (and *sb.*). Add: **1.** (Further examples.)
1847 J. D. Morell *Hist. View Philos.* (ed. 2) I. i. 95 To Logic, Hobbes devoted a considerable share of attention. The peculiarity of his logical system lies in the theory, that reasoning is merely a numerical calculation. **1905** B. Russell in *Mind* XIV. 490 The distinction of primary and secondary occurrences also enables us to deal with..the logical status of denoting phrases that denote nothing. **1922** E. P. Adams tr. *Einstein's Meaning of Relativity* i. 1 The object of all science, whether natural science or psychology, is to co-ordinate our experiences and to bring them into a logical system. **1939** *Mind* XLVIII. 304 To say that a term is of such and such a type or category is to say something about its 'logical behaviour', namely, about the entailments and compatibilities of the propositions into which it enters. **1966** W. V. Quine *Ways of Paradox* viii. 67 This condition is met by the usual logical languages, and presumably it can be met likewise by languages adequate to science in general.
b. *Computers* and *Electronics.* Of or pertaining to the logic (*logic *sb.* 3*) of computers and similar equipment; designed to carry out processes on electrical or other signals analogous to the processes of reasoning, deduction, etc., employed in (formal) logic; *logical element* = *logic element* (*logic *sb.* 3* b); *logical operation*: see *7; *logical shift*, a displacement of the digits of a sequence by a specified number of positions in a way that is not equivalent to multiplication by an integral power of the base; *esp.* a cyclic shift, in which digits taken from one end reappear, in the same order, at the other end.
In some of the uses below *logical* can equally well be regarded as having the sense of Logical *a.* 1.
1946 Burks, Goldstine, & von Neumann (*title*) Preliminary discussion of the logical design of an electronic computing instrument. (Rep. submitted to U.S. Army Ordnance Dept., PB 96703.) **1946** Goldstine & von Neumann *Princ. Large Scale Computing Machines* in J. von Neumann *Coll. Works* (1963) V. 23 The Memory Organ... In performing an operation (arithmetical or logical) it is usually necessary to store the quantities entering into it. **1950** [see *coincidence 7 b]. **1956** *IRE Trans. Electronic Computers* V. 132/2 One of the important properties of a digital computer is that it may be assembled simply and easily from a few well-chosen functional circuits. Each of these circuits represents a logical element that is useful to the system or logic designer in planning a computer. **1958** Gotlieb & Hume *High-Speed Data Processing* iii. 34 A concise and informative way of describing a computer is to draw its logical diagram which shows the paths and effects of the various signals through it. Logical diagrams are built up largely of gates. *Ibid.* v. 89 In a logical or cyclic shift the digits lost off one end of the number appear at the other end. **1962** *Newnes Conc. Encycl. Nucl. Energy* 671/2 Such a system is sometimes referred to as a logical switching system because the output responses are related by fixed rules to the inputs from the measurement channels. **1964** F. L. Westwater *Electronic Computers* iii. 49 The practical engineer must have the last word. Not infrequently, a design engineer will ask the logical designer to make alterations for various reasons. **1970** O. Dopping *Computers & Data Processing* vii. 117 In the logical shift, all bits take part and zeroes are shifted in at one end. This kind of shift is suitable for handling non-arithmetic information. **1971** J. H. Smith *Digital Logic* vii. 133 The reader using logical gating

should analyse his specifications to eliminate the trivial and reduce the circuit as far as possible.

2. (Earlier and additional examples.)

1588 A. Fraunce *Lawiers Logike* f. 120, I haue, for examples sake, put downe a Logicall Analysis of the second *Aegloge* in Virgill. **1828** Mill in *Westm. Rev.* IX. 144 Those who maintain, that to perform the logical analysis of an argument, in the manner pointed out by the syllogism, is not the best means of discovering whether it contain a flaw. **1958** G. J. Warnock *Eng. Philos. since 1900* ix. 120 The narrowly logical, context-neglecting manner adopted by the practitioners of 'logical analysis'.

7. Special collocations (see also sense *1 b): *logical addition*, the formation of a logical sum; *logical atomism* (see *atomism 1 b); *logical constant* (see quots. 1903, 1914); *logical construction*, an entity theoretically superfluous in that any statement referring to it can be replaced by an equivalent statement making no reference to it; *logical empiricism*, the name given to philosophical theories which replaced those of logical positivism (see quots. 1936, 1937); so *logical empiricist*; *logical fiction* = **logical construction*; *logical form*, the form, as distinct from the content, of a proposition, argument, etc., which can be expressed in logical terms; *logical grammar*, the rules of word-use in a proposition upon which its logical, as opposed to its purely grammatical, sense or meaning is held to depend; so *logical-grammatical* adj.; *logical implication*, implication which is based on the formal and not the material relationship between propositions; *logical machine*, an apparatus designed to facilitate logical calculations; also *transf.*; *logical multiplication*, the formation of a logical product; *logical operation*, an operation of the kind dealt with in logic (such as conjunction or negation); any analogous (non-arithmetical) operation on numbers, esp. binary numbers, in which each digit of the result depends on only one digit in each operand; *logical paradox* (see quot. 1967); *logical positivism*, the name given to the theories and doctrines of philosophers active in Vienna in the early 1930s (the Vienna Circle), which were aimed at evolving in the language of philosophy formal methods for the verification of empirical questions similar to those of the mathematical sciences, and which therefore eliminated metaphysical and other more speculative questions as being logically ill-founded; hence *logical positivist*; *logical product*, the conjunction of two or more propositions, or the intersection or two or more sets (written $p \wedge q$, $p \cap q$, $p.q$, pq, p and q); *logical structure*, the formal framework of logical rules to which a theory, language, proposition, etc., must conform in order to have truth-value; *logical subject*, the subject which is implied in a sentence or proposition, or which exists in the deep structure of a sentence; *logical sum*, the disjunction of two or more propositions, or the union of two or more sets (written $p \vee q$, $p \cup q$, p or q, $p+q$); *logical syntax* (see quot. 1934); *logical truth*, that which is true in logical or formal terms regardless of material meaning; *logical word*, a word of the type which gives logical context or form to a proposition but which, by itself, is non-representational and without meaning (see quot. 1940).

1868 C. S. Peirce in *Proc. Amer. Acad. Arts & Sci.* VII. 250 Let $a + b$ denote all the individuals contained under a and b together. The operation here performed will differ from arithmetical addition in two respects: 1st, that it has reference to identity, not to equality; and 2d, that what is common to a and b is not taken into account twice over... The process denoted by + ..I shall call the process of logical addition. **1903** B. Russell *Princ. Math.* ii. 18 From this point we can prove the laws of contradiction and excluded middle and double negation, and establish all the formal properties of logical multiplication and addition—the associative, commutative and distributive laws. **1970** O. Dopping *Computers & Data Processing* i. 23 In Boolean algebra..instead of the usual mathematical operations, there are certain logical operations; the most common ones of these are logical addition, logical multiplication, and negation. **1914,** etc. Logical atomism [see *atomism 1 b]. **1973** E. Gellner in Horton & Finnegan *Modes of Thought* 179 The..formal fact of the existence and nature of the mosaic's framework itself. This vision is shared by classical empiricism and by doctrines such as 'logical atomism'. **1903** B. Russell *Princ. Math.* i. 3 Pure Mathematics is the class of all propositions of the form 'p implies q', where p and q are propositions containing one or more variables, the same in the two propositions, and neither p nor q contains any constants except logical constants. And logical constants are all notions definable in terms of the following: Implication, the relation of a term to a class of which it is a member, the notion of *such that*, the notion of relation, and such further notions as may be involved in the general notion of propositions of the above form. **1914** —— *Our Knowl. External World* vii. 208 Such words as *or, not, if, there is, identity, greater, plus, nothing, everything, function*, and so on, are not names of definite objects, like 'John'..but are words which require a context in order to have meaning... 'Logical constants', in short, are not entities. **1922** tr. *Wittgenstein's Tractatus* 69 My fundamental thought is that the 'logical constants' do not represent. That the *logic* of the facts cannot be represented. **1958** G. J. Warnock *Eng. Philos. since 1900* x. 125 We thus get a distinction between so-called 'logical constants', the irreplaceable words on which the validity of general patterns of inference depends, and items of non-logical vocabulary. **1965** B. Mates *Elem. Logic* iv. 49 The logical constants occurring in ϕ are understood in the way usual among logicians (i.e., '∨' as standing for 'or', '—' for 'not', the universal quantifier for 'all', etc.). **1883** F. H. Bradley *Princ. Logic* 236 To show a new relation of elements in a logical construction is demonstration in the sense of reasoning. **1914** B. Russell *Our Knowl. External World* iv. 101 The only justification possible must be one which exhibits matter as a logical construction from sense-data. **1936** A. J. Ayer *Lang., Truth & Logic* iii. 73 The English State..is a logical construction out of individual people. **1956** J. O. Urmson *Philos. Analysis* iii. 36 If 'X' is an incomplete symbol then Xs are logical constructions. Thus if the expression 'the average man' is an incomplete symbol we may say that the average man is a logical construction. **1936** A. J. Ayer *Lang., Truth & Logic* 10 Our own logical empiricism to be distinguished from positivism. **1937** *Mind* XLVI. 345 Logical Empiricists are not attempting to be metaphysical, when they distinguish between language and reality. On the contrary, the distinction refers only to certain rules of usage for statements and modes of speech. Since we have investigated the rules of speech in empirical sciences, we are justified in calling our viewpoint 'Logical Empiricism'. **1966** J. J. Katz *Philos. of Lang.* iii. 16 The chief reason for this failure on the part of *logical empiricism*..and *ordinary language philosophy*..was that both were governed in their inferences..by an assumption about the nature of language. **1936** *Mind* XLV. 545 J. Somerville [in] 'The Social Ideas of the Wiener Kreis's International Congress' reflects on the Paris Congress 1935, [and] announces that the 'logical positivists' desire to repudiate Comte and to be henceforth known as 'scientific' or 'logical' empiricists. **1967** *Philos.* XLII. 293 A failure to work out in detail the consequences of logical empiricist principles for ethics. **1843** Mill *Logic* I. iii. v. 404 This tendency shows itself very visibly in the different logical fictions which are resorted to even by philosophers... Thus, rather than say that the earth causes the fall of bodies, they ascribe it to a *force* exerted by the earth, or an *attraction* by the earth. **1918** B. Russell in *Monist* XXVIII. 512, I believe that series and classes are of the nature of logical fictions. **1933** L. S. Stebbing *Mod. Introd. Logic* (ed. 2) 502 To say that the table is a logical fiction (or construction) is not to say that the table is fictitious..it is rather to deny that, in any ordinary sense, it is an object at all. *Ibid.*, 'Logical fiction' may be taken to be an unfortunate synonym for 'logical construction'. **1840** Mill in *Westm. Rev.* XXXIII. 266 A truth..to be believed in opposition to all that appears proof to the mere understanding; nay, the more to be believed, because it cannot be put into words and into the logical form of a proposition without a contradiction in terms. **1967** A. E. Blumberg in *Encycl. Philos.* V. 13/2 Logic confines itself to those arguments whose validity rests exclusively on the logical form of the statements composing them... There is still no fully satisfactory account of logical form. **1922** tr. *Wittgenstein's Tractatus* 55 A symbolism..which obeys the rules of logical grammar—of logical syntax. **1962** L. J. Cohen *Diversity of Meaning* iii. 81 The most obvious fault in the doctrine of logical grammar is that it suggests the conceptual study of meanings to be concerned with something that is timeless and unchanging. **1937** *Atlantic Monthly* CLIX. 49/2 The logical-grammatical construction involved is so commonly mishandled in current American speech and print. **1887** A. Seth *Hegelianism* v. 172 In the first sense..development means simply logical implication. **1904** B. Russell in *Mind* XIII. 209 It would seem that..logical implication is a simple notion, into whose composition the notion of terms does not enter. **1934** Cohen & Nagel *Introd. Logic* i. 9 (*heading*) Logical implication does not depend on the truth of our premises. **1870** W. S. Jevons in *Proc. R. Soc.* XVIII. 167 To explain the nature of the logical machine alluded to, it may be pointed out that the third of the fundamental Laws of Thought allows us to affirm of any object one or the other of two contradictory attributes. **1883** F. H. Bradley *Princ. Logic* 344 We shall discuss the Indirect Method, and with it the claims of the Logical Machine. **1943** *Mind* LII. 319 He is thus stymied at the outset, and being neither a logical machine nor an esthetic idiot he is likely to feel uncomfortable about it. **1868** C. S. Peirce in *Proc. Amer. Acad. Arts & Sci.* VII. 251 Let a, b denote the individuals contained at once under the classes a and b... If a and b were independent events, a, b would denote the event whose probability is the product of the probabilities of each. On the strength of this analogy..the operation indicated by the comma may be called logical multiplication. **1903, 1970** Logical multiplication [see *logical addition* above]. **1885** C. S. Peirce in *Amer. Jrnl. Math.* VII. 186, I prefer not to assign determinate values to **f** and **v**, nor to identify the logical operations with any special arithmetical ones. **1932** Lewis & Langford *Symbolic Logic* i. 7 It would have coincided with what we now know as symbolic logic. That is, it would have been an organon of reasoning in general, developed in ideographic symbols and enabling the logical operations to be performed according to precise rules. **1960** M. G. Say et al. *Analogue & Digital Computers* vii. 165 A zero input digit to this unit gives a one-digit output and a one-digit input gives a zero output. This is the logical operation 'not'. **1970** Logical operation [see *logical addition* above]. **1904** W. James *Ess. Radical Empiricism* (1912) i. 11 'Representative' theories of perception avoid the logical paradox, but on the other hand they violate the reader's sense of life, which knows no intervening mental image. **1954** I. M. Copi *Symbolic Logic* 332 These two kinds of paradoxes were first explicitly distinguished by F. P. Ramsey in 1926. Since then those of the first kind have been known as 'logical paradoxes'. **1967** J. van Heijenoort in *Encycl. Philos.* V. 45/2 In logic the word [*sc.* paradox] has taken on a more precise meaning. A logical paradox consists of two contrary, or even contradictory, propositions to which we are led by apparently sound arguments. **1931** Blumberg & Feigl in *Jrnl. Philos.* XXVIII. 281 To facilitate criticism and forestall even more unfortunate attempts at labelling this aspect of contemporary European philosophy, we shall employ the term 'logical positivism'. **1934** *Philos. Rev.* XLIII. 125 The logical positivism of the Vienna Circle..is based..upon this consideration of empirical meaning. **1968** M. Black *Labyrinth of Lang.* vi. 147 Logical Positivism has seen its best days. **1931** *Jrnl. Philos.* XXVIII. 291 The principle of causality is for the logical positivist not a categorical necessity of thought. **1967** J. Passmore in *Encycl. Philos.* V. 55/1 The logical positivists ordinarily took for granted the substantial truth of contemporary Science. Thus, it was a matter of vital concern to them when it became apparent that the verifiability principle would rule out as meaningless all scientific laws. **1868** C. S. Peirce in *Proc. Amer. Acad. Arts & Sci.* VII. 411 The numerical rank of a logical product depends on the identity or diversity..of parts of the factors. **1903** B. Russell *Princ. Math.* ii. 21 Most of the propositions of the class-calculus are easily deduced from those of the propositional calculus. The logical product or common part of two classes a and b is the class of x's such that the logical product of 'x is an a' and 'x is a b' is true. Similarly we can define the logical sum of two classes (a or b). **1955** A. N. Prior *Formal Logic* i. i. 7 We may also have a conjunction or logical product of more than two propositions, the analogy with arithmetical multiplication still holding. **1959** Logical product [see *logical sum* below]. **1871** A. C. Fraser *Life Berkeley* ii. 39, I have tried elsewhere..to explain the logical structure of the *Essay on Vision*. **1918** B. Russell in *Monist* XXVIII. 510 The first thing to do would be to discover the kinds of atoms out of which logical structures are composed. **1943** *Mind* LII. 26 Linguistic structure, though at times it conceals or distorts, has to be taken as capable of revealing logical structure, otherwise the study of logic would be impossible. **1970** L. J. Cohen *Implications Induction* i. 6 Many propositions are never hypothesised at all, even if alike in logical structure to those that are. **1898** *Mind* VII. 34 (*heading*) On the logical subject of the proposition. **1903** B. Russell *Princ. Math.* iv. 44 Every term..is a logical subject. **1933** [see *grammatical *a.* 1 b]. **1965** N. Chomsky *Aspects of Theory of Syntax* i. 23 In (8i) [*sc.* 'I persuaded a specialist to examine John'] the phrase 'a specialist' is the Direct-Object of the Verb Phrase and the logical Subject of the embedded sentence. *Ibid.* iv. 163 It seems that beyond the notions of surface structure..and deep structure (such as 'logical subject'), there is some still more abstract notion of 'semantic function' still unexplained. **1868** C. S. Peirce in *Proc. Amer. Acad. Arts & Sci.* VII. 411 The numerical rank of a logical sum depends on the identity or diversity..of the integrant parts. **1903** [see *logical product* above]. **1959** C. V. L. Smith *Electronic Digital Computers* ii. 32 Given two binary words..it is possible to generate a third word each bit of which is the logical sum or the logical product or, indeed, any Boolean function of the bits in the corresponding position of the given words. Operations of this sort may be called 'logical operations'. **1963** G. T. Kneebone *Math. Logic* ii. 53 In older accounts of symbolic logic, the terms 'logical sum' and 'logical product' are often used where 'disjunction' and 'conjunction' would now be preferred. There is indeed an analogy between the arithmetical operations of addition and multiplication. **1922** Logical syntax [see *logical grammar* above]. **1934** R. Carnap in *Philos. Sci.* I. 9 By the 'logical syntax' (or also briefly 'syntax') of a language we shall understand the system of the formal (i.e. not referring to meaning) rules of that language, as well as..the consequences of these rules. **1945** *Mind* LIV. 172 The weaknesses of the argument might be due in part at least to a certain poverty in the technical equipment of 'logical syntax'. **1970** L. J. Cohen *Implications Induction* i. 6 Some elementary principles in the logical syntax of experimental support. **1818–19** Coleridge *Philos. Lect.* (1949) ix. 276 This necessarily led men..to doubt whether a logical truth was necessarily an existential one, i.e. whether because a truth was logically consistent it must be necessarily existent. **1877** W. S. Jevons *Princ. Sci.* (ed. 2) viii. 153 Nothing is more certain than logical truth. **1943** *Mind* LII. 272 An exhaustive formulation of logical truth remains a worthy undertaking. **1940** B. Russell *Inquiry into Meaning & Truth* 20 We pass from the primary to the secondary language by adding what I shall call 'logical words', such as 'or', 'not', 'some', and 'all', together with the words 'true' and 'false' as applied to sentences in the object-language. **1958** S. E. Toulmin *Uses Argument* iv. 149 The validity of syllogisms being closely bound up with the proper distribution of logical words within the statements composing them. **1972** *Sci. Amer.* Sept. 82/3 Language performs this miraculous function largely through such little particles as 'if', 'when', 'not', 'therefore', 'all' and 'some', which have been called logical words because they account for the ability of language to formulate logical inferences (also known as syllogisms).

logically, *adv.* Add: Phr. *a logically perfect language*: a language in which the grammatical structure of sentences would be identical with their logical structure.

1918 B. Russell in *Monist* XXVIII. 520 In a logically perfect language the words in a proposition would correspond one by one with the components of the corresponding fact, with the exception of such words as 'or', 'not', 'if'. .. The language which is set forth in *Principia Mathematica*..aims at being that sort of a language that, if you add a vocabulary, would be a logically perfect language. **1922** —— in Wittgenstein *Tractatus* Introd. 7 He is concerned with the conditions which would have to be fulfilled by a logically perfect language. **1939** M. Black in Copi & Beard *Ess. Wittgenstein's Tractatus* (1966) 97 This view of the character of Wittgenstein's investigation..may have been suggested by Russell's own attempts to construct a logically perfect language. **1953** *Mind* LXII. 13 They have even gone so far as to suggest that philosophers should invent a logically perfect language as a substitute for ordinary speech. **1971** A. Flew *Introd. Western Philos.* xi. 388 One..of the many projects of Leibniz..was to develop a logically perfect language in which the truth or falsity of any proposition would be obvious from the symbols alone.

logicism (lǫ·dʒisiz'm). [f. Logic *sb.* + -ism.] The theory of Frege that a set of axioms for mathematics could be deduced from a primitive set of purely logical axioms, so that mathematics was essentially a part of logic.

1937 A. Smeaton tr. *Carnap's Logical Syntax of Lang.* v. § 84. 325 What should a logical foundation of mathematics achieve? On this question there are various views; the fundamental antithesis between them is particularly clearly brought out in two doctrines, *logicism*, which was founded by Frege (1884), and *formalism*, represented by Frege's opponents. (The designations 'logicism' and 'formalism' only appeared later.) **1970** A. E. Blumberg tr. *Stegmüller's Main Currents Contemp. German, Brit. & Amer. Philos.* viii. 327 The modern philosophy of mathematics is characterized by the fact that various schools have been formed to overcome the difficulties occasioned by the antinomies. The oldest of these schools is logicism and goes back to Frege. **1973** *Sci. Amer.* Apr. 103/2 Typically the choice is determined by one's degree of sympathy with one or another of three modern schools of mathematical thought: logicism, formalism and intuitionism.

logicist (lǫ·dʒisist), *sb.* and *a.* [f. Logic *sb.* + -ist.] **a.** *sb.* A (formal) logician; a mathematician who uses the methods or accepts the theory of logicism. **b.** *adj.* Of or pertaining to logicism. Also occas. **logici·stic** *a.*

a **1910** W. James *Some Probl. Philos.* (1911) v. 82 Hibben and the logicists seem to believe that conception, if only adequately attained to, might be all-sufficient. **1916** J. Dewey *Ess. Exper. Logic* 65, I am not questioning the right of the physicist, the mathematician, or the symbolic logicist to go ahead with accepted objects and do what he can with them. **1937** A. Smeaton tr. *Carnap's Logical Syntax of Lang.* v. § 84. 326 The sentence..can be derived with the help of the logicist system. **1952** S. C. Kleene *Introd. Metamath.* i. iii. 44 The logicistic definition of natural number now becomes predicative. **1956** E. M. Hutten *Lang. Mod. Physics* vi. 231 Our understanding of what statistical inference can do is obscured..by the logicist scheme of taking all hypotheses as universal implications, e.g. $(x)(fx \supset gx)$. **1973** *Sci. Amer.* Apr. 109/3 In the 'new math' the logicist theory that natural numbers are derived from the quantification of classes was at last put into educational practice.

logicize, *v.* Delete *rare* and (in sense 2) *nonce-use* and add further examples. (See also quot. 1919.) Hence **lo·gicized** *ppl. a.*, **lo·gicizing** *vbl. sb.*

a **1910** W. James *Mem. & Stud.* (1911) xv. 393 All these tricks for logicizing originality, self-relation, absolute process, subjective contradiction, will wither in the breath of the mystical fact. **1919** B. Russell *Introd. Math. Philos.* i. 7 Frege, who first succeeded in 'logicising' mathematics, *i.e.* in reducing to logic the arithmetical notions which his predecessors had shown to be sufficient for mathematics. **1924** C. K. Ogden tr. *Vaihinger's Philos. of 'As If'* iii. iv. 360 This creating, logicizing, arranging, falsifying, is the best guaranteed reality. **1937** [see *arithmetize *v.* 2 a]. **1957** J. Passmore *100 Yrs. Philos.* vi. 149 It is in the writings of G. Frege that the fundamental problems of a logicised mathematics first clearly emerged.

logico-. (Later examples.)

Used with adjs., sbs., and occas. with advbs. as second elements.

1858 A. De Morgan *On Syllogism* (1966) 89, I distinguish the two sides of logic as the *logico-mathematical* and the *logico-metaphysical*, frequently dropping the prefix *logico*. **1898** J. A. Hobson *John Ruskin* xiii. 319 No one..more subtly practised the vital as distinguished from the logico-mechanical method of teaching. *c* **1905** C. S. Peirce *Coll. Papers* (1933) IV. i. ix. 274 Dr. Georg Cantor, the great founder and *Hauptförderer* of the logico-mathematical doctrine of numbers. **1918** C. I. Lewis *Survey Symbolic Logic* p. v, Treatises..bristling with logico-metaphysical difficulties. **1922** D. Ainslie tr. *Croce's Aesthetic* (ed. 2) ii. v. 229 The same criticism..must have extended to the logico-grammarians of Port-Royal. **1930** T. Sasaki *On Lang. R. Bridges' Poetry* 3 The following different lines of approach are possible: (1) Logico-psychological [etc.]. **1931** J. Wisdom *Interpretation & Analysis* 13 The logico-analytic philosophers are constantly asking questions such as 'What does "This is a chair" mean?' **1935** *Mind* XLV. 263 It is only from a footnote that we learn that Pareto had realized the importance (and difficulty) of the 'theorem' that 'the logico-experimental truth of a theory and its social utility are independent'. **1937** *Mind* XLVI. 253 This book should prove stimulating to scientists, methodologists, and philosophers of all kinds, speculative, critical, or logico-positivist. **1938** C. Morris *Found. Theory of Signs* § 5. 92 The logico-grammatical structure of language. **1940** W. V. Quine *Math. Logic* 127 The atomic formulae formed by flanking '∈' with variables happen merely to be the ones appropriate to logico-mathematical matters. **1943** H. Read *Educ. through Art* iii. 70 What is now suggested, in opposition to the whole of the logico-rationalistic tradition, is that there exists a concrete visual mode of 'thinking'. **1946** *Mind* LV. 44 It refers only to a very small group of Logico-Analytic philosophers. **1949** *Mind* LVIII. 397 The latest logico-analysts avowedly make use of utterances that are as NonSensical as those of the metaphysician. **1951** *Mind* LX. 24 The Logico-positivist proposes not to *speak* of 'inference'. **1952** *Mind* LXI. 574 The great logico-philosophical revolution of the last half century, initiated by Frege and Russell. **1958** W. Stark *Sociol. of Knowl.* ii. viii. 320 If we study social reality as all reality should be studied, namely 'logico-experimentally', i.e. truly scientifically, we find..that human action is controlled by a set of drives. **1958** *Listener* 27 Nov. 879/1 The kind of thinking which I have called metaphysical thinking can lead to positive achievements, where both the system-philosophies and logico-linguistic investigations fail. **1970** J. N. Findlay tr. *Husserl's Logical Investigations* II. vi. viii. 826 The purely logico-grammatical laws which, as laws of complication and modification, distinguish the spheres of sense and nonsense. **1971** P. F. Strawson (*title*) Logico-linguistic papers.

lo·gily, *adv.* [f. Logy *a.*] In a dull or heavy manner.

1912 J. London *Son of Sun* viii. 326 The schooner,..from the weight of water on her decks, behaved logily.

logistic, *a.* and *sb.* Add: **A.** *adj.* **3. a.** *logistic curve*: also [after F. *logistique* (P.-F. Verhulst 1845, in *Nouv. Mém. de l'Acad. R. des Sci. et Belles-Lettres de Bruxelles* XVIII. 8)], a curve described by the equation $y=K/(1+Ae^{a-bt})$, where K, A, a, and b are constants, which approximates an exponential curve for small values of t, has a point of inflexion at $t=a/b$, and as t increases approaches $y=K$ asymptotically. Hence *logistic growth, law*, etc.

1925 G. U. Yule in *Jrnl. R. Statistical Soc.* LXXXVIII. 11 The logistic curve implies that, if we could plot the instantaneous percentage rate of increase of the population at any moment of time against the magnitude of the population, the resulting points should lie on a straight line, a line sloping downwards from left to right, since the rate of increase falls as the population increases. **1928** R. Pearl *Rate of Living* vii. 132 The growth of the stem follows a logistic curve. **1930** W. R. Inge *Christian Ethics & Mod. Probl.* v. 252 Professor Raymond Pearl, of Baltimore,..has evolved a theory that the growth of population follows what he calls a logistic curve, apparently independent of human volition. **1947** *Jrnl. R. Statistical Soc.* CX. 134 A most important..paper of Feller..on the application of Markoff processes to a series of population problems normally treated in a deterministic manner leading to exponential or logistic laws of growth. **1969** Sladen & Bang *Biol. of Populations* vi. 72 Pearl used the logistic equation of Verhulst, which is $dN/dt=rN(K-N)/K$. The first part, $dN/dt=rN$, is the exponential equation for growth. K is a constant concerned with realization of the potential, and N is the number of individuals in the population. **1974** *Encycl. Brit. Macropædia* XIV. 831/1 (*caption*) Logistic growth of a laboratory population of the small fruit fly.

d. Of or pertaining to mathematical or symbolic logic.

1918 C. I. Lewis *Survey Symbolic Logic* vi. 343 The logistic method is..applicable to any sufficiently coördinated body of exact knowledge. **1934** *Mind* XLIII. 101 First, he presents 'the basic calculus of exact logic by the logistic method'. **1963** H. B. Curry *Found. Math. Logic* i. 21, I shall discuss briefly the three principal varieties of higher-order logistic calculus.

4. Connected with or pertaining to logistics (cf. *logistics *sb. pl.*[2]).

1934 in Webster. **1957** *Listener* 14 Nov. 774/2 [Of local elections in Ethiopia] Everything had to be improvised; the logistic problems of this complicated terrain had to be solved as much as the psychological. **1958** *Times* 3 Nov. 11/7 When Montgomery, promised extra logistic support, fixed the date for Arnhem, Patton decided to get his forces so involved beyond the Moselle that Supreme Headquarters would find it impossible to find that extra support at his expense. **1971** *Sci. Amer.* Dec. 106/2 The Gombe Stream Centre, which arose from her work, is now a thriving institution with a dozen students and an entire little village of logistic and touristic support.

B. *sb.* **2.** (Later examples: cf. sense *A. 3 a.)

1925 G. U. Yule in *Jrnl. R. Statistical Soc.* LXXXVIII. 5, I have relegated to Appendix II some discussion of the mathematics of the curve, which, following Verhulst, we may term a 'logistic'. **1928** R. Pearl *Rate of Living* vii. 132 The seedling growth curves are slightly asymmetrical, but to a first approximation are sufficiently well graduated by the simple logistic $y=K/(1+e^{a+bx})$. **1974** *Nature* 3 May 12/3 In the absence of competition..the growth rate conforms to the logistic.

3. c. (usu. *sing.*) Mathematical or symbolic logic (see quot. 1918).

[**1905** *Philos. Rev.* XIV. 445 A logical renaissance must be noted... I give it the name of 'logistique' from an old word which appears to be revived. *Ibid.* 453 In the 'Logistique' (the revival of the word was recognized by the congress) the presence alone of MM. Peano, Couturat, [etc.]..was sufficient to guarantee the interest and importance of the questions treated.] **1918** C. I. Lewis *Survey Symbolic Logic* i. 3 Logistic would not have served our purpose because 'logistic' is commonly used to denote symbolic logic together with the application of its methods to other symbolic procedures. Logistic may be defined as *the science which deals with types of order as such*. It is not so much a subject as a method. **1933** *Mind* XLII. 117 Prof. Scholz writes throughout as an enthusiastic student of symbolic logic, or—to use the name more commonly employed on the Continent—Logistic. **1936** A. J. Ayer *Lang., Truth & Logic* iii. 88 The best-known example of such a symbolism [*sc.* artificial language symbolism] is the so-called system of logistic which was employed by Russell and Whitehead in their *Principia Mathematica*. **1956** A. Church *Introd. Math. Logic* (rev. ed.) I. 57 The word 'logistic'..originally meant the art of calculation or common arithmetic. Its modern use for mathematical logic dates from the International Congress of Philosophy of 1904. **1956** E. H. Hutten *Lang. Mod. Physics* ii. 34 There are still people who believe that Gödel's theorem represents the ultimate failure of logic and mathematics. But, to paraphrase Mark Twain, the reports of the early demise of logistic are greatly exaggerated.

logistical, *a.* Delete ? *Obs.* and add: **3. b.** = *logistic *a.* 3 d; of or pertaining to logicism.

1932 H. W. B. Joseph in *Mind* XLI. 439 And it would be consonant so far with the logistical doctrine that pure mathematics is just logic, or an outgrowth of logic. **1943** *Mind* LII. 264 The 'logistical thesis' that mathematics is reducible to logic. **1966** W. V. Quine *Ways of Paradox* viii. 64 (*heading*) A logistical approach to the ontological problem.

4. Pertaining to *logistics *sb. pl.*[2]; = *logistic *a.* 4.

1934 in Webster. **1957** *Economist* 7 Sept. 838/3 Soldiers who can, without any lessening of their primal martial qualities, handle their logistical and procurement problems. **1966** *Listener* 13 Jan. 65/2 You would..allow more time than six months for the 190,000 American troops to effectively dismember the 165,000 Viet Cong troops, an enemy which possesses all the strategical and logistical advantages. **1972** *Sci. Amer.* Feb. 77/3 Logistical problems such as transporting magnetic tapes and field equipment..proved to be as much of a challenge as technical considerations. **1975** *Listener* 9 Jan. 41/1 Sven Hedin in 1927..could show his journeys through East Asia, China and Mongolia with all the emphasis on the logistical dramas.

logistically (lǫdʒi·stikăli), *adv.* **1.** [f. Logistical *a.* + -ly[2].] In a logistic manner.

1932 Lewis & Langford *Symbolic Logic* v. 118 The logistically derived calculus of classes. **1933** *Mind* XLII. 32 We must hold constantly in mind that this logic is to be developed logistically.

2. [Cf. *logistics *sb. pl.*[2]] Connected with or from the point of view of logistics.

1956 R. Braddon *Nancy Wake* xviii. 212 A situation both logistically valuable and humanly comic. **1960** *Daily Tel.* 10 Oct. 24/6 The American official position has always been that it would be impractical, even logistically, to withdraw American troops from Germany and concentrate them, say, in Alsace Lorraine. **1966** M. R. D. Foot *SOE in France* x. 308 It turned out logistically impracticable. **1971** J. B. Carroll et al. *Word Frequency Bk.* p. xviii, Choosing *k* based on the length of individual samples was rejected as being unacceptably laborious and logistically impossible. **1971** *Physics Bull.* Dec. 726/1 If industrial work is to be done by universities, it must be set up and supported in such a way as to be logistically viable.

logistician (lǫdʒisti·ʃən). [f. Logistic *a.* + -ian.] One skilled in logic or logistics.

1932 *Mind* XLI. 433 Logisticians are the last persons who should quarrel with a notational convention whereby a syllogistic 'calculus' can be simplified. **1936** Ogden & Richards *Meaning of Meaning* (ed. 4) v. 95 Logisticians will only be logical when they admit that universals are an analogous convenience. **1937** *Mind* XLVI. 244 Prof. Scholz, who is editing the series of logistical studies to which this work belongs, has the rare distinction of being both an accomplished logistician and an erudite philosophical scholar. **1966** W. V. Quine *Ways of Paradox* viii. 65 Such conformity was the logistician's objective when he codified quantification.

logistics, *sb. pl.*[2] Add to def.: The organization of supplies, stores, quarters, etc., necessary for the support of troop movements, expeditions, etc.

1944 A. H. Burne *Art of War on Land* i. 10 All this must be superimposed on his knowledge of the material conditions—'the logistics' of the situation, as pointed out by Lord Wavell. *Ibid.*, Logistics, the science of moving and supplying troops. The term was invented by Baron de Jomini, but is seldom used in this country. **1947** D. S. Ballantine (*title*) U.S. naval logistics in the Second World War. **1947** Crowther & Whiddington *Science at War* 117 The Americans use the word 'logistics' to describe the technique of packing stores... It is derived from the French '*maître du logis*'. **1963** Mrs. L. B. Johnson *White House Diary* 27 Dec. (1970) 21 Bess Abell and some Filipino stewards arrived at the Ranch yesterday and got preparations under way for Chancellor Ludwig Erhard's visit—the vast logistics of deciding who was going to stay where. **1966** *Aviation Week & Space Technol.* 5 Dec. 22/1 The new plan is to provide, essentially, separate modules for each of at least three major experimentation areas—astronomy and biology, earth resources including meteorology, and orbital operations and logistics. **1971** *Daily Tel.* 8 May 8/6 He was responsible for the administration and logistics of the mine-sweeping fleet in the operation Neptune (the invasion of North-West Europe in 1944).

log-jam. [f. Log *sb.*[1] + Jam *sb.*[1]] **1.** An accumulation of logs in a river; a place where logs become jammed. Cf. *jam *sb.*[1] 1.

1885 E. L. Dorsey *Midshipman Bob* (1886) i. 73 His father got killed in a log-jam. **1897** Kipling *Five Nations* (1903) 39 Do you know that racing stream With the raw, right-angled log-jam at the end? **1900** *Jrnl. School Geogr.* (U.S.) Apr. 153 The breaking of a log jam or an ice gorge on one of our rivers. **1940** J. Buchan *Memory Hold-the-Door* v. 101 He could do what the lumberman does with a log-jam, and pick out the key log which, once moved, set the rest going. **1957** G. E. Hutchinson *Treat. Limnol.* I. i. 115 The successive formation of log jams.

2. *fig.* An obstruction or blockage; a delay; a deadlock. Cf. *jam *sb.*[1] 1 b.

1890 W. James *Princ. Psychol.* I. xi. 451 But at intervals an obstruction, a set-back, a log-jam occurs. **1907** *Springfield* (Mass.) *Weekly Republ.* 14 Feb. 8 The congressional log-jam which held back all legislature

nearly a week was finally broken Thursday afternoon. **1935** *Economist* 4 May 1009/1 A number of corporate refunding issues have been successfully offered.., leading the Secretary of the Treasury to affirm that the 'log-jam in the capital market' has at last been broken. **1951** M. Lowry *Let.* 5 June (1967) 245 There is a kind of log jam in my work. **1962** *Listener* 5 Apr. 597/2 Nothing is likely to break the Arab–Israeli log-jam until the Arabs achieve a greater measure of unity. **1965** *Ibid.* 20 May 737/2 Something may have happened to show us what Mr Ian Smith's electoral victory really means and how the political log-jam has begun to move. **1973** I. M. Sinclair *Vienna Convention on Law of Treaties* v. 139 Informal meetings among leading delegations failed to move the log-jam.

loglet (lǫ·glėt). [f. Log *sb.*[1] + -let.] A little log.

1914 W. De Morgan *When Ghost meets Ghost* ii. vi. 504 She brought a couple of young loglets to keep a little life in the fire.

logo (lǫ·go). Abbrev. of *logogram 2 c or *logotype 2. Also *attrib.*

1937 *Advertising & Selling* 1 Jan. 29 He wrote the first ad ever written for that new-fangled mechanical pencil called 'Eversharp'. Designed a logo for it, too. **1960** O. Skilbeck *ABC of Film and T.V.* 79 *Logo*, the layout of a sponsor's name, brand or slogan. **1967** T. Harknett *Two-Way Frame* xii. 92 An eight-wheeled trailer lorry and a five-ton van, both with the company logo on the side. **1969** *Guardian* 27 May 5/8 Almost every page carries that arch public relations device, an image-making logo. **1970** *Cabinet Maker & Retail Furnisher* 25 Sept. 628 (*caption*) A stylised version of the letter A is the logo used as a feature on Arkana's fleet of vans. The van body is in silver, with the cab and the logo in blue. **1971** *B.S.I. News* Dec. 4/2 With a single gold motif drawn from the Institution's crest and logo, these ties are now available. **1975** *Times* 13 Jan. 13/4 A national airline decided..to have their logo printed on the cover of book matches.

logocentric (lǫgose·ntrik), *a.* [f. Gr. λόγο-ς reason + *-centric.] Centred on reason.

1939 V. A. Demant *Relig. Prospect* v. 124 The connection of this biocentric dogma with a dogma of becoming is obvious. In what respects is the rational, logocentric tradition against which it revolts related to a true dogma of being? **1951** *Mind* LX. 575 Terminating in a logocentric predicament. **1952** [see *biocentric *a.*].

logogram. Add: **2. b.** *Philol.* A symbol or character used, alone or in combination, as the graphic representation of a whole word as a single letter.

1933 L. Bloomfield *Lang.* 287 The Egyptians..represented words not always by one symbol, but also by various arrangements of logograms. **1939** L. H. Gray *Foundations of Lang.* 360 Akkadian as written contains many Sumerian logograms. **1963** Bloomfield & Newmark *Ling. Introd. Hist. Eng.* ii. 35 Mesopotamian and Egyptian peoples developed a rebus technique by which signs for whole words (logograms) could be put together to form longer words.

c. *gen.* A symbol, as found in road-signs, advertising, &c., designed to represent in simple graphic form an object, concept, or attitude.

1966 *Sunday Times* 27 Feb. 11/2 Labour's original badge..has progressively turned into what the trade calls a logogram.

logographic, *a.* **2.** (Later example.)

1970 *Language* XLVI. 959 It now appears that the Proto-Indian script is a purely logographic script based on the so-called rebus principle.

logology. 2. (Later examples.)

1943 *Jrnl. R. Anthrop. Inst.* 5 Logology = the general study of linguistic elements of culture. **1961** K. Burke (*title*) The rhetoric of religion: studies in logology.

logopedics (lǫgopī·diks), *sb. pl.* [f. Gr. λόγο-ς word, speech, after Orthopædics, -ped-.] (See quot. 1951.) Also **logope·dia** [-ia[1]], in the same sense.

1923 Dorland *Med. Dict.* (ed. 12) 622/1 Logopedia, logopedics. **1951** S. D. Robbins *Dict. Speech Path. & Therapy* (1961) 67 *Logopedia, logopedics*, the study and correction of speech and voice defects and disorders as a general field of knowledge. **1960** F. T. Wien (*title*) Current problems in phoniatrics and logopedics. **1969** J. H. van Thal *Elem. Logopedics* i. 10 Logopedics, which is widely used in many countries..is more comprehensive ..than any other term in current use.

logophobia (lǫgŏfōu·biă). [mod.L., f. Gr. λόγο-ς word + -phobia.] Fear or distrust of words.

1923 [see *God-box* (*God *sb.* 16 a)]. **1933** H. R. Huse *Illiteracy of Literate* 48 The fear behind this logophobia is that secret names can be used in incantations. **1959** W. H. Mittins in Quirk & Smith *Teaching of English* iv. 126 The interest generated by such an approach might even help to overcome that language-resistance—the Americans call it 'logophobia'—which some teachers claim to be meeting more and more.

logopœia (lǫgŏpī·yă). [a. Latinized form of Gr. λογοποιία f. λόγος word + ποιεῖν to make + -ια abstract fem. ending.] (See quot. 1929.)

1929 E. Pound in *N.Y. Herald-Tribune* 20 Jan. xi. 5/4 Logopoeia, 'the dance of the intellect among words', that is to say, it employs words not only for their direct meaning, but it takes count in a special way of habits of usage, of the context we expect to find with the word... It holds the æsthetic content which is peculiarly the domain of verbal manifestation and can not possibly be contained in plastic or in music. **1934** —— *ABC of Reading* iv. 21 You still charge words with meaning mainly in three ways, called phanopoeia, melopoeia, logopoeia. **1957** N. Frye *Anat. Crit.* 244 The context that Ezra Pound has in mind when he speaks of the three qualities of poetic creation as *melopoeia, logopoeia*, and *phanopoeia*.

logorrhœa, logorrhea (lǫgŏrī·ă). [f. Gr. λόγο-ς word + ῥοία flow, stream (prob. after Diarrhœa).] Excessive volubility accompanying some forms of mental illness; also *gen.*, an excessive flow of words, prolixity. So **logorrhœ·ic, logorrhœ·tic** *adjs.*

1902 Baldwin *Dict. Philos. & Psychol.* II. 30/1 Logorrhea refers to the excessive flow of words, a common symptom in cases of mania. **1907** *Daily Chron.* 13 Feb. 7/4 In the case of a man suffering from the insanity known as logorrhea the ideas come rapidly tumbling over each other. **1935** *Punch* 5 June 662/2, I have invented logorrhœa—or, if you prefer it, logorrhage. Like pyorrhœa.., it afflicts three out of four. **1960** *Spectator* 21 Oct. 591 Protective shields against the prevailing logorrhoeic fall-out. **1965** *Listener* 14 Jan. 62/3 No one could, or would want to, surpass that logorrhoetic master, except himself. **1965** W. R. Brain *Speech Disorders* (ed. 2) v. 56 Patients with sensory or Wernicke's aphasia or jargon aphasia include logorrhoeic patients with abundant paraphasias and serious defects of comprehension. **1970** *Daily Tel.* 5 Feb. 6/4 We are left with a tedious tale of complicated intrigues written by an author suffering from acute logorrhoea. **1970** Hinsie & Campbell *Psychiatric Dict.* (ed. 4) 751/2 Also known as *logorrhea*..[tachylogia] is characteristic of the manic phase of manic-depressive disorder.

logotherapy (lǫgŏþe·răpi). *Psychol.* [ad. G. *logotherapie* (V. E. Frankl *Ärztliche Seelsorge* (1947)), f. Gr. λογό-ς reason + Therapy.] An existential type of psychotherapy which maintains that man's mental health depends on awareness of meaning in his life.

1948 *Amer. Jrnl. Psychotherapy* 685 To help his patients in their spiritual distress the author [*sc.* Frankl] uses the art of reasoning, i.e., 'logotherapy', and as his method is principally based on an explanation of the meaning of existence, he calls it 'Existential Analysis'. **1953** *Ibid.* VII. 10 To overcome psychologism we have introduced a psychotherapeutic method which we have called *Logotherapy*, an heuristic opposite to psychotherapy in its hitherto accepted narrow psychological sense. Logotherapy is to be understood as a therapy which derives from spiritual sources, and aims toward a spiritual goal. **1964** I. Lasch tr. *Frankl's Man's Search for Meaning* ii. 98 In logotherapy the patient is actually confronted with and reoriented toward the meaning of his life. *Ibid.* 99 According to logotherapy, this striving to find a meaning in one's life is the primary motivational force in man. **1968** Reinecke & Bailey tr. *Rudin's Psychotherapy & Relig.* ix. 188 We should not bypass Viktor Frankl's logotherapy, since this is the approach which most encompasses the spiritual dimensions of man. **1972** W. C. Coe *Challenges Personal Adjustment* vi. 141 V. E. Frankl..has developed what he calls 'logotherapy'. Its aim is to provide meaning..to man's existential being, uniqueness, and responsibility for self-actualization.

logotype. Add: **2.** = *logogram 2 c.

1957 *Archit. Rev.* CXXII. 421 These air outlet grills carry the company's loggotype [*sic*] (the three letters IBM which are the company's 'signature') in the form of magnetic plate, the position of which can be adjusted. **1968** *Heidelberg News* (Heidelberg Printing Machinery Co.) Sept. 2/2 It could concentrate its resources, as some printers have already done, on creating house styles, designing company logotypes, [etc.]. **1970** *Railway World* Apr. 166 A tatty sign affixed to the inside of the station windows (latterly in magnificent British Rail characterless logotype). **1970** *Sat. Rev.* (U.S.) 12 Sept. 94/1 The familiar *Life* logotype appears in the upper left corner. **1974** *Globe & Mail* (Toronto) 8 Feb. 12/5 The symbol was designed by Burton Kramer and Allan Fleming, authors of a number of other visual identikits including Canadian National Railways' CN logotype.

log-roller. 1. (Further examples.)

1897 [see *back-scratching *vbl. sb.*]. **1966** *Listener* 3 Mar. 324/1 Whether as editor,..impresario, log-roller, or friend, Ford seems to have been mixed up with almost every major writer of his time. **1968** *New Scientist* 5 Sept. 474/2 The same old professional log-rollers are going to say lightly-disguised variations of the thing they said to the same applauding audiences they met somewhere else last year. **1974** *Listener* 10 Jan. 54/3 D. G. Rossetti..would set Swinburne and the rest of his..practised log-rollers to promote his discovery.

log-rolling. 2. For *U.S. slang* read *colloq.* (orig. *U.S.*). Add further examples. Also *attrib.* or as *adj.*

1838 J. A. Quitman *Let.* 13 Dec. in J. F. H. Claiborne *Life & Corr. J. A. Quitman* (1860) I. 165 Tending to promote combinations and log-rolling schemes. **1860** S. Mordecai *Virginia* xxx. 303 But the log-rolling system of Virginia has diverted her energies from the completion of any one useful work. **1869** *Atlantic Monthly* Sept. 365/2 The log-rolling lobby generally exerted their powers upon objects which possessed a public character. **1889** G. B. Shaw *London Music 1888–89* (1937) 245, I received them with imprecations, having exhausted every form of words that logrolling amenity could take. **1914** W. B. Yeats *Responsibilities* 78 Log-rolling cranks and faddists. **1919** *New Statesman* 2 Aug. 437/2 There is no such thing as gratitude in politics—a fact which is perhaps the chief security we have against a universal orgie of log-rolling. **1929** A. Douglas *Autobiogr.* xxxiii. 222, I was involved in half-a-dozen controversies which covered matters of principle and the welfare of letters, as opposed to log-rolling and corrupt cliques. **1932** N. M. Butler *Looking Forward* vi. 87 This is no time..to permit log-rolling combinations of special interests to use public authority for their own benefit. **1957** J. S. Huxley *Relig. without Revelation* vi. 136 Politics would degenerate into a game of log-rolling. **1967** V. Nabokov *Speak, Memory* (ed. 2) xiv. 284 In their attitude toward literature they were curiously conservative; with them soul-saving came first, logrolling next, and art last. **1975** *N.Y. Times* 4 Mar. 33/3 In fact, a logrolling system, from which women rarely benefit, is the norm for faculty hiring.

Logudoro (lǫgudō·ro). The name of a town or area of Sardinia, used *attrib.* to designate the dialect or language used there. Hence **Lo:gudore·se, Lo:gudore·sian** *adjs.*

1849 J. W. Tyndale *Island of Sardinia* II. vi. 264 (*heading*) The Logudoro dialect. **1885** R. Tennant *Sardinia* v. 63 The rendering of the Lord's prayer in Sarde or Logudoro language, (so called to distinguish the Sarde proper from its northern and southern dialects) will give the best idea of the composite character of the language. **1946** Priebsch & Collinson *German Lang.* (ed. 2) i. 14 The Logudoresian dialect retains certain relict-words with affinities in Basque. **1952** *Archivum Linguisticum* IV. i. 92 Alghero, in Logodurese territory,..was an important centre of Catalan influence. **1954** Pei & Gaynor *Dict. Ling.* 190 The most important [Sardinian dialects]..are: Campidanese..and Logudorese (Logudoresian), spoken in the southern and central parts of the island. **1965** W. S. Allen *Vox Latina* i. 14 The Logudoro dialect of Sardinia.

logwood. Add: **3.** The extract of logwood used for colouring or dyeing. Also *attrib.*

1876 [see *chrome-black*]. **1880** *Encycl. Brit.* XIII. 80/1 Such an ink is costly..on account of the concentrated condition in which the logwood must be used. **1935** C. A. Mitchell *Documents & their Sci. Exam.* 56 Iron logwood inks have a greenish shade which gradually becomes black as the writing dries. **1966** J. S. Cox *Illustr. Dict. Hairdressing & Wigmaking* 91/1 Logwood... Used in the 19th cent. for dyeing hair. **1969** T. C. Thorstensen *Pract. Leather Technol.* xi. 173 Logwood has been extensively applied in the development of base colors in upper leathers, particularly calf.

logy, *a.* For *U.S.* read *N. Amer.* and add further examples.

1907 J. G. Millais *Newfoundland* 339 Logy, heavy, dull. Thus, a *logy* day. **1935** H. Davis *Honey in Horn* iv. 37 Ordinarily he could have out-wrestled her..but he was fagged and logy. **1955** *U.S. Bureau Amer. Ethnol. Bull.* No. 159. 277 Mrs. Murphy informed me that Ute medicine men placed a root (unidentified) in the mouth of an opponent's race horse to make it logy. **1955** W. Gaddis *Recognitions* ii. i. 291 And do you feel run down at the end of the day? that dull logy tired feeling that just seems to creep through you? **1973** E. Pace *Any War will Do* (1974) ii. 97 The heat, the flies, the logy ground swell.

Lohan (lō·hān). [Chin.] = *Arhat, Arahat.

1878 H. A. Giles *Gloss. Far East* 81 'Lohan cash' were cast in the reign of the Emperor K'ang Hsi, and were thus honourably named because believed to contain gold. **1880** J. Edkins *Chinese Buddhism* 5 Among [those]..who have gone deeper than the others into the profundities of Buddhist doctrine, are included those called..by the Chinese, P'usa and Lohan. **1905** *Daily Chron.* 26 Oct. 3/3 To call the Lohans or disciples of Buddha, 'Genii awaiting transformation into Buddhas' sounds as strange as if the twelve Apostles were described as awaiting transformation into Christs. **1936** *Discovery* Jan. 23/1 These T'ang figures and the great pottery Lohan, the Buddhist priest in contemplation, all live. **1964** M. Medley *Handbk. Chinese Art* 47 They [*sc.* Arhats] are called Lohan by the Chinese. **1971** L. A. Boger *Dict. World Pott. & Porc.* 201/2 The images of the Lohan are derived chiefly from the art work of one or two painters of the T'ang dynasty.

loiasis, var. *loaiasis.

‖ **loi-cadre** (lwa kadr'). *Fr. Pol.* [Fr.] A general outline law, the principles of which can be applied by the government in succeeding parallel situations.

1953 *Ann. Reg. 1952* 202 Eight votes of confidence. On the first—to empower the Government to draw up skeleton laws (*lois cadres*) reorganizing the nationalized industries. **1957** *Observer* 27 Oct. 15/6 It was a 'Loi-Cadre' for Algeria which provoked the collapse of the French government. **1967** *Economist* 29 Apr. 435/2 The French have long used and abused the technique of decree-laws and *loi-cadres* to overcome the difficulties of inefficient rule by the assembly. **1968** *Encycl. Brit.* I 624A/1 De Gaulle ignored a *loi-cadre*, laboriously passed by the French National Assembly, which provided for the setting up of regional assemblies in Algeria.

loid (loid). *Criminals' slang.* Also **'loid.** [Shortened f. Cellu)loid *sb.*] A celluloid strip used by thieves to force open a spring lock. Also *attrib.* Also as *v. trans.*, to break open (a lock) by this method; to let (oneself) in by this method. Hence **loi·ding** *vbl. sb.*

1958 M. Procter *Man in Ambush* xvi. 202 You said you could use a loid. Let's see you open that door. **1960** *Observer* 24 Jan. 5/5 Got yer stick (jemmy)? Got yer 'loid (celluloid strip for spring locks)? **1968** 'G. Bagby' *Another Day* vi. 107 What point..could there be in changing the cylinder..when..my visitor had managed entry by..'a loid job?'. He had worked a strip of heavy celluloid in over the lock tongue and pushed it back. *Ibid.* ix. 174, I loided myself into my apartment. **1968** *Observer* 10 Mar. 25/4 Mortice deadlocks with five or more levers, difficult to pick and impossible to loid. *Ibid.* 25/5 Doors are opened by picking, loiding, or using a false key. **1968** B. Turner *Sex-Trap* xiv. 134 'Have you got keys to all Creedy's places?' 'Beatty has. I use a loid myself.' He showed a tapered wedge of blank celluloid.

loin, *sb.* Add: **3.** *loin-rag* (= *loin-cloth*), *-steak*.

1929 D. H. Lawrence *Escaped Cock* ii. 49 He peeped round..adjusting his loin-rag. **1938** R. Graves *Coll. Poems* 184 Nor yet that brooding Hindu heat For which a loin-rag and a dish of rice Suffice until the pestilent monsoon. **1868** *Rep. Iowa Agric. Soc. 1867* 127 The reason..is the same that persons have for preferring loin-steaks to those cut from just aft of the horns.

loiner (loi·nəɹ). *slang.* [Origin uncertain.] An inhabitant of Leeds, West Yorkshire.

1950 M. Marples *University Slang* 180 *Loiner*,..possibly a corruption or mispronunciation of *oiner*. **1967** P. Ryan *How I became Yorkshireman* iii. 16 The lunchtime audience of Leeds loiners applauded vociferously. *Ibid.* ix. 59, I ran through the ranks of rumbling loiners and out into the eternal, grey twilight of Leeds.

loiter, *v.* Add: **1. a.** Freq. in legal phr. *to loiter with intent* (to commit a felony).

1891 *Act* 54 & 55 *Vict.* c. 69 § 7 The provisions [shall be] applied also to every suspected person or reputed thief loitering about or in any of the said places and with the said intent. **1899** C. Rook *Hooligan Nights* i. 16 You get lagged for loiterin' wiv intent to commit a felony or some dam nonsense like that. **1952** *Economist* 26 Jan. 207/3 Montgomery is always suspected of loitering with intent. **1957** [see *case *v.*[2] 5].

2. b. (Later example with *out*.)

1863 Lytton *Caxtoniana* I. 50 The little lake..on the banks of which I loitered out my schoolboy holidays.

‖ **lokal** (lokā·l). [Ger.] A local bar, a night-club.

1903 *Pop. Sci. Monthly* Dec. 126 The entire body, like a German scientific gathering, gravitated after adjournment to a summer garden or winter 'Lokal'. **1931** Wyndham Lewis *Let.* 7 Feb. (1963) 200 If Miss Hamilton insists upon..robbing the night *lokals* of Berlin of foreign custom, I at least will have nothing to do with it. **1947** L. Hastings *Dragons are Extra* vii. 157 Tourists were filling the *lokals* in the evenings. **1949** A. Wilson *Wrong Set* 170 Her later life of boites and lokals. **1969** A. Glyn *Dragon Variation* viii. 229 Drinking beer in a keller, singing German songs, drinking wine or apfelsaft in a Schwabing *lokal*, and singing quite different songs.

‖ **lokanta** (lokæ·ntă). [Turk.] In Turkey: a restaurant.

1954 M. Gough *Plain & Rough Places* ix. 142 The courtyard..is occupied, in summer-time, by the *lokanta*—the restaurant. **1964** D. C. Hills *My Trav. Turkey* xi. 120 We had hardly sat down in the *lokanta* when a gendarme captain swooped on me. **1969** 'A. Garve' *Ascent of D. 13* ii. 23 He descended the rock in time to lunch at a small *lokanta*, on watermelon, *shish kebab* and yoghurt. **1969** *Daily Tel.* 4 Jan. 15/3 There are many nearby *lokantas* where you can try Turkish food and wines.

lokoum, var. *locoum.

‖ **Lok Sabha** (lǫk sā·bă). [Skr. *lok* people + *sabhā* assembly, council.] The lower house of the Indian parliament.

1954 Binani & Rama Rao *India at Glance* (rev. ed.) 178 House of the People... Called Lok Sabha with effect from May 14, 1954. **1960** *Guardian* 24 Nov. 11/5 The Loksabha ..endorsed the Indian Government's..policies pursued during the current session of the UN. **1969** *Times* 13 Oct. (Indian Suppl.) 6/4 The fact that no Harijan or tribal member was elected to the Lok Sabha in the 1967 general election from a general seat is significant. **1971** *Femina* (Bombay) 2 Apr. 15/1 Mrs. Indira Gandhi announced the dissolution of the Lok Sabha. **1972** *Times of India* 28 Nov. 8/2 As soon as Mrs. Gandhi finished her statement in the Lok Sabha, the former minister for parliamentary affairs, Mr. Raghuramaiah, rose to seek some clarifications to assuage feelings in Andhra.

‖ **lokshen** (lǫ·kʃən), *sb. pl.* Also **lockschen, lockshan, lockshen.** [Yiddish, pl. of *loksh* noodle.] Noodles. Also *attrib.*, esp. *lokshen soup*.

1892 [see *farfel]. **1934** L. J. Greenberg *Jewish Chron. Cookery Bk.* 167 Thickly grease a pudding basin and sprinkle thickly with brown sugar; then put in a layer of lockshen. **1964** G. Sims *Terrible Door* xii. 67 Come on: the lockshan soup, the potato lutkas, are calling! **1969** *Coast to Coast 1967–68* 179 After the prayers Mrs Katzen served traditional fare; tonight it was lokshen soup, chopped herring, [etc.]. **1973** *Jewish Chron.* 20 July 23/3 He is not a man to decry lokshen soup.

Lolita (lǭlī·tă). The name of a novel (1958) and its main character by Vladimir Nabokov (1899–) about a precocious schoolgirl seduced by a middle-aged man, used to designate people and situations resembling those in the book. Also *attrib.* and *Comb.*

1959 *Encounter* Feb. 31/2 The melodrama turns this country of common routines into Lolitaland. **1960** B. Fretchman tr. S. de Beauvoir (*title*) Brigitte Bardot and the Lolita syndrome. **1960** *Spectator* 25 Nov. 843 A nymphomaniac launched into her life's work by a Lolita-ish experience with the man who wanted to marry her mother. **1964** C. Dale *Other People* v. 113 He drew back. .. This was June from next door, a sweet kid but a schoolgirl... Christ, he wasn't a Lolita type! **1967** H. Hunter *Case for Punishment* viii. 134 Jack Carter's our first assistant and he's absolutely head over heels in love with Jinnie Turner of the sixth form. A real Lolita affair. **1972** *Guardian* 25 Jan. 9/2 Louis Feraud..includes a group of dresses called schoolgirl frocks... Lolita lives again and one longs for the innocence of St Trinian's. **1975** *Listener* 6 Mar. 305/1 Chaplin had an uncontrollable infatuation with young girls... But his Lolita-like relationships in real life rarely matched the spiritual purity of love-on-the-screen.

lollapaloosa, etc.: see *lallapaloosa.

lollie, var. *lolly[1].

lollipop, *sb.* Add: **a.** (Earlier and later, incl. *attrib.*, examples.) Restrict *dial.* to sense in Dict. and add: now a sweet or water-ice on a stick.

1784 *London Chron.* 17–20 Jan. 72/3 She confessed..that a certain person..had enticed her to commit it [*sc.* the robbery], and given her sweetmeats, called lolly-pops. **1845** Thackeray *Legend of Rhine* ix, in *George Cruikshank's Table-Bk.* Sept. 193 Ask the youth whether the lollypop-shop does not attract him? **1944** W. de la Mare *Coll. Rhymes & Verses* 41 A bottle of lollipops loved by Bess Stood apart on a window shelf. **1953** C. T. Williams *Chocolate & Confectionery* vi. 104 The B.C.H. Hollow Sleeve Drop Roller Machine may be augmented by a special attachment for the production of lollipops. **1959** [see *brick *sb.*[1] 4]. **1959** N. Mailer *Advts. for Myself* (1961) 401 He is as pretentious as a rich whore, as sentimental as a lollypop. **1962** *Spectator* 13 Apr. 463/2 What will inevitably be known as the 'lollipop' tax. **1965** Hutton & Bode *Simple Sweetmaking* iv. 39 The lollipops have wax or cellophane paper twisted or tied round them. **1970** P. Villiard *Pract. Candymaking Cookbk.* v. 62, I am going to start your candymaking career off with some recipes for lollipops... You will need a supply of lollipop sticks.

b. Also, a showy or non-serious performance. Also *attrib.*

1952 J. Masefield *So Long to Learn* 203 Many of the speakers were of a kind that I called 'lollipop-speakers': they spoke every kind of verse as if it were a caramel to be sucked, without any glimmering of a notion that the words had any meaning. **1958** *Times* 16 Oct. 4/6 His [Beecham's] reluctance to offer 'lollipop' encores. **1959** *Punch* 16 Sept. 172/2 That will be no more than the 'lollipop' innings of a Compton, the once-in-a-while reappearance of a Cotton or a Kyle. **1972** *Times* 3 July 12/2 Colin Carr chose, mistakenly for a student, a lollipop, a Popper Polonaise, rather than good red meat, but it served to show off an enviable fluency on the instrument.

c. *attrib.*, spec. of or pertaining to a person using a circular sign on a stick to stop traffic so that children may cross the road. Less commonly (occas. in non-attrib. use), the pole bearing a disc used by such a person.

1959 *Courier & Advertiser* (Dundee) 28 Mar. 5/6 These old people, commonly called 'lollipop men', ..do a good job. **1960** S. Potter *Lang. in Mod. World* iv. 49 The traffic warden (or lollipop-man as the children affectionately call him) in his white overall makes himself prominent by raising his red disk on high: 'Stop, children crossing.' **1969** *Sunday Times* 9 Mar. 5 Top-hatted they stream from the school, one boy picks up the lollipop sign—which is hidden in a bush—and traffic is brought to a halt. **1969** *Courier-Mail* (Brisbane) 23 June 7/7 Civilian wardens to supervise school crossings..were used extensively overseas, and were known as 'lollipops' from the big discs on long poles they used to control vehicular traffic through school crossings. **1970** *Sunday Times* 25 Jan. 13 Mr. Blackmore, holding the lollipop that stops most traffic, said, 'That offside brake seems to be pulling.' **1971** *Daily Tel.* 27 July 13/1 (*heading*) Drivers must stop for lollipop men. *Ibid.*, Drivers stopped at school crossings by lollipop patrols must not proceed until the 'Stop, children crossing' sign has been taken away. **1972** *Sat. Titbits* 21 Oct. 20/1 Outside the school a 'lollipop lady' was holding up the traffic. **1972** *Times* 16 Dec. 12/2 We were on the lollipop patrol escorting the kids when they all scattered. **1973** *Daily Tel.* (Colour Suppl.) 16 Mar. 9/4 They watch the children in and out of school, something which on the mainland is done by a single elderly man with a 'lollipop stick'.

lollop, *sb.* Add: **2.** A trifling lazy person.

1896 in Farmer & Henley *Slang* IV. 223/2. **1919** H. L. Wilson *Ma Pettengill* iv. 125 Of course the poor lollop had never been able to think under any circumstances.

lolly[1]. Add: Also **lollie. a.** (Further, incl. Austral. and N.Z., examples.) Now usu. = *lollipop *sb.* a.

In Australia and N.Z. the usual sense is 'sweetmeat'.

1854 C. Spence *Clara Morison* II. ix. 102 Fanny ran away to the nearest lolly shop, and all her brothers and sisters followed her. **1860** C. M. Yonge *Friarswood Post Office* vii. 112 You may take your choice—gingerbread-nuts, or bits of cocoa-nut; or, what's jolliest, lollies with gin inside 'em! *Ibid.* x. 189 The children..bought all the 'lollies'. **1864** H. Weatherley *Treat. Art Boiling Sugar* 5 Whether they consist of the 'Loggets' or 'Cushies' of the Eastern part..the 'Humbugs' or 'Lollys' of the South..they each have their votaries. **1874** V. Pyke *Adventures G. W. Pratt* iii. iv. 78 One of them filled her lap with cakes and 'lollies' for the 'bairnie'. **1886** A. R. Butler *Glimpses Maori Land* iii. 42 A friend of mine goes to the stores..and says, 'I want some lollies.' **1898** 'H.' *Grain of Gold* ii. 4 The first part of every child's education..was to catch the old man by the coat tails and get a pat on the head and a 'lollie'. **1899** *Bulletin* (Sydney) 28 Jan. 11/3 A crowd of well-dressed women amused themselves by throwing lollies, bits of ginger, and small fruits into the mouths of a straw-hat push on the other side. **1911** E. M. Clowes *On Wallaby* iv. 136 Men, when they wanted to show their appreciation of her services, sent her a box of sweets—or lollies, as they are called out here. **1915** [see *chuck *v.*[2] 2 e]. **1935** G. Orwell *Clergyman's Daughter* ii. 131 London hawkers would come with baskets of doughnuts or water ices or 'halfpenny lollies'. **1936** C. R. Allen *Poor Scholar* i. 17 Even Herby fought down a longing for liquorice as they went by his favourite 'lolly' shop. **1955** *Times* 13 June 4/6 The 'real menace of sweets' to dental health was the constant sucking of lollies or other sweets, particularly at night. *Ibid.* 6 Aug. 8/4 A merry little cottage is painted in clear white and lolly pink. **1959** B. Comyns *Vet's Daughter* xii. 100 A scraggy woman with a fringe, dressed in lolly pink. **1961** *Coast to Coast 1959–60* 161 Anyway, Vivi, you look nice in those new shoes. Here's threepence for lollies. **1964** *Weekly News* (Auckland) 10 June 3/3 The most popular event was a giant lolly scramble. Hundreds of children spent a few minutes of furious activity seizing their share of about 7000 sweets tossed among them. **1969** M. Drabble *Waterfall* 87 He had acquired eleven wooden lollie sticks. **1969** *Landfall* XXIII. 99 What was an award for merit promises now to become a lolly-scramble. **1972** J. Wilson *Hide & Seek* i. 8 She looked longingly at the ice-cream van... 'Can I have a lolly?'

b. In Cricket, an easy catch. Also as *adj.* Cf. *dolly *a.* b.

1924 H. De Sélincourt *Cricket Match* v. 160 He..lofted it—a 'lolly' into the hands of point. **1960** J. Fingleton *Four Chukkas to Austral.* xv. 128 May did not last long, giving Benaud a lolly-catch at gully-slip.

c. *slang.* Money. Also *attrib.*

1943 M. Harrison *Reported Safe Arrival* 61 This 'ere bloke touches the Guv'ment fer a nice drop er lolly. **1962** *Spectator* 14 Feb. 194/2 Next year's Budget gives fistfuls of lolly away to everybody. **1965** *Listener* 28 Oct. 677/2 A young English con-man..stands to gain lots of lovely lolly. **1971** *Ink* 12 June 14/1 Bernie's salesmen were bringing in the lolly during the..boom. **1973** *Scotsman* 13 Feb. 8/3 The rank and file Bishops would be better off financially, if they accepted jobs as Principals in the Egg, Poultry and Potato Division of the Ministry of Agriculture... In terms of lolly they would be better off as Senior Officers in the Land, Drainage and Water Supply Division of the Ministry. **1973** G. Moffat *Deviant Death* ii. 14 There's only one person bringing in the lolly in that house.

lolly[2] (lǫ·li). *Canad.* [Shortened f. Loblolly. Cf. *E.D.D.*] (See quots.)

1792 G. Cartwright *Jrnl.* I. p. xii, *Lolly*, soft ice, or congealed snow floating in the water when it first begins to freeze. **1889** W. H. Withrow *Our own Country: Canada* 68 The distance to Cape Traverse is about nine miles, part solid ice, part drifting ice, part water, and sometimes a great deal of broken ice or 'lolly'. **1895** *Dialect Notes* I. 379 *Lolly*, ice and snow in the water along the shore... It is the hardest kind of thing to get the boat through. **1963** *Amer. Speech* XXXVIII. 299 *Lolly*, soft ice beginning to form in a harbor.

lollygag, var. *lallygag *v.* and *sb.*

Lolo (lōu·lōu). [Native name.] The name of an aboriginal people of south-western China, of a member of this people, and of their Tibeto-Burmese language. Also *attrib.* or as *adj.*

1736 R. Brookes tr. *Du Halde's Gen. Hist. China* I. 59 The Nation of the Lolos rul'd in Yun nan, and was govern'd by different Sovereigns. **1878** H. A. Giles *Gloss. Far East* 81 *Lolos*, wild hill tribes of Szechuan and Yün-nan. **1898** A. R. Colquhoun *China in Transformation* i. 24 It is clear there are but three great non-Chinese races in Southern China—the Lolo, the Shan and the Miao-tzu. **1901** E. H. Parker *China* i. 8 Among the mountains of north-east Yün Nan and south Sz Ch'wan, the powerful confederation of so-called Lolo tribes still maintain their independence. **1933** L. Bloomfield *Lang.* iv. 70 The other two groups [of *Tibeto-Burman*], *Bodo-Naga-Kachin* and *Lo-lo*, consist of lesser dialects. **1937** E. Snow *Red Star over China* v. 194 Moving..into Szechuan..[the Reds] soon entered the tribal country of warlike aborigines, the White and Black Lolos of Independent Lololand. **1948** D. Diringer *Alphabet* i. ix. 141 The languages of the Lolo-Mo-so group belong to the Tibeto-Burmese sub-family of the Tibeto-Chinese family of languages... Lolo is itself a sub-group of various languages, spoken by about 1,800,000 people in the south-western provinces of China. **1962** R. & D. Morris *Men & Pandas* iii. 47 Lieutenant J. W. Brooke was murdered in 1910 by the Lolos. **1974** *Encycl. Brit.* V. 567/2 The Tibeto-Burman branch of the Tibeto-Chinese language family falls into three groups: the Tibetan, the Yi (formerly called Lolo) and the Kachin.

loma (lōu·mă), *sb.*[2] *U.S.* (chiefly S.-Western). [Sp., f. *lomo* back, loin, ridge.] A broad-topped hill or ridge.

1849 *Picayune* (New Orleans) 4 May 2/3 [They] were riding quietly along the Loma Blanca, (white hill,) when they came suddenly upon a party of eight or ten Indians. **1863** *Ex. Doc. 37th U.S. Congress Sp. Sess. Senate* No. 1. 20 The new road is to follow the bottom at the edge of the *lomas*. **1923** C. F. Saunders *Southern Sierras Calif.* 75 All about are rounded hills, or lomas, rising in baldness. **1941** *Harper's Mag.* Oct. 498/1 Stand on the 'knoll' at Yucca Loma, drink in the desert, and then look down at your feet.

Loma (lōu·mă), *sb.*[3] and *a.* Pl. **Loma, Lomas.** [Native name.] **A.** *sb.* The name of a people inhabiting the border regions of Liberia, Sierra Leone, and the Republic of Guinea, and of their language. **B.** *adj.* Of or pertaining to the Loma or their language.

1957 *Encycl. Brit.* XX. 624/1 Above the plateau surface rise many mountain ranges, the highest of which are the Loma mountains. 1964 E. A. Nida *Toward Sci. Transl.* ix. 198 In a language such as Loma, spoken in Liberia, it is impossible to duplicate these sequences. 1969 *Liberian Studies Jrnl.* I. 25 The Kpelle, Loma, Bandi and Mende are included in the Southwestern Mande group. *Ibid.* 37 The people of eastern Lukasu and Yawiyasu have several linguistic similarities to the Loma people. 1970 D. Dalby *Lang. & Hist. Afr.* 112 The Loma syllabary (with a total of at least 185 characters) was devised in the 1930's.. reputedly inspired by a dream.

Lombard, *sb.*[1] and *a.* Add: **1. c.** The language of this people. Also *attrib.* or as *adj.*

1598 Florio *Worlde of Wordes* 3 How may we ayme at the Venetian, at the Romane, at the Lombard..at so manie, and so much differing Dialects..as be used and spoken in Italie? *Ibid.* 132/2 *Fio*.. In Lombard Italian for *Figlio*, a sonne, a childe. 1878, 1880 [see *Emilian *a.* and *sb.*]. 1936 G. F.-H. & J. Berkeley *Italy in Making* II. 353 All this he told me in Lombard dialect of which every word had to be translated into Italian by his son.

B. 1. b. *Lombard band* (see quots. 1959).

1936 A. W. Clapham *Romanesque Archit.* ii. 28 The so-called Lombard bands and wall-arcading..are distinctive of the first Romanesque style. 1959 *Chambers's Encycl.* I. 558/2 Shallow external pilasters cutting the wall-surface into bays and commonly called 'Lombard bands'. 1959 E. A. Fisher *Introd. Anglo-Saxon Archit. & Sculpture* 26 Lombard band ornamentation consisted of vertical pilaster strips of slight projection which divided a wall into bays.

Lombard-street. (Further examples.)

1815 *Pancratia* (ed. 2) 367, 9th [round]—Lombard-street to a China orange; Molineux was dead beat. 1821 P. Egan *Real Life London* I. vi. 83 Beat him hollow, it was all Lombard-street to a china orange. 1892 *Evening Standard* 9 Nov. 1/1 We describe the betting upon a moral certainty as being All Lombard-street to a China orange. 1974 *Times* 30 Nov. 10/5 If you didn't already know..then it's most of Lombard Street to a China Orange you'd never find out.

Lombardy. The name of a region of northern Italy, used *attrib.* in **Lombardy poplar,** to designate a columnar variety of poplar, *Populus nigra* var. *italica* (or *P. italica*), which was introduced from Italy to other countries. Also *absol.*

1766 [see Poplar 1 b]. 1797 S. Deane *Newengland Farmer* (ed. 2) 267 The Lombardy Poplar begins to be propagated in this country. 1799 Jane Austen *Let.* 17 May (1952) 62 The drawing-room window..commands a perspective view of the left-side of Brock Street, broken by three Lombardy poplars in the garden of the last house in Queen's Parade. 1882 [see Poplar 1 b]. 1917 L. M. Montgomery *Anne's House of Dreams* vii. 58 The Lombardies down the lane, tall and sombre. 1957 M. Hadfield *Brit. Trees* 151 The Lombardy poplar was once generally held to be a sport from the southern European black poplar. It is now said to be a true species. 1969 T. H. Everett *Living Trees of World* 94/1 The Berlin poplar, a tree of columnar growth and bright green foliage [is] believed to have the Lombardy poplar and *P. laurifolia* as its parents. 1974 A. Mitchell *Field Guide to Trees of Britain* 27 A narrow, columnar or spire-like tree, like a Lombardy poplar, looks much taller than it really is.

Lombrosian (lǫmbrōu·ziăn), *a.* [f. the name of Cesare *Lombroso* (1836–1909), Italian physician and criminologist + -ian.] Of or pertaining to Cesare Lombroso and to his theories of the physiology, psychology, and treatment of the criminal; also as *sb.*, an adherent or follower of Lombroso or his theories. Hence **Lombro·sianism**; **Lombro·sic** *a.*, of or pertaining to Lombrosianism.

1906 Joyce *Let.* 19 Aug. (1966) II. 151 He was ridiculing lombrosianism and antimilitarism. 1914 *Everyman* 16 Jan. 451/2 The 'conclusions' have been so recently published that Lombrosians and anti-Lombrosians must be..excused if they ask for time. 1922 G. B. Shaw in S. & B. Webb *Eng. Prisons under Local Govt.* p. lxii, The stigmata of the Lombrosic criminal. 1973 *Listener* 8 Nov. 624/1 'We don't go in for the Lombrosian type of nonsense of measuring skulls any more,' protest our present-day criminologists.

lomi-lomi. (Earlier and later examples.)

1850 W. Colton *Deck & Port* xi. 347, I was..determined to try..the bath and the 'lomi-lomi'. 1951 *Amer. Speech* XXVI. 23 Other common Hawaiian words are.. *lomilomi* (massage).

Lomongo: see *Mongo[1].

lomonosovite (lomǫ·nosǫvait). *Min.* [ad. Russ. *lomonosovit* (V. I. Gerasimovsky 1941, in *Dokl. Akad. nauk SSSR* XXXII. 498), f. the name of Michael *Lomonosov* (1711–65), Russian mineralogist: see -ite[1].] A phosphate, silicate, and oxide of sodium and titanium, $Na_5Ti_2(Si_2O_7)(PO_4)O_2$, found as dark brown triclinic crystals in the Kola peninsula, U.S.S.R.

1941 V. I. Gerasimovsky in *Compt. Rend. (Doklady) de l'Acad. des Sci. de l'URSS* XXXII. 498 The pegmatite patches..are composed of hackmanite..with a very considerable content of chinglusuite and lomonosovite (a sodium phosphate-titanium silicate). 1966 *Geochem. Internat.* III. 197 It is better to interpret lomonosovite as an inorganic clathrate of murmanite structure and sodium phosphate, with the possible formation of intermediate compounds between lomonosovite and sodium-poor lomonosovite.

London. Add: **London-bottled** *a.*, (of a wine) bottled in London; **London bridge,** a children's singing game; **London broil** *U.S.* (see quot. 1969); **London fog,** a dense fog once peculiar to London and large industrial towns; **London gin,** a dry gin; **London plane,** *Platanus* × *hispanica* (*P.* × *acerifolia*), a hybrid of *P. occidentalis* and *P. orientalis*, often planted as a street tree; **London shrinking,** a finishing process applied to fabric to prevent shrinkage; also **London-shrunk** *a.*

1959 *Times* 21 Sept. 13/2 London-bottled, it costs about 10s. 1972 *Guardian* 24 Feb. 11/2 The difference between London-bottled and château or domaine-bottled wine of the same vineyard..cannot be stated in hard terms. 1972 *House & Garden* Feb. 111/1 With London-bottled clarets starting at £1.50 and château-bottled at £2.50, the prices were not excessive. [1827 R. Thomson *Chronicles of London Bridge* 152 'Here follows the ancient Music to the Song and Dance of London Bridge is broken down.'.. 'A choice piece of simple melody, indeed,' said Mr. Postern,.. 'but you called it also a dance, Mr. Barbican; pray was it ever adapted to the feet, as well as to the tongue?'] 1894 A. B. Gomme *Traditional Games* I. 199 It is singular that the verses of this game [*sc.* 'Hark the robbers'], also enter into the composition of 'London Bridge is broken down'. It is probable, therefore, that it may be an altered form of the game of 'London Bridge'. 1909 *Encycl. Relig. & Ethics* II. 852/1 The singing game known as 'London Bridge' has many variants in the different localities where it is played, but fundamentally the theme is the same. 1939 F. Thompson *Lark Rise* ix. 159 Well-known games still met with at children's parties, such as 'Oranges and Lemons', 'London Bridge'. 1969 I. & P. Opie *Children's Games* viii. 235 A singing game such as 'London Bridge is Falling Down'. 1969 R. & D. De Sola *Dict. Cooking* 143/1 *London broil*, large flank steak broiled, then cut in thin slices diagonally across the grain for serving. 1973 E.-J. Bahr *Nice Neighbourhood* ii. 24 We cooked a London broil out on the grill and ate on the patio. 1974 *Columbia* (S. Carolina) *Record* 24 Apr. 14-B/1 Most steak buffs have their favorite cut, and this brings up something I haven't been able to figure out: the names they give steaks. I'm told that Britons had never heard of London broil until some Yank informed them what it was. 1830 M. Edgeworth *Let.* 8 Dec. (1971) 445 It is so very dark in a thick London fog that I can scarcely see what I write. 1887 [see Fog *sb.*[2] 2]. 1906 W. Marriott *Hints to Meteorol. Observers* (ed. 6) 67/1 *London fog*, the dry, gloomy, irritating fog peculiar to London and other large towns, aggravated by smoke. 1931 E. E. Cummings *Let.* 7 Jan. (1969) 120 A London fog struck the ville at the very moment of departure. 1972 E. Routley *Puritan Pleasures of Detective Story* iii. 36 Holmes..has moral status, and that makes one feel safer when one sets out with him into another London fog. 1920 G. Saintsbury *Notes on Cellar-Bk.* vii. 104 Gin, whether 'squareface' or London or Plymouth, [costs] not much more than half a crown. 1954 M. Sharp *Gipsy in Parlour* xxiv. 230 'Devon cider be a powerful brew,' said my Aunt Charlotte... 'London gin 's a sight worse,' retorted Clara. 1963 A. L. Simon *Guide Good Food & Wines* 721/1 *London Gins*, which differ according to the Distilleries responsible for them. 1970 *House & Garden* Nov. 126/2 The type of gin in most general use is London Dry... Though a few countries ..deem 'London' to indicate a geographical origin,..the term London Dry Gin is mainly accepted as indicating a type of gin. 1860 T. Rivers in *Gardeners' Chronicle* 21 Jan. 47/1 (*heading*) The London Plane trees. 1885 G. S. Boulger *Familiar Trees* 1st Ser. 23 Most of our London Plane-trees belong to an intermediate form. 1930 A. D. Webster *London Trees* 92 The Maple-leaved or London Plane stands first in the list of select trees for planting in towns. 1970 *Nature* 21 Mar. 1159/2 The London plane has proved extremely valuable as a tree that will endure the difficult environments of modern cities. 1957 *Textile Terms & Definitions* (Textile Inst.) (ed. 3) 61 *London shrinking*, a finishing process. 1940 *Chambers's Techn. Dict.* 509/2 London-shrunk (*Textiles*), a term used in the woollen and worsted trades to indicate that a fabric has been specially treated in order to prevent shrinkage during make-up and when worn. 1950 '*Mercury*' *Dict. Textile Terms* 323/2 *London shrunk*. The dry cloth to be shrunk is folded between an upper and lower layer of wet cloth. The cloth is then dried naturally, and afterwards pressed. 1968 J. Ironside *Fashion Alphabet* 239 *London-shrunk*, a process for shrinking wool fabric before tailoring.

Londonish (lɒ·ndəniʃ), *a.* [-ish[1].] Pertaining to or characteristic of London; exhibiting features or aspects of London.

1838 *Civil Engin. & Arch. Jrnl.* 154/2 A modest porch below, and a *neat* viranda—something of that sort, spruce and Londonish. 1852 Geo. Eliot *Let.* 23 June (1954) II. 37 Not that I don't like him..but I want nothing so Londonish when I go to enjoy the fields. 1922 *Sketch* 1 Nov. 194/3 A few mellow Cockney vowels to make us feel cosy —Londonish. 1925 W. Deeping *Sorrell & Son* vi. 56, I had been getting a little—Londonish—shall we call it. 1927 *Observer* 6 Nov. 9/4 The Cromwell-road is at once the most English and the most Londonish of our thoroughfares. 1956 M. Stewart *Wildfire at Midnight* iii. 32 You look a bit Londonish, if I may say so.

Londonization. (Further examples.)

1959 *Daily Tel.* 21 Nov. 15/2 Mr. Grimond..said Londonisation of the whole of the country must be prevented. 1962 *Rep. Comm. Broadc. 1960* (Cmnd. 1753) xiii. 139 The dangers of Londonisation are less than those of isolation.

Londony (lɒ·ndəni), *a.* [-y[1].] Suggestive of London or its characteristics.

1884 L. Troubridge *Life amongst Troubridges* (1966) 169, I thought him so very smart and Londony. 1907 D. O'Connor *Peter Pan Picture Bk.* 27 They made a chimney out of John's tall hat, which he had been Londony enough to bring with him. 1920 J. Galsworthy *In Chancery* I. x. 88 Rather pale she looked and Londony. 1949 D. Smith *I capture Castle* xiv. 263 A cool breeze was blowing in from the Park, smelling of dry grass and petrol—a most exciting, Londony smell. 1974 C. Milne *Enchanted Places* x. 69, I hated wearing overcoats in the country... Overcoats were Londony things.

lone, *a.* Add: **1. b.** *fig.* (Earlier example.) Also *attrib.*

1879 B. F. Taylor *Summer-Savory* xv. 122 In fact, in pretty nearly all his plays he had a 'lone hand'. 1888 Kipling *Barrack-Room Ballads* (1892) 118 A lone-hand raid of the rearmost cart. 1916 *Brit. Dominions Year Bk.* 1917 243 Lone-hand raids on Constantinople. 1922 Joyce *Ulysses* 455 This is a lonehand fight.

3. b. *lone star*, the single star on the state flag of Texas, hence called the *Lone Star State*. Also *Lone Star Stater*, a Texan.

1843 W. B. Dewees *Lett. from Early Settler Texas* (1852) 246 The lone star of Texas shall continue to wave proudly in the air as long as one brave Texan remains to defend it. 1845 *Congress. Globe* 28th Congress 2 Sess. App. 78/3 The 'lone star' has found a place upon the democratic banners. 1848 *Ibid.* 30th Congress 1 Sess. App. 973/1 Texas was then a 'lone star'. She is now one of thirty. 1860 *Ibid.* 5 Dec. 11/3 There is a clog in the way of the lone-star State of Texas in the person of her Governor. 1873 J. H. Beadle *Undevel. West* 805, I am proud to find him in honor and position among the 'Lone Star Staters'. 1873 Z. N. Morrell *Flowers & Fruits* (ed. 2) 20 Sam. Houston was then in Texas..intending..to set in motion 'a little two-horse republic under the Lone Star'. 1886 B. P. Poore *Perley's Reminisc.* I. 315 It took him only from February 28th to April 12th to conclude the negotiation which placed the 'Lone Star' in the azure field of the ensign of the Republic. 1909 'O. Henry' *Roads of Destiny* xvi. 267 The Lone Star State never yet failed to grant relief, [etc.]. 1943 B. House (*title*) I give you Texas: 500 jokes of the Lone Star State. 1971 *Times* 21 Sept. (Ireland Suppl.) 1/4 Two experts from Texas are using Cork as a base..appropriate, since co Cork has always had some of the aggressive independence of the lone star state.

c. *lone wolf*, (orig. *U.S.*) *fig.* (*a*) one who mixes little with others, keeps himself to himself; (*b*) a criminal who operates alone; also *attrib.* Hence (with hyphen) as *v. intr.*, to live, work, operate, etc., alone.

1909 F. H. Tillotson *How to be a Detective* 130 Occasionally the police run across Panhandlers known as 'lone wolves'—that is they do not mix with others of their class. 1927 *Dialect Notes* V. 454 *Lone wolf*, a bandit or house breaker who works without confederates. 1931 *Times Lit. Suppl.* 28 May 415/3 He was the 'lone wolf' of the campaign for federation. 1938 *Amer. Speech* XIII. 195 Lone-wolf v. 1938 E. Bowen *Death of Heart* ii. iv. 249, I am quite enough of a lone wolf as it is. 1944 R. F. Adams *Western Words* 93/1 *Lone-wolfing*, living alone, avoiding companionship of others. 1950 'S. Ransome' *Deadly Miss Ashley* iii. 35 He had been given hardly a dime's worth of information by the lone-wolf doctor. 1953 A. Baron *Human Kind* xvii. 121 They despised his ignorance, his vices and his pitiless lone-wolf philosophy. 1955 *Publ. Amer. Dial. Soc.* xxiv. x. 166 She is..a kind of *lone wolf* thief. 1955 *Times* 11 July 10/1 A 'lone wolf' terrorist. 1959 Anon. *Streetwalker* viii. 154 He's no lone wolf from Leeds or anywhere else. 1959 N. Mailer *Advts. for Myself* (1961) 408 The lone-wolf hope that we can begin to explore a little more. 1966 J. Philips *Wings of Madness* ii. iv. 131 You are in very serious danger if you try to lone-wolf it. 1966 G. Burnett *Dead Account* vii. 51 Remember what I said..no lone-wolfing, no withholding information. 1970 G. F. Newman *Sir, You Bastard* i. 19 An individualist to be watched unless he should develop into too much of a lone wolf. 1973 J. Rossiter *Manipulators* ii. 21 Detective Inspector De Moro..had given him a preliminary reprimand about lone-wolfing operations.

e. *lone pair* (Physical Chem.): a pair of electrons in the outer shell of an atom which are not involved in bonding.

1923 *Chem. & Industry Rev.* 2 Nov. 1051/1 A basic substance is one which has a lone pair of electrons which may be used to complete the stable group of another atom. 1964 J. W. Linnett *Electronic Struct. Molecules* ii. 31 In ammonia there are, therefore, three shared-pairs and one lone-pair.

6. b. More recently also in form *lone* (and *lones*).

1902 Kipling *Just So Stories* 197 They walked in the Wet Wild Woods by their wild lones. *Ibid.* 206 This is the picture of the Cat that Walked by Himself, walking by his wild lone through the Wet Wild Woods. 1908 *Westm. Gaz.* 28 May 2/4 The roads are dusty and dry When you walk 'em all by your lone. 1910 W. M. Raine *Bucky O'Connor* 21 But why for do they let a sick man like you travel all by his lone? 1917 W. J. Locke *Red Planet* vi. 75 After five minutes on my lones, I felt as if I should go off my head. 1941 W. de la Mare *Coll. Poems* 7 As she asks in her lone, This old, desolate crone.

lonely, *a.* Add: **1. b.** Colloq. or dial. phr. *on one's lonely(-o)*: on one's own; alone. Cf. Lone *a.* 6 b in Dict. and Suppl. (prob. infl. by *only*).

1919 D. H. Lawrence *England my England* (1922) 61 Oh, I'm going home by myself to-night—all on my lonely-O. 1924 'K. Mansfield' *Something Childish* 61 So you're on your lonely, missus? *a* 1930 D. H. Lawrence *Phoenix*

(1936) v. 594 A child was to be given a lump of soft clay and told to express himself, presumably in the pious hope that he might model a Tanagra figure or a Donatello plaque, all on his little lonely-o.

6. lonely-heart, a sentimental name for a friendless person; so **(Miss) Lonelyhearts,** a journalist who gives advice in a newspaper or magazine to people who are lonely or in difficulties; also *transf.* and *attrib.*; also **lonely-hearted** *a.* (and *absol.*).

1863 Lonely-hearted [in Dict.]. 1904 W. de la Mare *Henry Brocken* 200 Criseyde..the lonely-hearted. 1931 R. Campbell *Georgiad* i. 15 More lonely hearts are linked by the Reviews Than by the 'Link' or 'Matrimonial News'. 1933 'N. West' (*title*) Miss Lonelyhearts. *Ibid.* 14 Miss *Lonelyhearts* tells the story of a reporter,..detailed to write an agony column and answer daily the letters desperate with human misery addressed to his paper. 1938 G. Greene *Brighton Rock* i. i. 9 Come on over here, lonely heart. 1955 W. Gaddis *Recognitions* iii. ii. 749 Down the bar, the Big Unshaven Man was offered a job writing the lonely-hearts column for a newspaper in Buffalo. 1956 A. Wilson *Anglo-Saxon Att.* i. iii. 59 You're so busy being Miss Lonelyhearts to your public. 1958 M. Dickens *Man Overboard* ii. 31 Rose's weekly show was a toothsome mixture of soap opera and a Lonely-hearts column. 1959 'N. Blake' *Widow's Cruise* 29 He might just be the fulsome, pathetic lonely-heart he appeared to be. 1959 T. Griffith *Waist-High Culture* (1960) ii. 25 With a fellow lonely-heart,..he would drive out to remote lakes. 1959 *Listener* 28 May 924/2 The office of Connie, the girl who writes the 'Lonely Hearts' column. 1975 *Times* 1 Mar. 8/4 *Music Through Midnight*..the BBC's Miss Lonelyhearts spot. The other two nights there is *Contact*, a radio advice and counselling column.

loner (lōu·nəɹ). [f. Lone *a.* + -er[1].] A person who avoids company and prefers to be alone.

1947 *New Republic* 22 Dec. 7 Big John has decided to become a 'loner' for keeps. 1961 *Guardian* 26 Oct. 7/1 The American ex-patriates along the Seine..are what James Jones calls 'loners'. 1964 L. Linton *Of Days & Driftwood* xviii. 88 There are many 'loners' dotted around the coastal area, only accessible by boat. 1970 *Daily Tel.* (Colour Suppl.) 3 July 7 On course, as in private life, he is a loner, a man of few words who finds it impossible to chat and joke with the crowds. 1971 *Universe* 25 June 13/5 Mr Paisley is a bit of a maverick, a loner who won't be tied by party trappings. 1971 *Country Life* 29 July 278/1 A loner done to death by his fellows for stepping out of line. 1972 *Daily Mail* 30 Oct. 7/2 By nature I'm a loner.

lonesome, *a.* Add: **1. b.** *by* (or *on*) *one's lonesome*, all alone, without company or assistance. (Cf. *Lone *a.* 6 b). *colloq.*

1899 C. J. C. Hyne *Further Adventures Capt. Kettle* ii. 31 No, Kettle, if I'm to get well, some white man will have to go up by his lonesome for me, and square that witch doctor by some trick of the tongue. 1908 *Daily Chron.* 13 Aug. 5/7 Then, parting from him,..I went, all by my lonesome, along the Madeira Walk. 1920 B. Cronin *Timber Wolves* 125 'When I marry Amelia Peters,' says George, 'you can hit the trail on your lonesome.' 1953 H. Miller *Plexus* (1963) xii. 442 That evening I wandered off by my lonesome. 1973 G. Beare *Snake on Grave* ii. 10 One of Rommel's 88's had taken care of his old man somewhere in the Western Desert, and that had left Latch on his lonesome.

c. *lonesome for*.

1905 *Smart Set* Sept. 74 [He]..had become exceedingly lonesome for the nice young man. 1935 M. de la Roche *Young Renny* xi. 98 Bob has gone in to see Lizzie. She's feeling a bit lonesome for a sight of him.

long, *a.*[1] Add: **A. 1. c.** *to make a long arm* (examples); *the long arm of coincidence* (earlier and later examples); *to make a long nose* (later examples); *long in the tooth*: (orig. of horses) displaying the roots of the teeth owing to the recession of the gums with increasing age; hence *gen.*, old.

1852 Thackeray *Esmond* I. ii. 50 She was lean, and yellow, and long in the tooth; all the red and white in all the toyshops of London could not make a beauty of her. 1854 C. M. Yonge *Heartsease* I. ii. ii. 146 Rising, and making a long arm, he deposited them on the top of a high wardrobe. 1860 [see *Arm *sb.*[1] 2 b]. 1884 [see Arm *sb.*[1] 2 b]. *c* 1888 C. H. Chambers *Capt. Swift* (1902) ii. 29 I'm not safe here. This place is a hornet's nest. The long arm of coincidence has reached after me. 1895 G. B. Shaw *Our Theatres in Nineties* (1932) I. 229 Mr. Jerome..has discovered that in working the familiar but safe stage trick of *dénouement* by coincidence, the long arm cannot be too long. 1919 J. C. Snaith *Love Lane* xxi. 106 One of the youngest R.A.s on record, but a bit long in the tooth for the army. 1932 J. Conquest *Village Pompadour* xxv. 183 Long in the tooth, he escaped the traps laid by widow, débutante and free-lance. 1933 'R. Crompton' *William—the Rebel* x. 187 They merely made long noses at the Outlaws. 1936 W. S. Maugham *Cosmopolitans* 213 Go on... The long arm of coincidence was about to make a gesture. 1942 *R.A.F. Jrnl.* 27 June 20 Izzy Grant saw one [*sc.* a Gremlin]..making a long nose at him as he went into the ditch. 1951 'E. Crispin' *Long Divorce* xii. 141 'That's stretching the long arm of coincidence rather far.' 'It's pulling the damned thing right out of its socket.' 1957 J. Braine *Room at Top* xii. 124 A trifle long in the tooth, mark you, but she has style, real style. 1963 A. Huxley *Let.* 17 Nov. (1969) 964 Talk about the long arm of coincidence! The mail which brought your note..brought..at the same time a letter from Betty Wendel. 1972 *Sunday Express* 24 Dec. 2/5 To be honest I am getting quite long in the tooth and this is a method of bringing children into my Christmas. 1973 'B. Mather' *Snowline* vii. 83, I made a long arm for the telephone.

f. Also *long manure*.

1839 J. Buel *Farmer's Compan.* xx. 198 Great economy in dung may be effected by feeding these crops with the long manure of the yards and stables, instead of summer-yarding it.

5. b. *long purse*, one in which there is plenty of money; *long shillings*, good wages.

1809 M. L. Weems *Life F. Marion* iii. 26 Great Britain the nation of the longest purse in Europe. 1871 *Scribner's Monthly* II. 551 For longer purses there are hard woods in all combinations. 1910 'Saki' *Reginald in Russia* 105 The long arm, or perhaps one might better say the long purse, of diplomacy at last effected the release of the prisoners. 1910 *Chambers's Jrnl.* Sept. 603/2 There are 'long shillings' to be earned at the docks, but no easy ones; and the work is not only hard but dangerous. 1955 J. I. M. Stewart *Guardians* i. ix. 97 Lady Elizabeth's generalisation that here—in point of the long purse—was a particular in which Quail himself must lead any field.

c. *long suit*, *fig.* one's strong point.

1895 W. C. Gore in *Inlander* Dec. 114 *Long suit*, something one is familiar with or expert in. 1903 A. Adams *Log of Cowboy* xiv. 218 Young Pete..assured our foreman that the building of bridges was his long suit. 1916 E. V. Lucas *Vermilion Box* 26 Organizing has always been your long suit. 1923 U. L. Silberrad *Lett. J. Armiter* iv. 82 Charity's evidently your long suit. 1934 M. V. Hughes *London Child of Seventies* vi. 72 'Can you do simple long division?' 'Oh, yes, Dym,' said I hopefully, for that was my long suit. 1959 N. Coward *Look after Lulu!* ii. 68 Oh Lord! That's a teaser—arithmetic's never been my long suit.

d. *long chance*, one involving considerable uncertainty or risk.

1907 S. E. White *Arizona Nights* i. xiii. 191 He's plumb scared at the prospect of suffering anything, and would rather die right off than take long chances. *Ibid.* ii. iv. 262 He's one of those long-chance fellows. 1938 H. Nicolson *Let.* 17 Feb. (1966) 322, I do not think there is going to be a war yet. Not by a long chance. 1971 D. Eden *Afternoon Walk* ix. 125 It would be a long chance that the one I just saw was the same one.

7. a. Also used advb. without preceding *a* in Jamaican English (see also quot. 1961).

1942 L. Bennett *Jamaica Dial. Verses* 21 Me did tink me always hear sey Missis Queen bannish slavery lang time. 1961 F. G. Cassidy *Jamaica Talk* vi. 107 *Long time* means long ago ('Him gone long time'). 1971 *Jamaican Weekly Gleaner* 3 Nov. 5/1 Tams are also in (well, we did have that long time).

c. Colloq. phr. (orig. *U.S.*) *long time no see*, a joc. imitation of broken English, used as a greeting after prolonged separation.

1900 W. F. Drannan *31 Yrs. on Plains* (1901) xxxvii. 515 When we rode up to him [*sc.* an American Indian] he said: 'Good mornin. Long time no see you.' 1939 R. Chandler in *Sat. Even. Post* 14 Oct. 72/4 Hi, Tony. Long time no see. 1940 [see *hiya *int.*]. 1959 D. Beaty *Cone of Silence* viii. 105 'Hello, Clive.' 'Long time no see.' 1959 C. MacInnes *Absolute Beginners* 68 Hail, squire... Long time no see. 1971 D. E. Westlake *I gave at the Office* (1972) 164 'Hello, Arnold,' I said... 'Long time no see.'

8. (Earlier and later examples of *longest day*.)

1774 J. Andrews *Let.* 11 Aug. (1866) 340, I shall never get the idea out of my mind the longest day I have to live. 1836 A. H. Clough (*title*) The longest day. A poem written at Rugby School. 1911 H. S. Harrison *Queed* xxv. 321 You'd be a marked man to the longest day you lived. 1962 *Times* 27 Sept. 16/4 Mr. Darryl Zanuck's three-hour film, *The Longest Day*,..attempts to recapture some of the immensity of the D-Day operations.

9. c. *long on*: possessing a copious quantity of, having plenty of. Cf. *short on*, also *short of* (Short *a.* 18 e). orig. *U.S. slang.*

1913 Kipling *Divers. Creatures* (1917) 286 He was long on Kings. And Continental crises. 1929 W. R. Burnett *Little Caesar* iv. vi. 147 You're long on regard yourself, ain't you Rico? 1938 S. Chase *Tyranny of Words* vii. 78 Governor Lehmann, deficient in logic but long on human understanding, commuted the sentence. 1967 'H. Howard' *Routine Investigation* ix. 97 The battered Dodge may not have been long on looks, but it started first time. 1969 *Guardian* 22 Jan. 1/7 The new team is admittedly long on business management and short on statesmanship. 1973 *Good Food Guide* 429 Two inspectors describe it [*sc.* a restaurant] as long on gemütlichkeit and short on good cooking.

15. *long-exposure, -period, -range* (further examples), *-stay*; also **long-day,** (*a*) having a long working-day; (*b*) of plants, needing a long period of light each day before flowering.

1892 Long-day [in Dict.]. 1920 Garner & Allard in *Jrnl. Agric. Res.* XVIII. 559 It will be convenient to use the expressions 'long day' as meaning exposure to light for more than 12 hours and 'short day' as referring to an exposure of 12 hours or less. *Ibid.* 578 Hibiscus is a striking example of a long-day plant. 1947 *Sci. News* IV. 129 By and large, short day plants flower if they receive 8–9 hours of light a day, and long day plants flower if they receive 14–16 hours of light a day. 1966 G. E. Evans *Pattern under Plough* xiv. 146 Practical observations on a long-day plant, the lettuce. 1972 *Nature* 21 Apr. 407/1 It would be interesting to know whether other long day and short day plants exhibiting a photoperiodic response..behave in the same way as *Sinapis*. 1975 *Listener* 6 Mar. 319/1 The BBC requires Fitters (Shift-working)... The rate of pay is £62.42 p.w. for long-day shift working. 1902 *Chambers's Jrnl.* Nov. 706/2 A long-exposure survey of the whole heavens with one of the most modern photographic telescopes would indicate, I am convinced, no fewer than five hundred million stars. 1903 A. M. Clerke *Probl. Astrophysics* 348 The typical long-period variable is Mira Ceti. 1923 P. B. Ballard *New Examiner* 107 Long-period testing. 1968 R. A. Lyttleton *Mysteries Solar Syst.* iv. 109 The advent of the vast majority of comets, the so-called long-period comets, cannot be predicted. 1854 Long-range [see *asphyxiant *a.* and *sb.*]. 1932 J. Buchan *Gap in Curtain* i. 64, I..set myself..to a long-range forecast—what would be likely to happen on June 10th a year ahead. 1958 *New Statesman* 18 Jan. 59/1 Never mind your long-range missiles, Johnson said in effect. 1966 T. Pynchon *Crying of Lot 49* iii. 67 Incest or no, the marriage must be; it is vital to his long-range political plans. 1966 *Punch* 3 Aug. 186/1 According to the long-range forecast, there's a wettish month ahead. 1971 C. Bonington *Annapurna South Face* iii. 33 The Nepalese authorities..did not normally allow expeditions to use long-range radios in Nepal. 1952 C. P. Blacker *Eugenics* xi. 316 Long- and short-stay residential nurseries. 1970 *Guardian* 9 July 3/2 In France today..two thirds of the beds in mental hospitals are occupied permanently by long-stay patients. 1972 *Ibid.* 16 Feb. 7/3 Allegations of ill-treatment are confined to four of the long-stay wards. 1974 *Advocate-News* (Barbados) 19 Feb. 1/2 Meanwhile, during the year under review, a total of 189,000 'long-stay' visitors came to Barbados.

16. *long-barrelled, -billed* (earlier and later examples), *-descended, -grained* (also *-grain*), *-lashed, -leafed, -leaved* (further examples), *-rooted, -skirted, -sleeved* (later examples), *-spooned, -trousered.*

1902 *Daily Chron.* 20 Mar. 3/1 The rests for the long-barrelled muskets disappeared just at the beginning of the war. 1969 F. Wilkinson *Flintlock Pistols* 26 Pair of long-barrelled 17th century pistols of very fine quality. 1594 Barnfield *Affect. Sheph.* ii. ix. 13 (Arber), Wilt thou set springes..To catch the long-billd Woodcocke? 1822 J. Fowler *Jrnl.* (1898) 148 We thear for the first time seen the long Billed Bird;..the bill about one foot in length. 1970 S. Trueman *Intimate Hist. New Brunswick* xi. 144 Some seafowl, like the..long-billed curlew, had become either extremely scarce, or extinct over the years. 1847 Emerson *Poems* 53 He would come in the very hour.. And tell its long-descended race. 1866 Mrs. Gaskell *Wives & Daughters* I. xxiii. 260 Osborne was to do great things..marry a long-descended heiress. 1892 'Mark Twain' *Amer. Claimant* xix. 180 Every man is made up of hereditaries, long descended atoms and particles of his ancestors. 1831 J. M. Peck *Guide for Emigrants* ii. 156 The long grained Virginia corn is chiefly produced. 1970 'D. Halliday' *Dolly & Cookie Bird* iii. 33 You all meet over the trolleys with your long-grain rice sacks at the mainline Cash & Carry. 1974 *Times* 10 Jan. 10/1 With long grain rice, when correctly cooked, the grains remain separate. 1856 J. G. Whittier *Panorama* 128 A pleased surprise Looked from her long-lashed hazel eyes. 1913 C. Mackenzie *Sinister St.* I. ii. ii. 167 The long-lashed blue eyes and rose-leaf complexion. 1963 J. Fountain in B. James *Austral. Short Stories* 277 Her long-lashed eyes modestly lowered. 1819 E. Dana *Geogr. Sk. Western Country* 173 The long leafed pine is a stately tree, from 60 to 80 feet, clear of limbs. 1778 G. White *Let.* 3 July in *Selborne* (1789) ii. xli. 235 *Drosera rotundifolia*, round-leaved sundew. [*Drosera*] *longifolia*, long-leaved ditto. 1785 H. Marshall *Arbustrum Americanum* 83 Long-leaved Mountain Magnolia or Cucumber Tree. 1832 D. J. Browne *Sylva Amer.* 228 This invaluable tree is..called Long-leaved Pine, Yellow Pine, Pitch Pine and Broom Pine. 1942 W. de la Mare *Songs of Childhood* 86 The twilight rain shone at its gates, Where long-leaved grass in shadow grew. 1953 E. Sitwell *Gardeners & Astronomers* 3 The long-leaved planets in our garden-shed. 1964 W. L. Goodman *Hist. Woodworking Tools* 21 Inside the court there was a long-leaved olive tree. 1579 T. Lupton *Thousand Notable Things* ii. 28 If the disease be so long rooted. 1902 W. James *Varieties Relig. Experience* xi. 264 It costs, then, nothing..to renounce long-rooted privileges and possessions. 1960 *Farmer & Stockbreeder* 26 Jan. Suppl. 4/1 Long-rooted carrots only grow well in deep, sandy soil. 1821 M. Edgeworth *Let.* 19 Dec. (1971) 296 Very long skirted coat which he holds up often by tucking one hand under inside the bottom of the waist behind. 1921 D. H. Lawrence *Tortoises* 13 A gentleman in a long-skirted coat. 1974 R. Harris *Double Snare* xv. 105 Students..accompanied by their sandalled, long-haired, long-skirted birds. 1897 R. M. Gilchrist *Peakland Faggot* 95 Vignettes akin to those one sees on the porcelain faces of old Derbyshire 'long-sleeved clocks'. 1903 G. F. Abbott *Tale Tour Macedonia* 221 A long-sleeved black jacket. 1964 O. Coburn tr. *Braun-Ronsdorf's Wheel of Fashion* 263/1 The basque bodice, high-necked, long-sleeved and ever more tight-fitting. 1876 J. MacGregor *Rob Roy on Baltic* 286 A long, narrow, light racing-canoe, with a long-spooned paddle. 1964 *Seventeen* Jan. 46 Something very big in beach fashions—Petti's new long-trousered Surfer! 1967 A. Wilson *No Laughing Matter* iii. 362 The long-trousered suit that he had worn this holidays. 1974 I. Murdoch *Sacred & Profane Love Machine* 104 A thin long-trousered boy.

18. long-acting *a. Pharm.*, having effects that last a long time; **long and short stitch,** in embroidery, a flat stitch used for shading; **long-arm,** (*a*) a long-barrelled gun, as a musket, rifle, etc.; (*b*) a device used as an extension of the arm, e.g. a pole fitted with a hook, shears, etc., for lifting objects to, cutting branches, etc., at a height beyond the ordinary reach of the arm; freq. *attrib.*; **long Bertha** = *Big Bertha* (s.v. *Bertha[2]); **long blow** *Austral.* and *N.Z.* [*blow *sb.*[1] 1 c], a stroke of the shears in sheep-shearing which cuts away the fleece from rump to neck; **long bond** *Comm.* (see quot. 1948); **long card** *Contract Bridge* (see quots.); **long-case clock** = *grandfather's clock* [Grandfather 5], also ellipt. *long-case*; **long chain** *Chem.* [*chain *sb.* 5 g], a relatively large number of atoms (usu. of carbon) linked together in a line; freq. *attrib.* (usu. hyphenated); **long chair** = Chaise-longue; **long chalk** (see Chalk *sb.* 6 b); **long cist** *Archæol.*, a type of megalithic tomb having a long and narrow chamber to which there is direct entry; **long clothes** (earlier and later examples); also *fig.*; **long cross,** (*a*) *Printing* (see quot. 1884); (*b*) *Numism.*, a cross of which the arms extend to the outer circle on a coin; **long deal,** in card-playing (see quot.); **long drawer,** a drawer which extends the full width of a chest, wardrobe, etc.; **long dress,** a floor- or ankle-length dress, usu. worn as evening dress; **long ear,** a translation of the native name for a member of an extinct people which inhabited Easter Island and was distinguished by artificially lengthened ears; **long fallow** (see quots.); **long-fed** *a.* (see quot. 1969); **long Forties** *Naut.* (cf. Forty *sb.* 4); **long glass,** (*a*) a full-length looking-glass; (*b*) a drinking-glass approximately three feet long for holding a yard of ale (cf. Yard *sb.*[2] 9 c); **long grass,** used *gen.* of grass or grass-like growth, typical of certain areas in Africa, tall enough, for example, to conceal animals; **long green** *U.S. slang*, dollar-notes, money (cf. *green *sb.* 7 d); **long-haul** *attrib.* (see *haul *sb.* 1 c); **long-house,** (*b*) (earlier examples); also, a long dwelling-house in other areas, esp. a large communal village house in certain parts of Malaysia and Indonesia; **long ink** *Printing* (see quots.); **long john,** usu. in *pl.*, (*a*) a type of long, warm underwear; (*b*) a children's game; (*c*) (in *sing.*) a long coffee table; also *long John table*; (see also sense *18 c); **long jump,** also (with hyphen) as *vb.*; **long-keeping** *a.*, able to be kept for a long time; **long lady** = *farthing-candle* (Farthing *sb.* 5); **long legs** *W. Afr. colloq.* (see quot. 1971); **long-life** *a.*, remaining serviceable (quot. 1946 = remaining radioactive) for an unusually long time; **Long March,** *spec.* the year-long retreat of the Chinese Communists across south-western China during the period of Nationalist government; also (not always with capital initials) in other contexts; **long mirror** = *long glass (a); **long paddock** *Austral.* and *N.Z. slang* (see quot.); **long-persistence** *a.*, applied to a screen of a cathode-ray tube on which a spot remains luminous for a relatively long period after the electron beam has moved elsewhere; **long pull,** (*a*) *Printing*, in the operation of the hand-press, a pull on the bar almost to its fullest extent; (*b*) the practice in public houses of giving over-measure to attract custom; **long rains** [cf. Rain *sb.*[1] 2 b], in tropical countries, the rainy season; **long-room** (later examples); **long s,** a lower-case form of the letter s, printed ſ, no longer in general use; **long sauce** (see Sauce *sb.* 4 a); **long service,** (*b*) also used to denote a less specific period of service; **long ship,** (*b*) *Naval slang* (see quots.); **long-short,** (*a*) (earlier example); **long short story,** a short story (see *short *a.* 26) of more than average length, a novella; also **long-short** *ellipt.*; **long sight** (see also *sight *sb.*[1]); **long silk** *attrib.* of cotton, long-stapled; **long sleeve,** a sleeve which extends to the wrist; also (with hyphen) *attrib.*; **long slide** *Curling* (see quots.); **long-small,** a length of rod used in basket-making; **long song** (see quot.); **long-splice** *Naut.* (see quot. 1968); also as *vb.*; **long-staple** *a.* (earlier examples); also *ellipt.*; **long-stop** (see sense *18 d); **long-straw** *Thatching* (see quot. 1968); **long-straws,** the drawing of straws as a game; **long-sweetening,** (*a*) (earlier and later examples); **long-termer,** a person who is serving a long prison-sentence; **long-time** *a.*, also, extending for a long time into the future; requiring a long time; **long-timer** = *long-termer; **long twelves** *Printing*, a duodecimo (12mo) imposition scheme with the forme arranged in two rows of six long narrow type pages as opposed to three rows of four shorter and broader pages in standard 12mo schemes; **long verse** = *long-line 3; **long wave,** a wave of relatively long wavelength; *spec.* in *Broadcasting*, a radio wave with a wavelength longer than about one kilometre (but less than ten kilometres, in mod. use); freq. *attrib.* (usu. hyphenated); **long week-end,** a week-end holiday of more than the usual length; *fig.* the period between the wars of 1914–18 and 1939–45; **long-wool** (earlier examples of sense *b*).

1951 A. Grollman *Pharmacol. & Therapeutics* vi. 143 For prolonged mild sedation..small doses of a long-acting barbiturate are useful. 1971 D. Clark *Sick to Death* ii. 35 Sally was on long-acting insulin. That means she only had to inject twice a day. 1848 E. C. P. in C. H. Hartshorne *Eng. Medieval Embroidery* 121 'Long and short' stitch is employed for shading. 1960 [see *brick-stitch* (*brick *sb.*[1] 10)]. 1967 E. Short *Embroidery & Fabric Collage* iv. 91 The ones in existence depict the Buddha life size, the large areas of colour being filled in solidly in chain, satin, and long-and-short stitch. 1675 in *Public Rec. Colony of Connecticut* (1852) II. 270 Such Troopers as shall neglect to prouide themselues with long armes, viz. a carbin or muskett..shall be disbanded. 1952 Granville *Dict. Theatr. Terms* 113 *Long-arm*, a long wooden pole used for clearing borders and ceilings, that foul the lines in the flies. 1969 E. H. Pinto *Treen* 406 *Long-arms*, the trigger action, long-arm hook,..for removing objects from crowded windows, was a useful and necessary device in Victorian times... It is still made in modified form, to assist invalids in picking up objects which are otherwise out of reach. 1972 D. W. Bailey *Brit. Mil. Longarms 1815–65* 9 The barrels of military longarms were officially 'browned' from 1815. 1973 *Times* 14 July 12/1 Gang mowers and long-arm rotary cutters for roadside banks and verges. 1919 G. B. Shaw *Peace Conf. Hints* vi. 75 Within range of Long Bertha. 1929 H. B. Smith *Sheep & Wool Industry Austral. & N.Z.* (ed. 3) x. 77 The shearer now gets in a long blow with the machine, running from the britch end to the top of the neck. 1949 P. Newton *High Country Days* v. 49 Laying his sheep full length he swung into the 'long blow'—from rump to neck. 1952 J. Cleary *Sundowners* iii. 138 Paddy was beginning the longest cut, the 'long blow', from the flank to the top of the head. 1956 G. Bowen *Wool Away!* (ed. 2) iii. 36 If you see a shearer with a good long blow he is usually a good shearer. 1948 G. Crowther *Outl. Money* (ed. 2) ii. 73 Those that mature within five years are known as Short Bonds. Medium Bonds run from five to about twenty years, and all above that are Long Bonds. 1936 E. Culbertson *Contract Bridge Complete* i. 39 Low cards established from four-card or longer suits. They are called *long* cards. 1959 C. H. Goren *New Contract Bridge in Nutshell* (1960) 13 What is a long card? In the trump suit, long cards start at the fifth card. In a side suit, the fourth card is considered a long card. 1892 Long-case clock [see *grandfather 5]. 1899 F. J. Britten *Old Clocks & Watches* 320 Some of the earliest long-case clocks were liberally embellished with marqueterie. 1972 *Country Life* 9 Mar. 546/3 Strictly speaking, all clocks of this type should be called longcases, although since Victorian times they have been known to the general public as 'grandfathers'. 1972 B. Loomes *Yorks. Clockmakers* 10 Very few longcase clocks were made after 1860. 1930 *Biochem. Jrnl.* XXIV. 113 A peculiar long-chain fatty acid. *Ibid.* 114 The two long chains are connected to polar groups. 1951 *Sci. News* XXII. 98 The configuration in space of the long chains, formed by the linking together of successive amino-acids, which seem to be a common feature of all proteins. 1964 G. H. Haggis et al. *Introd. Molecular Biol.* ix. 216 Nucleic acids are long-chain molecules, and the individual units linked together to form these chains are called nucleotides. 1974 *Sci. Amer.* Mar. 72/3 The useful properties of a polymer depend almost entirely on the presence of long chains. 1891 Kipling & Balestier *Naulahka* (1892) vi. 54 It was full of white men..lying in the verandah in long chairs. 1929 E. Bowen *Last September* iv. 41 Help Uncle Richard in with the long chair. 1956 E. Ambler *Night-Comers* viii. 195 One of the long chairs was lying across the balustrade. 1925 V. G. Childe *Dawn European Civilization* xiii. 213 The long cists in North France, Belgium, Hessen, and Sweden have a holed-stone for the doorway. 1963 *Field Archaeol.* (Ordnance Survey) (ed. 4) 113 In Scotland, Wales, the Isle of Man, and the South-west of England burials in long cists which are coffin-like arrangements of stone slabs are frequently met with. 1819 Keats *Let.* 24 Sept. (1958) II. 215 A child in a[r]ms was passing by his chair..in the nurses a[r]ms—Lamb took hold of the long clothes saying 'Where, god bless me, Where does it leave off?' 1861 Dickens *Gt. Expect.* II. xvi. 254 He had just finished putting somebody's hat into black long-clothes, like an African baby. 1932 *Times Lit. Suppl.* 29 Sept. 676/3 Fibonacci, the first Christian writer to give a systematic exposition of the Hindu numerals, without which analysis might still have been in its long-clothes. 1683–4 J. Moxon *Mech. Exerc. Printing* (1962) 267 Then he [*sc.* the Press-man] Folds a sheet of the Paper he is to Work long-ways, and broad-ways, and lays the long Crease of it upon the middle of the Long-Cross. 1755 J. Smith *Printer's Gram.* 261 They [*sc.* compositors] lessen the Furniture on both sides the Long Cross, to enlarge the Bottom Margin. 1884 J. Gould *Letter-Press Printer* (ed. 3) 166/1 *Longcross*, the bar that divides a chase the longest way. 1924 *Southward's Mod. Printing* (ed. 5) I. xl. 246 (*caption*) Long cross. 1904 C. L. Stainer *Oxf. Silver Pennies* 50 Long cross voided, each limb terminating in crescent. 1972 *Oxf. Univ. Gaz.* CII. Suppl. No. 3. 50 A selection of 23 silver 'long-cross' pennies (1247–78) from a hoard found at Colchester. 1898 H. S. Canfield *Maid of Frontier* 86 It was what is termed a 'long deal', that is, no winning or losing card had slipped from the dealer's carelessly careful hands. 1810 E. Weeton *Let.* 11 May (1969) I. 261 You will find the necessary keys for the three long drawers. 1928 A. M. M. Douton *Bk. with Seven Seals* 19 They are in the top long drawer. 1975 *Country Life* 20 Feb. 426/2 The chest..has one long drawer..and below that are two deeper drawers. 1949 N. Marsh *Swing, Brother, Swing* iii. 36 She climbed into a long dress, six years old. 1954 J. Masters *Bhowani Junction* xxxii. 275 There's a dance to-night... Please come. Long dress. 1973 H. McCloy *Change of Heart* ix. 102 Long dress? Surely not in these days for a family dinner at home? 1891 W. J. Thomson in *Rep. U.S. Nat. Museum* (1889) 529 This unsatisfactory state of affairs was brought to an end..by a desperate battle, in which the 'long ears' had planned the utter annihilation of their enemies. 1919 K. Routledge *Mystery of Easter Island* xviii. 282 The Long Ears suddenly appear on the island at a much later time. 1958 T. Heyerdahl *Aku-Aku* xi. 353 The mayor..and his ancestors who had made the great statues on Easter Island called themselves long-ears. Is it not strange that they should bother to lengthen their ears so that they hung down to their shoulders? 1960 G. E. Evans *Horse in Furrow* x. 131 A bastard summer-land is so called to distinguish [it] from a true summer-land or long fallow. 1971 *World Archaeol.* III. 135 They [*sc.* the Tifalmin] practise long-fallow cultivation, clearing a patch of forest and abandoning it after two or three years, probably for fifteen years or more. 1909 *Daily Chron.* 12 Oct. 4/4 Long-fed beef, as fed by English farmers, cost 21s. 3d. 1969 Neumann & Snapp *Beef Cattle* (ed. 6) xi. 303 If cattle are fed finishing rations for 8 to 10 months, they are spoken of as 'long-fed' cattle. 1776 T. Pennant *Tour in Scotl. & Voy. Hebrides 1772* II. 145 Quantities of white-fish.. might be taken on the great sand banks off this coast. The long Fortys extend parallel to it. 1928 C. F. S. Gamble *Story N. Sea Air Station* xii. 183 The Grand Fleet was ordered to rendezvous in the 'Long Forties'; the Battle Cruiser Fleet to join farther south. 1843 C. Ridley *Let.* Nov. in *Cecilia* (1958) xii. 141 Little Matt..always gives himself a kiss in the long glass. 1883 J. Brinsley-Richards *Seven Yrs. at Eton* xxix. 322 There was a way of holding the long glass at a certain angle by which catastrophes were avoided. 1942 G. Mitchell *Laurels are Poison* xvii. 186 Have a look at yourself in the long glass. 1953 *Word for Word: Encycl. Beer* (Whitbread & Co.) 37/1 *Yard of ale*, known also as a long glass..held between 2½ and 3½ pints. 1858 E. H. D. Domenech *Missionary Adventures Texas & Mexico* iv. 276 The way of the long grass is not easy. 1863 *Macm. Mag.* Nov. 27/2 The long grass swarmed with hog-deer. 1912 D. Crawford (*title*) Thinking Black: 22 years without a break in the long grass of Central Africa. 1961 *Listener* 7 Sept. 346/1 This is the Africa of the 'long grass' such as *Hyparrhenia* and *Echinochloa* into which, half a century ago, one would have romantically disappeared. The areas of long-grass plains in the Sudan have to be experienced to be believed. 1964 C. Willock *Enormous Zoo* i. 14 Much of it is long-grass country. 1896 Ade *Artie* ix. 79, I never see him do a stroke of work, but he can always make a flash o' the long green. 1903 A. H. Lewis *Boss* xiv. 174 I'd naturally s'ppose that when you went ehy on th' long green, you'd touch th' old gentleman. 1946 S. Newton *Paul Bunyan* x. 63 We'll be there tomorrow afternoon with Napoleon and the long green. 1928 *Electrical Communication* VII. i. 62 Long Haul Single Channel Carrier Telephone Systems Connect Melbourne with..Victoria and South Australia. 1957 [see *haul *sb.* 1 c]. 1961 *Economist* 11 Nov. 575/2 The [airport] buildings themselves are expected to be adequate to meet all long-haul traffic until 1970. 1974 *Times* 5 Jan. 9/2 Other long-haul operators to be recommended..*Kenton Travel International*..offer a round-the-world trip starting at £795. 1975 *Islander* (Victoria, B.C.) 17 Aug. 3/3 Expenses of the long-haul teamsters who rested and stabled their horses there. 1751 C. Gist *Jrnls.* (1893) 51 They marched in under French Colours and were conducted into the Long House. 1753 G. Washington *Diaries* (1925) I. 50 We met in Council at the *Long House*. 1894 *Sarawak Gaz.* 1 May 67/1 The practice of herding together in 'long houses' prevents mental and moral improvement and hinders advance in gardening and planting and agricultural developement generally. 1905 *Chambers's Jrnl.* Oct. 714/1 The grim line of trophies hanging in the village long-house. 1912 Hose & McDougall *Pagan Tribes Borneo* I. iv, The Kenyah village frequently consists of a single long house. 1937 *Discovery* Sept. 257/1 Anga, our guide, who was head of the community on the opposite hill, invited us to visit his Longhouse... This dwelling place was cleverly constructed of bamboo and palm leaves (*atap*). 1949 B. A. St. J. Hepburn *Handbk. Sarawak* xix. 180 The 'long-house' system ensures that the individual incapacitated by illness or accident cannot be ignored or abandoned. 1961 *Listener* 9 Nov. 757/1, I came to my first long house after a journey of hours down the famous Rejan River, in the British territory of Sarawak in Borneo. 1965 C. Shuttleworth *Malayan Safari* ii. 32 The walls and roofs of the long-houses were built of palm leaves. 1966 G. E. Evans *Pattern under Plough* v. 72 The Welsh long-houses..with long sides and opposite doors providing a passage from side to side, and dividing the building roughly in two. 1971 *Lady* 15 July 88/3 The longhouse is an object lesson in community living. 1967 Karch & Buber *Offset Processes* 545 *Long ink*, ink that can be drawn out into a long thin string—such ink has considerable tack which will pull a plate clean and sharp. 1970 E. A. D. Hutchings *Survey of Printing Processes* 200 *Long ink*, an ink which will flow freely from a knife... Such an ink will, when dabbed on a finger and thumb, stretch out without breaking as these digits are drawn apart. 1943 T. R. St. George *C/o Postmaster* 12 Some odd garments affectionately known as 'longjohns'. 1961 A. Smith *East-Enders* ix. 156 In the living room there would be..a long John table, a small cocktail bar. 1961 *Sunday Express* 24 Sept. 20 'Longjohn' coffee table. 1962 W. Schirra in *Into Orbit* 47 We..stripped down to our long johns so that the technicians could plaster us all over with strips of wet paper. *Ibid.* 49 A series of waffle-weave patches on our long john underwear helps to keep the oxygen moving. 1964 *Spectator* 14 Feb. 217 The long john is a homely woollen undergarment of rustic provenience. 1969 J. Gardner *Founder Member* vii. 115 Boysie picked up the clothes... A suit of woollen long johns, a pair of heavy calf-length stockings. 1970 G. E. Evans *Where Beards wag All* xix. 219 The boys played *Long Johns* which they did as they walked along the road throwing the marbles ahead of them. 1971 *New Yorker* 4 Dec. 102/1 (Advt.), One-piece waist-to-toe Lightweight Long Johns with ribbed dress socks. 1972 *Guardian* 30 Nov. 15/6 (Advt.). Big Long John in opulent teak finish... This elegantly styled occasional table..47" × 17" overall width by 14½" high. 1934 R. Campbell *Broken Record* v.

116 An Impala..can long-jump thirty-seven feet without a run. 1963 *Times* 4 Feb. 3/4 He long jumped 26 ft. 10 in. for a world's best indoor performance. 1860 *Trans. Mich. Agric. Soc.* X. 229 That it is impossible to raise winter apples in the South, and that it is necessary to look to the North for a supply of long-keeping varieties. 1861 Mrs. Beeton *Bk. Househ. Managem.* 589 As late or long-keeping potatoes, the Tartan or Red-apple stands very high in favour. 1970 *Guardian* 6 June 12/4 Long-keeping cream..keeps longer than fresh cream but a shorter time than sterilised cream. 1896 Farmer & Henley *Slang* IV. 228/2 *Long-lady*, a farthing candle. 1953 A. Jobson *Househ. & Country Crafts* vii. 81 A farthing candle was known as a Long Lady. 1971 A. Kirk-Greene in J. Spencer *Eng. Lang. W. Afr.* 144 'Long legs' is a commonplace [in West Africa] for using influence in high places to secure a service. 1973 *Listener* 14 June 782/3 'Long leg' is a Nigerian colloquialism denoting corruption. 1928–9 *T. Eaton & Co. Catal.* Fall & Winter 245/2 The famous long-life Minerva Batteries. 1946 *Physical Rev.* LXX. 987/1 (*heading*) Long-life radio-iodine. 1966 *Daily Tel.* 7 Nov. 17/1 The association has already announced plans to buy a half-share in a Liverpool creamery and manufacture 'long life' milk for export to the Middle East and other tropical areas. 1969 *Guardian* 1 Aug. 7/6 Some bread is certainly tasteless but it's the prepacked sliced long-life cotton wool wadding that most people prefer. 1971 *Ibid.* 9 Aug. 7/5 Children may increasingly find themselves drinking longlife or dried milk next term. 1906 Kipling *Puck of Pook's Hill* 168 'Few people nowadays walk from end to end of this country.'.. 'The greater their loss. I know nothing better than the Long March when your feet have hardened.' 1937 E. Snow *Red Star over China* i. i. 19 The historic Long March of 6,000 miles, in which they crossed twelve provinces of China..and triumphantly emerged at last into a powerful new base in the North-west. *Ibid.* iv. vi. 180 The Long March..was begun in October 1934..the Red Army at last reached northern Shensi in October 1935. 1967 L. Deighton *Expensive Place* xxxvi. 217 The Long March meant the Nationalists killed two and a half million. 1970 *Guardian* 14 May 9/3 [Regis] Debray..overestimates the capacity of Latin Americans to envisage a 'long march'. Most Latin American revolutionaries..think mostly about the short, sharp blow that will lead to quick success. 1972 *Times* 23 Oct. 12/1 Mr Chou held a number of important posts and travelled widely before the Long March. *Ibid.* 27 Dec. 6/3 The MPLA had originally been based on the Congo (Zaire) but, after differences, had carried out the traditional 'Long March' (so beloved by revolutionaries when attempting to found a new state) to new bases in Zambia. 1973 *Times* 1 Oct. 6/5 The 'long march' on Besançon yesterday, organized by the leading French trade union organizations, proved a striking demonstration of the impact the six-month-old struggle of the Lip watch plant workers has had throughout France. 1869 L. M. Alcott *Little Women* (1871) II. i. 6 There were no.. long mirrors, or lace curtains in the little parlour. 1960 D. Lessing *In Pursuit of English* iv. 138 All her games were centred around the long mirror. 1933 L. G. D. Acland in *Press* (Christchurch, N.Z.) 4 Nov. 15/7 *Long paddock, the*. Slang for the road. People turn stock out on it, or travel them on it, to get cheap grazing. 1960 *Jrnl. Acoustical Soc. Amer.* XXXII. 1065 An instrument is described which extracts from the complex speech wave..information related to the subjective pitch of a sound. It then displays this information on the long-persistence screen of a revolving cathode-ray tube in such a manner that a continuous graph of pitch vs. time is obtained. 1966 D. G. Brandon *Mod. Techniques Metallogr.* v. 236 Visual observation of the image, even on a long-persistence screen, is very difficult at small probe diameters. 1683–4 J. Moxon *Mech. Exerc. Printing* (1962) 261 A long or a Soaking or Easie Pull, is when the Form feels the force of the Spindle by degrees, till the Bar comes almost to the hither Cheek of the Press, and this is also call'd a Soft Pull. 1770 P. Luckombe *Conc. Hist. Printing* 500 *Long pull* is when the bar of the Press requires to be brought close to the cheek to make a good impression. 1888 C. T. Jacobi *Printers' Vocab.* 77 *Long pull*, when the bar-handle of a press is pulled right over. 1901 *Contemp. Rev.* Mar. 355 The unlettered barmaid..tiring of handling the taps and the long-pull. 1909 *Daily Chron.* 30 Aug. 5/3 As the law stands magistrates have no power to stop the 'long pull'. 1964 *New Statesman* 21 Feb. 283/3 In 1921 a Licensing Act made this permanent—under the homely caption 'Long Pull Prohibited'. 1963 A. Smith *Throw out Two Hands* vi. 70 The plan was to take off from Zanzibar on January 1st... There was so much to be done before the long rains began. 1970 *Kenya Farmer* Feb. 3/2 If the long rains fail altogether—a most unlikely event—then a shortage would follow at the end of this year. 1841 *Knickerbocker* XVII. 458 In the long room of the Village Inn. 1870 J. K. Medbery *Men & Mysteries Wall St.* 22 A chamber is provided at the Exchange, where members may bargain with members at any hour throughout the day. This is known as the Long Room. 1962 S. Potter in L. Frewin *Boundary Bk.* 21 It is not the slightest use simply making vague references to the Long Room. 1808 C. Stower *Printer's Gram.* vi. 143 Since the very general introduction of round, in the room of long s's, many [type] cases have been made upon a plan different from the original ones. 1894 [see Serif]. 1914 A. E. Housman *Let.* 8 Mar. (1971) 410 His date..cannot be much earlier than 1800, as he seems not to use the long *s* except when *t* follows. 1960 G. A. Glaister *Gloss. Bk.* 241/1 It was not until John Bell's edition of Shakespeare in 1775 that the long s was generally discarded. 1830 William IV in W. A. Steward *War Medals* (1915) 348 Discharged soldiers receiving a gratuity for meritorious conduct shall be entitled to wear a medal having on one side the words For Long Service and Good Conduct, and on the other in relief the King's Arms. 1925 A. J. Toynbee *Survey Internat. Affairs 1920–23* ii. 109 Immediate legislation for the abolition of conscription and for setting up a long service army as provided in the Treaty. 1937 B. H. L. Hart *Europe in Arms* xii. 158 From that time onward the army became professional and long-service. 1916 M. T. Hainsselin *In Northern Mists* xvi. 63, I say, Padre, this is a pretty long ship, isn't it?.. Don't you know that 'a long ship' means one where it is a long time between drinks? 1946 J. Irving *Royal Navalese* 110 *Long ship*, a ship, or party, in which there is a long interval between..drinks. 1840 *Knickerbocker* XVI. 22 A buxom, rosy-cheeked girl, with a blue-striped long-short.. was busied around the fire-place. 1906 J. London *Let.* 15 Dec. (1966) 235 *The Times Magazine*..bought..one of my best long-short-stories. 1924 F. M. Ford *Let.* 18 Sept. (1965) 162 As for the novel: Hemingway..gave me the impression that it was a long-short story. 1942 'G. Orwell' in *Partisan Rev.* IX. 159 The paper shortage..may possibly bring back the 'long-short story', a form which has never had a fair deal in England. 1959 *News Chron.* 7 Oct. 8/5 Long shorts by Arthur Miller and Saul Bellow. 1870 J. Yeats *Nat. Hist. Commerce* ii. ii. 200 The long silk cotton of Algeria partakes at the same time of the character of the long silk staple of Georgia. 1814 Jane Austen *Let.* 9 Mar. (1932) II. 93 Mrs. Tilson had long sleeves too, & she assured me that they are worn in the evening by many. 1897 *Sears, Roebuck Catal.* 241/2 Ladies' Long Sleeve Vests..high neck and long sleeves with elastic ribbed cuffs. 1957 M. B. Picken *Fashion Dict.* 215/2 *Long sleeve*, sleeve which ends ½ inch below wrist joint. 1962 *Canada Month* Apr. 26/3 The west.. introduced the long slide now about twice the distance the old-style curler slides before launching his stone. 1969 R. Welsh *Beginner's Guide Curling* xii. 85 The delivery called the 'Slide' or the 'Long Slide' was introduced by Canadian curlers. 1912 T. Okey *Introd. Basket-Making* vii. 76 Some Luke, Long Small and Threepenny will be needed, and a few small two yearling sticks. 1953 A. G. Knock *Willow Basket-Work* (ed. 5) 9 The old trade names ..three feet, Tacks; four feet, Short-Small; five feet, Long-Small. 1856 *Chambers's Jrnl.* 28 June 402/1 An item in those streaming fathoms of verse technically known as 'long songs', in which as many as a hundred favourite ditties are sold for a penny. 1883 *Man. Seamanship for Boys' Training Ships R. Navy* (1886) 106 To form a long-splice with a piece of three and four-strand rope... Unlay the ends of the two ropes to the required distance [etc.]. *Ibid.* (*heading*) How do you long-splice a three or four-strand rope together? 1968 E. Franklin *Dict. Knots* 19 *Long splice*, a splice which has no apparent thickening of the rope at the points of joining. 1802 J. Simons *Let.* 15 Dec. in J. Steele *Papers* (1924) I. 341 Long Staple Cotton is in demand. 1843 *Knickerbocker* XXI. 39 It is here that the most valuable product of our country, the long staple cotton, is raised. 1867 *Harper's Mag.* Aug. 349/1 A bale of uplands cotton..demanding to be bought at the price of long-staple. 1947 *Agriculture* LIII. 448 (*heading*) Estimate no. 1. Long straw method (unclipped surface). 1963 *Times* 22 Apr. 9/2 For the customer it means a neater job and a roof that will last anything from 35 to 60 years instead of the 10 to 20 years of the traditional long-straw thatch. 1968 J. Arnold *Shell Bk. Country Crafts* xiii. 185 One notices that long-straw has a looser, more plastic appearance, compared with the stiff, 'close cropped', brush-like texture of reed. 1835 R. M. Bird *Hawks of Hawk-Hollow* I. ii. 33 Shall we sit down here, and play long-straws for sweethearts? 1714 in *N. Carolina Colonial Rec.* (1886) II. 132 Let who will go unpaid, Rum, long Sweet'n alias Molasses..must be had. 1936 M. Mitchell *Gone with Wind* xxi. 352 The sorghum used for 'long sweetening' did little to improve the taste. 1956 B. Holiday *Lady sings Blues* (1973) xviii. 142 Quite a few girls, especially long-termers in the joint, were lovers. 1970 Long-termer [see *fantasy *v.* 1 a]. 1927 Carr-Saunders & Jones *Survey Social Struct. Eng. & Wales* xx. 228 This is no indictment of the usefulness of long-time forecasts, because it is in any case impracticable to plan so far ahead. 1944 C. Sandburg in B. A. Botkin *Treas. Amer. Folklore* p. vi, A longtime book is this. One reading won't do for it. 1971 B. Malamud *Tenants* 13 Long voyage in a small room. There's a long-time book to finish. 1907 J. London *Road* (1914) v. 139, I know that the long-timers got more substantial grub, because there was a whole row of them on the ground floor in our hall. 1952 'J. Henry' *Who lie in Gaol* viii. 133 The long-timers are allowed to plant a few things in a plot there if they want to. 1770 P. Luckombe *Conc. Hist. Printing* 414 (*caption*) A sheet of long twelves. 1888 C. T. Jacobi *Printers' Vocab.* 77 *Long twelves*, a plan of imposition whereby the pages are laid down in two long rows of six pages. 1972 P. Gaskell *New Introd. Bibliogr.* 81 Long twelves (long 12°), when the sheet is folded once across the shorter side and five times across the longer, again making twelve leaves, twenty-four pages. 1871 H. Sweet in W. C. Hazlitt *Warton's Hist. Eng. Poetry* II. 3 Each long verse has four accented syllables. 1889 C. W. Kent *Elene* 8 The so-called 'long-verse' consists of two hemistichs. 1839 *Trans. Cambr. Philos. Soc.* VII. 95 Any particle *P* revolves continually in a circular orbit... The radius of this circle, and consequently the agitation of the fluid particles, decreases very rapidly as the depth *c* increases, and much more rapidly for short than long waves, agreeably to observation. 1895 H. Lamb *Hydrodynamics* viii. 276 Waves whose slope is gradual and whose length λ is large compared with the depth *h* of the fluid, are called 'long waves'. 1909 E. B. Titchener *Text-bk. Psychol.* i. 60 Let us take..a chart or projection of the solar spectrum, and let us work right through it, from the left or long-wave to the right or short-wave end. 1928 D. Brunt *Meteorol.* v. 38 The term 'low temperature radiation' [is frequently used] to denote the long-wave radiation of bodies at relatively low temperature. 1928 *Chambers's Jrnl.* Jan. 79/1 Many foreign long-wave stations have also been clearly heard with this set. 1963 *Meteorol. Gloss.* (Meteorol. Office) (ed. 4) 155 *Long wave*, in synoptic meteorology, a smooth, wave-shaped contour pattern on an isobaric chart with a wavelength of the order of 2000 km. .. Some four or five such waves..typically extend across a hemispherical chart. 1974 *Encycl. Brit. Macropædia* III. 311/1 [Broadcasting.] Long waves range from 30 to 300 kilohertz; medium waves from 300 kilohertz to three megahertz; and short waves from three to 30 megahertz. 1927 M. Kennedy (*title*) A long week-end. 1933 H. G. Wells *Shape of Things to Come* iii. § 8. 317 The old British institution of the *long week-end* flourishes. 1940 Graves & Hodge (*title*) The long week-end: a social history of Great Britain 1918–1939. 1944 Blunden *Cricket Country* xiv. 149 The name, 'The Long Week-End', has been devised to characterise the period between the two great wars. 1968 'E. Peters' *Grass Widow's Tale* ii. 24, I'm heading north..for a long week-end. 1859 *Trans. Illinois Agric. Soc.* III. 458 The Longwools attain to greater size and shear a larger fleece. 1877 J. Darby in J. Coleman *Sheep & Pigs Gt. Brit.* ii. v. 50 The best flocks of Devon Longwools are..derived solely from Leicester and Bampton—a most valuable cross in every respect... During the past fifteen years these sheep have been designated 'Devon Longwools'.

b. longspur, read: a bunting of the genus *Calcarius*, esp. *C. lapponicus*, the Lapland bunting; (later examples).

1917 T. G. Pearson *Birds Amer.* III. 22 Shore Larks that feed up and down the wintry seashore of New England and the middle states have also many Longspurs among them. 1953 D. A. Bannerman *Birds Brit. Is.* I. 314 While *Calcarius lapponicus* is the sole Palæarctic representative of the genus, the Nearctic fauna includes two other 'Longspurs'—as they are called in America. 1973 R. D. Symons *Where Wagon Led* i. iv. 45 The sun hit the top of the gray-green sage and the longspurs fluttered overhead.

c. longjohn, a tropical South American tree, *Triplaris surinamensis*, of the family Polygonaceæ; **long-leaf pine** *U.S.*, *Pinus palustris* (also *long-leafed, -leaved pine*: see sense *16); **long-pod** (later example).

1910 *Chambers's Jrnl.* Feb. 88/2 Impenetrable jungle, consisting mostly of chinchilla or sand box-trees, with now and then a sand-cocoa or a longjohn. 1969 S. M. Sadeek *Windswept & Other Stories* 30 We will build a nice house..with bamboo and longjohn. *a* 1816 B. Hawkins *Sk. Creek Country* (1848) 60 [On] the uplands to the south are the long-leaf pine. 1901 [see *laurel oak* (*laurel *sb.*[1] 6)]. 1904 T. E. Watson *Bethany* i. i. 8 Ours was just a plain house..of timbers torn from the heart of the long-leaf Georgia pine. 1969 T. H. Everett *Living Trees of World* 50/2 The longleaf pine, an open-headed kind that reaches 120 feet in height and has a natural range from Virginia to Florida and Mississippi is the most important timber tree of the south-eastern United States. 1972 *Country Life* 13 Jan. 104/4 A dependable way to ensure early pickings of broad beans..is to sow the seeds now..for which purpose I prefer longpod varieties.

d. long field (earlier example); also **long-fielder, -fieldsman; long-hop** (also in *Fives*), a ball which a player has ample time to hit after it bounces; **long-stop** (earlier example); also *fig.* a last resort, e.g. in an emergency; also (in literal sense) **long-stopper;** *long-stopping* vbl. sb. (earlier example).

1816 W. Lambert *Cricketer's Guide* (ed. 6) iii. 44 *Long field off side* this man should stand on the off side, between the middle wicket man and bowler at a considerable distance in the Field, so as to cover them. 1897 K. S. Ranjitsinhji *Jubilee Bk. Cricket* ii. 55 Nearly all good long-fielders take the ball, in catching, with their hands close to their bodies about chest-high. 1920 P. F. Warner *Cricket Reminisc.* xiv. 91 Never had he been a long-fielder. 1790 *Reading Mercury* 8 Mar. 3/3 He was the swiftest bowler and best long fieldsman at that time in the kingdom. 1900 A. E. T. Watson *Young Sportsman* 237 s.v. *Fives*, C..must above all avoid so returning it [*sc.* the ball] that it comes into the middle of the outer court as a long-hop. 1767 R. Cotton *Cricket Song* ix, in F. S. Ashley-Cooper *Hambledon Cricket Chron.* (1924) 184, I had almost forgot—they deserve a large bumper—Little George, the long Stop, and Tom Sueter, the Stumper. 1957 *Listener* 5 Sept. 349/1 Like all sorts of longstop laws we keep on the statute-book, but hardly ever use. 1962 *Punch* 11 Apr. 558/1 The National Assistance Board..is the long-stop of the Welfare State. 1973 *Times* 9 Nov. 1/5 The two uninvolved unions may..provide a long-stop cover. 1891 W. G. Grace *Cricket* x. 258 The most expert long-stoppers at the time when long-stop was even of more importance than the wicket-keeper. 1832 P. Egan *Bk. Sports* 348/2 Dick's shin-breakers stop'd them short In 'midst of their long-stopping!

B. 4. b. In the Morse code, a dash (opp. 'short'); a long buzz, etc., sounded as a signal.

1875 W. Thomson *Pop. Lect. & Addresses* (1891) III. 128 [It] renders quick and sure Morse signalling by longs and shorts impracticable. 1902 Kipling *Traffics & Discov.* (1904) 192 In longs and shorts, as laid down by.. Mr. Morse. 1916 [see *buzz *v.*[1] 9]. 1926 R. W. Hutchinson *First Course Wireless* 112 The key in the primary circuit enables the train of sparks to be continued for a long or a short period of time, thus producing the 'longs' and 'shorts', *i.e.* the 'dashes' and 'dots' of the Morse Code. 1943 F. J. Salfeld in *Penguin New Writing* XVII. 41 The alarm sounded: a series of urgent longs on the buzzer. 1948 'J. Tey' *Franchise Affair* x. 112 Do you know morse?.. I shall hoot the initials of your beautiful name on the horn... Two longs and three shorts. 1973 J. Drummond *Bang! Bang! You're Dead!* xxxviii. 134 A buzzer sounded..two longs, two shorts, another long.

7. (Earlier and later examples.)

1848 J. H. Newman *Loss & Gain* i. x. 71 'Reding ought to live here all through the Long,' said Tenby: 'does any one live through the Vacation, sir, in Oxford?' 1852 C. A. Bristed *Five Yrs. in Eng. Univ.* (ed. 2) 37 For a month or six weeks in the 'Long' they rambled off to see the sights of Paris. 1857 Mrs. Gaskell *Let.* 7 Dec. (1966) 490 Arthur Stanley..has just been spending the 'Long' at Moscow. 1861 D. G. Rossetti *Let.* June (1965) II. 406 Amateur workmen..offered on all hands, chiefly university men who stayed in Oxford that 'Long' for the purpose. 1863 G. M. Hopkins *Let.* 22 Mar. (1956) 15 The probability is I shall not see you for an age, unless we manage to meet in the Long. 1920 G. Saintsbury *Notes on Cellar-Bk.* x. 158 A mixture..first imparted to me..by a very amiable Dorsetshire farmer whom I met while walking from Sherborne to Blandford in my first Oxford 'long'.

8. b. *pl.* Long trousers. *colloq.*

1928 *T. Eaton & Co. Catal.* Spring & Summer 219/3 Flannel longs..boy's long trousers made from grey union flannel. 1947 D. M. Davin *Gorse blooms Pale* 57 His first suit of longs, all neatly pressed. 1954 'A. Garve' *Riddle of Samson* i. 15 A pair of grey flannel shorts that looked as though they'd been cut down from 'longs'. 1962 B. Harrisson *Orang-Utan* i. 37 They wanted to buy smart shorts (or, better still, longs), shirts and tie, a radio.

10. b. *pl.* Long-term stocks.

1964 *Financial Times* 12 Mar. 21/1 Partly reflecting technical influences, gilt-edged continued to gain ground, with the 'longs' closing up to $\frac{1}{16}$ better. 1969 *Daily Tel.* 16 Sept. 2 The 'longs' and undated stocks were particularly prominent and Treasury 6¾ p.c. 1995–98 rose a full point. 1972 *Times* 17 June 23/3 The 'longs'..closed 'uneasily steady', dealers said.

long, *adv.* Add: **1. b.** Also, *long as*, ellipt. for *so* (or *as*) *long as*.

1807 Wordsworth *To Small Celandine* in *Poems* I. 22 Long as there's a sun that sets Primroses will have their glory. 1938 G. Greene *Brighton Rock* I. i. 22 'It's all right,' he said, 'long as you are here.'

2. b. (Later examples.)

1849 Thackeray *Pendennis* I. xxv. 239 She fairly told Pen one day..that she felt herself breaking, and not long for this world. 1933 J. Masefield *Bird of Dawning* 43 He was shocked by the roaring wash of the water coming into the after hold. 'She's not long for this world,' he muttered. 1968 L. Goodman *Sun Signs* (1970) 193 These people either radiate incredible vitality or else complain that they're not long for this world.

7. (Later example.)

1887 W. Morris tr. *Homer's Odyssey* xii. 251 As the fisher sits on the headland with a rod that reaches long.

9. a. *long-awaited, -dead, -departed, -felt, -gone, -held, -lost* (as *sb.*), *-waited, -wearing.*

1914 *Times* 25 Aug. 6/4 The long-awaited battle is begun. 1974 *Melody Maker* 23 Mar. 19/3 The release of Jackson Browne's long awaited album 'For Everyman'. 1905 *Daily Chron.* 14 Nov. 3/4 The old Franciscan..mourned frantically for his long-dead brother. 1937 Auden in Auden & MacNeice *Lett. from Iceland* i. 21 Scribbling to a long-dead poet. 1974 A. Price *Other Paths to Glory* i. vii. 79 The two long-dead riflemen. c 1838 E. Brontë *Compl. Poems* (1941) 77 Old Hall of Elbë, ruined, lonely now;..Home of the departed, the long-departed dead. 1869 'Mark Twain' *Innoc. Abr.* xi. 102 Their long-departed owners seemed to throng the gloomy cells. 1952 R. Campbell tr. *Baudelaire's Poems* 19 Sweeping the far-off skylines with a gaze Regretful of Chimeras long-departed. 1862 A. Lincoln *Ann. Message to Congress* in *Evening Star* (Washington, D.C.) 1 Dec. 1/2 The judicious legislation of Congress..has satisfied..the long felt want of an uniform circulating medium. 1936 *Discovery* Mar. 83/2 To satisfy a long-felt want on the part of the serious student. 1950 W. de la Mare *Inward Compan.* 25 A happy house in that long-gone sunshine. 1943 D. Gascoyne *Poems 1937–42* 33 With long-held burning breath. 1960 R. W. Marks *Dymaxion World of B. Fuller* 117/2 Fuller's long-held theory that energy in gases evolves unique local patternings. 1853 Mrs. Gaskell *Cranford* xv. 308, I could no longer confirm her belief that the long-lost was really here. 1920 M. Beerbohm *And Even Now* 80, I was always in hope that when next the long-lost turned up..I should see him. 1928 *Publishers' Weekly* 16 June 2425 The long-waited reminiscences of the British Prime Minister. 1972 *Buenos Aires Herald* 4 Feb. 6/4 The long-waited inauguration in July of the Peligre Hydroelectric Dam. 1908 *Westm. Gaz.* 16 Apr. 4/2 Greasers are fitted everywhere to..add to the long-wearing life of the parts. 1963 *New Yorker* 23 Nov. 15 (Advt.), Our famous shirts..are made ..of exclusively woven, long-wearing materials. 1975 *Country Life* 2 Jan. 32/1 The engine is..low-revving and long-wearing.

long distance. A. 1. (See Long *a.*[1] 15.) (With hyphen.) Forming combinations used attrib. or as quasi-adj., esp. (*a*) of a telephone call; (*b*) of a race; (*c*) of a journey.

1884 *Whitaker's Almanack* 385/1 In America some remarkable trials of long distance telephoning have taken place. 1886 *Sci. Amer.* 2 Oct. 208/2 There is a popular belief that the long distance telephone is crowding the telegraph to the wall. 1887 [in Dict. s.v. Long *a.*[1] 15]. 1897 [see *centurion 3]. 1908 *Sat. Even. Post* 26 Sept. 15/3 A long distance line can only be used by one person at a time. 1919 [see *ceiling *vbl. sb.* 6 b]. 1923 [see *all-red* adj. (*all E. III. 13)]. 1925 W. J. Bryan *Mem.* 487 After the meal he made several long distance telephone calls. 1926 *Times* 6 May 3/4 The following other long-distance trains will also run, calling at the principal stations. 1929 *Daily Express* 7 Nov. 2/5 A Socialist member's resolution urging the nationalisation of railways and long-distance road transport was debated. *Ibid.* 11/3 The airship has now to compete with the flying boat as a long-distance craft. 1933 *Discovery* Apr. 131/2 The 'bypath' system gives faster service on local and long-distance calls. 1934 *Ibid.* Nov. 316/2 This railcar recently set up a world's record for long-distance running. 1934 [see *door-to-door* attrib. (*door 8)]. 1935 *Discovery* Dec. 352/1 Long-distance flight. 1935 G. Greene *England made Me* ii. 47 Put through any long-distance calls. 1959 A. Sillitoe (*title*) The loneliness of the long-distance runner. 1960 *Guardian* 13 June 3/3 Walking down a long-distance Russian train is apt to be..dull. 1961 *Times* 3 Oct. (Computer Suppl.) p. vi/4 Long-distance telephone calls. 1961 L. van der Post *Heart of Hunter* i. i. 25 The men sat with their heads bowed over arms clasped round their knees like long-distance runners from the race of their lives. 1968 *National Fisherman* Aug. 15-A/3 Formosa and South Korea started long-distance tuna fishing later than Japan. 1975 J. Rathbone *Kill Cure* ii. i. 53 A long distance call from Ankara to Istanbul involved time and trouble.

2. A long-distance telephone (call). Also as *adv.*, by long-distance telephone.

1904 'Mark Twain' *$30,000 Bequest* (1906) viii. 44 Aleck's imaginary brokers were shouting frantically by imaginary long-distance, 'Sell! sell!' 1920 Wodehouse *Jill the Reckless* (1921) xx. 295 Calling Izzy on the long distance. 1923 —— *Inimitable Jeeves* xv. 196 He became a sort of Voice Heard Off, developing a habit of ringing me up on long-distance. 1961 Webster, Called her up long-distance. 1969 A. Glyn *Dragon Variation* v. 147 He went straight to the telephone and called New York long-distance.

B. As *v. trans.* To make a long-distance telephone call to (a person); to report by means of such a call.

1945 *Time* 6 Aug. 79/1 Henry J. Kaiser last week long-distanced an old friend. 1950 *Newsweek* 30 Oct. 61 A UP staffer simply copied the story as it came in, went to an open phone, and long-distanced it to UP in San Francisco in a few seconds. 1965 E. Lacy *Double Trouble* viii. 79 My own family is down in the Bahamas, and I long-distanced them.

longer (lǫ·ŋgəɹ), *sb.*[3] *Canad.* (Atlantic Provinces). [f. Long *a.*[1] + -er[1].] A long pole or piece of timber used for fencing, a fishing stage, etc.

1772 G. Cartwright *Jrnl.* 17 Apr. (1792) I. 216 At noon I..searched the woods..where I found some good longers, and boat-hook staffs. 1837 *Times* (Halifax, Nova Scotia) 25 July 235/1 The skeleton of a man was found in the woods..by a man and boy who were cutting longers. 1878 *North Star* (St. John's, Newfoundland) 30 Mar. 3/2 On the afternoon of their death, the deceased.. left home for the woods to draw 'longers' across the pond. 1973 *Canad. Antiques Collector* Jan.–Feb. 13/2 Fences of longers made from spruce or marsh juniper meet and sometimes mingle with the hedgerows.

longeron (lǫ·ndʒěrǫn). [a. F. *longeron* stringer, beam, (longitudinal) member.] A frame member running lengthways along a fuselage.

1912 *Flight* 13 July 626/1 In front, the four longerons are assembled into a specially designed pressed-steel housing. 1915 W. E. Dommett *Aeroplanes & Airships* ii. 26 The framework has as its ground-work four members running fore and aft known as 'longerons'. 1931 J. E. Younger *Airplane Construction & Repair* viii. 130 Longerons are usually made of ash. 1962 F. I. Ordway et al. *Basic Astronautics* xi. 444 (*caption*) Space frame and longerons. 1966 E. V. Rickenbacker *Rickenbacker* (1968) vi. 107 The explosive bullet had hit the longeron just behind me.

longeur: see *longueur.

lo·ng-hair, *sb.* Also longhair. [f. Long *a.*[1] + Hair *sb.*] **1.** A cat with long fur. Also *attrib.*

1893 J. Jennings *Domestic or Fancy Cats* ii. 6 The several varieties which range under Long-hair embrace Persian, Angora, Chinese, Indian, French, and Russian. 1935 E. B. Simmons *Cats* xxviii. 143 The long-hair standard demands a body that is low on the legs. 1948 P. M. Soderberg *Cat Breeding* 78 Long-hairs need constant attention from the brush. 1958 *Listener* 28 Aug. 298/2 Pedigree cats are divided into two categories: long-hairs and short-hairs. 1972 M. Babson *Murder on Show* viii. 94, I was in the Cream and Blue-Cream Longhair aisle, cheering the lot of a lonely little Cream Longhair.

2. a. A 'brainy' person, an æsthete, an intellectual; also, a devotee of classical (as opp. to popular) music. (Freq. used contemptuously.) Also *attrib.*

1920 S. Lewis *Main St.* xxiii. 281 I'm surprised to find you talking like a New York Russian Jew, or one of these long-hairs! 1930 E. Ferber *Cimarron* xxiv. 378 These were the reformers—the long-hairs—fanatics. 1936 *Amer. Mercury* XXXVIII. x/2 *Long hair*, a symphony man. 1938 *Manch. Guardian Weekly* 2 Sept. 188/4 Very few [swing players] are 'long hairs' (people who like classical music). 1950 Blesh & Janis *They all played Ragtime* (1958) x. 208 Victor has withheld the records..because of fear of insulting the long-hairs. 1955 R. Blesh *Shining Trumpets* (ed. 3) xiii. 314 One can understand why Scott Joplin..could praise ragtime..and could heap fiery denunciation on its 'long-hair' detractors. 1957 O. Nash *You can't get there from Here* 84 So what do you want on yours—a lot of pinko longhairs, or red-blooded athletes and drum majorettes? 1958 Osborne & Creighton *Epitaph G. Dillon* ii. 57 Intense students, incompetent longhairs, and rather flashy deadbeats. 1960 'W. Haggard' *Closed Circuit* xvi. 192 All those longhairs advising us. 1967 *Listener* 16 Feb. 236/3 He planned to become a 'long hair' musician... He wanted to be a composer of symphonies.

b. A hippie, a beatnik.

1969 *Rolling Stone* 17 May 6/1 Would hippies and long-hairs sit on the youth commission? 1972 *Last Whole Earth Catalog* (Portola Inst.) 215/1 Long-hairs are doing new stuff with their bodies and nervous systems that occasionally needs medical attention or perspective. 1972 'R. Llewellyn' *Night is Child* (1974) i. 13 The noise of the band, a group of longhairs in..girlish duds, velvets, silks and falderals.

long-haired, *a.* (Stress variable.) Also long haired, longhaired. [f. Long *a.*[1] + Haired *a.*] Having long hair; *spec.* applied, at various times, (*a*) to Merovingians; (*b*) (freq. derog.) to æsthetes and intellectuals; (*c*) to cats with long fur; (*d*) to classical (as opposed to popular) music and musicians; (*e*) to beatniks and hippies. Sometimes without reference to the actual length of the hair: with or of intellectual or æsthetic pretensions. Spec. *long-haired chum* (see quot. 1890).

1552, etc. [in Dict. s.v. Long *a.*[1] 16]. 1872 Geo. Eliot *Middlem.* I. ii. xix. 340 Romanticism..was fermenting still..in certain long-haired German artists at Rome. 1881 W. S. Gilbert *Patience* i. 11 The peripatetics Of long haired aesthetics Are very much more to their taste. 1889 H. Weir *Our Cats* 16 Long-haired cats..are very diversified, both in form, colour, and the quality of the hair. 1890 Barrère & Leland *Dict. Slang* II. 27/2 *Long-haired chum* (tailors), a young woman, a young lady friend. 1914 C. Mackenzie *Sinister St.* II. xiii. 769 After a year with long-haired students I want a change. 1915 J. E. Patterson *Epistles from Deep Seas* 262 Goin' to have a 'long-haired chum', are we. 1917 'Taffrail' *Off Shore* 92 Some sort of a friendship, platonic or otherwise, with a 'long-haired' pal. 1922 S. Lewis *Babbitt* xiv. 184 The long-haired gentry who call themselves 'liberals'..and 'intelligentsia'... Irresponsible teachers and professors constitute the worst of this whole gang. 1935 *Vanity Fair* (N.Y.) Nov. 71/3 Straight or commercial musicians are often derisively called *salon-men* or *long-haired boys*. 1943 G. W. Willis *Tangleweed* xii. 174 It ain't a song. It's a composition. Long-haired. 1955 J. Cannan *Long Shadows* iii. 33 The long-haired fraternity hold the art of the camera in contempt—it rivals their daubs. 1959 C. MacInnes *Absolute Beginners* 157 Archeology, and long-haired music, and all those sorts of thing. 1959 *Times* 22 Nov. 5/6 Since the result of the general election..some of the long-haired boys in our movement have been holding inquiries and assessing blame for our defeat on everyone but themselves. 1962 J. M. Wallace-Hadrill (*title*) The long-haired kings. 1963 *Listener* 7 Feb. 264/2 It [sc. jazz] has begun an unwise flirtation with 'long-haired music'. 1965 M. Morse *Unattached* i. 59 Howard..found the atmosphere so sombre, long-haired and tedious that he quickly left. 1972 Ing & Pond *Champion Cats of World* 73 A definite breeding programme has begun in an endeavour to popularize once more the original long-haired cats. 1975 J. Symons *Three Pipe Problem* ii. 15 Sir Pountney was..opposed..to long-haired students, and to spineless intellectuals.

long-horn. [Long *a.*[1] + Horn *sb.*] **1.** A breed of beef cattle, orig. English, now common in the U.S., raised especially in the south-western states. Also *attrib.* and *transf.*

1834, etc. [in Dict. s.v. Long *a.*[1] 18 b]. 1901 W. A. White in *McClure's Mag.* Dec. 145/2 The picture of Tom Platt..standing at the head of a drove of wild-eyed human long-horns, as if to keep them from a stampede. 1903 A. Adams *Log of Cowboy* xxii. 353 There were lots of old long-horn cowmen living in the town. *Ibid.* 356 Some of those old long-horns didn't think any more of a twenty-dollar gold piece than I do of a white chip. 1949 O. Nash *Versus* 24 The big dog is to her Like a scarlet rag to a Longhorn. 1955 *Sci. News Let.* 22 Jan. 61/1 The longhorn first landed in this country when the animals were brought to America in 1521 by Gregorio Vallalobos, a governor-general sent to 'New Spain'. 1972 *Sat. Rev.* (U.S.) 6 May 20/3 Those are white-faced Herefords and black Angus... Don't see many longhorns anymore. 1974 *Sunday Times* (Colour Suppl.) 28 July 18 (*caption*) Longhorn Bull Hillpatrick at Cotswold Farm Park, near Cheltenham: a survivor from the Stone Age?

2. long-horn(ed) beetle = Longicorn *sb.*

1840 J. & M. Loudon tr. *Köllar's Treat. Insects* 15 The family of the long-horned beetles, (*Cerambycidæ*). 1894 *Insect Life* VI. 219 A Cerambycid or long-horned beetle; in pods of Enterolobium from Paraguay. 1936 *Forestry* X. 49 Longhorn beetles—Cerambycidae..are large, conspicuous, and easily recognized by their long horns or antennae. 1966 A. W. Lewis *Gloss. Woodworking Terms* 5 Longhorn beetle. Black and grey beetle about 1 in. long which sometimes attacks the softwoods used in buildings. 1968 R. D. Martin tr. *Wickler's Mimicry in Plants & Animals* viii. 86 The long-horned beetle *Erythrus rotundicollis* also resembles the same bug.

3. The long-eared owl, *Asio otus.*

1856 [in Dict. s.v. Long *a.*[1] 18 b].

4. long-horn(ed) grasshopper, a grasshopper of the family Tettigoniidæ (formerly called Locustidæ), having very long antennæ.

1893 *Insect Life* V. 271 The large Locustidæ, or long-horn grasshoppers, are very appropriately called 'cradlers' from the resemblance of the ovipositor of the female to a grain cradle. 1920 W. J. Lucas *Monogr. Brit. Orthoptera* 5 Locustodea (Long-horned Grasshoppers). 1972 L. E. Chadwick tr. *Linsenmaier's Insects of World* 80/2 In contrast with ordinary grasshoppers, most of which live on the ground, the long-horned grasshoppers do more climbing. 1972 Swan & Papp *Common Insects N. Amer.* 74 The tettigoniids, or longhorn grasshoppers, may be distinguished from the acridids, or grasshoppers, by their long filamentous antennae.

longifolene (lǫndʒifōu·līn). *Chem.* [f. mod.L. *longifol-ia*, specific epithet of the pine *Pinus longifolia* from which the oil was first isolated (f. Longi- + *-folia*, f. L. *folium* leaf) + -ene.] A colourless, somewhat viscous oil that is present in the turpentine oil from some species of pine and is a tricyclic sesquiterpene, $C_{15}H_{24}$.

1920 J. L. Simonsen in *Jrnl. Chem. Soc.* CXVII. 573 The sesquiterpene for which the name longifolene is proposed has..only been cursorily examined. 1962 E. L. Eliel *Stereochem. Carbon Compounds* x. 298 Among natural products, polycyclic bridged systems are exemplified by..the sesquiterpenoids longifolene and cedrene.

longitude. Add: Also with pronunc. (lǫ·ŋgitiūd).

longitudinal, *a.* and *sb.* Add: **A.** *adj.* **2.** (Further examples in *Geomorphol.*)

1822 Conybeare & Phillips *Outl. Geol. Eng. & Wales* p. xxiv, The longitudinal valleys are those which pursue a course parallel to the direction of the chains [of hills] which bound them. 1924 *Bull. Seismol. Soc. Amer.* XIV. 28 Tectonic earthquakes have also been divided into longitudinal and transverse earthquakes according as they are associated with the strike or transverse faults of a district. 1937 [see *concordant *a.* 4]. 1937 Wooldridge & Morgan *Physical Basis Geogr.* xxi. 354 The conception of truly longitudinal coasts as characteristic of the Pacific Basin breaks down on the coasts of Eastern Asia. 1971 C. R. Twidale *Structural Landforms* i. 11 Cross joints which cut vertically across the lineation or foliation of the rock; and longitudinal joints, which run parallel to such textural features in a vertical plane.

4. Involving information about an individual or group at different times throughout a long period (obtained by repeated examination or by eliciting recall on one occasion).

1949 B. J. Underwood *Exper. Psychol.* v. 117 Most researches in the past have been concerned with cross-sectional analyses of the organism, i.e., the capacity of the organism at the moment. When we study the influence of learning on these processes we are interested in the longitudinal aspects, i.e., how they developed. 1958 M. Argyle *Relig. Behaviour* vi. 59 These longitudinal studies can supply important supplementary information. 1962 *Arch. Gen. Psychiatry* VI. 328/1 Longitudinal studies of the urinary excretions of tryptamine and total indole-3-acetic acid..were made on 20 male schizophrenic patients. 1973 *Sci. Amer.* Sept. 35/3 Mathematical curves have been fitted with great success to measurements of individuals followed during the adolescent spurt... Such serial studies of individuals are called longitudinal, as opposed to the studies of populations called cross-sectional, in which each child is measured only once.

B. *sb.* **2. b.** *Aeronaut.* A longeron, esp. one in an airship.

1911 *Aero* May 38/1 There are three main longitudinals, which are of ash rebated for the reception of the diagonal cross-pieces. 1914 H. M. Buist *Aircraft in German War* i. 21 (*in figure*) Longitudinal. 1919 *Jane's All the World's Aircraft* 37a/1 Fuselage longitudinals and struts have sections of I-shape. 1950 *Gloss. Aeronaut. Terms* (*B.S.I.*) i. 50 *Longitudinal*, a girder on the outside of the hull structure running fore and aft. 1973 D. H. Robinson *Giants in Sky* ii. 31 The hull was identical with that of the LZ 2, the dimensions, the number of longitudinals and ring frames being identical with those of the ships lost at Kisslegg.

long knife. [f. Long *a.*[1] + Knife *sb.*] **1.** *N. Amer. Hist.* (Freq. *pl.*, and with capital initials.) A name given by North American Indians to white settlers, esp. of Virginia, or white soldiers. In Canada, *spec.* a citizen of the United States.

1774 J. R. Peyton *Let.* 10 Oct. in J. L. Peyton *Adventures my Grandfather* (1867) 143 The white troops, or 'Long Knives',..imagined no enemy was near. 1784 D. Boon in J. Filson *Discovery Kentucke* 62 The savages now learned the superiority of the long knife, as they call the Virginians, by experience. 1827 J. F. Cooper *Prairie* I. v. 135 If the Tetons lose their great chief by the hands of the Long-knives, old shall die as well as young! 1838 A. Jameson *Winter Stud. & Summer Rambles Canada* III. 55 A distinguished Pottowottomie warrior..was..a good friend to the Long-knives, (The Americans). *Ibid.* 142 The Indians gave the name of Cheemokomaun (Long Knives, or *Big Knives*) to the Americans at the time they were defeated by General Wayne..in 1795, and suffered so severely from the *sabres* of the cavalry. 1908 W. R. Nursey *Story Isaac Brock* xviii. 100 'My object,' said Brock, addressing the Indians, 'is to assist you to drive the "Long-knives" from the frontier.' 1959 R. Gant *World in Jug* 38 Paleface, hear me. Do not send pony soldiers and long knives against my nation. 1959 N. Sluman *Blackfoot Crossing* 33 Crowfoot says that the border—the medicine line—protects his people from the Long Knives. 1972 J. Mosher *Some would call it Adultery* iii. xvi. 147 Brought General Terry and his 'Long Knives', as the Indians called U.S. Cavalry, 'up across the border, Mounted Police or no'.

2. Phr. (*night of*) *the long knives*: a treacherous massacre (as, according to legend, of the Britons by Hengist in 472, or of Ernst Roehm and his associates by Hitler on 29–30 June 1934); hence used allusively of any similarly decisive or ruthless action.

The massacre of the Britons is described by Geoffrey of Monmouth in *Historia Regum Britanniæ* Bk. vi. xv.

1862 Borrow *Wild Wales* II. xx. 226 Hengist had commanded..that..each Saxon should draw his long sax, or knife,..and should plunge it into the throat of his neighbour... This infernal carnage the Welsh have appropriately denominated the treachery of the long knives. 1891 E. C. Brewer *Historic Note-Bk.* 531/2 *Long Knives* (*The Plot or Treachery of the*). This was a treacherous conference to which Geoffrey of Monmouth tells us the chief Britons were invited by Hengist at Ambresbury; others say by Vortigern. Beside each Briton a Saxon was seated, armed with a long knife; and, at a given signal, each Saxon slew the Briton seated by his side. Geoffrey tells us the signal was the utterance of these words: *Nemet oure Saxas*, and that the number massacred was 460. 1936 R. Olden *Hitler the Pawn* xvi. 378 The consequence is massacre, 'the night of the long knives', a St. Bartholomew's night. 1937 S. H. Roberts *House Hitler Built* ii. iii. 114 Such seem to have been the facts of this 'Night of the Long Knives' (the name given to it by Hitler and taken from one of the earliest marching songs of the Nazis). 1960 'W. Haggard' *Closed Circuit* iii. 29 Many would die in any night of the long knives. 1961 G. Thomas *Keep in Plays of Year* XXIV. 217 This is the night of the long knives and they all landed on Con. 1967 W. R. Manchester *Death of President* i. i. 93 His popularity margin with state voters was..wider than those of Connally or Kennedy-Johnson... He could be fairly confident of surviving the long knives of Austin. 1967 *Economist* 22 July 313/2 A government that can provide some degree of stability may be forgiven its day of the long knives—if the exercise is not repeated. 1968 J. Bingham *I love, I Kill* iv. 44 There was not the 'night of the long knives' feeling you get when a commercial management has ten thousand smackers at stake. 1968 *Sunday Times* 1 Dec. 11/5 The Long Knives flashed last week at Bush House. 1973 *Times* 18 June 1/7 They both lost their jobs last April 30, the night of the long knives when a whole series of officials suddenly resigned. 1973 R. Payne *Life & Death of Hitler* 274 The historical event known as the Night of the Long Knives took place in broad daylight.

long-legged, *a.* Add: **b.** Hence *long leg* (see quots.). *slang.*

1929 F. C. Bowen *Sea Slang* 85 *Long leg*, a big difference in the draught forward and aft in a sailing ship. 1961 F. H. Burgess *Dict. Sailing* 137 Any sailing vessel that draws a lot of water is said to have a long leg.

long-line, long line. 2. a. For def. read: A line of manuscript or type that runs across the page without columnar division. Also (with hyphen) *attrib.* Add further examples.

1914 E. A. Loew *Beneventan Script* xi. 289 The oldest extant Beneventan MSS. are written in long lines and not in two or more columns. 1939 W. H. P. Hatch *Princ. Uncial Manuscripts New Testament* 17 It was soon found that a wide column and a long line are more convenient for the reader. 1964 Dean & Legge *Rule of St. Benedict* p. xviii, The Douce Rule has thirty long lines to the page.

3. In Old English verse, two half-lines considered as a unit. Cf. G. *langzeile.*

1868 W. W. Skeat in Hales & Furnivall *Bp. Percy's Folio MS.* III. p. xxiv, There has been much discussion as to whether alliterative poems should be printed in couplets of short lines, or in long lines comprising two sections. 1877 H. Rehrmann *Essay concerning Anglo-Saxon Poetry* 9 All words of a long line are fit for alliteration that distinguish themselves in the verse by natural gravity or grammatical accent. 1929 W. E. Leonard in Malone & Ruud *Studies Eng. Philol. in Honor of F. Klaeber* 7 The law of the meter remains, an eight-beat long-line. 1970 *Rev. Eng. Studies* XXI. 133 The metrical division between two verses (hemistichs) of an alliterative long line.

lo·ngliner. Chiefly *N. Amer.* Also long-liner. [f. Long-line 1 + -er[1].] One who fishes with a long-line; a fishing vessel which uses long-lines.

1909 *Westm. Gaz.* 3 June 14/3 Dog-fish, these terrors to netsmen and long-liners. 1919 W. T. Grenfell *Labrador Doctor* (1920) x. 183 The Hearn long-liners and trawlers, who were just beginning their vast fishery in those waters. 1955 *Fishermen's Advocate* (Port Union, Newfoundland) 14 Jan. 10/5 Three new longliners are now under construction. 1959 *Globe Mag.* (Toronto) 12 Sept. 21/2 Groundfishing operations—draggers with their great, bottom-scraping nets, and longliners with multiple lines armed with hundreds of baited hooks—were the principal targets. 1969 *Guardian* 8 Mar. 7/5 Off this place the professional Japanese long-liners have taken swordfish in excess of a thousand pounds. 1974 *Nat. Geographic* Jan. 114/2 The boat is of the type still called a long-liner, from a time when the crews of such boats fished with lines and hooks. 1974 *Sci. Amer.* Mar. 119/1 The yellowfin-tuna fishery on the Pacific Equator, worked only recently by American and Japanese long-liners, is plainly disclosed in the Nantucketers' logbooks.

Hence **longlinerman,** a member of the crew of a longliner.

1955 *Fishermen's Advocate* (Port Union, Newfoundland) 14 Jan. 7/3 The courses, while aimed at longlinermen, were made available to all who wished to attend.

Longmyndian (lǫŋmi·ndiăn), *a. Geol.* [f. *Long Mynd* (see def.) + -ian.] Applied to a thick series of non-fossiliferous sedimentary rocks in the west Midlands, now believed to be of Pre-Cambrian age, whose main outcrop forms the hills of the Long Mynd in southern Shropshire (Salop). Also *absol.*

1888 C. Callaway in *Trans. Shrops. Archæol. & Nat. Hist. Soc.* XI. 239 We are able to determine the relative ages of the three most ancient rock-systems in the region, viz., the Malvernian, the Uriconian, and the Longmyndian... The great series which forms the Longmynd Hills was referred..to the Lower Cambrian; but until good evidence of its age is obtained, I have thought it better to use a local designation. *Ibid.* 240 The Longmyndian conglomerates are largely derived from the Uriconian. 1929 Evans & Stubblefield *Handbk. Geol. Gt. Brit.* iii. 20 The outcrop of the Longmyndian Rocks at Church Stretton forms the wild, moorland tract called the Longmynd. 1967 D. H. Rayner *Stratigr. Brit. Isles* ii. 45 The structures in the Longmyndian pediments..suggest strongly that the western part of the Wentnor outcrops are inverted. 1969 Bennison & Wright *Geol. Hist. Brit. Isles* iii. 67 The Longmyndian is in the form of a very large syncline.

long-neck. Add: **2. b.** In full, *long-neck clam.* An elongated, thin-shelled clam, *Mya arenaria*, found on the eastern coast of North America; = *long clam* (Long *a.*[1] A. 18 b).

1905 *Bull. N.Y. State Mus.* No. 71. 22 The soft or long-neck clam, Mya, is capable of locomotion only when very small. 1910 J. L. Kellogg *Shell-Fish Industries* 281 Mya, in different localities, is known as the clam, the soft clam, the long neck, long clam, squirt clam. 1923 D. K. Tressler *Marine Products of Commerce* 533 *Mya arenaria* commonly called soft clam, long clam, long neck..is found from South Carolina to the Arctic Ocean. 1970 R. Lowell *Notebk.* 200 File upon file, the beds of long-neck clams.

† **long nine.** *Obs. U.S.* [f. Long *a.*[1] 1 + Nine *sb.* 4 b.] A kind of cheap cigar.

1830 N. Dana *Mariner's Sk.* 213 (Th.), The fourfold row of long-nine-smoking beaux, that are regularly drawn up on Sunday forenoon in Market Square. 1835 *Harvardiana* I. 157 (Th.), He unfolded the wrapper; it contained two long-nine segars. 1851 *Yale Lit. Mag.* June 315 Pete had, as he always had after breakfast, a cigar in his mouth, a long nine. 1879 *Bradstreet's* 31 Dec. 3/3 Boys smoke 'long nines' while they still wear jackets.

‖ **longo intervallo** (lǫ·ŋgo intəɹvæ·lo). [Lat., lit. 'at a great distance'.] At some remove, though there is a gulf between them (of two persons, places, etc., being compared).

1693 Dryden tr. *Juvenal's Satires* (Dedication) p. iii, The most Vain, and the most Ambitious..have yielded the first place without dispute; and have been arrogantly content, to be esteem'd as second to your Lordship, and even that also, with a *Longo, sed proximi Intervallo.* 1890 W. James *Princ. Psychol.* II. xx. 227 Almost all subsequent progress has been made in Germany, Holland, and, *longo intervallo*, America. 1923 *Spectator* 17 Feb. 294/1 In casual passages *The Orissers* may seem dull, pompous and even ridiculous; but as a whole it has a strange power, a conviction and an intensity of imagination that mark it off from other novels and link it, *longo intervallo*, with *Moby Dick* and *Wuthering Heights.* 1935 J. C. Masterman *Fate cannot harm Me* ii. 44, I remember asking him once..who were his literary idols among the moderns. 'Well,' he said, 'Max Beerbohm of course..and then, but *longo intervallo*, P. G. Wodehouse.'

long-playing, *a.* [Long *adv.* 9 a.] That plays for a long time; *spec.* designating or pertaining to a microgroove gramophone record designed to be played at 33⅓ revolutions per minute. Cf. *L.P.* (*L 7).

1912 *Talking Machine News & Jrnl. Amusements* Nov. 529/1 The Petmecky Needle, which plays ten times, is the subject of an interesting little leaflet... Users interested in this long-playing needle should ask their dealers for full particulars. 1929 *Wireless Mag.* Oct. 252 (*heading*) Long-playing dance records. 1931 *Electronics* Dec. 236/1 Long-playing records... Already announced is a phonograph record that will play for 15 minutes. 1948 *Electronic Engin.* XX. 333 A new library of recorded music has been announced..which consists of a series of long-playing 10 and 12 in. records run at 33⅓ r.p.m. 1951 *Ann. Reg. 1950* 386 An important advance was made by the introduction of 'long-playing' records containing about 20 minutes of music in place of the 4½ minutes hitherto available. 1958 V. Bellerby in P. Gammond *Duke Ellington* ii. 151 With the advent of the Long Playing record he has shown himself ready to compose well outside his former 'three-minute' form. 1968 P. Oliver *Screening Blues* ii. 81 Singers who have been recently recorded under the more liberal circumstances of the long-playing era. 1972 *Gramophone* Oct. 834/2 (Advt.), Trade in your unwanted classical long-playing records..for guaranteed unplayed new records.

So **long-play** *a.* = *long-playing *a.*; also *ellipt.* as *sb.*, long-playing records collectively; **long-player,** a long-playing record.

1954 F. Ramsey (*title*) A guide to longplay jazz records. 1957 M. Goff (*title*) Short guide to Long Play. 1958 *Times* 19 Apr. 7/3 Already tape recording is hard on the heels of the long-play record and stereophonic reproduction promises to invade both. 1958 *New Statesman* 3 May 580/2 The Rank Organisation..has decided to start a record club..which will issue long-players. 1961 I. Murdoch *Bell* xiv. 187 I'll..put on my new long-player. 1962 A. Nisbett *Technique Sound Studio* 273 Long Play tape gives 50% more recording time on a spool than standard play. 1962 R. Douglas-Home *Sinatra* i. 8 The buying of long-players was (and still is) an expensive pursuit. 1968 *Blues Unlimited* Sept. 26 Phillips was introduced to long-play by Origin with two of his less impressive pieces. 1975 M. Kenyon *Mr Big* ii. 18 He chose a long-player of Carroll Gibbons and his Savoy Orpheans.

long run, long-run. Add: **2.** *Theatr.* A long period of being presented on the stage; a play or entertainment presented for a long period. Also *attrib.*

1714 [see Run *sb.*[1] 18]. 1883 D. Cook *On Stage* I. ix. 203 These are the days of 'long runs', when but one or two plays can be produced in a season. 1896 [see Run *sb.*[1] 18]. 1901 Beerbohm *Around Theatres* (1924) I. 320 The long-run system is often deplored on the ground that the mimes 'walk through' their parts. 1909 *Westm. Gaz.* 22 Apr. 2/3 Half the week is to be given to 'long-run' plays; the other half to new plays, revivals, and the classical drama. 1967 *Oxf. Compan. Theatre* (ed. 3) 195/2 Alfred Butt presented Laurette Taylor in *Peg o' My Heart.* This had a long run and was then transferred to the Globe.

B. *adj.* Taken or considered in the long run; = *long-term *a.*

1904 W. James *Meaning of Truth* (1909) iii. 89 Abstract truth, truth verified by the long run, and abstract satisfactoriness, long-run satisfactoriness, coincide. 1931 A. L. Rowse *Politics & Younger Generation* 7, I have had in mind the necessity of a long-run view of politics. 1946 J. S. Huxley *Unesco* i. 19 Long-run human progress. 1957 K. R. Popper *Poverty of Historicism* i. i. 6 There is no long-run uniformity in society on which long-term generalizations could be based. 1969 D. C. Hague *Managerial Econ.* ii. iii. 61 With many industrial goods, long-run costs will be constant or falling. 1975 *Times* 6 Jan. 12/6 We need..restraint from business and labour in which their longrun interests are elevated over..short-run interests.

long-running, *a.* [Long *adv.* 9 a.] Continuing for a relatively long period of time; *spec.* of a play: having a large number of consecutive performances.

1956 *Nature* 21 Jan. 142/1 This is precisely the case in a long-running crossing experiment with strains *H*..and *O*. 1968 G. C. Ramsey *Agatha Christie* iv. 50 It is..poetic justice that the most famous play in the English language should have supplied the title for the longest-running play in England. 1972 *Guinness Bk. Records* (ed. 19) 98/2 The longest-running musical show ever performed in Britain was *The Black and White Minstrel Show.* 1975 *Listener* 6 Feb. 166/2 A long-running and repetitive weekly discussion.

longshanks. Add: **1. b.** Hence applied generally to a tall or long-legged person, often as a term of derision.

1699 B.E. *New Dict. Canting Crew*, *Long-shanks*, long-legged. 1915 *Dialect Notes* IV. 206 There comes long-shanks across the fields. 1939 F. Thompson *Lark Rise* ii. 32 The two tamer children..would make a dash on their long stalky legs for their own garden gate followed by.. cries of 'Long-shanks! Cowardy, cowardy custards.' 1954 J. R. R. Tolkien *Fellowship of Ring* xi. 193 He was smoking a short black pipe. As they approached he took it out of his mouth and spat. 'Morning, Longshanks!' he said. 1959 I. & P. Opie *Lore & Lang. Schoolch.* ix. 169 In the following [terms] the chief emphasis is on height, 'Lofty' being the most popular nickname, followed by 'Longshanks'.

long-shore, *attrib. phr.* Add: Also longshore, 'longshore. **1. b.** *Physical Geogr.* Moving, taking place, or laid down more or less parallel to a shore.

1837 [in Dict., sense 1]. 1910 V. Cornish *Waves of Sea* vi. 179 When, at sea, the wind is obliquely on-shore there is not only a 'longshore current'.., but waves also break obliquely. Their effect to drive shingle along the shore is then obvious to the eye. 1952 R. F. Peel *Physical Geogr.* xv. 253 Longshore drift..is often the main factor controlling the supply of shingle to beaches. 1964 V. J. Chapman *Coastal Vegetation* viii. 193 Long-shore movement of beach material takes place..at the upper limit of the waves. 1968 R. W. Fairbridge *Encycl. Geomorphol.* 58/1 Currents from both flanks later build up the long-shore bar, both trough and bars being affected by long-shore currents, and the troughs are scoured by water escaping laterally behind the bars from the beach zone and longshore drift-currents bringing sand.

lo·ngshoring. [f. Long-shore *attrib. phr.* (*sb.*) + -ing[1].] The type of work done at a port; the occupation of a longshoreman.

1926 *Daily Colonist* (Victoria, B.C.) 1 July 7/1 Two men were injured while doing longshoring work on Monday. 1930 C. S. Johnson *Negro in Amer. Civilization* (1931) iv. 53 In a study of the conditions of the longshoring industry of the Chicago port in 1915 by Charles B. Barnes, 100 workmen were interviewed.

long shot. Also long-shot, longshot. [f. Long *a.*[1] + Shot *sb.*[1]] **1.** (In Dict. s.v. Long *a.*[1] 18.)

2. Something incredible or very unlikely; a far-fetched explanation; a wild guess; *spec.* a bet laid against considerable odds; = Outsider 1 b. See also Shot *sb.*[1] 9 d in Dict. and Suppl. Also *attrib.*

1867 [in Dict. s.v. Long *a.*[1] 18]. 1869 *Leisure Hour* May 326/1 He may also..learn to systematise his turf speculations, may know when it is prudent to 'back a jockey' or a long shot. 1906 'O. Henry' *Four Million* (1916) 33 A few long-shot winners at the New Orleans race-track. 1939 *Sun* (Baltimore) 30 Nov. 24/7 The long shots won the first and second races. 1955 *Sci. Amer.* Feb. 47/2, I made the first test very simple, because the whole idea seemed a long shot. 1970 *Globe & Mail* (Toronto) 26 Sept. 38/1 The other sign that fall has arrived is a sudden surge of long-shots from Western Canada speeding across the finish line in first place. 1971 *Publishers' Weekly* 23 Aug. 45/2 Since establishing his company nine years ago, Grossman has played hunches and longshots. 1975 *New Yorker* 20 Jan. 81/1 It was a day for long-shot players.

3. A cinema or television shot which includes figures or scenery at a distance; opp. *close-up. orig. *U.S.*

1922 *Sci. Amer.* Sept. 177/1 A quarter of a mile away, from which distance some of the so-called 'long shots' were filmed by the cameraman. 1930 *Sunday Times* 12 Oct. 4/3 Some of the close-ups are brilliantly clear and fascinating in detail, some of the long shots blurred and vague. 1934 *B.B.C. Year-Bk.* 58 The first television programme was transmitted,... New ways of using photo cells enabled artists to be followed from 'long shot' to 'close up' and vice versa. 1962 *Movie* Sept. 7/3 The last shot of the film is a longshot of a general advance. 1974 *Times* 19 Jan. 9/7 We got Griffith in long shot winding up like a clock gone haywire.

long-standing. 2. (Earlier and later examples.)

1814 Jane Austen *Mansf. Park* II. ix. 196 In spite of every long-standing expectation. 1848 Mill *Pol. Econ.* i. xi. 208 A long-standing and hereditary confidence in the safety of funds when trusted out of the owner's hands. 1975 *Times* 15 Jan. 14/1 The holing of the Japanese supertanker, Showa Maru, on a rock off Singapore last week is bound to reactivate sharply a long-standing conflict of views over control of the nearby Malacca Straits.

long-tail. Add: **1.** *spec.* a greyhound. Also *attrib.*

1876 *Coursing Calendar* 12 The former Duke of Hamilton..and others of their day, were followers of the 'long tails' on the very same ground. 1927 *Daily Express* 25 May 12 A little more foresight..might have made 'rag running' a very popular entertainment, commanding as much notice as the sport of long-tail racing. 1930 Billis & Kenyon *Pastures New* vi. 102 Some high-priced coursing dogs,—longtails as they were called—were brought into the colony.

b. Read: The long-tailed duck, *Clangula hyemalis.* (Later examples.)

1919 J. Masefield *Reynard* 93 Some longtails prinking. 1958 D. A. Bannerman *Birds Brit. Isles* VII. 141 A number of immature long-tails may be seen in the Channel from November onwards.

c. The white-tailed tropic bird, *Phaëthon lepturus.*

1905 *Chambers's Jrnl.* May 367/1 The tropic bird commonly called 'longtail'. 1960 J. Bond *Birds W. Indies* 22 White-tailed Tropicbird... Local names:.. Boatswain Bird; Long-tail.

4. long-tail pair *Electronics* = *long-tailed pair* (*long-tailed *a.* 3).

1946 *Electronic Engin.* XVIII. 298/2 Each of the amplifiers..embodies a 3-stage network, employing valves (EF. 50) in a long-tail pair connexion... The circuiting is such that the amplifiers respond to difference of signal potential between the input terminals, being several hundred times less sensitive to variations of the potential of the two terminals together with respect to ground. 1971 J. H. Smith *Digital Logic* iv. 70 Transistors T_1 and T_2 form a long tail pair where a fixed current is passed through R_1 into either T_1 or T_2, so that either one or the other transistor will conduct, but not both.

long-tailed, *a.* Add: **3. long-tailed pair** *Electronics*, a pair of identical valves (or transistors) with their cathodes (or emitters) connected together to a large resistor and usu. with their anodes (or collectors) connected to equal loads. Cf. *long-tail pair.

1947 *Electronic Engin.* XIX. 272/1 The type of circuit used is based on what is generally referred to as the 'long-tailed pair' and owes its inception to the late A. D. Blumlein. 1970 J. Shepherd et al. *Higher Electr. Engin.* (ed. 2) xxii. 713 Drift [in a d.c. amplifier] cannot be reduced by straightforward feedback since this reduces the gain in proportion; the technique of compensation may however be employed. In this method drift in one part of a circuit is balanced against drift in another part. The long-tailed or emitter-coupled pair circuit..is an example.

long-term, *a.* [f. Long *a.*[1] + Term *sb.*] Lasting for, pertaining to, or involving a relatively long period of time; maturing or becoming effective only after a long period. Also quasi-*advb.*

1908 *Daily Chron.* 24 July 1/6 The long-term men, who wore blue cotton overalls marked with the broad arrows, were in the rear. 1909 *Westm. Gaz.* 2 June 5/2 Mr. Fielding was able to place a 2½ per cent., long-term loan in London. 1937 *Discovery* June 178/1 A long-term programme of development. 1942 C. S. Lewis *Screwtape Lett.* xiv. 73 It is His long-term policy..to restore to them a new kind of self-love. 1956 C. Auerbach *Genetics in Atomic Age* 65 Thus each species has to strike a balance between the short-term requirement for a low frequency of mutation and the long-term requirement for an ample store of mutant genes. 1956 *Planning* XXII. 38 The more specific the investment the greater the need for forecasting, especially long-term. 1959 A. Lejeune *Crowded & Dangerous* xii. 134 The long-term future could look after itself. 1969 *Listener* 5 June 786/3 The difficulty in getting both politicians and the bureaucracy to think long-term, particularly on financial affairs. 1971 *Jrnl. Gen. Psychol.* LXXXV. 51 Short-term memory is required to act as..a cue to recall data from long-term storage. *Ibid.*, The effectiveness of extracting information from long-term memory may depend on short-term load. 1972 *Accountant* 23 Mar. 383/2 Valuation of long-term contracts. 1972 *Listener* 21 Dec. 854/2 Courses of action are followed without regard for their long-term consequences. 1974 *Times* 28 Dec. 9/6 Ordinary table wines..must not be kept long-term in the refrigerator.

Long Tom. Add: **1.** (Earlier example.)

1832 M. Scott *Tom Cringle's Log* xi, in *Blackw. Mag.* July 27/2 The long Tom must be a tearer to pitch its mouthful of iron this length.

2. (Earlier examples.)

1839 *Amer. Railroad Jrnl.* VIII. 98 The Long Tom.. consists merely of a trough. 1852 *Elora* (Ontario) *Backwoodsman* 17 June 2/4 The plough is a far more profitable instrument than 'the long Tom' or 'the rocker'.

4. *Austral.* A marine fish of the family Belonidæ.

1883 [in Dict., sense 3]. 1908 E. J. Banfield *Confessions of Beachcomber* i. iv. 154 The 'long tom' (*Zylosurus*, sp.) or alligator-pike, which shoots from the water. 1934 *Bulletin* (Sydney) 24 Jan. 20/2 The slender Long Tom must in future be known as *Lewinichthys ferox*... On the other hand the stout Long Tom is to be styled *Lhotskia macleayana.* 1965 *Austral. Encycl.* V. 362/2 Long toms or Needle-fish, members of the family Belonidae, in England and America called garfish... Long toms have the habit of leaping from the water, when either pursuing or being pursued.

5. *slang.* A particularly high-powered telephoto camera lens.

1968 'J. Welcome' *Hell is where you find It* iv. 58 There were Rolliefiexes, a Leica, a Long Tom for peeping, a couple of polaroids. 1968 'O. Mills' *Sundry Fell Designs* xiii. 142 Russ, grinning, remembered a Long Tom lens he had seen in the hands of one of the photographers. 1973 R. Busby *Pattern of Violence* x. 165 The long tom lenses and the barrels of the TV cameras peering down into the cleared arena.

Longton Hall (lǫ·ŋtən hǭl). The name of a house in Staffordshire where the first Staffordshire porcelain was manufactured; hence *attrib.* of the porcelain manufactured there. Also *ellipt.* Longton.

1757 *Public Advertiser* 4 Apr. 3/3 A Quantity of new and curious Porcelain or China, both useful and ornamental, of the Longton-Hall Manufactory, which has never been exposed to public View. 1885 C. Schreiber *Jrnl.* 27 Feb. (1911) II. 469 A certain Longton Teapot, which we had discovered among her china. 1925 W. W. Worster tr. *Hannover's Pott. & Porc.* III. xx. 533 A mode of decoration probably peculiar to Longton Hall is the application of designs like those on Battersea or Staffordshire enamels, in white enamel pigment on a strong blue ground. 1961 *Connoisseur* Dec. p. xiv (Advt.), A rare Longton Hall Vase finely painted with flowers. 1973 *Times* 14 Nov. 21/5 For Longton Hall a small melon tureen..made a record price at £4,000.

longuette (lǫŋge·t). [f. F. *longuette* somewhat long, longish.] A midi dress, a midi skirt.

1970 *New Yorker* 10 Oct. 167/1 A photographer hopped among them adding to the bulging dossier of *Women's Wear Daily* evidence of the arrival of its 'longuette'. 1971 *Time* 29 Mar. 29 Along Belgrade's Terazije, maxicoats and Longuettes, velvet knickers and leather gaucho pants abound.

longueur. Add: (Earlier and later examples.) Also in extended use, of music, etc.

The form *longeur* in quots. 1959 and 1970 is *erron.*

1791 H. Walpole *Let.* 26 May (1905) XIV. 437 Boswell's book is gossiping;.. but there are woful longueurs, both about his hero and himself. 1866 *Nation* (N.Y.) 16 Aug. 127/2 In what other writer than George Eliot could we forgive so rusty a plot, and such *langueurs* [sic] of exposition? 1892 I. Zangwill *Childr. Ghetto* I. 5 The terrible *longueurs* induced by the meaningless ministerial repetition of prayers already said by the congregation. 1950 'G. Orwell' *Shooting Elephant* 41 A kind of pattern ..survives the complications and the *longueurs* [in *King Lear*]. 1952 'M. Cost' *Hour Awaits* 251 This sense of impending pause in herself—this *longueur* of spirit. 1958 *Times* 22 Dec. 5/5 Should the recitalist be an imperfect executant, the audience must listen for much too long a period, with only one interval to alleviate the *longueurs.* 1959 *Guardian* 2 Nov. 7/1 One was aware of extensive stretches in which nothing historically worth while seemed to be going on—longeurs which the Director of History..would not have tolerated in his scenario. 1963 *Times Lit. Suppl.* 18 Jan. 40/3 Some of his own early experiences were clearly embedded: the longueurs of the middle-class Sunday, the deadly frustrations of office routine. 1970 *Times* 29 May 7/2 Despite the show's contradictions and longeurs, it does at least attempt something audacious and original. 1971 *Guardian* 18 Feb. 9/8 There are so many longueurs in most of the originals that one can hardly jib at Miss Wilson's frequent condensations. 1974 *Times* 6 Mar. 14/8 A perfect committee man, he would remain wholly silent—and even asleep—during the *longueurs* not unknown in university meetings.

longward(s (lǫ·ŋwǭɹd(z)), *adv.* [f. Long *a.*[1] + -ward, or -wards.] Towards longer wavelengths; on the long-wavelength side *of.*

1971 D. W. Sciama *Mod. Cosmol.* ix. 132 Because of the red shift the emission is spread out longwards of 21 cm. 1974 *Nature* 22 Feb. 513/1 Since the first report of a diffuse ultraviolet background, there have been several measurements attempted, both longward and shortward of Lyman α.

longyi (lǫ·ŋyī), var. Lungi. Cf. *lungyi.

1908 H. Jones *Let.* 2 May in H. J. W. Hetherington *Life & Lett. Sir H. Jones* (1924) ii. 203 The boxers, all naked except for their ornamental silk *longyis* tied up hard between their legs and around their middle. 1934 [see *Arakanese *a.* and *sb.*]. 1947 'N. Shute' *Chequer Board* 105 A green *longyi* wrapped around her waist and falling to her feet. 1957 R. Mason *World of Suzie Wong* ii. ii. 17 A Burmese woman..with her bright red *longyi* taut over her thighs. 1959 *Times* 10 Dec. 11/2 Burma, where men wear the *longyi.* 1971 *Nat. Geographic* Mar. 304/2 He wears the Burmese national costume, a wrap-around skirt called a *longyi*, derived from India, and a short formal jacket with three pockets and cloth buttons, derived from China.

Lonk (lǫŋk). [dial. var. of *Lank*, the first syllable of *Lancashire*: see *E.D.D.*] A large-sized variety of mountain sheep which originated in Lancashire or Yorkshire; the wool of this variety of sheep.

1863 in W. Fream *Youatt's Compl. Grazier* (1893) 473 If the Lonks be as hardy as they are good, they must be the most valuable sheep for the hills that we have at present. 1866 *Jrnl. R. Agric. Soc.* II. 367 The hill ranges of Yorkshire and Lancashire are believed to be the earliest home of the Lonks. 1911 *Chambers's Jrnl.* Dec. 778/2 The lonk is believed to have come originally from the Yorkshire hills. 1940 *Chambers's Techn. Dict.* 510/2 *Lonk*, wool that comes from the large type of mountain sheep of the same name, reared on the Lancashire and Yorkshire

moorlands. 1968 Fraser & Stamp *Sheep Husbandry* (ed. 5) ii. 110 The Lonk. Another horned and black-faced breed is native to a rather confined hill district of Lancashire and Yorkshire. It is a big sheep, handsome, with very clearly differentiated black-and-white markings on face and legs, strong-boned and active. 1972 J. Wainwright *Night is Time to Die* 7 The dark patch could be a sheep... Ye-es—it *could* be some stupid, wandering Lonk. 1974 *Times* 23 Feb. 14 Several farmers may turn out their Swaledale or Scottish Blackface, Herdwick or Lonk sheep onto one moor.

lonnin(g, var. Loaning *sb.*

Cf. the 14th–17th cent. examples with similar spellings in Dict. s.v. Loaning *sb.* and in *Dict. Older Sc. Tongue.*

1808 R. Anderson et al. *Ballads in Cumberland Dial.* 23 In dark winter neeghts i' the lonnins. 1849 F. T. Dinsdale *Gloss. Teesdale* 81 It's a lang lonnin that has nivver a turn. 1896 F. M. T. Palgrave *List Words Hetton-le-Hole* 29 *Lonning*, 'laning', i.e. lane... 'We find swiney up Mousely (Moorsley) lonen.'—Extract from boy's essay on Wild Flowers. 1933 *Times Lit. Suppl.* 6 Apr. 243/2 This is for those whose feet leave the road at the first chance and strike up through lonnins and intakes to untrodden ground. 1971 *Country Life* 9 Sept. 630/1 We came to the lonnin that turns away from the river and leads beside the church towards the dale road.

Lonsdale (lǫ·nzdēil). The title of Hugh Cecil Lowther (1857–1944), fifth earl of *Lonsdale*, used *attrib.* to denote any of various belts conferred upon professional boxing champions of the United Kingdom. Also *transf.* So *to give* (someone) *the Lonsdale* (slang), to dismiss, repudiate, 'throw out' (cf. *belt *sb.*[4]).

1910 *Daily Graphic* 18 Oct. 10/4 At the National Sporting Club..last evening, the contest was decided for the Bantam-weight Championship of Great Britain, the winner of which also holds the Lonsdale belt for that weight. 1914 *Boxing's Bk. Records* 9 (*caption*) Lonsdale Belt Holders. 1958 F. Norman *Bang to Rights* iii. 167, I can always give her the Lonsdale after a week or two. 1959 C. MacInnes *Absolute Beginners* 77 The Wiz was wearing a gladiator Lonsdale belt with studs on it. 1973 *Country Life* 26 July 249/1 The traditional Lonsdale Belt..with its enamel portrait of the 5th Earl.

lontar (lǫ·ntaɹ). [Malay.] = Palmyra; also, a manuscript written on leaves of this palm.

1820 J. Crawfurd *Hist. Indian Archip.* I. iv. 443 The *Lontar*, (*Borassus flabellifer*), the *Tar*, or Tal of Western India, grows abundantly in the Indian islands. 1935 I. H. Burkill *Dict. Econ. Products Malay Peninsula* I. 348 Lontar is also the best-known Malayan name. 1937 M. Covarrubias *Island of Bali* (1972) vii. 193 In Singaradja there is a library of these manuscripts,..where are preserved some splendid old *lontars* with illustrations... These are masterpieces of the art of illustration, with miniature pictures incised with an iron style on the blades of the *lontar* palm. 1952 'W. March' *October Island* i. 11 He moved majestically towards the four tall lontar palms. 1953 C. A. Gibson-Hill *Malay Arts & Crafts* § Hats & Dish-covers. Other fibres are employed, including strips from the Lontar Palm. 1961 P. Kemp *Alms for Oblivion* ix. 152 You know about *lontars*? Good. Well, in one important *lontar* is written, 'It is fair to lie to wifes [*sic*] and enemies.'

loo (lū), *sb.*[3] [Hind., f. Skr. *ulkā* flame.] The name given in Bihar and the Punjab to a hot dust-laden wind.

1888 Kipling *Phantom 'Rickshaw* 78 The *loo*, the red-hot wind from the westward, was booming among the tinder-dry trees. *a* 1936 —— *Something of Myself* (1937) iv. 98 A hot wind, like the *loo* of the Punjab. 1954 O. H. K. Spate *India & Pakistan* ii. 55 In the NW hot weather depressions generally take the form of violent dust-storms... Such dust-storms are distinct from the *loo*, a very hot dust-laden wind which may blow for days on end. 1965 E. Ahmad *Bihar* iv. 45 The hot scorching 'loo' winds of the Bihar plains during late April and May have an average velocity of 5–10 miles per hour. 1974 M. Peissel *Great Himalayan Passage* xi. 175 The Loo is caused by the hot expanding air of the Indian plains rushing into the cool hills.

loo (lū), *sb.*[4] [Etym. obscure.] A privy, a lavatory. Also *attrib.* and *Comb.*

A. S. C. Ross's examination of possible sources in *Blackw. Mag.* (1974) Oct. 309–16 is inconclusive: he favours derivation, in some manner that cannot be demonstrated, from *Waterloo.*

[1922 Joyce *Ulysses* 556 O yes, *mon loup.* How much cost? Waterloo. Watercloset. 1932 N. Mitford *Christmas Pudding* ix. 137 The absence in his speech of such expressions as 'O.K. loo'..'we'll call it a day'.] 1940 —— *Pigeon Pie* ii. 27 In the night when you want to go to the loo. 1943 C. Beaton in *Horizon* Jan. 37 They had dressed, teeth brushed, breakfasted, had visited the loo, and were on their precarious journey all in a question of fifteen minutes. 1944 Auden *For Time Being* (1945) 20 Between the bottle and the 'loo' A lost thing looks for a lost name. 1954 Koestler *Invis. Writing* xxxix. 419 The story of 'the loo-tank papers'..is another instance of the cloak-and-dagger atmosphere. 1955 G. Freeman *Liberty Man* ii. vi. 113 Johnnie, do take him to the loo, there's a good boy. 1957 P. Wildeblood *Main Chance* 57 The loo's on the landing, if you want to spend a penny. 1960 C. Mackenzie *Greece in my Life* 23, I think I should sigh for the old Grande Bretagne Hotel in spite of the squalor of the loo which was no paradise for dysentery. 1971 *Petticoat* 17 July 31/2 You can wait until he goes to the loo or, if he appears to have a bladder like an ox, send him to the kitchen for more coffee. 1972 *Guardian* 23 Feb. 18/5 Matching bathmats (£2.20)..and loo seat covers (£1.80 and £1.12). 1973 E. McGirr *Bardel's Murder* iv. 85 A neighbouring cat had come through the window and made away with the loo brush. 1974 *Observer* 28 Apr. 28/6 The loo rolls unfurling across the pitch.

loo, var. Lew *a.*[1] (Further examples.)

Several examples (–1881) in Dict. s.v. Lew *a.*[1]

1889 'M. Gray' *Reproach of Annesley* (ed. 5) vi. iii. 276 'Tis fine and loo here,.. and you med set down and hrest. 1892 H. C. O'Neill *Devonshire Idyls* 7 His house..was 'loo' from the cold north winds. 1906 *Daily Chron.* 16 Aug. 3/6 It is cool and pleasant to find a 'loo' corner on the Esplanade [in Penzance]. 1963 Nance & Pool *Gloss. Cornish Sea-Words* 109 *Loo*,.. (2) lee, sheltered out of a wind.

looard, var. Leeward *a.* (*sb.*) and *adv.*

1886 in H. Baumann *Londinismen* 100/1. 1910 'O. Henry' *Whirligigs* ii. 33 Morgan lived in a bamboo shack to 'loo'ard'. 1963 Nance & Pool *Gloss. Cornish Sea-Words* 109 *Looard*, leeward, is common sea-language.

looder, var. Lowder (in Dict. and Suppl.).

looey (lū·i). *N. Amer. slang.* Also looie, louie. [f. Lieu(tenant with pronunc. (lū-) + -y[6], -ie.] A lieutenant.

1916 *Rio Grande Rattler* 11 Oct. 5 The 'Looey' gaily saunters from his tent. 1920 J. Dos Passos *One Man's Initiation: 1917* vii. 86 Our louie's name's Duval. 1928 W. H. Upson *Me & Henry & Artillery* 120 Then the looeys started hollering to the sergeants. 1935 A. J. Pollock *Underworld Speaks* 73/1 *Looie*, a police lieutenant. 1942 M. Hargrove *See Here, Private Hargrove* lxiii. 182 How would you like a second looey's commission? 1967 I. A. Baraka in W. King *Black Short Story Anthol.* (1972) 119 Jimmy Lassiter, first looie. 1974 *Weekend Mag.* (Montreal) 27 Apr. 10/1 One scrap of the rarely-talked-about reality: after being a private 14 months, Angus was commissioned in the field as second looey.

loogan (lū·găn). *U.S. slang.* [Etym. unknown.] In derogatory use: a fellow, a 'fool'.

1929 D. Runyon in *Hearst's International* July 126/3 The poor loogan she is marrying will never have enough dough to buy her such a rock. 1932 J. T. Farrell *Young Lonigan* ii. 86 Bill's a loogin who always tries to wisecrack. 1933 'P. Cain' *Fast One* 200 There's Rose, with his syndicate behind him, and all the loogans he's imported from back East.

looie, var. *looey.

look, *sb.* Add: **1.** Phr. *if looks could kill* (or *slay*): used to denote an expression of hostility in a look.

1913 F. L. Barclay *Broken Halo* xxxvii. 372 If looks could slay, Margaret would not have left that room alive. 1922 F. Harris *My Life & Loves* I. ii. 37 When they let me up, I looked at Jones and if looks could kill, he'd have had short shrift. 1943 K. Tennant *Ride on Stranger* iv. 31 Just then she saw me..and if looks could have killed!

2. c. Also in pl., esp. in *from* or *by the looks* (*of*).

1883 R. Cleland *Inchbracken* iv. 28 It micht be e'en a bairn by the looks o' the bun'le. 1923 'B. M. Bower' *Parowan Bonanza* v. 54 You're just ahead of a big storm, by the looks, Mr. Rayfield. 1975 J. Symons *Three Pipe Problem* ix. 65 Acting doesn't pay too well from the looks of it.

e. Colloq. phr. *for the look of the thing*: for the sake of appearances.

1876 Trollope *Prime Minister* IV. xiii. 217 'I shall go down and vote for them of course,' said Mr. O'Mahony, 'just for the look of the thing.' 1910 'Saki' *Reginald in Russia* 63 He often wished, for the look of the thing, that people would sometimes burn candles at his shrine. 1924 M. Kennedy *Constant Nymph* xv. 212 Save for the look of the thing she had no particular wish for a reconciliation. 1940 L. H. Myers *Pool of Vishnu* iv. v. 191 Why are they willing to die, and to send those whom they love to their death, for the sake of—what shall I call it?—the look of the thing?

f. *Fashion.* With defining word(s): an appearance or effect indicated by the preceding word(s). Occas. without defining word (quot. 1973).

1938 [see *little-girl* attrib. (*little *a.* 14)]. 1939 *Vogue* 15 Apr. 2 (Advt.), Accent on That Fresh Young Look. 1940 *Mademoiselle* Mar. 56 Formold created for you that Tall Look of 1940. 1948 *Vogue* Mar. 41 The New Look has fined down... Length has crystallized into eleven inches from the ground for formal wear. 1966 *Listener* 3 Feb. 171/1 This year..the geometrical look is in. 1971 *Daily Mail* 3 Feb. 3 A specially-created hair-style to complete the hot-pants look. 1973 *Sunday Times* 28 Jan. 43/2 The clothes are created on them, and a whole look is painstakingly put together. 1974 *Country Life* 17 Jan. 107/1 The peasant look in knitwear.

look, *v.* Add: **1. a.** ¶ Colloq. phrases: *as quick* (or *soon*) *as look at you* (or *him*, etc.): very rapidly and readily; 'at the drop of a hat'; *not to know which way to look*: to be embarrassed; *not to look at* (someone): to find unattractive, to show no sexual interest in (someone); occas. in positive contexts.

1814 Jane Austen *Mansf. Park* I. v. 101 She came up to me..and talked and laughed till I did not know which way to look. I felt that I must be the jest of the room. 1817 M. Edgeworth *Harrington* xi. 247 Nor did I know well which way to look, when his lordship..asked Miss Montenero if she could possibly imagine that any such vulgar prejudices existed. 1861 C. M. Yonge *Young Step-Mother* xxviii. 420 Albinia did not know which way to look when all was ascribed to Mr. Kendal's great kindness to him. 1888 'R. Boldrewood' *Robbery under Arms* I. iii. 46 He was awful shook on Mad; but she wouldn't look at him. 1894 Somerville & 'Ross' *Real Charlotte* II. xxx. 78 There was no other woman here that signified except Miss Dysart, and it didn't seem likely she'd look at him. 1922 Joyce *Ulysses* 347 Give it to him too on the same place as quick as I'd look at him. 1926 I. Mackay *Blencarrow* xi. 105 Supposing Kathrine wouldn't look at him in any case? 1937 M. Sharp *Nutmeg Tree* xx. 266 If I married you I'd never look at another man so long as I lived. 1941 H. Nicolson *Diary* 8 July (1967) 177 He then stalks out of the room. We are left ashamed and wretched and do not know which way to look. 1946 E. O'Neill *Iceman Cometh* (1947) iii. 167 From what I've seen of 'em..they'd run over you as soon as look at you. 1955 J. I. M. Stewart *Guardians* iii. ii. 214 The novelists..constantly endow sensitive women with husbands whom, in fact, they would never *look at.* 1971 M. Torrie *Bismarck Herrings* xi. 156 Threatened to report you to the council as soon as look at you, she did. 1973 W. M. Duncan *Big Timer* iv. 29 You have to watch it nowadays. Jump you as soon as look at you. 1974 P. Dickinson *Poison Oracle* ii. 37 Will she even look at a male chimp? Doesn't she think she's human?

h. (Later examples.)

1947 A. Menen *Prevalence of Witches* ii. 36 Suddenly his eyes looked mischief again. 1956 H. Gold *Man who was not with It* (1965) vi. 53, I looked a question at her.

4. a. (Further examples.) *looky here* U.S. regional variant of 'look here'; also *look-a-here.* Also, *look who's* (or *what's*) *here*: see who (or what) is here.

1872 [see *gum-game* (*gum *sb.*[2] 9)]. 1876 'Mark Twain' *Tom Sawyer* x. 94 And besides, look-a-here—maybe that whack done for *him!* 1925 E. O'Neill *Desire under Elms* i. ii. 20 Looky here! Ye'd oughtn't t' said that, Eben. 1935 Z. N. Hurston *Mules & Men* i. iv. 95 'Looka here, folkses,' Jim Presley exclaimed. 'Wese a half hour behind schedule.' 1943 'C. Dickson' *She died a Lady* v. 38 Look here... Burn it all, all I was tryin' to do was see what she'd do flat out on an open road. 1945 A. Kober *Parm Me* 75 'Well, look who's here!' exclaimed the host. 1949 N. Marsh *Swing, Brother, Swing* iii. 48 'Well, well, well,' he said. 'Look who's here.' 1969 *Listener* 1 May 614/1 Look, my Bill doesn't include any blanket condemnation of unofficial strikes. 1971 *Black Scholar* June 54/2 Look here, Dr. Hare, I don't have a picture at this time. 1972 *Black World* Apr. 62 Lookahere Sammy... I'm glad to see ya. 1973 G. Sims *Hunters Point* xiii. 115 Look, we don't have to sit here. We could go down to the beach. 1974 *Daily Mail* 1 Oct. 8/5 *Look you!* Plaid Cymru protested to the BBC yesterday over the timing of its only party political broadcast.

6. a. *to look babies*: delete † and add later example.

1957 J. Braine *Room at Top* xi. 107, I could see my face in her pupils... 'You're looking babies,' she said.

d. For † *Obs.* read *rare, now dial.*

1879 *Boy's Own Paper* 18 Jan. 14/3 [The monkeys] both set to work and 'look fleas' in the hare's fur. 1961 F. G. Cassidy *Jamaica Talk* vii. 148 A very common usage makes *look* into a transitive verb meaning look for, gather: 'Arthur and I joined a group of boys to *look wood*'.

10. a. (Further examples.) Also, (*it*) *looks like*: it seems likely. *colloq.* (chiefly *U.S.*).

1910 W. M. Raine *Bucky O'Connor* 55 Your cook, Anderson, kid-napped the child, looks like to me. 1929 J. Buchan *Courts of Morning* 13, I admitted that it looked like it, and said that if Blenkiron had been captured by bandits..his captors had done the worst day's work of their lives. 1936 M. Mitchell *Gone with Wind* i. i. 10 Don't it look to you like she would of asked us to stay for supper? 1970 N. Marsh *When in Rome* v. 127 'Wouldn't it be a yell if..you were The Man?' 'Do I look like it?' 1972 G. Bromley *In Absence of Body* viii. 101 'And now I suppose you've got to find a replacement?' 'Looks like it.' 1973 *Guardian* 31 Jan. 4/7 Looks like your child's birthday is news again this year.

b. (Later example.)

1973 A. Broinowski *Take One Ambassador* ii. 21, I look like being in and out of the office a lot in the next few days.

18. d. *to look on* (or *to*) *the bright* (or *worst*, etc.) *side*: to regard or consider something with optimism (or dismay, etc.). Cf. Side *sb.*[1] 10.

a 1784 Johnson in E. Fuller *Thesaurus Quots.* (1941) 667/1 The habit of looking on the best side of every event is worth more than a thousand pounds a year. 1833 W. F. Hook *Let.* 9 Dec. in W. R. W. Stephens *Life & Lett. W. F. Hook* (1878) I. iv. 258, I am a bit of an optimist, I always look to the bright side of things. 1839 [see *bright *a.* 1 e]. 1848 J. Ruskin *Let.* 17 Mar. in M. Lutyens *Ruskins & Grays* (1972) xi. 98 My disposition is to look to the worst side of things and..I feared you were entirely ruined. 1852 Mrs. Stowe *Uncle Tom's Cabin* II. xxiv. 82 Well, of course, if you can look on the bright side, pray do. 1914 R. Frost *North of Boston* 69 But I don't count on it as much as Len. He looks on the bright side of everything. 1942 'P. Wentworth' *Pursuit of Parcel* xi. 51 Well, ducks, I shouldn't take on. Look on the bright side.

22. b. *to look towards a person* (later examples). Also, *to look at* (a person).

1880 Stevenson & Henley *Deacon Brodie* i. 24 Deacon, I looks towards you. 1890 B. L. Farjeon *Mystery of M. Felix* I. iii. 26 Mrs. Middlemore,..you're a lady after my own heart... Here's looking towards you. 1919 A. Noyes *Coll. Poems* I. 241, I looks to-wards you, Preston John, you've done us very proud! 1930 J. Dos Passos *42nd Parallel* i. 119 'Pard, have that on me.'..'Thanks, here's lookin at you.'

c. (Examples.)

1879 A. W. Tourgee *Fool's Errand* xliv. 330 There could be nothing looking towards marriage between us.

1903 A. T. HADLEY *Relations between Freedom & Responsibility* 15 A series of negotiations rather than discussions, looking toward compromise rather than toward mutual enlightenment. 1904 T. N. PAGE in *McClure's Mag.* 621 The South regarded jealously any teaching of the Negroes which looked toward equality. 1932 T. J. GRAYSON *Leaders & Periods Amer. Finance* xiii. 278 The thing to do was to take no precipitate action looking toward resumption.

30. look around. b. = *look round* (sense *42 c). Also, to search about *for*.

1883 'MARK TWAIN' *Life on Mississippi* xliii. 437 I'll look around a little, and if I can't do better I'll come back and take it. 1927 H. CRANE *Let.* 14 Mar. (1965) 290 I'm looking around for some sort of 'avocation'. 1974 'M. INNES' *Appleby's Other Story* x. 81 Upper servants are frequently left in residence as caretakers... If it happens at Elvedon, it will give you time to look around.

32. look back. e. (Later examples.)

1928 *Observer* 17 June 27/4 Since they adopted the bold experiment..of changing the date of their regatta.. Marlow Amateur Regatta has never looked back. 1936 'N. BLAKE' *Thou Shell of Death* i. 17 His origin is shrouded in mystery... Turned up suddenly in the R.F.C. early in the war, and never looked back. 1949 *Radio Times* 15 July 17/1 Jules Verne..wrote *Five Weeks in a Balloon*, scored an immediate success, and never looked back. 1973 *Times* 23 Apr. 4/7 The play ran into the war, and she has never looked back.

33. look down. e. (Earlier examples.)

1812 *Niles' Reg.* III. 45/2 Volunteer companies..are rolling to the frontiers, in force sufficient to look down opposition. 1837 *Knickerbocker* IX. 361 We're a free trader..and are forced to go well armed, to look down all resistance. 1838 J. F. COOPER *Homeward Bound* I. viii. 194 If the people cannot control and look down peculiarity ..one might as well live in a despotism at once.

37. look in. c. [After *listen in*, *LISTEN *v.* 2 e.] *intr.* To watch a television programme. *colloq.*

1927 *Pictorial Weekly* 5 Mar. 100/1 We shall then 'look-in' by wireless and see events and scenes at a distance. 1928 *Daily Sketch* 7 Aug. 11/1 The public..can 'listen-in' or 'look-in' to the transmissions. 1950 *Ann. Reg. 1949* 418 At the end of October there were..206,000 [television] sets licensed and..as many as a million people regularly looking-in. 1959 J. BOLAND *Operation Red Carpet* v. 67, I often look-in when he's on.

42. look round. c. (See quot. 1914.)

c 1869 TAYLOR & DUBOURG in M. R. Booth *Eng. Plays of 19th Cent.* (1973) III. 250 I've begged and prayed to him for time—only to look round. 1914 *Conc. Oxf. Dict.* Add., *Look round*, (esp.) examine the possibilities &c. with a view to deciding on a course. 1950 J. CANNAN *Murder Included* i. 8 Hugo..is out of the army and looking round; there was some talk of him starting a dairy herd. 1974 'R. TATE' *Birds of Bloodied Feather* iii. 74, I looked round for a job and found a modest occupation.

43. look through. d. *to look right* (or *straight*) *through* (a person): to pretend not to see (someone), to ignore (someone) deliberately.

1959 B. KOPS *Hamlet of Stepney Green* I. 9 Like me? He never even sees me. He looks straight through me. 1963 P. WILLMOTT *Evolution of Community* ix. 98 It's awful when they look right through you, because they think you're not as good as them. 1973 G. MITCHELL *Murder of Busy Lizzie* ii. 26 Clothilde's straight-laced mamma boycotted Eliza..and Clothilde..looked straight through the poor woman.

45. look up. e. (Earlier example.)

1806 R. COCHRANE *Let.* 6 Jan. in J. Steele *Papers* (1924) I. 461 One cause why it has been so low at this market was the scarsity of salt; our river is now full enough for Boats to run, I think the Article will look up.

47. Further collocations used attributively or as sbs.: **look-ahead**, an action of judging what can happen or is likely to happen in the (immediate) future; **look-and-say**, a method of teaching reading by identifying each word as a whole (as opposed to treating a word as a series of separate letters needing to be spelt); **look-around, -round** [cf. *to look (a)round*, senses 30, 42 (in Dict. and Suppl.)], an inspection, scrutiny, search; **look-through** *Papermaking* (see quots.).

1963 I. FLORES *Logic Computer Arithmetic* iv. 78 Another solution is to examine the inputs to a number of stages and, somehow, simultaneously predict the carry outputs for this group of stages. This is called the carry lookahead. *Ibid.* v. 83 Let us examine an adder which performs the carry function with a lookahead on several levels. 1967 A. BATTERSBY *Network Analysis* (ed. 2) xii. 210 The effect of a 'look ahead' decision rule is shown. 1973 *Sci. Amer.* June 93/3 Since the number of legal moves available to a player at each turn averages about 30, a full look-ahead to a depth of four would require examination of about 30^4, or 810,000 moves. 1909 B. DUMVILLE *Sci. of Speech* xii. 167 The books on school method usually mention three methods of teaching to read—the Alphabetic, the Look-and-Say, and the Phonic. 1964 M. CRITCHLEY *Developmental Dyslexia* iv. 16 Many have blamed the analytic, look-and-say, 'flash' or global systems of teaching—whereby the child learns to identify each word as a whole. 1973 *Guardian* 7 Mar. 5/2 The use of highly speculative Gestalt psychology as the theoretical basis for 'look and say' methods. 1947 *Ann. Reg. 1946* 157 Field-Marshal Smuts found time to fly to Berlin for what he described as a 'private look around' with no special..objectives. 1967 M. MCLUHAN *Medium is Massage* 10 'The Medium is The Massage' is a look-around to see what's happening. 1914 R. FROST *North of Boston* 65 We took one look round. 1932 J. BUCHAN *Gap in Curtain* ii. 97 He hoped, while in the country, to have a look round. 1937 E. J. LABARRE *Dict. Paper* 149/1 *Look-through*, a term applied to the appearance of paper when held to the light, thus disclosing the texture or formation. 1973 C. COHEN *Watermarks* (William Sommerville & Son Ltd.) 8/1 *Look-through*, the appearance of a sheet of paper when held up to the light: may be clear or mottled.

look-a-here: see *LOOK *v.* 4 a.

look-alike. *N. Amer.* [LOOK *v.* 9 + ALIKE *a.*] Something or someone that closely resembles another in appearance. Also *attrib.* or as *adj.*

1947 *Time* 1 Dec. 78 Lisle Maxwell Sanders..is often called 'Mr. Kieran' for his famed look-alike. 1949 J. ROEBURT *Tough Cop* xvi. 180 There were enough lookalikes in the flat photography of the period to puzzle me. 1961 M. BEADLE *These Ruins are Inhabited* (1963) ii. 31 Those look-a-likes, the two-shilling piece and the half-crown. 1969 L. GREENBAUM *Out of Shape* (1970) xxv. 175 Asher was flanked by his pregnant wife and a long-haired student. Except for the wife's stomach, the two women were look-alikes. 1972 *Islander* (Victoria, B.C.) 27 Feb. 10/3 This is the critical test, such misleading lookalikes as serpentine and its brother, bowenite, permitting no passage of light even through the finest sliver. 1974 *Publishers Weekly* 11 Feb. 62/3 He meets Gabrielle, Simone's look-alike sister, and falls in love all over again.

looker, *sb.* Add: **1. b.** (Further examples.) Also **looker-out**, in the book-trade, one who looks out wanted volumes from stock; **looker-upper** *colloq.*, one who looks something up.

1826 *New Monthly Mag.* XVII. 241, I have always casual lookers-in, and it is my cue..to keep..an open house. 1836 *Scottish Christian Herald* I. 286/2 Like the keepers of a puppet-show, to extort money from every looker-in. 1901 *Daily Chron.* 4 Dec. 9/2 (Advt.), Lookers-up (2 experienced); also several boys in beer factory. 1926 W. J. LOCKE *Old Bridge* II. viii. 134 The result..if sought, is there for the looker-round to behold. 1939 H. HODGE *Cab, Sir?* ii. 20 At the blind corners, where the separate sections are too far apart to keep in touch, there's a 'looker-out'. He waves up the cabs from one section to another. 1951 'J. TEY' *Daughter of Time* vii. 91 'Marta.. said you wanted something looked up.' 'And are you a looker-upper?' 1961 *Evening Standard* 14 July 19/4 Publisher has vacancy for warehouse-man to train as looker-out.

d. looker-in: a viewer of television. Also (now *rare*) *looker*.

The more usual word is *viewer*.

1927 *Pictorial Weekly* 5 Mar. 101 A speech which the 'looker-in' can actually see being delivered. 1928 *Daily Tel.* 30 Oct. 12/5 This afternoon 'lookers-in' will be given a chance of seeing the first still pictures to be publicly broadcast in this country. 1933 *Radio Times* 14 Apr. 72/2 *The First Television Revue*..should draw the majority of Britain's 'lookers' to their receivers. 1953 *Sunday Times* 25 Jan. 9/4 Producers should never allow themselves to be influenced by the knowledge that their audience contains many doggedly *literal* lookers. 1959 *Listener* 16 July 100/3 If the looker-in [of a televised church service] can only be a looker-on it would be better he did not watch.

3. A person, usu. a woman, of particularly pleasing appearance. *colloq.* (orig. *U.S.*).

1893 S. CRANE *Maggie* v. 41 The young men of the vicinity said, 'Dat Johnson goil is a puty good looker.' 1898 E. N. WESTCOTT *David Harum* 322, I was alwus a better goer than I was a looker. 1909 E. RICKERT *Beggar in Heart* 207 She isn't much of a looker—my missus has other points than looks. 1923 L. J. VANCE *Baroque* vii. 65 Just because daughter's a swell looker don't make father out an innocent. 1933 AUDEN *Witnesses* in *Listener* (Poetry Suppl.) 12 July p. ii, The days went by, he grew mature; He was a looker you may be sure. 1947 J. STEINBECK *Wayward Bus* vii. 80 She was a looker too—fine well-filled legs with rounded thighs. 1971 R. PARKES *Line of Fire* iv. 42 Bit of a looker... Otherwise..a ranking detective on a priority case, would hardly have bothered driving her home. 1973 *Washington Post* 5 Jan. 8/2 Sandra Archer, who plays the heroine from the Peace Corps..is such a looker that she can't help but make the Quest for Revolutionary Consciousness appear hopelessly glamorized.

looker (lu·kəɪ), *v. dial.* [f. the sb.] *trans.* and *intr.* To tend and guard (farm animals).

1887 PARISH & SHAW *Dict. Kentish Dial.* 95 *Looker*, to perform the work of a looker. 'John? Oh! he's *lookering*.' 1961 *John o' London's* 5 Oct. 400/3 In East Sussex a shepherd is still called a 'looker' and his occupation 'lookering'. 1962 R. JEFFRIES *Exhibit No. 13* x. 97 Jones was 'lookering' his bullocks. The ministry man was due.. and..the bullocks had to be treated with all possible care and attention.

look-in, *sb.* **2.** For def. read: *colloq.* An opportunity to take part in something, usually with a chance of success; a share of attention. Add later examples.

Quot. 1870 replaces that in Dict.

1870 *Bell's Life* 12 Feb. 3/6 If Fawcett imagines he has got a look in, Mullins will fight him for all the money he can get together. 1902 KIPLING *Traffics & Discov.* (1904) 27 We might even be able to give our Native Army a look in. 1905 *Official Guide Nat. Assoc. Professional Baseball Leagues* 58 With a team which never had a look-in for anything better than cellar championship..the club made money. 1911 G. B. SHAW *Getting Married* 244 We shall none of us have much of a look-in when Mrs. George comes. 1916 *Lit. Digest* 1 Jan. 7/2 Between Colonel Roosevelt and the diplomatic correspondence of this epoch the dictionary business is getting a look-in all right. 1916 'TAFFRAIL' *Pincher Martin* xvii. 330 It's time we had a look in at something. 1936 *Times Lit. Suppl.* 17 Oct. 837/1 The faithful, unmarried lover who never gets a look-in. 1964 *Word Study* Apr. 1/1 Nor would a quack bonesetter get a look-in at a position in a modern clinic. 1968 *Listener* 26 Sept. 390/1 An acknowledged modern artist gets a look-in at illustration 52.

looking, *vbl. sb.* Add: **1. d. looking-in** *Television* = *VIEWING *vbl. sb.* Also *attrib.*

1926 *Daily Herald* 31 Dec. 1/2 It is predicted that before many years have passed the family looking-in set will be as common in the home as is now the listening-in set. 1927 *Pictorial Weekly* 5 Mar. 101/2 'Looking-in' on Mr. Baird's apparatus is an interesting experience. 1951 *Ann. Reg. 1950* 414 From America came alarming reports of the craze for looking-in. 1957 R. HOGGART *Uses of Literacy* vi. 156 An undiscriminating looking-in, night after night, at T.V.

looking-forward. [f. LOOKING *vbl. sb.* + FORWARD *adv.*] The action of looking forward; an anticipation of future events.

1837 DICKENS *Let.* 3 Nov. (1965) I. 328 Anxious lookings-forward to the pleasure of your society. 1867 *People's Hymnal* 205/2 One the earnest looking forward, One the hope our God inspires. 1871 [see LOOKING *vbl. sb.* 1 b]. 1915 BEERBOHM *Lett. to R. Turner* (1964) 242 Well, dearest Reg, again all apologies..and all lookings-forward to later on. 1955 E. BOWEN *World of Love* xi. 221 Were there not those who said that everything *has* already happened, and that one's lookings-forward are really memories?

looking-glass. Add: **5. looking-glass image** *rare* = **mirror image*; **looking-glass world** (or **land**), a vision of the world as it would be if seen, reversed, through a looking-glass.

1929 A. HUXLEY *Do what you Will* 44 The professional Don Juan destroys his spirit as fatally as does the professional ascetic, whose looking-glass image he is. 1896 B. BERENSON *Italian Painters of Renaissance* (1930) II. ii. 63 He cannot persuade himself of the unreality of Looking-Glass Land until he has touched the back of the mirror. 1909 CHESTERTON *Tremendous Trifles* 234 Always the Kingdom of Heaven is 'at hand', and Looking-glass Land is only through the looking-glass. 1911 —— *Innocence of Father Brown* viii. 225 An unspeakable certainty that there was something still unexplained... Could not be fully explained by his fancy about 'looking-glass land'. 1871 'L. CARROLL' *Through Looking-Glass* xii. 218 You've been along with me, Kitty—all through the Looking-Glass world. 1963 *Daily Tel.* 15 Aug. 18/1 It is quite conceivable that there might exist a kind of looking-glass world, in which all matter is made up from anti-matter. 1967 'A. GILBERT' *Visitor* xii. 203, I felt as if I were in a Looking-Glass world where everything goes the wrong way.

lookit (lu·kit). *U.S. colloq.* [f. LOOK *v.* with arbitrary final element.] **a.** *int.* Listen! **b.** *v. trans.* Only in imp.: look at (something or someone).

1917 *Dialect Notes* IV. 396 *Look-at*, used among school children for *look!*.. Cf. *look-it* in Mass., Mich. 1925 T. DREISER *Amer. Trag.* (1926) I. xiv. 103 Oh, isn't that just the classiest, darlingest little coat you ever saw! Oh, do look at those sleeves... Lookit the collar. And the lining! And those pockets! 1926 *S.P.E. Tract* XXIV. 124 *Lookit*, listen to me. 1927 M. OSTENSO *Mad Carews* (1929) iv. 49 But lookit! Lookit the nice stockin's Mrs. Bowers made for ye. 1938 D. BAKER *Young Man with Horn* IV. iv. 242 'Lookit the jig-men,' Olga said. 'I thought you'd give us the go-by.' 1966 M. BREWER *Man against Fear* i. 8 Lookit my hair—whiter it gets every day. 1968 [see *JEEZ(E *int.*]. 1972 D. BLOODWORTH *Any Number can Play* xxiii. 231 Just get a load of that stuff, will you? It's not even killing the lousy jungle, lookit. It's bringing it alive!

look-out. Add: (Now usually stressed *loo·k-out.*) **2. a., b.** (Further N. Amer. examples.)

1888 *Century Mag.* Feb. 498/2 These lookouts or forerunners having returned, the herds are set in motion as early in the spring as may be. 1893 *Harper's Mag.* May 939/2 By each dealer's side sits the 'lookout'..lazily looking on in the interests of such fair play as is consistent with professional gambling. 1935 A. J. POLLOCK *Underworld Speaks* 73/1 *Look-out*, gambling house employee who observes the bets of players and the pay-off of dealers for regulatory purposes. 1955 J. S. GOWLAND *Smoke over Sikanaska* 16 The look-out had to be an expert woodsman, be able to read meteorological instruments, have an excellent degree of physical fitness and good eyesight. 1961 *Canada Month* 6 Oct. 42/3 However, the forestry people want money to buy greater preparedness through more lookouts, men, planes and equipment.

look-over. [f. LOOK *v.* 19.] An examination, a survey.

1909 R. A. WASON *Happy Hawkins* 183 Then I..took a stroll around to see that no one had been givin' us the look-over. 1916 'B. CABLE' *Action Front* 216, I want you to go down quickly and have a look over at the new ground. 1952 *Irish Digest* Feb. 8/2, I have myself treated one farmer..who was sent to me for a general 'look over'. This disclosed a malunited fracture of the ulna.

loo·k-see. *slang.* Also **looksee.** [Pidgin-like formation from LOOK *sb.* or *v.* + SEE *v.*] **1.** A survey; a tour of inspection, a reconnaissance; an investigation. Also *rare* (quot. 1926), appearance, looks.

1883 *Boy's Own Paper* 22 Dec. 185/1, I 'spec she just come here to makee look see how de people get on. 1906 J. LONDON *Let.* 25 Apr. (1966) 204 Would you care to have a look-see at it for publication in the magazine? 1908 *St. George's Rev.* I. 156 China..opium problem... It was my business to go out there and have what my John would call a 'look-see'. 1924 *Blackw. Mag.* Sept. 356/2, I sat up, and had a look-see. The ground sheet was crawling with scorpions. 1926 M. LEINSTER *Dew on Leaf* 82, I distrust the look-see of things. 1927 *Observer* 9 Oct. 22 We must be grateful to the B.B.C. for letting us have a 'looksee', as the Chinese say. 1928 *Sat. Even. Post* 4 Feb. 105/3 And I can take a look-see at what they're doing in aviation over there. 1939 J. PASCOE *Unclimbed N.Z.* v. 67 A hurried reconnaissance, or in Colonial argot, a 'look-see', disclosed that. 1942 C. BARRETT *On Wallaby* iii. 52 We had a looksee at Merre Gudda, which, the blacks say, is a haunted cave. 1943 C. H. WARD-JACKSON *Piece of Cake* 41 *A look-see*, a reconnaissance. Thus, 'Let's get some facts first; go down to the flights and take a look-see.' 1957 I. CROSS *God Boy* (1958) xv. 123 I'll wander up and have a look-see. 1967 B. COPPER *No Flowers for General* xi. 142 I'll have a looksee at the front. 1968 A. DIMENT *Bang Bang Birds* ii. 13, I took a long looksee through my ..binoculars.

2. (See quot.) *rare.*

1925 FRASER & GIBBONS *Soldier & Sailor Words* 147 *A look see*, a telescope, a periscope.

3. *attrib.*

1929 *Amer. Speech* V. 149 Several pidgin English terms are now accepted American slang:..'chow-chow' for food, and 'look-see man' for tourist or sightseer. 1971 M. TAK *Truck Talk* 101 *Look-see window*, a window in the rear of the sleeper that assists the driver in backing up by increasing his visibility. 1973 *Times* 17 May 27/2 As a result of his 'look-see' trip Lewis..came back home with around £100,000 worth of export orders. 1973 *Sci. Amer.* Oct. 114/1 'Look-see' diagrams that offer visual proof of complex algebraic formulas.

look-up (lu·kʌp). [f. vbl. phr. *to look up* (see LOOK *v.* 45).] **1.** A call, a visit. *rare.*

1855 D. G. ROSSETTI *Let.* 25 Nov. (1965) I. 278 Hughes ..gave them a look up about it. 1888 'R. BOLDREWOOD' *Robbery under Arms* I. xiv. 191 We foraged up Aileen's mare, and made it up to ride over to George Storefield's, and gave him a look-up.

2. The action of (or a facility for) looking something up in a dictionary, file, etc.; retrieval of information about items in an ordered collection. Freq. *attrib.*

1948 *Math. Tables & Other Aids to Computation* III. 157 Operations such as division, square root, table look-up, etc., where the required time cannot be predicted. 1958 A. D. BOOTH in *Aspects of Translation* 88 All that had been produced was a programme which would enable a computing machine to perform look-up operations which a human translator would perform with a dictionary. 1960 E. DELAVENAY *Introd. Machine Transl.* vi. 93 An appreciable amount of time will thus be saved in dictionary look-up. 1964 *Discovery* Oct. 55/1 The programme does this in several stages: (1) a dictionary look-up which provides information about parts of speech, [etc.]. 1967 COX & GROSE *Organiz. Bibliogr. Rec. by Computer* IV. 79 These citations are then found in the main file by a 'look-up' procedure. 1971 A. J. AITKEN in R. A. Wisbey *Computer in Lit. & Ling. Res.* 14 In addition, TLF also has a computer 'look-up' which in effect lists certain predictable collocations of certain common function words so that the computer can subdivide its examples according to these collocations.

looky here: see *LOOK *v.* 4 a.

loom, *sb.*[1] Add: **5*.** *Electr.* **a.** Flexible tubing which is fitted over the ordinary insulation of an electric wire to provide additional protection.

1917 A. L. COOK *Interior Wiring* xiii. 235 For wires carrying more than 300 volts or for damp places, flexible conduit or armored cable must be used. The flexible tubing used is sometimes called 'circular loom'. 1939 H. P. RICHTER *Pract. Electr. Wiring* xi. 158 Where wires cross each other, slip loom over both wires.

b. A group of parallel insulated wires bound together into a bundle; (see also quot. 1949).

1949 *Gloss. Aeronaut. Terms* (*B.S.I.*) II. 23 *Loom*, one or more cables pre-assembled for installation in an aircraft. 1962 *Which? Car Suppl.* Oct. 139/1 A wiring loom prevented the dipstick being removed or replaced easily. 1972 C. E. JOWETT *Electronic Engin. Processes* IV. vi. 141 The forming of looms should preferably be by means of plastic ties, at an approximate pitch of 25 to 38 mm.

6. *loom-room*; **loom-house** (earlier U.S. example); **loom-state** *a.*, of woollen fabrics, in the state in which they came from the loom, untreated.

1819 *Western Rev.* I. 303 The other two young women slept in a loom house adjoining. 1845 *Knickerbocker* XXV. 448, I went out to look at the loom-room. 1961 BLACKSHAW & BRIGHTMAN *Dict. Dyeing* 101 *Loomstate*, woven fabrics in the condition in which they come from the loom. For practical purposes the term is synonymous with *Grey* (adjective). 1972 *Times* 9 May 20/6 Lists of cotton 'grey' (loomstate) goods.

loon, *sb.*[2] Add: **1. b.** In phrases with *loon's* (see quots.). Also freq. *as crazy as a loon* (in reference to its actions in escaping from danger and its wild cry) and varr.; so, *as drunk as a loon*; *to hunt the loon* (see quot. 1880).

1830 *Kentucky Intelligencer* (Flemingsburg) 29 May 4 Patton informed me that McLaughlin had just gone from Elizabethville, and was 'drunk as a loon'. 1834 C. A. DAVIS *Lett. J. Downing* 42, I saw thru' it in a minute, and made it all as strait as a loon's leg. 1834 S. SMITH *Sel. Lett. J. Downing* 110 He begun to sing out like a loon for us to come and take him. 1840 C. F. HOFFMAN *Greyslaer* I. I. xi. 129 After tramposing for twenty-four hours on a stretch, with not even a loon's nap at the end of it. 1845 C. M. KIRKLAND *Western Clearings* 83 Why, you're both as crazy as loons! 1865 'MARK TWAIN' in Harte & 'Twain' *Sk. Sixties* (1926) 163 Our reserve..came filing down the street as drunk as loons. 1880 *Harper's Mag.* Dec. 31 Miss Lois had been hunting the loon with a handnet—a Northern way of phrasing the wearing of the willow. 1931 W. FAULKNER *Sanctuary* vi. 57 You're as crazy as a loon. 1934 J. T. FARRELL *Young Manhood* xxiv. 398 Jesus, I'm drunk as a loon. I'm drunk, Kelly. Drunk. 1951 *Publ. Amer. Dial. Soc.* xv. 66 Crazy as a loon.

c. *transf.* A crazy person; a simpleton.

Perhaps influenced by LOONY, LUNY *a.* and *sb.*

1885 'C. E. CRADDOCK' *Prophet Gt. Smoky Mts.* xii. 230 But ye air a smart man ter that loon, fur..he dunno he air a loon. 1918 C. SANDBURG *Cornhuskers* 99, I am a loon about the sea. 1945 *Coast to Coast 1944* 72 There we were, bottled up in camp because the loon in charge couldn't get the order signed for the trucks to leave.

2*. An early type of guided missile developed by the U.S. Navy.

1947 *Newsweek* 17 Mar. 64/2 The Navy also displayed a 'bat bomb' and a 'loon', two hitherto secret radio-controlled missiles. 1951 COGGINS & PRATT *Rockets, Jets, Guided Missiles & Space Ships* iii. [28/1] Later the [U.S.] Navy built its own improved version of the V-1, called the 'Loon'. It can be launched from a submarine far at sea in a surprise attack against a harbor or convoy. 1952 *Jane's Fighting Ships 1952–53* 410/1 'Loon' missile is contained in a watertight steel hangar and takes off from a ramp fixed to the submarine's deck.

loon (lūn), *sb.*[4] *colloq.* [f. the vb.] *pl.* A style of close-fitting casual trousers widely flared from the knees to the ankles. Also (in *sing.*) *attrib.*, as *loon pants, trousers.*

1971 *Melody Maker* 13 Nov. 50 (Advt.), New velvet and cord loon pants with 28" flare... New cotton drill loons and military trousers. 1972 *Guardian* 15 Aug. 17/3 As dead as tucked up loon pants. 1974 'D. CRAIG' *Whose Little Girl are You?* v. 94 They had bloody long-hairs in the police now and kids in loon trousers. 1974 D. WINSOR *Death Convention* xii. 98, I wriggled into a pair of brown velvet loons, dropped a cream lace tunic over them.

loon (lūn), *v.* [Etym. unknown.] *intr.* Esp. of young people: to spend one's leisure time in a pleasurable way, e.g. by dancing to popular music; to lie *about* or wander *about*. So **loo·ner,** one who loons; **loo·ning** *vbl. sb.* Cf. *LOON *sb.*[4]

1966 *Melody Maker* 30 July 8/6 The younger members of the MM staff spend a lot of time doing something called 'looning'. To judge by their general condition the next morning I gather this is what used to be known as 'raving'. 1969 *It* 4–17 July 12/2 It's sort of music essentially to loon about to. 1969 *Daily Tel.* 14 July 11 (*heading*) Long enough to loon in. *Ibid.*, A fashion designer..has just completed his first collection of clothes aimed purely for after work. He calls them 'looning' clothes. 1969 *Melody Maker* 13 Sept. 12/4 In the company of looners like Eric Burdon and Brian Auger, Zoot was the king looner. Zoot became a much beloved symbol of good fun and good time music. 1971 *It* 2–16 June 21/3 Children and the younger adults alike looning about in wonderful costumes.

loony, luny, *a.* and *sb.* Add: **c. loony bin** [*BIN *sb.* 7], a facetious term for a mental hospital; also *fig.* and *ellipt.*; **loony-doctor** *slang*, a doctor who treats mental illnesses, a psychiatrist.

1919 WODEHOUSE *My Man Jeeves* 195 If you're absolutely off your rocker, but don't find it convenient to be scooped into a luny-bin, you simply explain..it was just your Artistic Temperament. 1921 —— *Indiscretions of Archie* xxv. 303 Nine out of ten of them had views on Art which would have admitted them to any looney-bin, and no questions asked. 1938 J. PHELAN *Lifer* xix. 201 Left him behind us there, we did, stone balmy. Finished in a loony. 1942 N. STREATFEILD *I ordered Table for Six* 216 Mrs. Framley must have thought him fit for a looney bin. 1959 *New Statesman* 28 Mar. 434/3 In short, I do not want men to live for ever in a sort of global loony-bin. 1962 J. SYMONDS *Bezill* 68 Yes, Aunt Marion. She's locked up, you know, in the looney bin. 1974 *Times* 9 Feb. 10/4 (*heading*) In the looney bin. 1925 WODEHOUSE *Carry on, Jeeves!* vi. 139 Old Sir Roderick, who's a loony-doctor and nothing but a loony-doctor, however much you may call him a nerve-specialist. 1936 BENTLEY & ALLEN *Trent's Own Case* iv. 39 'Once, I remember, he said that the worst of these loony-doctors——.' 'Did he say "loony-doctors"?' 1960 WODEHOUSE *Jeeves in Offing* ii. 18 She's browsing with Sir Roderick Glossop, the loony-doctor.

loop, *sb.*[1] Add: **1. e.** *U.S.* The looped portion of a lasso. **f.** = NOOSE *sb.* 1.

1907 S. E. WHITE *Arizona Nights* I. v. 93 Some few whirled the loop, but most cast it with a quick flip. 1933, etc. [see *HONDA]. 1944 *Living off Land* vii. 138 Knots.. nooses (or loops). 1970 G. R. STEWART *Amer. Place-Names* 264/1 *Loop* Texas: so called because the post-master-to-be, at the time of suggesting a name, was playing with the loop of his lasso.

4. d. (ii) *Electr.* A point on an aerial at which the current or the voltage is a maximum.

1922 GLAZEBROOK *Dict. Appl. Physics* II. 1034/1 A standing wave forms on the simple antenna just as in an organ-pipe with stopped end, there being a node of current at the upper end and a loop at the earthed connection. 1928 [see *FEEDER 10 b]. 1968 *Radio Communication Handbk.* (ed. 4) xiii. 3/1 At positions of current loops, the current-to-voltage ratio is high and the wire will behave as a low impedance circuit.

e. Add to def.: Hence in *Electr.*, any complete circuit or path for a current.

1909 WEBSTER, *Loop*, a complete electric circuit. 1922 GLAZEBROOK *Dict. Appl. Physics* II. 659/1 If a stout wire be temporarily used to connect the two banks of plates inside, the condenser may be measured as a loop at telephonic frequencies. 1967 *Electronics* 6 Mar. 132/1 The feedback loop is formed by connecting the amplifier's inverting input (pin 2) to potentiometer. 1970 J. EARL *Tuners & Amplifiers* iv. 95 Small mains currents can flow in the loops formed by the several earths, and these can induce hum into the system.

f. Also, a similar path described by an aeroplane. (Cf. *LOOP *v.*[1] 6.) Phr. *to knock* [KNOCK *v.* 6 e] *for a loop* and varr.

1913 *Aeroplane* 25 Sept. 350/2 M. Pégoud succeeded in looping the loop completely. 1923 *Cosmopolitan* Apr. 84/1 It took Hurricane Sherlock just two boisterous rounds to smite 12-Punch O'Bernstein 'for a loop', as Hurricane put it. 1936 J. G. BRANDON *Pawnshop Murder* v. 46 Something had happened which had knocked even the imperturbable Wibley for the loop. 1968 J. WAINWRIGHT *Web of Silence* 126 Have you lost your marbles, Pewter?.. Have you gone for a complete loop? 1969 E. AMBLER *Intercom Conspiracy* (1970) vi. 110, I was really confused. That memorandum threw me for a loop. 1971 *Country Life* 18 Feb. 374/2 It was over Tewkesbury that I pointed my nose at the Mill and did my first loop. 1973 D. RAMSAY *Deadly Discretion* 153 That little charade of hers had knocked him for a loop.

g. *Skating.* A curve crossing itself, or any of several elaborations upon this.

1869 VANDERVELL & WITHAM *Syst. Figure-Skating* x. 187 The Large Loop. This being done entirely on one edge throughout, requires some medium degree of speed. 1901 *Encycl. Sport* IV. 370/1 Loops are of three kinds—the ordinary variety, the turn loop, and the bracket loop. 1935 *Encycl. Sports* 560/1 'Loop' is effected by over-balancing the body and recovering equilibrium by a quick turn of the foot. 1962 T. D. RICHARDSON *Art of Figure Skating* vii. 60 Loops are not skated in the normal positions..and what is more, they require an entirely different timing. 1973 *Times* 7 Feb. 15/8 Hoffmann was fourth in the rocker but skated the best loops.

h. A configuration in finger-prints.

1880 H. FAULDS in *Nature* 28 Oct. 605/1 The right ring-finger..has an oval whorl, but the corresponding left finger shows an open loop. 1894 'MARK TWAIN' in *Century Mag.* June 235 The bewildering maze of whorls or curves or loops which constituted the 'pattern' of a 'record' stand out bold and black. 1938 G. W. WILTON *Fingerprints* xvi. 78 Galton had experimented only with thumbprints, grouping his lineations into three classes of arches, loops and whorls. 1970 P. LAURIE *Scotland Yard* ix. 193 There are two basic finger-print patterns: loops, where the lines turn through two right angles, and triradii.

i. A slack length of film, flexible strip, or the like left between two mechanisms to allow for a difference between the supply and take-up motions, esp. (in cinematographic equipment) one between a sprocket that turns continuously and one that turns intermittently.

1912 F. A. TALBOT *Moving Pict.* vii. 70 A slight loop is made at either end of the gate. 1939 SPENCER & WALEY *Cinema To-Day* ii. 25 A flickering motion of these loops absorbs the difference between the steady feed of the sprocket wheels and the intermittent feed of the claw. 1962 G. A. T. BURDETT *Automatic Control Handbk.* ix. 54 When metal strip is being wound or reeled it is usual to form loops in the feed line to allow for flexibility and avoid undue tension.

j. = *loop aerial* (see *6).

1922 GLAZEBROOK *Dict. Appl. Physics* II. 1058/1 Before the advent of high amplification, it was impracticable to use single loops of manageable dimensions. 1936 *Jrnl. R. Aeronaut. Soc.* XL. 175 To obtain a full direction finding service, it is usual to instal an external circular loop which can be rotated and orientated by the operator when taking a bearing. 1966 *McGraw-Hill Encycl. Sci. & Technol.* I. 445/2 Use of small loops concealed within the set is a standard practice for broadcast receivers in areas where signal strength is high.

k. A length of film or magnetic tape whose ends have been joined to form an endless strip, so that continuous repetition of the recording is made possible (e.g. in rehearsing the synchronization required for dubbing a foreign-language sound track).

1931 K. F. MORGAN in L. Cowan *Recording Sound for Motion Pict.* x. 151 The machine is provided with a film elevator attachment for running a continuous loop of sound track. This attachment is particularly useful for the dubbing of continuous background sounds. 1951 R. SPOTTISWOODE *Film & its Techniques* xii. 355 It is most important [in dubbing]..not to upset the recording order of an emotional scene which has had to be broken down into several loops. 1959 HALAS & MANVELL *Technique Film Animation* xix. 216 Each pencil-test is shot on negative and projected in loops so that it can be viewed over and over again. 1962 A. NISBETT *Technique Sound Studio* 258 A tape loop may be used (*a*) to provide a repeated sound structure or rhythm (in radiophonics), (*b*) for an 'atmosphere' track where this is regular in quality, (*c*) in tape delay techniques. 1968 *Punch* 31 Jan. 153 At the Rank Organisation's Pinewood Studios they tell how Sophia Loren dubbed sixty-four loops in an hour and a half, against an average of ten or twelve loops an hour.

l. A sequence of control operations or activities in which each depends on the result of the previous one; *esp.* (more fully *closed loop*), one in which there is feedback, the result of a later operation being made to affect

one earlier in the sequence, usu. so as to maintain the output at a desired level.

1945 L. A. MACCOLL *Fund. Theory Servomechanisms* viii. 70 This procedure of adding feedback loops to internal parts of a servomechanism is employed frequently when it is necessary to take special steps to insure that the performances of those parts shall be accurate and reliable. 1948 BROWN & CAMPBELL *Princ. Servomechanisms* vii. 227 Figure..shows internal loop within a main closed loop. 1954 M. H. LAJOY *Industr. Automatic Controls* i. 9 The automatic washing machine which operates on a time basis and is not dependent upon whether or not the clothes are clean is an open loop system. 1962 F. I. ORDWAY et al. *Basic Astronautics* ix. 366 There are two basic control systems: the open-loop system and the closed-loop system. The open-loop system is familiar to us all. Examples of this are..a light switch, or the horn button on an automobile... In the closed-loop system a portion of the output is sensed and fed back to the input. The input is then altered to achieve a co-ordinated response at the output... A good example is the modern air conditioning system in which room air is fed back to the thermostat control. 1971 J. Z. YOUNG *Introd. Study Man* vii. 107 Certain cells in the hypothalamus..are very sensitive to slight rises in temperature above the normal (37°C). They then discharge nerve impulses that set in action the mechanisms that cool the body, such as sweating. This cools the blood and switches off the hypothalamus. In order to study such closed-loop feedback systems engineers use the device of 'opening' the loop. This has been done..by putting heating electrodes in the hypothalamus and arranging that they keep it at a constant temperature a few tenths of a degree above normal in spite of the cooling blood.

m. *Computers.* A sequence of instructions which is executed repeatedly (usu. with an operand that changes in each cycle) until some previously specified criterion is satisfied.

1947 GOLDSTINE & VON NEUMANN in J. von Neumann *Coll. Works* (1963) V. 86 When a simple induction takes place, C travels during each step of the induction over a certain path, at the end of which it returns to its beginning. Hence this path may be visualized as a loop. We will call it an induction loop or a simple induction loop. 1954 *First Gloss. Programming Terminol.* (Assoc. Computing Machinery) 12 *Loop*, the repetition of a group of instructions in a routine. 1955 R. K. RICHARDS *Arithmetic Operations in Digital Computers* xii. 359 Any one program may contain many loops which may interlock one another in a complex manner. 1964 C. DENT *Quantity Surveying by Computer* iii. 28 It will be seen that *s* in instruction 4 is reduced by unity at every cycle of the loop, by reason of instruction 7, and that eventually..the conditional jump instruction will decide that the number stored by instruction 5 is negative. Control then proceeds with the next sequence according to instruction 9. 1969 P. B. JORDAIN *Condensed Computer Encycl.* 294 The terminating condition test is usually the last instruction of the loop, but it may be anywhere. 1970 O. DOPPING *Computers & Data Processing* xiv. 215 Loops..are basic to the economy of automatic data processing. It costs a few dollars to have an instruction written, but the computer executes it at a price which may be only a fraction of a cent below what the same operation would cost if it were executed by a human being. The only way of amortizing programming is to arrange for most instructions to be repeated a great number of times.

n. *Nuclear Engin.* A system of pipes passing through or associated with a reactor that forms a closed circuit (under operating conditions).

1957 *New Scientist* 23 May 33/1 The chemical behaviour of such systems has to be investigated in a reactor by passing a pipe containing a portion of the system under investigation through or close to the reactor core. Such a pipe, with the necessary measuring equipment, is called a test loop. 1958 H. ETHERINGTON *Nucl. Engin. Handbk.* v. 142 To satisfy the demand for testing operation under conditions of high temperature and pressure in those reactors using a primary water loop, high-pressure loops have been developed. Questions of fuel stability, heat transfer, water chemistry, radiation-accelerated corrosion, fission-product leakage, and fuel stability under desired operating conditions may be answered in high-pressure water loops. 1974 *Times* 21 Jan. 15/2 The CEGB's first two plants..would be the 52nd and 53rd of their type—the Westinghouse four-loop design, in which a single reactor is linked to four steam-generating boilers. 1974 *Encycl. Brit. Macropædia* XII. 897/1 Pressurized water both cools the reactor and carries away the fission-produced heat through the primary system (or loop) to a heat exchanger. .. The pressurized water then circulates back through the reactor in a constant cycle. In the secondary loop (which is sealed off from the radioactive primary coolant water) the water boils and expands into steam.

o. A type of intra-uterine contraceptive device. Cf. *LIPPES LOOP.

1962 [see *LIPPES LOOP]. 1965 *Guardian* 28 May 6/3 The medical council has authorized the intra-uterine contraceptive device... The IUCD would nearly always prevent pregnancy... The 'loops'..cost only a few pence. 1967 *Times* 9 Oct. 4/4 [Pakistan] Although the 1970 target of 500,000 vasectomies and five million loop insertions might seem low,..the main object was to retard or halt the growing birth rate. 1971 *Petticoat* 17 July 6/4 There's one excellent method (the coil or loop) only suitable for someone who's already had a baby. 1974 *Guardian* 25 Mar. 10 Will the loop make your periods more painful? If you have comfortable periods before you have a loop fitted, you are unlikely to develop painful periods afterwards.

6. *loop-handle, -head, -lock, method, -net, road, system, way*; **loop aerial, antenna** *Radio*, an aerial consisting of one or more loops of wire; **loop film,** a loop (sense *4 k) of cinematographic film; **loop-line,** (*a*): for 3 c read 4 e; (examples); **loop pile** (see quot. 1963); **loop-stitch**: also as *v. trans.*, to connect or attach by means of loop-stitches; so **loop-stitching,** such work; **loop system,** a method of connecting electrical supply points (as lamp roses) by taking the wires to each point from terminals at its switch and at the previous supply point, instead of making a separate joint elsewhere in the circuit; **loop yarn** (see quot. 1940).

1913 *Year-Bk. Wireless Telegr.* 314 For the directive aerial, the writer had been employing a closed circuit or loop aerial, tuned with a condenser. 1966 J. P. HAWKER *Outl. Radio & Television* xxii. 367 The main advantage of the loop aerial is the two sharp null positions as the loop is rotated, occurring when the plane of the loop is parallel to the wave front. 1968 M. WOODHOUSE *Rock Baby* xiii. 131, I packed the D.F. set, the loop aerial, and the little transceiver into my rucksack. 1906 J. A. FLEMING *Princ. Electr. Wave Telegr.* iv. 280 (*caption*) Stationary potential oscillations set up on loop antennæ. 1932 F. E. TERMAN *Radio Engin.* xvi. 588 All practical direction-finding systems make use of a loop antenna. 1968 A. L. WEEKS *Antenna Engin.* ii. 56 Small loop antennas are frequently employed for low-frequency receiving applications. 1940 *Chambers's Techn. Dict.* 510/2 *Loop film*, the same as band film or cycle film. 1957 *Oxf. Pocket Bk. Athletic Training* (ed. 2) 9 Get a loop film taken of your technique at normal and slow motion speeds. 1949 W. F. ALBRIGHT *Archaeol. of Palestine* vi. 115 These craters were supplied with two tilted horizontal loop-handles. 1969 E. H. PINTO *Treen* 387 Some 19th-century planes have 17th-century type 'loop' handles. 1876 J. S. INGRAM *Centenn. Exposition* ix. 318 These were the larger and most important part of the exhibit, while the rest was made up of..prop nuts, loop heads, offsets and stay ends. 1859 G. A. SALA *Twice round Clock* 261 Then from the beginning of Italian opera in England, a grand trunk line extending to our days, I shunt off on to innumerable little branches and loop-lines. 1869 *Bradshaw's Railway Manual* XXI. 115 A loop line from Peterborough, through Boston and Lincoln, rejoining the main line at Retford. 1908 *Daily Chron.* 16 May 1/5 The loop-line railway linking up all the railway termini. 1956 *Railway Mag.* Nov. 745/1 The up main and goods loop lines were destroyed or heavily damaged for about 70 yd. 1970 *Ibid.* Oct. 585/2 Passenger trains were diverted over the loop line via Lochwinnoch which is rarely used by other than freight services. 1888 G. M. HOPKINS *Poems* (1918) 90 Then with loop-locks Forward falling..his twiny boots Fast he opens. 1901 L. W. WATERHOUSE *Conduit Wiring* 51 The wiring in this building has been carried out entirely on the 'loop' method, there being no joints in any of the wires or cables. 1869 *Game Laws Illinois* in *Fur, Fin & Feather* (1872) 175 That it shall be unlawful..to take or catch fish..by means of any seine, gill-net, tramel net, pike-net, or loop-net. 1924 R. BEAUMONT *Carpets & Rugs* vii. 262 The loop pile may wear flat or bare, but it remains part of the carpet structure. 1963 *Which?* Mar. 69/1 A loop pile carpet has closed loops while a cut pile has the top of the loop cut open. 1909 *Daily Mail* 5 Aug. 5/2 To construct loop-roads for fast motor traffic round villages. 1960 *New Left Rev.* July–Aug. 23/1 Loop roads for buses would penetrate some distance into the pedestrian precinct. 1963 *Times* 7 Mar. 7/4 Whether the town is to have a loop road or.. the High Street is to be widened. 1973 M. YORKE *Grave Matters* IV. ii. 76 He took the loop road that led away from the village. 1932 D. C. MINTER *Mod. Needlecraft* 189/1 In fig. 30 the patch edge..is loop-stitched to the paper. 1951 *Good Housek. Home Encycl.* 213/2 The wrong side of the garment is neatened by loopstitching the two raw edges together. 1969 *Jane's Freight Containers 1968–69* 544/1 One such hook and, at the other end, loop-stitching for permanent fixing to container sides. 1896 R. ROBB *Electr. Wiring* v. 119 The object of confining this construction to the loop system is to prevent joints being made in concealed places. 1925 G. A. WILLOUGHBY *House Wiring* ii. 73 The loop system of wiring does away with this possibility of fire, because the wires are looped from outlet to outlet and all joints are made within outlet boxes. 1929 *Times* 1 Nov. 18/3 Traffic proceeding towards London is being diverted at Hatton cross roads, *via* Cranford-lane to the Bath road and London (A.A. loop-way). *Ibid.*, A.A. loop-way signs. 1940 *Chambers's Techn. Dict.* 511/1 *Loop yarn*, a fancy yarn, with small loops; composed of three threads folded together, one of which is an effect thread and forms the loops, which are bound by another thread. 1957 SIMPSON & WEIR *Weaver's Craft* (ed. 8) xvi. 214 *Curl or Loop Yarn* is made by turning a comparatively thick thread around a much finer ground thread so as to form a succession of curls or loops along the surface of the yarn. 1964 [see *BOUCLÉ *a.*].

loop, *v.*[1] Add: **5.** (Earlier example.)

1837 J. KIRKBRIDE *Northern Angler* 3 Loop on the dropper-flies; the tail-fly should also be looped.

b. loop in. *trans.* (i) To connect into an electric circuit by the loop system.

1893 W. J. HOPKINS *Telephone Lines & their Properties* xiii. 203 It was the custom..to 'loop in' the several telephones, that is, to place them in series. 1899 W. P. MAYCOCK *Electr. Wiring* iii. 242 At A, three lamps *L* and switches *S* are 'looped in' to one fuse *F*. 1965 J. H. M. SYKES *Beginner's Guide Electr. Wiring* v. 113 Lighting circuits may be looped in, using three-plate ceiling roses, to avoid the necessity for joints.

(ii) To form (a wire) into a loop and insert it into a terminal.

1911 A. BURSILL *Princ. & Pract. Electr. Wiring* vi. 37 The wires must never be cut where they are looped-in. 1952 W. E. STEWARD *Mod. Wiring Pract.* 140 Frequently, the description of this [*sc.* looping-in] system given in books on wiring methods tends to create a false impression in the mind of the reader. From these descriptions it would appear that one length of cable is bared at intervals and looped in at switch and lighting terminals. In practice, when wiring in conduit, the two lengths of wire forming the loop are threaded in separately and the junction is made at the switch, light, or other terminal. 1967 G. A. T. BURDETT *Electr. Installations* 226 (*heading*) Looping-in the cable.

6. *to loop the loop*, to perform the feat of circling in a vertical loop, orig. on a specially prepared track (see LOOP *sb.*[1] 4 f), later in an aeroplane. Also *transf.*, *fig.*, and as *sb.*

Orig. a fairground phrase.

1902 *Strand Mag.* June 708 (*heading*) Looping the loop on a bicycle. *Ibid.* 708/1 'Looping the loop' in America has become even more popular than shooting the chutes. *Ibid.* 708/2 At first he could not induce the ball to loop the loop. 1903 G. BELL *Let.* 8 July (1927) I. 166 We went on a switchback that looped the loop... Hugo..was distinctly conscious of being upside down..for the fraction of a second. 1903 *Outing* XLII. 552/1 He knows how to win the steeplechase..and has been 'thrown out' for standing up in the loop-the-loops. [1908 A. BAZIN in *L'Aérophile* 15 May, Pourquoi pas 'looping the loop' tout de suite?] 1911 A. P. THURSTON *Elem. Aeronaut.* iii. 33 A glider..can be made to 'loop the loop', or follow any one of a number of curved paths. 1913 [see *LOOP *sb.*[1] 4 f]. 1922 A. S. EDDINGTON *Theory of Relativity* 3 The planets literally looped the loop in fantastic curves called epicycles. 1922 WODEHOUSE *Clicking of Cuthbert* ix. 209 A girl of such pronounced beauty that Ramsden Waters' heart looped the loop twice in rapid succession. 1935 [see *FALLING *ppl. a.* 5 b]. 1940 O. NASH *Face is Familiar* (1954) 10 It's pleasant to loop the loop, To daringly seize The flying trapeze With a cry of Allez-oop! 1960 B. KEATON *Wonderful World of Slapstick* (1967) 73 The climax of the act came when he started doing loop the loops, riding upside down. 1961 [see *AEROBATICS *sb. pl.*]. 1968 *Michelin Guide N.Y. City* 124 Coney Island..scenic railways, loop-the-loops and Ferris wheels compete with phantom trains, tunnels of love, sputniks.

7. *intr. Computers.* To execute a loop (*LOOP *sb.*[1] 4 m).

1958 GOTLIEB & HUME *High-Speed Data Processing* vi. 98 A common procedure is to use one sequence of instructions, cycling or looping through this sequence as often as required. 1969 P. B. JORDAIN *Condensed Computer Encycl.* 293 The ability to loop, and thus reuse instructions without duplicating them and wasting memory, is probably the single most important advantage gained by stored-program computers.

loop, *v.*[3] [f. LOOP *sb.*[2] Cf. LOOPED *ppl. a.*[2]] *trans.* To furnish with loop-holes.

1846 Z. TAYLOR *Let.* 9 Nov. in *N.Y. Morn. Express* (1847) 22 Jan. 2/3 The houses are of stone..all looped up for musketry.

‖ **loop** (lōᵘp), *int. S. Afr.* [Afrikaans, f. Du. imp. of *lopen* to walk.] A word of command to an animal to move forward.

1811 W. J. BURCHELL *Jrnl.* 18 June in *Trav. S. Africa* (1822) I. viii. 169 Philip mounted his seat,..with an animated voice calling out to the oxen, *Loop!* 1927 W. PLOMER *I Speak of Afr.* i. 40 'Loop!' he ordered in a loud voice. Shilling cracked his whip and shouted to the oxen. The voorlooper's head could just be seen through a forest of horns. 1937 F. B. YOUNG *They seek a Country* II. i. 162 'Ay, Blauwberg, would you?' (The long lash curled in the air like a salmon cast and stung the off-leader's muzzle.) '*Loop*, you devils, *loop!*'

looped, *ppl. a.*[1] Add: **5.** Intoxicated, drunk. *slang* (chiefly *U.S.*).

1934 in M. H. Weseen *Dict. Amer. Slang* xviii. 279. 1951 'M. SPILLANE' *Big Kill* ii. 46 The sap sounded half-looped. 1959 A. BAILEY *Making Progress* vi. 63 Slater..was almost looped on the Veuve Clicquot. 1962 J. POTTS *Evil Wish* ii. 28 Joe had gotten looped and called here. 1973 'R. MACDONALD' *Sleeping Beauty* xxxviii. 221 The message ..didn't come through too clear. She talked as if she was slightly looped.

looper[1]. Add: **1.** Also *attrib.*

1840 J. & M. LOUDON tr. *Köllar's Treat. Insects* III. 212 The most ruinous insect for fruit-trees is assuredly the green looper-caterpillar. 1932 E. STEP *Bees, Wasps, Ants* 184 A more striking..case is that of some 'looper' caterpillar,..from which a hundred *Microgaster* larvæ have broken out. 1964 V. B. WIGGLESWORTH *Life of Insects* x. 149 We have a whole series of insect larvae, of Noctuid moths, of 'looper' or Geometrid caterpillars, saw-fly larvae and others, all of which resemble pine needles.

3. *Aeronaut.* One who loops the loop, or who has done so; a machine specially adapted for looping the loop.

1914 *Aeroplane* 15 Jan. 63/1 Two more names have been added to the roll of loopers. *Ibid.* 12 Mar. 284/2 Mr. Hucks..first flew his two-seater, and later on the 'looper' at 700 feet, made one loop.

loop-in. [f. vbl. phr. *to loop in* (see *LOOP *v.*[1] 5 b).] A connection between two lengths of wire made at a terminal in the loop system. Usu. *attrib.*, as *loop-in system* = *loop system* (s.v. *LOOP *sb.*[1] 6).

1908 *Installation News* II. 86/2 In comparing the third method with the second, which is the usual method of loop-in wiring adopted, it will be found that there is a saving of wire in the new method. 1911 A. BURSILL *Princ. & Pract. Electr. Wiring* vi. 35 Fig. 18 shows four lamps connected in parallel on the loop-in system. 1960 E. L. DELMAR-MORGAN *Cruising Yacht Equipment & Navigation* xvii. 187 Even with the 'loop in' system one or at most two pairs of fuses will be sufficient. 1967 G. A. T. BURDETT *Electr. Installations* 224 To overcome the inherent dangers of using an ordinary three-plate ceiling rose as a looping-in point for the live feed an entirely new purpose-made loop-in ceiling rose has been introduced.

Ibid. 226 More recent patterns..have a fourth terminal which is intended for the earth continuity conductor loop-in.

looping, *vbl. sb.*[1] Add: (Later examples.)

1914 *Isis* (Oxf.) 28 Feb. 5/1 The 'looping' looked simplicity itself. A sudden plunge straight towards the earth ..a sudden reversal and a perpendicular climb, which gradually brought the airman upside down and..a second downward plunge. 1929 E. WILSON *I thought of Daisy* III. 123 'I'm not out for any looping,' said Daisy... 'I think I'll go home and go to bed.' 1964 E. BACH *Introd. Transformational Gram.* iii. 46 Care must be taken to ensure that unwanted recursion (looping) does not occur.

c. looping in (cf. *LOOP *v.*[1] 5 b), connection by the loop system.

1899 W. P. MAYCOCK *Electr. Wiring* iii. 243 The advantages of looping-in, as applied to the interconnection of switches and ceiling roses. 1902 F. C. RAPHAEL *'Electrician' Wireman's Pocket Bk.* 33 (*caption*) Simplex looping-in ceiling rose. 1930 —— *Electr. Wiring of Buildings* ii. 23 Looping in has become absolutely general, for not only does it save the danger of bad joints, but it is actually less expensive than to make joints. 1967 [see *LOOP-IN].

looping, *ppl. a.* Add: Also *looping caterpillar* = LOOPER[1] 1.

1875 HUXLEY & MARTIN *Course Elem. Biol.* x. 95 The polypes..are capable of crawling about by a motion similar to that of the looping caterpillar.

loopist (lū·pist). *rare.* [f. *LOOP *v.*[1] 6 + -IST.] = *LOOPER[1] 3.

1914 *Aeroplane* 15 Jan. 63/1 One of the latest loopists is M. Galtier, who on January 7th looped the loop at Chateaufort.

loopy, *a.* Add: **3.** *slang.* Crazy, 'cracked'.

1925 FRASER & GIBBONS *Soldier & Sailor Words* 147 *Loopy*, silly, daft. 1928 E. WAUGH *Decline & Fall* I. xiii. 145 He'll get off on a plea of insanity. Loopy, you know. 1942 'M. INNES' *Daffodil Affair* II. 40 One does not want a loopy colleague when embarked upon so distinctly rummy an investigation as the present. 1957 I. CROSS *God Boy* (1958) vi. 44 Honestly, the pair of them were looking at me as though I was loopy. 1961 P. WHITE *Riders in Chariot* iv. 88 'I will not waste my breath arguing with loopy Louie,' replied Mrs Jolley. 1970 *New Yorker* 15 Aug. 68/2 The wife..is mad neither in the sense of loopy nor in the sense of furious. 1973 A. HUNTER *Gently French* xiv. 125 He was loopy over her. If she said jump, he'd fall off a cliff.

loorie, var. *LOERIE.

loose, *a.* Add: **1. a.** *spec.* of horses etc.: allowed to run free in travelling or marching.

1845 J. C. FRÉMONT *Rep. Exploring Expedition* 10 A few loose horses, and four oxen..completed the train. 1843 *Oregon Hist. Soc. Q.* (1901) II. 191 About fifty wagons, with those who had large droves of loose cattle, now left. 1846 W. G. D. STEWART *Altowan* II. i. 41 The neighing of the loose troops, that ever and anon, broke forward to snatch the opportunity of browsing ere the crowd advanced. 1885 *Outing* VII. 21/2 All drove pack and loose animals before them.

l. Of money, cash, etc.: in relatively small denominations; in coins (as opp. to notes). So *loose change* (orig. *U.S.*), a quantity of coins kept or left in one's pocket, etc., for casual use.

1811 JANE AUSTEN *Sense & Sens.* I. xvii. 217 My loose cash would..be employed in improving my collection of music and books. 1827 A. SHERWOOD *Gazetteer Georgia* 112 It would be a kind of generous charity, to leave with the tavern-keepers..some of the loose change. 1872 E. EGGLESTON *End of World* 173 Unless he means to part with all his loose change. 1895 A. MACHEN *Three Impostors* 81 He never returned, but his watch and chain, a purse containing three sovereigns in gold, and some loose silver, with a ring..were found three days later. 1927 C. A. SIRINGO *Riata & Spurs* v. 54 That little burg saw the need of saloons and dance-halls to relieve the cowboy of his loose change. 1950 WILKINSON & FRISBY *They're Open* ii. 19 *Capacious left-hand pocket*,..in order that a large bulk of loose change may be carried. 1970 G. F. NEWMAN *Sir, You Bastard* viii. 245 He got rid of the loose change in his pocket. 1973 *Woman's Own* 4 Aug. 36 (Advt.), The clip-to coin section is just the right size for all your loose change.

m. *Gram.* Of certain syntactical elements: not essential to the meaning or construction, etc.

1932 E. KRUISINGA *Handbk. Present Day Eng.* (ed. 5) II. III. 235 The members of a loose group may be connected by other words or not... We distinguish *linked* groups and *unlinked* groups. 1933 O. JESPERSEN *Essent. Eng. Gram.* xxxiv. 357 A non-restrictive (or loose) clause, ..may be left out without injury to the precise meaning of the word it is joined to, as in 'The Prince of Wales, who happened to be there, felt sorry for the prisoners.' 1961 R. B. LONG *Sentence & its Parts* iii. 68 They [*sc.* subordinate interrogatives] function also as loose adjuncts... We went with Larry, *who knew everyone*. 1972 HARTMANN & STORK *Dict. Lang. & Ling.* 135 *Loose apposition*, a word or phrase used in apposition and often separated by sustained juncture in speech or by commas in writing.

3. a. (Later example.)

1908 *Animal Managem.* 17 The skin..when handled, should feel 'loose' and freely movable over the structures beneath.

9. loose back, a method of binding the spine of a book to make it open more easily; **loose body** *Med.* = *joint mouse* (*JOINT *sb.* 15); usu. *pl.*; **loose coupling** *Electr.* (see *COUPLING *vbl. sb.* 6 f (i)); **loose cover,** a detachable cover for a chair, couch, or car seat; also *attrib.*; **loose-fill, loose fill,** a type of house insulation (see quot. 1964); also *attrib.*; **loose head,** see *HEAD *sb.* 26 c; **loose-housing,** a method of housing cattle in winter in partly covered barns with access to a feeding area, in which the cows are not confined to a single stall; also *attrib.*; hence **loose-housed** *a.*; **loose scrummage, loose scrum,** in *Rugby Football*, a scrum formed by the players round the ball during play, and not ordered by the referee: opp. *set scrum(mage)*; hence, **loose scrummaging; loose smut,** a disease of cereals, esp. barley and wheat, caused by the fungus *Ustilago nuda*; **loose whale,** a whale which remains beside its harpooned mate.

1923 H. A. MADDOX *Dict. Stationery* 46 *Loose back*, also termed open or spring back. *Ibid.* 47 A loose back may be created by simply casing the book... The spring or loose back is actually bound into the book. 1956 *Bookman's Conc. Dict.* 277 *Spring Back*, an inner joint in a bookbinding which allows the book to open flat; known as Hollow or Loose Back. 1961 T. LANDAU *Encycl. Librarianship* (ed. 2) 268/2 *Open back*, a style of construction in which the cover is separated from the spine of the book by a special lining... Also called Hollow and Loose Back. 1886 H. MARSH *Dis. Joints* xv. 185 On examining his joint when the acute attack has gone off, the patient detects the loose body, and learns that it shifts its position. *Ibid.* 183 (*caption*) Specimens of the loose bodies found in the knee joint in Mr. Smith's case. 1952 [see *joint mouse* s.v. *JOINT *sb.* 15]. 1961 R. D. BAKER *Essent. Path.* xxi. 578 Portions of the damaged articular cartilage, or fragments of the thickened peripheral bone, break off and become loose bodies in the joint cavities. 1876 M. W. COOK *Tables & Chairs* i. 52 You may prefer to have your curtains, as well as the loose covers, of chintz. *Ibid.* 56 Nothing now-a-days looks so nice and ladylike, or is so economical as well-fitting loose covers. 1911 F. B. JACK *Woman's Bk.* 613/2 Loose covers are not much used nowadays, and, at the best, they soon get out of order and become shabby looking. 1929 *Radio Times* 8 Nov. 438/1 Odd Jobs about the House—II, A Few Hints on Loose Cover Cutting. 1936 R. LEHMANN *Weather in Streets* I. v. 104, I might keep her on for the sewing. She's very clever at loose covers. 1953 M. SHERIDAN *Furnisher's Encycl.* IX. iii. 414 A loose cover service may substantially increase the furnisher's business. 1959 *B.S.I. News* Nov. 20 Specifies..maximum foreign matter content for..loose cover cloths made from cotton. 1973 A. BROINOWSKI *Take One Ambassador* xii. 206 Comfortable upholstery, not bottomlessly soft, with well-cut linen loose covers. [1949 *Building Digest* IX. 305 The flat roof has..a loose vermiculite filling.] 1950 *Archit. Rev.* CVIII. 332/2 A 4 in. thickness of vermiculite loosefill for instance has the same thermal insulation as 24 times that thickness of concrete. 1956 *Good Housek. Home Encycl.* (ed. 4) 170/2 'Loose fill' which is poured or packed.. between the joists. 1964 J. S. SCOTT *Dict. Building* 198 *Loose-fill insulation*, insulating materials such as granulated cork, loose asbestos..vermiculite. Loose fill is placed between rafters or studs to increase the insulating value of a dry air space. 1969 *Daily Tel.* 16 Sept. 15 Most householders can climb into their lofts and lay mineral wool or glass fibre over the joists, or loose-fill between them. 1907 'OLD INTERNATIONAL' *Rugby Guide* 62 It was discovered that on the flank of the row where the ball came into the scrum there was a head overhanging the side of the scrum. This head was given the appellation of 'loose head'. 1917 [see *HEAD *sb.* 26 c]. 1927 [see *HOOKER[1] 6]. 1960 C. VENABLES *Instructions to Young Rugger Players* iii. 37 When the two scrummages are formed they pack down and, of course, the heads of the two front rows are interlocked. But, with three men in each row, this clearly means that one man in each row will have his head free. This is known as the 'loose head' and it is on that side that the scrum half will put the ball in, for the good reason that his own hooker will be nearer to the ball than the hooker on the other side. 1960 E. S. & W. J. HIGHAM *High Speed Rugby* xiv. 191 The *loose-head prop* puts the *inside* foot forward. *Ibid.* 195 We will refer to the hooker who has the loose head..as the 'loose-head hooker'. 1960 *Farmer & Stockbreeder* 5 Jan. 53/1 Half [the herd] is loose-housed and zero-grazed. *Ibid.* 12 Jan. 78/1 Three sides of the yard accommodating the 55 pedigree..Friesians loose-housed are filled by the covered lying shed. 1963 C. T. M. HERRIOTT tr. *Craplet's Dairy Cow* v. xvii. 374 Loose-housed animals are less nervous than those kept in byres. 1946 *Agric. Engineering* XXVII. 499/2 The loose-housing barn and milking parlor seemed to offer a possible improvement. 1948 *Pop. Bull. Washington Agric. Exper. Station* No. 190 (*title*) The loose housing and feeding of dairy herds. *Ibid.* 2 Loose housing is becoming popular. 1963 C. T. M. HERRIOTT tr. *Craplet's Dairy Cow* v. xvii. 369 Loose housing provides a well-compacted, thoroughly decomposed manure. *Ibid.*, Cows to be kept under a loose housing system can become accustomed to communal living. 1874 G. H. WEST *Rugby Union Football Ann.* 66 A light and very useful forward, especially in a loose scrummage. 1936 Loose scrummage [see *heel-back* (*HEEL *sb.*[1] 25)]. 1952 *Rugby Union Football* ('Know the Game' Series) (ed. 2) 26 A 'loose' scrum..is formed by..players closing round the ball when it is on the ground. 1958 PELMEAR & MORPURGO *Rugby Football* VIII. 339 Loose scrummaging (now sometimes known as 'rucking') was becoming the half-back's delight. 1960 E. S. & W. J. HIGHAM *High Speed Rugby* xiv. 201 It is possible to build up a loose scrum in two ways. 1890 *2nd Ann. Rep. Kansas State Exper. Station 1889* 213 The loose smuts are four closely allied species found on oats, wheat, and barley. 1909 *Bull. Bureau Plant Industry, U.S. Dept. Agric.* No. 152. 7 The loose smut..is easily distinguished from the covered smut by its earlier appearance, by its olive-green spore-mass.., and by the early shedding of the spores. 1924 *Jrnl. Agric. Res.* XXIX. 263/1 Formaldehyde and some of the organic mercury compounds have been found to control the loose-smut of barley in certain varieties. 1968 *Times* 16 Dec. 7/1 Loose smut in barley..has become a serious problem again because of the preponderance of susceptible varieties. 1903 F. T. BULLEN in *Strand Mag.* Nov. 539/1 All through the combat..the whalefishers will be closely beset by the 'loose' whale.

10. d. loose-ended *a.*, ended or finished off in a slack, untidy, or inconclusive way; also *fig.*; hence **loose-endedness; loose-footed** *a.* (later examples); **loose-knit** *a.*, connected in a tenuous or ill-defined way; not closely linked; **loose-lipped** *a.*, (*a*) loose-tongued; uninhibited in speech; (*b*) having full lips; **loose-mouthed** *a.* = **loose-lipped* adj.; **loose-skinned** *a.*, having skin wrinkled or hanging in folds.

1867 J. R. LOWELL in *Atlantic Monthly* Jan. 24 Loose-ended souls, whose skills bring scanty gold. 1937 *Times Lit. Suppl.* 15 May 379/1 The weaving of three themes through the tenuous and loose-ended plot. 1944 *Horizon* IX. 286 My purpose is to indicate..how we loose-ended mortals are dealt with. 1905 *Proc. Roy. Soc.* LXXV. 378 There was no slackness or loose-endedness about him either physically or intellectually. 1968 *Punch* 3 Jan. 4/2 The problem, which mightn't worry some people but had teased me for a fortnight by its sheer loose-endedness. 1878 J. H. BEADLE *Western Wilds* ii. 38 Every loose-footed man wanted to go. *Ibid.* xxviii. 442 Loose-footed young men erect a cabin, barely habitable in good weather. 1927 G. BRADFORD *Gloss. Sea Terms* 104/2 *Loose-footed*, a fore and aft sail not laced to (or without) a boom. 1948 R. DE KERCHOVE *Internat. Maritime Dict.* 433/2 *Loose-footed*, an expression used for denoting a fore-and-aft sail in which the foot is not laced to the boom. 1906 T. HARDY *Dynasts* II. I. vii. 42 As he shatters the moves of the loose-knit nations to curb his exploitful soul's ambitions. 1957 E. BOTT *Family & Social Network* iii. 94 Networks become loose-knit when people move from one place to another... If both husband and wife have moved considerably before marriage, each will bring an already loose-knit network to the marriage. 1963 *Times* 11 Mar. 3/5 The Welsh forwards performed doughty deeds individually, but were too loose-knit to hold their opponents, which was bad luck on D. C. T. Rowlands, whose most effective game thrives on a dominant pack. 1968 *Daily Tel. Mag.* 8 Nov. 27/4 The ARB team is loose-knit and embraces a cross-section of specialists. 1919 J. MASEFIELD *Reynard* 33 Loose-lipped with song and wine and revel. 1924 W. DE LA MARE *Ding Dong Bell* 70 Hook-nosed was I; loose-lipped. 1924 COMPTON MACKENZIE *Heavenly Ladder* xiii. 186 It was sad to see a young woman of thirty so loose-lipped and blowsy. 1928 *Daily Mail* 13 Aug. 5/1 Her mastery of what Sir William Watson has called the loose-lipped lingo of the streets. 1934 H. G. WELLS *Exper. Autobiogr.* II. ix. 679 If I were to put my reputation before my autobiographical rectitude, I think I should just let this little volume decay and char and disappear... Most of it is very loose-lipped indeed. 1934 W. B. YEATS *King of Great Clock Tower* 29 Had de Valera eaten Parnell's heart No loose-lipped demagogue had won the day. 1872 J. G. WHITTIER *Pennsylvania Pilgrim* in *Poet. Wks.* (1874) 447 We may trace How loose-mouthed boor and fine ancestral grace Sat in close contrast. 1931 W. FAULKNER *Sanctuary* xxii. 203 You'll know I ain't loose-mouthed. 1938 *Times Lit. Suppl.* 18 June 415/4 Were all Roman aristocrats loose-mouthed and pot-bellied? 1950 D. GASCOYNE *Vagrant* 53 He is apt to get oddly pedantic about the proprieties while even more loose-mouthed than ever. 1906 *Westm. Gaz.* 14 June 4/2 This old man had a full, loose-skinned face, with a comic mouth and forlorn eyes. 1909 *Ibid.* 5 Jun. 2/2 The sail heaved like a gigantic loose-skinned animal awakening. 1937 V. WOOLF *Years* 397 His swarthy wrinkled face.. always made her think of some loose-skinned, furry animal. 1941 BLUNDEN *Thomas Hardy* v. 110 Hands very white and soft and loose-skinned.

B. quasi-*sb.* and *sb.* **1. a.** *on* (or † *upon*) *the loose*: (*b*) of women: living by prostitution; (*c*) *gen.*, not tied down; not answerable to anyone.

1749 J. CLELAND *Mem. Woman Pleasure* II. 9 The giddy, wildness of young girls once got upon the loose. 1859 HOTTEN *Dict. Slang* 70 *On the loose*, obtaining a living by prostitution, in reality, on the streets. The term is applied to females only. 1879 *Roget's Thesaurus* 330 Impure; unclean &c...; on the streets, on the *pavé*, on the loose. 1890 BARRÈRE & LELAND *Dict. Slang* II. *Loose, on the*,..getting a living by prostitution. 1914 G. B. SHAW *Fanny's First Play* III. 201 Do you mean to say that you went on the loose out of pure devilment? 1935 N. ERSINE *Underworld & Prison Slang* 51 Ted is *on the loose*. 1949 PARTRIDGE *Dict. Underworld* 483/2 *On the loose*, obtaining a living by prostitution. 1951 E. PAUL *Springtime in Paris* i. 12 Just then Raoul was spending all his free time with the peace posters and Katya stayed at home. When the roles were reversed, and Katya was on the loose, no one knew precisely what she was about. 1958 *Times Lit. Suppl.* 30 May 293/2 A group of young Americans..some being genuinely on the loose or moving from job to job. 1970 V. CANNING *Great Affair* xvii. 319 There was a little mistiming at Sokota so your friend King Alfy is on the loose.

2. (Further examples.)

1894–5 *Rugby Union Football Handbk.* 11 'Offside' is still penalised in the loose, but not Solon himself..could define where a scrummage ends and the loose begins. 1922 'TOUCH FLAG' *Mod. Rugby Tactics* 49 Dangerous attacks frequently originate from chance openings in the loose. 1963 *Rugby World* Aug. 8/3 Wightman and Rogers impressed in the loose for England. 1974 *Country Life* 5 Dec. 1717/1 The All Blacks..were..gaining their expected supremacy at the line-out and in the loose.

loose, *v.* **4.** Delete ? *Obs.* and add later examples.

1916 'BOYD CABLE' *Action Front* 48 The artillery made a regular practice of loosing off a stated number of rounds per night. **1944** *R.A.F. Jrnl.* Aug. 286 Dropped our bomb-load..an' loosed off all our ammo.

b. (Examples with *off*.) Also *loosing off*.

1906 *Westm. Gaz.* 9 Mar. 4/1 The man for whom the whole of shooting is comprised in the gunning—in the 'loosing off', as he will call it. **1926** *Punch* 28 July 86/2 The bowler would acquire the trick of looking at one [wicket] while really he was loosing off at the other. **1928** Blunden *Undertones of War* ii. 8 The howitzer loosing off occasionally outside punctuated these amenities. **1946** J. Irving *Royal Navalese* 110 *To loose off*, to open fire.

12. Also in phr. *loose-all*, the signal to stop work given in the pits.

1911 D. H. Lawrence *White Peacock* vii. 485, I heard the far-off hooting of the 'loose-all' at the pits, telling me it was half-past eleven. **1913** —— *Sons & Lovers* ii. 30 Some men were there before four o'clock, when the whistle blew loose-all.

loose-leaf, *a.* Of a note-book, file, or the like: made to facilitate the insertion or removal of each leaf separately. Also as *sb.*

1902 *Accountant* 29 Nov. 1240/1 The difficulty he mentions is partly met by using a 'loose leaf' Ledger. **1907** *Daily Chron.* 6 Dec. 11/4 'Loose-Leaf' notebooks and diaries,..in which pages can be taken out or added at will, have already won a well-deserved popularity. **1917** H. B. Twyford *Purchasing & Storing* 409 A copy of every printed form used should be posted on a loose-leaf sheet. **1930** 'J. J. Connington' *Two Tickets Puzzle* xv. 232 Dr. Selby-Onslow nodded again: crossed the room to his desk; and pulled from a drawer a large loose-leaf volume. **1937** I. O. Evans *Cigarette Cards* 122 Sheets made up in this fashion could also be filed after the manner of the loose-leaves. **1955** W. Gaddis *Recognitions* ii. vi. 559 In the window was a large loose-leaf book, whose lined pages were filled in a cramped round hand. **1967** *Listener* 21 Sept. 370/2 The large loose-leaf notebook in which he wrote rehearsal notes. **1971** *Ann. Rep. Curators Bodl. Libr. 1969–70* 49 Transferring of the old handlists to typescript sheets in looseleaf binders continued. **1975** *BSI News* May 6/3 The third revision of BS 2782 *Methods of test for plastics* is being issued in a loose-leaf form.

loosely, *adv.* Add: **6.** *loosely-knit* adj. Cf. *loose-knit* adj. s.v. *Loose *a.* 10 d.

1935 Huxley & Haddon *We Europeans* i. 13 It [*sc.* group-sentiment] has spread beyond the family, the tribe, the loosely-knit federation of tribes to the yet more extensive aggregate, the nation. **1957** C. Hunt *Guide to Communist Jargon* xlvi. 153 It [*sc.* the Decree of the Central Committee] replaced the loosely-knit Union of Proletarian Writers by a single Union of Soviet Writers.

loosen, *v.* Add: **7.** *absol.* with *up*. **a.** To give money willingly, to talk freely, etc. *U.S. colloq.*

1908 K. McGaffey *Sorrows of Show-Girl* xi. 125 Loosen-up... You've got to donate for a couple of tickets to the annual benefit. **1911** G. S. Porter *Harvester* xx. 516 You're tight-mouthed... Loosen up! **1923** R. D. Paine *Comrades of Rolling Ocean* xi. 187 Somebody will have to loosen up to pay for the damage to my nervous system. **1922** C. Sandburg *Slabs of Sunburnt West* 6 Come across, kick in, loosen up. Where do you get that chatter? **1927** *Ladies' Home Jrnl.* 114 That is the first time he has ever loosened up. **1949** Wodehouse *Uncle Dynamite* i. 8 You will generally find women loosen up less lavishly than men.

b. In *Sport* or *Dancing*, to exercise the muscles before concentrated physical effort, to limber up. Also *loosening-up* vbl. sb. and ppl. adj.

1955 M. Gilbert *Sky High* xii. 165 The General came to a stop in the middle of his loosening-up exercises. **1956** R. Alston *Test Commentary* xvi. 139 Lindwall was given a couple of loosening-up overs. **1973** M. Russell *Double Hit* viii. 55 Make it an hour. I'll be twenty minutes loosening up... I'm after the exercise.

Looshai, var. *Lushai *a.* and *sb.*

loot, *sb.*² Add: **2.** *slang.* Money.

1943 Hunt & Pringle *Service Slang* 44 *Loot*, Scottish slang for money received on pay day. **1956** B. Holiday *Lady sings Blues* (1973) ii. 16 There was nothing to do except for Mom to go back slaving away as somebody's maid. In Baltimore she couldn't make half the loot she could up North. **1959** *Encounter* Oct. 73/1 MacInnes's teen-agers..are all economically self-supporting, in their own phrase, they've got the loot. **1968** J. Sangster *Touchfeather* xiii. 141 When you've got his sort of loot I don't suppose it matters. **1973** *Center City Office Weekly* (Philadelphia) 9 Oct. 3 Jefferson Medical College picks up $362,949 in loot for..continuing research in hemophilia.

loot (lūt), *sb.*³ *U.S. Mil. slang.* Shortened from Amer. pronunciation of Lieutenant 2.

1898 F. P. Dunne *Mr. Dooley in Peace & War* 11 R-run over an' wake up th' loot at th' station. **1918** *Stars & Stripes* 27 Dec. 7 He's a loot-colonel now and a D.S.C. **1921** B. Matthews in *S.P.E. Tract* v. 6 In the United States this officer is called the *lootenant*, which the privates of the American Expeditionary Force in France habitually shortened to '*loot*'. **1944** K. Levis in Murdoch & Drake-Brockman *Austral. Short Stories* (1951) 423 Some nitwit of a loot 's in charge that don't know a bush track from Pitt Street. **1948** J. G. Cozzens *Guard of Honor* 331 Don't thank the loot!

lop, *sb.*³ Add: **4. lop and top,** waste branches cut from timber trees, usually after the trees have been felled; **lop-stick** (earlier and later examples); also **lobstick.**

1892 Lop and top [in Dict., sense 1]. **1938** C. P. Ackers *Pract. Brit. Forestry* vi. 194 Lop and top may be overcome either by burning or by stacking it in 'trenches'. **1972** *Country Life* 30 Mar. 789/3 'Lopp and Topp'—the side and top branches—were the college property and if from ash or oak fetched 8s. to 9s. 6d. a load. **1821** N. Garry *Diary* 19 Aug. (1900) 149 After Dinner we observed that two of our Men had lopped away the Boughs and all the Lower Branches of two Trees leaving a Top. This is called a Lop-Stick. **1847** J. B. Nevins *Two Voy. Hudson's Bay* iv. 90 Two gentlemen were travelling a short time since, and lobsticks were cut for them. **1873** G. M. Grant *Ocean to Ocean* vii. 196 There is an old superstition that your health and length of days will correspond to your lobstick's. **1923** *Beaver* Aug. 421 To commemorate this great battle, three lobsticks were cut on each side of the river. **1949** *Argosy* Apr. 13 On the far side of the lake, if you must portage, use the 'lop-stick' mark. This is made by cutting all branches from one side of a tall tree which may be seen either from water or land. Its unnatural appearance attracts the eye; the side from which branches are cut indicates the direction of portage. **1964** *Islander* (Victoria, B.C.) 18 Oct. 1/12 There was a tradition among the Northern Indians that a lobstick honouring an individual would fall when its sponsor died.

lop, *v.*² Add: **2. b.** *to lop down*, to sit down, to lie down. *U.S. colloq.*

1839 C. M. Kirkland *New Home* ii. 17 Jist come in, and take off your things, and lop down, if you're a mind to. **1861** Mrs. Stowe *Pearl of Orr's Island* I. viii. 67 Ruby said she thought she'd just lop down a few minutes on the old sofa. **1892** F. P. Humphrey *New Eng. Cactus* 34 You'd best lop down on the lounge and get a nap.

4. *lop-brimmed* adj.

1901 S. E. White *Westerners* xvi. 131 His broad hat—straight-brimmed in a lop-brimmed camp—was pushed to one side.

lope, *sb.* Add: **2.** (Further examples.) Now used chiefly of people.

1809 M. L. Weems *Life F. Marion* xii. 108 He dashed off at a charging lope. **1910** C. E. Mulford *Hopalong Cassidy* xxi. 135 As he rode at an easy lope he kept a constant lookout for signs of rustling. **1953** R. Chandler *Let.* 15 Mar. in *R. Chandler Speaking* (1966) 28 Walks with a forward-leaning lope, huh? **1961** L. van der Post *Heart of Hunter* i. ii. 25 The others followed close on her heels with a strange stumbling lope. **1973** *Houston* (Texas) *Post* 14 Oct. (Spotlight Suppl.) 15/4 Somehow..Hero Hazard reaches the Sphinx..and evades the grisly Arab agents after a lope up and down the Great Pyramid.

loper¹. **3.** (Later examples.)

1952 J. Gloag *Short Dict. Furnit.* 323 Sliding rails which pull out from the carcase of a bureau to support the fall are called lopers. **1963** *Times* 20 Apr. 11/7 The bureau bookcase has continuously changed its outline, wood and fashionable finish according to the dictates of fashion, but apart from the invention, in this century, of the automatic opening loper, no basic improvement in it has been found possible between the reigns of Queen Anne and Queen Elizabeth II. **1971** *Canad. Antiques Collector* Mar. 10/1 The name 'pembroke' should be saved for a particular sort of small 4-legged dropleaf tables, usually with a drawer, the leaves supported on lopers (pull-out bars).

loper² (lōu·pəɹ), var. Looper². *S. Afr.*

1886 P. Gillmore *Hunter's Arcadia* iii. 18, I quickly substituted cartridges of lopers (buckshot) for the No. 3 that my chambers had previously contained. **1932** C. Fuller *Louis Trigardt's Trek* x. 120 They espied a fully armed Native and scared him away with a charge of *lopers*.

lopolith (lǫ·poliþ). *Geol.* [See quot. 1918 and -lith.] A large intrusive mass similar to a laccolith but having the base centrally sunken.

1918 F. F. Grout in *Amer. Jrnl. Sci.* CXCVI. 518 Professor John Barrell has suggested that as igneous forms they deserve a distinct name... The name proposed by the writer is 'lopolith' (from [Greek] λοπάς, a basin, a flat earthen dish, and λίθος, a stone). *Ibid.*, Pronunciation, lŏ'polith. **1962** [see *intrusive *sb.*]. **1972** M. P. Billings *Struct. Geol.* (ed. 3) xvi. 343 In plan, lopoliths are arcuate or circular.

Hence **lopoli·thic** *a.*

1959 *Econ. Geol.* LIV. 1208 Cooling of a lopolithic mass would proceed under the influence of two cooling gradients. **1968** R. W. Fairbridge *Encycl. Geomorphol.* 281/2 The second type [of intrusive dome mountain] is the complex laccolith (or lopolithic) intrusion with broad sill-type interfingering tongues of igneous material penetrating the surrounding sediments.

loppage. (Later example.)

1911 Fletcher & Kipling *School Hist. Eng.* ii. 32 Laws they made in the Witan, the laws of flaying and fine—Common, loppage and pannage, the theft and the track of kine.

loppiness. *rare.* [f. Loppy *a.*³] The quality of being loppy or choppy.

1908 *Daily Chron.* 10 Aug. 1/4 He complained that the loppiness of the water had taken the strength out of him.

loppy (lǫ·pi), *sb.* *Austral. slang.* [Prob. from Loppy *a.*¹] A handyman on rural stations.

1898 *Bulletin* (Sydney) 1 Oct. 14/3 A few more Western Queensland slang words... In a shearing shed: The boss is the 'finger', the shearers the 'brutes', the rouseabouts 'leathernecks', 'spoonbills', 'loppies' or 'Jacks'. **1933** *Ibid.* 8 Feb. 21 The loppies, who are meek and spry To shearers and the rest, Are perky chaps when I am by.

lop-sidedness. (Earlier and later examples.)

1843 H. James *Let.* 11 May in R. B. Perry *Tht. & Char. W. James* (1935) I. 46 Thought heaped up to topheaviness and inevitable lopsidedness. **1946** *Nature* 21 Dec. 890/2 He makes, incidentally, a powerful case for some attempt to redress the lopsidedness of scientific advance.

lop-stick: see Lop *sb.*³ 4 (in Dict. and Suppl.).

loquat. (Earlier and later examples.)

1820 *Trans. Hort. Soc. London* III. 229 You desire me [*sc.* Lord Bagot] to give you some information as to my mode of treating the Lo-quat. *Ibid.* 301 In 1813 ripe fruits of the Lo-quat were presented to the Horticultural Society by Lord Bagot. **1884** tr. *J. J. Rein's Japan* I. vii. 139 A Japanese fruit-tree, the Loquat (Eriobothrya japonica Lindl.), which the English have transplanted to their tropical and subtropical colonies. **1969** *Oxf. Bk. Food Plants* 104/2 Loquat..is also sometimes known as Japanese medlar... The yellow pear-shaped fruits are the size of crab apples and have a sweetish acid flavour.

loquence (lōu·kwĕns). *rare.* [ad. late L. *loquentia*: cf. Loquency.] Speaking; talk.

1677 T. Harvey tr. *Owen's Epigrams* 199 Thy Tongue is loose, thy Body close; Both ill: With Silence this, with Loquence that doth kill. **1886** R. F. Burton tr. *Arabian Nights* VIII. 346 When the Princess Miriam beheld Nur al-Din and heard his loquence and verse and speech, she made certain that it was indeed her lord Nur al-Din.

Loran (lōə·-, lǫ-ræn). orig. *U.S.* Also **loran.** [f. the initial letters of *lo*ng-*ra*nge *n*avigation.] A hyperbolic navigation system employing the difference in the times of arrival of pulsed radio signals from different stations. Freq. *attrib.*; also *ellipt.* for a Loran receiver.

[**1942** (*title*) Development of airborne receiver model LRN-1: report on project 191. (Radio Corporation of Amer. License Div. Lab. Rep. No. 207 (PB 32732), 9 Sept.).] **1943** (*title*) Development of Loran receiver trainer: report on project 191 (extension). (Radio Corporation of Amer. License Div. Lab. Rep. (PB 23321), 17 Mar.) **1945** *Tuscaloosa* (Alabama) *News* 18 Oct. 5 In the airplane, a loran receiver measures the difference in radio wave travel time in millionths of a second. **1945** [see *hyperbolic *a.* 2 b]. **1946** J. P. Baxter *Scientists against Time* ix. 151 The beauty of Loran for wartime use was that the ship or plane which used it emitted no signal that might give away its position. **1960** M. Sharcott *Place of Many Winds* i. 6 We passed Spring Island, where there is a Department of Transport Loran station. **1966** *McGraw-Hill Encycl. Sci. & Technol.* VII. 585 The distance at which reliable loran fixes are generally obtained is about 800 nautical miles over water from the pair of transmitting stations during daytime and about 1500 miles at night. **1972** K. Campbell *Thunder on Sunday* 11 'Get a Loran fix.'.. Peter Spence had his face glued into the rubber eye-piece of the Loran... He counted the jumping electric lines and the long number blips... He then transferred his eyes to the Loran map.

lorandite (lōə·rāndəit). *Min.* [ad. Hung. *lorandit* (J. Krenner 1894, in *Matemat. és Természett. Értesitö* XII. 473), f. the name of *Loránd* Eötvös, 19th-century Hungarian physicist: see -ite¹.] A sulphide of thallium and arsenic, $TlAsS_2$, found as scarlet monoclinic crystals.

1895 *Mineral. Mag.* XI. 32 Professor Krenner, of Budapest,..describes a mineral containing no less than 59·5 per cent. of thallium, to which he has given the name Lorandite. **1946** [see *crookesite]. **1957** *Contrib. Mineral. & Petrol.* XVI. 45 A new find of the thallium-arsenic-sulphosalt mineral lorandite in the Triassic dolomite of the famous Lengenbach quarry in Binnatal, Ct. Wallis (Switzerland) is described.

lord, *sb.* Add. **2.** *lord and master* (later examples). Also, a husband (now usu. *joc.*).

1665 R. Verney *Let.* 5 June in M. M. Verney *Mem.* (1899) IV. iv. 122 Peg Gardner saw your Lord and Master with some gentlemen in Parke. **1739–40** Richardson *Pamela* (1740) II. 251 Your lord and master came in very moody. **1816** Jane Austen *Emma* III. xvi. 300, I am waiting for my lord and master. **1864** C. M. Yonge *Trial* I. vii. 126 She was not going to be one of the womankind sitting up in a row till their lords and masters should be pleased to want them! **1922** Joyce *Ulysses* 639 The erring fair one begging forgiveness of her lord and master. **1961** [see *droit¹ 1 b]. **1975** P. Harcourt *Fair Exchange* ii. 121 'You're a Counsellor, a senior official..what advice would you give?'..'I can't see our lords and masters asking me.'

3. *Lord of the Manor*: rhyming slang for 'tanner' (sixpence, now equal to 2½ pence); also (*ellipt.*) *Lord*.

1839 H. Brandon *Poverty, Mendicity & Crime* 163/2 *Lord of the manor*, sixpence. **1882** *Sydney Slang Dict.* 5/2 *Lord of the Manor*, sixpence. **1933** *Times Lit. Suppl.* 16 Nov. 782/1 Twenty years ago you might hear a sixpence described as a 'Lord' meaning 'Lord of the Manor'; that is, a tanner. **1972** *Lebende Sprachen* XVII. 8/3 *Lord of the Manor*, tanner (*old sixpence*).

6. b. Also *lord* (or *lor'*) *lumme* (= lord love me).

1902 E. Nesbit *Five Children & It* ii. 61 'Lor' lumme,' said Billy Peasemarsh, 'if there ain't another on 'em!' **1903** J. London *People of Abyss* i. 8 Lord lumme, but it'll be the last I see af you if yer don't py me. *Ibid.* iv. 31 'Lor' lumme,' she laughed.

9. b. *House of Lords.* Also, a lavatory. *slang.*

1961 in Partridge *Dict. Slang* Suppl. 1139/1. **1967**

Listener 21 Dec. 802/2 In between you have the Business Man Jocular: 'I say, where's the geography, old son?' or 'When you need the House of Lords, it's through there.' **1969** J. Alexander *Where have All Flowers Gone?* i. 46 Half way up the stairs there was a lavatory... 'The House of Lords,' said Jake.

12. d. *Lord of the Flies*, a name for Beelzebub [tr. Heb. *ba'al-zĕbhūbh*: see Beelzebub]; also, used allusively of the book (1954) by William Golding (1911–) of the same name, in which a group of schoolboys marooned on an uninhabited tropical island revert to savagery and primitive ritual. (Use in quot. 1971 is *joc.*)

1931 A. Huxley *Cicadas* 37 But 'tis the shitten Lord of Flies Who with his loathsome bounties now fulfils On us their prayers. **1948** —— *Ape & Essence* (1949) 69 Such.. is the inscrutable justice of the Lord of Flies. *Ibid.* 90 The Lord of Flies, who is also the Blowfly in every individual heart. **1965** *Listener* 24 June 933/1 The ship's figurehead.., removed from its place by the children, becomes a kind of Lord of the Flies. **1967** J. Blackburn *Flame & Wind* ii. 72 Beelzebub—the Lord of the Flies—the Prince of the Hebrew devils. **1969** I. & P. Opie *Children's Games* 13 Such accounts..have..influenced educational practice..leading us to believe that a *Lord of the Flies* mentality is inherent in the young. **1971** *Guardian* 9 Feb. 8/4 Instead of the orgasmic moment, we arrive at Mr Neville's playtime: the lord of the flies.

13. c. Lord Derby, a large green and yellow-skinned variety of cooking apple, or the tree that produces it.

1876–85 Hogg & Bull *Herefordshire Pomona* II. plate 73 Lord Derby. The origin of this apple is unknown. It has probably been cultivated in the orchards for some years, without special notice. **1933** Hall & Crane *Apple* ii. 27 Some varieties, like Lord Derby, have a markedly upright habit. **1962** *Listener* 27 Sept. 495/1, I am thinking of cooking apples like..that good old-fashioned one, Lord Derby. **1966** C. R. Thompson *Pruning Apple Trees* ii. 194 There are a few varieties from which large apples are wanted early in the season, such as early Victoria and Lord Derby.

14. c. *Lord Muck*, a pompous self-opinionated man (see quot. 1966). *slang.*

1937 in Partridge *Dict. Slang* 539/1. **1955** J. Thomas *No Banners* xxix. 287 Hey, Lord Muck! May we have the honour of introducing ourselves? **1966** 'L. Lane' *ABZ of Scouse* 63 *Lord Muck of Muck Hall*, a bombastic person; a swollen-headed man who likes to assert his authority.

16. lord and lady (**duck**) *N. Amer.*, a pair of harlequin duck, *Histrionicus histrionicus*.

1770 G. Cartwright *Jrnl.* 29 July (1792) I. 20, I shot four eider ducks, and seven lords and ladies. **1835** E. Wix *Six Months Newfoundland Missionary's Jrnl.* (1836) 162, I had a fine view of a patch fox in my walk, saw several seals, and some of those very beautiful birds, called by the people of Newfoundland 'lords and ladies'. **1930** *Canad. Geogr. Jrnl.* I. 32/2 The Harlequin Duck..is known to trappers and prospectors in the far west as 'Lord and Lady Duck'. **1959** W. L. McAtee *Folk-Names Canad. Birds* (ed. 2) 16 Harlequin Duck [is also called] lord and lady (Usually in the plural, 'lords and ladies'. In allusion to the handsome plumage.)

Lord Lieutenant. Add: **1. c.** (Later examples.) The position is now mainly ceremonial, though he retains some of his former powers (see quots.), including the recommendation of persons for appointment as justices of the peace.

1962 W. O. Hart *Hart's Introd. Law Local Govt. & Admin.* ii. 64 With the disappearance of the militia the lord lieutenant has ceased to have a county force to command. His position has come to be largely one of great honour with no active duties to perform. **1963** *Times* 10 May 19/2 At the moment it was contemplated that there would be one lord lieutenant for the Greater London area, with a number of deputy lieutenants to assist him. **1963** K. B. Smellie *Hist. Local Govt.* iii. 42 The Lord Lieutenant and the Sheriff have mainly ceremonial duties. **1972** *Whitaker's Almanack 1973* 631 The duties of the Lord Lieutenant are to advise the Lord Chancellor as to the appointment of magistrates to the county bench... The Lord Lieutenant is usually a peer or baronet..and is often appointed *custos rotulorum* (keeper of the records).

lordosis. Add to def.: (occurring as a physical deformity).

b. A temporarily assumed posture, characteristic of some female mammals during mating, in which the back is arched downwards; the assumption of such a posture.

1941 *Endocrinology* XXIX. 411 Two males exhibited lordosis and hopping behavior typical of the sexually-receptive female [rat]. **1947** *Physiol. Rev.* XXVII. 243 Some female rats in heat display lordosis each time they are mounted by the male. **1965** *Science* 15 Jan. 306/3 Frequency of lordosis, ear-wiggling, and crouching (all indications of receptivity) were recorded. **1968** M. Diamond *Perspectives in Reproduction & Sexual Behavior* xx. 321 Normal males [*sc.* rats] rarely exhibit lordosis. **1971** *New Scientist* 13 May 370/3 The rats with no sense of smell were much more likely to show the typical lordosis (a posture where the belly is lowered and the rump stuck up in the air) that is normal in sexually receptive animals.

lordotic *a.* (examples, corresponding to sense *b of the sb.).

1971 *New Scientist* 13 May 370/3 With these animals removal of both olfactory lobes increased the likelihood of going into a lordotic posture. **1972** *Science* 5 May 519/1 After a 9-hour delay, the typical lordotic reactions and sticky estrous discharge had generally disappeared.

Lord's (lǭdz). Also **Lords.** The cricket ground at St. John's Wood, London, named after its founder Thomas *Lord* (1757–1832); now the headquarters of the Marylebone Cricket Club.

1799 *Times* 1 June 3/4 The colours..were presented..to the corps in Lord's cricket-ground. **1823** H. Bentley *Cricket Matches played by Mary-le-bone Club*, Lord's..26th August, 1793. Eleven of England against Twenty-two of the County of Middlesex. **1863** A. J. Munby *Diary* 10 July in D. Hudson *Munby* (1972) 167 Went to..Lord's to see..the Eton and Harrow cricket match. **1920** *Blackw. Mag.* 90/1 The Australians had won the toss at Lords... Our visitors had already on a Lords wicket pretty thoroughly extended a M.C.C. eleven. **1930** E. Waugh *Labels* 88 There was an interval, during which everyone..strolled about..as people do at Lord's in between the innings. **1959** *Listener* 19 Feb. 344/1 He loyally goes through the motions, military and social (Bank Picquet, Tower Guard, débutantes' balls, Ascot, Lord's), that are required of him. **1972** R. Wahn *Put Lock On It* ix. 99 When there's a Test on at Lords, radio stations cater to the thousands who sit up until dawn.

lordship, *sb.* **6.** Add after 'royalty': (in a mine or a book). (Earlier examples.)

1851 Blackwood *Let.* 18 Sept. in *Geo. Eliot Lett.* (1954) III. 160 Your warning voice about G.E.'s new novel keeps me uncertain. I incline to offer £900, or £1000 with a lordship not to begin until so many are sold. **1861** Stephens & Burn *Bk. Farm-Buildings* 171 The contractor will have Kingsbury agreeing.. **1885** J. Barrowman *Gloss. Scottish Mining Terms* 42 *Lordship*, rate per ton or other measure paid to the proprietor of minerals, royalty.

lordy, *int.* orig. *U.S.* = Lord *sb.* 6 c. Also *lordy me* (= lord help me).

1853 *Southern Lit. Messenger* Oct. 602/2 On the sofa..you sank down and bounded up and said Lordy! **1857** *Knickerbocker* L. 236 O Lordy me Sir! I'm so dreadful afeard you've both on you.. **1897** R. M. Stuart *In Simpkinsville* 155 Lordy, but it all takes my breath away. **1926** S. Anderson *Let.* 15 Apr. in R. L. White *S. Anderson/G. Stein* (1972) 52 It was a beautiful story, beautifully done. Lordy but that man can write. **1928** *Sat. Even. Post* 12 May 20/1 But seven hundred dollars and his pocket piece back again. Phew! Lordy! **1959** N. Mailer *Advts. for Myself* (1961) 198 Well, Lordy-me.. **1968** *Melody Maker* 30 Nov. 10/3 ..shouting 'Lordy, Lordy, yes'. **1973** H. Taylor *Serpent under It* (1974) xiv. 209 Lordy, ..if you pamper me like this you'll.. **1974** N. Freeling *Dressing of Diamond* 88 A kidnapping... And lordy, it's the child of a magistrate.

Lorelei (lǭ·rĕlai). Also 9 **Loreley.** [Name of a rock on the Rhine near Coblenz.] In German legend, a beautiful woman with long blonde hair who sat on the Lorelei rock and with her fine singing distracted boatmen, so that they drowned when their ships foundered on the rock. Also *transf.* Cf. Siren *sb.* 2.

1878 *Chambers's Encycl.* VIII. 745/1 The *Lorelei* of the Rhine is only a river-siren, though a more exquisite enchantress than ever Greek fancy conceived. **1920** 'O. Henry' *Strictly Business* (1917) ..motoring bacchanalians and diamond-hungry loreleis they are businesslike folk. **1937** N. Wainwright tr. *Dekobra's Madonna of Sleeping Cars* ix. 125 My little Lorelei had unfastened her hair, which fell in a golden cascade. **1966** 'A. Gilbert' *Fingerprint* iii. 45 Probably thought you were the original Lorelei. **1971** ..vii. 94 She..evolved into the most devastating Lorelei of Mayfair, with an utterly intriguing contempt for the men she could so easily fetter.

Lorentz (lǭ·rents). *Physics.* The name of H. A. Lorentz (1853–1928), Dutch physicist, used *attrib.* to designate various concepts and phenomena described by him, as **Lorentz (–FitzGerald) contraction** = *Fitzgerald contraction*; so **Lorentz-contract** vb. *trans.*; **Lorentz-covariant, -invariant** adjs., covariant, or invariant, under a Lorentz transformation; so *Lorentz-covariance*; **Lorentz transformation** [named in Fr. by H. Poincaré 1905, in *Compt. Rend.* CXL. 1505], the set of equations which in the special theory of relativity relate the space and time co-ordinates of one frame of reference to those of another moving rectilinearly with respect to the first; **Lorentz triplet,** a group of three spectral lines produced by the splitting of a single line with the frequency of the middle one by a magnetic field (the Zeeman effect, first interpreted by Lorentz).

¶ See also *Lorenz.

1908 *Sci. Abstr.* A. XI. 687 The equations for moving bodies, when subjected to a Lorentz transformation, are converted into the corresponding equations for the transformed quantities. **1916** *Monthly Notices R. Astron. Soc.* LXXVII. 155 The well-known Lorentz-contraction. **1920** [see *covariant B]. **1922** A. D. Udden tr. *Bohr's Theory of Spectra* ii. iii. 47 it follows immediately from this result according to the principle of correspondence that each fine structure component must be expected to split up into a normal Zeeman effect (Lorentz triplet). **1923** C. D. Broad *Sci. Thought* iv. 135 The Lorentz–Fitzgerald Contraction, if taken as a physical fact, affects all kinds of matter equally. **1955** R. Peierls *Laws of Nature* vi. 126 When one first hears of this 'Lorentz contraction' it sounds most artificial and unreasonable. **1955** W. Heisenberg in W. Pauli *Niels Bohr* 21 Such an interpretation..would destroy..just the obvious symmetry property of the theory of relativity, namely the Lorentz invariance, and it must therefore be considered inappropriate. **1956** W. Pauli *Ibid.* 35 Lorentz-invariant quantized field theories. **1958** ..189 The reason for this is that the field acting on an electron is Lorentz contracted. **1964** ..*Introd. Quantum Electrodynamics* iii. 23 The former is the more deep and can be made explicitly Lorentz covariant. **1966** P. Caws *Philos. of Sci.* xxii. 167 Einstein found the Lorentz transformations at hand when he needed them. **1970** G. K. Woodgate *Elem. Atomic Struct.* viii. 147 This so-called Lorentz triplet of one σ and two π lines is characteristic of the 'normal' Zeeman effect in electric dipole radiation. **1970** *Nature* 27 Oct. 272/3 Measuring scales..are assumed to be Lorentz-contracted by a factor $(1 - u^2/c^2)^{1/2}$.

Lorenz (lǭ·rents). *Physics.* Also (erron.) **Lorentz.** The name of L. V. Lorenz (1829–91), Danish physicist, used *attrib.* and in the possessive to designate the ratio $k/\sigma T$ (k = thermal conductivity, σ = electrical conductivity, T = temperature), which has approximately the same value for many metallic elements over a wide range of temperatures.

1922 Glazebrook *Dict. Appl. Physics* I. 459/2 They found that for seven pure metals..the value of Lorenz's constant $K/\sigma T$ was nearly the same over the range 18° to 100°C. .. *Materials* v. 122 The electron influence on the thermal conductivity of metals is normally far greater than the lattice contribution. When this is the case, the Lorentz constant holds reasonably well. **1966** Phillips & Williams *Inorg. Chem.* II. xix. 23 On the free-electron theory of metals it is predicted that there should be a direct relation between the thermal, k, and electrical, σ, conductivities... The Lorenz number, $K/\sigma T$, has the theoretical value of $2{\cdot}45 \times 10^{-8}$ watt ohm deg^{-2}.

lorgnette. (Further examples.)

1776 *Monthly Rev.* LIII. 536 Concerning Achromatic spying Glasses.. [we have thus translated the word *Lorgnettes*]. **1893** *Lett. Miss Riversdale* III. 120 Mr. Drummore moved on..with his lorgnette to his eye, scrutinising every dim.. **1899** G. A. Sala *Twice round Clock* 135 Surveyed, through powerful-lensed lorgnettes. **1958** H. Innes *Campbell's Kingdom* 75 To my astonishment she quizzed me through a gold lorgnette as I entered the room. **1970** V. Canning *Great Affair* vi. 104 My aunt..favoured Edwardian dress.., feathered hats, lorgnettes, [etc.].

lorie, var. Lory.

1848 H. W. Haygarth *Recoll. Bush Life Austral.* xii. 139 The lorie, with his splendid livery of blue and green.

loris. Add to etym.: [Used as a mod.L. generic name (E. Geoffroy Saint-Hilaire 1796, in *Mag. Encycl.* I. 48).] Substitute for def.: **a.** A small arboreal primate of the genus so called, distinguished by grey or black fur, large eyes, and thin limbs, and found in Sri Lanka (Ceylon) and southern India; also called *slender loris.* **b.** A larger primate of similar form but heavier build, belonging to the genus *Nycticebus* and found in south-east Asia; also called *slow loris.* (Further examples.)

1834 E. Griffith tr. *Cuvier's Animal Kingdom* I. 229 The Loris, commonly Lazy Monkeys. **1909** E. Prothero *Handy Nat. Hist. Mammals* iii. 73 The Slender Loris is a small animal only eight or nine inches long. **1967** J. R. & P. H. Napier *Handbk. Living Primates* .. The Slender Loris is not a particularly successful animal in captivity owing to delicacy and irascible temper. **1972** T. A. Vaughan *Mammalogy* vii. 119 Lorises occur in Africa south of the Sahara, in India, Ceylon, and Southeast Asia.

lorisid (lǭ·risid), *sb.* and *a.* [f. mod.L. family name *Lorisidæ*, f. generic name *Loris* (see above).] **A.** *sb.* A member of the family Lorisidæ, which includes lorises, pottos, and, in certain classifications, galagos. **B.** *adj.* Of or pertaining to this family.

1969 *Nature* 22 Nov. 821/1 Lorisid primates. *Ibid.*, The only lorisids which live in open country today are some of the galagines of Africa. **1970** *Ibid.* 25 July 356/1 The talonids of tarsiids, plesiadapids and lorisids. **1972** *Ibid.* 24 Mar. 180/2 Unlike the condition of the symphysis in living tarsiers, lorisids or lemurs..show distinct separation of the superior and inferior transverse tori.

lorisoid (lǭ·risoid), *sb.* and *a.* [f. mod.L. name of suborder *Lorisoidea*, f. generic name *Loris* (see above).] **A.** *sb.* A member of the suborder Lorisoidea, introduced by C. Tate Regan in 1930 and used in some recent classifications of primates. **B.** *adj.* Of or pertaining to this suborder.

1930 C. T. Regan in *Ann. & Mag. Nat. Hist.* VI. 385 The continent formed by the union of Africa and South America..was the original home of the Lemuroids and Lorisoids. *Ibid.* 387 In the Eocene there were in existence animals with many Lorisoid characters. **1953** W. C. O.

Hill *Primates* I. 101 Their [*sc.* dwarf-lemurs'] general affinities are certainly more lemuroid than lorisoid. *Ibid.* 105 Lorisoids tend to possess more rounded heads, with shorter faces, than the average lemuroid. **1965** *Punch* 21 Apr. 580/1 Dr. Manley's studies on social behaviour and reproductive patterns in lorisoid primates have been extended to include the Angwantibo. **1970** *Nature* 25 July 355/2 Such diverse groups of primates as the omomyids, most lemuroids, lorisoids, or the platyrrhines.

Lorraine (lǫrē·n). The name of a province in NE. France used *attrib.* in *Lorraine cross* = *cross of Lorraine* (Cross *sb.* 18). Also *cross Lorraine*.

1830 T. Robson *Brit. Herald* III. Gloss., *Cross patriarchal* or *double cross*, (French, *croix double*) composed of one piece in pale, and two traverse pieces... But French heraldists form their cross patriarchal somewhat different, and often call it a *croix Lorraine*. **1894** Cussans & Parker *Gloss. Terms Heraldry* 173 It is often blazoned as a *cross Lorraine*. **1920** Webster, Lorraine cross. **1970** *Guardian* 13 Nov. 1/1 Lorraine crosses, the symbol of the Gaullist Resistance.

Lorrainer (lǫrē·nəɹ). [f. *Lorraine + -er¹.] A native or inhabitant of Lorraine. Also *attrib.* Cf. *Lotharingian *sb.* and *a.*

1743 *Gentl. Mag.* Aug. 447/1 Of the Lorrainers..it is affirmed, that they are with Difficulty restrained from declaring in favour of their Sovereign. **1903** F. W. Maitland in *Cambr. Mod. Hist.* III. xvi. 574 The Lorrainers were not French. **1933** Kipling *Souvenirs of France* ii. 36, I love that imperturbable Lorrainer [*sc.* Poincaré]. **1966** M. R. D. Foot *SOE in France* xvi. 163 He was one of three brothers, barons of Lorrainer origin, landed gentry of the Limousin.

lorry, *sb.* (The form *lorry* is now usual.) Delete '*local*' and add: **1. b.** A large motor vehicle for carrying goods, etc., by road. Also *attrib.*

1911 *Encycl. Brit.* XVIII. 927/2 ..or lorry chassis. **1915** ..*Handbk.* (ed. 6) I. 11 Commercial motor vehicles, such as heavy motor lorries, are not specifically dealt with. **1945** *Morris Owner's Manual* ..Morris cars vans & lorries. **1950** *Amer. Speech* V. 274 American English has universally chosen *motor truck* and *truck* rather than *auto-truck* or the British *lorry*. **1955** *Times* 25 May 4/4 He spoke from a lorry on Waun Y Pound, a mountain top between Tredegar and Ebbw Vale. **1972** *Guinness Bk. Records* (ed. 19) 117 The world's largest lorry is the M-200 Lectra Haul built by Unit Rig and Equipment Co. of Fort Worth, Texas with a capacity of 200 tons.

3. *lorry driver, driving, load*; *lorry-borne* adj.; **lorry-bus,** a lorry used as a vehicle for public transport; also **lorribus**; **lorry-hop** *v.*, to hitch-hike by lorry; so **lorry-hopping** *vbl. sb.*; **lorry-jump** *v.* = *lorry-hop vb.; so **lorry-jumping** *vbl. sb.*; **lorry park,** an open space or lot reserved for the parking of lorries.

1937 H. L. Hart *Europe in Arms* xxiv. 323 The inability of these temporarily lorry-borne infantry to attack effectively after they had dismounted. **1935** *R.A.F. Jrnl.* Aug. 7 Our aircraft..have operated against..armoured formations and lorry-borne infantry. **1919** *Daily Mail* 12 June 4/4 In the welter of London's crowded streets we grasped at the relieving 'lorribus' (the converted Army lorry doing the duty as an emergency omnibus). **1963** *Economist* 23 Nov. 758/3 Ghana's ubiquitous lorry-buses..and mammy-wagons. **1926** *C.M.U.A. Jrnl.* Apr.–May 19 It sometimes happened that a lorry driver, arriving at noon, was told to go away and return next morning.. **1972** *Times* 20 Nov. 3/6 Lorry drivers will meet dockers' representatives to-morrow. **1937** D. L. Sayers *Busman's Honeymoon* viii. 166 Odd jobs of lorry-driving and taxi-work. **1973** J. Wainwright *Square Dance* 246 Shurrup, lad... Stick to lorry-driving. **1933** A. G. Macdonell *England, their England* xvii. 286 He walked a bit from Alton, and then lorry-hopped, in army fashion, as far as.. **1947** *Penguin New Writing* XXI. 98 An ex-convict..who lorry-hops across England in order to escape the police. **1925** Fraser & Gibbons *Soldier & Sailor Words* 143 *Lorry hopping* (or *jumping*), a familiar term at the Front for travelling by begging 'lifts' from passing transport vehicles. **1928** Blunden *Undertones of War* 102 By luck or judgment in lorry-hopping..one reached Boulogne. **1931** *Times Lit. Suppl.* 12 Nov. 899/3 She 'hitch-hiked', the American equivalent for 'lorry-hopping'. **1941** L. Hastings *Dragons are Extra* v. 101 He..lorry-jumped his way back to his own battery. **1965** *Guardian* 8 Apr. 4/5 The moral dangers of young girls 'lorry jumping' or frequenting roadside cafes. **1928** C. T. Brunner *Probl. Motor Transport* vi. 101 The expense of railway transport..could be reduced by packing the goods in containers..of a weight which would constitute a lorry load. **1972** *Times* 30 Nov. 23/4 Mr John Peyton..has actively stimulated opposition to the present EEC proposals for standardizing lorry loads. **1968** *Listener* 14 Mar. 335/1 Eight caravans..had been parked in a final gesture of despair in the middle of a Walsall Corporation lorry park. **1971** *Daily Tel.* 13 Sept. 3/6 Recommendations..to replace traditional transport cafés with a network of guarded lorry parks. **1974** *Ibid.* 9 Mar. 2/5 The removal of..goods..at a high security lorry park has led to a record number of lorries using it.

lorry (lǫ·ri), *v.* [f. the sb.] *trans.* To transport or convey by means of a lorry or lorries. Hence **lo·rried** *ppl. a.*

1920 *Blackw. Mag.* Jan. 125/1 They were 'lorried' to the Lys front. **1949** W. S. Churchill *Second World War* II. ii. xxiii. 416 Tanks..and artillery in front, and with lorried infantry in the centre. **1958** G. Donaldson *Shetland Life under Earl Patrick* 61 Only since such calls [at individual ports] ceased have all goods for places on the Mainland been landed at Lerwick and then lorried to their destinations. **1975** *Daily Tel.* 9 Sept. 12/6 There will be in 1 British Corps, three lorried infantry battalions.

lose, *sb.*² Add: *lose bet, game*, one in which the loser of the game wins the stakes.

1964 A. Wykes *Gambling* xi. 143 (*caption*) A 'lose' bet is that the shooter will throw a crap. **1971** *Jrnl. Gen. Psychol.* LXXXV. 268 High-risk bets are again more popular.

lose, *v.*¹ Add: **3. b.** Phr. *to lose one's nerve* (Nerve *sb.* 10): to become scared, uneasy; *to lose sleep over* (or *about*), *for*, etc., (something): to worry about (something) (usu. in negative contexts).

1912 *Chambers's Jrnl.* Nov. 739/1 There's nothing here to lose one's nerve about. **1934** G. B. Shaw *Too True to be Good* III. 86 When I was wounded and lost my nerve for flying, I became an army chaplain. **1942** H. C. Bailey *Dead Man's Shoes* ii. 79 'I'd like to know why you didn't tell me.' 'You told me not to lose any sleep over it.' **1944** 'N. Shute' *Pastoral* ii. 41 'I wasn't losing any sleep for them.' 'Those two have been at this for years.' **1959** [see *chance *sb.*]. **1959** N. Mailer *Advts. for Myself* (1961) 94 It's not the sort of thing I lose sleep over. **1967** J. Porter *Dover & Unkindest Cut of All* x. 109 Dover hadn't..any sleep over him.. 'You can't win 'em all,'.. **1970** *Guardian* 10 July 9/5 Stolid and conformist New Zealanders do not lose much sleep over the Negroes' troubles. **1974** *Ibid.* 18 Mar. 6/5 Although increasing restrictions on immigration..had been criticised..it is doubtful whether the immigrants themselves have lost much sleep over them. **1975** *Times* 14 Feb. 14/7 You have to get straight back on, or else you lose your nerve. The others are far more concerned with the loose horse than the girl lying face down in the dirt.

f. Also with *off*.

1872 *Rep. Vermont Board Agric.* II. 717, I think that tin buckets are preferable for catching sap to wooden ones, as they..have no hoops to lose off. **1906** *Dialect Notes* III. 145 I wanted to lose him off before we got to town.

j. *to lose a dinner* (or *a meal*): to vomit (what one has recently eaten).

1941 Baker *Dict. Austral. Slang* 44 *Lose a meal*, to vomit food. **1952** M. R. Kinnear *Swimming Pool* 40 ..

4. c. (20th-cent. examples with *in*.)

1902 *Chambers's Jrnl.* July 441/2 A bird does not gather speed when sailing in the air, as a falling stone would, neither does it lose in pace. **1903** *Q. Rev.* Oct. 413 As a consequence the work loses in freshness and even in clearness. **1949** *Harrap's Stand. French & Eng. Dict.* II. 218/2 The incident did not lose in the telling... To lose in value, in interest.

d. orig. *U.S.* *to lose out*: to be unsuccessful, to fail.

1858 H. Bushnell *Sermons for New Life* ix. 176 The child brought up a thief gets an infinite power of cunning..and loses out just as much in the power of true perception. **1909** 'O. Henry' *Roads of Destiny* iv. 66, I know you've lost out some by not having me to typewrite 'em. **1913** E. D. Biggers *Seven Keys to Baldpate* xiii. 165 But it's over, and you've lost out. **1926** C. Johnson *Negro in Amer. Civilization* (1931) xxvi. 396 Is it not true that the Negro female is losing out in the personal service? So often newspapers are specifying white in their want ads. **1938** E. I. Paul *Narrow St.* xvii. 133 Dindier made a bid for the premiership and lost out because Briand would not play ball with him. **1947** 'G. Orwell' *Eng. People* 38 When the Americans..add a preposition to every verb with a preposition that adds nothing to its meaning (*win out, lose out, face up to*, etc.). **1958** *Spectator* 10 Sept. 16/1 It has lost out to the new Soviet-type competition. **1963** S. Douglas *Years of Combat* v. 313 Traders might come whistling past one's ears, indicating all too clearly that..we had lost out in the preliminary tactical manoeuvrings. **1966** *Listener* 10 Mar. 327/2 It could be that both China and America are losing out to the Russians. **1971** *Guardian* 23 July 3/2 We are going to lose out unless the Government are prepared to do a tremendous public relations job for the tourist industry here. **1972** *Hansard* 10 July 15/2 Rep. Bella Abzug..lost out in her bid for a second term in Congress. **1973** *Times* 30 June 13/6 The monotonous proliferation of redundant prepositions in the ever more popular usages 'check up on', 'lose out on'..

e. (Of a clock, watch, etc.:) to become slow (Slow *a.* 12); to indicate a time earlier than the correct time. Also *trans.*, to run slow by the amount of (a specified period).

1861, 1870, 1917 [see *gain *v.*¹ 3 d]. **1955** *Oxf. Jun. Encycl.* VIII. 81/2 A pendulum clock with a steel rod loses ½ seconds per day for a rise of temperature of 10°F. **1972** *Which?* Aug. 244 At the same time each day the amount they had gained or lost was noted, and they were wound.

9. b. Delete † *Obs.* and add later examples. Esp. in phr. *you('re) lost me* = 'I failed to follow what you were saying'.

1961 L. Deighton *Ipcress File* vii. 42 'They have money..to investigate what they call "synthesised environment"'.. 'I said, 'You've lost me now—without trying.' **1967** H. Van Siller *Biltmore Call* 103 Frazer.. looked up, frowning. 'You've lost me. What do you mean, exactly?' **1970** R. Lewis *Wolf by Ears* i. 11 You will have to be a little more explicit in your statements. You've lost me.' .. 'Put simply, it's this way.' **1973** *Observatory* Oct. 162 You lost me at one stage.

loser. Add: **2. d.** *a bad, poor* (or *good*) *loser*: a person who loses with bad (or good) grace.

1892 A. W. Pinero *Magistrate* i. 28 I hate a bad loser, don't you, Guv? **1917** *Dialect Notes* IV. 326 *Loser* n., differentiated as good loser and bad loser. Not local. 'Those college boys are good losers and do not complain at defeat, but the academy boys..are sure bad losers.' **1931** D. L. Sayers *Five Red Herrings* xiii. 146 He was a bad loser. A slice off the tee..would put him off his game for the afternoon. **1947** 'G. Orwell' *Eng. People* 14 The admiration for a 'good loser'. **1951** [see *clobber *v.*²]. **1973** A. Mann *Tiara* xvii. 164 The British were always talking about 'being a good loser'. **1973** *Radio Times* 20–27 Dec. 10/1, I like to win at things. But..I'm a good loser.

e. An unsuccessful or incompetent person, a failure.

1955 *Amer. Speech* XXX. 304 *Loser*,..someone..hopeless. **1959** *Ibid.* XXXIV. 154 Those limited in ability or old-fashioned in dress or manners, [are] losers. *Ibid.*, The opposite of a B[ig] M[an] O[n] C[ampus] is a loser. **1972** *Melody Maker* 25 Nov. 53 (*heading*) Jiving K. Boots. The adventures of a loser musician. **1974** J. Stubbs *Painted Face* xiii. 180 Poor fellow... A born loser, every time.

4. *U.S. slang.* A convicted criminal, a person who has served a sentence in prison. So *two-time* (or *three-time*, etc.) *loser*, a person who has been in prison twice (or three, etc., times).

1912 D. Lowrie *My Life in Prison* xi. 127 T'day some four 'r five time loser'll drive up with a year. *Ibid.* xxvii. 337 He was a 'ten-time loser' the last time I saw him; *i.e.*, he had served nine previous terms at either San Quentin or Folsom. **1914** Jackson & Hellyer *Vocab. Criminal Slang* 56 *Loser* noun, current amongst prison habitues. An ex-convict... Examples: 'Three-time losers cop life in some states.' **1926** *Clues* Nov. 161/2 *Loser*, one who has served a prison sentence. **1931** *Amer. Speech* VII. 117 *Two-time loser*, a prisoner who has been convicted twice. 'He's a two-time loser; got the book the second time.' **1939** E. S. Gardner *D.A. draws Circle* (1940) xii. 213 He's a two-time loser. **1950** H. E. Goldin *Dict. Amer. Underworld Lingo* 129/1 *Loser*, anyone convicted of a felony. (A second felony conviction makes one a 'two-time loser'; a third conviction, a 'three-time loser', etc.) 'I'm a four-time loser on the next pinch (arrest).' **1973** *Houston* (Texas) *Chron. Texas Mag.* 14 Oct. 4/1 Bob, a three-time loser with a long line of busts and drug abuse..was sick of his life.

5. *Bridge.* **a.** A losing card.

1917 E. Bergholt *Royal Auction Bridge* (ed. 2) 56 The opponent will be able to make two tricks in that suit before Y has had a chance of discarding his two losers. **1921** F. Irwin *Compl. Auction Player* ii. 48 You hold five losers. That is a two-bid, no more. **1964** *Contract Bridge* ('Know the Game' Series) (ed. 2) 29/1 You should not play any trumps but utilize them to ruff losers. **1964** *Official Encycl. Bridge* 334/1 At no trump, all cards below the ace and not in sequence with it are possible losers.

b. loser-on-loser (see quot. 1964).

[**1929** *Auction Bridge Mag.* Jan. 328/1 (*title*) Discarding a loser on a loser...discarding a losing card on a loser.. may be the only play to secure the contract or game.] **1947** T. Reese *Reese on Play* xi. 162 Loser-on-loser play makes the hand against any defence... Some very complicated positions can arise after a loser-on-loser elimination. **1964** *Official Encycl. Bridge* 334/1 *Loser on loser*, the act of playing a card that must be lost on a losing trick in some other suit.

loseyite (lōu·zi,əit). *Min.* [f. the name of Samuel *Losey* (*c* 1833–1906), U.S. mineral collector + -ite¹.] A basic carbonate of zinc and manganese, $(Zn,Mn)_7(CO_3)_2(OH)_{10}$, found as bluish-white monoclinic crystals at Franklin, New Jersey, U.S.A.

1929 Bauer & Berman in *Amer. Mineralogist* XIV. 150 This mineral is named loseyite in honor of Mr. Samuel R. Losey, a native of Franklin and for many years, until his death, an ardent collector of the minerals occurring there. **1968** I. Kostov *Mineral.* ii. xi. 542 Loseyite is also monoclinic (*A2/a*), found as bluish white radiating bundles of lath-like crystals.

loss, *sb.*¹ Add: **4. c.** *to cut one's loss(es)*: to cease carrying on a losing transaction.

1912 *Q. Rev.* Jan. 287 It is now made the basis of the argument that England should 'cut her loss', and Ireland be sent adrift. **1926** T. E. Lawrence *Seven Pillars* (1935) viii. xciv. 519 Feisal had..made one of his lightning decisions to cut the loss; a wise decision, though it hurt us sorely. **1927** *Daily Express* 13 July 8/2 The only reasonable thing is for Great Britain to..cut her losses, and bring the whole matter to an end. **1939** 'G. Orwell' *Coming up for Air* ii. vii. 123 The trouble over Joe aged Father a great deal. To lose Joe was merely to cut a loss, but it hurt him. **1944** A. Bryant *Yrs. of Victory* ii. 39 Bonaparte..saw that he was beaten and, like the great man he was, cut his losses. **1969** *Listener* 28 Aug. 269/1 For this reason the CIA is cutting its losses, reducing its labyrinthine commitments.

5. c. (Examples.) *dead loss*: see *dead *a.* 30.

1940 W. Faulkner *Hamlet* i. i. 16 Major De Spain's barn taken fire and was a total loss. **1943** *Scrutiny* XI. 288 If you apply serious standards, then P. G. [Wodehouse] is a total loss.

d. *one person's loss is another's gain*: a semi-proverbial expression. Also with non-personal subject and complement.

1914 G. B. Shaw *Misalliance* 41 It civilizes them. And it uncivilizes us. Their gain. Our loss. **1925** *New Yorker* 22 Aug. 9/1 When the fighting was over she remained... Our loss was their gain. **1949** B. A. Botkin *Treas. S. Folklore* ii. i. 147 What is history's loss is folklore's gain. **1973** *Times* 22 Jan. 9/5 (*heading*) Newcastle's loss is Doncaster's gain.

10. *attrib.* and *Comb.* **loss leader** *Comm.*, an article put on sale at a non-profit-making

price in order to attract potential buyers of other articles; also *transf.*; hence **loss-leading** *vbl. sb.*; **loss-maker,** a business, etc., consistently working at a loss; **loss-making** *vbl. sb.*, the making of a loss (in business, etc.); also as *ppl. a.*, that makes a loss.

1922 Hayward & White *Chain Stores* vii. 109 Many chains have a fixed policy of featuring each week a so-called 'loss leader'. That is, some well known article, the price of which is usually standard and known to the majority of purchasers, is put on sale at actual cost to the chain or even at a slight loss..on the theory..that people will be attracted to this bargain and buy other goods as well. Loss leaders are often termed 'weekly specials'. **1942** H. Levy *Retail Trade Assoc.* xviii. 211 Prohibitions on 'loss leaders'. **1958** *Times* 15 Dec. 9/3 No doubt price cutting in individual lines often goes beyond the point where it is justified by reduction in direct cost: it is in part the loss-leader technique. People are attracted into a shop by some very low prices, and buy many other articles which give a handsome margin. **1969** *Daily Tel.* 7 Mar. 19 Sir Stanley Raymond, chairman of the Gaming Board, said yesterday he was convinced that Bingo was often used as a 'loss leader' to induce housewives into 'hard' gambling. In many cases Bingo provided only half of the takings in clubs. **1970** *Ibid.* 15 May 21/5 Some would like to see bank charges on personal accounts reduced or abolished, as a 'loss leader' to the existing personal customers. **1971** *New Scientist* 21 Jan. 102/1 The ranks of loss leaders and unrepeatable offers. **1964** *New Statesman* 28 Feb. 343/1 This concession was necessary to get any bill past his own back-benchers; and there are a few more..concessions..in particular a ban on loss-leading. **1968** *Daily Tel.* 12 Nov. 17/8 Only a madman or a company making a genuine attempt at loss leading would reduce rates. **1971** *Guardian Weekly* 23 Jan. 22 What happens when two companies, both lossmakers, merge into one? The answer, as often as not, is one big loss-maker. **1973** *Times* 24 Aug. 17/2 The company declined to give reasons for its withdrawals except to say that the bookstalls did not fit into its 'economic pattern' or plans for the future. It said the stalls were not lossmakers. **1971** *Sunday Times* 15 Aug. 41 He started by picking up a 40% stake in the lossmaking Carson's chocolate business in March 1964. **1974** *Times* 7 Mar. 19/6 The company has been loss-making since 1971.

lossless, *a.* Restrict † *Obs.* to sense in Dict. and add: **b.** *Electr.* Characterized by or causing no dissipation of electrical or electromagnetic energy.

1952 *Proc. IRE* XL. 1651/1 If..the system is lossless, then the transverse electromagnetic (TEM) mode can be propagated. **1962** Corson & Lorrain *Introd. Electromagn. Fields* xi. 357 Lossless dielectrics, good conductors, and low-pressure ionized gases. **1969** P. M. Chirlian *Basic Network Theory* i. 46 Inductors and capacitors are called lossless elements.

loss-proof (lǫ·sprūf), *a.* [f. Loss *sb.*¹ + Proof *a.* 1 b.] Guaranteed against loss, inflation, fluctuation in market value, etc.

1963 *New Yorker* 8 June 107 (Advt.), Travelers cheques ..Loss-proof and theft-proof. **1969** *Daily Tel.* 22 Nov. 14 (Advt.), The guaranteed loss-proof investment that *grows* despite Stock Market ups and downs.

lossy (lǫ·si), *a.* *Electr.* [f. Loss *sb.*¹ + -y¹.] Characterized by or causing dissipation of electrical or electromagnetic energy.

1948 H. A. Leiter in Smullin & Montgomery *Microwave Duplexers* ix. 378 Lossy cables are used..in order to isolate the oscillator from effects of mismatch in the unit under test. **1949** L. L. Langton *Radio-Frequency Heating Equipment* v. 76 A lossy capacitor. **1969** *Sci. Jrnl.* Dec. 44/3 At optical frequencies a metal transmission line structure would be very 'lossy' and only transparent dielectric materials such as glass can be considered. **1970** J. Earl *Tuners & Amplifiers* iv. 94 Most transformers are 'lossy' at the sub-bass end of the spectrum.

lost, *ppl. a.* Add: **1.** (Later examples of *lost soul*.)

1895 [see *booze *sb.* 1 c]. **1923** D. H. Lawrence *Ladybird* 224 The lost-soul look of the men. **1937** *Discovery* May 150/2 It emits a weird screaming wail like a lost soul.

2. (Earlier and later examples of *lost ball*.) *to get lost*: see *get *v.* 33.

1809 *Laws of Cricket* (rev. ed.), If lost Ball is call'd, the Striker shall be allowed four, but if more than four are run before lost Ball is call'd, then the Striker to have all they have run. **1967** *Laws of Cricket* ('Know the Game' Series) 15 If a ball in play cannot be found or recovered, any fieldsman may call 'Lost Ball', when 6 runs shall be added to the score.

6. lost cause, a cause (Cause *sb.* 11) that has failed or that is unlikely to succeed; *spec.* the cause of the South in the American Civil War (1861–65); **lost generation,** *spec.* that of the period of the 1914–18 war, a high proportion of whose men were killed in the trench warfare; also used more generally of any generation judged to have 'lost' its values, etc.; **lost property,** lost articles found but not claimed; so *lost property department, office*, an office dealing with (the disposal of) lost property; **lost river** *U.S.*, a river which disappears into the ground and re-emerges elsewhere; **lost rock** *U.S.*, a travelled boulder; **lost stone** *U.S.* = **lost rock*; **lost wax** = *cire perdue; **lost weekend,** one spent in dissolute living; also *transf.*

1865 M. Arnold *Ess. Criticism* p. xix., Oxford..home of lost causes, and forsaken beliefs, and unpopular names, and impossible loyalties! **1866** E. A. Pollard (*title*) The lost cause. **1901** 'Mark Twain' *Speeches* (1923) 231 You testify by honoring two of us, once soldiers of the Lost Cause. **1914** *Times Lit. Suppl.* 7 Aug. 378 Oxford has often been called 'the home of lost causes', or, as Mr. Cram puts it, 'of causes not lost but gone before'. **1933** C. Mackenzie (*title*) The lost cause. A Jacobite play. **1938** J. Betjeman *Oxf. Univ. Chest* v. 112 Wytham and Binsey are the less hackneyed of Oxford's lost causes on the edge of Oxford. **1940** C. F. Adams *And Sudden Death* xvii. 155 Why should I go round championing a lost cause? **1948** D. Wecter in J. G. Kerwin *Civil-Military Relationships in Amer. Life* 31 Their late adversaries, the United Confederate Veterans, licked their wounds and dwelt lovingly upon the Lost Cause. **1949** D. S. Freeman in B. A. Botkin *Treas. S. Folklore* p. viii, Perhaps every land that has the tradition of a Lost Cause builds its monuments in a certain sentimental determination and seeks through its memorials both to exemplify and to perpetuate its ideal. **1926** E. Hemingway *Sun also Rises* (*title-page*), 'You are all a lost generation.'—Gertrude Stein in conversation. **1930** R. Macaulay *Staying with Relations* iv. 57, I was nineteen when the affair [*sc.* the war] ended... I belong practically to the Lost Generation. **1939** C. Day Lewis *Child of Misfortune* ii. i. 146 'Ha,' Alec yelled. 'We're the Lost Generation.' **1951** E. Paul *Springtime in Paris* xi. 197 The era of the Lost Generation and the notorious expatriates. **1959** *Listener* 15 Oct. 616/2 Thomas Wolfe, F. Scott Fitzgerald, Sinclair Lewis, Ernest Hemingway, and Katherine Anne Porter—members of the so-called 'lost generation', strove and strayed on the left banks between the wars. **1969** [see *hippie, hippy *sb.* and *a.*]. **1970** D. T. Turner in Z. N. Hurston *Mules & Men* 11 New York was an exciting place for young black intellectuals and artists during the mid-Twenties. Afro-American culture had been rediscovered... The 'Lost Generation' danced wildly to jazz rhythms. **1844** A. W. Kinglake *Eothen* xxii. 340 The Governor..saw the value which I set upon the lost property. **1922** Joyce *Ulysses* 56 His lost property office secondhand waterproof. **1923** H. C. Bailey *Mr. Fortune's Practice* vi. 156 He was only calling on the lost property department to leave a lady's bag. **1941** V. Nabokov *Real Life S. Knight* (1945) x. 86 She fails to get the..job in the lost property office. **1959** *Manch. Guardian* 7 Aug. 12/2 A home-made double bass..will be included in a lost property sale. **1971** 'A. Garve' *Late Bill Smith* ii. 49 Everyone with any kind of problem was bringing it to Sue. She was harbour-master, postmaster, nurse, lost property office. **1843** 'R. Carlton' *New Purchase* I. ix. 58 Out come the mole rivers that have burrowed all this time under the earth, and which, when so unexpectedly found are styled out there,—'lost rivers'! And every district of a dozen miles square has a lost river. **1872** R. W. Raymond *Statistics of Mines* 197 The great 'lost river' which bursts out of the vertical side of the cañon of the Snake. **1831** J. M. Peck *Guide for Emigrants* ii. 136 Scattered over the surface of our prairies are large masses of rock, of granitic formation, roundish in form, usually called by the people *lost rocks*... These stones are denominated *boulders* in mineralogy. **1857** *Trans. Illinois Agric. Soc.* II. 347 Another curiosity is the boulders, or 'lost rocks', as they are frequently called, which are found on the surface of the earth in the middle and northern sections of Illinois. **1819** H. C. McMurtrie *Sk. Louisville* 29 (Th.), [Certain stones] in the Illinois and Missouri territories are denominated lost-stones, from their being strangers to the soil where they are found. **1933** H. F. Lenz *Alfred David Lenz Syst. Lost Wax Casting* 9 The modeling wax for casting in the 'cire perdue' or 'lost wax' method has several requirements. **1947** J. C. Rich *Materials & Methods of Sculpture* vi. 146 The 'lost-wax' or *cire-perdue* process is the traditional method of casting in bronze... The ancient Egyptians employed the lost-wax method, casting over ash cores. **1972** *Times* 28 Sept. 18/6 The bronzes were made by the lost-wax process in which the mould is destroyed. **1944** C. Jackson (*title*) Lost weekend. **1947** M. McCarthy in *Partisan Rev.* May–June 303 It was..comic that a man should have one name..for his wife..and another..for trips and 'lost' week ends. **1955** Koestler *Trail of Dinosaur* 56 He is the classic type who becomes addicted to the Communist drug, and never finds his way back from the lost week-end in Utopia. **1960** K. Amis *Take Girl like You* xix. 229, I quite expected to find you on the couch this morning, especially after your lost-weekend act. **1968** A. MacLeod *Dam* iii. 32 I'll have to go down to the pub and replenish the stock... And that will mean another bloody lost week-end. **1969** 'E. Lathen' *Come to Dust* iii. 34 He seems to feel that if Patterson emerges from some lost weekend, the press will seize on his connection with Neil Marsden. **1975** M. Kenyon *Mr Big* vii. 65 The *Express* used one paragraph headed Lost Weekend.

‖ **los von Rom** (lōs fǫn rōm). [Ger., = 'free from Rome'.] A slogan used by and applied to a movement which arose in Austria and Germany at the end of the 19th c., seeking to reduce the political influence of the Roman Catholic Church. Also used allusively of other policies of this kind.

1899 *Times* 8 Apr. 8/2 The really significant feature of the conversion of Herr Wolf, the Pan-Germanic member of the Reichsrath, and perhaps the most active promoter of the 'Los von Rom' agitation, is..his acceptance by the Evangelical community. **1920** W. T. Whitley in *Encycl. Relig. & Ethics* XI. 326/2 This Los von Rom movement seems..to have been anti-Slav, and reinforced Protestants quite as much, the tendency being to pave the way for these provinces to join the German Empire. **1923** G. M. Trevelyan *Manin & Venetian Revolution of 1848* ix. 165 Irritated by the Pope's Italian nationalism, the German Catholics might drift toward the position of *los von Rom*. **1957** *Oxf. Dict. Chr. Ch.* 822 *Los von Rom*... The Movement, which owed its name to the shout of a student at the Deutscher Volkstag at Vienna (1897), came to have its centre at Innsbruck. **1963** *Listener* 28 Feb. 360/1 None of the Five, not Dr Adenauer nor any other German, is likely to cry '*los von Rom*'.

lot, *sb.* **1. e.** Delete *rarely*, and add later examples.

1870 [see Throw *v.*¹ 41 e]. **1927** *Daily Express* 15 June 11/7 Mr. Patrick O'Maille..threw in his lot recently with Professor Magennis' party, Clainn Eireann.

2. d. Also in colloq. phr. *to be* or *to have had* (one's) *lot*.

1960 *Observer* 24 Jan. 5/2 When the bogies sus you and take your trousers to the forensic lab you've had your lot. **1961** J. Maclaren-Ross *Doomsday Book* i. vii. 77 Any trouble, mush, and it's your lot. *Ibid.* ii. v. 156 That's your lot, Marsh! This isn't one of your crime-plays, d'you hear. **1968** *Listener* 18 July 77/2 'Mummy, Tristram has just dropped your handbag into something rustic.' 'Right, that's your lot. Into the car, and we'll get back to civilisation.' **1973** *Times* 12 Dec. 2/7 Hate slogans scrawled on a blackboard. One read, 'Tina Wilson is going to be done over,' and another, 'Tina Wilson has had her lot tonight.'

6. a. (Earlier and later examples of *across* or *cross lots*.) Also *attrib.*, as *lot gate*. **b.** Also, land round a film studio where outside filming may be done.

a **1852** F. M. Whitcher *Widow Bedott Papers* (1883) xxii. 87 You see yer uncle and me went hum by the turnpike instid o' gwine cross lots. **1854** *Jrnl. Discourses* I. 83 [I dreamed that] I cut one of their throats from ear to ear, saying, 'Go to hell across lots.' **1858** J. R. Lowell *Two Gunners* in *Poetical Works* II. 125 Joe looked roun' And see (acrost lots in a pond..) A goose. **1887** *Scribner's Mag.* Dec. 735/1 He'd have had to foot it by the path 'cross lots... He's sold his hoss. **1928** 'R. West' *Strange Necessity* 205 The worst of making war, as of acting for the 'movies', is the amount of waiting around on the lot. **1929** W. Faulkner *Sartoris* iv. 281 Restless hounds waited for them at the lot gate. **1937** Partridge *Dict. Slang* 538/1 The most picturesque phrase of all is 'the lot', which is always used to describe the company's land surrounding the studio. It has been in use since the days when..hard-pressed pioneers rented vacant building lots. **1938** M. K. Rawlings *Yearling* xv. 153 As he clicked the lotgate, doves flew from the pines with a whistling of wings. **1948** 'N. Shute' *No Highway* vi. 151 Don't you ask me why, I'll tell you when I see you on the lot. **1966** *Listener* 15 Dec. 880/1 Many lots are devoted to the less arduous and expensive work of turning out films for television networks. **1969** in Halpert & Story *Christmas Mumming in Newfoundland* 174 Then commenced the chase, up lanes, 'across lots', down lanes.

8. b. (Later examples.)

1936 W. R. Titterton *Chesterton* i. ii. 30 If I quoted much of that marvellous essay I should have to quote the lot. **1956** H. & M. Williams *Plays of Year* XV. 178 It was to be a big wedding—the full treatment—Royalty—the lot. **1958** *Listener* 3 July 10/2 They are said to cure everything from rheumatism to ringworm, colic to snake-bite.. —the lot. **1961** [see *croak *v.* 5]. **1970** *Times* 7 Jan. 7/5 The death of his father..triggers off a crisis for him too, producing a temporary breakdown, dismissal from his job, separation from his wife, the lot. **1974** 'P. B. Yuill' *Bornless Keeper* vii. 65 They've searched the island twice —helicopters, dogs, the lot.

9. (Later examples.) Also with *adj.* as (*this, that*) *little lot*.

1898 J. D. Brayshaw *Slum Silhouettes* 141 Yus, it's a nobby little turn-out, ain't it?.. Mine? Lor' luv a duck! No, that's Sal Hogan's little lot. **1920** J. Mander *Story N.Z. River* v. 76, I just love you, lots. **1957** 'J. Wyndham' *Midwich Cuckoos* iii. 24 That there Miss Ogle ain't 'alf goin' to cop 'erself a basinful of 'Er Majesty's displeasure over this little lot. **1961** B. Crump *Hang on a Minute* 91 Have a bo-peep at this little lot, Jack, called Sam from the back of the truck. **1965** [see *buster 2 b]. **1968** *Listener* 1 Aug. 148/1 Mr Donoghue, like any good critic, doesn't mean us, he means you lot. **1975** P. G. Winslow *Death of Angel* 125 A group of lads she doesn't care about... Next stop Wormwood Scrubs, that little lot.

10. lot-jumper *U.S.*, one who appropriates another's lot.

1869 *Overland Monthly* III. 63 Then there had been a lot jumper's fight down at the end of the street. **1889** in J. B. Thoburn *Stand. Hist. Oklahoma* (1916) I. xx. 223 Gamblers, liquor dealers,.. lot-jumpers. **1931** G. F. Willison *Here they dug Gold* 241 Counterfeiters, lot-jumpers, mine-jumpers,.. and ruffians in general.

lotaustralin (lōutau·strālin). *Chem.* [f. mod.L. *Lot-us*, the generic name, + *austral-is*, the specific epithet (see Australian *sb.* and *a.*), of the lotus (Lotus 5) from which it was first isolated: see -in¹.] A toxic crystalline compound, $C_{11}H_{19}NO_6$, which occurs in *Lotus australis* and in white clover (*Trifolium repens*).

1938 Finnemore & Cooper in *Jrnl. Soc. Chem. Industry* May 165/1 The cyanogenetic constituent is in the main composed of the glucoside of the cyanohydrin of methyl ethyl ketone. This is new and it is proposed to call it lotaustralin. **1964, 1973** [see *linamarin].

Lotharingian (lōuþări·ndʒiăn), *sb.* and *a.* [f. *Lotharingia* (see below) + -an.] **A.** *sb.* A native or inhabitant of the ancient duchy of Lotharingia in northern Europe, situated between the Rhine and the Scheldt from Frisia to the Alps. **B.** *adj.* Of or pertaining to Lotharingia or its inhabitants, or to modern Lorraine (see quot. 1969). Cf. *Lorrainer.

1607 Topsell *Four-f. Beasts* 46 They [*sc.* beavers] are vsed by the Lotharingians and Sauoyens for meat allowed to be eaten on fish-daies. 1883 *Encycl. Brit.* XV. 5/1 By the treaty of Bonn (921) the Lotharingian duchy was ceded formally to France. *Ibid.*, The ever-restless spirit of the Lotharingians broke out into new commotions. 1909 J. H. B. Masterman *Dawn Mediæval Europe* xxiii. 216 Louis the German and Charles both fell upon his [*sc.* Lothair's] Lotharingian lands like birds of prey. 1959 *Chambers's Encycl.* VIII. 694/1 In a struggle with the turbulent Lotharingian magnates, the young king was killed. 1964 *Times Rev. Industry* Jan. 73/3 The river Our ..is situated within 100 miles of the Belgian coalfields, the Ruhr, the Lotharingian steel industry,.. and the Luxembourg iron industry. 1969 *Listener* 6 Feb. 165/1 In this new economic Europe, which hinges on the Rhine and the Rhône—the Lotharingian axis, as it's sometimes called—Brittany tends to feel even more isolated.

lotic (lōu·tik), *a. Ecol.* [f. L. *lōtus* washing (Needham & Lloyd *Life of Inland Waters* (1916) vi. 315).] Of fresh-water organisms or habitats, situated in rapidly moving water. Cf. *lenitic *a.*

1916 Needham & Lloyd *Life Inland Waters* vi. 363 The animals of lotic societies are mainly small invertebrates. 1931 R. N. Chapman *Animal Ecol.* xvii. 347 Along the shores of the Great Lakes there are lotic communities where the wave action is incessant, thus producing the fundamental conditions of a lotic environment and presenting an exception to the general statement that all lotic environments are streams. 1970 F. J. & W. B. Vernberg *Animal & Environment* ii. 44 Fresh-water systems are generally divided into two groups based on the activity or rate of movement of the water: (1) standing quiet (lentic) waters..; and (2) flowing (lotic) waters.

lotong (lotǫ·ŋ). Also lutung. [Malay.] A leaf monkey of the genus *Presbytis*, esp. *P. obscurus.*

1821 T. S. Raffles in *Trans. Linn. Soc.* XIII. 247 *Simia maura*? Linn. Lotong of the Malays. 1824 T. Horsfield *Zool. Res. Java* s.v. *Semnopithecus Maurus*, The name of the latter [monkey] is Lutung; but the Malays and Europeans apply this name to both species... In Sumatra the name of the 'Maure' is Lotong. 1839 T. J. Newbold *Pol. & Statistical Acct. Straits of Malacca* I. vii. 432 Of the genus Semnopithecus are..the Lotong, or Semnopithecus Maurus, of F. Cuvier; the kra, etc. 1903 J. L. Bonhote in *Fasciculi Malayenses: Zoology* I. 6 The lôtong is very generally distributed over those parts of the Peninsula that we visited. 1936 G. B. Gardner *Keris & Other Malay Weapons* v. 95 They have a curved iron handle forged on, resembling the tail of a *lotong* monkey. 1965 C. Shuttleworth *Malayan Safari* 10 A *lotong* (leaf monkey) chattered its alarm call near by.

lotsa (lǫ·tsă). Colloq. contraction of *lots of.*

1927 [see *hot *a.* 1 e]. 1945 A. Kober *Parm Me* 18 Mmm, is good the cake... You must use lotsa ecks [*sc.* eggs] fa such a cake. 1967 'P. Chambers' *Bad die Young* x. 146 We got lotsa time. 1969 in Halpert & Story *Christmas Mumming in Newfoundland* 88 The social drinkers soon left, even though the host was protesting that there was 'lotsa time to stay'. 1971 *It* 9–23 Sept. 4/2 The Notting Hill Carnival was lotsa fun for seven days and nights.

lotta (lǫ·tă). Also lotter. Colloq. contraction of *lot of.*

1906 E. Nesbit *Railway Children* viii. 175, I gets a lotter green paint and I paints her stem to stern. 1928 [see *bullshit 1]. 1944 C. Himes *Black on Black* (1973) 198 Lotta hustlers up there. 1945 A. Kober *Parm Me* 34, I hadda do a lotta talking to get her to come. 1965 'D. Shannon' *Death-Bringers* (1966) xiii. 164 A lotta people know my name who I don't know. 1969 *Coast to Coast 1967–68* 9 No, I'm a lotta things, but I ain't crazy. 1971 *Black World* Apr. 56 Lotta big talk, but when you get there nothin is happenin.

lotus. Add: **3. c.** The plant treated symbolically in Hindu and Buddhist thought; also, in Yogic exercises, a bodily position said to resemble the lotus blossom. Cf. *lotus gospel, pose*, etc. in *6.

1848 J. D. Hooker *Himalayan Jrnls.* Nov. (1854) I. x. 229 Low stone dykes, into which were let rows of stone slabs, inscribed with the sacred 'Om Mani Padmi om'.—'Hail to him of the lotus and jewel.' 1887 E. Arnold (*title*) Lotus and jewel, containing 'In an Indian temple', 'A casket of gems'..with other poems. 1949 S. Muzumdar *Yogic Exercises* 103 There are insuperable impediments because of which many will fail to master the Lotus. 1973 R. Rendell *Some lie & some Die* xviii. 183 Vedast..had taken up a Yoga position, a half-Lotus, on the floor.

6. (*Buddhism* and *Yoga*) *lotus gospel, pose, position, posture, seat, throne.*

1911 E. A. Gordon (*title*) The Lotus Gospel or Mahayana Buddhism and its symbolic teachings. 1966 'A. Hall' *9th Directive* ii. 17 He sat with his legs crossed under him in the Lotus pose. 1962 T. C. Lethbridge *Witches* v. 57 There are so many ways in which the Indian gods agree with those of Western Europe, even to their sitting in the 'lotus' position. 1964 I. Fleming *You only live Twice* i. 21 Since Bond had arrived in Japan he had assiduously practised sitting in the lotus position. 1968 *Guardian* 29 Feb. 8/4 Sitting in the lotus position..concentrating upon one's navel and repeating the mystic syllable, 'Om, Om'. 1884 Ram Chandra Bose *Hindu Philos.* vi. 177 It is called Padmásana (lotus-posture), and is highly beneficial in overcoming all diseases. 1965 W. Swaan *Jap. Lantern* xii. 136 Ideally, one should assume the cross-legged 'Lotus Posture' familiar from Buddha images. 1937 F. Yeats-Brown *Yoga Explained* ii. 71 (*caption*) The pupil is shown in the Lotus Seat, padmasana, but any comfortable position can be adopted. 1911 E. B. Havell *Ideals Indian Art* iii. 32 The whole spirit of Indian thought is symbolised in the conception of the Buddha sitting on his lotus-throne.

lotus-eating, *vbl. sb.* Add: Hence (as a back-formation) **lotus-eat** *v. intr.*

1898 W. J. Locke *Idols* v. 64 'What have you been doing with yourself all this time?' she said... 'Oh, Lotus-eating, generally,' he replied. 1911 —— *Glory of Clementina Wing* xv. 211 The week had evidently passed pleasantly for Quixtus... He was lotus eating. 1972 J. Potter *Going West* 60 Every dream fades. If he drank and lotus-ate, he would think. 1973 'B. Mather' *Snowline* xix. 231 Hippies..come here..from Bombay—lotus-eat for a month..and off they go.

louche. Delete *rare* and add to def.: Also, dubious, shifty, disreputable.

1873 G. H. Lewes *Diary* 16 Jan. in Geo. Eliot *Lett.* (1956) V. 368 The whole thing appeared *louche* and unpromising. 1905 G. B. Shaw *Lett. to G. Barker* (1956) 53 You could play Snobby. I want a slim, *louche*, servant-girl-bigamist, half-handsome sort of rascal. 1921 A. Huxley *Crome Yellow* xvii. 182 There had seemed to be something a little *louche* in the way she had suddenly found herself alone with Ivor. 1945 Auden *Sea & Mirror* ii. 46 A quick cold clasp now and then in some *louche* hovel. 1945 E. Waugh *Brideshead Revisited* 236, I knew of a louche little bar quite near here. 1959 P. H. Johnson *Humbler Creation* xlviii. 328 As if he were an unfrocked priest due for reception into the world of the *louche* and the lost. 1970 *Times* 7 Mar. (Saturday Suppl.) p. iv/6 There is plenty of marvellous delicate comedy and superbly *louche* menace. 1974 *Daily Tel.* (Colour Suppl.) 14 June 30/4 His louche greeting, 'Ladeez and Gentle*men*' was a byword among..BBC Light Programme listeners.

‖ **loucherbem** (lūʃɛbɛm). Also **loucherbème.** [f. Fr. *boucher* butcher.] The name given to a type of French slang formed by the transposing of certain letters and addition of others.

1937 J. Orr tr. *Iordan's Introd. Romance Lang.* iv. 360 The butchers of La Villette speak, or spoke, a slang called *loucherbem*, or *largongi*... The process..is a deliberate modification of an existing word. 1938 I. Goldberg *Wonder of Words* viii. 155 In French there is the *loucherbème* language, so called because its model of transformed words is *boucher*. 1939 L. H. Gray *Foundations of Lang.* 31 Only in a few instances do they actually change the form of words, as in the relatively recent French *loucherbème*. 1942 Partridge *Usage & Abusage* (1947) 160/2 *Gibberish* is applied..to the Loucherbem of the Paris butchers (*loucherbem* for *boucher* is itself Loucherbem).

Loucheux (lū·ʃö), *sb.* and *a.* [Canad. Fr., f. F. *louche* squint-eyed (see quot. 1828).] **A.** *sb.* **a.** A North American Indian people inhabiting the Yukon and Mackenzie River areas. **b.** The language of this people. **B.** *adj.* Of, pertaining to, or designating this people or their language.

1828 J. Franklin *Narr. Second Expedition Polar Sea* i. 23 The fact is, that Loucheux, or Squinter, was intended to convey the sense of the Indian name..Deguthée Dennee, which means, 'the people who avoid the arrows of their enemies, by keeping a look out on both sides'. *Ibid.*, The tribe of Indians whom Mackenzie calls the Quarrellers, but whom the traders throughout the fur country name the Loucheux. *Ibid.* 29 The Red River contributes its waters to the Mackenzie..and..is remarkable as being the boundary between the lands claimed by the Loucheux Indians and those of the Esquimaux. 1867 *Ann. Rep. Board of Regents Smithsonian Inst. 1866* 311 The physical characteristics of the Loucheux nation are..the same as those of the other aborigines of North America. *Ibid.*, The Loucheux language is a dialect of the Chepewyan. *Ibid.*, The Loucheux, though sunk in barbarism, are rather more intelligent than the other tribes. *Ibid.*, The Loucheux proper is spoken by the Indians of Peel's river. 1890 J. G. Frazer *Golden Bough* II. iii. 127 The Loucheux and Hare-skin Indians..are forbidden by custom to eat the sinew of the legs of animals. 1921 E. Sapir *Lang.* iv. 71 Such languages as Navaho,..Chipewyan, Loucheux. 1971 *Times* 22 Feb. (Canada Suppl.) p. v/1 In the north of the Yukon close by the Alaska border, a small band of Loucheux Indians have stood up to be counted: they do not want oil and gas exploration in their territory. *Ibid.*, The caribou move..through these flats, and the Loucheux take about 1,000 of them a year for food.

loud, *a.* Add: **6.** *loud-mouthed* adj. (further examples); *loud-talking* adj.; **loud-mouth** [f. the adj.], a person given to loud and self-assertive talk; hence as *vb.*, to talk in this manner, to bluster; **loud-mouthing** *vbl. sb.*

1934 J. O'Hara *Appointment in Samarra* iv. 111 He was a loud-mouth and a good one-punch fighter. 1940 J. B. Priestley *Postscripts* 17 You will find that the laziest loud-mouth in the workshop has suddenly been given power. 1950 A. Lomax *Mister Jelly Roll* (1952) v. 235 Morton, the old whorehouse pianist who seemed to be trying to loud-mouth his way back to big time. 1959 *Daily Mail* 31 Jan. 4/2 These 625 vain, devious loud-mouths..are our elected representatives. 1968 M. Richler *Cocksure* ix. 52 We are most decidedly not done for. My goodness, the last loudmouth to make that mistake was Hitler. 1867 Trollope *Last Chron. Barset* I. xli. 353 How, from the abjectness of his own humility..he would rebuke the loud-mouthed triumph of the bishop's wife. 1924 G. B. Shaw *St. Joan* i. i. 7 The self-assertive, loud-mouthed, superficially energetic, fundamentally will-less Robert. 1966 *Listener* 27 Oct. 609/1 The Surrey amateur with an Oxford background and the loud-mouthed crowd from the foundries suddenly coalesced. 1971 C. Bonington *Annapurna South Face* vii. 85 He was well built, very self-confident yet not loudmouthed. 1950 W. L. James in A. Dundes *Mother Wit* (1973) 431 Famous among those yet surviving names are 'corn field holler'..and 'loud mouthing'. 1973 *Daily Tel.* (Colour Suppl.) 26 Jan. 7/3 Money talks and there will be a lot of loud-mouthing before the future of the fourth channel is settled. 1848 Mrs. Gaskell *Mary Barton* I. i. 3 Merry and somewhat loud-talking girls. 1925 T. Dreiser *Amer. Trag.* (1926) I. i. ix. 56 He was too smooth and loud-talking.

loud, *adv.* Add: **1. d.** *out loud*, aloud, without restraint. Cf. Out-loud.

1821 M. Edgeworth *Let.* 22 Nov. (1971) 277 Lord Andover in the presence of Lord and Lady Suffolk and speaking *out loud* said 'Miss Dutton..be so good to walk with me into the library.' 1844 [see Out-loud]. 1881 Trollope *Ayala's Angel* III. xlix. 67 He knew it would make me laugh out loud. 1924, 1933 [see *cry *v.* 21 e]. 1939 G. B. Shaw *In Good King Charles's Golden Days* ii. 104 Even a dissolute court, as they say mine is—I suppose they mean a court where bawdy stories are told out loud instead of whispered. 1941 [see *cry *v.* 21 e].

e. Phr. *loud and clear*: (said, etc.) in a way that reduces or avoids confusion or misunderstanding; (esp. in radio or telecommunication) (heard or received) loudly and clearly; also *transf.*

1871 'L. Carroll' *Through Looking-Glass* vi. 133, I said it very loud and clear; I went and shouted in his ear. 1940 H. G. Wells *Babes in Darkling Wood* iii. ii. 256 I'd have to play the Star Spangled Banner, loud and clear, one hundred per cent. 1958 'Castle' & 'Hailey' *Flight into Danger* ix. 123 Hullo, Vancouver. 714 answering. Receiving you loud and clear. Over. 1959 *Listener* 19 Feb. 319/1 It seems desirable to say so loud and clear. 1962 L. Deighton *Ipcress File* vii. 43 Dalby looked up. 'You are receiving me?' 'Loud and clear,' I said. 1962 A. Shepard in *Into Orbit* 104, I also contacted..the chase planes—and heard them loud and clear. 1963 J. N. Harris *Weird World Wes Beattie* (1964) iii. 25 The fact is I remember that black Dodge loud and clear. 1970 C. Collingwood *Defector* iv. 33, I read you loud and clear. Now, suppose something does go wrong, what do I do? 1972 D. Lees *Zodiac* 90 They don't seem able to make up their minds whether to warn me off or knock me off but I do get the message loud and clear and..I'm going.

louderback (lau·dəɹbæk). *Geol.* [Named after G. D. *Louderback* (1874–1957), American geologist.] A cap of old lava on a tilted fault-block.

1930 W. M. Davis in *Bull. Geol. Soc. Amer.* XLI. 299 It seems to me highly appropriate that the lava sheets, which were thus spread unconformably on the Powell surface of the worn-down King mountains, and which now cover the back slopes of the tilted Gilbert blocks, should be called Louderbacks, after their discoverer. 1965 W. D. Thornbury *Regional Geomorphol. U.S.* xxiv. 473/2 Some of the range have lava caps, called by Davis (1930) louderbacks. 1968 R. W. Fairbridge *Encycl. Geomorphol.* 679/1 Since the original [lava] flow would normally have filled a valley, the louderback is usually also a form of inverted topography.

loud-hail, *v.* [Back-formation from next.] *trans.* and *intr.* To speak or call through a loud-hailer; to address (someone) through a loud-hailer; also *fig.* Also **loud-hailing** *ppl. a.* and *vbl. sb.*

1943 *Combined Operations, 1940–42* (Ministry of Information) xvii. 130 We 'closed' the 'Calpe', struggling with our loud-hailing equipment which remained resolutely silent. 1964 R. Petrie *Murder by Precedent* i. 19 I'd just left again and gone to Despatch when they loud-hailed me. 1965 *Listener* 3 June 837/1 Ku Klux Klan men, assorted torchbearers,.. and a loud-hailing director. 1969 C. Carfax *Silence with Voices* xviii. 129 She went back into the cabin after Dr Ford loud-hailed her. 1969 J. Elliot *Duel* i. ii. 43 'The extraordinary thing is ——' Harry loud-hailed so that people at the next table were forced to listen.

loud-hai·ler. [f. Loud *a.* 1 + Hailer.] A megaphone or other device for amplifying the voice, especially as used at sea. Also *transf.* and *attrib.*

1941 *Illustr. London News* 20 Sept. 366/3 Orders, signals and instructions..given over about one-eighth of a mile of water, orally, by the 'Loud Hailer' from Captain to Captain. 1943 *Combined Operations, 1940–42* (Ministry of Information) xvii. 131 'Maintain a smoke screen half a mile inshore and to the westward,' came over the loud hailer from the 'Calpe'. 1955 G. Freeman *Liberty Man* i. 2 Two metallic pipe-notes sounded through the loud-hailer system. 1956 J. Masters *Bugles & Tiger* xxii. 283, I heard a loud hailer calling my name. 1959 C. MacInnes *Absolute Beginners* 13 There was a loud-hailer echo up and down the flights. 1971 B. W. Aldiss *Soldier Erect* 241 They were allowed to use loud-hailers, even if the Japs would not trust them with rifles. 1971 *Daily Tel.* 16 Apr. 17/3 [He] alerted the Thames Coastguard..and then used a loudhailer to direct them to the highest point on the sandbank. 1973 C. Bonington *Next Horizon* xix. 264 John..ran out with the loud-hailer, shouting..the conventional form of greeting.

loudness. Add: Also, the (great or small) extent to which a sound is heard as loud. (Further examples.)

1937 *Ann. Reg. 1936* 65 It was suggested that an overall loudness of 90 phons at a lateral distance of 18 ft. should be the maximum permitted for a private car running on full throttle at 30 m.p.h. 1948 P. M. Morse *Vibration & Sound* (ed. 2) vi. 226 Corresponding to the physical quantities intensity and frequency are the physiological (or, rather, psychophysiological) quantities loudness and pitch. *Ibid.* 227 The bottom curve, for zero loudness, is the threshold of hearing, below which a sound ..is inaudible to the average person. 1959 *Chambers's Encycl.* XII. 728/1 It is important to distinguish between sound intensity—a physical quantity measurable in definite physical units; and loudness—the magnitude of the auditory sensation produced by that sound and therefore a subjective quantity depending upon personal judgements. 1961 G. A. Briggs *A to Z in Audio* 117 Although loudness depends primarily on actual intensity, it is also strongly affected by frequency. 1962 A. Nisbett *Technique Sound Studio* xi. 196 Equal loudness from the two voices will not generally be required.

2. *attrib.*, as *loudness level, scale*; **loudness control,** a device on an audio amplifier which corrects for the change in quality of reproduced sound at low volumes by boosting the bass (and often also the treble) relative to the middle frequencies, either combined with a volume control or as a separate on–off switch.

1961 G. A. Briggs *A to Z in Audio* 117 When the volume is turned down, music tends to sound thin and emaciated. The effect may be counteracted to some extent by bass and treble boost in the amplifier. A loudness control is a device which automatically applies bass and treble boost as volume is reduced. 1970 J. Earl *Tuners & Amplifiers* i. 18 An idea which is popular in Europe and America (though not often liked in Great Britain) is the so-called loudness control. *Ibid.*, Some amplifiers are equipped only with a loudness control, while others feature the ordinary (non-compensated) volume control which can be changed to loudness control action by means of a switch. 1975 *Gramophone* May 2026 (Advt.), Separate loudness control. 1938 *Bell Lab. Rec.* XVI. 213/1 The intensity level of the 1000-cycle reference tone was defined as the loudness level of this tone, and any other sound which is judged by listeners to be equally loud is said to have an equal value of loudness level. *Ibid.* 214/2 The unit to be used for intensity level measurements is the decibel, but the unit to be used for the equivalent loudness level measurements is the phon. 1934 *Discovery* Dec. 346/2 The number of phons and decibels coincides for the loudness scale applying to the frequency 1,000 cycles per second, but for no other frequency.

loud-speaker (laudspī·kəɹ). Also **loud speaker, loudspeaker.** [f. Loud *a.* + Speaker.] **1.** Any instrument for converting variations in an applied electric current or voltage (of appropriate magnitude and frequency) into corresponding sound waves that are able to be heard at a distance from the instrument.

Loud-speaking (see next) was formerly the usual term employed to refer to such instruments. Quot. 1884 is an isolated early example of *loud-speaker*.

1884 *English Mechanic* 26 Sept. 95/3 Before troubling to make or buy a 'loud speaker', 'Pathologist' might try the following dodge... Fix an ordinary 'toy' telephone over one ear... Fasten the end of the string from the 'toy' to the centre of the diaphragm of the Bell..and, at the middle of the string, hang a small weight..to keep it taut... The voice is almost..as clear as when using the Bell direct. 1920 *Telegraph & Telephone Jrnl.* VI. 111/2 It was quite remarkable how far and how distinctly it was possible to hear the talk from the loud-speakers. 1923 *Daily Mail* 1 Mar. 9 For each concert there will be seating accommodation..for 1,000 people, and to these the concert will be delivered by powerful loud speakers. 1924 *Wireless Weekly* 8 Oct. 745/1 Loud speakers of all kinds are an outstanding feature of the show. 1930 J. Buchan *Castle Gay* xiii. 208 Their ears were greeted by the bray of a loud-speaker to which the wives by their house-doors were listening. 1931 B. Brown *Talking Pict.* v. 117 In some form or other the dynamic or moving coil loud speaker, well known in radio, is..always used for sound picture work. 1943 J. B. Priestley *Daylight on Saturday* v. 27 They were now putting canned music over the loud speakers. 1957 L. F. R. Williams *State of Israel* ix. 158 Regulations strictly forbid..loudspeakers in their [*sc.* the polling stations'] immediate vicinity. 1970 J. Earl *Tuners & Amplifiers* iii. 76 Headphones..can nowadays give a subjective impression of quality, spaciousness and stereo effect equally as good as the best loudspeakers.

2. *attrib.*, as *loud-speaker enclosure, system, unit, van.*

1962 A. Nisbett *Technique Sound Studio* iv. 76 A manufacturer of loudspeaker enclosures invited members of the public to judge and compare the quality of three different stereo speaker systems. 1971 *Daily Tel.* (Colour Suppl.) 22 Oct. 57 (Advt.), Twin sealed loudspeaker enclosures, each containing 2 speakers (bass and treble). 1958 *Times Lit. Suppl.* 10 Oct. 583/2 The installation of loudspeaker systems in churches has rendered..manner even less important than matter in sermons than ever before. 1970 J. Earl *Tuners & Amplifiers* i. 25 A loudspeaker *system* is a box, usually called enclosure or cabinet, carrying a socket or couple of terminals at the rear with a grille at the front through which the sound comes. Such a loudspeaker system is scientifically designed in conjunction with the loudspeaker units inside. 1925–6 *T. Eaton & Co. Catal.* Fall & Winter 391 This Loud Speaker Unit when attached to your gramophone tone arm, makes the horn act as a loud speaker. 1970 Loudspeaker unit [see *loud-speaker system* above]. 1945 H. Nicolson *Let.* 3 July (1967) 474 One of my workers said something mean on the loud-speaker van. 1973 J. Drummond *Bang! Bang! You're Dead!* xlii. 144 A police loudspeaker van..pointed out that nothing could be gained by further resistance.

loud-speaking (lau·dspī:kiŋ), *a.* Also without hyphen (as one word). [f. Loud *adv.* + Speaking *ppl. a.*] Speaking loudly (in quot. 1855, *fig.*); *spec.* (the usual sense), capable of producing sound that can be heard at a distance; fitted with or employing a loud-speaker.

1855 [see Loud *adv.* 3]. 1879 *Telegraphic Jrnl.* VII. 112/2 The new telephone receiver of Mr. Edison is, undoubtedly, the greatest step..towards rendering the telephone a loud-speaking instrument. 1897 *Pearson's Mag.* July 51/1 This head-gear would hardly be necessary if a loud-speaking telephone were employed. 1921 *Sci. Amer.* Dec. 100/3 With a loud-speaking telephone on the living-room table and with a simple receiving set the members of the family can receive all the news..followed by a musical program. 1923 *Radio Times* 28 Sept. 14 A loud-speaking equipment that will enable all..to hear perfectly without headphones. 1951 *Catal. of Exhibits, South Bank Exhib., Festival of Britain* 82/2 Combined two-way loudspeaking intercommunication from bridge to selected points. 1960 *20th Cent.* Apr. 308 Loud-speaking cars toured hour by hour, giving names of known survivors.

louie, var. *looey.

Louis heel. Also **Louis Quinze heel.** [f. the name of Louis XV (1715–74), King of France, in whose reign it became fashionable.] A type of heel used on a woman's shoe (see quot. 1968).

1872 *Young Englishwoman* Dec. 651/1 She knows that it is impossible to walk gracefully with the high-pointed heel, and never wears any but the demi-talon Louis Quinze. 1901 F. H. Burnett *Making of Marchioness* i. ii. 30 'She's got Louis Quinze heels,' returned his Lordship. 1906 *Daily Chron.* 23 Feb. 6/5 From the pretty shop-girl..to the daughter of the upper classes the Louis heel is regarded amongst women as a pedestal of superiority to be appropriated for personal adornment. 1968 J. Ironside *Fashion Alphabet* 136 It is of any height or shape... Its character is defined by the sole material which..is attached to the breast (front) of the heel by adhesive... *But* the term 'Louis' is generally applied to a high heel with curved waist, flared at the base.

Louisianian (luˌīz-, luˌiziæ·niăn, luˌiziā·niăn), *a.* and *sb.* [f. *Louisiana* (see below), named after Louis XIV of France.] **A.** *adj.* Of or pertaining to the State of Louisiana at the mouth of the Mississippi. **B.** *sb.* A native or inhabitant of Louisiana.

1775 J. Adair *Hist. Amer. Indians* 240 The illustration of this may..shew our southern colonies what they may still expect from the masterly abilities of the French Louisianians. 1835 J. H. Ingraham *South-West* I. ix. 101 Americans; that is to say, Anglo-Americans as distinguished from the Louisianian French. 1854 C. E. A. Gayarré *Hist. Louisiana* (*French Domination*) I. 13 Is not this the very poetry of landscape, of Louisianian landscape? 1945 *Chicago Daily News* 26 July 2/3 The slow but straight-talking Louisianian is extremely popular with his junior officers. 1949 B. A. Botkin *Treas. S. Folklore* iv. i. 552 Marylanders grow lyrical over Brunswick stew..and Louisianians, over the superiorities of the Cajun and Creole cuisine. 1967 Mrs. L. B. Johnson *White House Diary* 5 Dec. (1970) 597 This is a warm, loving, family-type house..crowded to the limit with.. Louisianians.

Louis-Philippe (luiˌfilī·p). The name of Louis-Philippe, King of France from 1830 to 1848; used *attrib.* or *ellipt.* to designate the style of architecture, furniture, and interior decoration characteristic of his reign.

1908 A. Bennett *Old Wives' Tale* iii. 307 He minutely examined his mouth in the glass of the Louis Philippe wardrobe. 1919 W. S. Maugham *Moon & Sixpence* xi. 46 It was a very small room, overcrowded with furniture of the style which the French know as Louis Philippe. 1951 N. Mitford *Blessing* ii. iv. 183 To be able to tell at a glance whether an object was Louis XV or Louis Philippe, First or Third Empire. 1970 *Oxf. Compan. Art* 441/1 The slightly monotonous interiors..foreshadow the bourgeois interiors of Louis-Philippe. 1975 *Country Life* 6 Feb. (Suppl.) 31/2 Five piece gilt suite of the Louis Philippe period.

loukoum(i, varr. *locoum.

loulou (lū·lū). [Fr., f. *loup* wolf.] A nickname for a Pomeranian dog.

1894 G. du Maurier *Trilby* II. vi. 208 The same old couples petting the same toutous and loulous! *Ibid.* (*footnote*), Loulou—a Pomeranian dog. 1916 E. V. Lucas *Vermilion Box* ccv. 237 One of our sights for visitors here is a dog. A real dog, not a toutou or a loulou, but a great collie kind of thing.

lounge, *sb.* Add: **2. b.** The drawing-room of a private house; the public sitting-room in a hotel or institution. Also *transf.*

1881 [in Dict., sense 2]. 1908 *Daily Chron.* 13 Nov. 4/6 London hotels are extending their lounges. 1938 E. Bowen *Death of Heart* ii. i. 186 The sea..seemed an annexe of the livingroom... She learned later that Daphne called this the lounge. 1954 J. Betjeman *A Few Late Chrysanthemums* 94 It's ever so close in the lounge dear, But the vestibule's comfy for tea. 1965 M. Bradbury *Stepping Westward* iv. 181 Expressly for the purpose of hunting down Walker and bringing him to the English department faculty lounge, wherein the faculty were even now assembled. 1973 *Houston* (Texas) *Chron. Mag. People, Places, Pleasures* 14 Oct. 21/1, I am in a mobile lounge which is about to roll out from Dulles Airport to the supersonic Concorde. 1973 *Times* 8 Dec. 26/4 Burns Hotel... Licensed restaurant, bar lounge.

c. Ellipt. for *lounge-coat*, **-jacket, -suit.*

1893 J. Tomlin *Bond St. Syst. Cutting* 44 There are many different styles in which the Lounge is produced. 1905 *Daily Chron.* 16 Mar. 8/7 (Advt.), Good coat presser and baister for lounges and morning coats. 1928 *Tailor & Cutter* 29 Nov. 899/3 I'll take the lad; and..in six calendar months he will be able to make a lounge. 1968 A. A. Whife *First Course Gentlemen's Garment Cutting* (ed. 4) 69 The Reefer, Sports Jacket, Hacking Jacket..are designed on lines which originate in the basic structure of the Lounge.

3. (Earlier and later N. Amer. examples.)

1830 J. F. Watson *Ann. Philadelphia* 183 Formerly they had couches of worsted damask..in lieu of what we now call sophas or lounges. 1845 *Knickerbocker* XXV. 446 The hard-bottomed chairs were the same, and the lounge, and the tall mahogany clock. 1915 in B. A. Botkin *Treas. S. Folklore* (1949) iii. ii. 538 He stayed in the barn nights ..and slept on an ol' lounge he carried out from the house. 1972 E. Staebler *Cape Breton Harbour* xvi. 142 There is always some one sitting on the rocker or on the wooden lounge built into the corner.

4. *lounge-jacket, -wear*; also (sense *2 b) *lounge-diner* (also *lounge/diner*), *-hall*; **lounge bar,** a bar in a public-house which is furnished with the amenities of a lounge; **lounge lizard** *slang* (orig. *U.S.*), a man who spends his time idling in fashionable society, esp. in search of a wealthy patroness.

1937 *Hotel & Catering Management* Sept. 23/1 Smoke-room or lounge bars are obviously larger than cocktail bars, and provision has to be made for the serving of draught beers. 1971 'H. Calvin' *Poison Chasers* vii. 83 Two of the security men..came into the public bar, and the rest of the customers..went into the lounge bar. 1961 *Evening Standard* 3 July 23/4 (Advt.), Modern s/d house... Hall, lounge/diner, kitchenette. 1966 Lounge-diner [see *garden gnome* s.v. *garden *sb.* 6]. 1974 *Country Life* 7 Mar. (Suppl.) 21/1 Excellent Family House.. study/TV room, 31 ft. lounge/diner, breakfasting kitchen. 1910 *Bradshaw's Railway Guide* Apr. 1046 Lord Warden Hotel... Orchestra plays..in the beautiful lounge hall every evening. 1933 D. C. Peel *Life's Enchanted Cup* xi. 126 This house..contained..what house agents now call a lounge-hall. 1939 O. Lancaster *Homes Sweet Homes* 66 The luxury flat..is divided up..into a dining-room, drawing-room, lounge-hall, three bed, two bath, a kitchen and all the usual offices. 1887 E. B. Giles *Hist. Cutting in Eng.* i. 70 When repose is required [we use] the lounge jacket or dressing gown. 1899 R. Whiteing *No. 5 John St.* xxviii. 284 The billycock and the lounge jacket are, I think, my strong point. 1918 *Hatchet* 4 Apr. 39/1 (*caption*) Nautical lounge lizards. 1921 *Daily Colonist* (Victoria, B.C.) 10 Apr. 4/4 The delicate, poetic cast of features, and the misplaced eyebrow adorning the nether lip of these lounge lizards, denizens of dansants and cabarets. 1926 *Punch* 17 Nov. 534/1 Formal recognition of those firmly attached appendages of Society, the lounge-lizards. 1973 *Times* 29 Dec. 7/7 The £50 a week contract which.. lets her keep her lounge lizard husband, Queckett, in the manner to which he is accustomed, lacks conviction. 1969 *Sears Catal.* Spring/Summer 3 Now—and all through the summer—Sears has the answer to just what you're looking for in loungewear. 1974 *News-Palladium & Herald-Press* (Benton Harbor, Mich.) 8 May (Advt. Suppl.), See this utterly feminine loungewear in enchanting prints.

lounger. Add: **b.** An article of furniture or of dress designed to be used for relaxation.

1964 G. Sims *Terrible Door* xviii. 98 She wore..blue woollen stockings and highly polished, brown moccasin loungers. 1969 *Sears Catal.* Spring/Summer 3 Ankle length loungers with their own carefully constructed Bra sewn right in. 1969 A. Laski *Dominant Fifth* v. 182, I am sitting in a lounger under an umbrella. 1971 *Daily Tel.* 11 May 13/5 The best 'loungers' are those which allow the legs to be raised slightly above the body.

lounging, *vbl. sb.* Add: **b.** *lounging-chair* (later example), *-coat, robe, -room.*

1971 *Daily Tel.* 11 May 13/5 Now, this year, Vogue have a lounging chair which lets you get your head down. 1920 B. E. Stevenson *Gloved Hand* 11, I paused only to open my bag, change into a lounging-coat, and brush off the dust of the journey. 1960 *Harper's Bazaar* Oct. 87 A chiffon lounging coat printed with cinnamon flowers. 1908 *Sears Roebuck Catal.* 1117 (*heading*) Long kimonos or negligees and bath or lounging robes. 1945 R. Chandler *Let.* 13 Oct. in *R. Chandler Speaking* (1966) 44 The great man appeared.., clad in an expensive lounging robe. 1863 Mrs. Gaskell *Dark Night's Work* iv. 42 He used the study for a smoking and lounging-room principally. 1887 C. D. Warner *Their Pilgrimage* (1888) i. 2 In the spacious office and general lounging-room, sea-coal fires glowed in the wide grates.

loungy (lau·ndʒi), *a.* [f. Lounge *sb.* + -y[1].] Suggestive of lounges or lounging.

1911 Chesterton *Innocence of Father Brown* v. 124, I mean little, loungy men, who had just enough to live on and had nothing to do but lean about in bar-rooms and bet on horses.

‖ **loup** (lū), *sb.*[4] [Fr., lit. 'wolf'.] In full, *loup de mer.* The sea-bass, *Dicentrarchus labrax*, found off the coasts of western Europe and in the Mediterranean.

1766 Smollett *Trav.* I. xviii. 291 One of the best fish of this country, is called *Le Loup*, about two or three pounds

in weight; white, firm, and well-flavoured. 1938 E. Bowen *Death of Heart* i. iv. 76 Filing off wet third-class decks of lake steamers, choking over the bones of *loups de mer.* 1966 P. V. Price *France: Food & Wine Guide* 44 The *loup de mer*, so often found grilled over fennel sticks all over the south, is a sort of sea-perch or sea-bass. 1969 J. Elliot *Duel* i. v. 117 Keith..had lines out trolling for *rouget* and *loup de mer*. 1969 *New Yorker* 27 Sept. 120/2, I have seen more herbs thrown on the coals beneath, somewhat as fennel is blazed under a *loup*, the wolfish sea bass, at its last minutes on the grill.

loupe (lūp). [Fr.: cf. Loop *sb.*[4]] A small magnifier used by a watchmaker or jeweller.

1909 in *Cent. Dict.* Suppl. 1940 C. McCullers *Heart is Lonely Hunter* (1943) ii. ii. 105 He noticed the distorted look of Kelly's right eye as it appeared through his watchmaker's loupe. 1949 H. C. Weston *Sight, Light & Efficiency* iv. 122 Simple spectacle loupes are generally made to give a magnification of 2¼ or 2½. 1962 R. Webster *Gems* II. xxxiii. 598 It is now possible to discuss the various types of simple microscopes, such as the ordinary magnifiers with which jewellers are so familiar. These magnifiers, or loupes as they are sometimes called, must be considered with two points of view in mind. 1964 E. Bruton *Finsbury Mob* viii. 111 On an old kitchen table were a pair of jeweller's scales, a loupe,..various other bits of jewellers' paraphernalia, and a packet of detergent.

loup-garou. (Later examples.)

1939 H. M. Miner *St. Denis* vii. 136 The devil, in various forms, and *loup-garous* were abroad in the land and were a considerable problem. 1961 R. M. Patterson *Buffalo Head* ii. 69 Jerome, certain that the loup garou had come for him at last, backed closer to the fire.

lourie: see *lorie.

louse, *sb.* Add: **3. louse-borne** *a.*, of diseases: transmitted by lice.

1919 W. Byam et al. (*title*) Trench fever: a louse-borne disease. 1942 *Times* 21 Sept. 5/3 Typhus, which is louse-borne,..was overcome by active measures of disinfestation. 1964 M. Hynes *Med. Bacteriol.* (ed. 8) xxi. 322 *Trep. recurrentis obermeieri* is typical of the louse-borne disease. 1970 *Control of Communicable Dis. in Man* (Amer. Public Health Assoc.) (ed. 11) 275 (*heading*) Typhus-fever, epidemic louse-borne. 1974 Passmore & Robson *Compan. Med. Stud.* III. xii. 75/1 Louse-borne relapsing fever is a disease of cold weather.

louse, *v.* Add: **3.** With *up.* To infest with lice. orig. *U.S.*

1931 *San Francisco Examiner* 29 Jan. 34/4 Lousey, now fixed in Broadway actor jargon, is from small time troupers... The Maine tavern keeper who refused lodging to a repertoire company..explained: 'The last troupe loused up the beddin'.' 1931 *Gang World* Jan. 14 The precinct was fumigated yesterday, an' you ain't gonna louse it up again. 1955 R. P. Hobson *Nothing too Good for Cowboy* i. 15, I got loused up in that cabin once. 1968 *Listener* 9 May 601/2, I was occasionally loused-up myself, and people, rather than pass me, used to go on the other side of the road.

b. *slang.* To spoil, to mess up. Const *up.* Also **loused-up** *a.* orig. *U.S.*

1934 J. O'Hara *Appointment in Samarra* ii. 61 There's fifty bucks in it for you on account of lousing up your date. 1938 *Amer. Speech* XIII. 195 Louse up the show. 1948 *Sat. Even. Post* 25 Sept. 41 The hospital field is loused up enough. 1957 F. & R. Lockridge *Practise to Deceive* (1959) ii. 29 Of all the loused-up operations. 1958 E. Dundy *Dud Avocado* ii. i. 185 He said if I'm really serious about getting a part..the easiest way to louse it up would be to turn up with a hundred other people. 1959 *Tamarack Rev.* xii. 24 What a way to louse up this new magenta outfit... You'd think she'd spent her afternoon at a Yiddish tear-jerker. 1959 'S. Ransome' *I'll die for You* x. 119 Had a rough time getting her to come back... Damned if I'll let you louse me up now. 1967 [see next]. 1969 *N.Y. Rev. Bks.* 10 Jan. 38/2 It is safe to predict that President-elect Nixon will look for some outstanding public figure for this job, even though it may louse up the table of organization. 1972 *Human World* Nov. 48 If..he tries to sabotage his actions —he louses up a machine he is purporting to work, for example [etc.]. 1973 R. Ludlum *Matlock Paper* i. 7 A loused-up army record. 1975 *New Yorker* 5 May 115/1 The picture is a cheerfully loused-up reworking of the legend of King Arthur's Grail hunt.

louser. (See under Louse *v.*) Also, one who spoils things: used as a general term of abuse. Also **louser-up.**

1960 B. Moore *Luck of Ginger Coffey* iv. 86 You louser. ..What the hell do you know about love? All you want is to get up some woman's skirts. 1966 A. Prior *Operators* viii. 110 He entered her, cursing all women... The sluts, bags, lousers, slags. 1967 C. Cockburn *I, Claud* xxxv. 438 Hardly anyone can be packed off to some social equivalent of the Russian 'virgin lands' for lousing things up, because almost every louser-up can convincingly claim that he was not really responsible for the thing that happened. 1968 'N. Blake' *Private Wound* vii. 108 If any of you lousers interfere I'll plug him in the belly.

lousy, *a.* Add: **1. d.** 'Swarming' with; abundantly supplied with (money, people, etc.); full of. Const. *with. slang* (orig. *U.S.*).

1843 *Spirit of Times* 4 Mar. 7/3 He was lousy with money, and dared any man to face him. 1856 *Democratic State Jrnl.* (Sacramento, Calif.) 6 Oct. 2/3 The bed of the river is perfectly 'lousy' with gold. 1864 A. J. Munby *Diary* 15 July in D. Hudson *Munby* (1972) 199 Why Sir, these unfortunates are all over the place: the ground (he added with a gesture of disgust) is *lousy* with them. 1928 S. Vines *Humours Unreconciled* i. 13 The Totsuka Club was..in the words of Mr. Podler, 'just lousy with liars'. 1934 V. M. Yeates *Winged Victory* xix. 150 And if the Dover Patrol was costly in life, were not shipping magnates lousy with shekels? 1936 W. Holtby *South Riding* ii. i. 89 Leckton told me last month they threw in sixteen and a half couple of hounds and couldn't see a dog. Lost in thistles and willow herb—but lousy with foxes. 1956 R. Bradlow *Nancy Wake* xiii. 153 The town was lousy with Germans, she noted.

2. Delete 'Now *rare*' and add further examples. Also, inferior, poor, bad; ill; in low health or spirits.

1822 D. O'Connell *Let.* 22 May in *Corr.* (1972) II. 391 Perhaps to save some lousy postage you wrote across the letter. They stop all such letters in France. 1849 A. Gordon *Diary* 12 July in W. E. Woodward *Way Our People Lived* (1944) viii. 268, I wish I could never hear the word *lousy* again. I am willing to bet that Tommy Plunkett uses it fifty times a day, but he is no worse than the others. It is 'lousy' this and 'lousy' that. The rain is lousy, the trail is lousy, the bacon is lousy, and Gus Thorpe, losing in the card game, has just said that he has had a lousy deal. 1922 Joyce *Ulysses* 18 You come along with your lousy leer and your gloomy jesuit jibes. *a* 1930 D. H. Lawrence *Last Poems* (1932) 197 Oh great god of the machine What lousy archangels and angels you have to surround yourself with! 1932 N. Mitford *Christmas Pudding* xiv. 220, I still think it's lousy of you not to have taken me last night. 1933 'N. West' *Miss Lonelyhearts* (1949) 46, I felt swell before you came, and now I feel lousy. 1937 L. Bromfield *Rains Came* i. xxxix. 170 Life is so short and so lousy. 1950 'S. Ransome' *Deadly Miss Ashley* xv. 171 She felt too lousy to come to work. A foul cold, she said. 1959 J. Braine *Vodi* xi. 153 A dirty rotten little whore, who couldn't wait six months for her husband, not six lousy months before she went off with another man. 1959 R. Gant *World in Jug* 136 That sudden, nervous sweep which made him to my mind one of the lousiest drivers in Paris, city of lousy drivers. 1968 K. Weatherly *Roo Shooter* 23 You're not a bad bastard, Hunter,..in spite of your lousy cooking. *Ibid.* 74 'I got fifty-two last night,' he said... 'I only got a lousy twelve,' replied Hunter. 1973 P. Moyes *Curious Affair of Third Dog* xi. 148 A brisk, pretty, coloured nurse came in... 'Ah, you're awake... How do you feel?' 'Lousy,' said Henry.

B. As *adv.* 'Lousily', extremely. Chiefly *N. Amer.*

In quot. 1971 a mere intensive.

1932 *Amer. Speech* VII. 436 A man drunk is..'lousy drunk'. 1936 C. Day Lewis *Friendly Tree* xiv. 210 Well, not lousy-drunk. Just comfortable. 1971 D. Heffron *Nice Fire & Some Moonpennies* iv. 33 But she didn't lousy come to school that day! What a blow.

‖ **loutrophoros** (lūtrǫfǫ·rǫs). *Cl. Antiq.* [Gr. λουτρο-φόρος, f. λουτρόν water for a bath + -φορός bearing.] A tall two-handled vessel used for carrying water to the nuptial bath.

1896 C. H. Smith in *Catal. Greek & Etruscan Vases Brit. Mus.* III. 366 Beside her is a box or table on three legs terminating in lions' paws, on which stand a pyxis and an oinochoë with tall handle and neck (loutrophoros?). 1931 *Times Lit. Suppl.* 16 July 563/2 A splendid loutrophoros by Polygnotus. 1935 Richter & Milne *Shapes & Names Athenian Vases* 5 *Loutrophoros*,..tall vase with high, funnel-shaped neck, slender body, and flaring mouth. Water was brought in it from the fountain Kallirrhoe for the nuptial bath, and it was placed on tombs of unmarried persons. 1967 R. S. Folsom *Handbk. Greek Pott.* 157 The loutrophoros usually was an amphora, but sometimes the form appears as a hydria with three handles.

louver. Add: The form *louvre* is now usual in the U.K. and *louver* in the U.S.

4. Add to def.: Also used for other purposes, e.g. to deflect air issuing from an opening or to prevent the direct passage of light through it. Used in *sing.* in same sense; also, an individual slat or strip of such an arrangement. (Further examples.)

1833 J. C. Loudon *Encycl. Cottage, Farm & Villa Archit.* 1128/2 *Louvre*, see *Luffer-boards*. 1920 W. Neubecker *Pract. Sheet Metal Duct Construction* xiv. 155 (*caption*) Formation of movable louvre. *Ibid.*, When movable louvres are used they are pivoted on rods. 1923 *Man. Seamanship* (Admiralty) II. 282 An ideal system of ventilation for a small compartment. A small adjustable flap and louvre..is fitted in the supply trunk. 1933 Moyer & Fittz *Air Conditioning* ix. 236 The two-point thermostat *A* is located near the fresh-air intake louver. 1957 *Economist* 23 Nov. 710/1 Heat is distributed not only in conventional radiators but through louvres in the skirting board. 1966 D. F. Galouye *Lost Perception* xvi. 169 He had managed to..check on the intake louvre... The simple removal of four screws would unfasten the grating and provide access to the ventilation duct. 1968 *Autocar* 25 Jan. 30/2 The wheels splashed through puddles..and steam blew back from the bonnet louvres. 1972 *Sci. Amer.* June 127/3 The temperature probes can be housed in a box that has louvers large enough to ensure the free circulation of air.

5. *louvre door, screen, slat*; **louvre damper,** a louvre the inclination of whose slats may be varied to regulate the flow of air through them.

1920 W. Neubecker *Pract. Sheet Metal Duct Construction* xv. 160 (*caption*) Louvre dampers for large size ducts. 1953 N. W. Kay *Mod. Building Encycl.* 410/1 Louvre door. 1967 *Boston Sunday Herald* 26 Mar. iii. 5/7 (Advt.), Set up louver door. 1948 C. Isherwood in *Penguin New Writing* XXXIV. 127 Movable louvre-screens take the place of port-holes. 1969 *Sears Catal.* Spring/Summer 910 Louver slats fitted into 1⅛-in. thick frame.

louvered *ppl. a.* (also, esp. in the U.K., **louvred**) (further examples of sense b).

1898 F. W. Macey *Specifications in Detail* 264 Louvred doors are generally required for ventilation in various places. 1934 H. M. Vernon *Princ. Heating & Ventilation* ix. 170 A fresh-air shaft may be necessary, with a louvered opening above the level of the building. 1950 *Engineering* 7 Apr. 376/2 The long-distance transport of fruit and vegetables is being investigated..with special reference to the performance of a new type of louvred van. 1969 *Daily Tel.* 14 Jan. 15/4 Wardrobes..are better with louvred doors. 1972 *Gloss. Electrotechnical, Power Terms* (*B.S.I.*) iv. iii. 21 *Louvered ceiling*, luminaire system comprising a large installation of louvres above which are mounted the lamps.

lovally: see *loverly *a.*[2]

Lovat (lɒ·vǎt). The name of a place in Inverness-shire, used *attrib.* and *ellipt.* to denote a muted green colour, a tweed (suit), or another garment or material of this colour.

1907 *Daily Chron.* 21 Nov. 7/4 The famous 'Lovat' shades of tweed are very popular among lady motorists. 1911 *Daily Colonist* (Victoria, B.C.) 5 Apr. 2/5 (Advt.), Sweater Coats—New consignment, white, navy, lovat and camel hair, for ladies and gentlemen. 1914 C. Mackenzie *Sinister St.* II. iii. xi. 719 This world of light-green Lovat tweeds, of fashionable rusticity. 1922 J. Buchan *Huntingtower* i. 16 A most disreputable tweed suit..had once been what..is called a Lovat mixture, but was now a nondescript sub-fusc. 1930 H. Nicolson *Diary* 22 Feb. (1966) 42 The Prime Minister appears in Lovat plus-fours. 1937 D. L. Sayers *Busman's Honeymoon* iv. 87 Will you wear the Lovats or the grey suit? 1940 *Illustr. London News* CXCVII. 640 (Advt.), Overcoats... A wide choice of brown, fawns, greys or lovat. 1953 J. Trench *Docken Dead* vi. 81 He wore a shapeless suit of Lovat tweed. 1969 J. Wood *Three Blind Mice* i. 9 He wore..lovat-green socks.

love, *sb.*[1] Add: **1. e.** Also *love from*...

1785 Lady Newdigate *Let.* May in A. E. Newdigate-Newdegate *Cheverels* (1898) iv. 67 Love from all here Adieu. 1913 W. Owen *Let.* 19 Oct. (1967) 202 Love, Hopes, and Kisses from your own Wilfred. 1921 A. Huxley *Let.* 21 Nov. (1969) 205, I will telephone or write about both these dates. Love from Aldous. 1949 D. Smith *I capture Castle* xi. 188 Dear Cassandra, it was nice of you to write... Love from Neil.

4. *for love* (later example in weakened sense); *love at first sight*: the action or state of falling in love with someone whom one has never previously seen; *love's young dream*: the relationship of young lovers; the object of someone's love, a man regarded as the perfect lover.

c 1374 Chaucer *Troilus & Criseyde* ii. 667 This was a sodeyn love, how mighte it be That she so lightly lovede Troilus Right for the firste sighte; ye, pardee? *a* 1593 Marlowe *Hero & Leander* (1598) i. 175 Where both deliberat, the loue is slight, Who euer lov'd, that lov'd not at first sight? 1822 Hazlitt *Table-Talk* II. xvi. 354, I do not think that what is called *Love at first sight* is so great an absurdity as it is sometimes imagined to be. *a* 1834 Love's young dream [in Dict.]. 1839 [see *beat *v.*[1] 10]. 1868 W. Collins *Moonstone* I. vii. 91 You have heard of beautiful young ladies falling in love at first sight, and have thought it natural enough. 1898 J. K. Jerome *Second Thoughts* 155 The stout lady, now regarded as a would-be blighter of love's young dream, was hustled into the back seat. 1903 Kipling *Traffics & Discov.* (1904) 132 'Do you want a tow to Brixham?'..'What for?'..'For love; for nothing.' 1920 Galsworthy *Skin Game* i. 33, I don't mean any tosh about love's young dream; but I do like being friends. 1937 D. L. Sayers *Busman's Honeymoon* xv. 307 There now!.. If there ain't love's young dream a-comin' up the path. 1952 *Scrutiny* XVIII. 273 We know that what we have here is no drama of romantic love-at-first-sight. 1961 C. McCullers *Clock without Hands* iv. 89 In early youth, love at first sight, that epitome of passion, turns you into a zombie. 1966 L. Southworth *Felon in Disguise* viii. 121 It calls for a sweetheart act, a proper 'Love's Young Dream' routine. 1975 D. Bagley *Snow Tiger* xvi. 138 Don't you believe in love at first sight?

7. a. *for the love of Mike*: for goodness' sake! (A colloq. exclamation of exasperation or surprise, with no notion of the literal sense; prob. f. Mike *sb.*[4])

1922 Joyce *Ulysses* 763 O move over your big carcass out of that for the love of Mike listen to him. 1925 T. Dreiser *Amer. Trag.* (1926) I. i. ix. 57 For de love o' Mike, will you listen to dat, now. 1934 J. Brophy *Waterfront* i. 14 For the love of mike..shut those blasted windows. 1935 W. D. Hubbard *Thousandth Frog* i. 7 Dick could not repress an exclamation of astonishment. 'For the love of Mike. Look at them.' 1941 *Penguin New Writing* VIII. 91 Well, for the luvva Mike! 1942 *R.A.F. Jrnl.* 3 Oct. 11 Tired? Well for the love of mike! What about me? 1957 A. MacNab *Bulls of Iberia* xv. 181 For the love of Mike, let's hope he's brave.

g. *to make love*: now more usually, to copulate.

1950 M. Peake *Gormenghast* xxix. 173 One of the Carvers made love to her and she had a baby. 1966 Auden *About House* 15 Stocktaking, horseplay, worship, making love. 1967 B. Wright tr. *Queneau's Between Blue & Blue* xiv. 151 When you make love on a bunk,..the man has to bump his head. 1971 *Daily Tel.* 15 Jan. 17/1 Couples who make love frequently are more likely to have sons than those who do so less often.

9. a. (Later examples as a term of endearing address.)

1895 A. W. Pinero *Second Mrs. Tanqueray* iii. 104 Paula love, I fancied you and Aubrey were a little more friendly. 1966 *New Yorker* 29 Jan. 22/3 'Sit over here, love,' he said as another actress entered. 1967 *Listener* 5 Oct. 429/3 Lovely, loves, I loved it. And Alison here was hysterical, weren't you, Alison? 1975 J. Symons *Three Pipe Problem* iii. 25 But Sher. love, it was only a read-through. You don't expect me to *act*.

f. *love in disguise*: (see quots.).

1877 E. S. Dallas *Kettner's Bk. of Table* 282 Love in disguise is a calf's heart stuffed, then surrounded with forcemeat, next rolled in vermicelli, lastly deposited in a baking dish..and sent to the oven. 1958 W. Bickel tr. *Hering's Dict. Cookery* 451 *Love in disguise*, calf's heart, soaked in water, larded, boiled until tender, dried, coated with veal forcemeat, rolled in crushed raw noodles, roasted in butter in oven and basted frequently.

15. a. *love-allegory, -bed* (later examples), *-bite, -bond, -charm, -dance, -drug, -duel, -duet, -flight, -game, -look, -lyric, -magic, -marriage, -poet, -poetry, -secret, -sonnet, -talk, -theme.*

1933 R. Tuve *Seasons & Months* iv. 189 All this is to be found in the love-allegory of the *Golden Targe*. 1934 Dylan Thomas *18 Poems* 19 Invisible, your clocking tides Break on the lovebeds of the weeds. *a* 1963 S. Plath *Crossing Water* (1971) 33 Musky as a lovebed the morning after. 1749 J. Cleland *Mem. Woman Pleasure* II. 63 Then the turtle-billing kisses, and the poignant painless love-bites. 1903 H. Ellis *Stud. Psychol. Sex* III. 71 We may find references to love-bites in the literature of ancient as well as of modern times... In the Indian *Kama Sutra* of Vatsyayana a chapter is devoted to this subject. 1972 *Daily Tel.* 29 Jan. 3/1 Once I saw her sitting in class with a love bite on her neck. 1951 L. MacNeice tr. *Goethe's Faust* ii. v. 295 Rapture which yearns ever, Love-bond which burns ever. 1889 *Cent. Dict.*, Love-charm. 1935 *Amer. Speech* X. 119/2 Certain generic epithets.. have become so conventionalized that they too tell their tales... Love thief, love nest, love lure, love charm, love potion. 1948 B. G. M. Sundkler *Bantu Prophets S. Afr.* vi. 222 Various Native 'Chemist' shops sell..love-charms. 1911 J. A. Thomson *Biol. Seasons* II. 233 The long larval period of two or three years in the water, and the short aerial love-dance lasting for an evening or two. 1934 *Discovery* Nov. 309/1 For a few weeks they [*sc.* the termites] revive their ancestral free-living life in a mad love-dance. 1959 *Chambers's Dict.* Suppl., *Love-drug*, dagga. 1969 *Rolling Stone* 17 May 3/4 The new 'love drug', MDA (3,4-methylenedioxy-phenyl-iso-propylamine). 1880 'Mark Twain' *Tramp Abroad* 58 This was not a love duel, but a 'satisfaction' affair. 1932 R. Campbell *Taurine Provence* ii. 44 The great 'Lou Pouvenco'..bore a small fortune between his horns, until he was killed in a love-duel by a younger rival. 1946 *Essays & Stud.* XXXI. 105 Anyone who takes the trouble to get the score of Verdi's *Otello*, and compare the love-duet at the end of the first act, in particular Otello's solo passages, with the last but two and last but one paragraphs of the *Anna Livia* episode, will discover some very interesting similarities in phrasing. 1975 *Times* 12 Feb. 23/6 The dance of Discord and War..has to be reconciled by the love duet. 1908 E. J. Banfield *Confessions of Beachcomber* i. vii. 232 The love flight of the green and gold butterfly. 1936 *Brit. Birds* XXIX. 307 The love-flights of many species depend on a subtle change in the character of the wing-beat, most marked perhaps in the waders. 1925 F. Harris *My Life & Loves* I. 182, I waited a little while and then began the love game. 1973 B. Freemantle *Goodbye to Old Friend* iv. 54 He wondered if Anne were playing some odd kind of love game. 1637 S. Rutherford *Let.* 10 June (1891) clxxv. 331 Any little communion with Him [*sc.* Christ], one of His love-looks, should be my begun heaven. 1904 *Windsor Mag.* June 305/2 Do you think I don't know a love-look when I see it? 1856 *National Rev.* III. 372 The love-lyric..is probably the most intense expression of primitive passion. 1962 L. Durrell *Spirit of Place* (1969) 19 He had just published a sequence of love-lyrics called *Kingcup*. 1974 P. Dickinson *Poison Oracle* ii. 44 You get a basic story, but inside it you get dramatic sections and love lyrics. 1949 M. Mead *Male & Female* iii. 56 How the ..human sacrifice or love-magic fitted into the whole. 1850 Thackeray *Pendennis* II. xxi. 209 Look at your love-marriages... The love-match people are the most notorious of all for quarrelling afterwards. 1973 *Archivum Linguisticum* IV. 93 A love-marriage (as opposed to an arranged marriage). 1923 J. M. Murry *Pencillings* 224 Love poets are seldom the singers of happiness in love. 1965 *New Statesman* 16 July 87/2 Donne..has a title to be our greatest love poet. 1865 *Reader* 20 May 561/2 Claim passionate tenderness as especially feminine, and the inquiry is made whether all the best love-poetry in existence..has not been written by men. 1872 Geo. Eliot *Middlem.* III. v. xlvii. 75 Verifying in his own experience that higher love-poetry which had charmed his fancy. 1971 *Guardian* 14 July 11/2 The best of his [*sc.* Attlee's] poems are political ballads... The others (not so good) run even to love poetry. 1754 Richardson *Grandison* I. xxxvii. 265 And has he, can he have, *so many* Love-secrets, and yet..not let them transpire to such a sister? 1923 R. Graves *Feather Bed* 25 This meek ex-novice rifled Of her love-secrets? 1870 D. G. Rossetti *Let.* 26 Feb. (1965) II. 804 The love-sonnets are the preponderant portion. 1958 Blunden *War Poets* ii. 15 In the pre-war poems of Brooke something like a premonition can be seen recurring. A love-sonnet dated 1909 powerfully includes it. 1862 G. Meredith *Mod. Love* xxxiii. 65 My wife, read this! Strange love talk, is it not? 1968 'N. Blake' *Private Wound* v. 69 She used none of the experienced woman's verbal tricks to arouse me, none of the shameless, titillating, love-talk. 1938 R. Graves *Coll. Poems* p. xxi, With the love-theme went the old fear-theme, sharpened rather than blunted by the experiences of peace. 1957 Manvell & Huntley *Techniques Film Music* i. 21 Examples of original music by Griffith and Briel included a prominent love-theme (for the Little Colonel and Elsie Stoneman).

c. *love-lighted, -lit, -starved* adjs.

1785 T. Dwight *Conquest of Canäan* (1788) iii. 78 For earth too bright were these love-lighted fires! 1904 *Daily Chron.* 9 Feb. 5/2 Peering through the pale miracle of spring at his violets,..his blear eyes love-lighted. 1855 J. R. Lowell *In-doors Out* in *Uncoll. Poems* (1950) 107 No glimmering beacon's love-lit rays Will homeward guide the wand'rer's feet. 1948 Blunden *Shakespeare to Hardy* (1964) 208 Here she is in her father's garden, flowering, love-lit, awaiting the slow old Nurse. 1909 *Westm. Gaz.* 24 July 2/1 Love-starved young Keats hath cast his gift of clay. 1955 *New Yorker* 25 June 59/1 The heroine is a love-starved American secretary.

16. love-affair (earlier *sing.* and later *fig.* examples); **love beads,** a necklace of coloured beads worn as a symbol of universal love; **love-book,** (*b*): delete *nonce-use* and add earlier and later examples; **love comic,** a comic (sense *B. 2) in which the principal ingredient of the stories is love; **love-curl,** a lovelock, esp. on the forehead; **love draught:** delete † and add later examples; **love-hate,** (orig. a psychoanalytic) term used to describe ambivalent feelings of love and hate existing towards the same object; freq. *attrib.*; so as *vb.*; also *love-hatred*; **love-interest,** a theme or episode in a story, film, etc., of which the main element is the affection of lovers; **love-juice** restrict † to sense in Dict. and add: (*b*) an aphrodisiac; (*c*) a sexual secretion; **love-life,** relations between the sexes as they affect a particular person; **love-nest,** a secluded retreat for (esp. illicit) lovers; **love-object,** the object on which love is centred; **love-pass** = *love-passage*; **love-passage** (earlier example); **love-play,** wooing, caressing, spec. **foreplay*; also *fig.*; **love-scene** (earlier and later examples); **love-seat,** a special form of arm-chair (also, of sofa) designed for two occupants; **love-spoon,** a wooden spoon, sometimes with a double bowl, carved for presentation to one's intended wife; **love-up** [cf. *Love *v.*[1] 1 d] *slang*, an act of caressing, hugging, etc.

1862 *National Rev.* XIV. 220 They have suggested that some irregular love-affair was unprosperous. 1969 *Times* 25 Mar. 9/4 The crazy world of Erogenous Zones [*sc.* a play] is the result of 25-year-old Mike Stott's love affair with American strip cartoons. 1974 *Times* 4 Dec. 17 This century's love-affair with the motor-car. 1968 *Daily Colonist* (Victoria, B.C.) 20 June 1/5 Love beads, draped on him by Pierre Trudeau, adorn former Prime Minister Pearson at Toronto political rally. 1969 R. Lowell *Notebk. 1967–68* (1970) 217 Our love-beads Rattling together to show that we were young. 1973 'B. Mather' *Snowline* xiii. 155 Weirdo fringed shirts, headbands, love beads..as unsavoury a bunch of love children as I have ever seen. 1587 F. Clement *Petie Schole* 36 Bookeloue I say, but I meane not louebookes, which..be the enemies of vertue. 1936 C. S. Lewis *Allegory of Love* iv. 172 Hence those strange comings and goings in every medieval love-book. 1951 M. McLuhan *Mech. Bride* (1967) 151/2 It recently shifted a large section of its enterprises from murder to love comics. 1970 G. Greer *Female Eunuch* 214 The market is contested by..love comics and fotoromance. 1850 H. Melville *White Jacket* II. xxxvii. 240 Many sailors, with naturally tendril locks, prided themselves upon what they call love curls. 1926 T. E. Lawrence *Seven Pillars* (1935) lxxxvi. 479 In command was young Metaab, stripped to his skimp riding-drawers for hard work, with his black love-curls awry. 1841 Borrow *Zincali* I. ii. i. 228 The women.. dealing in love draughts and diablerie. 1906 *Westm. Gaz.* 27 Aug. 3/1 The love-draught which Tristram and Iseult drink together. 1925 J. Riviere et al. tr. *Freud's Coll. Papers* IV. 79 So the second antithesis, love–hate, reproduces the polarity pleasure–pain, which is bound up with the former. 1937 H. Nicolson *Diary* 16 June (1966) 302 Goering..has the love-hate complex of the average German bourgeois for England. 1950 E. J. Simmons *Dostoevsky* xix. 315 Versilov's..love-hate relations with Katerina which conclude with his mad attempt to murder her. 1967 A. Wilson *No Laughing Matter* ii. 216 She love-hates him enough to be unable to leave him. 1972 Ld. Robens *Ten Year Stint* ii. 15 My personal relationship with the men and their leaders was schizophrenic—a sort of love-hate relationship. 1951 H. Hatfield *Thomas Mann* iii. 36 The protagonists in *Two Friends*, a novella of the love-hatred between a responsible burgher and a ne'er-do-well, afford a certain parallel to Thomas and Christian Buddenbrook. 1961 *Times* 18 Mar. 11/4 The love-hatred of Isolde for Tristan. 1892 H. James *Notebks.* (1947) 129 There must be a 'love-interest'—which is one and the same with the other parts of the situation. 1938 R. G. Collingwood *Princ. Art* v. 84 The cinema, where it is said to be a principle accepted by almost every manager that no film can succeed without a love-interest. 1961 C. S. Lewis *Exper. in Crit.* iv. 38 The story of excitement or mystery usually has a 'love interest' tacked on to it. 1973 *Time Out* 2–8 Mar. 59/3 'Love-interest' rears its inept head just as the medical satire should show its teeth. 1896 Farmer & Henley *Slang* IV. 241/1 Love-juice. 1965 J. Gaskell *Fabulous Heroine* 59 The sheets smelt of linen instead of love-juice. 1968 L. Berg *Risinghill* 122 'What is "love-juice"?' 'The liquid produced in the vagina of a woman when she is sexually excited.' 1972 'R. Crawford' *Whip Hand* i. ix. 54 She was drugged by love-juices and on the brink of sleep. 1972 *Pussycat* XXXIII. lix. 7/2, I could feel his lovejuice so hot, trickling down into the start of my stomach. 1919 M. K. Bradby *Psycho-Anal.* v. 59 The character and development of the infantile love for father and mother will have an influence on the whole love-life of later years. 1934 'R. West' *Mod. Rake's Progress* 74 Ecclesiastics..called out to sanctify the love-life of our puny little George. 1959 A. Christie *Cat among Pigeons* viii. 89 Even Games Mistresses may have their love lives. 1972 T. Ardies *This Suitcase* xiii. 140 He's the guy who's trying to break up my love life. 1919 U. Sinclair *Brass Check* xi. 65 So before long we began to notice dark hints in the newspapers; such esoteric phrases as 'Sinclair's love-nest'. 1970 G. Greer *Female Eunuch* 154 Nobody knew of his love-nest. 1972 'H. Howard' *Nice Day for Funeral* ix. 124 Pamela and Frankie were sharing a love-nest at Lakeland Towers. 1923 J. T. MacCurdy *Probl. Dynamic Psychol.* xvi. 191 The 'sentiment of love'.. consists in identification with the love object. 1925 J. Riviere et al. tr. *Freud's Coll. Papers* IV. 45 In the choice of their love-object they have taken as their model not the mother but their own selves. 1960 C. Day Lewis *Buried Day* vii. 137 When it became apparent that..as a love-object, I myself was unsatisfactory, she started on dogs. 1967 M. E. Romm in C. W. Wahl *Sexual Probl.* 221 Fetishism is..the utilization in sex of a substitute for the love object. 1973 S. Fisher *Female Orgasm* xv. 437 Orgasm difficulties were observed to be linked to concern about the instability or potential loss of love objects. 1872 Hardy *Under Greenwood Tree* I. i. viii. 113 Good luck attended Dick's love-passes during the meal. He sat next Fancy. 1845 C. M. Kirkland *Western Clearings* 106 No one..had ever been able to ascertain whether there had actually been any 'love-passages' between them or not. 1911 Love-play [see *courting *vbl. sb.* 3]. 1944 T. Rattigan *While Sun Shines* ii. 226 You're both very much mistaken if either of you imagines that you're going to have twopence-worth of verbal loveplay with my fiancée on my telephone. 1963 A. Heron *Towards Quaker View of Sex* 55 Adult heterosexuality presents fewer problems where early love play is tolerated than where it is suppressed. 1964 L. Nkosi *Rhythm of Violence* 46 Lili: Jo, I don't want to play. Jojozi: [tries to kiss her.] Not even love-play? 1639 Massinger *Unnat. Combat* iii. iii. 181, I will bring you Where you..may see The love-scene acted. 1818 *Theatrical Inquisitor* XIII. 183 Love-scenes..which both French and English writers..regard as absolutely essential to their drama. 1932 R. Campbell *Taurine Provence* 37 Read his [*sc.* Shaw's] miserable love-letters (published) and his 'love-scene' between Caesar and Cleopatra. 1975 *Country Life* 6 Feb. 326/2 Intimate, tender love-scenes. 1904 Love-seat [see *double chair* (*double *a.* A. 6)]. 1915 F. W. Burgess *Antique Furnit.* 205 Such settees which closely resemble an adaptation of two single chairs, are commonly called 'love-seats'. 1970 *Canad. Antiques Collector* Dec. 21/1 A Victorian love seat Mr. Daniel saw being hauled away in a garbage truck. 1973 'D. Halliday' *Dolly & Starry Bird* x. 151 Johnson.. kissed her, and then..found a love seat and dropped there beside her. 1918 W. R. Butterfield in *Connoisseur* Aug. 191/1 At first,..love-spoons did not differ greatly from the wooden spoons in ordinary use in the household. 1968 J. Arnold *Shell Bk. Country Crafts* 193 The Welsh carvers ..produced a great deal of fine work, amongst which were the celebrated love-spoons. 1972 *Country Life* 20 Jan. 160/2 These [*sc.* stay busks] were rather in the manner of Welsh love-spoons and were made by young men for their intended marriage partners. 1953 'Caddie' *Sydney Barmaid* xxv. 136 Come on, wot about a little bit of a luv-up? 1968 M. Allwright *Roundabout* ix. 65 He looked so beaten by the world that I wanted to gather him in my arms on the spot and give him a good love-up.

b. love-bush, the Jamaican name for Dodder *sb.* 1.

1814 J. Lunan *Hortus Jamaicensis* I. 266 Cuscuta Americana... The negroes of Liguanea mountains call it love-bush. 1954 *Farmer's Guide* (Jamaica Agric. Soc.) 582 The common Love-bushes of Jamaica comprise about four species of *Cuscuta*. 1962 S. Wynter *Hills of Hebron* ii. 35 Pale yellow tendrils of the 'lovebush' wrapping themselves around the prickly arms of the cactus.

love, *sb.*[2] [Of obscure origin.] One of a set of transverse beams supporting the spits in a smoke-house for the curing of herring.

1865 [see Spit *sb.*[1] 4 d]. 1895 A. Patterson *Man & Nature on Broads* 44 A savoury bloater, fresh down from the 'loves', is engrossing our own attentions. 1962 Granville *Dict. Sailors' Slang* 73/2 *Loves*, wooden splines in a herring curing loft on which the fish are suspended to dry.

love, *v.*[1] Add: **1. d.** With *up.* To caress, fondle; to engage in love-play with. *colloq.* (orig. *U.S.*).

1921 J. Dos Passos *Three Soldiers* ii. iii. 83 You said you were goin' back and love up that goddam girl. 1928 *Dialect Notes* VI. 62 If a hillman [in the Ozarks] does admit that he loved a woman he means only that he caressed and embraced her—and he usually says that he *loved her up*. 1932 K. S. Prichard *Kiss on Lips* 167 Why don't you give her a hug..love her up a bit? 1957 J. Braine *Room at Top* xix. 166 If you love me up, I'll be as warm as toast. 1968 M. Allwright *Roundabout* ix. 59, I never meant any harm: it was just as if he was a puppy I was loving up.

2. b. (Earlier and later examples.)

c 1810 W. Hickey *Mem.* (1913) II. i. 10 Lord love your honour, to be sure I will. 1821 Scott *Pirate* I. i. 15 But, Heaven love you, Mr Mertoun, think what you are purposing. 1898 J. D. Brayshaw *Slum Silhouettes* 1 Mister Bloomfiel'? Lor' lummy! there ain't no misters 'ere. *Ibid.* 141 Mine? Lor' luv a duck! No, that's Sal Hogan's little lot. 1916 'Taffrail' *Pincher Martin* xii. 218 'Lord love us!..d'you mean to say'—Words failed him. 1922 Joyce *Ulysses* 176 Lord love a duck, he said, look at what I'm standing drinks to! 1934 T. S. Eliot *Rock* ii. 65 What's that? Lor-love-a-duck, it's the missus! 1938 'J. Bell' *Port of London Murders* xv. 247 ''Lor love us!' I says to meself. 'Something's up.' 1954 W. Sansom (*title*) Lord love us. 1955 M. Allingham *Beckoning Lady* iv. 55 Orf come 'is 'at, and lord luva duck!

d. *he loves me, he loves me not*, etc.: a formula used in divining-games. Also *transf.*

[1909 A. E. Gillington *Old Hampshire Singing Games* facing p. 1 Then they say 'David Bailey' (Boy's name), Do you love me? Yes, No, etc. till the skipping girl stops.

The two trilling the rope say—'Alma Bailey' (Girl's name), Do you love him? Yes, No, etc.] **1946** A. Uttley *Country Things* v. 64 He loves me. He don't. He'll have me. He won't. He would if he could, But he can't. **1959** I. & P. Opie *Lore & Lang. Schoolch.* xv. 339 Much energy and calculation is devoted to skipping through the alphabet... the following sequence being that used by an 11-year-old Portsmouth girl: Does he love me? Yes, no, yes, no... Will he marry me? Yes, no, yes, no. **1971** *Guardian* 10 July 11/2 Eric Lubbock's private game of 'he loves me, he loves me not' with press and politicians is coming to a blessed end.

e. Phrases. *an* (or *as*) *you* (or *thou*, etc.) *love me*, if (or since) you love me: used as an imprecation; *to love and leave you*: a formula of departure; *love them and leave them*, etc.: seduce and abandon women.

1818 Carlyle *Early Lett.* (1886) 148 Send a letter quickly, an thou love me. **1823** J. F. Cooper *Pioneers* I. i. 28 Natty—you need say nothing of the shot, nor of where I am going—remember, Natty, as you love me. **1885** R. Holland *Gloss. County of Chester* 212 *Love you and leave you*, a common saying when any visitor is going to take his departure. 'Well a' mun *love ye, and leave ye*.' **1917** 'S. Rohmer' *Si-Fan Mysteries* xxxv. 264 But in waiting for one who is stealthily entering a room, don't, as you love me, take it for granted that he will enter *upright*. **1930** *Amer. Speech* Dec. 92 Love 'em and leave 'em. **1938** W. G. Hardy *Turn back River* 33 Love 'em and leave 'em; that was the idea. **1946** K. Tennant *Lost Haven* (1947) xvi. 259, I wouldn't try to keep me if I was you... Love me and let me leave. **1960** K. Amis *Take Girl like You* ii. 36 I'm afraid I shall have to love you and leave you. **1967** J. Morgan *Involved* 11 'Dewi, I have to love you and leave you,' Frankie said. 'I'm supposed to be on duty.' **1975** H. McCutcheon *Instrument of Vengeance* vii. 123 'I have many interests.' 'But no girls? ..You just love them and leave them, no?'

loveable, etc.: see Lovable, etc.

love-bird. Add: **2.** A lover.

1911 *Maclean's Mag.* Nov. 39/2 Seems as if I'd lighted on a pretty nest of love-birds. **1949** A. Hynd *We are Public Enemies* iv. 121 Ma barged in on the love birds. **1974** J. Mitchell *Death & Bright Water* xx. 242 'Lovebirds, lovebirds,' Randy Blythe said. Callan sat up, one arm still round Helena.

loved, *a.* Add: **1. c.** *loved one*. (i) A beloved, a lover; *pl.*, one's family or relations. (ii) A dead relation (spouse, etc.). Freq. with capital initials.

1862 M. B. Chesnut *Diary* 9 June (1949) xiv. 240 How many, many of your friends and loved ones this scrap of paper may tell you have gone to their death. **1879** Geo. Eliot *Let.* 20 Feb. (1956) VII. 104, I want, if I can, to write a '*characteristik*' of my loved one—no memoir, but a brief sketch. **1906** Chesterton *Charles Dickens* viii. 187 To ask for the loved one, and then not to dare to cross the threshold. **1926** A. Huxley *Jesting Pilate* iv. 272 Lay the Loved Ones to rest in —Graveyard, the Cemetery Unusual. **1938** D. Castle *Do Your Own Time* iii. 35, I sat idly drawing designs on my writing-paper while nineteen heartsick men began writing to their loved ones. **1948** E. Waugh (*title*) The loved one. **1968** *Guardian* 17 Aug. 8/1 Montpellier has just acquired..Europe's first funeral parlour on the American model..the Loved Ones make their final appearances in rooms done up in sky blue, water green, pink, or beige. **1971** *Progress* (Cape Town) May 8/3 In many cases these people would have to travel more than two or three hundred miles to be with their loved ones. **1973** *Advocate-News* (Barbados) 22 Jan. 12/1 (Advt.). For a smaller charge, we do not only intimate the death of your loved ones in our Classified Columns we also give a lasting record of the event for posterity.

love-feast. **2.** and **3.** (Later examples.)

1870 [see *Ebenezer 1]. **1876** *Salem Republican* (Suisun, Calif.) 24 Aug. 2/1 A regular old-fashioned Democratic love-feast was engaged in by the many-scarred warriors of the party. **1882** *19th Cent.* Nov. 732 They who turn aside to attend a Ranters' Love-feast..must be wrong in the head. **1909** F. Calhoun *Miss Minerva* 96 She was always on hand at the Love Feast and the Missionary Rally. **1922** G. Edwards *From Crow-scaring to Westminster* iii. 35 One form of [Primitive Methodist] service was called a 'love-feast' at which small pieces of bread were taken round with water. **1943** N. Tennant *Ride on Stranger* iii. 28 Aunt Edith beamed over the love-feast. **1948** *Minneapolis Star* 17 Sept. 13/1 Senator Joseph H. Ball and Gov. Luther Youngdahl had a sort of love feast at the capitol Thursday. **1973** 'S. Harvester' *Corner of Playground* i. v. 47 Several of their customs had Egyptian overtones, including a 'love-feast'..close to the Pharaonic *agapes*.

love-in (lʌ·v-in). [*-in³.] A gathering for the purpose of establishing and enjoying love relationships.

1967 *Times* 28 Mar. 4/7 The 'love-in' in Elysian Park, Los Angeles, was equally odd but caused no more than a traffic jam. **1967** *Observer* (Colour Suppl.) 3 Dec. 12/1 The imitation 'Love-in' at the Alexandra Palace..attracted 'horrible yobs'. **1971** N. Butterworth *Flowers for Dead Witch* iii. 27 We had a few love-ins up on the cliffs.. We'd go for moonlight swims, and then dry off with a little horseplay.

lovekin (lʌ·vkin). Also **lovekins**. [f. Love *sb.* + -kin.] A lover.

1922 Joyce *Ulysses* 418 Reels off a credit. Lovey lovekin. **1925** T. Dreiser *Amer. Trag.* II. ii. xliii. 36 He'll be with me most of the time—the lovekins will.

Lovelace (lʌ·vlēs). The name of Robert Lovelace, a character in Richardson's *Clarissa Harlowe* (1747–8), used allusively for 'a seducer'.

[**1751** Richardson *Clarissa* (ed. 3) VIII. 294 Ladies..should rather prefer the honest heart of a Hickman..than the volatile mischievous one of a Lovelace.] **1812** Shelley *Let.* 11 June (1964) I. 305, I regard charges of resembling Lovelace with contemptuous indifference. **1850** Thackeray *Pendennis* II. ix. 92 If Arthur had been the most determined roué and artful Lovelace who ever set about deceiving a young girl, he could hardly have employed better means. **1944** S. Putnam tr. *E. da Cunha's Rebellion in Backlands* II. 211 A scandal in which a certain local bigwig, a police sergeant, was *magna pars*, the Lovelace of the episode.

lovelify (lʌ·vlifai), *v. rare.* [f. Lovel(y *a.* + -ify.] *trans.* To render lovely. So **lo·velified** *ppl. a.*, **lo·velifying** *vbl. sb.*

1897 G. B. Shaw *Our Theatres in Nineties* (1932) III. 73 Life, death, love and mankind are no longer themselves; they are glorified, sublimified, lovelified. **1935** *Punch* 17 July 65/1, I have heard of a 'Hairdresser and Beautician' who offers to 'lovelify' the ladies of Manchester; and, from New York of a Youthifying Beauty Cream. And here is a pretty piece from *The Windsor Magazine*:—'A hand-cream..has a lovelifying effect on hands roughened from gardening.'

lovely, *a.* Add: **3. a.** *absol.* or *sb.* A woman or girl of glamorous loveliness, esp. one who takes part in an entertainment or 'show'. Also *transf.*

1936 [see *privvett]. **1938** Auden & Isherwood *On Frontier* III. ii. 108 If..the working class prefers our larger and lovelier organs of enlightenment, which can afford snappier sports news..and bigger photographs of bathing lovelies. **1940** H. G. Wells *Babes in Darkling Wood* II. i. 143 Not for many years have I had that hungry craving for everything, give and receive, from another human being. I can't imagine the man. What a marvel, what a lovely! **1957** J. Braine *Room at Top* xvi. 150, I was taking Susan not as Susan, but as a Grade A lovely, as the daughter of a factory-owner. **1966** T. Pynchon *Crying of Lot 49* iii. 63 One of the girls, a long-waisted, brown-haired lovely in a black knit leotard. **1967** *Spectator* 3 Mar. 7/1 The six lovelies comprising Marie de Vere's Ballet Montparnasse. **1974** *Times* 22 Jan. 11 Gone are the memorable parade of velvet-faced classic lovelies, each indistinguishable from the other.

lover¹. Add: **2. c.** A pimp. *U.S. slang.*

1904 'No. 1500' *Life in Sing Sing* 250/1 *Lover*, a man who receives support from a prostitute. **1960** R. I. McDavid *Mencken's Amer. Lang.* 727 A pimp is a..McGimp, fish and shrimp, lover, Latin lover and many others.

d. As a form of address (to a lover, or casually). *colloq.* (orig. and chiefly *U.S.*).

1911 G. S. Porter *Harvester* x. 194 'Hello, lover!' cried Doctor Carey.. 'Are you married yet?' **1920** F. Scott Fitzgerald *This Side of Paradise* II. i. 209 *Rosalind*: Lover! Lover! I can't do with you, and I can't imagine life without you. **1959** N. Mailer *Advts. for Myself* (1961) 289 Maybe I wouldn't hear all the jazz you hear, lover, but I can't help it. **1966** D. Heyes..'Never (1964) v. 31 'You got it, lover,' the waitress said. **1966** L. Deighton *Billion-Dollar Brain* iv. 41 It's nothing like that, lover. I'm not going to get myself hurt. **1972** J. Burmeister *Running Scared* iv. 72 She swung her legs off the bed. 'How about some coffee, lover?'

4. lover boy, man *slang* (orig. *U.S.*), a lover, an attractive man, a woman-chaser; also used as a form of address; **lovers' lane**, a road or any other secluded place to which lovers resort; **lover's (lovers') quarrel**, a dispute between lovers; also *fig.*

1677 (*title*) The lovers quarrel: or, Cupids triumph. **1882** C. Lamb *Let.* 9 Jan. (1935) II. 384 Henceforth I interest all my thoughts in your commendable employment; look upon them as lovers' quarrels. **1842** Lytton *Zanoni* II. II. 137 The mysterious warning of Zanoni then suddenly occurred to him; he had forgotten it in the interest of his lover's quarrel with Viola. **1881** *Golden Gate Gaz.* (San Francisco) 26 Oct. 2/2 Sunday afternoon as a young lady and gentleman were promenading through 'Lovers' Lane' they were attacked by a ferocious dog. **1918** Harris *What Every Woman Knows* ... There is a romantically gloomy place called the wall of the village called the Lovers' Lane. **1941** *Times* 25 Aug. 4/1 In England, a lovers' quarrel took 50 years to cool off, finally got the knot tied. **1942** R. Frost *Witness Tree* 26 I had a lover's quarrel with the world. **1947** *News of World* 26 Jan. 3/6 Her..body..was found in a 'lovers' lane' on an empty building site. **1952** S. Ellin *Key to Nicholas Street* i. 14 Here was lover boy walking around with milady's key. **1958** M. Procter *Man in Ambush* xvi. 196 Lover boy's been talking to you. Take no notice. **1965** C. Williams *Man in Motion* ii. 26 He's a Lover Boy, one of these big, flashy, conceited types that has to..drive all the girls a break. **1961** H. S. Turner *Something Extraordinary* ii. 28 Clive..has never been able to stop himself calling him 'loverboy' in times of scorn. **1966** J. Pearl *Crucifixion P. McCabe* (1967) ix. 136 How would I know that I'd find Donna Lord parked in a lovers' lane at Briarwood Lake? **1968** L. Deighton *Only when I Larf* xvii. 226 'There's no hurry, loverman,' she said. **1972** F. Warner *Lying Figures* II. 9 Out on the prowl tonight, lover-boy? **1972** J. Brown *Chancer* xv. 208 Our arty friend, lover man, he was running a photography business. **1974** *Daily Tel.* (Colour Suppl.) 18 Jan. 34/4 Elisabeth Foster's body was found on January 3, 1972, under a hedge by a 'lovers' lane' near the picturesque village of Wrea Green.

lovering (lʌ·vəriŋ), *vbl. sb.* [f. Lover¹ + -ing¹.] Courting, fondling. Also *attrib.*

1884 in R. Lawson *Upton-on-Severn Words* 22. **1907** *Daily Chron.* 27 Nov. 3/3 Where the schoolboy demands gore in his books, she asks for 'lovering', as she calls it. *Ibid.*, Let him only think of the possibilities of that new 'lovering' school, where the headmaster kept a 'lovering table', at which dined the spoony couples. **1919** Galsworthy *Saint's Progress* I. ii. 20 Between these two young people no actual word of love had yet been spoken. Their lovering had advanced by glance and touch alone. **1922** C. Rae *Gate Curlew* xi. 178 She's nae objections to a bit of lovering behind her sister's coats. **1926** R. Macaulay *Crewe Train* ii. xv. 169 People like the lovering couple usually stayed out..for long hours on end.

loverly, *a.*ᵃ Also **lovally**. Repr. a Cockney pronunc. of Lovely *a.*

1905 [see *jellied a. 3]. **1937** in Partridge *Dict. Slang* 497/2. **1956** Lerner & Loewe *My Fair Lady* (1958) i. i. 27 All I want is a room somewhere, Far away from the cold night air.. Oh, wouldn't it be loverly? **1958** J. Wainwright *Web of Silence* 28 He 'ad the ackers—believe me—we 'ad a lovely time, I can tell that... A loverly job, it was.

lovership. (Earlier example, used as a form of address to a lover.)

1837 F. Trollope *Vicar of Wrexhill* III. xiv. 325 Your lovership must excuse me if I declare that it is my intention to accompany the young lady myself.

lovescape (lʌ·vskēip). *rare.* [f. Love *sb.* after Landscape.] A view or prospect of love.

1876 G. M. Hopkins *Wreck of Deutschland* xxiii, in *Poems* (1967) 59 With the gnarls of the nails in thee, niche of the lance, his Lovescape crucified. **1969** *Punch* 19 Feb. 286/1 It is finely elegiac and the townscapes and lovescapes are deeply felt.

loveward(s (lʌ·v,wɔəd(z)), *adv. rare.* [f. Love *sb.* + -ward or -wards.] Towards love.

1927 Joyce *Watching Needleboats at San Sabba* in *Pomes Penyeach*, I heard their young hearts crying Loveward above the glancing oar. **1957** G. M. Brown *Fisherman with Ploughs* 76 The children of the valley Drifted lovewards.

lovey-dovey (lʌvi,dʌ·vi), *sb.* and *a.* [f. Lovey + Dovie, dovey.] **A.** *sb.* = Lovey; also, brotherly love.

1819 [see Dovie, dovey]. **1904** *Daily Chron.* 26 Mar. 6/5 We will love, goodie as much as we can, lovey dovey. **1946** H. L. Mencken *In Life's Aug.* 46/2 And bring in a reign of peace, prosperity and lovey-dovey.

B. *adj.* Fondly affectionate; namby-pamby.

1886 J. Ashby-Sterry *Lazy Minstrel* 132 ..lovey-dovey by the blind which mamma prefers, but which I think looks lovey-dovey. **1900** H. Lawson *On Track* 63 Just as lovey-dovey talk is important to her and nonsense to you. **1967** 'D. Shannon' *Chance to Kill* (1968) vi. 91 You want to get all bound-minded and lovey-dovey with the dingos, O.K., but I got a right to feel how I like too. **1969** *Sunday Times* (Colour Suppl.) 19/3 There had been a fight. 'There was none of the usual lovey-dovey stuff between them.' **1973** 'H. Carmichael' *Too Late for Tears* viii. 99 This woman hadn't been all lovey-dovey with her husband... They'd been squabbling for months.

So **lo·vey-do·veyness** [-ness], a state of maudlin sentimentality.

1923 D. H. Lawrence *Stud. Classic Amer. Lit.* x. 210 He [sc. Melville] wanted the lovey-doveyness of perfect mutual understanding. **1968** *Punch* 29 May 788/3 The food pleasures, dear sports and general lovey-doveyness of family life.

loving, *ppl. a.* Add: **4. Comb.**, as *loving-heartedness*; *loving-hearted, -kind, -kindly* adjs.

1903 *Mary Dymock* I. v. vi. 33 In its early, loving-kindly days of gracious purpose. **1909** *Westm. Gaz.* 27 Feb. 4/3 The loving-hearted and most tender missionary who was head of the Conservatoire at Naples. **1909** R. Brooke *Coll. Poems* (1918) 99 Quiet and strange, and loving-kind, you sleep. **1926** *Contemp. Rev.* Feb. 226 All you may have been the orator, which never makes for loving-heartedness. **1960** *Clergy Rev.* Jan. 14 More's way, detached, peaceful and loving-kind, must have set him dreaming.

low, *a.* and *sb.* Add: **A.** *adj.* **2. f.** Also *Comb.*, as *low-back, -central, -front, -mid, -rising*, used chiefly attributively or quasi-adj.

1888 H. H. Palmer *Gram. Spoken Eng.* i. 13 Low-Rising: Nucleus-tone. **1932** J. J. Hogan *Outl. Eng. Philol.* 14, *ᴂ*: low-front. **1934** Webster, Low-back, low-central. **1944** T. H. Wetmore *in Studies in Speech & Drama in Honor of A. M. Drummond* 244 (*title*) The dialectal significance of the non-phonemic low-back vowel variants before R. **1952** Z. S. Harris *Methods Struct. Ling.* vi. 57 Low-rising intonation. **1962** *Amer. Speech* XXXVII. 163 The low-central free vowel /ɑ/. **1964** Low-front [see *high a. 4 b]. **1966** *Language* XLI. 306 /a/ low-back /ɔ/. *Ibid.*, In disyllabic words, Ngbaka shows only four sequences: high-mid, mid-high, mid-low, and low-mid. **1970** *Publ. Amer. Dial. Soc.* liv. 13 Sometimes.. it occurs as a retracted low-front or as a low-central monophthong.

h. Phr. *low to paper*: of type, of less than normal height.

[**1683–4** J. Moxon *Mech. Exerc. Printing* (1962) 346 Low against Paper.] **1888** *Encycl. Brit.* XXIII. 698/2 Types lower than the ordinary dimension are said to be *low to paper*, and if surrounded by higher types will not give a perfect impression. **1922** D. B. Updike *Printing Types* I. ii. 34 The standard height-to-paper is 0·918 inch.

Types exceeding or falling short of this measurement are termed respectively 'high-to-paper' and 'low-to-paper'.

7. c. (Later example.)

1921 *Chambers's Jrnl.* Aug. 533/1 He may feel that he is the superior in every way of some of the 'low white' with whom he comes into daily contact.

12. b. *low on*: deficient in, short of. *colloq.*

1966 *Listener* 23 June 926/3 Low on credibility however was *They Were So Few*. **1969** *Ibid.* 22 May 732/2 Her difficulty is that she is incapable of fulfilling herself, partly because she seems to be low on energy. **1974** J. Wainwright *Evidence I shall Give* xxxv. 200 He brewed instant coffee. He was low on sugar.

20. *low-altitude, -angle, -budget, -calorie, -consumption, -contrast, -cost, -density, -drag, -energy, -fat, -field, -flux, -impedance, -income, -intensity, -noise, -price, -rank, -rental, -risk, -status, -sulphur, -temperature, -tension* (later examples), *-value, -velocity, -voltage, -wage, -wattage, -wing*.

1938 R. W. G. Kingston in E. F. Norton *Fight for Everest 1924* 288 Low-altitude deserts. **1966** *Electronics* 3 Oct. 181 Doemler System GmbH last year made a successful low-altitude recovery with a parafoil that unlocks its wings for descent. **1971** C. Bonington *Annapurna South Face* iii. 34 He [sc. Kelvin Kent] had also organised four all-Gurkha porters, whose job it was to carry all the gear up to Base Camp. **1907** *Yesterday's Shopping* (1969) p. xiii/1 Low-angle Instrument Case. **1974** *Nature* 21 Mar. 300/1 Low-angle X-ray patterns from two hands of different rank. **1969** *Electronics* 14 Nov. 48 An antenna lacking the low-angle coverage..would not be able to communicate with the satellite unless the plane flew farther south. **1958** R. Sill in *Film Daily Yearbk. Motion Pict.* 985/1, 20th Century Fox..proclaimed that it has budgeted $65,000,000 for the production of 65 features this year, including 13 from..Regal Films, which specialize in low-budget films. **1962** B. Faulk *Brit. Broadcasting in Transition* vii. 129 Audiences seem to show no more enthusiasm for high-cost television than low-budget radio features. **1975** E. Bullins *Theme is Blackness* 129 Short, low-budget films that could be distributed nationally. **1969** *Lancet* 8 Dec. 1190/2 On low-calorie high-fat diets, fat and thin subjects developed the same levels of ketones in urine. **1969** *Sunday Times* 1 June 26 Apr. 28/1 Jars of low-calorie instant soup. **1960** *Farmer & Stockbreeder* 8 Mar. 81/1 The north-east, a low-consumption area. **1973** *Times* 12 Mar. (Italian Industry Suppl.) p. vi/1 Petrol has never been cheap in Italy, so low-consumption motors were developed there. **1966** G. D. Brandon *Mod. Techniques Metallogr.* ii. 14 The curves are for a 'soft', low-contrast, fast emulsion and a 'hard', high-contrast, slow emulsion are shown. **1951** H. L. Ickes *Diary* to Mar. (1953) I. 152 Obviously it wasn't a model of low-cost housing for people on the very lowest rung of the economic order. **1962** *Times* 17 Dec. (Canada Suppl.) p. xiii/7 Efficient and low-cost producers. **1969** *New Statesman* 22 Apr. 396/2 There is no cheaper way of running a radio station than setting a low-cost young man in a modern studio to talk about his breakfast and his fan mail. **1966** *New Left Rev.* May–June 70/2 Low density, two-storey three-bedroom housing. **1962** F. I. Ordway et al. *Basic Astronautics* iii. 78 Venus is [sc. the Moon] is a low-density world, there is little likelihood of extensive concentrations of metallic iron and nickel. **1966** *N.Y. Amsl. Sci.* CXV. 569 The resulting digital data, when stored on low-drag (reduced-pressure per inch) digital tapes. **1962** F. I. Ordway et al. *Basic Astronautics* ii. 48 An aerodynamic shroud..provides a low-drag housing for the entire vehicle. **1973** *Flying* Apr. 2/1 A low-drag high-speed wing. **1942** I. D. Stranathan *Particles of Mod. Physics* xii. 535 The mean free path may be longer, and the mean life correspondingly longer, for high energy mesotrons than it is for low energy mesotrons. **1966** *Farmer & Stockbreeder* 8 Mar. 149/2 A low-energy ration contained 750 cal per lb. **1969** *Times* 20 Sept. (Pure Food Suppl.) p. i/ Pies, burgers.. were rated lower than the standard the start of the experiment. **1972** J. Ball *Five Pieces Jade* xii. 150 He poured some cornflakes into a bowl, covered them with..low-fat milk. **1962** Corson & Lorrain *Introd. Electromagn. Fields* iii. 91 All that varies from one O' to another is the fact that sometimes the point lies in a high-field region, within a molecule, for example, and other times it lies in a low-field region. **1956** *Nature* 4 Feb. 202/1 The report of experimental facilities in [the nuclear reactor] DIDO may vary from 1:20 a day down to [2:0 a day for those in less-attractive low-flux positions. **1962** Simpson & Richards *Physical Princ. Junction Transistor* ii. 18 The device thus passes most of the emitter current from a low-impedance generator to a high-impedance load. **1952** *Ld. Lamb Village* xvi. 179 Roy Wilson may start in a low-income group, but if he can make a success, well, he's worth more than people who can't. **1971** *Guardian* 7 July 24/4 Up to 60,000 children in low income families have lost their right to free school meals. **1953** N. Tinbergen *Herring Gull's World* xiii. 209 Low-intensity copulations can be seen early in the season. **1964** P. Delattre in D. Abercrombie et al. *Daniel Jones* 53 Very low-intensity vowel. **1971** D. E. Westlake *I gave at the Office* (1972) 14 Marijuana was legal in the United States until 1935, and it was just a sort of low-intensity fact in American culture. **1961** *Times* 3 Apr. 6/4 Special 'maser' low-noise amplifiers will be used in the reception of the very weak signals from the satellites. **1861** H. Hudson *Martin Tupper* (1949) x. 121 Shepherd's mammoth low price clothing emporium, Chestnut street above Third. **1944** A. G. Hatcher in *Mod. Lang. Notes* LIX. 516, I have seen..low-price shoes. **1968** D. E. Allen *Brit. Tastes* viii. 204 Low-price catering. **1962** *Times* 8 Apr. 6/4 Special 'low-rank', 'low-rank' coals. [**1963** A. J. Moffat..] **1968** *Times* The decreased intensity of the diffuse scattering [of X-rays] at very low angles for the low-rank coals. **1965** *Times Lit. Suppl.* 25 Nov. 1057/3 Artistic low-rental areas. **1951** S. A. Stouffer in Parsons & Shils *Toward Gen. Theory Action* iv. v. 492 Low-risk and high-risk situations. **1956** J. Klein *Study of Groups* vii. 102 One thus obtains four types of sub-groups: high-status mobile, and non-mobile; low-status mobile, and non-mobile. **1974** J. Burnett *Useful Toil* i. 31 Farm labouring gradually sank during the course of the [nineteenth] century into a low-status occupation. **1968** G. J. Williams *Econ. Geol. N.Z.* xviii. 322/1 The coal is a low-sulphur lignite of lower grade than the lignites of Benbar and Mataura. **1909** *Installation News* III. 133/2 The low-temperature system of heating by means of convectors. **1957** R. Hoggart *Auden* iii. 32 He seems to be aiming not at a widely-acceptable demotic speech but at a low-temperature verse of intelligent observation and comment. **1948** *Penguin New Writing* XXXV. 93 Poetry of a kind which is at present particularly in need of rescue from oblivion: what may be called low-tension poetry. **1957** *Railway Mag.* Mar. 159/2 A cubicle containing the majority of the low-tension control equipment is housed in a compartment in the motor coach. **1962** A. Battersby *Guide to Stock Control* viii. 76 At the other end of the scale, spare parts should be in the low-value, long-residence class. **1956** *Nature* 21 Jan. 110/2 In areas where deep weathering or drift prevents difficult low-velocity-layer correction problems, the reflexion times seem to a known shallow marker-bed will give the necessary corrections for the deeper horizons. **1975** N. Luard *Robespierre Serial* vi. 45 A low-velocity handgun. **1922** Glazebrook *Dict. Appl. Physics* II. 748/1 On low-voltage alternating-current systems these [switches] consist of movable blades bridging two contacts. **1971** *Times* 19 Mar. 8 There had been a suggestion that rival import from the low-wage plants, like Japan. **1962** A. Wise *Death's Head* xi. 131 A few low-wattage bulbs. **1950** W. Faulkner *Hist. Dream* xix. 314 Erotic light..suffused the room, cast by the low-wattage lamp. **1959** *Listener* 9 Aug. 23 Rival 'wing cars' were just a temporary low-wing monoplane. **1950** Glos. *Aeronaut. Terms* 69 Low-wing monoplane, a monoplane in which the main supporting surfaces are at or near the bottom of the fuselage.

21. *low-ceiled, -eaved, -powered, -priced* (later examples), *-sided, -sized, -studded, -toned* (later examples), *-vaulted, -voiced* (later examples), *-waisted*.

1904 W. de la Mare *Henry Brocken* 106 The room in which we sat was low-ceiled and cheerful. **1905** *Westm. Gaz.* 20 May 5/3 Low-eaved houses, cobbled streets, and quiet squares. **1903** *Daily Chron.* 3 Aug. 3/7 A motor-car, however low-powered or slow it may be. **1935** R. Lehmann *Echoing Grove* xii Put a frame round the amorphous semi-transparent mass of human vapour and it will seem himself. **1972** G. Norris' *Spt. Cliff rings True* xv. 214 The drive led..to be done with a standard lamp. **1895** *Montgomery Ward Catal.* 5/1 This is one of the best low priced settings that we ever received. **1933** *Radio Times* 28 Sept. 22/2 Headphones specially low priced at 24/- per pair. **1874** J. W. Long *Amer. Wild-Fowl Shooting* 75 A small, low-sided boat..might be..dangerous on large waters. **1882** Noall & Milner *Glos. Cornish Dial.* 186 Low-sized, little, short of stature. **1907** Joyce *Let.* 1 Mar. (1966) II. 218 There were low-sized and sincerely..in. **1938** S. Beckett *Murphy* 10 A low-sized corpulent middle-aged woman. **1854** B. P. Shillaber *Life & Sayings Mrs. Partington* 121 The rooms..did not stand erect in the low-studded room. **1884** Low-studded [see Studded *ppl. a.* 6]. **1934** E. Wharton *Backward Glance* x. 231 It was a tiny garden patch, and a few steps brought us to the door of a low-studded cottage in a pea, graw was in the low-toned building. **1923** *Amer. Mercury* I. 81 Two *Fur Jacket* ..is..exceedingly low-toned. **1948** *B.B.C. Gloss. Broadcasting Terms* 18 Low-toned, manifesting a form of frequency distortion in which the lower audio-frequencies are accentuated. **1969** *Word* XXV. 155 Frequently derive..sequence low-low-toned. **1869** J. R. Lowell *Foot-Path* in *Under Willows* 225 Those angel stairways in my brain, That climb from these low-vaulted days To spacious sunshines far from pain. **1895** A. W. Pinero *Second Mrs. Tanqueray* ii. 63 She is a low-voiced, grave girl. **1953** 'N. Blake' *Dreadful Hollow* i. 18 He stared at her, a low-voiced, pleasant, anxious girl with kind eyes. **1923** A. Huxley *Antic Hay* ix. 124 Low-waisted jumper dresses. **1966** *Guardian* 27 July 6/4 Long, lean, low-waisted jumper top. **1974** *Country Life* 24 Oct. 1242/3 The long, low-waisted black jersey evening dress.

22. *low-burning*.

1966 E. Rickert *Reaper* 8 His mother sat by the low-burning peat. **1974** E. Hardwick *Seduction & Betrayal* 231 Miss Milman..is leading on..to a low-burning encounter with death.

23. *low-alloy a. Metallurgy*, containing a small proportion of alloying elements; **low-bush** *N. Amer.*, used *attrib.* to designate low-growing plants or their fruit, esp. **low-bush blackberry**, one of several species of *Rubus*; **low-bush blueberry**, a variety of *Vaccinium angustifolium*; **low-bush cranberry**, in Canada, *Viburnum edule*; **low comedy**, also *attrib.*; **low definition**: see *Definition 5 c; **low-fidelity** *a.*, characterized by an absence of 'high fidelity' (see *High fidelity); **low gear**: see *Gear *sb.* 7 b; **low grade** *Philol.* [Grade *sb.* 9 a], a reduced form, generally represented by [ə], in an ablaut series; **low-key** *Photogr.* (see quot. 1959); also *attrib.* and *fig.*, esp. muted, restrained, of modest ambition; so **low-keyed** *a.* (in *fig.* senses); **low-loader**, a lorry with a low deck (usu. without sides) to facilitate loading of heavy loads, esp. other vehicles; **low-loss** *a. Electr.*, characterized by or causing little dissipation of electric or electromagnetic energy; **low maple** *U.S.*, the mountain maple, *Acer spicatum*; **low-molecular** *a. Chem.*, having a low molecular weight; **low-neck**, a low-necked dress; **low-pass** *a.*, denoting a filter that attenuates components with a frequency greater than some cut-off frequency and passes components of lower frequency; **low profile**, (*a*) *attrib.* of a motor-vehicle tyre, having a relatively great width in proportion to its height; (*b*) (of or pertaining to) a low-keyed and unobtrusive policy, restrained or inconspicuous behaviour, etc.; **low quarters** (Quarter *sb.* 20 c) *pl. U.S.* (see quots. 1916, 1971); also **low-quarter** *attrib.*; **low-residue** *a.*, (of a meal or diet) designed to give rise to relatively little faeces and urine; **low-rise** *a.*, of a building: low, of one storey or few storeys; also *transf.*: as *sb.*, a low building; **low-yield** *a.*, producing little, giving a low return; *spec.* applied to a nuclear weapon having a relatively low explosive force.

1931 *Iron Age* CXXVIII. 1142/1 The low alloy steels were described [by H. J. French] as those containing usually no more than 6 or 7 per cent of the alloying metal. **1966** W. D. Hargreaves in D. L. Linton *Sheffield* 187 Rotherham produces about a million tons per year of carbon and low alloy steel. **1972** *Jrnl. Austral. Inst. Metals* XVII. 175/1 Where additional durability is required, a widely employed technique is to use a low-alloy 'marine grade' of steel. **1842** [J. C. Traill] *Let.* 18 Apr. in *Backwoods of Canada* (1836) 134 The low-bush cranberries are brought in great quantities by the Indians to the towns and villages. **1857** A. Gray *First Less. Bot. & Vegetable Physiol.* 212 R[ubus] *trivialis*, Michx. (Low-Bush Blackberry). **1891** J. M. Coulter *Man. W. Tex.* 124 Low bush blackberry. A southern blackberry, apparently common in eastern, southern, and western Texas. **1908** *Edmonton* (Alberta) *Jrnl.* 26 July 39/6 For the most part, everything hereabouts but low bush blueberries..and in the fall, low bush cranberries. **1969** E. Meuze *Two in Far North* i. 11 And sometimes find time to pick a handful of bright red low-bush cranberries. **1969** E. H. Pinto *Treen* 91 The harvesting of lowbush blueberries..begins in Maine in late July and August. **1849** *Theatrical Mirror* 17 Sept. 39 Whether in consequence of the non-arrival of the second low-comedy man, I would undertake the part in the last piece. **1886** in J. R. Ware *Passing Eng.* (1909) 171/2 The success of *Indiana* mainly depends upon the extravagant humours of the chief low-comedy merchant. **1906** E. Dyson *Factry'ands* xvii. 219 Feathers had some reputation as a yarn spinner. **1950** Wodehouse..a low-comedy..whose trick.. **1947** *Amer. Speech* XXII. 132 The relatively low-fidelity aluminum disks might not always record it clearly enough to interfere with established habits of transcription. **1962** Simpson & Richards *Physical Princ. Junction Transistors* xviii. 451 For low-fidelity communication links using portable equipment a narrow-band FM system..may be used. **1963** T. Pynchon *V.* vi. 135 Each anchor man had a walkie-talkie, tied in a common network to Zeitsuss's office and to a low-fidelity 15-inch speaker mounted on the ceiling. **1931** *Philol.* III. 265 Skt. i: Gr. ε-α, as low grade of Idg. (e.g.) on. **1907** *Mod. Philol.* V. 469 OE. *pæθan*, pre-Germ. *pʰpənō* may represent the low grade of a base *ᵖpen-* in *pent-to. **1895** *Leeds Mercury Suppl.* 4 Jan. (E.D.D.), Sarah were in a varry low kee. **1907** *Westm. Gaz.* 15 Jan. 14/2 There is not the slightest reason why we should not exploit this low-key type of picture. **1953** K. Reisz *Technique Film Editing* i. 65 The elegant low-key lighting (which is utterly unrealistic). **1958** N. Sharps *Dict. Cinematogr.* 110/1 Low-key, the term applied when a majority of the tones in the subject or image lie at the dark end of the grey scale. **1962** *Listener* 6 Dec. 934/1 Pleasant, low-key entertainment. **1972** *Guardian* 11 July 3/1 With the UDA building its barricades, low-keyed was the word around the Department. **1960** R. Davies *Voice from Attic* 136 Poets' bodies... **1969** *Times* 28 Dec. Is a trudge(?) .. **1969** *Daily Tel.* 13 Feb. 16/3 A fairly low-keyed inquiry. **1973** *Publishers Weekly* 30 May 40/1 A girl spending the winter recovering from a long illness, and her pet hamster are the elements in this low-keyed fantasy. **1927** *Times* (Weekly ed.) 1 Nov. 495/4 Two open coaches mounted on six-cylindered low-loader commercial chassis. **1960** *Guardian* 23 Nov. 11/4 Efforts were still being made to tow the low loader out of the hedge. **1973** J. Leason *Host of Extras* i. 23, I may have to carry them [sc. old cars] round to the power unit to save the cost of a low-loader all the way down to London. **1928** *Times* 31 Mar. 20/1 The effect of connecting up a valve amplifier with a highly efficient tuned circuit employing a coil of the low-loss type. **1946** *Nature* 9 Nov. 671/1 The oxides of iron combined with various proportions with a low-loss, non-magnetic binder, paraffin wax. **1962** Corson & Lorrain *Introd. Electromagn. Fields* xii. 418 This type of wave guide is found in almost all types of electronic equipment for use both at microwave frequencies where the cross-sectional dimensions of the propagation is a low-loss dielectric. **1785** H. Muhlenberg *Catal. Plant.* 3 Acer spicatum; *low maple*. **1856** [?] D. J. Browne *Sylva Amer.* 102 It is called Mountain Maple, from the dwarfish stature of the tree. **1897** G. B. Sudworth *Nomencl. Arborescent Flora U.S.* 282 *Acer spicatum*.. Mountain Maple.. Low Maple (Tenn.). **1946** *Nature* 28 Dec. 935/1 Crystallization phenomena in rubbers show certain similarities to, but also striking differences from, crystallization in ordinary low-molecular systems. **1866** Low-neck [in Dict., sense A. 1 c]. **1909** *Englishwoman* Apr. 319 Magazines with 'types of beauty'—in tights, ballet dancers' skirts or low-necks. **1922** Low-neck [see *bathing-suit* (*bathing *vbl. sb.* 2)]. **1929** J. H. Morecroft *Elem. Radio Communication* vii. 250 A simple calculation, say, of the low-pass type, will show the reason for the particular performance of filters. **1950** *Engineering* 17 Feb. 186/3

The model..consists essentially of a headpiece made up of successive layers of flexible material, felt and iron arranged so as to constitute a low-pass acoustic filter. Such a filter does not cause appreciable attenuation of the low and voice frequencies, but cuts out those frequencies which give rise to harmful shrillness. **1964** *Language* XL. 203 'Content-free speech' has also come to refer to doctored recorded corpora of speech which have been run through low-pass filters, so that features conveying affective qualities are retained while linguistic intelligibility is suppressed. **1967** *Guardian* 25 Sept. 3/8 A production Aston Martin..with..low profile tyres. **1970** *Ibid.* 24 Aug. 3/2 The Nixon doctrine of 'low profile' involvement, in other words a maximum of aid and a minimum of US troops. **1971** *Publishers' Weekly* 6 Dec. 19/2 Los Angeles..more than lives up to its reputation as a low-profile, seemingly endless sprawl with no center. **1972** *Times* 2 Aug. 14/5 Whittaker Hunt has been adopting an altogether lower profile. **1973** *Listener* 4 Jan. 7/3, I admire the brisk creativeness of American English. 'Low profile' is a perfectly vivid phrase for 'conciliatory demeanour'. **1880** 'Mark Twain' *Tramp Abroad* 275 He wore very low-quarter patent-leather shoes. **1916** *Dialect Notes* IV. 269 *Low-quarters*, Oxford shoes. 'He wears *low quarters* all the year round.' [New Orleans; Also N. Car.] **1971** *Current Slang* (Univ. S. Dakota) VI. 7 *Low quarters*, army dress shoes. **1962** A. Shepard in *Into Orbit* 98, I had a low residue breakfast of orange juice, a filet mignon wrapped in bacon, and some scrambled eggs. **1971** *New Scientist* 11 Mar. 551/2 Waste can be minimised by the careful planning of low-residue diets..but it will always present a problem [in space flight]. **1957** *Fortune* Sept. 213/2 What kind of people..prefer 'high-rise' apartment buildings, what kind of people prefer two-to-five-storey 'low-rise' houses? **1967** *London* (Ontario) *Free Press* 23 June 38/4 Hip Briefs for today's low-rise trousers. **1971** *Rand Daily Mail* (Home Owner) 27 Mar. 7/4 High rise for young couples or old folk, low-rise for couples with older children. **1972** *Sat. Rev.* (U.S.) 17 June 47/2 The finished units..were stacked 3-high..to form low-rise garden units. **1957** *Wall St. Jrnl.* 25 Jan. 1/3 Only low-yield nuclear tests will be conducted at the Frenchman's Flat Proving Ground... The announcement added high-yield devices (hydrogen bombs) are never tested in Nevada. **1959** *Times* 20 Oct. 19/3 Substantial two-way business in industrial shares partly reflected switching out of low-yield shares into higher yielding second rankers. **1972** *Sat. Rev.* (U.S.) 6 May 31/2 Localized Soviet attacks would be countered..by the Pentomic army equipped with a variety of sophisticated, low-yield nuclear weapons.

B. *sb.* **3. a.** *spec.* in E. Anglia, a hollow or valley between dunes; a pool left by the tide in such a hollow. (Further examples.)

1855 *Trans. Philol. Soc.* 33 [Norfolk words.] *Low*, a loch left by the tide on the shore. **1878** Miller & Skertchly *Fenland* ix. 291 The Tides..having a larger and deeper channel run with greater velocity into Lynn Deeps, and set westward through the lows in the sands into Boston Deeps, the Tide being about 20 minutes later there than in Lynn Deeps. **1929** *Jrnl. Ecol.* XVII. 138 Very characteristic of the Blakeney dunes are the 'lows'—narrow valleys between the dune-ridges, corresponding to the 'slacks' of the west-coast dunes. The lows differ from slacks in not being permanently moist..and in being liable to flooding by unusually high tides. **1964** V. J. Chapman *Coastal Vegetation* vi. 156 After flooding by the tide, water may remain in the low for some time. **1970** G. E. Evans *Where Beards wag All* xxi. 240 These men lived in these huts which they'd placed down there in the Low.

b. Also, an area of low gravitational field strength.

1966 *McGraw-Hill Encycl. Sci. & Technol.* 166/2 Gravity maps display highs and lows. **1972** *Nature* 3 Mar. 24/1 The main feature of the present map..is the large gravity 'low' in the Moray Firth.

c. A low point or minimum in price, temperature, numbers, or the like. Cf. *all-time* adj. (*All E. 13).

1911 *N.Y. Times* 20 Sept., Calumet and Hecla opened 17 points off at 373, which is 47 points above the established low of 1897. **1929** *Observer* 17 Nov. 3/4 The sharp rally..called the weighted average of eight leading industrial stocks up to 149·0 from the new low of 133·0. **1933** A. R. Longworth *Crowded Hours* viii. 133 In 1906 the finances of the Longworths, though they were on the way, had not reached their present 'low'. **1934** Wodehouse *Thank You, Jeeves* xiv. 203 But the heart was still sinking. And when I heard him snort emotionally in the darkness it touched a new low. **1953** —— *Performing Flea* 193 The population of Dormitory 309 fluctuated between a high of 68 and a low of 57. **1955** *Ann. Reg. 1954* 82 The death rate fell to a new low of 8·7 per thousand. **1964** *Financial Times* 3 Mar. 2/6 A new low for many years was reached in stocks of refined tin in London Metal Exchange official warehouses last week. **1968** *Globe & Mail* (Toronto) 17 Feb. 1/6 The predicted high for Elliot Lake today is 5 below; the low tonight 20 below. **1971** *Sci. Amer.* Sept. 107/2 The interior temperature of the house sometimes reached 80 degrees, and the nightly lows were seldom below 70. **1973** *Jrnl. Genetic Psychol.* CXXII. 186 Scores could range from a high of 64..to a low of 1.

5. = *low gear* (see *Gear *sb.* 7 b).

1934 in Webster. **1968** C. Nicole *Self Lovers* ii. 30 He descended the hill from his house in low, partly to minimise the potholes..and partly to enjoy the view. **1970** T. Hillerman *Blessing Way* xiii. 119 Put it in low and angle to the left. **1973** 'A. Hall' *Tango Briefing* viii. 96, I dragged the manual into low to kill the rest of the speed.

low, *adv.* Add: **5.** *low-cut, -flying, -growing* (later examples), *-hung* (later examples), *-paid* (also *absol.*), *-ranking, -slung, -trained* (also as pa. pple.), *-yielding*; **low-flung** *a. U.S. colloq.*, of low character or standing.

1897 *Sears, Roebuck Catal.* 203/3 Men's low cut canvas pumps. **1932** 'E. M. Delafield' *Thank Heaven Fasting* II. iii. 44 Lady Marlowe, superb..in her low-cut green satin, with an emerald tiara. **1962** P. Gregory *Like Tigress at Bay* iv. 40 Steve was forced to admit to himself that she presented a beautiful picture, sitting there in the spotlight in her low-cut, strapless gown. **1843** *Missouri Reporter* (St. Louis) 11 Apr. (Th.), Here we have a beautiful specimen of the dishonesty and low-flung slang of the clique. **1844** *Knickerbocker* XXIII. 506 Who wants a parcel of low-flung 'outside barbarians' to go in cahoot with us? **1853** J. G. Baldwin *Flush Times Alabama* 24 He..denounced Jefferson as a low-flung demagogue. **1861** *Oregon Argus* 28 Dec. (Th.), It would be impossible to attempt a controversy with such low-flung dogs. **1963** *Times Lit. Suppl.* 8 Feb. 95/2 The low-flying beginning of this book is much better than its high-flying philosophic end. **1906** *Westm. Gaz.* 26 May 11/3 The topmost bough of some low-growing tree or shrub. **1958** W. J. Stokoe *Caterpillars Brit. Moths* (new ed.) I. 299 Greater plantain (*Plantago major*) and other low-growing plants. **1916** D. H. Lawrence *Amores* 104 The low-hung lamps stretched down the road..as I hastened to meet The low-hung light of her eyes. **1924** E. Sitwell *Sleeping Beauty* xvi. 54 It seemed a low-hung country of the blind. **1964** W. L. Goodman *Hist. Woodworking Tools* 138 Comparatively low-paid labourers. **1974** *Times* 5 Dec. 4/4 The TUC had..recognized the low-paid as a special case. **1958** W. J. H. Sprott *Human Groups* ix. 152 The less popular members of the consistently low-ranking teams did not change their allegiance. **1931** *Morning Post* 21 Aug. 11/7 His low-slung car simply hurtled down the straight and was lost to view. **1965** G. McInnes *Road to Gundagai* ii. 31 A very full, low-slung muskmelon moon. **1973** 'R. MacLeod' *Nest of Vultures* ii. 30 Sleek Ferraris, opulent Rolls-Royces and an occasional low-slung Lamborghini. **1869** *Rep. Comm. Agric. 1868* (U.S. Dept. Agric.) 249 Low-trained hedges may be necessary where land is limited in area, and high in price. *Ibid.*, Evergreens or shrubs may be formed, trimmed, and low-trained a long time without pleaching. **1946** *Nature* 23 Nov. 762/2 The replacement of existing virus-infected, low-yielding clonal stocks of raspberries is a pressing necessity. **1968** *Economist* 2 Mar. 61/2 Town and City has always been a favourite low-yielding property share.

b. Also *low-flying* (so *low-fly* vb.), *-loading*.

1881 O. Wilde *Poems* 69 Young Mercury Low-flying to the dusky ford of Dis. **1949** *Happy Landings* July 8/1 One pilot, low flying over the Continent, had his aircraft blown-up. **1958** G. Dutton in B. James *Austral. Short Stories* (1963) 288 He low-flies through the unforeseeable complications of tree and rock. **1942** *Tee Emm* (Air Ministry) II. 78 The pilot who is recklessly daring..in properly authorised low-flying practices. **1945** *Ibid.* V. 33 There is..a fascination about low-flying. **1940** *Chambers's Techn. Dict.* 512/2 *Low-loading amplifier*, a more recent name for a loaded push-pull amplifier. **1962** *Times* 8 May 16/5 The luggage boot, with its flat, low-loading floor.

low, *v.*³ (Further examples of *'low*.)

1881 J. C. Harris *Nights with Uncle Remus* (1884) xvi. 81 Brer Fox 'low dat he know dat ar place same ez he do he own tater-patch. **1911** G. S. Porter *Harvester* xvi. 367 She 'lowed to make him a big man. **1938** [see *jes, jes']. **1948** G. V. Galwey *Lift & Drop* iv. 65 Wind's backed four points... We've got to 'low for it.

lowan (lōᵘ·ăn). *Austral.* [Aboriginal name.] = *mallee bird, fowl, hen* (Mallee²).

1861 H. W. Wheelwright *Bush Wanderings* iv. 63 The lowan is a plain dull-coloured bird, brownish black, a little larger than the common fowl. **1888** 'R. Boldrewood' *Robbery under Arms* II. ii. 23 The lowan (Mallee hen, they're mostly called) and tallagatta (brush turkey) were thick enough in some of the scrubby corners. **1939** *Nature* 12 Aug. 272/2 It [sc. Wyperfeld (Mallee) National Park, Victoria] is probably..the last home of that wonderful bird—the mound builder called the lowan, which incubates its eggs in the vegetative centre of mounds which it builds. **1963** [see *incubator-bird* (*incubator 5)].

low-boy. **2.** For *U.S.* read orig. *U.S.*, and add later examples.

1929 *Burlington Mag.* LIV. 313 The 'Lowboy'..is the pendant piece to the 'Highboy' and also has no counterpart in England. **1970** *Canad. Antiques Collector* May 11/2 Rare William and Mary lowboy of excellent proportions in burr-walnut and inlays of herringbone. **1973** *Country Life* 29 Mar. (Suppl.) 38/2 Queen Anne walnut two drawer lowboy..Height 2 ft. 3 ins. **1975** *Times* 6 Feb. 9/1 Hand made, hand finished traditional lowboy, in walnut.

lowbrow, low-brow (lōᵘ·brau), *sb.* and *a.* *colloq.* (orig. *U.S.*). [f. Low *a.* 1, in contrast with *highbrow, high-brow *sb.* and *a.*] **A.** *sb.* One who is not, or does not claim to be, highly intellectual or æsthetically refined.

1906 S. Ford *Shorty McCabe* iii. 64 The spaghetti works was in full blast, with a lot of husky low-brows goin' in and out. **1907** *Collier's* 30 Mar. 23/2 The overwhelming majority of Low Brows, who never read 'Peer Gynt'. **1914** 'High Jinks, Jr.' *Choice Slang* 15 *Low brow*, an uneducated person usually prone to exhibit his deficiencies. **1915** E. Wallace *Melody of Death* ii. 20 If you only knew how the low-brows are pitying you..you would not be posing for a picture of 'The Ruined Gambler'. **1926** *Glasgow Herald* 13 Mar. 4 The highbrows will be overwhelmed..: the lowbrows will be unmoved. **1927**, **1929** [see *broad-brow, broadbrow* (*broad *a.* D. 2)]. **1935** *Discovery* Sept. 281/1 If we can persuade the lowbrows and free-livers of the present day that there is something heroic, even sporting, in the cultivation of the higher regions of the mind, we shall have won more than half the battle. **1947** [see *classical *a.* 6 c]. **1953** [see *Aunt Edna* (*aunt 6)]. **1974** *Times* 30 Nov. 8/2 They considered themselves lowbrows and..preferred a rattling good yarn to all that highbrow twaddle.

B. *adj.* Of, pertaining to, or characteristic of a lowbrow; not highly, or not pretentiously, intellectual; unrefined, unæsthetic. Also quasi-*advb.*

1913 H. A. Franck *Zone Policeman 88* i. 28 With all its excellences it would be unjust to complain that the Zone 'Y.M.' is a trifle 'low-brow' in its taste. **1914** S. Lewis *Our Mr. Wrenn* v. 61 You ain't neither too highbrow or too lowbrow. **1923** *Spectator* 22 Sept. 391/2 Often the sole reason why he [sc. the man of genius] does not write 'low-brow' is because he cannot. **1928** *Collier's* 10 Nov. 30/1 The doctor who is sufficiently adventurous, or lowbrow, to visit a soda-fountain occasionally. **1934** S. R. Nelson *All about Jazz* i. 24 From the many attempts of our leaders to substitute..'symphonic syncopation', 'modern rhythm', and the like, it is obvious that the much-maligned and obvious 'low-brow' caste of the word has rendered it necessary for us to designate our music with something that is emphatically not—jazz! **1963** [see *highbrow, high-brow B. *adj.*]. **1974** *Country Life* 9 May 1126/2 Another enormous price was the £22,000 paid..for a mid-18th-century print of an actor by Toyonobu;..but, as all the subjects of such prints..seem to me to represent deplorably ham actors, I remain wholly unimpressed. Now I come to think of it, what a low-brow criticism.

absol. **1927** *Daily Express* 7 May 9/6 Our aim will be..to steer a course between the 'highbrow' and the 'low-brow' in music.

So **low-browism**, the condition of being low-brow; lack of interest in intellectual or æsthetic matters.

1931 A. Huxley *Music at Night* 222 The snobbery of culture..has now to wrestle with an organized and active low-browism. **1946** *Amer. Speech* XXI. 121 The Australian dislike of elegance and affectation leads to a general 'low-browism' of expression and a careless enunciation. **1957** R. Hoggart *Uses of Literacy* x. 255 At a time when it is so easy to be led into arrogant low-browism, some retain an idealistic love for 'things of the mind'. **1971** *Listener* 2 Dec. 763/3, I am not at all arguing for a populist lowbrowism.

low-browed, *a.* **1.** (Earlier, later, and *absol.* examples.)

1855 J. E. Cooke *Ellie* 71 The man, who was a coarse, low-browed fellow. **1905** *McClure's Mag.* May 20/1 Exactly the type of low-browed ruffian and professional thug that they were hiring over there. **1936** G. B. Shaw *Shaw on Theatre* (1958) 251 The simplicity of the lowbrowed.

low-country. Add: **1.** (Later examples.)

1797 *Last Advice of C. Pettigrew to Sons* (MS.) (D.A.E.), You may think it best to sell your possessions in this low country, and to move westwardly. **1823** E. James *Acct. Expedition Rocky Mts.* I. 38 Here commences the low country, which extends west to the Mississippi. **1828** W. Carr *Dial. Craven* (ed. 2) I. 302 *Low-country*, East Riding of Yorkshire, being, in general flat, particularly when contrasted with this mountainous district. **1874** 'H. Churton' *Toinette* x. 114 He came from somewhere down in the low country. **1963** R. I. McDavid *Mencken's Amer. Lang.* vi. 299 Low Country, especially in the South, denotes the Coastal Plain of South Carolina and Georgia, as distinguished from the Piedmont or Up Country. **1974** *State* (Columbia, S. Carolina) 15 Feb. 17-A/8 Sen. Strom Thurmond, R-S.C., said Thursday arrangements have been made for an additional 1.4 million gallons of gasoline for the South Carolina Low-country.

3. *S. Afr.* = *lowveld. Cf. *bushveld b.

1879 [see *bushveld]. **1929** D. Reitz *Commando* xiv. 126 Our road ran through the Sabi low country teeming with big game of all descriptions. **1930** *Official Year Bk. S. Afr.* No. 11. 18 The Low Country stretches from the Limpopo valley behind the escarpment past the eastern end of Zoutpansberg southwards to meet the South-Eastern region below the escarpment east of Carolina. **1947** J. Stevenson-Hamilton *Wild Life S. Afr.* xxiii. 189, I have weighed a good number of low-country leopards immediately after death.

lowder. Add: Also **looder**.

1881 *Contemp. Rev.* Aug. 190 The cure for this was to throw a fire-brand down the 'lighting-hole' in the 'looder'. **1899** J. Spence *Shetland Folk-Lore* 172 In a corner of the looder stood a toyeg..containing as much corn as would be a hurd o' burstin. **1910** *Old-Lore Misc.* III. i. 9 The table or bin on which the quern stands is called lúðr in Edda and looder in Orkney.

low down, *a.* and *adv.* Add: **a.** (Later examples.) **b.** (Earlier U.S. and later examples.)

In quot. 1850 in *b used in the geographical sense, and in quot. 1908 in a physical sense.

a. 1935 [see *growl *v.*³ 3]. **1959** 'F. Newton' *Jazz Scene* 291 The widespread practice of equating joy with height..and grief with depth... Thus the quality most desired in the old blues is that it should be *low-down* or *dragging*.

b. 1850 *Congress. Globe* 25 Apr. 821/1 The 'low down' Virginia Democracy had to yield to the western mountain Democracy. **1865** *Nation* (N.Y.) I. 586 His manners and conversation, showed him to be a good-deal above that class commonly called 'low-down, triflin' people', or poor white trash. **1869** *Overland Monthly* III. 130 There are the delusive 'kettlings', among the 'low-down' people. **1908** *Sears, Roebuck Catal.* 604/2 The closet..is a syphon jet bowl with a low down tank and seat. **1915** F. M. Hueffer *Good Soldier* III. i. 140 Low-down Bowery tough. **1916** 'Taffrail' *Pincher Martin* xii. 214 They considered that Pincher had played them a low-down trick in ignoring their charms and in going elsewhere for an object for his affections. **1928** *Variety* 15 Aug. 49/4 There were as many as 35 or 40 dark-skinned musicians on the stand, kidding around and giving their conception of low-down tunes. **1968** *Blues Unlimited* Sept. 10 John Lee Hooker comes here a lot, an' he plays some really lowdown things.

low-down, *sb. slang* (orig. *U.S.*). [f. the adj. or adv.] The fundamental, though not gene-

rally known, facts *on* (about) a person, situation, etc.; the 'inside story'.

1915 *San Francisco Call & Post* 2 Dec. 12 (*caption*) Aw, give us the low down on them, Bill. **1920** *Collier's* 15 May 57/2 He calls me back, and in about twenty minutes I have got the low down on Monsieur Kane Halliday. **1924** Wodehouse *Leave it to Psmith* ix. 184 Listen, Ed, while I slip you the low-down. **1930** *Punch* 5 Feb. 144/1 Our own book, Percival and I have decided, is to be called the Low-down on Taxi-drivers. **1930** Ade *Let.* 19 Dec. (1973) 149 You certainly shattered some of my early idols, although I think I had figured out the low-down on most of them before I read your book. **1935** Auden & Isherwood *Dog beneath Skin* I. ii. 34 The Old Man sent for me before I left. Wants me to get the low-down on the Dripping merger. **1939** *War Illustr.* 18 Nov. p. ii/3, I have often smiled in recollecting this bit of 'low down' on the exile of Doorn—possibly for the first time here divulged. **1946** R. Campbell *Talking Bronco* 13 To have the low-down from their cross-Fates, Predicting tons of human phosphates Imported here in flesh and bone. **1957** N. Micklem *Box & Puppets* v. 102, I was occasionally able to do him some slight service in return by giving him 'the low-down' on theologians about whom he was required to compose orations. **1959** G. Jenkins *Twist of Sand* iv. 73 I'll give the low-down, charts, position, damage and all the rest of it. **1973** M. Mackintosh *King & Two Queens* x. 147 One of his minions will..give me the official low-down on Fisher. Possible police record, etc.

lower, *a.* (*sb.*) and *adv.* Add: **A.** *adj.* **2.** (Examples of *lower income, middlebrow, middle-class, working-class*.)

1852 H. Martineau *Let.* 7 Apr. in R. K. Webb *H. Martineau* (1960) x. 301 Thinking men & women of the *lower middle*, & working classes. **1861** Mill *Repr. Govt.* iii. 67 A benefit of the same kind..is produced on Englishmen of the lower middle class by their liability to be placed on juries and to serve parish offices. **1881** Lady Monkswell *Diary* 26 Apr. in E. C. F. Collier *Victorian Diarist* (1944) 52 Presently an old lady, rather 'lower middle class', with grey hair & no cap, appeared and welcomed us. **1930** *Archit. Rec.* Feb. 113/1 The wages of the lower income groups will indeed have to be raised. **1937** *Daily Tel.* 15 Oct. 22/5 'American Dream' has many of the hall-marks of a lower-middlebrow best-seller. **1951** M. McLuhan *Mech. Bride* (1967) 35/2 Kindly human thoughts such as keep ordinary men in the lower-income brackets. **1968** D. Lawton *Social Class, Lang. & Educ.* ii. 7 The lower working class consists of groups traditionally outside the educational system. *Ibid.* v. 78 The typical lower working-class environment. **1971** R. Roberts *Classic Slum* vii. 105 Before 1914..lower-working-class men did not generally shave themselves, but patronised a barber twice a week. **1972** A. Davidson tr. *Moravia's Two of Us* i. 10, I was struck by his language, half courtly, half bureaucratic and, in any case, lower middle-class. **1973** *Guardian* 19 Apr. 4/4 Mario Biaggi, probably the most popular candidate for Mayor of New York among the city's lower income white population.

3. a. Delete † *Obs.* and add later examples.

1921 *Chambers's Jrnl.* 30 July 545/1 Appreciation of beauty..is that which most distinguishes the humans from their lowers. **1967** *Listener* 21 Dec. 802/1 If a man spoke rather loudly..keeping his vowels open, then he was an Upper. If he attempted this and just failed, then he was a Middle. If..his voice carried the flavour of the area in which he was born, then he was a Lower.

c. A lower plate of artificial teeth.

1878 C. Hunter *Mech. Dentistry* i. 7 For edentulous uppers or lowers plaster is employed. **1939** A. Thirkell *Before Lunch* vi. 162 'And how are the new lowers?'..Mrs. Pucken smiled broadly with a slightly seasick motion of her lower teeth. **1963** J. Osborne *Dental Mech.* (ed. 5) ix. 170 In function there is a tendency for upper dentures to be moved forwards and lowers backwards.

5. b. With duplication of *lower*: belonging in the lower reaches of a 'lower' class or grade.

1955 T. H. Pear *Eng. Social Differences* iii. 89 There are few serious impersonations of the lower-lower class. **1970** 'D. Craig' *Young Men may Die* ix. 68 We languish at an almost unbelievable and entirely unspeakable lower-lower-middle Costa Blanca resort.

B. *adv.* **b.** *Comb.*

1960 F. C. Stern *Chalk Garden* vii. 74 P[aeonia] *obovata alba*..is a lower-growing plant with rather smaller flowers. **1972** *Guardian* 24 Nov. 10/1 Lower-paid hospital workers are resorting to a series of unofficial strikes. **1972** *Times* 24 Nov. 2/7 Incomes standstill blow to lower-paid clergymen. **1975** *Times* 3 Jan. 2/3 Teachers should be looking after their lower-paid colleagues. **1975** *Times* 11 Jan. 7/6 Show a minimum by bidding Two Hearts or a near minimum by bidding Two Diamonds (a lower ranking suit).

lower, *v.* Add: **1. e.** To drink (beer or other liquor); to empty (a bottle or glass of liquor) by drinking. *colloq.*

1895 *Punch* 27 July 39/1 If you'd just seen me lower the beer. **1899** C. Rook *Hooligan Nights* iv. 63 Out comes a bloke wiv a razzo like 'arf a boiled beetroot... Looked as if you wouldn't like to pay for the 'arf of what 'e could lower. **1920** 'Sapper' *Bull-Dog Drummond* ii. 64 During the time that he took to drink a mild nightcap, Mr. Benton succeeded in lowering three extremely strong glasses of spirit. **1933** A. G. Macdonell *England, their England* vii. 113 The gallant Major..had already lowered a quart and a half of mild-and-bitter. **1962** 'L. Grex' *Terror wears Smile* ix. 143 He could lower a whole bottle of three-star brandy without batting an eye. **1974** P. Lovesey *Invitation to Dynamite Party* v. 56 He's more accustomed to lowering pints than lifting weights.

f. *to lower the boom*: to inflict a physical defeat *on* (someone), to treat someone severely, to put a stop to an activity. *N. Amer. slang.*

1950 *Western Folklore* Apr. 118 *Lower the boom*. In a fight, to knock out your adversary with one punch. 'I sure lowered the boom on him. He had to take down his pants to blow his nose.' **1951** *New Yorker* 30 June 21/1 Just as they were about to pawn my studs.., my patience evaporated and I lowered the boom on them. **1963** J. N. Harris *Weird World Wes Beattie* (1964) xv. 186 Wes had been borrowing from everybody and his brother, and the boys had lowered the boom on him. **1973** *Times* 14 July 5 Senator Inouye asked if President Nixon's actions, after being told all by Mr Dean, could be considered as 'lowering the boom'—the phrase used by the former Attorney General, Mr John Mitchell, in his testimony this week.

4. e. *Phonetics.* To replace (a sound) with an allophone or phoneme of lower tongue position (Webster).

1888 H. Sweet *Hist. Eng. Sounds* 21 In diphthongs of the (ij)-type there is a tendency to make the cleaving more distinct to the ear by divergence, the first element being lowered and retracted. **1927** E. V. Gordon *Introd. Old Norse* iv. ii. 255 Neither *i* nor *u* was lowered if *i* or *j* stood in the next syllable. **1959** A. Campbell *Old Eng. Gram.* 122, *ǣ* had been unrounded to *ē*, and *ȳ* unrounded and lowered to *ē*.

lowest, *adv.* (Later examples.)

1926 *Daily Chron.* 13 May 1/7 Any such agreement should..not adversely affect in any way the wages of the lowest-paid men. **1974** *Guardian* 24 Jan. 4/3 His scheme would help the lowest paid. **1975** *Times* 8 Mar. 7/7 East.. invited a switch to the lowest-ranking suit.

Lowestoft (lōᵘ·stǫft). The name of a town in Suffolk, used *attrib.* or *ellipt.* to denote a soft-paste porcelain made from 1757.

1790 E. Gillingwater *Hist. Acct. Lowestoft* ix. 423 It was agreed upon to open a subscription..in order to raise a sum sufficient for building a man of war..on which occasion the town of Lowestoft subscribed as follows:—Mr Walker, for the proprietors of the Lowestoft porcelain manufactory £10 10s. od. **1875** L. Troubridge *Life amongst Troubridges* (1966) 106 Two pretty old Lowestoft bowls. *Ibid.* 9 June 120 In the window was a most beautiful tea-set of old blue Lowestoft. **1912** C. Mackenzie *Carnival* xviii. 181 'But what a terrible teapot.'..'It's old Lowestoft.' **1968** R. H. R. Smithies *Shoplifter* (1969) 8 A small irregular pair of Georgian silver candlesticks flanked a bowl of armorial Lowestoft..the gold tracery glittering against the ultramarine glaze and the creamy porcelain. **1974** *Country Life* 6 June 1517/2 Rare Lowestoft coffee pot..circa 1776.

low fre·quency. [f. Low *a.* + Frequency 4.] **1. a.** A frequency (see Frequency 4 b in Dict. and Suppl.) having a relatively small number of cycles in a second; applied esp. to an electric current or voltage, an electromagnetic wave, or a sound wave. Abbrev. **L.F.*, esp. in radio and telecommunications, where it also refers specifically to electromagnetic waves of 30–300 kilohertz.

1900 M. A. Oudin *Stand. Polyphase Apparatus* xiv. 242 Transmissions of power are accomplished at a comparatively low frequency. **1928** *Daily Mail* 25 July 18/1 The best Gramophone and Wireless Sets at present in use fail to reproduce the low-frequency of bass notes in anything like proper proportion. **1933** *B.B.C. Year-Bk.* 384 Actually the microphone responds from very low frequencies up to about 8000 cycles per second. **1934** [see *high frequency 1 a]. **1943** C. L. Boltz *Basic Radio* vii. 121 In considering an imaginary capacitance, that of a tumbler switch,..we judged its capacitance in thousandths of a μμF. At low frequencies the reactance of such a switch is in perhaps millions of megohms. **1960** H. Carter *Dict. Electronics* 169 *Low frequency*. (1) A relative term used to distinguish waves or oscillations of a particular frequency or band of frequencies from those of higher frequencies. (2) More specifically applied to radio waves of frequencies between 30 and 300 kc/s, corresponding to the long wave range. (3) The term is also loosely applied to the audio-frequency signals in a radio receiver to distinguish them from the 'high frequency', i.e. the radio- and intermediate-frequency signals. **1962** A. Nisbett *Technique Sound Studio* iv. 78 The second harmonic [of the mains frequency] is generally the most serious component [of hum] except in certain types of hi-fi loudspeaker cabinet which provide a high output at low frequencies.

b. A low rate of occurrence, in space or time.

1935 [see *high frequency 1 b].

2. *attrib.* (usu. with hyphen). **a.** In sense *1 a.

1900 *Engineering* 28 Sept. 412/3 Low frequency induction telegraphy experiments..were not considered in this report. **1920** H. M. Dowsett *Wireless Telegr.* v. 118 Low-frequency amplification is the process applied to the signal current after its form has been altered by rectification so that it can affect a telephone. **1923** *Radio Times* 28 Sept. 36 Its volume is limited only by the amount of low frequency amplification employed in the receiving set. **1926** J. S. Huxley *Ess. Pop. Sci.* xvii. 191 The special sense-organs for perceiving low-frequency vibrations in water which, like a herring or any other fish, it [sc. the tadpole] carries on a 'lateral line' along its flank. **1934** *B.B.C. Year-Bk.* 385 The output from the microphone is taken to a two-stage low-frequency amplifier. **1941** *Electronic Engin.* XIV. 404 Low frequency response is expressed in the form of a curve showing variation in power output at different frequencies. **1964** W. Jassem in D. Abercrombie et al. *Daniel Jones* 344 The level of this low-frequency noise in the spectrum is 20–25 db below the peak at the higher frequencies. **1973** Boyd & Parkes *Dark Number* viii. 83 Some stimuli are so reverberant they just hum on as a low-frequency signal programmed into your unconscious.

b. In sense *1 b: occurring rarely, involving relatively few instances.

1957 *Publ. Amer. Dial. Soc.* xxviii. 78 Low-frequency verbs like *asseverate, reiterate, aver*, etc.,..would rarely be initials without *that*, or parentheticals. **1963** Ervin & Miller in J. A. Fishman *Readings Sociol. of Lang.* (1968) 79 The evidence..reveals a high degree of similarity in the existence of such sequences of high-frequency and low-frequency items. **1971** *Brit. Med. Bull.* XXVII. 19/1 Much of the risk to people from tribes where the frequency [of cancer of the penis] is high is lost when they move to low-frequency areas.

low-grade, *a.* [Low *a.* 20.] **a.** Of low or inferior quality.

1878 [see *high-grade *a.*]. **1879**, **1899** [see Low *a.* 20]. **1926** J. S. Huxley *Ess. Pop. Sci.* iv. 39 If the good types are now crossed together, a stock will be produced..not containing harmful recessive factors and therefore not continually producing a certain proportion of low-grade individuals. **1940** J. H. Jagger *English in Future* i. 12 Low-grade mental vocabularies consist of the auxiliary words common to all of us. *a* **1963** L. MacNeice *Astrol.* (1964) ii. 46 The Moon is especially responsible for rather low-grade persons. **1964** *Ann. Reg. 1963* 389 Going on from this, scientists found strains of bacteria which throve on the normally poisonous metal copper, and began research into the possibility of pumping solutions of such bacteria into cracks in rocks above low-grade copper ore. **1974** E. McGirr *Murderous Journey* 93 A cheap low-grade snooper.

b. Not high in amount or degree of intensity. Cf. Low *a.* 9.

1940 A. Huxley *Let.* 24 Apr. (1969) 452 There seems to be some kind of long-standing low-grade infection at work. **1950** A. L. Rowse *England of Elizabeth* p. viii, The accumulation of capital then—as against its erosion in a low-grade consumption now. **1961** *Lancet* 19 Aug. 402/2 Because of continuing low-grade colonic obstruction,..the left colon was excised.

low-headed, *a.* [Low *a.* 21.] **a.** Of trees: having a low crown of foliage.

1861 *Trans. Illinois Agric. Soc.* IV. 328 Plant dwarf, or dwarfed, low-headed cherries, only. **1869** *Rep. Comm. Agric. 1868* (U.S. Dept. Agric.) 201 The silk tree (*Albizzia julibrissin*) is a low-headed, spreading tree, possessed of the most graceful foliage.

b. *transf.* Favouring low-headed trees.

1865 *Trans. Illinois Agric. Soc.* V. 205 We have never.. been identified with the ultra low headed orchardists.

c. Of animals: carrying their heads low.

1932 W. Faulkner *Light in August* xiv. 311 The low-headed and eager dogs..began to bay.

lowland, *sb.* and *a.* Add: **A.** *sb.* **1.** *sing.* (Earlier U.S. examples.)

1841 Thoreau *Jrnl.* 8 Feb. in *Writings* (1906) VII. 207 Upland and lowland, forest and field have been ransacked. **1843** *Knickerbocker* XXII. 5 Everywhere, in lowland and highland,.. nothing is more evident than the ..degradation of the negro.

2. (Later examples.)

1938 Duke of Montrose in R. Bain *Clans & Tartans* 11 The Tartan as a dress properly belongs to the Highlands, and not to the Lowlands. **1961** C. R. MacKinnon *Highlands in Hist.* 95 Montrose had mustered his army at Blair Atholl, and decided to open his campaign in the Lowlands in order to encourage the king's supporters in the south.

b. *pl.* The Low Countries.

c **1685** in *Roxburghe Ballads* (Ballad Soc.) (1887) VI. ii. 421 And it is called the Sweet Trinity, And was taken by the false Gallaly, sailing in the Low-lands. **1923** G. B. Harrison *Shakespeare's Fellows* iii. 100 Between his service in the Lowlands and the success of *Every Man in his Humour*, 1598, he had tried acting. **1961** T. Henrot *Belgium* 28 Some fifteen Spanish grandees were named successively governors of the Spanish Low Lands.

low-level, *a.* [Low *a.* I and II.] Situated near or below ground level; *fig.* not advanced in skill, culture, etc.; low-ranking, unobtrusive, restrained.

1881 W. D. Hay *300 Years Hence* vii. 149 The Sahara Desert, or rather the low-level parts of it. **1908** *Chambers's Jrnl.* Sept. 647/1 The conformation of the barrier which it pierces compelled an abnormally long low-level tunnel. **1910** *Jrnl. R. Inst. Brit. Archit.* Apr. 472 The scheme [of Leonardo da Vinci] comprises a system of low-level streets for commercial purposes, and an upper residential stratum. **1914** W. Owen *Let.* 2 Feb. (1967) 232 Once fixed in a low-level Rut one is ever-after straightened there;—straightened intellectually and socially as surely as financially. **1923** Ogden & Richards *Meaning of Meaning* vii. 255 A useful low-level shorthand. **1930** *Times Educ. Suppl.* 11 Jan. p. i/2 We saw low-level sea ice. **1941** *Flight* 9 Jan. 23/1 During such low-level attacks it was easy to see what damage had been done, and there is a vivid report from an aircraft which roared over a main railway station at just below 100 feet. **1955** *Bull. Atomic Sci.* Jan. 5/1 Finally, the long-range genetic danger of exposure of mankind as a whole to low-level, but widespread and persistent radioactivity—the most ominous but least well understood of all dangers of the new age—is only beginning to be dimly perceived. **1955** T. H. Pear *Eng. Social Differences* xi. 264 An atmosphere of low-level excitement. **1964** E. Bach *Introd. Transformational Gram.* iv. 59 Thus a 'high-level' rule is a rule that is placed earlier in an ordered list than a 'low-level' rule. **1972** *Village Voice* (N.Y.) 1 June 53/2 Lee is obsessed with a woman who has let him down; Harry, who would rather sleep than hear the horny details, tells him, 'Obsession is low-level awareness.'

low-life, *a.* and *sb.* [Low *a.* 20.] **A.** *adj.* Coarse, disreputable, vulgar.

1794 [see Low *a.* 20]. **1861** M. B. Chesnut *Diary* 8 Dec. (1949) 171 Mr. Shuford he goes for low life things, hurting people's feelings. 1885 [see Low a. 20]. 1910 G. B. Shaw *Brieux: a Preface* 16 Servants, solicitors, and other low life personages. 1931 W. Faulkner *Sanctuary* xxvi. 265 A durn low-life Jew. 1939 *Time* 18 Dec. 21/1 There can be nothing very awful about even such ostentatiously 'low-life' dives as the Nut Club in Greek Street. 1959 F. Bull *I know Face* xi. 198 Soho low-life stories. 1964 *English Studies* XLV. 368 Could minor low-life characters speak at once so feelingly and with our own voice, to their mistress, at a tragic climax? 1972 *Listener* 10 Aug. 184/3 Low-life action and local vernacular.

B. *sb.* Pl. usu. **low-lifes.** A coarse, vulgar, or no-good person. (Esp. in Jewish use.)

1911 M. Glass *Potash & Perlmutter* i. 8 'If you think Pincus Vesell done me up good, Noblestone,' Potash said, 'you are mistaken. I got better judgment as to let a low-life like him get into me.' 1934 L. Golding *Magnolia St.* II. xiv. 467 Me, and what do you expect from such a low life? 1953 *Omaha* (Nebraska) *World-Herald* 25 Oct., Police Commissioner Nolan in opening a school of correct English for policemen yesterday, remarked that such words would 'refine the tastes' of the men, but his noble prospect left most of the city's lowlifes cold. 1955 J. Potts *Death of Stray Cat* (1960) xxii. 209 Why, that low-life! Stealing my stuff. 1959 H. Pinter *Birthday Party* (1960) III. 50 Keep an eye open for low-lives, for schnorrers and for layabouts. 1964 *Amer. Folk Music Occasional* i. 7 It would be an error to conclude that such songs are found only among the low-life. 1971 *Black World* Apr. 38/2 Our responsible leaders sniffed their disdain of the low-lifes and begged aristocrats for white-collar jobs.

lowly, *a.* Add: **2. b.** Of plants or animals, comparatively undeveloped.

1876 *City-Road Mag.* Jan. 44/2 There can be very little doubt that lowly forms can exist..at temperatures not much below 150° Fahr. 1886 A. Geikie *Class-Bk. Geol.* iv. 293 The progress of life from its earliest appearance in lowly forms of plant or animal has been continuous. 1922 *Q. Rev.* Apr. 328 The most conspicuous physical features in Europe..had no existence when these lowly organisms lived and died. 1927 Peake & Fleure *Apes & Men* 13 Birds first appear in the Jurassic system, while traces of lowly mammals have been found from the Trias onwards.

lowveld (lō͡u·felt, -velt). *S. Afr.* Also hyphenated and with capital initial(s). [ad. Afrikaans *Laeveld* low country.] The low-lying region of the eastern Transvaal and of Swaziland; also applied to corresponding regions of adjoining territories. Freq. *attrib.*

1878 A. Aylward *Transvaal of To-day* iii. 44 Bushveld and Lowveld are convertible terms. 1905 *Transvaal Agric. Jrnl.* Oct. 141 (*title*) Notes on the native flora and crops of the Lowveld of the Eastern Transvaal. 1929 J. Stevenson-Hamilton *Low-Veld* i The Transvaal Province of the Union of South Africa is divided into three zones of altitude known respectively as High, Middle, and Low Veld... It is not unusual to hear an altitude of three thousand feet referred to as 'Low-Veld'... The Low-Veld of the Transvaal lies approximately between the twenty-second and the twenty-sixth parallels of south latitude. 1931 *Discovery* Aug. 259/2 At Barberton [Transvaal], where problems connected with the improvement of cottons for low-veld areas are dealt with. 1949 *Cape Argus Mag.* 5 Nov. 9/3 The lowveld hills around Ungungundhlovu. 1959 *Cape Times* 4 Feb. 1/1 Mr. Ray Stockil yesterday announced his intention to retire as Leader of the Opposition in the Southern Rhodesia Parliament and to devote his attention to the economic development of the lowveld of the country. 1968 C. Burke *Elephant across Border* v. 189 A shallow ravine, covered in those lowveld succulents which are able to grow out of small patches of moss and cracks in the rock.

lox (lǫks, el ō͡u eks), *sb.*[1] Also **LOX.** [orig. f. *l*iquid *ox*ygen e*x*plosive; later interpreted as repr. liquid oxygen.] **a.** An explosive device which uses liquid oxygen as an oxidant (see quot. 1946).

1923 *Chem. Abstr.* XVII. 2646 Wood pulp as an ingredient renders LOXs less sensitive to detonation. 1933 *Mining & Metall.* XIV. 368/1 The research was to be centered upon the absorbent material used in the manufacture of LOX cartridges. 1946 G. J. Young *Elem. Mining* (ed. 4) v. 127 Lox consists of a canvas cartridge filled with granular carbonaceous material, moistened, which when soaked in liquid oxygen, removed from the soaking box, and placed in the drill hole can be detonated and is an effective blasting agent. 1966 R. Fordham *High Explosives & Propellants* v. 63 A type of explosive..which achieved popularity for a time and is still used in some countries, is the liquid oxygen explosive or LOX.

b. Liquid oxygen, esp. when used as a rocket propellant.

The 1923 example may have been intended in sense a.

1923 *Chem. Abstr.* XVII. 2646 The ballistic pendulum was found suitable for tests of LOX explosives. 1949 *News* (San Francisco) 3 Oct. 29/4 That liquid is called liquid oxygen, or 'L-O-X', as the missile men say. 1949 *Time* 6 June, The Viking burns alcohol and 'lox' in a single combustion chamber. 1959 *Adv. Space Sci. & Technol.* I. 193 Existing LOX systems probably will not function properly under weightlessness. 1962 *Flight International* XXXI. 168/2 The Buccaneer is the first Naval aircraft to use lox.

Hence as *v. trans.*, to fuel with liquid oxygen; **lo·xing** *vbl. sb.*

1961 A. B. Shepard in *Astronauts* Suppl. 4 The Redstone was loxed. 1962 J. Glenn in *Into Orbit* 187 One of the loxing valves stuck and the technicians had to shift a smaller valve to complete the loading process. 1970 R. Turnill *Lang. of Space* 72 Loading the lox into the launch vehicle's fuel tanks from a ground supply is referred to as loxing.

lox (lǫks), *sb.*[2] Pl. **lox, loxes.** [f. Yiddish *laks*, f. G. *lachs* salmon.] A kind of smoked salmon.

1941 S. Longstreet *Last Man around World* xii. 135 Listen, you lox-eater—until last year you thought herring was the only thing they took from the sea. 1950 F. Allen *Let.* Oct. in G. Marx *Groucho Lett.* (1967) 73 At the stage delicatessen..the lox is running good and the cream cheese is spreading easily. 1961 B. Malamud *New Life* (1962) 121 ... 1969 A. Glyn *Dragon Variation* ix. 293 He..folded himself on to a stool at the counter, and ordered coffee and a bagel with lox. 1972 *Times* 21 July 12/8 Where in Paris..can one enjoy a breakfast of beigles, lox (Yiddish for salmon) and cream cheese? 1973 *Daily Colonist* (Victoria, B.C.) 16 Sept. 35/5 The bagel..often is eaten with smoked salmon or lox from Nova Scotia.

loyal, *a.* Add: **3.** *spec.* **loyal toast,** a toast proposed and drunk (in the U.K. and British Commonwealth) to the monarch or (elsewhere) to some other important personage.

1799 *Times* 1 June 3/4 Many Loyal Toasts were given, and the day spent with great conviviality. 1835 [in Dict.]. 1856 [see *Highland a. 4]. 1930 M. H. Kelly *Spinifex* xi. 166 'Gentlemen'—Mainsala barked. 'The Loyal Toast!' 1972 *Guardian* 9 May 15/4 It was the most miserable meal I have ever presided over, and I had not the heart to propose the Loyal Toast at the end. 1974 *Jrnl.* Jan. 10/4 The time was come for toasts and speeches. After the Loyal Toast, Blackadder clipped his cigar.

loyalty. Add: **2. a.** *spec.* Of government employees.

1955 *Bull. Atomic Sci.* Apr. 132 Commissioner Murray, concurring in the result, found Dr. Oppenheimer a risk on grounds of 'loyalty' which he newly defined as obedience to security regulations. 1955 *Government Loyal & Disloyal* v. 230 Loyalty-security programs threaten to cripple the entire foreign service. *Ibid.* 231 Governments must guard themselves against harm and destruction, and there can therefore be no criticism of the objectives of loyalty investigations for government employees and scientists who work on government projects. 1964 Gould & Kolb *Dict. Social Sci.* 397 In the United States since the end of World War II, loyalty has been an important issue in politics... Under President Truman..there was an elaborate program for determining the 'loyalty' of servants of the U.S. government.

b. loyalty oath *U.S.*, an oath, usually mandatory, required of a prospective public employee or other person in which he swears to abstain from subversive activities.

1952 M. McCarthy in *Reporter* (N.Y.) 8 July 32/2 In order to..work for the Columbia Broadcasting System you have to sign a loyalty oath which is presented to you with your contract. 1965 L. Greenbaum *Out of Shape* xxii. 162 The older man had his contract terminated under a clause dealing with loyalty oaths. Apparently he was a card-carrier and hadn't said so. 1972 M. McCarthy *Birds of America* 230 Berkeley is horrible... You have to swear a loyalty oath... It's natural that the students would finally rebel. 1973 *Black Panther* 12 May 6/1 After signing a loyalty oath to the school, students are required to present two identification cards, one issued since November 1961, before gaining entrance to the campus.

Lozi (lō͡u·zi). [Native name.] **A.** *sb.* **a.** A Bantu people inhabiting Zambia. **b.** Their language. **B.** *adj.* Of or pertaining to the Lozi.

1948 M. Guthrie *Classification Bantu Lang.* 53 Nominals are used as members... in Lozi..*e silila* 'these are the knives'. 1949 M. Gluckman in M. Fortes *Social Struct.* 146 The Lozi..call them *Mambo*, 'the people of the West'. *Ibid.*, I worked through Lozi, a lingua franca which most Wiko men..speak and understand. *Ibid.* 151 The hostility of the Wiko..to their Lozi and Kwangwa neighbours. 1953 C. Cox in D. Hymes *Lang. in Culture & Society* (1964) viii. 249/1 These [languages] include..Lozi of Barotseland in Northern Rhodesia. 1959 G. D. Mitchell *Sociol.* 82 We shall be concerned with the Lozi system, which since British influence has undergone some modifications. 1967 *Economist* 11 Nov. 617/1 The..conference..saw Barotseland's Lozi leaders lose much of their status... Barotseland's development is lagging behind..because of..Lozi cultural indifference... Mr Kaunda could galvanise the Lozis into political co-operation. 1972 *Times* 18 Sept. 12/4 President Kaunda of Zambia has had to face..a Lozi challenge.

LSD[3] (e:lesdī·). Also (*rare*) **L.S.D.** [f. G. *lysergsäure-diäthylamid*, lysergic acid diethylamide.] Lysergic acid diethylamide (see *lysergic a.). Freq. with the number 25 appended.

[1947 W. A. Stoll in *Schweizer Arch. für Neurol. und Psychiatrie* LX. 279 Dieser Stoff ist das Lysergsäure-diäthylamid (LSD).] 1950 *Diseases Nervous Syst.* XI. 243/1 We believe that L.S.D. 25 is a drug which induces a controllable toxic state within the nervous system, that re-activates anxiety and fear with apparently just enough euphoria to permit recall of the provoking experiences. *Ibid.* 243 Supplies of L.S.D. 25 for this study were made available by the manufacturer, Sandoz Pharmaceuticals, New York City. 1955 *Sci. News Let.* 26 Feb. 135/1 LSD has been used by psychiatrists recently to bring on a mental disease state in healthy persons in the hope of learning more about how to treat or prevent real mental sickness. 1958 A. Huxley *Let.* 11 Jan. (1969) 843 A session with an RC psychiatrist, who had reluctantly submitted to taking LSD25. 1962 *Observer* 15 July 1/5 The American Medical Association has given a warning that use of the drug known as LSD-25, or lysergic acid, can lead to suicide, and that LSD-25 has joined the ranks of black-market drugs like marijuana and heroin. 1964 *Daily Tel.* 28 Mar. 9/3 The tablets are believed to be a solid form of LSD, lysergic acid diethylamide. They can be obtained in certain clubs and public houses in London and other big cities. 1964 D. F. Downing in M. Gordon *Psychopharmacol. Agents* I. xiii. 569 The effective oral dose in man amounts to 0·02–0·05 mg and (+)-LSD-25 is therefore 10^4 times as active as mescaline in causing a psychotomimetic effect and the most powerful psychotomimetic substance known. 1965 *New Scientist* 22 Apr. 225/1 Bizarre though the symptomatology may seem, the intensive research which has taken place in the 23 years since LSD 25 was discovered has failed to reveal any serious danger. 1966 *Daily Tel.* 31 Mar. 25 New York County Medical Society has called for severe penalties against the illicit manufacture, distribution and sale of the hallucination-producing drug known as LSD. It says that the drug is 'far more dangerous than heroin'. 1968 A. Hofmann in A. Burger *Drugs affecting Cent. Nervous Syst.* II. v. 205 In the case of a very favorable reaction to LSD, the normal feeling of the identity with the own self is weakened (depersonalization). The test person at the same time appears to himself as a stranger watching him and judging his actions coolly and critically. *Ibid.*, LSD produces a state of hypersuggestibility, which is characterized by an enormous sensitiveness to the influence of sensory and cognitive stimuli and which is of decisive importance for the use of LSD as an adjuvant to psychotherapy. 1973 E. Bullins *Theme is Blackness* 162 Not many of us smoked as much pot as before... Some of us dropped out and some were on LSD.

Ltd., abbrev. of Limited *ppl. a.* (sense 2 b).

1900 *Times* 19 Oct. 12/1 (Advt.), Employees of Bovril, Ltd.: are not eligible to compete. 1922 Joyce *Ulysses* 151, I am hastening to purchase the only reliable ink-eraser *Kansell*, sold by Hely's Ltd, 85 Dame Street. 1967 *Times Rev. Industry* Feb. 57 (Advt.), Dictation systems Dictaphone Company Ltd. 1973 *Guardian* 12 June 7/3 Chater & Scott Ltd.... Specialists in motoring books.

luau (lū·au). [ad. Hawaiian *lu'au*.] **a.** A party or feast with Hawaiian food and usually accompanied by Hawaiian entertainment; also *attrib.* **b.** A cooked dish of young taro leaves served with coconut cream and octopus or chicken.

1843 L. Smith *Diary* 3 Aug. in M. D. Frear *Lowell & Abigail* (1934) 168 The table..was..loaded with an immense quantity of food—pigs, turkies,..taro, luau, etc., etc. 1853 *Putnam's Mag.* II. 357 It was on a Saturday afternoon in the year 1852, that a merry party of us started forth to attend a luau or native feast, given by a chief who lived some miles from Honolulu. 1905 *Hartford* (Connecticut) *Courant* 4 Apr. 8/3 Luau is a square meal with roast pig and poi in it. 1951 H. Wouk *Caine Mutiny* 62 Willie and Keefer..full of beer and much whisky from the luau. 1957 *New Yorker* 19 Jan. 14 Choose from a range of restaurants..from quaint Japanese tea house to lively Hawaiian luau. 1960 *Ibid.* 26 Mar. 132 (Advt.), Luau dinners and after theater suppers. 1964 *Sun* 18 Aug. 20/1 In the old days, the luau was an expression of thanksgiving to the gods. Today, it is a feast for sight and palate. *Ibid.*, Guests sit cross-legged around the low luau table and eat with their fingers. 1966 *New Yorker* 22 Oct. 30 (Advt.), Charge everything from a surfboard to a luau on your American Express Credit Card.

lubber, *sb.* Add: **2. b. lubber-grasshopper** (earlier and later examples); **lubber-lift** *v.* (see quot.).

1877 *Field & Forest* II. 160 The 'Lubber' grasshopper [is the] large grasshopper *Romalia microptera*. 1962 Metcalf & Flint *Destructive & Useful Insects* (ed. 4) xiii. 377 Large and small, lubberly grasshoppers often invade cotton from near-by waste lands and destroy the plants... Lubber grasshopper, differential grasshopper. 1797 *Spirit of Farmer's Museum* (1801) 85 Our Democrats begin to muster, Rolling around an anxious eye, Some 'lubber lifting' power to spy. 1965 *Terms Forestry & Logging* (U.S. Dept. Agric. Bureau Forestry) 42 *Lubber lift*, to raise the end of a log by means of a pry, and through the use of weight instead of strength.

lube (liūb, lūb), *sb.* and *v.* Chiefly *N. Amer.* and *Austral.* Colloquial shortening of Lubricant *sb.*, Lubrication, Lubricate *v.* (cf. *hydro-lube). Freq. *attrib.*, as *lube bay, job, oil.*

1947 *Lubrication Engin.* Sept. 26 (*heading*) Lube spots. 1951 *Chem. & Engin. News* 24 Dec. 5456/1 (*heading*) Synthetic lube oil announced by Defense Department. 1956 *Ibid.* 3 Sept. 4247/2 (*heading*) Better lubes needed. 1957 Casamassa & Bent *Jet Aircraft Power Syst.* (ed. 2) xiv. 195/1 On the rear face are mounted the main lube and scavenge pump [etc.]. 1962 B. Malamud *New Life* (1962) 259 He had not lubed Levin's car. 1962 *Engineering* 10 Aug. 185/2 A reinforced plastic coating can be laid on without putting the tank out of commission if the products are crude or lube oils. 1964 *N.Z. Listener* 11 Dec. 5/1 ... passed the lube bay... Windy was so upset he got grease all over everything. 1966 ... *Wife* (1966) xv. 125 I'm just about done, I think, for the works—wash, lube job, tyre checks, gas—the usual. 1973 R. Hayes *Hungarian Game* liii. 323 The grease nipples were dry; the car needed a lube. 1973 *Tucson* (Arizona) *Daily Citizen* 22 Aug. 60/1 (Advt.), Clean and lube backing plates.

lubfish, var. Lobfish.

1817 Scott *Rob Roy* I. ii. 32 Stockfish—Titling—Cropling—Lub-fish.

lubra. Add: **b.** More generally: a woman. *Austral. slang.*

1966 W. S. Ramson *Austral. Eng.* 129 *Gin* and *lubra* are used of Aboriginal women and occasionally, though only in slang and never as a compliment, of women in general.

lubric, *a.* **3.** Delete †*Obs.* and add later example.

1909 J. Jusserand *Lit. Hist. Eng. People* III. 436 Here we have a Bellario, all virtue and sweetness..opposed to a series of lubric and ferocious monsters.

lubricate, *v.* Add: **2. c.** To grease the palm of; to bribe.

1928 *Daily Express* 12 July 1/1 He made specific charges. One was that taxicab proprietors have to 'lubricate' Scotland-yard before their taxicabs are passed for licensing.

lubricating, *ppl. a.* (Further examples.)

1867 *Amer. Jrnl. Sci.* XCIII. 346 The lubricating oils are of very low specific gravity. 1974 *Sci. Amer.* May 97/1 As lubricating oil circulates through a machine it picks up myriads of wear particles.

lubrication. Add: Also *attrib.*

1906 *Westm. Gaz.* 23 July 8/2 Motor-'buses, fitted with the new automatic lubrication appliance. 1907 *Ibid.* 13 Nov. 9/1 The lubrication system..are required to be fitted to all engine speeds. 1952 M. McLuhan *Mech. Bride* (1967) 124 He brings our 78 years of lubrication experience. 1975 D. *Bristow's Catal. Miscellany* 1/75 14 (*Operation manual*). The Singer Car..with lubrication chart loosely inserted.

lubricational (liūbrikē͡i·ʃənăl), *a.* [f. Lubrication + -al.] Of, pertaining to, or for lubrication.

1909 *Westm. Gaz.* 18 Nov. 4/2 An automatic lubricational oil pump is fitted at the end of the cam-shaft.

lubritorium (liūbritō͡ə·riə̆m). Chiefly *U.S.* Also **lu·britory.** [f. Lubri(cate *v.* + *-torium* as in *auditorium, sanatorium.*] A greasing bay in a service station; a service station.

1930 *Amer. Speech* V. 329 In New Philadelphia, Ohio, there is a 'lubritorium' which I presume is intended to assist automobiles to a proper degree of lubricity. In Ann Arbor, Michigan, there is a 'Lubritory' which serves the same purpose. 1942 Berrey & Van den Bark *Amer. Thes. Slang* §544 *Lubritory*, a gas station. 1954 *Encounter* June 13/1 The effort to 'professionalise' work has become the major means of giving one's job a badge of honorific quality... So the garage becomes the 'lubritorium'. 1963 *Punch* 24 Apr. 579/3 The sign outside a Glasgow service station: Lubritorium. 1969 M. Pei *Words in Sheep's Clothing* (1970) xxi. 209 'Lubritorium' is so much more learned a word than 'service station'. 1969 *West Australian* 5 July 71/5 (Advt.), Lubritorium and driveway attendant..required.

Lucanian (lukē͡i·niăn), *sb.* and *a.* [f. *Lucania*, name of a district of southern Italy, also called Basilicata.] **A.** *sb.* A member of the Lucani, a branch of the Sabelline race, inhabiting Lucania. **B.** *adj.* Of, pertaining to, or belonging to Lucania; *spec.* **Lucanian ox, cow** (see quots.).

1709 [see *Apulian a. and sb.]. [1797 *Encycl. Brit.* X. 314 *Lucæ boves* denoted elephants; first seen in Pyrrhus's wars in Lucania, whence the appellation (Pliny).] 1863 W. K. Kelly *Curiosities Indo-European Trad. & Folk-Lore* i. 5 The Romans gave the name of Lucanian *ox* to the elephant. 1878 J. R. King tr. *Cicero's Philippic Orations* XIII. v. 230 He..took possession of his estates in the Lucanian territory. 1879 Lewis & Short *Latin Dict.* 1079/1 *Luca bos*, Lucanian cow, for elephant (because the Romans first saw this animal in Lucania, in the army of Pyrrhus). 1907 A. E. Zimmern tr. *Ferrero's Greatness & Decline Rome* I. iv. 85 The Samnites and Lucanians, who were still under arms, sent ambassadors. *Ibid.* v. 89 Only the Samnites and the Lucanians, as being still in revolt, were to be excepted. 1949 *Oxf. Classical Dict.* 313/1 The Romans first encountered elephants ('Lucanian oxen') in Pyrrhus' army, but seldom used them in battle. *Ibid.* 514/2 Lucanian communities had an official known as *meddix*. 1957 *Encycl. Brit.* XIV. 456 The Social War, in which the Lucanians took part with the Samnites. 1960 *Times* 29 June 15/7 Slogans on the walls of a Lucanian town: 'We are Italians too.'

lucanid (lukē͡i·nid), *a.* and *sb.* [f. mod.L. family name *Lucanidæ*, f. generic name *Lucanus* (J. A. Scopoli *Entomologia Carniolica* (1763) 1), f. L. *Lūcānus* Lucanian, f. *Lucania* (see prec.).] **A.** *adj.* Of or pertaining to a stag-beetle of the family Lucanidæ. **B.** *sb.* A member of this family.

1925 A. D. Imms *Gen. Textbk. Entomol.* 512 Lucanid larvae inhabit the rotting wood of trees or their roots. 1932 J. S. Huxley *Probl. Relative Growth* vii. 208 A peculiarly interesting example..is provided by the Lucanid beetle *Cyclommatus tarandus*, with markedly heterogonic male mandibles. 1959 E. F. Linssen *Beetles Brit. Is.* II. 115 While not having the extraordinarily developed mandibles of male Lucanids, there is in the present family [*sc.* Scarabæidæ] a great complexity of 'outgrowth-horns'.

Lucas (liū·kās, ‖lūka). *Math.* The name of F. Édouard A. *Lucas* (1842–91), French mathematician, used *attrib.* to designate (*a*) the sequence of integers 1, 3, 4, 7, ..., formed in the same way as the Fibonacci numbers; (*b*) the sequences generated by the recurrence relation $u_{n+2} = Pu_{n+1} - Qu_n$ when $u_0 = 0$, $u_1 = 1$ (the Fibonacci numbers being a particular case corresponding to $P = 1$, $Q = -1$) and when $u_0 = 2$, $u_1 = P$, which are respectively defined by $u_n = (a^n - b^n)/(a - b)$ and $u_n = a^n + b^n$ ($n = 0, 1, 2, ...$), where a and b are the roots of $x^2 - Px + Q = 0$.

[1919 L. E. Dickson *Hist. Theory Numbers* I. xvii. 393 (*heading*) Recurring series; Lucas' u_n, v_n.] 1953 *Scripta Math.* XIX. 278 (*heading*) Linear expressions for the powers of Fibonacci and Lucas numbers. *Ibid.*, The *n*th term of the Lucas sequence 1, 3, 4, 7... 1954 *Duke Math. Jrnl.* XXI. 668 If any term of (*W*) vanishes, (*W*) is essentially the well-known Lucas sequence $L_n = (a^n - \beta^n)/(a - \beta)$. 1961 *Pacific Jrnl. Math.* XI. 385 It would be interesting to make a numerical study of several recurrences..to endeavor to find out whether the two Lucas sequences... 1966 Ogilvy & Anderson *Excursions in Number Theory* 164 The Lucas numbers satisfy the same recursion relation as the Fibonacci numbers, but have starting values $L_1 = 1$, $L_2 = 3$. 1972 F. Riesborn *Algebraic Numbers* i. 8 Prove: ... consecutive Lucas numbers are relatively prime...

Lucca (lu·ka). [The name of a city and province in northern Italy.] *Lucca lamb*, a variety of processed lambskin, used to make headwear; *Lucca oil*, a superior quality of olive oil.

1725 J. Steuart *Letter-Bk.* (1915) 238 A Chist of finest Cucca [*sic*] oil. 1734 C. Mortimer in W. Ellis *Mod. Husbandman* (1750) VII. II. 84 His Wife rubbed in..the Sallad-oil (which I had bought by the Name of Lucca-oil). [1861 S. Smith in Mrs. Beeton *Bk. Househ. Managem.* 244 Four times the spoon with oil of Lucca crown, And twice with vinegar.] *c* 1938 *Fortnum & Mason Price List* 52 (*caption*) Superfine Lucca oil. 1956 J. G. Links *Bk. Fur* ii. 52 Persian lamb..and all the old varieties of processed lamb (beaver, Tuscan,..Lucca and..others). 1962 —— *How to look at Furs* (rev. ed.) 108 The merino can be sheared less or more than it is in the case of beaver lamb. If it is sheared less, the result is often called by such names as Lucca lamb or Tuscan lamb (Italian lambs are often treated in this way). 1966 J. S. Cox *Illustr. Dict. Hairdressing* 92 *Lucca oil*, olive oil. 1970 *Kay & Co.* (*Worcester*) *Catal.* 1970–71 Autumn/Winter 287/1 Lucca Lamb Fur Hat. Get the feel of luxury with this fashionable..hat made in real Lucca lamb.

Hence **Lu·ccan** *a.* and *sb.*

1911 H. Gerard tr. *Maurel's Little Cities of Italy* I. I. vi. 63 Mirrors planted by the Luccan army on the towers of Asciano. *Ibid.*, The Luccans added a charming porch of elegant proportions. *Ibid.* 69 Matteo must have wandered often about the Luccan country. 1936 *Times Lit. Suppl.* 29 Aug. 690/2 He..finds the Pisan school Byzantine and the Luccan Roman. 1961 C. C. Bayley *War & Society in Renaissance Florence* i. 7 Florence emerged..with only a secondary prize, the former Luccan dependency of Pistoia. 1968 E. Hyams *Mischief Makers* iv. 51 Olives yielding an oil of almost Luccan quality.

Lucian. Add: **Lucianesque,** in a Lucianic style; **Lucianic** *a.* (later examples); **Lucianism,** admiration and emulation of Lucian.

1922 P. S. Allen *Erasmus* 6 The outcome of his thoughts ..was a Lucianic composition, *Moriae Encomium*. 1925 A. M. Harmon tr. *Lucian* IV. 111 The source and character of the reply contribute a truly Lucianic fillip of surprise. 1937 C. R. Thompson (*title*) Lucian and Lucianism in the English Renaissance: an introductory study. 1962 R. P. Adams *Better Part of Valor* IV. 48 Erasmus couples Lucianic irony and his own special form of wit. 1969 G. Holmes *Florentine Enlightenment 1400–50* iv. 112 *Momus* is the most substantial of his Lucianesque creations.

Lucianist[1]. Add: **b.** A student, admirer, or emulator of Lucian.

1940 C. R. Thompson *Translations of Lucian* i. 1 Erasmus..was the paramount Lucianist of the Renaissance. 1941 H. H. Hudson tr. *Erasmus's Praise of Folly* p. xix, Enough has been said, though more is available, to prove him a Lucianist. *Ibid.* p. xx, Rabelais..was both a Lucianist and an Erasmian. 1946 L. F. Dean tr. *Erasmus's Praise of Folly* 17 By 1506 he had become a thorough Lucianist.

Lucianist[2]. Add: (Later examples.)

1805 H. Adams *View Relig.* I. 187/2 *Lucianists*, so called from Lucianus, a disciple of Marcion. 1824 C. Buck *Theol. Dict.* (ed. 2) 313/2 *Lucianists*,..a sect so called from Lucianus,..a heretic of the second century, being a disciple of Marcion, whose errors he followed. *Ibid.*, There was another sect of Lucianists, who appeared some time after the Arians. They taught, that the Father had been a Father always,..even before he begot the Son, as having in him the power and faculty of generation. 1874 J. H. Blunt *Dict. Sects* 262 *Lucianists*, a section of the Marcionites, followers of Lucian... The particular tenet by which the Lucianists were distinguished from the Marcionites..was that, in the resurrection from the dead, neither the actual body nor the actual soul..would arise. 1882 H. M. Gwatkin *Stud. Arianism* ii. 31 Disciples of Lucian—Eusebius.., Menophantus.., and Leontius... These are all the Lucianists whom we can trace. *Ibid.* iii. 73 Eusebius himself was the ablest of all the Lucianists. 1903 J. F. Bethune-Baker *Introd. Early Hist. Christian Doctrine* xii. 163 The Lucianists thought that logic could settle everything. 1958 J. N. D. Kelly *Early Christian Doctrines* iii. ix. 230 He and Eusebius.., it is implied, were 'fellow-Lucianists', and Eusebius is elsewhere described as a disciple of Lucian. 1960 A. Bull tr. *Ricciotti's Age of Martyrs* v. ii. 265 The Origenists and the Lucianists were interested to see whether or not the unity of God..could be reconciled with their particular school.

Hence **Lucia·nic** *a.*

1882 H. M. Gwatkin *Stud. Arianism* iii. 72 We find him using the Lucianic creed. 1903 J. F. Bethune-Baker *Introd. Early Hist. Christian Doctrine* xii. 174 The Lucianic origin of the Creed has, however, been called in question in recent times. *Ibid.* 175 The Fourth Creed assigned to this Council, which might be Lucianic. 1962 *Catholic Dict. Theol.* I. 159 They could never hope to pass off the rambling and old-fashioned Lucianic creed as one that would now rally all theologians of the West.

luciferase (liusi·fĕrēͥz, -s). *Biol.* [ad. F. *luciférase* (R. Dubois 1887, in *Compt. Rend.* CV. 691): see next and *-ase.] Any enzyme which catalyses a reaction by which a specific luciferin produces light.

It was formerly believed that a single enzyme, luciferase, was common to all organisms which produce light.

1888 *Jrnl. R. Microsc. Soc.* 66 From the luminous parts of the animal the author [*sc.* R. Dubois] has succeeded in extracting two substances, the contact of which in the water, determines the appearance of the light. One of them was obtained in the crystalline state... It is soluble in water, and hardly soluble in alcohol; it may be called luciferine. The other body is an active albuminoid of the class of soluble ferments, and may be called luciferase. 1920 *Nature* 26 Aug. 843/1 The production of light by animals is due to the burning or oxidation of a substance called luciferin in the presence of an enzyme or catalyst called luciferase. 1952 [see *luciferin]. 1973 *Country Life* 9 Sept. 603/1 The luciferin, or light-producing material, in a glow-worm's body is oxidised and broken down, with the aid of an enzyme called luciferase, to release light energy with practically no wastage.

luciferin (liusi·fĕrin). *Biol.* Formerly also **-ine.** [ad. F. *luciférine* (R. Dubois 1887, in *Compt. Rend.* CV. 691), f. L. *lūcifer* light-bearing: see -in[1].] Any substance which is present naturally in an organism (such as the glow-worm) and which when oxidized in the presence of a specific enzyme (a luciferase) is capable of producing light.

It was formerly believed that one single substance, luciferin, was common to all organisms which produce light.

1888, 1920 [see *luciferase]. 1952 E. N. Harvey *Bioluminescence* p. xi, Luciferin and luciferase are general names used for these compounds manufactured by luminous animals, but it is probable that the luciferin or luciferase from a species in one group may be quite different chemically from that in another. 1954 Harvey & Tsuji in *Jrnl. Cell. & Compar. Physiol.* XLIV. 71 Luciferin may properly be defined as the oxidizable substance supplying molecules capable of absorbing enough excess energy to emit in the visible region. 1966 Johnson & Haneda *Bioluminescence in Progress* 10 Bacterial luciferin has become a matter of definition... $FMNH_2$ may be considered the 'luciferin' of this system. 1971 [see *luciferase].

Lucite (liū·səit). Also **lucite.** [f. L. *luc(i)-, lux* light + -ite[1].] A proprietary name for a solid, transparent plastic that is a methyl methacrylate resin; perspex.

1937 *Official Gaz.* (U.S. Patent Office) 6 July 9/1 E. I. DuPont de Nemours and Company. Lucite. For thermoplastic synthetic resin material, known as methyl methacrylate, in the form of sheets, rods, and tubes, and in powdered or granular form. 1939 *Reader's Digest* Feb. 81/1 Light flows through rods made of Lucite, a du Pont plastic, as water flows through a pipe. 1953 J. Y. Cousteau *Silent World* 89 Through the lucite windows [of the *Bathyscaphe*] the pilots could look into a landscape lighted by exterior floodlamps. 1973 R. Hayes *Hungarian Game* xv. 98 The system had two Lucite indicators, one red the other green.

luck, *sb.* Add: **3.** *as (good, ill) luck would have it*: by (good, ill) fortune; *best of Arab* (or *Welsh*) *luck*: cf. *British a. 6; *better luck another* (or *next*) *time*: an expression of encouragement to endure a disappointment; *devil's own luck*: uncannily good luck; *good luck to* (a person or thing): see *good luck; *just my* (or *his, our,* etc.) *luck*: typical of my (his, our) bad luck, or occas., of my good luck; (*one's*) *luck is in*, or *is/runs out*: luck is on one's side, or, one has come to the end of one's run of luck; *luck of the draw*: an expression of resignation to chance; *no such* (*good*) *luck*: the thing is beyond the power even of good fortune, unfortunately not; *push* (or *crowd, ride*) *one's luck*: to expect or count on an even better run of good fortune than one has had already; cf. *to push one's fortune* (Push *v.* 11 b); *you never know your luck*: you cannot be sure that luck won't change for the better, you may be lucky.

1598 Shakes. *Merry W.* III. v. 83 As good lucke would haue it..they conuey'd me into a bucke basket. 1680 A. Radcliffe in Rochester *Poems* 147 As luck wou'd have it in came Will. 1687 J. Philips tr. *M. de Cervantes's Don Quixote* I. I. ii. 9 As ill luck would have it, it happen'd to be upon a Friday Night. 1854 E. Twisleton *Let.* 29 June (1928) 213 They..asked us to a dinner a fortnight off, when as ill-luck would have it, we were engaged. 1855 F. W. Faber *Growth in Holiness* xvi. 297 When we fall we must rise again, and go on our way, wishing ourselves, after a Christian fashion, better luck another time. 1857 C. M. Yonge *Dynevor Terr.* II. xii. 176 'When you break down anywhere, send me a telegraph.' 'No such good luck,' sighed Clara. 1858 Trollope *Doctor Thorne* II. iv. 74 No harm had been done, and he might have better luck next time. 1862 W. Collins *No Name* I. I. xi. 150 All *he* ever said was 'better luck next time'. *c* 1874 D. Boucicault in M. R. Booth *Eng. Plays of 19th Cent.* (1969) II. 190 Well, as the divil's luck would have it, there was only ..a tailor's thimble, an' they couldn't get it full. 1891 E. Dowson *Let.* 30 June (1967) 205, I came here..& found as luck would have it a Rosière going on. 1892 R. L. Stevenson *Let.* 29 May (1911) IV. 54 No such luck; the ship delayed, and at last, about three, I had to send them home again. 1898 J. D. Brayshaw *Slum Silhouettes* 29 Well, yer never know yer luck; an' his was 'ard enuff, Gawd knows. 1901 Ade *Forty Mod. Fables* 156 Adams had a Run of Luck and he crowded it. 1903 G. B. Shaw *Man & Superman* I. 21 'She'll marry you.' 'No such luck, Jack!' 1907 —— *John Bull's Other Island* IV. 82 He has the divil's own luck, that Englishman, annyway. 1909 Galsworthy *Strife* I. 195 Just our luck, the men finding a fanatical firebrand like Roberts for leader. *a* 1911 D. G. Phillips *Susan Lenox* (1917) II. xxi. 494 Don't be a fool. Let's push our luck, now that things are coming our way. 1916 A. Bennett *These Twain* III. xx. 488 You never know your luck. If she'd been free I might have been fool enough to get married. 1923 *Brewer's Dict. Phr. & Fable* (new ed.) 689/1 *He has..the devil's own luck*, he is extraordinarily lucky; everything he touches turns to gold. 1926 J. Buchan *Dancing Floor* I. ii. 53 Something about the features..struck me as familiar. As luck would have it, it turned out to be Vernon. 1927 E. O'Neill *Marco Millions* I. iii. 45 Better luck next time. He'll learn! 1928 —— *Strange Interlude* VIII. 274 The damned radio has to pick out this time to go dead!.. Just my luck! *a* 1930 D. H. Lawrence *Phoenix II* (1968) 146 Gilbert at her side took step after step, and thought to himself his luck was out as regards women. 1938 R. D. Finlayson *Brown Man's Burden* 60 Just my luck to be caught in a thunderstorm. 1938 G. Greene *Brighton Rock* III. i. 99 'You ever come across this Kolley Kibber?' she asked. 'No such luck,' the barman said. 1956 J. Potts *Diehard* vii. 116 He paused. Was he pushing his luck too far? 1959 'J. Welcome' *Stop at Nothing* i. 11 He had never won the Derby and..had..announced that he would not die until he did. As he must by now be touching eighty.., this was pushing his luck pretty hard. 1963 A. Smith *Throw out Two Hands* vii. 79 If our luck was in we might hit Madagascar, but there is quite enough luck needed in ballooning without attempting to stretch it, and to hope for a landfall on a solitary island. 1966 'S. Harvester' *Treacherous Road* xix. 183 And the best of Arab luck to you, mate. 1966 M. R. D. Foot *SOE in France* x. 341 *Cinema-phono* had never been a lucky circuit, and its luck now ran out altogether. 1967 *Listener* 23 Nov. 667/2 As luck, or history, if you like, would have it, the Russian Revolution coincided with the spread..of wireless telegraphy. 1967 'J. Ashford' *Forget what you Saw* iii. 14 You never know your luck—one of these days we might actually set sail. 1967 M. Childs *Taint of Innocence* (1968) iii. 177 'It's so unfair.' 'Well, the luck of the draw.' 1969 'A. Garve' *Boomerang* iii. 132 'Anything else?' 'I don't think so—Except to wish you the best of Welsh luck!' 1970 N. Marsh *When in Rome* v. 127 Don't tell me you *are* the Lord Chamberlain... It would have been just my luck. 1972 M. Babson *Murder on Show* v. 58 My luck was in and I caught a taxi. 1973 'H. Howard' *Highway to Murder* iii. 37 Some say first impressions are best. Mine have been wrong as often as they've been right, so I guess it's the luck of the draw. 1974 I. Murdoch *Sacred & Profane Love Machine* 75 If I..was never heard of again. Thank you very much! No such luck! 1975 J. Symons *Three Pipe Problem* xv. 131 You ain't going to find no killer... I reckon this is some amateur riding his luck.

luck, *v.* Restrict *Obs.* exc. *dial.* to senses in Dict. and add: **1. c.** (Later examples.)

1946 Mezzrow & Wolfe *Really Blues* 376 Luck up on: get by luck, come into possession of unexpectedly. 1962 E. Lacy *Freeloaders* i. 9 This is the best writing pad I've ever lucked up on. 1971 *Black World* Apr. 56 All of em hopin to luck up on a few grand by hittin on me. 1973 *Black Panther* 29 Sept. 2/3 Riggs happened to luck up on a good hustle by attacking women's rights in tennis.

e. *to luck out* (U.S.), to achieve success or advantage by good luck in a difficult, testing, or dangerous situation.

1954 *Amer. Speech* XXIX. 303 He lucked out on the final examination... I just 'lucked out' on that shot... He really 'lucked out'. He didn't get any Saturday classes. 1967 *Boston Sunday Herald* 7 May VI. 8/4 If you luck out, good. If not, you've still got your original bookings. 1971 J. Ball *First Team* (1972) xxi. 317 Since we pretty much had to choose one from one..it looks as if we lucked out. 1972 J. Wambaugh *Blue Knight* (1973) ii. 22, I started making inquiries..and damned if I didn't luck out and get steered into a good job.

f. *to luck into*, to acquire by good fortune.

1959 *Time* (Atlantic ed.) 6 July 15/1 Loveless..lucked into booming revenues from old taxes as..Iowa expanded. 1966 S. Morrow *Moonlighters* (1967) ix. 98 He lucked into a prime location there. 1970 J. Potts *Affair of Heart* viii. 63 The rent was fantastically low; she had lucked into it a couple of years ago through an artist friend.

Lucky (lʋ·ki), *sb.*[3] [ellipt. f. *Lucky Strike*, a U.S. brand of cigarettes.] A Lucky Strike cigarette.

1934 A. Huxley *Beyond Mexique Bay* 153 If Luckies are to sell here, they will have to be put under the patronage of Saint Joseph. 1949 O. Nash *Versus* 139 As a Lucky Striker, Whenever I offer smart folk a Lucky [etc.]. 1953 A. Baron *Human Kind* 184 He fumbled for a packet of Luckies and offered Casey a fresh cigarette. 1963 'D. Rutherford' *Creeping Flesh* i. 63 I smoked six Luckies. *Ibid.* 80 A packet of Luckies, please. 1969 N. Freeling *Tsing-Boum* xxii. 160 Those French ones—no thanks. I only smoke Luckies, really.

lucky, *a.* Add: **1. e.** *lucky him* (or *you*, etc.): phrases expressing envy at another's good fortune.

1857 C. M. Yonge *Dynevor Terr.* II. xi. 171 She's..the finest figure in the whole county; lucky him who gets her. *Ibid.* xii. 176 Lucky you to have work at home, and to stay with it. 1893 W. Elwin *Let.* 4 May in E. Lutyens *Blessed Girl* (1953) x. 190 Lucky you to be mothered by Betty. 1965 A. Roudybush *Season for Death* (1966) xxviii. 165 'I'm going to the flicks.' 'Lucky you. God bless!' 1972 R. Hill *Fairly Dangerous Thing* I. iv. 36 'I'm busy every night but tonight.' 'Lucky you,' said Joe.

3. (Further examples.)

1905 A. Burvenich *Eng. Idioms* 219 To be born under a lucky star. 1911 C. E. W. Bean *'Dreadnought' of Darling* xxxv. 317 The Australian should probably thank Providence and his lucky star. 1920 W. J. Locke *House of Baltazar* xv. 187 It doesn't seem to be one of the House of Baltazar's lucky days. 1959 J. Braine *Vodi* xxv. 261 Thank your lucky stars to be well out of it. 1973 A. Behrend *Samarai Affair* xii. 115 Richardson wondered if this was going to be his lucky day.

7. lucky dip: see *dip sb. 1 h; **lucky dog,** someone considered by others to be lucky.

1841 S. Bamford *Passages in Life of Radical* (ed. 2) I. xxix. 175 They were a set of 'lucky dogs'... They escaped. 1844 [in Dict., sense 1]. 1922 Joyce *Ulysses* 634 You were a lucky dog if they didn't set the terrier at you directly you got back.

lucky-bag. Add: **1.** (Further examples.)

1891 *Confectioners' Union* IV. 530/1 (Advt.), Farthing Goods... Pick-me-up Lucky Bags. 1927 W. E. Collinson *Contemp. Eng.* 20 The term lucky bag is still used to denote a bag stuffed full of miscellaneous objects. 1931 A. Uttley *Country Child* iii. 37 There the paper bag lay, 'Monster Lucky Bag', with some others..on the counter, asking her to open it. 1968 'E. Peters' *Grass Widow's Tale* vii. 90 This trifle, hardly too big to have fallen out of a child's lucky-bag. 1973 *Dalesman* May 144/1, I well remember..the lucky bags which cost a ha'penny. These bags were pink, yellow or green and contained some sweets and a little novelty or toy.

2. (Earlier U.S. examples.)

1832 E. C. Wines *Two Yrs. in Navy* (1833) I. 55 All property that falls in his [*sc.* the master-at-arms'] way for which he cannot find an owner, is thrown into the 'lucky bag'. 1840 *Southern Lit. Messenger* VI. 233/2 Every man-of-war, you know, has her lucky bag, containing a little of every thing, and something belonging to everybody.

luculia (lukū·liă). [mod.L. (R. Sweet *Brit. Flower Garden* (1826) II. 145), f. *luculi swa* native name of *L. gratissima* in India.] A large deciduous shrub of the genus so called, belonging to the family Rubiaceæ, native to northern India and China, and bearing corymbs of fragrant pink flowers.

1826 R. Sweet *Brit. Flower Garden* II. 145 (*heading*) Delightfully fragrant Luculia. 1885 W. J. May *Greenhouse Managem.* (ed. 2) vi. 236 The best place for Luculias is in the beds or borders of a conservatory where they have plenty of room to grow. 1962 *Amat. Gardening* 21 Oct. (Suppl.) 32/3 The luculias need plenty of room and are best planted out in a well drained, peaty soil in the greenhouse.

Lucullan (liukʋ·lăn), *a.* [L. *Lucullanus, Lucullianus.*] = Lucullian, -ean *a.*

1857 Bostock & Riley tr. *Pliny's Nat. Hist.* VI. xxxvi. ii. 307 The largest of these columns, pillars of Lucullan marble..were erected in the atrium of Scaurus. *Ibid.* viii. 325 L. Lucullus was consul; the same person who gave its name, it is very evident, to the Lucullan marble. 1909 G. G. Ramsay tr. *Tacitus' Annals* II. xi. xxxii. 41 Messalina took herself off to the Lucullan Gardens. 1913 *Chambers's Jrnl.* 11 Jan. 87/1 Two thousand pounds for a really Lucullan feast as planned..by one who plans them constantly. 1935 *Times Lit. Suppl.* 11 Apr. 245/2 This is definitely a Lucullan book, of use only to the hostess with a chef and a large staff. 1935 J. M. Cobban *Senate & Provinces* iv. 116 Clodius and his fellows were acting at the direct instigation of the 'anti-Lucullan' party at Rome. *Ibid.* 123 Pompey..whom we have seen to be the prime force behind the anti-Lucullan movement. 1966 *Punch* 4 May 669/3 He gives us a Lucullan feast of stories, but my favourite is the tale of Mrs. Keppel.

Lucu·llic, *a.* = prec.

1904 J. McCabe tr. *Haeckel's Wonders of Life* xi. 98 The careful choice and preparation of savoury food..was just as important..as it is to-day in royal banquets or the Lucullic dinners of millionaires.

‖**lucus a non lucendo** (lū·kʋs a nǫn lūke·ndo, liū·kʋs ē͡i nǫn liūse·ndo). [L. phr., 'a grove, (so-called) from the absence of *lux* (light)', discussed by Quintilian in *De inst. oratoria* i. 6. 34.] A paradoxical or otherwise absurd derivation; something of which the essence or qualities are the opposite of what its name suggests. Also **lucus a non.**

1711 Addison *Spectator* 8 May, He composed an..Epic Poem..consisting of four and twenty Books, having entirely banished the Letter *A* from his first Book, which was called *Alpha* (as *Lucus a non lucendo*) because there was not an *Alpha* in it. *c* 1728 Swift in W. King *Dreamer* 88 And make his ignorance discerned, To get the name of council learned; (As *lucus* comes *à non lucendo*). 1749 Fielding *Tom Jones* III. VIII. iv. 172 This Sun, into which Jones was now conducted, was truly named as *Lucus a non lucendo*; for it was an Apartment into which the Sun had scarce ever looked. 1823 Byron *Don Juan* VI. lv. 240 Thus..has been shown 'Lucus a *non* Lucendo', *not* what *was*, But what *was not*; a sort of style that's grown Extremely common in this age. *Ibid.* XI. xxi. 54 Through Groves, so called as being void of trees, (Like *lucus* from *no* light;). 1845 R. Ford *Hand-bk. Trav. Spain* II. 941 *St. Jean de Luz*..is not a 'city of light', but of 'mud', and a Lutetia or lucus a non lucendo. 1848 J. R. Lowell *Fable for Critics* 69 'Illustrations'..are said to illustrate, because, as I view it, Like *lucus a non*, they precisely don't do it. 1904 W. F. H. King *Classical & Foreign Quotations* 186 To the *Lucus a non lucendo* principle, as it is called, are referred all such paradoxical derivations and descriptions which involve a contradiction in the mere stating of them. 1924 A. Huxley *Little Mexican* 13 Their philology was the picturesque *lucus a non lucendo*, bloody from by-our-Lady type. 1958 R. Liddell *Morea* III. 219 Was its name Hydraea (watery), a *lucus a non lucendo*—it is singularly waterless today.

luderick (lū·dərik, lʋ·drik). *Austral.* Also **ludrick.** [Aboriginal name.] A perciform herbivorous food fish, *Girella tricuspidata*, which has a dark-coloured back and silvery belly; also called blackfish, black bream or perch.

1898 E. E. Morris *Austral Eng.* 275/2 Luderick, or Ludrick, an aboriginal Gippsland name for a local variety of the fish *Girella simplex*. 1951 T. C. Roughley *Fish & Fisheries Austral.* 92 In 1947, when a conference of fisheries officials from all Australian States was held to discuss the question of the diversity of the names of fish in the various States, it was decided to adopt the Victorian name of 'luderick' for this fish throughout Australia... The luderick is an estuarine fish found in all Australian States, but it occurs in its greatest numbers in New South Wales, southern Queensland, and Victoria. 1969 *Man* (Austral.) Mar. 12/2 Over the seasons you get a variety of fish you'd be scratching to find anywhere else—drummer, luderick, [etc.].

Ludian (liū·diăn). Also **Lüd, Lude, Ludic, L'üdikš, Lüdish.** [f. Olonetsian *liüdi* (? ad. Russ. *ljudi*, people) + -an, -ian.] A language of the Finnish group of the Finno-Ugrian family of languages, used by a small number of speakers in the region of Olonets, now in the north-west part of the Russian Soviet Federative Socialist Republic.

1921 M. Stenbäck in T. Homén *East Carelia & Kola Lapmark* II. ii. 122 Among the Onega Wepsians... Those who speak Wepsian are said to speak *l'üdikš*. 1933 L. Bloomfield *Lang.* 68 The other languages of the Baltic branch [of the Finnish-Lapponic languages], *Carelian, Olonetsian, Ludian, Vepsian, Livonian, Ingrian*, and *Votian*, are far smaller, and some of them are near extinction. 1939 L. H. Gray *Foundations of Lang.* 369 The languages of the Uralic family are as follows: (1) *Finnish* group: Finnish proper.., Karelian.., Olonetzian, Ingrian, Lüdish, Vepsian, [etc.]. 1954 Pei & Gaynor *Dict. Ling.* 127 *Ludian*, a dialect (also called *Lüdish*) of the Finnish group of the Finno-Ugric (or Uralic) sub-family of the Ural-Altaic family of languages. 1955 B. Collinder *Fenno-Ugric Vocab.* p. ix, Lude.., spoken in the region west of lake Onega, is a kind of intermediary between Olonets and Veps. 1959 *Chambers's Encycl.* XIV. 200/2 Lüd, spoken to the north and west of Petrozavodsk. 1967 D. S. Parlett *Short Dict. Lang.* 81 *Ludian* 10 th[ousand] speakers in region of Olonets (north east of Leningrad on shore of Lake Ladoga) of a series of patois, basically northern Vepsian but with influences of Carelian, and forming a transition between the two. Impregnated with Finnish and Russian. 1974 *Encycl. Brit. Micropædia* V. 709/3 Ludic, a minor group of dialects spoken to the south of Karelian, is considered to be a blend of Karelian and Veps.

ludic (liū·dik), *a.* [ad. F. *ludique*, f. L. *ludĕre* to play.] Of or pertaining to undirected and spontaneously playful behaviour.

1940 Hinsie & Shatzky *Psychiatric Dict.* 323/2 This excess-energy must be expended (without purpose) in some way, most usually in play-activity, called *ludic activity*. 1969 P. L. Berger *Rumor of Angels* iii. 76 Ludic, or playful, elements can be found in just about any sector of human culture. 1971 D. G. Boyle *Lang. & Thinking in Human Devel.* vi. 65 Ludic play and symbolism are among the defining characteristics of humanity. 1972 *Times Lit. Suppl.* 3 Mar. 234/5 Poetry is sacred..for the same reasons that eroticism is sacred: it is ludic, that is to say useless. 1973 M. Amis *Rachel Papers* 207 My existence, too, was a prismatic web of mendacity—but for me it was..far more ludic, literary, answering an intellectual rather than an emotional need.

Ludlovian (lʋdlō͡u·viăn), *a. Geol.* [f. *Ludlovia*, med.L. name of Ludlow, town in Salop (Shropshire) in the vicinity of which are exposures of this series: see -ian.] Of, pertaining to, or designating the latest of the three divisions of the Silurian, preceding the Downtonian (or the second latest, if the Silurian is taken to include the Downtonian). Also *absol.*

1855 J. Phillips *Man. Geol.* vi. 104 This stage contains no less than seventy-eight species—the genera being mostly Wenlockian and Ludlovian. 1946 L. D. Stamp *Britain's Struct.* xii. 109 Though various divisions of the Silurian have been proposed the old three-fold one into: (3) Ludlow Series or Ludlovian; (2) Wenlock Series or Wenlockian; (1) Llandovery Series or Valentian; has stood the test of time and is most convenient. 1969 Bennison & Wright *Geol. Hist. Brit. Isles* vi. 121 There is..an almost perfect transition in the Welsh Borders from the Ludlovian into the Dittonian reflecting a gradually changing environment from marine to non-marine. *Ibid.* 120 *Monograptus scanicus* suggests a low Ludlovian age for these deltaic deposits.

ludo. (Later examples.)

1955 L. P. Hartley *Perfect Woman* xxxvi. 324 The four ludo-players identified her. 1973 *Times* 4 Oct. 2/5 Pedestrians feel like pawns on a vast spongy chess and ludo board. 1975 R. Player *Let's talk of Graves* iv. 108 Our dear Miss Grigg produced a new Ludo set and..a few crackers.

Ludolph (lū·dǫlf). [The name of *Ludolph* van Ceulen (1540–1610), who was born at Hildesheim (Germany), taught mathematics in the Netherlands, and calculated π to 35 decimal places.] *Ludolph's number*; the number π (see Pi *sb.*).

[1887 *Encycl. Brit.* XXII. 434/2 In Germany the 'Ludolphische Zahl' is still a common name for the ratio.] 1894 F. Cajori *Hist. Math.* 154 The value of π is therefore often named 'Ludolph's number'. 1959 *Webster's Biogr. Dict.* s.v. *Ceulen, Ludolph van*... Known for computations of the value of π (Ludolphian, or Ludolph's number).

Hence **Ludo·lphian** *a.*

1886 G. S. Carr *Synopsis Pure & Appl. Math.* I. II. 901/1 (Index), Ludolphian number. 1905 *Westm. Gaz.* 14 Oct. 3/1 The history of the search for this ratio, the Ludolphian number..is practically the early history of mathematics itself. 1953 H. Eves *Introd. Hist. Math.* iv. 92 Ludolph van Ceulen of Germany computed π to 35 decimal places by the classical method, using polygons having 2^{62} sides. He spent a large part of his life on this task and his achievement was considered so extraordinary that the number was engraved on his tombstone, and to this day is frequently referred to in Germany as 'the Ludolphian number'.

Ludwig's angina (lu·dvigz ændʒəi·nă). *Path.* [tr. mod.L. *angina Ludovici*, f. the name of W. F. von *Ludwig* (1790–1865), German surgeon, who described it in 1836.] Severe inflammation of the tissues of the floor of the mouth (usu. caused by streptococci).

[1875 *Brit. Med. Jrnl.* 25 Dec. 778/2 (*heading*) A note on Angina Ludovici. 1885 *Lancet* 26 Sept. 571/2 (*heading*) Two cases of angina Ludovici.] *Ibid.*, Ludwig's angina is a name applied to the condition known as submaxillary cellulitis. 1939 *Bull. Hist. Med.* VII. 1124 Ludwig's Angina is a comparatively rare disease despite the fact that the disease picture is fairly well known to all students of medicine. 1971 D. F. Mitchell et al. *Oral Diagn.* (ed. 2) xiv. 344 In this case, all of the spaces of the floor of the mouth (submandibular, sublingual, submental) are involved and the condition is called Ludwig's angina.

Ludwigsburg (lu·dvigzbʋ̈ɹg). The name of a town in Württemberg used *attrib.* or *absol.* to designate a variety of hard-paste porcelain made there from 1758 to 1824, characterized by its suitability for figure-modelling.

1863 W. Chaffers *Marks Pott. & Porc.* 185 Ludwigsburg, or Louisburg, called also Kronenburg porcelain. 1960 R. G. Haggar *Conc. Encycl. Cont. Pott. & Porc.* 188/1 (*caption*) Ludwigsburg porcelain miniature group. *Ibid.* 189/1 Ludwigsburg porcelain Chinese groups. *Ibid.* 273/1 Ludwigsburg models were reproduced at Amberg.. during the nineteenth century.

‖**lues Boswelliana** (liū·īz bǫzwelię̄ͥ·nă, -ā·nă). [f. Lues: cf. Boswellian *a.*, *-iana.] A disease of admiration; a biographer's tendency to magnify his subject.

1834 Macaulay in *Edin. Rev.* Jan. 508 Biographers, translators, editors,—all, in short, who employ themselves in illustrating the lives or the writings of others, are peculiarly exposed to the *Lues Boswellianæ*, or disease of admiration. 1928 [see *furor 4]. 1931 *Times Lit. Suppl.* 8 Oct. 765/3 Though he decidedly comes to praise Caesar, not to bury him, he avoids any excessive *lues Boswelliana*.

lueshite (liū·əʃəit). *Min.* [a. F. *lueshite* (A. Safiannikoff 1959, in *Bull. d. Séances, Acad. r. d. Sci. d'Outre-Mer* V. 1255), f. *Lueshe*, name of a locality north of Goma in eastern Zaïre where it was discovered: see -ite[1].] A black orthorhombic niobate of sodium, $NaNbO_3$.

1961 *Mineral. Mag.* XXXII. 966 Lueshite. 1968 I. Kostov *Mineral.* I. 250 The chief minerals [of the perovskite group] are perovskite ($CaTiO_3$), lueshite ($NaNbO_3$), and loparite ($NaCeTi_2O_6$).

luff, *v.* Restrict *Naut.* to senses 1 to 4 and add: **3. b.** To obstruct (an opponent's yacht which is attempting to pass to windward on the same course) by sailing one's own yacht closer to the wind.

1912 Heckstall-Smith & Du Boulay *Compl. Yachtsman* xiii. 299, I make it a fixed rule to luff every vessel that attempts to pass me to windward. 1960 E. Schiöttz *Pract. Yacht Racing* ix. 89 If a yacht clear astern sails between two yachts ahead she has no right to luff the windward one. 1965 *Sailing* ('Know the Game' Series) (ed. 2) 34/1 A yacht may luff a yacht clear astern or a windward yacht, until helmsman of windward yacht comes abreast of the mainmast of the leeward yacht.

5. *trans.* To alter the inclination of (the jib of a crane or derrick); to raise or lower in a vertical plane. Also with adverbs, as *luff in*, to raise (the jib), so moving the hook nearer to the operator; similarly *luff out*.

1913 H. Wilda *Cranes & Hoists* ii. 128 The lifting and lowering of the load is effected by means of multiple pulley blocks, the horizontal movement of the load by luffing the jib. 1922 H. H. Broughton *Electr. Handling of Materials* III. ii. 32 The load follows a horizontal path when luffing-in the jib. 1932 S. J. Koshkin *Mod. Materials Handling* v. 78 A derrick boom can be changed in inclination, or luffed, to give one of the components of motion to the load. *Ibid.* 80 The boom is luffed-in or out as little as possible. 1963 R. Hammond *Mobile & Movable Cranes* vi. 180 This arrangement reduces the power required to luff in the jib.

luffing *vbl. sb.* (further examples: cf. sense *5); so *luffing-in, -out*; **luffing crane,** a crane whose jib can be luffed in operation.

1913 *Engineering* 9 May 632/1 Most of these luffing-cranes are now fitted with some sort of compensating gear to keep the load approximately level while the luffing operation is being carried out. *Ibid.*, Greater acceleration in luffing is also obtained. 1922 H. H. Broughton *Electr. Handling of Materials* III. ii. 41 The design is exceedingly simple, the level path being obtained by the automatic paying-out or taking-in of the load rope during the operations of luffing in and luffing out respectively. 1963 R. Hammond *Mobile & Movable Cranes* vi. 181 The luffing motion control equipment consists of a negative-phase sequence panel, with contactors to give plain rotor-controlled 'luffing in' and 'luffing out' by the unbalancing of the stator voltages of the motor through an inbuilt transformer. 1967 *Courier-Mail* (Brisbane) 8 July 1/9 The cries of wharf laborer Derek Beuttel, 54, were not heard by fellow workmen because of the noise of three large luffing cranes.

‖**luftmensch, luftmensh** (lu·ftmenʃ). Pl. **luftmenschen.** [Yiddish, f. G. *luft* air + *mensch* person.] An impractical visionary.

1907 I. Zangwill *Ghetto Comedies* 240 The word 'Luftmensch' flew into Barstein's mind. Nehemiah was not an earth-man... He was an air-man, floating on facile wings. 1960 *Commentary* June 530/1 The revolutionary student, the nihilist-anarchists, the *Luftmenschen*. 1966 *New Society* 12 May 11/2 Americans, children of the soil, have become traders in air, advertising men, *luftmenschen*. 1968 L. Rosten *Joys of Yiddish* 212 The prototype of the luftmensh was one Leone da Modena,.. who listed his skills and cited no fewer than twenty-six professions... Why would so accomplished a man be classified as a *luftmensh*? Because out of all twenty-six professions.., he barely made a living. 1969 *Sat. Rev.* (U.S.) 26 Apr. 44/2 Menachem-Mendl, *luftmensch* extraordinary, a veritable Don Quixote of Eastern European commerce.

‖**Luftwaffe** (lu·ftvafə). [G., 'air-weapon'; cf. Lift *sb.*[1]] The German air force before, and until the end of, the 1939–45 war. Also *attrib.*

1935 *Times* 23 May 15/1 The armed forces are henceforth known collectively as the Wehrmacht (Defence Force) and consist of the Army (*Heer*), Navy (*Kriegsmarine*), and the Air Arm (*Luftwaffe*). 1939 *War Illustr.* 21 Oct. 169 The German Luftwaffe—literally 'air weapon' —has been built up by Field-Marshal Goering and General Milch into a formidable force both in size and quality. 1941 *Ann. Reg. 1940* 66 The repulse..broke down the legend of the invincibility of the German *Luftwaffe*. 1941 H. G. Wells *You can't be too Careful* v. ii. 244 The Luftwaffe he [*sc.* Goering] had launched. 1942 *Electronic Engin.* XV. 240 Generally the receiving valves are..similar to those used in German Luftwaffe sets. 1946 *R.A.F. Jrnl.* May 174 One Group alone has disposed of thousands of tons of Luftwaffe equipment. 1957 *Encycl. Brit.* I. 453/1 As soon as the arms limitation of the Versailles treaty had been repudiated, Germany embarked upon the creation of a great Luftwaffe.

lug, *sb.*[1] Add: **3. lug-pole** *U.S.* (= sense 1).

1773 *Mass. Gaz.* 4 Feb. (Suppl.) 1/2 A Defect in the Chimney, by Reason of the Wooden Lug-pole burning out. 1848 D. Drake *Pioneer Life Kentucky* (1870) v. 107 The tea kettle swung from a wooden 'lug pole'.

lug, *sb.*[2] Add: **2.** (Later examples.)

The statement in Dict. 'In Sc. the only word in use, *ear* being *obs.* exc. in combination' is no longer true. In Scotland as in England *lug* in this sense is *colloq.* or *joc.*

1908 *Old-Lore Misc.* I. VII. 270 Lang an last, da laird grippit him be da lug. 1916 'Taffrail' *Pincher Martin* ii. 28 Give 'im a clip under the lug! 1922 *Banffshire Jrnl.* 26 Sept. 6 An' hame-brewn sets th' lugs a crackin'. 1922 Joyce *Ulysses* 192 Has the wrong sow by the lug. *Ibid.* 261 Cowley's red lugs and Adam's apple in the door of the sheriff's office. 1939 —— *Finnegans Wake* 500 The snare drum! Lay yer lug till the groun.

3. d. *to put* (or *pile*) *on lugs*, to put on airs. *U.S. slang.*

1889 K. Munroe *Golden Days* xvii. 188 If you notice me..piling on any lugs..you just bump me down hard. 1896 Ade *Artie* vi. 54 The family did n't put on no such lugs in them days. 1903 A. Bennett *Leonora* iv. 106 American women..put on too much lugs, at any rate for an Englishman. 1905 *Springfield* (Mass.) *Weekly Republ.* 15 Sept. 12 Dr. Hall puts on no 'lugs', and is not above sitting on a cracker barrel in a country grocery for a chat with old acquaintances. 1920 S. Lewis *Main St.* 326 Oh, the lugs he puts on—belted coat, and piqué collar.

e. A demand for borrowed or exacted money. Esp. in phr. *to put the lug on*, to extort, to put pressure on. *U.S. slang.*

1929 D. Runyon in *Hearst's Internat.* Aug. 73/2 Why do you not put the lug on him? 1935 A. J. Pollock *Underworld Speaks* 84/2 *Out on the lug*, engaged in begging racket. *Ibid.* 93/1 *Put the lug on*, to borrow; to beat up a racketeer with blackjack or brass knuckles for muscling in on forbidden territory. 1936 *Kansas City* (Missouri) *Star* 15 Oct. 6/1 Indiana uses the 'Lug'. 1938 *Kansas City* (Missouri) *Times* 14 Feb. 1/6 The Democratic organization's lug on all city and county employees..for its campaign fund..has been started. 1940 *Topeka* (Kansas) *State Jrnl.* 26 Mar. 1/8 Shakedowns in Topeka are known to have ranged from $20 to $50 monthly, depending on the amount of illegal business done by the individuals on whom the lug was put. 1973 M. Truman *Harry S. Truman* vii. 129 My father also knew, from his inside contacts with Missouri Democrats, that the governor..was 'putting the lug' (to use Missouri terminology) on state employees to contribute to his campaign fund.

4. (Earlier U.S. examples.)

1835 J. Martin *New Gazetteer Virginia* 175 An eminent tobacco manufacturer of Richmond has offered the inhabitants of this district to take all of their tobacco, (lugs included) at $10 a hundred. 1851 *Southern Planter* (Richmond, Virginia) June 192/1 We quote lugs $5 25 to $7.

4*. Chiefly *N. Amer. slang.* Used contemptuously of a person: a lout, a sponger.

1931 *Broadway Brevities* 19 Oct. 2/1 Is his only sin the fact that he was born a lug? 1935 A. J. Pollock *Underworld Speaks* 73/2 *Lug*, an incapable person who frequently borrows small sums of money. 1936 R. Chandler *Black Mask* June 24/1 The girl snapped at me: 'Is this lug your partner?' 1952 *Landfall* VI. 265 Now there's your sermon!.. No, no, no. These lugs would never understand. 1953 K. Tennant *Joyful Condemned* xxix. 284 The big fellows slip through my fingers, leaving some tough lug to take the rap. 1968 B. Turner *Sex Trap* xi. 98 'The other lug's at June's,' Louis said. 1973 *Hansard* (Canada) 20 Mar. 2388/1, I will stand down when Mr. Speaker tells me to, not when you lugs tell me to. 1973 'B. Mather' *Snowline* x. 116 Any other names you can come up with?.. You don't owe these lugs anything.

5. lughole *dial.* and *colloq.*, ear-hole; **lug sole** *N. Amer.* (see quot. 1961).

1895 J. S. Fletcher *Wonderful Wapentake* 72 I'll come ..and pelt thi lughoil for tha. 1898 B. Kirkby *Lakeland Words* 96 Stuff thi lug-whols wi' woo. 1966 F. Shaw et al. *Lern Yerself Scouse* 20 *Is lugole*, his ear. 1973 *Times* 25 Aug. 10/8 A session with *Hello, Cheeky* is like being exposed to some noisy, rude and unstoppable urchin who wins you round or at least averts a skull-shattering clout about the lughole simply because he will go on regardless. 1961 Webster, *Lug sole*, a thick rubber sole that has deep indentations in a pattern designed to provide good footing and is used on sport and work shoes. 1970 *Toronto Daily Star* 24 Sept. 16/2 (Advt.), Heavy duty lug sole.

Luganda (liugæ·ndă). A language of the Bantu group, spoken in Uganda.

1889 R. P. Ashe *Two Kings of Uganda* p. ix, *Buganda* is the country itself, '*Muganda*' a native of the country, '*Baganda*' the plural of '*Muganda*', and '*Luganda*' is the language. 1902 *Encycl. Brit.* XXXIII. 541/2 The languages spoken in the Uganda Protectorate belong to the following stocks:—... *Bantu* (Lu-Ganda, [etc.]). 1933 L. Bloomfield *Lang.* iv. 67 The languages of the Bantu family,.. are very numerous,..; among the better known are *Luganda*, [etc.]. 1955 *Times* 6 July 8/3 Similar statements have been prevalent in Luganda newspapers and in the resolutions of some district councils outside Buganda. 1966 B. Kimenye *Kalasanda Revisited* 15 His Luganda was spoken with a sharp accent. 1971 *Guardian* 22 Apr. 10/3 He was brought up at Bombo in Buganda and speaks excellent Luganda.

Lugbara (lugbā·rā), *sb.* and *a.* Also **Logbara, Lugbware, Lugwari. A.** *sb.* **a.** A people inhabiting the border area of Uganda and Zaïre; a member of this people. **b.** The Sudanic language of this people. **B.** *adj.* Of or pertaining to this people.

1925 R. E. McConnell in *Jrnl. R. Anthrop. Inst.* LV. 439 The Lugwari Tribe of the West Nile District of Uganda does not seem to have been described, so I venture to record some notes. *Ibid.* 442 Though not aggressive, the Lugwari are a fearless and warlike people in defence. *Ibid.* 448 The Lugwari women very usually wear large clumsy iron rings around their ankles. 1932 C. G. & B. Z. Seligman *Pagan Tribes Nilotic Sudan* viii. 282 Rain-stones..said to be of Bari origin..taken from the Lugbware some years previously. 1933–5 A. N. Tucker in *Bull. Sch. Oriental Stud.* VII. 868 The Lugbara (also called 'Lugwari') live in the North-East Congo..and extend into the West Nile District of Uganda. 1935 Thomas & Scott *Uganda* v. 86 Languages of the Moru-Madi group [are spoken] by the Lugbara..and Lendu..of the West Nile. 1953 Trowell & Wachsmann *Tribal Crafts Uganda* i. 57 The Madi and Lugbara smoke their tobacco in a water-cooled pipe with a small clay bowl. 1959 *Chambers's Encycl.* XIII. 257/1 Sudanic languages... Sudan-Nile (Lugbara, Logo, Lendu, Baka in Belgian Congo and the Sudan). 1960 J. P. Crazzolara (*title*) A study of the Logbara, Ma'di, language, grammar and vocabulary. 1971 *Sunday Nation* (Nairobi) 11 Apr. 14/2 On January 25, 1971, a modestly educated Lugbara voice haltingly read out to the nation eighteen reasons why the army had taken over power. 1974 *Guardian* 26 Mar. 1 Amin's recent policy of cracking down on Lugbara officers. *Ibid.*, Ondoga was a Lugbara from the western Nile.

luge (lūʒ), *sb.* [Swiss dialect.] A sledge, of Swiss origin, of the bob-sleigh type. Also *attrib.*

1905 *Sci. Amer.* Suppl. 15 Apr. 24488/1 The 'luge' is a small sled peculiar to the Grisons. 1907 *Ladies' Field* 19 Jan. 278 On the ascent the luge flies straight up into the air. 1919 *Daily Mail* 10 Dec., (*caption*) A pair of winter holiday-makers in Switzerland enjoying a run on a Canadian luge. 1968 *Globe & Mail* (Toronto) 13 Jan. 41/4 Canada's luge team..will be announced Monday. The Canadian luge championships are scheduled Sunday.

Hence **luge** *v. intr.*, to toboggan on a luge. Also **lu·geing, lu·ging** *vbl. sb.*, **lu·ger.**

1907 *Ladies' Field* 19 Jan. 278 The gentle art of Luging... Les Avants is the most obvious place from which to luge... Caux is the only other [place] where lugers assemble in force. 1909 *Westm. Gaz.* 26 Jan. 5/2 Les Avants..has one of the finest natural luging courses in Europe. 1924 *Yorksh. Post* 28 Jan. 10/7 There is good curling, bob sleighing, and lugeing. 1927 *Daily Express* 28 Dec. 3/1 Youths..'luged' in a reckless, haphazard style.

Luger (lū·gəɹ). Also (erron.) **Lueger, Lüger.** The name of G. *Luger*, German firearms expert, used *attrib.* or *absol.* to designate a German type of automatic pistol. Cf. *PARABELLUM.

1904 W. B. WALLACE *Text Bk. Small Arms* 178 The Borchardt Leuger or 'Parabellum' automatic pistol belongs to Class I. 1912 H. M. RIDEOUT *William Jones* 139 For arms he carried only a Luger pistol. 1933 'P. CAIN' *Fast One* 229 Faber lifted the flap of the right side pocket, slipped a black Luger out onto the seat beside him. 1934 D. TEILHET *Talking Sparrow Murders* ii. 34 A short Lüger thudded to the floor. 1947 *Landfall* I. 264 He held a Lueger pistol. 1957 'D. RUTHERFORD' *Long Echo* viii. 159 He picked up the Luger and began to extract the magazine. 1961 *Sunday Express* 12 Mar. 17/8 Two middle-aged men..shoot at one another, one with a Luger and the other with a ·38. 1970 R. A. STEINDLER *Firearms Dict.* 171/2 s.v. *Pistole Parabellum*, The name of Georg Luger has been linked to the gun to such an extent that the gun is often simply called the 'Luger'. 1975 P. SOMERVILLE-LARGE *Couch of Earth* iii. 39, I would have expected..a Luger or a Walther.

luggage. Add: **4.** *luggage boot, -grid, -rack, -rest, -train* (earlier and later examples); **luggage locker**, a locker (sense 5) at a railway station, air terminal, etc., for use by passengers.

1972 *Country Life* 7 Dec. 1592/3 The luggage boot is..fairly well filled by the spare wheel. 1907 *Westm. Gaz.* 12 Feb. 4/2 A telescopic luggage-grid capable of taking heavy trunks. 1928 Luggage grid [see *GRID 9]. 1969 *Guardian* 21 Aug. 9/6 The dreary rows of luggage lockers. 1971 M. KELLY *25th Hour* i. 9, I drove to the main station..and..I put my case in a luggage locker. 1905 *Daily Chron.* 18 May 4/5 Holding on..to the luggage-rack in the narrow..boxes which serve for [train] compartments. 1928 *Chambers's Jrnl.* Jan. 21/2 Just below the luggage-rack is an indicator by which the passenger may control the amount of steam entering the radiator. 1973 C. WILLIAMS *Man on Leash* (1974) i. 5 The room was cool... He dropped the bag on a luggage rack. 1941 'N. BLAKE' *Case of Abominable Snowman* viii. 83 That luggage-rest..at the end of the bed. 1846 R. FORD *Gatherings from Spain* v. 49 A handful of opponents..may..burn the engines.., particularly smashing the luggage train. 1899 O. WILDE *Importance of being Earnest* III. 127 Apprised, sir, of my daughter's sudden flight.., I followed her at once by a luggage train.

lugger, *sb.*¹ Add: Also in *beef, ship lugger*.

1904 *Sun* (N.Y.) 5 Aug. 1 It was reported that beef luggers in all the cold storage plants were to be called out. *Ibid.* 11 Aug. 3 The men who are called ship luggers, and who load meat aboard the steamships.

lugger, *sb.*² (Earlier example.)

1757 *Gentl. Mag.* Jan. 45/2 On the 25th inst. a French lugger drove a vessel ashore at Hastings.

Lugol (lu·gǫl). [The name of Jean *Lugol* (1786–1851), French physician.] *Lugol's iodine, solution*, a solution of 5% iodine and 10% potassium iodide in water, which is used for the internal administration of iodine and as a biological stain.

1880 P. SQUIRE *Compan. Brit. Pharmacopœia* (ed. 12) 177 (*heading*) Lugol's solution. 1917 *Jrnl. Pharmacol.* IX. 363 The increase in resistance to Lugol's solution did not go beyond three times. 1948 L. MARTIN *Clin. Endocrinol.* iv. 83 If thyroidectomy be chosen for a case of thyrotoxic heart failure, it is best to secure the maximum improvement possible with strict rest in bed, digitalis, and the mercurial diuretics before giving Lugol's iodine. 1961 *Lancet* 23 Sept. 688/1, 30 minims Lugol's iodine were administered intravenously during the night. 1972 *Acta Path. & Microbiol. Scand.* A. LXXX. 185 The substance displays some properties similar to those of classic amyloid, such as..responsiveness to methyl-violet, iodine green and Lugol's solution.

lugubre. (Later example.)

1835 J. ROMILLY *Diary* 26 Feb. (1967) 69 In a fly..to drink tea with Mrs Clarkson..the affair most lugubre.

Lugwari, var. *LUGBARA *sb.* and *a.*

Luian, var. *LUVIAN *sb.* and *a.*

Luing (liŋ). [Name of an island in the Hebrides.] The name given to cattle evolved from a crossing of the beef shorthorn and Highland breeds. (See quots. 1970.)

1970 *Times* 8 July 2/4 The Luing (pronounced Ling) bred from the Highland and the Beef Shorthorn, confined itself to the new commercial cattle section... Some purists have queried whether it is..a breed within the strict meaning of the term. 1970 *Radio Times* 24–30 Oct. 51/1 The three Cadzow brothers..had to create the first new beef cattle breed for nearly two centuries—Luing cattle. To accomplish their task they turned the Hebridean island of Luing into a huge experimental ranch. 1971 *Country Life* 6 May (Suppl.) 32b, 87 arable acres support the hill flock and herd of pedigree Luing. 1972 *Field* 7 Dec. 1352 (*caption*) A new breed of beef cattle, the Luing. The breed was evolved by the Cadzow brothers..on the island of that name in the Inner Hebrides... The Luing is based on beef Shorthorn x Highland.

Luiseño (luisenyo). Also **San Luiseño.** [Sp., f. *San Luis Rey*, a mission established in S. California in 1798.] The name, orig. San Luiseño, given to a tribe of the Shoshone Indians; a member of this tribe or the language spoken by them. Also *attrib.* or as *adj.*

1858 *Daily Even. Bulletin* (San Francisco) 5 Nov., The true native Americans of the wild forests—such as the Yumas,..San Luiseños,..predominate. 1875 H. H. BANCROFT *Native Races Pacific States* I. iv. 460 The villages of the San Luiseños are in a section of country adjacent to the Cahuillas. 1884 H. H. JACKSON *Ramona* 244 In the Luiseño tongue that is Majel. 1965 R. F. SPENCER et al. *Native Americans* vi. 263/1 The Luiseño myth of their origins revealed a surprising loftiness of concept. *Ibid.* 264/2 These events have combined to substantially eliminate the Luiseño as an ethnic and cultural group. 1965 *Canad. Jrnl. Ling.* Spring 139 The Luiseno sub-branch of languages in South-western California.

Lukan, var. LUCAN *a.*

Lukanism (liū·kăniz'm). [f. *Lukan* LUCAN *a.* + -ISM.] A form of expression characteristic of St. Luke.

1919 S. C. CARPENTER *Christianity acc. Luke* 69 There are in the passage certain Lukanisms of style.

Lukanize (liū·kănəiz), *v.* [f. as prec. + -IZE.] *trans.* To invest with a Lucan character.

1919 S. C. CARPENTER *Christianity acc. Luke* 81 It would no doubt be possible to assert that he determined at all costs..to Paulinize and Lukanize the Master.

‖ **lukiko** (luki·ko). Also **lukiiko.** [Luganda, = audience-hall, council, levee.] A levee; the council or parliament of the Buganda people of Uganda.

1889 R. P. ASHE *Two Kings of Uganda* vi. 53 He, like us, has come to 'kika', that is to present himself at Mutesa's grand 'lukiko', or levee. 1902 *Encycl. Brit.* XXXII. 539/1 The native parliament or 'Lukiko' was reorganized and its powers were defined. [1904 G. R. BLACKLEDGE *Luganda-Eng. Vocab.* 51/1 *Lukiko*, audience room, council, levee.] 1955 *Ann. Reg.* 1954 122 It was suggested..that..the Lukiko should be given the opportunity to decide whether to recall the exiled Kabaka or to elect a new Kabaka, as a constitutional monarch. 1955 *Times* 3 Jan. 7/3 The compromise was not made public until after the Great Lukiko had debated the Namirembe reforms. 1969 *Times* 15 Sept. (Uganda Suppl.) p. ii/3 The Buganda Lukiko (Parliament) is not the only regional body to have suffered an eclipse. 1971 *Sunday Nation* (Nairobi) 11 Apr. 11/4 The Lukiiko..demanded that Obote's government should quit the soil of Buganda.

lulav (liū·lāv). Also **lulab, lulov.** [Heb. *lūlābh* branch.] (See quot. 1959.)

1892 I. ZANGWILL *Childr. Ghetto* I. 265 He bore to synagogue the tallest *Lulav* of palm-branches. 1893 *Ibid.* (ed. 3) 409 *Lulov*, palm-branch dressed with myrtle and willow, and used at the Feast of Tabernacles. 1959 D. D. RUNES *Conc. Dict. Judaism* 157/1 *Lulav*, palm frond, one of four species waved at Sukkoth services (others are citron, myrtle, willow). These plants are interpreted as representing various types of man. 1972 C. RAPHAEL *Feast of Hist.* iii. 72/1 An oil lamp..bearing symbols familiar in Jewish art, including..a *lulav* for Sukkot. 1973 *Jewish Chron.* 10 Aug. 18/5 Succot is a festival of thanksgiving for the harvest... What is the lulav for? *Ibid.*, The lulav represents the palm-tree and its dates are delicious. 1973 *Synagogue Light* Sept. 50/1 The required length of the stock of a Lulab, besides its upper leaves, should measure four hand-breadths.

Lullian, *a.* Add: (Later examples.)

1933 *Times Lit. Suppl.* 29 June 433/2 No records remain to us of the early Centenaries, but their nature can be safely deduced from that of the Lullian cult, which was practised for hundreds of years both in Majorca and on the mainland. 1946 E. A. PEERS *Fool of Love* vii. 108 He rifled the Lullian library of the College of Sapiencia. *Ibid.* 109 With Pasqual began a new orientation of Lullian studies.

Hence **Lu·llism**, the philosophy or beliefs of Raymund Lull or his followers.

1929 E. A. PEERS *Ramon Lull* xviii. 377 Ample evidence exists, supplied alike by the friends of Lullism and its enemies, as to the veneration with which its author was regarded. 1933 *Times Lit. Suppl.* 29 June 433/3 That Lullism could survive so long,..is an astonishing testimony to the hold which the Doctor Illuminate had upon his fellow-countrymen. 1946 E. A. PEERS *Fool of Love* vii. 109 At this very time Lullism was taking a fresh impetus both in Spain and abroad. 1954 T. MAYNARD *Long Road of Father Serra* (1956) ii. 27 Students from all over Spain were attracted to this main center of Lullism.

lulu (lū·lu). orig. *U.S. slang.* [Of obscure origin.] A remarkable or wonderful person or thing; freq. used ironically; also *attrib.*

1886 *Lantern* (New Orleans) 10 Nov. 6/3 Farrell's two baser was a lu-lu. 1896 ADE *Artie* ix. 76 Mebbe you think I ain't got a lulu of a head on me this morning. 1904 'O. HENRY' *Cabbages & Kings* xvii. 301, I smelt a million violets. She was a lulu. I told her I came in a private yacht. 1922 H. TITUS *Timber* iii. 38 *She's* a lulu though! 1940 R. STOUT *Over my Dead Body* xiii. 176 You certainly picked a lulu for an adopted daddy. 1946 J. IRVING *Royal Navalese* 111 *Lulu*, a lady of infinite allure but questionable character. 1963 *Economist* 9 Nov. 578/1 Its own rationale..of why it suddenly became obligatory for [oil] companies to consult governments before changing prices..might be described as a lulu performance. 1968 *Jazz Monthly* Oct. 19/2 A muddle unparalleled in the history of jazz record issues—and there have been some Lulus—seems to have taken place. 1969 *Islander* (Victoria, B.C.) 13 July 6/1 His first job was, to use modern parlance, a lulu—census taker for the Island's west coast. 1971 *Black Scholar* Sept. 43/2 Joyce, honey, when you goof, its a lulu. 1972 D. SALE *Love Bite* iv. 50, I do hope you're not scared of earth tremors... This one was a real lulu. 1974 *Evening News* (Edinburgh) 23 Nov. 8/4 There are some parts of a new book on spying that aren't fit to be printed... This one is a lulu. As long as two years ago, legal proceedings were initiated.

‖ **luluai** (lu·luai). *New Guinea.* [Native name.] A man appointed by the administration to be responsible for the maintenance of order in a village; a village headman.

1924 J. AINSWORTH *Rep. Administrative Arrangements Natives New Guinea* 17/1 in *Austral. Parl. Papers 1923–4* IV. 1819 A luluai or kukurai is the political head of the particular section of which he is either the hereditary or appointed headman. 1930 M. MEAD *Growing up in New Guinea* 372 *Luluai*, headman of village. 1937 *Official Handbk. New Guinea* IV. 302 For the purpose of local government, the Administration has appointed two native representatives in each village. The senior of these is called a '*luluai*', and the junior, his assistant, a '*tul-tul*'. The work of the *luluai* is that of a village headman. 1957 M. WEST *Kundu* v. 63 There is a luluai in every village appointed by the Kiap in Goroka. 1965 *Sunday Mail* (Brisbane) 10 Oct. 2 In the next village the headman, the 'luluai', had a wife who was dying of kuru. 1970 L. P. MAIR *Austral. in New Guinea* (ed. 2) v. 72 The native authorities were known..as luluais..and tultuls... The luluais are responsible for good order and control in the villages.

Lulworth (lɒ·lwəɹþ). The name of Lulworth Cove, Dorset, used *attrib.* in **Lulworth skipper** to designate a butterfly, *Thymelicus acteon*, of the family Hesperiidæ (cf. SKIPPER *sb.*¹ 2 c), first found there in 1833 by J. C. Dale.

1833 J. CURTIS *Brit. Entomol.* X. 442 (*heading*) The Lulworth Skipper. 1894 W. FURNEAUX *Butterflies & Moths* 199 The Lulworth Skipper..is a very local species. 1945 E. B. FORD *Butterflies* vii. 149 In England the Lulworth Skipper only flies within a mile or two of the sea. 1973 T. G. HOWARTH *South's Brit. Butterflies* 27 The Lulworth Skipper... The coloration of this butterfly is somewhat dingy.

lumbar, *a.* and *sb.*¹ Add: **A.** *adj.* **b.** Of, pertaining to, or performed on or within the spinal cord in the lumbar region.

1895 *Brit. Med. Jrnl.* Suppl. 27 Apr. 65/2 Fürbringer..recalls how Quincke, in 1891, at Wiesbaden, made known his method of lumbar puncture in cases of meningitis of various kinds. 1947 *Physiol. Rev.* XXVII. 253 The bitch exhibits normal estrous and mating despite complete lumbar section. 1963 *Lancet* 12 Jan. 116/1 In three infants the fundus oculi was examined, and in two a lumbar puncture was done.

lumber, *sb.*¹ Add: **3.** (Later examples.)

1928 *Chambers's Jrnl.* Feb. 119/1 Behind the lumber grand-stand, which..resembled every natural wooden grand-stand in the world, stretched a grass meadow. *Ibid.* 120/1 We found Miss J. and Miss N. in a home where the lumber had mellowed—featuring an entrancing tint. 1941 *Sun* (Baltimore) 15 Oct. 5/5 They take nuts and bolts out of packing cases, pick up broken and abandoned field telephone wire along the roadsides, whittle scrap lumber with penknives and produce workable Morse sending keys. 1945 J. J. MATHEWS *Talking to Moon* 66, I had lumber left over from the building of the chicken and pheasant houses. 1965 *Globe & Mail* (Toronto) 5 Jan. B5 A company that will manufacture prefabricated homes in the United Kingdom with Canadian lumber.

4. (sense 1) *lumber-cellar, -house* (earlier and later examples) (cf. LUMBER-HOUSE); *lumber-headed* (earlier example); (sense 3) *lumber-business, -king* (examples), *-merchant* (earlier example); **lumber baron** *U.S.*, a leading or wealthy timber merchant; **lumber-carrier,** (*b*) a vehicle for carrying lumber; **lumber-jack** (earlier and later examples); freq. *lumberjack* (unhyphenated); **lumber jacket** orig. *N. Amer.*, a warm jacket of the type worn by lumbermen; **lumber-mill** (earlier example); **lumber-port,** (*a*) a port-hole in the bow or stern of a ship for loading or unloading timber; (*b*) a seaport from which lumber is shipped; **lumber-raft,** a raft made of logs, boards, or the like; **lumber town** *U.S.*, a town chiefly engaged in the timber trade; **lumber-trade** (earlier and later examples); **lumber-wagon** *N. Amer.*, a springless wagon of a type used for hauling lumber or for general transport (see also quot. 1962); **lumber-yard** *N. Amer.*, a timber-yard.

1888 *N.Y. Life* 18 Feb. 27/2 One of the..lumber 'Barons' of Michigan. 1948 *Time* 29 Nov. 24/1 In many ways he seemed a throwback to the lumber barons, the cattle kings and the mining magnates who had ruled the West before him. 1792 J. BELKNAP *Hist. New-Hampshire* III. 211 Husbandry..is much preferable to the lumber business. 1896 *Vermont Agric. Rep.* XV. 79 Gov. Woodbury has spent years as superintendent of the Burlington branch of J. R. Booth's gigantic lumber business. 1928 *Collier's* 29 Dec. 5/4 On the left were rows of twenty-foot lumber piles, trams laid between them, and electric lumber carriers rolling on the trams. 1832 *Chambers's Edin. Jrnl.* 24 Mar. 59/2 Stone bottles..collected from all the lumber-cellars in the country. 1910 *Daily Chron.* 18 Jan. 3/4 A cramped and pokey lumber-cellar. 1818 T. G. FESSENDEN *Ladies' Monitor* 38, I would not wish your pedant lumber-headed. 1720 in A. McF. Davis *Tracts Currency Mass. Bay* (1902) 385 Hemp, Flax, Turpentine..to be stored up in the Lumber-house. 1899 H. B. CUSHMAN *Hist. Indians* 162 A lumberhouse and granary, each 18 x 20 ft. 1831 in E. C. Guillet *Valley of Trent* (1957) vi. 236 My misfortunes have been brought upon me chiefly by an incorrigible..race of mortals called *lumberjacks*, whom, however, I would name the Cossacks of Upper Canada. 1902 S. E. WHITE *Blazed Trail* 41 Typical native-born American lumber-jacks powerful in frame. 1972 *Daily Colonist* (Victoria, B.C.) 2 Feb. 16/5 She later became a lumberjack and spent three months in the woods. 1939 *These are our Lives* (Federal Writers Project, N. Carolina) 107 He was dressed in riding breeches and leather lumber-jacket. 1943 *R.A.F. Jrnl.* Aug. 16 They wore lumber jackets. 1952 S. KAUFFMANN *Philanderer* (1953) vii. 107 He went into the middle of the group and squatted next to Jake (he still remembered the smell of the woollen lumber-jacket). 1956 T. H. RADDALL *Wings of Night* 70, I pulled on my old lumber jacket and went out to do the firewood chore. 1968 J. IRONSIDE *Fashion Alphabet* 37 In its modern meaning, the lumber-jacket is very similar to an anorak. It is a short, straight jacket, reaching to the hips, with a centre-front fastening (usually zipped) and buckled at the sides to make it fit snugly. 1975 P. SOMERVILLE-LARGE *Couch of Earth* ix. 148 A black and red lumber-jacket, the sort American hunters wear. 1889 W. H. WITHROW *Our Own Country: Canada* 372 One of the great lumber-kings of the country. 1941 *Yankee* Dec. 19/3 They were..Anderton's lumber kings; so nobody minded their smelling strongly of horses, even in the Methodist basement. 1789 *Boston Directory* 181 Dillaway, Samuel, lumber-merchant. 1830 *Deb. Congress U.S.* 11 Mar. 606/2 You will not find..such constant, unceasing labor as in our lumber mills. 1838 *Yale Lit. Mag.* III. 76 The pirates had knocked out the lumber port, with the intention of sinking her [*sc.* a ship]. 1883 *Wheelman* I. 333 Calais [in Maine], the great lumber port of this part of the country. 1837 W. JENKINS *Ohio Gazetteer* 62 The Hockhocking river..furnishes..a downward navigation for flat boats and lumber rafts. 1898 Lumber-raft [in Dict.]. 1961 B. FERGUSSON *Watery Maze* v. 111 Irrawaddy lumber-rafts. 1880 *Harper's Mag.* Aug. 354/1 A cheerful little lumber town lying high among the hills. 1904 S. E. WHITE *Blazed Trail Stories* I. 3 The sawdust streets..of the lumber town were filled with people. 1972 R. NEELY *Sexton Women* (1974) ii. 12 She had been..brought up in a lumber town..near the Oregon border. 1689 in *Mass. Hist. Soc. Coll.* (1834) 3rd Ser. I. 98 They are supplied..with the lumber trade. 1732 COL. DUNBAR *Let.* 25 Aug. in *Calendar State Papers* (*Colonial Ser.*) (1939) 201 The undertaker for the masting has and does carry on a greater lumber trade than any man in N. Engld. 1963 *Canada Month* Nov. 22/1 The lumber trade furnished employment for thousands of lumber jacks, river drivers, and sailors. 1831 S. STODDARD *Diary* 30 Nov. in *Mich. Hist. Mag.* (1927) XI. 472 Breakfast swallowed we stepped into our next rig, which was a lumber wagon drawn by two very good horses. 1887 C. D. WARNER *Their Pilgrimage* (1888) xiii. 288 At this season one meets them [*sc.* the hop-pickers] on all the roads driving from farm to farm in lumber wagons. 1902 E. BANKS *Autobiogr. Newspaper Girl* 1 Gathered about the little village station in hard-seated lumber-wagons. 1961 *Edmonton* (Alberta) *Jrnl.* 24 July 11/7 They forded the North Saskatchewan River in a lumber wagon. 1962 *Amer. Speech* XXXVII. 270 *Lumber wagon*, an old, broken-down automobile, particularly one that rides rough. 1786 *Maryland Jrnl.* 4 Apr. (Th.), Lumber-yard, at the head of Baltimore Bason. 1851 C. CIST *Sk. Cincinnati in 1851* 207 Connected with the machinery is a lumber yard. 1961 W. E. GREENING *Ottawa* 108 The district close to the lumberyards was full of waterfront dives. 1973 C. WILLIAMS *Man on Leash* (1974) iv. 52 Lew was..running a lumberyard and building supply here.

lumber, *sb.*² Restrict † *Obs.* to senses 1 a and 2, and add: **1. b.** *to be in lumber* (further examples); also, to be in trouble.

1963 J. PRESCOT *Case for Hearing* viii. 125 My poor old dad was in and out of lumber all his life. 1965 A. PRIOR *Interrogators* xi. 202 We're out on a limb hoping for a confession, and if we don't get it we're in dead lumber. 1967 'M. CARROLL' *Begotten Murder* iv. 104 It rather looks to me as if someone is trying to get Susan in lumber. 1972 L. HENDERSON *Cage until Tame* vi. 43 I've got to keep at it. Break my bloody leg or something stupid like that and I'm in lumber.

3. *slang.* A house or room; *spec.* one where stolen property is hidden; a house used by criminals.

1753 J. POULTER *Discoveries* (ed. 2) 33 They pike up the Prancers, that is, go up Stairs, and fisk the Lumbers, that is, search the Rooms. *a* 1790 H. T. POTTER *New Dict. Cant & Flash* (1795) 40 *Lumber*, a house convenient for the reception of swindlers, sharpers, and cheats. 1800 G. PARKER *Life's Painter* xiv. 117 Have you any body in the lumber behind the bar? *Ibid.* xv. 140 *Lumber*, a room. 1846 *Swell's Night Guide* 34 The polka is greatly in favour with the femmes of this lumber. *Ibid.* 74 His long room, or 'slanging lumber', is the scene of many a choice spree and downey movement. 1923 S. T. FELSTEAD *Underworld of London* iii. 108 The proprietor of the 'lumber', where stolen property is stored pending a suitable buyer, also wants his whack. 1938 F. D. SHARPE *Sharpe of Flying Squad* xiv. 151 Her husband was taken into custody at a 'lumber' (hide-out for stolen property) in Walthamstow. 1950 R. FABIAN *Fabian of Yard* xxxiv. 206 *Lumber*, address used by a prostitute for her profession only.

lumber, *v.*¹ Add: **2.** (Later U.S. examples in special senses.)

1855 J. E. COOKE *Ellie* 207 Keeping the footman lumberin at the knockers on both sides o' the street. 1890 *Dialect Notes* I. 65 'Listen how he lumbers', said of a deep-mouthed dog's barking when he has treed a 'coon or 'possum. 1904 T. WATSON *Bethany* (1920) 165 And he himself did not always know what he had on his mind until he pushed back his specs, and began to 'lumber' [= hold forth].

lumber, *v.*² Add: **1. a.** (Further examples.) Now usu., to leave (someone) *with* something unwanted or unpleasant; to get (someone) into trouble or difficulties; freq. *pass.*

1924 E. WALLACE *Room 13* i. 9 'If they lumbered you with the crime, it was because you was a mug,' said Lal complacently. 'That's what mugs are for—to be lumbered.' 1951 A. BARON *Rosie Hogarth* III. iv. 180, I suppose you're afraid... Of getting lumbered, eh? 1958 T. HALL in P. Gammond *Decca Bk. Jazz* xix. 233 Poor old Don Rendell..got really lumbered. He left his clarinet with Gee's with the proviso that it would be forfeited if the trousers and windcheaters weren't returned by the following Monday. Needless to say, they weren't. 1961 SIMPSON & GALTON *Four Hancock Scripts* 35/2 Every time I travel on a train I get lumbered with a carriageful of the most miserable-looking bunch of face-aches you've ever seen in your life. 1964 G. DAVIS *Friday before Bank Holiday* i. 11, I want to realise on the cottage..but I'm lumbered unless I can find another home for Fiddler. 1968 J. LOCK *Lady Policeman* xii. 113, I tell him I'm lumbered for court in the morning.

3. b. *trans.* To go over (ground) cutting the timber on it.

[1831 *Trans. Lit. & Hist. Soc. Quebec* II. 269 His intention..was to clear land and lumber some.] 1851 J. F. W. JOHNSTON *Notes N. Amer.* I. 52 We clean up two or three acres every year of the lumbered land (land from which the timber has been cut). 1871 R. L. DASHWOOD *Chiplo-quorgan* v. 60 This part of the country has been 'lumbered', being too difficult of access. 1900 *U.S. Dept. Agric. Yearbk.* 365 The cut-over lands..which..have been lumbered heavily, not only for timber but also for fuel. 1971 *Lebende Sprachen* XVI. 9/2 This valley was lumbered in 1955. We lumbered more than a million acres last year.

lumber, *v.*³ Add: Also, to arrest, imprison.

1882 *Sydney Slang Dict.* 6/1 *Lumber*, to take or carry away to the lock-up. 1931 *Police Jrnl.* Oct. 501 Did the detective (busy) arrest (lumber) Jack? 1953 K. TENNANT *Joyful Condemned* ii. 17 Don't you worry about the police. If there's a warrant out for you..they'll lumber you sooner or later. 1961 B. CRUMP *Hang on a Minute* 136 We were sneaking into the church to bunk down last night when the johns lumbered us. 1970 M. KENYON *100,000 Welcomes* iv. 30 We're pros—twice in twelve years I've been lumbered... Only twice in twelve years screwing.

lumbriculus (lɒmbri·kiūlɒs). [mod.L. (A. E. Grube 1844, in *Archiv für Naturgeschichte* X. 207), f. mod.L. *Lumbricus* (cf. LUMBRICUS) the name of a genus of earthworms + *-ulus*.] An aquatic, oligochæte worm of the genus so called, resembling an earthworm.

1901 [see *DIVISION 1 f]. 1927 HALDANE & HUXLEY *Animal Biol.* ix. 172 The same (multiplication) would have been true if instead of a Planarian you had cut up the little polyp Hydra, and almost the same with the smaller Annelid worms such as Lumbriculus. 1963 R. P. DALES *Annelids* viii. 154 *Lumbriculus* has been found capable of regenerating the head twenty-one times in succession.

lumbrous, *a.* (Earlier example.)

1836 J. HILDRETH *Dragoon Campaigns Rocky Mts.* I. iii. 26 Our lumbrous vessel heavily groped her way through the waters.

lumen. Add: **2.** [First adopted, in Fr., by A. Blondel 1894, in *La Lumière électrique* 7 July 10.] A unit of luminous flux (now incorporated into the International System of Units), equal to the flux emitted by a point source of intensity one candela (formerly, one candle) into a solid angle of one steradian.

The total flux emitted by a source of one candela is 4π (= about 12·57) lumens.

1898 *Astrophysical Jrnl.* VII. 300 Luminous current is defined as the rate at which luminous energy is emitted by a point-source through a solid angle of one steradian. Unit: The luminous current of one candle, i.e., of one Hefner lamp. Name: 'Lumen'. Proposed by L. Weber. 1937 *Times* 13 Apr. p. iv/2 Electric discharge lighting..has been recently installed to give an even and intense illumination of 1,120,000 lumens so that any flaw or defect in the panels being produced may be detected at once. 1953 AMOS & BIRKINSHAW *Television Engin.* I. vii. 125 The efficiency of electric lamps is usually expressed in lumens per watt... For gas-filled lamps the efficiency increases steadily with the rating and is approximately 10 lumens per watt for 25 W lamps, 13 lumens per watt for 100 W lamps and 17·5 lumens per watt for 1 kW lamps. Thus a 100 W lamp radiates a total of 1,300 lumens and is equivalent to a source of $1{,}300/4\pi$, i.e., approximately 100 candle-power. 1974 *Which?* Mar. 88/3 The long-life bulbs gave a light output of about 600 lumens after one hour of life.

3. Special Comb.: **lumen-hour,** the quantity of light corresponding to a flux of one lumen radiated for one hour; similarly **lumen-second.**

1925 *Trans. Illuminating Engin. Soc.* (U.S.) XX. 630 The Lumen-hour is the unit of quantity of light. It is equal to a flux of one lumen continued for one hour. 1975 *Times* 21 Mar. 20/3 The public wanted a certain package of candlepower (expressed in lumens); the industry assured certain fixed wattages, and then aimed at the minimum cost per lumen-hour for those wattages. 1930 tr. *L. P. Clerc's Photogr.* ii. 7/2 The unit of exposure is the lumen-second or candle-metre-second. 1958 *Newnes Compl. Amat. Photogr.* x. 110 The light output of a flash-bulb is measured in lumen seconds.

‖ **lumen siccum** (liū·men si·kɒm). [L., = dry light.] The dry light of rational knowledge or thought.

1605 BACON *Adv. Learn.* II. f.48 But this same *Lumen siccum*, doth parch and offend most mens watry and soft natures. 1819 COLERIDGE *Philos. Lect.* (1949) xiii. 374 Must there not be some power, call it with Lord Bacon the '*lumen siccum*'; or 'the pure light', with Lord Herbert;..that stands in human nature but in some participation of the eternal and the universal by which man is enabled to question, nay to contradict, the irresistible impressions of his own senses, nay, the necessary deductions of his own understanding? 1946 *Mind* LV. 285 Taylor's intellect was no *lumen siccum*, but was always strongly personal in its approach and attitude.

lumeter (liū·mītəɹ). [f. L. *lūm-en* light + -METER.] = *LUXMETER.

1911 *Trans. Optical Soc.* XII. 104 Messrs. R. & J. Beck, Ltd., the makers of the 'Lumeter' instrument. 1923 L. C. MARTIN *Colour* 178 The lumeter reads in foot-candles, that is to say the brightness which a perfectly reflecting and diffusing surface would have if illuminated by so many foot-candles. 1947 *Brit. Jrnl. Psychol.* XXXVIII. 89 The apparatus was calibrated for intensity of illumination (as this would appear to the subject's eye) by placing a lumeter in the position of the tube T.

lumichrome (liū·mikrōum). *Chem.* [ad. G. *lumichrom* (P. Karrer et al. 1934, in *Helv. Chim. Acta* XVII. 1010), f. L. *lūmi(n-, lūmen* light + Gr. χρῶμ-α colour.] 6,7-Dimethylalloxazine, $C_{12}H_{10}N_4O_2$, a crystalline compound that is formed by ultra-violet irradiation of riboflavin in acidic solution and shows a sky-blue fluorescence.

1935 *Chem. Abstr.* XXIX. 798 Straw-yellow crystals of lumichrome..are obtained by irradiating solns. of lactoflavin with sunlight. 1949 *Proc. Soc. Exper. Biol. & Med.* LXX. 585/2 The presence of lumichrome crystals in the media containing riboflavin after growth of the tubercule bacillus indicated that the loss of the vitamin was due to conversion to lumichrome. 1962 S. UDENFRIEND *Fluorescence Assay in Biol. & Med.* xiv. 460 Fresh milk and cream emit greenish-yellow fluorescence which is due mainly to riboflavin. When exposed to daylight for appreciable periods of time or irradiated in any other way, the fluorescence turns from yellow to blue. The latter is characteristic of lumichrome. 1972 *Plant Physiol.* XLIX. 991/2 The present report establishes the formation of lumichrome, presumably from either free or bound riboflavin, by the action of light on etiolated shoots of corn and oats and on yeast cells.

Lumière (liūmiɛ̄ᵊɹ). *Photogr.* The name of the brothers Auguste (1862–1954) and Louis (1864–1948) *Lumière*, French photographers, used *attrib.* to denote a process of colour photography invented by them involving a colour screen consisting of a glass plate coated with a mixture of starch grains dyed in the primary colours which was placed in front of the panchromatic emulsion on exposure and for viewing; also applied to plates used in this process. Cf. *AUTOCHROME *a.* and *sb.*

1907 *Westm. Gaz.* 24 Aug. 14/2 In some respects the new Lumière process is old. *Ibid.* 20 Sept. 4/3 Even in the case of the Lumière plate a keen eye can detect the red and blue grains upon the lighter portions. 1920 *Chambers's Jrnl.* Apr. 238/2 Twelve volumes, illustrated by 1260 plates of his [*sc.* Burbank's] most beautiful creations, reproduced in colour from original *lumière* plates. 1955 G. R. SHARP tr. *Lorelle's Colour Bk. Photogr.* 199 In the Lumière process a transparent film base carries a mosaic pattern of microscopic coloured filters which is coated with a fine grain panchromatic emulsion. 1966 LACOUR & LATHROP *Photo Technol.* xv. 194/2 To produce the Lumiere Autochrome plate, starch grains were pulverized and one third dyed blue, one third dyed green and the remaining grains dyed red. The colored grains were remixed and spread in a very thin layer on a glass plate.

lumiflavin (liūmiflēi·vin). *Chem.* [f. L. *lūmi(n-, lūmen* light + *FLAVIN 2.] 6,7,9-Trimethylisoalloxazine, $C_{13}H_{12}N_4O_2$, a crystalline compound that is formed by ultra-violet irradiation of riboflavin in alkaline solution and shows a yellow-green fluorescence.

1934 *Chem. Abstr.* XXVIII. 2036, 1000 l. whey yields 170 mg. pure lumiflavin. 1946 *Jrnl. Biol. Chem.* CLXII. 96 Lumiflavin can either inhibit or stimulate the use of riboflavin or FAD by *Lactobacillus casei*, depending upon the relative amounts of lumiflavin present. 1956 *Jrnl. Biochem.* (Tokyo) XLIII. 643 To estimate the fluorescence of flavin separately from similar fluorescent substances in living body, the procedure of converting flavins to lumiflavin is of excellent one [*sic*]. 1962 S. UDENFRIEND *Fluorescence Assay in Biol. & Med.* vii. 243 Riboflavin can also be converted to the more highly fluorescent derivative, lumiflavin, by photodecomposition in alkaline solution, the formation of lumiflavin involving cleavage between the ribitol and flavin ring.

luminaire (liū·minēᵊɹ). orig. *U.S.* [Fr.; see LUMINARY *sb.*] An electric light and its fittings; such a lighting unit. Cf. LUMINAIR.

1921 *Trans. Illuminating Engin. Soc.* (U.S.) XVII. 249 The matter of a generic term for 'lighting unit' was considered by the Committee... The suggested term 'luminaire' was deemed to be the most acceptable... It is not a coined word but is a term already in use in the French language. 1925 M. LUCKIESH in Cady & Dates *Illuminating Engin.* ix. 363 In school lighting..luminaires should not be spaced farther apart than a distance of 1·5 times their elevation above the desk-tops. 1933 *Trans. Illuminating Engin. Soc.* (U.S.) XXVIII. 274 A luminaire is a complete lighting unit consisting of a light source, together with its direct appurtenances, such as globe, reflector, refractor, housing and such support as is integral with the housing. 1940 *Chambers's Techn. Dict.* 513/2 *Luminaire*,..a term sometimes used in America to denote an electric-light fitting. 1960 *How TV Works* iv. 24 The ceiling of a large studio such as Studio 6 in Manchester..may be fitted out with up to a hundred different lights, or 'luminaires' as they are more often called. 1971 *Daily Tel.* 21 Aug. 10 The British Standards Institution's committee responsible for electric light fittings has announced that in the interests of international standardisation it recommends the use of the word 'luminaire' instead of 'lighting fittings' in future. 1971 *Bahamian Rev.* Nov. 32/2 The 'Night Guard' luminaire is turned on automatically at dusk and switched off automatically at dawn, providing night-long security and safety. 1972 HENDERSON & MARSDEN *Lamps & Lighting* xix. 341 Luminaires can be divided into two categories: those which are essentially decorative and those which are essentially functional. Decorative luminaires are usually in the form of an assembly of decorative components around a light source, ranging from a simple pendant to the large..prestige chandelier. 1973 *Times* 1 Feb. 22/8 It is preposterous for BSI to force *luminaires* upon us.

Luminal (liū·minæl), *sb. Pharm.* Also **luminal.** [prob. f. L. *lūmin-, lūmen* light (as a rendering of PHEN-) + *-AL².] A proprietary name of phenobarbitone.

1912 *Trade Marks Jrnl.* 20 Mar. 402 Luminal... A medicine for human use as a sedative and hypnotic. The Bayer Company, Limited,..Manchester; manufacturers. 1928 *Daily Express* 7 Dec. 12 Gave him five grains of luminal. He'll sleep for a few hours and wake up penitent. 1937 *Times* 6 Sept. 7/5 For the defence, it was suggested that the shock of the collision, together with the use of luminal (a narcotic), would account for the symptoms. 1951 'E. CRISPIN' *Long Divorce* iv. 37 As a doctor, she had remedies to hand—barbitone, nembutal, luminal. 1960 W. G. LENNOX *Epilepsy* II. xxvi. 860 Luminal (Winthrop-Stearns) is phenobarbital that costs more.

luminance. Add: **2.** *Physics.* The amount of luminous flux emitted by unit area of a source into unit solid angle (the objective analogue of subjective brightness).

1950 JENKINS & WHITE *Fund. Optics* (ed. 2) vii. 104 To distinguish it from the visual sensation of brightness, it is usually termed the luminance in the technical literature, but..we shall use the commoner name brightness, with the understanding that..the photometric quantity..is meant. 1952 *Electronics* Nov. 208/2 Luminance is a purely photometric quantity. Use of this name permits brightness to be used entirely with reference to the sensory response. 1957 V. J. KEHOE *Technique Film & Television Make-Up* viii. 95 As to..the reception of a color telecast on a monochromatic receiver, only the luminance portion will be supplied to the single electron gun in the monochrome tube to produce a black-and-white version of the signal. 1968 *Sci. Amer.* Nov. 8/2 Consider two areas that have the same luminance but one of which is seen against a dark background while the other is seen against a light background. The former will appear to be brighter. 1970 *Jrnl. Gen. Psychol.* LXXXII. 111 Midway between these red apertures a yellow spot, three minutes of arc in diameter and 3·1 ftL in luminance, provided the fixation. 1972 *Sci. Amer.* June 101/1 Measurements of a photograph of the vase..showed that the luminance of the moon was 15 foot-lamberts and the space one moon diameter below was 20 foot-lamberts. The contour effect is so strong that the apparent brightness of the two areas is just the reverse of the objective luminance.

luminarism (liū·minăriz'm). [-ISM, after LUMINARIST.] The art or doctrine of the luminarists.

1903 *Edin. Rev.* Oct. 373 We shall probably hear less of Turner as the pioneer of impressionism, luminarism, and pre-Raphaelitism. 1953 *New Republic* 6 July 18 In our own art the progression from linear to formal, formal to coloristic, and finally to the complete luminarism of the 1870s has been a slow process.

luminism (liū·miniz'm). [f. L. *lūmin-, lūmen* light + -ISM.] = *LUMINARISM.

1905 *Sat. Rev.* 11 Feb. 174 And now, stated in general terms, what is the principle of Monet's luminism? 1920 *Glasgow Herald* 27 Aug. 4 The world of art has given us Futurist, Cubist, Vorticist, luminism and others. 1927 F. J. MATHER *Hist. Mod. Painting* 365 The theoretical Cubists merely turned upon the prevailing Expressionism the criticism which Cézanne had applied to Luminism. 1970 R. LOWELL *Notebk.* 245 One misses Emerson drowned in luminism.

luminize (liū·minəiz), *v.* [f. L. *lūmin-, lūmen* light + -IZE.] *trans.* To make luminous; to apply a luminous substance to.

1958 *Oxford Mail* 9 July 1 Home radiation hazard in luminizing clocks. 1959 *Times* 30 Apr. 6/5 A Swiss firm was using strontium to luminize their watches. 1967 *Health Physics* XIII. 613/2 Where large numbers of the same type of dial are to be luminised, the paint is often applied by machine.

Hence **lu·minized** *ppl. a.*, **lu·minizing** *vbl. sb.*; **lu·minizer,** one who luminizes objects.

1958 *Times* 20 June 11/7 Luminizing belongs to the daunting group of hazards associated with radiation. 1958 *Ann. Rep. Chief Inspector of Factories on Industr. Health 1957* 36 in *Parl. Papers 1958–9* (Cmnd. 558) XIII. 183 The case of a female luminiser, aged 42, is of special interest. 1959 *New Scientist* 19 Nov. 980/2 The clock industry has its own luminizing regulations. 1967 *Health Physics* XIII. 613/1 It was possible to make body radioactivity measurements on almost all the personnel employed on luminising in these workshops. *Ibid.* 613/2 Radium luminous compound is usually obtained by the luminiser in the form of a powder. *Ibid.* 614/1 In some cases the rest periods are spent on work with luminised articles.

luminol (liū·minǫl). *Chem.* [f. L. *lūmin-, lūmen* light + -OL.] A pale yellow crystalline bicyclic hydrazide, $C_8H_7N_3O_2$, which gives a blue luminescence when oxidized in alkaline solution and is used in the determination of oxidizing agents and metal ions.

1934 E. H. HUNTRESS et al. in *Jrnl. Amer. Chem. Soc.* LVI. 241/1 It is the purpose of the present note to offer an improved method for the synthesis of this material [*sc.* 3-aminophthalhydrazide], which in the interest of simplicity we have long referred to as 'Luminol'. 1965 *Analytical Biochem.* XII. 309 This assay [for hematin iron] is based on the luminescence of alkaline luminol in the presence of hematin iron and hydrogen peroxide. 1970 R. P. WAYNE *Photochem.* iv. 120 The oxidation of luminol..in alkaline solution by ferricyanide..is one of the best-known chemiluminescent reactions. 1974 *Nature* 25 Jan. 193/2 Using a ⁶⁰Co-γ-source, irradiated saccharides give a bright blue light (≈4,460 Å) when they are dissolved in luminol solution.

luminophore (liū·minŏfōᵊɹ). Also **-phor.** [f. L. *lūmin-, lūmen* light + -O + -PHORE.] **a.** A luminescent substance.

1907 *Jrnl. Chem. Soc.* XCII. II. 419 The addition of sodium or potassium sulphate to the mixture intensified both this green phosphorescence and that due to the luminophore added. 1930 *Chem. Abstr.* XXIV. 785 (*heading*) Contribution to the preparation of luminophores. 1950 H. W. LEVERENZ *Introd. Luminescence of Solids* v. 147 Luminescent materials in general (luminophors)..have the additional property of being able to convert part of the absorbed primary energy into emitted luminescence radiations whose spectral characteristics are determined almost entirely by the luminophor. *Ibid.*, The generic term *luminophor* is subclassified into fluorophors..and phosphors. 1968 *Proc. Internat. Conf. Luminescence, Budapest 1966* I. 1290 Precipitates of uniform size..are demanded especially by the luminophore industry.

b. A group of atoms in a molecule which is considered to be responsible for its luminescence.

1910 *Encycl. Brit.* X. 72/1 H. Kauffmann..suggested that the property [of fluorescence] is due to the presence of at least two groups. The first group, named the 'luminophore', is such that when excited by suitable aetherial vibrations [it] emits radiant energy; the other..acts with the luminophore in some way or other to cause the fluorescence. 1949 P. PRINGSHEIM *Fluorescence & Phosphorescence* v. 392 The strong absorption of visible light which is characteristic of dyes was ascribed to the presence of certain unsaturated groups, such as the azo group —N=N—, the ethylene group —HC=CH—, and the carbonyl group >C=O, which were called chromophors. These chromophors were at first supposed also to be the carriers of fluorescence as 'luminophors'.

luminosity. Add: **1. b.** The effectiveness of light of any particular wavelength in producing the sensation of brightness when perceived.

[1888 *Amer. Jrnl. Sci.* CXXXVI. 359 While..the luminosity of any spectral ray increases proportionately to the heat in this ray, and indeed is but another manifestation of the same energy..there is..a failure to recognize how totally different optical effects may be produced..according to the wave-length.] 1898 *Astrophysical Jrnl.* VII. 303 That particular property of any color which determines its value as an illuminant is called its 'luminosity'. 1936 *Proc. R. Soc.* A. CLV. 664 The relative luminosity (visibility) in the red region has been measured..as far as 770 mμ. In this paper an account is given of measurements up to 900 mμ. At this wavelength the relative luminosity is one-sixty millionth of that at 556 mμ. 1950 F. H. ADLER *Physiol. Eye* xx. 645 The luminosity of any one wave length is usually compared to that of 550 mμ. 1966 C. W. WILMAN *Seeing & Perceiving* xiv. 116 Curve *Ph* in Fig. 78 represents the relative luminosity (brightness) of various parts of the spectrum.

c. *Astr.* The intrinsic brightness of a heavenly body (as distinct from its apparent brightness, diminished by distance); the rate of emission of electromagnetic radiation (visible or invisible) within any part of the spectrum.

1906 *Astrophysical Jrnl.* XXIII. 248 Prevailing opinion..admits the presence in the heavens of at least a few stars of extraordinary intrinsic brilliancy... stars having a luminosity exceeding that of the Sun by ten-thousand fold or more. 1924 H. DINGLE *Mod. Astrophysics* vi. 74 By brightness we mean—at present, at any rate—the apparent brightness of a star, which will be determined by the star's distance from us as well as by its intrinsic luminosity. 1930 R. H. BAKER *Astron.* ix. 369 The apparent magnitude of a star relates to its brightness as we observe it, depending on its real brightness, or luminosity, and on its distance. 1974 *Nature* 1 Mar. 34/2 We have..assumed a thermal spectrum, with an electron temperature of 2×10^8 K, in deriving an upper limit to the X-ray luminosity of the source. *Ibid.*, A 3σ upper limit was obtained to the source luminosity of $4{\cdot}0\ (\pm 1{\cdot}0) \times 10^{32}$ erg s⁻¹ for the energy range 0·5 to 1·5 keV.

3. Special Comb.: **luminosity curve,** a graph showing how emitted energy, or perceived brightness, varies with wavelength; **luminosity function** *Astr.*, a function giving the number or proportion of heavenly bodies with an absolute magnitude equal to, or greater than, any chosen value.

1886 ABNEY & FESTING in *Phil. Trans. R. Soc.* CLXXVII. 425 By successive alterations in the distance of the comparison-light other pairs of points in the spectrum are determined until the limits of the visible spectrum are reached. The curves of intensities of different parts of the spectrum plotted from these observations will be found to be fairly smooth. This curve we call the 'luminosity curve'. *Ibid.* 452 We..determined the luminosity-curve of a candle. 1900 *Astrophysical Jrnl.* XI. 220 (*heading*) Determining the luminosity curve of the solar spectrum. 1937 *Nature* 6 Mar. 409/2 The maximum of this luminosity curve was at 510 mμ. This was a mean value for forty-eight young observers. 1941 J. D. COBINE *Gaseous Conductors* xiii. 514 The luminosity curve for an incandescent lamp is shifted slightly toward the long wave lengths compared with sunlight. 1951 *Proc. IRE* XXXIX. 1143/2 The luminosity curve expresses the relation..between the luminance and the radiance of spectrum colors. *Ibid.* 1144/1 Based on the new definition of the lumen, and on the new International Temperature Scale, the peak value of the luminosity curve corresponds to 680 lumens per watt. 1924 *Astrophysical Jrnl.* LIX. 13 The comparison..is easily made with the aid of what may be called the apparent luminosity function..which expresses the frequencies of absolute magnitudes among the stars brighter than m_0. 1958 *Ibid.* Suppl. III. 211 Distances of galaxies are correlated with their apparent magnitude; however the uncertainty of the luminosity function of galaxies complicates and weakens the statistical treatment.

luminous, *a.* Add: **1.** (Further examples.)

1877 W. H. BALMAIN *Brit. Pat. 4152* 7 Nov., My invention relates to a method of rendering paints, varnishes, whitewashings, and temperings luminous, and consists in the introduction into ordinary paints, varnishes, or washes of a phosphorescent substance, by which means the object to which the paint, or varnish, or wash is applied is made visible in the dark and more or less capable of imparting light to other objects. 1926 R. W. LAWSON tr. *Hevesey & Paneth's Man. Radioactivity* xxv. 194 These radioactive 'luminous substances' differ from the phosphorescent substances formerly used, in that they do not require previous illumination. 1929 *Jrnl. Amer. Med. Assoc.* 9 Feb. 466/1 (*heading*) Occupational poisoning in manufacture of luminous watch dials. 1953 KIRK & OTHMER *Encycl. Chem. Technol.* X. 658 Of increasing commercial importance are the luminous pigments which produce fluorescent and phosphorescent effects of value in interior decoration, for direction signs and safety devices, television picture tubes, special military applications, and the like. 1958 *New Statesman* 28 June 846/3 The sins of the hitch-hiking undergraduates were drainpipes, luminous socks, and frying sausages in their bedrooms. 1969 R. F. LANG tr. *Henglein's Chem. Technol.* 546 Radium salts are used..for manufacturing luminous paint mixed with ZnS and CaS. 1975 K. BARCLAY tr. *Orum's Whipping-Boy* xi. 82 My alarm clock has a luminous dial.

d. *luminous efficiency*, (of radiant energy) the ratio of the luminous flux to the total flux; (of a light source) the ratio of the luminous flux to the total flux emitted by, or power supplied to, the source.

1902 *Encycl. Brit.* XXVIII. 89/1 The luminous efficiency of any source of light, that is to say, the percentage of rays emitted which affect the eye as light compared with the total radiation, is dependent upon its temperature. In an ordinary oil lamp the luminous rays do not form much more than 3 per cent of the total radiation. 1927 H. N. RUSSELL et al. *Astron.* II. xvi. 529 The luminous efficiency of sunlight is obtained by dividing the total luminous flux by the total radiant flux..and comes out 100 lumens, or about 8 candles per watt. 1961 CARNT & TOWNSEND *Colour Television* iii. 55 Objects moving across the screen may..leave a yellow trail..as the red and green phosphors continue to glow after the blue phosphor has ceased... The luminous efficiency of these phosphors is low, only a few per cent of the electrical energy being converted into radiant energy... Ten or so lumens per watt can be obtained at 20 kV.

lumirhodopsin (liū:mirodǫ·psin). *Biochem.* Also **lumi-rhodopsin.** [f. L. *lūmi(n-, lūmen* light + *RHODOPSIN.] An orange intermediate that is formed when rhodopsin is bleached by light and changes spontaneously to metarhodopsin.

1950 G. WALD et al. in *Science* 17 Feb. 180/1 On exposing the solution to light at these low temperatures, the maximum shifts about 5 mμ toward the blue, rising about 5% in height in cattle rhodopsin, falling about this amount in frog rhodopsin, still with little change in shape. This is the light reaction. We shall call its product lumirhodopsin. 1975 *Nature* 3 Jan. 56/2 Lozier and Stoeckenius have observed [in *Halobacterium halobium*] four spectrally distinct intermediates analogous to the rhodopsin, prelumirhodopsin, lumirhodopsin and metarhodopsin of invertebrates.

lumisterol (liūmi-stērǫl). *Biochem.* [ad. G. *lumisterin* (A. Luttringhaus 1931, in *Chemiker-zeitung* 12 Dec. 956/2), f. L. *lūmi(n-, lūmen* light + *-sterin* after CHOLESTERIN, *ERGOSTERIN, with altered ending (see *-STEROL).] A steroid alcohol, $C_{28}H_{44}O$, which is a stereoisomer of ergosterol and occurs as an intermediate when this is converted to vitamin D_2 by ultra-violet irradiation and warming.

1932 *Nature* 20 Feb. 277/1 In lumisterol the crystals are fine needles with *b* as needle axis. 1949 W. H. EDDY *Vitaminology* iii. 46 Lumisterol can be converted into calciferol and may form with it an addition compound consisting of one part lumisterol and one part calciferol.. This addition compound..was called vitamin D_1. 1967 W. R. BUTT *Hormone Chem.* xi. 235 It has been established that acetate is the primary carbon source of cholesterol, there being probably about thirty separate reactions involved in the pathway through squalene and lumisterol to cholesterol. 1970 R. A. MORTON *Fat-Soluble Vitamins* 187 The compound originally called vitamin D_1 was later found to be a mixture of ergocalciferol (D_2) and lumisterol, a photodecomposition product of D_2.

lumme (lɒ·mi), *int.* Also **lummy.** A corruption of (*Lord*) *love me.*

1898 J. D. BRAYSHAW *Slum Silhouettes* 146 Wot! Pay for me death? Oh, lummy! not me. 1921 H. WILLIAMSON *Beautiful Yrs.* xviii. 205 He's an awful liar, you know, but can't he half climb a tree! Lumme, much better'n I can! 1934 P. ALLINGHAM *Cheapjack* xxii. 297 You've forgotten all about sister Alice! Lumme, you must get 'er something. 1935 D. L. SAYERS *Gaudy Night* xvii. 370 'Lor' lumme!' I says, 'there's old Winderpane gawn.' 1942 *R.A.F. Jrnl.* 3 Oct. 33 Two years ago..they set London on fire... Lumme, that was a night. 1963 *Times* 4 Mar. 5/2 A pitch which has evoked from Trueman the classic comment: 'Lumme! A green dusty.'

lummox (lɒ·məks). *dial.* and *U.S.* Also **lommocks, lommox, lummicks, lummux,** etc. [Of obscure formation. Goes with the dial. verb *lummock* to move heavily or clumsily.] A large, heavy, or clumsy person; an ungainly or stupid lout.

a 1825 R. FORBY *Vocab. E. Anglia* (1830) II. 201 Look o' yin great lummox, lazing and lolloping about. 1854 A. E. BAKER *Gloss. Northamptonshire Words* I. 402 A great fat lommocks. 1854 'DOW JR.' *Patent Sermons* IV. 149 (Th.), Man in his original state is little more than a big lummux of a baby. 1857 J. G. HOLLAND *Bay Path* 381 (Th.), I hope you'll leave somebody else to home besides this lazy lummox. 1893 W. K. POST *Harvard Stories* 186 Well, don't you be such a lazy lummox. 1919 H. L. WILSON *Ma Pettengill* iii. 92 Oswald is a big fair-haired lummox that sings tenor in the Presbyterian choir. 1924 F. HURST (*title*) Lummox. 1934 [see *GORMLESS *a.*]. 1952 J. STEINBECK *East of Eden* xvii. 172 Those great lummoxes would chew a little thing like you to the bone. 1953 K. TENNANT *Joyful Condemned* vi. 54 There's that settee... This big lummox..can sleep on that. 1957 J. BRAINE *Room at Top* iv. 45 The big lummox standing possessively beside her. 1971 T. KILROY *Big Chapel* vi. 123 And that other lummox backing him up good-oh!.. A fine pair of cross-fire merchants the two of them! 1973 A. GARNER *Red Shift* 55 You big lummox,.. He's talking about the Irish.

lump, *sb.*¹ Add: **1. a.** (Earlier example of *lump of sugar.*) Also *ellipt.*, = *lump of sugar.*

1728 E. SMITH *Compleat Housewife* (ed. 2) 213 When 'tis fine draw it into dry Bottles, and put a Lump of Sugar into every Bottle. 1899 H. JAMES *Awkward Age* III. xi. 101 Sugar?—isn't that the way to say it? Three lumps? *a* 1916 'SAKI' *Toys of Peace* (1919) 24 Little friendly questions about weak or strong tea, how much, if any, sugar, milk, cream, and so forth. 'Is it one lump? I forgot.' 1922 H. S. WALPOLE *Cathedral* I. v. 85 No, I'm afraid I don't—thank you, Mrs. Sampson. One lump, please.

f. = *lump work* (LUMP *sb.*¹ 8). Of persons: those who contract to do work 'in the lump', i.e. for a lump sum.

[*a* 1852 H. MAYHEW *London Labour* (1861) II. 330/1 The first man who agrees to the job takes it in the lump, and he again lets it to others in the piece.] 1902 *Eng. Dial. Dict.* III. 691/1 You can do it either by the day or lump. 1969 *Daily Mail* 3 Sept. 2/1 He then spelt out exactly how the thousands of 'labour only' sub-contractors—'The Lump' in building trade slang—deprive the Treasury of more than £3 million a year. 1970 *Daily Tel.* 27 May 11/1 In the building industry, the outgoing government argues, this employment of a quarter of a million men—known as 'the Lump' because it accepts a lump sum for its work and attends to its own tax and social insurance problems—has resulted in widespread tax and National Insurance evasion. 1972 *Times* 21 Nov. 21/2 They attribute much of the confusion in the building 'jungle', as they often call it, to the operations of the 'lump', the growing number of labour-only sub-contractors and 'self-employed'. 1973 *Guardian* 22 Feb. 9/5 A Bill which would prohibit 'lump' labour in the building and construction industry was given a formal first reading in the Commons... There had been a definite increase in the lump in the past 10 years. 1974 *Shelter News* Easter 3/2 Some companies already party to the agreement admit a limited use of lump labour and argue very convincingly that they have little choice if they are to meet completion dates.

g. *U.S. slang.* A parcel of food given to a tramp or vagrant. Cf. Eng. dial. *lump*, a luncheon (see *E.D.D.*).

1912 D. LOWRIE *My Life in Prison* ix. 105, I noticed he had a lump (lunch) with him. 1914 JACKSON & HELLYER *Vocab. Criminal Slang* 56 *Lump*, current chiefly amongst yeggs, hobos, and the indigent. A donation of victuals intended for consumption outside the house. 1926 J. BLACK *You can't Win* vi. 67 She'll give you a sit-down for yourself, chances are, but bring back a 'lump' for us. 1931 'D. STIFF' *Milk & Honey Route* xiv. 161 It may be that he has the boy along only to wash his clothes or to bum his lumps. 1967 K. ALLSOP *Hard Travellin'* xviii. 214, I met a husky burly taking of his rest And he flagged me with a big lump and a can.

h. *pl.* Hard knocks, scolding.

1935 *Jrnl. Abnormal Psychol.* XXX. 363 *Lumps, get the*, to [be] beaten up. 1949 [see *FOOT *sb.* 29]. 1970 J. H. GRAY *Boy from Winnipeg* 32 My father would sit and take his verbal lumps, saying nothing. 1971 B. MALAMUD *Tenants* 130 Now I take my lumps, he thought. Maybe for not satisfying Mary.

4. b. (Earlier example.)

1849 N. KINGSLEY *Diary* (1914) 53 The farther north we get the more our anxiety is increased, as those big lumps are not quite eradicated from our minds yet.

5. d. (Later examples.)

1923 H. G. WELLS *Men like Gods* I. viii. 143 We shall all be..judged in a lump. 1934 G. B. SHAW *Too True to be Good* Pref. 9 The unqualified assertion that the rich, in a lump, are miserable.

7*. *slang.* The workhouse (see also quot. 1933).

1874 HOTTEN *Slang Dict.* 219 *Lump*, the workhouse; also called the Pan. 1898 J. D. BRAYSHAW *Slum Silhouettes* 146 If Sal can't bury me, the 'Lump''ll have to. 1933 'G. ORWELL' *Down & Out* xxxii. 236 These..are some of the cant words now used in London... The lump—the casual ward. 1972 G. F. NEWMAN *You Nice Bastard* 347 *In the lump*, in the workhouse.

8. *lump-lac, -tobacco*; **lump-sugary** *a.*, suggestive of lump-sugar.

1815 KIRBY & SPENCE *Introd. Entomol.* I. x. 317 In this country..it is distinguished by the names..Lump-lac when melted and made into cakes. 1873 *Beeton's Dict. Commerce* s.v. *Lac*, Lump lac is the deposit [of lac] formed into cakes. 1909 *Chambers's Jrnl.* Sept. 585/2 The body has a dry, lump-sugary appearance. 1851 C. CIST *Sk. Cincinnati in 1851* 244 Charles Bodmann..manufactures lump tobacco.

lump, *sb.*² Add: **1.** (Later examples.)

1844 *Knickerbocker* XXIV. 471 We discussed the merits of dun-fish,..lump, halibut,..and trout. 1867 [see *BAGATY, BAGGETY]. 1969 A. WHEELER *Fishes Brit. Isles & N.-W. Europe* 345 (*heading*) Lumpsucker (Sea Hen, Hen-fish, Lump).

2. lump-fish, lump sucker: later examples.

1969 H. HORWOOD *Newfoundland* 223 A lumpfish that I weighed on a pier-head in Conception Bay a few years ago went over thirty pounds. 1972 *Country Life* 30 Nov. 1541/3 You can afford to entertain with the real thing [*sc.* Sevruga caviar] and not pass off Danish Lumpfish roe (dyed black) as a substitute. 1974 *Observer* (Colour Suppl.) 15 Dec. 76/2 The lumpfish, known also as the cock- or hen-paddle on account of the thick crest shaping its back, is a creature of character. *Ibid.*, There is lumpfish caviare, which comes dyed black, and pearly, in small glass pots from Iceland and Denmark. 1959 A. HARDY *Fish & Fisheries* x. 193 Another surprise in this first haul was a lumpsucker *Cyclopterus lumpus* which I had previously decided to leave out of the book, thinking it to be an entirely coastal species... Its body is covered with little protuberances giving it a somewhat toad-like appearance; but it is bright with a pink hue on its lower parts. 1974 *Observer* (Colour Suppl.) 15 Dec. 76/2 They [*sc.* the female lumpfish] swim off leaving the males in charge, who cling to the rock by means of a suction disc between the pelvic fins—hence yet another name, lumpsucker.

lump, *v.*³ Add: **2. a.** *spec.* in *Taxonomy*: To classify (plants and animals) without using minute variations as a basis for the establishment of a large number of different species or genera. Cf. LUMPER *sb.* 3 (in Dict. and Suppl.).

1893 [in Dict.]. 1945 A. YOUNG *Prospect of Flowers* xx. 151 Our Village Schoolmistress carries lumping to an extreme degree. 1962 MACKWORTH-PRAED & GRANT *Birds S. Third Afr.* I. p. xi, It will be noted that we have not followed the modern trend of 'lumping' species and that we keep specific rank in this work for more birds than is usual nowadays. 1973 *Nature* 30 Mar. 353/1 The general absence of subgenera and species groups [in Chiarelli's classification of primates], combined with some tendency to 'lump' has meant the virtual disappearance of certain significant distinctions.

6. (Influenced by LUMPER *sb.* 1.) To act as a lumper, to load or unload cargoes. Hence, (*colloq.*) to carry or shift (something heavy) *about*. Also **lu·mping** *vbl. sb.*

1890 BARRÈRE & LELAND *Dict. Slang* II. 33/2 (Thieves), to *lump* the lighter, to be transported. In this case to *lump* signifies to load. 1905 *Westm. Gaz.* 10 Oct. 10/1 He..soon had the squad of irregulars at hard work 'lumping' as heartily as any gang of dock labourers. 1911 'KIWI' *On Swag* 14 For a month or so [I] was lumping on the wharf at the Spit. 1925 A. B. ARMITAGE *Cadet to Commodore* vi. 43, I earned sixteen shillings a week by 'lumping' in the docks. 1946 K. TENNANT *Lost Haven* (1947) xiii. 201 'I promised her a salmon.' He felt a fool lumping the great thing about.

lumped, *ppl. a.* Add: **2.** *Electr.* (Containing impedances or circuit elements) localized at a particular point or points, rather than distributed uniformly throughout part of a circuit.

1912 A. E. KENNELLY *Applic. Hyperbolic Functions to Electr. Engin. Probl.* iii. 35 This semi-angle cannot be correct, because it is based on a lumped leakance instead of a distributed leakance as assumed. 1948 E. G. BOWEN *Text-bk. Radar* xiii. 439 The units are connected by a low impedance coaxial cable which..is equivalent to a lumped capacity at video frequencies. 1967 *Electronics* 6 Mar. 163/1 Factors that combine to make analysis of crosstalk difficult are:.. The distributed, rather than lumped, nature of the inductances and capacitances of the circuits involved. 1973 B. KINARIWALA et al. *Linear Circuits & Computation* i. 4 We will confine ourselves to the discussion of lumped circuits—that is, circuits composed entirely of lumped elements.

lumpenproletariat (lɒ·mpənprōulitēᵊ·riăt). [a. G. *lumpenproletariat* (K. Marx 1850, in *Die Klassenkämpfe in Frankreich* and 1852, in *Der achtzehnte Brumaire des Louis Bonaparte*), f. *lumpen*, rag (*lump* ragamuffin: see LUMP *sb.*¹) + *proletariat* (see PROLETARIATE, -AT).] A term applied, orig. by Karl Marx, to the lowest and most degraded section of the proletariat; the 'down and outs' who make no contribution to the workers' cause. So **lu·mpenproleta·rian** *a.* and *sb.* Also **lu·mpen** *a.*, boorish, stupid, unenlightened, used derisively to describe persons, attitudes, etc., supposed to be characteristic of the lumpenproletariat; also *ellipt.* or as *sb.*

1924 H. KUHN tr. *Marx's Class Struggles France* I. 38 The financial aristocracy, in its methods of acquisition as well as in its enjoyments, is nothing but the reborn *Lumpenproletariat*, the rabble on the heights of bourgeois society. 1936 R. GESSNER *Some of my Best Friends are Jews* xxii. 306 No bandits dared attack her desperate inhabitants, the rough and dangerous Jews who had become lumpen-proletarians under the Czar. 1937 F. BORKENAU *Spanish Cockpit* i. 15 A not unimportant number of these 'lumpenproletarian' elements have joined the anarchist movement. 1940 'G. ORWELL' *Inside Whale* 132 The lumpen-proletarian fringe..composed partly of genuine artists and partly of genuine scoundrels. 1942 *New Statesman* 17 Oct. 255/1 He [*sc.* Hitler] mixed with the *Lumpen-proletariat*, the nomadic outcasts in the no-man's-land of society. 1944 KOESTLER in *Horizon* Mar. 167 Thus the intelligentsia..becomes the Lumpen-Bourgeoisie in the age of its decay. 1948 J. STEINBECK *Russ. Jrnl.* (1949) ix. 220 This journal will not be satisfactory either to the ecclesiastical Left, nor the lumpen Right. 1949 A. WILSON *Wrong Set* 57 Like called to like. The Colonel's lady and Lily O'Grady were both 'lumpen' under the skin. 1958 *Time* 28 June 35/3 At his cinematic best a shaggy lumpen-proletarian helplessly meshed in the woof of modern life..Charlie Chaplin. 1963 D. MACDONALD *Against Amer. Grain* i. 58 The spoofs of Dada have now become the serious offerings of what one might call the lumpen-avant-garde. 1969 R. BLACKBURN in Cockburn & Blackburn *Student Power* 186 Regis Debray has suggested that the lumpen-bourgeoisie of Latin America substitute police vigilance for an authentic class consciousness. 1970 *New Yorker* 18 July 22/2 Cope, a well-known proponent of lumpen aesthetics, had met the challenge with cool authority. 1971 'P. KAVANAGH' *Triumph of Evil* (1972) ii. 19 The rightist reaction of the white *lumpenproletariat* is easily imagined. Their instinctive response is racist and anti-intellectual. 1972 *Times Lit. Suppl.* 6 Oct. 1202/4 The *lumpenbourgeoisie*, behind a variety of leaders, is sick of dissent—student demonstrations, anti-apartheid, pornography, drug-taking, immigration, strikes, and crime. 1972 *Times* 14 Nov. 14/8 The underlying antipathy..towards the *lumpen* masses. 1973 *Ann. N.Y. Acad. Sci.* CCXI. 128 Fortunately for a lumpen linguist like me, my job doesn't require exercise of the vernier virtuosity that one sees in Linguistic Atlas maps and synopses. 1974 *Black Panther* 27 Apr. 13/2 The outlaw and the lumpen will make the revolution. The people, the workers, will adopt it. 1974 *Listener* 2 May 598/2 These institutions are likely to be manipulated only by a comparatively favoured group in society..leaving a real *lumpen* element at the bottom.

lumper. Add: **3.** *Taxonomy.* Substitute for def.: A taxonomist who is unwilling to use minute variations as a basis for the establishment of a large number of different species or genera. (Later examples.)

1945 A. YOUNG *Prospect of Flowers* xx. 151 Botanists are divided into two classes, 'splitters' and 'lumpers', 'splitters' being those who split plants into a large number of species and sub-species, while 'lumpers', impatient of minute distinctions, are inclined to lump them together. 1967 A. W. JONES *Introd. Parasitol.* xxix. 419 The more conservative taxonomists, called irreverently 'lumpers', defended established categories from attack by the radical 'splitters'. 1972 *Sci. Amer.* Nov. 60/2 One can use the work of many different taxonomists, without regard to whether they are 'lumpers' or 'splitters' in their method of classification, as long as the work is self-consistent.

lumpers (lɒ·mpəɹz), *sb. pl. slang.* [f. *lump sum*: cf. *-ER⁶.] A lump sum paid as compensation for loss of employment.

1960 *Economist* 23 July 352/2 They all give the impression that they remain happiest when a colonial official takes his 'lumpers' (lump sum compensation) and retires for good. 1963 *Punch* 17 July 77/2 The civil servants..deprived..of their career, have the consolation of their 'lumpers'—..up to £10,000. 1965 *New Society* 11 Nov. 11/2 The settler saw the official receive the 'golden handshake' or 'lumpers'.

lumpless (lʌ·mplės), *a.* [f. LUMP *sb.*^{1} + -LESS.] Having no lumps.

1908 *Daily Chron.* 3 Mar. 8/1 As soon as the ingredients are fairly worked into a lumpless, creamy whole, stop beating.

lumpy, *a.* Add: **1. b.** Also of broken weather.

1928 *Sat. Even. Post* 10 Mar. 8/1 'Had good weather?' 'Lumpy weather all the way.'

2. Applied to a person.

1926 A. BENNETT *Lord Raingo* I. xxxviii. 216 The fair but lumpy young woman silently left the room. **1928** E. O'NEILL *Strange Interlude* I. 14 Pretty vicious face under caked powder and rouge..lumpy body. **1934** *Times Lit. Suppl.* 8 Mar. 162/2 In the early nineties I was looking at Rubens's 'Rape of the Sabines' in the National Gallery, when a British workman beside me remarked, 'Lumpy lot to lift, eh?' **1959** I. & P. OPIE *Lore & Lang. of Schoolch.* ix. 168 The unfortunate fat boy..is known as..lumpy. **1973** *Observer* 5 Aug. 18/7 Lumpier and mousier, the girl grooms socialised affectionately with their ponies.

luna. 3. (Earlier and later examples.)

1869 *Amer. Naturalist* II. 679/2 Luna moth. **1876** *Field & Forest* II. 72 Mr. Rodgers..gives the history of the Luna moth. **1948** *Natural Hist.* Dec. 451/1 The Hercules moth is a close relative of the well-known Luna moth. **1972** SWAN & PAPP *Common Insects N. Amer.* 267 Luna Moth: *Actias luna*... Wings are delicate green, shading to pale gray, with transparent eyespots... Larva..resembles the Polyphemus larva except that Luna has yellow lateral stripe and no oblique stripes on the side. **1974** A. DILLARD *Pilgrim at Tinker Creek* x. 159 Luna moths are those fragile ghost moths, fairy moths, whose five-inch wings are swallow-tailed, a pastel green bordered in silken lavender.

lunabase (liū·nābeis). *Astr.* [f. L. *lūna* moon + BASE *sb.*^{1} (as the sb. corresponding to the adj. *basic*, in the petrographic sense).] The lunar maria or lowlands (the dark-coloured regions as seen from the earth).

Orig. proposed for the rock composing them (see quot. 1944).

1944 J. E. SPURR *Geol. applied to Selenology* II. iv. 20 The differently colored formations which largely make up the marking of the lunar 'map' have, in the Imbrian study (volume I), been interpreted as probably corresponding to two main types of terrestrial rock: the light-gray 'uplands' as siliceous, the dark mare as basic or basaltic... In future descriptions the light-colored rocks may be called lunarite. The dark-gray mare..may be called imbase... In the mare formations south of the Mare Imbrium, a later, darker phase of presumably basic rock..may safely, and briefly, be called novabase (later lunar basaltic rock). And the inclusive term for both would be lunabase (lunar basaltic rock). **1966** *Earth Sci. Rev.* I. 231 Spurr (1944) gave the name 'lunarite' to the more highly reflective parts [of the moon's surface], and the name 'lunabase' to the darker parts. Large areas of lunarite are referred to as 'continents', and specific areas of lunabase are called 'maria'. **1967** *New Scientist* 2 Feb. 263/2 The mountains [on the moon] end with steep terminal slopes of 20° or so, which cannot be due to erosion as a result of melting by subsequent lavas from the mare region (the dark lunabase).

lunanaut: see *LUNARNAUT.

Luna Park. The name of an amusement centre on Coney Island, Brooklyn, New York; also *transf.*, any such entertainment park. Also as *v. intr.*

1911 H. HAWTHORNE *New York* viii. 53 Luna Park and Dreamland are large sections of Coney..within whose gates are real fairylands of plaster palaces. **1921** A. HUXLEY *Crome Yellow* ii. 8 The lights of Luna Park. **1930** J. COLLIER *His Monkey Wife* iii. 33 The Fun Fairs and Luna Parks of the carnal capitals of Europe. **1936** F. CLUNE *Roaming round Darling* xxiv. 244 Over a by-wash, then, Luna-parking down a big dipper, belly half-way up the other side. **1956** M. DUGGAN *Immanuel's Land* 90 Up at Tibidabo in the empty luna-park an aluminium aeroplane turns slowly at the end of a long pole. **1957** A. HEPBURN *Compl. Guide N.Y. City* (new ed.) 109 Biggest and best of the amusement centers are Steeplechase, Luna Park, Dreamland. **1967** G. GREENE *May we borrow your Husband?* 10 As squalid as a closed fun-fair with Lunar [*sic*] Park boarded up.

lunar, *a.* and *sb.* Add: **A.** *adj.* **1.** (Further examples.)

1958 *Observer* 17 Aug. 1/6 It was new moon on Friday, and the 'lunar probe' must be launched in the next two or three days or postponed for a month. **1971** *Sci. Amer.* Oct. 49/3 Type *D* material consisted of miscellaneous fines smaller than a centimeter in diameter, material sometimes called lunar soil. **1972** *Science* 2 June 1014/3 It is appropriate to define the base of the 'lunar crust' at the discontinuity at 65 km.

b. lunar orbit, (*a*) the orbit of the moon around the earth; (*b*) an orbit around the moon; **lunar window,** a launch window for a mission to the moon.

1728 *Chambers's Cycl.* II. 578/1 The transverse Diameter of the Lunar Orbit. **1834** [in Dict., sense 1]. **1968** *Guardian* 28 Dec. 9/1 The Lunar Landing Module..will shuttle two astronauts from lunar orbit, down to the moon's surface. **1969** *Daily Mail* 14 Jan. 1/2 For us, lunar orbit was the busiest time of the flight. **1969** *New Scientist* 17 July 114/2 They are inserted into an elliptical lunar orbit varying between 60 and 170 nautical miles above the Moon's surface. **1969** *Guardian* 15 Jan. 18/2 A lunar window exists for Russia at the moment. **1970** N. ARMSTRONG et al. *First on Moon* ii. 33 The time of launch..had been chosen with great care. The date and the hour had been fixed to take full advantage of the so-called 'lunar window'.

c. Employed in or relating to travel to or from the moon or on its surface; *lunar* (*excursion*) *module*, a module designed to take an astronaut from an orbiting spacecraft to the moon's surface and back (abbrev. *LM, LEM* (*L 7)).

1962 *Daily Tel.* 17 July 17/6 If anything should go wrong with the lunar space-craft and the astronauts found themselves unable to take off, from the moon, unmanned supply vehicles would be sent. **1962** *New Scientist* 19 July 123 At a hundred miles or so from the lunar surface retro-rockets will slow the craft into a lunar orbit. It will then be split in two, to form a mother craft and a lunar excursion module. **1965** *Punch* 20 Jan. 84/1 The great Apollo 'lunar project' by which the Americans, bless them, still hope to land a man, alive, on the Moon in 1970. **1966** *Electronics* 3 Oct. 134 The Lunar Excursion Module of the Apollo program has to descend safely as well as take off and rendezvous with the command module for the long trip home. **1967** *Technology Week* 20 Feb. 16 (*caption*) Full-size mockup of Apollo Telescope Mount..is based on Apollo Lunar Module and is being used for placement of instruments. **1969** *Daily Mail* 14 Jan. 5/2 We were to test the lunar module, the vehicle which will land on the Moon, in a high Earth orbit. **1970** N. ARMSTRONG et al. *First on Moon* i. 20 By 1962..the future programs were well designed and the lunar mission was going to become a reality. **1971** *New Scientist* 3 June 574/1 Special wire mesh wheels have been built for America's Lunar Roving Vehicle which..is to drive around the lunar surface during the Apollo 15 mission in July. **1972** *Daily Tel.* 24 Apr. 1/4 They climbed into the lunar rover to drive 3·3 miles to North Ray Crater. **1975** S. JOHNSON *Urbane Guerilla* v. 181 Neil Armstrong at last stepped down from the lunar module *Eagle* on to lunar soil.

B. *sb.* **2. b.** *colloq.* A look.

1906 GALSWORTHY *Man of Property* i. 21 Now and then he would level his umbrella and take a 'lunar', as he expressed it, of the varying heights. **1938** N. MARSH *Artists in Crime* v. 65 Let us take what used to be called a 'lunar' at the case. **1950** 'J. GUTHRIE' *Is this what I Wanted?* iii. 71 Charles took a lunar.

lunarite (liū·nărəit). *Astr.* [f. L. *lūna* moon + -ITE^1.] The lunar uplands (the light-coloured regions as seen from the earth).

Orig. proposed for the rock composing them (see quot. 1944).

1944, 1966 [see *LUNABASE]. **1967** *New Scientist* 2 Feb. 263/3 In a few places the lunarite [lava] flows seem to dip under the lunabase of the surrounding plain, indicating that the lunarite and lunabase flows may at an early time have been contemporaneous.

lunarnaut (liū·nāɹnǭt). Also **lunanaut.** [f. LUNAR *a.* + *-naut*, after *aeronaut, astronaut*.] One who travels or has travelled to the moon.

1965 *Guardian* 30 Aug. 12/1 The moon men (or lunarnauts, as they will doubtless come to be called) must be able to park their spacecraft. **1966** *Electronics Weekly* 27 July 29/2 (*heading*) Lunanauts will land with a reassuring 'bump'. **1969** *Scottish Daily Mail* 13 Jan. 2/6 Lunarnaut Frank Borman warned today that space flights are 'chancy'. **1971** *Daily Tel.* 30 Jan. 3/3 The Command and Service Module orbit the Moon while the lunarnauts carry out experiments and walk on its surface.

lunarscape (liū·nāɹskēip). [f. LUNAR *a.* + SCAPE *sb.*^{3}] A picture or view of the moon's surface; the lunar landscape.

1965 *Newsweek* 25 Jan. 89 No one knows in detail what the lunarscape is like. **1966** *Punch* 27 Apr. 636/1 Superb blending of the Op and the Pop, comprising pointillist lunarscape in Coke bottle-tops. **1967** *Time* 25 Aug. 38 Lights flashed everywhere, bounding off the Day-Glo lunarscapes along the wall. **1970** *Guardian Weekly* 15 Aug. 12 The moon began to show out over the North Sea... Through my powerful binoculars I was able to study its face. The lunarscape looked like some great contoured globe. **1971** *New Scientist* 22 Apr. 224/1 These [paintings] include some dramatic lunarscapes.

lunate (liū·nēit), *sb. Archæol.* [f. the adj.] A small prehistoric stone (usu. flint) artifact which was probably used as an arrow-head and has an elongated half-moon shape with the straight edge unworked and the curved edge sharpened by chipping.

1932 *Jrnl. R. Anthrop. Inst.* LXII. 261 A fair proportion of lunates and other microliths showed a peculiar retouch... The back of the implement is not blunted in the ordinary way, but is trimmed obliquely from both surfaces, the result being to make it sharp instead of blunt. **1949** W. F. ALBRIGHT *Archaeol. of Palestine* iii. 59 The Natufian was a thorough-going microlithic culture, consisting largely of flint blades and points, most typical of which is the so-called lunate, a crescent or arc-shaped blade, probably used to tip reed arrows. **1960** K. M. KENYON *Archaeol. in Holy Land* ii. 36 Most characteristic of all are the lunates, very fine little flakes with a straight edge and a crescent-shaped back.

lu·nately, *adv. rare.* [f. LUNATE *a.*] In a crescent form.

1872 H. C. WOOD *Contrib. Hist. Fresh-Water Algæ N. Amer.* 109 Cells..more or less lunately curved.

lunatic, *sb.* Add: **1. c. lunatic fringe,** a minority group of adherents to a political or other movement or set of beliefs; also *attrib.*; **lunatic soup** *Austral.* and *N.Z. slang*, alcoholic drink.

1913 T. ROOSEVELT *Hist. as Lit.* 305 There is apt to be a lunatic fringe among the votaries of any forward movement. **1936** *Economist* 4 Jan. 3/1 Dr. Townsend has now left all his rivals behind in competition for the votes of the lunatic fringe. **1945** 'G. ORWELL' *England your England* (1953) 56 The lunatic fringe even contrived to be simultaneously pro-Russian and pro-Nazi. **1951** E. PAUL *Springtime in Paris* iii. 59 Most of the paintings are imitations of other paintings, either conscious or unconscious, ranging from primitives and false primitives to the latest lunatic fringe. **1953** *Manch. Guardian Weekly* 21 May 15/4 Five [letters]..refer to his 'plug-ugly tactics', the 'lunatic fringe' brand of anti-communism. **1958** *Times* 9 Sept. 9/3 The lunatic fringe of society, to which only too many irresponsible hoaxers here and in America belong, finds the telephone its best friend. **1968** *Guardian* 12 Dec. 9/1 Lunatic-fringe Utopians call for something entirely new. **1969** *Observer* 23 Nov. 25/3 Views like these, as anyone who has listened to a few conversations on the verandahs of suburban Kingston will recognise, aren't lunatic fringe. **1970** *Ibid.* 11 Oct. 32/3 Both sides..have what you might call lunatic fringes; on the middle-aged right there are those who would lynch anyone with long hair, on the young left..there are those who refuse to state their case rationally. **1973** D. RAMSAY *Deadly Discretion* 175 Antique shops were magnets for the lunatic fringe. **1933** *Bulletin* (Sydney) 6 Sept. 42/1 Lunatic soup, as the few fellows about who knew him as Darkie called the brandy he drank. **1941** BAKER *Dict. Austral. Slang* 45 *Lunatic soup*, cheap red wine.

lunation. 2. Delete † *Obs.* and add later example.

1953 A. C. CLARKE *Prelude to Space* xxii. 115 If..there's a last-minute hold-up, launching will be delayed.., at the most, thirty-six hours. After that we'll have to wait for the next lunation—that is, for four weeks... We're anxious to land in daylight.

lunch, *sb.*^{2} Add: **2.** (Later examples.) Now in common use for LUNCHEON 2. Also, a light meal at any time of the day.

1936 *S.P.E. Tract* XLV. 183 In several..instances a word has been liberated in America from the restrictions that limit its application in England; for example..*lunch*. **1950** E. A. MCCOURT *Home is Stranger* (1951) viii. 124 At midnight the womenfolk hurried out to the hall to prepare lunch. **1951** *Good Housek. Home Encycl.* 540 *Lunch*, the midday meal..may cover anything from the sandwich lunch taken by some office and factory workers to the fairly elaborate formal meal given when guests are entertained at a public function or private party. **1960** S. PLATH *Colossus* (1967) 20, I open my lunch on a hill of black cypress. **1963** L. DIACK *Labrador Nurse* I. vii. 37 After a 'lunch' (i.e. a snack), all hands..would set to work. **1965** J. S. GUNN *Terminol. Shearing Industry* I. 36 *Lunch*, a light snack which was taken at the mid-morning or mid-afternoon break... Lunch was food taken at any time away from main meals, even in the evening. **1968** *New Society* 22 Aug. 265/2 Though the U still have lunch (not dinner) in the middle of the day and U-dogs still have their dinner then, U-children have changed; they no longer have mid-day dinner, in the nursery, but have lunch with their mothers.

b. *out to lunch*: insane; stupid, unaware; socially unacceptable. *N. Amer. slang.*

1955 *Sci. Digest* Aug. 33/1 'Out to lunch' refers to someone who, in other years, just wasn't 'there'—and he is told immediately to 'Get with it!' **1959** *She* May 21/2 *Out to lunch* has nothing to do with social life but implies one is not in the groove. **1966** *Toronto Daily Star* 16 June 74/2 A girl who would be attracted to Bud's mean streak and bad temper must be a little out to lunch. **1974** *Melody Maker* 13 July 13/5, I think he's out to lunch. He's blown out—completely.

3. *lunch-bell, -box, -break, -cake, -can, -counter, -date, -hour, -house, -money, -pail, -party, -room, -stand, -tin, -wagon.*

1875 MRS. STOWE *We & Neighbors* v. 67 The ringing of the lunch bell interrupted the conversation. **1864** *Rep. Comm. Patents 1862* (U.S.) I. 158 *Improved Lunch Box*... This invention consists of an arrangement of dishes, cups, etc., arranged within a case for the use of travellers. **1921** *Daily Colonist* (Victoria, B.C.) 23 Oct. 6/1 (Advt.), Folding lunch boxes. **1970** G. GREER *Female Eunuch* 233 Her house is ideally a base which her tired-warrior hunter can withdraw to..while he..is prepared by laundry and toilet and lunch-box for another sortie. **1960** *News Chron.* 14 June 6/6 Half of today's secretaries are married women and rely on the lunch break to do..their shopping. **1971** C. STORR *Thursday* xii. 133 She timed herself to reach the [building] site just before the lunch break. **1886** M. L. DODS *Handbk. Pract. Cookery* (new ed.) 219 Lunch Cake,..Bake in a moderate oven for one hour and a quarter. **1901** *Daily Colonist* (Victoria, B.C.) 27 Oct. 7/2 Our buns, scones and lunch cakes are the acme of perfection. **1951** *Good Housek. Home Encycl.* 540/2 *Lunch cake*, a fairly plain, substantial fruit cake. **1897** R. M. STUART *In Simpkinsville* 14 They'd give him biscuits out o' their lunch-cans. **1869** *Demorest's Young Amer.* Oct. 460 (*caption*) Eating at a small lunch counter or eating-saloon. **1904** G. S. FULLERTON *Syst. Metaphysics* xv. 242 To obtain a sandwich from the woman at the lunch-counter. **1934** *Archit. Rev.* LXXVI. 159/1 (*caption*) Lunch counter with recessed stainless steel front. **1960** *New Statesman* 26 Mar. 435/2 Possibly President Eisenhower will be a shade or two happier when a coloured Georgian and a white Georgian, seated on adjacent stools, have pie and coffee at a Woolworth lunch-counter in Atlanta—and possibly he may not. **1970** *Times* 23 Mar. 13/3 A nostalgic echo of those days when it seemed that all

we had to do to achieve integration was to sit down at enough lunch counters together. **1933** *Radio Times* 14 Apr. 95 A tragic lunch date. **1968** *Listener* 25 July 103/3 A message came regretting he could not keep our lunch date, because he must be sure to cast his vote. **1908** KIPLING *Lett. of Travel* (1920) 129 Canadian Clubs..assemble their members during the mid-day lunch-hour. **1909** H. G. WELLS *Ann Veronica* xiv. 302 She waited in the laboratory at the lunch-hour. **1929** *Radio Times* 8 Nov. 426/1, 1.15–2.0. A lunch hour concert. **1959** J. O'DONOVAN *Visited* xviii. 116 Edith sat by the phone all the lunch-hour. **1846** *Knickerbocker* XXVIII. 558 The following parody was found inscribed on the newspaper-board of a 'lunch-house' in Saint Louis, Missouri. **1902** G. H. LORIMER *Lett. Merchant* viii. 108 One of those fellows..goes around and makes the boys give up their lunch money to buy flowers. **1955** M. GILBERT *Sky High* x. 144 Sixpence a week and their lunch money. That's all they get. **1891** M. E. WILKINS *New Eng. Nun* 44 Matilda came in her voluminous alpaca, with her tin lunch-pail on her arm. **1926** E. HEMINGWAY *Torrents of Spring* x. 71 He set down his lunch-pail. **1964** M. GALLANT in R. Weaver *Canad. Short Stories* 2nd Ser. (1968) 77 There was an unbridgeable gap..between the girl whose father went off to work with a lunch pail and the daughter of a man who ate..in the company cafeteria. **1884** F. M. CRAWFORD *Amer. Politician* I. iii. 43 At a lunch party..they sat and talked about pictures. **1936** H. NICOLSON *Let.* 19 Feb. (1966) 244, I gave a lunch-party. **1961** NEW ENG. BIBLE *Luke* xiv. 12 When you give a lunch or dinner party. **1830** *N.Y. Mercantile Advertiser* 16 Aug. 4/6 His Breakfast, Lunch and Dining rooms are capacious and comfortable. **1919** Lunch room [see *HOLE *sb.* 7 b]. **1958** *New Statesman* 1 Feb. 143/1 By the end of the novel, Flem..having by trickery acquired a half-interest in a Jefferson lunch-room has taken himself off to the county seat. **1887** C. B. GEORGE *40 Yrs. on Rail* v. 79 Superintendent Johnson..noticed this lunch-stand, with its modest, yet appetizing display. **1950** *N.Z. Jrnl. Agric.* Aug. 191/1 Lunch tins which can be painted attractive colours are easily washed, inexpensive, and light and convenient to carry. **1894** *Life* 4 Oct. 215/1 'That, my dear,' responded Adalbert, 'is a lunch wagon.' **1959** N. MAILER *Advts. for Myself* (1961) 69 Inside, out of the rain, the lunch wagon was hot and sticky.

luncheon. Add: **2. b.** *U.S.* Applied to a late supper.

1903 *Boston Even. Transcript* 3 Oct. 5 At this table, from 9 o'clock until midnight, a bountiful standing luncheon was served continuously.

3. *luncheon-basket* (earlier and later examples); **luncheon-car,** on a railway train, a restaurant-car where luncheons are provided; also *attrib.*; † **luncheon-dinner** = *lunch-dinner* (LUNCH *sb.*^{2} 3); **luncheon meat,** a type of pre-cooked meat containing preservatives; **luncheon voucher,** a money voucher given to employees which is exchangeable for meals at certain restaurants.

1859 QUEEN VICTORIA *Jrnl.* 7 Oct. in D. Duff *Victoria in Highlands* (1968) 159 Our pony carried the luncheon baskets. **1958** J. CANNAN *And be a Villain* i. 15 Ah, the Victorian railways!..the luncheon baskets you wired ahead for. **1903** A. BENNETT *Leonora* vii. 203 The express, with its two engines, its gilded luncheon-cars, and its post-office van. *Ibid.* 204 Catching the luncheon-car attendant by the sleeve. **1909** *Westm. Gaz.* 8 Sept. 2/1 A new luncheon-car express, starting from King's Cross at 1.5 p.m. **1970** *Country Life* 31 Dec. 1293/1 On the next stage north there was a luncheon car, and at The Mound he ate his lunch. **1819** M. EDGEWORTH *Let.* 4 Mar. (1971) 178 After luncheon-dinner I finished at Oxford. **1882** W. D. HAY *Brighter Britain!* I. ii. 25 There are three common meals—breakfast, luncheon-dinner, and dinner-supper. **1945** 'R. CROMPTON' *William & Brains Trust* ix. 166 Although it meant opening her last remaining tin of Luncheon Meat. **1953** J. HUNT *Ascent Everest* IV. xii. 155 Over our bacon and, possibly, eggs or fried luncheon meat. **1957** E. CRAIG *Collins Family Cookery* 814 Dice luncheon meat and use as a filling for bread. **1960** A. E. BENDER *Dict. Nutrition* 76/2 *Luncheon Meat*..must have a meat content not less than 80%. **1955** *Evening Standard* 28 Oct. 15/3 (*heading*) Doorman/timekeeper for staff and goods entrance... Pension scheme, welfare fund, luncheon vouchers, etc. **1966** A. LA BERN *Goodbye Piccadilly* v. 48 Their wallets contained more luncheon vouchers than treasury notes. **1973** *Times* 24 Jan. (Security Conf. Printing Suppl.) p. i/3 The notes in our wallets, stamps for letters..luncheon vouchers..are a few of the little pieces of paper so essential to modern life.

luncheo·nette. orig. *U.S.* [LUNCHEON 2 + -ETTE.] A small restaurant or snack bar serving light lunches.

1924 *Public Opinion* 11 July 31/2 Luncheonettes supply icecream soda and a ham sandwich. **1930** J. O. DAHL (*title*) Soda fountain and luncheonette management. **1939** C. MORLEY *Kitty Foyle* viii. 74 Sparta's, a Greek candy and luncheonette a block farther up Main. **1959** *New Statesman* 24 Oct. 534/3 They sleep late, meet on the corners, drift in and out of the poolrooms and the luncheonettes. **1969** *New Yorker* 20 Sept. 38/2 A luncheonette, where a couple of young men in shirt sleeves sat hunched over the counter eating..noodle soup. **1972** *Daily Tel.* 10 Apr. 9/8 France is to have 100 quick-lunch restaurants *à l'Anglaise*... These luncheonettes will all be along France's main highways. **1974** D. RAMSAY *No Cause to Kill* I. 44 She didn't care for luncheonette food.

lunching, *vbl. sb.* [f. LUNCH *v.*] The action of taking lunch. Also *attrib.*

1920 R. L. ALSAKER *Maintaining Health* 271 Lunching before going to bed is a bad habit. **1968** *Economist* 18 May 74/2 One can see bloated central staffs, outside directors who are appositely called 'lunching directors' and cannot control their management colleagues in any way.

lu·nchless, *a.* [f. LUNCH *sb.*^{2} + -LESS.] Having had no lunch; without lunch.

1904 KIPLING *Traffics & Discov.* 339, I found myself stranded, lunchless, on the sea-front. **1920** GALSWORTHY *In Chancery* III. i. 244 Sitting lunchless in the hall of his hotel. **1958** E. NEWBY *Short Walk in Hindu Kush* ii. 17 In my lunchless state I envied them.

lu·nch-time. [f. LUNCH *sb.*^{2} + TIME *sb.*] The time at which lunch is eaten. Also *attrib.*

1859 GEO. ELIOT *Let.* 10 Oct. (1954) III. 180 He can't take us wrongly any day either at 1 o'clock, (lunch-time) or at half past 5 (dinner). **1866** G. M. HOPKINS *Lett. to R. Bridges* (1955) 5 He..left me at lunch-time. **1890** [see LUNCH *sb.*^{2} 3]. **1909** H. G. WELLS *Ann Veronica* xiv. 290 Capes came into the laboratory at lunch-time. **1929** *Radio Times* 8 Nov. 389/2 The lunch-time programme arranged by Mr. Christopher Stone. **1957** MANVELL & HUNTLEY *Technique Film Music* 235 Kisenga, an African musician..is seen playing this work in a reconstruction of the war-time National Gallery lunch-time concerts. **1963** *Times* 22 May 9/5 We'll be at the coast by lunchtime and the children can't wait for a swim and neither can I. **1968** J. BINGHAM *I love, I Kill* xv. 222 Have me thrown out, boyo. If you hurry up, it'll make the lunchtime editions. **1971** D. CRYSTAL *Ling.* i. 11 There is at least one teachers' common-room..where problems of etymology..provide the normal, lunch-time gossip.

‖ **lundum** (lu·ndʊm). [Pg.] A primitive Portuguese song and dance, from which the *fado* probably developed.

1936 R. GALLOP *Portugal* xi. 252 The *lundum*..shared the affections of the Lisbon populace from the last quarter of the eighteenth to the middle of the nineteenth century. The *lundum* came to Portugal from Brazil... It reached Brazil from the west coast of Africa. **1957** R. CAMPBELL *Portugal* ix. 191 It was this crossing of the primitive sensual ferocity and black, wailing misery of the negro *lundum*..that gave us the *fado*. **1957** [see *belly-dancing* s.v. *BELLY *sb.* 17]. **1969** S. BRADFORD *Portugal & Madeira* 27 With them [*sc.* Angolan slaves] went memories of their native dances, one of them being the lundum, a dance of Congolese origin.

lunette. Add: **15.** *Physical Geogr.* A broad shallow mound of wind-blown material built up along the leeward side of a lake basin, esp. in arid parts of Australia, and typically having a crescent shape with the concave edge of the crescent along the lake shore.

1940 E. S. HILLS in *Austral. Geographer* III. VII. 15 (*title*) The lunette, a new land form of aeolian origin. *Ibid.* 15/1 Along the eastern shores of almost every lake and swamp in the plains of northern Victoria there occurs a crescentic ridge of silty clay or clay 'loam'... It is..proposed to designate them by a new term—*lunette*. **1942** C. A. COTTON *Geomorphology* (ed. 3) xx. 275 Dust captured from the air during gales that produce dust storms is brought down by spray whipped up from lakes, so that crescentic mounds of loamy material of this origin grow up immediately to leeward of the lakes. Being rarely more than 20 or 30 feet high these broad mounds are not conspicuous unless they rise from very level plains, as is the case in south-eastern Australia, where there are many examples of such land-scape forms, there termed lunettes. **1957** G. E. HUTCHINSON *Treat. Limnol.* I. i. 127 The most important type of wind action in forming lake basins..is deflation or wind erosion. The clearest evidence of this process is provided by those cases in which the deflated material is piled up as a curved mound of sand or *lunette*..along the lee shore of the depression.

lung. Add: **6. a.** *lung cancer, function.* **b.** *lung-breathing, -bursting* adjs.

1907 *Westm. Gaz.* 1 June 16/3 The complete proof of this evolution of the lung-breathing four-footed creatures of the earth from purely aquatic forms has been lost. **1949** *Oxf. Jun. Encycl.* II. 359/2 If the larval form [of the Axolotl] is kept..it will gradually turn into the mature, lung-breathing salamander. **1971** S. CAVELL *World Viewed* vii. 41 Baudelaire's..lung-bursting inflation of Delacroix. **1973** C. BONINGTON *Next Horizon* xxi. 286 The last length of rope..was the most strenuous of all, taking two hours of lung-bursting effort to reach the top. **1926** *Jrnl. Amer. Med. Assoc.* 17 July 147/1 A diagnosis of endothelioma has been made frequently in primary lung cancers. **1953** *Newsweek* 25 May 60 Dr. Alton Ochsner..believes that lung cancer..'is unquestionably due to the carcinogenic effect of cigarette smoking'. **1975** 'G. BLACK' *Big Wind* ii. 39 When she was still a deb..lung cancer was still diagnosed as galloping consumption. **1966** *Lancet* 24 Dec. 1386/1 Systematic lung-function studies were not carried out in these patients.

7. lung book, a lamellate respiratory organ found in spiders, scorpions, and certain other arachnids; cf. *book-lung* (*BOOK *sb.* 18); **lung-fish** (later example); **lung fluke,** a parasitic trematode flatworm of the genus *Paragonimus*; also *attrib.*; **lung snail,** a snail of the order Pulmonata (see PULMONATE *sb.*).

[**1861** J. BLACKWALL *Hist. Spiders Great Brit.* I. 4 The internal organs of respiration in connection with the anterior pair of stigmata present the appearance of membraneous sacs formed by lamellæ applied to one another like the leaves of a book.] **1881** E. R. LANKESTER in *Q. Jrnl. Microsc. Sci.* XXI. 541 The lamellæ of the Scorpion's lung-book. **1932** BORRADAILE & POTTS *Invertebrata* xv. 447 The spiders, at least, have passed through a primitive lung-book stage from which they have not all emerged. In fact they show all the stages of replacement of lung books by tracheae. **1971** *Nature* 12 Feb. 455/1 The species [sc. *Micrathena gracilis*, a spider] possesses a well-developed stridulatory organ with a file on the cover of the lung book (the respiratory organ). **1968** A. S. ROMER *Procession of Life* viii. 165 The dipnoans owe their popular name of lungfishes to the fact that, except for two ray-finned fishes..they are the only living fishes to possess these air-breathing structures. **1900** STILES & HASSAL in *16th Ann. Rep. Bureau Animal Industry, U.S. Dept. Agric.* 560 (*title*) The lung fluke (*Paragonimus westermanni*) in swine and its relation to parasitic hemoptysis in man. **1931** *Jrnl. Amer. Vet. Med. Assoc.* LXXVIII. 229 (*title*) Lung flukes of the genus Paragonimus in American mink. **1937** *Discovery* Feb. 34/2 The lung-fluke disease, or paragonim[i]asis, of which they [*sc.* mitten crabs] are a carrier in China, does not really threaten Europe as yet. **1970** *Black's Vet. Dict.* (ed. 9) 516/1 Lung flukes attack cats, dogs, pigs, and man in the Far East and the United States. **1909** *Westm. Gaz.* 26 June 15/2 The land and most of the freshwater snails belong to the lung snails, the gills being reduced to a mere vestige.

Lung-ch'üan (luŋtʃuā·n). The name of a district in the province of Chekiang, China, used to designate a type of Chinese celadon ware produced mainly during the Sung dynasty (A.D. 960–1279).

1904 E. DILLON *Porcelain* v. 63 Lung-chuan ware was made during Sung times. **1936** *Burlington Mag.* Jan. 9/1 The lovely green celadons of Lung-ch'üan... Soon it may be possible to identify the celadons made at the different factories of the Lung-ch'üan district. **1960** H. HAYWARD *Antique Coll.* 62/1 Typical [of Celadon wares] are..the much-exported Lung-chüan celadons of the Sung and Ming dynasties. **1971** L. A. BOGER *Dict. World Pott. & Porc.* 56/1 The characteristic Sung Lung-ch'üan celadon has a body that approaches a white porcelain in character, but its main glory rests in the radiant, light bluish-green glaze.

lunge, *sb.*^{2} **2.** (Earlier example.)

1845 J. J. HOOPER *Taking Census* in *Some Adventures Simon Suggs* 155 That was a most unfortunate lunge I made into that hole in the river.

lunge, *sb.*^{3} Substitute for def.: Either of two large North American freshwater fishes, *Salvelinus namaycush*, a char or lake trout found in northern lakes, or *Esox masquinongy*, a pike found in the Great Lakes. (Earlier and later examples.)

1851 *Vermont Laws* 49 Such person or persons shall forfeit and pay..the sum of one dollar for each trout or lunge so taken. **1857** *Porter's Spirit of Times* 11 Apr. 86/3 The lower end of the lake..is supplied with the large cat-fish,..Oswego, black, longe, great-bass, pike-perch, perch, &c. **1902** *Jrnl. Amer. Folk-lore* Oct. 246 *Longe* or *lunge*, a common abbreviation of *muskelunge* (*maskalonge*) among English-speaking people in the region about the Great Lakes. **1953** *Canad. Geogr. Jrnl.* XLVII. 17/1 Recently, thanks to government hatcheries, 'lunge has been added to the menu.

lungful. Add: (Later examples.) *spec.* a quantity of inhaled cigarette-smoke.

1942 *R.A.F. Jrnl.* 3 Oct. 33 The little man accepted a cigarette.., drawing down a lungfull of smoke. **1964** 'E. LATHEN' *Accounting for Murder* (1965) xiv. 128 He sucked a healing lungful of smoke. **1973** A. HUNTER *Gently French* ii. 20 He pulled in a contemptuous lungful. **1973** P. MALLOCH *Kickback* xxix. 154 Hold your breath all the time. You don't want to get a lungful.

‖ **lung-gom-pa** (lu·ŋgɒmpa). Also **lung-pa.** [Tibetan.] A Tibetan monk who has the mystical power of walking many miles at great speed without stopping.

1931 A. DAVID-NEEL *With Mystics & Magicians in Tibet* vi. 200, I met the first *lung-gom-pa* in the Chang thang of Northern Tibet. *Ibid.*, The feat expected from the *lung-gom-pa* is one of wonderful endurance rather than of momentary extreme fleetness... True *lung-gom-pas* must be very rare. **1937** *Times Lit. Suppl.* 7 Aug. 579/1 He claims to have seen *lung-gom-pas* on two occasions. These are hermits who are said to have gained the power of travelling a hundred miles a day on their own feet. **1952** E. MOSSBACHER tr. *Maraini's Secret Tibet* iii. 51 *Lung-pa*, ..the 'wind-men'—monks who, after years of extreme asceticism and strenuous preparation, succeed in freeing themselves almost completely from the weight of the human frame and are therefore able to travel hundreds of miles in a single day. **1954** W. NOYCE *South Col* iii. 39 On the boat we read two books about Buddhism. I read of 'tumo'..and 'lung-gompa', the art of going into a trance and travelling many miles at incredible speeds.

lungy, *a.* Add: **b.** Coming from the lungs.

1909 *Westm. Gaz.* 21 Apr. 2/1 As the armed companies turned this corner of the narrow road a lungey Oriental cheer..saluted each. **1935** 'R. CROMPTON' *William—the Detective* ix. 211 Sounds lungy to me.

lungyi (lʊ·ŋyī), var. LUNGI. Cf. *LONGYI.

1906 W. DEL MAR *Romantic East* i. 8 The lungyi..whenever the wearer can afford it, is of thin silk and is simply a square of about five feet..put on like a petticoat and folded in over the right hip. **1948** *Amer. Speech* XXIII. 228/2 *Longyi, lungyi*, Burmese skirtlike attire which falls to the ankles. The national dress, worn by both sexes. **1959** *Times* 19 Mar. 14/6 The Burmese in their gaily coloured *lungyis*.

lunik (liū·nik). *Astronautics.* Also **Lunik.** [f. L. *lūn-a* moon + *-NIK, after *sputnik*, or ad.

Russ. *lúnnik* (similarly f. Russ. *luná* moon).] Any of a series of Russian spacecraft sent to or close by the moon.

1959 *Daily Tel.* 5 Jan. 16/6 The Russians had a word for the moon rocket soon enough: Lunik. Unlike Sputnik the word was not official. *Ibid.*, The only vocal demonstrations..were from the generation still young enough to qualify for..trips..aboard Luniks. **1959** H. NICOLSON *Let.* 14 Sept. (1968) 370, I hate the lunik for having bumped into the radiant moon. **1961** *Ann. Reg. 1960* 388 The Russians attempted no further 'lunik' shots. **1964** *Yearbk. Astron. 1965* 70 The Russians se[n]t up their Moon probe Lunik III, which represented the greatest triumph in space research up to that moment. [**1966** *Ann. dell'Istituto Univ. Orient., Sezione Slava* IX. 213 From the evidence presented above the fact seems to emerge that, even if *lunnik* was coined in Moscow on the morning of 3 January 1959, it was simultaneously fabricated in the West.] **1967** *Punch* 28 June 937/1 The Luniks were soon joined by American Surveyors and Lunar Orbiters and by the end of 1966, by no stretch of poetic imagination could the moon be described as companionless.

lunk (lʊŋk). *colloq.* (orig. *U.S.*) [Abbrev. of LUNKHEAD.] A slow-witted, unintelligent person.

1867 *Harper's Weekly* 25 May 330/2 They're tigers, you thick-headed lunk. **1907** J. MASEFIELD *Tarpaulin Muster* iv. 70 He's dead all right... He might ha' known..going alone among them Indians... None but a red-headed runt'd have been such a lunk as to try it. **1931** F. HURST *Back St.* II. xxvi. 231 What a lunk he continued to be! **1955** W. GADDIS *Recognitions* I. vii. 229 Most artists have a great lunk of a man they trail around with them, they never know what to do with him, he gets drunk, he gets into trouble. **1975** *New Yorker* 24 Mar. 93/3 He looks incredulous, as if he couldn't figure out how he got turned into such a lunk.

lunker (lʊ·ŋkəɹ). *N. Amer. colloq.* [Origin unknown.] An animal, esp. a fish, which is an exceptionally large example of its species; a 'whopper'. Also *attrib.*

1912 *Dialect Notes* III. 582 Isn't that calf a lunker. **1920** *Outing* July–Aug. 197, I said that I caught trout in a tin pan, and here's the proof. This old lunker of a rainbow gave me a bath. **1947** *Sports Afield* Dec. 21/1 A bronzed lunker came out of the shadowy depths and smashed the pigskin. **1968** *Globe & Mail* (Toronto) 3 Feb. 33/5 The area has been dubbed lunker country and anglers find it's more than just a slogan. **1972** *Angling Times* 6 Apr. 14/3, I shall be going out..in search of these lunker bass.

lunkhead. Delete '*colloq. U.S.*' and substitute '*colloq.* (orig. *U.S.*)'. Add earlier and later examples.

1884 'MARK TWAIN' *Huck. Finn* xxii. 225 So the duke said these Arkansaw lunkheads couldn't come up to Shakespeare. **1908** *Daily Chron.* 23 July 3/2 Now do you see, you lunkhead? **1934** WODEHOUSE *Right Ho, Jeeves* v. 52 A lunkhead capable of mucking things up as Gussie had done. **1951** E. PAUL *Springtime in Paris* ix. 165 They are not all lunkheads or mountebanks. **1966** *Punch* 7 Dec. 868/1 The poor lunkhead's concerns soon get lost under all the modelling and backlighting.

lunokhod (liū·nǒkɒd, -χɒd). *Astronautics.* Also (as the proper name of individual vehicles) with capital initial. [a. Russ. *lunokhód*, f. *luná* moon + *-khod*, suffix denoting something that travels (f. *khodít'* to go).] A type of Russian self-propelled, radio-controlled vehicle for transmitting information about the moon as it travels over its surface.

1970 *Guardian* 18 Nov. 1/2 Russia is likely to try to bring its moon crawler Lunokod-1 back to earth. **1973** *Nature* 30 Nov. 241/2 The thickness of the regolith decreased as the lunokhod approached the rille and the lip consisted of a rock 'border' with boulders of at least 1 to 2 m. *Ibid.*, Magnetometer experiments, a new feature of Lunokhod-2.

lunula. Add: **5.** *Archæol.* A gold, crescent-shaped, neck ornament found in archaeological sites of the Early Bronze Age.

1719 J. HARRIS *Hist. Kent* I. II. 249/1 Many also of the *Lunulae* were found here. **1773** *Archæologia* II. 37 The small circular plates at the extremities of the Lunula. **1867** *Archæol. Jrnl.* XXIV. 197 In another remarkable discovery of golden relics.., namely the two *lunulæ* found at Padstow.., the precious deposit was likewise accompanied by an object of bronze. **1911** *Encycl. Brit.* II. 353/1 The flat, crescent-shaped, diadem-like objects called 'lunulae', which are..characteristic of Ireland. **1939** G. CLARK *Archaeol. & Society* v. 133 If the distribution of crescentic jet necklaces and the gold 'lunula' ornaments are together plotted on the map they will be found to coincide closely with the distribution of food vessels.

Luo (lū·o), *sb.* and *a.* Also **Luoh, Lwo. A.** *sb.* **a.** The name of an East African people in Kenya and the upper Nile valley; a member of this people. **b.** The Nilotic language of this people. **B.** *adj.* Of or pertaining to the Luo or their language.

[**1905** C. ELIOT *E. Afr. Protectorate* viii. 148 Whereas the villages in the north are surrounded with mud walls, those of the Ja-luo are protected by a thick-set hedge of euphorbias and aloes.] **1911** *Encycl. Brit.* XV. 565/1 *Jur*, the Dinka name for a tribe of negroes of the upper Nile valley, whose real name is Luoh, or Lwo. **1942** *E. African Ann. 1941–2* 17/1 The Luo are a Nilotic negro tribe of agriculturalists living in the hot fertile country east of Lake Victoria. **1957** W. M. HAILEY *Afr. Survey* (rev. ed.) iii. 98 There are language committees concerned with the Kikuyu and Luo languages and an increasing volume of vernacular literature is being produced. **1964** C. WILLOCK *Enormous Zoo* v. 85 A mixed horde of Kikuyu, Luo and one Turkana appeared armed with pangas. **1968** *New Scientist* 12 Dec. 599/3 The Luo, the second largest tribe in Kenya after the Kikuyu. **1968** Y. R. CHAO *Lang. & Symbolic Syst.* 99 Dinka and Luo..have almost 1 million speakers each. **1969** *Listener* 24 July 100/1 Mboya was the only Luo leader of stature still in the party hierarchy. **1970** *Guardian* 6 June 9/6 Josphat shouted something in Luo which I did not hear. **1970** *Language* XLVI. 402 Some of the so-called prefixing forms of Luo are almost isomorphic semantically with their suffixal counterparts in Tarascan.

lupinosis (liūpinōu·sis). [f. LUPINE, LUPIN *sb.* + -OSIS.] Poisoning of animals, esp. sheep, after ingestion of lupines, either that caused by the presence of lupine alkaloids in the lupines, or (and now usu. spec.) that caused by toxins produced by a fungus of the genus *Phomopsis* growing on lupines.

1899 E. V. WILCOX in *Bull. Montana Agric. Exper. Stat.* No. 22. 39 The symptoms have become so well known and are so constant and uniform that the disease caused by the lupine poisoning has been called lupinosis. **1905** MOUSSU & DOLLAR *Dis. Cattle, Sheep, Goats & Swine* II. vii. 242 The symptoms of lupine poisoning are so well known in Europe that chronic lupine poisoning has been given the name lupinosis. **1928** W. C. MILLER *Black's Vet. Dict.* 571/2 In Europe, by far the greater number of cases [of poisoning by lupines] are of a chronic type, which results in the production of a train of symptoms to which the name 'lupinosis' has been given. **1943** *N.Z. Jrnl. Agric.* LXVII. 83/3 There was considerable swelling about the head, and the skin came off the ears and the nose and also along the top of the head. These symptoms are identical with those described as occasionally occurring in cases of lupinosis. **1961** *Jrnl. Austral. Inst. Agric. Sci.* XXVII. 62/1 In 1880, 14,138 of a total number of 240,000 sheep in one district in Pomerania died of lupinosis. **1966** *Brit. Vet. Jrnl.* CXXII. 508 Lupinosis of sheep was shown to be due to the grazing of dead standing lupin roughage which had been exposed to rain after drying off in the spring. **1967** *Adv. Veterinary Sci.* XI. 86 Two distinct forms of injury are recognized in animals ingesting one or more of the several parts of lupines. The first of these is due to the pharmacologic activity of the bitter principle (a variable mixture of alkaloids) and has become variably known as lupine poisoning, alkaloidal poisoning, American lupinosis, or lupine madness. The second is an icteric disease caused by a hepatotoxin which was designated..as 'ictrogen'... The disease was called 'lupinosis', or sometimes more specifically, 'European lupinosis'. **1972** *Mycologia* LXIV. 316 Typical signs of lupinosis were produced experimentally in Merino ewes fed either naturally infected lupines or pure cultures of the fungus [sc. *Phomopsis leptostromiformis*] on autoclaved white lupine seeds.

lupulone (liū·piŭlōun). *Chem.* Also **-on** (-ɒn). [ad. G. *lupulon* (W. Wöllmer 1916, in *Ber. d. Deut. Chem. Ges.* XLIX. 781), f. mod.L. *lupulus*, used either as the specific epithet of the hop, *Humulus lupulus*, or † as a name for the genus *Humulus* (cf. L. *lupus* hop): see -ONE.] A bitter, crystalline, cyclic ketone, $C_{26}H_{38}O_4$, that is an important constituent of hops and has strong antibiotic activity.

1919 *Chem. Abstr.* XIII. 495 The so-called α- and β-bitter acids..are more appropriately termed 'humulone' and 'lupulone'. **1937**, etc. [see *HUMULONE]. **1967** *Chem. Rev.* LXVII. 26/1 While lupulone crystallizes from the soft resin of continental European hops, British and American hops yield cohumulone.

lupus. Add: **4. b.** Used in various mod.L. (or sometimes Englished) collocations to designate various forms and manifestations of *lupus vulgaris* or to designate various other skin diseases: **lupus erythematosus** [tr. F. *lupus érythémateux* (Casenave 1850, in *Gaz. des Hôpitaux* 27 July 354/3): see ERYTHEMATOUS *a.*], a disease which is now considered to be manifested in two related forms, that of chronic discoid *lupus erythematosus*, which usu. involves only the skin and causes scaly red patches to form esp. on the face, and that of systemic *lupus erythematosus*, which produces a similar skin condition but involves the connective tissues generally and is attended by widespread symptoms of illness, esp. fever, malaise, and arthralgia; **lupus vulgaris,** a tuberculous disease of the skin, characterized by the formation of brownish nodules; = LUPUS 4.

1852 *Med. Times & Gaz.* XXVI. 141 (*heading*) Treatment of lupus exedens. **1857** *Lancet* 1 Aug. 116/2 (*heading*) Horrible deformity from lupus vorax. **1860** *Boston Med. & Surg. Jrnl.* LXII. 462 In lupus vulgaris it [*sc.* a liquid remedy] is seldom used, but in lupus erythematosus it may be considered a true specific. **1878** *Lancet* 13 July 35/1 (*heading*) Clinical lecture on disseminated follicular lupus. **1883** *Med. Chron.* (Baltimore) I. 271 (*heading*) Notes of a case of erythematous lupus complicated by the tubercular syphiloderm. **1947** *Sci. News* IV. 106 Tuberculosis can attack not only the lungs but also most other parts of the body—bones and joints, kidneys, glands, brain, and skin. In the last mentioned the disease is known under the Latin name of Lupus vulgaris. **1966** WRIGHT & SYMMERS *Systemic Path.* II. xxxix. 1547 In systemic lupus erythematosus the skin manifestations are merely part of a much more widespread disorder. **1974** PASSMORE & ROBSON *Compan. Med. Stud.* III. xxxi. 47/2 Lupus vulgaris most commonly begins in childhood.

‖ **lur**^{1} (lūəɹ). Also **lure** (lūəɹ). Pl. **lurer, lures, lurs.** [Da., Norw., and Sw.] A Bronze Age musical instrument of the horn family found in Scandinavia.

1876 STAINER & BARRETT *Dict. Mus. Terms* 275/2 *Lures*, ancient Scandinavian trumpets. **1879** GROVE *Dict. Mus.* I. 56/2 There is a Swedish instrument of this kind called *Lure*. **1955** *Times* 24 May 5/5 The *lur* is still occasionally used in Denmark. **1961** C. W. MONK in A. Baines *Mus. Instruments* xi. 277 Some examples of the Danish Bronze Age *lurs*, cast in the shape of a mammoth tusk, have mouthpieces astonishingly like modern melodic brass instruments. **1968** G. JONES *Hist. Vikings* I. i. 18 The long, slender, gracefully curved lurs or trumpets. **1970** BRAY & TRUMP *Dict. Archaeol.* 136/1 Lurer come from the peatbogs of Denmark and are almost invariably found in pairs.

Lur^2 (lūəɹ). A member of an aboriginal people inhabiting Luristan in western Iran. Chiefly in *pl.*

1845 *Encycl. Metrop.* XXIII. 272/2 The Lurs are divided into the Pīsh-kūh..and Pusht kūh. **1909** *Daily Chron.* 18 Feb. 5/1 Colonel Bell formed a favourable opinion of the Lurs as a whole. **1965** B. SWEET-ESCOTT *Baker St. Irreg.* iii. 89 Most of the Persian troops had not been paid for several weeks, and..when the surrender had taken place.., they had promptly sold their rifles to the neighbouring Lurs and Kurds.

Lurex (liū·reks). Also **lurex.** The proprietary name of a type of yarn which incorporates a metallic thread; also, fabric made from this yarn.

1945 *Official Gaz.* (U.S. Patent Office) 16 Oct. 360/1 The Dobeckmun Company, Cleveland, Ohio... Lurex. For yarn and thread comprised either in whole or in part of laminated film having a metallic appearance, including gimp comprised of a filament of yarn spirally overwrapped with coated or laminated foil which has been slit to a narrow width, said overwrapping imparting a metallic appearance. **1958** *Spectator* 12 Dec. 853/1 The party from the Camberwell Jazz Club, dressed in Italian-cut jackets shot with lurex, were not enthusiastic. **1965** *Guardian* 31 Mar. 15/2 Far away in the Outer Hebrides, the crofter weavers have found pleasure as well as profit in weaving 'Lurex' threads into tweeds of an altogether new lightness and beauty. **1967** *Spectator* 4 Aug. 139/3 The curtain rises on Donald Pleasance, in lurex dressing-gown,..kneeling before a marble urn and listening to Verdi. **1973** *Guardian* 18 June 15/2 Plain snakeskin Lurex blousons.

Lurgi (lūə·ɹgi, lʊ̄·ɹgi). The name of the *Lurgi* Gesellschaft für Wärmetechnik m.b.H., of Frankfurt, W. Germany, used *attrib.* to denote a method of gasification suitable for low-grade coal such as lignite by reaction with steam and oxygen at high pressure; so *Lurgi gas, plant.*

1934 ROBERTS & JENKNER *Internat. Coal Carbonization* xi. 249 In the Lurgi process the heat transference occurs directly by means of hot gaseous products of combustion, which pass through the coal. **1950** D. A. TOWNEND in D. H. Bangham *Progress Coal Sci.* xxvi. 436 Lurgi gas from brown coal has the following composition. **1950** WARING & FOSTER in Foster & Lund *Econ. of Fuel Gas from Coal* v. 81 Table 17 gives..the average analyses of the gas from the Lurgi plant at Bohlen. *Ibid.* 83 The economic advantage of the Lurgi process is that gasification of a cheaper, low-rank fuel is possible and that a comparatively high-Btu gas is delivered under pressure without the expense of carburetion with oil. **1973** *Nature* 7 Dec. 326/2 The well established Lurgi process is used to gasify the coal almost completely and a gas of low calorific value..is produced. *Ibid.*, The Lurgi process converts almost all the coal into gas, unlike the system used to manufacture town gas which produces coke and tar as well. **1974** *Sci. Amer.* Mar. 20/2 In some places Lurgi plants made a gas that was distributed as city gas.

lurgy (lʊ̄·ɹgi). Also **lurgi.** Usu. in phr. *the dreaded lurgy.* A fictitious, highly infectious disease invented and made a byword by the Radio Goons (*GOON 4).

1954 *Radio Times* 9 Nov. 20/3 The Goon Show... Poor Arnold Fringe is suddenly stricken with the Dreaded Lurgi... Within a few days Lurgi has claimed nine thousand victims. **1969** I. & P. OPIE *Children's Games* ii. 75 (*heading*) The dreaded lurgy. **1971** *It* 15–29 July 5/3 The youth of Australia have been saved once more from the dreaded lurgy, marijuana. **1974** H. MACINNES *Climb to Lost World* ix. 149, I was beginning to feel weak and knew that I had caught the dreaded swamp lurgy.

luringly (liūə·riŋli), *adv.* [f. LURING *ppl. a.* + -LY^2.] In a luring or enticing manner.

1897 J. L. ALLEN *Choir Invisible* xvi, This second image..drawing always nearer, summoning him more luringly. **1961** R. CRAFT *Diary* 31 Oct. in Stravinsky & Craft *Themes & Episodes* (1966) II. 203 He talks luringly of the Favors of a houri.

Luristan (lūəristā·n). The name of a district in western Iran used *attrib.* to designate the engraved bronze articles and castings, of the twelfth century B.C., found in the region.

[**1935** L. BINYON *Spirit of Man in Asian Art* iv. 105 Persian art, as we see it in the bronzes of animals recently discovered in Luristan, propagated its motives of design among the outlying, unsettled, still nomad tribes.] **1961** *Times* 6 June 22/7 (Advt.), A pair of unique Luristan bronze cheekpieces from a bit. **1970** *Oxf. Compan. Art* 830/1 Scientific excavations..have been held to show (though it is still unproven) that the Luristan bronzes are the handwork of the Medes, an Indo-European people who..began to infiltrate into Persia at about this period. **1971** *Ashmolean Mus. Rep. of Visitors 1970* 16 *Purchased*: seventeen 'Luristan' and 'Amlash' bronzes, Iran, late 2nd to early 1st millennium B.C.

lurk, *sb.*^{1} Add: **2.** (Earlier example.)

c **1842** *Exposure of Impositions practised by Vagrants* 5 Persons who go on this lurk, generally represent themselves as Captains or Masters of merchant ships which have been wrecked, and they have, of course, lost all their property.

b. Chiefly *Austral.* and *N.Z. slang.* A scheme, 'dodge', plan of action, ruse (not necessarily implying fraud). (See also quot. 1941.)

1916 C. J. DENNIS *Songs Sentimental Bloke* 125 *Lurk*, a plan of action. **1918** [see *JERRY *v.*]. *c* **1926** [see *FREEZE *sb.*^{1} 1 (i)]. **1938** *Observer* 13 Nov. 11/3 *Dart*, a scheme or racket; *lurk*, ditto. **1941** BAKER *Dict. Austral. Slang* 45 *Lurk*, a 'dodge', scheme, racket. 2. A hanger-on, an eavesdropper or sneak. **1953** A. UPFIELD *Murder must Wait* xvii. 149 Thanks a lot for the antidote... It is one hell of a good lurk. **1961** *N.Z. Listener* 17 Mar. 3/1 The Navy..is just a good lurk for those who want to delve deep into the public purse. *Ibid.* 15 Sept. 29/2, I suspect Barry Crump found he was on a good advertising lurk when he was interviewed for *Book Shop*. **1966** B. COOPER *Drown him Deep* xx. 165 She was a very rich girl indeed, and Hilary, with considerable influence over her, might well be on to a very good 'lurk'. **1967** C. DRUMMOND *Death at Furlong Post* xv. 182 If the hounds were out, the lurk was to get ahead of them and go to earth.

c. *Austral.* and *N.Z. slang.* A job.

1916 C. J. DENNIS *Songs Sentimental Bloke* 20, I found 'er lurk Wus pastin' labels in a pickle joint. *Ibid.* 125 *Lurk*, a regular occupation. **1958** R. STOW *To Islands* 126 'What's your lurk, mate?' 'Me? Stockman on a mission.' **1965** *Telegraph* (Brisbane) 2 June, O'Grady's current lurk is holidaying as an unpaid deck hand on the South Molle cruise ship Crest while he absorbs sunlight and material for a new book.

4. *slang.* A hiding-place; a 'hang-out'.

1906 E. DYSON *Fact'ry 'Ands* viii. 97, I come out frim me lurk, 'n' went over ther ground. **1924** *Chambers's Jrnl.* 20 Sept. 683/2 Why did the old beggar come to this secret lurk in the East End and disguise himself? **1972** *N.Z. News* 26 Jan. 4/1 The first bar I saw was the Cockney's Pride—a completely new lurk for me. **1974** J. GARDNER *Return of Moriarty* 32, I met her in a servant's lurk.

lurk, *v.* Add: **5.** [Perh. connected with LURCH *v.*^{2}] In pa. pple.: beaten, lost (in a game of chance) (see also quot. 1929). *slang.*

1917 M. T. HAINSSELIN *Grand Fleet Days* xx. 172 'What-Ho!'..said the Admiral, 'not a bad idea at all! Let's have a garden...' '——' said the Watch-keepers, in the sheltered seclusion of the wardroom, knowing full well that they would be lurked for the digging. **1929** F. C. BOWEN *Sea Slang* 87 *Lurked, to be*, to be ordered to do some unpleasant job without a chance of avoiding it. **1938** C. MORGAN *Flashing Stream* III. 222 Four straight aces. Good enough? You're lurked, Sandford. **1946** J. IRVING *Royal Navalese* 111 The man who 'cuts' for drinks and loses is 'lurked for the round'.

lurker^1. **2.** (Earlier and later examples.)

c **1842** *Exposure of Impositions practised by Vagrants* 4 Lurkers are persons who go about with briefs, containing false statements of losses by fire, shipwrecks, accidents, &c. **1925** H. LEVERAGE in *Flynn's* IV. 869/2 *Lurker*, a swindler. **1973** G. BUTLER *Coffin for Pandora* i. 27, I knew the lingo. A macer was a cheat or a sharper and a lurker was a man with a story of hard luck to tell.

lurkingly, *adv.* (Later example.)

1929 R. B. C. GRAHAM *Thirty Tales & Sk.* 178 They eyed the women just as a starving dog looks at a butcher's shop, sideways and lurkingly.

lu·rkingness. [f. LURKING *ppl. a.* + -NESS.] The quality of lurking.

1912 GALSWORTHY *Inn of Tranquility* 51 The mist..seemed to have in its sheer silence a sort of muttered menace, a shuddery lurkingness.

lu·rkman. *Austral. slang.* [f. LURK *sb.*^{1} + MAN *sb.*^{1}] (See quot.)

1945 BAKER *Austral. Lang.* vii. 138 We are..originators of the following terms for various sharpers, tricksters and others who live by their wits: *spieler*..*lurk man*..and *amsterdam*.

lurrier (lʊ·riəɹ). [f. LURRY *v.* + -ER^1.] An operative in textile-printing (see quot. 1897).

1897 C. F. S. ROTHWELL *Printing Textile Fabrics* 34 The lurrier brings the colour required from the colour shop, the pieces and back-greys from the stock room, and also does any odd jobs required by the printer. **1921** *Dict. Occup. Terms* (1927) § 399 *Lurrier, lurryman*, colour carrier.

Lusatian (liusēi·ʃiăn), *sb.* and *a.* [f. med.L. *Lusatia* + -AN.] **A.** *sb.* A native or inhabitant of Lusatia, name of a former region of eastern Germany between the Elbe and the Oder; = WEND *sb.* I. **b.** The West Slavic language spoken in Lusatia. **B.** *adj.* Of or pertaining to Lusatia or its inhabitants.

1555 R. EDEN tr. *Martyr's Decades of Newe Worlde* 290 The Slauon tounge..vsed of..the Bohemians, Lusacians, Silesians, Morauians, [etc.]. **1862** R. G. LATHAM *Elem. Compar. Philol.* 766 Lusatian language. **1877** A. H. KEANE tr. *Hovelacque's Sci. of Lang.* 275 The Sorbian, or Sorabian, called also Wendic, or Lusatian comprises two distinct varieties, High and Low Sorbian. *Ibid.* 276 About the middle of the sixteenth century the Lusatian territory was twice as extensive as at present. **1921** *19th Cent.* May 894 We need only except Lessing, who was a Lusatian. **1933** L. BLOOMFIELD *Lang.* iv. 60 One of these, *Lusatian* (*Wendish, Sorbian*), survives as a speech-island of some 30,000 persons in Upper Saxony. **1949** *Archivum Linguisticum* I. I. 89 In view of the scarcity of Lusatian literature it is to be regretted that no selections have been supplied. **1972** W. B. LOCKWOOD *Panorama Indo-European Lang.* 158 Slavonic in Lusatia—it may be termed Sorbian or Lusatian as well as (Lusatian) Wendish—falls into two divergent dialect groups.

lush, *sb.*^{2} Add: **1. c.** A habitual drunkard, one addicted to drink.

1890 J. A. RIIS *How Other Half Lives* (1891) xix. 221 The first long step in crime taken by the half-grown boy..is usually to rob a 'lush', i.e., a drunken man who has strayed his way. **1899** ADE *Doc Horne* i. 1 'My uncle didn't think so,' remarked the lush. *Ibid.* iv. 39 The drinking man, often mentioned as the lush. **1945** J. STEINBECK *Cannery Row* xxix. 124, I don't like to leave the place without a man. Some lush might get smart and the kids couldn't handle him. **1958** J. & W. HAWKINS *Death Watch* (1959) 135 She took a drink now and then, but she wasn't a lush. **1958** *Spectator* 14 Feb. 210/1 Some high-class Hollywood bitches and lushes. **1972** D. DELMAN *Sudden Death* (1973) iii. 94 He's a drunk, ain't he?.. He's a lush. And a lush is a lousy security risk.

2. lush-head, -hound, a drunkard; **lush-roller, -worker,** one who steals from drunks.

1925 H. LEVERAGE in *Flynn's* IV. 869/2 *Lush-roller*, one who robs drunken men. **1930** Lush-worker [see *GLOM *v.*]. **1935** G. INGRAM *'Stir' Train* ii. 31 He's a 'lush-hound' and I knew he must be a coward. **1945** L. SHELLY *Jive Talk Dict.* 29/1 *Lush head*, chronic drinker. **1946** MEZZROW & WOLFE *Really Blues* (1957) Dedication, To all the junkies and lushheads in two-bit scratchpads. **1948** MENCKEN *Amer. Lang.* Suppl. II. 682 A *creep-joint* or *panel-house* is one in which patrons are robbed, a *roller* or *mush-worker* is a girl who robs them, and a *lush-worker* is one who specializes in drunks. **1957** *Amer. Speech* XXXII. 278 *Zoot suit* meaning flashy clothes, and *lushhead* or *lush* for drunkard are no longer considered good jazz lingo, though they are or were in common nonjazz usage.

lush, *a.*^{1} Add: **2. d.** Also, luxurious; of a woman: sexually attractive.

1939 *Punch* 8 Nov. 517/1 Business-men from neutral countries should be met with red-carpeted gangways and military bands, and passed in lush motor-cars from one feast to the next. **1942** [see *BINT *sb.*^{2}]. **1958** *Economist* 8 Nov. 497/2 The egg board's lush new London headquarters.

lush, *v.*^{2} Add: **3.** With *up*. **a.** *intr.* To get drunk. **b.** *trans.* To ply with drink, to make (a person) drunk. **c.** *trans.* To provide with a luxurious standard of living.

1926 MAINES & GRANT *Wise-Crack Dict.* 10/2 *Lush up*, to get drunk. **1927** K. NICHOLSON *Barker* 150 *Get lushed up*, become intoxicated. **1927** *Punch* 25 May 573/1 He at once enclosed a bit of tennis-lawn for them [*sc.* rabbits] as a *manège*..and altogether lushed them up to the good things of this world. **1933** WODEHOUSE *Mulliner Nights* ii. 48 If I lush this cat up satisfactorily, shall I not be in a position later on to make a swift touch? **1952** W. R. BURNETT *Vanity Row* (1953) xv. 107 Mr. Hobart got so lushed up... He was spilling drinks down the front of his shirt. **1959** R. GANT *World in Jug* 39 By that time Andy Mendoza had got himself lushed up and started careening around the set playing a slow drag. **1960** *News Chron.* 9 Apr. 3 (*caption*) You are lushed up on the good life and are convinced that we never had it so mechanised and marvellous. **1961** WODEHOUSE *Ice in Bedroom* i. 8, I see you're lushing up the dumb chums.

Lushai (lu·ʃai), *a.* and *sb.* Also **Lhooshai,** etc. [Native name.] **A.** *adj.* Of or pertaining to a mountainous region in India between Burma and Bangladesh, its inhabitants, or the language spoken by them. **B.** *sb.* **a.** A native or inhabitant of this region. **b.** The Tibeto-Burman language spoken there.

1862 C. U. AITCHISON *Coll. Treaties India* I. 77 On the southern frontier of Cachar lies the territory of the Lhooshai Kookees... Timber merchants are..in the habit of employing Lhooshais in felling the trees in their forests. **1868** *Ann. Indian Administration* XII. 86 Mr. N. T. Davey remarks on the difficulties..likely to result in an attempt to explore..the hilly tract lying between Cachar and Chittagong. These hills are inhabited by the Looshais. **1873** E. BALFOUR *Cycl. India* III. 499/1 In the beginning of 1871, the Looshai made a prolonged raid on the North-East Provinces of British India, but were driven back. **1874** T. H. LEWIN *Progressive Colloq. Exercises in Lushai Dial.* 3 The Lushai dialect is in fact the *lingua frânca* of the country. The clan-name Lushai probably means 'the decapitators', being derived from 'lú' a head and 'shá' or 'shát' to cut. **1876** W. W. HUNTER *Statistical Acct. Bengal* VI. 59 The Lushâis or Kukis are a powerful and independent people, split up into different clans. **1887** C. A. SOPPITT *Short Acct. Kuki-Lushai Tribes* i. 2 The Lushai people have only been known to us within comparatively recent years. *Ibid.* 4 In Hill Tipperah there are three tribes, named Paitu, Omroi, Korêng, all nearly connected with the Lushais. *Ibid.* 79 (*heading*) Comparison of the dialects of the Kuki Lushai tribes and of the language known as Lushai. **1890** KIPLING *Barrack-Room Ballads* (1892) 17 We've chivied the Naga an' Looshai, we've give the Afreedeeman fits. **1910** *Encycl. Brit.* I. 773/1 The inner line formerly maintained along the Lushai border has since 1895 been allowed to fall into desuetude, but Lushais visiting Cachar are required to take out passes from the superintendent of the Lushai hills. **1915** *Encycl. Relig. & Ethics* VIII. 198/1 The Lushais are a superstitious people, and believe firmly in witchcraft. **1959** P. C. CHOUDHURY *Hist. People of Assam* iv. 95 The Central Chin group..includes the Lushâis and other allied peoples. **1972** *Language* XLVIII. 476 The most useful Kukish language for comparative purposes has hitherto been Lushai. **1972** W. B. LOCKWOOD *Panorama Indo-European Lang.* 227 To the south of these places [*sc.* Manipur and Tripura], in the southern tip of Assam, the local language is Lushai with 250,000 speakers. **1974** *Encycl. Brit. Micropædia* VI. 396/3 Lushai villages traditionally were situated on the crests of hills or spurs and..were fortified by stockades. *Ibid.*, The Kuki clans have been largely absorbed by the Lushai, adopting the Lushai customs and language.

lu·sher. [f. LUSH *v.*^{2} 2.] One who is excessively self-indulgent, especially one who drinks excessively.

1895 G. MEREDITH *Amazing Marriage* II. xxxi. 80 The suspicion cast on the dreary lusher was the wife's wild shot at her husband. **1914** *Dialect Notes* IV. 110 Once all the politicians, nearly, were lushers. **1928** M. C. SHARPE *Chicago May* 288/1 *Lusher*, lone drinker.

lushy, *a.*^{1} Add: Also as *sb.*, a drunkard. *U.S.*

1944 *New Yorker* 8 July 28/2 All our horn blowers were lushies. **1945** L. SHELLY *Jive Talk Dict.* 13/2 *Lushie*, a drunkard. **1946** MEZZROW & WOLFE *Really Blues* (1957) 94 The lushies didn't even play good music.

Lusitanian, *a.* Add: **2.** *Biol.* Of plants or animals, having their origin in south-western Europe, esp. Portugal.

1907 R. F. SCHARFF *European Animals* v. 88 Almost all the members of that South-western, or Lusitanian, element in our fauna have a discontinuous range, which is a sure indication of great antiquity. **1927** PEAKE & FLEURE *Hunters & Artists* 38 It is a remarkable fact that the west of the Spanish peninsula and the south-west corner of Ireland share several peculiar plants and animals, such as the winter strawberry, the London pride, the great spotted slug, and a peculiar lacustrine shellfish. This so-called 'Lusitanian' association may be supposed to have lived on the Atlantic coast-lands of the Ice Age. **1935** *Discovery* Nov. 318/1, I look upon it [*sc.* the field cricket] as clearly a member of that group which Scharff has called Lusitanian. **1970** *Watsonia* VIII. 93 Several of the main phytogeographical elements in the British flora, including the Lusitanian, are represented in this section.

Luso- (lū·so). [f. *Lusitania* = Portugal.] In *Comb.*: of Portugal, Portuguese.

1951 SMITH & MARCHANT *Brazil* v. 146 The Brazilian Negroes are blending physically and culturally with the national types, mostly Luso-Brazilian ethnically. **1957** R. CAMPBELL *Portugal* 20 The Luso-Spanish sieges of Numantia. **1958** *Archivum Linguisticum* X. I. 30 The Luso-Hispanic frontier. **1964** *New Statesman* 3 Apr. 514/1 Portugal's assertion of complete Luso-African unity. **1969** J. MANDER *Static Society* ii. 79, I have stressed the unity that Latin America possesses in virtue of its Luso-Hispanic inheritance. **1973** *Black World* Sept. 19/1 [Jorge de Lima] taught Luso-Brazilian literature at the Federal and National Universities.

lusting, *vbl. sb.* (Later examples.)

1895 KIPLING *Seven Seas* (1896) 82 The lying, and the lusting, and the drink. **1909** *Westm. Gaz.* 18 Aug. 2/3 Thou art grim with the lusting of gain.

lustre, *sb.*^{1} Add: **1. e.** In ceramics, the surface sheen produced by glazing; the material used for glazing. Also *ellipt.*, = *lustre ware* below. Hence *lustre-glazed, -painted* adjs.

1829 S. SHAW *Hist. Staffordshire Pott.* x. 227 The Lustre of our day is a good red clay body, with a fine brown glaze; upon which is laid, for Gold Lustre, a very thin coating of a chemical mixture containing a small quantity of Gold in solution. **1892** J. R. & F. KIDSON *Hist. Notices Leeds Old Pott.* 87 The Agate ware made at the Leeds Pottery was contemporary with the earliest makes of their Silver Lustre. **1897** *Sears, Roebuck Catal.* 681/2 Luster band, open Meakin's English-ware... The decorations are of a heavy luster band and a flower sprig in luster which resembles gold very closely. **1939** *Burlington Mag.* May 227/2 A lustre-glazed tile made at Valencia in the late fifteenth century. **1961** *Antiquaries Jrnl.* XLI. 9 A lustre-painted bowl of Malaga ware in the Staatliche Museen, East Berlin. **1969** G. LEWIS *Collector's Hist. Eng. Pott.* xvi. 157 The newly introduced technique of electro-plating made the silver lustre less popular and production ceased. **1973** *Country Life* 11 Nov. 1049/1 Morris's philosophy extended as much to..rush-bottomed chairs as to gold lustre.

7. *lustre bowl, china, jug, mug, tea-pot, tile*; **lustre ware** (further examples); *spec.* pottery which is given a metallic lustre by the application of a metal oxide to its surface; also *transf.*

1908 *Sears, Roebuck Catal.* 359/4 Iridescent luster bowl made of the most select Bavarian china. **1952** M. Laski *Village* viii. 136 The furniture..consisted of..a painted pine corner cupboard with..some genuine old lustre china inside. **1908** J. M. Synge *Lett. to Molly* (1971) 276 Look round in Galway for lustre jugs or Irish curios. **1971** 'E. Ferrars' *Stranger & Afraid* i. 16 A good deal of clutter, in the way of Staffordshire teapots, lustre mugs, Bohemian glass and photographs. **1935** N. Mitchison *We have been Warned* I. 107 Miss Waterhouse would now be giving.. strong tea to her weaving class, pouring steadily from a beautiful lustre tea-pot. **1943** D. Welch *Maiden Voy.* vi. 46, I was only talking about the lustre teapot... It's shiny stuff—like metal, only china. **1933** *Burlington Mag.* Nov. 224/1 These spiral scrolls and leaves are familiar..in the thirteenth century lustre tiles. **1875** E. Meteyard *Wedgwood Handbk.* 31: After the commencement of the present century lustre-wares were generally made throughout the Potteries. **1938** *Times Lit. Suppl.* 17 Sept. 595/4 The household of the meek Jewish tailor and the home of the jolly publican are shining pieces of cockney lustreware. **1961** *Antiquaries Jrnl.* XLI. 1 A large cover of Hispano-Moresque lustreware, imported from Malaga. **1971** *Canad. Antiques Collector* Jan. 17/1 English lustre ware affords a fascinating variety of colour, charm and decoration for the collector.

luteal (liū·tiăl), *a.* [f. L. *lūte-us* yellow (in mod.L. *corpus luteum*: see *corpus 2) + -al.] Of or pertaining to the *corpus luteum*.

1927 *Amer. Jrnl. Anat.* XL. 211 The inhibitory effect of luteal tissue upon follicular development does not explain why certain animals have an anoestrum. **1932** E. Allen *Sex & Internal Secretions* xvi. 794 The lutenizing [*sic*] factor will not cause the luteal change in the ovary of an immature animal. *Ibid.* 795 During the luteal phase the so-called 'B' factor is utilized. **1939** S. R. M. Reynolds *Physiol. Uterus* iii. 69 The action of the luteal hormone. **1966** *Ann. Rev. Physiol.* XXVIII. 60 Prolactin may stimulate luteal function in the ferret. **1969** K. W. McKerns *Gonads* iii. 72 (*heading*) Factors affecting luteal activity in intact rabbits.

lutecium, var. *lutetium.

luteinization (liū:tīnəizēi·ʃən). *Physiol.* [f. Lutein + -ization.] The formation of lutein in the cells that remain of the Graafian follicle after expulsion of the ovum, during which process the follicle is converted into the *corpus luteum*; the formation of a *corpus luteum*.

1929 *Anatomical Rec.* XLIII. 239 (*heading*) The origin of the corpus luteum in the rat as indicated by studies upon the luteinization of the cystic follicles. **1931** *Amer. Jrnl. Physiol.* XCVII. 292 It [*sc.* the luteinizing hormone] does, however, cause luteinization of the follicles which are produced by the gonad stimulator. **1967** G. S. Richardson *Ovarian Physiol.* ii. 22 Increased follicular growth without luteinization or corpus-luteum formation. **1968** Passmore & Robson *Compan. Med. Stud.* I. xxxvii. 18/2 Following ovulation and subsequent luteinization of the follicle, oestrogen synthesis is supplemented by the production of progestagens.

Hence **lu·teinize** *v. trans.*, to cause (a tissue associated with the *corpus luteum*) to form lutein; **lu·teinized** *ppl. a.*; **lu·teinizing** *ppl. a.* and *vbl. sb.*; *luteinizing hormone*, a glycoprotein hormone secreted by the adenohypophysis which in the female helps to induce ovulation and brings about the formation of the *corpus luteum*, and in the male promotes the secretion of androgen by acting on the Leydig cells of the testis (abbrev. LH (*L 7)).

1929 *Anatomical Rec.* XLIII. 242 The cells of either the granulosa or the theca have become luteinized. **1931** *Amer. Jrnl. Physiol.* XCVII. 291 (*heading*) The gonad stimulating and the luteinizing hormones of the anterior lobe of the hypophesis [*sic*]. *Ibid.* 294 This preparation.. does luteinize the follicles produced by the gonad stimulating fraction. *Ibid.*, The water soluble preparation contains primarily the gonad stimulating hormone with little or none of the luteinizing substance. *Ibid.* 295 Stimulation to luteinizing activity. **1935** *Endocrinology* XIX. 45 The ovary of the infantile rat can be luteinized to a much higher degree than was found in the above experiments with urine. **1966** *Ann. Rev. Physiol.* XXVIII. 61 Heavily luteinized ovaries. **1969** K. W. McKerns *Gonads* x. 279 Some 95 percent of the ovarian tissue consisted of luteinized cells. **1970** *Sci. Jrnl.* June 47/1 Three gonadotrophic hormones are produced by the anterior section of the pituitary gland: the mammotrophic hormone, prolactin, the follicle stimulating hormone (FSH) and the luteinizing hormone (LH). **1970** *Nature* 26 Sept. 1344/2 In the cyclic female albino rat, a release of pituitary luteinizing hormone (LH) occurs on the afternoon of proestrus.

luteo-. Add: **b.** Also before a vowel **lute-.** Used as the combining form of *corpus luteum* (*corpus 2), as in **luteo-lysis** *Physiol.* [*lysis 3], degeneration of the *corpus luteum*, such as occurs when the discharged ovum is not fertilized; so **luteoly·tic** *a.* [*-lytic], bringing about or sufficient to bring about luteolysis; **luteo·ma** *Path.* [*-oma], an ovarian tumour consisting of cells resembling those of the corpus luteum.

1961 J. W. Everett in W. C. Young *Sex & Internal Secretions* (ed. 3) I. viii. 537/1 By luteolysis we shall refer to corpus luteum regression in any of its manifestations. **1965** *Endocrinology* LXXVI. 1218/1 The luteolysis induced by estrogen. **1971** *Nature* 23 Apr. 526/1 We have measured the output of prostaglandin from the uterus in one condition which is associated with premature luteolysis, uterine distension. **1974** *Ibid.* 10 May 176/2 In many species luteolysis is dependent upon the presence of the uterus and an intact utero-ovarian vasculature. **1961** *Recent Progress Hormone Res.* XVII. 122 This experiment.. argues against the possibility that the luteolytic effect of progesterone is due to a direct action on the corpus luteum. **1965** *Endocrinology* LXXVI. 1214/1 Particular attention was focused on establishing the luteolytic dose of estradiol cyclopentylpropionate. **1971** *Nature* 23 Apr. 526/1 Prostaglandin $F_{2\alpha}$..has a luteolytic action in various species including the guinea-pig. **1931** R. J. E. Scott *Gould's Med. Dict.* (ed. 3) 749/1 *Luteoma*, a tumor developing from the corpus luteum. **1947** *Radiology* XLIX. 280/1 Figure 10 shows a mixed luteoma and tubular adenoma. **1966** R. W. Evans *Histol. Appearances Tumours* (ed. 2) xxviii. 666 There exists a rare though well-recognized ovarian tumour which, because of its debatable nature and genesis, has been variously and confusingly called: adrenocorticoid tumour or hypernephroma of ovary, virilizing lipoid-cell tumour, luteoma, androgenic hilus-cell tumour and masculinovoblastoma.

luteotrophic (liū:tiotrōu·fik), **-tropic** (-trōu·pik, -trǫ·pik), *a.* *Physiol.* [f. *luteo- b + *-trophic, *-tropic.] Maintaining or capable of maintaining the *corpus luteum* in being during pregnancy and thus counteracting luteolysis; *luteotrop(h)ic hormone*, such a hormone which in rats and mice is produced by the adenohypophysis and which is probably identical with the hormone prolactin of other animals, including man (though attempts to demonstrate that prolactin is luteotrophic in these other animals have failed) (abbrev. LTH (*L 7)).

1941 E. B. Astwood in *Endocrinology* XXVIII. 310 In speaking of a substance which will maintain the functioning of formed corpora lutea it has been found convenient to use the term 'luteotrophic' substance or 'luteotrophin'. **1955** *Jrnl. Endocrinol.* XIII. 19 Cutuly (1941) found that luteotrophic hormone stimulates corpora lutea of hypophysectomized rats sufficiently to allow implantation. **1961** *Recent Progress Hormone Res.* XVII. 120 The luteinizing hormone..may be luteotropic in animals other than rats and mice. **1964** H. H. Cole *Gonadotropins* i. 1 We know nothing of other possible pituitary luteotropic hormones. **1964** *Endocrinology* LXXV. 625/1 Such extracts have been shown to be luteotrophic in hypophysectomized rabbits. **1968** Passmore & Robson *Compan. Med. Stud.* I. xxxvii. 16/2 The factors which control the maintenance and regression of the corpus luteum are not understood, but during pregnancy a placental gonadotrophin exerts a powerful luteotrophic effect.

Hence **lu:teotro·phin, -tro·pin,** any substance exerting a luteotrophic effect; *spec.* = **luteotrop(h)ic hormone*.

1941 [see *luteotrophic *a.*]. **1945** *Index-Catal. Library Surg.-General's Office, U.S. Army* 4th Ser. IX. 1323/1 Luteotropin: see Pituitary hormone. **1961** *Recent Progress Hormone Res.* XVII. 120 LH may have the dual role of causing ovulation and of acting as a luteotropin. **1961** W. C. Young *Sex & Internal Secretions* (ed. 3) I. viii. 530/2 The term luteotrophin was proposed by Astwood (1941) to refer to a substance that maintains function of corpora lutea, in distinction to substances that cause them to form. It is now conceded that the substance described in that paper was probably the lactogenic hormone... Although lactogen seems to be the hypophyseal luteotrophin in rats, such is not necessarily true for all species... Nevertheless, the expression luteotrophin in the generic sense continues to be desirable. **1965** *Endocrinology* LXXVI. 1218/1 It therefore appears..that a hypophysial luteotropin is only necessary for a short period after ovulation.

lutetium (liuti·ʃiŏm, -siŏm). *Chem.* Also lutecium. [ad. F. *lutécium* (G. Urbain 1907, in *Compt. Rend.* CXLV. 761), f. F. *Lutèce*, L. *Lutet-ia* (see Lutetian *a.*): see -ium.] A rare metallic element that is the heaviest member of the lanthanide series and forms colourless salts in which it is trivalent. Atomic number 71; symbol Lu.

1907 *Jrnl. Chem. Soc.* XCII. ii. 956 (*heading*) A new element: lutecium, resulting from the decomposition of Marignac's ytterbium. **1929** *Encycl. Brit.* XIV. 490/2 Lutecium occurs along with ytterbium, erbium, etc., in the minerals gadolinite, euxenite, xenotime, etc. It is separated from the other members of the group by the fractional crystallization of the bromates, lutecium bromate being the most soluble passes into the mother-liquor. **1939** *Physical Rev.* LVI. 21 The half-life of the natural radioactivity of lutecium..is..$7 \cdot 3 \pm 2 \times 10^{10}$ years for Lu^{176}, which probably is the active isotope. **1952** *Chem. & Engin. News* 7 July 2843/2 The table of international atomic weights, as published by the 'Comptes Rendus' of the New York meeting of the Union for 1951.. shows changes in the weights of seven elements... Lutetium replaces lutecium, to agree with the original Latin spelling Lutetia..instead of the French Lutèce. **1973** J. J. Lagowski *Mod. Inorg. Chem.* xvi. 604 Lutetium, the first element which follows the lanthanides, starts the last *d*-series, but it forms a M^{3+} cation with a filled 4*f* orbital. Thus, lutetium on the basis of its atomic and ionic radii also appears to be a member of the lanthanide series.

luthier. (Later examples.)

1932 *Times Lit. Suppl.* 4 Feb. 68/2 That [knowledge] which comes from the long practice of the craft of the *luthier*. **1951** H. M. Robinson *Cardinal* ii. 183 With two luthiers in the house, a violin ought to get finished once in a while. **1974** *Daily Colonist* (Victoria, B.C.) 1 Sept. 16/8 Dunn learned to make the 16th-century instruments while sharing a shop with Canadian luthier Ray Nurse and now makes about 20 a year.

lutite (lʊ·təit). *Geol.* Formerly also **-yte.** [f. L. *lut-um* mud + -ite¹ 2 b.] = Pelite. So **luta·ceous** *a.* [-aceous] = Pelitic *a.*

1904 A. W. Grabau in *Amer. Geologist* XXXIII. 242 The third texture [*sc.* that of 'rock flour or impalpable powder'], finally, may be designated by the term *lutaceous*, (from *lutum*, mud), and for consolidated rocks of this type the term *lutyte* may be used, irrespective of chemical composition. Pelyte has been used particularly for argillaceous rocks of the group. **1941** F. H. Lahee *Field Geol.* (ed. 4) 785 These terms [*sc.* psephite, psammite, pelite] are derived from the Greek. The equivalent terms, derived from the Latin and preferred by some geologists, are rudite.., arenite.., and lutite. **1959** W. W. Moorhouse *Study of Rocks in Thin Section* xviii. 334 The clastic sediments are classified according to size as rudytes (rudaceous), which are conglomerates, arenytes (arenaceous) which are sands, and lutytes (lutaceous), which are shales and muds. **1970** *Nature* 14 Mar. 1068/2 A sediment core..reveals the top 20 cm to be a greyish-brown fossiliferous lutite.

Lutomer (liū·toməɹ). Also Ljutomer. The Slovene name for the region of Slovenia in which wine is produced; used *attrib.* for the wine produced there.

Before the creation of Yugoslavia in 1918 both the region (then part of Austria) and the wine were known by their German name *Luttenberg*.

[**1851** C. Redding *Wines* (ed. 3) xi. 288 The Luttenberg wines of Lower Styria are among the first.] **1954** 'Bon Viveur' *A.B.C. of Wine Drinking* 74 *Lutomer Riesling*, sweetish—definitely—fairly sturdy and with a slight bouquet. **1965** *Sun* 23 Jan. 6/6 The favourite white, Lutomer Riesling, that great party standby. **1970** L. W. Marrison *Wines for Everyone* ix. 122 One has the feeling that in those restaurants where carafe wines are available, the white is more likely to be lutomer riesling than graves or mâcon. **1971** J. Jeffs *Wines of Europe* 499 In a tasting of 1958 wines in 1964, the best was undoubtedly a lutomer traminer which..was very well balanced.

lutulent, *a.* (Later *literary* example.)

1922 Joyce *Ulysses* 377 Exterior splendour may be the surface of a downwardtending lutulent reality.

lutz (luts). [prob. f. name of Gustave *Lussi* (1898–), Swiss figure skater, who invented it.] A jump in ice-skating in which the skater takes off from the outside back edge of one skate and lands, after a complete rotation in the air, on the outside back edge of the opposite skate.

1938 D. Cummings *Figure Skating* xv. 83 Similar to the flip jump is the *Lutz*; but it is more difficult. **1940** M. Y. Vinson *Advanced Figure Skating* vi. 175 A really fine lutz ..is the ambition of every jumper. **1959** *Times* 12 Mar. 4/4 The title therefore was not awarded, in spite of Jones's splendid free skating programme, which included a fine.. lutz. **1964** *Times* 3 Feb. 3/5 There was a momentary unsteadiness in landing the double lutz, but all her other jumps were splendidly executed. **1976** *Times* 19 Jan. 9/6 The rather perilous landing of a double lutz.

luv (lʊv, luv). Spelling used to represent an affectionate, dialectal, or colloquial, etc., occurrence of the word *love*: esp. freq. as a term of address.

1898 J. D. Brayshaw *Slum Silhouettes* 35 They say as luv is blind. **1909** R. Brooke *Let.* 21 Aug. (1968) 174 Boxer's in Luv. **1957** J. Braine *Room at Top* vi. 52 T'lad's cum to enjoy hisen, 'aven't you, luv? **1966** 'L. Lane' *ABZ of Scouse* 63 *Luv*, ironical term of affection applied to the opposite sex. Often used by waitresses as in *what'll yer 'ave luv?* **1968** A. Clarke *Darkened Room* x. 126 The nurses called me 'Luv' or 'Dear'. **1972** G. Bell *Villains Galore* xiii. 205 Watch that money, luv! It's not safe there.

Luvian (lū·viăn), *sb.* and *a.* Also Luwian, Luian. [f. G. *Luvisch*, *Luvier* from *Luvia*, name given to part of Asia Minor: see E. Forrer in *Mitteilungen Deut. Orient-Gesellschaft* (1921) LXI. 20–39: see -ian.] **A.** *sb.* **a.** A member of an Anatolian people contemporary with the Hittites, known from cuneiform inscriptions. **b.** The language of the Luvians. **B.** *adj.* Of or pertaining to the Luvians or their language.

[**1923** H. R. Hall in Buckler & Calder *Anatolian Studies* 168 Although until we have the cuneiform texts before us it is quite impossible to control the work of Forrer..yet, whatever we may think at present of his elaborate analysis of the eight languages which he thinks the Hittites or their subjects spoke,..'Ur-Luvisch', 'Luvisch',..and so on, yet the rough historical results.. can no doubt be accepted without demur.] **1924** *Cambr. Anc. Hist.* II. xi. 253 More [tablets], however, are couched in some six native allied dialects, according to the latest decipherers.., who agree in regarding the dialects as Indo-European... To the six dialects they give the names Kanesian, Luvian, [etc.]. **1933** C. D. Buck *Compar. Gram. Greek & Latin* 15 Closely related to the cuneiform Hittite are the hieroglyphic Hittite and Luwian. **1934** Webster, *Luian*,..an ancient language of the Hittite empire. **1939** L. H. Gray *Foundations of Lang.* xi. 324 There are also Hittite inscriptions in pictographic or hieroglyphic characters... Their language may be akin to that seen in the thus far scanty fragments of *Luian* or *Luvian*, closely related to Hittite. **1952** O. R. Gurney

Hittites i. 18 Other Indo-European dialects (Luwian, Palaic, Lycian, and 'Hieroglyphic Hittite') established themselves in other parts of Anatolia. **1961** L. R. Palmer *Mycenaeans & Minoans* 26 The possibility..that the predecessors of the Greeks were Luvians from western Asia Minor should please both philologists and archaeologists. **1963** —— *Interpretation Mycenaean Greek Texts* 339 The word [*sc.* *a-ja-me-na*] may be a Luvian loan-word (*a-ja* 'do, make') in the sense 'wrought'. This possibility is strengthened by the fact that *kuwana-* occurs in Luvian and is held to be the source of Greek κύανος. **1966** J. Puhvel in Birnbaum & Puhvel *Anc. Indo-European Dial.* 238 This linguistic division of the substrata is indirectly discernible in the divergent Hattic, Kaneshite, and Luwian pantheons within the hospitality of the state cults of the Hittite empire... 'Hieroglyphic Hittite' is in reality a dialectal form of Luwian ('East' or 'Late' Luwian). **1973** K. A. Kitchen in D. J. Wiseman *Peoples of Old Testament Times* 67 Goliath (*Golyat*) is claimed as a dissimilated form of a Walwatta.., from a Luvian base *walwi/a*.

Luwian, var. *Luvian *sb.* and *a.*

lux (lʊks), *sb.* *Physics.* Pl. lux. [L., = 'light'.] A unit of illumination (now incorporated into the International System of Units) equal to the illumination of a surface all of which is one metre from a uniform point source of light of unit intensity (now one candela), i.e. (as now defined) one lumen per square metre.

1889 *Engineering* 13 Sept. 313/1 Mr. Preece was not.. fortunate with his unit of surface illumination, the 'lux' being the illumination produced by a lamp of one Carcel power at a distance of one metre, which is practically the same illumination as that afforded by an English standard candle at a distance of 1 ft... It is probable, sooner or later, that such a unit will be adopted. **1897** *Jrnl. Inst. Electr. Engin.* XXVI. 638 It was proposed by the Committee of the International Congress of Geneva that M. A. Blondel should present a report, with..a complete system of defining the dimensions and units of photometry... This report was submitted for examination of a commission..and was accepted... Lux = Lumen/Square metre. **1911** A. P. Trotter *Illumination* ii. 17 Sir William Preece ..adopted the Carcel-metre, and he showed that it was equal to a standard candle of 1·058 foot. At the Paris Electrical Congress of 1889 he adopted the name 'lux' for this... The unit of illumination produced by a bougie-décimale at a metre has been called a bougie-metre, and the name 'lux' was revived at the Geneva Congress of 1896 and was applied to this unit... This lux is, roughly, one-twelfth of a foot-candle, or about one-fourteenth of Preece's lux. The illumination produced by a Hefner at a metre has also been called a 'lux', and it seems probable that the International candle at a metre may be called an International lux. **1927** *Forestry* I. 87 Experiments carried out..in a light intensity of 48,000 lux. *Ibid.*, The lux is a light intensity produced by a light of one candle-power at a distance of one metre. Forty-eight thousand lux is roughly equivalent to full sunlight at noon in mid-summer in Germany. **1953** Amos & Birkinshaw *Television Engin.* I. 279, 1 lux = 1 lumen per square metre = 0·0929 foot-candle. **1970** *Nature* 26 Dec. 1349/2 On many days during the growth of these plants light intensities reached 80,000 lux.

luxe. 1. Delete † *Obs.* and add later examples. Also *attrib.* or quasi-*adj.*

1908 *Westm. Gaz.* 28 July 11/3 We possess..a special department for hiring out luxe carriages. *Ibid.*, One of our Charron luxe cars fills all the essentials of a gentleman's private car. **1932** R. Fry *Let.* 26 Apr. (1972) II. 667 Epidauros..was all on a grand scale and tremendous *luxe* in the architecture. **1959** *N.Y. Times* 8 Nov. i. 98 (Advt.), The incomparable beauty and luxe of silk. **1961** *Ibid.* 26 Nov. i. 91 (Advt.), The great revival of the luxe little mesh evening bag. **1967** L. Deighton *London Dossier* 124 Jermyn Street..is a thoroughfare resonant with *luxe*. **1968** L. Durrell *Tunc* iii. 94 For tonight the Pera hotel.. will enable you to rest. There is every luxe. **1974** *New Yorker* 3 June 98/2 He looks round his palace of a house with sniffy and quite unfair resentment, considering its comfort and luxe.

Luxembourgeois (lʊ-ksēmbuɹʒwā), *a.* Of or pertaining to Luxemb(o)urg or to its inhabitants. Also as *sb. pl.*, the natives or inhabitants of Luxemb(o)urg.

1905 T. H. Passmore *In Further Ardenne* v. 121 If its effects upon conduct be any test of a religious system, Luxembourgeois Catholicism comes out brightly, for in the matter of honesty and chastity the people are resplendent. **1951** T. Edwards *Belgium & Luxembourg* vii. 80 A playground for both Luxembourgeois and Belgians. **1971** *Guardian* 8 June 11/4 Though the Luxembourgeois run the hotel, all have English as a working language. **1973** *Sunday Times* 21 Jan. 24/6 Many Luxembourgeois aristocrats have long ago departed.

Luxembourger (lʊ·ksēmbɜ̄ɹgəɹ). A native or inhabitant of Luxemb(o)urg.

1913 G. Renwick *Luxembourg* i. 38 During the two previous reigns German influence was supreme at Court. ..But..that is so no more, and to be a Luxembourger is not now a drawback. **1942** *R.A.F. Jrnl.* 16 May (recto rear cover), Luxembourgers paid tribute to fallen R.A.F. pilots and were punished for their gesture. **1951** W. J. Taylor-Whitehead *Luxembourg* i. 24 The Luxembourgers work hard, and they work long (though sometimes irregular) hours. **1956** B. Miles *Attic in Luxembourg* i. 3 Redouté, the rose-painter, was a Luxembourger. **1972** R. Mayne *Europeans* vi. 164 Some European officials.. smiled at Luxembourgers' accents in French or German. **1973** *Guardian* 23 Mar. 15/7 The Luxembourgers have been saying that they will sign, only if the bank has its headquarters in their capital.

Luxemburgisch (lʊksēmbɜ̄·rgiʃ). Also Luxembourgish. [f. *Luxemb(o)urg* + -ish¹.] = *Letzeburgesch.

1957 *B.B.C. Handbook* 251, 1940... 30 Nov. Luxembourgish broadcasts (as part of Belgian Service) began. **1961** R. E. Keller *German Dial.* 10 These two dialects—or languages—are the daily medium of *all classes* of 300,000 people in the case of Luxemburgisch and of over three million people in the case of *Schwyzertütsch*. **1965** [see *Letzeburgesch].

luxmeter (lʊ·ksmītəɹ). Also luxometer. [f. *lux *sb.* (+ -o) + Meter *sb.*³] An instrument for measuring the luminance (brightness) or illuminance (illumination) of a surface.

1910 H. M. Hobart *Dict. Electr. Engin.* II. 319/2 *Luxmeter, photometric.* This is a photometer of the 'cat's-eye type'. **1912** W. C. Clinton tr. *Bloch's Sci. of Illumination* v. 101 A further development..in the direction of greater portability and lightness has been made in the luxometer. **1915** Gaster & Dow *Mod. Illuminants* vii. 261 Another small instrument of the surface-brightness type, the 'Luxometer', has recently been introduced. **1932** *Proc. Internat. Illumination Congr., 1931* I. 215 Two types of luxmeter have been studied: the luxmeter with a scale of illuminations, and the luxmeter with a single comparison illumination, controllable either by the distance of the lamp or by the interposition of a photometric wedge. **1967** E. Chambers *Photolitho-Offset* vii. 85 Meters can be purchased for measuring the number of Lux (luxmeters) or [*printed* of] foot-candela (foot-candela meters).

luxury. Add: **7.** *attrib.*, as *luxury coach, cruise, duty, edition, flat, liner, shop, tax, trade. Comb.*, as *luxury-loving* adj.

1936 O. Lancaster *Progress at Pelvis Bay* 61 As many as fifty or sixty..luxury coaches leave daily from the square. **1941** N. Coward *Australia Visited* i. 5 Society ladies returning home exhausted after a lightning luxury cruise. **1971** E. Candy *Words for Murder Perhaps* iv. 50 The drug pusher, the criminal classes. You don't get that sort of person in good hotels or on luxury cruises. **1904** *Hansard Commons* 19 Apr. 564 An article of luxury which might very fairly pay a luxury duty. **1930** T. E. Lawrence *Lett.* (1938) 709 We must do our best to get the whole of the luxury edition placed before the date of publication. **1937** *Night & Day* 9 Dec. 12 Luxury flat. A story by K. K. Busvine. **1939** O. Lancaster *Homes Sweet Homes* 66 The luxury flat..is divided up with fiendish ingenuity into a dining-room, drawing-room, lounge-hall, three bed, two bath, a kitchen and all the usual offices. **1958** *New Statesman* 5 Apr. 439/2 A luxury-flat developer ..instructed his architects to do neo-Georgian. **1934** G. B. Shaw *Too True to be Good* Pref. 11 The rich tourists in the palace hotels and luxury liners. **1947** L. MacNeice *Dark Tower* 50, I followed you—but not on a luxury liner. Mine was a cargo boat. **1959** N. Marsh *Singing in Shrouds* iii. 38 A luxury liner and organised fun would be more his cup-of-tea. **1899** 'Mark Twain' in *Forum* Mar. 32 We are..the most luxury-loving people on the earth. **1927** H. Waddell *Wandering Scholars* i. 25 It was an apple-orchard in blossom to his luxury-loving, exquisite and peaceful soul. **1962** H. R. Loyn *Anglo-Saxon Eng.* ii. 69 The gold coinage, papyrus, fine stuffs and spices point to the existence of a luxury-loving aristocracy. **1935** N. Mitchison *We have been Warned* iv. 349 Turning..to Mayfair, they began to pass the discreet and rather impressive luxury shops. **1973** S. Jackman *Guns covered with Flowers* vi. 74 The luxury shops and the department stores. **1904** *Hansard Commons* 19 Apr. 564 The luxury tax at that time was 6*d.* **1974** R. Jeffries *Mistakenly in Mallorca* xiii. 122 The large cars on national or tourist plates whose owners were not forced to pay the luxury taxes that all Spaniards had to. **1905** *Westm. Gaz.* 9 Jan. 3/1 Their action deprives of employment persons who were..employed in luxury trades. **1966** G. Greene *Comedians* i. iii. 84, I can see great possibilities of improvement. My mother was not catering for the luxury-trade.

ly, abbrev. *langley.

1959 S. L. Hess *Introd. Theoret. Meteorol.* ix. 131 The rate of receipt of solar energy is nearly constant at the equator where it varies between 790 and 895 ly day⁻¹. **1963** G. L. Pickard *Descriptive Physical Oceanogr.* v. 51 One langley (ly) is one gram calorie per square centimetre.

lycænid (ləisī·nid), *sb.* and *a.* [f. mod.L. family name *Lycænidæ*, f. the generic name *Lycæna* (J. C. Fabricius 1807, in *Mag. für Insektenkunde* VI. 85), f. Gr. λύκαινα she-wolf.] **A.** *sb.* A butterfly of the family Lycænidæ, which includes the blues and the hairstreaks. **B.** *adj.* Of or pertaining to this family.

1892 W. L. Distant *Naturalist in Transvaal* 68, I have often mistaken it for a large Lycænid. **1913** *Oxf. Univ. Gaz.* 4 June 950/1 A beautiful series of 84 Lycaenid butterflies from the Nicobar Islands. **1972** L. E. Chadwick tr. *Linsenmaier's Insects of World* 197/1 (*caption*) The hairstreaks, a large group of the lycaenids, spend most of the time in bushes and trees. **1974** *Nature* 1 Mar. 16/3 Eliot's paper is designed to evoke thought and curiosity about his lycaenid butterflies.

lycaon (likēi·ǫn). [mod.L. (S. Brookes in E. Griffith et al. tr. *Cuvier's Animal Kingdom* (1827) V. 151), f. Gr. λυκάων, L. *lycāon* a wolf-like animal.] A wild dog of the monotypic genus so called, found in Africa south of the Sahara; the African hunting dog.

1827 E. Griffith et al. tr. *Cuvier's Animal Kingdom* V. 151 (*heading*) Burchel's Lycaon. **1915** Roosevelt & Heller *Life-Hist. Afr. Game Animals* I. viii. 265 The skull of *Lycaon* is easily recognizable from that of a wolf or a jackal. **1945** C. L. B. Hubbard *Observer's Bk. Dogs* 187 *Lycaon* is the most peculiar and interesting of all dog-like carnivores native to Africa. **1958** *Listener* 16 Jan. 102/2 The four-legged animal which deals out more destruction than almost any other is the lycaon, or wild dog.

Lycaonian (likēiōu·niăn), *a.* and *sb.* Also Lykaonian. [f. L. *Lycaonia*, Gr. Λυκαονία Lycaonia + -an.] **A.** *adj.* Of or pertaining to ancient Lycaonia in southern Asia Minor, its inhabitants, or the language spoken by them. **B.** *sb.* **a.** The language of Lycaonia. **b.** A native or inhabitant of Lycaonia.

1582 Bible (Reims) *Acts* xiv. 10 And the multitudes when they had seen what Paul had done, lifted vp their voice in the lycáonian tongue. **1890** W. M. Ramsay *Hist. Geogr. Asia Minor* 392 When, in 361–2, it was found advisable to divide further the large province of Isauria, all the Lykaonian cities were taken from it and from Pisidia. **1893** —— *Church in Roman Empire* ii. 57 Greek then, and not Latin or Lykaonian, would be the common language of these two classes of the population. **1911** [see *Iconian *a.* and *sb.*]. **1926** *Public Opinion* 25 June 582/1 The two gods to whom the Lycaonians..were accustomed to pray. **1957** *Encycl. Brit.* XIV. 511/2 The Lycaonians were to a great extent independent of the Persian empire... The mention of the Lycaonian language in the Acts of the Apostles..shows that the native language was spoken by the common people at Lystra as late as A.D. 50.

lyceal (ləisī·ăl), *a.* [f. Lycée + -al.] Of or pertaining to the French Lycées or similar establishments.

1904 G. S. Hall *Adolescence* I. 345 Marro tabulated the conduct of 3,012 boys in gymnasial and lyceal classes in Italy from eleven to eighteen years of age.

‖ **lycéen** (liseaṅ). [Fr.: see Lycée.] A pupil at a *lycée* in France.

1883 H. James *Little Tour in France* (1885) xiii. 90 Little pale-faced *lycéens*..who are about to be restored to those big educative barracks. **1909** *Daily Chron.* 23 Sept. 4/4 One is naturally led to inquire what steps are taken to ensure the discipline of the Lycéens out of school. **1971** G. Steiner *In Bluebeard's Castle* iv. 85 The Victorian public-school boy, the *Gymnasiast* or *lycéen*. **1974** *Guardian* 26 Mar. 4/4 A background of marching lycéens and charging police, not to speak of protesting school masters.

lyceum. Add: **4. b.** *lyceum bureau, hall, lecture, lecturer, lecturing, system* (earlier example).

1924 I. S. Cobb *Kansas* ii. 20 Fate, personated by the booking agency of a lyceum bureau, decreed that I should jump out of the Teutonic comforts of St. Louis. **1831** *Mass. Private & Special Statutes* 4 Mar., They are hereby made a corporation, by the name of Lyceum Hall,..for the purpose of affording means..for the prosecution of literary and scientific studies [etc.]. **1837** H. Martineau *Society in America* I. i. iii. 61, I attended another Lyceum lecture in Massachusetts. **1922** L. Mumford in H. E. Stearns *Civilization in U.S.* 6 The Lyceum lecture..was taken as a soporific rather than a stimulant. **1844** *Knickerbocker* XXIV. 294 The remark of a lyceum lecturer upon matrimony. **1881** *Harper's Mag.* Mar. 628/2 During the days of his lyceum lecturing, no man was more popular [than Dr. Chapin] upon the platform. **1843** 'R. Carlton' *New Purchase* I. 174 The common school system, the lyceum system.

6. (With capital initial.) The name of a theatre near the Strand in London, used *attrib.* to denote a performance characteristic of those given at this theatre, esp. of the melodramatic type associated with Henry Irving; also *transf.*

1898 G. B. Shaw *Let.* 29 Jan. in C. St. John *Ellen Terry & Bernard Shaw* (1931) 294 Henry [Irving]..is as much behind the times now as Pinwell's and Fred Walker's and Mason's pictures are (I always call them the Lyceum school). **1898** —— *Plays Pleasant & Unpleasant* p. xix, Popular entertainments like Gounod's opera or the Lyceum version, in which poetry and philosophy are replaced by romance. **1901** —— *Three Plays for Puritans* p. xi, I found that the whole business of stage sensuousness, whether as Lyceum Shakespear, musical farce, or sham Ibsen, finally disgusted me. **1936** 'N. Blake' *Thou Shell of Death* xiii. 229 Tones that would have done credit to a Lyceum melodrama. **1964** 'A. Gilbert' *Knock knock, who's There?* i. 21 This wasn't a cosy pub..it was all set for a Lyceum melodrama.

Lycian (li·siăn), *sb.* and *a.* [f. L. *Lycia*, Gr. Λυκία Lycia + -an.] **A.** *sb.* **a.** A native or inhabitant of ancient Lycia in south-west Asia Minor. **b.** The language and script used in Lycia. **B.** *adj.* Of or pertaining to ancient Lycia, its inhabitants, or the language and script used by them.

1598 Chapman tr. *Homer's Seaven Bookes of Iliades* 45 Sarpedon had the Lycians charge. **1607** Topsell *Four-f. Beasts* 115 These beasts are..plentifull..in Lycia:..the Lycian Roes doe neuer goe ouer the Syrian Mountaines. **1718** Pope tr. *Homer's Iliad* xvi. 570 Two sounding Darts the Lycian Leader threw. **1839** C. Fellows *Jrnl. Excursion Asia Minor* viii. 229 On the side of the tomb..under two lines of the peculiar characters of this town, (perhaps Lycian,) is a group of figures. **1841** —— *Acct. Discoveries Lycia* vii. 168 The tedious, but..useful occupation, of copying the Lycian inscription. **1845** [see *bilingual *a.* 2]. **1883** *Encycl. Brit.* XV. 92/2 It [*sc.* Lycia] was..inhabited from a very early period by a distinct people,

known to the Greeks as Lycians. *Ibid.* 94/1 A few of these inscriptions are fortunately bilingual, in Greek and Lycian. **1913** H. R. Hall *Anc. Hist. Near East* vi. 270 There is a reference to Lycian pirates in another letter. **1933** [see *Carian *sb.* and *a.*]. **1941** [see *Ilian *a.* (*sb.*)]. **1948** D. Diringer *Alphabet* 464 The Lycian alphabet..is certainly of Greek origin. **1951** J. B. Bury *Hist. Greece* (ed. 3) viii. 337 Cimon..constrained the Lycian communities to enrol themselves in the confederacy of Delos. **1972** R. Meiggs *Athenian Empire* xvii. 307 The Lycians also had ceased to bring their tribute. *Ibid.*, The Carian and Lycian coasts were being used by pirates. **1973** *Times* 1 Oct. 17/6 Lycian is an Indo-European language and belongs to the Luwian sub-group of Anatolian. *Ibid.*, It would be important to know in which of the two Lycian dialects the inscription is written.

lycid (li·sid), *a.* and *sb.* [f. mod.L. family name *Lycidæ*, f. generic name *Lycus* (J. C. Fabricius *Mantissa Insectorum* (1787) I. 163), f. Gr. proper name Λύκος, L. *Lycus*: see -id³.] **A.** *adj.* Of or pertaining to a beetle of the family Lycidæ. **B.** *sb.* A beetle of this family.

1932 J. S. Huxley *Probl. Relative Growth* vii. 237 See.. Mjöberg's discussion (1925) on the larviform females of certain Lycid beetles. **1934** Webster, *Lycid n.* and *adj.* **1962** *New Scientist* 6 Dec. 577/1 The cerambycid beetle *Elytroleptus ignitus* is a mimic of the lycid beetles *Lycus loripes* and *L. simulans*... They are typical cases of Batesian mimicry, the cerambycids being palatable to predators but gaining protection by looking like the distasteful lycids. **1972** L. E. Chadwick tr. *Linsenmaier's Insects of World* 165/1 There are a great many mimetic forms that resemble protected beetles (such as the already-mentioned lycids).

lycomarasmin (ləi:komăræ·smin). *Chem.* Also † **lyco-marasmine.** [ad. G. *lyco-marasmin* (Plattner & Clauson-Kaas 1945, in *Experientia* I. 196), f. mod.L. *lyco-persici* varietal or specific epithet (taken as gen. of *lycopersicon*: see *lycopersicin) + Gr. μαρασμ-ός withering + -in¹.] A phytotoxic dipeptide, $C_9H_{15}N_3O_7$, which contains glycine and aspartic acid residues and was isolated from *Fusarium bulbigenum* var. *lycopersici*, the fungus which causes tomato wilt.

1945 *Experientia* I. 196 On hydrolysis lyco-marasmine yields glycine, aspartic and (probably) pyruvic acid and ammonia. **1956** *Ann. Rev. Microbiol.* X. 361 Lycomarasmin, therefore, causes iron deficiency in the stem and iron excess in the leaves. **1972** S. A. J. Tarr *Princ. Plant Path.* xiii. 236 Lycomarasmin is a dipeptide which brings about wilting and yellowing of tomato cuttings, and its phytotoxicity is increased in the presence of iron.

lycopene (ləi·kopīn). *Chem.* [f. *lycop(in + -ene.] A polyunsaturated hydrocarbon, $C_{40}H_{56}$, which is a red carotenoid pigment present in tomatoes and many berries and fruits.

1935 *Jrnl. Biol. Chem.* CXII. 424 In an investigation of the plastid pigments of marsh dodder,..α- and β-carotenes, γ-carotene, lycopene, and rubixanthin have been isolated. **1950** *Genetics* XXXV. 209 Tomatoes which contain principally *beta*-carotene differ from those which contain mostly lycopene by a single incompletely dominant gene. **1972** Goodwin & Mercer *Introd. Plant Biochem.* xi. 279 The conversion of phytoene into coloured carotenoids involves a stepwise desaturation to neurosporene and lycopene which are then converted into the cyclic carotenoids.

lycopersicin (ləi:kopɜ̄·ɹsikin). *Chem.* [f. mod.L. *lycopersic-on*, the name of the genus to which the tomato belongs, f. Gr. λύκο-ς wolf + περσικ-ός peach (f. Περσίς: see Persian *a.* and *sb.*): see -in¹.] **a.** = *lycopene.

1913 B. M. Duggar in *Washington Univ. Stud.* I. 23 Since the pigment derived from the tomato has now been demonstrated to be distinct from carotin.., and since it was first described in the tomato, it seems well to suggest that, in view of the inappropriateness of the earlier terms, lycopersicin should be adopted instead.

b. = *tomatin.

1945 G. W. Irving et al. in *Science* 6 July 9/1 We have obtained from the expressed juice of Pan American tomato plants..a preparation which..possesses marked fungistatic activity toward Fol. This antibiotic agent, which will be designated 'lycopersicin', occurs throughout the mature plant. **1946** *Jrnl. Bacteriol.* LII. 601 In a recent publication..this substance was referred to as 'lycopersicin'. Inasmuch as it has since been learned that the term 'lycopersicin' was once used..as a synonym for 'lycopene', the red pigment of the tomato, the designation of the antibiotic agent has been changed to 'tomatin' to avoid possible confusion.

lycopin (ləi·kopin). *Chem.* [f. mod.L. *lycopersicon* (see *lycopersicin) + -in¹.] = *lycopene.

1903 C. A. Schunck in *Proc. R. Soc.* LXXII. 174 Believing that this substance has not been isolated before, or if it has, has been mistaken for carotin, I venture to apply to it the name *Lycopin*. **1965** Bell & Coombe tr. *Strasburger's Textbk. Bot.* 43 Over seventy different carotinoids are known. Of these we need refer here only to the carotin of the carrot.., to the lycopin of the tomato and other red fruits, and to the yellow violaxanthin.

Lycra (ləi·kră). Also lycra. A proprietary name of an elastic polyurethane fibre and fabric used esp. for underwear and swimming costumes.

1958 *Official Gazette* (U.S. Patent Office) 18 Nov. TM 87/1 E.I. du Pont de Nemours and Company... Lycra. For synthetic fibers and filaments for generalized use in the industrial arts. First use Feb. 3, 1958. **1959** *Trade Marks Jrnl.* 8 Apr. 380/2 Lycra... Raw or partly prepared natural or synthetic fibres; and untwisted and unspun filaments;..E. I. Du Pont De Nemours and Company. **1961** *Harper's Bazaar* Sept. 103 Lycra, the ultimate in lightweight power. **1963** *Guardian* 8 Nov. 12/4 Proofed.. lycra jersey ski pants. **1968** *Vogue* 15 Apr. 23/2 This.. pantie-corselette..of softest lycra power net. **1972** *Country Life* 18 May 1275/3 Beachwear in newest Lycra material, light but giving corselette support.

lyctus (li·ktŏs). [mod.L. (J. C. Fabricius *Entomologia Systematica* (1792) I. ii. 502), f. Gr. Λύκτος, L. *Lyctus* name of a city in Crete.] A wood-boring beetle of the genus so called; a powder-post beetle. Also *attrib.* Hence **ly·ctid** *sb.*, a beetle of the family Lyctidæ; as *adj.*, of or pertaining to this family.

1917 *U.S. Dept. Agric. Farmers' Bull.* No. 788 (*title*) Powder-post damage by *Lyctus* beetles to seasoned hardwood. **1926** E. O. Essig *Insects Western N. Amer.* 438 The European lyctus..can be separated by the single row of large round shallow punctures on the elytra. **1936** *Discovery* Feb. 42/1 The Powder-Post or Lyctus beetles are menacing timber-yard pests. **1958** *Times Rev. Industry* May 18/3 Death watch beetles and furniture beetles..are joined by..the lyctus beetle. **1959** E. F. Linssen *Beetles Brit. Is.* II. 63 The dust made by Anobiid larvae is granular, that of the Lyctid grubs is a fine powder. **1963** N. E. Hickin *Insect Factor in Wood Decay* v. 134 Lyctid beetles are small.

Lycurgan (ləikɜ̄·ɹgăn), *sb.* and *a.* Also Lycurgean, Lycurgian. [f. L. *Lycurgus*, Gr. Λυκοῦργος traditional lawgiver and founder of the Spartan constitution, dated in antiquity variously to the ninth and eighth centuries B.C.] **A.** *sb. rare.* An adherent of Lycurgus or his methods. **B.** *adj.* Of, pertaining to, or characteristic of Lycurgus, or the constitutional innovations attributed to him; harsh, severe.

1584 W. Allen *True Defence Eng. Catholiques* vii. 149 By the meanes of such Lycurgians as this, we haue in England new lawes against al claime of iurisdiction spiritual or temporal, that can be made by anie person whosoeuer, borne out of the Realme. **1846** G. Grote *Hist. Greece* II. ii. vi. 454, I incline to adopt the opinion of Thucydides as to the time at which the Lycurgean constitution was introduced at Sparta. **1934** [see *die-hardism]. **1956** A. Toynbee *Historian's Approach to Relig.* 242 The Lycurgean régime at Sparta was exceptional. **1962** *Listener* 30 Aug. 323/2 The Lycurgan training for public service enriched Greek 'paedeia'. **1970** *Oxf. Classical Dict.* (ed. 2) 1007/2 Proposals to revive the strictness of the Lycurgan training..were obstructed by the ephors.

lyddite, *v. rare.* [f. the sb.] *trans.* To destroy, wreck, etc., by the explosion of lyddite.

1906 F. Campbell *Dearlove* 78 She was pleased they had not dynamited or lyddited him.

lydite (li·dəit). *Min.* Also 9 lydit. [f. Lyd(ian *a.* and *sb.* + -ite¹.] = *Lydian stone* (s.v. Lydian *a.* 3).

1816 R. Jameson *Syst. Mineral.* (ed. 2) I. 192 Lydit. **1861** H. W. Bristow *Gloss. Mineral.* 223/2 Lydian Stone ..Lydite. **1907** *Q. Jrnl. Geol. Soc.* LXIII. 31 It is by no means a pure sand; there are no 'lydite' or other pebbles. **1948** *Proc. Prehist. Soc.* XIV. 129 Brown and purple quartzite cobbles, pieces of vein quartz and of lydite. **1968** *Mineral. Abstr.* XIX. 333/1 Six chemical analyses of phyllites and clay slates are listed; these are in zeolitic and greenschist facies and contain grapholite-bearing lydites.

lye, *sb.* Add: **4.** *lye-cask, hominy, -leach.*

1821 W. B. Dewees *Lett. from Early Settler Texas* (1852) 20 Our subsistence was principally upon..a kind of lye hominy seasoned with hickory nut kernels. **1843** 'R. Carlton' *New Purchase* I. ix. 63 A lie-cask, or rather, an inverted pyramidical box to contain ashes. **1847** J. O. Halliwell *Dict. Archaic & Provincial Words* Lie-leach. **1854** M. J. Holmes *Tempest & Sunshine* xv. 202 Now be keerful and not run afoul of the plaguey lye leech! **1919** J. P. Dunn *Indiana* II. 1170 A woman situated like Mrs. McCoy, in her Indian boarding school, with no food but lye hominy in the house..'degraded her soul' by cooking lye hominy. **1948** E. N. Dick *Dixie Frontier* 290 Lye hominy was made by soaking the whole grains of corn in lye water to remove the hulls.

lye, *sb.*², var. Lie *sb.*² 4.

1855 Ogilvie *Imp. Dict.* Suppl., *Lye*, a term employed, in *railway lang.*, to denote the sidings or short offsets from the main line, into which trucks may be run for the purpose of loading or unloading. **1901** *Daily Record* (Glasgow) 31 Aug. 3 A boy..was accidentally killed at the lye of South Renfrew Station on Thursday night.

lying, *vbl. sb.*¹ Add: **3. lying-in-state,** (of the corpse of a public figure) being on display for public tribute before burial.

1923 W. de la Mare *Riddle* 241 Positive constellations of candles as if for a Prince's..lying-in-state. **1947** M. Field *Boys' & Girls' Film Bk.* 67 When he died thousands of people went to his lying-in-state. **1972** *Whitaker's Almanack 1973* 565/2 The Duke of Windsor's lying-in-state took place in St. George's Chapel, Windsor.

lying-in. b. (Further examples.)

1823 J. Constable *Let.* 24 Aug. in *Corr.* (1964) ii. 282 Miss Cookson is on a visit with another lying in sister. **1861** D. G. Rossetti *Let.* 20 Apr. (1965) II. 396 Dr. Babington, head of the Lying-in Hospital. **1912** *Q. Rev.* July 60 A slight increase in the ratio of lying-in claims to the number of members... A lying-in benefit of 30s. **1964** D. Owen *Eng. Philanthropy* i. ii. 50 The years 1749–65 saw the founding of..the British Lying-In (1749), the City of London Lying-In (1750),..these in addition to the Lying-In Charity,..established in 1757. **1975** *Country Life* 2 Jan. 50/3 Villages which had a mutual aid or 'lying-in society'.

Lyle gun (ləil gʊn). [Named from its inventor D. A. *Lyle* (d. 1937) of the U.S. Ordnance Dept.] A cannon, invented in 1877, designed to project a rope from the shore to a stranded ship to facilitate salvaging and rescue operations.

1880 *Ann. Rep. U.S. Life-Saving Service* 109 The keeper ..then proceeded to charge the Lyle gun. **1899** *Ibid.* 471 The sample submitted is designed for use with the 2½-inch Lyle gun. **1911** *Encycl. Brit.* XV. 606/2 The keeper fires a line over the wreck with the Lyle gun, a small bronze cannon..having an extreme range of about 700 yards. **1927** G. Bradford *Gloss. Sea Terms* 106/2 *Lyle gun*, a life-saving gun designed to shoot a projectile with a line attached. This line..establishes a connection with a stranded ship by which an endless line and tail block is hauled off and afterwards a hawser. **1944** H. Norby *Questions & Answers* vi. 128 No part of Lyle gun equipment is to be used for any other purpose.

Lylian (li·liăn), *a.* Also **Lylyan.** [f. the name of John *Lyly* (*c* 1554–1606), English dramatist and novelist.] Of, pertaining to, or having the characteristics of John Lyly or his works.

1923 E. K. Chambers *Eliz. Stage* III. 36 The plays of the Lylyan school..illustrate very precisely, on the side of staging, that blend of the classical and the romantic tempers which is characteristic of the later Renaissance. **1928** E. A. Gerrard *Eliz. Drama* ii. i. 145 The scene between Lucilla and the Enchanter is purely Lylian. **1959** *Times* 1 Sept. 11/2 It was Lyly who opened the door for Shakespeare, and when he was forced to abandon the Lylian view he abandoned comedy. **1962** G. K. Hunter *John Lyly* iv. 164 The approach to Alexander is underlined for us in a typically Lylian way. *Ibid.* iv. 218 With typical Lylian balance the action returns to the point from which it started.

Lyman (ləi·măn). *Physics.* The name of Theodore *Lyman* (1874–1954), U.S. physicist, used *attrib.* to designate a series of lines (individually designated alpha, beta, etc.) discovered by him in the ultraviolet part of the spectrum of atomic hydrogen, with wave numbers represented by the formula $R(1 - 1/m^2)$ (where R is the Rydberg constant and $m = 2, 3, \ldots$), the first line of which has a wavelength of 121·6 nanometres.

1922 *Encycl. Brit.* XXXII. 559/2 The formulæ for the hydrogen series are as follows:—Lyman series: $\nu = N/1^2 - N/m^2$ $(m = 2, 3, \ldots)$. Balmer series: $\nu = N/2^2 - N/m^2$ $(m = 3, 4, \ldots)$. Paschen series: $\nu = N/3^2 - N/m^2$ $(m = 4, 5, \ldots)$. **1929** J. K. Robertson *Introd. Physical Optics* xviii. 359 Lines of the Lyman series arise from electron drops from outer orbits to the innermost or normal orbit, for which $k = 1$. **1959** *Sunday Times* 5 Apr. 8/3 The sun itself has been photographed in the extreme ultra-violet (the Lyman-alpha) region of the spectrum. **1967** W. R. Hindmarsh *Atomic Spectra* ii. 13 Ritz..proposed his well-known combination principle in 1908; he recognized that the wave-numbers of the lines could always be represented as a difference between two terms... It was with the use of this principle that the Paschen, Lyman, Brackett and Pfund series in the spectrum of hydrogen were predicted ..before they were known experimentally. **1968** G. M. B. Dobson *Exploring Atmosphere* (ed. 2) viii. 158 Another possibility is that the very strong radiation which is given out by hydrogen in the sun, at a wavelength of 1216 Ångstrom units (known as the Lyman alpha line) produces the *D* region [of the ionosphere].

lymph. Add: **5. lymph node,** any of several small rounded gland-like structures of the lymphatic system, which are disposed along the course of the lymph vessels and which are responsible for removing foreign bodies from the lymph stream and for producing lymphocytes and antibodies; a lymph gland.

1892 *Proc. N.Y. Path. Soc. 1891* 65 (*heading*) Large-celled indurative hyperplasia of the lymph nodes. **1925** *Jrnl. Amer. Med. Assoc.* 28 Feb. 669/2 The cervical, axillary, epitrochlear and inguinal lymph nodes were moderately enlarged. **1955** *Sci. News Let.* 18 June 393/2 Antibodies, he found, are formed in lymph nodes, better known to the layman as glands. **1961** H. A. Skinner *Origin Med. Terms* (ed. 2) 259/2 All of the early writers seem to have adhered to the term [lymph] gland..and the term has been continued, although there is no indication of any gland function. This was a matter of vigorous debate at the anatomical congress at Basle [in 1895]..and Toldt especially advocated use of the term 'lymph node'. On the last ballot his suggestion was voted down by a large majority. In the Nomina Anatomica adopted at Paris in

1955 the term node was approved. **1972** J. W. Shields *Trophic Function Lymphoid Elem.* xi. 71 The lymph nodes are characteristically oriented to receive substrate more or less directly from the peripheral tissues.

lymphadenopathy (li:mfædĭnǫ·păþi). *Med.* [f. Lymph + Adenopathy.] Disease of the lymph nodes.

1920 *Bull. Johns Hopkins Hosp.* XXXI. 413/2 An acute infection associated with lymphadenopathy and lymphocytosis. **1973** *Nature* 27 July 206/2 The pertinent findings on physical examination were marked gingival hypertrophy, lymphadenopathy, petechiae of the lower extremities, and hepatomegaly.

lymphangiography (li:mfændʒi,ǫ·grăfi). *Med.* [f. Lymph + Angiography.] A technique or procedure for demonstrating and examining the lymph vessels *in vivo* by injecting a contrast medium into them and examining them with X-rays; an examination by this technique.

1941 Dorland & Miller *Med. Dict.* (ed. 19) 830/1 *Lymphangiography*, the roentgenologic visualization of lymphatic vessels following the injection of contrast medium. **1952** *Clin. Sci.* XI. 13 (*heading*) Lymphangiography in man. **1961** *Radiology* LXXVI. 179/1 Lymphangiography, the radiographic demonstration of the lymphatic system by intralymphatic injection of contrast material, has opened a new field of investigation. **1973** J. F. Meaney et al. *Complications & Legal Implications Radiologic Special Procedures* ix. 92 Lymphangiography is usually considered a relatively simple procedure. *Ibid.*, Many outpatients having lymphangiographies do not have their temperatures recorded.

Hence **lympha·ngiogram**, a radiograph taken by this technique; **ly:mphangio·grapher**, one who carries out or employs lymphangiography; **lympha:ngiogra·phic** *a.*; **lympha:ngiogra·phically** *adv.*

1955 *Brit. Med. Jrnl.* 16 Apr. 941/1 (*caption*) Normal lymphangiogram: 6 ml. of 70% diodone was injected and the films were exposed one minute later. **1964** *Clin. Radiol.* XV. 346/1 We have encountered only one serious reaction during our series of fifty-five lymphangiograms. **1967** A. Rüttimann *Progress Lymphology* 127 (*heading*) Lymphangiographic evaluation of lymphoma. **1968** *Proc. Conf. Lymph & Lymphatic Syst.* vii. 154 Kinmonth and associates reported the largest series of patients studied lymphangiographically. **1973** J. F. Meaney et al. *Complications & Legal Implications Radiologic Special Procedures* ix. 99 Every lymphangiographer should be aware of the potential danger.

lymphangitis. Add: Also **-angiitis.** (Earlier and later examples.)

1842 Dunglison *Dict. Med. Sci.* (ed. 3) 425/2 Lymphangeitis. **1889** *Syd. Soc. Lex.* Lymphangiitis. **1960** R. A. Runnells et al. *Princ. Vet. Path.* xxii. 691/1 Epizootic lymphangitis is a chronic suppurative infection of horses, mules, and donkeys caused by the fungus *Zymonema farciminosum*. **1966** Wright & Symmers *Systemic Path.* I. xxvi. 803/2 Elephantiasis of the penis and scrotum... Repeated attacks of lymphangiitis and the presence of a sufficient number of worms may, in the course of time, block so many lymph channels that chronic oedema develops.

lympho- (li·mfo), comb. form of Lymph 3, used in numerous biological and medical terms, as **ly·mphoblast** *Biol.* [-blast], any cell which is a precursor of a small lymphocyte; so **lymphobla·stic** *a.*; **ly:mphoblasto·ma** *Path.* [*-oma], malignant proliferation of lymphoblasts; **lymphoge·nic** [*-genic], **lympho·genous** [*-genous], (*a*) producing lymph or lymphocytes; (*b*) arising in, produced by, or disseminated via the lymphatic system; **ly·mphokine** *Immunol.* [f. Gr. κινεῖν to move], any of various soluble substances released by lymphocytes following activation by contact with an antigen which are thought to be involved in cell-mediated immunity but to lack the antigen-specificity of antibodies; **lymphope·nia** [*-penia], reduction in the number of lymphocytes in the blood; **lymphopoiesis** (-poi,ī·sis) [*-poiesis], the formation of lymphocytes; so **lymphopoie·tic** *a.*

1909 *Cent. Dict.* Suppl., Lymphoblast. **1935** *Jrnl. Amer. Med. Assoc.* 7 Sept. 765/2 Lymphoblasts approach the structure of myeloblasts. **1962** *Lancet* 27 Jan. 206/2 In the early stages of the production pathway, the lymphocytes are large and possess an intensely basophilic cytoplasm; these are the cells which have often been called lymphoblasts. **1905** *Ibid.* 12 Aug. 465/2 Dr. Jones concluded,..secondly, that apparently a lymphoblastic marrow was usually accompanied by an increased production of erythroblasts, but that, on the other hand, a leucoblastic marrow was not associated with increased erythroblastic production. **1961** *Ibid.* 5 Aug. 291/2 In 2 cases biopsy of a lymph-gland was undertaken and reported as lymphoblastic lymphoma. **1920** *Jrnl. Urol.* IV. 137 (*title*) Lympho-blastoma (lympho-sarcoma) of the prostate. **1926** *Jrnl. Amer. Med. Assoc.* 17 Apr. 1185/1 Lymphoblastoma, a name considered by some as synonymous with malignant lymphoma..may include lymphatic leukemia (lymphocytic and lymphoblastic), aleukemic lymphatic leukemia (pseudoleukemia, aleukemic lymphadenosis), lymphocytoma, Hodgkin's disease (lymphogranuloma), lymphadenoma, lymphomatosis, lymphosarcoma, round cell sarcoma, leukosarcoma and lymphadenosarcoma. **1970** S. D. Kobernick tr. *Masson's Human Tumors* ii. x. 353 The structure is very simple: a chaotic mixture of small cells with round nuclei, larger in lymphoblastomas than in lymphocytomas. **1901** Lymphogenic [see *lymphomatosis]. **1968** *Proc. Conf. Lymph & Lymphatic Syst.* viii. 173 Experimental animals with lymphogenic encephalopathy have decreased response to pain. **1889** *Syd. Soc. Lex.*, *Lymphogenous*, producing lymph. **1909** *Practitioner* Nov. 656 The unilateral distribution of renal tuberculosis..is explained by lymphogenous infection. **1935** N. P. Sherwood *Immunol.* iii. 59. The four avenues [of spread of infectious agents] commonly mentioned are surface spread, dissemination by way of the lymphatics (lymphogenous), blood stream (hematogenous) and direct extension to adjacent tissues. **1968** *Proc. Conf. Lymph. & Lymphatic Syst.* viii. 169 (*heading*) Lymphogenous encephalopathy. **1969** D. C. Dumonde et al. in *Nature* 4 Oct. 38/1 In the guinea-pig these four phenomena are mediated by cell-free soluble factors, which are generated during interaction of sensitized lymphocytes with specific antigen, but which are expressed without reference to immunological specificity. The generic term 'lymphokine' is suggested to describe this group of biological activities. **1973** *Ibid.* 2 Mar. 22/2 When T cells are activated by antigen, they proliferate..but they..do not become antibody-secreting cells. They do, however, secrete a variety of non-antigen-specific factors ('lymphokines') such as migration inhibition factors (MIF), chemotactic factors, cytotoxic factors and mitogenic factors, at least some of which presumably play a role in cell-mediated immune responses, for which T cells are primarily responsible. **1974** *Sci. Amer.* Apr. 36/2 Transplantation antigens and other foreign material ..stimulate the production and release of 'sensitized' lymphocytes, or effector cells... In addition to killing graft cells directly the sensitized lymphocytes secrete a variety of chemical agents called lymphokines, some of which act directly on the foreign cells and some of which attract other leukocytes..which digest damaged cells and cell fragments. **1909** *Cent. Dict.* Suppl., Lymphopenia. **1921** *Lancet* 10 Dec. 1205/2 The blood content passes from exhibiting a lymphopænia to a lymphocytosis. **1964** L. Martin *Clin. Endocrinol.* (ed. 4) i. 35 A leucocytosis of 10–15,000 per c.mm. is usual with a polymorph excess, lymphopenia and eosinopenia. **1918** Stedman *Med. Dict.* (ed. 5) 569/1 *Lymphopoiesis*,..the formation of lymphocytes. **1968** Passmore & Robson *Compan. Med. Stud.* I. xxvii. 2/2 In the lymphoreticular organs the reticulum cells..also differentiate into stem cells which divide and mature to form lymphocytes in the process known as lymphopoiesis. **1915** J. E. R. McDonagh *Biol. & Treatm. Venereal Dis.* xv. 144 (*heading*) Syphilis of the lympho- and haemopoetic [*sic*] system. **1966** M. W. Elves *Lymphocytes* iii. 71 The lymphopoietic role of the thymus.

lymphocyte. Substitute for def.: A kind of small leucocyte which has a single round nucleus and little or no granulation in the cytoplasm, constitutes about a quarter of the total leucocytes in the blood stream, is found in large numbers in the lymph nodes and other lymphoid tissue, and is a major agent in most immunological processes. *Large lymphocytes*, which constitute a small proportion of the total number of lymphocytes, are larger than and apparently the precursors of the *small lymphocytes*, which constitute the bulk of them. (Earlier and later examples.)

1890 *Jrnl. Morphol.* IV. 108 It is quite generally agreed that the origin of the white corpuscles of the blood is to be found in the lymph leucocytes, or lymphocytes, to borrow a convenient term... The lymphocytes are characterized by a vesicular nucleus, usually with a nucleolus and a scanty reticulum, and by a very small protoplasmic envelope. **1891** *Rep. Lab. R. Coll. Physicians Edin.* III. 118 If this view of the nature of lymphocytes is accepted, we at once realise the futility of the attempts..to establish distinctions between the leucocytes found in different situations,—the white blood corpuscles, the wandering cells in the tissues, the lymphocytes of the lymph, lymph-glands and spleen, the thymus-cells, the marrow-cells, &c. **1905** A. Stengel tr. *Ehrlich's Dis. Blood* 71 At the introduction to this chapter it was pointed out that a retrograde movement in hematology is in progress which attempts to establish the derivation of all the white blood-cells from the lymphocytes. **1913** Gulland & Goodall *Blood* ii. vii. 58 Large lymphocytes have the same general characters as the small lymphocytes. **1962** *Lancet* 26 May 1098/2 The blood-picture was:..eosinophils 2%, lymphocytes 11%, monocytes 1%, [etc.]. **1968** Passmore & Robson *Compan. Med. Stud.* I. xxvii. 4/2 Lymphocytes, or more properly small lymphocytes, to distinguish them from their immediate precursors, the medium and large lymphocytes, are formed in the cortex of lymph nodes by the maturation and division of lymphoblasts and lymphoid stem cells. **1970** *Courier-Mail* (Brisbane) 28 Jan. 30 Unfortunately this desensitisation technique seems to affect only a proportion of the recipient's lymphocytes: the remainder attack the transplant.

So **ly:mphocyto·ma** (pl. **-o·mas, -o·mata**) *Path.* [*-oma], malignant proliferation of lymphocytes; any condition characterized by this; a tumour composed of lymphocytes; **ly:mphocyto·matous** *a.*

1908 Osler & McCrae *Syst. Med.* IV. 827 The hyperplastic and neoplasm-like conditions of the lymph nodes depending upon an overgrowth of cells of the type of lymphocytes..are here brought into one general class, the lymphocytomata. This includes all the lymphocytomatous tumors classed variously as lymphoma, lymphadenoma, lymphosarcoma, pseudoleukæmia, adenia, lymphadenomatosis, lymphomatosis, etc. **1920** J. E. R. McDonagh *Venereal Dis.* iv. 55 (*heading*) Syphilitic lymphocytomata. **1949** *Year Bk. Radiol.* 406 The medical profession has believed that no attempt at treatment of lymphoid tumors (lymphosarcoma, lymphocytoma, Hodgkin's disease and macrofollicular lymphoma) will be successful.

lymphœdema (limfĭdī·mă). *Path.* Also lymphedema. [f. Lymph + Œdema.] Œdema resulting from obstruction of lymph vessels or lymph nodes.

1889 *Syd. Soc. Lex.*, Lymphœdema. **1961** *Lancet* 19 Aug. 408/1 The XO female was noted to have peripheral lymphœdema and neck-webbing at birth. **1968** *Proc. Conf. Lymph & Lymphatic Syst.* vii. 153 In 1892, Milroy reported congenital, bilateral lymphedema of the legs in twenty-five of ninety-two members of one family.

Hence **lymphœde·matous** *a.*

1934 *Arch. Internal Med.* LIV. 614 Lymphedematous swelling..is present at birth. **1962** D. I. Abramson *Blood Vessels & Lymphatics* xxiv. 733 The protein content of tissue fluid from the lymphedematous area was similar to that of the serum.

lymphogranuloma (li:mfogrænĭulōu·mă). *Path.* [f. *lympho- + Granuloma.] Used, usu. with mod.L. or Eng. adjs., to designate any of three diseases (see quot. 1958), esp. *lymphogranuloma venereum* (or *lymphogranuloma inguinale*), a venereal disease, esp. of the tropics, caused by micro-organisms of the group Chlamydiæ, and manifested esp. as inflammation, followed by suppuration and breakdown, of the lymph nodes and lymph vessels, particularly in the inguinal region.

1924 J. Schaumann in *Brit. Jrnl. Dermatol.* XXXVI. 515 When in November 1914, I described the disease I proposed designating by the anatomico-clinical name of benign lymphogranuloma, I had already detected the essential characteristics, which justified its classification amongst lymphadenic affections. **1932** DeWolf & Van Cleve in *Jrnl. Amer. Med. Assoc.* 24 Sept. 1065/1 They [*sc.* Durand, Nicolas, and Favre] chose the name 'subacute inguinal lymphogranulomatosis' for the disease, a rather unfortunate designation because of the confusion to which it may lead. We have chosen the shorter designation, 'lymphogranuloma inguinale', as this name is becoming more widely used by European writers. **1958** R. W. Raven *Cancer* II. xxiv. 473 Few medical terms can have such different meanings in different countries as lymphogranuloma. When used without qualification, in English-speaking countries, 'lymphogranuloma' usually refers to 'lymphogranuloma inguinale' (Nicolas–Favré disease); in most European countries it implies Hodgkin's disease ('lymphogranuloma malignum') but in Scandinavia it refers to sarcoidosis ('lymphogranuloma benignum'). **1960** J. Marshall *Dis. Skin* xxv. 630 (*heading*) Hodgkin's disease (malignant lymphogranuloma of Paltauf–Sternberg). **1967** A. C. Allen *Skin* (ed. 2) xi. 423/1 At different stages in the evolution of the information on the disease, stress was placed on the involvement of one organ over others, as indicated by the application of the terms 'uveoparotid fever', 'osteitis tuberculosa multiplex cystica', 'lupus pernio', and 'benign lymphogranuloma'. **1974** Passmore & Robson *Compan. Med. Stud.* III. xiii. 13/2 Lymphogranuloma venereum. Alternative names for this disease are lymphogranuloma inguinale and tropical bubo. It is caused by *Chlamydia* or *Bedsonia*.

lymphogranulomatosis (li:mfogrænĭŭlomătōu·sis). *Path.* [f. *lymphogranulomat-*, taken as stem of prec. + -osis.] Used, usu. with mod.L. or Eng. adjs., to designate any of three diseases, viz. *lymphogranulomatosis benigna*, sarcoidosis; *lymphogranulomatosis inguinalis* [tr. F. *lymphogranulomatose inguinale* (Durand et al. 1913, in *Bull. et Mém. Soc. méd. d. Hôp. de Paris* XXXV. 274)], lymphogranuloma venereum; *lymphogranulomatosis maligna*, Hodgkin's disease.

1911 Dorland *Med. Dict.* (ed. 6) 470/1 *Lymphogranulomatosis*..Hodgkin's disease. **1915** J. E. R. McDonagh *Biol. & Treatm. Venereal Dis.* xlvi. 550 This case..would probably have been called lymphogranulomatosis on the Continent, and lymphadenoma or Hodgkin's disease affecting the skin in this country. **1926** *Arch. Dermatol. & Syphilol.* XIV. 36 In January, 1913, Nicolas, Durand and Favré published a report concerning a series of cases of lymphatic enlargement of the groins which they named 'lymphogranulomatosis inguinalis'. **1932** [see *lymphogranuloma]. **1935** *Brit. Jrnl. Dermatol.* XLVII. 225 The symptomatic triad of benign lymphogranulomatosis. *Ibid.*, I attributed the diabetes to the presence of lymphogranulomatosis benigna in this organ. *Ibid.* 227 The polynuclear leucocytosis in malign lymphogranulomatosis (Hodgkin's disease). **1971** Brunson & Gall *Concepts of Dis.* xxv. 925/1 In the United States and Great Britain, Hodgkin's disease is regarded generally as a malignant neoplasm; in Germany it is considered to be a peculiar inflammatory process and is designated by the term *lymphogranulomatosis*. **1973** R. B. Scott *Price's Textbk. Pract. Med.* (ed. 11) ix. 874 Sarcoidosis. *Synonyms.* Lymphogranulomatosis benigna [etc.].

lymphography. Add: **2.** A technique similar to lymphangiography but including demonstration and examination of lymph nodes.

1935 Dorland & Miller *Med. Dict.* (ed. 17) 776/1 *Lymphography*, roentgen visualization of the lymph vessels and nodes after the injection of a contrast medium. **1969** *Brit. Med. Jrnl.* 6 Dec. 579/1 Lymphovenous shunts have been detected in patients by lymphography with Ultrafluid Lipiodol. **1973** J. F. Meaney et al. *Complications & Legal Implications Radiologic Special Procedures*

ix. 96 Fraimow et al. demonstrated the physiologic effects of lymphography on the lungs.

lymphomatosis (limfomătōᵘ·sis). *Path.* [f. *lymphomat-* (taken as stem of mod.L. LYMPHOMA) + -OSIS.] Any of various diffuse neoplastic or hyperplastic disorders originating in lymphoid tissue.

1900 DORLAND *Med. Dict.* 368/2 *Lymphomatosis*, general lymphatic engorgement. **1901** *Encycl. Medica* VII. 196 Among them [*sc.* synonyms of lymphadenoma] are adenia, lymphadenia, lymphogenic diathesis, lymphadenosis, lymphoma, lymphomatosis, pseudo-leukemia, lymphosarcoma, and the non-committal Hodgkin's disease. **1933** *Jrnl. Exper. Med.* LVIII. 254 Lymphomatosis (lymphoid leukosis) of chickens is identical with the similar disease of mammals. **1947** *Radiology* XLIX. 354/1 Terminal changes were mainly of two types: generalized atrophy (premature aging) and mediastinal lymphomatosis. **1961** *Brit. Vet. Jrnl.* CXVII. 323 Much confusion has resulted from the use of the term 'visceral lymphomatosis', which some workers understand to be a genuine lymphoid leucosis, whilst others use it to designate the visceral lesions associated with fowl paralysis. **1965** *New Scientist* 17 June 800/2 The three main leukaemias of poultry —lymphomatosis, erythroblastosis and myeloblastosis. **1970** JUBB & KENNEDY *Path. Domestic Animals* (ed. 2) I. iv. 386/1 All primary tumours of the lymphoreticular tissues are malignant... These tumours occur in all domestic animals... Very many names have been applied to them depending on the pattern of growth, the presence or absence of 'leucaemic' changes in the circulating blood, and the judgement of the observer on what was the chief cell involved. In the following discussion we shall include them all in the designation lymphomatosis.

lynchet. Now the usual form of LINCHET. Delete '*dial.*' and add: **2.** (Later *attrib.* example.)

1917 J. MASEFIELD *Old Front Line* 42 The line of the lynchet-top merges into the slope behind it.

b. *Archæol.* A cultivation terrace. Also *attrib.*

1796 *Gentl. Mag.* LXVI. 822/1 On the declivities of the elevated and chalky tracts of Wiltshire, Dorsetshire, and other counties, there very frequently occurs a beautiful assemblage of *terraces*, mostly horizontal, and rising in a continued series like the steps of Egyptian pyramids... These, which are commonly arable,..are popularly called *lynchets*... They are generally regarded in the neighbourhood as the offspring of human exertion in remote ages, to facilitate and extend the dominion of the plough. **1869** D. MACKINTOSH *Scenery Eng. & Wales* IV. ii. 89 Many terraces are still cultivated but..there is..a general desire to plough down the 'lynchets' (as they are locally called), and..formerly their number was much greater than at present. **1908** A. H. ALLCROFT *Earthwork of Eng.* ii. 40 All but the very summits of the highest Downs were early ploughed, and the lynchets must in many cases be of mediæval, if not of Saxon date. **1953** R. J. C. ATKINSON *Field Archaeol.* (ed. 2) 19 When lynchets (cultivation terraces) are photographed facing a setting sun, the sloping faces of the terraces..will reflect more light than the surrounding ground. **1954** M. BERESFORD *Lost Villages* ix. 297 On the valley sides are lynchet-like terraces which look as if they mark where the ploughs moved. **1968** J. ARNOLD *Shell Bk. Country Crafts* 13 Celtic farmers increased the areas of their arable land by excavating some of the hillsides and making terraces or lynchets.

Hence **ly·nchetted** *ppl. a.*, of land: cultivated in this way.

1928 *Antiquity* June 171 To the south and west of the lynchetted area lies what is known as 'The Druid's Circle'. **1933** *Ibid.* VII. 494 The..rarity in Cumbria of the lynchetted form of settlement. **1954** S. PIGGOTT *Neolithic Cultures* ii. 33 Certain types of settlement, characterized by huts within irregular lynchetted areas, may be Neolithic in date.

lynching, *vbl. sb.* (Earlier examples. Also *lynching-bee.*)

1836 D. CROCKETT *Exploits & Adventures Texas* vii. 103 This is what we call Lynching in Natchez. **1837** *Southern Lit. Messenger* III. 648 The outrages of the borderers, the frontier law of 'regulation' or 'lynching', which is common to new countries all over the world, are ascribed to slavery. **1900** *Congress. Rec.* 31 Jan. 1369/1 They have sometimes had 'lynching bees',..they have sometimes lynched men for murder, for arson, for rape. **1903** C. T. BRADY *Bishop* ix. 172, I don't join no more lynchin'-bees. **1943** *Christian Cent.* 1 Dec. 1/2 Evidently there is a widespread and growing fear lest the United Nations..let loose in Europe what might turn out to be little less than a gigantic lynching bee.

Lynch law. Add: Now usu. **lynch law.** (Earlier and later examples.)

The particulars supplied by Ellicott, together with other evidence, clearly establish the fact that the originator of Lynch law was Captain William Lynch (1742–1820) of Pittsylvania in Virginia. According to Ellicott, 'this self-created judicial tribunal was first organised in the state of Virginia about the year 1776'; an article in the *Southern Lit. Messenger* (1836) II. 389 gives the date definitely as 1780.

1811 A. ELLICOTT in C. V. Mathews *A. Ellicott* (1908) 220 Captain Lynch just mentioned was the author of the Lynch laws so well known and so frequently carried into effect some years ago in the southern States in violation of every principle of justice and jurisprudence. **1902** J. LONDON *Daughter of Snows* 284 It's lynch law, you know, and their minds are made up. They're bound to get me. **1963** *Times* 18 Apr. 9/1 It smacks more of lynch law than the reasoned kind of wisdom one hopes for in our elected officials. **1974** *Times* 28 Nov. 8/7 Mr Jeremy Thorpe.. urged..that the hijackers of the British Airways VC10 should not be handed over to the 'lynch law' of the Palestine Liberation Organization.

Judge Lynch (earlier example); **lynch mob,** a mob intent on lynching.

1838 H. MARTINEAU *Retrospect of Western Travel* II. 87 A distant Lynch mob was outraging..a free and innocent citizen. **1840** W. G. SIMMS *Border Beagles* 248 The murmurs began to close with the ominous inquiry after that venerable border magistrate, Judge Lynch. **1972** B. GARFIELD *Line of Succession* (1974) III. 255 If he pulls through alive you'll just have to hold off a lynch mob.

Lynch-like, *a.* [f. *Lynch*: see LYNCH LAW.] Characteristic or suggestive of Judge Lynch.

1837 R. M. BIRD *Nick of Woods* I. 221 Since Stackpole, having endured the penalty for stealing him, considered himself as having a legal, Lynch-like right to the animal, which no one could dispute.

lynch-man. [Cf. LYNCH LAW.] One of the early administrators of lynch law.

1811 A. ELLICOTT in C. V. Mathews *A. Ellicott* (1908) 221 The Lynch-men associated for the purpose of punishing crimes in a summary way without the tedious and technical forms of our courts of justice. *Ibid.* 222, I should not have asserted it as a fact had it not been related to me by Mr. Lynch..together with several other Lynch-men as they are called.

Lyngby (li·ŋbi). Also **Lingby.** [See quot. 1964.] Used *attrib.* or *ellipt.* to designate a mesolithic culture of the Baltic area or its artefacts (see quots.).

1925 V. G. CHILDE *Dawn European Civilization* i. 11 The..Lingby culture..belongs to the end of the ice-age or the very beginning of the Ancylus period. **1936** J. G. D. CLARK *Mesolithic Settlement of N. Europe* ii. 70 The tanged flake from Christiansund..is often spoken of as 'a Lyngby point'. *Ibid.* iii. 79 Our knowledge of the Lyngby culture apart from the reindeer antler objects is slight. **1951** A. COATES *Prelude to History* x. 253 To the east several cultures, the Ahrensburg..in north Germany, Lyngby in Jutland and Swiderian in north Poland, derive from the East Gravettian of the upper palaeolithic. **1964** W. L. GOODMAN *Hist. Woodworking Tools* 12 Their earliest tools, made of reindeer antler, with the stump of the brow tine sharpened to a cutting edge, are known as 'Lyngby axes', from the site at Norre-Lyngby, in Denmark, where they were first found.

lyngorm: see *LINGWORM.

lynn, U.S. var. of LINN[2] (in Dict. and Suppl.). Also *attrib.*

1787 W. SARGENT in *Mem. Amer. Acad. Arts & Sci.* IX. 158 Lynn,..a light white wood very proper for finishing the inside of dwelling houses. **1796**, etc. [see LINN[2]]. **1819** E. DANA *Geogr. Sk. Western Country* 84 Sugar maple, black and white walnut,..lynn, sycamore, cotton wood. **1819** E. EVANS *Pedestrious Tour* 299 Here are the lynn tree, gum tree, [etc.]. **1839** in *Trans. Mich. Agric. Soc.* (1855) VI. 263 The table lands are mostly timbered with the varieties of oak, beech, maple, lynn, hickory. **1849** E. CHAMBERLAIN *Indiana Gazetteer* (ed. 3) 170 The other forest trees..are ash, walnut,..lynn, [etc.]. **1886** *Harper's Mag.* June 58/2 Ropes are made of lynn bark.

lyochrome (ləi·ŏkrōᵘm). *Biochem.* [ad. G. *lyochrom* (Ellinger & Koschara 1933, in *Ber. d. Deut. Chem. Ges.* LXVI. B. 317), f. Gr. λύ-ειν to loosen + -o + Gr. χρῶμ-α colour.] = *FLAVIN 2.

1933 *Chem. Abstr.* XXVII. 2167 The name lyochrome is proposed for pigments of this class. **1938** [see *INTRAVITAL *a.*]. **1964** WAGNER & FOLKERS *Vitamins & Coenzymes* iv. 48 The term 'lyochrome' was rejected, and the designation 'flavin' was adopted for these growth-promoting water-soluble pigments. **1974** *Encycl. Brit. Macropædia* IV. 922/2 (*heading*) Flavins (lyochromes).

Lyonnais (lī,ǫnē·, ‖ li,onę), *sb.* and *a.* Also **Lyon(n)ese** (lī,ǫnī·z); (fem.) **Lionnoise, Lyonnaise, Lyonoise.** [Fr.] **A.** *sb.* A native or inhabitant of the city of Lyons (Fr. *Lyon*), or the former province of Lyonnais, in eastern France; these people collectively; also, the French dialect of this area.

1653 [see *ANGEVIN *a.* and *sb.*]. **1768** STERNE *Sentimental Journey* II. 201 The maid was a Lyonoise of twenty, and as brisk and lively a French girl as ever moved. **1777** P. THICKNESSE *Year's Journey* II. xlii. 78 Two harangues made by the Emperor Claudius in the senate, in favour of the Lyonoise. **1867** C. M. YONGE *Pupils of St. John* xv. 249 To Irenæus and his Lyonnese such falsehood was the next thing to apostasy. **1868** J. S. NORTHCOTE *Celebrated Sanctuaries of Madonna* iv. 134 The devotion of the Lyonnese..clung rather to the crypt and the image of St. Pothinus. *Ibid.* 142 Suchet was a Lyonnese by birth. *Ibid.* 146 Nor must we..fail to notice the last ornament presented to the sanctuary by the piety of the Lyonnese. **1955** J. THOMAS *No Banners* xvii. 156 Fernande continued in her archaic *Lyonnais*: 'Joe's been arrested, too.' **1960** L. DURRELL *Spirit of Place* (1969) 327 No traveller will repeat these words to a Lyonnais without provoking the passionate cry: 'Unjust!' **1966** *Observer* 30 Oct. 5/2, I saw between 15,000 and 20,000 *Lyonnais* in the square. *Ibid.*, The *Lyonnaises* peered into an English bathroom in the Ideal Home Exhibition House. **1972** *Guardian* 11 Mar. 15/1 The nineteenth century writer, Edouard Aynard, described the Lyonnese as 'a northern race astray in the south'.

B. *adj.* Of, pertaining to, or characteristic of Lyons or its inhabitants; *spec.* designating either of two styles of book-binding associated with Lyons.

1801 C. WILMOT *Let.* 13 Dec. in *Irish Peer* (1920) 17, I have issued forth betimes, under the conduct of a Lyonese Laquais de Place. **1867** C. M. YONGE *Pupils of St. John* xiv. 224 The whole..day..was spent by this Lyonnese multitude in baiting these..men. **1893** S. T. PRIDEAUX *Hist. Sk. Bookbinding* i. 46 The Lyonnese binders..used very fine stamps. **1894** H. P. HORNE *Binding of Bks.* ii. 77 The use of azured tools..became a characteristic of Lyonese bindings. **1928** E. P. GOLDSCHMIDT *Gothic & Renaissance Bookbindings* I. 317 These bindings are generally referred to as being in the 'Lyonnese' style. It is quite possible that a good many may come from Lyons, but I have never found any proof for this. **1960** E. A. LOWE *Eng. Uncial* 3 The collection of Concilia in the sixth-century Paris MS. Lat. 12097 written mostly in half-uncial has many pages in almost contemporary uncial. Experts consider the collection Lyonese. **1960** L. DURRELL *Spirit of Place* (1969) 328 The first steam-powered boat was Lyonnais in conception and design. De Jouffroy of Lyons built it. **1966** M. R. D. FOOT *SOE in France* viii. 215 He made firm friends with two Lyonnais business men. **1972** R. COBB *Reactions to French Revolution* ii. 51 The 29 Mai..would *never* be forgotten,.. another example no doubt of Lyonnais particularism.

b. *Cookery.* Designating food, esp. sliced potatoes, cooked or served with onions, or with an onion sauce. Freq. placed after the sb., and as *à la Lyonnaise.*

1846 C. E. FRANCATELLI *Mod. Cook* 8 Lyonnaise Sauce. Peel..onions... Slice them... Fry them... put them into a small stewpan, with..brown sauce. [**1846** A. SOYER *Gastronomic Regenerator* 470 (*heading*) Pommes de Terre à la Lyonnaise.] **1866** H. ST. CLAIR *Dainty Dishes* 130 Potatoes à la Lyonnaise..are very good to eat with cutlets. **1877** E. S. DALLAS *Kettner's Bk. of Table* 359 *Lyonnese Potatoes* (Pommes-de-terre à la Lyonnaise).— These are cooked potatoes combined with cooked onions. **1939** A. SIMON *Conc. Encycl. Gastron.* I. 34 *Lyonnaise*, Sauce. A white wine and onion sauce. **1960** E. DAVID *French Provincial Cooking* 228 What is to be served after these sausage dishes?.. No dish requiring potatoes after the Lyonnais one. **1974** *Times* 2 May 8/1 Lyonnaise potatoes... Fried potatoes made with blanched raw potatoes... Sauté potatoes with onion.

Lyons (ləi·ənz, lioṅ). Also **Lyon.** [F. *Lyon* Lyons, a city in France.] Used *attrib.* to designate various products associated with the city (see quots.); also *absol.*

1765 STERNE *Tr. Shandy* VII. xxxi. 162, I could.. sometimes not so much as see even a Lyons-waistcoat, but this remembrance..would present itself. **1778** F. BURNEY *Evelina* I. xvi. 100 We found her..very busy in wiping her negligee..as she said it was a new Lyon's silk. *Ibid.* III. xxi. 234 'Twill be a most excellent opportunity to shew off her best Lyons' silk. **1851** MAYHEW & BINNY *1851* xvi. 160 The jewels and the tapestry, and the Lyons silks, are not now the sole objects of attraction. **1855** DICKENS *Dorrit* (1857) I. i. 6, I can cut my bread..like Lyons sausage. **1869** 'MARK TWAIN' *Innoc. Abr.* (1870) xxx. 244 Lyons velvets rank higher in America than those of Genoa. **1872** *Young Englishwoman* Dec. 646/1 Lyons grosgrains and moire silks. **1874** W. CROOKES *Pract. Handbk. Dyeing & Calico-Printing* II. iii. 195 (*heading*) Aniline blues... *Lyons blue* (Girard and De Laire). **1895** *Brit. Warehouseman* Feb. 13/2 A triple-pleated blouse of purple Lyons velvet was worn. **1902** S. J. WEYMAN *In Kings' Byways* I. 161 The hiding-places..had been.. contrived to hold runlets of Nantz and bales of Lyons. **1908** A. BENNETT *Old Wives' Tale* III. vi. 385 She bought ..coffee, Lyons sausage, dried prunes. **1914** F. W. ATACK tr. *Wahl's Manuf. Org. Dyestuffs* III. xvi. 171 The monophenyl- and monotolyl-derivatives [of Pararosaniline] occur..in the dyestuffs known as Phenyl violet,..Lyons blue, etc. **1924** A. T. DE MOUILPIED *Quest for Colour* iv. 20 Summary of the more important discoveries in the history of synthetic organic dyestuffs... 1860 Bleu de Lyon—a spirit soluble blue derived from Magenta. [Discoverer] Girard and De Laire. **1952** 'P. WENTWORTH' *Brading Collection* xi. 66 Great ladies..had bought changeable silks there, and fine Lyons velvet. **1959** D. OGRIZEK *France Observed* 211 The contribution local artists have made towards enhancing the prestige of Lyons silk has always been recognised. **1960** E. DAVID *French Provincial Cooking* 225 Randall & Aubin..sell a coarsely-cut garlic-flavoured sausage which is not unlike the Lyon sausage. **1971** *Colour Index* (Soc. Dyers & Colourists) (ed. 3) V. 5573 Commercial name..Lyon Blue..C. I. Solvent Blue 3.

lyophile (ləi·ŏfəil), *a.* [ad. G. *lyophil* (Freundlich & Neumann 1908, in *Zeitschr. f. Chem. und Ind. d. Kolloide* III. 81/2), f. Gr. λύ-ειν to loosen; see -PHIL, -PHILE.] **1.** *Physical Chem.* = *LYOPHILIC *a.*

1915 W. W. TAYLOR *Chem. Colloids* i. 7 The term *lyophile* has been applied to those systems in which there is a marked affinity between the phases and *lyophobe* to the others. **1927** H. S. VAN KLOOSTER tr. *Kruyt's Colloids* i. 10 Emulsoids, or lyophile colloids,..form disperse systems in which the properties of the dispersing phase are considerably modified by the disperse colloid. **1941** R. J. HARTMAN *Colloid Chem.* xviii. 358 The terms lyophobe ('hatred' toward external phase) and lyophile ('love' toward external phase) were suggested by Freundlich.

2. *Biol.* and *Med.* Also **lyophil** (-fil). Of, pertaining to, or employing lyophilization; lyophilized.

1934 S. MUDD et al. in *Jrnl. Immunol.* XXVI. 341 (*heading*) The preparation, properties and applications of lyophile serum proteins and complement. **1949** *Jrnl. Bacteriol.* LVII. 575 The lyophil process has been shown to be a successful method for the preservation of yeast cultures in a viable state. **1958** *Angiology* IX. 189/2 The condenser

system of the lyophile apparatus should be cleaned and dried before re-use. **1961** *Phytopathology* LI. 259/1 (*heading*) Observations on lyophil preservation and storage of Pythium species.

lyophilic (ləiŏfi·lik), *a.* *Physical Chem.* [f. *LYOPHIL(E *a.* + -IC.] Of a dispersed colloidal phase: having an affinity for the dispersion medium; not readily precipitated out by small quantities of electrolyte. Also applied to sols containing such a phase, which generally have a lower surface tension and a higher viscosity than the dispersion medium and which give gels on evaporation or cooling.

1911 [see *LYOPHOBIC *a.*]. **1938** H. L. HIND *Brewing* I. vi. 111 The complex carbohydrates and proteins belong to the class of lyophilic colloids, which is much the more important class in brewing. **1940** GLASSTONE *Text-bk. Physical Chem.* xiv. 1210 Substances which most readily form lyophilic sols are those of high molecular weight; each particle, therefore, consists of a small number of molecules and in some instances possibly of only one large molecule. **1959** [see *LYOPHOBIC *a.*]. **1969** G. D. PARFITT *Dispersion of Powders in Liquids* iii. 81 Solutions of macromolecules and association colloids are of the lyophilic type and form spontaneously when the components are brought into contact.

lyophilization (ləiǫ:filəizēⁱ·ʃən). *Biol.* and *Med.* [f. *LYOPHIL(E *a.* + -IZATION.] = *FREEZE-DRYING *vbl. sb.*

1938 *Jrnl. Allergy* X. 3 The chief advantage of lyophilization lies in the means it offers to preserve antibodies and other labile serum components..over extended periods of time. **1949** E. P. ABRAHAM et al. in H. W. Florey et al. *Antibiotics* II. xv. 638 Solvent transfer and lyophilization still form the basis of all commercial penicillin production, processes. **1958** W. C. FRAZIER *Food Microbiol.* xvi. 214 Milk and other liquid dairy products may be dried by lyophilization. **1967** *Oceanogr. & Marine Biol.* V. 164 Lyophilization is probably the gentlest way to dry a tissue and is essential if delicate materials are to be preserved.

lyophilize (ləiǫ·filəiz), *v.* [f. *LYOPHIL(E *a.* + -IZE.] *trans.* To subject to lyophilization; to freeze-dry.

1938 *Jrnl. Allergy* X. 3 In each instance a negative Kahn reaction was obtained on the 'immune' serum before it was lyophilized. **1949** A. G. SANDERS in H. W. Florey et al. *Antibiotics* II. xvi. 684 Spores from these fungi can be lyophilized, that is, dried from the frozen state, without impairing their vitality. **1973** LANDHUIS & IDE in J. A. Capella et al. *Corneal Preservation* xii. 141 Tissue lyophilized without pretreatment in glycerin became hard, took longer to rehydrate, and healed more slowly.

Hence **lyo·philized** *ppl. a.*; also **lyo·philizate** [after *filtrate, precipitate*, etc.], a substance or material which has been lyophilized; **lyo·philizer,** an apparatus for carrying out lyophilization.

1938 *Jrnl. Allergy* X. 9 Patients to be treated with lyophilized immune serum were selected. **1963** *Problemy Gematol. i Pereliv. Krovi* VIII. vi. 62 Both homogenates and lyophilizates may be used as a source of erythroplastin in experimental research and clinical laboratory tests for hemostasis. **1967** *Oceanogr. & Marine Biol.* V. 163 In a lyophilizer the water is sublimed away during a drying period lasting several hours but the tissue is at zero degrees as long as ice is present. **1971** *Times* (Suppl.) 27 July p. iii/6 Problems concerning skin replacement in burns have been tackled..where the preparation of lyophilized (freeze dried) and deep frozen skin has been successful. **1972** *Nature* 14 Apr. 347/1 The lyophilizate was then redissolved to give a final concentration of 5 × SSC. **1972** IZAK & LEWIS *Mod. Concepts in Hematol.* 76 Lyophilization was performed in a Virtis batch lyophilizer.

lyophobe (ləi·ŏfōᵘb), *a.* *Physical Chem.* [ad. G. *lyophob* (Freundlich & Neumann 1908, in *Zeitschr. f. Chem. und Ind. d. Kolloide* III. 81/2), f. Gr. λύ-ειν to loosen: see -PHOBE.] = *LYOPHOBIC *a.*

1915 [see *LYOPHILE *a.* 1]. **1927** H. S. VAN KLOOSTER tr. *Kruyt's Colloids* i. 10 The suspensoids, or lyophobe colloids ..are systems in which most of their physical properties differ only slightly from their dispersion media. **1941** [see *LYOPHILE *a.* 1].

lyophobic (ləiŏfōᵘ·bik), *a.* *Physical Chem.* [f. *LYOPHOB(E *a.* + -IC.] Of a dispersed colloidal phase: not having an affinity for the dispersion medium; readily precipitated out by small quantities of electrolyte. Also applied to sols containing such a phase, which generally have a similar surface tension and viscosity to those of the dispersion medium, and which on evaporation or cooling give solids which cannot readily be reconverted into sols.

1911 H. FREUNDLICH in *Chem. News* 22 Sept. 140/1 Colloidal solutions stand between these two extremes. One class, distinguished from coarse suspensions only by the ultramicroscopic dimensions of their particles, are termed 'Suspension-colloids' or 'Lyophobic Sols'. These include colloidal metals, sulphides, and many hydroxides. 'Emulsion-colloids', or 'Lyophilic sols', which include albumin, gelatin, starch, &c., approach more nearly to the true solutions. **1959** K. J. MUSELS *Introd. Colloid Chem.* viii. 181 If we consider a paraffinic solvent,..the alcohol group is now lyophobic and the hydrocarbons are lyophilic. **1968** *Physics Bull.* Nov. 377/1 The stability of lyophobic colloids, like a quartz suspension or gold sol, is due to the competition between electrostatic forces of repulsion.. and the van der Waals attraction.

lyotrope (ləi·ŏtrōᵘp), *a.* *Physical Chem.* [ad. G. *lyotrop* (H. Freundlich *Kapillarchemie* (1909) 54), f. Gr. λύ-ειν to loosen + τροπή turn, turning.] = *LYOTROPIC *a.* 1.

1915 W. W. TAYLOR *Chem. Colloids* i. 8 Among the reactions in which this lyotrope influence has been recognised are the following: the catalysis of esters, the inversion of cane sugar, the setting of gelatine, and the heat-coagulation of albumin. **1965** *Biol. Abstr.* XLVI. 480/2 Hofmeister's lyotrope salts modify the molecular structure of erythrocyte stroma protein.

lyotropic (ləiŏtrǫ·pik, -trōᵘ·pik), *a.* *Physical Chem.* [f. *LYOTROP(E *a.* + -IC: see *-TROPIC.] **1.** Associated with the change of internal pressure of a solution from that of the solvent which is caused by the solute. *Lyotropic series*, a series in which ions are arranged in order of their lyotropic effects, esp. their ability to cause precipitation of a lyophilic sol.

1924 W. A. PATRICK in H. S. Taylor *Treat. Physical Chem.* II. xx. 1285 Those properties of solutions or liquids that are in causal relation to the forces of molecular attraction are termed lyotropic. We thus have two great groups of properties of solutions, the colligative, and the lyotropic. **1925** G. BARGER tr. *Freundlich's Elem. Colloid Chem.* 27 The characteristic series of cations and anions so obtained..are called the lyotropic series. **1926** H. S. HATFIELD tr. *Freundlich's Colloid & Capillary Chem.* 49 We can therefore distinguish from the group of properties which are comprehended in the van't Hoff theory a second independent group, which are due to the change of the internal pressure by the solute. These properties will be described as lyotropic; the influence of the solute upon the surface tension, the compressibility, the solubility of difficultly soluble substances, and others will be called lyotropic influences. **1937** *Chem. Rev.* XX. 170 The numbers obtained in the case of salting-out experiments with lyophilic colloids play an important part in other lyotropic phenomena. *Ibid.* 179 Since the heat of hydration of ions depends on the electric field which surrounds them, the lyotropic effects are beyond doubt caused by the different electric field strengths of the ions. **1956** *Bacteriol. Rev.* XX. 51/2 Attempts have been made to correlate the toxicity of various salts to the lyotropic..series. **1970** *Biochemistry* IX. 2802/1 We have adopted an approach to the elucidation of lyotropic mechanism in which the effects of related perturbants on a standard collagen-buffer system have been compared.

2. Of a mesophase: having its phase transitions readily effected by a change of concentration. Also applied to the mesomorphism exhibited by such a mesophase.

1933 *Trans. Faraday Soc.* XXIX. 1008 (*heading*) Lyotropic mesomorphism. **1966** *Molecular Crystals* II. 55 Lyotropic systems are also thermotropic and it is the co-operative action of the temperature and of the solvent which enables them to pass successively from the solid crystalline state to the liquid crystalline state and to the isotropic liquid or dissolved state. **1969** *Analytical Chem.* Nov. 26A/3 Liquid crystals are divided into two major groups. One of these is identified as thermotropic, indicating that the class is prepared by heating... Lyotropic liquid crystals constitute the second major group and are prepared by mixing two or more components. In two-component systems involving water the second component is generally an amphiphile. Lyotropic systems can be large in number and varied in composition. **1974** P. G. DE GENNES *Physics Liquid Crystals* i. 5 For all the systems listed in this section, the transitions are induced most easily by changing the concentration of rods rather than the temperature; for this reason they are commonly called lyotropic. *Ibid.* 6 Depending on which..conditions hold, amphiphilic compounds may be lyotropic or thermotropic.

lyotropy (ləiǫ·trŏpi). *Physical Chem.* [f. *LYOTROP(E *a.* + -Y[3].] The change of internal pressure brought about in a solution by the solvent or solvents; the mechanism by which this is brought about.

1927 H. S. VAN KLOOSTER tr. *Kruyt's Colloids* xvii. 237 Lyotropy, therefore, is probably not a hydration pure and simple, but one that orients the dipoles of water. **1937** *Chem. Rev.* XX. 169 (*heading*) Quantitative lyotropy.

lype (ləip). *Sc.* Also **lipe.** [See LIPE *sb.*[1]] Part of the roof of a mine rendered dangerous by faults in the rock.

1835 *Trans. Highland Soc.* X. plate ix (*caption*) Lipes or 'glazed backs'. Dislocations or 'hitches'. **1883** W. S. GRESLEY *Gloss. Terms Coal Mining* 163 *Lypes*,..irregularities in the *roof* indicating danger from *falls*. **1912** G. CUNNINGHAM *Verse* 70 'Boot up steps and doon steps, and veezes and lypes. **1920** A. H. FAY *Gloss. Mining & Mineral Industry* 411/1 *Lype* (Scot.), an irregularity in the mine roof... A projecting rock in a mine roof that may fall at any time. Usually in the plural, and sometimes spelled Lipe. **1952** B. HOLMAN *Behind Diamond Panes* 82 All the care in roof support did not prevent a 'fall' because of a 'greasy lipe'.

lyric, *a.* and *sb.* Add: **A.** *adj.* **1.** *lyric drama, lyric stage*: examples.

1825 W. AYRTON *Let.* 5 Apr. in J. Ebers *Seven Yrs. King's Theatre* (1828) 255 Signor Tramezzani,..one of the finest singers and actors that ever graced the lyric stage, took the character of Guglielmo. **1842** *Ainsworth's Mag.* I. 183 Still we hail with undiminished delight the lyric drama. **1877** G. B. SHAW *How to become Mus. Critic* (1960) 22 The most exciting situation in lyric drama—the duet in the fourth act of Les Huguenots. *Ibid.* 28 It requires a faculty for light comedy, which is almost unknown on the lyric stage. **1938** *Oxf. Compan. Mus.* 526/2 *Lyric Drama*, another name for opera, covering all kinds. The term is applied not so much to any particular work as to the whole class—i.e. opera as distinct from the spoken play. **1957** *Oxf. Compan. Theatre* 589/1 Beethoven's..solitary contribution to the lyric stage, 'Fidelio' (1805). **1958** A. JACOBS *New Dict. Mus.* 218 *Lyric drama*, occasional synonym for opera (especially in French, as *drame lyrique*); hence also *the lyric stage*, i.e. the operatic stage.

B. *sb.* **4.** The words of a popular song; freq. *pl.* Also *attrib.*, as *lyric-writer*.

1876 STAINER & BARRETT *Dict. Mus. Terms* 276/2 *Lyric*, poetry or blank verse intended to be set to music and sung. **1927** *Melody Maker* Aug. 759/3 On July 8 Edgar Leslie, the prolific and most successful lyric writer in America, arrived in London. **1933** *Punch* 16 Aug. 180/3 The gramophone plunged fervently into that lyric called 'I've Got a Date with an Angel'. **1934** C. LAMBERT *Music Ho!* IV. 272 The lowbrow poet—the type of writer who in the nineteenth century produced 'Champagne Charlie' and now produces revue lyrics. **1938** *Oxf. Compan. Mus.* 526/2 Another well-known poet constantly advertises himself in the British musical press as 'Lyric Author...2,000 songs...not one failure to give great pleasure'. **1946** E. O'NEILL *Iceman Cometh* (1947) II. 132 They all join in a jeering chorus, rapping with knuckles or glasses on the table at the indicated spot in the lyric. **1958** *Times* 2 Aug. 7/4 Teenagers in Minneapolis, believing that the words of some 'pop' songs can encourage juvenile crime, have.. 'opened a nation-wide "better lyrics" contest'. **1967** *Listener* 3 Aug. 130/1 Having introduced a new sound in the music, they saw that they had next to change the type of lyric. **1968** *Ibid.* 7 Nov. 610/1 According to Mick Farren, lyric-writer of the Deviants: 'Pop music is..the last free medium.' **1972** *Jazz & Blues* Sept. 12/1 The banality of the lyrics. **1973** *Listener* 19 Apr. 522/1 The bo' weevil fugues..in blues lyrics.

lyricist. Add: **b.** A writer of lyrics (*LYRIC *sb.* 4).

1909 *Pall Mall Gaz.* 12 Apr. 4/3 No doubt it could be paralleled in the works of our own music-hall 'lyricists'. **1932** *Amer. Speech* VII. 243 Scores of ways of saying 'I Love You' have been invented by the lyricists of popular songs. **1963** *Movie* Apr. 11/3 Music by Joseph Kosma. Lyricist: Jacques Prevert. English lyrics by Johnny Mercer, sung by Nat King Cole. **1971** *Daily Tel.* 11 Oct. 15 'Jesus Christ, Superstar', a 'folk opera' depicting the last seven days of Christ,..is making a fortune for its composer and lyricist.

lysate (ləi·zēⁱt). *Biol.* [f. LYS(IS + *-ate*, after *filtrate, precipitate.*] A solution or preparation containing the products of lysis of cells, esp. bacterial cells.

1922 *Brit. Med. Jrnl.* 19 Aug. 298/2 The other point is the question of the effect of dispersion of the lysate—that is, the bacterial bodies in the emulsion used. **1954** *Sci. News* XXXIII. 91 This process of bacterial destruction is called 'lysis'; the resulting liquid is a 'lysate', and may contain over a hundred thousand million bacteriophage particles per cubic centimetre. **1973** *Nature* 12 Jan. 96/1 These antibodies gave strong reactions with ultrasonic lysates of both normal and CML leucocytes.

lyse (ləiz), *v.* *Biol.* [Back-formation from *LYSIS 3: cf. *analysis/analyse, catalysis/catalyse*, etc.] **1.** *trans.* To bring about lysis of (a cell, etc.).

1927 *Brit. Jrnl. Exper. Path.* VIII. 121 It [*sc.* an active phage] lysed certain laboratory strains of coliform bacilli. **1947** *New Biol.* II. 75 Fleming made the observation that the small bacterial colonies close to the mould colony were being 'lysed' (dissolved). **1967** *New Scientist* 13 July 97/1 It will burst open or lyse the offending bacteria. **1970** *Nature* 16 May 594/1 They lysed algal cells by treatment with penicillin.

2. *intr.* To undergo lysis (sense *3).

1933 *Jrnl. Infectious Dis.* LII. 272 The tendency of placental blood to lyse was found to be overcome by this use of hypertonic salt solution. **1971** *Nature* 22 Jan. 272/2 Many bacteria lyse when growth of the cultures ceases. *Ibid.* 26 Nov. 231/2 When leaves of these two species were slowly dried.., chloroplast lamellae and mitochondrial cristae often disintegrated or lysed while tonoplasts.. usually remained intact.

So **lysed, ly·sing** *ppl. adjs.*

1922 *Brit. Med. Jrnl.* 19 Aug. 290/1 The filtrate contains the bacteriophagic principle as active as the lysed non-filtered culture. **1924** *Jrnl. Bacteriol.* IX. 401 It is established that, though non-lysing, they carry *some* lytic agent. **1929** *Jrnl. Path. & Bacteriol.* XXXII. 41 The type of resistant colony is related to these characteristics of the lysing phage. **1934** *Biol. Rev.* IX. 338 Phage particles just liberated from a lysing bacterium are more active. **1949** H. W. FLOREY et al. *Antibiotics* I. i. 36 The lysates were known as Sentocym preparations, with a prefix to indicate the type of lysed organism which they contained.

Lysenkoism (ləise·ŋko,iz'm). [f. the name *Lysenko* (see below) + -ISM.] Belief in or advocacy of the views of the Russian agronomist T. D. Lysenko (1898–1976), who opposed modern genetics and advocated neo-Lamarckian views and who for a time achieved great influence in Soviet Russia. Cf. *MICHURINISM.

1948 *Discovery* Nov. 325/1 The opponents of Lysenkoism did not remain silent. **1962** *New Scientist* 11 Jan. 101/2 It would mean that if genes can be offset by a non-hereditary factor, that very factor could in effect become hereditary... This, surely, would be Lysenkoism, or an aspect of it. **1969** I. M. LERNER tr. *Medvedev's Rise & Fall T. D. Lysenko* vi. 136 The period of absolute domination of Lysenkoism in Soviet biology and agronomy was relatively short-lived. **1974** *Nature* 11 Oct. 558/1 A convinced anti-Lysenkoist, whose career suffered considerably during the years of Lysenkoism, Astaurov became a focus for the anti-Lysenkoist reaction.

Hence **Lyse·nkoist,** one who believes in or advocates the views of Lysenko; also as *adj.*

1949 *Jrnl. Heredity* XL. 197/1 The Lysenkoist offensive is a two-pronged attack. **1950** *New Biol.* VIII. 6 The argument between Lysenkoists and geneticists is in the tradition of those great controversies which periodically ruffle, sometimes convulse, the progress of a science. **1969** *Nature* 13 Sept. 1182/1 He describes the total failure of.. attempts to breed cows with high butter-fat by hybridization with Jerseys by Lysenkoist principles. **1973** *New Yorker* 24 Sept. 119/1 In 1971..Lysenkoists were still accusing classical geneticists of failure to achieve practical results.

lysergic (ləisə̄·ɹdʒik), *a.* *Chem.* [f. *lys* (in HYDROLYSIS) + ERG(OT *sb.* + -IC.] *lysergic acid*: **a.** A crystalline tetracyclic compound, $C_{16}H_{16}N_2O_2$, containing the indole nucleus and occurring in two enantiomorphic forms, the dextrorotatory one being produced by the hydrolysis of ergot alkaloids.

1934 JACOBS & CRAIG in *Jrnl. Biol. Chem.* CIV. 547 A new substance was obtained..which possessed both acid and basic properties and which we have named lysergic acid. **1961** *New Scientist* 2 Nov. 278 The alkaloids of ergot are all derivatives of lysergic acid, an indole compound. **1964** D. F. DOWNING in M. Gordon *Psychopharmacol. Agents* I. xiii. 572 Lysergic acid derivatives with biological activity are..known to occur both in fungi and higher plants. The most active hallucinogenic substance (LSD-25) has, however, been obtained only by synthesis from (+)-lysergic acid. **1969** T. A. BAN *Psychopharmacol.* xviii. 347/1 Lysergic acid diethylamide, or, as commonly referred to, LSD_{25}, is a synthetic product prepared from lysergic acid.

b. Used *ellipt.* for *lysergic acid diethylamide*, or (less commonly) any derivative of the acid.

1952 *Jrnl. Mental Sci.* XCVIII. 314 High doses of glucose raising the blood sugar above 200 mgm. inhibit the action of lysergic acid (LSD. 25). **1954** *Brit. Jrnl. Psychol.* XLV. 274 The visions produced by such drugs as mescaline and lysergic acid. **1955** *Sci. Amer.* June 39/1 The creation of experimental psychoses with lysergic acid thus opens the way to studying treatments and the nature of mental illness. **1957** J. S. HUXLEY *Relig. without Revelation* vii. 168 Mescalin and lysergic acid induce remarkable intensifications and modifications of consciousness, including often a sense of transcendence of self. **1958** *Sunday Times* 1 June 7/5 Lysergic acid should be tried for its extraordinary property of resurrecting childhood. **1962** [see *LSD[2]]. **1964** *New Statesman* 28 Feb. 347/3 (Advt.), Author seeks objective confirmation of subjective mystical experience under Lysergic Acid (LSD 25). **1969** T. A. BAN *Psychopharmacol.* xviii. 348/1 Other lysergic acids. Besides LSD_{25} there are at least eight other lysergic acid derivatives with psychotomimetic properties. **1972** *Daily Tel.* (Colour Suppl.) 11 Aug. 12/3 The acids, like the barbiturates and amphetamines, are synthetically produced, and the best known is lysergic acid, LSD.

Hence **lysergic acid diethylamide** (dəi,e·þil-, dəi,ĕþəi·lămǝid), the diethylamide of lysergic acid (in which the group $—N(C_2H_5)_2$ replaces the hydroxyl group of the acid), an extremely powerful synthetic hallucinogen which can produce profound changes in perception (esp. vision) and mood, with psychotic symptoms resembling those of schizophrenia, and which has been given orally in psychotherapy, usu. as the water-soluble tartrate, a colourless odourless powder; also called *LSD[2], *LYSERGIDE.

1944 *Chem. Abstr.* XXXVIII. 1502 With $NHEt_2$ condensation gave *d*-lysergic acid diethylamide. **1950** *Diseases Nervous Syst.* XI. 241/1 Occasionally, it was observed that patients were able to verbalize the repressed components of their conflicts during a toxic delirium. This led us to consider various drugs that might induce a transitory delirious state. It was during this search that the Sandoz Company called to our attention and made available d-lysergic acid diethylamide (L.S.D. 25). **1955** *Sci. News Let.* 2 July 4/1 LSD, short for lysergic acid diethylamide, is a chemical that produces hallucinations and delusions in healthy persons like those in mental sickness. **1965** *New Scientist* 22 Apr. 224/3 Lysergic acid diethylamide (LSD 25) and mescaline are..often called 'psychotomimetic', because some of the symptoms which they produce create certain resemblances to schizophrenic illness. **1967** *Economist* 7 Oct. 20/1 The coming thing is lysergic acid diethylamide (LSD). **1969** [see sense a above]. **1971** *Nature* 29 Jan. 347/1 Small doses (~ 100 μg) of lysergic acid diethylamide (LSD) taken orally cause hallucinations and distortions of visual perception in man. *Ibid.* 16 July 191/1 Lysergic acid diethylamide (LSD-25) is believed to cause chromosome abnormalities.

lysergide (ləisə̄·ɹdʒəid). *Pharm.* [f. *LYSERG(IC *a.* + AM)IDE.] = **lysergic acid diethylamide.*

1965 *Jrnl. Amer. Med. Assoc.* 11 Jan. 93/1 Lysergic acid diethylamide..is the most potent of the hallucinogenic drugs. and is also referred to as LSD-25, lysergide, and delysid. **1968** M. SHEPHERD et al. *Clin. Psychopharmacol.* viii. 188 Lysergide has been administered to patients suffering from many types of mental illness. *Ibid.* 190 The dangers of lysergide therapy have been repeatedly stressed.

lysin. Delete entry in Dict. and see next and *LYSINE.

lysin (ləi·sin). *Biol.* Also †**-ine.** [ad. G. *lysine* (W. Kruse 1893, in *Beiträge zur path. Anat. und zur allgemeinen Path.* XII. 339), f. *lysis* LYSIS: see -IN[1].] Any substance (as a bacteriolysin or hæmolysin) which is able to lyse cells; *spec.* an antibody with this ability.

1900 A. C. JONES tr. *Fischer's Struct. & Functions Bacteria* xvii. 168 To formulate a theory of immunity that shall not be lost in clouds of hypothesis is at present impossible. The alexines, antitoxines, lysines, and antilysines, that the wordy research of the last few years has given us, are at present quite unknown. **1902** *Brit. Med. Jrnl.* 5 Apr. 845 Some immune serums appear to exercise an agglutination of the red cells immediately before the lysin action. *Ibid.* 12 Apr. 920 [The first-mentioned poisons] as well as..the lysin of cholera belong to the lysin group. **1922** *Brit. Jrnl. Exper. Path.* III. 259 Gengou ..has described a bacteriolytic substance which he has found in extracts of leucocytes. He found, however, that the lysin was absorbed by saturation with the microbe. **1954** A. WHITE et al. *Princ. Biochem.* xxiv. 658 If the cells are lysed, the antibodies are lysins. **1966** *McGraw-Hill Encycl. Sci. & Technol.* VII. 637/2 Lysins vary in the range of host species whose cells they will attack and in their requirements for accessory factors for lysis; the immune lysins are strictest in their requirements.

lysine (ləi·sīn). *Chem.* Also †**-in.** [ad. G. *lysin* (E. Fischer 1891, in *Arch. f. Anat. u. Physiol., physiol. Abtheil.* 269), prob. f. *lysatinin* LYSATININE: see -INE[5].] An amino-acid, $COOH·CH(NH_2)·(CH_2)_3·CH_2NH_2$, which is probably a constituent of all proteins.

1892 *Jrnl. Chem. Soc.* LXII. 1500 Not lysine, but lysine carbonate..was obtained. **1897** [see LYSATININE]. **1919** Lysin [see *HISTIDINE]. **1946** *Nature* 7 Sept. 349/2 Destruction of lysine by heat has been observed in milk, casein and oat protein. **1964** *Economist* 4 July 70/3 Lysine, a chemical that has great potentialities as a protein-forming additive for animal feedstuffs. **1965** LEE & KNOWLES *Animal Hormones* ii. 33 The structural formula of vasopressin is the same in all mammals except in the pig and the hippopotamus, where the amino acid arginine is replaced by lysine. **1973** *Daily Colonist* (Victoria, B.C.) 29 Sept. 28/1 Lysine is one of the amino acids that is an essential component of protein in human nutrition.

lysis. Add: **3.** *Biol.* [perh. derived from the suffix *-lysis* in *bacteriolysis, hæmolysis* (see *-LYSIS 2).] The disintegration or dissolution of cells or cell organelles; *esp.* the dissolution of bacterial cells brought about by bacteriophage.

1902 *Jrnl. Exper. Med.* 17 Mar. 4 That complete agglutination has no effect upon subsequent solution (lysis) of the corpuscles will be shown when treating of the latter phenomenon. **1922** *Brit. Med. Jrnl.* 19 Aug. 296/2 The Twort phenomenon and the d'Herelle phenomenon are identical. They are two different aspects of..the transmissible lysis of bacteria. **1922** *Brit. Jrnl. Exper. Path.* III. 258 The lysis takes place with dead as well as with living bacteria. **1937** *Jrnl. Immunol.* XXXII. 1 (*heading*) A natural hemolysin from the rat producing nuclear lysis of chicken erythrocytes. **1940** *Jrnl. Gen. Physiol.* XXIII. 643 (*title*) The growth of bacteriophage and lysis of the host. **1970** *Nature* 11 July 138/1 Certain antibiotics can cause physical disintegration or 'lysis' of cells when added to growing cultures of sensitive bacteria.

-lysis (lisis). A word-forming element [f. Gr. λύσις a loosening, parting] in many technical terms, primarily denoting decomposition, disintegration, dissolution.

1. In words in which the first element indicates the agent; e.g. (in *Chem.*) ELECTROLYSIS (*c* 1840), *HYDROGENOLYSIS; (in *Biol.*) *BACTERIOLYSIS (sense 1), *BIOLYSIS.

2. In words in which the first element indicates the substance or object affected; e.g. (in *Chem.* and *Biochem.*) *FRUCTOLYSIS, *GLYCOLYSIS (1892); (in *Biol.*) *BACTERIOLYSIS (sense 2), *HÆMOLYSIS (cf. *LYSIS 3). **b.** In a few terms in *Surg.* *-lysis* denotes surgical detachment of a part indicated by the first element, as **cardiolysis.*

3. In words in which the first element indicates some other characteristic; e.g. (in *Biol.*) *AUTOLYSIS, *HETEROLYSIS (sense 1); (in *Chem.*) *HETEROLYSIS (sense 2), *HOMOLYSIS. (CATALYSIS (1836) was adopted directly from Gr. κατάλυσις dissolution.)

lyso- (ləi·so), a comb. form of LYSIS (sense *3), used to form various biological and biochemical terms, as **lysole·cithin** *Biochem.*, any compound obtained from a lecithin by the hydrolytic removal of one of the two fatty acid groups.

1923 LEVENE & ROLF in *Jrnl. Biol. Chem.* LV. 743 Thus the product was a mixture of several substances, which we propose to name lysolecithin and lysocephalin. Inasmuch as lecithins differ in the nature of their saturated fatty acids, it is probable that there exist several lysolecithins. **1972** *Sci. Amer.* Jan. 88/3 Lecithin, like the bile salts, is synthesized by the liver, and after it is secreted into the small intestine an enzyme secreted by the pancreas converts lecithin to lysolecithin. This substance breaks up cells by removing the lipids from their membranes.

lysogenesis (ləisodʒe·nėsis). *Bacteriology.* [f. *LYSO- + -GENESIS.] Orig., the production of lysis or a lysin; in mod. use = *LYSOGENY.

1901 *Lancet* 19 Oct. 1031/1 Agglutination..is not concerned in lysogenesis, which is a function of the action of immune body to addiment alone. **1902** *Encycl. Brit.* XXVI. 68/1 It has been completely established that in this phenomenon of lysogenesis [*sc.* dissolution of bacterial cells] there are two substances concerned, one specially developed or developed in excess, and the other present in normal serum. **1932** *Jrnl. Path. & Bacteriol.* XXXV. 855 The use of *B. sanguinarium* as an indicator of lysogenesis arose out of previous work..on the characteristics of a lysogenic enteritidis strain. *Ibid.* 862 The phenomenon of lysogenesis. **1951** *Jrnl. Bacteriol.* LXII. 293 (*heading*) Studies on lysogenesis. I. The mode of phage liberation by lysogenic *Escherichia coli*. *Ibid.*, The stable association of a bacteriophage with a bacterial strain, known as lysogenesis, has received scarce attention. **1968** A. L. SMITH *Microbiol. & Path.* (ed. 9) xxvi. 278/1 This kind of phage is referred to as a prophage, and this condition of mutual tolerance is termed lysogeny or lysogenesis.

lysogenic (ləisodʒe·nik), *a.* *Biol.* [f. *LYSO- + *-GENIC.] Pertaining to, or capable of producing or undergoing, lysis (sense *3); *spec.* applied to a bacterium which, without being attacked by a phage, can lyse and liberate phage, an effect which is due to the ability of such a bacterium to carry, through an indefinite number of generations, a phage which is integrated with the bacterial genome and normally replicates synchronously with it but which may, under a suitable stimulus, become detached from the bacterial genome, replicate rapidly, and lyse the bacterial cell.

1899 [see *immune body* s.v. *IMMUNE *a.* (*sb.*) 3 b]. **1902** *Encycl. Brit.* XXVI. 68/1 The first of these is the lysogenic action, which consists in the production of a change in the corresponding bacterium whereby it becomes granular, swells up, and ultimately may undergo dissolution. *Ibid.*, Lysogenic action is not confined to the case of bacteria, but obtains also with other organised structures. **1911** STEDMAN *Med. Dict.* 499/1 *Lysogenic*, relating to the formation of lysins. **1929** *Austral. Jrnl. Exper. Biol. & Med. Sci.* VI. 277 (*heading*) Observations on a permanently lysogenic strain of *B. enteritidis gaertner*. **1929** J. H. DIBLE *Recent Adv. Bacteriol.* iv. 73 Of greater interest are certain strains which, whilst themselves resistant, have incorporated with them the lytic principle, which remains present through future subcultures. These are the 'lysogenic' strains of Bordet. **1930** F. M. BURNET in *Syst. Bacteriol.* (Med. Res. Council) VII. xxxix. 494 There are some bacterial strains that it is impossible to deprive of lysogenic power. **1934** *Biol. Rev.* IX. 346 The ability of certain bacterial strains, so-called lysogenic bacteria, to liberate phage during their growth. **1953** [see *INDUCTION 9 d]. **1953** *Cold Spring Harbor Symp. Quant. Biol.* XVIII. 71/1 It is possible that the phage population is heterogeneous, containing lytic, lysogenic, etc. phages. **1965** *New Scientist* 21 Oct. 167/2 Strains of bacteria carrying prophage DNA are described as 'lysogenic'. **1970** PASSMORE & ROBSON *Compan. Med. Stud.* II. xviii. 101/2 The prophage is normally inhibited from replication in the lysogenic state by the presence of virus-specified repressor substance that controls the operon for the structural gene. **1971** LECHEVALIER & PRAMER *Microbes* xli. 430/1 Such phages (e.g., coliphage λ) are called lysogenic viruses.

Hence **lysogeni·city, lyso·geny,** the state or property of a bacterium of being lysogenic; the phenomenon of lysis of a lysogenic bacterium with the subsequent release of phage.

1932 *Jrnl. Path. & Bacteriol.* XXXV. 857 Double lysogenicity, *i.e.* the production of two distinct phage types by a single strain, is relatively common. *Ibid.* 857 In certain important bacterial groups lysogenicity is a normal characteristic. **1934** *Biol. Rev.* IX. 347 It was found that another group of Salmonellas..did show the existence of a common type of lysogenicity. **1954** *Genetics* XXXIX. 429 (*heading*) Segregation of lambda lysogenicity during bacterial recombination in *Escherichia coli* K12. **1956** *Nature* 14 Jan. 92/1 The occurrence of lysogeny in *SU*298 was demonstrated. **1962** *Lancet* 19 May 1054/2 Little is known about the mechanism of induction of tumours by viruses, and..it seems clear that nothing directly akin to bacteriophage lysogeny occurs. **1970** PASSMORE & ROBSON *Compan. Med. Stud.* II. xviii. 101/2 When a bacteriophage infects a sensitive bacterium, it can either replicate and lyse the host cell..or it may be incorporated in the host genome in a state of lysogeny.

lysogenization (ləisǫ:dʒĕnəizēⁱ·ʃən). *Bacteriology.* [f. *LYSOGEN(IC *a.* + -IZATION.] The process by which a bacterium acquires a phage which becomes stably integrated into its genome; the establishment of the lysogenic state.

1953 *Cold Spring Harbor Symp. Quant. Biol.* XVIII. 71 (*heading*) Studies on lysogenization in *Escherichia coli*. **1970** *Nature* 17 Jan. 226/1 The parallels between the lysogenization of bacteria by bacteriophage and the transformation of cells in tissue culture by DNA tumour viruses are too compelling to ignore.

Hence **lyso·genize** *v. trans.*, (of a bacteriophage) to become stably integrated into the genome of (a bacterium); also *absol.*; **lyso·genized** *ppl. a.*; **lyso·genizing** *vbl. sb.*

1953 *Cold Spring Harbor Symp. Quant. Biol.* XVIII. 71/1 There do occur, however, *lambda* mutants that lysogenize poorly or not at all. *Ibid.*, Different strains of artificially lysogenized bacteria. **1957** *Virology* III. 42 (*heading*) Mutations in a temperate bacteriophage affecting its ability to lysogenize *E. coli.* **1961** *Jrnl. Molecular Biol.* III. 399 (*heading*) A mutation affecting the DNA content of bacteriophage lambda and its lysogenizing abilities. **1971** *Nature* 5 Nov. 12/1 The Paris group..set about isolating mutants of this phage which by lysogenizing CRT46 cells can cause the cells to replicate at 42° C.

lysosome (ləi·sŏsōᵘm). *Biol.* [f. *LYSO- + *-SOME[4].] A cytoplasmic cell organelle widely found in animal tissues which contains hydrolytic enzymes enclosed in a membrane.

1955 C. DE DUVE *et al.* in *Biochem. Jrnl.* LX. 615/2 For practical purposes, it is proposed to refer to these granules as lysosomes, thus calling attention to their richness in hydrolytic enzymes. **1970** *Sci. Jrnl.* Aug. 15/4 The invader is met by the lysosome, engulfed and destroyed by the enzymes.

Hence **lysoso·mal** *a.*

1957 *Symp. Soc. Exper. Biol.* X. 59 All the lysosomal enzymes..were recovered to a large extent in the parenchymal cells. **1965** *Listener* 18 Mar. 404/1 Amoebae or other protozoa take particles into food vacuoles, release lysosomal enzymes into the vacuoles and so digest the particles.

lysozyme (ləi·sŏzəim). *Biochem.* [f. *LYSO- + *EN)ZYME.] Any of various similar enzymes of relatively low molecular weight which are widely found in animal and plant tissues and secretions and which are capable of hydrolysing a particular mucopolysaccharide found in the cell walls of certain Gram-positive bacteria and hence of lysing such bacteria.

1922 A. FLEMING in *Proc. R. Soc.* B. XCIII. 306 In this communication I wish to draw attention to a substance present in the tissues and secretions of the body, which is capable of rapidly dissolving certain bacteria. As this substance has properties akin to those of ferments I have called it a 'Lysozyme', and shall refer to it by this name throughout the communication. **1934** J. H. PARSONS *Dis. Eye* (ed. 7) x. 144 The tears are not a good culture medium, and though they contain traces of lysozyme, they cannot be regarded as actively bactericidal. **1949** H. W. FLOREY et al. *Antibiotics* I. p. v, No consideration has been given to antibacterial substances of animal origin, such as lysozyme. **1962** FLORKIN & MASON *Compar. Biochem.* III. viii. 413 *Ficus* lysozyme is different from the egg white enzyme. **1965** PEACOCKE & DRYSDALE *Molecular Basis Heredity* xii. 144 Similar studies have been initiated on the genetically controlled structure of the lysozyme produced in bacteria infected by normal and mutant T4 bacteriophage.

lytic (li·tik), *a.* [ad. Gr. λυτικ-ός able to loose.] **1.** *Med.* Of, pertaining to, or causing a lysis (sense 2).

1889 *Syd. Soc. Lex.*, *Lytic*, of, or belonging to, a loosing or dissolving. **1907** *Practitioner* Apr. 500 It boots little whether that increase in heat [during fever] is due to an excessive production, as the 'genetic' protagonists proclaim, or an inadequate dissipation, as their 'lytic' opponents contend. **1962** *Lancet* 26 May 1123/2 Ice placed inside the oxygen tent and cold sponging are remarkably efficient in lowering the temperature of these infants to 86–88°F (31°C)... In our experience 'lytic' agents are not required.

2. *Bacteriology.* Pertaining to or causing lysis (sense *3); *lytic cycle*, the sequence of events that takes place from the infection of a bacterium by a virulent phage to lysis of the bacterium and the release of progeny phage.

1902 *Jrnl. Exper. Med.* 17 Mar. 281 Only when the lytic serum is very fresh will solution be effected. **1922** *Brit. Jrnl. Exper. Path.* III. 259 The lytic substance, lysozyme, is very stable. **1925** C. H. BROWNING *Bacteriol.* ix. 214 He concluded that the agent causing this solution or lytic action was a living virus. **1946** *Nature* 23 Nov. 745/1 The lytic effect of certain myxobacteria upon the true bacteria (eubacteria) has been known for some years. **1963** *Biol. Abstr.* XLI. 1198/2 Studies on control of enzyme induction and initiation of lytic cycle in a lysogenic bacterium. **1968** R. RIEGER *Gloss. Genetics & Cytogenetics* 34 The introduction by infection of the genetic material of virulent phages into a susceptible host results invariably in the death and dissolution of the cell with the release of (100–10000) new phages ('lytic cycle'). **1972** *Nature* 24 Mar. 144/3 It is much more likely that the tails kill sensitive cells either by injecting lytic enzymes or by acting at the cell surface.

-lytic (li·tik), ending of adjs. corresponding to sbs. that end in *-LYSIS (some of the earliest examples of which, ANALYTIC, CATALYTIC, correspond to Gr. originals in -λυτικός: cf. *LYTIC *a.*).

lytically (li·tikăli), *adv.* *Bacteriology.* [f. *LYTIC *a.*: see -LY[2].] (By infection) with a lytic phage.

1967 *Jrnl. Virol.* I. 917/2 The enzyme..in lytically infected cells is dependent on virus genome for its development. **1969** A. M. CAMPBELL *Episomes* i. 5 Such specialized transduction is not observed with lytically grown phage. **1972** *Jrnl. Molecular Biol.* LXX. 68 The sequences of SV40 RNA which appear only late in lytically infected cells.

lyxoflavin (liksoflēⁱ·vin). *Biochem.* Also **-flavine.** [f. *LYXO(SE + *FLAVIN 2.] A yellow crystalline flavin, $C_{17}H_{20}N_4O_6$, which is a lyxose derivative and occurs in heart tissue.

1949 *Arch. Biochem.* XXII. 64 We have succeeded in isolating L-lyxoflavine from the human heart muscle. This compound was obtained in the form of monoclinic crystals. **1953** *Proc. Soc. Exper. Biol. & Med.* LXXXII. 590/2 Relatively large amounts of lyxoflavin are required for such growth stimulation, and since available evidence indicates that the compound does not occur naturally, it cannot be considered a vitamin. **1963** J. P. LAMBOOY in Florkin & Stotz *Comprehensive Biochem.* XI. ii. 31 Lyxoflavin stimulates a small increase in the growth rate of rats, chicks, and pigs when adequate amounts of riboflavin are present in the diet.

lyxose (li·ksōᵘz,-s). *Chem.* [a. G. *lyxose* (Fischer & Bromberg 1896, in *Ber. d. Deut. Chem. Ges.* XXIX. 581), f. *xylose* XYLOSE by reversal of the first syllable.] A crystalline pentose sugar, $C_5H_{10}O_5$, which differs from xylose in the configuration of the carbon atom adjacent to the aldehyde group, is rare in nature, and is obtained synthetically by degradation of the calcium salt of galactonic acid.

1896 *Jrnl. Chem. Soc.* LXX. I. 348 The lactone can be reduced with 2½ per cent. sodium amalgam at 0° to the sugar, lyxose. **1961** *Tetrahedron* XV. 80 This is the first time that L-lyxose..has been found in a natural product [*sc.* curamycin].

M

M

M. Add: **II. 4*.** Further symbolic uses in science.

a. In *Physics M* is used to designate the series of X-ray emission lines of longer wave-length than the *L*-series obtained by exciting the atoms of any particular element (cf. *L 6* a); these arise from electron transitions to the atomic orbit of third-lowest energy, with principal quantum number 3, which is thus termed the *M-shell*, and electrons in this shell *M-electrons*. **M-capture,** the capture by an atomic nucleus of one of the *M*-electrons.

1911 C. G. Barkla in *Phil. Mag.* XXII. 408 From the similarity of the behaviour of all the elements we must then admit..the possibility of further series M, N, &c. 1923 H. L. Brose tr. *Sommerfeld's Atomic Struct. & Spectral Lines* iii. 145 Electronic transitions that end in the M-shell, furnish differences of energy that correspond to emissions of lines of the M-series. 1924 *Phil. Mag.* XLVIII. 722 Krypton..has 2 K, 8 L, 18 M, and 8 N electrons. 1924 R. W. G. Wyckoff *Struct. Crystals* ii. 72 The M-lines have been observed from the elements from dysprosium..to uranium. 1934 H. E. White *Introd. Atomic Spectra* xvi. 306 Like the *K* and *L* series lines, the *M* series lines follow nearly straight lines on a Moseley diagram. 1968 *Physical Rev.* CLXVI. 944/2 The predicted exchange-overlap correction..is even larger for *M* capture than for *L* capture. 1970 E. P. Bertin *Princ. & Pract. X-Ray Spectrometric Analysis* i. 27 An electron that has enough energy to expel a *K* electron obviously can also expel any *L* or *M* electron.

b. In *Physics m* and *M* denote magnetic quantum numbers, corresponding to the component of an angular momentum (often indicated by a subscript) in some physically distinguished direction (usu. that of a magnetic field). [Introduced by A. Landé 1921, in *Zeitschr. f. Physik* V. 233.]

m is usually used for a single particle, and *M* for an assemblage of particles.

1923 *Jrnl. Optical Soc. Amer.* VII. 415 The numbers at the left show the magnetic quantum number *m* characterizing each level. 1926 *Proc. R. Soc.* A. CXII. 80 It is necessary also..to take into consideration the effect of the orientation of orbits with respect to an imaginary magnetic field, and such orientation involves a fourth quantum number *m*, which in turn is itself composite and is equal to the vector sum of two subsidiary quantum numbers m_a and m_s. 1962 H. D. Bush *Atomic & Nucl. Physics* ii. 38 It is necessary to characterize electrons in atoms by four quantum numbers n, l, m_l, m_s. 1967 W. R. Hindmarsh *Atomic Spectra* I. vi. 72 We can also perform a simple calculation for the case where the Zeeman splitting is large compared with the hyperfine structure but still small compared with the fine structure. In this case the quantum numbers M_J and M_I (which defines the *z*-component of the nuclear spin) are well defined, but not *J* and *I*.

5. M. = member, as in M.B.E., Member(ship) of the Order of the British Empire, M.C., Members of Congress (examples), M.I.A.E., Member of the Institute of Automobile Engineers, M.I.C.E., Member of the Institution of Civil Engineers, M.I.E.E., Member of the Institution of Electrical Engineers, M.I.M.E., Member of the Institute of Mechanical Engineers, Member of the Institute of Mining Engineers, M.I.Mech. E., Member of the Institution of Mechanical Engineers, M.I.Struct.E., Member of the Institution of Structural Engineers, M.J.I., Member of the Institute of Journalists, M.L.A., Member of the Legislative Assembly, M.L.C., Member of the Legislative Council, M.P.P., (in Canada) Member of Provincial Parliament, M.R.C.S. (examples), M.V.O., Member of the Royal Victorian Order.

m- (Chem.) = Meta- 6 b; M = middling: of paper, showing slight imperfections; M. or m. = million (£300 m. = three hundred million pounds); M = morphine.

M.A., mental age; M.A.D., MAD, magnetic anomaly (or airborne) detector (or detection); M. and D., medicine and duty, marked on a serviceman's sick report when he is feigning illness; M. and V., meat and vegetable(s); M.A.P., Ministry of Aircraft Production; M.A.S.H. (*U.S.*), Mobile Army Surgical Hospital; M.A.T.S. (*U.S.*), Military Air Transport Service; M.C., Military Cross (established 1915); M.C.C., Marylebone Cricket Club, the governing body of English cricket; the official title of touring teams generally deemed to represent England; Mcf, mcf, etc., a thousand cubic feet; so MMcf(d), etc., a million cubic feet (per day); MCP, male chauvinist pig; M.C.P.A., 2-*m*ethyl-4-*c*hloro*p*henoxy*a*cetic-acid (salts and esters of which are used in sprays as herbicides); M.C.R., middle common room (in the University of Oxford); M.C.U. (*Photogr.*), medium close-up; M.D., Managing Director; m.d., mental, or mentally, deficient or defective; M.D., Musical Director; M-day, mobilization day; m.e. (*Bibliogr.*), marbled edges; M.E., medical examiner; M.E., Middle English (see English *sb.* 1 b); mF (also mf, etc.), microfarad; M.F., m.f., motherfucker; M.F., M.G., machine-finish(ed), -glazed (*machine 9 c, 10); M.F.N., most favoured nation; M.F.V., m.f.v., motor fleet vessel; motor fishing vessel; M.G., machine-gun; M.G. (*Building*), make good; M.G.B. [Russ. *Ministérstvo Gosudárstvennoĭ Bezopásnosti*], Ministry of State Security; M.G.C., machine-gun company; m.g.d., million gallons per day; M.G.M., Metro-Goldwyn-Mayer (a film company); freq. *attrib.*, esp. to denote a roaring lion used as a symbol by this company; MHD, magnetohydrodynamic(s); M.H.W., mean high water; M.I., Military Intelligence (followed by numerals which indicate departments); M.I.5, M.I.6, sections of Military Intelligence which (until 1964) dealt with matters of state security; cf. **D.I.*; M.I., Mounted Infantry; MIA, missing in action; MICR, magnetic ink character recognition; mip, mean indicated pressure; MIRV, multiple independently targeted re-entry vehicle (a type of missile); hence as *vb. trans.*, to equip with multiple independently targetable warheads; so *MIRVed* ppl. adj., *MIRVing* vbl. sb.; M.I.T., Massachusetts Institute of Technology; M.K.S., m.k.s., metre-kilogramme-second; M.L., motor launch; M.L.D., MLD, minimum lethal dose; M.L.F., multilateral force; M.L.W., mean low water; M.M., Military Medal (established 1916); mm., millimetre (examples); m.m.f., M.M.F., magnetomotive force; M.M.P.I., Minnesota Multiphasic Personality Inventory; M.O., mass-observation; M.O., medical officer; m.o., modus operandi; M.O., m.o. (*Chem.*), molecular orbital; M.O., money order; M. of I., M.O.I., Ministry of Information; M.O.H., Medical Officer of Health; M.O.L., manned orbiting (or orbital) laboratory; M.O.S. (*U.S. Mil.*), Military Occupational Specialty; MOS(T), metal-oxide-semiconductor (transistor); M.O.T., MoT, Ministry of Transport; also *ellipt.*, a Ministry of Transport test to establish the roadworthiness of a motor vehicle; also *attrib.*; m.p., M.P., melting point; M.P., military police(man); m.p.g., miles per gallon; m.p.h., miles per hour; so as *v. intr.* (nonce), to travel; M.Q. (*Photogr.*), metol-hydroquinone; M.R.A., Moral Rearmament, *Buchmanism; M.R.B.M., medium range ballistic missile; mRNA († M-RNA), messenger RNA; M.R.P. [Fr. *Mouvement Républicain Populaire*], Popular Republican Movement, the Christian Democratic Party of France under the Fourth Republic; M.S. (*Photogr.*), medium shot; M.S., minesweeper; M.S., morphine (sulphate); MS, multiple sclerosis; MSG, monosodium glutamate; MSH, melanocyte-stimulating hormone; M.S.I. [It. *Movimento Sociale Italiano*], Italian Social Movement, the Fascist Party of Italy under the Republic; M.T., machine translation; M.T., motor transport; M.T.B., motor torpedo boat; M.T.C., Mechanized Transport Corps; M.T.I., m.t.i., moving-target indication (a radar system that gives prominence to moving objects); M.V.D. [Russ. *Ministérstvo Vnútrennikh Del*], Ministry of Internal Affairs, replacing the N.K.V.D.; M.Y.O.B., mind your own business. Also *M and B, *Ms², *musa².

1889 G. M'Gowan tr. *Bernthsen's Text-bk. Org. Chem.* xvii. 329 The xylene of coal tar consists of a mixture of the three isomers, *m*-xylene being present to the extent of 70 to 85 p.c. 1929 L. A. Coles *Introd. Mod. Org. Chem.* xxvii. 327 *m*-Dinitrobenzene is always prepared by the nitration of benzene in two stages. 1968 R. O. C. Norman *Princ. Org. Synthesis* xi. 368 The t-butyl group is removed by reaction with more of the starting *m*-dialkylbenzene. 1971 *Nomencl. Org. Chem.* (I.U.P.A.C.) (ed. 3) A. 18 The position of substituents is indicated by numbers except that *o*- (*ortho*), *m*- (*meta*) and *p*- (*para*) may be used in place of 1,2-, 1,3-, and 1,4-, respectively, when only two substituents are present. 1894 *Amer. Dict. Printing & Bookmaking* 354/1 *M paper*, paper which is not up to the highest standard of the manufacturer. 1937 E. J. Labarre *Dict. Paper* 170/1 M paper is that which is not up to the first sorting, but in which the imperfections are trivial. 1948 *Words into Type* 545 M's or M paper. Paper not up to the standard quality. 1934 Webster, *M*,..mega- (million). 1955 *Times* 3 May 10/3 (*heading*) $1M. declined. *Ibid.* 9 May 8/2 $28m. for aid to Spain. *a* 1912 W. T. Rogers *Dict. Abbrev.* (1913), *M.*, morphin. 1914 Jackson & Hellyer *Vocab. Criminal Slang* 60 *M*, or *Morph*, used by morphine fiends. Sulphate of morphia. 1922 M (= morphine) [see C (*C III. 3)]. 1935 A. J. Pollock *Underworld Speaks* 74/1 *M.*, morphine (a white alkaloid derived from opium). 1953 W. Burroughs *Junkie* xii. 125 When I have an H or M shooting habit I am non-sociable. 1940 R. S. Woodworth *Psychol.* (ed. 12) iv. 111 Of the two measures, MA and IQ, which is the better index of intelligence? 1960 J. B. Carroll in Saporta & Bastian *Psycholinguistics* (1961) 340/1 There has been much interest in the relation between language maturity and measures of intellectual functioning such as MA or IQ. 1952 *Coronet* May 78/1 Scientists..thought, a compasslike magnetic device could be made to respond to even as small a body of metal as a submarine... U.S. Navy engineers tried dangling the tiny magnetic element on the end of a hundred-foot cable, through which electrical impulses traveled to a recording instrument in the plane. It worked, and MAD (magnetic air-borne detector) was soon helping our Navy send U-boats to the bottom. 1967 *Sunday Mail Mag.* (Brisbane) 22 Jan. 6/3 The long metal shape of a sub shows up distinctly as MAD passes over the hull. 1968 A. Hine *Magnetic Compasses & Magnetometers* xi. 308 The purpose of M.A.D. is to find small irregularities in the general pattern of the Earth's magnetic field, which are associated with ferro-magnetic deposits of rock and oil-bearing strata. 1917 A. G. Empey *Over Top* 299 *M. and D.*, what the doctor marks on the 'sicker' or sick report when he thinks Tommy is faking sickness. 1919 W. H. Downing *Digger Dial.* 33 *M. & D.*, medicine and duty. A familiar sick-parade slogan. 1935 G. Blake *Shipbuilders* 256 If that wound's not healed by to-morrow it's M. and D. for you. 1925 Fraser & Gibbons *Soldier & Sailor Words* 148 *M and V*, a familiar expression for the tinned meat and vegetable ration. 1944 *R.A.F. Jrnl.* Aug. 260 We are given..a hot tin of M. & V. (Meat and Vegetable: you pour it into your mess tin and eat it with a spoon). 1944 A. Jacob *Traveller's War* 273 The spearhead of the Eighth Army will eat a Christmas dinner of tinned meat and vegetable stew (the famous 'M. and V.'), biscuits and tinned fruit and tea. 1972 A. Neave *Flames of Calais* iii. 34 By nightfall my troop was cooking M & V beneath the plane trees of the market place. 1942 Partridge *Dict. Abbrev.* 60/1 M.A.P. 1946 *Happy Landings* (Air Ministry) July 1/2 Returning to England as Controller of Research and Development at M.A.P. 1953 *Economist* 14 Nov. 505/2 The Ministry of Aircraft Production. The problems that will confront the Atomic Energy Corporation have a family likeness..to those of MAP. 1950 *Army Information Digest* Dec. 51 Critical cases are flown by..helicopter direct to Mobile Army Surgical Hospitals (MASH). 1970 *Monthly Film Bull.* (Brit. Film Inst.) July 140/2 *M-A-S-H*, U.S.A., 1969 Director: Robert Altman... Hawkeye Pierce, Duke Forrest and Trapper John McIntyre arrive to join the 4077th Mobile Army Surgical Hospital in Korea, and are kept busy operating on wounded men sent back from the front lines. 1955 R. J. Schwartz *Compl. Dict. Abbrev.* 109/1 MATS. 1958 *Times* 24 July 9/7 M.A.T.S...has to have up-to-date aircraft capable of carrying freight as well as troops. 1917 *Illustr. London News* 30 June 759/1 The five classes of the Order [of the British Empire] are:..4. Officers (O.B.E.), 5. Members (M.B.E.). 1936 *Discovery* Sept. 292/1 Major A. B. Klein, M.B.E. 1955 *Times* 8 July 15/2 Mr. Stace has recently been honoured with the M.B.E. and we are all extremely gratified that such an honour should have been conferred on another stalwart of the industry. 1972 *Times* 6 Dec. 32/3 In proud and ever loving memory..of Capt. John Henry Brunt, V.C., M.C...and his beloved father, T. H. Brunt, M.B.E. 1832 *Boston Even. Transcript* 29 June 2/2 Two bundles were lately received at the Opeloasas post office, franked 'H. A. Bullard, M.C.' which contained two Marseilles vests for children in that parish. 1904 *N.Y. Even. Post* 23 Sept. 5 John Wesley Gaines, M.C., made a careful study some years ago of the evils of a President's eligibility to reëlection. *a* 1917 M.C. [see *D.S.O.* (*D III. 3)]. 1969 S. Mays *Fall out Officers* viii. 51 He's only got one eye and a lump shot aff his knee; both shot aff when he got his M.C. 1862 *Lillywhite's Cricket Scores* I. 128 Neither the M.C.C. books or 'Bentley' gave the name of the ninth batsman on the M.C.C. side. 1972 *Times* 6 Dec. 11/1 MCC's outcricket tightened up after an alarming start. 1960 V. B. Guthrie *Petroleum Products Handbk.* III. 43 *Mcf*, 1,000 cu ft. *Ibid.* xvii. 18 *MCF*, abbreviation for thousand cubic feet. 1975 *Economist* 30 Aug. 8/3 Someone has interpreted mcf as million cubic feet whereas according to American practice this means 1,000 cubic feet. 1971 *Publishers' Weekly* 1 Nov. 22 MCPs, you should know by now, are Male Chauvinist Pigs, an epithet that has grown so common it is now abbreviated. 1946 G. Blackman in *Jrnl. Ministry of Agric.* LIII. 17 These names..are unlikely to be remembered, so that the abbreviations MCPA and DCPA can perhaps serve better. *Ibid.*, White charlock, corn buttercup and shepherd's needle are best destroyed with MCPA. 1958 *New Biol.*

XXVI. 45 The selective growth regulating herbicides such as M.C.P.A...and 2,4-D..are the most commonly used herbicides on pastures and the buttercup species differ in their sensitivity to these. 1971 *Arable Farmer* Feb. 12/2 MCPA, discovered during the war and still widely used alone or in mixture for cereals. 1966 *Rep. Comm. Inquiry Univ. Oxf.* II. 483 For those who enjoy the college atmosphere and who belong to a college which has a MCR *and* which is usefully situated in relation to their living accommodation or place of work, this can largely supply their need. 1959 W. S. Sharps *Dict. Cinematogr.* 110/2 *M.C.U.*, abbreviation for Medium Close-up. 1955 R. J. Schwartz *Compl. Dict. Abbrev.* 110/2 *MD*..; Managing Director. 1963 *Times* 14 May p. i/4 (Advt.), Do give this a little thought and perhaps you might have a word with the Managing Directors of some of the subsidiary companies. .. Have you talked to the M.D.s about it? 1942 Partridge *Dict. Abbrev.* 61/1 *M.D.*,..mentally deficient. 1968 'L. Black' *Outbreak* xiv. 137, I just don't see..how the m.d. ward got infected. 1971 *Oz* May 5/1 He is classified in Orange County..as MDSO (mentally defective sexual offender). 1927 *Melody Maker* Apr. 397/1 It would be useless for any M.D. to call a rehearsal and run through music *before* seeing the 'subject' he has to accompany. 1967 *Stage* 2 Mar. 3/2 Ernest Woodhouse the MD of Creswell Colliery Band. 1937 *Reader's Digest* Aug. 97 M-day in America. 1970 N. Armstrong et al. *First on Moon* i. 18 That was M day, and a good one to remember. *a* 1912 W. T. Rogers *Dict. Abbrev.* (1913) 123/1 *m.e.* (book), marbled edges. 1952 J. Carter *ABC for Bk.-Collectors* 12 *M.e.*, marbled edges. 1961 Webster, *ME*,..medical examiner. 1968 H. Waugh *30 Manhattan East* (1969) 120 I'll give your boss the M.E.'s verdict as soon as I get it. 1874 H. Sweet in *Trans. Philol. Soc. 1873–4* 526 The word *grèèt* [in the eighteenth century] = M.E. *grèèt* (O.E. *greât*) is an example of exceptional retention of the older *èè*. 1927 *Englische Studien* 10 Nov. 74 The investigations which have been published hitherto with a view of classifying the ME. dialects by the aid of place-name material. 1972 M. L. Samuels *Ling. Evol.* v. 85 The plural *-eþ* had been replaced by *-en* in the Midlands, and early ME texts from East Anglia show this feature as their only distinction of number. 1914 E. A. Dawe *Paper* 157/2 (*index*) Machine finish (M.F.). 1965 S. C. Gilmour *Paper* (ed. 2) xix. 218 M.F. papers are much used for printing formes of type with line blocks. 1892 C. T. & W. H. Jones *Telegr. Connections* 15/2 On one end of the condenser will usually be found a stamp like this: 2·5 M.F. The microfarad (for which M.F. stands) is the one-millionth part of a farad. 1911 L. W. Bishop *Wireless Operators' Pocket-Bk.* v. 62 The highest capacity of a variable condenser should not be over ·004 mf. 1940 *Chambers's Techn. Dict.* 544/2 *mF.*, *μF.*, abbrev. for micro-farad. 1958 *Times Rev. Industry* May 32/1 Single phase 26V current in conjunction with a 1·5 mf tuning capacitor. 1969 C. M. Rodgers *Songs of Blackbird* 38 None of us can relax until the last m.f.'s Been done in. 1971 B. Malamud *Tenants* 165 The blacks have to murder you white MF's for cripplin our lives. 1973 S. Henderson *Understanding New Black Poetry* 44, I am not speaking merely of words like 'nigger' and 'the big M.F.', as Ron Welburn calls it. 1942 Partridge *Dict. Abbrev.* 62/1 *M.F.N.*, most favoured nation. 1961 *Times* 16 May 15/7 Consequently, the M.F.N. concept is not helpful. 1948 *Hansard Commons* 8 Mar. 971/1, I hope I shall be able to satisfy the hon. Gentleman with the figures for..the number of fishery protection vessels. We have now.. eight ships plus two M.F.V.s. 1949 P. F. Anson *Scots Fisherfolk* 84 The fleet has also been increased by the addition of a large number of Admiralty-built, diesel-motor vessels (M.F.V.s), originally used by the Navy, but since allocated to fishermen. 1973 A. MacVicar *Painted Doll Affair* xiii. 148 They're at sea in that m.f.v. 1974 'M. Hebden' *Pride of Dolphins* II. ix. 188 They were moving up astern of a big grey-painted vessel..an ex-naval M.F.V. 1914 E. A. Dawe *Paper* iv. 25 The paper passing round this heated cylinder is dried, and glazed on one side, hence the term M.G., or machine-glazed paper. 1937 E. J. Labarre *Dict. Paper* 170/2 Characteristic of machine glazed papers (abbr. M.G.) is that they are only glazed on one (the under) side, the other being in the (rough) condition in which it comes from the wet end of the machine. 1965 S. C. Gilmour *Paper* (ed. 2) xix. 236 Litho posters are generally printed on M.G. paper. 1915 D. O. Barnett *Let.* 8 July 207 I'm going to be M.-G. officer. 1969 S. Mays *Fall out Officers* xx. 154 Sergeant Yardley of M.G. Squadron. 1940 *Chambers's Techn. Dict.* 544/2 *M.G.* (Build.), abbrev. for make good. 1950 Webster Add., M.G.B. 1974 T. P. Whitney tr. *Solzhenitsyn's Gulag Archipelago* I. I. iv. 145 The MGB wasn't interested in the truth and had no intention of letting anyone out of its grip once he was arrested. 1917 W. Owen *Let.* 12 Feb. (1967) 433 We have a Canadian,..various M.G.C.'s, a S.W.B. 1963 F. D. Fawcett *Cycl. Initials & Abbrev.* 97/2 *MGC*, Machine Gun Corps, disbanded 1922. 1955 *Times* 7 July 1/5 (Advt.), The extent of the supply works to be supervised and maintained include the following... Ultimate maximum output of finished water 21·5 m.g.d. 1935 Ade *Let.* 27 June (1973) 186 The sad facts in regard to *The County Chairman* are that M.G.M. bought the talking rights and recently sold them to Fox. 1952 S. Kauffmann *Philanderer* (1935) iv. 68 What we want is some kind of gimmick or slogan or handle for the public to tie on to with this thing... Like 'Ask the man who owns one'. Or the M.G.M. lion. A trade-mark. 1974 W. Garner *Big enough Wreath* xvi. 250 He'd allowed his *id* an MGM roar. 1974 P. M. Hubbard *Thirsty Evil* i. 14 Jimmy was my agent... We had not reached the point yet where he had to haggle with M.G.M. over the film-rights. 1960 *Aeroplane* XCIX. 837/2 The latter included ion-drive, solar propulsion, plasmajet propulsion, photon-drive, and MHD (Magnetohydrodynamic) propulsion. 1965 *New Scientist* 3 June 652/2 By passing the exhaust gases from the MHD generator into a conventional steam-raising plant, an overall thermal efficiency of 55 per cent might be achieved. 1974 *Nature* 1 Mar. 89/3 In the post-war years Ferraro did much to encourage research in MHD. [1920 *Tide Tables* (Admiralty, Hydrographic Dept.) I. p. xxi (*heading*) M.H.W.S...M.H.W.N.] 1923 N. Davey *Stud. Tidal Power* iv. 63 Taking the M.H.W. and M.L.W. areas of the estuary at Salcombe..the half-tide outflow system will utilise a mean average of 49% of the maximum valume of water above M.L.W.S. level. 1964 V. J. Chapman *Coastal Vegetation* ii. 19 It is possible to regard the littoral as extending from mean high water (M.H.W.) to mean low water (M.L.W.); alternatively, it can be defined as reaching from extreme high water mark (E.H.W.M.) to extreme low water mark (E.L.W.M.). Most of the early ecological work is based upon one of these two definitions. 1939 D. Wheatley *Sixty Days to Live* vi. 58 'It's a matter which may affect the welfare of the whole nation. I really mean that'. 'Well, if you put it that way. Is this M.I.5, or something?' 1940 H. Innes *Trojan Horse* ii. 32 The search of a representative of M.I.5. 1963 V. Gielgud *Goggle-Box Affair* xiii. 126 He ultimately opted for Polish nationality, promising apparently to do occasional jobs for M.I.6 on the side. 1964 M.I.5 [see D.I. (*D. III. 3)]. 1965 B. Sweet-Escott *Baker St. Irreg.* iii. 85 M.I.9..had the job of helping allied prisoners of war to escape. 1901 Kipling *Five Nations* (1903) 164 (*title*) M.I. 1929 J. Buchan *Courts of Morning* II. i. 177 They've first-rate cavalry, but indifferent M.I. 1946 *Newsweek* 6 May 36/2 From D Day until May 6, 1945..more than 10 per cent of the total casualties were listed as MIA. 1970 *Sunday Mail* (Brisbane) 8 Feb. 5/1 Diana O'Grady has groped through 34 months as the wife of an MIA (missing in action). *a* 1912 W. T. Rogers *Dict. Abbrev.* (1913) 125/1 *M.I.A.E.*, Member of the Institute of Automobile Engineers. 1935 *Discovery* Sept. 276 H. Warren, M.I.E.E., M.I.A.E. 1885 *List of Subscribers, Classified* (United Telephone Co.) (ed. 6) 89 *Engineers* (Civil)... Lowe J. E., M.I.C.E., M.I.M.E. (of Bolling & Lowe). 1966 MICR [see *magnetic *a.* 1]. 1970 O. Dopping *Computers & Data Processing* iii. 62 MICR readers do not utilize the information contained in the vertical distribution of ink. They analyze only the horizontal distribution. *a* 1912 W. T. Rogers *Dict. Abbrev.* (1913) 125/2 *M.I.E.E.*, Member of the Institute of Electrical Engineers. 1937 *Discovery* Apr. p. xxviii/1 Professor John Hollingworth, M.A., D.Sc., F.C.G.I., M.I.E.E. *Ibid.*, Professor Dempster Smith, M.B.E., M.Sc.Tech., M.I.M.E. 1909 *Cent. Dict.* Suppl., M.I.Mech.E. 1961 Webster, mip. 1962 *Engineering* 23 Nov. 662/2 A series of marine engines of various sizes, all having the high mip pressure of 10·4 kg per sq. cm. 1967 *Technology Week* 20 Feb. 12/3 A program to demonstrate the feasibility of a solid propellant Post Boost Control System (PBCS) to deliver Multiple Independent Re-entry Vehicle (MIRV) payloads is expected to be funded shortly. 1968 *N.Y. Times* 3 Nov. IV. 7 MIRVing the Polaris system allows a dozen warheads to be fitted to a single Poseidon missile. 1970 *Time* 20 Apr. 22 Each MIRVed rocket is capable of carrying a number of warheads. 1974 L. Deighton *Spy Story* xii. 119 Can they retarget the MIRVs before launching? 1974 *Times* 5 Mar. 7/3 The multiple independently targeting technology for warheads (Mirvs) is the most striking Soviet advance adduced in the report. 1974 *Nature* 6 Dec. 431/1 Proposals for SALT–2 now include multiple warheads in the list of constraints,..presumably because it is considered economically desirable not to go the whole way on MIRVing. 1975 *Sci. Amer.* Jan. 48/1 The tentative agreement represents an incremental advance for arms control..in that for the first time it includes strategic bombers and limits at least the number of missiles that can be MIRVed. 1937 *Discovery* Mar. p. xx/1 W. B. McKay, M.Sc.Tech., M.I.Struct.E. 1950 J. D. MacDonald *Brass Cupcake* (1955) xii. 124 Very distinguished Boston type. Taught mathematics at M.I.T. 1954 D. Riesman *Individualism Reconsidered* v. 303 The M.I.T. students who brought us together dreamed up the whole idea, then found the means to implement it. 1968 Chomsky & Halle *Sound Pattern Eng.* p. x, The general point of view that underlies this descriptive study is one that several of us have been developing for more than fifteen years, at M.I.T. and elsewhere, at first independently, but increasingly as a joint effort. 1909 *Cent. Dict.* Suppl., MJI. 1935 *Engineering* 26 July 95/1 This new system has been designated the Giorgi-M.K.S. system, and in it the practical electrical units are essential constitutional elements in one-to-one relation, with the result that the numerical conversion factors need no longer be..memorised. 1963 Jerrard & McNeill *Dict. Sci. Units* 15 The Giorgi or M.K.S. system attracted little attention until about 1935 but after this interest in them [*sic*] increased and in 1948 the 9th International Conference on Weights and Measures adopted the M.K.S. definition as their definition of the ampere. 1968 M. S. Livingston *Particle Physics* xii. 208 In the rationalized mks system of electromagnetic units, charge becomes a fourth fundamental quantity. 1942 Partridge *Dict. Abbrev.* 63/1 *M.L.*,..motor launch. 1945 'N. Shute' *Most Secret* i. 4 The *Raumboot* is rather like our own M.L., isn't it? 1897 *Medicine Hat* (Alberta) *News* 8 Apr. 4/2 Another of our M.L.A.'s is to join his confreres in the new Western Eldorado, the marvellously rich Kootenay. 1933 *Bulletin* (Sydney) 6 Sept. 12/3 The combination of the guillotine and the division system is being developed to a stage when the main duty of an M.L.A. will be to sit on a bench and be counted like a sheep. 1970 *Globe & Mail* (Toronto) 26 Sept. 8/2 Indeed the new talk about Fraser power has been met with instant avowals of undying opposition from some of the Government's own MLAs in the Fraser Valley. 1971 *Sunday Australian* 8 Aug. 7/2 He is an old family friend and 'retainer' of Norman Smith, Labor MLA for the Mt Isa area from 1940 to 1960. 1849 *Niagara* (Ontario) *Chron.* 25 Oct. 3/2 J. Leslie, M.L.C. 1930 L. G. D. Acland *Early Canterbury Runs* ii. 29 He..was a member of the old Provincial Council and an early M.L.C. 1971 *Hindustan Times Weekly* (New Delhi) 4 Apr. 3/4 Two MLAs and one MLC of Uttar Pradesh..joined the Congress (N) today. 1901 M.L.D. [see *addiment]. 1928 L. E. H. Whitby *Med. Bacteriol.* xx. 198 One M.L.D. is that amount of toxin which, on subcutaneous inoculation, will with certainty cause the death of a 250 gm. guinea-pig within four days. 1961 N. G. Pandalai *Textbk. Bacteriol.* (ed. 2) xii. 245 While the M.L.D. of atropine for an adult person is 130 mg., of strychnine 30–40 mg., and of cobra venom 4·4 mg., the M.L.D. of the crude tetanus toxin is only 0·2 mg. or even less. 1973 White & Tinbury *Essent. Immunol. & Microbiol.* II. vii. 120 The ultimate measured effect of toxin is the MLD. 1963 *Times* 10 June 10/5 Thus the first and perhaps most important objection to the multilateral force (M.L.F.) has been largely removed. 1904 *Amer. Jrnl. Sci.* XVII. 335 The datum plane used..is mean low water—M.L.W.—at the Battery as used by the Department of Docks and Ferries. 1923, 1964 M.L.W. [see *M.H.W.* above]. 1933 M.M. [see *Minnie², minnie]. 1971 D. Niven *Moon's a Balloon* v. 63, I caught a glimpse of the medal ribbons of the D.C.M. and M.M. on his chest. 1878 *Jrnl. R. Microsc. Soc.* I. 355 A quarter of a century ago..Harting proposed..micro-millimetre... Thus we had *m.* for the metre, *mm.* for the millimetre, and *mmm.* for the micromillimetre. 1897 *Sears, Roebuck Catal.* 583/1 Pin Fire Pistol Cartridges... In sizes 7 M-M is 32 caliber, 9 M-M is 38 caliber, 12 M-M is 44 caliber. 1971 *Guardian* 16 Dec. 1/2 The police did not identify the weapon but said that the ammunition was 9 mm. 1973 G. G. Spalding in P. Hepple *Outlook for Natural Gas* iv. 55 (*heading*) Estimated well deliverability (MMCFD). 1974 *Petroleum Rev.* XXVIII. 792/2 The field came on production in September 1971 and is currently producing 560 MMcfd. 1975 *North Sea Background Notes* (Brit. Petroleum Co.) 19 Under this agreement, BP undertook to deliver at least 50 million cubic feet of gas a day for a period of 15 years. Later revisions to the contract have resulted in a present commitment of 168 mmcfd as the basic contract rate. 1893 R. M. Walmsley *Wormell's Electr. in Service of Man* II. 392 The number of lines..which a magnetomotive force (M.M.F.) can set up depends not only on its own magnitude but on the reluctance..of the path provided. 1962 *Newnes Conc. Encycl. Electr. Engin.* 467/1 The m.m.f. per metre length of path is *H*.., the value of *H* determining the magnetic flux density $B = \mu H$ at the point, where μ..is the absolute permeability of the medium. 1946, etc. MMPI [see *Minnesota]. 1948 *Psychol. Bull.* XLV. 402 This would seem to indicate that..the MMPI scales do not actually differ from each other as they are supposed to. 1970 *Jrnl. Gen. Psychol.* LXXXIII. 70 Noting the discrepancy between behavior and test data (in this instance, the MMPI), Peterson..also concluded that the manifestations of psychosis in the MMPI..were instances of inaccurate diagnosis. 1939 Madge & Harrisson *Britain, by Mass-Observation* i. 10 Through M-O you can already listen-in to the movements of popular habit and opinion. 1971 *Guardian Weekly* 10 Apr. 18/1 The MO reports now held by Sussex University have been cleverly worked over. 1916 F. M. Ford *Let.* 19 Dec. (1965) 80 The M.O. who has just sounded my poor old lungs again says I am to be sent to Nice. 1924 *Army Q.* Oct. 138 Stop a minute—give this chit to the M.O. 1944 *Living off Land* v. 97 Day's travel from the nearest M.O. 1974 S. Gulliver *Vulcan Bulletins* 23 You have to get your ideas sorted out not to spend your time in the MO's queue asking for tranquillisers. 1955 R. J. Schwartz *Compl. Dict. Abbrev.* 114/3 *MO*,..method of operation (*modus operandi*). 1956 'E. McBain' *Cop Hater* (1958) vii. 66 It was possible that the two deaths were unrelated..but not very probable. The m.o. was remarkably similar. 1974 R. Edwards *Dixon of Dock Green* 108 His m.o. was to pull two or three jobs in a line and then fade from the scene. 1937 *Trans. Faraday Soc.* XXXIII. 1481 The *m.o.* solution depends essentially upon the solution of a one-electron problem. 1947 *Q. Rev.* I. 151 The energy of a M.O. is lowest..when the component atomic orbitals overlap one another as much as possible. 1968 R. O. C. Norman *Princ. Org. Synthesis* ii. 35 In the lower-energy MO, termed the σ1s bonding orbital.., there is an accumulation of charge in the region between the nuclei. 1909 J. Joyce *Let.* 4 Sept. (1966) II. 246, I received your M.O. for £3.5.0. 1940 H. Nicolson *Diary* 3 Aug. (1967) 104, I am feeling very depressed by the attacks upon the Ministry of Information... Since the M. of I. should be an offensive instrument, its value to our war-effort will be diminished by this constant sniping from the rear. 1942 E. Waugh *Put out more Flags* 169, I will say for the Ministry of Information they were uncommon civil. Not at all like they are here. At the M. of I. they were never too busy to see one. 1909 *Cent. Dict.* Suppl., MOH. 1911 G. B. Shaw *Doctor's Dilemma* Pref. p. lxxii, When one of the first-rate posts becomes vacant, all the leading M.O.H.s compete for it. 1961 *Lancet* 19 Aug. 440/1 She worked as assistant M.O.H. and inspector of midwives in Manchester. 1939 *War Illustr.* 18 Nov. p. iii/2 M.O.I., Ministry of Information. 1957 J. Braine *Room at Top* x. 100 A MOI poster. 1963 *Aviation Week & Space Technol.* 16 Dec. 30 As described by Defense Secretary McNamara, the system, called MOL for manned orbiting laboratory, will consist largely of hardware already under development. 1965 *Sci. News Let.* 2 Jan. 6/2 The most immediate USAF manned space project is the Manned Orbital Laboratory (MOL), scheduled for some time between 1967 and 1969. In the MOL two men will spend 30 days in orbit around the earth, studying both outer space and each other's reactions to it. 1969 *Times* 13 June 7/6 The cancelled Manned Orbiting Laboratory (M.O.L.) of the U.S. Air Force. 1955 R. J. Schwartz *Compl. Dict. Abbrev.* 115/2 *MOS*,..Military Occupational Specialty. 1969 I. Kemp *Brit. G.I. in Vietnam* iii. 43 'What's your M.O.S.?' he barked. 'Operations and Intelligence Assistant, First Sergeant.' 1964 R. D. Lohman in *Semiconductor Products* May 31 The metal-oxide-semiconductor (MOS) transistor is a new semiconductor device which combines many advantages of vacuum tubes and bipolar transistors. *Ibid.*, A typical n-channel MOS. 1967 Millman & Halkias *Electronic Devices & Circuits* xiv. 384 There are two types of field-effect transistors, the junction field-effect transistor (abbreviated JFET, or simply FET) and the insulated-gate field-effect transistor (IGFET), more commonly called the metal-oxide-semiconductor (MOS) transistor (MOST or MOSFET). 1971 *New Scientist* 8 July 77/2 With the very high-precision masking now used in MOS techniques the space between metallised paths can be 0·0003 in or less. 1973 *Sci. Amer.* Aug. 54/2 Today virtually all desk calculators and all pocket calculators are designed around MOS circuits. 1955 R. J. Schwartz *Compl. Dict. Abbrev.* 115/2 *MOT*,.. Ministry of Transport. 1968 *Listener* 13 June 787/3 The annual MoT test could incorporate a check on noise, and any vehicle which had become noisier..than its original design limits would fail the test. 1971 *Exchange & Mart* 15 July 67 (Advt.), Spot cash!! For any make of Car or Van in any condition e.g. Failed M.O.T. and Damaged Vehicles. 1972 A. Aird *Automotive Nightmare* ii. 41 The inclusion of other items, and in particular corrosion, in the MOT test would also significantly increase the repair demand. 1973 J. Morris *Age of Arthur* 178 The roads are shown as classified by the Ministry of Transport (MoT). 1885 *Phil. Mag.* XX. 513, 2 [exceptions] are due to the m.p. of carbon tetrabromide, CBr_4, being too high. 1947

Sci. News IV. 153 A polisher using a powder of oxamide (M.P. 417° C.)..did not produce any effect on speculum metal (M.P. 745° C.) or copper (M.P. 1,083° C.). 1917 A. G. Empey *Over Top* 300 M.P., Military Police. Soldiers with whom it is unsafe to argue. 1967 *Coast to Coast 1965–66* 22 A bloody M.P. I suppose, that's what you were. 1972 D. E. Westlake *Cops & Robbers* (1973) vi. 91 What I really wanted was to drive a tank, but I wound up an (M.P. 1931 *Daily Express* 28 Apr. 2/1 Petrol consumption ..averages 20–26 m.p.g. 1955 *Times* 30 Aug. 11/5 The petrol consumption is claimed to be at the rate of 100 m.p.g. on long journeys and 90 m.p.g. in town. 1973 *Country Life* 20 Sept. 802/3 The engine..is very thirsty for its size, giving me 14 mpg overall and nearer 12 mpg in town driving. 1909 *Cent. Dict.* Suppl., m.p.h. 1935 *Discovery* Oct. 293/2 Speed was increased by only two m.p.h. 1942 *R.A.F. Jrnl.* 16 May 9 The qualities of this type of aircraft are judged..by its m.p.h., manoeuvrability and ascending speed. 1960 Wodehouse *Jeeves in Offing* xix. 185, I deposited Upjohn at the 'Bull and Bush' and started m-p-h-ing homeward. 1975 *Guardian* 20 Jan. 7/3 The car with..a top speed of 107 mph, is aimed initially at the American market. 1826 *Colonial Advocate* (Toronto) 9 Feb. 2/4 John J. Lafferty, Esq. M.P.P. was called to the chair. 1970 *Toronto Daily Star* 24 Sept. 35/1 Lewis, MPP for Scarborough West. *a* 1912 W. T. Rogers *Dict. Abbrev.* (1913) 129/1 *M.Q.*, Metol-Quinol (Developer). 1940 *Chambers's Techn. Dict.* 561/1 *M.Q.*, an abbrev. for metol-quinol or metol-hydroquinone developers. 1939 *Nation* (N.Y.) 5 Aug. 135/1 M.R.A. and Hollywood were made for each other. 1949 *New Statesman* 15 Oct. 422/1 The success of M.R.A...is so brilliant that one cannot understand why the world is still in such a mess. 1969 *Listener* 3 July 19/3 Less than 1 per cent of our members are associated with MRA. 1960 *Acronyms Dict.* (Gale Research Co.) 126 MRBM... Medium Range Ballistic Missile. 1961 *Ann. Reg. 1960* 167 Opposition to giving M.R.B.M.s to NATO was partly emotional. 1848 Thackeray *Pendennis* (1849) I. ii. 10 A professional friend, M.R.C.S. 1936 *Discovery* June 182 Winifred de Kok, M.R.C.S., L.R.C.P. 1961 M-RNA [see *messenger RNA* (*messenger 7)]. 1965 *Jrnl. Molecular Biol.* XI. 187 A very small proportion of rapidly labeled RNA can be identified as M-RNA in the cell cytoplasm. 1965 *Ibid.* XIV. 257 [They] have presented evidence for the attachment of ribosomal particles to mRNA while the RNA is still attached to its DNA template. 1970 Ambrose & Easty *Cell Biol.* iv. 129 When the code has been transcribed from DNA on to mRNA, the latter leaves the nucleus, passing through the nuclear membrane into the cytoplasm. 1946 *Ann. Reg. 1945* 172 The Socialist Party (S.F.I.O.) and Christian Democrats (M.R.P.) were in favour of *Yes*. 1958 *Spectator* 30 May 676/3 Feelings within the MRP were stiffening as in the Socialist Party. 1953 K. Reisz *Technique Film Editing* iv. 87 *M.S.* Father, daughter and doctor. 1923 *Man. Seamanship* (Admiralty) II. 177 The executive officer of a minesweeping vessel should first ascertain the allowed establishment of M/S stores. 1953 W. Burroughs *Junkie* (1972) v. 52, I was getting sick and wondered if I would get home to the M.S. I had stashed in my apartment. 1969 R. R. Lingeman *Drugs from A to Z* 176 When used by addicts, morphine is most commonly in the form of a salt, e.g., morphine sulfate (the origin of M.S., a slang term for morphine), which is soluble in water and hence injectable. 1955 *Sci. News Let.* 14 May 311/3 The search for twins with MS, or multiple sclerosis, has yielded 33 identical sets so far, but the National Multiple Sclerosis Society would like to locate about 350 more, fraternal as well as identical. 1959 *Observer* 29 Mar. 8/4 Best results are obtained by dissolving M.S.G. in the cooking liquor and adding towards the end of cooking. 1969 *New Scientist* 6 Mar. 505/2 Every packet soup, fish-finger and chicken croquette contains a dose of MSG to..'wake up all the flavour *nature* put in your food'. 1953 MSH [see **melanocyte-stimulating hormone*]. 1958 MSH [see *melatonin]. 1965 Lee & Knowles *Animal Hormones* x. 128 As in amphibians, the elasmobranch skin colour is controlled by the level of MSH. 1955 M. Grindrod *Rebuilding of Italy* viii. 79 The MSI had..by 1952..built up some sort of status for itself. 1959 *Engineering* 6 Feb. 184/2 MT is discussed in two general chapters, on dictionary searching, the 'stem-ending' method of analysis, identification of idioms and a possible method for resolving multi-meanings. 1966 Y. Bar-Hillel in *Automatic Transl. of Lang.* (NATO Summer School, Venice 1962) 20 On the other hand, the number of research groups which have taken up MT as their major field of activity is still on the increase. 1917 'Contact' *Airman's Outings* 167 The Squadron Commander meets us... 'Seen anything?' he asks. 'Fourteen trains and some M.T.,' I reply. 1946 *R.A.F. Jrnl.* May 174 There are W.A.A.F. clubs,..W.A.A.F. on M.T. 1947 L. Hastings *Dragons are Extra* i. 25 He collected second-hand cars... Tinkering them up, and practising the same sort of cannibalism that long after became an M.T. necessity in modern war. 1938 *Jane's Fighting Ships* 140 A number of the M.T.B's of the Thornycroft type are still in existence. 1944 *R.A.F. Jrnl.* Aug. 265 Cargo vessels, with M.T.B.'s weaving among them, leaving long white wakes. 1955 'N. Shute' *Requiem for a Wren* 128 Left Gosport in an M.T.B. 1942 Partridge *Dict. Abbrev.* 65/2 MTC. 1944 M.T.C. [see *Croix de Guerre]. 1956 *Electronic Engin.* XXVIII. 15 A further important feature of this equipment is the m.t.i. (moving target indicator) system, which is more effective in removing unwanted clutter and permanent echoes, and more stable in operation than any other previous type. 1966 *McGraw-Hill Encycl. Sci. & Technol.* VIII. 620/1 MTI is almost a necessity when moving targets are being sought over a region from which the ground clutter echoes are very strong. 1949 Koestler *Promise & Fulfilment* 262 At the bottom of their hearts they know quite well what the M.V.D. has in store for them. 1959 *Listener* 4 June 996/3 One falls into the hands of the M.V.D. 1902 *Encycl. Brit.* XXXI. 339/2 The Royal Victorian Order..consists of.. knights commanders (K.C.V.O.), commanders (C.V.O.), and members of the fourth and fifth classes (M.V.O.), the distinction between these last divisions lying in the badge and in the precedence enjoyed by the members. 1915 *Dialect Notes* IV. 246 *N.o.y.b.*, none of your business. Also *m.y.o.b.*, mind your own business. 1951 P. Branch *Lion in Cellar* xvi. 184 'Who are you?'..'M.Y.O.B.' 1972 J. Wilson *Hide & Seek* i. 8 'I had to go and collect something,' Alice mumbled. 'What?' 'M.Y.O.B.' 'Don't be cheeky.' 1974 L. Meynell *Fairly Innocent Little Man* xi. 140 'M.Y.O.B.,' Hooky said, 'if you know what that means.' 'At school we say it means Mess Your Own Breeches.'

b. Abbrev. for Master: M.A.A., Master-at-Arms; M.B.A., Master of Business Administration; M.C. (earlier and later examples; also as *v. trans.* and *intr.*); M.Ed., Master of Education; M.Litt. (*magister litterarum*), Master of Letters; M.R., Master of the Rolls; M.S. (*U.S.*), M.Sc., Master of Science.

1916 'Taffrail' *Pincher Martin* vii. 117 'Petty Officer William Weatherley,' the M.A.A. went on, 'requests hextension o' leaf till two p.m. on Monday.' 1934 Webster *M.B.A.*, Master in, or of, Business Administration. 1965 *New Statesman* 26 Nov. 828/3 Even those few firms that give lip service to the idea of business schools are unwilling to regard a good MBA degree as representing more than one or two years of 'relevant experience'. 1790 E. Sheridan *Jrnl.* 5 Jan. (1960) 192 It was Tyson's Benefit, and as he is my acquaintance, independent of being M:C:, it was but decent that at least one of us should appear there. 1812 *Dramatic Censor 1811* 470 They are no more..co-equal with what they were, than the M.C. of a watering-place is on a par with the late Earl of Chesterfield in mind and manners. 1937 A. Upfield *Mr. Jelly's Business* viii. 74 Almost seventy people waiting for the M.C. to announce the first dance. 1938 *New Yorker* 22 Oct. 23/3, I m.c'd and they had a couple of kids from the local dancing school doing tap. 1950 Webster Add., *m.c.*, *MC*, or *M.C.*..v.i. & t. 1954 E. Warner *Trial by Sasswood* (1955) x. 185 He was to m.c. an assemblage in the market compound. 1968 *Listener* 5 Sept. 307/1 With his mutton-chop whiskers John Peel appropriately looks like an MC at an Edwardian music hall. 1934 Webster, *M.Ed.*, Master of Education. 1937 *Discovery* June p. xlviii/1 G. P. McHugh M.Sc., Ph.D. (Lond), M.Ed. (Dunelm). 1909 *Cent. Dict.* Suppl., M.Lit. 1955 R. J. Schwartz *Compl. Dict. Abbrev.* 113/3 *M Litt*, Master of Letters. 1895 *Funk's Stand. Dict.*, M.R. 1964 *Mod. Law Rev.* XXVIII. III. 274 Lord Denning M.R. said that Silverthorne was liable. 1909 *Cent. Dict.* Suppl., M.S. 1955 *Sci. Amer.* Sept. 36/3 He was born in Miami, Fla., and has a B.S. and an M.S. in physics from the University of Florida. 1909 *Cent. Dict.* Suppl., M.Sc. 1936 *Discovery* May 156 Julius Grant, Ph.D., M.Sc., F.I.C.

d. m = Milli-, as in mA, milliampere(s); mg, mgm, milligram(s); ml, millilitre(s); mm (see sense 5 a in Dict.); mrad, millirad(s).

1896 T. E. Herbert *Electr. in Applic. to Telegr.* iii. 27 1/1000 ampere or 1 m.a. 1927 *Wireless World* 16 Nov. 670/1 The..three valves..draw a high-tension current totalling 2 or 3 mA. 1970 *Which?* June 183/2 Even at maximum volume most transistor radios will not use more than 100 mA. 1894 J. Walker tr. *Ostwald's Man. Physico-Chem. Measurements* iii. 39 The correction is 0·00014, i.e. 0·14 mg. per gram. 1951 *Good Housek. Home Encycl.* 339/2 Its vitamin C content is low—1 mg. per oz. 1968 *Listener* 28 Nov. 703/1 The good effects of last year's Road Safety Act, making it an offence to drive with a level of alcohol in the blood of more than 80 milligrammes per 100 millimetres [*sic*], may be wearing off. This may well be because 80mg/100ml is such a high level. 1909 *Cent. Dict.* Suppl., *mgm.*, an abbreviation of *milligram*. 1939 *Nature* 11 Mar. 442/1 It is..possible to carry out exact estimations of carbon..using 2–3 mgm. of material. 1892 G. Collar *Notes on Metric System* 7, 10 millilitres (ml.) make 1 centilitre (cl.). 1968 ml [see *mg* above]. 1961 *New Scientist* 11 May 297/1 The average exposure in the course of radiological procedures was only of the order of 20 mrad annually, which is far less than the 100 mrad from natural background sources to which we are subjected. 1970 mrad [see *dose *sb.* 1 b].

e. M = Mega- b, as in MeV (also Mev, etc.), million electron volt(s); MHz, megahertz; MW, megawatt(s).

1934 *Physical Rev.* XLVI. 1109/2 The spectrum.. indicates gamma-ray lines of roughly 2, 4, 5·5 and 7 m.e.v. *Ibid.*, (*heading*) Energy in MEV. 1955 *Bull. Atomic Sci.* May 171/3 U-238 does not undergo fission with neutrons below 1·1 Mev. 1964 Heisenberg in *Cambr. Rev.* 24 Oct. 48/1 A mass difference up to 400 MeV. 1951 *Physica* XVII. 213 Measurements have been carried out on paramagnetic resonance absorption in iron ammonium alum at a frequency of about 9200 MHz. 1970 *E. Afr. Standard* (Nairobi) 2 Jan. 11/7 (Advt.), B.B.C. listeners can also hear programmes in English for East Africa between 7.30 and 8.45 p.m. on 15·42 MHz. 1956 *Proc. CERN Symposium* I. 22/2 Production of the flutter fields by pole-face windings..would require ~ a few MW of power per pole face. 1962 *Newnes Conc. Encycl. Electr. Engin.* 324/1 Fuel-fired steam power stations..are now built for outputs up to 1,000 MW or more.

f. Designating a motorway. Also *fig.* and *Comb.*

1959 *Times* 10 Nov. 6/5 Two people were injured in an accident on the London–Birmingham motorway, the M.1, yesterday. 1963 *Listener* 10 Jan. 100/2 Some of the by-ways of sound radio—by-ways to me but no doubt a positive M1 to thousands of others—can be rewarding. 1964 *Daily Tel.* 3 Mar. 21/1 (*heading*) M-drivers 'can stop if drowsy'. 1966 'A. York' *Eliminator* iv. 72, I would have thought you'd have wanted to try her on the M. 1973 *Guardian* 26 May 8/5 Bartholomew's new Motorway Atlas..is a useful tool for those who go up and down the Ms all the time.

'm. = Madam *sb.*

1864 J. S. Le Fanu *Uncle Silas* I. iv. 37 He bowed gravely, with a: 'Yes, 'm—shall, 'm.' 1933 E. A. Robertson *Ordinary Families* vii. 140 Her submissive 'Yes 'm' and 'No, miss'. 1945 in B. A. Botkin *Treas. S. Folklore* (1949) II. iv. 352 Mr. Linktum come down. Yes'm, Mr. Abe Linktum and his partner, Horace Greeley, comed down.

m'. = My *poss. adj.*

1712 [see My *poss. adj.* 1 β]. 1837 J. F. Cooper *England* II. xii. 51, I take it, the polite way of pronouncing this word is by a sort of elision—as m'horse, m'dog, m'gun. *Ibid.*, I think more noble peers, however, said '*me* lurds', than 'm' lurds'. 1852 Dickens *Bleak Ho.* (1853) i. 4 'Mr. Tangle,' says the Lord High Chancellor... 'Mlud,' says Mr. Tangle. 1907 R. Brooke *Let.* 12 Aug. (1968) 99 M'uncle is changing his house or face or wife or something. 1908 [see *house *sb.*¹ 4 c]. 1914 [see *æsthete]. 1942 G. Greene *Brit. Dramatists* 33 Lady Teazle 'm'ludding' and flirting a fan. 1945 E. Waugh *Brideshead Revisited* I. ii. 47 A series of harrowing interviews with m'tutor. 1967 *Guardian* 31 July 5/1, I remember him playing the part of a prosecuting counsel..full of patronising 'M'luds'. 1968 A. Diment *Gt. Spy Race* vii. 103 Why, chappie in m'tutors, doing four years in the Scrubs. 1970 O. Norton *Dead on Prediction* v. 84 Not evidence, m'dear. Is it? 1972 J. Porter *Meddler & her Murder* iii. 39 Allow me to carry the tray for you, Eve, m'dear.

ma, *sb.* Add: **a.** Also a familiar shortening of (or substitute for) Mrs. Also applied *colloq.* to a middle-aged or elderly woman, esp. one in authority.

1932 J. Cary *Aissa Saved* xxxvii. 196 You come now, ma; she very sick. 1943 K. Tennant *Ride on Stranger* iii. 27 'What-ho, Ma,' came a howl from the kitchen. 'Coming.' Beryl appeared wiping her hands. 1951 J. Common *Kiddar's Luck* xi. 161 My father called them the Ma gang, because they had a habit of always alluding to each other as Ma This and Ma That. *Ibid.*, It was becoming fairly regular to see Ma McGrewin, or Ma Smailes, or Ma Forbes sitting boldly by the fireside. 1966 Wodehouse *Plum Pie* v. 124 'Did Ma Purkiss make a speech?' 'Yes, Mrs. Purkiss spoke.' 1974 M. Babson *Stalking Lamb* xvi. 120 'Here, Ma,' Aaron said, 'I've brought company.'

b. *Austral.* (With capital initial.) Popular name for New South Wales. So *Ma State*.

1934 *Bulletin* (Sydney) 24 Jan. 25/1 The Cabbage Gardeners will have to be licked outright if Ma is to have a hope. 1945 Baker *Austral. Lang.* 187 *New South Wales*, New South, the Ma State or Ma. 1949 *Geogr. Mag.* Feb. 373 *New South* and *The Ma State*, New South Wales (oldest state in the Commonwealth).

ma. (mēi), *a.* Colloq. abbrev., in some schools, of *major* = the elder of two namesakes.

1932 Wodehouse *Louder & Funnier* 12 Faber *ma* was so annoyed by his snorts and chuckles that he hit him over the head with a *croisson* or small French roll. 1963 *Times* 5 June 14/3 There was gentler satire from McGavin ma. with John Betjeman's 'Village Inn'.

maa (mā), *v.* [Echoic, in imitation of the sound made by a sheep or goat. Cf. Mae *v.*] *intr.* To bleat. Hence as *sb.* and **maa·ing** *vbl. sb.*

1827 Darley *Sylvia* II. i. 34 It will make me *ma-a* like a he-goat on a rock-top when he misses the beard of his charmer. 1886 J. Stewart *Twa Elders* 147 The boys would maa and bleat. 1922 Joyce *Ulysses* 162 All are washed in the blood of the lamb, bawling maaaaaa. 1928 *Blackw. Mag.* Mar. 324/1 Poor old goat!..His caperings were fantastic, his *maaings* continuous. 1940 N. Marsh *Surfeit of Lampreys* (1941) xvii. 268 'M-a-a-ah,' said Rattisbon with a formidable and sheep-like cry.

maalesh, var. *maleesh *int.*

maar (māɹ). *Geol.* Pl. ‖ maare, maars. [G. dial., 'crater-lake'.] **a.** (Usu. with capital initial.) One of the craters or crater-lakes of the Eifel district in Germany. **b.** Any volcanic crater which does not lie in a cone and was formed by a single explosive event (and is usu. occupied by a lake).

1826 *Edin. Jrnl. Sci.* V. 152 The craters of this country [*sc.* Prussia] have nearly, without exception, become bodies of water, or *Maare*, as they are called by the natives. *Ibid.* 154 To the S.E. of Steffler, lies a small *maar*, or crater-lake. *Ibid.* 156 The water in the three lakes appears to stand at the same level... One only, the Schalkenonchrener maar, has any visible outlet. 1882 A. Geikie *Text-bk. Geol.* 560 Occasionally, as in some of the Maare of the Eifel, these non-volcanic fragments constitute most of the débris. 1895 *Bull. Philos. Soc. Washington* XII. 251 The maars are of still rarer occurrence, and represent the antithetic phase of volcanism. 1933 R. A. Daly *Igneous Rocks & Depths of Earth* viii. 161 Maars are relatively flat-floored craters of explosion at vents that are either coneless or else provided with inconspicuous cones. *Ibid.*, Reck has recently published..fine illustrations of maars in Abyssinia. 1968 R. W. Fairbridge *Encycl. Geomorphol.* 681/1 Maare sometimes resemble calderas. 1971 *Nature* 30 July 330/2 The lake occupies the crater of a maar—a volcanic eruption crater lying below the general level of the surrounding country and surrounded by a low tuff ring.

maas (mās). *S. Afr.* Also **amaaz, amas, amasi.** [f. Zulu (*pref.*) *ama-* + *si* milk.] Thickened sour milk.

1833 S. Kay *Trav. Caffraria* I. v. 121 Their general diet extremely simple. This ordinarily consists of milk, which ..they invariably use in a sour curdled state. It is called *amaas*, and rendered thus thick and acidulous by being kept in leathern sacks or bottles. 1857 *Cape Monthly Mag.* I. 289 The men are not allowed to drink any amasi or thick sour milk from a kraal of which they may think of courting a girl. 1882 W. R. Ludlow *Zululand & Cetewayo* vii. 73 Maas, which is the chief food of the Zulus, where there are large herds of cattle, is most delicious and nourshing food. 1946 J. Hertslet *Bantu Folk Tales* v. 61

A wether had been killed and a large piece of its meat was put with the maas and beer. 1952 S. Cloete *Curve & Tusk* (1953) iii. 33 He ate..maas, the sour thickened milk of the indigenous people. 1953 R. Campbell *Mamba's Precipice* ii. 24 They solidified the milk in calabashes and called it *Maas*. 1967 O. Walker *Hippo Poacher* 18 He provided them with *amasi* or sour milk.

maasbanker (mā·sbaːŋkər, mɒsbɒ·ŋkəɹ). Also **maasbancker, maasbank, marsbanker, masbanker, massbanker.** [Afrikaans, f. Du. *marsbanker* Mossbunker.] The South African name for the scad or horse mackerel, *Trachurus trachurus*.

1727 J. G. Scheuchzer tr. *Kæmpfer's Hist. Japan* I. I. xi. 136 *Adsi* is the *Maasbancker* of the Dutch. 1843 J. C. Chase *Cape Good Hope* II. 169 Maasbank..June..Like Mackarel, but stronger, not always wholesome. 1853 L. Pappe *Synopsis Edible Fishes Cape Good Hope* 25 *Caranx Trachurus* Lacep. (Maasbanker; Bastard Mackerel.).. Caught in winter at both ends of the Colony. Its flesh is well flavoured and wholesome. 1902 *Trans. S. Afr. Philos. Soc.* XI. 215 The Maasbanker..is identical with the Maasbanker (*Caranx trachurus*) of Holland, and indeed is to be found almost everywhere within the temperate and tropical zones of both hemispheres. 1930 C. L. Biden *Sea-Angling Fishes of Cape* vii. 146 The English horse-mackerel is the masbanker of the Cape (*Trachurus trachurus*). 1947 *Cape Times* 2 May 9 Maasbankers continue to be plentiful. 1953 *Ibid.* 20 May 4/4 The Government has limited the number and capacity of plants which may be used for the manufacture of fish meal and fish oil from pilchards and massbankers. 1959 G. Jenkins *Twist of Sand* ii. 43 We'd had a fair haul of pilchard, stockfish and maasbanker. 1970 *Cape Times* (S.A. Fishing Rev.) 28 Oct. 2/2 The purse seine boats..hunt the shoals of pilchard, mackerel, maasbanker and anchovy along the West Coast. 1972 L. G. Green *When Journey's Over* (1973) xii. 148 Nelson saw snoek and *marsbankers* in boxes and barrels or hanging up to dry.

Maastrichtian, var. *Maestrichtian *a.*

‖ **maat** (māt). *S. Afr.* [Afrikaans = Mate *sb.*² 1.] A companion, partner, friend.

1812 W. Burchell *Trav. S. Afr.* 24 July (1822) II. 466 Another chieftain who was his *maat* (partner, or agent). *Ibid.* 555 Thus a Hottentot..when he visits Litakun.. goes directly to the house of his correspondent, whom he calls his *maat* (a Dutch word identical with 'mate'). 1827 G. Thompson *Trav. S. Afr.* (ed. 2) I. xii. 238 Many of the Bechuanas selected *maats* or comrades..from among their allies. 1932 C. Fuller *Louis Trigardt's Trek* x. 124 We drank it as my *maats* had just come to my wagon.

mabela (mabī·la). *S. Afr.* Also **mabbele.** [ad. Zulu *i(li)bele*, pl. *amabele*.] Indian millet, *Sorghum vulgare*, or the meal or porridge made from it. Also *attrib.*

1824 W. J. Burchell *Trav. S. Afr.* II. xviii. 586 The plant resembles... Maize or Indian corn... The Bichuanas call it *mábbĕlĕ*..and are fond of chewing the stalk. 1939 A. W. Wells *S. Afr.: Planned Tour* xx. 228 (*caption*) Natives in a valley of mealies and mabela corn near the Caledon River. 1946 *Cape Argus Mag.* 14 Dec. 4/5 He.. set before him on a stool a bowl of mabela, a flat scone toasted on the coals and a calabash bowl of milk.

mac¹. Add to def.: Also a familiar form of address used to any stranger.

1937 Partridge *Dict. Slang* 503/1. 1962 L. Deighton *Ipcress File* xxv. 158 'Make on the feet, mack,' he said. 1963 *Landfall* XVII. 14 You'd only to hear my father..to know where he stood, solid for intolerance, mac, but solid. 1965 *Sc. Nat. Dict.* VI. 169/2 *Mac*,..in colloq. use applied in Eng. to anyone known or thought to be a Scotsman, in Scot. common as a fam. form of address to any stranger. 1968 *New Yorker* 16 Mar. 42 For the last time, Mac—we don't have any mead. 1973 J. Wainwright *Pride of Pigs* 128 The bouncer..tapped him on the shoulder and said: 'Hey, mac.'

mac, var. *mack *sb.*⁵

macabre, *a.* Add: **1.** (Later examples.)

1870 C. M. Yonge *Caged Lion* ix. 166 It is the Danse Macabre... It was invented as a warning to those of sinful life. 1938 *Oxf. Compan. Mus.* 251/1 *Danse macabre*. The idea of Death as a dancer, or as a fiddling inciter to the dance, is very ancient. 1955 *Times* 19 May 3/7 This *danse macabre* could only have been written by a poet. 1966 *Listener* 17 Nov. 746/3 An *Allegretto* in G minor—a waltz-like *danse macabre*—which quotes the opening of the first cello concerto. 1974 H. Waugh *Parrish for Defence* (1975) i. 5 The hours before dawn belonged to the souls of the dead... The restless whirling of their frenzies would be the *Danse Macabre*.

2. Hence as *sb.*; also **maca·brely** *adv.*

c 1920 T. E. Lawrence *Lett.* (1938) facing p. 233 It's just struck me that there's all the elements of a macabre in the passage which R. G. censored. 1948 F. R. Leavis *Great Tradition* i. 19 The unfortunate *macabre* of the cab-journey. 1961 A. Wilson *Old Men at Zoo* ii. 107 The black suit she wore, against which her white face..stared out so macabrely. 1964 *Economist* 24 Oct. 365/2 This is where [it] ..might become most macabrely relevant. 1968 D. Francis *Forfeit* xiii. 163 What he said..was macabrely at variance with the way he said it.

macadamia (mækădēi·miă). [mod.L. (F. von Mueller 1858, in *Trans. Phil. Inst. Victoria* II. 72), f. the name of John *Macadam* (1827–65), Scottish-born chemist, who was secretary of the Philosophical Institute of Victoria at that time.] An evergreen tree of the Australian genus so called, of the family Proteaceæ, esp. *Macadamia ternifolia*, the Queensland nut, which bears white-fleshed nuts in hard shells. Also *attrib.*

1904 J. H. Maiden *Forest Flora New South Wales* I. 217 Mr. Betche and I described a *Macadamia*..from Camden Haven, N.S.W. 1929 M. D. Frear *Our Familiar Island Trees* 64 Perhaps the best of these [recent introductions to Hawaii] is the Macadamia Nut, sometimes called the Queensland Nut from its native habitat. 1949 S. A. Clark *All the Best in Hawaii* xx. 207 Macadamia nuts call for special pause and explanation, for mainlanders seem rarely to have heard of them... A man named E. W. Jordan brought some of these nuts from Tasmania to Honolulu about the year 1890. *Ibid.* 208 Some smooth-leaf varieties of macadamia produce smaller and less sweet nuts. As a delicacy..macadamia nuts, salted and bottled, have hardly three decades of history... Several Honolulu concerns ship bottled macadamias anywhere. 1955 W. Gaddis *Recognitions* II. i. 295 She returned to her desk..took a macadamia nut from the jar. 1963 *Economist* 14 Dec. 1198/3 Oils are available..from..macadamia nuts. 1969 *Coast to Coast 1967–68* 4 He could hear the familiar morning sounds through that hush,..the sow's demanding squeals, the butcher-birds and magpies in the macadamia-trees. 1970 *Daily Colonist* (Victoria, B.C.) 23 Dec. 4/1 Economies [made by United Air Lines] have ranged from cancelling routes..and ceasing to serve macadamia nuts—a Hawaiian delicacy—with drinks.

macaroni. Add: **8*.** An Italian. *slang.*

1845 [see *frog¹ 3 b]. 1901 'L. Malet' *Hist. R. Calmady* v. x. 461 You don't suppose I mean to stand here till the second anniversary of the Day of Judgment, watching your blithering chicken-shanked macaronies suck rotten oranges, do you? 1942 E. Paul *Narrow St.* xxix. 266 'Cut the throats of the macaronis,' Madame Absalom said. She disliked Italians slightly more than the rest of the human race. 1946 D. Hamson *We fell among Greeks* viii. 91 They dropped us practically on to the Italian garrison at Karpenísi... Doug was playing hidey-ho with a couple of macaronis, taking potshots round bushes at each other.

8.** Nonsense, meaningless talk. *slang* (chiefly *Austral.*).

1924 Lawrence & Skinner *Boy in Bush* iii. 46 Yes. Jam, macaroni, cockadoodle. We're plain people out here-aways, not mantle ornaments. 1941 Baker *Dict. Austral. Slang* 45 *Macaroni*, nonsense, foolishness. 1945 —— *Austral. Lang.* vi. 128 *Macaroni*..and *borak* cover the same meaning of misleading chatter. 1965 J. von Sternberg *Fun in Chinese Laundry* (1966) iv. 67 What is flashed from the projector overhead will be the same old macaroni.

9. (sense 1) *macaroni pudding, soup*; **macaroni cheese,** a savoury of which the principal ingredients are macaroni and cheese.

[1769 E. Raffald *Experienced Eng. Housekeeper* xii. 261 To dress Macaroni with Permasent Cheese... Boil it.., pour it on a Plate, lay all over it Permasent Cheese toasted.] 1877 Trollope *Is he Popenjoy?* (1878) I. i. 2 It is as though one were asked to eat boiled mutton after woodcocks, caviare, or maccaroni [macaroni, 1877 *serial publ.*] cheese. 1934 A. Ransome *Coot Club* iii. 40 Tell her we won't be late. Macaroni cheese to-night. 1972 B. Nilson *Pears Bk. Light Meals* xii. 184 Macaroni cheese with ham. 1861 Mrs. Beeton *Bk. Househ. Managem.* xxvii. 654 *Sweet macaroni pudding*... Put the macaroni, with a pint of the milk, into a saucepan with the lemon-peel. 1963 N. Heaton *Puddings* II. 56 *Macaroni Pudding*. .. When cool, add the beaten egg and the sugar. 1845 E. Acton *Mod. Cookery* (ed. 2) i. 11 *Maccaroni soup*. Throw four ounces of fine fresh mellow maccaroni into a pan of fast-boiling water. 1949 H. Smith *Master Bk. Soups* xv. 198 *Thick Macaroni Soup*. Prepare 3 pints of good gravy. .. Garnish with 6 ozs. macaroni cooked in salted water.

Macartney. Add: **b. Macartney rose,** an evergreen white-flowered climbing rose, *Rosa bracteata*, introduced in 1795 by Lord Macartney.

1811 *Curtis's Bot. Mag.* XXXIV. 1377 Macartny's [*sic*] Rose... Native of China, whence it was introduced by Lord Macartny [*sic*], on his return from his embassy to that country. 1837 [in Dict. s.v. Rose *sb.* 3]. 1908 H. H. Robbins *Our First Ambassador to China* p. viii, The design on the cover of the book is one adapted from a sprig of the original Macartney rose, growing in the North of Ireland. 1955 C. C. Hurst in G. S. Thomas *Old Shrub Roses* ix. 93 So many different species have been concerned in the nineteenth and twentieth centuries... Macartney Roses from *R[osa] bracteata* [etc.]. 1974 *Country Life* 21 Mar. 631/1 The Macartney Rose, *Rosa bracteata*, belongs essentially to late summer..an exquisite creature with large white blooms and golden stamens.

Macassar. Add: **1. Macassar ebony,** the dark-coloured wood of *Diospyros celebica* and related species from Celebes and the Andaman Islands.

1889 G. S. Boulger *Uses of Plants* vii. 177 Those [sticks] most commonly cut out of the solid log are Oak.. home-grown; Ceylon or Macassar Ebony..; and the Palmyra Palm. 1920 A. L. Howard *Man. Timbers of World* 76 (*heading*) Ebony, macassar... This wood is imported in large billets and round logs... The colour ranges from dark brown to black, and a large proportion of the logs are streaked with yellow or yellowish brown, some very handsomely figured pieces being occasionally found. 1936 *Nature* 9 May 790/2 Counter front..in finely figured macassar ebony. 1947 J. C. Rich *Materials & Methods of Sculpture* x. 288 Macassar ebony, or Coromandel,..is considered by many wood carvers to be inferior to African ebony. 1972 *Handbk. Hardwoods* (Building Res. Establishment) (ed. 2) 68 In Macassar ebony pale to medium brown zones contrast with black wood.

2. Also **Macasar, Makasar, Makassar, Mangkasar, Mungkasar.** A native or inhabitant of Macassar; the name of their language. Also *attrib.* or as *adj.*

1808 *Asiatick Researches* X. 194 The dialect of Mungkásar or Macassar, the bravest and most renowned of the Búgís tribes, differs..from the Búgís proper... I have formed a short radical vocabulary of both the Búgís and Mungkásar [languages]. 1816 T. S. Raffles *On Malayu Nation* 125 To the collection that has already been made of the various laws and usages of the Malays, Sumatrans, Javanese, Bugis, Macasars and Sulus, may be added the compendium of the Muhammedan law of inheritance. 1817 —— *Hist. Java* II. xi. 162 By the aid of the Dutch..the Makásar chief was driven from his post. 1840 J. Brooke *Jrnl.* 14 Feb. in R. Mundy *Narr. Events Borneo & Celebes* (1848) I. vii. 106 From his residence at Tetiagi..he transmits information to the Makassar government. 1886 *Misc. Papers relating to Indo-China* (Straits Branch R. Asiatic Soc.) I. xii. 108 The descriptive catalogues of the extensive Bugis and Makassar literature..may now be consulted. 1911 *Encycl. Brit.* V. 598/1 The Macassars are well-built and muscular, and have in general a dark-brown complexion. 1968 *Ibid.* V. 132/2 They [*sc.* the Toradja tribe] came under a Bugis and Makasarese civilizing influence... The Bugis and Makasars..came into touch with Hindu culture in southern Celebes..and later were converted to Islam. *Ibid.* XIV. 654/1 Gaily coloured, plaited basketry is an outstanding feature of Makasarese markets... The people, the Mangkasaras or Makasars, are a branch of the Malay people, similar to but not identical with the Bugis, who inhabit the same region.

Macassarese (mǎkæsǎrī·z). Also **Makas(s)arese.** **a.** The Macassar people. **b.** Their language. Also *attrib.* or as *adj.*

1880 [see *Dyak]. 1948 D. Diringer *Alphabet* vii. 430 The Macassarese and the Buginese..are the most important and the most advanced peoples of the island [*sc.* Celebes]... Macassarese is spoken nowadays in all the districts from Balu Kumba to Segere. 1959 *Chambers's Encycl.* III. 211/1 The peoples of the south (Buginese and Macassarese) are Moslems. 1968 [see *Macassar 2].

McBurney's point (mǎkbɒ̄·ɹnīz). *Surg.* and *Anat.* Also **McBurney point.** [f. the name of Charles *McBurney* (1845–1913), U.S. surgeon, who described it in 1889.] A point on the surface of the abdomen situated along a line from the umbilicus to where the anterior superior spine of the right ilium can be felt and at a distance of 1½ to 2 inches from this spine, which point normally lies directly above the appendix and is the point of maximum tenderness in appendicitis.

1890 L. A. Stimson in *N.Y. Med. Jrnl.* 25 Oct. 449/2, I found him..very weak, with marked tenderness and slight deep induration at a point midway between the umbilicus and the right anterior superior spine of the ilium... For the sake of brevity, and as a proper recognition of the value of this symptom, I shall speak of this point as 'McBurney's point'. *Ibid.* 450/1 The McBurney point was well marked. 1910 *Practitioner* Feb. 261 In peritonitis the pain is situated over the inflamed viscus, *e.g.* at McBurney's point in appendicitis. 1972 *Sci. Amer.* Aug. 122/2 Draw a line from the navel to the frontal protuberance of the right hipbone. Place your finger about one and a half inches from the bony end. That is McBurney's point.

macca (mæ·kă). [f. Macaw².] **1.** A Jamaican name for the palm *Acrocomia sclerocarpa*, distinguished by its prickles; hence, used for the prickles of other plants and animals. Also *attrib.*

1873 [implied in sense *2]. 1910 Anderson & Cundall *Jamaica Anancy Stories* 31 De man dat hab on boot must go befo' so' mash macca. 1946 E. N. Burke *Stories told by Uncle Newton* I. 17 There was soft grease—applied warm —for prickles ('macca' to us children) in the fingers or toes. 1956 J. Hearne *Stranger at Gate* xvii. 131 They went through the bush..the dry *macca* crunching beneath their shoes. *Ibid.* xxxi. 239 If I have to walk a hundred mile on macca thorn, I am going to be dere. 1961 F. G. Cassidy *Jamaica Talk* i. 7 A thoroughly Jamaican word ..is *macca*, which now means any kind of prickle, thorn, bur, or sharp spine on plants or animals... It is used in many combinations too, like *macca breadfruit, macca yam, macca fern*.

2. macca-fat, a Jamaican name for the fruits of certain oil-yielding palms, esp. those of the genus *Acrocomia* and the oil palm, *Elæis guineensis*; **macca-fat palm,** one of these trees. Cf. *macaw-fat* (Macaw²).

1873 C. J. G. Rampini *Lett. from Jamaica* 71 Clumps of wild ginger, Macca-fat palms,—surely the most graceful of all that graceful tribe. 1961 F. G. Cassidy *Jamaica Talk* i. 7 The *macaw tree* [is]..also called the *macca-fat palm* in Jamaica. *Ibid.*, The two most prominent features of these trees [are]—the fruits (not dates, but a similar nut-like berry with an edible rind, the *macca-fat* of today) and the thorns.

Maccabean, -æan (mækăbī·ăn), *a.* and *sb.* [f. *Maccabee + -an.] **A.** *adj.* Of or pertaining to Judas Maccabæus or the Maccabees. **B.** *sb.* = *Maccabee.

1821 R. Laurence *Bk. of Enoch* 207 Between the period of the captivity, and the rise of the Maccabæan dynasty. 1840 J. H. Howlett *Metrical Chronol.* (ed. 3) 16 Under the Maccabaean princes, what did the Jews become? 1845 [see *Maccabee]. 1890 Church & Seeley (*title*) The hammer, a story of Maccabean times. 1920 *Q. Rev.* July 4 His [*sc.* Disraeli's] party reaped the benefit of his Maccabean courage. **1952** GERTH & MARTINDALE tr. *Weber's Anc. Judaism* IV. xiv. 360 The high priest Onias, who in the confusion of Maccabean party struggles had escaped to Egypt..had not scrupled to build a temple there. **1973** [see *HASMONEAN *sb.* and *a.*].

Maccabee (mæ·kăbī). *Jewish Hist.* [ad. L. *Maccabæ-us*, Gr. *Μακκαβαῖος*, the epithet of the Jewish patriot Judas.

'The source of the name is uncertain, but it is most natural to connect it with *maqāb*, "hammer".' 1883 *Encycl. Brit.* XV. 130/2.]

A supporter or successor of Judas Maccabæus, the leader of a religious revolt against Antiochus IV, 175–164 B.C., as recorded in the Books of the Maccabees.

1375 BARBOUR *Bruce* I. 465 Thai was lik to the Machabeys, That, as men in the bibill seys [etc.]. *a* **1420** WYCLIF *1 Macc.* (*heading*) Here biginnith the firste book of Machabeies. **1550** *Briefe & Compendiouse Table, Concordaunce Bible* Title-page, The thirde boke of the Machabees. **1614** RALEGH *Hist. World* I. II. x. 377 Then Modin the Natiue Citie of the Macchabees. **1702** R. L'ESTRANGE tr. *Josephus' Works* 1058 Flavius Josephus, his discourse of the Maccabees. **1845** *Encycl. Metrop.* IX. 643/2 Hence, all who fought under that standard were called Maccabees, or Maccabeans. **1920** H. F. HENDERSON *Relig. in Scotl.* i. 16 The patriotism of the Jew, especially in the age of the Maccabees, was an ardent passion that enabled him to overcome enemies four times his number. **1956** W. R. FARMER (*title*) Maccabees, Zealots, and Josephus. An inquiry into Jewish nationalism in the Greco-Roman period.

transf. and *attrib.* **1865** *Sunday at Home* 194/2 But at length one of the Maccabee princes..subdued them. **1883** *Encycl. Brit.* XV. 131/1 The Vatican [MS.] does not contain the Maccabee books. **1959** *Listener* 20 Aug. 292/2 The Maccabees of this small but resolute movement..were the Aaronsohn family.

Maccabiah (mækăbī·ă). [f. *Maccabi*, the name of an Israeli sporting organization + *-ah* Heb. adj. ending.] In full, *Maccabiah Games*. A sporting contest, modelled on the Olympic Games, for Jewish athletes, and held at four-yearly intervals in Israel.

[**1962** *New Jewish Encycl.* 299/1 The major project of the Maccabi is the organization of Maccabiads, or 'Jewish Olympics'.] **1963** J. COMAY *Introducing Israel* viii. 129 The Maccabiah, an 'Olympic Games' for Jewish athletes.. takes place in the huge Ramat Gan Sports Stadium. **1969** *National Herald* (New Delhi) 29 July 10/3 The kindling of the Maccabiah flame was followed by the release of hundreds of doves. **1973** *Black Panther* 11 Aug. 12/3 Israel has declared that athletes from Rhodesia who took part in the so-called Maccabiah Games had no official status.

McCarthyism (măkā·ɹþi,iz'm). [f. the name of Joseph R. *McCarthy* (1908–57), U.S. senator + -ISM.] The policy of hunting out (suspected) Communists and removing them from Government departments or other positions. Hence **McCa·rthyist** *a.*, of or pertaining to McCarthyism; **McCa·rthyite**, one who gave support to such a policy; also *attrib.* and *transf.*

1950 *N.Y. Post* 5 Apr. 44/1 To call McCarthyism a fascist atmosphere would be descriptive enough. **1951** [see *big lie* (*BIG *a.* B. 2)]. **1952** *N.Y. Times* 26 Oct. I. 70/7 McCarthyism breeds fear, suspicion and unrest. It turns neighbor against neighbor and makes every American a suspected traitor until and unless he can prove his innocence by McCarthyite standards. **1952** *San Francisco Examiner* 18 Sept. 3/6 The McCarthyites unquestionably have an appeal, Senator Lehman conceded, since they come before the people as super-patriots. **1953** *Ibid.* 23 July 20/8 The editorial denounced a House committee investigation of Communistic clergymen... The 'Daily Worker' called it 'the new McCarthyite drive against the integrity and independence of the church'. **1954** *Encounter* Aug. 10/1 For the unlucky rest of us, McCarthy and McCarthyism—the proper noun and the common one—are familiar to excess. **1954** RORTY & DECTER *McCarthy & Communists* 118 They are not so much *pro*-McCarthyites as anti-anti-McCarthyites (opponents of all who oppose McCarthy). **1955** H. READ *Grass Roots of Art* (rev. ed.) iv. 85 We are at once struck by the fact that whatever the political ideology may be—socialist, fascist, communist, or MacCarthyist—..all express one opinion about art. **1957** *Times Lit. Suppl.* 27 Dec. 782/2 Has this book been written to add fuel to McCarthyist flames that have long been dying away in the United States? **1959** *Ibid.* 6 Nov. p. xix/3 The whole academic community is only now emerging from the shock of the assault on the academies that was one of the most sinister results of 'McCarthyism'. **1960** *20th Cent.* June 578 People who were not active McCarthyites..were none the less sources of strength for him. **1971** *Guardian* 19 Jan. 10/2 Once people start.. wondering about their neighbours, a McCarthyite infection could take hold. **1973** *Ibid.* 18 June 4/4 A tradition of paranoia out of which grew McCarthyism and many earlier witch-hunts. **1973** *Times Lit. Suppl.* 6 July 765/1 The tendency is timeless—and by no means sinister or subversive even in McCarthyite terms.

‖ **macchia** (ma·kiă). [Corsican It.] = *MAQUIS 1.

[**1868** J. A. SYMONDS *Let.* 4 May (1967) I. 805 The sea shore [of Corsica]..is a tangle of sweet & splendid flowers. .. The whole air is fragrant with multitudinous scents wafted from these 'macchi', as the natives call them.] **1924** A. HUXLEY *Let.* 3 Dec. (1969) 238 The mountain is covered with woods of cork trees and a kind of macchia or bush of fragrant shrubs. **1936** *Nature* 2 May 735/2 Such characteristic native vegetation as the beautiful 'fijnbos' (the macchia or maquis of the south-west Cape) may be irretrievably damaged by fire. **1962** *Times* 24 Mar. 11/4 A cottage lost in the *macchia*. **1966** C. MACKENZIE *My Life & Times* V. 133 One looked down..into the *macchia* of Ventrosa.

McCoy (măkoi·). In the colloq. phr. *the real McCoy* (or *Mackay, McKie*): the 'genuine article', the real thing.

Its origin remains uncertain: see, for example, *Amer. Speech* (1958) XXXIII. 297 f.

1883 R. L. STEVENSON *Lett. to C. Baxter* (1956) 123 For society, there isnae sae muckle; but there's myself—the auld Johnstone, ye ken—he's the real Mackay, whatever. **1922** *Collier's* 7 Oct. 26/2 'At's the real McCoy you got there, brother!..Comes right down from Canada!' **1930** *Amer. Mercury* Dec. 456/2 *McCoy*, genuine liquor. 'This is McCoy. You can't fake Quebec wrappers.' **1934** M. ALLINGHAM *Death of Ghost* v. 68 There's something very attractive about the real McKie when you meet it. **1935** R. C. WOODTHORPE *Shadow on Downs* xii. 298 It looks like the real Mackay. A lovely colour. **1940** *Penguin New Writing* I. 73 'You look the real McCoy, now,' he said. **1942** C. BARRETT *On Wallaby* v. 103 It's the real Mackay we've got, Chas.; didn't it spring out of the wall? **1949** N. MARSH *Swing, Brother, Swing* vi. 140 If I could pick my work I'd be in an outfit that went for the real mackay. **1952** W. R. BURNETT *Vanity Row* (1953) v. 43 A real gentleman... Not a phony... He was the real McCoy. **1959** L. BERNSTEIN *Joy of Mus.* (1960) 167 The operetta score..was musically elaborate, closer to opera, even containing finales, with everybody *singing* his way through the plot, vocalizing different sentiments at the same time —the real contrapuntal McCoy. **1972** *Guardian* 17 Feb. 10/3 Sadler's Wells is playing host to the regal offspring Royal Ballet, and not, please note, a second eleven but the real Macoy.

mace, *sb.*¹ Add: **5. mace-reed** = REED-MACE.

1901 G. MEREDITH *Reading of Life* 126 A hundred mares, all white! their manes Like mace-reed of the marshy plains Thick-tufted, wavy.

Mace (mēⁱs), *sb.*⁵ In full, *Chemical Mace*. The proprietary name of a chemical preparation used as a disabling weapon by being sprayed at a person's face; also *attrib.* Hence (usu. with small initial) as *v. trans.*, to attack with this liquid.

1966 *Official Gaz.* (U.S. Patent Office) 22 Nov. TM 174, SN 233,338. General Ordnance Equipment Corp., Pittsburgh, Pa. Filed Nov. 26, 1965. Chemical Mace. The word 'Chemical' is disclaimed apart from its use with the mark. For Non-Explosive Defensive Weapons in the Nature of Tear Gas Packaged in Aerosol Containers. **1966** *Law & Order* June 50/3 The development..of the Chemical Mace, a liquid, long range, selective, tear gas projector is an event of considerable importance to law enforcement. **1967** *N.Y. Times* 3 Aug. 17/2 The gas is called Mace and it comes in a small black aerosol container like a hair spray can. **1968** *Listener* 9 May 595/3 A new anti-riot chemical called Mace. Sprayed like tear gas it turns the victim temporarily blind. **1968** *Sun* (Baltimore) 18 Sept. A.14/6 Scores of innocent adult bystanders.. were clubbed, maced, arrested. **1969** E. AMBLER *Intercom Conspiracy* (1970) vii. 143 The stuff was..some kind of chemical Mace or nerve gas. **1970** W. WAGER *Sledgehammer* (1971) xvi. 103 Williston raised the can of Mace, aimed for his enemy's eyes and nose. **1971** *Harper's Mag.* Sept. 63 He responded to foul and abusive language by Macing an old woman's hot dog. **1972** J. ROSSITER *Rope for General Dietz* i. 16 The two pen-shaped aerosols of Mace chemical gas I invariably carried. **1973** *Black Panther* 25 Aug. 2/2 They feel that because they are in power they can call ten or 20 pigs for just one man, mace him, beat him up and call him profane names.

Macedo- (mæ·sīdo), combining form of MACEDONIAN *a.*¹ in the names of dialects spoken in Macedonia, an area in the central Balkans (see *MACEDONIAN *a.*¹ and *sb.*¹), as *Macedo-Bulgarian*, *-Illyrian*, *-Rumanian*, *-Rumanic*.

1861 MAX MÜLLER *Lect. Sci. of Lang.* v. 182 This Romance language is spoken in Wallachia and Moldavia, and in parts of Hungary, Transylvania, and Bessarabia... It is divided by the Danube into two branches: the Northern or Daco-romanic, and the Southern or Macedo-romanic. **1880** A. H. SAYCE *Introd. Sci. of Lang.* II. vii. 120 The Danube divides it [*sc.* Rumanian] into two branches, the northern or Daco-Rumanic, and the southern or Macedo-Rumanic, the latter of which abounds with Albanian and Greek words. **1908** T. G. TUCKER *Introd. Nat. Hist. Lang.* 240 A southern division [of Rumanian]..is formed by the *Macedo-Roumanian* (or *Vlach*) of parts of Roumelia, Macedonia, and Thessaly. **1937** J. ORR tr. *Iordan's Introd. Romance Ling.* ii. 95 The current forms in Macedo-Rumanian. *Ibid.* iii. 267 The five Macedo-Rumanian, the two Megleno-Rumanian, and the two Istro-Rumanian localities. **1948** L. SPITZER *Linguistics & Lit. Hist.* 3 Meyer-Lübke would quote..Macedo-rumanian. **1949** ENTWISTLE & MORISON *Russ. & Slavonic Lang.* vii. 370 It is the pronunciation current in Macedo-Bulgarian dialects. *Ibid.* 374 The forms in *-m* are widely distributed in Bulgarian dialects, and are universal in Macedo-Bulgarian. **1959** G. NANDRIS *Handbk. Old Church Slavonic* I. 20 It [*sc.* epenthetic *l*] is dropped as a rule in Macedo-Bulgarian. **1960** R. AUTY *Ibid.* II. 10 The researches of Vatroslav Oblak showed the affinities of O[ld] C[hurch] S[lavonic] with present-day Macedo-Bulgarian dialects. **1965** G. Y. SHEVELOV *Prehist. Slavic* 611 There are also two small peripheral areas with a tendency to self-isolation from the other Slavs..: West Baltic and Macedo-B[ul]g[arian]. **1966** E. G. POLOMÉ in Birnbaum & Puhvel *Anc. Indo-European Dial.* 70 Pisani preferred to consider the former as Macedonian,..without, however, connecting *siiri-* with 'Macedo-Illyrian' *Σίρρας*. **1966** H. BIRNBAUM *Ibid.* 161 We can..assume here the existence of certain C[ommon] Sl[avic] dialect groups... These dialect groups will be shown to correspond to East Slavic.., Lekhitic.. Sorbian.., Czechoslovak.., Sloveno-Serbocroatian..,and Macedo-Bulgarian. **1967** D. S. PARLETT *Short Dict. Lang.* 103 Macedo-Rumanian (or Arumanian), in Albania, Thessaly, Macedonia. **1972** W. B. LOCKWOOD *Panorama Indo-European Lang.* 45 Macedo-Rumanian or Aromunian..is the language of the Aromuni, the Rumanians of the south.

macédoine (masedwan). [Fr., f. *Macédoine* Macedonia, with reference to the diversity of peoples in the Macedonian empire of Alexander the Great] **1.** Mixed fruit or vegetables cut up into small pieces. Also *attrib.*

1846 C. E. FRANCATELLI *Mod. Cook* 32 White macedoine of vegetables. *Ibid.*, Garnish of brown macedoine. Prepare the vegetables for the *Macedoine* according to the directions given in the preceding recipe. **1855** E. ACTON *Mod. Cookery* (rev. ed.) xxiii. 453 Jelly of two colours, with *macedoine* of fruit. **1894** L. HERITAGE *Cassell's New Universal Cookery Bk.* 67 Soup with Macédoines. *Ibid.* 1127 Macédoine of Fruits. **1895** 'M. RONALD' *Century Cook Bk.* 378 Macédoine salad. **1960** *Good Housek. Cookery Bk.* (rev. ed.) 313/2 Macédoine of fruit.

2. *fig.* A medley or mixture of unrelated things.

1820 H. LUTTRELL *Advice to Julia* 18 Such is the tattle of our beaus. These simple elements compose..The *Macedoine* of London-talk. **1857** G. C. MUNDY *Our Antipodes* (ed. 4) 10 Now for a *macédoine* of advertisements, word for word as entered. **1925** *Glasgow Herald* 3 Nov. 10 Europe contains as many different types of features,..as India. But there are few parts of Europe which present such a mosaic, or macédoine, of contrasts side by side. **1974** *Times* 30 Apr. 8/5 Ann Buck designs in what I call the *macedoine* style, with all sorts of bits and pieces of fabric welded into a garment.

Macedon. For 'Macedonia' read 'ancient Macedonia' and add later examples of sense 2.

1940 A. H. M. JONES *Greek City* i. 8 As king of Macedon he [*sc.* Alexander] could draw as he wished on the resources of Macedonia. **1969** A. TOYNBEE *Some Probl. Greek Hist.* IV. ii. 426 The invaders found the entire manpower of Macedon arrayed against them. **1969** 'M. RENAULT' *Fire from Heaven* (1970) vi. 275 In Athens, the marble tablet which witnessed the peace with Macedon had been torn down, in formal declaration of war. **1972** F. W. WALBANK *Polybius* i. 26 Polybius describes with amazement Demetrius of Phalerum's prophecy of the future downfall of Macedon.

Macedonian, *a.*¹ and *sb.*¹ **A.** *adj.* For 'a country north of Greece' read 'an ancient country north of Greece; now, a geographical area in the central Balkans lying astride the frontiers of southern Yugoslavia, northern Greece, and southwestern Bulgaria; also, the name of a province in Yugoslavia' and add further examples.

1946 [see B. c below]. **1958** *Listener* 4 Dec. 912/2 The first Macedonian dictionary is being prepared, there are newspapers in the Macedonian language, and Skopje has acquired a brand-new Macedonian university. **1974** *Encycl. Brit. Micropædia* VI. 442/1 Yugoslav Macedonia contains the great majority of the Macedonian people.

B. *sb.* **a.** (Later examples.)

1844 [see *EPIROT]. **1897** W. E. GLADSTONE *Let.* 19 Jan. in *Macedonian Question* (1902) 3 Why not Macedonia for the Macedonians as well as Bulgaria for the Bulgarians and Servia for the Servians? **1902** N. BUXTON *Ibid.* 44 There is, moreover, a certain enmity in Bulgaria towards the Macedonians. **1935** S. CHRISTOWE *Heroes & Assassins* vii. 116 The Macedonians took as their motto 'Liberty or Death'. **1950** E. BARKER *Macedonia* ii. 23 After King Alexander instituted his dictatorship in January 1929.. the Macedonians began to settle down and to accept Yugoslav rule passively. **1966** D. DAKIN *Greek Struggle in Macedonia* xii. 307 There seems also to have been some conflict between the National Greeks and the Greek Macedonians over the question of leadership. **1971** PALMER & KING *Yugoslav Communism & Macedonian Question* iv. 64 Evidently there was no attempt by the Macedonians to proclaim an independent or autonomous state, as the Croatians did. **1974** *Encycl. Brit. Micropædia* VI. 442/1 In the Greek province are Slavic speakers, Macedonians, and Pomaks, or Bulgarian-speaking Muslims.

b. The language of ancient Macedonia, recorded in fragmentary remains, and usually regarded as a variety of Greek within the Indo-European family.

1556 [in Dict., sense A]. **1933** C. D. BUCK *Compar. Gram. Greek & Latin* 14 Languages for which Illyrian origin is claimed or disputed are Venetic and Messapian in ancient Italy, Macedonian, and Albanian. Macedonian, that is, the native speech of the Macedonians as distinguished from the Attic *κοινή* which they came to adopt as their official language, is known from proper names and rather numerous glosses. **1939** L. H. GRAY *Foundations of Lang.* 330 The relation of Macedonian to Greek is uncertain. **1966** E. G. POLOMÉ in Birnbaum & Puhvel *Anc. Indo-European Dial.* 70, *σιβύνη* and *σιγύνη* are both occasionally ascribed to Macedonian by scholiasts and lexicographers. **1972** W. B. LOCKWOOD *Panorama Indo-European Lang.* 6 Since Macedonian was in contact with Illyrian and Thracian, borrowings from these languages could account for the exotic strain... It is to be assumed that the Macedonian dialect (or language) succumbed to Attic Greek.. during the Hellenistic Age.

c. The modern language of the province of Macedonia in Yugoslavia and adjacent areas of Bulgaria and Greece. Also *attrib.* So *Macedonian-speaking*.

1946 R. CAPELL *Simiomata* II. 85 A Macedonian nationalist, one Gotchi, has been leading a band of Macedonian-speaking Andartes. **1955** R. JAKOBSON *Slavic Lang.* (ed. 2) 15 Still preserved is the nasal component in ..some border dialects of Slovenian and Macedonian. **1964** M. PARTRIDGE *Serbo-Croatian* 13 Together with Bulgarian, Macedonian and Slovene, Serbo-Croatian belongs linguistically to the southern branch of the Slavonic group. **1972** W. B. LOCKWOOD *Panorama Indo-European Lang.* 162 The greatest concentration of Macedonian speakers was found in that part of the province which passed to Serbia. In 1945, this area became the Macedonian Constituent Republic of Yugoslavia with Macedonian as its official language. The number of speakers exceeds one million. **1974** *Encycl. Brit. Micropædia* VI. 442/2 Macedonian became the official language of the People's Republic of Macedonia, an autonomous area in Yugoslavia, when it was established in the 1940s.

Macedonian, *sb.*² (Earlier example.)

1559 [see *EUNOMIAN *sb.* (*a.*)].

Macedonic (mæsīdǫ·nik), *a.* [a. L. *Macedonic-us*, Gr. *Μακεδονικ-ός* Macedonian.] = MACEDONIAN *a.*¹

1859 E. MASSON tr. *Winer's Gram. New Testament Diction* I. 33 The previously distinct dialects..were blended into a popular spoken language, with a predominance of the Macedonic variety.

maceral (mæ·sĕrăl). *Geol.* [f. L. *mācer-āre* MACERATE *v.* + -AL, after MINERAL *sb.*] Any of the microscopic structural constituents of coal.

1935 M. C. STOPES in *Fuel in Sci. & Pract.* XIV. 11/1 To..construct an acceptable petrological classification, therefore, a prime need became obvious, viz., a word to cover all petrological units seen in microscopic sections of coals, as distinct from the visible units seen in hand specimens... I now propose the new word 'Maceral' (from the Latin, macerare, to macerate)... The word 'macerals' will, I hope, be accepted as a pleasantly sounding parallel to the word 'minerals', conveying the suggestions of the fundamental difference between them. *Ibid.*, The concept behind the word 'macerals' is that the complex of biological units represented by a forest tree which crashed into a watery swamp and there partly decomposed and was macerated in the process of coal formation, did not in that process become uniform throughout but still retains delimited regions optically differing under the microscope, which may or may not have different chemical formulae and properties. These *organic* units, composing the coal mass I propose to call *macerals*, and they are the descriptive equivalent of the *inorganic* units composing most rock masses and universally called *minerals*. **1970** *Nature* 11 July 194/2 The petrographic components (macerals) of coals possess different stability ranges.

macerate (mæ·sĕrēⁱt), *sb.* [f. the vb., after *precipitate*, *filtrate*, etc.] A product obtained by maceration.

1961 in WEBSTER. **1974** *Nature* 27 Sept. 294/2 In scanning electron microscopical (SEM) investigations of Precambrian sedimentary rocks, the risk of contamination during the preparation of rock macerates is extremely high. **1975** *Ibid.* 20 Mar. 184/2 Nitrate production was depressed when *Hyparrhenia* root macerate was added to soils.

macerator. Add: **c.** A pulping machine. *U.S.*

1912 *Publishers' Circular* 12 Oct. 503 Then the macerator, the greatest consumer of contemporary literature, takes them [*sc.* books] to its bosom.

Mach (măk, mæk, ‖ māχ). The name of Ernst *Mach* (1838–1916), Austrian physicist and philosopher: **a.** Used, usu. *attrib.*, to designate certain concepts associated with his work on aerodynamics, as **Mach('s) angle**, the angle between a generator of the Mach cone and its axis; **Mach cone**, a cone that extends backwards from a body moving at supersonic speed and coincides with the shock wave it produces, separating a region affected by the motion (inside the cone) from a region unaffected by it (outside the cone); **Mach('s) number**, the ratio of the relative speed of a body and a fluid to the speed of sound at the same point; so *Mach one, two* (or *1, 2*), etc., a speed corresponding to a Mach number of one, two, etc.

1930 DOUGALL & DEANS tr. *Ewald's Physics Solids & Fluids* v. 262 During a small interval of time τ, a point source of disturbance expands to a sphere of radius $c\tau$, the centre of which has moved through a distance $q\tau$. The cone touches this sphere, so that $\sin a = c\tau/q\tau = c/q$...$a$ is called Mach's angle. **1933** *Proc. R. Soc.* A. CXXXIX. 307 The Mach cone $x = y$ and $z = 1$ would satisfy the required conditions. **1937** DODGE & THOMPSON *Fluid Mech.* xiii. 370 The ratio V/c will be found to appear in all flow problems where compressibility is an important factor. It is known as Mach's number..in honor of the Austrian scientist. **1938** *Jrnl. R. Aeronaut. Soc.* XLII. 194 The part of the wave at a distance from the shell moves normal to itself with the speed of sound and this fact determines the angle it makes with the direction of motion of the shell (the 'Mach angle'). **1947** *Time* 8 Sept. 76/2 During both flights it reached 'mach ·828'. **1948** 'N. SHUTE' *No Highway* 34 It was diving at round about Mach unity, and the wings came off. **1953** *Sci. News* XXIX. 93 The non-dimensional quantity $M = V/a$ is defined as the 'Mach Number' of the flow. **1957** *Spaceflight* I. 51/1 In a rocket exhaust where the gases are moving at 6,000 ft./sec.,..the velocity of sound is 2,000 ft./sec., so the Mach No. is 3. **1967** *Technology Week* 23 Jan. 29/2 (Advt.), McDonnell testing and development facilities range from man-rated space chambers to Mach 28 wind tunnels.

b. Used, usu. *attrib.*, to designate an optical illusion first investigated by Mach (in *Sitzungsber. der K. Akad. der Wissensch.* (*Math.-Natur. Cl.*) (1865)), in which a place where the spatial rate of variation of surface brightness abruptly increases or decreases (as at the inner and outer edges of an indistinct shadow) may appear extra dark or bright to an extent that cannot be accounted for simply in terms of the objective variation in brightness.

1932 W. M. DEANS tr. *Pohl's Physical Princ. Mech. & Acoustics* i. 4 Mach's bands have led to much trouble in the carrying out of physical observations. **1936** *Brit. Jrnl. Psychol.* XXVII. 103 The same generalizations can be made about the Mach effect on a rotating colour wheel. **1965** GRAHAM & BROWN in C. H. Graham et al. *Vision* xvi. 474/2 Mach rings with spatial variations in color. **1965** F. RATLIFF *Mach Bands* ii. 43 The Mach bands appear at once in the shadow cast on a piece of white paper by the edge of a card held under a fluorescent desk lamp, which provides an extended source of light. Covering the ends of the lamp, which usually are not uniformly bright, somewhat enhances the effect. **1970** —— in Cohen & Seeger *Ernst Mach* 32 The Mach bands are one of the most compelling of all the visual 'illusions', and have been mistaken for objective physical phenomena.

machærodont, *a.* Add: Also **machairodont**. Also as *sb.*

1889 NICHOLSON & LYDEKKER *Man. Palæont.* (ed. 3) II. 1448 The extinct Machærodonts or Sabre-toothed Tigers. **1925** C. R. EASTMAN tr. *Zittel's Text-bk. Palaeont.* III. 75 A possible derivation of both Felines and Machairodonts may be from Dinictis. **1973** *Nature* 3 Aug. 311/2 A minimum of twenty-three large mammal species were represented, including at least five extinct forms—a large baboon (?*Simopithecus*), a sabretooth cat (machairodont), [etc.].

machair (maχēᵊ·ɹ). *Sc.* Also **machaire, machar, machir, machirr.** [Gael. *machair*, Ir. *machaire*.] A flat or low-lying coastal strip of arable or grassland usually overlying shell sand. Also *attrib.*

1684 A. SYMSON in A. Mitchell *Geogr. Coll. relating to Scotl.* (1907) II. 86 These three parishes last described.. are commonly called the Machirrs or Machirrs of Whithern, which word Machirrs, as I am informed, imports white ground, and indeed those parishes, contain by far much more arable and white land, than up in the Moors, though the parishes there be much larger. **1878** *Q. Jrnl. Geol. Soc.* XXXIV. 848 Benbecula..has only one hill; and if we except the 'Machair', as the 'good land' along the west coast is called, all the rest of the island consists of low-lying moor, bog, and lake, with long shallow inlets of the sea straggling in. *Ibid.* 849 Adjoining the sandy shores are the delightful 'machairs', with their wealth of bright colour; while inland from the 'machairs' stretch the brown sombre peat and moorland. **1899** *Blackw. Mag.* Feb. 423/2 The burial-ground..occupied a little knoll in the middle of the 'machar', close to the sea (*machar* is the fine sweet pasture or links lying along the shore). **1924** *Glasgow Herald* 15 Mar. 4 In Highland glens or by the machirs of the Western Isles a crone..will still be consulted as if she were the Delphic sibyl. **1930** J. BUCHAN *Castle Gay* xvii. 271 The machars, yellowing with autumn, stretched for miles before him. **1955** F. F. DARLING *West Highland Survey* 21 There is a sufficiency of shell among the sand to encourage a fairly typical *machair* flora. *Ibid.* 51 The introduction of rabbits in the nineteenth century has gone a long way towards ruining the agricultural potential of the island, as these animals have created several small deserts on the *machair*. **1958** *Irish Times* 7 June, The term 'machaire' is used by English-speakers here to denote 'coastal strips of pasture land'. **1971** *Country Life* 24 June 1606/1 Her parents complained of attempts to put holiday caravans and a public lavatory on the machair. **1973** *Stornoway Gaz.* 3 Mar. 6/2 We are appealing to any reader (rugby enthusiast or not) who might be able to suggest (or lend) any reasonably flat area of ground—a stretch of machair or croftland—within, say, five miles of Stornoway.

mâche (maʃ). Also **mache**. [Fr.] = CORN-SALAD.

1830 [see *lamb's lettuce* s.v. LAMB *sb.* 7 b]. **1961** *Harper's Bazaar* June 84/2 Other salad greens..watercress, and French *mache*. **1962** S. COMBES *Dict. Cuisine French* 44/2 *Mâche*, lamb's lettuce, corn-salad. **1964** *Harper's Bazaar* Sept. 125/2, I like a green salad..lettuce or *mâche* or endive.

macheer (mătʃīᵊ·ɹ). *Western U.S.* Also **machere**. [Corruptly ad. Sp. *mochila*.] A leather flap attached to a saddle. Also *attrib.*

1847 *Calif. Star* (San Francisco) 21 Aug. 2/3 [A] man declares his *macheres* (saddle cover) was stolen from under him, although seated upon his horse and his horse in motion. **1853** G. D. BREWERTON *Overland with Kit Carson* (1930) 49 Our saddles were of the true Mexican pattern, wooden trees covered with leathers called *macheers*. *a* **1861** T. WINTHROP *John Brent* (1883) v. 45 Showers shrank his buckskins and soaked the macheers of his saddle to mere pulp. **1873** J. MILLER *Life amongst Modocs* 50 The Prince unfastened his cloak from the macheers behind my saddle. **1927** C. M. RUSSELL *Trails plowed Under* 166 I've seen bronc riders use an old macheer saddle with a Texas tree.

‖ **macher** (ma·χəɹ). *U.S.* [Yiddish, f. G. *macher* maker, doer.] A man of importance, a bigwig; a braggart. Often *derogatory*.

1930 *Amer. Speech* VI. 126 A *Macher*..from the German word... Literally translated the word means 'maker' and idiomatically..is used derisively, referring to a braggart. **1964** S. BELLOW *Herzog* 87 He's a fine fellow... Not like that *macher*, Alexander. Always some scandal about him. **1964** W. MARKFIELD *To Early Grave* (1965) ii. 29 A man who sat on every board and committee, a powerful *macher* in Jewish communal affairs, who had passed around word that he needed a speech-writer. **1969** D. S. DAVIS *Where Dark Streets Go* (1970) i. 10 His father's a big *macher* in the union. **1973** *Jewish Chron.* 2 Feb. 19/3 It doesn't matter a tinker's cuss whether you amend the constitution to call the chairman president, macher or grand panjandrum.

machete (mătʃe·tĭ). Now the more usual form of MATCHET.

1832 M. R. MITFORD *Lights & Shadows Amer. Life* III. 215 The monteros drew their *machetes*, the sharp broadswords they usually carry about with them. **1854** [see MATCHET]. **1956** H. G. DE LISSER *Cup & Lip* v. 66 A dozen black men rushed forward, two with upraised machetes. **1958** J. COPE *Golden Oriole* xxiii. 131 On the far side of the wooden building an old man hired by Chipi was slowly chopping with a machete at the tropical growth. **1962** S. WYNTER *Hills of Hebron* ii. 29 The shadow of a man flung across the dirt track as he stalked along, one hand swinging free, the other with his machete held at the ready. **1973** *Black World* Sept. 12/2 The blade of the machete has a deeper significance in the song of the cane cutter.

Machian (mā·kiăn), *a.* and *sb.* [f. *MACH + -IAN.] **A.** *adj.* Of, pertaining to, or characteristic of Ernst Mach (see *MACH) or his theories or ideas. **B.** *sb.* A follower or adherent of Mach or his ideas.

1927 D. KVITKO tr. *Lenin's Materialism & Empirio-Criticism* i. 69 We shall confine ourselves to those opinions which reveal the 'subjective' ignorance of our Machians. *Ibid.* ii. 85 Your statement..is utterly fallacious, and follows only from your Machian position. **1938** A. FINEBERG tr. *Lenin's Materialism & Empirio-Criticism* i. 82 Karl Pearson, the English Machian, who avoids all philosophical artifices. *Ibid.* iii. 169 James Ward...does not controvert Mach, but..utilises the entire Machian trend in physics in his fight against materialism. **1941** *Mind* L. 82 The Machian positivists of Peirce's day denied that there are 'really' any natural laws. **1958** *Victorian Stud.* I. 253 Russell's own theory of knowledge underwent an increasingly Machian development from *Our Knowledge of the External World*..to the *Analysis of Mind*, in which a full-blooded neutral monism of Mach's variety is expounded. **1963** K. R. POPPER *Conjectures & Refutations* vi. 173 In our own day essentialism has been dethroned; a Berkeleian or Machian positivism or instrumentalism has..become fashionable. **1974** *Nature* 11 Jan. 99/2 Relation (8) may be looked upon either as an equipartition between the gravitational energy and the energy of expansion according to Einstein's equations or as a Machian relation determining the gravitational constant in terms of the matter content of the Universe.

‖ **machila** (măʃī·lă). Also **machilla**. [Pg., perh. f. Tamil *macil*, *mañcil* stage in a journey, f. Hindi *manzil*, f. Arabic.] A conveyance, usually for one person, used in Africa, consisting of a hammock slung between two poles, carried by two or more bearers.

1884 MRS. M. A. PRINGLE *Towards Mountains of Moon* vii. 89 The Portuguese [in Quilliman] go from house to house in a sort of palanquin, called here a *machilla* (pronounced *masheela*). **1897** H. H. JOHNSTON *Brit. Cent. Afr.* iv. 91 We then started for Kotakota, Jumbe's men insisting on carrying me in a machilla. **1900** GROGAN & SHARP *From Cape to Cairo* 168 [He] obtained..a team of boys to carry me in a machila to the highlands of Kivu. **1944** J. C. HEENAN *Cardinal Hinsley* 64, I have now travelled in Africa in every kind of conveyance—ox carts, glorified perambulators, machilla. **1952** S. CLOETE *Curve & Tusk* (1953) iv. 38 We will carry you, Lord, we will carry you in a machila, a hammock.

b. *attrib.*, as *machila-carrier*, etc.

1900 GROGAN & SHARP *From Cape to Cairo* vi. 57 [He] utilized the bandsmen when off duty as machila-carriers. **1906** R. C. F. MAUGHAM *Portuguese E. Afr.* 14 Machilla-travelling on the Frontier.

machinability (măʃī:năbi·lĭti). [f. MACHIN(E *v.* + ABILITY.] The ability to be cut by machine tools.

1921 *Glasgow Herald* 22 Sept. 5 The influence of lead on the greater possibility of liquation from gun metal castings, the easier machinability, expected minimised corrosion and improvement for bearings. **1940** [see *GRAPHITIZER]. **1941** F. T. SISCO *Mod. Metall. for Engin.* x. 159 It is impossible to evaluate the machinability of ferrous or non-ferrous materials by a single laboratory test. **1961** *Jrnl. Iron & Steel Inst.* CXCVII. 171/1 The machinability of nodular cast iron is compared with that of steels of similar microstructure and mechanical properties. **1968** A. KOBAYASHI in *Encycl. Polymer Sci. & Technol.* VIII. 349 The machinability of polyester cast resin is relatively poor.

machinable (măʃī·năb'l), *a.* [f. MACHIN(E *v.* + -ABLE.] Capable of being cut by machine tools.

1917 *Sci. Amer. Suppl.* 16 June 374/2 If well handled in the cupola the product will be strong and machinable. **1930** *Engineering* 23 May 676/2 The resistor of a new electric furnace..is a new machinable alloy of iron with several other metals. **1951** WOLDMAN & GIBBONS *Machinability & Machining of Metals* i. 1 The most machinable metal is the one which will permit the fastest removal of the greatest amount of material per grind.

machine, *sb.* Add: **1. a.** *machine-for-living(-in)* [tr. F. *machine à habiter* ('Le Corbusier' *Vers une Architecture* (1923) p. ix)], a house. Also in imitative phrases.

1931 A. HUXLEY *Music at Night* 217 In Le Corbusier's phrase, a house is a 'machine for living in'. **1934** —— *Beyond Mexique Bay* 132 Le Corbusier himself could hardly have done the trick better: King's is the perfect machine-for-praying-in. **1960** R. W. MARKS *Dymaxion World of B. Fuller* 22/1 The house was actually the world's first tangible embodiment of what one French architect hopefully designated as a 'Machine-for-Living'. **1966** 'J. MELVILLE' *Nell Alone* vii. 75 The whole house was..a machine for Mrs Richier to live in.

b. Short for *bathing-machine* (earlier and later examples).

1788 E. SHERIDAN *Jrnl.* (1960) iv. 114, I went down to the bathing House where I found a great Number of Ladies and Gentlemen waiting to take their turn in the Machines. **1825** E. WEETON *Jrnl.* 14 June (1969) II. 384 Southport..is sadly exposing..and the modest complain much, gentlemen's and ladies' machines standing promiscuously in the water! **1870** G. MEREDITH *Lett.* (1970) I. 426 We have a flat sandy shore, and you see half a dozen fat men at a time scampering out of the machines.

e. A motor car. *U.S.*

1901 *McClure's Mag.* XVIII. i. 21/2 His assistant crouching at his feet out of range of the swift-flying currents of air produced by the mad flight of the machine. **1912** *Collier's* 21 Sept. 37/2 Leslie, Lanagan, and I hurried in the chief's machine to the Swanson home. **1915** *Sat. Even. Post* 3 Apr. 62/2 The reliability of the machine was so amazing that, in seven years of business, not a single breakdown had been reported. **1919** *Ibid.* 25 Jan. 45/3 As I neared my own house I slowed the machine.

3. Delete † *Obs.* and add later examples.

1941 G. MARX *Groucho Lett.* (1967) 48 They will wind up at Las Vegas playing the machines. **1962** *Gloss. Terms Automatic Data Processing* (*B.S.I.*) 92 *Tabulator* (*accounting machine*), a machine which reads data from a medium ..and produces lists, tables or totals. **1968** *Times* 11 Oct. 8/2 Thorpe has analysed the fish-calls of 40 sandwich terns by means of a sound spectrograph, a machine which analyses sounds, in terms of their pitch and loudness and produces a graphical representation of the sound. **1970** *Washington Post* 30 Sept. B.14/1 The stoves, the refrigerators and other machines. **1973** *Black Panther* 14 Apr. 6/1 The Visiting Room containing tables and chairs and machines for snacks and soft drinks.

4. b. A bicycle (earlier and later examples); a flying-machine, an aircraft; a mechanical printing press; a fire-engine (U.S.); a typewriter; a calculating machine or computer. Also short for *shearing-machine* (Austral. and N.Z.).

1659 T. ST. SERF tr. *Cyrano de Bergerac's Govt. World in Moon* sig. D3v, I caused a Machine to be made of Iron.. and being well seated in the seat, I cast my Magnetique Bowl into the Air. **1679** R. HOOKE in *Phil. Coll. R. Soc.* I. 15 A Machin newly Invented for Flying in the Air. **1751** R. MORRIS *Narr. Life J. Daniel* xii. 170 He had brought his machine to absolute perfection, and..had been making an experiment for flying in it. **1809** *Nicholson's Jrnl. Nat. Philos.* Nov. 172 It may be of some amusement to some of your readers to see a machine rise in the air by mechanical means. *Ibid.* 173 The little machine is completed. **1825** T. C. HANSARD *Typographia* 699 The machine being put in motion, the paper which is to be printed is laid upon a board, passed through, and receives the impression on one side. *Ibid.* 700 One machine was to perform the work of eight presses. *Ibid.* 712 My machine has had a trial of six months: its ordinary speed is..at an average of two thousand impressions, or one thousand perfected sheets per hour. **1832**, etc. [see *calculating machines* s.v. *CALCULATING *vbl. sb.*]. **1833** *Penny Mag.* Monthly Suppl. Nov.–Dec. 508/2 One thousand perfect copies..could only have been daily produced at one press by the labour of two men. The machine produces sixteen thousand copies. *c* **1848** B. A. BAKER *Glance at N.Y.* (*c* 1857) 9 I've made up my mind not to run wid der machine any more. **1859** BARTLETT *Dict. Amer.* (ed. 2) 259 *Machine*, the name for a fire-engine among the New York 'b'hoys'. **1871** *Porcupine* 29 July 276/1 Tantalising the toll-bar keepers on a 'machine'. **1891** W. D. HOWELLS in 'Twain' & Howells *Mark Twain-Howell's Lett.* (1960) II. 639 The machine with which this letter is written is a Hammond. **1892** A. POWELL *Southward's Pract. Printing* (ed. 4) ii. 11 Presses..are of two kinds, (a) hand presses and (b) mechanical presses. The latter are, in England, usually called 'machines' or 'printing machines'. *Ibid.* xlviii. 429 The one-side, single-cylinder machine, which is generally referred to when a cylinder machine, or, indeed, 'a machine', is mentioned among printers. **1899** *Northern Times* (Golspie, Sutherland) 22 June 1/1 (Advt.), Splendid cycles... Boys' machines at £5 10s. **1900** F. M. FORD *Let.* Oct. (1965) 10 My dear Galsworthy, Excuse my writing by machine; Christine at this moment monopolizes the only pen there is in the house. **1909** *Aeronautics* Dec. 151 Any machine— 'plane or dirigible. **1915** *Southward's Mod. Printing* (ed. 3) II. i. 1 In the printing office the hand press is spoken of as the 'press' and the machine press as the 'machine'.. The press can be worked by hand power only; the machine may be driven by steam, gas, or other motive power. **1919** 'BOYD CABLE' *Old Contemptibles* viii. 124 He paid more attention now to watching for enemy machines, and never failed..to rush his pilot to a machine and into the air if a German was reported in sight. **1927** H. CRANE *Let.* 30 Mar. (1965) 295 I'm *so* unhappy without a machine. Hope I get my new one soon. **1940** *Bulletin* (Sydney) 10 Jan. 16/1 'Anyone,' [Old Harry] declared, 'could put up tallies with machines. With the tongs now—'..Harry produced half a dozen pairs of tongs and some sheep. **1944** 'N. SHUTE' *Pastoral* iv. 89 The machine before them opened out and trundled down the runway. **1946** V. S. GANDERTON in H. Whetton *Pract. Printing & Binding* x. 114/2 Work on the principle of matching the new sheet to the machine, not the machine to the sheet. **1946** *Ann. Computation Lab. Harvard Univ.* I. Foreword, Harvard University's need for a machine such as the IBM Automatic Sequence Controlled Calculator has long been a matter of discussion. **1954**, etc. [see *machine code*, *instruction*, etc., in sense *10]. **1956** G. BOWEN *Wool Away!* (ed. 2) i. 9 Relative quality of workmanship between blades and machines is a debatable point. **1957** D. D. MCCRACKEN *Digital Computer Programming* ii. 14 The first two digits [of the instruction]..tell the machine what to do. **1960** N. R. SCOTT *Analog & Digital Computer Technol.* v. 170 Instructions to the machine consist of combinations of these addresses with code numbers designating the arithmetical operations to be performed. **1970** E. A. D. HUTCHINGS *Survey of Printing Processes* iv. 58 The inking unit is situated at the top of the machine, above the type-bed. **1970** O. DOPPING *Computers & Data Processing* xix. 306 A company which changes computers normally changes to a machine which is considerably faster than the old one. **1972** *Daily Tel.* 5 Aug. 9/3 For us, it was back to our bicycles. We stacked our machines in the back of the car and set off for gently contoured Norfolk.

e. *slang.* The penis; a condom (see also quot. 1896).

1749 J. CLELAND *Mem. Woman Pleasure* I. 200 Coming out with that formidable machine of his, he lets the fury loose. **1785** GROSE *Dict. Vulgar T.*, *Machines*,..See *Cundum*. *c* **1863** PHILO CUNNUS *Festival of Passions* II. 12, I then seized his stiff machine in my grasp. **1896** FARMER & HENLEY *Slang* IV. 262/2 *Machine*,..1. The female *pudendum*... 2. The *penis*.

f. *N.Z. colloq.* A totalizator.

1889 G. P. WILLIAMS in A. E. Woodhouse *N.Z. Farm & Station Verse* (1950) 26 What a lot [of money] you left behind in the 'machine'. **1891** WILLIAMS & REEVES *In Double Harness* 8 When racing was developed by the aid of the 'machine'. **1900** J. SCOTT *Tales Colonial Turf* 218 The bookmakers would not pay 30–1 as the machine is doing.

6. b. [A Gallicism.] A painting of large size.

1932 R. FRY *Characteristics French Art* III. 62 He was too poor in spirit ever to try, himself, to paint one of the big machines which made one an historical painter. **1965** *Listener* 28 Oct. 672/1 The small pictures and the *machines* appear to be the different sides of the same coin.

8. For *U.S.* read orig. *U.S.*, and add further examples. Also *transf.*

1884 *Tit-Bits* 28 June 164/3 The Business Machine was furious. He said that [etc.]. **1941** *Ann. Reg. 1940* 281 Britain unaided could not hold out against the spectacular German machine. **1942** *R.A.F. Jrnl.* 18 Apr. 13 It dared to oppose the Nazi war machine. **1948** P. D. WHITTING in M. Beloff *Hist.* xvi. 338/1 Hitler, backed by a finely organized propaganda machine, could rouse the German nation to frenzied hatred of one country after another. **1965** *New Statesman* 7 May 706/1 To some people who have observed the scope and method of operations at Transport House, the question is not simply one of finding new custodians to mind the party machine. It is whether that machine ought not to be taken apart and entirely reconstructed. **1972** *Guardian* 28 Oct. 13/4 The Labour machine had failed to pick up..the magnitude of the swing towards Cyril Smith. **1973** *Black Panther* 20 Oct. 17/3 The 'Miracle Mets'..surprised everyone by..swamping the 'Big Red' Cincinnati machine in the National League playoffs.

9. a. (sense 3) *machine aesthetic, art, form, sculpture*; (sense 4) *machine house, part, -power, -process*; (sense 4 b) *machine embroidery, lace, stitch*; (sense 8) *machine candidate, party, power*; *machine-like* adj. (earlier and later examples).

1945 H. READ *Coat of Many Colours* lxvii. 320 (*heading*) Machine aesthetic. **1967** *Listener* 8 June 745/3 The Bauhaus is one thing, and the machine aesthetic..is another. **1973** *Times* 8 Aug. 10/4 In Léger's writings of the Twenties, it is not so much speed as the mass-produced object and the 'machine aesthetic' which occupies his attention. He saw the mass-produced object as the surfacing of an anonymous natural beauty of man-made forms independent of the self-conscious architect-designer. **1945** H. READ *Coat of Many Colours* lxvii. 323 What the critics of machine art object to..is not the fact of standardization, but rather the failure to reproduce certain qualities which they regard as essential to art. **1959** —— *Conc. Hist. Mod. Painting* vi. 213 [The Bauhaus] established for the first time a course in basic design that could serve as a training for the machine art of an industrial civilization. **1950** *Economist* 9 Dec. 1004/1 The two machine candidates in New York City. **1960** G. LEWIS *Handbk. Crafts* 66 The word 'machine', perhaps, makes this sort of embroidery sound dull and mechanical, but in actual fact machine embroidery is decorative, exciting and creative, and has the added advantage that it is relatively quick to do. **1973** 'E. FERRARS' *Foot in Grave* vi. 104 I'll write and I'll paint, and I'll take up machine embroidery. **1909** W. R. SORLEY *Interpretation of Evolution* 29 Instinct..impresses the machine-form upon portions of the external world, as in the bird's nest or beaver's dam. **1955** P. HERON *Changing Forms of Art* 70 They are the crystallized thoughts of an inventor, but one who is aware of the beauty of the machine-forms which come to him out of the blue. **1808** J. STEELE *Let.* 31 Aug. in *Papers* (1924) II. 562, I bought them [*sc.* steelyards]..last winter for the use of my Machine house. **1913** J. VAIZEY *College Girl* xxvi. 360 'A neck arrangement', composed of the cheapest of machine lace. **1698–1712** SHAFTESBURY *Philos. Regimen* (1900) 114 Machine-like to be moved and wrought upon, wound up and governed exteriorly, as if there were nothing that ruled within or had the least control. **1932** E. BOWEN *To the North* xv. 151 Machine-like efficiency is not, she had been given to understand, compatible with high intelligence. **1944** *Horizon* Feb. 97 What society wants is the machine-part which does the job. **1972** *Sci. Amer.* June 122/3 A distributor of bearings and similar machine parts. **1858** *N.Y. Daily Tribune* 1 Nov. 7/6 Both of these alleged swindlers are prominent members of the 'Masheen' party of the First Ward. **1924** *Army Q.* Oct. 38 The replacement of muscle-power by machine-power is the cardinal fact in every department of material life. **1937** B. H. L. HART *Europe in Arms* xxiii. 312 As was inevitable, machine-power overcame an ill-equipped opponent. **1951** S. SPENDER *World within World* v. 284 The new phase of domination, and threat by machine-power politics. **1968** *Brit. Med. Bull.* XXIV. 189/1 The first industrial revolution is largely the history of enlisting machine power in the performance of many thousands of tasks. **1935** *Burlington Mag.* July 48/2 One familiar with machine-processes. **1970** *New Scientist* 12 Mar. 513 The ultimate comment on technology came from the American artist Tinguely who built machine-sculptures that could be exhibited only once—because they destroy themselves. **1934** WEBSTER, Machine stitch. **1964** *McCall's Sewing* vii. 98/2 Use a fine machine-stitch..and a fine machine needle.

c. *machine-driller, -knitter, -printer*; *machine-darning, -knitting, -moulding, -printing* (earlier and later examples), *-production, -riveting, -switching*; *machine-coated, -finished, -generated, -glazed, -knitted, -printed, -processable, -readable, -set, -tooled* adjs.; *machine-darn, -knit, -mould* vbs.

1963 R. R. A. HIGHAM *Handbk. Papermaking* ix. 228 To the papermaker, the term *machine-coated* signifies a paper which has been coated on the paper machine as an integral part of the papermaking process. To the printer, however, machine-coated means a class of paper. **1932** D. C. MINTER *Mod. Needlecraft* 182/2 This type of tear may also be machine-darned. *Ibid.* 177/2 Machine-darning is suitable for table-linen. **1967** E. SHORT *Embroidery & Fabric Collage* i. 17 (*caption*) Motif on net with machine darning and cut work. **1906** *Westm. Gaz.* 11 Jan. 3/1 The wages of machine-drillers on the surface are 10s. a day. **1892** W. W. GREENER *Breech-Loader* 52 The machine-made and machine-finished gun may be distinguished: First, by [etc.]. **1960** *Gloss. Paper, Stationery* (*B.S.I.*) 17 *Machine-finished* (*M.F.*) *paper*, paper treated mechanically on a paper-machine to obtain a smoother and more uniform appearance on both sides than on the unfinished paper. **1973** S. JENNETT *Making of Bks.* (ed. 5) xi. 182 Machine-finished Papers (or M.F.) have the normal finish of the paper-making machine. The surface is moderately smooth and shiny, but not glossy. **1961** F. KAUFMAN *Electronic Data Processing & Auditing* vii. 117 The loss of a machine-generated decision is surely no worse than the failure of careless or overburdened people to make such decisions. **1914** Machine-glazed [see *M.G.* s.v. *M 5]. **1959** *Gloss. Packaging Terms* (*B.S.I.*) 66 *Machine glazed* (*M.G.*) *paper or board*, paper or board which has had one side made smooth and glossy by drying on a heated, polished metal cylinder, forming part of the drying section of the machine. The other side remains relatively rough. **1962** F. T. DAY *Introd. to Paper* iv. 44 The M.G. high-speed single-cylinder paper making machine illustrated here is a standard type of equipment employed in the mill for making thin machine glazed papers. A popular name for the M.G. or cylinder machine is the 'Yankee'. **1927** T. WOODHOUSE *Artificial Silk* 83 Enormous lengths were machine-knitted into hose and half-hose. *Ibid.* 79 The utilization of artificial silk yarn for hand-knitted and machine-knitted articles. *Ibid.* 83 If a machine-knitter does not wind the yarns in his own mill, he can have them supplied in the form of bottle bobbins. **1886** *Family Friend* Jan. 87/1 Machine-knitting. **1927** T. WOODHOUSE *Artificial Silk* 86 In machine-knitting several courses are formed simultaneously. **1922** *Encycl. Brit.* XXX. 36/1 By 1915–6 cast-iron cylinders were cast from metal patterns and machine-moulded. **1888** *Lockwood's Dict. Mech. Engin.* 217 *Machine-moulding*, ..embraces the moulding of wheels and ordinary work by the aid of special machines. **1949** F. BOWERS *Princ. Bibliogr. Descr.* x. 355 It seems necessary for the purposes of descriptive bibliography to draw a chronological line after which the methods of description for machine-printed books will in general hold. **1963** *Times Lit. Suppl.* 26 Apr. 312/1 Machine-set and machine-printed books. **1909** *Westm. Gaz.* 24 Sept. 8/1 An old man..described as a machine-printer. **1825** T. C. HANSARD *Typographia* 714 Machine printing will..be only applicable to works of extensive sale. **1892** A. POWELL *Southward's Pract. Printing* (ed. 4) liii. 467 (*heading*) Some difficulties in machine printing. **1972** P. GASKELL *New Introd. Bibliogr.* 260 Inks for machine printing differed little from those for the hand-press period. **1967** COX & GROSE *Organiz. Bibliogr. Rec. by Computer* VII. 185 The B.N.B. machine-processable records. **1971** J. B. CARROLL et al. *Word Frequency Bk.* p. ix, Machine-processable data for lexicography. **1898** J. A. HOBSON *John Ruskin* ix. 217 The 'driving' tendency of modern machine-production. **1931** L. WATT *Future of Capitalism* iv. 42 This 'dilemma' of technological unemployment (unemployment resulting from the development of machine-production) would..face any form of economic organisation. **1961** *Times* 30 Oct. (Computer Suppl.) p. ix/6 The three basic types of machine-readable document. **1971** *Computers & Humanities* V. 301 To collect in machine-readable form a million-character corpus of modern vernacular literature. **1888** *Lockwood's Dict. Mech. Engin.* 217 *Machine-riveting*, riveting performed by a single application of steady pressure at the same instant upon the tail and the head of a rivet. **1908** KIPLING *Lett. of Travel* (1920) 154 The brittle pulp-paper, the machine-set type, are all as standardised as the railway cars of the Continent. **1967** KARCH & BUBER *Offset Processes* iii. 47 The layout man may choose to use proofs (also called proof-press prints). These proofs may be hand-set, machine-set or both. **1922** GLAZEBROOK *Dict. Appl. Physics* II. 834/1 Several types of machine switching or automatic exchange systems have been devised, but in each of them the principle is to move the terminal of the calling line to that of the called line, which is fixed. **1950** J. ATKINSON *Herbert & Procter's Telephony* II. i. 1/1 The idea of automatic or machine switching is by no means new. **1962** *Times* 3 Mar. 11/3 Hook-making..is a high-speed machine-tooled operation.

10. machine age, a name given to an era notable for its extensive use of mechanical devices; also *attrib.*; **machine code**, a code (CODE *sb.*¹ 3 b or *3 c) prepared by or for the use of a machine; *spec.* = **machine language*; so; *machine-coded* adj.; **machine finish**, a moderately smooth finish that paper has after leaving the machine on which it was made (see quots. 1937, 1960); **machine-hours**, hours during which a machine operates; **machine instruction** *Computers*, an instruction (*INSTRUCTION 4 c) in a machine language; **machine language** *Computers*, a language (*LANGUAGE *sb.* 1 d) that a particular computer can handle or act on directly, without further translation; **machine-oriented** *a. Computers*, (of a computer language) devised in the light of the requirements of a particular kind of computer; **machine-pistol**, a submachine-gun; **machine proof** = *press-proof* (PRESS *sb.*¹ 16 b); **machine-room** *Printing*, the room in which printing presses are operated; **machine-shop** (earlier U.S. example); **machine-time**, time during which a computer is in use; **machine translation**, translation by a computer; **machine word** see *WORD *sb.*; **machine-work**, (*b*) (examples).

1922 L. MUMFORD in H. E. Stearns *Civilization in U.S.* 11 These buildings..shall embody all that is good in the Machine Age. **1934** H. READ *Art & Industry* I. i. 6 Has he [*sc.* the artist] any function in a machine-age society? **1967** SINGHA & MASSEY *Indian Dances* i. 36 Shaivism itself, under the impact of the new materialistic machine age had lost its religious fervour. **1954** *First Gloss. Programming Terminol.* (Assoc. Computing Machinery) 4 *Computer code* (*Machine code*), the code representing the operations built into the hardware of the computer. **1958** G. GREENE *Our Man in Havana* I. iv. 45 Of course it's [a book-code] not so hard to break as a machine-code. **1971** LOWE & HIDDEN *Computer Control in Process Industries* v. 108 The machine language, or machine code,..is the repertoire of instructions for the basic operations that the central processor is designed to perform. **1964** F. L. WESTWATER *Electronic Computers* ix. 144 A special routine called a 'compiler'..produced an efficient machine-coded program from the pseudo-code. **1907** CROSS & BEVAN *Text-bk. Paper-Making* (ed. 3) x. 270 The mill or machine finish is one which can be varied within wide limits. **1937** E. J. LABARRE *Dict. Paper* 170/1 Machine finish is the surface of the paper (1) as it leaves the last drying cylinder of the paper machine; (2) as it leaves the calenders immediately following the paper-machine. **1960** G. A. GLAISTER *Gloss. Bk.* 245/1 *Machine-finish*, paper made smooth, but not glossy, by receiving the normal finish of a Fourdrinier paper-making machine which completes its process by passing the paper over heated drums and through steel calendering rolls. These smooth the surface to the required degree. **1921** ECCLESTON & ROBINSON *Business Costs* 377 Direct labor and overhead... Machine Hours 30. **1966** A. BATTERSBY *Math. in Managem.* vii. 173 For the sake of simplicity, we may choose to use a measure such as 'idle machine-hours', on the grounds that a reduction in idleness will automatically bring down operating costs. **1956** Machine instruction [see *LANGUAGE *sb.* 1 d]. **1970** Machine instruction [see *MICROPROGRAMMER]. **1949**, etc. Machine language [see *LANGUAGE *sb.* 1 d]. **1967** A. HASSITT *Computer Programming* ii. 41 There are a series of programs..which accept Fortran statements as data and produce machine language statements as output. *Ibid.*, Although there are many different machine languages, many concepts are common to all of these languages. Some of the common ideas are binary arithmetic, index registers, memory addresses, [etc.]. **1968** *Brit. Med. Bull.* XXIV. 192/1 The user prepared his program in a.. computer language..which the computer itself translated into its own basic machine language. **1967** D. WILSON in Wills & Yearsley *Handbk. Managem. Technol.* iii. 47 These programs are often referred to as problem-oriented languages, as opposed to the lower-level assembly or autocoder languages which are more commonly used at present and are machine-oriented. **1970** O. DOPPING *Computers & Data Processing* xiv. 227 When problem-oriented programming languages..are used instead of simple machine-oriented languages, programming time is often reduced drastically. **1940** *Illustr. London News* CXCVI. 786 (*caption*) Much has been heard of the machine-pistols used by Nazi parachutists. **1962** *Spectator* 1 June 710/1 A Police State that tries to stop runaway schoolboys with machine-pistol fire. **1973** M. WOODHOUSE *Blue Bone* xiv. 153 A man with red hair and a machine pistol. **1951** S. JENNETT *Making of Bks.* vi. 88 The machine proof..is pulled immediately before the forme goes on the press, or while it is actually on the press. **1961** T. LANDAU *Encycl. Librarianship* (ed. 2) 233/1 *Machine revise*, a proof printed when the forme is on the printing machine... Also called machine proof. **1833** *Penny Mag.* Monthly Suppl. Nov.–Dec. 510/1 We will conduct our readers to Mr. Clowes's printing establishment, where there are more printing machines at work than at any other office in the world... Upon entering the machine-room the stranger will naturally feel distracted by the din of so many wheels and cylinders in action. **1904** *Brit. Printer* Feb. 6/2 One of the strong points of the establishment—its machine-room accommodation—is examined. **1946** V. S. GANDERTON in H. Whetton *Pract. Printing & Binding* x. 128 (*caption*) A fine example of a modern letterpress machine room. **1972** P. GASKELL *New Introd. Bibliogr.* 294 The machine-room overseer, an important man who ran the hand-press department as well as the machine-room. **1827** *Aurora* (Philadelphia) 25 July 1/3 A Machine Shop, from 60 to 70 feet long and 20 feet wide, two stories high. **1968** *New Scientist* 12 Sept. 548 In my environment the majority of user time, not machine time but user time, is spent in writing and running short programs. **1973** *Computers & Humanities* Mar. 198 The operation that consumes most machine time is the verification of Rule II, where we test that each sentence is contained in the union of at most three others. **1949** W. WEAVER in Locke & Booth *Machine Transl. of Lang.* (1955) 20 Mr. Max Zeldner, in a letter to the *Herald Tribune* on June 13, 1949 [*published* June 26], stating that the most you could expect of a machine translation of fifty-five Hebrew words which form the 23d Psalm would start out Lord my shepherd no I will lack [etc.]. **1956** *Nature* 7 Jan. 1/1 Dr. Booth is optimistic that even the problems of machine-translation of literary work may prove less complex than they at present appear. **1960** E. DELAVENAY *Introd. Machine Transl.* 123 Machine translations today are still very imperfect. **1968** J. LYONS *Introd. Theoret. Ling.* iv. 159 The automatic analysis of written texts for the purpose of machine-translation. **1861** B. HEMYNG in H. Mayhew *London Labour* (1862) Extra vol. 222/1 She then supported herself and her child by doing machine-work for a manufacturer. **1867** A. D. WHITNEY *Summer in L. Goldthwaite's Life* i. 10 No machine-work, but all real dainty finger-craft.

machine, *v.* Add: **2. a.** Also with *in*.

1894 J. E. DAVIS *Elem. Mod. Dressmaking* 47 Tacking is not strong enough to hold sleeves well to the arm-hole for machining-in.

b. *absol.* To manufacture things by machinery.

1916 H. G. WELLS *Mr. Britling* I. i. 16 They had standardised and machined wholesale, while the British were still making the things one by one.

4. a. Delete *nonce-use* and add later example. **c.** To render mechanical; to treat as if machinery.

1916 F. M. FORD *Let.* Sept. (1965) 72 The French Press ..continues to blaze and coruscate about my gifts... Of course these salvos are a little machined by the French Govt. **1916** H. G. WELLS *Mr. Britling* I. ii. 67 The reality of life is adventure, not performance. What can be ruled about can be machined. **1919** J. L. GARVIN *Econ. Found. Peace* 183 As they drilled under arms or machined their Socialism. **1959** *Listener* 19 Nov. 868/2 'The new poets,' Apollinaire wrote in 1917, 'will one day machine poetry (*machiner la poésie*) as the modern age has machined the world.'

machine-gun, *sb.* (see MACHINE *sb.* 10.) Add earlier and later examples. Also *fig.*, *attrib.*, and *Comb.*

Some more recent types of machine-gun can be fired from the hip and are adaptable for firing single shots.

1870 *Jrnl. R. United Service Inst.* XIV. 504 Machine Guns: The 'Gatling Battery'—The Agar and Claxton Guns —The French and Montigny Mitrailleurs. **1882** *Army & Navy Mag.* Dec. 195 Machine-guns, which have an effective range not differing materially from that of small arms, would..have to advance with infantry. **1906** *Westm. Gaz.* 4 Oct. 2/3 Motor-bicycles, those machine-gun terrors of the road. **1909** *Ibid.* 9 Dec. 5/1 Experiments carried out with the machine-gun fitted aeroplane. **1915** R. W. CAMPBELL *Private Spud Tamson* xix. 288 Z-r-r-p—Z-r-r-p—Z-r-r-p spat the machine-gun batteries behind the little knolls. **1919** 'BOYD CABLE' *Old Contemptibles* ix. 146 The steady postman's-knock *rat-tat-tat* of machine-guns. *Ibid.* xvi. 255 The rifle and machine-gun fire rose again. **1928** H. LASKI in *Holmes–Laski Lett.* (1953) II. 1077 They cross-examined me with machine-gun rapidity. **1937** KOESTLER *Spanish Testament* v. 104 Militiamen were frequently obliged to smoke out the machine-gun nests set up in the monasteries. **1941** *Ann. Reg. 1940* 154 [France] sent..5,000 machine-gun rifles. **1955** S. SPENDER *Coll. Poems 1928–53* 99 Machine-gun anger quickly scythed the grasses. **1970** *Toronto Daily Star* 24 Sept. 3/3 They put their machine-gun nests on roofs and in the rooms and gardens of private houses. **1972** 'E. LATHEN' *Murder without Icing* (1973) iii. 30 Convulsions on one level or another could be expected with..machine-gun rapidity. **1973** *Times* 14 Nov. 8/1 The Israelis..have reinforced their machinegun posts beside the road.

So **machine-gun** *v. trans.*, to turn a machine-gun on, to fire at with a machine-gun; **machine-gunning** *vbl. sb.*; **machine-gunner**, one who operates a machine-gun. Also *fig.*

1915 R. BROOKE *Let.* 29 Jan. (1968) 658 There's a good chance of my going to Hythe for a fortnight to learn machine-gunning. **1915** 'I. HAY' *First Hundred Thousand* xix. 279 The machine-gunner is a more or less accepted nuisance by this time. **1917** 'CONTACT' *Airman's Outings* 185 Other guerilla work is done by craft which..machine-gun whatever worth-while objects they spot. **1918** E. M. ROBERTS *Flying Fighter* 38 Whatever they saw was sure to be machine-gunned. **1930** J. B. PRIESTLEY *Angel Pavement* ii. 76 They sweep, lash, and machine-gun the streets with rain. **1942** *R.A.F. Jrnl.* 2 May 5 Our work was mainly bombing and machine gunning, to weaken the will of the enemy. *Ibid.* 16 May 21, I..maintained my course to machine-gun the decks. **1955** *Times* 25 June 6/3 He said that a conspiracy was hatched last November for a military revolt... The aim was to murder the President of the Republic through air bombardment or land action and to intimidate civilians by air bombing or machine-gunning the streets. **1967** B. PATTEN *Little Johnny's Confession* 17 The last young thigh..machineguns you down. **1970** T. HUGHES *Crow* 24 The face Of a machine-gunner a long burst not long enough. **1972** *Daily Tel.* 1 June 2/8 This was followed by a machine-gunning of the Spanish Embassy on December 3. **1973** D. LEES *Rape of Quiet Town* vi. 102, I aimed at one of the machine-gunners.

machi·neless, *a.* [f. MACHINE *sb.* + -LESS.] That does not use or does not require a machine or machines. Spec. *machineless waving* (see quot. 1966); also *ellipt* as *sb.*.

1909 in WEBSTER. **1926** *United Free Ch. Miss. Rec.* May 201/1 It was a machineless age, the world moved at a slow pace. **1933** *Catholic Med. Guardian* Apr. 90 Catholic Law Associations..advocate the return to machineless handicrafts. **1942** *Horizon* July 58 One of Thelma's ladies was settin' over yonder..gitting a machineless. **1966** J. S. COX *Illustr. Dict. Hairdressing* 93/1 *Machineless waving*, (1) Hot permanent waving in which the required heat is either generated chemically in a pad applied to the hair, or by means of a pre-heated metal clamp. (2) Cold Permanent Waving.

machi·ne-wash, *v.* [f. MACHINE *sb.* + WASH *v.*] *trans.* To wash in a washing-machine. So **machi·ne-wa·shable** *a.*

1960 *Farmer & Stockbreeder* 26 Jan. (Suppl.) 4/3 Some drip-drys can be successfully machine-washed. **1962** *N.Y. Post* 9 Oct. 54 Stretch-ever elastic made without rubber so you can machine-wash it. **1963** *New Yorker* 8 June 87 Guaranteed machine-washable. **1970** *Guardian* 28 May 11/5, I doubt if you would really machine-wash a child's walking shoe. *Ibid.* 16 June 13/2 In Welsh peasant cotton, machine washable. Jacket with inset waistband. **1975** *Country Life* 13 Mar. 673/3 Machine-washable continental quilts.

machinofacture (măʃī:nofæ·ktiŭɹ, -tʃiŭɹ). [f. MACHINE *sb.* + -O + FACTURE.] The making of articles by machine; mechanization.

1903 L. F. WARD *Pure Sociol.* 26 The invention of tools, instruments, utensils, missiles, traps, snares, and weapons comes under this head, crowned by the era of machinofacture, artificial locomotion, and electric intercommunication. **1928** E. & C. PAUL tr. *Marx's Capital* xiii. 403 An organised system of working machines which are one and all set in motion by the transmitting mechanism from a central automaton, constitutes the fully developed form of machinofacture.

machir(r), varr. *MACHAIR.

Machism (mā·kiz'm, -χ-). [f. name of Ernst *Mach* (see *MACH) + -ISM.] The theories of Ernst Mach, esp. his concept of empiriocriticism.

1927 D. KVITKO tr. *Lenin's Materialism & Empirio-Criticism* i. 69 In philosophical literature, writers of various tendencies have long since discovered the chief sin of Machism in spite of all its covering. **1949** P. FRANK *Mod. Sci. & its Philos.* x. 194 A logical contradiction exists only between a metaphysically conceived Machism, which is then subjective idealism, and a metaphysically conceived materialism, which accepts only matter as having existence. **1968** *Listener* 14 Nov. 648/3 Lenin was particularly incensed at Machism, and wrote his *Materialism and Empiriocriticism* against Mach and Avenarius. **1972** J. T. BLACKMORE *Ernst Mach* xiv. 214 If we may define 'pure Machism' as a phenomenalism hostile to atomism and theoretical physics, [etc.].

machismo (mătʃī·zmo). [Mexican-Sp., f. *mach(o* masculine + *-ismo* -ISM.] The quality of being *MACHO; male virility, masculine pride. Also *attrib.*

1948 B. GRIFFITH *Amer. Me* 50 Machismo makes a boy swear big round oaths as a youngster. **1962** *Listener* 12 Apr. 649/1 The Mexican obsession with the concept of *machismo*, of masculinity. **1967** MCCORMICK & MASCAREÑAS *Compl. Aficionado* iii. 74 To bite the cape like an animal while folding it is false machismo. **1969** J. MANDER *Static Society* i. 56 The exaggerated masculinity, the famous *machismo*, of the Mexican. **1969** A. MARIN *Rise with Wind* v. 63 The bus driver..drove with Latin *machismo* through a tangle of narrow streets. *Ibid.* x. 125 A eunuch was without *machismo*, that peculiar combination of pride and virility that was the essence of manliness. **1971** *Guardian* 20 July 11/8 The contradictions, hypocrisy, and 'machismo' ideal in Puerto Rican culture. **1973** GAGNON & SIMON *Sexual Conduct* (1974) viii. 251 In the prison, toughness may substitute for intercourse as a measure of *machismo*.

Machmeter (mā·k-, mæ·kmītəɹ, mā·χ-). *Aeronaut.* Also **Mach meter**. [f. *MACH + METER *sb.*³] An air-speed indicator that reads directly in Mach numbers.

1947 *Jrnl. R. Aeronaut. Soc.* LI. 734/1 An instrument known as a Mach meter, which indicates the Mach number of flight at any height, has already been developed and is now being fitted in certain high-speed aircraft. **1948** 'N. SHUTE' *No Highway* xii. 305 What was the Machmeter showing? **1958** 'P. BRYANT' *Two Hours to Doom* 118 He watched the Machmeter carefully as it moved up from point nine. **1964** J. E. D. WILLIAMS *Operation of Airliners* xx. 320 The indication of the Mach meter is not..instantaneously responsive to changes of engine thrust. **1975** L. J. CLANCY *Aerodynamics* iii. 32 A Machmeter..depends on measurements of total and static pressure by the pitot-static tube.

macho (matʃo, mæ·tʃo), *sb.* and *a.* orig. *U.S.* [Mexican-Sp. *macho*, a male animal or plant; *adj.*, masculine, vigorous.] **A.** *sb.* A man; *spec.* a 'tough guy'; also, manliness, virility; an impression of this. **B.** *adj.* Ostentatiously or notably manly and virile.

1928 *Nation* 29 Feb. 233/2 The Machos (Americans) have taken El Chipote. *Ibid.* 11 Mar. 288/1 Here was I in their midst, a Macho Yankee Gringo, yet treated with consideration. **1951** *Sat. Rev. Lit.* 22 Sept. 15/2 In the Continente, men were supposed to be *machos*—males—and the women were supposed to bear their children, besides keeping their houses, and awaiting with patience their returns from the beds of their mixed-breed mistresses, or the battlefield. **1959** N. MAILER *Advts. for Myself* (1961) 19 Every American writer who takes himself to be both major and *macho* must sooner or later give a *faena* which borrows from the self-love of a Hemingway style. *Ibid.* 418 'Man, you can take care of yourself,' he said with glee. 'I don't know about that,' I answered, obeying the formal minuet of the *macho*. **1964** *Punch* 25 Mar. 444/1 A quality much prized in Mexico called *macho*, namely 'masculinity, virility'. **1964** S. BELLOW *Herzog* 157 Provided that he remain *macho* she would listen with glistening eyes. *Ibid.* 186 A prince of the erotic Renaissance, in his *macho* garments. **1968** T. HOWARD *Black Light* xix. 163 The medical practitioner had to salvage a little of that all-important Latin-American *macho*—or manliness. **1972** *Publishers' Weekly* 23 Oct. 40/2 Reveals the *macho* of the sport for what it is: gridiron Darwinism, young athletes psyched out of their skulls. **1972** *New Yorker* 2 Dec. 159/3 And so we have separate cultures—black-*macho* movies and white-*macho* movies, equally impoverished, equally debased. **1975** *Ibid.* 14 July 65/1 She [*sc.* Greta Garbo] played opposite Clark Gable once, and the collision, though heated, didn't quite work; his macho directness—and opacity—reduced her from passionate goddess to passionate woman.

machree (măkrī·). *Ir.* Also **Machree, ma chree, mochree.** [Ir.-Gaelic *mo chroidhe* (of) my heart, my dear.] My dear! Often in phr. *Mother Machree.*

1829 G. GRIFFIN *Collegians* (ed. 2) II. xx. 92 Coax him, *ma chree, ma lanuv*, to gi' me the price o' the whiskey. *Ibid.* xxiv. 195 Oh, ma chree, m'asthora... What ails you? **1831** S. LOVER *Legends & Stories Ireland* p. xxiii, *Machree*, my dear. *Ibid.* 89 They wint to the wine-cellar, but, jew'l machree, they soon run back into his room. **1866** MRS. GASKELL *Wives & Daughters* II. xvi. 160 Thanks to you, little Molly—cuishla ma chree, pulse of my heart. **1918** N. MUNRO *Jaunty Jock* 159 'I could live on the berries.., I love them!' 'Doubtless, mochree,' would he answer her, laughing. **1918–19** *T. Eaton & Co. Catal.* Fall & Winter 389/2 Vocal Records... Mother Machree (Tenor). **1952** E. O'NEILL *Moon for Misbegotten* (1953) 132 In a minute you'll start singing 'Mother Machree'! **1970** M. KENYON *100,000 Welcomes* iv. 32 'You ever been to Ireland, Louie?'..'Every year... Ol' Mother Machree. And see the Micks get stoned.' **1973** *Listener* 1 Mar. 277/2 The Celtic Revival, born as it was of romantic nationalism—Mother Machree, Kathleen ni Houlihan and the whole pack of Gaelic Norns started calling to us.

‖ **macht-politik** (ma·χt,pǫlītīk). Also **macht politik, machtpolitik.** [G., f. *macht* power, strength + *politik* policy, politics.] Power politics; strength as a potential factor to use in gaining a desired result.

1916 J. W. HEADLAM in Prince von Bülow *Imperial Germany* (rev. ed.) p. xxiii, Let us consider these things purely as a balance-sheet of loss and gain according to the undiluted principles of 'Macht-Politik'. **1940** *Horizon* May 326 Our leaders display not the qualities of *Machtpolitik*, but the caution of the peaceful negotiator. **1958** *Times Lit. Suppl.* 11 Apr. 190/4 Indeed, the uniformly cynical *Machtpolitik* practised by every leading actor in the drama is profoundly depressing. **1964** P. MEADOWS in I. L. Horowitz *New Sociol.* II. 456 A space age revolution.. in which the technological *Machtpolitik* of space conquest is also serving as a relatively moral equivalent for war. **1969** J. MANDER *Static Society* vi. 179 Naked *machtpolitik* ..calls countervailing forces into being. **1973** *Times* 26 Oct. 20/1 The Chinese, in a gesture of fastidious distaste, abstained from the Security Council resolution on the reasonable grounds that although it was in effect a blatant example of Soviet-American *macht politik* it would be perverse to vote against it simply on that account.

Machzor (mā·χzǭɪ). Also **Machsor, Mahzor.** Pl. **-im.** [Heb. *maḥzor* a cycle.] In Jewish liturgy, the name for the books of prayers and readings used on various occasions in the liturgical year; usu. used *spec.* for the book of prayers for use at festivals.

1864 *Chambers's Encycl.* VI. 155 Later, the term [sc. *Siddurim*] was restricted to the weekday ritual, that for festivals being called *Machsor* (Cycle). **1883** *Encycl. Brit.* XV. 292/2 The *Maḥzor*, meaning prayer-book, is capable of division..into..the Smaller and the Larger. The Smaller Maḥzor contains ordinary prayers,..with the poetical insertions and lessons from the Pentateuch and the Prophets... The Larger Maḥzor..embodies the ordinary prayers..for the whole year, and the lessons..for all feasts and fasts and..other occasions. **1891** M. FRIEDLÄNDER *Jewish Relig.* 391 The *Machzor*..or Prayer-book for the Holy-days contains numerous additions to the ordinary prayers. **1892** I. ZANGWILL *Childr. Ghetto* I. 276 Three sets of Machzorim, or Festival Prayer-Books. **1893** —— *Ghetto Tragedies* 66 She bent her eyes downwards on her neighbour's *Machzor*. The woman immediately pushed the prayer-book more towards Rebecca. **1973** *Synagogue Light* Sept. 42/2 The festival prayer book, the Machzor, is used during the High Holidays because many additional prayers not recited throughout the year are read on these days. **1974** *Encycl. Brit. Micropædia* VI. 505/3 *Maḥzor*.., originally a Jewish prayer book arranged according to liturgical chronology and used throughout the whole year... *Maḥzor* has come to mean the festive prayer book, as distinguished from the Jewish prayer book (Siddur) used on ordinary sabbaths and on week-days.

McIntosh (mæ·kintǫʃ). Also **MacIntosh, Mackintosh.** The name of John *McIntosh* (b. 1777), Canadian farmer, used *attrib.* in **McIntosh Red** to designate a red-skinned variety of eating apple or the tree producing it, first developed on his farm. Also *absol.*

1878 *Canad. Horticulturalist* Mar. 42 Winter apples; here my list will be small, but I think reliable: Talman's Sweet, Pomme Grise, American Golden Russet, and McIntosh Red. **1908** *Busy Man's Mag.* Feb. 89 The First McIntosh Red Tree in Canada. **1910** L. WOOLVERTON *Canad. Apple Grower's Guide* 190 McIntosh [is] a very fine dessert apple for early winter use. **1932** [see *DELICIOUS *a.* 2 b]. **1933** M. DE LA ROCHE *Master of Jalna* xii. 136 He extracted a large, perfectly shaped MacIntosh Red from his pocket. **1965** MRS. L. B. JOHNSON *White House Diary* 15 Jan. (1970) 221 Our best brands are Northern Spies and MacIntosh Reds. **1969** *New Yorker* 20 Dec. 76/2 Already on our six acres..we have planted..a MacIntosh apple. **1970** *Globe & Mail* (Toronto) 28 Sept. 5/2 (*caption*) Alison Loates..found a bright-red McIntosh was her fancy.

mack, *sb.*[4] (Later examples.)

1926 MAINES & GRANT *Wise-Crack Dict.* 11/1 *Mac*, man who lives off the earnings of a woman. **1931** G. IRWIN *Amer. Tramp & Underworld Slang* 125 *Mac*, a pander; a lover or associate of lewd women. No doubt from the French word for this class, 'maqereau'.., although the shorter word has been in use in America for years. **1950** BLESH & JANIS *They all played Ragtime* (1958) ii. 39 The dapper, foppish 'macks'..in their Stetsons, box-back coats, and St. Louis 'flat' shoes got their gambling stakes from the girls. **1972** T. KOCHMAN *Rappin' & Stylin' Out* 243 'Pimp', or 'mack man'..a person of considerable status in the street hierarchy, who, by his lively and persuasive rapping ('macking' is also used in this context), has acquired a stable of girls to hustle for him and give him money. **1973** *Washington Post* 21 Apr. D 7/2 Now comes 'The Mack', a movie about the rise and fall of a sweet pimp named Goldie.

mack (mæk), *sb.*[5] Also **mac.** A common abbrev. of MACKINTOSH 2 and 3.

1901 'R. ANDOM' *Troddles & Us & Others* xxi. 245 It rained pretty steadily... Murray and Wilks, having left their 'macs.' behind, were constrained to spend one solid day loafing about the..inn premises. **1902** *Captain* VII. 468/1 Who said you might wear my mackintosh?.. Suppose you give it up... Buck up. It looks like rain... Mack up, please. I want it. **1917** A. G. EMPEY *From Fire Step* 170 In front of the door stood an officer in a mack (macintosh). **1923** *Daily Mail* 12 Feb. 12 (Advt.), Girls' Mack Capes with Hoods... All guaranteed waterproof. **1929** GALSWORTHY *Roof* iii. 54 'Have we got to dress?'.. 'No; bung on your mack and shoes.' **1963** *Times* 15 May 14/6 Small knots of mac-clad farmers. **1973** A. BEHREND *Samarai Affair* viii. 85 Richardson slipped on his mack and went round to India buildings. **1974** D. WINSOR *Death Convention* ii. 10, I had dug my hands in my mac pockets.

Mack (mæk), *sb.*[6] [f. *Mack*, name of manufacturer.] The proprietary name of several types of heavy vehicle, as lorries, tractors, etc. Also *attrib.*

1913 *Hand Bk. Gasoline Automobiles* (U.S. Automobile Board of Trade) 165 (*caption*) Mack 2-ton truck. **1921** *Official Gaz.* (U.S. Patent Office) 4 Jan. 192/1 International Motor Company, New York, N.Y... Mack.. [for] Motor-Trucks. Claims use since about Oct. 13, 1911. **1930** J. DOS PASSOS *42nd Parallel* v. 401 Charley..started to help clean the parts of the carburetor of a Mack truck. **1948** PARTRIDGE *Dict. Forces' Slang* 114 *Mack*, a ten-ton lorry. (Army.) **1959** E. K. WENLOCK *Kitchin's Road Transport Law* (ed. 12) 85/1 Mack type N.M. heavy artillery tractors may be used for snow ploughing, or grit or salt distribution on frosty roads, notwithstanding their excess width. **1962** *Times* 27 Nov. 13/3 A fleet of 'Macks'—snow-shifters. **1971** M. TAK *Truck Talk* 103 *Mack*, a popular, economical and long-wearing make of tractor, the one best known to the public... The Mack engine has a characteristic sound. .. The Mack trademark is a bulldog. **1972** MARSHALL & BISHOP *Lorries, Trucks & Vans* 123 No other name can surely be more autonomous with the US truck industry than Mack. 'Built like a Mack' is still a phrase used by Americans to signify something solid or sturdy. **1973** *Trade Marks Jrnl.* 13 June 1141/2 Mack... Commercial motor road vehicles and parts and fittings therefor.. Mack Trucks, Inc..., Allentown, State of Pennsylvania, United States of America; manufacturers.

Mackay: see *McCOY.

mackayite (mākəi·əit). *Min.* [f. the name of John W. *Mackay* (see quot. 1944²) + -ITE¹.] A green hydrated tellurite of ferric iron, perhaps $Fe_2(TeO_3)_3.xH_2O$.

1944 FRONDEL & POUGH in *Amer. Mineralogist* XXIX. 217 The new mineral described here under the name mackayite was found in specimens from two deposits in Goldfield, Nevada. *Ibid.* 218 At the instance of Mr. C. D. Woodhouse, the writers propose the name mackayite for the species after John W. Mackay (1831–1902), a mining operator who in a few years amassed a great fortune on the Comstock Lode, Nevada. The name is intended to recognize Mackay's financial endowment of the School of Mines of the University of Nebraska. **1968** I. KOSTOV *Mineral.* 518 Mackayite is tetragonal.., found as short prismatic green crystals.

McKenney (māke·ni). [f. the name of W. E. *McKenney* (1891–1950), who popularized it.] The name given to a suit preference signal in Bridge, devised by the American player Lavinthal in 1934. Freq. as *McKenney convention.*

1952 I. MACLEOD *Bridge is Easy Game* xiii. 158 You should study and play what is called the 'McKenney' suit preference signal. *Ibid.* 159 East must make a 'McKenney' by throwing the Jack of Diamonds calling for a Spade switch. There is a danger that a McKenney signal may be confused with the ordinary demand for the suit to be continued. **1959** *Listener* 20 Aug. 298/3 Many partnerships, therefore, have a rule, 'No McKenney at trick 1'—McKenney being an alternative name for a suit preference signal. **1964** *Official Encycl. Bridge* 341/2 *McKenney*, standard term in Great Britain for the suit preference signal... McKenney's support..caused the convention to be called in European countries the McKenney convention. **1965** *Listener* 3 June 842/1 The convention known as the McKenney convention is a form of suit-preference signalling, publicized by McKenney, but invented by another American player, Hy. Lavinthal. It differs from other suit-signalling conventions in as much as the signals relate to suits other than that played.

mackerel¹. **1. a.** For '*Scomber scomber*' substitute '*Scomber scombrus*' and add: Also used for other fishes of similar appearance belonging to the family Scombridæ. (*Austral.* and *N.Z.* examples.)

1838 J. S. POLACK *New Zealand* I. ix. 322 The *pātiki*.. is equally excellent with the European fish [sole], as are also the *mackarel*, of which there are several varieties. **1843** E. DIEFFENBACH *Trav. N.Z.* II. 209 *Scomber loo*.. (*Scomber scombrus*, Solander, Pisc. Austr., p. 31.) Solander observed this mackerel in Queen Charlotte's Sound. **1886** R. A. SHERRIN *Handbk. Fishes N.Z.* 61 In season mackerel are often found between Cape Colville and the Great Barrier. **1951** T. C. ROUGHLEY *Fish Austral.* 96 Common mackerel (Slimy mackerel—*Pneumatophorus australasicus*). The common mackerel (of the family Scombridæ) is found in all Australian States and occurs in great numbers, particularly round the southern half of the continent where it inhabits ocean waters no great distance from the coast. **1960** DOOGUE & MORELAND *N.Z. Sea Anglers' Guide* 250 Common mackerel..is the mackerel of English-speaking countries... Other names: *Pneumatophorus japonicus*; southern mackerel, English mackerel, frigate mackerel; tawatawa (Maori).

c. Phr. *Holy Mackerel*, an exclam. expressing wonder or astonishment.

1899 ADE *In Babel* (1903) 111 Hot? Holy sufferin' mackerel! Me pushin' up the lid..to get a little fresh air. **1944** T. RATTIGAN *While Sun Shines* II. 218 Holy mackerel! A Duke! **1958** 'J. BROGAN' *Cummings Report* xviii. 189 Holy mackerel! What a way to run an army! **1961** *Amer. Speech* XXXVI. 40 *Holy Mary* is probably the idea underlying *holy Moses* and *holy mackerel*.

4. *mackerel fleck*; **mackerel shark,** substitute for def.: a shark of the family Lamnidæ, esp. a mako shark (*MAKO¹) or the porbeagle, *Lamna nasus* (examples); **mackerel-skied** *a.*, having, or characterized by, a mackerel-sky.

1940 R. GIBBINGS *Sweet Thames run Softly* xx. 182 Tall nimbus clouds reared their heads towards the mackerel flecks in the upper air. **1819** *Plough Boy* I. 135 The revenue cutter brought in two very strange fish, found eating a dead horse, supposed to be mackerel sharks. **1959** A. HARDY *Fish & Fisheries* iv. 73 The inclusion here [*sc.* in the British list] of the mako or mackerel shark, *Isurus oxyrinchus*, is a surprise. **1971** *Islander* (Victoria, B.C.) 17 Oct. 6/1 A..fisherman..found a drift bottle inside a mackerel shark. **1921** W. DE LA MARE *Mem. Midget* xxxviii. 255 One mackerel-skied afternoon, Mrs. Monnerie and I and Susan were returning across the Park.

mackereling, *vbl. sb.* Add: Also **mackerelling.** (Earlier and later examples.)

1856 J. REYNOLDS *Peter Gott* xiv. 167 When fishermen can make as much in a few weeks at mackerelling, as in as many months at codfishing,..they prefer to run their chance in the former. *Ibid.*, Mackerelling is often called a lottery. **1880** *Harper's Mag.* Sept. 510/1 Among the rest are two of the singular 'porgy steamers' turned to mackereling. **1881** S. P. McLEAN *Cape Cod Folks* iii. 62, I was going mackerellin' with ye myself that time. **1952** *Gloucester* (Mass.) *Daily Times* 20 Mar. 8/7 His present seiner is the Jean and Patricia which stays with mackereling only.

b. A mackerel-like effect.

1866 G. M. HOPKINS *Jrnls. & Papers* (1959) 139 Scud spots etc, and some very faintly made out mackerelling. .. Bright, with mackerelling now and then. **1883** —— *Let.* 21 Dec. in Hopkins & Dixon *Corr.* (1955) 165, I have further noticed streamers, fine ribbing or mackerelling, and other..textures.

McKie: see *McCOY.

Mackinaw. Add: Also **Macinac, Macinaw, Mackina,** and with lower-case initial. Also, a heavy woollen cloth, now usu. with a plaid design; *pl.*, garments made of this cloth. **Mackinaw (boat),** (*b*) a schooner-rigged boat formerly used on the Great Lakes; **Mackinaw coat** (or **jacket**), a thick double-breasted jacket; also *ellipt.*; **Mackinaw shirt,** a plaid woollen shirt; **Mackinaw skiff** = *Mackinaw* (*boat*) (in Dict.); **Mackinaw trout** (further examples); also, a North American char, *Cristivomer namaycush.*

1812 J. G. LUTTIG *Jrnl. Expedition Upper Missouri* (1920) 54 The Mackina Boat took 5 hunters to the Island. **1822** L. CASS *Let.* 4 Oct. in *Wisconsin State Hist. Soc. Coll.* (1911) XX. 287 The heavy Mackinac blankets are almost impervious to the rain, and are universally worn by the Indians in this quarter. **1826** T. FLINT *Recoll.* 102, I have seen a Mackinaw skiff, carrying five tons, which came from the lakes into the Chicago of Michigan. **1836** in *Mass. Hist. Soc. Proc.* (1892) 2nd Ser. VII. 276 Covering, a cotton counterpane, a sheet.., besides my own great coats and green Mackinaw. **1840** *Southern Lit. Messenger* VI. 604/1 The celebrated Mackinaw trout, so called after the town, near which they are found, is generally caught by the hook. **1841** *Western Herald & Farmers' Mag.* (Sandwich, Ontario) 31 Mar. 3/2 They have also a large

assortment of blankets..of the real Mackinaw. **1842** *Southern Lit. Messenger* VIII. 586/2 A party of six..had occasion..to ascend the Missouri, in a Mackinaw [boat], with the purpose of trading. **1872** Mackinaw blanket [see *blanket coat* s.v. *BLANKET *sb.* 7]. **1887** *Rep. U.S. Comm. Fisheries* 24 At Duluth, Minnesota, the mackinaw boats average about 32 feet in length. **1900** *Atlantic Monthly* LXXXV. 102/1 It is then the woodsman dons his Mackinaw jacket. **1902** J. LONDON *Daughter of Snows* 316 He was interrupted by a warm-complexioned man clad in faded mackinaws. **1902** S. E. WHITE *Blazed Trail* 16 They all wore heavy blanket Mackinaw coats. *Ibid.* 375 A tall..individual dressed in a faded mackinaw and a limp slouch hat. **1912** J. SANDILANDS *Western Canad. Dict.* 28 *Mackinaw*, a heavy woolen cloth much in favour among lumberjacks. A lumberjack speaks of his thick winter jacket as his mackinaw. **1916** H. KEPHART *Camping & Woodcraft* I. 147, I usually discard the sweater in favor of a mackinaw shirt. **1920** S. LEWIS *Main St.* 230 He had given up..wearing red mackinaws in lumbercamps. **1930** J. DOS PASSOS *42nd Parallel* I. 101 A young man, his head and ears huddled into the collar of a mackinaw. **1938** E. HEMINGWAY *Fifth Column* (1939) 479 He wore..a mackinaw shirt. **1941** L. D. BALDWIN *Keelboat Age* 50 The Mackinaw boat in use on the Missouri was an adaptation of the flatboat and of the Mackinaw skiff. **1945** R. W. SERVICE *Ploughman of Moon* 157, I couched on the floor, lying on a buffalo-robe and wrapped in a mackinaw blanket. **1956** J. S. GOWLAND *Sikanaska Trail* 44 A guard in plain clothes brought along my mackinaw jacket. **1961** E. HUNTER *Mothers & Daughters* I. 9 She had hardly ever seen him without his hooded Mackinaw. **1961** *Vancouver Sun* 17 Aug. 23/1 The laker (mackinaw trout) is a record. Largest Canadian sport-caught lake char..is an 87 pounder. **1964** *Atlantic Advocate* July 77/1 William M...was operating the mill and producing a large red and black check design for Mackinaw or cruiser cloth. **1968** R. F. ADAMS *Western Words* (ed. 2) 187 The Mackinaw was a flat-bottomed boat with a pointed prow and a square stern... A large Mackinaw was as much as 50 or 60 feet long. **1971** J. McDOUGALL *Parsons on Plains* i. 1 My first recollections are of stumps, log heaps,..bateaux, Mackinaw boats. **1973** D. MACKENZIE *Postscript to Dead Let.* 9 He was..wearing a red-and-grey mackinaw over a shirt without a tie. **1973** J. RYDER *Trevayne* (1974) xxxiv. 264 A man in a mackinaw coat and a fur cap.

mackinawite (mæ·kinǭ,əit). *Min.* [f. *Mackinaw*, the name of a mine in Snohomish County, Washington + -ITE¹.] A black tetragonal iron sulphide.

1962 H. T. EVANS et al. in *Program Ann. Meetings Geol. Soc. Amer.* 47A Recently, another iron sulfide mineral, mackinawite, very similar in physical properties to valleriite, but containing no copper.., has been identified from the Mackinaw Mine, Snohomish County, Washington. **1969** *Amer. Mineralogist* LIV. 1190 Information about the occurrence, optical properties, microhardness and cell-size is given for a nickelian mackinawite (Fe 38·1%, Co 3·3%, Ni 18·7%) from Vlakfontein in the Transvaal. *Ibid.*, This mackinawite was examined in reflected light in both air and oil. It shows the very strong pleochroism and anisotropy characteristic of this mineral. **1970** *Nature* 15 Aug. 700/1 Modern surface sediments containing H_2S are often coloured black by fine grained, iron monosulphide minerals such as mackinawite, $Fe_{1+x}S$, and greigite, Fe_3S_4.

mackintosh. Add: **2.** Now freq. used to designate any type of rain-proof coat.

1895 *Montgomery Ward Catal.* 296/3 Men's Double Texture Mackintosh Box Coats, double breasted, all wool black tricot.., fancy plaid lining, silk velvet collar. **1897** *Sears, Roebuck Catal.* 188/2 Ladies' Single Military Cape Mackintosh. Made from extra fine all wool Scotch mixed cheviot, with fine, soft dressy finish. **1956** A. S. C. ROSS in N. Mitford *Noblesse Oblige* 30 *Burberry* and *raincoat* are of the same genre, *macintosh* or *mac* being normal.

mackintoshed (mæ·kīntǫʃt), *a.* Also **macintoshed.** [f. MACKINTOSH + -ED².] Wearing, or protected by, a mackintosh.

1904 J. VAIZEY *More about Pixie* xxiv. 257 The cloaked and mackintoshed figures..seemed all black, all the same. **1927** *Daily Mail* 30 June 9/4 Mackintoshed enthusiasts occupied them [*sc.* boats]. **1962** *Punch* 21 Nov. 751/1 Macintoshed ruthless plainclothesmen. **1963** *Times* 25 Apr. 6/1 They sauntered self-consciously off, mackintoshed and blackbooted, to mingle with the unsuspecting crowd.

Mackintosh Red: see *McINTOSH.

mackintoshy (mæ·kintǫʃi), *a.* [f. MACKINTOSH + -Y¹.] Of, belonging to, or suggestive of mackintoshes; characterized by or given to wearing mackintoshes.

1939 'G. ORWELL' *Coming up for Air* IV. vii. 279 The gas-bill and the school-fees, and the mackintoshy smell. **1941** E. BOWEN *Look at Roses* 253 'They're mackintoshy sort of people,' she said.

McLeod (măklau·d). Also erron. **Macleod.** [Name of Herbert *McLeod* (1841–1923), English scientist, who invented the instrument in 1874 (*Phil. Mag.* XLVIII. 110).] *McLeod gauge*: a type of mercury-in-glass manometer for the absolute measurement of low pressures, in which a fixed large volume of the gas to be measured is compressed by the mercury into a small volume and the resulting pressure found from the height of the column of mercury it supports, the desired pressure then being obtained by multiplying this by the ratio of the smaller to the larger volume in accordance with Boyle's law.

1880 J. E. H. GORDON *Physical Treat. Electr. & Magn.* II. xxxv. 83 The apparatus is in connection with a McLeod gauge, by means of which pressures to 0·00005 millim. can be determined. **1923** *Proc. Cambr. Philos. Soc.* XXI. 505 The residual pressure of the gas in the box was measured by a Macleod gauge. **1971** *Sci. Amer.* Aug. 114/2 Although the McLeod gauge indicates absolute pressure and is universally used for calibrating other instruments, it is inconvenient to operate.

maclock, var. *MUCKLUCK.

McLuhanism (măklū·ăniz'm). The social ideas of the Canadian writer H. Marshall *McLuhan* (1911–80), such as that the effect of the introduction of the mass media is to deaden the critical faculties of individuals. Hence **McLuhane·sque** *a.*; **McLu·hanite** *a.* and *sb.*, of or pertaining to, an adherent of, McLuhan; **McLu·hanize** *v. trans.*, to convert to McLuhanism; to render in a manner typical of McLuhan.

1967 *Punch* 4 Oct. 520/1 Most of us 20th-century electric circuitry villagers by now have got at least the drift of McLuhanism. **1967** *Spectator* 10 Nov. 571/2, I am leaving on one side all McLuhanite arguments that books are on the way out. **1967** *Observer* 10 Dec. 24/3 A suitably McLuhanesque fanfare. **1968** *Listener* 4 Jan. 8/1 Then there is Father Walter Ong, whom Martin Dodsworth once McLuhanised into 'the Ong with the numinous prose'. *Ibid.* 6 June 722/3 The demolition of fashionable McLuhanite myths about television has been another of the year's benefits. *Ibid.* 751/1 It was really a kind of McLuhanesque exercise in the depiction of the influence of media on people's lives. **1968** *Punch* 17 July 100/3 Mr. Turner and director Peter Hammond had bravely attempted to McLuhanise the intractably literary. **1970** *Times* 31 Mar. (Australian Suppl.) p. vi/8 The thesis of Australia in the seventies as a model McLuhanized society is fervently denied by publishers. **1971** *Guardian* 14 Jan. 7/1 McLuhanites and Orwellians are likely to block our view of their masters' arguments. *Ibid.* 1 Nov. 9/1 One occasionally wonders whether the image—as in some McLuhanesque lens—is not more important than its object. **1972** *Ibid.* 8 Feb. 4/6 It is on the basis of evolution ..and not instant McLuhanism that the Swiss delegation has opened negotiations at Brussels. **1973** *Ibid.* 19 Feb. 8/6 Designers on a newspaper—whatever McLuhanesque horror next?

maclura (măklʲūə·ră). [mod.L. (T. Nuttall, 1818), f. the name of William *Maclure* (1763–1840), American geologist.] A deciduous North American tree, *Maclura pomifera*, of the genus so called, belonging to the family Moraceæ and bearing an inedible fruit resembling an orange; the Osage orange or bow-wood. Also *attrib.*

1818 T. NUTTALL *Genera N. Amer. Plants* II. 233 Maclura. (Bow-wood, Yellow-wood.) **1857** *Trans. Illinois Agric. Soc.* II. 222 A few Maclura hedges are growing. *Ibid.* 302 Mr. Tisdell has two hedges of the maclura growing on his farm. **1858** J. A. WARDER *Hedges & Evergreens* 21 The division of the prairies into twenty-acre lots, by the dense hedges of maclura. *Ibid.* 52 The cost of the maclura hedge.

M'Naghten rules (măknǭ·t'n). Also **McNaghten, MacNaughten, Macnaughton,** etc., **rules** (or **case,** etc.). [Named after Daniel *M'Naghten* who was tried for murder in 1843 and acquitted on a plea of insanity.] Name applied to the answers given in the House of Lords in 1843 after the trial of Daniel M'Naghten for the murder of Sir Robert Peel's secretary, Edward Drummond, when five questions were put respecting crimes alleged to have been committed by persons suffering from insanity; subsequently used as criteria for judging an accused person's responsibility for his actions.

[**1843** *Times* 14 Mar. 4/3 Last night the Lord Chancellor ..brought forward the circumstances of M'Naughten's trial for the consideration of the Peers... He proceeded to caution their Lordships against supposing that, even if the verdict in M'Naughten's case should appear to have been given upon faulty or inconclusive evidence, it would be necessary to alter the law upon the subject.] **1892** D. H. TUKE *Dict. Psychol. Med.* I. 318/1 Juries are never, at the present day, charged strictly in conformity with the McNaghten rules. [**1902** *Encycl. Brit.* XXIX. 491/2 The answers [by Chief Justice Tindal] to these questions are commonly called 'The rules in Macnaughton's case' and they still nominally contain the law of England as to the criminal responsibility of the insane.] **1958** B. HAMILTON *Too Much of Water* xii. 271 So long as the McNaughton rules run, I don't think he's got a *chance* of Broadmoor. **1959** JOWITT *Dict. Eng. Law* II. 1122/2 In *R. v. McNaghten* or *M'Naghten* or *Macnaughten* (1843)..a discussion took place in the House of Lords upon the direction to the jury..and as a result a series of questions was put to the judges. The answers of the majority constitute the 'McNaghten Rules', and have been accepted as laying down the law as to insanity with reference to criminal responsibility. **1959** *Listener* 24 Sept. 481/1 Insanity in this context was until lately very narrowly defined by the McNaghten rules. **1962** *Lancet* 1 Dec. 1148/1 Giving the background to present-day legal views on diminished responsibility, the McNaughten rules, uncontrollable impulse, and psychopathy. **1968** N. WALKER *Crime & Insanity in Eng.* vi. 105 It is difficult to find a clear statement of the medical objections to the M'Naghten Rules at this date. *Ibid.* 107 Mr Justice Bray explained the M'Naghten Rules to the Gloucester jury. **1971** *Reader's Digest Family Guide to Law* 773/1 The McNaughten Rules can be applied to all offences, but they are rarely used in anything but murder cases. **1973** *Times* 24 May 12/7 What emerges clearly from the Canadian trial is the immense benefit to both doctors and lawyers in *this* country from the subsidence of the McNaghten rules and the introduction of the plea of diminished responsibility.

Mâcon (mā·kon). [Name of a city in the department of Saône-et-Loire, France.] A wine of Burgundy, produced in the district around Mâcon.

[**1699** M. LISTER *Journey to Paris* 161 The White Wines of value are those of *Mascon* in *Burgundy*.] **1863** T. G. SHAW *Wine* viii. 258 A bottle of the homely Mâcon, instead of the elegant Beaujolais, had to be called for. **1908** CHESTERTON *Man who was Thursday* viii. 155 Draining a glass of Mâcon. **1963** *Spectator* 1 Mar. 275, I chose a Macon from the wine-list and subsequently paid £1 for it. **1967** A. LICHINE *Encycl. Wines* 336/2 A slightly lesser amount [of wine] will be sold simply as Mâcon with no other indication of source. **1972** H. W. YOXALL *Enjoyment of Wine* xv. 117 A glass or two of cool, light, dry wine—a young moselle, a muscadet, a mâcon blanc..—is a first-rate aperitif.

macon (mēi·kən). [f. M(UTTON + B)ACON.] During the 1939–45 war: mutton salted and smoked like bacon.

1939 *Daily Express* 22 Nov., *Macon* has now been adapted by other newspapers as a name for mutton bacon. This is only the latest of many words and phrases originally coined in this office which have later been used generally. **1939** *News Rev.* 30 Nov. 15 Macon, the Scottish dish which may eke out any wartime shortage of bacon. Macon is mutton cured in the same way as bacon. **1939** *Times* 7 Dec. 10/4 Macon was introduced..at the Savoy Hotel yesterday... Mr. Cecil Rodd, introducing macon.. said he did not pretend to know how the word macon came into being; it just happened. **1968** *Punch* 7 Feb. 177/2 The Ministry of Food then stood in for Mrs. Beeton, instructing them how to..work wonders with such unlikely raw materials as macon (bacon made from mutton, children)..and..coelacanth.

Maconochie (măkǫ·nŏki). *colloq.* [The name of the makers, *Maconochie* Brothers, of London.] **1.** Meat stewed with vegetables and tinned, esp. as supplied to soldiers on active service; a tin of such meat. Also *Maconochie ration.*

1901 'LINESMAN' *Words by Eyewitness* i. 9 The hungry shelterers trooped back to their 'dixies', and wasted not a thought upon them [*sc.* shells], until the 'Maconochie' had vanished. **1915** F. H. LAWRENCE in T. E. Lawrence *Home Lett.* (1954) 689 We have found several tins of Maconochie Rations & also Jam in the house. **1917** 'I. HAY' *Carrying On* viii. 220 How would a Maconochie apiece suit your boys? **1940** 'GUN BUSTER' *Return via Dunkirk* II. xix. 225 He manages to scrape together two tins of Maconochie (stew), a tin of cold potatoes,..and some 'issue biscuits'. **1954** W. FAULKNER *Fable* 61 His company commander was shaving out of a Maconochie tin.

2. *joc. slang.* The stomach. Also *Maconochie Cross, Medal* (see quots.).

1919 W. H. DOWNING *Digger Dial.* 32 *Maconochie*,.. stomach. *Ibid.*, *Maconochie Cross*, Military Cross. *Maconochie Medal*, Military Medal. **1925** FRASER & GIBBONS *Soldier & Sailor Words* 149 *Maconochie*..was also sometimes used for stomach, *e.g.*, 'He got hit in the Maconochie'. **1965** BROPHY & PARTRIDGE *Long Trail* 147 *Maconochie Medal*, the Military Medal; *Maconochie Cross*, the Military Cross.

‖ **macoumère** (makumę̄r). *West Indies.* Also **macomère, macoumere.** [French Creole.] A god-mother, or the mother of one's god-child; more generally applied to any female friend of a family, or as a derogatory term for an old woman or gossip.

1942 H. C. GORDON *West Indian Scenes* xix. 217 *Macomère Crab*, who was a kind soul, complied with her request. **1952** S. SELVON *Brighter Sun* vi. 95 And is why yuh looking at me so, *macoumère*? **1960** *Tamarack Rev.* XIV. 22 This time so, they ain't bothering about the other people in the queue at all, they mauvalanging and bursting out in some loud kya-kya laugh like them macoumere in the market in..Trinidad.

macrergate (mækrə̄·ɪgēit). *Ent.* Also **macroërgate** [ad. G. *makroergate* (E. Wasmann 1895, in *Biol. Centralbl.* XV. 606), f. Gr. μακρ-ός large + ἐργάτης worker.] A large worker ant.

1901 W. M. WHEELER in *Amer. Naturalist* XXXV. 879 The larger macroërgates are nearly eight times as large as the normal workers. **1910** —— *Ants* vi. 97 The *macrergate* is an unusually large worker form which in some species is produced only in populous or affluent colonies. **1915** H. ST. J. K. DONISTHORPE *Brit. Ants* 40 The macrergate is larger in stature than the normal worker of the species in question.

macro (mæ·kro), *sb.* [f. the prefix MACRO-.] **1.** *Computers.* A macro-instruction. Freq. *attrib.*

1959 *Jrnl. Assoc. Computing Machinery* VI. 132 The built-in system macro instructions in SCAT presently consist of (1) two macros for generating the..standard entry and exit from subroutines, (2) a set of debugging macros [etc.]. **1961** LEEDS & WEINBERG *Computer Programming Fund.* iv. 103 The assembly program can be instructed to recognize this instruction sequence through the use of a single form called a macro-instruction... Recognition of the 'macro' form causes the assembly program to assemble the corresponding sequence of one or more machine instructions. **1963** L. SCHULTZ *Digital Processing* xi. 240 To avoid having to write the groups repeatedly, the programmer..instructs the machine to establish a table in which it stores the group of instructions. To identify the group, he assigns a code name, called a macro name. *Ibid.*, The programmer writes one instruction (a macro) and the machine 'generates' several. **1966** *New Scientist* 27 Oct. 162/3 In the ICT command language it is possible to define new commands in the language in terms of those already defined. An example of this 'macro' facility is given here... The effect is to allow the user to condense a long and complicated job description into a shorthand phrase of his own choice. The use of 'macros' in this way makes it extremely easy for the user to set up his own conventions. **1970** O. DOPPING *Computers & Data Processing* xix. 324 As an advantage of automatic coding, it is sometimes pointed out that the macros and subroutines are tested and error-free, in contrast to home-made subroutines. **1972** J. J. DONOVAN *Systems Programming* iv. 111 In employing a macro, the programmer essentially defines a single 'instruction' to represent a block of code. For every occurrence of this one-line macro instruction in his program, the macro processing assembler will substitute the entire block.

2. *Photogr.* (See *MACRO- 2 c.)

macro-. Add: **1. b.** *macrochromosome, macrovegetation*; **ma·croform** (see quot.); **ma·crofossil** *Palæont.*, a fossil discernible to the naked eye; **ma·cro-instruction** *Computers*, an instruction in a programming or source language which is equivalent to a specified set of ordinary instructions in an object language (which may be the source language or machine language); freq. as two words (cf. *MACRO *sb.* 1); **macro·phagous** *Zool.* [-PHAGOUS], feeding on relatively large pieces of food; **macrotri·chium** *Ent.* [Gr. θρίξ, τριχ- hair], usu. in pl. **macrotri·chia,** in certain insects, the larger hairs on the body, esp. those on the surface of the wings.

1905, 1969 Macrochromosome [see *microchromosome* s.v. *MICRO- 1]. **1967** *Anglo-Amer. Catal. Rules: Brit. Text* 267 *Macroform*, a reproduction large enough to be easily read by the unaided eye. Used in contradistinction to 'microform'. **1937** G. D. HANNA in Camp & Hanna *Methods in Paleont.* II. 79 Under the term macrofossils it will be desirable to treat together many of the larger groups of invertebrates such as the Mollusca, Brachiopoda, Echinodermata, Bryozoa, and Crustacea... All forms more than 10 mm. in diameter will be treated as macrofossils. **1969** BENNISON & WRIGHT *Geol. Hist. Brit. Isles* ii. 34 The larger fossils (macrofossils) of the Chalk are largely neritic. **1974** A. HUXLEY *Plant & Planet* ii. 9 Tertiary and Cretaceous sediments in Venezuela and Borneo..yielded a detailed pollen record but virtually no macro-fossils. **1959** Macroinstruction [see *MICRO-INSTRUCTION]. **1959, 1961** [see *MACRO *sb.* 1]. **1961** *Times* 3 Oct. (Computer Suppl.) p. x/5 Through the use of these macro-instructions the computer can be given the semblance of being a much more powerful machine than it really is. **1970** O. DOPPING *Computers & Data Processing* xix. 312 By introducing macro instructions in the source language, the designer can bring about the same ease of programming as could be achieved by giving the computer a more powerful operation list than it really has. But naturally, one does not get the same advantages in terms of economy of memory space and computer time as would be obtained if the more powerful instructions were really built into the machine. **1972** Macro instruction [see *MACRO *sb.* 1]. **1949** I. F. & W. D. HENDERSON *Dict. Sci. Terms* (ed. 4) 240/1 Macrophagous... Feeding on relatively large masses of food; *opp.* microphagous. **1951** M. ABERCROMBIE et al. *Dict. Biol.* 130 All land animals are macrophagous. **1964** *Oceanogr. & Marine Biol.* II. 396 He [*sc.* A. R. Fontaine] finds both microphagous and macrophagous mechanisms are used in this species [sc. *Ophiocomina nigra*]. **1934** C. H. CURRAN *Families & Genera N. Amer. Diptera* 487 Macrotrichia—The larger microscopic hairs on the surface of the wings. **1957** RICHARDS & DAVIES *Imms's Textbk. Ent.* (ed. 9) I. 12 Setae or Macrotrichia are commonly known as hairs and each arises from a cup-like pit or alveolus. *Ibid.* 38 Macrotrichia or true setae..are found on the main veins and their branches,..less frequently on the wing-membrane. **1970** E. F. RIEK in *Insects of Australia* (Commonwealth Sci. & Industr. Res. Organization) xxxvii. 875/2 Hooked macrotrichia may occur over most of the fore margin of the hind wing [in Hymenoptera]. **1958** *Jrnl. du Conseil Internat. Explor. de la Mer* XXIV. 32 Samples of macrovegetation were collected in the spring. **1960** *Oikos* XI. 183 (*heading*) Subaquatic macrovegetation in Ösbysjön, Djursholm.

d. **macrocra·nial** *a.*, having a long skull; macrocephalic; **macropha·llic** *a.*, having a large phallus.

1902 *Biometrika* Aug. 462 Dolichocephaly and chamaecephaly in both races are associated with macrocranial characters. **1907** *Practitioner* Aug. 318 The population of the south-west of Scotland, exclusive of Glasgow, is long-headed or macrocranial. **1970** F. SNOWDEN *Blacks in Antiquity* 23 The circumcised and macrophallic Ethiopians. **1972** *Sunday Times* 23 Apr. 43 Commercial porn.. with its inevitable distortions, brutalised women and macrophallic faceless men.

e. Terms in which *macro-* indicates subject-matter treated on a larger scale, or more comprehensive phenomena or levels of treatment, than is implied either by the word to which *macro-* is attached or by the corresponding term beginning *micro-*, as *macro-planning, -sociology* (hence *-sociological* adj.); *macro-historical* adj.; **ma·croclimate,** the general climate of a relatively large area; so **ma:cro-clima·tic** *a.*; **ma:cro-evolu·tion,** evolutionary change over a long period, leading to the appearance of new groups of plants or animals; **ma·crophysics,** the part of physics that is concerned with bodies and phenomena on a macroscopic scale; so **macrophy·sical** *a.*

1939 *Geogr. Jrnl.* XCIII. 463 As in all local climates strong departure from the macroclimate develops in the climates of cities in calm cloudless weather. **1967** M. J. COE *Ecol. Alpine Zone Mt. Kenya* 59 This remarkable contrast between the micro- and macroclimates of these areas is expressed even more clearly by Mani. **1971** J. Z. YOUNG *Introd. Study Man* xxxviii. 544 The British..move with less protection in their damp but equable macroclimate. **1939** *Ecology* XX. 30 Microclimate..makes it possible for smaller groups of preclimax vegetation to persist as relict colonies long after macroclimatic changes have shifted formations. **1946** S. A. WILDE *Forest Soils* iii. 21 Macroclimatic soil-forest zones. **1939** *Ann. Reg. 1938* 369 The gene theory was applied increasingly to problems of macro-evolution. **1940** R. GOLDSCHMIDT *Material Basis of Evolution* iii. 8 Microevolution..has been used..for evolutionary processes observable within the span of a human lifetime as opposed to macroevolution, on a geological scale. **1972** *Listener* 3 Aug. 138/2 There may be a different mechanism for grand heroic macro-evolution and for ordinary humdrum everyday evolution at species or sub-species level. **1955** *Bull. Atomic Sci.* Feb. 42/2 There are some who profess to see in matters of culture, in matters precisely of the arts and sciences, a certain macro-historical pattern, a grand system of laws which determines the course of civilization and gives a kind of inevitable quality to the unfolding of the future. **1902** MANN & MILLIKAN tr. *Drude's Theory of Optics* p. vii, Pure electromagnetic experiments lead to conclusions in what may be called the domain of macrophysical properties only. For the explanation of optical dispersion a hypothesis as to the microphysical properties of bodies must be made. **1956** E. H. HUTTEN *Lang. Mod. Physics* v. 179 Two micro-physical events are connected in a different way from two macro-physical events. **1960** *Times* 15 Mar. 2/2 A study of the relationships between the macro-physical properties of soils and the physical chemistry of their colloidal constituents. **1909** *Cent. Dict.* Suppl., Macrophysics. **1936** *Discovery* Mar. 96/2 Thirteen unsolved problems are listed, ranging from those concerning the creation of the universe..to the structure of the atom, from macrophysics and astronomy *via* biology..back to physical science again. **1956** *Nature* 18 Feb. 321/1 Mr. F. H. Ludlam..pointed out that the partitioning (into micro-, macro- and synoptic-physics) of cloud studies.. must not be made too rigid. **1966** *Economist* 11 June 1206/1 To mention only a few [contributions], largely concerned with macro-planning: welfare-theoretical problems and planner's and consumer's sovereignty are the theme of Joan Robinson. **1973** *Financial Times* 26 Mar. 16/4 The macro-planning functions of the metropolitan county (mainly transport and town and country planning, together with fire and police, consumer protection, [etc.]). **1951** R. FIRTH *Elem. Social Organiz.* i. 18 If the distinguishing feature of the anthropologist is micro-sociological technique, his theory is macro-sociological. **1970** *New Society* 5 Feb. 232/3 The approach, the initiated will recognise, is essentially macrosociological. **1941** G. GURVITCH in *Jrnl. Philos.* XXXVIII. 486 Macrosociology is the study of the world of groups and of global societies, each of which are microcosms of the forms of sociality. **1958** W. STARK *Sociol. of Knowl.* i. 20 It might be useful..to call the one the macrosociology of knowledge, because it fixes its attention on the inclusive society.

2. Combs. in which *macro* can be regarded as a separable element having adjectival force, with the meaning: macroscopic, large-scale; overall, comprehensive; large. Hence used as an independent word not preceding the sb. it qualifies.

1934 WEBSTER, *Macro* adj. = *macro-*. **1937** *Mind* XLVI. 327 Quantum mechanics, moreover, claims applicability not only to micro-processes; it also contains laws of mechanics which are applicable to macro-processes. **1954** *Gloss. Terms Iron & Steel* (*B.S.I.*) I. 19 *Macro etch test*, etching with acids or other reagents to reveal the macro structure, flow lines and/or defects. **1957** C. DAY LEWIS *Poet's Way of Knowl.* 14 Until recently the scientist has sought for general laws governing macro-events, and has drawn from them inferences about individual events. **1959** K. R. POPPER *Logic Sci. Discovery* II. viii. 196 Observable physical effects are interpreted as 'macro laws'. **1961** G. H. ORCUTT et al. *Microanalysis Socioecon. Syst.* p. xv, Given the possibility of experimentation at both micro- and macrolevels, it has been possible for the physical sciences to achieve great successes at both levels without the necessity of predicting aggregate behavior on the basis of knowledge about microbehavior. **1962** J. T. MARSH *Self-Smoothing Fabrics* xv. 263 This type of staining test..does, however, demonstrate irregularity of distribution on a macro-basis as distinct from a micro-basis, and provide useful evidence of uneven impregnation during padding, or of resin migration during drying. **1966** *McGraw-Hill Encycl. Sci. & Technol.* XI. 143/2 The instruments and apparatus used in microanalytical work are to some extent miniature replicas of macro instruments and apparatus. **1967** COX & GROSE *Organiz. Bibliogr. Rec. by Computer* v. 107 Before I move from the macro-level of the overall design of a MARC service to the minutiae of record data content and file organization. **1967** D. WILSON in Wills & Yearsley *Handbk. Managem. Technol.* iii. 47 The logic for the individual programs is expressed in terms of macro and micro block-diagrams. The macro block-diagrams show the main logic for a particular program and may be prepared by the systems analyst; the micro block-diagrams show the detailed logic for program coding and will be prepared by the programmer. **1970** *Guardian Weekly* 18 Apr. 15 Revolutions, including the Russian one, made the mistake of insisting that the macro-change, of the whole system, should come before micro-change, of the life style. **1971** *New Scientist* 21 Jan. 100/1 Undoubtedly a number of serious macro-problems face the Ulster community. *Ibid.* 100/2 In the absence of a knowledge of these macro-changes, systems will continue to evolve on the basis of year-to-year crisis decisions. **1971** *Good Motoring* Sept. 9/2 As vehicle speeds rise from 25–30 mph to 60–70 mph, however, tyre grip in wet weather diminishes unless there is also 'macro' or large-scale roughness of the surface. **1973** A. E. WILKERSON *Rights of Children* 307 These issues..reflect the necessity for a macro-approach to child welfare. **1974** *Times Lit. Suppl.* 8 Mar. 242/5 The forms and methods of economic management, both micro and macro.

b. *Chem.* Of or pertaining to macroanalysis. Cf. *MACRO-SCALE.

1933 *Biochem. Jrnl.* XXVII. 434 The present micro-method for urea in 0·2 cc of blood compares very favourably with the comparatively macro-method of Van Slyke and Cullen. **1938** ARTHUR & SMITH *Semi-Micro Qualitative Analysis* I. 3 Dealing with samples of only 3 to 5 mg. (compared with the usual 0·5 to 1·0 g. macro sample), the micro method consumes much less time for reactions.. and in many other ways proves superior to the macro methods. **1946** MELDRUM & DAGGETT *Textbk. Qualitative Analysis* i. 2 The two..generally applicable methods of qualitative analysis of inorganic materials are the macro and the semimicro methods. **1974** [see *MICRO- 8 b].

c. *Photogr.* Denoting apparatus or procedures used in macrophotography, as **macro lens** (also *ellipt.* as *macro*), a lens suitable for taking photographs unusually close to the subject.

1956 *Focal Encycl. Photogr.* 688/1 A camera with back focusing is a great advantage in macro work. **1961** L. A. MANNHEIM tr. *Croy's Camera Close Up* 85 The maximum aperture of a macro-lens depends..on its design. *Ibid.* 88 Macro-exposures. Exposures in close-up work have to allow for..the subject and its lighting..and the scale of reproduction. **1968** GAUNT & PETZOLD *Pict. Cycl. Photogr.* 425/2 The main requirement for a macro-camera is adequate lens extension. **1971** *Amat. Photographer* 13 Jan. 80/1 (Advt.), 50 mm f/3·5 Macro. **1974** *Ibid.* 29 May 8/1 The picture on the left could only have been taken with a macro lens. *Ibid.*, It's impossible to combine macro and telephoto in one lens.

macroanalysis (mækro,ænæ·lĭsis). *Chem.* [f. *MACRO- 1 e + ANALYSIS.] Quantitative analysis of samples of the size for which the older chemical techniques were usually developed, commonly 0·1–1 gramme.

1938 MELDRUM & FLOSDORF *Qualitative Analysis Inorg. Materials* I. 7 Microanalysis may be applied when the amount of material available..is very small... Many of the reactions are essentially the same as those used in the systematic macroanalysis. **1955** [see *MICRO- 8 b]. **1966** *McGraw-Hill Encycl. Sci. & Technol.* XI. 143/1 Methods of quantitative analysis vary with the amount of sample taken and of constituent being determined. A macroanalysis (decigram analysis) uses a sample of about 0·1–0·5 gram.

macrobiotic, *a.* and *sb.* Add: **a.** *adj.* **2.** Of or pertaining to a Zen Buddhist dietary system intended to prolong life, comprising pure vegetable foods, brown rice, etc. Also as *sb.*, a follower or adherent of this system; **macrobiotics** *sb. pl.* (const. as *sing.*), the use or theory of such a dietary system.

1936 L. P. WEAVER tr. *Székely's Cosmos, Man & Society* III. ii. 207 (*heading*) Macrobiotics, the optimal and omnilateral correlations of longevity. **1965** W. DUFTY *You are All Sanpaku* (Eng. version of *Y. Sakurazawa's Zen Macrobiotics*) 178 Macrobiotics is not the kind of vegetarianism which is merely sentimental. **1968** *New Yorker* 10 Feb. 22 Macrobiotic. That means she'll eat fish. **1969** *Listener* 17 July 79/3 Macrobiotics is a life science discovered in Japan by Georges Ohsawa and it's based on the two pillars of philosophy and diet—the staple food being brown rice. **1970** M. SPARK *Driver's Seat* 48 Rice, unpolished rice is the basis of macrobiotics... It is a cleansing diet. Physically, mentally and spiritually. **1970** *Time* 16 Nov. 59 Most macrobiotics, as Ohsawa's devotees call themselves, try to follow his other nine diets, which are graduated from six to minus three to include increasing amounts of fish and vegetables—organically grown—along with brown rice. **1971** *Courier-Mail* (Brisbane) 10 May 9/4 Macrobiotic nibblers of seeds, bean curd, kale, other seaweeds and brown rice, are spreading their message of love and peace by feeding. **1971** *Daily Tel.* 18 Oct. 4/6 The macrobiotic diet, a rigid system based on vegetables and cereals..is denounced..by the American Medical Association as bad for health and even potentially lethal. **1972** *Guardian* 1 Dec. 11/2 Right now she is into the Zen macrobiotic brown-rice bit. **1973** 'P. REID' *Harris in Wonderland* vii. 55 Supper ..was macrobiotic, leeks and brown rice on a wooden plate. **1975** *Listener* 1 May 589/3 Macrobiotics crumbling to death in the name of good health.

macrocarpa (mækrǒkā·ɪpă). [mod.L., specific epithet of *Cupressus macrocarpa* (T. Hartweg 1847, in *Jrnl. Hort. Soc.* II. 187), f. Gr. μακρό-ς large + καρπός fruit.] An evergreen

tree, *Cupressus macrocarpa*, native to the Monterey peninsula of California and widely cultivated elsewhere, esp. as a fast-growing hedge or wind-break; = *Monterey cypress.* Also *attrib.*

[**1866** 'SENILIS' *Pinaceæ* 73 It [*sc.* large-coned cypress] is to be found in two forms. *Macrocarpa*, when raised from seed, has a distinct, continuous, and erect leader with the side branches regularly disposed and gracefully drooping.] **1905** *Daily News* 12 Feb. 4/6 The Bird only moved from a macrocarpa tree to the cypress and lime trees. **1922** *N.Z. Jrnl. Agric.* 20 Mar. 176 The macrocarpa is very rapid in its growth. **1935** *Discovery* Dec. 375/1, I have found macrocarpa hedges a very popular breeding place for various species. **1937** 'R. HYDE' *Persephone in Winter* 42 The faint wet tang of macrocarpa leaves. **1948** J. BETJEMAN *Sel. Poems* 78 Those macrocarpas still survive the gales They must have had last winter. **1950** *N.Z. Jrnl. Agric.* Feb. 163 Good, well-kept shelter belts of macrocarpa. **1959** A. McLINTOCK *Descr. Atlas N.Z.* p. xiii, Plantations and shelter belts of introduced (exotic) trees, mainly radiata pine and macrocarpa, break up the farmlands. **1966** G. W. TURNER *Eng. Lang. Austral. & N.Z.* iv. 82 Children..grew up among the pine and macrocarpa hedges or bleak unsheltered paddocks which replaced the bush.

macrochoanite (mæ:krokō̆u·ănəit). *Palæont.* [f. MACRO- + Gr. χοάν-η funnel + -ITE¹ 2 a.] A nautiloid cephalopod once included in a sub-order Macrochoanites, in which each septal neck reaches back into the preceding one. Hence **ma:crochoani·tic** *a.*, (having septal necks) characteristic of a macrochoanite.

1883 A. HYATT in *Proc. Boston Soc. Nat. Hist.* XXII. 260 We, therefore, divide the Ellipochoanoida into the true Microchoanites,..and the Macrochoanites. *Ibid.*, In the more complicated forms of Goniatitinæ, while the young are quite generally macrochoanitic, the later larval stages and the adults are universally short funnelled. **1964** C. TEICHERT in R. C. Moore *Treat. Invertebr. Paleont.* K 39/1 Macrochoanitic necks are found only in the Endocerida. *Ibid.* 95/1 Hyatt distinguished the Bactritidae as 'Macrochoanites'.

macroco·smically, *adv.* [f. MACROCOSMIC *a.*: see -LY².] In relation to the macrocosm.

1881 [see *MICROCOSMICALLY *adv.*]. **1939** C. WILLIAMS *Descent of Dove* 128 Considered macrocosmically or microcosmically.

macrocyclic (mækrosəi·klik), *a.* [f. MACRO- + CYCLIC *a.*] **1.** *Bot.* Of a rust fungus: having a long life cycle.

1926 ARTHUR & KERN in *Mycologia* XVIII. 90 The vegetative body is either long-cycle (macrocyclic), consisting of two unlike and discontinuous generations, or short-cycle (microcyclic), consisting of one continuous generation. *Ibid.* 91 We are indebted to Dr. G. Lagerheim, of Stockholm, for the suggestion of the terms macrocyclic and microcyclic which seem to us to be very appropriate and satisfactory. **1929** [see *ÆCIUM]. **1950** E. A. BESSEY *Morphol. & Taxon. Fungi* xii. 396 Rusts possessing such a cycle are called macrocyclic or long-cycle rusts. **1970** J. WEBSTER *Introd. Fungi* 376 It is also believed that forms with shorter life cycles arose..from macrocyclic ancestors.

2. *Chem.* Containing or being a ring composed of a relatively large number of atoms.

1947 R. L. WAKEMAN *Chem. Commercial Plastics* xi. 225 Macrocyclic esters can thus be formed, containing even as many as 16 or more members in the ring. **1960** [see *MACROLIDE]. **1964** M. HYNES *Med. Bacteriol.* (ed. 8) x. 137 These substances, as well as erythromycin, have a macrocyclic lactose ring, and are sometimes termed macrolides. **1972** BANTHORPE & CHARLWOOD in A. A. Newman *Chem. Terpenes* vii. 377 The macrocyclic compound cembrene.. is probably derived from GGPP.

Hence (as a back-formation) **ma·crocycle,** a macrocyclic compound or molecule.

1956 *Nature* 14 Jan. 70/1 Synthesis of azaporphins and related macrocycles. **1971** *Ibid.* 17 Sept. 183/1 Several [antibiotics] have structures made up of two types of chemical entity; for example..the macrolides (macrocycles attached to carbohydrate residues). **1974** *Chem. Rev.* LXXIV. 351/2 Christensen and coworkers have discussed for several classes of macrocycles their unique ion binding properties.

macrocyte. Add: Hence **macrocytic** (-si·tik) *a.*, typical or characteristic of a macrocyte; characterized by macrocytes; **macrocytosis** (-səitō̆u·sis) = MACROCYTHÆMIA.

1893 DUNGLISON *Dict. Med. Sci.* (ed. 21) 656/2 *Macrocytosis*, condition in which the red corpuscles are increased in size, becoming macrocytes. **1930** DAVIDSON & GULLAND *Pernicious Anæmia* xi. 237 We would describe the condition as a macrocytic hæmolytic anæmia. **1930** M. M. WINTROBE in *Proc. Soc. Exper. Biol. & Med.* XXVII. 1072 The first and most obvious group [of anemias], which can be called macrocytic, includes the cases of pernicious anemia, sprue, and a case of pernicious anemia of pregnancy. **1932** *Jrnl. Exper. Med.* LVI. 551 Evidence that a normal red corpuscle could change into a macrocytic or a microcytic form. **1947** *Radiology* XLIX. 289/2 Macrocytic anemias appeared in certain animals on chronic exposure to doses in the range under discussion. **1962** *Lancet* 12 May 1004/2 Clinical and hæmatological examinations..showed anæmia with..macrocytosis, and reticulocytosis. *Ibid.* 1 Dec. 1142/2 The peripheral blood is not strikingly macrocytic. **1968** J. H. BURN *Lect. Notes Pharmacol.* (ed. 9) 79 In pernicious anaemia there is enough haemoglobin but too few red cells. The red cells are swollen (macrocytic) and hyperchromic.

ma:cro-econo·mics, *sb. pl.* (usu. const. as *sing.*). Also **macroeconomics.** [f. MACRO- + ECONOMIC B. *sb.* 2 c.] The science or study of the economy as a whole. Opp. *MICRO-ECONOMICS *sb. pl.* So **ma:cro-econo·mic** *a.*

1948 *Econometrica* XVI. 309 (*heading*) Some conditions of macroeconomic stability. **1948, 1949** [see *MICRO-ECONOMICS *sb. pl.*]. **1957** *Economist* 5 Oct. 36/3 The macroeconomic prophets, having seen their forecasts repeatedly falsified, have lost something of their confidence. **1961** G. ACKLEY *Macroecon. Theory* i. 4 Macroeconomics deals with economic affairs 'in the large'. It concerns the over-all dimensions of economic life. **1966** *Listener* 3 Mar. 312/2 Even within the field of what we call macro-economics—the study of what determines the over-all level of activity and other aspects of the economy as a whole—there have been considerable elements of continuity. **1969** *Daily Tel.* 1 Nov. 10 The Government may be difficult to challenge in matters of macro-economics or high finance. When it comes to the practical mechanics of daily life, however, in which a convenient coinage matters a great deal, the public is the best judge. **1971** *New Scientist* 15 Apr. 147/1 The pundits argue about the overall, macro-economic effects of the budget. **1975** *Times* 16 July 15/5 The wonderful world of established macro-economics—no serious unemployment, no intolerable inflation, no controls.

macroërgate, var. *MACRERGATE.

macrogametophyte (mækrogæ̆mī·tofəit). *Bot.* [f. MACRO- + *gametophyte* (s.v. GAMETE).] = *MEGAGAMETOPHYTE.

1931 *Bot. Gaz.* XCII. 23 (*heading*) Development of the macrogametophyte and embryo of *Daucus carota*. **1938** G. M. SMITH *Cryptogamic Bot.* II. vii. 193 The developing embryo of *Selaginella* always pushes deeply into the macrogametophyte by means of a suspensor.

macroglobulin (mækroglǫ·biŭlin). *Biochem.* [ad. G. *makroglobulin* (attributed to Pedersen and Waldenström by Waldenström 1948, in *Schweiz. Med. Wochenschr.* 25 Sept. 928/2): see MACRO- b and GLOBULIN.] Any of the immunoglobulins of very high molecular weight (about 1,000,000 or more).

1952 J. WALDENSTRÖM in *Adv. Internal Med.* V. 408 Pedersen compared this normally occurring macromolecule with the pathologic so-called 'macroglobulin' found in the disease described by us as macroglobulinemia. **1961** [see *macroglobulinæmia* below]. **1967** *Times* 21 Nov. 3/6 As well as these small antibody molecules, the body produces much larger antibodies, called macroglobulins.

So **ma:croglobulinæ·mia** [ad. G. *makroglobulinämie* (J. Waldenström 1948, in *Schweiz. Med. Wochenschr.* 25 Sept. 928/2)], an excess of macroglobulins in the blood; esp. (more fully *Waldenström's macroglobulinæmia* or *syndrome*) a disease similar to myelomatosis but less often fatal; **ma:croglobulinæ·mic** *a.*

1949 K. PEDERSEN in *Jrnl. Franklin Inst.* CCXLVIII. 570 In a few rare cases of essential macroglobulinemia.., a very high concentration of the 20 S component has been found. **1961** *Lancet* 5 Aug. 289/2 While there can be no doubt that Waldenström's syndrome exists, it seems that such increases in macroglobulins are not confined to a single disease entity. The term macroglobulinæmia will therefore be used in this paper to refer to the presence in the serum of increased amounts of macroglobulins, regardless of the clinical associations. *Ibid.* 290/1 On paper electrophoresis, macroglobulinæmic sera yield patterns which are indistinguishable from those found in multiple myeloma. **1971** D. HAWKINS in S. O. Freedman *Clin. Immunol.* vi. 179 Patients with Waldenstrom's [*sic*] macroglobulinemia may have proteinuria and even nephrotic syndrome.

Ma:crolepido·ptera, *sb. pl.* *Ent.* [f. MACRO- + LEPIDOPTERA *sb. pl.*] A collective term for the larger butterflies and moths.

1882 W. F. KIRBY *European Butterflies & Moths* p. iii, The present work is designed to provide entomologists.. with a comprehensive illustrated guide to the study of European Macro-Lepidoptera. **1907** R. SOUTH *Moths Brit. Isles* 1st Ser. 6 Quite a number of the species included in that division [*sc.* Micro-lepidoptera] are actually larger than the many kinds that were placed in the other contingent styled Macro-lepidoptera. **1946** *Nature* 12 Oct. 498/2 A well-worked group of insects such as the Macrolepidoptera affords admirable material for faunistic comparisons. **1955** E. B. FORD *Moths* p. xv, The Macro-Lepidoptera are an artificial assemblage, consisting of an arbitrary selection of certain families, and parts of families, without scientific validity. **1972** L. E. CHADWICK tr. *Linsenmaier's Insects of World* 226/2 These families [*sc.* Microlepidoptera] consist of very small species, yet they are in no way separated from the others (the 'Macrolepidoptera').

macrolide (mæ·krŏləid). *Pharm.* [ad. Polish *makrolid* (Z. Katula 1958, in *Postepy Hig. i Med. Dosiviadczalnej* XII. 491): see MACRO- and LACTIDE.] Any of a class of antibiotics containing macrocyclic lactone rings.

1960 *Biol. Abstr.* XXXV. 4202/2 From the fermentation products of various strains of *Streptomyces* there have been isolated several dozen antibiotics, whose common characteristic is a macrocyclic lactic ring. These antibiotics are called macrolides. **1967** [see *ERYTHROMYCIN]. **1968** *New Scientist* 28 Mar. 678/2 Tylosin is a macrolide antibiotic, and organisms resistant to it are often cross-resistant to other macrolides, such as erythromycin, oleandomycin, and spiromycin.

macromolecule (mæ:kromǫ·lĭkiŭl, -mō̆u·lĭkiŭl). *Chem.* [f. MACRO- + MOLECULE.] †**a.** A group of chemical molecules in a crystal bound together in a characteristic shape, which was once believed to account for the symmetry of the crystal. *Obs.*

1886 G. J. STONEY in *Rep. Brit. Assoc. Adv. Sci. 1885* 989 In iron pyrites, FeS_2, the hemihedral form which is characteristic of this mineral..can be traced on from the chemical molecule FeS_2 through a macromolecule which includes six of these as sub-molecules, and which is connected with the other similar macromolecules in a regular way.

b. [ad. G. *makromolekel* (Staudinger & Fritschi 1922, in *Helv. Chim. Acta* V. 788).] A molecule composed of a very large number of atoms and having a high molecular weight (e.g. a molecule of a polymer, a protein, or a nucleic acid).

1935 *Nature* 19 Oct. 626/1 The now generally accepted view that the polymers and condensates must be regarded as constituted of macromolecules formed by the polymerization or condensation of single units. **1957** *New Biol.* XXIII. 66 Proteins are macromolecules normally built up from hundreds or thousands of individual amino acid molecules, and possessing molecular weights in the range of 10,000 to several millions. **1959** *Sunday Times* 5 Apr. 8/6 The macromolecules that are thought to have been the progenitors of living matter. **1969** *Sci. Jrnl.* Nov. 15/3 The two key macromolecules of living things, proteins and nucleic acids, principally play the parts of catalyst and information carriers respectively.

So **ma:cromole·cular** *a.*, of, pertaining to, or consisting of a macromolecule or macromolecules.

1931 *Chem. Rev.* VIII. 409 The physical properties of rubber..indicate..that it is macromolecular. **1958** *New Scientist* 26 June 256/2 We know how the macromolecular mechanism of the chromosomes reshuffles the genes when the sperm and the eggs are formed. **1964** G. H. HAGGIS et al. *Introd. Molecular Biol.* iv. 80 The helix is emerging today as a fundamental feature of macromolecular structure at several levels. **1970** *New Statesman* 20 Mar. 406/3 Otto Wichterle, director of Macromolecular Chemistry in Prague.

macronutrient (mæ:kroniū·triĕnt). *Plant Physiol.* [f. MACRO- + NUTRIENT *sb.*] Any of the chemical elements (as potassium, nitrogen, calcium, sulphur, phosphorus, or magnesium) which are normally taken up by plants as inorganic salts and which are required for growth and development in relatively large amounts (rather than trace amounts).

1942 *Bot. Gaz.* CIII. 651 An extensive experiment in plant nutrition was designed in order to examine the effects of varying concentrations of macro-nutrient elements in the nutrient medium on the ascorbic-acid content of tomato fruits. **1970** *Nature* 25 July 376/1 The lack of macronutrients especially N, P and K, and sometimes micronutrients.

macropædia (mæ:kropī·diă). [f. MACRO- + Gr. παιδεία learning.] The main section of the 15th edition of the *Encyclopædia Britannica* (published in 1974) in which information is presented in the form of extended articles. (Cf. *MICROPÆDIA, *PROPÆDIA.) Hence **macropæ·dic** *a.*

1974 *Times* 12 Jan. 12/1 Finally, there is macropaedia, 19 volumes of substantive essays ranging the world of learning, with articles from 750 to 250,000 words each. Twelve articles are book length. *Ibid.* 16 Jan. 14/4 The *Encyclopaedia Britannica*..proclaimed the imminent publication of a new edition..the macropaedia, supplying knowledge in depth, with 19 volumes. **1974** *Times Lit. Suppl.* 11 Oct. 1120/4 Some sorts of item, those needing more than 750 words but less than full-scale Macropaedic attention,..get Procrustean treatment.

macrophage. Add: Hence **macropha·gic** *a.*

1904 *Brit. Med. Jrnl.* 10 Sept. 562/1 The lymph glands and other macrophagic organs. **1971** *Nature* 5 Feb. 412/2 A sparse scattering of cells, presumably macrophagic or inflammatory.

ma:crophoto·graphy. [f. MACRO- + PHOTOGRAPHY.] Photography in which objects are reproduced larger than or at their actual size but without the degree of magnification that use of a microscope would give.

1889 E. J. WALL *Dict. Photogr.* 114 *Macro-photography*, a term used to denote the enlargement of the negative. **1940** A. L. M. SOWERBY *Wall's Dict. Photogr.* (ed. 15) 438 *Macro-photographs*, term applied to photographs of small objects reproduced at or about natural size. Macrophotography occupies a position intermediate between ordinary photography, in which objects are much reduced, and photo-micrography, in which objects are shown greatly enlarged. **1958** *Newnes Compl. Amat. Photogr.* xiv. 147 Macrophotography, showing all the colour variations of a natural subject in close-up, is a fruitful field. **1964** *Times Rev. Industry* Apr. 42/2 Representatives..demonstrating photomicrography, oscillography, macrophotography and general industrial photography to present or potential Polaroid users in industry. **1967** *Pix* 6 May 24 It might be said that photomacrography (or macro-photography) is that area of close-up photography which lies between what can be done by a simple dioptre lens (close-up attachment lens) and what requires the attachment of the camera to a microscope (i.e., photomicrography).

So **ma:cropho·tograph**, a photograph produced by macrophotography.

1900 Dorland *Med. Dict.* 370/1 *Macrophotograph*, an enlarged photograph. **1933** *Burlington Mag.* Jan. 15/2 A wealth of information is recorded by the X-rays, macrophotographs, micro-photographs and details. **1940** [see above]. **1973** *Sci. Amer.* Feb. 63/2 (Advt.), Macrophotographs (formerly up to 10 times actual size) can now be made up to 14·8 times actual size.

macropsia (mækrǫ·psiă). *Ophthalm.* [f. Macro- + *-opsia*, as in Megalopsia.] = Megalopsia.

1890 Billings *Med. Dict.* II. 96/1 *Macropsia*, a condition of vision in which objects appear abnormally increased in size. **1899** [see Micro- 3]. **1961** A. Huber *Eye Symptoms in Brain Tumors* i. 24 The usual hallucinations and the phenomena of micropsia and macropsia actually bring us to the symptomology of disturbances of the higher visual functions.

macro-scale (mæ·kroskē̆l). Also **macro scale, macroscale.** [f. Macro- + Scale *sb.*[3]] A large or macroscopic scale; *spec.* in *Chem.*, the scale of macroanalysis.

Macro is freq. apprehended as an adj. qualifying *scale* (cf. *macro- 2).

1931 J. W. Brown in C. A. Mitchell *Recent Adv. Analytical Chem.* II. xv. 304 Developments in the application of general macro-scale chemical methods of qualitative and quantitative analysis to amounts of material 50, 100 or 1,000 times smaller. **1941** J. H. Reedy *Elem. Qualitative Analysis* (ed. 3) 2 Formerly chemical analysis was carried out on a 'macro' scale, using considerable amounts of material. **1964** N. G. Clark *Mod. Org. Chem.* xxiv. 496 Quantitative analysis was originally designed to be performed on a sample of 0·4–1·0 g for each element determination (only carbon and hydrogen are estimated on the same sample); this is called the macro-scale. **1968** C. A. Doxiadis *Between Dystopia & Utopia* 52 Also, we should not forget that projecting in the macro-scale..is easier than in the micro-scale. It is easier to predict where the future population will settle in one generation than what type of house, or dress, a certain lady is going to like next year. **1970** *Interior Design* Dec. 767/1 Nevertheless, changes will occur, of course, but on a micro- not macro-scale. **1972** *Physics Bull.* Nov. 668/1 A polymer crystallized from the melt will show, on a macroscale, random crystalline orientation.

macroscopic, *a.* Add: Also *fig.*, general, comprehensive, concerned with large units.

1931 M. Dobb in W. Rose *Outl. Mod. Knowl.* II. xiv. 623 Those macroscopic, as distinct from microscopic, issues of the economic order. **1960** E. Delavenay *Introd. Machine Transl.* 132 Macroscopic study concentrates on large-scale aspects of phenomena—for instance macroscopic linguistics lies bearing on very general statistical rules of language. **1963** *Listener* 10 Oct. 536/2 The problem is to explain *macroscopic individuality* (in common usage) in terms of microscopic non-individuals, rather than the other way round. **1964** I. L. Horowitz *New Sociol.* 3 The rationalists, or 'macroscopic' tendency, concerned with developing 'general theories' of human behavior.

macrosegment (mæ·krosegmĕnt). *Linguistics.* [f. Macro- + Segment *sb.*] A continuous unit of speech between two pauses, with a single intonation.

1958 C. F. Hockett *Course in Mod. Ling.* iv. 38 The stretch of material spoken with a single intonation is called a *macrosegment*. *Ibid.* 41 Though the center of an intonation is by definition the most prominent syllable in the macrosegment, it need not carry the highest pitch. **1963** [see *microsegment]. **1964** K. L. Pike in D. Abercrombie et al. *Daniel Jones* 430 The total phonemic phrase (i.e., the macrosegment including the sum of units between primary juncture, ..) in unemotional speech has an overall intonation contour. **1965** *Language* XLI. 244 A major segment (..Hockett's macrosegment) consists of a series of pitch levels terminated by a major juncture. **1971** *Ibid.* XLVII. 739 In the nucleus column for macrosegment and mega-segment, a O symbol occurs with no explanation given.

macroseism (mæ·krosaiz'm). *Geol.* [f. Macro- + Seism.] A major earthquake; in mod. use (*rare*), any earthquake, as opposed to an imperceptible earth tremor (cf. *microseism).

1903 *Sci. Amer.* Suppl. 2 May 22855 Prof. Milne pointed out the distinction which exists between macroseisms, or large earthquakes, and microseisms, or small earthquakes. The former he described as world-shaking disturbances. **1907** *Jrnl. Geol.* XV. 401 Great confusion exists because of the different uses of the terms 'macroseism' and 'microseism', as well as the adjectives derived from them. The usage here is that of both Milne and de Montessus, which makes 'macroseism' apply to the greater disturbance on the ground. **1924** *Bull. Seismol. Soc. Amer.* XIV. 29 Various useful compounds [of *seism*] have been suggested, such as ..*macroseism*, de Montessus, 1907, for an ordinary sensible earthquake. **1972** *Gloss. Geol.* (Amer. Geol. Inst.) 424/1 *Macroseism*, a syn. of *earthquake*, as opposed to *microseism*.

Hence **macrosei·smic** *a.*, of or pertaining to a macroseism or (in mod. use) those effects of an earthquake that are perceptible without the aid of instruments; **macrosei·smically** *adv.*

1903 *Nature* 9 July 235/1 This is probably true for other phases of motion, and it has also been shown to exist for macro-seismic disturbances. **1907** *Jrnl. Geol.* XV. 408 Macroseismic origins. **1938** *Nature* 1 Oct. 624/1 The region over which the shock was felt macroseismically extended as far as the island. **1940** *Ibid.* 6 Jan. 14/1 On the basis of the macroseismic data, the accompanying sketch map showing the isoseismal lines has been constructed. **1947** K. E. Bullen *Introd. Theory Seismol.* xv. 252 Macroseismic data..usefully supplement the data obtained from seismographs. **1973** *Nature* 17 Aug. 384/2 Fairly complete pictures of what are now termed the macroseismic aspects of Britain's larger earthquakes.

macrosmatic, *a.* Delete '*rare*' and substitute for def.: *Zool.* Having well-developed olfactory organs. Also *fig.* (Add earlier and later examples.)

1890 W. Turner in *Jrnl. Anat. & Physiol.* XXV. 106, I propose..to arrange the Mammalia in relation to the development of the olfactory apparatus into three groups:—(a) Macrosmatic, where the organs of smell are largely developed, a condition which is found..in the majority of mammals. [Etc.] **1894** *Proc. Zool. Soc.* 9 *Echidna*..is, to use Turner's nomenclature, 'macrosmatic'. **1924** *Jrnl. Comparative Neurol.* XXXVII. 318 Even macrosmatic animals like dogs and fishes locate odorous substances by random seeking reactions, not by direct orientation. **1962** *Science Survey* III. 260 Cats and dogs, most of the predators, rodents and deer, and many others are called macrosmatic because a large part of their nasal labyrinths are covered with a special olfactory epithelium. **1968** *Times* 5 Oct. 20/7 [Orwell's] a macrosmatic writer tracking down the stench of hypocrisy or the gangrene of intellectual treachery. **1971** *Nature* 16 Apr. 432/1 Groddeck argued that man is as macrosmatic as the dog.

macrospore. a. Substitute for def.: = *megaspore* (s.v. *mega-). (Add further examples.)

1955 G. M. Smith *Cryptogamic Bot.* (ed. 2) II. x. 281 The heterospory was pronounced [in the fossil fern *Archaeopteris*], the macrospores having a diameter about ten times that of microspores. **1965** [see *megaspore* s.v. *mega-]. **1974** *Mycopathologia et Mycologia Applicata* LIII. 56 Some fungi..produce two spore states which are similar in type but differ in size. The large septate spores are frequently referred to as macrospores and the smaller, nonseptate spores as microspores. The term macrospore is also used infrequently as synonymous with chlamydospore... This latter usage seems undesirable.

macrosporophyll (mæ:krospǫ·rǭfil). *Bot.* [f. Macro- + *sporophyll* (s.v. Sporo-).] = *megasporophyll* (s.v. *mega-).

1888 *Encycl. Brit.* XXIV. 130/2 Carpel = leaf bearing macrosporangia (macrosporophyll). **1955** G. M. Smith *Cryptogamic Bot.* (ed. 2) II. vii. 201 According to the nature of their sporangia, the sporophylls are called macrosporophylls and microsporophylls.

ma·crostructure. Also **macro-structure.** [f. Macro- + Structure *sb.*] Large-scale or overall structure; *spec.* the structure of a metal that is visible (on polished and etched surfaces) to the naked eye or under low magnification.

1920 *Glasgow Herald* 18 Dec. 4 The lecturer spoke of microscopical methods, of macro-structure, and dealt with the variation of physical properties of steel. **1930** *Engineering* 14 Mar. 357/1 Macrostructure of cast alloys. **1960** E. H. Gombrich *Art & Illusion* xi. 365 Van Gogh's own [accent] can be forged with relative ease. But then his swirling lines still belong to the macrostructure of his style. **1964** *English Studies* XLV. (Suppl.) 180 Bogard also rejects the possibility of Shakespeare having influenced the macro-structure of Webster's tragedies.

b. *concr.*

1956 *Nature* 14 Jan. 81/1 The protein and its associated copper bonds..are stable or protected in highly organized macrostructures such as mammalian melanin granules.

So **ma:crostru·ctural** *a.*

1893 [see *microstructural *a.*]. **1963** J. Wiesenfarth *Henry James & Dramatic Analogy* i. 15 One of the principles of James's dramatized novel is that it should always represent. Such macrostructural devices as scenes and pictures..are integral elements by which this ideal is realized. **1964** *English Studies* XLV. (Suppl.) 188 It is strange that no one appears to have taken much interest in the macro-structural relations between his plays and those of other Elizabethan and Jacobean dramatists. **1974** B. Jessop *Traditionalism, Conservatism & Brit. Political Culture* ii. 42 This requires macrostructural analysis as well as survey analysis.

macula. Add: **1.** *Anat.* Any of various structures which have the appearance of a spot; *spec.* the *macula lutea* (see below).

1901 *Phil. Trans. R. Soc.* B. CXCIV. 74 Fundus oculi (right eye) of the Lemurine Dourouculi. The macula is present, but the macula ring has disappeared. **1932** S. Zuckerman *Social Life Monkeys* x. 153 Since the eyes of the lower mammal are usually set more to the sides than to the front of the head..a specially sensitive area or macula is not developed. **1952** *Sci. News* XXIII. 73 When an object is fixed, the eyes are directed towards it in such a way that it is focused on the central spot of the retina—the area for acute vision, known as the macula. **1964** [see *crista].

2. *Anat.* and *Path.* Used in various mod.L. collocations, as **macula densa** [(K. W. Zimmermann 1929, in *Zeitschr. f. mikrosk.-anat. Forschung* XVIII. 529), f. L. *densus* thick, dense], a small mass of cells of uncertain function closely associated with the juxtaglomerular cells; **macula lutea** [L. *luteus* yellow], an oval, yellowish area near the centre of the retina, where visual acuteness is most pronounced; the yellow spot.

1836–9 R. B. Todd *Cycl. Anat. & Physiol.* II. 530/1 Its bottom..presents a sieve-like spot, *macula cribrosa*. **1848** Dunglison *Dict. Med. Sci.* (ed. 7) 370/1 *Foramen centrale et limbus luteus retinæ*; the central foramen and yellow spot of the retina; discovered by Sömmering. Macula lutea. **1857** *Ibid.* (rev. ed.) 560/2 *Maculæ albæ*, white spots, seen on serous membranes..and which appear to be the result of previous inflammatory action. **1942** *Lancet* 3 Oct. 394/2 Zimmerman [*sic*] first recognised the different appearance of the cells of the distal tubule most closely applied to the afferent arteriole in its juxtaglomerular portion. In this part, the ordinary tubular cells seem aggregated, and become in some species much higher and more columnar. Because of this grouping together of nuclei, and the impression of increased density of epithelial cells produced, Zimmerman called this the macula densa. **1952** Macula densa [see *juxtaglomerular *a.*]. **1962** *Gray's Anat.* (ed. 33) 1302 The pyramid and adjoining part of the elliptical recess [of the internal ear] are perforated by a number of holes (macula cribrosa superior). **1967** G. M. Wyburn et al. *Conc. Anat.* viii. 206/2 In the centre of the retina is the yellow spot, the macula lutea. **1968** Macula densa [see *juxtaglomerular *a.*].

macular, *a.* Add: *spec.* of or pertaining to the *macula lutea*.

1909 M. Greenwood in L. Hill *Further Adv. Physiol.* 397 If there is a good deal of macular pigmentation the mixed light undergoes selective absorption. **1932** S. Zuckerman *Social Life Monkeys* x. 166 Monkeys have hands and, what the handed prosimiæ lack, macular vision to guide their manipulations. **1961** R. D. Baker *Essent. Path.* iv. 49 In Tay-Sachs disease..lipids are collected in ganglion cells..and in the macula of the retina (macular star).

maculate, *ppl. a.* Add: Now chiefly *lit.* and *poet.* (Later examples.)

1919 T. S. Eliot *Sweeney among Nightingales* in *Poems*, The zebra stripes along his jaw Swelling to maculate giraffe. **1932** W. Faulkner *Light in August* (1933) xiii. 300 Leaning in the window, breathing the hot still rich maculate smell of the earth. **1964** C. S. Lewis *Discarded Image* vii. 161 In Shakespeare's *Lucrece* we need to know fully who the 'spotted princess' (219–28) is: Tarquin's Reason, rightful sovereign of his soul, now maculate. **1965** E. Bishop *Questions of Travel* 22 House, open house.. Darkened and tarnished By the warm touch Of the warm breath, Maculate, cherished, Rejoice!

maculate, *v.* Add: Now *rare*. (Later examples.)

1737 A. Baxter *Inquiry Human Soul* (ed. 2) II. 202 *Lucretius* tells us maculating dreams accompany youth. *a* **1945** E. R. Eddison *Mezentian Gate* (1972) ii. 21 That were to maculate the purity of your own proper nature.

maculature. Delete † *Obs.*⁻⁰ and add: **2.** *Engraving* (see quots.).

1904 *Burlington Mag.* V. 70 One of these [impressions of the Hundred Guilder Plate]..is a 'maculature', an impression on a sheet of ordinary paper passed over the plate to remove the ink. **1914** *Brit. Mus. Guide Processes of Engraving* 52 A maculature is another form of weak impression. A copper plate needs to be inked between each impression. Sometimes a second impression is taken from the plate before re-inking, as a means of extracting the remainder of the ink from the lines. This is called a maculature.

maculopapule (mæ:kiŭlǫpæ·piūl). *Med.* [f. Macul(a + -o + Papule.] A maculopapular lesion. So **ma:culopa·pular** *a.*, having characteristics of both a macule and a papule; characterized by such lesions.

1900 Dorland *Med. Dict.* 370/2 Maculopapule. **1905** Gould *Dict. New Med. Terms* 247/1 Maculopapular. **1912** H. French *Index Differential Diagn. Main Symptoms* 424 If a macule takes on a slight degree of elevation it is sometimes styled a maculo-papule. *Ibid.* 528 If the lesions, originating as erythematous macules, do not take on the full character of papules, they are said to be maculo-papular. **1928** C. P. Emerson *Physical Diagn.* iii. 82 There may be some maculopapules or true papules on the palms or soles. **1962** *Lancet* 12 May 998/2 He had a scattered maculopapular rash over the trunk. **1968** *Biol. Abstr.* XLIX. 4986/2 The scales are grayish in appearance in chronic squamous maculopapular dermatoses.

macumba (măku·mbă). Also **makumba.** [Pg.] A religious cult-practice of the Negro population of Brazil characterized by sorcery, ritual dancing, and the use of fetishes. Also *attrib.*

1939 *Peabody Bull.* Dec. 8 The frenzy of a *macumba* or the tropical sensuality of a son. **1941** *Survey Graphic* Mar. 181 The religious *macumba* or *candomblé* found in the morros (the hills) combines Catholicism with African and Indian magic rituals. **1948** H. Mielche *From Santos to Bahia* vii. Once a year..for four days the city surrenders to the spell of Makumba. **1951** Smith & Marchant *Brazil* v. 145 Brazilian *macumbas* and *candomblés* are undergoing rapid changes. There is a curious fusion..with other religions and cults, especially with Catholicism and Spiritualism. **1963** *Guardian* 4 June 7/1 Some of the most striking scenes are depictions of the macumba rites. **1964** *Listener* 6 Aug. 211/1 They had penetrated a *macumba* temple in Rio de Janeiro. **1969** J. Mander *Static Society* vii. 211 The Brazilian North-East, where Voodoo or *makumba* cults flourish. **1971** *Daily Colonist* (Victoria, B.C.) 23 May 33/3 The people of Bahia are..devotees of Macumba, that unique north east Brazilian blend of voodoo and Christianity.

Macushi (măkū·ʃi). Also **Macusi. a.** A Carib Indian people inhabiting Guyana (formerly

British Guiana) and Brazil; a member of this people. **b.** The language of this people. Also *attrib.* or as *adj.*

1881 *Encycl. Brit.* XII. 828/2 In British Guiana the Carib tribes are the Ackawais and Caribisi of the coast and forest regions, the Arecunas and Macusis of the savannah region. **1934** E. Waugh *Handful of Dust* v. 273 She Macushi woman. All these people Macushi people. *Ibid.* 274, I wish I could speak Macushi. **1934** —— *Ninety-two Days* iii. 78 There was a large Indian village. These were sophisticated Macushis who were in constant contact with the ranches. *Ibid.* 89 He was at work on a translation of the scriptures into Macushi. **1974** H. MacInnes *Climb to Lost World* iv. 57 A Macusi Indian..had been taken to England by some missionaries.

macushla (măku·ʃlă). *Anglo-Ir.* [f. Ir. *mo* my + *cuisle* vein, pulse (of the heart). Cf. *acushla.] (My) dear heart; darling. (Used as a term of address.)

1887 W. B. Yeats *Dawn-Song* in *Irish Fireside* 5 Feb. 83/3 Wake, *ma cushla*, sleepy-headed. **1918–19** *T. Eaton & Co. Catal.* Fall & Winter 369/2 Vocal Records... Macushla (Tenor). **1946** A. Seton *Turquoise* viii. 115 Sure, and ye're a foine figger of a woman yourself, macushla. **1950** O. Nash *Family Reunion* (1951) 89 In a word, Macushla, There's a scad o' things that to make a house a home it takes.

mad, *sb.*[2] (Earlier and later U.S. examples.)

1834 in J. S. Bassett *Southern Plantation Overseer* (1925) 65, I will be darnde if I can do anything with them and they all ways in the mads. **1867** W. L. Goss *Soldier's Story* xiv. 258 The Colonel has got his mad up. **1878** E. B. Tuttle *Border Tales* 50 A grizzly will stand in the middle of the road, growling and getting his mad up. **1916** H. L. Wilson *Somewhere in Red Gap* ii. 57 She kept her mad down better. She set there as nice and sweet as a pet scorpion. **1950** J. D. MacDonald *Brass Cupcake* (1955) iii. 25 When I want a personality course, friend, I'll go to someone who hasn't a mad on at the world. **1973** M. & G. Gordon *Informant* xxxiii. 128 Well, thanks a lot! I go through hell for you and you take your mad out on me.

mad, *a.* Add: **1. b.** (Earlier examples.)

1605 Shaks. *Lear* II. iv. 289 O Foole, I shall go mad. **1782** Cowper *Poems* I. 314 What! hang a man for going mad? Then farewell British freedom. **1795–1804** W. Blake *Vala* i. in *Compl. Writings* (1972) 265 Thou wilt go mad with horror if thou dost Examine thus Every moment of my Secret hours. **1839** in *Amer. Speech* (1956) XL. 130 O dear, I shall go mad, My husband is so crazy.

fig. **1901** G. B. Shaw *Three Plays for Puritans* Pref. p. xxix, Besides, I have a technical objection to making sexual infatuation a tragic theme. Experience proves that it is only effective in the comic spirit..but..to worship it, deify it, and imply that it alone makes our lives worth living, is nothing but folly gone mad erotically. **1914** —— *Parents & Children* in *Misalliance* p. cii, The sort of Rationalism which says to a child 'You must suspend your judgment until you are old enough to choose your religion' is Rationalism gone mad. **1923** L. W. Reese *Wild Cherry* 22 The weather has gone mad with white. **1949** T. Rattigan *Playbill* 56 The lighting for this scene has gone mad.

4. c. (Examples.)

1776 [see *music-mad* s.v. Music sb. 12 b]. **1825** H. Wilson *Mem.* I. 41 One of her new admirers, who, being flute-mad, and a beautiful flute player, was always ready. **1848** [see *woman-mad* adj. s.v. Woman sb. 7]. **1904**, etc. [see *man-mad* adj. s.v. *man sb.[1] 20]. **1943** E. M. Almedingen *Frossia* ii. 58 Look at all this promiscuity... They have all gone sex-mad. **1946** K. Tennant *Lost Haven* (1947) 8 All the family were horse-mad. **1974** 'P. B. Yuill' *Hornless Keeper* xiii. 129 Perhaps you can save her from a sex-mad rabbit and win her undying love.

5. (Further examples.)

1887 F. Francis *Saddle & Mocassin* 111 The more he studied it [*sc.* the bill] the madder he got. **1902** W. James *Varieties Relig. Experience* xi. 264 He can't 'get mad' at any of his alternatives; and the career of a man beset by such an all-round amiability is hopeless. **1925** E. Wallace *King by Night* viii. 32 Don't get fresh with that girl of mine... You just get mad at her. **1939** [see *chisel v.[1] 4]. **1956** M. Duggan in C. K. Stead *N.Z. Short Stories* (1966) 90 Are you mad at me? Simpson asked. **1962** H. Hood in R. Weaver *Canad. Short Stories* (1968) 2nd Ser. 210 'Why is Daddy mad?' said Deedee. '*I'm not mad*!' **1973** *Black World* June 57/1 Gloria mad at me three days now.

8. Also (*as*) *mad as a cut snake* (Austral.), † *Ajax*, *a hornet* (U.S.), *a meat axe* (chiefly Austral. and N.Z.), *a wet hen*.

1588 Shaks. *L.L.L.* iv. iii. 7 By the Lord this Loue is as mad as Aiax, it kills sheepe. **1607** Chapman *Bussy d'Ambois* iii. 468 Murther market folkes, quarrell with sheepe, And runne as mad as Aiax. **1732** T. Fuller *Gnomologia* 140 Love is as mad as Ajax; it kills Sheep, so it kills me. **1823** J. Doddridge *Dialogue of Backwoodsman & Dandy* in *Logan* 42 Every body that was not ax'd was mad as a wet hen. **1855** T. C. Haliburton *Nat. & Hum. Nat.* I. 85, I feel as mad as a meat axe. **1902** W. N. Harben *Abner Daniel* 54 The Colonel is as mad as a wet hen about the whole thing. **1919** Mencken *Amer. Lang.* 80 In the familiar simile, *as mad as a hornet*, it [*sc.* the word *mad*] is used in the American sense. **1923** Wodehouse *Inimitable Jeeves* xviii. 249 My uncle will be as mad as a wet hen when he finds out that he has been fooled. **1927** *Amer. Speech* II. 360 He was as mad as a hornet when he heard how the election went. **1932** 'W. Hatfield' *Ginger Murdoch* 30 'But you're mad!' said Mick, 'mad as a cut snake!' **1946** J. Fountain in *Coast to Coast 1945* 252 The cow's mad—mad as a meat-axe! **1951** S. Mackenzie *Dead Men Rising* 203 'Mad as a cut snake,' Johnson said admiringly, 'and there's not a better feller in the whole camp.' **1963** *Moderna Språk* LVII. i. 10 *As mad as a cut snake*: 'mad' is used in the sense of 'angry', and the phrase means 'extremely angry'. **1970** D. M. Davin *Not Here, Not Now* v. iii. 263 She's mad as a meataxe anyway about the whole idea. **1971** *Wall St. Jrnl.* 22 July W. 1/4 The chicken farmers of Quebec..are as mad as, well, a wet hen.

9. *mad-afraid, -keen* adjs.; **mad minute** *Army slang*, a minute of rapid rifle-fire or frenzied bayonet-practice (see quots.); **mad money** *colloq.*, money for use in an emergency; *spec.* (see quot. 1922); **mad scientist**, a scientist who is mad or eccentric, esp. so as to be dangerous or evil: a stock figure of melodramatic horror-stories; freq. *attrib.*

1895 Kipling *Seven Seas* (1896) 90 When the steers are mad-afraid. **1949** A. Christie *Crooked House* xvi. 126 She's mad keen on this detecting stuff. **1974** L. Lamb *Man in Mist* xiii. 88 Derek Boots was not exactly the type to join us here... I was not so mad keen on him. **1917** A. G. Empey *Over Top* 298 *Mad minute*, firing fifteen rounds from your rifle in sixty seconds. A man is mad to attempt it, especially with a stiff bolt. **1942** in Baker *Austral. Lang.* (1945) viii. 155 The *mad minute*, bayonet drill. **1945** C. H. B. Pridham *Superiority of Fire* vi. 57 By 1914, many men in each regiment could exceed even twenty rounds in the 'mad minute'. **1964** C. Falls in S. Nowell-Smith *Edwardian England* xiv. 537 Reservists and young soldiers alike could shoot steadily and accurately at a relatively slow rate for long periods, or in emergency fire what they called their 'mad minute'. **1965** Brophy & Partridge *Long Trail* 147 *Mad minute*, a newspaper phrase for British rapid fire during the Retreat from Mons... The name was also applied to the frenzied minute spent charging down the assault course, bayoneting straw-filled dummies, representing enemy soldiers. **1922** *Dialect Notes* V. 148 *Mad money*, money a girl carries in case she has a row with her escort and wishes to go home alone. **1933** Partridge *Slang To-day & Yesterday* v. 283 *Mad money*, return fare, it being very generally believed by the New Zealand troops..that every English girl infallibly carried her return fare in case her soldier friend became *mad*, i.e., acted with an excessive freedom of manner. **1943** J. Steinbeck *Once there was War* (1959) 136 He has a nest egg or mad money. **1961** L. Deighton *Ipcress File* x. 61, I think he grabs an S.I. now and again when he needs some mad money. **1970** 'D. Shannon' *Unexpected Death* (1971) ix. 135, I haven't even a dime of mad money with me, hope I don't need it. **1972** O. Sela *Bearer Plot* i. 15, I reached for the wad of notes Keith kept as mad money. **1940** 'N. Blake' *Malice in Wonderland* III. xviii. 282 A sort of mad-scientist motive for the whole series of outrages. **1963** 'G. Bagby' *Murder's Little Helper* (1964) iv. 36 The whole idea smacked too much of some mad-scientist fable out of a comic strip. **1972** B. Turner *Sodom's Women* ix. 82 He would have passed for the mad scientist in one of those films which star giant insects.

mad, *v.* **1.** (Later U.S. examples.)

1873 M. Holley *My Opinions* 249 At the same time it madded some of the Republicans. **1893** 'O. Thanet' *Stories Western Town* 31, I madded him first; I was a fool. **1916** H. L. Wilson *Somewhere in Red Gap* vi. 268, I think to find him all madded up and mortified; but he's strangely cheerful for one who has suffered. **1924** W. M. Raine *Troubled Waters* vi. 59 O' course, it ain't that any of them's afraid to mad that crazy gunman, Tait.

Madagascan (mædăgæ·skăn), *a.* and *sb.* [irreg. f. *Madagascar* (see next) + -an.] **A.** *adj.* Of or pertaining to Madagascar (now the Malagasy Republic). **B.** *sb.* A native or inhabitant of Madagascar.

1886 *Ibis* 135 The alternative hypothesis..that the Madagascan and Columbian species [of Snipes] have changed. **1890** *Cent. Dict.*, *Madagascan*,..a native or an inhabitant of Madagascar. **1953** L. D. Stamp *Africa* xx. 511 Every Madagascan farm has a few pigs. **1973** *Listener* 20 Sept. 369/1 The Madagascan revolt, in which eighty thousand were killed. **1973** *Country Life* 20 Dec. 2122/2 The Japanese..have developed a real yen for it [*sc.* Beaujolais]. So have the Madagascans, the Tahitians and the Nationalist Chinese.

Madagascar (mædăgæ·skăɹ). The former name of the Malagasy Republic, a large island off the east coast of Africa, used *attrib.* in special collocations, as **Madagascar cat,** the ring-tailed lemur, *Lemur catta*; **Madagascar** (clove) **nutmeg, manna** (see quots.); **Madagascar periwinkle,** a tropical plant (*Vinca rosea*) with white or rose-coloured flowers.

1900 H. A. Bryden *Animals Afr.* 12 Some of these curious lemurs, which are usually known as 'Madagascar Cats'. **1866** Lindley & Moore *Treas. Bot.* I. 28/2 One species, *A*[*gathophyllum*] *aromaticum*, grows in Madagascar... The fruit..encloses a kernel of an acrid caustic taste, known as Madagascar clove nutmegs. **1889** *Cent. Dict.*, *Dulcitol*,..is commercially obtained from an unknown plant in Madagascar, and in the crude state is called *Madagascar manna*. **1821** M. Browne *Diary* (1905) 104 There were..myrtles, beautiful campanulas, geraniums, Madagascar periwinkles, etc.

Madagascarian (mædăgæskɛ̄ə·riăn), *a.* [f. prec. + -ian] = *Madagascan *a.* Also **Madaga·scarene, Madaga·scrian** *sbs.* = *Madagascan *sb.*

1824 M. A. Hedge *Radama* iv. 78 The first order is usually composed of those termed the white Madagascrians. **1856** C. Nordhoff *Merchant Vessel* xix. 246 The natives..are mostly black, the descendents of Madagascarenes. **1875** *Encycl. Brit.* III. 758/2 Madagascar, the Comoros, and the widely-scattered Mascarene Islands constitute a fifth Subregion,..and for this we may most reasonably use the name 'Madagascarian'. **1893** A. Newton *Dict. Birds* II. 347 Those [genera] belonging to the insular or Madagascarian Subregion.

‖**madal** (madā·l). Also **madāla, maddale.** A double-headed drum used in Nepal and eastern India. Cf. *mridangam.

1914 A. H. F. Strangways *Mus. Hindostan* ix. 228 The *maddale* (*mṛdanga*-shape) hollowed out of a tree. **1954** J. Masters *Bhowani Junction* 329 My Birkhe beat cheerfully on a madal, which is a deep and narrow Gurkha drum. **1960** S. Prajñānanda *Hist. Devel. Indian Mus.* i. 7 The different musical instruments of folk-music like *ekatāra*,..*maddāla*,..etc. bear testimony to..the cultural taste and outlook of the peoples of Bengal. **1969** *Illustr. Weekly India* 27 July 33/1 They assemble..and dance—both men and women—to the accompaniment of the *madal* (the Santali drum). **1971** K. Kent in C. Bonington *Annapurna South Face* App. H. 314 Most boys [in Nepal] learn to play the flute and *madal*—a double-ended drum, with one end slightly smaller than the other.

madam, *sb.* Add: **3. c.** (*d*) A brothel-keeper; cf. *madame 4. (*e*) Nonsense, humbug. *slang.*

(*d*) *a* **1911** D. G. Phillips *Susan Lenox* (1917) I. xxi. 393 The madam fixes things so that every girl always owes her money. **1912** T. Dreiser *Financier* xlvi. 510 In a few moments the 'madam', as the current word characterized this type of woman, appeared. **1918** C. Garnett tr. *Dostoevsky's White Nights* 117 Before me was standing a person with a stupid smile, the 'madam' herself. **1926** J. Black *You can't Win* iv. 30 The following week I called at Madam Kate Singleton's... In a minute the madam came down. **1959** N. Mailer *Advts. for Myself* (1961) 279 A rather remarkable woman who had been the madam of a whorehouse. **1960** Auden *Homage to Clio* 85 Henry Adams Was mortally afraid of Madams: In a disorderly house He sat quiet as a mouse. **1964** *Punch* 30 May p. xiii/1 Barbara Stanwyck as Lesbian madam of New Orleans brothel. **1975** *Daily Tel.* (Colour Suppl.) 18 July 7/1 The oldest girl is a woman, maybe Czechoslovakian, maybe the madam.

(*e*) **1927** E. Wallace *Feathered Serpent* xvii. 218 'I was getting a hundred quid for this job..and I couldn't turn him down.' 'The usual "madam"!' sneered the inspector. 'It's not "madam", Mr. Brown,' said Jerry earnestly, 'though I admit it sounds as likely as cream in skilly; but it's true.' **1932** A. Gardner *Tinker's Kitchen* 284 Madam = made up story; flattery. **1936** J. Curtis *Gilt Kid* ii. 18 'What did the old boy say?' 'Just the usual madam.' **1965** *Sunday Times Mag.* 11 July 21 Both sides are expert with the madam—a form of kidology which seems to come naturally to most Merseysiders. **1973** J. Wainwright *Touch of Malice* 130 It was not the sort of place conducive to putting over a spot of old madam. The normally glib flannel tended to stick in his throat and the guff and eyewash hadn't enough elbow-room to..sound..feasible.

4. (*b*) **madam-shop** [Madam sb. 3], a small shop which sells ready-to-wear clothes for the fashionable woman of mature taste.

1952 *Times Lit. Suppl.* 28 Nov. 778 The juggling with couture dressmakers..and owners of 'madam shops' in order to keep them contented with their share of reporting. **1963** *Harper's Bazaar* Feb. 20/3 The sight of two assistants in one of those little madam shops fitting girdles on to plastic models. **1965** *Guardian* 16 July 8/5 Boutiques are nothing new in Manchester. There is a stalwart 'madam shop' tradition; little shops for women of substance. **1966** H. W. Yoxall *Fashion of Life* viii. 68 The contemporary Madam Shops, so aptly named,..are enjoying a new popularity. **1967** *Guardian* 30 Mar. 5/1 The madam shops... Shops catering for women over 25 who want to dress in fashion, and can afford to.

madame. Add: **4.** = *madam *sb.* 3 c (*d*).

1871 *N.Y. Herald* 29 July 6/2 The Madame..sent her to an infamous den in Forsyth Street kept by a Mrs. Hines. **1922** V. Woolf *Jacob's Room* ix. 171 Only Madame herself..had about her that leer, that lewdness. **1934** A. Woollcott *While Rome Burns* 157 Visiting the local Maison Tellier..and taking the madame and all her girls out duck-shooting. **1961** M. Jones *Potbank* xxvi. 111 Her behaviour..made me think of a *madame* in the more discreet sort of brothel. **1969** G. Greene *Trav. with my Aunt* i. viii. 83 There was a discipline in the old-time brothels. The madame..played a rôle similar to that of the headmistress of Roedean. **1975** *Sunday Times* 12 Jan. 37/2 She plays the part of the notorious Xaviera Hollander, New York call-girl and *madame*.

madarosis. (Further examples.)

1902 Gould & Warren *Internat. Text-bk. Surg.* (ed. 2) II. xxvii. 864 The affection may go on until nearly all the eye-lashes are lost and the lids left bald—madarosis. **1956** 'H. MacDiarmid' *Stony Limits & Scots Unbound* 39 Nor.. can we..Shut our eyes despite their madarosis of the sun. **1972** A. Sorsby *Mod. Ophthalm.* (ed. 2) II. xx. 557 (*caption*) Leprosy showing lagophthalmos and madarosis of brows and lashes.

madbrain. A. *sb.* Delete † *Obs.* and add later example.

1876 G. Meredith *Beauch. Career* II. xvi. 285 He began to think her lost beyond hope, embarked for good and all with the madbrain.

madcapery (mæ·dkæpĕri). Also **madcappery.** [f. Madcap *sb.* + -ery.] The behaviour of a madcap; mischievous or reckless conduct.

1905 D. Sladen *Playing Game* xii. 139, I wondered what madcapery Rich had been up to. **1966** *Punch* 20 April 586/2 Seeing the two plays in such quick succession ..does provide an opportunity to follow in one long sweep the two stages of Prince Hal's progress from madcappery to maturity.

maddening, *ppl. a.* Add later examples of the sense 'irritating, annoying, vexatious'.

1896 A. Beardsley *Let.* c 25 Oct. (1971) 188 Dent must be simply maddening. **1925** N. Coward *Vortex* ii. 55 'Have you only one set, Florence?'..'Yes, isn't it maddening? Clara promised to bring hers down but forgot.' **1944** —— *Middle East Diary* 36 So many of my Naval friends are here and it's maddening that I shall have no time to go and visit them. **1947** A. Huxley *Let.* 19 Jan. (1969) 565 There will still be revisions to do on the screen play..maddening work, resembling jig saw puzzles rather than literature.

madder, *sb.*[1] Add: **4.** **madder-bleach,** a special method of bleaching cotton; **madder-print,** madder-printed cloth or cotton (*Cent. Dict.* 1890).

1909 L. A. Olney *Textile Chem. & Dyeing* II. 58 The Madder Bleach... In calico printing..where a particularly clear and white ground is desired this form of bleach is used.

maddery (mæ·dəri), *a.* *Nonce-wd.* [f. Madder *sb.*[1] + -y[1] 2.] = Madderish *a.*

1873 G. M. Hopkins *Note-bks. & Papers* (1937) 186 Its dewlaps and bellyings painted with a maddery campion-colour.

madding. Add: **1.** Esp. in phrase *far from the madding crowd* (see quot. 1749 in Dict.), a conventional phrase denoting a secluded place removed from public notice.

1874 Hardy (*title*) Far from the madding crowd. **1889** J. K. Jerome *Three Men in Boat* i. 9, I..suggested that we should seek out some retired and old-world spot, far from the madding crowd, and dream away a sunny week among its drowsy lanes. **1944** F. Clune *Red Heart* 14 People..far from the madding crowds, west of the Darling River. **1952** G. Sarton *Hist. Sci.* I. xvi. 397 He [*sc.* Plato] did not want to teach in the streets and markets, but on the contrary in a place that was sufficiently distant from the madding crowd and secluded.

mad-doctor. For def. read: A physician who specializes in disorders of the mind; a psychiatrist. (Add further examples.)

1852 Dickens & Wills *Curious Dance round Curious Tree* in *Househ. Words* 17 Jan. 385/1 Nothing was too wildly extravagant, nothing too monstrously cruel, to be prescribed by mad-doctors. **1877** J. M. Granville *Care & Cure of Insane* I. 2 It must never be forgotten that the so-called 'mad doctors' have been the first to press this truth on the profession. **1890** (*title*) Mad doctors by one of them, being a defence of asylum physicians. **1972** C. Achebe *Girls at War* 9 That humble practitioner who did the miracle became overnight the most celebrated mad-doctor of his generation.

made, *ppl. a.* Add: **3. b.** Of bills of exchange: (see quots.).

1868 E. Seyd *Bullion & Foreign Exch.* 89 The foregoing Foreign Bills of Class 2 are called *drawn* Bills, being usually negotiated from the Drawer direct to a London Foreign Banker; but where such drafts are made in the Country, and sent up to a correspondent in London, who then negotiates the same with his own Indorsement on them, they are called *made* Bills. *Ibid.* 90 Bills drawn abroad and payable abroad, but negotiated in the United Kingdom, are also *made* Bills.

5. b. *made to measure*: see Measure *sb.* 2 a; also (usu. with hyphens) *attrib.*; *made to order*: see Order *sb.* 24 c; also (usu. with hyphens) *attrib.*; so *made-to-order-ness*, the state or condition of being made to order.

1960 *Sunday Express* 24 July 12/6 Made-to-measure tweed skirt. **1973** *Guardian* 26 Feb. 3/3 Good lighting is made-to-measure lighting..the result of applying a lighting engineer's expertise to your particular office problem. **1974** *Country Life* 3–10 Jan. 58/1 Made-to-measure..corsets, brassieres, maternity foundation. **1921** *Daily Colonist* (Victoria, B.C.) 3 Apr. 9/1 (Advt.), Ladies' and gents' smart made-to-order spring suits made with particular care to every detail. **1947** 'G. Orwell' *England your England* (1953) 8 There was the made-to-order stuff which I produced quickly. **1973** *Amer. Speech* 1969 XLIV. 277 Or the phrase used in a television advertisement for 'Sentry—the made-to-order insurance'. **1973** R. Stout *Please Pass the Guilt* (1974) ii. 11 He sat..in his made-to-order chair. **1923** *Glasgow Herald* 8 Nov. 4 There is an air of cynical made-to-order-ness about the second [poem].

7. c. *to have* (*got*) *it made*, to be sure of success; to have it easy, to have no more obstacles to overcome. *colloq.* (orig. *U.S.*).

1955 in *Amer. Speech* May 118. **1960** J. Updike *Rabbit, Run* (1961) 108 Say. You really think you have it made. **1961** J. Heller *Catch-22* (1962) vi. 51, I had it made, I tell you. Fifty grand a year I was knocking down, and almost all of it tax-free. **1967** A. Diment *Dolly Dolly Spy* vi. 85 The money you earn is fantastic... You've got it made, mate. **1968** —— *Gt. Spy Race* iii. 45 She had.. big, well-proportioned hips. I tell you, if the derrière gets with-it again this bird had it made. **1972** 'H. Howard' *Nice Day for Funeral* iii. 49 This was the kind of set-up half the dames I know would've given their back teeth for. I had it made. **1974** *Times Lit. Suppl.* 8 Mar. 227/2 The abstentions of 1972 were due not to this disillusionment but to an overwhelming conviction that Mr Nixon had it made, so why take time..to go out and vote?

9. *made-over* (examples); **made-up,** (*e*) of articles of trade, ready-made, not made to measure; also **made-up tie,** a tie, esp. a bow-tie, with a fixed bow or knot; (*f*) of stakes, arranged after the original programme of races is drawn up; (*g*) of a book, with its deficiencies made good by the insertion of a leaf, etc., from another copy of the same edition.

1912 R. A. Wason *Friar Tuck* xxx. 208 When Jim came back he was a made-over man, and everyone asked him if he had religion. **1916** J. E. Wells *Man. Writings Middle Eng.* 294 A copy, and perhaps a somewhat made-over copy of an earlier text. **1929** W. K. Gregory *Our Face from Fish to Man* II. 153 Even the most imposing human faces are but made-over fish traps. **1967** A. Lewin *Unaltered Cat* I. ii. 24 A place very like her own, a made-over brownstone only two blocks away. **1725** M. W. Montagu *Let.* c 10 June (1966) II. 53, I wish you would lay out part of my Money in a made up Mantua and petticoat. **1849** *Theatrical Programme* 16 July 55 (Advt.), The immense patronage they have received this season in their made-up skirt-rooms. **1876** *Coursing Calendar* 110 In the made-up stakes for puppies Mr. Farmer's brace..made a good display. **1883** *Graphic* 14 Apr. 395/4 (Advt.), A full assortment of made-up articles of the best and most suitable description. **1895** *New Rev.* June 631 It is an odious fact that this country spends about a million and a half a year in the purchase of made-up clothes from Germany. **1913** C. Mackenzie *Sinister St.* I. II. ii. 171 The boys..bought made-up bow-ties of purple and pink that were twisted round the stud with elastic. **1913** R. B. McKerrow in *Trans. Bibliogr. Soc.* XII. 303 The cruder sort of made-up copies that one often finds in the second-hand market. **1929** J. Buchan *Courts of Morning* 25 Sandy in a greasy dress suit and a made-up black tie. **1952** J. Carter *Bks. & Bk.-Collectors* (1956) vi. 195 The..language of book-collectors..is thickly encrusted with jargon... 'Made-up,' says one, with a sniff. 'Q6 is a cancel as usual,' says another. **1972** M. Gilbert *Body of Girl* v. 52 He was wearing a blue suit,..a flannel shirt and a made-up bow tie.

10. Colloq. phr. *made of money*, extremely rich, very wealthy.

1849 D. W. Jerrold (*title*) A man made of money. **1855** Mrs. Gaskell *North & South* I. xii. 143 'I shall order horses.' 'Nonsense, John. One would think you were made of money.' **1876** Trollope *Prime Minister* III. xv. 254 You're living here in a grand house, and your father's made of money. **1895** Mrs. H. Ward *Story B. Costrell* ii. 35 You don't care, not you!—one 'ud think yer were made o' money. **1918** C. Mackenzie *Early Life Sylvia Scarlett* II. ii. 292 He asked if I thought he was made of money and could buy top-hats like matches. **1967** E. Coxhead *Thankless Muse* iv. 103 'Then why don't we keep it?' 'You think Clare's made of money, I suppose?' **1975** C. Egleton *Skirmish* x. 104 Book him into a hotel..but nothing fancy, we're not made of money.

Madeira[1]. Add: **2. c.** **Madeira sauce,** a rich brown sauce made with Madeira and served with braised or roast meats.

1872 R. C. Smith *Madeira & its Associations*, The lively Frenchman..dwells upon the virtues of delicious Madeira. He offers at once to despatch his grandfather in Madeira sauce. *A la Sauce Madère on mangerait son grandpère.* **1877** *Cassell's Dict. Cookery* 57/1 *Beef, Lumber, in Madeira Sauce*..Prepare a sauce with brown stock or broth, some butter, flour, cayenne pepper, salt, pepper, and a glass of Madeira. **1946** G. Millar *Horned Pigeon* xix. 305 She followed this with kidneys and a Madeira sauce. **1956** L. Diat *French Cooking for Home* 43 Boiling spoils the flavour of Madeira Sauce.

madeleine (mæ·dlē̆in). Also **madeline.** [F., prob. f. name of *Madeleine* Paulmier, 19th-c. French pastry-cook.] A (kind of) small rich cake baked in a shell-shaped tin. Sometimes (with allusion to Proust, quot. 1922) taken as typical of something that strongly evokes memories or nostalgia. Also, in English cooking, a kind of baked pudding or small fancy cake.

1845 E. Acton *Mod. Cookery* xviii. 473 (*heading*) Madeleine puddings. (To be served cold.) **1846** C. E. Francatelli *Mod. Cook* 404 Madeleines..are made with the same kind of batter as Genoese cakes, to which currants, dried cherries, candied peel or angelica, may be added. **1902** G. H. Ellwanger *Pleasures of Table* vi. 169 Dumas tells the story of the excellent cake called madeleine, an entremets which all who have been in France will remember. **1922** C. K. S. Moncrieff tr. *Proust's Swann's Way* I. 61 And suddenly the memory returns. The taste was that of the little crumb of madeleine..my aunt Léonie used to give me, dipping it first in her own cup of.. tea. **1939** O. Lancaster *Homes Sweet Homes* 42 The flavour of Bordeaux pigeon summons with all the completeness of Proust's tea-soaked madeleine an unforgettable cloud of Mons. Doré's angels hovering over the Colosseum. **1948** *Good Housek. Cookery Bk.* 577 Dip a cherry in a little jam, place on top of each madeleine and put on 2 leaves of angelica. **1958** *Spectator* 1 Aug. 174/1 We all have our little fragment of *madeleine* that brings back a dearly remembered but half-forgotten past. **1960** E. David *French Provincial Cooking* 32 At the little town of Commercy originated the small, fragile, shell-shaped cakes called madeleines so beloved of French children... (How the English madeleine, a sort of castle pudding covered in jam and coconut, with a cherry on the top, came by the same name is something of a mystery.) **1962** *Punch* 21 Mar. 462/1 It may be said—if it may then Proust has certainly said it—that the intensity of nostalgic emotion has little to do with the quality of the material which evokes it. (I dare say that madeleine cakes are very nasty.) **1972** 'M. Innes' *Open House* I. ii. 12 He might have been Proust's Marcel, hard upon imbibing the displeasing little sopped *madeleine* which brought his childhood flooding back to memory.

‖**madérisé** (maderize), *a.* [Fr.] Of wine, affected with maderization (see below). So **ma·derize** [F. *madériser*] *v. intr.*, to become *madérisé*.

1941 Schoonmaker & Marvel *Amer. Wines* 219 The French also describe a white wine which has been too long in barrel..as *madérisé*, or 'madeiraed'. **1950** O. A. Mendelsohn *Earnest Drinker* xx. 189 When a white wine starts to go a brownish colour and gets a curious..musty smell about it, it is called *Madérisé* and means that the wine is on the road downhill. **1958** A. L. Simon *Dict. Wines* 105/2 *Madérisé*, the polite French word to use to describe the bottle stink of a wine which has been kept too long. **1961** W. E. Massee *Wines & Spirits* 10 A white table wine that has maderized is undrinkable. **1972** H. W. Yoxall *Enjoyment of Wine* v. 39–40 White wine deepens its tone with age and becomes what they call maderisé, though not all white wines that are so darkened acquire the musty taste of a truly maderized wine.

maderization (mădiᵊrəizē̆i·ʃən). [ad. F. *madérisation* (also used), f. Madeira[1] 2 + -ization.] A brown discoloration in white wines often appearing after overlong or unsuitable storage.

1951 R. Postgate *Plain Man's Guide to Wine* iv. 76 The only noticeable change is an occasional darkening of colour, called *madérisation*. **1952** H. W. Allen *White Wines & Cognac* ii. 53 Still without a hint of *madérisation* and golden as sunlight. **1952** A. Lichine *Wines of France* i. 7 A characteristic called maderization after Madeira wine, whose colour they take. **1959** W. James *Word-bk. Wine* 113 *Maderization*, a flaw in white wines caused by their absorption of too much oxygen during vinification or maturation; the wine turns rusty in the bottle, and takes on a brown colour and a musty, flat, taste; when much advanced there may be some resemblance to a very poor madeira. **1971** *Times Lit. Suppl.* 12 Mar. 297/2 Such a bouquet suggests maderization, and I cannot believe that a lover of German wines would enjoy this very much. **1973** *Vogue* 15 Apr. 153/3 Old champagne can be great. It goes brown with age, and turns sweeter and heavier—a condition known as 'maderization'.

madhead[2]. After '*Obs.*' add 'exc. *dial.*'

1959 *New Statesman* 26 Dec. 904/3 He'd quietened down now though, and butter wouldn't melt in his mouth, he said..after all his stories about him being a madhead. **1959** I. & P. Opie *Lore & Lang. Schoolch.* ix. 170 Red heads attract a barrage of nicknames..glow-worm, mad head [etc.].

madhouse. Add: **a.** (Later examples.)

1916 G. B. Shaw *Androcles & Lion* Pref. p. lxiii, One person in every five dies in a workhouse, a public hospital, or a madhouse. **1922** Joyce *Ulysses* 358 The nobleman with the foreign name..had to have her put into a madhouse, cruel only to be kind. **1929** F. N. Hart *Hide in Dark* v. 182 It seems fairly essential to get at what facts are available..if some of us aren't to wind up in a madhouse. **1955** G. Willans *Fasten your Lapstraps!* i. 23 There is a dull sound of barley sugar being crunched and gum chewed—the whole place..is like a Victorian madhouse. **1971** J. Namier *Lewis Namier* xiii. 234 To L insanity was man's ultimate degradation; a madhouse, however well appointed, was hell.

b. *fig.* A scene of uproar or confusion bewildering to the onlooker.

1919 G. B. Shaw *Heartbreak Ho.* III. 95 Is this England, or is it a madhouse? **1929** H. Crane *Let.* 7 Feb. (1965) 335 This City [*sc.* Paris], as you know, is the most interesting madhouse in the world. **1946** E. O'Neill *Iceman Cometh* (1947) III. 160 God, I'm glad I'm leaving this madhouse! **1956** R. Braddon *Nancy Wake* xiv. 155 The Moulins railway junction was a mad-house of torn and tangled lines and shattered rolling stock. **1973** *Radio Times* 26 Apr. 48/1 They [*sc.* chefs] roast and stew and bake in a kind of madhouse of shouted commands, cancelled orders and frayed tempers.

Ma·dison. orig. *U.S.* [Origin unknown.] A group dance, popular in the 1950s and 1960s.

1962 *Listener* 27 Dec. 1074/2 Girls in traditional Kimono dancing the twist or the Madison. **1966** *Punch* 25 May 768/2 We found the Bishop and Dolly Girl an hour later. She was teaching him the rudimentary technique of The Madison. **1968** M. & J. Stearns *Jazz Dance* 4 Before and during Presley's initial success, the first wave of dances became popular. Group dances such as the Madison and the Birdland arrived first. **1969** C. Booker *Neophiliacs* vii. 177 In the same month of October [1962]..the unknown Beatles first joined the fashionable beat of the Madison in the charts.

Madison Avenue (mæ·disən æ·vĭniu). The name of a street in New York City, which is the centre of the American advertising business; hence *allusively*, (American) advertising generally, the advertising business; *collect.*, American advertising agents. Also *attrib.*

1955 H. Kurnitz *Invasion of Privacy* (1956) v. 39 A tall, lean young man..dressed in the dark grey and neat stripes of a Madison Avenue advertising executive. **1957** J. Blish *Fallen Star* ii. 33 We had..Madison Avenue gossip..for lunch. **1958** J. K. Galbraith *Affluent Society* xviii. 200 The violent mores of Hollywood and Madison Avenue. **1959** N. Mailer *Advts. for Myself* (1961) 404, I kept expecting him to go Madison Avenue, I was certain he would sell out sooner or later. **1959** V. Packard *Status Seekers* (1960) i. 12 Madison Avenue has been busily trying to understand our tastes and buying behaviour. **1960** *Guardian* 15 Oct. 6/5 An Eisenhowerish vocabulary which..sometimes has more of a Madison Avenue ring. **1961** [see *Harley]. **1970** *Times* 24 Aug. 15/5 Britain's Madison Avenue widely predicted that after a decent interval, *Ad Weekly* would also attempt something spectacular. **1972** *Jazz & Blues* Sept. 9/2 We know what's wanted. I had something to offer Madison Avenue.

madly, *adv.* Add: **2.** Colloq. uses. **a.** Passionately, fervently (cf. quot. 1590 in Dict.).

1767 Boswell *Let.* 30 Mar. (1924) I. 108, I was so madly in love as to think of marrying her. **1880** O. C. Stone *Few Months New Guinea* iv. 55 The natives..seem madly fond of *kuku* [*sc.* tobacco], and would pawn their very clothes for it if they wore any. **1888** W. S. Gilbert *Trial by Jury* 12, I love him—I love him—with fervour increasing, I worship and madly adore. **1924** A. Christie *Man in Brown Suit* ii. 19, I could never marry a man unless I loved him madly. **1959** [see *beetle *sb.*[2] 2 b]. **1974** M. Cecil *Heroines in Love* v. 151 Upper-class heroines.. were never 'in love', always 'madly in love'.

b. Extremely, very, 'awfully'.

1888 H. James *Reverberator* II. i. 16, I was not madly impatient to see you married. **1902** G. Bell *Lett.* (1927) I. vii. 130 It is a madly interesting place. **1935** N. Marsh *Enter Murderer* ii. 31 She's madly keen on criminology. **1937** C. Connolly in L. Russell *Press Gang!* 79 Hubert said Balliol was perfect for case-histories like mine, but I realised I should find it madly ungay. **1945** N. Mitford *Pursuit of Love* ix. 73 It's madly wearing to the optic nerve centres. **1954** J. B. Priestley *Magicians* i. 20 Mavis..brightened up like a touched-off firework and was at once, in her own phrase, 'madly gay'. **1967** [see *draggy *a.* b]. **1974** *Radio Times* 3 Jan. 5/4 Mr Williams's facial mobility is *madly* impressive.

mado (mā·do). *Austral.* [Aboriginal name.] A small marine fish, *Atypichthys mado* (or *A. strigatus*), found in southern Australian and northern New Zealand waters.

1898 E. E. Morris *Austral Eng.* 277/2 *Mado*, a Sydney fish, *Therapon cuvieri*, Bleek; called also *Trumpeter-Perch*. **1906** D. G. Stead *Fishes of Australia* 134 The mado is a handsome little fish, having alternate brown and yellow longitudinal stripes along the body. **1958** *Austral. Encycl.* V. 458/2 *Mado*, a small fish (*Atypichthys mado*) very common round wharf-piles and in inlets along the coasts of the southern half of Australia. **1960** Doogue & Moreland *N.Z. Sea Anglers' Guide* 229 Mado..yellow with longitudinal brown streaks..Distribution: North Island south to about East Cape. Also occurs in Australian waters.

Madonna. Add: **4.** *Madonna-like* adj.; **Madonna blue,** a shade of deep blue; **Madonna lily** (earlier and later examples).

[**1930** Maerz & Paul *Dict. Color* 198/2, 1917 Madonna.] **1932** A. Christie *Peril at End House* vii. 85 She was wearing a gown of Madonna blue. **1935** *Times* 2 Oct. 17/4 A Madonna-blue lining. **1971** N. Freeling *Over High Side* II. 110 Twilight had fallen, of a pure madonna blue. **1850** Mrs. Gaskell *Let.* 24 Jan. (1966) 101 She was a madonna-like person with a face..full of thought and gentle love. **1895** G. B. Shaw *Let.* 20 Mar. (1965) 502 You must..be sweet..and Madonna like. **1909** M. Diver *Candles in Wind* I. iv. 40 The soft brown hair..giving a Madonna-like air of purity to the oval face. **1966** 'R. Standish' *Widow Hack* vi. 65 She has large, generous features which compose themselves in a Madonna-like serenity. **1877** E. S. Phelps *Story of Avis* ix. 172 An exquisite motion which an artist..would not have wasted..on anything less than a Madonna lily. **1963** W. Blunt *Of Flowers & Village* 173 My whole room is scented by a great pot of Madonna lilies. **1974** *Country Life* 21 Mar. 632/1 A child in a bonnet creeps shyly down the path, dwarfed by the Madonna lilies.

Madras. Add: **1.** Also, **Madras cotton,** cotton fabric produced in Madras, *esp.* the brightly checked or striped cottons the colours of which run together in laundering. Also **Madras** *ellipt.*

1890 L. Hearn *Two Years in French West Indies* 216 The making-up of the Madras into a turban is called 'tying a head'. **1897** *Sears, Roebuck Catal.* 216/3 French Madras is a fine soft finished fabric with the colors woven through. .. Light colors, plaids or stripes. **1921** *Daily Colonist* (Victoria, B.C.) 22 Oct. 7/2 (Advt.), 50-Inch Colored Madras at $1.50 a Yard..this Madras will make beautiful side curtains. **1942** J. Hoye *Staple Cotton Fabrics* vii. 148 When made in fancy weaves and of fine yarns, they are known as Madras ginghams. **1962** L. Deighton *Ipcress File* xxviii. 177 A bright Madras jacket. **1964** Hollen & Saddler *Textiles* (ed. 2) 153/2 The imported Indian madras, today, is handwoven with white warp and colored filling 'guaranteed to bleed'. **1968** H. Waugh *Con Game* iii. 34 Three pairs of men's shoes and one wrinkled Madras jacket. **1972** *Vogue* Feb. 86 Madras checked safari jacket, unlined seersucker.

Madrasi (mădræ·si), *a.* and *sb.* Also **Madrassi, -assee.** [Urdu *Madrasī*, f. *Madras* the city in southern India.] **A.** *adj.* Of or pertaining to Madras. **B.** *sb.* A native or inhabitant of Madras.

1878 *Chambers's Jrnl.* Feb. 115/1 English, after the rickety fashion of a Madrassee, Sam spoke fairly enough. **1879** H. Hartigan *Stray Leaves* 2nd Ser. 129 While ruminating, a Madrasi servant came out. **1921** *Contemp. Rev.* Sept. 291 'Western civilisation,' said an eminent Madrasi the other day, 'has led to war.' **1924** *Blackw. Mag.* Aug. 227/2 The officer commanding..Madrasi Christians, will not admit this. **1971** R. Russell tr. *Ahmad's Shore & Wave* v. 49 His partner, a solemn-looking, bald Madrasi, was suffering all this in silence.

madrigalesque (mædrigāle·sk), *a.* [f. Madrigal *sb.* + -esque; cf. F. *madrigalesque* f. It. *madrigalesco*.] Having the features or characteristics of madrigals.

1911 *Encycl. Brit.* XVII. 295/2 Long afterwards we occasionally meet with the word again, when a 17th or 18th century composer sets to some kind of accompanied singing a poem of madrigalesque character. **1924** W. H. Hadow *Music* iv. 99 A pleasant light comedy set to madrigalesque music with a real sense of characterization.

madrilene (mædrilī·n, -lē̆·n), *sb.* Also **madrilène.** [ad. F. (*consommé à la*) *madrilène*, f. Sp. *Madrileño, -leña* of Madrid.] A clear soup which is usually served cold (see quot. 1907).

1907 G. A. Escoffier *Guide Mod. Cookery* xiii. 224 *Consommé a la madrilène*. Add four oz. of raw tomato and one oz. of capsicum to the consommé per every quart of the latter. Mix these ingredients with the clarification, and serve as cold as possible. **1931** J. Berjane *French Dishes for Eng. Tables* i. 13 Consommé Froid aux Tomates, or Madrilène. **1952** S. Kauffmann *Philanderer* (1953) xi. 183 He found a lot of food in the icebox and had fixed some madrilène and a big salad and some fruit cup by the time she arrived. **1955** E. Bowen *World of Love* v. 96 Terence..began to spoon up liquidly-jellied madrilene. **1956** A. L. Simon *Wine & Food Menu Bk.* 182 Consommé madrilène..should be rather highly seasoned and its colour should be not more than just a maiden's blushing pink. **1964** A. Launay *Caviare & After* 135 *Madrilène* (*à la Madrilène*), dishes flavoured with tomato juice or clear soups flavoured with tomato juice and usually served chilled.

Madrilenian (mædrilī·niăn), *a.* and *sb.* Also **Madrileñan, Madrilene, Madrilenean.** [f. Sp. *Madrileño, -leña* of Madrid, the capital of Spain.] **A.** *adj.* Of or pertaining to Madrid. **B.** *sb.* A native or inhabitant of Madrid.

1841 Borrow *Zincali* I. II. ii. 241 The Madrilenian Gypsy women are indefatigable in the pursuit of prey. **1846** R. Ford *Gatherings from Spain* xviii. 254 This bit of land was taken possession of by a worthy Madrilenian. **1873** Browning *Red Cott. Nt.-Cap* I. 42 Father Miranda, goldsmith of renown: By birth, a Madrilene. **1882** F. R. McClintock *Holidays in Spain* 20 The carriages and horses of the Madrilenian high-life are as good as anything to be seen in Hyde Park. **1889** M. S. Van de Velde *Cosmopolitan Recoll.* II. iv. 135 One of the unsavoury apartments is occupied by Señor de Castellan—a gentleman, a scholar, and an officer in the navy—who was once the most conspicuous figure in the Madrilene world of fashion. **1909** *Daily Chron.* 2 July 3/1 A far better opportunity..of seeing the inner life of a Madrilenean family of distinction. **1921** J. B. Trend *Pict. Mod. Spain* 65 Some critics have accused Baroja of distorting the Madrileñan character. **1957** A. MacNab *Bulls of Iberia* xii. 129 Vicente Pastor... A Madrilenian born and bred,..had the dry, dominating, Castilian style in the ring. **1972** *Times* 11 May (Spain Suppl.) p. vii/2 Madrilenians take their bullfighting seriously. **1974** *Times* 2 May 17/2 Madrid's residents are crowded... Few Madrilenians live in one-family homes.

Madrileño (mædrilē̆·nʸo). Fem. **Madrileña** (-lē̆·nʸă). [Sp.] = *Madrilenian *sb.*

1832 E. C. Wines *Two Yrs. in Navy* (1833) I. x. 307 The tragedian was a *Madrileno*, as nearly all the actors on the Spanish stage are. **1866** Mrs. W. P. Byrne *Cosas de España* I. viii. 180 The Madrileños and Madrileñas were already sunning themselves in the bright morning air. **1903** L. Williams *Toledo & Madrid* II. i. 117 Both were *Madrileños*, and taught the sciences at Toledo and Granada. **1932** E. Hemingway *Death in Afternoon* v. 50 Madrileños love the climate. **1943** E. A. Peers *Spain in Eclipse* I. iv. 85 'Han pasado, han pasado..' shouted crowds of Nationalist Madrileños. **1950** G. Brenan *Face of Spain* v. 101 Mr. Washbrook is a New Englander..his wife is a handsome and vigorous Madrileña. **1965** C. D. Eby *Siege of Alcázar* (1966) i. 28 Swarms of Madrileños of the lower classes were arguing along the curbs. **1975** *Times* 19 Apr. 12/4 Do as the Madrileños do: punctuate your perambulations at the various bars.

Madurese (mædiŭrī·z), *a.* and *sb.* [f. *Madur*(*a* + -ese.] **A.** *adj.* Of or belonging to Madura, an island lying off the north-east coast of Java. **B.** *sb.* **a.** A native or inhabitant of Madura. **b.** The Austronesian language of Madura. Also **Madu·ran.**

1817 T. S. Raffles *Hist. Java* I. ii. 59 The Madurese.. display a more martial and independent air..than the natives of Java. *Ibid.* viii. 359 In the provinces east of Surabáya, the language partakes much of the Madurese. **1853** J. R. Logan in *Jrnl. Indian Archipelago* VII. 33 Anam heads are common in eastern Java and especially amongst the..Madurans. **1878** R. N. Cust *Sk. Mod. Lang. E. Indies* ix. 137 The Javanese is the chief Language of the island of Java... Its Character..is used by the Sundanese, Balinese, Madurese, and the people of Lombok. *Ibid.* 138 The Madurese..has one Dialect, the Sumanap. **1880** *Encycl. Brit.* XIII. 607/1 The limits of the Madurese area are not so easily given. **1933** L. Bloomfield *Lang.* iv. 71 The languages of the great islands of the East, such as..Maduran. **1969** *Language* XLV. 685 Although there is still much of Madurese morphology which requires further description, this study goes far beyond anything done for Madurese heretofore.

maduro (madū·ro). [Sp., = ripe, Mature *a.*] A dark-coloured cigar.

1889 L. Friedlander *Tobacconist* (ed. 5) 28 M.. Maduro—very dark brown. **1908** S. Ford *Side-Stepping with Shorty* vii. 106 The spots of him that you could see.. was the colour of a twenty-cent maduro cigar. **1939** C. Graves *Cigars & Man* 16 The two remaining darker shades, Colorado-Maduro and Maduro are seldom met with in England. **1957** C. Mackenzie *Sublime Tobacco* xiv. 240 '1 pound full Cabañas'... These may have been Maduro, the darkest of all, which are never seen to-day.

maduromycosis (ma·durǝmǝikōᵘ·sis). *Path.* [mod.L., f. Madur(a + -o + Mycosis.] A chronic destructive infection of the foot (rarely of other parts) that is accompanied by many discharging sinuses and is caused by various actinomycetes and fungi; also called *Madura foot* when appropriate.

1916 Chalmers & Archibald in *Ann. Trop. Med. & Parasitol.* X. 170 But confusion may arise between the terms 'Mycetoma' and 'True Mycetoma', and, therefore.. we suggest the word 'Maduromycosis' instead of True Mycetoma. **1937** *Nature* 27 Feb. 377/2 Maduromycosis, a fungus disease affecting the legs of human beings. **1961** R. D. Baker *Essent. Path.* ix. 213 Maduromycosis or mycetoma is an infection of an extremity, usually the lower, as the result of a puncture wound.

maemae, var. *maimai.

maestrale (mǝistrā·le). Also **maestral, maestro** (mǝi·stro). [It. *maestrale*, f. L. *magistrāl-is*, f. *magister* Master *sb.*[1]] The name of a wind experienced in the Mediterranean (see quot. 1944). Cf. Mistral.

1766, 1813 [see Mistral]. **1902** *Encycl. Brit.* XXX. 622/1 In summer a north-west 'trade' wind, the Maestro occurs in the Adriatic. **1920** *19th Cent.* Aug. 288 The waves ..on the beaches of Corfu..come in with a pleasant surge when the northerly *maestro* is blowing. **1944** *Italy* (Geogr. Handbk. Ser. B.R. 517, Admiralty, Naval Intelligence Div.) I. v. 415 The *maestro* (or *maestrale*), although bearing the same name as the mistral of the Rhône valley, is not to be confused with it. The name is given to NW. winds in the Adriatic, and NW., N., and NE. winds in Liguria and Tuscany. In the west the maestrale is a winter wind (Genoa), but is less cold and dry than the mistral proper. In the Adriatic it is a summer wind. **1967** D. S. Walker *Geogr. Italy* (ed. 2) iii. 224 The influence of persistent winds (especially the *maestrale*) is apparent in the tortured shapes of the trees.

‖**maestria** (maˌe·striă). [It.] Skill, mastery.

1876 Stainer & Barrett *Dict. Mus. Terms* 279/1 *Maestria* (It.), skill, address, authority. **1921** R. Fry *Let.* 15 Mar. (1972) II. 505 He [*sc.* Derain]'s becoming more and more Baroque... It may be dangerous with his incredible *maestria*. **1928** *Daily Express* 16 Jan. 4 Sir Charles..can occasionally be persuaded to render the 'Volga Boatmen's Song', which he sings with incomparable maestria. **1938** *Daily Tel.* 25 July 9/1 While admiring the maestria of this piece of writing by Mr. Shaw I find in it..several inaccuracies. **1955** Taylor & Kerr *Mus. Lovers' Encycl.* 633/1 Maëstria.

Maestrichtian (māstri·χtiăn), *a.* *Geol.* Also **Maastrichtian.** [ad. F. *maestrichtien* (A. Dumont 1849, in *Bull. de l'Acad. R. d. Sci.*, etc., *de Belgique* XVI. II. 360), f. *Maestricht* (now *Maastricht*), the name of a city in SE. Holland; see -ian.] Of, pertaining to, or designating a division of the Upper Cretaceous in Europe that is now regarded as a stage lying next below the Danian. Also *absol.*

1885 A. Geikie *Text-bk. Geol.* (ed. 2) 834 In the Cotentin, a limestone with *Baculites anceps, Scaphites constrictus*, and other fossils has been paralleled with the Maestricht chalk (Maestrichtian sub-stage). **1931** Gregory & Barrett *Gen. Stratigr.* xi. 171 The Senonian deposits in the N.E. of France and Belgium were those of a shallow sea, such as the phosphatic chalks of the Campanian and the limestone 'tuffau' of the Maastrichtian. **1967** D. H. Rayner *Stratigr. Brit. Isles* x. 337 Maastrichtian chalk is rare in Britain but small remnants of the lower zone (*Belemnitella lanceolata*) are known from the Norfolk coast and north-eastern Antrim. **1971** *Nature* 19 Feb. 553/2 Although disputed by some, the latest Cretaceous and earliest Tertiary (Maestrichtian and Danian stages) have generally been recognized as a brief interval during which many organisms underwent a severe attenuation in diversity or became extinct. **1975** *Ibid.* 6 Mar. 50/1 By about 70 Myr ago (late Cretaceous: Maastrichtian) there seems to have been a gap large enough to have prevented east–west migration of marine reef colonies.

maestro. Add: **a.** Also with pronunc. (mǝi·stro). Pl. **maestri, maestros.** (Further examples.)

1947 *Penguin Music Mag.* IV. 44 The maestros on the rostrum. **1952** J. K. Sherman (*title*) Music and maestros.

b. *transf.* A master or leader in any art, profession, etc.

1938 M. Allingham *Fashion in Shrouds* xv. 238 The old *maestro* allows beautiful suspect to slip through nicotine-stained fingers. **1942** Berrey & Van den Bark *Amer. Thes. Slang* § 629 Coach; trainer; manager.. *maestro*. **1945** R. Chandler in *Atlantic Monthly* Nov. 51/3 There is no reason to expect from the anonymous toilers of the screen a quality which we are very obviously not getting from..the sulky maestri of the little magazines. **1945** S. Lewis *Cass Timberlane* (1946) xix. 114 Reverence for jazz and familiarity with such contemporary maestri as Benny Goodman and Pee-wee Russell. **1952** Phillips & Reese *Bridge with Mr. Playbetter* xxix. 123 'Perhaps, if you lead a low Diamond originally,' answered the maestro. **1953** *Proc. Geologists' Assoc.* LXIV. 139 That maestro of Highland tectonics, Sir Edward Bailey. **1958** *Listener* 2 Oct. 537/3 The 'Ted Ray Show' came back with the maestro in full hilarity. **1960** V. Nabokov *Invitation to Beheading* xix. 193 We beseech you, be calm, maestro. If something was not just right, it was the result of an oversight. **1964** L. Nkosi *Rhythm of Violence* 64, I deman' an explanation why my bottle is being impounded... Ah, the maestro needed a shot in the arm too! **1971** *Sunday Nation* (Nairobi) 11 Apr. 26/1 Anyone willing to part with a work of the maestro is asked to contact the Society.

maestro: see *maestrale.

‖ **maestro di cappella** (maˑstro di kapeˑlă, məiˑstro —). *Mus.* Also **maestro di capella.** [It., lit. 'master of the chapel'.] = *Kapellmeister* (see Kapelle); choir-master; musical director, conductor.
1724 *Short Explication Foreign Words in Musick Bks.* 44 *Maestro*, is Master. Thus, *Maestro de capella*, is Master of the Chapel Musick or Master of Musick only, meaning thereby one of the first Rank. 1774 'J. Collier' *Mus. Trav.* 16 All the musicians of Britain..together with every *Maestro di Capella* in Italy. 1880 J. H. Shorthouse *John Inglesant* xxiii. 316 The elder, whose name was Giacomo Andria, was maestro di capella of one of the churches. 1947 C. Gray *Contingencies* 98 In 1850, Raimondi became *maestro di cappella* at St. Peter's in Rome. 1964 *Conc. Oxf. Dict. Opera* 238/2 The *Maestro di cappella* was orig. the equivalent of the German Kapellmeister..but today the term is used only with reference to religious music.

Maeterlinckian (mēⁱtəɹliˑŋkiăn, māˑtəɹ-), *a.* [f. *Maeterlinck* (see below) + -ian.] Of, pertaining to, or having the characteristics of Maurice Maeterlinck, Belgian author (1862–1949), or his writings.
1895 G. B. Shaw *Our Theatres in Nineties* (1932) I. 189 The Maeterlinckian treatment of Pelléas et Mélisande. 1904 W. L. Courtney *Devel. M. Maeterlinck* 35 It is hardly a characteristic example of the Maeterlinckian drama. 1909 *Times Lit. Suppl.* 22 Apr. 150/2 Conflicting with the spirit of mystery and fascinating Maeterlinckian unreality. 1933 *Ibid.* 5 Oct. 657/2 The old haunted house of the Harveys on the Quay, with its Maeterlinckian terrors for a sensitively imaginative child. 1974 P. de Vries *Glory of Hummingbird* (1975) xiii. 173 On the one hand, flocks of Maeterlinckian bluebirds fresh descended from heaven; on the other, a passel of pubescent near-villains queuing up for release from hell.

Mae West (mēⁱ west). *slang* (orig. *R.A.F.*). [f. the professional name of an American film actress and entertainer (1892–), with reference to her curvaceous figure (see quot. 1941).] An inflatable life-jacket, orig. issued to R.A.F. men in the war of 1939–45, later in more general use where the risk of drowning is involved.
1940 *Reader's Digest* May 31/2 The aviators have adopted amusing novelties. For example..*Mae West* for a life jacket. 1941 *N.Y. Times Mag.* 27 July 21/2 One can understand..why an airman's life-belt should be a 'Mae-West'. It..gives the wearer a somewhat feminine figure. 1942 *R.A.F. Jrnl.* 16 May 33 A second more determined pull opened it [*sc.* the parachute] at about 400–500 feet. The pilot did not inflate his Mae West. 1945 *Daily Mirror* 15 Aug. 3/3 McGarvey discarded his Mae West and swam to the vessel. 1952 T. J. Mulvey *These are your Sons* v. 100 When you are flying over the hills of Korea, Mae Wests are a poor substitute for parachutes. 1958 M. Dickens *Man Overboard* x. 152 I'm glad I wasn't in your submarine, Ben. I'd have worn a Mae West all the time. 1971 *Daily Tel.* 16 Jan. 21/8 He was burned about the face and floated in his 'Mae West' until rescued by the Air-Sea Rescue Service. 1974 *Times* 4 Sept. 1/8 One of them had a Mae West on with the words, 'Morning Cloud' written across it.

‖ **mafeesh** (măfīˑʃ), *a.* and *int.* Also **mafish, mefeesh.** [ad. colloq. Eastern Arab. *mā fī-š* there is nothing.] (See quot. 1925.)
1855 R. F. Burton *Pilgrimage to El-Medinah & Meccah* I. i. 11 When a little boy, presuming that the occasion might possibly open the hand of generosity, looked in my face and exclaimed 'Bakhshish', he obtained in reply 'Mafish'; which convinced the bystanders that the sheep-skin contained a real sheep. 1897 A. Conan Doyle *Tragedy of Korosko* (1898) viii. 225 'But what..about the three ladies?' The black soldier shrugged his shoulders. 'Mefeesh!' ['Mafeesh!', 1897 serial publ.] said he. 'One of them is old, and..there are plenty more women if we get back to Egypt.' 1916 C. J. Dennis *Moods of Ginger Mick* (1918) 135 *Mafeesh*, finish; I am finished. 1919 W. H. Downing *Digger Dial.* 32 Mafish.., finish, finished. 1924 Kipling *Debits & Credits* (1926) 318 As soon as he reached the place it was *mafeesh* with him, as usual... You never noticed him. *Ibid.* 320 House, two ricks an' stable *mafeesh*, the big glasshouse with every pane smashed. 1925 Fraser & Gibbons *Soldier & Sailor Words* 150 *Mafeesh*, (*Arabic*). Dead. Done with. Finished. Used colloquially everywhere on Eastern Fronts... It had other meanings: 'I can't', 'I know', 'Get out', 'Go to hell', [etc.]. 1931 T. R. G. Lyell *Slang* 497 *Mafeesh*, old lad—nothing doing! I gave away the last of them a week ago.

mafficker. (Example.)
1910 *Blackw. Mag.* July 9/2 The 'mafficker' may hereafter come within sight of the enemy.

mafia. Add: Now usu. with pronunc. (mæˑfiă, maˑfiă). Delete ‖ and (from def.) 'erroneously'. Also in the U.S. and elsewhere, and *transf.* (Further examples.) Hence also **mafioˑsa** (fem.), **maˑfiaist,** a member or supporter of the mafia; **maˑfiaism,** the doctrines or practices of the mafia. Also *attrib.* and *transf.*
1924 A. Christie *Poirot Investigates* iii. 92 Suddenly they learn that one of these secret societies, the Mafia, or the Camorra..is on their track. 1927 *Daily Tel.* 22 Nov. 7 Thuggism meant an end to human life; Mafiaism poisoned every department of it. 1948 *Oxf. Jun. Encycl.* I. 135/2 Through a secret society called the Mafia, they [*sc.* the Czechs and Slovaks] gave help to the Allies against the Austrians and Germans. 1948 E. L. Irey *Tax Dodgers* (1949) vii. 145 His gunmen shot the Mafiaists out of the top rung of the underworld. 1959 *Times Lit. Suppl.* 9 Oct. 575/1 Mr. Maxwell also prints statements by members of the ruling classes and outsiders: a nun, a priest, a semi-*mafioso* doctor, a Tuscan *carabiniere*. 1965 J. Wainwright *Death in Sleeping City* i. 75 She was born into the Mafia. She was a Mafiosa... He was a lousy Mafioso—because he wasn't all rotten. 1967 *Times* 28 Feb. (Canada Suppl.) 29 Their Mafia-conscious compatriots south of the border. 1967 *Listener* 20 Apr. 529/2 He [*sc.* J. F. Kennedy] was constantly personalizing the office he held, filling the White House with his 'Irish mafia',..extending the influence of the Kennedy court to cover the whole range of American cultural and intellectual activity. 1969 C. Drummond *Odds on Death* v. 118 No law enforcement agency had ever learned to cope with *mafiosi*. 1970 *New Yorker* 3 Jan. 44 The composers' Mafia, with its dedication to atonality and the production of new noises, holds no terrors for him. 1970 E. Tidyman *Shaft* (1971) vi. 92 If a Mafia don was breaking his kid into the business today, he would break him in through the Harvard Business School. 1971 M. McCarthy *Birds of America* 263 I'm prepared for attacks, naturally, from the academic Mafia. 1971 P. Ziegler *King William IV* xiii. 162 It was the turn of the Ultra Tories. This mafia of malcontents..pledged themselves to destroy the Duke [of Wellington]. 1972 'J. Ripley' *My Word you should have seen Us* 88 The F.B.I. know Joseph Colombo as a Capo Mafioso. 1973 *Guardian* 16 Feb. 13/8 British Brussels is..split on the issue of the TUC's participation in Europe. The Labour Party mafia is oversensitive about the TUC. 1974 *Sumter* (S. Carolina) *Daily Item* 24 Apr. 12A/4 Another top Mafioso, Aniello Dellacroce, just finished an income-tax sentence in the Atlanta pen. A few days before his release, another Mafioso leader, Frank Valenti, was locked behind federal bars.

mafic (mæˑfik), *a.* *Min.* [f. Ma(gnesium + L. *f-errum* iron + -ic.] Pertaining to, containing, or designating the dark-coloured minerals of igneous rocks, which are predominantly ferromagnesian in character. Opp. *felsic *a.*
1912 W. Cross et al. in *Jrnl. Geol.* XX. 561 We suggest the term..*mafic* for the group of modal ferromagnesian minerals of all kinds. 1920 [see *femic *a.*]. 1926 [see *ankaramite]. 1969 *Nature* 20 Dec. 1153/1 The lavas are intruded by ultramafic and mafic bodies and dyke swarms. 1970 *Ibid.* 31 Oct. 413/1 The African rifts are sites of under-saturated, predominantly mafic volcanics erupted in comparatively small volume from single vents. 1971 I. G. Gass et al. *Understanding Earth* v. 83/1 The oceans, with their thin (about 5 km) mafic crust.

mafioso: see Mafia in Dict. and Suppl.

mafish, var. *mafeesh *a.* and *int.*

‖ **ma foi** (ma fwa), *int.* [Fr., lit. 'my faith'.] On, upon my word (see Word *sb.* 15 a). Cf. † Mafey *int.*
c 1400 *Brut* 17 'Ma foy,' quod þe fader, 'y may no more axen.' 1778 F. Burney *Evelina* I. xiv. 76 '*Ma foi*, Sir,' answered she, '..none of my acquaintance is in town.' 1791, 1842 [see Mafey *int.*]. 1868 C. M. Yonge *Chaplet of Pearls* II. xxxvi. 159 [He] had the civility to give me a guide and an escort... *Ma foi!* I believe they were given to..take me by indirect roads. 1906 Baroness Orczy *I will Repay* xvii. 197 'A splendid combination, *ma foi!*' said Merlin. 1928 A. Christie *Mystery of Blue Train* xxvii. 221 This time, *ma foi*, I thought we had got him.

mafoo (māˑfū). Also **ma-fu.** [ad. Chinese *ma-fu*, f. *ma* horse + *fu* servant, labourer.] A Chinese groom, stable-boy, or coachman.
1863 G. Fleming *Trav. Manchu Tartary* ii. 16 Ma-foo, as he was soon christened, from his occupation as horse-keeper. 1880 W. Gill *River of Golden Sand* II. viii. 310, I at first rode a hired pony, and my new Ma-Fu walked on in front leading the grey. 1890 Barrère & Leland *Dict. Slang* II. 37/2 *Mafoo* (pidgin), horse-boy, groom. 1924 *Blackw. Mag.* Feb. 232/2 Let them send word to the mafoo to have the ponies saddled. *Ibid.* Aug. 265/2 He borrowed Cantegril's famous two-horse barouche, all complete, with its variegated *mafoos*. 1939 'A. Bridge' *Four-Part Setting* xxiii. 313 They dismounted and left the mafoos and ponies in an open grassy space.

mag (mæg), *sb.*⁵ *Astr.* Also **mag.** (with point). Abbrev. of *magnitude*(*s*) (Magnitude 3 a).
1840 *Mem. R. Astron. Soc.* XI. 283 (*heading*) Relative order of mag. as observed. 1851 *Monthly Notices R. Astron. Soc.* XI. 187 A double star..appearing to the naked eye as a bright 4 mag., and designated by A as 3·4. 1918 *Astrophysical Jrnl.* XLVII. 264 The absolute magnitudes of the bulk of the A stars lie within a range of 2·6 mags. 1972 *Nature* 22 Dec. 439/2 The distribution of galaxies brighter than 16 mag is..not especially isotropic.

mag (mæg), *sb.*⁶ Also **mag.** (with point). Colloq. abbrev. of Magneto.
1920 *Blackw. Mag.* Nov. 562/2 Having wrestled for an hour with the mags., they were eventually induced to give forth reluctant sparks. 1930 [see *differential *a.* and *sb.* B. 4]. 1943 *R.A.F. Jrnl.* Aug. 11, I could see her eyes intent on the rev counter as she cut out first one mag, then the other. 1958 'Castle' & 'Hailey' *Flight into Danger* i. 16 Each engine has two magnetos... In the run-up each engine in turn is opened to full throttle and each of the mags tested separately. 1973 E. Arnold *Proving Ground* (1974) iv. 61 Sayers pulled off the throttle and cut off the mag.

mag (mæg), *sb.*⁷ Also **mag.** (with point). Colloq. abbrev. of Magnesium, often used for *magnesium alloy.*
1969 C. Trickey *Building & Racing 850 Mini* i. 10/2 While, naturally, mag. alloy wheels are highly desirable from other points of view, the steel ones are obviously the best choice for a formula which is trying to keep costs to a minimum. 1970 *Telegraph* (Brisbane) 8 Dec. 14 Many 'mag' wheels sold as extras for hotted-up cars are made of cast aluminium. They cost about $40 each. Genuine magnesium alloy wheels, as used on racing cars, cost about $120 each. 1973 D. J. Abodaher (*title*) Mag wheels and racing stripes.

Magar (māˑgăɹ). Also **Muggur.** **a.** A member of one of the tribes of western Nepal, of Mongol origin and noted for their prowess in fighting. **b.** The language spoken by this tribe. Also *attrib.* or as *adj.*
1811 W. Kirkpatrick *Acct. Kingdom Nepal* vii. 220 Besides the Sanskrit,..the principal vernacular languages of this country are, the Purbutti, the Newar, the Dhenwar, the Mugger, [etc.]. 1835 B. H. Hodgson in *Jrnl. Bengal Asiatic Soc.* II. 219 Both Gorungs and Magars still maintain their own vernacular tongues, Tartar faces, and careless manners. *Ibid.* 221 The attachment of the Magars to the house of Gorkha is but recent, and of no extraordinary or intimate nature. 1859 R. G. Latham *Descriptive Ethnol.* II. xii. 409 Gorkha, so far as it is not other than Hindu, seems to have been chiefly Magar. 1877 L. H. Morgan *Anc. Society* III. vi. 373 If the organizations..of the Magars of Nepaul..were examined upon the original evidence, it is highly probable that they would be found exactly analogous to the Iroquois tribes. 1911 *Encycl. Brit.* XIX. 379/2 The Gurkhalis, Magars, and Gurungs together comprise the military population. 1925 T. Longstaff *This My Voyage* v. 93 Besides speaking Khaskura..he was familiar with the peculiar dialects of the Magars and Gurungs, two of the most redoubtable of the fighting clans. 1971 C. Bonington *Annapurna South Face* iv. 47 Our porters were recruited in the main from two of the major hill tribes, the Magars and the Gurungs. 1971 J. Pemble *Invasion of Nepal* i. 9 Lévi thought that the Mongolian Magar tribesmen were the earliest inhabitants of the western hills.

magazine, *sb.* Add: **5. c.** = *magazine programme.
1936 *Radio Times* 30 Oct. 88/2 'Picture Page'. A Magazine of Topical and General Interest. 1949 *Ibid.* 15 July 12/2 Music Magazine. A weekly review. 1953 *Ann. Reg. 1952* 360 The establishment of the radio magazine *New Soundings*. 1957 *B.B.C. Handbk.* 153 *Family affairs*: a weekly magazine for mothers with children. 1975 *ITV Evidence to Annan Committee* 46 A thirty-minute news magazine is taking shape for 6 pm transmission.
6. b. (Later examples.)
1915 'I. Hay' *First Hundred Thousand* vii. 77 Pumpherston graciously accepted the charger of cartridges.., rammed it into the magazine, adjusted the sights,..and fired his first shot. 1919 'Boyd Cable' *Old Contemptibles* xvii. 277 Carruthers..took a box of cartridges from a niche in the wall, and proceeded to recharge his magazine. 1964 H. L. Peterson *Encycl. Firearms* 255/1 This turret system was revived many years later as a practical magazine for the Lewis machine gun.
d. Also, in a camera, projector, etc.
1889 *Judge* (U.S.) 22 June 180/2 Every operator can develop and print his own negatives and refill his magazine. 1958 *Amat. Photographer* 31 Dec. 3/2 (Advt.), The Hanomatic slide changer is complete with a plastic magazine holding 36 slides. 1964 C. Willock *Enormous Zoo* v. 77 John Buxton used up one magazine of film and then reloaded with terrible precision. 1967 H. M. R. Souto *Technique Motion Pict. Camera* i. 13 The first mechanism has the task of drawing the unexposed film (or raw stock) from the storage chamber, called a *magazine*, and after exposure, driving it into a similar magazine.
7. (sense 5 b) *magazine rack, -reader, table*; (sense 6 b) *magazine slot*; **magazine cover,** the cover, freq. pictorial, of a magazine; **magazine programme,** a periodical broadcast programme comprised of varied items of entertainment linked together as a single series (see quot. 1941); **magazine rights,** the rights of publishing matter in a magazine; **magazine section,** a section included in some newspapers the contents of which resemble a magazine; **magazine story,** a story written for publication in a magazine.
1938 *Toronto Daily Star* 30 Dec. 12/6 Famous Hollywood Glamour Girls. Magazine cover models. 1942 E. Paul *Narrow St.* xvi. 124 Mireille was not the most attractive, from the magazine-cover standpoint. 1951 M. McLuhan *Mech. Bride* (1967) 120/2 The feminine images of our ads and magazine covers. 1941 *B.B.C. Gloss. Broadcasting Terms* 18 *Magazine programme*, programme made up of miscellaneous items (e.g. talks, interviews, musical acts), loosely related one to the other by a compère or by other means of presentation. 1970 *Times* 23 Feb. 25/3 B.B.C. Newcastle..will have its own budget which will be sufficient to allow the production of another 30-minute weekly magazine programme. 1972 P. Black *Biggest Aspidistra* iii. iv. 175 Godfrey Bazely, a Midland Region broadcaster..loved the world of farming... The BBC gave him a new magazine programme aimed at farmers and their families. 1917–18 *T. Eaton & Co. Catal.* Fall & Winter 416/1 Morris Chair... Paper and magazine rack under arm. 1955 E. Bowen *World of Love* ii. 42 One or two ruched taffeta cushions and a magazine-rack..survived from her few attempts to bring the room into line with her ideas. 1969 *House & Garden* Apr. 160/2 Magazine rack in Afrormosia. 1833 Mill *Let.* 24 Sept. in *Works* (1963) XII. 179 They would not be attractive to the bulk of Magazine-readers. 1882 W. James *Will to Believe* (1897) 109 Thousands of innocent magazine readers lie paralyzed and terrified in the network of shallow negations which the leaders of opinion have thrown over their souls. 1909 *Westm. Gaz.* 14 July 11/2 In America 'magazine rights'

did not necessarily mean publication by instalments. The term was used to distinguish magazine rights from newspaper syndicate rights. 1959 N. Mailer *Advts. for Myself* (1961) 158 Sam throws the Magazine Section away... Sam is enraged at editorial dishonesty. 1969 *Listener* 30 Jan. 148/1 Leavis did not apologise that his terms of reference should be the Robbins Report and Harold Wilson and the magazine sections of the English Sundays. 1910 Kipling *Land & Sea Tales* (1923) 178 The tiny twenty-two cartridge had dropped into the magazine-slot. 1858 J. A. Froude *Let.* 17 Jan. in J. W. Cross *George Eliot's Life* (1885) II. viii. 4, I had made acquaintance with 'Janet's Repentance', and had found there something extremely different from general magazine stories. 1885 C. M. Yonge *Nuttie's Father* II. ii. 23 The hero of many a magazine story. 1932 Q. D. Leavis *Fiction & Reading Public* i. iii. 47 The magazine story is almost without exception a commercial article. 1942 Magazine story [see *exclusive A. *adj.* 9]. 1966 H. Roth *Button, Button* (1967) i. 15 A small, locked safe..unnoticeable..because the top was extended to make it look like a magazine table. 1967 A. Diment *Dolly Dolly Spy* xi. 145 The magazine table caught them neatly behind the naked knees and..they overbalanced.

magazinedom (mægăzīˑndəm). [f. Magazine *sb.* 5 b + -dom.] The world or sphere of magazines.
1890 *Review of Reviews* I. 9/1 Such a guide to magazinedom as you propose to establish would be extremely useful. 1902 *Tablet* 22 Sept. 448 It is the very romance of magazinedom. 1907 *Daily Chron.* 19 Feb. 3/2 It is true that magazinedom is very crowded, but there is always room at the top.

magaziny. Add: (Later examples.) Also **magaziney.**
1938 *New Statesman* 8 Jan. 56/2 Nor is *The Best Poems* negligible. It is magaziney, but contains respectable work by De la Mare [etc.]. 1941 *Scrutiny* IX. 379 He [*sc.* Mr. Plomer] attempts also pseudo-ballads made to the Auden formula, which are amusing in a magaziny way. 1961 *Times Lit. Suppl.* 23 Jun. 385/1 Of these fourteen pieces at least eight are almost defiantly magaziny. 1966 *Punch* 9 Mar. 362/3 The violence seems somehow bookish and the literary chat magaziney.

Magdalenian (mægdālīˑniăn), *a.* and *sb.* *Archæol.* Also **Madelainean, Madelenian.** [ad. F. *magdalénien* (G. de Mortillet, *c* 1867), f. the place-name *La Madeleine* in the department of Dordogne, France: see -an.] **A.** *adj.* Of or belonging to the Lower Palæolithic culture represented by remains found at La Madeleine. **B.** *sb.* A man or woman of this culture.
1885 A. Geikie *Text-bk. Geol.* (ed. 2) vi. 914 Those [deposits] which contain well-finished implements associated with carved bone and ivory, as at the caves of La Madeleine (Périgord), have been called Magdalenian. 1896 A. H. Keane *Ethnol.* 87 Madelenian or Third Cave Age. 1904 [see Solutrian, Solutrean *a.*]. 1911 W. J. Sollas *Anc. Hunters* 323 The Magdalenians were quite capable of making respectable buckles or fibulæ. 1920 *Q. Rev.* Oct. 376 We are moved by a greater number of artistic qualities than were the Magdalenian reindeer men. 1956 H. Read *Art of Sculpture* ii. 28 This highly stylized figure anticipates the schematic character of the later Magdalenian figures. 1963 *Field Archaeol.* (Ordnance Survey) (ed. 4) 21 From the period of the Magdalenians who were hunting and making their wonderful cave paintings in France round about 20,000 b.c...there is very little in Britain... A few typical Magdalenian objects have been found in Mendip caves. 1973 *Listener* 10 May 606/2 The Magdalenian hunters of southern Europe fifteen thousand years ago invented the harpoon.

mage. 2. Delete † *Obs.* and add later example.
1877 Smith & Wace *Dict. Christian Biogr.* I. 477/2 The author of that superstition was Masdec,..a mage, who gathered the credulous around him.

Magen David (māgeˑn dāvīˑd). Also **Mogen David** (mǫˑgən dǫˑvid). [Heb. *māghēn Dāwīdh* shield of David, f. *Dāwīdh* David, king of Judah and Israel, d. circa 973 b.c.] A six-pointed star symbolizing Jewishness or Zionism.
1904 *Jewish Encycl.* VIII. 251/2 *Magen Dawid* ('David's Shield'), the hexagram formed by the combination of two equilateral triangles. 1934 *Encycl. Jewish Knowl.* 324/2 *Magen David*... During the World War it was used..as equivalent to the Red Cross, such organizations described themselves as the 'Red Magen David'. 1964 W. Markfield *To Early Grave* (1965) vi. 109 During the Sinai campaign she beat two Egyptians to death with a *Mogen David* she herself welded together from a pair of captured gasoline drums. 1972 C. Raphael *Feast of Hist.* (dust-jacket), The 'square' Magen David on the jacket appears in some Jewish manuscripts from late-fifteenth-century Portugal. 1975 *Jewish Chron.* 7 Feb. 21/5 Can anyone explain why Jewish greetings cards should be so expensive?.. Why should the addition of a..Magen David make them so much dearer than their non-denominational counterparts?

mageship (mēⁱˑdʒ,ʃip). [See -ship.] The position or function of a mage.
1875 E. Dowden *Shakspere: his Mind & Art* i. 37 Prospero must forever have remained somewhat apart and distinguished from other Dukes..by virtue of the enchanted island and the marvellous years of mageship.

magged, *a.* Add: **b.** Irritated; exhausted, jaded.
1839 F. Trollope *Fragment* in *Dom. Manners Amer.* (ed. 5) 272 Mrs. Rapp... We shall be right down magged without her... Angelina... It won't convene for me to be mixing doe cakes..all day. 1903 *Eng. Dial. Dict.* IV. 7/2 I'm quite magged with my day's glanin'.

Maggid (māˑgid). Also **Magid, m-.** Pl. **Maggidim.** [Heb. *maggīdh* narrator.] An itinerant Jewish preacher.
1892 I. Zangwill *Childr. Ghetto* I. 35 The central place of honour which befits a Maggid. 1902 H. Hapgood *Spirit of Ghetto* x. 289 He was sitting opposite an old 'magid', or wandering preacher. 1941 G. G. Scholem *Major Trends Jewish Mysticism* ix. 329 Rabbi Baer the *Maggid*, or popular preacher. 1960 L. P. Gartner *Jewish Immigrant* iv. 105 Hirsch Dainow (1833–1877), a recently arrived Maggid of some note.

maggie. Add: **2. b.** *dial.* and *Austral.* A magpie.
1825 J. T. Brockett *Gloss. North Country Words* 131 *Maggy*, a magpie. 1878 *Zoologist* Sept. 332/1 Magpie. *Piet*; *Maggie*. 1934 *Bulletin* (Sydney) 19 Sept. 21/2 Out maggie went backwards through the door, arguing every inch of the ground. 1965 *Jrnl. Lancs. Dial. Soc.* Jan. 7 *Maggie*: Tunstall, Burton, Hornby, Bentham, [etc.].
4. In full, *Maggie Ann* (also *Maggy Anne*). Margarine. *colloq.*
1933 Partridge *Slang To-day & Yesterday* 385 *Maggie Ann*. Margarine: C20. 1959 I. & P. Opie *Lore & Lang. Schoolch.* ix. 163 Margarine or 'marg' is 'Maggy Anne'. 1971 D. Lees *Rainbow Conspiracy* vi. 91 Sam never paid him enough to put maggy on his bread.

maggoty, *a.* Add: **2*.** *Austral.* and *N.Z.* Also **maggotty.** Angry, bad-tempered, esp. in phr. *to go maggoty*, to lose one's temper.
Dialectal evidence of this use is presented in *Eng. Dial. Dict.*
1919 W. H. Downing *Digger Dial.* 33 *Maggotty*, angry. 1936 F. Sargeson *Conversations with Uncle* 24 There was a shearer who used to go maggoty if a lamb wouldn't sit still. 1951 D. Stivens *Jimmy Brockett* 31, I didn't need to, but I shaved every day and my old man made me maggotty by asking me one day, 'Do you shave up or down?' 1959 D. Forrest *Last Blue Sea* 74 He's down there..going maggotty about doctors and Japs and boongs.

maghemite (mæˑghīməit). *Min.* [f. Mag(netite + *hem-atite* (var. Hæmatite) + -ite¹.] A modification of ferric oxide, Fe_2O_3, belonging to the spinel family, which is found as brown isometric crystals, is highly ferromagnetic, and is principally an alteration product of magnetite.
1927 P. A. Wagner in *Econ. Geol.* XXII. 846 The interesting form of ferric oxide under discussion appears to be identical in atomic structure with magnetite, but has the chemical composition and the physical properties of hematite... It is desirable that it should have a distinctive designation... I would suggest *maghemite*. 1944 *Trans. Geol. Soc. S. Afr.* XLVI. 30 Maghemite is the commonest and most widely distributed mineral in the Bushveld magnetic iron ores. 1962 W. A. Deer et al. *Rock-Forming Min.* V. 73 Maghemite is metastable, and inverts to haematite (α-Fe_2O_3) on heating.

Maghribi (magriˑbī), *sb.* and *a.* Also **Maghrabee, Maghrabi, Maghrebi.** [Arab. *maġribī*, lit. 'western'.] **A.** *sb.* **a.** A native or inhabitant of the Maghrib, a region of north-western Africa, including Morocco, Algeria, and Tunisia. **b.** The Arabic spoken in this region. **B.** *adj.* Of or pertaining to the Maghribi or their language.
1840 E. W. Lane tr. *Arabian Nights' Entertainments* II. xviii. 600 The people of Western Africa... Called by the Arabs 'El-Maghrib'. This name is generally given to the districts of Northern Africa west of Egypt. The inhabitants of those parts are called 'Maghrabees'. 1855 R. F. Burton *Pilgrimage to El-Medinah & Meccah* I. viii. 228 We lay down upon the sand, to rest among a party of Maghrabi pilgrims travelling to Suez. 1898 A. J. Butler tr. *Ratzel's Hist. Mankind* III. v. v. 196 The Islamite world witnessed a great struggle for supremacy between..the Maghrebin in the west, the Mashrikin in the east. 1948 D. Diringer *Alphabet* ii. iv. 272 Kufic gave rise to a number of varieties,..in northern (known as Maghribi, or western) and central Africa. 1969 *Word* XXV. 116 In many varieties of Maghribi Arabic when it co-occurs with a sibilant in the same word.

Maghzen (maˑχzən). Also **Maghasen, Makhzen, -an.** [Arab. *makzan*.] The Moroccan government; the dominant official class in Morocco; irregular Algerian horsemen in the service of France. Also *transf.*
1854 J. R. Morell *Algeria* xxi. 392 Immediately that the tribes were subdued, the victors required them to supply irregular horsemen, called *makhzen* or *goum*, to attack the refractory. 1874 tr. *G. Rohlfs's Adventures Morocco* ix. 193 Every day at midday on his return from the Maghasen (palace of the Sultan and seat of government) I was called. 1894 G. Montbard *Among Moors* 169 A knot of soldiers of the Maghzen are plunging their horses into the stream. 1901 *Chambers's Jrnl.* 6 July 504/1 Laraiche, once offered to the British Maghzen (Government) in return for help against the enemies of the Sultan. 1930 *Economist* 24 May 1179/1 This loan of 62,500,000 francs was originally contracted by the Maghzen of Morocco. 1973 R. E. Dunn in Gellner & Micaud *Arabs & Berbers* 92 At the 'national' level it influenced foundation of the *makhzan*'s policies towards the Saharan region, especially in relation to the southward expansion of French Algeria.

magic, *sb.* Add: **1. d.** *like magic*: without any apparent explanation; with incredible rapidity. (Cf. Like *a.*, etc. B. 1 b.)
1857 *Knickerbocker* Jan. 98 Broiled chicken and oysters ..disappeared from before us like magic. 1900 *Congress. Rec.* 9 Jan. 704/2 Germany's Chinese trade is increasing like magic.
4. magic-man, a magician, sorcerer; also *fig.*
1905 *Westm. Gaz.* 6 Apr. 3/2 Disease was thought to be a visitation of supernatural wrath, to be appeased by offerings to the priests and magic-men of the time. 1923 R. Graves *Whipperginny* 51 Time and Space, folly's wonder, Three-card shufflers, magic-men! 1959 Halas & Manvell *Technique Film Animation* v. 62 Magic-men, mummers and actors wore masks in earlier times to typify ..the farcical, the comic, the eccentric, the pathetic, the tragic and the insane in human portraiture.

magic, *a.* Add: **2.** *magic box*: applied *colloq.* to various, esp. electronic, devices; *magic carpet*: a legendary carpet on which a person could be transported wherever he wished; also *transf.* and *fig.*; *Magic Marker*: the proprietary name of an instrument, consisting of a tube of quick-drying ink and a felt-tipped pen, used for marking objects; also *magic marker*.
1897 Kipling *Capt. Cour.* ix. 189 From San Diego to Sixteenth Street, Chicago, let the magic carpet be laid down. Hurry! oh, hurry! 1909 H. G. Wells *Tono-Bungay* iv. i. 434, I had to come off my magic carpet and walk once more in the world. 1931 *Times Lit. Suppl.* 20 Aug. 625/2 His Magic Carpet is a book of travels, by means of which he is transported into lands that he is fated never to see. 1935 A. J. Pollock *Underworld Speaks* 74/2 *Magic box*, crooked farobank box in which 53 instead of 52 cards are manipulated by dealer. 1935 J. Hargan *Gloss. Prison Lang.* 5 *Magic-box*, coil arrangement which starts any car—without the aid of key. 1936 *Daily Tel.* 15 Aug. 15/5 (*heading*) Nicknames of the freight expresses... The Magic Carpet is not from Arabia, but Kidderminster, bringing fine weaves to London's floors. 1945 *Daily Mirror* 15 Aug. 4/2 Most sensational development was the 'Magic Box', which gave our pilots a picture of the ground beneath them even though it was hidden by darkness and cloud. 1959 *20th Cent.* Nov. 325 Both radio receivers and television sets have been known and called for brief periods 'magic boxes'. 1960 *Guardian* 22 Mar. 10/4 The job of the 'magic boxes' in the wheelhouse is to take some of the chance out of trawling. 1960 D. Lessing *In Pursuit of English* vi. 204 The yapping..of the..puppies distracted Flo from her magic box. 1960 L. Meynell *Bandaberry* viii. 127, I had suddenly an immense desire to get on a magic carpet and float away..to some carefree place. 1956 *Official Gaz.* (U.S. Patent Office) 4 Dec. TM5/2 Speedry Products, Inc., Richmond Hill, N.Y... *Magic Marker*. Applicant disclaims the term 'Marker' apart from the mark as a whole. For Felt Nib Marking Pens Comprising Small Containers for Such Ink, Equipped with Caps and having Felt Nibs at their Ends for Marking. First use on or about Sept. 1, 1952. 1962 L. Deighton *Ipcress File* xxxii. 209 A slim green file; on the cover it said 'Henry' in magic-marker lettering. 1964 'S. Forbes' *Long Hate* (1966) ii. 16 Block letters. Magic Marker, unquestionably. Children used Magic Markers in school these days. 1971 R. Russell tr. *Ahmad's Shore & Wave* vii. 59 Nur Jahan felt as though she were on a magic carpet soaring towards the sky. 1971 C. Bonington *Annapurna South Face* App. B 255 Magic markers 20. 1973 'J. Patrick' *Glasgow Gang Observed* xiii. 120 With chalk, magic markers, but most often with..aerosol spray paint, the pitch was marked out with slogans. 1974 D. Mackenzie *Zaleski's Percentage* iv. 94 That restaurant's his magic carpet. If they take it away from him, he'll never stop falling.
3. *magic circle*: (*b*) the title of a society of conjurers; (*c*) a circle used in magic as a protection against evil; (*d*) a small group of people who are privileged to receive confidential information, make important decisions, etc.
1912 *Magic Wand* Aug. 377/2 (*heading*) 'The Magic Circular'.—A monthly review of the magic art, issued for private circulation only amongst members of the Magic Circle. 1934 Webster, *Magic circle*, 1. a. A circle drawn by a magician about any person or place, within which demons raised by incantations were believed to have no power. b. A clearly defined place, group, or the like, entrance into which is regarded as desirable, pleasing, etc. 1955 *Radio Times* 22 Apr. 32/3 All this week magicians from all over the world are gathering to celebrate the Golden Jubilee of the Magic Circle. 1965 *Listener* 10 June 859/1 Kaiser wrote a series of plays centred on the love of two people. The pattern is throughout the same: the lovers are enclosed in a kind of magic circle which isolates them from the rest of the world. 1971 G. Mitchell *Lament for Leto* ii. 51 He..was an amateur conjuror and.. was 'on the fringe of the Magic Circle'. 1973 *Jrnl. Genetic Psychology* CXXII. 168 The complete, circular form of the Ufos is reminiscent of the mandala or 'magic circle': i.e., totality. 1974 E. Ambler *Dr. Frigo* i. 65 Could he have concluded belatedly that letting me into the magic circle had been a mistake..? 1974 *Times* 13 Nov. 16/7 The Tories..would be making a profound mistake if they were to take the right to decide their leader out of the exclusive hands of the elected MPs and put it back into those of some kind of new-furbished Magic Circle.
4. *magic (chain-)stitch* (see quot. 1900).
1900 Day & Buckle *Art in Needlework* 41 A playful variation upon chain-stitch..is effected by the use of two threads of different colour... The light thread disappears, and comes out again to the left of the dark one... This 'magic stitch'..is to be found in Persian, Indian, and Italian Renaissance work. 1934 M. Thomas *Dict. Embroidery Stitches* 34 *Chain stitch—chequered*, also known as *magic stitch* and *magic chain stitch*. 1957 M. B. Picken *Fashion Dict.* 329/2 *Magic Chain-stitch*... Chain-stitch worked with two threads of different colors in one needle.

5. *Nuclear Physics.* Applied to a set of numbers which correspond to nuclei of exceptional stability when either the number of protons or the number of neutrons in it is equal to one of the set (now taken to be 2, 8, 20, 28, 50, 82, 126, and perhaps 184); hence applied to nuclei containing such a number of protons or neutrons; *doubly magic*, containing a magic number of protons and also a magic number (not necessarily the same) of neutrons. Occas. extended to an analogous set of numbers (see quot. 1974) for electrons in atoms.
1949 O. Haxel et al. in *Physical Rev.* LXXV. 1766/1 A simple explanation of the 'magic numbers' 14, 28, 50, 82, 126 follows at once from the oscillator model of the nucleus. 1956 *Nature* 28 Jan. 159/1 The current work on fast-neutron capture..refines the earlier studies by D. J. Hughes of the variation of cross-section with atomic weight which..throws more light on the magic-number nuclei. 1969 *Physics Lett.* XXVIIIB. 544/2 Since the nucleon shells at $Z = 114$, $N = 184$ are not as 'magic' as those at $Z = 82$, $N = 126$, we would not expect an unusually small capture cross section for $^{294}110$. 1969 *Nature* 27 Dec. 1253/2 In calcium-40 which, like lead-208, is doubly magic and so a suitable isotope for nuclear structure calculations, the protons have been found to bunch together towards the centre of the nucleus. 1971 *Ibid.* 12 Feb. 451/2 One complication is that for heavier nuclei, the magic numbers probably differ for neutrons and protons. 1971 *Physics Bull.* Dec. 711/2 Much interest has been aroused by the prediction that the nucleus ($A = 298$, $Z = 114$) may be a doubly magic nucleus.., which could involve an island of stability in the mass region around 300. 1974 *Encycl. Brit. Micropædia* VI. 484/1 The magic numbers for atoms are 2, 10, 18, 36, 54, and 86.

magic (mæˑdʒik), *v.* [f. the sb.] *trans.* To transform, make, etc., (as if) by magic; also *to magic* (something) *away*, to cause, as if magically, to disappear.
1906 Kipling *Puck of Pook's Hill* 304 There was Oak and Ash and Thorn enough in that year-end shower to magic away a thousand memories. 1909 L. M. Montgomery *Anne of Avonlea* xxiv. 282, I actually have a half-guilty feeling as if I had 'magicked' it [*sc.* a storm] up. 1925 W. de la Mare *Miss Jemima* 31, I discovered, as if the Fairy Creature herself had magicked it there..a large hay-wain. 1952 A. Grimble *Pattern of Islands* v. 104 A steel hook bought from a trade-store could only be magicked once, as a finished article. 1957 J. Frame *Owls do Cry* ix. 38 It was to have these things sitting in their head, like a charm, to magic away the drudgery. 1966 *New Statesman* 25 Nov. 792/2 Children, who are still capable of being magicked, will, I hope, get it for Christmas. Though it's not a children's book. 1972 *Times* 16 Feb. 13/3 We cannot magic them away. 1973 D. Francis *Slay-Ride* iv. 56 He magicked some huge open sandwiches on about a foot of French loaf.

magical, *a.* **3.** *magical circle*: also = *magic circle (sense *c*).
1915 *Encycl. Relig. & Ethics* VIII. 322/1 The famous constituent of the mediaeval magical circle.
Hence also **magicaˑlity,** magical power or quality.
1924 W. J. Locke *Coming of Amos* iv. 43 An untouched cheque-book of whose magicality he was innocently certain.

magic eye. [f. Magic *a.* + Eye *sb.*¹] **a.** A miniature cathode-ray tube used as a tuning indicator on a radio receiver, or to indicate the correct adjustment of other electrical equipment.
Magic Eye is registered as a proprietary name in the U.S.
[1936 *Official Gaz.* (U.S. Patent Office) 6 Oct. 21/1 RCA Manufacturing Company, Inc.,..*Magic eye* for radio receiving sets of the type equipped with cathode ray tubes for resonance indication and accessories and parts thereof.] 1937 *Night & Day* 26 Aug. 28/1 Should one..plump for the [wireless] set with a 'mystic eye', or are those which boast a 'magic eye' or an 'electric eye' to be preferred? 1939 *Proc. IRE* XXVII. 631/2 The ground wave alone was introduced into a receiver which had a 'magic eye'. With this wave, the 'eye' remained exactly fixed in deflection. 1962 A. Nisbett *Technique Sound Studio* v. 100 A 'magic eye' is a useful guide, but should not be relied on unless the recordist is quite sure of what its indications mean in terms of the particular programme material. 1968 *Radio Communication Handbk.* (ed. 4) ii. 20/2 Magic-eye tubes are often used as voltage indicators in a.f. and r.f. measuring equipment and also widely as indicators of signal level for tape recorders.
b. A photo-electric cell or similar electrical device used for identification, detection, or measurement.
1938 *Sun* (Baltimore) 22 Feb. 20/3 Eleven pairs of 'magic eyes'..have counted approximately 7,000,000 motor vehicles during the last year. 1945 *Nature* 15 Sept. 320/2 The aids to the bombing of invisible targets which were variously called 'H_2S', the 'gen-box', 'Mickey' and the 'Magic Eye'. 1958 *Times* 2 May 7/2 The large radio antenna carried on the superstructure of the Victorious.. was a form of 'magic eye'. 1962 *Daily Tel.* 26 Oct. 17/7 The 'magic eye' records the wagon numbers of freight vehicles moving at speeds up to 100 mph. 1974 *Country Life* 6 June 1449/1 An automatic 'magic eye' lets in exactly the right amount of light every time you want to take a picture.

magic lantern. Add: Also *fig.*
1775 *Morning Chron.* 29 May 1 (Advt.), At Marylebone Gardens, To-morrow..will be presented The Modern Magic Lantern..being an attempt at a Sketch of the Times in a variety of Caricaturas. 1840 C. Fox *Jrnl.* 21 July (1972) 100 We are all shadows in the magic lanthorn of Time. 1933 S. Spender *Poems* 28 Where magic-lantern faces skew for greeting.
magic-lantern *v.* Delete (*nonce-wd.*) and add further example.
1935 W. Empson *Poems* 29 All those large dreams by which men long live well Are magic-lanterned on the smoke of hell.

magico- (mæˑdʒiko). Combining form of Magical *a.* with other adjs. as *magico-erotic, -oriental, -profane, -religious.*
1915 *Encycl. Relig. & Ethics* VIII. 258/2 These are of a magico-erotic nature, and, like similar rites among savages, are founded on the belief that the ghost can cause fruitfulness, or perhaps may incarnate himself in the barren woman who performs the rite. 1930 *Times Lit. Suppl.* 27 Nov. 1004/3 As against the 'magico-oriental' view of the image maintained by the Iconoclast stands the Platonism of the image-worshipper. 1941 *Jrnl. R. Anthrop. Inst.* LXXI. 85/2 Numerous acts and ceremonies..which the European..would place under the category of magico-profane, magico-religious or superstition. 1908 *Man* VIII. 46 The magico-religious ideas and practices of savage and proto-historic man. 1922 *Nature* 29 Apr. 540/2 The complex and inexorable system of magico-religious *gennas*. 1967 C. L. Wrenn *Word & Symbol* 17 Anglo-Saxon magico-religious arts.

Magid, var. *Maggid.

Maginot (‖ maʒino, mæˑdʒinǒᵘ). The name of a French minister of war, André *Maginot* (1877–1932), used to designate the line of fortifications (*Maginot Line*) built before the war of 1939–45 along the north-eastern borders of France, and in which the French placed excessive confidence. Also *transf.* (designating other similar lines of defence) and *fig.*
1936 *Times* 30 Oct. 15/4 The Maginot Line. M. Daladier ..inspected the recently completed frontier fortifications between Maubeuge and Valenciennes. 1937 A. Vagts *Hist. Militarism* xii. 457 The American coast is the Maginot Line of the United States. 1938 *Nation* (N.Y.) 23 July 78 Little Steel's Maginot Line against unionism. *Ibid.* 83/2 Along the border the Czech Maginot Line is artfully concealed. 1940 *Economist* 27 July 118/1 Sea-power is no Maginot Line to lull us again into lethargy. 1940 L. B. Namier in *19th Cent.* Nov. 468 Her [*sc.* France's] system of alliances in East-Central Europe was as imperfect and as deceptive as her Maginot line later on. 1947 C. S. Lewis *Miracles* xiv. 146 The belief that nothing but Nature exists and that if anything else did she is protected from it by a Maginot Line. 1972 *Times* 14 Nov. 17/2 The Dow has still to close above what is jestfully known on Wall Street as the 'Maginot Line' of 1,000. 1973 B. Freemantle *Goodbye to Old Friend* i. 14 He..lifted out his tray containing pens..and set it..at the head of the blotter. My Maginot Line, he thought. Behind the tray, I'm safe. 1975 *Guardian* 27 Jan. 4/3 Casemate 35/3, a key heavy gun emplacement in the Maginot Line.
Hence *attrib.* and *Comb.*, as **Maginot-minded** *a.*, obsessed with the inviolability of the Maginot Line; *Maginot(-Line) complex, mentality*, etc.: indicating an obsession with defence (of the status quo, etc.) as an attitude of mind.
1940 *Economist* 13 July 37/1 The efficacy of blockade is a mirage, a delusion comparable to the Maginot-complex that bemused and enervated France. 1942 *Ann. Reg. 1941* 79 People began to wonder whether the Government had not become 'Maginot-minded'. 1955 *Bull. Atomic Sci.* Mar. 81/1 Maginot-line mentality..denotes..a one-sided preoccupation with putting all reliance on a single strategy. 1962 *Listener* 24 May 893/1 Defence is only relevant to my present argument as evidence of the Maginot attitude of the old and Great Powers. 1962 *Observer* 1 July 7/1 There is also the danger that a national shelter programme could lead to a Maginot Line psychology. 1973 *New Society* 13 Sept. 632/2 The NGA has been attacked by the non-craft unions for having a 'Maginot line mentality'.

magisterially, *adv.* Add: **1.** (Later examples.)
The dominant sense now is 'authoritatively'.
1963 *Times* 4 Mar. 14/2 M. Planchon, who wrote every word of the text and was responsible for every step of the choreography, rarely stumbles and magisterially weaves the disparate threads into one homogeneous pattern. 1972 *Times* 15 June 7/3 These dons, judges and headmasters.. who moved so magisterially within their cathedral closes, [etc.]. 1975 *Gramophone* Oct. 633/3 The new Karajan reading is certainly magnificent, magisterially played by the finest orchestra in the world.

magistery. Add: **5.** = Magisterium 2.
1899 C. B. Pallen tr. *Sarda y Salvany's What is Liberalism?* xxxii. 105 The Church alone possesses supreme doctrinal magistery in fact and in right.

magistral, *a.* **1. a.** (Later example.)
1942 Partridge *Usage & Abusage* (1947) 5, I obtained permission from..eminent scholars..to quote at length from their magistral works.

magistrate, *sb.* Add: **3.** *magistrates' court*: a court for the trial of minor offences and small civil cases and for the preliminary hearing of more serious cases.
1867 T. W. Saunders *Practice in Magistrates' Courts* (ed. 3) i. 1 The various kinds of magistrates' courts—petty and special sessions—quarter sessions—clerk to the justices. *Ibid.*, It is..the design of the following pages to treat of the *practice* of the magistrates' courts as we find these courts established. 1904 J. F. & B. E. Crump *Magistrate's Pocket Manual* xii. 47 There is perhaps even greater need to observe the spirit of these instructions in the Magistrates' Courts, because..rough and uncultivated manners..make so great a strain on the patience and self-control of the Bench. 1952 *Act 1 Eliz. II* c. 55 § 38 Where a person is taken into custody for an offence without a warrant and is retained in custody, he shall be brought before a magistrates' court as soon as practicable. 1959 Jowitt *Dict. Eng. Law* II. 1341/2 *Petty sessional court*, a court of summary jurisdiction consisting of two or more justices..now known as the magistrates' court. 1965 *New Statesman* 10 Dec. 920/2 The police courts, reborn as magistrates' courts in 1952, are still police courts not only to those who never frequent them but also to the old lags who frequent them most.
4. *Sc. slang.* In full, *Glasgow magistrate.* A herring.
1833 *Chambers's Edin. Jrnl.* 2 Nov. 314/2 My neighbour, thinking it absurd to mince such a matter as a *Glasgow Magistrate*, handed up a whole one to the chairman. 1874 Hotten *Slang Dict.* 177 *Glasgow Magistrate*, a salt herring. When George IV. visited Scotland, a wag placed some salt herrings on the iron guard of the carriage belonging to a well-known Glasgow magistrate, who made one of a deputation to receive his Majesty. 1890 Barrère & Leland *Dict. Slang* II. 38/1 *Magistrate* (Scotch slang), a herring. 1895 J. Nicholson *Kilwuddie* (ed. 4) 119 Ham's unco dear, sae, if ye like, we's hae a 'magistrate'. 1950 *Scots Mag.* Dec. 171 Herring were cured there by Walter Gibson, a merchant of Glasgow and Provost of that city in 1688, and it is perhaps because of Provost Gibson that salt herring acquired their nickname of 'Glasgow Magistrates'.

Maglemose (mæˑgləmǒᵘsə). [The Danish place-name *Maglemose* (great moss) near Mullerup on the west coast of Sjælland.] Used *attrib.* to designate the Mesolithic culture of northern Europe represented by bone implements and microliths found at Maglemose.
1915 W. J. Sollas *Anc. Hunters* (ed. 2) 544 The Maglemose industry is widely distributed around the Baltic. 1921 M. C. Burkitt *Prehistory* xii. 155 These..formed the Maglemose culture along the shores of the Baltic. 1931 *Times Lit. Suppl.* 8 Jan. 23/1 Azilian and Maglemose man. 1932 *Antiquity* VI. 218 It proves..that during the Maglemose period, there was fresh water at a spot now many feet below the sea. 1948 A. L. Kroeber *Anthropol.* (rev. ed.) xvi. 668 For the first time we find European habitation sites along beaches (Ertebølle and Asturian phases of Mesolithic culture), river mouths (Azilian), on lake and bog shores (Maglemose).

Maglemosian (mæglǝmǒᵘˑsiăn), *a.* and *sb.* Also -ean. [f. prec. + -ian.] **A.** *adj.* = prec. **B.** *sb.* A person of the Maglemosian culture; the culture itself.
1918 in Webster Add. 1921 M. C. Burkitt *Prehistory* xiii. 163 The first culture found is the so-called Maglemosean. 1925 V. G. Childe *Dawn European Civilization* i. 9 The Maglemosians had axes and adzes and their command over nature was thereby enormously extended. 1943 J. & C. Hawkes *Prehist. Brit.* i. 26 The Maglemosians probably crossed the area now covered by the North Sea as fishers and fowlers. 1949 W. F. Albright *Archaeol. of Palestine* iii. 62 If we allow the Tahunian a time spread of between one and two thousand years (say about 6000–4500 b.c.) it would have been roughly contemporary with the late Maglemosean. 1954 S. Piggott *Neolithic Cultures* i. 14 The Star Carr discoveries in Yorkshire have shown that cultures of Maglemosean affinities were established in what is now eastern England at least as early as the beginning of Boreal times. 1970 *Guardian* 5 Aug. 18/4 An almost complete skeleton of an elk..unearthed near Blackpool..belongs to the early mesolithic period which began around 8000 b.c... This new find is the first evidence that this maglemosian culture extended to what is now the North-west of England.

maglev (mæˑglev). Also **mag-lev.** Abbrev. of *magnetic levitation* (see *levitation 1 c).
1973 *Financial Times* 7 Sept. 25/7 Krauss-Maffei expects to be running a 20-passenger 'maglev' vehicle capable of speeds up to 250 mph later this autumn. 1973 *Sci. Amer.* Oct. 18/3 Two entirely different approaches to magnetic levitation (which is often shortened to 'maglev') are being pursued. 1974 *Globe & Mail* (Toronto) 16 May 4/4 The Canadian program is aimed at the development of mag-lev vehicles which will travel at 300 miles an hour, carrying about 100 passengers each.

magma. Add: **3. a.** In mod. use: A hot, fluid or semi-fluid material beneath the earth's crust from which igneous rocks are believed to be formed by cooling and solidification and which erupts as lava. (Further examples.)
1944 A. Holmes *Princ. Physical Geol.* iii. 28 A volcano

is essentially a rift or vent through which magma (molten rock material highly charged with gases) from the depths is erupted at the surface. 1955 *Sci. News Let.* 19 Mar. 187/3 The liquefied material forms a fluid mass, called magma, that is lighter than the overlying rocks and tends to rise at an opening. Magma is called lava when it reaches the earth's surface. 1971 I. G. Gass et al. *Understanding Earth* xxi. 310/2 At the close of volcanism, the magma in these chambers will also solidify to coarse-grained rocks having the same composition as the lavas which previously were being erupted.
5. *fig.*
1928 R. A. S. Macalister *Archæol. of Ireland* iv. 219 A conglomerate of lazy abbreviations..studded in a repulsive magma of exotic gibberish. 1933 H. G. Wells *Shape of Things to Come* iii. §1. 261 The need for a planned 're-nucleation' in the social magma that arose out of this dissolution.
6. *attrib.*, as (sense 3) *magma reservoir.*
1909 A. Harker *Nat. Hist. Igneous Rocks* ii. 36 At the depth at which we suppose a large magma-reservoir to be situated the conditions would be quite different. 1971 I. G. Gass et al. *Understanding Earth* xxi. 303/1 Because these volcanoes are large and their volcanic history is long, the magma reservoirs beneath them..have fractionated, giving rise to diversified lava assemblages.
magmatic *a.* (examples).
1890 in *Cent. Dict.* 1903 Geikie *Text-bk. Geol.* (ed. 4) II. 808 (*heading*) Magmatic ores. 1910 *Encycl. Brit.* III. 513/2 The modifications of the granite are ascribed to magmatic segregation (chemical and physical processes which occasioned diffusion of certain components towards the cooling surfaces). 1933 *Geogr. Jrnl.* LXXXI. 332 Magmatic movement, however caused, has been a factor in the formation of the Rift Valleys. 1957 G. E. Hutchinson *Treat. Limnol.* I. i. 33 Intermittent volcanic activity.. repeatedly drew off material from the magmatic reservoir. 1971 I. G. Gass et al. *Understanding Earth* i. 24/1 Quartz and nepheline if brought into contact at magmatic temperatures react to give albite.

magmatism (mæˑgmătiz'm). *Geol.* [f. L. *magmat-*, *magma* (see Magma) + -ism.] **a.** The theory advocated by magmatists.
1948 *Q. Jrnl. Geol. Soc.* CIII. 140 Dr. Reynolds 'tolls the knell' of the passing theory of 'magmatism' as an all-embracing explanation of the origin of all plutonic rocks. *Ibid.* 141 A new era in petrogeny, an era in which 'magmatism' will become incorporated with 'transformism', plutonic processes will become closely linked with metamorphic processes, and petrogenesis..will become an integral part of a wider science of geochemistry.
b. Motion or solidification of magma; magmatic activity.
1952 H. Ramberg *Orig. Metamorphic & Metasomatic Rocks* 3 (*heading*) Natural boundary between metamorphism and magmatism. 1970 Clifford & Gass *Afr. Magmatism & Tectonics* p. vii, Studies..to illustrate the relationship between two important geological processes—magmatism and tectonics. 1974 *Nature* 19 Apr. 650/1 Seismicity, tectonism and magmatism are now concentrated near plate margins.

magmatist (mæˑgmătist). *Geol.* [f. as prec. + -ist.] One who believes that many granitic rocks, or plutonic rocks in general, were formed from magma.
1947 *Geol. Mag.* LXXXIV. 209 Much of the disagreement..arises from a belief on the part of the Magmatists that Hutton proved the magmatic origin of granite. 1947 *Q. Jrnl. Geol. Soc.* CII. 443 He agrees with the 'magmatists' that 'the now classical views of Sederholm, put forth in explanation of the pre-Cambrian granite phenomena of Finland, are reasonable and may be applied..to other Archaean formations, such as the Canadian Shield'. 1957 H. H. Read *Granite Controversy* iii. 154 The inability of the magmatists to interpret the plain field-facts arises.. from the inertia of a century of belief in the magmatic origin of all granitic rocks.

‖ **magna cum laude** (mæˑgnă kʊm lǭˑdi, mæˑgnă kum lauˑdi), *phr.* Chiefly *U.S.* [L., 'with great praise'.] With great distinction: designating a degree, diploma, etc., of a higher standard than the average (though not the highest). Also *transf.* and *fig.*
1900 *Dialect Notes* II. 13 A few words or phrases of direct Latin importation used at some of the older institutions...more commonly *cum laude*, *magna* or *summa cum laude* for the degree of honor attained in studies. 1933 Balmer & Wylie *When Worlds Collide* i. 22 He was graduated from Harvard with a *magna cum laude*. 1963 M. McCarthy *Group* v. 102 He was angry because Helena had failed to get *magna cum laude*, when a lot of the Jewish girls had. 1973 *Physics Bull.* Feb. 116/3 From 1911 to 1914 he was an 1851 Exhibition research scholar in physics, and spent the period in Göttingen taking special advanced courses in a variety of subjects and carrying out research on *Die Struktur des Gels der Kieselsäure*, for which he was awarded the degree of PhD 'magna cum laude'. 1973 E. Pace *Any War will Do* (1974) i. 47 The minute I heard poor Ollie shout, I knew you had passed. And when the counter splintered, you got magna cum laude.

magnaflux (mæˑgnăflʊks). Also **Magnaflux.** [f. *magna-*, taken as comb. form of Magnetic *a.* + Flux *sb.*] A method of testing metal parts for internal or surface defects by magnetizing the metal and observing the pattern assumed by a magnetic powder that is applied to it (either directly, or in oil that is used as a bath or sprayed on the metal). Usu. *attrib.*
1938 *S.A.E. Jrnl.* May 36/1 Another method of magnetic testing..involves the use of Magnaflux in which the parts to be inspected are first longitudinally or circularly magnetized... When properly magnetized, defects result in local flux leakage. Magnaflux powder, being of high magnetic permeability..adheres along the lines of flux leakage. 1940 Crump & Maul *Our Airliners* vii. 139 Every piece is measured and checked for wear and many of the parts go to the magnaflux tank. 1946 *Jrnl. R. Aeronaut. Soc.* L. 535/2 Japanese designers considered that all important parts should be hardness-tested after heat treatment; that welded primary structures should be magnaflux tested; and casting radiologically examined. 1963 Jones & Schubert *Engin. Encycl.* (ed. 3) 783 Although the Magnaflux method of inspection has been universally adopted throughout the aircraft industry it is by no means confined to it.

magnalium (mægnēⁱˑliəm). [f. Magn(esium + Al(umin)ium.] A light aluminium-based alloy containing some magnesium.
1900 *Motor-Car World* I. 90/1 Magnalium is a new alloy of aluminium with from ten to thirty per cent. of its weight of magnesium. 1910 *Encycl. Brit.* I. 708/2 This alloy, under the name of magnalium, is coming into use for small articles in which lightness and rigidity have to be combined. 1948 J. J. Stevens tr. *A. von Zeerleder's Technol. Aluminium* ii. 20 The binary aluminium-magnesium alloys, which were developed in 1899 by Mach under the name 'Magnalium', exhibit rising mechanical properties with rising magnesium content... After 1910 these alloys fell out of favour owing to low corrosion resistance.

‖ **magna mater** (mæˑgnă mēⁱˑtəɹ, mæˑgnă māˑtəɹ). [L., lit. great mother.] A mother-goddess, a fertility deity; also *transf.*
1728 [see Great A. *adj.* 12 b]. 1845 *Encycl. Metrop.* XVII. 481/2 *Cybele*, otherwise known as Ops, Rhea, Vesta,..Idæa Mater, Magna Mater, Mater Deorum, Bona Dea, and Tellus. 1896 [see *fieldstone]. 1919 J. Buchan *Mr. Standfast* xxi. 356 There was a strange cult in the ancient world, the worship of *Magna Mater*—the Great Mother. 1962 'M. Innes' *Connoisseur's Case* viii. 92 She was a kind of *magna mater* whose true sphere was a teeming nursery. 1974 *Times Lit. Suppl.* 13 Sept. 977/3 In Frieda [von Richthofen]'s case this was a straightforward matter. Like other *magnae matres* of the time (Mr. Green discusses Alma Mahler, Isadora Duncan, Lou Andreas-Salome and Mabel Dodge Luhan), she sought to enhance life by erotic love.

magnateship (mæˑgnēⁱtʃip). [See -ship.] The dignity or position of a magnate.
1916 W. J. Locke *Wonderful Year* i. 4 The vast, original Margett had retired..to county magnateship. 1937 *Sat. Rev. Lit.* 24 Apr. 14 The Ball Brothers glass jar magnateship.

magnesian, *a.* Add: **2.** *Min.* [See *-ian 2.] Of a mineral: having a (small) proportion of a constituent element replaced by magnesium.
1930 W. T. Schaller in *Amer. Mineralogist* XV. 571/2 Magnesium-magnesian. 1951 C. Palache et al. *Dana's Syst. Min.* (ed. 7) II. 503 Cuprian melanterite and pisanite are obtained from solutions containing added copper sulfate; and magnesian and other compositional variants may be similarly obtained by appropriate additions. 1968 I. Kostov *Mineral.* ii. ix. 498 Kirovite a magnesian variety [of melanterite].

magnet. Add: **4.** *magnet-drawn* adj.
1923 R. Graves *Whipperginny* 43 She was magnet-drawn by his least wish. 1952 R. Campbell tr. *Baudelaire's Poems* 67 When to a cherished cat my gaze Is magnet-drawn.

magnetic, *a.* and *sb.* Add: **A.** *adj.* **1.** (Selected further examples; see also the sbs., and sense *5).
1926 *Jrnl. Iron & Steel Inst.* CXIV. 112 The concentrate from the magnetic separation is always mixed with much water, and cannot on that account be directly used. 1964 F. L. Westwater *Electronic Computers* iv. 65 If we now suppose that we have another electro-magnet which we can call a 'reading' magnet or magnetic head as it is more usually termed, when the magnetised spot of wire passes under the reading head there will be a change of magnetic flux through the coils on the head. 1964 T. W. McRae *Impact of Computers on Accounting* i. 6 The current vogue is to use magnetic storage devices which employ the principle of magnetic hysteresis or retained magnetism. The more popular of these devices are: 1. Magnetic tape. 2. Magnetic drum. 3. Magnetic disks. 4. Magnetic cards. 5. Magnetic core (i.e. ferrite core). 1966 A. Battersby *Math. in Managem.* viii. 200 An alternative to optical scanning is to use characters printed in magnetic ink. The micr system (Magnetic Ink Character Recognition) is mainly used by banks, and the specially designed characters are now becoming a familiar feature of our cheque-books. 1967 D. Wilson in Wills & Yearsley *Handbk. Managem. Technol.* iii. 42 Many data processing installations are now planning to use magnetic disks or magnetic cards for the first time. 1970 *Sci. Amer.* Feb. 29/2 By 1972, if not sooner, it should be economic to replace the standard magnetic-core memory of the computer (the type of memory in which tiny magnetic rings of ceramic are threaded on a matrix of thin wires) with a series of LSI memory modules. 1971 J. H. Smith *Digital Logic* vi. 126 A variation of the drum is the magnetic disc, very much like a long playing record, but again coated with magnetic material. A whole stack of these discs are normally assembled to give a reasonably fast memory of very large capacity.
5. Special collocations (see also sense 1 in Dict. and Suppl.): **magnetic anomaly,** a local deviation from the general pattern of the earth's magnetic field; **magnetic bottle,** a magnetic field that confines a plasma inside it to a restricted region; **magnetic brake,** a friction brake that is actuated magnetically; **magnetic bubble,** a small, mobile region of reverse magnetization in a very thin sheet of magnetic material in which the magnetization is perpendicular to the sheet and predominantly in one direction; **magnetic drum** *Computers*, a cylinder that can be rotated and has a magnetizable outer surface on which data can be recorded on circular tracks by means of a set of fixed heads (one opposite each track); **magnetic induction** (see *induction 10 b); **magnetic lens,** (a device producing) a magnetic field capable of focusing a beam of charged particles; **magnetic memory,** (*a*) a dependence of the magnetic state of a body on its previous magnetic history; (*b*) *Computers*, a memory (sense *2 d) that employs the magnetic properties of bodies; **magnetic mine,** a submarine mine that is detonated by the approach of a magnetized body such as a ship; **magnetic mirror,** (*a*) a magnetized surface that reflects light; (*b*) a magnetic field that causes approaching charged particles to be reflected; **magnetic moment** (see *moment *sb.* 8 c); **magnetic quantum number,** the quantum number *m* (see *M 4* b); **magnetic resonance accelerator** = *cyclotron; **magnetic stripe** (see *stripe *sb.*³); **magnetic tape,** tape (now usu. of plastic) coated or impregnated with a magnetic material, or made of magnetic material, for use as a recording medium.
1899 *Jrnl. Brit. Astron. Assoc.* IX. 134 (*heading*) Magnetic anomalies in Russia. 1929 Eve & Keys *Appl. Geophysics* ii. 38 Three auxiliary magnets of different strengths are usually carried in order to provide for the measurement of large magnetic anomalies and thus extend the range of measurement. 1971 *Nature* 5 Feb. 374/2 The magnetic evidence for seafloor spreading is based on the pattern of the linear magnetic anomalies which lie over and to the sides of mid-oceanic ridges. 1957 *Sci. Amer.* Aug. 80/2 Since no conceivable material will withstand the temperature of these [thermonuclear] reactions, the hope is to contain them in a 'magnetic bottle' that will take advantage of the plasma's response to electromagnetic forces. 1967 *Science Year* 341 No one yet knows how to confine a hot plasma in a 'magnetic bottle' long enough for the energy liberated by relatively infrequent thermonuclear reactions to overcome the inevitable loss from electromagnetic radiation and from plasma escaping from the bottle. 1922 Glazebrook *Dict. Appl. Physics* II. 323/1 In magnetic brakes..an electromagnet..is held just above the rail to which it is attracted upon excitation. 1970 *Nature* 10 Oct. 114/1 Magnetic bubbles, small stable regions of reverse magnetization in a highly anisotropic ferrimagnetic sheet, were described by Dr F. C. Rossol. 1971 *Sci. Amer.* June 78/1 A promising alternative..exploits a new technology in which data bits are stored in the form of magnetic 'bubbles' moving in thin films of magnetic material... The bubbles are stable over a considerable range of conditions and can be moved from point to point at high velocity. 1950 W. W. Stifler *High-Speed Computing Devices* xiv. 304 A rotating magnetic drum..is a static storage system, according to this definition, because data recorded on the drum are static with respect to the surface of the drum. 1955 *Sci. Amer.* June 98/2 For the larger, intermediate-speed memory of a computer the favorite device at present is the magnetic drum. 1968 *Brit. Med. Bull.* XXIV. 192/1 Different users require.. different combinations of back-up storage (magnetic tape units,..magnetic drum units, and so on). 1919 *Phil. Mag.* XXXVIII. 709 Such a 'magnetic lens' is not of much immediate value as the magnetic spectrum of positive rays is very complex. 1966 *McGraw-Hill Encycl. Sci. & Technol.* VIII. 33/2 Magnetic lenses are employed as condensers, objectives, and projection lenses in magnetic electron microscopes, as final focusing lenses in the electron guns of cathode-ray tubes, and for the selection of groups of charged particles of specific velocity in velocity spectrographs. 1887 *Jrnl. Soc. Telegr. Engin.* XVI. 523 No matter how treated, a piece of soft iron has a 'magnetic memory'. 1947 L. B. Loeb *Fund. Electr. & Magn.* (ed. 3) xvii. 242 A continuous curve is retraced that never restores the magnetic material back to its zero value. This behavior can be ascribed to a sort of 'magnetic memory' of the substance. 1957 *Economist* 30 Nov. 797 The electronic valves, transistors and magnetic memories used in the modern computer have been developed from knowledge gained from other spheres of electronic applications. 1939 *News Rev.* 30 Nov. 13/1 We already know the secrets of the magnetic mine. 1939 *War Weekly* 1 Dec. 185/1 The new magnetic mine is said to be in use by the Germans at the present time. 1974 *Encycl. Brit. Micropædia* VI. 910/2 Three common types of naval mine are in use, classed by their method of detonation: magnetic mines, pressure mines, and acoustic mines. 1894 *Jrnl. Inst. Electr. Engin.* XXIII. 448 (*heading*) Researches on the reflection of polarised light from magnetic mirrors. 1952 R. F. Post in *Classified Conf. Thermonucl. Reactors held at Denver June 28, 1952* (AEC Rept. Wash.-115) 83 A second possible attack on the problem of containment of a linear discharge at its ends is through the use of what might be called an 'enhanced magnetic mirror'. The ordinary 'magnetic mirror' effect has been known for some time. 1956 L. Spitzer *Physics of fully Ionized Gases* i. 11 The constancy of the magnetic moment has the immediate result

that gyrating particles will tend to be reflected from regions of increasing magnetic field... Such a reflecting region may be called a 'magnetic mirror'. **1966** *McGraw-Hill Encycl. Sci. & Technol.* X. 388/1 The 'mirror machine' (invented by R. F. Post) allows confinement [of the plasma] in a tube with ends 'plugged' by magnetic mirrors... A magnetic mirror is merely a localized region where the magnetic field is made much stronger than average, so that charged particles tend to be reflected as they approach the mirror. **1923** Magnetic quantum number [see *M 4* b]. **1961** POWELL & CRASEMANN *Quantum Mech.* x. 355 Since the single electron in these atoms has zero angular momentum in the ground state, the magnetic quantum number must be $m = 0$. **1935** Magnetic resonance accelerator [see *CYCLOTRON]. **1968** F. KERTESZ *Lang. Nucl. Sci.* (Oak Ridge Nat. Lab. TM 2367) 17 A number of the accelerators..have names ending in '...tron', recalling the generic term *cyclotron* or magnetic resonance accelerator. **1937** *Bell Syst. Techn. Jrnl.* XVI. 165 (*heading*) Sound recording on magnetic tape. **1942** *Jrnl. R. Aeronaut. Soc.* XLVI. 68 This paper described an electro-magnetic method of producing and controlling reverberation by the use of a magnetic tape recording system. It consists of recording a sound pattern magnetically on steel tape. **1957** *Economist* 30 Nov. 807/2 The computer, in turn, produces the instructions on magnetic tape which is read by the control apparatus of the cutting machine. **1958** *Listener* 16 Oct. 605/1 Stereophonic magnetic tape recordings have been available for some time. **1967** C. BERNERS-LEE in Wills & Yearsley *Handbk. Managem. Technol.* i. 6 It may well be that the data already exists on magnetic tape in a computer installation.

B. *sb.* **3.** (Further example.)

1971 *Nature* 26 Nov. 187/2 Gravity and magnetics have theorems which limit the amount of interpretation which can be applied to observations of potentials and in general geophysicists work within the confines of these restrictions.

4. *pl.* Magnetic devices or materials.

1965 *IEEE Trans. Magnetics* Mar. 3/1 The name became Magnetics Group, and its scope:..treatment of all matters in which the dominant factors are the fundamental developments, design, and certain applications of magnetic devices. **1971** *Physics Bull.* Nov. 646/1 The size of the magnetics market however does not hold any reassurances for research.

5. *pl.* Magnetic properties or phenomena collectively.

1972 *Nature* 22 Dec. 438/1 They have recorded bathymetry, magnetics and gravity along a profile over a ridge striking N 50° W. **1974** *Ibid.* 4 Jan. 9/1 A new geophysical model for the Red Sea is presented in Fig. 5... The synthetic magnetics is compared with the observed. The fit of the anomalies for the older crust is not quite so good as for the Gulf data.

magnetician. (Examples.)

1838 *Ann. Electr. Magn. & Chem.* II. 53 Had Mr. Clarke, the 'magnetician', known that fact, he might have saved himself the trouble. **1846** C. FOX *Jrnl.* 13 Oct. (1972) 168 Mineral veins, however, a quite different case; infinite scope therein for Papa and all Magneticians. **1897** *Proc. R. Soc.* LX. p. x, His first paper on that subject..was completely overlooked by magneticians. **1939** *Geogr. Jrnl.* XCIII. 157 Gilbert the magnetician. **1966** *McGraw-Hill Encycl. Sci. & Technol.* IV. 20/1 One of these [components], called variation by the navigator and magnetic declination by the magnetician, is the angle between magnetic and geographic meridians. **1970** *Nature* 29 Aug. 982/2 He found that magneticians were far readier to collect than to interpret data, and set himself to analyse geomagnetic variations.

magnetization. Add: **2.** Special Comb.: **magnetization curve,** a graph of magnetic induction against magnetic field strength in the same region.

1890 J. HOPKINSON in *Proc. R. Soc.* XLVIII. 2 A magnetisation curve is all that I have obtained free from doubt; the sample was heated and its magnetisation determined at various temperatures for a force of 0·50. **1968** C. G. KUPER *Introd. Theory Superconductivity* v. 86 Figure (5.5) shows a typical magnetization curve for a small cylinder in a transverse field.

magneto, *sb.* Add: *spec.* one in an internal-combustion engine employing spark ignition.

1902 J. E. HUTTON in A. C. Harmsworth et al. *Motors* viii. 154 The magneto is driven off the main shaft of the engine..by means of a chain. **1929** A. HUXLEY *Let.* 22 Feb. (1969) 307 At Albengo on the Italian riviera, our magneto suddenly gave out. Short circuit, something fused. **1959** *Motor Man.* (ed. 36) vi. 169 Special equipment is needed to develop the high tension required for ignition... There are two systems. One, commonly known as coil ignition, takes in low tension current from the battery and converts it to high tension. The other depends upon a magneto-electric machine—a magneto, for short—which is driven by the engine and generates high-tension current. **1971** *Arable Farmer* Feb. 49/1 On external types of magneto..the points are normally accessible by removing the end cover.

b. *attrib.*, as **magneto ignition,** in some internal-combustion engines, ignition by means of a voltage generated by a magneto that is driven by the engine.

1902 J. E. HUTTON in A. C. Harmsworth et al. *Motors* viii. 159 The magneto ignition system, which is applied to all the new model motor-engines of M. Mors, consists of a magneto-electric rotary machine, combined with a series of mechanical contact-breakers. **1905** *Westm. Gaz.* 12 Dec. 4/2 The all but universal use of magneto-ignition. *c*1933 E. MOLLOY *Mod. Motor Repair* III. 877/1 The distance of the gaps..should be adjusted as accurately as possible to between ·015 to ·018 in.; if the engine is fitted with magneto ignition. **1966** *McGraw-Hill Encycl. Sci. & Technol.* VIII. 52/2 The higher cost of magneto ignition is not warranted in modern automobiles, where storage batteries are required for other electrically operated equipment.

magneto-. Add: *magneto-induced* adj.; **magne:toca·loric** *a.* [ad. F. *magnétocalorique* (Weiss & Piccard 1918, in *Compt. Rend.* CLXVI. 352)], applied to the reversible change of temperature that accompanies a change in the magnetization of a paramagnetic or ferromagnetic material; **magne:toca·rdiogram** *Med.*, a record of the variations in the magnetic field of a patient's body that occur as a result of the beating of the heart; so **magne:toca·rdiograph,** an instrument used to make such records; **magne:tocardiogra·phic** *a.*, **-cardio·graphy**; **magne:toche·mistry,** the branch of science concerned with the relation between magnetism and chemical phenomena, molecular and atomic structure, etc.; so **magne:toche·mical** *a.*; **magne:to-explo·der,** a magneto-electric apparatus for firing an explosive charge; **magne:to-io·nic** *a.*, of or pertaining to the joint effect of a magnetic field and ionized gas (e.g. in the ionosphere) on the propagation of radio waves; **magne:to-mecha·nical** *a.*, pertaining to the interrelation of magnetic and mechanical properties, esp. the magnetic moment and angular momentum of an atom or particle; *spec.* applied to the ratio of these quantities (or its reciprocal); **magneto-optics,** substitute for def.: the branch of physics which deals with the optical effects of magnetic fields; (further examples); **magne:toresi·stance,** dependence of the electrical resistance of a body on an external magnetic field; **magne:toso·nic** *a.*, pertaining to or designating a type of magnetohydrodynamic wave that has two speeds of propagation (both functions of the magnetic field strength and the speed of sound in the fluid), can travel in any direction relative to that of the field, and is characterized by a displacement of the fluid in any direction in the plane defined by the directions of propagation and of the field except the direction normal to the former; **magne:tosta·tics** [after ELECTROSTATICS], the branch of physics dealing with unchanging magnetic fields; so **magne:tosta·tic** *a.*, **-sta·tically** *adv.*; **magne·totail** [TAIL *sb.*[1]], the broad, elongated part of the magnetosphere that extends from the vicinity of the earth in a direction away from the sun; **magne:totellu·ric** *a.*, pertaining to or designating a technique for investigating the electrical conductivity of the earth by measuring simultaneously fluctuations in its magnetic and electric fields at the surface; hence **magne:totellu·rics** [-IC 2], the branch of geophysics concerned with this.

1921 *Sci. Abstr.* A. XXIV. 724 (*heading*) Magnetocaloric effect. **1937** J. W. T. SPINKS tr. *Herzberg's Atomic Spectra & Atomic Struct.* vi. 211 The magnetocaloric effect [in paramagnetic substances] is so small at room temperature that it cannot be observed. **1965** A. H. MORRISH *Physical Princ. Magn.* iii. 83 Since for a normal paramagnetic salt $(\partial M/\partial T)_H$ is negative, an increase in the field produces a heating and conversely decreasing H gives rise to a temperature drop. This is often called the magnetocaloric effect. The important application of this effect is in the production of temperatures below 1°K. **1963** BAULE & MCFEE in *Amer. Heart Jrnl.* LXVI. 95/1 The electromotive forces of the heart set up currents in the torso which in turn produce magnetic fields. These fields are exceedingly small... Nevertheless, they may be detected, as is shown by the magnetocardiogram. **1971** *New Scientist* 10 June 631/2 As well as being more convenient, a magnetocardiogram (MKG) is faster to take than an electrocardiogram. **1963** *Amer. Heart Jrnl.* LXVI. 96/2 The magnetocardiograph offers the potentiality of detecting otherwise 'silent' components of the electromotive forces of the heart. **1967** *N.Y. Times* 6 May 33 Dr. Cohen said..it is..too early to assess the long-range potential of the magnetocardiograph. **1970** *Amer. Heart Jrnl.* LXXIX. 231/2 Since the electromotive surface representing the activation boundaries lies in the heart, electrocardiographic or magnetocardiographic lead fields need only be known within this region. **1967** *Bull. Exper. Biol. & Med.* LXIV. 1024 Magnetocardiography is also valuable as an addition to electrocardiography. **1911** *Jrnl. Chem. Soc.* C. II. 367 (*heading*) Magneto-chemical researches on the atomic structure of the halogens. **1943** P. W. SELWOOD *Magnetochem.* iii. 58 The dimerization of free radicals and of other paramagnetic molecules is one of the most fruitful fields for magnetochemical research. **1972** *Indian Jrnl. Chem.* X. 726/2 A magnetochemical study of these derivatives has also been carried out to get some information about the nature of bonding of the metal ions with the polyphosphate chain. **1914** *Chem. Abstr.* VIII. 2648 (*heading*) Magneto-chemistry. **1937** *Nature* 20 Mar. 489/1 By magnetochemistry is meant the examination of the problems of chemical structure in the light of magnetic measurements and modern magnetic theory. **1969** *Jrnl. Leeds Univ. Union Chem. Soc.* XI. 42 The earliest significant work on magnetochemistry was probably by Faraday who devised a method for measuring susceptibility which is still used today. **1973** *Nature* 31 Aug. p. xi (Advt.), Magnetochemistry, especially the study of metal–metal interactions in paramagnetic clusters. **1869** *Chambers's Jrnl.* Apr. 271/2 A magneto-exploder.. was shewn, which will fire a fuse, and consequently a cannon. **1908** *Installation News* II. 149/2 Water-tight bells and magneto exploders for blasting purposes. **1871** *Eng. Mechanic* 3 Feb. 480/1 He is referring to a galvanic, and not a magneto-induced current. **1925** APPLETON & BARNETT in *Electrician* 3 Apr. 398/1 The same theory (for which we propose the name magneto-ionic theory) has also been independently put forward by Messrs. Nicholls and Schelling ('Nature', March 7th, 1925). **1932** E. V. APPLETON in *Jrnl. Inst. Electr. Engin.* LXXI. 645/2 It is highly probable that interference between the various magneto-ionic components of singly- and multiply-reflected downcoming waves is partly responsible for resultant intensity fading. **1973** *Physics Bull.* May 291/3 A first start is made to the problem of the transfer of radio emission in a pulsar magnetosphere by applying the method of magnetoionic theory. **1925** *Sci. Abstr.* A. XXVIII. 611 Magneto-mechanical anomaly of the atom. **1930** Magneto-mechanical [see *GYROMAGNETIC *a.* 1]. **1950** W. FINKELNBURG *Atomic Physics* iii. 153 The magnetic moment of paramagnetic atoms is due to the electron rotating in its orbit or to the spin of the electron. The so-called magnetomechanical parallelism enables us to distinguish between these two contributions to the atomic magnetism. **1953** [see *GYROMAGNETIC *a.* 1]. **1958** CONDON & ODISHAW *Handbk. Physics* IV. viii. 136/1 Because of the magnetomechanical interactions.., internal stresses due to cold-working, impurities, precipitates, etc., contribute a spatially varying component of the short-range energy. **1913** P. ZEEMAN *Res. Magneto-Optics* p. xi, To the memory of this sage [*sc.* Faraday]—..the pioneer in magneto-optics as in so many things, I have ventured to dedicate this volume. **1960** *Physical Rev. Lett.* IV. 357 (*heading*) Photon momentum effects in the magneto-optics of excitons. **1930** L. W. MCKEEHAN in *Physical Rev.* XXXVI. 949 The magneto-resistance effect in non-ferromagnetics is only measurable with accuracy in very intense magnetic fields. **1961** *Engineering* 22 Dec. 823 One of the first practical devices to use the phenomenon of magnetoresistance is a voltage regulator. **1966** *New Scientist* 3 Feb. 286/2 The magnetoresistance effect in semiconductors, which has recently been shown to exist at frequencies well up into the microwave region. **1962** W. B. THOMPSON *Introd. Plasma Physics* v. 84 There is remarkably little evidence for the propagation of magnetosonic waves through a plasma. *Ibid.* 91 In the magnetohydrodynamic case, there are three possible parametrizing velocities C_s, C_A and the magnetosonic speed $C_T = \surd(C_A^2 + C_s^2)$. **1971** *Nature* 13 Aug. 443/3 The Crab pulsar..seems to be associated with a series of wisps of gas... Wisps 2, 3 and 4 have been generated by the magnetosonic waves which this motion of Wisp 1 produces. **1893** *Notices Proc. R. Inst. Gt. Brit.* XIII. 348 Magnetostatic screening by soft iron would follow the same law as electrostatic screening, if the magnetic susceptibility of the iron were infinitely great. **1950** *Physical Rev.* LXXV. 156 The domain patterns..represent configurations of low magnetostatic energy. **1969** M. A. UMAN *Lightning* 101 The magnetostatic field present during a lightning discharge is directly proportional to the discharge current. **1960** *Physical Rev. Lett.* V. 47/1 The figure of 10^{-8} erg/cm³ represents..a solar wind flux sufficient to compress the geomagnetic field magnetostatically to $\sim 9R_e$ if we utilize the usual model. **1897** A. G. WEBSTER *Theory Electr. & Magn.* ix. 353 (*heading*) Parallel treatment of electrostatics and magnetostatics. **1952** E. G. RAMBERG tr. *Sommerfeld's Electrodynamics* I. 40 In most treatments the analogy between electrostatics and magnetostatics is emphasized. **1968** J. C. ANDERSON *Magnetism* i. 10 In magnetostatics it is convenient to define a magnetic pole strength, q, analogous to electric charge. **1971** *Daily Tel.* 4 Aug. 4/3 This region's outlines become distorted by a varying stream of electrified atomic particles from the Sun called solar wind. It sweeps part of the magnetosphere..into a tear-drop shape which scientists would now refer to as the Earth's magnetotail. **1973** *Nature* 9 Mar. 79/2 The multiple satellite mission will help unravel the spatial and temporal variations of the radiation belt, the magnetopause, the bow shock and magnetotail. **1953** L. CAGNIARD in *Geophysics* XVIII. 605 (*heading*) Basic theory of the magneto-telluric method of geophysical prospecting. **1967** L. R. ALLDREDGE in Matsushita & Campbell *Physics of Geomagn. Phenomena* I. I. ii. 55 Smith et al...describe instrumentation for magnetotelluric experiments for which simultaneous measurements of geomagnetic micropulsations and telluric currents are made. **1971** *Nature* 3 Sept. 13/1 To learn something about the behaviour of the upper mantle..Tammemagi and Lilley..have made four magnetotelluric soundings at 150 km intervals roughly along the latitude of Canberra. **1960** *Jrnl. Geophysical Res.* LXV. 4202/1 Useful field results have not been numerous in the literature, and field magnetotellurics is still in the developmental stage. **1967** A. T. PRICE in Matsushita & Campbell *Physics of Geomagn. Phenomena* I. II. iii. 295 There is, at present, much activity in the theory and practice of magnetotellurics, which should greatly help the task of unravelling and interpreting many geomagnetic phenomena.

magnetogyric (mæɡnī:tədʒəi·rik), *a. Physics.* [f. MAGNETO- + Gr. γῦρ-ος ring, circle + -IC.] **1.** Pertaining to or exhibiting the Faraday effect (*FARADAY 1).

1904 A. SCHUSTER *Introd. Theory Optics* xii. 272 All substances turn the plane of polarization when they are traversed by light in the direction of a magnetic field. They become therefore 'magneto-gyric'. *Ibid.* 290 A simple and rational connexion between the Zeeman effect and magneto-gyric properties.

2. Applied to the ratio of the magnetic moment of an atom or particle to its angular momentum.

1965 M. G. SCROGGIE *Electron in Electronics* ix. 204, γ_s, the magnetogyric ratio for spin, is different from the orbital γ, being practically twice as large. **1967** CONDON & ODISHAW *Handbk. Physics* (ed. 2) VIII. ix. 113/1 The term $g_N(e/2Mc)$ is often replaced by γ, the magnetogyric ratio; hence, $\mu = \gamma\hbar I$. **1972** *Science* 27 Oct. 364/1 As long as **H**₁ is applied, **M** precesses..at a frequency given by the Larmor relation $\omega_R = \gamma H_1$ (in radians per second) where γ is the magnetogyric ratio of the nucleus. **1974** *McGraw-Hill Yearbk. Sci. & Technol.* 314/2 Compared to H¹, the C¹³ nuclei are much weaker magnets. The magnet strength (magnetogyric ratio γ) is only about ¼ that of protons.

magnetohydrodynamic (mæɡnī:təhəidrodəinæ·mik), *a.* [f. MAGNETO- (repr. *electromagnetic*) + HYDRODYNAMIC *a.*] = *HYDROMAGNETIC *a.* Abbrev. MHD (see *M 5).

1943 [see *HYDROMAGNETIC *a.*]. **1950** H. ALFVÉN *Cosmical Electrodynamics* iv. 76 This coupling between mechanical and electromagnetic forces produces a type of wave motion, called magneto-hydrodynamic waves. **1960** *Times* 14 Apr. 2/2 (Advt.), Central Electricity Generating Board. ..Teams are being formed to work on the following projects:—..Magnetohydrodynamic generation. Fundamental research on (relatively) low temperature plasmas and their interactions with magnetic fields [etc.]. **1961** *Engineering* 16 June 817/3 The magnetohydrodynamic prospect, direct generation by passing hot gas through a magnetic field. **1962** F. I. ORDWAY et al. *Basic Astronautics* vi. 287 In a binary system, the angular momentum of the main star seems to have been transferred to the orbital motion of the second star. The means by which the angular momentum was shuffled off is not entirely clear, but magnetohydrodynamic or viscous effects could have played the key role. **1971** I. G. GASS et al. *Understanding Earth* vi. 94/2 Runcorn has re-examined the possibility of exciting the wobble [of the rotating earth] by the effects of magnetohydrodynamic turbulence in the fluid outer core.

Hence **magne:tohydrodyna·mical** *a.*, **-dyna·mically** *adv.*

1953 L. H. ALLER *Astrophysics* ix. 354 Alfvén suggested a theory of sunspots based on magneto-hydrodynamical waves. **1959** *Jrnl. Geophysical Res.* LXIV. 1220/1 The entire medium can..be stirred up without requiring any magnetohydrodynamical work. **1962** W. B. THOMPSON *Introd. Plasma Physics* iv. 58 The possibility of producing a magnetohydrodynamically confined plasma has stimulated an enormous research effort in plasma physics.

magnetohydrodynamics (mæɡnī:təhəidrodəinæ·miks), *sb. pl.* (const. as *sing.*). [f. prec.: see -IC 2.] The more usual name for *HYDROMAGNETICS *sb. pl.*

1950 *Sci. Amer.* Oct. 39/3 Among these [theories] one can mention the magneto-hydrodynamics of charged particles proposed by the Swedish cosmologist H. Alfvén. **1958** *Observer* 7 Sept. 13/5 'Magnetohydrodynamics', a branch of theoretical physics which combines classical theory of the behaviour of water and of magnetic fields,..is becoming one of the most elaborate and enthralling branches of science. **1973** M. HARWIT *Astrophysical Concepts* vi. 194 Magnetohydrodynamics..tells us that the presence of a force, such as a gravitational or electrostatic force, acting normal to the magnetic field, can produce a drifting motion.

Hence **magne:tohydrodyna·micist,** one who studies magnetohydrodynamics.

1955 *Sci. Amer.* Feb. 41/1 How the corpuscles travel through space is still in dispute. The Alfvén school of 'magneto-hydrodynamicists', for instance, maintains that the Chapman–Ferraro theory is basically incorrect, and that the solar corpuscles, as well as cosmic rays, are accelerated by magnetic and electrical fields in space.

magneton (mæ·ɡnėtɒn). *Physics.* [ad. F. *magnéton* (P. Weiss 1911, in *Compt. Rend.* CLII. 189), f. *magnéti-que* MAGNETIC *a.*: see *-ON[1].] Any of several units of magnetic moment used in atomic and nuclear physics: *Bohr magneton* [*BOHR], a unit that arises naturally in Bohr's theory of the atom, equal to $eh/4\pi m$ (where e and m are the charge and mass of the electron, and h is Planck's constant, in S.I. units), i.e. about $9 \cdot 27 \times 10^{-24}$ joule per tesla; *nuclear magneton,* a unit used for expressing nuclear magnetic moments, defined analogously to the Bohr magneton with the mass of the proton replacing that of the electron and equal to about $5 \cdot 05 \times 10^{-27}$ joule per tesla; *Weiss magneton* [named after P. Weiss (see above)], a disused unit (equal to approximately one-fifth of a Bohr magneton) arrived at as the highest common factor of the measured magnetic moments of the atoms of various elements.

At first often regarded as a 'particle' or 'atom' of magnetism.

1911 *Sci. Abstr.* A. XIV. 191 The author [*sc.* P. Weiss] is led to postulate the existence of a common constituent of magnetic materials—the elementary magnets of [*sic*] which he terms *magnétons,* and which are to be found in the atoms of all magnetic substances. *Ibid.* 243 (*heading*) Size of the magneton deduced from the coefficients of magnetisation of solutions of iron salts. **1915** *Proc. Physical Soc.* XXVII. 430 A further comparison with experimental determinations is rendered possible by assuming that the magnetic moment of an electron revolving with the unit of angular momentum is equivalent to 5 magnetons. **1917** R. W. HUTCHINSON *Adv. Textbk. Magn. & Electr.* II. 313 This fundamental magnet he calls the 'magneton'. Calculation shows that iron, nickel and magnetite contain 11, 3, and 7 of these magnetons per molecule respectively. **1919** *Nature* 6 Nov. 227/2 The suggestion, first proposed by Weiss, that there exists a natural unit of magnetism called the magneton, analogous in some respects to the atom of electricity, still lacks definite proof. **1923** Bohr magneton [see *BOHR]. **1927** N. V. SIDGWICK *Electronic Theory of Valency* 208 The Bohr magneton..is almost exactly five (4·96) times the Weiss magneton. **1932** J. H. VAN VLECK *Theory Electr. & Magn. Susceptibilities* ix. 228 Although many molecules are still found to have moments which are integral multiples of the Weiss magneton within the experimental error, this is probably fortuitous, for there are many reasons for believing the Weiss magneton phenomenon to be spurious. **1935** *Physical Rev.* XLVII. 745/2 Using the value 3/2 for the spin of K³⁹ and 0·0152 for $\Delta\nu$ we get 0·39 nuclear Bohr magnetons for the magnetic moment of the K³⁹ nucleus. *Ibid.* 801/1 We obtain a value of 3·14 nuclear magnetons for the magnetic moment of the Li⁷ nucleus. **1962** H. D. BUSH *Atomic & Nucl. Physics* iii. 60 For all nuclei, including the proton the magnitude of the moment is not an integral multiple of the nuclear magneton. **1973** *Nature* 26 Jan. 239/1 Quantum fluctuations of the electromagnetic field close to the particle [*sc.* a free electron] imply an apparent increase in its magnetic moment beyond the value of one Bohr magneton implied by the unmodified Dirac theory.

magnetopause (mæɡnī·təpɔ̄z). [f. MAGNETO- + PAUSE *sb.*] The outer limit of a magnetosphere.

1963 SONETT & ABRAMS in *Jrnl. Geophysical Res.* LXVIII. 1243/2 We call the region discussed in this paper the magnetopause (the transition region between the magnetosphere proper and the interplanetary medium). **1969** *Physics Bull.* Feb. 54/1 The third spacecraft..is intended to investigate the 'solar wind' of low energy protons and electrons together with the geomagnetic cavity or magnetopause, where these particles are just in balance with the earth's magnetic field. **1971** I. G. GASS et al. *Understanding Earth* xviii. 259/2 The solar wind protons are reflected at the magnetopause, and so never enter the magnetosphere.

magnetophone. Add: **2.** An early form of moving-coil microphone.

1922 *Encycl. Brit.* XXXII. 528/1 The behaviour of microphones in this respect was often unsatisfactory, and telephone earpieces, or magnetophones, were frequently substituted for them. This results in diminished sensitivity, but the binaural effects are much improved. **1929** J. A. RATCLIFFE *Phys. Princ. Wireless* v. 73 A microphone of quite a different type is much used at the present day for broadcasting. It is known as the magnetophone, and consists of a small coil of very light aluminium wire attached to the back of a freely moving diaphragm. The coil is placed between the poles of an electro-magnet. **1931** *B.B.C. Year-Bk.* 445/2 The magnetophone is relatively insensitive, and its output requires considerable amplification. **1962** *Listener* 8 Nov. 765/3 Two of the earliest [*sc.* BBC microphones] are to be seen..: Round-Sykes 'magnetophone', [etc.].

3. Also **-phon.** [a. G. *magnetophon.*] A tape recorder. (Used chiefly as a rendering of the German or with reference to German machines.)

1946 *Wireless World* Feb. 53/1 The equipment, known as the magnetophone, uses a thin plastic tape, about 0·25 in. wide, coated with a metallic powder film. **1949** *Electronic Engin.* XXI. 124/3 Sound recorders of the magnetophone type make use of two oscillatory currents in the recording head. **1953** F. SPANDÖCK in E. G. Richardson *Technical Aspects Sound* I. xv. 410 After the Second World War magnetophones were built in many countries in many different forms. **1954** *Sci. Abstr.* B. LVII. 245/1 The various tests carried out on magnetophones, microphones and amplifiers are enumerated. **1974** *Encycl. Brit. Macropædia* XVII. 55/2 The Magnetophon Company of Germany succeeded in producing a magnetic film-coated tape that constituted a low-cost recording medium. During World War II the magnetophon, using the coated tape, served the Nazi propaganda organizations well.

magnetoplumbite (mæɡnī:toplʌ·mbəit). *Min.* [ad. G. *magnetoplumbit* (G. Aminoff 1925, in *Geol. Fö. i Stockholm Förh.* XLVII. 289), f. Gr. μάγνητ-, μάγνης MAGNET + L. *plumb-um* lead: see -ITE[1].] A strongly magnetic greyish black oxide of manganese, lead, titanium, and ferric iron occurring as acute dipyramidal crystals.

1926 *Chem. Abstr.* XX. 1194 (*heading*) A new oxide mineral from Långban, magnetoplumbite. **1944** C. PALACHE et al. *Dana's Syst. Min.* (ed. 7) I. 728 Adelsköld, .. (1938), gives *C6/mmc* as the space group of an artificial material supposedly the same as magnetoplumbite... The supposed composition $PbFe_{12}O_{19}$, by analogy with so-called β-alumina, proposed by Adelsköld (1938) is not indicated by analysis. **1968** I. KOSTOV *Mineral.* II. iv. 259 Magnetoplumbite has a perfect {0001} cleavage.

magnetosphere (mæɡnī·tosfiəɹ). [f. MAGNETO- + SPHERE *sb.*] The region surrounding the earth or a heavenly body in which its magnetic field is effective and prevails over magnetic fields due to other causes (in the case of the earth not spherical but much elongated on the side away from the sun).

1959 T. GOLD in *Jrnl. Geophysical Res.* LXIV. 1219/1 The region above the ionosphere in which the magnetic field of the earth has a dominant control over the motions of gas and fast charged particles..is known to extend out to a distance of the order of 10 earth radii; it may appropriately be called the magnetosphere. **1962** *Flight Internat.* LXXXI. 222 Instead of distinct inner and outer [radiation] belts, the picture now appears to be of one large region in which particles of different characteristics are trapped. Different types of particles have been observed at 1½ Earth radii (from the centre of the Earth), three radii, four radii (the older concept of the outer belt) and between six radii and the outer edge of the magnetosphere, as the region is now known. The outer limit varies from day to day but is at approximately 8–12 Earth radii. **1967** *New Scientist* 26 Jan. 198/1 Like the Earth, Jupiter is expected to have a 'magnetosphere' with a 'tail' pushed away from it by the solar wind. **1971** I. G. GASS et al. *Understanding Earth* xviii. 260/1 The Van Allen belts are zones of stably trapped particles within the magnetosphere. **1971** *Nature* 13 Aug. 443/3 Much work needs to be done to investigate the kinds of plasma instabilities that can occur in the magnetospheres of neutron stars.

Hence **magnetosphe·ric** *a.*

1961 *New Scientist* 19 Jan. 167/2 What have been termed 'magnetospheric waveguides' may provide new methods of radio communication between the northern and southern hemispheres. **1969** *Physics Bull.* June 223/1 Instabilities..occur in both laboratory and magnetospheric and interplanetary plasmas. **1974** *McGraw-Hill Yearbk. Sci. & Technol.* 277/2 The so-called whistlers in the kilohertz frequency range, as well as the long-period waves,..have become important tools in magnetospheric exploration both from the ground and from satellites.

magnetostriction (mæɡnī:tostri·kʃən). [f. MAGNETO- + L. *strictiōn-em* drawing or pressing together (f. *stringere* to draw together, draw tight).] A dependence of the state of strain of a body (and hence its dimensions) on its state of magnetization. Freq. *attrib.*

1896 *Phil. Mag.* XLI. 454 (*heading*) On the effects of magnetic stress in magnetostriction. **1926** *Jrnl. Iron & Steel Inst.* CXIII. 657 A reciprocal relation between magneto-striction and the effect of stress on magnetisation was investigated experimentally. **1937** *Nature* 18 Dec. 1068/2 A magnetostriction oscillator has been developed which produces intense audible vibrations of frequencies of 8000 cycles per second, capable of fracturing glass. **1959** H. BARNES *Oceanogr. & Marine Biol.* 74 The magneto-striction effect depends upon the fact that when certain metals, notably nickel, are placed in a varying magnetic field they undergo mechanical changes; nickel itself contracts in an increasing and expands in a decreasing magnetic field. As with the mechanical changes brought about by the piezo-electric effect, these also may be transmitted to a medium and, if they are sufficiently rapid, ultrasonic waves are produced. **1966** C. R. TOTTLE *Sci. Engin. Materials* vi. 138 Iron crystals expand along the direction of magnetization and contract at right angles to it, whereas nickel contracts along the magnetization direction and expands perpendicular to it. This phenomenon is known as magnetostriction.

Hence **magne:tostri·ctive** *a.*, of, exhibiting, or employing magnetostriction.

1911 *Chem. Abstr.* V. 1550 (*heading*) A study of the Joule and Wiedemann magnetostrictive effects in steel tubes. **1939** *Nature* 11 Mar. 416/1 Pierce and the late J. H. Vincent independently developed similar oscillators based upon the magnetostrictive properties of iron and nickel and their alloys. **1951** *Electronic Engin.* XXIII. 16 The maximum dimensional change is obtained when the magnetostrictive element is excited at its natural resonant frequency. **1964** T. W. MCRAE *Impact of Computers on Accounting* i. 6 The Ferranti 'Pegasus' introduced the so-called magnetostrictive delay line store, whereby a mechanical stress wave passing along a wire represents a bit.

magnetron (mæ·ɡnėtrɒn). *Electronics.* [f. MAGNET(IC *a.* + *-TRON.] A diode with a cylindrical anode surrounding a coaxial cathode in which the flow of electrons is controlled by a magnetic field applied parallel to the axis, and now usu. designed to produce microwave pulses of high power.

1924 *Sci. Monthly* XVIII. 650 As the negatively charged grid cuts off the current of the three-element tube, so an external magnetic field will also do it in the two-element tube... Such a device, called a magnetron, was shown. **1945** *Times* 15 Aug. 2/1 In July 1940, Professor J. T. Randall, of Birmingham, produced a magnetron which was the first high-power generator of centimetric waves in the world. The magnetron remains the heart of every modern Radar equipment. **1958** *Engineering* 7 Feb. 174/3 The machine [*sc.* a linear accelerator] is of the travelling-wave type, using radio-frequency power generated by a magnetron valve operating at a frequency of 3,000 Mc/s. **1971** *New Scientist* 1 Apr. 36/1 Outred is developing microwave cavity sources using very high-powered (kilowatts) magnetrons for the infrared spectroscopy programme.

magnitude. Add: **2. c.** The intrinsic size of an earthquake or underground explosion (as distinguished from the intensity of its effects at any particular place), usu. expressed by a number that is a logarithmic function of the maximum resulting seismometric deflection adjusted to allow for distance.

1935 C. F. RICHTER in *Bull. Seismol. Soc. Amer.* XXV. 1 In the course of historical or statistical study of earthquakes in any given region it is frequently desirable to have a scale for rating these shocks in terms of their original energy, independently of the effects which may be produced at any particular point of observation. On the suggestion of Mr. H. O. Wood, it is here proposed to refer to such a scale as a 'magnitude scale'. *Ibid.* 2 The requirements of research..call for some estimate of the magnitude, in the sense here used, of each important shock in the California region. **1947** K. E. BULLEN *Introd. Theory Seismol.* xiv. 234 The smallest earthquakes

reported felt are of magnitude 1·5;..those of magnitude 4·5 are capable of causing slight damage near the epicentre. **1959** B. F. HOWELL *Introd. Geophysics* ix. 125 Values of magnitude calculated at different observatories, using different seismometers at one observatory, or even different phases (body or surface waves) on the same seismogram, may be different. **1971** *Daily Colonist* (Victoria, B.C.) 4 Feb. 1/2 The quake had a magnitude of 6·7 on the Richter scale. **1972** *Sci. Amer.* Jan. 14/3 Underground explosions in the megaton range can have a body-wave magnitude of 6·5 to 7.

3. a. (Further examples.) Now regarded as a number on a continuous scale representing the negative logarithm of the brightness, such that a decrease of five magnitudes represents a hundred-fold increase in brightness and a decrease of one magnitude an increase of 2·512 times.

Before decimal or fractional numbers were used there was an intermediate stage illustrated by quots. 1796, 1846 (and also 1851).

1796 W. HERSCHEL in *Phil. Trans. R. Soc.* LXXXVI. 168, I..am for instance, denotes that a star so marked is between the first and second magnitude. 2.1m signifies the same thing, with an intimation that the star so marked is nearly of the second magnitude, but partakes still something of the lustre of a star of the first order. **1846** J. F. W. HERSCHEL in *Mem. R. Astron. Soc.* II. 444, I shall extend the examination to all stars of the 3rd and 4th magnitudes, those of the 5th however not included. **1851** *Monthly Notices R. Astron. Soc.* XI. 157 The differences..clearly show that widely different *scales* of magnitude have been adopted. *Ibid.*, This triple star..is designated by Argelander..as of magnitude 5·4 (or about 4·7). **1897** D. P. TODD *New Astron.* xvi. Even the surpassing brilliancy of the sun can be indicated on the same scale; the number −25·4 expresses his stellar magnitude. **1930** R. H. BAKER *Astron.* 1. 20 The magnitude of the brightest star, Sirius, is −1·6; Canopus is −0·9. **1967** C. M. HUFFER et al. *Introd. Astron.* xvii. 248/2 The photoelectric photometer, under the best conditions and by averaging several observations, can make measures down to 0·001 magnitude in accuracy.

magnolia. Add: **b*.** The colour of magnolia blossom, usu. a shade of pale pink. Hence *attrib.*, passing into *adj.*, of the colour of magnolia blossom.

1931 *Daily Chron.* 21 May 6/1 A gown of pale magnolia-pink satin. **1963** *Times* 25 Apr. 6/6 For the wedding yesterday Princess Alexandra chose..a traditional dress ..fine cotton lace, faintly tinted with magnolia. **1971** R. FULLER in *Listener* 28 Oct. 583/1 My study has been repainted 'Magnolia'.. Magnolia shows bright enough, if really Only off-white. **1974** 'S. HARVESTER' *Forgotten Road* xii. 137 The pale blue cotton shirt had been replaced by one of a droopy magnolia tint.

magnolious (mæɡnō·liəs), *a. slang.* [Humorously f. MAGNOLIA + -OUS.] Magnificent, splendid, large. Hence **magno·liousness** [see -NESS], the fact or quality of being magnolious.

1865 G. A. SALA *My Diary in Amer.* I. 114 But..she might be the sheriff's daughter..., accustomed to go out on Sundays with a 'magnolious' parasol and a 'spangiferous' crinoline. **1913** *Dialect Notes* IV. 19 *Magnolious,* very fine, splendid. 'Used in Wyoming. 'How do you like my suit?' 'It's *magnolious.*' **1921** JOYCE *Lett.* (1957) I. 173 With many thanks again and wishes for your magnolious expansiveness. *Ibid.* 175 Best wishes for your continued magnoliousness. **1942** BERREY & VAN DEN BARK *Amer. Thes. Slang* § 29/5 *Magnificent,*..magnolious, phenomenal, red-hot.

magnon (mæ·ɡnɒn). *Physics.* [f. MAGN(ETIC *a.* or MAGN(ETISM + *-ON[1].] The quantum or quasiparticle associated with a spin wave in a magnetic material.

1941 *Jrnl. Physics* (Moscow) IV. 358/1 Such magnetic excitations..will be called..'magnons' (this name was suggested by L. Landau). **1971** *Nature* 18 June 424/1 Light scattering research has been revolutionized by the introduction of the laser... Scattering from magnons or spin waves has furthered understanding of antiferromagnetic materials and their phase transitions. **1973** *Sci. Amer.* Jan. 91/1 Phonons and magnons are typical examples of quasiparticles that are bosons.

magnoperate, *v.* Add: **3.** *intr.* (*nonce-use.*) To act in a grand manner.

1926 J. AGATE *Contemporary Theatre 1925* 18 Everything about Richard is magnificent... He may be said not to act, but to magnoperate.

Also (as *nonce-wds.*) **magnopera·tion,** a great 'operation', a *magnum opus*; **magno·perator,** a great 'operator'; **magno·perous** *a.*, operating in a grand manner.

1928 BEERBOHM *Seven Men & Two Others* (1950) 230 Mr. Nat Heinz, the famous 'Firsts Agent', had recently come over from New York... I wrote at once a respectful note to this magnoperator. **1930** *Times* 28 Mar. 15/4 Not until the very last volume did Sir John Fortescue seek the help of a 'devil' in any part of his 'magnoperation'. **1939** JOYCE *Finnegans Wake* 15/37 The shadow of the huge outlander,..magnoperous, had balked at the bar of a rota of tribunals.

magnox (mæ·ɡnɒks). *Nuclear Engin.* Also **Magnox.** [f. *magn*esium *no ox*idation.] Any of various magnesium-based alloys containing a small proportion of aluminium that were developed for the containers of the fuel in nuclear reactors. Freq. *attrib.*

1953 *Technical Memorandum Min. of Supply, Div. of Atomic Energy (Production), Res. & Devel. Branch, Metall. Lab.* 30 Dec. 1 A magnesium alloy 'Magnox E'. **1955** *U.K. Atomic Energy Authority Rep.* SCS-M-301 (*title*) The superpurity aluminium (aluminium) and beryllium in magnox alloys. **1957** *Financial Times Ann. Rev. Brit. Industry* 15/3 Materials able to stand up to much higher temperatures than..the present metallic uranium fuel and Magnox cans. **1964** *Economist* 2 June 913/1 Corrosion of magnox components in a reactor. **1970** *Nature* 30 May 833/2 The magnox programme..was never expected to lead to the cheapest generation of electricity by the magnox systems are unduly conservative. **1971** *Nature* 28 May 233/1 The UK had then just announced plans for its first nuclear power programme based on magnox.

magnum (mæ·ɡnəm), *a.* [Cf. MAGNUM.] (Freq. with capital initial.) Of a cartridge: adapted so as to be more powerful than its calibre suggests. Of a gun: designed to fire such cartridges. Also *absol.*

Magnum is registered as a proprietary name in the U.S.

1939 *Official Gaz.* (U.S. Pat. Office) 26 Mar. 728/2 Smith & Wesson, Inc., Springfield, Mass. Filed Feb. 20, 1935. *Magnum* for Revolvers. Claims use since Dec. 28, 1934. **1937** J. S. SHARPE *Compl. Guide Handloading* xx. 266/1 A Magnum load..is anything which is 'stepped up' in power and should not be treated by the handloader. **1962** *Amer. Speech* XXXVII. 267 A ·357 Magnum bullet. **1970** *New York* 26 Nov. 50/1 A picture of Huggins and his Magnum on the floor of UCLA. **1971** V. CANNING *Firecrest* xiii. 188 Guards who carried twelve-bore shotguns and ·375 Magnum rifles. **1973** *Black Panther* 15 Mar. 6/3 It was a magnum type revolver bullet that killed Carr. **1973** C. C. NONTE *Firearms Encycl.* 159 While magnum cartridges are legion in the rifle field, there are currently only three used in handguns.

Magnus effect (mæ·ɡnəs efekt). [Named after Heinrich G. *Magnus* (1802–70), German scientist, who first described it.] The effect of rapid spinning on a cylinder or sphere moving through a fluid in a direction at right angles to the axis of spin, which results in a sideways force at right angles to both the direction of motion and the axis of spin and towards the side where the peripheral motion of the body is in the opposite direction to its overall motion.

1921 tr. *Cranz & Becker's Handbk. Ballistics* I. 314 A quantitative theory of the deviation of a sphere under the influence of the adhering air (Magnus effect) was attempted ..by Tait. **1925** *Flight* 8 Jan. 14/1 It does not appear to be until quite recently that the idea of applying the Magnus effect has occurred to anyone. **1943** R. C. BINDER *Fluid Mech.* xi. 245 The Flettner rotor, which employs the Magnus effect, has been applied to the propulsion of marine vessels. Vertical cylinders are extended some distance above the deck. Each cylinder is rotated about its axis by a small motor, and an air force is produced for moving the craft. **1973** *Sci. Amer.* Apr. 121/2 The drag of the turbulent wake and the sidewise forces of the Magnus effect (the same ones that govern the behaviour of the flow pattern around a spinning ball) are carefully considered. They are at the heart of baseball, tennis, golf and high-quality soccer.

Magosian (məɡōu·siān), *a. Archæol.* [f. the place-name *Magosi* in Uganda + -AN.] Of, pertaining to, or designating a stone-age culture in Uganda.

1932 WAYLAND & BURKITT in *Jrnl. R. Anthrop. Inst.* LXII. 369 The Magosian Culture of Uganda... Magosi, an outpost of the King's African Rifles in Karamoja (North-East Uganda), is situated 160 miles due north of the Uganda Railway at Tororo and about 60 miles to the west of the..frontier of the Kenya Colony... The discovery of the Magosian Culture. **1946** *Ibid.* LXXVI. 1. 60/1 The Magosian culture was first noticed at Sawmills..by Jones in 1924. At that time..it was not even given a name. **1959** J. D. CLARK *Prehist. S. Afr.* vi. 150 We find a similar link between the Middle and the Later Stone Ages. These are the various forms of the Magosian culture and they are now placed in the Second Intermediate stage. This culture is named from the small rock cistern in north-eastern Uganda, excavated by Wayland. **1968** *Encycl. Brit.* II. 12/2 In the final stages of the Middle Stone Age, known as the South African Magosian, microlithic elements appear, just as in the case of east Africa. **1970** BRAY & TRUMP *Dict. Archaeol.* 129 The sequence continued into Sangoan, followed by Early Middle Stone Age (Lupemban) industries related to those of the Congo, then Magosian.

magpie. Add: **3. a.** and **b.** (Later examples.)

1917 G. W. E. RUSSELL *Politics & Personalities* iv. ix. 157 The most hideous of all known costumes—the episcopal 'Magpie'—costs £100. *Ibid.* 160 Carrying with his own apostolic hands the sacred appliances of Mitre or Magpie. **1920** P. DEARMER *Ornaments of Ministers* (new ed.) plate 29 (*caption*) The..figures of the two bishops well illustrate the 'magpie' dress. **1923** *Brewer's Dict. Phr. & Fable* (new ed.) 700/2 Formerly bishops were humorously or derisively called magpies because of their black and white vestments.

8. Add: *magpie-minded* adj.

1955 G. A. N. LOWNDES *Brit. Educ. Syst.* iii. 49 Satisfying his curiosity..with unrelated snippets of knowledge which may lead to his becoming magpie-minded.

b. *attrib.* and *quasi-adj.*: with allusion to the acquisitiveness, curiosity, etc., of the magpie.

1908 M. WILMOT *Russ. Jrnls.* (1934) xi. 371 He deplores the magpie-mindedness which possibly rides the language of the Country. **1901** G. B. SHAW *Caesar & Cleopatra* II. 114 He maintains an air of magpie keenness and profundity. **1936** P. FLEMING *News from Tartary* III. ii. 116 All our actions..were closely scrutinized..by the Chinese with ill-concealed amusement and a magpie curiosity. **1940** *Proc. Prehist. Soc.* VI. 120 Beaker people ..showed a magpie acquisitiveness for other people's chattels. **1953** R. LEHMANN *Echoing Grove* ii. 32 Where does all this stuff come from? she said... But Mother hoarded, didn't she? **1962** G. K. HUNTER *John Lyly* iv. 140 Lyly has..a magpie talent..of grabbing the magpie taste for classical motifs. **1974** 'J. LE CARRÉ' *Tinker, Tailor* xv. 153 That whole magpie collection of tattered hotel junk.

c. Used *attrib.* of something with black and white colouring.

1885 KIPLING *Phantom 'Rickshaw* (1889) 9 My eye was arrested by the sight of four *jhampanies* in 'magpie' livery... Was it not enough that the woman was dead.. .and that the black and white servitors re-appeared? **1933** *Daily Mail* 28 Mar. 11 Magpie Millinery. Black and white millinery is popular at all seasons. **1932** *Daily Tel.* 25 Apr. 4/5 Black and white is to continue its long reign this summer. Charming examples of this smart, practical 'magpie' fashion will be found at most inexpensive prices. *Ibid.* (Advt.), Ermine that imparts the fashionable magpie effect. **1940** H. J. MASSINGHAM *Field Fellowship* ii. 33 Moreton Old Hall in Cheshire is the conventional example of the magpie style. *Ibid.*, We find the wood-workmanship of the Avon Valley carried to the utmost point of elaboration and complexity... This is the country of the 'magpie' timber-framing. **1971** J. B. GOOD *Opal Terminal.* 26 *Magpie,* black and white patch formed together.

magslip (mæ·ɡslip). [said to be f. MAG(NETIC *a.* + **slip*(-RING.)] A kind of electric motor designed so that, when it is operating as a receiver, the angular position of its rotor is maintained the same as that of the rotor of a similar motor electrically connected to it and operating as a transmitter; *orig.* applied to a particular design of receiver in which the rotor is a simple magnetic member without windings polarized by a fixed coil and moved by a three-phase stator.

1947 *Electronic Engin.* XIX. 259/1 Small instruments of the magslip or selsyn type, in which shaft position corresponds to the angular datum of the input quantity. **1950** *Engineering* 29 Dec. 557/3 The forces developed by the gyroscopes are..small and therefore have to be amplified. This is achieved by means of a sensitive hydraulic-motor unit developed by the magslips, the signal being transmitted from the gyroscope to the hydraulic unit by a system of 'magslips'. The magslip is, of course, well known as a device for the electrical transmission of angular displacements. **1954** *Jrnl. Brit. Interplanetary Soc.* XIII. 281 Running up the axis of the life compartment and passing through the nose is the coelostat tube topped by a semi-elliptical mirror which can be rotated and followed by a servo magslip motor thus allowing viewing in all directions. **1959** E. G. & A. T. BURDETT *Automatic Control Handbk.* iv. 11 Magslips were developed for use in the Royal Navy for transmission of bearing, range and elevation between observing instruments, calculating tables, searchlights and guns. **1967** W. HEALEY *Princ. Automatic Control* xiv. 263 Some confusion existed over terminology, but military specifications have led to standardization in terms of voltages, frequencies, etc., so that older units such as Selsyns and Magslips are similar in principle but different in size.

magsman. Add: **2.** *Austral.* A story-teller, raconteur.

1944 W. E. HARNEY *Taboo* (ed. 3) 56 We were discussing dreams... It let him go with his complexes and repressions, but I was itching to have my say, for I too am a good 'magsman'. **1965** — *To Ayers Rock & Beyond* v. 45 We would sit around the camp-fire to sing songs or recite poems of the past..and in this way we had the warm school of oral teaching was the starting point with many a bush poet and magsman, such as I, who kept up the yarning into later days. **1967** *Telegraph* (Brisbane) 8 Apr. 4/1 Hardy..became the official yarn-spinning champion of Australia today. He won the magsman's championship in Darwin.

‖**magtig** (ma·χtiχ), *int. S. Afr.* [Afrikaans, shortened form of *allamagtig* ALMIGHTY *a.*] An exclamation of astonishment, awe, etc.

1891 B. MITFORD *Weird of Deadly Hollow* xii. 77 Magtig! It's all through the *vrouw-mensche.* Whoever heard of bringing *vrouw-mensche* out springbok-hunting? **1925** W. PLOMER *Turbott Wolfe* iii. 141 'Magtig!' he said, 'supposing [some dim—] 'Jy?'' **1947** 'J. Speak of Afr.' 1. 53 Magtig! but he's a strong man! **1943** 'H. KNIGHT' *Covenant* (1944) ii. iii. 69 A hundred and fifty! Magtig, that's a lot to lose. **1948** *Sunday Times* (Johannesburg) 17 Oct. 21 It only Oom Piet were a reasonable man. But, magtig, what's the good of talking that? **1973** S. CLOETE *Company with Heart of Gold* 126 Magtig, what is a farm if there is no one to leave it to?

Magyar, *sb.* and *a.* Add: **B.** *adj.* **2.** *Dress-making.* Of or pertaining to a style of blouse, bodice, etc., in which the sleeves are cut in one piece with the main part of the garment. Also *ellipt.* as *sb.*

1911 *Daily Colonist* (Victoria, B.C.) 5 Apr. 24/1 (Advt.), Evening gown of grey chiffon..with low neck and magyar sleeves. **1922** *Home Chat* 13 Apr. 128 Just take away the lace insertion, and you get a quite plain little Magyar of white muslin. **1924** H. J. REEVE *Elem. Dress Pattern-Making* 37 The Empire yoke may be cut Magyar type to the hips. **1935** *Prairie Fr.* (Leeds) Jan. 21 Adapting the Magyar Pattern. **1923** *Daily Mail* 29 June 15 A tight-fitting bodice which buttons down the back, the bolero bodice, and cross-over magyar. **1952** *Woman's Jrnl.* June 73/3 Asta's interpretation is in white fleecy wool, with..magyar sleeves and immensely wide cuffs. **1958** V. CLIFFE *Making your own Clothes* iii. 42 It is not at all difficult, especially if you begin with what is known as a Magyar blouse. **1964** M. BRAY *More Dress Pattern Designing* xiii. 150 (*heading*) The kimono or magyar block. **1968** W. CLARK *Dressmaking Techniques for Trade Students* 33 The magyar is not to be confused with a Dolman sleeve which, being cut in one with the bodice, is closer fitting. **1979** H. STANLEY *Modelling & Flat Cutting for Fashion* vii. 93 Variations of style in category (b) are limited to the Kimono (or Magyar) and to styles with gussets of differing shapes.

mah (mā, *unstressed* mă), var. MY *poss. adj.* Common in written Black English. Cf. *ma* s.v. MY *poss. adj.*

1933 J. M. BREWER in *Publ. Texas Folklore Soc.* XI. 101 Yuh kin sow in yuh half of yuh wants to. **1935** Z. N. HURSTON *Mules & Men* I. i. 31 Nobody don't want to hear you..mah sense to. **1936** C. HIMES *Black on Black* (1973) 168 Mah belly feels lak mah throat wuz cut. **1953** S. A. BROWN in *A. Dunbar's Mecca* II. ..Whar wuz settin' in de do' wid mah pipe. **1966** E. BULLINS 'Theme is Blackness (1973) 10 Now, brothers and mah good sisters, now are we really honest? **1966** M. THELWELL in *A. Chapman New Black Voices* (1972) 241 Ah kno-o-ows Ah's gotta be on mah way ..Ah'm gittin' mah peace in mah heart she would. **1968** *Esquire* Apr. 86/1 'Who are so blessed as to have had bestowed upon them at birth the lifetime gift of soul, these are the most communicative and meaningful sounds'.. the familiar 'mah' instead of 'my'.

Mahabharata (mahabā·ratā). Also **Mahābhārata, Mahabharat.** [Skr., 'the great history of the Bharata dynasty'.] An ancient Hindu epic.

1784 W. HASTINGS *Let.* 4 Oct. in C. Wilkins tr. *Bhăgvăt-Gēētā* (1785) 6 The Măhābhărăt contains the genealogy and general history of the house of Bharat, so called from Bhurrut its founder. *Ibid.* 11 The Măhābhărăt..is said to consist of more than one hundred thousand metrical stanzas, of which he [*sc.* Wilkins] has at this time translated more than a third. **1788** *Asiatick Researches* I. 351 The next in celebrity [to the *Ramayana*], if it be not superior to it in reputation for holiness, is the *Mahabharat* of Vyāsa. **1795** C. WILKINS (*title*) The story of Dooshwanta and Sakoontala extracted from the Mahābhārata, a poem in the Sanskreet language. **1863** M. WILLIAMS *Indian Epic Poetry* 16, I come now to the Mahā-bhārata or Great Bhāratsthi, that is, the great poem which describes the descendants of Bharata. **1894** J. C. OMAN *Gt. Indian Epics* II. i. 91 Of the one hundred thousand verses of the Mahabharata not more than a fourth part is concerned with the main story of the epic. **1901** E. W. HOPKINS *Gt. Epic of India* ii. 63 The later Rāmāyaṇa..unquestionably betrays acquaintance with the Mahābhārata. **1921** E. J. THOMAS *Song of Lord* 22 In Book VI of the Mahābhārata begins the account of the great battle, probably a real event of prehistoric times. **1948** S. RADHAKRISHNAN *Bhagavadgītā* 14 The authorship of the *Gītā* is attributed to Vyāsa, the legendary compiler of the *Mahābhārata.* **1967** SINGER & MASSEY *Indian Dances* II. viii. 88 Plays were written with themes from other sacred epics such as the *Mahabharata,* the *Shiva Purana* and the *Bhagavad Purana.* **1972** T. HOLROYDE *Indian Mus.* ii. 77 Musical instruments had already been mentioned in the Upanishads (800 to 600 B.C.) and the *Mahābhārata* (c. 500 to 400 B.C.). **1972** M. SHEPPARD *Taman Indera* 69 The patrons of the Javanese and Malay shadow plays seem always to have preferred the Mahabharata, with its plethora of gods and heroes. **1973** J. A. B. VAN BUITENEN *Mahābhārata* I. p. xiii, The central story of *The Mahābhārata* takes its matter from the legitimacy of the succession to the kingdom of Kurukshetra in northern India.

mahaila (mahai·la). Also **mahailah, maheila, maheileh, mahela(h), mehala.** [App. f. Arab *safina mahila* ship treated with bitumen; cf. Arab. *mahl(u)* liquid pitch.] A large river sailing-boat used in Iraq.

1904 H. V. GEERE *By Nile & Euphrates* vi. 86 On the Shatt el-Arab many varieties of native craft are to be seen, *dhows, mehalas.* **1916** T. E. LAWRENCE *Home Lett.* (1954) 319 The bellam is the passenger boat..and the Mahaila the cargo boat. Mahailas sit on the water like the scooped rind of a melon-rind. **1922** *Blackw. Mag.* Jan. 70/2 They were..waiting for a mahaila to pass on her way upstream. **1952** E. F. DAVIES *Illyrian Venture* iii. 41 The river was..full of bellams and mahelas with their lateen rig were sailing downstream. **1969** R. MILLAR *Kut* i. 5 Downstream ..improvised a mock fleet from the few available river steamers and dragged mahailas (small river dhows).

Mahal. Add: **4.** The name of a type of coarse-woven carpet made in villages near Arak, Iran. Also *attrib.*

1911 G. G. LEWIS *Pract. Bk. Oriental Rugs* xiii. 201 Mahal is but a trade name. **1913** W. A. HAWLEY *Oriental Rugs* 286 If the field has concentric medallions, the rug may be a..Mahal. **1931** A. U. DILLEY *Oriental Rugs & Carpets* iv. 121 With us Mahal means the lowest grade of Sultanabad weaving. **1967** K. LARSON *Rugs & Carpets of Orient* 104/1 *Mahal,* made in: the districts of Mahallat, Mushkabad and Dulakhoo. **1972** P. L. PHILLIPS tr. *Formenton's Oriental Rugs & Carpets* 150 Mahal carpets are easily recognizable because of their large knots and their soft colours.

Mahamad (mā·hāmād). Also **Ma'amad** and with small initial. [mod.Heb. *ma'amad.*] The body of trustees ruling a Sephardic synagogue; freq. in *pl. gentlemen of the Mahamad.* Also *attrib.*

1831 *Ascamot, or Laws & Regulations Jewish Congregation* v. 37 The institution of the *Mahamad* in this congregation is very ancient... The *Mahamad*..is composed of the four *Parnassim* and a *Gabay,* elected annually. **1893** I. ZANGWILL *Childr. Ghetto* (ed. 3) I. 8 The blue-blooded Dons, 'the gentlemen of the Mahamad', who ruffled it with swords and three-cornered hats... **1904** *Jewish Encycl.* VIII. 259 The mahamad exercised complete control of the congregation a despotic control which degenerated into a sort of police supervision. **1933** *Public Opinion* 27 Aug. 161/1 The gentlemen of the Mahamad had sent a call to members of the Spanish and Portuguese Jews' Congregation of London to assemble at their ancient burial-ground in Mile-end Road. **1964** *New Judaea* May 125/2 The Sephardim were to learn that..the Haham..would brook no trespass by the Mahamad. **1971** *Encycl. Judaica* XI. 638/1 One of the characteristic features of *ma'amad* policy was that on certain days they called upon the *ma'amad* held in power. ..

Maharana (mahā·rā-nā). [Hindi, f. *mahā* great + *rānā* (dial. var. of *rājā*) RAJA(H).] = MAHARAJA(H), *spec.* in the title of the Maharana of Udaipur.

1823 J. MALCOLM *Mem. Cent. India* I. viii. 342 The old chief immediately unbuckled his sword, which..he laid at the feet of the Maha Rana.. **1908** *Imp. Gaz. India* XXIV. 90 The ruling chief..bears the title of Maharana..... **1909** ... **1941** ... **1971** *Femina* (Bombay) 3 Apr. 63/2 A former palace of the Maharana of Udaipur, the Lake Palace Hotel is presently run by the Maharana himself.

Maharanee. Add: Also **maharani.** (Earlier and later examples.)

1855 H. H. WILSON *Gloss. Judicial & Revenue Terms* 318/1 *Mahárání*,..the principal wife of a Rájá, or queen in her own right: applied also to courtesy to Hindu ladies of rank, although not of princely dignity. **1924** J. BUCHAN *Ld. Minto* v. 286 The Maharanee gives her sights to him from behind her purdah. **1938** W. J. COMPTON *Marie Gosh of Bustar* i. 3 The late Chief, Maharani Prafulla Kumari Devi, at her accession in 1922..was the last legitimate survivor of one of the most ancient royal families of northern India. **1973** R. RUSSELL tr. *Ahmad's Shore & Wave* ix. 110 God knows how many maharanis' hearts he had conquered.

Maharashtri (mahā·rā·ʃtri). Also **Mahārāṣṭrī.** [Skr. *Mahārāṣṭrī,* f. *Mahārāṣṭra* Great Kingdom.] The Prakrit language of the Maharashtra region of India, the modern descendant of which is Marathi (MAHRATTI).

1801 H. T. COLEBROOKE in *Asiatick Researches* VII. 227 The Mahárāshtra or Mahrátta is the language of a nation which has..become..the principal... **1880** A. F. R. HOERNLE *Compar. Gram. Gaudian Lang.* p. xvii, There are in reality only two varieties of Prákrit. One includes the Sauraseni and the (so-called) Mahárāshtri. **1900** R. B. JOSHI *Comprehensive Marathi Gram.* 6 Those who cannot be derived from the Mahárāshtri through the Apabhramsa form of that Prakrit language. **1960** *Encycl. Brit.* ... XXXII... **1971** *Journ.*... the Mahārāshṭrī... expressed in the Bhāgavata... **1973** ...

Maharashtrian (mahā·rā·ʃtriən). Also **Mahrashtrian.** [f. *Maharashtra* + -IAN.] **A.** *adj.* Of, pertaining to, or characteristic of Maharashtra. **B.** *sb.* A native or inhabitant of Maharashtra. Cf. *MAHARATTI.

1957 *Ann. Reg.* 1956 10 Maharashtrian demonstrations took place outside Parliament House in Delhi. **1958** A. TOYNBEE *East to West* xxxii. 99 But how is one to decide whether Bombay city is to be Maharashtrian while its capital is Gujerati? **1965** P. ROBINSON *Pakistani Agent* i. 12 Maharashtra..a slightly built Maharashtrian in his late twenties. *Ibid.* viii. 106 The brasiers, which were in the Mahrashtrian style. **1969** *Capital* (Calcutta) 27 Feb. 358/1 In Bombay, the Shiv Sena, in its attempt to drive out all non-Maharashtrians from the city, committed atrocities on them. **1970** *Hindustan Times Weekly Rev.* (New Delhi) Nov. 12/1 He is an 'invader' in Bombay just as the Maharashtrian is an 'invader' in Ahmedabad. **1971** *Hindustan Times Weekly Rev.* (New Delhi) 4 Apr. p. iv/5 A Maharashtrian girl married to a Sarasvat..felt that 'marriage is mostly give and take'. **1971** *Femina* (Bombay) 30 Apr. 7/1, I was sitting and chatting with a couple of young Maharashtrians.

maharee, mahari, varr. *MEHARI.

Maharishi (mahā·riʃi). Also **Maharshi.** [Skr. *mahārṣi,* f. *mahā* great + *ṛṣi* inspired sage.] The title of a Hindu sage or holy man. Also in more general use: (the title of) a popular leader of spiritual thought or opinion. Cf. *GURU.

1785 C. WILKINS tr. *Bhăgvăt-Gēētā* 144 *Măhărshees,* great saints, of whom there are reckoned seven, who were at the creation produced from the mind of Brahmā. **1810** E. MOOR *Hindu Pantheon* 94 The terms *Devarshi, Rajarshi, Maharshi,* are nearly synonimous with *Rishi*; meaning saint, deified saint, great saint, or great sage. **1855** H. H. WILSON *Gloss. Judicial & Revenue Terms* 318/1 Maharshi, ...**1966** C. F. LUTES in Mahesh Yogi *Sci. of Being* (rev. ed.) 13 It was in Madras, in 1958, that Maharishi founded the Spiritual Regeneration Movement. **1967** *Guardian* 1 Jan. 8/1 The Dial Press..have published as 'A Brahmin Book' McLuhan *hot & cool,* the new 'critical symposium' on Maharishi McLuhan. **1970** R. PLATT *Pushbutton Butterfly* (1972) i. 107 Testimonials of faith and gratitude to Guru Maharishi Viparina. **1972** *Guardian* 1 Nov. 10/6 The Beatles..turned to the Maharishi, but he let them down.

Mahatma. Add: **1.** (Further examples.)

1855 *Daily News* 14 Feb. 5/2 Teacups are found by Mahatmas where no teacups should have been, unless they were either unaccountably lost or surreptitiously introduced. **1909** (see *MAHAT* a.). **1911** A. H. NETHERCOT *First Five Lives of Annie Besant* v. 295 (*heading*) The Chela of the Mahatmas.

2. A title indicative of love and respect conferred on revered persons in India; applied freq. to Mohandas Karamchand Gandhi (1869–1948).

1931 *Manch. Guardian* 11 Aug. 8/2 The great soul in the insignificant-looking body is the distinguishing mark of the Mahatma. **1938** C. LEALDE *Gost & Kali* v. 295 His fanatical devotion..deterred the ambition of great names concealed here that he was an intimate friend of them all: of the Mahatma, of Nehru, of Subhas Bose. **1949** H. N. BRAILSFORD in H. S. L. Polak et al. *Mahatma Gandhi* viii. 96 On 9 January, 1915, ..Gandhi landed in Bombay... Soon after..in a public meeting he was given the title of Mahatma. **1949** H. N. BRAILSFORD ... the title of 'Mahatma'.. of which the literal meaning is 'great soul'. It is the custom among Indians to bestow such distinctions on leaders whom they love and admire. c **1949** P. YOGANANDA *Autobiogr. Yogi* xliv. 392 Mr. Desai led us..to the writing-room where, cross-legged, sat Mahatma Gandhi... He never refers to himself as 'Mahatma'. He was none the more humble, and witty, protests about the title. **1973** J. FERGUSON *Politics of Love* iv. 104 This is why Mahatma Gandhi said 'If you Christians rely on soldiers for your safety, you are denying your own doctrine of the Cross.'

Mahayana (mahayā·na). [Skr. *mahāyāna,* f. *mahā* great + *yāna* vehicle.] The 'Great Vehicle', a name given to the more general form of Buddhism practised in northern Asia. Cf. *HINAYANA. Also *attrib.* So **Mahaya·nian** *a.*, **Mahaya·nism, Mahaya·nist** *a.* and *sb.*, **Mahayani·stic** *a.*

1868 [see *HINAYANA]. **1883** MAX MÜLLER *India* iii. 87 The Northern conquerors..seem to have made a kind of compromise with Buddhism, and to have patronized that compromise, or to an amalgamation of Saka legends with Buddhist doctrines, that we owe the so-called Mahâyâna form of Buddhism. **1891** W. W. ROCKHILL *Land of Lamas* 105 A curious preservation of the Mahâyânist doctrine of the *Kâyatraya.* **1907** [see *HINAYANA]. **1907** D. T. SUZUKI *Outl. Mahayana Buddhism* 21 The idea is distinctly Mahayanistic. **1927** A. HUXLEY *Proper Stud.* v. 187 In the first century of our era..the Mahayana or Great Vehicle was created, and Buddhism became an entirely new religion. **1945, 1951** [see *HINAYANA]. **1956** A. TOYNBEE *Historian's Approach to Relig.* i. vi. 89 The Mahayanian Buddhists in China. **1961** J. MASTERS *Road past Mandalay* xii. 162 They worshipped God according to the rites of the Mahayana and Hinayana. **1971** *Sun* (Ceylon) 17 Sept. 5/6 This impact of Japan and Mahayana Buddhism on him opened a further avenue for his visionary zeal. **1973** *Times* 14 Apr. (Nepal Suppl.) p. ii/3 Although every variation of the belief that Buddha Sākyamuni survive, the favourite moulds of Mahāyāna art were the five Celestial or Dhyāni Buddhas and the Bodhisattvas, whose images became the centres of devotional cults.

Mah Jong, mah-jong (mā dʒɒ·ŋ). Also **mah jongg, mah jongh,** etc., and as one word. [Chinese *ma-ch'iao* (Shanghai dial. *-tsiang*) sparrows (f. *ma* hemp + *tsiang* small birds), the name of the game.] An old Chinese game resembling certain card games, introduced into Europe and America in the early 1920s. The 136 or 144 pieces used in the game are known as tiles, and they are divided into five or six suits. The object of the players, usually four in number, is to build a hand containing four sets of three tiles and a pair. Also *attrib.*

1922 R. E. LINDSELL (*title*) Ma-Cheuk or Mah-Jongg. **1923** CHIANG LEE *Mah Jong* 7 Mah Jong, or Mah Tsiong (Sparrows), as it is pronounced in the city of Ning Po where it is supposed to have originated, has long been in vogue in China as a card game for about eight centuries. **1923** *Daily Mail* 23 June 6 There will be..demonstrations of Mah Jong, the wonderful Chinese game which threatens to oust Bridge. **1923** JOYCE *Let.* 6 Nov. (1966) III. 82, I have prepared a nice intricate Mah Jongg puzzle for myself. **1931** *Times Lit. Suppl.* 7 May 365/1 Particularly interesting are the monographs..dealing with China's methods of civil warfare (hireling warriors fixing harmless fusillades in the intervals of their mah jongg). **1963** J. KIRKUP *Tropic Temper* 63 Two old men were playing mah-jong. **1967** *Guardian* 16 Aug. 6/6 Following the traditional 'Cleansing of the Tiles' the North Cotswold Mah-Jongg Canton started another season last night at the Slow-on-the-Wold Hotel. **1971** *Daily Tel.* 17 Aug. p.i/1 ... **1972** E. T. RIESMAN *Conversations in Japan* 116 They described the favorite pastime, mah-jongg, still popular with many. **1973** K. HOWERS *Hong Kong* 246 The interests of these 'unions' centre on..

MAHLERIAN

mahjong. **1972** *Straits Times* (Malaysian ed.) 22 Nov. 21/7 The $10,000 hold-up of mahjong players at the Keong Kee sawmill's recreation club.

Hence **Mah Jong** *v. intr.*, to complete one's hand at the game of Mah Jong.

1923 J. BRAY *How to play Mah Jong* 12 Each player in turn draws a tile and discards one in place of this until some player completes his hand; i.e. 'Mah Jongs'. *Ibid.* 21 None of the following tiles in the hand assist so far in Mah Jonging (completing the hand)..and may be discarded. **1924** M. C. WORK *Mah-Jongg Up-to-Date* I. 32 The Chinese word 'Woo' (meaning peace) has been extensively used by many American writers in the same sense as it is used by Chinese players; viz., to Mah-Jongg, to win the game, to go out. **1933** E. TALBOT *Mah-Jong made Easy* 31 Sometimes a player collects tiles that will give a small mah-jong to oneself and discards those of high value that give another player the opportunity of making a big hand, even if he does not mah-jong. **1966** 'HAN SUYIN' *Mortal Flower* I. 24 Everyone..shopping, tea-ing, mahjonging or donkey-riding.

Mahlerian (māliə·riăn), *a.* and *sb.* [-IAN.] **A.** *adj.* Of, pertaining to, or characteristic of the Austrian composer Gustav Mahler (1860–1911) or his music. **B.** *sb.* An admirer or adherent of Mahler; an exponent of Mahler's music. Also **Ma·hlerish** *a.*, resembling the classical-romantic style of Mahler; **Ma·hlerite** *sb.* = *MAHLERIAN *sb.*

1947 D. NEWLIN *Bruckner, Mahler, Schoenberg* v. 130 The typically Mahlerian direction..sets the stage for a mysterious march-like beginning, likewise typically Mahlerian. **1948** *Penguin Music Mag.* VI. 132, I am no ardent Mahlerite. **1959** *Listener* 27 Aug. 333/2 He [sc. a conductor] was a great Mahlerian. *Ibid.* 10 Dec. 1052/3 Almost a Mahlerish work. **1962** *Times* 2 Feb. 13/2 Mahlerites are likely to owe their allegiance to the prewar gramophone records conducted by Dr. Bruno Walter. **1962** *Listener* 25 Oct. 695/1 We hear an eclectic, youthful style penetrated by what we recognize as the unmistakable and emerging Mahlerian voice. **1963** *Times* 7 Mar. 8/3 Emotionally the music lacks the concentrated intensity of anguish to win it Mahlerian popularity. **1971** *Guardian* 17 Apr. 8/1 Does he then consider that composer's [sc. Shostakovitch's] expiatory fifth symphony better music than his weird Mahlerian fourth? **1974** *Times* 18 Apr. 7/7 Blaukopf has heeded the lesson and his essay has a proper Mahlerian sharpness to it.

mahogany. Add: **3. b.** *colloq.* = BAR *sb.*[1] 28.

1936 N. COLLINS *Trinity Town* i. 18 From the moment Mr. Primrose appeared behind his own mahogany and superseded the barmaid, he dominated everything. **1955** *Punch* 4 May 557/1 Every interval sees twenty orchestral players with their elbows on the mahogany and off-duty singers in full fig on high stools being selectively sweet about their rivals.

7. mahogany birch *U.S.* (examples); **mahogany flat** *slang*, a bed-bug.

1813 H. MUHLENBERG *Catal. Plant.* 88 *Betula lenta*.. soft birch,..black birch,..sweet birch,..or mahogany birch. **1908** N. L. BRITTON *N. Amer. Trees* 256 The cherry birch, or Black birch,..is also called Sweet birch and Mahogany birch. **1864** HOTTEN *Slang Dict.* 176 *Mahogany flat*, a bug. **1942** BERREY & VAN DEN BARK *Amer. Thes. Slang* § 120/60 *Bedbugs*,..flats, mahogany flats. **1952** *New Biol.* XIII. 86 The abdomen [of the bed-bug] when empty is flat (hence the name 'mahogany flat'). **1959** SOUTHWOOD & LESTON *Land & Water Bugs* 188 Bedbug. Also known as 'wall lice', 'mahogany flats' and 'crimson ramblers' these insects have been carried all round the world. **1967** B. J. BANFILL *Pioneer Nurse* v. 61 Until two months ago we had only a log shanty. Somehow the Mahogany Flats took over and we had to burn it.

mahoohoo, var. *MOHOOHOO.

mahorka (māhǫ·ɹkă). Also **makharka, makhorka.** [Russ. *makhórka* shag.] A coarse tobacco smoked in Russia mostly by soldiers and peasants. Also *attrib.*

1902 R. N. BAIN tr. *Tales from Gorky* ix. 263 The chums were going along the high-road smoking *makharka* cigars of their own manufacture. **1949** I. DEUTSCHER *Stalin* i. 22 The circles met in the workers' own overcrowded slum-dwellings and filled the air with the biting smoke of *makhorka*. **1951** KOESTLER *Age of Longing* iv. 68 The tall old man with the dignified stoop, the mahorka smell and the sparkling, close-set eyes. *Ibid.* vi. 94 Old Arin sat down on the foot of the bed, smoking his mahorka cigarettes. **1962** *Listener* 2 Aug. 182/2 The sights and sounds of St Petersburg—..the smell of Mahorka tobacco and of the droshkies. **1964** *Sunday Mail Mag.* (Brisbane) 22 Nov., Villagers in the Caucasus have been smoking a mossy mixture known as mahorka.

Mahsud (mā-sū·d). Also **Mahsood.** A member of one of the principal tribes of Waziristan in north-west Pakistan, noted for their bellicosity. Also as *adj.*

1873 E. BALFOUR *Cycl. India* (ed. 2) V. s.v. *Wazira*, The Wazeeri are divided into three great divisions, or Usman kheyl, Ahmedzye, and Mahsood. **1882** H. B. ROWNEY *Wild Tribes India* II. i. 120 The Wuzeerás... are a bold and ferocious people... The principal tribal divisions among them are: the Máhsuds, Áhmedzyes, Othmánzyes, and Bithunnees. **1895** *Blackw. Mag.* July 79 The Mahsud headmen solemnly bound themselves to renounce looting the Powindah caravans. **1898** H. W. MILLS *Pathan Revolt N.-W. India* (ed. 3) iv. 22 Even after the Mahsuds had attacked Wano, Waziristan was not permanently occupied. **1909** *Daily Chron.* 2 June 1/7 The Mahsuds, who have been causing trouble on the Indian frontier, are said to have secured 10,000 rifles. **1920** *Blackw. Mag.* Oct. 445/1 You take the road to Mahsudland. *Ibid.* 457/2 The half-veiled, unfriendly, grey Mahsud hills. **1924** [see *BRAHUI *sb.* and *a.*]. **1957** R. HILTON *N.-W. Frontier* x. 133 The 20th Punjabis..hustled the mob of Mahsuds out of the camp at the points of their bayonets. **1963** *Listener* 10 Jan. 96/2 A column was out against refractory Mahsuds.

mahwa. Add further variant spellings: **moa, mohua, mohur, mohwa, mowra, mowrah.**

1. Substitute for def.: A large tree, *Madhuca latifolia*, native to India and belonging to the family Sapotaceæ, which bears fleshy, edible flowers and is used as a timber-tree; also, a smaller, related tree, *Madhuca butyracea*, the Indian butter tree, native to Nepal, whose seeds yield an oily, butter-like substance. Also *attrib.* (Further examples.)

1889 G. S. BOULGER *Uses of Plants* III. 139 The Mahwa, besides its saccharine flowers, yields 33 per cent. of a butter from its seeds, used in India as food, and now imported for soap and candle making. **1907** B. M. CROKER *Company's Servant* xii. 108 A sickly tamarind and two frail gold mohur trees. **1920** *Nature* 1 Apr. 147/1 The possibility of utilising the mowra flowers of India for the purpose [of manufacturing alcohol] is discussed... They are used by the natives as a foodstuff, and especially for the preparation by fermentation of an alcoholic liquor called daru or mohwa spirit. **1922** W. SCHLICH *Man. Forestry* (ed. 4) I. II. 134 Mention should be made of the flowers of..the 'mohwa' tree, the corolla of which is eaten fresh or dried, or distilled into intoxicating liquor. **1934** 'G. ORWELL' *Burmese Days* ii. 20 Gold mohur trees like vast umbrellas of bloodred bloom. **1951** *Dict. Gardening* (R. Hort. Soc.) III. 1229/2 *M[adhuca] latifolia*. Mahwah or Moa... The fl[owers] are edible and the wood valuable. **1954** J. MASTERS *Bhowani Junction* xxvii. 234 There was a gold mohur tree in bloom, its branches leaning out. **1964** R. PERRY *World of Tiger* x. 142 The oak-like mohua tree..carries a tremendous weight of flowers—a hundredweight or more on a single tree. **1971** *Illustr. Weekly India* 18 Apr. 35/1 Toddy and mohua liquor are also sometimes drunk.

3. mowrah meal, the dried and powdered residue of the seeds of the mahwa tree, after the oil has been extracted, used as a pesticide to kill worms in turf.

1939 R. B. DAWSON *Pract. Lawn Craft* xxv. 180 Mowrah meal..does not maintain the turf in worm-free condition for longer than about two years. **1948** I. G. LEWIS *Turf* xii. 121 The residue of an Indian bean after the oil has been extracted for commercial purposes, Mowrah Meal is a material which is entirely safe to use. **1973** R. GROUNDS *Perfect Lawn* viii. 73 Mowrah meal applied in summer or autumn..will effectively reduce the worm population.

Mahzor, var. *MACHZOR.

mai (mai). Shortened form of MATAI.

1831 G. BENNETT in *London Med. Gaz.* 12 Nov. 184/2 *Podocarpus Species*... This tree, the Mai or Matai of the natives of New Zealand, is an unpublished species of Podocarpus. **1845** [see *HINAU]. **1855** R. TAYLOR *Te Ika a Maui* 440 Matai, mai.., a tree with a fine thick top, and leaf much resembling that of the yew. **1874** LINDLEY & MOORE *Treas. Bot.* II. Suppl. 1315/1 Mai (N. Zeal.), *Podocarpus spicata*. **1946** *Jrnl. Polynesian Soc.* LV. 145 Matai, a tree (*Podocarpus spicatus*), black-pine: often shortened to māi.

maid, *sb.*[1] Add: **8.** *maid service.*

1951 *N.Y. Times* 28 Oct. VIII. 19 (Advt.), Shamrock apartment hotel... Maid service. **1968** *Globe & Mail* (Toronto) 17 Feb. 49/1 (Advt.), Prince Carlton Hotel Ltd. Rooms. Maid service, telephones. **1969** 'O. BLEECK' *Brass Go-Between* (1970) vi. 69 In addition to daily maidservice, the Adelphi offered a restaurant and bar. **1971** *Country Life* 23 Dec. 1814/4 [Villa]..Sleep 10, c.h.w., refrigerator, maid service. **1973** H. NIELSEN *Severed Key* v. 60, I wonder if this place provides maid service.

maid, *v.* Add: **1.** (Later examples.)

1936 M. MITCHELL *Gone with Wind* iv. 63 My Prissy been maidin' fo' Miss India fo' a year now. **1958** V. P. JOHNS *Servant's Probl.* i. 11 During the two months I've been maiding for Mr. Atterbury, one or the other of them have been in every day.

3. *trans.* To wait on (a person) as a maid.

1909 R. HICHENS *Bella Donna* xxi. 228, I must learn to maid you. **1929** 'R. OKE' *Frolic Wind* v. 83 It was, of course, certainly untrue that she had a fourth to maid her. **1934** A. CHRISTIE *Murder on Orient Express* III. iv. 219 Susanne..used to look after my clothes and maid me. **1949** C. H. B. KITCHIN *Cornish Fox* xi. 164 What would happen to the Colonel after the wedding?.. Upton wouldn't be at all pleased if Delia continued to 'maid' her father from Southview.

Hence **mai·ding** *vbl. sb.*

1900 [see MAID *v.* 1]. **1921** *Chambers's Jrnl.* 1 Jan. 73/1 If you require 'maiding', you tell the lady of the bureau of your floor, and she supplies you with an attendant. **1955** 'C. H. ROLPH' *Women of Streets* vi. 78 Maiding to a prostitute is a definite job.

maiden. Add: **A.** *sb.* **1. c.** *the answer to a maiden's prayer*, an eligible bachelor. Also *transf.*

1935 *Mademoiselle* Aug. 15 Here, you Freshmen, Seniors, et al, is the answer to a maiden's prayer. **1957** J. FLEMING *Maiden's Prayer* II. 109 You're the answer to a maiden's prayer, dear heart. No need for you to do a stroke of work, you can marry money and live the life of a gentleman. **1971** J. BRUNNER *Honky in Woodpile* xi. 83, I was still in college. Thought he was the greatest.. answer to a maiden's prayer! **1974** A. PRICE *Other Paths* II i. 112 You're the answer to a maiden's prayer.

9. b. A strawberry plant bearing its first crop. Also *attrib.*

1928 *Daily Express* 28 May 5/3 The 'runners' are laid from the 'maidens' or last year's [strawberry-]beds. **1974** *Times* 13 July 11/3 Another advantage of growing only maiden strawberries is that it gives us one more crop to work into our programme of crop rotation... I have now decided..to grow only maidens—that is, to take one crop off a plant and then discard it.

c. Short for *maiden bell* (see sense *B. 4 f).

1909 *Daily Chron.* 1 Oct. 7/5 The High Wycombe 'tenor'..thus issues proudly from the Whitechapel Foundry a 'maiden'.

10. a. *maiden-catching, -eyed, -folded, -furled, -hued* adjs.; **maiden-thought** *poet.*, Keats's term for the stage of human development after 'the infant or thoughtless Chamber', one of innocent, untarnished hope.

1957 AUDEN & KALLMAN *Magic Flute* I. i. 28 Had I a maiden-catching net, Fair maids by dozens I should daily get. **1930** J. MASEFIELD *Wanderer of Liverpool* 24 The maiden-eyed morning. **1916** D. H. LAWRENCE *Amores* 100 Then lets her black hair loose, the darkness fall About her from her maiden-folded bands. **1876** G. M. HOPKINS *Wreck of Deutschland* xxxiv, in *Poems* (1967) 62 The heaven-flung, heart-fleshed, maiden-furled Miracle-in-Mary-of-flame. **1913** E. F. BENSON *Thorley Weir* i. 21 The dog-rose spread its maiden-hued face skywards. **1818** KEATS *Let.* 3 May (1958) I. 281 We no sooner get into the second Chamber, which I shall call the Chamber of Maiden-Thought, than we become intoxicated with the light and the atmosphere. **1954** L. MACNEICE *Autumn Sequel* 22 The Customs Office of Maiden-thought.

b. maiden's wreath, a perennial herb of the genus *Francoa*, bearing pink or white flowers in spikes or racemes; cf. *bridal wreath* (*BRIDAL *a.* 2 c).

1893 W. ROBINSON *Eng. Flower Garden* (ed. 3) 419/1 Francoa (Maiden's Wreath).—Chilian plants of the Saxifrage family... They are rather tender. **1908** G. JEKYLL in *Colour Flower Garden* xiii. 116 Maiden's Wreath (*Francoa ramosa*) is a plant for many uses. The foliage, though sparing in quantity, is distinct and handsome. The long flower-stems are flung out with a kind of determination of character. **1952** A. G. L. HELLYER *Sanders' Encycl. Gardening* (ed. 22) 198 Francoa (Maiden's Wreath; Bridal Wreath).

B. *adj.* **2.** *maiden name* (earlier and later examples).

1689 S. SEWALL *Diary* (1878) I. 305 Visited Cousin Powers, and Cous. Lapworth, whose maiden name was Ann Lee. **1862** BORROW *Wild Wales* III. v. 41, I asked her her maiden name. **1922** JOYCE *Ulysses* 366 Her maiden name was Jemima Brown And she lived with her mother in Irishtown.

4. b. (Earlier *Cricket* example.)

1851 J. PYCROFT *Cricket Field* iv. 58 In point of style the old players did not play the steady game with maiden overs as at present.

f. In Bell-founding: (see quots.).

1901 H. E. BULWER *Gloss. Technical Terms Bells* 2 *Maiden bell*, a bell that requires no tuning after it comes from the mould. **1910** *Encycl. Brit.* III. 688/1 The metal is then boiled and run molten into the mould... When extricated it ought to be scarcely touched and should hardly require tuning. This is called its maiden state. **1912** H. B. WALTERS *Church Bells Eng.* ii. 47 Sometimes a whole peal used to be turned out so nearly correct that no tuning was needed; such bells were known as a 'maiden peal'.

5. d. (Earlier example.)

1596 SHAKS. *1 Hen. IV*, v. iv. 132 Full bravely hast thou flesht thy Maiden sword.

maiden aunt (mē¹·d'n ānt). [f. MAIDEN *a.* 1 + AUNT.] An unmarried aunt. Hence **mai:den-au·ntishly** *adv.*, in a manner characteristic of maiden aunts; primly.

1709 PRIOR *Henry & Emma* in *Poems on Several Occasions* 244 The ancient Maiden Aunt. **1711** [see AUNT 1 a]. **1758** [see MAIDEN *sb.* and *a.* B. 1]. **1847** C. M. YONGE *Scenes & Characters* xiii. 158 That worst of plagues, a prying maiden aunt. **1917** T. S. ELIOT *Prufrock* 33 Miss Helen Slingsby was my maiden aunt. **1928** *Observer* 22 July 7 Camberley..was rather like a shocked maiden aunt, who had been forced to look on at something not quite 'nice'. Maiden-auntishly she took up again the knitting of the Franco-German War. **1938** W. S. MAUGHAM *Summing Up* 176, I..ask myself whether in another forty years the bright young things of current letters will appear as jejune as do now their maiden aunts of The Yellow Book. **1975** *Harper's & Queen* May 137/3 As for aged relations, maiden aunts in whose name so much is suffered, Miss Charlton hasn't got any.

maidenhair. Add: **5. b.** A woman's pubic hair.

1928 D. H. LAWRENCE *Lady Chatterley* xv. 265 That's where to put forget-me-nots, in the man-hair, or the maiden-hair. *a* **1930** —— *Last Poems* (1932) 14 The dim blotch of black maidenhair like an indicator, Giving a message to the man.

maidenhead[1]. Add: **1.** Also = HYMEN[2] 1. (Further examples.)

1592 SHAKS. *Rom. & Jul.* I. i. 23, I will bee civill with the Maids, and cut off their heads..the heads of the Maids, or their Maiden-heads. **1867** R. G. LATHAM *Dict. Eng. Lang.*, *Maidenhead*,..Virginity..the hymen, or virginal membrane. **1928** F. W. S. BROWNE tr. *T. H. van de Velde's Ideal Marriage* II. iv. 57 Within this space is the sexual orifice... In maidens this is closed by the *hymen*.. or virginal membrane, popularly called 'maidenhead'.

maiden's blush. Add: **1.** Also *maiden-blush rose.*

1861 C. M. YONGE *Stokesley Secret* ii. 35 The standard maiden-blush rose. **1928** D. H. LAWRENCE *Let.* 14 Nov. (1962) II. 1100 Those maiden-blush roses.

3. Either of two Australian trees with pinkish wood, *Sloanea australis*, of the family Elæocarpaceæ, or *Euroschinus falcatus*, of the family Anacardiaceæ.

1884 A. NILSON *Timber Trees New South Wales* 54 *E[chinocarpus] Australis*.—Maiden's Blush.—A beautiful tree, sometimes attaining a height of 150 feet. **1965** *Austral. Encycl.* I. 174/2 It [sc. *Euroschinus falcatus*] is variously known as pink poplar, ribbonwood, Donnelly's cedar and most generally perhaps as maiden's blush, though this last vernacular is now retained for a very different tree, *Sloanea australis*. *Ibid.* III. 364/2 *S[loanea] australis* (maiden's blush or bush alder) is a small tree with showy flowers, found in the coastal brush forests from Illawarra in southern New South Wales to Queensland.

4. *Austral.* (See quots.)

1941 BAKER *Dict. Austral. Slang* 45 *Maiden's blush*, ginger beer and raspberry. **1961** PARTRIDGE *Dict. Slang* Suppl. 1177/1 *Maiden's blush*, ginger beer and raspberry cordial: Australian. **1966** G. W. TURNER *Eng. Lang. Austral. & N.Z.* vi. 116 *Maiden's blush* (a drink, either of port and lemonade or rum and raspberry).

maiding: see under *MAID *v.*

maidless (mē¹·dlės), *a.* [f. MAID *sb.*[1] + -LESS.] Not having or without a maid-servant.

1892 C. M. YONGE *That Stick* I. xi. 117 My Lady could not well be allowed to go maidless. **1909** *Daily Chron.* 19 Aug. 7/3 The clever housekeeper knows the value of saying to the cook—or family, if maidless—'It is going to be so hot that we'll have cold meals.' **1926** *Public Opinion* 23 Apr. 410/3 The maidless mother is the chief victim of the storm. **1973** *Sci. Amer.* June 10/1 My outside interests are those one acquires in raising four children (a geologist, an astronomer, a mathematician and a violinist) and running a busy, exciting, maidless household.

maid of honour. Add: **2.** (Further examples.)

No longer restricted to Richmond, Surrey.

1942 MRS. BELLOC LOWNDES *Let.* 19 Nov. (1971) 235 There were lobster patties, and queer looking Maid of Honour cakes. **1960** *Good Housek. Cookery Bk.* (rev. ed.) 394/1 Maids of Honour..3 oz. butter, 2 eggs,..1 oz. blanched almonds..puff pastry, [etc.]. **1966** J. DOS PASSOS *Best Times* (1968) i. 11, I had so many little cheesecakes called Maids of Honor that I felt a bit sick.

3. The principal unmarried female attendant on the bride at a wedding; a bridesmaid. Also *fig.*

1895 in *Funk's Stand. Dict.* **1906** M. E. FREEMAN *Light of Soul* 348 Lily asked Maria to be her maid of honor. She planned to be married in church. **1911** H. S. HARRISON *Queed* xxi. 267 A victoria containing two lovely young girls sponsor and maid of honor for South Carolina. **1922** JOYCE *Ulysses* 321 The maids of honour..sisters of the bride, wore very becoming costumes. **1974** 'R. B. DOMINIC' *Epitaph for Lobbyist* x. 90 The maid of honour and the bridesmaids passed by... Betty Jo appeared on the arm of her father..floating in a cloud of white.

‖ **maiko** (məi·ko). [Jap.] A girl who is being trained to become a geisha.

1904 R. J. FARRER *Garden of Asia* xii. 111 They are but Maiko—geisha so young and untried as to be beneath Japanese consideration. *Ibid.* 113 She rises.., flanked by the two Maiko. **1938** BUSH & KAGAMI *Japanalia* 69/1 Young apprentice geisha are called *Maiko* in Kyōto and *Hangyoku* in Tōkyō. **1966** P. S. BUCK *People of Japan* (1968) vii. 93 Young geisha, or *maiko*, usually dance with some gaiety. **1971** *Vogue* Dec. 92/1 An apprentice Geisha is a Maiko—a girl usually between seventeen and twenty.

mail, *sb.*[3] Add: **2.** Also (orig. *U.S.*) without article.

1873 J. H. BEADLE *Undevel. West* xxii. 441, I think this office gives us three times as much mail as that at Salt Lake. **1913** *U.S. Official Postal Guide* July 12 The postage on fourth-class mail may be prepaid by..ordinary postage stamps. **1941** *Men Only* Sept. 12 Forwarding mail is another job that usually falls to the Mess Secretary.

b. For *U.S.* read orig. *U.S.* (Earlier and later examples.)

a **1844** M. C. FIELD in S. F. Smith *Theatr. Apprenticeship* (1846) 204 He walks as if he had the missing mail in his pocket and an extra to issue immediately. **1873** T. B. ALDRICH *Marj. Daw* 163, I go over to K— for my mail. **1941** *Men Only* Sept. 12 Although the addressee has made half a dozen moves since leaving the Mess, he expects his mail to reach him without delay. **1953** *Manch. Guardian Wkly.* 3 Sept. 7/4 Mr. Lattimore..had his personal mail forwarded to the White House.

c. Used as the name of a newspaper.

1789 (*title*) The Evening Mail. **1823** (*title*) Waterford Daily Mail. **1896** (*title*) The Daily Mail. **1896** LLOYD GEORGE *Family Lett.* (1973) 108 There are excellent reports in the Mail & S.W. Daily News of Saturday's meeting at Barry. **1922** JOYCE *Ulysses* 242 Look here Martin, John Wyse Nolan said, overtaking them at the *Mail* office. **1975** *Times* 10 Apr. 17/4 The tremendous financial support that the *Daily Mail* has received..from its readers for its Vietnam Orphans Fund.

4. b. *mail boat* (earlier and later examples), *-boy, carriage, -hack, matter* (later example), *-plane, room* (later examples), *-van, wagon* (earlier examples); *mail-carrying* adj.; **mail-bag** (earlier example); **mail-box** (examples of *b*); **mail-car,** (*c*) *Austral.* and *N.Z.*, a motor vehicle used for the conveyance of the mail (and also sometimes of passengers); **mail-catcher** (earlier example); **mail-contract,** a contract for the conveyance of postal matter; **mail cover** *U.S.*, the monitoring of all mail sent to a specified address; **mail-day,** the day on which mails are dispatched or received; **mail drop** [*DROP *sb.* 17 d], a place where mail may be left to be collected by another person; **mail order** orig. *U.S.*, an order for goods sent to a business house by post; hence as *v. trans.*; also **mail-order firm, house,** one transacting business mainly by post; so *mail-order business, catalogue*; **mail-rider** (earlier and later examples); **mail-road** (earlier example); **mail run** *Austral.* and *N.Z.*, = *mail-route*; **mail-slot** *U.S.*, = *letter-slit*; **mail stage** (later examples); **mail-time,** the time mails take to pass between two places; **mail truck,** a motor vehicle used for the conveyance of the mail.

1812 *Theatrical Inquisitor* I. 273 The majority of readers..ramble through books as post-boys ride through towns..and..can tell you as little of the contents as those who carry the mail-bags can of the letters. **1795** in R. Putnam *Mem.* (1903) 397 It has been suggested to me that the mail boats are much too heavy for pushing with the requisite speed. **1933** L. A. G. STRONG *Sea Wall* 1 He did not even heed the mailboat, as she glided gracefully in..to the harbour. **1872** *Rep. Comm. Patents 1870* (U.S.) II. 751/1 In a mail-box, the arrangement herein shown and described..for the purpose of guiding and holding the mail matter. **1922** M. B. HOUSTON *Witch-Man* xix. 260 She stopped now at the foot of Little Glory to look in the mail-box. Always she brought him his mail now. **1973** 'H. HOWARD' *Highway to Murder* x. 128 One of the mail boxes said Miss Thorpe's flatlet was on the third floor. **1842** in J. S. Buckingham *E. & W. States Amer.* II. 118 [I] saw descending the hill..the mail-boy on his horse at full speed. *a* **1861** T. WINTHROP *John Brent* (1883) xvii. 159 Jake Shamberlain aint a hog, and his mail boys aint of the pork kind. **1874** *Congress. Rec.* 15 Apr. 3099/1 Hitherto seven [officers] were known as mail-boys and the others as mail-messengers. **1907** N. MUNRO *Daft Days* i. 3 The tune of the mail-boy's song. **1942** C. ASTON in *N.Z. New Writing* I. 55 Martin heard the mailcar go past. **1945** C. MANN in Murdoch & Drake-Brockman *Austral. Short Stories* (1951) 261 Sent by the mail-car up to Town. **1947** 'A. P. GASKELL' *Big Game* 48 She had returned from town..in the mailcar. **1860** J. G. HOLLAND *Miss Gilbert's Career* x. 166 The Crampton line of public travel and mail carriage was only one of the many tributaries to the great trunk lines. **1909** *Westm. Gaz.* 1 June 8/3 The various lines of passenger and mail-carrying steamers. **1875** *Chicago Tribune* 18 Sept. 5/3 The Post-Office Department has introduced the use of a 'mail-catcher'. *a* **1861** T. WINTHROP *John Brent* (1883) vi. 50 His ranch is down the valley, towards Pravo. He owns half the United States mail contract. **1965** *N.Y. Times* 24 Feb. 26 When a person is subjected to a mail cover, the Post Office records the name and address of anyone sending mail to him, as well as the postmarking and the class of mail. **1974** *Daily Tel.* 29 Jan. 17/4 The FBI began its investigation of Miss Paton as a result of a 'mail cover' on the New York headquarters of the party. **1855** J. EVANS *Let.* 15 June in G. N. Jones *Florida Plantation Rec.* (1927) 131 People will be there [sc. at the post office] every Mail day. **1926** [see *BACK BLOCKS, BACKBLOCKS *sb. pl.*]. **1933** B. WILLOUGHBY *Alaskans All* 102 Everywhere in the land mail-day is the most longed-for, the most important day the Alaskan knows. **1972** *Time* 17 Apr. 45/1 It is hiring a full-time employment counselor to help them find new jobs if they are fired, and even has a special mail drop to receive anonymous tips. **1973** *Sat. Rev. World* (U.S.) 4 Dec. 46/2 Today's students traveling abroad use American Express offices as a mail drop. **1909** 'O. HENRY' *Roads of Destiny* x. 165 One afternoon Jimmy Valentine..climbed out of the mail-hack. **1906** *Churchman* 10 Nov. 724 All mail matter for the secretary of the convention should be addressed to [etc.]. **1867** *Commercial & Financial Chron.* V. 26/2 Mail and telegraph orders will receive our personal attention. **1875** *Chicago Tribune* 8 July 6/3 Few buyers were present and the 'mail order' business also was light. **1897** *Sears, Roebuck Catal.* 579 There was a time when the consumer paid what was asked... [This] is radically changed,—changed by the Mail Order Business. **1906** S. E. SPARLING *Introd. Business Organiz.* 318 The mail order is based almost exclusively upon circular advertising, and while the mail-order firms employ general publicity, they rely almost exclusively upon circularization in developing and holding the trade. **1916** H. L. WILSON *Somewhere in Red Gap* v. 175 Hetty Daggett..orders this [skirt] by catalogue,..from the mail-order house in Chicago. **1928** *Collier's* 29 Dec. 7/1 Wasn't it my suggestion that marriage be turned over to the mail-order houses? **1930** J. B. PRIESTLEY *Angel Pavement* iv. 167 He read all the advertisements in his newspaper, which specialised on Saturdays in the mail-order business. **1968** 'N. BLAKE' *Private Wound* iv. 64 A brand-new suite of furniture which looked as if it had been mail-ordered out of a catalogue. **1969** *Times* 7 Nov. 14/7 The sophisticated stores—even the better mail order catalogues—want to know what is happening. **1972** *Guardian* 23 Feb. 11/3 Allcraft..mail-order a candle-making kit. **1926** *Daily Colonist* (Victoria, B.C.) 2 July 1/5 Vessel expected to arrive here on July 12—Mr. Wells and Mr. Evans then to leave on mail plane. **1931** C. KELLY *U.S. Postal Policy* vii. 137 At the great airports may be seen the alert, efficient, young eagles who pilot the mail planes. **1940** *Jrnl. R. Aeronaut. Soc.* XLIV. 743 The design of a high-speed mailplane for overnight trans-Atlantic services. **1801** in C. Cist *Cincinnati in 1841* (1841) 177 The mail-rider..from the upper route. **1846** *Knickerbocker* XXVI. 52 The mail-rider sank down apparently through the solid ground with his horse and saddle-bags. **1944** F. CLUNE *Red Heart* 81 The far-western mail-rider who, once a fortnight, jogged over the red-soil plains. **1818** in H. B. Fearon *Sk. Amer.* 430 About three miles from the great mail road to Cincinnati. **1911** *Daily Colonist* (Victoria, B.C.) 7 Apr. 7/2 Came before the police magistrate yesterday to answer to the charge of wrecking the mail room of the steamer, throwing mail bags over board, and generally making things interesting. **1968** *Punch* 21 Feb. 270/3 The Mail Rooms (I believe you call them Dispatch Rooms)..were crowded. **1946** M. TRIST in Murdoch & Drake-Brockman *Austral. Short Stories* (1951) 418 A mail run used to be a mail run in those days. **1961** B. CRUMP *Hang on a Minute* 88 [He was] doing the Whenuaroa mail-run in a flash new truck. **1957** J. KEROUAC *On Road* (1958) 183 He peeked down through her mail-slot. **1821** T. NUTTALL *Jrnl. Trav. Arkansa* i. 35 On the morning of the second of October 1818, I took my departure from Philadelphia in the mail stage. **1834** *Southern Lit. Messenger* I. 181, I took my seat in the mail stage, and travelled three hundred miles without once going to bed. **1912** *Chambers's Jrnl.* Jan. 5/2 The mail-time between that town [sc. Villa Rica] and London will be reduced from thirty days to about eighteen. **1961** L. VAN DER POST *Heart of Hunter* I. vii. 112 Catch the fortnightly mail truck to Ghanzis. **1963** A. LUBBOCK *Austral. Roundabout* 44 Twice a week, the mail-truck from Bourke brought out their stores and fresh vegetables. **1909** *Chambers's Jrnl.* June 343/2 Mail-vans in large numbers..are now being driven by mechanical power. **1959** *Times* 2 Apr. 10/4 A Post Office spokesman said that the mailvan had previously collected three bags from a firm in Bell Street. **1831** *Boston Even. Transcript* 23 June 2/1 We..were carted thence in the mail waggon to Sandwich. **1871** E. EGGLESTON *Hoosier Schoolmaster* (1872) xvii. 135 You can git on the mail-wagon that passes there about five o'clock.

mail, *v.*[4] For *U.S.* read orig. *U.S.* Add later examples.

The usual word in the U.K. is still *post*.

1948 H. T. MOORE in D. H. Lawrence *Lett. to B. Russell* 8 Lawrence is telling Lady Ottoline that Russell, who has mailed him the synopsis of his lectures..still needs to break away. **1968** M. RICHLER in R. Weaver *Canad. Short Stories* 2nd Ser. 170 Once more Mervyn mailed off his novel.

mailabi·lity. *U.S.* [f. MAILABLE *a.*] The quality or fact of being mailable.

1883 *U.S. Official Postal Guide* Jan. 664 Mailability of Doubtful Matter. **1903** *Publishers' Circular* 3 Mar. 275/3 As the card does not bear on the address side the words 'United States of America', its mailability is not affected by my circular of the 16th ult. **1944** *Newsweek* 29 May 72 Manhattan's postmaster, Albert Goldman, doubtful as to the mailability of 'Strange Fruit'. **1968** *U.S. Postal Bull.* LXXXIX. No. 20226, 2/1 (*heading*) Mailability of concealable firearms.

mailed, *a.* **1.** Delete † *Obs.* and add later example.

1922 JOYCE *Ulysses* 305 With his mailed gauntlet he brushed away a furtive tear.

2. *mailed fist*, (a threat of) armed force or superior might.

1897 [in Dict.]. **1898** *19th Cent.* Jan. 164 Japan is a foe who will not be terrified by the mailed fist of Germany. **1898** *Review of Reviews* Mar. 214 Prince Henry of the mailed fist has by this time reached his destination. **1920** M. BEER *Hist. Brit. Socialism* II. III. i. 16 The first rude contact with the mailed fist brought him back to the sober realities of life.

mailer[2]. Add: **4.** *S. Afr.* A person who purchases liquor from a bottle-store and resells it to an illicit liquor dealer or shebeener.

1950 *Cape Times Week-end Mag.* 17 June 5 As soon as the bottlestore opens, the *mailer* is there. He gets his regulation two bottles and takes this to the shebeen. Then he goes to another bottlestore for a further two bottles. And so he goes on the whole day. **1950** [see *DOP *sb.*[2] 2]. **1959** *Cape Argus* 14 Nov. 2/9 When we stopped the delivery the shebeens arranged for mailers to get the liquor for them.

mailing, *vbl. sb.*[2] Add: **mailing list** orig. *U.S.*, a register of addresses to which goods and postal matter may be sent; **mailing shot** *Advertising*, material dispatched to potential customers as part of an advertising campaign.

1909 *Daily Chron.* 12 Oct. 4/5 In the States there are 600,000 farmers on the mailing list. **1928** *Publishers' Weekly* 26 May 2201 It is proposed that the booksellers of the country place in this Clearing House..duplicates of their mailing lists. **1961** *Lebende Sprachen* VI. 70/1 Office furniture, machines and supplies..mailing list. **1968** *Heidelberg News* (Heidelberg Printing Machinery Co.) Sept. 3/3 The primary aim of a mailing shot is to obtain sales leads. So include a reply paid card and only as much information as is needed to provoke interest. **1969** *Times* 5 Nov. 21/2 A gift offer of a beauty pack with a new bathroom cleaner, real shillings back in an off-pack offer for the bleach, mailing shots, and trade press advertising. **1972** *Times* 30 Nov. 23/5 They decided to go ahead with a 10,000 mailing shot, coupled with..advertisements. **1973** *Sat. Rev. Society* (U.S.) May 68/2 Ask to be put on the mailing list for Selected U.S. Government Publications.

‖ **maillot** (mayo). [Fr. (13th c. in Robert, but the undermentioned senses are not recorded in Fr. before the 19th c.), lit. 'swaddling clothes'; prob. alteration of *maillol*, *maille* mesh, mail (see MAIL *sb.*[1]).] **1.** Tights.

Romance etymologists mention, but have not verified the existence of, the Maillot referred to in quot. 1936.

1888 H. JAMES in *Cent. Mag.* Apr. 872 The hungry conjurer, the gymnast whose *maillot* is loose, have something of the glamour of the hero. [**1936** A. HASKELL

Prelude to Ballet 101, 1801..An important development of costume affecting the dance is the invention of tights by Maillot, costumier of the Paris Opera.] **1939** in WEBSTER Add.

2. A tight-fitting, usu. one-piece, swimming costume.

1928 *Daily Express* 30 June 5/3 Dressmakers have prepared for these holiday-makers most delightful bathing outfits, and mannequins walk in maillots, bathing caps and wraps. **1928** *Daily Mail* 9 Aug. 3/4, I have heard of a wonderful eelskin maillot. **1960** 'J. & E. BONETT' *No Grave for Lady* v. 79 Lumina, in a white maillot, was..wincing as her bare feet met the sharp stones. **1974** F. SELWYN *Cracksman on Velvet* II. 80 She was dressed in the latest Parisian *maillot*... It was hardly to be imagined that so fragile a costume would survive a single immersion in sea-water.

3. A jersey, 'top'.

1948 A. WAUGH *Unclouded Summer* iv. 65 At a gala evening where the men would discard their *maillots* for starched linen. *Ibid.* 78 His heart was thudding against the cotton of his *maillot*. **1955** *Times* 29 July 9/4 This is the *maillot jaune*, the yellow jersey which the rider, who is so far ahead on aggregate, is privileged to wear as the emblem of leadership. **1955** D. BARTON *Glorious Life* vi. 74 In a not quite clean maillot and a seersucker skirt.

maimai (məi·məi·). *N.Z.* Also **maemae, mai mai, mimi.** [Alteration of Austral. Aboriginal *mia-mia* MIA-MIA.] A makeshift Maori shelter of sticks, grass, etc. (see quots. 1863, 1873). Hence, in more recent use, a duckshooter's hide or stand.

1863 S. BUTLER *First Year in Canterbury Settlement* v. 72 The few Maories that inhabit this settlement..always go on foot, and we saw several traces of their encampments—little *mimis*, as they are called—a few light sticks thrown together, and covered with grass, affording a sort of half-and-half shelter for a single individual. **1873** J. H. H. ST. JOHN *Pakeha Rambles through Maori Lands* ix. 153 In the days of bush fighting it used to be a common occurrence at the end of a day's march, when the maemae's had been knocked up by the side of a stream, to see three or four of the men gravely set to work with pannikin ..and 'wash' for a prospect. **1963** *Weekly News* (Auckland) 15 May 27 W. Porter operated successfully from a well-made maimai in the shallows of the Waikato River. **1966** J. K. BAXTER *Pig Island Lett.* 43 He needs the maimai's breast of shade.

main, *sb.*[1] **8.** Add to def.: The pl. *mains* is freq. used, esp. *attrib.* and in *Comb.*, in a collective sense: the public supply, esp. the electricity supply.

1929 *Radio Times* 8 Nov. 433/1 Faultless Radio, coupled to an all-mains system of current supply, operating..without mains hum. **1930** *Morning Post* 18 Aug., With any good receiver, costing from about £12 for a battery-operated model, to £30 for a mains-model, several foreign stations may be regularly well received. **1936** *Discovery* July 203/1 It is still safer to switch off the current at the mains. **1959** *Times* 26 Aug. 5/4 A transistor radio receiver will soon be a serious challenge to the mains-driven sound radio receiver. **1962** *Which?* Aug. 261/1 The voltage will probably be fixed—a mains voltage of 200 to 250 volts. **1968** *Listener* 22 Aug. 239/1 At the airport there were lights—not from the mains, but from a generator. **1969** *Soviet Weekly* 13 Sept. 12 You rent a cottage with mains water and a gas cooker for 32 to 38 roubles a month. **1971** 'H. CALVIN' *Poison Chasers* xii. 161 There was an old-fashioned mains radio on the sideboard, and I switched it on. **1975** *Daily Tel.* 13 June 13/4 Villages on high ground at the end of the mains system have been temporarily without supplies, the South West Water Authority announced.

main, *a.* Add: **1. e.** (Later example.)

In context *arch.*

1922 JOYCE *Ulysses* 339 And he answered with a main cry: *Abba! Adonai!*

11. main beam, (*a*) *Building* (see quot. 1940); (*b*) the undipped beam of the headlights of a motor vehicle; **main chancer** [*CHANCER *sb.*], an opportunist, one who has an eye to the main chance; **main course, dish,** one of a number of substantial dishes in a large menu; the principal dish of a meal; also *fig.*; **main crop,** the chief crop, excluding the early and late varieties or sections; also *attrib.* (usu. as one word); **main drag** (see *DRAG *sb.* 3 e); **main frame** *Computers* (see quots. 1964, 1970); **main man** *U.S. slang* (see quot. 1970); **main plane** *Aeronaut.*, a principal supporting surface of an aircraft (as distinguished from a tail plane); also *mainplane*; **main range** *Austral.* and *N.Z.*, the principal ridge of a chain of mountains; **main sequence** *Astr.*, in the Hertzsprung–Russell diagram of stellar magnitude against spectral type or decreasing surface temperature, a continuous band of star types extending from the upper left of the diagram (hot, bright stars) to the lower right (cool, dim stars) to which most of the stars in the neighbourhood of the sun belong; freq. *attrib.* (usu. hyphenated); **main squeeze** *U.S. slang*, an important person; also (with pun on SQUEEZE *sb.* 2 b) a man's principal woman friend (see also quot. 1941).

1940 *Chambers's Techn. Dict.* 522/1 *Main beam*, in floor construction, one of the principal beams transmitting loads direct to the columns. **1964** *Which?* Apr. 47/2 Our drivers particularly liked the headlamps of the *BMC Bluebird* and *Commer* on main beam and also found their dipped beam better than the other caravans. **1967** *Autocar* 28 Dec. 2/2 The little trio of warning lights—red for ignition, green for blinkers and blue for main-beam—are set..into the walnut of the facia. **1940** 'N. BLAKE' *Malice in Wonderland* I. vii. 95 There was a terrific row on the local council... Beauty-snobs versus main-chancers. **1974** *Publishers Weekly* 5 Aug. 57/1 Candid but never mean-spirited, O'Brien comes across as a deeply dedicated party man who is far more than a main-chancer. **1889** A. FILIPPINI *Table* 21 French dinners are generally served in three main courses, viz., *Relevés, Entrées*, and *Rotis*; all the rest are considered side courses. **1936** E. CRAIG *Woman, Wine & Saucepan* i. 17 If the courses are few, choose one wine, red or white, according to what you are serving as the main course. **1967** *Guardian* 22 July 4/5 As low as 5s for a main course (meat balls, potato salad). **1970** 'M. UNDERWOOD' *Silent Liars* II. xii. 134 How many more main courses to come? **1782** MAWE & ABERCROMBIE *Every Man his own Gardener* (ed. 9) 119 Onions or leeks for the main crop should be sown the beginning or middle of this month. **1859** R. THOMPSON *Gardener's Assistant* 249 The main crop of the Long Horn, Altrincham, and other large sorts [of carrots] for winter use, should be sown [etc.]. **1877** S. HIBBERD *Amateur's Kitchen Garden* 49 The second early and main crop sorts [of peas]. **1883** *Culture of Vegetables & Flowers* (Sutton & Sons) 147 Potatoes for main crops should now be got in. **1908** *Daily Chron.* 2 Oct. 3/5 The 'White City' potato..is confidently expected to surpass all the maincrop varieties hitherto produced. **1942** *R.A.F. Jrnl.* 18 Apr. 8 There should be full scale raking down of land and sowing of maincrop Carrots, Onions.. and Lettuce. **1962** *New Scientist* 12 Apr. 31/2 Arran Pilot, an early potato, which develops more rapidly than main-crop varieties. **1971** *Arable Farmer* Feb. 35/2 Maris Piper. Early maincrop. **1957** A. MACNAB *Bulls of Iberia* viii. 81 Three modes: the high spectacular curtain-raisers, the low dominating 'benders', and the main dish or *natural* passes with the breast pass as their natural complement. **1972** J. BALL *Five Pieces Jade* x. 118 A small, intimate meal was waiting in basic Japanese style... There was no evident main dish as in Western dining. **1964** *Gloss. Data Processing & Communication Terms* (Honeywell Inc.) 27/1 *Main frame*, (1) the central processor of the computer system. It contains the main storage, arithmetic unit and special register groups... (2) All that portion of a computer exclusive of the input, output, peripheral and in some instances, storage units. **1970** A. CHANDOR et al. *Dict. Computers* 245 *Main frame*. Originally implied the main framework of a central processing unit on which the arithmetic unit and associated logic circuits were mounted, but now used colloquially to refer to the central processor itself. **1973** *Ann. N.Y. Acad. Sci.* CCXI. 282 The development of small main-frame computers called 'minicomputers'. *Ibid.* 283 Their chief advantages are small size, low price—as low as $2500 for a main frame. **1974** *Sci. Amer.* Apr. 79/1 The laboratory station mainframe has the essentials built-in (power supply, logic state indicators and programmers, and pulse sources to provide active stimulus for the student's circuits). **1967** I. A. BARAKA in W. King *Black Short Story Anthol.* (1972) 126 'Hey, man, I saw that d' fagit Bobby Hutchens down in the lobby with a real D.C. queer.'.. 'Hey man you cats better cool it... You talkin' about Ray's main man. You dig?' **1970** C. MAJOR *Dict. Afro-Amer. Slang* 79 *Main man*, favorite male friend; one's hero. **1910** R. FERRIS *How it Flies* xx. 464 *Main plane*, the principal supporting surface of an aeroplane. In the biplane, or the multiplane type, it denotes the lowest surface, unless some other is decidedly larger. **1913** A. E. BERRIMAN *Aviation* iii. 21 The presence of the two main planes as the distinguishing feature of biplanes. **1946** *Happy Landings* (Air Ministry) July 5/1 The aircraft.. was seen..minus the port outer mainplane and engine. **1973** *Nature* 14 Sept. 95/1 A further important observation was of a 20 cm ball which appeared at a height of about 50 cm over the trailing edge of the mainplane of an aircraft in flight. **1888** 'R. BOLDREWOOD' *Robbery under Arms* II. iii. 52, I say,..we haven't made any mistake—crossed over the main range and got back to the coast, have we? **1950** *N.Z. Jrnl. Agric.* Jan. 19/1 The barren and unproductive land consists of bare mountain tops, and native bush areas on the slopes of the main ranges. **1971** *N.Z. Listener* 19 Apr. 56/4 A *main range* is the major backbone of any individual group of mountains. [**1926** A. S. EDDINGTON *Internal Constitution of Stars* vii. 151 The three stars belong to what is now called the 'main series' running from types *O* and *B* down the dwarf series to type *M*.] **1927** P. DOIG *Outl. Stellar Astron.* I. i. 10 The dwarf branch is now more frequently referred to as the 'main sequence', a name due to Eddington. **1962** F. I. ORDWAY et al. *Basic Astronautics* vi. 284 A new star not yet hot enough to initiate thermonuclear reactions obtains its luminosity from gravitational contraction. Later, as the star heats up, thermonuclear reactions commence; and it joins the main sequence. **1966** *McGraw-Hill Encycl. Sci. & Technol.* XIII. 106/2 Stars that do not belong to the main sequence fall into two main groups: red giants..and white dwarfs... Modern theoretical work indicates that red giants and white dwarfs began their stellar careers as normal main-sequence stars and owe their distinctive properties to evolutionary changes incident on the exhaustion of nuclear energy sources. **1896** ADE *Artie* vii. 63, I went in and asked the main squeeze o' the works how much the sacque meant to him. **1927** D. HAMMETT in *Black Mask* Feb. 12/2 Vance seems to be the main squeeze. **1941** J. SMILEY *Hash House Lingo* 36 *Main squeeze*, hostess. **1970** H. E. ROBERTS *Third Ear* 9/2 *Main squeeze*, a man's closest woman friend.

main (mē¹n), *v.* [f. *main road* (MAIN *a.* 8 b).] **1.** *trans.* To convert into a main road.

1927 *Daily Tel.* 7 June 11/3 The widening and 'maining' of the road leading to the Royal Hotel corner. **1930** *Jrnl. Town Planning Inst.* XVI. 102/1 It is reported that the process of 'maining' roads has been steady and continuous. **1969** A. BIRD *Roads & Vehicles* iii. 40 The new county councils were made responsible for maintaining all 'main' roads in their counties, though it was left to them to determine which should be 'mained'.

2. *slang.* To insert (heroin or a similar drug) into a vein; to mainline.

1970 *Time* 16 Mar. 17 All my friends were on heroin. I snorted a couple of times, skinned a lot, and after that I mained it. **1972** J. BROWN *Chancer* v. 69 The bastard, he mained me. I said to skin it, but he mained it. First time. **1973** *Daily Mail* 3 Apr. 19/4 *Maining*, injecting straight into the vein.

Maine (mē¹n), *sb.*[2] *U.S.* **a.** The name of the State of *Maine* used in *Maine* (*liquor*) *law*, a law forbidding the manufacture or sale of intoxicating liquors; hence applied to similar laws (see quot. 1897).

1852 *Lantern* (N.Y.) I. 119/2 Does the Maine Law, in abolishing the use of liquors, include the Cotton-Gin? **1853** *Illustr. London News* 26 Nov. 438/1 The Earl of Harrington approves of a Maine law for England. **1855** *Knickerbocker* XLV. 479 They have the Maine Law down below. **1860** W. L. SARGANT *Robert Owen & his Social Philos.* xix. 216 Nor had he had recourse to a Maine law. **1864** T. H. NICHOLS *40 Yrs. Amer. Life* I. 76 But drunkenness becoming common,..spirits were banished, the apple orchards cut down,..'Maine Laws' were finally passed. **1871** *Scribner's Monthly* I. 673 Its special suggestiveness resides in the fact that it originates with the friends of the Maine law. **1897** *Encycl. Social Reform* 1107/1 A prohibitory law was passed in Maine in 1846, and in 1851 a more stringent one, including the provision for the seizure and destruction of intoxicating liquors (known as the 'Maine Law'..) was enacted... Vermont in 1852, New Hampshire in 1855, and Connecticut in 1854, passed the Maine law. **1960** R. K. WEBB *H. Martineau* xi. 346 She saw the Maine Law..as the exercise of the will of a democratic people.

b. *Maine law man*: a prohibitionist.

1858 *Leisure Hour* 3 June 352/2 Upon the extensive mines..are large numbers of teetotallers and Maine Law men.

mainland. Add: **1. c.** *Canad.* That part of British Columbia on the mainland of Canada, as opposed to Vancouver Island.

1921 *Daily Colonist* (Victoria, B.C.) 26 Mar. 5/4 Vancouver Island has done a great deal to support the Mainland in the past, and is doing so at present. The Mainland is no good to us.

d. The continent of Australia, as opposed to Tasmania.

1934 T. WOOD *Cobbers* xiii. 164 They are tied to Australia—'the mainland', they call it... 'The mainland' is an object of suspicion, envy, and dislike. **1944** *Living off Land* v. 102 So far as malaria is concerned, there's not much of it on the mainland. **1958** C. KOCH *Boys in Island* 15 He lived in an island. At six years old he knew about that. He heard it at school, and knew about it from Uncle Charlie's talk about the Mainland, Australia. **1968** G. DUTTON *Andy* 200 Soon as I blew from the boob, I headed down for the Old Tassy... I'll get over on the Mainland soon, or up the islands.

e. *N.Z. colloq.* The South Island.

1949 *Journeys* XXXIV. 56 I'm sure South Islanders are right in claiming that they live on the 'mainland'. **1965** *Weekly News* (Auckland, N.Z.) 10 Feb. 3/1 Greater numbers of adventurous North Islanders than ever before have crossed the seas this summer to the 'mainland'.

2. b. Special combs., as **mainland China,** the People's Republic of China, as opposed to Taiwan; hence *mainland Chinese.*

1967 *Guardian* 14 July 5/3 She lived in Hongkong and.. visited Japan, Korea, and what is known as 'Mainland China'. **1972** *Korea Herald* 17 Nov. 2/2 (*caption*) Chiao Kuanhua, first vice foreign minister of mainland China is greeted Tuesday by David Scotfox on behalf of Alec Douglas-Home at London airport. **1973** J. GOODFIELD *Courier to Peking* iii. 42 Teh Chang..went to mainland China... He was from Hong Kong. **1975** *Times* 3 Mar. 14/5 Mainland China is about to fight back against..the degenerate Hong Kong cult of violence, *Kung fu.* **1966** C. ISRAEL *Hostages* 38 World War III won't begin..until everyone has a chance to see what happens when the Mainland Chinese take their seat in the General Assembly. **1972** W. GARNER *Ditto, Brother Rat!* xxi. 159 Chinese, right? Mainland Chinese, too, I'd say.

mainlander (further examples).

1941 BAKER *Dict. Austral. Slang* 45 *Mainlander*, a person living on the Aust. continent, in contrast to one living in Tasmania or an island dependency. **1966** *Tribune* (Palmerston North, N.Z.) 1 Apr. 8/7 It's great what a pause in the wrong place does to the meaning of a phrase. Judging by a news announcer on Tv. the other night we can now lay claim to being Mainlanders. The news item doesn't matter but in the script were the words 'Palmerston, North Otago'. **1969** *Times* (Taiwan Suppl.) 9 Dec. p. iii/2 Taiwan's population today is 13.6m.—10.1m. native-born Taiwanese and 3.5m. mainlanders. **1971** *Guardian* 5 Nov. 21/3 If the mainlanders do not deliver on the current promises, the Taiwanese will be further disillusioned.

ma-in-law, colloq. abbrev. of MOTHER-IN-LAW.

1961 in PARTRIDGE *Dict. Slang* Suppl. **1971** R. RENDELL *One Across* v. 45 We've got some old tab coming here. ..Pal of my ma-in-law's. **1971** J. TYNDALL *Death in Lebanon* x. 181 So Ma-in-law has been quick to take her revenge. **1974** *People's Jrnl.* (Inverness & Northern Counties ed.) 29 June 4/2 My ma-in-law used to say.. 'Fierce is the light which beats upon the throne.'

main line. [f. MAIN *a.* + LINE *sb.*[2] 26 b.] **1.** The principal line of a railway.

1841 [see LINE *sb.*[2] 26 b]. **1865** [see MAIN *a.* 8 b]. **1880** *Harper's Mag.* July 196/2 The main line will soon have reached the Rio Grande Valley. **1959** *Chambers's Encycl.* XI. 488/2 The first lengthy main line, the Grand Junction, was opened in 1837. **1970** O. S. NOCK *Rail, Steam & Speed* i. 17 A 'Gladstone' gliding effortlessly down one of the 1 in 264 banks of the Brighton main line.

b. *transf.* A principal route, connection, conduit, family, etc. *U.S.*

1845 *Hunt's Merchant's Mag.* XIII. 127 The main line of state works of Pennsylvania extends from the city of Philadelphia..to the city of Pittsburg. **1925** *Amer. Speech* I. 136/1 The cable that hauled the logs was called 'the main line'. **1945** S. J. PERELMAN *Crazy like Fox* 124 Why not follow the example of glamorous Mrs Barney Kessler, socially prominent matron of the Main Line. **1952** *Sunday Times* 3 Feb. 5/4 'Main line'..in America..is widely applied to important roads or conduits or arteries of one kind or another.

c. A large or principal vein, into which drugs can readily be injected; hence, an intravenous injection of drugs; the act or habit of making such an injection. *slang* (orig. *U.S.*).

1933 *Amer. Speech* VIII. II. 27/2 The *main line* is the blood vessel into which the injection is made. **1938** *Amer. Speech* XIII. 182/2 From my dropper I'll shake the dust, From my spike I'll scrape the rust, And my old main-line I'll bust. **1953** W. BURROUGHS *Junkie* (1972) iii. 33, I began shooting in the main line to save stuff and because the immediate kick was better. **1959** 'E. MCBAIN' *Pusher* iv. 36, I gave him a snort... He got on mainline a couple of weeks later. **1968** 'J. WELCOME' *Hell is where you find It* xiv. 165 What about the purple hearts? Gone on main line yet? **1970** *Sunday Times* 8 Mar. 1/2 She was taking big doses of heroin..by mainline-injection into a vein—every four hours.

2. *attrib.* or quasi-*adj.* (often with hyphen).

a. Of or pertaining to the main line of a railway.

1879 E. J. SIMMONS *Mem. Station Master* (1974) viii. 124, I couldn't take charge of this main-line station if you gave me all the world. **1884** R. PIKE *Railway Adventures & Anecdotes* 93 A gentleman..got into a first-class carriage ..with the intention of proceeding home by one of the main line down trains. **1926** *Daily Express* 11 May 1/1 Great improvements in the main line train services continue to be made throughout the country. **1958** *New Statesman* 11 Oct. 484/2 Wuhsi..apparently sees a foreigner rarely although it is a large main-line city. **1969** M. GILBERT *Etruscan Net* I. i. 13 At main line stations they have a nose for these things.

b. *transf.* and *fig.* Principal; occupying a principal or important position; also, 'middle-of-the-road', 'mainstream'.

1941 BLUNDEN *Thomas Hardy* iii. 42 Here he wrote one of his main-line novels. **1941** *Time* 20 Jan. 77 High life in Philadelphia's Main Line society. **1958** *Listener* 13 Nov. 775/1 Asimov's two books..in addition to being main-line science fiction, are also 'whodunits' in the classic tradition. **1963** *Guardian* 29 Apr. 9/1 She is from a wealthy main-line Philadelphia family. **1965** P. FREUND *Spymaster* 114 His family, though not Main Line, was acceptable. **1968** *Punch* 11 Sept. 378/3 The enthusiasm of American main-line politicians for TV is said to be waning. **1972** *Times* 15 Apr. 16/4 The 'mainline Christian churches', say these young folk, have sold out to the 'establishment'.

c. Of, pertaining to, or characterized by the intravenous injection of drugs. Also *fig.*

1938 *Amer. Speech* XIII. 187/2 *Main-line shooter*, a vein-shooter. **1944** R. A. MOORE *Textbk. Path.* xlvii. 522 On occasion, as in 'main-line' drug addicts, the fungus may be introduced directly into the blood. **1957** J. KEROUAC *On Road* (1958) 173 A big shot of heroin in the main-line vein. **1969** *Daily Tel.* 29 Jan. 21/7 He was allowed a daily prescription of three grains—18 pills—of heroin. These were dissolved and injected directly into the bloodstream, which was known as the 'mainline' method. **1969** *Punch* 26 Feb. 295/3 Flattery is the mainline drug of the theatre—there is no one connected with that ridiculous institution who does not need his fix. **1972** *Times* 10 Apr. 10/4 As a mainline addict of BBC news..it has taken a week of comparing news on the two channels..to push me into the ranks of the deserters. **1974** G. JENKINS *Bridge of Magpies* iii. 46 Then you'll be like that poor bugger coming off now in the boat. Started smoking grass. Grew the stuff..in a potty in his cottage... He's on to mainline stuff now.

main-line, *v. slang* (orig. *U.S.*). Also (without hyphen) **mainline.** [f. prec.] *trans.* and *intr.* To inject (a drug or drugs) intravenously. So **main-liner; main-lining** *vbl. sb.*

1934 *Detective Fiction Weekly* 21 Apr. 113/2 The addict who shoots stuff into the veins is said to be a gutter, or mainliner. **1938** *Amer. Speech* XIII. 187/2 If you can main-line a cube of that stuff, it is on the house. **1950** R. CHANDLER *Let.* 18 May in *R. Chandler Speaking* (1966) 79, I don't think any writer could think up an expression like 'mainliner' for a narcotic addict who shoots the stuff into a vein. **1951** *N.Y. Times* 27 June 19/1 An intravenous injection of the drug..is called 'mainlining'. **1959** A. BESTIC *Girl Outside* xiii. 213 She took to the needle. She's what they call a mainliner. **1959** 'E. MCBAIN' *Pusher* viii. 77 Snorting?.. Skin pops?.. Larry, Larry, are you mainlining? **1964** B. ELLIS *I came back from Hell* iii. 24 She was only skin-popping in those days (injecting her arms), not mainlining (injecting her veins). **1967** M. M. GLATT et al. *Drug Scene* v. 58, I don't know how much longer I would have lasted before I would have tried to mainline. *Ibid.* 59 Peter was revolted by the sight of someone else 'mainlining hard drugs' yet, whilst under the influence of hashish, he thought that he might go on to try mainlining for a bigger kick. **1970** N. MARSH *When in Rome* v. 126 The big leap. Pothead to main-liner. **1971** *Nature* 5 Nov. 14/3 The serious possibilities were raised of teratogenicity, tachycardia (especially if main-lining ever started), brain damage, [etc.]. **1972** M. PEREIRA *Singing Millionaire* vii. 71 He made himself a fix..and he mainlined it. **1974** D. GRAY *Dead Give Away* ii. 23 What about the ones that can't cope in the rat race?.. The meths drinkers, the mainliners?

mainpast (mē¹·npɑst). *Law. Obs.* exc. *Hist.* [ad. AF. *meynpast*, f. med.L. *manupastus*, f. L. *manū* abl. of *manus* hand + *pāstus*, pa. pple. of *pāscere* to feed (cf. PASTURE *sb.*).] A man's household; a domestic; a dependant.

Anglo-Norman examples are cited in *Mod. Lang. Notes* (1932) June 375–6.

1865 F. M. NICHOLS tr. *Britton* I. I. ii. 13 Let the coroner inquire of whose tithing or whose mainpast such fugitive was. **1891** MAITLAND & BAILDON tr. *Court Baron* (Selden Soc.) 53 Thy son who is thy mainpast entered the lord's garden. **1895** POLLOCK & MAITLAND *Hist. Eng. Law* I. II. iii. 555 The head of a household answers for the appearance in court of the members of his household, his servants, his retainers, those whom his hand feeds, his *manupastus* or *mainpast*. **1909** W. S. HOLDSWORTH *Hist. Eng. Law* III. ii. 295 We can see traces of the older principles under which he was held to be liable in the rule which made him responsible for the doings of his household or 'mainpast'. **1961** R. F. HUNNISETT *Medieval Coroner* iv. 64 They [sc. the coroners] had to inquire in whose tithing or mainpast the outlaw had been and enrol it, in order that it might be amerced at the eyre for his flight.

mainprize, *v.* For '*Obs.*' read '*Obs.* exc. *Hist.*' and add later examples.

1865 S. DOBELL *Life & Lett.* (1878) II. iii. 272 Not Moses only or Elias, But Heaven mainprized, and every standing saint Astonied into marble. **1895** POLLOCK & MAITLAND *Hist. Eng. Law* II. II. ix. 582 If a man was arrested he was usually replevied (*replegiatus*) or mainprised (*manucaptus*), that is to say, he was set free so soon as some sureties (*plegii*) undertook (*manuceperunt*)..his appearance in court. **1904** M. BATESON *Borough Customs* (Selden Soc.) I. 99 If distress be delivered by pledge or mainprise of any one, if he who is..mainprised does not come to justify himself..let his..mainpernours be distrained to produce him.

main stem. [MAIN *a.* 8 b.] The principal stem; also *transf.* and *fig.* in various (chiefly *U.S. slang*) senses, as a main street, main line of a railway, pre-eminent person, etc.

1832 *Amer. Railroad Jrnl.* I. 804/2 The western fork.. connects it with the main stem. **1900** ADE *Fables in Slang* iv. 24 To grow up and be the Main Stem, like Mr. Jeffries. **1900** 'FLYNT' & WALTON *Powers that Prey* x. 250 Investigations that have been begun in 'the main stem'. **1907** J. LONDON *Road* (1914) vii. 218 The kids began 'battering' the 'main-stem'. **1923** D. H. LAWRENCE *Birds, Beasts & Flowers* 42 Each single twig, Each one setting off straight to the sky As if it were the leader, the main-stem, the forerunner. **1928** [see *FOLD *v.*[1] 5]. **1931** [see *DRAG *sb.* 3 e]. **1941** J. SMILEY *Hash House Lingo* 36 *Main stem*, principal street of a city. **1945** E. T. WALLACE *Barington* 62 The hog bounced us around and finally got back on the main stem. **1959** *Times Lit. Suppl.* 6 Nov. p. xxii/1 London audiences were listening to a play called *Broadway*.. typical of drama at that time in and around the Main Stem.

main stream. Also **mainstream, main-stream.** [MAIN *a.* 8 b.] The principal stream or current (of a river, etc.). Also *transf.* and *fig.*, the prevailing direction of opinion, fashion, society, etc.; *spec.* of jazz: see quot. 1960[2]. Also *attrib.* Hence **mai·nstreamer,** a musician, etc., who is in the 'main stream' of his profession.

1667 [see MAIN *a.* 8 b]. **1831** CARLYLE in *Foreign Q. Rev.* VIII. 355 But after Luther's day, the Didactic Tendency again sinks to a lower level; mingles with manifold other tendencies; among which, admitting that it still forms the main stream, it is no longer so pre-eminent, positive, and universal, as properly to characterize the whole. **1865** M. ARNOLD *Ess. in Crit.* 171 Byron and Shelley will be long remembered..for their..Titanic effort to flow in the main stream of modern literature. **1938** F. M. FORD *Let.* Oct. (1965) 302 The very considerable influence that Mr. Pound..exercised on literary mainstreams. **1952** *New World Writing* Apr. 33 Negro artists, in moving out into the mainstream of American culture, should gain a sense of solidarity with both the national and the general world of art. **1957** S. DANCE in S. Traill *Concerning Jazz* 55 Mainstream jazz, typified by musicians like Basie, Ellington..Armstrong. **1957** *Jazz News* Apr. 2/3 Rising like a Pheonix [*sic*] from the dying embers of the British modern jazz world is a musical form that has been tagged 'Mainstream'. **1958** 'E. CRISPIN' *Best SF Three* 9 Mainstream fiction..has been almost uniformly catatonic in its withdrawal from environment. **1960** W. NAYLOR *Silver Birch Anthol.* 7 Ask him a question on a comparatively minor issue and time and time again he will use it as a tributary through which to return to the mainstream of his philosophy. **1960** *Sunday Times* 11 Sept. 37/1 Very broadly you can break up jazz asymmetrically into the big, simple, driving noise of the traditionalists and the smaller, sophisticated, elegiac sound of modern jazz. In between, there is a discernible third group of 'mainstream' enthusiasts. **1961** *New Left Rev.* July–Aug. 42/2 Humph is authentic mainstreamer: he's been influenced by everything in jazz up to the moment. **1966** R. ELLISON in A. Chapman *New Black Voices* (1972) 407 The main stream of American literature is in me, even though I am a Negro, because I possess more of Mark Twain than many white writers do. **1969** *Listener* 26 June 904/3 Gesualdo.. belongs less to the mainstream of music than to one of its smaller and more wayward canals. **1973** C. BONINGTON *Next Horizon* xxii. 297 A climber who had always been on the outside of the mainstream British climbing scene. **1974** *Listener* 10 Jan. 54/2 Dick Taverne..possesses a rigid habit of mind, so far quite alien to mainstream British politics.

Main Street, main street. [MAIN *a.* 8 b.] **a.** The principal street of a town, esp. in the U.S.; freq. used as a proper name.

In quots. 1598–1743 preceded by an article and not a special collocation.

[**1598** FLORIO *Worlde of Wordes* 327/2 *Rióne*, a maine streete, a high way. **1687** S. SEWALL *Diary* (1878) I. 183 At night a great Uproar and Lewd rout in the Main Street. **1698** *Rec. East Hampton, N.Y.* (1887) II. 434 Ten Acres of Land..[bound] on the west with the Maine Street of the said Town. **1717** in *Narragansett Hist. Reg.* (1884–5) III. 279 In or through the Main street, called Hope street, in this town [sc. Bristol, R.I.]. **1741–3** [see MAIN *a.* 8 b].] **1810** F. CUMING *Sk. Tour Western Country* 194 Main street, parallel to Water street, is one hundred feet wide. **1817** S. BROWN *Western Gazetteer* 92 Main street presents to the traveller as much wealth, and more beauty than can be found in most of the Atlantic cities. **1855** *Knickerbocker* XLVI. 328 Louisville is an imposing, wealthy city. Main-street, in its entire extent would do honor to any metropolis in America. **1888** 'R. BOLDREWOOD' *Robbery under Arms* II. iii. 54 Go on down Main Street (the first street in a diggings is always called Main Street). **1889** W. D. HOWELLS *Hazard of New Fortunes* I. I. xi. 106 You know the kind of street Main Street always used to be in our section—half plank-road..and the rest mud-hole. **1892** KIPLING *Lett. of Travel* (1920) 8 Every..thing, is reported, digested, discussed, and rediscussed up and down Main Street [in Vermont]. **1961** L. MUMFORD *City in Hist.* iii. 74 It was rather a Broad Ways..it served as the classic 'Main Street'. **1968** *Michelin Guide N.Y. City* 78 Before the Revolution..Greenwich Street, which then bordered the Hudson and its warehouses, was 'Main Street'. *Ibid.* 82 Mulberry Street is the Italian 'Main Street': colorful shops offer national specialities. **1973** A. BROINOWSKI *Take One Ambassador* vii. 86 You should see this station in rush hour. It's like Main Street Christmas Eve.

b. Used allusively, esp. since the publication in 1920 of Sinclair Lewis's novel *Main Street*, as a symbol of mediocrity, parochialism, or materialism in small-town life. Also *attrib.*

[**1855** *N.Y. Tribune* 31 Dec. 4/4 It has risen to its present position of bloated arrogance and swaggering insolence by the liberal and unstinting patronage it has received from the full purses and free hands of Eastern men in Main street and elsewhere.] **1916** 'B. M. BOWER' *Phantom Herd* i. 5 You'll have to let me weed out some of these Main Street cowboys. **1920** S. LEWIS (*title*) Main Street. **1931** *Times Lit. Suppl.* 9 Apr. 282/3 He dislikes uniformity, mass-control, Main-street and Rotarian ideals. **1948** *Time* 6 Dec. 20/3 Harry Truman has never lost his great respect for Marshall, nor is he unmindful of the prestige and authority Marshall carries on Main Street as well as in Moscow. **1972** B. GARFIELD *Line of Succession* (1974) II. 126 The President..was an amalgam of liberal traditions..and the values of Main Street. **1973** T. H. WHITE *Making of President 1972* (1974) ix. 231 From faculty club to student union, from bar to parlor, from Wall Street to Main Street, all wanted out of Vietnam.

Hence **Main Streeter, Mai·nstreeter,** a typical inhabitant of a small American town; one who shares the values of Main Street (sense *b). Also **main-street** (*N. Amer.*) *v. intr.*, to campaign in main streets during an electoral campaign; hence **main-streeting** *vbl. sb.*

1934 WEBSTER, Main Streeter. **1945** *Sat. Rev.* (U.S.) 6 Oct. 8/2 His books started some Americans laughing at others and made it possible for people to realize that somebody else was a Main Streeter. **1947** *Time* 27 Jan. 4/3 It has raised the hope of this Mainstreeter from Podunk to its highest ebb since the era of Wendell Willkie's 'One World'. **1966** *Maclean's Mag.* 1 Jan. 31 Though she [sc. Olive Diefenbaker] refuses to speak in public she mainstreets better than The Chief [sc. John Diefenbaker]. **1967** *Canad. Ann. Rev. 1966* 133 Duff Roblin, whose rural mainstreeting, recorded bagpipe accompaniment, and unflagging oratory had won him his fourth personal victory in a row. **1971** *Time* 14 June 20 Boston has witnessed a merry binge of mainstreeting, leafletting and parties with some of the excitement of a mayoral election. **1974** *Globe & Mail* (Toronto) 5 June 8/3 New Democratic Party Leader David Lewis traded polemics with Parti Quebecois supporters in Montreal's working-class St. Jacques district yesterday before doing some mainstreeting not far from the area where he lived as a young immigrant half a century ago.

maintain, *v.* Add: **9. c.** To give a drug to (an individual, esp. a drug addict) in maintenance doses so as to sustain a particular therapeutic effect. Cf. *MAINTENANCE 7 c.

1957 *Jrnl. Amer. Med. Assoc.* 14 Dec. 1970/1 The current discussion..can be reduced to the desirability and feasibility of treating the addict as a total person,..withdrawing drugs.., or of maintaining him on an appropriate amount of drugs if it is determined he cannot be successfully cured of his addiction. **1971** *Nature* 22 Oct. 558/1 The recipient dogs were anaesthetized with intravenous thiopentone and maintained with halothane, nitrous oxide and oxygen through an endotracheal tube. **1972** *Science* 26 May 881/1 In April 1971, a program was inaugurated to maintain at least 20,000 addicts on methadone in New York State alone.

maintained school (see quots.).

1944 in *Parl. Papers 1943–4* (Cmd. 6523) (*title*) Principles of government in maintained secondary schools. **1960** *Where?* iii. 15/1 *Maintained school*, a school maintained by a local education authority, including county, voluntary aided, and voluntary controlled schools. **1966** *Rep. Comm. Inquiry Univ. Oxf.* I. 74 It is also likely that the swing to science has played some part in raising the rate of recruitment from maintained schools. **1973** *Guardian* 25 Apr. 13/4 The 12-man governing board of Sacred Heart [a Catholic maintained school].

maintainability (mēɪn-, mɛntēɪ·nābi·lɪti). [f. MAINTAINABLE *a.*; see *-BILITY.] The quality of being easily maintained; capability of being maintained.

1943 *Sci. Amer.* June 250 Engineering..determines the serviceability and maintainability of new types of equipment. **1971** *New Scientist* 3 June 574/3 Every project should have goals for reliability..and 'maintainability'. **1974** *Daily Colonist* (Victoria, B.C.) 9 Oct. 3/2 It demonstrates the maintainability of the Candu reactor system.

maintenance. Add: **3.** Freq. *attrib.*

1884 *List of Subscribers* (London & Globe Telephone Co.), In case of unsatisfactory service..send complaint in writing..to London and Globe Telephone and Maintenance Company, Limited. **1942** *Aeronautics* May 40/1 The public does not hear very much about the Maintenance Command of the Royal Air Force. **1957** *Ann. Reg.* 1956 67 A prolonged ban on overtime working by the maintenance men. **1958** *Listener* 23 Oct. 650/1 Maintenance engineers at London Airport vote unanimously to continue their strike. **1970** *E. Afr. Standard* (Nairobi) 23 Jan. 18/2 (Advt.), Applicants..should have at least five years' experience, either in a supervisory capacity or as a maintenance engineer in sole charge of a plant. **1974** *Times* 7 Mar. 1/1 Maintenance workers will be allowed into the pits during the weekend.

7. b. (Further examples.) Also *maintenance order*: a court order, in the case of a broken marriage, compelling the husband to pay the wife a regular fixed sum for her maintenance.

1803 G. COLMAN *John Bull* IV. 42 I'll settle a separate maintenance upon ould mother Brulgruddery [*sc.* his wife]. **1866** *Act* 29 & 30 *Vict.* c. 32 § 1 It shall be lawful for the Court to make an order on the husband for payment to the wife during their joint lives of such monthly or weekly sums for her maintenance..as the Court may think reasonable. **1907** *Act* 7 *Edw. VII* c. 12 § 1 The court may ..make an order on the husband for payment to the wife ..for her maintenance. **1920** *Act* 10 & 11 *Geo. V* c. 33 § 10 The expression 'maintenance order' means an order other than an order of affiliation for the periodical payment of sums of money towards the maintenance of the wife or other dependants of the person against whom the order is made. **1960** M. SPARK *Bachelors* vi. 88 But there's usually a question of maintenance orders. I distinctly recall his being described as a bachelor. **1971** *Reader's Digest Family Guide to Law* 265/2 In many cases a wife will be left, after a separation, with insufficient money for her own day-to-day expenses. Her need for maintenance will be even more urgent if she is taking care of any children of the marriage... She can apply to the local magistrates' court or to a divorce court for a maintenance order. **1973** *Times* 14 Mar. 20/4 To qualify for maintenance a wife does not have to ask for a divorce.

c. The action of providing (a person) over a period of time with doses of a drug sufficient to maintain its effect on the body while usu. being less than the dose given initially; usu. *attrib.*, as *maintenance dose*.

1936 STEDMAN *Med. Dict.* (ed. 13) 319/2 *Maintenance dose*, the dose given in a protracted case of illness, to keep the patient under the influence of the drug after this has been attained by the initial dosage. **1963** *Brit. Pharmaceutical Codex* 683 Dose [of propylthiouracil]. Controlling dose, 200 to 600 milligrams daily; maintenance dose, 50 to 200 milligrams daily. **1965** *Jrnl. Amer. Med. Assoc.* 23 Aug. 648/1 Addicts coming to a maintenance program usually fear that physicians will not prescribe enough medication. **1971** *Nature* 29 Jan. 290/2 The committee believes that the British system of maintenance, in which an addict is supplied with daily doses of heroin, and the methadone maintenance treatment,..both have their advantages and drawbacks.

‖ **maintien** (mañtiañ). [Fr.] Bearing, deportment.

1849 GEO. ELIOT *Let.* 4 Oct. (1954) I. 313 Those happy souls..who do really effect much good, simply by their calm and even *maintien*. **1889** G. B. SHAW *London Music 1888–89* (1937) 73 Josabeth had the *maintien* of the French stage in a degree that would have enraptured A. B. Walkley.

Maioli (māyǫ·li). The name of Thomas Mahieu (fl. 1549–72), latinized as *Maiolus*, French book-collector and secretary to Catherine de Médicis, used *attrib.* to designate a French style of book-binding with elaborate gold tooling, used for some of the books in his library. Also *absol.*

[**1837** 'J. A. ARNETT' *Inquiry into Nature & Form of Bks. of Ancients* vi. 142 Another patron, of the name of Maioli, is well known, from his bindings, though of his personal history no traces are left.] **1890** *Catal. Exhib. Recent Bk.-Bindings 1860–9* (Grolier Club) 37 [Book] Brown morocco..Sides decorated with a Maioli design. **1894** W. S. BRASSINGTON *Hist. Art of Bookbinding* xiii. 178 (*caption*) Maioli binding, Italian, early sixteenth century. **1928** E. P. GOLDSCHMIDT *Gothic & Renaissance Bookbindings* I. 105 Both the Groliers and the Maiolis were made by the same binders and gilders at the same time. **1961** J. P. HARTHAN *Bookbindings* (ed. 2) 12 The 'atelier au trèfle' which supplied bindings to Grolier and Henri II is probably the source also of these fine Maioli bindings.

maiosis, maiotic, obs. varr. *MEIOSIS 3, *MEIOTIC *a.* 2.

‖ **maire**[2] (mɛ̄r). [Fr.; see MAYOR.] A mayor; the chief municipal officer of a French town or of one of the arrondissements or districts of Paris.

1790 H. WALPOLE *Let.* 30 Aug. (1905) XIV. 289 At Marseilles..a Monsieur Cazalet..had been invited to dine with the *maire*! **a 1861** A. H. CLOUGH *Poems* (1862) 139 Their *maire*, he said, could neither write nor read. **1900** C. M. YONGE *Modern Broods* xiii. 121 She gabbled away most eloquently to the Maire, almost as fluently as a born Frenchwoman. **1955** *Times* 22 Aug. 8/6 As a record of continuous office-holding it is overshadowed by that of Edmond Mathis, maire of Ehuns in the Haute-Saône from 1878 to 1953. **1964** C. MACKENZIE *My Life & Times* III. i. 31 Monsieur Boeuf the *maire*, who looked exactly like his name. **1974** S. SHELDON *Other Side of Midnight* ii. 55 Let's get married by some *maire*.

‖ **mairie** (mɛ̄·ri). [Fr., f. *maire* mayor (see prec.).] In France, a town hall; a public building housing the municipal offices of a town or arrondissement and often also serving as the official residence of the mayor.

1864 DICKENS *Mrs. Lirriper's Legacy* i, in *All Year Round* Extra Christmas No., 1 Dec. 6/2 The Major went down to the Mairie. **1896** E. DOWSON *Let.* 27 Apr. (1967) 359 In his official capacity I cannot see the maire..but I call daily at the mairie & at his house. **1925** E., O., & S. SITWELL *Poor Young People* 13 Outside the mairie of this country place. **1935** W. FORTESCUE *Perfume from Provence* 82 The mysteries in the Mairie having been performed, the procession..went into the church, where the Marriage Mass was celebrated. **1966** N. FREELING *King of Rainy Country* 96 A real French town hall..with grandiose pillars outside to support the dignity of the Republic. **1974** *Times* 16 Apr. 5/1, I found M. Jarrot at the *mairie* of Montceau-les-Mines.

maison. Delete *Sc. Obs.* and add later examples. Now usu. in the sense of a business (esp. a fashion) house or firm.

1922 JOYCE *Ulysses* 165 He went on by la Maison Claire. **1932** 'E. M. DELAFIELD' *Thank Heaven Fasting* I. ii. 32 The tall, yellow-headed assistant from the Maison Leroy in Sloane Street..twisted the front into the mass of her hair. **1935** A. CHRISTIE *Death in Clouds* xiii. 138 I'll easily get what I want from Henri's or the Maison Richet. **1966** J. S. COX *Illustr. Dict. Hairdressing* 94 *Maison*... A mode of address used by many high-class hairdressers of the Victorian and Edwardian periods... The intention was to convey to the customer that the firm had a knowledge of French hair styles... An example was *Maison Stephens et Cie.*, Queen's Road, Clifton. **1968** *Listener* 29 Aug. 280/2 The *spécialité* of this particular *maison* is the primal bourgeois myth, the Dick Whittington legend brought up to date. **1971** C. BRAHMS *Sound of Murder* ix. 112 In gold letters on a faded label were the words 'Maison Petronella'. **1971** D. AYERST *Guardian* xxiii. 330 Dresses by 'Maison N'Importe' and all the rest of it.

2. *Comb.* **maison close** [lit. 'closed house'], a brothel; **maison de couture,** a fashion house; **maison de passe** [lit. 'house of passage'], a brothel (see quot. 1967); **maison de santé** [lit. 'house of health'], a nursing home; also, *euphem.*, a home for the mentally sick; **maison tolérée** [lit. 'tolerated house'], a licensed brothel.

1939 E. AMBLER *Mask of Dimitrios* viii. 126 Your *maison close* must have proved disappointing. The inevitable Armenian girls, of course. **1950** *Landfall* IV. 174 The squalid harlot in the *maison close*. **1961** *Guardian* 10 Feb. 9/4 Toulouse-Lautrec..became..an habitué of the *Maisons Closes* where the girls regarded him as a sort of ..mascot. **1927** E. HEMINGWAY in *Scribner's Mag.* Apr. 359/2 The American lady had bought her own clothes for twenty years now from the same maison de couture [etc.] in the Rue Saint Honoré. **1933** A. CHRISTIE *Lord Edgware Dies* xii. 107 The name of Lao Tse..would suggest to her a prize Pekingese dog, the name of a famous *maison de couture*. **1964** A. ADBURGHAM *Shops & Shopping* xxi. 247 Paul Poiret..described..a visit paid to his *maison de couture* by Margot Asquith. **1970** P. J. PETERSEN *Baby, it's Cold Inside* 53 The rigors of squiring her through a score of *maisons de couture*, jewelers', and millinery shops were..unimaginable. **1960** B. MARSHALL *Divided Lady* xxix. 99 The hotel we drew up at eventually reminded me of a *maison-de-passe* I once knew in Barcelona. **1967** L. DEIGHTON *Expensive Place to Die* ix. 71 It's not a brothel... It's a *maison de passe*..a place that people go to when they already have a partner. **1841** POE *Murders in Rue Morgue* in *Tales* (1845) 142 Some raving maniac escaped from a neighboring *Maison de Santé*. **1859** *Times* 5 Feb. 5/6 (Advt.), Maison de Santé, or Residence for Invalids. **1910** W. J. LOCKE *Simon* xxiv. 337 He seemed to be happy enough at the maison de santé. **1940** JOYCE *Let.* 3 Nov. (1966) III. 496, I had arranged for the reception at a Swiss *maison de santé*. **1942** 'A. BRIDGE' *Frontier Passage* vii. 125 We don't want him shut up in a *maison de santé* for spy mania. **1893** G. B. SHAW *Let.* 12 Dec. (1965) 412 A woman of bad character, the keeper of two *maisons tolérées* in Brussels. **1927** *Observer* 17 Apr. 20/2 Agreement on that point [*sc.* the abolition of licensed houses of prostitution decreases the..traffic] greatly strengthens the hands of the League in working for the abolition of the *maison tolérée*. **1970** K. CHESNEY *Victorian Underworld* x. 343 Throughout her existence in a *maison tolérée*,..she was entirely in the hands of the bawd to whom she was committed.

maisonnette. Add: **2.** (Usu. in the form *maisonette.*) A part of a residential building which is let separately, usu. distinguished from a flat by not being all on one floor.

1912 *Chambers's Jrnl.* Feb. 144/2 Flats or maisonettes, such as those upon Anne's Mansions, Westminster, London. **1919** T. S. ELIOT *Whispers of Immortality* in *Poems*, Grishkin has a maisonette. **1923** A. HUXLEY *Antic Hay* ix. 129 'It's a dream delightful little maisonette,' she explained. 'Full of awful things. We had to take it furnished.' **1957** *New Yorker* 5 Oct. 171/2 Le Corbusier wisely utilized the type of apartment the English call the maisonette, with a narrow private stairs to connect upper and lower rooms. **1959** *Times* 28 May 13/5 To-day one sees maisonettes advertised as such but also covering a range of separate living accommodation from the Rotch porch of the converted house to the modern slab block of flats. **1961** L. MUMFORD *City in Hist.* Note to plate 51, The London County Council's Roehampton estate, with its mixture of tower and slab, of flats and maisonettes. **1970** *Times* 6/6 Maithili is the dialect of the old country... **1975** J. SYMONS *Three Pipe Problem* iii. 22 Why did these houses agents insist upon using not the word rooms, nor apartment, nor even flat, but the atrocious *maisonette*?

Maithili (mai·tili). The name of a dialect of Bihari, one of the Indo-Aryan group of languages; also, the name of its script.

1881 G. A. GRIERSON *Introd. Maithili Lang. of N. Bihar* i. 5 The Maithilī character is nearly the same as Bangālī, differing only in one or two letters. *Ibid.* ii. 8 Words derived direct from the Sanskrit, which were originally neuter, become masculine in Maithili. **1928** *Funk's Stand. Dict.* (1946) 743 Many lyrics in Maithili script date from the 15th century. **1948** D. DIRINGER *Alphabet* ii. vi. 365 The Oriya, Maithili and Early Manipuri characters seem..to be somewhat connected with the Bengali script. **1964** D. C. SWANSON in Householder & Saporta *Probl. Lexicogr.* i. 74, I have been unable to elicit any 'diminutive' suffixes from speakers of Maithili. **1968** *Punch's Stand. Atlas* 606/1 Maithili is the dialect of the old country of Mithila or Tirhut, famous from ancient times for its learning.

‖ **Maitrank** (moi·traŋk). [G., f. *Mai* May + *trank* drink, beverage.] = *May-drink* (MAY *sb.*[2] 3 a).

1856 GEO. ELIOT *Jrnl.* in J. W. Cross *George Eliot's Life* (1885) II. viii. 42 Delicious *Mai-trank*, made by putting the fresh *Waldmeister*—a fragrant plant with a small white flower..into mild wine, together with sugar. **1897** G. DU MAURIER *Martian* 273 You might sit for sixpence in a pretty garden and drink coffee, beer, or Maitrank. **1899** G. JEKYLL *Wood & Garden* vi. 60 The pretty little Woodruff..revives memories..of Maitrank; that best of the 'cup' tribe of pleasant drinks, whose flavour is borrowed from its flowering tips.

‖ **maître** (mɛ̄·tr'). [Fr., = master.] **1.** *slang.* Also **maître d'**, **maître de** (mɛ̄·tr' də or di). Used *ellipt.* for *MAÎTRE D'HÔTEL 2.

1899 R. WHITEING *No. 5 John St.* xviii. 188 But the viands! Some of them, as the maître is so good as to tell me..come straight from Paris. **1953** P. ARMEE *House* xi *not Home* (1954) x. 16 Hat-check girls, waiters, the maître d', some were grinning. **1959** ..Capital was happy to Harry. **1959** *Guardian* 12 Dec. 3/1 The maître de (they do not bother about the 'hotel' bit) ushered..parties..to tables. **1967** C. DRUMMOND *Death at Furlong Post* iv. 56 A very large steak..made with great blobs of butter by a spivvy little maître with a spirit stove. **1969** A. GLYN *Dragon Variation* vii. 208 The maître d', who knew her well, greeted her and led her to her usual table on the north side. **1972** M. CATTO *Sam Cabanova* iv. 74 With the help of the maître who looked after the cellar they chose a fabulous Château d'Yquem. **1973** W. MCCARRY *Delaal* iii. 161 I've just brought over a new maître de, a young Italian fellow who was running the place [*sc.* a restaurant]. **1974** *Publishers Weekly* 17 June 59/1 The sort of lady who is instinctively escorted to corner tables by *maîtres d'*.

2. The title and form of address of a French lawyer.

1883 C. M. YONGE *Stray Pearls* I. xvi. 211 Technically he [*sc.* a lawyer] was only Maître Dargent, and his mother and sisters have been called Mademoiselle. Monsieur and Madame were much more jealously limited to nobility than..now. **1910** BARONESS ORCZY *Lady Molly of Scotland Yard* vi. 148, I hear from Maître Vendôme that he has safely received my letter. *Ibid.* 165 He was met by his late client's lawyer, Maître Vendôme. **1957** D. DU MAURIER *Scapegoat* xviii. 253 The lawyer bowed. 'Oh you have the kindness, Maître, to see that the notification of the death is sent to the newspapers.' **1971** *Guardian* 15 Sept. 2/6 Maître Nicolet then said in court that the centre of the European gang was in France.

3. a. maître d'armes (mɛ̄·tr' dārm) [lit. 'master of arms'], a fencing instructor.

1896 C. A. HENTY *Through Russian Snows* vii. 143 The veterans were always ready to give him lessons with the sabre or rapier in addition to those he received from the *maître d'armes* of the regiment. **1933** *Times Lit. Suppl.* 24 Mar. 220/2 He will by then be lost in a maze of technical terms, and he will be fortunate..to have a Sturm *maître d'armes* to turn the uninitiated to. **1956** J. D. AYLWARD *Eng. Master of Arms* ix. 110 He was ..abreast of the progress made during the seventeenth century by the great French *maîtres d'armes*. **1961** F. C. AVIS *Sportsman's Gloss.* 194 *Maître d'Armes*, a master of arms or fencing instructor.

b. maître de ballet (mɛ̄·tr' də balɛ) [lit. 'ballet-master'], originally one who composed and superintended the production and performance of a ballet; now, a trainer of ballet dancers; hence **maîtresse de ballet**, the female counterpart.

1838 T. CREEVEY *Let.* in *Creevey Papers* (1903) II. iii. 64 Yesterday I spent a very amusing hour with Sefton at the Opera House, seeing the *maître de ballet* manoeuvre about 50 *figurantes* for the approaching new ballet of *Alfred*. **1828** J. EBERS *Seven Yrs. King's Theatre* xii. 351 An amiable and able individual..has experienced the com-

mon fate of mortality; I mean M. Boisgerard, the second Maître de Ballet. **1845** *Morning Post* 9 Apr. 5/5 *Maîtres de ballet* have to begin a new course of study. **1913** A. E. JOHNSON *Russ. Ballet* 3 But though his [*sc.* Noverre's] ambition as a *maître de ballet* outran his perceptions as an artist,..he initiated and firmly established a new form of art. **1915** M. E. PERUGINI *Art of Ballet* xxxiii. 278 A wise choice was made in the selection of the late Madame Katti Lanner as *maîtresse de ballet*. **1936** *Times* 1 Sept. 11/3 Under its *maître de ballet*..it has reached international standards. **1970** *Times* 27 May 11/5 Hu's fight master is a former Peking *maître de ballet*.

maître d'hôtel. Add: **2.** A hotel manager; now usually the manager of a hotel dining-room, a head waiter.

1891 R. H. SAVAGE *My Official Wife* vi. 43 The attentive *maître d'hôtel* flew past us and threw open the door of a splendid apartment. **1907** E. GLYN *Three Weeks* i. 17 Her red wine the *maître d'hôtel* poured into her glass himself. **1923** M. ARLEN *These Charming People* 85 The agreeable and polished M. Kisotto, prince of *maîtres d'hôtels*, chanced by our table. **1950** T. S. ELIOT *Elder Statesman* i. 9 The *maître d'hôtel*. And the waiters all seem to be your intimate friends. **1973** D. MILLER *Chinese Jade Affair* xvii. 157 The 'canard aux navets' was being served in a liberal fashion by the Maître d'Hotel himself.

3. *attrib.* (See quots.)

1845 E. ACTON *Mod. Cookery* (ed. 2) iv. 107 *French Maître d'Hôtel, or Steward's Sauce.* Add to half a pint of rich, pale veal gravy..salt, minced parsley, and lemon-juice. **1861** MRS. BEETON *Bk. Househ. Managem.* x. 223 *Maître d'Hôtel Butter*, for putting into Broiled Fish just before it is sent to Table. *Maître d'Hôtel Sauce (Hot)*, to serve with Calf's Head, Boiled Fish, and different Fish. **1951** *Good Housek. Home Encycl.* 344/1 Simply prepared dishes garnished with *maître d'hotel* butter. **1969** *House & Garden* Dec. 64/2 *Maître d'hôtel*. If given this name, a dish will probably be fairly quickly and simply cooked— and garnished with *maître d'hôtel* butter. **1969** A. LAWS *Fish* ..with or without a fresh herb and melted butter in some form or other. When used to describe butter, it means that the butter has been seasoned, mixed with parsley and lemon juice worked into it and is then put to get cold so that it can be made into pats. **1974** *Times* 6 Nov. 14/8 My salmon steak with maître d'hotel butter was good.

‖ **maîtresse** (mɛtrɛ·s). [Fr., mistress.] Used in phrases, as **maîtresse en titre** [lit. 'mistress in name'], an official or acknowledged mistress; **maîtresse femme**, a strong-willed or domineering woman.

1830 THACKERAY *Catherine* in *Fraser's Mag.* July 104/1 We had a great mind to make..Mrs. Catherine *maîtresse en titre* to Mr. Alexander Pope. **1853** C. BRONTË *Villette* III. xxxii. 83 Ah! I know you well..cette *maîtresse-femme*, my cousin Beck herself. **1876** C. M. YONGE *Three Brides* xxxvii. 179 There are four kinds of wives—the crowned queen, the dead-weight, the *maîtresse femme*, and the helpmeet. **1899** W. J. LOCKE *Great Pandolfo* xii. 240 She had been the *maîtresse en titre* of one of his friends. **1920** S. HUNTER *Ambrose* Romance *Lenebenus* ii. 42 At first Leo did not dare to marry her: though he took her to the Palace as an acknowledged *maîtresse en titre*. **1946** C. L. THOMSON *Jane Austen* ii. 50 The wonderful heroine [of *Corinne*], a *maîtresse femme* like Mme de Staël herself. **1951** *Times Lit. Suppl.* 5 July 544/1 She tells of..dwarf, and cicisbeo, and *maîtresse en titre*. **1956** G. B. SHAW *Millionairess* Pref. 105 In the humblest cabin that contains a family you may find a *maîtresse femme* who rules in the household by a sort of divine right. **1960** O. MANNING *Great Fortune* v. 65 Sophie, whose aggressive manner had deceived Guy into thinking her a *maîtresse femme*. **1970** E. SHANNEY *Elizabeth* ii. 44 Her..was a *maîtresse femme*. **1953** M. STEEN *Anna Fitzalan* ii. 72 For two and a half years Anna fulfilled the functions of a nurse, as well as those of a *maîtresse en titre* of Evan's household. **1964** *Guardian* 26 Oct. 16/8 An accredited speech-writer is, like a 'maîtresse en titre,' an offence to public morals. **1973** D. CHANDLER *Mainbeaux* I. Arabella [Churchill] was..combining the roles of maid-of-honour to the Duchess and *maîtresse-en-titre* to the Duke, to whom she bore several children. **1973** E. HYAMS *Final Agenda* vii. 98 The widow was a *maîtresse femme* whose place in the hierarchy was three steps higher than that of her..

Maitreya (maitrē·yā). [Skr., f. *maitrī* friendship.] In Buddhist theology, the name of the Buddha who will appear in the future; also, an artistic representation of this future Buddha.

1889 M. MONIER-WILLIAMS *Buddhism* iv. 135 Gautama is the fourth Buddha of the present age... He is to be followed by the fifth Buddha, Maitreya [*sc.* a name meaning 'full of love towards all beings']. **1923** A. B. KEITH *Buddhist Philos.* xvii. 269 We have in Buddhaghosa's own authority for the advent of Maitreya. **1953** E. J. THOMAS *Hist. Buddhist Thought* xiii. 168 As a part of this doctrine of a succession of Buddhas arose the belief in a future Buddha, Metteyya, or Maitreya. **1962** *Guardian* 23 Dec. 11/3 A seated Maitreya in gilt bronze, Silla—early seventh century. **1972** P. M. BARTLS *Korea* 142/1 The figures are monks, images of Maitreya, the Buddha of the future. **1972** D. BLOODWORTH *Any Number can Play* iii. 17 The gross ..was strikingly resembled the Maitreya who was the Buddha who come to herald an age of happiness.

maize. Add: **3.** (Further examples.) Freq. denoting a colour of cloth or dress-material.

1838 E. GROSVENOR *Diary* 9 June in G. Huxley *Lady Elizabeth & the Grosvenors* (1965) iii. 28 Succeeded in finding..a maize and silver dress. **1853** MRS. GASKELL *Cranford* xiii. 244 If a happy sea-green could be met with, the gown was to be sea-green: if not she inclined to maize. **1858** GEO. ELIOT *Scenes Clerical Life* I. ii. 31 Maize is a colour that decidedly did *not* suit his complexion. **1923** *Daily Mail* 15 June 6 (Advt.), Crêpe de Chine..several shades, including: Pale Pink..Maize, Lemon. **1970** *Cape Times* 28 Oct. 19/3 (Advt.), Rambler Rogue, 1970. 5500 miles, maize with cognac trim. **1975** *New Yorker* 26 May 54/1 (Advt.), Our all silk honan tie... Navy, red, brown, maize, rust, bottle green.

4. *maize-yellow*.

1867 C. T. DAVIS *Manuf. Leather* (ed. 2) 607 Light ochre yellow, maize yellow to dark gold ochre yellow.

Maje·stic, *sb.* [f. the adj.] A variety of potato, producing light-skinned, kidney-shaped tubers.

1917 J. W. SANDERS *Bk. Potato* (ed. 3) xvii. 99 *Majestic*, tubers, oval-shaped; hardy, strong grower, enormous cropper; fine quality and appearance. **1926** R. N. SALAMAN *Potato Varieties* xxvi. 263 Majestic has attained great popularity, and today is one of the most widely grown field crops throughout Great Britain. **1955** R. GAULT *Vegetables* x. 103 *Main Crop.* Majestic, a heavy cropping white kidney, with shallow-eyed large tubers. **1963** *Times* 22 Apr. 2/6 Majestic, which takes up over 50 per cent of the maincrop acreage, was introduced in 1912.

majesticalness. (Later example.)

1909 *Westm. Gaz.* 19 Apr. 1/3 The majesticalness of this mere bundle of human genius and human sorrows.

majesty. Add: **b.** *His Satanic Majesty* (examples). Also *his sable majesty* (see *SABLE *a.*).

1751 RICHARDSON *Clarissa* (ed. 4) IV. 126 To be sure, Jack, she means to do great despight to his Satanic Majesty in her hopes of reforming me. **1922** JOYCE *Ulysses* 315 The priests and bishops of Ireland doing up his room in Maynooth in his Satanic Majesty's racing colours.

majlis (ma·dʒ,lis). Also **madjlis, majlas, majliss, medjlis, mejliss, mejlis,** [Arab. *majlis*.] An assembly or council; *spec.* the national assembly in Iran; also, a reception-room.

1822 G. F. LYON *Narr. Trav. N. Afr.* vi. 261, I found Mukni sitting in the greatest agitation, pale, and alone in the Mejlis, or Court of his Castle, and apparently not welcome to me. **1830** J. L. BURCKHARDT *Notes on Bedouins & Wahábys* 59 At his evening assemblies (*majlis*), everybody sat down where he could find a convenient place. **1855** R. F. BURTON *Pilgrimage* II. xv. 36 The Shaykh..led us into the Majlis, which was swept and garnished with all due apparatus for the forthcoming reception ceremony. **1903** *Hist. Cape Yes* ..1876 The pinching machine now used in several major league training camps... **1909** A. HAMBRO ..it was not like the Reign of Europe's moderns... The Majlis... [*garbled*] **1913** *Encycl. Brit.* XXI. 196/2 By a rescript dated the 5th of August [1906] Muzzaf-ud-Din Shah gave his assent to the formation of a national council (*Majlis i shora i milli*). **1930** *Glasgow Herald* 29 Nov. 11/3 The British Government expects the Egyptians to ratify the Anglo-Persian agreement. **1955** *Lit. Digest* 27 June 19/2 Premier Riza Khan replied in the Majlis that..he would proclaim Persia a republic. **1940** F. STARK *Winter in Arabia* 160 There in the Mansab's majlis they sit. **1955** *Times* 22 June 11/3 The lead..has been given by the Majlis, where a strong group of Government supporters has been formed. **1964** *Ann. Reg.* 1963 296 The Majlis was opened on 6 October, thus bringing to an end a period of some two and a half years during which the Constitution had been suspended. **1973** *Daily Colonist* (Victoria, B.C.) 21 Mar. 28/1 [Bahrain] There is an air of desert democracy mixed with a certain noblesse oblige about the majlis, the formal Arab court session in which the ruler sits on his throne. **1973** *Nat. Geographic* Feb. 227/2, I found him presiding over the majlis, his open meeting attended by petitioners and friends each day [in Oman]. **1974** *Christian Science Monitor* 5 Dec. 38/2 Colorful Gulf dhows sail past concrete skyscrapers outside the windows of the Majlis.

major, *sb.*[1] Add: **1. c.** An officer in the Salvation Army.

1907 G. B. SHAW (*title*) Major Barbara. **1970** C. BRAMWELL-BOOTH *Catherine Booth* xvi. 377 The Army Mother gave colours to..Major Rose Clapham and her assistants who were to establish the work in South Africa. **1975** *Times* 30 Apr. 9/3 Major Eva den Hartog, of the Dutch Salvation Army, has spent the last 17 years at the disaster centres of the world.

d. Major Mitchell *Austral.* = *Leadbeater's cockatoo* (*LEADBEATER[1]).

1898 E. E. MORRIS *Austral Eng.* 280/1 *Major Mitchell*, vernacular name of a species of Cockatoo, *Cacatua leadbeateri*... The bird was named after the explorer, Major (afterwards Sir Thomas) Mitchell, who was Surveyor-General of New South Wales. The cry of the bird was fancifully supposed to resemble his name. **1911** C. E. W. BEAN *'Dreadnought' of Darling* xvii. 169 We came across a flock of 'Major Mitchells',..white and pink, and exceedingly pretty when they fly and show the whole expanse of their pink feathers. **1933** A. MOORHEAD *Rum Jungle* iii. 40 Two of the Major Mitchells lighted on my shoulder. **1966** *Times* 28 Mar. (Austral. Suppl.) p. xvi/5 The Major Mitchell or pink cockatoo.

e. (See quot. 1919.)

1919 R. BLATCHFORD *My Life in Army* xvi. 52 The 'major' was 'major' then. **1919** *Athenæum* 25 July 664/1 'Major' as..abbreviation of the Sergeant Major is applied to N.C.O. **1925** FRASER & GIBBONS *Soldier & Sailor Words* 150 *Major*, *the*, the usual name among N.C.O.'s generally for the Sergeant Major.

major, *a.* and *sb.*[2] Add: **A.** *adj.* **I. 1. a.** *major road*, esp. in phr. *major road ahead* used on traffic signs (until 1975).

1930 *Highway Code* 7 It is the duty of a driver on a minor road when approaching a major road to *go dead slow*. **1937** R. F. BROAD *Motor Driving made Easy* (ed. 6) v. 61 The *Halt at Major Road Ahead* sign..enforces a definite standstill upon road users. **1950** *N.Y. Times* 20 Apr. 1/5 Railroad firemen and engineers called a strike against four major roads for Wednesday. **1967** L. MEYNELL *Mauve Front Door* xi. 137 Sidney Street runs into Falloden Street at right angles; there isn't any sort of traffic sign ('Halt' 'Major Road Ahead' or anything like that). **1968** *Highway Code* 14 You must..be ready to stop at the line marked by the road at the 'Major Road Ahead' sign. **1971** A. DRAPER *Death Penalty* v. 37 Hudson drove the car at full speed, not even bothering to halt at the major road ahead.

c. Also *ellipt.* in other sports, for *major penalty*, *score*, etc.

1951 *Football Record* (Melbourne) 8 Sept. 18 They opened with four behinds, and then rattled on sixteen majors. **1964** *Kingston* (Ontario) *Whig-Standard* 28 Dec. 9/7 Kingston took 21 penalties and Gananoque sat out 14 sentences, including a misconduct to James McGlade for incurring his second major of the game. **1969** *Sun-Herald* (Sydney) 13 July 48/1 Goals by Brian Younge and Ray Wilson seemed to put the seal on the match for Hawthorn until a last-minute burst by the Blues took them to within a straight goal, with two majors by Bletbyn. **1970** *Toronto Daily Star* 24 Sept. 17/5 Peter Murray of Jarvis raced 93 yards for a major, and then..three plays, quarterback Mike Chrzan sneaked from three yards in the fourth, Lang converting. **1970** *Globe & Mail* (Toronto) 25 Sept. 33/3 Parkdale quarterback Rich Chudziak tossed a 38-yard pass to Wally Lytwynec, who ran 30 yards for the major.

d. In Bridge. *major suit*: spades or hearts (tricks taken by the chosen suit these are trumps score more points than when clubs or diamonds are trumps). Also *ellipt.*

1916 R. F. FOSTER *Auction Bridge for All* vii. 27 High cards, two sure tricks, are just as necessary in the major suits as in the minor suits. **1927** M. C. WORK *Contract Bridge* ii. 11 Many find it easier to remember 30 for Minors, 30 for Majors and 35 for No Trump. **1958** *Listener* 9 Oct. 572/3 It would give an exaggerated picture to reverse the following One Diamond with Two of a major. **1973** *Country Life* 13 Dec. 2046/1 Many pairs nowadays bid Stayman, which is an inferior contract... It scores less than the major suit slam. **1974** *Times* 27 Apr. 10/5 After opening a strong No Trump without a four-card major, I rarely know how much interest to read into my partner's second response.

e. *N. Amer. major league*, the highest division of teams in baseball, etc. Also *attrib.*

1942 *Sun* (Baltimore) 3 Apr. 1876 The pitching machine now used in several major league training camps. **1947** *Partisan Rev.* XIV. 258 He conceded his best years to all of from himself by means of his devotion to professional sports, major league baseball especially. **1963** C. ANDERSON *Young Sportsman's Guide Baseball* ii. 17 A visit to a minor league park. **1971** In a score of minor league players had participated in the American Legion sponsored Leagues. **1968** *Globe & Mail* (Toronto) 17 Feb. 13 (Advt.), Start with the finest beaches in Florida, add great fishing,..major league baseball, greyhound and horse racing. **1971** H. SEYMOUR *Baseball* II. p. vi, 1..scored unofficially..for two major-league clubs.

B. *sb.*[2] **6.** orig. *U.S.* In some universities, a subject to which special attention is given during a certain period of study. Also, this subject seen as a qualification. Also, a student thus specializing.

1890 in T. W. Goodspeed *Hist. Univ. Chicago* (1916) 142 A subject taken as a major requires four and a half hours classroom work or lecture work a week. **1907** *Columbia Univ. Catal.* Mar., Open only to students taking a major in the Department of English. **1928** [see *ELECTIVE *sb.* 2]. **1948** *Democrat* 5 Aug. 1/7 Mr. Gray will receive the Bachelor of Science degree with a major in business administration. **1949** *Daily Ardmoreite* (Ardmore, Okla.) 23 Jan. 6/6 McLaurin is a Negro education major at the University of Oklahoma who is receiving instruction in a separate classroom. **1953** *Manch. Guardian Weekly* 24 Dec. 14/3 Courses for the 'English major'. There are courses in phases of English literature designed largely..for those who are 'majoring' in English. **1959** *Times* 8 Sept. 3/2 (Advt.), Royal Australian Air Force. Education Officers required with Majors in Maths or Physics. **1961** *New Statesman* 3 Mar. 338 In English Studies the 'major' could be English literature. **1964** L. NKOSI *Rhythm of Violence* ii. 18, Sarie is a Social Science major; she's done a..study of African languages among Afrikaner students.. Sarie is a Social Science 'major'. **1970** *Brit. Jrnl. Social Psychol.* LXXXVII. 91 These female students were psychology majors served as experimenters. **1972** *N.Y. Times* 3 Nov. 22/4 The regulars are unhappy..that the McGovern organizers ..have reached into the district, bypassing..a shaggy-headed political-science major at Brooklyn College who sometimes teaches the neighborhood in dungarees, [etc.]. **1973** *Black World* Mar. 62 He is presently a theater-arts major at Los Angeles Community College.

7. A large company, organization, etc.

1968 *Wall St. Jrnl.* 24 Apr., Says a Columbia Pictures Corp. screen writer: 'You can see what Ross is doing today will be done by the majors tomorrow.' **1968** *Blues Unlimited* Sept. 29 Apart from Goldstar, the major outputs were from the majors in Hollywood. **1973** *Time* 25 June 85/2 In some cases, according to industry-contract report, Robert Killian, 'the majors are taking over the choice locations, putting up giant 20-pump stations with 24-hr. service and are replacing the small dealers'. **1974** *ITV 1974* 10/2 The [Independent Television] Authority.. created a system made up of several large so-called 'network companies' (sometimes also called the 'majors') and a number of smaller 'regional companies' (sometimes called the 'minors'). ..The five major companies are also responsible of network programmes to be used by the whole service. **1974** *Spartanburg* (S. Carolina) *Herald* 24 Apr. A2/2 Of the four majors reporting so far, only Exxon showed a decline from the fourth-quarter profits, while Amoco's net rose 81 per cent.

major, *v.* Add: **3.** *intr.* Of a university student: to take, or qualify *in*, a major course of study; to specialize *in*. orig. *U.S.*

1924 P. MARKS *Plastic Age* xvii. 185 Having decided to major in English, he found that he was required to take a composition course the second half of his sophomore year. **1927** *Brit. Weekly* 1 Sept. 470/1 It is a thesis for the Th.D. degree, for which he has already stood the examinations, 'Majoring' in Greek New Testament at Louisville. **1929** *Publishers' Weekly* 20 July 252 Bulbous-headed adolescents who have majored in English descend in shoals. **1955** W. GADDIS *Recognitions* i. i. 7 As a youth in a New England college he had..majored in classical poetry. **1958** *Times Lit. Suppl.* 5 Sept. 502/2 He majored in forgery and illegal entry inside several specially recommended prisons. **1966** *Listener* 25 Feb. 290/1 A student who in the School of Social Studies 'majors'—Sussex seems to revel in this translatlantic noun-verb—in history also reads various 'contextual' subjects. **1971** *Progress* (Cape Town) May 2/1 He majored in Afrikaans at Witwatersrand University. **1973** *Daily Tel.* 12 June 21/2, I am studying A levels in geography, English and art, with a view to majoring in geography. **1973** *Jrnl. Amer. Orient. Soc.* XC. 83/1 For both B.A. and M.A. Eliot majored in literature.

Majorcan (mădʒǭ·ɹkăn, măjǭ·ɹkăn), *sb.* and *a.* Also 7 **Majorkine.** [f. *Majorca* the name of one of the Balearic Islands + -AN.] **a.** *sb.* A native or inhabitant of Majorca. **b.** *adj.* Of or pertaining to Majorca. Cf. *MALLORCAN *sb.* and *a.*, *MALLORQUIN *sb.* and *a.*

1697 H. MAUNDRELL *Jrnl.* 27 Apr. in *Journey from Aleppo* (1703) 121 We..were very courteously receiv'd by the Guardian, Father Raphael, a Majorkine by birth. **1848** E. J. SABINE tr. *A. von Humboldt's Cosmos* II. 147 One maritime people after another, Phoenicians,.. Majorcans, and Spaniards, made successive efforts to penetrate onwards in the Atlantic Ocean. **1858** *Leisure Hour* 18 Feb. 103/1 This Majorcan driver said he had always thought a Protestant meant 'somebody who does not accept the Bible'. **1873** *Genl. Mag.* Feb. 129 The family of Napoleon I. was originally native of that island i.e. Majorca. **1876** C. T. BIDWELL *Balearic Islands* 109 Myrtle-covered points, in the use of which the Majorcans display natural skill. *Ibid.* 111 Majorcan carpets and mattings cover the floors. **1935** J. A. FRASER *Spain & W. Country* ix. 106 The few following treaties of the great Majorcan [*sc.* Ramon Lull]. **1964** *New Statesman* 14 May 766/3 There is some farcical reporting of Majorcan dialogue. **1967** *Punch* 16 Nov. 767/1 The Majorcans believe in being 'formal' as the Spaniards also do. **1973** *Times* 1 June 13/7 A Majorcan official remarked that the British went about as though they owned the country.

major-domo. Add: **c.** *U.S.* In south-western states, an overseer on a farm or ranch; also, the water-master or official in charge of irrigation in New Mexico.

1834 in *Calif. Hist. Soc. Q.* (1929) VIII. 228 Four rancheros, each one made up of an Indian village, a house for the *mayordomo* directing it, [etc.]. **1836** D. B. EDWARD *Hist. Texas* ix. 291 Having thus glanced at the Major domo, we shall take a peek at his Locum tenens. **1902** F. H. NEWELL *Irrigation in U.S.* 107 He is usually known as the 'water master' or 'ditch-rider'; or, in Spanish-speaking communities as *majordomo*. **1920** J. HART *Vigilante Girl* xiv. 195 Arthur's chair was taken to the drawing-room. They found the major-domo and a group of *vaqueros* waiting. **1948** F. JOHNSTON *Lost & Living Cities Calif. Gold Rush* p. ii, The precious metal was found..by Francisco Lopez, majordomo of the Mission San Fernando, in Placerita Canyon.

majorette (mĕ³dʒɔre·t). orig. *U.S.* [f. MAJOR *sb.*[1] + -ETTE.] = *DRUM-MAJORETTE.

1941 *San Francisco Examiner* 20 Apr. (Amer. Weekly Section) 4/1 During the past few years the drum major.. has given way to the so-called majorette. **1951** M. MCLUHAN *Mech. Bride* (1967) 122/1 The majorette is no man beater. **1959** *Times* 20 Nov. (Canada Suppl.) p. xx/4 Each club has a band and a corps of bare-legged prancing girls called majorettes. **1960** *Wanganui* (N.Z.) *Photo News* 13 Feb. 20 (caption) Coach of the Wanganui Majorettes. **1973** *Sat. Rev. Society* (U.S.) May 64/1 A bubbly ex-majorette named Linda.

majoritarian (mădʒɒrɪtē³·riăn), *a.* [f. MAJORITY + *-arian* as in *libertarian*, etc.] Governed by or believing in decision by a majority; supporting the majority party. Also as *sb.*

1918 J. BUCHAN *Nelson's Hist. War* XX. 118 Early in June came the delegation of the German Majority Socialists, which included—besides Scheidemann, the Majoritarian leader—that Hermann Müller who, on the eve of the declaration of war, had invited the French Socialists to vote against war credits. **1957** *Britannica Bk. of Year* 512/1 Some new words arose from politics and world affairs... Majoritarian was an adjective meaning 'ruled by the beliefs of the majority'. **1964** GOULD & KOLB *Dict. Social Sci.* 404/2 The idea of party responsibility..aims at facilitating the practice of majoritarian democracy. **1965** *Listener* 21 Jan. 113/3 The fateful divide in the pre-revolutionary Russian Social-Democratic movement between the Mensheviks (i.e., minoritarians) and the Bolsheviks (i.e., majoritarians)..went far deeper than a purely political argument. **1971** P. WORSTHORNE *Socialist Myth* viii. 196 The idealist disillusion with majoritarian democracy. **1972** *Mod. Law Rev.* XXXV. 1. 73 In exercising their discretion the courts have normally adopted a posture which is both 'legalistic' and 'majoritarian'..majoritarian in the sense that the views of the majority have been treated as the most probative.. factor.

Hence **majorita·rianism**, belief in, or the existence of, rule or decisions by a majority.

1961 in WEBSTER. **1968** A. LIJPHART *Politics of Accommodation* viii. 125 In short, the rule is majoritarianism tempered by the spirit of concurrent majority. **1969** *Internat. Encycl. Social Sci.* IX. 537/2 Opponents of strict majoritarianism advance numerous arguments. First, they point out that the majoritarian principle might be used to destroy the conditions of its own existence, such as freedom of association and expression. **1975** *N.Y. Times* 22 Mar. 27/3 The public interest defined as the greatest happiness for the greatest number..has hardened into an increasingly imperious 'majoritarianism'.

majority. Add: **2.** Also *transf.*

1909 *Westm. Gaz.* 28 Sept. 3/1 The majority celebration of the pneumatic tyre promises to be the biggest trade function on record. **1930** R. M. FOSS *Let.* 25 May (1965) 324 If it [*sc.* the book] did not go on I'd be prepared to guess that it will become almost a historical novel. **1966** *Listener* 7 July 10/1 It is sad that, as it approaches its majority, this organisation should have run into deep waters.

3. *absolute majority* (examples).

1888 J. BRYCE *Amer. Commonwealth* I. v. 55 Neither Polk in 1844, nor Taylor in 1848,..had an absolute majority of the popular vote. **1971** G. K. ROBERTS *Dict. Political Analysis* 120 In politics..the number of votes constituting a majority is equal to 50 per cent plus one of votes cast (sometimes referred to as an 'absolute majority').

7. *majority-rule, -vote*; **majority calling,** in Bridge (see quot. 1964); **majority carrier** *Electronics*, a charge carrier of the kind carrying the greater proportion of the electric current in a semiconducting material (i.e. an electron in n-type and a hole in p-type material); **Majority-Socialist,** one who, after the division of the German Socialist party during the 1914–18 war, acted with the larger party; **majority verdict,** the verdict of the majority of a jury; also *transf.*

1947 *Daily Express* 28 June 2/7 The system of 'majority calling' at auction bridge, the adoption of which in this country was first advocated by the 'Evening Standard'. **1929** *Laws Contract Bridge Portland Club* 6 While, at the Portland Club, 'value calling' is invariably played at 'Auction', 'majority calling' is being used at 'Contract'. **1964** *Official Encycl. Bridge* 343/1 *Majority calling*, the principle by which the rank of suits was changed so that a lower level, regardless of scoring value. **1951** W. SHOCKLEY et al. in *Physical Rev.* LXXXIII. 151/2 The density of minority carriers is much smaller than the density of majority carriers. **1960** *Electronics* 29 Apr. 78/1 The intrinsic semiconductors are made into extrinsic..semiconductors by their majority-carrier densities. **1964** RICHARDS *Physical Princ. Junction Transistors* v. 100 Using equation (2.11) with the assumption that the conductivity is of minority carriers..is that of majority carriers for..density determines, we obtain γ_x = .. **1969** D. R. B. TUCKER *Instead of Bk.* 169 Rule is evil, and..it is none the better for being majority rule. **1950** *Scribner's Encycl. Soc.* 29/1 When such deep and permanent cleavages are absent, majority rule seems more likely to secure more equal freedom than any other, at least if free discussion, thought. **1963** *Times* 1 May 12/7 A Majority-Socialist of course..the majority vote was supposed. **1968** J. R. PENNOCK in *Internat. Encycl. Social Sci.* IX. 536 The term 'majority rule' needs to be discussed in two quite different meanings. In its most widely held form, the term means the principle of decision-making within a specified group. At its simplest, the rule requires that the votes of each member shall be counted as equal to that of every other and that no vote or division by a minority may override that of a majority. By extension, majority rule is sometimes contrasted with any rule requiring that decisions be unanimous or by any number larger than a simple majority. **1973** *Guardian* 10 Dec. 4/5 The Smith regime..is proving..how slender is the hope of orderly progress towards majority rule. **1974** *Five Tables in a Land* 49..The Conservatives...

1928 J. BUCHAN *Hist. War* XX. 125 The great governing parties of the Empire... The Conservatives on the extreme right and the Minority Socialists on the extreme left, were the Catholic Centre, the Radicals, the National Liberals, and the Majority Socialists. **1923** E. A. ROSS *Russ. Soviet Republic* 20 At this time [*sc.* 1918] a Soviet was formed in Berlin, to which members of the 'Majority' Socialists..sent as delegates. **1960** S. DELMER *Weimar Germany* i. 6 Ebert and his 'Majority' Socialists called on him [*sc.* Prince Max] to hand over power to them. **1966** *Westm. Gaz.* 7 Apr. 5/2 The jury..sent in to court to inquire if the parties would accept a majority verdict. **1966** H. KALVEN et al. *Amer. Jury* xxxvi. 461 Numerous states allow for a majority verdict..in civil cases. **1973** *States allow for* ...**1972** *Times* 22 Apr. 3/5 If a jury cannot agree,..the judge may accept a majority verdict. **1973** *Listener* ..

makak, var. *MOCOCK.

‖ **makan** (ma·kan). *Malaysia.* [Cf. Malay *makan* to eat; *makanan* food.] Food.

1927 R. J. H. SIDNEY *In Brit. Malaya To-Day* 231 His stomach may become distended with a number of weird mixtures..from the School *makan* shop. **1963** J. KIRKUP *Tropic Tempter* 101, I wander back to the hotel for my makan. (Cold turkey.) **1971** *Carry Singapore in your Pocket* (Singapore Tourist Promotion Board) (ed. 3) 29 Watch a meal of noodles being fried in 5 minutes at the 'makan' (food) stall.

‖ **makara** (mɛ·kărā). Also **Makara.** [Skr.] A mythical Hindu sea-animal, variously represented in Indian art; the equivalent of Capricorn in the signs of the zodiac.

1873 E. BALFOUR *Cycl. India* (ed. 2) III. M 58/2 Kama ..bears on his banner the fish Makara, an aquatic monster something like the sign of the zodiac Capricornus. **1901** A. C. GIBSON tr. *Grünwedel's Buddhist Art India* ii. 57 The sea-elephant, Makara, a creature formed of the fore-part of an elephant with the body and tail of a fish..has been retained everywhere in Indian art. **1934** *Burlington Mag.* Oct. 164/1 But essentially India is the Makara, a sea-monster among the rest, whose figure, since the days of King Asoka, has played a part in Indian art; it is seen on toranas of all times; ..it followed the spread of Indian art to the islands, ..whither it even ornamented Tibet. **1962** A. K. COOMARASWAMY *Introd. Indian Art* (ed. 2) ix. 42 Decorative work with *makaras* and other marine monsters [is] represented. **1971** *Illustr. Weekly India* 11 Apr. 67/2 (*December 22 to Jan. 19*) *Capricorn* (*Makara*). Frankness with friends may make them..a good ally. **1974** J. O'HARA..

Makassar, Makas(s)arese: see *MACASSAR, *MACASSARESE.

make, *sb.*[2] Add: **2. a.** *spec.* with implied reference to the manufacturer or source of manufacture; = BRAND *sb.* 6.

1909 WEBSTER, *Make*,..often referring to quality or grade of a manufactured article; as, whose *make* is it? **1937** *Discovery* Feb. 61/2, I tested the records on four Jersey cows of the same make. **1957** J. BETJEMAN *Continual Dew* 5 Talk of sports and makes of cars in various bogus Tudor bars. **1975** 'D. CRAIG' *Dead Liberty* xix. 208 It would be better not to park the same..vehicle in that street again... He wished now he had not mentioned the make of his car.

8. *on the make* (earlier and later examples); also, intent on winning someone's affections; seeking sexual pleasure; improving, advancing, getting better.

1869 J. R. BROWNE *Adventures Apache Country* 507 'Oh, you're on the make, are you?' 'Why, yes, to be candid, I'd like to make fifty thousand or so.' **1874** HOTTEN *Slang Dict.* 221 Any one is said to be 'on the make', who goes too high a price for his goods, or endeavours in any way to take advantage. **1918** BARRIE *What Every Woman Knows* iii. 55 There are few more impressive sights in the world than a Scotsman on the make. **1929** J. P. MCEVOY *Hollywood Girl* (1930) 41 Jimmy..said Bruce was a hot lot: on the make—he wouldn't have wasted that much time on me. **1934** J. O'HARA *Appointment in Samarra* (1935) vii. 222 It's the first time I ever knew of you going on the make for some dame. **1955** *Times* 11 May 7/3, I think we are on the make again, that the tide was..the tide is running in our favour. **1967** N. FREELING *Strike Out* 97 He struck me always as a guy on the make. **1973** 'A. BLAISDELL' *Crime by Chance* vii. 126 You mean he was on the make? Yes let's say that. **1973** N. M. DUNCAN *Big Timer* xxii. 150 Riordan was on the make. He'd found out something he could use. **1974** R. ADAMS *Shardik* ii. 46 Insinuating, daredevil, and with the ease of one familiar and obsequious, of a presuming servant on the make.

10. = *DECLARATION 8 b (see also quot. 1964).

1901 J. B. ELWELL *Bridge* 13 In considering a heart make, the dealer should be influenced by the general strength of his hand and by the number of honours he holds in the trump suit. **1905** R. F. FOSTER *Compl. Bridge* 316 The declaration is often called the make. **1964** *Official Encycl. Bridge* 343/1 *Make*..As a noun, it means a successful contract, but usually a hypothetical one in the post-mortem: 'Five diamonds would have been a make'. **1974** *Country Life* 17 Jan. 98/1 Four Spades is a make, but Five Clubs is one trick short.

11. A (sexual) conquest; *spec.* a woman of easy virtue. Cf. *MAKE *v.*[1] 65 *slang* (orig. *U.S.*).

1942 BERREY & VAN DEN BARK *Amer. Thes. Slang* § 439/1 Woman of easy morals..make. **1952** *Landfall* V. 98 'A widow's an easy make,' he said, 'you pedal and let her steer.'

12. An identification of, or information about, a person or thing from police records, finger-prints, etc. *slang* (orig. *U.S.*).

1950 in WENTWORTH & FLEXNER *Dict. Amer. Slang* (1960) 332/1 We got a make on his prints. **1959** 'E. McBAIN' *Pusher* vii. 62 Couldn't get a make on the torn cigarette stub. **1966** 'L. EGAN' *Detective's Due* (1966) vi. 71, I think this is too good to be true, but we've got a make on the prints from Ibid. To the D.M.V., just came through the wire, owner's name plate number... It belongs to a fifty-five two-door Ford. **1967** W. PRE *Protectors* ix. 82 We've got a make on Beth Pages. She's on the paper. **1969** R. M. SMITH *Ransom* iv. 175 We got a make on the Chevy.. Stolen last week. **1973** *Daily Colonist* (Victoria, B.C.) 6 Sept. 4/3 For years authorities have been trying to get a better 'make' on the grocery clerk, although he picked him up a licence in a nearby state after theirs has been in place.

13. *make and mend*: the action of making and repairing clothes; spec. *Naut.*, a period set apart for seamen to repair their clothes; hence, a period of leisure; a half-holiday; also *attrib.* and as vb.

1884 W. D. HOWELLS *Rise S. Lapham* (1885) i. 5 She cooked, swept, washed, ironed, made and mended from daylight till dark. **1899** *Navy & Army Illustr.* 14 Oct. 107 (*caption*) Thursday has been in the Navy, ever since King William IV..first instituted the practice, generally observed as 'make and mend' day. **1903** KIPLING *Traffics & Discov.* (1904) 154 I've gone to my tea kettle gentlemen to breakfast, and then we'll make and mend clothes till dinner-time, the Serjeant said. **1926** 'TAFFRAIL' *Pincher Martin* xv. 273 Th' navy's 'avin' its make an' mend, an' it's wot I calls a regular 'oliday. **1932** FRASER & GIBBONS *Soldier & Sailor Words* 150 *A make and mend*, a naval holiday. In

old days..usually on a Thursday. Now-a-days..the weekly half holiday (often transferred to a Saturday) continues to be known as a 'Make and Mend'. **1935** *New Survey London Life* IX. iv. xiii. 423 On Tuesday afternoon we do make-and-mend class. **1942** F. G. HACKFORTH-JONES *On the-One-One* xii. 153 Saturday was regarded as a make-and-mend day. **1943** *Sunday Times* .. (Women's Group on Public Welfare) iii. 60 Any 'Make and Mend' organisation concentrates upon re-clothing the people of a permanent service. **1955** *Times* 12 July 9/6 They.. moved on to their 'sun-porch', usually a clearing in the forest where they spend their time leisurely make and mend.

make, *v.*[1] Add: **1. d.** (Later example.) For *slang* read *colloq.*

1934 T. S. ELIOT *Rock* i. 13, 'e was a fine fellow on one side of 'is face. **1973** ..

3. b. *to be made for each other* (or *one another*): to be such as to harmonize perfectly or form an ideal combination; to be ideally suited: usu. of a specified man and woman.

1751 RICHARDSON *Clarissa* (ed. 3) III. 328 Her features are all harmony, and seem to be made for one another. **1857** DICKENS *Let.* Sept. (1938) II. 887 Poor Catherine and I are not made for each other, and there is no help for it. **1937** J. N. MCILWAITH *Kinsmen at War* xvii. 179 Dear Sally—they were made for one another. **1971** D. CLARK *Sick to Death* iv. 83 When two people—what's the popular phrase?—are made for each other.

29. b. (Later examples.) Also, in milder sense, to 'acquire', manage to get. (See also quots. 1926, 1953.)

1914 D. O. BARNETT *Lett.* (1915) 13 The company made a day of it, but it was claimed almost at once, so we made it in good time. **1926** J. BLACK *You can't Win* ix. 93, I was an expert house burglar..carefully choosing only the best homes... I 'made' them in the small hours of the night. *Ibid.* xv. 203 Better wait till night if you want to make a train. **1928** KIPLING *Limits & Renewals* (1932) 53 I'm making a temp'ry collar and lead of flatters. **1936** J. G. MILLAR *Horned Pigeon* ii. 16 Skinner and I were both good at what the Army calls 'making things', which is what the civilian would call 'fiddling' or even 'stealing' things. *Ibid.* Potatoes and onions I had 'made' from the food-dumps round Alexandria. **1953** W. BURROUGHS *Junkie* (1973) iii. 34 Taking junk hidden by another junkie is known as 'making his stash'. **1954** *Ibid.* 156 *Making Cars*... Breaking into parked cars and stealing the contents.

39. Delete † *Obs.* and add N. Amer. examples. (See also *Canadianisms*.)

1856 J. REYNOLDS *Peter Gott* 43 Since the last war with England, the fishermen from the States have not been allowed to make fish upon the shores of the British Provinces. **1909** E. C. ROBINSON *In Unknown Land* 30 A fisherman comes here to 'make' fish, not to catch them. **1923** *Canad. Fisherman* 243/1 It is a new departure for Newfoundland green fish to be taken to Europe to be dried. **1973** E. S. MORISON *European Discovery Amer.: Northern Voy.* xiv. 491 In our times 'making fish', as the Maine people call curing cod, haddock, and pollock in the sun, is simple enough..the fish..are headed, gutted, split, and washed fish in brine for two or three days, then rinsed, and spread them on home-made flakes to be cured in the sun. **1974** *Nat. Geographic* Jan. 122/1 Many south coast men [in Newfoundland] have to 'make' fish— that is, preserve their own.

46. a. (Further examples.) Phr. *to make one's day*: to render a day delightful or redeem it from routine, dullness, or banality. Similarly, *to make one's evening*, etc.

1903 A. BENNETT *Leonora* vi. 168 The conductor..told her that she had simply made the show. **1909** F. BARCLAY *Rosary* xxiv. 252, I knew her presence made my day and her absence meant night. **1935** WODEHOUSE *Luck of Bodkins* xi. 189 To see him sit there..with his head bandaged and all those blisters..made his day. **1942** *Punch* 14 Oct... **1946** *Observer* 21 Feb. ..**1955** J. G. ..

1942 T. RATTIGAN *Flare Path* i. 113 We've got some nice cheerful news for you boys..That would make his evening. **1953** X. FIELDING *Stronghold* 120 For Ioanna, the evening had been made. **1958** *Listener* 16 Oct. 623/2 What made the evening for many people was the interview with Mr. Khrushchev in the conference itself. **1969** *Tea & 4 Aug.* 2/4 I'll be driving along, it made my day. **1971** D. MCLAVE *Question of Negligence* (1973) xvii. 141 Get another card..did he say it made your day? I said that, too. **1973** 'D. SHANNON' *Spring of Violence* (1974) vi. 107 'Have you made the gun?' 'Right off. It's a Hi-Standard revolver.'

51. f. *U.S. Underworld slang.* To recognize or identify (a person, etc.). Cf. *MAKE *sb.*[2] 12.

1906 A. H. LEWIS *Confessions of Detective* 122 You wouldn't have come within a block of him. In the language of the Central Office, he'd have 'made' you. **1930** D. HAMMETT *Maltese Falcon* (1930) 62 Somebody's made him. **1932** J. M. SULLIVAN *Criminal Slang* 16 *Make someone*, to identify or recognize. **1950** HOLLEY & HUFFER *Vocab. Criminal Slang* 57 You had better ring up (disguise) so he won't make you. **1955** *Publ. Amer. Dial. Soc.* xxiv. 81 The victim then can possibly identify, or make, 'the thief. **1971** 'O. BLEECK' *Thief who painted Sunlight* (1972) xiv. 127 I'm a pretty good tail. You didn't make me. **1973** 'D. SHANNON' *Spring of Violence* (1974) vi. 107 'Have you made the gun?' 'Right off. It's a Hi-Standard revolver.'

53. c. *to make believe* (further examples).

1902 G. B. SHAW *Mrs. Warren's Profession* Pref. p. xxiii, Let's make believe we do not believe: we make-believe. **1904** KIPLING *Traffics & Discov.* 334, I did all that I could; I made believe I'd brothers behind. **1925** S. LEWIS *Martin Arrowsmith* xxxiii. 380, I make believe read my French books. **1951** G. GREENE *End of Affair* i. v. 37 As long as I could make-believe that he was really there. **1974** 'J. WAINWRIGHT' *Hard Hit* xii. 173 He's making believe.

f. *to make do*: to manage *with* (what is available, esp. an inferior substitute). Also *absol.*, esp. in phr. *to make do and mend*: to repair for continued use (cf. *MAKE *sb.*[2] 13); also as *sb.* *phr.* See also *MAKE-DO.

1899 E. WHARTON *Greater Inclination* 89 She had.. accepted it [*sc.* marriage] as a provisional compensation,.. she had made it 'do'. **1927** *Observer* 28 Aug. 16/6 The listener who was content to receive only the programmes of his own local station..could make do with a very simple and inefficient form of direct-coupled tuning arrangement. **1934** J. MARSHALL *Andromeda* xii. 156 She had always had to make do with... **1938** *Technology* Jan. 375/4 We may put all our energy into making them into.. **1941** H. W. *Observer* 20 Jan. ..**1941** ..making do, and making do.. **1946** ..a perfect ... **1952** ..and good will. **1962** ... H. WOODHOUSE *Rock Baby* vii. 31 Rasmussen had to make do with two oranges and only half a dozen rounds of toast because I was there.

65. b. (Further examples.) Also *fig.*, to achieve, accomplish, reach. Freq. *to make it*: to arrive at a certain place; to succeed in traversing a specified distance; to achieve a desired object; to be successful; *spec.* to achieve sexual intercourse.

1885 [see *BULLOCKY *a.* 2]. **1905** R. BEACH *Pardners* (1912) ii. 46 'We'll make it over into Mexico without any trouble,' he said. **1912** R. A. WASON *Friar Tuck* xxvi. 187 Badger-face tried to raise himself on his elbow, but he couldn't quite make it. **1916** H. L. WILSON *Somewhere in Red Gap* i. 25, I hurried him like to set out and dress and make the party. **1924** E. HEMINGWAY *In Our Time* (1926) ix. 121 'He looked like him or the bull and then he finally made it.' **1928** [see *BEST *adv.* 3 b]. **1935** L. A. G. STRONG *Seven Arms* 46 No matter if they can't make it in one day, sleeping out won't hurt them. **1938** G. GREENE *Brighton Rock* ..I ..so 'Will you be there?' 'No,' Hale said. 'I can't make it.' **1951** N. M. GUNN *Well at World's End* ii. 122 Ay! Don't you think we'll make it? **1955** *Pampiona* that day and asked if we were game. **1955** W. GADDIS *Recognitions* I. vii. 166 But I have to make it with her. She said. **1957** J. KEROUAC *On Road* III. vii. 225 They went in a party to lie in broad daylight..and there, he made Lucille... **1959** *Listener* 25 June.. **1962** ..he never made it... **1967** 'E. BRIGGS' ..makes the dinner, I'll make it. The latest that he has a lot of... Rebellious wanted.. **1968** ..'You make it with me?'..**1969** *Guardian* 23 Jan. 10 Jayne is interested in sex.. **1971** ..made it with a girl. **1973** ..the truth is Buchanan never made it as President. **1973** *Times Lit. Suppl.* 16 Nov. 1389/2 He finally makes it with long-desired Rachel.

c. To be successful in advances to (a member of the opposite sex); to win the affection of; *spec.* to persuade (a person) to consent to sexual intercourse; to seduce. *slang* (orig. *U.S.*)

1918 H. C. WITWER in *Amer. Mag.* June 110/3 Look at that big stiff tryin' to make the dame! **1921** *Sat. Even. Post* 1 Oct. 17/1 She give up trying to make me and got off at Albany. She was a good looker but I have no time for gals that tries to make strangers on a train. **1928** ANDERSON & STALLINGS *What Price Glory* 1, in *Three Amer. Plays* (1926) 7 Gosh! I guess even Lippy could make a kid in the rear.. the Lincoln he's the most popular. ...This broad I was trying to make insisted on putting up her hair. **1953** R. LEWIS *Mamrage* xii. 215 The guys..think they can make you P.D.Q. [*sc.* pretty damn quick] even if they're old and fat. **1949** ... **1950** ... **1951** ...

1955 ..*Parker's Bar* ..**1962** ...**1963** ... E. GOFFMAN *Where Action Is* 200 James Bond makes the acquaintance of an unattainable girl and then rapidly makes the girl.

66. *to make time* (*with*) (a person): to make advances to (a member of the opposite sex); to court or flirt with; also = sense 65 c above. *N. Amer. slang.*

1934 G. S. LORIMER *Stag Line* iv. 122 'You can't make any time with me,' I said giving him a lazy smile. 'I've got a date.' **1953** W. BURROUGHS *Junkie* (1972) xiv. 141 At another table two young men were trying to make time with some Mexican girls. **1962** E. LUCIA *Klondike Kate* viii. 171 The parties and going out at night. They made time with some kid until Marcia's age. **1971** E. BULLINS in W. King *Black Short Story Anthol.* (1972) 82 Say, are ya makin' much time with this bitch?

There was a guarded flash in his eyes. **1971** 'D. SHANNON' *Ringer* (1972) viii. 140 Frankly, he'd have liked to make time with that girl, but she'd turned up her nose at him. **1973** D. HUGHES *Along Side Road* (1974) ii. 6 Which'll he did it if he wanted to make time with her, eh?

70. *to make as if* (later examples).

1919 V. WOOLF *Night & Day* iv. 51 Denham..made as if he were tearing up handfuls of grass..from the carpet. **1967** 'M. YOURCENAR' *Coup de Grace* 86 With her fourth bundle she looked like a discharged servant-girl; shifting it from one arm to the other she made as if to escape. **1971** *Leader* (Durban) 7 May 9/2 The girls want to look like boys and the boys make as if they're girls.

b. *to make like*: to pretend (to be), to behave like or as if; to imitate. *colloq.* (orig. *U.S.*). Cf. sense 73 a in Dict.

a 1881 S. LANIER *Poems* (1892) 179 Then he..madelike he neither had seen nor heard. **1928** J. PETERKIN *Scarlet Sister Mary* x. 100 Any man might be good if he made like he was good. **1939** J. STEINBECK *Grapes of Wrath* xxiii. 396 This rich fella..he makes like he's poor. **1951** P. FRANKAU *Winged Horse* iii. 277 Couldn't make like me when he saw me and said he knew I'd be..ready for the new life. **1957** ..Next time around she made like a big shot and started the ball rolling by buying me a drink. **1973** R. WARE ..'Make like a tree and leave' ..riding on N.Z.' 20/3 On a large wave it is also possible to spread your arms and legs and 'make like' a kite in the surf. **1968** M. WOODHOUSE *Rock Baby* xxiii. 227 Well, you are making much noise, - For a guy who was making like an Iceberg.

72. (Later examples.)

1910 BELLOC *Verses* 3 The tide is making over Arun Bar. **1956** C. WILLOCK *Death at Flight* viii. 104 We shall hold this into a platform..in order to give us a little extra height when the tide makes.

b. Of ice: to form. *N. Amer.*

1784 N. WEBSTER in E. E. Ford *Notes Life N. Webster* (1912) I. 88 Cold; ice makes in the river. **1817** *Montreal Herald* 8 Feb. 2/5 The ice was making so rapidly that we quickly gained the safety of the place. **1888** J. MCDOUGALL *George Millward McDougall* 174 As soon as the snow falls and ice makes, dogs and hunt..fish..are at a great advantage..as they become the means of transport for the most part. **1890** *N.Y. Tribune* 12 Dec. 3/3 Several good guides..will await him in an attempt to reach Kadiak Island by crossing Alaska Peninsula before the ice makes. **1933** R. MERRICK *Fear North* ii. 16 Any day, he making everywhere. **1971** J. MCDOUGALL *Parsons on Planes* iv. 31 After ice makes, the fish freeze almost as soon as you take them out of the water.

82. make with —. d. [tr. Yiddish *mach mit.*] To bring into operation; to use, affect; to concern oneself with, play (orig. *U.S.*).

1940 J. O'HARA *Pal Joey* 131 The poor man's Bing Crosby is still making with the throat here in Chi. **1943** *Amer. Speech* XVIII. 46 To make with the eyes, to flirt with the eyes. **1949** ..to give forth a Bronx cheer'. *Ibid.* Come, Baby, make with your hands. **1959** W. BURROUGHS *Naked Lunch* 80 A.J., surrounded and fighting against ..with the hog-call. **1962** L. DEIGHTON *Ipcress File* xviii. 114 Dalby had..changed into a red Hawaiian shirt. 'You're making with the native costume' I said. **1971** E. MAINS' *Never let Up* xv. 149 Don't be a Smart Aleck. Make with the alcohol. I haven't got all night. **1973** 'H. HOWARD' *Nice Day for Funeral* iii. 34 Make with the feet into your bathroom. **1973** D. LEES *Zodiac* 188 Well people like Zodiac make with the dreams you have to listen. **1974** 'A. GILBERT' *Nice Little Killing* i. 6 Make with the coffee, hot, sugar.. You're embarrassing Jim. Time his place closed.

86. make down. b. (Example.)

1877 'S. TYTLER' *Childhood 100 Yrs. Ago* i. 19 Old clothes of their seniors carefully kept and 'made down' for their descendants.

91. make out. c. (*b*) Delete 'Chiefly *U.S.*' and add further examples; *to make it out* (example).

1861 W. COLLINS *Tom Tiddler's Ground* iv, in *All Year Round Extra Christmas No.* 39 Cobblers and tinkers and sons and farm-labourers who couldn't make it out in the city and then..**1925** *Bruce Carne* 167 *Action* 7, I took a fancy to seein' how the engines made out under war conditions. **1932** W. G. MCADOO *Crowded Yrs.* xi. 55 Without my knowledge..he thought I should not make it out. **1931** WODEHOUSE *Hot Water* x. 170 'How did you make out?' 'Okay. I'm in the château.' **1937** N. M. AYSCATT *Murder in Midsummer* vii. 64 Could we not try out his make out? **1950** ..**1958** ..**1961** ..is..made out this big yard and all, but we haven't..the ..work..**1942** J. DILL in W. S. Churchill *Second World War* (1951) IV. i. 17 ..**1968** ..**1971** ..**1966** ..

(c) *spec.* To gain sexual satisfaction; to have sexual intercourse (*with*). *slang* (orig. *U.S.*). Cf. *MAKING *vbl. sb.*[1] 10 b.

1939 J. BAIRD *Waste Heritage* vii. 99 Oh, say, how'd you make out with Hazel? **1961** *Times* 27 Apr. 17/2 The ill-defined object of being 'made out' sexually and emotionally with different girls. **1962** *Amer. Speech* XXXVII. 30 When I was young, if one 'made out', his accomplishment was a good night kiss; now it is usually implied by either *to neck* or *to pet*, or both.

92. make over. d. Delete 'Now only *U.S.*' and add further examples.

1903 H. JAMES *Ambassadors* viii. 111 The new quantity was represented by the fact that Chad had been made over. **1928** R. MACAULAY *Keeping Up Appearances* ix. 94 Feeling..in need of restoration, she..had a small port. That's better, she agreed with herself..makes you feel quite made over. **1936** 'J. TEY' *Shilling for Candles* xvi. 180 She..never recognized Chris... She'd heard that they made you over in Hollywood. Perhaps that was it. **1958** *Technology* Jan. 375/4 Had we better do what we can to 'make over' the traditional grammar school for an age of scientific industry? **1972** *Daily Tel.* (Colour Suppl.) 8 Sept. 26/1 A sewing machine stands on a table with a limp-looking dress, in the process of being 'made over', hanging on a nearby chair.

96. make up. 1. (*b*) (Earlier and later examples.) Also, of a woman: to apply cosmetics to (one's face). Hence **made-up** *a.*

1808 *Monthly Pantheon* I. 346/1 Yes, she produces a good effect!—she's well *made up*! *a* **1817** JANE AUSTEN *Persuasion* (1818) IV. x. 222 Morning visits are never fair by women at her time of life, who make themselves up so little. If she would only wear rouge. **1926** GALSWORTHY *Silver Spoon* III. vi. 258 Marjorie Ferrar stepped into the Box, not exactly nervous, and only just 'made-up'. **1930** V. SACKVILLE-WEST *Edwardians* i. 15 She was heavily but badly made-up, with a triangle of red on either cheek. **1945** S. LEWIS *Cass Timberlane* (1946) xlvii. 333 Small white wool socks..to be worn with bare legs that were made-up to look tanned. **1968** A. MUNRO in R. Weaver *Canad. Short Stories* 2nd Ser. 295 The women's faces in the room, made up some time before, have begun to show the effects of heat.

(*c*) (Further examples.)

1901 C. MORRIS *Life on Stage* v. 26 Of course when you are making up for a character part you go by a different rule. **1931** *Daily Tel.* 21 May 13/3 She came to a car in which one girl was smoking a cigarette and two others were busily 'making up'.

ma:ke-and-brea·k. Also without hyphens. [f. the vbs. as used in *to make contact* (MAKE *v.*[1] 8), *to break contact* (BREAK *v.* 29): see CONTACT I c.] **a.** The alternate making and breaking of electrical contact. Freq. *attrib.*

1857 *Chambers's Jrnl.* 22 Aug. 121/1 The transmission of the current having..to be made through a make and break key of metal. **1875, 1898** [see MAKE *sb.*[2] 9]. **1904** A. B. F. YOUNG *Compl. Motorist* (ed. 2) 197 The time of firing is controlled by a simple 'make-and-break' commutator placed on the half-time shaft. **1911** *Encycl. Brit.* XXVI. 520/2 In Squier and Crehore's 'Synchronograph' system 'sine waves' of current, instead of sharp 'makes and breaks', are employed for transmitting signals. **1938** *Archit. Rev.* LXXXIII. p. lviii, A semi-rotary quick make-and-break switch. **1958** *Newnes Compl. Amat. Photogr.* 120 To charge a large capacitor from accumulators requires a more complex power pack which includes a make-and-break coil, transformer and rectifier.

b. An apparatus for automatically making and breaking electrical contact.

1908 *Daily Chron.* 7 Nov. 3/3 The 'make and break' or commutator..is..frequently placed in an inaccessible position. **1911** A. B. SMITH *Mod. Amer. Telephony* xxv. 689 The circuit consisted of an induction coil..and mechanical 'make and break' that gave 10,000 sparks per second. **1929** *Times* 5 Nov. 7/5 The distributor and make-and-break are easily seen. **1970** K. BALL *Fiat 600, 600D Autobook* xi. 138/2 Failure of the horn to operate may be due to..dirty contacts on the horn make-and-break.

make-belief. Delete '*rare*' and add further examples.

1909 *Westm. Gaz.* 20 Aug. 4/1 The contempt with which all things are remembered, to which cling suggestions of the theatrical and the make-belief. **1960** J. RAE *Custard Boys* II. xix. 212 He had been my guide through the days of make-belief and blinkered memory. **1971** *Daily Tel.* 11 Oct. 9/1 That great reservoir of make-belief which lies hidden in so many children.

make-believer, -believing. (Further examples.)

1907 G. B. SHAW *Major Barbara* Pref. 149 The romantic make-believer lay outside the pale of sympathy in literature. **1909** *Times Lit. Suppl.* 18 Mar. 105/2 A little make-believing girl, not unlike..the make-believing child of the 'Child's Garden of Verses'. **1954** T. S. ELIOT *Confid. Clerk* I. 40 It begins as a kind of make-believe And the make-believing makes it real. **1957** L. MACNEICE *Visitations* 17 Coral islands of first love Where makebelieving boy and girl Assume the music of the spheres.

make-do (mē̆·k,dū̆). [f. *to make do*: see *MAKE *v.*[1] 53 f.] A makeshift; a temporary expedient. Also *attrib.* or quasi-*adj.*, characterized by makeshift methods. So *make-do-and-mend* (cf. *MAKE *v.*[1] 53 f).

1895 *Dialect Notes* I. 372 These 'ere make-dos are no 'count. **1916** KIPLING *Tales of 'Trade'* 13 The full tale of their improvisations and 'make-do's' will probably never come to light. **1923** *Daily Mail* 12 Feb. 4 When..prices steadily mounted to their peak, thousands of careful housewives adopted what was known as a 'make-do' policy. **1944** in R. W. Zandvoort et al. *Wartime English* (1957) 151 This garment is being made in the Make-do-and-Mend Department of the store from an old costume. **1948** *Times* 19 Aug. 5/2 Large schemes..must be ruled out; make do and mend must take their place. **1961** L. MUMFORD *City in Hist.* xiv. 420 The spirit of make-shift and make-do too often prevailed.

make-ready. 2. Delete *U.S.* and add further examples. Also *attrib.*

1915 *Southward's Mod. Printing* (ed. 3) II. iii. 39 As a preventive against slurring through baggy tympans as much of the make-ready as is practicable should be done at the back of the forme. **1946** *N. & Q.* 9 Mar. 93/1 Most early books are marked by bad paper, brownish ink, poor typography, lack of care in 'make-ready', and crude illustrations. **1959** *Times* 14 Jan. 12/4 A saving of 50 hours' make-ready time. **1967** KARCH & BUBER *Offset Processes* ii. 24 Makeready can be mechanical..with the 3 M makeready machine. **1970** E. A. D. HUTCHINGS *Survey of Printing Processes* 200 Make-ready, specifically, all operations carried out on the production machine from receipt of the job to the start of the production run.

makeshift. Add: **2.** Also *transf.*, of a person. Quot. 1848 may belong to sense 1.

1848 [see WIRE-PULLER]. **1912** [see *GAWD]. **1920** H. CRANE *Let.* 22 Dec. (1965) 50, I..hope that I have not been relinquished as one of Akron's temporary 'make-shifts' or 'reliefs'. **1951** E. PAUL *Springtime in Paris* ix. 162 Dr. Thiouville..completed his studies under the De Gaulle régime and the first few Middle-of-the-Road make-shifts.

make-up. Add: **1.** *spec.* The character or temperament of a person (cf. quot. 1821 in Dict.). Also *transf.*

1898 [see *GET *v.* 73 b]. **1908** W. MCDOUGALL *Introd. Social Psychol.* iv. 117 Constitutional conditions..that exert a constant influence..responsible for much in the mental make-up of the adult. **1928** *Morning Post* 23 Mar. 15 M. Poincaré..works alone, and has nothing of the people's tribune in his make-up. **1942** *R.A.F. Jrnl.* 13 June 5 The lack of conscience in the make-up of the Japanese. **1965** G. N. GARMONSWAY in Bessinger & Creed *Medieval & Ling. Stud.* 140 He becomes the prototype of the more ruthless type of Germanic hero, with a touch in his make-up of the berserk or ruffian. **1975** *Country Life* 6 Feb. 343/1 Dogs possessing faults in their make-up..are liable to pass them on to their offspring.

b. The balancing of accounts at the end of a certain period; cf. MAKE *v.*[1] 96 j. Also *attrib.*

1952 *Economist* 21 June 839 Electricity repayments by B.E.A. completed by the date of the May make-up. **1975** *Guardian* 20 Jan. 12/2 On the latest make-up date, the banks' acquisitions of interest-bearing liabilities were..higher.

2. b. Also used by women generally.

1932 'N. SHUTE' *Lonely Road* vii. 135 She had made herself look quieter than before. It may have been that she had less make-up on. **1959** J. BRAINE *Vodi* xxi. 229 Her mother would..put on a little makeup. The lipstick and rouge seemed to give the large face..not less but more dignity. **1974** *Listener* 7 Nov. 603/3 We've all got our 'titfers' on, four layers of make-up, every jewel we can find.

d. The action or process of 'making up' with cosmetics, etc.

1890 *Cent. Dict.*, *Make up*,..the preparation of an actor for impersonating the character assigned to him, including dress, painting and altering the appearance of the face, etc. **1922** J. ERSKINE *Coll. Poems* 120 If this world be a stage, what hours we give To tedious make-up in the tiring-room. **1930** *Punch* 8 Jan. 47/2 One 'make-up' each morning, says Popkin, is plenty. **1951** *Good Housek. Home Encycl.* 20/2 It is a good idea to have also a magnifying mirror for shaving or make-up. **1974** *Country Life* 28 Nov. 694/3 A cleanse and make-up from £2.50, a make-up lesson £5.

e. A make-up room; the place where an actor or the like is 'made up'; the work of such a place. *colloq.*

1960 *Guardian* 5 May 9/4, I..lost no time phoning Groucho, who was 'in make-up'. **1967** M. SHULMAN *Kill 3* III. viii. 146 'She's been to make up?' 'Only a bit of powder. We thought the van look was best.' **1972** *New Statesman* 28 Apr. 554/3 There are still important women in BBC TV. They have four of the top 70 jobs, to do with planning, education and children's programmes. And Head of Make-up, of course. **1973** *Listener* 20 Dec. 838/3 They've got a new girl in make-up here and I'm not sure whether she has caught the Ronnie Corbett look.

4. b. *colloq.* or *dial.* Something (esp. food) made up from odds and ends. Also *attrib.*

1854 A. E. BAKER *Gloss. Northamptonshire Words* II. 6 *A make-up dinner*, a dinner composed of scraps and remnants. **1877** J. M. NEILSON *Poems* 48 The treacle mak'-up on the candyman's stand. **1924** 'L. MALET' *Dogs of Want* v. 129 The sort of refreshments caterers supply at a public entertainment like this—left-overs and make-ups, from the local restaurants.

5. b. Replacement of the water lost from a boiler or the like by evaporation, leakage, etc.; water added for this pupose.

1930 *Engineering* 10 Jan. 56/1 Others..treat the entire boiler feed water, including condensate and make-up, [etc.]. **1971** *Sci. Amer.* May 71/2 The amount of water required for 'makeup' in a large cooling tower is considerable. **1972** R. G. KAZMANN *Mod. Hydrol.* (ed. 2) iv. 133 A 500 megawatt plant will use 500,000 gallons of water an hour for makeup purposes.

6. (sense 2 a) *make-up girl, man* (later examples); (sense *2 b) *make-up kit*; (sense *2 d) *make-up mirror*; (sense 3) *make-up copy, editor, hand, man*; **make-up bed**, a bed that can be set up temporarily.

1911 A. BENNETT *Hilda Lessways* IV. i. 276, I quite forgot about the make-up bed for Florrie. **1960** G. A. GLAISTER *Gloss. Bk.* 249/1 *Make-up copy*, i.e., a set of folded sheets and plates put in the correct order as a pattern. **1903** J. RALPH *Making of Journalist* xiii. 153 Around the walls are the desks..that of the night or 'make-up' editor, and that of the managing editor. **1972** H. EVANS *Editing & Design: Newsman's English* i. 8 In the composing room ..an editorial man watches... He may be called a production editor, a make-up editor or a stone sub or stone editor. **1957** *Times* 28 Aug. (Radio & T.V. Suppl.) p. xix/2 Ordeal by Television..under the impersonal inspection of..camera-men, make-up girls. **1971** 'W. HAGGARD' *Bitter Harvest* vii. 74 A make-up girl had powdered his bald patch... Now he sat..in the sweaty studio. **1949** R. HOSTETTLER *Technical Terms Printing Industry* 122/2 Make-up hand. **1955** A. HUXLEY *Genius & Goddess* 51 Ruth..squandered a year's accumulated savings on a make-up kit and a bottle of cheap perfume. **1922** S. A. CUNEO *From Printer to President* 43 It was necessary for the owner to be a typesetter, a make-up man, a press feeder, in fact an all 'round printer. **1959** Make-up man [see **goose pimples*]. **1973** W. MCGIVERN *Reprisal* (1974) 116 A part in a picture..a make-up man to fuss over her. **1964** *Harper's Bazaar* Nov. 99 The make-up mirror..in black opaline and ormolu, £27. **1966** H. NIELSEN *After Midnight* (1967) vi. 86 A make-up mirror on the bedside table.

make-work (mē̆·k,wǝ̄ɹk). orig. *U.S.* [f. MAKE- + WORK *sb.*] Work or activity of little or no value devised mainly to keep someone busy. Also *attrib.*

1937 *Political Q.* VIII. 608 A nation-wide organization of relief work, including..genuine public works and a host of make-work projects of questionable merit. **1947** *Yale Law Jrnl.* Dec. 182 Community development subordinated to make-work and pump-priming. **1957** [see *cost-accounted* ppl. a. s.v. *COST *sb.*[2] 6]. **1972** *Guardian* 13 Sept. 12/1 Still the old make-work continues, the mailbags and the excessive cleaning, which..occupy almost a quarter of the prison population.

Makhzen, -an: see *MAGHZEN.

‖ **makimono** (makimǭ·no). Also **emakimono** [Jap. *e* picture, painting], **makemono**, (with hyphen) **maki-mono**. [Jap., something rolled up, a scroll.] A Japanese scroll containing pictures, usually with explanatory writing, to form a continuous narrative, designed to be examined progressively from right to left as it is unrolled.

1882 G. A. AUDSLEY *Ornamental Arts Japan* I. plate 16 (*caption*), She appears to be rising out of the mist at her feet bearing a *makimono* covered with writing. **1884** SATOW & HAWES *Handbk. for Travellers Cent. & N. Japan* (ed. 2) 554/2 *Maki-mono*, a long roll graphic or pictorial. **1889** S. BING in *Artistic Japan* III. 161 All the paintings.. are executed..on rolls of silk or paper, which are known in the Japanese language as *Kakémono* (that is to say, a thing to hang up) when the subject is upright, and under the word *Makimono* when it stretches itself out in a horizontal manner. **1901** L. HEARN *Jap. Misc.* 243 The shrine contained a *makémono*, or scroll, inscribed with the spirit-names of many ancestors. **1911** L. BINYON *Flight of Dragon* x. 81 The makimono, or continuous long scroll-paintings of landscapes, admirably fulfilled the aim of Taoist art. **1926** A. WALEY tr. *Sacred Tree* 31 The first painted makimonos..were regarded merely as a succession of topographical records, joined together more or less fortuitously. **1958** Y. YASHIRO *2,000 Yrs. Jap. Art* iv. 161 The finest expression and the fullest development of *Yamato-e* are the *e-makimono* (scroll-paintings). **1966** J. ROSENFIELD tr. *Noma's Arts Japan* vi. 122 These long horizontal paintings, the *emakimono*, are sometimes said to be peculiarly Japanese in form, but similar scrolls have been made by the Chinese as well.

making, *vbl. sb.*[1] Add: **8. b.** *pl.* Paper and tobacco for rolling a cigarette. *N. Amer., Austral.*, and *N.Z. colloq.*

1905 'O. HENRY' in *Everybody's Mag.* Dec. 817/1 He took out his 'makings' and rolled a cigarette. **1907** S. E. WHITE *Arizona Nights* I. ix. 161 'Well,' agreed Rogers, 'pass over the "makings" and I will.' **1910** R. W. SERVICE *Trail of '98* (1911) I. iv. 15 'Got the makings?' 'No, I'm sorry; I don't smoke.' **1912** *Collier's* 21 Dec. 23/2 A revelation to any man who doesn't know a real 'makin's' cigarette. **1930** J. DEVANNY *Bushman Burke* 26 He grinned and took out the 'makings'. **1940** *Amer. Speech* XV. 213/1 The day before payday, the camp's 'smoking' has become scarce and 'rollings' or 'makings' are at a premium. **1941** *Coast to Coast* 207 We rolled our makings, lit them and stood about waiting for someone to start talking about something. **1949** S. P. LLEWELLYN *Troopships* 7 The normal Kiwi..his clothes in a heap beside him, the 'makings' handy. **1963** B. PEARSON *Coal Flat* iv. 70 Rogers offered him a cigarette but McKenzie shook his head and pulled out the makings in a way that seemed to suggest that tailor-mades were a stigma of middle-class ideology. **1963** H. GARNER in R. Weaver *Canad. Short Stories* (1968) 2nd Ser. 30, I gave McKinnon enough money to get me a package of makings.

10. b. making-out, the action of the vbl. phr. *to make out* (*MAKE *v.*[1] 91 c (*c*)).

1957 F. KOHNER *Gidget* vii. 76 'No sweaty hands, no making out in drive-in movies.' 'Making out?' 'My God, Larry, where've you been living. I guess you still call it necking.' **1964** C. NICOLAI *Murder in Fine Arts* xxviii. 248 He had a violent aversion to necking or 'making out' as the current expression had it. **1967** 'J. H. ROBERTS' *February Plan* I. iv. 113 She had been..putting him off when it came to making out (how many evasive phrases were there for the act of sexual intercourse?). **1969** E. GOFFMAN *Where Action Is* 157 Adult males may define a female as an object to initiate a sexually potential relationship with... This action is sometimes called 'making out'. In our society there are special times and places set aside for making out.

making-up. Add: (Further examples.)

1808 C. STOWER *Printer's Gram.* 467 As soon as the first person brings him his matter, he [*sc.* the clicker] counts off the number of lines,..then gives him another taking of copy, and proceeds with the making up [of the type pages]. **1816** JANE AUSTEN *Emma* I. xii. 206 She hoped they might now become friends again... Making-up indeed would not do. She certainly had not been in the wrong. **1827** L. T. REDE *Road to Stage* 49 My readers will call to mind the excellent 'making up' of Mr T. P. Cooke, in the Monster, 'Frankenstein'. **1838** DICKENS *Let.* 15 Nov. (1965) I. 453 On the other side I forward you a copy of the making-up. **1841** W. SAVAGE *Dict. Art of Printing* 470 When a compositor is in a companionship he has composed his copy to within the quantity of a page of the work, he gives the overplus of his copy after having composed his own last page, to him who is composing the copy that follows his matter... This is called giving the making-up. **1898** E. C. MADDISON *Speculation on Stock Exchange* 25 The process known as 'making-up'..arrival of the..contango-day.. **1905** J. C. LINCOLN *Partners of Tide* iv. 74 His worst quarrel with Gus and her friend, Clara Hopkins—the quarrel that lasted two weeks without a making up—came about because the new member refused to tell what the initials 'stood for'. **1913** *Trans. Bibliogr. Soc.* XII. 303 There seems no reason to suppose..making-up of copies in the early part of last century.

b. (Earlier and later examples.)

1875 *Encycl. Brit.* I. 95/1 The price at which such transactions are adjusted is the 'Making-Up' price of the day. **1969** *Jane's Freight Containers 1968–69* 195/1 Three rail sidings are available for 'making-up' trains.

‖ **makkoli** (ma·kǒli). [Korean.] A popular alcoholic beverage in Korea.

1972 P. M. BARTZ *S. Korea* 73/1 Domestic production [of wheat] is one-quarter the volume of imports and is used mainly for noodles and *makkoli*, the common alcoholic beverage. **1972** E. FODOR *Japan & E. Asia* 477 These are both cloudy and light-tan colored and are known together as makkoli.

mako[1] (ma·ko). [Maori.] In full, **mako shark**. A large blue shark of the genus *Isurus*, esp. *I. oxyrinchus*; also called mackerel shark.

1727 J. G. SCHEUCHZER tr. *Kæmpfer's Hist. Japan* I. i. 133 Mako never exceeds three or four fathom in length. **1848** R. TAYLOR *Leaf from Nat. Hist. N.Z.* 24 Mako, a shark, peculiar..the teeth..as ear ornament. **1871** F. W. HUTTON *Fishes N.Z.* 77 *Lamna Glauca*. Tiger Shark. Mako... The shark from which the Maoris obtain the teeth which they decorate their ears. **1936** ..wore pieces of mako tooth.. **1938** *Times Lit. Suppl.*..land waters.. He was a very big Mako shark built to swim as fast as the fastest fish in the sea. **1959** *Manch. Guardian*..The sharks come to Cornwall..blue sharks, makos, porbeagles. **1964** A. WHEELER *Fishes Brit. Isles & N.-W. Europe* 53/1 The mako has no commercial value beyond its angling appeal. **1970** *E. Afr. Standard* (Nairobi)..19/4 The big fish..was the capture of Kenya's first mako shark. **1973** *Shooting Times & Country Mag.* 27 May 17/1 The four British species of shark are blue, mako, porbeagle and thresher.

mako[2], var. *MAKOMAKO[2].

Makololo (makǒlō·lo), *sb.* and *a.* Also **Makalolo, Ma Kololo. A.** *sb.* A Negro people of Africa now living in Zambia near the junction of the Zambesi and Kafue rivers; a member of this people. **B.** *adj.* Of or pertaining to this people.

1857 D. LIVINGSTONE *Missionary Trav. & Res. S. Afr.* ix. 179 Mamochisane..said she would never consent to govern the Makololo so long as she had a brother living. *Ibid.* x. 204 The Makololo chiefs pride themselves on eating with their people. **1867** [see *SAMBOK *sb.*]. **1871** J. MACKENZIE *Ten Yrs. North of Orange River* ii. 34 Both the Matebele and Makololo countries were far beyond the territories of the Transvaal. **1871** C. M. YONGE *Pioneers & Founders*..There was no choice but to have recourse to the Makololo, and thus let loose one set of savages against the other. **1904** A. H. KEANE *Boer States* vii. 101 The Makololo and other Basuto..founded a powerful State.. **1923** ..*Louis Trigardt's Trek*..The foundation of the Ma Kololo nation. **1947** ..The largest group of houses..was at Linyanti, in Makololo territory. *Ibid.*..Among the Bakuena, and the Makololo..there was no despotic cruelty.

makomako[1] (mā·komā·ko). *N.Z.* [Maori.] = *KORIMAKO.

1848 R. TAYLOR *Leaf from Nat. Hist. N.Z.* 8/2 Makomako, a bird.. **1873** W. L. BULLER *Hist. Birds N.Z.*..*Anthornis Melanura* (Bell-Bird)... Native names.. Makomako, Korimako.. Kopara. Of the above names, Korimako is most generally used by the northern and Makomako by the southern tribes. **1875** W. E. ATKINSON *Let.*..in *Richmond-Atkinson Papers* (1960) II. 399, I have also the mako mako, silver-eye, [etc.]. **1911** *Encycl. Brit.* XIX. 626/2 The tui and makomako rank high as songsters. **1966** [see *KORIMAKO].

makomako[2] (mā·komā·ko). *N.Z.* Also mako. [Maori.] A small New Zealand tree, *Aristotelia serrata* (or *A. racemosa*), of the family Elæocarpaceæ, which bears clusters of small pink flowers and dark red berries; also called wineberry.

1848 R. TAYLOR *Leaf from Nat. Hist. N.Z.* 20/2 *Mako*, a tree; the bark used as a black dye. *Makomako*, a tree (*Friesia racemosa*). **1861** A. S. ATKINSON *Jrnl.* 24 Mar. in *Richmond-Atkinson Papers* (1960) I. 696 Just before the fireplace were a few makos which they had been sleeping on. **1866** LINDLEY & MOORE *Treas. Bot.* II. 92/1 The berries of *Aristotelia racemosa*, the Mako-Mako of the natives, are eaten. **1883** J. HECTOR *Handbk. N.Z.* (ed. 3) 130 Mako, a small handsome tree, six to twenty feet high, quick-growing, with large racemes of reddish nodding flowers. **1949** P. BUCK *Coming of Maori* (1950) II. ix. 239 The spears were formed of..rods of light *mako* wood. **1966** *Encycl. N.Z.* II. 670/1 (*caption*) *Aristotelia serrata*, makomako or wineberry, is a small tree of graceful appearance.

Makonde (măkǭ·ndi), *sb.* and *a.* **A.** *sb.* A Bantu-speaking people of Tanzania and Mozambique; a member of this people. **B.** *adj.* Of or pertaining to this people.

1911 *Encycl. Brit.* I. 330/2 Bantu negroids... Ma-konde. *Ibid.* III. 359/1 The archaic *Makonde* or *Mabiha* [language] of the lower Ruvuma, and the coast between Lindi and Ibo. **1948** D. DIRINGER *Alphabet* 26 The Makonde people, an important tribe of Tanganyika Territory. **1973** *Times* 17 Feb. (Mozambique Suppl.) p. iv/3 (Advt.), The Makondes were more given to tattooing. **1974** *Times* 2 May 5/2 Mr Simon Makada, a Makonde..and a Frelimo youth leader. **1974** *Afr. Encycl.* 317/3 The Makonde are traditionally an agricultural people... In Mozambique many Makonde and Makua people are involved in war against the Portuguese administration.

makoré (makǒre·). Also **makora, makori.** [Native name in West Afr.] A large West African tree, *Tieghemella* (or *Mimusops*) *heckelii*, of the family Sapotaceæ; also, the dark red-brown wood produced by it.

1915 J. H. HOLLAND *Useful Plants Nigeria* III. 422 *Dumoria Heckeli*... Vernac[ular] names... Mako, Makoré, Makerou. **1936** *Nature* 9 May 791/1 Skirting and dado in specially cut logs of Makoré with small marquetry banding in the Makoré doors of maple burr. **1947** J. C. RICH *Materials & Methods of Sculpture* x. 292 Makore is also known as Makori, Makora, and African cherry. The wood is beautifully figured. **1972** *Handbk. Hardwoods* (Building Res. Establishment) (ed. 2) 121 Marine plywood of makoré is also widely used in boat-building.

maktuk, var. *MUKTUK.

makuk, var. *MOCOCK.

makumba, var. *MACUMBA.

makuta: see *LIKUTA.

makutu (măkū̆·tu). *N.Z.* [Maori.] Sorcery, witchcraft; a magic spell. Also as *v. trans.*, to bewitch.

1846 J. WHITE *Jrnl.* (MS.) 1 If anyone..offended us.. we should try to kill them by makutuing them to see if makutu was true. **1905** W. B. *Where White Man Treads* 41 Another decimating tool in the hands of spite and revenge was makutu (witchcraft)... Let one fear that a makutu spell had been laid upon him,..he would..die! **1938** R. D. FINLAYSON *Brown Man's Burden* 70 She was sure someone had put a makutu spell on her and the evil spirit was eating her life away. **1943** N. MARSH *Colour Scheme* i. 25 'I'll tell my great-grandfather..and he'll *makutu* you.' 'Kid-stakes! Nobody's going to put a jinx on me.' **1963** *Evening Post* (Wellington, N.Z.) 16 Dec., The true tohunga whiawhia, who had gained complete mastery of the powers of makutu, was able to project his spirit in such a manner as to kill men and women. **1971** *N.Z. Listener* 29 Mar. 11 A number of modern case histories of patients whose illness or death has been connected with makutu.

Mal, var. *MALER *sb.* and *a.*

mal-. Add: *malperformance.*

1938 R. G. COLLINGWOOD *Princ. Art* xii. 283 It is the malperformance of the act which converts what is merely psychic (impression) into what is conscious (idea). **1964** M. CRITCHLEY *Developmental Dyslexia* ix. 60 Dyslexics show only a mild tendency towards a malperformance of higher order right-left orientation exercises.

Malabar. Add: **1.** An inhabitant of the Malabar coast.

1582 N. LICHEFIELD tr. *Lopez de Castaneda's Hist. Conquest E. Indias* f. 37[v] They asked of the Malabars which went with him what he was? **1681** R. KNOX *Hist. Relation Ceylon* IV. ix. 159 This Plain is..inhabited by Malabars, a distinct People from the Chingulayes. **1867** C. J. BOYLE *Far Away* 72 Started on foot up the gorge, our bags on the shoulders of Malabars. **1931** M. YEO *St. Francis Xavier* xii. 155 One night his faithful Malabar..woke to the sound of blows. *Ibid.*, The Malabar told his tale.

2. The language spoken on the Malabar coast.

1801 T. MUNRO in G. R. Gleig *Life T. Munro* (1830) I. 322 From Miliserum to the Chandergeery river no language is understood but the Malabars of that coast. **1837** T. BACON *First Impr. Hindostan* I. 99 He was compelled to fall back upon his only two words of Malabar. **1872** [see TAMULIC *a.*]. **1931** M. YEO *St. Francis Xavier* x. 124 Tamil, the pre-Aryan language of southern India, which Francis, Portuguese-fashion, calls Malabar.

3. A kind of handkerchief (see quots.).

1882 CAULFEILD & SAWARD *Dict. Needlework* 340 *Malabars*, cotton handkerchiefs, printed in imitation of Indian handkerchiefs, the patterns of which are of a peculiar and distinctive type, and the contrasts of colour brilliant and striking. **1957** M. B. PICKEN *Fashion Dict.* 218/2 *Malabar*, cotton handkerchief printed in brilliant colors and designs of East Indian type.

4. *attrib.*

1696 J. OVINGTON *Voy. Suratt* 213 Many of their Women by their usual Custom..have gain'd the Name of Malabar Quills. **1778** Malabar language [see TAMIL]. **1872** tr. *St. Francis Xavier's Let.* in H. J. Coleridge *Life & Lett. St. Francis Xavier* II. 73 Enrico Enriquez ..writes and speaks the Malabar tongue.

Hence **Malabar-ese** *a.* [-ESE], of or pertaining to the Malabar coast or its inhabitants; **Malaba·rian** *a.* and *sb.*; **Malaba·ric** *a.* and *sb.*; **Malaba·rish** *a.* and *sb.*

1709 A. W. BOEHM *Propagation of Gospel in East* 17 This place (*sc.* Tranquebar) is altogether stocked with Malabarian Heathens. *Ibid.* 28, I will set down here..the Malabarick Letters. *Ibid.*, I caused..The Lord's Prayer.. to be put into Malabarick. **1717** J. T. PHILIPPS *Acct. People Malabar* i. 2 According to the Malabarish way of Reasoning. *Ibid.* 8 The chief Reasons why the Malabarians refuse to embrace the Christian Religion. *Ibid.* 53 There are some things..which are very Malabarick. **1718** ZIEGENBALG & GRUNDLER *Let.* in R. Millar *Hist. Propagation of Christianity* (1731) II. viii. 305 We catechise..every Friday in Malabarian. **1788** Malabarian [see TAMIL]. **1808** C. STOWER *Printer's Gram.* Index, Malabaric alphabet. **1922** *Blackw. Mag.* May 611/1 The Moplahs..are the descendants of Arab fathers and Malabarese mothers. **1922** O. JESPERSEN *Language* iii. 65 He..was also the first to see that the Dravidian (by him called Malabaric) languages were totally different from Sanskrit.

ma:labso·rption. *Physiol.* [f. MAL- + ABSORPTION.] Imperfect absorption (of food material by the body).

1932 DORLAND *Med. Dict.* (ed. 16) 739/2 *Malabsorption*, disorder of normal nutritive absorption; disordered anabolism. **1962** *Lancet* 2 June 1168/2 The rickets of coeliac disease is commonly attributed to malabsorption of vitamin D. **1974** PASSMORE & ROBSON *Compan. Med. Stud.* III. xix. 58/1 The term malabsorption syndrome is applied to a number of conditions that affect the gut or digestive glands and which result in disturbance of the absorption of one or more nutrients.

Malacca. Malacca cane (earlier and later examples).

1844 *Ainsworth's Mag.* VI. 501 He..struck the pavement..with the brass-shod point of his Malacca cane. **1846** A. J. H. DUGANNE *Daguerreotype Miniatures* 7 He held..a musk cigar, and twirled in his right..a unique and delicate Malacca cane. **1963** J. KIRKUP *Tropic Temper* 83 There were some stalls selling..the famous Malacca Canes, stout swinging-sticks with curved and decorated handles. **1965** C. SHUTTLEWORTH *Malayan Safari* vi. 88 Malacca cane grows in clumps in the jungle.

malachite. Add: **2. malachite kingfisher**, a small blue, green, and red kingfisher, *Corythornis cristata*, found in Africa south of the Sahara; **malachite sunbird**, a blue and green sunbird, *Nectarinia johnstoni*, found in parts of southern and eastern Africa.

[**1875** *Ibis*..Malachite-crested Kingfisher. This charming little Kingfisher is found invariably frequenting the small streams and ditches close to Durban.] **1903** A. C. STARK *Birds S. Afr.* III. 82 The Malachite Kingfisher is found throughout the greater part of Africa south of the Sahara. **1906** *Daily Chron.* 20 Feb. 7/7 A great rarity is a Malachite sun bird, a most graceful.. formed bird with a bill as long as its body and as fine as a piece of wire. **1947** J. STEVENSON-HAMILTON *Wild Life S. Afr.* xxxiv. 300 The beautifully coloured malachite kingfisher (*Corythornis cristata*)..likes to sit on an overhanging branch. **1966** E. PALMER *Plains of Camdeboo* xi. 297 The male malachite sunbird..green and gold. **1971** *Country Life* 28 Oct. 1127/2 Tiny malachite kingfishers, brilliant red and blue, darted along the tall rushes.

malacon (mæ·lăkǫn). *Min.* Also † **malacone.** [ad. G. *malakon* (T. Scheerer 1844, in *Ann. d. Physik* LXII. 436), f. Gr. μαλακόν, neut. of μαλακός soft.] A soft brown altered form of zircon.

1854 DANA *Syst. Min.* (ed. 4) II. ii. 196 Malacone. **1868** *Ibid.* (ed. 5) 275 The following tetragonal zircon-like minerals are probably altered zircons... Malacon. **1921** *Mineral. Abstr.* I. 173 The malacon of Madagascar shows a higher degree of radioactivity than is accounted for by its content of thorium. **1954** *Amer. Jrnl. Sci.* CCXXXIX. 306 The third type of [zircon] is characteristic of the younger granite in the Oxford House Area, and of the late pre-Huronian and Huronian granites of the Lake Superior region. This variety is malacon, a form of zircon with relatively low refringence, weak birefringence and a dirty, altered appearance. **1968** I. KOSTOV *Mineral.* II. v. 472 Malacon is a metamict zircon.

maladaptation. Add: Hence **malada·ptive** *a.*; **malada·ptively** *adv.*

1931 *Brit. Jrnl. Psychol.* Jan. 334 Prof. Fisher begins his study of the abnormal by considering..how alternative the responses of the normal individual often are, when he is faced with a difficult situation. *Ibid.* Oct. 132 Both images and symbols function maladaptively [in mental defectives]. **1957** P. LAFITTE *Person in Psychol.* v. 58 Three modes of adaptive and eight modes of maladaptive behaviour that are used in response to frustration. **1969** *Sci. Jrnl.* Nov. 70/3 Long fingers are..generally maladaptive in terrestrial primates. **1971** *Sci. Amer.* Sept. 132/3 Some aspects of what we have called progress or social evolution may be maladaptive. **1974** *Times Lit. Suppl.* 30 Aug. 923/1 Man's instincts..have become seriously maladaptive. *Ibid.* 923/5 The hypertrophied human brain has become as maladaptive as the dinosaur's armour.

‖ **malade imaginaire** (malad imaʒinɛ̄r). [Fr. (after the title of Molière's play, 1673); cf. MALADE.] A person suffering from an imaginary illness.

1818 LADY MORGAN *Florence Macarthy* IV. v. 202, I trust you will not think I am playing the *Malade Imaginaire*, when I assign indisposition as an excuse for my absence. **1888** *Athenæum* 15 Dec. 811/1 A *malade imaginaire*..who sacrifices her daughter to her own comfort. **1911** G. B. SHAW *Doctor's Dilemma* Pref. p. xxviii, Doctors ..nurse the delusions of the *malade imaginaire*. **1958** M. STEWART *Nine Coaches Waiting* v. 60 Enough pills and boluses to satisfy the most highly-strung *malade imaginaire*. **1974** E. AMBLER *Dr. Frigo* II. 80 He's..inclined to worry about his health. If you mean do I think him a malade imaginaire, no I don't.

‖ **maladif** (maladīf), *a.* [Fr.; cf. MALADIVE *a.*] = MALADIVE *a.*

1481 [see MALADIVE *a.*]. **1859** QUEEN VICTORIA *Let.* 8 Jan. in R. Fulford *Dearest Child* (1964) 157, I thought he was too maladif to marry. **1886** *Portfolio* 51/2 Standing in the corner..a fragile, *maladif* little figure, with a disproportionately impressive head. **1911** D. H. LAWRENCE *Let.* 30 Dec. (1962) I. 90 You'll find it [*sc.* the story], perhaps, thin—*maladif*. **1950** D. GASCOYNE *Vagrant* 55 For a maladif, mandarin-miened, mauve melody-man.

maladive, *a.* (Later examples.)

1907 R. BROOKE *Let.* Oct. (1968) 112 Lippo Lippi's girl-Madonna's, with the maladive sensuality of their tired faces. **1940** *Horizon* Mar. 171 There is the road to New-church-on-Arrow, up which Kilvert..used to stride to see his sweet but *maladive* Emmeline. **1953** *Economist* 6 June 641 The most dangerous trend in post-war Britain..the tendency..to indulge in maladive self-attention.

maladju·sted, *a.* [f. MAL- + ADJUSTED *ppl. a.*] Inadequately adjusted; exhibiting or characterized by psychological maladjustment.

1885 [see MAL-]. **1930** R. S. WOODWORTH *Psychol.* (ed. 8) xiii. 564 Ways of helping maladjusted individuals. **1931** E. WILSON *Axel's Castle* iv. 93 Only thirty at the time of his death, Tristan Corbière had been an eccentric and very maladjusted man. **1953** R. LEHMANN *Echoing Grove* 32 Where did she get this magpie streak from? Can she be maladjusted? **1957** P. LAFITTE *Person in Psychol.* vi. 75 Persistence is characteristic of the maladjusted person, who cannot change because he is unable to learn by getting to grips with his situation. **1970** R. C. CARSON *Interaction Concepts of Personality* vi. 203 The families from which maladjusted persons emerge.

maladjustment. Add to def.: *spec.* in *Psychol.*, unsuccessful adaptation to one's social environment. (Further examples.)

1909 E. T. DEVINE in *Fabian News* Dec. 94/1 Not poverty and not punishment explains the misery of our modern commercial and industrial communities, but rather social maladjustment. **1923** J. S. HUXLEY *Ess. Biologist* i. 55 Whatever overstress and maladjustment the complexity of modern civilization has brought with it. **1930** R. S. WOODWORTH *Psychol.* (ed. 8) xiii. 565 This.. picture of the situation in which the maladjustment arose is of value to the troubled adult. **1949** E. JENKINS *Six Criminal Women* 166 Cook did not take to the road through want, but rather from what the psychiatrists call 'maladjustment'. **1965** J. R. WITTENBORN in B. B. Wolman *Handbk. Clin. Psychol.* 1039/1 Evidence of periodic or constant maladjustment of varying degree from early life. **1970** R. C. CARSON *Interaction Concepts of Personality* vi. 203 Processes that appear to be implicated in the acquisition of maladjustment within the family.

† **malaëra·tion.** *Med. Obs.* [f. MAL- + AERATION.] Imperfect aeration (of the blood).

1843 R. J. GRAVES *Syst. Clin. Med.* xx. 245 Her countenance exhibits no proof of malaëration of the blood. **1908** *Practitioner* Mar. 321 Such cyanosis may..indicate lack of propelling power in the circulation, with consequent malaëration.

malagas (mæ·lăgæ·s). *S. Afr.* Also 8 **malagos; malagash,** Afrikaans **malgas** (malχa·s). [Afrikaans, f. Du. *mallegas*, a. Pg. *mangas*(*-de-velludo*), 'velvet-sleeves', the wandering albatross.] The Cape gannet, *Morus* (or *Sula*) *capensis*.

1731 G. MEDLEY tr. *Kolb's Present State Cape Good-Hope* II. 143 As soon as the *Malagos* 'spies a Fish under her, she pops her Head nimbly into the Water. **1906** STARK & SCLATER *Birds S. Afr.* IV. 17 *Sula capensis*. Malagash. *Ibid.* 19 The Malagash is found along the coasts of South Africa. **1936** E. L. GILL *First Guide S. Afr. Birds* 186 Malagas, Malgas, Cape Gannet; *Morus capensis*. This is the gannet of the South African guano islands. **1944** M. DE B. NESBITT *Road to Avalon* (1949) xxi. 163 The malagas seabird..is prolific here. **1964** A. L. THOMSON *New Dict. Birds* 330/1 The Cape Gannet (or Malagash) *S*[*ula*] *capensis* (with dark tail), breeding off South Africa and migrating northwards. **1970** MCLACHLAN & LIVERSIDGE *Roberts's Birds S. Afr.* (ed. 3) 21 Cape Gannet. Malgas. *Morus capensis*... These birds may be picked up at some distance by their conspicuous whiteness. **1974** *Eastern Province Herald* 7 Nov. 22 There are vast colonies of malgas (gannets) and other birds which live off the shoals of fodder fish in our waters.

Malagueña (mælăge·nyă). [Sp. (see MALAGA).] **1.** A woman of Malaga.

1845 R. FORD *Hand-bk. for Travellers Spain* I. 352/1 Teresita, the daughter, is a pretty specimen of a *Malagueña*. **1868** H. O'SHEA *Guide to Spain & Portugal* (ed. 3) 312/2 The Malagueñas are considered to be the prettiest women in all Spain.

2. (Also with small initial.) A Spanish dance resembling the fandango; also, a song accompanying this dance.

1883 GROVE *Dict. Mus.* III. 599/2 The dance-songs—*fandangos, rondeñas*, and *malagueñas*,..have..more symmetry and more animation. *Ibid.*, Songs and dances often derive their names from the provinces or towns in which they are indigenous; thus..*malagueña* from Malaga. **1922** J. HERGESHEIMER *Bright Shawl* (1923) 69 The ceaseless playing of the guitars, strains of jotas and malagueñas. **1925** *Blackw. Mag.* Jan. 80/2 Hansson sang a queer malagueña which he had heard often in Andalucia. **1968** K. BIRD *Smash Glass Image* ix. 113 The next dance was slow, nostalgic. A *malagueña* perhaps? **1973** *Times* 6 Jan. 14/4 One typical dance [in the Canaries], the *malagueña*, reveals its peninsular origin in its name.

malaky, var. *MALARKEY.

malamute (mæ·lămiut). Also **malemut(e).** [The name of an Eskimo tribe living on the Alaskan coast.] A large grey or black and white dog with a thick coat, pointed ears, and a plumed tail curling over its back, belonging to the spitz breed so called, which first developed in Alaska. Also *attrib.*

[**1874** *Ann. Rep. Board of Regents Smithsonian Inst.* 27 Maglemut.] **1898** *Klondike Nugget* (Dawson, Yukon Territory) 20 July 3/2 After appointing a committee to keep stray Malamoots and donkeys off the diamond, the game was called. **1898** *Yukon Midnight Sun* (Dawson, Yukon Territory) 1 Oct. 2/5, I have the first man to find in the country who owns a dog, it matters not whether he is malamute, husky,..or little mongrel. **1908** *St. Nicholas* Mar. 387/2 Few pure malamutes..are now employed in the mail service. **1913** *Outing* Feb. 521/1 Sitting among a litter of Malamute puppies. **1922** G. C. F. PRINGLE *Tillicums of Trail* 84, I first picked on a big grey-muzzled malemute named Steal as a likely fellow to lead... All malemutes are born thieves, some men think. **1934** *Times Educ. Suppl.* 15 Sept. p. iv/2 These Arctic or sub-Arctic dogs—huskies, malamutes, Samoyeds..are wonderful in their strength, powers of endurance, and stoutness of heart. **1948** *Chicago Tribune* 23 Feb. IV. 1/6 Along with the Eskimos will be a representation of the Malemute sled dogs that provide the only motive power available to Eskimos. **1952** M. K. WILSON tr. *Lorenz's King Solomon's Ring* x. 117 The purest wolf-dogs that exist are certain breeds of Arctic America, particularly the so-called malemuts, huskies etc. **1955** *Reader's Digest* (Austral.) Dec. 122/2 This malemute was bitten by a mad red fox. **1962** M. E. MURIE *Two in Far North* I. vi. 51 Long strings of Malemutes lope down the streets. **1971** F. HAMILTON *World Encycl. Dogs* 569 The Malamute must give an impression of great strength and tremendous propelling power.

malaprop, *sb.* and *a.* Add: Also as *v. intr.*, to utter a malapropism; *trans.*, to make a malapropism of (a word).

1959 I. JEFFERIES *Thirteen Days* v. 67 The usage was so wild that I thought he was malapropping. **1970** M. TRIPP *Man without Friends* iii. 27 She malaprops any word of more than two syllables.

malaria. Add: **c.** *malaria-carrying* adj.

1916 *Brit. Mus. Econ. Ser.* IV. 8 This group [of mosquitoes]..includes also several of the malaria-carrying *Anopheles*. **1946** *Nature* 10 Aug. 202/2 The experimental infection of laboratory-hatched larvæ of the malaria-carrying *Anopheles gambiæ* Giles with a fungus of the genus *Cœlomyces* Keilin.

malariologist. Delete *rare* and add further examples.

1944 *Living off Land* v. 115 Sir Ronald Ross, of the Indian Medical Service..became the world's foremost malariologist. **1966** *New Scientist* 18 Aug. 374/3 Malariologists are seeking to add to the established methods for controlling the disease. **1971** *Nature* 5 Mar. 65/1 For many years he was a 'wet-foot' malariologist in some of the most difficult countries in the world.

malariology (mălē̆ᵊriǫ·lŏdʒi). [f. MALARI(A + -OLOGY.] The scientific study of malaria.

1925 *Science* 9 Oct. 320/2 In the field of malariology Dr. Carter has long held..the same position of preeminence that he enjoyed in relation to yellow fever. **1936** *Discovery* Nov. 353/1 Instruction in malariology was given to a number of medical practitioners. **1967** D. F. CLYDE *Malaria in Tanzania* p. x, Britain provided..grants with which the research units in Dar es Salaam and Tanga were enabled to lay the scientific foundations of modern malariology in this country. **1974** *Nature* 1 Feb. 318/3 His fleeting acquaintance with the discipline of malariology and with the realities of malaria control.

malarkey (mălā·ɹki). *slang* (orig. *U.S.*). Also **malaky, malarky, mullarkey.** [Origin unknown.] Humbug, nonsense, foolishness.

1929 J. P. MCEVOY *Hollywood Girl* vii. 102 It's a wonder you notice me, I told him. That's a lot of malaky, says he. **1930** *Variety* 29 Oct., The song is ended but the Malarkey lingers on. **1934** *Esquire* Dec. 49/3 Daughter of Mrs. Sally Alden, father unknown! What malarkey! All hooey, even protected by the official records of a friendly republic. **1938** *Down Beat* Mar. 5/4 We've got to say to the recording companies..'Cut out the Mullarkey and give us some down-home stuff!' **1945** J. STEINBECK *Cannery Row* xiii. 55 He knew God damn well the story was so much malarky. **1958** *Sunday Times* 20 Apr. 31/1, I will only give you the politician's malarky about imponderables and changing circumstances. **1960** G. M. WILSON *It rained that Friday* xi. 107 Somebody's passed the word round that the island's haunted. I told them it was a lot of malarkey. **1963** J. MITFORD *Amer. Way of Death* iii. 139 The malarkey that surrounds the usual kind of funeral. **1964** *Punch* 23 Dec. 964/3 Any mullarkey from ratting to potato picking. **1965** *Sunday Mail Mag.* (Brisbane) 25 Aug. 5/3 Here was a man who didn't give you the old malarky. **1971** [see *FLOPSY BUNNY]. **1973** *Observer* 25 Mar. 12/2 Tall stories..of rattlesnakes bringing up a nestful of baby robins,..or some such malarkey.

malathion (mælăþǝi·ǫn). [The substance is manufactured from diethyl *mal*eate (an ester of MALEIC acid) and a *thio*-acid (see s.v. THIO-).

The name orig. proposed was *malathon* (see quot. 1953[1]). Later *malathion* was registered in the U.S. both as a generic (i.e. non-proprietary) name and as a trade-mark of the American Cyanamid Co.]

An organophosphorus insecticide which is relatively harmless to plants and animals; in commercial preparations it is a brownish liquid with a strong smell of garlic and is also used in the form of a powder.

1953 *Official Gaz.* (U.S. Patent Office) 14 Apr. 322/2 American Cyanamid Company... *Malathion* For Insecticide. **1953** *Substitution of Malathion for Malathon as a Coined (Generic) Name for the Insecticidal Chemical O,O-Dimethyl Dithiophosphate of Diethyl Mercaptosuccinate* (Interdepartmental Comm. Pest Control, Bureau Entomol. & Plant Quarantine, U.S. Dept. Agric.), On January 30, 1952 the Interdepartmental Committee on Pest Control approved the name 'malathon' as coined name for O,O-dimethyl dithiophosphate of diethyl mercaptosuccinate. Because of difficulty encountered in the trademarking of the name selected at first the commercial sponsor, American Cyanamid Co., decided to change the name to 'malathion'. The American Chemical Society and the American Medical Association are agreeable to the change. Malathion has been registered with the Trade-Mark Division and released for general use. On March 27, 1953, the Interdepartmental Committee on Pest Control approved the name 'malathion' as a coined (generic) name for the chemical in question. **1955** *Sci. News Let.* 3 Sept. 157/1 The insecticides DDT, EPN, heptachlor and malathion have been tested in granular forms. **1960** *Farmer & Stockbreeder* 19 Jan. 119/3 However, as a cure [for red mite in woodwork], it is best to use a B.H.C. or malathion preparation. **1964** *Which?* Apr. 114/1 The organophosphorus compounds, such as dimethoate, malathion, menazon, are mainly poisonous to sucking insects, such as greenfly. **1968** M. PYKE *Food & Society* viii. 119 It is reckoned that whereas 0·5, 0·06, 9 and 1·2 micrograms of DDT, lindane, malathion and carbaryl are ingested for each kilogram of human body weight, the least amounts of these pesticides likely to be harmful are 10, 12, 14 and 20 micrograms respectively.

Malawi (mālā·wi), *a.* and *sb.* [Name of a Central African state, formerly Nyasaland.] **A.** *adj.* Of or pertaining to Malawi or its inhabitants. **B.** *sb.* A native or inhabitant of Malawi. Also **Mala·wian** *sb.* and *a.*

1960 *Economist* 15 Oct. 252/1 Dr Banda's Malawi Congress party. **1963** *Ibid.* 6 July 23/2 His [*sc.* Dr. Banda's] Malawians have lost from the absence of a Nyasa delegation. **1964** *Times* 6 July (Suppl.) 1/2 He [*sc.* Dr. Banda]..has invited the Prime Minister of Southern Rhodesia, Mr. Ian Smith, to the Malawi independence celebrations. **1964** *Observer* 12 July 13/2 The brand-new Malawi flag with a red rising sun. **1965** *Guardian* 4 Oct. 9/6 A brawl at a nightclub between a young British bank clerk and a Malawian. **1966** *Listener* 5 May 637/2 In Rhodesia..Malawis are regarded as the best employees. **1970** *Times* 26 Mar. (Malawi Suppl.) p. i/6 As for Rhodesia, it is inconceivable that the Malawians should adhere strictly to sanctions. *Ibid.* p. iv/4 At Fort Johnston we.. acquired the company of a Malawian student hitching a ride home to see his parents. **1971** *Rand Daily Mail* 4 Dec. 1/4 The tiny Malawi Defence Force. *Ibid.* 3/8 Terrorists might enter Malawi purely to embarrass the Malawian Government. **1972** *Whitaker's Almanack 1973* 756/2 In 1970 a 70-mile line was opened..linking the Malawi rail system with the Mozambique network. **1973** *Guardian* 17 Apr. 12/2 Dr. Banda..believed passionately that grammar was what the Malawians most needed.

malaxation. Add: **c.** A form of massage.

1887 D. MAGUIRE *Art of Massage* (ed. 4) 46 Malaxation is the same movement [as pétrissage], differing only on account of the flat of the hand being applied with more or less strength before bringing the fingers together to exercise the pétrissage. **1961** *Brit. Med. Dict.* 870/1 *Malaxation*, pétrissage, one of the movements carried out in massage, in which the muscles are squeezed in the masseur's hands and rolled and pressed; kneading.

Malay, *sb.* and *a.* Add: **A.** *sb.* **1. b.** *spec.* in S. Afr. One of the Muslim community of Cape Town and adjoining districts (see quots. 1944 and 1972). In full, *Cape Malay*.

1785 tr. *Sparrman's Voyage to Cape of Good Hope* I. 12 The burial grounds of the Chinese and free Malays that live at the Cape. **1812** A. PLUMPTRE tr. *Lichtenstein's Trav. S. Afr.* I. i. ii. 28 This practice receives great encouragement from the natural inclination that the slaves, particularly the Malays, have to music. **1844** J. BACKHOUSE *Narr. Visit Mauritius & S. Afr.* v. 82 The religion of the False Prophet was introduced into the Colony by the importation of Malacca slaves, by the Dutch. Hence the terms Malay and Mahomedan became synonymous in the Colony. **1944** I. D. DU PLESSIS *Cape Malays* i. 1 Strictly speaking 'Malay' stands for that section of the local Muslim community in which the descendants of Malay slaves and political exiles are to be found. **1972** *Standard Encycl. S. Afr.* VII. 145/2 Coming from Java and the neighbouring Indonesian islands, the Cape Malays belong essentially to the Javanese and Balinese section of the Malay race.

B. *adj.* **c.** Of, pertaining to, or characteristic of the Cape Malays.

1881 *Cape Monthly Mag.* Mar. 192 A Malay hat. **1953** DU PLESSIS & LÜCKHOFF *Malay Quarter* 9 A Malay wedding. **1974** *S. Afr. Panorama* Mar. 7 The oldest of its two gables, dated 1800, is..reputedly the work of a Malay craftsman.

Malayalam (malāyā·lăm). Also 9 **Malayalma, Malayalim.** [Native name; cf. next.] A Dravidian language, closely related to Tamil, spoken in southern India. Also *attrib.* or as *adj.*

1837 H. HARKNESS *Anc. & Mod. Alphabets Pop. Hindu Lang.* p. i, Under the head of Malayalma are also included the Characters of..Tuluva. **1839** F. SPRING (*title*) Outlines of a grammar of the Malayalim language. **1864** A. J. ARBUTHNOT (*title*) Malayalam selections. **1902** [see *DRAVIDIAN A. *adj.*]. **1908** T. G. TUCKER *Introd. Nat. Hist. Lang.* 136 *Malayâlam*, occupying a comparatively small area of the west or Malabar coast south of Mangalore. **1911** *Encycl. Brit.* XVII. 466/1 *Malayalam*, a language of the Dravidian family, spoken on the west coast of southern India. **1960** *Universe* 14 Apr. 24/2 Malayalam, a vernacular language of Southern India, was used for the first time as the liturgical language in an ordination in the Syro-Malabar rite at Ernakulam. **1971** *Leader* (Durban) 7 May 8/4 He wrote in Malayalam, which is the mellifluous, if impossibly difficult language of Kerala... His directness and his Malayalam-accented English were always irresistible.

Malayali (malāyā·li). Also **Malayalee.** [Native name, f. Dravidian *mala* mountain + *āḷ* possess.] A member of a Malayalam-speaking people inhabiting the state of Kerala on the Malabar coast of southern India. Cf. *MALAYALAM. Also *attrib.*

1856 R. CALDWELL *Compar. Gram. Dravidian Lang.* 491 The higher castes are styled 'Hindus', or else 'Tamilians', 'Malayâlis', &c., according to their language and nation. **1889** L. J. FROHNMEYER *Progressive Gram. Malayalam Lang.* p. xii, The Malayalees have not a few Folk Songs. *Ibid.* xvi. 133 The Malayali..prefers to divide in two or three activities, what Europeans would express by one Verb. **1928** *Chambers's Jrnl.* XVIII. 244/1 Chetwynd managed to hold the Malayalees together for another three days. **1961** P. SPEAR *India* vii. 91 The southern Dravidian people were articulated in four groups..the Tamils,..the Telugus or Andhras,..the Malayalis along the west coast in the modern state of Kerala; and the Kanaras. **1971** *Hindustan Times Weekly Rev.* (New Delhi) 4 Apr. p. iv/5 We celebrate both Malayali and Bengali festivals.

Malayan, *a.* and *sb.* Add: **B.** *sb.* **3.** During the existence of the Federation of Malaya (from 1948 until 1963), an inhabitant of Malaya (regardless of race or creed).

1954 V. BARTLETT *Rep. from Malaya* ii. 23 A Malayan is anybody—be he Malay, Chinese, Tamil, Eurasian or European—whose home is in Malaya, whereas a Malay is a member of a large race of whom about 3,000,000 live in Malaya, and more than 70,000,000 in Java, Sumatra and the other islands of Indonesia. **1958** G. MIKES *East is East* 114 A *Malayan* is an inhabitant of Malaya, irrespective of race or creed. **1961** L. D. STAMP *Gloss. Geogr. Terms* 304/2 Since the independence of Malaya..Malayan has a political significance as a citizen of Malaya.

Malayanize (mălē̆ᵊ·ănǝiz), *v.* [f. MALAYAN *a.* and *sb.* + -IZE.] *trans.* To make Malayan in character or composition. So **Mala:yaniza·tion.**

1955 *Times* 24 Aug. 5/5 With regard to Malayanization of the Civil Service, there were good men among Malayan members of the Civil Service. **1956** *Times* 22 Nov. 9/5 The Government [of Singapore] would begin to put its Malayanization scheme into effect next January. **1960** *Times* 25 Nov. 13/2 Singapore had better get ahead with 'malayanizing' its university. **1965** R. MCKIE *Company of Animals* i. 15 'Malayanization'—the replacement of Europeans in government services by local-born people. **1971** N. BARBER *War of Running Dogs* iii. 46 Everything would change with independence and Malayanization. **1972** *Accountant* 5 Oct. 420/2 It was contended for the taxpayer that there was no evidence..that he had accepted any variation of the agreement because of the policy to Malayanize the staff.

Malayic (mălē̆ᵊ·ik), *a. rare.* Also 8 **Malaic.** [See -IC.] = MALAYAN *a.*

1723 R. MILLAR *Hist. Propagation Christianity* II. viii. 478 Sermons..in the Malaic Tongue. **1890** D. G. BRINTON *Races & Peoples* 230 The Malayic stock.

Malayo-. (Further examples.)

1876 *Encycl. Brit.* V. 288/1 Their [*sc.* the Macassars'] language..belongs to the Malayo-Javanese group. **1879** A. H. KEANE in A. R. Wallace *Australasia* 607 Papûans proper in the centre; Malayo-Papûans in the Indian Archipelago. **1880** A. H. SAYCE *Introd. Sci. of Lang.* II. viii. 188 The agglutinated adjuncts..may be almost wholly dispensed with, as in Malayo-Polynesian. **1887** A. FEATHERMAN *Social Hist. Races Mankind* II. i. 251 The Malayo-Melanesians are the most important branch of the Melanesian stock. **1896** A. H. KEANE *Ethnol.* 285 The Negroid Malayo-Malagasy peoples of Madagascar. *Ibid.* 331 Semi-cultured and rude Malayo-African populations. *Ibid.* 333 The Philippine half-castes may be roughly classed as..Malayo-Indonesians, Malayo-Europeans, and Malayo Chinese. **1911** WEBSTER, Malayo-Negrito. **1933** [see *AUSTRIC *a.*]. **1934** PRIEBSCH & COLLINSON *German Lang.* I. ii. 35 A Creolized language, in this case a blend called Malayo-Portuguese, which was used by the whites in dealing with slaves.

Malaysian, *a.* Add: Of or pertaining to the Federation of Malaysia (formed in 1963 from the states of Malaya, Sabah, Sarawak, and (until 1965) Singapore).

1962 *Times* 9 Jan. 9/3 English is to remain as an 'international language' in the proposed Malaysian Federation. **1963** *Daily Tel.* 12 Sept. 21/7 The agreement covers the question of sovereignty for Singapore and arrangements as between the Singapore Government and the Malaysian Government for emergency security measures. **1963** *Guardian* 16 Sept. 9/4 The Malaysian Government is fully aware that recognition is not a right but a privilege. **1971** *Ibid.* 22 Feb. 3/1 About ninety bills face the Malaysian Parliament when it re-opens tomorrow after 21 months. **1973** J. M. WHITE *Garden Game* 97 A crumpled fistful of Malaysian money.

B. *sb.* A native or inhabitant of the Malay archipelago (quot. 1625) or of Malaysia.

1625 PURCHAS *Pilgrimes* I. III. xiv. 321, I cannot imagine what the Hollanders meane to suffer these Maleysians, Chinesians, and Moores of those countries. **1955** *Times* 15 June 9/6 The population of the Malay States comprises roughly 50 per cent. Malaysians, 40 per cent. Chinese, 10 per cent. others. **1965** *Guardian* 24 July 8/6 His efforts to overrun the Malaysians in Borneo. **1970** *Daily Tel.* 16 June 2/8 (*heading*) Malaysian jailed for blackmail. **1974** 'G. BLACK' *Golden Cockatrice* v. 93 You are as good a Malaysian as I am a Taiwanese.

malaysianite (mălē̆ᵊ·ziănǝit). *Geol.* [f. MALAYSIAN *a.* + -ITE[1].] Any tektite from the tektite field of the Malay peninsula.

1940 *Pop. Astron.* XLVIII. 45 Malaysianites tend to show irregular and heavily etched surfaces. **1964** *New Scientist* 16 Jan. 160/2 The widespread Australasian strewn field..consisting of the australites, the javanites, the malaysianites, billitonites and philippinites.

maldesce·nded, *a. Med.* [f. MAL- + DESCENDED *ppl. a.*] Of a testis: not having descended all the way into the scrotum from the abdominal cavity during development of the fœtus, or having descended ectopically.

1908 *Ann. Surg.* XLVIII. 321 (*heading*) Treatment of the undescended or maldescended testis. **1933** *Post-Graduate Med. Jrnl.* IX. 248 There is a widespread.. belief that if a young boy with a mal-descended testis is left alone the organ will drop..at puberty. **1936** H. BAILEY *Dis. Testicle* ii. 11 A maldescended testis is one which cannot be made to touch the bottom of the scrotum. **1962** *Lancet* 19 May 1059/1 A few weeks ago we attempted to evaluate the dangers of malignancy arising in undescended and maldescended testes.

maldesce·nt. *Med.* [f. MAL- + DESCENT.] The state (of a testis) of being maldescended; incomplete or ectopic descent.

1908 *Ann. Surg.* XLVIII. 324 The different types of maldescent of the testis. **1948** L. MARTIN *Clin. Endocrinol.* viii. 159 In cryptorchidism the testes remain in the abdomen, in maldescent they lie in the inguinal canals or unduly high in the scrotum. **1954** G. I. M. SWYER *Reproduction & Sex* vi. 65 Sometimes maldescent is present on one side only, the testicle on the other side being situated in the normal position. **1964** *Arch. Dis. Childhood* XXIX. 607/1 Any testis not in the scrotum, that is, 4 cm. or less from the pubic crest in infants and boys is undescended. If it is more than this distance but not as low as it should be it is incompletely descended. The terms mal-descent and imperfect descent have no useful meaning.

maldistribution (mæ:ldistribiū̆·ʃǝn). [f. MAL- + DISTRIBUTION.] Faulty distribution. Hence **ma:ldistri·buted** *ppl. a.*

1895 *U.S. Dept. Agric. Farmers' Bulletin* XXXIII. 22 This ruinous state of affairs is not attributable to overproduction, but to maldistribution. **1928** *Britain's Industr. Future* (Liberal Industr. Inquiry) III. xix. 243 We agree with the Socialists in thinking that a distribution so uneven as is at present found is maldistribution. **1931** *Times Trade & Engin. Suppl.* 24 Jan. 431/2 We are suffering from maldistribution of gold. **1933** *Planning* 6 June 1 Unemployment has grown..to a world-wide problem of maldistributed leisure. **1935** *Discovery* Jan. 14/1 Poverty persists and populations become maldistributed. **1957** *Economist* 23 Nov. 694/2 The chronic maldistribution of wealth. **1969** *Listener* 6 Mar. 318/1 The story which began in the wilderness ends..in automated (if maldistributed) affluence. **1971** *Daily Tel.* 3 July 12 Summer is the time when maldistribution of business at the start of the session catches up with the Lords.

Maldivian (mǫ̆ldi·viăn), *sb.* and *a.* Also 9 **Maldivan. A.** *sb.* A native or inhabitant of the Maldive Islands in the Indian Ocean; their language. **B.** *adj.* Of or pertaining to the Maldive Islands or their inhabitants.

1836 *Trans. Bombay Geogr. Soc.* I. 77 The Maldivans have a written, as well as an unwritten law. **1841** *Jrnl. R. Asiatic Soc.* VI. 42 (*title*) Vocabulary of the Maldivian language. **1959** *Times* 19 Mar. 6/3 He was able to learn something of the Maldivian language. **1970** *Trans. Philol. Soc.* 148 The Maldivian numerals are lexically related to the Indic numeral systems. **1972** *Guardian* 15 Mar. 13/5 The Maldivians are hopefully extending the runway. **1972** *Language* XLVIII. 464 The only language closely related to Sinhalese is Maldivian, spoken in the Maldive Islands and on the island of Minicoy, which is part of India.

‖ **mal du siècle** (mal dü syɛkl'). [Fr.] World-weariness, weariness of life, deep melancholy because of the condition of the world.

In French with specific reference to the early 19th-c. Romantic poets.

1926 A. HUXLEY *Essays New & Old* 38 The *mal du siècle* was an inevitable evil. **1932** *Times Lit. Suppl.* 16 June 439/1 We have the sentimental and practical romanticism, the *mal du siècle*. **1957** G. SMITH *Friends* ii. 36 A bijou house in Chelsea, premature obesity, *mal du siècle* and incipient gout. **1963** R. WELLEK in N. Frye *Romanticism Reconsidered* 109 Little has been accomplished by calling familiar states of mind—*Weltschmerz, mal du siècle*, pessimism—'negative' romanticism.

male, *a.* and *sb.*[2] Add: **A.** *adj.* **1. e.** Also used, not jocularly, in referring to professions that are usually considered to be predominantly female, as *male midwife, model, nurse, prostitute*.

1878 TROLLOPE *John Caldigate* (1879) II. iv. 45 He took the child very gently... He had already assumed for himself the character of being a good male nurse. **1915** MRS. BELLOC LOWNDES *Diary* 12 Mar. (1971) 58 It is very difficult to get the right type of man to be a male nurse. **1948** A. KINSEY et al. *Sexual Behavior in Human Male* vi. 216 Some male prostitutes ejaculate five, six, or more times per day with regularity over long periods of years. **1961** S. BAKER *Visual Persuasion* ii. 47/2 Male model agencies and actors' guilds always have a few personalities on hand. **1972** 'W. HAGGARD' *Protectors* vi. 71 The man in the chair..was kept alive by two male nurses. **1974** *Mother & Baby* Feb. 44 (*heading*) Would you like a male midwife? **1974** H. WAUGH *Parrish for Defence* (1975) xix. 321 He is presently serving as a male prostitute for other males. **1975** *Times* 11 Mar. 4/8 It would be wrong to rule out male midwives as a matter of principle. **1975** 'G. BLACK' *Big Wind* vii. 114 He looked like a male model in one of those ads for expensive men's knitwear.

2. a. male-sterile *a.*, (of a hermaphrodite plant) incapable of producing fertile pollen; so **male-sterility.**

1921 *Jrnl. Genetics* XI. 269 (*heading*) Male-sterility in flax, subject to two types of segregation. *Ibid.* 271 In the breeding work it was not thought necessary to emasculate the male-sterile flowers. **1946** *Nature* 21 Sept. 422/2 Further tabulation of male-sterility genes in varieties of the onion is given. **1959** *New Biol.* XXVIII. 75 Certain willow-herb crosses which produce male-sterile progeny.

B. *sb.* **4.** Comb. *male climacteric, menopause, pill, supremacist* (also *attrib.*), *supremacy*; *male-determining, -dominated* adjs.; **male chauvinism**, an attitude attributed to men of excessive loyalty to members of the male sex and of prejudice against women; hence **male chauvinist**, one who adopts this attitude; also *attrib.*, esp. in phr. *male chauvinist pig*; **male impersonator**, a female who impersonates a male on the stage.

1970 *Time* 17 Aug. 23 European women have accepted their lot much more readily than their American counterparts. Recently, however, growing numbers..have launched their attack on male chauvinism. **1973** O. LANCASTER *Littlehampton Bequest* 84 Their marriage has always been a completely unselfish relationship, both taking an active part in the struggle against Imperialism, Neo-Colonialism and Male Chauvinism. **1970** *New Yorker* 5 Sept. 27/1 Hello, you male-chauvinist racist pig. *Ibid.*, Repent Male Chauvinists. **1972** *Southerly* XXXII. 75 The male chauvinist aspects of mateship have come in for considerable discussion since the spread of women's liberation critiques. **1972** *Punch* 1 Mar. 289/1, I know, I know; me male chauvinist pig, you Jane. But the exercise has finally woken me up to ask—why should there be separate magazines for men and women at all? **1974** J. HELLER *Something Happened* 333, I enjoy fucking my wife. She lets me do it any way I want. No Women's Liberation for her. Lots of male chauvinist pig. **1966** G. B. MAIR *Kisses from Satan* vii. 78 I've got a male climacteric and I don't like it. **1972** *Daily Colonist* (Victoria, B.C.) 8 Feb. 2/1 Men may experience a certain slackening in sexual interest that may be related to what medical literature calls 'male climacteric' (a retrogression of the sex glands). **1931, 1957** Male-determining [see *FEMALE B. *sb.* 3]. **1958** W. J. H. SPROTT *Human Groups* 65 The home..is in certain important respects male-dominated. **1964** W. MCCORD in I. L. Horowitz *New Sociol.* xxv. 434 The city is a male-dominated area (more men enter the city seeking jobs than women). **1973** I. SINGER *Goals of Human Sexuality* i. 33 She submits to the relationship as a way of conforming to the demands of a male-dominated society. **1895** STUART & PARK *Variety Stage* 222 Serio-comics, sisters, dancers, male impersonators, and ballad and character vocalists. **1930** *Bulletin* 14 May 5 That popular male impersonator and pantomime principal boy, Miss Nora Delany. **1972** *Times* 29 Sept. 16/8 To the connoisseurs of the music hall there have been only two great male impersonators, one of whom was Vesta Tilley, the other Hetty King. **1949** ERNST & LOTH *Sexual Behaviour & Kinsey Report* viii. 93 Another reason which has sometimes been given is that many men reach in middle age a climacteric which is dubbed a male menopause. **1963** E. LANHAM *Monkey on Chain* xii. 195 You think dirty, you act dirty—like a dirty old man. What is it? Male menopause? **1971** J. WAINWRIGHT *Last Buccaneer* II. 187 Middle-aged men get odd ideas. It's a sort of male menopause—a change of life. **1966** *New Statesman* 27 May 767/1 Techniques like the 'male pill', tying of the spermatic cords..are in use or under investigation. **1971** *Black Scholar* Dec. 7/2 Excepting the woman's role as caretaker of the household, male supremacist structures could not become deeply embedded in the internal workings of the slave system. **1973** J. JONES *Touch of Danger* xxviii. 167 You just don't want to understand Jane... She's a threat to all you male supremacists. **1908** CHESTERTON *Man who was Thursday* i. 12 Most of the women were of the kind vaguely called emancipated, and professed some protest against male supremacy.

Male, var. *Maler *sb.* and *a.*

malease. Restrict † *Obs.* to sense 2 and add later examples of sense 1.
1923 *S.P.E. Tract* xiii. 35 All French loan-words that are sufficiently naturalized to be considered as English should be treated as English... We should be glad if London editors would..print as follows:..*malease* for *malaise.* **1929** R. Bridges *Testament of Beauty* i. 231 [St.] Francis climbed—rather his gentle soul had learn'd From taste of vanity and by malease of the flesh. **1930** *Times Lit. Suppl.* 4 Dec. 1022/2 Yet her achievement was muted by excess of ardour and by a subtle, fundamental malease. **1938** L. P. Smith *Unforgotten Yrs.* vi. 165 My tutor gave it unusual praise, in which praise I was conscious of the mingling of a curious malease.

maleate (mălī·ĕit). *Chem.* [f. Male(ic *a.* + -ate[4].] A salt or ester, or the anion, of maleic acid.
1853 H. Watts tr. *Gmelin's Hand-bk. Chem.* VIII. 153 The Salts of Maleic acid, the Maleates, are all soluble in water. **1925** *Jrnl. Amer. Chem. Soc.* XLVII. 1073 First, association occurs between permanganate ion or molecule with maleate or fumarate ion. **1970** R. W. McGilvery *Biochem.* xi. 220 Fumarate is a *trans* compound (maleate is the corresponding *cis* isomer), and hydrates to yield the L-enantiomorph.

maledictive (mælĭdi·ktiv), *a.* [f. L. *maledict-* (see Malediction) + -ive.] Characterized by cursing or curses; uttering maledictions.
1865 S. Ferguson *Lays of Western Gael* 54 Daily in their mystic ring They turn'd the maledictive stones. **1905** K. Meyer *Cáin Adamnáin* p. vii, A poem on the maledictive psalms selected by Adamnan. **1922** Joyce *Ulysses* 326 Seats of learning and maledictive stones.

maleesh (mā·lī∫), *int.* Also **ma(a)lesh.** [ad. colloq. Eastern Arab. *mā 'ale-š* no matter.] No matter! never mind! Also as *sb.*, indifference, slackness.
1913 'S. Rohmer' *Mystery of Dr Fu-Manchu* iv. 44 '*Ma 'alesh!*' came her soft whisper; 'but I am afraid to trust you—yet.' **1919** W. H. Downing *Digger Dial.* 33 *Maleesh,* it doesn't matter. **1925** Fraser & Gibbons *Soldier & Sailor Words* 151 *Maleesh*: Never mind. It doesn't matter. Used colloquially on Eastern Fronts. **1946** *Happy Landings* (Air Ministry) July 12/1 The manpower problem will solve itself in time, but there is no excuse for the *maleesh* attitude. **1947** *Landfall* I. 162 Not so much of the *ma lesh!* **1958** L. Durrell *Mountolive* xvi. 315 *Ma-a-lesh!* Let it be forgiven! Nothing avails our grief! **1971** *Guardian* 22 June 3/3 The general air of 'malesh', Arabic for 'never mind', or 'forget it', continues to hang over the police force.

maleic, *a.* Add: Also † **malæic. maleic anhydride,** the anhydride, $C_4H_2O_3$, of maleic acid which is a crystalline compound that forms addition compounds with substances containing conjugated carbon–carbon double bonds; *maleic* (*anhydride*) *value,* a measure of the number of conjugated double bonds in a substance (e.g. an oil) obtained by reaction with maleic anhydride.
1857 W. A. Miller *Elem. Chem.* III. v. 335 Hydrated malæic acid when maintained in a state of fusion at 300°, is converted into a crystalline mass of fumaric acid. If distilled by a temperature suddenly elevated to 460°, malæic anhydride is formed. **1898** *Jrnl. Chem. Soc.* LXXIV. i. 177 Maleinoid chlorobromosuccinic acid is prepared by saturating with chlorine a solution of bromine in chloroform, adding maleic anhydride, again passing chlorine into the liquid, which is then exposed to bright sunlight. **1936** Ellis & Jones in *Analyst* LXI. 814 The calculation of the 'maleic anhydride value' (M.A.V.) or, more shortly, 'maleic value' is given... At the moment the term 'maleic value' is suggested for this figure in preference to 'diene value'. **1944** H. G. Kirschenbauer *Fats & Oils* v. 46 The theoretical Maleic Anhydride Value of trielaeostearin is 87·2. **1950** K. A. Williams *Oils, Fats & Fatty Foods* (ed. 3) iii. 116 The maleic anhydride value is given by the expression: 12·692 × ml. of alkali used ÷ weight of sample taken in grams. **1964** N. G. Clark *Mod. Org. Chem.* xix. 389 A purely aliphatic compound, maleic anhydride, an intermediate for alkyd resins..is obtained from benzene by passing its vapour, mixed with air, over a vanadium pentoxide catalyst at 450°. **1972** *Materials & Technol.* IV. xiii. 465 The peculiar structure of maleic anhydride, in which a C=C double bond is in conjugation with a C=O double bond, makes it the most important and widely used dienophilic component in the Diels–Alder reaction with 1,3-dienes to produce non-aromatic six-membered ring compounds.

maleinoid, obs. var. *Malenoid *a.*

‖ **mal élevé** (mal ɛlve), *adj. phr.* Also fem. **mal élevée.** [Fr., lit. 'badly brought up'.] Of a person: bad-mannered, ill-bred. Of an object or situation: lacking in refinement.
1878 H. James *Europeans* II. v. 198 Even that *mal-élevée* little girl..makes him do what she wishes. **1924** A. D. Sedgwick *Little French Girl* iii. v. 263 It was odd..but not *mal élevé.* **1949** E. Bowen *Encounters* p. xi, If I was *mal élevée*..so were my betters. **1964** A. Wilson *Late Call* 24 She echoed her mother's horror at all things *mal élevées.* **1974** E. Ambler *Dr. Frigo* i. 34 The wives of certain French officials have expressed disapproval [of her]. *Mal elevée* is their verdict.

malemut(e, varr. *Malamute.

maleness. a. Delete † (*obs.*) and add later examples; also = virility.
1925 *New Yorker* 27 June 17/2 Her strongest reason was the fascination of St. Mawr's terrific 'maleness'. **1929** D. H. Lawrence *Pansies* 126 So it is with Englishmen. They are all double roses And their true maleness is gone. **1937** *Discovery* Oct. 315/1 His [sc. the stag's] maleness reigns supreme. **1975** A. Fraser *Whistler's Lane* xi. 173 His jacket smelt of horses and tobacco and general maleness.

malenoid (mæ·lĕnoid), *a.* *Chem.* Formerly also **malei·noid.** [f. Malein(ic *a.* + -oid.] Resembling maleic acid in having a *cis* configuration in geometrical isomerism.
1895 [see *fumaroid *a.*]. **1898** [see *maleic *a.*]. **1907** [see *cis- 3]. **1938** [see *fumaroid *a.*]. **1964** *Internat. Encycl. Chem. Sci.* 709/2 *Malenoid form,* the *cis* form of geometrical isomerism.

Maler (mā·ləɹ), *sb.* and *a.* Also **Mal, Male, Moler, Muler.** [Native word = 'hillmen', ult. f. Dravidian *mala* mountain.] **A.** *sb.* **a.** A Dravidian people living in the Rajmahal hills of northern India; a member of this people. **b.** The language spoken by this people (also called *Malto, *Rajmahali). **B.** *adj.* Of or pertaining to the Maler people or their language.
1811 F. Buchanan *Jrnl. Survey Bhagalpur* (1930) 133 The Not Pahariyas..eat and intermarry with the Moler... Many of the men speak the Moler language. *Ibid.* 138 The interpreter tribe is here called Desi Moler... Many of them speak the Muler language, and they eat and can intermarry with the Muler of the hills. **1853** J. R. Logan in *Jrnl. Indian Archipelago* VII. 50, I infer that the Male and Kol resemble the coarser Binua tribes of the Malay Peninsula more than the Burmans. **1872** E. T. Dalton *Descriptive Ethnol. Bengal* viii. 264 The Málers were the first of the aboriginal tribes of Bengal that were prominently noticed by the officers of the East India Company. **1873** E. Balfour *Cycl. India* III. s.v. *Male,* The Male or Rajmahali are described as mostly very low in stature, but stout and well proportioned. **1885** G. C. Whitworth *Anglo-Indian Dict.* 262/1 *Rájmaháli,* the name of a tribe called also Pahári. Also the name of their language, which is Dravidian. These people call themselves Maler. **1892** H. H. Risley *Tribes & Castes Bengal: Ethnogr. Gloss.* II. 54 In 1782..one of the archers murdered a Málé woman, and in order to punish this..Mr. Cleveland proposed the formation of a district tribunal. **1906** G. A. Grierson *Ling. Survey India* IV. 446 In former days the Maler made frequent raids on the plains. **1915** *Encycl. Relig. & Ethics* VIII. 344/2 *Māl, Mālē, Māl Pahāriā,* a non-Aryan tribe, containing various groups... The inter-relations of the North and South groups have not been clearly ascertained, but they seem to be, to a large extent, pure Dravidians, and those in the Rājmahāl Hills in Bengal are closely allied to the Orāons. **1930** C. E. A. W. Oldham in F. Buchanan *Jrnl. Survey Bhagalpur* 138 Here we find Bhuiyās still speaking the Maler or Malto language. **1938** S. S. Sarkar *Mālērs of Rajmahal Hills* i. 8 The dress of the womenfolk of the Mālērs and the Orāons differs greatly from one another. *Ibid.* 15 The Mālēr country is bounded by the Ganges in the north. **1963** L. P. Vidyarthi *Maler* i. 11 To a Maler, the pleasure or wrath of spirits and supernatural powers controls the nature of happenings in the Maler family, village,..field or..forests.

‖ **malerisch** (mā·ləri∫), *a.* [G., 'painterly', f. *maler* painter.] After H. Wölfflin's use (1915 *Kunstgeschichtliche Grundbegriffe*): of or pertaining to a manner of painting characterized more by the merging of colours than by the more formal linear style; painterly.
1933 *Burlington Mag.* Dec. 269/2 Ruskin somewhere draws a distinction between drawing with a brush and painting with a brush—a distinction which in our time has been further elaborated by Heinrich Wölfflin. In Wölfflin's sense, the English water-colourists are always linear, and never malerisch. **1937** *Ibid.* Oct. 168/1 The most splendid examples of Raphael's draughtsmanship... They are '*malerisch*'. **1955** *Times* 2 Aug. 10/2 This is..a resuscitation of the great malerisch tradition, which was rejected by those who thought they were following Cézanne. **1961** *Times* 1 Mar. 15/3 The *malerisch* flow and swirl of the paint itself.

malfunction (mælfʊ·ŋk∫ən), *sb.* [f. Mal- + Function *sb.* 3.] Faulty functioning. So **malfu·nction** *v. intr.*; **malfu·nctioning** *vbl. sb.* and *ppl. a.*
1928 E. Bagby *Psychol. of Personality* xi. 154 Conditions of this sort are usually found to be *symptoms* of a malfunctioning of fundamental bodily processes. **1941** Johnson & Haven *Automatic Arms* 128 The slide on an automatic pistol may fail to remain open after the last shot, due to a malfunction of the catch... The use of the term 'malfunction' conveys nothing unless we know *what malfunctioned.* **1957** *New Scientist* 9 May 38/2 As an ophthalmic optician I was entertained by 'Geminus' (25 April) on his visits to his optician and his own experiments at correcting the malfunctions of his extra-ocular muscles. **1957** *New Biol.* XXII. 83 They [sc. psychosomatic disorders] come to the attention of patient and physician through some malfunctioning of an organ. **1958** *Times* 14 Mar. 10/1 Through 'mechanical malfunctioning', a B-47 bomber aircraft dropped what is described as 'an unarmed nuclear device'. *Ibid.* 17 Oct. 15/2 Eyes whose tear-ducts were malfunctioning because of the absence of gravity. **1961** *New Scientist* 7 Dec. 617/1 Some tiny valves in the spaceship's plumbing system failed to function properly... This malfunction has been brushed off in the newspapers. **1966** T. Pynchon *Crying of Lot 49* i. 14 A dented, malfunctioning version of himself. **1972** *Daily Tel.* 22 Jan. 17/2 An investigation was being made to discover why the warm-air heating system at the house..had malfunctioned. *Ibid.*, Recommendations would be made to the house builders when the cause of the malfunction had been discovered. **1973** *Sci. Amer.* Oct. 69/3 An instrumental malfunction at the crucial time can spoil the results of months of preparation.

malgas, var. *Malagas.

‖ **malgré.** Add: Freq. followed by a pronoun, as **malgré lui** (lüi), in spite of himself or herself; also **malgré elle, eux, moi,** etc., in spite of herself, themselves, myself, etc. Also **malgré tout,** despite everything.
1830 C. C. F. Greville *Mem.* (1874) II. xi. 38 This was my first dinner at Dudley's, brought about *malgré lui* by Lady Glengall. **1843** Thackeray *Confessions of G. Fitz-Boodle: Mr. & Mrs. Berry* in *Fraser's Mag.* Mar. 359/2 We obliged the Frenchman to drink *malgré lui.* **1852** Geo. Eliot *Let.* 24–25 July (1954) II. 50, I congratulate you on your ability to be cheerful *malgré tout.* **1883** G. Bloomfield *Reminisc.* II. xiii. 45 They say that, *malgré eux,* they cannot yield. **1904** Wodehouse *Gold Bat* xiii. 142 He could imagine their feelings when the prodigal strolled into their midst—an old Wrykinian *malgré lui.* **1929** A. Huxley *Holy Face* 26 That too, too merry laughter of clergymen who want to prove that, *malgré tout,* they can be good fellows. **1932** *Mind* XLI. 24 He [sc. Callicles] stands revealed as a moralist *malgré lui.* **1933** *Times Lit. Suppl.* 26 Jan. 56/2 The Bulpington as a warrior *malgré lui,* whose conduct in action might..have brought him to the execution squad. **1943** Wyndham Lewis *Let.* 15 Aug. (1963) 363 As you know I am here malgré moi; but I attempt to put to some serious use my prolonged immersion in this N. American civilisation. **1950** A. Huxley *Let.* 11 Aug. (1969) 629 One likes him, *malgré tout.* **1963** *Economist* 28 Dec. 1318/1 These central bankers..have created, almost *malgré eux,* a code of practice. **1970** *Guardian* 11 Dec. 10/6 The weight of the acting fell mostly on Anna Cropper as the fertility goddess *malgré elle.*

mali, var. Mallee[1].
1908 C. Sorabji *Between Twilights* xii. 150 While my *Mali* and I did tidyings in the Garden, I spoke to him gently about the plant. **1934** 'G. Orwell' *Burmese Days* ii. 20 A nearly naked mali, watering-can in hand. **1971** *Hindustan Times Weekly* (New Delhi) 4 Apr. 10/7 The groundsmen and malis were being paid from the University Sports Fund.

Mali (mā·li), *a.* and *sb.* Also **Melle, Melli. A.** *adj.* Of or pertaining to Mali, an ancient empire (of the 13th and 14th centuries) and a modern republic (founded in 1960) in west Africa. **B.** *sb.* A native or inhabitant of Mali. So **Ma·lian** *a.* and *sb.*
1906 H. H. Johnston *Liberia* II. xxvii. 896 The Melli..empire..arose and flourished in Nigeria before the foundation of Timbuktu. **1911** *Encycl. Brit.* XVII. 565/1 It is not known by whom the Melle (Mali) state was founded... The first king whose name is preserved was Baramindana, believed to have reigned from 1213 to 1235. **1960** *Times* 28 Mar. 6/1 The Mali folklore is less unfamiliar to Parisians than that of its sister republic. **1960** *Daily Tel.* 24 Aug. 1/1 The Mali force..was due to be withdrawn from the Congo. *Ibid.*, Irish troops of the U.N. were to have taken over from the Malis. **1960** *Economist* 31 Dec. 1367/2 The Malians themselves (or Soudanese, as they used to call themselves) have only lately emerged from black Africa's first working experiment of a sovereign federation. **1967** S. Knight *Window on Shanghai* xv. 65, I especially like the beautifully robed Malians... They are all members of the Malian government. **1970** P. Oliver *Savannah Syncopators* 106 Mali lutes, Senegal *kora.* **1973** *Times* 15 Oct. 8/8 Police today charged a 31-year-old Malian with illegal entry into Italy after the deaths yesterday of three young Africans believed to have been involved in illegal traffic of African labourers to Europe. **1974** *Black World* Feb. 56 West African elements, in particular the Malians, were installed in America before the arrival of the Europeans.

Malibu, malibu (mæ·libū). Chiefly *Austral.* and *N.Z.* [The name of *Malibu* Beach in California.] In full, *Malibu board.* A short lightweight surf-board.
1962 *Austral. Women's Weekly* Suppl. 24 Oct. 3/3 *Malibu,* type of surfboard made from foam, balsa, or fibreglass and under 10 ft. long. **1965** *N.Z. Listener* 17 Dec. 4/4 Australia had its first look at the Malibu board when film star Peter Lawford took one 'down under' in 1954. **1969** *Times* 25 July 5/2 Worried by the number of thefts of malibu boards, police have issued leaflets in the surfers' own language. **1970** *N.Z. News* 8 Apr. 16/3 A. Griffin won both the ski and malibu board races.

malic, *a.* Add: Applied to enzymes whose substrate is malic acid, as *malic dehydrogenase; malic enzyme* (see quot. 1951).
1937 H. Tauber *Enzyme Chem.* vii. 170 Malic dehydrogenase may be prepared by washing frog or ox muscle with *M*/15 phosphate buffer. **1948** *Jrnl. Biol. Chem.* CLXXIV. 997 There might be a mechanism for fixation of CO_2 by pyruvate other than that involving the 'malic' enzyme. **1951** *Physiol. Rev.* XXXI. 86 Enzymes catalyzing the reversible oxidative decarboxylation of *l*-malic acid are widely distributed. These enzymes are distinct from malic dehydrogenase, which catalyzes the reversible oxidation of *l*-malic acid, and are referred to as 'malic' enzymes. The majority of the known malic enzymes function specifically with TPN as coenzyme. **1972** *Biochem. Genetics* VII. 303 Polymorphic variation of the human mitochondrial malic enzyme was detected in Caucasians and Negroes.

malihini (mălihī·ni). [Hawaiian *malihini* stranger.] In Hawaii, a stranger, a newcomer; a beginner, a novice.
1914 *Outing* LXIV. 26/1 A couple of canoes are launching..where the neophytes, the 'malihini', learn their first lessons in riding the rollers. **1966** *Surfer* VII. 39 Hawkshaw rode tandem with two malihinis (beginners) on his shoulders. **1967** J. Severson *Great Surfing* Gloss., *Malihini,* a newcomer to the Hawaiian Islands. Newcomers to the sport of surfing are also referred to as malihinis.

Malines (mălī·n). [The name of a town (also called Mechlin) in Belgium.] **1.** In full *Malines lace* = Mechlin lace. In full *Malines* (also *Maline*) *net,* a net made for millinery or veils.
1833 Carlyle *Sart. Res.* I. iv. in *Fraser's Mag.* Nov. 590/2 To Teufelsdröckh the highest Duchess is respectable..but nowise for her pearl bracelets, and Malines laces. **1842** Thackeray *Fitz-Boodle's Confessions* in *Fraser's Mag.* June 713/1, I thought she would not let the evening pass without talking of her Malines lace. **1850** *Harper's Mag.* Aug. 431/2 Another pattern is..embroidered and trimmed with *malines.* **1897** *Sears, Roebuck Catal.* 318/1 A 27 inch Maline net veiling..in black, brown, cream or navy blue. **1909** *Daily Chron.* 4 Mar. 4/5 Lady Anne Lambton was the wearer of Malines net, the skirt panelled with lace. **1911** J. Masefield *Jim Davis* iii. 30 It contained a dozen yards of very beautiful Malines lace. **1912** *Queen* 13 Apr. 613/1 This coat..had long sleeves slit up at the wrist, and edged with white net bordered with Malines. **1964** Mrs. L. B. Johnson *White House Diary* 12 Mar. (1970) 89, I dressed early this morning, in solid black, with my mantilla draped over my maline hat, and set out..for the Requiem Mass for King Paul.
2. A Belgian breed of the domestic fowl. Also *attrib.*
1906 E. Brown *Races Domestic Poultry* viii. 117/1 The birds which are known as *poulets* and *poulardes de Bruxelles* are entirely Malines. **1910** *Encycl. Poultry* II. 295 The Malines fowl is of the Asiatic type, and is large in size... There are two varieties—the Coucou and the white. **1938** T. Newman *Princ. & Pract. Poultry Husbandry* i. 23 The Maline..was the result of a cross between the Langshan and Campine, and the Antwerp Brahma.

Malinke (mäli·ŋke). The name of a people of western Africa and their language. Also *attrib.* or as *adj.* Cf. *Mandinka *sb.* and *a.*
1883 *Encycl. Brit.* XV. 475/2 *Mandingoes,* otherwise known as *Wangawara, Malinkes,* or *Wakore*..are one of the most widely distributed and important peoples of Western Africa to the north of the equator. **1911** F. W. H. Migeod *Lang. W. Afr.* I. i. 33 Among the tribes whose languages furnish evidence of the extent of Mandingo influence are the Malinke, Bambara, Susu, [etc.]. *Ibid.* viii. 202 Malinke, Susu, and Soninke take suffixes to the verb. **1913** *Ibid.* II. xxi. 332 The Malinke held in the past a less leading position than many other branches of the Mandingos. **1950** D. Jones *Phoneme* xxviii. 206 Variphones consisting of r-like and l-like sounds would also appear to occur in some languages of Africa, for instance, in Bambara and Malinke. **1970** P. Oliver *Savannah Syncopators* 106 Lobi and Malinke xylophones. *Ibid.* 112 *Malinke,* extensively distributed tribe occupying large areas of Senegal, Guinea and western Ivory Coast. **1970** *Western Folklore* XXIX. 241 Among the Kissi and Malinké of Liberia..*banga* has the specialized meaning of a children's instrument of several strings.

Malinois (malinwa·). [Fr., f. *Malinois* of or from Malines in Belgium.] A wire-haired variety of the Belgian Sheepdog.
1947 C. L. B. Hubbard *Working Dogs of World* 104 The Malinois forms one of the three main groups of Belgian Sheepdogs... The breed takes its name from the district of Malines, where it probably originated. **1948** B. Vesey-Fitzgerald *Bk. Dog* 564 The Malinois represents the wire-haired Belgian Sheepdog. **1971** F. Hamilton *World Encycl. Dogs* 56 The Malinois is short-haired,..generally fawn in colour, with dark tips to the hairs on back and sides, giving a 'blackened' or shaded effect, and with a black mask.

Malinowskian (mæ:linǫ·f,skiăn), *a.* [f. the name of Bronislaw Kasper *Malinowski* (1884–1942), Polish-born anthropologist + -an.] Of, pertaining to, or characteristic of Malinowski and his works.
1954 *Amer. Anthropologist* Oct. 915 A fascinating personality—a Malinowskian anthropologist and a classical Confucian scholar rolled into one. **1955** Homans & Schneider *Marriage, Authority & Final Causes* 15 We may also call it Malinowskian functionalism. **1956** L. A. Fallers *Bantu Bureaucracy* 129 It is rather in the nature of a 'mythical charter', in the Malinowskian sense. **1960** *Guardian* 19 Feb. 6/2 Malinowskian methods of field investigation. **1968** *Listener* 9 May 614/3 Interminable guest-shows communicated nothing (though even that laughing, wisecracking phatic stuff has its Malinowskian place).

mal-inse·rtion. *Anat.* [f. Mal- + Insertion.] Abnormal insertion (Insertion 3).
1904 E. H. Fenwick *Handbk. Clin. Electr.-Light Cystoscopy* xxi. 439 Some form of ureteric obstruction—either an acute bend of the ureter, due to adhesion or mal-insertion in movable kidney,..or a stone impacted low down in the ureter. **1964** S. Duke-Elder *Parsons' Dis. Eye* (ed. 14) xxix. 460 It occurs chiefly in cases of congenital origin—probably mal-insertion of the muscles—but has been met with after interference with the pulley of the superior oblique in frontal sinus operations.

Malkite, var. Melchite.
1909 A. S. Lewis *Codex Climaci Rescriptus* p. xiii, As this version was adopted by the Malkite Church, it cannot be older than the Council of Chalcedon (A.D. 451).

mall[1]. Add: **4.** Now usu. pronounced (mæl). **b.** (Further examples.) Also, a shopping-precinct.
1914 Joyce *Dubliners* 24 All the branches of the tall trees which lined the mall were gay with little light green leaves. **1963** *Observer* 15 Sept. 23/6 The central paved avenue, or 'mall' [in a shopping-centre], wider than any street, with booths in the middle. **1969** *Daily Tel.* 19 Aug. 13/2 Basically, the housewife is demanding more comfort and convenience in shopping. This means covered and heated 'malls' with car-parking facilities adjacent. **1974** *Economist* 21 Dec. 47/1 The developers have discovered an even more potent device for generating sales: the rigidly controlled 'shopping environment' of the enclosed malls... Woodfield Mall, near Chicago..includes 235 stores in a roofed-over area of 191 acres. **1975** *Times* 19 Feb. 4/8 Malls smoke danger. Special fire precautions are needed in single-storey shopping malls.

Mallaby-Deeley (mæ:lābi,dī·li). [f. Harry *Mallaby-Deeley* (1863–1937), an English clothing manufacturer.] A cheap suit of clothes (see quots.). Also *transf.*
1920 *Punch* 10 Mar. 196 Oo, Lumme! Wot price Reginald in 'is Mallaby-Deeleys? **1937** *Evening Standard* 5 Feb. 3/1 In 1920 he [sc. Mallaby-Deeley] became the leader of cheap mass-production tailoring... A 'Mallaby-Deeley' soon became another name for a suit. **1959** R. Postgate *Every Man is God* xxiii. 218 'You don't know what a Mallaby Deeley is?' said his one-time Major... David Roddman learned that a Mallaby Deeley was a cheap mass-produced suit invented by a smart M.P. Ex-officers, out of work, for the use of. **1962** Granville *Dict. Sailors' Slang* 74/2 *Mallaby-Deeley's,* plain clothes worn by Naval officers and men on shore leave.

‖ **mallam** (mæ·lăm). [ad. Hausa *mālam(i)* (often used as a title).] A learned man, scribe, teacher.
1932 J. Cary *Aissa Saved* viii. 44 The white mallam put his hand on her head. *Ibid.* x. 53 She went to all the priests and mallams and juju men for a hundred miles round and commanded them to give the baby medicine. **1936** —— *Afr. Witch* xxii. 329 The old *mallam* went into his own room, which was the late Emir Aliu's office. **1939** —— *Mister Johnson* 82 He has been described..as a ham-fish or *mallam* judge—that is, an office keeper. **1965** A. Nicol *Truly Married Woman* 88 'What foolishness is this?' the Mallam said sternly.

mallard. Add: **2. c.** *mallard call, decoy, duck, -shooting.*
1852 R. Glisan *Jrnl. Army Life* (1874) ix. 102 A fine mallard duck suddenly flew up. **1874** J. W. Long *Amer. Wild-Fowl Shooting* xiv. 186 Morning and evening mallard-shooting. *Ibid.* xix. 214 They decoy exceedingly well to mallard decoys, and come readily to the mallard call.

mallee[2]. Add: (Later examples.) **mallee root** *slang* (see quot. 1941).
1911 E. M. Clowes *On Wallaby* ix. 249 On the Wimmera Plains is massed the dwarf eucalyptus known as the Mallee Scrub, the roots of which make such ideal firewood. **1936** I. L. Idriess *Cattle King* ii. 10 They passed through the mallee belt, then out on to plain country. **1941** Baker *Dict. Austral. Slang* 45 *Mallee root,* a prostitute. **1944** A. Russell *Bush Ways* xxii. 104, I think of the mallee scrubs and red gum fringes of the lower Murray River. **1958** E. O. Schlunke *Village Hampden* 127 'They're going to lynch you, Regerson,' Harry told him, grinning. 'You'd better take to the mallee before they come for you.' **1963** *Times* 12 Mar. p. x/7 The mallee fowl is one of the megapodiidae or mound-building birds of the dry interior. **1966** G. Durrell *Two in Bush* v. 166 Mallee scrub consists of a small species of eucalyptus between six and twenty feet high, and in places the trees grow very close together, their branches entwining and forming a continuous canopy. **1973** *Sun-Herald* (Sydney) 26 Aug. 85/2 (*caption*) A mallee fowl on its mound.

mallein (mæ·li,in), *v.* [f. the sb.] *trans.* To inoculate (a horse or mule) against glanders.
1915 *Punch* 4 Aug. 101/2 All mules on joining units will in future be malleined.

mallet, *sb.*[1] Add: **1. f.** *Mus.* A light hammer used for playing the vibraphone, xylophone, or similar instrument.
1930 *Melody Maker* Jan. 69/2 You should use at least three different 'hardnesses' of mallets for solo playing. **1968** *Ibid.* 23 Nov. 18 Having started with four mallets right from the beginning, I found myself playing the instrument in piano style.
6. mallet finger *Med.*, a condition in which a finger is permanently flexed at the distal joint, usu. resulting from a blow to the tip of the extended finger, which ruptures the extensor tendon; a finger so affected; **mallet-headed** *a.*, (*b*) applied to a chisel made to be struck with a mallet; (*c*) having a head shaped like that of a mallet.
1894 Gould *Dict. Med.* 472/2 *Mallet finger,* a deformity of a finger characterized by deficient extension or undue flexion of the terminal phalanx. **1934** Key & Conwell *Managem. Fractures* xx. 718 (*heading*) Drop or mallet finger (baseball finger). **1956** *Jrnl. Amer. Med. Assoc.* 21 July 1135 (*heading*) New technique for treatment of mallet fingers and fractures of distal phalanx. **1967** *Punch* 29 Mar. 458/3 Less heroically, women can contract Mallet Finger by 'tucking the bed-clothes under the mattress when bed-making'. **1906** E. Johnston *Writing & Illuminating* xvii. 396 The chisels are either *Hammer-headed* or *Mallet-headed.* **1909** *Westm. Gaz.* 26 Oct. 5/2 The announcement made by the Rules of Golf Committee on the subject of mallet-headed clubs will have far-spreading consequences.

Malling (mǭ·liŋ). The names of two villages, East and West *Malling,* in Kent, used to designate: **1.** English pottery of the late sixteenth century in the form of tin-enamelled jugs, one of which was found at West Malling; = *tiger-ware.*
[**1903** *Times* 11 Feb. 10/6 The West Malling Elizabethan jug..will be sold [on]..February 19th.] **1933** W. B. Honey *Eng. Pott. & Porc.* iii. 36 A 'Malling jug' of 1550..was sold for five hundred and sixty guineas. **1968** A. Ray *Eng. Delftware Pott.* xi. 91 The 16th century 'Malling' jugs are the earliest tin-glazed pottery.
2. A rootstock for fruit trees developed at the East Malling Research Station. Also *attrib.*
1927 *Gardeners' Chronicle* 19 Mar. 199/2 Present-day planters have the advantage of being able to buy trees on the East Malling stocks. *Ibid.* There is a very strong demand amongst market growers for dessert Apples on Jaune de Metz (Type IX), the most dwarfing of the Malling selections. **1928** A. H. Hoare *Eng. Grass Orchard* 73 It is good news for orchard planters that it has been found possible to select and propagate by layering the most desirable of the seedling crabs and the very vigorous types of 'Paradise' stocks, as they are called. These are known as Malling Types XII., XIII. and XVI., and Bristol No. 5... They are not yet plentiful. **1936** H. V. Taylor *Apples Eng.* p. iii (Advt.), Selected trees on Malling stocks. **1966** A. G. Brown in *Fruit Present & Future* (R. Hort. Soc.) 18 East Malling has continued to work with rootstocks and has recently introduced a new dwarfing stock Malling 27. **1973** *Pears Encycl. Gardening: Fruit & Vegetables* 23 Trees on Malling IX, which are planted closer than those on other rootstocks..will ensure the largest weight of fruit from a small garden, in the quickest possible time.

Mallorcan (mălǭ·ɹkăn, mălʸǭ··), *sb.* and *a.* [ad. Sp. *Mallorca* Majorca + -an.] = *Majorcan *sb.* and *a.* Also, the language of the Mallorcans.
1868 H. O'Shea *Guide to Spain & Portugal* (ed. 3) 341/1 The inhabitants, especially the Mallorcans, are an honest, interesting..people. *Ibid.* 342/2 There is great similarity between Mallorcan and the Languedocian patois of Montpellier. *Ibid.* 343/1 The learned Scaliger expatiated on the excellence of the Mallorcan pottery. **1927** N. L. Duryea *Mallorca* i. 6 Oriental eyes brood in Mallorcan faces. **1934** *Scottish Geogr. Mag.* L. 136 A great proportion of the Mallorcan trees are exceedingly old. **1937** W. Fortescue *Sunset House* ix. 162 A little red Mallorcan sausage. **1954** T. Maynard *Long Road of Father Serra* (1956) i. 13 Not all Mallorcans were so pious as the Serras. *Ibid.* 17 The beauty of the Mallorcan scene. **1972** A. Lowe *Catalan Vengeance* 140 The *Carta Catalana,* the commercial map of oriental trade drawn in 1375 by the Mallorcan Jew Abraham Cresques.

Mallorquin (mălǭ·ɹkin, mălʸǭ··), *sb.* and *a.* Also -quine. [Sp. *Mallorquín,* f. *Mallorca* Majorca: see *Majorcan *sb.* and *a.*] = *Majorcan *sb.* and *a.* Also, the language of the Mallorquins.
1839 *Penny Cycl.* XIV. 339/1 In character the Mallorquines somewhat resemble the Catalans. **1868** H. O'Shea *Guide to Spain & Portugal* (ed. 3) 342/2 The 'Mallorquin' is a corruption of the Catalan dialect. **1869** H. W. Bates *Illustr. Trav.* I. 267/2 The women of Soller are..by no means so plain in their appearance as the majority of the Mallorquine females. **1875** *Encycl. Brit.* III. 277/2 A railway is in course of construction from Palma by Inca to Alcudia, and the stock is all held by Mallorquins. **1926** R. Macaulay *Crewe Train* i. i. 4 As..his knowledge of the Mallorquin idiom improved..he discovered that Mallorquins are almost the kindest and most sociable people in the world. **1929** D. H. Lawrence *Let.* 26 May (1932) 806 The Mallorquin servants cooked Spanish rice over a fire. **1952** [see *Ibizan *a.*]. **1954** T. Maynard *Long Road of Father Serra* (1956) i. 2 They have their own language, Mallorquin, even if this is no more than a dialect of Catalan. **1973** *Times* 3 Mar. 26/2 *Mallorquin,* the local dialect of the native Catalan language, runs a poor second. **1974** R. Jeffries *Mistakenly in Mallorca* iii. 24 A Mallorquin inlaid sideboard. *Ibid.* v. 34 She was not..a Mallorquin or a Spaniard.

malnou·rished, *a.* [f. Mal- + Nourished *ppl. a.*; cf. Malnutrition.] Suffering from or provided with insufficient nutrition or nourishment; undernourished.
1928 *Daily Tel.* 11 Dec. 17/1 He [sc. Lord Eustace Percy] had emphasised the importance of local authorities making adequate provision for mal-nourished children. **1963** *Economist* 29 June 1348/1 Half the world's population is still hungry or malnourished or both. **1972** *Lancet* 22 July 143/1, 200 randomly selected malnourished children under the age of five years.

malnou·rishment. [f. Mal- + Nourishment.] = Malnutrition.
1932 *N.Y. World-Telegram* 16 June 22/7 The inroads of starvation, diseases and malnourishment. **1936** E. P. O'Donnell *Green Margins* 222 Many of them were tuberculous from malnourishment. **1969** N. W. Pirie *Food Resources* iii. 99 There will be protein malnourishment unless food, averaged over a day, contains a reasonable percentage of protein.

mal-observation. (Earlier and later examples.)
1843 Mill *Logic* II. v. iv. 387 It is mal-observation, when something is not simply unseen, but seen wrong; when the fact or phenomenon..is mistaken for something else. **1895** A. Lang in R. M. Dorson *Peasant Customs* (1968) II. 444 If he denies it, if he says with Lord Kelvin that hypnotism is all imposture and malobservation, I am silenced. **1922** *Encycl. Brit.* XXXII. 199/2 Repression of the truth would have to be added as a third to mal-observation and forgetfulness, as a very subtle source of error in testimony to the occurrence of the supernormal.

maloca (mälōᵘ·kă). Also 9 **malocca.** [Pg., a large hut, f. Amer. Sp., raid, attack, f. Araucanian *malocan* to fight (Webster).] A large hut in certain Indian settlements in South America.
1860 Mayne Reid *Odd People* 122 The Mundrucus build the *malocca*..only in their case it is not used as a dwelling, but rather as a grand arsenal, a council-chamber,..and, if need be, a fortress. *Ibid.* 47 Three or four hundred individuals not unfrequently assemble under the roof of a single *malocca.* **1944** S. Putnam tr. *E. da Cunha's Rebellion in Backlands* ii. iii. 175 The ruins of the *malocas,* or native villages. **1951** J. C. Fennessy *Sonnet in Bottle* iii. v. 81 The village consisted only of three large malocas, with palm-thatched roofs reaching nearly to the ground. **1962** N. Maxwell *Witch-Doctor's Apprentice* xiv. 171 In the Colombian Amazon, a maloca is the traditional dirt-floored, round or oval Indian house whose walls are made of the same thatch as its roof.

malocclu·sion. *Dentistry.* [f. Mal- + Occlusion.] Faulty occlusion (of the teeth).
1888 J. N. Farrar *Treat. Irregularities Teeth* I. lv. 563 (*heading*) Mal-occlusion of the anterior teeth.—Treatment. **1937** *Times* 22 Nov. 9/4 A constant stream of visiting dentists..inspected their malocclusions. **1968** *New Scientist* 12 Sept. 538/2 Malocclusion of the jaws, with resulting pyorrhoea, was the pharaohs' biggest dental problem.

malonate (mæ·lǫnĕit). *Chem.* [f. Malon(ic *a.* + -ate[4].] The anion, or an ester or salt, of malonic acid.
1862 W. A. Miller *Elem. Chem.* (ed. 2) III. v. 367 The malonates of potash and ammonia are deliquescent, but uncrystallizable. **1898** J. Wade *Introd. Study Org. Chem.* xxv. 153 Ethyl malonate, $CH_2(COOEt)_2$, is a heavy, colourless liquid, which has a pleasant aromatic odour, and boils at 198°. **1951** A. Grollman *Pharmacol. & Therapeutics* xxix. 676 Both malonate and succinate of sodium are absorbed only slowly from the intestine, and act as saline cathartics. **1962** *Times* 1 May 4/1 (Advt.), They'll be available later in an attractive bound volume scented with perfumes based on certain allyl esters such as salicylate and malonate. **1962** H. A. Krebs in A. Pirie *Lens Metabolism Rel. Cataract* 351 The effects of malonate on the oxidation of succinate.

malonic, *a.* Add: *malonic ester,* the diethyl ester, $CH_2(COOC_2H_5)_2$, of malonic acid, which is a liquid widely used in synthesis, as of carboxylic acids RR′C(COOH)$_2$ or RR′CH·COOH by alkylation with alkyl halides (*malonic ester synthesis*).
1881 *Chem. News* 28 Jan. 47/1 (*heading*) Syntheses by means of malonic ester. **1906** J. J. Sudborough *Bernthsen's Text-bk. Org. Chem.* (rev. ed.) x. 238 This so-called 'malonic ester' synthesis is an important method for the preparation of higher dibasic acids. **1968** R. O. C. Norman *Princ. Org. Synthesis* x. 330 The use of malonic ester considerably increases the versatility of this general method [of amino-acid synthesis]. **1971** N. Allinger et al. *Org. Chem.* xxiv. 665 In the malonic ester synthesis, diethyl malonate is converted by sodium ethoxide into its sodium salt, which is then allowed to react with an alkyl halide.
Hence **ma·lonyl,** the radical —CO·CH$_2$·CO— derived from malonic acid by removal of the two hydroxyl groups.
1889 G. M'Gowan tr. *Bernthsen's Text-bk. Org. Chem.* 282 Barbituric acid, malonyl urea, $C_4H_4N_2O_3$, is a dibasic acid. **1931** *Jrnl. Chem. Soc.* 273 The Friedel–Crafts reaction has now been extended to the coupling of malonyl chloride..with methyl ethers of resorcinol. **1970** R. W. McGilvery *Biochem.* xvi. 317 The butyryl group can be displaced onto the neighboring cysteinyl group by an incoming malonyl group. **1971** *Nomencl. Org. Chem.* (I.U.P.A.C.) (ed. 2) C. 313 Malonyl (preferred to propanedioyl) —CO·CH$_2$·CO—.

Malo-Russian (mēilo,rʊ·∫ăn), *sb.* and *a.* [f. Russ. *Malorossiya* Little Russia, or ad. *Malorós(s), -rús* or *Malorossiyánin* Little Russian.] **A.** *sb.* A member of the Little Russian race inhabiting the south of Russia; their language. **B.** *adj.* Of or pertaining to the Little Russians, Ruthenian.
1862 [see Ruthenian *sb.* 2]. **1880** W. R. Morfill *Russia* 74 The Malo-Russian is very rich in Skazki (national tales) and in songs. *Ibid.* 75 The Malo-Russian philologists. **1923** E. A. Ross *Russ. Soviet Republic* 58 Between Great Russia and the Black Sea live the Ukrainians or Little Russians (Malo-Russians).

malpais (malpa,ī·s). *U.S.* [Sp., f. *malo* bad + *país* country, region.] Rugged or difficult country of volcanic origin.
1844 G. W. Kendall *Narr. Santa Fé Expedition* II. 384 We had crossed the *mal pais,* or bad country, as it is called. **1847** G. A. F. Ruxton *Adventures Mexico & Rocky Mts.* xi. 79 The tract of country known as the *Mal Pais,* a most interesting volcanic region. **1896** *Nat. Geogr. Mag.* VII. 295 Bench mesas may be classified by structure into bolson mesas, stream-terrace mesas, talus-fan mesas, and malpais mesas. **1918** C. E. Mulford *Man from Bar-20* xvii. 178 Slipping on the treacherous malpais and loose stones. **1942** Kearney & Peebles *Flowering Plants Arizona* 17 Some of these are limestone areas, whereas others are volcanic, locally known as 'mallapy' (*mal pais*).

malpresentation. (Earlier and later examples.)
1852 *Lancet* 2 Oct. 299/2 The proportion of mal-presentations is increased in cases in which labour comes on at the full term. **1962** R. H. Smythe *Anat. of Dog Breeding* v. 104 There is always the possibility of mal-presentation or malformation of an individual puppy.

malrotation (mælrotēⁱ·∫ən). *Med.* [f. Mal- + Rotation.] Faulty or abnormal rotation of a part of the body, esp. of the intestines during development.
1932 *New Eng. Jrnl. Med.* CVI. 280 At operation a distended stomach was found... There was no evidence of mal-rotation. **1961** *Lancet* 9 Sept. 598/1 There was also malrotation and severe mesenteric lymphadenitis.

malt, *sb.*[1] Add: **3.** (Further examples.) Now usu. = *malt whisky.*
1925, etc. [see *ball *sb.*[4]]. **1967** A. Diment *Dolly Dolly Spy* iv. 39 He..poured himself a big shot of the finest Malt. **1972** A. Ross *London Assignment* 32, I had put plenty of Malvern water into the malt, figuring to make it last. **1974** *Times Lit. Suppl.* 11 Jan. 38/5 Ross Wilson ignores the religious divisions of the trade ('malts' are the Catholic whiskies). **1975** *Scottish Field* Jan. 13/2 Hugh MacDiarmid, a whisky drinker for 60 years, expounds on the great malts.
4. a. *malt whisky* (further examples).
1968 *Scottish Field* Nov. 56 (Advt.), Chivas Regal is blended with the softest Glenlivet malt whiskies. **1970** Simon & Howe *Dict. Gastron.* 389/1 About a hundred years ago it became the custom to blend straight malt whisky with neutral grain whisky.
b. *malt* (*tea-*)*loaf.*
1901 B. S. Rowntree *Poverty* viii. 288/2 Malt loaf. 3d. **1969** E. Gébler *Shall I eat you Now?* 98 They drank tea and ate slices of a sticky malt tealoaf.
5. malt shop *U.S.*, a shop where malted milk is sold.
1943 D. Baker *Trio* ii. 88 She was out of the gate and into a little section of bright lights—college clothes,..malt shops. **1949** *Time* 26 Sept. 25/1 The word was relayed through the drive-ins, malt shops and garages.

Malt (mǫlt), *sb.*[2] *slang.* = Maltese *sb.* 1 a.
1959 C. MacInnes *Absolute Beginners* 40 A sickly simper,..that she turned on like a light for the two beefo Malts. **1969** *Sunday Mail* (Brisbane) 16 Mar. 10 A new and frightening power is suddenly emerging in the Sydney underworld. It is called by those who know about it 'The Malts'. Police fear it is taking the form of a Maltese 'Mafia'. **1971** *Guardian* 6 July 11/6 The 'Malts', as one used to refer to them, or sometimes 'yellow-bellied bastards'.

maltase. Substitute for def.: Any enzyme which hydrolyses maltose and other α-glycosides. (Add earlier and later examples.)
1890 *Jrnl. Chem. Soc.* LVIII. 998 The author starts with the assumption that the diastase of malt is composed of a mixture of two enzymes—maltase and dextrinase. **1946** *Thorpe's Dict. Appl. Chem.* (ed. 4) VII. 481/2 Maltase also hydrolyses..α-glycosides but not β-glycosides. **1964** N. G. Clark *Mod. Org. Chem.* xvi. 327 Hydrolysis with dilute hydrochloric acid yields two molecular proportions of glucose, and the glycosidic linkage is established as α by the fact that similar hydrolysis occurs with the enzyme maltase (which is specific for α-glycosides). **1968** A. White et al. *Princ. Biochem.* (ed. 4) xviii. 387 Maltases IV and V account for about two-thirds of the total maltase activity.

malted, *ppl. a.* **2.** (Earlier and later examples.)
1887 *Official Gaz.* (U.S. Patent Office) XLI. 358 Trade-Marks..Food preparation for infants and invalids.—Horlick's Food Company, Mount Pleasant and Racine, Wis.,..'The words *Malted Milk* and the letters "M.M."' **1920** [see *beezer 2]. **1930** Belloc *New Cautionary Tales* 41 He gave him medicine by the pail And malted milk, and nutmeg ale. **1960** A. E. Bender *Dict. Nutrition* 82/1 Malted milk—liquid separated from mash of ground barley malt and wheat flour mixed with milk and the mixture dried. **1974** *Times* 16 Mar. 11/6 My palate has reverted to childhood and adores American food—sandwiches, malted milk,..jello.

malted (mǫ·ltĕd), *sb.* A drink of malted milk (Malted *ppl. a.* 2).
1945 *Newsweek* 8 Jan. 84 The couple..went into a Walgreen drugstore for ice cream. There, high above the ads for malteds, they noticed Vedovelli's paintings. **1949** A. Miller *Death of Salesman* ii. 117 Let's go downstairs and get you a malted. **1953** *Manch. Guardian Weekly* 13 Aug. 7 The soda fountain serves such luxuries as candy and malteds. **1969** Gish & Pinchot *Lillian Gish* xiv. 202 Order her a malted and a cheese sandwich.

Maltese, *a.* and *sb.* Add: **A.** *adj.* **2. Maltese cross,** (*b*) *Philately,* name given to the post-mark used on British postage stamps from their introduction in 1840 until 1844; (*c*) *Cinemat.* = *Geneva mechanism*; **Maltese dog,** substitute for def.: = sense *B. 4 (later example); **Maltese terrier** = sense *B. 4.
1881 Philbrick & Westoby *Postage & Telegr. Stamps Gt. Brit.* 291 Several obliterating marks are found: 1. The Maltese cross, or *croix patée,* as then in use. **1909** J. G. Hendy *Hist. Postmarks Brit. Isles 1840–76* 9 The Maltese Cross or *croix patée* obliterator—the latter is really the correct term to apply to it from a heraldic point of view—varied a good deal in shape in different places. **1917** C. N. Bennett *Guide to Kinematogr.* ix. 127 Modern Maltese cross movements are descendants of the 'Geneva stop' by which overwinding of watches is prevented. **1934** J. B. Seymour *Stamps Gt. Brit.: Line-Engraved Issues 1840–53* 9 The design generally used for obliterating the British stamps from 1840 to 1844 has been called, incorrectly, a Croix Patée or Maltese Cross; it is neither one nor the other. But it is so well known as the Maltese Cross that it is deemed advisable to call it so for the purpose of this work. *Ibid.* 102 The stamps from the first forty plates [sc. of the one penny red imperforate] are usually obliterated with the Maltese Cross. **1949** P. C. Litchfield *Guide Lines to Penny Black* 16 Stamps with the red Maltese Cross are more likely to have come from one of the early plates. **1953** Maltese cross [see *Geneva[2] b]. **1963** *Gt. Brit.: Specialised Stamp Catal.: Queen Victoria* (Stanley Gibbons Ltd.) 23 The Maltese Cross is a valuable plating aid for collectors of the Penny Red imperf. Its presence on a stamp is strong evidence that that stamp is from a plate put to press before 1845. **1913** V. Shaw *Encycl. Kennel* 130 Many people persist in calling the delicate little Maltese dog a terrier, whereas he is nothing of the kind, but a fragile member of the toy family. **1880** H. Dalziel *Brit. Dogs* III. viii. 430 Whether the dog we now call a Maltese terrier be a descendant more or less pure from the breed Strabo wrote of, it is now impossible to say. **1954** M. K. Wilson tr. *Lorenz's Man meets Dog* iv. 42 They [sc. two large dogs] nearly pulled a Maltese terrier in half.
B. *sb.* **4.** A very small, long-coated, white dog of the breed so called, formerly known as *Maltese dogs* or *Maltese terriers* (sense A. 2 in Dict. and Suppl.).
1867 Mrs. F. Lehmann *Let.* 22 Jan. in Geo. Eliot *Lett.* (1956) IV. 336 The Lewes's were enchanted with Chang. They say he is a real Maltese. **1950** A. C. Smith *Dogs since 1900* 318 The Maltese. This, one of the oldest of the toy breeds in Europe, at one time had 'Terrier' added to its name. **1972** *Country Life* 10 Feb. (Suppl.) 21/1 'Floriana Maltese'..good straight white coats, with black points.

malthoid (mæ·lþoid). *Austral.* and *N.Z.* [f. Maltha + -oid.] The proprietary name of a bituminous material made from wood fibre and used as a roof- or floor-covering or for covering other surfaces.
1936 *Patent Office Jrnl.* (N.Z.) 3 Dec. 367/1 *Malthoid.* Pabco Products (Australia), Limited... Paints, varnishes, coatings, and paint compounds, [etc.]. **1959** *Numbers* IX. 9 The malthoid roof. **1960** B. Crump *Good Keen Man* 76 When all the boards were on [the hut] we covered the whole structure with malthoid and tacked it into place. **1960** J. Fingleton *Four Chukkas to Austral.* 7 An odd sort of trial upon a malthoid pitch.

Malto (mæ·lto), *sb.* [Native word = 'language of the Maler': see *Maler *sb.* and *a.*] A Dravidian language spoken by the Maler people living in the Rajmahal hills of northern India. Also called *Rajmahali. Also as *adj.*
1884 E. Droese *Introd. Malto Lang.* p. i, Malto is the language of one of the aboriginal races of India who call themselves Maler *i.e.* men, and go among their Aryan neighbours by the name of Paharias (Hill people). **1906** G. A. Grierson *Ling. Survey India* IV. 446 Malto is almost exclusively spoken in the Rajmahal Hills in the north-east of the Sonthal Parganas... Malto is the name used by the people themselves in order to denote their language. The word simply means 'the language of the Maler', and *maler* in Malto means 'men' and is the name the people apply to themselves... It is..probable that Malto like Malayālam is derived from the common Dravidian *mala,* mountain, so that the original meaning of 'maler' would be 'hillmen'. *Ibid.* 447 Malto does not possess a literature of its own... The Malto language very closely agrees with Kurukh. **1938** S. S. Sarkar *Mālērs of Rajmahal Hills* i. 18 The Mālērs speak *Mālto.* **1955** T. Burrow *Sanskrit Lang.* viii. 387 The Dravidian languages Kurukh and Malto are preserved even now in Northern India, and may be regarded as islands surviving from a once extensive Dravidian territory. **1963** L. P. Vidyarthi *Maler* iv. 57 The Maler..call themselves 'Male' in their native Malto language which means hill-man. **1972** W. B. Lockwood *Panorama Indo-European Lang.* 224 Malto, the vernacular of 100,000 tribesmen, is found in the Rajmahal Hills.

maltol (mǫ·ltǫl). *Chem.* [a. G. *maltol* (J. Brand 1894, in *Ber. d. Deut. Chem. Ges.* XXVII. 810): see Malt *sb.* and -ol.] A crystalline compound, $C_6H_6O_3$, which occurs in larch bark and chicory and is formed by roasting malt; 3-hydroxy-2-methylpyran-4-one.
1894 *Jrnl. Chem. Soc.* LXVI. i. 271 Beer, prepared from caramel malt, gives a violet coloration with ferric chloride; this is not due to salicylic acid, but to the presence of a compound termed maltol, which is distinguished from salicylic acid by giving no reaction with Millon's reagent. **1945** *Jrnl. Amer. Chem. Soc.* LXVII. 2276 Hydrolysis of streptomycin chloride with *N* sodium hydroxide...yields a weakly acidic substance, m.p. 161–162°, which has been characterized as maltol. **1972** *Sci. Amer.* Mar. 16/2 In recent years the use of maltol, which can intensify or modify the flavor of preserves, desserts, fruit, soft drinks and foods generally high in carbohydrates, has expanded greatly.

maltreater (mæltrī·təɹ). [f. Maltreat *v.* + -er[1].] One who maltreats or ill-uses.
1902 O. Wister *Virginian* iv. 57 A maltreater of hawsses [horses]. **1906** B. von Hutten *What became of Pam* i. ii, Tyrants,..drunkards, maltreaters [of women]. **1965** M. Spark *Mandelbaum Gate* iii. 55 Much as the timid spinsters of the old days, while abroad, would be moved to violence against the maltreater of the donkey.

‖ **malum in se** (mæ·lǝm in sī). Pl. **mala in se.** [med.L.] Something intrinsically evil or wicked. Also as *adj. phr.*
1623 J. Williams *Let.* 30 Aug. in J. Hacket *Scrinia Reserata* (1693) i. 157 But to grant a Pardon even for a thing that is *malum in se.* **1811** *Edin. Rev.* Feb. 275 That corruption is not..a *malum in se,* as Mr Windham has been pleased to assert. **1826** *Ibid.* Feb. 331 Whether there be..any solecism which is a *malum in se,* as distinct from a *malum prohibitum.* **1856** *Sat. Rev.* 26 July 290/1 It is the bargaining intention, and this is *malum in se.* The *malum prohibitum* is the contract. **1893** [see *hypertension 1 a]. **1959** Jowitt *Dict. Eng. Law* II. 1130/1 *Mala in se,* acts which are wrong in themselves, such as murder, as opposed to *mala prohibita.* **1961** *Times* 18 Mar. 12/6 That part of the criminal law which covered offences *mala in se.*

malva (mæ·lvă). [L. *malva* mallow; adopted by Linnæus in his *Genera Plantarum* (1737) 308 as the name of a genus.] = Mallow.
[**1548** W. Turner *Names of Herbes* sig. E3 Malva is called in greeke Malache, in englishe a Mallowe or a Mallo.] **1883** W. Robinson *Wild Garden* (ed. 3) xiv. 150 Some of the Malvas are..vigorous-growing plants. **1959** W. Strache *Forms & Patterns in Nature* 35 (*caption*) Malva flower. **1966** E. Palmer *Plains of Camdeboo* xvii. 277 The early settlers... used..the Malva to bathe sore eyes.

Malvern (mǫ·lvəɹn). The name of a town in Hereford and Worcester, used *attrib.* and *ellipt.* to designate alkaline mineral water obtained from springs there.
1757 J. Wall *Exper. & Observations Malvern Waters* (ed. 2) App. 68 She was last Year perswaded to try Malvern Water,..and..the Water has been constantly sent to London all the Winter for her Use. **1907** *Yesterday's Shopping* (1969) 27 Malvern water... Bots. 6/0 [per doz.]. **1930** A. Bennett *Imperial Palace* lxiv. 525 She watched over his flowers and his Malvern water. **1965** *New Statesman* 14 May 753/1 One of the lackeys will carry a 10-day supply of Malvern water. **1967** *Observer* (Colour Suppl.) 28 May 38/2 Malvern. The one British mineral water that is sold on a national scale. **1974** J. Stubbs *Painted Face* iv. 70 A picnic basket, complete from cold capon..to Malvern water.

Malvernian (mǫlvɜ̄·miăn), *a.* *Geol.* [f. the name of the *Malvern* Hills, a range in England between Hereford and Worcester: see -ian.] Of, pertaining to, or characteristic of the Malvern Hills; *spec.* (*a*) applied to a Pre-Cambrian series of plutonic rocks that forms most of the hills; also *absol.*; (*b*) applied to a north–south orientation like that of the hills.
1879 *Q. Jrnl. Geol. Soc.* XXXV. 654 The contained pebbles..consist of quartzite, quartz, gneiss, mica-schist, red felspar, and granitoid rock. This assemblage strongly suggests derivation from the Malvernian series represented at Primrose Hill. **1899** *Proc. Geologists' Assoc.* XV. 426 The strike of the axes of the main folding of the country [around Stourbridge] is first north and south, in continuation of the main axes of the Malvern ridge (Malvernian folding), but at Abberley Hill it swings round almost at right angles to this direction, the axes now striking from west to east (Hercynian or Mercian folding). **1948** L. J. Wills *Palaeogeogr. Midlands* ii. 7 The distribution of the Malvernian away from Malvern itself is quite unknown. **1969** Bennison & Wright *Geol. Hist. Brit. Isles* iii. 68 The Malvernian is composed predominantly of metamorphosed igneous rocks, diorites and gabbros. *Ibid.* x. 248 (*heading*) Structures of Malvernian trend.

malversate (mælvɜ̄·ɹsĕit), *v.* [Back-formation from Malversation.] *trans.* To use (funds) for purposes other than those for which they were intended.
1881 S. Butler *Family Lett.* (1962) 198 If I malversated these funds or lost them, my co-trustee might come down on the company for having regarded the order of one trustee to the disregard of a joint order. **1928** G. B. Shaw *Intelligent Woman's Guide Socialism* 417 The public school (meaning a very exclusive private school malversating public endowments).

Malvi (mā·lvi). Also **Malwi.** A dialect of Rajasthani.
1883 *Census of Central Provinces 1881* (India Census) II. 91 (*table*) Languages and dialects amalgamated therein... Urdû, Mâlvi, Nimâri, [etc.]. **1901, 1957** [see *Jaipur]. **1968** B. Walker *Hindu World* I. 586 Other Rājasthāni dialects are *Mārwāri*.., *Jaipuri*.., *Mālvi*.., *Mewāri*.., *Udaipuri.*

‖ **mal vu** (mal vü). [Fr.] Held in very poor esteem, or in no esteem, looked down upon. Also *attrib.*
1904 H. O. Sturgis *Belchamber* iii. 42 There is nothing so easy as to get rid of me. I am horribly *mal vu* by the authorities. **1958** L. Durrell *Mountolive* ii. 62 He is..rather an old-fashioned reactionary in his outlook, and is consequently rather *mal vu* by his brother craftsmen. **1958** *Times Lit. Suppl.* 21 Nov. 672/3 Her daughter Elsie's marriage to a clever but socially *mal vue* [*sic*] research scientist.

‖ **mamaku** (ma·maku). *N.Z.* [Maori.] A tree fern, *Cyathea medullaris,* or the starchy food formerly prepared from its pith.
1846 C. Heaphy *Jrnl.* 15 May in N. M. Taylor *Early Travellers in N.Z.* (1959) 232 Encamped in order to collect and bake *mamaku,* the stem of the fern tree, which was to constitute our provisions. **1882** W. D. Hay *Brighter Britain!* II. 152 The inner stems of mamaku or tree-fern. **1905** W. B. *Where White Man Treads* 16, I have forgotten 'mamaku' (the tree fern)—fanning itself in the mid-heat of summer. **1949** P. Buck *Coming of Maori* (1950) i. iii. 34 Tree fern pith (*mamaku*), and fern root (*aruhe*) were placed before them. **1966** *Encycl. N.Z.* II. 785/1 Mamaku or black tree fern (*Cyathea medullaris*) was an important source of starch food... The pith [of the trunk] was steamed for about two days..and the resulting sago-like substance could be eaten cold or dried for future use.

‖ **mamaliga** (məməli·gə). [Rumanian *mămăligă.*] = Polenta c; maize porridge: a staple food in Rumania.
1878 J. W. Ozanne *Three Years in Roumania* 121 Maize forms the staple food of the lower classes. Every day a portion is boiled for the use of the family. It is called *mamaliga,* and is usually eaten alone; a little milk or a piece of salt fish being, however, sometimes added. **1925** J. A. Hammerton *Countries of World* XXXIV. 3425/1 The peasants live for the most part on mamaliga. **1970** *Sat. Rev.* (U.S.) 12 Sept. 107/3 In Rumania, unfermented grape juice or must is sold at wooden stalls..with slices of hot *mamaliga,* the national dish of of cornmeal mush.

mamaloi (ma·ma,lwa). Also **mamanloi.** Pl. **mamaloi, mamalois.** [ad. Haitian Creole *mamalwa,* f. *mama* mother + *lwa* *loa[2].] A voodoo priestess.
1884 S. St. John *Hayti, or Black Republic* v. 194 The Haytians have corrupted the compounds Papa Roi and Maman Roi into Papaloi and Mamanloi. **1935** R. A. Loederer *Voodoo Fire in Haiti* The priests are called 'Papaloi' by their followers, and the priestesses 'Mamaloi', both words being Creole corruptions of 'Papa Roi' and 'Mama Roi'.

‖ **mama-san** (mama,san). Also **mamasan.** [Jap., f. *mama* mother + *san* honorific title.] In Japan and the Far East, a matron in a position of authority: *spec.* one in charge of a geisha-house; the mistress of a bar.
1949 L. H. Crockett *Popcorn on Ginza* xvi. 218 Mama-sans of some of the better Tokyo houses become well informed on all sorts of political intrigue. **1958** G. Mikes *East is East* 56 '*Mamasan*' is not surprised if someone wishes to linger late in one of those tiny little Japanese houses where the parties take place. **1962** A. Campbell *Heart of Japan* iv. 113 The mama-san wanted us all to go home... A geisha party isn't supposed to go on more than three hours. **1968** *Manch. Guardian Weekly* 4 Jan. 10 We were visited by two enormous Sumo wrestlers, with their equally large Mama-San. **1971** *Nat. Geographic* Oct. 571/2 The representative bars have an institutional figure called 'mama-san', a matronly, amiable, garrulous, and straightforward woman who permits the sailors to fraternize with any of a covey of young ladies.

mamba (mæ·mbă). Also 9 **momba.** [ad. Zulu *imamba.*] A large venomous African snake of the genus *Dendroaspis* (family Elapidæ), esp. the *green mamba* (*green *a.* 12 b), *D. angusticeps,* and the black mamba, *D. polylepis.* Also *attrib.*
1862 J. S. Dobie in *S. Afr. Jrnl.* (1945) 44 He called it a green mamba and was about 9 feet long. **1878** P. Gillmore *Great Thirst Land* xxix. 346 A black mamba—a description of snake common in Natal, and reported to be very deadly. **1897** J. Bryce *Impressions S. Afr.* iii. 23 The black *momba,* which is nearly as large as a rattlesnake, is a dangerous creature. **1910** J. Buchan *Prester John* iv. 80 A black mamba might appear out of the tangle. **1912** F. W. Fitzsimons *Snakes S. Afr.* (new ed.) 194 There are two varieties of the Mamba. One is vivid leaf-green, the other is olive or brownish-black. **1921** *Chambers's Jrnl.* 26 Feb. 203/2 It was a mamba snake, eight feet long. **1939** S. Cloete *Watch for Dawn* 62 With the bite of a mamba or a ringhals the heart sometimes beat even after life had gone. **1958** *Listener* 23 Jan. 154/1 Black mambas..have the unattractive habit of lurking in trees and dropping down on to the unwary. **1965** A. Nicol *Truly Married Woman* 96 A poisonous one, a mamba, its middle still rippling slowly. **1969** A. Bellairs *Life of Reptiles* I. v. 191 The deadly mambas (*Dendroaspis*) have fangs which are relatively a little shorter and thinner than those of the average cobra.

mambar, var. *Mimbar.

mambo (mæ·mbo). [Amer. Sp., 'prob. fr. Haitian creole (voodoo priestess)' (Webster).] **1.** A kind of rumba, a ball-room dance (and its music) of Latin-American origin.
1948 *Call-Bulletin* (San Francisco) 17 Sept. 11/1 Tony De Marco predicts the new dance fad will be 'The Mambo', which was introduced..last week. A zingier form of rhumba. **1950** *Newsweek* 4 Sept. 76 The difference between the rumba and the mambo is the difference between the regular foxtrot and the jitterbug. **1951** R. Chandler *Let.* 19 Apr. in *R. Chandler Speaking* (1966) 213, I doubt if he knows the new dance called a mambo, because it seems to be only recently discovered or developed. **1952** *Down Beat* 25 Jan. 2/2 Prado, who says he introduced the mambo in Mexico City in 1948, claims it is merely Afrocuban rhythm with a dash of American swing. **1955** *Caribbean Q.* IV. ii. 102 Some St. Lucians..patronize the several clubs which have jukeboxes stocked largely with mambos of the Perez Prado school. **1957** J. Kerouac *On Road* (1958) 93 Mambo blasted from jukeboxes. **1959** *Listener* 24 Sept. 473/1 Innumerable folk-dance sources, ranging from the square dance to Latin American mambos and the rest. **1964** W. G. Raffé *Dict. Dance* 299/1 *Mambo* (Haiti), ritual dance of Voodoo; an initiation ceremonial which follows the ancient *Danse Shalame* (Salome) in its symbolic-realism. 'Mambo' is the official name for the chief priestess. **1966** *Crescendo* Dec. 27/1 Only bossa novas and mambos seem to be played with anything like enthusiasm by these bands. **1972** *Village Voice* (N.Y.) 1 June 4/3 Now the pop-music business, having scraped the hillbilly barrel and blown the froth off the mambo craze, has taken over R[hythm]-and-B[lues].
2. A voodoo priestess.
1964 R. Severn *Blood & Gold* x. 102 She must be a *Mambo*.., a *Voodoo* priestess. **1964** [see sense 1 above]. **1966** *Daily Tel.* 11 Aug. 18/6 The author and his herbalist friend became close companions with a girl who was often thus possessed, but had to be initiated and trained by an expert sorceress or *mambo.*

mamillary, *a.* Add: **2.** *mamillary body* [prob. tr. mod.L. *corpus mamillare*], either of a pair of small white hemispherical structures lying side by side between the tuber cinereum and the posterior perforated substance in the interpeduncular fossa on the ventral surface of the brain.
1832 J. Quain *Elem. Anat.* (ed. 2) viii. 642 A thin lamella of white substance,..called *locus perforatus,*..is triangular, the sides being formed by the crura cerebri, the base by the mammillary bodies. **1921** Tilney & Riley *Form & Functions Cent. Nervous Syst.* xxxiv. 613 The mammillary body serves as a relay station in the olfactory tract. **1942** O. Larsell *Anat. Nervous Syst.* xviii. 249 The mammillary bodies are two elevations..just rostral to the posterior perforated substance. **1967** G. M. Wyburn et al. *Conc. Anat.* vii. 190/2 The depressed area between the peduncles, the interpeduncular fossa, has the following structures from behind forwards: the posterior perforated substance, the paired mamillary bodies, the tuber cinereum and infundibulum..and the optic chiasma.

mamma[1], mama. Add: **c*.** *Mam(m)a mia!* [It., lit. 'mother mine!']: an exclamation expressing surprise or astonishment.
1848 Mrs. Gaskell *Mary Barton* II. iii. 47 The little hopeless stranger..spoke in some foreign tongue, with low cries for the far distant 'Mamma mia!' **1969** G. Greene *Trav. with my Aunt* i. vii. 72 There came a strange grating noise. '*Mamma mia,*' the nurse said, 'what's that?' **1971** R. Falkirk *Chill Factor* x. 94 'It rain every day?' asked the Italian. 'Every day,' I said. '*Mama mia,*' he said.
d. *mamma's boy,* a boy who has been pampered and spoiled; one who is excessively timid; also applied to a man. *colloq.*
1850 C. M. Yonge *Henrietta's Wish* xiii. 183, I would not give a farthing for Fred if he was always to be the mamma's boy you would make him. **1861** —— *Young Step-Mother* xxiii. 336 It might be no great harm if Maurice were a tame mamma's boy. **1929** W. Faulkner *Sound & Fury* 141 Yah..go on then, mamma's boy. If he goes swimming he'll get his head wet and then he'll get a licking. **1967** A. M. Stein *Executioner's Rest* vi. 100 Who was this Erridge? He was mama's boy, and if you don't know the kind of trouble mama's boy tourists go looking for when they're loose on the shores of the Mediterranean, you just haven't travelled.

mammaplasty (mæ·măplæsti). *Surg.* Also **mammo-.** [f. Mamma[2] + -plasty.] Alteration of the shape or size of a breast by plastic surgery. Hence **mammapla·stic** *a.*
1938 *Revue de Chirurgie Structive* VIII. 39 (*heading*) Review of 80 cases of mammaplasty. **1947** H. May *Reconstructive & Reparative Surg.* xvii. 397 (*heading*) Mammaplastic procedures in the female. **1968** *N.Y. Times* 4 Apr. 26 He said the mammoplasty in the first case involved the transplant of adipose (fatty) tissue. **1969** C. Allen *Textbk. Psychosexual Disorders* (ed. 2) xii. 296 The procedures included..construction of a pro-vagina and mammoplasty.

mammer, *v.* (Later examples of *mammered* ppl. a.)
1861 T. Hughes *Tom Brown at Oxf.* III. vii. 127, I be that mad wi' myself, and mammered, and down, I be ready to hang myself. **1883** W. H. Cope *Gloss. Hampshire Words* 56 *Mammered,* perplexed. **1888** B. Lowsley *Gloss. Berks. Words* 108 *Mammered,* amazed, confused, puzzled.

mammering, *vbl. sb.* **1.** and **2.** (Later examples.)
1941 E. R. Eddison *Fish Dinner* (1972) vii. 108 Arquez, seeing..Clavius wounded and in a mammering whether to fly or fight: threw another chair. **1945** A. L. Rowse *West-Country Stories* 17 The innumerable mammering of bats' voices. **1947** M. Lowry *Let.* 13 Aug. (1967) 151, I am all of a doodah. I am in a mammering.

mammo- (mæ·mo), comb. form of Mamma[2], used in various medical and biological terms, as **mammoge·nic** *a.* [*-genic], stimulating the growth of the breasts, esp. at puberty; of or pertaining to this activity; so **ma·mmogen,** any substance which has or is supposed to have mammogenic activity; **mammoge·nesis,** the stimulation of the growth of the breasts, esp. at puberty; **mammo·graphy** *Med.* [-GRAPHY], a technique or procedure for diagnosing and locating abnormalities of the breasts by means of X-rays; an examination by this technique; hence **ma·mmogram, ma·mmograph**, a radiograph taken by this technique; **mammogra·phic** *a.*

1940 *Endocrinology* XXVII. 892 Only mammogen 1 (the duct growth factor) is present in ether-soluble extracts of the AP. Work is now progressing on the concentration of mammogen 2, the lobule-alveolar growth factor, and also on the perfection of a suitable technic for the assay of this hormone. **1958** *Proc. R. Soc.* B. CXLIX. 306 Oestrogen may resume its role as a mammogen. **1971** COWIE & TINDAL *Physiol. Lactation* iii. 115 Although the concept of *specific* mammogens of pituitary origin was never widely accepted..the 'Mammogen' theory rightly centred attention on the role of the anterior pituitary in mammogenesis. **1958** *Proc. R. Soc.* B. CXLIX. 312 The placenta may contribute as much to mammogenesis as the pituitary and ovaries combined. **1971** COWIE & TINDAL *Physiol. Lactation* iii. 117 Only in the rat and mouse have such detailed analyses of the hormones concerned in mammogenesis been made. **1938** *Proc. Soc. Exper. Biol. & Med.* XXXVII. 608 This new pituitary principle will be called the 'mammogenic hormone'. **1940** *Endocrinology* XXVII. 888 (*heading*) Evidence for the presence of a second mammogenic (lobule-alveolar) factor in the anterior pituitary. **1958** *Proc. R. Soc.* B. CXLIX. 304 Mammogenic activities of the ovarian hormones. **1971** COWIE & TINDAL *Physiol. Lactation* iii. 122 There is..a vast literature on the mammogenic potencies of ovarian steroids. **1937** N. F. HICKEN in *Surg., Gynecol. & Obstetr.* LXIV. 594/1 The procedure utilizes contrast fluids which are injected directly into the milk ducts, thus giving an accurate roentgenographic pattern of the ductal and secretory system of the mammary gland. The terms 'mammography' and 'mammograms' have been coined to describe these examinations. *Ibid.*, Gentleness, persistence, and patience are requisite in obtaining good mammograms. *Ibid.* 594/2 These mammographic visualizations afford an excellent method of studying both the physiological and pathological changes of the breast. **1969** *New Scientist* 13 Mar. 570/3 Preliminary designs for a commercial version of the system for producing mammograms. **1968** *N.Y. Times* 25 Mar. 25 This image Dr. Wolfe and many other radiologists agree is far clearer and more detailed than film mammographs. **1970** *Daily Tel.* 23 Oct. 17/4 The mammograph is another X-ray machine. Each breast is X-rayed individually, rather as though it were under a microscope. **1973** M. J. BRENNAN in Holland & Frei *Cancer Med.* xxvii. 1776/2 Combined clinical and mammographic screening detected breast cancer at a rate of 2·72 per 1,000 initial examinations. **1969** *New Scientist* 8 May 276/1 Thermography can detect presymptomatic breast cancers even when they are not palpable nor shown up by mammography. **1971** *Daily Colonist* (Victoria, B.C.) 26 Nov. 2/2 If you are still worried, you could have a mammography done—an X-ray technique commonly used in women with breast lumps that have to be identified.

mammoplasty, var. *MAMMAPLASTY.

mammoth, *sb.* and *a.* Add: **B.** *adj.* (Earlier and later examples.)

Freq. in American usage before 1850. The reference in quots. 1802[1] and 1803 is to a large cheese presented to Jefferson.

1802 *Port Folio* (Philadelphia) II. 31 (Th.), A baker in this city offers Mammoth bread for sale. **1802** *Balance* (Hudson, N.Y.) 19 Oct. 331 (Th.), No more to do with the subject than the man in the moon has to do with the mammoth cheese. **1803** J. DAVIS *Trav. U.S.A.* ix. 329 Its extraordinary dimensions induced some wicked wag of a federalist to call it the Mammoth Cheese. **1813** *Niles' Reg.* IV. 32/2 The Mammoth bank bill passed the senate this day on a third reading. **1924** W. R. INGE *Lay Thoughts* (1926) III. ii. 192 The new journalism, with its 'mammoth combines', is good business, but bad democracy. **1956** *Hansard Commons* 10 May 1450/2 The coal industry today is having to undertake this mammoth reorganisation because of the failures of hon. Members opposite in the years between the wars. **1966** AUDEN *About House* 39 He offered Mammoth-marrow And, perhaps, Long Pig. **1974** *Economist* 21 Dec. 65/1 Britain's mammoth current account deficit.

mammotrophic (mæmotrǭu·fik), **-tropic** (-trǭu·pik, -trǫ·pik), *a. Physiol.* [f. *MAMMO- + *-TROPHIC, *-TROPIC.] Stimulating or having the ability to stimulate the breasts, *esp.* in relation to growth or milk secretion.

1935 *Proc. Soc. Exper. Biol. & Med.* XXXII. 1049 The hypophyseal mammotropic hormone. **1942** *Ibid.* LI. 308 (*heading*) The direct mammotrophic action of lactogenic hormone. **1970** *Sci. Jrnl.* June 47/1 All males normally produce small quantities of mammotrophic hormone, but to what end remains a mystery. **1973** *Nature* 8 June 349/2 To further define the relationship between the mammotrophic and sebotrophic hormones we have studied the effect of suckling on sebum secretion.

So **mammotro·phin, -tro·pin** = *PROLACTIN.

1935 *Proc. Soc. Exper. Biol. & Med.* XXXII. 1049 (*heading*) Detection of mammotropin in the urine of lactating women. **1952** [see *lactogen* s.v. *LACTO-]. **1958** *Proc. R. Soc.* B. CXLIX. 309 By adding mammotrophin (*MH*) to these steroids, incomplete lobulo-alveolar growth was induced. **1961** *Lancet* 7 Oct. 792/1 The demonstration of mammotrophin activity.

mammy[1]. Add: **1. b.** (Earlier example.)

1826 DISRAELI *Viv. Grey* I. I. iii. 14 'Mammy-sick!' growled Barlow *primus*.

2. (Earlier example.)

1837 *Southern Lit. Messenger* III. 744/1 [Aged Negro domestics] were greeted always by the kind appellatives of 'daddy and mammy'.

mammy[2]. *W. Afr.* [Of obscure origin.] Used *attrib.*, as **mammy boat, mammy chair**, a (wicker) basket or chair used on ships for conveying persons to and from surf-boats on the West African coast; **mammy-cloth** (see quot. 1971[1]); **mammy lorry, wagon**, a small open-sided vehicle in West Africa; **mammy trader**, a market woman in West Africa.

1904 *Chambers's Jrnl.* 3 Dec. 15/1 You may elect to travel over the side in the 'mammy-chair', a huge barrel with part of its side cut away slung in the air by the steam-winch. **1909** MOORE & GUGGISBERG *We Two in W. Afr.* 16 So I found myself sitting in the 'Mammy chair', an ordinary basket-chair with ropes slung to the arms and back, ..and in a moment I was whisked off the deck, swung over the side at the end of a long derrick, [etc.]. **1920** *Blackw. Mag.* June 848/1 A mammy chair was lowered, and we made the usual undignified ascent to the main deck. **1928** *Daily Express* 27 Jan. 6/3 A 'mammy' boat, which is simply a wicker basket with seats, is slung over the ship's side by crane. **1935** L. G. GREEN *Great Afr. Mysteries* xv. 192 The 'mammy-chair' is like a swingboat at a fair; just a wooden box with two seats facing each other. **1957** *Times* 17 Dec. 9/6 As we drove on to the ferry a 'mammy-waggon' full of them [*sc.* Ghanaian students] was pushed on behind us. **1959** *Times* 9 Nov. (Ghana Suppl.) p. iv/7 Coordination of the activities of the vigorous 'mammy' traders. **1959** *Listener* 31 Dec. 1156/2 The car-park [in Ghana] with its taxis and mammy-lorries. **1961** G. GREENE *Burnt-Out Case* I. i. 7 When there were no European visitors there were always the old women,..their bodies wrapped in mammy-cloths. **1961** *Listener* 2 Nov. 697/1 The mammy-wagons are the friendly little open-sided buses which ply, always crowded, between towns and villages, and which have their names painted in bright colours. **1962** *Ibid.* 22 Feb. 335/1 A mammy-lorry had stopped just outside, and was disgorging its contents. **1965** W. SOYINKA *Road* 19 Goes into the mammy-waggon stall through hidden entrance up-stage. **1970** P. OLIVER *Savannah Syncopators* 59 (*caption*) Hausa women... One wears a 'mammy-cloth' printed with portraits of Queen Elizabeth the Second and Prince Philip. **1971** A. BAILEY *In Village* (1972) vii. 56 One man I knew used to dress in the evenings in a mammy cloth, the colorful cotton robe the Africans swaddled themselves in. **1971** *Reader's Digest* (U.S. ed.) Oct. 30/1 [In Jiddah] West African 'mammy traders' hawk cocoa beans, salves for arthritis and gaily colored cloth.

mamootie (mæ·mūti). Also **mammotie, mamooty, mam(m)oty, mamuty, mometty.** [ad. Tamil *mammaṭṭi*, altered form of *maṇveṭṭi*, f. *maṇ* earth + *veṭṭi* spade.] A digging tool used mainly in India, shaped like a hoe with the blade at an acute angle to the helve.

1782 C. SALMON in G. W. Forrest *Sel. Lett. Govt. India* (1890) III. 855 He marched..with two battalions of sepoys, leaving the cavalry..who were ordered to make a show of entrenching themselves by digging with mamuties. **1852** F. A. NEALE *Narr. Residence Siam* viii. 138 By means of a mometty or hatchet..this fellow dug..a reservoir. **1858** P. L. SIMMONDS *Dict. Trade Products* 238/1 *Mammotie*, a road hoe, used in Ceylon. **1881** W. W. GREENER *Gun* 587 For shooting on the plains nothing but a mamoty, a crowbar, a mallet, and a couple of axes are required, and these can all be had in India. **1920** *Blackw. Mag.* Oct. 467/2 A company of pioneers each man with pick or shovel or mamootie slung in leather slings on his back. **1934** 'G. ORWELL' *Burmese Days* (1935) vi. 83 He..swung his mamootie aloft again and hacked at the dry ground. **1940** *Nature* 20 July 91/2 The scraping is effected with an ordinary *mammoty* (like a pointed spade with a recurved handle), removing about ½ in. of soil.

mampus (mæ·mpǝs). *dial.* [Of obscure origin.] A great number, a crowd.

c **1730** *Dorset Vocab.* in *N. & Q.* (1883) 21 July 45/1 A mampus, multitude. **1857** *Gloss. Provincial Words Dorset* 6 *Mampus*, a great number. **1880** HARDY *Trumpet-Major* I. ii. 32 The mampus of soldiers that have come upon the down. **1904** —— *Dynasts* I. II. iv. 46 What a mampus o' folk it is here to-day!

‖ **mamur** (mamūǝ·ɹ). [Arab. *ma'mūr*.] An Egyptian official governing a district.

1836 E. W. LANE *Account* I. iv. 152 The whole of Egypt is divided into several large provinces, each of which is governed by an 'Osma'nlee (or a Turk); and these provinces are subdivided into districts, which are governed by native officers, with the titles of *Ma-moo'r* and *Na'zir*. **1899** A. S. WHITE *From Sphinx to Oracle* 108 Shortly afterwards I received the Mamur, who was accompanied by the officer commanding the police. **1900** G. BELL *Let.* 2 May (1927) I. 86 The Mamur (the Sultan's land agent). **1901** F. W. FULLER *Egypt* 259 Each mamurieh is under the charge of a Mamur, or Sub-Governor, who has in most cases been selected from among the native officers of the Egyptian army. **1902** *Encycl. Brit.* XXVII. 691/1 The provincial police is under the direction of the local authorities, the *mudirs* or governors of provinces, and the *mamurs* or district officials. **1958** L. DURRELL *Mountolive* ii. 43 Today, with British rule, the Copt is debarred from holding the position of Governor or even of *Mamur*—the administrative magistrate of a province. **1962** E. E. EVANS-PRITCHARD *Ess. Social Anthropol.* vii. 145 A Zande Prince went through the ceremony of blood-brotherhood with a *mamur* (Egyptian or Sudanese official) to whom he was bitterly hostile.

mamzer. Delete † *Obs.* and add later examples. Also **momser, momza, momzer.** Pl. **mamzeri·m.** Also in extended uses as a term of abuse or familiarity.

1955 H. KURNITZ *Let.* Dec. in G. Marx *Groucho Lett.* (1967) 250 When the momzerim in the aisle seats have finished clubbing me I will come to California. **1960** WENTWORTH & FLEXNER *Dict. Amer. Slang* 342/1 *Momzer, momser*, one who borrows frequently, or who expects much attention and many favors; a sponger... Any disliked person; a bastard. **1963** 'R. L. PIKE' *Mute Witness* (1965) ii. 23 Johnny Rossi? The west-coast hood?.. And we're keeping *momsers* like that alive, now? **1964** W. MARKFIELD *To Early Grave* (1965) iv. 86 That mass man, that totalitarian type, that *momza*. **1968** L. ROSTEN *Joys of Yiddish* 254 *Momzer*..is often used to describe a very bright child, a clever or ingenious person, a resourceful, gets-things-done, corner-cutting type. **1970** L. M. FEINSILVER *Taste of Yiddish* 47 *A Mamzer*..(a bastard, a thief; hence a dirty crook). **1972** *Guardian* 10 Oct. 4/2 The problem of Mamzerim, the Jewish bastard outcasts. **1973** *Jewish Chron.* 19 Jan. 1/4 Hanoch and Miriam Langer are a brother and sister declared by a rabbinical court in Petach Tikva more than seven years ago to be *mamzerim* (children of an adulterous union). *Ibid.* 2 Feb. 23/4 The husband could not be given permission to remarry in an Orthodox synagogue, so he married in a Reform synagogue. Incidentally, any child from this marriage is not a mamzer.

man, *sb.*[1] Add: **2. b.** With a qualifying word, applied to prehistoric types of man, as *Cro-Magnon Man, Neanderthal Man, Peking Man*, etc. (see under the qualifying words).

4. c. Colloq. phr. *to separate* (or *sort out*) *the men from the boys*: to distinguish which persons in a group are mature, manly, expert, etc.

1962 J. BRAINE *Life at Top* ii. 39 Every day one was tested, the men were separated from the boys. **1967** *Listener* 16 Feb. 239/1 It is always quite a pleasure to see some really professional jazz players and hear their sounds and one of the best examples is *Jazz Goes to College*..which certainly sorts out the men from the boys. **1968** *House & Garden* May 36/4 The Dry Martini..is a drink that will quickly separate the men from the boys and the girls from their principles. **1972** *Listener* 17 Aug. 201/3 The ability to memorise a whole [chess] game within a few minutes is..what separates the men from the boys.

e. (Further examples.) Now in more general use as a form of address to both men and women, as a meaningless expletive, or as an interjection expressing surprise, delight, deep emotion, etc.; esp.: (*a*) in *S. Afr.*; (*b*) among Negroes; (*c*) among jazz musicians and enthusiasts.

1874 A. BATHGATE *Colonial Experiences* ix. 110 Man! they all looked as if they had been shot, and would hardly believe me. **1896** H. A. BRYDEN *Tales S. Afr.* i. 29 'Man!' he said, 'if I had not not been *shamed* into following you, I would never have come across that place.' **1901** M. FRANKLIN *My Brilliant Career* (1966) ii. 7 Here,..with splendid roads, mail thrice weekly, and a railway platform only eight miles away, why, man, my fortune is made! **1933** *Metronome* Aug. 23 Trum's greeting was in the Negro dialect he usually employed: 'Man! How is you?' **1952** M. TRIPP *Faith is Windsock* i. 21 The Jamaican stood up and stretched himself. 'Glad you woke me, man,' he said. **1958** *Observer* 14 Dec. 7/7 The coloured boy..doesn't like the English. 'I'm nineteen, man, and when I go into a public house, they all look at me.' **1958** *Star* (Johannesburg) 17 Dec. 17/1 His second play captures a South African atmosphere in flashes—his policeman begins almost every sentence with 'Man!' and ends it with 'Hey!' **1960** *Monthly Rev.* May 27 Negroes habitually call each other 'man' in reaction to a lifetime of being addressed by white folk as 'boy'. **1961** 'B. WELLS' *Day Earth caught Fire* ix. 145 'Cut that out, man,' the beatnik said. **1966** *Evening Standard* 1 Feb. 8/1 'I like to speak the truth, man,' he said addressing me [*sc.* a woman] with the universal Jamaican appellation of 'man'. **1969** C. F. BURKE (*title*) God is beautiful, man. **1971** *Black Scholar* Jan. 43/2 Man, this would make these fighters so mad. **1971** *Black World* June 54/2 Hey, only the squares, man, only the squares have it to keep. **1972** *New Nation* (Singapore) 25 Nov. 8/5 Close-cropped Charlie continued: 'Oh, you live in Bukit Timah... I used to visit my uncle there, man.' **1973** *Caribbean Contact* Feb. 4/3 'No, man,' the St. Lucian replies.

h. *man to man* (further examples). Also as *adj.* or *adv. phr.* (freq. hyphenated), straightforward(ly), frank(ly). Hence (*nonce-wds.*) **man-to-man-ness, man-to-mannish** *a.*

1901 E. A. ROSS *Social Control* 29 So long as the struggle is man to man..a conscience is a handicap. **1902** E. NESBIT *Five Children & It* ix. 239 The best 'man to man' tone at his command. **1908** *Daily Chron.* 11 Aug. 4/5 Another form of suasion is the 'man-to-man' talk. **1940** N. MARSH *Surfeit of Lampreys* (1941) x. 141 Mike gave Alleyn a man-to-mannish look. **1943** J. B. PRIESTLEY *Daylight on Saturday* xxi. 167 'e remembered..an' stood me a pint. Now that's what I call man-to-man style. **1948** *Dublin Rev.* II. 11 The [Chinese] character for 'jen' is made up of 'man' and 'two'... I translate it 'man-to-man-ness'. **1952** S. KAUFFMANN *Philanderer* (1953) ix. 146 Perry spread his palms wide. 'Why don't you ask her?' Confidentially, man-to-man. 'I'd like to do it just to show her our gratitude for that week-end.' **1955** *Times* 5 Aug. 7/3 Talks man to man can serve the great international peace settlement. **1958** E. HYAMS *Taking it Easy* 49 Bachelor had a brisk, man-to-man style [of writing] which was immensely popular. **1973** 'M. INNES' *Appleby's Answer* xv. 128 A confidential and man-to-man note will be in order. **1974** A. PRICE *Other Paths* I. iii. 30 We must talk man-to-man now—if you would leave us for a moment, Mrs. Mitchell.

k. *spec.* One's representative or envoy *in* a specified place.

1958 G. GREENE (*title*) Our man in Havana. **1961** I. FLEMING *Thunderball* xiv. 154 Seems to me your President is taking all this a bit more seriously than his Man in Nassau. **1963** L. DEIGHTON *Horse under Water* vi. 33 'Welcome to Gibraltar,' said Joe MacIntosh, our man in Iberia. *Ibid.* xxxix. 154 Through the twilight 'our man in Cardiff' lifted a finger at the crooked castle of Caerphilly.

m. *man John*: delete 'nonce-substitutions' and add earlier examples.

1838 [see JOHN 1 b]. **1845** DISRAELI *Sybil* III. vi. vi. 214 My missus says that not a man John of them is to be seen.

p. (*a*) *spec.* One who is a specialist in or is expert in a specified subject; a supporter or adherent of a specified person; one who favours a specified product. See also *cruelty man.

1921 T. S. ELIOT *Let.* 26 Oct. in *Waste Land Drafts* (1971) p. xxii, I have wondered whether he is quite the best man for me as he is known as a nerve man and I want rather a specialist in psychological troubles. **1938** A. HUXLEY *Let.* 12 Apr. (1969) 435 The cancer-man..had hit on something very fundamental. **1958** *Listener* 25 Sept. 449/2, I am therefore wholeheartedly a Galbraith man... Professor Galbraith is the first writer to attempt a systematic economic analysis of such a society. **1960** *Ottawa Citizen* 25 Mar. 1/8 Pearkes 'Bomarc Man' Despite U.S. Cutbacks... Defence Minister Pearkes today denied the Bomarc-B missile is a 'dead pigeon' and affirmed he still retained full confidence in it as an air defence system. **1964** *College Composition & Communication* XV. iv. The program of English studies was in the hands, not of the wise dean who encouraged the variousness of my undergraduate program, but of several gentlemen called 'the Malory man', 'the Restoration drama man', and 'the *Paradise Lost* man'. **1969** *Listener* 24 July 102/1, I had been the *Daily Telegraph*'s main German man for some time. **1973** J. WAINWRIGHT *Touch of Malice* 89, I wouldn't have thought..you were a beer man. **1974** *Country Life* 25 Apr. 1008/1 A wonderful cruising ground for the yachtsman, be he a power or sail man. **1974** P. LOVESEY *Invitation to Dynamite Party* ii. 25 'Is it the hard stuff that you've taken a fancy to?' 'Not me, Sarge. I never was a whisky man.'

r. *every* (or *each*) *man for himself*: applied to a situation in which each person is preoccupied with his own safety or advancement.

c **1386** CHAUCER *Knight's Tale* (1894) 1182 At the kinges court, my brother, Ech man for him-self. c **1515** BARCLAY *Ecologues* (1928) I. 1009 Eche man for him selfe, and the frende for all. **1562** J. HEYWOOD *Dialogue of Proverbs* II. ix, in *Works* sig. k4 Every man for him selfe, and god for us all. **1629** T. ADAMS *Workes* 422 That by-word, Euery man for himselfe, and God for vs all, is vncharitable, vngodly; and impugneth directly the end of euery good calling, and honest kinde of life. **1825** J. NEAL *Bro. Jonathan* III. xxix. 90 'Forward! Forward!' were the cries, on every side of our hero. 'Forward! forward! every man for himself!' **1922** H. WALPOLE *Cathedral* I. i. 14 They had been troublous times. It had been every man for himself.

s. *the man* (also *the Man*): a person in authority; such persons collectively; *spec.* (*a*) a prison governor; (*b*) a policeman or detective; the police; (*c*) one's employer, 'boss'; (*d*) (*Negro slang*) a white man; white people collectively; (*e*) a drug-pusher. *U.S. slang.*

1918 G. M. BATTEY *70,000 Miles in Submarine Destroyer* (1919) 302 Any body in authority is 'the man'. **1928** R. FISHER *Walls of Jericho* 306 *The man*, designation of abstract authority. He who trespasses where a sign forbids is asked: 'Say, biggy, can't you read *the man's* sign?' **1933** *Amer. Speech* VIII. III. 29/2 Unuther goddam word out o' you-all, an' Ah'll send ev'ry goddam one of ye up t' the Man. **1953** W. BURROUGHS *Junkie* (1972) ix. 87 When I first hit New Orleans, the main pusher—or 'the Man', as they say there—was a character called Yellow. *Ibid.* 159 '*The Man*', junk seller. 'The Man' is a New Orleans expression, and can also refer to a Narcotics Agent. **1962** *Amer. Speech* XXXVII. 270 *Man*, a policeman. A word used by teen-age drivers. 'When I heard the siren, I knew it was the Man.' **1962** J. BALDWIN *Another Country* (1963) II. ii. 243 One of the musicians came to the doorway, and said, 'Ida, honey, the man says come on with it if you coming.' **1963** *N.Y. Times* 18 May 12/2 A well-educated Negro said today: 'The demonstrations, I think, suggested to "The Man" that tokenism won't make it and that he has to come to grips with the problem right now.' 'The Man', in Negro parlance, is the white man. **1965** *Times Lit. Suppl.* 25 Nov. 1035 *Man* is the title by which one Negro addresses another... *The Man* is the way in which he speaks of the enemy, of the white. **1968** *Blues Unlimited* Nov. 8 They were sharecroppers but the last crop put them so deeply in debt with 'the man' that they plan to quit farming. **1970** *Guardian* 3 Nov. 10/1 'The Man is repressive. The Man is fascist...' To the bombers and kidnappers the Man is authority. He is every policeman. He is President Nixon. He is Prime Minister Trudeau. **1972** J. MILLS *Report to Commissioner* 106, I heard you were the man, and if I really believed you were the man, you'd be dead now. **1972** *Guardian* 12 Aug. 9 Rus is not Uncle Tomming it around Harlem with 'the Man'. He has brought a foreign visitor. **1973** *Black World* Nov. 92/2 Their writing mainly concerns the street life—the pimp, the junky, the forces of drug addiction, exploitation at the hands of 'the man'. **1974** R. THOMAS *Porkchoppers* xi. 98 They'll be on my back for telling them something that they don't think the man needs to know.

18. *man of sense* (earlier example); **man of action**, a man whose life is characterized by physical activity or deeds rather than by thoughts and ideas; **man of destiny**, a man looked upon as an instrument of destiny; *spec.* Napoleon I; **man of distinction**, a person who is distinguished in his looks, manners, and bearing; **man of the moment**: see *MOMENT *sb.* I c; **man of the people**, a man who comes from or identifies himself with the common people; a working-class man.

1597 Man of Action [see ACTION 1 a]. [illegible]

19. (Further examples.) [illegible]

1932 W. FAULKNER *Light in August* i. 23 She traverses the ranked battery of maneyes and enters the store. **1923** R. GRAVES *Whipperginny* 25 And frozen music dulls their need Of drink and man-flesh greed. **1918** D. H. LAWRENCE *New Poems* 43 The man-life north-imprisoned, shut in the hum of the purpled steel. **1967** A. BATTERSBY *Network Analysis* (ed. 2) xv. 271 If the senior executives of a company really feel that the risk is great, they might do better to shelve network analysis and continue with existing methods until better man-management relations have been forged. **1971** *Mod. Law Rev.* XXXIV. vi. 680 The principle of an employer's responsibility for the results of 'teamwork by his team' goes to the centre of all thinking upon the techniques of organisation of groups and 'man-management' in large organisations. **1971** C. BONINGTON *Annapurna South Face* i. 7 The smaller the party, the closer you can get to the mountains uncluttered by all the ..complications of man-management caused by a larger expedition. **1888** G. M. HOPKINS *Poems* (1967) 105 Squadroned masks and manmarks treadmire toil there. **1962** W. STEGNER *Wolf Willow* III. ii. 148 'What's a windigo?' 'What the Crees used to call an Injun that made use of man-meat.' **1971** B. MALAMUD *Tenants* 207 The sweaty youths, holding long spears, leaped and yelped as they danced, their man-meat swinging like grapes in bunches in their loincloths. **1928** BLUNDEN *Jap. Garland* 16 Clawtendrils reach, man-monsters glare. **1934** DYLAN THOMAS *18 Poems* 27 The growing bones, the rumour of manseed Within the hallowed gland. **1971** G. M. BROWN *Fishermen with Ploughs* 4 Hoof-fast Njal bore his manseed wombfurled waveward. c **1879** G. M. HOPKINS *Poems* (1967) 82 Breathing bloom of a chastity in mansex fine. **1888** *Ibid.* 105 Manshape, that shone Sheer off, disseveral, a star, death blots black out. **1922** JOYCE *Ulysses* 368 Perhaps they get a man smell off us. **1938** M. K. RAWLINGS *Yearling* xi. 113 I'm jest lonesome for boy-smell and man-smell. **1961** R. P. HOBSON *Rancher takes Wife* viii. 118 She [*sc.* a cow moose] would trot over to the timber, lick her calf to its feet, and lead it off—away from the contaminated area of man-smell. **1682** J. BUNYAN *Holy War* (1905) 190 For here lay the excellent wisdom of him that builded Mansoul, that the Walls could never be broken down. **1929** R. BRIDGES *Testament of Beauty* IV. 1391 From this dilemma..Man-soul made glad escape in the worship of Christ.

b. *man-friend*. Cf. sense 22 a below.

1736 LD. HERVEY *Mem.* (1952) 182 The King's letter to the Queen about this affair was an extraordinary one, asking her, as he would have done a man-friend, what she thought of all this business. **1893** *Ladies' Home Jrnl.* Apr. 39/1 There is no impropriety in a man friend writing to you without having asked your permission. **1922** D. H. LAWRENCE *Fantasia of Unconscious* xi. 204 The woman..will have a new man-friend, if nothing more.

d. *man-picker, -tamer, -tracker; man-carrying, -lifting, -loving, -stopping, -tracking* vbl. sbs. and ppl. adjs.; *man-shy* adj.

1909 *Westm. Gaz.* 25 Feb. 4/1 The latter..achieved for the first time in history a man-carrying machine propelled by its own motive power. **1961** *Times* 28 Apr. 3/6 The placing in orbit of a man-carrying satellite. **1846** *Pract. Mechanic* June 235/1 The man-lifting engine erected at United Mines, Gwennap..for lowering and raising the miners. **1899** *Rep. Brit. Assoc. Adv. Sci.* 948 Man-lifting kites might be used instead of a balloon. **1927** C. L. M. BROWN *Conquest of Air* 122 By 1905 Cody had attracted official attention to his man-lifting kites. **1895** W. JAMES in *Internat. Jrnl. Ethics* Oct. 8 The old warm notion of a man-loving Deity. **1939** W. B. YEATS *Last Poems* 17 Man-picker Niamh leant and sighed By Oisin on the grass. **1931** F. D. DAVISON (*title*) Man-shy. **1899** *Kynoch Jrnl.* Oct.–Nov. 15/2 The manstopping powers of the..bullets appear to be considerable at short ranges. **1905** *Ibid.* Apr.–June 96 Man-stopping bullets are not allowed. **1973** *Guardian* 17 Mar. 13/2 Both guns increase the 'man-stopping' effectiveness of police marksmen. **1922** JOYCE *Ulysses* 526 Master! Mistress! Mantamer! **1931** D. RUNYON *Guys & Dolls* (1932) i. 32 Nip and Tuck go back to Georgia..with a big reputation as man-trackers. *Ibid.* i. 17 They is man-tracking bloodhounds from Georgia.

c. *man-filled, -gripped, -measured, -powered, -propelled* ppl. adjs. Also *MAN-MADE *a.* [illegible]

f. *man-centred, -faced, -headed, -high* (later examples), *-smelling, -stale, -tall, -visaged* adjs. [illegible]

20. man-boy, a youth, an immature man; **man-crazy** *a.* = **man-mad* adj.; **man-hair**, a man's pubic hair; **man-hungry** *a.*, desirous of a man; **man-in-space** orig. *U.S.*, an astronaut; freq. *attrib.*; **man-load**, a load such as one man can carry; **man-machine**, (*b*) a man who acts mechanically; an automaton (cf. quot. 1904 s.v. 22 a below); (*c*) used as *adj.* to denote a relationship or communication between a man and a machine; **man-mad** *a.*, madly desirous of a man or men (cf. *man-wood* adj. in Dict.); **man-mountain** (further examples); now applied esp. to a wrestler; **man-pack, manpack**, a compact package containing equipment or supplies, designed to be carried easily by one man; also *attrib.*; **man-power**, (*a*) the power or agency of man in work; (*b*) used as a unit of power or rate of working; (*c*) the number of persons available for a purpose, esp. for military service; also *attrib.*; **man-riding** *Mining*, the vehicular transport of miners underground; freq. *attrib.*; so **man-rider**, a train of vehicles designed for this; **man-rope** (**knot**) (see quot.); **manself** [after HIMSELF *pron.*] *nonce-wd.*, man's self; **man-shift**, (the amount of work done in) a single shift worked by one man; **man-size(d)** *a.*, of the size of a man; large, full-size; large enough to occupy, suit, or satisfy a man; **man-starved** *ppl. a.*, suffering from the lack of a man or men; desirous of a man; **man-strength** = **man-power* (*c*); **man-tailored** *a.*, of women's clothes: tailored after the fashion of men's clothes; **man-wise** *adv.* and *pred. a.*, in the fashion or way of men; in respect of a man; concerning individual men.

1927 W. DEEPING *Kitty* ix. 115 This husband of hers, this man-boy, what was he? **1945** S. SPENDER *Citizens in War* iv. 35 Tolerance defeated the religious man-boy who expected at least to be persecuted. **1923** G. ATHERTON *Black Oxen* xlii. 262 The young women would say, in their nasty slang, that I was probably man-crazy. **1967** H. VAN SILLER *Biltmore Call* 128 Never liked the woman... Man crazy and two-faced. **1972** A. AMIN tr. *Ahmad's No Harvest but Thorn* xi. 115 Let them call her bold... That Jeha was man-crazy. **1928** Man-hair [see *MAIDENHAIR 5 b]. **1951** J. CORNISH *Provincials* 220 Douglas fell..to a girl described by Bunty as twenty-eight and strictly man-hungry. **1959** *Daily Tel.* 21 Nov. 1/2 Capsule recovery is a crisis point in our man-in-space programme. **1962** *Amer. Speech* XXXVII. 43 Before April 12, 1961, the concept expressed by *cosmonaut*..was rendered..by such terms as ..*man-in-space*. **1968** *Economist* 28 Dec. 11/2 The man-in-space programme has earned itself enemies in powerful reaches of the scientific establishment. **1878** H. M. STANLEY *Through Dark Continent* I. xiv. 376 Suna commanded his Katekiro to make up 300 man-loads of hoes and old iron. **1951** M. MCLUHAN *Mech. Bride* (1967) 18/1 There is a wide range of mental states engendered in the same man–machine relationship. **1953** R. LEHMANN *Echoing Grove* 240 He was now an automaton, a man-machine. **1967** COX & GROSE *Organiz. Bibliogr. Rec. by Computer* IV. 99 This is what is meant by man–machine dialoguing. **1904** Man-mad [in def. of *man-wood* adj. in Dict.]. **1928** R. MACAULAY *Keeping up Appearances* vii. 62 If a woman went on that way about men you'd call her man-mad. **1959** *News Chron.* 28 Aug. 6/3 The man-mad duchess..picks up the young peasant painter. **1942** BERREY & VAN DEN BARK *Amer. Thes. Slang* § 429/1 Corpulent person,..*man mountain*. *Ibid.* § 707/2 *Man mountain*, a large wrestler. **1956** J. SYMONS *Paper Chase* xxx. 240 He was trying desperately to establish some kind of grip that would move this man mountain. **1960** E. W. HILDICK *Jim Starling & Colonel* xii. 105 'Are you—er—Mr Thimble?'.. 'Well, I'm not Man-Mountain Dean.' **1972** J. MOSEDALE *Football* viii. 105 A man-mountain for his time, he played tackle for the Giants and Packers. **1967** *Electronics* 6 Mar. 52/2 The jungle..absorbs and deflects transmissions from man-pack radio sets. **1970** *Islander* (Victoria, B.C.) 8 Feb. 5/3 Transportation was all by man-pack, pack-horse, canoe and raft. **1862**, etc. Man-power [in Dict., sense 19 a]. **1893** *Eng. Mechanic* Dec. 332/2 Maxim's early trials gave ..about 1 lb. per man-power. **1917** W. S. CHURCHILL in *World Crisis 1916–18* (1927) II. 378 It is not possible to settle the question of man-power without a clear idea of the plan of campaign. **1919** *Brit. Manufacturer* Nov. 25/1 One of these man-power ploughs, adapted for small holdings and for use on terraced land, is driven by a cable. **1926** A. BENNETT *Lord Raingo* xii. 60 Biggest piece of political camouflage ever attempted, the Man-Power bill is. **1972** *Accountant* 28 Sept. 386/2 The popular vogue of 'manpower planning'..must be more competently compiled where new relevant facts and figures are fed back from the human resource accounting system. **1972** *Daily Tel.* 28 Nov. 16 A great deal of manpower was being wasted before the merger. **1967** *Gloss. Mining Terms* (*B.S.I.*) x. 11 *Manrider*, a manriding train. **1950** H. F. BANKS in E. Mason *Pract. Coal Mining* II. xxxv. 530/2 Where locomotives and large mine cars are in use, they serve also as a means of man riding. *Ibid.* 531/1 Fig. 78 illustrates one type of man-riding car. **1952** T. BRYSON *Mining Machinery* (ed. 3) xiv. 381 The train..comprises six man-riding carriages with seating accommodation for 18 persons, two brake cars.., and one tool car. **1972** (*title*) Specification for cars for manriding in mines (B S.I.). **1973** *Times* 2 Mar. (Wales Suppl.) p. iii/3 The coal board is now concentrating on increasing 'manriding'—trains from pit bottom to coalface—wherever the geology makes it possible. **1883** *Man. Seamanship for Boys' Training Ships R. Navy* (Admiralty) (1886) 121 A double-wall, double-crowned.. is used for man-ropes, stopper-knots, &c., also called a man-rope knot. **1880** G. M. HOPKINS *Note-bks. & Papers* (1937) 271 He..would raise man..infinitely above manself to the divine justice. **1930** *Engineering* 3 Jan. 23/3 Variations in the standard of living..are reflected in the average output per man-shift. **1946** *Nature* 17 Aug. 226/2 In 1938..the Ruhr coal miner produced 30·5 cwt. per man-shift; in the same year the British miner produced 23 cwt. per man-shift. **1913** R. W. SERVICE *Rhymes of Rolling Stone* 103 The man-size mountains palisade us round on every side. **1928** *Publishers' Weekly* 16 June 2441 Since writing is a man-size job, he would have his hands full. **1963** *House & Garden* Mar. 130/1 (Advt.), Fourpenn'orth of Phurnacite 'eggs' will heat enough water for a man-size bath. **1975** *Times* 13 Jan. 8/1 We would..drink whisky out of mansize glasses. **1920** J. GREGORY *Man to Man* (1921) xvii. 209 He was trying to hide a pair of man-sized feet behind his table. **1920** S. LEWIS *Main St.* xxiii. 277 A regular man-sized pair of pants. **1954** 'N. SHUTE' *Slide Rule* vii. 167 These woodworkers, accustomed as they were to man-sized factories, regarded Airspeed Ltd as a joke. **1972** F. DURBRIDGE *Bat out of Hell* v. 174 Thelma..blew her nose on a tiny handkerchief which seemed ridiculous in her man-sized hand. **1961** *Times* 24 Oct. 13/2 Beautiful, man-starved women. **1931** *Ann. Reg. 1930* I. 23 The United States was greatly adding to its man-strength. **1922** JOYCE *Ulysses* 153 Molly had that elephantgrey dress with the braided frogs. Mantailored with selfcovered buttons. **1970** *Women's Wear Daily* 23 Nov. 31 (Advt.), Man-tailored jeans by Anvil Brand are now available. **1901** Man-wise [see *groupwise* adv. s.v. *GROUP *sb.* 6]. **1930** D. H. LAWRENCE *A propos Lady Chatterley* 39 Men experience the great rhythm of emotion man-wise, women experience it women-wise.

b. With a period of time, as *man-day, -hour, -minute, -month, -week, -year*, a day, hour, etc., of one person's work (or life). Also in extended use. Cf. *man-load, -shift* (sense 20 above).

1925 E. O. SHEBBEARE in E. F. Norton *Fight for Everest, 1924* III. vi. 364 Each case would contain a sufficient ration of all kinds of food for a given number of 'men-days'. **1957** M. SWAN *Brit. Guiana* I. iv. 83 The doctor told me that before the conquest malaria accounted for about 55,000 hospital man-days a year. **1972** D. HASTON *In High Places* vi. 74 As for food... Some dried meat, bacon, nuts, chocolate and a few hot drinks per man-day. **1972** M. D. PAPAGIANNIS *Space Physics & Space Astron.* 280 In less than 10 years from the first manned flight we have accumulated more than a year of man-days in space. **1917** *Anti-Submarine Rep.* (Admiralty, R. Naval Air Service) Dec., Many air stations have large complements, and a comparison of the number of 'man hour's' work carried out with the the hours of flying on anti-submarine service will be found of interest. **1924** H. H. EMSLEY *Factory Costing* 76 The number of productive 'man-hours' in each department during each 'cost period'. **1951** R. A. KNOX *Stimuli* III. xxii. 131 Experts are uncomfortably reckoning out what chances..we are throwing away in terms of man-hours. **1971** *Nature* 29 Oct. 625/1 Countless man-hours have been spent on 'measuring' such allegedly useful entities as 'IQ' and 'reading age'. **1973** D. WESTHEIMER *Going Public* ix. 133 How many man-hours do you spend actually executing an assignment? **1934** WEBSTER, Man-minute. **1947** CROWTHER & WHIDDINGTON *Science at War* ii. 111 The actual working-time was only 53 man-minutes. **1969** J. ARGENTI *Managem. Techniques* vii. 42 This package cuts out nearly all the hard, highly skilled man-months of programming. **1973** *Computers & Humanities* VII. 190 The preparation of the double transcription of the Vulgate text took seven man-months. **1945** *Times* 19 Jan. 2/4 About 1,300,000 man-weeks remained to be consumed by March 31 if the labour force remained the same. **1961** *Jrnl. Oil & Colour Chemists' Assoc.* XLIV. 296 With proper planning fifty paints can be fully investigated in approximately two man weeks. **1928** A. S. EDDINGTON *Nature Physical World* ix. 180 We must describe the amount of humanity in it [*sc.* Great Britain] as 400 million man-years. **1956** J. MASTERS *Bugles & Tiger* ii. 71 All my family's hundreds of man-years of work now focused on me. **1969** J. ARGENTI *Managem. Techniques* vii. 41 Linear Programming..might take two man-years to develop an 'LP Model' of a company. **1973** *Times* 12 Feb. (Anchor Project Suppl.) p. i/5 This level is likely to be little more than half the output a man-year that will be expected from single, integrated works of more than seven million tonnes capacity.

21. man's man, a man whose qualities are appreciated by other men; a man who is popular with other men; so **man's woman** (see also quot. 1905).

1897 G. DU MAURIER *Martian* 214 He had been essentially a man's man hitherto, in spite of his gay light love for lovely women; a good comrade par excellence, a frolicsome chum, a rollicking boon-companion, a jolly pal! **1902** J. BUCHAN *Watcher by Threshold* v. ii. 277, I know he's supposed to be a man's man..but I'm honest enough to own to detesting him. **1960** Man's man [see *DIG *v.* 6 c]. **1972** P. NEWTON *Sheep Thief* v. 39 He had had comparatively little to do with women. Essentially a man's man he had been wrapped up in his calling. **1905** A. BENNETT *Sacred & Profane Love* II. vi. 191, I was at last a man's woman. I had a protector. **1930** W. S. MAUGHAM *Writer's Notebk.* (1949) 232 She is what is called a man's woman, and exchanges chaff across the dining-room with the other guests. **1968** E. MCGIRR *Lead-Lined Coffin* iii. 108 Jeanie's a man's woman, you know, not a woman's woman.

22. a. (Further examples.)

1904 *Westm. Gaz.* 22 Nov. 12/1 Nowadays the Kaiser's men-machines take the best part of the pavement from anyone whom they may safely elbow off. **1907** A. C. BENSON *Diary* 31 Jan. (1926) 157 Yet I do not squabble with my men-friends. **1910** F. BARCLAY *Rosary* ii. 10 But of men friends she had many. **1923** D. H. LAWRENCE *Birds, Beasts & Flowers* 172 Men-peasants from jungle villages dancing and running with sweat and laughing. **1952** S. SPENDER *Shelley* ii. 9 Hogg..was the first in a series of rather earth-bound men-friends. **1973** J. BURROWS *Like an Evening Gone* i. 12 How many attendant men friends?

c. men-only *a.*, designating a place, etc., restricted for the use of men.

[**1935** (*magazine-title*) Men only. **1955** H. W. ALLEN in *Stag Party with 'Men Only'* 12 The *Sporting Times*, affectionately known from the colour of its paper as the *Pink 'Un*, that spicy and distinctly *Men Only* weekly of Victorian days.] **1965** 'S. RANSOME' *Alias his Wife* vi. 57 Without discouraging women as customers, it had become a men-only place. **1967** N. FITZGERALD *Affairs of Death* ii. 34 It's a men-only bathing place. **1972** G. BEARE *Bee Sting Deal* x. 123 There was one hotel..rigidly dry and rigidly men-only.

23. Men's Liberation, Men's Lib. orig. *U.S.* [after **Women's Liberation, Women's Lib.*], a movement aimed at freeing men from traditional views of their character and role in society; **men's room** orig. *U.S.*, a lavatory for men; also (ellipt.) *men's*; **men's wear, menswear**, clothes for men.

1957 S. BECKETT *All that Fall* 24 *Mrs. Rooney*. 'Where were you all this time?' *Mr. Rooney*. 'In the men's.' **1972** R. LOCKRIDGE *Preach No More* iii. 43 This friend of his had..gone to the men's, or something. **1970** *New Yorker* 19 Dec. 101 We recently read in the *Times* about a group called Men's Liberation, Inc., whose aim is to free men from their traditional role of 'all-powerful provider' and embolden them to cry, complain, feel sorry for themselves, and change their minds. The members of Men's Lib say they are tired of 'having to prove our masculinity twenty-four hours a day' and believe that if their cause should prevail 'outmoded concepts would disappear in the face of reality'. **1971** *N.Y. Times* 17 June 41 To the Women's Liberation Movement the sex-role distinctions mean pressure to conform to separate and unequal rewards for being a woman. To the Men's Liberation Movement they mean pressure to conform to the 'ego ethic'—or conform to perform. **1972** 'J. MELVILLE' *Ironwood* ix. 154 Three months of Tessa and they [*sc.* her male assistants] must have been heading for freedom and Men's Lib. **1929** 'E. QUEEN' *Roman Hat Mystery* iii. 31 Search the lounge downstairs. The men's room, the ladies' room. **1947** AUDEN *Age of Anxiety* (1948) ii. 51 Malin excused himself and went to the men's room. **1962** J. WAIN *Strike Father Dead* 138 He's in the men's room. He's been wanting to go all evening, but as long as you were playing he didn't want to miss a note. **1947** *Partisan Rev.* XIV. 364 Seymour's taste in clothes and men's wear was loud, practically spectacular. **1972** P. FLOWER *Cobweb* II. 58 Among the heaps of flowers a wreath from the Menswear Dept. **1973** D. ROBINSON *Rotten with Honour* 35 [He] looked at Hale as if Hale were modelling menswear. **1974** *Radio Times* 28 Feb. 4/3 The family textile and menswear business.

man, *v.* Add: **4.** Also with *up*; *spec.* to supply with the full number of workers required.

1947 *Times* 6 Feb. 5/5 Must industries be fully 'manned up' rather than 'manned'? Must the strong, simple transitive verb..become as obsolete in England as it appears to be in America? **1947** *Hansard Commons* 18 Dec. 1895/1 Mining and agriculture are the two most vital industries which we must man-up. **1951** E. A. G. ROBINSON in D. N. Chester *Lessons Brit. War Econ.* iii. 52 Their most urgent jobs were not in fact manned-up as they would have liked. **1955** *Times* 20 May 11/6 The need to 'man up' undermanned industries 'in the national interest' might justify higher wages. **1969** *Times* 6 Mar. 23/1 'We're manned up again,' came a voice of relief... Superstitious members..now have the satisfaction of seeing their numbers rise from 13 to 20.

mana (mā·nă). [Maori.] Power in general, authority, prestige; *spec.* in primitive religion, an impersonal supernatural power which can be associated with people or with objects and which can be transmitted or inherited.

1843 E. DIEFFENBACH *Trav. N.Z.* II. III. ix. 371/2 Mana—command, authority, power. **1855** R. TAYLOR *Te Ika a Maui* 279 The natives..feel..that with the land, their *mana*, or power, has gone likewise. **1858** *Richmond-Atkinson Papers* (1960) I. 367 The most loyal reverence for the Queen's name and 'mana'..[is] to be found in Ngapuhi. **1877** R. H. CODRINGTON *Let.* in Max Müller *Lect. Orig. & Growth Relig.* (1878) 54 There is a belief in a force altogether distinct from physical power, which acts in all kinds of ways for good and evil, and which it is of the greatest advantage to possess or control. This is Mana. a **1910** W. JAMES *Some Probl. Philos.* (1911) i. 17 What made things act was the mysterious energy in them, and the more awful they were, the more of this *mana* they possessed. **1920** *Times Lit. Suppl.* 29 Apr. 264/2 Notions of the type of mana or orenda are of 'a nascently philosophic order'. **1937** R. H. LOWIE *Hist. Ethnol. Theory* xii. 204 It is not merely spirits and deities that loom as sacred, but also the impersonal force Melanesians call 'mana'. **1951** R. FIRTH *Elem. Social Organiz.* vii. 215 The data accumulated on the character and functions of.. principles of mana and taboo..have enabled large regions of human religious experience to be mapped out on a comparative basis. **1959** *Chambers's Encycl.* IX. 47/2 It is this rudimentary concept of a mystic, quasi-impersonal force connected with anything mysterious and arresting that finds expression in *mana*. **1965** *Listener* 2 Dec. 920/2 Warhol has always provided a good example of the kind of *mana* which emanates from certain chosen individuals in modern society. **1975** H. MCCLOY *Minotaur Country* v. 53 He has *mana*... The thing most people think they're talking about when they say charisma.

attrib. and *Comb.*

1924 W. B. SELBIE *Psychol. Relig.* 208 A fearful cringing before some mysterious mana-charged object. **1937** *Brit. Jrnl. Psychol.* Oct. 197 Their religious beliefs centre round a concept of magic of the 'mana' type i.e. vague, abstract, impersonal power. **1949** KOESTLER *Promise & Fulfilment* II. v. 274 He has a *mana*-circle in Tel Aviv.

manage, *v.* **11. b.** (Earlier examples.)

1854 M. L. CHARLESWORTH *Ministering Children* iv. 47 Rose was..wondering how William would manage about getting some logs for Mercy's fire. **1873** 'S. COOLIDGE' *What Katy Did* xi. 195 I've been thinking how we are to manage about the housekeeping. **1895** 'G. MORTIMER' *Like Stars that Fall* viii. 108 'How will you manage about your hair?' 'I shall cut it short, I think.'

managed, *ppl. a.* Add: **2. managed bond** (see quot. 1972); **managed currency**, a currency system which is not tied to the gold standard but is regulated by the government of the country concerned; **managed economy**, an economy in which the framework and general policies are regulated or controlled by the government.

1923 J. M. KEYNES *Tract on Monetary Reform* 166 This is what is meant by saying that gold has 'intrinsic value' and is free from the dangers of a 'managed' currency. *Ibid.* 167 Gold itself has become a 'managed' currency. **1924** W. F. SPALDING *Eastern Exchange Currency & Finance* (ed. 4) p. v, India and one or two other Eastern nations have found defects in a 'managed' currency. **1929** G. D. H. COLE in *New Leader* 8 Nov. 16/3 A plea for a 'managed currency'..a currency that is regulated not automatically by the supply of gold, but deliberately. **1941** G. N. CLARK *Holland & War* 9 In September 1936 ..France and Switzerland gave up the gold standard and Holland too went over to a managed currency. **1969** *Adv. Sci.* XXVI. 64/2 The first objective of a managed economy..was originally stability. *Ibid.* 66/2 In considering the problems of a managed economy it is well to begin by recognizing that management is by no means the same as control. [**1971** *Investors Chron.* 2 Apr. 22 (Advt.), It's called the Abbey Selective Investment Bond, and it's issued and managed by Abbey Life Assurance.] **1972** *Observer* 8 Oct. 18/2 In theory a managed bond is an ideal investment. The investor buys units in a special bond fund which the managers can invest in property, shares or fixed-interest stock according to their expert appraisal of the investment scene.

management. Add: **1. c.** *spec.* The administration of a commercial enterprise. Also in phrases designating specific methods of business administration, as *management by exception, by objectives*.

1906 DICKSEE & BLAIN (*title*) Office organisation and management. **1962** A. BATTERSBY *Guide to Stock Control* x. 95 They [*sc.* the junior staff] are the instruments of 'management by exception', filtering out the run-of-the-mill cases from the few unusual ones which call for the personal attention of the Materials Controller. **1962** H. O. BEECHENO *Introd. Business Stud.* ii. 17 Trade union development and legislation on conditions of work, plus more enlightened management have eradicated most forms of unfair treatment of employees. **1965** H. I. ANSOFF *Corporate Strategy* (1968) 9 Management of a business firm is a very large complex of activities which consists of analysis, decisions, communication, leadership, motivation, measurement, and control. *Ibid.* iii. 36 Their potential pervasiveness is such that objectives have been used as a basis for an integrated view on the entire management process which has become known as 'management by objectives'. **1968** JOHANNSEN & ROBERTSON *Managem. Gloss.* 81 *Management by Objectives*, setting targets within an organisation..as a basis for achieving greater efficiency and providing motivation and an incentive to managers. **1969** J. ARGENTI *Managem. Techniques* 146 Management by Exception..simply says, 'don't tell someone if everything is O.K.—only tell him if something has gone wrong'. **1970** *Lebende Sprachen* XV. 11/2 Operational research is an integral part of good management.

6. *spec.* That group of employees which administers and controls an industry in contradistinction to the labour force in that industry or in industry in general. (Further examples.)

1912 F. W. RAFFETY *Modern Business Pract.* II. II. i. 197 Management..deals with the three sources of force or energy—capital, labor and land... The management should never lose sight of their important influence. **1940** W. TEMPLE *Hope of New World* i. 61 If there is to be tension at all, let it be between the financial interests of Shareholders and the productive interests of Management and Labour in co-operation. **1959** [see *BOARD *sb.* 17]. **1965** H. I. ANSOFF *Corporate Strategy* (1968) i. 20 Management from top to bottom continually seeks to improve efficiency, to cut costs, to sell more, to advertise better. **1968** *Lebende Sprachen* XIII. 4/2 It has been part of management philosophy to encourage and support the self-development of its employees. **1969** J. ARGENTI *Managem. Techniques* 150 In every company there are two or three dozen absolutely vital ratios that must be watched and checked by the management.

7. *management accounting, consultancy, consultant, game, science.*

1963 BROAD & CARMICHAEL (*title*) A guide to management accounting. **1967** *Oxford Computer Explained* 32 *Management accounting*, the provision of a continuous and up-to-date check and control on income and expenditure. **1958, 1966** Management consultancy [see *CONSULTANCY]. **1972** *Accountant* 19 Oct. 487/1 The rapid development of management consultancy services has resulted in the profession extending its activities far beyond its traditional role of auditor. **1961** *Times* 14 July 2/2 Management consultants..must also have..experience in personnel management. **1969** *Times* 6 Mar. 23/2 This brings the number of management consultants on the commission up to three. **1964** M. ARGYLE *Psychol. & Social Probl.* x. 133 There has been a swing away from traditional teaching methods towards group discussion, joint report-writing by syndicates, case-studies and the management game. **1965** H. I. ANSOFF *Corporate Strategy* (1968) ix. 148 Management science can offer only partial assistance to the decision maker. **1969** *Times* 2 May 34 (Advt.), Post-graduate studentships in management sciences.

managerial, *a.* Add: In more recent use esp. of a manager of a commercial enterprise.

1912 F. W. RAFFETY *Modern Business Pract.* II. II. i. 204 Management..involves not only the forces of production but careful consideration of the results to be obtained... It is this latter purpose which distinguishes the man with managerial ability from the purely technical man. **1924** J. STAMP *Stud. Current Probl. Finance & Govt.* ii. 43 It [*sc.* the Excess Profits Duty Act] conferred quite a considerable number of discretions to the Commissioners of Inland Revenue, such as..the amount to be allowed as managerial remuneration. **1940** H. G. WELLS *Babes in Darkling Wood* IV. iii. 379 They would soon draw plenty of recruits and assistants from the managerial class in the big trade machines. **1941** J. BURNHAM *Managerial Revolution* vi. 70 The theory of the managerial revolution asserts.. the following: Modern society has been organized through a..set of major economic, social, and political institutions which we call capitalist... At the present time these institutions..are undergoing..transformation... Within the new social structure a different social group..—the managers—will be the dominant..class. **1955** P. F. DRUCKER *Pract. Managem.* xxii. 268 Opportunities for participation that will give him [*sc.* the worker] a managerial vision. **1958** *Listener* 30 Oct. 685/1 The salaried middle class of professional, technical, and managerial people. **1973** *Amer. Speech 1969* XLIV. 268 Managerial positions are unquestionably the highest occupational level of the three.

Hence **manage·rialism**, belief in or the art of conducting or planning business or other enterprises by the use of managerial techniques. Also **manage·rialist**, an adherent of managerialism; also *attrib.* or as adj.

1946 'G. ORWELL' *James Burnham* 5 According to Burnham..'managerialism' has reached its fullest development in the U.S.S.R. *Ibid.* 6 He describes the New Deal as 'primitive managerialism'. [Burnham did not use the word *managerialism* in *The Managerial Revolution*.—Ed.] **1952** K. R. POPPER *Open Society* (ed. 2) I. 4 A lapse into totalitarianism (or perhaps into 'managerialism'). **1965** H. I. ANSOFF *Corporate Strategy* (1968) iii. 40 A 'managerialist' point of view came into being which..subjected the microeconomic theory to thorough criticism. *Ibid.*, The managerialists have offered a number of substitute explanations of the behaviour of the firm. **1966** *Harper's Mag.* June 67 Many remained caught in the irrelevancies of such questions as whether the Soviet Union was a 'degenerate workers' state' or a 'managerialist bureaucracy'. **1970** *Guardian Weekly* 14 Feb. 8 No doubt some cynics and managerialists will deride them [*sc.* local city councils] as 'mere talking-shops'. **1973** *Human World* Feb. 7 The anxiety Mr Maddox senses in the environmentalists comes to no more than a yearning for a universal poeticized prosperity. This they find compatible with a beneficent world-wide managerialism.

managing, *ppl. a.* Add: **3.** *managing clerk* (CLERK *sb.* 6 b); *managing director* (DIRECTOR 1 b); *managing editor* (earlier U.S. example).

1799 *Times* 1 June 1/3 (Advt.), Wanted..a managing clerk; he must understand well the Business of Conveyancing. **1861** MILL *Repr. Govt.* v. 91 Even a joint-stock company has always in practice, if not in theory, a managing director. **1865** A. D. RICHARDSON *Secret Service* 18, I found the Managing Editor in his office. **1866** MRS. H. WOOD *Elster's Folly* II. x. 230 The lawyer laughed. 'Hopkins did not know you; and strangers are generally introduced to..our managing clerk.' **1922** JOYCE *Ulysses* 484 They are immediately appointed to positions of high public trust..as managing directors of banks. **1940** W. TEMPLE *Hope of New World* i. 61 The great Managing Directors tend to rule the Boards of which they are at once both members and servants. **1972** M. GILBERT *Body of Girl* xiv. 128 She started here as my secretary. Then..she took on a number of jobs herself. I suppose you'd have called her a managing clerk by that time. **1972** J. MCCLURE *Caterpillar Cop* iv. 47, I would have you know that the managing director of your paper is a personal friend of mine. **1974** L. MEYNELL *Fairly Innocent Little Man* xiv. 180, I had a good job with a law firm. In time I daresay I would have become managing clerk.

‖ **manaia** (manəi·ă). [Maori.] A motif in Maori carving with a bird-like head and a human body.

1896 A. HAMILTON *Art Workmanship of Maori Race* 12 The thin board-like central piece with a human figure between two *pitau* spirals is called *Manaia* by the Arawas. **1910** J. COWAN *Maoris of N.Z.* xii. 165 Some of the wall slabs are carved into fantastic figures of fabulous water-monsters..others represent the mythical creatures known as the *manaia* and *wheku* with bird-like beaks and snaky tails all coiled in endless spirals. **1916** [see *bird-headed* (*BIRD *sb.* 8)]. **1924** E. BEST *Maori* II. xxi. 574 A common design in the carved work of superior houses and elevated storehouses is that known as the *manaia*..a figure composed of a long, slim body, a birdlike head, and an indefinite number of legs. **1959** FREEMAN & GEDDES *Anthropol. in South Seas* 117 The important symbol in Maori carving termed the *manaia* I believe is basically avian in origin. **1963** T. BARROW *Life & Work Maori Carver* 23 The most fascinating [mystery] is the origin and nature of a small creature called a *manaia*. There are many explanations; that it represents a malignant spirit, that it is merely a human image seen in profile, that it is a bird that attacked man in ancient times, or that it represents the *mana* or spiritual power of the human figure it is with. **1966** *Encycl. N.Z.* II. 410/2 Apart from the naturalistic figure, every type of full-faced figure has a *manaia* to match... The head of the *manaia* can, in each case, be recognised as half of the head of the appropriate matching figure divided down the middle of the face.

‖ **mañana** (mæn^y^ā·nă), *adv.* and *sb.* [Sp., tomorrow, morrow, f. *cras mañana* lit. tomorrow early, f. pop.L. **maneana* early, f. L. *māne* in the morning.] **A.** *adv.* Tomorrow, on the day after today. **B.** *sb.* Tomorrow, the day next after the present. Often taken as a synonym of easy-going procrastination as said to be found in Spanish-speaking countries: the indefinite future. Phr. *land* (or *kingdom*) *of mañana*: sometimes applied *spec.* to Mexico.

1845 R. FORD *Hand-bk. for Travellers Spain* I. ii. 144 Andalucia... Nowhere will the stranger hear more frequently those talismanic words which mark national character..the Manaña [*sic*], [etc.]. **1879** J. W. BODDAM-WHETHAM *Roraima & Brit. Guiana* xviii. 208 With an Indian the morrow is as indefinite a period as 'mañana' with the Spaniards. **1885** *Harper's Mag.* Jan. 217/2 Is Cedar Keys just on the borderland of that vast region known as the kingdom of *Mañana*? **1889** E. RIPLEY *From Flag to Flag* 165 Their *mañana* never came, never was intended to come. **1903** A. ADAMS *Log of Cowboy* 138 Flood had had years of experience in dealing with Mexicans in the land of *mañana*, where all maxims regarding the value of time are religiously discarded. **1910** *Daily News* 27 Sept. 4 The 'manana' boys didn't work quick enough for him, so he put the crew on. **1927** D. H. LAWRENCE *Mornings in Mexico* 59 *Mañana*, to the native, may mean to-morrow, three days hence, six months hence, and never. **1938** 'G. ORWELL' *Homage to Catalonia* i. 12 A promise that there should be machine-gun instruction *mañana*. Needless to say *mañana* never came. **1961** *Guardian* 23 Jan. 6/7 'Mañana' is a concept that exists far outside the Spanish world. **1973** C. BONINGTON *Next Horizon* xiv. 201 The elaborate etiquette and principle of mañana (leave everything until tomorrow, in the absolute confidence that tomorrow will never arrive), which dominates all dealings in South America. **1973** *Nat. Geographic* May 658/1 With a dispatch that put the lie to those who claim this [*sc.* Mexico] is the land of *mañana*, the orders went out to build a subway.

mananosay (mænănǭu·sĕi). *U.S.* Also **mananose, maninose.** [prob. ad. Algonquian name.] The soft-shell clam, *Mya arenaria*.

[**1709** J. LAWSON *New Voy. Carolina* 162 Man of Noses are a Shell-Fish commonly found amongst us.] **1843** J. E. DEKAY *Zool. N.Y.* V. 240 *Mya arenaria*..in some districts ..still retains its ancient aboriginal appellation of *Maninose*. **1859** BARTLETT *Dict. Amer.* (ed. 2) 84 The Soft Clam or Manonosay (*Mya arenaria*), obtained from the shores of tidal rivers by digging one or two feet in the loose sand. It has a long, extensible, cartilaginous snout, or proboscis, through which it ejects water; whence it is also called Stem-clam. **1870** *Putnam's Monthly Mag.* May 525/1 Even to the toothsome Manonosays that squirted water up through the sand what time the tides were out. **1895** *Sun* (N.Y.) 30 July 9/1 Mananosay, maninose (Maryland), man-of-noses (North Carolina), names for the round clam, from an Algonquian word meaning 'shellfish that one gathers by hand'. **1967** L. S. TAWES *Coasting Captain* xiii. 443, I used to take my launch and go fishing, sometimes digging mananoses.

manasseite (mānæ·se,əit). *Min.* [f. the name of E. *Manasse* (see quot. 1941) + -ITE[1].] A hydrated carbonate and hydroxide of magnesium and aluminium that is polymorphous with hydrotalcite and occurs as soft, foliated, whitish masses having a pearly lustre and a greasy feel.

1941 C. FRONDEL in *Amer. Mineralogist* XXVI. 310 These facts indicate with reasonable certainty that the mineral is identical in composition with hydrotalcite. The name manasseite is proposed for this mineral in honor of Ernesto Manasse (1875–1922), Italian chemist, mineralogist and petrographer,..who made important contributions to our knowledge of hydrotalcite and pyroaurite. **1962** W. A. DEER et al. *Rock-Forming Min.* V. 63 Spinel may alter to talc, mica, serpentine, or corundum: Struwe (1958) has also recorded its alteration to diaspore..and manasseite, $Mg_6Al_2CO_3(OH)_{16}.4H_2O$.

mancala (mæ·nkă·la). Also **mankalah, munkalah.** [ad. colloq. Arab. *manḳala*, f. *naḳala* to move.] A board game, originally Arabic but now common throughout Africa and Asia, played by two players on a special board (see quot. 1952) the object of which is the capture of the opponent's pieces. Also used as a generic term for regional variations of the basic game.

[**1813** J. GALT *Lett. from Levant* xxix. 242, I saw there today a game... The Idriots call it Mandoli, or the Almonds, and it is played at a board by two persons.] **1836** E. W. LANE *Acct. Manners & Customs Mod. Egyptians* II. iv. 46 One of the games most common among the Egyptians is that of the *munkalah*. **1877** *Encycl. Brit.* VII. 726/2 They [*sc.* the Egyptians] are acquainted with.. other games, among which is one peculiar to themselves, called Mankalah, and played with cowries. **1952** H. J. R. MURRAY *Hist. Board-Games* vii. 158 Anthropologists use the term *mancala* for any similar game played on a board in which the pattern..usual for board-games is replaced by two, three, or four rows of holes deep enough to contain a number of pieces at the same time. **1969** R. C. BELL *Board & Table Games* II. iv. 72 Ba-awa played by the Twi people of Ghana, is one of the simpler forms of mancala and appears to be extremely ancient.

Manchegan (mantʃē·găn), *a.* and *sb.* [ad. Sp. *manchego, -ga*: see -AN.] **A.** *adj.* Of or pertaining to La Mancha, a region (formerly a province) of central Spain. **B.** *sb.* A native or inhabitant of La Mancha.

[**1779** H. SWINBURNE *Trav. Spain* xxxvi. 319 The Manchegos have a pretty song about these eyes of the Guadiana.] **1841** BORROW *Zincali* I. II. ii. 245 Conde instantly unclasped one of those terrible Manchegan knives which are generally carried by contrabandistas. **1846** R. FORD *Gatherings from Spain* xxi. 296 The Manchegan bull, small, very powerful, and active, is considered to be the original stock of Spain. *Ibid.* xxiii. 326 Sancho, a true Manchegan,..maintained that for a *zapateo*, a knocking of shoes, none could beat him. **1897** A. F. JACCACI *On Trail of Don Quixote* iv. 96 It is arid, savage La Mancha which makes the Manchegan peasants shy, taciturn, and sombre. **1959** R. CROFT-COOKE *Quest for Quixote* i. 63 He was proud of being a Manchegan, a country-bred Spaniard. **1966** H. YOXALL *Fashion of Life* xxv. 239 Seven glasses were set in front of each of us, with a small piece of Manchegan cheese between each glass.

Manchester. Add: **1. Manchester terrier**, a small, short-coated, black and tan terrier of the breed so called, once particularly popular in the Manchester area. Also *absol.* Cf. *Black and tan* (BLACK *a.* 14).

1894 R. B. LEE *Hist. & Descr. Mod. Dogs Gt. Brit. & Ireland: Terriers* iv. 75 The Kennel Club acknowledged it as the 'Manchester' terrier, as well as by its own name of the black and tan. **1943** 'C. DICKSON' *She died a Lady* v. 37 A so-called Manchester terrier sprang on the front of the chair. **1971** F. HAMILTON *World Encycl. Dogs* 461 The only acceptable color for a Manchester Terrier is black-and-tan. *Ibid.*, Manchesters are very attractive little dogs equally at home in town or country.

b. (Further examples.)

1885 [see *BISMARCK 1]. **1957** *Encycl. Brit.* II. 828/2 Bismarck brown (Manchester brown), prepared by the action of nitrous acid on *m*-phenylene diamine, contains triaminoazobenzene.

Manchesterian (mæn,tʃe·stiə·riăn), *a.* and *sb.* Also 8 **Manchestrian.** [f. MANCHESTER + -IAN.] **A.** *adj.* Of or pertaining to Manchester. **B.** *sb.* An inhabitant of Manchester; also, one of the Manchester School of politicians. Cf. *MANCUNIAN *sb.* and *a.*

1778 J. WEDGWOOD *Let.* 3 Mar. (1965) 218 Nothing but half a score Highland, Manchestrian, and Liverpool regiments amongst us will raise their malignant spirits again. **1821** *Kaleidoscope* 3 July 423/3 Professing myself to be a plain Englishman and a Manchesterian. **1837** *Times* 5 Oct. 3/2 The engine went its way to Whitmore with the Liverpool passengers..and returned to fetch the unlucky Manchesterians. **1879** W. T. ARNOLD *Let.* 15 Aug. in D. Ayerst *Guardian* (1971) xxi. 206, I like the work very much, and strange to say like Manchester or at all events the Manchesterians. **1897** *Essays in Liberalism* 70 'Sordid inhuman wretch', 'brutal Manchesterian', are the terms applied to those who demonstrate the national loss of wealth which must result from the substitution of 'Fair' for Free Trade.

Hence **Mancheste·rianism** = MANCHESTERISM.

1897 *Essays in Liberalism* 33 A sneer at Cobden, a contemptuous allusion to Manchesterianism and the 'dismal science'.

Manchesterize (mæ·n,tʃestərəiz), *v.* [f. MANCHESTER + -IZE.] *trans.* To make representative or typical of Manchester. Hence **Ma:nchesteriza·tion.**

1925 E. T. SCOTT *Let.* 27 Dec. in D. Ayerst *Guardian* (1971) xxix. 452 The policy, both as to popularising and Manchesterising the paper..will be determined outside the Room. **1964** *Economist* 19 Dec. 1370/3 They do not want to see their [cotton] industries 'manchesterised'. **1965** *Ibid.* 23 Oct. 416/3 The [cotton] industry has argued ..that the unrestricted entry of imported textiles from low-cost Commonwealth countries is largely to blame for the state of the industry. (The Continentals have a word for it: 'manchesterisation'.) **1971** D. Ayerst *Guardian* xxix. 452 The policy of 'Manchesterisation' continued.

Manchu (mæn,tʃū·), *sb.* and *a.* Also **Manchew, Manchoo, Manchow, Mantcheou, Mantchoo, Mantchu,** etc. [Manchu, 'pure', the name of a tribe descended from the Nü-chên Tartars.] **A.** *sb.* **1.** A member of a Tungusic race inhabiting Manchuria, which conquered China in 1644 and was the ruling class until the Revolution in 1912.

[**1655** A. Semedo *Hist. China* III. 292 He used to cry out in their Language *Hoo Manzu*.] **1697** L. Le Compte *Mem. Journey through China* I. i. 17 One of the Petty Kings of the Eastern Tartary..whose subjects called Mouantchéou ..entered the Province of Leauton with a numerous Army. **1736** R. Brookes tr. *Du Halde's Gen. Hist. China* IV. 86 The places belonging to the Mantcheoux have Mantcheou Names. **1759** *Universal Hist., Mod.* IV. 278 Here the present empire of the eastern Tatars, or Manchews..had its beginning. **1821** G. Staunton tr. *Narr. Chinese Embassy* 152 The Mantchoos and the Mongals bear a great resemblance to each other. **1883** S. W. Williams *Middle Kingdom* (new ed.) I. i. 44 The Manchus are an agricultural or a hunting people. **1940** E. Pound *Cantos* lviii. 77 Tai Tsong..Forbad manchus marry their sisters.

2. The language of the Manchus.

1822 G. Staunton *Misc. Notes China* 95 Table of Contents of a Chinese and Mantchoo-Tartar Dictionary. **1888** H. E. M. James *Long White Mountain* 132 Yet, so wonderful are the ways of men, the Court and the people alike are now abandoning Manchu for the cumbrous and barbarous Chinese. **1920** *Contemp. Rev.* Apr. 526 Ferdinand Verbiest ..to please Kang-hi had learnt Manchu.

B. *adj.* Of or pertaining to the Manchus, their country (Manchuria), or their language.

1736 R. Brookes tr. *Du Halde's Gen. Hist. China* IV. 90 A great Number of Mantcheou Mandarins. **1770** W. Guthrie *New Geogr. Gram.* 472 The Chinese went to war with the Manchew Tartars. **1844** C. Fox *Jrnl.* 22 Aug. (1972) 153 They gave him a hymn to translate into the Manchow language. **1848** S. W. Williams *Middle Kingdom* II. 562 Out of a Manchu population of four thousand ..not more than five hundred survived. **1882** *Encycl. Brit.* XIV. 96/1 Tobacco..grown in the province [*sc.* Manchuria] being greatly prized throughout the Chinese empire under the name of 'Manchu leaf'. **1898** *19th Cent.* Jan. 163 The Manchu dynasty is the cement that holds the heterogeneous components of the Chinese Empire together. **1948** N. Whymant *China Manual* i. 16 The Manchu leader..mounted the throne and in 1644 the Ch'ing or Manchu dynasty began. **1972** T. Shabad *China's Changing Map* ii. 45 The criteria used by the Chinese Communists in distinguishing a Manchu ethnic group are not entirely clear. **1972** *Mainichi Daily News* (Japan) 6 Nov. 17/6 The authors have no acquaintance with Korean or Manchu grammar.

Manchurian (mæn,tʃū·riăn), *a.* [f. *Manchuria*, the country of the Manchus, now a dependency of China + -an.] **1.** Of or pertaining to Manchuria.

1876 A. R. Wallace *Geogr. Distribution Animals* I. 220 Japan and North China, or the Manchurian Sub-region. **1899** J. F. Fraser *Round World on Wheel* xxxi. 395 In five minutes down swooped several Manchurian officers. **1911** *Encycl. Brit.* XVII. 554/1 Eventually a Manchurian convention was arranged between China and Russia. **1937** *Discovery* Jan. 12/2 A candle copied from an old Manchurian pattern. **1972** W. B. Lockwood *Panorama Indo-Europ. Lang.* 154 The Manchurian languages are dispersed over an immense tract of the Soviet Far East.

2. Manchurian crane, a crane found in eastern Asia, *Grus japonensis*; **Manchurian ermine** (see quot.); **Manchurian roe, sika, wapiti,** local races of deer, *Capreolus capreolus bedfordi*, *Cervus nippon mantchuricus*, and *C. canadensis xanthopygus*; **Manchurian tiger,** a subspecies of the tiger, *Panthera tigris longipilis*, found in Manchuria and Siberia, and distinguished by its large size and shaggy fur; also called Siberian tiger.

1869 *Proc. Zool. Soc.* 628 Mantchurian Crane. *Grus montignesia.* **1898** R. Lydekker *Deer of all Lands* 102 The Manchurian wapiti is said to be smaller than the (?typical) American race. *Ibid.* 115 (*caption*) Buck and Doe of Manchurian Sika in winter pelage. *Ibid.* 231 (*heading*) The Manchurian Roe—*Capreolus Manchuricus.* **1899** R. Ward *Rec. Big Game* (ed. 3) 458 Lastly, we have the Manchurian tiger (*F. tigris longipilis*), characterised by its large size, heavy build, short limbs, and the great length and thickness of the fur. **1931–4** J. D. D. La Touche *Handbk. Birds E. China* II. 298 The Manchurian Crane breeds in Manchuria, Corea, and Eastern Siberia, and passes through Japan on migration. **1957** M. B. Picken *Fashion Dict.* 219/1 *Manchurian ermine*, fur of Chinese weasel. **1964** A. L. Thomson *New Dict. Birds* 162/1 Another species that has become alarmingly scarce is the Manchurian Crane *G[rus] japonensis*, a large white bird with black wings, a dark grey face, and a broad streak of the same colour running downwards at either side of the neck. **1964** R. Perry *World of Tiger* i. 8 A Manchurian tiger does in fact tire quickly and rest often when traversing deep snow. **1972** G. K. Whitehead *Deer of World* v. 74 The Manchurian Wapiti is similar to the North American animal. *Ibid.* 77 The principal distribution of the Manchurian Sika deer is in the central and southern parts of Manchuria. [*Ibid.* 84 Throughout the greater part of the Korean Peninsula the Roe is the Chinese or Manchurian race, *C[apreolus] c[apreolus] bedfordi.*]

Manchu-Tungus (mæn,tʃū·,tu·ŋgus). [f. *Manchu *sb.* and *a.* + *Tungus.] Name given to a language family comprising Manchu and Tungus.

[**1933** L. Bloomfield *Lang.* iv. 69 The Tungusé-Manchu family lies to the north of the Mongol, dividing Yakut from the rest of the Turco-Tartar area.] **1955** *Times* 15 Aug. 7/3 There are the peoples, often semi-nomad, who are in contact with civilization but only slightly affected by it: Indians and mestizos in Bolivia and Peru, the many groups, amounting in 1941 to nearly 25,000,000 souls, whom the Republic of India classifies as adivasis, the Paleoasiats and the Manchu-Tungus of the Soviet Far East. **1956** Jakobson & Halle in Saporta & Bastian *Psycholinguistics* (1961) 349/2 In Manchu-Tungus and in Paleosiberian languages. **1964** *Language* XL. 301 Coordinate with Turkic, Mongolian, and Manchu-Tungus.

mancia (mæ·ntʃia). [It.] A gratuity, a tip.

1951 [see *have *v.* 15 d]. **1963** T. Pynchon *V.* xiv. 409 Guides: there to do any bidding, to various degrees of efficiency, on receipt of the recommended baksheesh, pourboire, mancia, tip.

Mancunian (mæn,kiū·niăn), *sb.* and *a.* [f. L. *Mancunium* Manchester + -an.] **A.** *sb.* A native or inhabitant of Manchester. **B.** *adj.* Of or pertaining to Manchester.

1904 H. Beswick *Last Karkawbar* 134 'Th' Owd Rivvur' —as some old Mancunians dub the Irwell. **1908** *Westm. Gaz.* 22 Oct. 2/3 How strangely provincial—may we even say Mancunian?—is the very recent theory that Mr. Cobden invented Free Trade. **1926** *Glasgow Herald* 2 Oct. 8 In the Manchester docks..lies the real secret or the industrial trick, as the Mancunians choose to phrase it, which is at the foundation of the city's greatness. **1931** *Daily Tel.* 6 Jan. 10/3 The Mancunians who wish to play bowls on Sunday might surely be allowed their simple pleasure. **1947** H. Miles tr. *Maurois's Disraeli* (ed. 2) II. iv. 124 A truly Mancunian rain had drowned the enthusiasm. **1963** *Times* 2 Feb. 9/3 Any middle-aged Mancunian.. can remember posters drawing attention to celebrity concerts. **1973** P. Geddes *Ottawa Allegation* ii. 18 Still the Mancunian accent, true, yet fading fast. **1973** *Guardian* 16 Mar. 13/2 Michael Croft, the beaming Mancunian who founded the National Youth Theatre.

mand (mænd). [Final element of *com)mand*, *de)mand*, etc.] B. F. Skinner's term for an utterance aimed at producing an effect or result, etc. Cf. *tact.

1957 B. F. Skinner *Verbal Behavior* II. iii. 35 The term 'mand' has a certain mnemonic value derived from 'command', 'demand',..and so on... A 'mand', then, may be defined as a verbal operant in which the response is reinforced by a characteristic consequence and is therefore under the functional control of relevant conditions of deprivation or aversive stimulation. **1959** *Anthropol. Ling.* I. i. 41 It is interesting to speculate how far the program for the acquisition of mands and tacts will account for all verbal behavior. **1968** D. Lawton *Social Class, Lang. & Educ.* iv. 56 For Skinner, language behaviour is an example of learning by operant conditions... Requests, demands or commands (mands) tend to be reinforced by satisfaction of needs. Another kind of utterance is termed a '*tact*', which is a response to a situation rather than a response to a need (e.g. 'this apple is red'). **1972** *Language* XLVIII. 482 Beneath the linguistically questionable trappings (cf. Chomsky 1959) of mands, tacts, and echoic responses—..is there a brilliance which linguists in general have been prevented from seeing? **1973** *Archivum Linguisticum* IV. 52 'The ethnography of communication'—contains the item 'mands'.

mandala (mæ·ndălă). [Skr. *maṇḍala* disc, circle.] A symbolic representation of a magic circle usually with symmetrical divisions and figures of deities, etc., in the centre, used by Buddhists in meditation and found in many cultures as a religious symbol; *spec.* in *Jungian Psychol.*, an image of a similar magic circle visualized in dreams and symbolizing the dreamer's striving for unity of self and completeness. Also *attrib.*

1859 Max Müller *Hist. Anc. Sanskrit Lit.* i. 218 The division of the Sanhitâ which is adopted in the Bṛihaddevatâ, is that of Maṇḍalas, Anuvâkas, and Sûktas. **1882** *Encycl. Brit.* XIV. 228/1 Their practical belief..busied itself almost wholly with obtaining magic powers (*Siddhi*), by means of..magic circles (*Mandala*). **1927** W. Y. Evans-Wentz *Tibetan Bk. Dead* 136 (*caption*) The Great Mandala of the Knowledge-Holding and Wrathful Deities. **1931** C. F. Baynes tr. *Wilhelm & Jung's Secret of Golden Flower* 97 For the most part, the *mandala* form is that of a flower, cross, or wheel, with a distinct tendency toward four. *Ibid.*, I have come across women who did not draw mandala symbols but who danced them. **1933** E. J. Thomas *Hist. Buddhist Thought* xv. 193 The great spell.. should be inscribed in a circle (maṇḍala) made of certain substances with appropriate divisions and figures. **1938** C. G. Jung *Psychol. & Relig.* iii. 96 Historically..the mandala served as a symbol in order to clarify the nature of the deity philosophically. *Ibid.* iv. 109 Since modern mandalas have..close parallels in ancient magic circles, in the centre of which we usually find the deity, it is evident that in the modern mandala man—the complete man—has replaced the deity. **1941** A. Huxley *Grey Eminence* iii. 66 An elaborate circular diagram, curiously like one of those symbolic *mandalas*, into which the Buddhists contrive to cram such a wealth of doctrinal significance. **1949** K. Raine *Pythoness* 17, I piece the divine fragments into the mandala Whose centre is the lost creative power. **1966** P. White (*title*) The solid mandala. **1973** *Jrnl. Genetic Psychol.* CXXII. 168 The complete, circular form of the Ufor is reminiscent of the mandala or 'magic circle': i.e., totality. **1974** *Time* (Canada) 18 Mar. 56/1 The appetite of the young for religious experience is leading along exotic paths these days—demons and gurus, mandalas and myths.

Mandan (mæ·ndăn), *a.* and *sb.* Also **Mandane, Mandano, Mandon.** [ad. Dakota *Matani.*] **A.** *adj.* Of, pertaining to, or designating a Siouan Indian people of North Dakota. **B.** *sb.* This people; a member of this people; also, their language.

1794 in *Mass. Hist. Soc. Coll.* (1810) 1st Ser. III. 24 The tribes of Indians which he passed through, were called the Maskego Tribe..Mandon tribe, Paunees, and several others. **1806** M. Lewis in *Deb. Congress U.S.* (1852) 9th Congress 2 Sess., App. 1065 The bridlebits and blankets I have seen in the possession of the Mandans. **1806** [see *berdache]. **1831** J. M. Peck *Guide for Emigrants* 21 At the Mandan villages, 1600 miles from the Mississippi, it [*sc.* the Missouri] is said to be nearly as wide..as at St. Charles. **1877** L. H. Morgan *Anc. Society* II. vi. 158 In intelligence and in the arts of life the Mandans were in advance of all their kindred tribes. *Ibid.* III. iii. 440 In Mandan my brother's wife is my wife. **1911**, etc. [see *Hidatsa]. **1933** L. Bloomfield *Lang.* iv. 72 The Siouan family includes many languages, such as..Winnebago, Mandan, Crow. *Ibid.* xvii. 283 A Mandan Indian sent the ..picture to a fur-trader. **1971** S. E. Morison *European Discovery Amer.: Northern Voy.* iv. 86 Eventually the Mandan become the most favored tribe.

mandarin¹. Add: **1. b.** (Earlier and later examples.)

1791 Boswell *Johnson* I. 5 From a man so still and so tame..conversation worth recording could no more be expected, than from a Chinese mandarin on a chimney-piece. **1845, 1855** [see Niddle-noddle *v.*].

c. *transf.* A person of much importance, a great man. Often used *colloq.* of Government officials, leading politicians or writers, etc.

1907 *National Rev.* Aug. 838 Our Parliamentary Mandarins are ineffably shocked at the impiety of an independent Radical. **1908** *New Age* 6 June 112/2 The chams, lamas, and mandarins of London letters are doubtless devising adjectives for it [*sc.* a book]. **1919** B. Ruck *Disturbing Charm* II. ix. 234 If you let it get known..that you've got a view like that, you'll have some of the Mandarins snaffling that office of yours for themselves. **1925** Fraser & Gibbons *Soldier & Sailor Words* 151 The Mandarins of the War Office. **1947** *Oxf. Univ. Handbk.* 261 If he is an athlete of any distinction, the mandarins of his particular game will know all about him long before he arrives. **1961** *Listener* 9 Nov. 782/3 Mr. Thody, who has written a book about Sartre no less brilliant than his two books on Camus, makes comparisons between the two *mandarins*. **1971** P. Loraine *Photographs have been Sent* v. iii. 164 The Medical Mandarins maintained stony silence.

2. (Later examples.) Also *transf.*

1959 V. Cronin *Pearl to India* xiii. 188 Some philologists claimed Mandarin to be derived from Hebrew. **1963** *Listener* 17 Jan. 140/1 BBC Mandarin, or Announcers' English, was devised as a refined product, based on so-called 'southern educated' English. **1964** *Amer. Speech* XXXIX. 26 Home's own writings..are written in a kind of middle-class international Victorian Mandarin which defies analysis. **1971** K. Hopkins *Hong Kong* 235 Cantonese is very much the predominant language but there are minorities who speak..Mandarin.

4. mandarin language (earlier example); **mandarin blue** (see quot. 1949); **mandarin coat** (see quot. 1957); **mandarin collar,** a narrow collar standing up from a close-fitting neckline.

1912 *Home Chat* 13 Apr. 112/2 In flamingo red, Mandarin blue or wood-violet mauve linen. **1949** *Dict. Colours Interior Decoration* (Brit. Colour Council) 17/1 *Mandarin blue*, a descriptive name for one of the blues specially produced for China by British dyers at the beginning of the twentieth century. **1911** *Daily Colonist* (Victoria, B.C.) 22 Apr. 7/1 (Advt.), Mandarin Coats. In exquisite hand-embroidered silks and silk lined. **1957** M. B. Picken *Fashion Dict.* 67/1 *Mandarin*, long, loose, richly embroidered silk coat with wide sleeves. **1972** *Vogue* June Special 94 Mandarin coat and slit dress of matching print. **1952** C. Howard *Compl. Dressmaking* 104 The collar is a mandarin collar. [**1953** *News Chron.* 2 June 7/1 You can spot him [the Sultan of Selangor] by his yellow knee-length baju (jacket) with mandarin-type collar.] **1967** G. B. Mair *Girl from Peking* vi. 69 I'll go Chinese. Gold and blue brocade with high mandarin collar and short sleeves. **1971** P. Loraine *Photographs have been Sent* III. i. 91 A white silk suit with a mandarin collar. **1697** L. Le Compte *Mem. Journey through China* I. v. 134, I acquainted her that I spake the Mandarin language..which they constantly use at the Court.

b. *attrib.* or quasi-*adj.*, in *transf.* sense of 'superior, esoteric, "highbrow"'; applied esp. to literary productions or style: 'ornate, refined; high-flown' (often in derogatory use).

1916 H. G. Wells *Mr. Britling* I. i. 16 The conservative classes whose education has always had a mandarin quality—very, very little of it, and very old and choice. **1947** J. Hayward *Prose Lit. since 1939* vii. 47 If literature..is not to become the arcane cult of a mandarin class, it must impose its values. **1952** *Times Lit. Suppl.* 29 Aug. (Suppl.) p. xii/2 The influence that delivered the English novel from Mr. Bennett and Mrs. Brown turned out to be D. H. Lawrence and not his mandarin contemporaries. **1959** *Ibid.* 1 May 257/2 In reaction from Victorian pomposness, bureaucratic jargon, *fin de siècle* poetic prose, and 'mandarin English', a go-as-you-please attitude has crept into the language. **1962** *Listener* 13 Sept. 406/2 The conventionally acceptable accents and Mandarin prose we learn at school. **1973** *Times Lit. Suppl.* 6 July 787/2 M

Bourdieu's style has, from the first paper onwards, been growing increasingly mandarin. *Ibid.* 787/3 The mandarin observer, 'freed from the constraints and urgencies of practice'.

mandarin². **1.** (Earlier and later examples.)

1771 J. R. Forster tr. *Osbeck's Voy. China* I. 307 Here are two sorts of China oranges (*Citrus sinensis*). The first is that called the *Mandarin-orange*, whose peel is quite loose. **1926** H. H. Hume *Cultivation Citrus Fruits* xxix. 477 The other mandarin or kid-glove oranges are attacked [by citrus rust] but not often severely. **1969** *Oxf. Bk. Food Plants* 86/2 Tangerine (*Citrus reticulata*). The fruit is also known under several other names, one of which is 'mandarin', denoting its origin in the Far East.

mandarinship. (Earlier example.)

1697 L. Le Compte *Mem. Journey through China* I. i. 12 Executioners..ready to bind and cudgel whom his Mandarineship should think fit.

‖ **mandat** (mandā). [a. F. *mandat* (see Mandate *sb.*).] **1.** A paper money (in full *mandat territorial*) issued by the French Revolutionary Government from 1796 to 1797, replacing the Assignat.

1858 P. L. Simmonds *Dict. Trade Products* 238/2 *Mandats*, a national paper-money, issued..in France to replace the assignats which had become wretchedly depreciated. **1902** J. R. Moreton-Macdonald in Carlyle *French Revolution* (ed. C. R. L. Fletcher) III. 339 The Directory.. destroyed the die of the *Assignats*, but immediately replaced them by *Mandats Territoriaux* which rapidly followed the same course. **1930** S. E. Harris *Assignats* ix. 225 The Assignats were to be exchanged for Mandats which were to be supported by a special security. **1955** E. Pound *Section: Rock-Drill* lxxxviii. 45 Willing to see a currency of hard money..as France since the time of mandats and assignats. **1974** *Encycl. Brit. Micropædia* I. 594/3 In early 1796 the assignats were replaced by the mandats territoriaux (land warrants) at the rate of one mandat for 30 assignats. The failure of the mandat to gain public confidence forced the Directory to return to a metallic currency (Feb. 4, 1797).

2. In France, a money-order.

1896 E. Dowson *Let.* c 15 Mar. (1967) 345, I enclose a mandat for the price & postage, & shall be much obliged if you can get them [*sc.* newspapers] & post them here. **1939** E. Ambler *Mask of Dimitrios* xii. 251, I received here a letter from him enclosing a *mandat* for the three thousand francs.

mandate, *sb.* Add: **2.** (Earlier example.)

1501 Douglas *Pal. Hon.* (c 1553) II. sig. F1 Submyttand me..ȝour plesour and mandate till obeysyng.

4. (Further examples.)

1886 *Hansard Commons* 9 Apr. 1244, I am perfectly aware that there exists in our constitution no principle of the mandate... But..I maintain that there are certain limits which Parliament is morally bound to observe, and beyond which Parliament has morally not the right to go in its relations with the constituents. **1936** R. C. K. Ensor *England, 1870–1914* i. 25 Gladstone (who ten months earlier had been telling the queen that his work was done, his mandate exhausted, and he himself in need of a long rest) declared on 24 January [1874] his intention of dissolving parliament. **1968** *Daily Progress* (Charlottesville, Virginia) 11 July C14/4 We need to win only 36 [seats], which I am sure we can do, and that will be an overwhelming mandate for Scottish freedom.

b. *spec.* A commission issued by the League of Nations (1919–1946) authorizing a selected power to administer, control, and develop a territory for a specified purpose; the territory so allocated. Also *attrib.*

1919 *League of Nations Charter* Art. xxii, The character of the mandate must differ according to the stage of the development of the people, the geographical situation of the territory, its economic conditions, and other similar circumstances. **1920** *Glasgow Herald* 7 Aug. 9 It will still be necessary for the Council to set up a permanent Mandate Commission. **1921** *Spectator* 12 Feb. 189/2 The draft mandates for Palestine and Mesopotamia, which are to come before the Council of the League of Nations on February 21st, were published unofficially last week. *Ibid.* 2 Apr. 419/1 They apparently look upon mandate-making as a kind of old-fashioned diplomacy. **1924** *Brit. Weekly* 30 Oct. 98/4 The Mandate Section of the General Secretariat of the League of Nations. **1937** F. P. Crozier *Men I Killed* xii. 277 Are the British Mandates..a success? **1946** *Ann. Reg. 1945* 166 The Trusteeship System, replacing the Mandate System of the League, will cover a wider range of backward territories. **1972** *Whitaker's Almanack 1973* 951/2 Syria, which had been under French mandate since the 1914–18 war, became an independent Republic during the 1939–45 war. **1974** *Encycl. Brit. Micropædia* VI. 557/3 Both the territories and the authority to administer them were called mandates. Among them were Britain's mandates in Palestine and Tanganyika. *Ibid.* 558/1 The Mandate System was replaced by the UN Trusteeship System in 1946.

c. doctor's mandate: a mandate from the people empowering the government to take extreme measures in the national interest.

1931 *Times* 7 Oct. 14/1 Mr. MacDonald would issue a manifesto as the head of the National Government appealing for what is called a 'doctor's mandate'. **1961** I. Jennings *Party Politics* II. vii. 291 The Government appealed to the people on a 'doctor's mandate'. **1965** A. J. P. Taylor *Eng. Hist. 1914–45* x. 324 MacDonald asked for 'a doctor's mandate'—a blank authority for the National government to do whatever they could agree on. **1973** *Times* 17 Dec. 15/1 Some of the substantial voices that now call for a doctor's mandate from the people stand close enough to the Prime Minister for him to feel the full cogency of their persuasions. **1974** *Observer* 13 Jan. 1/3 The Prime Minister's main demand would be for a 'doctor's mandate' to enable him to take the measures he considers necessary to make the pay-and-prices policy effective.

mandate, *v.* Add: **3.** To assign (territory) under a mandate of the League of Nations. Cf. *mandate *sb.* 4 b. So **manda·ted** *ppl. a.*

1919 J. M. Keynes *Econ. Consequences Peace* 248 The Mandated States should be compelled to adhere to this Union for ten years. **1920** *Glasgow Herald* 7 July 11 The Island of Nauru in the Pacific (which is mandated to the British Empire). **1922** *Weekly Dispatch* 5 Nov. 8 We were authorised to raise local native forces to protect the mandated area. **1922** *Times Lit. Suppl.* 23 Nov. 756/3 The result of the late war has been to eliminate Germany from the map, her territories being mandated to the British and other nations. **1944** J. S. Huxley *On Living in Revolution* xii. 119 Crown colonies, protectorates, condominiums, mandated territories of various categories. **1958** A. R. Radcliffe-Brown *Method in Social Anthropol.* I. iii. 90 Cadets who are selected for the administration of the Mandated Territory are sent to the territory for one or two years to make acquaintance with the kind of life and work they will have. **1970** *Internat. & Compar. Law Q.* XIX. 218 When this section was enacted, New Guinea was a Mandated Territory of the League of Nations.

4. To give a mandate to, to delegate authority to (a representative, group, organization, etc.). Freq. as **manda·ted** *ppl. a.*, permitted to act on behalf of a group, etc., approved by means of a mandate.

1958 *Spectator* 20 June 191/1 A delegate conference was called, and garages invited to mandate their representatives to vote for or against continuance [of a strike]. **1967** *National Observer* (U.S.) 3 July 13 Mr. Reagan must raise the money to pay off that deficit and to pay for mandated new programs. **1968** Smith & Zurcher *Dict. Amer. Politics* (ed. 2) 231 *Mandated expenditure*, an expenditure which a State requires a municipal government to make, often from locally collected funds, and often without reimbursement from State funds. **1969** D. Widgery in Cockburn & Blackburn *Student Power* 126 Universities with a strong and democratic union came to Council with an elected delegation fully briefed and mandated on all issues by general meetings. **1972** *Daily Tel.* 29 Apr. 14 The [union] delegates are elected by, and frequently mandated by, those members who attend branch meetings. **1973** *Black World* May 35/2 The Committee mandated its current chairman..to visit all O.A.U. member states. **1973** *Black Panther* 21 July p. B, Their annual salary increases exceed the nationally-mandated rate. **1974** *Daily Tel.* 25 May 6/5 Mr Thorne, a member of the national executive, said he was mandated to vote for industrial action.

mandatory, *a.* and *sb.* Add: **A.** *adj.* **a.** (Further examples.)

1966 *Listener* 15 Dec. 880/1 Mr. George Brown..puts Britain's case for mandatory sanctions to Security Council. **1972** *Soviet Weekly* 26 Feb. 1 Each delegation has one vote, and only unanimous decisions are mandatory. **1972** *Incorporated Linguist* XI. 36 For further education, postgraduate training is not mandatory, but there is much to be said for it.

c. *spec.* Designating a power or state in receipt of a mandate from the League of Nations, or the system of rule by mandate. Cf. *mandate *sb.* 4 b.

1921 *First Assembly* (League of Nations Union) 260 The Commission shall examine the annual reports of the Mandatory States and advise the Council as to the execution of the terms of the Mandates. **1922** *Encycl. Brit.* XXX. 509/1 Under the Peace of Versailles a new form of colonial possession came into being... The 'mandatory' system was..evolved. **1930** [see *incident *sb.*¹ 1 b]. **1946** *Ann. Reg. 1945* 167 The compromise eventually adopted required the Mandatory Powers to express their willingness to place the territories for which they were responsible under the Trusteeship system. **1950** M. Hay *Foot of Pride* ix. 269 'It will be interesting,' wrote C. R. Ashbee, civil adviser of the mandatory administration, in 1922, 'to watch..the inevitable failure of the Rutenberg scheme.' **1958** *Listener* 21 Aug. 273/1 An underground organisation [in Palestine]..directed against Britain, the mandatory power. **1974** *Encycl. Brit. Micropædia* VI. 558/1 Exercise of the mandates was supervised by the League's Permanent Mandates Commission, but the commission had no real way of enforcing its will on a mandatory power.

B. *sb.* **b.** *spec.* A power or state in receipt of a mandate from the League of Nations to administer and develop a territory (in quot. 1927, the territory assigned in this way).

1919 *League of Nations Covenant* Art. xxii, The wishes of these communities must be a principal consideration in the selection of the Mandatory. **1927** *Daily Express* 24 May 3 A memorandum issued by the Arab Executive (Nationalists)..accuses Great Britain of ignoring the covenant principle to assist mandatories to become self-governing. **1928** *Manch. Guardian Weekly* 8 June 443/3 He adopts quite frankly the very contestable position that sovereignty over the mandated territory belongs solely to the mandatory. **1936** V. Margueritte *League Fiasco* ii. 83 The Mandates Commission, consisting of nine members, would have nothing to do but examine the annual reports submitted by the mandatories. **1937** H. F. Angus *Probl. Peaceful Change Pacific Area* iii. 155 It was assumed that whichever of these countries acted as mandatory would do more for the interests of native populations than had been done by their predecessor.

M and B (em ănd bī·). Also **M & B,** and with points. A registered trade mark of the *M*ay & *B*aker Co. Ltd., applied to pharmaceutical preparations marketed by them, esp *M & B 693*, sulphapyridine tablets.

A symbol incorporating the letters M and B was registered as a trade mark in 1935 (*Trade Marks Jrnl.* (1935) 30 Oct. 1349).

1938 *Lancet* 28 May 1210/2 For experimental infections in mice the effective dose of M. & B. 693 varies from 0·25 mg. per g. to 2 mg. per g. **1943** J. B. Priestley *Daylight on Saturday* i. 3 With ultra-violet rays and radiant heat and *M. & B. 693* to be had for the asking. **1951** W. S. Churchill *Second World War* (1952) V. 373 The admirable M and B, from which I did not suffer any inconvenience, was used at the earliest moment, and after a week's fever the intruders were repulsed. **1953** *Trade Marks Jrnl.* 18 Feb. 139/1 M & B... Chemical products used in industry, science and photography; [etc.]... May & Baker Limited, Dagenham, Essex; manufacturing chemists. **1957** F. S. Taylor *Hist. Industr. Chem.* xvi. 250 Sulphapyridine (M & B 693) revolutionized the treatment of pneumonia. **1968** *Listener* 19 Dec. 815/3 One pocket stuffed with quinine and the other with M and B tablets.

Mande (mā·ndē), *sb.* and *a.* = *Mandingo *sb.* and *a.*

1883 R. N. Cust *Sk. Mod. Lang. Afr.* I. xi. 179 The Mande-nga occupy a mountainous Region... The final syllable is a Suffix, which conveys the meaning of the people themselves, while their language should properly be called Mande. *Ibid.* 186 The Mende are Pagans and turbulent. Care must be taken to distinguish the Mende from the Mande. *Ibid.* 186 They [*sc.* the Vei] belong to the Mande Cluster. **1911** *Encycl. Brit.* XVII. 564/2 Delafosse divides the Mandingo group linguistically into three main sections: (1) the *Mande-tamu*, (2) the *Mande-fu*, and (3) the *Mande-tā*, according as they use for the numeral 10 the root *tamu*, *tā* or *fu*. *Ibid.* 565/1 The manati was the totem of the Mande group. **1930** C. G. Seligman *Races Afr.* iii. 59 The Mandingo—more correctly the Mendi or Mande—constitute one of the most important groups of French Senegal. **1952** Westermann & Bryan *Lang. W. Afr.* ii. 31 The Mande languages are spoken over a vast area extending from the Atlantic coast to the Black Volta. **1955** P. Strevens *Papers in Lang.* (1965) ix. 114 Within the area..are spoken Temne and Mande languages in Sierra Leone, [etc.]. **1968** G. Jackson *Let.* 6 Mar. in *Soledad Brother* (1971) 152 The oldest language is one spoken in Africa: Mande. **1970** P. Oliver *Savannah Syncopators* 112 *Mandingo.* Mande-speaking peoples of which the Malinke are the largest.

Mandingo (mændi·ŋgo), *sb.* and *a.* Also **Manding, Mundingo. A.** *sb.* **a.** A large group of Negro peoples of the upper Niger in West Africa; a member of these peoples. **b.** The language or languages of these peoples. **B.** *adj.* Of or pertaining to these peoples.

1623 R. Jobson *Golden Trade* 27, I take my beginning from the mouth of the River, whereat our first entrance, we find the Black men called Mandingos. **1757** [see *Ibo *sb.* 1 a]. **1798** *Proc. Afr. Assoc.* 4 The inhabitants are chiefly Mandingoes, and seem to be a well disposed and peaceable race. *Ibid.* 9 Of the other wild animals in the Mandingo countries, the most common are the hyæna, the panther, and the elephant. *Ibid.* 13 By the Niger, is here undoubtedly meant, the river of Senegal; which in the Mandingo language is called *Bafing*, or the Black river. **1888** L. A. Smith *Music of Waters* 329 Another very rhythmical air is the following Mandingo one. **1925** P. Radin tr. *Vendryes' Lang.* II. ii. 112 A West African language, Mandingo, distinguishes *a fa* 'his father', from *a-ta kursi* 'his breeches'. **1930** C. G. Seligman *Races Afr.* iii. 60 The typical Mandingo are described as tall and slender in build. **1936** G. Greene *Journey without Maps* II. i. 99 Some Mandingo traders whom he caught smuggling goods over the border from French territory. **1948** *Caribbean Q.* I. i. 11 Of these 101 were Mandingoes from Senegambia and the upper Niger. **1969** *Times* 19 July 9/5 Many of these slaves were..conversant with the two main languages of Senegambia: Wolof and Mandingo. **1972** *Times* 29 June 16/2 The term 'Manding' embraces a number of West African peoples who speak related forms of the same language and share a similar culture. The Manding language, in a variety of dialects.., is spoken by some 10 million people. *Ibid.* 16/5 The pious Manding who made a famous pilgrimage to Mecca in 1324. *Ibid.*, The Manding have always been natural travellers.

Mandinka (măndi·ŋkă), *sb.* and *a.* = *Malinke.

1957 M. Banton *W. Afr. City* iv. 62 Tribes largely resident in Freetown or the Colony area, for example, Kru, Sherbro and Mandinka. *Ibid.* vii. 122 The Kissi..speak a separate language similar to that of the Bulom but now much influenced by Mandinka. *Ibid.* 127 There are about 125,000 Mandinka in West Africa, from Senegal to the Ivory Coast.

mandolin, -ine. Add: **2.** *transf.* A kitchen utensil fitted with cutting blades and used for slicing vegetables. Usu. spelt *mandoline*.

1951 E. David *French Country Cooking* 19 A vegetable slicer which goes by the charming name of *Mandoline.* **1959** *Times* 16 Nov. 15/4 Slice the peeled potatoes evenly and thinly (a slicing device known as a *mandoline* makes this task a matter of moments). **1961** *Spectator* 25 Aug. 270 With the aid of that blessed instrument called a mandoline the cucumber is thinly and evenly sliced. **1969** *Daily Tel.* 16 Jan. 17 Arrange a chopping board, sharp knife, grater and cucumber slice (or mandoline) around the colourful basket of raw vegetables. **1975** *Habitat Catal.* 68 Mandolin... Wood frame stainless steel slicer and crinkle cutter.

mandor (mæ·ndǭɹ). Also **mandore, mandur.** [Malay *mandor* (*mandur*), ad. Pg. *mandador* one who gives orders.] A foreman or overseer in Malaysia or Indonesia. Cf. *kangany.

1889 S. J. Hickson *Naturalist in North Celebes* 65 The coolies were under the supervision of two mandūrs or foremen. **1926** *Blackw. Mag.* Apr. 508/1 A Malay 'mandor' is told that at a certain time on that day he must bring so many men. **1928** *Ibid.* Apr. 473/1, I remember a Malay mandor of mine laughing aloud when he was told that another Malay and two children had died..from exposure. **1958** Ginsburg & Roberts *Malaya* ix. 333 In the managerial class the labor foremen, *kanganies* or *mandors*, are a combination of labor boss and patriarch depending on personality factors and on the strength of the trade-union organization. **1962** B. Harrisson *Orang-Utan* iii. 119 We heard..through one of the road engineers..that one of his Malay mandors had been asked to look for a baby gibbon which somebody wanted to keep as a pet. **1965** C. Shuttleworth *Malayan Safari* iii. 41 An Indian *mandor* (foreman) was taken from the veranda of his house whilst sleeping. **1969** K. S. Sandhu *Indians in Malaya* iii. 114 *Mandurs* (overseers)..and railway porters were classified as skilled workers and allowed to return to Malaya [from India]. **1970** T. Lilley *Projects Section* xv. 192 Raja Gopal's house..stands..about a hundred yards from the labour lines..of which Raja Gopal is..the Mandore in charge. **1972** *Sunday Times* (Kuala Lumpur) 30 Apr. 3/2 Tin mine mandore Yaacob bin Abdul Wahab had to master more than the course.

Mandrax (mæ·ndræks). *Pharm.* A proprietary name for tablets containing methaqualone and diphenhydramine hydrochloride, used as a sedative.

1963 *Trade Marks Jrnl.* 10 Apr. 485/1 Mandrax... Hypnotics, being pharmaceutical preparations and substances. Roussel-Uclaf,..Paris. **1967** *Scottish Med. Jrnl.* XII. 63/1 Mandrax..is a hypnotic preparation, which has been actively marketed in Great Britain since Autumn 1965... It has become commonly prescribed as a non-barbiturate hypnotic. **1970** L. Leech in *Drug Dependence* (U.S. Nat. Inst. Mental Health) 22 Another factor of importance has been the increased use of non-barbiturate hypnotics, particularly methaqualone and diphenhydramine in the form of Mandrax, by young people within the 'pill scene'. **1972** *Police Rev.* 17 Nov. 1505/1 He admits taking Mandrax tablets obtained on prescription.

mandur, var. *mandor.

mandy, Mandy (mæ·ndi). Colloq. abbrev. of *Mandrax (*tablet*).

1970 *Daily Tel.* 8 Sept. 2/3 Dr Tylden says that hypnotic tablets of methaqualone and antihistamine known as 'Mandies' to addicts, had been mentioned to her by youngsters from all parts of Britain... The favourite mixture was four 'Mandies' and half a pint of cider, which could lead to sudden unconsciousness. **1971** *Frendz* 21 May 11/2 Avoid dealing while tripping on Acid, Speed or Mandies—you'll goof on the action. **1973** *Daily Tel.* 11 July 2/8 Addicts, who call the white tablets [of methaqualone] 'mandies', 'mainline' by crushing the tablets and injecting themselves.

mane, *sb.* Add **1. a.** *fig.* (Later examples.)

1893 F. Thompson *Hound of Heaven* in *Poems* 49 To all swift things for swiftness did I sue; Clung to the whistling mane of every wind. **1927** Joyce *Flood* in *Pomes Penyeach*, A waste of waters ruthlessly Sways and uplifts its weedy mane. **1936** R. Campbell *Mithraic Emblems* 38 The World put down its lovely mane.

4. *mane-flinging* ppl. adj.

1945 P. Larkin *North Ship* 19 As some vast seven-piled wave, Mane-flinging, manifold, Streams at an endless shore.

‖ **maneaba** (mane·abā). [Native name.] In the Gilbert and Ellice Islands, a meeting-house.

1944 G. H. Eastman *Front Line Islands* 5 Landing at Nui Island we proceeded as our custom was to the public *maneaba* (meeting-house), where the Resident Commissioner spoke to the people. *Ibid.* 10 Our people at Hull Island..have recently erected a large new *maneaba*, which is now used as school house and for women's meetings and various other community gatherings. **1952** A. Grimble *Pattern of Islands* ii. 58 Every Gilbertese village of any size had its own maneaba, or speak-house, in those days. **1970** A. Coates *Western Pacific Islands* viii. 63 In the absence of any king, the wise men conducted affairs in their council house—*manéaba*—their jurisdiction extending just as far as was acceptable, which usually meant to the limits of the land occupied by the clans whose senior members had a reserved place in the *manéaba*, which is somewhat like a hereditary parliament, with the important—and very Pacific—exception that all decisions must be taken on the basis of unanimity. **1974** *Nat. Geographic* Dec. 753/2 Once ashore, I was escorted to the large meetinghouse, the *maneaba*, with thatched eaves that stood only four feet above the ground and a roof that soared upward to a crisscross of massive beams a full forty feet overhead.

man-eater. Add: **2.** (Further examples.)

1922 Blunden *Bonadventure* xvi. 95 To sleep there was to be slowly suffocated, let alone the folly of sleeping among man-eaters [*sc.* mosquitoes]. **1957** R. Campbell *Portugal* iv. 61 Aulus Fedeus had another [moray]..a man-eater.

4. *fig.* Of a person (see quots.). *colloq.*

1906 E. Dyson *Fact'ry 'Ands* xi. 136 To Spats' Beauties she was always Porline or The Man-Eater. **1928** A. Huxley *Point Counter Point* xiv. 264 Marjorie isn't the only bore. Nor Lucy the only man-eater. **1929** F. C. Bowen *Sea Slang* 88 *Man Eater*, a particularly tough officer under sail. **1944** T. Rattigan *While Sun Shines* II. 209 Aw, she's no man-eater. You don't get real man-eaters this side of the Atlantic. **1968** D. Gray *Died in Red* xx. 122 'She 's pretty, you said?'.. 'Very, sir.' 'And a man-eater?' 'I'd say so, yes, sir.' **1974** J. Montgomerie *Implosion* xiii. 97 A womaniser, to use an old-fashioned term. Was a woman ever described as a maniser? No, but I'd heard the designation man-eater.

man-eating *ppl. a.* Also *fig.*

1954 T. S. Eliot *Confid. Clerk* II. 61 Between a couple of man-eating tigers like you and Lizzie, he's got to have protection. **1958** *Times Lit. Suppl.* 26 Dec. 749/4 He develops an obsession for a fearful man-eating television actress. **1959** *Times* 28 May 15/5 *The Affair in Arcady* has everything—a man-eating young heroine..a sinister step-father, [etc.]. **1974** K. Benton *Craig & Tunisian Tangle* xi. 158 That sadistic bitch... She's got her man-eating eye on you.

maned, *ppl. a.* **c.** (Later examples.)

1924 R. Campbell *Flaming Terrapin* iv. 68 A fierce train, maned like a ramping lion With smoke and fire, thunders on rolling iron. **1925** E. Sitwell *Troy Park* 66 Whinnying, neighed the maned blue wind.

manège. Add: **2. b.** *transf.* and *fig.*

1825 T. Moore *Mem. Life R. B. Sheridan* II. xxi. 493 Had his talents, even then, been subjected to the *manège* of a profession. **1955** E. Pound *Classic Anthol.* IV. 214 In the cars' manège war-skill appears. **1973** *Times* 12 Jan. 10/6 She has given up even attempting the fouettés in the ballroom scene; but if the manège she substituted was slightly tame, Nureyev made up by the electrifying speed of his pirouettes.

maneton (mantoṅ). *Aeronaut.* [F. *maneton* crank-pin.] (See quot. 1949.)

1919 *Gloss. Aeronaut. Terms* (R. Aeronaut. Soc.) 44 *Maneton*, the small end of the crankshaft of a rotary engine. **1939** *Times* 30 Mar. 9/3 One simple device for assembling the maneton end of the crankshaft. **1949** *Gloss. Aeronaut. Terms (B.S.I.)* II. 11 *Maneton*, the detachable short end of a crankshaft in a rotary or radial engine.

ma·n-folk. *poet.* [Man *sb.*¹] People, human beings, men.

1875 W. Morris tr. *Virgil's Æneid* XII. 825 Let not that manfolk shift their tongue, or cast their garb aside. **1887** —— tr. *Homer's Odyssey* III. 252 Amid other dwellings of manfolk. *Ibid.* I. 393 Of all that befalleth manfolk dost thou deem it the evillest thing?

Manga. 1. (Earlier example.)

1824 W. Bullock *Six Months' Residence Mexico* xvii. 216 An elegant manga or cloak, of velvet, fine cloth, or fine figured cotton.

manganese. Add: **3. manganese purple** = *manganese violet*.

1937 *Burlington Mag.* Dec. 277/2 A large jar of early Florentine maiolica..of the class painted in manganese—purple and green, the only pigments at that time known to the maiolica painter. **1963** *Times* 25 May 11/5 An unpleasant manganese-purple tinge.

manganoan (mæŋgănǭu·ăn), *a.* *Min.* [f. Mangan(ese + *-oan.] Of a mineral: having a (small) proportion of a constituent element replaced by bivalent manganese.

1930 W. T. Schaller in *Amer. Mineralogist* XV. 571/2 Manganese—manganoan. **1944** C. Palache et al. *Dana's Syst. Min.* (ed. 7) I. 777 Mossite is..reported also (manganoan) from Yinnietharra, Western Australia. **1968** I. Kostov *Mineral.* II. ix. 498 Luckite is a manganoan melanterite.

manganous, *a.* **b.** Delete 'with its lower valency' and substitute 'with a valency of two'.

mange, *sb.*¹ Add: **3. mange mite,** a parasitic mite of the family Sarcoptidæ, causing mange in various mammals.

1873 A. S. Packard *Our Common Insects* xi. 125 (*caption*) Mange mite. **1911** *Encycl. Brit.* XXVIII. 13/1 The dermatozoa..are lice, fleas, ticks, acari or mange mites. **1950** *N.Z. Jrnl. Agric.* Jan. 68/1 Two types of mange mites affect pigs. **1962** Metcalf & Flint *Destructive & Useful Insects* (ed. 4) xx. 973 When hogs are scratching and rubbing vigorously and their hair is standing erect,.. it is probable that the animals are infested with mange mites.

mangeao (mæŋge,ɑu·, mæŋgī·o). Also 9 **mangi, mangiao,** 9—**mangeo.** [Maori.] A New Zealand tree, *Litsea calicaris*, of the family Lauraceæ, with tough, light brown wood.

1848 R. Taylor *Leaf from Nat. Hist. N.Z.* 20/2 *Mangiao*, a tree; the ash of this country. **1867** J. D. Hooker *Handbk. N.Z. Flora* II. 766/1 Mangeao, *Tetranthera calicaris*. **1873** in E. E. Morris *Austral Eng.* (1898) 282/1 Mangi—remarkably tough and compact, used for ship-blocks and similar purposes. **1882** W. D. Hay *Brighter Britain!* II. 187 The perfumy mangiao. **1960** [see *five-finger 1 d]. **1963** Poole & Adams *Trees & Shrubs N.Z.* 42 L[itsea] *calicaris*... Mangeao. Tree reaching 15m. Bark smooth, dark greyish brown... L[owland] Forest. North of North Island to lat. 38°.

manger (mē̆i·ndʒəɹ), *v.*² *rare.* [f. Manger *sb.*¹] *trans.* To fasten (an animal) to a manger.

1905 W. Holman Hunt *Pre-Raphaelitism* II. 72 An old ram mangered by a halter.

mange-tout (manʒtū). [Fr., lit. 'eat-all'.] In full, *mange-tout pea*. A variety of pea producing pods which are eaten with the seeds they contain; = *sugar-pea* (Sugar *sb.* 5 c).

1928 *Daily Express* 18 July 5/2 It [*sc.* a variety of pea] is largely grown in France, where it goes by the name of 'mange-tout'..and is eaten, shell and all. **1951** *Good Housek. Home Encycl.* 591/1 The seeds [*sc.* peas] are contained in a green pod, which is not usually eaten (except in the case of the sugar or mangetout variety). **1966** *Sunday Times* (Colour Suppl.) 27 Nov. 50/3 A bowl of *mangetout* peas. **1970** *Harrod's Summer Food News* 5/1 Whether you need..a pound of potatoes, or..mange tout peas, the quality is of the best at Harrods. **1972** *Country Life* 4 May 1121/3 From the French we are learning to enjoy the *Mangetout* or sugar pea. **1974** *Observer* (Colour Suppl.) 15 Sept. 12/3 Asked with impious briskness precisely when to sow mange-touts, divide artichokes or prune plums.

mangi, var. *mangeao.

mangle, *sb.*³ Add: **b. mangle-board** [Da. *manglebræt*], a board with which linen and cotton may be pressed and smoothed.

1892 E. Rowe *Hints on Chip-Carving* iii. 47 The border ..may be seen on a mangle-board from Jutland, dated 1708. **1928** *Daily Express* 22 June 12/6 The exhibits include various examples from Denmark, Sweden, Norway and Holland... Dates on the mangleboards go back as far as 1590.

c. *fig.* A bicycle. *Austral. slang.*

1941 Baker *Dict. Austral. Slang* 45 *Mangle*, a bicycle. **1965** G. McInnes *Road to Gundagai* viii. 122 'Where's the grid?' 'My bike!' 'Yeah, the old mangle.'

mangosteen. Add: **4.** A name used in Barbados for the jujube, *Zizyphus mauritiana*.

1750 G. Hughes *Nat. Hist. Barbados* v. 134 The Dunk-tree, or Mangustine. This is a middle-sized Tree. **1859** Bartlett *Dict. Amer.* (ed. 2) 263 *Mangosteen*, in Barbadoes this name is given to the Jujube (*Ziziphus jujube*). **1965** E. G. B. Gooding et al. *Flora Barbados* 272 (*heading*) *Zizyphus mauritiana*..*Z. jujuba* (L.).. Mangustine.. Shrub or small tree, 2–4·5 m., young stems tomentose.

mangy, *a.* **3.** (Later examples.)

1896 J. K. Snowden *Web of Old Weaver* x. 127, I cannot see that it much benefits any man to tell him all these mangy quaverings. **1922** Joyce *Ulysses* 735 We have to be thankful for our mangy cup of tea. *Ibid.*, The old mangy parcel he sent at Xmas. **1930** R. Campbell *Poems* 10 The poet wags his mangy stump of rhyme. **1942** Berrey & Van den Bark *Amer. Thes. Slang* § 30/4 Poor; mean; contemptible..*mangy*.

Manhattan (mænhæ·tăn). Also **manhattan.** [The name of the island on which the older part of New York is situated.] Used *ellipt.* for **Manhattan cocktail,** a cocktail made of vermouth and whisky with a dash of bitters.

1890 B. Hall *Turnover Club* 16 This order was as follows ..the Actor, 'A Manhattan cocktail'. **1895** C. D. Warner *Golden House* v. 43 He and old Fairfax sipped their five-o'clock 'Manhattan'. **1906** *Mrs. Beeton's Bk. Househ. Managem.* xlix. 1511 *Manhattan.* Ingredients: ½ a wineglassful of vermuth, ½ a wineglassful of whisky, 30 drops of green syrup, 10 drops of Angostura bitters, 6 drops of Curaçoa, [etc.]. **1909** [see *Bronx 1]. **1937** M. Sharp *Nutmeg Tree* viii. 84 If I could have one [a cocktail], I'd have a Manhattan. **1955** W. Gaddis *Recognitions* II. iv. 474 The dumpy woman was drinking a manhattan. **1963** C. D. Simak *They walked like Men* xii. 70 The bartender set my drink before me and began fixing her Manhattan. **1972** D. Bloodworth *Any Number can Play* x. 80 Larry returned with soda-water for Max and a manhattan for himself.

Manha:ttane·se, *sb.* and *a.* [f. prec. + -ese.] **A.** *sb.* A native or inhabitant of New York or Manhattan Island; usu. *collect.*; also, the dialect of New York. **B.** *adj.* Of or pertaining to New York or Manhattan Island.

1828 J. F. Cooper *Notions Amer.* I. 200 The New Yorkers (how much better is the word Manhattanese!) cherish the clumsy inconvenient entrances. **1842** —— *Wing-and-Wing* I. 11 Hundreds collected on the spot, which, in Manhattanese parlance, would probably have been called a battery. *Ibid.* 193 This gentleman was an American, and a native Manhattanese; his near relatives, of the same name, still residing in New York. **1844** *Knickerbocker* XXIII. 586 The sound of India-crackers and the pleasant smell of lobsters is already perceptible to the senses of the awakening Manhattanese. **1875** W. Whitman in *Gentl. Mag.* Dec. 706, I was Manhattanese, friendly, and proud. **1904** *Forum* (N.Y.) Jan.–Mar. 410 'Her Own Way'..brings together a number of highly piquant Manhattanese types of to-day, sketched with captivating drollery. **1908** S. Ford *Side-Stepping with Shorty* vi. 91 He drops the imitation society talk that he likes to spout, and switches to straight Manhattanese. **1909** *Nation* (N.Y.) 9 Sept. 238/3 Perhaps the most amusing thing in the book is an interpolated story based on a difference of opinion between New Englanders and Manhattanese on the subject of doughnuts and crullers. **1942** Berrey & Van den Bark *Amer. Thes. Slang* § 181/5 *Manhattanese*, language peculiar to New Yorkers.

manhood. 7. manhood suffrage (earlier examples).

1859 Disraeli in *Hansard Commons* 31 Mar. 1245 Why, Sir, I have no apprehension myself that if you had manhood suffrage tomorrow the honest, brave, and good-natured people of England would resort to pillage, incendiarism, and massacre. **1864** *Q. Rev.* CXVI. 262 If Mr Gladstone is generally supposed to have taken up the battle-cry of manhood suffrage, he has only himself to thank for it. **1867** John 1st Earl Russell *Let.* 27 Mar. in B. & P. Russell *Amberley Papers* (1937) II. 24 Dizzy must know that, & I believe means 'Manhood suffrage'.

‖ **mani²** (mā·ni). [Tibetan *maṇi*, f. Skr. *maṇi* precious stone (as in the jewel-lotus prayer *om maṇi padme hum* 'Oh the lotus-jewel, Amen').] In full, *mani wall*. A Tibetan 'prayer wall', covered with stones piously inscribed (see also quots.)

1863 E. Schlagintweit *Buddhism in Tibet* xiii. 196 Mani, originally a Sanskrit word meaning 'a precious stone',..is used to designate walls of about six feet in height and four to eight feet in breadth. **1882** *Encycl. Brit.* XIV. 197/2 It [*sc.* the palace at Lé] is surrounded by poplar plantations, with manis and ch'hordtens beyond. *Ibid.*, 'Mani', a long stone wall, several feet wide, running along the roadside, covered with loose stones deposited by the passers-by, inscribed with the prayer or ejaculation, 'Om mani padme hom'. **1925** B. Beetham in E. F. Norton *Fight for Everest, 1924* I. viii. 171 The mani-walls (prayer-stones) had once been of an unusually imposing nature. **1952** H. W. Tilman *Nepal Himalaya* iv. 39 The longest mani wall I have ever seen—nearly three hundred yards of it... On each side are flat stones with carved Buddhas or religious texts for the benefit of passers-by. **1953** Tsung-Lien Shen & Shen-Chi Liu *Tibet* v. 118 The approach to a village or monastery is marked by a *Mani* wall or by a pagoda. **1959** *Times* 23 May 7/6 The people here are Buddhist and there is abundant evidence of their faith—*mani* (or prayer) walls.

mania. 1. Add to def.: Now a more general term for a condition of over-excitement and restless activity, usually denominated according to its severity (see quot. 1971). Cf. **hypermania*, *hypomania.

1925 J. Riviere et al. tr. *Freud's Coll. Papers* IV. 164 The most remarkable peculiarity of melancholia, and one most in need of explanation, is the tendency it displays to turn into mania accompanied by a completely opposite symptomatology. **1969** Ullmann & Krasner *Psychol. Approach Abnormal Behavior* II. xxi. 417/1 A person may gradually progress from hypomania to delirious mania over time, or the delirious mania may, in rare instances, be emitted without a 'warm-up' period. **1971** T. Roberts *Handbk. Psychiatric Nurses* II. 62 *Mania.* There are four main types:—Hypomania... Acute mania, producing a wild frenzied aggressive attack of excitement and over activity. Chronic mania sometimes called Scott's mania. Hypermania: delirious or Bell's mania. **1973** *Sci. Amer.* Sept. 117/3 Mania is manifested by psychic elation, increased motor activity, rapid speech and the quick flight of ideas. The stigmata of depression are melancholia, the slowing of thought, unusual thought content (for example overwhelming guilt over imagined transgressions and delusions of rotting away), motor retardation, sleep disturbances and preoccupation with bodily complaints.

manic (mæ·nik), *a.* [f. Mania: see -ic.] Of, pertaining to, affected with, or resembling mania. Also as *sb.*, a person affected with mania.

1902 *Buck's Handbk. Med. Sci.* (rev. ed.) V. 124/1 She showed a typical picture of a manic excitement with great exhilaration, flight of ideas, and distractability. **1921** R. M. Barclay *Kraepelin's Manic-Depressive Insanity & Paranoia* 54 The conspicuous contrasts between manic and depressive attacks. **1922** R. S. Woodworth *Psychol.* xi. 259 In the excited insane condition known as 'mania' or the 'manic state', the patient is excessively distractible. **1957** P. Lafitte *Person in Psychol.* vi. 75 The manic hopelessly flees the devil wherever he goes. **1957** *Observer* 6 Oct. 18/7 That manic transatlantic zing which is part of the fascination of the Marx Brothers. **1964** *New Statesman* 6 Mar. 367/1 Entirely faithful as expositor, Dr. Segal brings to bear her own contributions, particularly in regard to what she calls manic reparation.

2. *Comb.* **manic depression,** the condition of manic-depressive illness; **manic-depressive** *a.*, characterized by or affected with alternating periods of elation and mental depression; also *fig.*; hence as *sb.*, a person so affected.

1958 M. Argyle *Relig. Behaviour* ix. 108 The upper and middle classes have higher rates for manic depression and neurosis. **1964** —— *Psychol. & Social Probl.* vi. 78 Schizophrenia is thought by some to be inherited by two or more recessive genes, manic depression via two or more dominant ones. **1973** *Nature* 16 Feb. 480/1 A disproportionate number of patients with schizophrenia or manic-depression were born in the early months of the year. **1902** A. R. Defendorf *Clin. Psychiatry* 282 Manic-Depressive Insanity. This term is applied to that mental disorder which recurs..throughout the life of the individual and in which a defective hereditary endowment seems to be the ..prominent etiological factor. **1931** E. Wilson *Axel's Castle* iii. 92 The modern psychologists would probably diagnose him as introverted, narcissistic and manic depressive. **1934** P. Bottome *Private Worlds* xxi. 206 A gland experiment which would help them to get a certain physical reaction from manic depressives. **1943** J. S. Huxley *Evolutionary Ethics* ii. 13 Complete alternation of conflicting moods is exemplified in manic-depressives. **1959** R. Graves *Coll. Poems* 315 Have seen two parallel red-ink lines drawn Under their manic-depressive bank accounts. **1961** J. Heller *Catch-22* (1962) xxvii. 298 Greed depresses you. Crime depresses you. Corruption depresses you. You know, it wouldn't surprise me if you're a manic-depressive! **1961** L. Mumford *City in Hist.* xv. 520 It was the manic-depressive rhythm of the market, with its spurts and stoppages, that made the large urban center so important to industry. **1970** *Jrnl. Gen. Psychol.* LXXXIII. 76 A study of the neurotic tendencies shown in dementia praecox and manic depressive insanity. **1973** *Sci. Amer.* Sept. 117/3 Manic-depressive psychosis is marked by severe disturbances in mood that are self-limited in time but are recurrent and frequently cyclic.

Manichæistic (mæ:niki,i·stik), *a.* Also **Manicheistic.** [f. as Manichæist: see -istic.] Of the nature of Manichæism.

1924 O. Lodge *Making of Man* ii. 49 A Manichæistic conception of existence can never have been really satisfying. **1932** *Times Lit. Suppl.* 3 Mar. 148/3 With something of Manicheistic fury Miss Leader tracks down the sense of physical abjection and inferiority.

Manichæism, Manicheism. Add: Also *transf.*

1894 W. Archer *Theatr. 'World' 1893* 224 The time is past for the elementary Manicheism on which *The Tempter* is based. **1959** B. & R. North tr. *Duverger's Political Parties* (ed. 2) p. xv, The elementary Marxist opposition of middle class to working class will also often be used... There are many more shades of social stratification than this rough manicheism suggests.

‖ **manicou** (maniku). [Fr.] An opossum of the genus *Marmosa*, found in Central and South America.

1953 P. L. Fermor *Violins of St.-Jacques* 28 The most remarkable [pets] were a family of manicous, the mother of which carried her dozen young about by twisting her tail parallel to her spine in order that they might loop their own round it and secure their positions on the parent back. **1968** E. Lovelace *Schoolmaster* xi. 171, I will go late and headlight in the forest for manicou.

manicure, *sb.* Add: **2.** attrib. *manicure case, department, girl, parlour, scissors, set, stick.*

1895 *Montgomery Ward Catal.* 255/3 Combination Toilet and Manicure Case..contains complete set of 12 white Florence toilet and manicure fittings. **1967** 'K. O'Hara' *Unknown Man* ii. 10 There was a chip out of one nail, so that she would need the manicure case. **1887** *Mod. Society* 23 July 749/2 A young lady employed in the manicure department of one of our most popular establishments. **1922** S. Lewis *Babbitt* xviii. 219 Young women who had recently been manicure girls. **1946** *Vogue* June 35/1 The manicure-girls stare contemptuously. **1912** *Collier's* 9 Nov. 11/3 Jack Zelig was a daily patron of a manicure parlor. **1893** *T. Eaton & Co. Catal.* Spring & Summer 35/3 Scissors, manicure, 25 c. **1913** J. Webster *Daddy-Long-Legs* 194 Sawing off picture wire with manicure scissors. **1952** M. Norton *Borrowers* ii. 17 Homily.. unhooked the blade and handle of half a pair of manicure scissors from a nail on the wall. **1973** H. Nielsen *Severed Key* i. 14 Keith..located the cosmetic case... 'Women usually carry a manicure scissors in these things,' he said. **1897** *Sears, Roebuck Catal.* 435/2 Manicure and Buttonhook set, fancy, Solid Sterling Silver handles. **1945** E. Bowen *Demon Lover* 72 A manicure-set in a purple box; all the objects..lay in grooves on white velvet. **1962** *Guardian* 5 Dec. 6/4 A really complete manicure set..for 22s 6d. **1909** E. Banks *Mystery F. Farrington* 162 Orange-wood manicure sticks.

manicure, *v.*, **manicured,** *ppl. a.* Add: Also *transf.* and *fig.* **manicuring** *vbl. sb.* (earlier example).

1892 G. & W. Grossmith *Diary of Nobody* xviii. 230 I'm going in for manicuring. It's all the fashion now. **1922** M. Arlen *Piracy* III. vi. 192 Even her soul was manicured. c **1926** 'Mixer' *Transport Workers' Song Bk.* 97 You would make a good oil-painting if your face was manicured. **1936** F. Clune *Roaming round Darling* viii. 68 The only prisoner in sight was a 'lifer' manicuring the garden. **1957** J. Braine *Room at Top* i. 10 Big houses with drives and orchards and manicured hedges. **1964** M. McLuhan *Understanding Media* II. xxii. 217 This shift from the mobile open road to the manicured roots of suburbia may signify a real change in American orientation. **1972** *Times* (Jamaica Suppl.) 7 Aug. p. vi/3 Manicured and well-watered fairways of tourist golf courses. **1973** A. MacVicar *Painted Doll Affair* xiv. 164 A tractor-powered gang-mower was manicuring the last of the fairways. **1975** *New Yorker* 19 May 115 (Advt.), Your day may be occupied..testing the exquisitely manicured golf course.

‖ **maniéré** (manyere), *a.* [Fr.] Affected or characterized by mannerism; = Mannered *a.* 3.

1743 H. Walpole *Let.* 14 Aug. (1857) I. 263 The Sasso Ferrati you sent me..is not so *maniere* [*maniéré*, in *Corr.* (1955) XVIII] as the Dominichin. **1764** S. Foote *Patron* I. 2 But the figures are all finish'd alike. A maniere, a tiresome sameness throughout. **1813** A. Romilly in S. H. Romilly *Romilly-Edgeworth Lett.* (1936) 64 Her little maniéré airs are with difficulty supported for a few hours. **1839** Thackeray in *Fraser's Mag.* June 749/1 You will see here a large drawing..which will shew at once how clever that young artist is, and how weak and maniéré. **1908** *Westm. Gaz.* 14 Nov. 19/1 The suit for outdoor wear delights in a daring *maniéré* simplicity.

‖ **manière criblée** (manyɛ̄r kr*i*ble). [Fr., f. *manière* Manner *sb.*¹ + *criblé* Cribbled *ppl. a.*] The cribbled style of engraving. Cf. Cribbled *ppl. a.* and *criblé *sb.* (and *a.*)

1903 *New Internat. Encycl.* XI. 834/2 Manière criblée. Probably the oldest process of engraving upon metal for the purpose of printing. It derives its name from the white dots with which the dark ground of the print is covered, resembling the holes of a sieve. **1960** H. Hayward *Antique Coll.* 174/1 *Manière criblée*, the name is derived from a group of dots made on the plate with a punch to break up otherwise black areas of background. Prints of this type were made in the late 15th and early 16th cent., especially in Florence. **1965** Zigrosser & Gaehde *Guide to Collecting Orig. Prints* iv. 50 Those early prints known as dotted prints (*manière criblée, Schrotblatt*), made with various goldsmith's punches on metal.

manifest, *sb.* **2.** Delete † *Obs.* and add later examples.

1915 A. Huxley *Let.* Dec. (1969) 87 Meanwhile all is forgiven and forgotten if you subscribe to the *Palatine*.. vide multicoloured manifest thus conceived, which you had better distribute. **1922** *Glasgow Herald* 12 Apr. 11 The annexation itself had been proclaimed by a personal manifest of the Emperor King.

3. Also *transf.* Also, a similar list of freight or passengers carried by a train or aeroplane. Also, a fast freight train (chiefly *U.S.*).

1869 'Mark Twain' in *Buffalo Express* 21 Aug. 1/3 The doctor is not done taking inventory. He will make out my manifest this evening. **1872** in I. M. Tarbell *Hist. Standard Oil Co.* (1904) I. 286 The party..covenants and agrees..to make manifests or way-bills of all petroleum or its products, transported over any portion of the railroads. **1929** *Amer. Speech* IV. 342 *Manifest*, a fast merchandise freight train. **1931** G. Irwin *Amer. Tramp & Underworld Slang* 127 *Manifest*, a fast freight train, from the 'manifest' of the goods carried. **1934** *Amer. Ballads & Folk Songs* 24 The manifest freight Pulled out on the stem behind the mail. **1936** Mencken *Amer. Lang.* (ed. 4) 582 A fast freight is a *manifest* or *red-ball*. **1956** W. A. Heflin *U.S. Air Force Dict.* 315/1 *Manifest*, a document that lists in detail the passengers or other items carried in one aircraft for a specific destination. **1959** R. Collier *City that wouldn't Die* ix. 143 Watching Schied initial the bomb manifest and hand it down to the chief armourer. **1969** *Jane's Freight Containers 1968–69* 128/1 Daily containers move in expedited Piggy-back trains and transcontinental manifest trains. **1971** *Sunday Australian* 8 Aug. 11/2 He came to my office with..samples of passenger manifests.

manifest, *a.* Add: **1. b.** *manifest destiny* (often written with capital initials), *Polit.* 'The doctrine of the inevitability of Anglo-Saxon supremacy. A phrase used by those who believed it was the destiny of the United States or of the Anglo-Saxon race to govern the entire Western Hemisphere' (*D.A.E.*). Also *transf. U.S.*

1845 J. O'Sullivan in *U.S. Mag. & Democratic Rev.* July & Aug. 5 Our manifest destiny to overspread the continent allotted by Providence for the free development of our yearly multiplying millions. **1856** *Spirit of Times* 13 Dec. 235/2 He was a 'manifest destiny' man. **1858** *Economist* 6 Feb. 139/2 What reasons can be shown for the manifest destiny to whip Africans which do not apply with equal force to prove the manifest destiny to subdue half-caste Spaniards into conformity with their will? **1867** J. R. Lowell *Biglow Papers* 2nd Ser. p. vii, The incarnation of 'Manifest Destiny', in other words, of national recklessness as to right or wrong. **1927** J. W. Pratt in *Amer. Hist. Rev.* XXXII. 795 One can hardly read a work on the history of the United States in the two decades before the Civil War without meeting the phrase 'manifest destiny', widely used as a convenient statement of the philosophy of territorial expansion in that period. **1959** *Listener* 18 June 1048/2 The long and stern struggle by which Canada had maintained its separateness against American 'Manifest Destiny'.

c. *manifest function*. Sociol. (See quots.)

1949 R. K. Merton *Social Theory* I. i. 51 *Manifest functions* are those objective consequences contributing to the adjustment or adaptation of the system which are intended and recognised by participants in the system; *Latent functions*, correlatively, being those which are neither intended nor recognised. **1957** M. Banton *W. Afr. City* ix. 168, I shall describe..how the companies fulfil their explicit functions of providing mutual aid and entertainment—what R. K. Merton has termed the manifest functions.

manifold. Add: **C.** *sb.*¹ **4. b.** Short for *manifold-paper*.

1897 B. Stoker *Dracula* xvii. 229, I began to typewrite from the beginning of the seventh cylinder. I used manifold, and so took three copies of the diary. **1926** *Paper Terminol.* (Spalding & Hodge) ii. 16 *Manifold*, slightly waxed tissue or other thin interleaving paper made for employment with carbon paper. Also an extremely thin typewriting made for the multiplication to as many as ten or twelve carbon copies of typewritten letters. **1954** *Ibid.* 38 *Manifold*, papers similar in character to Bank, although thinner. They range in substance from 16½ × 21 in. 4½ lb. 500's to 8 lb. 500's and are used when a large number of carbon copies is required. **1967** Karch & Buber *Offset Processes* xi. 485 *Manifold*, thin, strong, [for] duplicate copies, sales books, etc.

5. Substitute for '(See quot.)': Any pipe that splits into a number of branches; *spec.* the one running from the carburettor of an internal-combustion engine to the cylinders, and the one from the silencer to the cylinders. (Earlier and later examples.)

a **1884** Knight *Dict. Mech.* Suppl. 579/2 *Manifold*, the chambers with no. ›les into and from which the pipes of a radiator lead. **1919** [see *exhaust-manifold* s.v. *exhaust *sb.* 3]. **1948** *'Motor' Manual* (ed. 33) v. 99 The induction manifold of a modern engine generally is heated by the exhaust. This..can be easily and neatly arranged in most engines because the exhaust manifold usually is close to the inlet manifold. **1961** *Economist* 30 Dec. 1306/1 Ford provides only the carburettor, inlet manifold and camshaft for the Classic [motor car]. **1971** *Sci. Amer.* Sept. 222/3 All gases are admitted through needle valves to a manifold that connects to the laser.

MANIFOLDER

manifolder (mæ·nifōᵘldəɹ). [f. MANIFOLD *v.* + -ER¹.] A machine for multiplying copies of a document, etc., or a person using this. Also used of a typeface (see quot. *c* 1961).
1903 in *Funk's Stand. Dict.* Suppl. 1911 WEBSTER (citing G. H. Putnam), He seems to have added to his employment that of a manifolder and seller of manuscripts. *c*1961 *Imperial Type Faces* (Imperial Typewriter Co.), Pica Gothic type is an exceptionally good manifolder.

manifolding, *vbl. sb.* (s.v. MANIFOLD *v.*). Add: Also *concr.*, = MANIFOLD *sb.*¹ 5 (in Dict. and Suppl.).
1938 *Times* 9 Aug. 8/7 To adjust the tappets it would be best to lift the manifolding. 1963 BIRD & HUTTON-STOTT *Veteran Motor Car* 52 Internal manifolding and clean-cut architectural appearance were all in marked contrast to most contemporary engines.

manilla², manila. Add: **2.** Also, = *manilla paper* (now used of various papers of a light yellow-brown colour: see quots.).
1926 *Paper Terminol.* (Spalding & Hodge) ii. 16 *Manilla*, a superfine tough quality of wrapping and label paper made from Manilla hemp. The term is now loosely applied to cheap imitations made from wood pulp. 1954 *Ibid.* 38 *Manil(l)a*, a coloured paper..not necessarily containing Manila fibre. 1959 *Gloss. Packaging Terms* (*B.S.I.*) 67 *Manilla*, an imitation manila made in a wide variety of qualities. The term is now generally used to indicate the characteristic colour.
b. The light yellow-brown colour of manilla paper.
1934 in WEBSTER. 1954 P. HIGHSMITH *Blunderer* (1956) viii. 71 He put on old manila trousers.

manipulability (măni:piŭlăbi·lĭti). [f. MANIPULAB(LE *a.*: see *-BILITY.] The quality of being manipulable.
1947 *Partisan Rev.* Sept.–Oct. 473 Every idea is judged in terms of its political manipulability. 1957 B. F. SKINNER *Verbal Behavior* v. 124 The increasing separability and manipulability of response elements in a minimal unit repertoire is a step toward ideal conditions. 1957 L. EISELEY *Immense Journey* (1958) 6 As I tapped and chiseled there in the foundations of the world, I had ample time to consider the cunning manipulability of the human fingers. 1967 *Punch* 11 Oct. 546/2 Doctors are no more gullible than the rest of us, but they are a captive audience, and the extent of their manipulability can be observed on a firm's local sales charts. 1971 J. B. CARROLL et al. *Word Frequency Bk.* p. vi, Only the Kučera-Francis work is of sufficient size and manipulability to yield a substantial word list and citation base for lexicography. 1974 *Time* 7 Jan. 60/3 His recommendations that society change its 'system of production, ownership and consumption' depend on faith in man's manipulability and desire to change.

manipulable, *a.* Delete *rare* and add later examples.
1957 B. F. SKINNER in Saporta & Bastian *Psycholinguistics* (1961) 233/1 Suffixes such as *-ness* or *-hood* are usually readily manipulable as separate elements in composing new terms. 1963 *Punch* 2 Jan. 21/1 A swiftly manipulable key-board. 1965 J. FEINBERG in M. Black *Philos. in Amer.* 154 It would be an oversimplification.. to identify 'the cause' of an infelicitous condition with *any* manipulable necessary condition. 1971 *Nature* 12 Feb. 445/3 As well as being easy to learn and formally manipulable, [computer languages]..should be thought-provoking. 1971 *Human World* Nov. 14 Is sovereignty something that can be arranged like this? Is it so infinitely manipulable?

manipulatable (măni·piŭlēⁱ:tăb'l), *a.* [f. MANIPULAT(E *v.* + -ABLE.] = MANIPULABLE *a.* (in Dict. and Suppl.).
1934 in WEBSTER. 1964 E. A. NIDA *Toward Sci. Transl.* ix. 204 One must expect to find numerous and subtle distinctions, which cannot easily be reduced to readily manipulatable rules. 1970 *Computers & Humanities* IV. 327 These three considerations are separable and individually manipulatable.

manipulate, *v.* Add: **3. b.** *Finance.* To cause (stocks) to rise or fall by affecting the market in other ways than those arising out of ordinary business; to influence (the market) in such ways.
1870 J. K. MEDBERY *Men & Mysteries Wall St.* 188 The stock..was most admirably manipulated, until it finally touched 152. 1903 S. S. PRATT *Work of Wall St.* 147 A market is rigged when it is manipulated.
4. *trans.* To stimulate (the genitalia); also *refl.*, to masturbate.
1949 O. SCHWARZ *Psychology of Sex* ii. 31 The large majority of these children simply manipulate their genitals in a purely playful manner. 1953 A. C. KINSEY et al. *Sexual Behavior Human Female* v. 161 The female may stimulate them [*sc.* the breasts] with her hand..while she simultaneously manipulates her genitalia. 1965 MASTERS & JOHNSON in J. Money *Sex Research* iv. 107 It is a rare woman who directly manipulates the clitoris, or, if she does, maintains this type of stimulative activity for any significant length of time. 1971 'V. X. SCOTT' *Surrogate Wife* 134 At one point he began manipulating himself.
manipulated *ppl. a.* (example).
1903 S. S. PRATT *Work of Wall St.* xi. 147 A deal is the operation resulting from a secret combination or agreement among Wall Street men to effect a certain purpose, generally of a manipulated character in the market.

manipulation. **4.** Add examples of use in finance.
1888 *Nation* (N.Y.) 9 Aug. 107/2 Manipulation signifies a common understanding and design on the part of a clique of operators to raise or depress values in order to get other people's money. 1908 *Westm. Gaz.* 26 Aug. 2/2 The opportunity for market manipulation is obvious.

manipulative, *a.* (Later examples.)
1909 *Westm. Gaz.* 8 Sept. 11/4 The powerful manipulative interests are watching events closely. 1947 M. M. LEWIS *Lang. in Society* i. 24 The manipulative and the declarative are the twin incentives by which the development of language is fostered in the child, and remain the essential functions of language in society. 1966 *New Statesman* 22 Apr. 581/2 The cause was won by the politicians' desire to appear manipulative about almost any issue, regardless of its suitability for manipulation. 1968 R. KYLE *Love Lab.* ix. 125 Lillian had just completed a manipulative session and was wearing only a wrapper. 1971 *Nature* 30 Apr. 578/2 For they are also clearly preadaptive to efficient use of the hands as manipulative organs, in such activities as food gathering and transport. 1972 *Maclean's Mag.* Mar. 61/1 Spicer is strong, attractive and manipulative. 'He's good at persuading people to do things they might not want to do.'
Hence **mani·pulativeness.**
1949 M. MEAD *Male & Female* ii. 26 Greater awareness does not always mean greater manipulativeness. 1975 *N.Y. Times* 4 Mar. 31 You pity their failings—their determination to 'advance' their children as if life were a relay race,..their manipulativeness and insularity.

manipulator. Add: **1. c.** One who controls the price of stocks by specially contrived methods.
1888 *Economist* 17 Nov. 3/3 If the people are apathetic it is utterly useless for the manipulators to undertake to push prices up. 1903 S. S. PRATT *Work of Wall St.* 146 A professional may or may not be a manipulator, but a manipulator is always a professional. *Ibid.* 256 By false tips and matched orders or wash sales the manipulators endeavored to establish fictitious quotations for their stocks. 1904 *N.Y. Tribune* 15 May 4 Manipulators desperately endeavoring to bring back recessions which will permit them to 'get even'.
2. f. A device for handling radioactive material, operated by remote control from behind a protective shield.
1952 *Nucleonics* Nov. 41/1 They have been named master-slave manipulators because all the seven degrees of freedom of the tongs are slaved to the single master handle. 1955 *Reactor Handbk.: Engin.* (U.S. Atomic Energy Comm.) 859 A general-purpose manipulator is considered to be a remotely-controlled mechanical arm capable of gripping diverse objects... All movements of the manipulator arm are controlled by a human operator. .. The manipulator may have a lower or higher load capacity than the human arm. 1958 *New Scientist* 24 Apr. 21/1 A manipulator believed to be the only one manufactured in Europe that is operated by electrical remote control will be displayed at next week's Hanover Fair... The power manipulator is equipped with interchangeable grasping units..and can lift and position objects weighing up to 750 lbs. 1962 *Newnes Conc. Encycl. Nucl. Energy* 488/2 These manipulators fall into three classes: 1. Handling tongs and sphere units... 2. Master-slave manipulators... 3. Rectilinear power manipulators. 1969 *IEEE Trans. Nucl. Sci.* XVI. 594/2 The geometry of the Brookhaven AGS and of other synchrotrons requires a larger degree of remoteness than is possible with mechanically-connected master-slave manipulators.

Manipuri (mænipūᵊ·ri), *a.* and *sb.* [Native name.] **A.** *adj.* Of or pertaining to Manipur, a state in the region of Assam in north-east India; *spec.* used of a style of dancing (also *ellipt.*). **B.** *sb.* The people of Manipur; a member of this people; also, the Tibeto-Burman language of the Manipuri.
1906 E. A. GAIT *Hist. Assam* viii. 195 The Manipuri Raja was mindful of the services rendered him a few years previously. 1911 [see *KHASI]. 1918 W. PETTIGREW *Tāngkhul Nāga Gram. & Dict.* 4 As in Manipuri, there are two noticeable changes in intonation in TN, a high and low tone to a great number of words which to the ignorant sound the same. 1927 *Blackw. Mag.* June 817/1 The Manipuris..became such strict Hindus as even to 'out-Brahman' the Brahmans. 1948 D. DIRINGER *Alphabet* II. vi. 365 The Oriya, Maithili and Early Manipuri characters seem also to be somewhat connected with the Bengali script. 1957 G. B. L. WILSON *Dict. Ballet* 178 *Manipuri*, in Hindu dancing, the style of north-east India. The dances are in the form of dance-drama (supported by dialogues and songs) and many dancers participate. It is mainly lyrical and lighter than the other Indian styles. 1968 *Encycl. Brit.* XIV. 790/2 About 60% of the population [of Manipur] speak Manipuri. *Ibid.* XV. 118/2 *Meithei*, also known as Manipuri, the dominant population of the valley of Manipur. 1969 *Femina* (Bombay) 26 Dec. 53/1 Kishori also left an imprint as a noted Manipuri dancer in the early dance dramas produced by the Indian National Theatre. Her rhythmic grace is still remembered. 1971 *Ceylon Observer* (Mag. Ed.) 19 Sept. 4/8 (Advt.), New silver motive white Manipuri saree.

‖ **ma nishtana** (mā niʃtā·nā). *Judaism.* Also **mah nishtan(n)ah.** [Heb., 'Why (is this night) different (from all other nights)?'] The opening words of the four questions in the Passover Haggadah, traditionally asked by the youngest member of a Jewish household on Seder Night; hence used to designate this part of the Passover celebrations.
1902 *Jewish Chron.* 18 Apr. 20/1 After the sanctification (*Kiddush*) with which our Haggadah begins..the four questions *Ma nishtana* are read. *Ibid.* 25 Apr. 20/2 In olden times the children were provided with small nuts and chestnuts to give more zest to the question, 'Ma nishtana'—'why is this night different from all other nights?' 1904 *Jewish Encycl.* VI. 141/2 Its initial words, 'Mah Nishtannah', are used as the name of the Haggadah, as in the question: 'What has Korah..to do in the Mah Nishtannah?' 1960 *Jewish Chron.* 8 Apr. 27/5 There was for me, the youngest boy, the important task of relearning the Ma Nishtana. 1972 C. RAPHAEL *Feast of Hist.* iii. 90 The spirit of *ma nishtana*—the Four Questions in the Haggadah.

manism (mēⁱ·niz'm). [f. MANES *sb. pl.* + -ISM.] The worship of the *manes* or shades of the dead; ancestor-worship. Hence **mani·stic** *a.*
1904 G. S. HALL *Adolescence* II. xii. 179 Culture developed through the four stages of manism, when each [*sc.* the moon, sun, planets, etc.] was regarded as only the sensuous object it seemed, [etc.]. 1933 J. D. UNWIN *Sexual Regulations & Human Behaviour* i. 1 Uncivilised peoples ..Manistic—these do not erect temples but they pay some kind of post-funeral attention to their dead. 1937 A. HUXLEY *Ends & Means* xv. 311 All human societies are in one or another of four cultural conditions: zoistic, manistic, deistic, rationalistic. 1950 *Funk's Stand. Dict. Folklore* II. 673/1 Manism, the general term for the worship of the spirits of the dead, and, more specifically, for ancestor worship, is widespread in the world. 1956 E. E. EVANS-PRITCHARD *Nuer Relig.* xiii. 311 Many such origins have been propounded: magic, fetishism, manism, animism.

manitoka (mænitōᵘ·kă). *S. Afr.* Also **manotoka.** [Perhaps a name fabricated by the botanist P. Macowan.] The South African name of a large shrub, *Myoporum insulare*, of the family Myoporaceæ, native to Australia and bearing small white flowers followed by edible blue berries.
1906 F. BLERSCH *Handbk. Agric. S. Afr.* xiii. 267 Hedge shrubs and trees... Manotoka (*Myoporum insulare*). 1948 H. V. MORTON *In Search of S. Afr.* ii. 48 The old buildings and houses of former occupants..were standing roofless and deserted in a jungle of manitoka trees. 1956 *Cape Times* 1 Mar. 8/6 The English myrtle hedge..is far more interesting, is greener and less likely to dry out. This can also be said of the manitoka and tecoma hedges. 1973 M. PHILIP *Caravan Caravel* 19 A tarred road circled the middle of the thickly grassed park, and the caravans were standing in rows on four different levels, with hedges of fleshy, narrow-leaved manitokkas [*sic*] bushing out between the caravan sites.

mankalah, var. *MANCALA.

mankey, var. *MANKY *a.*

man-killer. Add: Also *transf.* and *fig.*
1929 F. C. BOWEN *Sea Slang* 89 *Man-killer*, a hard-working sailing ship in which accidents were frequent. Some ships earned extraordinary reputations in that way. 1957 R. CAMPBELL *Portugal* v. 86 They rechristened her [*sc.* a mare], Garota, or 'dear little girlie-wirlie', in order to ..give the impression that this man-killer was as sweet-tempered and innocent as she looked. Actually she had already killed two men. 1964 MRS. L. B. JOHNSON *White House Diary* 24 July (1970) 183 This was a man-killer of a day for Lyndon.

manky (mæ·ŋki), *a. local.* Also **mankey.** [f. MANK *a.* + -Y¹.] Bad, inferior, defective; dirty.
Possibly influenced by Fr. *manqué*.
1958 F. NORMAN *Bang to Rights* iii. 124 He would have to have all his teeth out as it seems that they were all mankey. 1971 B. W. ALDISS *Soldier Erect* 121 Have you chucked out that dirty manky beer you poisoned me with last time I came? 1973 A. GARNER *Red Shift* 14 That's your manky palate, lad. The dressing and the wine have to balance. 1973 'J. PATRICK' *Glasgow Gang Observed* 231 *Clatty*, 'stourie', dirty, 'manky'. 1974 *Jrnl. Lancs. Dial. Soc.* XXIV. 28 [Westhoughton]. *Manky*, naughty.

manless, *a.* Add: **1. b.** Of a woman: lacking the company of men, *spec.* having no husband or suitor. Hence **ma·nlessness,** the state or condition of being manless.
1924 *Public Opinion* 7 Nov. 460/2 We find girls robbed of wholesome excitements..by the loneliness and manlessness of their lives. 1942 E. PAUL *Narrow St.* xix. 161 Like so many manless women, Thérèse was extraordinarily sensitive and hostile to the masculine touch. 1962 K. ORVIS *Damned & Destroyed* xxv. 189 Fear of living out her life as a manless woman. 1970 O. NORTON *Dead on Prediction* i. 5 I'd just been going through a pretty manless period. 1972 *Guardian* 16 Feb. 13/2 The 24-year-old, man-less office worker.

manling. Delete † *Obs.* and add later examples.
1895 KIPLING *Second Jungle Bk.* 184 A Manling with a knife threw stones at my head. 1922 A. S. M. HUTCHINSON *This Freedom* 276 Her baby boy, her tiny manling.

man-made (mæ·n,mēⁱd), *a.* Also (occas.) as two words and (more freq.) **manmade.** [f. MAN *sb.*¹ + MADE *ppl. a.*] Made by man; made or devised by human effort, i.e. not existing in nature; artificial: applied esp. to fibres and fabrics manufactured from chemicals of natural or synthetic origin. Also *ellipt.* and *fig.*

a 1718, 1839 [in Dict. s.v. MAN *sb.*¹ 19 e]. 1921 W. DE LA MARE *Mem. Midget* xxxiii. 237 There was nothing man-made in Fanny; and if there are women-shaped mermaids I know what looks will be seen in their faces. 1939 G. B. SHAW *In Good King Charles's Golden Days* I. 29, I tell you that from the moment you allow this manmade monster called a Church to enter your mind your inner light is like an extinguished candle. 1948 *Sci. News* VII. 86 When..the country is small and the civilisation old, one finds that the landscape is eventually almost entirely man-made. 1955 *Times* 5 July 16/3 Man-made fibres, by which I mean those made by the viscose and acetate processes and the various synthetic fibres. *Ibid.* (Suppl.) p. ii/1 The first man-made forest of such a size to be used for large-scale paper-making will provide the raw material for the 300-acre combined pulp, newsprint and timber mill. 1960 *Farmer & Stockbreeder* 1 Mar. 62/1 Yet another Welsh valley is to be flooded. In time, another man-made lake. 1966 *McGraw-Hill Encycl. Sci. & Technol.* V. 242/1 Rayon is a man-made fibre but not a synthetic fiber. Nylon is a synthetic fiber. 1968 *Daily Tel.* 4 Nov. 2/1 Still no let-up is visible in the prodigious progress of man-made fibres... This year's man-made production at the ninth-month mark was more than 26 p.c. above last year's level at the same stage. *Ibid.*, Man-mades now have a quarter of the total world fibre output. 1969 D. C. HAGUE *Managerial Econ.* II. iv. 90 A large proportion of output in a modern economy is accounted for by..large firms. In the United Kingdom, this happens in the..manmade fibre industries. 1969 *New Yorker* 12 Apr. 85/1 In order to avoid capturing one of their own microflora or gathering any of the other manmade contaminants on the surface of the moon, the astronauts will have to take the sample for the exobiologists from underground. 1974 *Country Life* 3–10 Jan. 54/2 Coats and dresses in ultrasuede, a man-made, washable material.

manna¹. Add: **9. manna-gum,** a species of *Eucalyptus*, *E. viminalis*.
[1834 G. BENNETT *Wanderings in N.S.W.* I. xvi. 319 The elegant drooping manna-trees..were numerous.] 1884 A. NILSON *Timber Trees New South Wales* 74 *E*[*ucalyptus*] *viminalis*—Manna Gum;..Flooded Gum.—An elegant tree, attaining a height of 150 feet and a diameter of 8 feet. [1887 Manna-drooping Gums: see *GUM *sb.*² 5]. 1937 *Discovery* Dec. 364/2 The Greater Gliding 'Possum..feeds also on several other eucalypts, particularly manna-gum and long-leafed box. 1940 H. ELLIS *My Life* iv. 102 The curious variety of manna gum I knew only there [*sc.* in Australia]. 1961 *Coast to Coast 1959–60* 60 Before me was a track running between walls of manna-gum, casuarina, tea-tree in full blossom, and wedding-bush even whiter.

mannan (mæ·næn). *Chem.* [f. *MANN(OSE + *-an*, after *GLUCOSAN.] Any of a group of polysaccharides that are composed chiefly of mannose residues and occur widely in plants, esp. as reserve foods.
1895 *Jrnl. Chem. Soc.* LXVIII. II. 128 (*heading*) Mannan as a reserve material in the seeds of Diospyros Kaki. 1931 [see *FRUCTOSAN]. 1965 *New Scientist* 6 May 381/1 The cell-walls of the [coffee] bean consist of a mannan which is an insoluble polysaccharide..and protein.

manned, *ppl. a.* (s.v. MAN *v.*). Add examples relating to aviation and space travel.
1906 *Nature* 8 Nov. 35/1 (*title*) The first 'manned' flying machine. 1907 *Ibid.* 4 Apr. 538/2 During the course of the last few years very rapid strides have been made in investigating the upper air by means of manned and unmanned balloons and kites. 1936 G. HEARD *Exploring Stratosphere* 18 As it was obviously impossible to obtain sufficient data by manned balloons, small sounding-balloons were released bearing a light cage containing self-recording instruments. 1946 H. HARPER *Dawn of Space Age* 4 It is intended to develop manned and instrument-carrying rockets capable of being projected in and beyond the earth's atmosphere. 1957 *Observer* 28 July 6/3 In the present phase of manned bombers, which will last into the mid-1960s, the United States Strategic Air Command has some 1,500 B47 medium bombers capable of attacking the Soviet Union from bases around its periphery. 1960 *John o' London's* 7 Apr. 395/1 Manned entry into interplanetary space is inevitable. 1968 *Times* 16 Oct. 8/7 The probe seemed to be a trial for an imminent manned flight to the moon. 1971 *Guardian* 3 July 1/3 The indication of a technical failure rather than a human failure..should be encouraging both to the Russian and the US manned space programmes.

mannequin (mæ·nɪkin, mæ·nəkwin). [Fr.] A woman (or occas. a man) employed in the showrooms of dress-makers, costumiers, etc., to wear and show off garments. Also, a model of a human figure for the display of clothes, etc. Also *attrib.*
1902 *Pall Mall Mag.* XXVII. 119 Another salon.. ornamented with tall mirrors in which were reflected the slender elegant figures of several mannequins, most of them exceedingly pretty and all arrayed in magnificent dresses. 1919 BEERBOHM *Seven Men* 11 Then came 'Stark: *A Conte*', about a midinette who, as far as I could gather, murdered..a mannequin. 1924 [see *film super* s.v. *FILM *sb.* 7 c]. 1927 *Sunday Express* 7 Aug. 3/6 Bogus mannequin schools.., of which there are several in London, promise to train girls to become expert mannequins. 1930 *Daily Express* 6 Oct. 13/5 Autumn Mannequin Parades will be held on Tuesday and Wednesday this week. 1939 M. B. PICKEN *Lang. Fashion* 97/2 *Mannequin*.., model of human figure for display of garments, hats, furs, etc. 1940 D. MCCARTHY *Drama* 217 The *mannequin* show illustrates the 'moral' and commercial exploitation of sex interest. 1951 M. MCLUHAN *Mech. Bride* (1967) 99/1 Her mannequin past is in the way. 1960 S. BECKER tr. *Schwarz-Bart's Last of Just* III. iii. 108 The mannequin and its one unscrewed foot. 1966 J. S. COX *Illustr. Dict. Hairdressing* 95/1 *Mannequin*.., an artificial head used to display a hair style, or on which to create a hair style, or work hair for a postiche.
Hence **ma·nnequining** *vbl. sb.*, **ma·nnequinism,** the business of mannequins.
1927 *Sunday Express* 15 May 5/3 'Mannequining is a serious business now,' said the head of a mannequin school to me. 1928 *Daily Express* 2 June 4/4 More and more..distinguished women..have recently joined the ranks of teachers of mannequinism.

manner, *sb.*¹ **2. b.** Delete *Obs.* and add later examples.
1907 G. B. SHAW *How he lied to her Husband* 142 Henry places himself on guard in the manner of a well taught boxer. 1967 E. SHORT *Embroidery & Fabric Collage* iii. 60 The coverlet tells the story of Tristan, in a series of scenes showing different incidents, in the manner of a strip cartoon.

mannerism. Add: **a.** *spec.* (Freq. with capital initial.) Applied to a style of art which originated in Italy *c* 1520 and preceded the Baroque, characterized by stylistic exaggeration in figure and composition, etc.
Apart from the generalized use already illustrated in the Dict., the term has been used (*a*) of the followers of Michelangelo, Raphael, etc., as Bronzino, Pontormo, the Zuccari, etc. (mainly a 19th-c. use), (*b*) from *c* 1920 under the influence of M. Dvořák and W. Friedländer applied without pejorative associations to Italian art, in particular, from *c* 1520 to *c* 1590. Sense (*b*) is first found in English in the 1930s.
1845 A. JAMESON *Mem. Early Italian Painters* II. vii. 203 Those faults which have rendered many of his [*sc.* Parmigiano's] works unpleasing, by giving the impression of effort, and of what in art is called mannerism. 1851 tr. *Kugler's Schools of Painting in Italy* (ed. 2) II. v. ix. 469 Many of the painters in question would, fifty years earlier, have done great things; now they fell into repulsive mannerism. 1891 tr. *J. Adeline's Art Dict.* 249 Mannerism may be defined as *manner* in a bad sense. Qualities of treatment which when moderately displayed mark individuality of style, when carried to excess and too often repeated degenerate into mannerism. 1937 E. K. WATERHOUSE *Baroque Painting in Rome* 1 From 1535 until 1590 the history of painting in Rome is the history of Mannerism. 1940 A. BLUNT *Artistic Theory in Italy 1450–1600* v. 76 Those, like Dolce and Aretino..unable to follow Michelangelo as he moved on into Mannerism. 1943 *Art Bulletin* Mar. 87/2 The word mannerism..is currently used either to designate the period between the High Renaissance and the Baroque or else as a name for the anti-classical movement in sixteenth-century art. 1947 D. MAHON *Stud. Seicento Art & Theory* 59 Though the relatively recent understanding of Mannerism as a quite different artistic language from Baroque has somewhat curtailed the original scope of the latter (involving the secession of such figures as Tintoretto and El Greco) the use of the expression Baroque as a general term in its widest sense (such as Gothic), seems to have come to stay. 1961 J. D. ROSENBERG *Darkening Glass* (1963) v. 88 What we study as Mannerism he [*sc.* Ruskin] dismissed as vapidity and affectation. 1962 *Listener* 12 July 54/1 It is to move out of the serene and classical harmony of High Renaissance portraiture into the contrivance of the style that is known as Mannerism. 1965 G. SHEPHERD in P. Sidney *Apology for Poetry* 66 In the history of the pictorial arts, Lommazo and Zuccaro are spoken of as theorists of a phase of late mannerism. 1972 F. GARVIE tr. *Wundram's Art of Renaissance* 42 Architecture in the age of Mannerism played a major part historically by overcoming the static in favor of the movemented.

mannerist. Add: *spec.* An exponent or adherent of Mannerism in art (see prec.).
1695, 1716 [in Dict.]. 1833 J. CONSTABLE in C. R. Leslie *Mem. Life J. Constable* (1843) xii. 135 A certain set of painters who, having substituted falsehood for truth, and formed a style mean and mechanical, are termed mannerists. 1845 A. JAMESON *Mem. Early Italian Painters* II. x. 250 In the middle of the sixteenth century Italy swarmed with painters: these go under the general name of the mannerists, because they all imitated the manner of some one of the great masters who had gone before them. 1864 R. N. WORNUM *Epochs of Painting* 303 Hosts of copyists and mannerists arose,..with a mania for representing the naked human figure, [who] sacrificed almost every beauty, quality, and motive, to the paramount desire of anatomical display. 1907 B. BERENSON *N. Italian Painters of Renaissance* 156 The Mannerists, Tibaldi, Zuccaro, Fontana, thus quickly give place to the Eclectics. 1926 [see *ACADEMISM 2]. 1951 A. HAUSER *Social Hist. Art* I. v. 388 The antitheses of 'Gothic' and 'Renaissance'..are still..irreconcilable in the outlook of the mannerists. 1956 A. HUXLEY *Adonis & Alphabet* 229 There is not the slightest reason to believe that Catholic fervour was less intense in the age of the Mannerists than it had been three generations earlier.
2. In appositive use, passing into *adj.*
1934 R. WITTKOWER in *Art Bulletin* XVI. 216 The Laurenziana belongs to a..group of buildings arranged on similar principles, common between 1520 and 1580/90 and to be called Mannerist. 1939 *Handbk. Drawings & Watercolours Dept. Prints & Drawings Brit. Mus.* 38 The leading figure of this mannerist movement, which is largely occupied in elaborate decorative schemes in palaces and churches, was Francesco Salviati. 1944 *Archit. Rev.* XCVI. 187 The author is, it seems, of the generation to which what we now define as Mannerist is nothing but a late phase of the Renaissance. 1964 *English Studies* XLV. 98 The transition from the ambiguities of Mannerist expression to that of Baroque realism. 1972 *Guardian* 17 Nov. 12/3 It was the influence of Raphael that informed the Mannerist artists whose work clusters round that of the giants.

mannerize, *v.* (Later examples.)
1910 A. C. BENSON *Silent Isle* xv. 103 Tennyson..became solemn, mannerised, conscious of responsibility. 1956 *Archit. Rev.* CXIX. 161/2 The forms which have developed from the rationale and the initial ideology of the modern movement are being mannerized and changed into a conscious imperfectionism.

manness. Add: **2.** = MALENESS.
1921 W. J. TURNER *Music & Life* p. ix, You are not concerned with their..brains..but with something which we may call their 'manness'. 1947 J. STEINBECK *Pearl* (1948) v. 64 The quality of woman..could cut through Kino's manness.

mannie. Add: (Further, incl. non-*Sc.*, examples.) In Jewish use: see quot. 1909.
The Jewish use is perhaps a contraction of *Emanuel*.
a 1689 W. CLELAND *Poems* (1697) 105 A puffie cheek'd red bearded mannie. 1909 J. R. WARE *Passing Eng.* 173/2 *Manny* (Jewish E. London), term of endearment or admiration prefixed to Jewish name, as 'Manny Lyons'. 1914 JOYCE *Dubliners* 103 My little man! My little mannie! 1916 'BOYD CABLE' *Action Front* 262 'My manny here was good enough,' said Macalister, 'to tell me he wouldna' bandage my eyes.' 1922 D. H. LAWRENCE *England, my England* 154 There's too much of the little mannie about him. 1971 M. WEST *Summer of Red Wolf* vi. 125 What do you want me to pack for the poor mannie?

manning, *vbl. sb.* Add: **1.** Also, in more recent use: the action of furnishing a factory, an industry, etc., with men.
1955 *Times* 22 Aug. 7/4 He refers to the dockers' insistence on full manning, with the implication that this is a restrictive practice.
3. (Further examples.)
1953 *Britannica Bk. of Year* 638/2 Other compound nouns were *manning requirements* (the number of men required by an industry). 1962 *Listener* 1 Mar. 377/1 On the labour side there are..rules about manning-scales (that is, how many men to a job). 1975 *Times* 14 Apr. 12/2 We put them there, those too many men... Manning levels were never changed. 1975 *Times* 5 June 18/4 The stoppage is by 80 door hangers and welders, who are protesting about management plans to reduce manning scales.

mannitol. Add def.: A sweet crystalline hexahydric alcohol, $CH_2OH(CHOH)_4CH_2OH$, known in three optically isomeric forms, which is found in many plants (such as sugar cane, celery, and larch) and is used in aqueous solution in kidney function tests. (Now the usual name.) (Earlier and later examples.)
1879 *Jrnl. Chem. Soc.* XXXVI. 1034 Mannityl-hexsulphuric acid is obtained as an uncrystallisable liquid by dissolving mannitol in sulphuric monochloride. 1939 *Jrnl. Amer. Chem. Soc.* LXI. 761/2 To prepare the *l*-glyceraldehyde in an analogous way from *l*-mannitol was much more difficult, since the *l*-mannitol..is not yet obtainable commercially. 1951 A. GROLLMAN *Pharmacol. & Therapeutics* xxxii. 750 Mannitol, a hexahydric alcohol sugar is filtered through the glomeruli but is neither reabsorbed nor excreted by the tubules. 1957 *Technology* July 187/2 Mannitol, which also comes from seaweed, is used by the chemical, explosive, pharmaceutical and electrical industries. 1970 H. MCLEAVE *Question of Negligence* (1973) xxvii. 208 'Intravenous saline and mannitol,' he murmured. His registrar injected the drug to shrink the brain and relieve the compression.

Mannlicher (ma·nliχəɹ). [f. the name of *Mannlicher* (see below).] A firearm of a type invented by Ferdinand Ritter von Mannlicher (1848–1904), esp. a type of sporting rifle. Also *attrib.*
The umlauted *ä* in the Kipling example is erroneous.
1884 in H. Cholmondeley-Pennell *Smoke of Battle* (ed. 2) 16 (Advt.), The Mannlicher Rifle is the one of all the specimens of Repeating Arms..which seems to us the best arm for our soldiers. 1896 T. F. FREMANTLE *Notes on Rifle* xi. 175 The ·256 Mannlicher is capable of making very fair shooting at these long ranges. 1902 KIPLING *Traffics & Discov.* (1904) 20 My gun was too good, too uniform—shot as close as a Männlicher [Mannlicher, 1902 magazine publ.] rifle. 1922 J. BUCHAN *Huntingtower* xiv. 267 McGuffog, who was a marksman, was also given a sporting Mannlicher. 1935 E. HEMINGWAY *Green Hills Afr.* (1936) I. ii. 46 There was the short-barrelled explosion of the Mannlicher. 1970 I. V. HOGG *Military Pistols & Revolvers* 41/2 The Mannlicher uses a form of delayed blow-back action.

mannose (mæ·nōᵘs, -ōᵘz). *Chem.* [a. G. *mannose* (Fischer & Hirschberger 1888, in *Ber. d. Deut. Chem. Ges.* XXI. 1805), f. *mann(it)ose* MANNITOSE.] A crystalline sugar, $C_6H_{12}O_6$, which is known in three optically isomeric forms which are epimers of those of glucose, and whose dextrorotatory form is obtained by the hydrolysis of mannans.
1888 *Jrnl. Chem. Soc.* LIV. 934 Mannose..is obtained by dissolving the above phenylhydrazone in four parts of hydrochloric acid..cooled with ice and salt. 1911 [see *EPIMER]. 1949 M. A. JENNINGS in H. W. Florey et al. *Antibiotics* II. xxxi. 975 The series included strains known to vary in such properties as..fermentation of carbohydrates (mannite, trehalose and mannose). 1968 A. SOLS in F. Dickens et al. *Carbohydrate Metabolism* I. iii. 71 Mannose can be efficiently metabolized by most tissues because it is a very good substitute for hexokinase.

‖ **mano** (mā·no). *Anthropol.* [Sp. *mano* hand.] A primitive stone implement, held in the hand and used for grinding cereals and other foodstuffs.
1901 *Ann. Rep. Board of Regents Smithsonian Inst. 1899* 37 The grinding-stone concordantly changes from a simple roller or crusher to a mano (or muller), and finally to a pestle, at first broad and short, but afterwards long and slender. 1911 W. K. MOOREHEAD *Stone Age N. Amer.* II. xxvii. 103 The stones used on these [mortars] are flat, or oval water-worn stones and not finished, like mano stones common to the Cliff Dweller country. 1944 G. C. VAILLANT *Aztecs Mexico* (1950) i. 35 The flat grinding-stones and mullers, still used in Mexico and called *metates* and *manos*, prove that the people relied on corn as their principal food. 1959 E. TUNIS *Indians* 119/2 The grinding was done by rubbing the grains across it with another stone, the *mano*, held in the hands. 1960 C. WINICK *Dict. Anthropol.* 342/2 *Mano*, a cylindrically shaped grindstone slightly tapered at both ends. It was held in the hand (whence its name handstone) and used as the upper stone in milling. 1964 A. D. KRIEGER in Jennings & Norbeck *Prehist. Man in New World* 32 The most important new trait, however, is that of food-grinding with stone implements: basin-shaped milling stones and manos. 1971 [see *INDIGENIZATION]. 1972 *Sci. Amer.* May 89/3 From the Maya Mountains came the metamorphic rock used to make not only axe heads of stone but also the *manos*, or stone rollers. 1974 *Encycl. Brit. Macropædia* XI. 936/2 They [*sc.* villagers of the Ocós and Cuadros phases of Meso-American civilization] were productive corn farmers as well, raising a small-eared race of maize called Nal-Tel, which their wives and daughters ground on *metates* and *manos* and cooked in globular jars.

‖ **manoao** (ma·noau). Also **monoao.** [Maori.] A New Zealand evergreen tree, *Dacrydium kirkii*, of the family Podocarpaceæ; formerly applied to another species of *Dacrydium*, *D. colensoi*; also called *silver pine*.
1867 J. D. HOOKER *Handbk. N.Z. Flora* II. 766/1 Manoao, *Dacrydium Colensoi*. 1889 T. KIRK *Forest Flora N.Z.* 191 (*heading*) The manoao. *Ibid.* 192 The wood of the manaoa [*sic*] is of a light-brown colour. 1950 *N.Z. Jrnl. Agric.* Feb. 115/2 The open pumice country, clothed in a tangled mass of manuka and manoao. 1951 *Post-Primary School Bulletin* (Wellington, N.Z.) V. XII. 274 The manoao (*Dacrydium kirkii*)—a tree related to the rimu but having its young leaves more than an inch long and its old leaves only an eighth of an inch long. 1963 POOLE & ADAMS *Trees & Shrubs N.Z.* 26 Monoao. Tree reaching 25m. Bark light brown.

Manoeline, -lino, varr. *MANUELINE *a.*

manœuvrability (mănū:vrăbi·lĭti). Also (*U.S.*) **maneuverability.** [f. *MANŒUVRABLE *a.*] Capacity for being manœuvred.
The spelling with medial *-e-* in quots. 1942 and 1954 is unusual.
1923 *Rep. & Mem. Aeronaut. Res. Comm.* No. 851. 1 (*heading*) The comparison of the manœuvrability of aeroplanes by the use of a cinematograph camera. 1927 *Daily Express* 24 Sept. 8 Much will depend on the skill of the pilots in taking the corner, and also on the manœuvrability of the machines. 1930 C. J. STEWART *Aircraft Instruments* p. xvii, The manœvrability and the adequacy of the various control surfaces of an aircraft. 1942 *Tee Emm* (Air Ministry) II. 85 A compromise between stability and manœuvreability would not be difficult. 1954 *Encounter* June 13/2 The European national economies..are peculiarly dependent upon world trade balances, and the area of manœuvreability is limited. 1968 *Encycl. Brit.* II. 874/1 Large steering gear ratios make high-speed manœuverability more difficult. 1972 *Drive* Summer 88/1 We were glad of the boat's manoeuvrability when the current increased under the bridge at Arundel.

manœuvrable (mănū·vrăb'l), *a.* [f. MANŒUVRE *v.* + -ABLE.] Capable of being manœuvred, used esp. of aircraft and motor vehicles.
The spelling with medial *-e-* in quot. 1942 is unusual.
1921 *Aeronautical Jrnl.* XXV. 520 Getting off across wind..is only attempted with relatively high-powered or manœuvrable aeroplanes. 1942 *Tee Emm* (Air Ministry) II. 86 The aircraft will become more stable, but it will be less manœuvreable. 1972 *Motor* 8 Apr. 21/1 (Advt.), There are other things that we've done to our cars which would make any size of car more manoeuvrable. 1973 J. DRUMMOND *Bang! Bang! You're Dead!* xxiv. 118 'What do you need?' 'A reasonable strike-force. Manoeuvrable.'

manœuvre, *v.* Add: **2. b.** Earlier examples of **manœuvring** *ppl. adj.*
1801 M. EDGEWORTH *Belinda* II. xvi. 122 This manœuvring lady represented this report as being universally known and believed, in hopes of frightening her niece into an immediate match with the baronet. 1822 —— *Let.* 9 Feb. (1971) 346 Mrs. Lock was one of the manoeuvring mothers who wanted to draw him in for her daughter.

manœuvring, *vbl. sb.* (Earlier example of *attrib.* use.)
1814 JANE AUSTEN *Mansf. Park* I. v. 93 Speaking from my own observation, it [*sc.* marriage] is a manœuvring business.

man-of-war. Add: Also **man-o'-war. 2. c.** (Earlier and later examples.)
1774 J. ANDREWS *Let.* 30 Dec. (1866) 79 Partaking of the extreem ill qualities of a soldier as well as that of a man-of-war's man. 1931 *Times Lit. Suppl.* 19 Feb. 124/4 'Matelot' is undoubtedly the French for sailor, but we are not told that it is the English man-of-war's-man's name for himself.
3. *man-of-war bird.* Delete † and add later examples.
1906 W. L. SCLATER *Birds S. Afr.* IV. 495 The 'Cape Sheep', 'Great Albatros', 'Man of War Bird' and 'Goney' are all names which are sometimes applied to this bird [*sc.* the wandering albatross]. 1949 M. LOWRY *Let.* May (1967) 178, I think he [*sc.* Melville] is confusing it with his man-of-war bird, the frigate bird. 1952 E. HEMINGWAY *Old Man & Sea* 30 He saw a man-of-war bird with his long black wings circling in the sky. 1962 *Times* 6 Apr. 7/2 (Advt.), Some of the country's most spectacular birds live here [*sc.* in the Florida Everglades]. The snowy egrets, cranes, water turkeys, man-o-war birds, scarlet ibis, [etc.].
6. Used *attrib.* to designate a boy's garment resembling that worn by a sailor, a sailor suit. *Obs.* exc. *Hist.*
1883 in L. de Vries *Victorian Advts.* (1968) 49 Man-o'-war suit. Complete 10/9. 1911 *Daily Colonist* (Victoria, B.C.) 5 Apr. 2/5 (Advt.), Stylish summer hats for little boys and girls... Duck man-o-war hats tams. 1922 JOYCE *Ulysses* 341 His little man-o'-war top and unmentionables were full of sand. 1965 CUNNINGTON & BUCK *Children's Costume in Eng.* 183 There was the man-o'-war suit..complete with lanyard, knife and good conduct stripes.

‖ **manoir** (manwār). [a. F. *manoir* MANOR.] A French manor-house; a country house built in this style.
1853 C. BRONTË *Villette* II. xviii. 136 This house..is rather a manoir than a chateau. 1885 A. EDWARDES *Girton Girl* I. v. 107 The look of the old manoir was cheery. 1926 V. HUNT *Flurried Yrs.* I. 76 We would walk among high hedges..to this or that old ruined *manoir* that would not let. 1954 V. SACKVILLE-WEST *Let.* 31 Aug. in H. Nicolson *Diaries & Lett.* (1968) 264 We can think back on that lovely country with the poplars..and the castles and the *manoirs*. 1968 F. WHITE *Ways of Aquitaine* II. 32 The château of Meillant is a perfect example of the high Renaissance *manoir*.

‖ **manoletina** (manolĕtī·nā). [Sp., f. *Manolet(e*, the professional name of the Spanish bullfighter Manuel L. R. Sánchez (1917–47) + *-ina*.] In bullfighting, a decorative pass popularized by Manolete, in which the muleta is held behind the back in the left hand. Also known as the *orteguina*.
1952 J. MARKS *To the Bullfight* v. 67 We may watch him [*sc.* the matador], at short range, ignoring it [*sc.* the bull] in imitation of *Manolete*'s most imperious yet least estimable manner, as it rushes by under his right arm and the muleta that skims its back in a *manoletina*. 1957 A. MACNAB *Bulls of Iberia* xv. 193 Three *manoletinas* (very exciting with such big horns, which come out under man's armpits each time). 1959 V. J.-R. KEHOE *Aficionado!* xvi. 201/2 This pass is called an *orteguina* or, more commonly today, a *manoletina*—after Domingo Ortega who revived it out of an old school, and Manolete who popularized and refined it to a high degree... it is another adorno. 1961 —— *Wine, Women & Toros!* x. 129/1 His faenas include ..*manoletinas* looking at the crowd and then—surprise—he kills very well! 1971 J. LEIBOLD *This is the Bullfight* xvii. 204 The manoletina is probably the most popular of all the decorative passes and is utilized by bullfighters of every degree of experience. The manoletina is always performed as a right-handed pass with the sword extending the cloth and while it is a very spectacular pass it is not nearly as dangerous as it appears. 1974 *Aiken* (S. Carolina) *Standard* 18 Apr. 4–c/4 Then the grandstand falls silent as Diaz prepares to execute a 'manoletina', a pass rarely seen in bullfighting. The matador must not turn his head as the bull comes pounding across the sand, but only feels the rush of wind as the animal charges by.

manometer. Add: Hence **mano·metry,** the use of manometers.
1923 GLAZEBROOK *Dict. Appl. Physics* III. 191/1 (*heading*) Medium-precision manometry. 1961 *Lancet* 12 Aug. 349/1 Articles on..manometry.., chromatography, and electrophoresis.

manometrically (mænome·trikăli), *adv.* [f. MANOMETRIC, -METRICAL *adjs.*: see -LY².] In a manometric way; by means of a manometer.
1899 *Nature* 2 Feb. 321/1 The arc light cannot be used manometrically, nor..is it probable that the magnesium flame could be thus employed. 1936 *Biochem. Jrnl.* XXX. 2319 Acetoacetic acid was usually determined manometrically by the aniline citrate method. 1974 PASSMORE & ROBSON *Compan. Med. Stud.* III. xix. 97/1 Propulsive activity [of the colon] is more difficult to study manometrically and detect radiographically.

manool (mænōᵘ·ǫl). *Chem.* Also **manoöl.** [a. G. *manool* (Hosking & Brandt 1935, in *Ber. d. Deut. Chem. Ges.* LXVIII. 1311): see *MANOAO and -OL.] A bicyclic diterpenoid alcohol, $C_{20}H_{34}O$, which occurs in the oil of *manoao* wood and is used as a base for perfumes.
1935 *Chem. Abstr.* XXIX. 6591 The red resin extd. with alc. from *D. biforme* yields..a neutral light yellow viscous oil.., giving on fractional distn. about 90% of a very viscous colorless liquid (II) solidifying to crystals of the compn. $C_{20}H_{34}O$... II is therefore a bicyclic diterpene alc. with 2 double bonds, for which the name *manoöl* is proposed. 1969 *N.Z. News* 23 July 4/3 Manool, an ambergris-like product which is extracted from trees that take hundreds of years to grow, may provide New Zealand with a new export market.

manor. Add: **3.** (Further example of *fig.* use.)
1874 A. J. MUNBY *Diary* 20 Apr. in D. Hudson *Munby* (1972) 366, I was 'struck all of a heap' by seeing..a picture of Wigan wenches working at brow. What right had this artist to poach on my manor..?
e. *slang.* A police district; a local unit of police administration. Also *transf.*, one's home ground, one's own particular territory.
1924 S. SCOTT *Human Side Crook & Convict Life* vii. 107 There are straight crooks and crooked crooks on the 'Manor' of a detective, and he gets to know them apart. 1928 E. WALLACE *Gunner* xii. 93, I wouldn't advise you to break in on Gennett's 'manor'—he's rather touchy, and he's got charge of the case. 1945 M. ALLINGHAM *Coroner's Pidgin* xxiii. 202 Do you realize who the Coroner is for those parts?.. It's Montie Forster's manor. 1959 *Observer* 1 Mar. 10/1 This 'manor'—a tenement neighbourhood in North London—is theirs by right of birth and conquest. 1961 C. WITTING *Driven to Kill* 24 If anything's happened to Pearce, it's in my manor, so I'm interested. 1962 R. COOK *Crust on its Uppers* iv. 47 'Then they whipped him down to the nick on the hurry-up.' 'Which manor?' 'The local nick.' *Ibid.* 49 'Where to for the next one?' says he. 'Well off the manor, if you like.' 1970 P. LAURIE *Scotland Yard* 287 Every non-policeman knows that detectives call their working-area their 'manor' or 'patch'... What they actually said was 'ground'. 1974 *Times* 23 Nov. 2/6 Mr Buck is trying to alter that situation by appealing to everyone in his 'manor' (with a population one million higher than that of Northern Ireland) to dial 999 at the slightest sign of suspicious activity.

manorial, *a.* (Later example of **manorializing** *vbl. sb.*)
1918 *Eng. Hist. Rev.* Jan. 70 The king..must forestall the lord's manorializing tendency by adding these thegns and freemen to his own estates.

mano:rializa·tion. [f. MANORIAL *a.* + -IZATION.] The process of making or becoming manorial. Also **mano·rialize** *v. trans.*, 'to conform or subject to the tenure of the manorial system' (Webster 1909).
1896 C. M. ANDREWS in R. H. I. Palgrave *Dict. Pol. Econ.* II. 684 Economic life had become manorialised but not feudalised. 1907 *Q. Rev.* July 147 Varieties occurred in the process of manorialisation. 1927 J. J. HOGAN *Eng. Lang. in Ireland* 15 Manorialization was in full progress at the beginning of the thirteenth century in Leinster. 1939 *Antiquity* XIII. 22 Villages which were but partially manorialized.

manorship. Delete † *Obs.* and add later example.
1920 *Public Opinion* 6 Aug. 127/3 Both names..are.. deeply associated with the neighbourhood of Old Jordans, its homesteads, and churches and manorships.

manostat (mæ·nostæt). [f. *mano-* (in MANOMETER) + *-STAT.] Any device for automatically maintaining a constant pressure in an enclosed space. Hence **manosta·tic** *a.*
1900 *Jrnl. Physical Chem.* IV. 546 (*heading*) On a manostat. 1923 GLAZEBROOK *Dict. Appl. Physics* III. 191/2 The majority of manostats are designed on the broad principle that whenever the pressure departs from the desired constant value, the manometer itself, through the change in liquid level in one of its limbs, automatically actuates either a supply or removal of gas..or else some equivalent means of restoring the desired pressure. 1936 *Jrnl. Amer. Chem. Soc.* LVIII. 1703/1 The new manostatic technique. 1966 *Encycl. Industr. Chem. Analysis* I. 617 As a general rule, manostats controlling four fixed pressures of 10, 50, 100, and 200 Torr, and accurately adjusted to these pressures, provide a range of operating pressures quite adequate for every distillation situation. Below 50 Torr, pressure regulation is unsatisfactory when mercury is used as the manostatic fluid.

manotoka, var. *MANITOKA.

‖ **manque** (maňk). [Fr., f. *manquer* to fail, lack: see quot. 1903.] In roulette, the name of one section of the cloth covering the numbers 1 to 18; a bet placed on this section.
1850 *Bohn's Handbk. Games* 348 The manque wins, when the ball enters a hole numbered eighteen, and all those under that number. 1885 CAPT. PALMER *Monte Carlo & Rouge et Noir* 8 The space for Manque is so marked on the table, and the ball stopping in any of the compartments containing any number up to 18 makes Manque win. 1903 'L. HOFFMANN' *Card & Table Games* (ed. 3) 626 If he places his money on *Manque* (so called because in this event the ball 'fails'..to fall into a higher number than 18), he is considered to wager that the ball will fall into one of the numbers from 1 to 18 inclusive. 1923 L. H. DAWSON *Hoyle's Games Modernized* II. 279 To bet on both *Passe* and *Noir* or *Rouge* and *Manque* at the same time, two separate stakes would be required. 1953 T. KING *21 Games of Chance* 31 In addition to the central strip of numbers, there are outer spaces marked *Passe, Manque, Pair, Impair, Noir* and *Rouge*. 1969 J. BINSTOCK *Casino Administration* vi. 72 This [*sc.* roulette] is played on a table with 37 numbers from zero to 36, the so-called 'even chance' bets being *pair* and *impair*, *passe* and *manque*, *rouge* and *noir*.

‖ **manqué** (maňke), *a.* Fem. **manquée.** [Fr., pa. pple. of *manquer* to miss, be lacking.] After its noun: that might have been but is not, that has missed being. Also occas. *pred.*
1778 F. BURNEY *Diary* Aug. (1904) I. i. 54 Dr. Johnson's favourite is Mr. Smith. He declares the fine gentleman *manqué* was never better drawn. 1895 G. B. SHAW in W. Archer *Theatr. 'World' 1894* p. xxvi, A villain if you like..a kicked, cuffed, duped pantaloon by all means; but a hero *manqué*, never. 1898 C. R. ASHBEE *Cellini's Treatises on Goldsmithing & Sculpture* p. xi, There is about his figures always something *manqué*. 1913 tr. *Gleizes & Metzinger's Cubism* i. 15 People have tried to present Cézanne as a sort of genius *manqué*. 1927 *Sat. Rev.* 17 Sept. 370/2 'The History of Anthony Waring' is a poem manqué. 1942 W. STEVENS *Let.* 29 July (1967) 4/4 That such a person [*sc.* a poet] is to be visualized as..an eccentric or a person somehow manqué is nonsense. 1948 F. R. LEAVIS *Great Tradition* ii. 61 Casaubon..is an intellectual *manqué*. 1960 V. BRITTAIN *Women at Oxf.* v. 86 Published memories of Miss Maitland..suggest that she was a hospital matron *manquée*. 1962 *Listener* 12 July 73/2 A clever, unhappy young writer *manqué*. 1974 *Listener* 26 Sept. 408/3 Was Lady O a courtesan manquée?
b. In other uses: defective, spoilt, missing, lacking, etc.
1773 H. WALPOLE *Let.* 27 Mar. (1904) VIII. 262 Dr. Goldsmith's *She Stoops to Conquer*... The author's wit is as much *manqué* as the lady's. 1793 F. BURNEY *Let.* 3 May (1972) II. 98 Our party was manqué in every way;—I came early, but with a head ache; your melancholy Note did not relieve it. 1841 THACKERAY in *Fraser's Mag.* June 724/1, I never yet had a good dinner in my life at Véfour's; *something* is always *manqué* at the place. 1876 C. M. YONGE *Womankind* i. 6 The single woman ceases to be *manquée*, and enjoys honour and happiness. 1881 H. JAMES *Notebks.* (1947) 31 He used to talk to me about Spain, about the East..till it seemed to me that life would be *manquée* altogether if one shouldn't have some of that knowledge. 1894 R. FRY *Let.* 27 Mar. (1972) I. 158 Millais..is the most *gifted* man we ever had, but somehow he's *manqué*, never done what he might have done. 1940 W. STEVENS *Let.* 9 Aug. (1967) 362 Thus, one's chords remain *manqué*; still there they are.

man-rate (mæ·nrēⁱt), *v.* [f. MAN *sb.*¹ + RATE *v.*¹] *trans.* To make (a rocket, spaceship, etc.) suitable for manned flight; to certify as safe for manned flight. So **man-rated** *ppl. a.*; **ma·n-rating** *vbl. sb.*
1963 E. STUHLINGER et al. *Astronautical Engin. & Sci.* iv. 68 Manrating is a new term. To manrate a space vehicle means to make it suitable for carrying a manned payload. 1967 *Technology Week* XX. 23 Jan. 29/2 (Advt.), McDonnell testing and development facilities range from man-rated space chambers to Mach 28 wind tunnels. *Ibid.* 30/1 (Advt.), Their man-rated lunar descent engine is one of a family of small throttleable engines. 1968 *New Scientist* 21 Mar. 631 After it [*sc.* a new Soviet booster] has been man-rated, however, and used in orbital rendezvous and docking missions, the USSR will have achieved a major platform for further advance. 1971 *Sci. News Let.* 19 June 416 The general question of the reliability of unmanned spacecraft has always been a thorny problem for NASA. Manned spacecraft have redundant systems—back-up systems in case one fails. This method, called 'man-rating a spacecraft', is costly.

mansard. Add: **c.** *Comb.*, as *mansard-roofed* adj.
1887 J. E. TAYLOR *Tourist's Guide Suffolk* 31 The Tower Ramparts, where the red-tiled, mansard-roofed cottages have been built on the very top. 1915 E. ATKINSON *Johnny Appleseed* 80 From there he saw the white mansard-roofed mansion.
Also **ma·nsarded** *a.*
1903 *Westm. Gaz.* 11 Feb. 2/1 Handsome little hôtels, *mansarded* and *œil de bœuf'd*. 1951 W. SANSOM *Face of Innocence* viii. 100 Each pantiled or mansarded or beamed façade. 1962 *Listener* 5 Apr. 592/1 Why should we have to suffer tall mansarded roofs?

manscape (mæ·nskēⁱp). [f. MAN *sb.*¹, formed in imitation of LANDSCAPE; see also SCAPE *sb.*³] A view or picture of a sea of faces in a crowd. (Fanciful.)
1927 *Sunday Express* 24 Apr. 1/1 A manscape of a hundred thousand souls is a moving spectacle. 1948 *Archit. Rev.* CIV. 11/1 But this is landscape. What of manscape?

manse, *sb.* Add: **3.** Also in phr. *son* (*bairn, child, daughter*) *of the manse*: the son (daughter) of a Protestant minister, *esp.* in the Church of Scotland.
1855 MRS. OLIPHANT *Lilliesleaf* III. ix. 116 To think that this was our Mary, a bairn of the Manse. 1903 G. W. BALFOUR in M. C. Balfour *From Saranac to Marquesas* p. xix, One of the few survivors left of the happy company of 'children of the manse'. 1962 *Glasgow Herald* 8 Oct. 7 Dr McIntyre..is a son of the manse, always a good step on the high road to fortune in Scotland. 1965 *Listener* 9 Sept. 372/2 Albert Schweitzer, a son of the manse, was born an Alsatian and brought up in the village of Günsbach. 1976 A. RICHARDS *Former Miss Merthyr Tydfil* 52 Dorothea was a daughter of the manse, a history graduate.

-manship, *suffix.* [f. MAN *sb.*¹ 4 p + -SHIP after CHURCHMANSHIP, CRAFTSMANSHIP, SPORTSMANSHIP, etc.] Used with prefixed sb. (occas. vb.) to denote skill in a subject or activity, esp. now so deployed as to disconcert a rival or opponent.
This traditional terminal element underwent a profound change of meaning after 1947 under the influence of *GAMESMANSHIP. Cf. also *brinkmanship, *LIFEMANSHIP, *oneupmanship.
1821 [see *bullmanship* s.v. BULL *sb.*¹ 11 a]. 1880 [see BUSHMANSHIP]. 1894 *Pall Mall Gaz.* 23 Oct. 4/3 Parisiennes continue to witch the world with noble bikemanship in their graceful kilted knickerbockers. 1909 M. B. SAUNDERS *Litany Lane* I. iii. 34 Otherwise the allegory—and the good-humoured clubmanship—sufficed. 1925 L. O'FLAHERTY *Informer* vi. 78 We can imagine him perfecting himself in the arts of gunmanship, deceit, [etc.]. 1939 *Amer. Speech* XIV. 80/1 Professor Kenneth B. Haas ..inserted a short paragraph concerning 'Consumer Vocabulary' in an article entitled 'Buymanship as an Economic Prophylaxis'. 1950 *Sunday Times* 9 July, Many gamesmen find a specious field for the exercise of their knowledge in the allied craft of queuemanship. 1951 C. D. MILNER *Dolomites* 81 Many fine climbers who were.. developing British cragmanship. 1959 *Evening Standard* 13 June 4/3 His hobbies..include farming, motoring.. and general do-it-yourself-manship. 1962 *Economist* 28 July 338/2 Connoisseurs of conferencemanship will be happy to find resolutions calling for the abolition of indirect taxation, [etc.]. 1973 *Nature* 24 Aug. 526/1 He has some useful and pointed things to say on 'grantsmanship'.

mansion, *sb.* **3. e.** (Earlier and later examples.)
1876 A. TROLLOPE *Prime Minister* II. iv. 67 He had been to look at a flat,—a set of rooms,—in the Belgrave Mansions, in Pimlico. 1892 A. W. PINERO *Magistrate* II. 105 *Messiter.* Where at, sir? *Vale.* Albert Mansions, Victoria Street. 1955 *Times* 15 July 10/3 The process happens more often in the case of a mews, a yard, or a court, and is almost frequent in the case of a 'mansions'. 1972 *Mainichi Daily News* (Japan) 7 Nov. 6/6 The earnings of the real estate division, including those obtained through sales of mansions (high class apartments) and lots for villas, will increase by 33 per cent.

mansion-house. b. (Earlier examples of U.S. sense.)
1679 *Public Rec. Colony of Connecticut* (1859) III. 42 He shall build upon his sayd accomodations a good sufficient mansion house. 1745 E. KIMBER *Itinerant Observations Amer.* (1878) 37 A Negro Quarter, is a number of Huts or Hovels, built at some distance from the Mansion-House. 1812 *Niles' Reg.* III. 9/2 The majority then retired to the Mansion house. 1837 W. JENKINS *Ohio Gazetteer* 162 A large and elegant Mansion house has been erected on the ground with numerous smaller cottages and out buildings. 1844 in C. Cist *Cincinnati Misc.* (1845) I. 68/1 The mansion house of E. S. Haines..and various single buildings are observable for their fine appearance.

manslaughter (mæ·nslǭtəɹ), *v. colloq.* [f. the sb.] *trans.* To kill (a person) without malice aforethought.
1920 R. MACAULAY *Potterism* III. ii. 133, I had left the house morally certain that Arthur Gideon had murdered (or anyhow manslaughtered) Oliver Hobart. 1922 A. A. MILNE *Red House Mystery* ix. 81 'Murdered him?' 'Well, manslaughtered him, anyway.'

manslaughterer. Add: (Later examples.) Also, one who commits manslaughter (sense 2).
1885 A. EDWARDES *Girton Girl* I. iii. 71 As a physician, I consider him a manslaughterer. 1912 E. A. PARRY *What Judge Saw* xvii. 297, I regarded the doctor as a manslaughterer at the time. 1965 *Lancet* 20 Nov. 1070/2 With cyclists, children, and old people safely tucked in bed, I do not have to feel a potential manslaughterer just because I sit behind the steering-wheel.

‖ **manso** (ma·nso). [Sp.] A meek, tame, or cowardly person or animal. Also *attrib.* or as *adj.* Cf. **Indio manso*.
Used esp. of 'cowardly' bulls in *Bullfighting*.
1836, 1860 [see *INDIO]. 1912 A. CONAN DOYLE *Lost World* viii. 115 'Yes sir, war drums,' said Gomez, the half-breed. 'Wild Indians, bravos, not mansos; they watch us every mile of the way; kill us if they can.' 1932 E. HEMINGWAY *Death in Afternoon* 298 *Manso*, tame, mild and unwarlike; a bull which does not have the fighting blood is *manso*, as are also the steers called *cabestros* when they are trained. 1952 J. MARKS *To the Bullfight* v. 58 He [*sc.* the president] has also a green and a red one [*sc.* handkerchief], which he holds over the edge of the box:..the red to condemn a cowardly *manso* to the stigma of black banderillas. 1957 A. MACNAB *Bulls of Iberia* vii. 73 Since 1951, in Spain the 'firing' of *manso* bulls has been symbolical only;..the sticks have no fireworks. 1967 MCCORMICK & MASCAREÑAS *Compl. Aficionado* ii. 30 He could pick out the *mansos* with eighty per-cent accuracy, which is better than most toreros can do.

mansonia (mænsōᵘ·niă). [mod.L. (J. R. Drummond 1905, in *Jrnl. Linn. Soc. Bot.* XXXVII. 260), f. the name of F. B. *Manson* (fl. 1905), a forester in Burma who collected the first specimens of *M. gagei*.] A large tree of the genus so called, belonging to the family Sterculiaceæ, esp. the West African *Mansonia altissima* or the hardwood obtained from it. Also *attrib.*
[1934 W. D. MACGREGOR *Silviculture Mixed Deciduous Forests Nigeria* iii. 29 *Mansonia* seedlings 4–9" in height appear to be 1932 regeneration.] 1936 *Nature* 6 June 954/2 The Mansonia (*Mansonia altissima*)..occurs in the deciduous forests of west tropical Africa. 1958 *Archit. Rev.* CXXIV. 191/2 The floors are of mansonia wood blocks, the staircase and fittings of sapele and the ceilings gaboon plywood. 1969 B. J. RENDLE *World Timbers* I. 134 Mansonia was introduced to the world market from Nigeria, as a substitute for walnut, in the 1930s.

manta. 2. Substitute for def.: [Amer. Sp. *manta*, adopted as a generic name by E. B. Bancroft in 1829: see quot.] In full, **manta ray.** A very large ray (RAY *sb.*²) of the genus so called, found in tropical seas; also called devil-fish. (Later examples.)
1829 E. N. BANCROFT in *Zool. Jrnl.* IV. 454 The Manta has, I believe, been generally supposed to belong to the Ray family. 1905 D. S. JORDAN *Guide to Study of Fishes* I. 448 The devil rays or mantas of the Tropical seas, *Manta* and *Mobula* being the most specialized genera. 1958 *Listener* 14 Aug. 247/2 The huge manta rays..opening their gills to let in the tiny cleaner-fish. 1972 *Islander* (Victoria, B.C.) 9 Apr. 7/1 A pair of manta rays making black silhouettes in the midday sun.

mantel, *sb.* Add: **3. d.** *mantel-clock* (earlier and later examples), *-mirror*; **mantel-place** *southern U.S.* = MANTELPIECE.
1870 W. M. BAKER *New Timothy* 25 The mantel-clock strikes six sharp insisting blows as she exclaims. 1963 *Times* 18 May 6/5 Fabre paid £1,200 for a bronze and ormolu elephant mantel clock, also in the Caffieri manner. 1865 MRS. STOWE *House & Home Papers* 86 Now come the great mantel mirrors for four hundred [dollars] more. 1842 W. G. SIMMS *Last Wager* in *Gift 1843* (Philadelphia) 286 You have a very singular ornament for your mantle-place.

mantelshelf. Add: *transf.* (Later examples from *Mountaineering*.) Also *attrib.*
1941 C. F. KIRKUS *Let's Go Climbing!* iv. 60 Half-way up is a ledge..with no holds for some distance above... This is known as a mantelshelf. 1955 M. E. B. BANKS *Commando Climber* xii. 230 A little higher he was stopped by a rock wall capped by an icy mantelshelf. 1963 'G. CARR' *Lewker in Norway* viii. 161 A firm right foothold, a smooth 'mantelshelf' movement, and he was up and standing on the rock bridge. 1968 P. CREW *Encycl. Dict. Mountaineering* 81 The mantelshelf technique has three main steps as follows: a pull up to raise the body as high as possible; changing one arm, and then the other, into a press-up position; cocking one leg up onto the ledge and slowly standing up.

man-tiger. (Later examples.)
1872 *Nature* 29 Feb. 343/1 The analogous belief in man-hyenas, man-tigers, &c. 1941 J. MASEFIELD *Gautama* 35 Man-tigers dragged the white form nearer As victim on that stone to lie.

Mantinean (mæntinī·ăn), *sb.* and *a.* Also 6 Mantynyan, 9 **Mantineian.** [f. Gr. Μαντίνεια, L. *Mantinea* Mantinea.] **A.** *sb.* A native or inhabitant of the ancient city of Mantinea in Arcadia. **B.** *adj.* Of or pertaining to Mantinea or its inhabitants.
1550 T. NICOLLS tr. *Thucydides' Hystory* [*Peloponnesian War*] v. fol. 145 That the Argiues shulde make allyance wyth the Lacedemonyans, forsakinge the same, whyche they had concluded wyth the Athenyans, the Mantynyans and the Elyans. 1629 HOBBES tr. *Thucydides' Peloponnesian War* v. 335 A thousand Athenians..being come after the Battell to ayde the Mantineans. 1753 W. SMITH tr. *Thucydides Hist. Peloponnesian War* II. v. 180 Three thousand heavy-armed Eléans, as auxiliaries to the Mantinéans, came up. 1808 W. MITFORD *Hist. Greece* II. xvii. 263 The Tegeans, and all those other Arcadians who had not, with the Mantineians, renounced the Lacedæmonian alliance. 1846 G. GROTE *Hist. Greece* II. II. viii. 586 Sparta..drove back the Mantineians within their own limits. 1877 L. H. MORGAN *Anc. Society* II. xiv. 349 Bachofen has collected and discussed the evidence of female authority..among the..Mantineans. 1951 J. B. BURY *Hist. Greece* (ed. 3) 327 The Mantineans..were never ready to join hands with their Tegeate neighbours. *Ibid.*, The new synoecism of the Mantinean villages. 1968 V. EHRENBERG *From Solon to Socrates* vii. 284 Sparta.. made..an expedition to free the Parrhasians in Arcadia from Mantinean rule. 1970 *Oxf. Classical Dict.* (ed. 2) 644/2 After the peace of 387 the Spartans obliged the Mantineans to dismantle their walls and live in villages.

mantistic (mænti·stik), *a.* [f. Gr. μάντις prophet + -ISTIC.] Pertaining to divination or prophecy; prophetic, mantic.
1876 A. WILDER *Knight's Symbolical Lang. Anc. Art & Mythol.* 141 An idea of peculiar spiritual or mantistic qualities supposed to be peculiar to the female sex.

mantle, *sb.* Add: **2. b.** (Later examples: cf. sense 10*.)
1949 A. G. TANSLEY (*title*) Britain's green mantle. 1962 *Listener* 1 Mar. 376/1 Since..the upper part of the atmospheric mantle is rich in carbon dioxide, advanced life-forms on the surface of Venus seem unlikely.
10*. One of the three major layers composing the earth, extending from the bottom of the crust (at a depth of about 30 km.) for about 2,900 km. to the boundary with the region beneath (the core), and differing from the crust and the core in physical properties (esp. density) and in chemical composition.
Orig. not distinguished from the crust (see quots. 1940, 1955).
1940 R. A. DALY *Strength & Struct. Earth* i. 1 The earth contains a spheroidal core... The rest of the planet, beneath ocean and atmosphere,..may be distinguished as a whole by the name 'mantle'. *Ibid.* 21 At the depth of about 2,900 kilometers..there is..an interface or rapid transition between the earth's silicate mantle and its 'iron' core. 1955 *Sci. Amer.* Sept. 58/3 All the earth outside the core is called the mantle. The whole of the mantle (apart from the oceans and pockets of magma in volcanic regions) is now known to be essentially solid. 1958 *Nature* 13 Sept. 692/1 The Mohorovičić discontinuity is the boundary between the mantle and the assorted surface rocks of the earth, and it marks a very sharp change in the velocity with which earthquake waves travel. 1962 *Listener* 30 Aug. 304/1 Geophysicists are preparing to drill a hole six kilometres deep through the earth's crust to the mantle. 1969

New Yorker 12 Apr. 96/2 If the structure of the moon proves to be similar to the earth's—that is, to have a core, a mantle, and a crust—then geologists may be able to learn a good deal about the structure of the earth. **1973** *Sci. Amer.* Apr. 24/2 Outside the liquid outer core is a mantle of solid rock some 2,900 kilometers thick, which approaches to within 40 kilometers of the earth's surface under the continents and to within 10 kilometers under the oceans. The thin rocky skin surrounding the mantle is the earth's crust. No drill has penetrated the earth's crust deeper than a few kilometers.

11. mantle-cavity, the space between the mantle (sense 6) and the body of a mollusc or brachiopod; **mantle fibre** *Cytology*, any spindle fibre which is attached to a chromosome.

1853 *Phil. Trans. R. Soc.* CXLIII. 37 It [*sc.* the heart] lies parallel to the rectum, with the auricle forwards at the base of the mantle-cavity. **1958** *New Biol.* XXV. 100 The visceral mass (of Cuttlefish) is closed all round by a muscular mantle in such a way that below and to the sides of it there is a cavity—the mantle cavity. **1896** E. B. Wilson *Cell* ii. 74 The daughter-chromosomes are dragged apart solely by the contractile mantle-fibres, the central spindle-fibres being non-contractile. **1920** L. Doncaster *Introd. Study Cytol.* iii. 31 In some animals these fibres are said to be of different thickness, those attached to chromosomes being thicker and called 'mantle fibres' (from their appearance at a later stage). **1966** D. M. Kramsch tr. *Grundmann's Gen. Cytol.* iii. 149 We are dealing with a central spindle, on whose exterior aspect the chromosomal spindle fibres are attached as mantle fibres.

Mantoux (mantū, mæ·ntū). *Med.* The name of Charles *Mantoux* (1877–1947), French physician, used *attrib.* with reference to a method, introduced by him in 1908, of testing for past or present tuberculous infection by intradermal injection of diluted tuberculin.

1931 R. J. E. Scott *Gould's Med. Dict.* (ed. 3) 764/1 Mantoux test. **1932** *Observer* 28 Feb. 11/2 The Mantoux test has a definite though limited application in the diagnosis of tuberculous disease. **1956** *Nature* 25 Feb. 367/2 A study was made of the interrelationships of the Mantoux and Lepromin reactions. *Ibid.* 368/1 The Mantoux-negative individuals. **1971** D. Lambert in C. Bonington *Annapurna South Face* App. G. 290 Immunity [from tuberculosis] is tested first by the Mantoux Test, and this is examined after three days.

mantra. Add: (Later examples.) Also *attrib.* and *fig.* Also, a holy name, for inward meditation.

1900 W. Maxwell in W. W. Skeat *Malay Magic* v. 116 Numerous *mantra*, or charms..are in use among the Pawangs. **1931** G. B. Sansom *Japan* iii. xii. 222 Recitation of the formulae known as *mantra*. **1956** E. Wood *Yoga Dict.* 93 *Mantras*, forms of speech which carry a material effect upon the mind, emotions or body, or even on things... The scriptures contain many great *mantras*, of which the greatest is the word *Om*. *Ibid.*, If these [sounds] are arranged by some person competent in this matter, words..and sentences of power can be formed. Such a person was called a mantramaker (*mantrakāra*) in ancient India. **1962** Brahmacharini Usha *Ramakrishna-Vedanta Wordbk.* 47 *Mantra*,..also mantram... The particular name of God... The mantra, regarded as one with God, represents the essence of the guru's instruction to his disciple, who is enjoined to keep it sacred and secret, and to meditate on the aspect of God which it symbolizes for the rest of his life. **1964** V. S. Naipaul *Area of Darkness* x. 249 The Tibetans..were suffering because they had forgotten the *mantras*, charms, which might have repelled their enemies. **1965** *New Statesman* 16 Apr. 616/3 He is less happy when defending the mantra, or holy names on which one must meditate. **1967** Singha & Massey *Indian Dances* iv. 56 The priest lit the sacred fire, recited the mantras (sacred texts), and tied the tali round the girl's neck. **1969** *Enactment* (Delhi) Nov. 18/1 You used to chant his name like a *mantra*. **1971** *Daily Tel.* 19 Aug. 10/4 People who ever have dinner in English country hotels know the *mantra*, or holy formula: 'Coffee will be served in the lounge.' **1972** *Language* XLVIII. 174 Secondly, the Brahmanas cannot be regarded far removed in time from the mantras, and therefore their authors might not render a particular mantra word.. which..was not altogether unintelligible in their times. **1972** *Village Voice* (N.Y.) 1 June 76/3 The birds were placed in cardboard boxes..and rushed to a building where teams of hippies, chanting the Hare Krishna mantra, worked into the dawn of the next day to save their lives. **1973** *Times* 22 Jan. 20 The Maharishi uses a technique based on a mantra, a sound which each meditator is given to intone inwardly.

man-trap. *transf.* and *fig.* (Further examples.)

1857 W. Chandless *Visit to Salt Lake* ii. xi. 330 The planks (of the streets) worn out and broken through, leaving large holes, popularly known as 'man-traps'. **1903** G. B. Shaw *Man & Superman* iii. 121 You know better than any of us that marriage is a mantrap baited with simulated accomplishments. **1922** Joyce *Ulysses* 425 You never seen me in the mantrap with a married highlander. **1929** W. Faulkner *Sartoris* (1932) v. ii. 365 Some new kind of mantrap [*sc.* an aeroplane] that flies fine—on paper. **1965** H. Sheppard *Dict. Railway Slang* 7 *Man trap*, catch points to prevent unauthorised entry from siding. **1974** P. M. Hubbard *Thirsty Evil* i. 9 She was no man-trap, but she did not miss much.

man-trap, *v.* [f. the sb.] *trans.* To beset with man-traps. Also *fig.* and as **ma·ntrapping** *ppl. a.*

1911 J. London *Son of Sun* (1913) iv. iv. 159 Besides, the runs are all man-trapped—you know, staked pits, poisoned thorns, and the rest. **1952** Dylan Thomas *Let.* 8 Oct. (1966) 378 Every lane was mantrapped for me. *c* **1953** —— *Ibid.* 416 And eel up wheezily..from all the claws and bars and breasts of the mantrapping seabed. **1957** A. Clarke *Later Poems* (1961) 61 Discharge, excrete, their centuries, Man-trapped in concrete.

‖ **mantri** (mæ·ntri). Also 8 **mantree.** [Hindi, f. Skr. *mantri*, f. *mantrín* wise, eloquent, skilled in sacred texts or spells, f. *man* to think; cf. Mantra.] **a.** In India, etc.: a minister, counsellor. **b.** In Indonesia (including the former Netherlands East Indies): a minor official or subordinate functionary vested with some authority.

1783 W. Marsden *Hist. Sumatra* 287 The officers next in rank to the *Sultan* are called *Mantree*, which some apprehend to be a corruption of the word *Mandarin*, a title of distinction amongst the Chinese. **1814** T. S. Raffles *Substance of Minute on Land Rental on Island of Java* 218 In each division there shall be fixed a station of police, to which shall be appointed a competent officer, with such number of inferior *Mantris, Peons*, &c. as shall be deemed necessary. **1873** E. Balfour *Cycl. India* III. 146/2 *Mantra*, Sans. Counsel, hence mantri, Sans. a counsellor. **1896** Carey & Tuck *Chin Hills* I. i. 3 The elders of the village, called..by the Lushai officers 'Kharbari' and 'Mantri', surround the person of the Chief. **1917** Coomaraswamy & Duggirala tr. *Mirror of Gesture* 15 *The Ministers* (mantri)—Those who shine as royal ministers are men of their word. **1965** I. Southall *Indonesia* xxiii. 227 If the villager ever does get to the hospital, he will find an outpatient clinic with a mantri in charge. It is the duty of this mantri to treat all simpler problems and refer the difficult to the doctor.

Mantuan. Add: **A.** *adj.* (Earlier and later examples.)

1538 King Henry VIII *Epistle* sig. B2 We moste hartely desyre you, that ye wol vouchesafe, to rede those thynges, that we wrote this laste yere, touchynge the Mantuan Councille. **1940** G. F.-H. & J. Berkeley *Italy in Making* III. viii. 141 The Mantuan volunteers..gave a very creditable account of themselves.

B. *sb.* (Later examples.)

1842 K. H. Digby *Mores Catholici* xi. i. 6 Mezentius is never named by the great Mantuan without the epithet 'contemtorque Deûm'. **1953** T. F. Murray tr. *Dal-Gal's Pius X* v. 91 He had not spent long among the Mantuans before his people..were charmed by his meekness and won by his charity.

manual, *a.* Add: **1.** Also, as opposed to *automatic*, applied to hand-operated devices, systems, etc.

1921 *Conquest* Jan. 124/1 Consider what happens in an ordinary 'manual' telephone exchange when a subscriber makes a call. **1923** *Bell Syst. Techn. Jrnl.* Apr. 56 The system most commonly employed today for connecting [telephone] subscribers' lines together is the so-called 'manual' system; that is, a system in which operators.. make the actual connections. **1959** *Gramophone* Sept. 82 (Advt.), The 'Conquest' Automatic Record Changer for stereophonic and monophonic play... Provision for manual operation. **1966** *McGraw-Hill Encycl. Sci. & Technol.* XIV. 43/2 There are two general types of transmission, manual shift and automatic. Manual-shift transmissions..are used in conjunction with a clutch.

Manueline (mæ·niuĕləin), *a.* Also **Manoeline, Manoelino, Manoellian, Manuelline.** [f. the name of *Manuel* I (b. 1469, reigned 1495–1521), King of Portugal.] Of or pertaining to a style of Portuguese architecture developed during the reign of Manuel I and characterized by its ornate elaborations of Gothic and Renaissance styles.

1908 W. C. Watson *Portuguese Archit.* x. 145 Nearly all these churches and palaces were built or added to in that peculiar style now called Manoelino. *Ibid.* 147 The Jesus College at Setubal..is the best example in the country of a late Gothic church modified by the addition of certain Manoelino details. **1911** *Encycl. Brit.* XXII. 144/2 In architecture the name of King Emanuel was given to a new and composite style (the Manoeline or Manoellian), in which decorative forms..were harmonized with Gothic and Renaissance designs. **1931** S. Sitwell *Spanish Baroque Art* ii. 50 There is always an imaginary Orient at the back of their ideas... In the Manoeline there are Moghul, Persian, Moroccan influences. **1937** *Archit. Rev.* LXXXII. 129/2 The Manoeline style of architecture is as distinct from what is termed Italian 'Renaissance' as Romanesque is from Gothic... In this respect by 'Manoeline' is meant Portuguese Renaissance. **1948** G. Kubler *Mexican Archit. 16th Cent.* II. viii. 382 In Portugal, the period before *ca.* 1525 was dominated by Manueline ornament. **1960** J. Lees-Milne *Baroque in Spain & Portugal* ii. i. 144 The Manoeline style—if style it may be called—has a very positive bearing upon the last phase of Portuguese Baroque. **1968** *Encycl. Brit.* III. 271/2 The Capellas Imperfeitas ('unfinished chapels') are one of the most marvellous examples of Manuelline architecture.

manufacturing, *ppl. a.* (Earlier and later examples.)

1722 Defoe *Jrnl. Plague Year* 257 The Manufacturing Trade in England suffer'd greatly. **1789** H. More *Lett.* (1925) 165, I have written to different manufacturing towns for a [school] mistress. **1892** *Rep. Vermont Board Agric.* XII. 140 As these manufacturing centers increase in size, so do the farm lands in like ratio increase in value. **1942** *Short Guide Gt. Brit.* (U.S. War Dept.) 7 The great 'midland' manufacturing cities of Birmingham, Sheffield and Coventry. **1959** *Chambers's Encycl.* IX. 51/1 The manufacturing industries of Manchester are much more varied than in a specialized manufacturing town.

manuka. (Earlier and later examples.)

1832 G. Bennett in *London Med. Gaz.* 18 Feb. 750/1 This tree..is probably a species of Leptospermum. It is found abundantly at New Zealand,..and is named Kaetatowa, or Manuka, by the natives. **1882** W. D. Hay *Brighter Britain!* II. 195 The Manuka or Manukau..is the 'ti-tree' of settlers. **1920** *Nature* 22 July 667/1 The most abundant foods [for trout] were the green manuka-beetle,..the larvae of caddis-flies..and a small mollusc. **1936** 'R. Hyde' *Passport to Hell* ii. 47 The brown and white manuka blossoms. **1948** *Coast to Coast 1947* 76 Johnny put up a manuka shelter around the pool. **1959** *Tararua* XIII. 47 We may find on the topographical maps such New Zealand terms as..'teatree' in Auckland, 'manuka' further south. **1960** *New Scientist* 4 Aug. 330/1 Sheep farmers were beginning to get worried by manuka, which showed signs of investing large tracts of country. And then, suddenly the coccid appeared and started to destroy the weed.

manure, *sb.* Add: **1.** Also, other substances, esp. various chemicals, used as fertilizers.

1794 R. Kirwan in *Trans. R. Irish Acad.* V. i. 137 The substances principally used as manures, are chalk, lime.. gypsum, [etc.]. **1824** J. C. Loudon *Encycl. Gardening* (ed. 2) 243 Saline and calcareous substances form the principal fossil manures. **1904** T. W. Sanders *Roses & their Cultivation* 70 It is of no use applying manure that has been allowed to decay naturally, because nearly all the essential salts have been washed out or evaporated. **1922** Joyce *Ulysses* 698 James W. Mackey..agent for chemical manures. **1951** *Dict. Gardening* (R. Hort. Soc.) III. 1248/2 Manures may be classified into Organic Manures which are substances of animal or vegetable origin, and Inorganic Manures which are of mineral origin. **1971** L. D. Hills *Grow your own Fruit & Vegetables* ii. 27 There is a clear distinction between organic and inorganic manures and fertilizers.

3. *manure-spreader.*

a **1884** Knight *Dict. Mech.* Suppl. 580/2 *Manure Spreader*, a cart having a bed of traveling slats..to distribute the load while the vehicle is moving over the surface of the ground. **1915** J. London *Let.* 26 Jan. (1966) 445 My plan still holds of using litter-carriers to dump manure..into..waiting manure-spreaders. **1943** J. S. Huxley *TVA* vi. 49 He insisted on buying a mechanical manure-spreader and using it with a tractor. **1969** K. M. Wells *Owl Pen Reader* i. 46 The uphill clatter of Farmer Jim's manure-spreader as he drove it over the stubbles.

Hence **manu·r(e)y** *a.*, splashed or littered with manure.

1890–3 E. M. Taber *Stowe Notes, Lett. & Verses* (1913) 29 The stable-yard repulsive, muddy and manury. **1932** *Sunday Express* 3 July 17/4 Many's the time I've seen him all mud and manurey.

manuscript. B. *sb.* **1. b.** Add to def.: Also applied to an author's typed copy.

1967 R. A. Waldron *Sense & Sense Devel.* vi. 116 *Manuscript* is no longer something necessarily 'written by hand' but is usually the author's original copy of the text of a printed book, etc. **1967** *Anglo-Amer. Catal. Rules: Brit. Text* 267 *Manuscript*, a writing made by hand. (Original typescripts are generally treated as manuscripts in libraries.) **1967** *Listener* 2 Mar. 282/3 Perhaps you are going to transcribe the tape-recording into typewritten manuscript.

man-woman. a. Delete † *Obs.* and add later examples.

1920 D. Lindsay *Voy. Arcturus* xviii. 240 'What do you call men-women?' 'Persons of mixed sex, like yourself.' **1975** P. G. Winslow *Death of Angel* iv. 109 The new man-woman, emerging from the chaos that is matter.

Hence **man-womanly** *a.*, having the characteristics of both sexes.

1929 V. Woolf *Room of one's Own* 148 It would be well to test what one meant by man-womanly.

Manx, *a.* and *sb.* **B.** *sb.* **2.** Add to def.: Now extinct. (Further example.)

1970 B. M. H. Strang *Hist. Eng.* ix. 402 One IE language (Manx) has become extinct since the Second World War.

Manxwoman (example).

1974 J. Mann *Sticking Place* viii. 132 I'm a Manxwoman.

manxome (mæ·ŋksŏm), *a.* *poet.* *nonce-wd.* [Invented word; cf. -some[1] (as in *fearsome, gruesome, loathsome*, etc.).] ? Fearsome.

1871 'L. Carroll' *Through Looking-Glass* i. 22 Long time the manxome foe he sought.

many, *a.* and *sb.* Add: **A.** *adj.* **5. e.** Also used of something that is repeated to excess.

1941 H. L. Mencken *Newspaper Days* (1942) xii. 193 The poor old man..nursing a hangover from a Bar Association banquet, had thrown in one too many quick ones, and so got himself plastered. **1956** A. Wilson *Anglo-Saxon Attitudes* i. ii. 43 Some people have made one imaginative leap too many and show little sign of being able to return to the realm of reason.

6. a. *many-angled* (later example), *-antlered, -centuried, -corridored, -faceted, -fingered, -hearted, -minded, -mooded, -mouthed* (later example), *-pleated, -splendoured, -twined, -volumed* adjs. Also *many-dimensional* adj.

1892 W. B. Yeats *Countess Kathleen* iii. 57 Heaven's many-angled star reversed. **1930** Blunden *Poems* 48 Bronze noonlight domes the dim blue gloom Where

many-antlered oaks immure A hush. **1848** J. R. Lowell *Columbus* in *Poems* 2nd Ser. 11 Many-centuried shade Of some writhed oak. **1931** Blunden *To Themis* 53 Beneath the accustomed dome Of this chance-planted, many-centuried tree. **1859** Tennyson *Vivien* in *Idylls of King* 132 The myriad-room'd And many-corridor'd complexities Of Arthur's palace. **1905** W. James in *Mind* XIV. 196 Satisfaction is a many-dimensional term that can be realized in various ways. **1933** A. N. Whitehead *Adventures of Ideas* xi. 242 And space is many-dimensional. *a* **1963** C. S. Lewis *Poems* (1964) 102 The many-dimensional timeless rays. **1909** *Daily Chron.* 7 Sept. 4/4 A many-faceted diamond. **1947** *Mind* LVI. 291 The many-faceted problem of perception soon came to dominate the epistemological scene. **1909** E. Pound *Personae* 36 All tremulous beneath the many-fingered breath. **1935** W. Empson *Poems* 19 Crossing and doubling, many-fingered, hounded. **1882** in *Eng. Dial. Dict.* (1903) IV. 33/1 (s. Dev.), He was always many-hearted [= 'soft-hearted']. **1904** W. de la Mare *Henry Brocken* 193 Yonder fine many-hearted poplar. **1895** *Funk's Stand. Dict.*, *Many-minded*, showing changes of mind; changeable in opinion; fickle; versatile. **1932** W. B. Yeats *Words for Music* 37 Even Cicero And many-minded Homer were Mad as the mist and snow. **1920** *19th Cent.* Aug. 272 To know him [*sc.* Stephen Phillips] was to realise how many-mooded and complex a man he was. **1935** C. Day Lewis *Time to Dance* 50 You shall recall one open as the day, Many-mooded as the light above English hills. **1917** D. H. Lawrence *Look! We have come Through!* 18 The spouse all full of increase Moiled over with the rearing of her many-mouthed young. **1927** —— *Mornings in Mexico* 81 The many-pleated, noiseless mountains of Mexico. *a* **1907** F. Thompson *Kingdom of God* in *Sel. Poems* (1908) 131 'Tis ye, 'tis your estrangèd faces, That miss the many-splendoured thing. **1962** *Sunday Times* (Colour Suppl.) 25 Nov. 29 (*title*) The many-splendoured fisherman. **1971** *Nat. Geographic* Oct. 548/1 The Ocean Terminal and nearby Harbour Centre in Kowloon offer a dazzling promenade past 150 or more stores selling jewelry, watches, ginger jars, television sets.., pearls, and other many-splendored items. **1909** E. Pound *Personae* 35 What should avail me the many-twined bracelets? **1857** J. R. Lowell *Orig. Didactic Poetry* in *Atlantic Monthly* Nov. 112 Many-volumed thunder. **1927** W. B. Yeats *Senate Speeches* (1961) 138 This many-volumed ancient history.

b. (ii) *attrib.* phrases consisting of *many* with a sb. in sense 'having, consisting of, many of the things named', as *many-course, -electron, -interest, -particle, -volume, -word.*

1955 D. Chapman *Home & Social Status* xi. 172 The many-course dinner with wines. **1929** *Trans. Faraday Soc.* XXV. 672 We use this system of energy levels for the many-electron problem just as was done in atoms. **1970** G. K. Woodgate *Elem. Atomic Struct.* i. 5 In many-electron atoms the electrostatic interaction with the nucleus is summed over all the electrons. **1955** M. Gluckman *Custom & Conflict in Afr.* v. 135 They [*sc.* rituals] are inappropriate in the family, our single many-interest group. **1955** W. Pauli *Niels Bohr* 135 The states formed in the reaction are states of a many-particle system. **1941** *Mind* L. 141 That these naïve beliefs [*sc.* about ambiguity] are false is easy to realise in this age of many-volume dictionaries. **1924** R. M. Ogden tr. *Koffka's Growth of Mind* v. 329 Transference from a one-word to a many-word sentence is carried out. **1940** A. H. Gardiner *Theory of Proper Names* ix. 29 In some of my many-word names (e.g. Edgar Allan Poe) the coherence of the parts is much slighter than in others. **1973** A. Quinton *Nature of Things* v. 127 Many-word sentences have to be used to guard against the misunderstandings.

c. many-body, pertaining to or involving three or more bodies or particles; applied *spec.* to the problem of predicting their positions and motions at any future time given their present values and the way the bodies interact; **many–many** *a.* (see quots.); **many–one** *a.*, applied to a correspondence or relation such that two or more members of one set are associated with or related to each member of a second set; hence **many–oneness**; **many-valued** *a.* [cf. G. *mehrwertig*] *Philos.*, 'possessing three or more truth-values in place of the customary two of truth and falsehood' (Webster 1961); **many-worded** *a.*, of a term or description involving the use of several words.

1927 Fisher & Hartree tr. *Born's Mech. Atom* iv. 248 The analytical difficulties of the many-body problem. **1962** W. B. Thompson *Introd. Plasma Physics* i. 2 The microscopic dynamics of a plasma must be understood as a study in many-body physics. **1964** L. Wilets *Theories Nucl. Fission* iv. 56 The first quantitative attempt to apply many-body techniques to finite nuclei was performed by Brueckner, Lockett, and Rotenberg... The results of the infinite many-body problem were used to obtain the K-matrix as a function of density. **1922** W. E. Johnson *Logic* II. vii. 156 Here the denominating correlation is not one–one but many–many, and yet the names and the things happen to be numerically equal. **1933** *Mind* XLII. 53 In this third use of 'term' the total situation located by 'Tom fears Francois' is not two-termed but many-termed, and the total situation located by 'England fears France' many-many-termed. **1965** *Language* XLI. 44 Transformational relations are one–one, but expansion relations are many–many. **1910** Whitehead & Russell *Principia Math.* I. ii. 438 Thus many–one relations are the converses of one–many relations. *Ibid.* 575 The transition from series generated by one-one or many–one relations of consecutive terms to series generated by transitive relations of *before* and *after*. **1936** *Jrnl. Philos.* 17 Dec. 706 The relation of tokens to their type..is consistently many–one: a type can have many tokens, but a token only one type. But the relation of type to word is many–many. A word can be represented by many types; but so can a type represent many words. **1955** A. N. Prior *Formal Logic* 279 A 'many–one' relation, $Cls\to 1$, is an *R* such that if *x* is an *R* of any given thing then it is not an *R* of anything else.. The null relation is many–one for the same sort of reason as it is one–many. **1959** E. M. Patterson *Topology* (ed. 2) iv. 82 A homomorphism between a group G_1 and a group G_2 is a many–one transformation. **1971** *Language* XLVII. 8 This overwhelming predominance of many–one mapping over one–many, as we move from semantics to phonetics, can then be seen as further evidence of language's directionality. **1966** S. Beer *Decision & Control* xvii. 441 The effect of that Act..was to provide a richer mapping, to reduce the many–oneness of the homomorphic transformation. **1934** *Philos. of Sci.* I. 118 A many-valued system of logic is a code of inference which endows propositions with truth values intermediate between true and false. **1936** *Mind* XLV. 273 His probability logic is a many-valued logic. **1963** R. Dickerson in H. W. Baade *Jurimetrics* 63 The potential usefulness of many-valued logic for building mathematical models helpful in dealing with problems of vagueness. **1965** *Philosophy* XL. 172 Languages based on a many-valued logic are artificial constructs. **1969** N. Rescher (*title*) Many-valued logic. *a* **1832** Bentham *Chrestomathia* App. 9. § 2 in *Works* (1843) VIII. 188/1 In the many-worded appellative, part of speech, the word *part* is instructive. **1843** Mill *Logic* I. i. ii. 30 A mixed term belongs to the class of what have been called many-worded names. **1901** A. Sidgwick *Use of Words in Reasoning* v. 143 Description..is more often than not many-worded. **1957** G. Ryle in M. Black *Importance of Lang.* (1962) 151, I am still not quite sure why..every possible grammatical subject of a sentence, one-worded or many-worded, stands to something.

manyatta (mænyæ·tā). Also **manyat.** [Masai.] Among certain African peoples, particularly the Masai, a group of huts forming a unit within a common fence.

1905 A. C. Hollis *Masai* 292 Meat may not be eaten in the manyat, or warriors' kraals. **1921** *Blackw. Mag.* Jan. 118/1 The Manyatta, a rambling collection of dome-shaped huts surrounded by a straggling zeriba, seemed to have settled down for the day. **1964** *Listener* 30 July 163/2 He returned to the family *manyatta*—a thorn-fenced cluster of low mud hovels shared with the beasts. **1971** *E. Afr. Standard* (Nairobi) 13 Apr. 9/1 A gang of Borana tribesmen attacked a Samburu manyatta at Shaba Hills.

man-year: see *man *sb.*[1] 20 b.

manyfold (me·nifōuld), *adv.* Also **many-fold, many fold.** [a mod. re-formation from Many *a.* + -fold (after *three-, fourfold*, etc.), etymologically identical with Manifold, though this was rarely used in this sense (Manifold *adv.* 1 b).] In the proportion of many to one.

1879 [see -fold]. **1916** D. S. Kimball in *Mod. Business* (Alexander Hamilton Inst.) VII. ii. vi. 293 If this difference is reflected many-fold in the product of a large and costly machine the difference in output might very well be marked. **1920** S. S. Huebner *Marine Insurance* p. v, The tonnage of vessels under the American Flag was, within a brief period, increased many fold. **1961** M. Hynes *Med. Bacteriol.* (ed. 7) x. 126 Urinary infections with resistant bacteria may respond to antibiotics that are concentrated many-fold by renal excretion. **1969** *Sci. Jrnl.* 36/1 If any nation can find..all the major ingredients to produce unlimited supplies of energy, fertilizer, fresh water and other necessities, then the useful part of the Earth will be increased manyfold.

many-headed, *a.* **b.** (Further example.)

1934 Wodehouse *Right Ho, Jeeves* xv. 181 We might have been a rather oversized greyhound and a somewhat slimmer electric hare doing their stuff on a circular track for the entertainment of the many-headed.

manyogana (manyō·ganā). Also **manyokana.** [Jap., f. *Manyō*(*shū*) 'collection of a Myriad Leaves', name of an 8th-cent. anthology of Japanese poetry + *-gana* combining form of *kana (phonetic) letters, script.] A system of writing in use in Japan in the 8th century, found *spec.* in the Manyōshū, in which Chinese characters are used to represent Japanese sounds.

1868 J. J. Hoffmann *Japanese Gram.* 6 The running-hand form was used in the old Japanese Bundle of Poems ..*Man-you-siu* or the Collection of the Ten Thousand Leaves, compiled about the middle of the eighth century. The first *Kana*-form was, consequently, called *Yamato-kâna*.., the other *Man-you-kâna*. **1909** tr. *S. Okuma's Fifty Yrs. of New Japan* II. i. 2 We also used these [Chinese] characters merely as symbols for our own sounds. This latter method..we find..generally used in our old works like the '*Kojiki*' and the '*Manyōshū*', whence these symbols came to be called the *Manyō-kana*. **1928** G. B. Sansom *Hist. Gram. Japanese* i. 23 The name of this anthology was the *Manyōshū*, or 'collection of a Myriad Leaves', and the characters thus used were known as *Manyōgana*. **1934** S. Yoshitake *Phonetic System of Ancient Japanese* i. 7 It was that great philologist Motowori Norinaga who first discovered how strictly certain Man-yō-gana were differentiated. **1948** *Introd. Classic Japanese Lit.* (Kokusai Bunka Shinkokai) (1956) p. x, The choice of subject for ballads ranged much more widely..encouraged by the popularization of new methods of writing, by the introduction of the Japanese syllabary (*kana*) in the primitive form known as *man'yō-gana* (*Man'yōshū* style *kana*) where the Chinese ideographs were used with their phonetic value. **1951** J. K. Yamagiwa in Reischauer & Yamagiwa *Transl. Early Japanese Lit.* 277 In the eighth and ninth centuries, abbreviations and simplifications of the *Man'yōgana* resulted in the creation of the two syllabic scripts, the *katakana* and the *hiragana*. **1959** *Chambers's Encycl.* VIII. 58/2 Two [writing] systems gradually developed: one an elaborate distortion of Japanese into Chinese forms (*kambun*); the other a phonetic adaptation of Chinese characters for the reproduction of Japanese sounds (*manyōgana*). **1965** D. Keene *Manyōshū* p. xviii, The so-called '*Manyō-gana*' are the Chinese characters which were commonly used as phonograms in the *Manyōshū*, from which the present system of *kana* was evolved. **1974** *Canad. Jrnl. Linguistics* XIX. 217 Taken together, the *ongana* and *kungana* comprise the *man'yōgana*, the eighth-century precursors of the later *kana* syllabaries.

manzanilla. Add: **2.** A variety of olive, distinguished by small thin-skinned fruit.

1911 *Daily Colonist* (Victoria, B.C.) 13 Apr. 6/1 (Advt.), Your Easter Sunday menu will demand the luscious, healthful Olive. We carry a big stock, all kinds..Manzanillas—Small Olives. **1964** *Economist* 8 Aug. 547/2 Special varieties [of olive]..like the Manzanilla and the Goodal in Spain. **1974** *Observer* (Colour Suppl.) 8 Sept. 66/3 The country around Seville in Spain is green olive country, both for the manzanillas and the huge queen olives.

manzanita. Add: Also **manzanito.** (Earlier and later examples.)

1848 E. Bryant *California* xviii. 236 We have met occasionally with a reddish berry called by the Californians manzanita (little apple). **1869** C. L. Brace *New West* xi. 138 Around her were..dishes of the manzanita seed. **1888** B. Harte *Drift from Redwood Camp* in *Phyllis of Sierras* 227 A bent manzanito-bush..flew back against his breast. **1918** C. E. Mulford *Man from Bar-20* xvii. 178 He pushed through matted thickets of oak brush and manzanito. **1928** A. Bierce *Can Such Things Be?* i. 8 Unable in the darkness to penetrate the thickets of manzanita and other undergrowth. **1971** *Black Scholar* Apr.–May 47/2 Our job is to clear out 10 miles of oak trees, manzanita shrubs and sagebrush on both sides of the highway. **1972** *Village Voice* (N.Y.) 1 June 76/1 A sign next to a luxuriant manzanita plant welcomes you to God's land.

manzello (mænze·lo). [Origin uncertain.] A musical instrument resembling a soprano saxophone.

1962 *Melody Maker* 7 July 7/3 The manzello, which he uses frequently in solos, sounds rather like a soprano while the stritch is somewhere in the tenor range. **1966** *New Statesman* 4 Nov. 677/1 On manzello he [*sc.* Roland Kirk] echoes some of Sidney Bechet's canary-yellow flights. **1969** *Daily Tel.* 15 Feb. 15/1 Kirk's solo ability on flute, tenor saxophone, manzello and stritch.

Mao (mau). [f. *Mao* Tse-Tung; see *Maoism.] Used *attrib.* of a simple style of clothing based on dress in Communist China, as *Mao cap, collar, jacket, trousers*, etc.

1967 *Guardian* 1 Aug. 5/5 A close-fitting standing collar to which Paris is now giving the name Mao. **1968** *N.Y. Times* 22 May 50 'Extravagances' such as the Mao jacket. **1968** *Punch* 24 July 107/2 Out, apparently, are Mao caps, Guevara beards, Maharishi gowns and Zapata moustaches. **1969** A. Sinclair *Last of Best* ix. 231 Silk Mao jackets. **1972** *Daily Tel.* 3 Mar. 14 The blue-clad men and women, all in Mao jackets and trousers. **1973** *Sunday Express* (Trinidad) 1 Apr. (Suppl.) 7/3 From the East was Madame Sung Chih-Kuang, 50, wife of the first Red Chinese Ambassador to Britain, who wore a grey cotton Mao suit. **1974** J. Mitchell *Death & Bright Water* iii. 19 Mao boiler suit to show he was one of the righteous, but hand-sewn and made of silk.

b. Mao 'flu, Hong Kong influenza (see *Hong Kong 2).

1968 *Guardian* 14 Dec. 1/1 Hongkong or Mao 'flu now sweeping the United States, is likely to reach Britain next month or in February. **1973** 'G. Ashe' *Life for Death* v. 49 He's had the damned Mao flu and it's put him right out.

Maoism (mau·iz'm). [f. the name of *Mao* Tse-Tung (1893–), Chairman of the Central Committee of the Chinese Communist Party + -ism.] The Marxist-Leninist theories of Mao Tse-Tung developed and practised in China. Hence **Mao·ist** *sb.*, a follower of these theories; also *attrib.* or as *adj.*, of or pertaining to these theories.

1951 B. I. Schwartz *Chinese Communism & Rise of Mao* xiii. 189 The essential features of Maoism. *Ibid.* 190 Another peculiar feature of the Maoist strategy..is the preference for 'border area' bases. **1961** *Listener* 23 Nov. 886/2 Maoist Marxism. **1962** *Ibid.* 18 Jan. 112/1 There are several kinds of communism in the world—Titoism, Khrushchevism, Maoism in China. **1964** *New Statesman* 17 Apr. 592/3 The Maoists are blind not to see this and to identify their cause with a nostalgia for the Stalin epoch. **1967** *Guardian* 15 May 6/1 The demonstrators have been conducting their affairs in Maoist style—little red books of quotations and all. **1968** *Listener* 28 Nov. 736/1 The awful children..are now revealed as a tiny Maoist cell at the heart of the family. **1969** *N.Y. Rev. Bks.* 16 Jan. 6/2 Many of the good facets of Maoism, such as participation in social and political life and the linking of practice and theory in education. **1970** G. Greer *Female Eunuch* 22 The most telling criticisms will come from my sisters of the left, the Maoists, the Trots. **1970** *New Scientist* 19 Mar. 539/2 Neither Tories nor Labourites can be blamed for sounding like Maoists, desperately looking for a Little Red Book. **1971** *Peace News* 5 Nov. 5/1 His blend of religious fervour and Maoist rhetoric does not endear him either to pacifists or to liberals. **1973** *Listener* 2 Aug. 147/2 This year Mao will be 80, so we may assume that the Maoist era is coming to its end.

Maoize (mau·əiz), *v.* [f. *Mao* (see *Maoism) + -ize.] *trans.* To imbue with the doctrines of Maoism. So **Maoiza·tion.**

1970 *Guardian* 18 Sept. 11/2 The guerillas' growing ideological dependence on China..could lead to the 'Maoisation' of the Middle East. **1971** *Time* 12 July 22 Mao demolished..others who did not share his own mystical concept of the revolution. He hoped to replace them with freshly radicalized, totally Maoized youth who would be prepared to spend their lives in permanent struggle.

maomao (mau·mau). *N.Z.* [Maori.] A blue-skinned marine food fish, *Scorpis violaceus* or *S. æquipinnis*, found in New Zealand and Australian waters.

1886 R. A. A. Sherrin *Handbk. Fishes N.Z.* 67 The delicious little maomao may be caught at the Rurima Rocks in immense quantities. **1949** P. Buck *Coming of Maori* (1950) ii. viii. 215 The longest bag net..was used to catch *maomao* (*Scorpis violaceus*), a deep-sea fish which travelled in shoals. **1966** *Encycl. N.Z.* II. 408/2 The blue maomao also occurs in Australia, where it is known as the hardbelly.

Maori, *sb.* (*a.*) Add: **1. b. Maori bug** (see quot. 1966); **Maori dog,** a type of dog, which is now extinct, first introduced to New Zealand by the Maoris; also in *fig.* use, *cunning as a Maori dog*, a phrase of vulgar abuse; cf. *kuri. **c. Maori oven** = *copper Maori; **Maori P.T.** *N.Z. slang*, taking it easy and doing nothing.

1944 *Mod. Jun. Dict.* (Whitcombe & Tombs) 251 *Maori bug*, a strong-smelling, dark-coloured beetle. **1959** *Numbers* IX. 8 We..shook Maori bugs out of the blankets. **1966** *Encycl. N.Z.* I. 269/1 *Bug, Maori* (*Platyzosteria novae-zelandiae*). Maori bug is the commonly accepted name for the largest endemic cockroach of New Zealand... This species is capable of liberating a characteristic, unpleasant odour when disturbed. **1947** D. M. Davin *For Rest of our Lives* 103 Cunning as a Maori dog, you know him. **1950** *N.Z. Jrnl. Agric.* Feb. 136/1 Pastoralists..[in the early days] had their problems, too. Maori dogs were troublesome. **1953** M. Scott *Breakfast at Six* iii. 30 'The professional charmer,' jeered Larry... 'Cunning as a Maori dog,' supplemented Sam vulgarly. **1966** *Encycl. N.Z.* I. 491/2 The Maori dog (kuri)..was probably introduced during the period of the Great Migration (c. 1350 A.D.)... It became extinct some years after the arrival of the European settlers. **1849** Maori oven [see *go-ashore* s.v. *go *v.* VIII]. **1905** [see *copper Maori]. **1961** Partridge *Dict. Slang* Suppl. 1179/1 *Maori P.T.*, 'taking it as easily as possible, i.e. resting when one should be undertaking physical training'..: New Zealand soldiers': 1939–45. **1966** G. W. Turner *Eng. Lang. Austral. & N.Z.* vi. 135 In New Zealand the word *Maori* sometimes enters into slang with contemptuous connotation, but there is a good-natured tolerance, even tinged with envy for an admirable adjustment to life's problems, in a term like *Maori P.T.*.. to mean lying down and doing nothing. **1967** *Listener* 2 Mar. 299/3 *Maori P.T.* is New Zealand *dolce far niente*. **1969** *Pocket Oxf. Dict.* (ed. 5) 1035/2 *Maori P.T.*, (sl.) loafing, doing nothing.

3. *black Maori, white Maori* (N.Z.) (see quots.).

1883 *Illustr. Guide Dunedin* 169 (Morris), Tungstate of lime occurs plentifully in the Wakatipu district, where from its weight and colour it is called White Maori by the miners. **1965** G. J. Williams *Econ. Geol. N.Z.* xiii. 190/2 This hard ferro-manganese material forms the pebbles known to the early alluvial miners as 'black Maori' (as contrasted with 'white Maori'—scheelite).

Maoridom (mauᵊ·ridəm). Also **Maori-dom.** [f. Maori *sb.* (*a.*) + -dom.] The Maori world; Maori culture.

1882 W. D. Hay *Brighter Britain!* I. 19 In the very heart of recent Maori-dom. *Ibid.* 278 Auckland city people know little or nothing of Maoridom. **1955** W. J. Phillipps *Maori Carving Illustr.* 8 The tapu system of ancient Maoridom. **1974** *N.Z. Listener* 20 July 11/2, I told him the *Listener* should have an authentic Maori voice speaking in it, a voice that was recognisable throughout Maoridom.

Maoriland (mauᵊ·rilænd). [f. Maori *sb.* (*a.*) + Land *sb.*] A name for New Zealand. **Ma·orilander,** a white man born in New Zealand.

1863 F. E. Maning *Old N.Z.* i. 15, I prepared to make my *entrée* into Maori land in a proper and dignified manner. **1881** *Every Boy's Annual* 657/2 Our goose-wing pen bears us lightly down upon the Maori-land. Captain Abel Tasman, a Hollander, in 1642 discovered New Zealand. **1884** K. Nicholls (*title*) The King Country, or Explorations in New Zealand: a narrative of 600 miles of travel through Maoriland. **1896** *Melbourne Argus* 22 July 4/8 (Morris), Always something new from Maoriland! **1896** *Melbourne Punch* 9 Apr. 233/2 (Morris), Norman is a pushing young Maorilander. **1915** *Morning Post* 16 June 9/6 The Maorilanders gave 'Hakas' till their voices failed them. **1933** *Bulletin* (Sydney) 4 Jan. 11 A Maoriland youth.

Maoriness (mauᵊ·rinės). [f. Maori *sb.* (*a.*) + -ness.] = *Maoritanga.

1963 *Times* 6 Feb. (Suppl.) p. ii/5 The failure of humanitarianism when confronted with Pakeha land hunger and Maori determination to hold fast to their *Maoritanga* (Maoriness) bred conflict. **1974** *N.Z. Listener* 20 July 10/3, I have a curious kind of status which derives from my work, from the fact of my Maoriness,—such as it is.

‖ **Maoritanga** (mauᵊrita·ŋā). [Maori.] The culture, traditions, and heritage of the Maori people; the individuality of the Maori; Maoriness.

1940 I. L. Sutherland *Maori People Today* v. 176 What then..is meant by *Maoritanga*? It means an emphasis on the continuing individuality of the Maori people, the maintenance of such Maori characteristics and such features of Maori culture as present day circumstances will permit. **1948** N. Smith *Maori People & Us* ii. 37 He [*sc.* the Maori] is very sensitive about the honour and standing of his tribe, his ancestry, and his *Maoritanga*. *Ibid.* xi. 218 An improved mode of living among the Maori people can be developed side by side with their *Maoritanga*. **1966** G. W. Turner *Eng. Lang. Austral. & N.Z.* viii. 171 Should the Maori preserve his separate identity, his *maoritanga*, or should he become Europeanized as much as possible? **1968** B. Biggs in E. Schwimmer *Maori People in Nineteen-Sixties* 76 Maori are remaining distinctively Maori without their language, often regarded as the *sine qua non* of Maoritanga. **1971** *N.Z. Listener* 8 Mar. 46 Pakehas who try to give Maoris a bad conscience about their knowledge of Maoritanga. **1974** *Ibid.* 20 July 10/4 The old man did the speaking and it gave me considerable pleasure to hear these men, these korouas, steeped in their Maoritanga, steeped in the tribal values and tradition of Tuhoe.

Mao tai (mau tai). Also **Mao Tai, mao-tai, Mao-T'ai, Maotai.** [*Maotai*, name of a town in south-west China.] A mellow and strong pot-stilled spirit produced in Maotai.

1965 O. A. Mendelsohn *Dict. Drink* 211 Maotai, Chinese strong pot-stilled..spirit as celebrated in the Far East as is cognac in the Western world. **1967** D. Bloodworth *Chinese Looking Glass* xxiii. 239 The famous Chinese white wines—Mao-T'ai, Mei-Kuei-Lu, Pai-kan..—have the mule-kick of a bath-tub gin. **1970** *Observer* 10 May 8/4 A glass of *mao-tai*, the potent Chinese equivalent of vodka. **1972** *Guardian* 22 Feb. 24/2 Mr Nixon..consumed little Mao Tai, a potent Chinese drink distilled from sorghum. **1973** *Times* 13 Sept. 16/6 Mao tai, the Chinese national spirit, had increased almost ten-fold in price on the Hongkong market since President Nixon set his lips to it.

map, *sb.*[1] Add: **1. e.** *fig.* In recent phrases: (*a*) *off the map*: out of existence; into (or in) oblivion or an insignificant position; of no account; obsolete; also (with hyphens) as attrib. phr.; (*b*) *on the map*: in an important or prominent position; of some account or importance; in existence (see also quots.); so *to put on the map*: to establish the position or vogue (of someone or something).

(*a*) **1904** W. H. Smith *Promoters* ii. 54 When she [*sc.* Carthage] wouldn't let up, the only thing left was to wipe her off the map. **1911** R. D. Saunders *Col. Todhunter* vii. 99 A good set-to is the best way..to put a stop to quarrelin'. It just wipes the whole thing off the map. **1914** *Grand Mag.* Jan. 429/2 [He] had been so harried by the Federal officers that he had faded off the map. **1915** *War Illustr.* II. 328/1 One of the curious off-the-map incidents of the war was brought to notice the other day. **1922** *Tatler* 6 Sept. 386/1 A man who owns a lot of coaches.. said that the big stuff..was off the map as far as he was concerned. **1924** W. M. Raine *Troubled Waters* xiii. 143 'Anything new, Matson?'..'Don't forget we've been off the map 'most three weeks.' **1928** *Weekly Dispatch* 13 May 2/6 Cochineal insects, except for making tinctures to colour jellies, are practically off the map today. **1938** E. Bowen *Death of Heart* i. i. 19 An off-the-map, seedy old family friend. **1973** E. Lemarchand *Let or Hindrance* xi. 131 We're a bit off the map up here.

(*b*) **1913** C. E. Mulford *Coming of Cassidy* viii. 122 Cowan had just put Buckskin on th' map by buildin' th' first shack. **1916** *Munsey's Mag.* June 146/2 'The Fortune Hunter', the play that put Winchell Smith on the dramatists' map. **1918** E. Pound *Let.* 4 June (1971) 138 He [*sc.* Henry James] certainly has put America on the map. Given her a local habitation and a name. **1919** Wodehouse *Damsel in Distress* vii. 93 What I mean to say is, you are on the map. You have a sporting chance. **1924** W. M. Raine *Troubled Waters* xix. 205 Didn't know you knew I was on the map. You're sure honouring me. **1926** A. Bennett *Lord Raingo* i. xii. 60 Some say if there's two members of the War Cabinet, it isn't Andrew Clyth and Tom Hogarth—it's Andrew Clyth and Andrew Clyth... But that isn't so. Tom's on the map all right. **1934** *B.B.C. Year-Bk.* 74 Weekly Chamber Concerts..further helped to put the Hall 'on the map'. **1944** F. Clune *Red Heart* 2 The war has put the Red Heart on the map. **1971** *Daily Tel.* 27 Nov. 12 Nepal is very much on the tourist map today, with many tours to the Far East including a few days there. **1973** *Times* 24 Apr. (São Paulo Suppl.) p. i/7 The exhibition was so successful that in one weekend São Paulo put Brazil firmly on to the export map.

f. A diagram representing the spatial distribution of anything or the relative positions of its components.

1881 *Phil. Trans. R. Soc.* CLXXI. 653 The research.. on a method of photography by which the least refrangible end of the solar spectrum could be mapped has reached such a stage that it seems desirable that I should..present a map of the solar spectrum between wave lengths 7600 and 10,750. **1926** *Encycl. Brit.* III. 622/1 An excellent photographic map of this spectrum has been given by T. R. Merton. **1950** W. L. Bragg *Rev. Recent Adv. X-Ray Analysis* ii. 41 (*caption*) Electron density map of the phthalocyanine molecule (left) and key to the structure (right). **1973** *Nature* 21–28 Dec. 509/1 The ability to record activity in many cells at once would allow the construction of a detailed map of the functional connections within a ganglion.

g. *Genetics.* A diagram which represents the linear order and relative distance apart of the known genes of (part of) a chromosome.

1915 T. H. Morgan et al. *Mechanism Mendelian Heredity* iii. 64 In the construction of the chromosome maps shown in the frontispiece the distance taken as a unit is that within which 1 per cent. of crossing over will occur. **1935** *Genetics* XX. 317 (*heading*) Cytological and crossover maps. **1935** L. H. Snyder *Princ. Heredity* xv. 160 (*caption*) Linkage map for *Drosophila melanogaster*. **1939** *Jrnl. Genetics* XXXIX. 335 There is now evidence for the existence of seven sex-linked genes, and for these he has furnished a tentative map. **1954** *Adv. Genetics* VI. 1 (*heading*) Map construction in *Neurospora Crassa*. **1970** Ambrose & Easty *Cell Biol.* x. 338 (*caption*) Linear genetic map of *Drosophila* showing the four linkage groups corresponding to the four chromosomes.

2. (Examples.)

1782 Cowper *Hope* in *Poems* I. 171 He draws upon life's map a zig-zag line, That shows how far 'tis safe to follow sin. **1899** W. E. H. Lecky (*title*) The map of life.

d. *slang.* A person's face.

1908 [see *hang *v.* 25 f]. **1922** Wodehouse *Clicking of Cuthbert* ix. 205 The portrait..was that of a man in the early thirties... 'What a map!' exclaimed the young man. **1935** —— *Luck of Bodkins* xv. 178 It's mostly a case of having a map that photographs well. **1936** J. Curtis *Gilt Kid* xiv. 144 What d'you want to sit there staring at me for? I'm not a bloody oil-painting. You ought to know my map by now.

3*. *Math.* = *mapping *vbl. sb.* 2.

1949 [see *inclusion 3]. **1966** Sze-Tsen Hu *Introd. Gen. Topology* ii. 27 Continuous functions will be called mappings or maps. *Ibid.* 28 A map $f:X\to Y$ from a space *X* into a space *Y*. **1971** G. Glauberman in Powell & Higman *Finite Simple Groups* i. 8 The main tool in investigating this property is the transfer homomorphism of *G* into *S/S'*. Unfortunately, we do not have time to define this map.

4. *map-board, -case, -light, paper, reference, -roller; map-drawing* vbl. sb.; *map-like* adj. (examples); **map-fire,** artillery-fire in which maps are used for laying the guns; **map-net** = *graticule 2; **map projection** = Projection *sb.* 7; **map-reading** *vbl. sb.*, the inspection and interpretation of a map; so (as a back-formation) **map-read** *v.*, to consult and interpret a map; **map square,** one of several squares (sense 6) drawn on a map for ease of reference.

1947 D. M. Davin *Gorse blooms Pale* 192 The G.I. and the A.D.C. vaulted from their perch in the back of the jeep and then bent over it again to get their map-boards. **1916** H. G. Wells *Mr. Britling* i. v. 181 He..turned over the map in the map-case beside him, and tried to find his position. **1940** 'Gun Buster' *Return via Dunkirk* ii. xx. 250 The A.C.P.O. and I had indulged ourselves in two expensive rain-proof map-cases. **1948** W. S. Churchill *Second World War* I. ii. xxii. 365 A few feet behind me, as I sat in my old chair, was the wooden map-case I had had fixed in 1911, and inside it still remained the chart of the North Sea. **1943** J. S. Huxley *Evolutionary Ethics* i. 6 This business of map-drawing. **1971** *Guardian* 25 Feb. 2/4 Israel's formal reply to Cairo's offer is..likely to include the first tentative attempt at 'map drawing' since the six-day war. **1922** *Encycl. Brit.* XXX. 252/2 The precision with which 'map fire' could be carried out. **1963** *Times* 13 Mar. 10 Standard equipment includes two-speed screen wipers and washers—operated from the steering wheel—anti-dazzle mirror, reversing lights, map light and a heating and ventilation system. **1966** L. Cohen *Beautiful Losers* (1970) i. 92 His cufflink gleamed in the maplight. **1920** E. Sitwell *Wooden Pegasus* 60 Our map-like cheeks are painted red. **1935** Auden & Isherwood *Dog beneath Skin* 11 Meadows where browse the Shorthorn and the maplike Frisian. **1932** J. W. Cameron *Maps & Map-Work* iii. 30 A map projection is any definite system of drawing meridians and parallels, the network of lines thus formed being called a *map-net* or *graticule*. **1954** Fisher & Lockley *Sea-Birds* p. xvi, One of these is on a mapnet invented by the late Professor C. B. Fawcett and is used with his permission and that of the Royal Geographical Society. **1942** H. A. Maddox *Dict. Stationery* (ed. 2) 68 *Map paper*, a specially made smooth cartridge or strong printing paper—opaque, strong and free from atmospheric influence. **1963** R. R. A. Higham *Handbk. Papermaking* vii. 202 Chart and map papers. The best grades are produced from rag pulps, although sulphite and sulphate mixtures are also used. **1890** Map-projection [see Trapeziform *a.*]. **1905** [see Projection *sb.* 7 b]. **1961** L. D. Stamp *Gloss. Geogr. Terms* 308/1 *Map-projection*, the representation of part or whole of the spheroidal surface of the earth on a plane-surface. **1960** *Lebende Sprachen* V. 166/2 The pilot is flying too high and too fast to map-read his way across country. **1965** D. MacKenzie *Lonely Side of River* 176 She map-read accurately, giving him plenty of time to respond to her directions. **1968** F. White *Ways of Aquitaine* xii. 161 When one looks at a really fine map.. it is easy for anyone who can map-read to see the whole shape of the landscape. **1919** H. Shaw *Text-bk. Aeronaut.* xviii. 208 *Map Reading.*—It is above all necessary for the pilot of a machine to be able to understand a map. **1937** *Discovery* June 192/1 Map-reading classes. *Ibid.*, More maps and yet more maps, is the author's demand..; and more map-reading, too, we must add. **1971** M. McCarthy *Birds of America* 10 At a very early age he became a whizz at map-reading. **1954** J. Masters *Bhowani Junction* i. vi. 53 This is the map reference the Wimpy gave..where it spotted those men. **1955** E. Waugh *Officers & Gentlemen* i. ix. 107 The assault of the island was rehearsed..scrambling inland to objectives which in Mugg were merely map-references, but, in the Mediterranean, were gun-emplacements. **1969** M. Pugh *Last Place Left* iii. 18, I.. then gave him some of the facts and the map references. **1851** C. Cist *Sk. Cincinnati in 1851* 245 Shade and map-rollers, turning in ivory, done in a superior style. **1917** 'Contact' *Airman's Outings* 272, I looked overboard to make certain of the map square.

map, *v.*[1] Add: **1. e.** *trans.* To establish the relative positions, or the spatial relations or distribution, of (the components of).

1881 [see *map *sb.*[1] 1 f]. **1950** *Adv. Genetics* III. 117 With linked loci which are closely spaced the recombination value *y* may itself be used as a conventional measure of distance..in which to map the chromosome. **1969** *Times* 13 Feb. 10/3 Professor Jacobson has inverted an eye from young toads..and then mapped the connexions made between the retina and the brain. **1974** *Sci. Amer.* Mar. 122/3 Conformational analysis is the completion of the old program for mapping molecules in space. *Ibid.* 94/2 One laborious method for mapping wave forms consisted in noting a series of voltmeter readings and the corresponding angular positions of the alternator shaft.

f. *Math.* To associate with each element of (a set) one or more elements of another set in accordance with a mapping (sense *2); occas., to associate (an element) similarly. Const. *into, to*; also const. *onto* for a certain kind of mapping (see *on to, onto *prep.*).

1939 M. H. A. Newman *Elem. Topology of Plane Sets of Points* i. 12 A (1, 1)-correspondence is set up between the set of all positive integers, *I*, and the set of positive even integers, *E*, by mapping *n* of *I* on 2*n* of *E*. *Ibid.* iii. 57 A circle can be mapped continuously on a square region. **1941**, etc. [see *homomorphically *adv.*]. **1965** Patterson & Rutherford *Elem. Abstr. Algebra* ii. 57 The correspondence is one-one and maps G_1/H onto Im *f*. **1971** *Sci. Amer.* Aug. 94/3 The symbolism $f:S\to\mathbf{R}$ expresses the fact that *f* causes each point of *S* to be assigned a value in **R**; the expression is usually read as '*f* maps *S* to **R**' or, more formally, as '*f* is a function from *S* to **R**'.

2. map out. a. (Later example.)

1910 *New Mag.* Nov. 204/2 A new country was mapped out by those two men.

b. (Further examples.)

1853 Mrs. Gaskell *Ruth* I. xii. 262 Don't let us perplex ourselves with endeavouring to map out how she should feel, or how she should show her feelings. **1955** *Times* 9 May 10/1 The suggestion was made..to set the ball rolling and map out the times in which later discussion should proceed. **1972** *Daily Tel.* 15 Nov. 1/1 A political plan mapped out by President Lanusse.

3. *intr.* To fall into place on a map or plan.

1893 Harkness & Morley *Theory of Functions* 338 Show that..lines parallel to the axes map into unipartite Cartesians.

Mapai (mapəi·). [mod. Heb.] A Left-wing party in the State of Israel. Also *attrib.*

1949 Koestler *Promise & Fulfilment* ii. v. 275 As he does not belong to the inner circle of Mapai,..he can't get a job. **1950** Theimer & Campbell *Encycl. World Politics* 236 The largest party is the Labour Party known as Mapai..similar in programme to West European labour parties. **1956** *Ann. Reg.* 1955 283 A State trial which had prejudicial consequences for Mapai. **1964** *Ann. Reg.* 1963 297 He had settled in Palestine in 1924 and become editor of a daily Hebrew newspaper of the Mapai Party. **1971** W. Laqueur *Dict. Politics* 268 The most powerful party in the *Knesset* is the Israel Labour Party made up of *Mapai*, [etc.].

Mapam (mapā·m). [mod. Heb.] A political party of the far Left in the State of Israel. Also *attrib.*

1950 Theimer & Campbell *Encycl. World Politics* 236 Mapam..based on the communal agricultural settlements which are a feature of Israel. **1955** *Times* 9 May 9/3 The election should show the relative strength of the..Mapam. **1956** *Ann. Reg.* 1955 284 The extreme Left, Mapam,.. increased their representation. **1972** *Guardian* 27 Mar. 4/2 A meeting between Mrs Meir and Mapam leaders.

mapau (mā·pau). *N.Z.* Also **mapou.** [Maori.] A name for several New Zealand trees, esp. *Myrsine* (or *Suttonia*) *australis*, of the family Myrsinaceæ, an evergreen bearing clusters of white flowers and black berries. Cf. *matipo.

1868 *Trans. N.Z. Inst.* I. 37 White Mapau or Piripiriwhata (*Carpodetus serratus*)... Red Mapau (*Myrsine Urvillei*). **1882** W. D. Hay *Brighter Britain!* II. 198 The Mapau..affords good material for fencing. **1889** T. Kirk *Forest Flora N.Z.* 75 By the settlers it is frequently called 'black mapou' on account of the colour of the bark. **1949** P. Buck *Coming of Maori* (1950) i. iv. 57 The canoe also brought..a *mapau* (*Myrsine urvillei*) named Ateateahenga to be used in planting ceremonies. **1951** *Dict. Gardening* (R. Hort. Soc.) IV. 2057/2 S[*uttonia*] *australis*. Mapou. Evergreen tree 10 to 20 ft., or a tall shrub. **1963** Poole & Adams *Trees & Shrubs N.Z.* 166 M[*yrsine*] *australis*... Mapou.

maphrodite, aphetic f. Hermaphrodite *a.* 4 b.

1849 N. Kingsley *Diary* (1914) 33 She is to appearance a bark or maphrodite brig. *Ibid.* 35 Was routed early this morning to see a ship close along side... she is a Maphrodite Brig.

maple. Add: **1. b. hedge maple,** *Acer campestre.*

1906 *Westm. Gaz.* 13 Nov. 12/2 Every lane is aflame with hedge-maples. **1957** M. Hadfield *Brit. Trees* 376 Common Maple. *Acer campestre* Linnaeus. Hedge maple, field maple. Usually a small tree..but also seen in hedge-rows as a pollard or, owing to repeated cutting, a shrub.

2. b. The colour of maple.

1853 *Heal & Son Catal.* 60/1 Wardrobes, japanned maple, or any colour for gentlemen's use. **1926–7** *Army & Navy Stores Catal.* 297/2 Oil Varnish stains..in the following colours..Mahogany, Rosewood, Maple, Satinwood. **1967** [see *duck's egg c].

3. *maple forest, timber; maple-leaved, -timbered* adjs.; **maple beer,** a beverage made from maple sap; **maple candy,** a sweet made from maple sap; **maple leaf,** a representation of the leaf of the maple tree (as an emblem of Canada); **maple molasses** (examples); **maple pea,** a variety of garden pea with wrinkled seeds; = *maple rouncival*; also *absol.*; **maple sugar** *N. Amer.* (earlier and later examples); **maple swamp,** a swamp in which maple is the prevailing tree; **maple syrup** (examples); hence **maple-syruping** *vbl. sb.*; **maple syrup (urine) disease,** a rare condition which is usu. fatal at a very early age or (if the infant survives) leads to mental deficiency, and is caused by the absence of an enzyme which decarboxylates various metabolites of the amino-acids leucine, isoleucine, and valine, so that these substances are present in high concentrations in the blood and urine and impart a characteristic smell of maple syrup to the latter.

1788 *Amer. Museum* IV. 350/1 *Maple beer.*—To every 4 gallons of water (while boiling) add a quart of maple melasses. **1857** 'Porte Crayon' *Virginia Illustr.* i. 23 The table was spread with the best in the house—cold bread and meat..maple beer. **1973** L. Russell *Everyday Life Colonial Canada* viii. 103 Spruce beer, made from the tender twigs of that tree, and maple beer, from the late, weak sap fermented with hops. **1840** *N.Y. Mirror* 4 Apr. 37/2 Your great dealers in Newtown pippins and maple candy. **1879** *Morning Chron.* (Halifax, Nova Scotia) 2 July 1/8 The average boy and a good sized lump of maple candy, form the materials from which we might deduct self-evident conclusion regarding the facility with which attachments are formed in early life. **1975** *Times* 22 Apr. 6 The Prince of Wales samples maple candy while visiting a sugar camp near Ottawa. **1840** *Knickerbocker* XVI. 267 A small and beautiful lake [with]..a rich tract of maple forest on one side. **1910** Kipling *Rewards & Fairies* 146 Still autumn sets the maple-forest blazing. **1860** *Trans. Lit. & Hist. Soc. Quebec* IV. 20 The Mayflower..I am told is the emblem of Nova Scotia, as the Maple leaf is of Canada. **1900** [in Dict.]. **1964** *Globe & Mail* (Toronto) 29 Oct. 1/9 A single maple leaf on a white field flanked by red bars will be recommended to the House of Commons today as Canada's new flag. **1967** *Canadian Antiques Collector* Jan. 25/1 At least three different Maple leaf forms have been found in the pressed glass patterns of Canada. **1785** H. Marshall *Arbustrum Americanum* 77 Maple-leaved Liquidamber Tree or Sweet Gum. **1813** H. Muhlenberg *Catal. Plant.* 32 Maple-leaved Mealy Tree (*Viburnum acerifolium*). **1930** Maple-leaved [see *London plane]. **1804** T. G. Fessenden *Orig. Poems* 29 The lips of my charmer are sweet, As a hogshead of maple molasses. **1863** 'G. Hamilton' *Gala-Days* 225 A land flowing with maple molasses and sugar. **1897** C. Durand *Reminisc.* 83 We made our household sugar, and luscious maple molasses, not mixed as it is now too often with water and common Muscovado sugar. **1733** W. Ellis *Chiltern & Vale Farming* xxxii. 219 The Maple is a larger and sweeter Pea for the Hog. **1744** —— *Mod. Husbandman* Mar. vi. 53 (*heading*) The Nature and Culture of the common and rouncival Maple Pea. **1960** *Farmer & Stockbreeder* 23 Feb. 4/2 For maples interest is mostly restricted to seed lots. **1969** *Times* 14 Jan. 8/6 One pigeon, encouraged by occasional rewards of maple peas fed to it..learnt to glide in a stationary position in a wind tunnel so that its performance could be measured. **1720** *Phil. Trans. R. Soc.* XXXI. 27 Maple Sugar is made of the Juice of Upland Maple, or Maple Trees that grow upon the Highlands. **1852** A. Cary *Clovernook* 74 Everyday in winter she used to feed them [*sc.* the bees] maple-sugar if she had it. **1885** 'Mark Twain' in *Century Mag.* XXXI. 195/2 We occupied an old maple-sugar camp. **1907** *Springfield* (Mass.) *Weekly Republ.* 9 May 16 The Holyoke canoe club opened the river year with a maple-sugar eat at their club-house. **1921** *Daily Colonist* (Victoria, B.C.) 30 Mar. 3/6 Up to the present the weather has not been favourable for the manufacture of maple sugar in the district of Quebec. **1931** W. Cather *Shadows on Rock* v. i. 204 The country people had been coming..bringing maple sugar, spruce beer. **1969** E. H. Pinto *Treen* 94/1 Maple Sugar Moulds...early 19th-century, carved wood moulds for maple sugar, from Quebec Province, Canada. **1974** *Country Life* 3–10 Jan. 28/3 Maple sugar was..often the only source of sugar available to pioneers in the backwoods. **1667** *Early Rec. Providence, Rhode Island* (1894) V. 317 Standing on the west Side of a Maple Swampe. **1789** J. Morse *Amer. Geogr.* 143 One species generally predominating in each soil has originated the descriptive names of..maple, ash and cedar swamps. **1855** *Knickerbocker* XLVI. 225 Cutting hoop-poles in the maple swamps. **1849** in *Glimpses of Past* (Missouri Hist. Soc.) (1933) I. 5 At the different houses they received sugar, coffee, lard, candles, flour, maple syrup, [etc.]. **1885** *Outing* Oct. VII. 77/1 A moment later, all smoking and puffy and swimming in maple sirup, it disappears. **1905** Calkins & Holden *Art of Mod. Advertising* 113 Maple-sirup is a product to which justice has never been done. [**1954** J. H. Menkes et al. in *Pediatrics* XIV. 462 (*heading*) A new syndrome: progressive familial infantile cerebral dysfunction associated with an unusual urinary substance... A characteristic feature of their illness was the passage of urine with an odor strikingly similar to that of maple syrup. **1957** *Amer. Jrnl. Dis. Children* XCIV. 571/2 (*heading*) Maple sugar urine disease.] **1959** *Brit. Med. Jrnl.* 10 Jan. 90/1 (*heading*) 'Maple syrup urine disease.' An inborn error of the metabolism of valine, leucine, and isoleucine associated with gross mental deficiency. *Ibid.* 92/2 A disease which may be related to 'maple syrup disease'. **1967** Mrs. L. B. Johnson *White House Diary* 11 June (1970) 527 We went in to breakfast.., including blueberry pancakes, and, naturally, Vermont maple syrup. **1968** Passmore & Robson *Compan. Med. Stud.* I. xi. 23/1 Maple syrup disease, so rare as to be a clinical curiosity, is an inborn error of metabolism. **1969** *New Scientist* 3 July 10/1 Prevention of postnatal brain damage by dietary treatment has been reported in a number of other inborn errors of metabolism. Examples include galactosaemia, tyrosinosis, maple syrup urine disease and possibly homocystinuria. **1973** J. Drummond *Bang! Bang! You're Dead!* xv. 39 A pretty girl, with a small cute body the colour of maple syrup. **1975** *Budget* (Sugarcreek, Ohio) 20 Mar. 3/8, March 17—Very damp on the outside again. Ideal weather for maple syruping. **1845** C. M. Kirkland *Western Clearings* 3 He had purchased fine farming land and maple timber. **1849** *Ex. Doc. 31st U.S. Congress 1 Sess. House* No. 5. II. 631 At 7½ a.m., went over good maple-timbered land to corner.

maple tree. (Further examples.)

1660 *Early Rec. Warwick, Rhode Island* (1926) 322 Bounded by a mapell tree on the Northwest corner. **1700** *Early Rec. Providence, Rhode Island* (1893) IV. 139 The maple tree is the north east Corner bound. **1810** F. A. Michaux *Hist. Arbres Forestiers de l'Amérique Septentrionale* I. 28 Red flowering maple,..Swamp maple,..Soft maple..[ou] Maple tree..dans Pensylvanie, la Virginia, et l'Ohio. **1867** J. N. Edwards *Shelby* xx. 337 To send the blood coursing through his veins like the sap in the maple-trees. **1974** *Country Life* 3–10 Jan. 28/1 One of the earliest..signs that the winter freeze is coming to an end is the sight of buckets hanging on the trunks of maple trees.

mapling (mēi·pliŋ), *sb.* [f. Maple.] A ripple-like figure in wood, characteristic of maple.

1909 W. Bateson *Mendel's Princ. Heredity* 144 A damasked pattern showing where the mapling would have been if the plant had been a coloured one.

mappable (mæ·păb'l), *a.* [f. Map *v.*[1] + -able.] That may be represented on or by a map.

1920 A. S. Eddington *Space, Time & Gravitation* vi. 106 Hurdle-counts will no longer be accurately mappable on a plane sheet of paper, because they do not conform to Euclidean geometry. **1969** Bennison & Wright *Geol. Hist. Brit. Isles* i. 14 A bed of particular lithology, and mappable as a recognizable stratigraphic unit. **1974** *Nature* 8 Feb. 344/2 In area 105 a mappable disconformity has been recognised at a stratigraphic level just above the KBS Tuff.

Hence **mappabi·lity.**

1969 *Proc. Geol. Soc.* Aug. 145 Vertical and lateral boundaries should be based on the lithological criteria that provide the greatest unity and practical utility (i.e. essentially, mappability).

mapping, *vbl. sb.* Add: **b.** *Genetics.* The making of a genetic map (*map *sb.*[1] 1 g); the process of determining the chromosomal position of a gene in relation to other genes.

1935 L. H. Snyder *Princ. Heredity* xv. 153 (*heading*) The mapping of chromosomes. **1965** *Genetics* LI. 157 (*heading*) Mapping of temperature-sensitive mutants in bacteriophage T5. **1967** *Jrnl. Molecular Biol.* XXVII. 163 (*heading*) Genetic mapping in *Bacillus subtilis*. **1969** A. M. Campbell *Episomes* i. 7 Of special interest was the mapping of the Hfr character itself. **1970** Ambrose & Easty *Cell Biol.* x. 334 (*heading*) Gene mapping with *Drosophila*.

2. *Math.* A correspondence by which each element of a given set has associated with it one element (occas., one or more elements) of a second set. (Some topologists use *transformation* in this sense and *mapping* in the narrower sense of a continuous transformation.)

1931 H. P. Robertson tr. *Weyl's Theory of Groups & Quantum Mech.* iii. 110 A mapping or correspondence *S*.. is determined by a law which associates with each point *p* of the field a point *p'* as image. **1935** [see *homomorphic *a.* 2]. **1958** G. T. Whyburn *Topological Analysis* ii. 24 A continuous transformation will be called a mapping. **1964** W. J. Pervin *Found. Gen. Topology* i. 10 Other terms for mapping are 'function', 'transformation', and 'operator'. **1966** F. M. Hall *Introd. Abstr. Algebra* I. ix. 156 If the mapping θ is between the object space *A* and image space *B* we write $\theta: A\to B$ (a mapping θ of *A* into *B*). **1968** M. Bruckheimer et al. *Math. for Technol.* ii. 17 An example of a one-many mapping is the mapping $a\to\pm\sqrt{a}$ of the set of non-negative real numbers to the set of all real numbers. **1971** *Nature* 17 Dec. 396/2 The ideal transducer..neither stores nor dissipates energy; viewed as a mapping from input space, (e_1,f_1), to output space (e_2,f_2), $e_1f_1+e_2f_2=0$.

‖ **maquereau** (makəro). Pl. **-eaux.** [Fr.] = Mackerel[2].

In quot. 1920 the form with final *-o* is idiosyncratic.

1898 A. M. Binstead *Pink 'Un & Pelican* xi. 256 His head was pounded like a Hamburg steak, and sixty-five dollars rolled into the basket, principally contributed by the females who had been accustomed to support the maquereau during his life. **1919** C. Mackenzie *Sylvia & Michael* i. 21 The officer..called him a *maquereau*. **1920** E. Pound *Hugh Selwyn Mauberley* 14 Bewildered that a world Shows no surprise At her last maquero's Adulteries. **1922** F. M. Ford *Let.* 15 Aug. (1965) 144 It is really too much to expect the very mildest of men to aid in sharpening a pen whose sole occupation is the describing of himself as, let us say, maquereau. **1933** 'G. Orwell' *Down & Out* x. 78, I counted the number of times I was called *maquereau* during the day. **1939** E. Ambler *Mask of Dimitrios* xi. 214 The clientèle began to change... We had ..more *maquereaux* and fewer gentlemen, more *poules* and fewer *chic* ladies. **1971** D. Wallis *Bad Luck Girl* i. iii. 30 No girl..goes on ship these days. They..lead you to some place where her *maquer*[*e*]*au* can slug you.

maquette (mǎke·t). [Fr. (1752), ad. It. *macchietta* speck, little spot, dim. of *macchia* spot, f. *macchiare* to spot, stain, f. L. *maculāre*;

cf. MACULATE *v.*] A small preliminary model, in wax or clay, etc., or a preliminary painted sketch, from which a work in sculpture is elaborated. Also *transf.* and *fig.*

1903 *Athenæum* 24 Jan. 122/3 M. J. B. E. Detaille has, after a long delay, executed four *maquettes*, each comprehending three large panels. **1926** W. J. Locke *Stories Near & Far* 78 The maquette or model in clay. **1951** H. Read *Meaning of Art* (ed. 3) 240 The sculptor's maquette, or model, was reproduced, generally by other hands, either by casting in bronze, or by being reproduced to scale by mechanical methods in marble. **1958** *Times* 8 Oct. 6/4 One might describe his art as a prolonged *maquette* for some ultimate synthesis or other. **1965** Zigrosser & Gaehde *Guide to Collecting Orig. Prints* ii. 16 In some instances, the artist has actively collaborated in the adaptation of his own maquette, or sketch, by working on the plates or stones, and by 'proving' and approving the color separations. **1970** *Country Life* 31 Dec. 1260/3 This was the oil maquette for the figure of the wife on the Westminster Abbey monument to the poet Nicholas Rowe. **1972** P. Marks *Collector's Choice* iii. 181, I don't know why you had to go out and buy those Carpeaux maquettes last week. **1973** *Times* 11 Apr. 6/4 (*caption*) A maquette of Henry Moore's 'Family Group'.

‖ **maquillage** (makiyɑ̄ʒ). [Fr., f. *maquiller* to make up one's face, f. OF. *masquiller* to stain, alt. of OF. *mascurer* to darken.] The action of applying make-up to one's face; also, make-up, cosmetics; also *transf.* Hence **maquilla-ged** *a.* made up.

1892 *Ladies' Home Jrnl.* Dec. 8/2 All this is..thrown away upon the devotees of maquillage. **1921** M. Sadleir *Privilege* vi. 81 It was a relief to find Anthony innocent, at least, of maquillage. **1929** D. H. Lawrence *Pansies* 55 The caged mind..leaves a rind Of maquillage and pose and malice to shame the brutes. **1938** E. Ambler *Cause for Alarm* iii. 47 The edge of a heavy and clumsily applied maquillage. **1941** 'R. West' *Black Lamb* (1942) II. 417 Doubtly dazzling with the radiance of a Slav blonde and an ample maquillage. **1957** S. Gainham *Cold Dark Night* ix. 49 Her voice did not rise and there seemed no change in the beautifully maquillaged face. **1959** R. Graves *Coll. Poems* 304 Confirming hazardous relationships By kindly maquillage of Truth's pale lips. **1971** *Daily Tel.* 25 Sept. 7/3 Plastering the players all over with visual and aural *maquillage* in the form of endlessly changing lights and phonic effects.

maquis (ma·ki). Also **macquis**, (erron.) **maqui**. [Fr., 'brush-wood, scrub', ad. Corsican It. *macchia* thicket, *MACCHIA, f. L. *macula* spot.] **1.** The dense scrub characteristic of certain Mediterranean coastal regions, esp. in Corsica, often used as a refuge by fugitives. Also applied to similar areas of scrub or brushwood elsewhere.

1858 T. Forester *Rambles Corsica & Sardinia* viii. 65 There are also..several lagoons on the coast, of which the Stagna di Biguglia, near which we turned off into the *maquis*, is the largest. **1900** 'H. S. Merriman' *Isle of Unrest* xxii. 246 It is..usual for a man to take to the macquis the moment that he finds himself involved in some trouble. **1906** Conrad *Mirror of Sea* xlii. 259 Dominic's brother had to go into the *maquis*, into the bush on the wild mountain-side. **1913** A. Huxley *Let.* 30 July (1969) 55 The maquis..is thicker than anything I've ever seen in Europe: oak, hazel, maple, wild cherry and a pine or two in the more open spaces. **1923** *Nature* 24 Feb. 268/1 These types of vegetation, like the Mediterranean 'maquis', develop in regions of winter rains and long dry summers. **1932** *Forestry* VI. 171 The nature of the undergrowth—dense shrubby maquis. **1945** G. Millar *Maquis* xi. 226 Boulaye met us with a long face in the Maquis. **1957** J. Braine *Room at Top* xv. 136 The village ends abruptly.. beyond it is nothing but the moors..and a solitary farmhouse a mile west. That too has a military air; the moors are Gilden's maquis and behind its walls are planned the sudden raid, the ambush. **1962** E. Hyams *Wild Garden* 167 The Jerusalem Sage..may be used to form a ground-smothering maquis in conjunction with cistuses, lavenders, rosemary and other sun-loving shrubs. **1973** *Daily Tel.* 13 Oct. 9/6 We were..standing in a hillside garden in the Alpes Maritimes. Around us spread the maquis for unending miles. **1973** *Islander* (Victoria, B.C.) 9 Feb. 15/3 Most of the vegetation [in Corsica] is not forest as we know it..but a tangled growth called maquis.

2. *transf.* A secret army of patriots in France during the German occupation in the war of 1939–45, so named from their being conceived as hiding in country of this kind; also, a member of this army. Also in extended use (applied to similar groups) and *attrib.*

1944 *Ann. Reg. 1943* 230 The patriots of the 'maquis' fought with admirable courage. **1944** *New Statesman* 27 May 346/2 The Italian *Maquis*, whose guerrilla operations are now being directed and financially encouraged by the Allied Command. **1944** *Hutchinson's Pict. Hist. War* 12 Apr.–26 Sept. 523 (*caption*) Field-Marshal Montgomery stops his car to have a word with a Maquis guard. **1945** W. S. Churchill *Victory* (1946) 7 The tributes which are paid to the heroic French and Belgian Maquis. **1951** N. Annan *Leslie Stephen* iv. 160 Victorian thinkers..appeared to be unaware of the spiritual maquis burrowing underground in Europe. **1965** B. Sweet-Escott *Baker St. Irreg.* vii. 205 If it had still been considered important to raise a maquis in Hungary, it might well have been worth exploring. **1966** M. R. D. Foot *SOE in France* xi. 365 A large group of maquis known to exist in the Auvergne mountains.

3. *attrib.* and *Comb.*

1937 *Brit. Birds* Sept. 98 Small parties were seen..flying about the maquis-covered slopes. **1944** *Times* 24 Apr. 5/6 The *maquis* country is usually thought of as confined to Haute-Savoie. It is in fact much more extensive. **1971** *Homes & Gardens* Sept. 80/1 Panoramic vistas of maquis-covered foothills merging into the stark blue inland peaks of over 2,500 feet. **1974** J. Thomson *Long Revenge* ii. 22 The Maquis radio operator..contacted us.

Hence **maquisard** (makizɑ̄r), a member of the maquis. Also *transf.* and *attrib.*

1944 *Times* 24 Apr. 5/7 Often enough the *maquisards* have had to fight against heavy odds... Today..the situation is better. Arms are reaching the *maquis* in growing volume. **1945** G. Millar *Maquis* v. 85 The greyheaded Maquisard, a local railway worker in peace-time. **1959** *Economist* 3 Jan. 23/2 He has always insisted that, except for a tiresome band of *maquisards* in the eastern extremity of the island [Cuba], everything was under control. **1963** N. Freeling *Gun before Butter* ii. 203 During the war presumably, he had got acquainted with the local Maquisard formation, doubtless. **1967** 'E. Peters' *Black is Colour* viii. 152 The moustache that would have done credit to a Corsican *maquisard*. **1973** *Times* 18 Jan. 19/6 The German onslaught..was disastrous to the maquisards who stayed and fought.

mar, *sb.* Add: **3.** *Comb.*: **mar resistance**, resistance to loss of gloss by abrasion; so **mar-resistant** adj.

1941 L. Boor in *Mod. Plastics* Sept. 80/2 Depending on the type of plastic under consideration,..various methods of evaluating the optical degradation of the test areas may be used... The particular aspect of hardness of which this test is a measure has been given the tentative designation 'mar resistance'. **1969** C. O. Hasson in W. R. R. Park *Plastics Film Technol.* iv. 87 The gloss of the marred area is measured and compared to the unabraded film to determine the mar resistance. **1973** *Washington Post* 13 Jan. C3/4 (Advt.), Seven-drawer kneehole desk..with mar-resistant top.

Mar[2] (mɑ̄r). [Aramaic.] An honorific title for saints and higher clergy, chiefly in the Nestorian and Jacobite churches.

1694 M. Geddes tr. *Hist. Church of Malabar* 11 Their Bishop at that time was Mar Joseph, who..had been brought thither by Mar Audixa, Patriarch of Babylon. **1864** G. B. Howard *Christians of St. Thomas* iv. 154 In the year 1653 Archdeacon Mar Athanasius Stephanos made his appearance,..claiming to be the rightful Metropolitan of the Christians of St. Thomas. **1892** G. M. Rae *Syrian Church in India* xix. 304 The first native of Malabar that ever received consecration..was Mar Athanasius Matthew. *Ibid.* 309 The pretensions of Mar Koorilos were finally disposed of; Mar Dionysius resigned his dignity..; Mar Athanasius was by royal proclamation declared metropolitan. **1933** *Encounter Rev.* LI. 390 Fr Hugh Connolly was editing a number of Texts and Studies consisting of a translation of certain Syriac liturgical homilies of Mar Narsai, a fifth century Nestorian theologian. **1956** N. Zernov *Christian East* ix. 103 In 1665 a Monophysite bishop, Mar Gregorius, arrived in South India from Palestine. *Ibid.* 105 The Bishops Mar Mathew Athanasius and Mar Thomas Athanasius formed a separate Church, now known under the name of the Mar Thoma Church. **1964** P. F. Anson *Bishops at Large* 27 They have given themselves impressive ecclesiastical titles,..Hierarch, Mar, Metropolitan.. and so on.

maquis (ma·ki), *sb.* Add: **2. b.** Also *absol.*

marc, *sb.* Add: **2. b.** Also *absol.* c **1420**, **1542** [see MARE *sb.*[1] 46 b]. **1853** Mrs. Gaskell *Ruth* III. i. 43 For the present we should neither meddle nor mar in Ruth's course of life. c **1865** G. M. Hopkins *Poems* (1967) 121 So be it; I must maim and mar. **1889** [see MAKE *v.*[1] 46 b].

mar-. *mar-feast* (further examples) (*arch.*).

1897 W. Morris tr. *Homer's Odyssey* XVII. 446 What God brings this pain, this mar-feast, hath hither sent to us? **1922** Joyce *Ulysses* 710 Superannuated bailiff's man, marfeast, lickplate, spoilsport, [etc.].

marabou[1]. Add: **3.** (Further *attrib.* examples.)

1973 *Times* 10 Oct. 12/6 An attractive bridge coat..in antiqued silk velvet finished with a marabout collar. **1975** *Lady* 1 May 881/2 Marabou cap and muff.

marabunta (mærəbʌ·ntə). [Native name.] In Guyana, a name used for several social wasps. Also *attrib.*

1883 E. F. im Thurn *Among Indians Guiana* v. 130 The wasps vary much more in general appearance and size, some of them being large and beautifully coloured insects. The forest-dwelling social species are indiscriminately called by the colonists marabuntas. **1888** H. Kirke 25 *Yrs. Brit. Guiana* iii. 76 A large brown wasp, called a marabunta, builds his pretty paper combs under the eaves and galleries of our houses; his sting is severe. **1928** C. W. Beebe *Jungle Peace* (1919) vi. 81, I listened to the buzzing of a marabunta wasp. **1934** E. Waugh *Handful of Dust* iv. 267 A marabunta had left a painful swelling on his hand. **1965** J. Carew *Black Midas* i. 12 Marabunta built clay nests in the thatch. *Ibid.* vi. 141 Tonic and Woody Sam removed spiders, cobwebs, marabunta nests..from the roof.

maraca (mærɑ̄·kə). Also **maracca**. [Pg. *maracá*, prob. from Tupi (Webster).] Usu. *pl.* A Latin-American percussion instrument made from the dried shell of a gourd or other material with beans or beads, etc., inside to produce a rattling sound; a gourd rattle. Also *attrib.*

1824 H. E. Lloyd tr. *Spix & Martius's Travels in Brazil* II. iv. 226 The *Maracá*, a longish gourd shell, filled with maize, fastened to a handle, with which in their dancing, they make a rattling as with castanets. **1928** *Vanity Fair* (N.Y.) Nov. 72 A fashionable evening event along the Havana water-front is a concert by black boys with their primitive African instruments, the bongo, timbales, guiro, maracas, and claves. **1933** *Punch* 4 Jan. 6/3 It [sc. the rumba] may be recognized by a number of taps produced in rapid succession... They are made by a couple of maracas in the orchestra... Maracas..are the dried husks of the calabash fruit. **1952** See *ILLUSTRA-NIA *a.* and *sb.*] **1954** J. Steinbeck *Sweet Thursday* ii. 16 An orchestra took shape—two guitars, a guitar, rhythm and maraca men. **1960** *Melody Maker* 23 July 10/5 Alan Haven mentioned maracca sticks and jingle-sticks. **1971** 'E. Anthony' *Tamarind Seed* i. 13 A coal-black barman ..rattling a shaker as if he were playing the maraccas. **1972** *Jazz & Blues* Sept. 51 The almost African rhythms with much maracca shaking.

marae. Add: Now usu. with pronunc. (mæ·rəi). The sense in Dict. is now only *Hist.* (Later examples.) Now, among Polynesian peoples, a space in front of a meeting-house or among the houses of a village, set apart for social functions. Also *fig.*

1877 Ranken in *Jrnl. R. Anthrop. Inst.* VI. 236 The *marais*, or terraced enclosures for sacred purposes, are exactly like those of Mexico and Peru. **1905** W. B. *Where White Man Treads* 57 Every order and tribal regulation had to be decided in meeting-house convened, and proclaimed in the marae [village green] to the whole people. **1910** J. Cowan *Maoris N.Z.* xxxii. 338 We fall in on the lawn—the marae or parade-ground up to the village marae. **1920** C. G. Seligmann *Melanesians Brit. New Guinea* xx. 223 Each local group of each clan has..at least two *marae* [sic], which serves as the meeting place for the men of the local group, and is their clubhouse in the fullest sense of the term. **1924** R. W. Williamson *Social & Pol. Syst. Cent. Polynesia* I. 350 This *malae* was, I presume, a *malae* or open space where meetings..were held, as in Samoa and Tonga. *Ibid.* II. xv. 60 In some islands the *marae* or *malae* was merely an open space..along with a special wooden house erected in it. **1934** *Nature* 10 Nov. 740/2 The turtle.. was the food of the gods [in Tahiti], eaten only by chiefs and priests at the *marae*. **1937** *Man Colour Scheme* x. 166 They were hangin' about the Marae in groups. **1949** P. H. Buck *Coming of Maori* (1950) iv. 18 On coming to New Zealand, it is a curious fact that the two fundamental features of the central Polynesian temples were not combined but remained as distinct entities. Thus the open court, distinguished by the term *marae*, is retained as a secular feature in front of the tribal or family meeting houses. **1959** A. McLintock *Descr. Atlas N.Z.* 72 In the country proper [i.e., districts] the Maori has retained his traditional mode of life which is symbolised in the *marae*..It is a literal sense the term means the open courtyard in front of the communal meeting house; today, however, it embraces all aspects of community life—community buildings, tribal gatherings, church activities, and other marae functions. **1966** Tindale & Lindsay *Rangatira* iii. 32 The central stone of Taputapuatea,..the marae temple, floored with slabs of basalt. *Ibid.* x. 93 They entertained their guests on the open marae ground among the houses. **1965** *Weekly News* (Auckland) 1 May 6/6 Dominating the marae was a cross 40 feet high illuminated with coloured lights. **1969** *Islander* (Victoria, B.C.) 6 July 5/2 The design [of a hotel in Tahiti] is reminiscent of ancient Tahitian outdoor worshipping temples (marae). **1974** *N.Z. Listener* 20 July 10/3 There was a group being welcomed on to the marae as I arrived.

marage (mɑ̄·rèidʒ), *v. Metallurgy.* [f. MAR-[TENSITE + -AGE *v.*] *trans.* To allow (a maraging alloy) to cool slowly in the air so that it develops great strength without significant changes in dimensions as a result of the transformation of austenite to martensite and subsequent age-hardening of the martensite.

1962 *Trans. Amer. Soc. Metals* LV. 524/2 Specimens were initially maraged 3 hr. at 900°F., then were maraged for various times at a second temperature. **1963** *Engineering* 24 May 715 When homogenised and maraged [sc. steel] is claimed not to be notch sensitive despite its exceptional strength.

So **ma·raged** *ppl. a.*, **ma·raging** *vbl. sb.*, esp. in *marag(e)ing alloy*, *steel*, steel that has been or may be hardened by maraging, generally containing up to 25 per cent nickel and smaller amounts of titanium, cobalt, molybdenum, or other elements.

1961 *Engineering* 24 Mar. 407/3 The original American release says that the steel develops its high strength by means of a remarkably easy heat treatment involving age-hardening of martensite. For this they have invented a term 'mar-aging'. **1962** *Trans. Amer. Soc. Metals* LV. 61/2 Maraged hardness increased linearly [sic] as the product, cobalt times molybdenum, increased. *Ibid.* 520/2 In annealing of 18% Ni maraging steel, temperatures in the range 1100 to 1300 F should be avoided to prevent retained austenite. **1968** R. Kurkas *Physical Metall.* from Ibid..

marahuana, marajuana, varr. *MARIJUANA, MARIHUANA.

Maranao (mæ·rɑ̄nau). Also **Maranaw**. [ad. Maranao *Maranâw*, f. *ranaw* lake.] **a.** A Moro people inhabiting the province of Lanao del Sur and parts of central Cotabato province in the island of Mindanao (in the Republic of the Philippines), and some areas of northern Borneo. **b.** The Austronesian language of these people.

1957 [see *ILLANO]. **1962** H. C. Conklin in J. A. Fishman *Readings Sociol. of Lang.* (1968) 419 Pronoun systems in Tagalog, Ilocano,..Maranao,..and some other Philippine languages. **1963** H. Otfelius *Islands of Pleasure* 126 There are also the island Moros. The Maranao and Maguindanaw are farmers... Because their main home is located around Lake Lanao on the island of Mindanao the Maranaos are commonly called the Lake Moros. **1968** J. Kirkup *Filipinescas* viii. 154 The noble movements.. portrayed a Maranaw warrior. **1969** J. Cockcroft *Philippines* 55 About 87 different languages and dialects are spoken in the Philippines... Tagalog, spoken in Manila,..Maranao, in Lanao. **1974** *Encycl. Brit. Micropædia* VI. 595/1 Like other Filipino Muslims, the Maranao differ markedly from the Christians. **1975** *Language* LI. 365 At the opposite extreme is a language like Maranao, in which there seems to be a specific Voice morpheme for each possible realization of a verb's subcategorization feature.

Maranatha. (Further examples.)

1846 W. F. Hook *Church Dict.* (ed. 5) 398 *Maranatha* could not be any species of ex-communication, but only a reason for pronouncing *Anathema* against those who express their hatred against Christ, by denying Him to be the Messiah. **1938** C. S. Lewis *Out of Silent Planet*.. **1926** A. Chambers tr. *Amounchy's Mysticism & Eastern Church* vi. 121 And, finally, the closing cry quivering with joyous awe: Maranatha! ('Come, our Lord!'). This appeal in Aramaic takes us back to the earliest period of the primitive Church in Jerusalem. **1961** *New Engl. Bible* I *Cor.* xvi. 22 *Marana tha*—Come, O Lord! **1964** E. A. Nida *Toward Sci. Transl.* viii. 170 Such expressions as 'Abba Father', 'Maranatha', and 'baptized into Christ' could be used with no explicit knowledge of their origin.

maraschino. Delete ‖ and add: **1. b.** Also, **maraschino cherry**, a cherry preserved in real or imitation maraschino.

1905 'O. Henry' in *N.Y. World Mag.* 22 Jan. 2/1 The world seemed no larger than the red spot on a Maraschino cherry in a table d'hote grapefruit. **1918–19** T. *Eaton & Co. Catal.* Fall & Winter 389/1 Luscious Whole Red Maraschino Cherries in a snowy-white ice-cream, coated with fine chocolate. **1961** [see *CRÈME 1 b]. **1964** J. Drummond *Welcome, Proud Lady* xxi. 96 Maraschino cherries, olives, sliced lemon, salted nuts were arranged in small crystal bowls.

marasmus. Delete ‖ and add to def.: esp. in undernourished children. (Further examples.)

1900 W. G. Thomson *Pract. Dietetics* (ed. 2) viii. 464 Marasmus is a form of starvation occurring chiefly in artificially fed infants, but also in those at the breast, in whom there is great wasting of the muscular and other soft tissues. **1951** R. W. B. Ellis *Dis. in Infancy & Childhood* vii. 234 Marasmus (Infantile atrophy). This is a condition of extreme and chronic malnutrition, and whilst it often arises simply from underfeeding, it may also be due to a variety of other causes, e.g. congenital syphilis ..or parasitic infection. It is not therefore a disease *sui generis* but a clinical picture of which it is first necessary to discover the cause before treatment can be satisfactorily undertaken. **1968** Meneghello & Rizzardini in A. Dorfman *Child Care in Health & Dis.* iii. 42 Almost all children in Chile suffer from so-called calorie-protein malnutrition, or marasmus, and this form of malnutrition is the major problem in Chile... The other form, protein malnutrition or kwashiorkor, is the most common type in other countries of the region. **1971** *Sci. Amer.* Oct. 54/3 They have advanced the understanding of the starvation disease called marasmus, which is increasing in many developing countries because mothers are giving up prolonged breast-feeding and their infants are not receiving an adequate substitute diet during a critical time in development.

Maratha, var. MAHRATTA.

1858 [see MAHRATTA 3]. **1886** Kipling *Departmental Ditties* (ed. 2) 7 Rajah Runtum..Roused his Secretariat to a fine Maratha fury. **1913** E. M. Forster *Let.* 1 Jan. in *Hill of Devi* (1953) 12 Cooked rakishly over one eye was a Maratha Turban of scarlet and gold. **1925** S. N. Sen *Administrative Syst. Marathas* (ed. 2) p. vii, The Marathas ..engaged in a life and death struggle. **1942** *R.A.F. Jrnl.* 13 June 4 The Maratha heirs of Sivaji. **1971** R. Russell tr. *Ahmad's Shore & Wave* i. 10 The Mughal Empire fell and the Marathas overspread the land. **1973** *Times* 16 Apr. 12/1 A Maratha Hindu for the city and Gujarati for the dominant Maratha caste to the south has been raised.

Marathi. Delete 'obs.' and add examples.

1827, 1868 [see MARATHI]. **1835, 1880** [see *KANARESE *a.* and *sb.*]. **1908** T. G. Tucker *Introd. Nat. Hist. Lang.* 127 Marāthī in the north-west Deccan. **1932** N. S. Vaswani *Administrative Syst. Marathas* (ed. 2) 2 Next in importance, are the *bakhars* or Marathi prose chronicles. **1933** L. Bloomfield *Lang.* iv. 63 Such great languages as Marathi. **1949** G. Orwell *Note-Bk.* in *Coll. Ess.* (1968) IV. 512 We've always had difficulty with the Marathi newspapers. **1953** E. M. Forster *Hill of Devi* 41 He could recall his father speaking Marathi and praying in a Marathi temple. **1958** A. Toynbee *East to West* xxxii. 99 If the city [sc. Bombay] had been merged in a separate Marathi-speaking state, [etc.]. **1967** 'W. Haggard' *Conspirators* ii. 20 His parents had spoken Marathi, beautiful Brahmin's Marathi. **1971** *Illustr. Weekly India* 25 Apr. 21/2 The personal feelings of eminent Marathi poet P. S. Rege. **1971** *Femina* (Bombay) 30 Apr. 16/2 (Advt.), In four languages: English, Hindi, Gujarati and Marathi. **1971** H. R. F. Keating *Inspector Ghote goes by Train* ii. ix. 211 He ordered in Hindi. He ordered in Marathi. He ordered in crude English.

marathon (mæ·rəþǫn). Also **Marathon**. [The place-name *Marathon* (Gr. Μαραθών) on the north-east coast of Attica: see MARATHONIAN *a.*] **a.** The name first given on the occasion of the revived Olympic Games held in Athens in 1896 to a long-distance foot-race (now usu. of 26 miles 385 yards), with allusion to the run of Pheidippides at the time of the battle at Marathon in 490 B.C., as recorded in Herodotus and later sources. Also *attrib.*

Herodotus records that Pheidippides ran from Athens to Sparta to secure aid before the battle; the race first instituted in 1896 was based on a later less sound tradition that Pheidippides ran from Marathon to Athens with news of the Persian defeat.

1896 *Fortn. Rev.* June 950 We now come to the great glory of the Greeks—the victory in the Marathon Race. **1905** *Programme of Olympic Games Athens* 1906 3 Flat Races... Marathon Race, 42 kilometers. From Marathon to Athens on the road. **1908** T. A. Cook *Olympic Games* 82 The whole of Hellas seemed concentrated at Athens to see the result of the great Marathon Race in the stadium. **1908** *Westm. Gaz.* 22 July 1/3 There are two things which no one who wishes to win the Marathon can ever afford to forget. **1936** *Discovery* Feb. 48/2 Never would a Marathon runner or a racing car. **1955** I. F. N. Petre *In Long Run* viii. 71 On July 21st..the A.A.A. Marathon Championship was due to be run at Birmingham on a roughly circular out-and-home course. **1964** H. Watman *Encycl. Athletics* 108/1 Owing to the disparity in the nature, if not distance, of various courses, marathon times should not be taken too seriously. *Ibid.*, The only woman on record to have completed a full marathon is Dale Greig. **1966** J. Hopkins *Marathon* 11 British runners are to be found among the leaders in almost every international Marathon.

b. In extended uses, applied to other long-distance races or competitions calling for endurance. Also *attrib.* (and as quasi-*adj.*).

1908 *Daily Chron.* 5 Nov. 1/2 A competition..under a title of 'The Marathon' was decided last night... It was intended that the contestants should..peel a quarter ton of potatoes. **1928** *Daily Express* 2 July 11/5 The dance marathon here ended at midnight, nine couples left in the competition stepping together after twenty days of continuous dancing. **1932** G. Morton *Mystery of Hermit's End* ii. 19 The latter was taking intensive training for marathon swimming. That form of sport had caught on. **1932** *Times Lit. Suppl.* 26 May 393/1 Its friend Cap Bridges, the marathon swimmer. **1968** *Radio Times* 28 Nov. 8/3 The London-Sydney Motor Marathon. *Progress Reports*. **1969** *New Yorker* 20 Dec. 64/2 One of the phenomena of America's Depression days was the marathon dance. **1975** *Ibid.* 10 Mar. 100/2 It's an album of pictures: fine dancing, marathon dancing, the Lindy.

c. Other *transf.* and *fig.* uses. Esp. an event or activity of long duration.

1915 'Bartimeus' *Tall Ship* x. 185 That was a bit of a Marathon, wasn't it? He measured the distance across the lawn with a humorous eye. **1951** *N.Y. Herald-Tribune* 29 Nov. 1 The House of Commons ended a marathon... after sitting through a marathon session of 20 hours and 20 minutes. **1952** C. Day Lewis *Italian Visit* iii. 39 After a marathon walk through the Vatican galleries. **1958** W. Gaddis *Recognitions* II. vii. 613 There's been somebody tagging around after me all day, this marathon walker, I met him in a bar. **1971** *Britannica Bk. of Year 1970* 72/2 *Marathon*, *adj.*: a group session in which members remain together for an extended period (24 to 48 hours) and interact openly and responsively so as to increase self-understanding. **1972** *Daily Tel.* 7 June 2/3 The question whether a judge and special assessors should replace the conventional jury in complicated and marathon criminal cases.

Hence **ma·rathon** *v. intr.*, to run as in a marathon race (*rare*); **ma·rathoning** *ppl. a.* (*rare*).

1920 *Chambers's Jrnl.* Aug. 519/2 Do I have to marathon ten miles and back? **1964** H. Watman *Encycl. Athletics* 107/2 He [sc. Kolehmainen] won by less than 13 sec...the closest result in Olympic marathoning history.

marble, *sb.* Add: **4. b.** Phr. *to pass in one's marble* and varr., to die, to give up. *Austral. slang.*

1908 *Austral. Mag.* 1 Nov. 1250 Instead of dying you can 'chuck a seven', 'pass in your marble', or 'peg out'. **1918** A. Wright *Over Odds* 102 'I suppose the old pot knew y'old man before he passed in his marble,' ventured Dick. **1924** *Truth* 27 Apr. 6 *Throw in the marble*, to relinquish. **1951** D. Stivens *Jimmy Brockett* 304 I'm not going to pass in my marble just yet.

c. Phr. *to make one's marble good*: to make a good impression (on a person), to ingratiate oneself, to improve one's position. *N.Z.* and *Austral. slang.*

Quot. 1938 illustrates a similar *S. Afr.* use.

1936 *Mixer Transport Workers' Song Bk.* 31 Some taps the boss before they join,..By this they make their marble good. **1938** A. M. Brown *Sh. Africa*.. **1941** Baker *Dict. Slang* (1961) *Suppl.* 1179/1 A word I have heard used in the Cape [Province], mostly from people attending Rhodes University College, Grahamstown, is *marble*. Examples are: 'His marble is high'—he is 'well-in' (with such-and-such a person). 'He is polishing his marble with so-and-so'=he is trying to ingratiate himself. **1944** J. H. Fullarton *Troop Target* iii. 26, I was making my marble good. **1947** D. M. Davin *Gorse blooms Pale* 206 The crowd.. wanted to see if he could make his marble good with us. **1963** D. Crick *Martin Place* 223 Take my tip, if you wanter make your marble good: say nothing.

5. c. (See quots.) Also **marble crust**.

1894 *Cowral Winter Sports* No. 12/1 *Marble*, a snow-crust formed by alternate freezing and thawing. Found on Southern slopes. **1920** F. Lunn *Ski-ing* 49 *Frozen* xvii. 90 Snow *Marble* crust is so slippery that it is almost impossible to obtain a purchase on it with the skis. **1969** H. Hiller *Ski* xiv. 185 *Marble* crust looks like its name and is formed by the wind. The snow is of high altitude in early winter.

6. c. As (false) transl. of F. *meubles*: furniture, movables, personal effects; 'the goods'. *slang.*

1864 Hotten *Slang Dict.* 176 *Marbles*, furniture, moveables; 'money and marbles', cash and effects. **1867** Trollope *Claverings* II. vi. 67 She won't get any money from me, unless I get the marbles for it. **1896** Farmer & Henley *Slang* IV. 280/1 *Marbles*,..furniture; moveables. *Money and marbles*=cash and effects. [From Fr. *meubles*.] Hence, any substantial good or thing.

d. *pl.* Mental faculties; brains; common sense. *N. Amer. slang.*

1927 *Amer. Speech* II. 360 *Marbles, doesn't have all his* (verb phrase), mentally deficient. 'There goes a man who doesn't have all his marbles.' **1953** J. Pollock *Underworld Speaks* 75/1 *Marbles*, the brain. **1957** M. Millar *Soft Talkers* i. 7 She's a fattish little bugger with some of her marbles missing. **1958** Wodehouse *Cocktail Time* xvii. 148 Do men who have got all their marbles go swimming in lakes with their clothes on? **1967** M. L. Roby *Cat & Mouse* i. 19 He ain't right in the head. Lost his marbles, mister? **1970** J. Wainwright *Take-Over Men* i. 8 You lost your goldarn marbles? **1973** *Ottawa Jrnl.* 6 Feb. 9/4 'I still have most of my marbles,' he said cheerfully. **1973** R. Parkes *Guardians* xi. 204 Crazy bastard... I think he's blown his marbles.

7. b. *marble-hearted* adj. (earlier and later examples).

1605 Shakes. *Lear* I. iv. 283 Ingratitude! thou Marble-hearted Fiend. **1927** R. Graves *Poems (1914–26)* 203 Not marble-hearted but your own true love.

8. a. *marble chips*, *saw* (earlier and later examples); **c.** *marble-built*, *-flagged*, *-slabbed*, *-stoppered* adjs.; **d.** *marble-still*, *-tall* adjs.

1791 W. Blake *French Revolution* 1, in *Compl. Writings* (1972) 138 Shall this marble built heaven become a clay cottage. **1** 1908 H. Crane *Let.* 19 Aug. (1965) 173 Examining pebbles and cinders and marble chips through the telescope. **1946** *Happy Landings* (Air Ministry) July 2/2 White stone or marble chips spread out and rolled into the macadam surface of the roads. **1928** in *Nature* 1 Dec. W. B. Yeats *Wanderings of Oisin* 53 A marble-flagged, pillared room. **1839** *Hist. Dict. Arts* 801 The marble saw is a thin plate of soft iron, continually supplied..with water and the sharpest sand. **1890** *Cent. Dict.*, *Marble-saw*,..a machine for cutting marble... Such machines will cut a block of marble into several slabs simultaneously, or can be arranged to cut out pyramidal blocks, or to shape a cylinder or a frustum of a cone. **1864** *Harper's Mag.* Dec. 40/1, I had a snug marble-slabbed brick house. **1933** R. Crompton *William—the Rebel* viii. 162 Lay his catch upon the marble-slabbed hat-stand in the hall of the inn. **1904** W. de la Mare *Henry Brocken* xiii. 168 He stood, thus, marble-still. **1972** *Country Life* 30 Nov. 1481/3 The screw-topped or marble-stoppered lemonade bottles of long ago. **1938** Belloc *Sonnets & Verse* 196 The Islands have received it, marble-tall.

9. marble bone *Path.* [tr. G. *marmorknochen*], (*a*) (also *pl.*) = *OSTEOPETROSIS; also called *marble bone(s) disease*; (*b*) an affected bone in a person with osteopetrosis; **marble cake** orig. *U.S.*, a cake made of light and dark sponge, having a mottled appearance suggestive of marble; **marble orchard** *U.S. slang*, a cemetery; **marble-player**, **marbles-player**, one who plays the game of marbles; **marble-top** *usu. attrib.*, designating a piece of furniture the top of which is covered with marble; also **marble-topped** adj.; **marble town** *U.S. slang* = *marble orchard*.

1922 *Arch. Surg.* V. 462 In 1921, Schultz discussed the nature of the disease of marble bones (Albers-Schönberg). **1922** *Jrnl. Amer. Med. Assoc.* 2 Dec. 1955/2 A patient..was found, on roentgen-ray examination, to have a pathologic fracture as the result of a rather obscure bone condition which has been termed osteosclerosis fragilis generalisata, Marmorknochen (marble bone), or Albers-Schönberg disease. **1947** *Brit. Jrnl. Radiol.* XLIII. 73 Marble bone disease is due to..an unknown agent which damages the bone-forming blastema at the beginning of the second period of development of each individual bone. *Ibid.* 73 Fractures in marble bones, for the most part, do not splinter. **1961** R. D. Baker *Essent. Path.* xxi. 560 In osteopetrosis (Albers-Schönberg's disease; marble bones) the bones are abnormally hard and thick, but are easily fractured. **1973** Forfar & Arneil *Textbk. Paediatrics* xxiii. 1525/1 (*heading*) Albers-Schönberg disease (osteopetrosis, marble bones disease). **1871** Mrs. T. J. V. Owen *Illinois Cook Bk.* 202 Marble Cake... White part... Three teacupsful white sugar,..Dark part... Three teacupsful

brown sugar, One teacupful molasses, [etc.]. **1878** N. A. Donnelley *Lakeside Cook Bk.* 29/1 Marble Cake. **1903** K. D. Wiggin *Rebecca* xxvi. 290 She began to stir the cake. **1971** M. McCarthy *Birds of America* 74 My husband used to like a marble cake. **1929** M. A. Gill *Underworld Slang* 8/1 *Marble orchard*, cemetery. **1940** J. Cain *Mildred Pierce* 155 You'll get your names in the marble orchard. **1973** R. Beck *Trick Baby*.. *Ten Lost Years* x. 110 A couple more punches and it would have been the marble orchard for him. **1950** A. Bennett *Clayhanger* I. i. 9 Six men playing the noble game of rinkers... They were celebrated marble-players. **1955** *Publ. Amer. Dial. Soc.* xxiii. 7 Marble players are not imaginative as far as their terminology is concerned. **1959** I. & P. Opie *Lore & Lang. Schoolch.* xi. 228 Young marbles players..easily become prey to strange thoughts. *Ibid.*, Marbles..players are addicted to charms. **1883** *Head & Son Catal.* Sept. 2025 Hall Tables... St. Ann's Marble Top. **1891** 'O. Thanet' *Otto the Knight* 60 [She was] a woman of property,..owning two marble-top bureaus and a sewing-machine. **1963** *Homes & Gardens* Feb. 1 Marble top coffee table, 36" × 15", £35 10s. 0. **1849** Marble-topped [see LONGFELLOW adv.]. **1854** Miss Cooper *French Life*.. in *Fraser's Mag.* Apr. 433/2 The 'guéridon' (round, marble-topped table)..the cafe-frequenter's *article* in a French drawing-room. **1886** 'Mark Twain' *Let.* 7 Aug. (1920) 257 They never used a stove, but cooked their meals on a marble-topped table. **1934** J. C. Wood *Captain's Deathbed* (1950) 181 There are marble-topped tables at the corner. **1959** W. Golding *Free Fall* iv. 85 When we were sitting at the marble-topped table my plans began to come apart. **1971** *M. Lee Dying for Fun* xlii. 203 Would he have to change the décor of his flat,.. these marble-topped café tables? **1973** *Times* 18 Sept. 14.. The tomb..was thought to have been placed under one of the marble-topped tables in..the [hotel] lobby. **1945** L. Shelly *Jive Talk Dict.* 29 *Marble town*, a graveyard. **1970** C. Major *Dict. Afro-Amer. Slang* 80 *Marble town*, (1940's) a cemetery.

marbleize, *v.* (Further, including U.K., examples.)

1909 Galsworthy *Fraternity* vii. 60 She had before her two little books. One of these was bound in marbleized paper. **1974** *State* (Columbia, S. Carolina) 3 & 4 Mar. G 13/6 She was more than happy to pay the extra fabrics and tees on two dozen heavy gold-and-white marbleized candles.

marbling, *vbl. sb.* Add: **2. b.** In meat: the quality or state of having the lean streaked with thin layers of fat. Cf. MARBLED *ppl. a.* 2 c.

1935 W. H. Tomhave *Meats & Meat Products* xvi. 201 Marbling is always present in ribs, loins, and chucks. **1939** *Daily Express* 7 Nov. 5/4 There are certain indications by which the official graders can detect clearly whether or not there is 'marbling' in the beef while it is in the side. **1963** *New Yorker* 22 June 41 Marbling is the most important thing in a steak. **1972** T. McHugh *Time of Buffalo* xxiii. 231 Despite the greater cost, buffalo meat is not overpriced, for its marked leanness entails less fatty marbling and less trimming.

marbling (mɑ̄·zbliŋ), *ppl. a.* [-ING[2].] = MARBLED *ppl. a.* 2 c.

1958 *Times* 8 Dec. (Suppl.) p. viii/3 Tender beef can be procured from older animals that are well-finished, i.e., in which there has been considerable deposition of 'marbling' fat between the muscle bundles. **1971** *Country Life* 9 Sept. 643/3 The older mature animals, if well finished can still provide meat of a high quality, due in part to the intimate penetration of 'marbling fat' into the deep muscle tissue.

marc. Add: = *marc brandy*.

1934 O. Sitwell *Let.* 15 Aug. in Mrs. Belloc Lowndes *Diaries & Lett.* (1971) 126 We settled them with a glass each of their beloved marc. **1949** Mrs. Belloc Lowndes *Merry Wives Westminster*.. G. Millar *Horned Pigeon* xvi. 252 Ramon and Alban sat inside, drinking marc. *Ibid.* 253 Ramon ordered three more marcs. **1956** J. Baldwin *Giovanni's Room* ii. 44, I ordered black coffee and a cognac, a large one. Giovanni was far from me, drinking *marc*.

marcato (mɑ̄ɹkɑ̄·to), *adv.* and *a. Mus.* [It., pa. pple. of *marcare*, to mark, accent, of Gmc. origin: cf. MARK *v.*] (With each note) emphasized. Also *transf.*

1840 Busby & Hamilton *Dict. 3,000 Mus. Terms* (ed. 3) 109 *Marcato*, a term implying a strong and marked style of performance. **1842** J. F. Warner *Universal Dict. Mus. Terms* 56/1 *Marcato*, marked, distinguished, rendered prominent, as e.g. Ben marcato, well marked, in a clearly marked, distinct manner... This term is sometimes used over such passages or individual notes, or the melody is to be well marked and distinguished. **1922** Joyce *Ulysses* 214 He read, *marcato*. **1962** *Times* 6 Dec. 17/7 Mr. Spaul got exciting results with some 'marcato' arrangements.

Marcel (maɹse·l), *sb.* Also with small initial. [f. the name of François *Marcel* Grateau (1852–1936), French hairdresser who invented the method.] In full, *Marcel wave*. A kind of artificial wave of the hair produced by using heated curling-tongs. Also *fig.*

1895 in N. & Q. (1941) 6 Sept. 129/1 (Advt.), Experts in the Marcel and Last Vienna Wave. **1908** *Smart Set* Sept. 86/1 And when she 'comes to', her Marcelle wave is straight as a shad. **1909** 'O. Henry' *Roads of Destiny* 61 Man, what do you suppose she did? Loosened up that Marcel wave in the surf at Coney. **1926** *Glasgow Herald* 25 Sept. 9 It began to rain... Many a beautiful marcel was sacrificed to save a masterpiece of millinery. **1930** R. Macaulay *Staying with Relations* xvi. 226 Little marcel waves lapped at the Eugenia's white sides as they lay at anchor in the San José harbour. **1934** E. Sitwell *Aspects Mod. Poetry* i. 21 Mr. Austin Dobson, and his Marcel Waves, the wriggling, giggling horrors of his frisettes and other imitations of French forms. **1964** L. Hairston in J. H. Clarke *Harlem* 285 The waves in my hair done up..! called Sonny for an appointment; I had to have a marcel!

b. *Comb.*, as **Marcel-waved**, **-waver**, **-waving**.

1914 'O. Henry' *Gentle Grafter* ix. 72 A combination steak beater, shoe horn, marcel waver. **1923** *Chambers's Jrnl.* Sept. 568/1 She could..[bring into] view one of her exquisitely marcel-waved hairs. **1925** *Daily Tel.* 13 May 20/5 (Advt.), Expert Marcel Waver and Manicurist. **1929** marcel wave waving. **1927** *Daily Express* 30 Nov. 11/3 Miss Aylwin has been earning her living..doing shingling and marcel waving. **1932** [see *finger-waving* vbl. sb.]. **1968** *Times* 30 Jan. 9/7 Your head is frizzed or Marcel-waved into a neat pompadour. **1974** *Daily Tel.* 1 Aug. 11/2 This marcel waving... He has been..

Marcel, *v.* [f. prec. sb.] *trans.* To wave (hair) in the 'Marcel' fashion. Also *fig.*

1906 B. von Hutten *What became of Pam* x. 71 A gentleman who marcelled herself in an Oxford Street shop. **1906** Kipling *Debits & Credits* 155 The wind marcelled the waters. **1910** *Daily Express* 10 Dec. 11/1 Her unbobbed hair is marcelled in broad waves from a high forehead. **1931** E. Paul *Springtime in Paris* iv. 81 Wood and figures may be dyed, marcelled or manoeuvred like veils around a dancer. Behind them, the makeshoes a three. **1974** *Observer Mag.* 26 Nov..

marcel, *v.* [f. Marcel *v.*...]

1909 'O. Henry' *Options* 103 A stone house with an engraving of an idol with marcelled hair, playing a flute, over the door. **1921** F. Courtenay *Physical Beauty* xi. 11 You will usually find the finger..comb sets (mounted on springs) which will give it. **1923** *Daily Colonist* (Victoria, B.C.) 17 Jan. 6/3 (Advt.), Swan-Marcello perfect marcelling. Home appointments made. **1925** *Daily Express* 14 Oct. 6 Women's heads in the late Roman period..are represented not only with distinct marcelling, but also with elaborate jewellery to emphasise the waves. **1930** R. Macaulay *Staying with Relations* xvi. 226 The marcelling special sea. **1933** L. C. Douglas *Forgive us our Trespasses* (1937) xii. 250 Victor..calmly continued his monologue in a marcelled, affected baritone. **1936** 'Mark Twain'..? **1938** D. Baker *Young Man with Horn* (1939) ii. 60 Even the marcelling hadn't gained any real ground. **1971** G. Wheeler *Epitaph for Mister Wynn* (1972) xxvi. 150 The halo of silver hair..was in marcelled good order.

marcescent, *a.* (Later examples.) **marcescence** (later example).

1904 J. C. Willis *Man. Flowering Plants* (ed. 2) 77 If it [sc. the perianth] remains unwithered round the fruit, persistent, withered, marcescent, enlarged as in Physalis, ... **1966** *Acta Phytogeographica Suecica* XLIX. 31 *Marcescent* ... Around the stem..there is a dense insulating mantle of marcescent dry leaves. **1974** *Encycl. Brit. Macropædia* XIV. 2/1 The stems of forest species of Giant Lobelia are usually bare of marcescent foliage. **1974** *New Phytologist* LXXIII. 481 Mature pachycauls..exhibit several characteristics often associated with pachycaul trees of other regions, e.g. wide pith, presence of starch-sheath, unbranched axis and marcescence.

March, *sb.*[1] Add: **2. a.** *March-bloom*; *March-hatched* adj.

1881 G. M. Hopkins *Poems* (1967) 67 Look! March-bloom, like on mealed-with-yellow sallows! **1921** F. M. Ford *Let.* 15 July (1965) 133 March hatched cockerels. **1960** *Farmer & Stockbreeder* 5 Jan. 101/1 My February- and March-hatched pullets started to lay.

b. *March fly*, (*a*) *U.S.*, a dark-coloured, hairy fly of the family Bibionidæ; (*b*) *Austral.*, a blood-sucking horse-fly of the family Tabanidæ.

1895 J. H. & A. B. Comstock *Man. Study of Insects* xix. 450 They [sc. Bibionidæ] are most common in early spring, which has suggested the name March-flies; but some occur later in the season. **1907** W. W. Froggatt *Austral. Insects* 294 They [sc. Tabanidæ] are popularly known in Australia as 'March Flies'; in England and America they usually go by the name of 'Horse or Gad Flies'. **1968** E. J. Banfield *Confessions of Beachcomber* I. vii. 221 The sluggish March-fly, which, with the vicious mosquito, are at last... **1953** *Tasmanian (Hobart)*.. **1948** Dale Collins *Ord.* 127 A vicious March fly with threatening buzz dived at her. **1970** Collins & McAlpine in *Insects of Australia* (Commonwealth Sci. & Industr. Res. Organization) xxiv. 701/2 Tabanidae (March flies, Horse flies)... In Europe and North America, the term 'March fly' is applied to the Bibionidae. **1973** Swan & Papp *Common Insects N. Amer.* 603 March flies appear in large numbers early in the spring. **1973** *Islander* (Victoria, B.C.) 12 Aug. 2/1 Many a March-fly-bitten manege picnic is perhaps the..in which..worse..

march, *sb.*[2] Add: **4.** *march shire*.

1917 *Eng. Hist. Rev.* Oct. 483 The Warden..had simply taken over certain duties hitherto discharged by the sheriff in the March shires.

march, *sb.*[4] Add: **1. a.** Also, a procession organised as an expression of (esp. political) dissent, or to draw attention to a particular problem, etc. Cf. *hunger-march.

1908, etc. [see *hunger-march* s.v. *HUNGER *sb.* 4 c]. **1962** *Ann. Reg. 1961* 105 In May the United States Committee for non-violent action organized a 'march' from San Francisco to Moscow. **1965** 'Bread Winner'.. to present a formal petition to Parliament. **1962** [see *ALDERMASTON]. **1970** K. Giles *Death in Church* ii. 29 He's a crank and a fire-brand... Not a 'march' goes by without him carrying a banner. **1973** *Black World* Nov. 42/2 Perhaps the most arresting phenomenon of the last decade has been the sudden thrust of the thousands of 'sit-ins' and 'marches' by Afro-Americans in the South... [etc.] has been the obligation and opportunity of the entire world to re-evaluate the contributions of the Black American to world society.

d. *march past*. (Examples of non-military use.)

1904 O. W. Campbell *Shelley & Unromantics* ii. 10 Telling them [sc. Shelley's friends] over is like calling for a march past of the Farce. **1959** P. H. Gibbs *March Guardian Whly.* 10 Aug. 10/1 The march past of the pilgrims in the Grand Place of Ypres. **1969** *Times* 5 June 3/8 The teams (of 28 swimmers) risked a premature ducking by a carefully rooted march-past.

2. d. (Earlier examples.)

[1775 Burke *Speech moving Conciliation with Colonies* 14 The march of the human mind is slow.] **1818** Keats *Let.* 3 May (1931) I. 157 It proves there is really a grand march of intellect. **1821** Lady Morgan *Italy* I. 67 Impediments are now thrown in the march of mind... To retrograde, not to go forward in the march of mind.

3. b. *to steal a march* (further examples).

1771 Smollett *Humph. Cl.* I. 127 She yesterday wanted to steal a march on poor Liddy. **1844** Dickens *Chuzz.* Apr. I.. *Godey's Lady's Bk.* Sept. 133/1 He evidently intended to steal a march upon me, and smuggle a fine picture to New York, under my very nose. **1950** T. S. Eliot *Cocktail Party* ii. 93 He's quite triumphant Because he thinks he's stolen a march on her.

march, *v.*[2] Add: **1. f.** To engage in a protest, etc., march (see *MARCH *sb.*[4] 1 a).

1967 *Freedomways* VII. 102 Where is the Federal Government today as civil rights workers in Louisville face screaming mobs, throwing rocks and bottles at them as they peacefully march to end housing discrimination? **1969** *New Yorker* 14 June 76/3 He has repeatedly been asked to march and picket. **1972** *Times* 19 Oct. 5/7 (*heading*) Doctors march in Vienna.

‖ **märchen** (mɛ̄·ɹçən). Also **Märchen**. [G., fairy-tale, f. MHG. *merechyn* short verse narrative, f. OHG. *mâri*, MHG. *mære* news, tale (f. OHG. *mâri*, MHG. *mære* famous, known, *a.*) + the dim. suffix OHG. *-chîn* -KIN.] A folk-tale or story. Also *attrib.*

1871 L. M. Alcott *Good Wives* x. 130 You and I will read these pleasant little Märchen together. **1885** *Athenæum* 22 Aug. 230/2 The Punjaub tales..are, naturally, rather modern and civilised..there is no Servian and Romaic Märchen. **1920** A. Lang in *Folk-Lore* XII. 355 My friend..starts from the idea so common to the *Märchen*, of the person whose 'soul', 'life', or 'strength', is secreted in some animal, plant, or external object. *Ibid.* (1904) 'M. W.', 'There is no doubt..that the story of the shadowy Anglian king Offa, blended with *märchen* elements, was well known in England in the time of Cynewulf. **1928** W. W. Lawrence *Beowulf & Epic Trad.* 166 Study of märchen has made it clear that the situations at least that they throw upon sophisticated literature. **1930** H. L. Lorimer *Homer & Monuments* viii. 515 Outside the apologoi the poet of the *Odyssey* prefers to eliminate elements of the märchen type. **1963** Brown & Foote *Early Eng. & Norse Stud.* 159 The Norse riddle-contest... *Ibid.*, Oriental märchen-motives. **1964** C. S. Lewis *Discarded Image* vi. 135 The Gnomes are closer to the Dwarfs of *märchen*.

marcher[2]. Add: **1. b.** *spec.* A person who takes part in a protest, etc., march (see *MARCH *sb.*[4] 1 a).

1908, **1922** [see *hunger-marcher* s.v. *HUNGER *sb.* 4 c]. **1933** C. Day Lewis *Magnetic Mountain*.. our marchers, Shy of hoped, perhaps, for..a feeling of community with his fellow-marchers. **1962** *Guardian* 19 Sept. 1/4 We observe its [sc. wave's] direct and unquestioned influence on Aldermaston marchers and other protest... **1962** *Times*.. several..protesting marches... **1960** *Guardian* 28 Oct. 7/4 Many bright remarks about French politics, anti-bomb marchers... **1967** *Listener* 11 Oct. 470/3 About 4,000 and marchers who protested beside the Holy Loch. **1969** *New Yorker* 14 June 78/2 He had marched with the Selma marchers.

marching, *vbl. sb.* Add: **d.** *marching money* (later examples); **marching order** (earlier and later examples); also *fig.*

1941 Baker *Dict. Austral. Slang* 46 *Marching money*, money to travel. **1962** R. Cook *Cruise on Uppers* (1964) vi. 31 There were no ways—hardly marching money even. **1780** W. Heath *Let.* 31 July in *Mass. Hist. Soc. Coll.* (1905) 7th Ser. V. 93 George Washington has put the army in light marching order. **1837** *King's Regulations Army* 92 General Officers are to cause the Troops..to be frequently paraded, and exercised at least once a week in Heavy Marching Order. **1848** A. H. Clough *Let.* 26 Feb. in T. Arnold *N.Z. Lett.* (1966) 78, I anticipate considerable trouble in getting any Constitution into marching Order. **1850** J. J. Hooper *Horse Guards* 24 In many garrisons the greater part of the time thus employed is in heavy marching order. **1918** L. E. Ruggles *Navy Explained* 127 If the stew is covered over with a crust..they call it heavy marching order. **1937** R. de Hernhold.. **1963** *Times* 29 Oct. 8/5 The paratroops went out... in marching order... **1968** J. Thomson *Long Revenge*... Lucas *Dauphin*.. the Britain will have to abandon its heavy marching order. **1929** Duhamel's *Pasquier Chron.* 88 The barometer, too, nearly had its marching orders, and then neither of us would have known whether it was going to be wet or fine. **1961** S. Chaplin *Day of Sardine* viii. 136 He was never any good to you. It makes no difference to me if you give him his marching orders—just as long as we shift him..he'll have gone... **1974** M. Innes' *Appleby's Other Story* x. 79 As you should be kept on him, to my mind. Given his marching orders, he ought to be.

marching, *ppl. a.*[2] Add: **1. b.** *marching band*, a band that marches; *marching girl* (*Austral.* and *N.Z.*), a girl trained to march in formation, a drum-majorette.

1953 *Here & Now* (N.Z.) July 9 Not for a long time have I observed such a symptom of our malaise as the business of 'marching girls'. **1954** *N.Z. Listener* 10 Dec., Preceded, of course, by marching girls to provide just that touch of military pageantry so dear to us all. **1955** Keenaway & Gascoyne *Pict. Hist. Austr.* 1. 7/3 There is a..music in the distinctly Australian practice of the marching bands and of ragtime. **1969** *Atlantic Monthly* July 51/2 Marching bands and their formations.. **1966** *Austr.* 'Seeing them at..' **1961** .. 2 (*caption*) In gay, colorful uniforms marching girls were cheered by city crowds as they strode through city streets during the Australia Day celebrations. **1974** *Fiji Times* 3 July (*caption*) Marching girls parade through Lautoka during the Salusalu Festival at the weekend. **1974** P. de Vries *Glory of Hummingbird* i. 4 The sousaphone tuba he played in the school's marching band.

Ma·rcobrunner. Also **Marcobrunn**, **Markbrunner**, **Markobrunn**, **Markobrunner**. [G.,f. the name of a vineyard in the Rheingau.] A Rhenish white wine.

1833 C. Redding *Hist. Mod. Wines* ix. 190 Marcobrunner, the growth of the Rheingau, Laubenheimer... **1845** J. Forster *Life* (1872)... Markbrunner... **1862** G. Towey *Wine & Wine Countries* 202 The landlord, in considering his.. *Steinberger, Marcobrunner*, a fine flavour. **1862** C. Tovey *Wine & Wine Countries* 201 The landlord..Marcobrunner at 24s. **1936** 'Vitis' Ger. Ambrosia xix... .. best wines in Germany. **1848** H. Redding.. **1874** *Encycl. Brit.* XXIV. 828/2 The Rhine wines of the first quality are Johannisberger, Steinberger, Marcobrunner... **1935** C. Morgan.. Assmannshauser, etc. **1961** J. Liebig.. **1958** H. A. Vachell *Some Happenings* xvi. 261 Let us drink your health in some Marcobrunner. **1962** L. Deighton... **1933** E. Waugh... excellent bottle of Marcobrunner followed by a glass of his own particular old brandy. **1969** A. Lichine *Encycl. Wines* 433/2 The alternative [name] Marcobrunner will be found on the labels of two estates,..which own the best parts of the district. *Ibid.* 434/2 Markobrunn, sometimes written Marcobrunn.

Marconi. Insert after 'inventor': , Guglielmo, Marchese *Marconi* (1874–1937).

2. Used *attrib.* to designate a type of rig used on sailing vessels, or the various parts of such a rig.

1912 *Yachting Monthly* XIII. 356/2 Istria, Mr. Allom's 'Marconi' boat, promptly installed herself as 'the yacht of the year. **1915** *Ibid.* XIX. 151/1 Mr. Charles Nicholson, whose 'Marconi' boat, Istria, built in 1912, created such a sensation. **1916** *Rudder* 422/1 In the coming races for the Marblehead Race (Dunlap-Rowell) ... by the Marconi rig. **1921** *Ibid.* 28 [They] will carry 125 square feet of canvas in one sail on a so-called Marconi mast. **1940** 'O. Marty' *Wandering Years* 145 That eighty-foot Marconi mast was held in place by a network of steel wires... **1960** R. Y. F. Loomis 'Hobjaw' *Story* ii. 14, I expressed to Rigg my preference for a fast, marconi-rigged, wind-ward working cutter. **1962** E. F. Jacobs *Yachting*... **1965** *Yachting World* ..marconi... at the masthead of a tall curved mast with an almost vertical ... a leg-of-mutton or Bermuda rig. **1976** 'Motor Boating & Sailing' Jan. 136/1 This little cruiser has a very tall Marconi mainmast... *Ibid.* Marconi or gaff configuration.

Also **marco·nigraph** *v. trans.* and *intr.*, to send a message by marconigraphy (*to*); **marco-nigraphing** *vbl. sb.*; **marconi-graphy**, Marconi's system of radiotelegraphy; radio. (All now *disused*.)

1903 *Nature* 23 Apr. 583/1 The history of the series of inventions and discoveries which have culminated in Transatlantic Marconigraphy. **1907** *Daily Chron.* 27 Sept. 5/2 The Lusitania was marconigraphed at 5.30 p.m. yesterday 300 miles west of Fastnet. **1909** E. Stanton-Porter *Girl of Limberlost* xxiv. 453 Hi... I want you..I'll tele..Marconigraph, anything. **1912** R. Brooke *Let.* (1968) 284 It is..deeper and quieter than the telephone, though not so clever as marconigraphy.

marconi (mɑ̄ɹkǭ·ni), *v. Disused.* [f. the sb.] **1.** *trans.* and *intr.* To send a message by radio (*to*).

1912 C. N. & A. M. Williamson *Heather Moon* i. iv. 47, I marconied her we had gone off on the high seas to come. **1919** *Times* 25 June 13/6 Messages were signalled to the coastguards requesting them to marconi to the Fleet.

2. *trans.* To transmit by radio. Also *fig.*

1908 *Isle of Man Weekly Times* 12 Sept. 9/4 An author sometimes dreams of the ideal actress who shall 'Marconi' across the footlights the puppet he has given life to. **1922** *Glasgow Herald* 2 Nov. 6/2 These figures represent a code which can be wired, cabled, or marconied anywhere. **1926** P. Bottome *Old Wine* xi. 127 Marconied the news..across space.

Marconist (mɑ̄ɹkǭ·nist). *Disused.* [f. MARCONI + -IST.] The operator of a Marconi radiotelegraphy system.

1900 in *Encycl. Dict.* (1904) Suppl. s.v., Then the Marconists began to pull the enemy's leg by sending what our Marconist calls 'Rot'. **1934** *Jrnl. R. Aeronaut. Soc.* XXXVII. 628 On board of the machine, an increase of the crew to probably three pilots, two Marconists and one mechanic.

marcottage (maɹkǫtɑ̄·ʒ, mɑ̄ɹkǫ·tèdʒ). *Hort.* Also **marcotting**. [Fr. *marcottage* layering.] A method of propagating trees or shrubs in which a ring of bark is removed from a branch, and the wound covered with a thick layer of soil, moss, etc., into which new roots grow, before the plant is cut from its parent. Hence (as a back-formation) **marcot**, the wound from which new roots grow, or the new plant formed. Also as *v. intr.*, to use this method of propagation.

1927 P. M. Sisguera in *Philippine Agriculturist* XV. 63 Marcottage is the term applied to the process of reproduction by which a branch of a plant is 'girdled' and meanwhile wrapped with soil or other media in order to induce rooting while still attached to the mother plant. A marcot is the branch so treated... Planting marcots instead of seeds has several advantages. *Ibid.* 64 Time for marcotting. It is preferable to marcot during the rainy season. **1934** [see *air-layering* s.v. *AIR *sb.* III. II]. **1955** R. C. M. Wright *Plant Propagation* xv. 99 'Air layering' 'Chinese layering' or 'marcotting' is believed to have been used by gardeners for thousands of years. **1968** R. J. Garner *Grafter's Handbk.* (rev. ed.) iii. 60 Marcotting..is a very ancient method of propagation. Ibid. 98 It is..variously known as air layering, Chinese layering, circumposition, or, in India and elsewhere, as gootee. *Ibid.*, An ingenious method of securing a continuous supply is to suspend a vessel containing water above the marcot. **1959** Hartmann & Kester *Plant Propagation* xiv. 406 The ancient Chinese method, or a variation of it, consisted of plastering a ball of clay or other soil mixture about the region of the girdle, then surrounding it with moss. **1969** *Gloss. for Landscape Work* (*B.S.I.*) v. 12 Marcotting. A form of air layering in which a soil-based rooting medium is used.

Marcusian (maɹkiū·ziǎn), *a.* and *sb.* [f. the name of Herbert *Marcuse* (1898–), American writer.] **A.** *adj.* Of, pertaining to, or connected with Marcuse or his political views. **B.** *sb.* One who holds the political views of Marcuse.

1968 *Economist* 20 July 25/3 In Italy, as elsewhere, it has brought into the open a younger generation that is more articulate than the mass and can variously and very articulate about being a maoist, trotskyist, castrist, approximately be classified as maoist, trotskyist, castrist, marcusian and so forth. **1968** *Listener* 31 Oct. 594/2 (*heading*) Marcusians. *Ibid.*, The boring old proletariat is unrepresented in this Marcusian world. **1969** *Guardian* 8 Sept. 9/6 Marcuse, and the hippie generation who represent Marcusian premature utilitarianism, would be a social subset with the sensory, the imaginative and the quietist. **1969** *Parallel Ost.* 58 A Marcusian analysis of the developed world shows the system beginning to crumble from 'within'. **1970** *Guardian* 16 July 7/4 The beginning of the 'Marcusian scene' here.. **1971** *Times* 29.. 'the Marcusian' view.. **1972** ... on Jan. 10/1 It will be asked whether the Marcusian or Marcusian infection is beginning to take effect in this country.

mard (mɑ̄ɹd), *a.* = MARRED *ppl. a.* **b.** *dial.*

1903 in *Eng. Dial. Dict.* IV. 30/1. **1967** E. Lawrence *White Peacock* vii. 493 The little devils are soft, mard-soft. **1913** — *Love Poems* 53 Eh, tha'rt a mard-'arsed kid. **1963** A. *Lewis Dict.*.. "Marcus" ...x. 187 Elsewhere the weak one may carry the label 'mard'.

mard (mɑ̄ɹd), *v. dial.* [Cf. prec. and MAR *v.* 3 c.] To 'spoil' (a child).

1874 in *Eng. Dial. Dict.* (1903). **1911** D. H. Lawrence *White Peacock* vii. 493 She hardened 'em till they were soft.

Mardi gras (mɑ̄ɹdi grɑ̄). [Fr., lit. 'fat Tuesday'.] Shrove Tuesday; the last day of carnival, esp. in France. In U.S. esp. as celebrated in New Orleans. Also *attrib.*

1699 M. Lister *Journey to Paris* (ed. 3) 177 My Lord Ambassador was at a Ball at Monsieur de Montargis mardy gras. **1728** J. Lucas tr. *Cheyne's..* Also *Mardi-Gras* *Day*; 1729 Yesterday was *Mardi Gras*—the last day of the folly. **1848** H. Greville *Diary* 8 Mar. (1883) I. 236 Two Mardi Gras: 'Pretty' 'The heady ..carnival. *Ibid.* and is celebrated in England. ... **1879** 'Mark Twain' *Life on Mississippi* xvi. **1883** New Orleans. 'Mardi-Gras' is a gaiety which becomes the Mardi-Gras festivities. **1900** *Ann Fabian* in *Slang* 148 His Father was too Serious a Man to get out with the New Orleans Mardi Gras Clubs. **1909** 'O. Henry' *Options* 184 The reception they were going to put up for the Mardi Gras. **1931** B. Gibbons..could take like an afternoon tea in Bury St Edmunds with a curate's aunt. **1924** *Blackw. Mag.* Nov. 709/2 There are those to whom Mardi-Gras is yet a religious festival. **1931** *Times Lit. Suppl.* 9 Apr. 285/3 The Mardi Gras irresponsibility..excuses the daring pranks of a naughty youth. **1931** *Christ. Let.* 12 Dec. (1965) 392 It isn't any sort of Mardi-Gras thing of that with the Indians express, despite the flamboyant colors of their costumes. **1945** *Sat. Even. Post* 15 May. 14 Mardi Gras in Coney. **1972** *Guardian* 29 Dec. 12/1 Russell Harty took an 'Aquarius' [TV] team to Mardi Gras, the great New Orleans Shrove Tuesday freak-out.

mardy (mɑ̄·ɹdi), *a. dial.* [f. *MARD *a.* + -Y[1].] 'Spoilt', sulky, whining. Also as *sb.*, a spoilt child.

1903 *Eng. Dial. Dict.* IV. 34/1 A boy who cries with a..called by his fellows 'a mardy baby'. **1913** D. H. Lawrence *Sons & Lovers* vii. 127 'Now, Miriam,' said Maurice, 'you come an' 'ave a go.' 'No,' she cried, shrinking back. 'Ha! baby. The mardy-kid!' said her brothers. **1915** — *Rainbow* i. 12 Young Tom, whom he called a mardy baby. **a 1930** — *Phoenix II* (1968) 170 As for Harold, he was right. He was very respectable and a bit of a mardy, perhaps..but he was all right. **1959** J. Braine *Vodi* ii. 27 'Don't be so bloody soft, man,' Tom said. 'Don't be such a mardy.' **1961** *Guardian*.. 'You're mardy. You're dead mardy.' **1961** J. I. M. Stewart *Man who won Pools* 93 'I don't want you as a nipper, you Pots.' **1975** D. Clark *Premedicated Murder* v. 169 'Don't you get all mardy about it if you like,' said Green, unabashed.

mare. Add: **1. c.** (Later examples.)

1590 Shaks. *Mids. N.* III. ii. 463 The man shall have his mare again. **1922** Joyce *Ulysses* 231 She's a gamey mare and no mistake. **1925** C. W. Ogle in *Caribbean Anthol. Short Stories* 43 Forget her, Joe boy! Bah! These mares give me the creeps.

mare[4] (mɑ̄·re, mɑ̄·ri, mǣ·ri). *Astr.* Pl. **maria** (mɑ̄·riǎ), occas. **mares**. [L., = 'sea': used in 17th-c. L. works (e.g. J. Hevelius *Selenographia* (1647) vi. 226), the proper names (which are still current) given to the various regions were taken into Eng. often without translation.] Any of the extensive areas of flat land ('seas') on the surface of the moon, which appear dark and were once thought to be seas; also, any of the dark areas visible on Mars.

1878 R. Turner *View of Heavens* 11 The Oceans, Seas, and Lakes, are, ... **1866** *Monthly Notices R. Astron. Soc.* XX. 52 The plains or *maria*, which appear dark... **1883** J. Herschel..upon ... **1900**... These mountains ... **1908** *Maria* ... **1929** Sir J. Jeans *Universe around Us*... **1931** E. Moore *Old Moon* 167 Maria..are only the level and, as at present..levels. **1946** *Nature* 28 ... **1959** ... **1962** *Encycl. Brit.* XV. 803 The sunlit *maria*.. **1964** ... **1969** *Times* 21 July..the *maria*..'Seas'.. Mare Tranquillitatis... **1970** *Nature* 6 June 925/1 The form and magnitude of the lunar maria.. **1974**..

marechal. (Later examples of *maréchale.*)

1863 Mrs. Gaskell *Her First Floor went to Crowley Castle* in *All Year Round* Extra Christmas No., 3 Dec. 16/2 Her hair ..daintily powdered and scented with maréchale. **1905** *Smart Set* Sept. 113/1 They were of satin..the yellow, ylang-ylang, myrrh, and maréchale.

‖ **maréchal** (mareʃal). Also 7 **Marishal**, 8 **Mareschall**, *Sc.* **Marechal**. [Fr.] The French word for 'marshal' or 'field-marshal' used occas. in English.

1699 M. Lister *Journey to Paris* (ed. 3) 202 The Marischal his Father..embraced me, and wished me.. **1745** M. W. Montagu *Let.* 27 Jan. (1966) II. 349 It is reported here that the taking of Marechall Bell Isle will make the residence of France very unsafe to all the English. **1783** T. Blaikie *Diary Scotch Gardener* (1931) 187 At St Germains the Mareschal de Noailles... **1906** A. J. Buchanan *Mr. Standfast* xx. 355 They will not come. **1926** *Ibid.* For ..maréchal..saw in a marshal.. **1934** Locke *Mountebank* xiv. 170 Were British Generals real, like French Generals, Lyautey and Joffre?

Maréchal Niel (mareʃal nīl). Also anglicized **Marshal Niel**. [Fr., f. the name of Adolphe *Niel* (1802–69), Marshal of France: see prec.] A climbing Noisette rose of the variety so called, introduced in 1864, and bearing large, well-formed, fragrant, yellow flowers.

1864 *Gardeners' Chron.* 17 Dec. 1209/1 First-class Certificate has been given to a Tea Rose called Maréchal Niel, produced by M. Eugène Verdier, of Paris... This variety appears to be one of the noblest of the yellows. **1867** T. Rivers *Rose Amateur's Guide* (ed. 9) 136 The rose to which I allude is Marshal Niel. **1888** H. Kirke 25 *Yrs. Brit. Guiana* iv. 81 Gardens in British Guiana are as a rule, disappointing... Roses, except the strong tea-scented varieties, such as Maréchal Niel, do not flower successfully. **1899** O. Wilde *Importance of being Earnest* ii. 67 *Algernon*. Might I have a button-hole first?..*Cecily*. A Maréchal Niel? [sic]. **1905** *Smart Set* Sept. 157/2 You had a great bowl of Maréchal Niels on the piano. **1928** [see *GLOIRE DE DIJON]. **1955** G. S. Thomas *Old Shrub Roses* ix. 82 'Cloth of Gold' was the parent of the famous 'Maréchal Niel'. **1971** N. Young *Compl. Rosarian* v. 76 Chromatella (also known as Cloth of Gold) became the parent [in 1864] of Maréchal Niel which was the most esteemed of all yellow roses until well into this century.

‖ **mare clausum** (mɑ̄·ri klǭ·zəm). [L., closed sea, from the title of a Latin work

(1635) by John Selden (1584–1654), English jurist, written in answer to *Mare liberum* (1609) by Grotius.] A sea under the jurisdiction of a particular country.

This term and *mare liberum* originated during the struggle between England and the Netherlands in the 17th century.

1652 M. Nedham tr. *Selden's Of Dominion of Sea* sig. g1 Mare Clausum is the Sea possessed in a private manner, or so secluded both by Right and Occupation, that it ceaseth to bee common. 1849 J. Allen *Navigation Laws Gt. Brit.* iv. 34 Although at one time the professed admirers of Mare Liberum, the Dutch began to consider that Mare Clausum possessed more substantial charms. 1856 *Newsp. & Gen. Reader's Compan.* I. § 1013 The Yellow Sea, which for ages has been, with few exceptions, a *mare clausum*, is now a *mare liberum* to all the world. 1911 *Encycl. Brit.* XVII. 698/2 *Mare clausum* and *mare liberum*..in international law, terms associated with the historic controversy which arose out of demands on the part of different states to assert exclusive dominion over areas of the open or high sea. 1949 *Canad. Jrnl. Pol. & Social Sci.* XV. 344 The old theory of mare clausum, under which coastal waters were sometimes fixed at sixty miles, one hundred miles, two days' journey, etc., from the shore, and under which claims were also made by agreement between two or more countries over an ocean common to them. 1973 [see *MARE LIBERUM].

Maree, var. *MARIA *sb.* and *a.*

Marek (ma·rek). The name of Dr. Josef *Marek*, Hungarian veterinary surgeon, used in the possessive in **Marek's disease** to designate *fowl paralysis* (*FOWL *sb.* 5 c), first described by him in 1907.

1961 *Brit. Vet. Jrnl.* CXVII. 332 It is suggested that this complex be termed Marek's disease, a term which has some precedence and which implies a disease and not a pathological entity. 1964 *Black's Vet. Dict.* (ed. 7) 353/1 Fowl paralysis itself (Marek's disease) has of recent years become more common. 1970 *Poultry Bull.* Dec. 12 The control of Marek's Disease by vaccination is at present very topical. 1972 *Country Life* 5 Feb. 284/1 No sooner had a vaccine for Marek's Disease been found than Fowl Pest swept through our poultry flocks.

|| **mare liberum** (mā·ri ləi·bĕrŏm). [L., free sea, from the title of a Latin treatise (1609) by Hugo Grotius (1583–1645), Dutch jurist.] A sea open to all nations. Cf. *MARE CLAUSUM.

1652 M. Nedham tr. *Selden's Of Dominion of Sea* sig. a2 This People [the Netherlanders]..carried out their design ..by..a daily intrusion upon the Territorie by Sea, that in time they durst plead and print *Mare Liberum*..to defie the Dominion of England over the Sea. 1806 J. Randolph in *Abridgm. Deb. Congress U.S. 1789–1856* (1857) III. 428/2 Again: Is the *mare liberum* any where asserted in this unnamed book, that free ships make free goods? No, sir; the right of search is acknowledged; that enemy's property is lawful prize, is sealed and delivered. 1849, etc. [see *MARE CLAUSUM]. 1973 *Times* 9 May 16/2 Should we discover that we too might benefit from wider zones of national jurisdiction, we should after all only be returning to the philosophy prevailing before Grotius' idea of *Mare Liberum* took hold, to John Selden's equally respectable *Mare Clausum*.

maremma. Add: **2.** *attrib.*, as *maremma sheep-dog*.

1924 A. Huxley *Little Mexican* 283 The old gentleman ..taking his big maremma dog for a walk [in Tuscany]. 1945 C. L. B. Hubbard *Observer's Bk. Dogs* 101 Maremma Sheepdog. 1948 —— *Dogs in Brit.* iii. xvii. 202 Common in central Italy from Tuscany to the Abruzzi, the Maremma Sheepdog is also called the Cani da Pastor Maremmani, Abruzzi Sheepdog and Maremmes Sheepdog. The breed is centuries old and has been bred pure by Tuscan farmers for herding sheep and cattle; lately the race has been in more general use, particularly as a guard dog. 1972 *Country Life* 2 Nov. 1186/3 Maremma (Italian) Sheep Dog.

Marengo (mare·ngo). [See def.] The name of a village in northern Italy, the scene of Napoleon's victory over the Austrians in 1800, used in the name of the dish *chicken, fowl, poulet à la Marengo*, said to have been served to Napoleon after the battle of Marengo.

1861 Mrs. Beeton *Bk. Househ. Managem.* 464 (*heading*) Poulet à la Marengo... Fowl à la Marengo. 1877 E. S. Dallas *Kettner's Bk. of Table* 121 (*heading*) Chicken à la Marengo,—the chicken ..is fried in oil. 1959 *Good Food Guide* 169 Chicken Marengo and tournedos.

mareogram (mæ·ri,ogræm). [f. L. *mare* sea + -o + -gram.] A graphical record of variations in sea level.

1904 *Publ. Earthquake Investigation Comm. Foreign Lang.* (Japan) XVIII. 14 The following table..gives the mean monthly values of the distance between the sea surface and the datum line in the mareogram at each of the two places. 1931 C. Davison *Japanese Earthquake of 1923* xvi. 104 The mareogram at Toba is reproduced in Fig. 25. Here, the first movement occurred at 6.57 p.m., the maximum amplitude (of 4 ft. 1¼ in.) being reached with the fifth wave at 3.37 p.m. 1949 Gutenberg & Richter *Seismicity of Earth* 95/1 Tsunamis are frequently recorded on mareograms written by tide gauges.

Hence **mareogra·phic** *a.*

1939 *Jap. Jrnl. Astron. & Geophysics* XVII. 121 The tsunamis that were associated with earthquakes..left conspicuous records on some of our mareographic stations.

1956 *Jrnl. Earth Sci. Nagoya Univ.* IV. 5 Some of these after-shocks were strong,..but no tunamis were observed, even on the mareographic records.

Marezine (mæ·rĕzīn). *Pharm.* Also **marezine.** [f. *mare-* + PIPER)AZINE.] 1-Methyl-4-α-phenylpiperazine hydrochloride, $C_{18}H_{22}N_2$.HCl, used in the form of tablets as an anti-emetic, esp. for the prevention and treatment of motion sickness. In the *British Pharmacopœia* called *cyclizine hydrochloride*.

Marezine is registered as a proprietary name in the U.S.

1952 *Official Gaz.* (U.S. Patent Office) 23 Dec. 1007/1 Burroughs Wellcome & Co... *Marezine* for medicinal preparations intended for use in the prevention or treatment of allergies, motion sickness, nausea and vomiting in pregnancy, or nausea due to other causes. 1954 *Jrnl. Pharmacol. & Exper. Therap.* CXII. 297 (*heading*) Pharmacologic properties of cyclizine hydrochloride (marezine). 1965 *Economist* 6 Nov. 615/2 The Food and Drug Administration announced last week that three anti-histamines—Bonine, Marezine and Cyclizine—which are used to combat nausea, including morning sickness, must carry a warning on their labels that, if taken by a pregnant woman, they may harm her unborn child. 1969 *Guardian* 21 July 9/2 Each [astronaut]..carries..45 milligrams of marezine for motion sickness.

marezzo (mære·tso). [It., f. *marezzare* to water (silk), marble.] A kind of artificial marble (see quots.). More fully *marezzo marble*.

1876 *Encycl. Brit.* IV. 508/1 Marezzo marble is made of cement mixed with fibre for strength and to resist a blow. 1901 *Notes on Building Construction* (ed. 5) III. iii. 250 The basis of Marezzo marble, as well as of Scagliola, being plaster of Paris, neither of them is capable of bearing exposure to the weather. *Ibid.*, The Artificial Marble now manufactured in London is made on the same principle as the Marezzo, but differs from it in the character of the cement used. 1958 F. S. Merritt *Building Construction Handbk.* xii. 9 *Marezzo*, an imitation marble formed with Keene's cement to which colors have been added. 1964 J. S. Scott *Dict. Building* 201 *Marezzo marble*, an artificial marble like *scagliola*, which differs from it mainly in having no chips of added coloured material. When precast, it is cast on a smooth sheet of plate glass or slate to give a polished surface.

marg, var. *MARGE².

margarine. Add: Also commonly with pronunc. (mādʒărī·n). For def. read: A substance made from edible oils and meat fats with water or skimmed milk, used as a spread on bread, etc., and as a cooking fat. (Further examples.)

1890 [see FILLED *ppl. a.* 1 b]. 1907 *Act 7 Edw. VII* c. 21 § 13 For the purposes of the Sale of Food and Drugs Act and this Act the expression 'margarine' shall mean any article of food, whether mixed with butter or not, which resembles butter and is not milk-blended butter. 1960 A. E. Bender *Dict. Nutrition* 79/1 Margarine... Compulsorily fortified in many countries (and voluntarily in most) with vitamins A and D, so that nutritionally equal to butter. 1963 *Which?* July 211/1 Margarine is made from a selection of vegetable oils, sometimes whale or fish oil, and often lard, usually with skimmed milk, and water. 1975 *Listener* 16 Jan. 80/3 Sheep's head..mixed..with minced onion and margarine.

b. *margarine works*; **margarine-cheese** (see quots.).

1899 *Act 62 & 63 Vict.* c. 51 § 25 The expression 'margarine-cheese' means any substance, whether compound or otherwise, which is prepared in imitation of cheese, and which contains fat not derived from milk. 1902 *Encycl. Brit.* XXV. 93/2 From America cheese has come into the English market, made from skim-milk which has again been provided with fatty matter, generally emulsified margarine—hence the term 'margarine cheese' or 'filled cheese'. 1909 *Chambers's Jrnl.* Jan. 24/1 Margarine works are equipped with cooling machinery.

Hence **margari·ne** *v. trans.*, to smear or spread (bread) with margarine; **margarined** *ppl. a.*

1918 *Punch* 15 May 315 She knows which side her bread's margarined. 1924 Galsworthy *White Monkey* iii. iv. 243 'Well,' he said, over their cocoa and margarined bread: 'I must see Mr. Mont, that's certain.' 1960 D. Storey *This Sporting Life* ii. i. 159 She..began to margarine the bread.

Margarita (măĭgărī·tă). Also **margarita.** [f. a woman's name.] **1.** A Spanish wine.

1920 G. Saintsbury *Notes on Cellar-Bk.* ii. 21 My cellars ..have seldom for fifty years been without a certain 'Margarita'. 1924 *Trade Marks Jrnl.* 17 Dec. 2876 Very superior very pale dry sherry. Margarita... Sherry wine, the produce of Spain. John Harvey & Sons, Limited,.. Bristol; wine and spirit merchants.

2. A cocktail made with tequila and citrus fruit juice.

1965 O. A. Mendelsohn *Dict. Drink* 212 *Margarita*, mixed drink made from tequila and citrus juice, drunk from vessel whose rim has been dipped in salt. 1969 *Sat. Rev.* (U.S.) 8 Feb. 32/3 You'll learn how to make..mixed drinks, ranging from Margaritas and Mai-Tais to Irish Coffee and Moscow Mules. 1969 *Guardian* 4 July 7/4 Serve as an appetizer with..if you happen to have both limes and *tequila*—margaritas. 1969 A. Glyn *Dragon Variation* ix. 288 She handed the telephone back to the bartender and sipped her margarita. 1971 C. Fick *Danziger Transcript* (1973) 44 Her husband moved past us into..the bar... 'I'm dying for a margarita.'

marge² (māɪdʒ), **marg** (maɪg, maɪdʒ). Colloq. abbrevs. of MARGARINE, as pronounced (māɪdʒărī·n) or (mā·ɪgărīn).

1922 Joyce *Ulysses* 150 Potatoes and marge, marge and potatoes. 1933 'G. Orwell' *Down & Out* xxiv. 179 'No butter, only marg,' she said, surprised. 1937 E. Garnett *Family from One End Street* iv. 82 Fetch me a bit of marg, Lily Rose—I can't waste butter on him. 1939 N. Monsarrat *This is Schoolroom* viii. 197 Living on the dole.. bread-and-marge, and meat once a month. 1960 J. Betjeman *Summoned by Bells* vii. 70 In quieter tones we asked in Hall that night Neighbours to pass the marge. 1974 W. Foley *Child in Forest* ii. 161 The remembered slice of marge-spread toast.

margin, *sb.* Add: **2. b.** *spec.* Profit(s), profit margin (further examples).

1851 C. Cist *Sk. Cincinnati in 1851* 319 We are in the enjoyment of a clear margin, of at least, half a cent per pound, over our eastern neighbors. 1890 *Congress. Rec.* 25 June 6499/2 When an employer feels that his margin is slipping away from him, the first thing done is to scale down the price of wages. 1940 *Economist* 31 Aug. 291/1 Fixed margins, which were established..to check an incipient boom, are now..the subject of some controversy. 1971 *Daily Tel.* 21 Oct. 23 (*heading*) Doubled margins work wonders for Spillers. 1972 W. A. Pantin *Oxf. Life* iv. 48 The estates allotted to the support of chairs were in some cases left in the hands of the university, which paid out of them the appropriate salary to the professor, and sometimes made a profit out of the margin. 1972 *Times* 16 Oct. 4/6 Wheat..is grown as much for the straw as for its £57 an acre gross margin. *Ibid.* 4/7 Bulbs ..give easily the biggest margin to the acre.

c. (Earlier and later examples.)

1848 W. Armstrong *Stocks* 10 The purchaser then hands over this margin to the person with whom he hypothecates the Stock. 1870 J. K. Medbery *Men & Mysteries Wall St.* 62 Seven per cent a year is generally allowed on all margins advanced by customers. 1880 *Harper's Mag.* Oct. 782/2 All speculated, but they did not speculate on margins. 1934 S. S. Huebner *Stock Market* xxiii. 434 If the customer purchases stock worth $100 and deposits $20 as margin,..he has a margin of 20 per cent. 1971 *Investors Chronicle's Beginners, Please* (ed. 5) 310 The latter [*sc.* the broker] makes an interest charge..and receives from his client..sufficient to maintain an adequate 'margin' to protect the broker against fluctuations in the value of the collateral.

e. *margin of safety*: an allowance made for safety, *spec.* a number equal to the factor of safety (*FACTOR *sb.* 8) minus one.

1888 R. S. Ball *Exper. Mech.* (ed. 2) ii. 33 So great a margin of safety is necessary on account of the jerks and other occasional great strains that arise in the raising and the lowering of heavy weights. 1905 *Trans. Inst. Naval Archit.* XLVII. ii. 203 One hears, both in the lecture theatre and drawing office, much talk of 'margins of safety'. 1941 N. H. Anderson *Aircraft Layout* vii. 179 If the margin of safety comes out a negative quantity, the design is not satisfactory; in all cases a positive margin of safety is required. 1968 F. K. Teichmann *Fund. Aircraft Structural Analysis* 78 In general, the designer of aircraft structures selects structural elements with a margin of safety of zero. This does not mean that safety is not built into the craft..since the loads, or load factors actually imposed on the craft are determined rather precisely.

5. margin call, a demand by a broker that an investor deposit further cash or securities to guarantee the margin (sense 2 c) on his investment; **margin clerk**, one who records dealings in margins (sense 2 c); **margin release, marginal release**, the mechanism on a typewriter which enables the typewriting to run into the set margin; hence **margin release key**, the key on the machine which operates this mechanism.

1961 R. E. Badger et al. *Investment Princ. & Pract.* (ed. 5) iv. 84 If the market value of the securities in an account declines, the broker issues a 'margin call' to the customer, requesting the deposit of additional cash or securities to protect against loss on the loan. 1963 Phillips & Lane *Personal Finance* vii. 180 If a stock declines in market price, a margin call may be issued by the broker who must require the required margin coverage at current market prices, else he must sell the stock to protect the interests of the creditors who have supplied the previously borrowed funds. 1972 *N.Y. Law Jrnl.* 14 Nov. 2/2 The deficit arose after defendant was unable to meet margin calls. 1886 *Harper's Mag.* July 213/2 Speculative sales of land..are also made in the Call Room ..the caller of provisions and margin clerk presiding. 1973 *N.Y. Law Jrnl.* 30 July 4/6 The regulation should be..in the language of the brokers and margin clerks who must work under it. 1913 *Pitman's Commercial Encycl. & Dict. Business* IV. 1620/2 Note what a wonderful improvement there is..in all the minor parts, such as..margin release keys. 1914 *Pitman's Commercial Self-Educator* I. 163/1 Closely connected with these Marginal Stops is the Margin Release, which in the latest models takes the form of a key marked 'Marginal Release' or 'M.R.' 1939 F. A. Faunce *Secretarial Efficiency* ii. vi. 87 Notice the location of..the backspacer, the margin release, and the ribbon shift. 1962 J. B. Priestley *Margin Released* p. viii, Typewriters..have a key labelled 'Margin Release', frequently needed by hasty and careless typists like myself. 1967 J. Harrison *Secretarial Duties* (ed. 3) viii. 106 When erasing, move the carriage fully to the left or right, using the margin-release key.

margin, *v.* **4.** (Later examples.)

1902 H. L. Wilson *Spenders* xxxiv. 408 As the stock fell, the banks requested the brokers to margin up their loans, and the brokers, in turn, requested Percival to margin up his trades. 1973 *N.Y. Law Jrnl.* 30 July 4/2

Once a margin account has been properly margined, the regulation imposes no further requirements respecting the status of the account in the absence of a subsequent transaction.

marginal, *a.* and *sb.* Add: **A.** *adj.* **2.** (Further examples.)

1892 *Photogr. Ann.* 229 The lenses are of special optical glass, constructed with the nicest precision of curvature, so maintaining good marginal definition. 1925 J. Laird *Our Minds & their Bodies* ii. 31 For scientific purposes the marginal or borderline cases are usually the most instructive. 1934 H. C. Warren *Dict. Psychol.* 159/1 *Marginal contrast*, an accentuated type of simultaneous contrast, which occurs in regions close to the boundary between two contrasting areas. 1950 tr. *Mountaineering Handbk.* (Assoc. Brit. Members Swiss Alpine Club) ix. 88 Marginal crevasses..run diagonally upwards and towards the centre of the glacier. 1957 R. G. Collomb *Dict. Mountaineering* 52 The changing course of a glacier causes marginal crevasses (i.e. splits at the edges)..; the changing angle of slope over which it flows causes marginal crevasses and transverse crevasses in mid-stream. 1962 Blake & Trott *Periodontology* v. 49 Marginal gingivitis involves only the free gingiva. 1962 M. L. Haselgrove *Photographers' Dict.* 141 (*caption*) A beam parallel to the principal axis will not be brought to a focus at the principal focus; marginal rays will converge to *M* while axial rays converge to *A*. 1965 G. Y. Shevelov *Prehist. of Slavic* 74 As in many other cases this columnal final stress was replaced by the marginal final stress.

c. *Psychol.* Of, on, or pertaining to the edge of the field of consciousness.

1894 C. L. Morgan *Introd. Compar. Psychol.* i. 17 We.. pay attention solely to focal consciousness, omitting all reference to the great body of marginal subconsciousness. 1899 W. James *Talks to Teachers* ii. 18 The expressions 'focal object' and 'marginal object', which we owe to Mr. Lloyd Morgan, require, I think, no further explanation. 1903 [in Dict., sense 3]. 1927 W. E. Collinson *Contemp. Eng.* 108 The field of consciousness with its focal and marginal presentations.

3. Add to def.: Freq. in *Econ.*; esp. of or pertaining to goods produced and marketed at a small margin of profit; *spec.* in phrases *marginal cost, man, utility*. Cf. quots. 1887 and 1890 in Dict. (Further examples.)

1890 A. Marshall *Princ. Econ.* I. p. x, The term 'marginal' increment I borrowed from von Thünen, and it is now commonly used by German economists. 1909 J. A. Hobson *Industr. Syst.* v. 109 So with the case of the 'marginal shepherd', the tenth man whom a farmer calculates it is just worth his while to employ because he can get him for the price of twenty sheep a year, and he will just save that number by his work. *Ibid.*, Marginal productivity of labour..the productivity of any single man, 'marginal' or other. 1920 A. Marshall *Princ. Econ.* (ed. 8) vi. i. 517 The farmer's interests are equally served by hiring 10 or 11 men; but..the eleventh man (supposed to be of *normal* efficiency) is the marginal man, when the markets for labour and sheep are such that one man can be hired..for the price of 20 sheep. 1925 J. M. Keynes in A. C. Pigou *Memorials A. Marshall* 22 It undoubtedly gave Jevons priority of publication as regards the group of ideas connected with 'marginal' (or, as Jevons called it, 'final') utility. 1930 *Economist* 28 June 1441/1 The proposals were based upon marginal cost on full-time running. 1931 *Encycl. Social Sci.* V. 366/1 In 1871.. W. Stanley Jevons coined the phrase final utility, Carl Menger spoke of marginal utility and Léon Walras used the term rarity. 1934 *Punch* 17 Oct. 424/1 People babble cheerfully about the Marginal Man, the law of Diminishing Returns and so on. 1957 *Times* 21 Dec. 5/2 The weather at Idlewild international airport was reported to be 'marginal', and the captain had to ensure that sufficient fuel remained to divert to an alternative base. 1958 *Listener* 25 Sept. 447/1 The ordinary American citizen enjoys amenities which not even the rich enjoyed a century ago, and marginal output consists of goods that by any standards are patently inessential. *Ibid.* 448/1 Marginal production in affluent America today..is in general much less urgent..than was marginal production 100 years ago. 1963 *Times* 16 Apr. 13/1 At any point of time there are some who would buy less coal if the price went up, either economizing in fuel or switching to some alternative. We economists call these the 'marginal' users. 1969 J. Argenti *Managem. Techniques* 155 Marginal cost calculations can be extremely complex in a large multi-product company... Unless marginal costing is used, the answer given to the question 'will it pay us to meet this order or should we turn it down?' will be wrong. 1973 *Black Panther* 8 Sept. 12/3 In Southern Appalachia, Black Appalachians are 'marginal' (just above the poverty level).

b. Applied to land, ore, etc., barely worth developing. Also applied to a person working such land, etc. (Cf. MARGIN *sb.* 2 a, quot. 1863.)

1910 P. H. Wicksteed *Common Sense of Pol. Econ.* ii. vi. 571 'Marginal land'..is not land..considered with reference to the volume of supply. 1935 *Economist* 30 Nov. 1095/2 Huge quantities of marginal ore which, unattractive at the old price of gold, are attractive at the new. 1943 J. S. Huxley *TVA* i. 7 Reafforestation and the proper use of marginal lands. 1944 *Ann. Reg. 1943* 283 Policy..of aiding the thousands of inefficient marginal farmers to raise their standards of farming. 1954 M. Beresford *Lost Villages* x. 346 Derbyshire... There seems to have been surprisingly little retreat of settlement from marginal lands in the Peak. 1975 *Times Lit. Suppl.* 2 May 471/3 The bringing back into production of marginal agricultural land..formed part of his [*sc.* Keynes's] vision.

c. *Sociol.* Of an individual or social group: partly belonging to two differing societies or cultures but not fully integrated into either.

1928 R. E. Park in *Amer. Jrnl. Sociol.* May 881 (*heading*) Human migration and the marginal man. *Ibid.*, One of the consequences of migration is to create a situation in which the same individual..finds himself striving to live in two diverse cultural groups. The effect is to produce an unstable character... This is the 'marginal man'. *Ibid.* 893 It is in the mind of the marginal man—where the changes and fusions of culture are going on—that we can best study the processes of civilization. 1937 E. V. Stonequist (*title*) The marginal man. 1957 V. W. Turner *Schism & Continuity in Afr. Society* iv. 108 Sandombu was from many points of view..an atypical, marginal man in Mukanza Village. 1963 *Rev. Eng. Stud.* XIV. 258 A homosexual, he regarded himself as a marginal man, out of step with society, yet contemptuous of its hypocritical standards. 1964 Gould & Kolb *Dict. Social Sci.* 407/1 Sociologists wrote of the second generation immigrants as the most distinctively marginal group, measured by their relatively high index of crime. 1964 R. D. Hopper in I. L. Horowitz *New Sociol.* 324 The Creole marginal group constituted about 10 per cent of the population at the time of the revolution. 1964 L. A. Costa Pinto in *Ibid.* 471 In a marginal society it is possible to find strong support..for quite opposite decisions.

d. Of minor importance, small, having little effect; usu. const. *to*.

1929 *New Statesman* 1 June 232/1 For the 'marginal' voter—he who is at the point of indifference whether he comes to the poll or not—is unlikely..to be induced to do so by any really important consideration. 1954 M. Beresford *Lost Villages* vii. 261 The lands marginal to medieval corn-growing. 1955 *Times* 11 June 9/6 Most of the changes are..shifts of emphasis rather than reversals of previous policy; they are important but they are marginal. 1959 *Times* 14 Jan. 3/6 There is no major writer who uses the stage as his preferred medium of creation... Mr. Graham Greene and Mr. Angus Wilson, for example, still seem marginal to the drama. 1959 *Times Lit. Suppl.* 29 May 321/3 Twenty-six changes of punctuation, of which four or five seem..to be improvements, fourteen almost certainly wrong, and seven or eight marginal. 1964 [see *MARGINALITY]. 1969 *Listener* 16 Jan. 92/3 The lack of a character with which we can identify—..the soldiers.. remain uncharacterised and marginal—soon stills our wish to be emotionally involved.

e. *Pol.* Pertaining to a constituency, etc., in which an election or issue is likely to be closely contested and the majority very small. Also *ellipt.* as *sb.*

1951 *Times* 25 Oct. 6/3 (*heading*) Marginal seats... A significant feature of today's General Election polling is the substantial number of constituencies which may be described as marginal, and where the result of the voting is problematical. 1955 *Times* 7 June 7/4 A marginal constituency is one where the retiring member, having won the last election by a small majority, stands a sporting chance of being defeated at the next one. 1960 Butler & Rose *Brit. Gen. Election 1959* xi. 135 Despite the concentration on marginal seats, the Conservatives managed to put on a full-scale campaign almost everywhere. 1965 *Listener* 24 June 921/1 It was difficult for him to give a lead on this particular topic directly after such a marginal vote in his favour. 1966 *New Statesman* 25 Feb. 246/1 For Labour MPs in 'marginals'..it means that their perilous positions could be secured. 1970 *Guardian* 16 June 10/4 Three of the four Labour-held marginals theoretically at risk would be lost only on a swing against Labour of between 4 and 5 per cent. 1974 *Times* 13 Feb. 4/6 Redistribution can make a safe seat marginal.

4. *Stock-broking.* Pertaining to, of the nature of, margins (sense 2 c).

1870 J. K. Medbery *Men & Mysteries Wall St.* 59 Nor is there any dissimilarity between the conditions of purchase in complete and in marginal transactions. *Ibid.* 62 The broker..demands of his customer either solid deposit of money or stocks, or marginal deposit of money. 1930 *Economist* 22 Mar. 653/1 Dominion and colonial stocks.. attracted only the 'marginal' business of the market. 1936 *Ibid.* 15 Feb. 368/2 The Great Western, whose 'marginal' security is an ordinary stock. 1938 J. B. Williams *Theory Investment Value* iii. 21 To the marginal investor, it will be indifferent whether he invests in stock or in promissory notes. 1954 B. Graham *Intelligent Investor* ii. 34 Marginal trading—a potent cause of financial ruin to many—has been held within strict limits and at times suspended entirely.

B. *sb.* **2.** *Zool.* A feather on the edge of a bird's wing.

1887 *Proc. Zool. Soc.* 347 They [*sc.* feathers along the posterior border of the wing] are best termed marginals (*tectrices marginales*). 1898 F. E. Beddard *Struct. & Classification Birds* 9 The patagium is mainly filled up with several rows of feathers, which are collectively termed the marginals.

marginalism (mā·ɪdʒinăliz'm). *Econ.* [f. MARGINAL *a.* + -ISM.] An economic analysis which gives prominence to marginal factors in the economy. Hence **ma·rginalist** *a.*, of or pertaining to marginalism; also *sb.*, an adherent of marginalism.

1926 J. A. Hobson *Free-Thought in Social Sci.* iii. 118 The recent extension of Marginalism treats 'doses' as infinitesimal quantities, applies them to the demand as well as the supply side of the economic equation..and to all economic activities. *Ibid.* 119 Our modern Marginalists commit a..mistake in affecting to treat economic material in general as being..other than it actually is. 1929 —— *Wealth & Life* ii. iii. 119 The development of a marginalist doctrine, representing the movements of minutely divisible units of capital and labor into businesses and trades. 1951 J. O. Kamm *Econ. Investment* ii. 33 (*heading*) Marginalism and investment policy. 1956 R. F. Harrod in A. Pryce-Jones *New Outl. Mod. Knowl.* 470 Such were the Austrian school of marginalists and their counterpart in England, Jevons. 1969 P. Anderson in Cockburn & Blackburn *Student Power* 220 Alfred Marshall, father of marginalist economics. 1973 K. Menger in Hicks & Weber *Carl Menger* iii. 38 Austrian marginalism and mathematical economics. *Ibid.*, The Austrian marginalists and the mathematical economists agree in most of their fundamental economic views.

marginality (mādʒinæ·liti). [f. MARGINAL *a.* + -ITY.] The quality or state of being marginal (in various senses of the adj.).

1929 *Pitman's Econ. Educator* III. 1352/1 The exchange value of commodities determines their marginal utility and itself expresses their marginal cost. The notion of marginality in this way sheds light upon the economic conflict. 1951 M. S. Rix *Investment Arithmetic* xii. 117 Priority percentages are also sometimes used to compare the gearing, or marginality, of one company's equity with another. 1955 *Times* 21 June 11/7 The region now has 28 Conservative to 15 Labour marginal seats. Conservatives also have the larger share of 'marginality' in the South and West. 1961 S. R. Herman in J. A. Fishman *Readings Sociol. of Lang.* (1968) 508 The choice of language at the various stages in the above case history reflects the passage from a position of marginality..to a position of adjustment as a member of Israel society retaining the elements of identity derived from socialization in another culture. 1963 L. Trilling in N. Frye *Romanticism Reconsidered* 96 And the converse of what explains Nietzsche's relative marginality explains Dostoevsky's position at the very heart of the modern spiritual life. 1964 R. D. Hopper in I. L. Horowitz *New Sociol.* 19 Marginality here means that, in pre-revolutionary societies, there is formed a group that is marginal to the structure of political power and social prestige. 1971 *World Archaeol.* III. 146 All of these facts combine to form a general picture of marginality in resources. 1974 R. Jessop *Traditionalism, Conservatism & Brit. Polit. Culture* ii. 42 A marginal [social] class..is more likely to become class-conscious than one that is relatively secure and unexploited. Marginality in turn can be due to adverse market conditions, to technological change, to political action, to bad luck, or some other factor.

marginally, *adv.* Add: **a.** (Later example.)

1970 M. Jones *Ducal Brittany* ii. 47 The Breton version of the homage, transcribed..at the back of the *Livre des Osts*, is marginally glossed as simple homage.

b. Round or about the margin or edge of anything; in a way that is close to the limit or margin; by a small margin, slightly.

1909 H. G. Wells *Tono-Bungay* iii. ii. 250 Wandering marginally through distinguished gatherings, I would catch the whispers: 'That's Mr. Ponderevo!' 1931 *Brit. Jrnl. Psychol.* Jan. 226 Attention is directed to locating the proper box for response rather than to picking up a card (because this is such a very common process and everyone can do it 'sub-consciously' or 'marginally'). 1965 *New Statesman* 30 Apr. 672/1 The situation in Britain is only marginally less alarming. 1972 *Accountant* 17 Aug. 191/1 The law will be applied strictly..where schemes differ marginally from the prescribed conditions. 1974 'A. Garve' *File on Lester* xxxvi. 128 The papers are terrible this morning—the *Star* marginally the worst, but only marginally.

margination. Add: **2.** Annotation with marginal notes. *rare.*

1874 Swinburne *Let.* 13 July (1959) II. 308, I quite agree with you that the Homeric margination ought to be most carefully preserved.

maria, *pl.* of *MARE⁴.

Maria (ma·riă), *sb.* and *a.* Also **Maree.** **A.** *sb.* A member of a jungle-dwelling Dravidian people of central India; also, the native (Dravidian) language of this people. **B.** *adj.* Of or pertaining to this people or their language.

1827 R. Jenkins *Rep. Territories Rajah of Nagpore* ii. 33 The wildest of these Gonds are the Marees. They generally go in a state of complete nudity, and even their women have no covering, but aprons of leaves. *a* 1863 S. Hislop in G. Smith *Stephen Hislop* (1888) 22 Moria Gonds..are more civilized than the Márias. 1863 *Sel. Rec. Govt. India Foreign Dept.* No. 39. 14 The language in this talook is Teloogoo and Maria. The population is composed mainly of Marias and Telingas. 1882 H. B. Rowney *Wild Tribes India* i. i. 5 The Máree Gonds..live in the wildest parts of the province of Nágpore. 1938 W. V. Grigson *Maria Gonds of Bastar* iii. 49, I divide the so-called Marias into two divisions, the Hill Marias of the Abujhmar mountains, and the Bison-horn Marias. 1944 I. Singh *Gondwana & Gonds* ii. 22 The Marias formerly roamed about in a state of nature and then..adopted leaves and barks as their garments. 1962 *Listener* 29 Nov. 893/1 Every young member of Indian jungle tribes like the Maria. 1971 [see *GOND *sb.* and *a.*].

|| **mariachi** (mariā·tʃi). [Mexican Sp. *mariache*, *mariachi*.] A group of itinerant Mexican folk musicians; also, a member of such a group. Also *attrib.*

1941 *Time* 19 May 97 There are stacks of records by the omnipresent Mexican street bands, the *mariachis*. 1948 'P. Quentin' *Run to Death* xiii. 106 The *mariachis* had started to play. Over the frenzied twang of guitars a deep, chesty baritone was extolling Guadalajara. 1964 *Listener* 16 July 90/3 The *mariachis* personify all the robust colour of Mexico... They have the looks and the swagger of brigands and the dress-sense of toreadors, and they make jarring, irresistible music in the convulsive patterns of Latin-American rhythms. 1966 [see *CHARRO]. 1967 S. Blanc *Rose Window* (1968) viii. 79 The breeze carried snatches of *mariachi* music from the terrace where the guests were dining. 1970 W. Apel *Harvard Dict. Mus.* (ed. 2) 526/2 Characteristically Mexican is the *mariachi*,

the typical band that serves to entertain people in cafés and at village and country dances and celebrations. 1971 *Islander* (Victoria, B.C.) 19 Dec. 2/2 A mariachi band was ordered and two bottles of tequila to make their evening's work more pleasurable.

|| **mariage blanc** (mariaʒ blaň). [Fr., lit. 'white marriage'.] An unconsummated marriage.

1926 *Irish Statesman* 13 Nov. 234 It is improbable, and yet quite possible, this story of a *mariage blanc*. 1931 *Times Lit. Suppl.* 17 Sept. 702/2 One must admire the tact with which Mr. Church handles the painful theme of the full-blooded Norah's revolt against the frustration of a *mariage blanc*. 1958 *Observer* 23 Feb. 14/2 We get a hint that Evan's and Cherry's is really a *mariage blanc*. 1958 G. Mitchell *Spotted Hemlock* xi. 214 'The autopsy revealed that the girl was a virgin.'..'*Un mariage blanc?* Good Lord!' 1975 *Listener* 21 Aug. 253/3 Opal..suggested a *mariage blanc* between Natalie and Bosie that would have enabled Opal and Bosie to have a lasting liaison.

|| **mariage de convenance** (mariaʒ də koňvənaňs). [Fr., lit. 'marriage of convenience'; cf. CONVENIENCE *sb.* 6.] A marriage arranged or contracted from motives of convenience (sense 6) or expediency. Also *fig.* Cf. *marriage of convenience* (*MARRIAGE 8).

1854 Thackeray *Newcomes* I. xxviii. 275 What the deuce does a *mariage de convenance* mean but all this. 1864 Trollope *Small House at Allington* I. ix. 83 I'm only a half sort of lover, meditating a mariage de convenance to oblige an uncle. 1912 I. Nitobé *Jap. Nation* vi. 163 The sorrows of *mariage de convenance* in Europe. 1923 J. S. Huxley *Ess. Biologist* vii. 295 Some men are pragmatic and utilitarian in regard to Truth; by others she is worshipped as fanatically as any goddess. So some men deliberately make *mariages de convenance*. 1955 *Bull. Atomic Sci.* Mar. 97/3 History suggests that the evil partnership between Communist China and the Soviet Union may yet develop the brittle quality of a mariage de convenance. 1957 L. Durrell *Justine* iv. 235 Someone trapped into a *mariage de convenance*. 1974 M. Cecil *Heroines in Love* iv. 91 The fashionable world, with its ill-bred ideas about money and *mariages de convenance*.

mariahuana, var. *MARIJUANA, MARIHUANA.

Marial (mēə·riăl), *a.* [F. *marial* = MARIAN *a.*¹ 1; cf. MARIAL².] = MARIAN *a.*¹ 1.

1904 *Catholic Herald* 23 Dec. 8/4 On the initiative of the Archbishop of Leopol [*sic*], a grand Marial Congress was organised and held in his Cathedral city. 1952 D. M. J. Langdon *Our Lady of La Salette* v. 30 Montserrat in the mountains of Spain, and hundreds of other centres of Marial devotion. 1966 *Approaches* Nov. 123 The Marial Congregations for men, which are now actively organising their lay apostolate in all fields of public life.

Marian, *sb.*² Add: **2.** (Later examples.)

1969 A. Fraser *Mary Queen of Scots* ii. xxii. 433 The castle of Edinburgh, so long held by Kirkcaldy and Maitland on behalf of the Marians..was at last effectively besieged. *Ibid.*, Maitland..had died a loyal Marian. 1974 G. Donaldson *Mary Queen of Scots* v. 122 Kirkcaldy and Maitland did not emerge as open Marians until much later and were at first numbered among 'secret favourers of the Queen'. *Ibid.* 134 Balfour of Pittendreich became a Marian, though he was not at Langside.

3. An English Catholic of Queen Mary's reign (1553–8).

1868 H. H. Milman *Ann. S. Paul's Cathedral* xi. 256 The Primate Pole was in his grave, Heath of York a strong Marian. 1899 F. W. Maitland *Coll. Papers* (1911) III. 130 Canon MacColl laboured under the misfortune of knowing something about the votes that these Marians gave in Parliament. 1904 W. H. Frere *Eng. Church 1558–1625* ii. 23 The champions of the Marians..were to dispute with the champions of the exiles. 1955 C. Read *Mr. Secretary Cecil* iv. 101 He abandoned flight and decided to throw in his lot with the Marians.

Marianne (maria·n). Also in anglicized form (*rare*) **Mary Anne.** [Fr.] The name of a Republican Secret Society formed in France after the coup of 1851 to restore a Republican Government; hence a familiar name for the Republican form of Government and, by extension, a personification of the French Republic.

1870 Disraeli *Lothair* I. xi. 100 Our refreshment at council is very spare..but we always drink one toast... It is to one whom you love and whom you have served well. Fill glasses, brethren, and now 'To *Mary-Anne*'. 1890 Barrère & Leland *Dict. Slang* II. 45/2 'La Marianne' ..was the name of a secret Republican Society in France. 1933 Kipling *Souvenirs of France* ii. 53, I asked a friend, an Alsatian General, whence the flood of material had come. 'From Marianne,' was the reply. 'She has all sorts of things like these in her stocking—when she needs them.' 1934 H. G. Wells *Exper. Autobiogr.* II. ix. 747 (*caption*) Marianne asks Dadda to tell her all about it. 1940 —— *Babes in Darkling Wood* iv. iii. 374 Nor will Marianne be in a position to act the vindictive hostess this time... France and Britain had their chance of making a world peace in 1918, and they muffed it. 1958 H. J. Greenwall *When France Fell* ii. i. 66, I went to the *Marianne* office in the Avenue des Champs Elysées. 1962 *Listener* 12 July 57/2 Had Daumier wished to treat a similar theme we can be almost certain he would have selected Marianne or the Gallic cock as an impersonal national emblem. 1971 *Guardian* 3 Apr. 3/5 There is no official model for the bust of Marianne, the incarnation of the Republic, found in every Mairie in France.

Mariavite (mēə·riăvəit). [Pol. *Mariawita*, f. L. phr. *qui Mariæ vitam imitantur*.] A member of a Polish Christian sect which flourished in the early 20th century; also *attrib.*

[1906 *Daily Chron.* 23 Apr. 5/5 The church at Leschno.. had been occupied by the sect of Marianites. 1906 *Times* 28 May 5/3 The *Osservatore Romano* publishes to-day an Encyclical..condemning the doctrine of the so-called Marianisti, and definitely suppresses the Marianisti congregation.] 1906 *Tablet* 9 June 897/2 Some priests..had founded, without permission from their lawful superiors, a kind of pseudo-monastic society, known as the *Mariavites* or *Mystic Priests*. *Ibid.* 898/2 The so-called Mariavite priests who are in good faith may no longer be led astray. 1949 D. Attwater *Catholic Encycl. Dict.* 306 *Mariavites*, a sect in Poland originating in 1906 with a number of apostate clergy. 1957 *Oxf. Dict. Chr. Ch.* 857/2 *Mariavites*, a Polish sect, founded in 1906 by J. Kowalski, a priest of Warsaw, and Felicia Kozlowska, a Tertiary sister,..on their excommunication from the RC Church.

|| **mari complaisant** (mari koňplezaň). [Fr.] A husband who tolerates his wife's adultery.

1898 W. J. Locke *Idols* xv. 211 You are three outstanding people... Hang it all—the *mari complaisant*—and Merriam is the last man in the world—it beats me altogether. 1933 *Times Lit. Suppl.* 5 Oct. 673/1 Without being the *mari complaisant* he has hitherto stood aside, probably seeing much more than he could admit. 1937 Mrs. Belloc Lowndes *Diary* 20 Jan. (1971) 145 The unfortunate Mr Simpson was..regarded as *un mari complaisant*. 1950 D. Ames *Corpse Diplomatique* ix. 72, I don't understand English women. In France there is such a thing as a *mari complaisant*, but I've never heard of an *amant complaisant*! 1958 L. Durrell *Justine* i. 29 She was reputed to have had many lovers, and Nessim was regarded as a *mari complaisant*. 1972 P. M. Hubbard *Whisper in Glen* xii. 115 If you see him as the *mari complaisant*, you'd better think again... If he knew I was carrying on with you..he'd fling me out.

mariculture (mæ·rikɒltiŭɪ, -tʃəɪ). [f. L. *mari-*, *mare* sea + CULTURE *sb.*] The cultivation of the resources of the sea, esp. of fish for food. So **maricu·ltural** *a.*, of or pertaining to mariculture; **maricu·lturist**, one who engages in or specializes in mariculture.

1903 *Science* 9 Oct. 461 Such maps would be purely agricultural and maricultural, dependent upon the harvests of the land and sea. 1909 *Cent. Dict.* Suppl., *Mariculture*, the development of the resources of the sea, especially with respect to food-fish; coined in distinction from *agriculture*. 1969 *Sci. Jrnl.* Dec. 30 The man who studies agriculture stands in a different place from the man who studies mariculture. There are other differences between the two fields but not so basic as the need for the mariculturist to swim. 1970 *New Scientist* 20 Aug. 378/1 The first step in an effort to use the deep ocean water for mariculture, the generation of electrical power and the production of fresh water. 1971 *McGraw-Hill Yearbk. Sci. & Technol.* 23 If we are very optimistic and assume for a moment that the yield of the traditional fishery will not change, mariculture could perhaps lead to an increment in the crop produced from the sea from the present 1% to 1.1%. 1973 *Daily Colonist* (Victoria, B.C.) 12 Apr. 12/4 Some areas should be left wild; others should be utilized for recreation; others might be developed and employed for mariculture; and still other areas might be exploited or used as dumping grounds. 1974 *Victorian* (Victoria, B.C.) 5 Apr. 2/1 The federal-provincial governments have allotted $38,000 to two companies involved in oyster mariculture in a cost-sharing program to improve and develop the oyster industry in British Columbia.

Marie (mā·ri). Also **marie.** A female Christian name used *attrib.* of a type of plain sweet biscuit.

1878 *Official Guide & Album Cunard Steamship Co.* 186/3 (List of Biscuits) Marie. 1888 *Mrs. Beeton's Bk. Househ. Managem.* ii. 31/2 Biscuits..Marie..1s. 9d. per tin. 1906 E. Nesbit *Railway Children* ii. 27 There were biscuits, the Marie and the plain kind. 1919 D. Ashford *Young Visiters* vi. 50 He swollowed his tea and eat a Marie biscuit hastilly. 1947 E. Taylor *View of Harbour* xii. 203 A jug of cocoa and a plate of Marie biscuits. 1966 B. Kimenye *Kalasanda Revisited* 43 Though they sipped their tea, they simply could not bring themselves to touch a marie biscuit. 1972 C. Drummond *Death at Bar* i. 10 The old tea lady..slopped the fluid over on to the marie biscuits in the saucer.

Marie Antoinette (ma·ri antwane·t). The name of the Austrian queen consort (1755–1793) of Louis XVI of France, used *attrib.* to designate various styles of dress, decorative art, etc., characteristic of her or her reign (see quots.).

1925 F. Scott Fitzgerald *Great Gatsby* v. 110 We wandered through Marie Antoinette music-rooms. 1950 '*Mercury*' *Dict. Textile Terms* 337 *Marie Antoinette*, a curtain, having appliqué sprays, flowers and leaves of cord and tape. 1960 Cunnington & Beard *Dict. English Costume* 133/1 *Marie Antoinette skirt*,..a day skirt with 7 gores, 1 in front, 2 on each side, and 2 behind, box-pleated. 1966 J. S. Cox *Illustr. Dict. Hairdressing* 96/1 *Marie Antoinette chignon*, a full globular low dressed chignon. *Marie Antoinette coiffure*, hair styles similar to those worn by Marie Antoinette.

Marie Louise (ma·ri lu,ī·z). The name of *Marie Louise* (1791–1847), second wife of Napoleon I, used to designate a variety of yellow-skinned pear. Also *attrib.*

1817 *Trans. Hort. Soc.* II. 406 A box of Pears, sent to the Society by Dr. Van Mons, of Brussells, was examined. .. They are stated, by Dr. Van Mons, to be seedlings raised by himself: 1, Napoleon..; 2, Marie Louise: in shape and size resembling a small Bon Chrétien; yellow, with bright red next the sun; very rich, and ripe about the middle of October. 1820 *Ibid.* IV. 519 Specimens of the Marie Louise Pear were received. 1860 R. Hogg *Fruit Manual* 200 Marie Louise... One of our very best pears. Ripe in October and November. 1894 'Mrs. Alexander' *Choice of Evils* I. ii. 34 Not even the Mervyn gardens could produce finer pears than the solitary 'Marie Louise' tree. 1929 E. A. Bunyard *Anat. Dessert* 104 Marie Louise, cannot be overlooked,..being named after that most acquiescent of Empresses, Napoleon's second wife, and, like her, it needs the support of a strong arm to stay its languishing growth. 1958 [see *DOYENNE¹].

marigold. Add: **8.** The colour of the marigold flower; hence *attrib.* passing into *adj.*, of this colour, bright yellow.

1774 Marigold colour [in Dict., sense 7]. 1839 [see *café au lait* (*CAFÉ 3)]. 1872 G. M. Hopkins *Let.* 22 Mar. (1956) 55, I am jaundiced all marigold under the eyes. 1916 D. H. Lawrence *Amores* 49 Cluck, my marigold bird, and again Cluck for your yellow darlings. 1923 *Daily Mail* 28 Feb. 1/3 (Advt.), Shades of..Jade, Silver Grey or Marigold. 1975 *Country Life* 20 Mar. 744/2 Their clothes are matched to a limited number of colours...there are cornflower/cool cornflower, mint/cool mint, and marigold/cool marigold.

marijuana, marihuana (mærihwā·nă). Also mara-, maria-; mariguan(a. [Amer. Sp.] **1. a.** A preparation of the hemp plant, *Cannabis sativa* (see *CANNABIS), for use as an intoxicating and hallucinogenic drug; usu. applied to a crude preparation of the dried leaves, flowering tops, and stem of the plant that is generally smoked.

1894 *Scribner's Mag.* May 596/2 [The] 'toloachi', [and] the 'mariguan',..are used by discarded women for the purpose of wreaking a terrible revenge upon recreant lovers. 1918 *Jrnl. Amer. Med. Assoc.* 21 Dec. 2094/1 The symptoms mentioned..as being produced by smoking Mara Huiwane or marajuana are similar to those produced by the mescal plant. 1923 W. Smith *Little Tigress* 102 The cockroach is unable to stagger around any more because he has no more marijuana to smoke. *Marijuana* is a form of drug that brings false heart to the user. 1927 *Amer. Speech* III. 37 The tobacco of the Turkish coffee-houses has been replaced in Mexico by marihuana, the native variety of hashish. 1928 *Daily Express* 11 Oct. 3 What is Marijuana? A deadly Mexican drug, more familiarly known as 'Mary Jane', which produces wild hilarity when either smoked or eaten. 1935 J. Steinbeck *Tortilla Flat* xiv. 246 His eyes were as wide and pained as the eyes of one who smokes marihuana. 1939 [see *DAGGA¹]. 1944 R. A. Moore *Textbk. Path.* lix. 657 After long-continued use of marihuana there is mental deterioration. 1952 M. McCarthy *Groves of Academe* (1953) ii. 19 On one occasion, even, marijuana had been smoked on the steps of the gymnasium. 1960 N. Polunin *Introd. Plant Geogr.* ix. 267 Cannabis consumption as hashish, marijuana, etc., causes states of ecstasy and stupefaction. 1968 P. Oliver *Screening Blues* ii. 83 The song *Weed Smoker's Dream* referred to the hallucinations induced by marijuana. 1968 *Times* 19 Dec. 4/6 Basic findings are that.. subjects who have not smoked marihuana before do not have strong subjective experiences, even after strong doses.

b. *fig.*

1948 *Sat. Rev.* (U.S.) 19 June 24/3 One of the most cogent reasons for the success of the comics, is namely, that they are the marijuana of parents as well as of their offspring. 1965 *New Statesman* 10 Dec. 916/2 The politician who gave the Beatles the MBE..knows that if he suddenly cut off about 20 million young addicts from their supply of aural marijuana the withdrawal effects could produce unpleasant political side-effects.

2. The hemp plant; = *CANNABIS 1.

1907 A. B. Lyons *Plant Names* (ed. 2) 511 (Cannabis) a. Add Syn[onym] Marihuana (Mex.). 1934 R. E. Carradini *Narcotics & Youth Today* 1 In this classification [of narcotic drugs] should be included..*Cannabis Indica*, popularly known as Indian hemp or hashish, and the native plant, *mariahuana*. 1934 *Jrnl. Amer. Med. Assoc.* 21 July 212 On appeal to the Supreme Court of Utah, Navaro contended that the term 'mariguana', as used in the statute, signified a plant, not a drug. 1959 *Encounter* Oct. 56/2 The land where the marihuana grows. 1973 *Daily Tel.* 4 July 4 Nepal is to ban the growing of marijuana and poppies and the production of cannabis and opium.

3. *attrib.* and *Comb.*, as *marijuana addict, cigarette, habit, -smoker, -smoking* vbl. sb. and ppl. adj., *weed*.

1936 *Amer. Speech* XI. 12 Argot of narcotic addicts... *Twister*,..one who rolls his own twists, or marajuana cigarettes or, by extension, a marajuana addict. 1938 *Manch. Guardian Weekly* 2 Sept. 188/3 Some [swing players] are 'mugglers' (Marijuana addicts). 1933 *Jrnl. Amer. Med. Assoc.* 28 Oct. 1398/1 The first person to be convicted in Hillsborough County [in Florida] of selling marajuana cigarets..was sentenced..to a year in the county jail. 1966 T. Pynchon *Crying of Lot 49* iii. 64 Leonard the drummer..produced a fistful of marijuana cigarettes. 1953 W. Burroughs *Junkie* (1972) x. 107, I told them you were sick and that you needed a shot of morphine, and they said, 'Oh, we thought it was just a question of a marijuana habit.' 1951 E. Paul *Springtime in Paris* xi. 202 Narcotics addicts and marihuana smokers are infiltrating the cafés. 1935 *Amer. Mercury* Aug. 426/1 Such practices as adultery, marihuana smoking, gambling, ..and rowdyism seldom worry the investigator. 1957 P. Frank *Seven Days to Never* iii. 102 She married..a marijuana-smoking drugstore cowboy. 1958 *Times* 3 Dec. 6/6 Unauthorised clubs..were a great moral danger to young people. The marihuana weed was smoked.

marimba. Substitute for def.: A kind of deep-toned xylophone, originating in Africa and consisting of wooden keys on a frame with a tuned resonator beneath each key. Hence, a modern orchestral instrument evolved from this. Also *attrib.* (Further examples.)

1923 A. H. Newbold *Bamboo Curtains* 27 You and I on a polished floor, And a big marimba band. 1924 S. Sitwell *13th Cæsar* 58 Loud is the marimba's note Above these half-salt waves. 1927 *Melody Maker* Aug. 732/3 (Advt.), The Rage To-Day.—'Foote' Xylophones and Marimbas. 1934 A. Huxley *Beyond Mexique Bay* 119 The visitor of 1933 may think himself lucky if he hears a *marimba* orchestra discoursing tangos. 1957 Wodehouse *Over Seventy* iii. 42 You can dance nightly to the strains of somebody's marimba band. 1958 A. Jacobs *New Dict. Mus.* 226 Milhaud wrote a concerto (1947) for marimba and vibraphone. 1961 *Guardian* 22 May 5/5 Her company of 75 dancers, singers, and marimba players presented us with authentic reconstructions of the Aztec ritual dances. 1970 J. Blades *Percussion Instruments* xvi. 425 In *Vision of St. Augustine* Tippett writes for marimba.

marimbaphone (mări·mbăfōun). [f. prec. + Gr. φωνή sound, voice.] = prec.

1921 *Daily Colonist* (Victoria, B.C.) 9 Oct. 23/2 The instrumentation will be as follows: 25 Violins, 6 Pianists.., 4 Marimbaphones, [etc.]. 1923 *Weekly Dispatch* 11 Mar. 5 The marimbaphone or Mexican xylophone..depends for its depth of tone on a number of tubes of varying lengths filled with liquid. 1926 A. Huxley *Jesting Pilate* iv. 263 Organ chimes, giant marimbaphone, vibraphone. 1962 *Listener* 22 Nov. 885/3 A marimbaphone and six chimes.

marina¹. Add: **2.** A dock or basin with moorings for yachts and other small craft. Also *attrib.* orig. *U.S.*

1935 *Yachting Monthly* LIX. 223/2 Fees for keeping a yacht in a municipal 'marina' (trick name for basin) are modest. Yachts are tied up in slips in these marinas. 1959 *Manch. Guardian* 15 Aug. 5/2 Everywhere in the United States the outboard motor, the cabin cruiser, the 'marina' (a little dock for pleasure craft) are to be seen. 1960 *Sunday Times* 3 Apr. 37/5 But what about entering and leaving this well-dammed basin, or marina—the American word? 1961 *Times* 28 Apr. 6/3 (*heading*) Hayling Island marina plan to be pressed. 1963 R. I. McDavid *Mencken's Amer. Lang.* 264 Spanish may share with Italian the credit for the current vogue of *marina*, as a de luxe designation for a yacht basin. 1969 *Daily Tel.* 11 Jan. 19/3 It will include two hotels, flats, 'boatels', yacht clubs, restaurants, two public houses and a 'marina drome'. 1971 *N.Z. Listener* 25 Oct. 7/1 In 1928 boat-owners added another word to their vocabulary, 'marina'. Coined in the United States to describe a new kind of luxury boating facility..it has altered the style and living of an increasing number of New Zealanders.

Marina³ (mărī·nă). [f. the name of H.R.H. Princess *Marina*, Duchess of Kent (1906–68).] Used *attrib.* in *Marina green*, a shade of green.

[1934 *Times* 29 Nov. (Suppl.) p. iii/2 The going away *ensemble* [of Princess Marina] will be in almond-green tweed.] 1935 *Discovery* Aug. 231/1 There is a popular shade of green at present in vogue, termed 'Marina green'. 1935 *Times* 18 Nov. 21/4 Cleverly designed dinner gowns in satin and lace in rust, Marina green, and royal blue. 1958 M. Dickens *Man Overboard* iii. 46 That sickly colour which was once foisted on to a loyal public as Marina green when the Duchess of Kent was married.

marinate, *v.* Add: **1. a.** (Further examples.) Also *absol.*

1960 *Times* 18 Jan. 15/5 Fillet [steak] is exceptionally tender and should not need to be marinated or beaten. 1970 *Daily Tel.* (Colour Suppl.) 30 Oct. 62 It is..absolutely in order to make the 'gamey' flavour less pungent by marinating.

marinated, *ppl. a.* **2.** (Further example.)

1969 *Daily Tel.* 14 Jan. 15 Spicy marinated pork.

marine, *a.* and *sb.* Add: **A.** *adj.* **1.** *marine band*: a geological horizon containing fossils of marine origin situated between horizons of freshwater origin.

1920 W. Gibson *Coal in Gt. Brit.* xxii. 277 With the exception of the marine fauna of Skipsey's Marine Band and of that above the Slaty Band Ironstone, the invertebrate fauna is not of much variety. 1939 Raistrick & Marshall *Nature & Orig. Coal Seams* ii. 33 The remains of fresh-water or marine shells are less common, forming when they occur in quantity 'mussel bands' and 'marine bands'. 1969 Bennison & Wright *Geol. Hist. Brit. Isles* ix. 224 The marine horizons (called marine bands) are sometimes of great lateral extent.

6. marine biology, the study of plants and animals living in the sea; hence **marine-biological** *a.*; **marine biologist**, one engaged in this study; **marine iguana**, a large lizard, *Amblyrhynchus cristatus*, found on the Galapagos Islands; **marineland**, a type of zoological garden, designed to exhibit and preserve marine animals.

1884 in *Jrnl. Marine Biol. Assoc.* (1887) I. 22 Report of the Foundation Meeting of the Marine Biological Association. 1959 D. A. Bannerman *Birds Brit. Is.* VIII. 119 The expedition..was mainly to investigate some of the marine-biological and agricultural problems of the Tristan da Cunha group. 1963 T. A. Sebeok in J. A. Fishman *Readings Sociol. of Lang.* (1968) 30 Aristotle was also the first marine biologist who accurately classed the dolphin with the mammals. 1887 *Jrnl. Marine Biol. Assoc.* I. 1 It is not proposed to limit the contents of this journal to formal reports..but to include within its pages..brief records of observations relating to the marine biology and fisheries of the coasts of the United Kingdom. 1936 *Discovery* June 166/2 M. Paul Lemoine..referred to his [*sc.* Sir Robert Mond's] work on behalf of marine biology. 1967 *Oceanogr. & Marine Biol.* V. 111 Of all the different facets of marine science, marine biology is perhaps the oldest in India. [1839 Darwin *Jrnl.* in Fitzroy & Darwin *Narr. Voy. H.M.S. Adventure & Beagle* III. xix. 453 (*heading*) Marine lizard feeds on sea-weed.] 1924 C. W. Beebe *Galapagos* v. 111 The sea or marine iguana is as good a name as could be desired. 1962 C. J. & O. B. Goin *Introd. Herpetology* viii. 136 *Amblyrhynchus cristatus*, the Marine Iguana of the Galápagos Islands, feeds on marine algae at low tide. 1972 *Country Life* 6 Apr. 842/3 In addition to the giant tortoises..there are the marine iguanas. 1963 *Times* 17 Apr. 12/5 Southsea Castle committee agreed today to open negotiations with Billy Smart's Circus for the proposed establishment of a marineland in the castle grounds. 1968 *N.Z. News* 16 Oct. 6/2 Complete plans and specifications have been drawn up and a tender accepted for a marineland on Windsor Reserve at Devonport, Auckland.

B. *sb.* **4. c.** Phr. *tell that to the marines* (earlier and later examples): an expression of incredulity. Cf. quot. 1892 s.v. HORSE-MARINE².

1806 J. Davis *Post Captain* v. 29 He may tell that to the marines, but the sailors will not believe him. *c* 1829 D. Jerrold in M. R. Booth *Eng. Plays of 19th Cent.* (1969) I. 179 No palaver; tell it to the marines. What, tacking and double tacking! Come to what you want to say at once. 1902 J. Conrad in *Blackw. Mag.* Dec. 802/2 'You shall get nothing from me, because I have nothing of mine to give away now.' 'Tell that to the marines!' 1928 *Times* 21 July 17/5 He said that I should..most likely be shot. I ventured to suggest that he should tell that to the Marines. 1933 E. O'Neill *Ah, Wilderness!* (1934) iv. ii. 145 And I suppose you just sat and let yourself be kissed! Tell that to the Marines! 1944 W. S. Maugham *Razor's Edge* vii. 279 '*A d'autres, ma vieille*,' I replied, which I think can best be translated by: 'Tell that to the marines, old girl.' 1967 D. Francis *Blood Sport* xiv. 172 'When this is over you can sleep for a fortnight.' 'Yeah?' he said sarcastically. 'Tell it to the marines.'

6. b. marine blue, a dark blue, the colour of the uniform worn by the Royal Marines; freq. *ellipt.* as *marine*.

1873 *Young Englishwoman* IV. 78/1, I should suggest.. one polonaise of black cashmere..and one of deep marine blue vigogne. 1895 *Montgomery Ward Catal.* 3/1 Cashmere... All the fashionable colors..black, marine blue, light blue. *Ibid.* 12/3 Plain China silk..navy, marine, gray. 1910 *Encycl. Brit.* VIII. 746/2 The following is a list of the more important basic colours derived from coal-tar.. marine blue, indoïne blue, [etc.]. 1925 E. Sitwell *Poor Young People* 2 The colours most in favour are marine Blue,..myrtle green. 1930 Maerz & Paul *Dict. Color* 168/1 In the early nineteenth century there became popular in the general textile field a color adapted from the uniforms of sailors. This color seems first to have been called *Marine Blue*, followed shortly by the name *Navy Blue*. The color was long dyed with indigo. 1934 *Times* 30 Nov. 20/2 The Queen of Denmark's dress and..coat were both of deep marine-blue velvet. 1961 H. E. Bates *Now sleeps Crimson Petal* 43 The marine blue thorns of sea-thistle were touched with sepia rose.

|| **marinera** (marinē·ră). [f. Sp. *marinero*, *-a* marine, seafaring.] = *CUECA.

1926 D. L. Joseph in J. F. Dobie *Rainbow in Morning* (1965) 54 Play for him a *Marinera* While they whip him. 1964 W. G. Raffé *Dict. Dance* 305/1 Of Spanish extraction, the *Marinera* came to Peru from Chile and was originally called *Cueca Chilena*, or *Chilena*, the name being changed to *Marinera* during the 19th century when Peru and Chile were at war (1879–1883). The new name honoured the Peruvian Navy. 1973 K. Benton *Craig & Jaguar* v. 53 They even succeeded in making me dance a *marinera* with one of the—er—very nubile young women. 1974 *Encycl. Brit. Macropædia* I. 669/1 This uncertain major–minor tonality pervades the Andean music. It appears in the *chinguinada* for church festivals, with a bass ostinato (repeated melodic-rhythmic pattern); in the mestizo *yaraví* serenade and in the mestizo *marinera* couple dance, with its metre alternating ⁶⁄₈ and ¾ patterns.

marinescape (mărī·nskēip). [f. MARINE *a.* and *sb.* + SCAPE *sb.*³, after LANDSCAPE.] A picturesque view or prospect of maritime scenery. Cf. SEA-SCAPE.

1928 *Daily Express* 13 Jan. 11 The Mediterranean was like a sheet of sapphire... Everywhere the marinescape was dotted with rowing boats and other vessels. 1969 J. W. Mavor *Voy. Atlantis* v. 112 It was the most awesome marinescape that I had ever seen.

marine store(s. Add: **1. b.** (Earlier example.)

c 1829 D. Jerrold in M. R. Booth *Eng. Plays of 19th Cent.* (1969) I. 184 Go into the mercantile line—take a shop for marine-stores.

2. (Earlier example.)

1837 Barham *Legend H. Tighe* in *Ingol. Leg.* (1840) 1st Ser. 158 In Ratcliffe Highway there's an old marine store.

3. (Further examples.)

1836 Dickens *Sk. Boz* 1st Ser. I. 91 The marine-store dealer at the corner of the street. 1844 *Ainsworth's Mag.* VI. 112 A marine-store keeper of the pilfered orts and ends of literature.

Mariological (mēəriolǫ·dʒikăl), *a.* Also with small initial. [f. MARIOLOG(Y + -ICAL.] Of or pertaining to Mariology. So **mariolo·gically** *adv.*

1954 *Theology* LVII. 90 A mariological interpretation. *Ibid.*, Interpret the text mariologically. 1967–8 H. R. McAdoo in Clark & Davey *Anglican/R.C. Dialogue* (1974) vi. 93 There are..similarly differences as between Roman Catholics and Anglicans, and Anglicans would add Mariological definitions. 1969 E. L. Mascall in A. Richardson *Dict. Christian Theol.* 208/1 By a small majority it was decided by the [Second Vatican] Council not to issue a separate Mariological document. 1971 *Catholic Dict. Theol.* III. 264/1 The most notable event in the Mariological deliberations of the Vatican Council was the vote taken on 29 October 1963 to decide whether the schema *de Maria matre ecclesiae* should be taken as a final chapter of the decree on the Church or as a decree by itself. 1974 *Encycl. Brit. Macropædia* XI. 563/1 The extravagances of Marian cult and Mariological thought that have sometimes tended to go beyond the limits both of biblical foundation and of ecclesiastical regulations.

marionettish (mæriŏne·tiʃ), *a.* Also **marionetish.** [f. MARIONETT(E + -ISH¹.] Suggestive of a marionette in appearance or movement.

1921 E. O'Neill *Emperor Jones* v. 184 There is something stiff, rigid, unreal, marionettish about their movements. 1930 *Observer* 18 May 26/2 'Intimate Snapshots' was fresher and rounder in conception than in execution... I thought the liftman unclear in his explanation, and the journalists marionetish. 1963 A. Ross *Australia 63* x. 196 If fast, short-pitched bowling induces an initial, marionettish jerkiness he soon settles down to more convincing methods. 1969 E. Wilkins *Rose-Garden Game* viii. 192 Their marionettish aspect encourages the impression that at any moment the gilded mandorla might begin to turn on its axis and spin.

marione·ttist. Also **marionnettist.** [f. MARIONETT(E + -IST.] One who operates marionettes.

1918 *Marionnette* (Florence) Jan. 98 The Marionnettist has put his dolls in their box...his wife is counting the pennies. 1924 *Glasgow Herald* 10 May 8 It was here the man of so many quarrels..fought the mock duel with.. the marionettist's monkey. 1936 R. S. Sibbald *Marionettes in North of France* i. 44 M. Delannoy..gives a list of the marionettists who applied to the city of Lille..for permission to show their dolls.

Mariotte (mæ·riŏt). Also erron. **Marriotte.** The name of Edme *Mariotte* (*c* 1620–84), French physicist, used in the possessive and *attrib.* to designate apparatus he devised and a principle he enunciated: **Mariotte('s) bottle** or **flask**, a bottle with an outlet near the bottom and an adjustable glass tube passing through a cork in the neck, which if filled to above the bottom of the tube gives a flow of constant head equal to the height of the bottom of the tube above the outlet; **Mariotte's law**, Boyle's law (see LAW *sb.*¹ 17 c (*b*)); **Mariotte's tube**, a U-tube having one arm short and sealed at the end and the other elongated and open to the air.

1845 E. West tr. *Peschel's Elem. Physics* I. v. iii. 220 On the same principle Mariotte's flask acts. *Ibid.* iv. 228 In such a case it is more convenient..to use..a Mariotte's tube.., which being graduated according to Marriotte's [*sic*] law, will show the amount of pressure. 1898 J. C. P. Aldous *Elem. Course Physics* 171 (*heading*) Boyle's (or Marriotte's [*sic*]) law. *Ibid.* 172 (*heading*) Boyle's (or Mariotte's) tube. 1930 Dougall & Deans tr. *P. P. Ewald's Physics Solids & Fluids* v. 217 The experiment is conveniently carried out by using the so-called Mariotte's bottle. 1966 *McGraw-Hill Encycl. Sci. & Technol.* II. 313/1 The phenomenon was discovered independently by Edme Mariotte about 1650 and is known in Europe as Mariotte's law. 1974 *Nature* 26 Apr. 798/1 Continuous flow of fluid..was maintained relatively constant by means of gravity flow from a Mariotte flask. 1975 Williams & Wilson *Biologist's Guide to Princ. & Techniques Pract. Biochem.* iii. 93 This can be overcome by the use of a Mariotte flask which will keep the operating pressure constant.

|| **mariposa** (maripo·să). [Sp., lit. 'butterfly'.] In *Bullfighting*, a movement in which the bullfighter draws the bull by flapping the cape behind his own back. Also *fig.*

1932 [see *FAROL]. 1936 R. Campbell *Mithraic Emblems* 143 If you would lepidopterize These mariposas that I fling. 1957 A. MacNab *Bulls of Iberia* xv. 169 The bull.. is attacking what looks like a giant butterfly... The butterfly is Estudiante himself, and he is doing the famous *quite* of the *mariposa*... The matador holds the cape behind him,..as he holds out his arms sideways to form the 'wings' with the cape... As the bull turns and goes for one wing, he drops his arm and that wing folds up. 1967 McCormick & Mascareñas *Compl. Aficionado* iii. 92 Like all beginning novilleros, Domingo longs to learn the adornos first, the *revolera*, the *serpentina*, the *mariposa* with the cape.

mariposa lily. Add: Also **mariposa tulip.** Substitute for def.: A bulbous plant of the genus *Calochortus*, belonging to the family Liliaceæ, and native to the Pacific coast of North America and Mexico. Also *absol.* (Earlier and later examples.)

1868 A. Wood in *Proc. Acad. Nat. Sci. Philadelphia* 169

This splendid flower [sc. *Calochortus venustus*]..has long been known to the native Californians by the name of *Mariposa* (Spanish for butterfly). 1869 J. Muir *Jrnl.* 9 June in *My First Summer in Sierra* (1911) 43 The grasses and flowers in glorious array,..mosquitoes, Mariposa tulips, lupines. 1920 B. M. & R. Rice *Pop. Stud. Calif. Wild Flowers* 27 It would probably be more correct to say Mariposa Tulip than Mariposa Lily, for botanists place them in the tulip family. 1925 J. L. Cotter *Culture of Bulbs* 118 The name Calochortus now embraces the Mariposa Lily proper, and the Star Tulip. 1934 S. E. White *Folded Hills* 198 Here also were..mariposas. 1971 F. M. Synge *Collins Guide to Bulbs* (ed. 2) 71 It is in..the Mariposa lilies or Mariposa tulips that we chiefly find the most outstanding horticultural plants [of the genus *Calochortus*]. *Ibid.* 73 The leaves of this section are..more conspicuous than those of the Mariposas.

mariposite (mæripōᵘ·zait, -soit). *Min.* [f. *Maripos-a*, the name of the county in California, U.S.A., where it was first found + -ite¹.] A variety of muscovite that contains a relatively high proportion of silica and up to one per cent. chromic oxide and is found as green or greenish-yellow crystals.

1868 B. Silliman in *Proc. Calif. Acad. Sci.* III. 381 The mineral probably new, and must be referred to the mica section of an hydrous silicate. Should it..prove to be new, I would suggest the name *Mariposite* as an appropriate name for it, as it was on the Mariposa estate that it first attracted my attention, and where it exists in great abundance. 1916 *Bull. U.S. Geol. Survey* No. 610. 139 A comparison of the two analyses of mariposite made by Hildebrand with Penfield's analysis of alurgite shows that the two minerals are probably identical. 1946 *Amer. Mineralogist* XXXI. 1 Chrome micas, classed variously as fuchsite, mariposite, and chromiferous muscovite are of wide distribution throughout gold-bearing districts of the Canadian Shield. 1965 G. J. Williams *Econ. Geol. N.Z.* x. 149/2 The majority of economic occurrences of fuchsite (and mariposite) seem to have a direct ultramafic association.

maritime, *a.* and *sb.* Add: **A.** *adj.* **6.** **maritime pine,** the cluster pine, *Pinus pinaster*, a southern European tree distinguished by cones in clusters, often planted in coastal areas to bind sandy soil.

1894 H. M. Ward *Laslett's Timber & Timber Trees* (ed. 2) xxxi. 349 The Cluster Pine..much used in the south and west of France, where it is known as the Maritime Pine from the extensive planting on the coasts, yields a highly resinous reddish wood. 1914 W. J. Bean *Trees & Shrubs Hardy in Brit. Is.* II. 187 Cluster Pine, Maritime Pine..As its common name implies, admirably adapted for maritime localities. 1966 T. H. Everett *Living Trees of World* 34/1 Of all two-needled pines, the species with the longest leaves,..is the cluster pine or maritime pine..a southern European tree that thrives in coastal areas.

B. *sb.* **1. b.** In *pl.*, with capital initial. The eastern provinces of Canada adjoining the Atlantic Ocean (Nova Scotia, New Brunswick, and Prince Edward Island).

1926 *Daily Colonist* (Victoria, B.C.) 24 July 1/2 Continuing his Maritime tour, Premier Meighan is in another here tonight again stressed his policy for the Maritimes. 1934 W. M. Whitelaw (*title*) The Maritimes and Canada before Confederation. 1938 L. M. Montgomery *Anne of Ingleside* viii. 58 The July night was unreasonably cold as a summer night in the Maritimes sometimes is. 1947 G. Taylor *Canada* xvi. 385 The three small provinces linked as the Maritimes..together only amount to 1 per cent. of the area of the Dominion. 1956 C. R. Fay *Life & Labour in Newfoundland* viii. 134 Schooners come from the Maritimes with much needed provisions and lumber. 1960 *Times* 12 Nov. (Canada Suppl.) p. vi/6 The salty Maritimes or the wide-open Prairies. 1970 M. H. Orkin *Speaking Canad. Eng.* vii. 186 Nova Scotia in *The Mayflower Province*, since..New Brunswick as *The Loyalist Province*. Together with Prince Edward Island, they are collectively known as the *Maritime Provinces* or, more familiarly, the *Maritimes*.

‖ **Marivaudage** (marivodaʒ). [Fr., f. the name of P. C. de *Marivaux* (1688–1763), novelist and dramatist.] The expression of affected language and exaggerated sentiment in the style of Marivaux; an overdone attempt at refinement, affectation.

1765 H. Walpole *Let.* 19 Nov. in *Corr.* (1948) XIII. 144 Crébillon is entirely out of fashion, and Marivaux a proverb: *marivauder*, and *marivaudage* are established terms for being prolix and tiresome. 1882 G. Saintsbury *Short Hist. French Lit.* iv. ii. 410 All the work of Marivaux, dramatic and non-dramatic, is pervaded more or less by a peculiarity which at the time received the name of Marivaudage. This peculiarity exists partly in the sentiment, and partly in the phraseology. The former is characteristic of the eighteenth century, disguising a considerable affectation under a mask of simplicity, and the latter (sparkling with abundant, if somewhat precious wit) is ingeniously constructed to suit it and carry it off. 1894 Beerbohm in *Yellow Bk.* II. 284 The qualities that I tried..to travesty—paradox and marivaudage. 1930 *Mod. Lang. Rev.* XXV. 71 *Marivaudage*..to quote Faguet..'consists much more in analysing to excess a just thought than in decking out to excess an empty one'. 1959 *Oxf. Compan. French Lit.* 455/2 The term *marivaudage*, coined from his name, is used to signify the analysis of the delicate sentiments of the heart and the subtle, affected style used by Marivaux to this end in his comedies. 1969 *Observer* 12 Jan. 26/5 It [sc. Lesbianism] is a tricky subject, poised on the brink of either tiresome *marivaudage* or tasteless titillation.

mark, *sb.*¹ Add: **7. d.** (Earlier and later examples of *slang* sense.) Also, *a soft* or *easy mark* (cf. *easy *a.* 13 b), a person who is easily persuaded or deceived (*slang* (orig. *U.S.*)); also (*Austral. slang*), *a good* (or *bad*) *mark* (see quot. 1941).

1845 R. Howitt *Impressions Australia Felix* 233, I heard it casually from the lips of apparently respectable settlers as they rode on the highway, 'Such and such a one is a good mark!'—simply a person who pays his men their wages, without delays or drawbacks; a man to whom you may sell anything safely. 1860 F. & J. Greenwood *Under a Cloud* II. xiv. 332 'There's a mark!' exclaimed one to the other, looking towards the spot where Hatcher was standing. 1883 J. Greenwood *Tag, Rag & Co.* iii. 24 Publicans..are usually the unfortunate tradesmen fixed on as a mark. 1896 Ade *Artie* xvi. 150 He was the wise guy and I was the soft mark. *Ibid.* xviii. 173 When that kind of a mark comes in they..get ready to do business. 1904 G. H. Lorimer *Old Gorgon Graham* 288 He was too easy a mark to succeed in Wall Street. 1922 Joyce *Ulysses* 75 Didn't catch me napping that wheeze. The soft touch. Soft mark. 1929 'E. Queen' *Roman Hat Mystery* xiii. 187 For a shrewd man, he certainly was an easy mark for the wiseacres. 1941 Baker *Dict. Austral. Slang* 46 *Mark, good (or bad)*, a general term of approval (or disapproval) for a person. 1956 H. Gold *Man who was not with It* (1965) i. 6, I floated down..to kick the smaller mark where it would tell on him. 1962 J. Ludwig in R. Weaver *Canad. Short Stories* (1968) 2nd Ser. 255 He was repulsive, old, a mark, a fool—his nerve went. 1966 'J. Hackston' *Father clears Out* 93, I was a bit nervous and an easy mark to bamboozle. 1968 B. Turner *Sex Trap* xv. 148 You thought I might like to turn myself into an easy mark! A square with a loud voice and a shopping-list? Marriage! 1971 *Friends* 21 May 11/2 In 90% of all rip-offs the naive mark hands over his capital to some total stranger who never returns. 1973 E. McGirr *Bardel's Murder* ii. 31 In the twenties it was the Yanks who was the suckers, but now..it's us who are the marks.

11. g. Also *fig.* in phr. *full marks* used as an expression of considerable praise or commendation. Also *top marks*.

1934 D. L. Sayers *Nine Tailors* ii. v. 175 That is well observed. Yes, Bunter, you may have full marks for that. 1941 H. Nicolson *Let.* 27 Mar. (1967) 154 It is a fine show and I give Ronnie Campbell full marks. 1945 W. de la Mare *Scarecrow* 22 'Full marks, my dear,' said Mr. Bosover, squeezing her hand. 1966 *Listener* 29 Sept. 473/3 Full marks for the deadly opening sentence. 1973 *Times* 2 June 9/4 Full marks as always, to this hard-up, persevering, resourceful and imaginative museum. 1973 J. Thomson *Death Cap* xiii. 177 He gave Holbrook top marks for finding such a delightfully sharp and acid old lady.

12. c. Add examples with prefixed figure representing a limit or total, or an approximation of this.

[1860 Thackeray *Roundabout Papers* vi, in *Cornh. Mag.* Aug. 256 The *Cornhill Magazine*..having sold nearly a hundred thousand copies, he (the correspondent) 'should think forty thousand was now about the mark'.] 1929 *Star* 21 Aug. 18/2 Dennis Brothers' 1s. shares can usually be regarded as a reasonable purchase under the £3 mark. 1936 *R.A.F. Jrnl.* May 170 The membership already exceeds the twenty-five hundred mark. 1948 D. Ballantyne *Cunninghams* iii. 15 Around the forty mark. 1965 N. Paul *Cine-Photogr.* ii. 19 There is a tremendous range of models on the market today, especially in the 8 mm. range, where they can..soar up to the £300 mark. 1973 *Times* 14 Apr. (Nepal Suppl.) p. ii/9 If the present trend continues, the 25 million mark may be reached before the end of the century.

d. Also *Austral. Rules Football* (see quots. 1968). Cf. *mark *v.* 15 d.

1965 *Sun-Herald* (Sydney) 4 July 51 Geelong Rover, Barry Farman, won a trophy for the best mark taken by his team. 1967 *Ibid.* 16 Apr. 67 Ryan's first kick struck the back of the Collingwood man on the mark, but Perkins ruled he had moved back over the mark. 1968 Eagleson & McKie *Terminol. Austral. Nat. Football* ii. 22 *Mark*,..the catching of the ball on the full directly from the kick of another player. *Ibid.* 23 *Mark*,..the spot at which a player caught the ball ('took a mark'), and behind or over which he must make his kick. 1969 *Sun-Herald* (Sydney) 13 July 48/1 Close took a beautiful one-handed mark, his fourth for the game,..but kicked only a point.

e. *Athletics.* A line drawn to indicate the starting-point. Also in phrases (lit. and fig.), as *on your mark*(*s*), (*get set, go*), the instructions given to competitors at the start of a race; *to be first off the mark*: to gain an initial lead over one's opponents; *to be quick off* (or *on*) *the mark*: to lose no time in starting, to waste no initial advantage; *to be slow off the mark*: to start slowly or sluggishly; to waste time in starting; *to get* (or *be*) *off the mark*: to start (well); *to overstep the mark*: to infringe the rules by placing one's foot over or beyond the mark; so *fig.*, to go beyond a fixed limit or standard, to 'go too far'; cf. the phr. *overshoot the mark* (Overshoot *v.* 2 b, and Mark *sb.*¹ 7 a, e); *to toe the mark* (see Toe *v.* 2).

1887 M. Shearman *Athletics & Football* 65 It requires, however, much skill and practice not to 'take off' before the mark [in jumping]. *Ibid.* 198 Nothing was said until the men got upon their marks. 1905 *Pearson's Mag.* Sept. 290/2 He..beat his field by a yard or two off the mark. 1912 E. H. Ryle *Athletics* 91 This method..assists a runner to keep steady on the mark while awaiting the report of the pistol. 1917 E. Pound *Let.* 12 Aug. (1971) 114, I have been a bit slow getting the *Little Review* off the mark. 1919 E. P. Oppenheim *Strange Case J. Thew* I. xiii. 112 'Did he make any trouble?' 'He had no chance..I was first off the mark.' 1921 *Daily Colonist* (Victoria, B.C.) 24 Mar. 13/3 It was felt that the Parent-Teachers' Associations were in some cases overstepping the mark in regard to requests. 1928 *Observer* 5 Feb. 23/5 When you really wish to get going you have a second, a third and a top speed change which will get the car off the mark, in the old phrase, in an inspiriting manner. 1929 F. N. Hart *Hide in Dark* i. 13 All set? On your mark! 1930 L. W. Olds *Track Athletics* iv. 26 On Your Marks... Get Set... Go or Gun. 1931 *Oxford Mail* 29 Aug. 8/3 G. Fisher and L. Rogers were quickly off the mark, 20 runs being scored in the first ten minutes. 1934 P. G. Wodehouse *Right Ho, Jeeves* i. 9 Get off the mark. 1936 ... public is at a loss. 1938 R. Warner *Professor* ii. 227 Now he appeared like a person who fears that he has in some way overstepped the mark, has involuntarily wounded another's feelings, or alluded to some subject that were better left unmentioned. 1943 *Endeavour* Apr. 74/2 The dog-breeder and poultry farmer were..quick off the mark in using the new therapy. 1946 B. Marshall *George Brown's Schooldays* v. 22 On your marks, go. The thing was to look... 1947 'G. Orwell' *Eng. People* 24 With all their political ignorance the English people will often show surprising sensitiveness when some small incident seems to show that 'They' are overstepping the mark. 1954 G. Mitchell *On your Marks* xv. 172 'On your marks,' said the starter. The swimmers were poised and ready. 'Go!' 1958 J. Wain *Contenders* v. 112 Ned was off the mark at once... Blue Seal Pottery seemed to become a vogue overnight. 1970 Marlow & Watts *Track Athletics* ii. 18 It is recommended that the shoulders should be forward when the athlete settles into the blocks on the command 'On your marks'. 1971 *Morning Star* 19 Oct. 2 The Institute of Teachers and the Association of Science Teachers were quick off the mark, followed by the more orthodox national trade unions. 1972 R. Adams *Watership Down* xxvi. 208 The pursuit was a bit slow off the mark..because poor old Bugloss wasn't there to give the orders. 1972 W. A. Pantin *Oxf. Life* iv. 50 Dr. William Buckland..was very quick off the mark with putting in plans for the future use of the building. 1973 *South Wales Echo* 30 June (Suppl.) 1/2 Then the whole Burke Special team can get off the mark. 1973 *Guardian* 8 Oct. 10/2 Beatrix Potter..wrote about Tommy Brock (the badger) and Mr. Todd (the fox) living on terms of toleration—until Tommy badly overstepped the mark.

13. b. (Earlier and later examples.)

a 1847 T. Chalmers in Bartlett *Dict. Amer.* (1860) 262 Men..called out to make a mark on the world. 1854 *Harper's Mag.* Sept. 561/1 There was a time when Jacob made his mark upon the stock-brokers and money-changers of that monetary locality. 1905 L. Strachey *Characters & Commentaries* (1933) ii. iv. 34 Whether he himself might not have made his mark in politics..is a subtle speculation. 1921 R. Macaulay *Dangerous Ages* xii. 208 Why dream wistfully of doing one's bit, making one's mark, in a world already as full of bits..as a kaleidoscope? 1942 C. Bell *Landmarks 19th Cent. Painting* 147 They have not failed to leave a mark on history. 1952 A. J. Cronin *Adventures in Two Worlds* i. 15 In medicine..or some other field, I believe that you will make your mark. 1973 *Observer* (Colour Suppl.) 5 Aug. 18/2 Woolton..making his mark as Civilian Boot and Shoe Controller in World War I. 1974 *Radio Times* 3 Jan. 49/4 Pop music has produced many trendsetters—we pick the ones who made their mark in the 70s.

e. *Telecommunications.* Each of a succession of marks on paper whose relative duration and separation are used to convey information in telegraphy; hence any kind of signal that conveys information by its presence or absence (rather than by its magnitude). Opp. *space*; hence **mark-space** *attrib.*

1837 *Amer. Jrnl. Sci. & Arts* XXXIII. 187 (*Describing Morse's telegraph*) To find the spaces..count the points at the bottom of each line. 1859 T. P. Shaffner *Telegr. Manual* xxxiv. 469 The length of the mark or of the space upon the ribbon paper will be in proportion to the length of the contact made with the key. 1892 C. L. Buckingham in C. F. Brackett et al. *Electr. in Daily Life* 143 In 1846 Bain proposed to employ perforated strips of paper to effect automatic transmission in connection with an electro-chemical process for recording, in which marks upon a moving strip of paper are made by electrolysis attending the passage through it of the current. 1938 *Admiralty Handbk. Wireless Telegr.* II. R. 59 When the key is pressed, the valve oscillates and we have the mark period. *Ibid.*, During the mark, the oscillating circuit constitutes the load. 1953 *Electronics* July 183/2 Mark-Space Phototubes. As the perforated tape passes through the keying head, pulses of light from the exciter lamps pass through the holes and alternately strike the tone-on (mark) and tone-off (space) phototubes. The phototubes conduct and produce alternate pulses of current. 1958 P. E. K. Donaldson et al. *Electronic Apparatus for Biol. Res.* xv. 234 The circuit performs oscillations with a mark-space ratio of approximately unity, the waveforms having the form of Figure 15.14. 1969 J. J. Sparkes *Transistor Switching* iii. 52 A circuit capable of producing rectangular waves with very sharp rising and falling edges, at mark-space ratios of between 1:1 and 500:1..is shown. 1972 *Sci. Amer.* Dec. 16/3 The computer's input element is a photoelectric paper-tape reader that handles 1,000 'marks' (punched holes) per second.

15. b. Also, freq. followed by a numeral, a designation of the stage of development in design and construction of a manufactured product or piece of equipment, as a weapon, an aeroplane, etc. Abbrev. Mk. Also *transf.* and *fig. Mark 1*: first class (*slang*).

1888 *Treat. Mil. Small Arms & Ammunition* 52 Enfield Revolver Pistol, Mark II. 1899 *Kynoch Jrnl.* Oct.–Nov. 12/1 Despite the unfortunate failures at Edinburgh and

Bisley of the bullet known as Mark IV. 1914 *Times Bk. Navy* vii. 99 The gun has progressed through successive stages, or 'marks' as they are technically known. Marks 3, 4, 5, and 5w were 'built' guns. 1916 T. E. Lawrence *Let.* 22 Dec. (1938) 214 Will you please wire for 4 locks complete for Gun Maxim Converted Mark II to be sent to Yenbo? 1926 L. Nason *Chevrons* 120 This is going to be a real, old issue, Mark I scrap. 1942 W. S. Churchill in *Second World War* (1951) IV. 768 There should be no difficulty in sparing 1,000 tanks and 1,000 anti-tank and A.A. guns. No doubt older marks might form the bulk. 1946 *R.A.F. Jrnl.* May 180 No fewer than 26 different 'Marks' were built and put into service... In accordance with Service custom they merited and received the bestowal of a special Mark number. 1948 'N. Shute' *No Highway* i. 7 The Mark I model [*sc.* an aeroplane] which went into production first had radial engines. 1959 *Listener* 19 Mar. 514/1, I trust the Ministry of Supply is making sure that all senior naval, military, and air force officers are supplied with Mark II* crystal balls. 1973 P. Johnson in Johnson & Gale *Highland Jaunt* i. i. 13 Time's winged chariot is hurrying near, powered by the latest mark of internal combustion engine. 1973 J. Wainwright *Devil you Don't* 5 It was a great car—a Jag Mark II. 1974 *Economist* 21 Dec. 29/3 They are calling him [*sc.* Helmut Schmidt] Kissinger Mark II. 1975 J. Symons *Three Pipe Problem* xv. 129 For his white clients Riverboat often played the role of American Negro, Mark One.

22. *a mark at*, a good hand at.

1881 *Punch* 3 Dec. 263/2 Till my chummy Scholard Mike, who's a mark at A.B.C., Read me Littler's little tale.

23. (sense 11 g) *mark-sheet, system*; **mark sensing,** the process by which a machine 'senses' or reads data in the form of electrically conducting marks made by hand on cards with a lead pencil; so **mark-sense(d** *a.*, marked in this way and for this purpose.

1965 *Math. in Biol. & Med.* (Med. Res. Council) i. 10 They used 'mark-sense' cards for the details of the investigation and for the results of calculations. 1971 *Computers & Humanities* VI. 16 A newer approach employs mark-sense cards upon which musical symbols have been printed. This allows the person doing the encoding to accomplish it in a single step, quietly, and without the need for cumbersome equipment. 1964 T. W. McRae *Impact of Computers on Accounting* i. 14 Figure 1.8 illustrates a mark sensed card. The mark sensed positions are in columns 70–78 and the holes punched from these..are in columns 31–33. 1959 E. M. Grabbe et al. *Handbk. Automation, Computation, & Control* II. v. 19 Mark sensing uses special electrical conducting marks placed in various positions on punched cards. 1970 O. Dopping *Computers & Data Processing* iii. 55 Information can also be recorded in a card with a pencil, and in a subsequent operation, 'mark sensing', the markings are translated to punched holes in the cards. 1961 *Oxf. Local Exam.* (*Latin, Paper 1*) p. 3 The total mark..is to be halved before it is entered on the marksheet. 1964 *Oxf. Mag.* 12 Mar. 254/1 Now that the dust has settled on the marksheets of the January Part II examination. 1885 *Encycl. Brit.* XIX. 754/2 He devised an ingenious system of recording the convicts' daily industry by marks... The mark system had already been tried with good results in Ireland.

mark, *sb.*² Add: **4. b.** Also, now the name of the unit of German currency.

1911 E. Sykes *Banking & Currency* (ed. 3) viii. 61 The Imperial Bank of Germany can issue notes to the value of 550,000,000 marks against the deposit of securities. 1932 J. W. Angell *Recovery of Germany* (rev. ed.) ii. 19 Mark prices fell... In..1921..two things happened which definitely sealed the fate of the old mark currency. *Ibid.* 21 Everyone who was unfortunate enough to possess mark currency was in danger of having it lose half its value in his pocket overnight. *Ibid.* 56 The mark notes, bank deposits, and securities became practically worthless. 1933 E. Roll *Spotlight on Germany* i. 39 The function of the mark as a medium of exchange had disappeared almost as completely as its ability to form a standard of value. *Ibid.* 41 The mark was finally stabilized at one million-millionth of its pre-war value. 1964 M. McLuhan *Understanding Media* (1967) ii. xiv. 156 The depreciation of the citizen went along with that of the German mark.

mark, *v.* Add **2. c.** *to mark down* (earlier and later examples); hence *marked-down* ppl. adj., also *ellipt.* as *sb.*; cf. *mark-down *sb.*; *to mark up*: to label (goods) with a higher price; to raise in price.

1859 *N.Y. Herald* 5 Jan. (Advt.), Mark every article Way Way Way down To some price which will make it.. Sell and go quick. 1870 *Amer. Naturalist* III. 3 The prices of venison and other game was so far 'marked up' that gold..was charged for salmon. 1902 G. H. Lorimer *Lett. Merchant* iv. 52 The clerks all knocked off their regular work and started in to mark up prices. 1908 *Sat. Even. Post* 24 Oct. 13/1 At another store there is a marked-down sale of parasols. 1910 'Saki' *Reginald in Russia* 43 They have turned instead to the muddy lanes and cheap villas and the marked-down ills of life. 1913 F. H. Burnett *T. Tembarom* vi. 67 A remnant of crimson stuff secured from a miscellaneous heap at a marked-down sale at a department store. 1923 E. O'Neill *Moon of Caribbees* 18 Don't you boys forget to mark down cigarettes or tobacco or fruit, remember! Three shillin's is the price. 1942 M. Schlauch *Gift of Tongues* (1943) vi. 135 'Marked-downs' (overheard in a department store for 'reduced dresses'). 1962 E. Godfrey *Retail Selling & Organization* xvi. 164 A minority of firms..have abandoned them [*sc.* fixed profit margins] in favour of marking-up each item on its merits. 1965 A. Christie *At Bertram's Hotel* xii. 113 She also availed herself of some marked down lines in furnishing fabrics. 1974 *Times* 22 Oct. 12/6 Exclusive goods can be marked up to carry a full, and I suspect..very high, profit margin.

e. *to mark off* (Engin. and Shipbuilding): to mark (an object) with lines to serve as a guide for subsequent cutting, machining, alignment, etc.; to represent (a dimension or detail) on an object in this way.

1894 W. J. Lineham *Text-bk. Mech. Engin.* vi. 185 If the work is too large to mark-off on a table it should be levelled, and all lines be drawn by reference to an ideal horizontal or vertical plane. 1895 J. Donaldson *Drawing & Rough Sketching for Marine Engineers* (ed. 6) 87 When the forging of a crank shaft, or other part of an engine, is delivered..it is essential..that it be accurately marked off for machining. *Ibid.*, The throw of the vertical crank is marked off from the horizontal line. 1925 F. J. Drover *Marine Engin. Repairs* xix. 120 The cheeks are marked off for thickness. 1966 J. H. Dixon tr. *Dormidontov's Shipbuilding Technol.* i. 15 Hull details are marked off from the full-size or scaled down lofting data and from working drawings. *Ibid.* vi. 127 When the metal is being marked off, the outlines of the components are drawn full size, with an indication of the allowances and tolerances for machining.

f. To mark the ears of (a lamb, or less commonly a calf); also, to dock and geld. Cf. *marking *vbl. sb.* 1 a. *Austral.* and *N.Z.*

1933 *Press* (Christchurch, N.Z.) 4 Nov. 15/7 *Mark*, to ear-mark. Now a frequent euphemism for *cut and tail*. 1941 Baker *Dict. Austral. Slang* 46 *Mark*,..to geld lambs. 1950 *N.Z. Jrnl. Agric.* Sept. 197/1 Lambs are usually marked at between 1 to 6 weeks of age; 3 weeks is probably the ideal age.

6. a. Also with *down*.

1894 'R. Andom' *We Three & Troddles* xxii. 210 The demon dyspepsia had marked him down.

d. Also with *out from*.

1888 Mrs. H. Ward *R. Elsmere* I. viii. 222 The even gentleness which always marked her out from others. *Ibid.* III. xxxii. 34 The reasoning faculty..which marks us out from the animal. 1905 T. E. Harvey *Rise of Quakers* ii. 10 Her son's serious ways, by which he was marked out from his other brothers and sisters.

7. f. *to mark down*: to make a note of; to set down in writing.

1881 J. Fothergill *Kith & Kin* II. ix. 241 More than one matron then present had silently marked him down.. in her book of 'eligibles'. 1936 *Discovery* May 145/1 An exhaustive search was not feasible, but 51 nightingales were marked down in Devon.

g. In phr. *to mark* (*someone's*) *card* [Card *sb.*² 6 e]: (from the practice in horse-racing) to provide (someone) with information; to tip (someone) off; to 'put (someone) right'. *slang.*

1961 Partridge *Dict. Slang* Suppl. 1179/2 *Mark* (*someone's*) *card*, to give him the information he needs; to put him right: barrow-boys': since ca. 1945. Ex the racecourse. 1962 R. Cook *Crust on its Uppers* ii. 34 They'd marked my card there was a new dance-hall been opened over at Peckham. 1970 P. Laurie *Scotland Yard* 291 *Mark X's cards, to*, to brief X discreetly. 1970 G. F. Newman *Sir, You Bastard* 282 The third was to phone the insurance assessor and mark his card.

13. Delete 'Now *poet.*' and add later examples.

1832 Disraeli *Contarini Fleming* II. iii. v. 241, I looked up, I marked the tumultuous waving of many torches. 1849 C. Brontë *Shirley* II. xii. 278 She smiled, well pleased to mark the delight of her pupil. 1860 Thackeray *Four Georges* i, in *Cornh. Mag.* July 5 Lift up your glances respectfully, and mark him eyeing Madame de Fontanges. 1893 Kipling *Many Inventions* p. viii, The children wise of outer skies Look hitherward and mark A light that shifts. 1901 G. B. Shaw *Caesar & Cleopatra* v. 199 Rufio, satisfied, nods at Cleopatra, mutely inviting her to mark that. 1922 C. Bell *Since Cézanne* 11 They babble in the Burlington Fine Arts Club—where nobody marks them. 1972 'M. Innes' *Open House* ii. xi. 99 He paused to mark this [*sc.* a compliment] going home.

15. d. *Austral. Rules Football* (see quot. 1968). Cf. *mark *sb.*¹ 12 d.

1968 Eagleson & McKie *Terminol. Austral. Nat. Football* ii. 23 *Mark*, catch the ball on the full directly from the kick of another player. 1969 *Sun-Herald* (Sydney) 13 July 48/1 Hudson marked 35 yards out, but a bad kick fell well short of the goal.

16. Also, *mark you* (or *me*, etc.). Cf. *look you* s.v. Look *v.* 4 a.

1837 E. Howard *Old Commodore* I. v. [*sc.* vi] 210 Mark you me, Mr. Alsop, mark you me. We have done our duty, sir. 1862 Mrs. H. Wood *Channings* II. ix. 137 'Impudence,' shortly answered Yorke. 'Mark you, Miss Channing! I have not done with you.' 1867 A. J. Evans *St. Elmo* xvii. 232 It remains to be seen whether a grand success is not destined to crown it. Mark you! The grapple is not quite over. 1903 G. B. Shaw *Man & Superman* i. 23 For mark you, Tavy, the artist's work is to shew us ourselves as we really are. 1922 Joyce *Ulysses* 139 It was the speech, mark you, the professor said, of a finished orator. a 1953 E. O'Neill *Touch of Poet* (1957) iii. 94 Provided, mark you, that you and your daughter sign an agreement I have drawn up. 1966 *Listener* 17 Feb. 240/1 Well, mark you, I wouldn't be the inveterate optimist I am, if I didn't live in the belief that the year of my life is yet to come.

Markan (mā·ɹkăn), *a.* [f. *Mark* + -an.] = Marcan *a.*

1909 B. W. Bacon *Beginnings Gospel Story* p. xxviii, Let the reader simply subtract mentally the Markan element from Luke. 1926 V. Taylor *Behind Third Gospel* v. 130 Again the Markan stripe is characterized by unity of subject-matter. 1930 A. G. Hebert tr. *Brilioth's Eucharistic Faith & Pract.* i. 5 An energetic analysis of the texts leads him to the conclusions that the Markan narrative is alone authentic and original, [etc.]. 1971 *Novum Testamentum* XIII. 187 In general the changes made by Matthew and Luke to their Markan source reveal primarily a Matthean and Lukan redaction history and not a Markan one.

mark-down (mā·ɹk,daun), *sb.* [f. *to mark down* s.v. Mark *v.* 2 c.] A reduction in price; an article the selling price of which has been reduced. Also *attrib.*

1880 in *Dict. Amer. Eng.* (1942) III. 1486/2 The success of our mark-down sales. 1962 *Economist* 2 June 891/1 Specialists found it difficult to absorb shares without drastic markdowns in price. 1962 E. Godfrey *Retail Selling & Organization* ii. 16 Goods are bought in specially or are mark-downs from other sections of the store. 1972 *Oxford Times* 27 Oct. 9 To show a false mark-down price in a sale is a definite offence. 1974 *Oxford Mail* 10 Jan. 2/1 (Advt.), Gigantic genuine mark-down Fashion Sale.. Headington.

marked, *ppl. a.* Add: **1. b. marked cheque** (see quots. 1907 and 1951); **marked transfer** (see quots.).

1896 R. H. I. Palgrave *Dict. Pol. Econ.* II. 695/2 Marked cheques are payable in the first clearing of the next day, and cannot be refused for any reason. 1907 W. G. Cordingley *London Commercial Dict.* 115 *Marked cheque*, a cheque marked by the banker on whom it is drawn, stating that it is 'good' for the amount named upon it. 1930 Hutchinson & Lovell *Short Dict. Legal, Commercial & Econ. Terms* 77 *Marked cheques*,..cheques initialled by a banker as an indication that he holds sufficient balance to meet such cheque when presented. 1951 R. W. Jones *Thomson's Dict. Banking* (ed. 10) 395 *Marked Cheque*. Marking cheques takes one of two forms —marking for clearing purposes and marking at the request of the drawer. 1901 W. G. Cordingley *Dict. Stock Exchange Terms* 59 *Marked Transfers*. The company..then issue a fresh Certificate to him for the part unsold, and endorse the Transfer that they hold Certificates to cover the number of shares sold, when the Transfer is said to be 'Marked' or 'Certified'. 1961 Webster, *Marked transfer*, an instrument for transferring a portion of the shares of a stockholder's certificate after being certified as good by a proper official on the London stock exchange.

c. Of a linguistic construction, form, etc.: distinguished or determined by a particular feature; distinguished as intrinsically unnatural (see also *marker 7; opp. **unmarked*).

The corresponding Russian word was used by Trubetzkoy before Bloomfield: see R. Jakobson *Selected Writings* (ed. 2, 1971) I. 734 ff.

1933 L. Bloomfield *Lang.* xvi. 268 A phrase consisting of the preposition *to* and an infinitive expression belongs to the special form-class of *marked infinitive phrases*, whose function differs from that of unmarked infinitive expressions. 1953 C. E. Bazell *Ling. Form* ii. 19 The Latin marked construction here can be uniquely derived from the unmarked in every case. 1959 R. Quirk in Quirk & Smith *Teaching of English* i. 44 It seems likely that this structural notion of marked and unmarked members of opposing pairs can be applied cautiously but with profit to many of the binary choices in usage at more complex linguistic levels than the lexical. 1963 *Canad. Jrnl. Ling.* VIII. 92 Whenever a demonstrative system is reduced to a binary one, the first member which can be correlated with the first person always remains and thus becomes the marked member. 1965 *Amer. Speech* XL. 175 There are certain marked forms [of the verb] which are past and future respectively. 1968 Chomsky & Halle *Sound Pattern Eng.* 402 Certain aspects of this general problem can be dealt with if we incorporate the Praguian notion of 'marked' and 'unmarked' values of features into our account in some systematic way, and if we then revise the evaluation measure so that unmarked values do not contribute to complexity. *Ibid.* 404 A major difference between the Praguian conception of markedness and our own is that in the former the marked coefficient of a feature was assumed always to be + and the unmarked coefficient always −. 1972 M. Shapiro in *Language* XLVIII. 345 In my concept and utilization of the terms marked and unmarked, the former is to be understood as denoting relative complexity and differentiatedness, the latter as denoting the absence of these relative attributes.

markedness. Add: *spec.* in *Linguistics*: the quality or state of being marked (*marked *ppl. a.* 1 c).

1968 [see *marked *ppl. a.* 1 c]. 1968 *Language* XLIV. 714 The notion of markedness as developed within Praguean phonology..allows phonological segments to be differentiated. 1972 *Sci. Amer.* Sept. 78/3 Any syntactic structure is a member of a transformational chain and any two partially synonymous constructions display an interrelation of markedness and unmarkedness. 1973 A. H. Sommerstein *Sound Pattern Anc. Greek* iii. 82 The Markedness theory holds that phonological features are not in general symmetrical: that for each feature in each environment, one of its possible values is more normal, more 'unmarked', than the other.

marker. Add: **1. b.** One who records prices on the stock exchange.

1870 J. K. Medbery *Men & Mysteries Wall St.* 21 The 'marker' or black-board clerk writes off the prices upon the tablet.

f. (Later examples.)

1907 S. E. White *Arizona Nights* i. viii. 149 In the meantime the marker was engaged in his work. First, with a sharp knife he cut off slanting the upper quarter of one ear [of a calf]. 1928 *Collier's* 29 Dec. 6/2 There was a lull in the stream of lumber. The marker turned for a look at the order board.

j. *U.S.* In surveying: a person who makes the marks on trees to indicate boundaries or lines of survey.

1743 *New Jersey Archives* (1883) 1st Ser. VI. 154 You are to employ..an assistant surveyor..& also proper

chainbearers & markers. 1843 *Amer. Pioneer* II. 379 In running the back line of the survey..I was about one hundred yards in advance of the chainmen and marker.

3. b. A monument, memorial stone, etc., marking a place of special interest.

1906 *Springfield* (Mass.) *Weekly Republ.* 15 Feb. 16 The committee appointed to investigate the matter of a marker for the Washington elm reported in favor of a granite marker. 1959 A. G. Woodhead *Study of Greek Inscriptions* iv. 45 The commonest types of monument were the small cylindrical marker (*columella*), seldom more than two feet high, [etc.].

c. A flare, distinctive sign, or object of any kind used as a guide to a pilot of an aircraft seeking a particular area, obstruction, etc.; *spec.* a flare dropped from an aeroplane to illuminate or mark a target. Also with defining word, as *ground marker, sky-marker*, etc. (see these sbs.).

1936 M. B. Garber *Mod. Mil. Dict.* 195 *Marker*, a symbol, letter, or figure on the ground, visible from aircraft, by means of which the operators are able to determine their position. 1944 *Times* 22 Mar. 4/5 The leading aircraft then dropped markers right across the middle of the target. 1944 R. C. K. Ensor *Miniature Hist. War* iv. 53 Accuracy was also greatly improved by the system of 'pathfinders' dropping 'markers' to guide the rest. 1951 *Gloss. Aeronaut. Terms* (B.S.I.) iii. 24 *Markers*, objects of approved shape or colour indicating specific areas and obstructions.

4. (Earlier and later examples.)

1888 *Congress. Rec.* 12 Dec. 202/2 The waving of the bloody shirt would not have been a marker. 1904 W. H. Smith *Promoters* xxv. 366 What little I've told you isn't a marker to other things he said.

5. a. In various games such as bridge or whist, a scoreboard or card, or other implement used to record the score. (See also *bridge-marker*, *bridge *sb.*² c.)

1907 *Army & Navy Stores Catal.* 198/2 Whist marker, inlaid, pair 2/7. *Ibid.* 375/2 Playing card table..with drawer divided to take cards and bridge scoring blocks, whist cards and markers. 1960 R. C. Bell *Board & Table Games* I. vi. 157 Little bone sticks marked with..dots are used to keep the score... Usually each player [at Mah Jong] starts with..2 markers with 10 red dots [etc.].

b. *U.S. slang.* A promissory note; an I.O.U.

1887 F. Francis *Saddle & Mocassin* xii. 225 Before half the deal was over, the whole bank of checks was gone, and Cuff was giving markers for hundreds as hard as he could go it. 1931 D. Runyon *Guys & Dolls* (1932) vii. 147 Now I am going to pay my landlady, and take up a few markers here and there, and feed myself up good. *Ibid.* xiii. 279 He is willing to take Charley's marker for a million if necessary to get Charley out.

6. *Genetics.* Any allele (usu. one which is easily recognized phenotypically and whose gene has been located on a specific chromosome) which is used in genetic experiments to identify a chromosome or to locate less well-known genes on a genetic map.

1930 *Genetics* XV. 219 There is a certain expected amount of crossing over between the two mutant genes treated as markers of a given point. 1938 *Ibid.* XXIII. 291 In order to obtain Notch deficiencies, normal males carrying yellow (*y*—0·0) as a marker..were mated with females homozygous for cherry (w^{ch}—1·5) and wavy (*wy*—40·7). 1940 *Nature* 10 Aug. 199/2 It is..possible to recognise, by means of the markers used, what combination of chromosomes is present in the pseudo-backcross progeny. 1961 *Proc. Nat. Acad. Sci.* XLVII. 378 (*heading*) Transformation studies on the linkage of markers in the tryptophan pathway in *Bacillus subtilis*. 1969 A. M. Campbell *Episomes* ii. 22 Phage resulting from DNA infection is distinguished from progeny of the helper phage itself by genetic markers.

7. *Linguistics.* A word, affix, etc., which distinguishes or determines the class or function of the form, construction, etc., with which it is used. Also *attrib.*

1933 L. Bloomfield *Lang.* xvi. 269 Our determining adjectives, our prepositions, our co-ordinating conjunctions, and our subordinating conjunctions, may be viewed as markers. 1953 C. E. Bazell *Ling. Form* v. 67 *And* [in 'hot and cold'] may be described as a 'marker', after which no question of constituents arises. 1960 E. Delavenay *Introd. Machine Transl.* vii. 110 Those strictly linguistic markers sought in the sentence by the programmers of automatic translations. 1961 R. B. Long *Sentence & its Parts* iii. 65 The marker pronouns and adverbs are not themselves usable in declaratives ordinarily, but the marker verb forms are. 1964 *Amer. Speech* XXXIX. 53 The German plural marker *-er* is historically a derivational suffix. 1968 P. M. Postal *Aspects Phonol. Theory* i. 13 By the 'markers' of a linguistic level I refer to the formal structures which the rules of that level assign to sentences as part of their total structural description. 1972 Hartmann & Stork *Dict. Lang. & Ling.* 137/1 Markers may indicate the category of a linguistic unit at any level of analysis. 1973 *Archivum Linguisticum* IV. 32 The remaining forms are sufficiently distinguished for person, i.e. in the absence of specific person markers, by differing marks of gender-cum-number.

8. *attrib.* and *Comb.*, as (sense *3 c) *marker burner; marker flag, light*; **marker beacon** (see *beacon *sb.* 6 d); **marker bomb,** a bomb emitting a coloured light, dropped in an air-raid to serve as a point of direction.

1929 Marker beacon [see *beacon *sb.* 6 d]. 1934 *Jrnl. R. Aeronaut. Soc.* XXXVIII. 828 When a pilot had to maintain height, to observe instruments to ensure that he did not get into difficulties, and had also to listen for the marker beacon. 1935 *Times* 19 Oct. 9/5 The installation will consist of a main beacon at one corner of the aerodrome and two marker beacons 300 yards and 3,000 yards away respectively. 1971 *Gloss. Electrotechnical, Power Terms* (*B.S.I.*) iii. vi. 24 *Marker beacon*, in aviation, radio beacon which radiates a signal to define an area above the beacon. 1944 *Hutchinson's Pict. Hist. War Oct. 1943–Apr. 1944* 84 (*caption*) Different stages in the lighting up of a target by a marker bomb. 1947 *Shell Aviation News* No. 113. 7/3 These marker burners would give a clear patch in bad weather that would provide the pilot with actual visual contact with the ground. 1945 *Penguin New Writing* XXIII. 9 Dan buoys appeared..their marker flags drowned and bedraggled. 1959 *Listener* 12 Feb. 278/2 Among the maze of wheel tracks and hoof marks, I noticed a red marker flag. 1943 *Chambers's Techn. Dict.* Suppl. 964/1 *Marker light* (Signalling), an indicating light on a signal post, to indicate the position or aspect of the main signal should its light have failed. 1960 O. Skilbeck *ABC of Film & T.V.* 82 *Marker light*, device giving synch. between picture and sound film by simultaneous photographic exposure instead of the Clapper board. Outmoded by the use of Magnetic film for Direct recording. 1971 M. Tak *Truck Talk* 104 *Marker lights*, the small lights that serve to outline a truck's length and width at night and in bad weather.

market, *sb.* Add: **1, 3.** (U.S. examples.) Also, now, = *supermarket.

1888 *Ladies' Home Jrnl.* June 2/2 Cherries..should be avoided when..sold in city markets. 1895 *Funk's Stand. Dict.*, *Market*,..a private store for the sale of provisions; as, a meat-*market*. 1905 *Delineator* Mar. 498/3, I patronize a small market..kept by an old Frenchman and his son. They are experts on cutting and trimming meat. 1911 *Woman's Home Compan.* Apr. 4/2, I have used inadequately filtered water, uninspected milk and shopped in markets where inspection of sanitary conditions was never dreamed of. 1967 'D. Shannon' *Rain with Violence* (1969) i. 19 She's pretty sure Mrs. Gerner usually shopped at the nearest market up on Marengo. 1972 *Jrnl. Social Psychol.* LXXXVII. 78 Markets were chosen which served a relatively large number of black customers.

1. b. *to go to market*: also *fig.*, to behave in an angry manner, to become angry. *Austral.* and *N.Z.*

a 1925 F. S. Anthony *Follow Call* (1936) 17 Peter came home drunk once every week, and made his poor wife milk the herd of twenty-four cows by herself; and then about 8 p.m. he'd arise from the sofa and go to market because the poor woman hadn't cooked a hot tea for him. 1945 Baker *Austral. Lang.* 121 A man in a temper is said..*to go to market*.

7. a. *to be in* (or *on*) *the market*: also *fig.*

1908 M. Diver *Great Amulet* x. 115, I don't feel called upon..to advertise the fact that I am not..'on the market'. 1929 *Amer. Speech* V. 123 A widow who was ready for another husband was said to be..'in the market'. 1936 L. C. Douglas *White Banners* x. 212 She was in the market for diversion. 1955 'A. Gilbert' *Is she Dead Too?* vii. 136 When I'm in the market for trouble of that kind I'll tip you the wink. 1955 *Times* 30 Aug. 3/1 This..is an important match with third place in the County table still in the market.

b. *black market* (see as main entry), *buyer's market* (see *buyer 3), *common market* (see *common *a.* 21), *seller's market* (see *seller).

c. In Horse Racing, the kind or amount of business done in bets, the state of betting.

1886 Earl of Suffolk & Berks. et al. *Racing & Steeple-Chasing* v. 85 On arriving at the rails, which separate the private stands' enclosure from the ring, he finds the market well set. 1897 E. H. Cooper *Mr. Blake of Newmarket* xxvi. 255 'I've missed the market!' My friend..explained..that he had not got the best bet against the horse which he might have got. 1972 J. Mitchell *Betting* ii. 34 Because of..the shortness of the period..for which the betting market on a race is active, racecourse bookmakers keep their betting as simple as possible.

10. a. *market-boat* (earlier examples), *-boy, -net, -people, -sloop, survey, -wagon*; also *market-made* adj.

1780 *New Jersey Archives* (1914) 2nd Ser. IV. 401 Mrs. Roker, and one other woman, were going in a market boat from Philadelphia. 1853 'P. Paxton' *Stray Yankee in Texas* 278 [He] bought a market-boat, and tried trading upon the bayou. 1863 A. D. Whitney *Faith Gartney's Girlhood* v. 44 The market-boys, and the waiters, and the confectioners' parcels. 1947 Auden *Age of Anxiety* (1948) ii. 44 You will soon Not bother but acknowledge yourself As market-made, a commodity Whose value varies. 1922 Joyce *Ulysses* 219 She passed out with her basket and a market net. 1696 *Mass. Bay Acts & Laws* 162 Hucksters and Traders of the Town, shall not..Buy of any of the Market People there. 1830 J. F. Cooper *Water Witch* I. xi. 259 The rogues will pass the pennant like innocent market people. 1873 'Vieux Moustache' *Boarding-School Days* 36 He had been a hand on a New York market-sloop. 1885 *Outing* VII. 206/2 A big market-sloop came along bound west. 1967 C. Berners-Lee in Wills & Yearsley *Handbk. Managem. Technol.* 4 Do the conclusions of this market survey stand up? *Ibid.* 13 Useful programs for planning advertising campaigns have been written, based partly on market survey data. 1973 J. Goodfield *Courier to Peking* vii. 92, I don't think any market surveys have been done. 1802 *Deb. Congress U.S.* (1851) 18 Mar. 1027 In the state of New Jersey five hundred and forty two [of the carriages taxed] are..principally market-wagons. 1895 C. D. Warner *Golden House* i. 9 Here and there [was] a lumbering market-wagon from Jersey.

b. market basket (earlier examples); **market economy,** a system of economy which is subject to free competition; **market gardener** (earlier and later examples); **market gardening,** keeping a market garden; **market hunter,** one who hunts game for the market; so **market-hunting** *vbl. sb.*; **market mammy** *colloq.*, an African woman stallholder; **market master** (examples); **market money,** money for buying things in a market (see also quot. 1891 in sense 10 a in Dict.); **market potential** (see quots.); **market reporter,** one who records the market rates of goods or stocks; **market research,** the systematic investigation of the demand for particular goods, a branch of marketing research; also *attrib.* and as vb. (to carry out market research); so *market-researched* ppl. adj., **market researcher,** a person engaged in this activity; **market shooter,** one who shoots game for the market; **market socialism,** an economic system in which a country's resources are publicly owned but production is geared to the private customer; hence **market-socialist** *attrib.*; **market square,** an open square in which a town market is held; also *fig.*; **market stall** (earlier example); **market value** (earlier examples); **market work,** also, the work connected with selling goods in a market.

1807 *Salmagundi* 27 June 247 Particular description of market-baskets, butchers' blocks and wheel-barrows. 1853 Dickens *Bleak Ho.* xxxiv. 333 Her market-basket.. is a sort of wicker well with two flapping lids. 1951 R. Firth *Elem. Social Organiz.* iii. 117 The other frame of organization consists of a market economy where a man lives largely by selling his produce or his labour in competition with others. 1966 *Listener* 24 Nov. 755/2 A market economy is an arrangement by which people, by their purchases, make clear what they want or do not want. Their market behaviour, in turn, is an instruction to producers as to what they should produce or not produce. 1968 *Ibid.* 4 July 7/2 In the market economy of classical capitalism the interests of the working class ran clean counter to those interests which relied upon the smooth working of the economic system. 1972 *Accountant* 5 Oct. 409/1 In a market economy, in which prices of goods and services are ultimately determined by the forces of supply and demand, any intervention by government in the form of fixing or holding prices and incomes cannot for long be effective. 1826 *Westm. Rev.* V. 20 The fruit and vegetables of a market gardener, on which his subsistence depends. 1832 *Chambers's Edin. Jrnl.* 7 Apr. 76/2 An industrious but rather unsuccessful market-gardener..from Kirkcaldy. 1963 *Times* 22 Feb. 10/3 Sir Thomas Playford, orchardist and market gardener, has all the assurance even in the uncertain world of politics of a man who knows that he will reap what he sows. 1875 *Encycl. Brit.* I. 384/1 *Market Gardening*... The growth of fruits and of culinary vegetables is in various parts of Great Britain an important department of farming. 1958 Hayward & Harari tr. *Pasternak's Dr. Zhivago* i. vii. 218 It is not for the sake of market gardening that we are going all this enormous distance. 1874 J. W. Long *Amer. Wild-Fowl Shooting* 185 Blue-winged teal..are much sought for by market-hunters. 1940 M. B. Trautman *Birds Buckeye Lake, Ohio* 170 According to former market hunters and old sportsmen, the Eastern Least Bittern was the most numerous transient and summering species between 1860 and 1900. 1897 *Outing* XXX. 293/1, I had little dreamed that Michigan would ever so far forget herself as to encourage market-hunting in preference to sportsman like methods. 1962 A. Lejeune *Duel in Shadows* vii. 91 There were two Africans, a small man.. and his wife, fat as a Market Mammy. 1966 *Punch* 21 Sept. 457/1, I stared at..this old market-mammy who had lost her 'touch' and was clinging so stubbornly to the last shreds of her trading reputation. 1972 *Daily Tel.* 21 Jan. 4/3 About 7,000 white-clad market mammies—women stallholders—with drummers and horn blowers yesterday demonstrated through Accra in support of last week's military coup against the Busia regime. 1851 C. Cist *Sk. Cincinnati in 1851* 87 A city treasurer, a marshal, a wharf and three market masters are elected. 1913 J. W. Sullivan *Markets for People* 104 The Pennsylvania markets usually get along with a single market-master. 1633 G. Herbert *Church Porch* in *Temple* 13 Think heav'n a better bargain, then to give Onely thy single market-money for it. 1868 *Putnam's Mag.* Jan. 40/2 Strawberries are down to ten cents a box..but you didn't leave a cent of market-money. 1962 S. Strand *Marketing Dict.* 436 *Market potentials*, determining the marketability of a product. 1969 J. M. Rathmell *Managing the Marketing Function* v. 200 *Market potential* is an expression of a market's absorption of a total industry's production in units or dollar sales... The major operational value of market potential is its usefulness in determining spatial rather than temporal objectives. 1854 B. J. Taylor *Jan. & June* 83 And so, as Market Reporters have it, 'we have movements to note'. 1926 *Market Res. Agencies Guide to Publ.* (U.S. Dept. Commerce) p. iii, With large sums of money being spent for market research..the necessity of having an inventory of accomplishments becomes obvious. 1927 *Ibid.* p. iv, Market research is the study of all problems relating to the transfer of goods from producer to consumer, involving relationships and adjustments between production and consumption, preparation of commodities for sale, their physical handling, wholesale and retail merchandising, and financial problems concerned. 1931 R. Simmat *Market Res.* i. 7 Reputable agencies will not accept the advertising for a product which market research has revealed to be unsuitable. 1937 *Discovery* Feb. 47/2 Another field for psychological investigation in many parts of the Special Areas is in market research, where methods of sale and customers' psychology need to be examined side by side. 1951 M. McLuhan *Mech. Bride* (1967) 31/2 As market-research tyranny has developed, the object and ends of human consumption have been blurred. 1956 *Planning* XXII. 19 Individual pub-

lishers and individual booksellers continue to do what they can for themselves in the way of market research and sales promotion. 1964 M. Argyle *Psychol. & Social Probl.* xiii. 162 One of the earliest pieces of market research discovered that coffee had an unfavourable image of being connected with tiredness and strain. 1965 *Spectator* 1 Jan. 3/1 The *Sunday Times*..claimed that £100,000 was to be spent on market research. 1969 P. Dickinson *Pride of Heroes* 58, I commissioned a little firm in Chicago to market-research the American idea of what English beer ought to taste like. 1970 *New Scientist* 1 Jan. 4/2 A 'hard sell' is necessary and this must be preceded by market research. 1973 *Times* 1 Feb. 8/2 (Advt.), Some market research experience would be an advantage. 1964 *Economist* 17 Oct. 276/3 The 1800 is as much a challenge to orthodox market-researched motoring as the Mini was five years ago. 1951 M. McLuhan *Mech. Bride* (1967) p. vi/2 Today the tyrant rules not by club or fist, but, disguised as a market researcher, he shepherds his flocks in the ways of utility and comfort. 1969 J. Argenti *Management Techniques* 157 Market Researchers rely heavily on the analysis of statistics—a notoriously difficult subject—and on answers to questionnaires—also notoriously unreliable. 1975 *Country Life* 2 Jan. 51/2 Editors have been convinced by accountants, market-researchers and computers that stories don't help to sell their products. 1880 *Golden Days for Boys & Girls* 6 Mar. 3/2 He knew very well that the ambitious and high-spirited Oscar was not a market-shooter from choice. 1897 *Outing* XXX. 293/2 The market-shooter, with no dogs to take care of, can sneak through the known haunts of the quail. 1965 *Listener* 15 Apr. 547/2 The new methods involve a form of market socialism, on the Yugoslav model... Production would be largely guided by the market. 1969 *Guardian* 3 July 12/5 The effect of the Czechoslovak events has been to discredit, in the eyes of the official [Russian] Establishment, the idea of 'market socialism'. 1972 D. Collard *Prices, Markets & Welfare* iv. 41 Where preferences are private but the ownership of resources public we have a market-socialist system. *Ibid.* xiii. 119 In the early 'thirties..Lange, Lerner and Dickinson..suggested a series of devices known as 'market socialism'. 1794 *Mass. Hist. Soc. Coll.* 1st Ser. III. 254 Besides the lower floor of Faneuil hall being used as a flesh market, a number of stalls are erected on Market square..and let to the market men. 1836 D. B. Edward *Hist. Texas* 148 A block shall be designated for a market square. 1963 *Times* 14 Feb. 8/5 The scientist who refuses contact with the people and opts for the Ivory Tower rather than the Market Square is a traitor to himself and to humanity. 1827 Drake & Mansfield *Cincinnati in 1826* vi. 55 The Revenue of the Corporation is derived; From..Rent of Market-stalls. 1691 Locke *Money* in *Wks.* (1824) IV. 99 According to the market value. 1791 *Deb. Congress U.S.* (1834) 1st Congr. App. 1993 The rapid increase that has taken place in the market value of the public securities. 1887 H. H. Jackson *Between Whiles* iv. 226 Donald liked slow cruising and the market-work best.

market, *v.* Add: **3.** To 'trade *on*', to take advantage of.

1906 Hardy *Dynasts* II. i. i. 10 These cloaked visitors of every clime That market on your magnanimity To gain an audience.

marketeer. Add: **3.** A supporter of Britain's entry into the Common Market (cf. *common *a.* 21).

1962 *Listener* 15 Nov. 799/2 The Marketeers within the Labour Party. 1969 *Guardian* 29 Aug. 16/8 The TUC General Council is trying to avoid an embarrassing public debate on the Common Market... Pro- and anti-Marketeers united in angry condemnation of Mr. Jenkins's attempts to force the issue on to the agenda. 1970 *Times* 4 May 8/2 Mr. Douglas Jay..leads his all-party group of anti-Marketeers, under the banner of the new Common Market Safeguards Campaign. 1971 *New Scientist* 27 May 522/1 The Paris summit has made the marketeers a little bolder. 1975 *Times* 25 Feb. 14/2 A keynote of the marketeers' case is..that Britain is in Europe and that the decision is..whether to come out.

marketing, *vbl. sb.* Add: **3.** *attrib.* and *Comb.*, as (sense 1) *marketing advantage, agent, conference, day, director, policy, problem, revolution, strategy, survey*; (sense 2 a) *marketing bag, basket*; **marketing mix** (see quot. 1969); **marketing research,** the systematic study of all the factors involved in marketing a product (see quot. 1963); so **marketing researcher.**

1919 A. Marshall *Industry & Trade* App. J. 800 The low grade industries which congregate in London owe comparatively little to the marketing advantages which are to be found there. 1951 M. McLuhan *Mech. Bride* (1967) 92/1 The marketing agents reciprocate by using still more applied science. 1934 T. S. Eliot *Rock* ii. 65 Enter Mrs. Ethelbert with marketing bag, hilariously. 1925 W. de la Mare *Two Tales* 20 A small, old, spectacled lady with a large marketing-basket, was..issuing out from behind the shop. 1970 *Brit. Printer* June 57 The BFMP's fourth marketing conference in London served to emphasise the urgent need for printers to understand the term 'marketing' let alone employ its principles. 1619 Marketting day [in Dict., sense 1 a]. 1961 *Times* 14 July 2/2 Marketing Director to organize and direct the Sales Organization. 1969 D. C. Hague *Managerial Econ.* III. xiv. 293 One of the major issues in all business is how to arrive at the correct 'marketing mix'—the correct balance of product design, price, advertising and other promotional expenditure, spending on the sales force, and so on. 1972 *Lebende Sprachen* XVII. 46/1 *Marketing mix.* 1930 G. R. Collins *Marketing* v. 78 The process of formulating marketing policies should recognise the fact that consumers' demands are the origins of economic impulses. 1920 P. T. Cherington *Elem. Marketing* 3 Increase of the scale of production alone introduces maladjustments between producing and consuming conditions with regard to quantity, quality, time, and place, which it becomes part of the marketing problem to correct. 1951 E. S. Bradford *Marketing Res.* p. xi, Marketing research is coming to be recognized as essential to a successful development of marketing. 1956 G. R. Collins *Marketing* (rev. ed.) v. 78 Marketing research..must..have for its major objective the minimizing of errors in marketing judgment. 1963 *Gloss. Managem. Terms* (Brit. Inst. Managem.) (Typescript), The distinction between market research and marketing research is important but is not yet as widely used and understood in the U.K. as in the U.S.A. One is a study of the market for the product and the other is a study of the marketing of the product. 1967 G. Wills in Wills & Yearsley *Handbk. Managem. Technol.* 175 Marketing research began in the second decade of the century as a commercial tool for the collection of facts for a better direction of sales effort. *Ibid.* 178 The analysis of territorial sales potential..is within the marketing research function of the business. *Ibid.* 179 Any marketing problem has three certain dimensions which marketing manager and researcher ignore at their peril—time, cost, and resources. 1963 *Times* 5 June 17/4 Today we [*sc.* the British] are lagging behind in a..transition that might be described as the Marketing Revolution. 1969 D. C. Hague *Managerial Econ.* iii. xiii. 288 Dr Rothschild was impressed by the ease with which military terminology insinuates itself into discussions of oligopoly. We talk of price wars, sales campaigns, marketing strategies, industrial espionage and so on. 1974 *Times* 18 Feb. 12 Many hoteliers now regard conferences as an integral part of the marketing strategy for their establishments. 1941 'Balbus' *Reconstruction & Peace* 58 Marketing surveys and campaigns for increasing consumption.

market-place. Add: Also *fig.*

1958 J. Barzun in Caplow & McGee *Academic Marketplace* p. v, The observers' point of departure was the current problem of mobility within the academic marketplace. 1975 *Times* 10 Apr. 16/1 Pension schemes..tied to the earnings-related inequalities of the market-place.

marking, *vbl. sb.* Add: **1. a.** (*Austral.* and *N.Z.* examples.) Cf. *mark *v.* 2 f.

1891 R. Wallace *Rural Econ. Austral. & N.Z.* ii. 43 Lamb-tailing and marking were on at this time—the end of June. 1946 F. Davison *Dusty* iv. 45 He handled lambs at marking time..as gently as if he owned them. 1950 *N.Z. Jrnl. Agric.* Sept. 197/1 Lamb marking consists of three different operations: the placing of a distinguishing mark in one or both ears, the amputation of tails, and the castration of male lambs.

c. With advs., freq. in *attrib.* uses.

1726 [in Dict., sense 1 a]. 1894 W. J. Lineham *Text-bk. Mech. Engin.* vi. 183 (*heading*) Marking-off, machining, fitting and erecting. 1909 *Westm. Gaz.* 24 Mar. 10/3 A marking-down process was also indulged in in anticipation of selling by speculators and profit-taking. 1917 [see *assembly 1 b]. 1948 R. de Kerchove *Internat. Maritime Dict.* 451/2 Marking-off platers work to drawings supplied from the drawing office and information given from the mold loft. 1964 S. Crawford *Basic Engin. Processes* i. 36 A typical marking-out table is shown. *Ibid.* 39 This mild-steel clamp is machined to length, width, and thickness prior to marking-out. 1966 J. H. Dixon tr. *Dormidontov's Shipbuilding Technol.* v. 115 A template workshop..fitted with optical marking apparatus and a marking-off table.

d. *Stock Exchange.* The recording of prices at which bargains are made.

1903 *Pitman's Business Man's Guide* 305/1 *Marking*,.. on the Stock Exchange this signifies the recording of the prices at which actual business has been done in any security between the hours of eleven and three. 1930 *Economist* 22 Mar. 653/1 Dealings begin, the marking of bargains is effected in one or other of the Stock Exchange Daily Lists, [etc.]. 1934 F. E. Armstrong *Bk. Stock Exchange* iii. 52 The marking of bargains in the *Official* and *Supplementary Lists* is..the record of business actually transacted, and provides an index to an established price level. 1959 *Times* 7 Apr. 10/6, 20,451 separate bargains were recorded. This is the highest total reached since the 'marking' of bargains was inaugurated. 1971 T. G. Goff *Theory & Pract. Investment* v. 53 A symbol beside a marking may indicate that there is some unusual feature about the underlying deal.

4. *marking brush, pencil*; **marking stitch** (earlier example).

1855 'Q. K. P. Doesticks' *Doesticks, what he Says* xxii. 198 A dry-goods box with a marking brush sticking out of the top of it. 1882 M. H. Foote *Led-Horse Claim* xvii. 228 He..swept, with one stroke of his marking-brush, a black circle around the figures. 1886 Marking pencil [see *B, BB, BBB* (*B III)]. 1961 *Lebende Sprachen* VI. 70/1 Marking pencil. 1861 Mrs. Stowe *Pearl of Orr's Island* I. xvii. 152, I was going to begin and teach her some marking stitches.

markka (mā·ɹkā). Also **marka.** [Finn.] **a.** The principal monetary unit of Finland, equal to 100 penniä. **b.** The name of the coin equal to this amount. (Cf. Mark *sb.*² 4 b.)

1903 *Finland by the 'Nord' Steamers* 11 The standard coin is the Marka, equal to 10d. of our money, divided into 100 Penni. 1911 G. Renwick *Finland Today* 336 Finnish coinage presents no difficulties, the Finnish markka (plural, markkaa) being equal to the French franc. 1963 *Economist* 1 June 925/3 Some forest owners are speculating on a rise in prices, possibly via a devaluation of the [Finnish] markka. 1972 *Daily Tel.* 3 Aug. 19/4 Your pound's worth in London... Finland..9·95 markkas.

Markov (mā·ɹkǫf). *Math.* Also **Markoff.** [The name of Andrei Andreevich *Markov* (1856–1922), Russian mathematician, who investigated such processes.] *Markov process*: any stochastic process for which the probabilities, at any one time, of the different future states depend only on the existing state and not on how that state was arrived at. *Markov chain*: a Markov process in which there are a finite or countably infinite number of possible states or in which transitions between states occur at discrete intervals of time; also, one for which in addition the transition probabilities are constant (independent of time). Also *Markov property*, the characteristic property of Markov processes.

1939 *Jap. Jrnl. Math.* XVI. 47 (*heading*) Markoff process with an enumerable infinite number of possible states. 1942 *Trans. Amer. Math. Soc.* LII. 37 Then $p_{ij}(t)$ can be considered a transition probability of a Markoff chain: A system is supposed which can assume various numbered states, and $p_{ij}(t)$ is the probability that the system is in the *j*th state at the end of a time interval of length *t*, if it was in the *i*th state at the beginning of the interval. 1950 W. Feller *Introd. Probability Theory* I. xv. 337 A Markov process is the probabilistic analogue of the processes of classical mechanics, where the future development is completely determined by the present state and is independent of the way in which the present state has developed. *Ibid.* 337 A definition of the Markov property. 1953 J. L. Doob *Stochastic Processes* v. 170 A Markov chain is defined as a Markov process..whose random variables can (with probability 1) only assume values in a certain finite or denumerably infinite set. The set is usually taken, for convenience, to be the integers 1,..., *N* (finite case) or the integers 1, 2,... (infinite case). *Ibid.* 186 The problem of card mixing is a good example of the application of Markov chains. 1953 J. B. Carroll *Study of Lang.* iii. 85 A Markoff process has to do with the different 'states' into which a phenomenon can get, and the statistical probabilities which govern the transition of the phenomenon from one state to another. 1956 *Nature* 4 Feb. 207/1 The most simple case is when all the atoms of the assembly are supposed to have no volume and no interactions (such as in an ideal gas). In that case it can be treated as a Markov process. 1960 Kemeny & Snell *Finite Markov Chains* ii. 25 A finite Markov chain is a finite Markov process such that the transition probabilities $p_{ij}(n)$ do not depend on *n*. 1962 J. Riordan *Stochastic Service Syst.* iii. 28 The simplest infinite-server system is unique among its fellows in the possession of the Markov property that future changes are independent of the past. 1966 S. Karlin *First Course in Stochastic Processes* ii. 27 A discrete time Markov chain $\{X_n\}$ is a Markov stochastic process whose state space is a countable or finite set, and for which $T = (0, 1, 2, \ldots)$. *Ibid.*, The vast majority of Markov chains that we shall encounter have stationary transition probabilities. 1968 P. A. P. Moran *Introd. Probability Theory* iii. 140 Thus a Markov chain observed in the reverse direction of time will be a Markov process. However, it will not in general be a Markov chain because the observed transition probabilities will not be independent of *t*. 1973 *Manch. Sch. Econ. & Social Stud.* XLI. 401 (*heading*) A Markov chain model of the benefits of participating in government training schemes.

Markovian (mā·ɹkōᵘ·viăn), *a.* [f. prec. + -ian.] Having the Markov property.

1950 W. Feller *Introd. Probability Theory* I. xv. 338 Examples of non-Markovian processes... Let $X^{(n)}$ equal 1 or 0 according to whether the *n*th drawing [from the urn] results in a black or a red ball. The sequence $\{X^{(n)}\}$ is *not* a Markov process. 1965 N. Chomsky *Aspects of Theory of Syntax* ii. 89 A derivation involving only phrase structure rules (rewriting rules) has a strict 'Markovian' character. That is, in a derivation consisting of the successive lines $\sigma_1, \ldots, \sigma_n$.., the rules that can be applied to form the next line σ_{n+1} are independent of $\sigma_1, \ldots, \sigma_{n-1}$ and depend completely on the string σ_n. 1972 *Nature* 3 Mar. 20/2 To render the process Markovian, we have assumed exponential distributions of division times for both normal and abnormal cells.

Marks and Sparks (mā·ɹks ænd spā·ɹks). Also **Marks.** Colloq. name for the merchandising company of Marks and Spencer Ltd. or any of the stores owned by this firm. Freq. *attrib.* of clothes bearing the firm's trademark.

1951 Partridge *Dict. Slang* (ed. 4) Add. 1106/2 *Marks*, a Marks & Spencer store. 1964 *Sunday Times* 5 Apr. 41/5 In their Marks and Sparks' woollies and living in what looks like a remarkably nice housing estate, Topsy and Tim clearly stand in for classless society. 1966 A. Prior *Operators* i. 8 His Marks and Sparks shirt. 1971 'D. Halliday' *Dolly & Doctor Bird* xi. 141, I sat there in my Marks and Sparks knickers and brassière. 1972 R. Quilty *Tenth Session* 104 The girls get just about all their odds and ends in Marks and Sparks or the bargain basement at C & A.

marksman. **5.** (Earlier example.)

1887 M. Shearman *Athletics & Football* 57 The starter is helped by a 'marksman', who places the men on the scratch.

ma·rk-up. [f. *to mark up* s.v. *mark *v.* 2 c.] The amount added by a retailer to the cost price of goods to cover overhead charges and provide profits.

1920 P. I. Carthage *Retail Organization & Accounting Control* ii. 44 The mark up of merchandise refers to the percentage or the amount of money added to the cost in order to obtain the selling price. 1931 J. F. Pyle *Marketing Princ.* xx. 518 *Mark-up* is..the difference between the marked price of the merchandise, at retail, and the cost of merchandise bought. Thus the mark-up represents the gross margin that is *hoped for*. 1936 *Economist* 22 Feb. 422/1 The 'mark up' (retailers' margin) is evidently adequate to provide relatively stable earnings on a stabilised capital investment. 1957 *Times* 19 Aug.

9/2 Shops often add to the price which they pay for their merchandise to manufacturers and wholesalers a percentage 'mark up' of 30 or 40 per cent. **1965** *Spectator* 22 Jan. 116/3 A big, new market for poor-quality materials and an even larger mark-up where the garments are supplied ready-made. **1972** M. WOODHOUSE *Mama Doll* xiii. 175 The estimated value of the items on the inventory is somewhere around fifty-five million pounds... Allowing a hundred per cent markup on sale, that would make their wholesale value some twenty-seven million pounds. **1974** *Times* 7 Feb. 20/8 The exorbitant mark-up imposed by European booksellers.

marl (māɹl), *sb.*[4] [Reduced form of MARBLED *ppl. a.* 2.] A yarn made from two different coloured threads twisted together so as to produce a mottled effect; the fabric produced from this yarn. Usu. *attrib.*

1892 *Queen* 5 Mar. p. xi (Advt.), Ladies write for Patterns of the entirely new designs in..Marls, Tweeds,..and Beiges. **1922** *Daily Mail* 18 Dec. 1 (Advt.), Knitted sports suit in rich Marl mixtures and plain colours. **1926** *Illustr. Official Jrnl.* (Patent Office) 20 Oct. 1668 Spinning marl or multi-ply yarns. **1968** E. GALE *From Fibres to Fabrics* iv. 45 Two marl threads are sometimes twisted together to form one yarn. **1970** A. M. COLLIER *Handbk. Textiles* iv. 88 A marl yarn is produced by combining two yarns of uneven thickness, with uneven rates of delivery so that there is excess material at intervals.

marlberry. Substitute for def.: A small, evergreen tree, *Ardisia paniculata*, of the family Myrsinaceæ, bearing clusters of pink flowers and dark red berries, and found in the West Indies and tropical America. (Examples.)

1884 C. S. SARGENT *Rep. Forests N. Amer.* 100 *Ardisia Pickeringia* Marlberry-Cherry. **1897** G. B. SUDWORTH *Nomencl. Arborescent Flora U.S.* 316 *Icacorea paniculata*. Marlberry. **1917** *Rep. Board of Regents Smithsonian Inst.* 384 In addition to these are the paradise tree or bitterwood; soapberry tree;..marlberry; [etc.]. **1924** J. A. THOMSON *Sci. Old & New* v. 27 Even the names transport us into a land of pure delight—the paradise tree, the myrtle-of-the-river, the marlberry, and the bois-fidèle.

marler (mā·ɹləɹ). [f. MARL *v.*[2] + -ER[1].] A marline-spike.

1929 J. MASEFIELD *Hawbucks* 15 A strong sheath knife with a marler at the back.

marlin (mā·ɹlin), *sb.*[2] [App. abbrev. of MARLINSPIKE, from the shape of the beak.] A large, marine, game fish belonging to the genera *Makaira* or *Tetrapterus* of the family Istiophoridæ, having the upper jaw elongated to form a beak.

1917 *Vanity Fair* (N.Y.) July 62/1 The marlin—*Tetrapterus mitsukurii*—is sometimes called the Japanese swordfish. **1923** D. K. TRESSLER *Marine Products of Commerce* 737/1 The marlin (*Tetrapterus mitsukurii*)..is also called spikefish. **1926** *Glasgow Herald* 24 June 4/5 The extract concerning the loss of a black marlin swordfish, estimated to weigh 600 lb., is one of the most dramatic passages. **1931** *Times Lit. Suppl.* 16 Apr. 301/1 Since then marlin have been taken, on heavy tackle, of 372 lb. **1937** *Discovery* Nov. 357/1 One [swordfish], a black marlin, has been recorded in New Zealand weighing 976 lbs. **1940** *Nature* 6 Apr. 555/2 The black marlin differs from the striped and the blue marlin in the fact that the pectoral fin, when adducted, remains in the horizontal position, whilst in the others it can be brought up flat against the side of the body. **1959** *Angling Times* 13 Mar. 6/3 Madeira has them: the white marlin and the blue marlin. **1970** M. SLATER *Caribbean Cooking* 11 *Marlin*, one of the most sought-after game fish, found in rocks and reefs.

marm (māɹm, mæm). [Var. of MA'AM, freq. in U.S. writers.] **1.** = MA'AM 1.

1837 R. M. BIRD *Nick of Woods* I. 120 Say the word, marm. **1845** C. M. KIRKLAND *Western Clearings* 17 'Massy no, marm!' said Jane, with a giggle. *Ibid.* 18 No, Marm; but—this 'ere is something about the team, I guess.

2. = MA'AM 2.

1837 [see MA'AM 2]. **1887** M. E. WILKINS *Humble Romance* 107 Marm Lawson was not a duchess; but she was Marm Lawson. The 'Marm' itself was a title. *a* **1922** T. S. ELIOT *Waste Land Drafts* (1971) 59 line 50 The pleasant violin At Marm Brown's joint, and the girls and gin.

3. = MA'AM 3.

1865 S. HALE *Lett.* (1919) 16 The silk is seven dollars a yard, and the marm that makes it asks a great deal. **1872** *Congress. Globe* 42nd Congress 2 Sess. App. 632/3 It will be seen that in the great race thus far the English and American marms are about 'nip and tuck'.

b. Used for 'mother'. (Also in address.)

1835 J. F. COOPER *Monikins* II. i. 35 He could scare one by threatening to tell his *marm* how he behaved. **1838** — *Eve Effingham* II. xvi. 168 Who taught you to call me marm!..Say 'ma' this instant. **1845** S. JUDD *Margaret* I. 37 'Has your marm got that done?' asked Martha Madeline. **1890** S. O. JEWETT *Strangers & Wayfarers* 9 You've got real nice features, like your marm's folk.

marmalade, *sb.* Add **3. c.** quasi-*adj.*: of the colour of marmalade (so *marmalade-coloured* adj.).

1926 S. T. WARNER *Lolly Willowes* III. 184 Jim was..a mottled marmalade cat. **1938** K. HALE (*title*) Orlando, the marmalade cat. **1951** 'C. CARNAC' *It's her Own Funeral* iv. 38 James, the marmalade cat, was sitting disapprovingly outside. **1957** *Times Lit. Suppl.* 15 Nov. p. xvii/2 Miss White's account of how a fierce little Siamese and an unsnubbable marmalade kitten learnt to live amicably together. **1961** *Guardian* 20 Jan. 9/7 A magnificent dark marmalade-coloured Persian cat. **1965** G. MCINNES *Road to Gundagai* ii. 29, I faced a crowd of blonde giants with marmalade fuzz on their chests. **1972** J. AIKEN *Butterfly Picnic* iii. 55 The local marmalade-coloured rock. **1973** 'E. PETERS' *City of Gold & Shadows* iii. 47 A very austere dress in a dark russet-orange shade that touched off the marmalade lights in her eyes.

ma·rmalade, *v.* [f. the sb.] *trans.* To spread with marmalade. Hence **ma·rmaladed** *ppl. a.*

1967 D. PINNER *Ritual* x. 107 David..selected a piece of toast, marmaladed it, and munched it. **1968** C. NICOLE *Self Lovers* i. 18 Brice marmaladed toast. **1975** J. WOOD *North Kill* vii. 106 Eating steamed haddock along with marmaladed toast.

marmalady, *a.* Delete *rare*[−1] and add later examples.

1920 JOYCE *Let.* 3 Jan. (1957) I. 135 *Nausikaa* is written in a namby-pamby jammy marmalady..style. **1960** P. COLERIDGE *Running Footsteps* 59 A clipped marmaladey moustache.

marmite (marmīt). [F. *marmite* pot or kettle: see MARMIT.] **1. a.** An earthenware cooking-vessel; a stockpot.

1805 C. JAMES *New Mil. Dict.* (ed. 2), *Marmite*,..porridge-pot, kettle; a machine in which soldiers boil their victuals. **1882** C. M. YONGE *Unknown to Hist.* II. xxi. 294 The French suite, every one of whom liked to have his own little arrangements of cookery, and to look after his own *marmite* in his own way. **1919** B. RUCK *Disturbing Charm* I. xi. 107 The door into the huge French kitchen stood open, giving a glimpse of marmites, burnished copper pans, crocks, and five-decker cookers. **1955** L. WOOLLEY *Alalakh* vi. 218 'Marmite', Type 154, of normal cooking-pot ware, black clay reddened on surface. **1960** *Home & Garden* Oct. 150/2 A range of ovenware..5s. 6d for a marmite.

b. *slang.* A bomb or shell resembling a pot.

1915 G. ADAM *Behind Scenes at Front* 48 The graves in the churchyard have been torn open by 'marmites'. **1919** [see *BLACK MARIA 2].

2. (mā·ɹmait.) (Properly with capital initial.) The proprietary term for an extract made from fresh brewer's yeast.

1907 *Yesterday's Shopping* (1969) 18/1 Marmite (Vegetable Ext.)... Marmite Bouillon, in tubes. **1920** *Trade Marks Jrnl.* 25 Aug. 1588 *Marmite*... A Concentrated Culinary Preparation, being an Article of Food. The Marmite Food Extract Co., Limited. **1923** *Nature* 12 May 626/2 In the prevention of beri-beri the addition of oatmeal and dhall to the British ration, the addition of marmite, and later, the issue of bread containing 25 per cent. of atta, were found valuable. **1925** D. H. LAWRENCE *Let.* 17 Dec. (1962) II. 871 So Sonya will never cook us another goose, only marmite pie and nut-cutlet. **1928** R. MACAULAY *Keeping up Appearances* i. 5 Doctors..prefer that one should breakfast..on marmite, Ry-Vita biscuits, and an apple. **1934** *Nature* 20 Oct. 623/2 The yeast extract 'Marmite' has long been recognised as a source of the vitamin B complex. **1947** T. H. WHITE *Mistress Masham's Repose* i. 7 She had..brown eyes the colour of marmite, but more shiny. **1966** A. E. LINDOP *I start Counting* xviii. 208 He knocked me up a Marmite sandwich.

maro. (Later examples.)

1833 A. SMITH in M. D. Frear *Lowell & Abigail* (1934) 63 Many others wear nothing but a narrow strip of tapa about their loins, called a *maro*. **1860** MAYNE REID *Odd People* 213 A coarser and scantier pareu is to be seen among the poorer people,..and not unfrequently this is only a mere strip wrapped around the loins; in other words, a 'malo', 'maro', or 'maso'—as it is indifferently written in the varied orthography of the voyagers. **1898** [see TAPA]. **1969** R. T. WILCOX *Dict. Costume* (1970) 201 *Malo*, the Hawaiian man's girdle or loincloth. Originally of tapa cloth which was made from tree bark, but now of cotton dyed in brilliant colors.

marocain (marŏkēi·n). [f. F. *maroquin*: see MAROQUIN *a.* and *sb.*] A dress fabric of ribbed crêpe. Also, a garment made from marocain.

1922 *Glasgow Herald* 28 June 8/6 For her going away dress the bride had a three-piece suit of grey wool marocain. **1922** *Tatler* 30 Aug. 354/2 An evening frock of black crêpe marocain. **1923** *Weekly Dispatch* 14 Jan. 15 A new evening model..had a trailing skirt with narrow draperies of black panne, while the upper part..was in white marocain. **1926** C. SIDGWICK *Sack & Sugar* xi. 126, I wore an embroidered marocain. **1957** M. B. PICKEN *Fashion Dict.* 88/1 *Marocain*.., ribbed crepe of silk or wool or combination.

maroodi (mārū·di). Also **maroudi.** [ad. Arawak *marodi*.] In Guyana, any of several birds included in species of guan belonging to the genera *Penelope* or *Pipile*.

1825 C. WATERTON *Wanderings S. Amer.* 23 The forest contains an abundance of..maams, maroudis and waracabas. **1876** C. B. BROWN *Canoe & Camp Life Brit. Guiana* 345 Maroodies of two kinds (the common and white-headed)..were also numerous. **1883** E. F. IM THURN *Among Indians Guiana* 62 Now and then a maroodi (*Penelope*) cried shrilly from among the trees. **1922** *Blackw. Mag.* Apr. 535/1 A quail and a maroodi rewarded their zeal.

maroon, *sb.*[1] Add: **3.** Used as an air-raid warning, etc., in the war of 1914–18.

1918 *Flying* 6 Feb. 90/1 Clearly, the authorities ought to have posted notices..explaining that the maroons are warnings to take cover. **1918** *Daily Mirror* 12 Nov. 2/1 London went wild with delight when the great news came through yesterday... Bells burst into joyful chimes, maroons were exploded, bands paraded the streets, and London gave itself up wholeheartedly to rejoicing. **1934** E. WHARTON *Backward Glance* xiii. 358 Four years of war had inured Parisians to every kind of noise connected with air-raids, from the boom of warning maroons to the smashing roar of the bombs.

maroon, *sb.*[2] **2.** (Further examples.)

1785 in *South Carolina Hist. & Geneal. Mag.* (1912) XIII. 188 On Monday we form a maroon party to visit some saw mills. **1838** C. GILMAN *Recoll. Southern Matron* xxxii. 223 Feeling the necessity of refreshment, we alighted for a while beneath a tree by the roadside, for a *maroon*.

maroon, *v.* Add: **2. b.** *transf.* To place or leave in a position from which one cannot escape. Also *fig.*

1910 *N.Y. Even. Post* 6 Jan. (Th.), Train No. 4.., due here from Los Angeles on January 1, is marooned in the desert. **1912** *Ibid.* 15 July 1/7 Rescue parties found dazed families..marooned on roofs. *Ibid.*, The torrent rushed..through the [station] yard,..marooning several hundred passengers. **1916** W. OWEN *Let.* 19 June (1967) 395, I am marooned on a Crag of Superiority in an ocean of Soldiers. **1946** *Sun* (Baltimore) 10 Aug. 4/1 It comes out for..direct assistance and encouragement to farmers marooned in declining or unproductive lines. **1973** *Jewish Chron.* 29 June 16/2 Marooned in the decaying house, she hears voices and sees the ghosts of the family. **1974** *Sunday Tel.* 7 July 26/2 Living a few miles out [of a city] is all very well in itself, but it often involves two cars—one for an otherwise marooned wife and family.

marooner. Add: † **4.** = MAROON *sb.*[2] 1. *Obs. U.S.*

a **1738** W. BYRD *Hist. Dividing Line* in *Writings* (1901) 37 We were told that on the South Shore, not far from the Inlet, dwelt a Marooner that Modestly call'd himself a Hermit.

‖ **maror** (mārō·ɹ). Also **moror.** [Heb. *mārōr*.] A dish of bitter herbs eaten as part of the Jewish Passover *seder*.

1893 [see *CHAROSET(H)]. **1903** W. ROSENAU *Jewish Ceremonial Institutions* viii. 122 A dish is placed, on which are put three unleavened cakes... On the top of them are put..the 'Charoseth' (a mixture of scraped apples and almonds, 'Moror' (bitter herbs), parsley and salt water... The bitter herbs, usually consisting of horseradish, stand, on account of their pungent taste, for the hard work of the Israelites in Egypt. **1905** *Jewish Encycl.* XI. 146/1 None has done his duty on that night [*sc.* Passover] until he has given voice to the three words 'pesaḥ' (pascal lamb), 'maẓẓah' (unleavened bread), and 'maror' (bitter herb). **1959** [see *CHAROSET(H)]. **1968** A. MATTS tr. *Sperling's Reasons for Jewish Customs* 188 The best choice for *maror* is a piece of horseradish, because horseradish has a particularly sharp and bitter taste. **1972** C. RAPHAEL *Feast of Hist.* ii. 37 Passover..was instituted with three binding ceremonies: (1) *Pesach*—a special sacrifice at the Temple... (2) *Matzah* (unleavened bread)... (3) *Maror* (bitter herbs)—eaten at the Seder to remind us of the bitterness of the Egyptian slavery.

maroudi, var. *MAROODI.

‖ **marouflage** (maruflāʒ), *sb.* *Art.* [f. F. *maroufler* to attach (a painted canvas to a wall) with *maroufle*, a strong adhesive (also, a lining or layer of such adhesive): see -AGE.] **1.** The act or process of pasting a painted canvas to a wall, traditionally using an adhesive made of white lead ground in oil. Hence **marouflage** *v.*, **marouflaging** *vbl. sb.*

1883 J. W. MOLLETT *Illustr. Dict. Art & Archæol.* 205/2 *Marouflage*, a method of house-painting in France, upon a lining of prepared canvas fixed upon the surface to be decorated. **1909** *Cent. Dict.* Suppl., *Marouflage*, this device allows mural paintings to be in oil colors, and also allows the work to be done at a distance from the building for which it is intended. **1934** H. HILER *Notes Technique Painting* iv. 265 *Marouflage* as the different methods of gluing the canvas to the wall are called, was very popular throughout the nineteenth century, and most modern European wall decorations were executed by this method. *Ibid.*, A paste of rye flour, to which two or three heads of garlic were added, was the commonest way of marouflaging on the Continent. **1969** R. MAYER *Dict. Art Terms & Techniques* 235/1 Traditionally, the adhesive used in marouflage is commercial white lead in oil.

2. [Transf. sense of *maroufle* a lining, layer (of glue): see prec. sense.] In ironwork, a piece of leather or other material used as a backing to show off decoration.

1957 R. LISTER *Decorative Wrought Ironwork* 55 This decoration was called *Marouflage*. Examples of door plates backed with leather are to be seen in King's College Chapel, Cambridge. **1960** H. HAYWARD *Antique Coll.* 175/2 *Marouflage*, a scarlet cloth backing to pierced lock plates.

marque[2] (māɹk). [Fr., = mark, sign.] = *MARK *sb.*[1] 15 b. A model or brand esp. of motor vehicle.

1906 C. JARROTT *Ten Yrs. Motors & Motor Racing* x. 174 Half-way to Bordeaux, and out of the first four cars two were Dietrich—this seemed a good record for the *marque*. **1956** *Road & Track* Oct. 5 (Advt.), The marque of Mercedes-Benz. *Ibid.* 14 The firm and marque has not actively participated in competition. **1958** *N. & Q.* Feb. 86/1 'Marque' is surely a recent borrowing from the French

language where it is in general use to denote a particular type of variety of a product, e.g. Frigidaire, Hoover, etc... The channel whereby the word has passed from French to English is doubtless international motor-racing. **1960** *Times* 4 July (Suppl.) 1/1 Sixty years is long..also in the reign of a branded product—particularly in the reign of a *marque* of cigarette. **1972** *Daily Tel.* 28 Jan. 6/8 The Renault 5 has all the comfort associated with the marque, despite its lack of size. **1973** *Times* 5 Oct. (Safety Suppl.) p. iv, But it should be emphasized that both marques are in the executive/luxury category and that to make a small family saloon as safe as a Mercedes or a Volvo would involve problems of weight and cost that are probably insuperable.

‖ **marquesa** (maɹkē·zā). [Sp.; cf. MARCHESA.] In Spain: a marchioness.

1846 R. FORD *Gatherings from Spain* xviii. 249 We well remember the death of a kind and venerable Marquesa at Seville. **1966** H. YOXALL *Fashion of Life* vii. 63 The *marquesa* of the great neighbouring house was interested. **1968** A. BROWN *Slay me Suddenly* ix. 131 A *marquesa* may have dignity because she belongs to one of the oldest lineages in Spain.

Marquesan (maɹkēi·săn, -z-), *a.* and *sb.* [f. *Marquesas* Islands in the Pacific + -AN.] **A.** *sb.* One of the Polynesian aborigines of the Marquesas Islands; also, the language spoken by them. **B.** *adj.* Of or pertaining to the Marquesas Islands.

1799 J. WILSON *Missionary Voy. S. Pacific* 138 Several of the Marquesans were continually plaguing the captain to take them to Otaheite. **1837** *Evangelical Mag.* Aug. 395 The spiritual enemies, which the Marquesan islanders have to encounter. **1845** J. COULTER *Adventures Pacific* xii. 155 Any liability to capsize is counteracted by the outrigger that all the Marquesan canoes have attached to them. *Ibid.* xiv. 211 Eight or ten nuts (commonly known as the candle-nut, from their emitting a bright flame, and being used by Marquesans as a substitute for candles) are strung on a piece of reed. **1896** R. L. STEVENSON *In South Seas* I. ii. 13 The alarms and sudden councils of Marquesan chiefs. **1919** F. O'BRIEN *White Shadows South Seas* iii. 28 Hence descended the Marquesans, vikings of the Pacific, in giant canoes, and sprang upon the fighting men of the Tahitians. *Ibid.* ix. 96 The Catholic chants sung thus in Marquesan took on a wild barbaric rhythm that thrilled the blood. **1936** V. A. DEMANT *Christian Polity* xi. 192 In regard to..Oceania..mention need only be made of..the effigies of chiefs among the Marquesans. **1969** J. H. VANCE *Deadly Isles* (1970) xvi. 118 A massive white-haired Marquesan with a face carved from teak. **1974** *Listener* 21 Nov. 679/2 The pathetic dependence of the Marquesans on imported foodstuffs. *Ibid.*, Curiosity about ancient Marquesan culture.

marquis. Add: **5.** A North American variety of spring wheat. Also *attrib.*

1906 C. SAUNDERS in *Bull. Canad. Dept. Agric.* No. 57. 29 *Chelsea* and *Marquis* are new cross-bred sorts produced at the Central Experimental Farm. **1924** J. A. THOMSON *Sci. Old & New* xliii. 253 Marquis is a hard, red spring wheat with excellent milling and baking qualities; it is now the dominant spring wheat in Canada and the United States. **1936** DENTON & LORD *World Geogr. for Canad. Schools* 85 Dr. Charles E. Saunders, the Dominion cerealist, had been experimenting for several years, trying to produce a new wheat which would ripen earlier than Red Fife. In 1903 his efforts were crowned with success. The new wheat was named 'Marquis'. **1960** D. E. BUBLITZ *Life on Dotted Line* 51 Marquis wheat was chosen as their crop that first year. **1965** I. REEKIE *Along Old Melita Trail* x. 150 Marquis wheat, developed from a cross of Red Fife, was released in 1911.

marquise. Add: **3. b.** *Archit.* (See quot. 1891.)

1891 tr. *J. Adeline's Art Dict.* 251/2 The term *marquise* is applied to a light roof which projects from the façade of a building. It is generally placed over a flight of steps. On the outside of theatres marquises of considerable length are not infrequently to be seen. Almost invariably they have a glass roof. **1904** B. VON HUTTEN *Pam* v. i. 237 A moment later, she stood in the door, under the little 'marquise'. **1924** 'L. MALET' *Dogs of Want* i. 7 [She] stood under the glass marquise, at the top of the flight of steps. **1930** A. BENNETT *Imperial Palace* vii. 28 On the steps under the marquise she took off her cloak.

marquisette (maɹkize·t). [Fr., dim. of *marquise* MARQUISE.] (See quots. 1968.)

1908 *Tatler* 6 May (Suppl.) p. iv, A simple house gown of black marquisette. **1909** *Daily Chron.* 12 Apr. 7/1 The soft marquisettes and satin-faced foulards. **1913** [see *POURREAU]. **1928** *Times* 9 May 11/3 A picture gown of fine gold embroidered marquisette, over peach georgette. **1930** *Times* 13 Mar. 11/6 Lovely ankle-length frocks in flowered and plain chiffons, lace, satins, and marquisettes. **1957** *New Yorker* 5 Oct. 110/1 Its dazzling décolletage veiled in black marquisette. **1968** E. GALE *From Fibres to Fabrics* xii. 132 Some marquisettes are not woven, but are produced on a simple type of lace machine. **1968** J. IRONSIDE *Fashion Alphabet* 240 *Marquisette*, a leno-weave gauze fabric made of cotton, rayon, silk or synthetic fibres.

Marrano. (Later examples.) Also *attrib.*

1941 G. G. SCHOLEM *Major Trends Jewish Mysticism* vii. 240 The beginning of the persecution of the Jews in Spain and the appearance of Marrano Judaism after 1391. **1942** L. B. NAMIER *Conflicts* 133 More logical were those who attempted amalgamation: but even this, as a mass movement, merely produces Marranos. **1960** L. P. GARTNER *Jewish Immigrant* i. 16 The mid-seventeenth century, when the earlier flight of Marranos from Spain and Portugal was no longer important. **1973** *Jewish Chron.* 2 Feb. 3/5 The 90 members of the tribe near the town of Pachuca..claim that some of their ancestors were Marranos (baptised Jews suspected of secret adherence to Judaism) who were burnt at the stake in Mexico during the 16th century. **1973** *Publishers Weekly* 3 Sept. 44/1 The term 'Marranos' originally denoted those Spanish Jews who, during the Inquisition, or long before it, converted to Catholicism but continued secretly to practice their Jewish faith; loosely it denotes all secret Jews.

Marranism (further example). Also **Marra·nic** *a.*, **Marra·noism.**

1924 *Glasgow Herald* 25 Sept. 6 In most Latin American countries, the arrivals find no organized Jewish life—only Marranoism. **1941** G. G. SCHOLEM *Major Trends Jewish Mysticism* vii. 254 The descendant of a Marranic family. *Ibid.* viii. 311 The conception of a voluntary Marranism.

marri (mæ·ri). [Aboriginal name.] A Western Australian red gum tree, *Eucalyptus calophylla*, or its timber.

1920 R. GRIMWADE *Anthography of Eucalypts* plate 75 *E[ucalyptus] calophylla*. Marri. Western Australian Red Gum... A tall, sturdy tree, well distributed over the coastal regions of Western Australia. **1934** W. F. BLAKELY *Key to Eucalypts* 84 *E. calophylla*, R.Br. 'Marri' or 'Red Gum'... A medium-sized to large tree. Bark rough, flaky, persistent on trunk and branches. **1938** A. W. D'OMBRAIN *Gallery of Gum Trees* 48 In Western Australia ..there are two distinctive and beautiful red flowering gums, the Scarlet Flowered Gum,..and..the much larger Pink Flowered Marri. **1963** *Times* 12 Mar. (Austral. Suppl.) p. v/7 Marri and gimlet and grey-barked tuart, and..other valuable timbers. **1969** T. H. EVERETT *Living Trees of World* 269/1 The marri or red gum (*E. calophylla*) has its home in southwestern Australia. It becomes 150 feet tall with a trunk 5 feet in diameter... Its flowers, in large decorative terminal clusters, are cream-colored or pink; they produce abundant nectar. This is a good ornamental and shade tree. Its lumber is used for boxes and light construction.

marriage. Add: **8.** *marriage contract* (examples), *-day* (later examples), *-hall, manual, market* (earlier and later examples), *mart, -tie* (later examples), *-vow* (later examples); **marriage broker,** (*b*) in cultures in which arranged marriages are the norm, one who arranges marriages for a fee; **marriage bureau,** an agency which arranges introductions with a view to marriage; also *fig.*; **marriage certificate,** a copy of the record of a legal marriage which is given to the contracting parties; **marriage counselling, guidance,** the giving of advice on problems connected with marriage, usu. as a form of social service; also **marriage counsellor, marriage guidance counsellor; marriage of convenience** = *MARIAGE DE CONVENANCE; **marriage payment** *Anthropol.*, payment of a traditional kind made in many tribal societies to a bride or her parents by the bridegroom or his parents; cf. *BRIDE *sb.*[1] 5; **marriage rate,** the ratio of the number of marriages per year to the population (usually expressed per thousand); **marriage-ring** (earlier and later examples).

1892 I. ZANGWILL *Childr. Ghetto* I. 18 The same mould covers them all—..the marriage-brokers repose with those they mated. **1932** L. GOLDING *Magnolia Street* I. ii. 32 She didn't marry again, though she was only in the middle thirties when he died and the marriage-brokers got busy. **1968** L. ROSTEN *Joys of Yiddish* 325 A wisecrack defines a *shadchen* as 'a marriage broker who knows the perfect girl for you—and married the wrong girl himself'. **1942** OLIVER & BENEDETTA *Marriage Bureau* 19 How much better it would be if there were an organization that could arrange the actual match-making and see that suitable people met each other. And this was my idea for the Marriage Bureau. **1953** 'H. JENNER' *Marriage is my Business* i. 19 The story has often been told of the two exdebs and how they coped with opening a marriage bureau. **1960** N. EPTON *Love & English* VI. v. 355 The marriage bureau..lists *all* your assets for the calm perusal of registered clients. *Ibid.* 356 This modern matchmaker, the marriage bureau, is middle-class and eminently respectable. **1972** *Accountant* 5 Oct. 411/3 Its aim would be the establishment of a 'marriage bureau' for the smaller businesses of Europe. **1848** MRS. GASKELL *Let.* 11 Nov. (1966) 62, I wish I had five sisters, who were bound to love me by their parents' marriage certificate. **1860** E. EDEN *Semi-Attached Couple* II. xviii. 225 He had seen too much of life, to believe in these sudden discoveries of marriage certificates. **1911** G. B. SHAW *Doctor's Dilemma* III. 61 She carried her marriage certificate in her face and in her character. **1945** MENCKEN *Amer. Lang.* Suppl. I. 475/1 *Marriage lines* is confined to the vulgar. On higher levels *marriage certificate* is used. **1967** 'G. NORTH' *Sgt. Cluff & Day of Reckoning* xi. 101 Brittle where it had been folded, the marriage certificate held his gaze. **1824** MILL in *Westm. Rev.* I. 537 Plato is represented as exceedingly wicked, for having expounded, in his Republic, the footing upon which he thought that the marriage contract could most advantageously be placed. **1961** NEW ENG. BIBLE *Matt.* i. 19 Joseph desired to have the marriage contract set aside quietly. **1945** Marriage counselling [see *COUNSELLING *vbl. sb.*]. **1959** HERBERT & JARVIS (*title*) A modern approach to marriage counselling. **1970** *Guardian* 31 July 18/6 In the 30 years the marriage counselling movement has been in existence the lack of specific privileges for marriage counsellors has never come to a showdown. **1946** Marriage counsellor [see *COUNSELLOR 1 b]. **1970** G. GREER *Female Eunuch* 19 The revolutionary woman must know her enemies, the doctors, psychiatrists,..marriage counsellors. *a* **1640** MASSINGER et al. *Old Law* (1656) v. 69 As I am Lord of the day (being my marriage day the second) I doe advance bonnet. **1852** M. ARNOLD *Tristram* I, in *Empedocles on Etna* 115 That her lord and she Might drink it on their marriage day. **1907** W. B. YEATS *Deirdre* 36 Although her marriage-day had all but come. **1935** *Time* 13 May 28/2 Old Stone Church is one of an increasing number whose pastors run 'marriage guidance bureaus'. **1945** *Times* 25 July 5/5 Groups of people in the provinces have recently been setting up their own local marriage guidance councils. **1948** D. R. MACE *Marriage Counselling* iii. 18 Marriage guidance..falls into three natural divisions. The first is education for marriage... The second is marriage preparation..intensive preparation of engaged couples... The third is marriage counselling..offering help to married people who are confronted with difficulties. **1965** HALL & HOWES *Church in Social Work* iii. 63 Started in 1938 and revived and reconstituted on a national basis in 1943, the National Marriage Guidance Council..quickly became established. **1967** O. WYND *Walk Softly, Men Praying* xii. 186 About all that was available to me was the kind of noise that might have come from a junior apprentice marriage-guidance counsellor. **1970** G. GREER *Female Eunuch* 17 Women are not happy even when they do follow the blue-print set out by sentimental and marriage guidance counsellors and the system that they represent. **1924** W. J. LOCKE *Coming of Amos* xvi. 204 Of the marriage in the bleak marriage-hall of the Mairie, she remembered little. **1936** H. M. & A. STONE (*title*) A marriage manual. **1965** MASTERS & JOHNSON in J. Money *Sex Research* iv. 109 If the suggestions of the marriage manuals are followed, the male develops the concept that he is to find the clitoris and stay with it. **1969** 'J' *Sensuous Woman* (1970) xix. 121 Don't let any marriage manual talk you into missing that moment when he explodes in you. **1973** I. SINGER *Goals of Human Sexuality* vi. 149 The fifteen popular marriage manuals..all treat sexual response as if it were work. **1850** *Punch* 3 Aug. 54/2 If..the Duke of Cambridge were permitted to take his coronet into the home marriage-market. **1942** E. BOWEN *Seven Winters* 46 The ulterior designs of the marriage market. **1971** R. RUSSELL tr. *Ahmad's Shore & Wave* iv. 45 The oldest, it was true, was married to a prosperous consultant engineer, but the other two were still in the marriage market. **1823** BYRON *Don Juan* XII. xlvi. 7 The Smithfield Show Of vestals brought into the marriage mart. **1972** *Village Voice* (N.Y.) 1 June 70/3 For all our talk about being liberated, most of my friends are too shy, or perhaps just too snobby, to show up at marriage marts. **1711** Marriage of convenience [see CONVENIENCE *sb.* 6]. **1949** G. B. SHAW *Buoyant Billions* IV. 53 The proportion of happy love marriages to happy marriages of convenience has never been counted. **1972** C. DRUMMOND *Death at Bar* ii. 64 Reg and June are a bit peculiar in that it was simply a marriage of convenience... They went their own way. **1975** *Listener* 16 Jan. 84/3 The co-ordinators had..paired each of the piano sonatas with each of the string quartets. Was this just a marriage of convenience..for numerical rather than musical reasons? **1924** W. H. RIVERS *Social Organisation* iii. 46 Marriage by exchange..may co-exist with marriage by purchase, and ..in some cases, it is only a means of avoiding..the marriage payments. **1951** *N. & Q. Anthropol.* (ed. 6) II. iv. 111 In either case the settlement of a marriage-payment.. may be the custom. **1957** V. W. TURNER *Schism & Continuity in Afr. Society* ix. 265 He..affected to take umbrage because the marriage-payment..he had made for her in cloth was not returned. **1963** W. J. GOODE *World Revolution & Family Patterns* iv. 167 The marriage payment or groom service points to the great concern with the legal possession of children. **1859** Marriage rate [see *birth-rate* (*BIRTH *sb.*[1] 13)]. **1891** G. B. LONGSTAFF *Stud. in Statistics* iii. 14 Speaking generally, the birth-rate corresponds to the marriage-rate, but with marked exceptions. **1970** W. D. BORRIE *Growth & Control World Population* iv. 62 There was a suggestion of a considerable degree of rationality in the reactions of these populations to the food situation, with marriage rates rising in good times and falling in bad times. *a* **1631** DONNE *Jeat Ring Sent* in *Poems* (1633) 292 Marriage rings are not of this stuffe. **1878** O. WILDE *Ravenna* 5 A moon of fire Round-girdled with a purple marriage-ring. **1913** J. MASEFIELD *Daffodil Fields* 82 She dropped her marriage-ring upon the table. **1865** G. M. HOPKINS *Note-bks. & Papers* (1937) 54 The Frenchman said the marriage-tie was in every case a bad thing. **1951** R. FIRTH *Elem. Social Organiz.* ii. 59 The marriage-tie is made the basis of service as a cook. **1935** B. MALINOWSKI *Coral Gardens* II. VI. v. 234 Whether the marriage vows are treated as a sacrament or as a mere legal contract. **1965** A. CHRISTIE *At Bertram's Hotel* vii. 79 The walls of Jericho seemed to be a symbolical way of referring to a certain lady's marriage vows... They tumbled down.

marriageability. (Examples.)

1895 in *Funk's Stand. Dict.* **1910** *Encycl. Relig. & Ethics* III. 345/2 The karōḥ (lit. 'seclusion'), or declaration of marriageability of girls, is celebrated with great solemnity. **1944** G. B. SHAW *Everybody's Pol. What's What?* viii. 61 Equality of income and complete touchability and marriageability exist at present at different levels in the trades, the professions, the ranks, and the classes. **1974** *Sci. Amer.* Sept. 113/1 According to Oriental astrology, girls born that year may murder their husbands, which tends to reduce their marriageability.

marriage-bed. (Later examples.)

1913 D. H. LAWRENCE *Love Poems* 5 The low-built shed Where hangs the swallow's marriage bed. **1922** JOYCE *Ulysses* 458 He felt it his mission in life to urge me, to defile the marriage bed, to commit adultery at the earliest possible opportunity.

married, *ppl. a.* Add: **1.** *married couple*, a husband and wife; often in contexts where they are acting jointly as domestic servants.

a **1817** JANE AUSTEN *Persuasion* (1818) III. viii. 147 Admiral and Mrs. Croft..among the married couples. **1871** *Monthly Packet* Oct. 388 The grandmother had at that time as lodgers a married couple and a single man. **1890** W. BOOTH *In Darkest Eng.* II. vi. 209 We shall be confronted with married couples who..demand that we should provide for them lodgings. **1962** *Times* 22 Oct. 1/3 Mr. and Mrs. G. K. require a married couple—chauffeur house-parlourman and cook-housekeeper. **1965** J. S. GUNN *Terminol. Shearing Indust.* II. 35 s.v. *Twins*. Two shearers who travel and work together, neither being prepared to 'take a pen' unless his mate gets one too. These inseparable mates are sometimes referred to as a 'married couple'. **1971** C. WHITMAN *Death Suspended* iv. 80 The staff..consisted of..a married couple employed as manager and housekeeper. **1972** *Listener* 21 Dec. 861/2 Up-and-coming young married couples who..are the people on whom the future of society depends. **1973** *Times* 7 Apr. 26/3 (Advt.), Young married couple needed. See domestic situations.

b. *fig.* (Later example.)

1973 *Times* 23 May 19/1 The..happily married Paternoster pedestrian precinct and the extended churchyard to the north-east.

3. Special collocation: *married print* Cinemat., a positive film carrying both pictures and a sound track.

1953 K. REISZ *Technique Film Editing* 276 From this cut negative and the negative of the re-recorded composite sound-track, a married print is prepared which is ready for projection to cinema audiences. **1959** F. CHAGRIN in Halas & Manvell *Technique Film Animation* xix. 238 Sound and vision have equal importance; they are equal partners, aiming at an ideal married-print status and preparing for it scrupulously by synchronizing their moods and their movements. **1965** P. WAYRE *Wind in Reeds* ix. 135 From this original the first 'married' print was produced with an optical sound track down one side of the film.

B. *sb.* A married person; also *collect.* Freq. in phr. *young marrieds*.

1890 A. JAMES *Diary* 30 Nov. (1964) 159 The married, however, thro' their ignoble state are doomed to shatter all the ideals of the soaring spinster. **1934** J. O'HARA *Appointment in Samarra* i. 8 The kids home from boarding-school and college, and the younger marrieds, most of whom she knew by their first names, and then the older crowd. **1938** *Time* 5 Dec. 8/3 No representative letter can be chosen from such heterogeneous applications—from men and women, from youngsters, young marrieds, middleaged, and oldsters. **1939** [see *CUT-IN *sb.* 1]. **1958** *Times* 22 Nov. 7/7 These teeming 'young marrieds' are very different from those less numerous in the Depression, a generation ago. **1959** *House & Garden* June 8 Smart young marrieds go 'natural' with Irish Linen. **1963** A. HERON *Towards Quaker View of Sex* i. 11 Among the married, faithfulness may be achieved by 'working to rule'. *Ibid.* 57 Free heterosexual relations between the young unmarried or between the married. **1965** *New Statesman* 30 Apr. 672/2 Wealthy parents, who are subsidised by the bachelors and the childless married. **1970** K. GILES *Murder Pluperfect* ii. 14 Victorian marrieds, of a certain class, used to sleep around. **1972** J. CASSON *Lewis & Sybil* ix. 192 Our life together as young naval marrieds was to begin in Alexandria.

Marriotte, erron. var. *MARIOTTE.

Marrism (mā·riz'm). [f. the name of N. Ya. *Marr* (1865–1934), Russian linguist and archæologist + -ISM.] The linguistic theories advocated by Nikolai Yakovlevich Marr, in which language is regarded as a phenomenon of social class rather than of nationality; the advocacy of such theories. Hence **Ma·rrist** *a.*

1950 *Archivum Linguisticum* II. 116 The quintessence of Marrism, he [*sc.* B. Serebrennikov] tells us, resides in four fundamental theses, viz. the four elements as the source of the world's vocabularies, linguistic growth by hybridisation, stadial evolution, and the semantic transformation of primitive totem-names. *Ibid.* 118 It is he [*sc.* V. Vinogradov] who summarises the three existing attitudes to Marr among Soviet scholars, viz. (1) that Marrism is Marxism in linguistics, [etc.]. *Ibid.* 8 Her [*sc.* R. Sor's] articles are perhaps the first reasoned statement of the Marxist standpoint that is not at the same time Marrist. **1960** [see *HYBRIDIZATION 2]. **1963** V. KIPARSKY in *Current Trends in Ling.* I. 94 From 1925 to World War II, when the only officially accepted linguistic school in Soviet Russia was Nikolaj Marr's 'Japhetology', later simply called 'Marrism', there was no interest..in Slavic languages. **1965** H. KUČERA in *Language* XLI. 118 The rehabilitation of comparativist techniques—another by-product of the 1950 renunciation of Marrism. *Ibid.* 125 The reader will find many brief references to the Marrist controversy of 1950. **1966** B. COLLINDER in Birnbaum & Puhvel *Anc. Indo-European Dial.* 199 Marrism, which was officially encouraged in Russia for political reasons, has raged as a kind of Asiatic flu in some European universities west of the Iron Curtain.

marrite (mā·rait). *Min.* [f. the name of John Edward *Marr* (1857–1933), British geologist + -ITE[1].] A sulphide and arsenide of lead and silver, $PbAgAsS_3$, which is found as grey monoclinic crystals in dolomite at Lengenbach, Switzerland.

1904 [see *LENGENBACHITE]. **1905** R. H. SOLLY in *Mineral. Mag.* XIV. 76 The colour of marrite is lead-grey to steel-grey, but the crystals are usually tarnished with iridescent colours... The name marrite is proposed in honour of Dr. John Edward Marr, F.R.S., of Cambridge. **1967** *Zeitschr. für Kristallogr.* CXXV. 459 Marrite, a rare sulfosalt from the Binnatal, Switzerland, has been shown to have composition $PbAgAsS_3$. The phase is monoclinic, space group $P2_1/a$.

‖ **marron glacé** (maroṅ glase). [F. *marron* chestnut + GLACÉ *a.*] A sweetmeat consisting of a chestnut preserved in syrup.

1871 *Leisure Hour* 23 Sept. 600/1 'Marrons glacées' appeared at the confectioners'. **1875** *All Year Round* 2 Jan. 279/2 Old women are..preparing chestnuts for their ..destinies as 'marrons glacés', &c. **1880** G. A. SALA *Amer. Revisited* (1882) I. 57 Did my stock-in-trade comprise a few *marrons glacés*, so much the better for my youthful patroness. **1904** A. BENNETT *Great Man* v. 45 'A pound of marrons glacés.'..'What are they?'..'Taste.'.. 'It's like chestnuts,' Harry mumbled through the delicious brown frosted morsel. **1906** KIPLING *Actions & Reactions* (1909) 210 Oranges, bits of banana, and *marron[s] glacés*. **1925** C. SIDGWICK *Humming-Bird* xviii. 188 Behind the loaded counter the proprietor and his assistants were weighing chocolates and marrons glacés. **1967** 'L. BLACK' *Two Ladies in Verona* ii. 18 A low table garnished with.. an enormous box of marrons glacés. **1973** *Times* 1 Dec. 14/1 Marrons glacés. 14 oz. tins. 2 for £5.20... Marrons in syrup. Large whole chestnuts in extra heavy syrup.. 18 oz. tins. 3 for £5.25.

marrow, *sb.*[1] Add: **5.** *marrow-freezing* adj. (examples); **marrow kale** = *marrow-stem* (*kale*); **marrow oil,** a dressing for the hair (cf. *MARROWFAT 2); **marrow scoop** = *marrow-spoon*; **marrow-stem (kale)** = *CHOU MOELLIER (see quot. 1925); also *marrow-stemmed kale*.

1847 C. BRONTË *Jane Eyre* I. xv. 295 A marrow-freezing incident enough. This was a demoniac laugh. **1900** *Sketch* 21 Feb. 191/2 A play..of a rather marrow-freezing kind. **1872** R. HOGG in *Jrnl. Roy. Hort. Soc.* III. 174 Marrow Kale..is the *Chou moellier* of the French, a form of the Jersey kale which produces a long, thickly swollen stem like a gigantic cigar, the swollen part being filled with a mass of tender pith. **1855** F. DUBERLY *Let.* 29 Jan. in E. E. P. Tisdall *Mrs. Duberly's Campaigns* (1963) iv. 124, I want you to write to Savory & Moore, Bond St. for 2 Large Bottles Marrow Oil, 4 Bottles of Bandoline. **1864** TROLLOPE *Can You forgive Her?* I. xl. 310 All his apparatus for dressing,—his marrow oil for his hair, [etc.]. **1969** E. H. PINTO *Treen* 79 The marrow scoop and toothpick are missing. **1970** *Canad. Antiques Collector* Mar. 23/1 There were numerous objects..such as soup ladles..marrow scoops. **1972** *Collector's Guide* Aug. 10/2 (Advt.), Fine Marrow Scoop (crested), by Wm. Chawner, 1763, £40. **1920** *Conquest* Apr. 256/3 In place of grass it is possible to grow crops such as cabbage, vetches, rape, marrowstem kale, [etc.]. **1925** MALDEN & NISBET in W. G. R. Paterson *Farm Crops* II. 191 The Marrow-stem Kale is what is known as a 'variety-hybrid'. That is, it is the result of crossing two distinct varieties—the kohl-rabi and the Thousandhead Kale, each a variety of *Brassica oleracea*. *Ibid.* 194 Mildews and moulds do less damage to the marrow stem than to turnips. **1960** *Farmer & Stockbreeder* 8 Mar. 131/2 Marrow-stem kale goes fibrous and woody very quickly after Christmas, and loses most of its feeding value. **1961** I. MOLNAR *Man. Austral. Agric.* 214 The Marrow Stemmed Kale (chou mollier) is a kale with a swollen axis. **1972** D. H. ROBINSON *Fream's Elem. Agric.* (ed. 15) x. 281 Marrowstem Kale is capable of producing very large quantities of greenstuff greatly relished by stock.

marrow, *sb.*[2] **1.** (Later examples.)

1935 A. J. CRONIN *Stars look Down* I. ii. 20 His dad had gone with the marrows in his set. *Ibid.* xxii. 205 Jack Reedy..and his marrow, Cha Leeming, worked their shift in the Scupper Flats.

marrowfat. Add: **2.** *N. Amer.* A tallow-like substance prepared by boiling down marrow.

1717 J. KNIGHT *Jrnl.* 16 Aug. (1932) 164 These Indians gave me..2 pretty bigg bladders of Marrow fatt. **1791** in *Publ. Champlain Soc.* (1934) XXI. 528 They broke the Bones of the Buffalo & made marrowfatt. **1841** G. CATLIN *Lett. on N. Amer. Indians* I. 116 'Marrow-fat' is collected by the Indians from the buffalo bones which they break to pieces, yielding a prodigious quantity of marrow, which is boiled out and put into buffalo bladders. **1846** R. B. SAGE *Scenes Rocky Mts.* viii. 79 Marrow-fat, an article in many respects superior to butter. **1888** *Century Mag.* XXXVI. 898/1 Then he slicked his hair with marrow-fat from a horn. **1947** *Beaver* Dec. 21 Clean Marrow fat rendered down like white butter.

†**3.** *U.S. slang.* (See quot.) *Obs.*

1903 A. B. HART *Actual Govt. Amer. Conditions* 75 The 'marrow-fat' fraud consists in a voter's putting in more than one ballot, while the clerk puts down fictitious names to cover the extra ballots.

marrowsky (mārau·ski). Also **marouski, Marowsky, morowski, mowrowsky.** [Asserted to have been derived from the name of a Polish count, doubtfully identified with Count Joseph Boruwlaski. See *N. & Q.* 13th Ser. I. 331, 437, 467.] **a.** A variety of slang, or a slip in speaking, characterized by transposition of initial letters, syllables, or parts of two words. Also *marrowsky language*.

1863 R. NICHOLSON *Autobiogr. Fast Man* xviii. 200 Fanny King, or as Bill Leach, in the interesting language called *Marouski*, termed her, Kanny Fing. **1883** G. A. SALA *Living London* 491 The vocabulary of Tim Bobbin, Josh Billings,..and the 'Marowsky' language.

b. An instance of this.

1923 in *N. & Q.* 27 Oct. 331/2 In my childhood..an old cousin used to entertain me with what we now call *spoonerisms*, but which she termed *morowskis*. **1962** V. NABOKOV *Pale Fire* 185, I remember one perfect evening when my friend sparkled with quips, and marrowskies, and anecdotes.

Hence **marrow·skyer,** one who uses marrowsky language or makes marrowskies in his speech; **marrow·skying** *vbl. sb.*, the intentional or accidental transposition of initial letters, etc.

1860 HOTTEN *Dict. Slang* (ed. 2) 173 s.v. *Medical Greek*. At the London University they have a way of disguising English..which consists in transposing the initials of words... This disagreeable nonsense is often termed *marrowskying*. **1912** *Brit. Med. Jrnl.* 22 June 1443 It would be interesting if 'marrowskyers'' blunders could also be classified. *Ibid.*, All actors live in dread of 'marrowskying', that curious transposition of syllables. **1922** O. JESPERSEN *Lang.* viii. 150 'Marrowskying' or 'Hospital Greek' transfers the initial letters of words, as *renty of plain*.

Marrucinian (mærūsi·niăn), *sb.* and *a.* Also **Mar(r)ucine, Marucian, Marusian.** [f. L. *Marrucini* + -AN.] **A.** *sb.* **a.** A member of an Oscan-Umbrian people living near Teate in ancient Eastern Italy. **b.** The language of this people. **B.** *adj.* Of or pertaining to this people or their language.

1578 tr. *Appian's Auncient Historie of Romanes Warres* I. 25 The Marsians, the Malinians, the Vestinians, the Marucians,..and the Samnites, whiche people before had euer bin enimies, and hurtfull to the Romaines. *Ibid.* 31 A little Marusian aunswered hym, and kylled hym. **1600** P. HOLLAND tr. *Livy's Rom. Hist.* VIII. 302 Every one of them, namely, the Marsians, the Pelignians and Marucines, were in feats of armes comparable and egall to the Samnites every way. *Ibid.* XXVII. 660 Hee dispatched likewise messengers before, through the territories of the Larinates, Marrucines, Ferentines, and Pretutians. **1601** —— tr. *Pliny's Nat. Hist.* II. lxxxiii. 39 Medowes and olive rowes.. in the Marrucine territorie. **1863** W. P. DICKSON tr. *Mommsen's Hist. Rome* III. IV. vii. 235 The small but hardy confederacies in the Abruzzi—the Pælignians, Marrucinians, Frentanians, and Vestinians. *Ibid.* 237 The territory of the Pælignians, situated in the centre of the Marsian, Samnite, Marrucinian, and Vestinian cantons. **1888** J. WRIGHT tr. *Brugmann's Elem. Compar. Gram. Indo-Germanic Lang.* I. 9 Of the Volscian, Picentine, Sabine, Aequiculian, Vestinian, Marsian, Pelignian and Marrucinian dialects we have only very scanty remains. **1933, 1939** [see *MARSIAN *sb.* and *a.*]. **1966** M. S. BEELER in Birnbaum & Puhvel *Anc. Indo-European Dial.* 51 The so-called minor dialects such as Paelignian, Volscian, and Marrucinian. **1974** R. A. HALL *External Hist. Romance Lang.* 48 The Sabellic dialects..: Sabine..; Paelignian, Marrucinian, Vestine, and Marsic in the central Apennines east of Rome; and Volscian.

marry, *v.* Add: **1. c.** *marry up* (earlier example). Also *fig.* (const. *with*).

1698 J. COLLIER *Short View Immorality Eng. Stage* iv. 154 This Spark..makes a lucky Hand on't at last, and marries up a rich Lady. **1942** *Tee Emm* (Air Ministry) II. 97 They already have had the idea of taking photos of the countryside and marrying them up with similar areas on a map. **1958** *Times* 15 Feb. 7/3 Now that poverty-stricken Jordan is married up with a rich relation. **1970** BAKEWELL & GARNHAM *New Priesthood* iv. 44 If a writer..is not ripe for TV drama, then we would certainly sit down with him, pull his script to bits, send it away, come back, try again, try it on an editor, marry them [*sic*] up with an editor if possible.

2. b. With *off*.

1860 QUEEN VICTORIA *Let.* 18 Dec. in R. Fulford *Dearest Child* (1964) 292, I do wish somebody would go and marry her off—at once. **1865** M. C. HARRIS *St. Philip's* viii. 59 If the young girls did not mind being.. finally married off to some of her *protégés*, it was all very well. **1894** V. HUNT *Maiden's Progress* i. 2, I sincerely hope you will be married off before *I* come on, or I shall have no peace. **1908** *Smart Set* June 14/1 Mr. Hardcastle was insisting upon marrying off Aunt Ella to Señor Dominguez y Aguirra. **1973** 'E. FERRARS' *Foot in Grave* viii. 150 I've worked quite hard at trying to get you married off.

3. b. To obtain (something) by getting married. Esp. on phr. *to marry money*: to marry a rich spouse.

1858 TROLLOPE *Doctor Thorne* I. xiv. 297 He must marry money, or he will be a ruined man. **1911** L. J. VANCE (*title*) Marrying money. **1914** JOYCE *Dubliners* 98 There'll be no mooning and spooning about it. I mean to marry money. **1934** *Harrap's Stand. French & Eng. Dict.* I. 329/1 *Épouser*..*une grosse dot*, to marry money. **1936** HERZOG & BLOOAH *Jabo Proverbs from Liberia* 180 If you marry a beautiful woman, you marry trouble.

5. a. (Later examples.)

1908 *Daily Chron.* 19 Aug. 4/7 By 'marrying' or blending the wine from different vineyards he discovered that there resulted an effervescing wine. **1919** KIPLING *Years Between* 63 With a thin third finger marrying Drop to wine-drop domed on the table. **1925** BELLOC *Mr. Petre* vi. 150 A genius who could marry the commonest tricks to unheard rapidity and daring. **1942** L. BENNETT *Jamaica Dial. Verses* 44 Pumpkin? yuh wi Haffe buy cho-cho wid i, Dem married, same like 'ow cassada Married to yampi. **1952** M. KERR *Personality & Conflict in Jamaica* 59 Soap and rice are popular gifts, as the shopkeepers in the country tend to 'marry them'. **1958** *Times Lit. Suppl.* 3 Oct. 567/4 It is always something of a problem with a heavily illustrated book to know how best to marry a long text with the illustrations relating to it. **1960** O. SKILBECK *ABC of Film & TV* 82 *Marry*, to combine related mute and sound negatives in printing, or different picture negatives in optical work. **1963** *Times* 24 Apr. 12/7 Rolls-Royce believe that their success with a combination of silica and aluminium opens up the possibility of 'marrying' different types of ceramics and metals. **1967** CASSIDY & LE PAGE *Dict. Jamaican Eng.* 294/1 *Marry*, to couple something in short supply (and much in demand) with something not much wanted, so that one must purchase the latter in order to have the former. **1969** *Daily Tel.* 11 Nov. 6/7 Scientists are 'marrying' strains, such as the

Hongkong virus, to other influenza viruses that are known to reproduce and spread faster.

d. *Stockbrokers' slang.* To set (one transaction against another).

1931 *Economist* 10 Oct. 675/1 The shareholders..can do nothing, for the Trust only agrees to 'marry' buying orders for clients against selling orders. **1959** *Ibid.* 21 Mar. 1099/1 The brokers in these shares then find it convenient to 'marry' the buying and selling orders.

6. a. Also *to marry into*: to enter (a family, etc.) by marrying; to obtain by getting married. (See quot. 1662 in Dict.) Also *to marry well*: to have a successful marriage (in terms of harmony, material gain, or social standing).

1604 [see WELL *adv.* 6 c]. **1871** GEO. ELIOT *Middlem.* (1872) I. I. iv. 61, I wish you to marry well; and I have good reason to believe that Chettam wishes to marry you. **1876** C. M. YONGE *Womankind* xxxi. 281 Mothers.. cherish wishes that their girls may 'marry well'; *i.e.* richly. **1899** O. WILDE *Importance of Being Earnest* I. 39 You can hardly imagine that I and Lord Bracknell would dream of allowing our only daughter..to marry into a cloak-room, and form an alliance with a parcel. **1904** H. O. STURGIS *Belchamber* i. 10 They both married, and married what is called 'well', while many of their fairer..sisters were left ungathered on the stem. **1905** A. BURVENICH *Eng. Idioms* 224 *Marry*... To—into a fortune; *épouser une fortune, een rijk huwelijk doen*. **1923** R. MACAULAY *Told by Idiot* I. xiii. 50 What chances does a girl want, except to marry well? **1931** *Daily Express* 21 Sept. 19/1 He married into money, so that I find it difficult to credit that his contemplated return is for the purpose of restoring his balance at the bank. **1933** R. H. LOWIE in *Encycl. Social Sci.* X. 146/2 No Masai of good standing marries into a blacksmith's family. **1946** [see *BROUHAHA]. **1963** *Listener* 14 Feb. 274/1 A socialist politician who has married into a farm. **1974** 'J. MELVILLE' *Nun's Castle* i. 19 Among the other luxuries life provided for Lady Dorothy were kindred on the eastern seaboard of the United States... Probably she had someone married 'well' into the Kremlin if the truth were told.

c. *fig.* (Later examples.) Also, (of wine, etc.) to mature. Now esp. *to marry up with*, to link with.

1960 T. MCLEAN *Kings of Rugby* xi. 170 Risman began to bear to the left to marry up with the outside backs. **1963** *Harper's Bazaar* Aug. 26 Blended..the whisky must now 'marry' another year before bottling. **1969** *New Yorker* 27 Sept. 120/2 The juices are the foundation for a generous gravy, in which blanched vegetables like carrots, onions..are heated and then allowed to marry for an hour or more. **1969** [see *GAZPACHO]. **1971** C. BONINGTON *Annapurna South Face* App. C. 257 Even though there were indications that the British mask used on Kanchenjunga and the American mask used on Everest were superior in design, I preferred to keep things as simple as possible, eliminating any risk of parts failing to marry up.

d. *to marry above, below* or *beneath*, (oneself): to marry a person of higher or lower social position.

1721 J. KELLY *Compl. Coll. Scottish Proverbs* 252 Marry above your Match, and you get a Master. **1741** RICHARDSON *Pamela* (1742) III. 4 For who will offer to reproach me for marrying, as the World thinks, below me. *Ibid.* IV. 246 A Man, who..marry'd beneath him. **1860** A. J. MUNBY *Diary* 21 Feb. in D. Hudson *Munby* (1972) 51 Saw Mrs Lock (herself considerably reduced in the world, being in fact a *lady* who married beneath her). **1876** TROLLOPE *Prime Minister* III. xx. 336 He had married much above himself in every way. **1895** A. W. PINERO *Second Mrs. Tanqueray* IV. 160 All my family have chucked me over... Jus' because I've married beneath me.

e. *to marry out*: to marry a person of a different clan, group, religion, etc.

1842 *Southern Lit. Messenger* VIII. 331/2 'Marrying out', as the Friends call one of a different faith, is regarded by them with especial horror. **1923** C. ROTH *Hist. Marranos* xii. 316 His son, Jacob Israel Bernal, married out of the faith and left the community. **1954** E. E. EVANS-PRITCHARD *Inst. Primitive Society* vi. 68 The rule of exogamy—the rule of marrying-out—which required that a man must not marry inside a defined set of his own kin. **1957** V. W. TURNER *Schism & Continuity in Afr. Society* ix. 287 Most women marry out at least once. **1959** F. M. WILSON *They came as Strangers* III. iv. 212 When his daughter married a Gentile..he never reproached her for marrying 'out'. **1964** E. HUXLEY *Back Street New Worlds* ii. 20 Jewish women stick much closer to their own kind, with only between one-tenth and one-fifth marrying 'out'. **1973** *Jewish Chron.* 19 Jan. 23/4 The smugness..may be one of the reasons for the apathy among Orthodox Jews which leads to marrying out.

marrying, *vbl. sb.* Add: **1.** *transf.* and *fig.* Also **marrying-in, -out.**

1917 B. GRIMSHAW *Nobody's Island* v. 61 Such greens, such blues, such marryings of both, not brush and colour, nor pen and ink, could paint. **1940** *Chambers's Techn. Dict.* 528/1 *Marrying*, the process of lashing poles together in scaffold erection. *Ibid.*, *Marrying* (*Cinema*), the printing of the mute negative and the negative sound-track on the release print. **1952** *Economist* 8 Nov. 401/1 The marrying of reactors with power generators. **1964** GOULD & KOLB *Dict. Social Sci.* 239/2 Endogamy and exogamy are processes of marrying-in and marrying-out. **1973** *Times Lit. Suppl.* 6 July 775/2 A drama of 'marrying out' (a painful process which one does not have to be Jewish to experience).

Mars. Add: **2*.** The proprietary name of a chocolate-covered bar with a toffee-like filling. Usu. in form *Mars bar*.

1932 *Trade Marks Jrnl.* 5 Oct. 1273 *Mars*,..a sweetmeat. Mars Confections Limited. **1943** *Penguin New Writing* XVI. (verso front cover, Advt.), Mars are made from the finest available materials—including chocolate.., glucose.., separated milk. **1948** C. DAY LEWIS *Otterbury Incident* ix. 112 Toppy gave the errand boy half a Mars bar he had in his pocket. **1963** L. DEIGHTON *Horse under Water* xi. 45 He did that twenty years ago when you were wearing a Micky Mouse gas-mask and saving your coupons for a Mars bar. **1973** J. WHITE *Norfolk Child* 176 But there were peppermints,..and, of course, Mars. We..cut them into slices to make them last longer; first we bit the toffee off the top of the slice, then we nibbled the chocolate round the edges, and last of all..the delicious honey-sweet centre. **1973** C. BONINGTON *Next Horizon* iii. 57 'I've got a can of sardines and two Mars Bars,' I said. **1973** *Times* 20 Sept. 18/8 Getting an overweight patient to regulate his eating habit (using a knife and fork to eat a Mars Bar).

Marsa, Mars(e, varr. MAS, MASSA. Also with small initial.

Freq. in Black English contexts.

1874 'H. CHURTON' *Toinette* xxviii. 301 He gave Toinette to young Marse Geoffrey. **1880** A. W. TOURGÉE *Bricks without Straw* 408 The old time 'Marse' was now almost universally used, and few 'niggers' presumed to speak to a white man..without removing their hats. **1884** 'MARK TWAIN' *Huck. Finn* xxxviii. 391 Why, Mars Tom, I doan' *want* no rats. **1890** *Century Mag.* Nov. 65 Suit you, marsa? **1901** B. T. WASHINGTON *Up from Slavery* 12, I recall the feeling of sorrow among the slaves when they heard of the death of 'Mars' Billy'. **1909** 'O. HENRY' *Roads of Destiny* ii. 39 Marse Robert robbing the bank! *Ibid.* xxi. 350 I'm not going back on Mars' Jeff. **1949** [see *LINE *sb.*[2] 17 b]. **1950** R. AMES in A. Dundes *Mother Wit* (1973) 488/1 The old time darky's..love for 'ole marse' and 'ole mist'ess'. **1955** E. POUND *Classic Anthol.* I. 8 Thaar's where ole Marse Shao used to sit.

Marseillais, *a.* (*sb.*). Add: **1.** *adj.* (Later examples.)

1896 C. M. YONGE *Release* II. xxi. 294 Rumours reached us of the arrival of the Marseillais tigers. **1966** M. R. D. FOOT *SOE in France* xi. 377 He..sank himself without effort into the marseillais underworld.

2. *absol.* and *sb. pl.* **a.** (Later examples.) Also in *sing.*, an inhabitant of Marseilles.

1939 E. AMBLER *Mask of Dimitrios* i. 12 The other guests were a very noisy pair of Marseillais. **1951** E. PAUL *Springtime in Paris* ix. 159 Auvergnats, Catalans, Basques, Normans, Bretons, Marseillais. **1966** M. R. D. FOOT *SOE in France* ix. 248 The marseillais Basset and his Burgundian mate, Jarrot.

Marseilles. Add: **3.** Applied *attrib.* to a type of pottery produced in Marseilles during the seventeenth and eighteenth centuries.

1870 C. SCHREIBER *Jrnl.* (1911) I. 101 Bought..cover of Marseilles ware,..two Marseilles plates with Chinese figures. **1960** R. G. HAGGAR *Conc. Encycl. Cont. Pott. & Porc.* 302/2 The subsequent history of Marseilles faience is one of great trade expansion and the multiplication of factories... The death blow was given to the industry by the increased demand for earthenware of English type, and the French Revolution. **1974** *Times* 29 Oct. 19/4 A fine Marseilles *Veuve Perrin* soup tureen.

marsh. Add: **1. b.** *local.* A meadow; a stretch of grassland near a river or the sea.

1787 W. MARSHALL *Rural Econ. Norfolk* I. 320 The upper sides [of the fens] being frequently out of the water's way, afford a proportion of grazable land: hence, probably, they are provincially termed marshes. **1852** L. A. MEREDITH *My Home in Tasmania* I. 163 (Morris), A marsh here is what would in England be called a meadow, with this difference, that in our marshes, until partially drained, a growth of tea-trees..and rushes in some measure encumbers them; but, after a short time, these die off..and a thick sward of verdant grass covers the whole extent. **1892** J. E. TAYLOR *Tourist's Guide Suffolk* (ed. 2) 23 Some of the larger river-valleys, such as the Stour, Orwell, Deben, Alde, Blyth, and Waveney, have these soils laid down in grass, forming rich and well-known grazing land, or 'marshes' as they are locally termed.

4. a. marsh hay, hay made from marsh grasses; **marsh spot,** a deficiency disease of garden peas, caused by a lack of manganese.

1742 W. ELLIS *Mod. Husbandman* (1750) IV. 101 Curing coarse Marsh-Hay. **1839** *Cultivator* VII. 33 The common marsh hay is no better than the 'bog meadow hay' of the east. **1852** *Trans. Mich. Agric. Soc.* III. 132 They feed well at the straw stack and thrive on marsh hay. **1895** A. PATTERSON *Man & Nature on Broads* 50 A sack of sweet 'mesh' hay an' a blanket or tew to tuck yerself in. **1924** *Beaver* Sept. 460/2 There is every indication of a poor [hay] crop... The Indians who depend upon the marsh hay have the same cry. **1961** J. W. ANDERSON *Fur Trader's Story* iii. 24 Marsh hay was used at these posts to feed the cattle. **1931** J. BRYCE in *Essex Farmers' Jrnl.* X. 71/1 Marsh-Spot disease of peas..is carried in the seed. *Ibid.*, Marsh-Spot is usually brought on to a farm in the seed peas. **1934** *Jrnl. Ministry of Agric.* Dec. 833 From time to time complaints are raised concerning a defect in pea seeds to which the name Marsh Spot has been given. The term appears to have originated from the fact that the trouble is most common in seeds from crops grown in low-lying marsh land... The defect..shows itself mainly in the form of a dark, often slightly sunken spot..on the face of each seed-leaf. **1959** *New Biol.* XXX. 91 Diseases such as..'marsh spot' of peas are caused by the low availability of manganese in the soil.

b. marsh fritillary, a tawny butterfly, *Euphydryas aurinia*, formerly called the *greasy fritillary* (GREASY *a.* 9); **marsh quail** (earlier and later examples); **marsh tacky** *U.S.*, a small pony bred in marshy districts; **marsh treader** *U.S.*, a water bug of the family Hydrometridæ.

1890 C. W. DALE *Hist. Brit. Butterflies* 202 The Marsh Fritillary varies more generally than any other British butterfly. **1958** *Listener* 24 July 125/2 A series of observations on the Marsh Fritillary butterfly. **1750** J. BIRKET *Some Remarks Voy. N. Amer.* (1916) 32 Killd some Squirrels and some very pretty birds called Marsh quails something bigger than a field fare. **1935** J. C. LINCOLN *Cape Cod Yesterdays* 137 The Cape Cod market gunner of yesterday also shot upland birds for the market—quail, partridge, snipe,..'marsh quail' and an occasional pheasant. **1838** C. GILMAN *Recoll. Southern Matron* xix. 131 An accident happening to my horse, I was obliged to hire one of the little animals called 'marsh tackies' to carry me over a creek. **1937** D. C. HEYWARD *Seed from Madagascar* 118 He could..gallop his 'marsh tackey' through thickets so dense that a rabbit could scarcely get through them. **1895** J. H. & A. COMSTOCK *Man. Study of Insects* xiv. 124 The Marsh-treaders, family Limnobatidæ. **1902** L. O. HOWARD *Insect Bk.* 282 The Marsh Treaders (*Hydrometridæ*). **1972** SWAN & PAPP *Common Insects N. Amer.* 115 Marsh treader; *Hydrometra martini*.

c. marsh grass, pennywort, rosemary (later examples).

1868 *Putnam's Mag.* May 592/1 Clumps..begin to make their appearance above the reeds and tall marsh-grass. **1936** D. MCCOWAN *Animals Canad. Rockies* xix. 169 His supper was often of coarse marsh grass. **1972** *Islander* (Victoria, B.C.) 23 Jan. 6/1 Acres of marsh grass where animals and birds feed. **1885** *Outing* VII. 179/1 All the ground about is carpeted with the light-green leaves of the marsh-pennywort. **1960** *Oxf. Bk. Wild Flowers* 46/2 Marsh Pennywort or White-rot (*Hydrocotyle vulgaris*). This small, creeping or floating plant..differs from other Umbellifers in having undivided round leaves, with the leaf stalk attached at the centre of the leaf blade. **1861** MRS. STOWE *Pearl Orr's Island* I. x. 82 'Marsh rosemary is a very excellent gargle.' Said Mr. Sewell. [**1960** *Oxf. Bk. Wild Flowers* 118/2 Bog Rosemary or Marsh Andromeda (*Andromeda polifolia*). This small evergreen shrub up to 1 foot high has leaves which are shiny above and covered with a white bloom below.]

Marsh[2]. The name of James *Marsh* (1794–1846), British chemist, used in the possessive and *attrib.* to designate a sensitive test for the presence of arsenic in a substance (published by him in 1836), in which the substance is subjected to the action of nascent hydrogen so that any arsenic in it forms the gas arsine, which may be detected by decomposing it (e.g. by passing it through a heated glass tube) and looking for a dark stain of arsenic (distinguishable from a similar stain given by antimony by its solubility in sodium hypochlorite solution).

1855 J. SCOFFERN *Elem. Chem.* 479 Arseniuretted hydrogen thus employed, as a means of removing and discovering arsenic, is called Marsh's test, from its inventor, Mr. Marsh. **1930** D. L. SAYERS *Strong Poison* xx. 265 In a small apartment usually devoted to Bunter's photographic work..stood the apparatus necessary for making a Marsh's test of arsenic. **1968** *Materials & Technol.* I. ix. 399 Very minute quantities of arsenic can be detected by means of the Marsh test.

marshal, *sb.* Add: **3. d.** Designating an officer of high rank in the Royal Air Force, as *Marshal of the Royal Air Force, Air Chief Marshal, Air Marshal, Air Vice-Marshal*; also †*Marshal of the Air* (obs.).

1919 *Times* 4 Aug. 12/6 His Majesty..has approved of new titles for the commissioned ranks of the Royal Air Force. These are..Marshal of the Air, Air Chief-Marshal, Air Marshal, Air Vice-Marshal... It will probably be some time before we have a Marshal of the Air, as at present there is no officer of the rank of either Air Chief-Marshal or Air Marshal. *Ibid.* 7 Aug. 14/6 Air Vice-Marshal Trenchard.. made the Air Force become the powerful and formidable fighting machine that it was. *Ibid.* 19 Aug. 10/4 The King has been pleased to approve the promotion of Air Vice-Marshal Sir H. Trenchard, K.C.B., to Air-Marshal. **1922** *Man. Seamanship* (Admiralty) I. 11 Marks of Rank... R.A.F. Marshal of the Air. R.N. Admiral of the Fleet. **1943** W. S. CHURCHILL *End of Beginning* 141 Most of the air-marshals, the leading men in the Air Force, think little of dive bombers. **1947** *Whitaker's Almanack* 461 (*heading*). Marshals of the Royal Air Force. **1968** *Who's Who* 1331/2 (*heading*) Harris, Marshal of the Royal Air Force Sir Arthur Travers.

marshalate. Add: **1. b.** The order of Marshal in the French army.

1945 R. HARGREAVES *Enemy at Gate* 80 Commanders.. ear-marked..for ultimate promotion to the Marshalate of France. **1973** *Times Lit. Suppl.* 16 Nov. 1390/4 A slighter but handsomely produced and welcome book on the marshalate. *Napoleon's Marshals* is partly [etc.].

2. *U.S.* The office of a marshal (cf. MARSHAL *sb.* 9 a).

1954 W. FAULKNER *Fable* 159 After some months under the threat of the full marshalate, he compromised with his father on the simple deputyship.

Marshall (mā·ɹʃăl). The name of George C. *Marshall* (1880–1959), Secretary of State in the U.S.A. from 1947 to 1949, used *attrib.* to designate a plan initiated by him in 1947 to supply financial assistance to certain Western European countries to further their recovery

219/2 The possibility of a 'Marsokhod' seems to have reached the stage of constructive planning... The lunokhods are controlled from Earth—a Mars vehicle would have to be self-controlling at least as far as such vital commands relating to its motion as 'start' and 'stop' are concerned.

martelé (mā·ɹtəle), *a.* *Mus.* [Fr., pa. pple. of *marteler* to hammer.] = *MARTELLATO *a.*
1876 in STAINER & BARRETT *Dict. Mus. Terms* 281/1. **1961** D. D. BOYDEN in A. Baines *Mus. Instruments* vi. 124 The new bow..was more powerful, and its strength and the quicker take-up of the hair permitted *sforzando* effects such as *martelé*, which had been rare previously. **1962** *Punch* 28 Nov. 804/3 The martelé string chords of the first subject.

‖ **martellato** (mɑːɹtɛlā·to), *a.* *Mus.* [It., pa. pple. of *martellare* to hammer.] Lit. 'hammered'; said of notes which are heavily accented and left before their full time has expired. Also *transf.* Hence as *adv.* and *sb.*
1876 in STAINER & BARRETT *Dict. Mus. Terms* 281/1. **1928** G. B. SHAW *Let.* 6 Feb. in *B. Shaw & Mrs. Campbell* (1952) 267 Such music as your [dramatic] pupils may have must come from within them; all you can teach them is the value of your..*martellato molto*. **1928** *Daily Express* 27 Aug. 3/2 The notes in small type preceding the long B. flat..must be played *martellato* and with clearness. **1931** G. JACOB *Orchestral Technique* v. 50 The martellato in the left-hand part of the original. **1959** *Listener* 8 Oct. 596/2 One was put instantly in mind of William Walton's *martellato* style. **1969** *Times* 7 Nov. 13/1 His *martellato* octaves sounding like sharp cracks of a gun. **1973** *Times* 28 Mar. 11/8 Verne Reynolds's Sonata, composed last year and proving to be no more than an impersonal essay in virtuoso keyboard devices, all sound and no real fury, despite an excess of *martellato* in the outer movements.

martemper (mā·ɹtempəɹ), *v.* *Metallurgy.* [f. MAR(TENSITE + TEMPER *v.*] *trans.* To treat (steel) so as to reduce its tendency to crack or distort by quenching rapidly to a temperature just above that at which martensite begins to form, allowing the temperature to equalize throughout, and then cooling slowly. So **ma·rtempered** *ppl. a.*, **ma·rtempering** *vbl. sb.*
1943 B. F. SHEPHERD in *Iron Age* 28 Jan. 50/1 Martempering is a term used to describe a heat treating operation whereby martensite is produced with a minimum of residual hardening strains. **1943** —— in *Metal Progress* July 105/2, I have coined the term 'martempering' to describe this operation of hardening by cooling quickly. **1947** *Steel Processing* XXXIII. 103/2 Steel was more susceptible to cracking when martempered than when treated by the more conventional method. **1953** AITCHISON & PUMPHREY *Engin. Steels* iii. 115 The properties of a martempered steel are in no way inferior to those of the same steel quenched and tempered in the traditional manner. *Ibid.* 116 Martempering is unlikely to supplant the direct quenching of normal, commercial steels. **1956** *Jrnl. Iron & Steel Inst.* CLXXXIII. 447/1 Warpage was least on martempered specimens. **1968** E. R. PETTY *Physical Metall. Engin. Materials* xi. 206 Quench stresses and attendant distortion or cracking may be reduced by..an interrupted quench (martempering).

Martenot, shortened form of *ONDES MARTENOT.
1931 *Electronics* July 18/1 As an example of a very fully developed [electronic musical] instrument..the 'Martenot' may be cited, named after its inventor. **1932** *19th Cent.* May 606 The Martenot instrument..looks like a small portable harmonium, from which connexions run to one or more loud-speakers. *Ibid.* 607 Anyone already able to play the piano can make a fair showing on a Martenot after a few hours' practice. **1936** *Discovery* July 222/1 The Martenot instrument... Honegger, Milhaud, Ibert are among the composers who have scored for it, and Stokowski has made use of it in his orchestra. **1938** *Oxf. Compan. Mus.* 287/2 The instrument known as Ondes Musicales..or the Ondium Martenot, or the Martenot..is purely melodic.

martensitic (mɑːɹtenzi·tik), *a.* *Metallurgy.* [f. MARTENSIT(E + -IC.] Pertaining to or containing martensite; resembling the structure or mode of formation of martensite.
1909 in *Cent. Dict.* Suppl. **1915** *Chem. Abstr.* IX. 437 A martensitic structure is produced in 0·28% C steel without increase in hardness by heating to white heat and then cooling slowly. **1937** *Discovery* May 153/2 Carbon is mainly responsible for the characteristic martensitic mode of hardening. **1950** J. OSBORNE *Dental Mech.* (ed. 3) xii. 222 Two types of stainless steel find application in dentistry, Martensitic and Austenitic steels. **1967** A. H. COTTRELL *Introd. Metall.* xvii. 265 An important example of a shear transformation occurs when steel is quench-hardened. The F.C.C. austenitic structure then transforms by shear, with some dilatation, into martensite, the distorted form of B.C.C. iron... In recognition of this, shear transformations are often referred to as martensitic transformations.
Hence **martensi·tically** *adv.*, in a martensitic manner.
1972 *Physics Bull.* Dec. 712/1 The difficult areas of x ray crystallography, such as in the study of martensitically transformed materials,..can be opened up by an electron microscope technique.

Martha (mā·ɹþă). **a.** The name of the sister of Mary and Lazarus in Bethany; hence used with allusion to Luke x. 40, 41 for one much concerned with domestic affairs. In Christian allegory a symbol of the active life, opp. *MARY 4.
a **1225** *Ancrene Wisse* (1962) f. 112a/8 Husewifschipe is marthe dale. Marie dale is stilnesse & reste of alle worldes noise. **1907** [see *MARY 4]. **1907** M. HALL *Woman's Trek from Cape to Cairo* ix. 133 My host was equally busy..while his wife, 'Martha like', busied herself in seeing a chicken cooked. **1909** *Westm. Gaz.* 30 July 2/1 The domestic servant problem..could never have become just what it is had this generation of mistresses followed in the steps of that ancestry which accorded the Marthas of the homes a certain amount of appreciation for services rendered. **1910** S. R. CROCKETT *Dew of their Youth* II. xviii. 140 You are of the tribe of Marthas, Jen, and you certainly work hard enough for everybody. **1938** E. GOUDGE *Towers in Mist* x. 277 She belonged to that noble army of Marthas who cook the dinners that the Marys gobble up to keep them going between their visions and their dreams. Grace was the best kind of Martha. **1953** G. GREENE *Living Room* I. i. 8 In this household she is obviously the anxious Martha; the weaker character intent on carrying out orders. **1962** J. D. SALINGER *Franny & Zooey* 158 For all I know, I may be a little jealous... It may very well be that I hate like hell to play Martha to someone else's Mary. **1962** *Friend* 3 Aug. 951/1 The Marthas and Marys are both essential. There is no division between the Mystical and the Practical. **1967** MRS. L. B. JOHNSON *White House Diary* 10 Aug. (1970) 553 The 'Mary' and the 'Martha' in my life have an eternal war.
b. Martha-coloured *a.*, used *transf.* (see quot. 1936) by the poet Edith Sitwell.
1924 E. SITWELL *Sleeping Beauty* xiv. 50 The Martha-coloured scabious Grew among dust as dry as old Eusebius. **1936** —— *Sel. Poems* Pref. 25 'The Martha-coloured scabious'..is the result of a personal memory... As a child, I had a nursery maid called Martha, who always wore a..gown..exactly the colour of a scabious. **1937** —— *I live under Black Sun* II. x. 296 The Martha-coloured scabious waved aimlessly.

Martha Gunn (mā·ɹþă gʊn). *Pottery.* (See quot. 1957.) Also *attrib.*
1922 R. K. PRICE *Astbury, Whieldon & R. Wood Figures & Toby Jugs* 86 Another rare Ralph Wood Toby jug in the collection is Martha Gunn. **1957** MANKOWITZ & HAGGAR *Conc. Encycl. Eng. Pott. & Porc.* 141/2 *Martha Gunn*, a female Toby jug, modelled in the likeness of Martha Gunn (1727–1815) the celebrated Brighton bathing-woman who is alleged to have dipped the Prince of Wales in the sea. **1960** R. G. HAGGAR *Conc. Encycl. Cont. Pott. & Porc.* 135/2 A large number of jugs called *Jacquelines*..are comparable to the English 'Martha Gunn' toby jugs made by Ralph Wood and others.

martial, *a.* and *sb.* Add: **A.** *adj.* **1. b.** *martial art* (usu. *pl.*), any of various fighting sports or skills mainly of Japanese origin, such as judo, karate, and kendo.
1933 *Official Guide to Japan* (Japanese Government Railways) p. clxxxvi, Contests [of kendo] take place nowadays at the annual meetings of the Butoku-kai, or Association for Preserving the Martial Arts, in Kyōto. **1955** E. J. HARRISON *Fighting Spirit of Japan* (ed. 2) x. 97 Of that branch of Japanese esoterics which belongs to what may generically be styled *bujutsu*, literally 'martial arts', though the Japanese terminology has a far wider and more comprehensive scope than its English equivalent, I may justly claim to know something. **1966** [see *KENDO]. **1968** [see *KUNG-FU]. **1974** *Isle of Wight County Press* 23 Nov. 31 Mr. Singleton, who holds a Kendo black belt, a brown belt in Karate, and has just taken up Ju-Jitsu, said he had no intention of 'cashing in' on the current martial arts boom.

Martian, *a.* and *sb.* Add: **A.** *adj.* **1. b.** Also *transf.*
1950 A. HUXLEY *Themes & Variations* 256 Man's Martian aggression against himself. *Ibid.*, The Martian forces of overpopulation and erosion. **1953** *News Chron.* 2 June 1/1 Step by step, in Martian clothing, the two figures move forward, pursuing their race against time and the mountain in the slowest of slow motion.
B. *sb.* (Earlier and later examples.)
1883 W. S. LACH-SZYRMA *Aleriel* III. iii. 109 He..brought with him another Martian, differently attired. **1901** G. B. SHAW *Three Plays for Puritans* Pref. p. xvii, He feels concerned about the..destruction of the human race by the Martians. **1949** [see *EARTHLING *sb.*[2] 1]. **1951** R. BRADBURY *Silver Locusts* 108 From the back of the machine a Martian with melted gold for eyes looked down at Tomás. **1967** MRS. L. B. JOHNSON *White House Diary* 13 May (1970) 519 What will I remember about Camp David? The two 'Martians' in their silver suits..firefighters on hand for any emergency.

martin[1]. Add: **3. martin-box** *U.S.* (earlier and later examples); **martin bug**, a blood-sucking bug, *Oeciacus hirundinis*, whose principal host is the house-martin; **martin-cage**, a cage for holding martins; **martin-house** = *martin-box*.
1828 *Farmer's Almanac 1829* (Wendell, Mass.) sig. F3v Whose house is that with white capped chimnies, black sashed windows, and a nice little marten [*sic*] box just an epitome of the State House? **1854** B. F. TAYLOR *Jan. & June* 60 A martin-box of a cottage scuds round the corner of the Meeting House. **1871** MRS. STOWE *Oldtown Fireside Stories* 108 Your questions tumbles over each other thick as martins out o' a martin-box. **1946** J. STUART *Tales Plum Grove Hills* 85, I know what Mom is thinking when she looks at the martin boxes. **1923** E. A. BUTLER *Biol. Brit. Hemiptera-Heteroptera* 322 By most authors the martin bug is considered to be generically distinct from the rest of our British Cimicina, and is referred to the genus *Oeciacus*. **1935** *Brit. Birds* XXVIII. 278 The martin bug is rarer and resembles the much-disliked bed-bug which infests human habitations. **1959** SOUTHWOOD & LESTON *Land & Water Bugs Brit. Isles* vii. 187 The martin bug is recorded from most English counties south of a line from the Wash to the Bristol Channel. **1844** *Knickerbocker* XXIII. 442 Like a superannuated old man with a martin-cage upon his crooked back. **1826** 'A. SINGLETON' *Lett.* 74 The tasteful slave makes, perchance a martin-house, by erecting a high pole having a number of hollow calabashes hung around the top of it. **1835** R. M. BIRD *Hawks of Hawk-Hollow* I. iii. 44 Here's..the identical old Folly, with..the pot in the chimney, and the martin-house on a pole. **1854** B. P. SHILLABER *Life & Sayings Mrs. Partington* 101 He heard Gruff scolding Ike for throwing snowballs at his new Martin-house. **1884** 'C. E. CRADDOCK' *In Tennessee Mts.* I. 40 There was a lofty martin-house whence the birds whirled fitfully.

Martin[4]. Add: **4.** *Martin ware, Martinware*, a type of brown, salt-glazed, freq. elaborately modelled pottery, made by the Martin brothers in the late 19th and early 20th centuries. (See quot. 1897.)
1897 F. LITCHFIELD *Chaffers's Marks Pott. & Porc.* (ed. 8) 870 *Martin Ware.* The four brothers Martin carry on a small pottery at Southall. The initiator of the business was Robert Wallace Martin... He was joined by his three brothers..to produce the ware which now bears their name... A great point with the Martins is that the decoration of a specimen is never repeated. **1918** E. CLARKE *Story of My Life* xxxi. 411 On the top of the low book-shelves stand a few choice bronzes..and some fine specimens of my favourite Martin-ware. **1922** J. F. BLACKER *ABC Eng. Salt-Glaze Stoneware* xxv. 219 The best period of Martin ware. **1937** *Times Lit. Suppl.* 25 Sept. 696/3 The four brothers Martin responsible for the production of the nineteenth and early twentieth century pottery known as Martinware. **1957** MANKOWITZ & HAGGAR *Conc. Encycl. Eng. Pott. & Porc.* 142/1 A large collection of Martinware is exhibited at the Public Library and Museum, Southall, Middlesex. **1966** G. A. GODDEN *Illustr. Encycl. Brit. Pott. & Porc.* 214 (*caption*) A selection of Martinware birds with loose heads. **1973** *Times* 16 Feb. 21/5 A Martinware bird of exceptional size was sold at Sotheby's..for £1,250.

martin[5]. (Examples.)
1856 P. THOMPSON *Hist. Boston* 714 A twin-heifer is called a martin, and is said to be incapable of bearing young. **1903** *Rep. Kansas State Board Agric. 1901–2* II. 211 Pure-bred steer, spayed or martin heifer, two years old and under three.

martineta (mɑːɹtinē·tă). [Amer. Sp., prob. f. Sp. *martinete* night heron.] A species of tinamou, *Eudromia elegans*, found in southern Argentina. Also *attrib.*
1872 W. H. HUDSON in *Proc. Zool. Soc.* 546 The Martineta frequents the elevated tablelands. **1892** W. R. KENNEDY *Sporting Sk. S. Amer.* 7 The 'Martinetas', or large Partridge (*Rhynchotus rufescens*), are generally found in pairs, and if one rises the other is probably not far off. **1925** L. E. ELLIOTT in J. A. Hammerton *Countries of World* VI. 4140/2 On the great undulating plains [of Uruguay] are thousands of..the double quail or martineta. **1961** G. DURRELL *Whispering Land* iv. 94 The commonest bird we saw was undoubtedly the martineta, a species of tinamu. It is plump, a partridge-shaped little bird, about the size of a bantam. Its plumage is a rich array of autumn browns, speckled and streaked with golds, yellows and creams... On its head there is an elongated crest of dark feathers, which curves like a half-moon over its head. **1964** A. L. THOMSON *New Dict. Birds* 822/2 The Martineta Tinamou *Eudromia elegans* is earth-coloured and adorned with a long slender crest which the bird when excited carries directed forwards.

Martini[1]. Substitute for def.: In full, *Martini-Henry (rifle)*. A rifle used in the British army from 1871 to 1891, combining a breech-mechanism invented by Friedrich von Martini with a ·45-calibre barrel devised by Benjamin Tyler Henry (see *HENRY[1]). So *Martini-Henry carbine*. (Add further examples.)
1890 KIPLING *Barrack-Room Ballads* (1892) 11 We sloshed you with Martinis. *Ibid.* 48 When 'arf of your bullets fly wide in the ditch, Don't call your Martini a cross-eyed old bitch. **1892** R. CHURCHILL *Men, Mines & Animals S. Afr.* vii. 115 A good magazine is probably a better weapon than a Martini–Henry. **1900** [see *DANE GUN]. **1915** OMMUNDSEN & ROBINSON *Rifles & Ammunition* iii. 59 Another arrangement..was to make the grooves deeper at the breech than at the muzzle. The principle was adopted for the Enfield and retained in the Martini–Henry. **1962** H. L. PETERSON *Bk. Gun* 182/1 From 1871 until after the adoption of smokeless powder, the Martini served as the principal British single-shot weapon. **1967** ROBERTS & WATERS *Breech-Loading Single-Shot Match Rifle* i. 28 Martini–Henry rifles chambered for the various large British hunting cartridges were..popular in Africa and India.

Martini[2] (mɑːɹtī·ni). [f. the name of *Martini* and Rossi, Italian wine-makers.] The proprietary name of a type of vermouth; a cocktail consisting of gin and vermouth; *dry Martini*, such a cocktail containing more gin than vermouth, sometimes with the addition of orange bitters. Also *attrib.*
1894 *Puck* (U.S.) 28 Nov. 238/2 (Advt.), The Club Cocktails Manhattan, Martini, Whisky, [etc.]. **1896** *Crescent* (Brooklyn, N.Y.) 1 Aug. 11/2 As he sipped his martini and inhaled its seductive bouquet, a far-away look came into his baby-blue eyes. **1903** ADE *People you Know* 20 It was an actual mystery to him that any one could dally with a Dry Martini while there was a Hydrant on every Corner. **1906** *Mrs. Beeton's Bk. Househ. Managem.* xlix. 1511 Martini cocktail. **1909** [see *BRONX 1]. **1919** C. MACKENZIE *Sylvia & Michael* v. 203 What are you going to have? Ordered a Martini here the other day. **1930** WODEHOUSE *Very Good, Jeeves!* ii. 57 A nicely-balanced meal, preceded by a couple of dry Martinis, washed down with half a bot. **1953** D. A. EMBURY *Fine Art of Mixing Drinks* 36 Martinis, Manhattans, and other cocktails containing wine can be stirred with a rod or long spoon. *Ibid.* 106, I have already referred to the Martini as the most perfect of apéritif cocktails. **1958** *Times Lit. Suppl.* 19 Dec. 733/4 The sham, artificial life of all-night martini parties. **1962** 'E. MCBAIN' *Like Love* (1964) vii. 100 A tray with a Martini shaker and two iced Martini glasses. **1967** A. LICHINE *Encycl. Wines* 542/1 The Italian Cinzano may be red or white, sweet or fairly dry—as also may Martini and Gancia. **1968** [see *GIBSON[2]]. **1972** J. MOSEDALE *Football* vii. 100 People tend to think of us professionals as guys who spend all their time around the swimming pool with a blonde in one hand and a martini in the other... A lot of us don't like martinis.

Martiniquan (mɑːɹtinī·kăn), *a.* and *sb.* Also **Martinican, Martiniquian, Martiniquien.** [f. *Martinique* (see below) + -AN.] **A.** *adj.* Of or pertaining to Martinique, an island of the West Indies. **B.** *sb.* A native or inhabitant of Martinique. So **Martiniquais** *a.* and *sb.* [Fr.].
1890 L. HEARN *Two Yrs. French W. Indies* 358 In Louisiana an almost similar dish is called *jimbalaya*: chicken cooked with rice. The Martiniquais think it..a delicacy. **1891** W. F. HUTCHINSON *Under Southern Cross* vi. 58 The Cuban becomes twice a Spaniard, a Barbadian three times an Englishman, and a Martiniquien four times a Frenchman. *Ibid.* 69 (*caption*) A fair Martiniquienne. **1907** A. E. ASPINALL *Pocket Guide W. Indies* xiv. 255 St. Pierre..was for all the world like a small French provincial city, with its cabarets and cafés, at the tables of which the Martinicans passed their leisure hours. **1926** A. BELL *Spell of Caribbean Islands* xvi. 256 The outstanding features of Martinican life. **1942** N. SMITH *Black Martinique* (1943) iii. 26 A wild-eyed youth.., brown-skinned, a true Martiniquais, came bursting toward us through the crowd. **1953** *Caribbean Q.* III. i. 28 Trinidadian patois has more faithfully retained truly Creole words which the Martiniquan has lost. *Ibid.* 29 The Martiniquan humour of M. Gratiant. **1955** *Ibid.* IV. i. 13 Another example of this poem of solidarity with Africans is the Martiniquais Paul Niger's *Je n'aime pas l'Afrique*. **1964** *Sat. Rev.* (U.S.) 10 Oct. 75/1 Lafcadio Hearn..wrote so poignantly of the 'little' Martiniquans. **1967** D. LOWENTHAL in R. D. Abrahams *Positively Black* (1970) iii. 75 A black or colored man from Jamaica or Martinique is simply a Jamaican or Martiniquian. **1970** J. BROWN *Un-melting Pot* vii. 99 '*Peau noir, masque blanc.*' The old Martiniquan saying sums up as well as any the central predicament of West Indian man. **1972** F. WARD *Golden Islands Caribbean* ii. 47 (*caption*) The Creole costume remained in vogue with Martinican women. *Ibid.* 48 A Martinican is unequivocably [*sic*] and irrevocably a Frenchman. **1973** *Nation* (Barbados) 23 Dec. 32/4 The Barbadians..will be allowed to fish freely in Martiniquan waters. *Ibid.* 32/5 Martiniquans do not eat flying fish. **1975** *New Yorker* 31 Mar. 54/3 We're all here by accident, and a Haitian and a black American could just as easily have been a Martinican or a Jamaican.

martyr, *sb.* Add: **4. a.** (Further examples.)
c **1862** E. DICKINSON *Poems* (1955) II. 417 The Martyr Poets—did not tell—But wrought their Pang in syllable—..The Martyr Painters—never spoke. **1870** *Brewer's Dict. Phr. & Fable* 558/2 *The Martyr King*, Charles I. of England, beheaded January 30th, 1649. **1876** G. M. HOPKINS *Wreck of Deutschland* xxi, in *Poems* (1967) 58 Thy unchancelling poising palms were weighing the worth, Thou martyr-master. **1908** L. DAWSON *Nicknames & Pseudonyms* 194 *Martyr King*.., Charles I..King of England. *Ibid.* 195 *Martyr President*.., Abraham Lincoln. **1939** R. CAMPBELL *Flowering Rifle* IV. 117 The Press then bills him as a martyr-hero. **1942** BERREY & VAN DEN BARK *Amer. Thes. Slang* § 184/14 The Martyr President,..Uncle Abe, *Abraham Lincoln*.
b. *martyr-habit, -spasm*; **martyr complex**, an exaggerated desire to sacrifice oneself for others and to have the sacrifice recognized.
1931 *Times Lit. Suppl.* 22 Jan. 58/4 In Rachel, we are asked to believe they fix a 'martyr-complex', or give her a taste for unnecessary self-sacrifice. **1902** W. JAMES *Varieties Relig. Experience* iv. 98 The 'misery-habit', the 'martyr-habit', engendered by the prevalent 'fearthought'. **1916** BLUNDEN *Harbingers* 66 Marble writhed to martyr-spasm.

martyrion (mɑːɹti·riǫn). Also in L. form **martyrium.** Pl. **martyria.** [Gk. μαρτύριον, L. *martyrium* martyr's tomb or shrine.] = MARTYRY 2.
1711 J. BINGHAM *Origines Ecclesiasticæ* (ed. 2) III. VIII. i. 110 (*heading*) Why some Churches called *Martyria*. *Ibid.*, Such [Churches] as were built over the Grave of any Martyr, or called by his Name to preserve the Memory of him, had usually the distinguishing title of *Martyrium*, or *Confessio*, or *Memoria* given them. **1884** ADDIS & ARNOLD *Cath. Dict.* 553/1 The name 'martyrium' (μαρτύριον)..at first meant the church built over a martyr's remains. **1931** D. ATTWATER *Catholic Encycl. Dict.* 328/1 *Martyrion*, a primitive name for the place of burial of a martyr, for a church built over a martyr's tomb, and then for any church. **1931** *Times Lit. Suppl.* 23 July 584/1 The second is the largest church at Korykos and perhaps a 'martyrion'. **1960** *Antiquity* XXIV. 131 In the course of the 4th century a Christian cemetery grew up round this martyrion. **1965** *Virtue's Catholic Encycl.* II. 659/2 The original name for a church was a *martyrium*, since it was often built upon the actual site of martyrdom. **1974** *Times Lit. Suppl.* 8 Nov. 1267/1 Those of Syria and Mesopotamia were largely intended as cathedrals (rather than 'martyria').

proprietors who live in a little Mas. **1932** R. Campbell *Taurine Provence* 48 The 'Mas' is a mixture between a manor and a huge farm-house. **1942** 'A. Bridge' *Frontier Passage* xii. 221 The peasant's rough-and-ready method of defending his *mas*. **1964** F. White *West of Rhone* iv. 51 The *mas*, the farms, were devoted to stock-raising. **1966** 'R. Standish' *Widow Hack* vi. 69 The villa was a well-faked *mas provençal*, but instead of the conventional rounded Roman [roof] tiles, they were flat.

Masai (mā·sai, măsai·). [Bantu.] **A.** *sb.* **a.** A pastoral people of mixed Hamitic stock inhabiting parts of Kenya and Tanzania; a member of this people. **b.** The Nilotic language of this people. **B.** *adj.* Of or pertaining to this people or their language.

1857 Erhardt & Krapf (*title*) Vocabulary of the Enguduk Iloigob, as spoken by the Masai-tribes in East Africa, in two parts (Masai–English, and English–Kimasai, with a song of the Masai). **1873** [see *Kavirondo]. **1894** [see *Kikuyu]. **1898** E. Clodd in R. M. Dorson *Peasant Customs* (1968) I. 384 Among the Masai..while it is bad form to kiss a lady, it is *comme il faut* to spit on her. **1902** *Encycl. Brit.* XXV. 140/1 The Masai east of Lake Victoria..are undoubtedly Hamites, mixed..with negro elements. **1933** L. Bloomfield *Lang.* iv. 67 Among the languages of this region..we may mention..Nuba,..Dinka, and..Masai. **1947** *E. Afr. Ann.* 1946–7 41/2 The Samburu..are closely associated with the Masai and are called from the Masai word 'Samburer' which means 'butterfly'. **1955** *Times* 12 July 9/7 The Mau Mau gangs must be leading a life not altogether unlike that which their grandfathers led in the struggles with the Masai over 60 years ago. **1962** *Listener* 6 Sept. 343/2 No one lived there and even Masai cattle seldom came. **1967** M. J. Coe *Ecol. Alpine Zone Mt. Kenya* 1 The retention in the Masai language of many words that refer to Mt. Kenya. **1969** *Reporter* (Nairobi) 13 June 36/4 This new spirit of the Masai is exemplified by 700 Masai families who migrated from their motherland some years ago. **1973** *Sunday Express* 11 Mar. 23/2 She has just come back from Kenya.., where she has been photographing the Masai.

Masarwa (măsā·ɹwă). *S. Afr.* [Native name.] **A.** *sb.* The name given to the Bushmen distributed over the Northern Kalahari desert. **B.** *adj.* Of or pertaining to this group.

1871 J. Mackenzie *Ten Yrs. North of Orange River* viii. 128 The other subject race is that of the Bushmen, called Barwa by the Bechuanas in the south, and Masarwa by those in the north of the country. **1896** H. A. Bryden *Tales S. Afr.* ii. 42 You may never..see a Masarwa Bushman..who does not show marks of fireburn upon the nether limbs. **1905** G. W. Stow *Native Races S. Afr.* xiv. 265 Bushman speaking peoples as the Masarwa. **1928** E. H. L. Schwarz *Kalahari & its Native Races* viii. 172 The term Masarwa..is used for all Bushmen in the Northern Kalahari. *Ibid.* 174 There were several Masarwa women about the palace precincts. **1931** [see *game-land* (*game *sb.* 16 a)]. **1971** *Sunday Times* (Johannesburg) (Mag. Section) 28 Mar. 3/3 The people of Dilepe..are not prepared to receive a Masarwa in their midst.

mascal, var. Mescal.

1850 W. R. Ryan *Personal Adventures Upper & Lower Calif.* I. 193 The usual beverages, besides water, are wine, *mascal*, and *aguardiente*. **1887** F. Francis *Saddle & Mocassin* iv. 68 The Mexican..reduced himself to a state of complete intoxication with *mascal*.

mascara (mæskā·ră). Also (*rare*) mascaro. [It. *mascara, maschera*, Sp. *máscara* Masker *sb.*[1]] A preparation for colouring the eyelashes and eyebrows.

1890 *Cent. Dict.*, *Mascaro*, a kind of paint used for the eyebrows and eyelashes by actors. **1922** I. & H. Klumph *Screen Acting* 59 You will have your own make-up..cold cream, mascara for the eyelashes. **1927** N. Martin *Constant Simp* (1928) vi. 101 No black globules of mascara hung from the tips of her long lashes. **1930** H. Rubinstein *Art Feminine Beauty* xvi. 250 My mother had brought us up to use a little powder on our faces but would have been horrified at the thought of rouge or lipstick, or mascara for the eyes. **1938** M. Allingham *Fashion in Shrouds* xvi. 267 Her long eyelashes thick with mascara. **1958** E. Dundy *Dud Avocado* I. vii. 121 My eyes, which stung and swam from the running mascara. **1966** T. Pynchon *Crying of Lot 49* v. 125 She knew she looked terrible—knuckles black with eye-liner and mascara from where she'd rubbed.

Hence **masca·ra** *v. trans.*, to treat with mascara. So **masca·raed** *ppl. a.*

1935 S. Lewis *It can't happen Here* xxxviii. 407 A lady with mascaraed eyelashes. **1938** N. Marsh *Death in White Tie* iii. 36 Our hostess's mascaraed eyes. **1944** W. S. Maugham *Razor's Edge* ii. 68 They had the same heavily mascaraed eyelashes, the same brightly painted lips, the same rouged cheeks. **1961** M. Howard *Surgeon's Dilemma* viii. 247 Adelia, frantically mascaraing her eyelashes.

‖ **mascaret** (maskare). [Fr. (16th c.), f. Gascon *mascaret* spotted cow, f. *mascara* (cf. Pr. *mascarar*, OF. *mascurer*, F. *mâchurer*) to daub, to black the face: app. arising from the resemblance of the tidal bore to the movement of running cattle.] A tidal bore in an estuary in France.

1841 W. A. Brooks *Treat. Improvement Navigation of Rivers* xi. 127 The mascaret of the Dordogne consists of two, three, and sometimes four lofty waves, which follow each other rapidly, damming back the whole breadth of the river. **1900** P. Dearmer *Highways & Byways in Normandy* xii. 315 At high tides there is the spectacle of the famous *mascaret* or bore. **1902** *Encycl. Brit.* XXXII. 508 The tide begins to make itself felt at Poses, 11 miles above Elbeuf, and between Caudebec and Villequier the *mascaret*, or bore, has its greatest development. **1959** *Listener* 5 Feb. 248/1 The *mascaret* is to the Seine what the bore is to the Severn. **1971** S. E. Morison *European Discovery Amer.: Northern Voy.* viii. 256 The navigation of the Seine is more hazardous..and even dangerous when a tidal bore, the *mascaret*, sweeps up the river unexpectedly.

‖ **mascaron** (maskaroṅ, mæ·skărǫn). [Fr., = grotesque mask (1603 in Robert), ad. It. *mascherone*, f. *maschera* Mask *sb.*[3]]. In decorative art: a grotesque face or mask (cf. Mask *sb.*[3] 4 a).

1664 Evelyn tr. *Fréart's Parallel Antient Archit. with Mod.* Pref. 3 They produce nothing save Mascarons, wretched Cartouches, and the like idle and impertinent Grotesks. **1883** J. W. Mollett *Illustr. Dict. Art & Archaeol.* 207/1 *Mascaron*,..the face of a man or animal employed as an ornamentation for decorating the keystones of arches or vaults, or the stones of an arch, &c. **1926** R. Glazier *Man. Hist. Ornament* (ed. 4) 11. 169 The Italians from about 1480 favoured elaborate borders [in MSS.]..composed of classical columns,..mascarons, etc.

mascon (mæ·skǫn). *Astr.* [f. *mass concentration*.] One of the concentrations of denser material thought to exist under some lunar maria, discovered as a result of the variations they produce in the speed of an orbiting satellite; also, a similar region on another planet.

1968 Muller & Sjogren in *Science* 16 Aug. 680/1 The Urey–Gilbert theory of lunar history has predicted such large-scale high-density mass concentrations below these maria, which, for convenience, we shall call mascons. **1969** *Times* 22 Feb. 7/7 Exactly what the mascons are has puzzled selenologists since the discovery last year, and one explanation is that they are the remnants of enormous meteorites. **1969** *Sci. Jrnl.* May 61/1 The so-called mascons extend laterally for 50 to 200 km, are about 5 km thick and lie at a depth of between about 50 and 100 km. **1971** *Nature* 26 Nov. 168/2 The following findings were announced... Gravitational anomalies which may be caused by martian mascons. **1973** J. Strong *Search the Solar Syst.* x. 104 Density anomalies, or 'mascons' on the Moon were located by their disturbance of the lunar orbiters.

mascot. Delete '*slang*' and add earlier and later examples.

1881 Farnie & Reece *Mascotte* I. 4 Ah! blest their lot whom fate shall send A true Mascotte, a fairy friend!.. Luck's his for ever! **1918** E. M. Roberts *Flying Fighter* 328 Each man had his own little fetish. It was known as the pocket-piece or mascot. In some cases it might be a dice or a playing-card... In other cases it might be a locket, then again a medal, while many of us carried little wooden dolls. **1921** R. Hichens *Spirit of Time* vi. 106 'She says you brought her luck.'..'I didn't know I was a mascot,' he remarked. **1955** *Halsbury's Statutory Instruments* II. 2276 No mascot shall be carried by a motor vehicle..where it is likely to strike any person with whom the vehicle may collide. **1975** *Times* 6 June 17/4 Christie's ..launched a new collecting field yesterday with a sale of motoring mascots, the figurines mounted on car bonnets.

attrib. **1898** *Speaker* 31 Dec. 783/2 His mascot snakes that were kept, buried in flannel, in an oblong glass-topped box.

ma·scotism. [-ism.] = next.

1924 W. J. Locke *Coming of Amos* viii. 94 If there is anything in luck, in mascotism or anti-mascotism, one might have reason to believe that Amos..had the most maleficent of influences.

mascotry (mæ·skǫtri). [f. Mascot + -ry.] Attachment to or belief in mascots; the use of mascots.

1916 *Church Times* 7 Apr. 332/4 Mascotry,..the growing superstition with regard to mascots. **1923** *Sunday at Home* May 504/3 There is a worse aspect of mascotry than mere folly,..for mascotry is, in its essence, simply idolatry.

masculinism (mæ·skiŭliniz'm). *rare.* [f. Masculine *a.* + -ism.] **a.** Tendency to masculine physical traits in a woman. **b.** Advocacy of the rights of men.

1895 [see *infantilism]. **1916** H. Ellis *Essays in War-Time* viii. 88 The advocates of Woman's Rights have seldom been met by the charge that they were unjustly encroaching on the Rights of Man. Feminism has never encountered an aggressive and self-conscious Masculinism.

masculinist (mæ·skiŭlinist). [f. Masculine *a.* + -ist.] **1.** An advocate of men's rights, opp. *feminist *sb.* Cf. *hominist.

1918 V. Woolf *Writer's Diary* (1953) 6 He [sc. Milton] was the first of the masculinists. **1967** *Times* 13 Oct. 8 No militant masculinists stooped to conquer; indeed almost the only men to be seen on the premises, apart from the staff, were male journalists invited to help celebrate the occasion.

2. A person of the female sex who adopts or affects characteristics or qualities usually thought of as masculine.

1928 *Daily Express* 11 July 5/3 The suffragists modelled their appearance, their manners, and their education on that of boys and men... They were, in short, thorough-going masculinists.

3. *attrib.*

1951 R. Campbell *Light on Dark Horse* xvii. 257 Bloomsbury still awaits its 'masculinist' Messiah.

masculinization (mæ:skiŭlinəizē·ʃən). [f. next: see -ization.] The action of *masculinize *v.* (in either sense); the process of becoming masculine; also, a masculine state or condition.

1895 H. James *Notebks.* (1947) iv. 196 There is a big comprehensive subject in the *avènement*, or rather in the masculinization of women. **1922** F. H. A. Marshall *Physiology of Reproduction* (ed. 2) xv. 696 (*caption*) Masculinisation of guinea pig. **1924** *Glasgow Herald* 24 Dec. 4 The masculinisation of woman deprives society of a 'precious asset'. **1927** *19th Cent.* Aug. 266 We find him throwing all his authority against the masculinisation prevalent in our schools. **1952** Srb & Owen *Gen. Genetics* xiv. 287 Freemartins are sterile, and show considerable masculinization in a number of traits. **1970** *Sci. Jrnl.* June 50/2 They are of course normal genetic and anatomical females on whom the male sex hormones from their adrenals have imposed a false masculinization. **1971** *Daily Tel.* (Colour Suppl.) 10 Dec. 18/2 In the production of millions of children a year, it is not surprising that occasionally nature's complex technology should break down to produce an imbalance of hormones with masculinisation of the female foetus or feminisation of the male.

masculinize (mæ·skiŭlinəiz), *v.* [f. Masculin(e *a.* + -ize.] *trans.* **a.** To render masculine or more masculine in nature, form, or character.

1912 Mme Moret tr. *A. Moret's Kings & Gods Egypt* i. 29 She even tried to change the very name she had received at her birth and to masculinise it by omitting the feminine ending. **1927** C. C. Martindale *Relig. of World* 19 The first way in which Brahmā could be '*thought*', was, as sufficiently 'masculinised' (Brahmā) to be a god, supreme, yet to that extent specified.

b. *Biol.* To induce male sexual characteristics in.

1924, 1962 [see *feminize *v.* 2]. **1971** *Daily Tel.* (Colour Suppl.) 10 Dec. 21/3 The amount of androgen required for normal masculinizing of the brain is greater than that required to masculinize the genitalia.

Hence **ma·sculinized** *ppl. a.*, **ma·sculinizing** *ppl. a.* and *vbl. sb.*

1927 *Daily Express* 28 Sept. 8/7 The second type of masculinised female is the politically-minded woman. **1928** *Sunday Express* 19 Aug. 5 'The Cinderella Man'.. was a masculinised version of the old story. **1936, 1939** [see *androgen]. **1961** M. F. A. Montagu *Genetic Mechanisms in Human Dis.* xii. 113 Completely masculinised genetic females are almost common, forming something like 1 in 40 of the masculine population. **1970** *Nature* 3 Oct. 94/2 In the glow-worm *Lampyris* the apical cells of the testis produce a masculinizing hormone.

masculinoid (mæ·skiŭlinoid), *a.* [f. Masculin(e *a.* + -oid.] Masculine (but not male); of male form or appearance.

1921 [see *adrenal *a.* and *sb.* A]. **1923** J. S. Huxley *Ess. Biologist* iv. 145 The 'masculinoid' woman (to use the current jargon) tends physically also to be less feminine.., while the 'feminoid' man shows the reverse tendency.

Masdeu (masdǿ). [See quot. 1851.] A sweet firm-bodied wine of a dark colour and mellow flavour produced at a vineyard in the South of France.

1851 C. Redding *Hist. Mod. Wines* (ed. 3) vi. 154 This wine is not a factitious French port from the harbour of Cette... It is a genuine production called Masdeu, from the vineyard which produces it, between Perpignan and Collioure. **1903** *Whitaker's Almanack* 453 Of wines imported in casks the following are the usual measurements: Pipe of Port or Masdeu = 115 gallons, [etc.]. **1958** A. L. Simon *Dict. Wines* 108/1 *Masdeu*, red, fortified wines of Roussillon, very popular in England in early Victorian days. Also written *Masdu* and *Masdieu*. **1965** O. A. Mendelsohn *Dict. Drink* 214 *Masdeu*, French (Roussillon) red dessert wine.

masdevallia (mæsdĭvæ·liă). [mod.L. (H. Ruiz Lopez & J. Pavon *Floræ Peruvianæ et Chilensis Prodromus* (1794) 122), f. the name of Jose *Masdevall* (d. 1801), Spanish physician and botanist + -ia[1].] An epiphytic orchid of the large genus so called, belonging to the family Orchidaceæ and native to the cool, mountainous regions of South America.

1845 *Curtis's Bot. Mag.* LXXI. 4164 Windowed Masdevallia... is one of the very curious productions of nature. **1896** F. H. Woolward *Genus Masdevallia* 1 The first *Masdevallia* known to science was..discovered by the Spanish botanists, Ruiz and Pavon, in the Andes of Peru, during their residence in that country from 1777 to 1794. **1911** *Encycl. Brit.* XX. 172/1 *Masdevallia* is common in cultivation and has often brilliant scarlet, crimson or orange flowers. **1970** B. & W. Rittershausen *Pop. Orchids* ix. 211 The flower [of *Masdevallia muscosa*] is of a translucent yellow colour, the lip being somewhat large for Masdevallias.

mase (mēiz), *v.* [Back-formation from next, the ending *-er* being treated as the suffix -er[1] of agent nouns.] *intr.* To *lase, esp. in the microwave part of the spectrum. Hence **ma·sing** *vbl. sb.*

1962 *Engineering* 2 Feb. 191 The device 'mases' 30 per cent of the time. **1962** *New Scientist* 1 Mar. 486 One knows when the alignment is good enough only by the fact that the device starts to 'mase'. **1964** *Encycl. Sci. Suppl.* (Grolier) 308/1 Many substances have been made

to 'mase' and 'lase'—that is, to behave like masers or lasers. **1966** Smith & Sorokin *Laser* vii. 369 At this point the stage was nearly set for the actual achievement of an injection maser. The main remaining question was: in what sort of a structure could masing action be achieved? **1973** *Nature* 21/28 Dec. 468/1 Since none of the CH lines have so far been seen in absorption, irrespective of the nature of the source where they have been found, one is led to believe that all transitions are masing weakly.

maser (mēi·zəɹ). [An acronym: see quot. 1955[2].] A laser, esp. one that emits microwaves.

The first masers emitted microwaves. Later ones emitted in other parts of the spectrum, and these were also called masers for a time until *laser* came to be adopted as the general name for all such devices.

1955 *Sci. News Let.* 5 Feb. 83/1 Scientists can, for the first time, generate microwaves of extremely high frequency by tapping directly the energy of molecules, Dr. Charles H. Townes of Columbia University's physics department reported in New York. His device for doing so is known as the 'maser'. *Ibid.*, Work on the maser began three years ago. **1955** Gordon, Zeiger, & Townes in *Physical Rev.* 15 Aug. 1264/1 The device utilizes a molecular beam in which molecules in the excited state of a microwave transition are selected. Interaction between these excited molecules and a microwave field produces additional radiation and hence amplification by stimulated emission. We call an apparatus utilizing this technique a 'maser', which is an acronym for 'microwave amplification by stimulated emission of radiation'. **1958** Schawlow & Townes in *Ibid.* 15 Dec. 1940 (*heading*) Infrared and optical masers. **1960** *Times* 17 Oct. 2/2 (Advt.), Development work on existing long range programmes in: solid-state millimetre and optical masers, parametric amplifiers, [etc.]. **1960, 1962** [see *laser[2]]. **1962** *Engineering* 2 Feb. 190/2 The basic type of maser, from which the principle of *optical* masers is derived, is a quite recent electron device... Research later extended the upper frequency range of this technique of low-noise signal amplification, until it eventually reached the visible region—giving rise to the optical maser, otherwise known as the laser. **1967** T. P. Melia *Introd. Masers & Lasers* iv. 34 The power output of the ammonia maser is very small (about 10^{-10} watt) but the frequency stability is of the order of 1 in 10^{12} over a 1 minute period and 1 part in 10^{10} over very much longer periods. **1968** *Nature* 30 Mar. 1237/2 (*heading*) Is interstellar hydrogen capable of maser action at 21 centimetres? **1969** S. G. & H. Lipson *Optical Physics* xi. 369 A maser is, of course, a very-narrow-band amplifier; it is because of its narrow band that the noise-level achieved is very low, since there is a theoretical minimum noise-level per unit band-width. **1972** *Sci. Amer.* Sept. 136/3 The greatest contributor..to the receiver's ability to handle Telstar's faint signals (about 10^{-12} watt) was a maser amplifier that combined very high gain with extremely low noise.

mash, *sb.*[1] Add: **3. d.** *slang.* Mashed potatoes; esp. in the phr. *sausage(s) and mash*.

1904 A. E. W. Mason *Truants* xxii. 213, I..go into a public-house..and have a sausage and mash and a pot of beer. **1939** W. S. Maugham *Christmas Holiday* i. 7 They could drop in..and eat kedgeree and sausages and mash. **1973** 'H. Carmichael' *Too late for Tears* vi. 85 He.. ordered sausage and mash with peas. **1974** *Woman* 4 May 18/1 Sizzling sausages and tasty mash.

mash, *sb.*[3] Add: **2.** (Earlier and later examples.) Also, *to make* (or *have*) *a mash* (*on*).

1884 E. W. Nye *Baled Hay* 135 Two Laramie girls on horseback yanking a fly drummer along the street..because he tried to make a mash on them. **1909** 'O. Henry' *Roads of Destiny* iv. 61, I certainly seemed to have a mash on her. **1912** D. Canfield *Squirrel-Cage* III. xxix. 319, I thought it would be fun to tease Paul about the mash you made on old What's-his-name.

Comb. **mash note,** a love-letter.

1890 B. Hall *Turnover Club* 134 He is greatly afflicted by that dreadful bane of fine-looking actors, yclept the 'mash note' in the profession. **1899** *Chicago Record* 7 Jan. 4/6, I was writin' mash notes to myself. **1913** R. W. Chambers *Gay Rebellion* i. 6 That poem seemed to deal a direct blow at this suffragette strike. Several women subscribers sent in mash notes. **1930** W. R. Burnett *Iron Man* 85 He gets mash notes by the ton. **1970** *New Yorker* 14 Nov. 87/2 (Advt.), A pen that roared through the Twenties and Thirties writing checks, letters, autographs ..jazz and mash notes.

mash, *sb.*[5] Eng. and U.S. dial. variant of Marsh.

1671 *Coll. S. Carolina Hist. Soc.* (1897) V. 336 About ye rivers mouth & up the river beyond ye mashes. **1840** *Knickerbocker* XVI. 210, I reckon you won't get nothing for him without you turn him out on the mash. **1843** 'R. Carlton' *New Purchase* I. ix. 58 They had been sufficiently fortunate as to..learn the nature of 'mash land'. **1876** W. Whitman *Specimen Days* (1882–3) 94 The sedgy perfume..reminded me of 'the mash' and south bay of my native island. **1892** P. H. Emerson *Son of Fens* iii. 23, I went back to the sheep... I used to drive 'em down to mash along with the cows.

mash, *v.*[4] *U.S.* Var. of Mesh *v.* 3 b.

1850 *Rep. Comm. Patents 1849* (U.S.) 155 What I claim as new..is..the shaft H, with the pinions I, mashing into racks H. **1860** Bartlett *Dict. Amer.* (ed. 3) 265 In machinery, one wheel is said to *mash* into or with another, i.e. to 'engage' with it.

‖ **masha** (mæ·ʃă). [Hindi *māshā*.] An Indian weight, equivalent to about fifteen grains troy or eight rattis, originally based on the weight of a seed of the bean *Phaseolus mungo* var. *radiatus*.

1848 J. H. Stocqueler *Oriental Interpreter* 148/2 Masha, a weight of fifteen grains troy. Used by native goldsmiths and jewellers, and in the native evaluation by assay of the precious metals. **1855** E. Acton *Mod. Cookery* (rev. ed.) xxxii. 612 Five eggs, two tolahs of milk, one masha of salt, two mashas of cayenne pepper. **1969** *Daily Tel.* 1 Oct. 19/8 Customers had their purchases weighed in tolas, mashas and ruttees by a Pakistani shop-keeper.

Masham (mæ·ʃăm). The name of a small town in the northern part of Yorkshire, used to designate a breed of sheep produced by crossing Wensleydale or Teeswater rams with Blackface or Swaledale ewes.

1951 A. Fraser *Sheep Husbandry* (ed. 2) ii. 68 The Wensleydale..is pre-eminently a crossing breed... It provides sires to cross with the hill breeds of sheep indigenous to the North of England... The cross lambs thus produced are variously known as 'Mashams'—Half-breds—or Greyfaces, and are very useful lambs for feeding on low ground. **1971** *Farmers' Weekly* 19 Mar. 43/3 Certain wool types such as Blackface..have met a better market demand than..Masham.

mashed, *ppl. a.* Add: Hence as *sb.*, mashed potatoes (esp. in the phr. *sausage and mashed*). *slang.*

1923 in J. Manchon *Le Slang*. **1926** G. B. Shaw *Translations & Tomfooleries* 225 One [*sc.* a public-house] had a placard up 'Sausage and Mashed'. **1934** T. S. Eliot *Rock* i. 40 Restaurants where you can get..sausage and mashed or toad-in-the-'ole for twopence. **1963** H. Garner in R. Weaver *Canad. Short Stories* (1968) 2nd Ser. 52, I ate a lunch of meat pie, mashed and gravy.

mashie-niblick. *Golf.* [f. Mashie, mashy + Niblick.] An iron club combining the features of the mashie and the niblick, now called the number 7 iron.

1907 W. M. Butler *Golfer's Manual* 5 The marriage of different pairs of clubs has produced such implements as mashie-niblicks and putting-cleeks. **1912** H. Vardon *How to play Golf* vii. 146 Nowadays tools called mashie-niblicks are popular. **1929** Wodehouse *Mr. Mulliner Speaking* vi. 196 How bitterly he regretted now those raking drives, those crisp flicks of the mashie-niblick of which he had been so proud ten minutes ago. **1931** *Times* 28 Feb. 4/6 A little bullet-headed mashie niblick..was also highly efficient. **1964** J. Stobbs *A.B.C. of Golf* 61 Out have gone..the tried and proved regular favourites... Among these..the mashie niblick, a large round-shaped-faced lofted mashie (modern equivalent the 7-iron). **1972** T. Scott *Club Golfers' Handbk.* 127 *Mashie-niblick*, another name which has been replaced by a number.

Mashona (măʃǭu·nă), *sb.* and *a.* Pl. Mashona, Mashonas. [Native name.] **A.** *sb.* **a.** A group of Bantu peoples inhabiting parts of Rhodesia and Mozambique; a member of one of these peoples. **b.** Any of the languages of these peoples. **B.** *adj.* Of or pertaining to these peoples.

1835 A. Smith *Diary* (1940) II. 222 The Bakalaka and the Mashoona speak a different language to the Bechuanas. **1846** H. H. Methuen *Life in Wilderness* vi. 143 To the north-east the land becomes mountainous, and the Mashona, a tribe possessing guns, inhabit it. **1893** A. M. Hartmann (*title*) A grammar of the Mashona language. **1894** —— (*title*) English–Mashona dictionary. **1895** G. W. H. Knight-Bruce *Memories Mashonaland* ii. 41 The Mashona..are very fond of meat, but they hardly ever get it. To give him meat is the one way that will nearly always ensure a wild Mashona working well. **1896** *Scientific African* 78 (Pettman), The careful workmanship displayed in the crude musical instrument known as the Mashona piano. **1900** A. H. Keane *Boer States* vii. 101 Mashonas..and several other smaller groups for whom there is no collective national or racial designation, except the all-comprehensive 'Kaffre'. **1911** *Encycl. Brit.* XVII. 837/2 The name Mashona has been derived from the contemptuous term *Amashuina* applied by the Matabele to the aborigines owing to the habit of the latter of taking refuge in the rocky hills with which the country abounds... The Mashona..are in general a peaceful, mild-mannered people. **1911** J. G. Frazer *Golden Bough: Magic Art* (ed. 3) I. vi. 352 The king of the Matabeles..prays..: 'O great spirits.., I thank you for having granted last year to my people more wheat than to our enemies the Mashonas.' **1928** C. Bullock *Mashona* iii. 12 The word Mashona is used *faute de mieux*; nor need we look for its etymological derivation, because it has none. It is simply one of those British bowdlerisations..a race name to cover numerous clans. **1936** J. Buchan *Island of Sheep* iv. 68 They are Mashonas and are timid as rabbits. **1959** *Chambers's Encycl.* II. 107/2 In the Shona (Mashona) group [*sc.* of Bantu languages] there are about 40 dialectal forms. **1972** *Police Rev.* 24 Nov. 1521/1 Cecil John Rhodes..began negotiating..for rights to explore the mineral wealth of the regions east of the royal kraal at Bulawayo, at that time inhabited by the Mashona tribe.

mashwa (ma·ʃwă). Also machwa, mashua, mashuwa. [Marathi, 'fishing-boat', f. Skr. *matsya* fish.] A kind of small open boat (see quots.).

1885 G. C. Whitworth *Anglo-Indian Dict.* 184/1 The machwás of different ports vary slightly in build. They are considered to be amongst the swiftest sailing-vessels known; they are very sharp in the bows, with hollow keel, well rounded in the stern, and the mast slopes a little forward; these vessels are chiefly made of teak, and cost from Rs. 300 to Rs. 400; they last about forty years. **1906** H. W. Smyth *Mast & Sail in Europe & Asia* ix. 310 A smaller class of Zanzibar vessel is the *mashuwa*, or open fishing-boat worked by the Arab of the neighbourhood. They are generally dilapidated little crafts with the usual long bow and transom-stern. **1929** *Ibid.* (ed. 2) x. 359 The term Mashwa, or Mashuwa on the African coast, is applied apparently to any small round-stern vessel, meaning originally a fishing-boat. The larger seagoing craft all carry a small 'mashwa' on deck. **1942** *Mariner's Mirror* XXVIII. 21 The *Mashwa* or *Machwa*. The larger Arab fishing boats, the *mashwas*, are built after the design of the *sambuk* and the *jehazi* in their respective home ports. At Zanzibar they are employed also for the transport of coral rag and firewood. All are fitted with a single mast and are undecked save at the ends; the stern is of the square transom type; the tack of the sail is extended by means of a bowsprit as in the *jehazi*. **1948** R. de Kerchove *Internat. Maritime Dict.* 453/1 *Mashuwa, mashwa*, generic name in the Persian Gulf and in Southern Arabia for an open ship's boat with straight or curved stem and transom stern, usually propelled by oars, but also rigged with one mast on occasion... This name also refers to a small open boat with raking stem and rounded stern of Deccan used for fishing and local trading on the Gujarat coast between Bombay and Cam-Bay.

mask, *sb.*[3] Add: **1. b.** Also = *gas mask.

1915 H. W. Wilson *Great War* IV. 331 A Highlander wearing a mask. **1918** *Ibid.* XI. 454 French soldiers wearing the masks, fitted with goggles and respirators, that rendered them immune to noxious gases. *Ibid.* 455 Special masks were devised for the horses.

4. b. Also, the head-skin of any 'game'; the face or muzzle of a dog.

1894 C. Phillipps-Wolley et al. *Big Game Shooting* II. xv. 417 Peel off the whole mask from the antlers downwards to the muzzle. **1928** C. S. Stockley *Big Game Shooting* 88 Skins..should..be..hung on a frame to dry, the mask being filled with dry grass or paper. **1945** C. L. B. Hubbard *Observer's Bk. Dogs* 214 A light Cairn may have a dark mask. **1972** *Daily Tel.* 8 Apr. 17/4 Six more dog pelts, all complete with masks—the head of the dog—have been found..at Nuneaton... Mr Horner, whose firm exports between 40,000 and 50,000 dog skins a year, commented: 'I am baffled as to why someone should prepare pelts with masks.'

g. A cosmetic preparation, a face-pack.

1931 H. G. Wells *Work, Wealth & Happiness of Mankind* I. v. 221 She has her face put under a 'mask', an affair of beaten-up eggs and other ingredients which tightens on the face. **1934** M. Verni *Mod. Beauty Culture* I. v. 28/2 The foundation for a general type of mask is fuller's earth made into a paste with liquid. *Ibid.*, If the operator has neither given nor received a mask treatment before, she should give herself one. **1955** C. Hart *Handbk. Beauty* 29 Masks serve the purpose of stimulating the circulation, thus improving your complexion color... You can buy a mask ready-made or concoct one at home. Buttermilk all by itself is a good mask; so is oatmeal..so is an egg. *Ibid.*, Cover your hair.., because the mask stuff is sticky.

h. *Electronics.* In the manufacture of microcircuits, a thin surface layer that is removed in parts so as to permit selective modification of the underlying material.

1956 *Bell Syst. Techn. Jrnl.* XXXV. 25 For the emitter, a film of aluminium approximately 1,000 Å thick was evaporated onto the surface through a mask which defined an emitter arc of 1 × 2 mils. **1957** *Jrnl. Electrochem. Soc.* CIV. 549/1 A SiO_2 surface layer provides a selective mask at high temperatures against the diffusion of some donors and acceptors into Si. **1967** *Electronics* 6 Mar. 60/2 Faster IC's can be built with the dimensional resolutions possible in present diffusion masks. **1973** *Sci. Amer.* Apr. 65/2 Masks can be made from various types of material, for example insulators such as silicon dioxide,..or metals such as aluminum, molybdenum and gold. To expose the regions in which ion implantation is desired the mask is removed by chemical etching.

5. mask jug, a jug with a lip or front shaped like a face; also *masked jug*.

1910 Downman & Gunn *Eng. Pott. & Porc.* (ed. 5) 58 (*caption*) Masked jug in Fulham brown stoneware of the 18th century. **1963** *Times* 1 May 15/5 A Worcester yellow-ground mask jug fell to Tilley at £700. **1970** *Canad. Antiques Collector* Jan. 30/1 Mask jugs, in which a face, or even a figure, formed the shape of the front of the jug, opposite the handle, had been made for centuries in most European countries.

mask, *v.*[4] Add: **1. c.** To provide with a gas mask.

1916 *War Illustr.* IV. 607 Machine-gun section masked, ready for the enemy. **1918** H. W. Wilson *Great War* XI. 455 French soldier with one of the French army dogs, both masked against enemy gas attack.

2. g. To disguise the real character of or diminish the effect of.

1922 W. G. Kendrew *Climates of Continents* 18 The south-east trades..are now at their greatest strength, but sea breezes mask them on the immediate littoral. **1931** *Times* 18 Feb. 6/5 It was this late stroke, masked, into the left corner which defeated the game of Joshua Crane.

h. Of a sound or other object of perception: to diminish or prevent the perception of (another stimulus, usu. one affecting the same sense).

1923 *Physical Rev.* XXI. 706 When the masking tone is loud it masks tones of higher frequency better than those of frequency lower than itself. **1949** McCord & Witheridge *Odors* xvii. 190 The opportunity for masking objectionable textile odors by..impregnating the goods with a definitely noticeable perfume seems to be negligible. **1960** *Lang. & Speech* III. 160 The low-frequency noise masks the voicing of a consonant. **1966** *Jrnl. Exper. Psychol.* LXXII. 233/1 Backward masking refers to the power of certain stimuli, the masking stimuli, to disrupt or mask the processing of other stimuli, the target stimuli, which have been presented *earlier*. **1973** D. L. Richards *Telecommunication by Speech* iii. 113 Room noise that masks

speech from the other station will make conversation more difficult. **1974** *Encycl. Brit. Macropædia* XVI. 555/1 The substantially greater intensity of one odour may mask another.

i. *Chem.* To prevent (a substance or ion) from taking part in a certain reaction by causing it to undergo another preliminary reaction.

1934 in Webster. **1936** *Industr. & Engin. Chem.* (*Analytical Ed.*) 15 Nov. 409/1 A characteristic example in the analysis of anions is that..in which sulfite is masked by the addition of formaldehyde. **1970** D. D. Perrin *Masking & Demasking Chem. Reactions* i. 2 Silver is masked by ammonia against precipitation as the hydroxide or chloride.

masked, *ppl. a.*[2] Add: **2. c.** Of a sound or other stimulus (see *mask *v.*[4] 2 h, *masking *vbl. sb.*[2] 2 b).

1924 *Physical Rev.* XXIII. 270 The frequency of the masked tone is indicated on each curve. **1965** T. S. Littler *Physics Ear* vii. 134 We usually determine the level at which a sound has to be maintained in order to be just audible in the presence of another. This is known as the masked threshold of the masked or maskee sound in the presence of the masking or masker sound.

d. *Chem.* Of a substance or molecular group: prevented from taking part in a certain reaction by being in a bound form, esp. as a result of another preliminary reaction.

1932 *Discovery* Mar. 96/2 The first traces of the masked iron indicative of haemoglobin appear in the nuclei of certain cells. **1940** R. E. Oesper tr. *Feigl's Specific & Special Reactions for Qualitative Analysis* iv. 49 This fact was used..to release the cationically masked sulfate groups in the green complex chronic sulfate solutions. **1974** *Angewandte Makromolekulare Chemie* XXXIV. 111 Tertiary aliphatic amines containing masked methylol groups, e.g. methylolalkylether.., prove efficient catalysts in the production of polyurethane foams.

maskelynite (mæ·skĕlinəit). *Min.* [ad. G. *maskelynit* (G. Tschermak 1872, in *Sitzungsber. d. K. Akad. d. Wissensch.* (*Math.-Natur. Cl.*) LXV. 131), f. the name of Nevil Story-*Maskelyne* (1823–1911), English mineralogist): see -ite[1].] A colourless aluminosilicate of calcium and sodium which has a composition near to that of andesine and is found in some kinds of meteorite.

1875 E. S. Dana in J. D. Dana *Syst. Min.* (ed. 5) App. II. 37 Maskelynite. **1921** *Bull. Geol. Soc. Amer.* XXXII. 402 The name maskelynite..was first given by Tschermak to an isotropic constituent of the habit and composition of a labradorite feldspar occurring in the stone of Shergotty. The name has since been extended to cover a colorless interstitial substance, sometimes quite isotropic and sometimes faintly doubly refracting, of like appearance and habit, and a common if not universal constituent of the chondritic meteorites. **1971** *Physics Bull.* Oct. 578/2 It appears that intense shock disorders the lattice to produce the isotropic mineral maskelynite.

masking, *vbl. sb.*[2] **2.** Delete *Photogr.* and add to def.: Obscuring or covering (wholly or in part) by the interposition or overlaying of something. Also *masking-out*. (Add further examples.)

1933 E. Molloy *Mod. Motor Repair* III. 980/1 Where a line is required or a clear finished edge, this is obtained by masking. *Ibid.*, Any upholstery will require masking..if the whole car is going to be sprayed. **1949** C. E. Chapel et al. *Aircraft Maintenance & Repair* xi. 361/1 Figure 6 illustrates the masking of an airplane wing in order that the wing may be finished in two colors. **1951** *Electronics* Feb. 109/1 In color photography a technique known as 'masking' has been developed in which a negative image is combined with a positive image in printing, thus effectively introducing negative light in the final result. **1956** *Bell Syst. Techn. Jrnl.* XXXV. 334 If part of the work is coated with an insulating varnish, electrolytic etching will take place only on the uncoated surfaces. This technique, often called 'masking', has the limitation that the etching undercuts the masking if any considerable amount of material is removed. **1967** E. Chambers *Photolitho-Offset* iv. 39 Additional equipment should include an airbrush, for use in connection with continuous tone negatives and positives complete with frisket-paper for masking-out and vignetting. **1972** L. Lamb *Picture Frame* i. 16 Adhesive tape was used for temporary masking-out so that a colour could be given a straight, hard-edge.

b. The action of one stimulus in diminishing the sensitivity of a subject to another (cf. *mask *v.*[4] 2 h).

1923 *Physical Rev.* XXI. 706 For tones introduced into the same ear, except when the frequencies are so close as to produce beats, the masking is greatest for tones nearly alike. **1949** McCord & Witheridge *Odors* xvii. 186 Masking is an effect produced at the point of sense perception in the respiratory passages together with the final interpretation in the brain and does not constitute an aromatic correction of the aerial components. **1962** A. Nisbett *Technique Sound Studio* x. 170 A second reason for holding effects back..is the masking that is caused by even moderately quiet background sound. **1967** *Science* 15 Sept. 1335/3 Visual backward masking consists in the retroactive interference with the perception of one visual stimulus, the target, by another visual stimulus, the mask, closely following the target in time. **1974** *Encycl. Brit. Macropædia* IX. 243/2 Some dentists use auditory analgesia (a 'masking' of pain by sound).

c. *Chem.* The prevention of a substance or ion from taking part in a certain reaction by causing it to undergo another preliminary reaction.

1936 F. Feigl in *Industr. & Engin. Chem.* (*Analytical Ed.*) 15 Nov. 409/1 The masking of certain reactions of mercury, copper, nickel, and cobalt by cyanide and thiocyanate has been known for a long time. **1963** P. W. West in E. W. Berg *Physical & Chem. Methods of Separation* xviii. 343 The term *masking* was introduced many years ago by Prof. Fritz Feigl, who utilized this technique in making specific or highly selective spot tests when the available reagents were general. **1971** *Nature* 19 Mar. 194/1 Generally speaking 'masking' has come to mean the prevention of the normal reaction of an ion upon addition of a reagent with which it reacts.

4. *Comb.* **masking board** *Photogr.*, the board on which the printing paper is fixed in an enlarger; **masking frame** *Photogr.* (see quot. 1940); **masking tape,** adhesive tape used for masking, esp. to protect certain areas of a surface before applying paint.

1958 *Newnes Compl. Amat. Photogr.* 319 The paper is removed from the masking board, and the next negative lined up in position. **1972** G. L. Wakefield *Exposure Control in Enlarging* viii. 110 As a good masking board for the enlarger easel is expensive..it is worth making one. **1940** F. J. Mortimer *Wall's Dict. Photogr.* (ed. 15) 441 *Masking frame*, an accessory used with vertical enlargers to hold the paper flat and provide a white border. **1958** *Newnes Compl. Amat. Photogr.* 318 The sheet of bromide paper is then placed in the masking frame. **1972** G. L. Wakefield *Exposure Control in Enlarging* viii. 107 It had been intended to build the microammeter and associated electronics into a masking frame. **1936** Brimm & Boggess *Airplane & Engine Maintenance* 282 After the numbers or letters have been outlined by pencil, they should be blocked in with masking tape. **1962** *Which? Car Suppl.* Oct. 138/1 Marks had been left by masking tape on the right rear body panel. **1971** D. E. Westlake *I gave at the Office* (1972) 53 Someone had fairly recently put strips of masking tape over some sort of company name on the doors [of the trucks].

masking, *ppl. a.*[2] (Further examples.)

1923 [see *mask *v.*[4] 2 h]. **1969** T. C. Thorstensen *Pract. Leather Technol.* viii. 115 In the study of masking agents it was established very early in chrome tanning research that the anions could be listed in order of increasing affinity for the chromium tanning complex. **1970** D. D. Perrin *Masking & Demasking Chem. Reactions* i. 1 A masking reagent..is one that lowers the concentration of a free metal ion or a free ligand to such a level that certain of its chemical reactions are prevented. **1970** *Jrnl Gen. Psychol.* LXXXII. 37 An exhaust fan provided masking noise.

masochism. Now usu. with pronunc. (mæ·sŏkiz'm). Substitute for def. A form of sexual perversion in which a person finds pleasure in abuse and cruelty from his or her associate (cf. Sadism). Recently applied more generally to a form of perversion in which a sufferer derives or is believed to derive pleasure from pain or humiliation. (Later examples.)

1930 W. R. Inge *Christian Ethics & Mod. Probl.* iii. 109 Modern psychology has invented the word masochism for that perversion of the sex-instinct which takes the form of self-torture. **1943** H. Read *Politics of Unpolitical* ii. 18 Masochism is the unconscious impulse to dissolve oneself in the power of another person and to participate in his annihilating power. **1963** A. Heron *Towards Quaker View of Sex* 71 *Masochism*, obtaining pleasure (usually sexual) through suffering pain, subjugation or humiliation. **1974** *Listener* 21 Nov. 679 That Johnson had masochistic fantasies..was first mooted 20 years ago... That Johnson's masochism did not stop at fantasy seems to be indicated by..entries in Mrs Thrale's diary.

masochist (mæ·sŏkist). [f. as Masochism + -ist.] One who is given to masochism.

1895 tr. *M. Nordau's Degeneration* v. i. 538 Masochists or passivists..clothe themselves in a costume which recalls, by colour and cut, feminine apparel. **1917** C. R. Payne tr. *Pfister's Psychoanal. Method* 78 Very many tormentors of animals are..sadists, consequently also, more or less masochists. **1957** *London Mag.* Jan. 75 Psychologically her affinities were with the more heroic type of masochist, such as Saint-Exupéry, T. E. Lawrence, and Orwell. **1974** 'J. Melville' *Nun's Castle* iv. 88 'Are you telling me Lynnet was some sort of masochist?'..'I think she associated love and pain.'

masochistic (mæsŏki·stik), *a.* Also (*erron.*) masso-. [f. prec. + -ic.] Of, pertaining to, resembling or characterized by masochism.

1904 G. S. Hall *Adolescence* II. 112 Women may acquire a Massochistic love of violence and pain for the ideal of pleasure. **1928** *Music & Lett.* July 125 The almost masochistic melancholy of the average fox-trot. **1958** *Essays & Stud.* XI. 65 The very energy of the style is masochistic—a tormenting awareness of its own impotence to do, or change, anything. **1959** *Times* 30 Sept. 13/5 Eighteenth-century patrons seemed to take a masochistic delight in seeing themselves lampooned on the stage. **1974** [see *masochism].

Hence **ma:sochi·stically** *adv.*, in a masochistic manner.

1936 *Times Lit. Suppl.* 18 July 597/2 Minor Canons speak of themselves masochistically as 'mere' minor canons. **1947** S. J. Perelman *Westward Ha!* (1949) vii. 87 Masochistically allowing chickens to peck at our bare tootsies.

Mason (mē̆i·s'n), *sb.*[3] orig. *U.S.* [f. the name of John *Mason*, who was granted the patent for such jars in 1858.] *Mason jar*: a wide-mouthed glass jar with an airtight screw top widely used in home bottling; also *Mason fruit-jar*.

1885 *N.Y. Weekly Tribune* 6 Aug. 13/2 The Illinois Agricultural Society calls attention to the fact that Mason fruit-jars have been sent to that State packed in straw foul with Canada thistle. **1888** L. Hargis *Graded Cook Bk.* 472 Quince and apple butter... Put a little of the mixture in a plate and invert, if it adheres the butter is done. Fill Mason jars and seal. **1947** *Nat. Geogr. Mag.* June 822/1 His annual cherry crop amounted to one forlorn little cherry which he covered with a mason jar. **1950** *N.Z. Jrnl. Agric.* Jan. 51/3, 2-pint mason jar feeders..were placed in the group of hives. **1972** H. C. Rae *Shooting Gallery* ii. 80 Landscapes evolved in his mind, like smoke in a Mason jar. **1973** J. Jones *Touch of Danger* xxvii. 162, I began to swish the martini mixture in the big Mason jar.

Mason and Dixon. *U.S.* = *Mason's and Dixon's line*: see Line *sb.*[2] 17 b. Now usu. Mason-Dixon.

1779 in W. B. Reed *Life & Corr. J. Reed* (1847) II. 134 The Virginia gentlemen offer to divide exactly the 40th degree with us... Perhaps we [of Penna.] would be as well off with Mason and Dixon's line continued. **1834** C. A. Davis *Lett. J. Downing* 36 And he tell'd me Georgia would go for me, arter the Gineral, as soon as any north of mason and dickson. **1843** *Knickerbocker* XXII. 185 The writer, who dwelleth near Mason and Dixon, descants upon the awful climate. **1948** *Downers Grove* (Illinois) *Reporter* 21 Oct. 3/2 Two Dixiecrats, out of their element north of the Mason-Dixon, nevertheless registered their convictions on the States' Rights issue. **1969** I. Kemp *Brit. G.I. in Vietnam* vii. 155 You're crazy, Limey! You couldn't keep me here for all the girls south of the Mason-Dixon Line. **1975** *Times* 20 Aug. 12/3 The tortured condition of the Jewish middle-class intellectual seems..the dominant subject of novels written north of the Mason-Dixon line.

Also *transf.*

1948 *Sat. Even. Post* 30 Oct. 124/4 This same little band has long since adopted three war-ruined families in the old country—two in Bizonia, one on the other side of Europe's bitter Mason and Dixon's line. **1970** *New Yorker* 22 July 4/3 Sub-Mason-Dixon dining.

Masonite[3] (mē̆i·sŏnəit). Also masonite. The proprietary name of a type of fibreboard made from wood-fibre pulped under high steam pressure.

1926 *Official Gaz.* (U.S. Patent Office) 23 Nov. 802/1 Mason Fibre Company, Laurel, Miss. Filed July 24, 1926. *Masonite. Particular description of goods.*—Fiber Board, Insulating Board, Composite Construction Board, Synthetic Lumber, or Artificial Lumber. **1933** *Sci. Amer.* Jan. 39/1 Floors will be of Masonite cushioned flooring. **1948** *Archit. Rev.* CIII. 270 (*caption*) The walls are covered with masonite sheeting fixed with broken recessed joints and slightly rounded corners, and painted grey-white. **1950** in W. Schack *Man. Plastics & Resins* 279 *Masonite* (trade name) refers to a series of headboards having considerable structural strength and resistance to abrasion and moisture. **1970** *Jrnl. Gen. Psychol.* LXXXII. 58 A five-digit counter, a speaker, two lights, and two standard Gerbrands rat levers were mounted on the lower panel (the unused lever was concealed by a masonite mask). **1970** *Southerly* XXX. 186 His blunted thumb jerked over the hump of his shoulder, pierced the masonite lining, pierced the weatherboards, rode the thirty odd miles of train-tracks and arrived in Sydney. **1973** *Jrnl. Genetic Psychol.* CXXII. 19 The stimuli used for pretraining and oddity training were multidimensional common household items and toys mounted on 4 × 4 inch brown-masonite squares.

masonry, *sb.* Add: **3.** Also *transf.*

1926 J. Black *You can't Win* xv. 202 George..took no pains to hide himself, feeling sure that the masonry of the road..would protect him against the common enemy—the law.

b. (See quot. 1896.) *slang.*

1841 Lytton *Night & Morning* (ed. 2) II. III. viii. 178, I was one of them, and know the masonry. **1896** Farmer & Henley *Slang* IV. 289/1 *Masonry*, secret signs & pass-words.

mass, *sb.*[1] Add: In the use of Roman Catholics freq. with pronunc. (mɑs). **7. mass-time** (earlier and later examples); **mass rock** *Hist.*, a rock at which persecuted Irish Catholics would gather to celebrate the mass in secret.

1914 W. P. Burke *Irish Priests in Penal Times* p. vii, The 'Mass Rock'.., the 'Priests' Hollow',..and many a similar name..are witnesses..to a hunted priesthood. **1932** H. Concannon *Blessed Eucharist in Irish Hist.* xix. 389 He was often placed on the summit of a high rock, to signal the approach of the 'priest hunters', while around a great stone, 'a Mass rock',..the parishioners were assembled for Mass. **1933** Father Augustine *Ireland's Loyalty to Mass* x. 166 Carraig an Aifrinn, (the Mass Rock). **1959** D. D. C. P. Mould *Peter's Boat* vi. 79 There are, in fact, an enormous number of known Mass rocks in this country, and that is to be expected, for they were in use whenever persecution was hot, and in fact, come almost into living memory. **1483** in *N. & Q.* (1973) Apr. 125/1 His Highnes wyth a sword in his hand berying yt upright all the masse tyme. **1845** 'C. Malone' in S. J. Brown *Poetry of Irish Hist.* (1927) 258 At Mass-time once I went to play. **1922** Joyce *Ulysses* 279 And once at masstime he had gone to play... A boy. A croppy boy.

mass, *sb.*[2] Add: **4. a.** Also freq. in pl. (Further examples.)

1958 *Observer* 18 May 16/4 If you weren't a poor little rich girl, or very pretty or exciting socially, there were still masses of fun to be had. **1974** A. Morice *Killing with Kindness* viii. 56 I'm sure you've got masses to do.

e. (Earlier example.)

1873 E. Atkinson tr. *Helmholtz's Pop. Lect. Sci. Subjects* 1st Ser. 101 Although we are not usually clearly conscious of these beating upper partials, the ear feels their effect as a want of uniformity or a roughness in the mass of tone.

g. *Psychol.* (See *apperceiving mass, *apperception mass.)

1907 W. James in *Jrnl. Philos.* 18 July 397 My statements may seem less obscure if surrounded by something more of a 'mass' whereby to apperceive them.

10. *attrib.* and *Comb.*, passing into *adj.* Examples, of which a selection is listed below, are very numerous in the 20th century. Used to mean: of, involving, composed of masses of people (or things) or the majority of people (or a society, group, etc.); done, made, etc., on a large scale.

mass-appeal, art, audience, behaviour, circulation, communication (hence *mass communicator*), *consciousness, consumer* (hence *-consuming, consumption*), *cult, culture, deportation, education, emotion* (hence *emotional* adj.), *entertainment, fear, grave, hypnosis* (hence *-hypnotized* adj.), *hysteria, immigration, literacy, migration, mind* (hence *-minded* adj., *-mindedness*), *movement, murder* (hence *mass murderer*), *party, persuasion, propaganda, psychology, public, society, suggestion, suicide, unemployment*; *mass-made* ppl. a.; *mass-advertising, -buying, -merchandising* (also *-merchandised* ppl. a., *-merchandiser*), *-selling, -thinking* vbl. sbs. **b.** spec. *Mil.*, as *mass attack* (also as vb.), *formation, raid*; *mass-bombing* vbl. sb. **c.** *Physics*, as *mass-flow, transport*.

1958 *New Statesman* 20 Dec. 881/2 Not all the spice goes into the thick wedges of Christmas pudding; and surely, as we sink deeper and deeper in the treacle of mass-advertising, we deserve this one tiny life-saver? **1944** L. MacNeice *Christopher Columbus* 14 To assert..that all art should have mass-appeal is like asserting that all mathematics should be 'for the million'. **1962** *Rep. Comm. Broadcasting 1960* 16 in *Parl. Papers 1961–2* (Cmnd. 1753) IX. 259 It has been said that in fact people watch these items; that the justification lies precisely in the fact that they are mass-appeal items. **1962** S. Strand *Marketing Dict.* 441 In marketing, a product is said to have *mass appeal* when it is desired by the multitudes in all income levels. **1938** *Current Hist.* Feb. 54 (*caption*) A Mass Art. **1959** M. McCarthy *Sights & Spectacles* p. xiv, Some of the editors felt that the theatre was not worth bothering with, because it was neither a high art, like Art, nor a mass art, like the movies! **1964** Hall & Whannel *Pop. Arts* iii. 68 Mass art often destroys all trace of individuality and idiosyncrasy which makes a work compelling and living. **1970** I. C. Jarvie *Towards Sociol. of Cinema* vii. 99 Cinema is in a real sense a mass art, not only in that it appeals to the mass audience, but also in that the individual is lost in the darkness of the continuous performance. **1947** Crowther & Whiddington *Science at War* 101 When convoys of merchant ships were mass-attacked by U-boats in 1942, they were liable to suffer heavy losses. **1959** *Chambers's Encycl.* XIV. 733/1 On 8 August a mass attack was launched on a convoy near the Isle of Wight. **1960** L. L. Snyder *The War 1939–45* vi. 117 August 6, 1940. From his country home..Goering issued orders for the first great mass attack on England. **1938** *Time* 21 Nov. 53/1 Hitchcock pictures..are often too intricately built and written to appeal to mass audiences. **1957** R. Hoggart *Uses of Literacy* vi. 147 Genuine controversy..alienates, divides and separates, the mass-audience. **1957** P. Martineau *Motivation in Advertising* xiv. 169 The 'mass-audience' housewife. **1967** M. McLuhan *Medium is Massage* 22 The mass audience..successor to the 'public'. **1974** *B.B.C. Handbk.* 261/2 Opera, so obviously appropriate to television but with a limited appeal to the mass audience. **1940** Harrisson & Madge *War begins at Home* i. 21 When hundreds of those replies show similar attitudes, we know we are on to something really important in terms of mass behaviour. **1941** E. C. Shepherd *Mil. Aeroplane* 4 Anti-aircraft fire can..break up the formations so that mass bombing or pattern bombing becomes impossible. **1929** Mass buying [see *mass selling* below]. **1950** B. Schulberg *Disenchanted* (1951) v. 61 Occasionally appearing in mass circulation magazines with stories increasingly ordinary. **1957** F. Williams *Dangerous Estate: Anat. Newspapers* xviii. 284 The tabloids and mass-circulation Sunday newspapers. *Ibid.* 291 The great mass-circulation newspapers.. command their millions. **1963** *Times Lit. Suppl.* 26 Apr. 298/4 The mass-circulation magazines. **1941** Beals & Brody *Lit. Adult Educ.* v. 139 Concentration of power in the agencies of mass communication is vicious. **1946** J. S. Huxley *Unesco* ii. 25 The spread of information through all media of Mass Communication—in other words, the press, the cinema, the radio and television. **1954** J. B. Priestley *Magicians* ii. 47 Mass communications become stronger in their effects every year. **1968** P. Oliver *Screening Blues* 9 Mass communication through recording, and later through radio, spread the culture of the blues beyond the local definition until the whole Negro world was its habitat. **1971** B. Mafeni in J. Spencer *Eng. Lang. W. Afr.* 100 The great potentialities of the language [*sc.* Nigerian Pidgin] as a medium of mass communication. **1965** *Punch* 9 June 858/1 One way and another Lord Thomson is quite a mass communicator. **1967** *Ibid.* 4 Jan. 23/2 The delusion you are suffering from usually attacks columnists, commentators and other mass communicators. **1922** D. H. Lawrence *Aaron's Rod* x. 139, I want to get myself awake, out of it all—all that mass-consciousness. **1947** Koestler in *Partisan Rev.* XIV. 141 Structural changes in economy without functional changes in mass-consciousness, must always lead to a dead end. **1931** A. Huxley *Music at Night* iv. 208 A man who has no interest in the things of the mind..is the ideal consumer, the mass consumer. **1971** G. Steiner *In Bluebeard's Castle* iv. 87 The informational energy required by a mass-consumer society is being transmitted pictorially. **1954** *Encounter* Mar. 5/2 Those who accept conformity do not challenge the existence of a mass-producing, mass-consuming society, even though they refuse its values. **1969** R. B. Fuller *Operating Man. Spaceship Earth* viii. 117 As we study industrialization, we see that we cannot have mass production unless we have mass consumption. **1963** *Spectator* 14 June 783 Mr. Macdonald's target is 'mass culture' and its nauseating child 'Midcult', which, if anything, is worse than 'Masscult'. **1971** M. W. Young *Fighting with Food in Massim Society* i. 11 Cargo beliefs are still held by many individuals, however, and a fresh catalyst could well provoke another mass cult. **1939** *Life* 24 July 65/2 The State of Texas..has never been properly recognized for its contributions to U.S. mass culture. **1957** R. Hoggart *Uses of Literacy* i. 23 We are moving towards the creation of a mass culture. **1957** *Economist* 5 Oct. 45/1 Whether the reader turns to Shakespeare or to Mr Erle Stanley Gardner for solace may depend less on his psyche and the state of 'mass culture' than on the thorny economics of the book business. **1975** *Listener* 9 Jan. 40/2 The mass culture which is everyone's hope..depends upon everyone sharing what has previously been the preserve of a privileged few. **1952** C. P. Blacker *Eugenics* 146 Since these words were written, we have witnessed the inhumanities of concentration and extermination camps, of labour camps and mass deportations, which have scarcely been surpassed by previous tyrannies. **1955** Koestler *Trail of Dinosaur* 160 The mass-deportations in the 'thirties produced new waves of the plague. **1927** A. Huxley *Proper Stud.* 113 The ordinary system of mass education. **1940** Graves & Hodge *Long Week-End* iv. 50 The terms 'thriller' and 'shocker'..had been in use since the Eighties—an early by-product of mass-education. **1964** C. Barber *Ling. Change Present-Day Eng.* ii. 20 The influence of mass-media and of mass-education. **1927** A. Huxley *Proper Stud.* 183 Periodic revivalism and the evoking of great mass-emotions on such occasions as pilgrimages provide the necessary emotional excitement. **1961** A. O. J. Cockshut *Imagination of Charles Dickens* v. 78 Mingling with the crowd he is untouched by mass emotions. **1963** *Economist* 3 Aug. 445/2 Everyone interested in political, mass emotional, historical causes should read it. **1933** *Radio Times* 14 Apr. 71/1 A music-hall or other large centre of mass-entertainment. **1960** *20th Cent.* Dec. 557 The huge metropolitan machines of mass entertainment. **1932** H. Nicolson *Public Faces* xii. 324 There is only one human emotion stronger than mass-hatred, and that is mass-fear. **1952** Pankhurst & Holder *Wind-Tunnel Technique* i. 31 The rate of mass flow per unit cross-sectional area (the mass velocity) is given by [etc.]. **1957** *Times* 4 Oct. 11/7 In the by-pass engine a given thrust is obtained by a greater mass-flow of air at reduced velocity, the engine being both bigger and heavier than the 'straight jet'. **1917** A. G. Empey *Over Top* 299 *Mass formation*, a close order formation in which the Germans attack. **1948** E. Pound *Pisan Cantos* (1949) lxxvii. 49 By getting me onto the commission To inspect the mass graves at Katin. **1972** *New Statesman* 28 Jan. 109/3 The communists..murdered several thousand political prisoners, whose bodies were later discovered..in three separate mass graves. **1974** *Times* 4 Sept. 8/1 Turkish troops today uncovered more bodies from the mass grave at the Turkish Cypriot village of Maratha. **1951** M. McLuhan *Mech. Bride* (1967) 10/2 Power, glitter and mass hypnosis engendered by regular ranks. **1967** G. Playfair *Prodigy* iii. 82 There was..an element of mass-hypnosis in Bettymania, as there is in all such manifestations of idolatry. **1946** R. Campbell *Talking Bronco* 61 Mass-hypnotized, dinned drunken by the tireless Mechanic repetition of the wireless. **1934** A. Huxley *Beyond Mexique Bay* 276 The mindless mass hysteria of howling mobs. **1954** *Economist* 22 May 611/1 Yet Mr Graham does not produce mass hysteria, although he is certainly dramatic. **1969** Milgram & Toch in Lindzey & Aronson *Handbk. Social Psychol.* (ed. 2) IV. xxxv. 507 Such diverse phenomena as 'collective excitement, social unrest.., riots.., mass hysteria, [etc.]'. **1973** *Times* 7 Dec. 19/4 There is an argument for introducing rationing in London where some form of mass hysteria seems to have occurred. **1949** Koestler *Promise & Fulfilment* vii. 69 The next transformation came after the First World War when mass immigration started in earnest. **1973** *Times* 5 Dec. 18/2 Israel was founded on..mass immigration as the dynamic of growth. **1937** C. Madge in C. Day Lewis *Mind in Chains* 147 A fact of to-day which is so near to us that it is hard for us to see it is mass-literacy. **1967** *Economist* 15 Apr. p. i/3 The Franklin Book Programs Inc.. uses all kinds of techniques from mass-literacy and mass-distribution schemes to training programmes. **1934** T. S. Eliot *Rock* i. 46 Those whose souls are choked and swaddled In the..new winding sheets of mass-made thought. **1959** V. Packard *Status Seekers* (1960) i. 3 They can dine on mass-merchandised vichyssoise. *Ibid.* 4 Everybody could enjoy the good things of life—as defined by mass merchandisers. **1962** S. Strand *Marketing Dict.* 442 *Mass merchandising*, a large-scale selling method, appealing to the population as a whole with convincing advertising. **1972** R. H. Buskirk et al. *Concepts of Business* xii. 169 The mass-merchandising specialty store.. frequently calls itself a discount house. **1901** E. A. Ross *Social Control* xxix. 396 The common perils of war or mass migration may call for stricter corporate discipline. **1935** Huxley & Haddon *We Europeans* ix. 273 Mass-migration and military conquest. **1932** H. J. Massingham *World without End* x. 264 A man or woman who deserts his or her distinctive colour of being and joins the mass-mind. **1936** G. M. Young *Victorian Eng.* xxviii. 158 The propagandist, the advertiser and all other agents of the mass-mind. **1955** Koestler *Trail of Dinosaur* III. 242 An unexpected mutation of the mass-mind may occur. **1964** M. McLuhan *Understanding Media* xi. 107 Creating the paradox of the 'mass mind' and the mass militarism of citizen armies. **1934** Webster, Mass-minded. **1942** D. Powell *Time to be Born* (1943) xii. 280 The masses of women..too mass-minded in their ambitions to be even faintly understood by her. **1939** H. J. Massingham *Countryman's Jrnl.* xxxiii. 143 It is good for a man..to converse with his own spirit or what mass-mindedness has left of it. **1961** *Ann. Reg. 1960* 430 The first of these was Eugene Ionesco's *Rhinoceros*, a savage satire on mass-mindedness. **1897** *Amer. Jrnl. Sociol.* Nov. 344 They are cultivating mass sympathies and drilling themselves in mass movements. **1927** D. H. Lawrence *Let.* 11 July (1932) 684 They [*sc.* the Germans] are capable of mass-movement. **1935** *Discovery* Oct. 292/1 Occasional mass-movements, like those of the lemmings of Scandinavia, were not to be confused with true migration. **1939** *Encycl. Brit. Bk. of Year* 19/2 In the United States there is taking place what may roughly be termed a mass movement towards Adult Education. **1967** H. Arendt *Orig. Totalitarianism* (new ed.) x. 313 The decisive differences between nineteenth-century mob organizations and twentieth-century mass movements are difficult to perceive. **1931** S. Nearing (*title*) War: organized destruction and mass murder by civilized nations. **1945** R. A. Knox *God & Atom* vi. 79 Democracy labours for breath, when the power of mass-murder is concentrated..in the hands of a few. **1958** *Times Lit. Suppl.* 13 June 334/3 His solution for the whole problem of mass-murders is 'family-love'. **1967** H. Arendt *Orig. Totalitarianism* (new ed.) xii. 421 The bulk of the armed SS served at the Eastern front where they were used for 'special assignments'—usually mass murder. **1952** B. Wolfe *Limbo* (1953) vii. 80 Look at this baby-faced mass murderer. **1960** *Sunday Express* 25 Dec. 10/2 A reprieved mass-murderer in Dartmoor. **1947** 'G. Orwell' *Eng. People* 22 The Communist Party [in Britain]..has never shown signs of growing into a mass party of the kind that exists in France. **1954** B. & R. North tr. *Duverger's Pol. Parties* I. ii. 63 The mass-party technique in effect replaces the capitalist financing of electioneering by democratic financing. **1956** C. W. Mills *Power Elite* xiii. 310 The increased means of mass persuasion that are available. **1960** J. B. Priestley *Lit. & Western Man* ii. 16 The time of vast urban masses and all our techniques of mass persuasion. **1950** Koestler et al. *God that Failed* 65 Independent of the mass-propaganda methods of the Munzenberg enterprises. **1967** H. Arendt *Orig. Totalitarianism* (new ed.) viii. 264 Entirely new and complicated forms of mass propaganda were adopted by all parties. **1900** *Amer. Jrnl. Sociol.* Jan. 521 This is the reason why ethnology is finding its most promising developments today in the line of ethnic or folk-psychology, which is only a cross-section of mass-psychology. **1916** B. Russell *Princ. Social Reconstruction* ii. 60 Even the few questions which are left to the popular vote are decided by a diffused mass-psychology. **1929** Chesterton *Thing* xxii. 166 The abandonment of individual reason, in favour of press stunts and suggestion and mass psychology. **1937** Koestler *Spanish Testament* II. 308, I could wish that everyone who talks of mass psychology should experience a year of prison. **1960** H. Read *Forms of Things Unknown* iv. xi. 179 A resort to mass psychology is evasive. **1938** *Times Lit. Suppl.* 17 Sept. 598/1 The mass-public does not want opinion, it wants news. **1960** *New Left Rev.* May–June 65/2 Personalities..with an existence quite different in tone from the mass public. **1939** *War Weekly* 25 Oct. 1313/1 First night mass raid on London... On the Tuesday night, a night mass air attack was tried for the first time. **1929** *Publishers' Weekly* 19 Oct. 1928/1 Our shop, like other small shops, is not geared for mass selling or mass buying. **1968** *Listener* 29 Aug. 285/3 There is now hardly a significant publication, from the weekly reviews to the mass-selling dailies, which does not have equity in one or other of the programme-contracting companies. **1948** T. S. Eliot *Notes Def. Culture* ii. 40 This gives it [*sc.* 'bourgeois' society] a difference in kind from the aristocratic society which preceded it, and from the mass-society which is expected to follow it. **1950** *Antioch Rev.* X. 382 Mass society is no more stable under dictatorship than it is under democracy. **1957** F. Williams *Dangerous Estate: Anat. Newspapers* xviii. 285 The straits to which an industrial mass-society has brought millions of its members. **1967** G. Steiner *Lang. & Silence* 69 The possibility that.. certain elements in the technological mass-society..have done injury to language is the underlying theme of this book. **1901** E. Ross *Social Control* 148 In public opinion there is something which is not praise or blame, and this residuum is mass suggestion. **1920** W. McDougall *Group Mind* ii. 42 A proposition which voices the mind of the crowd..and so comes with the power of a mass-suggestion. **1924** W. B. Selbie *Psychol. Relig.* 116 We have here to reckon with the influence of mass suggestion. **1947** *Mind* LVI. 60 There can be mass suggestion, there are methods of group persuasion, but reasoning is carried on by each of us not only for himself but by himself. **1937** M. Covarrubias *Island of Bali* (1972) vii. 199 The famous krisses of the kings of South Bali taken by the Dutch as war booty at the time of the great mass-suicide of Den Pasar in 1906. **1959** *Manch. Guardian* 9 July 6/6 A.. policy of mass suicide masquerading as a policy of defence. **1924** *Public Opinion* 30 May 528/3 Our modern saints of co-operative mass-thinking. **1958** R. Williams *Culture & Society* III. 298 Mass-thinking, mass-suggestion, mass-prejudice would threaten to swamp considered individual thinking and feeling. **1953** *Phil. Trans. R. Soc.* A. CCXLV. 535 The mass-transport velocity can be very different from that predicted by Stokes on the assumption of a perfect, non-viscous fluid. **1957** G. E. Hutchinson *Treat. Limnol.* I. v. 347 The small deepwater waves of negligible amplitude..produce virtually no mass transport of water. **1974** *A.I.Ch.E. Jrnl.* XX. 88/1 One of the potential advantages of slurry reactors is reduction of intraparticle mass transport resistances. **1937** 'G. Orwell' *Road to Wigan Pier* viii. 153, I wanted to see what mass-unemployment is like at its worst. **1965** J. L. Hanson *Dict. Econ.* 387 Mass unemployment is the most serious type since it is due to a general deficiency of demand. **1974** *Times* 5 Nov. 15/3 The monetarists fear that some unemployment may be the necessary price of avoiding mass unemployment.

d. mass action, (*a*) *Chem.*, the effect which the concentration of a reactant has on the rate of a chemical reaction; (*b*) the action of a mass of people; **mass balance** *Aeronaut.*, a state in which inertial coupling between the angular movement of a control surface and other degrees of freedom of the aircraft is eliminated, so avoiding flutter of the surface; also, a mass attached to a control surface to bring about such a state; also as *v. trans.*; so **mass-balanced** *ppl. a.*, **mass-balancing** *vbl. sb.*; **mass concrete,** concrete which is not reinforced; **mass defect,** a deficiency of mass; *spec.* in *Nuclear Physics*, the sum of the masses of the constituent particles of a nucleus, as free individuals, less the mass of the nucleus (a quantity which effectively represents the binding energy needed to disperse the particles of the nucleus); **mass distribution,** the distribution of goods in bulk; **mass-effect,** (*a*) (see quot. 1902); (*b*) *Metallurgy*, the effect of size and shape in causing different rates of cooling, and so different hardnesses, in different parts of an object following heat treatment; (*c*) (usu. in *pl.*) a total or 'grand' effect; (*d*) an effect due to or dependent on mass; **mass-energy,** (*a*) the property of which mass and energy are regarded as different but interconvertible manifestations, being related by the equation $E=mc^2$ (propounded by Einstein in *Ann. d. Physik* (1905) XVIII. 641), where E is the energy equivalent of a mass m and c is the speed of light; (*b*) *attrib.*, relating to (the equivalence of) mass and energy; **mass man,** a hypothetical average man; one typical of mass society, characterized by a lack of individuality and a tendency to be manipulated by stereotyped ideas from the mass media; **mass market,** the market for mass-produced goods; also (with hyphen) as *vb.*; hence **mass-marketed** *ppl. a.*, **mass-marketing** *vbl. sb.*; **mass meeting** (earlier and later examples); also *transf.* and *fig.*; **mass noun,** a noun which in common usage lacks a plural form (opp. **count-noun*); **mass number** *Nuclear Physics*, the total number of protons and neutrons in an atomic nucleus; **mass phenomenon** (see quot. 1968); **mass-point** *Physics*, an entity conceived as having mass and (like a geometrical point) occupying a position but lacking spatial extension; **mass radiography,** radiography of the chests of large numbers of people by a quick routine method; **mass-ratio,** the ratio of the masses of two things; *spec.* in *Aeronaut.*, the ratio of the mass of a rocket with full fuel tanks to that of the same rocket with empty fuel tanks; **mass-reflex** *Physiol.*, (in patients who have suffered gross injury to the spinal cord) a reflex which may involve all parts of the body innervated from the part of the spinal cord below the lesion; **mass spectrograph,** a type of mass spectrometer (in the broader sense) in which deflected ions are made to strike a photographic plate so as to produce a photographic mass spectrum; hence *mass-spectrographic* adj.; **mass spectrometer,** any instrument in which material in a vacuum is ionized and the resulting ions are formed into a beam, separated according to the ratios of their mass to their net electric charge (e.g. by deflecting them in a magnetic field or accelerating them in an electric field), and then detected; *esp.* one in which the detection is done electrically rather than photographically; so **mass spectrometry;** also *mass-spectrometric* adj., *-spectrometrically* adv.; **mass spectroscope** (*rare*), an instrument for producing a mass spectrum; a mass spectrometer (in the broader sense); **mass spectroscopy,** the art of using the mass spectrometer or mass spectrograph; that branch of science which involves the use of these instruments; so *mass-spectroscopic* adj.; **mass spectrum,** a record obtained with a mass spectrometer or mass spectrograph, in which ions from a sample material are represented as dispersed according to their mass-to-charge ratio; **mass transfer** *Chemical Engin.*, movement of one substance through or into another on a molecular scale; **mass unit** = *atomic mass unit* (s.v. *ATOMIC *a.* and *sb.* A. 1); **mass wasting** *Geomorphol.*, movement of rock, soil, fallen snow, or the like under the influence of gravity; **mass-word** = **mass noun*. See also *MASS MEDIUM, *MASS OBSERVATION, *MASS PRODUCTION.

1891 G. M'Gowan tr. *E. von Meyer's Hist. Chem.* 461 Berthollet..deduced precisely the opposite from his own assumption—that mass-action comes into play in chemical processes. **1903** H. C. Jones *Princ. Inorg. Chem.* xxviii. 346 If these crystals are redissolved in more water, and the solution evaporated to crystallization, the neutral salt will separate, showing a further splitting off of sulphuric acid due to the mass action of the water. **1924** W. B. Selbie *Psychol. Relig.* 73 It is not only that mass action has a marked effect upon the will, but that [etc.]. **1958** R. Williams *Culture & Society* III. 298 The derived ideas have arisen from urbanization..from the working class, mass-action. **1970** Passmore & Robson *Compan. Med. Stud.* II. iii. 1/2 Whatever the nature of a receptor, its reaction with a drug is presumably chemical and can be described by the Law of Mass Action. **1974** *Daily Tel.* 17 Sept. 6/3 He went on to justify the disruption of academic life during student campaigns, and said that mass action in support of student policies would go ahead. **1931** *Flight* 11 Sept. 906/1 The control surfaces have been mass-balanced in order to reduce the risk of flutter. **1935** Pippard & Pritchard *Aeroplane Struct.* (ed. 2) xii. 238 A bob weight used to give mass balance to an aileron should be placed as far outboard of any point of wing support as possible. **1959** J. L. Nayler *Dict. Aeronaut. Engin.* 168 The mass balance may be a single lump of metal or be distributed along the span of the control surface and connected to it by a series of links. **1966** F. G. Irving *Introd. Longitudinal Static Stability Low-Speed Aircraft* viii. 82 Full mass-balance..may not be required as an anti-flutter measure on slow-speed aircraft. **1934** *Jrnl. R. Aeronaut. Soc.* XXXVIII. 76 Aeroplanes employing mass-balanced ailerons. *Ibid.*, Those [aeroplanes] in which the anti-flutter device of mass-balancing is incorporated. **1947** C. F. Toms *Introd. Aeronaut.* v. 195 There can arise cases in which dynamic mass-balancing with a single mass is not possible and two masses are required. **1972** T. H. G. Megson *Aircraft Struct.* xiii. 468 The position of the inertia axis may be adjusted by a redistribution of wing weight, a process known as mass-balancing. **1930** *Engineering* 25 July 101/2 After the completion of the mass concrete foundation, the reinforcement was erected for the columns. **1924** J. G. A. Skerl tr. *Wegener's Orig. Continents & Oceans* xi. 160 Subterranean mass-defect and mass-excess above the sea-level..mutually counterbalance, isostasy thus prevailing in mountain masses. **1927** F. W. Aston in *Proc. R. Soc.* A. CXV. 510 There would have been no loss of energy, that is mass defect, in the latter [*sc.* alpha particles] to represent the binding forces holding the four particles together. **1962** H. D. Bush *Atomic & Nucl. Physics* iii. 69 According to the principle of the equivalence of mass and energy, the mass defect is equivalent to the energy necessary to separate the constituent particles: $\Delta E=c^2.\Delta M$. This relation shows that a mass defect of one a.m.u. is equivalent to a binding energy of 931×10^6 eV (931 MeV). **1936** J. F. Pyle *Marketing Princ.* (ed. 2) xviii. 575 Mass distribution, the corollary of mass production, is made possible through the effective use of sales-promotional methods and devices. **1951** C. W. Mills *White Collar* I. ii. 26 What good would your mass production be without our mass distribution? **1967** *Economist* 15 Apr. p. i/3 One of the most effective spearheads of American book publishing..uses all kinds of techniques from mass-literacy and mass-distribution schemes to training programmes. **1902** Hillebrand & Penfield in *Amer. Jrnl. Sci.* CLXIV. 217 The alkalies and lead play so small a rôle, and the remaining constituents so prominent a part in the complex chemical molecules, that the latter control or dominate the crystallization by virtue of what may be called their mass-effect. **1925** *Jrnl. Iron & Steel Inst.* CXII. 473 (*heading*) Initial temperature and mass effects in quenching. **1934** Webster II. 1510/2 Arranged in or involving masses, as a mass effect. **1936** *Mind* XLV. 226 It may be true that macroscopic phenomena involve—or, if we like to say so, are mass-effects of—large quantum numbers. **1939** Carpenter & Robertson *Metals* II. xvi. 1362 To discover whether duralinium was subject to mass-effect in the same way as steel, Teed quenched 5 in. diam. and 0·564 in. diam. bars of the same material from 490°C. into water. **1945** H. D. Smyth *Gen. Acct. Devel. Atomic Energy Mil. Purposes* iv. 35 It was found that the total absorption of neutrons by such spheres could be expressed in terms of a 'surface' effect and a 'mass' effect. **1947** A. Einstein *Mus. Romantic Era* xvi. 248 The rhythmic piquancy and the mass-effects of Auber. **1968** D. R. Cliffe *Technical Metall.* v. 112 The actual cooling rate depends upon several factors, including the diameter of the bar or thickness of section (i.e. mass effect). **1970** G. K. Woodgate *Elem. Atomic Struct.* ii. 28 This is an example of isotope shift arising from the normal mass effect. **1935** *Proc. R. Soc.* A. CXLIX. 415 The data..afford strong evidence of the validity of the laws of conservation of mass-energy, and of momentum, in some atomic transmutations. **1938** Einstein & Infeld *Evolution of Physics* 208 According to the theory of relativity, there is no distinction between mass and energy... Instead of two conservation laws we have only one, that of mass-energy. **1942** J. D. Stranathan *'Particles' of Mod. Physics* ix. 374 According to the mass-energy equivalence concept, the rest mass of an electron represents an energy $V=..0{\cdot}511\times10^6$ EV. **1968** M. S. Livingston *Particle Physics* i. 6 The released energy [in fission and fusion] comes from the excess mass which is transformed into energy... We can compute the energy release through the Einstein equivalence relation $E=mc^2$. This mass energy was stored in the heavy nuclei and in the very light nuclei in the form of excess mass at the time our galaxy was formed. **1928** A. Huxley *Point Counter Point* v. 76 They were armed to protect individuality from the mass man, the mob. **1945** H. Read *Coat Many Colours* lxxi. 348 Screwed-up tissue papers, cigar butts, all the characteristic droppings of Mass-man. **1966** D. Jenkins *Educated Society* ii. 70 They will become increasingly dependent on the public media..on the way to becoming mass-men. **1973** *Black World* May 8/1 Black community newspapers and talk shows on Black radio stations are other avenues for reaching our mass man. **1959** *Times* 9 Mar. (Britain's Food Suppl.) p. vii/7 The housewife who has an outside job now forms an important factor in the mass-market. **1959** *Time* 16 Mar. 61/1 An ultrasonic dishwasher for the home is already technically feasible but too expensive to mass-market. **1962** *Guardian* 7 Feb. 8/1 The mass-market furniture designers. **1971** *Engineering* Apr. 41/3 For which their whole mass-market philosophy had prepared them. **1960** *New Left Rev.* Sept.–Oct. 3/2 Mass-marketed commodities. **1945** B. Nash *Developing Marketable Products* ii. 23 This relationship between the large-volume manufacturer, the many different users, and the mechanisms of mass marketing is typified by the activities of every manufacturer using mass-marketing methods. **1733** B. Lynde *Diary* (1880) 39 Our mass meeting at which the village intend to urge their being a township. **1851** A. O. Hall *Manhattaner* 4 We steamed..by mass meetings of democratic looking logs and snags. **1880** A. E. Housman *Let.* 10 May (1971) 20 They were chairing Harcourt from the station to a mass-meeting at the Martyrs' Memorial. **1904** Wodehouse *Gold Bat* viii. 90 They would never run the risk involved in holding mass-meetings in one another's studies. **1960** R. Campbell *Coll. Poems* III. 73 In the mass-meeting of the waves. **1970** *Daily Tel.* 17 Apr. 2 Their 10-man executive committee would be calling a mass meeting of the 2,000-strong student body. **1933** L. Bloomfield *Lang.* xii. 205 Mass nouns never take *a* and have no plural. **1963** *Language* XXXIX. 209 The mass noun *blood* in the singular takes *the* and *some* but not numerical quantifiers. **1965** [see *COUNT *sb.*^{1} 9]. **1970** *Archivum Linguisticum* I. 24 There is no special class of mass-nouns in Portuguese marked by the absence of metaphony. **1923** F. W. Aston in *Phil. Mag.* XLV. 945 These integers are provisionally called 'mass-numbers'. The mass-number may be taken to represent the number of protons in the atom. **1946** *Electronic Engin.* XVIII. 153/2 The separation of the uranium isotope of mass number 235..from that of mass number 238..constituted one of the major investigations in the development of the atomic bomb. **1962** H. D. Bush *Atomic & Nucl. Physics* iii. 63 A nuclide is indicated by the chemical symbol with the mass number as a superscript and atomic number as a subscript, the latter often being omitted. For example, $_7N^{14}$ or N^{14} refers to one of the isotopes of nitrogen ($Z=7$), namely the one which has mass number 14. **1936** J. R. Kantor *Objective Psychol. Gram.* iii. 32 Comparative grammar deals with auditory-vocal mass phenomena. **1954** R. W. Brown in G. Lindzey *Handbk. Social Psychol.* II. xxiii. 833/2 Social scientists have not been content to define mass phenomena as collective misbehavior. **1967** H. Arendt *Orig. Totalitarianism* (new ed.) ix. 277 Statelessness, the newest mass phenomenon in contemporary history. **1968** A. F. C. Wallace in *Internat. Encycl. Social Sci.* X. 55/2 'Mass phenomenon' signifies that class of social event in which a large number of people at the same time behave in a way which constitutes a notable interruption of their routine, socially sanctioned role behavior. **1911** J. Ward *Realm of Ends* xii. 255 The mass-points of the modern physicist.. Leibniz held to be only phenomenal. **1956** E. H. Hutten *Lang. Mod. Physics* vi. 243 Newtonian mechanics..treats only of such phenomena as can be described in terms of a few concepts, e.g. masspoint, force, etc. **1971** *Amer. Jrnl. Physics* XXXIX. 484/2 Consider a finite one-dimensional mass-point lattice in which nearest neighbors are joined by massless ideal springs. **1943** *Electronic Engin.* XVI. 108 Equipment for mass radiography installed in a large industrial concern for examination of the workers. **1954** E. Jenkins *Tortoise & Hare* x. 114 These mobile X-ray units for mass radiography. **1971** *Brit. Med. Bull.* XXVII. 93/2 Of the people screened by mass radiography..a large number have disabilities that warrant referral to a medical practitioner. **1946** A. S. Eddington *Fund. Theory* ii. 32 (*heading*) Mass-ratio of the proton and electron. **1949** W. Ley *Conquest of Space* (1950) i. 26 The mass-ratio is 2·7:1—that is,..the rocket at take-off weighs 2·7 times as much as its empty hull, machinery and payload. **1958** *New Scientist* 9 Jan. 19 To obtain an intercontinental range of 2,500 miles requires a mass ratio of 12:1, to achieve a satellite orbit requires a ratio of 25:1, and to escape to the Moon needs a ratio of 100:1... The German V2 had a mass ratio of only 3·2:1. **1968** R. A. Lyttleton *Mysteries Solar Syst.* i. 42 The Earth and Mars can be paired together satisfactorily as components resulting from a single rotationally unstable planet since their mass-ratio is as high as 9:1. **1917** Head & Riddoch in *Brain* XL. 233 It is evident, therefore, that under certain conditions the spinal cord below the level of the lesion may show signs of diffuse reflex activity. Scratching the sole of the foot may not only evoke a flexor spasm, but may cause premature evacuation of the bladder and an outburst of excessive sweating. This we have spoken of as a 'mass-reflex'. **1970** M. Hollander tr. *Monnier's Functions Nervous Syst.* II. xii. 242 A mass reflex is an irradiation phenomenon consisting of flexion of the legs with visceromotor reflexes: sweating, micturition and defecation. **1920** F. W. Aston in *Phil. Mag.* XXXIX. 611 The Positive Ray Spectrograph or, as it may be more conveniently termed, Mass-Spectrograph. **1945** H. D. Smyth *Gen. Acct. Devel. Atomic Energy Mil. Purposes* x. 109 Kellex, whose mass spectrograph methods of isotope analysis were sufficiently advanced as to become of great value to the project, as in analyzing samples of enriched uranium. **1963** J. B. Farmer in C. A. McDowell *Mass Spectrometry* ii. 8 The primary, although not the exclusive, task of mass spectrographs is..mass comparison... Mass spectrometers are primarily useful for work involving abundance determination. **1935** *Proc. R. Soc.* A. CL. 253 The masses which we have suggested give a much closer fit with the observed transmutation data than do the mass-spectrographic values. **1957** G. E. Hutchinson *Treat. Limnol.* I. iii. 215 The more modern mass spectrographic determinations..indicate very clearly that..the D and O^{18} contents [of water] vary concomitantly. **1932** *Physical Rev.* XL. 429 A new type of mass spectrometer..in which no magnetic fields are used. **1955** *New Biol.* XVIII. 66 It is possible by using a mass-spectrometer to determine the proportions of heavy nitrogen, N^{15}, and the normally more abundant isotope, N^{14}, in a given sample. **1963** Mass spectrometer [see *mass spectrograph* above]. **1966** A. J. Ahearn *Mass Spectrometric Analysis of Solids* i. 2 In the past, the term mass spectrometer was reserved for electrical detection. However, the term mass spectrometry is now the generally accepted designation for this mass analysis technique regardless of the type of detection employed. **1969** Price & Williams *Time-of-Flight Mass Spectrometry* p. ix, There are now about 300 time-of-flight mass spectrometers in use throughout the world. **1966** D. G. Brandon *Mod. Techniques Metallogr.* iii. 124 We shall also consider the possibility of ion-bombarding a specimen to sputter a thin layer from the surface, followed by a mass-spectrometric analysis of the atoms in this layer. *Ibid.* 172 The surface of the sample is bombarded by positive ions, and the sputtered ions are collected and analysed mass-spectrometrically. **1952** (*title*) Mass spectrometry (Inst. Petroleum, London). **1974** *Nature* 29 Mar. 458/1 The first fraction..was evaporated to dryness and subjected to mass spectrometry... Molecular ions at m/e 420 and m/e 448 are consistent with the product being a mixture of palmitic and stearic aldehyde 2,4-dinitrophenylhydrazones. **1958** H. E. Duckworth *Mass Spectroscopy* 204 Mass spectroscopes. **1963** J. B. Farmer in C. A. McDowell *Mass Spectrometry* ii. 7 The term mass spectroscope is applied to any device which has the ability to separate gaseous ions according to mass–charge ratio. **1938** R. W. Lawson tr. *Hevesy & Paneth's Man. Radioactivity* (ed. 2) xix. 178 Mass-spectroscopic investigations. **1968** M. S. Livingston *Particle Physics* vi. 120 The results of mass-spectroscopic measurements have been summarized from time to time to deduce by least-squares analysis the best values of atomic masses. **1926** R. W. Lawson tr. *Hevesy & Paneth's Man. Radioactivity* xviii. 135 Not until the method of mass-spectroscopy had been devised and developed was it possible to establish that many of the ordinary elements are..mixed elements. **1958** H. E. Duckworth *Mass Spectroscopy* i. 10 For many years photographic plates had served as the standard ion detectors in mass spectroscopy, but in the early 1930's these began to be supplanted by systems of electrical detection. **1920** F. W. Aston in *Phil. Mag.* XXXIX. 611 (*heading*) The mass-spectra of chemical elements. *Ibid.* 625 A positive ray spectrograph capable of giving a focussed mass-spectrum is..described. **1966** *Encycl. Industr. Chem. Analysis* II. 465 Mass spectra are related in a direct and simple way to molecular structure, thereby providing a unique and characteristic spectrum for each molecule which can be vaporized in the ion source. **1937** W. H. Walker et al. *Princ. Chem. Engin.* (ed. 3) xiv. 447 The mass transfer from the main body of the gas to the interface can be visualized as meeting two resistances in series, that of the turbulent main body of gas and that of the gas film. **1962** J. C. Wright *Metall. in Nucl. Power Technol.* ix. 169 Mass transfer between ferrous constructional materials and a sodium coolant circuit is accelerated if the oxygen content exceeds 30 p.p.m. by weight. **1971** R. Hardbottle tr. *Grassman's Physical Princ. Chem. Engin.* ix. 553 A typical feature of mass transfer is the slowing down of transfer of the migrating components by components not participating in mass transfer. **1942** Pollard & Davidson *Appl. Nucl. Physics* v. 74 If a positron is emitted it must be treated as costing 0·0011 mass unit or 1 Mev extra. **1953** Barnett & Wilson *Inorg. Chem.* iv. 28 There has been a loss of 0·0184 mass units. **1966** Gucker & Seifert *Physical Chem.*(1967) ii. 25 Prior to 1961 two different mass units were used: the physical mass unit.. and the chemical mass unit. **1951** *Ohio Jrnl. Sci.* LI. 299 (*heading*) Mass wasting, classification and damage in Ohio. **1968** R. W. Fairbridge *Encycl. Geomorphol.* 697/2 Frequently the immediate cause of mass wasting can be related directly to changes in shearing stress brought about by (1) increase in the weight of materials, (2) withdrawal of support, or (3) earth tremors. **1914** O. Jespersen *Mod. Eng. Gram.* II. v. 115 Words which represent 'uncountables'..are here called mass-words; they may be either material..such as *silver, quicksilver, water, butter,*..or else immaterial, such as *leisure, music, traffic, progress,* [etc.]. **1935** *Jrnl. Eng. & Germ. Philol.* XXXIV. 429 Masswords (like *gold, embers, knowledge*). **1954** Pei & Gaynor *Dict. Ling.* 133 *Mass-word*, Jespersen's term for words denoting concepts, properties or things which ordinarily cannot be separated into distinct component units.

Mass, *sb.*^{4} Var. Mas, *Mars(e.

1837 *Southern Lit. Messenger* III. 174 Mass Phil been very uneasy about you.

Massa. For def. read: Representing *master* in the written form of Negro speech. Also **massa.** (Add further examples.)

1821 P. Egan *Life in London* II. v. 289 But the black *slavey*, who is entering the room, is singing out, '*Massa*, you ought to be shamed.' **1935** in Z. N. Hurston *Mules & Men* (1970) I. i. 25 It was night and Ole Massa sent John, his favorite slave, down to the spring to get him a cool drink of water. **1953** *Caribbean Q.* III. III. 157 As one writer put it:..It has ever been their practice, to prevent the extension of useful practical knowledge, other than the hewing of cane fields, and feeding the mill, to the Negro lest he may one day rise to the rank of an Overseer, and supersede Massa Busher (Overseer). **1973** *Sunday Express* (Trinidad & Tobago) 8 Apr. 13/3 Day after day he sat glued to the t.v. and behaved generally as though he didn't know 'massa day is done'. **1973** J. Ryder *Trevayne* (1974) xi. 94 'I'll set up interviews.' Bonner looked at Andrew and laughed. '*Your* interviews, massa.'

massage, *sb.* Add: Now usu. with pronunc. (mæ·-). (Further examples.) Also *fig.*

1907 *Yesterday's Shopping* (1969) 538/3 Massage Rollers, for the face..13/0. **1929** H. Walpole *Hans Frost* II. iii. 143 Then with that by now practised and customary spiritual massage she set to work on herself. **1963** E. H. Edwards *Saddlery* xxii. 168 A convenient and efficient massage pad..can be made from leather stuffed with hay. **1966** J. S. Cox *Illustr. Dict. Hairdressing* 97/1 *Massage oil*, an oil which is applied to the scalp to aid the process of massage and increase its beneficial effect. **1969** E. H. Pinto *Treen* 17 The massage roulette..is English, of mahogany, and was advertised for sale between 1880 and 1890. The three rollers revolve independently. **1973** *Times* 8 Aug. 7/1 At present £52m is spent annually on pills and low calorie foods, then there are health hydros, sauna baths, massage parlours, exercising machines, gymnasiums and slimming clubs.

b. Used *euphem.* for: sexual activity.

The expressions *massage establishment, parlour,* etc., used as signs outside a building, are frequently alleged to mean 'brothel' (see quots.).

1913 *Collier's* 25 Jan. 7/1 Along with them go the announcements of 'massage parlors' (an all-too-obvious euphemism), free whiskies, and other agencies of public injury. **1921** E. Weekley *Etym. Dict. Mod. Eng.* 1416 *Stew*^{1}..Brothel... The public hot-air baths acquired a reputation like that of our massage establishments some time ago. **1966** *Economist* 3 Sept. 894/2 The American authorities..allow their free-spending men to attract the usual proliferation of bars, night-clubs and 'massage parlours'. **1971** *Sunday Tel.* (Brisbane) 16 May 42/5 A sheriff's deputy herds a group of women arrested in raids last month on 22 'massage parlors' in Los Angeles. **1972** D. Bloodworth *Any Number can Play* ix. 66 The imitation opium den and the model massage parlour. **1972** *Screw* 12 June 33/3 (Advt.), A totally dominant English massage by a lovely female, in your home or office. **1972** *Times* 4 Nov. 9/1, I noticed a few discreet establishments that looked as unauthentic, or as authentic according to your viewpoint, as New York's massage parlours. **1973** *Publishers Weekly* 8 Jan. 66/2 There is a difference between what a massage used to be, and what it has come to be a euphemism for lately. **1973** *Listener* 5 Apr. 439/2 The bars and massage parlours are..to re-open. But where are the clients now?.. The Americans have gone. **1974** *Publishers Weekly* 29 Apr. 45/1 The 'massage parlors' which have mushroomed in America's big cities. *Ibid.*, Massage parlor prostitutes in New York. *Ibid.*, 'Massage parlor' prostitution is considered by many to be 'glamorous' therapy.

massage, *v.* Add: Now usu. with pronunc. (mæ·-). Also *transf.* and *fig.*. Hence **massa·ging,** *vbl. sb.*

1924 R. Hichens *After Verdict* III. § 1. 314 Arabs were washing sacks below the dam and massaging them with persistently stamping bare brown feet. **1931** *Times Lit. Suppl.* 30 Apr. 340/2 'Shellacking', 'massaging'.., euphemisms which cover the long ordeal of beating, pounding..by which the detective learns the truth. **1936** F. Clune *Roaming round Darling* viii. 68 In the outer courtyard were several straw-hatted inmates in cover-alls massaging the lawns. **1966** *Aviation Week & Space Technol.* 5 Dec. 23/2 The interim has permitted an overall massaging of the initial reports, which resulted..in the revision that last week amounted to a major change in design philosophy for post-Apollo. **1974** *Globe & Mail* (Toronto) 23 Sept. 7/5 The Japanese government's efforts to massage the press coverage of the visit have been substantial. It offered to pay the fares of Canadian journalists to go to Tokyo for the interview and is bringing 30 Japanese journalists with the Prime Minister.

massager (mæ·sāʒəɹ). [f. Massage *v.* or *sb.* + -er^1.] **a.** One who practises massage, = Masseur, Masseuse.

1921 E. G. Lowry *Washington Close-Ups* 85 He is..a politician to his finger tips and a strong josher; a real handshaker and an elbow massager.

b. A massaging machine.

1940 *Life* 4 Nov. 54 Wooden barrel massager with rollers like clothes wringers is designed for action on the hips and buttocks. **1969** E. M. Brecher *Sex Researchers* (1970) x. 291 Dr. Dickinson was responsible..for the introduction into American gynecological practice of the electrical vibrator or massager. This device, applied to the mons and near the clitoris, produces intense erotic stimulation. **1969** 'J' *Sensuous Woman* (1970) iii. 29 Since the machine is advertised as a facial massager, you can purchase it without embarrassment.

Massagetæ (mæsagī·təi). Also **Massagetes.** [L. *Massagetæ*, Gr. Μασσαγέται, perh. f. native name *Masakata* Great Sakas.] An ancient Scythian people that lived to the east of the Caspian Sea.

1601 Holland tr. *Pliny's Nat. Hist.* VI. xvii. 123 The principall nations of Scythia, bee the Saræ, Massagetæ, [etc.]. **1709** I. Littlebury tr. *Herodotus' Hist.* I. I. 134 The Massagetes resemble the Scythians in their Habit and Way of Living. **1874** G. Grote *Hist. Greece* III. II. xvii. 329 The Scythians, originally occupants of Asia, or the regions East of the Caspian, had been driven across the Araxês, in consequence of an unsuccessful war with the Massagetæ. **1911** J. G. Frazer *Golden Bough: Magic Art* (ed. 3) I. i. 15 The Spartans, Persians, and Massagetae sacrificed horses to him [*sc.* the sun]. **1926** *Cambr. Anc. Hist.* IV. i. 15 Cyrus died fighting... His opponents were the Massagetae, a savage race who occupied the great plain to the East of the Caspian. **1967** C. A. Robinson *Anc. Hist.* (ed. 2) xxi. 410 The first business of the Greek king of Bactria was to protect this land, the gateway of Iran, from the semibarbarism of the north, and in particular to guard against the strong confederacy of the Massagetae, who were massed across the Oxus and Jaxartes rivers. **1969** J. Conway tr. *Bengtson's Greeks & Persians* xv. 318 Alexander was unable to capture him [*sc.* Spitamenes], but the Scythians beyond the Jaxartes, the Massagetae, to whom Spitamenes had fled, cut off his head and sent it to Alexander as a gift. **1973** G. Widengren in D. J. Wiseman *Peoples Old Testament Times* xiii. 320 When Cyrus..again turned to the east, he ultimately suffered a defeat in a battle against some nomadic northern people of Iranian origin, either..the Massagetae ..or the Danae.

Massaliot(e: see *Massiliot *sb.* and *a.*

massasauga. Delete ‖ and for Latin names substitute: *Sistrurus*, esp. *S. catenatus*. Also *attrib.* (Earlier and later examples.)

1835 C. Bradley in *Ohio Archaeol. & Hist. Soc. Publ.* (1906) XV. 257, I learned that my conquered enemy was the massassagua, the Michigan rattlesnake. **1941** *Amer. Midland Naturalist* XXV. 659 (*title*) Habit and habitat studies of the massasauga rattlesnake. **1954** R. C. Stebbins *Amphibians & Reptiles Western N. Amer.* 464/1 The massasauga is usually unaggressive, but the venom is dangerous.

massecuite (mæskwī·t). Also **masse cuite.** [Fr., lit. = cooked mass.] In sugar-making, the juice of the sugar-cane after concentration by boiling.

1882 *Spons' Encycl. Industr. Arts* v. 1927 The *massecuite* is quickly let out of the pan into a tank. **1886** *Harper's Mag.* June 88/2 After the *masse-cuite* has left the pan, the crystallization is completed by cooling. **1921** [see *box-hand* (*BOX *sb.*^{2} 24)]. **1925** *Chambers's Jrnl.* June 368/2 This dark-looking mass of semi-liquid substance is called masse-cuite. **1968** *Encycl. Brit.* XXI. 379/2 The cycle ..varies from 2 minutes for very high-test massecuites to 15 minutes or more for low grades.

massed, *ppl. a.* Add: **1.** (Earlier example.) spec. *massed entry*: see quot. 1964.

1881 'Mark Twain' *Prince & Pauper* ix. 84 The massed world on the river burst into a mighty roar of welcome. **1955** W. W. Greg *Shakes. First Folio* v. 157 In *The Winter's Tale*..there are massed entries in all but two scenes, but some individual entries are repeated later on. **1964** F. Bowers *Bibliogr. & Textual Crit.* v. i. 137 A massed entry is the notation at the start of a scene of the entrance in a group of all the characters, although some will make their entrances later at different points in the scene.

2. In reference to carved or written inscriptions, having the words arranged to form a solid column of lettering.

1906 A. E. R. Gill in E. Johnston *Writing & Illuminating* xvii. 392 In that space set out the Inscription, either 'Massed' or 'Symmetrical'. **1940** G. Hewitt *Lettering* xvi. 163 Italics..are not often successful when massed.

3. massed practice, training, trials *Psychol.*, a method of conditioning or training in which practice is concentrated with hardly any rest between repetitions.

1938 *Jrnl. Exper. Psychol.* XX. 201 The method of distributed practice should, therefore, yield relatively less accumulated inhibition than the method of massed practice. **1940** Hilgard & Marquis *Conditioning & Learning* vi. 148 Spaced practice favors conditioning over massed practice. **1949** B. J. Underwood *Exper. Psychol.* xi. 366 Several studies have shown that conditioning takes place more rapidly by distributed than by massed trials. *Ibid.* 631 *Massed training*, training in which trials are given in rapid succession with little if any time intervening between successive units of work. **1953** C. E. Osgood *Method & Theory Exper. Psychol.* III. xii. 504 The empirical law that 'distributed practice is superior to massed practice' has been found to hold under nearly all conditions. **1960** H. J. Eysenck *Behaviour Therapy & Neuroses* i. 17 The extinction of 4 tics in a female patient by..repeated voluntary repetition of the tic by massed practice. **1969** W. Mayer-Gross et al. *Clin. Psychiatry* (ed. 3) 186 The optimal technique was massed practice..followed by prolonged rest to permit the dissipation of reactive inhibition.

Massic (mæ·sik), *a. Obs.* exc. *Hist.* [ad. L. *Massic-us* the name of a mountain in Campania.] Designating an ancient wine produced in Campania, Italy. Also *ellipt.*

1638 T. Hawkins tr. *Horace's Odes* I. i. 19 Some others use Old Massique wines to ply. **1653** B. Holyday tr. *Horace's Odes* I. i. 19 In Massic Wines some bouze their time away. **1751** J. Stirling tr. *Horace's Odes* I. i, in *Horace's Works* I. II. 2 Cups of old Massic. **1833** C. Redding *Hist. Mod. Wines* 8 The Falernian..grew upon the volcanic Campania near Naples, where also the Massic was produced. **1920** *Punch* 15 Sept. 209, I raise my cup of massic Not to the earlier but the later 'classic'. **1961** H. W. Allen *Hist. Wine* v. 10 Massic appears to mean, not an individual wine, but any really fine Campanian wine. Virgil, Horace and Martial all use it in this rather general sense. **1965** O. A. Mendelsohn *Dict. Drink* 214 *Massic*, a wine from Campania, Italy, mentioned in classical writings.

massif. a. Delete '?' and '(*obs.*)' and add later examples.

1966 *New Statesman* 13 May 701/2 Martin proposes to remove everything in favour of his uniform *massif* of offices. **1973** *Times* 9 Jan. 12/7 There must, dear God, be something between slums and concrete massifs.

massify (mæ·sifəi), *v.* [f. Mass *sb.*^{2} + -ify.] *trans.* To form into a mass, *spec.* a mass society. Also **massifica·tion; ma·ssified** *ppl. a.*

1954 W. G. Peck *Christian Econ.* i. 20 We shall further massify and denature our people. *Ibid.* x. 162 He has been calling for security through the further 'massification' of his society. **1961** R. Williams *Long Revolution* III. 359 What the Americans call the 'massification' of society can only happen..if a majority of the people whom they regard as 'the masses' accept this version of themselves. **1966** D. Jenkins *Educated Society* ii. 59 The Nazis carried to its ruthlessly logical conclusion the attitude of those who wish to massify society for their own ends. *Ibid.* iv. 176 [A] society suffering from a dull uniformity..is because people have become massified or because those in positions of privilege have become listless. **1967** D. Cooper *Psychiatry & Anti-Psychiatry* ii. 45 Human motivation in courts of law and also in massified political judgements. **1969** G. S. Jones in Cockburn & Blackburn *Student Power* 42 Although there has been a substantial increase in the number of students, there has been nothing like the same process of massification that has occurred in the USA or Japan.

‖ **mässig** (mę·siχʸ). *Mus.* [G.] = Moderato *adv.*

1884 F. Niecks *Conc. Dict. Mus. Terms* 171 *Mässig* (Ger.), Moderate. **1938** *Oxf. Compan. Mus.* 545/2 *Mässig* (Ger.), (a) 'Moderate', 'moderately'; (b) 'in the style of' (e.g. *Marschmässig*, 'in march style').

Massiliot (mæsi·liǫt), *sb.* and *a.* Also **Massaliot, Massaliote, Massilian.** [f. L. *Massilia*, Gr. Μασσαλία Massalia.] **A.** *sb.* A native or inhabitant of Massalia (or Massilia, mod. Marseilles), a Greek colony founded *c* 600 B.C. to the east of the mouth of the Rhône on the Mediterranean coast of southern France. **B.** *adj.* Of or pertaining to Massalia or its inhabitants.

1601 P. Holland tr. *Pliny's Nat. Hist.* III. iv. 55 Townes in the other parts..namely, Agatha, in times past belonging to the Massilians. **1856** Grote *Hist. Greece* XII. II. xcviii. 614 In an age when piracy was common, the Massaliot ships and seamen were effective in attack and defence. *Ibid.* 615 Except the Phenicians and Carthaginians, these Massaliots were the only enterprising mariners in the Western Mediterranean. **1873** G. W. Kitchin *Hist. France to 1453* I. ii. 23 The whole seaboard from Var to Rhone was given to the Massiliots. **1898** A. Hassal *Jervis's Hist. France* (rev. ed.) i. 7 Pompey was beaten, and Cæsar's hand fell heavily on the Massiliots. **1932** *Times Lit. Suppl.* 8 Sept. 617/2 The voyage of the Massilian Pytheas. **1945** *Proc. Prehist. Soc.* XI. 56 Massilian coins have been obtained from the archaeological strata. **1949** *Oxf. Classical Dict.* 543/2 Massiliot seamen played an important part in the Second Punic War. **1957** *Encycl. Brit.* XIV. 965/1 The Massaliots made their way by prudence and by vigilant administration of their oligarchical government. *Ibid.*, In the 4th century B.C. the Massaliot Pytheas visited the coasts of Gaul. **1963** Cary & Warmington *Anc. Explorers* (rev. ed.) ii. 36 Syracusan and Massilian traders could join hands in the ports of Rome. **1964** J. Boardman *Greeks Overseas* v. 227 In the sixth century the Marseilles route is vouched for by the Massaliot and Phocaean pottery found at Mt Lassois. **1970** *Oxf. Classical Dict.* (ed. 2) 654/1 Massaliote ships helped Rome in the Second Punic War.

massily, *adv.* Delete '† *Obs.*' and add later example.

1922 E. R. Eddison *Worm Ouroboros* vii. 75 Benches of green jasper massily built and laden with velvet cushions.

Massim (mæ·sim). Also **Misima.** The eastern Papuo-Melanesian people of the Territory of Papua New Guinea; also, a member of this people; the language of this people. Also as *adj.*

[**1889** E. T. Hamy in *Revue d'Ethnogr.* VII. 504 La plupart des Mélanésiens de la mer d'Entrecasteaux sont aujourd'hui appelés collectivement *Massims*.] **1893** S. H. Ray in *9th Internat. Congr. Orientalists* II. 760 Of these, the Misima is the same as the New Guinea Melanesian, whilst the Nada forms..show the pronouns usual as possessive suffixes. **1894** A. C. Haddon *Decorative Art Brit. New Guinea* 184 Following Dr. E. T. Hamy, I adopt the term Massims as a collective name for the mixed people inhabiting the archipelagoes off the south-east of New Guinea... The term Massim..originally arose from an imperfect knowledge of the island of Misima, but I employ it solely as a convenient conventional name. **1910** C. G. Seligmann *Melanesians Brit. New Guinea* 9 The most characteristic cultural feature of the Massim is the existence of a peculiar form of totemism with matrilineal descent. **1936** R. Firth *Art & Life New Guinea* 22 The Massim area, in the south-east, is particularly famous for the wealth of carving with which its spatulæ are embellished. **1961** B. A. L. Cranstone *Melanesia* 42 The prow-boards of Massim canoes..are always carved by specialists. **1971** M. W. Young *Fighting with Food in Massim Society* iii. 54 Unlike in many areas of the Massim, men do not attribute malevolent supernatural powers to women.

massive, *a.* Add: **2. a.** (Further examples.) Now freq. used in weakened senses, 'large, great, far-reaching'.

1958 [see *AIR-LIFT 2]. **1958** G. Kennan *Russia, Atom & West* v. 77 The Marxists claim, of course, that colonialism invariably represented a massive and cruel exploitation of the colonial peoples. **1958** *New Statesman* 6 Sept. 306/2 The tendency to make *massive* a substitute not merely for *enormous, immense* and *huge* but even for *large* and *extensive* as applying to all sorts of phenomena, social, financial, political and psychological. **1963** *Daily Mail* 25 Feb. 16/6 Cowdrey again showed massive good form to make 86 in 163 minutes. **1965** R. & D. Morris *Men & Snakes* ix. 201 We required a massive sample of the population. **1973** *Word 1970* XXVI. 119 He died suddenly of a massive heart attack.

e. massive retaliation, name given to a military strategy, inaugurated in the U.S.A. by J. F. Dulles in the 1950s, which uses the threat of a punitive response with thermonuclear weapons to deter aggression.

[**1954** *N.Y. Times* 13 Jan. 2/3 [Speech of J. F. Dulles] Local defense must be reinforced by the further deterrent of massive retaliatory power.] **1954** *Times* 29 Mar. 5/5 The plan of 'meeting any aggression with massive retaliation in places of our choosing'. **1955** *Bull. Atomic Sci.* Jan. 29/1 We believe that the announced concept of massive retaliation contains the elements of adequate military support to attain the expansive aims of communism. **1959** *Times Lit. Suppl.* 25 Sept. 542/2 Mr. Dulles's 'massive retaliation' speech of January 1954..could hardly be read in the Soviet Union otherwise than as a threat. **1971** E. Luttwak *Dict. Mod. War* 128/1 Massive retaliation was never central US policy, and was in any case associated with the diplomatic technique of brinkmanship.

3. Also *Zool.*, applied to organisms which are compact in structure.

1888 Rolleston & Jackson *Forms Animal Life* (ed. 2) 250 There appear to be two fresh-water Sponges in Great Britain... The former is branched, the latter massive and lobate.

massively, *adv.* (Later examples.)

1959 *Times Lit. Suppl.* 20 Feb. 97/2 The road of elaborate verbal analysis of a poem which the trained reader immediately and massively *feels* to be wrong. *Ibid.* 27 Feb. 113/3 We must have massively responded..to a poem..before we can usefully talk about it. **1962** *Listener* 29 Mar. 540/1 The Administration was never crystal-clear on exactly how we would massively retaliate with nuclear weapons. **1974** *Physics Bull.* Apr. 132/3 Past CEGB technological choices and commercial practices have resulted in massively unproductive investments.

massivity (mæsi·viti). [f. Massive *a.* + -ity.] The fact or condition of being massive.

1908 W. H. Dawson *Evolution Mod. Germany* 13 Everywhere one sees the worship of massivity, the striving after crude, imposing effects—in the modern monuments, the public buildings, the bridges. **1921** S. Graham *Europe* xiv. 186 A mighty stone structure, of great height and massivity. **1953** E. Barker *Age & Youth* iv. 72 But New College had a massivity, if I may coin that word—an invisible massivity which matched the visible physical massivity of its chapel and hall and cloisters and tower—that captured my spirit and stretched my powers.

Mass-John: see Mas 2.

massless, *a.* Add: (Further examples.) Also **ma·sslessness.**

1947 *Mind* LVI. 50 Consider a weight of mass *m* hanging downwards under gravity on a (massless) elastic strand. **1963** K. W. Ford *World of Elementary Particles* 115 One necessary condition for masslessness seems to be the absence of charge. Every charged particle has mass, although not every neutral particle is massless. **1971** *Guinness Bk. Records* (ed. 18) v. 70/1 Of all sub-atomic concepts only the neutrino calls for masslessness. **1972** *Sci. Amer.* June 53/1 Massless uncharged particles that interact so little with other particles that the outer layers of the sun and solid bodies such as the earth are virtually transparent to them.

mass medium (mæ:s mī·diəm). [f. Mass *sb.*^{2} + Medium *sb.*] A medium of communication (such as radio, television, newspapers, etc.) that reaches a large number of people; usu. in pl. *mass media.*

The *pl.* form *mass media* is sometimes erroneously construed as *sing.*

1923 S. M. Fechheimer in N. T. Praigg *Advertising & Selling* v. 238 (*title*) Class appeal in mass media. *Ibid.*, The several million readers of a big mass medium. G. Snow in *Ibid.* 240 Mass media represents the most economical way of getting the story over the new and wider market in the least time. **1942** D. Waples *Print, Radio & Film* 19 Radio (which, according to German propaganda theory, is the 'real' mass medium). **1946** J. S. Huxley *Unesco* ii. 58 The media of mass communication—the somewhat cumbrous title (commonly abbreviated to 'Mass Media') proposed for agencies, such as the radio, the cinema and the popular press, which are capable of the mass dissemination of word or image. *Ibid.* 60 The use of the mass media to foster education, science and culture... Regarded from this angle, the mass media fall into the same general category as the libraries and museums—that of servicing agencies for man's higher activities. **1957** R. Hoggart *Uses of Literacy* ii. 27 A great deal has been written about the effect on the working-classes of the modern 'mass media of communication'. **1964** M. Argyle *Psychol. & Social Probl.* vii. 101 Propaganda over the mass media may be intended to change behaviour, beliefs or feelings, or all three. *Ibid.* xiii. 157 Mass media such as TV, radio, newspapers and posters. **1966** N. Tucker (*title*) Understanding the mass media. **1968** P. Oliver *Screening Blues* 3 For the Negro in Tuscaloosa, Alabama..the blues record afforded the first real opportunity of contact through a mass medium with others of his social status. **1971** *New Scientist* 18 Mar. 610/1 The department's trendy, mass-media sounding title is a historical accident. **1974** *Black World* Dec. 5/1 Nevertheless, the mass media continues to emphasize the attitudes, beliefs and rhetoric of a minority of the Black community.

mass observation (mæ:s ǫbzəɹvē·ʃən). [f. Mass *sb.*^{2} + Observation.] The study and record of the social habits of people (taken in the mass); also (with capital initials) the name of an organization established for this purpose. Also *attrib.* So **mass-observationist, -observer; mass-observe** *v. trans.*

1933 L. Bloomfield *Lang.* ii. 38 The observer..by this mass-observation, gives us a statement of the speech-habits of a community. **1937** Madge & Harrisson *Mass-Observation* 8 If after reading the pamphlet you should wish to co-operate by becoming a Mass-Observer, send a card. *Ibid.* i. 10 A group of people started *Mass-Observation*, which aims to be a scientific study of human social behaviour, beginning at home. **1937** Jennings & Madge (*title*) May the twelfth: Mass-Observation day-surveys 1937. **1938** *Times* 10 Mar. 15/4 Nor are they mass-observationists..testing the movements of public opinion with cold curiosity. **1939** 'N. Blake' *Smiler with Knife* xvi. 233 'I thought..you wur one of them Mass Observers...' 'No. I'm just one of the mass-observed.' **1948** J. Betjeman *Coll. Poems* (1958) 229 The Mass-Observer with the Hillman Minx. **1951** 'A. Garve' *Murder in Moscow* ii. 34 He's supposed to have an assignment..but he spends all his time doing his own private mass observation. **1967** G. Wills in Wills & Yearsley *Handbk. Managem. Technol.* 184 The two other major quantification techniques are the consumer survey..and the mass-observation study. **1970** *Guardian* 10 Dec. 8/4 The Mass Observers..mass observed blind drinkers. **1974** *Daily Tel.* (Colour Suppl.) 19 Apr. 23/3 The Mass-Observation method..is the absolute opposite of interviewing.

mass-priest. (Later examples.)

1902 J. Buchan *Watcher by Threshold* II. 109 The family changed its faith, and an Episcopal chaplain took the place of the old mass-priest in the tutoring of the sons. **1922** Joyce *Ulysses* 379 She said that he had a fair sweet death through God His goodness with masspriest to be shriven, holy housel and sick men's oil to his limbs.

mass production. [f. Mass *sb.*^{2} + Production.] The production of manufactured articles in large quantities by a standardized process. Also *transf.* and *attrib.*

1920 *Teacher's World* 19 May 283/2 (Advt.), Mass production. High class..chairs. **1922** Turner & Wood *Man. Up-to-Date Organisation* 110 Mass Production is a continuous replica of a standardised master-pattern or design. **1923** J. M. Scott-Maxwell *Costing & Price-Fixing* 31 Factories on a mass production basis will have all material most carefully specified, examined, and tested. **1929** W. Deeping *Roper's Row* xxxii. 353 The new world of smudged faces, mass-production man in the making. **1930** G. B. Shaw *Apple Cart* I. 25 Don't forget our racing motor boats and cars, sir: the finest on earth, and all individually designed. No cheap mass production stuff there. **1944** W. Temple *Church looks Forward* x. 82 The problems due to mass-production press upon us. **1946** J. S. Huxley *Unesco* i. 18 A mass-production system can indirectly destroy creative initiative and aesthetic appreciation. **1947** J. Hayward *Prose Lit. Since 1939* 26 Literary and historical biography was already a flourishing industry. What particular..forces were concerned in its mass-production cannot be investigated here. **1948** [see *conveyor belt* s.v. *CONVEYER 4 c]. **1962** H. O. Beecheno *Introd. Business Stud.* x. 86 Mass-production methods have called for mass-selling methods. **1967** M. McLuhan *Medium is Massage* 50 Printing..provided the first uniformly repeatable 'commodity', the first assembly line—mass production.

Hence **ma:ss-produ·ce** *v. trans.* and *intr.*, to manufacture (articles) by mass production; also **ma:ss-produ·ced** *ppl. a.*; **ma:ss-produ·cer; ma:ss-produ·cing** *ppl. a.*; **ma:ss-pro·duct.**

1922 Joyce *Ulysses* 661 They drank..Epps's massproduct, the creature cocoa. **1923** *Daily Mail* 22 Jan. 5 All cars made in the United States are not necessarily mass-produced. **1929** E. Gill *Art-Nonsense* 318 The ordinary person..says, what is to hinder him from having..mass-produced pots and pans designed by artists to look artistic? **1929** A. Huxley *Do what you Will* 90 The mass-producers will do their best to make everybody more and more prosperous. **1930** —— *Vulgarity in Lit.* vii. 45 Asceticism is not popular in this mass-producing age. **1938** *New Statesman* 23 July 143/2 Trades which mass-produce luxury or semi-luxury goods. **1940** *Life* 9 Dec. 84/2 U.S. now mass-produces. **1950** M. Hay *Foot of Pride* iii. 74 St. Vincent Ferrer and other preachers sent thousands of panic-stricken people to the font where a priest mechanically turned them into Christians. Theologians..decided that these mass-produced Christians had been validly baptized. **1954** *Encounter* Mar. 5/2 A mass-producing, mass-consuming society. **1958** Mass-produce [see *BATTERY 12* c]. **1967** E. Short *Embroidery & Fabric Collage* iii. 68 There is no way of mass-producing patchwork. **1973** *Country Life* 15 Mar. 718/1 This is our 18th Century style bedroom suite. Not for everyone because it is not mass-produced.

massula (mæ·siŭlă). *Bot.* [mod.L., dim. of L. *massa*: see Mass *sb.*^{2}] (*a*) In heterosporous ferns of the genera *Azolla* and *Salvinia*, the tissue surrounding the maturing microspores; (*b*) in certain orchids, a cluster of pollen grains developed from a single cell.

1856 J. S. Henslow *Dict. Bot. Terms* (ed. 2) 105 Massula (a little lump). One of the smaller fragments which ..compose the pollen mass in Orchidaceæ. **1882** S. H. Vines tr. *F. G. J. von Sachs's Text-bk. Bot.* II. 454 In the microsporangia [of *Azolla*] the mucilage looks like a large-celled tissue, and forms from two to eight separate clumps (*Massulæ*), each of which encloses a number of microspores. In some species..these massulæ have their surfaces covered with hair-like appendages. **1894** S. H. Vines *Students' Text-bk. Bot.* I. III. 410 In *Azolla* the microspores are likewise embedded in this substance, but in more than one group or massula according to the species. **1895** —— *Ibid.* II. III. 564 The pollinium may consist..of larger groups of cells, termed *massulæ* (e.g. Orchis). **1938** G. M. Smith *Cryptogamic Bot.* II. ix. 358 The functional macrospore lies in the large massula at the base of a sporangium. *Ibid.* 360 A microspore remains embedded within a massula during the entire course of its development into a gametophyte. **1959** Wirth & Withner in C. L. Withner *Orchids* v. 157 In some genera, such as *Peristylus*, there is a variation on the normal mode of development. All the cells resulting from the divisions of an archesporial cell remain attached to each other, eventually forming a small cluster of tetrads known as a massula. **1965** Bell & Coombe tr. *Strasburger's Textbk. Bot.* 582 *Azolla* is of interest for the arrangements which ensure fertilization. After liberation from the microsporangium the sixty-four microspores are aggregated into 5–8 roundish floating balls (massulae) formed of the frothy periplasmodium. Arising from the surface of each massula are a number of stalked, barbed hooks.

massy, Eng. and U.S. dial. var. of Mercy *sb.* 4.

1817 A. Royall *Lett. from Alabama* (1830) ix. 22 Massy upon me! **1867** W. F. Rock *Jim an' Nell* 31 Law! massy, Jim. **1884** J. C. Egerton *Sussex Folk & Ways* 41 'Massy!' she said, 'the girls nowadays don't know naun about work!' **1905** *Dialect Notes* III. 17 *Massy sakes! sakes alive!* interj. All feminine exclamations. **1944** H. Wentworth *Amer. Dial. Dict.* 380/1 Massy = mercy.

mast, *sb.*^{1} Add: **3.** Also used in various special senses (see quots.).

1910 *Flying in Bournemouth* (Souvenir Wallisdown Flying Whit-Week) 13 (*caption*) Mast with supporting wires. **1914** R. Stanley *Text-bk. Wireless Telegr.* xiv. 189 The aerial to be supported by ten tubular steel masts each 300 feet high. **1918** E. S. Farrow *Dict. Mil. Terms* 371 *Mast*, in aeronautics, the upright part, usually extending upward from the center of a monoplane for support of controls and guy or truss wires; a vertical upright in the main or supplementary planes. **1920** [see *HORN *sb.* 21 h (i)]. **1924** *Harmsworth's Wireless Encycl.* II. 1409 *Mast*, term used in wireless work to describe generally any structure used to raise and support the aerial wires. **1931** D. Rose *J. de la Cierva's Wings of Tomorrow* 110 A sort of whipping action of the rotor blades [of a helicopter] which jerked at the mast as they turned in their circle. **1956** W. A. Heflin *U.S. Air Force Dict.* 317/1 *Mast* n., 1. A spar for the support of an antenna. 2. A rotor mast... 3. In certain early airplanes, any of various vertical members used to hold guy wires or struts. **1966** *Listener* 1 Dec. 799/2 The German pavilion..is made of a semi-transparent plasticized fabric supported on a network of cables slung between a very few slender steel masts.

mast, *sb.*^{2} **4. mast year** (add definition, and earlier and later examples): a year in which forest trees produce a good crop of fruit.

1744 W. Ellis *Mod. Husbandman* (1750) VI. II. 89 Under these Trees, the Hogs generally get Pork in a Maste-Year. **1927** *Forestry* I. 28 The rarity of full mast years and heavy weed growth are often held responsible for the failure of most attempts at natural regeneration of oak. **1953** H. L. Edlin *Forester's Handbk.* ii. 30 But with many kinds [of trees] full seed crops are only found at intervals of three or four years—the *mast* years.

‖ **mast** (mɑst), *sb.*^{6} [Pers.] (See quots.)

1819 W. Ouseley *Trav. East* I. vi. 268 Here also was abundance of *mâst*,..coagulated milk or clotted cream, slightly sour, which when diluted with water forms *ab i dûgh*..a beverage in warm weather equally grateful and salubrious. **1933** *Discovery* Sept. 284/1, I was given *mast* or sour milk [in Persia]. **1963** *Times* 6 Feb. 12/6 Quantities of tea served in glasses and drunk, at any rate by the Iranians, with lots and lots of sugar and the local yoghourt, known as *mast.* **1968** C. Roden *Bk. Middle Eastern Food* 60 Yoghourt is an essential part of the Middle Eastern diet... In Iran today it is known as *mâst.*

mastaba(h. Add: **2.** (Further examples.) Also *attrib.* and *Comb.*

1883 *Encycl. Brit.* XX. 122/2 They have been extended upwards and around into a great stepped mass of masonry .., the successive faces of which rise at the characteristic mastaba angle of 75°. **1931** G. A. Reisner *Mycerinus* i. 3 Some of the mastaba tombs also belonged to members of the royal family. **1932** *Times Lit. Suppl.* 26 May 389/1 The concession of the area of that pyramid and of large areas in the vast mastaba-field of Giza. **1936** *Ibid.* 13 June 501/1 Around the three great pyramids..are grouped.. the *mastaba*-tombs of the nobles. **1952** E. B. Garside tr. *Ceram's Gods, Graves, & Scholars* II. x. 120 A mastaba is an oblong structure with sloping sides containing cult rooms and connected by a shaft with a burial chamber in the rock beneath. **1964** W. L. Goodman *Hist. Woodworking Tools* IV. i. 160 The only tool shown in use on mastaba reliefs. *Ibid.*, This method of constructing the cabin or house with overlapping vertical boards may explain why the walls of the early mastaba tombs..show the deep recesses.

mastectomy (mæste·ktŏmi). *Surg.* [f. Gr. μαστ-ός breast + *-ectomy.] Excision or amputation of a breast.

1923 in Stedman *Med. Dict.* (ed. 7). **1935** *Jrnl. Med. Assoc. Georgia* XXIV. 119 (*heading*) The seasponge as a postoperative dressing following radical mastectomy. **1973** M. J. Brennan in Holland & Frei *Cancer Med.* xxvii. 1777/1 Since skillful biopsy followed by a short wait before definitive treatment does not penalize results, there is no reason for immediate mastectomy when the frozen section does not give certain diagnosis.

master, *sb.*^{1} Add: **I. 4. b.** *his* (or *her, my,* etc.) *master's voice*, a catch-phrase, originating from the trade name of a gramophone record company, denoting, freq. ironically, the voice of authority.

[**1910** *Trade Marks Jrnl.* 5 Oct. 1604 His Master's Voice. ..Talking machines, talking machine needles, talking machine records and other talking machine accessories,.. The Gramophone Company, Limited,..London,..manufacturers.] **1922** M. A. von Arnim *Enchanted April* xviii. 297 'Francesca!' shouted Briggs. She came running... 'Her Master's Voice,' remarked Mr. Wilkins. **1922** Joyce *Ulysses* 465 My master's voice. **1969** 'H. Pentecost' *Girl with Six Fingers* (1970) I. iv. 68 'He—he'd hesitate to do anything that would displease Angela.' 'Not when he hears his master's voice,' said Jericho. **1972** R. Quilty *Tenth Session* 42 Bank's personal secretary..announced.. that the doctor would like to see Angus in his office... 'His master's voice,' he informed Norman.

9*. a. An original disc († or cylinder) with grooves cut by a stylus during recording. **b.** (Also *master matrix.*) A disc with ridges in place of grooves that is made from the plating of an electroplated original and is used as a stamper or (more usually) to make a 'mother'.

1904 S. R. Bottone *Talking Machines & Records* 69 With the master running in the phonograph, the trained ears of the specialists enable them to detect the most minute imperfections. **1908** *Daily Chron.* 29 Oct. 7/1 A special room is devoted to recording, or making the master from which copies are to be taken. **1918** H. Seymour *Reproduction of Sound* 68 The original master is recorded by means of a feed thread... This is thereupon electrotyped—that is, a metallic negative is grown upon it, and this is called the master matrix. A small number of casts in wax are made from this... These are similarly electrotyped, and become the working matrices. **1935** H. C. Bryson *Gramophone Record* vi. 130 (*caption*) Stripping copper masters from the wax. *Ibid.* 142 The wax is positive, the copper master is negative, the mother is positive, the stamper is negative, and the record produced from it is therefore positive and will play on a gramophone. **1952** Godfrey & Amos *Sound Recording & Reproduction* v. 139 Only one master can be prepared from a wax original.., but more than one master can sometimes be made from a lacquer-coated disk. **1962** A. Nisbett *Technique Sound Studio* 259 The master has ridges instead of grooves. In half processing (when only a small run is required) the master is also used as a stamper. **1972** *Jazz & Blues* Oct. 29/2 The final batch of King Jazz recordings were done on 12" masters.

c. (See quot. 1958.)

1930 W. Desborough *Duplicating & Copying Processes* iii. 21 The master is produced in negative or mirror form by placing the face of a special hekto carbon at the back of the sheet of paper on which the master is to be made. **1957** G. W. W. Stevens *Microphotogr.* xi. 213 The time and care expended on a master negative give it considerable value ..therefore..it is advisable to make at least two masters once the camera has been set up. **1958** T. Landau *Encycl. Librarianship* 204/2 *Master*, the original plate or stencil in duplicating processes from which copies are made. In photocopying, the negative from which a positive print is made. **1971** R. Busby *Deadlock* iv. 39 Could you make that four copies?.. You'll need two for the index apart from this master.

II. 15. *the master* (or *Master*): further examples (see quot. 1870 in Dict.); *old master* (further examples).

1869 'Mark Twain' *Innoc. Abr.* 260 Who painted these things? Why, Titian..Raphael—none other than the world's idols, the 'old masters'. **1889** G. B. Shaw in *Hawk* 13 Aug. 172/2 Behind the Master's house is the Master's grave; for Wagner..is 'buried in the back garden, sir, like a Newfoundland dog'. **1908** *Pop. Mechanics* Nov. 741/2 A Munich artist and inventor has devised an electric machine by which the famous paintings of the 'old masters' can be renewed. **1937** H. G. Wells *Camford Visitation* v. 58 All his fancied novelties of criticism were fully foreseen by the Master [*sc.* Karl Marx]. **1959** *Times* 20 Aug. 9/5 In calling the anonymous hero of *The Aspern Papers* 'H. J.' I may have misled a few people into thinking that I have attempted to make 'the publishing scoundrel' into a portrait of 'The Master'. **1962** *Listener* 14 June 1043/3 The style, a pastiche of Nabokov's..is a style less functional, more baroque, than the Master's.

b. (Further examples.) *The Old Masters*: the name given to the annual Winter Exhibition of the Royal Academy from the name of the first exhibition (1870), 'Exhibition of the Works of the Old Masters'.

1824 J. Fisher *Let.* 18 Jan. in C. R. Leslie *Mem. Life J. Constable* (1845) viii. 130, I am shut up in lodgings here,—the walls covered with *old masters.* **1861** G. Meredith *Let.* 19 Nov. (1970) I. 111 He is in new chambers full of pictures, Old Masters, we hear. **1870** *Times* 3 Jan. 5/1 Exhibition of old masters at the Royal Academy. The noble new rooms of the Royal Academy..the place of exhibition of a selection from the finest works of the old masters in English private galleries, opened..for the public today. **1880** Geo. Eliot *Let.* 5 Feb. (1956) VII. 248 If you had come to town you would have liked to see ..the Old Masters at the Academy but perhaps you may yet come before they are removed. **1890** G. B. Shaw *London Music in 1888–89* (1937) 284 My ticket for the Press view at the Old Masters on Friday! **1908** *Outlook* 19 Dec. 848/2 The disappointment at the prospect of no Old Masters exhibition this winter is barely mitigated by the *menu* of the banquet provided by the McCulloch collection. **1968** S. C. Hutchison *Hist. R. Acad.* xiii. 141 The Old Masters Exhibition of 1879 was conceived on a much larger scale than its forerunners and occupied nine galleries. **1975** *Times* 23 June 12 The value of Old Masters is perhaps more subjectively determined than in any other field... The Old Master market has been very strong in the 1970s.

c. *spec.* in Chess. Cf. **chess-master*, *GRAND MASTER 3*.

1852 H. Staunton *Chess Tournament* p. l, To reward the ability of first-rate masters throughout the world, they offered..prizes. **1910** *Encycl. Brit.* VI. 104/1 The terms *master* and *amateur* are not used in any invidious sense, but simply as designating, in the former case, first-class players, and in the latter, those just on the borderland of highest excellence. **1969** *Times* 25 Jan. 17/7 There are three types of F.I.D.E. master title: international grand-master, international master, and international woman master. **1973** *Daily Tel.* 25 Apr. 16/3 After six of the 11 rounds of the Birmingham International Chess Tournament, Tony J. Miles, 18, the Birmingham schoolboy, still leads the mixed concourse (one grand master, several international masters, candidate masters and others). **1973** *Sci. Amer.* June 93/2 Kalme is a senior master with a rating of 2,455 on the International Chess Federation scale, on which 2,200 denotes a master and Fischer's rating is 2,785.

IV. 22. *young master* (later examples).

c **1874** D. Boucicault in M. R. Booth *Eng. Plays of 19th Cent.* (1969) II. 201 Did you see the young master? **1923** Wodehouse *Inimitable Jeeves* i. 9 He had been clearing away the breakfast things, but at the sound of the young master's voice cheesed it courteously.

b. Prefixed, with disparaging implication, to the name of an adult.

1885 C. M. Yonge *Nuttie's Father* II. ix. 105 I'm not going to have a *tête-à-tête* with Master Mark. **1959** *Listener* 22 Jan. 155/2, I never liked Jinnah... I thought a great deal of his ambition was for Master Jinnah rather than anything else. **1975** T. Heald *Deadline* vii. 151, I think you'll find Master Wimbledon will do very well at anything if he thinks it will further his career. He's an exceedingly ruthless and ambitious young man.

V. 24. d. *master carpenter* (later examples), *clockmaker.*

1824 R. Humphreys *Mem. J. Decastro* 76 He calls for his scene-painter, composer, master-carpenter, property-man. **1933** *Burlington Mag.* Nov. 239/2 A master-clockmaker of the class of Thomas Tompion. **1936** *Ibid.* Aug. 54/1 At nineteen he was..a master-carpenter and a painter. **1964** W. L. Goodman *Hist. Woodworking Tools* II. vii. 78 The wide-ranging migrations of the master-masons and master-carpenters of the Middle-Ages.

e. (Further examples.)

1930 *English Jrnl.* XIX. 628 The Minnesota *Quarterly* is at present running a series of tales about a master criminal. **1932** Wodehouse *Louder & Funnier* 72 The psychology of the Master Criminal is a thing I have never been able to understand. **1939** T. S. Eliot *Old Possum's Pract. Cats* 33 He's the master criminal who can defy the Law. **1968** D. M. Smith *Mod. Sicily* lv. 505 That notorious character Don Calogero Vizzini,..master criminal and boss of Villalba,..made a fortune out of wartime shortages. **1975** L. Deighton *Yesterday's Spy* vi. 54 Champion was some kind of master spy.

f. *Chess.* Designating play or players of the highest class at national or international level. (Cf. *15 c.)

1894 J. Mason *Princ. Chess* 186 In the case of master players, a slight initial error..will permeate the remainder of the game. **1938** P. W. Sergeant *Championship Chess* 191 The *Benoni Counter-Gambit* had for years been out of favour in master-play. **1958** *Listener* 13 Nov. 803/1 It is really only in master chess..that knowledge of the latest variations and finesses becomes important. **1959** *Ibid.* 5 Nov. 794/1 You need not go as slow as the sixteen moves an hour customary in master practice.

25. a. *master bathroom, bedroom, dish, list, pattern* (also *fig.*), *star, switch, tape.*

1959 *Sunday Times* 3 May 19/4 The kitchen and master bathroom. **1926** *New Republic* 7 Apr. p. iii (Advt.), Large living-room and dining-room opening on court, four master bedrooms. **1937** M. Hillis *Orchids on your Budget* ii. 35 Crisp white argentine curtains in the master bedroom. **1945** Nelson & Wright *Tomorrow's House* ii. 12/2 The discussion continued in the privacy of the master bedroom. **1972** *Country Life* 5 Oct. (Suppl.) 32d/2 Master bedroom suite with dressing room, bath and shower rooms. **1960** Auden *Homage to Clio* 26 Doomed to observe Beauty peck at a master-dish. **1962** Y. Malkiel in Householder & Saporta *Probl. Lexicogr.* 9 Fragmentary master-lists of the items. **1967** *N.Y. Times* (Internat. ed.) 11–12 Feb. 9/6 Master list of investment and speculative stocks—key lists for both the conservative and more speculative investor. **1948** H. Hall *Home Dress-Making Simplified* xxxiii. 311 The safe plan, when remodeling, is to fit a master pattern in muslin... This pattern is then used as a guide for fitting the section when cutting. **1960** R. Lister *Decorative Cast Ironwork* 230 *Master pattern*, a pattern from which a production work is cast. **1962** W. Nowottny *Lang. Poets Use* vi. 138 It is Eliot's peculiar insight..to have..expressed the longing for the master-pattern that will free us from the fret of a world in which there are too many equipollent patterns. **1904** W. de la Mare *Henry Brocken* vi. 62 The master-stars shone earlier here. **1907** A. Hay *Introd. Course Continuous Current Engin.* xvi. 284 A small switch which is arranged to control simultaneously a number of larger switches is spoken of as a master switch. **1943** *Gloss. Terms Electr. Engin.* (*B.S.I.*) 52 *Master switch or circuit-breaker*, a switch or circuit-breaker to which the operation or function of one or more other switches is subservient. **1975** P. Organ *House on Cheyne Walk* i. 1 The night is going. Soon someone somewhere will throw a master switch..and the lights will go out..all over London. **1954** Master tape [see *FILE *sb.*^{2} 4 b]. **1962** *Times* 5 July 15/6 In a disc-cutting room at Decca's London studios, a master tape of a recent recording of music by Prokofiev and an ordinary commercial pressing of a disc made from it were started simultaneously. **1967** *Lebende Sprachen* XII. 137/2 *Master tape*, original recording by informant of a language laboratory drill. **1973** *Daily Tel.* (Colour Suppl.) 12 Oct. 29/2 The signals..from the master tape are then fed into the recording head that drives the cutting stylus.

b. *master-coup, -dream, -fulcrum, plan, -power, -voice.*

1939 R. Campbell *Flowering Rifle* VI. 148 The cynic master-coup of propaganda. **1928** Blunden *Jap. Garland* 21 Sleep's master-dream there stands alone: The tower of East and West! **1957** T. Hughes *Hawk in Rain* 11 And I..strain towards the master-Fulcrum of violence where the hawk hangs still. **1935** Master plan [see *FREEWAY]. **1939** H. M. Lewis *City Planning* xii. 128 A master plan.. may be defined as a general plan for the future layout of the city. **1957** L. F. R. Williams *State of Israel* 77 The bulk of the population is being systematically spread out in accordance with a master plan drawn up by the Government. **1960** *Washington Post* 16 Nov. A.16 The Friends, who have retained consultants to study the Zoo's problems and form a master plan, found that the institution has almost never received adequate funds. **1971** B. de Ferranti *Living with Computer* ix. 82 He therefore ..gives them names so that they can be filed and described on the master plan. **1821** Hazlitt *Table-Talk* I. v. 96 His mind grappled with that which afforded the best exercise to its master-powers. **1910** W. de la Mare *Three Mulla-Mulgars* xiii. 174 Of its calm authority the master-voice said, 'So shall it be.'

29. master-batch, a concentrated mixture employed in the production of synthetic rubbers and plastics; hence as *v. trans.*, to mix in a master-batch; **masterbatching** *vbl. sb.*; **master-brain,** = *MASTER-MIND *sb.* b; **master**

card, (*a*) *Bridge* (see quots.); (*b*) a record card which summarizes the information recorded on a number of other cards; **master class,** (*a*) the most powerful or influential class in society; (*b*) a class receiving instruction from a 'master' (a person of distinguished skill), esp. in *Music*; **master clock,** a clock which transmits regular pulses of electricity for controlling impulse dials or computer operations; **master-craftsman,** a craftsman thoroughly conversant with his trade; one who employs workmen; also *transf.*; hence *master-craftsmanship*; **master number** = *matrix number* (*MATRIX 7); **master oscillator,** an oscillator used to produce a constant frequency, esp. the carrier frequency of a radio or television transmitter; **master race,** a race of people considered to be pre-eminent in greatness or power; *spec.* the Germans or 'Aryans' (see *ARYAN *sb.* 2) regarded as a superior people (cf. *HERRENVOLK) during the Nazi period; **master rod,** in a rotary or radial engine, a rod which connects one of the pistons to the crankshaft and carries the wrist pins to which the link rods are connected; **master-scene, shot** *Cinematography*, (see quots.); **master–slave manipulator,** a type of manipulator (sense *2 f) which reproduces at the handling end the positions and motions of the operator's fingers.

1937 H. BARRON *Mod. Rubber Chem.* vii. 79 Certain fundamental ingredients such as accelerators, antioxidants and sulphur are always added in small quantities... It is now common practice to add them as a 'masterbatch'. 1953 N. L. CATTON *Neoprenes* 32 A convenient method of making sure that mixing operations are carried out in the elastic phase is through the use of concentrated masterbatches. 1959 *Times* 13 Mar. 10/2 The Cariflex range consists of hot and cold polymers, oil masterbatches, carbon black masterbatches and hot and cold latices. 1964 *Amer. Speech* XXXIX. 272 Most stocks..in a tire are mixed in two stages. The masterbatch is the first stage, in which all the rubbers and pigments are mixed except for the curing agents. 1953 *Industr. & Engin. Chem.* May 1059/2 Copolymers of butadiene and styrene..have been extended with various rosin-type acids in a manner similar to that employed in extending with petroleum oils (latex masterbatching). 1971 CROWTHER & EDMONDSON in C. M. Blow *Rubber Technol. & Manuf.* viii. 269 It is generally accepted that masterbatching improves the physical properties of those compounds where a high degree of carbon black dispersion must be achieved. *Ibid.* 270 The curing ingredients themselves may be masterbatched. 1923 E. WALLACE *Missing Million* xiv. 112 The master-brain who took his pick of the cleverest criminals at large. 1905 R. F. FOSTER *Compl. Bridge* 316 *Master card*, the best left in play of any suit which has already been led. 1937 *Times* 13 Apr. p. iv/3 A..tabulator, which ..can collate, record and analyse 900 production facts a minute, as well as produce master cards summarizing the figures it tabulates. 1961 T. LANDAU *Encycl. Librarianship* (ed. 2) 236/2 *Master card*, a main entry in a card catalogue bearing tracings indicating all the added entries, [etc.]. 1964 COHEN & BARROW *A.B.C. of Contract Bridge* 282 *Master card*, a card that can be beaten only if it is trumped. 1970 O. DOPPING *Computers & Data Processing* iv. 72 If the machine has only one card feed, data are transferred from a master card to the following cards which are often called detail cards. 1861 MILL *Repr. Govt.* ii. 38 Slavery is..corrupting to the master-class when they once come under civilized influences. *c* 1926 'MIXER' *Transport Workers' Song Bk.* 47 Note the constant drop in wages..Endorsed by every rag-time press That the master-class command. 1930 J. DOS PASSOS *42nd Parallel* I. 104 Big Bill talked about..sticking together in the face of the masterclass. 1963 *Ann. Reg. 1962* 446 Pablo Casals conducting a master class. 1965 *Listener* 10 June 873/2 Master-class exercises in comparative semantics and applied imagery. 1970 *Times* 26 Feb. 16/1 Old singers don't forget—they teach. And the best of them conduct master classes. 1973 *Times* 17 Apr. 32/4 Crash course on British Portraiture... Contact: Master Classes. 1973 *Ms.* Nov. 46/3 Ms. Rosenberger has worked out a variety of exercises like the one she suggested to the tense master-class student. 1904 Master clock [see *JOURNEYMAN 3 b]. 1922 G. L. OVERTON *Clocks & Watches* x. 108 If the master clock should have gained or lost at all, the arrangements provided for putting it right again are such that all the connected dials are at the same time automatically set to the right time. 1951 Master clock [see *impulse clock* s.v. *IMPULSE *sb.* 5 b]. 1967 H. JACOBOWITZ *Electronic Computers made Simple* xiii. 261 Unless the computer is asynchronous.., all its operations are controlled by a series of equally spaced timing pulses from a master clock oscillator. 1937 G. B. BROWN *Arts in Early Eng.* VI. II. xi. 215 We are in almost complete ignorance as to the personalities of the master craftsmen who created the schemes of enrichment. 1952 C. DAY LEWIS tr. *Virgil's Aeneid* II. 36 That master-craftsman of crime, Ulysses. *Ibid.* VIII. 177 Now there is need for Your strength, your speediest work and your master-craftsmanship. 1969 *John Edwards Mem. Foundation Q.* V. IV. 146 Such documents are of great importance to the discographer who is trying to assemble the usual discographic data of master number, recording date..and release number. 1973 in B. Holiday *Lady sings Blues* 208 In the era of 78 rpm records most companies assigned a number to each recording they made for the sake of handy reference, called the 'master' number. 1928 STERLING & KRUSE *Radio Manual* viii. 305 The tuning of this type of circuit consists fundamentally of setting the master oscillator at the desired wavelength and then resonating the antenna circuit for maximum antenna current. 1951 W. E. PANNETT *Radio Installations* vii. 177 The quartz crystal..has largely superseded valve circuits as a master oscillator for controlling the frequency of a transmitter. 1971 M. G. SCROGGIE *Found. Wireless & Electronics* xv. 244 The function of what is usually called the master oscillator is to define the frequency of the station at least as accurately and constantly as the now very narrow international regulations require. 1942 J. S. HUXLEY in *South Atlantic Q.* XLI. 354 The [German] nation is formidably united behind its own ideal of an 'Aryan' Master-Race. 1945 D. THOMSON et al. *Patterns of Peacemaking* iii. 118 Her expansionism has been due to a triple convergence of forces: the economic greed and lust for power of her ruling classes, the primitive and tribal nationalist and imperialist sentiments of her people, and hatred of the foreigner which takes the form of theories of Japanese being a 'master-race'. 1949 D. MACARDLE *Children of Europe* xiv. 212 Boys and girls who had been impregnated with the teaching of Nazi schools..shared the overwhelming pride of the 'Master Race'. 1953 P. ABRAHAMS *Return to Goli* vi. 215 Imperceptibly, a 'master race' mentality is in the making [in Kenya]. 1974 *Times* 21 Oct. 3/4 To suggest that I [*sc.* Sir Keith Joseph] was arguing for a 'master race'..is..very wrong. 1922 *Encycl. Brit.* XXX. 36/2 Connecting-rods of rotary and radial engines consist usually of one master rod, ball or roller-bearinged, with the big-end enlarged to form circular lugs to secure wrist pins carrying the plain or auxiliary type of rod of the remaining cylinders. 1928 Master rod [see *link rod* s.v. *LINK *sb.*^{1} 7]. 1946 J. W. VALE *Aviation Mechanic's Engine Manual* iii. 87 Constructing the crankshaft in sections permits use of the one-piece master rod. 1960 O. SKILBECK *ABC of Film & T.V.* 83 *Master scene*, usually an establishing shot; the basic scene from which a sequence is edited by intercutting or adding closer shots. A master scene script is an optional stage in writing..in which only the above appear. 1960 D. WILSON *Television Playwright* 16 The plays in this book are printed in 'Rehearsal Script' form, that is to say, the manuscript as the author wrote it. In the Cinema this would be called a 'Master-scene Script'. 1953 K. REISZ *Technique Film Editing* 280 *Master shot*, single shot of an entire piece of dramatic action taken in order to facilitate the assembly of the component closer shots of details from which the sequence will finally be covered. 1959 P. BULL *I know Face* v. 94 We started..with a 'master-shot', which means that the actors go through the entire scene from rather far off and only portions of the shot are likely to be used in the final film. 1972 R. MANVELL *Internat. Encycl. Film* 356 *Master shot*, the main shot of a complete piece of dramatic action, which facilitates the assembly of the component shots of which it will finally be composed. 1952, etc. Master–slave manipulator [see *MANIPULATOR 2 f]. 1960 *Times* 24 May 20/4 Closed-circuit stereoscopic television combined with a light-duty master–slave manipulator.. enables an operator to see the articles which he is moving in a solid radiation-proof container.

VI. 30. As a suffix forming compounds (mostly proprietary names) designating articles, appliances, etc., that are held to be supreme or superior in the field denoted by their first elements.

1935 *Vogue* 1 Dec. 110 Rainbow Matchmaster. 1937 *Amer. Speech* XII. 265 Then there are the *Buffet-Master*, a 'combination grill and waffle iron', *Menu-master*, *Kookmaster*, and *Grill Master*, all electric grills. 1947 J. BERTRAM *Shadow of War* 334, I glanced across the lighted interior of the big Skymaster. 1955 M. MCCARTHY *Charmed Life* (1956) ii. 52 The combination waffle iron and sandwich grill, the roto-broiler, the mixmaster. 1960 *Times Lit. Suppl.* 1 July 414/2 The shops were filled with goods and buyers..in the houses washers, dryers..Mixmasters. 1972 M. CRICHTON *Terminal Man* I. v. 41 Robert Morris was sitting in the hospital cafeteria..when his pagemaster went off... He..went to the wall phone to answer his page.

masterate (mā·stərět). [f. MASTER *sb.*^{1} + -ATE^1; cf. DOCTORATE *sb.*^{1}] The degree or dignity of a master (see MASTER *sb.*^{1} 13).

1902 *Science* 17 Oct. 612 The masterates should, of course, be permanent, and should not involve financial relations with the Institution. 1961 *Press* (Christchurch, N.Z.) 23 Mar. 10/8 There was not enough incentive for the best students to proceed to masterate degrees. 1966 *New Statesman* 16 Sept. 411/2 (Advt.), The Department [in Massey University, New Zealand] offers courses leading to the masterate level in Arts and to the Diploma in Education. 1971 *Commonwealth Universities Yearbk.* 1408 There is no uniformity among either universities or faculties as to the stage at which honours are awarded but they are frequently deferred to the masterate.

master-key. (Further *fig.* examples.)

1821 HAZLITT *Table-Talk* I. xvi. 376 Affectation is the master-key to both. *a* 1930 D. H. LAWRENCE *Phoenix II* (1968) 101 And he, she knew to her anguish and mortification, he was still the master-key to almost all life, for her. 1934 *Planning* II. XL. 2 Self-government for industry may prove to be the master key to many of our..problems.

masterless, *a.* Add: **1. c.** *transf.* Of unknown authorship or provenance.

1899 A. LANG *Homeric Hymns* 6 The conventional attribution of the Hymns to Homer..is merely the result of the tendency to set down 'masterless' compositions to a well-known name. 1903 *Library* IV. 397 It has become possible to assign to..Peter von Olpe a small group of books hitherto masterless.

ma·ster-mi·nd, *sb.* [MASTER *sb.*^{1} 24, 25 b.] **a.** An outstanding or commanding mind or intellect; a person with such a mind. Also *transf.*

1720 [in Dict. s.v. MASTER *sb.*^{1} 25 b]. 1821 HAZLITT *Table-Talk* I. ix. 198 He shews the marks of a great moving intellect, so that we trace the master-mind, and can sympathise with the springs that urge him on. 1839 POE *William Wilson* in *Gift 1840* (Philadelphia) 235 The despotism of a master mind in boyhood over the less energetic spirits of its companions. 1841 —— *Colloquy Monos & Una* in *Graham's Mag.* Aug. 52/2 At long intervals some master-minds appeared, looking upon each advance in practical science as a retro-gradation in the true utility. 1857 DICKENS *Dorrit* II. xxiv. 529 The master-mind of the age ..became mute again. 1909 WODEHOUSE *Mike* xxxii. 183 You can't expect two master-minds like us to pig it in that room downstairs. 1930 H. G. WELLS *Autocracy of Mr. Parham* III. viii. 220 This immediate personal recognition of the new régime by the master mind of Britain. 1949 E. JENKINS *Six Criminal Women* 11 Hers was one of the earliest master-minds in advertising. 1961 *Observer* 19 Feb. 5/4 An electronic master-mind called the 'Honeywell Data Centre'.

b. *spec.* Such a mind, or a person, directing a criminal enterprise.

1872 TROLLOPE *Eustace Diamonds* (1873) III. lxix. 203 The police thought that I had been the master-mind among the thieves. 1913 'D. D. CARTER' in *Hearst's Mag.* July 137 (*title*) The master mind. 1920 'SAPPER' *Bull-Dog Drum.* v. 121 A gang of international criminals..controlled by a master-mind. 1936 E. AMBLER *Dark Frontier* v. 80 From now on, he, Carruthers, would be the master mind. 1959 [see *crime-writing* s.v. *CRIME *sb.* 4]. 1960 *Observer* 24 Jan. 5/2 These were recognised prop-men or putters up of jobs, what the mugs called master minds.

Hence **ma·ster-mi·nd** *v. trans.*, to be the master-mind behind (an enterprise, a crime, etc.); to plan and direct; also occas. *intr.*; so **ma·ster-mi·nding** *vbl. sb.* and *ppl. a.*

1941 *Time* 26 May 24/2 The *Telegram* hired a series of detective storytellers to mastermind the Hess case. 1941 *True Detective* June 18/3 Often suspected but never convicted of masterminding some of the plots hatched in his grog shop. 1945 *Sun* (Baltimore) 30 Nov. 8 Baseball owes him nothing much but an opportunity to retire from the strenuous side of the competition and take over the job of master-minding. 1949 R. CHANDLER *Little Sister* xxi. 151 Somebody had to master-mind this deal. 1956 W. H. WHYTE *Organization Man* (1957) vi. 75 What they mean by sales is..master-minding the work of those who do the helping. 1957 'B. BUCKINGHAM' *Boiled Alive* xxxiii. 249 She master minded it all right, but..Pepe is not without blame. 1973 *Guardian* 29 June 15/7 A ruthless master-minding conspirator. *Ibid.* 15/8 It was possible he was.. capable of master-minding a vast conspiracy. 1973 A. HUNTER *Gently French* viii. 74 Quarles has master-minded for several of the gangs. 1974 D. RAMSAY *No Cause to Kill* II. 132 Hy Goldman..was..sceptical of her master-minding potential.

masterpiece. Add: (Further examples of orig. sense.)

1940 *Burlington Mag.* Sept. 93/2 Who could have ordered a cabinet made in this, apparently, needlessly expensive way? I think the only solution is that it must have been made as a 'master-piece', a test of workmanship to gain admission to a trade guild. 1951 *Oxf. Jun. Encycl.* VII. 143/2 The apprentice..was also at times required to provide a sample of his work, or 'masterpiece'. 1965 *Listener* 11 Nov. 761/1 The Mud Bath was another young man's 'masterpiece' (I am using the word 'masterpiece' in its original sense—as a test of a young painter's professional skill).

1. c. *colloq.* A person, thing, or action to be admired for being remarkable, singular, etc.

1906 KIPLING *Puck of Pook's Hill* 249 Ah, he was a masterpiece!..he never winked an eyelid. 1933 W. S. MAUGHAM *Sheppey* I. 20 It was a masterpiece the way I kidded him. 1963 N. MARSH *Dead Water* (1964) vii. 178 She spotted it [*sc.* sex] everywhere... She was a masterpiece. *Ibid.* viii. 223 He'm a masterpiece for holding his liquor.

master-stroke. 1. (Further examples.)

1690 in E. Waller *Poems* II. Pref. sig. A5, Some Painters will hit the chief Lines, and master strokes of a Face so truly, that [etc.]. 1867 EMERSON *May-Day* 108 And in their vaunted works of Art, The master-stroke is still her part.

mast-fed, *a.* [MAST *sb.*^{2}] Fed on mast.

1566 [in Dict s.v. MAST *sb.*^{2} 4]. 1843 'R. CARLTON' *New Purchase* II. lvi. 246 It was *mast* fed, i.e. fed on acorns and beech nuts. 1851 C. CIST *Sk. Cincinnati in 1851* 284 Much the larger share of this, is..made of mast-fed and still-fed hogs.

mast-head, *sb.* Add Now usu. **masthead** (without hyphen). **1. c.** In newspapers, journals, etc., the title, colophon, motto, etc., printed in some conspicuous position, usually immediately preceding the editorial matter or at the top of the first page. orig. *U.S.*

1838 *Hennepin* (Illinois) *Jrnl.* 22 Dec. 1/1 Many of our Whig friends..were anxious that the Journal should.. carry Whig colors at the mast-head. 1923 O. G. VILLARD *Some Newspapers* iii. 43 The vision of its purpose..which it now daily carries under its 'mast-head' on the editorial page. 1932 P. VAN D. STERN *Introd. Typogr.* 198 *Masthead*, the heading on the first page of a newspaper. 1937 *Amer. Speech* XII. 13 About 35 dailies use the word *Telegram* in their masthead. 1952 S. KAUFFMANN *Philanderer* (1953) ix. 141 She was a young, bright, snub-nosed, bobbed-hair cartoon character, to appear in the masthead and all institutional advertising. 1959 *Manch. Guardian* 22 Aug. 4/1 The masthead on page one will follow the style adopted..above our principal comment of the day. 1967 *Bucks Examiner* 3 Feb. 1/7 This week the 'Bucks Examiner' has a new masthead—the name given by newspaper men to the title at the top of page one. 1973 *Guardian* 28 Feb. 13/3 Varsity, the Cambridge University newspaper.. will appear next term under the mast head of its brasher rival Stop Press.

3. masthead cutter, -sloop (see quot. 1961);

masthead genoa, a Genoa jib which is attached to the topmost part of the mast.

1949 *Yachting Monthly* LXXXVII. 235/2 Her rig is a masthead cutter—that is to say the working jib sets to the masthead. 1961 F. H. BURGESS *Dict. Sailing* 144 *Masthead cutter* or *sloop*, a cutter or sloop whose fore stay, on which the luff of the foresail is set, reaches up to the masthead. 1962 J. A. S. RUSSELL in *Roving Commissions 1961* 147 Her masthead cutter rig, with the mast well inboard and a boomed staysail, is snug and versatile. 1958 *Yachting World Ann.* 115 *Springtide* has 900 sq. ft. in her working rig of main, mizzen, and boom staysail. The area may be increased by 400 sq. ft. with the masthead genoa. 1967 J. HOWARD-WILLIAMS *Sails* vii. 102 A bending mast..will mean abandoning ideas of a masthead genoa. 1954 D. PHILLIPS-BIRT *Rigs & Rigging of Yachts* 49 The masthead sloop—the sloop in which a headsail may be set on a forestay from the masthead. 1958 *Yachting World Ann.* 102/2 For some reason, yachtsmen appear shy of the masthead sloop rig; yet it is difficult to see what a yacht like this would gain by the lower fore-triangle.

mastika (mæstī·kă). Also **mastica, masticha.** [ad. mod.Gr. μαστίχα mastic.] = MASTIC *sb.* 6.

1926 P. M. SHAND *Bk. Wine* x. 264 Chios is also the home of the Mastic plant, whence a horrible liqueur, very popular in the Levant, called Chio-Mastica-Raki..is made. 1930 E. WAUGH *Labels* vi. 148 He said if one ever drank *mastika* one returned to Greece. 1966 'S. HARVESTER' *Treacherous Road* ix. 85 They were drinking mastica, the local brandy. 1967 A. LICHINE *Encycl. Wines* 346/2 *Mastika* (or *Masticha*), a favourite Greek aperitif, made on the island of Chios, from a brandy base with gum mastic added. 1968 L. DURRELL *Tunc* iii. 114 He poured me a glass of fiery mastika. 1969 *Listener* 2 Jan. 31/2 The aniseed-flavoured mastika. 1974 *Times* 21 Feb. 14 The guide maintained that, while Bulgarians were indeed virile, this was due more to drinking slivovice and mastika than to anything they ate.

mastitis. Add to def.: (in man or other mammals). (Further examples.)

1950 *N.Z. Jrnl. Agric.* Mar. 265/1 Milk from cows suffering from mastitis may be dangerous to public health. 1970 *Black's Vet. Dict.* (ed. 9) 531/2 When penicillin became available for veterinary use, it began to change the bacteriological picture of mastitis in our dairy herds.

ma·stman. *U.S.* [MAST *sb.*^{1} 1.] (See quot. 1890.)

1839 C. F. BRIGGS *Adventures Harry Franco* I. xxii. 236 All hands call him dismal Jerry, except Mike, the mast man, and he calls him Sergeant Longshanks. 1890 *Cent. Dict.*, *Mastman*, a seaman stationed at a mast in a man-of-war to keep the ropes clear and in order. 1901 A. M. KNIGHT *Mod. Seamanship* 382 Caution the mastmen to keep fast the weather sheets until the yards are down.

mastodon. *fig.* (Later examples.)

1851 C. CIST *Sk. Cincinnati in 1851* 187 There are five other tubs, which in the aggregate, contain as much as the great mastodon just described. 1936 [see *BLUE-PRINT, BLUEPRINT *sb.* 1 a]. 1965 *Listener* 2 Sept. 337/2 It is high time we..began to inquire more coolly how we can expect this mastodon in our midst to behave.

mastoidectomy (mæstoide·ktŏmi). *Surg.* [f. MASTOID *a.* and *sb.* + *-ECTOMY.] Any operation for the relief of inflammation of or within the mastoid process (esp. within the cavities, i.e. the mastoid cells and mastoid antrum, inside this process), as by penetrating into and cleaning out these cavities; excision of the mastoid process.

1898 *Laryngoscope* IV. 365 But the now perfected mastoidectomy, even with its tympanic communication, established under the best antiseptic measures,..left more to be desired. 1909 *Practitioner* Nov. 695, I have of late reserved the operation of mastoidectomy for cases of mastoid abscess in the acute stage with local signs. 1956 *A.M.A. Arch. Otolaryngol.* LXIII. 248/1 The surgery included both mastoidectomies (endaural and postauricular approach) and fenestrations.

masturbate, *v.* For def. read: To produce an orgasm by stimulation of the genitals, not by sexual intercourse. (Add further examples.) Also *fig.* and as *v. trans.*, to cause (another person) to have an orgasm by stimulation of his or her genitals.

1934 R. CAMPBELL *Broken Record* vi. 125 A bankclerk has to masturbate his mind all day: when he comes home he has no strength to mount the muse. 1964 H. MONTGOMERY HYDE *Hist. Pornography* i. 19 He [*sc.* Pepys] finished the book the same evening, and we now know that it excited him to masturbate. 1966 L. COHEN *Beautiful Losers* (1970) I. 36 She wanted to see me masturbate for the last time. 1968 M. COURTENAY *Sexual Discord in Marriage* vi. 61 He..had masturbated with heterosexual fantasies. 1970 B. W. ALDISS *Hand-Reared Boy* 62 During this miserable period I masturbated myself for consolation, and Ann also did it to me. 1974 *Daily Tel.* 3 Oct. 3/1 Clients at the parlour..paid £15-plus to be masturbated by young women in various stages of undress.

masturbator, masturbatory: further examples. Also **masturba·torily** *adv.*

1868 *Index Expurgatorius of Martial* 90 Titius was a masturbator. 1924 J. RIVIERE et al. tr. *Freud's Coll. Papers* I. 100 The tendency to anxiety in masturbators. 1932 *Jrnl. Amer. Med. Assoc.* 24 Dec. 2201/2 Mechanical masturbators. 1948 A. C. KINSEY et al. *Sexual Behavior Human Male* II. v. 170 Many pre-adolescents take a good many years to discover masturbatory techniques. 1960 *Spectator* 3 June 809 The prose background to his [*sc.* Swinburne's] violent, cerebral, masturbatory poetry. 1963 *Listener* 24 Jan. 163/2 To be sexually attractive was one thing, but to become a sexual fetish was another. Marilyn Monroe became a pair of lips, a walk, a set of numbers, 38-24-36. The artificiality of the unresolved tension I have described finds a parallel here in this ersatz, masturbatory sex stimulus. 1967 B. BROSS *Pleasures of Love* vii. 98 These magazines are being bought by masturbators, male ones and lesbians. 1973 S. FISHER *Female Orgasm* xii. 338 If a woman wants to escape restriction and explore the new she is particularly positive toward masturbatory gratification. 1973 M. SEYMOUR-SMITH *Guide Mod. World Lit.* 225 One senses Williams hanging masturbatorily over his nasty, midcult images of evil.

masturbation. For def. read: The action or practice of masturbating; deliberate erotic self-stimulation. (Add further examples.) Also *attrib.* and *fig.* Also **mutual masturbation,** stimulation of the genitals of one person by another in order to produce an orgasm without sexual intercourse.

1897 H. ELLIS *Stud. Psychol. Sex* I. 117 The sexual relationship rarely goes beyond close physical contact, or at most mutual masturbation. 1924 J. RIVIERE et al. tr. *Freud's Coll. Papers* I. 90 Neurasthenics whose potency has already been seriously diminished by masturbation. 1929 D. H. LAWRENCE *Pornogr. & Obscenity* 32 The masturbation self-enclosure produces idiots. 1948 A. C. KINSEY et al. *Sexual Behavior Human Male* III. xiv. 497 As more usually employed, the word 'masturbation' refers to any self stimulation which is deliberate and designed to effect erotic arousal. *Ibid.* xxi. 616 Mutual masturbation between two males may be dismissed, even by certain clinicians, as not homosexual. 1949 M. MEAD *Male & Female* xiii. 271 The attempts of some child specialists to break down the masturbation taboo. 1951 R. CAMPBELL *Light on Dark Horse* xxiii. 346 The liberaloid mentality that they had acquired from reading the masturbations of ..Rousseau. 1954 W. FAULKNER *Fable* (1955) 59 That sort of masturbation about the human race people call hoping. 1961 R. F. C. HULL tr. *Jung's Coll. Works* IV. 212 She had induced a number of girls of her own age to perform mutual masturbation. 1967 E. LEA in R. E. L. Masters *Sexual Self-Stimulation* 325 The Dresden Criminal Museum..houses an elaborate..masturbation machine that the user operates with a foot pedal. 1968 *Listener* 14 Mar. 350/3 It contains..many, many accounts of adventures with women, masturbation phantasies, [etc.]. 1968 M. COURTENAY *Sexual Discord in Marriage* vi. 68 He became worried about the possible harmful effects of masturbation. 1972 *Daily Tel.* 28 Nov. 17/3 Lines such as 'I like to play with my ding-a-ling' and 'Most of all with your ding-a-ling' are intended as deliberate stimulation to self and mutual masturbation. 1974 *Ibid.* 3 Oct. 3/3 For the £15 massage—with masturbation—the girls received £3 commission. *Ibid.*, He had first set up massage facilities.., but clients had wanted masturbation.

masurium (măzĭū̆ə·riŏm). *Chem.* [mod.L. (W. Noddack et al. 1925, in *Sitzungsber. d. Preuss. Akad. d. Wissenschaften* 409), f. G. *Masur-en* name of a region in NE. Poland + -IUM.] A name proposed for the element of atomic number 43 (later named *TECHNETIUM), which was claimed to have been discovered spectroscopically in certain platinum ores.

1925 *Glasgow Herald* 16 June 6 These new elements have been named by their discoverers 'Masurium', after the Masurian Lake region, and 'Rhenium', after the Rhineland. 1941 *Chem. & Industry* 11 Oct. 729/2 Japanese investigators have studied the decay curves of the chemically separated masurium fraction from molybdenum after bombardment with slow neutrons and they distinguished the two masurium isotopes 99 and 101. 1947 [see *ILLINIUM]. 1962 J. H. WHITE *Inorg. Chem.* xxv. 480 Technetium..has been called masurium, and it has not yet been isolated from natural minerals.

masut, var. *MAZUT.

mat, *sb.*^{1} Add: **1. e.** *N.Z.* A type of cloak or cape worn by the Maori (cf. quot. 1777 under sense 1 a in Dict.); also used allusively to refer to the Maori way of life.

1807 J. SAVAGE *Some Acct. N.Z.* viii. 50 The dress of the natives consists in a mat finely wove of the native flax. 1832 A. EARLE *Narr. Residence N.Z.* (1966) 59 They were clothed in mats, called Ka-ka-hoos. 1840 W. DEANS *Let.* 29 Mar. in J. Deans *Pioneers of Canterbury* (1937) 23 Two New Zealanders, clad in a native mat. 1849 W. T. POWER *Sketches in N.Z.* xvii. 146 New habits are rapidly modifying the old ones... In throwing off the mat and the blanket, they also dispense with shark oil and red ochre. 1874 J. C. JOHNSTONE *Maoria* i. 16 The rough *pureki*..when seen upon the men in the canoes which boarded the first vessels that visited the Island, was not inappropriately called 'a mat', and the ugly name came to be applied to any description of garment worn by the Maoris. 1905 W. B. *Where White Man Treads* 54 He is a warrior; and at any moment may cast off his mat and defend his privileges. 1947 'A. P. GASKELL' *Big Game* 92 She must have been somewhere to a Maori High School and then come back to the mat. 1970 D. M. DAVIN *Not Here, Not Now* III. vi. 197 All a man can do is go back to the mat and cry, or laugh.

f. *Bowls.* = FOOTER *sb.*^{1} 4.

1892 J. BROWN *Man. Bowling* (ed. 2) 69 The mat shall be placed by the lead of the party who lost the previous head. 1910 *Encycl. Brit.* IV. 347/2 The bowler delivers his bowl with one foot on a mat or footer, made of india-rubber or cocoanut fibre, the size of which is also prescribed by rule as 24 by 16 in. 1959 *Times* 12 Aug. 4/6 Their No. 3..went to the mat. 1962 *Bowls* ('Know the Game' Series) 4 At the beginning of the first end the mat is placed lengthwise on the centre line of the rink, the back edge to be four feet from the ditch.

g. Phr. *on the mat*, orig. in army use (see quot. 1919); in trouble with some authority. Cf. *on the carpet* (s.v. *CARPET *sb.* 1 b).

1898 *Pearson's Mag.* Oct. 372/2 The sergeant..shouts with military brevity: 'On to the mat, John Smith.' [*ante*, Close to the medical officer's desk is a thick padded carpet about a yard square.] 1917 A. G. EMPEY *Over Top* 302 *On the mat*, when Tommy is haled before his commanding officer to explain why he has broken one of the seven million King's regulations for the government of the Army. His 'explanation' never gets him anywhere unless it is on the wheel of a limber. 1919 *Athenæum* 1 Aug. 695/1 'He 's on the mat' means the same [as 'He's for the high jump']; the pre-war orderly room was furnished with a piece of carpet, in the exact centre of which the accused stood. 1925 FRASER & GIBBONS *Soldier & Sailor Words* 154 *Mat, on the*: up for trial. In trouble. 1935 A. J. POLLOCK *Underworld Speaks* 92/2 *Put on the mat*, a thorough questioning, usually by the police. 1949 J. R. COLE *It was so Late* 62 Then I was on the mat again. Now it seems a wonder I kept out of trouble as long as I did. 1973 J. THOMSON *Death Cap* x. 136 Mrs Holbrook had been given the impression that she was on the mat in front of her husband's superior officer.

h. A piece of padded material, canvas, etc., used as a floor covering in gymnastics, wrestling, etc. Hence *fig.*, in phr. *to go to the mat*: to engage in a struggle or controversy; to argue.

1903 P. LONGHURST *Wrestling* I. 5 Ordinary gymnasium mats covered with canvas or sail-cloth form the best surface for this style of wrestling. 1924 WODEHOUSE *Leave it to Psmith* i. 28, I..heard..you and Aunt Constance going to the mat about poor old Phyllis. 1937 D. ALDIS *Time at her Heels* i. 26 She just didn't have time at the moment to call him in and go to the mat with him about it. 1950 *Oxf. Jun. Encycl.* IX. 489/1 In this type of wrestling a 'fall' is gained by bringing the opponent's two shoulders simultaneously into contact with the mat. 1967 V. L. DREHMAN *Head over Heels* i. 3 Wide mats must be used for the learner in tumbling. 1970 *New Yorker* 12 Dec. 131/1 These senators felt that the President had handed them two lemons, had gone to the mat for his choices when he didn't have to.

2. b. (Earlier examples.)

1800 M. EDGEWORTH *Parent's Assistant* (ed. 3) V. 32 These here half dozen little mats, to put under my dishes. 1852 MRS. GASKELL *Let.* Dec. (1966) 217 The little ones had worked mats, & gathered flowers &c &c for her dressing-room.

3. a. (Later examples.)

1898 POUND & CLEMENTS *Phytogeogr. Nebraska* iii. 53 This group is composed of acaulescent or low-stemmed plants which grow aggregated into dense cushion-like mats. 1916 B. D. JACKSON *Gloss. Bot. Terms* (ed. 3) 224/1 *Mat*, a closely intertwined vegetation, with roots and rhizomes intermixed. 1930 *Forestry* IV. 70 A 'mat' of actively growing mycelium of the parasite was present in the watering flasks of many of the cultures. 1971 *Nature* 30 Apr. 599/1 At Kariba..where the lake took 4–5 yr to fill, extensive mats of *Salvinia auriculata* and *Pistia stratiotes* accumulated over much of the lake. 1972 *Science* 27 Oct. 403/1 Microbial mats occur in Yellowstone at temperatures up to about 70°C. 1973 *Nature* 4 May 12/1 The meeting concluded with a field trip to the algal mats at the north shore of the Great Salt Lake.

8. a. *mat-house, -shed* (examples), *-skirt, -work.* **c.** *mat-forming, -making* (examples). **d. mat-man** *slang*, a wrestler.

1951 *Dict. Gardening* (R. Hort. Soc.) IV. 1878/1 [*Saxifraga*] Section 2. Hirculus. Mat-forming plants with undivided, deciduous, oval leaves. 1971 D. BARTRUM *Rock Gardens* vii. 162 The 'Cheddar Pink' is another very tough mat-forming plant that doesn't mind being walked upon when not in flower. 1898 W. C. SCULLY *Between Sun & Sand* 18 (Pettman), On either side of it stood respectively, a mat-house and a square tent. 1854 THOREAU *Walden* 283 Might not the basket, stable-broom, mat-making, corn-parching, linen-spinning, and pottery business have thrived here? 1890 LD. LUGARD *Diary* 21 Feb. (1959) I. ii. 111 The Banga reed grows by the river here, but very poorly, and useless for building and mat-making. 1923 *N.Y. Times* 11 Feb. I. II. 1/4 (*heading*) Navy matmen on top. In a finely contested wrestling match..the Naval Academy won. 1930 *Ibid.* 16 Feb. XI. 3/5 It was..the first time in four seasons that the Midshipmen have been able to take the measure of the matmen from South Bethlehem. 1942 BERREY & VAN DEN BARK *Amer. Thes. Slang* § 707/2 Matman. 1968 *Globe & Mail Mag.* (Toronto) 17 Feb. 8/3 He became one of the best known mat men in Canada. 'Wrestling always fascinated me,' he says now. 1971 *Soviet Weekly* 8 May 14 A popular group exercise among the matmen. 1926 M. LEINSTER *Dew on Leaf* I. iii. 41 Warehouses, mat-sheds, and hovels. 1939 'A. BRIDGE' *Four-Part Setting* ii. 8 To sit in a mat-shed on the sand and drink cocktails. 1908 *Daily Chron.* 15 Aug. 1/6 A Maori chief..saying..he was to fasten the native mat-skirt about his body. 1944 *Horizon* Jan. 48 P.T. exercises ..mat work, track work.

mat, *sb.*^{2} **3.** Substitute for def.: A sheet of cardboard placed on the back of a print or drawing and then covered by a mount which forms a margin round the area of the print; also used for the mount itself. Cf. MOUNT *sb.*^{2} 3 a. Also *attrib.* (Add later examples.)

1909 F. WEITENKAMPF *How to appreciate Prints* xiii. 291 Sometimes mat and mount are fastened together on all four sides, forming what is known as a 'sunk mount'. 1932 —— *Quest of Print* xii. 270 Some collectors place a sheet of celluloid, cellophane, or similar material..over the print

and under the mat. 1965 ZIGROSSER & GAEHDE *Guide to Collecting Orig. Prints* vii. 100 Quality of Mat Board. Only 100 percent rag-fiber mat stock is to be used. 1967 *Boston Sunday Globe* 23 Apr. B. 58/7 Sometimes prints come with a mat (white space around the art), but the framed picture is much better if the mat is made from a mat board. 1973 F. TAUBES *Painter's Dict. Materials & Methods* 149 Made, as a rule, from cardboards of various colours, mats serve in the framing of watercolors and all kinds of prints... The width of a mat..should be equal on top and at the sides but somewhat greater at the bottom, or the picture will have a tendency to 'droop'. 1974 P. HIGHSMITH *Ripley's Game* v. 50 He needed more mat paper.

mat, *sb.*^{5} Colloq. abbrev. of MATINÉE.

1914 G. ATHERTON *Perch of Devil* I. viii. 55 Although Mr. Compton won't take me to any balls, there are the movin' pictures and the mats—matinées. 1940 *Amer. Speech* XV. 204/2 *Mats*, matinees.

mat, *sb.*^{6} Abbrev. of MATRIX 4.

1923 M. V. ATWOOD *Country Newspaper* 20 Just a word should be added about matrices, or 'mats' as they are always called. 1937 E. J. LABARRE *Dict. Paper* 172/1 This matrix or 'mat' is then baked and used for making a metal plate for flat or roll printing. 1942 F. BROWN *Angels & Spaceships* (1955) 38 The cost of getting special Linotype mats cut would be awfully high. 1967 V. STRAUSS *Printing Industry* v. 225/2 Matrices, called mats by the industry, are intermediate elements in the production of stereotypes but they are also independent items of commerce. *Ibid.* 226/1 During mat-making an intaglio replica of the original relief material is produced. 1975 *Printing Historical Soc. Newslet.* No. 28. 3 A few large display matrices of the Caslon series (original founders' mats) are offered to PHS members.

mat (in Cinematography): see *MATTE^3.

-mat (mæt), *suffix.* [abbrev. of *-MATIC *suffix.*] A terminal element in sbs. (usu. proprietary names) denoting a device that works automatically, as *ROTISSOMAT, or a business that contains automatic or self-service equipment, as *AUTOMAT, *LAUNDROMAT. orig. *U.S.*

1935 [see *-MATIC *suffix*]. 1951 *Amer. Speech* Oct. 166 Probably the 'Automat', a self-serving institution supplying food, gave special impetus to the extension of the suffix *-mat*. A 'Mail-O-Mat' is available in the Reception Section of the Pentagon building at Washington. 1971 *Amat. Photographer* 13 Jan. 78/1 (Advt.), Nikkormat FTN f2 Nikkor, £80.

mat (mæt), *v.*^{4} [f. *MAT *sb.*^{2} 3.] *trans.* To mount a print on a cardboard backing, or to provide it with a border. So **ma·tted** *ppl. a.*; **ma·tting** *vbl. sb.*

1965 ZIGROSSER & GAEHDE *Guide to Collecting Orig. Prints* vii. 100 The collector or owner who will have to rely on commercial framers to have his prints matted is cautioned to insist on the following points. *Ibid.* 104 If the matted print belongs to a study collection and is stored horizontally, hinging at the side is preferable because it makes for easier and safer handling. 1967 *Boston Sunday Globe* 23 Apr. B. 58/7 Oil paintings, on the other hand, do not take glass or matting. 1968 P. NUTTALL *Picture Framing* ii. 24 Different papers can be used to surface the mounting (matting) boards before the window is cut.

Matabele (mætăbī·li). Also 9 **Matabeli, Matabili.** [Native name.] **1.** A people of Zulu stock living in Rhodesia; also, a member of this people. Also *attrib.* or as *adj.* Cf. *NDEBELE. (See also quot. 1925.)

1823 R. MOFFAT *Jrnl.* 27 May in I. Schapera *Apprenticeship at Kuruman* (1951) iv. 84 Several men from the Barolongs had just passed them on their way to Mahumapeloo to request their assistance..to endeavour to make the Matabeles retreat. 1835 A. SMITH *Diary* 16 June (1940) II. 75 Most of the people appeared to be the original inhabitants of the country, having a different look to the Matabeli. 1839 W. C. HARRIS *Wild Sports S. Afr.* iv. 22 The country of Moselekatse, king of the Abaka Zooloos, or Matabili. *Ibid.* xiv. 111 Numerous Matabili villages. 1854 R. MOFFAT *Matabele Jrnls.* (1945) I. 319 Scarcely a child is to be seen in a real Matabele town. 1894 *Jrnl. Anthrop. Inst.* XXIII. 83 As soon as a Matabele is dead his relations tie the corpse in a blanket. 1896 F. C. SELOUS *Sunshine & Storm Rhodesia* viii. 64, I should not be at all surprised to see it stated that the rebellion was caused by the inhuman behaviour of the white men in Rhodesia, who, it will be said, were in the habit of shooting down the poor, meek, inoffensive Matabele. 1919 H. H. JOHNSTON *Compar. Study Bantu & Semi-Bantu Lang.* I. v. 798 Tebele (Sindebele). This dialect of the Matabele (Amandebele) Zulus [etc.]. 1925 A. WERNER in J. A. MacCulloch *Mythol. all Races* VII. II. vii. 246 There are various tribes of ogres having only one arm and one leg, while others..have not this peculiarity. The Basuto call the former class of beings 'Matabele'—probably from having come to look on their dreaded enemies, the Zulu tribe of that name, as something scarcely human. 1936 *Discovery* June 172/1 The Xulu-Xosa (including the so-called 'Kaffirs', and the Matabele). 1967 T. O. RANGER *Revolt in S. Rhodesia* i. 38 While the Company made use of the missionaries for its own ends, employing them to obtain a favourable press for the Matabele war, Company officials themselves shared these missionary beliefs. 1972 *Police Rev.* 24 Nov. 1521/1 Cecil John Rhodes..began negotiating with Lobengula, king of the Matabele.

2. Matabele ant, a large, black, stinging ant, *Megaponera foetens*, found in Rhodesia.

1924 E. STEP *Go to Ant* xiii. 193 One of these [stink ants] is the Stink Ant proper (*Paltothyreus tarsatus*) and the other is the Matabele Ant (*Megaponera foetens*). 1932 *Discovery* July 225/2 The big black Matabele ant, *Megaponera foetens*, often marched through our camp in military formation. 1953 S. H. SKAIFE *Afr. Insect Life* xxii. 361 The Matabele ant, *Megaponera foetens*, is one of the largest and best-known members of this sub-family [*sc.* the Ponerinæ].

matador. Add: **1. b.** *fig.*

1930 R. CAMPBELL *Adamastor* 8 The matador of truth, he trails his scorn Before their lowered horns and bloodshot eyes. 1944 AUDEN *For Time Being* (1945) 118 George, you old matador, Welcome back to the Army.

c. Used *attrib.* to designate garments resembling those worn by a matador. Cf. TOREADOR C.

1959 M. SHADBOLT *New Zealanders* 217 She was..clad in black, in tight sweater and matador slacks. 1960 *Tamarack Rev.* XIV. 137 The girls dress mostly in matador pants and bright blouses. 1962 C. ROHAN *Delinquents* 144 She used to sit on the top of the front steps, clad in her matador pants and skin-tight sweater.

‖ **matadora** (mætădǭ·ră). [Sp.] A woman matador.

1955 *People* (Austral.) 11 Jan. 20/2 In Mexico the girls are invading the bullrings. Pretty matadoras with flashing swords and whirling red capes are taking much of the limelight. 1956 P. MCCORMICK *Lady Bullfighter* i. 30 The words..are particularly welcome when someone substitutes the word '*matadora*' for '*señorita*'. 1968 C. CINTRON *Torera: Mem. Bullfighter* II. 70 Since they barely tolerated the idea of a gringa rejoneadora, they wouldn't even consider the possibility of an American matadora. 1974 *Publishers Weekly* 24 June 54/2 The cast includes a nervous novice matador, an ace matador,..an aspiring gypsy matadora.

matagouri. Add: (Also with pronunc. (-gau·ri).) Also **matagory, -gowry**; = *IRISHMAN 2. Also *attrib.* (Later examples.)

1934 *Bulletin* (Sydney) 16 May 39/1 Her [*sc.* an old ewe's] long wool became entangled in the thorns of a stunted matagory bush. 1939 C. BRASCH in A. E. Currie *Cent. Treas. Otago Verse* (1949) 100 Alone on the parched rise, inhuman matagowry Dry-green and fibrous. 1958 *Landfall* XII. 18 My hands were thorned with the matagauri from pulling off the finer neck wool. 1962 J. FRAME *Edge of Alphabet* xxx. 164 He remembers his last visit up Central..where the matagouri, the 'spiked plant that does not cry' grows on the hills among the snow-grass. 1963 *Times* 6 Feb. p. vii/3 What were once harsh, unlovely *matagouri* flats are now green with lucerne.

Mata Hari (ma·ta hā·ri). [f. Malay *mata* eye + *hari* day.] The name taken by Margaretha Gertruida Zelle (1876–1917), used *fig.* to signify a beautiful and seductive spy; also *attrib.* and as *vb.*

1936 E. WAUGH *Waugh in Abyssinia* iii. 96 Patrick's spy..was soon known to the European community as Mata Hari. 1947 'N. BLAKE' *Minute for Murder* vii. 163 He's got it into his thick head that I'm a male Mata Hari. 1948 'P. QUENTIN' *Run to Death* iii. 25 Why don't you come out from behind the Mata Hari and tell me the truth? 1962 M. CARLETON *Dread Sunset* vii. 122, I somehow can't see Miss Gantry as Jaspar's Girl Friday, let alone a Mata Hari. 1963 H. SLESAR *Bridge of Lions* (1964) i. 19 'She?' Shortlake grinned. 'Ah. Mata Hari stuff.' 1970 T. LILLEY *Projects Section* xix. 257 She was beginning to look haggard; the strain of Mata-Hari-ing was sure to tell. 1973 'A. HALL' *Tango Briefing* xiv. 183 You wouldn't have to frisk this pint-sized Mata Hari: you could see she was armed half a mile away.

‖ **matai** (mata,*i*). [Samoan.] In a Samoan extended family, the person who is chosen to succeed to a chief's or orator's title and honoured as the head of the household.

1928 M. MEAD *Coming of Age in Samoa* (1929) iv. 40 The *matai* exercises nominal and usually real authority over every individual under his protection, even over his father and mother. 1934 F. M. KEESING *Mod. Samoa* i. 30 In general the *matais*, or titled men,..engage mostly in ceremonial activities while the work as the white man views it is done by the rest of the community. 1956 F. M. & M. M. KEESING *Elite Communication in Samoa* ii. 17 The community is made up of a series of households of 'extended family' or 'multiple family' type, each under the authority of a 'head' or '*matai*'. 1974 *Sunday Advocate-News* (Barbados) 16 Dec. 21/2 The matais usually deal out justice to law breakers in the Samoan spirit of 'forgive and forget' before the matter ever reaches the ears of the authorities. 1974 *Encycl. Brit. Macropædia* XVI. 205/2 Within the villages, close kinship ties have traditionally bound individuals into a sternly collectivist society. Elected family leaders called *matai* (titled chiefs or orators) form village or district councils to administer group affairs.

matapo, var. *MATIPO.

Matara, var. *MATURA.

matata (ma·tātă). *N.Z.* [Maori.] = *fernbird* (*FERN *sb.*^{1} 2 b).

1835 W. YATE *Acct. N.Z.* (ed. 2) ii. 60 Matata—A small dusky-coloured bird, with a white and brown spotted breast. 1863 A. S. ATKINSON *Jrnl.* 29 Sept. in *Richmond-Atkinson Papers* (1960) II. 64 The korimakos in the bush and the larks and matatas in the open about us. 1882 [see *fern-bird* (*FERN *sb.*^{1} 2 b)]. 1966 R. A. FALLA et al. *Field Guide to Birds N.Z.* 205 Fernbird. *Bowdleria punctata.* Other names: Matata, Tataki Thrush.

Matawila, var. *METAWILEH.

match, *sb.*^{1} Add: **7.** (Earlier example of use in Cricket.)

1700 *Post Boy* 30 Mar. 2/1 A Match at Cricket, of 10 Gentlemen on each side, will Be Play'd.

12. match ball, a ball of the quality and dimensions specified by the laws of the game; also, in lawn tennis, a ball that may decide a match; **match-book,** (*b*) *Cricket* (see quot. 1934); **match-card** *Cricket* (see quot. 1934) (= *score-card* (a)); **match-fit** *a.*, in good physical condition for a match; also *transf.*; hence **match-fitness; match-play, player** (add examples of lawn tennis usages); so *match-playing* vbl. sb.; **match-point,** (*a*) the state of a game when one side or player needs only one point to win the match; also, the point itself; (*b*) in *Bridge*, a unit used in scoring in tournament play; so *match-pointed* ppl. a.; **match race,** a race run as a competition; **match-winner,** one who, by his skill, makes a major contribution to the winning of a match; hence **match-winning** *ppl. a.*

1849 in 'Bat' *Cricketer's Manual* (Advt.), Dark's and Duke's Match Balls. 1895 KIPLING *Day's Work* (1898) 344 In the black jersey..of the First Fifteen, the new match-ball under his arm. 1927 *Daily Express* 5 July 2/1 They.. retrieved two successive match balls in the third set. 1928 *Daily Tel.* 26 June 11/7 On the tenth game, the American had three match balls. The first time he smashed out of court. 1934 W. J. LEWIS *Lang. Cricket* 8 A match ball is one of superior quality for match-play, of the size and weight specified in the Laws. 1961 *Amer. Speech* XXXVI. 44 With a lob Fraser lifted the deciding matchball out of bounds. 1969 M. BRADY *Lawn Tennis Encycl.* 129 Match Balls. The following championships were won after the winner was match point down. 1845 W. DENISON *Cricketer's Compan.* p. iv, The Compiler feels it necessary to offer his thanks to the Presidents and Secretaries..for ..allowing him to make extracts from their match-books. 1900 W. A. BETTESWORTH *Walkers of Southgate* II. vii. 253 In the [Harrow] Sixth Form match-book Mr. Tremlett's score..is given as follows. 1934 W. J. LEWIS *Lang. Cricket* 161 *Match book*, a book containing the scores or tabulated records of the matches played by a club or an eleven. 1901 H. BLEACKLEY *Tales of Stumps* iv. 96 Match-cards with the 'order of going in' had been printed and eagerly purchased by the spectators. 1908 *Westm. Gaz.* 20 Oct. 11/2 The Jam Sahib of Nawanagar, who, however his name may be printed on the match-cards, will always be known to the cricket-loving public. 1934 W. J. LEWIS *Lang. Cricket* 162 *Match card*, a card giving the names of the players in the order of going in to bat, and a summary of the score up to the time of issue. 1960 V. JENKINS *Lions Down Under* xii. 179 Terry Davies, Tony O'Reilly and Jeff Butterfield..were still far from match-fit. 1962 *Times* 1 Aug. 3/1 J. G. Willcox..is still not match-fit. 1967 *Listener* 24 Aug. 241/2 It took me about three months of being back in the theatre to feel match fit. 1960 V. JENKINS *Lions Down Under* xvi. 251 He..failed to reach match-fitness in time. 1961 *Times* 1 July 3/1 It was more than the flesh and blood of a man below physical and match-fitness could stand. 1920 W. T. TILDEN *Art Lawn Tennis* 67 Match play, where both men are in the same class as tennis players, resolves itself into a battle of wits and nerve. 1909 E. H. MILES *Lessons Lawn Tennis* (ed. 3) 50 The ideal match player. *Ibid.*, Hints on Match-playing. 1921 A. W. MYERS *20 Yrs. Lawn Tennis* 48 In the end he won the match, Dixon, after lazily reaching match point some eight or nine times,..retiring at two sets all. 1928 *Daily Express* 22 June 1/6 The British doubles team held match point twice in the fourth set with Eames serving. 1936 E. CULBERTSON *Contract Bridge Complete* xxxiv. 371 In duplicate Contract with match-point scoring, the unit of play is not a game or a rubber, but an individual deal. 1940 *Ibid.* (ed. 2) xxxiv. 369 Top-score (7 match-points) went to the North-South pair that fulfilled a six-spade contract. 1955 *Times* 6 July 8/5 The British [bridge] team continued to disappoint ..although they beat Finland to-day by 53 match points. 1965 *Listener* 30 Dec. 1091/1 Except at match-point scoring. It can be argued that in this method of scoring One No Trump is almost certain to give East-West a very good score. 1969 *New Yorker* 14 June 61/1 Pasarell can drive Graebner out of his mind, because he sometimes waits until Graebner has him at or near match point. 1973 *Country Life* 13 Dec. 2048/1 This was a good match-point result, for many pairs reached Six Diamonds. 1958 *Listener* 30 Oct. 709/2 Match-pointed Pairs. Game All. Dealer West. 1974 *Guardian* 25 Mar. 24/7 Match-pointed pairs events require an entirely different approach from team competitions. 1804 M. CUTLER in W. P. & J. P. Cutler *Life & Corr. M. Cutler* (1888) II. 172 It was a match race of two two-year-old colts for $1,000. 1854 W. G. SIMMS *Mellichampe* xviii. 157 If by..a match-race on foot with an Indian runner, I could do the creature a service, I could go to work cheerfully. 1874 B. F. TAYLOR *World on Wheels* 105 The train..ran a match race with a train on the Michigan Central, and reached Chicago twenty-five minutes ahead. 1948 *Chicago Daily News* 1 Nov. 13/3 He whipped Sir Barton, a 4-year-old, in a memorable match race. 1955 *Amer. Speech* XXX. 22 The term match or match race in common acceptance encompasses any specially arranged two-horse race. R. [Rules of the N.Y. State Racing Commission] 159, however, applies the term to such a race only when the track management has added 'no money or other prize'. 1961 F. C. AVIS *Sportsman's Gloss.* 166/2 *Match Race Conditions*, those normally applied to a cycle race, and affecting the number of competitors; their positions; the start, etc. 1908 *Westm. Gaz.* 21 Aug. 3/1 If you..should encounter as keen a match-winner as yourself [at golf]. 1964 J. MERCER *Great Ones* x. 69 At Wembley nothing happened to upset matters. Reg

Lewis was undoubtedly our match-winner. 1908 *Westm. Gaz.* 21 Aug. 3/1 And if you, match-winning disciple, find yourself placed in this delicate position [etc.]. 1909 *Ibid.* 12 June 16/1 In the indefatigable Mr. Brearley they have a match-winning bowler on fast wickets. 1958 *Times* 20 Sept. 2/6 Without Wardle they have no match-winning spinner. 1960 V. JENKINS *Lions Down Under* xii. 173 The accepted New Zealand pattern for match-winning rugby. 1972 G. ROSS *Hist. Cricket* i. 16 Mynn was the greatest match-winning cricketer the game produced before W. G. Grace.

II. The state of being matched. **11*.** *Electr.* An equality of impedance between two coupled devices (cf. *MATCH *v.*^{1} 5 c).

1931 *Proc. IRE* XIX. 725 By introducing capacitative elements,..a match can be obtained. 1952 D. D. KING *Measurement at Centimeter Wavelength* ii. 51 No assurance of match exists without prior knowledge of the impedance to be matched. 1962 SIMPSON & RICHARDS *Physical Princ. Junction Transistors* v. 89 The condition of conjugate match can be obtained quite readily when tuned transformer coupling is used as in many radio-frequency applications.

match, *sb.*^{2} Add: **5.** *match-point*; **match-book,** a 'book' containing (safety) matches; **match-box,** (*c*) *slang* a very small house; also *attrib.*, esp. in phrases *match-box skirt* (see quot. 1968), *match-box toy*, a toy small enough to fit into a match-box; **match-pot,** a small vessel for holding matches; **match-stand,** a stand for holding matches; **match-stick,** also (*b*) *slang*, a nickname for a thin person; (*c*) *attrib.*, esp. designating simple drawings in short straight lines.

1951 C. ARMSTRONG *Black-Eyed Stranger* (1952) iv. 31 The old match-books in the gutter. 1966 H. WAUGH *Pure Poison* (1967) xxiii. 140 The counter..boasted a ledger..a postcard rack and a basket of matchbooks advertising the motel. 1968 MRS. L. B. JOHNSON *White House Diary* 7 Feb. (1970) 628 One table asked me for my autograph and I wrote it for them on White House matchbooks. 1905 *Pearson's Mag.* July 103/2 Few fingers are too old and few too tiny to help the match-box maker. 1920 JOYCE *Let.* 29 Sept. (1957) I. 148 Got notice to quit this match-box and am..looking for a flat. 1946 A. J. CRUSE (*title*) Match-box labels of the world. 1962 *Harper's Bazaar* Oct. 134/2 A matchbox skirt. A flipover belt ties on the hips. 1964 *New Statesman* 10 Apr. 579/1 Lesney Products (matchbox toys) has reported an accelerated increase in exports, turnover and profits. 1968 J. IRONSIDE *Fashion Alphabet* 45 *Match-box*, a skirt given a squared box-like look by outside stitched seams back and front, introduced by Dior. 1969 E. H. PINTO *Treen* 211/1 A large range of German 'matchbox' wooden toys was made at that time [1930]. 1929 W. FAULKNER *Sartoris* v. 351 The thick cables along the veranda eaves would be budding into small lilac match-points. 1856 J. C. ROBINSON *Inventory of Objects Mus. Ornamental Art* 37 Wedgwood Match-pot... Pair of Match-pots. 1882 *Hamilton Palace Collection Catal.* No. 600 A two-handled Chinese Vase and Cover, of rock crystal, with a matchpot at the side. 1873 *Young Englishwoman* July 357/2 This match-stand is made of pasteboard, covered with velvet. 1909 *Chambers's Jrnl.* Aug. 506/1 One lot tells us of '4 silver-mounted match-stands..and a quantity of imitation jewellery'. 1959 I. & P. OPIE *Lore & Lang. Schoolch.* ix. 169 Thin people inspire almost as many names and jokes as fat people..match-stick (sometimes abbreviated to 'matchy'), needles, [etc.]. 1959 HALAS & MANVELL *Technique Film Animation* 14 In France, Emile Cohl began to make his little white match-stick figures jump about against a black background as early as 1908. 1963 *Times* 22 May 3/3 One man mastered it, little Harmer, of the matchstick legs and the cool footballer's brain. 1966 *Guardian* 22 Apr. 6/5 Matchstick men —taught by adults, copied by infants—can be death to child art. 1967 J. WAINWRIGHT *Worms must Wait* i. 5 They were wasted, emaciated men... They were match-stick men... They were men who should have died, but who refused to die.

match, *a.* Add: **1. match dissolve** [cf. *DISSOLVE *v.* 7 b] *Cinemat.* (see quots. 1959, 1970); also *matched dissolve*; **matchmark** (see quots.).

1953 K. REISZ *Technique Film Editing* II. xi. 172 When diagrams *are* used, it is most important to make exactly clear what they refer to... This is accomplished most simply by the matched dissolve. 1959 W. S. SHARPS *Dict. Cinematogr.* 110/1 *Match dissolve*, the overlapping of two *shots* so that, because of the identical positions of their subjects, only one person or object appears to be seen about the point of overlap. 1970 W. WAGER *Sledgehammer* (1971) ix. 39 Match dissolve. That's a film term for a standard motion-picture transition, say, from the face of a clock in a police chief's office to the face of another clock in the senator's bedroom. 1918 WEBSTER Add., *Match-mark*.., a mark placed on the contiguous separable parts of any device to aid in the proper reassembling of any of those parts. 1962 *Gloss. Terms Glass Industry* (*B.S.I.*) 41 *Match mark*, a line or seam on glassware formed at the join of two mould parts.

match, *v.*^{1} Add: **5. c.** *Electr.* To equalize (two coupled impedances) so as to bring about the maximum transfer of power from one to the other; to make (a device) equal in effective impedance *to*.

1929 E. MALLETT *Telegr. & Telephony* vii. 162 Where the impedances cannot be matched transformers may be introduced to give the same effect. 1931 *Proc. IRE* XIX. 725 At high frequencies a transformer consisting of primary, secondary and mutual inductances cannot be constructed to match a generator effectively to a resistive load. 1938 Q. P. HARNWELL *Princ. Electr. & Electromagn.* iv. 114 When resistances are matched, half the power developed is delivered to the load and half is lost in the source. 1959 R. L. SHRADER *Electronic Communication* xiii. 371 The requirement is to match a 4-ohm speaker to a 4,000-ohm power tube. 1966 R. C. HANSEN *Microwave Scanning Antennas* III. ii. 186 The sinuous feed and couplers..terminated in radiating elements which are matched to free space.

8. b. *to match up to*: to equal; to be comparable with.

1958 *Times* 18 Sept. 13/3 But Blanche can match up to Emily, indeed she surpasses her in the end. 1958 *Listener* 13 Nov. 800/2 This musical image..cannot match up to the breadth and immediacy of 'Mars'. 1964 M. GOWING *Britain & Atomic Energy* v. 163 Britain's manpower resources did not match up to her programmes.

matchableness. (Later example.)

1857 C. M. YONGE *Let.* 1 Oct. in C. Coleridge *C. M. Yonge* (1903) viii. 214 A most perfect marriage..as to the matchableness of the two people.

matched, *ppl. a.* Add: **3.** *matched orders*: the name given to systems of manipulation on the Stock Exchange, which involve artificial treatment of orders to buy and sell. Also occas. *sing.* Also *matching orders.* orig. *U.S.*

1903 S. S. PRATT *Work of Wall St.* 146 The Syndicate may be washing sales by matched orders through curb brokers in order to market watered stock. 1908 *Westm. Gaz.* 26 Aug. 2/2 An order to buy is given to one broker and an order to sell to another, and an arrangement is made that the two brokers shall only deal with one another... The opportunity for market manipulation is obvious, and 'matched orders' have led to grave scandals. 1920 A. C. PIGOU *Econ. of Welfare* II. vi. 178 It should be observed that the device of 'matched orders' may be made difficult by a rule forbidding offers and bids for large amounts of stock on the terms 'all or none'. 1930 J. E. MEEKER *Work of Stock Exchange* (rev. ed.) xvi. 455 A.. subtler evil consists of 'matching orders'... Instances where there is danger of matched orders..deserve brief explanation. 1941 DICE & EITEMAN *Stock Market* (ed. 2) 467 When a trader gives one broker orders to buy a given stock and another orders to sell a like amount of stock in a manner to either advance or depress the price artificially, such a trader is said to be giving matched orders. 1951 G. L. LEFFLER *Stock Market* xx. 327 Artificial market activity was accomplished..through matched orders and ..without matched orders... Matched orders were in violation of state law and the rules of the New York Stock Exchange... The same effect as the matched order could be obtained without its actual use. 1964 P. WYCKOFF *Dict. Stock Market Terms* 163 *Matched orders.* 1. Orders to buy or sell a particular stock which are placed simultaneously with different brokers by the same person... 2. Buy and sell orders matched legitimately by the specialist in a stock in order to arrange an opening price as closely as possible to the previous close.

4. Designed to match in colour, style, etc.; matching.

1972 A. ROUDYBUSH *Sybaritic Death* (1974) vii. 70 Clare's status-conscious matched luggage. 1974 P. DE VRIES *Glory of Hummingbird* iii. 49 A matched set of luggage. 1975 P. G. WINSLOW *Death of Angel* vi. 142 He presented her with a set of matched luggage..for the weekends.

mate, *sb.*^{2} Add **1.** (Further, esp. Austral., examples.)

1845 C. GRIFFITH *Present State of Port Philip* 79 Two [bushworkers] generally travel together, who are called mates; they are partners, and divide all their earnings. 1890 'R. BOLDREWOOD' *Miner's Right* 136 We have been firm friends and true mates all this time. 1901 M. FRANKLIN *My Brilliant Career* i. 3 Daddy's little mate isn't going to turn Turk like that, is she? 1908 E. J. BANFIELD *Confessions of Beachcomber* I. v. 174 With a mate he had been for many months, bêche-de-mer fishing, their station ..a lonely islet in Whitsunday Passage. 1911 C. E. W. BEAN '*Dreadnought' of Darling* xxxv. 311 Perhaps the strongest article in the out-back code is that of loyalty to a mate. 1942 C. BARRETT *On Wallaby* iv. 75, I told my mates some of these facts on returning. 1966 *Observer* 17 Apr. 30/1 A 17-year-old boy..said, 'I haven't got a real mate. That's what I need.' 1968 K. WEATHERLY *Roo Shooter* 109 Old Sam, born and reared in the bush, a good mate and bushman. 1973 *Parade* (Melbourne) Sept. 34/1 An obelisk in the Jewish section of the Melbourne General Cemetery records the names of those who fought for Australia in the 1914 War. Many of them trained in the Faraday Street School cadets. They assimilated the lessons of patriotism and were great mates.

b. (Austral. examples.)

1852 R. CECIL *Diary* 31 Mar. (1935) 36 When the diggers address a policeman in uniform they always call him 'Sir', but they always address a fellow in a blue shirt with a carbine as 'Mate'. 'Mate' is the ordinary popular form of allocution in these colonies. 1862 A. POLEHAMPTON *Kangaroo Land* 99 A man, who greeted me after the fashion of the Bush, with a 'Good day, mate'. 1974 *Sydney Morning Herald* 14 Feb. 7, I asked a station attendant (attired in a dirty open-necked shirt and trousers, recognizable only by a dirty cap) if the train was the North-West Mail. 'I wouldn't have a clue, mate,' was the reply.

d. Also *to be mates with.*

1880 H. LAPHAM in D. M. Davin *N.Z. Short Stories* (1953) 57 At this time I was mates with a young fellow called Jim Smith, a good enough lad as a mate, and would do just as big a day's labour as any man.

e. (See quot. 1904 and cf. 4 b.)

1881 H. & C. R. SMITH *Isle of Wight Words* 21 *Meyat*, a mate; the carter's assistant. 1884 J. C. EGERTON *Sussex Folk & Ways* ii. 26 [A] carter-boy credited with the following..advice to his father, whose 'mate' he was. 1904 GOODCHILD & TWENEY *Technol. & Sci. Dict.* 384 *Mate*, an assistant or subordinate who assists a more skilled workman. 1951 *Engineering* 9 Mar. 296/3 It is claimed by the strikers that a new electrician's mate..should resign. 1963 *Times* 9 Mar. 9/5 The Scottish chimney sweep..always has a mate.

3. d. *spec.* A point on tramway lines which is cast solid and pairs or 'mates' with the movable tongue or switch on the other rail; an 'open' or 'fixed' point. orig. *U.S.*

1909 in WEBSTER. 1922 *Glasgow Herald* 3 Oct. 8 The weight of the inserts varies from about 100 to 300 lb., depending on the angle of the crossing or mate.

mate, *v.*^{2} Add: **1.** Delete ? *Obs.* and add later examples.

1874 SWINBURNE *Bothwell* I. i. 4, I might sleep well and laugh and walk at ease, With none to mate me. 1891 W. MORRIS *Poems by Way* 129 *Fig-tree.* I who am little among trees In honey-making mate the bees.

6. a. *intr.* Of an engineering part: to make a good or proper fit *with*.

1909 in WEBSTER. 1956 S. PARKER *Drawings & Dimensions* vii. 52 When considering the application of 'Go' and 'Not Go' gauges..the questions 'Will *like* mate with *like*?' and 'How much force may be used when gauging?' inevitably arise. 1959 *Motor Manual* (ed. 36) v. 132 The end of the axle hub is threaded and extends outside the wheel to take a large eared nut, which is coned internally to mate with a cone on the edge of the wheel hub. 1967 *Electronics* 6 Mar. 15/3 (Advt.), Push-pull coupling mates easily.

b. *trans.* To fit or join *with* or *to*.

1959 *Nat. Geographic* Feb. 159/1 High on skeletal service towers, we watched engineers mate, or couple, rocket stages and gingerly install the payloads containing custom-built miniaturized instruments. 1962 V. GRISSOM in *Into Orbit* 119 On 1 July the capsule was taken from the hangar..to be mated to the Redstone. 1970 N. ARMSTRONG et al. *First on Moon* iii. 55 By mid-May it was time to 'mate' the electrical systems of the rocket boosters with those of the spacecraft.

mated, *ppl. a.*^{2} Add: **b.** Fitted or fitting together.

1958 V. H. WATSON *Rover Cars* xiv. 230 Wash and examine all parts, renew as necessary, bearing in mind that the crown wheel and pinion are 'mated' parts and must be replaced as such.

matelot (mæ·tlo). [Fr., sailor; cf. *MATLO(W.] **1.** A sailor. *Naut. slang.*

1911 'GUNS' & 'THEELUKER' *Middle Watch Musings* 8 You've just got two old and respected matelots well scrubbed for nuffing. 1915 *Blackw. Mag.* July 89/1 We invite the matelots to lie on their backs on the upper-deck. 1947 *Landfall* I. 283 What time we had left was spent on fruitless errands for the Pommie matelots. 1955 *Times* 10 Aug. 3/3 The Navy was there once more and so were the two matelots of Monday. 1970 [see *JOCK^1 1 b]. 1974 *Listener* 21 Nov. 672/1 Our screen matelots..should be as reticent as..Captain Horatio Hornblower.

2. Applied to a shade of blue.

1927 *Daily Tel.* 26 Apr. 13/5 (Advt.), Fawn, Rose, .. Almond, Matelot. 1928 *Daily Express* 27 July 5/4 A lovely deep blue known as 'matelot'.

mater. 1. (Later examples.)

1905 *Sci. Amer.* 12 Aug. 120/2 The Astrolabe of Regiomontanus... *Mater* is the name given to the large disk divided into degrees. 1974 *Ibid.* Jan. 101/1 Each plate being engraved on both sides and all being stacked in the mater, or main body, of the astrolabe.

Mater Dolorosa (mē̆ⁱ·tər dǫlǭ·rōᵘ·să). [med. L., lit. 'sorrowful mother'.] A title of the Virgin Mary, emphasizing her role in the Passion of Christ; a representation, in painting or sculpture, of the Virgin Mary sorrowing. Also *transf.*, a woman who has the attributes of the sorrowful mother.

The term probably originated in the medieval Latin hymn beginning 'Stabat mater dolorosa Iuxta crucem lacrimosa'.

1800 J. DALLAWAY *Anecdotes of Arts in England* 516 He has a mater dolorosa and a boy playing on a lute by Guido. *c* 1869 TAYLOR & DUBOURG in M. R. Booth *Eng. Plays of 19th Cent.* (1973) III. 308 Crosses her hands on her bosom, à la Mater Dolorosa. 1872 GEO. ELIOT *Middlem.* IV. VIII. lxxx. 283 Dorothea's face..had the pale cheeks and pink eyelids of a *mater dolorosa*. 1888 *Cycl. Painters & Paintings* III. 220/2 Mater Dolorosa.. By Murillo, Madrid Museum... Companion to Ecce Homo. 1892 *Servite Manual* (Servite Fathers) 200 (*heading*) Via Matris Dolorosæ or, The Stations of the Way of the Seven Dolours of Mary. 1895 *Yellow Bk.* IV. 280 A vast cape that might have enshrouded the form of a Mater Dolorosa hung by the side of a jauntily-striped Langtry-hood. 1904 G. GRONAU *Titian* vii. 147 We can follow exactly the history of this picture and of its companion, the 'Mater Dolorosa', which Titian sent off to Charles V in 1555. 1917 M. G. SEGAR *Some Minor Poems Middle Ages* xxi. 48 Old English poems on the Mater Dolorosa are always wonderfully full of feeling. 1923 M. INNES *Ten Florentine Painters* 117 This group of the 'Mater Dolorosa' and her dead Son is called a *Pietà*. 1936 R. LEHMANN *Weather in Streets* IV. ii. 398 Mrs. Cunningham stood in the double doorway..receiving with a smile of the lips, but not of the hollowed *mater dolorosa* eyes, congratulations upon Amanda. 1938 *Times Lit. Suppl.* 11 June 415/3 Patricia is the picture of the *mater dolorosa* of all time. 1942 G. D. CARLETON *Mother of Jesus* xiv. 68 The Mother of Jesus is indeed the Mater Dolorosa, the Sorrowful Mother. 1951 L. MACNEICE tr. *Goethe's Faust* I.

120 In a niche on the wall is an image of the Mater Dolorosa. 1955 W. Smith tr. Mieggé's Virgin Mary viii. 171 All this conforms with..medieval piety..and the more recent veneration of the *Mater dolorosa*.

material, *a.* and *sb.* Add: **A.** *adj.* **2.** (Further examples.)

1685 tr. *Arnauld & Nicole's Logic* III. xiii. 65 The truth of a Consequence..is only propounded conditionally, and separated from the material Truth, as I may so say, of what it contains. 1883 F. H. Bradley *Princ. Logic* 471 If 'material' is a name for what transcends mere 'concepts' and commits itself to truth, then of course all logic must be material. 1889 J. Venn *Princ. Empirical Logic* Pref., In such a province..as that of Material or Inductive Logic the case is very different. 1936 *Mind* XLV. 442 Wisdom..gives the name *Material Analysis* to the substitution, for sentences about wholes, of sentences mentioning the individual parts of the whole, where the parts are of the same order of ultimacy (as previously defined) as the whole. *E.g.*, material analysis of sentences about awe will be in terms of fear and admiration. 1937 A. Smeaton tr. *Carnap's Logical Syntax of Lang.* IV. 237 We will..assign to the material mode of speech any sentence which is to be interpreted as attributing to an object a particular property, this property being quasi-syntactical, so that the sentence can be translated into another sentence which attributes a correlated syntactical property to a designation of the object in question. 1946 *Mind* LV. 321 If we follow the material logicians in holding that universal propositions are existential as to individuals also [etc.].

b. *material implication*: a relationship which holds between two propositions, irrespective of content, save only when the first is true and the second false.

1903 B. Russell *Princ. Math.* ii. 14 How far formal implication is definable in terms of implication simply, or material implication as it may be called, is a difficult question. 1932, etc. [see *FORMAL *a.* and *sb.* 1 d]. 1965 Hughes & Londey *Elem. Formal Logic* iii. 17 A material implication is always true when its antecedent is false, and also when its consequent is true. That is, a false proposition materially implies any proposition; and a true proposition is materially implied by any proposition. 1974 *Encycl. Brit. Macropædia* XI. 40/2 Reading $p \supset q$ as 'If p, q'..is made convenient by the fact that 'if' is often used in English in a 'material-implication' sense (someone who asserted 'If it isn't raining, it's snowing' would usually be held to have spoken truly *unless* the antecedent were true but the consequent false; i.e., unless it was neither raining nor snowing).

3. e. Philos. *material object, thing*: an object considered as a physical existent independent of consciousness; hence *material objectness*, the state of existing as a material object.

1605 Bacon *Adv. Learning* I. i. § 3 View and inquiry into these sensible and material things. 1649 tr. *Descartes' Discourse* 59 Imagination..is a particular manner of thinking on materiall things. 1713 Berkeley *Three Dialogues* 61, I wou'd, therefore, fain know, what Arguments you can draw from Reason, for the Existence of what you call *real Things*, or *material Objects*. 1737 A. Baxter *Inquiry Human Soul* (ed. 2) II. 204 He supposes that from the surfaces of all material things there are continually flying off thin membranes. 1865 Mill *Exam. Hamilton's Philos.* i. 6 We at first limit ourselves to the case of physical, or what are commonly called material objects. 1899 W. James *Talks to Teachers* vii. 58 The result of all this is that intimate familiarity with the physical environment, that acquaintance with the properties of material things, which is really the foundation of human consciousness. 1912 B. Russell *Probl. Philos.* iv. 58 Common sense regards tables and chairs and the sun and moon and material objects generally as something radically different from minds. 1920 A. N. Whitehead *Concept of Nature* ii. 43 Thus colour is not part of the reality of the material object. 1932 H. H. Price *Perception* ii. 52 'Material-objectness' cannot be defined without mention of it. 1933 *Mind* XLII. 291 On the phenomenalistic view of material-objectness, there is evidence that material objects exist. 1940 A. J. Ayer *Found. Empirical Knowl.* i. 1 It does not normally occur to us that there is any need..to justify our belief in the existence of material things. 1941 *Mind* L. 282 The puzzle is this: if sense-data are all that we are directly aware of in perception, how have we ever acquired the concept of 'material-objectness' at all? 1959 J. L. Austin *Sense & Sensibilia* (1962) ii. 7 'Material thing' is not an expression which the ordinary man would use. *Ibid.* x. 107 The material-object language *must somehow* be 'reducible' to the sense-datum language. 1964 *Philos. Rev.* LXXIII. 324 A corpse, like a material object, is non-conscious, rather than merely unconscious.

4. c. *material culture*: the physical objects (tools, articles of domestic and religious use, dwelling-places, etc.) which give evidence of the type of culture developed by a social group.

1929 *N. & Q. Anthropol.* (ed. 5) 187 The study of the artefacts and material culture of a people should not be viewed solely from their material aspects. *Ibid.* 188 The investigator of socio-religious matters may find that he cannot get information, and he will then find that a study of material culture provides him with a convenient avenue of approach. 1931 *Encycl. Social Sci.* IV. 622/1 Material equipment of culture is not, however, a force in itself... Material culture requires a complement less simple, less easily catalogued or analyzed, consisting of the body of intellectual knowledge. 1937 R. H. Lowie *Hist. Ethnol. Theory* iii. 27 Evolution is a positive fact in material culture. 1971 *World Archaeol.* III. 119 Most items of material culture will have been removed.

d. Gram. *material noun*, = **mass noun*.

1892 [see *class-noun* (*CLASS *sb.* 10)]. 1892 H. Sweet *New Eng. Gram.* I. 56 When a material noun is used to express an individual object of definite shape, it is no longer a material noun, but a class-noun. 1925 Grattan & Gurrey *Our Living Lang.* xviii. 110 A material-noun is a word which stands for the whole mass of matter possessing the qualities implied by the word, or for an indefinite quantity of that matter—for example, *water, iron, veal, butter*. 1969 R. Kingdon *Palmer's Gram. Spoken Eng.* (ed. 3) II. 61 Common nouns..are subdivided into Material nouns and Class nouns. Material nouns..name substances.

B. *sb.* **6*.** Preceded by a qualifying word, as *officer material*, a person who has (or persons who have) qualities thought of as suitable for an officer; similarly *headmaster material*, etc.; also with non-animate qualifier, as *football material*, a person or persons potentially suitable as footballers; *grammar-school material*, a person or persons suitable for admission to a grammar school.

1892 *College Index* (Auburn, Alabama) Nov. I. i. 23 He still kept a sharp lookout for football material. 1927 *Officers Training Corps Gaz.* Apr. 59/1 The sorely needed officer material caused by the early casualties. 1946 *Amer. Speech* XXI. 238 *Officer material*, an enlisted man short in mentality. 1951 H. Wouk *Caine Mutiny* xi. 123, I regard both those men as excellent officer material. 1964 'E. Peters' *Flight of Witch* i. 9 Tom Kenyon, confident, clever and ambitious, was obvious headmaster material. 1968 L. Berg *Risinghill* 86 The inspector who denounced the school the first time was a man who was interested in grammar-school 'material'. 1969 M. Pugh *Last Place Left* xxvii. 194 It was difficult recruiting men.. 'Some of them, well, they're not leadership material.' 1969 B. Weil *Dossier IX* ii. 4 His secretary came in. She was a tall fair girl with a tendency to stride. Wren officer material, Asher thought. 1970 'J. Melville' *New Kind of Killer* i. 8 Her report..was, on the whole, a favourable one: 'This officer is good promotion material.' 1971 D. Eden *Afternoon Walk* viii. 109 Aren't all top executives ulcer material?

7. *materials technology*, *material(s testing*; **material(s control** (see quot. 1959); hence *materials controller*; **material(s handling**, the movement and storage of materials in a factory; **material(s science**, that branch of science which treats of the structure and properties of materials, esp. in so far as they are relevant to their usefulness and potential applications; so *material(s scientist*.

1918 C. E. Knoeppel *Organization & Administration* xv. 252 The material control sheet takes care of pieces ordered, pieces rough, pieces in progress, and pieces finished. 1938 W. B. Cornell *Business Organization* xvi. 318 The other important duties of the standards and methods section include..development of stores and material control and production control methods. 1959 *Gloss. Terms Work Study* (B.S.I.) 33 *Material control*, procedures and means by which the correct quantity and quality of materials and components are made available to meet production plans. 1962 A. Battersby *Guide to Stock Control* 123 The Materials Control office calculated that the increase in the first-grade stock would have to be 200 items to preserve the same risk level as before. *Ibid.* x. 90 The Finance Director..will then need a subordinate who can translate his general decisions on policy into detailed instructions. The subordinate would be the Materials Controller: this title is better than the more usual 'Stock Controller' because, as we have seen, he controls the *flow* of materials rather than the stocks themselves. 1921 E. T. Elbourne *Factory Administration & Cost Accounts* (new ed.) 807 A form of production service having special reference to material handling and custody. 1932 S. J. Koshkin *Mod. Materials Handling* i. 2 It is of the greatest importance that the materials-handling methods and devices should be sufficiently worked out at the time the plant is designed so as to make them an integral part of the design. 1966 A. Battersby *Math. in Managem.* vii. 163 A recent paper dealing with steel fabrication showed that an extra crane costing £15,000 would not be needed in a new materials-handling system. 1961 (*title of periodical*) Progress in materials science. 1966 P. Feltham *Deformation & Strength of Materials* p. v, One of the effects of the unprecedented advance in the synthesis and use of new materials in the last few decades..has been the widespread introduction of 'materials science' into curricula of university courses. 1974 *New Scientist* 9 May 349/1 (Advt.), Applicants should have a degree in physics or a material science discipline. 1974 *Nature* 16 Aug. p. vii (Advt.), A vacancy exists for a Materials Scientist to study basic factors concerned with lubrication, wear, fatigue and corrosion of plastics and metals used in the manufacture of orthopaedic implants. 1962 *Technology* June 129/1 The course on materials technology will draw on the methods of physics, chemistry, metallurgy and engineering. 1972 N. J. Parratt *Fibre-Reinforced Materials Technol.* p. ix, Much of industry is concerned with making materials or with turning them into useful hardware, so that it cannot safely ignore any advances in materials technology. 1924 *Trans. Amer. Soc. Mech. Engin. Index* 124/2 (*heading*) Materials testing. 1950 *Chem. Engin. Progress* XLVI. 110/3 Materials Testing Reactor which..will be used in studies of material to be employed in building reactors. 1969 R. F. Lang tr. *Henglein's Chem. Technol.* 315 The so-called technical laboratories..carry out the routine analyses..as well as material testing.

materialistical, *a.* (Further example.)

1973 C. & R. Milner *Black Players* v. 115 There's too much paranoia, too much greed... And suspicion, and too much materialistical fear.

‖ **materia prima** (mātiə·riă prəi·mă). [L., 'first matter'; cf. Gr. ἡ πρώτη ὕλη (see Hyle).] = *first matter* (Matter *sb.*[1] 6 c).

1551 S. Gardiner *Explication True Catholique Fayth* 137 If we agree with the Philosophers that there is (*Materia prima*)..the same (*Materia prima*) beyng as it were (substancia) that altereth not. [*a* 1586 Sidney *Apol. Poetrie* (1868) 55 The quiddity of..*Prima materia* will hardely agree with a Corslet.] 1603 Holland tr. *Plutarch's Morals* xvi. 229 The substance or matter that hath neither forme nor any colour, which they call *Materia prima*, is a subject capable of all formes. 1665 J. Locke *Let.* in P. King *Life of Locke* (1829) 20 Poor *materia prima* was canvassed cruelly. *Ibid.*, The young monks..dispute as eagerly for *materia prima*, as if they were to make their dinner on it. *a* 1721 Prior *Locke & Montaigne* in *Dialogues of Dead* (1907) 240 When he [*sc.* Descartes] gave Us his subtil matter, he only new Christened Aristotles *Materia prima*. 1905 W. James in *Jrnl. Philos.* 25 May 281 'Pure experience'..was the name I gave to the *materia prima* of everything. 1961 *Times* 28 Dec. 11/4 A continuum of active or quiescent energy which..would reintroduce a *materia prima* to the satisfaction of metaphysicians.

‖ **mater lectionis** (mēi·təɪ lekti,ōu·nis). *Gram.* Pl. **matres lectionis.** [mod.L., lit. 'mother of reading'.] A letter which has the function of a diacritical mark; *spec.* in Hebrew writing, a sign indicating a vowel sound.

1846 B. Davies tr. *Gesenius's Hebrew Gram.* I. i. § 8 The vowel-letters are also called by grammarians, *matres lectionis* (since they partly serve as guides in reading the unprinted text). 1925 P. Radin tr. *Vendryes's Lang.* v. i. 328 In the Greek alphabet the principle of the *matres lectionis* was skilfully used for the creation of a special sign for each vowel. 1962 Davidson & Mauchline *Introd. Hebrew Gram.* (ed. 25) 13 Thus ו would come in time to stand for long *o* and *u*, י for long *e* and *i*. ה likewise came to have a vocalic usage and the three letters are often termed vowel-letters or vocalic consonants (also *matres lectionis*).

maternity. Add: **3.** *maternity leave*; **maternity benefit**, a welfare service payment made to women, under specified conditions, after the period of confinement; **maternity home** = *maternity hospital*; **maternity jacket** *slang* (see quot. 1925); also *ellipt.* as *maternity*.

1911 *Q. Rev.* Oct. 561 When both husband and wife are insured, they are to be entitled, in case of the wife's confinement, to both sickness and maternity benefit. 1911 [see *BENEFIT *sb.* 3 d]. 1945 *Release & Resettlement* (H.M. Govt.) xii. 40 The benefits of the National Health Insurance Scheme include..maternity benefit. 1957 Brudno & Bower *Taxation in U.K.* 523/1 Maternity benefits. 1970 *National Insurance: Maternity Benefits* (Dept. Health & Social Security, NI 17A) I. i. 6 There are two maternity benefits—maternity grant and maternity allowance. Both benefits may be paid for a confinement provided the conditions are satisfied. *Ibid.* III. xxiii. 18 Any question about contributions, including whether the contribution tests for maternity benefit..are satisfied, is reserved for the Secretary of State for Social Services. 1922 *Encycl. Brit.* XXX. 652/1 The supervision of midwifery, including the establishment of maternity homes. 1926 *Act* 16 & 17 *Geo. V* c. 32 § 12 The expression 'maternity home' means any premises used or intended to be used for the reception of pregnant women or of women immediately after childbirth. 1967 *Encycl. Brit.* XV. 66/1 Increasing grants have been made by some governments for the construction of free maternity homes or maternity wards in general hospitals. 1925 Fraser & Gibbons *Soldier & Sailor Words* 154 *Maternity jacket*, the name given the double breasted tunic, worn formerly in the Royal Flying Corps. 1958 'N. Shute' *Rainbow & Rose* ii. 66 A very young man in the double-breasted 'maternity' jacket of the Royal Flying Corps. *Ibid.* iii. 91 She insisted on giving me a buttonhole..but my maternity hadn't got a buttonhole to put it in. 1970 *Guardian* 21 May 13/2 At present, maternity leave is based on the custom and practice of either individual companies or sections of industry. 1973 *Times* 17 Mar. 2/2 Mrs Christine Page..said: 'Because of the economic uncertainty of some families, we are asking for this maternity leave to be negotiated.'

b. Used *attrib.* of garments spec. designed for pregnant women.

1893 *Ladies' Home Jrnl.* Apr. 28/2 Summer maternity gowns. 1907 *Dress* Jan. p. xx, My Maternity corsets are recommended by physicians everywhere. 1951 I. Shaw *Troubled Air* x. 164 She was wearing one of her shapeless tent-like maternity dresses. 1962 M. Drabble *Summer Bird-Cage* vi. 92 She would wear pretty maternity dresses and be an excellent mother. 1963 'G. Bagby' *Murder's Little Helper* (1964) ii. 14 What kind of a burglar is it that swipes maternity clothes? *Ibid.*, She hadn't splurged on maternity garments. 1964 M. Drabble *Garrick Year* i. 11 Three years of child-bearing and modelling maternity clothes. 1966 *Guardian* 5 Aug. 8/4 The wide coats and dresses have the ease of maternity wear. 1971 B. Malamud *Tenants* 211 Under a short maternity skirt her stomach is in flower. 1971 C. Storr *Thursday* vi. 63, I can just see myself in maternity smocks for the next ten years.

mateship. Add: (Austral. examples.)

1905 H. Lawson in B. Stevens *Bush Ballads* (1910) 76 The College Wreck..Tramps West in mateship with the man Who cannot write his name. 1930 W. H. Hancock *Australia* x. 199 Thwarted individualism found consolation in the gospel of mateship. 1960 J. Fingleton *Four Chukkas to Austral.* xviii. 158 Benaud knew, too, that he had to cultivate mateship. 1962 *Guardian* 14 Aug. 6/5 A beery sense of 'mateship' among the men. 1968 *Courier-Mail* (Brisbane) 20 July 2/2 When I was..Prime Minister ..he gave me not only his personal co-operation, understanding and loyalty but also his mateship, often in the most difficult circumstances, even when we disagreed.

matey. (Further examples.)

1859 H. Kingsley *Recoll. G. Hamlyn* II. xiv. 277, I took him for a flash overseer, sporting his salary, and I was as thick as you like with him. And 'Matey,' says I, (you see I was familiar, he seemed such a jolly sort of bird),

'Matey, what station are you on?' 1909 S. Watson *Wops the Waif* i. 2 'I say, Tickle matey, wot's all them a-readin' of on that bill over there?' interrupted Wops. 1916 'Taffrail' *Pincher Martin* ii. 16 A gray, cigar-shaped vessel lying in a dry dock, with dockyard 'maties' swarming on board her. 1931 Kipling *Limits & Renewals* (1932) 197 Polyphemus damned back at Jemmy like a Chatham matey. 1958 E. Hyams *Taking it Easy* I. ii. 72 The dockyard 'mateys' who had been sent us to do the work were all Bretons from the Arsenal at Brest. 1973 J. Drummond *Bang! Bang! you're Dead!* xxv. 86 Right, matey, 'oo told you? 1974 'M. Hebden' *Pride of Dolphins* III. iii. 232 Many of the lights in the control room were missing, probably stolen by dockyard mateys.

matey (mēi·ti), *a.* [f. Mate *sb.*[2] + -y[1].] Like a mate or mates; friendly and familiar (*with*); sociable, companionable. Hence **ma·teyness** = *MATINESS.

1915 T. Burke *Nights in Town* 50 You are all so—what is the word?—matey, isn't it? Yes, that's the note of the London [music] hall—mateyness. 1919 Wodehouse *Damsel in Distress* xv. 172 After the game he took me off to his cottage and gave me a drink... We got extremely matey. 1926 *Contemp. Rev.* June 682 The *British Worker*, an equally deplorable organ produced by *intelligentsia* trying to be 'matey'. 1929 W. Deeping *Roper's Row* xxi. § ii, Elizabeth would..want to be matey with people. 1946 Wodehouse *Joy in Morning* i. 4, I..continued to tut-tut a bit at having missed the young pipsqueak, with whom my relations had always been of the matiest. 1958 *Times Lit. Suppl.* 20 June 347/3 When the implied mateyness is further established by such otiose asides..one stirs uncomfortably to ask why a writer who has so much that is interesting to say should waste our time with these lazy transcriptions. 1967 *Listener* 24 Aug. 244/2 Continuous pop and advertisements, presented with that bogus transatlantic mateyness which apparently 'sends' teenagers. 1974 V. Gielgud *In Such a Night* ix. 82 She would throw a party. And in the ensuing haze of drink and mateyness grievances would..disappear.

math[1]**.** (Later examples.)

1917 Mother St. Jerome *Garden of Life* 18 You feel as you lie in the math The watching unseen of his eyes. 1968 R. E. Zupko *Dict. Eng. Weights & Measures* 103 *Math*,.. in Herefordshire equal to approximately 1 acre..or to the amount of land that a man could mow in a day.

math[2]**.** Add: (Earlier and later examples.) Also **matha.**

1828 H. H. Wilson in *Asiatick Researches* XVI. 103 The disciples, who are domesticated in the several *Maths*, profess also perpetual celibacy. *a* 1851 M. M. Sherwood *Hist. Little Henry & his Bearer* (1866) 84 This sahib..had himself built a mutt. 1877 M. Williams *Hinduism* 224 *Yatis*, monks or ascetics,..often congregate in Maṭhas or 'monasteries'. 1913 'A. Avalon' tr. *Tantra of Great Liberation* p. xvii, Kedarnath... A *matha* and temple dedicated to Shrī Sādāshiva. 1956 R. Redfield *Peasant Society & Culture* iii. 81 Some sweet-voiced, gifted expounder sitting in a temple, *mutt*, public hall or house-front.

math[3] (mæþ). *U.S. colloq.* = *MATHS.

Math. is used as an abbreviation in written English in the U.K. but not in speech, the normal form being *Maths*.

1890 in *Cent. Dict.* 1895 W. C. Gore in *Inlander* Nov. 64 *Math.* n. Mathematics. 1899 J. London *Let.* 24 Oct. (1966) 62 She's well up in the higher math. 1916 [see *my guess is* (*GUESS *sb.* 1)]. 1938 I. Goldberg *Wonder of Words* x. 200 He [*sc.* a student] says *math* and *ec.* 1961 C. Winston *Hours Together* (1962) vii. 139 There was Morton Kersh, with the math book propped against the milk bottle on the kitchen table. 1971 *Black World* June 80/2 T. J. failed Math and English. 1973 *Jrnl. Genetic Psychol.* CXXIII. 163 The follow-up results with the math-science boys. 1974 *State* (Columbia, S. Carolina) 8 Mar. 19-B/5 Parents who have children in a Title I reading, math or kindergarten program are also invited to attend.

mathematical, *a.* and *sb.* Add: **A.** *adj.* **1. b.** *mathematical model*: see *MODEL *sb.* 2 e.

e. *mathematical linguistics*: a branch of linguistics concerned with the application of mathematical models and procedures to the analysis of linguistic structure; so *mathematical linguist*; *mathematical logic*: logic that is mathematical in its method, using symbols and following definite and explicit rules of derivation; modern logic; symbolic logic; so *mathematical logician*; *mathematical philosophy*: that branch of philosophy concerned with the nature of mathematics.

1951 *Language* XXVII. 221 There is a growing cleavage between the mathematical linguists, or metalinguists, and the physical linguists, whom I should call just plain linguists. 1956 J. Whatmough *Lang.* xi. 220 The development of mathematical linguistics is opening a new field of inquiry, and may rightly be expected to bring greater order into a subject, which..has been and still remains chaotic. 1961 C. Mohrmann et al. *Trends European & Amer. Ling.* 21 In his report to the 1957 Congress, Professor Joshua Whatmough mentioned the initiation of the Seminar in Mathematical Linguistics at Harvard University two years previously—the first appearance of the subject in an academic curriculum. 1964 E. Bach *Introd. Transformational Gram.* vii. 145 At several universities.. courses in 'mathematical linguistics' are offered. 1968 J. Lyons *Introd. Theoret. Ling.* ii. 71 What one might refer to loosely as 'mathematical' linguistics is now a very important part of the subject. 1972 Hartmann & Stork *Dict. Lang. & Ling.* 137 Mathematical linguistics may be said to begin with the counting of linguistic units such as phonemes, graphemes, or vocabulary items. 1858 A. De Morgan *On Syllogism* (1966) 78 This *mathematical* logic.. will commend itself to the educated world. 1880 J. Venn in *Princeton Rev.* 248 What with the logicians who hate mathematics, and the mathematicians who despise logic, a theory of so-called mathematical logic does not find many friends. 1908 B. Russell in *Amer. Jrnl. Math.* XXX. 222 (*title*) Mathematical logic as based on the theory of types. 1940 W. V. Quine (*title*) Mathematical logic. 1941 O. Helmer tr. *Tarski's Introd. Logic* ii. 18 Logic..has undergone a complete transformation with the effect of assuming a character similar to that of the mathematical disciplines; in this new form it is known as mathematical or deductive or symbolic logic. 1967 J. van Heijenoort *From Frege to Gödel* p. vii, Mathematical logic is what logic, through twenty-five centuries and a few transformations, has become today. 1850 A. De Morgan *On Syllogism* (1966) 58 There is no occasion for the mathematical logician to pay the least deference to the Christian followers of Aristotle; the master himself was a mathematician. 1883 F. H. Bradley *Princ. Logic* 360, I may suggest to the mathematical logician that [etc.]. 1903 B. Russell *Princ. Math.* iii. 457 But now, thanks mainly to the mathematical logicians, formal logic is enriched by several forms of reasoning not reducible to the syllogism. 1879 W. James *Coll. Ess. & Rev.* (1920) 141 Clifford's..chapters on the 'Philosophy of the Pure Sciences'..form as luminous an introduction to mathematical philosophy as was ever written. 1897 B. Russell in *Mind* VI. 112 The problems of mathematical philosophy. 1901 *Mind* X. 30 It is..imperative, in the interests of mathematical philosophy, to supply the defect. 1919 B. Russell (*title*) Introduction to mathematical philosophy.

mathematicism (mæþĭmæ·tisiz'm). *Philos.* [f. Mathematic *a.* + -ism; cf. G. *mathematizismus*.] The opinion that everything can be described ultimately in mathematical terms, or that the universe is fundamentally mathematical.

1933 *Mind* XLII. 107 Russell..is dismissed on the grounds that his subjectivism rests on a surreptitious basis of realism, that his mathematicism ignores qualities, and that his neutral monism leaves no room for mind in its indefeasible quality of awareness. 1962 L. J. Cohen *Diversity of Meaning* i. 19 Forgotten principles, like the mathematicism of Plato's *Timaeus*.

mathematicization (mæþĭmæ:tisəizēi·ʃən). [f. Mathematiciz(e *v.* + -ation.] = *MATHEMATIZATION.

1952 F. X. Meehan in *Proc. Amer. Catholic Philos. Assoc.* XXVI. 19 Philosophy became for them a mathematicization of nature under the impetus given by Descartes and Newton. 1962 *Technology* May 12/3 The 'mathematicization' of science and technology.

mathematico-. (Further examples.)

1857 *Encycl. Brit.* XIII. 578/2 Now the mathematico-logical theories tend..to convert logical study into a mere cramming of the memory with formulae. 1865 Mill *Auguste Comte* 119 The sciences, all..except the mathematico-astronomical couple are still..in a very early stage. 1890 W. James *Princ. Psychol.* II. xxxviii. 669 Take any other mathematico-mechanical theory and it is the same. 1908 C. S. Peirce in *Monist* XVIII. 463 This would be very far from establishing the idea of certain mathematico-logicians that a line consists of points. 1937 [see *CENTRIFUGE *sb.*]. 1940 C. L. Hull et al. (*title*) Mathematico-deductive theory of rote learning. 1965 *Math. in Biol. & Med.* (*Med. Res. Council*) IV. 135 We feel more secure when we have the guiding influence of a mathematico-physiological model in our experiments. 1973 *Nature* 6 July 60/2 Economic theorists take fairly naturally to the set-theoretic and mathematico-logical underpinning of mathematical methods.

mathematization (mæ:þĭmătəizēi·ʃən). [f. Mathematiz(e *v.* + -ation.] Mathematical treatment; the state of being mathematized.

1936 W. F. R. Hardie *Study in Plato* v. 38 In the *Philebus* Plato refers to the discovery of the musical intervals (17d), and pays tribute to the inspiration derived from this example of successful mathematization. 1952 G. Sarton *Hist. Sci.* I. xvi. 416 His [*sc.* Plato's] mathematization of political thought. 1956 E. H. Hutten *Lang. Mod. Physics* ii. 37 There is no part of physics that can be said to be completely formalised as a system. At best, we find a 'mathematization'. 1962 L. J. Cohen *Diversity of Meaning* i. 19 He may too easily ignore what was contributed to the mathematization of science by medieval influences like Roger Bacon's stress on the need for a strictly mathematical treatment of optics. 1973 *Sci. Amer.* Dec. 101/1 In the Copernican system the planets are harmoniously ordered out from the sun so that the shorter the period, the closer the distance. It is this pattern that opened the way to the mathematization and mechanization of the universe.

maths, colloq. abbrev. Mathematics *sb. pl.*

1911 W. Owen *Let.* 14 Sept. (1967) 81 The Answers to Maths. Ques. were given us all this morning. 1917 *Wireless World* Sept. 385 Extremely 'rusty' in 'maths'. 1931 [see *DUB *sb.*[6]]. 1956 A. S. C. Ross in N. Mitford *Noblesse Oblige* 32 The U equivalent is *master*, *mistress* with prefixed attribute (as *maths-mistress*). 1960 M. Spark *Bachelors* x. 155 I've got a pile of homework to do. Maths papers.

-matic (mæ·tik), *suffix.* [f. Auto)matic *a.*] A terminal element in words (usu. proprietary names) designating devices which work automatically or mechanically, as *Hoovermatic*, *traffomatic*. orig. *U.S.*

1935 *Amer. Speech* Feb. 35/1 In Providence, Rhode Island *traffomatic* signals turn red lights to green as cars approach them. *Adjustomatic* and *Ceomatic* also occur, and there is a *Nickel-mat* restaurant. 1938 S. Chase *Tyranny of Words* xiv. 173 Dial-a-matic. 1941 *Word Study* Nov. 7/1 This year bids fair to be the *-matic* year of all years. 1957 *Journal des Traducteurs* II. 50 Admen all over the continent have coined hundreds of magic brand names with this productive adaptation suffix, exploiting to the full the suggestion of 'minimum effort' it connotes..: *Accumatic* watch, *Ajustomatic* pipe, *Coffeematic* percolator. 1959 *Amer. Speech* Oct. 237 English trade names.. such as *Hoovermatic* (an automatic washer and dryer) or *Sensimatic* (a Burroughs bookkeeping machine). 1966 L. Cohen *Beautiful Losers* (1970) III. 238 The Bowl-a-Matic habitually divided every strike between First and Second Player regardless of who or how many threw. 1971 *Amat. Photographer* 13 Jan. 92/2 (Advt.), Pakmatic.., Easymatic,.. Instamatic. 1972 *House & Garden* Feb. 79 'Brush-o-matic', to clean upholstery.

‖ **matière** (matiẹ̄r'). [Fr.] The quality given to his pigment by an artist. Also *fig.*

1915 W. H. Wright *Mod. Painting* xiv. 320 De Vlaminck has a rich and compelling *matière* and an art sense which is almost coquettish. 1921 R. Fry *Lett.* (1972) I. 62 In the *Nature Morte*..he [*sc.* Picasso] has gained a certain plastic quality by the extraordinary contrasts of *matière*. 1936 *Burlington Mag.* July 35/1 Though the *matière* is still heavy, the paint is applied in parallel oblique strokes. 1960 *Times* 12 May 10/5 Dubuffet has experimented with the potentialities of *matière* in his paintings. 1963 *Guardian* 29 Feb. 9/2 Kossoff struggles with his grotesquely sensual matiere like a Vietnamese trooper slurping in the mud of a rice paddy. 1971 *Guardian Weekly* 7 Aug. 19 Lecture Four. The matière of verse demonstrated further in Wallace Stevens; who isn't my poet.

Matilda (mă·tĭ-ldă). *Austral. slang.* Also **matilda.** [A female Christian name.] = Swag *sb.* 10. So *to walk* or *waltz Matilda*: to carry one's swag, to travel the road (cf. *WALTZ *v.*); *Matilda-waltzer*, a tramp, traveller.

1893 *Bulletin* (Sydney) 18 Nov. 20/3 A swag is not generally referred to as a 'bluey' or 'Matilda'—it is *called* a swag. *Ibid.*, No bushman thinks of 'going on a wallaby' or 'walking Matilda'..he goes on the track. 1916 J. B. Cooper *Coo-oo-ee* i. 3 Somehow things appear different to a man on a coach to a swaggie padding the hoof along the same road with 'Matilda' slung from his shoulders. 1917 A. B. Paterson *Waltzing Matilda* in *Saltbush Bill* (1924) 24 And he sang as he looked at his old billy boiling, 'Who'll come a-waltzing Matilda with me?' *c* 1926 'Mixer' *Transport Workers' Song Bk.* 139 That night he packed up his 'Matilda'. 1933 *Bulletin* (Sydney) 3 May 20/2 Matilda-waltzers with one black (or brown) shoe or boot are an everyday sight. 1934 *Ibid.* 12 Dec. 25/2 He chucks Matilda into a corner and is rummaging for tucker when Old Dave comes back. 1962 Marshall & Drysdale *Journey among Men* xii. 116 It was a beautiful place for a camp. It was good to stretch out in the arms of Matilda at the end of the day, and slowly smoke a cigarette. 1965 J. S. Gunn *Terminol. Shearing Industry* II. 29 In the days of itinerant workers and wanderers the swag had many names, including 'matilda', 'bundle', [etc.].

matildite (mă·ti-ldəit). *Min.* [a. It. *matildite* (A. D'Achiardi *I Metalli* (1883) I. 136), f. the name of the *Matilda* mine (near Morococha in the department of Junín) in central Peru: see -ite[1].] A sulphide of silver and bismuth, $AgBiS_2$, occurring as brittle, grey to black, orthorhombic crystals that are opaque with a metallic lustre.

1892 E. S. Dana *Dana's Syst. Min.* (ed. 6) 115 Matildite. 1900 *Mineral. Mag.* XII. 313 A description is given of 128 mineral species found in Japan... The occurrence of the rare mineral matildite is recorded. 1969 *Canad. Mineralogist* IX. 655 A characteristic feature of matildite in most of its occurrences from widely separated regions of the world is its intergrowth in galena which form [*sic*] Widmanstätten-like structures.

matily (mēi·tili), *adv.* [f. *MATEY *a.* + -ly[2].] In a friendly, familiar, or companionable fashion.

1973 *Daily Tel.* 10 Mar. 17/1 A squirrel chattered matily at me as I climbed the stile. 1974 *Listener* 21 Nov. 674/3 *The Hammers*..is jolly, warm-hearted,..matily naive.

matinée. Add: **1.** (Further examples.) Also one in a cinema.

1858 *Boston Even. Gaz.* 6 Mar., A dramatic matinee is a novel idea. 1879 *Dramatic Notes* 23, I have given 379 *Matinées*, equal to one year and a quarter of night performances. 1973 P. Evans *Bodyguard Man* i. 13 A child raised on Saturday matinée soap-operas at the local cinema.

2. A woman's lingerie jacket.

1896 *Woman's Life* 11 July 178/2 The pretty summer matinée of white cambric or lawn that accompanies this petticoat is made with full bell sleeves edged with lace. 1908 A. Bennett *Old Wives' Tale* III. iv. 338 She sat up and managed to drag her *matinée* from a chair and put it round her shoulders. 1960 Cunnington & Beard *Dict. Eng. Costume* 134/2 *Matinee* 1851.., a hooded pardessus made of jaconet or muslin and worn outdoors over a morning dress.

3. *attrib.* and *Comb.*, as *matinée actor, girl, public, ticket*; **matinée coat**, a baby's short outer garment; **matinée hat**, a lady's hat worn at matinées; **matinée idol**, a handsome actor of a type supposed to be especially attractive to matinée audiences.

1895 W. Archer in *World* 13 Feb. 25/2 He will learn next to nothing from matinée actors and audiences. 1929 *Treasure Cot Catal.* Nov. 31 Very pretty Matinee Coat in

good quality Crepe-de-chine. 1957 W. Rodwell *Cutting & Designing Juvenile Outerwear* I. 2 Basic draft for infant's matinée coat... This draft is for the earliest type of coat worn by children. 1899 H. Van Dyke *Fisherman's Luck* 101, The matinée girl is not likely to have a very luminous or truthful idea of existence floating around in her pretty head. 1894 *Million* 10 Feb. 223/2 The disappearance of the *matinée* hat would be seriously felt by the writer of jokes. 1898 [in Dict.]. 1905 G. B. Shaw *How to become Mus. Critic* (1960) 256, I once, in Drury Lane Theatre, sat behind a *matinée* hat decorated with two wings of a seagull. 1902 'H. McHugh' *Back to Woods* vii. 107 Mrs. John..you should be proud of this matinee idol husband of yours. 1909 'O. Henry' *Roads of Destiny* xiii. 210 Being conspirators from the cradle and matinee idols by proclamation. 1924 J. Buchan *Three Hostages* iv. 54 He's not in the least the ordinary matinée idol. He is.. adored by women and also liked by men. 1952 Granville *Dict. Theatr. Terms* 116 A perfect example of the matinée idol was the late Owen Nares. 1973 *Times* 26 Mar. 12/5 In the heyday of the matinee idol he had commanded.. a sober, serene and loyal admiration. 1919 F. Hurst *Humoresque* 206 It was just their always playing in the same pictures, and that silly matinée public..got to linking their names together. 1904 'A. Dale' *Wanted: a Cook* 208 What do you say to mentioning matinée tickets once a week?

ma·tiness. [f. *MATEY *a.* + -ness.] Friendly quality or character.

1928 *Evening News* 28 Dec. 8/3 There is the same expansive geniality, the same note of unassumed 'matiness'. 1950 *Landfall* IV. 345 The defensive bar-room and back-fence matiness of New Zealand life.

mating, *vbl. sb.*[1] Delete † *Obs.* and add later *attrib.* examples.

1908 *Westm. Gaz.* 12 Aug. 7/3 Mackenzie ran into a mating net with Gunsberg in a lively game. 1968 S. Morrison *Chess* vi. 52 Black would reply 10..K–R3 and..escapes the mating net.

mating, *vbl. sb.*[2] Add further *attrib.* examples.

1888 W. Day *Horse* xviii. 239 The management of mares at the mating season, and during the period of gestation, is a subject requiring more than mere passing notice. 1905 E. Wharton *House of Mirth* II. xiii. 500 The blind motions of her mating instinct. 1919 T. S. Eliot *Hippopotamus* in *Poems*, At mating time the hippo's voice Betrays inflexions hoarse and odd. 1936 *Brit. Birds* XXIX. 307 The 'mating rite' is performed only on the nest. 1936 *Discovery* Oct. 307/2 During the mating season, its call is startling. 1938 *Proc. Amer. Philos. Soc.* LXXXIX. 412 As conjugation takes place usually or only between individuals of diverse types, these types will be referred to as mating types. 1953 D. A. Bannerman *Birds Brit. Isles* I. 78 The mating-call..can be heard all this time. 1960 G. Blanchet *Search in North* xi. 142 The nesting season..was already ended, the mating songs were no longer heard and the young..on the wing. 1962 *Times* 30 Mar. 15/4 The mating-call or whistle of the juvenile 'wolf'. 1973 *Country Life* 29 Nov. 1848/2 The condition of the ewes by the end of mating time should be at least maintained for the first half of pregnancy.

mating, *ppl. a.* (Further examples, corresponding to *MATE *v.*[2] 6.)

1941 N. H. Anderson *Aircraft Layout* iv. 99 Dural mating parts should not be used in bearings. 1967 M. Chandler *Ceramics in Mod. World* iv. 129 Yet another method is to fire the rings separately, then grind the mating surfaces and join them with an epoxy resin adhesive.

matipo (mæ·tipo). *N.Z.* Also **matapo.** [Maori.] Either of two New Zealand evergreen trees, *Pittosporum tenuifolium*, of the family Pittosporaceæ, which is also called black matipo and bears clusters of purple flowers, or *Myrsine australis*, the red matipo (see *MAPAU).

1866 M. A. Barker *Let.* in *Station Life N.Z.* (1870) 94 Varieties of matapo, a beautiful shrub, each leaf a study, with its delicate tracery of black veins on a yellow-green ground. 1879 J. B. Armstrong in *Trans. N.Z. Inst.* XII. 329 The tipau or matipo, Pittosporum tenuifolium.. makes the best ornamental hedge I know of. 1921 H. Guthrie-Smith *Tutira* vii. 51 Later, appeared slender matapo (Pittosporum tenuifolium). 1960 N. Hilliard *Maori Girl* II. xiii. 152 Henry commented on the legs of the girls: 'Too thin! Like *matipo*.'

matless (mæ·tlės), *a.* [f. Mat *sb.*[1] + -less.] Not furnished with a mat or mats.

1880 J. Ross *Hist. Corea* x. 318 The dead body..is not, like the Chinese, put on a matless floor.

matlo(w (mæ·tlo). *slang.* [Phonetic ad. F. *matelot* sailor.] A sailor. Cf. *MATELOT.

1903 Kipling *Traffics & Discov.* (1904) 58 Simultaneous it hits the Pusser that 'e'd better serve out mess pork for the poor matlow. 1908 *Westm. Gaz.* 31 July 2/1 Evolutions and exercises to keep the modern 'matlow' busy and happy. 1914 'Bartimeus' *Naval Occasions* xxii. 200 In less formal surroundings..he is wont to refer to himself as a 'matlow'. 1916 'Taffrail' *Pincher Martin* ii. 16 Matloes, Pincher, the same as you an' me. 1946 *News Chron.* 27 June 2 (R.N. Advt.), It's a good life and a good job, too! I'm a Matlo—Able Seaman to you. 1974 S. E. Morison *European Discovery of America: Southern Voyages* xiii. 316 Magellan was always a sailor's sailor, and *os rudos marinheiros*, as Camoëns called the common 'matlow' of that era, always stood by him in his contests with officers.

matoke (mato·ke). [Local name.] A preparation of the flesh of bananas, used as food in Uganda; also, the fruit itself. Also *attrib.*

1959 N. W. Simmonds *Bananas* ix. 261 In Uganda.. the unripe fruit is peeled, wrapped in banana leaves and.. steamed... The product is called 'matoke' and it was the basic food of the Baganda people. 1962 *Lancet* 5 May 942/1 The excretion of 5-hydroxyindolacetic acid (5-H.I.A.A.) increases after the ingestion of matoke bananas, a normal article of diet of the Baganda people of Uganda. 1966 B. Kimenye *Kalasanda Revisited* 17 The gleaming vehicle..was only on the road when they needed fresh supplies of matoke from the market. *Ibid.* 33 Kibuka had to resort to collecting matoke peelings in an old bucket. 1969 *Reporter* (Nairobi) 16 May 39/1 For large areas of Uganda the staple food crop is bananas, or as it is known locally, matoke.

matra (mā·tră). [Skr.] In Indian music, a beat, or a subdivision of one, within a rhythmic phrase.

1898 B. A. Pingle *Indian Mus.* (ed. 2) v. 161 The number of Mátrás in a Tála is fixed. 1971 *Shankar's Weekly* (Delhi) 4 Apr. 24/3 The tabla player has to go through a cycle of matras, each of which is divided into a number of beats. 1972 P. Holroyde *Indian Mus.* v. 199 Tala (= palm of the hand) is a complex organization of rhythms or differing beats..in multiple groupings or thekas which are again subdivided by the individual components of the beat, the fractional matras. *Ibid.* 200 Matras are the smaller individual beats, each matra being determined in length by the pace of the overall rhythm.

‖ **matraca** (matrā·kă). Also **mattraca.** [Sp.] In Spain: a kind of mechanical wooden rattle used instead of church bells on Good Friday.

1910 C. B. Luffmann *Quiet Days in Spain* i. 14 A singular contrivance in the high towers takes the place of bells on Good Friday; this is the 'mattraca'—three long boxes of heavy wood arranged round a spindle, with several roughly fashioned hammers with rings in their handles, through which a rod is run to keep them in place. At short intervals during the whole of Good Friday the mattraca is turned round, grindstone fashion, and creates a most awful din... As I can find no clue to the origin of the mattraca, I fall back on the belief that it was intended to inform the multitude of the building of, or nailing of Christ to, the cross. 1974 S. E. Morison *European Discovery of America: Southern Voyages* v. 92 Holy Week in Seville... The supreme Passion on Good Friday when one heard the clacking of the *matraca* in place of cheerful bells.

matral (mæ·trăl, mēi·trăl), *a. rare.* [f. *MATR(I- + -al.] = *MATRICENTRED *ppl. a.*

1956 [see *MATRICENTRED *ppl. a.*].

matri- (mæ·tri, mēi·tri), used, esp. in *Anthropol.* and *Sociol.*, as the combining form of L. *māter* (*mātr-is*) mother, in various words denoting aspects of social organization defined by relationship through women. Some examples are given below as main words. Cf. also Matriarch, Matriarchal, etc., in Dict.

Matric. Add: Also **matric.** (Later examples.)

1936 Auden & Isherwood *Ascent of F6* (1937) I. i. 18 The Crowthers' pimply son has passed Matric. 1937 A. S. Neill *That Dreadful School* ii. 38 My staff and I have a hearty hatred of all examinations, and to us the Matric. is anathema. But we cannot refuse to teach children their Matric. subjects. 1948 [see *INTER., INTER]. 1965 W. Soyinka *Road* 83 To have served in Burma was to have passed your London Matric. 1967 D. P. Carew *Many Years, Many Girls* vii. 135 Sixth formers took the School Leaving Certificate... Matric could be achieved on this same exam. 1971 *Sunday Express* (Johannesburg) 28 Mar. (Home Jrnl.) 14/2 My daughter is writing Matric this year. 1971 R. Russell tr. *Ahmad's Shore & Wave* iv. 31 Their parents were only concerned with seeing that somehow or other they passed their matric. 1971 *Southerly* XXXI. 100 Not to mention the Matric next term. 1972 *Nature* 1 Dec. 267/2 He got a first in the London Matric, and became a 'temporary laboratory assistant' at Bradford Technical College.

matric (mēi·trik), *a. Math.* [f. L. *mātrīc-*, stem of *mātrix* Matrix.] Of or pertaining to a matrix or matrices.

1921 *Proc. Nat. Acad. Sci.* VII. 84 The notion of a norm or numerical value of a complex quantity.., as it arises in algebra, has a more or less immediate generalization to more extensive matric systems. 1941 *Mind* L. 273 The total matric algebra of order sixteen over the field of complex numbers. 1952 *Electronic Engin.* XXIV. 264 The application of matric theory to networks containing thermionic valves. 1958 R. V. Andree *Sel. Mod. Abstr. Algebra* v. 107 The Pauli matrices form a closed set under matric multiplication. 1972 *Computer Jrnl.* XV. 228/2 Pairwise interactions..may be used to define a distance function without requiring the qualitative data themselves to constitute a matric space.

matricentred (mætrise·ntəɹd), *ppl. a.* [f. *MATRI- + Centred, centered *ppl. a.*] Centred on the mother. Hence **matrice·ntric** *a.*, **matricentri·city.**

1956 R. Firth *Two Stud. Kinship in London* 41 We might speak then of South Borough kinship as being *matri-centred* or *matral*. 1957 M. Banton *W. Afr. City* xi. 205 The composition of tribal and Creole matricentric households does not differ significantly. 1957 V. W. Turner *Schism & Continuity in Afr. Society* p. xix, Another consequence of virilocal marriage making for instability in residential structure is the measure of autonomy it confers on the matricentric family. *Ibid.* iii. 76 The transition from matricentricity to matriliny as the basis for local groupings must always be hazardous. 1963 J. J. Honigmann *Understanding Culture* v. 85 Matricentric households are a 'natural' solution that arises in any social system..when people occupy a very low ranking, economically hazardous position.

matriclan (mæ·triklæn). [f. *MATRI- + Clan *sb.*] A matrilineal clan.

1950 D. Forde in Radcliffe-Brown & Forde *Afr. Syst. Kinship & Marriage* 309 All members of a matriclan within a village..regard one another as kin. 1957 *Jrnl. R. Anthrop. Inst.* LXXXVII. 92 But wealth, that is money and livestock, is vested in the matriclan... The matriclan, holding only moveable property, is a non-localized descent group. 1959 G. D. Mitchell *Sociol.* 69 Tribal offences include sexual relations between a man and a woman of the same matri-clan.

matriculability (mătri:kiŭlăbi·lĭti). [f. Matricul(ate *v.* + -ab(le + -ility.] Ability or fitness to matriculate.

1927 *Cambr. Univ. Reporter* 11 Oct. 142 The name of every candidate for matriculation on November 1 or 2, together with evidence of matriculability..must be sent to the Registry. 1963 *Statutes & Ordnances Univ. Cambr.* Suppl. 710 The Matriculation Registration Form when completed, together with any necessary evidence of matriculability, shall be submitted to the Registry.

matriculand (mătri·kiulæ:nd). [f. Matricul(ate *v.* + *-and[2].] = Matriculant.

1975 *Times Lit. Suppl.* 23 May 556/3 In the sixteenth century Salamanca had about 5,000–7,000 matriculands a year.

matrifocal (mæ·trifōukăl), *a. Sociol.* [f. *MATRI- + Focal *a.*] Applied to a family in which the mother is left with the responsibility for and authority over the household; mother-centred. Hence **matrifoca·lity**, the condition of a family which depends on the mother.

1952 *Internat. Afr. Inst. Memorandum* XXVI. 12 The matrifocal cells of the compound family. 1956 R. T. Smith *Negro Family Brit. Guiana* ix. 221 We maintain that the matri-focal system of domestic relations..can be regarded as the obverse of the marginal nature of the husband-father rôle. 1969 O. Lewis in D. P. Moynihan *On Understanding Poverty* vii. 198 For example, matrifocality, a high incidence of consensual unions, and a high percentage of households headed by women, which have been thought to be distinctive characteristics of Caribbean family organization or of Negro family life in the United States, turn out to be traits of the culture of poverty. 1969 J. & S. Baratz in T. Kochman *Rappin' & Stylin' Out* (1972) 11 This assumption left social scientists with no other alternative than to wrongly describe..the matrifocal family unit so prevalent in lower-class black society as 'evidence of male emasculation'.

matrilateral (mætrilæ·tĕrăl), *a.* [f. *MATRI- + Lateral *a.*] Of or pertaining to relationship involving the mother's brother or sister; used esp. of cross-cousin marriage.

1951 *Jrnl. R. Anthrop. Inst.* LXXXI. 24/2 Matrilateral cross cousin marriage (father's sister's son—mother's brother's daughter). 1957 V. W. Turner *Schism & Continuity in Afr. Society* viii. 254 Ndembu men appear to marry their patrilateral and matrilateral cross-cousins with equal frequency. 1963 *Brit. Jrnl. Sociol.* XIV. 65 From the point of view of this hypothetical individual, whom anthropologists for many years have called *Ego*, his non-lineage kin are *patrilateral* and *matrilateral* as the case may be. 1971 *World Archaeol.* III. 191 It is probable that this alliance..continues today due to prescribed matrilateral cross-cousin marriage.

matriline (mæ·triləin). [f. *MATRI- + Line *sb.*[2] 24.] The matrilineal line of descent.

1957 *Jrnl. R. Anthrop. Inst.* LXXXVII. 11 As Evans-Pritchard's analysis of the distribution of bride-wealth shows..there is explicit recognition of both the bride's patriline and matriline. 1963 W. J. Goode *World Revolution & Family Patterns* v. 221 The marriage of a mother's brother's daughter to a father's sister's son repeats the relationship of deference and respect which is due from the matriline. 1965 I. M. Lewis in M. Banton *Relevance of Models for Social Anthropol.* 109 The functional implications of descent are often much more significant than whether descent is traced in the patri- or matri-line. 1971 R. Needham *Rethinking Kinship & Marriage* i. 20 It is conceivable, and may in one alleged case be so in fact, that a terminology composed of matrilines should govern the affairs of a society that was preponderantly patrilineal. 1972 P. Laslett *Household & Family in Past Time* 22 The daughter, not the son, brings her working class spouse into the family circle... This..occurs within a matriline, not a patriline.

matrilineage (mætrili·nĭĕdʒ). [f. *MATRI- + *Lineage 2 c.] Matrilineal lineage.

1949 M. Fortes *Social Struct.* 60 Membership of the matrilineage is *ipso facto* membership of a widely dispersed exogamous clan. 1951 B. Z. Seligman in *N. & Q. Anthropol.* (ed. 6) II. ii. 89 A matrilineage consists of all the descendants through females of a single ancestress. 1957 *Jrnl. R. Anthrop. Inst.* LXXXVII. 93 It is the group, a dispersed, basic matrilineage, which defines the extent of the lateral inheritance of wealth. 1957 V. W. Turner *Schism & Continuity in Afr. Society* iii. 80 We regard a *minimal* matrilineage as a group consisting of the descendants through women of a common grandmother,

a *minor* matrilineage as the matrilineal descendants of a common great-grandmother. 1963 W. J. Goode *World Revolution & Family Patterns* iv. 194 The father is increasingly able to assert new rights against the traditional rights of the matrilineage.

matrilineal (mætrili·nĭăl), *a.* [f. *MATRI- + Lineal *a.* 2.] Of, pertaining to, or based on (kinship with) the mother or the female line; recognizing kinship with and descent through females.

1904 N. W. Thomas in *Man* LIII. 84 Mr. Hill-Tout argues that totemism originates in a patrilineal just as much as in a matrilineal state of society. 1906 —— *Kinship Organisations & Group Marriage Austral.* 30 Membership of a phratry depends on birth and is taken *directly* from the mother (matrilineal descent) or father (patrilineal descent). 1914 W. H. R. Rivers *Kinship & Social Organisation* ii. 40 The people of Buin still practice matrilineal descent. 1921 *Edin. Rev.* July 163 His discovery of matrilineal institutions in Europe. 1946 *Nature* 28 Sept. 457/2 It is clear that the ancient Chamorro had an elaborate social organisation with matrilineal clans and village chiefs whose power was based mainly on inherited wealth and monopolies. 1951 E. E. Evans-Pritchard *Social Anthrop.* ii. 29 According to Bachofen, there was first everywhere promiscuity, then a matrilineal and matriarchal social system, and only late in the history of man did this system give way to a patrilineal and patriarchal one. 1969 *Times* 22 Oct. (Ghana Suppl.) p. vii/3 The women have legal rights which would be the envy of any Englishwoman. They own property absolutely and the matrilineal system—inheritance through the women—means that the children belong to the mother's family.

Hence **matrili·neally** *adv.*

1907 *Athenæum* 20 Apr. 477/1 The method of reckoning descent matrilineally. 1937 *Nature* 20 Feb. 328/1 The husband in such a group, finding property to descend matrilineally, would try to provide for his portionless sister's family. 1964 Gould & Kolb *Dict. Social Sci.* 367/1 In a patrilineally organized society a man must also have kinship relations through his mother, and mutatis mutandis in a matrilineally organized society.

matrilinear (mætrili·nĭăɪ), *a.* [f. *MATRI- + Linear *a.*] = *MATRILINEAL *a.*

1910 *Manch. Guardian* 8 Aug. 5 It is said that they [*sc.* the Choctaw Indians] were governed by chiefs who succeeded by matrilinear descent. 1913 B. S. Phillpotts *Kindred & Clan* 275 Kindreds organized on matrilinear or on patrilinear lines. 1926 *Contemp. Rev.* Apr. 528 That these powers are related to a previous system of matrilinear descent there can be no doubt. 1950 Darlington & Mather *Genes, Plants & People* v. 57 The historic change from matrilinear to patrilinear inheritance has also led to important conflicts.

matriliny (mæ·trilini). [f. *MATRI- + Lin(e *sb.*[2] + -y[3].] The observance of matrilineal descent and kinship.

1906 N. W. Thomas *Kinship Organisations & Group Marriage Austral.* ii. 19 We may now examine the relation of matriliny to the seat of authority in the family. 1927 [see **bien entendu*]. 1937 *Nature* 20 Feb. 328/1 Though most of the tribes of Assam are now patrilineal, yet an earlier stratum of matriliny should not be overlooked. 1955 [see *father-right* (*FATHER *sb.* 12)]. 1965 I. M. Lewis in M. Banton *Relevance of Models for Social Anthropol.* 109 The lumping together of societies on the basis of patriliny or matriliny alone can only lead to confusion.

matrilocal (mætrilōu·kăl), *a.* [f. *MATRI- + Local *a.*] Applied to the custom in certain social groups for a married couple to settle in the wife's home or community.

1906 N. W. Thomas *Kinship Organisations & Group Marriage Austral.* iii. 30 When the husband removes and lives in his wife's group the marriage is *matrilocal*. 1927 *Contemp. Rev.* July 84 The clan is held together by matrilocal marriage. 1933 *Times Lit. Suppl.* 8 June 390/2 Marriage is matrilocal, or, in other words, the man goes to live with his wife. 1949 M. Fortes *Social Struct.* 70 Households with female heads can be..'matrilocal', that is, made up of a woman and her dependants by marriage and motherhood. 1956 R. Pieris *Sinhalese Social Organization* VI. i. 203 In the *binna* (matrilocal) marriage, the husband lived in his wife's parental home. 1956 J. Whatmough *Lang.* 250 But the Apache 'great family', with matrilocal residence after marriage, has its special terminology. 1966 Nye & Berardo *Emerging Conceptual Frameworks in Family Analysis* ii. 26 The extended family is a group founded on kinship and locality, and resulting from the rules of patrilocal or matrilocal marriage.

Hence **matriloca·lity**, the custom of matrilocal residence; **matrilo·cally** *adv.*, in a matrilocal manner.

1935 *Geogr. Jrnl.* LXXXVI. 262 The increasing amount of patrilocal marriages instead of the former pure matrilocality of the people. 1938 *Jrnl. R. Anthrop. Inst.* LXVIII. 301 The Hopi also transmit functions, such as priesthood and office, from male to male, without renouncing their matrilocality. *Ibid.*, Matrilocality here means, fundamentally, the succession of female descendants each one of whom remains her whole life in one spot. 1941 *Sudan Notes & Rec.* XXIV. 55 In a few cases the Shilluk settle matrilocally. 1963 *Economist* 13 Apr. 157/2 A mass of theorising about 'matrilocality' and the Cockney Mum.

matrimonial, *a.* and *sb.* Add: **A.** *adj.* **3. b.** *matrimonial agency, bureau* = *marriage bureau* (see *MARRIAGE 8); *matrimonial agent*, one who works in a matrimonial agency.

c 1888 C. E. Morland *Matrimonial Agency* (1902) 3, I have always expressed myself shocked at the indelicacy of marriages arranged by Matrimonial Agencies. *Ibid.* 4 A letter of introduction from a certain matrimonial agent. 1890 W. Booth *In Darkest Eng.* II. vi. 233 *A matrimonial bureau*... In London at the present moment how many hundreds..of young men and young women..are practically without any opportunity of making the acquaintance of each other! 1895 [see *CO-ED *a.*]. 1908 A. Bennett *Buried Alive* ii. 51 I've been thinking for years of getting married again... And what is there except a matrimonial agency? 1917 N. Douglas *South Wind* ii. 18 He..resembles a broken-down matrimonial agent. 1925 E. H. Young *William* xxxvi. 308, I hope the matrimonial bureau will be a great success. 1966 K. White *Lett. from Gourgounel* 40 He approaches Matrimonial Bureaux to try and find a wife. 1967 C. Watson *Lonelyheart 4122* iii. 33 A matrimonial bureau. Well, what else could be meant by 'Handclasp House. Are you weary of the solitary path?' *Ibid.* v. 46 He has the traditional English middle-class attitude to matrimonial agencies. Terribly *infra dig.*

matrix. Add: The normal pronunc. of *matrices* is now (mēi·trisīz).

4. c. A (positive or negative) copy of an original disc recording that is used in the making of other copies; *spec.* one used as a stamper.

1904 S. R. Bottone *Talking Machines & Records* 86 A copper matrix is first made from it [*sc.* the original disc] by electrotyping... From the copper matrix thus produced any number of duplicates can be moulded out of ebonite by hot pressure. 1918 H. Seymour *Reproduction of Sound* 175 A stamper is a working matrix for pressing records, and as such is merely a duplicate of the master matrix. *Ibid.* 310 To get an exact replica of a master matrix, an intermediate process is necessitated, in the production of a 'mother' matrix, which is a facsimile in metal of the original record..from which the subsequent negatives can be obtained. 1922 O. Mitchell *Talking Machine Industry* vi. 69 Several matrices are formed, which are then nickel-plated, polished and receive a strong backing of heavy steel as a support. They are then ready for the presses. 1929 *Melody Maker* Apr. 375/2 For sale. Matrices by leading U.S.A. Records Company, containing latest and biggest American successes. 1935 H. C. Bryson *Gramophone Record* vi. 142 Negative matrices, i.e., master and stamper, may be made to reproduce by means of a special needle, the end of which is split into a tiny fork which just fits over the ridge. 1952 [see *MOTHER *sb.*[1] 10*]. 1964 P. J. Guy *Disc Recording & Reproduction* vi. 76 When the third shell—which is called the Matrix or Stamper—is stripped off, its ridges are nickel plated.

d. *Photogr.* A dyed print in relief used for transferring colour to a final colour print.

1947 *Van Nostrand's Sci. Encycl.* (ed. 2) 917/2 Dyed relief films or differentially hardened films are generally called matrices when used in an imbibition process. 1957 R. W. G. Hunt *Reproduction of Colour* v. 44 The matrix film is..then washed with hot water to leave a hardened gelatin relief image. 1970 C. B. Neblette *Fund. Photogr.* xxii. 312 After drying, the films, now termed matrices, are dyed in the proper colours.

6. b. *spec.* in *Logic.* An array of symbols representing truth-values, giving the result of all possible assignments of truth-values to components of a propositional form or proposition; = **truth-table*. Also, that part of a truth-table which is an array of the total truth-possibilities (see quot. 1965). Also, a set of basic truth-tables for a particular system of logic (see quot. 1973). Also *attrib.*, as *matrix method*, etc.

1914 C. I. Lewis in *Jrnl. Philos., Psychol. & Sci. Methods* XI. 600 The matrix algebra for implications is useful as an instrument for investigating the interrelations of necessity, truth, possibility, falsity. 1932 —— & Langford *Symbolic Logic* vii. 201 That any principle, expressible in the symbols of the system, holds or does not hold can be determined by investigating its truth-status for all combinations of the truth-values of the elements. This is the matrix method. 1955 A. N. Prior *Formal Logic* 243 In the other matrices, the corner values are the usual ones for the corresponding two-valued operators. 1965 Hughes & Londey *Elements of Formal Logic* iii. 21 The array of possible combinations of truth-values of the variables (always set out on the left) will be called the matrix. 1973 J. J. Zeman *Modal Logic* v. 89 What we refer to may be called the matrix method... A truth-value system, or matrix, may be thought of as a set of tables, one for each of the primitive operators of the system, which may be used in computing a 'truth table' for any wff of the system.

c. *Logic.* An expression that would become a statement if its variables were replaced by constants, i.e. by names of individuals or classes or statements, as appropriate; = either **propositional function* or **statement-form*, depending on which type it is; *esp.*, in predicate calculus, a quantifier-free part of a formula (see quots. 1954, 1971).

1908 B. Russell in *Amer. Jrnl. Math.* XXX. 238 Then p/a, which we will call a matrix, may take the place of a function. 1910 Whitehead & Russell *Principia Math.* I. 262 Let us give the name of matrix to any function, of however many variables, which does not involve any apparent variables. 1932 Lewis & Langford *Symbolic Logic* ix. 267 Such functions as $f(x)$ and $p \vee q$..we may call matrices. 1954 I. M. Copi *Symbolic Logic* ix. 298 In any *wff* in prenex normal form $(Qx_1)(Qx_2)..(Qx_n)\,G$ the group of quantifiers $(Qx_1)(Qx_2)..(Qx_n)$ is the prefix and the quantifier-free formula G is the matrix. 1971 G. Hunter *Metalogic* 252 A formula A is in prenex normal form iff it is of the form $Qv_1..Qv_nB$, where each Q is either $\wedge$ or $\vee$, $n \geq 0$, B is a wff, and no quantifiers occur in B. The B..is called the matrix of A, and the part of A (if any) that precedes the matrix is called the prefix.

d. *Computers.* An interconnected array of diodes, cores, or other circuit elements that has a number of inputs and outputs and somewhat resembles a lattice or grid in its circuit design or physical construction.

1948 *Gloss. Computer Terms* (Mass. Inst. Technol. Servomechanisms Lab. Rep. R–138) 7 *Matrix switch*, a multi-position switch used in computers for decoding binary numbers. By mixing the output of flip-flops holding a binary number in an array of crystal rectifiers or resistors, it permits selection of one or a group of output lines. 1952 *RCA Rev.* June 185 The selected core will be magnetized in the desired direction while all other cores in the matrix will remain unaffected. The read-out is obtained by applying read-in current pulses. 1955 R. K. Richards *Arithmetic Operations in Digital Computers* iii. 75 In other applications the matrix is used to 'gate' an external signal (such as a series of pulses) onto one of a multiplicity of signal lines. 1969 P. B. Jordain *Condensed Computer Encycl.* 315 Such a diode matrix will encode decimal into binary. 1970 O. Dopping *Computers & Data Processing* x. 136 Cores are usually built together to form square or possibly rectangular matrices in which each matrix contains as many cores as there are words in the memory bank. Most matrices nowadays contain (64 × 64 =) 4,096 cores.

e. *Television* and *Broadcasting.* A circuit designed to accept a number of inputs and produce outputs that are linear combinations of them in different proportions. Freq. *attrib.*

1953 *Proc. IRE* XLI. 842/1 We show the pickup system giving information in the form of three channel voltages... All three of these voltages are fed to the inputs of three separate matrix or mixing units. 1954 *Ibid.* XLII. 201 (*heading*) Matrix networks for color TV. *Ibid.* 201/2 In a 3-tube video matrix with R_L and r_b equal to 5000 ohms, the gain per tube will be one third of the gain of a one-tube amplifier. 1961 Carnt & Townsend *Colour Television* vii. 185 The matrix circuits, in effect, solve the three equations $R' = Y' + 0{\cdot}96\,I' + 0{\cdot}62\,Q'$, $G' = Y' - 0{\cdot}27\,I' - 0{\cdot}65\,Q'$, $B' = Y' - 1{\cdot}11\,I' + 1{\cdot}70\,Q'$, and derive their name from the matrix notation used in algebra as a form of shorthand for writing and solving such equations. 1970 *Jrnl. Audio Engin. Soc.* XVIII. 627/1 (*caption*) One form of resistive matrix employed in four-channel stereo receiver or adapter to provide electrical separation of front and rear signals in left and rear stereo channels if transmission matrix of Fig. 4 is utilized at FM broadcast transmitter. 1971 D. J. Seal *Mazda Bk. Pal Receiver Servicing* i. 7 Two colour difference signals (R–Y and B–Y) are fed from the decoder to the matrix, where they are combined with the luminance signal (Y) to form the three primary colour signals red, green and blue. 1973 *Sat. Rev. Arts* (U.S.) 1 Apr. 49/2 *Matrix system*, the quadraphonic disc system in which four signals are fed into a circuit (matrix) that mixes (encodes) them into two signals that are inscribed on the walls of a record groove... If the disc is played on a properly equipped quad system, the signals are fed into a complementary matrix, which restores (decodes) them back to the original four, and the disc is heard quadraphonically.

7. *matrix algebra*; **matrix mechanics** *Physics*, a form of quantum mechanics developed by W. Heisenberg in which the operators corresponding to physical co-ordinates (position, momenta, etc.) are represented by matrices with time-dependent elements; **matrix number**, a number assigned by a record company to a matrix in the manufacture of gramophone records; **matrix printer**, a printer in which each printed character is made up of dots printed by the tips of small wires selected out of a rectangular array; **matrix sentence** *Linguistics* (see quot. 1967[1]).

1930 Ruark & Urey *Atoms, Molecules & Quanta* xvii. 577 (*heading*) The laws of matrix algebra. 1969 D. C. Hague *Managerial Econ.* i. 24 With complex decision problems..calculus and matrix algebra may have to be used. 1926 P. A. M. Dirac in *Proc. R. Soc.* A. CXII. 666 In Heisenberg's matrix mechanics it is assumed that the elements of the matrices that represent the dynamical variables determine the frequencies and intensities of the components of the radiation emitted. 1966 *McGraw-Hill Encycl. Sci. & Technol.* VIII. 181/2 Matrix mechanics is disadvantageous for obtaining quantitative solutions to actual problems; because it is concisely expressed in a form independent of special coordinate systems, however, matrix mechanics is advantageous for proving general theorems. 1974 G. Reece tr. *Hund's Hist. Quantum Theory* xiii. 171 First Heisenberg showed in the language of matrix mechanics that two identical coupled systems always behaved like the two oscillators. 1965 G. Melly *Owning-Up* xi. 129 He was a record collector who knew the matrix number..of every record in his immaculately filed collection. 1968 Leadbitter & Slaven *Blues Records 1943–66* 7 All matrix numbers where known are given for completeness sake, even though they may only be the issue number with A or B suffixes or numbers allocated by pressing plants. 1958 Gotlieb & Hume *High-Speed Data Processing* iii. 59 IBM has matrix printers for use with the 702 and 705 data processors. 1964 T. W. McRae *Impact of Computers on Accounting* i. 17 The 'matrix' printer and the 'chain' printer can reach speeds of up to 1200 lines a minute. 1964 Matrix sentence [see *CONSTITUENT B. 5]. 1967 D. Steible *Conc. Handbk. Ling.* 77 *Matrix sentence*, in transformational grammar, a basic sentence, known in traditional grammar as an independent clause, into which other structures may be embedded. 1967 *Word* XXIII. 338 The ultimate fate of an included-sentence construction, of course, is to become so thoroughly embedded in the matrix sentence as to be no longer distinguishable. 1968 *Canad. Jrnl. Ling.* XIII. 83 Infinitive phrases have a wide range of function within the matrix sentence. 1972 HARTMANN & STORK *Dict. Lang. & Ling.* 138 Matrix sentences often coincide with what is known in traditional grammar as main clauses.

b. Applied to precious stones (see quot. 1909).

1909 *Cent. Dict.* Suppl., *Matrix-gem*, an opal, turquoise, ruby, or other gem intimately mixed with the matrix material and cut with it. 1921 *Brit. Mus. Return* 157 in *Parl. Papers* XXVII. 651 A suite of specimens of sapphire ..comprising two matrix specimens.

matrix (mēi·triks), *v.* [f. prec. sb. (Orig. formed as the vbl. sb.)] *trans.* To combine (signals) in different proportions so as to obtain one or more linear combinations of them.

1969 CARNT & TOWNSEND *Colour Television* II. iii. 106 Matrixing *R–Y* and *B–Y* produces *G–Y*. 1971 D. J. SEAL *Mazda Bk. Pal Receiver Servicing* i. 7 The decoder accepts the PAL chrominance signal.., demodulates it and produces two colour difference signals to be matrixed and amplified by the colour difference amplifiers. 1972 H. F. OLSON *Mod. Sound Reproduction* ix. 188 The four inputs *LF*, *LR*, *RF*, and *RR*, representing the left front, left rear, right front, and right rear respectively, are matrixed and encoded to the left, *L*, and right, *R*, channels of the conventional two-channel stereophonic recording system. Each of the two channels *R* and *L* contains a mixture of the four channels in such a manner that the four channels can be reconstituted in reproduction by means of the decoder and matrix. 1972 *Gramophone* Jan. 1412/1 The extra channels of information intended for feeding to the back loudspeakers are simply mixed (or matrixed) with the front channel signals prior to the cutting of the disc.

So **ma·trixed** *ppl. a.*, **ma·trixing** *vbl. sb.*

1951 *Proc. IRE* XXXIX. 1158/2 All that is required is the addition (or subtraction..) of fractions of the camera currents to derive the currents required to control the receiver. [*Note*] This operation has been called 'matrixing' in some discussions. The analogy to the matrix operation, which is used in vector algebra to change from one system of co-ordinates to a second, is obvious. 1955 J. W. WENTWORTH *Color Television Engin.* iv. 125 Matrixing can be used to improve color fidelity only when camera signals are produced by linear devices; if the camera signals are nonlinear functions of red, green, and blue, the linear equations of the matrix process cannot be worked out properly unless the matrix circuit is preceded by linearity correctors. 1971 D. J. SEAL *Mazda Bk. Pal Receiver Servicing* v. 89 Since the luminance signal (–Y) is applied to all three cathodes, the tube performs the final matrixing, and the three gun currents are proportional to the original red, green and blue separation signals from the colour cameras. 1972 H. F. OLSON *Mod. Sound Reproduction* ix. 189 The performance, as exemplified by..spatial effects, of a two-channel-to-four-channel coded and matrixed system is inferior to that of four discrete channels. 1972 *Observer* (Colour Suppl.) 22 Oct. 54/4 Not all the matrix systems are compatible with each other, though a machine like the Pioneer QX-8000 *is* capable of dealing with all forms of matrixed four-channel.

matroclinous (mætrokləi·nəs), *a. Biol.* [f. L. *mātr-*, *māter* mother + Gr. κλίν-ειν to lean + -OUS.] Resembling the female rather than the male parent; involving or possessing a tendency to inherit a character or characters from the female parent only. So **ma·trocliny**, matroclinous inheritance.

1913 *Jrnl. Exper. Zoöl.* XV. 587 Morgan has explained the case of criss-cross inheritance on the ground that the *X*-chromosome is the carrier of all sex linked factors. The sons are matroclinous because they receive their unpaired chromosome directly from the mother, and must show all the sex linked characters which she showed. 1917 *Genetics* II. 247 But even if one hybrid should be judged to indicate patrocliny in one cross, the same hybrid appears in the reciprocal cross where it would be a case of matrocliny for the same hybrid. 1925 C. C. HURST *Exper. in Genetics* xxxviii. 535 Gametic chromosomes, male and female, are either equal with 7–14–21 or 28 each, or unequal and matroclinous with a maternal bias in the ratio of 1·5–2–3–4 or 5:1. 1939 *Nature* 14 Jan. 81/2 To account for the purely matroclinous inheritance, the following explanation is suggested. The eggs of *M. formosa* are fertilized by the other two species, but the paternal chromosomes remain inactive. 1961 A. MÜNTZING *Genetic Res.* xxvi. 257/1 F. von Wettstein..observed that this matrocliny was clearly noticeable also in [the] F_2. 1968 R. C. KING *Dict. Genetics* 149 In *Drosophila* the daughters produced by attached-X females are matroclinous in terms of their sex-linked genes.

matron. Add **1. d.** A female dog or horse used for breeding.

1931 A. C. SMITH *About our Dogs* vi. 67 When the matrons have been purchased..the question of finding mates for them is uppermost. 1948 C. L. B. HUBBARD *Dogs in Brit.* xxi. 365 Brood matrons were sent to Welsh Hound studs. 1966 *Telegraph* (Brisbane) 22 Jan. 5/1 King's Daughter could have been an above average stud matron. 1971 *Country Life* 11 Nov. 1267/1 The breeding of hounds ought not to be merely a question of appearance and dash... They should also be the offspring of stallions and matrons of time-honoured residence in the kennels.

4. b. matron of honour, a chief bridesmaid who is married.

1903 *N.Y. Tribune* 20 Sept., Her only attendant, as matron of honor, wore pale blue crepe de chine. 1948 *Daily Oklahoman* (Okla. City) 9 June 13/3 The bride and I wonder whether I should be called matron of honor or maid of honor. 1961 A. BARNES *Mod. Wedding Etiquette* vii. 61 There is never more than one matron of honour and she wears, not the usual youthful style bridesmaids' dresses, but an afternoon gown with hat and gloves. Her duties are exactly the same as those of the chief bridesmaid. 1963 *Listener* 14 Mar. 473/3 The groom's brother ..has to listen to the indignant and delightfully comic complaints of the matron-of-honour. 1969 EDWARDS & BEYFUS *Lady Behave* 268 If the bridesmaid is married she is known as a Matron of Honour.

matronize, *v.* **1.** (Earlier example.)

1741 S. RICHARDSON *Familiar Lett.* 187 Childbed matronizes the giddiest Spirits.

matso, var. *MATZAH.

matsu. Substitute for def.: A local name for several pine trees, especially the two native to Japan, *Pinus densiflora*, the Japanese red pine, and *P. thunbergii*, the black pine, both valuable ornamental and timber trees. (Earlier and later examples.)

1727 J. G. SCHEUCHZER tr. *Kæmpfer's Hist. Japan* I. I. 118 The common Fir, which they call *Matzuoki*, will come to the age of a thousand. 1884 tr. *J. J. Rein's Japan* I. vii. 151 The Aka-matsu or red pine..and the Kuro-matsu or black pine..are the commonest conifers in the country. 1916 E. H. WILSON *Conifers & Taxads Japan* 25 The Japanese Red Pine, or Aka-matsu, is very widely distributed in Kyushu. *Ibid.* 27 This Black Pine, or Kuro-matsu, is found in Japan. *Ibid.* 28 Another name for this pine [*sc.* Black Pine] in Japan is O-Matsu (Male Pine). 1923 DALLIMORE & JACKSON *Handbk. Coniferæ* 466 *Pinus Thunbergii*,..Black Pine,..Kuro-matsu. 1938 D. T. SUZUKI *Zen Buddhism & its Influence on Jap. Culture* II. iii. 247 The Japanese species of the pine known as *matsu* generally spreads its branches irregularly and the trunk is gnarled. 1965 J. OHWI *Flora Japan* 115/2 *Pinus thunbergii*... Kuro-matsu, O-matsu. 1971 S. ELIOVSON *Gardening Jap. Way* 86 Pine (*Matsu*). This is the symbol of longevity, being hardy and evergreen during cold winters.

‖ **matsuri** (mætsu·ri). [Jap.] A solemn celebration or festival held periodically at every Shintō shrine in Japan in order to deepen the consciousness of the gods in the daily lives of the worshippers.

1727 J. G. SCHEUCHZER tr. *Kæmpfer's Hist. Japan* I. III. 223 It is a custom which obtains in all cities and villages, to have two such Matsuri's celebrated every year with great pomp and solemnity in honour of that God, to whose more particular care and protection they have devoted themselves. 1841 *Manners & Customs of Japanese* iii. 65 Many and..various peculiarities belong to the *Matsuri* festival. 1883 E. G. HOLTHAM *Eight Yrs. Japan* viii. 194 The Kiyōto matsuri, specially connected with the 'Gion' quarter of the city, inhabited by singing and dancing girls and such like,..was well worth seeing. 1928 F. A. LOMBARD *Outl. Hist. Jap. Drama* iii. 62 *Matsuri*..have from ancient times in Japan been occasions of great popular rejoicing. 1964 *Asia Mag.* 20 Sept. 9 (*caption*) The winding festival or *matsuri*.

matt, *a.* Now the usual spelling, esp. in *Photogr.*, of MAT *a.*

1876 *Encycl. Brit.* V. 170/1 Water gilding..is finished either 'matt' or burnished... Matt-work is protected with one or two coats of finish-size. 1892 *Photogr. Ann.* II. 76 The metal separates in a matt grey form. 1925 *Morris Owner's Manual* 79 The operation is finished when the valve face has a clean, even, matt-surfaced ring around it. 1933 *Archit. Rev.* LXXIII. 265 To treat plastic materials without humbug so as to bring out their..pleasant matt texture. 1944 L. MACNEICE *Springboard* 32 The matt-grey iron ship, Which ought to have been the Future. 1953, 1958 [see *EGG-SHELL c]. 1967 E. CHAMBERS *Photolitho-Offset* iii. 27 Some smooth, semi-matt prints give fair results, but, in general, sepia tone prints, matt-surface prints and hand-coloured prints should be avoided.

matte[3] (mæt). *Cinemat.* Also **mat.** [Fr.] A mask (MASK *sb.*[3] 4 e) used to obscure or shade (part of) the image shown. Also *attrib.*

1938 *Motion Pict. Sound Engin.* (Acad. Motion Pict. Arts & Sci.) 39 A more recent procedure is to use a W-type mat which reduces the track width from the center as well as from both..edges. 1948 MENCKEN *Amer. Lang.* Suppl. II. 699 *Matte shot*, a film made with a section blocked out, to be filled later on another set. 1949 W. H. OFFENHAUSER *16-Mm. Sound Motion Pict.* xii. 382 A traveling matte..is merely a film interposed at the gate between the light source and the image-bearing film that alters the light intensity in the manner of a neutral density film. *Ibid.* 549/2 Matte rolls are a pair of film rolls used as light modulators. 1959 W. S. SHARPS *Dict. Cinematogr.* 110/2 *Matte box*, a box fitted to the front of a camera to hold mattes. The matte box frequently combines the functions of a lens hood and filter holder as well. 1960 O. SKILBECK *ABC of Film & TV* 84 *Matte*, a specially photographed Mask which..leaves blanks which can be correspondingly filled with something else at a second printing; e.g. adding a false Background as though seen through a window. 1972 L. D. GIANNETTI *Understanding Movies* 73 Mattes are used to block out certain areas of the real scene where the animated drawings will appear in the finished print.

matte, var. *MATT *a.*

1897 [see *GLOSSY *a.* a]. 1909 *Chambers's Jrnl.* Oct. 684/1 Taken altogether, it cannot compare with the 'matte' appearance of the flatted enamel which is so extensively used in French decoration. 1969 *Sears Catal.* Spring/Summer 6 Uniquely styled in matte-finished vinyl. 1973 *Sci. Amer.* Oct. 128/1 For those who wish a definite opaque matte appearance, talc is cut to a third, and rice starch or particularly fine precipitated chalk is added to improve absorption.

matter, *sb.*[1] Add: **26. matter wave** *Physics*, a de Broglie wave (see *DE BROGLIE).

1930 RUARK & UREY *Atoms, Molecules & Quanta* xxi. 722 When electrons impinge on polycrystalline metal surfaces the fraction scattered at an angle θ with the normal to the surface does not decrease uniformly as θ increases... The results seemed likely to remain unexplained, until Einstein discussed de Broglie's matter waves, in 1924 and 1925. 1972 *Sci. Amer.* Oct. 106/3 The bombarding particles have the properties not only of particles but also of waves. Such 'matter waves' are called de Broglie waves (after Louis de Broglie, who first suggested that the wave-particle duality observed for electromagnetic radiation might also exist for matter). 1974 G. REECE tr. *Hund's Hist. Quantum Theory* xi. 141 The intuitive theory of matter waves and of fields of matter is modified to the point where there is room for the idea of particles.

matter, *v.* Add: **2.** Examples with personal subject.

1909 H. W. C. NEWTE *Sparrows* xl. 505 With your appearance and talents you should be a great social success with people who matter. 1926 G. B. SHAW *Translations & Tomfooleries* II. 49 Well, what about Edith? Doesn't she matter? 1933 E. A. ROBERTSON *Ordinary Families* x. 233 Wives still matter politically, if a man's to be offered a comfortable Conservative seat.

mattery, *a.* **1.** Delete 'Now *rare*' and add later examples.

1955 W. W. DENLINGER *Compl. Boston* I. 152 Small mattery spots on the inside of the thigh..are often the first indication that a dog is suffering from distemper. 1961 S. CHAPLIN *Day of Sardine* i. 8 This was a sick kitten. Thin as a rake, mattery eyes, scabby eyes.

Matthew Walker. *Naut.* In full *Matthew Walker knot*, a multi-stranded rope end knot, prob. named after its originator.

[1808 D. LEVER *Young Sea Officer's Sheet Anchor* 5 Matthew Walker's knot.] 1841 R. H. DANA *Seaman's Manual* vii. 37 A Matthew Walker knot,..unlay the end of a rope [etc.]. 1856 C. NORDHOFF *Man-of-War Life* xii. 212, I practised..and was soon master of long and short splices, manrope knots, turks-heads, and Matthew Walkers. 1860 *All Year Round* 28 July 382/1 'Which knot?' asked Toby... 'Matthew Walker, spritsail-sheet, stopper or shroud?' 1883 *Man. Seamanship for Boys' Training Ships R. Navy* (Admiralty) (1886) 105 A Matthew Walker knot is used for the standing part of the lanyards of lower rigging. 1953 *People* (Austral.) 26 Aug. 47/3 Mason will talk by the hour about Matthew Walker knots, bell ropes ..and cross-pointing if he encounters a pair of sympathetic ears. 1961 F. H. BURGESS *Dict. Sailing* 144 *Matthew Walker*, a knot, single or double, used to make a collar on a rope; the strands may be relaid when formed (a development of the wall knot). 1968 E. FRANKLIN *Dict. Knots* 20 *Matthew Walker* and *Double Matthew Walker*, probably the best and most useful of the multistrand rope end knots. It is claimed that the unknown Matthew Walker is the only man to have a knot named after him. 1971 I. IMRIE tr. *Svensson's Handbk. Seaman's Ropework* 79 The Mathew Walker knot on a three-stranded rope is the simplest of all knots with a rope's own end.

mattraca, var. *MATRACA.

mattress[1]. Add: **3. b.** *U.S.* A bed of sugar-cane. Hence **mattress** *v. trans.*, to form (sugar-cane) into 'mattresses'.

1829 A. SHERWOOD *Gazetteer Georgia* (ed. 2) 255 The stacks or banks in which seed cane is preserved during winter, are called mattresses. 1833 B. SILLIMAN *Man. Sugar Cane* 12 They are cut near the ground, and carted to the vicinity of the fields where they are to be planted; being formed..into long beds about fifteen feet wide, which are called *mattresses*. 1850 *Rep. Comm. Patents: Agric. 1850* (U.S. Dept. Agric.) 423 It was..supposed that the cane was spoiled in the mattress by the continued warm weather after it was mattressed.

4. b. mattress antenna, array = *BILL-BOARD, BILLBOARD 2, also *ellipt.*; **mattress-jig** *slang*, sexual intercourse.

1947 D. G. FINK *Radar Engin.* iv. 247 The power efficiencies of properly adjusted mattress arrays vary from 80 to 95 per cent. *Ibid.* 249 The gain of a paraboloid is generally lower than that of a mattress of the same area expressed in square wavelengths. 1950 Mattress antenna [see *BILL-BOARD, BILLBOARD 2]. 1961 R. L. MATTINGLY in H. Jasik *Antenna Engin. Handbk.* xxv. 26 A common transmitter-receiving antenna can be implemented with a mattress array. 1896 FARMER & HENLEY *Slang* IV. 290/2 *Mattress-jig*, copulation. 1922 JOYCE *Ulysses* 418 Smutty Moll for a mattress jig.

Matura (mā·tʊrā). Also **Matara.** [Name of a town (now called *Matara*) in Sri Lanka.] *Matura diamond*: a colourless variety of zircon used as a gem. Cf. JARGON *sb.*[2]

1880 *Encycl. Brit.* XIII. 532/1 The Singalese variety [of pale jargoons], found chiefly at Matura, has been termed 'Matura diamond'. 1936 L. J. SPENCER *Key to Precious Stones* xvii. 189 Much of the colourless zircon..is known as 'Matura diamond'. 1962 R. WEBSTER *Gems* I. vii. 118 The colourless stones..were..called by the misnomer 'Matara diamonds', a term which should now be forgotten.

maturation. Add: **5. a.** Also in *Psychol.*, the physical growth which, together with learning, leads to full development. Also *transf.* Hence **matura·tional** *a.*, of or pertaining to maturation.

1921 *Psychol. Rev.* May 196 The process runs along parallel with the process of maturation and it is not clear in any case just what is contributed by heredity and what is due to learning. 1929 A. GESELL in C. Murchison *Found. Exper. Psychol.* 651 (*heading*) Maturational correspondence in identical twins. *Ibid.* 658 Maturational factors preserve his native endowment. 1938 R. S. WOODWORTH *Exper. Psychol.* xxix. 764 Maturation consists largely in development of the ability to learn. 1943 C. T. MORGAN *Physiol. Psychol.* vii. 124 Many experiments have been carried out on the general problem of the relative importance of maturation and learning. 1953 *New Biol.* XIV. 27 Clearly this 'setting' of the perceptual and behavioural pattern in the third and fourth years must have as its base important maturational changes in the physiology and anatomy of the brain. 1958 K. LOVELL *Educ. Psychol. & Children* vii. 103 In studying the effects of maturation on learning in children, use has been made ..of identical twins. 1961 J. BERKO in Saporta & Bastian *Psycholinguistics* vi. 372/1 Throughout childhood, girls are perhaps from a maturational point of view slightly ahead of the boys who are their chronological age mates. 1968 M. BUNGE in Lakatos & Musgrave *Probl. Philos. Sci.* 134 Scientific Research can pass through several phases of maturation, the degree of maturity attained depending on the depth and the logical organization of the ideas involved. 1970 D. S. WRIGHT et al. *Introducing Psychol.* iv. 70 There is still some value in contrasting maturation and learning... This is simply a convenient way of classifying different antecedent conditions of the single process of development.

b. Add to def.: *spec.* of their gametes. (Further examples.)

1896 E. B. WILSON *Cell* 338 *Maturation*, the final stages in the development of the germ-cells. More specifically, the processes by which the reduction of the number of chromosomes is effected. 1972 BALIN & GLASSER *Reproductive Biol.* v. 319 (*heading*) Maturation of epididymal spermatozoa.

7. *attrib.*, as **maturation division** *Biol.*, either of the two divisions of meiosis.

1896 E. B. WILSON *Cell* v. 185 It is plain that the nature of the maturation-divisions can only be approached through a study of the origin of the tetrads. 1966 *Chromosoma* XIX. 99 (*heading*) The maturation divisions of the parthenogenetic stick insect *Carausius morosus* Br.

mature, *a.* Add: **1. d.** Of a soil: having a fully developed profile. Of a soil profile or its parts: fully developed.

1926 C. F. MARBUT in Tansley & Chipp *Study of Vegetation* vii. 131 Mature soils..owe their essential characters to the nature of the climate in which they are developed. 1927 [see *IMMATURE *a.* 2 e]. 1954 W. D. THORNBURY *Princ. Geomorphol.* iv. 76 A mature soil profile exhibits well-developed horizons. 1971 E. A. FITZPATRICK *Pedology* vii. 249/2 Marbut placed great emphasis on mature freely drained soils.

2. d. *mature student*: an adult who undertakes a course of study at a later age than normal.

[1924 L. G. E. JONES *Training of Teachers in Eng. & Wales* xiv. 352 The sprinkling of more mature students is a great help.] 1953 C. A. RICHARDSON et al. *Educ. of Teachers in Eng., France & U.S.A.* iii. 59 The mature student is one who..wishes to embark on a course of training as a teacher at an age considerably later than the normal age of entry to college. 1969 H. C. DENT *Educ. Syst. Eng. & Wales* (ed. 4) x. 205 From 1962–63 the Minister of Education ceased to give State scholarships (except to mature students). 1969 *Guardian* 16 Sept. 5/3 The formation of a new union for mature students has given fresh hope to people who are taking up teachers' training courses late in life. 1970 *St. Hilda's College* (*Oxf.*) *Rep. 1968–69* 21 One mature student was accepted to read for the B.A. in English. 1971 *Mod. Law Rev.* XXXIV. 652 A better solution to the problem of democratising recruitment would be to facilitate the attendance at law school of mature students.

6. b. Of a progressive cataract: characterized by complete opacity.

1850 [see *IMMATURE *a.* 2 d]. 1904 L. W. FOX *Dis. Eye* xii. 310 Operations should not be performed on both eyes at the same time, even though both cataracts are mature. 1970 A. H. KEENEY *Ocular Exam.* ix. 143/1 A mature cataract is one that has developed complete opacification throughout.

maturely, *adv.* **3.** Delete *rare*[-1] and add later examples.

1902, 1965 [see *DISSECTED *ppl. a.* 2 b]. 1965 M. SPARK *Mandelbaum Gate* iv. 108 The rest were Arabs and Jews, most of whom were maturely sixteen years of age and upward.

matzah (ma·tsa). Also **matso** (ma·tso), **-za, -zho, -zo, -zot, -zoth, mazzot, -oth, motso, -za, mozza.** Pl. **matzoth, -os.** [Yiddish *matse*, f. Heb. *maṣṣāh*.] (A wafer of) unleavened bread, eaten by orthodox Jews during the Passover.

1846 *Jewish Manual, or Pract. Information Jewish & Mod. Cookery* p. xiv, *Matso*, Passover cakes. 1871 E. LEVY *Jewish Cookery Bk.* 32 Soak two matzos, or crackers, in cold water. 1891 M. FRIEDLÄNDER *Jewish Relig.* II. 380 The head of the family..who reads the [Passover] Service has before him on the table..three unleavened cakes (*matsoth*). 1893 I. ZANGWILL *Childr. Ghetto* (ed. 3) xxii. 202 Now is the national salutation changed to 'How do the *Motsos* agree with you?' 1904 *Jewish Encycl.* VIII. 393/1 The eating of mazzot during the seven days of the Passover festival is intended to recall the hurried departure from Egypt. 1922 JOYCE *Ulysses* 79 Something like those mazzoth: it's that sort of bread: unleavened shewbread. 1927 *Daily Express* 16 Apr. 9 The command concerning unleavened bread, known as 'Matzos', will be observed. 1950 G. MIKES *Milk & Honey* 152 On the feast day..matzot is eaten instead of bread. 1952 S. SPENDER *Learning Laughter* xi. 159 This night we eat *matza* and no leavened bread. 1963 *Ann. Reg. 1962* 374 In March the Soviet Government banned the making of *matzot* (unleavened bread) required for Passover observances. 1965 M. SPARK *Mandelbaum Gate* ii. 33 The unleavened bread, crisp *matzko* that made crumbs everywhere, was uncovered. 1970 L. M. FEINSILVER *Taste of Yiddish* iii. 323 Even the Air Force..has a pretty good idea of what matzo is, for a New Jersey woman reported an Unidentified Flying Object as looking like a big matzo. 1973 *Jewish Chron.* 2 Feb. 23/2 This particular hechsher only applies to matzot manufactured in Britain.

attrib.

1846 *Jewish Manual, or Pract. Information Jewish & Mod. Cookery* i. 9 *Matso soup*... Take half a pound of *matso* flour. 1906 *Mrs. Beeton's Bk. Househ. Managem.* lv. 1574 Take up a little of the motza paste..roll lightly in the motza meal. 1921 *Dict. Occup. Terms* (1927) 433 *Matzo* (*or matzoth*) *baker*..expeditiously removes circular matzo bread i.e. Jewish passover bread. 1953 W. P. MCGIVERN *Big Heat* ix. 111 He liked Jewish food..mozza ball soup. 1970 S. J. PERELMAN *Baby, it's Cold Inside* 208 The exhibit you see here..was first believed by the coroner to be a fossilized matzo ball. 1971 M. MCCARTHY *Birds of America* 36 She should be making *gefüllte* fish and matzoth balls. 1973 *Times* 3 Feb. 13/5 You can try the kneidlach soup (with matzo-meal dumplings).

mauby (mō^u·bi). Also **mawby.** [See MOBBIE, MOBEE.] In Barbados, a drink made from the diluted extract of the bark of a tree or shrub of the genus *Gouania* or related plants; cf. MOBBIE, MOBEE 1.

1790 J. B. MORETON *Manners & Customs West India Islands* 105 Cool drink or mauby is a delicious nectar to them in the morning. 1954 S. M. SADEEK *Windswept & Other Stories* (1969) 25 He drank, at first sipping the tall mauby slowly. 1957 I. BAYLEY in F. A. Collymore *Notes for Gloss. Words & Phr. Barbadian Dial.* 56 Mauby is made from bark, boiled in water for about an hour, and concentrated to a basic extract. When making the drink, some of the stock is diluted with water; sugar and vanilla essence are added... The result is a bitter-sweet drink, with a frothy head, popular for quenching the thirst, and for 'cooling' the blood. 1958 J. CAREW *Black Midas* i. 17 One Friday Tanta Moore invited him in for a drink of mauby. 1962 *Listener* 16 Aug. 248/2 A man stands beside his handcart dispensing mawby or snowball, which consists of shaved ice covered with scarlet syrup. 1965 *Ibid.* 8 Apr. 523/3 Constant chatter is punctuated with the high-pitched, rather doleful cry of the 'mauby' girl: she.. carries a two-gallon tank on her head, with a tap on it that she manipulates to dispense cold tea to her customers; a special kind of cold tea, brewed from the bark of the mauby plant. 1968 E. LOVELACE *Schoolmaster* i. 8 Fruit wines from cashew and guava, and sorrel and mauby. 1973 *Advocate-News* (Barbados) 15 Dec. 8/1 In short, it is like serving mauby and ginger beer in a cocktail glass. 1974 *Sunday Advocate-News* (Barbados) 10 Mar. 14/1 Formerly, mauby women walked the streets with their load on their heads, and a bench in hand. But today, some of them drive or are driven.

maucherite (mau·χ-, mau·k-, mau·ʃərəit). *Min.* [ad. G. *maucherit* (F. Grünling 1913, in *Centralbl. f. Min., Geol. u. Paläont.* 225), f. the name of W. *Maucher* (1879–1930), mineral dealer of Munich; see -ITE[1].] A nickel arsenide, approximately $Ni_{11}As_8$, occurring as brittle, reddish grey tetragonal crystals that are opaque with a metallic lustre and tarnish on exposure.

1913 *Jrnl. Chem. Soc.* CIV. II. 516 Maucherite, a new nickel mineral from Thuringia... This new mineral, which was at first mistaken for rammelsbergite, occurs.. in veins..in the copper-shales..at Eisleben, Thuringia. 1940 *Mineral. Mag.* XXV. 570 A physical, chemical, and preliminary structural study of maucherite..from Eisleben, Thuringia, and Sudbury, Ontario, and temiskamite ..from Elk Lake, Ontario, shows that these minerals.. represent a single mineral species. 1963 *Mineral. Abstr.* XVI. 263/1 Syntheses from nickel and arsenic were carried out by means of dry thermal procedures... At arsenic vapour pressure, Ni_3As_2, maucherite, niccolite, rammelsbergite or paramelsbergite [*read* pararammelsbergite], are formed depending on the temperature.

maul, *sb.*[1] **4.** (Later examples.)

1960 *Times* 7 Dec. 17/1 The ability to start attacks from line-outs and loose mauls. 1973 *Times* 9 Feb. 11/2 We did not do well in the rucks and mauls against Wales.

‖ **Maulana** (maulā·nā). Also **Maulanah, Mawlana, Mulana.** [Arab. *maulānā* our Lord: cf. MOOLVEE and MULLAH.] A title given to a learned Muslim.

1832 H. H. WILSON *Relig. Sects of Hindus* in *Asiatick Researches* XVII. 297 Maulana Rum observes—'What is the world? forgetfulness of God, not clothes, nor wealth, nor wife, nor offspring.' *a* 1846 G. OUSELEY *Biogr. Notices Persian Poets* 112 The author of this sublime poem on Divine Love and the Súfi Philosophy, was Muláná Jeláluddin. *Ibid.* 116 The Muláná was sixty-nine years of age when he closed his earthly career in Koniah. 1855 H. H. WILSON *Gloss. Judicial & Revenue Terms* 335/2 *Mauláná*,..the title of a person of learning or respectability, teacher, doctor: in the Maratha countries, the usual designation of the Mohammadan village schoolmaster. 1888 *Encycl. Brit.* XXIII. 656/1 Ottoman literature may be said to open with a few mystic lines, the work of Sultán Veled, son of Mauláná Jelál-ud-Dín, the author of the great Persian poem the *Mathnawí*. 1895 *Jrnl. Asiatic Soc. Bengal* LXIII. III. 54 When sixteen years of age, he studied under Maulánáh 'Abd-ul-'Azíz of Delhi. 1902 E. A. GAIT *Rep. Census India 1901* VI. iv. 174 Other reformers were spreading the doctrines of the Patna School, the most successful of whom was Mauláná Karámat A'li of Jaunpur. 1936 *Encycl. Islām* III. 417/2 Allāh is often called *Mawlānā* 'our Lord' in Arabic literature. *Ibid.* 418/2 Peace was negotiated..by Mawlānā Ḥamza. 1962 *Economist* 9 Mar. 889/1 The maulanas want..the kind of obscurantist stranglehold on the [Pakistani] nation..the president is determined to resist. 1975 *Bangladesh Times* 18 July 2/1 Maulana Abdur Rab.. breathed his last recently at his village home due to heart attack... The late maulana was renowned for his social services, community developments and preaching the ideals of Islam.

mauler. Add: **2.** *slang.* A hand; a fist. Cf. MAULEY.

1820 'W. T. MONCRIEFF' *Mod. Collegians* 18 The Old Gentleman wishes to know how to put his maulers in action. 1936 'F. GERALD' *Millionaire in Memories* vii. 193 To a young fellow who fancied himself with his 'maulers' the proposition was attractive..plenty of fighting. 1960 S. H. COURTIER *Gently dust Corpse* xii. 170 Wait till I get my maulers on Lew Boston. 1973 J. ROSSITER *Manipulators* viii. 92 You keep your big maulers off this.

maum, mawm, varr. MA'AM or MAM[1]. *U.S.*

1826 A. ROYALL *Sk. Hist., Life & Manners U.S.* 121 Yes mawm, no mawm. 1835 A. B. LONGSTREET *Georgia Scenes* 110 'Aunt' and 'mauma', or 'maum', its abbreviation, are terms of respect commonly used by children, to aged negroes. 1881 *Harper's Mag.* Apr. 728/2 Maum Dulcie, is my habit ready? 1928 J. PETERKIN *Scarlet Sister Mary* ii. 14 Mary had grown up in Maum Hannah's old house in the Quarters. 1950 *Publ. Amer. Dial. Soc.* xiv. 46 Maum' Dinah.

mauma, maumer, maumie, varr. *mamma*, = MAMMY 2.

1835 W. G. SIMMS *Partisan* 141 You have told me nothing of old mauma. 1838 C. GILMAN *Recoll. Southern Matron* xv. 101 Who does not remember his youthful Christmas; the reiterated charge to his *maumer* to awaken him first? 1881 *Harper's Mag.* Apr. 737/2 Maumie, how I should like to see a Hoodoo meeting! 1890 *Ibid.* July 232/1 Respectable colored 'maumas', ample of girth, in spotted white aprons. 1895 *Century Mag.* May 155/2 Only a few, a very few, of the faithful old 'maumers' and loyal house- and body-servants remain. 1950 *Publ. Amer. Dial. Soc.* xiv. 46 *Mauma*,..name by which elderly Negro women were called by the children of the family with which they were connected.

Mau Mau (mau·mau). [Kikuyu.] Name of an African secret society originating among the Kikuyu, having as its aim the expulsion of European settlers and the ending of British rule in Kenya. Also *attrib.*

1950 *E. Afr. Standard* 16 June A/5 We have arrived at a state of dissatisfaction and insecurity, bitterness and delusion which has given rise to the formation of such societies as the 'Dini ya Mswambwa' and the 'Mau Mau Association'. 1952 L. S. B. LEAKEY *Mau Mau & Kikuyu* xi. 96 The Kikuyu regard the Mau Mau Association as nothing more than the old Kikuyu Central Association under another name. 1952 *N.Y. Times* 26 Oct. I. 3/2 Large groups of natives fled today to lofty mountain hideouts to escape the Government drive..against Mau Mau terrorists, who have sworn death to the white man. 1955 *Times* 13 July 6/5 Further evidence of the infiltration of Mau Mau into tribes other than the Kikuyu, Embu, and Meru was revealed here to-day. 1960 F. D. CORFIELD *Hist. Survey Orig. & Growth Mau Mau* 78 On 21st September [1948] the Director of Intelligence and Security reported..that a new movement, the *Mau Mau*..had appeared in Naivasha. 1971 *Sunday Nation* (Nairobi) 11 Apr. 12/5 Events beyond human control—famine, Mau Mau, the changing environment.

Hence **mau-mau** *v. U.S. slang*, to threaten, terrorize; so as *sb.*; **mau-mauer; mau-mauing** *vbl. sb.*

1970 T. WOLFE *Radical Chic & Mau-mauing Flak Catchers* (1971) 97 Going downtown to mau-mau the bureaucrats got to be the routine practice in San Francisco. *Ibid.* 124 In public you used the same term the whites used, namely, 'confrontation'. The term '*mau-mauing*' was a source of amusement in private. The term *mau-mauing* said, 'The white man has a voodoo fear of us, because deep down he still thinks we're savages. Right? So we're going to do that Savage number for him.' *Ibid.* 148 And it was here that Bill Jackson proved himself to be a brilliant man and a true artist, a rare artist of the mau-mau. 1971 *Harper's* June 9 His [*sc.* Norman Mailer's] demonstration of the inadequacies and distortions of Kate Millett's *Sexual Politics* is convincing and indicates that the English Department of Columbia University had been mau-maued by that termagant of Women's Lib. 1974 *Maclean's Mag.* Dec. 108/1 We've had a belly full of Indian rhetoric. After a threat is made for the fifth, twelfth or fortieth time, the excitement wears off; boredom becomes contempt when we find out that each hollow boast brings more government money tinkling into the Indian organizations' pockets. American writer Tom Wolfe calls the technique 'mau-mauing'; it's no accident that the most successful mau-mauer, Harold Cardinal, runs one of the strongest Indian organizations in Canada.

maumet. Add: **2. c.** A baby, child. *dial.*

1932 S. GIBBONS *Cold Comfort Farm* v. 69 Far from those that loves her and cowdled her in their bosoms when she was a mommet. 1949 *Antiquity* XXIII. 42 The poor mommet whose father was 'in trade' is discernible even on its tomb as 'a really vulgar babe'.

maund, *sb.*[2] **1.** (Later examples.)

1909 *Chambers's Jrnl.* Oct. 665/2 The import of dyeing materials into Kashmir in one year was: Indigo, fourteen maunds, or one thousand one hundred and twenty pounds. 1955 *Times* 12 May 12/5 Throughout the Himalaya the unit of weight for barter is the load a man can

carry—a maund, or 80 pounds. 1969 *Commerce* (Bombay) 26 July 188/1 Raw jute arrivals in the last week remained static at the previous level of 30,000 maunds a day. 1972 *Nat. Geographic* Oct. 532/2, I harvested about 33 maunds an acre. At a fraction more than 82 pounds to the maund that is roughly 2,700 pounds. 1975 *Bangladesh Observer* 21 July 7/3 On July 4 Bangladesh Rifles personnel chased an alleged smuggler near the border, but he fled away leaving one maund thirty seers of fertilizer.

maunderingly (mǭ·ndəriŋli), *adv.* [f. MAUNDERING *ppl. a.*[2] + -LY[2].] In a maundering manner; inconsequently.

1909 I. ZANGWILL *Melting-Pot* III. 172 *David.* (..he picks up the violin, and as his fingers draw out the broken string he murmurs) I must get a new string. (He resumes his dragging march toward the door, repeating maunderingly) I must get a new string. 1915 F. M. HUEFFER *Good Soldier* IV. vi. 282 Edward..believed maunderingly that some essential attractiveness in himself must have made the girl continue to go on loving him. 1955 J. SLOTKIN in *Perspective* Autumn 152 He raised his head maunderingly, like a jackal from a lion's kill no longer warm.

maundful. [f. MAUND *sb.*[1]] The amount contained in a maund.

1828 W. CARR *Dial. Craven* (ed. 2) I. 314 *Maund-ful*, a basket full. 1924 *Chambers's Jrnl.* 20 Sept. 673/2 Two maundfuls at a time, the pilchards are tipped out of the baskets.

maundy. Add: **4. maundy money,** now usu. distributed by the reigning monarch at Westminster Abbey.

1967 *Everyman's Encycl.* VIII. 293/1 In 1833..the dole was replaced by a money payment ('Maundy money') distributed by the sovereign in person or through a royal almoner.

maureeyah, var. *MOYA.

Mauritanian (mǫritēi·niăn), *a.* and *sb.* Also **Mauretanian.** [f. *Mauretania* (see below) + -AN.] **A.** *adj.* Of or pertaining to the ancient country of Mauretania in North Africa, or the modern independent republic of Mauritania on the west coast of Africa, formerly a French colony. **B.** *sb.* A native or inhabitant of Mauritania. Also **Maurita·nic** *a.*

1594 MARLOWE & NASHE *Dido* IV. iv. 215 And will my guard with Mauritanian darts, To waite vpon him. 1607 TOPSELL *Four-f. Beasts* 247 That same herbe (called *Doronicum*)..[is] highly esteemed among the Arabians[,] Graecians, and Mauritanians. *Ibid.* 461 These also are the Epethets of lions, wrathfull, maned..stout, great, Masilian, Mauritanian, Parthian, [etc.]. 1652 J. TAYLOR *Rule of Holy Dying* (ed. 2) iii. 95 Masinissa the Mauritanian. *a* 1680 EVELYN *Diary* an. 1645 (1955) II. 467 There were kept in it two Eagles, a Crane, a Mauritanian Sheepe, a stag, and sundry foule. 1776 [see MOOR *sb.*[2] 1]. 1895 [see **Afro-European* adj. and sb.]. 1911 'I. HAY' *Safety Match* xiv. 221 Algiers—that curious combination of Mauretanian antiquity and second-rate French provincialism. 1959 *Times* 7 Jan. 11/7 The Mauritanian trader is a common sight in the markets. 1964 *Times* 29 Sept. 10/7 An attack by a gang of white youths on Mr. Youssouf Gueye, the First Secretary of the Mauritanian delegation. 1967 *Oceanogr. & Marine Biol.* V. 450 Two regions of the Atlantic Ocean, namely the Lusitanian..and the Mauritanian extending from Gibraltar to Cape Blanc, and including the Canary Islands and Madeira. 1968 *Encycl. Brit.* XIV. 1124/2 Mauretanian history really begins with the Punic Wars... Mauretanians were good military material and were used by the Romans in both Africa and other parts of the empire. 1972 *Guardian* 8 Feb. 13/1 A couple of Mauritanians selling silver spoons.

Mauritian (mǫri·ʃăn), *sb.* and *a.* [f. MAURITIUS + -AN, -IAN.] **A.** *sb.* A native or inhabitant of the island of Mauritius in the Indian Ocean. **B.** *adj.* Of or pertaining to Mauritius.

[1835 J. JEREMIE *Recent Events at Mauritius* i. 5 The Mauriciens are to set an example 'au monde entier'.] 1865 *Chambers's Jrnl.* Sept. 546/2 To Mauritians..this droll little Mauritian journal..may very probably appear a most influential exponent of public opinion. 1884 H. W. LITTLE *Madagascar* xiii. 280 Cheap rum..is often used.. for the purchase of bullocks for the Mauritian markets. 1896 B. BURLEIGH *Two Campaigns* 60 Fiery Mauritian rum. *Ibid.* 111 The place was under a Mauritian named Hemming. 1918 L. HUXLEY *Life J. D. Hooker* II. xxviii. 12 The Mauritian Herbarium was to go back, as soon as the assistant had revised the lists. 1926 *Chambers's Jrnl.* Apr. 212/1 The Mauritian Creole is a crafty fellow. 1963 *Guardian* 26 Feb. 4/2 A young Mauritian said he would now shave off his beard. 1966 M. R. D. FOOT *SOE in France* xi. 380 The Mayer brothers,..Mauritians by origin. 1967 R. I. MCDAVID in G. V. Bobrinskoy *Lang. & Areas* 86 A viable language in its own right—like.. Haitian or Mauritian Creole. 1971 *Leader* (Durban) 7 May 1/2 Officials of the sports body..have already had negotiations with Mauritian football authorities. 1972 *Guardian* 8 Mar. 13/3 Ramgoolam would have to decide whether to..find a Mauritian for the top job, or (swallowing Mauritian pride..) find a suitable Englishman. 1973 *Times* 5 Mar. (Mauritius Suppl.) p. i/3 Warmed by the sun and Mauritian friendliness we relaxed. 1974 *Times* 14 Mar. 16/4 The Mauritians invited me to their binge.. celebrating the sixth anniversary of their independence.

Maurya (mau·riă). [Skr., f. the name of Chandragupta *Maurya*, who founded the dynasty.] The name of a dynasty that ruled northern India from 321 to *c* 184 B.C.; a member of this dynasty. Also *attrib.* Hence **Mau·ryan** *a.*

1870 *Jrnl. R. Asiatic Soc.* IV. 96 The celebrated Mauryan king..was not only sovereign of the valley, but of the whole of Northern India. *Ibid.* 122 (*heading*) Maurya dynasty, 130 years. 1886 *Encycl. Brit.* XXI. 272/2 The oldest hitherto known specimens of Indian writing are five rock-inscriptions, containing religious edicts in Pāli..issued by the emperor Aśoka..of the Maurya dynasty. 1910 *Ibid.* XIV. 399/1 After Asoka the Mauryas dwindled away, and the last of them..was treacherously assassinated in 184 B.C. 1935 *Times Lit. Suppl.* 4 July 434/4 The interest that the Mauryan State took in preventing famine and in relieving distress appears to have been more vigorous than in modern days. 1956 A. TOYNBEE *Historian's Approach to Relig.* vii. 90 In the age of the Maurya Emperor Açoka. 1959 *Chambers's Encycl.* VI. 553/2 About 185–175 King Demetrius conquered the Mauryan empire in north India up to Patna on the middle Ganges. 1968 I. W. MABBETT *Short Hist. India* v. 61 The Mauryan dynasty, like the rest of them, eventually became a line of kings jostling with others, a principality in the north-east. 1973 *Times* 14 Apr. (Nepal Suppl.) p. ii/2 The oldest monuments found on Nepalese soil are contemporaneous with the Mauryan empire of North India.

mauryah, var. *MOYA.

‖ **mauvais coucheur** (movę kūʃör). [Fr., lit. 'bad bedfellow'.] A difficult, uncooperative, or unsociable person.

1959 *Times* 19 Nov. 15/2 Brecht at best was a *mauvais coucheur*. 1960 C. P. SNOW *Affair* iv. 45 The moral roughneck, the *mauvais coucheur*, often seemed to her to have a dignity and elevation not granted to the rest of us. 1963 *Times Lit. Suppl.* 26 Apr. 297/3 Goddard was a man of talent but he was obviously a *mauvais coucheur*.

‖ **mauvaise langue** (movęz lāṅg). [Fr., lit. 'bad tongue'.] An evil or venomous tongue; a vituperative gossip; a scandal-monger.

1888 C. M. YONGE *Beechcroft at Rockstone* II. xiii. 14 The foreboding that the *mauvaises langues* would get hold of it. 1924 A. D. SEDGWICK *Little French Girl* IV. i. 304 Marigold is a wretched gossip, and worse. She's a *mauvaise langue*; I would not trust her story. 1936 'R. WEST' *Thinking Reed* i. 18 In another moment that couple of *mauvaises langues* would have something to wag about. 1939 C. BEATON *My Royal Past* ix. 101 A lady..with a very *mauvaise langue*. 1974 *New Society* 11 Apr. 58/2 We are told that all Europe is at a standstill, until it knows who has been elected president (and those who are *mauvaise langue* will add that if it is Chaban-Delmas, there will be a further delay whilst he acquaints himself with foreign affairs).

‖ **mauvais pas** (movę pā). *Mountaineering.* [Fr., lit. 'bad step'.] A place that is dangerous or difficult to negotiate.

1816 SHELLEY *Let.* 25 July (1964) I. 500 The one [mule] which I rode fell, in what the guides call a *mauvais pas*. 1843 J. D. FORBES *Trav. through Alps of Savoy* iv. 67 The spot has acquired the name of the *Mauvais Pas*, which it bears more frequently than its proper one of La Roche de Moré. 1871 L. STEPHEN *Playground of Europe* v. 315 As a mere gymnast upon the mountains..in overcoming *mauvais pas* of all descriptions—the guide is incontestably superior. 1907 G. D. ABRAHAM *Compl. Mountaineer* iv. 56 The leader can retie himself on the end and scramble safely up the *mauvais pas*. 1940 F. S. CHAPMAN *Helvellyn to Himalaya* vii. 166 We decided to return once more to the couloir..thus short-circuiting the *mauvais pas* we had seen in our reconnaissance. 1971 N. TENNENT *Islands of Scotl.* i. 30 The northern section of the ridge..has some fine situations, including a tricky descent on the Iorsa side to the 'impassable' 'mauvais pas'.

‖ **mauvais quart d'heure** (movę kār dör). [Fr., lit. 'bad quarter of an hour'.] A short period of time which is embarrassing and unnerving; a brief but unpleasant experience.

1864 M. B. CHESNUT *Diary* 8 Jan. (1949) 353 General Preston rushed in to the breach... But Mr. Willie Mountford had his *mauvais quart d'heure*. 1871 M. ARNOLD *Friendship's Garland* 95 You must needs have..a *mauvais quart d'heure*. 1883 LD. SALTOWN *Scraps* II. iii. 77 My modesty was severely tried, and I do not remember to have often spent a more *mauvais quart d'heure*, which was actually about the length of time that my martyrdom endured. 1897 J. MCCABE *Twelve Yrs. in Monastery* iv. 64 We gave him many a *mauvais quart d'heure* by running to the door when we saw his shadow near it, and chasing him through the convent. 1905 E. M. FORSTER *Where Angels fear to Tread* ii. 47 He takes it so well. But you must have had a *mauvais quart d'heure*. 1920 *Punch* 18 Feb. 138/3 Last week's programme had its *beaux moments*, but it had also at least two *mauvais quarts d'heure*. 1924 P. C. WREN *Beau Geste* I. i. 25 At the end of ten minutes, a very *mauvais quart d'heure*, I beckoned the Sergeant-Major. 1960 *Guardian* 21 Sept. 8/6 The long-term effects of a transitory *mauvais quart d'heure*. 1965 *Economist* 5 June 1125 John Kennedy had his *mauvais quart d'heure* between April and June, 1961.

mauvais sujet. (Earlier and later examples.)

1793 T. FREMANTLE in *Wynne Diaries* (1952) xix. 251 Find the surgeon to be a mauvais subjet [*sic*], talk much to him. He crys, and promises to amend. 1813 A. ROMILLY *Let.* 12 Aug. in S. H. Romilly *Romilly–Edgeworth Lett.* (1936) 56 The youngest son, who was a very mauvais sujet, has just been killed in a duel. 1825 H. WILSON *Mem.* I. 179, I will lay my life, you two desperate mauvais sujets came here together! 1897 E. A. BARTLETT *Battlefields Thessaly* viii. 171 These *mauvais sujets* had maltreated their own countrymen and women. 1945 R. HARGREAVES *Enemy at Gate* 56 Rome had failed to save one of its children, and even if that *mauvais sujet* had been guilty of recusancy, in the offspring's downfall the parent was no less undone. 1953 *Essays in Crit.* III. 217 The things in my criticism that he shakes his head over..endorsing the charges of..'bad form' and '*mauvais sujet*', *are* the effort to win that recognition. 1975 A. CHRISTIE *Curtain* v. 46 The *mauvais sujet*—always women are attracted to him.

mauve, *a.* Add: **b.** *Comb.* Also with other colours, as *mauve-pink, -red.*

1899 B. W. WARHURST *Colour Dict.* 44 Lilac-rose is a dull rose..which..as a tint will appear as..Mauve-pink. 1917 D. H. LAWRENCE *Look! We have come Through!* 57 Their mauve-red petals on the cloth. 1952 A. G. L. HELLYER *Sanders' Encycl. Gardening* (ed. 22) 92 *Catalpa Fargesii*..with its fine var. *Duclouxii* with large mauve-pink flowers.

maven (mēi·vən). *U.S. colloq.* Also **mavin, mayvin.** Pl. **-im.** [ad. Heb. *mevin* understanding.] An expert or connoisseur.

1965 *Hadassah News Letter* Apr. 30 (Advt.), Get Vita at your favorite supermarket, grocery or delicatessen. Tell them the beloved *Maven* sent you. It won't save you any money: but you'll get the best herring—Vita. 1968 L. ROSTEN *Joys of Yiddish* 223 *Mavin* was recently given considerable publicity in a series of newspaper advertisements for herring tidbits. 'The Herring *Mavin* Strikes Again!' proclaimed the caption. The picture showed an empty jar. A real advertising *mavin* must have thought that up. 1969 *Time* 12 Sept. 78 Much of the credit for the Cinderella publishing story goes to Robert Gottlieb, then the editorial genie in residence at Simon & Schuster, now the mavin at Alfred Knopf. 1970 L. M. FEINSILVER *Taste of Yiddish* iii. 323 Canada Dry has been touting its product as 'Maven's Choice' in American Jewish weeklies, where Switzerland Emmentaler cheese announces itself with: 'Calling all Mayvinim!' 1972 *Publishers Weekly* 10 July 22/1 If Shawn is not exactly a boxing *maven*, he knows even less about baseball. 1973 *Milwaukee Jrnl.* 4 Mar. V. 4/1 Miss Decter (no Ms., please) is one of the culture 'mavins' of New York, to use a newly chic term borrowed from the Hebrew. 1973 *N.Y. Times* 8 July X. 31/4 Mama, who had managed to support herself by becoming a local real estate *maven*, negotiated the purchase.

maverick, *sb.* [Etym. note: read 'Samuel A. *Maverick* (1803–1870), a Texas cattle-owner who left the calves of his herd unbranded'.]

1. (Earlier and later examples.)

1867 in J. G. MCCOY *Hist. Sk. Cattle Trade* (1940) 83 The term maverick which was formerly applied to unbranded yearlings is now applied to every calf which can be separated from the mother cow. 1942 S. KENNEDY *Palmetto Country* 223 Most likely candidates for rustling are unbranded calves and cattle, called 'mavericks' in the West. 1974 G. JENKINS *Bridge of Magpies* xv. 229 Cowpunchers riding herd and trying to rope the most bloody-minded maverick that ever cut loose on the plains of Texas.

2. a. Add to def.: An independent person; an individualist; also applied *spec.* in the U.S. to a politician who will not affiliate with a regular political party. Also *attrib.* in sense 'independent, unattached'.

1886 *Calif. Maverick* (San Francisco) 13 Feb. 4/1 People would say, 'He holds maverick views', meaning that his views were untainted by partisanship in the matter. 1901 *McClure's Mag.* Dec. 147 Occasionally they found a maverick legislator, or traded for one. 1948 *Manch. Guardian Weekly* 8 Jan. 4/1 A few maverick liberals. 1948 *Chicago Daily News* 11 June 16/7 One Republican Senator, and not by any means a conspicuous maverick, pointed out that the Senate might have acted. 1954 *Manch. Guardian Weekly* 30 Dec. 8/3 His fellow 'maverick' ..also appeared to have gone into not uncomfortable shadows. 1957 *Times Lit. Suppl.* 8 Mar. 1/3 A contrast of Maverick with Movement verse will illustrate this... To Mavericks, Movement verse expresses 'antagonism towards sensibility and sentiment'. 1957 *Oxf. Mag.* 17 Oct. 22/2 The story is not just of local boy making good, but also, and more significantly, of maverick making friends. 1963 *Listener* 17 Jan. 115/1 Neither the novelists of the central tradition of English writing, nor the great Mavericks of the nineteenth century, whose strong apprehension of evil forced them to break through the domestic web in which they wished to remain, had any clear theological pattern in which to embody their sense of evil. 1968 P. OLIVER *Screening Blues* 18 Maverick lines that move from blues to blues are given new rhymes and new meanings by their juxtaposition with other ideas, while they retain the quality of surprise. 1973 *Radio Times* 4–10 Aug. 8 Ruth Inglis met the 77-year-old maverick zoologist at his Oxford base.

maverick, *v.* (Examples.)

1883 in *Amer. Speech* (1958) May 141 The Indians stole them..and the Texans 'mavericked' the unbranded. 1948 J. K. ROLLINSON *Wyoming Cattle Trails* 139 The artful practice of burning or working over brands was resorted to, with honest mavericking as a side line.

b. *intr.* To stray or wander like a maverick.

1910 W. M. RAINE *Bucky O'Connor* 203 It hadn't penetrated my think-tank that this was your hacienda when I came mavericking in.

mavrodaphne (mævrodæ·fni). [mod.Gr., f. late Gr. μαῦρος dark (Gr. ἀμαυρός) + δάφνη laurel.] A dark-red sweet Greek wine made from the grape of the same name.

1911 in WEBSTER. 1935 A. L. SIMON *Wines & Liqueurs from A to Z* 36 *Mavrodaphne*, a sweet dessert wine from Patras (Greece). 1945 E. WAUGH *Brideshead Revisited* I. ii. 52 Neither the mixture of wines, nor the Chartreuse, nor the Mavrodaphne Trifle..explains the distress of that

hag-ridden night. 1958 R. LIDDELL *Morea* II. ii. 52 The sweet wine, *Mavrodaphne*..reminds me too often of the Looking-glass precept: 'Come fill up your glasses with treacle and ink.' 1960 M. STEWART *My Brother Michael* vi. 82, I sipped the wine. It was dark as mavrodaphne. 1972 *Times* 10 June 9/1 Mavrodaphne, the only Greek wine apart from retsina to be well-known outside its homeland, comes from the Peloponnese.

mavrone (mǎvrō^u·n), *int.* Anglicized form of Irish *mo bhrón* my grief (f. *brón*), used as an exclamation of sorrow.

1892 W. B. YEATS *Countess Kathleen* iii. 64 The treasure-room is broken in—mavrone, mavrone The door stands open, and the gold is gone. 1922 JOYCE *Ulysses* 197 He wailed:—And we to be there, mavrone, and you to be unbeknownst. 1939 —— *Finnegans Wake* II. 232 Stop up, mavrone, and sit in my lap.

maw, *sb.*[1] **3.** (Earlier and later *fig.* examples.)

1818 KEATS *Endymion* II. 272 Weary, he sat down before the maw Of a wide outlet. 1931 F. HURST *Back St.* xxxi. 277 The small vestibule where stood a row of these iron men of chance. Occasionally one of the guests, feeding coins into these metal maws, staked her. 1955 *Times* 5 Aug. 9/7 The grim realities of economics are rapidly driving the stubbornly 'independent' man into the insatiable maw of large combines.

mawby, var. *MAUBY.

Mawlana, var. *MAULANA.

mawley, var. MAULEY.

maw-wormy, *a.* [f. MAWWORM[2].] = *maw-wormish* adj.

1885 in J. R. Ware *Passing Eng.* (1909) 174/2 Without being mawwormy, I fail to see why a wreath should be presented to any man who makes a business of giving opera.

maxi- (mæ·ksi), combining form of MAXIMUM 5 denoting things, esp. articles of clothing, which are very long or large of their kind. Also as *sb.* Cf. *MIDI-, *MINI-.

1961 *Spectator* 30 June 965 If you get into a maxi-taxi after midnight with a companion, there is 3/3d on the clock before you start. 1966 *Times Educ. Suppl.* 24 June 2015/2 There will be Lady X in Rutland realizing with a gasp of horror that she is wearing the same maxi-skirt as Lady Y. 1966 *Daily Tel.* 24 Oct. 11/1 The maxi-bag must be the next thing. And by the natural sequence of bag fashion psychology, the manufacturers are now designing bags on briefcase lines. 1967 *Punch* 3 May 624/1 The clampdown in Greece..may shut up those tireless critics, in our own land of the free, of mini-skirt girls and maxi-hair boys, now banned by the Greek Minister of the Interior. 1967 *Evening Standard* 7 Dec. 9/1 You're in a black crêpe maxi-skirt... If everyone is wearing maxi's, you're all right. 1968 *Daily Tel.* 18 Jan. 15/3, I doubt if we'll see the Queen in a real maxi yet awhile, but the belted emerald maxi coat sketched was certainly the Royal dressmaker's newest looking garment in the collection he showed yesterday. 1968 *Guardian* 2 May 7/4 Here are expensive clothes from 30 to 70 guineas. They are not for youngsters who wear maxis for kicks. 1970 R. LOWELL *Notebk.* 167 The girl's maxi-coat, Tsar officer's, dragged the snow. 1970 *Time* 16 Nov. 67 To relieve the sterile monotony of nurses' uniforms, Fashion Designer Pierre Cardin recently unveiled three new creations at a London showing. Two of his designs—nunlike wimples with white maxidresses—were harmless affairs that might make ward nurses look functional if not fashionable. 1971 'V. X. SCOTT' *Surrogate Wife* 230 He was wearing a great big olive green maxicoat. 1971 *Ink* 12 June 1/2 'Have a whiff' was released by Pye on a maxi-single (two tracks on each side). 1973 S. B. JACKMAN *Guns covered with Flowers* vi. 98 Marija slipped off her tweed maxi-coat. 1975 *Daily Tel.* 4 Jan. 16 This figure is quite impressive until it is compared with the 1974–75 budget of £1,325 million, which means that it is only a 1½ per cent. mini-reduction in what promises to be a maxi-budget.

maxim, *sb.*[1] Add: **3.** *spec.* Used of precepts of morality or prudence occurring in Old English verse.

1883 H. KENNEDY tr. *B. ten Brink's Hist. Eng. Lit.* I. 64 The poet is fond of beginning a new maxim or a chain of them with the second half of a verse. 1892 S. A. BROOKE *Hist. Early Eng. Lit.* II. 277 The *Gnomic Verses*..consist of folk-proverbs, maxims, short descriptions of human life, [etc.]. 1902 *Jrnl. Eng. & Germ. Philol.* IV. 477 The *Wife's Complaint*..closes with a general maxim deduced from the sad experiences of the once happy couple. 1966 S. B. GREENFIELD in E. G. Stanley *Continuations & Beginnings* 143 The diction of secular gnomes or maxims.

maximal, *a.* (Later examples.)

1955 *Bull. Atomic Sci.* Apr. 107/2 The tradition of an insistence on maximal loyalty—the tradition of hyper-patriotism—is an old one in the United States, and it becomes very demanding in periods of crisis. 1958 E. WINTER in Middleton & Tait *Tribes without Rulers* 140 What is termed the maximal lineage consists of all the descendants in the male line of a single eponymous ancestor. 1960 P. SUPPES *Axiomatic Set Theory* viii. 244 *A* is a maximal chain in *B* if and only if *A* is a chain in *B* and there is no chain *C* in *B* such that $A \subset C$. 1963 W. V. QUINE *Set Theory* x. 231 There are notable equivalents of the axiom of choice... One is Hausdorff's law that every partial ordering harbors a maximal ordering. 1964 E. MENDELSON *Introd. Math. Logic* 10 A maximal ideal is a proper ideal which is included in no other proper ideal. 1971 D. CRYSTAL *Ling.* 199 The investigation of grammatical models in terms of maximal units. *Ibid.* 201 The sentence is the maximal unit of grammatical analysis. 1972 *Language* XLVIII. 264 The present delayed publication is intended especially to permit me to acknowledge that before today the only adequate publication of the principle of maximal redundancy has been in the Presidential Address of December 1969.

maximally *adv.* (later examples).

1969 *Daily Tel.* 11 Jan. 5/3 The answer, as with the short-term gain.., is to tax yourself maximally. In this case you would assume that £700 would be left to you after 30 p.c. had been deducted. 1970 *Nature* 30 May 804/1 The enzyme was assayed in maximally activated conditions. 1971 *Jrnl. Gen. Psychol.* LXXXIV. 88 It has been found that the same manners of input timing that maximally increase brightness also act to affect hue and desaturation maximally. 1971 D. CRYSTAL *Ling.* 175 Phonology deals with sounds and contrasts between sounds only within the context of some language (maximally, in any language). 1973 *Nature* 2 Mar. 68/1 Histamine is synthesized maximally in the hypothalamus.

maximalism (mæ·ksimăliz'm). [f. MAXIMAL *a.* + -ISM or ad. Russ. *maksimalizm*.] The policy or theory of a 'maximum' programme of some kind. Cf. *MAXIMALIST.

1920 *Glasgow Herald* 11 May 10 Bologna is undoubtedly to-day one of the chief Italian strongholds of Maximalism. 1951 N. GORODETZKY *St. Tikhon Zadonsky* ix. 189 The tendency of many Russians to seek extremes, maximalism, paroxysms, even in spiritual matters. 1967 J. M. CAMMETT *Antonio Gramsci* iv. 88 Thereafter, the break between *Ordine Nuovo* and maximalism widened, eventually contributing heavily to the Socialist schism at Livorno. 1967 C. SETON-WATSON *Italy from Liberalism to Fascism* xii. 524 Maximalism..provided only revolutionary talk as a substitute for revolution.

Maximalist (mæ·ksimălist). Also **maximalist.** [f. MAXIMAL *a.* + -IST or ad. Russ. *maksimalist*, f. L. *maximum*, or ad. F. *maximaliste*.] A member of the more extremist 'fraction' of the Russian Socialist-Revolutionary Party which split off from the main body of the party in 1904 and which used and advocated terrorist methods. Later regarded as a translation of Russ. *bol'shevik* and used as an alternative name for a Bolshevik. Also, a member of any similar group outside Russia. Also *attrib.* or as *adj.*, of or pertaining to a policy or theory of maximum demands (of some kind specified in the context).

1907 I. ZANGWILL *Ghetto Comedies* 408 'Ah, you're a Maximalist,' said the beadle. 'No, I am only a Minimalist.' 1909 *Westm. Gaz.* 23 Dec. 7/4 He is said to have joined the 'Maximalists' in 1907. 1921 tr. *Trotsky's Defence of Terrorism* ix. 173 Plekhanovists, Maximalists, Anarchists... Absolutely all the 'shades of Socialism'. 1921 *Glasgow Herald* 14 Oct. 9 Forty out of 100 Deputies are Reformists..headed by some of the best brains in the country, like Signore Turati..and Modigliani. To them are opposed an equal number of Maximalists. 1933 M. EASTMAN tr. *Trotsky's Hist. Russ. Revolution* II. xiii. 309 While on the right the 'democracy' was competing with the Bolsheviks, on the left too there were the anarchists, the Maximalists, the Left Social Revolutionaries, trying to crowd them out. 1933 *Mind* XLII. 181 The rule..of always acting so as to produce the greatest amount of good, which latter rule the 'Maximalists' say we ought always to follow. *Ibid.*, The 'Maximalist' Theory (this designation suggests quantity better than 'Optimific') concentrates generally upon the consideration, not of individual situations, but of what happens when a general practice, custom or institution is dropped and replaced by another. 1954 *Ann. Reg. 1953* 153 There were the 'maximalists'..who wanted a real surrender of sovereignty, and the 'minimalists'. 1955 D. W. TREADGOLD *Lenin & his Rivals* IV. xi. 212 They were bitterly denounced by orthodox SR's, who took fright..at 'their own reflection in the Maximalist mirror'. 1962 R. R. ABRAMOVITCH *Soviet Revolution* i. 30 This concept shocked the Russian Marxists..they considered it a betrayal of scientific Socialism, a reversion to the old, utopian, Bakuninist, maximalist-anarchist ideas. 1966 *Economist* 5 Nov. 568/1 We have moved from a position of protest to one of responsibility and have put behind us the old maximalist dreams and illusions. 1967 C. SETON-WATSON *Italy from Liberalism to Fascism* xii. 511 The [Italian Socialist] party elected a 'maximalist' executive at its Rome congress in September 1918 and proclaimed the dictatorship of the proletariat to be its goal. 1969 D. M. SMITH *Italy* (ed. 2) VII. xxvii. 216 The maximalists were made strong and uncompromising by the belief that history was on their side.

maximality (mæksimæ·liti). [f. MAXIMAL *a.* + -ITY.] The property of being maximal.

1964 A. P. & W. ROBERTSON *Topological Vector Spaces* i. 4 This would contradict the maximality of *μ*. 1971 D. GORENSTEIN in Powell & Higman *Finite Simple Groups* i. 102 But $[Q_1, Z] = Q_1$, so $Q_1 = Q$ by the maximality of *Q*.

maximed, *a.* [f. MAXIM *sb.*[1]] Expressed as or in a maxim.

1883 J. C. VAN DYKE *Bks. & how to use Them* i. 19 There is another maximed truth in this connection..: 'Knowledge is a two-edged sword.'

maximin (mæ·ksimin), *sb.* and *a.* [f. MAXI(MUM + MIN(IMUM *sb.* and *a.*, after *minimax*.] The largest of a set of minima; usu. *attrib.* (passing into *adj.*), *spec.* designating a strategy that maximizes the smallest gain that a participant in a game or other situation of conflict can guarantee himself. Cf. *MINIMAX *sb.* and *a.*

1954 BLACKWELL & GIRSHICK *Theory of Games & Statistical Decisions* i. 27 ξ_0 is called a maximin strategy for player I. 1957 LUCE & RAIFFA *Games & Decisions* iv. 81 A strategy which dictates choices *c* and *g* on moves 2 and 4, respectively, cannot be maximin for this game. 1958 *Engineering* 21 Mar. 369/2 Player A will thus determine the smallest value in each row and choose the largest of these minima, the 'maximin'. 1968 G. OWEN *Game Theory* vii. 140 Though it is of course impossible to determine how a person will act.., we can nevertheless set a minimum to the amount that a player will accept for himself. This is the amount that he can obtain by unilateral action, whatever the other player does. This is, of course, the maximin value of the game for that player. 1970 BEVERIDGE & SCHECHTER *Optimization* xii. 656 This result represents the mini-max theorem of von Neumann and Morgenstein [*sic*] (1953), which states that *A*'s expected gain is at least *P*, and *B*'s expected loss is at worst *L*, where $P = L$... The value of *P* (and *L*) lies between the row maxi-min and the column mini-max. 1972 J. RAWLS *Theory of Justice* § 26. 152 The maximin rule tells us to rank alternatives by their worst possible outcomes.

maximize, *v.* Add: **1. a.** (Further examples.) Also *spec.* in *Economics*.

1943 E. R. WALKER *From Econ. Theory to Policy* v. 93 He [*sc.* the businessman]..is frequently engaged in exploratory action, designed to discover methods of maximizing profit. 1968 *Listener* 28 Mar. 403/1 The evidence is not inconsistent with the hypothesis that peasants act as if they maximised profit. 1970 *Jrnl. Gen. Psychol.* LXXXII. 15 The adversarial system of justice, which demands that each litigant do whatever is necessary to maximize the probability of a favorable result. 1970 T. LUPTON *Managem. & Social Sci.* (ed. 2) i. 11 So arranging spells of work and rest that output would be maximized. 1971 *Daily Tel.* 12 June 20/7, I wish to maximise my after tax income. Do you consider an investment in Guaranteed Income Bonds is a safe way of doing this? 1971 J. Z. YOUNG *Introd. Study Man* xxv. 361 Difficulties of adjustment will be maximized by ignorance and uncertainty. 1972 *Jrnl. Social Psychol.* LXXXVI. 57 These groups were selected so as to maximize the variety of occupational areas included.

Hence **ma·ximized, ma·ximizing** *ppl. adjs.*

1920 A. S. PRINGLE-PATTISON *Idea of God* (ed. 2) 356 Eternal, not in the sense of a maximized consciousness of time, but as an apprehension different in type. 1927 N. P. WILLIAMS *Ideas of Fall & Orig. Sin* 395 The maximising and minimising versions of the Fall-Doctrine. 1949 *Mind* LVIII. 195 The concept of a maximising individual, the famous 'Economic Man', has arisen. This maximising assumption is usually held to be the fundamental assumption of Economics. 1968 *Listener* 28 Mar. 403/1 Time enough, if the peasant is already a maximising Economic Man in the Henry Ford mould.

3. *intr.* To attain a maximum value.

1972 *Physics Bull.* Oct. 587/3 Photons of energy *hν* greater than satisfies equation (1) suffer absorption while at some energy $h\nu_{max}$ the gain maximizes. 1973 [see *MINIMIZE *v.* 3]. 1974 *Globe & Mail* (Toronto) 21 Oct. 4/3 If emissions were curtailed now, the resultant ozone destruction would maximize around 1990 and would remain significant for several decades.

maximum. Add: **4.** (Later example.)

1941 R. R. PALMER *Twelve who Ruled* x. 239 The Maximum of September 29 caused trouble from the start.

5. b. Also in *Comb.*, as *maximum-security* (used *attrib.*).

1966 *Punch* 28 Dec. 962/2 The family wing of every prison enjoyed maximum security arrangements. 1969 E. AMBLER *Intercom Conspiracy* (1970) ii. 27 The special maximum-security passes that would be needed to gain admittance. 1972 *Guardian* 29 Jan. 9/5 Albany, the new maximum security prison on the Isle of Wight.

maxixe (maʃī·ʃə, mæksī·ks). [Pg.] A round dance of Brazilian origin resembling the two-step.

1914 *Maclean's Mag.* Nov. 82/2 The Maxixe comes to us from Brazil. 1922 C. BELL *Since Cézanne* xx. 227 Sportsmen at the bar who like a fox-trot or a maxixe. 1925 *Chambers's Jrnl.* July 427/1 They sang..and danced the *maxixe* until cockcrow! 1954 *Ballet Ann.* VIII. 102/1 Ready to seek consolation..to the rhythm of the *maxixe*. 1969 F. RUST *Dance in Society* x. 82 In the early months of 1914, the general craze for the tango in England helped to popularize another South American dance—the Brazilian maxixe.

Maxwell[1] (mæ·kswel). The name of Charles *Maxwell*, 19th-century English soldier and explorer, used in the possessive to designate **Maxwell's duiker,** a small brown West African antelope, *Cephalophus maxwelli*, brought back from Sierra Leone by him.

[1827 E. GRIFFITH et al. tr. *Cuvier's Animal Kingdom* IV. 267 Maxwell's Antelope. (*A. Maxwellii.*) A specimen somewhat inferior in size was brought home from Sierra Leone by Colonel Charles Maxwell.] 1905 SCLATER & THOMAS *Bk. Antelopes* I. 182 Maxwell's Duiker appears to extend from Senegal and Gambia all along the west coast of Africa to the mouths of the Niger. 1960 *Times* 29 Sept. (Nigeria Suppl.) p. xxi/4 The little Maxwell's duiker..is everywhere abundant.

Maxwell[2] (mæ·kswel). *Physics.* [The name of James Clerk *Maxwell* (1831–79), Scottish physicist.] **1.** Used in the possessive and *attrib.* to designate various concepts originated by him, as **Maxwell('s) demon,** a being imagined by Maxwell as allowing only fast-moving molecules to pass through a hole in one direction and only slow-moving ones in the other direction, so that if the hole is in a partition dividing a gas-filled vessel into two parts one side becomes warmer and the other cooler, in contradiction to the second law of thermodynamics; **Maxwell('s) distribution**, the distribution of molecular velocities predicted by Maxwell's law, the number with a velocity between v and $v+dv$ being proportional to $\exp(-\frac{1}{2}mv^2/kT)v^2dv$ (where m is the mass of a molecule, k is Boltzmann's constant, and T is the absolute temperature); **Maxwell('s) equation**, each of a set of four linear partial differential equations (first proposed by Maxwell in 1864) which summarize the classical properties of the electromagnetic field and relate space and time derivatives of the electric and magnetic field vectors, the electric displacement vector, and the magnetic induction vector, and also involve the electric current and charge densities; usu. *pl.*; **Maxwell('s) law**, a law in classical physics giving the probabilities of different velocities for the molecules of a gas in equilibrium.

1879 W. Thomson *Pop. Lect. & Addresses* (1889) I. 137 Clerk Maxwell's 'demon' is a creature of imagination.., invented to help us to understand the 'Dissipation of Energy' in nature. **1885** *Science* 31 July 83/1 (*heading*) Maxwell's demons. **1956** E. H. Hutten *Lang. Mod. Physics* iv. 152 It would require a Maxwell demon..to select the rapidly moving molecules according to their velocity and concentrate them in one corner of the vessel. **1971** *Sci. Amer.* Sept. 182/2 Maxwell's demon became an intellectual thorn in the side of thermodynamicists for almost a century. The challenge to the second law of thermodynamics was this: Is the principle of the increase of entropy in all spontaneous processes invalid where intelligence intervenes? **1899** R. E. Baynes tr. *Meyer's Kinetic Theory of Gases* 370 If the number of particles is limited..Maxwell's distribution cannot exist at every moment, but will occur with exactness only when the changing states which succeed each other in the course of a sufficiently long period are all taken into account together. **1955** Friedman & Weisskopf in W. Pauli *Niels Bohr* 138 The spectrum of neutrons and protons emitted from nuclei bombarded with neutrons of 14 Mev or with protons of similar energy fits approximately the predicted Maxwell distribution of an evaporating compound nucleus. **1907** *Sci. Abstr.* A. X. 1295 The principle of relativity in conjunction with Maxwell's equation leads to the conclusion that the inertia of a body changes in a quite determinate manner with its energy-content. **1962** Corson & Lorrain *Introd. Electromagn. Fields* iii. 101 If the symmetry of the [electrostatic] field is simple and if the charge density ρ is zero, as it often is, we can usually integrate the Maxwell equation $\nabla \cdot \mathbf{D} = \rho$ to find the displacement vector **D**. **1964** E. A. Power *Introd. Quantum Electrodynamics* i. 4 Maxwell's equations are not invariant under Galilean invariance and thus are not valid in all inertial frames. Historically this was a most important result leading to special relativity. **1899** R. E. Baynes tr. *Meyer's Kinetic Theory of Gases* iii. 48 That this extension of Maxwell's law to compound molecules is admissible was first recognised by Boltzmann. **1943** Margenau & Murphy *Math. Physics & Chem.* xii. 432 The Maxwell law for the distribution of velocities in an ideal gas.

2. (Usu. written **maxwell.**) The unit of magnetic flux in the C.G.S. system, equal to the flux through an area of one square centimetre normal to a uniform induction of one gauss.

In the International System of Units the unit of magnetic flux is the weber (= 10^8 maxwells).

1900 *Nature* 30 Aug. 414/1 The Commission proposes to assign to the unit of magnetic flux, of which the magnitude will be subsequently defined, the name of Maxwell. **1924** A. Still *Elem. Electr. Design* iv. 68 It is desired to estimate the total flux in maxwells carried by a closed circular iron ring. **1959** R. L. Shrader *Electronic Communication* iii. 72 The gauss is the flux density in maxwells per square centimeter.

Maxwell-Boltzmann (mæ·kswel,bǫ·ltsmăn). *Physics.* The names of J. C. *Maxwell* (see prec.) and L. *Boltzmann* (see *Boltzmann), used (in some cases as an alternative to *Maxwell* or *Boltzmann* alone) to designate concepts arising out of their work on the kinetic theory of gases.

1901 O. Lodge in *Phil. Mag.* II. 241 Any rotational energy possessed by a dumbbell about its longitudinal axis could have no influence on smooth collisions, and accordingly could not be transferred or altered in amount; therefore such rotation ought not to be included in the partition of energy within the meaning of the Maxwell-Boltzmann law when properly stated. **1927** R. C. Tolman *Statistical Mech.* iv. 53 In a system containing a large number of molecules appreciable deviations from the Maxwell-Boltzmann distribution will have a very small probability of occurrence. **1951** D. Bohm *Quantum Theory* i. 16 For a perfect gas..we obtain the familiar Maxwell-Boltzmann distribution of velocities. **1968** M. S. Livingston *Particle Physics* iii. 58 In classical mechanics the distribution of thermal velocities and energies among the molecules of a gas, in thermal equilibrium due to the exchange of energy in random impacts, is described by Maxwell-Boltzmann statistics.

Maxwellian (mækswe·liăn), *a.* *Physics.* Also **maxwellian.** [f. *Maxwell[2] + -ian.] Of, pertaining to, or originated by J. C. Maxwell; in accordance with Maxwell's theory.

1886 *Electrician* 26 Mar. 386/2 The Maxwellian stress. **1914** L. Silberstein *Theory of Relativity* ii. 48 In using the Maxwellian stress..in his theory, Lorentz considers it ..as a system of 'merely fictitious tensions'. **1939** *Brit. Jrnl. Psychol.* XXIX. 253 The light from an electric bulb, rendered nearly parallel by a condenser, filled a large lens at whose principal focus the eye was placed. This lens appeared uniformly illuminated by the 'Maxwellian view'. **1958** H. J. Gray *Dict. Physics* 318/1 The Maxwellian view refers to the method of making a lens apparently flooded with a uniform brightness: a real image of a source of light is formed by a lens in the pupil of the eye. An extended area of bright white light or coloured light is produced and has wide application in photometry and colorimetry. **1968** R. A. Lyttleton *Mysteries Solar Syst.* i. 32 For a whole group of individual stars, these relative speeds are distributed rather like the maxwellian distribution of velocities for the particles of a gas.

May, *sb.*[3] Add: **5. a.** (sense 1) *May-glad* adj., *-hope*, *-mess*; *May-time* (earlier, later, and *attrib.* examples).

1911 E. Pound *Canzoni* 4 No poppy in the May-glad mead. *a* **1889** G. M. Hopkins *Poems* (1967) 38 May-hope of our darkened ways! **1877** —— *Poems* (1967) 67 Look, look: a May-mess, like on orchard boughs! *c* **1350** *Will. Palerne* (1881) 823 Alle freliche foules þat on þat frip songe, for merþe of þat may time þei made moche noyce. **1633** P. Fletcher *Purple Island* xii. lxxxii. 179 More fruitfull then the May-time Geminies. **1930** T. S. Eliot *Ash-Wednesday* 14 The broadbacked figure..Enchanted the maytime with an antique flute. **1963** A. Clarke *Coll. Plays* 161 Who will wash them by the river's edge, Hang them unseen upon as white a hedge In Maytime?

c. May-blob, the marsh marigold, *Caltha palustris*; also applied to other plants (see *Eng. Dial. Dict.*); **may-haw** (examples); **may-pop** (earlier example).

1863 May-blob [see *granny's nightcap]. **1881** S. Evans *Evans's Leicestershire Words* (new ed.) 192 *May-blob*, the marsh-marigold. **1908** *Pacific Monthly* XX. 94/2 Could they or their children after them pick out a May-blob from a May-pop? **1916** D. H. Lawrence *Amores* 38, I can smell the gorgeous bog-end, in its breathless Dazzle of may-blobs. **1960** *Oxf. Bk. Wild Flowers* 4/2 Marsh Marigold (*Caltha palustris*)... The plant has many other names such as Kingcup and May Blobs. **1868** *Amer. Naturalist* II. 468 They [*sc.* deer] visit the ponds in which the May-haw grows, the fruit of which is juicy with the flavor of the apple. **1938** M. K. Rawlings *Yearling* xi. 112 He concentrated on light bread and mayhaw jelly. **1851** *De Bow's Rev.* XI. 49 May Pop, Passion Flower, is also abundant here.

may (mēi), *sb.*[4] [f. May *v.*[1]] An utterance of the word 'may'; a possibility.

1897 G. Saintsbury *Flourishing of Romance* ii. 30 These 'mays' are not evidence. **1935** G. K. Zipf *Psycho-Biol. of Lang.* (1936) 303 From the great number of.. *mays* of today are taken those which are to belong to the matrix of *musts* of tomorrow.

may, *v.*[1] Add: **B. 9. d.** In † *be as be may*, *be that* (or *it* or *this*) *as it may*, *that is as may be*, and similar expressions: whether that is so or not, that may well be so: phrases used to indicate that a statement or act, etc., is perhaps true or right from one point of view but not from another, or that there are other factors to be taken into consideration.

c **1386** Chaucer *Man of Law's Tale* (1894) 1012 Be as be may, ther was he at the leste. **1470–85** Malory *Works* (1967) I. 73 Be as hit be may. **1530** [see May *v.*[1] 8 d (*b*)]. **1593** Shaks. *3 Hen. VI* 1. i. 194 But be it as it may: I here entayle The Crowne to thee. **1796** [see May *v.*[1] 9 a]. **1820** R. Southey *Let.* 17 Nov. in *N. & Q.* (1975) Sept. 400/1 Be that as it may, I wish you would let me know what books of mine you have not received from Longmans. **1834** M. Edgeworth *Tour in Connemara* (1950) i. 10 There goes a story, you know that no woman must ever appear at Ballinasloe Fair... Be this as it may, we were suffered to drive very quietly through the town. **1875** Trollope *Way we live Now* I. xlix. 311 'Good news?' she asked... 'That's as may be,' he said. **1883** A. Dobson *Old-World Idylls* 211 Rose kissed me to-day. Will she kiss me to-morrow? Let it be as it may, Rose kissed me to-day. **1910** Galsworthy *Motley* 168 'Yu'le tak' the ole 'arse then?' 'That's as mebbe—waal, gude naight.' **1928** F. Hurst *President is Born* 14 Be that as it may, the circle of giving in Centralia was anything but a vicious one. **1935** G. Heyer *Death in Stocks* iv. 41 That's as may be, and if it's true you couldn't say but what it's a judgment. **1939** G. M. Gathorne-Hardy *Fourteen Points & Treaty of Versailles* 14 The Fourteen Points..has been described, by Mr. Winston Churchill, as 'certainly an accommodating document'. Be that as it may, it was presumably what the President had actually meant. **1949** F. Maclean *Eastern Approaches* I. viii. 133 If the authorities..had received no instructions regarding my journey it could only be due to a most regrettable omission... To this he answered that this was as it might be; but without explicit instructions.. he could not allow me to remain on Chinese territory. **1949** H. Pakington *Young W. Washbourne* vi. 51 'But if it was used as a sitting-out place it wouldn't be secluded,' said Mrs. Harbottle. That was as it might be, retorted Mrs. Wilkins. **1958** *Economist* 1 Nov. 387/2 There have been reports of some exchange of views between these two formidable figures. Be that as it may, the manner in which their views have been made known shows a wide and characteristic difference. **1975** T. Heald *Deadline* ii. 23 'I shall have to liaise with the police.' 'That's as may be,' said Lord Wharfedale.

Maya (mā·yă, moi·(y)ă, mēi·ă), *sb.*[2] and *a.* Also 9 **Maye.** [Sp.] **A.** *sb.* **a.** A member of an ancient Indian people of Yucatan and Central America; these people collectively. **b.** The language of this people. **B.** *adj.* Of, pertaining to, or designating this people.

1825 J. Conder *Mod. Traveller Mexico & Guatimala* I. 191 The number of these [Mexican] languages exceeds twenty, of which fourteen have grammars..tolerably complete; viz. the Mexican or Aztec, the Otomite,..the Maye or Yucatan. **1832** J. Bell *Syst. Geogr.* V. 584 In the space between the Rio del Norte and the American frontier, are the following tribes: 1. The *Tawakenoes*, on the *Rio Brassos*, 200 warriors, 800 souls;..5. *Mayes*, at the mouth of the Guadaloupe river, 200 warriors, 800 souls. **1845** *Trans. Amer. Ethnol. Soc.* I. 252 *K* has in the Maya a different sound from our *c* before *a, o, u.* **1875** H. H. Bancroft *Native Races Pacific States* II. 117 This Maya culture. *Ibid.* 118 Yucatan was occupied in the sixteenth century by the Mayas proper. **1877** [see *Chibcha]. **1914** T. A. Joyce *Mexican Archæol.* viii. 202 Both divisions of this people originally spoke Maya. *Ibid.* 203 The Maya language as a whole exhibits certain points of similarity to that of the Mixtec and Zapotec. *Ibid.* xi. 282 Of the land system among the Maya we know very little. **1928** T. Gann *Discoveries Cent. Amer.* 89 The old man..knew no word of any language but Maya. *Ibid.* 206 In none of them [*sc.* cities] is the Maya arch found. **1959** E. Tunis *Indians* 21/2 The Maya of Mexico were the only Indians who achieved a written language, with symbols that stood for individual words. **1974** *Nat. Geographic* Nov. 661 In Guatemala's lofty highlands..Maya Indians dwell amid the cloud-ripping hulks of dead and dormant volcanoes. *Ibid.*, Here live Maya-speaking Indians in villages that garland the lakeshore with a litany of saints' names.

Hence **Ma·yan** *a.*, of, pertaining to, or designating the Mayas; *sb.*, one of the Mayan people; their language; **Ma·yanist**, an expert in, or student of, Mayan culture.

1889 S. Hale *Mexico* viii. 82 The Mayan legends..tell of nothing but wars and conquests, struggles and defeats. **1911** *Encycl. Brit.* XVIII. 335/2 These Pipils..migrated into territories previously occupied by an older race of Mayan origin... The easternmost limit of prehistoric Mayan civilization, on the Pacific coast of Central America is Fonseca Bay. **1923** H. I. Priestley *Mexican Nation* ii, In the modern oil region along the Pánuco River and the coast, was a segregated group of Mayans. **1926** *Blackw. Mag.* Nov. 647/2 This man..was of pure Maya blood, and was filled..with old Mayan lore and tradition. **1933** E. Pinchon *Viva Villa!* x. 135 The craftsmanly and aristocratic Mayans and Tehuans of the south. **1950** *Caribbean Q.* II. ii. 27 Later that year Dr. Linton Satterthwaite, Associate Curator of the Pennsylvania University Museum and a Mayanist of note, was put in touch with the writer. **1961** T. Proskouriakoff in L. Deuel *Conquistadors without Swords* (1967) xxvii. 384 We Mayanists spend an inordinate amount of time deciphering half obliterated hieroglyphic texts. **1962** L. Kemp tr. *Leon-Portilla's Broken Spears* iii. 31 By the time Cortes ransomed him from the natives eight years later he spoke Mayan fluently. **1965** B. Ross *Mexico* i. 21 They asked him where he had obtained the cloth and the pottery? The Mayan said that it came from a land not very far away. **1968** *Listener* 11 July 53/3 They describe the attempts of Olson, newly arrived in Yucatan, to come to terms with Mayan culture. *Ibid.* 54/1 He takes off for Yucatan unprepared for his task, learning Mayan as he goes, and from inadequate books. **1973** *New Yorker* 24 Mar. 116/2 He got the word in Mayan for 'road' or 'journey'. *Ibid.* 116/3 One can thus understand the interest among Mayanists when..it was announced that a wholly unknown fourth codex had come to light. **1973** *Listener* 7 June 746/1 The Mayans..invented essentially the same way of writing large numbers as a sequence of digits that we use.

may-be, maybe, *adv.* Delete *arch.* and *dial.* and add later examples. Also phr. *and I don't mean maybe*: I am positive (*colloq.*).

1897 Kipling *Capt. Cour.* vii. 147 Don't want nothin', 'less, mebbe, an anchor that'll hold. **1926** Maines & Grant *Wise-Crack Dict.* 5/1 And I don't mean maybe, jazz form of putting one's self on oath. **1930** Joyce *Anna Livia Plurabelle* 23 I'll make it worth your while. And I don't mean maybe. **1931** E. Linklater *Juan in Amer.* v. v. 401 When a guy..tries to get fresh..a girl's justified in giving him the razz—and I don't mean maybe! **1933** E. Caldwell *God's Little Acre* viii. 110 We've sunk a hole twenty feet since this morning, and I don't mean maybe, either. **1953** *Manch. Guardian Weekly* 20 Aug. 7 Except maybe a guy had been there two years. **1968** B. Foster *Changing Eng. Lang.* i. 30 Some years ago a song popular in this country bore the title 'Maybe', and on consulting the *O.E.D.* one finds this word described as archaic and dialectal! Yet by now it is in everyday use in Britain. **1968** *New Society* 29 Aug. 305/1 The *Oxford English Dictionary* describes 'maybe' as archaic and dialectal. But within the last generation it has been re-instated in England, obviously as a result of its popularity in the American vocabulary, where 'perhaps' is a very rare bird. Yet all this time 'maybe' (or 'mebbe') has been thriving in northern England, Scotland and Ireland. **1975** *New Yorker* 13 Jan. 36/1, I wanted to think maybe she was different now.

May-butter. (Later U.S. example.)

1812 J. J. Henry *Accurate Acct. Heroes Campaign against Quebec* 23 We gave salted pork, and they returned two fresh beaver tails, which when boiled, renewed ideas, imbibed with the May-butter of our own country.

May-day[1]. Add: **c.** *attrib.*, *spec.* of political processions, celebrations, etc., on the first day of May.

1906 [see *communism 1 b]. **1930** I. Low *His Master's Voice* x. 120 Scaffolding began to appear in the Red Square for May Day celebrations. **1939** [see *hunger-march]. **1973** *Listener* 9 Aug. 184 In 1890, during the first of the big May Day demonstrations, we read of trade-union branches assembling from every quarter of London.

May-day[2]. Also **Mayday, mayday.** [Phonetic repr. of F. *m'aider* imper. inf. 'help me!', or shortening of *venez m'aider*.] An international radio-telephone signal of distress. Also *transf.* and *attrib.*

1927 *Internat. Radio Telegraph Convention* 51 Rules apply to the radio telephone distress call which consists of the spoken expression mayday, (corresponding to the French pronunciation of the expression 'm'aider'..). **1929** *Times* 18 June 16/1 The pilot, after wirelessing the S.O.S. of the Air Service 'May Day'..endeavoured to return to Lympne aerodrome with the power still at his disposal. **1930** *B.B.C. Year-Bk.* 399 *Aircraft and Wireless*... In case of distress, due to engine failure over the sea, the word 'Mayday'—equivalent to the S.O.S. used by ships—transmitted through the microphone, will summon immediately all possible help. **1951** O. Berthoud tr. *Clostermann's Big Show* i. 30 If you can't get back to the coast, bale out after calling 'May Day' on frequency D. **1962** *Listener* 1 Mar. 370/2 With the onset of lambing time the farmers' 'Mayday' signal begins to hum along the network of telephone wires. **1962** *Sunday Express* 5 Aug. 1/2 Her first 'Mayday' distress message came soon after midday. **1971** *Islander* (Victoria, B.C.) 16 May 13/1 When her topmast chains snapped, she radioed a mayday signal. **1971** *Daily Tel.* 23 Aug. 1/5 The radio operator sent a Mayday distress call before the sea rushed into the cabin.

may-duke. (Later U.S. examples.)

1841 *Knickerbocker* XVII. 154 The air is impregnated with the fragrances..of the blossoming may-dukes. **1874** *Rep. Vermont Board Agric.* II. 359 This variety, and the.. May Duke, Late Duke, and other Dukes,..are hardly less hardy than plums.

Mayfair (mēi·fēə̆ɹ). [May *sb.*[3] + Fair *sb.*[1]] **a.** A fair held in May, esp. that held annually from the 17th century until the end of the 18th century in Brook fields near Hyde Park Corner. **b.** The district of London, very fashionable since the 19th century, between Oxford Street and Piccadilly, occupying the site of the old fairground. Also as quasi-*adj.*

1701 B. Fairfax in *Tatler* (1786) I. Notes 418, I wish you had been at May-fair, where the rope-dancing would have recompensed your labour. **1709** *Tatler* 24–26 May, The Crowd of the Audience are fitter for Representations at May-fair, than a Theatre-Royal. **1748** H. Walpole *Let.* 3 Sept. (1903) II. 336 A chosen committee waited on the faithful pair to the minister of May-fair. **1754** *Connoisseur* 17 Oct. 227 Catalogue of Males and Females to be disposed of in Marriage to the Best Bidder, at Mr. Keith's Repository in May Fair. **1848** Thackeray *Van. Fair* li. 453 Yesterday, Colonel and Mrs. Crawley entertained a select party at dinner at their house in May Fair. **1874** H. C. Pennell (*title*) The Muses of Mayfair: selections from *vers de société* of the nineteenth century. **1933** J. Buchan *Prince of Captivity* ii. i. 146 Clothes slightly astray from the conventions of Mayfair. **1953** A. Christie *Pocket Full of Rye* i. 8 Miss Grosvenor..wailed in a voice whose accent was noticeably less Mayfair than usual.

c. *attrib.* and *Comb.*

1752 H. Walpole *Let.* 27 Feb. (1903) III. 85 They were married with a ring of the bed-curtain..at Mayfair chapel. **1843** Carlyle *Past & Present* iii. ix. 252 Patent-Digester, Spinning-Mule, Mayfair Clothes-Horse. **1866** M. Mackintosh *Stage Reminisc.* vi. 74 So elegant and comfortable that even luxurious west-enders might have fancied themselves at home in their own May-Fair drawing-rooms. **1940** N. Marsh *Surfeit of Lampreys* (1941) ii. 31 The twins were saying..that..the only thing..was for them to turn crooks and be another lot of Mayfair boys. **1943** I. Brown *Just Another Word* 24 'She got terribly akimbo'..a species of Mayfair slang for what was earlier called 'high horse'. **1957** R. W. Zandvoort *Handbk. Eng. Gram.* v. ii. 225 It's such a bore, don't you know. In the..example (which really represents 'Mayfair' English) the tag practically ceases to be a question. **1962** J. D. Salinger *Franny & Zooey* 155 He was still for a moment. Then, in an almost unintelligibly thick Mayfair accent: 'I'd rather like a word with you, Miss Glass.' **1967** K. Giles *Death in Diamonds* i. 21 He's insolent, smooth and tough, very hip in his talk. Like the Mayfair boys in the thirties. **1970** C. Drummond *Stab in Back* iv. 88 They 'ad a saying when we were young about 'Mayfair Boys', meaning gentlemen crooks.

Hence **May·fairish** *a.*, of the nature or character of Mayfair.

1938 G. Allighan *Sir John Reith* iv. 235 The B.B.C. organisation..is too isolated from the common people. There is a West End outlook, a Mayfairish idea of the elite, an everlasting implication of superiority. **1967** [see *ideal home].

mayo (mēi·o), colloq. abbrev. of Mayonnaise.

1960 Wentworth & Flexner *Dict. Amer. Slang* 335/1 *Mayo*, mayonnaise. Common lunch-counter use since *c* 1930. **1969** L. Hellman in *Atlantic* Apr. 118 Run down to the corner and get me a ham and cheese on rye and tell them to hold the mayo. **1971** *New Yorker* 10 July 20 We were sitting at a luncheonette counter the other day, just about to bite into a b.l.t. down, with mayo, when a familiar voice addressed us from an adjacent stool.

Mayologist (māyǫ·lŏdʒist). [f. *Maya *sb.*[2] + -ologist (see -ology).] A student of Maya antiquities.

1926 *Glasgow Herald* 18 May 9 Mr. Joyce, perhaps the ablest of living Mayologists, has been sent..on a mission of inspection.

mayonnaise. For etym. read: [F. *mayonnaise*, also *magnonaise*, *mahonnaise*, the latter being prob. fem. of *mahonnais* of Port *Mahon*, capital of Minorca, taken by the duc de Richelieu in 1756.] Also with defining word, as *egg*, *fish*, *lobster*, *salmon mayonnaise*.

See quots. s.v. *lobster[1] 5.

1861 Mrs. Beeton *Bk. Househ. Managem.* 225 For a fish Mayonnaise, this sauce may be coloured with lobster-spawn, pounded. **1910** *Encycl. Brit.* VII. 74/2 The mayonnaise (originally *mahonnaise*) is ascribed to the duc de Richelieu. **1975** D. Bloodworth *Clients of Omega* xvi. 153 The partners..began to eat salmon mayonnaise off the altar.

2. *Contract Bridge* = *goulash 2.

1927 M. Work *Contract Bridge* 138 *Mayonnaise*, old name of Goulash.

Hence **mayonnai·sed** *a.*

1968 C. Drummond *Death & Leaping Ladies* i. 7 A hearty trencherman himself, a victor over many a mayonnaised lobster. **1972** P. A. Whitney *Listen for Whisperer* vii. 130 An array of tiny mayonnaised shrimps.

mayoress. 2. Delete '*nonce-use*' and add earlier and later examples. *U.S.*

1884 *Chicago Tribune* 5 May 4/3 She cannot under the laws monopolize the holding of nice special offices like Mayoress. **1947** *Chicago Sun* 4 Nov. 16/5 The dish..will be delivered..by Mrs. Margaret Craig, mayoress of Twickenham.

3. A person appointed to assist a female mayor.

1932 *Times* 10 Nov. 8/6 Mrs. Emily F. George, the Labour Mayor of Bermondsey, has appointed her daughter..to be Mayoress.

Mazahbi, var. *Mazhabi. **Mazal tov**, var. *mazel tov.

Mazatec (mæ·zātek), *sb.* and *a.* **a.** (A member of) an Indian people inhabiting northern Oaxaca in southern Mexico. **b.** The language of this people. Also *attrib.* or as *adj.* Hence **Mazate·can** *a.*

1909 Webster, *Mazatec*, one of a tribe of Zapotecan Indians of northeastern Oaxaca, Mexico. They raise silk and are noted for gorgeous silken fabrics. **1949** E. A. Nida *Morphol.* (ed. 2) II. 24 Mazatec, a language of Mexico. **1957** *Encycl. Brit.* V. 138/2 *Mazatec* (Guerrero, Puebla, Oaxaca; includes *Trique* and *Chocho*). **1958** J. Berry in J. A. Fishman *Readings Sociol. of Lang.* (1968) 751 One of the Mazatec dialects. **1962** D. H. Hymes in *Ibid.* 111 The Mazatecs of Mexico. *Ibid.* 129 Neighboring dialects may differ, as when one group of Mazatec abstract the tones of their language for a whistled code, while the Soyaltepec Mazatec do not. **1963** *Times* 27 April 9/7 Land where heron live at peace with primitive Mazatecan Indians. **1967** S. C. Gudschinsky *How to learn Unwritten Lang.* ii. 19 Transitive relationship is expressed in Mazatec by a transitive verb followed by a noun object and a noun subject. *Ibid.* 22 Mazatec possessive phrases are quite different from English. **1974** *Encycl. Brit. Micropædia* VI. 728/1 The Mazatec language is most closely related to Chocho, Ixcatec, and Popoloca. *Ibid.*, The Mazatec are Roman Catholic with syncretistic elements.

Mazbi, var. *Mazhabi.

Mazdaist (mæ·zdă,ist). [f. as Mazdaism + -ist.] An adherent of Mazdaism.

1920 in Webster. **1932** *Antiquity* VI. 277 Where Mazdaists and Christians were found together, there must certainly have been adherents of the Manichæan creed. **1941** *Burlington Mag.* Aug. 59/1 Mohammedans..Mazdaists..Christians, Buddhists and perhaps Jews as well had their own temples.

maze, *sb.* Add: **4. a.** *spec.* in *Psychol.*, a device, consisting of a correct path concealed by blind alleys, used to study human and animal intelligence and learning. Also *attrib.* and *Comb.*

1901 W. S. Small in *Amer. Jrnl. Psychol.* XII. 228 The process of learning the way through this maze is adequately described as a gradual establishment of direct associations by profiting by chance experience. **1914** F. A. C. Perrin (*title*) An experimental and introspective study of the human learning process in the maze. **1921** *Lancet* 19 March 597/2 The..Porteous maze tests which, testing as they do the foresight, the capacity to plan, the practical judgment and concentration of the child, supply a marked lack of the Binet scale. **1940** *Brit. Jrnl. Psychol.* Jan. 191 Many maze workers have noted the fact of variability in performance from day to day. **1951** G. Humphrey *Thinking* viii. 257 Verbal instruction improves the score in stylus-maze running. **1958** *Spectator* 8 Aug. 201/1 The obsessional maze-running experiments of the American rat-psychologists. **1964** M. Argyle *Psychol. & Social Probl.* v. 67 [Areas in which delinquents differ from non-delinquents] Impulsiveness, weakness of 'ego-control', tendency to cut corners in maze tests.

‖ **mazel tov** (ma·zel tŏv, tǫf). Also **mazal tov, mazzel tov**, etc. [ad. mod.Heb. *mazzāl ṭōḇ* good luck, f. Heb. *mazzālōth* pl., constellations.] As a salutation: good luck, congratulations.

1862 *Once a Week* VII. 192/2 Whereupon all present wish him *mazal tov* (good speed), and the ceremony is at an end. **1892** I. Zangwill *Childr. Ghetto* I. 37 There was a hubbub of congratulation ('*Mazzoltov, Mazzoltov*', 'Good Luck'). **1932** L. Golding *Magnolia St.* iii. iv. 520 The air pattered like hail with their good luck wishes..*Mazel tov, mazel tov!* **1957** L. Stern *Midas Touch* iii. xx. 149 '*Mazeltof*, father,' he said. **1959** B. Kops *Hamlet of Stepney Green* ii. ii, Going to have another one?..Mazeltoff, darling. **1970** *New Yorker* 19 Sept. 32/1 'Uri is going to be a father.' Yigael..said, 'He told me. *Mazel tov*.' **1972** 'H. Howard' *Nice Day for Funeral* x. 138, I said, 'Mazeltov. I hope you're both very happy.' **1973** *Jewish Chron.* 2 Feb. 19/3 Mother and baby..are..doing well. A hearty mazzeltov to them!

Mazhabi (mɒ·zhabi). *India.* Also **Maz(ah)bi, Muzbi, Muz(hu)bee.** [Hindi, f. Arab. *mazhab* religion.] A convert to the Sikh religion from Islam, *spec.* in the Punjab; a converted Chuhra or member of the sweeper caste.

1849 J. D. Cunningham *Hist. Sikhs* 379 *Muzhubee*, converts from Mahometanism are so called. **1858** R. Temple *Let.* 25 May in Yule & Burnell *Hobson-Jobson* (1886) 464/1 To the same destination (Delhi) was sent a strong corps of Muzhubee (low-caste) Sikhs, numbering 1200 men, to serve as pioneers. **1908** Kipling *Lett. of Travel* (1920) 162 A fair sprinkling of Punjabis—ex-soldiers, Sikhs, Muzbis, and Jats—are coming in on the boats. **1917** *Encycl. Relig. & Ethics* IX. 608/2 Such converts may form new castes, like the Mazbi or Mazhabi Sikhs, who were by origin Chūhṛās, or scavengers, outside the Hindu pale. **1923** *Chambers's Jrnl.* 20 Jan. 113/1 He was not going to give his daughter in marriage to a *Muzbee*. **1965** Sher Singh *Sansis of Punjab* ix. 188 If a Sansi takes food from a Choohra, Mazahbi, Chamar or other lower castes, he is outcast and fined.

mazout, var. *mazut.

mazuma (māzū·mă). *U.S. slang.* Also **mazume.** [Yiddish.] Money, cash.

1904 G. V. Hobart *Jim Hickey* i. 15 We're a sad bunch ..when we haven't a little mazume in the vest pocket. **1906** 'O. Henry' *Four Million* (1916) 131 Burn a few punk sticks in the joss house to the great god Mazuma. **1907** C. E. Mulford *Bar-20* viii. 90 When th' mazuma is divided up it won't buy a meal. **1913** —— *Coming of Cassidy* xii. 191 'What's this?' he demanded.. 'Money,' replied Hopalong. 'It's that shiny stuff you buy things with. Spondulix, cash, mazuma.' **1926** *Amer. Speech* May 456/1 How many of those using the word *mazuma* know its meaning? As originally used by the Jewish people it is '*m'zumon*' and is a Chaldean word meaning in literal translation the 'ready necessary'. It is employed in the Talmud which is written in Chaldean and not Hebrew. **1941** *Amer. Mercury* May 615/1 When you die, you have to leave your mazuma behind. **1943** W. H. Chase *Sourdough Pot* xvi. 97 A sign tacked on a tent pole, informed the public, that 'No Credit—You must produce the mazuma. We are not in business for our health.' **1972** *Times Lit. Suppl.* 29 Sept. 1154/3 Likewise piling up its mazuma by legerdemain.

mazut (māzū·t). Also **masut, mazout.** [Russ. *mazút*, ad. Arab. *makhzulat* refuse, waste.] The viscous liquid left as residue after the distillation of Russian petroleum, used as fuel oil and a coarse lubricant.

1897 *Chambers's Jrnl.* 19 June 393 Masut, the new substitute for coal. **1907** *Times Engin. Suppl.* 16 Oct. 4/2 Among liquid fuels which may be employed that known as 'masut' will help to supply the increasing demand. **1924** J. A. Hammerton *Countries of World* II. 1277/1 'Crude oil'..is refined into benzine, petrol and kerosene, the refuse or 'mazout' being consumed for common fuel and furnaces. **1924** *Blackw. Mag.* Feb. 152/1 They caught the sickly sweet smell of half-burnt 'Mazut' fuel. **1951** R. Campbell *Light on Dark Horse* xx. 288 The oil, or mazout, has also spoilt the quality of the red-mullet. **1974** P. Highsmith *Ripley's Game* v. 47 He tackled with broom and dustpan the exterior of the pipes and the floor around their *mazout* furnace.

mazzel tov, var. *mazel tov.

Mazzinian (mætsī·niăn), *sb.* and *a.* [f. the name of Giuseppe *Mazzini* (1805–72), Italian patriot and revolutionary + -an.] **A.** *sb.* An adherent of Mazzini. **B.** *adj.* Of, pertaining to, supporting, or resembling Mazzini or his policy. So **Mazzi·nianism, Mazzi·nism**, the principles or methods of Mazzinians; **Mazzi·nist** = *Mazzinian A.

1850 J. Miley *Hist. Papal States* III. viii. iv. 644 They ..have loaded the memory of Gregory XVI., just as much as the Mazzinians, with every species of libellous invective. **1860** *Illustr. London News* 7 Jan. 18/3 Even the Mazzinists claim him at times as their own. **1861** tr. *A. Bresciani's Jew of Verona* p. iv, Misrepresentation which had been so assiduously thrown around the recent events by the Mazzinian press. **1862** *Dublin Rev.* Feb. 203 The universal dread which Mazzinism inspired. **1866** H. E. Manning in S. Leslie *Life H. E. Manning* (1921) xiii. 195 Mazzinianism and Fenianism are one in principle. **1875** P. K. O'Clery *Hist. Italian Revolution* v. 165 The Mazzinian programme was the establishment of a single republic... The literary propaganda of Mazzinianism. *Ibid.* 187 They were no less revolutionists than the Mazzinians. **1932** G. F.-H. Berkeley *Italy in Making* I. xiv. 209 Aurelio Saffi, the well-known Mazzinian, who lived in Forlì. **1937** *Times Lit. Suppl.* 1 May 322/4 Meredith's attack..was a satire of social extravagances tinged with Mazzinian idealism. **1965** C. Hibbert *Garibaldi & his Enemies* i. x. 140 Louis Napoleon could not free himself so easily..from the indiscretions of the Mazzinians. *Ibid.*, The Mazzinian *L'Italia e Popolo* of Genoa. **1967** C. Seton-Watson *Italy from Liberalism to Fascism* xi. 422 The militant irredentists from Istria and Dalmatia..cared nothing for the Mazzinian dream of Italo-Slav friendship.

mazzot, mazzoth, varr. *matzah.

‖ **mbongo** (m,bǫ·ŋgo). *S. Afr.* Also **imbonga, imbongo, mbonga, mbongi.** [Zulu.] An official who sings the praises of the (Zulu) king; hence applied to any flatterer of a high personage or institution.

1839 W. C. Harris *Wild Sports S. Afr.* xiv. 116 We.. were accordingly preparing to start when a herald, called, in the Matabili language, *Imbongo*, a proclaimer of the king's titles, suddenly made his appearance outside the kraal. **1871** in T. Baines *Northern Goldfields Diaries* (1946) III. 687 His *Imbonga*, or court flatterer..recounted the battles of the Matabele. **1945** *Cape Times* 29 May, Municipal mbongo... The Mayor made quite a good case for the appointment of an official who would be a sort of professional praiser. **1948** *Cape Argus* 16 Sept. 7 Government supporters have danced around their Ministers like dutiful, adoring and disciplined mbongos. **1957** *Ibid.* 15 June 1/7 They were..a lot of mbongas to hymn the praises of the incompetent Nationalist hierarchy. **1973** *Drum* 8 Mar. 53 The mbongis..climb up his ancestral tree singing the praise poems of Vorster's illustrious forebears.

Mbret, var. *Mpret.

Mc. All words of Scottish or Irish origin beginning with *Mc* (e.g. *McCarthyism, *McCoy) are placed alphabetically as if spelt *Mac-*.

Mdlle: see Mademoiselle 1.

Mdme: see Madame 1.

me, *pers. pron.* Add: **4.** (Later examples.)

1814 Jane Austen *Let.* 2 Mar. (1932) II. 92, I am to call upon Miss Spencer: Funny me! **1895**, etc. [see *little me* s.v. *little *a.* 13]. **1961** 'P. Dennis' (*title*) Little me: the intimate memoirs of that great star..Belle Poitrine. **1973** 'D. Halliday' *Dolly & Starry Bird* viii. 111 'As Timothy would say, silly me,' Johnson said in a voice as hard as his bifocals.

5. For 'Now chiefly *arch.* and *poet.*' read 'Freq. in U.S. colloq. usage', and add further examples.

1821 W. Sewall *Diary* (1930) 75 Purchased me some linens. **1874** *Rep. Vermont Board Agric.* II. 512 In 1861 I built me a horse barn, twenty-eight by forty. **1916** 'B. M. Bower' *Phantom Herd* iii. 42 I'm going to make me one. **1972** '*Gramophone*' *Pop. Record Catal.* Dec. 129/2 (*song-title*) I'm gonna get me a gun. **1974** M. Hastings *Dragon Island* xv. 129 I'll grab me the first-aid box and..see how my patients are doing.

6. (Later examples.)

1903 *Dialect Notes* II. 320 *Me and you*,..almost universal for 'you and I'. **1905** *Westm. Gaz.* 11 Nov. 3/1, I can foresee..that unless me and Ellen advise you, you'll become simply ——. **1936** Mencken *Amer. Lang.* (ed. 4) ix. 457 Him and me are friends. **1966** 'J. Hackston' *Father clears Out* 50 Me an' me mate's eyes was dancin' out of our 'eads.

7. c. Now somewhat more commonly used intensively like French *moi*.

1923 *Dialect Notes* V. 244, I am not going to-day, *me*. **1963** *Listener* 20 June 1041/3 Me, I like fighting, too.

9. With the verb *to be*: suited to or representative of my tastes, ability, personality, etc.; appropriate for me; my real self.

1899 J. London *Let.* 29 July (1966) 47 This is me all the time and all over. **1905** A. Bennett *Sacred & Profane Love* i. v. 83 But that poor little book isn't *me*... I shall never write another like it. **1925** R. Hall *Saturday Life* vii. 79 Nothings, just nothings, they didn't count; this is the thing that's *me*. **1938** J. Cary *Castle Corner* 557 The house has got to be contemporary, it's got to be art..and it's got to be me. **1949** R. Chandler *Little Sister* xxviii. 209, I don't like the script... It just isn't me, if you know what I mean. **1957** P. Wildeblood *Main Chance* 56 Cardigans are not..particularly *me*. **1963** F. T. Visser *Hist. Syntax Eng. Lang.* I. iii. 240, I like this dress, it's me!

II. 10. Used *colloq.* and *dial.* (also *Austral.* and *N.Z.*) as a poss. adj.

Treated in Dict. s.v. My *poss. adj.*, e.g. in 19th-cent. representations of Irish speech.

1862 G. Meredith *Let.* 23 June (1970) I. 152 B. Wyse came the other day..and hoped for forgive *miss*: 'Me deer Mardith', etc.! **1901** M. Franklin *My Brilliant Career* x. 84 Now it's your turn, me fine lady. *Ibid.* xxvi. 220 A couple of letters..stuffed in me pocket. **1911** W. Owen *Let.* 20 Sept. (1967) 83 Love to Mary and me brethren twain. **1946** K. Tennant *Lost Haven* (1968) i. 20 Me own mother sent me word he wants to see me! *Ibid.* ii. 39 Me granddaughter's coming to stay with me. **1960** G. Slatter *Gun in my Hand* ii. 21 An' it ran like a hairy goat an' I did me chips. **1966** F. Shaw et al. *Lern Yerself Scouse* 20 Me dollypegs, my legs... *Me webs*, my feet. **1968** K. Weatherly *Roo Shooter* 21 'Cost you a night's shooting, me backside,' roared the driver. **1972** *Southerly* XXXII. 6 Me motorbike's out at the station. **1973** *Sunday Express* (Trinidad & Tobago) 1 Apr. (Suppl.) 13/1 Ah on de road, Putting out me hand Like a mas in a band.

meach, var. *meech *v.*

meachin(g), varr. *meeching *ppl. a.*

‖ **mea culpa** (mēi·ă ku·lpă, mī·ă kɒ·lpă). [L., lit. 'through my own fault'.] A phrase from the prayer of confession in the Latin liturgy of the Church; used *lit.* and *transf.* as an exclamation of repentance, and as the name of such an exclamation; also *mea maxima culpa*. See Confiteor.

c **1374** Chaucer *Tr. & Cr.* (1894) ii. 525 Now, *mea culpa*, lord! I me repente. **1602** W. Watson *Decacordon* 40 Shall lay their hands a little heavier on their hearts with *Mea maxima culpa*. **1818** Lady Morgan *Florence Macarthy* IV. v. 208 Mingled a broken *ave-maria* and *mea-culpa*, in utter consternation and superstitious fear. **1891** E. Dowson *Let.* 7 Feb. (1967) 184 Today, mea culpa, mea maxima culpa, I have done nothing. **1922** Joyce *Ulysses* 279 He beat his hand upon his breast, confessing: *mea culpa*. **1948** R. Butler *Words of Mass* i. 10 The striking of the breast at the *mea culpa* is an emphatic recognition of guilt and expression of sorrow. **1958** *Times* 17 Oct. 17/1 Eisenstein made a public *mea culpa* at the time in the form of an open letter to the Committee. **1960** [see *guilt-complex]. **1974** 'D. Shannon' *Crime File* (1975) v. 79 *Mea culpa*... We all do stupid things sometimes.

mead[1]. Add: **c. mead-bench, -hall**: later examples.

1959 A. G. Brodeur *Art of Beowulf* 16 A mead-bench is a seat in a royal hall, where the dispensing of good drink symbolizes the warm relationship between lord and retainer. **1903** L. F. Anderson *Anglo-Saxon Scop* 36 Personal valour and prowess on the field of battle, courage.., hardihood..., these were the all-absorbing topics of conversation in the mead-hall. **1968** E. B. Irving *Reading of Beowulf* v. 242 The clustering of the clan family in the lighted mead-hall.

meadow, *sb.* Add: **4. a.** *meadow-down*, *-farmer*.

1877 G. M. Hopkins *Poems* (1967) 71 Meadow-down is not distressed For a rainbow footing it. **1742** W. Ellis *Mod. Husbandman* II. iii. 109 There are two Sorts of Farmers, who carry on this Business..*viz.* the Grass, or Meadow-farmer, and the Plough-farmer. **1884** R. Jefferies *Life of Fields* 139 The meadow-farmers, dairymen, have not grubbed many hedges.

b. meadow brown (butterfly), for Latin name substitute '*Maniola jurtina*' and add examples of absol. use; **meadow-hen** (earlier examples); **meadow-lark** (earlier and later N. Amer. examples).

1930 *Times Educ. Suppl.* 4 Oct. p. iv/4 The meadow-brown's heavy, indolent flight. **1974** *Lady* 1 Aug. 169/1 Butterflies abound..from the innumerable brown ringlets, 'gate-keepers', speckled woods and meadow browns, to various beauties like the common blues. *a* **1841** W. P. Hawes *Sporting Scenes* (1842) I. 18 The principal inhabitants are gulls, and meadow-hens. **1863** 'G. Hamilton' *Gala-Days* 97 You know you didn't scare a little meadow-hen. **1775** B. Romans *Conc. Nat. Hist. E. & W. Florida* 114 Meadow larks, fieldfares, rice birds, &c. are very frequently had. **1948** H. Jacobs *We chose Country* 161 Birds were everywhere, first killdeers, making a din in the fields at dusk, then meadowlarks, caroling in the morning sun. **1963** G. H. Thomson *Crocus Country* xx. 131 The bird we loved most..was the meadow lark. **1969** N. W. Parsons *Upon Sagebrush Harp* vii. 42 There were many birds, but the meadowlark moved me most.

c. *meadow pea* (examples), *trefoil* (earlier example); **meadow lily**, a common lily of the eastern U.S., *Lilium canadense*; also, formerly used for lily-of-the-valley, *Convallaria majalis*.

1832 W. D. Williamson *Hist. State Maine* I. 125 [We have] two varieties of meadow-lilies,..May-lily, or 'lily of the valley'; and nodding-lily. **1894** *Jrnl. Amer. Folk-Lore* VII. 102 *Lilium Canadense*,..meadow lily, nodding lily. **1946** E. Hodgins *Mr. Blandings builds his Dream House* (1947) viii. 104 When the bluebells and the columbine faded, the meadow lilies and the wild geranium took up the torch... August was well along. **1896** G. Henslow *How to Study Wild Flowers* 98 *Lathyrus pratensis*, Meadow Pea. This genus resembles vetches, but has fewer leaflets. **1960** *Oxf. Bk. Wild Flowers* 22/2 Meadow Vetchling or Meadow Pea (*Lathyrus pratensis*). Although this Pea has thin, rather weak stems, it may reach up to 3 feet in height by scrambling over other plants. **1629** J. Parkinson *Parad.* i. lxxix. 341 The Medica's are generally thought to feede cattell fat much more then the Medow Trefoile, or Clauer grasse.

meady (mī·di), *a. rare.* [f. Mead[1] + -y[1].] Resembling or suggestive of mead.

1887 G. Meredith *Ballads & Poems Tragic Life* 102 Yellow flamed the meady sunset. **1900** *Sydney Mail* 31 Mar. 777 If you simply let it stand it will usually work up into the meady beverage.

meal, *sb.*[1] Add: **3. a.** *meal-bag*, *-drift*; *meal-dusty*, *-white* adjs.

1644 *Essex County, Mass. Probate Rec.* (1916) I. 46 Too meal baggs. **1738** *New Hampsh. Probate Rec.* (1914) II. 622 He knows of no meal Bag that his son had but what he borrowed of him. **1876** *Wide Awake* 72/1 She was bundled up so you would hardly have known her from one of the meal-bags. **1968** E. R. Buckler *Ox Bells & Fireflies* ii. 34 Father puts three basketsful [of potatoes] into each meal bag. **1877** G. M. Hopkins *Poems* (1967) 70 What lovely behaviour Of silk-sack clouds! has wilder, wilful-wavier Meal-drift moulded ever and melted across skies? **1951** W. de la Mare *Winged Chariot* 40 Meal-dusty polls, glossed plumage. **1938** —— *Memory* 95 This meal-white snow.

meal, *sb.*[2] Add: **2. b.** *meals-on-wheels*: a service, usually provided by a women's voluntary organization, whereby meals are taken by car to old people, invalids, etc. Also *attrib.*

1961 *Times* 17 Jan. 7/1 Of those who get meals-on-wheels, less than 10 per cent have adequate meals on week days when there is no delivery. **1961** *Guardian* 19 Apr. 2/5 The need for an extension of the meals-on-wheels schemes as a contribution to the care of the home-bound. **1966** 'O. Mills' *Enemies of Bride* xii. 106 Having just completed a tour with a Meals-on-Wheels van to less fortunate old ladies, she was dressed in a green, shirt-waister dress of the Women's Voluntary Services. **1966** 'K. A. Saddler' *Gilt Edge* i. 22 I'm on National Assistance. I get Meals-on-Wheels twice a week. **1970** J. Fleming *Young Man, I think you're Dying* ii. 25 On weekday mornings the meals-on-wheels service would bring hot food for the invalid.

d. *to make a meal of* (further examples); also, to treat in an over-fussy or laborious manner.

1961 C. Willock *Death in Covert* iv. 93 Dyson..was making a meal of everything. He had carefully paced the distance... He had stuck sticks in the ground. **1968** *Guardian* 22 Aug. 7/2 When university people get on to fashion they make a meal of it. **1970** *Ibid.* 5 Dec. 8/5 He had as one of his guests Inia Te Wiata, whose name has been pretty familiar to radio listeners for some years now but Mr Murray still made a meal, if not a light snack, of promoting it. **1970** *N.Z. News* 21 Jan. 16/4 New Zealand scored 325 for six wickets declared—everyone except M. G. Burgess and G. E. Vivian making a meal of University bowlers. **1971** M. Polland *Package to Spain* vii. 95 His small sharp face was full of righteous outrage... Henry was making a meal of it. **1972** J. Eastwood *Henry in Silver Frame* xxiii. 191, I wouldn't want the gutter-press to make a meal of me.

4. *meal-break*; **meal ticket** *orig. U.S.*, (*a*) a ticket entitling a person to a meal; (*b*) *fig. slang*, a source of income or livelihood (esp. a husband or wife regarded as such).

1958 *Guardian* 20 Aug. 1/1 The central committee's advice to busmen..was to insist on full meal breaks. **1971** 'H. Calvin' *Poison Chasers* ix. 112 Ronnie Samson had to stop me to let the boys off for a meal break. **1870** O. Logan *Before Footlights* 248 The rather scrubby party who occasionally purchases..a 'Meal Ticket', and thus gets entrance to the festive dining hall. **1899** 'J. Flynt' *Tramping with Tramps* 395 *Meal-ticket*, a person 'good' for a meal. **1912** *Collier's* 23 Nov. 38/2 I've been doin' a lot for her—a regular meal ticket an' ticket agency for her. **1926** [see *good *a. adj.* 1 f]. **1929** T. Wolfe *Look Homeward, Angel* (1930) xxvii. 377 She can't bear to give him up... He's her meal-ticket. **1939** Ade *Let.* 7 July (1973) 214 The play remained at the Garden until the following summer and next year it was being played by three companies. It turned out to be my meal ticket. **1972** 'H. Howard' *Nice Day for Funeral* iii. 40 He was her meal-ticket. Why should she want him sent to the pen? **1972** *Lebende Sprachen* XVII. 34/1 US meal ticket—BE [*sc.* British English] luncheon voucher. **1973** *Jewish Chron.* 2 Feb. 23/5 Our young women do not look for 'meal-ticket' marriages.

meal, *v.*[2] **1.** (Earlier and later examples.)

1628 O. Feltham *Resolves* (ed. 3) lxxvi. 236 With Earthen Plate, *Agathocles* (they say) Did vse to meale. **1891** H. C. Bunner *Zadoc Pine* 201 A lodging-house for those who 'mealed' at the hotel. **1918** *Jrnl. Friends' Hist. Soc.* 7 John Lecky..generally arranged to meal at Friends' houses. **1960** A. Powell *Casanova's Chinese Restaurant* 113 'Doesn't Carolo ever eat himself?'.. 'He often meals with us as a matter of fact.'

mealie. Add: (Further *attrib.* examples.) Cf. *mielie.

1855 W. C. Holden *Hist. Natal* x. 282 The consumption of mealie (maize) meal in D'Urban has increased. **1884** E. P. Mathers *Glimpse of Gold Fields* 29 At one of our camps I tried to get some mealie porridge made. **1925** P. Smith *Little Karoo* 18 Just outside the door was the worn mealie-stamper, cut out of a tree-trunk and shaped like an hour-glass, in which the mealies were pounded into meal. **1928** R. Campbell *Wayzgoose* i. 27 Your notice boards like mealie-stems are stripped. **1929** D. Reitz *Commando* xiv. 129 Rations were chiefly game and mealie-meal. **1944** M. de B. Nesbitt *Road to Avalon* (1949) xiv. 113, I sit on the mealie planter, control the levers and see that the hoppers are well filled as they drop mealies in the furrows. **1956** N. Gordimer *Six Feet of Country* 9 The girl catching her stockings on the mealie-stooks. **1962** *Cape Times* 13 June 1/4 A Harvard aircraft..crashed in a mealie land near Dunnottar. **1971** *Sunday Times* (Johannesburg) 28 Mar. 12/5, I made all sorts of things like mealie bread to make a bit of money for my poor old husband to buy a piece of industrial ground. **1973** *Times* 28 Nov. 8/7 Mealie-meal, the staple food, was..in short supply. **1974** G. Jenkins *Bridge of Magpies* iii. 51 A slovenly breakfast of half-burnt mealie-meal porridge and boiled penguin eggs.

2. mealie-cob, a corn cob; **mealie-cob worm**, the caterpillar of *Heliothis armigera*, a noctuid moth; **mealie-pap** = *mieliepap* (*mielie).

1859 R. J. Mann *Natal* 137 (Pettman), The young mealy-cob is generally preferred to bread. **1932** S. Zuckerman *Social Life Monkeys* xii. 195 The meally cobs and fruit it [*sc.* chacma] plunders from cultivated lands. **1911** J. D. F. Gilchrist *S. Afr. Zool.* 150 (Pettman), The Mealie-cob worm..does extensive damage to mealies, peas, tomatoes, and lucerne. **1880** E. F. Sandeman *Eight Months in Ox Waggon* xxvii. 273 Guinea fowls..form a very relishing change from the never-varying *menu* of bók or mealie pap. **1902** J. H. M. Abbott *Tommy Cornstalk* 28 A few Kaffir transport drivers..are boiling their 'mealie-pap' in three-legged pots. **1903** J. Y. F. Blake *West Pointer with Boers* iii. 49 We had to come down to straight mealie pap (corn meal mush), and fresh beef. **1922** J. Buchan *Bk. Escapes* vi. 120 For food he had to trust to mealie-pap at Kafir kraals. **1966** D. Varaday *Gara-Yaka's Domain* xii. 135 Mealie-pap—maize-porridge—the bush substitute for bread was cooking.

mealy, *a.* Add: **2.** *mealy pudding* = *white pudding* (White *a.* 11 e).

1914 F. B. Jack *Cookery for every Household* 517/1 *Mealy Puddings.* 1 lb. oatmeal. ½ lb beef suet... When the puddings are required, toast them a few minutes in front of the fire. **1946** F. M. McNeill *Recipes from Scotland* 65 *White* or *Mealie Puddings*... The puddings will keep for months if hung up. **1951** *Good Housek. Home Encycl.* 550/1 *Mealy Pudding*, an oatmeal pudding which is served with grilled sausages, bacon, herrings, etc.

4. b. mealy-bug. Substitute for def.: a scale insect of the family Pseudococcidæ, esp. one of the genus *Pseudococcus*; (further examples).

1927 *Chambers's Jrnl.* Aug. 502/1 The mealybug, so called from its white waxy or mealy coating, belongs to the big family of scale insects known as the Coccidae. **1953** S. H. Skaife *Afr. Insect Life* xi. 122 Mealy bugs are small insects, the largest being only about one-eighth of an inch long, pink or purplish in colour but appearing to be white because their bodies are covered with a waxy powder. **1971** *Daily Colonist* (Victoria, B.C.) 5 May 27/5 The Comstock Mealybug, also known as the Pseudococcus Comstocki, is a serious threat to citrus, sugar beets, grapes, apples and other crops.

mealy-mouth (mī·li,maụ-þ). [Mealy *a.* 8; cf. Mealy-mouthed *a.*] **1.** A mouth which never utters plain terms; a soft, indirect, or reticent manner of speaking; hence a mealy-mouthed person. Also ironically: see quot. 1941. *slang.*

1600, 1828, 1862 [see Mealy *a.* 8]. **1898** B. Kirkby *Lakeland Words* 100 *Mealy-Mooth*,..eny body 'at's mealy-moothed's o' that sooart. **1919** J. C. Snaith *Love Lane* xlvii. 254 It was no use having a divided mind, it was no use having a mealy-mouth. **1941** J. Smiley *Hash House Lingo* 37 *Mealy mouth*, fault finding customer. **1942** *Scrutiny* X. 400 Iudushka, the chief character, is the apotheosis of the mealy-mouth.

2. *dial.* Applied to various birds, as the willow warbler, *Phylloscopus trochilus*, or the whitethroat, *Sylvia communis*.

1885 C. Swainson *Provincial Names & Folklore Brit. Birds* 26 Willow Warbler..Mealy mouth (Craven). **1961** *Countryman* LVIII. 468 A missel thrush is a 'greybird'.. the whitethroat a 'mealymouth' [in S. Pembrokeshire].

mean, *sb.*[2] Add: **10. g.** *means-end(s)* (used *attrib.*): of or pertaining to the ways of achieving a result considered together with the result.

1933 *Psychol. Rev.* XL. 60 The 'means-end-readiness' Tolman defined as a certain selectivity as regards stimuli, and as regards the responses..to such stimuli. **1951** R. Firth *Elem. Social Organiz.* iii. 85 Structural change implies that there was some imperfection in the previous means-ends schedule of a substantial number of members of the society. **1958** *Listener* 5 June 931/2 We must look at the contexts where the means-end model *is* appropriate. .. The second means-end context is that of making or producing things. We mix the flour in order to make a cake. **1963** A. Kaplan in P. A. Schilpp *Philos. R. Carnap* 840 If there were a final end it could not have the derived cognitive meaning of a means-end implication. **1965** H. I. Ansoff *Corporate Strategy* ii. 24 The two sets of rules have a means-ends relationship; objectives set the goals, and strategy sets the path to the goals. **1965** *Language* XLI. 80 Aiming towards a means-ends model of language.

15. means test, an official inquiry into an applicant's private resources, determining or limiting a grant or allowance from public funds; also (with hyphen) as *vb.*; hence **means-testable** *a.*; **means-tested** *ppl. a.*; **means testing** *vbl. sb.*

1930 *Economist* 7 June 1263/2 We should not cavil greatly at the principle of granting, on the basis of a means test, maintenance allowances for children compelled to attend school. **1935** *Planning* II. 11 A system of Transitional Payments was introduced..subject, however, to a household 'Means Test' administered by the local Public Assistance Committees. **1940** *Manch. Guardian Weekly* 23 Feb. 153/3 Non-contributory pensions, to which alone the household means test was to apply, had been conditioned by the applicant's means ever since 1908. **1957** *Times* 31 Dec. 11/5 Provision should be made in the new scholarship scheme for giving adequate assistance to students who would be debarred under 'means tests' from accepting scholarships offered by the State and other sources. **1963** *Economist* 1 June 882/2 All university awards are means-tested now. **1966** *Ibid.* 29 Jan. 388/1 The genuinely poor long-term sick should be generously treated, preferably through generous means-tested national assistance. **1970** *Guardian* 29 Oct. 13, 1971 is going to be a record year for means-testing... Millions of new means-test forms will have to be designed. .. More people..will..make the sudden move from means-tested exemption from Health Service charges to paying the full rate. **1970** *Daily Tel.* 12 Dec. 2/6 Many of them are now living on fixed private incomes or with the aid of supplementary benefits which are means-tested. **1972** *Times* 30 Sept. 15/4 The poorest being supported by means testable benefits. **1973** *Times* 13 Feb. 16/7 Means-testing over a wide range of social benefits has been introduced on a scale unprecedented since the war.

mean (mīn), *sb.*[3] *colloq. rare.* [*sb.* use of Mean *a.*[1] 5.] A mean person; = *meany.

1938 E. Bowen *Death of Heart* ii. iv. 241 You *are* a mean, Dickie!

mean, *a.*[1] Add: **2. f.** *U.S. colloq.* In low spirits or poor state of health; poorly, not quite well.

1857 'Dow, Jr.' *Dow's Patent Sermons* 1st Ser. 7 As mean..as a rooster in a thundershower. **1911** H. S. Harrison *Queed* vii. 90 Mebbe you could do better writing and harder writing if only you didn't feel so mean. **1911** J. F. Wilson *Land Claimers* i. 21 'Feel pretty mean,' the packer asked him kindly.

3. a. Of domestic animals or things in general: poor in quality or condition; comparatively worthless; unpleasant, disagreeable.

1817 in *Trans. Illinois State Hist. Soc.* 1910 148 Hogs in this Country are the meanest that I have ever seen... I do not believe you ever see half so mean hogs as we have here. **1823** W. Faux *Memorable Days Amer.* 219 The horses here are nearly all mean,..dwarfish things. **1842** C. M. Kirkland *Forest Life* I. 140 You've had a pretty mean time, I reckon. *a* **1890** in Barrère & Leland *Dict. Slang* (1890) II. 49/1 The night was dark and stormy, about as mean a night as was ever experienced in Washington. **1936** Ade *Let.* 29 Apr. (1973) 192 We arrived home on the 17th without mishap and almost immediately ran into mean weather including one snow fall which completely covered the ground. **1948** 'N. Shute' *No Highway* ix. 245 Eight handles it [*sc.* a coffin] had, for carrying, but gee, that was a mean load. **1973** *Kingston* (Ontario) *Whig-Standard* 11 Aug. 7/5 That was just about the meanest electric storm I ever sat through.

5. b. (Later examples.)

(*a*) **1841** 'Dow, Jr.' *Short Patent Sermons* 78 [One girl] thought me real mean for uttering such super-diabolical sentiments. **1891** R. T. Cooke *Huckleberries* 14 It would be awful mean of me to leave you here alone.

(*b*) **1862** R. H. Newell *Orpheus C. Kerr Papers* 1st Ser. ii. 21, I see he felt powerful mean, so I walked up to him. **1884** 'Mark Twain' *Huck. Finn* xli. 421 [She] tucked me in, and mothered me so good I felt mean.

d. *colloq.* (orig. *U.S.*). Remarkably clever, adroit, etc.; excellent; formidable.

1920 H. C. Witwer in *Collier's* 15 May 6/3 Everything was jake until K. O. Krouse shook a mean dice and win $28 from Battlin' Lewis on the way to Toledo. *Ibid.* 57/2 You never heard tell of Kane Halliday?.. The big..foot-ball star, the weights thrower..what they call a round-about athalete? *You* know, one of them bimbos which flings a wicked spear and hurls a mean hammer and that there stuff, get me? *Ibid.* 62/2 Your wonder child may pack a mean wallop. **1924** *Ladies' Home Jrnl.* Feb. 21/1 'That Lucy Layman sure does shake a mean foot,' began Edmond airily. **1931** D. Runyon *Guys & Dolls* (1932) iii. 64 She swings a very mean skillet, and gets me up some very tasty fodder. **1963** *Economist* 23 Nov. 775/2 Mr Ronnie Scott..plays a mean saxophone. **1968** *Globe & Mail* (Toronto) 13 Jan. 34/7 He blows a mean trumpet and sings well, too. **1973** *Observer* 9 Sept. 30/3 Does a mean goulash, taught him by his grandmother and perfected in Hungary. **1973** *Listener* 25 Oct. 578/2 Jack Palance smokes a mean cigar in *Oklahoma Crude*.

7. *mean-faced* adj.

1918 Mrs. Belloc Lowndes *Out of the War?* 211 Thin, mean-faced, yet sharply intelligent-looking man. **1953** E. S. Grenfell in C. K. Stead *N.Z. Short Stories* (1966) 69 A shiftless, mean-faced fellow.

B. *adv.*[1] (Later U.S. example.)

1861 O. W. Norton *Army Lett.* (1903) 26 Virginia has acted meaner than South Carolina.

mean, *a.*[2] and *adv.*[2] Add: **A.** *adj.* **7. a.** *mean square*: the (arithmetical) mean of the squares of a set of numbers; *mean-square deviation* (*mean-square error*, † *error of mean square*): the mean of the squares of the differences between a set of numbers and some fixed number; usu. identical with the variance; formerly, † the square root of this (usu. identical with the standard deviation).

1845 *Encycl. Metrop.* II. 460, k' and k'' are approximately found by taking the actual mean, and the actual mean square, of a large number of observed durations. Again, the mean risk of error in estimating a single duration is $\cdot 39894\sqrt{[(k''-k'^2)\div s]}$. **1894** Error of mean square [see *deviation 2 d]. **1956** A. A. Townsend *Struct. Turbulent Shear Flow* iii. 44 The contributions to the mean-square rate of strain of the eddies larger than $1/k$ come mostly from values of $E(k')$ for k' near k. **1968** R. A. Lyttleton *Mysteries Solar Syst.* vii. 245 This regrouping of the equations of condition..results in a further reduction of the number of unknowns at the price..of increasing the mean-square error associated with each equation. **1970** S. Brandt *Statistical & Computational Methods* vi. 83 A sum of squares..divided by the number of degrees of freedom is called the mean square or more explicitly the mean-square deviation... Its square root (which has the dimension of the measured quantity, i.e. the mean) has the lengthy name root-mean-square deviation.

mean, *v.*[1] Add: **1. f.** *to mean well* (used ironically).

1813 Jane Austen *Pride & Prej.* III. v. 108 Perhaps she *meant* well, but, under such a misfortune as this, one cannot see too little of one's neighbours. **1889** F. Anstey in *Granta* 17 May 9/2 Still, with all their presumption, they meant well, poor fellows! **1910** R. Brooke *Let.* 9 Jan. (1968) 206 He is a silly man... Yet he means well. **1961** G. Smith *Business of Loving* ix. 206 Felix obviously meant well. **1973** S. B. Jackman *Guns covered with Flowers* x. 159 He smiled apologetically, 'He means well'. Stevens grinned. 'And you can't say worse than that about anyone.'

2. b. (Earlier and later examples.)

1854 Thackeray *Newcomes* I. xxix. 287 What the devil do you mean about your Chimène and your Rodrigue? **1930** G. B. Shaw *Apple Cart* i. 16 What do you mean? Isn't it what I have always said?

c. (*if*) *you know, see, understand, what I mean*, i.e. 'have I made it clear?'

1846 G. E. Jewsbury *Sel. Lett. to Mrs. Carlyle* (1892) 203 There would be a want of reverence in it, if you understand what I mean. **1968** *Guardian* 24 Apr. 9/8 If I thought..he was going to back-chat me like he does now ..I'd half-kill him now, you know what I mean? **1968** *Listener* 30 May 711/3 A braying Brooklyn accent studded with 'You know what I means'. **1974** *Sunday Times* 20 Jan. 12/4 [He'll] be only too keen to get back to his boat, if you see what I mean.

3. b. Of a person or thing: to be of some account or importance, to 'matter' *to* (someone); to be a source of benefit, or an object of regard, affection, or love *to* (someone).

1888 Mrs. H. Ward *R. Elsmere* II. iii. xxvi. 279 It was only by a great effort that he could turn his thoughts from the Squire, and all that the squire had meant to him during the past year. **1912** *Red Mag.* 1 Mar. 515/1 It came over me how much she meant to me and how hard a wrench it was going to be to live along without her. **1914** *Times Lit. Suppl.* 8 Jan. 12/1 Comprehension of what Lady Gregory has meant to him and to others who worked with her. **1922** Joyce *Ulysses* 356 He would never understand what he had meant to her.

c. *pass.* To be destined (by providence); to have special significance.

1897 Kipling *Captains Courageous* viii. 169 It couldn't have been *meant*. It was only the tide. **1956** M. Stewart *Wildfire at Midnight* i. 16 So *handy* having that address. It's as if it were *meant*. **1962** M. Allingham *China Governess* (1963) xii. 151 That's miraculous! That's what people mean when they say a thing is 'meant'. **1974** I. Murdoch *Sacred & Profane Love Machine* 239 When I need you, you are here. You must see how *meant* it all is.

mean, *v.*[3] Restrict † *Obs. rare* to senses in Dict. and add: **3.** *trans.* [f. Mean *a.*[2] 7 or *sb.*[2] 8.] To calculate the arithmetical mean of. Also with *up*.

1882 W. J. L. Wharton *Hydrogr. Surveying* 210 We need not mean up each column of times. *Ibid.* 213 When working several sets, calculate them simultaneously as far as this, and mean the results. *a* **1888** P. F. Shortland *Naut. Surveying* (1890) 64 The permanent errors will destroy each other in the results of all..observations so meaned.

meander, *v.* Add: **4.** *U.S.* To pass or travel deviously along or through (a river, etc.).

1821 in *Missouri Hist. Soc. Coll.* (1906) II. 61 We still continued meandering the Arkansas. **1831** J. O. Pattie *Personal Narr.* 13 We crossed the Missouri..and meandered the river as far as Pilcher's fort. **1839** Z. Leonard *Adventures* (1904) 69 We separated, each party to meander the rivers that had been respectively allotted to them. **1839** in *Trans. Mich. Agric. Soc.* (1856) VII. 360 Branches of Swan creek meander this track in such manner as to facilitate drainage.

me-and-you. *slang.* [Jocular adaptation of colloq. pronunc. (mī·niū) of Menu.] = Menu 2.

1932 P. P. *Rhyming Slang* 23 Me and you..Menu. **1943** N. Marsh *Colour Scheme* iv. 68 Come on, Beautiful. Let's have a slant at the me-and-you.

meanie, var. *meany.

meaning, *vbl. sb.*[1] Add: **6.** *attrib.* and *Comb.*, as *meaning-analysis*, *-area*, *-change*, *-content*, *-relation*, *-relationship*, *-unit*; *meaning-bearing*, *-carrying* adjs.

1936 Wirth & Shils tr. *Mannheim's Ideology & Utopia* i. 46 There are..vast possibilities of precision in the combination of meaning-analysis and sociological situational diagnosis. **1966** C. G. Hempel *Philos. of Nat. Sci.* viii. 103 The characterization of these substances by their molecular structure is arrived at, not by meaning analysis, but by chemical analysis. **1958** C. Rabin in *Aspects of Translation* vii. 124 The object in question has to be assigned either to the meaning-area of *watch* or to that of *clock*. **1957** N. Chomsky in Saporta & Bastian *Psycholinguistics* (1961) 266/2 We have counter examples to the suggestion.. that morphemes be defined as minimal meaning-bearing elements. **1972** *Archivum Linguisticum* III. 83 Generativists have attempted to remove all meaning-bearing elements from transformations into the deep structure. **1965** H. A. Gleason *Ling. & Eng. Gram.* 192 We might have had the meaning-carrying verb *washed*. **1954** F. G. Cassidy *Robertson's Devel. Mod. Eng.* (ed. 2) ix. 232 The various types of meaning-change. **1968** M. Black *Labyrinth of Lang.* vii. 163 A reasonably adequate account of the nature of such meaning-changes..is bound to be very complicated. **1946** Priebsch & Collinson *German Lang.* (ed. 2) ii. iii. 234 In several Indo-European languages are types of compounds in which the meaning-content or 'determination' is left unexpressed. **1948** *Mind* LVII. 80 Each separately is intelligible by itself and has its own meaning-content independently of the other. **1934** Priebsch & Collinson *German Lang.* ii. iii. 208 From the point of view of the meaning-relation of their component words compounds fall into several well-marked classes. **1963** J. Lyons *Structural Semantics* ii. 28 The only meaning-relation that is relevant in the case of phonemes is sameness and difference of meaning. **1961** R. B. Long *Sentence & its Parts* i. 11 Meaning relationships are obviously varied. **1966** *English Studies* XLVII. 256 We find the same meaning-relationship between the adjunct clause and its verb. **1938** I. Goldberg *Wonder of Words* xvi. 323 The smallest meaning-unit of a word is called a semanteme. **1962** W. Nowottny *Lang. Poets Use* vii. 162 What we call the 'same word' is not a single meaning-unit.

meaningful, *a.* Add: (Later examples.) Also, amenable to interpretation; having a recognizable purpose or function; *spec.* in *Logic*, resulting from the application of the rules of a language or sign system; able to function as a term in such a system.

1922 *Times Educ. Suppl.* 29 Apr. 200/2 The pupils see at once that their studies are meaningful. **1934** Cohen & Nagel *Introd. Logic* ix. 185 It has been said that to the two alternatives *true* and *false*, there is a third, the *meaningless*... The question of what constitutes a meaningful expression is a large one. 1940 W. V. QUINE *Math. Logic* iv. 164 Under Russell's scheme an abstraction prefix..can be applied outright to any meaningful formula. 1942 T. C. POLLOCK *Nature of Lit.* ix. 192 If the charge is made that a particular work..is..'meaningless' ..a good critic may be able to discriminate privately.. that it is..highly meaningful. 1952 C. P. BLACKER *Eugenics* ix. 203 It is a different matter to test the tests and to prove them to be meaningful and workable. 1953 *Mind* LXII. 8 We often ask what a word means, but we do not ordinarily ask whether a word is meaningful or not. If it were not meaningful we would not call it a word. 1954 *Essays in Crit.* IV. 349 Tragedy affirms a cosmos of which man is a meaningful part. 1959 B. WOOTTON *Social Sci. & Social Path.* iii. 92 Meaningful conclusions are, moreover, inhibited by the inadequacy of the available evidence. 1970 O. DOPPING *Computers & Data Processing* i. 18 We can add the information contents from different parts of the message and obtain a meaningful result from the addition. 1971 *Sat. Rev.* (U.S.) 18 Dec. 59/1 All of us..see a need to be related in a meaningful way to the black experience. In our case, this means teaching black people. 1973 *Times* 9 Feb. 24/4 Federation met again last month to try to get meaningful talks going again on a new disputes machinery. 1973 *Physics Bull.* May 281/3 The phoneme is the smallest meaningful unit of sound a listener can perceive.

meaningfully, *adv.* (Later examples.)
1937 *Mind* XLVI. 385 The finitist must hold..that of 'the sequence of time intervals' also, it cannot meaningfully be supposed that it denotes an infinite extension. 1955 *Sci. Amer.* Aug. 84/2 We cannot meaningfully look for any further sense of 'rational'. 1961 J. MCCABE *Mr. Laurel & Mr. Hardy* (1962) vi. 122 He must arrange the sequence of action clearly, meaningfully, dramatically.

meaningfulness (mī·niŋfŭlnės). [f. MEANINGFUL *a.* + -NESS.] The fact or quality of being meaningful.
1919 R. H. FISHER *Outside of Inside* 105 In mystical trance he discerned the meaningfulness of the third heaven. 1922 J. Y. SIMPSON *Man & Attainment of Immortality* xiv. 320 The exquisite sensitivity of their minds to the meaningfulness of his language. 1935 *Mind* XLIV. 426 'Redness' is a concept which has application in experience, and this is all that is required for meaningfulness. 1940 W. V. QUINE *Math. Logic* iii. 147 The meaningfulness of an expression—the eligibility of an expression to occur in statements at all, true or false—is a matter over which we can profitably maintain control. 1941 A. D. WOOZLEY in T. Reid *Ess.* p. xxix, He is far more interested in..universals as related to words—*i.e.* in the meaningfulness of language. 1951 *Trans. Philol. Soc.* 43 Here also the fundamental argument is meaningfulness. 1953 *Essays in Crit.* III. 107 Verification of meaningfulness in literature must lie, in the last analysis, outside literature. 1959 *Cambr. Rev.* 30 May 567/2 Whether or not the verifiability criterion is a criterion of the meaningfulness of a statement or of the scientific character of a statement, the defects which it has in either case can be reproduced *quid pro quo* for the falsifiability criterion. 1959 H. GARDNER *Business of Crit.* II. iii. 131 Our first step towards making it [*sc.* a play or poem] meaningful to us is to be aware of the meaningfulness of the images to men of its own day. 1971 *Human World* Nov. 40 The question of the meaningfulness of religious utterances.

meaningless, *a.* Add: (Later examples.) Also, not 'meaningful' (cf. *MEANINGFUL *a.*).
1890 W. JAMES *Princ. Psychol.* I. xvi. 676 He learned lists of meaningless syllables by heart. 1934, 1942 [see *MEANINGFUL *a.*].

mean while, meanwhile. Add: **A.** *sb.* **1.** (Later examples.)
1908 H. G. WELLS *First & Last Things* III. viii. 123 The organized state..has not arrived..and in the meanwhile they must act like its anticipatory agents. 1960 G. SANDERS *Mem. Professional Cad* II. viii. 170 In the meanwhile the boat..evidently decided to end it all. 1972 *Listener* 1 June 705/2 In the meanwhile sanctions would be continued.
2. After '*Obs.*' add 'exc. *arch.*'
1922 JOYCE *Ulysses* 381 This meanwhile this good sister stood by the door.

meany (mī·ni). *colloq.* Also **meanie.** [f. MEAN *a.*^{1} + -Y^6, -IE.] A mean-minded or stingy person.
1927 H. C. BROWN *In Golden Nineties* iii. 107 It was whispered by some old meanies that many of the five-foot floral offerings were purchased by the actors themselves and sent to the theatre with fictitious names attached. 1928 J. P. MCEVOY *Showgirl* xiii. 212/1 This old meany.. last night..found himself, much to his aged surprise, in the Klaw Theatre. 1936 L. C. DOUGLAS *White Banners* xiv. 315 Colonel Livingstone was an old meanie. 1951 J. B. PRIESTLEY *Festival at Farbridge* II. i. 34 He was at heart, she felt, a cunning old meanie. 1974 *Times* 15 Feb. 4/4 A bunch of local 'baddies' reinforced by 'meanies'.

mearing: see MERING *vbl. sb.* in Dict. and Suppl.

measle, *sb.* **1.** (Later examples of *sing.*)
1924 GALSWORTHY *White Monkey* I. iii. 21 Fleur knew how catching the word was; it would run like a measle round the ring. 1948 MENCKEN *Amer. Lang. Suppl.* II. 383 False singulars, made by back formation, are numerous, *e.g.*,..*measle*, nor are they confined to the untutored.

measle, *v.* Add: **4.** *fig.* To be full of or teem *with* (objectionable things). *rare.*
1856 C. READE *Never too Late* II. xxv. 245 All this..in thieves' cant, with an oath or a nasty expression at every third word. The sentences measled with them.

measly, *a.* **4.** (Earlier and later examples.)
1864 M. E. BRADDON *Henry Dunbar* II. xi. 212 The audacity to offer a measly hundred pounds or so for the discovery of a great crime! 1905 *Dialect Notes* III. 14 *Measly*,..poor. 'I don't want that *measly* stuff.' 1919 E. O'NEILL *Ile* in *Moon of Caribbees* (1923) 13 Did you ever hear o' me pointin' s'uth for home with only a measly four hundred barrel of ile in the hold? 1973 J. PORTER *It's Murder with Dover* vi. 55 Ten measly years in the nick doesn't worry anybody. 1974 *Sunday Tel.* 9 June 34/4 A spineless exhibition by the early Yorkshire batting—they have mastered only a measly five batting points all season—put them on the rack yet again.

measurability. Delete '*rare*' and add further examples.
1955 W. PAULI *Niels Bohr* 71 Measurability of electromagnetic fields. 1957 *Times Lit. Suppl.* 18 Oct. 620/2 A chapter on 'Illusionism and Perspective' deals with..measurability in quattrocentro compositions. 1969 *Nature* 14 June 1033/1 Though a physical quantity is considered primarily an attribute of the object under investigation, the need for measurability implies that the characteristic features of the relevant instruments of observation are an integral part of the quantity. 1975 *Times* 26 May 6/2 The main cause of this domination is measurability. Money provides an incontrovertible unit of measurement.

measure, *sb.* Add: **2. a.** *made to measure* (later example). Also *transf.* and *fig.*: fashioned to fulfil specified requirements; appropriate for a particular purpose; chiefly *attrib.*
1928 *Punch* 11 July p. xix/1 (Advt.), It is to give all the benefits of made-to-measure shoes..that these Lotus Bespoke Models have been created. 1937 *Evening News* 5 Feb. 8/2 We shall have made-to-measure houses..with provision for adding rooms as families grow. 1958 *Listener* 9 Oct. 563/2 The new knowledge..may ultimately enable us to produce made-to-measure mutant viruses. 1959 *Ibid.* 2 Apr. 608/1 The Boat Race..made to measure for television.
16. *spec.* Applied to Old English verse.
1774 R. HENRY *Hist. Gt. Brit.* II. 432 The kinds and measures of their [*sc.* Saxon and Danish poets'] verses. 1802 J. SIBBALD *Chron. Scottish Poetry* IV. p. lviii, In the same kind of measure [as the Fragment of the genuine Cædmon] are almost all the popular rhymes which still continue to be repeated by children in their ring-dances. 1873 H. MORLEY *First Sk. Eng. Lit.* 20 There is one measure for Beowulf, Cædmon's Paraphrase, and all subsequent First English poems. 1877 H. REHRMANN *Ess. Anglo-Saxon Poetry* 10 A foot or measure is made up ..of one accented syllable and its connected unaccented syllable, or syllables. 1942 J. C. POPE *Rhythm of Beowulf* 44 According to Heusler, alliteration must introduce the first measure in the second half-line.

measure, *v.* Add: **2. c.** *transf.* and *fig.* uses.
1824 R. HUMPHREYS *Mem. J. Decastro* 156 Mr. Cross.. when he wrote a character for a person, measured the extent of his genius the same as a tailor does the body for a garment, and was generally very successful in making a *good* fit. 1859 B. JERROLD *Life D. Jerrold* vi. 94 The pig was to be measured for his part. *a* 1890 in Barrère & Leland *Dict. Slang* (1890) II. 49/1 He had been measured for a funeral sermon three times, he said, and had never used either one of them. 1896 FARMER & HENLEY *Slang* IV. 296 *To have been measured for a new umbrella*..(American), —(1) To appear in new but ill-fitting clothes; whence (2) to pursue a policy of doubtful wisdom. 1942 BERREY & VAN DEN BARK *Amer. Thes. Slang* § 117/16 Be buried,.. be measured for a new overcoat.
4. b. Also, to be comparable *with*.
1904 G. PARKER *Ladder of Swords* vi. 61 Her words for the great cause had measured well with her deeds. 1907 *Smart Set* Mar. 126/2 The dog moved a cat-like step forward, making up the interval, and the man made a mental note that its single stride measured with his.
c. *fig.* *to measure up to*: to be equal (in ability, etc.) to; to match (cf. *match up to* s.v. *MATCH *v.*^{1} 8 b); to have the necessary qualifications for. orig. *U.S.*
1910 *N.Y. Even. Post* 16 Dec. 8 A man should be found for Senator who in ability and character will measure up to the just demands of such a situation. 1930 J. W. JOHNSON *Black Manhattan* xiv. 177 These plays..made a high demand on the versatility of the company... The demand was fully measured up to. 1931 G. T. CLARK *Leland Stanford* xii. 405 Stanford..had in mind the problem of selecting some one who measured up to his ideal. 1958 *Spectator* 7 Feb. 181/2 The later period, however, does not measure up to the earlier in the originality or interest of its political literature.

measured, *ppl. a.* Add: **2. a.** *measured mile*: a distance of one mile carefully measured, esp. such a distance used for determining the speed of a ship; also *attrib.*
1608 J. CHAMBERLAIN *Let.* 15 July (1939) I. 262 A gentleman wan a great wager for riding five measured miles..twenty times over in less than five howres. 1666 [in Dict.]. 1720 *Britannia Depicta, or Ogilby Improved* (title-page), Wherein are..Engraven, All ye Cities.. scituate on or near the Roads, with their respective Distances in Measured and Computed Miles. 1843 *Pract. Mechanic* June 358/2 The *Mermaid* then..performed the measured mile at the rate..of above thirteen miles through the water per hour. 1901 *Trans. Inst. Naval Archit.* XLIII. 23 The whole idea of the design has been, not so much to procure a very high measured-mile speed for a few hours, as to assure a good, continuous ocean speed. 1950 *Engineering* 10 Feb. 147/1 Measured-mile posts were in existence at Hartley..prior to 1884.

measurely, *adv.* (Later U.S. example.)
1878 J. H. BEADLE *Western Wilds* xiv. 222 It is measurely free from winter storms.

measurer. **1.** (Examples of special use.)
1636 *Early Rec. Dedham, Mass.* (1892) III. 36 We doe order yt all high wayes..he orderly set out by our Measurer. 1641 *Rec. Colony & Plantation New Haven* (1857) I. 51 Bro: Pecke chosen measurer for the towne to fill and strike all the corne. 1706 *Rec. Early Hist. Boston* (1882) VIII. 37 Allexander Seers, Samuell Bridge,..to serve as Measurers of board, Timber, and Plank. 1827 DRAKE & MANSFIELD *Cincinnati in 1826* vi. 51 The council have power to appoint..Measurers of wood and coal. 1841 in C. Cist *Cincinnati in 1841* (Advt.), George Warren, Measurer of Stone-work, Birch-work, and Plastering. 1916 BLUNDEN *Harbingers* 24 The binman found the measurer pleased, For hops were clean and work was through. 1972 *Classification of Occupations* (Dept. of Employment) III. 408/2 Workers in this group weigh and otherwise measure materials, goods and products,..for example:..Measurers.

measuring, *ppl. a.* **b.** **measuring-worm** (earlier and later examples).
1843 J. E. DEKAY *Zool. N.Y.* VI. 41 It walks after the manner of some caterpillars called Measuring worms. 1903 W. J. HOLLAND *Moth Bk.* 323 The larvæ, which are commonly known as 'measuring-worms', 'span-worms', or 'loopers', have the power in many cases of attaching themselves by the posterior claspers to the stems and branches of plants, and extending the remainder of the body outwardly at an angle. 1939 DUNCAN & PICKWELL *World of Insects* x. 172 Certain large measuring-worms, of the family Geometridae, have the ability, when danger threatens, to stiffen themselves out at angles to the stem on which they have been feeding or crawling, and thereafter for several minutes or hours to resemble perfectly bare and lifeless twigs. 1956 W. R. BIRD *Off-Trail in Nova Scotia* viii. 215, I saw some of those long 'loopers' or 'measuring worms' that I hadn't seen since a boy.

meat, *sb.* Add: **1. c.** (Later examples.)
1853 MRS. GASKELL *Cranford* xv. 296 After that she acknowledged that 'one man's meat might be another man's poison'. 1902 J. CONRAD *End of Tether* xiv, in *Youth* 370 One man's poison, another man's meat. 1905 A. BURVENICH *Eng. Idioms* 240 It is nuts to him; meat and drink to him, viz. the very sort of thing he likes. 1914 G. B. SHAW *Misalliance* 17 Whats one woman's meat is another woman's poison. 1929 J. B. PRIESTLEY *Good Companions* I. iii. 82 She had a trick of repeating phrases, raising her voice the second time, that had been meat and drink to mimics at Washbury for years. *a* 1930 D. H. LAWRENCE *Phoenix* (1936) 701 In the free, spontaneous self, one man's meat is truly another man's poison. And therefore you *can't* draw any average..unless you are going to poison everybody. 1937 PARTRIDGE *Dict. Slang* 515 *Meat, the nearer the bone the sweeter the*, a..low catchphrase applied by men to a thin woman. 1939 F. THOMPSON *Lark Rise* i. 20 In spite of their poverty and the worry and anxiety attending it, they were not unhappy, and, though poor, there was nothing sordid about their lives. 'The nearer the bone the sweeter the meat', they used to say. 1939 N. MARSH *Overture to Death* xxii. 254 I'm no psycho-analyst, but I imagine she'd be meat and drink to any one who was.
3. Also, *local U.S.*, confined to certain types of meat, usu. pork.
1832 J. K. PAULDING *Westward Ho!* I. 124 Nothing is called meat in these parts but salt pork and beef. 1845 C. M. KIRKLAND *Western Clearings* 93 Venison is not 'meat' to be sure, in our parlance; for we reserve that term for pork, *par excellence*. 1891 *Fur, Fin & Feather* 182 A bearskin is worth $5 to him..besides, he likes the flesh if meat (i.e. pork) is 'skeerse'. 1902 *Dialect Notes* II. 239 *Meat*, bacon always understood. 1903 *Ibid.* 320 *Meat*,..pork. Not often applied to beef, mutton, etc. 1927 *Ibid.* V. 469 *Meat*,..ham; —used only of the hog.
c. (*b*) Also, matter of importance or substance; the gist or main part (of a story, situation, etc.). Cf. MEATY *a.* 1 b.
1901 KIPLING *Kim* xv. 390 At evening time..she won to the meat of the matter, explained low-voicedly by the lama. 1937 *Jrnl. R. Aeronaut. Soc.* XLI. 1025 There was so much real meat in this paper that it was impossible to enter into any long discussion about it. 1942 *Tee Emm* (Air Ministry) II. 129 Delving into detail, digging out the meat, and giving your advice. 1951 in M. McLuhan *Mech. Bride* (1967) 36/4 It is not only full of meat, but so interestingly written that I am going to loan it around the store. 1955 *Bull. Atomic Sci.* June 226/3 But the real meat of the book is in the depiction of the moral conflicts keenly felt by these men. 1960 *Times* 23 June 3/4 This makes the meat of Wimbledon. 1970 *Nature* 12 Sept. 1092/2 Shift registers..perform the meat of a computer calculation.
d. *transf.* Delete example in Dict. and add:
1872 'MARK TWAIN' *Roughing It* I. 357 Come along—you're my meat *now*, my lad. 1907 S. E. WHITE *Arizona Nights* I. vii. 136 'Whew!' I whistles, 'That's a large order—But I'm your meat.' 1917 A. G. EMPEY *From Fire Step* xvi. 103, I gleefully fell in with the scheme, and told Cassell I was his meat. 1922 A. BENNETT *Let.* 14 Nov. (1966) I. 319 It was not everybody's meat, but it was in my opinion somebody's meat. 1922 E. O'NEILL *Hairy Ape* (1923) iii. 29 Say, dis is a cinch! Dis was made for me! It's my meat, get me! 1942 W. E. JOHNS *Biggles sweeps Desert* v. 66 The way you snaffled my Hun! I call that a bit thick... He was my meat, absolutely, yes by Jingo. *Ibid.* 67 'You were jolly nearly his meat,' Biggles pointed out, coldly. 1956 B. HOLIDAY *Lady sings Blues* (1973) iv. 46 'I Cried' was my damn meat, just like 'Rhythm' was Lester's. 1972 *Dict. Contemp. & Colloq. Usage* (Eng.-Lang. Inst. Amer.) 19 *Meat*,..one's field of interest;..as: Math, that's my *meat*.

e. *coarse slang.* The penis; the female genital organs; the human body regarded as an instrument of sexual pleasure; a prostitute.
1595 GOSSON *Pleasant Quippes* sig. B2 That you should coutch your meat in dish, And others feele, it is no fish. 1597 SHAKES. *Henry IV: Part Two* (1623) II. iv. 83/1 Away you mouldie Rogue, away; I am meat for your Master. 1611 L. BARRY *Ram-Alley* v. sig. H4v Faith take a maide, and leaue the widdow, Maister. Of all meates I loue not a gaping oyster. 1664 T. KILLIGREW *Parson's Wedding* v. ii, in *Comedies & Tragedies* 142 Your bed is big enough for two, and my meat will not cost you much. 1860–1 W. WHITMAN *Leaves of Grass* (ed. 3) 301 Feeling with the hand the naked meat of his own body. 1923 J. MANCHON *Le Slang* 192 *A bit of meat*, une putain. 1967 G. DAVIS in W. King *Black Short Story Anthol.* (1972) 341 Maxine's mother was never home on Saturday mornings, so I kept Maxine's three younger brothers outside while Teddy slipped the meat to her in the bedroom. 1970 G. GREER *Female Eunuch* 265 It would be unbearable, but less so, if it were only the vagina that was belittled by terms like *meat*. 1971 B. MALAMUD *Tenants* 31 I'm not saying I don't appreciate her company, especially when my meat's frying, but not when I have something I got to write. *Ibid.* 136 Sam wanted the brothers to beat up on you and crack your nuts for putting the meat to his bitch. *Ibid.* 143 I got you in bed with nothin on you You gonna eat my meat. 1971 *Black Scholar* Sept. 36/1 She was in his arms..and gabbing his erect meat. 1973 D. BARNES *See Woman* (1974) 94 I've tried the white meat..so I can understand why you might be wondering if the dark meat isn't better.
f. The centre (of a cricket bat, of the head of a golf club, etc.), esp. in phr. *to hit* (a ball) *on* or *with the meat*. *slang.*
1909 *Westm. Gaz.* 15 Jan. 4/2 If you did not take the gutta-percha ball right in the middle of the club (right 'on the meat', according to the modern abominable phrase) it declined to go at all. 1922 WODEHOUSE *Clicking of Cuthbert* ix. 203 You think..that lovely woman loses in queenly dignity when she fails to slam the ball squarely on the meat? 1925 *Country Life* 11 July 48/2 It is easy to drive a lob bowler..on the 'meat' or drive of the bat. 1959 R. FULLER *Ruined Boys* II. ix. 143 Wilkes hit the second ball of the over with the meat of the bat. 1963 *Times* 28 Jan. 4/3 It was apparent that here was the severest and purest hitter in the game at the pinnacle of his form, tuning up as though the ball were tied to the meat of his racket by a string of elastic. 1974 *Guardian* 6 Aug. 23/2 Kitchen was well held, full off the meat, by Younis at forward short leg in Arnold's first over.
5. a. *meat can, -platter, -tin, -trough, -tub* (earlier example), *vat.* **b.** *meat-chopper, -eating* sb. and adj. (examples), *-freezer, -freezing* (examples of *attrib.* use), *-packer, -packing, -producer, -producing* adj., *rationing, -tenderizer.* **d.** similative, as *meat-faced, -pink* adjs.
1897 *Outing* XXX. 284/1 For active service..the two regiments would need to be supplied with..meat cans. 1868 *Mich. Agric. Rep.* VII. 348, 1 lightning meat chopper. 1922 JOYCE *Ulysses* 302 In the course of the argument..meatchoppers..were resorted to. 1956 *Amer. Speech* XXXI. 87 Back-formations from nouns in *-er* indicating agents, such as..*meat chopper*. 1853 *Trans. Mich. Agric. Soc.* IV. 154 The Americans are notoriously a meat eating people. 1905 *Vegetarian Messenger* Jan. 14 Vegetarianism *v.* meat-eating. 1921 G. B. SHAW *Back to Methuselah* II. 77 One of his sons invented meat-eating. 1939 DYLAN THOMAS *Map of Love* 24 Sewing a shroud for a journey By the light of the meat-eating sun. 1958 A. R. RADCLIFFE-BROWN *Method in Social Anthropol.* I. v. 115 The two chief meat-eating birds. 1922 JOYCE *Ulysses* 33 The meatfaced woman, a butcher's dame. 1909 *Daily Chron.* 2 Nov. 5/4 Australian globe-trotters, meat-freezers, financiers. 1908 *Westm. Gaz.* 14 Dec. 2/1 The meat-freezing works employ over 3,000 men. 1909 *Chambers's Jrnl.* Jan. 23/1 Argentina..had in 1884 the first meat-freezing works established. 1903 E. JOHNSON *Amer. Railway Transportation* 131 The large meat-packers..own their own cars. 1921 *Daily Colonist* (Victoria, B.C.) 7 Apr. 8/1 The strike situation between the meat packers and their employees remained unchanged this morning. 1973 *Guardian* 23 Feb. 18/1, I finally got a job as a meat packer on £18 a week. 1873 *Rep. Iowa Agric. Soc. 1872* 175 The panacea for all these ills is to be found in tanneries..meat packing and curing houses, [etc.]. 1891 J. J. FLINN *Chicago* 330 Meat packing is the oldest of Chicago's industries. 1907 G. B. SHAW *Major Barbara* Pref. 179 Who chips a corner of the veneering from the huge meat packing industries of Chicago. 1974 *Daily Colonist* (Victoria, B.C.) 28 July 16/8 Meatpacking companies have reaped most of the benefits of the present subsidy while producers received very little. 1939 AUDEN & ISHERWOOD *Journey to War* i. 31 Hairy, meat-pink men. 1863 'G. HAMILTON' *Gala-Days* 75, I decided upon a meat-platter. 1916 *Daily Colonist* (Victoria, B.C.) 2 July 7/1 (Advt.), Large Meat Platters, regular value 60 c. July Clearance each 35 c. 1909 *Westm. Gaz.* 14 Dec. 2/1 The consumer is called upon to pay £5,600,000, in order that..the Colonial meat-producer may receive the negligible gift of £350,000. 1932 J. S. HUXLEY *Probl. Relative Growth* vi. 202 Animals cannot show their full potentialities as meat-producers if kept in unfavourable nutritive conditions. *Ibid.* iii. 90 Meat-producing animals. 1918 *Times* 7 Feb. 3/1 Should the currency coupon become the basis of meat rationing, it is probable that [etc.]. 1969 *New Yorker* 27 Sept. 122/2 Unite, slaves of the steam kettle and the shakers of meat tenderizer and MSG. 1970 M. SLATER *Caribbean Cooking* 12 The pulp [of pawpaw] contains pepsin and the skin and seeds are useful as meat tenderisers. 1889 *Century Mag.* Apr. 909/2 They say that he sometimes fills an old meat-tin with water in anticipation of a long march. 1902 E. BANKS *Autobiogr. Newspaper Girl* 64, I would have been capable of going into the street and knocking down any little butcher's boy who refused peaceably to deliver up to me the contents of his wooden meat-trough. 1779 E. PARKMAN *Diary* (1899) 171 We are unhappily low in ye Meat Tub. 1847 *Rep. Comm. Patents 1846* (U.S.) 310 The mode by which I obtain a vacuum in meat vat A. for curing meat.

6. **meat-ant**, a name used in Australia for an ant of the genus *Iridomyrmex*, esp. *I. detectus*, a large copper-coloured ant; **meat ball** (see quot. 1970); also *fig.*; also used *attrib.* to designate a type of landing system for aircraft (*U.S. slang*); **meat-block**, a block of wood on which meat is cut up; **meat-breakfast**, a breakfast that includes a meat dish; **meat card**, a card entitling the holder to a ration of meat; **meat coupon**, one of the coupons of which a meat card is made up; **meat cube**, a small cube of concentrated meat extract; **meat-grinder**, a mincing machine; also *fig.*; **meat-head** *slang* (chiefly *U.S.*), a stupid person; so **meat-headed** *a.*, stupid; **meat hog**, a hog intended for food (*U.S., rare*); **meat-hook**, (*a*) in Dict., sense 5 a; (*b*) an arm or hand *slang*; **meat-house**, (*a*) a house in which meat is hung; (*b*) *dial.*, 'a larder; *fig.* a house where a liberal allowance of good food is given' (E.D.D., *q.v.*); (*c*) *slang*, a brothel; **meat jelly**, a jelly prepared from meat; **meat loaf** = *LOAF *sb.*^{1} 2 e; **meat man**, (*c*) (examples); (*d*) a man responsible for supplying meat to a camp; **meat-market**, (*c*) *slang*, a rendezvous for prostitutes, homosexuals, etc.; **meat rack** = **meat-market* (c); **meat ticket**, (*a*) *Mil. slang*, an identity disc; (*b*) = **meal ticket* (b); **meat tool** *coarse slang*, the penis; **meat train**, the men, horses, etc., conveying meat to a party; **meat-wagon** *slang* (chiefly *U.S.*), (*a*) an ambulance; (*b*) a police van, 'black Maria'; (*c*) a hearse; **meat-works** chiefly *Austral.* and *N.Z.*, an establishment where meat is processed and packed; also, a slaughterhouse.
1907 W. W. FROGGATT *Austral. Insects* 95 Most of them [*sc.* ants of the genus *Iridomyrmex*] are small, except our 'Mound Ant', sometimes known as the 'Meat Ant', *Iridomyrmex detectus*, which is the commonest and most widely distributed ant in Australia. 1952 *Coast to Coast* 101 The ant-lion had seized the meat-ant by one leg. 1970 BROWN & TAYLOR in *Insects of Australia* (Commonwealth Sci. & Industr. Res. Organization) xxxvii. 958/1 The meat ants (*Iridomyrmex* spp.) can be a serious pest around homes and food-processing plants. *c* 1838 C. MATHEWS in M. R. Booth *Eng. Plays of 19th Cent.* (1973) IV. 138 You must be content with pâté or forced meat ball. 1877 *Cassell's Dict. Cookery* 413/1 *Meat Balls, Minced, Fried.*—Take some roast mutton, some chestnuts, and neck of veal boiled in water with salt and vinegar. 1941 P. GALLICO in *Sat. Even. Post* 14 June 110/2 As for that revolting meat-ball, I never wish to see him again. 1957 *Economist* 31 Aug. 688/2 The mirror reflects a bright light astern and upward into a beam which the pilot follows straight to a landing by keeping the 'meatball' light precisely centred in the mirror. 1957 *New Yorker* 5 Oct. 82/3 I'll polish off a meatball sandwich. 1960 *Aeroplane* XCIX. 65/1 The equipment evaluated by the F.A.A. included the U.S.A.F. 'Meat Ball' system, the U.S. Navy mirror system. *Ibid.* 401/1 The pilot aligns the 'meatball', or blob of reflected light, in the centre of the deck-landing mirror. 1962 *Flight Internat.* LXXXII. 100/1 A steady descent at 250kt to 1,000 ft..would be followed by a further descent to 600 ft, with the talkdown continuing until the pilot picked up the 'meatball' light of the mirror sight. 1969 R. AIRTH *Snatch!* iv. 36 He looked a very tough meatball. 1970 SIMON & HOWE *Dict. Gastron.* 262/2 *Meat balls*, any combination of meat, raw or cooked, shaped into balls. 1838 E. FLAGG *Far West* II. 59 Mr. W...was on the stump, in shape of a huge meat-block at one corner of the market-house. 1857 GEO. ELIOT *Scenes Clerical Life* (1858) I. 105 The unpleasant circumstances..together with heavy meat breakfasts, may..have contributed to his desponding views. 1910 *Bradshaw's Railway Guide* Apr. 1149 *Hotel Bedford—Paris*... Room, meat breakfast, electric light.., from 6s. 1870 *Food Jrnl.* 1 Dec. 622 The restaurateurs are compelled to ask for their customers' 'meat card'. 1918 *Times* 6 Feb. 8/2 (*heading*), The London meat card. *Ibid.* 25 Feb. 9/5 Only three coupons each week of a meat card can be used for butcher's meat. *Ibid.*, You must not tear off meat coupons yourself. This duty rests with the retailer. 1919 'I. HAY' *Last Million* 97 'Got my meat coupons?' They shook their heads... 'Better have bacon and eggs,' announced Hebe. 'They're not rationed.' 1951 *Good Housek. Home Encycl.* 392/2, 1 meat cube, or 1 tsp. meat extract. 1971 *Guardian* 19 May 8/2 Make stock with a meat cube, sieve in left-over vegetables and you have an economical soup. 1951 M. MCLUHAN *Mech. Bride* (1967) 128/1 That man..whose school training wins him the privilege of getting at once into the technological meat grinder. 1969 R. & D. DE SOLA *Dict. Cooking* 150/1 *Meat grinder*, utensil or attachment for grinding meat, usually provided with a variety of cutting blades for different grinds. 1970 W. BURROUGHS JR. *Speed* vii. 145 I'd be sucked right into the actual meatgrinder of another reel. 1972 *Listener* 17 Aug. 199/1 The North Vietnamese.. managed to tie down South Vietnam's strategic reserve.. in a gruesome 'meat-grinder'. 1945 H. I. PHILLIPS *Private Purkey's Private Peace* i. 8 'Harriet!' mocked a battered buddy. 'Honey bunch! Lissen to the meathead.' 1967 *Boston Globe* 18 May 27/3 It seems to this meat-head that the building of a branch of the state university..gives all of us a great chance to upgrade some area which needs a little up-grading. 1971 *Newsweek* 29 Nov. 52/1 Archie Bunker, the middle American hero of 'All in the Family'.. sees himself menaced by a rising tide of spades,..meat-heads,..fags and four-eyes. 1949 W. R. BURNETT *Asphalt Jungle* (1950) vi. 72 Some meat-headed tart. 1856 G. N. JONES *Florida Plantation Rec.* (1927) 169, I doe not see but verry few of the shoats that I turned out for meat hogs this year. 1919 W. H. DOWNING *Digger Dial.* 33 *Meat-hook*, arm. 1932 *Amer. Speech* VII. 334 *Meat hooks*, hands. 1945 L. SHELLY *Jive Talk Dict.* 29 *Meat hooks*, the hands. 1805 W. CLARK in Lewis & Clark *Orig. Jrnls. Lewis & Clark Expedition* (1905) III. 284 Had the Meet house covered and the Meat all hung up. 1831 J. M. PECK *Guide for Emigrants* II. 126 Around it [*sc.* the cabin] are put a meat or smoke house [etc.]. 1862 R. HENNING *Let.* 28 Aug. (1966) 95 The woolshed, meathouse..and dwelling-house are so very well put up. 1896 FARMER & HENLEY *Slang* IV. 296/2 *Meat-house*, a brothel. 1906 *Dialect Notes* III. 146 *Meat-house*,..smoke-house; a wooden shed in which smoked and preserved meats are kept. 1923 J. MANCHON *Le Slang* 192 *Meat-house*,.. maison publique. 1937 PARTRIDGE *Dict. Slang* 515/1 *Meat-house*, a brothel. 1381 *Forme of Cury* (1780) xxxvi. 103 For to make mete Gelee that it be wel chariaunt. 1865 MRS. STOWE *House & Home Papers* 248 Those fine, clear meat-jellies which form a garnish..palatable to the taste. 1965 V. HOLLAND tr. *A. Escoffier's Ma Cuisine* 67 Beef tea and meat jelly for invalids. 1932 E. CRAIG *Cooking with E. Craig* 58 (*heading*) Banana and meat loaf. 1939, etc. Meat loaf [see *LOAF *sb.*^{1} 2 e]. 1831 R. COX *Adventures Columbia River* II. x. 222 The meat-men did not return until nine this morning..but at eleven the hunters' signal drew us to the shore, and the meat-men were despatched. 1910 *Dialect Notes* III. 445 *Meat-man*,..butcher, or driver of a butcher's wagon. 1920 *Ibid.* V. 83 *Meat man*, for butcher. 1896 FARMER & HENLEY *Slang* IV. 296/2 *Meat-market*,..any *rendezvous* of public women. 1937 PARTRIDGE *Dict. Slang* 515/1 *Meat-market*, a rendezvous of harlots. 1957 J. OSBORNE *Entertainer* i. 18 Every tart and pansy boy in the district are in that place... It's just a meat-market. 1973 *Amer. Speech 1970* XLV. 58 *Meat market* n., street on which homosexuals gather, cruise, and pick up tricks. 1972 B. RODGERS *Queens' Vernacular* 132 *Meat rack*, outdoor setting..where homosexuals gather to parade their wares. 1919 W. H. DOWNING *Digger Dial.* 33 *Meat Ticket*, (or Dead Meat Ticket)—an identification disc. 1925 FRASER & GIBBONS *Soldier & Sailor Words* 154 *Meat ticket*, the identity disc. 1929 *Papers Mich. Acad. Sci., Arts & Lett.* X. 307 *Meat-ticket*, wrist-tag for purposes of identification; 'dog-tag'. 1936 P. BOTTOME *Level Crossing* xvi. 192 He'd better not try to settle too much on my meat ticket! 1948 PARTRIDGE *Dict. Forces' Slang* 117 *Meat ticket*, identity disc. 1971 B. MALAMUD *Tenants* 88 What do you do..with your meat tool? You got no girl, who do you fuck other than your hand? 1845 J. C. FRÉMONT *Rep. Exploring Expedition* 234 The meat train did not arrive this evening, and I gave Godey leave to kill our little dog. 1925 J. H. TABER *Story of 168th Infantry* I. xvi. 189 By this time all of the old members of the regiment..cheerfully referred to the ambulance as the 'meat wagon'. 1939 *Forum* (N.Y.) July 42/1 He must have pulled his rip cord because he woke up in the meatwagon. 1942 BERREY & VAN DEN BARK *Amer. Thes. Slang* § 81/13 *Hearse*, meat cart, -crate, or wagon. 1943 R. CHANDLER *Lady in Lake* (1944) xvi. 95 Murder-a-day Marlowe, they call him. They have the meat wagon following him around to follow up on the business he finds. 1954 *Britannica Bk. of Year* 637/1 The meat-wagon was the police-van in which the criminal rode to captivity. 1956 S. LONGSTREET *Real Jazz* 7 The band would march out behind the meat-wagon, black plumes on the hearse horses. 1964 *Listener* 31 Dec. 1055/2 The bogeys..bundle us into the back of a meat-wagon. 1971 'E. MCBAIN' *Hail, Hail, Gang's all Here* i. 15 We'll need a couple of meat wagons. The minister and two other people were killed, and..there're a lot of injured. 1973 'H. HOWARD' *Highway to Murder* xiii. 153 She hadn't deserved to become a parcel of broken flesh and bone in the meat wagon. [1895 T. A. COGHLAN *Wealth & Progress New South Wales* I. 367 All the cattle killed, except 27,891 treated in the meat-preserving works, were required for local consumption.] 1934 WEBSTER, *Meat works.* 1948 V. PALMER *Golconda* ii. 9 Driving them across country to the meatworks at Wyndham. 1960 B. CRUMP *Good Keen Man* 46 The hut stank like a meat works. 1968 *Times* 23 Jan. (Austral. Suppl.) p. xiv/3 In one year before the meatworks opened a large property shot 10,000 scrub bulls (not branded or castrated).

meater. Restrict † *Obs.* to sense in Dict. and add: **2.** *rare.* One who eats (butcher's) meat; a meat-eater.
1920 *Contemp. Rev.* Dec. 819 The 'meater' lives at higher pressure and exhausts his energy quicker than the non-meat-eater.

meatless, *a.* Add: **2.** (Further examples.) *spec.* Of foods specially prepared or supplied for vegetarians: containing no butcher's meat.
1909 C. H. SENN (*title*) Meatless fare and lenten cookery. 1909 *Daily Chron.* 2 Nov. 7/3 Even that anomaly—to most people—'the meatless meal' is included. 1969 *Daily Tel.* 9 Oct. 16/4 The era of meatless meat, chickenless chicken and cheeseless cheese is already dawning in America. 1972 *Guardian* 8 June 7/5 Some of them [*sc.* countries] had their meatless days. *Ibid.* 31 July 2/4 The city's three big department stores..decreed meatless menus.
Hence **mea·tlessness.**
1918 *Punch* 27 Mar. 206/2 If he [*sc.* our butcher] were removed we should be plunged into absolute meatlessness.

me·bbe, me·bby, *colloq.* and *dial.* variants of MAY-BE, MAYBE (see *E.D.D.*).
1844 'J. SLICK' *High Life N.Y.* I. vi. 65 Mebby I'll mention where I got them. 1886 H. BAUMANN *Londinismen* 106/2 Mebbe. 1906 W. CHURCHILL *Coniston* I. x. 119 Mebbe I be, and mebbe I hain't. 1910 R. BROOKE *Let.* 8 June (1968) 240 Jacques'll be here about then, mebbe. 1948 E. POUND *Pisan Cantos* (1949) lxxxi. 110 A vacant lot where you'd occasionally see a wild rabbit Or mebbe only a loose one. 1961 S. CHAPLIN *Day of Sardine* vii. 148 'We're goin' to the pictures and mebbe afterwards we'll fill up with some fish and chips. 1972 *Last Whole Earth*

Catalog (Portola Inst.) 152/1 I read in *Newsweek* or somewhere that this is the New Renaissance—Mebbe so.

mebos (mī·bǫs, *Afrikaans* mē·bǫs). *S. Afr.* Also **meebos.** [Afrikaans, prob. ad. Jap. *umeboshi*, a dried, preserved plum.] A confection made from apricots dried, flattened or pulped, and preserved in salt and sugar.
1793 tr. *C. P. Thunberg's Trav. Europe, Afr. & Asia* III. 120, I saw several kinds of fruit, the produce of this country [*sc.* Japan], either dried or preserved in yeast, in a mode which is, I fancy, only practised at Japan or China. The fruit that was only dried, such as plumbs and the like, was called *Mebos*. 1899 *Answers* 18 Nov. 7/2 The best sort of preserve is called *mebos*, and is made of stoned and sun-dried apricots, flattened out, and pickled with salt and sugar. 1912 *Northern Post* 27 Sept. (Pettman), I have now come to the conclusion that our old navigators became acquainted with this delicacy in Japan, learned to like it, and afterwards at the Cape attempted to imitate it, but used the fruit of apricot trees.., and that the word *Meibos* or *Meebosje* had its origin in [Jap.] *Umeboshi*. 1939 S. CLOETE *Watch for Dawn* iii. 39 He had mebos and raisins and sun-dried peaches. 1954 M. KUTTEL *Quadrilles & Konfyt* i. 9 Have a jar of mebos handy on board..in case of sea-sickness. 1959 J. COLLIER *Stellenbosch Revisited* ii. 45 A delicacy called 'mebos' which consisted of apricots dried, flattened and treated with salt and sugar. 1974 *Eastern Province Herald* 27 Nov. 37 Mevrou Van Niekerk..fed them on mebos and honey cakes in her big farm kitchen.

Mebyon Kernow (me·byǫn kǝ·ɹno). [Cornish.] 'Sons of Cornwall', the name of a Cornish party of independence.
1962 *Rep. Comm. Broadcasting 1960* 224 in *Parl. Papers 1961–2* (Cmnd. 1753) IX. 259 We note..a submission entered by Mebyon Kernow (Sons of Cornwall) advocating a service of broadcasting for Cornwall. 1963 *Guardian* 8 Apr. 3/6 Mebyon Kernow, the Cornish home rule movement. 1967 *Ibid.* 5 Dec. 3/2 Mebyon Kernow (Sons of Cornwall) are to put forward candidates in the next general election. 1975 *Isis* 17 Oct. 9/3 Mebyon Kernow (The Sons of Cornwall), the nationalist organisation, has never demanded autonomy but the more reasonable 'self-government in domestic affairs' within the United Kingdom.

mecamylamine (mekămai·lămīn). *Pharm.* [f. ME(THYL + *CAM(PHANE + -YL + AMINE.] A potent ganglion-blocking drug, 3-methylaminoisocamphane, $C_{11}H_{21}N$, which is used in treating hypertension.
1955 *Sci. News Let.* 29 Oct. 281/1 A nerve-blocking drug called mecamylamine. 1956 *Arch. Internal Med.* XCVII. 561/1 Such side effects as constipation were just as prominent with the small completely absorbed doses of mecamylamine as with the relatively large poorly absorbed doses of other ganglion-blocking agents. 1968 L. GYERMEK in A. Burger *Drugs Affecting Cent. Nervous Syst.* I. iv. 290 Mecamylamine depresses the vasomotor center independently of its ganglionic blocking action. 1971 *Nature* 17 Sept. 207/1 Previous work in this laboratory with monkeys trained to puff cigarette smoke showed that mecamylamine, a nicotinic-blocking agent, reduced their smoking. 1974 M. C. GERALD *Pharmacol.* viii. 159 Ganglionic blocking agents such as mecamylamine (Inversine)..prevent cholinergic transmission at the autonomic ganglia.

Meccano (mi·kā·no). Also **meccano.** The proprietary name of a set of metal pieces, nuts, bolts, etc., and tools, specially designed for constructing small models of buildings, machines, or other engineering apparatus; any portion of such a set. Also *attrib.* and *fig.*
1907 *Trade Marks Jrnl.* 23 Oct. 1893 Meccano... Constructional toys. Frank Hornby,..Liverpool; manufacturer of constructional models. 1908 (*title*) Meccano (mechanics made easy): manual of instructions for the whole series of models. 1924 H. DE SÉLINCOURT *Cricket Match* iii. 56, I shall make a prison cell of meccano, and pretend you're locked..inside. 1927 *Sunday at Home* Aug. 675/1 There was meccano in the goldfish bowl. 1928 *Television* May 20/1 String, cardboard, and pieces of rough wood with Meccano parts;..all combined to make the television machine. 1928 'R. WEST' *Strange Necessity* ii. 199 The complete meccano set for the mind that is in The First Men in the Moon. 1930 J. B. PRIESTLEY *Angel Pavement* vi. 307 It seemed only yesterday when he was.. putting the Meccano set by the boy's bedside. 1934 *Amer. Speech* IX. 167/1 An economy may be made for the foreigner by presenting to him the elementary words only, thereby substituting a linguistic Meccano for the expense of ready-made models. 1945 J. BETJEMAN *New Bats in Old Belfries* 45 Pressed 'neath the box of his Meccano set. 1957 *New Scientist* 25 Apr. 9/2 The vocabulary is largely a 'Meccano' vocabulary; it is built up from bits and pieces which can be stuck together in twos, threes and fours. 1960 S. R. RANGANATHAN *Colon Classification* (ed. 6) 13 The Meccano feature makes it necessary to give, in addition to the unit-schedules, a set of Rules for constructing Class Numbers with the aid of the unit-schedules. 1965 *Listener* 23 Sept. 464/3 In crisp Technicolor images and prose painstakingly constructed on a Meccano-like principle. 1970 J. A. HOWARD *Aerial Photo-Ecol.* xi. 122 The mechanical or spider templates are provided by 'meccano' type strips. 1972 R. PERRY *Fall Guy* iii. 48 The suspension bridge was still standing, its graceful lines making Sydney harbour bridge look like a clumsy piece of Meccano. 1972 *Guardian* 14 Oct. 3/2 The Eiffel Tower..is quite a piece of Meccano: there are more than 18,000 structural components in the 985 ft high tower. 1973 *Ibid.* 10 Feb. 10/2 He's the sort of uncle who'd give you a Meccano set for Christmas and then stay all afternoon and tinker with it. 1974 *Trade Marks Jrnl.* 30 May 939/2 Meccano Multikit... Toys and playthings, all sold in kit form. Meccano Limited,..Liverpool,..manufacturers and merchants.

mech (mek), colloq. abbrev. of MECHANIC *sb.* 3.
1951 PARTRIDGE *Dict. Slang* (ed. 4) 1107/2 *Mech*, mechanic; esp. in the old *air mech* of the R.F.C. and the current *flight mech* of the Air Force: coll.: since ca. 1912. 1968 J. SANGSTER *Touchfeather* xvi. 191 Bud carries a heater, the mechs don't. 1973 A. HUNTER *Gently French* iii. 23 Hanson called over a mech. The mech started it for us and drove it out.

mechanic, *a.* and *sb.* Add: **B.** *sb.* **2.** *Mechanics' institute*: see also INSTITUTE *sb.*^{1} 4.
d. One who cheats at gambling games, *spec.* cards; a card-sharp. *U.S.* and *Austral. slang.*
1909 in *Cent. Dict.* Suppl. 1944 D. RUNYON in *Collier's* 12 Feb. 12/4 What I must know is are you a mechanic at gin? 1949 J. SCARNE *On Cards* i. 1 To the average card player the man who makes his living by cheating at cards is a sharper, sharp or shark, but to card hustlers and house men he is known as a *mechanic*. *Ibid.* ii. 10 Some mechanics keep two fingers curled around the long edge of the deck and two around the short upper edge. 1953 BAKER *Australia Speaks* v. 121 *Mechanic*, a person who cheats at cards, especially a professional card sharp (Americans use *mechanic* for a dishonest player at faro). 1966 K. GILES *Provenance of Death* v. 146 The games were straightish, no real mallarky, though Jack himself was a 'mechanic' with the cards if necessary. 1966 *Daily Tel.* (Colour Suppl.) 30 Sept. 27/2 As croupier..always on guard for the sharps—the mechanics. 1970 R. & J. PATERSON *Cranberry Portage* vii. 40 He was what other gamblers called a 'mechanic', an expert manipulator of playing-cards.

mechanical, *a.* and *sb.* Add: **A.** *adj.* **3. e.** *mechanical* (*wood*) *pulp* (see quot. 1926).
1888 CROSS & BEVAN *Text-bk. Paper-Making* vi. 105 Mechanical Wood Pulp.—A very large quantity of pulp is used in the commoner kinds of paper, such as cheap news, etc., which is obtained by disintegrating wood by mechanical means alone. 1890 A. WATT *Art of Paper-Making* x. 113 Mechanical wood pulp is also used in a moderate degree. 1926 *Paper Terminol.* (Spalding & Hodge) ii. 17 *Mechanical wood*, the lowest grade of wood pulp prepared by the purely mechanical process of grinding. 1936 *Economist* 8 Feb. 304/1 Of sulphate as well as of mechanical pulp, both annual production and stocks have been sold out. 1937 [see *GROUND *ppl. a.* 4]. 1953 *Economist* 26 Sept. 883/1 Mechanical pulp, from which newsprint is made, is about 10s. a dry ton higher [than chemical pulp]. 1963 R. R. A. HIGHAM *Handbk. Paper-Making* v. 129 Mechanical pulp is an important cheap grade used in the manufacture of papers such as newsprint, wallpaper, [etc.].
f. *mechanical twin* (Metallurgy): a twinned crystal produced by mechanical deformation; so *mechanical twinning*.
1913 *Engineering* 10 Oct. 510/3 There is now good reason to doubt that mechanical twinning ever occurs in metals. 1923 GLAZEBROOK *Dict. Appl. Physics* V. 344/1 These internal stresses cause internal straining of the metal, which in turn causes the formation of numerous mechanical twins. 1935 G. E. DOAN *Princ. Physical Metall.* iii. 77 In mechanical twinning, each atom moves a certain distance relative to the neighboring plane. 1966 *McGraw-Hill Encycl. Sci. & Technol.* VIII. 293/1 Typical phenomena observable are..effects of deformation—strain markings,..mechanical twins, and microcracks. 1966 W. J. McG. TEGART *Elem. Mech. Metall.* v. 121 Another important mode of deformation is that of mechanical twinning.
5. a. *mechanical equivalent of heat*: see EQUIVALENT *sb.* 2 c.
8. c. *mechanical advantage* (of a machine): the ratio of the load to the force applied to the machine; *mechanical zero* (see quot. 1971^2).
1894 W. J. LINEHAM *Textbk. Mech. Engin.* ix. 481 The first is the principle of virtual velocities, and the second mechanical advantage. 1962 A. NISBETT *Technique Sound Studio* v. 94 A reversed-reading instrument is used: it has its mechanical zero at the right-hand end of the scale and is deflected back to the scale zero (its 'electrical zero') by a steady current. 1971 B. SCHARF *Engin. & Its Lang.* ix. 68 To calculate the efficiency of a machine we divide the real mechanical advantage by the ideal mechanical advantage and multiply the result by 100%. 1971 *Gloss. Electrotechnical, Power, Terms* (*B.S.I.*) I. iv. 9 *Mechanical zero*, equilibrium position which the index will approach when the measuring element..is de-energized.
10. *Comb.*, as *mechanical-minded* adj.
1947 J. STEINBECK *Wayward Bus* xvii. 223, I suppose people really think they are mechanical-minded.
B. *sb.* **1.** (Later examples.)
1963 *Times* 20 Apr. 4/1 The play scene is omitted and the rest of the mechanicals, a couple of fellows boldly designated on the programme as 'Bottom's Friends', are left deliberately vague. 1968 *Listener* 1 Feb. 148 Frivolity seems to me the only, precarious excuse for this novel—a let-out, for instance, for treating the income-earning group as characters and the rest as 'mechanicals'.
3. *Printing.* The artwork and 'copy' as finally assembled. Also *transf.*
1967 V. STRAUSS *Printing Industry* xi. 744/2 At its simplest, the mechanical is a piece of artists' illustration board, somewhat larger than the final size of the printed piece. To this board are attached, by cementing or pasting .., a number of line images, all in the same focus and, of course, of inspected quality. This board bears, furthermore, all notations that will enable it to serve as the blueprint of the job. 1967 KARCH & BUBER *Offset Processes* iii. 47 This final assembly of reading matter and artwork for the printer is called a *mechanical*. 1967 *Britannica Bk. of Year* (U.S.) 66 Many regulations, particularly in relation to TV, prevented advertisers from using international campaigns if the basic mechanicals—artwork, films, and so on—were not produced in Italy by Italians. 1973 *Publishers Weekly* 12 Mar. 38 The layout [of an advertisement] was changed at the last minute, and the mechanical bearing [the publisher] Quadrangle's name either was not replaced, or it fell off.

mechanicalization (mĭkæ:nikăləizē·ʃən). [f. MECHANICALIZE *v.* + -ATION.] The being or becoming mechanical in character or in means of operation; *esp.* in military terminology.
1922 *Glasgow Herald* 5 Jan. 4 A record of experimental progress towards mechanicalisation of wireless telegraphy in as simple and portable a form as possible. 1924 *Westm. Gaz.* 26 Mar., Colonel Rudkin..proved himself an expert on mechanicalisation—the new word to indicate army tendencies. 1924 *Times Lit. Suppl.* 16 Oct. 654/4 The mechanicalization of the army. 1926 *Glasgow Herald* 27 Aug. 11 Military minds turn to 'mechanicalisation', an ugly word but an indication of the abolition of what Tommy Atkins terms 'foot-slogging'. 1927 *Sunday Times* 6 Mar. 20/3 The real benefits that result from a mechanicalisation of industry.

mechanicalize, *v.* Add: (Later examples.) Hence **mecha·nicalized** *ppl. a.*
1924 *Army Q.* Oct. 39 The first step suggested is that the divisional transport should be mechanicalized. 1926 *Glasgow Herald* 9 Sept. 9 One of the problems studied.. was the landing of a mechanicalised force on an open beach in the face of opposition.

mechanically, *adv.* Add: **7.** *Comb.*, as *mechanically-minded* adj.
1922 *Guardian* 19 May, Any mechanically-minded person can make a simple receiving set for a pound or two. 1937 B. H. L. HART *Europe in Arms* xvii. 231 Some doubt must remain..as to the ability of horse-minded soldiers to become mechanically minded. 1972 G. DURRELL *Catch me a Colobus* ii. 45 Oscar the orang-utan,..the most mechanically-minded of all the apes.

mechanicism. Delete *Obs. rare*$^{-1}$ and add further examples.
1962 W. STARK *Fund. Forms of Social Thought* II. xii. 176 A fresh high-water mark of mechanicism was reached in the eighth decade of the nineteenth century. 1971 *Archivum Linguisticum* II. 115 But if to avoid the limitations of Kuryłowicz's method, we have to admit that a form passes arbitrarily from one function to another, not taking into account the facts of polarization, attraction, etc., then we fall into mechanicism.

mechanicist (mĭkæ·nisist). [f. MECHANIC *a.* and *sb.* + -IST.] = MECHANICALIST.
1934 *Mind* XLIII. 248 'Mechanical' is a highly ambiguous term, and many who call themselves mechanicists would not deem it inconsistent with their principles to recognise in the molecule a different unit from the atom, with different properties. 1936 *Nature* 4 Apr. 559/2 Haldane asserts that the real universe is a universe of personality and the manifestation of God... No doubt materialists and mechanicists would take exception to some of the arguments. 1939 *Ibid.* 8 July 52/2 Controversies about the philosophy of animate Nature..keep alive the old feud between vitalists and mechanicists. The latter favour the view that life is a by-product of blind processes of dead matter.

mechanico-. Add: *mechanico-acoustic, -material* adjs.; **mechanico-morphic** *a.* = *MECHANOMORPHIC *a.*; **mechanico-morphism** = **mechanomorphism*; **mechanico-physical** *a.*, of or pertaining to the philosophy which explains all phenomena as the outcome of the physical laws of the motions and interactions of matter.
1964 Y. R. CHAO in D. Abercrombie et al. *Daniel Jones* 39 The mechanico-acoustic set-up for recording speech on wax masters. 1920 D. H. LAWRENCE *Touch & Go* 9 The strike situation..is a mechanico-material struggle, two mechanical forces pulling asunder from the central object, the bone. 1937 *Mind* XLVI. 176 We wish to stress here what we conceive to be the inadequacy of all 'mechanico-morphic' representations of concrete becoming in nature. 1935 J. MURPHY tr. *Schrödinger's Sci. & Human Temperament* Introd. 19 We must..abandon the mechanical structure. We must turn to the statistical concept... In other words, Schrödinger pleads for the abandonment of what may be called mechanico-morphism in the pursuit of natural science,..the casting aside of all models and the wholesale employment of mathematical formulas in their stead. 1890 W. JAMES *Princ. Psychol.* II. xxviii. 666 The modern mechanico-physical philosophy..which..includes the nebular cosmogony, the conservation of energy, the kinetic theory of heat and gases, etc.,..begins by saying that..the only laws [are] the changes of motion which changes in collocation bring.

mechanics. Add: **3.** *transf.* The procedural or operational details (*of* something).
1925 E. B. WILSON *Cell* (ed. 3) ii. 172 This fact is fundamentally important for the mechanics of mitosis. 1930 *Writer* Sept. 265/2 Is the playwright perpetually making references in his dialogue to the mechanics of his craft? 1932 *Punch* 13 Jan. 55/1 Miss E. B. C. Jones is well up in the mechanics of psycho-analysis. 1940 *Punch* 10 Apr. 407/3 It is a pity the mechanics of the story take up so much of the time. 1960 L. PICKEN *Organization of Cells* viii. 373 (*heading*) The mechanics and energetics of changes in cell shape. 1974 *Nature* 22 Mar. 280/3 Speakers at current meetings in the field generally give little information on how their results are actually obtained, that is, the pure mechanics of the business.

mechanism. Add: **2.** (Later examples.) Also used, esp. *Psychol.*, of the means or agency by which mental processes and bodily actions are caused to take place (cf. quot. 1885 in Dict.).
1910 G. N. CALKINS *Protozool.* i. 29 Bodies closely associated with the mechanism of nuclear division and of locomotion. 1913 J. S. HALDANE *Mechanism, Life & Personality* I. 9 Descartes, in his writings about the nervous system,..suggested nervous mechanisms. *Ibid.* II. 58 The real difficulty for the mechanistic theory is that we are forced..to postulate that the germ-plasm is a mechanism of enormous complexity and definiteness, and..that this mechanism..can divide and combine with other similar mechanisms. 1921 H. C. MILLER *New Psychol. & Teacher* 161 This mental mechanism [*sc.* the complex] lies at the root of all bias, all injustice, and all inability to think clearly. 1924 *Brit. Weekly* 28 Aug. 471/3 He will see strange recesses in human personality and unsuspected mechanisms fashioning religious beliefs. 1929 K. S. LASHLEY (*title*) Brain mechanisms and intelligence. 1941 *Psychosomatic Med.* July 227/1 The adaptive mechanisms by means of which the organism strives to achieve this goal. *Ibid.* 233/1 By 'cognitive field' or 'practical insight' we mean a mechanism capable of registering and integrating stimuli. 1958 *Amer. Jrnl. Psychiatry* Sept. 204/2 The brain, unlike a machine into which any mechanism can be set, appears to have mechanisms of its own. 1964 COFER & APPLEY *Motivation* xi. 573 Conceptions of learned drive basically assert that responses which produce strong stimuli are the mechanisms of such drives. 1968 M. BUNGE in Lakatos & Musgrave *Probl. Philos. Sci.* 128 Something mediating between inputs and outputs, i.e. a mechanism triggered by the inputs and which has the required outputs. 1972 *Physics Bull.* Mar. 141/1 Furthermore some mechanism must be found for judging the quality of the work done by the chief scientist and the controller.
6. Delete † *Obs.* and add earlier and later examples.
1690 LOCKE *Hum. Und.* I. iii. 22 Thereby making Men no other than bare Machins... And upon that ground they must necessarily reject all Principle of Vertue, who cannot put Morality and Mechanism together. 1902 BALDWIN *Dict. Philos. & Psychol.* II. 59/1 In biology: mechanism is opposed to vitalism, and in more recent controversy to neo-vitalism. 1909, 1917 [see *FINALISM 2]. 1956 O. L. ZANGWILL in A. Pryce-Jones *New Outl. Mod. Knowl.* 170 *Mechanism* has sought to account for all behaviour in terms of the quasi-automatic activities of the central nervous system.

mechanist. **3.** Delete † *Obs.* and add later examples. Also as *adj.*
1913 J. S. HALDANE *Mechanism, Life & Personality* I. 6 The constant controversies..between mechanists and non-mechanists. 1919 G. B. SHAW *Heartbreak House* p. xv, Their mechanist theory taught them that medicine was the business of the chemist's laboratory, and surgery of the carpenter's shop. 1925 C. D. BROAD *Mind & its Place* 43 One feels that the disputes between Mechanists and Vitalists are unsatisfactory. 1931 *Brit. Jrnl. Psychol.* Oct. 137 Those whom their opponents call 'vitalists' can see in the relationship an attribute of something living. Those whom their opponents dub 'mechanists' can see in the relationship [between association processes and neuronic excitation] a property they attribute to mechanism. 1965 *Listener* 19 Aug. 286/2 Pain due to emotional disturbance..arises in the mind (or in the brain if you tend to be a mechanist).

mechanistic, *a.* Add: (Later examples.) Also, pertaining to or holding mechanical theories in psychology and linguistics.
1915 B. HOLMES in *Chicago Med. Recorder* Mar. 2 (*heading*) The Mechanistic view of Dementia Precox. 1923 W. MCDOUGALL *Outl. Psychol.* i. 30 All the varieties of psychology which propose..to replace the hypothesis of a mind, a soul, a self, [etc.].., by that of a brain or a bodily organism working on strictly mechanical or physical principles..may be conveniently classed together as mechanistic psychologies. 1924 W. B. SELBIE *Psychol. Relig.* 278 On the negative side they have adduced evidence of a kind which makes a merely mechanistic explanation of the universe impossible. 1933 L. BLOOMFIELD *Lang.* 33 The *materialistic* (or, better, *mechanistic*) theory supposes that the variability of human conduct, including speech, is due only to the fact that the human body is a very complex system. 1939 *Ann. Reg. 1938* 301 The Government's revenue would be so great that extinction of the debt would be automatic. This 'mechanistic' theory was severely ridiculed. 1952 J. DREVER *Dict. Psychol.* 163 *Mechanistic theory*,..the interpretation of psychological processes on a mechanical basis, and denial of the reality or efficacy of ends and purposes. 1963 MARX & HILLIX *Syst. & Theories Psychol.* II. ix. 203 One man, Gustav Fechner, seems to have shared the mechanistic-romantic conflict with Freud. 1966 M. PEI *Gloss. Ling. Terminol.* Mechanistic theory,..an approach to language and linguistics based on objective methodology in recording and classifying language phenomena. 1967 R. A. WALDRON *Sense & Sense Devel.* ix. 201 The rather more austere and mechanistic tendency of twentieth-century linguistics has given such phraseology a quaint, old-fashioned air. 1968 M. BUNGE in Lakatos & Musgrave *Probl. Philos. Sci.* 128 The first approach can be called phenomenological or global, the second mechanistic or atomistic.

mechani·stically, *adv.* [f. MECHANISTIC *a.*] In a mechanistic manner; on mechanistic principles.
1923 W. MCDOUGALL *Outl. Psychol.* vi. 189 One attempt to explain mechanistically this fundamental type of profiting by experience has been widely accepted. 1925 C. D. BROAD *Mind & its Place* ii. 77 The attempts which have been made..to treat mental phenomena mechanistically. 1935 *Mind* XLIV. 88 If you show that mental activity has independent causal powers, and that it works purposively and not mechanistically, surely [etc.]. 1947 J. C. RICH *Materials & Methods of Sculpture* xi. 350 Metal negative molds..are used commercially for mechanistically manufacturing papier mâché reproductions in large quantities. 1961 *N. & Q.* June 237/2 Mr. White's method is usually to utter a few simple principles about the art of fiction.. and then test them rather mechanistically against his author's four novels. 1971 *Nature* 25 June 495/2 This situation looks unlikely to change, unless there is more collaboration in the design of mechanistically meaningful experiments. 1974 FRITH & MCLAUCHLAN in R. K. Harris *Nucl. Magn. Resonance* III. xii. 393 CIDNP is normally most useful mechanistically when performed in the high field of a normal spectrometer.

mechanization. (Later examples.)
1915 A. W. GOUGH *God's Strong People* 32 A tyranny built..on a degradation and mechanization of the personal power. 1928 *Daily Mail* 7 Feb. 7/5 The 'mechanisation' of the army. 1937 *Daily Express* 11 Jan. 10/2 Mechanisation has come to the farmer. 1954 [see *AUTOMATION]. 1960 *Post Office Electr. Engineers' Jrnl.* LIII. II. 75/2 The mechanization of the trunk service, involving the installation of automatic trunk exchanges..is now well advanced.

mechanize, *v.* **1.** Add to def.: To change (an industry, etc.) to a mechanical form of working; to provide with machines; spec. *Mil.*, to equip with mechanical weapons and vehicles, as tanks, armoured cars, etc.
1942 E. WAUGH *Put out More Flags* i. 21 The yeomanry ..had recently been mechanized, in the sense that they had had their horses removed; few of them had ever seen a tank. 1952 *Oxf. Jun. Encycl.* VI. 286/2 The spreading of farmyard manure and other winter jobs have been mechanized. 1957 *Encycl. Brit.* II. 412/2 The Germans had broken abruptly with the past by mechanizing their artillery.

mechanized, *ppl. a.* Add further examples, esp. in *Mil.* sense: of, pertaining to, equipped with, or using mechanical vehicles and weapons.
1928 *Daily Mail* 8 Feb. 7/5 The fast two-seater tanks of the mechanised army. 1939 *Punch* 8 Nov. 504/1 Their plan..was to attack..the Maginot line..along the whole extent from North to South, at the same time executing a turning mechanized movement through..Switzerland. 1941 *Picture Post* 3 May 19/1 Mechanised warfare and careful preparation have again defeated flesh and blood. 1941 *Punch* 16 July 59/2 Enemy mechanized units penetrated our defences. 1942 *Times* (Weekly ed.) 7 Jan. 4 The conscience of humanity is still a factor of which even the lords of mechanized war may have to take account. 1942 *R.A.F. Jrnl.* 16 May 9 A vital means of countering the German Panzer and mechanised troops. 1968 R. M. OGORKIEWICZ *Design & Devel. Fighting Vehicles* i. 30 Some of the new ideas were explored in a series of experiments which began with the Experimental Mechanised Force assembled in 1927 on Salisbury Plain. 1970 MACKSEY & BATCHELOR *Tank* 33/1 It is one of the paradoxes of early mechanised warfare that the technologists not only produced the machines but also foresaw their tactical possibilities ahead of the soldiers.

mechano- (me·kăno), repr. Gr. μηχανο-, comb. form of μηχανή MACHINE *sb.*, and used in the senses (*a*) 'machine(s)', as in MECHANOLOGY, (*b*) 'mechanical (and)', as in MECHANOTHERAPY, *MECHANOCHEMICAL *a.* Also **me:chanoca·loric** *a. Physics* [*CALORIC *a.*], applied to the phenomenon by which a flow of the superfluid form of liquid helium (helium II) generates a difference in temperature between the sending and receiving ends; **me:chano-ele·ctric, -ele·ctrical** *adjs.*, pertaining to or producing a conversion of mechanical movements into corresponding electrical effects; also **me:chano-electro·nic** *a.* (*rare*).
1939 MENDELSSOHN & DAUNT in *Nature* 29 Apr. 720/1 The mechano-caloric effect is evidently the reverse of the 'fountain phenomenon', for whereas the latter means that the setting up of a temperature difference results in a flow of liquid helium II, the mechano-caloric effect shows that a flow of liquid helium II is accompanied by a development of heat (or cold). 1964 *New Scientist* 24 Sept. 773/1 Shortly after the discovery of this 'thermo-mechanical' effect the complementary 'mechano-caloric' phenomenon was observed. 1958 *Jrnl. Gen. Physiol.* XLI. 1245 (*heading*) The sites for mechano-electric conversion in a pacinian corpuscle. *Ibid.* 1263 The simplest interpretation of our results would seem to be that the entire mechano-electric transducer lies in the membrane of the non-myelinated ending. 1974 *Nature* 1 Mar. 69/2 Isometric tension was measured with two RCA 5734 mechano-electric transducers. 1961 M. L. GAYFORD *Acoustical Techniques & Transducers* vi. 279 Trying to obtain a perfect electrical copy of the groove excursion by the mechano-electrical groove-following means provided by a gramophone pick-up. 1974 *Nature* 20–27 Dec. 728/1 The muscle fibre was..mounted horizontally with one of its tendons hooked to the anode of a RCA 5734 mechano-electrical transducer tube. 1950 *18th Internat. Physiol. Congr. Copenhagen* 171 The movements of the membrane ..are communicated to the plate of a mechano-electronic transducer (R.C.A. 5734), a triode which by means of a pivoted anode translates mechanical movements into electric current variations.

mechanochemistry (me:kănoke·mistri). [f. prec. + CHEMISTRY.] The study or phenomenon of mechanochemical processes.
1928 P. M. TRAVIS *Mechanochemistry* 3 This new science of mechanical dispersion, involving the principles of physical chemistry, I am calling 'Mechanochemistry', because it involves dispersion or deflocculation by mechanical means rather than by chemical. 1951 *Jrnl. Polymer Sci.* VII. 407 A simplified model which discloses the principles of the mechanochemistry of polyelectrolytes will be shown here. 1961 *New Scientist* 2 Mar. 548/3 In the USSR ..the field of investigation has been aptly named 'mechanochemistry', in line with 'electrochemistry' and 'photochemistry'. 1972 *Science* 3 Nov. 460/2 A complete understanding of the sliding filament model for generating flagellar bending waves must include an understanding of the mechanochemistry of the active sites at the molecular level.
Hence **me:chanoche·mical** *a.*, pertaining to both mechanics and chemistry; *spec.* able to convert chemical energy directly into mechanical energy; **me:chanoche·mically** *adv.*
1949 W. GARNER *Textile Laboratory Manual* i. 14 (*heading*) Mechano-chemical methods for single fibres. 1949 A. KATCHALSKY in *Experientia* V. 320/1 In the phenomena discussed here the chemical ionization energy is transformed directly into mechanical energy. We may therefore regard them as the first synthetic examples of mechanochemical systems. 1961 *New Scientist* 2 Mar. 549/2 The action of an internal combustion engine causes substantial changes in lubricating oil viscosity due to mechanochemical breakdown of the polymer being sheared between the narrow clearances. *Ibid.* 550/3 Mechanochemically initiated polymerization has been found to occur with a large variety of polymerizable substances using substrates other than rubber. 1964 *Oceanogr. & Marine Biol.* II. 198 Takata suggested that in *Acetabularia* there may be an ATP sensitive protein that forms a mechanochemical system controlling protoplasmic streaming. 1972 *Science* 3 Nov. 460/2 The basic mechanochemical process by which energy from ATP dephosphorylation is used to produce mechanochemical work may be very similar to the active sliding process. 1974 *Nature* 31 May 475/2 Exergonic band cleavage would therefore occur in the mechanochemically coupled step.

mechanomorphic (me:kănomǭ·ɹfik), *a.* [f. *MECHANO- + Gr. μορφή form + -IC.] Having the form or qualities of a machine or mechanism; of or pertaining to the Deity regarded as a mechanical force. Hence **me:chanomo·rphism**, the concept of something (esp. the Deity) as mechanomorphic.
a 1885 D. D. WHEDON *Ess., Rev. & Discourses* (1887) 265 A still more curious notion of dignity..rejects the *anthropomorphic* and substitutes therefor a *mechanomorphic* theory. 1926 B. H. STREETER *Reality* i. 2 Materialism pictures the Universe as an Infinite Machine; this by analogy may be called mechanomorphism. *Ibid.* 22 Mechanomorphic materialism. 1937 A. HUXLEY *Ends & Means* ix. 124 They have felt the effects of mechanomorphism. 1950 —— *Themes & Variations* 197 Cubists, who liked to paint machines or to represent human figures as though they were parts of machines..representation of human beings in a mechanomorphic guise.

mechanoreceptor (me:kănorĭse·ptǭɹ). *Physiol.* Also **mechano-receptor.** [f. *MECHANO- + RECEPTOR.] A sensory receptor which responds to mechanical stimuli, such as pressure changes resulting from touch or sound.
1927 J. H. PARSONS *Introd. Theory Perception* i. 7 Receptors which are found in animals may be classified into three groups..chemo-receptors, mechano-receptors, and radio-receptors. *Ibid.* 9 Mechano-receptors can be divided into two classes, contact and distance receptors. 1951 *Sci. Amer.* Jan. 17/2 Among these mechano-receptors are the receptor organs for rotatory and translatory motion in the inner ear, the receptors responsible for the pressure sense of the skin, the muscle spindles imbedded in all muscles that fix and move bodily masses, and the so-called Pacinian or Vater's corpuscles found throughout the connective tissues. 1970 *Nature* 28 Mar. 1263/2 These passive movements are perceived by mechanoreceptors of the pedicel.
Hence **me:chanorece·ption**, the process by which a mechanical stimulus is converted by a mechanoreceptor into a nervous impulse; **me:chanorece·ptive** *a.*, capable of performing mechanoreception.
1951 *Jrnl. Physiol.* CXV. 16 The mechanoreceptive area was localized by touching the tongue with a thin wire while listening to the pressure impulses in the loudspeaker. 1958 *Jrnl. Gen. Physiol.* XLI. 1249 (*heading*) Mechanoreception after excision of capsular structure. 1968 R. T. VERRILLO in D. R. Kenshalo *Skin Senses* vii. 139 (*heading*) A duplex mechanism of mechanoreception. 1974 *Nature* 26 Apr. 740/2 Neuromas in man must..contain mechanoreceptive endings since they are always tender when gently palpated.

meclozine (me·klǒzin, -īn). *Pharm.* Also **meclizine.** [f. ME(THYL + C(H)LO(RINE *sb.* + PIPERA)ZINE.] A piperazine derivative, $C_{25}H_{27}N_2Cl$, which is an antihistamine drug used mainly as an anti-emetic, esp. in preventing motion-sickness, and is usu. given as the dihydrochloride, a white crystalline compound.
Meclozine is the name in the *British Pharmacopœia*, *meclizine* that in the *Pharmacopeia of the U.S.A.*
1955 *Jrnl. Pharmacol. & Exper. Therap.* CXV. 483 From a perceptual-motor viewpoint there would seem to be no valid reason why operating personnel could not take advantage of the beneficial effects which cyclizine, meclizine and promethazine have upon motion sickness. 1957 *Brit. Pharmaceut. Codex 1954* Suppl. 39 Meclozine

hydrochloride has the properties of the anti-histamine drugs. **1959** *Which?* July 69/1 The 1 mg. of hyoscine protected over 80 per cent of the subjects from sea-sickness, and..this dose, for one hour, was significantly more effective than the cyclizine and meclozine. **1963** *Times* 27 May 18/4 (Advt.), The thalidomide tragedy did not directly affect your Company, but one of our most widely prescribed products, which contains meclozine, is for use in pregnancy sickness... Near the end of the year meclozine itself came under criticism. **1965** *New Scientist* 4 Nov. 328/2 Anti-histamines, containing meclizine, cyclizine and chlorcyclizine must now carry a strong warning label. **1971** J. J. Burns in B. N. La Du et al. *Fund. Drug Metabolism & Drug Disposition* xvii. 361 Although large doses of chlorcyclizine and meclizine can induce malformations in laboratory animals, no evidence of harm to the human fetus has been reported.

Mec Vannin (mek væ·nin). [Manx, lit. 'sons of (the Isle of) Man'.] Name of a Manx nationalist party.

1966 *Isle of Man Weekly Times* 7 Jan. 8/2 (*heading*) Mec Vannin... Manx Nationalists could not be fully convinced by the argument that..the banks and the houses ..are still Manx..even though their ownership and administration is increasingly passing out of native hands. *Ibid.* 8/3 The man who comes to open a shop, or a boarding-house or a factory, which will be run by himself in the Island, is a candidate for Mec Vannin. **1974** *Times* 8 Apr. 4/6 Mec Vannin..intends to field candidates for all the Tynwald seats at the next general election.

med.[1], **med** (med). (*a*) Abbrev. of Medic *sb.* or Medical *sb.* 1; (*b*) abbrev. of Medical *a.*; (*c*) abbrev. of Medicine *sb.*[1]

(*a*) **1851** [see Medic *sb.*]. **1853** in Farmer & Henley *Slang* (1896) IV. 298/1 Take..Sixteen interesting meds, With dirty hands and towzeled heads. **1899** A. H. Quinn *Pennsylvania Stories* 19 The Meds waited till the visitors were opposite them. **1942** [see (*b*) below].

(*b*) **1933** Partridge *Slang To-day & Yesterday* III. iii. 190 Of the 'plural' variety [of abbreviation] is *med. lab.*, medical laboratory. **1942** Berrey & Van den Bark *Amer. Thes. Slang* § 529/2 Doctor; physician,..*med*,..*med man*. *Ibid.* § 529/5 Medical student,..*med*. *Ibid.* § 530/9 Medical college,..*med college* or *school*. *Ibid.* § 534/1 *Med business*, the medical profession. **1955** in *Amer. Speech* (1956) XXXI. 233 The law students began the boress by storming the Medical Building singing about the evils of the med students. **1974** *Spartanburg* (S. Carolina) *Herald* 18 Apr. A3/2 The lawmakers from Charleston County, site of the existing med school, are opposed to the move.

(*c*) **1942** Berrey & Van den Bark *Amer. Thes. Slang* § 531/8 Medicine; drugs, *dope*, *med*. *Ibid.* § 831/22 *Courses of study*... Medicine, *med*. **1962** 'E. McBain' *Like Love* (1964) xi. 158 A page out of a textbook, elementary stuff, we had this in first-year med. **1973** H. McCloy *Change of Heart* vii. 71, I want to study psychiatry when I get through med.

Med[2] (med). Abbrev. of Mediterranean *sb.* 1. Also *attrib.*

1948 in Partridge *Dict. Forces' Slang*. **1955** G. Freeman *Liberty Man* I. i. 14 We went all round the Med., Istanbul, Capri, Gib. **1960** 'S. Harvester' *Chinese Hammer* ii. 26 He is off-course over the Med heading for Turkey. **1962** *Punch* 3 Oct. 486/1 Hoardings are turning Med-blue. **1971** *Guardian* 6 July 11/5 The summer cruise of the Med Fleet was a grand affair.

‖ **médaillon** (medaiyoṅ). [Fr., lit. = medallion.] A small, flat, round or oval-shaped cut of meat or fish.

1900 [see *escalope]. **1921** W. J. Locke *Mountebank* xvi. 207 A *médaillon de veau périgourdine*, a superimposition of toast, *foie gras*, veal and truffles. **1964** L. Joyce-Cowen *Million Menus* 90a Lobster médaillons with Gruyère cheese soufflé topping. **1971** *Guardian* 4 June 11/4 Medaillons of Pork Tenderloin.

medal, *sb.* **1.** Delete '† *Obs.*' and add later examples.

1922 Joyce *Ulysses* 622 A pious medal he had that saved him. **1972** *Sci. Amer.* Feb. 57/2 (Advt.), Medals are recognized as a beautiful medium for artistic expression, as will be exemplified in this fine art series.

2. b. Freq. with a defining word indicating the degree of excellence attained, as in *gold medal* (see *gold[1] 1 b), *silver medal* (see *silver *sb.* and *a.* IV. 21), *bronze medal*.

1852 C. Fox *Let.* 29 Sept. in *Jrnls.* (1972) 207 They have not hesitated to give Anna Maria two bronze medals. **1908** T. A. Cook *Olympic Games* i. 16 For all these gold, silver, and bronze prize medals have been allotted. **1973** P. O'Donnell *Silver Mistress* vii. 133 That may not be a hundred percent right, but do I get a bronze medal for trying?

4. medal chief *U.S.*, an Indian chief who received a medal from the Colonial or U.S. authorities; **medal ribbon,** a ribbon of specific colours and design for attaching a particular medal or for wearing without a medal; **medal round** *Golf*, a round of medal-play.

1772 D. Taitt *Let.* in N. D. Mereness *Trav. Amer. Colonies* (1916) 518 [Letter] To the Great and Small medal chiefs. **1813** *Niles' Reg.* V. 270/2 At this moment a medal chief of the Choctaw nation is soliciting to be employed. *a* **1816** B. Hawkins *Sk. Creek Country* (1848) 27 He is one of the great medal chiefs. **1909** *Westm. Gaz.* 4 Oct. 1/3 The medal-ribbon which a soldier tore off his tunic. **1944** *R.A.F. Jrnl.* Aug. 286, I did not recognize his medal-ribbon, so asked what it was. **1971** *Daily Tel.* (Colour Suppl.) 16 July 5/3 The steward..looked too distinguished with his medal ribbons to be tipped. **1898** W. A. Morgan '*House*' *on Sport* I. 180 It does not at all follow that the man who can play a match can also play a medal round. **1909** *Westm. Gaz.* 27 Aug. 12/3 It was arranged to play a medal round in the forenoon.

medallist. 3. (Later examples.)

1968 *Times* 18 Oct. 12/1 Tommie Smith, together with the bronze medallist, John Carlos, has made sporting history. **1973** M. Mackintosh *King & Two Queens* xvi. 220 Don't be a heroine. I didn't choose an Olympic gold medallist.

medarsa, var. Madrasah.

1920 *Blackw. Mag.* Dec. 750/1 The 'universities'—'Medarsas'—of Fez and Marrakesh..are now open once more to the Christian visitor.

meddlesome, *a.* Add: *Meddlesome Matty* (or *Mattie*): a nickname for a meddlesome person (allusively, from quot. 1814).

1814 A. & J. Taylor *Orig. Poems* (ed. 11) II. 3 (*title*) Meddlesome Matty. [Not in 1805 ed.] [**1923** D. H. Lawrence *Kangaroo* xi. 230 Jaz is a meddlesome-Patty.] **1927** *Times* 17 Aug. 11/5 My warning was addressed to those who would make of the League 'a kind of international Meddlesome Matty'. **1938** A. G. Macdonell *Autobiogr. of Cad* xxi. 259 Every reformer finds his obstructionists... In both cases Meddlesome Matties were to blame. **1960** D. Holman-Hunt *My Grandmothers & I* iv. 92 Good gracious, child, what a meddlesome matty you are.

Mede, *sb.* **1.** (Further examples.)

c **1384, 1526, 1880** [see *Elamite *sb.* and *a.*]. **1930** J. D. Duff tr. *Rostovtzeff's Hist. Anc. World* (ed. 2) viii. 123 The Iranian tribes of Medes and Mannai..grew stronger by degrees, and their pressure upon Assyria steadily increased. **1969** J. Conway tr. *Bengtson's Greeks & Persians* i. 5 Cyaxares the Mede seized the last, ephemeral Assyrian kingdom. *Ibid.*, Greek tradition mentions the Medes and the Assyrians in one breath.

medersa, var. Madrasah.

1923 G. Casserley *Algeria To-Day* iii. 58 There is a finely-built Medersa or theological college for Mahommedans. **1935** *Times Lit. Suppl.* 3 Jan. 4/3 President of a *medersa*—that is, an Arabic University—and commandant of the neighbouring fort of Gouridam. **1968** *Vogue* 15 Apr. 124/2 Fès, the ancient university city of Morocco ..this crowded ancient place of mosques and medersas.

medevac (me·dīvæk). *U.S.* Also **medivac.** [f. Med(ical *a.* + Evac(uation.] A military helicopter for transporting wounded soldiers to hospital. Hence **Medevac, Med-Evac** *v. trans.*, to transport by medevac.

1966 [see *heli-]. **1967** *Harper's Mag.* Jan. 77 The two wounded Aid Men continued to crawl about and administer care. There would be no medevac; there was no landing zone for it. **1969** *Time* 28 Nov. 23 At My Lai, Ridenhour reported, one soldier shot himself in the foot so that he would be Medevacked out of the area. **1973** *Washington Post* 13 Jan. A. 3/3 You look at an NLF (National Liberation Front) soldier who can't get Med-Evac'ed in 20 minutes. **1973** *Maclean's Mag.* July 17/3 While McCoy attended his patients, and the nurses and Thomas stood by, Logozar and Hartwell debated who would fly the medivac.

medi- (mī·di). *Zool.* = Medio-.

1903 *Amer. Anthropologist* V. 627 The meditemporal [fissure] consists of a segment in the cephalic region of the lobe, 6 cm. in length. *Ibid.* 631 The medifrontal [fissure] springs from the orbitofrontal.

media. Add: Esp. as pl. of *medium *sb.* 5. Also erron. as sing. in same sense (cf. *mass medium note). Also *attrib.* or as *adj.*

1923 [see *mass medium]. **1927** *Amer. Speech* III. 26 It was finally decided to allot a definite media to each member. *Ibid.*, One of the best advertising medias in the middle west. **1929** E. O. Hughes *Outl. Advertising* v. 82 The advertising *media* to which reference will be made.. are newspapers, journals, magazines and such-like printed publications. **1958** *Times Lit. Suppl.* 5 Dec. 698/5 The media which appeal to our visually conscious age call for organizing ability as well as individual talent. **1962** M. McLuhan *Gutenberg Galaxy* 246 Is not the essence of education civil defence against media fall-out? **1966** K. Amis in *New Statesman* 14 Jan. 51/3 The treatment of *media* as a singular noun..is spreading into the upper cultural strata. **1966** *Economist* 10 Dec. 1166/1 In any recession, media buyers, never a very adventurous lot, stay with the safest bets in any advertising medium. **1968** *Sun* (Baltimore) 4 July A16/3 Then there is 'media coordinator', that is, a tape recorder operator. **1969** T. Parsons *Politics & Social Struct.* III. x. 250 Different media (or often the same media) carry qualitatively different content. **1971** *Daily Tel.* 4 Nov. 17/4 Decrees laid down in 1965 formally forbid direct contact between the [Dutch] Royal House and the publicity media. **1971** *Radio Times* 18 Nov. 82 The media have an ambiguous relationship with the radical left. **1972** *Guardian* 16 May 16/6 McGovern..occasionally stages a 'media event' which uses the unwitting people as props for TV news footage. **1972** *Times* 19 July 13/6 Miss Allen seems to be under the impression that the media is confined to newspapers. **1973** *Black World* Mar. 80 The theme of the poem deals with the white backlash, a media term describing the hostile reactions of white folks to civil-rights activities. **1973** 'R. Macdonald' *Sleeping Beauty* i. 9 'You from a media?' 'No, I'm just a citizen.' **1974** *Listener* 28 Feb. 260/1 The impartial experts and media-men. **1975** *Atlantic Monthly* Jan. 29/2 'Media' is itself a code name for a stereotype. *Ibid.*, Enemies of journalism..refer to the profession as the 'media' in disdain.

medial, *a.* **2.** *medial line*: delete † (*obs.*) and add later example; also *medial area* (see quot.).

1908 T. L. Heath tr. *Euclid's Elements* III. x. 50 A medial straight line..is so called because it is a mean proportional between two rational straight lines commensurable in square only. *Ibid.* 55 It is in the Porism that we have the first mention of a medial area. It is the area which is equal to the square on a medial straight line.

median, *a.*[1] and *sb.*[1] Add: **A.** *adj.* **1. c.** *median strip*, a strip of ground, paved or landscaped, dividing a street or highway. *N. Amer.*

1954 J. C. Ingraham *Mod. Traffic Control* ii. 31 Choice of the type of median strip—solid concrete, grass and curbing or grass alone. **1967** *Boston Sunday Herald* 7 May 1/1 A fence erected on the median strip to discourage the road-crossings. *Ibid.*, The need for the closing of dangerous, life-taking crossovers in the median strip. **1968** *Globe & Mail* (Toronto) 3 Feb. 31/4 A street of exceptionally broad sidewalks and a median strip for trams. **1971** *Fremdsprachen* XV. 67/1 The city of Lawndale, Calif., recently installed nearly two miles of man-made turf on its traffic median strips.

3. a. (Earlier and later examples.) *median dose*, a dose which is effective in half of those receiving it.

1882 F. Galton in *Rep. Brit. Assoc. Adv. Sci. 1881* 245 The Median, in height, weight, or any other attribute, is the value which is exceeded by one-half of an infinitely large group, and which the other half falls short of. **1929** Kelley & Shen in C. Murchison *Found. Exper. Psychol.* 838 Some investigators have often preferred the median to the mean as a measure of central tendency. **1947** *Radiology* XLIX. 302/2 Most of the following data were obtained after median lethal and lower doses of radiation. **1973** *Daily Tel.* 30 Jan. 7/1 The Statistical Office report shows the median wage for Northamptonshire as £1,260.

B. *sb.* **3.** Substitute for def.: A line drawn from a vertex of a triangle to the midpoint of the opposite side. (Earlier example.)

1883 *Encycl. Brit.* XVI. 15/1 If *a*, *b*, *c* be the three sides of a triangle, and *α*, *β*, *γ* the three medians, i.e., the lines drawn from the angles to the middle points of the opposite sides.

Median, *a.*[2] and *sb.*[2] Add: **B.** *sb.* **2.** The language of ancient Media, a dialect related to Old Persian; = *Medic *sb.*[2]

1813 [see *Kurd]. **1841** R. G. Latham *Eng. Lang.* I. i. 3 The Ossetic, a language spoken by an insulated tribe of Mount Caucasus, and a supposed remnant of the Ancient Median, is Indo-European. **1848** *Trans. R. Irish Acad.* XXI. II. 241 In Median..sounds were sometimes confounded. *Ibid.* 244, I..observed some Median words transcribed in one of the inscriptions, and a few other words that, though altered, appeared to be of Persian or Median origin. **1908** T. G. Tucker *Introd. Nat. Hist. Lang.* 189 It has, however, been argued strongly that the Avestic language is in reality Old Median. **1939** L. H. Gray *Foundations of Lang.* 32 From the New Testament ..we know that in the first century A.D. Parthian, Median, Elamite, Cappadocian, Pontic, Phrygian, Pamphylian, Cretic.., and Arabic were spoken. **1950** R. H. Kent *Old Persian* I. i. 6/2 Among the less known Old Iranian languages the most important was Median.

mediate, *a.* Add: **2. c.** Also *Psychol.* (See quot. 1897.)

1897 C. H. Judd tr. *Wundt's Outl. Psychol.* iii. 239 'Mediate recognition'..consists in the recognition of an object, not through its own attributes, but through some accompanying mark or other. **1912** *Amer. Jrnl. Psychol.* Jan. 106 The cases of true mediate association found in these experiments.

mediate, *v.* Add: **5.** (Further examples.) Spec. in *Psychol.*, to bring about (a result) by acting as a mediating agency between an idea, intention, etc., and its realization; to act as such a mediator.

1931 G. Stern *Meaning & Change of Meaning* xiv. 388 The comprehension of *Buchmacher* was mediated by the English word. **1942** Cofer & Foley in *Psychol. Rev.* XLIX. 522 Behavioral generalization in children was found to be mediated by verbal responses. **1957** B. F. Skinner in Saporta & Bastian *Psycholinguistics* (1961) 67/2 Instead of going to a drinking fountain, a thirsty man may simply 'ask for a glass of water'... The consequences of such behavior are mediated by a train of events no less physical or inevitable than direct mechanical action. **1958** B. Bernstein in J. A. Fishman *Readings Sociol. of Lang.* (1968) 227 The working-class child has to translate and thus *mediate* middle-class language structure through the logically simpler language structure of his own class to make it personally meaningful. **1968** M. Bunge in Lakatos & Musgrave *Probl. Philos. Sci.* 128 [A] theory that takes the risk of hypothesizing something mediating between inputs and outputs. **1971** Sutherland & Mackintosh *Mechanisms Animal Discrimination Learning* I. ii. 5 The behaviorists in general have denied the existence of the process of selective attention in animals except where the process could be mediated by overt orientation responses. **1972** *Jrnl. Social Psychol.* LXXXVII. 129 Of particular interest to the present investigators was the extraction of the dimensions actually used by individuals in mediating similarity. **1974** *Nature* 1 Mar. 73/1 Cholera enterotoxin mediates specific biochemical events in both intestinal and non-intestinal tissues by stimulating adenyl cyclase and cyclic AMP. **1974** *Author* Summer 89/2 People wanted direct, authentic communication that had not been mediated by sub-editors or script writers. The tape recorder made this new authenticity possible.

mediated, *ppl. a.* Add: Also *Psychol.*, arrived at by mediation; involving mediation (*mediation 3 b).

1942 Cofer & Foley in *Psychol. Rev.* XLIX. 514 A theoretical analysis of mediated generalization as a specific conditioned response phenomenon. **1950** Keller & Schoenfeld *Princ. Psychol.* v. 161 (*caption*) Mediated generalization in the conditioning of the galvanic skin response to verbal stimuli. **1971** A. Paivio *Imagery & Verbal Processes* ix. 319 Mediated transfer studies generally involve the learning of two or more lists.

mediating, *ppl. a.* Add: **2. b.** *Psychol.*, interposing between an idea, intention, etc., and its result; acting as a mediator (sense *6).

1953 C. E. Osgood *Method & Theory Exper. Psychol.* III. ix. 401 Certain stimulus patterns..are variably associated with systems of mediating reactions. **1963** W. W. Grings in M. H. Marx *Theories Contemp. Psychol.* xxxi. 517 The lack of a clear-cut basis for deducing the nature of the mediating response. **1966** W. N. Runquist in J. B. Sidowski *Exper. Methods Psychol.* 503/2 A common mediating association ('army-ocean', mediated by 'navy'). **1968** K. Duncker in Wason & Johnson-Laird *Thinking & Reasoning* ii. 34 The final form of a solution is typically attained by way of mediating phases of the process. **1971** Sutherland & Mackintosh *Mechanisms of Animal Discrimination Learning* IX. ii. 309 The nature of the 'mediating processes' possibly available to children but not to animals.

mediation. Add: **3. b.** *Psychol.* (See quot. 1934.) Also *attrib.*

1912 *Amer. Jrnl. Psychol.* Jan. 102 The occurrence of associations whose mediation does not come into consciousness in any recognizable manner is certainly an interesting and well-attested phenomenon. **1934** H. C. Warren *Dict. Psychol.* 162/1 *Mediation*, the interposition of one or more ideas or acts between an initial stimulus or idea and a given end result whose genesis is under investigation. **1953** C. E. Osgood *Method & Theory Exper. Psychol.* III. ix. 395 Short circuiting enters into all behavior, and its most important role lies in formation of those representational mediation processes. **1963** Ervin & Miller in J. A. Fishman *Readings Sociol. of Lang.* (1968) 91 While there is overlap..with American studies in verbal mediation, the experimental questions have been quite different. **1970** *Jrnl. Gen. Psychol.* LXXXIII. 53 Osgood..developed it [*sc.* the Semantic Differential] and rationalized it in terms of mediation theory. **1971** A. Paivio *Imagery & Verbal Processes* ix. 320 The experimental and language-habit approach to the investigation of mediation paradigms.

mediational (mīdiē·ʃənăl), *a.* [f. Mediation + -al.] Of or pertaining to mediation; mediating.

1951 Parsons & Shils *Toward Gen. Theory Action* III. i. 280 One can assume, within reasonable limits (without, that is, detailed psychophysiological, 'mediational' investigation), what potential 'perceptions' are possible. **1953** C. E. Osgood *Method & Theory Exper. Psychol.* III. ix. 404 Mediational modification is indicated..by a single asterisk. **1956** *Scottish Jrnl. Theol.* IX. 228 The leading mediational theologian [*sc.* J. Baillie] of our time. *Ibid.* 234 That supreme sensitivity and mediational quality which have always marked his thought and outlook. **1964** *Language* XL. 211 'Operant' is one of those psychological terms (together with 'latency' and 'mediational') which recur frequently in psycholinguistics. **1968** M. Black *Labyrinth of Lang.* vii. 158 Causal theories..insist upon the importance of the 'mediational process' induced by perceiving the sign. **1971** *Jrnl. Gen. Psychol.* LXXXV. 125 A mediational process involving colour cues.

mediator. Add: **6.** That which effects a transition between one stage and another; *spec.* in *Psychol.*, that which acts as an agent in mediation (sense *3 b). Also *attrib.*

1953 C. E. Osgood *Method & Theory Exper. Psychol.* III. ix. 402 The self-stimulation produced by mediators. *Ibid.* 404 (*caption*) A change in the instrumental sequence elicited by a mediator. **1965** *Language* XLI. 139 Such words as *istorik*, *istorija*.., which appear already in Old Russian literature, are justly regarded by her as direct borrowings from Greek or Latin, and not as mediated by other languages... On the other hand, the more detailed data obtained on foreign words have enabled Mrs. Worth to correct and to define more precisely the role of the mediator languages, especially that of Polish. **1967** *Word* XXIII. 14 Among the numerous 'mediator' languages suggested so far those based on the symbols of mathematical logic are of especial interest to modern interlinguistics. **1969** [see *E.S.T.* (*E. III)]. **1970** N. E. Spear in M. R. D'Amato *Exper. Psychol.* xii. 604 S's consequential use of potential mediators for learning. **1970** *Jrnl. Gen. Psychol.* LXXXIII. 5 Rhine also furthers his analysis to the affect-arousal components that certain verbal mediators may eventually elicit. **1971** A. Paivio *Imagery & Verbal Processes* ix. 319 Potential mediators are either inferred from association norms or are 'built in' experimentally.

‖ **media vuelta** (mē·diă vue·ltă). Also **media-vuelte.** [Sp., lit. 'half turn'.] In bull-fighting, a method of killing the bull by approaching from behind (see quot. 1962).

1932 [see *banderilla *v.*]. **1934** R. Campbell *Broken Record* viii. 183 Don Simao uses a *media-vuelte* in sticking in his banderillas from horseback; that is, he takes the bull while it is turning, citing it from behind. **1962** B. Conrad *Encycl. Bullfighting* 163/1 *Media vuelta*... A way of placing *banderillas* in which the *torero* comes at the bull from behind, forcing the animal into a *media vuelta* as the man approaches. **1967** McCormick & Mascareñas *Compl. Aficionado* iv. 121 Once in a great while..you will see an aging matador, green with fear, killing *a la media vuelta*.

medic. B. *sb.* Delete '*Obs.* exc. as U.S. college slang' and add further examples.

1895 W. C. Gore in *Inlander* Nov. 64 *Medic*, a student in the medical department. **1902** W. W. Hall *Appl. Relig.* I. 22 For ages medics have been laying down rules for the regimen of diseased people. **1925** Fraser & Gibbons *Soldier & Sailor Words* 154 *The Medics*, R.A.M.C. [Royal Army Medical Corps]. **1945** *Finito! Po Valley Campaign* 61 (*caption*) Our medics treat German wounded. **1968** Mrs. L. B. Johnson *White House Diary* 17 Jan. (1970) 618 He was a Medic and, although twice wounded, he had gone on..to save life after life, crawling across the battlefield. **1970** *Oxford Mail* 21 Feb. 12/3 (*heading*) The 'medics' are fast movers. **1973** *Black Panther* 3 Mar. 6/2, I was a member of the U.S. Army from 1967 to 1970. During the Tet offensive of 1968, I served as a combat medic with the 101 ABN. Div. at Hue, South Vietnam. **1974** *Evening Standard* 1 Mar. 48/3 Dr Brian Warren, Mr Heath's personal physician, called to see him at Downing Street—but as a friend, not as a medic.

Medic (mī·dik), *a.*[2] and *sb.*[2] [ad. L. *Mēdic-us*, Gr. Μηδικός Median.] **A.** *adj.* = Median *a.*[2]

1888 G. Bertin *Abridged Gram. Lang. Cuneiform Inscriptions* 81 Medic Grammar. **1889** *Jrnl. Anthrop. Inst.* Aug. 31 The Medic language is not the same as the Akkadian, though in syntax and in vocabulary it presents a very marked connection. **1912** H. G. Rawlinson *Bactria* ii. 25 A treaty was made between the rival nations, and ratified by a marriage between the Medic king and a Lydian princess. **1933** C. Dawson *Enquiries Relig. & Culture* ii. 80 Assyria was at the height of her power..but her pitiless militarism was gradually ruining..Asia and prepared the way for a new series of barbarian invasions—Cimmerian, Scythian and Medic.

B. *sb.* = *Median *sb.*[2] 2.

1894 A. H. Sayce *Primer of Assyriology* ii. 20 The script and language of Elam—sometimes, but incorrectly, called Scythian, Medic, or Protomedic.

Medicaid (me·dikēid). Also **medicaid.** [f. Medic(al *a.* + Aid *sb.*] Name given in the United States to a scheme making available state and federal funds for the use of persons judged to require assistance with medical expenses, and provided for under Title XIX of the Social Security Act, 1965. Also *attrib.*

1966 *Life* 29 July 4/1 The word we are going to be hearing a lot more is something called 'Medicaid'—also known as Title 19. **1967** *Economist* 2 Dec. 960/2 The states have until 1970 to set up their Medicaid schemes, but 33 have already done so, attracted by the generous federal contribution. **1972** *Daily Colonist* (Victoria, B.C.) 6 Jan. 1/7 Nearly $1 billion in medicaid payments 'went down the drain' in New York City through medical malpractice. **1973** *Interfaces* May 44 The launching of the Medicare and Medicaid programs..created a need for new technical words. **1974** *Spartanburg* (S. Carolina) *Herald* 22 Apr. B4/6 If you can be paid you will also be eligible for Medicaid provided through the South Carolina Department of Social Services.

medical, *a.* and *sb.* Add: **A.** *adj.* **1. a.** (Further examples.)

1753 *Phil. Trans. R. Soc.* XLVII. 399 Every body has consider'd what the Italians printed..upon the subject of medical electricity, as too hasty a publication. **1776** T. Pennant *Tour in Scotl. & Voy. Hebrides 1772* II. 247 The number of medical students are now annually reckoned at about 300. **1777** T. Cavallo *Compl. Treat. Electr.* x. 287 *Medical Electricity*, a practical method of applying Electricity to the human body, when affected with different distempers. **1814** G. J. Singer *Elem. Electr.* III. iv. 286 Mr. Partington, whose experience as a medical electrician is considerable. **1835** F. D. Maurice *Let.* 6 Oct. in J. F. Maurice *Life F. D. Maurice* (1884) I. xiii. 186 If I could get any influence over the medical students, I should indeed think myself honoured. **1864** Mrs. Gaskell *Wives & Daughters* (1866) I. iii. 26 He went very steadily to work ..advertising in medical journals, reading testimonials, sifting character and qualifications. **1878** W. James in R. B. Perry *Tht. & Char. W. James* (1935) II. IV. liii. 29 The degrading sophistries of medical materialists. **1885** *Househ. Words* 20 June 155 (Farmer), Medical students have liberally assisted in the formation of slang, their special department thereof being known as medical Greek. **1902** W. James *Varieties Relig. Experience* i. 13 Medical materialism finishes up Saint Paul by calling his vision on the road to Damascus a discharging lesion of the occipital cortex, he being an epileptic. **1935** *Economist* 7 Sept. 456/2 The [National Health Insurance] scheme provided for the payment of cash benefit in time of sickness, for medical benefit..and sanatorium benefit. **1945** *Release & Resettlement* (H.M. Govt.) xii. 40 The benefits of the National Health Insurance Scheme include medical benefit (i.e. free treatment by an insurance doctor and medicine). **1966** I. Jefferies *House-Surgeon* ii. 20 'I know you've *graduated*,' she interposed, 'but at heart you're still a medical student.' **1975** 'M. Duke' *Death of Holy Murderer* i. 15, I put aside the medical journal I'd been trying to read.

d. medical board [Board *sb.* 8 b], a body of medical men responsible for the medical examination of soldiers, the maintenance of public health, etc.; hence *medical-board* vb. trans., to refer for consideration by a medical board (*rare*); **medical certificate,** a certificate from a doctor, attesting the state of a person's health, etc.; **medical examiner,** (*a*) a doctor who carries out an examination for physical fitness; (*b*) *U.S.*, a medically qualified public officer whose duty is to investigate deaths that occur under unusual or suspicious circumstances, to perform post-mortems, and sometimes to initiate inquests; hence **medical examination,** an examination to establish the degree of a person's physical fitness, etc.; **medical garden** (earlier example); **medical hall** *Ireland*, a pharmacy, a chemist's shop; **medical jurisprudence,** the law as it relates to the practice of medicine; also = *forensic medicine*; **medical officer,** a doctor appointed by a public authority or company to attend to matters relating to health (in Gt. Britain in the public sphere the post of 'medical officer of health' was abolished on 1 Apr. 1974); **medical register,** a register, in Britain kept by the General Medical Council, of all doctors legally in practice; so *medical registration*; also *attrib.*; **medical school,** (a school or faculty of) a college or university in which medicine is studied.

1814 *Niles' Reg.* VI. 36/2 A Medical Board is now sitting at the city of Washington, by order of the secretary of war. **1843** *Times* 29 Nov. 4/3 Two distinct medical boards have declared themselves satisfied as to his sound state of health. **1922** *Encycl. Brit.* XXXI. 898/2 Every medical board paper was a kind of forecast. **1947** L. Hastings *Dragons are Extra* iii. 63 He was then medical-boarded and sent to the nursing-home in England. **1973** 'R. MacLeod' *Burial in Portugal* i. 18 The last medical board had cut his army pension, reckoning his back was improving. **1838** Dickens *Nickleby* (1839) xv. 131 A very little more and it [*sc.* a back comb] must have entered her skull. We have a medical certifiket that if it had, the tortershell would have affected the brain. **1853** C. Schreiber *Jrnl.* (1952) 12 The medical certificate would be quite sufficient, and..without his appearing the fellowship was safe. **1914** Kipling *Let. of Travel* (1920) 210 One could eat..in one's cabin without a medical certificate from the doctor. **1971** J. Clegg *Dict. Social Services* 96 A medical certificate (on the official form) must be sent to the local social security office. **1973** E. McGirr *Bardel's Murder* ii. 35 'Did you get a medical certificate?' 'That and pain killers.' **1974** *Times* 23 Sept. 2/3 She had attempted suicide five years ago..but she had medical certificates that said she was now all right. **1898** F. de H. Hall (*title*) Medical examination for life assurance. **1916** *Brit. Med. Jrnl.* 26 Aug. 300/2 (*heading*) Medical examination of recruits. **1935** S. G. Leigh *Guide to Life Assurance* (ed. 4) viii. 92 Many people have an aversion from the medical examination. **1848** *Risks of Life Assurance* 5 Dr. Ingleby..then the medical examiner for the Norwich Office. **1877** *Acts & Resolves Gen. Court of Massachusetts 1877* 580 The governor shall nominate..able and discreet men, learned in the science of medicine, to be medical examiners. **1889** Pollock & Chisholm *Med. Handbk. Life Assurance* p. viii, Plain instructions for the guidance of the medical examiner, to afford him a ready handbook of reference on insurance rules. **1928** *Bull. Nat. Research Council* LXVI. 13 Only in New York City and Boston, where the medical examiner's system prevails, is the bulk of the work done in the properly equipped morgues. **1937** T. A. Gonzales et al. *Legal Med. & Toxicol.* ii. 9 The coroner has investigative and judicial powers; the medical examiner can conduct only an investigation. **1973** *Times* 5 June 6/4 No official ruling on the cause of death has yet been given by the medical examiner, the equivalent of the coroner. *a* **1706** J. Evelyn *Diary* an. 1658 (1955) III. 217, I went to see the Medical Garden, at Westminster, well stored with plants, under Morgan, a very Skillful Botanist. **1922** Joyce *Ulysses* 669 Purchased by him at 10.20 a.m. on the morning of 27 June 1886 at the medical hall of Francis Dennehy. **1938** J. Cary *Castle Corner* 279 Ah, ye dirty devil, and what sort of a drip are ye to be dropped in a medical hall. **1788** S. Farr (*title*) Elements of medical jurisprudence. **1845** [see Forensic *a.*]. **1937** T. A. Gonzales et al. *Legal Med. & Toxicol.* p. vii, The section on legal medicine is concerned with the principal pathologic conditions encountered in cases of forensic importance,.. and medical jurisprudence or that part of the law which deals with the practice of the physician. *Ibid.*, The subjects which are considered a part of medical jurisprudence, such as the corpus delicti,..malpractice, insanity, and insurance. **1835** *Times* 2 Nov. 7/1, I have only heard of two unions, in which the number of medical officers is equal to the number employed before the formation of these unions. **1841** Dickens *Let.* 29 Oct. (1969) II. 414 With regard to the Medical officers it will be best to say that I am going out of town. **1860** F. Nightingale *Notes on Nursing* i. 10, I have known a medical officer keep his ward windows hermetically closed. **1916** W. Owen *Let.* 10 Feb. (1967) 379 The Medical Officer says I should get them removed. **1964** *Times* 22 Feb. 6/2 Medical officers of health should be more closely concerned with the functions of their local authority that had a bearing on health in any way. **1974** *Daily Tel.* 1 Apr. 6/6 Regional hospital boards, executive health councils and local authority departments with health responsibilities will disappear. So too will medical officers of health, some of whom will now become 'community physicians'. **1886** *Act* 49 & 50 *Vict.* c. 48 § 14 The medical register shall contain a separate list of the names and addresses of the colonial practitioners. **1926** *Brit. Med. Jrnl.* II. 430/1 A few years ago the Government of the Irish Free State announced that it had decided to establish a separate Medical Register. **1973** *Times* 26 July 2/4 Dr —— was suspended from the medical register yesterday for professional misconduct in prescribing drugs for other than bona-fide treatment. **1886** *Lancet* 10 July 85/2 The great importance of due medical registration. **1926** *Brit. Med. Jrnl.* II. 322/2 A decision on the medical registration question has not yet been reached. **1765** in J. Carson *Hist. Med. Dept. Univ. of Pennsylvania* (1869) 55 The institution of Medical Schools in this country has been a favorite object of my attention. **1841** *Southern Lit. Messenger* VII. 550/2 We had lighted upon the University, in the act of organizing an extensive

Medical School. **1909** J. S. Fletcher in H. Greene *Crooked Counties* (1973) 228 He was anxious to see one or two experiments..being carried on in some of the medical schools. **1973** *Guardian* 25 Jan. 15/6 The shortage of doctors in some regions is directly related to the absence of a regional medical school.

B. *sb.* **3.** A medical examination for fitness.

1917 'Taffrail' *Sub* i. 27 It was neither the *x* and *y* part of the business nor the 'medical' which caused me qualms. It was the dreaded interview. **1938** N. Marsh *Death in White Tie* x. 103 He wanted me to go to Edinburgh to take my medical... I wanted to go to Thomas's. **1946** H. Wayne *Two Odd Soldiers* i. 7 The Recruiting Officer welcomed and encouraged us..and told us to present ourselves on the following Tuesday for our 'medical'. **1968** *Listener* 8 Feb. 187/3 Mr Scannell's adolescence appeared to be no more remarkable than a hundred others. He coughed at his RAF medical; he spouted Yeats to a literal-minded girl. **1973** *Times* 1 June 5/6 (*heading*) Medicals for drivers urged.

Medicare (me·dikēᵊɹ). Also **medicare.** [f. Medi(cal *a.* + Care *sb.*[1] 4.] **a.** Name given in the United States to a scheme of health insurance for the elderly, provided for under Title XVIII of the Social Security Act, 1965; cf. *Medicaid. **b.** Name of a similar scheme in Canada. Also *attrib.*

1962 *Economist* 16 June 1099 Ailing Medicare. **1963** *Times* 25 Feb. (Canada Suppl.) p. vi/6 The Canadian medicare problem must be seen against Canada's geographical, economic and political background. **1963** *Punch* 16 Oct. 577/3 The cost of providing Medicare.. would be less than the annual cost of burying them. **1965** *New Statesman* 6 Aug. 177/1 Last week he [*sc.* President Johnson] signed the medicare bill. **1968** *Globe & Mail* (Toronto) 13 Feb. 1/7 Under the federal medicare legislation starting July 1, Ottawa will pick up half the average per capita cost. **1970** *Ibid.* 28 Sept. 17/1 A meeting of the Quebec Cabinet in Montreal tonight to discuss the possibility of passing emergency legislation to implement medicare. **1971** *Maclean's Mag.* Oct. 65/3 However he discharges his other campaign promises from abolishing deterrent fees for users of Medicare to renegotiating the pulp-mill deal, [etc.]. **1971** *Optometry Today* (Amer. Optometric Assoc.) 22 At present, optometrists..cannot provide optometric services to American senior citizens under the Medicare program. **1973** [see *Medicaid]. **1974** *Globe & Mail* (Toronto) 2 Mar. 8/3 Saskatchewan, which introduced medicare amid stormy and turbulent summer days in 1961, is embarking on a denticare programme.

Medici (me·ditʃi), *a.* [It. surname: cf. Medicean *a.*] = Medicean *a.*, esp. *Medici collar* (see quot. 1968); *Medici lace* (see quot. 1969), *Medici porcelain* (or *china*), a type of porcelain produced under the patronage of Francesco de' Medici; *Medici print*, *reproduction*, etc., one produced by the Medici Society, Ltd., London. Also *ellipt.*

1799 N. W. Wraxall *Mem. Courts* II. xxi. 114 A broad Medicis of Dresden lace surrounded her bosom and shoulders. **1868** J. Marryat *Hist. Pott. & Porc.* (ed. 3) xiv. 456 (*caption*) Brocca of Medici Porcelain, 15 inches high. **1873** *Young Englishwoman* May 234/1 The Medici's collarette..is..becoming only to those ladies who have a slender swan-like neck. **1890** *Amer. Mail Order Fashions* (1961) 13 [Drawers] Cambric, ruffle of Medici lace, 75 c. **1900** F. Litchfield *Pott. & Porc.* ii. 18 It was of soft paste, is extremely rare, and is known as 'Medici china'. **1902** Jourdain & Dryden *Palliser's Hist. Lace* (rev. ed.) iv. 56 The Medici collars were supported by fine metal bars called 'verghetti'. **1906** *Burlington Mag.* Oct. 47/1 If their quality can be maintained the *Medici* prints now being issued by Messrs. Chatto and Windus, of which the first three plates now lie before us, will undoubtedly represent a great advance upon anything which has yet been done in colour reproduction. **1908** Beerbohm in *Around Theatres* (1924) II. 405 Nor was I embarrassed by the hoop and Medici collar that Dalila wore. **1936** *Catal. Medici Prints* 3 This Series has now attained a world-wide reputation under the title of 'Medici Prints'. **1940** *Burlington Mag.* May p. x/2, Medici porcelain. *Ibid.*, A 'Medici' plate, marked with the letter F and the dome of the Cathedral of Florence. **1949** F. Towers *Tea with Mr. Rochester* 24 The study with its Axminster carpet and Medici prints and rows of encyclopædias. **1957** J. Braine *Room at Top* vi. 60, I was..looking at the Medici reproduction of Olympe. **1960** R. G. Haggar *Conc. Encycl. Cont. Pott. & Porc.* 306/1 With one exception, all the known pieces of Medici porcelain are painted in blue, or blue and manganese purple. **1963** A. Gernsheim *Fashion & Reality* II. 79 Many capes had a small Medici collar. Such upstanding collars were worn on evening dresses, 1888–92..and fur Medici collars on coats and jackets during the winter of 1889–90. **1964** M. Clive *Day of Reckoning* iv. 42 There was a gap between those large, pale, repellant Victorian Arundel prints and the neat Medici reproductions which took their place. **1968** J. Ironside *Fashion Alphabet* 51 *Medici collar*, a stiffened collar usually of lace, fan-shaped; high at the back of the neck and tapering away at the sides. **1969** R. T. Wilcox *Dict. Costume* (1970) 206/1 *Medici lace*, French bobbin-lace similar to Cluny lace, but of finer thread. Woven closely and open in intricate pattern with one edge scalloped.

medicine, *sb.*[1] Add: **2. f.** *to take one's medicine*, to submit to or endure something disagreeable; to learn a lesson; *a dose*, *taste*, etc., *of one's own* (*kind of*) *medicine*, repayment or retaliation in kind; 'tit for tat'.

1865 A. D. Richardson *Secret Service* v. 75 The leaders refused to take their own medicine. **1894** P. L. Ford *Hon. Peter Stirling* xxvii. 150 'He snubbed me..,' explained Miss De Voe, smiling slightly at the thought of treating Peter with a dose of his own medicine. **1903** *N.Y. Times* 21 Sept., Canada can do nothing—she must take her medicine and make the best of it. **1904** 'O. Henry' *Cabbages & Kings* xvii. 299 You go back and take your medicine like a man. **1939** T. S. Eliot *Family Reunion* I. i. 18 Make him feel that what has happened doesn't matter. He's taken his medicine, I've no doubt. **1941** V. Perdue *Singing Clock* (1945) viii. 52 It was only fair for them to get a taste of their own medicine. **1961** L. van der Post *Heart of Hunter* I. vii. 111 Giving him some of his own medicine, I said: 'How very charming of you!' **1961** C. Willock *Death in Covert* xii. 217, I set the spring-guns, sir... I'm willing to take my medicine for that... If it's found out..it'll make the other things look black for me. And them I did not do. **1968** E. Gaines in A. Chapman *New Black Voices* (1972) 102 He ain't the first one they ever beat and he won't be the last one, and getting in it will just bring you a dose of the same medicine.

4. Also *colloq.* in phr. *bad medicine*, something or someone sinister or ill-fated.

1825 G. Simpson *Jrnl.* in *Fur Trade* (1931) 136 Some of them have it that I am one of the 'Master of Life's Sons' sent to see 'if their hearts were good' and others that I am his 'War Chief' with bad medicine if their hearts were bad. **1869** *Harper's Mag.* Jan. 151/2 Will Comstock was sure that it was bad 'medicine' (luck) to camp on the Stinking Water. **1945** P. Cheyney *I'll say she Does!* ii. 38 I'd like to talk to him. He's bad medicine. **1964** 'E. Peters' *Flight of Witch* i. 25 'Did you know that outcrop of rock is known locally as the Altar?'..'So that's it,' he said. 'Just bad medicine.' **1969** 'J. Morris' *Fever Grass* xxiii. 214 When people like me..get together we can be bad medicine fer anybody from outside. **1973** P. O'Donnell *Silver Mistress* xiii. 216 Momma's gotta go redundant. She's a great kid..but she's gonna be bad medicine on the run.

6. a. *medicine bottle* (earlier example), *cabinet*, *chest* (earlier and later examples), *cupboard*; **medicine ball,** a stuffed leather ball which is thrown and caught to provide exercise; **medicine glass,** a small drinking-glass graduated for use in measuring medicines; **medicine show** *N. Amer.*, a travelling show, in which entertainers attract customers to whom medicine can be sold.

1895 *Crescent* (Brooklyn, N.Y.) 1 Nov. 14/1 While Charlie Notman opines that the 'gym' needs more medicine ball, the few inoffensive ones now on hand are kicking for more 'gym'. **1903** W. L. Savage in *Athletics & Outdoor Sports for Women* 49 The illustrations below show two of the methods for passing medicine balls. **1930** *Bulletin* 13 Feb. 8/2 The Prince of Wales..instead of indulging in ..tennis and quoits, preferred to devote the time after tea to throwing the medicine ball. **1965** G. McInnes *Road to Gundagai* x. 156 Down the middle of the hall three teams were competing at medicine ball. **1974** J. Heller *Something Happened* 314 It was a relay race, and he was ten yards ahead,..carrying a heavy medicine ball. **1852** Dickens *Bleak Ho.* (1853) v. 35 Quantities of dirty bottles: blacking bottles, medicine bottles, [etc.]. **1899** *Montgomery Ward Catal.* 576/2 Medicine cabinet, made of oak. **1922** S. Lewis *Babbitt* i. 14 Above the set bowl was a.. medicine cabinet. **1955** W. Gaddis *Recognitions* II. vii. 572 Esther set off with her to the bathroom, where they interrupted someone who was looking through the medicine cabinet. **1974** M. Babson *Stalking Lamb* xiv. 93 The sleeping tablets..in the bathroom medicine cabinet. **1731** in E. Singleton *Social N.Y. under Georges* (1902) 85 A very fine Medicine Chest with great variety of valuable Medicines. **1957** P. Kemp *Mine were of Trouble* i. 6, I remember a bulky 'medicine chest' which seemed to contain chiefly iodine, quinine and cascara. **1907** *Yesterday's Shopping* (1969) 131/3 Fumed oak Medicine Cupboard..4/7. **1966** A. E. Lindop *I start Counting* xiii. 151, I fled to the bathroom. I yanked open the medicine cupboard so hard that the mirror came off its hinges. **1853** Medicine glass [see *bed table* s.v. *bed *sb.* 19]. **1938** H. Asbury *Sucker's Progress* 355 Minnie and Colorado Charley then organized a medicine show with which they traveled through Mexico and Central America. **1958** P. Gammond *Decca Bk. Jazz* ix. 117 Thelonious Monk, passing through with a travelling medicine show, was once heard as an uncomplicated swing stylist there. **1962** E. Lucia *Klondike Kate* 7 They travelled with medicine shows, carnivals, [etc.]. **1970** P. Oliver *Savannah Syncopators* 96 With the demise of the work song, the blues became the song vehicle to accompany labour and, in earlier years, for the medicine show or the barber shop.

b. *medicine bag* (earlier and later examples), *bundle*, *lodge* (earlier and later examples), *pipe*, *sack*; **medicine line,** a name given by American Indians to the border between Canada and the United States; **medicine man** substitute for def.: a magician or shaman among American Indians and other peoples (earlier and later examples); hence *colloq.*, a doctor (cf. sense 2 a in Dict.); also *transf.* and *fig.*; **medicine murder,** murder committed to obtain parts of the body for 'medicine'; ritual murder; **medicine wolf** *U.S.* = Coyote; **medicine woman,** an Indian woman dealing in magic.

1797 C. Chaboillez *Jrnl.* in B. C. Payette *Northwest* (1964) [I] made him consent to go for his Medicine Bag. **1801** A. Henry *Jrnl.* in E. Coues *New Light Hist. Greater Northwest* (1897) I. 162 The fellow came accordingly with his drum and medicine bag. **1865** Milton & Cheadle *N.W. Passage by Land* iv. 66 The chief..dressed in a spangled shirt, a cap covered with many-coloured ribbons, and an elaborately-worked medicine-bag, rose and made an oration. **1971** J. McDougall *Parsons on Plains* xix. 180 With their medicine-bags in hand they stood like statues. **1936** *Canad. Geogr. Jrnl.* XII. 98/2 Wherever they went they carried with them a medicine pipe and bundle upon the back of a milk-white steed. **1952** *Beaver* Sept. 27 Medicine bundles, both personal and tribal, were of great importance. **1969** *Ibid.* Summer 49/1 Each member had his own medicine bundle containing various items. **1910** A. L. Haydon *Riders of Plains* 95 The Indians..called the International Boundary the 'Medicine Line', assuming that in the absence of any agreement between the two Governments relative to this crime, they were perfectly safe on one side of the line with regard to what had been done on the other. **1913** L. V. Kelly *Range Men* 143 In their own tongue they [*sc.* Indians] called it the 'medicine line', and were very well pleased with the condition of affairs. *Ibid.* 162 The Canadian Indians were not entirely disappointed at the results of their own forays south of the 'medicine line'. **1962** W. Stegner *Wolf Willow* II. vii. 96 By that time Crow and Gros Ventre and Sioux and Blackfoot and Assiniboin would already know that the 'Medicine Line', as they called it, was something potent in their lives. **1970** *Beaver* Winter 28/1 In 1801 he was.. just across the North Dakota medicine line, the magical boundary between Canada and the States. **1973** R. D. Symons *Where Wagon Led* I. vii. 110 The various ranch outfits would make up big trail herds and move them across the Medicine Line. **1808** C. Mackenzie in L. R. Masson *Les Bourgeois* (1889) I. 354 The women were directed to go into the woods for branches to cover the Medecine lodge. **1901** F. H. Giddings *Inductive Sociol.* 207 Religious Societies—..In North American Indian tribes, they are known as Medicine Lodges. **1944** *Beaver* June 35 At the end of the medicine lodge inside, Moanday erects two poles, a cross-bar between. At the foot of each pole he lays the body of a dog he has killed. **1801** A. Henry *Jrnl.* in E. Coues *New Light Hist. Greater Northwest* (1897) I. 162 An Indian who pretended to be a medicine man was employed by Maymiutch to cure his sick brother. **1890** E. Dowson *Let.* 10 Oct. (1967) 170 In spite of my rooted aversion to the genre I shall have to call in a medicine-man. **1898** G. B. Shaw *Doctors' Delusions* (1932) 107 Out of sheer credulity as to the infallibility of the medicine man, we are drifting into a legal procedure which relieves them from all necessity to gain our confidence by the good they do us. **1922** Joyce *Ulysses* 16 She bows her old head to a voice that speaks to her loudly, her bonesetter, her medicine-man. **1939** J. Dillard in A. Dundes *Mother Wit* (1973) 278/2 Medicine men..falsify their experience..in order to qualify for their craft. **1947** Auden *Age of Anxiety* (1948) iii. 72 The medicine men who keep this body Politic free from fevers, Cancer and constipation. **1961** *Listener* 20 Apr. 683/2 He [*sc.* a foreign correspondent] has, like a medium or a circus medicine man, simply to go into a trance to pronounce what '*the*' American people feel about it all. **1965** R. & D. Morris *Men & Snakes* ii. 46 A double-sexed two-headed snake named Sachan..was the typical emblem of the Near Eastern and Mediterranean medicine man. **1966** B. Kimenye *Kalasanda Revisited* 92 The medicine man at Nakivubo bus park had a fresh supply of snuff. **1974** Wodehouse *Aunts aren't Gentlemen* iii. 17, I was a bit early for my appointment, and was informed on arrival that the medicine man was tied up for the moment with another gentleman. [**1947** *Times* 25 Oct. 3/5 Seventeen Africans were sentenced to death in two cases..here today for committing murder for 'medicine'.] **1952** *Basutoland 1951* (H.M.S.O.) ix. 57 During 1951 fourteen trial cases..were heard by the High Court. Two of these cases dealt with the crime known as Medicine Murder. **1966** *New Statesman* 25 Feb. 265/1 A chapter..on the abominations of 'medicine murder'. **1833** G. Catlin *N. Amer. Indians* (1841) I. 111 At that hour.., with medicine-pipes in his hands and foxes tails attached to his heels, entered Mah-to-he-hah (the old bear). **1971** J. McDougall *Parsons on Plains* xii. 110 First, the oldest conjuror took the big medicine pipe with the long stem. *a* **1831** J. Smith *Jrnl.* in M. S. Sullivan *Trav. J. Smith* (1934) 5 You observe at the door [of the Indian Lodge] three straight and handsome poles set up in a triangular form and joined together at the top, on which is suspended the medicine sack of the owner, consisting of such things as he fancies to possess a certain undefined charm. **1883** 'Mark Twain' *Life on Mississippi* 613 See my medicine-sack and my war club tied to it. **1837** W. Irving *Capt. Bonneville* II. xv. 147 This little, whining, feast-smelling animal, is..called among the Indians the 'medicine wolf'. **1846** [see *cayeute, cayote]. **1860** E. J. Lewis in *Colorado Mag.* (1938) XV. 30 Went up town and saw a young grizzly bear, a young swift or medicine wolf much resembling a fox. **1834** *Knickerbocker* IV. 372 The mother evinced her sagacity, as a diviner or medicine woman. **1836** *Ibid.* VIII. 152 It was at the wigwam of an old Indian 'medicine-woman' that I stopped.

medick. (Later examples.)

1912 W. Somerville *Agric.* iii. 69 Species standing wide apart, like beans and medick, have no mutual interest in any particular organism. **1960** *Oxf. Bk. Wild Flowers* 20/2 Spotted Medick or Calvary Clover... The leaflets are much bigger than those of other Medicks.

medico. Add: **1.** (Later examples.) Pl. **medicoes, medicos.**

1915 A. Conan Doyle *Valley of Fear* I. iv. 53 The medicos will have a word to say before we finish. **1923** A. Bennett *Riceyman Steps* I. x. 46 Dr. Raste would sometimes say with a dry, brief laugh, 'we medicoes'. **1943** A. Huxley *Let.* 7 May (1969) 489 One can only.. pray that the army doctors aren't merely pumping him full of toxic sulfanilamide..which is what so many medicos do. **1951** E. Paul *Springtime in Paris* xv. 284 Leaving Hortense to the medicos, young and old, I went with Busse to the Gare de Lyon. **1971** *Lancet* 23 Oct. 919/1 We really need a new word for doctor. Perhaps the Common Market will insist on a common term—medico perhaps? **1973** *Nature* 6 Apr. 377/2 The twenty thousand or so scientists, engineers, medicos and so on on the staff of British universities.

medico-. Add: *medicosocial* adj.

1961 *Lancet* 29 July 259/1 (*heading*) Medicosocial aspects. **1967** *Times* 20 Feb. 13/4 Everything is to be gained by the widest publicity being given to these serious medico-social problems affecting all sections of the community.

medina (mīdī·nă). Also **Medina.** [Arab., 'town'.] The non-European section of a North African town.

1906 D. Sladen *Carthage & Tunis* II. i. 330 Inside, these great houses of the Medina are just as stately. **1935** G. Gorer *Africa Dances* I. vii. 76 The dispossessed negroes went to live in the Medina, a geometrical ghetto of one-room stone huts. **1950** R. Landau *Invitation to Morocco* iii. 28 Leaving the native *medina* surrounded by its ancient walls untouched, he erected outside it an entirely new town. **1961** J. Anthony *About Tunisia* i. 28 My favourite walks were in the residential quarters high in the Medina. **1972** *Country Life* 13 Jan. 86/2 The crowded Medinas of Islamic cities. **1972** W. McGivern *Caprifoil* (1973) x. 174 He never made the mistake of pretending he understood Arabs. He had grown up with them..listened in the *medinas* to the rise and fall of the wise men's prayers.

Medinal (me·dinăl). *Pharm.* Also **medinal.** A proprietary name for the sodium salt, $C_8H_{11}N_2O_3Na$, of barbitone; it is a hypnotic of similar action to barbitone.

1908 *Trade Marks Jrnl.* 12 Aug. 1319 Medinal... Medicines for human use. Chemische Fabrik auf Actien (Vorm. E. Schering),..Berlin, Germany, manufacturers. **1922** *Daily Mail* 16 Dec. 7 The verdict was Death from an Overdose of Medinal taken by Misadventure. **1946** J. B. Priestley *Bright Day* vii. 217 As I undressed I chewed two tablets of medinal... I knew that medinal acts slowly with me, but it was all I had there. **1965** J. Pollitt *Depression & its Treatment* v. 66 As these patients are already retarded it is rarely necessary to give tranquillisers or sedatives during the day, but barbitone sodium (Medinal) at night is often most effective in preventing early waking.

‖ **medio** (me·dio). Also **medio real.** [Sp. *medio* half.] An obsolete Mexican coin, also used in Cuba, worth half a *real fuerte* or 6¼ *centavos*.

1824 J. R. Poinsett *Notes on Mexico* 141 A *medio*, the sixteenth of a dollar. **1844** G. W. Kendall *Santa Fé Expedition* II. xii. 239 One of the lads [handed] each of the unfortunate prisoners a medio. **1845** T. J. Green *Jrnl. Texian Expedition* 246 One medio would buy a leaden rivet. **1859** R. H. Dana *To Cuba* 47 As there is no coin in Cuba less than the medio, 6¼ cents, the musicians get a good deal or nothing. **1897** T. A. Janvier *Mexican Guide* vii. 81 Medio real = 6¼ centavos. **1909** 'O. Henry' *Roads of Destiny* xxi. 358 I've got one Chili Dollar, two *real* pieces, and a *medio*.

medio-. Add: **1.** *medio-laterally* adv.

1971 *Nature* 20 Aug. 542/2 The nucleus is crescent shaped in cross section and extends 1·4 mm rostrocaudally and mediolaterally.

2. In *Philol.*, as **medio-palatal** *a.*, articulated with the tongue against the middle part of the hard palate; **medio-passive** *a.*, of the voice of a verb, intermediate between active and passive (cf. Middle *a.* 4 a); also as *sb.*; hence *medio-passivization*.

1902 E. W. Scripture *Elem. Exper. Phonetics* xvii. 297 Roof articulations are indicated by the names..pre-, medio-, postpalatal. **1942** Bloch & Trager *Outl. Ling. Analysis* 15 Different points of articulation are designated by the terms prepalatal, mediopalatal, and postpalatal, indicating that the front [of the tongue] touches or approaches respectively the anterior, the middle, or the posterior part of the hard palate. **1949** R.-M. S. Heffner *Gen. Phonetics* vi. 148 In Midwest American pronunciations of English there is an open *r* sound, which is formed by raising the middle region of the dorsum of the tongue towards the mediopalatal junction between the hard palate and velum. **1962** Chavarria-Aguila & Penzl in Householder & Saporta *Probl. Lexicogr.* IV. 241 A contrasting pair of voiced and voiceless medio-palatal fricatives occur. **1921** E. Sapir *Lang.* iv. 74 Of the seven suffixes..-*o*(*ht*)- indicates activity done for the subject (the so-called 'middle' or 'medio-passive' voice of Greek). **1933** E. H. Sturtevant *Compar. Gram. Hittite Lang.* vi. 250 Hittite has a medio-passive, which has the same uses as the Gk. middle voice. It most frequently represents the subject as acting upon or in respect of himself. **1934** Priebsch & Collinson *German Lang.* II. vi. 296 Germanic retained in Gothic an inflected medio-passive present from Indo-European. **1952** O. R. Gurney *Hittites* vi. 118 The verb has two voices—active and medio-passive. **1968** *Encycl. Brit.* XI. 558/1 Besides the active there survives a medio-passive voice [in Hittite] characterized in the present by a somewhat loose -*r*(*i*) attached to the old middle endings. **1972** *Language* XLVIII. 393 A reflexive—hence, by ergative criteria, mediopassive—form of the verb. *Ibid.* 395 In keeping with the ergative focus, the object (non-ergator) is retained formally in an intransitive, mediopassive form, there being no external agent of the action described. **1973** A. H. Sommerstein *Sound Pattern Anc. Greek* ii. 13 The infinitive ending in all medio-passive senses is /-sthay/. **1975** *Language* LI. 97 Ambiguity between reflexive and medio-passive constructions is possible (e.g. 'I got out' vs. 'I cut myself'); but in practice this is not much of a problem, since many verbs do not undergo medio-passivization.

Mediterranean, *a.* and *sb.* Add: **B.** *sb.* **1. b.** Also, pertaining to the lands or countries in or around the Mediterranean Sea; spec. *Mediterranean anæmia* or *disease*, thalassæmia, esp. thalassæmia major; *Mediterranean climate*, the climate of lands around the Mediterranean Sea, characterized by hot, dry summers and mild, wet winters; also applied to any similar climate in other regions, as parts of California, Chile, South Africa, and southern Australia; *Mediterranean pine* = *maritime pine* (*maritime *a.* 6), or Stone-pine.

1760 G. Washington *Diaries* (1925) I. 145 Planted 4 nuts of the Mediterranean Pine in the Pen. **1896** *Jrnl. R. Microsc. Soc.* 324 (*heading*) Influence of the Mediterranean climate on plants. **1908** R. de C. Ward *Climate* v. 124 The sub-tropical belt is exceptionally wide... The fact that the Mediterranean countries are so generally included in this belt has led to the use of the name 'Mediterranean climates'. **1924** *Scottish Geogr. Mag.* XL. 150 (*heading*) The Mediterranean climates of Eurasia and the Americas. **1933** Mediterranean pine [see *coronilla]. **1936** Whipple & Bradford in *Jrnl. Pediatrics* IX. 279 (*heading*) Mediterranean disease—thalassemia (erythroblastic anemia of Cooley). *Ibid.*, This interesting disease presents three important abnormalities together characterizing a syndrome which may be designated 'Mediterranean disease'. *Ibid.* 292 The clinical diagnosis was Mediterranean anemia, pericardial effusion, bronchopneumonia terminal. **1953** A. Smith *Blind White Fish in Persia* iii. 51 Truly the Elburz mountains divide the arid heat of Tehran from the Mediterranean climate of the Caspian. **1954** *Blood* IX. 648 The term Mediterranean anemia is used in this paper to encompass the entire group of diseases characterized by microcytosis, hypochromia, ovalocytosis, anisocytosis and poikilocytosis, and, most prominently, targeting, together with certain familial hereditary patterns and clinical features. It is recognized that this group covers the spectrum from the most benign to the most severe forms clinically, and the range from hypochromic polycythemia to the severe Cooley's anemia hematologically. **1961** R. D. Baker *Essent. Path.* xviii. 494 In thalassemia (Cooley's or Mediterranean anemia) there is an inherited anomaly of the red blood corpuscles consisting of flat 'target' cells. **1969** Neuberger & Cahir *Princ. Climatol.* vi. 81 Climates having dry summers and wet winters as a result of the shifts of the subsidence belts are said to be 'Mediterranean' or 'California' climates. **1971** F. A. Ward *Primer of Haematol.* vi. 54 If,..in Mediterranean anaemia, it can be shown that excessive destruction of red cells is taking place almost exclusively in the spleen, then splenectomy will relieve but, of course, not cure the condition.

3. A racial type found especially in countries bordering on the Mediterranean sea; a person of this racial type. Also *attrib.* passing into *adj.* (Cf. *Eurafrican *a.* and *sb.*)

1888 C. Morris *Aryan Race* i. 13 The hair of the Mediterraneans is not so long or so cylindrical in section as in the Mongolians. **1899, 1910** [see *Eurafrican *a.* and *sb.* 1]. **1921** *19th Cent.* May 896 It would be difficult to deny that the latinised 'Mediterraneans' are the most finely tempered peoples of Europe. **1921** *Man* CVII. 180 The brown dolichocephals called Mediterraneans. **1928** [see *Eurafrican *a.* and *sb.* 1]. **1935** Huxley & Haddon *We Europeans* iv. 137 The Nordic, Eurasiatic, and Mediterranean types which are now scattered through the European population. *Ibid.* vi. 172 The Mediterranean type is much more widely distributed than the Mediterranean area. **1939** C. S. Coon *Races of Europe* iv. 83 Some Mediterraneans were probably white skinned, and others brown. *Ibid.*, The Mediterranean group seems to be of purely *sapiens* ancestry, without Neanderthaloid or other mixture. **1959** *Chambers's Encycl.* XI. 433/1 Caspians.. are taller and less glabrous than the European Mediterraneans. **1960** J. Comas *Man. Physical Anthropol.* ix. 602 This is the Dinaric race, which seems to be a Mediterranean type brachycephalized by some non-Mediterranean agency. **1962** C. S. Coon *Orig. Races* (1963) i. 19 Races like the..Mediterranean, East Baltic, and Dinaric, which loom large in the Europe-centred literature of anthropology, are neither subspecies nor, in a strict sense, local races. **1971** J. C. King *Biol. of Race* vi. 114 For the European populations one began to hear of Nordic, Alpine, Mediterranean..and God alone knows how many other races and subraces. **1974** N. Freeling *Dressing of Diamond* 11 He didn't look the part... A round Mediterranean head, and coarse black hair... A slightly Slav face. **1974** I. Murdoch *Sacred & Profane Love Machine* 288 He ..looked into the big dark eyes..Mediterranean eyes, African eyes.

Mediterra·neanize, *v.* [f. Mediterranean *a.* + -ize.] *trans.* To make Mediterranean in character or attributes. So **Mediterra:neaniza·tion.**

[**1896** T. Common tr. *Nietzsche's Case of Wagner* iii. 9 *Il faut méditerraniser la musique*: I have reasons for using this formula. The return to nature, to health, to gaiety, to youth, and to *virtue*!] **1915** *Nation* (N.Y.) 6 May 485/3 The Greeks are Mediterraneanized Slavs whose only heritage is a language which Demosthenes could neither have pronounced nor understood. **1921** *19th Cent.* May 894 The ancient Alpine race has been mediterraneanised, latinised, slavonised, and teutonised in Europe, and the Teuton in his turn has undergone mediterraneanisation, latinisation and slavonisation. **1947** N. Cardus *Autobiogr.* III. 244 He mediterraneanises Wagner, to use Nietzsche's term.

medium, *sb.* and *a.* Add: **A.** *sb.* **5.** *spec.* of newspapers, radio, television, etc., as vehicles of mass communication. Also *attrib.* and in *pl.* (see *media).

1967 M. McLuhan (*title*) The medium is the massage. **1968** *Listener* 28 Mar. 394/3 My recent visits to the theatre, together with my colour television set, have convinced me that McLuhan has it wrong. The medium *impedes* the message.

7. For 'gas-jet' in def. substitute 'source of light'. (Later example.)

1933 P. Godfrey *Back-Stage* vii. 90 'Two more floods up-stage, Bill,' says the stage-manager. 'What mediums, sir—amber or pink?'

9. c. (Earlier example.)

1777 in *Essex Inst. Hist. Coll.* (1906) XLII. 319 There cargo is Salt..37 bales, cases, hhds of mediums [etc.].

10. A medium-dated security.

1968 [see *medium-dated* adj., sense B. 3 below]. **1974** *Daily Tel.* 25 May 20/6 The popularity among high taxpayers of low coupon mediums such as Treasury 3 p.c. 1979 is not all that difficult to comprehend. **1975** *Times* 25 Apr. 24/7 'Shorts', after being ⅛ point higher at one stage, ended mixed... 'Longs' and 'mediums' were up to ¾ point up.

B. *attrib.* and *adj.* **1. d.** Of sherry, wine, etc., having a flavour intermediate between dry and sweet. So *medium dry*, *medium sweet* adj. phrs.

1906 *Hatch, Mansfield Price List* 11 First Quality, Extra or Medium Dry. *Ibid.* 20 Light, Medium Sweet. **1933** H. W. Allen *Sherry* iii. 46 This medium wine is likely to find its way into a blend of Amoroso. **1951** R. Postgate *Plain Man's Guide to Wine* iii. 57 Their labels contain nothing more informative than 'Best South African Sherry: Medium Dry', or some such phrases. **1960** I. Jefferies *Dignity & Purity* i. 18, I expect you'd like sherry, wouldn't you? Medium? **1961** *Twining Bros. Wine List* 2 Amontillado Rico (Medium Dry)... Ancient Browne Rednutt (Medium Sweet). **1969** *Guardian* 13 Feb. 9/1 The medium dry white wine. **1972** A. Hunter *Vivienne* viii. 101 The waiter..remembered the Major's buying Mrs. Selly a drink, a medium sherry. **1972** *Country Life* 23 Mar. 673/3 One can make it [*sc.* mead] very dry or quite sweet... I intended to make mine medium sweet. **1973** J. Porter *It's Murder with Dover* v. 45 He sipped his medium sweet cider. **1974** 'A. Gilbert' *Nice Little Killing* iv. 61 Maybe a Dubonnet or a medium sherry—spirits never.

e. The designation of meat cooked between 'well done' and 'rare'. So *medium done*, *medium rare* adj. phrs. (cf. Rare *a.*[2] b).

1939 P. K. Newill *Good Food* iv. 72 Beef..medium.. 22–25 [minutes per pound]. **1953** J. & M. Roberson *Meat Cookbk.* ii. 47 Beefsteaks..medium 1 in. 7 min. on each side. **1968** L. O'Donnell *Face of Crime* (1969) vii. 104 His own steak was just as ordered: medium rare and delicious. **1972** 'E. McBain' *Sadie when she Died* iv. 45 Carella ordered prime ribs, medium rare. **1972** *House & Garden* Feb. 111/1 Steak au poivre was ordered *bleu*..but arrived medium done. **1975** M. Kenyon *Mr Big* xix. 176, I never saw myself..tellin' Jeeves to do the steak medium.

3. *Comb.* **a.** *medium-haul*, *-heel*, *-range*, *-rise*, *-term*, *-weight* adjs. **b.** *medium-powered*, *-priced* adjs.; **medium-dated** *a.* (see quots. 1958 and 1968).

1948 *Financial Times* 5 May 1/6 Among medium-dated stocks War Loan 3½ per cent was a good spot. **1953** *Economist* 25 July 287/3 In the gilt-edged market interest at first was concentrated upon the 'shorts' and the medium dated issues. **1958** 'Nedlaw' *Your Guide to Stocks & Shares* III. 78 *Medium dated*, a gilt-edged security having more than five years but less than ten years to run to its final maturity or redemption date. **1968** P. A. S. Taylor *Dict. Econ. Terms* 70 *Medium-dated*, '*mediums*'. Securities with a life of between five and fifteen years. **1963** *Punch* 4 Sept. 352/3 The new medium-haul Trident. **1965** 'W. Haggard' *Hard Sell* iii. 31 SAGA was building a medium-haul aircraft. **1974** *Times* 14 Mar. 5/4 The new dishes now being served on all medium-haul flights. **1973** A. Roy *Sable Night* iii. 24 Dark red suède medium-heel shoes. **1963** Bird & Hutton-Stott *Veteran Motor Car* 41 Georges Richard himself was more interested in the medium-powered, medium-priced car. **1895** *Montgomery Ward Catal.* 15/3 The most satisfactory medium-priced cottons on the market. **1972** E. Hargreaves *Fair Green Weed* v. 67 They were booked in at a medium-priced hotel. **1943** *Sun* (Baltimore) 24 Aug. 2/6 But it could be of real value to medium and short-range planes. **1974** 'F. Clifford' *Grosvenor Square Goodbye* II. 166 The provision of a medium-range aircraft at Heathrow. **1968** *Guardian* 19 June 3/3 'Medium-rise' housing—flats going up to only four or five stories. **1972** *Times* 17 June 8/8 Aaron Wallis lived in Battersea, in a medium-rise block of no great character. **1958** *Spectator* 15 Aug. 216/3 A high credit rating for medium-term loans. **1965** *McGraw-Hill Dict. Mod. Econ.* 321 A medium-term forecast made in July, 1963 could cover the period from January, 1965, through 1967. **1969** *Times* 5 May 22/2 The Midland Bank Finance Corporation..specialises in medium-term credit. **1971** *Jrnl. Gen. Psychol.* LXXXIV. 243 Span memory is distinct from medium-term memory in the case of semantic material. **1895** *Montgomery Ward Catal.* 282/3 Men's medium weight, natural wool color undershirts. **1964** *McCall's Sewing* I. 11/1 Select medium-weight fabrics which drape nicely and add roundness. **1968** M. S. Livingston *Particle Physics* i. 5 Medium-weight nuclei are the most tightly bound and the most stable.

c. *attrib.* and *Comb.* examples of sense 8 b of the sb.

1872 Schele de Vere *Americanisms* 245 A Circle is held for Medium Developments and Spiritual Manifestations at Bloomfield-street every Sunday. **1886** W. James in *Proc. Amer. Soc. Psychical Res.* 105 Her pupils contract in the medium-trance. **1919** J. M. Keynes *Econ. Consequences Peace* iii. 37 Mr. Lloyd George's unerring, almost medium-like, sensibility to every one immediately round him. **1972** A. Ford *Life beyond Death* i. 49 'I'm willing to go along with this stuff,' a medium baiter said one time, 'if you'll positively guarantee to bring me Socrates.'

d. Special collocations: **medium bomber,** a bomber intermediate between the heavy and the light; **medium close-up** *Cinematogr.*, a cinematographic or television shot intermediate between a medium shot (see below) and a close-up; also called *medium-close shot*; **medium frequency,** an intermediate frequency (of oscillation); *spec.* in *Broadcasting*, a frequency of a medium wave, viz. one between 300 kilohertz and three megahertz; **medium shot** *Cinematogr.*, a cinematographic or television shot intermediate between a close-up and a long shot; **medium wave** *Broadcasting*, a radio wave with wavelength between a hundred metres and a kilometre (see quot. 1929 for former limits); freq. *attrib.* (usu. hyphenated).

1935 *Flight* 22 Aug. 204a/2 The specialized light bomber ..may..be supplanted eventually by the very fast medium bomber. **1938** *Encycl. Brit. Bk. of Year* 161/1 A medium bomber can carry enough incendiary bombs to start 150 separate simultaneous fires. **1956** *U.S. Air Force Dict.* 321/2 *Medium bomber*,..currently (1956), a bomber having a gross weight, including bomb load, of between 100,000 and 250,000 pounds..; a medium bomber is thought of as having medium range, and as being best used at medium altitudes, as well as having a medium gross weight. **1971** E. Luttwak *Dict. Mod. War* 45/2 'Medium' bombers and 'light' bombers retain a residual role in 'tactical' situations. **1957** Manvell & Huntley *Technique Film Music* ii. 34 Long track, mostly in medium or medium-close shot, with one large pull-back during the market-place scene. **1933** A. Brunel *Filmcraft* 147 Scene 168. Medium close up. The slave-driven father looks desperately round again. **1948** E. Lindgren *Art of Film* v. 82 A medium close-up of his body on which shadows of the prison bars form a pattern. **1950** E. E. Brodbeck *Handbk. Basic Motion-Pict. Techniques* 103 *Medium close up*... Such a scene shows the most vital part of the subject plus some of that part's surrounding area. **1969** W. Rutherford *Gallows Set* vi. 77 David in medium close up came on the monitor screens. **1920** *Whittaker's Electr. Engineer's Pocket-Bk.* (ed. 4) 348 The result..has been the adoption of two frequencies, a medium frequency for general power and lighting, and a low frequency for systems supplying rotary converters. **1946** *Happy Landings* (Air Ministry) July 9/1 The radio compass, when tuned to any medium frequency, was seriously affected by thunderstorms... The compass pointer did not point towards the M/F station selected. **1966** *McGraw-Hill Encycl. Sci. & Technol.* I. 363/2 European medium-frequency (mf) broadcasting channels are assigned at 9-kc intervals rather than the 10-kc intervals used in the Western hemisphere. **1933** A. Brunel *Filmcraft* 150 Scene 488. Medium shot. The waiter pouring out champagne into the glasses on Pauline's table. **1937** *Discovery* Nov. 330/2 The [television] cameras are also fitted with lenses of suitable focal length for getting long-shots, medium-shots and close-ups as desired. **1953** K. Reisz *Technique Film Editing* ii. 81 In a long-lasting shot (33) Bill slows down as he approaches the camera and finally comes to rest in medium shot. **1966** H. P. Manoogian *Film-Maker's Art* vi. 219 In traditional editing the scene is established, usually in a long shot, and as the action within the scene proceeds a number of medium shots and close-ups are taken to relate that action. **1975** *New Yorker* 26 May 32/3, I think we can get around the full lotus by having a stunt man do it, and using medium and long shots in a half-lit room. **1928** *Wireless World* 7 Nov. 626/2 It will..be best to concentrate on maximum efficiency in medium-wave transformers. **1929** *Jrnl. Inst. Electr. Engin.* LXVIII. 22/2 The definitions were fixed as follows [by the International Radio Technical Committee]... Medium waves: kilocycles/sec 1500 to 100; m 200 to 3000. **1938** D. H. Surgeoner *Aircraft Radio* iii. 23 Medium waves are..suitable for both communications and direction-finding and these can be transmitted over reasonable distances. **1961** *Radio Times* 6 Apr. 9/1 The Network Three transmitters, both medium-wave and VHF, will be used for one channel..and the BBC's television sound transmitters for the other. **1974** Medium wave [see *long wave* s.v. *LONG a.¹ 18].

mediumistic, *a.* (Earlier and later examples.)

1868 *Harper's Mag.* July 193/2, I should think they were both highly mediumistic. **1920** D. H. Lawrence *Lost Girl* xv. 344 The terror, the agony, the nostalgia of the heathen past was a constant torture to her mediumistic soul. **1941** L. A. G. Strong *John McCormack* xiv. 217 That almost mediumistic quality which enabled the singer to surrender to the song and yet keep his own individuality. **1954** A. Huxley *Let.* 5 Dec. (1969) 717 Huene's reactions were wholly aesthetic; Gerald's mainly verbal and mediumistic—with other personalities talking through him from a variety of mental levels. **1974** *Times Lit. Suppl.* 27 Sept. 1046/3 What relation do they bear to mediumistic trances?

mediumly (mī·diŭmli), *adv.* [f. Medium *a.* + -ly².] Moderately; to a medium or average extent.

1909 *Nation* (N.Y.) 16 Dec. 598 An excellent anthology, for old as well as for the 'mediumly' young. **1958** *Times Lit. Suppl.* 30 May 301/4 Well-known names and personalities even to the mediumly informed. **1961** A. Wilson *Old Men at Zoo* iv. 213 I'm wedged as hell and mediumly happy there.

medivac, var. *medevac.

medjelis, medjliss, varr. *majlis.

Medo- (mī·do), combining form of Mede *sb.* (L. *Mēdus*, Gr. Μῆδος), used parasynthetically with terms denoting other peoples or countries, esp. *Medo-Persian* adj.

1769 J. Brown *Dict. Bible* II. 381/2 The three kingdoms..conquered by Cyrus and his Medo-Persian troops. **1835** J. B. Robertson tr. *Schlegel's Philos. of Hist.* I. vii. 272 The Assyro-Babylonian empire which preceded the Medo-Persian. **1869** G. Rawlinson *Man. Anc. Hist.* II. 9 A revolution in the Medo-Persian capital. **1875** *Encycl. Brit.* I. 604/2 The name for God in Medo-Scythic is *Annap.* **1899** *Captain* Apr. 96/2 There are certain Medo-Persian laws pertaining to a successful stamp collection. **1938** *Times Lit. Suppl.* 1 Jan. 14/2 For a long period the technique of American journalism has been almost Medo-Persian in its rigidity.

medulloblastoma (mĭdɒ:loblæstōᵘ·mă). *Path.* Pl. **-blastomas, -blastomata.** [f. Medull(a + -o + Blast(o- + *-oma.] A malignant tumour of the central nervous system that usually occurs in the cerebellum of children.

1925 Bailey & Cushing in *Arch. Neurol. & Psychiatry* (Chicago) XIV. 193 We shall adopt for our cases the term *Medulloblastoma* as appropriate to the particular tumors to which we independently wish to call attention. **1926** —— *Classification Tumors of Glioma Group* 118 Medulloblastomas..are rapidly growing, soft, malignant tumors which not infrequently, when they reach the leptomeninges, disseminate themselves through the subarachnoid spaces. **1974** Passmore & Robson *Compan. Med. Stud.* III. xxxiv. 75/1 Medulloblastomata usually arise from the inferior vermis and involve the flocculus and nodule. Hence the presentation is often difficulty in walking, truncal ataxia and a tendency to sudden falls.

mee (mī). [ad. Chinese *mien* flour, noodles, dough, prob. via the Hokkien pronunc. (mīⁿ).] A Chinese dish popular in Malaysia, consisting basically of noodles, with a variety of other ingredients.

1935 A. Dixon *Singapore Patrol* xiii. 107 We were confronted by a large dish piled with a startling mixture of spaghetti, bamboo shoots, sliced prawns, and tiny cubes of pork. The concoction was well seasoned with garlic, chillies, and soya-bean sauce. Its name..was *mee*. *Mee* is a favourite dish of the Chinese, and is on sale at all hours of the day and night. **1955** P. Anderson *Snake Wine* II. vi. 163 A Chinese cook tosses a handful of bean-sprouts, a bundle of pale yellow *mee*. **1963** J. Kirkup *Tropic Temper* viii. 82, I had some nice fried mee at a Chinese stall. **1965** C. Shuttleworth *Malayan Safari* v. 72 They sit.. wrestling manfully with knives and forks over English lamb chops, with their stomachs clamouring for Chinese *mee* or *kuay teow*. **1966** D. Forbes *Heart of Malaya* ii. 31 One could get a plate..of the noodles called *mee*, for a few cents.

meech, *v.* Chiefly *U.S.* Also **meach.** [Dial. var. of Miche *v.*] (Examples of sense 2 of Miche *v.*)

1857 'Dow, Jr.' *Dow's Patent Sermons* 1st Ser. 203 When you fall short of the object for which you jump, you go meechin off, like a cat that has missed her mouse. **1833** S. Smith *Life & Writings J. Downing* 176 The old man hauled in his horns and meeched off looking shamed enough. **1902** H. L. Wilson *Spenders* xxxi. 366 I'd hate to have you come meachin' around after that stock has kited.

meeching, *ppl. a.* Also **meaching, meachin, me(e)chin.** [var. of Miching *ppl. a.*; now dial. and the preferred spelling in the U.S.] Skulking, furtive; mean.

1610 Beaumont & Fletcher *Scornf. Lady* (1616) v. i. 9 Sure shee has some Meeching raskall in her house. **1792** *Mass. Spy* 22 Mar. 1/1 There is a kind of meaching souls in the world. *a* **1800** *Spirit of Farmer's Museum* (1801) 287 We observed, however, that he had lantern jaws and a meaching look. **1836** T. C. Haliburton *Clockmaker* 1st Ser. (1837) xv. 140 Father goes up to him, lookin as soft as dough, and as meechin as you please. **1844** 'J. Slick' *High Life N.Y.* II. 219, I gin her hand a leetle mechin shake. **1869** *Harper's Mag.* 193/2 Of old the contrast between the Southerner's proud self-assertion and the Northerner's meeching humility was inexpressibly mortifying to every thoughtful inhabitant of the free States. **1884** W. D. Howells *Rise S. Lapham* (1885) ix. 153 I'm not going to have you do anything that will make you feel meeching afterward. **1944** H. Wentworth *Amer. Dial. Dict.* 383/1 *Meech*,..1926, Maine, 'He was so meechin' that butter wouldn't melt in his mouth.' Common.

meek, *a.* Add: **4.** *meek-faced, -mild, -swarded* adjs.

1871 J. G. Whittier *Miriam* 29 That lean, fierce priest ..Meek-faced, barefooted. **1933** W. de la Mare *Fleeting* 92 Meek-faced, they snuff the air. **1951** —— *Winged Chariot* 23 Meek-mild as chickweed. **1953** C. Day Lewis *Italian Visit* I. 24 Immortal landscape of a day..Meek-swarded, comely pastoral.

meemies (mī·miz), *sb. pl. slang.* [Origin obscure.] In full, *screaming meemies*. **1.** Hysterics; a hysterical person. (See also quot. 1927.)

1927 *New Republic* 9 Mar. 72 The following is a partial list of words denoting drunkenness now in common use in the United States... To have the screaming meemies [etc.]. **1942** *Ibid.* 19 Jan. 78 If the mother gets the screaming meemies every time the siren growls, this deep and damaging fear is reflected and associated with all sirens, and all talk of bombs. **1946** Mezzrow & Wolfe *Really Blues* (1957) xvii. 323 The jangled nerves, the reefer flights, the underworld meemies. **1970** *N.Z. Listener* 12 Oct. 13/3 Thunderclap. Gets the screaming meemies. I couldn't help hearing her shouting, in the next room. **1973** Gagnon & Simon *Sexual Conduct* (1974) v. 151 We've got a nice circle of friends that aren't a bunch of screaming meemies.

2. Also *sing.* **meemie, mimi.** (See quots.)

1944 *Infantry Jrnl.* Sept. 23 In Sicily it was the Nebelwerfer mortar, nicknamed the 'woof-woof' or 'screaming mimi'. **1945** *N.Y. Herald Tribune* 4 Feb. 3 The nebelwerfer is a six-barreled projector firing six-inch rockets... The rockets, not very accurate, are variously called 'screaming meemies' and 'moaning Minnies', but, like most Army slang terms, these names are also applied to other enemy explosives. **1945** *Sat. Rev. Lit.* 3 Nov. 7 The Screaming Meemie is a German multi-barreled rocket-mortar (so named for the sound it makes going off).

meer, var. *Mir *sb.*¹

meet, *sb.* Add: **1. b.** *slang.* A meeting; an assignation or appointment, esp. a meeting with a supplier of drugs; a meeting-place, esp. one used by thieves.

1879 *Macm. Mag.* Oct. 503/1 At six I was at the meet (trysting-place). **1889** 'Mark Twain' *Lett.* (1917) II. 512 We'll manage a meet yet. **1893** L. W. Moore *His Own Story* xxxvii. 460 Where he could see the 'meet'. *Ibid.* 461 He made a 'meet' for the following day. **1916** C. J. Dennis *Songs Sentimental Bloke* 23, I dunno 'ow I 'ad the nerve ter speak, An' make that meet wiv 'er fer Sundee week! **1929** *Chicago Tribune* 11 Oct. 14/3 The [drug-] peddler takes his stock to a point on the street or possibly a pool hall. The place where he meets his customer is called a 'meet' or a 'stand'. **1930** *Amer. Speech* VI. 118 Coast dental meet set for July 8–12. **1938** D. Runyon *Take it Easy* xv. 291 He finally arranges a personal meet with the Judy. **1944** [see *grouse *a.*]. **1955** W. Gaddis *Recognitions* II. v. 490 I'm going to make a meet, he answered..—I'm going out to meet a passer, to hand this stuff over to him. **1962** K. Orvis *Damned & Destroyed* xxv. 183 Your man made a meet... No drugs changed hands. **1967** J. Morgan *Involved* 26 I've made the meet with Alfie Stride for twelve-thirty. **1970** G. F. Newman *Sir, You Bastard* vi. 180 Manso was considering trying to make a meet with you.

2. a. *Geom.* A point, line, or surface of intersection.

1893 J. W. Russell *Pure Geom.* 156 The meets of opposite sides of a hexagon..inscribed in a conic are collinear. *Ibid.* 236 Given five points on each of two conics, to construct the conic which passes through the four meets of these conics and also touches a given line. **1958** A. Barton *Introd. Coordinate Geom.* viii. 147 (*heading*) Meet of two tangents. *Ibid.* x. 213 Find the equation of the lines joining *O* to the meets of $4x - 3y = 10$ and $x^2 + y^2 + 3x - 6y - 20 = 0$.

b. *Algebra.* The intersection of two or more sets; also, the infimum of two or more elements of a lattice.

1933 G. Birkhoff in *Proc. Cambr. Philos. Soc.* XXIX. 441 Let *Π* be any collection of subalgebras S_k... By the meet $\Delta(\Pi)$ of *Π* we mean the set of elements in every S_k of *Π*. **1933** H. F. Baker *Princ. Geom.* VI. ii. 70, [*h*] and [*k*] have a common space [*m*], which we may call their meet. **1938** T. G. Room *Geom. of Determinantal Loci* i. 6 The join of two spaces is defined as the space of least dimension which contains all the points of each of them, and the meet (or intersection) of the two spaces as the space of least dimension containing all points common to both of them. **1965** D. E. Rutherford *Introd. Lattice Theory* i. 3 We frequently call the l.u.b. of a subset the union of the elements which compose the subset, and correspondingly we call the g.l.b. of the subset the intersection or meet of its elements. **1972** A. G. Howson *Handbk. Terms Algebra & Analysis* xv. 76 Every two elements of *P* have a meet and a join, e.g. $12 \wedge 30 = 6$, $3 \vee 5 = 15$, and so *P* is a lattice.

meet, *v.* Add: **2. e.** *more than meets the eye*: greater significance than is at first apparent.

1853 'P. Paxton' *Stray Yankee in Texas* 308 There might be more in it than at first met the eye. **1906** Galsworthy *Man of Property* III. viii. 368 There's more here, sir, however,..than meets the eye. I don't believe in suicide, nor in pure accident, myself. **1911** Beerbohm *Zuleika D.* xvi. 247 Quick in more than meets the eye, John. Spiritually quick. You saw me putting on my hat; you did not see love taking on the crown of pity. **1923** Wodehouse *Inimitable Jeeves* ii. 21 'There is more in this than meets the eye,' I said. 'Why should your uncle ask a fellow to lunch whom he's never seen?' **1943** F. W. Crofts *Affair at Little Wokeham* ii. 21 There's more in most things than meets the eye.

g. Naut. *to meet (her, the ship)*: see quots. 1776 and 1948.

1776 W. Falconer *Universal Dict. Marine* (Phr. French Marine), s.v. *Rencontre!* The order to the helmsman, to meet the ship, right the helm, or put it towards the opposite side, in order to check the ship's sheer. **1841** R. H. Dana *Seaman's Manual* 183 If the order is..'Ease her!' 'Meet her!' or the like, the man should answer by repeating..the order. **1856** C. Nordhoff *Merchant Vessel* 233 The obstinate craft takes a mighty, almost resistless sweep to the other side, and 'meet her', is the cry, while poor Jack tugs desperately at the heavy-moving wheel. **1902** B. Lubbock *Round the Horn* 187 Occasionally he says sharply, 'Meet her! Meet her!' and sometimes he jumps to the wheel and gives us his powerful aid in grinding it up and down. **1948** R. de Kerchove *Internat. Maritime Dict.* 461/1 *Meet her*, an order given to the helmsman to shift the rudder so as to check the swing of the vessel's head in a turn. **1968** H. F. Chase *Boatswain's Manual* (ed. 3) xi. 255 If given too much wheel,..her head may start to fall off to port. When this is about to happen the helmsman will 'meet her' by putting the wheel to starboard for a few moments.

4. Freq. used in the imperative as part of a formula of introduction.

1920 C. E. Mulford *Johnny Nelson* vi. 37 'Meet th' Doc, Nelson,' said Dave. Johnny turned. 'Glad to meet you, Doctor.' **1926** A. A. Thomson (*title*) Meet Mr. Huckabee. **1933** 'Hay' & 'Armstrong' *Orders are Orders* II. 46 Waggermeyer (*breezy as ever*) Good morning, boys! Meet Miss Marigold, my secretary and continuity girl. **1961** T. Hughes (*title*) Meet my folks!

12. Sc. *to meet in with*, to encounter (a person); = sense 4.

1825 Jamieson Suppl., *To meet in wi'*, to meet with. S. B. **1828** D. M. Moir *Mansie Wauch* xi. 96, I..advised him to take a step in at his leisure to St. Mary's Wynd, where he would meet in with some merchants in scores. **1878** R. Cuddie *Corstorphine Lyrics* 18, I met my auld frien' Tam. Wha had met in wi' some guid chiel' and tasted half a dram. **1889** R. L. Stevenson *Master of Ballantrae* ix. 247, I was not always as I am today; nor (had I met in with a friend of your description) should I have ever been so.

13. *to meet up with*: to overtake or fall in with; to meet, encounter; to become acquainted with. *colloq.* (orig. *U.S.*). Also *absol.*, *to meet up*.

1837 A. Sherwood *Gazetteer Georgia* (ed. 3) (Provincialisms), *Met up with*, for overtook. **1889** K. Munroe *Golden Days* ix. 96 They'd meet up with you somewheres along Coloma way. **1905** *Dialect Notes* III. 87 He started before I did, but I met up with him before we got to town. **1919** F. Hurst *Humoresque* 297 Tell him his little Sid is here with thirty minutes before she meets up with the show on the ten-forty. **1935** [see *get *v.* 66 c]. **1949** G. Davenport *Family Fortunes* I. i. 10 It was better than staying at home and they would probably meet up with friends. **1955** E. Hillary *High Adventure* vi. 104 We..there met up with Evans and his two Sherpas. **1959** *N.Z. Listener* 12 June 5/4 It was there that I met up with Lowry again. **1972** D. Haston *In High Places* xi. 116 Everyone met up on Pokhara airstrip. **1972** *Daily Tel.* 13 May 20/8 The Sun Life is a very competitive office and would seem to meet up with your requirements. **1973** C. Bonington *Next Horizon* v. 92, I met up with Mick Burke in the camp site.

14. Comb. *meet-the-people*: phr. used *attrib.* to describe a tour, etc., made by an eminent person to meet members of the general public (in quot. 1943 of a hamburger with 'popular' appeal). orig. *U.S.*

1943 *Amer. Speech* XVIII. 148 *Meet-the-People burger* (musical comedy 'Meet the People' Los Angeles 1940). **1959** *Economist* 30 May 820/2 Their 'meet-the-people' tours of factories and shipyards. **1959** *Punch* 16 Sept. 160/2 His colleagues mysteriously disappeared during a 'meet the people' tour of the Isle of Wight. **1975** *Times* 2 July 6/2 The Belgian Prime Minister..continued his meet-the-people tour in Edinburgh yesterday.

meeterly, *a.* and *adv.* **b.** *adv.* (Further example.)

1847 E. Brontë *Wuthering Heights* I. xiii. 321 They's a pack uh corn i' t' corner, thear, meeterly clane.

meeting, *vbl. sb.* Add: **7.** *meeting bonnet, clothes, coat, -day, gown, hat, -time*; **meeting seed** (see quot. 1851).

1867 J. R. Lowell *Biglow Papers* 2nd Ser. p. lxxix, Her new meetin'-bunnet Felt somehow thru' its crown a pair O' blue eyes sot upon it. **1775** in O. E. Winslow *Amer. Broadside Verse* (1930) 141/2 He got him on his meeting clothes. **1867** 'T. Lackland' *Homespun* I. 63 The 'meetin clothes' of the children are laid away for another week, and the old ones got out again. **1887** M. E. Wilkins *Humble Romance* 139 An thar was Israel in his meetin coat, an' me in my best gown. **1644** *Early Rec. Portsmouth, Rhode Island* (1901) 32 It is..ordered that the businesse of such metinge dayes shal be specified. **1686** S. Sewall *Diary* (1878) I. 146 Mr. Moodey exercises at our House being our meeting-day. **1776** T. Pennant *Tour in Scotl. & Voy. Hebrides 1772* II. 364 After three market-days or meeting-days within the town of Halifax. **1856** M. J. Holmes *'Lena Rivers* 30 Nobody'd think any better of them for being rigged out in their very best meetin' gowns. **1887** M. E. Wilkins *Humble Romance* 300 Hatty in her meeting-gown of light-brown delaine, and her white meeting-hat..was not pretty. **1851** *Knickerbocker* XXXVIII. 372 (Th.), Some people call it 'caraway' and 'aniseseed', but we call it 'meetin'-seed', 'cause we cal'late it keeps us awake in meetin'. **1889** R. T. Cooke *Steadfast* xxxix. 414 Mothers of young families distributed fragrant bunches of dried 'meetin' seed' among their flocks. **1905** E. U. Valentine *Hecla Sandwith* 25 [He] sat contentedly munching 'meeting seed' which Molly Tucker..had given him. **1940** E. Early *New England Sampler* 319 In old New England gardens there grew three plants called *Meetin' Seed*—Fennel, Dill, and Caraway. **1639** *Rec. Colony & Plantation New Haven* (1857) I. 26 On the Lords Day in the meeting time. **1881** Mrs. Stowe *Sam Lawson's Oldtime Fireside Stories* 200 We were in disgrace, we boys; and the reason of it was this: we had laughed out in meeting-time!

meeting-house. Add: **2.** (Earlier and later examples.)

1632 *Rec. Cambridge, Mass.* (1901) 4 Every person.. shall [be]..within [the] meeting-house in the Afternoone. **1634** *Rec. Watertown, Mass.* (1894) I. 1 The charge of the Meeting House shal be gathered by a Rate justly levied. **1910** *Dialect Notes* III. 445 *Meeting house*,..church. Older generation. **1959** *Amer. Speech* XXXIV. 9 Around the turn of the nineteenth century, Baptists,..and others began dropping the term *meetinghouse* and replacing it with *church*.

b. *Polynesia.* A tribal hall (see quot. 1949).

1865 L. Andrews *Dict. Hawaiian Lang.* 144/1 *Ha-le-ha-la-wai*, s. *Hale* and *halawai*, to meet; assemble. A meeting house; a synagogue; a place of meeting. **1897** A. Hamilton *Maori Art* (1901) II. 112 *Whare-matoro*, a large meeting house. **1944** D. Stewart in D. M. Davin *N.Z. Short Stories* (1953) 267 The young Maori..led me over to the meeting-house, a long, low, gusty barn of a building. **1949** P. Buck *Coming of Maori* (1950) III. iv. 374 The meeting houses formed the social focus of the tribe, hence they were generally named after tribal ancestors. When the people assembled within its walls for tribal discussions, the orators were justified when they said, 'We have gathered together within the bosom of our ancestor.' The carved meeting houses were a source of pride to the people and they gave an atmosphere to the village that nothing else could equal. **1960** N. Hilliard in C. K. Stead *N.Z. Short Stories* (1966) 239 There I go walking into the meeting-house with my shoes on. **1974** *N.Z. Listener* 20 July 10/4 They are clustered, the kuia, around the tent in which the body lies alongside the meetinghouse.

3. *meeting-house chamber, ground, land, lot, post, rate, yard* (earlier example).

1651 *Official Rec. Springfield, Mass.* (1898–9) I. 200 The above mentioned bargain about the meeting house chamber. **1689** S. Sewall *Diary* (1878) I. 286 Paid 40£..for the Releases of Meetinghouse Ground. **1690** *Ibid.* 334 Mrs. Judith Winthrop's Deed of the Meeting-house Land in Boston. **1735** *New Hampsh. Probate Rec.* (1914) II. 523, I give to my son..a lot of land lying in the Meeting house lot. **1647** *Rec. Watertown, Mass.* (1894) I. 11 A wrighting shall be sett upon the meting-house-post, to give Warning [etc.]. **1656** *Ibid.* 48, 2ly [that] ye give acompt of the meeting howse rate. **1712** *New Hampsh. Probate Rec.* (1907) I. 687 Northerly on the fence by the metinge house yard.

mefeesh, var. *mafeesh *a.* and *int.*

‖ **méfiance** (mefiãns). [Fr.] Mistrust.

1876 H. James *Let.* 8 Feb. in R. B. Perry *Tht. & Char. W. James* (1935) I. 366 Touches that are too raffiné, words and phrases that are too striking, or too complete, inspire him with an instinctive *méfiance*. **1921** D. H. Lawrence *Sea & Sardinia* vii. 285 Partly it is barbaric *méfiance*. **1938** E. Bowen *Death of Heart* II. vi. 280 'Oh, did you,' said Dickie, with a certain *méfiance*.

Meg³: see *Megger.

mega-. Add: **me·ga-city,** a very large city; **megaco·lon** *Path.* [Colon¹], gross dilatation and hypertrophy of the colon; a colon in this condition; **megaka·ryocyte** (also **-caryo-**) *Biol.* [Karyo- + -cyte], any of the giant cells with large multilobar nuclei which are found in small numbers in normal blood marrow and which are believed to give rise to blood platelets by their fragmentation; so **me:gakaryocy·tic** *a.*; **me·ga-machine,** a social system dominated by technology and functioning without regard for specifically human needs; **mega-millionaire,** a multi-millionaire; **megaphyllous** *a. Bot.* [Gr. φύλλον leaf], having large leaves; **me·garipple** *Geol.*, an extensive undulation of the surface of a sandy beach or sea bed that is typically tens of metres from crest to crest and tens of centimetres in height, but may be much larger; **me·gashear** *Geol.* [Shear *sb.*²], a trans-current fault in which the displacement is very large (of the order of a hundred kilometres); **megaspore,** substitute for def.: the larger of the two kinds of spores in heterosporous cryptogams; later, extended to the homologous structure in seed plants (i.e. the immature embryo sac); (further examples); **megasporophyll,** substitute for def. of senses (*a*) and (*b*): any leaf or modified leaf which bears megasporangia; **megate·chnics,** the extensive mechanization of a society with a highly developed technology; **megavi·tamin** *attrib.*, based upon the administration of large doses of vitamins.

1968 *Harper's Mag.* Feb. 61 The noisy, ugly, chaotic, increasingly dangerous and ever-spreading mega-cities. **1975** *N.Y. Times* 20 Aug. 37/2 Or, as the Mailer–Breslin platform said, 'New York will become the first insane asylum of the megacity.' **1906** Dorland *Med. Dict.* (ed. 4) 419/2 Megacolon. **1908** *Practitioner* Sept. 459 True congenital idiopathic megacolon, or Hirschsprung's disease. **1949,** etc. Megacolon [see *Hirschsprung]. **1890** W. H. Howell in *Jrnl. Morphol.* IV. 118, I shall speak of them hereafter as megakaryocytes, or large nucleated giant cells. **1938** H. Downey *Handbk. Hematol.* I. vii. 449 (*heading*) Blood platelets and megacaryocytes. **1966** *Lancet* 24 Dec. 1416/1 There was virtual absence of erythroid precursors in the bone-marrow, with normal myeloid series and megakaryocytes. **1938** H. Downey *Handbk. Hematol.* I. vii. 482 Many of the blood platelets of the peripheral blood were rather large megacaryocytic fragments. *Ibid.*, Wuyts produced a marked megacaryocytic reaction in rabbits. **1972** *Nature* 7 Apr. 293/3 Erythroid, granulocytic or megakaryocytic cells. **1967** L. Mumford *Myth of Machine* i. 12 Cosmic order was the basis of this new human order. The exactitude in measurement, the abstract mechanical system, the compulsive regularity of this 'megamachine', as I shall call it, sprang directly from astronomical observations and scientific calculations. **1967** *Harper's Mag.* Oct. 110 The megamachine was 'invisible' because its tens of thousands of interacting parts were human. **1970** L. Mumford in *New Yorker* 31 Oct. 85 What is needed to save mankind from the megamachine—or whatever controls the megamachine—is to displace the mechanical world picture with an organic world picture, in the center of which stands man himself. **1973** *Physics Bull.* Jan. 5/2 The sophisticated industrial megamachines of the present century are also based on system centred technology, and the troubles they have led us into are only too clear. **1968** *Time* 8 Mar. 21 Nelson Aldrich Rockefeller, 59, a megamillionaire via the Rockefellers, a political patrician through the Aldriches. **1973** *Observer* 12 Aug. 11/1 He has managed to reach the near top of the mega-millionaire league table. **1904** *Science* 21 Oct. 529/1 The pteridophytes..may be disposed according to the prevalent size of their leaves in a series, leading from microphyllous to megaphyllous types. **1909** D. H. Scott in A. C. Seward *Darwin & Mod. Sci.* 203 A large proportion of the higher plants are microphyllous in comparison with the highly megaphyllous fern-like forms from which they appear to have been derived. **1953** L. M. J. U. van Straaten in *Geol. en Mijnbouw* XV. 3/1 Initial stages of transverse megaripples covered with the common small scale current ripples are frequently seen on the large, sandy tidal flats. **1968** *New Scientist* 18 Apr. 113/1 The strange features that have come to be known as megaripples—regularly formed giant undulations, measured by echo sounders, that straddle the ocean floor with distances of three to four miles between crest and crest. **1954** S. W. Carey in *News Bull. Geol. Soc. Austral.* July 1 The known megashears, where orogens are displaced hundreds of km., were next examined. **1971** *Nature* 2 July 23/1 We regard these two [American] plates as distinct entities separated mainly by the Cayman–Puerto Rico megashear. **1900** *Jrnl. R. Microsc. Soc.* 605 The division of the megaspore of *Erythronium* is..essentially the same as in *Lilium philadelphicum*. **1965** K. Esau *Plant Anat.* (ed. 2) xviii. 563 The ovule developing from the placenta of the ovary is the seat of formation of the megaspores (or macrospores). **1967** L. Mumford *Myth of Machine* ix. 189 When all the components, political and economic, military, bureaucratic and royal, must be included, I shall usually refer to the 'megamachine': in plain words, the Big Machine. And the technical equipment derived from such a megamachine thence becomes 'megatechnics'... At its inception no inferior chief could organize the megamachine and set it in motion. **1967** *Harper's Mag.* Oct. 108 Under the impulsion of unprecedented 'megatechnics'—'nuclear energy, supersonic transportation, cybernetic intelligence, and instantaneous distant communication'—the far-flung settlement patterns of Megalopolis are resistlessly expanding in many parts of the world, transforming man and the earth. **1970** *New Yorker* 10 Oct. 76/2 The idea of universal mechanization (megatechnics) was established in the megamachine of Egypt. **1970** L. Pauling *Vitamin C & Common Cold* 70 The use of very large amounts of vitamins in the control of disease has been called megavitamin therapy. Megavitamin therapy is one aspect of orthomolecular medicine. It is my opinion that in the course of time it will be found possible to control hundreds of diseases by megavitamin therapy. **1972** *Daily Colonist* (Victoria, B.C.) 20 Feb. 19/2 Dr. Abram Hoffer, a Saskatchewan psychiatrist who pioneered megavitamin therapy for schizophrenia. **1975** *Nature* 14 Aug. 529/3 Efforts of the FDA to regulate megavitamin promotion, however, were set back by a court decision and by the passage last year of a bill in the US Senate that specifically prevents the FDA from classifying high-potency vitamin preparations as drugs.

b. *megabar* [*bar *sb.*⁶ 1, 2], *-bit* [*bit *sb.*⁴], *-byte, -curie, -dalton* [*dalton²], *-hertz, -newton, -parsec, -rad, -volt* (examples; hence *megavoltage*), *-watt*; also **me·gabuck** *colloq.* [*buck *sb.*⁸], a million dollars; **me·gacorpse,** a million dead bodies, a term used in estimating the possible effects of nuclear warfare; **me·gadeath,** the death of a million persons, as a unit in estimating the possible effects of nuclear warfare; **megaunit** (me·găyūnit) *Biol.* and *Med.*, a million international units.

1903 Richards & Stull *New Method determining Compressibility* 43 The pressure of a megadyne per square centimeter would be called a megabar. **1925** J. Joly *Surface-Hist. Earth* iii. 55 The megabar is one million dynes per sq. cm. It is nearly one atmosphere. **1969** *New Scientist* 9 Jan. 81/3 The pressure needed to produce metallic hydrogen may well be less than a megabar. **1957** *Electronics* 1 Oct. 163 (*heading*) High-speed computer stores 2·5 megabits. **1972** *Sci. Amer.* Sept. 139/2 A broadcast-quality color-television signal in digital code calls for 90 megabits per second. **1946** *Picture Post* 7 Dec. 10/1 Atomic research is so expensive that American scientists have ceased to use the dollar as their unit. They have laughingly coined the term 'megabuck'—one megabuck equals a million dollars. **1952** *Galaxy* June 16/2, I had already pencilled in a tentative campaign in the budget well under a megabuck. **1968** *Amer. Anthropologist* LXX. 608/2 He certainly had no megabuck research grant. **1973** *Nature* 12 Jan. 86/1 The diamond project was not cheap (I think I first heard the word 'megabuck' in connexion with diamond synthesis). **1973** Megabyte [see *kilobyte* s.v. *kilo- b]. **1958** *Tuscaloosa* (Alabama) *News* 4 Sept. 4/5 The eeriest new word coined in the space age is 'Megacorpse'. **1968** *Economist* 13 Apr. 29/2 Dr. Kahn, a controversial figure best known for his calculations on thermonuclear war and his invention of the term 'megacorpse', has begun to broaden the institute's scope. **1947** *Radiology* XLIX. 326/1 The amount of radioactivity from these fission products with moderately long half-lives was in the range of hundreds of megacuries. **1957** *New Scientist* 10 Oct. 28/3 Large amounts of radioactivity can be measured in megacuries: one megacurie is the equivalent of one [metric] ton of radium. **1960** *New Biol.* XXXI. 121 The weight of 1 per cent of the particle [*sc.* T2 phage] is 1·2 Md. [*Note*] Md = megadaltons or 1,000,000 molecular weight units. **1973** *Sci. Amer.* Apr. 22/2 The molecular weight of the *E. coli* chromosome is 2,500 megadaltons. **1953** *Birmingham* (Alabama) *News* 21 June E3/1 He does not deal in numbers of atomic bombs or precise methods of delivery, in kilotons or megadeaths. **1959** *New Statesman* 21 Nov. 693/3 Mr Krushchev's announcement that a single Soviet factory is producing 250 megadeath weapons a year is a timely reminder of the risks of delay. **1962** R. E. Lapp *Kill & Overkill* viii. 100 '55 megadeaths' does not sound as bad as 55 million Americans dead. **1971** *Islander* (Victoria, B.C.) 5 Dec. 15/4 The brain that was good enough to produce the skilled hunter was also good enough to produce huge empires, noble causes to die for, vast armies, and megadeath. **1941** *Chem. Abstr.* XXXV. 2159 (*heading*) Molecular changes following irradiation with hertzian waves of a frequency of 1875 megahertz. **1966** *Electronics* 3 Oct. 171 Transatlantic airliners will

communicate with the satellite on the 118 to 136 Megahertz band. **1971** D. W. Sciama *Mod. Cosmol.* iv. 52 The revised 3C catalogue gives a virtually complete list of the [radio] sources between declinations −5° and +90° that are brighter than 9 flux units at 178 megahertz. **1970** *Sci. Jrnl.* June 16/1 It is possible to transform maraging steels with a high content of embrittling components..into fine wire having a tensile strength of about 5200 meganewtons per square metre. **1975** *Physics Bull.* Apr. 165/1 Traditionally, standard forces up to meganewtons are produced on deadweight machines. **1933** *Proc. Nat. Acad. Sci.* XIX. 1001 At the distance of three megaparsecs, the largest galaxies in the Virgo group have linear diameters of approximately six kiloparsecs. **1938,** etc. Megaparsec [see *Hubble]. **1973** *Physics Bull.* Nov. 674/1 There is evidence for an intergalactic [magnetic] field of order 10^{-13} T which seems to be uniform over scales of several thousand megaparsecs. **1958** *Times Rev. Industry* July 25/3 For many applications the megarad (one million rads) is more suitable. **1960** A. Charlesby *Atomic Radiation & Polymers* iv. 65 A reactor running at 100 megawatts power output could provide 30 megarads to 1 ton of material. **1953** *Economist* 31 Jan. 307/2 During last summer, British production [of penicillin] fell below the 1951 weekly average of 1,216,000 mega units. **1970** *New Scientist* 19 Mar. 543/1, 2.4 megaunits of long acting benzathine penicillin can maintain a treponemicidal blood and tissue level for three weeks or more. **1924** C. R. Underhill *Magnets* i. 11 The volt per microcoulomb or the megavolt per coulomb. **1957** *Technology* July 181/1 A research physicist at work on the 450 megavolt synchro-cyclotron. **1961** *Lancet* 16 Sept. 616/2, 3 patients with a recurrent infiltrating neoplasm after megavoltage therapy were included. **1961** D. W. Smithers in Tanner & Smithers *Tumours Oesophagus* xxi. 265 The next step forward came with the introduction of megavoltage [X-ray] apparatus. **1900** Webster, Megawatt. **1955** *Times* 16 July 6/4 The first atomic stations of the Central Electricity Authority will have two nuclear reactors each, together providing a net output of electricity of 100 to 200 megawatts. **1969** P. W. McDaniel in D. Z. Robinson et al. *Nucl. Energy Today & Tomorrow* (1971) II. iv. 210 In 1962 the largest U.S. power reactor was 180 megawatts.

megacycle (me·găsəik'l). [f. Mega- b + Cycle *sb.*] **a.** One million cycles (of an oscillation or other periodic phenomenon). **b.** *ellipt.* One million cycles per second; = *megahertz* (*mega- b).

1928 Sterling & Kruse *Radio Manual* 45 The frequency of 1,000,000 cycles can then be expressed as 1,000 kilocycles or 1 megacycle. **1936** A. Hund *Phenomena High-Frequency Syst.* ii. 56 It may be assumed from the theory of the ionized layer that for waves below 10 m in length (above 30 megacycles/sec) the sky wave passing toward this layer will never be returned. **1955** *Radio Times* 22 Apr. 3/1 The BBC decided..to build..short-range V.H.F. stations operating on frequencies around 90 megacycles per second. **1959** [see *kilocycle]. **1971** C. Bonington *Annapurna South Face* iii. 38 Kelvin plunges into details of frequencies, megacycles and all the other gibberish of technical experts.

megagametophyte (megăgæmī·tofəit). *Bot.* [f. Mega- + *gametophyte* (s.v. Gamete).] A gametophyte that develops from a megaspore; a female gametophyte.

1933 *Amer. Jrnl. Bot.* XX. 217 (*heading*) Morphology of the megagametophyte and the embryo sporophyte of *Isoetes lithophila*. **1964** H. J. Dittmer *Phylogeny & Form Plant Kingdom* xxi. 487 A single megagametophyte may contain several archegonia with ripe eggs.

megalithic, *a.* Add: **2.** Special collocations, as **megalithic fathom,** a name given to a measure of length equal to 5·44 ft., used in the construction of certain British megalithic monuments; **megalithic yard,** half a megalithic fathom (2·72 ft.).

1961 A. Thom in *Math. Gazette* XLV. 83 The unit of length used was the same from Land's End to John o' Groats... This unit was 5·44 ft... Half of this, 2·72 ft., might be called the Megalithic Yard, but whether this or the Megalithic Fathom (5·44 ft.) was the length of the standard rod carried about the country it is not possible to say. **1962** —— in *Jrnl. R. Statistical Soc.* A. CXXV. 243 In an earlier paper (Thom, 1955) it was shown that in a statistically significant number of cases the diameters of stone circles were multiples of 5·44 ft. This might be called the megalithic fathom. *Ibid.* 246 About 55 per cent. of all circles have the diameter an even number of fathoms. In setting out a circle it is the radius rather than the diameter which has to be measured out on the ground, so that where the diameter is an odd number of fathoms, that is, in about 45 per cent. of all circles, the constructors used a half fathom. Thus for our present purpose it seems better to take as the unit a length of half a fathom (2·72 ft) and for convenience call it the megalithic yard. **1967** —— *Megalithic Sites in Britain* v. 36 We first demonstrate that there is a presumption amounting to a certainty that a definite unit was used in setting out these rings. It is proposed to call this the Megalithic yard (MY). Two of these might be called the Megalithic fathom... It will appear that the Megalithic yard is 2·72 ft and so the Megalithic fathom is 5·44 ft. **1975** *Country Life* 16 Jan. 134/2 The megalithic yard of 2·72 ft that Professor Thom has recently so convincingly shown to have been used by the builders of Stonehenge, Avebury, Carnac.

megalo-. Add: **me·galopod** *a.* and *sb.* [Gr. πούς foot], (*a*) *adj.* having large feet, megalopodous; (*b*) *sb.* a person with large feet.

1909 *Cent. Dict.* Suppl., Megalopod, *a.* and *n.* **1951** Auden *Nones* (1952) 40 The basalt Tombs of the sorcerers shatter And their guardian megalopods Come after you pitter-patter

megalomaniac, *a.* (Examples.)

1899 *Pall Mall Gaz.* 14 Feb. 2/3 A megalomaniac world is always apt to regard a waistcoat-pocket community as a joke. **1929** W. J. Locke *Ancestor Jorico* 29 They had to attribute the great fortune to the megalomaniac dreams of a dying man. **1974** J. Pope Hennessy *R. L. Stevenson* xii. 226 A project emanating from Fanny's now megalomaniac brain.

megalopolis (megălǫ·pǫlis). [f. Gr. μεγαλο- great + πόλις city; see Megalo-, -polis.] Used (freq. with capital initial) as a designation of a very large city or its way of life; also, the practice of building large cities. Also *attrib.* Cf. † Megapolis.

The normal Eng. pl. (after Metropolis) would be *megalopolises*, but the irregular formation *megalopoli* occurs occasionally.

1832 Webster, Megalopolis. **1869** M. Collins *Ivory Gate* II. 211 Paul and his wife are back in the precincts of Megalopolis. **1945** *Archit. Rev.* XCVII. 5/1 Normal impulses, suppressed and frustrated in Megalopolis, return in the form of collective aggressions. **1951** M. McLuhan *Mech. Bride* (1967) 38/2 Megalopolis is both humanly and economically wasteful. **1959** *Economist* 25 Apr. 330/1 The dream of a massive shift of the American economy to the Great Lakes, and the growth of a vast new megalopolis along their shores. **1964** *Economist* 12 Sept. 1010/2 'Megalopolis'—or excessive urbanisation resulting from modern technology. **1968** C. A. Doxiadis *Between Dystopia & Utopia* 4 The city turned into a metropolis, which has now grown into a megalopolis. **1969** *Daily Tel.* 29 May 16/3 Los Angeles..is a vast megalopolis of over 10 million, not so much a city as an endless sprawling suburb without a centre..and criss-crossed with multi-lane freeways. **1970** *New Scientist* 5 Feb. 259/1 From the mud hut level of Indian villages and towns to the concrete towered megalopoli of Europe and North America. **1972** *Listener* 9 Mar. 293 Big money is unlikely to be made except perhaps in a very few megalopolis centres.

So **megalopo·litan** *sb.* and *a.*, (*a*) *sb.* an inhabitant of a megalopolis; (*b*) *adj.* of or pertaining to a megalopolis or the way of life characteristic of large cities.

[**1633** E. Grimstone tr. (*title*) The History of Polybius the Megalopolitan.] **1926** *Brit. Weekly* 29 July 355/3 Follows the winter season, with its dawning of 'megalopolitan civilization', its extinction of spiritual creative force. **1930** E. Waugh *Labels* i. 15 After the exaggerated cleanliness and sparkle of the preceding country, this exaggerated sombreness and squalor, called up..all the hatred and weariness which the modern megalopolitan sometimes feels towards his own civilisation. **1946** *Theology* XLIX. 94 The report deplores the sprawl of houses but not that of megalopolitan culture. **1957** R. Hoggart *W. H. Auden* 15 Auden's interest is in men in urban societies, in men now living through their perennial moral and metaphysical problems in megalopolitan settings. **1971** *Daily Tel.* 30 Nov. 10/4 The same international, megalopolitan forces, rolling over everything in their path, threaten Welsh and English alike. **1974** *Listener* 10 Jan. 53/2 Megalopolitan life brings with it anonymity and bewilderment.

Meganthropus (megæ·nþrǫpŭs). [mod.L. (G. von Koenigswald, 1942), f. Gr. μεγα- great + ἄνθρωπος man.] A large fossil hominid of the genus so called, esp. *Meganthropus palæojavanicus*, fragments of whose remains were first discovered by G. von Koenigswald in Java in 1941. Also *attrib.*

1942 G. von Koenigswald *Let.* 15 Jan. in *Anthropol. Papers Amer. Mus. Nat. Hist.* (1945) XL. 16/1 *Meganthropus* is a newly discovered fossil hominid perhaps related to *Australopithecus*. **1944** *Science* 16 June 480/2 The *Meganthropus* jaw is much too large and massive. **1959** J. D. Clark *Prehist. S. Afr.* iii. 73 It has been suggested that the *Meganthropus* form living in Java at this time is another early hominid of the Australopithecine group. **1967** M. H. Day *Guide to Fossil Man* (ed. 2) 238 (*caption*) The Meganthropus II mandibular fragment.

megaphone. Add: **2.** Also *attrib.* and *fig.*

1905 'O. Henry' *Trimmed Lamp* (1907) 180 The megaphone man roars out at you to observe the house of his uncle. **1909** *Daily Chron.* 7 June 5/1 They felt sure that the British Press were not the paid megaphones of financial buccaneers. **1951** M. McLuhan *Mech. Bride* (1967) 138/1 Is there any role left for the individual in a world of collective megaphone personalities? **1960** V. Nabokov *Invitation to Beheading* xvi. 162 The director..examined a piece of paper, and in a megaphone voice addressed Cincinnatus. **1968** *Listener* 20 June 803/1 Tell Joanna she must stop being a megaphone for a negative tendency in a parent–child participating democracy.

megaphone (me·găfōᵘn), *v.* [f. the sb.] *intr.* and *trans.* To speak or utter (as) through a megaphone. Hence **me·gaphoned, me·gaphoning** *ppl. adjs.*

1901 *Daily Colonist* (Victoria, B.C.) 30 Oct. 1/7 The Cufic..megaphoned the lightship asking to be reported. **1901** R. C. Lehmann *Anni Fugaces* 70 The air grows blue with loud reproaches Hurled at the crews by megaphoning coaches. **1912** J. H. Moore *Ethics & Educ.* 97 Long ago she [*sc.* a dog] and her associates were accustomed to megaphone to each other in this way. **1920** *Glasgow Herald* 21 Apr. 8 The captain megaphoned an invitation to come on to the flag deck. **1927** H. G. Wells in *Sunday Express* 2 Oct. 12/7 The impressive gatherings.., the megaphoned and broadcast speeches. **1963** A. Smith *Throw out Two Hands* iv. 46 Aldermanic individuals had to be given captive ascents on a long rope while they megaphoned their impressions to the gaping faces down below. **1967** *Punch* 26 July 121/2 The compere's megaphoned voice could be heard from the concert hall.

me·gaphonist. [f. Megaphone 2.] One who speaks through a megaphone.

1906 'O. Henry' *Four Million* 203 'What's eatin you?' demanded the megaphonist. **1949** *Box Office* 5 Nov. 20 Elliot Nugent is the megaphonist and William H. Wright the producer. *Ibid.* 21 As the month began, no megaphonist had been set for 'Jet Pilot'.

megapolis. (Later example.)

1855 I. C. Pray *Mem. J. G. Bennet* 450 Capital alone is wanting to make this city in point of influence..the megapolis of the world.

Megarian, *a.* Add: Also applied to a type of bowl of the Hellenistic period, usu. hemispherical and with relief ornament.

[**1874** A. Dumont *Peintures Céramiques* vi. 50 On leur a donné en Grèce le nom de Mégare, parce qu'elles se rencontrent..sur le territoire de cette ville.] **1905** H. B. Walters *Hist. Anc. Pott.* I. xi. 499 A series of vases known as Megarian or Homeric bowls, of hemispherical form, without handles. **1960** R. M. Cook *Greek Painted Pott.* vii. 215 The so-called 'Megarian' bowls, commercially the most successful relief ware of the Greeks, extend through most of the Hellenistic world and age. The shape is roughly hemispherical with or without a low ring foot. **1961** *Oxf. Univ. Gaz.* 10 Mar. 832/2 A 'Megarian' bowl with figures of Pan and (?) Heracles and Auge.

‖ **megaron** (me·gărǫn). [Gr. μέγαρον hall.] The great central hall of a type of house characteristic esp. of the Mycenæan period. Also *attrib.*

1877 *Architect* 4 Aug. 54/2 Palace of Ulysses at Ithaca. .. Within was another portico and a stately doorway, admitting to the *megaron*; this was a large apartment, unfloored, but lofty, roofed, and used as a dining hall by the men. **1906** *Ann. Brit. Sch. Athens* XII. 253 A fixed central hearth was introduced into the megaron to suit the needs of a more rigid climate than the original Africo-Aegean one. **1907** T. D. Seymour *Life Homeric Age* vi. 188 The great hall or *megaron* is the centre of the life of the household—not unlike the baronial halls of the old English castles. **1928** C. Dawson *Age of Gods* iii. 58 They brought with them a new type of house, with a pillared porch opening into a single large room or hall. This is the so-called Megaron house, which was characteristic of the early Greeks. **1950** H. L. Lorimer *Homer & Monuments* i. 6 A form of the megaron house, rectangular, with a single entrance to the main room and a more or less central hearth standing clear of the walls. **1963** *New Scientist* 9 May 301 Evidence which strongly suggests that the 'megaron' was an Anatolian rather than a Greek invention.

megascopically, *adv.* (Examples.)

1894 H. S. Washington *Volcanoes Kula Basin* 21 Megascopically it is much more frequent and prominent than either the augite or the olivine. **1909** J. P. Iddings *Igneous Rocks* I. II. iii. 445 Coarse-grained rocks..must be measured megascopically on the surface of proper-sized specimens. **1938** A. Johannsen *Descr. Petrogr. Igneous Rocks* IV. 377 Megascopically, laths of black pyroxene up to 2 cm. in length are seen. **1943** *Amer. Mineralogist* XXVIII. 499 The material under the botryoidal surface is also megascopically crystalline. **1970** K. C. Jackson *Textbk. Lithology* iv. 214 Epidote..is difficult to recognize megascopically because of small grain size and masking by associated micaceous minerals.

megasea (megæ·siă). [mod.L. (A. H. Haworth *Saxifragearum Enumeratio* (1821) 6), f. Gr. μέγας large, in reference to the large leaves of the plants.] A perennial herb of the genus formerly so called; = *Bergenia.

1886 [see *Bergenia]. **1914** G. Jekyll *Colour Schemes for Flower Garden* (ed. 3) 96 The Megaseas persist the whole year round. **1962** R. Page *Educ. Gardener* I. iv. 135 The bergenias which we used to call saxifrage or megasea are handsome through the twelve months of the year with their bold leathery leaves.

megastructure (me·găstrɒ:ktiŭı), *sb.* [f. Mega- + Structure *sb.* 5.] A massively large construction or complex, esp. one consisting of many buildings. Hence **me:gastru·ctural** *a.*

1965 *Life* 24 Dec. 146 Other planners foresee skeletal Megastructures of enormous, light-weight 'space frames' —grid-works towering high in the air. **1967** *Listener* 6 July 12/2 Those temporary heavens of exhibition and transport —the Victorian megastructures—..have acquired a permanence in people's minds out of all proportion to the functions they originally housed. *Ibid.* 20 July 73/3 The megastructural schemes by Corbusier's followers in Japan. **1968** *Daily Tel.* (Colour Suppl.) 13 Dec. 30/1 Rudolph's design..truly qualifies as a 'megastructure', a city-within-a-city. **1973** *Observer* (Colour Suppl.) 2 Dec. 33/3 An early attempt to blend housing, commerce and townscape in one huge building—the lumpish description, 'urban megastructure', has been much used.

megathermic (megăþə·ımik), *a.* [f. Megatherm + -ic.] Pertaining to, connected with, or consisting of megatherms.

1903 W. R. Fisher tr. *Schimper's Plant Geogr.* 226 The megathermic flora is already perceptibly impoverished.

megaton (me·gătɒn). [f. Mega- b + Ton¹.] **a.** A unit of explosive power, equal to that of one million tons of T.N.T. Freq. *attrib.*

1952 *N.Y. Herald-Tribune* 18 June 23/7 The first true super-bomb to be detonated is expected to have a power of two megatons. **1955** *Times* 30 May 7/7 The Pacific tests in 1952 and 1954 showed that a hydrogen bomb equivalent to 10 to 15 million tons of T.N.T. (commonly called 10–15 megatons) dropped on the centre of an ordinary city would cause total destruction within a circle of four to five miles. **1957** *Oxford Mail* 20 Aug. 1/5 Britain, he said, was concentrating on ballistic missiles because aircraft were unable to have 100 per cent success against an enemy equipped with megaton bombs. **1959** *Listener* 2 Apr. 613/3 The force of each detonation is believed to have been in the kiloton rather than the megaton range. **1962** *Newnes Conc. Encycl. Nucl. Energy* 48/1 A Bikini-type megaton bomb would destroy completely most ordinary buildings in an area of 84 square miles. **1965** *New Statesman* 21 May 810/2 Six megatons were used in the whole of the Second World War. **1972** [see *magnitude 2 c].

b. *fig.*

1957 M. Shulman *Rally round Flag, Boys!* (1958) vi. 74 A broth of a woman..filled with energy in the megaton range. **1963** *Daily Tel.* 9 July 1 Another British spy scandal of 'megaton proportions' was forecast yesterday. **1969** N. Cohn *AWopBopaLooBop* (1970) ix. 86 Music splintering and feet shuffling, butts twitching by the megaton.

Hence **me·gatonnage,** explosive power of nuclear weapons, as expressed in megatons.

1963 *Economist* 23 Nov. 746/2 Nato is probably much closer to equality with the Warsaw alliance, in numbers of footsloggers as well as in nuclear megatonnage, than used to be thought. **1967** *Listener* 9 Feb. 186/2 The Russians have not been very explicit..about doubling the megatonnage of their warheads. **1971** *Guardian* 3 Aug. 11/5 The deliverable megatonnage of nuclear weapons. **1971** *Nature* 24 Dec. 493/2 The Soviet Union has already substantially overtaken the United States in numbers of ICBMs, in megatonnages, and in underground tests of large nuclear devices.

Megger (me·gəı). The proprietary name of apparatus designed esp. for measuring electrical insulation resistance. Also **megger.** Also **Meg³,** a type of Megger.

1903 *Trade Marks Jrnl.* 2 Sept. 962 Megger... Electric testing apparatus..and scientific instruments..Evershed and Vignoles, Limited,..London,..manufacturers. **1920** *Whittaker's Electr. Engineer's Pocket-Bk.* (ed. 4) 282 *Evershed's Megger*. The Megger insulation testing set contains in one case a hand-driven magneto generator and a direct reading moving coil ohmeter. **1923** *Nature* 13 Jan. 63/2 The 'Meg' insulation tester.., a remarkably light and cheap megger running to 10,000 mgo. which should prove a boon to linemen. **1923** [see *insulation tester*]. **1924** *Trade Marks Jrnl.* 28 May 1189 Megger... All goods included in class 6 [Machinery of all kinds, and parts of machinery, except agricultural and horticultural machines]. Evershed & Vignoles, Limited,..London,.. electrical engineers. **1928** *Naval Electr. Manual* (Admiralty) I. ix. 204 The 'Megger' is a direct-reading ohmmeter for the measurement of high resistances, such as the insulation resistance wiring, etc. **1930** *Engineering* 31 Jan. 129/3 A demonstration model of their megger earth tester. **1936** *Discovery* July 220/2 This instrument [*sc.* the Timber Humidity Meter]..has the appearance of a Megger. **1942** *Trade Marks Jrnl.* 11 Nov. 468/1 Meg... Electrical apparatus for testing insulation resistance and electrical apparatus for testing earth resistance, each comprising an electric generator and an electric measuring instrument. Evershed & Vignoles Limited,..London,..electrical and mechanical engineers. **1942** *Tee Emm* (Air Ministry) II. 84 Always remember the importance of megger-testing the insulation. **1950** *Engineering* 14 Apr. 404/1 The same firm were showing a new pattern of Megger insulation tester. **1957** G. Clark *Archæol. & Society* (ed. 3) ii. 63 By using a 'Megger' Earth Tester, and measuring resistivity at intervals, it is possible to contour degrees of resistivity and so to detect areas of ancient disturbance, in the form of ditches.

Megillah (mĕgi·lă). Also † **meghillah, megilla(h), megille.** Pl. **megil(l)oth, megil(l)a(h)s.** [Heb. *megillah* roll, scroll.] **a.** Each of five books of the Old Testament, namely S. of S., Ruth, Lam., Eccles., and Esther, appointed to be read by adherents of the Jewish faith on certain feast days; freq. with particular reference to the book of Esther, read at the feast of Purim. Also, a copy of any one, or all, of these books.

1650 E. Chilmead tr. *Leon Modena's Hist. Rites of Jews* III. x. 165 They read..the whole book of Esther; which they have..in a long Roll:..and this they call..*Meghillah*, that is to say, *Volumen*, a Volume, or Roll. **1652** E. Bargishai *Brief Compendium Vain Hopes of Jews Messias* 17 In the same *Talmuth* and book *Megilla*..saith *Chennina*, when the *Messias* shall appear, then shall God ..shew great honours to the Jews. **1818** *Jewish Preceptress* 85 After the usual service, the Megilla, or book of Esther, is read. **1836** *Nethiboth Olam* (*Old Path*) 26 Feb. 27/1 We request the attention of our readers to the reason given why the reading of the Megillah is more important than any of the commandments. **1857** C. D. Ginsburg *Song of Songs* p. vii, The following is an exposition of the first of the five books called Megilloth. *Ibid.* 2 This Song is the first of the..five Megiloth, or books which are annually read in the Synagogues. **1926** 'R. Learsi' tr. *Ash's Kiddush Ha-Shem* 89 Young men drew artistic designs in colors for..*Megillahs*. **1932** A. Z. Idelsohn *Jewish Liturgy* 391 Mĕgilla chanting. **1957** *Oxf. Dict. Chr. Ch.* 882/1 *Megilloth* (Heb..., 'rolls'), the name given to five Books in the OT, all of them in the third and latest section of the OT canon known as the 'Hagiographa', which were read by the Jews on certain feast days. **1968** L. Rosten *Joys of Yiddish* 230 *Megillah*... Hebrew: 'scroll'. **1.** *Megillah* usually describes the Book of Esther..; also the Book of Ruth. (There are five *megillahs* in all.) **1973** *Synagogue Light* Sept. 12/2 We read in Megilas Ruth [etc.]. **1974** *Jewish Chron.* 13 Sept. 17/1 A megilla in an engraved casket..was presented to the Brighton and Hove Hebrew Congregation.

b. *slang.* With allusion to the length of the Megillah: a long, tedious, or complicated story; freq. in phr. *a whole Megillah* (Yiddish *a gantse Megillah*).

1957 L. M. Feinsilver in *Chicago Jewish Forum* Summer 228/1 *A gantse Megillah* or 'a whole *Megillah*' has been thrown around by a number of TV personalities..presumably with little idea of the origin of the phrase. **1968** L. Rosten *Joys of Yiddish* 230 *Megillah*... 2. Anything very long, prolix; a rigmarole... 3. In popular parlance: Anything complicated, boring, overly extended, fouled up. **1968** *Punch* 1 May 626/1 Feeding all the *megillah* to the papers about his family of Irish Polacks who came over with the Pilgrim Fathers. **1970** L. M. Feinsilver *Taste of Yiddish* i. 28 The *Megille*, or scroll, of Esther which is read aloud on Purim, and which takes some time to read, gave rise to the humorous description of a long story as *a gantse megille* (a whole *megille*). **1970** S. Sheldon *Naked Face* (1971) x. 110 'Do you know the most peculiar thing about this whole megillah?' queried Moody thoughtfully.

mehari (mĕhā·ri). Also **maharee, mahari, maherry, mehara, meheri,** etc. [F. *méhari*, f. Algerian Arab. *mehrī*, Class. Arab. *mahrī*, of Mahra, a province in South Arabia.] An Arabian, single-humped camel, used for riding.

1738 T. Shaw *Travels or Observations relating to Barbary & Levant* 240 That species of the Camel-kind, which is known to us by the name of the *Dromas* or Dromedary, is here called *Maihāry*; though it is much rarer in Barbary than in the Levant. **1821** G. F. Lyon *Narr. Trav. N. Afr.* i. 23 A small Kafflé passed us, consisting of ten or twelve camels, and amongst them one or two Maherries. **1826** Denham & Clapperton *Narr. Trav. & Discoveries in N. & Cent. Afr.* i. 22 A marauding party was sent out to plunder some maherhies. *Ibid.*, Nine camels of the maherhy species were brought in. **1848** J. Richardson *Travels in Great Desert of Sahara* I. iii. 87 The two races, the coast-camel, and the Maharee or desert-camel... The most fierce and dominant was the Maharee. **1854** J. R. Morell *Algeria* 480 The mahari supports fatigue better than the camel, and never betrays an ambuscade. **1857** H. Barth *Trav. N. & Cent. Afr.* I. viii. 178 Riding my own méheri, I was quite at liberty to go before or fall behind. **1863** J. Hutton tr. *Daumas's Horses of Sahara* II. 342 Fetoum remounted her *mahari* and gave the signal for retreat. **1904** R. Hichens *Garden of Allah* II. viii. 113 'He who smokes the keef is like a Mehari with a swollen tongue,' he rejoined. **1921** M. W. Hilton-Simpson *Among Hill-Folk Algeria* 120 In the great desert he rides the 'mehari', or trotting camel. **1923** G. Casserly *Algeria To-Day* 235 A Targui's *mehara* (riding camel) is usually white, is speedy and full of endurance. **1926** *Contemp. Rev.* Oct. 425 Captain Domèvre of the Spahis..after being severely wounded in action with his *méhari*, is transferred to the military intelligence section at Beyrout.

Meiji (mēi·dʒi). [Jap., 'enlightened government'.] The name given to the period of the rule of the Japanese emperor Mutsuhito (1868–1912), which was marked by the modernization and westernization of Japan. Freq. *attrib.*

1873 E. M. Satow tr. *Kinsé Shiriaku: Hist. Japan* 125 The chronological period was also changed to Meiji (Enlightened Government), and an imperial proclamation was published making it a rule for all time that there should be only one chronological period for each reign. **1894** D. Murray *Japan* xv. 378 The year-period, which from January 1865, had borne the name of Keiō, had been changed to *Meiji* (Enlightened Peace), and was fixed to begin from January, 1868. **1901** C. Lowe tr. *A. von Siebold's Japan's Accession to Comity of Nations* p. vi, The writer..felt particularly called upon to contribute to an appreciation of the magnificent achievements of the statesmen and diplomatists of the *Meiji* Era. **1931** I. Nitobé *Japan* iv. 225 The two great reforms of Japanese history—those of the Taika and Meiji eras. **1936** J. A. B. Scherer *Three Meiji Leaders* i. 1 Hirobumi Ito..became Japan's first Premier under the new Meiji government. *Ibid.* xiv. 133 The three Meiji leaders differed widely from one another, but each was of a classic grandeur. **1957** *Encycl. Brit.* XII. 967/1 During the first years of the Meiji period, when western ideas were being adopted in the newly reopened Japan, the European style of painting was more cultivated than any other. **1961** A. M. Craig *Chōshū in Meiji Restoration* i. 6 Chōshū and Satsuma, han whose samurai were subsequently to carry out, as the dominant center of the new Meiji government, the revolution of the early Meiji period. **1972** M. Kochan tr. *P. Akamatsu's Meiji 1868* II. ii. 258 As far as government accounts in the early years of Meiji can be known, nine and a half million *yen* are said to have been devoted to military expenditure in the period between November 13, 1871, and December 31, 1872. **1974** G. Jenkins *Bridge of Magpies* xiii. 206 My grandfather..lived at the time of the Meiji Revolution which made Japan into a modern state.

‖ **mein Gott** (məin gɒt), *int.* [G., = my God.] Used as a typically German exclamation corresponding to *My God* (God 7).

c **1838** C. J. Mathews in M. R. Booth *Eng. Plays of 19th Cent.* (1973) IV. 130 Oh mine Got, mine Got, when I tink of zat bewitching Miss Parker! **1867** Trollope *Claverings* I. xix. 239 'Yes, yes;—mein Gott, yes,' said Schmoff. **1891** Kipling *Life's Handicap* 261 Mein Gott! I would sooner collect life red devils than liddle monkeys. **1924** 'Sapper' *Third Round* vi. 142 And think—five, ten minutes more and I also to pieces would have been blown. *Mein Gott!* it makes me sweat. **1968** R. Sawkins *Snow along Border* iv. 33 Oh, mein Gott, I've just remembered something odd.

‖ **mein Herr** (məin hēr). [G., = my lord.] Used in a jocular or ironic tone in addressing a German man; so as *sb.*, a male German.

1922 Joyce *Ulysses* 202 The gross virgin who inspired *The Merry Wives of Windsor* let some meinherr from Almany grope his life long for deephid meanings in the depth of the buckbasket. **1967** [see *Herr]. **1975** J. Blackburn *Mister Brown's Bodies* xxii. 185 Very clever, *mein Herr*, though the real prize will excape you.

meinie. 5. (Later example.)

1970 J. Wain *Winter in Hills* iv. 361 A speech was now made by McAlister, the Scotch poet in whose ostensible support this meinie was assembled.

meiobenthos (məi:obe·nþǫs). [f. Gr. μείων smaller + *benthos.] The section of the benthos that includes animals neither small enough to be grouped with the microfauna nor large enough to be grouped with the macrofauna. So **mei:obe·nthic.**

1942 M. F. Mare in *Jrnl. Marine Biol. Assoc.* XXV. 519 A new terminology is needed, and these groups are here designated the macrobenthos, meiobenthos, and microbenthos. The macrobenthos is equivalent to the macrofauna of the bottom... The meiobenthos here comprises the fauna of intermediate size, such as small crustacea (copepods, cumaceans, etc.) small polychaetes and lamellibranchs, nematodes and foraminifera. The microbenthos comprises all the small organisms. **1967** *Oceanogr. & Marine Biol.* V. 522 The number of individuals given by Kiseleva has the great merit of taking into account the meiobenthos. **1969** *Nature* 15 Nov. 720/2 Nematodes are among the commonest meiobenthic animals. **1971** *Ibid.* 28 May 260/1 (*heading*) Deficiency of gravity corers for sampling meiobenthos and sediments. **1973** *Ibid.* 30 Mar. 324/1 Where hydrodynamic conditions and bottom configuration are so subdued that meiobenthic bioturbation becomes a locally prominent sedimentological process.

meiofauna (məi·ofǭnă). [f. Gr. μείων smaller + Fauna.] = *meiobenthos. So **mei:ofau·nal** *a.*

1967 M. L. Wass in Olson & Burgess *Pollution & Marine Ecol.* vi. 272 Benthic animals under stress may be meiofaunal in size. *Ibid.*, An extensive study has been done..on a group of protozoan meiofauna. **1973** *Nature* 30 Mar. 323/2 Ostracods and nematodes were the most widely distributed and active members of the meiofaunas. *Ibid.* 324/1 Data confirming..the complete disappearance of some meiofaunal groups, in sediments of the deeper seafloor.

meiosis. Add: Pl. **meioses. 3.** *Biol.* (Formerly also **maiosis.**) The division of a diploid cell nucleus into four haploid nuclei, which offsets a doubling of chromosome numbers at a subsequent fertilization and normally comprises a reduction division (meiosis I) followed by an equational division (meiosis II); commonly used to include also the accompanying division of the cytoplasm.

1905 Farmer & Moore in *Q. Jrnl. Microsc. Sci.* XLVIII. 489 We propose to apply the terms Maiosis or Maiotic phase to cover the whole series of nuclear changes included in the two divisions that were designated as Heterotype and Homotype by Flemming. **1907** *Rep. Brit. Assoc. Adv. Sci.* 689 There is reason to believe that a sorting of the chromosomes, analogous to that seen in meiosis, takes place in the third division of the ascus. **1925** E. B. Wilson *Cell* (ed. 3) vi. 576 Meiosis brings about two additional results... One is to establish new haploid combinations of the original maternal and paternal chromosomes in the germ-cells... Not less important is the reorganization of the chromosomes, individually considered, that takes place during meiosis, by means of 'crossing-over'. **1934** L. W. Sharp *Introd. Cytol.* (ed. 3) xvi. 254 Meiosis involves two nuclear divisions but only one chromosomal division. **1949** Darlington & Mather *Elem. Genetics* i. 34 Since crossing-over can probably take place between any two chromomeres, it will be rare indeed for two identical haploid nuclei to be produced from different meioses. **1971** *Sci. Amer.* Aug. 56/3 Without mitosis there could be no meiosis, the type of cell division that gives rise to eggs and sperm.

meiotic (məi,ǫ·tik), *a.* [f. Gr. μειωτικ-ός diminishing (see Meiosis).] **1.** *Rhet.* That represents things as less than they really are; characterized by meiosis or litotes.

1907 *Westm. Gaz.* 17 June 2/2 Is there not..a good deal to be said for the meiotic method [of portraying the Caesars] preferred by Sir Lawrence Tadema? **1915** *Oxf. Mag.* 18 June 38/2 We have occasionally mentioned in mild meiotic terms that the Oxford roads do not wholly satisfy our ideal of perfection.

2. *Biol.* (Formerly also **maiotic.**) That is characterized by meiosis (sense *3); of, pertaining to, or occurring at meiosis.

1905 [see *meiosis 3]. **1913** W. Bateson *Mendel's Princ. Heredity* (rev. ed.) 270 This differentiation will come about at this reduction, or meiotic division, as it is called. **1952** Srb & Owen *Gen. Genetics* vii. 119 The period between the two meiotic divisions is designated by the special term interkinesis. It may be rather long, but typically it is very short. **1962** *Lancet* 15 Dec. 1270/1 Three types of meiotic non-disjunction have been recognised. **1973** *Nature* 5 Oct. 261/2 For spermatozoa and meiotic preparations, ripening testes were removed.

MEI P'ING

Hence meio·tically *adv.*
1676 R. MEGGOTT *Sermon preached on St. Paul's Day* 6 This is that which according to the Hebrew way of speaking is here meiotically expressed by *It is not good.* 1959 *Times* 7 Oct. 4/1 To put it meiotically, the match was rather one-sided. 1966 *Times* 28 Mar. (Austral. Suppl.) p. xiv/5 A *y* chromosome might..be attached to another meiotically driven chromosome. 1968 R. RIEGER *Gloss. Genetics & Cytogenetics* 274 Thus, the zygotic (diploid) chromosome number is meiotically reduced to the gametic (haploid) number. 1973 *Nature* 5 Oct. 259/1 When the brandspore germinates its nucleus divides meiotically to re-establish the haploid sporidial phase.

‖ **mei p'ing** (mē̆i piŋ). Also **mei ping** and with capital initials. [Chin., lit. = prunus vase.] A Chinese porcelain vase with a narrow neck designed to hold a single spray of flowers.
1915 R. L. HOBSON *Chinese Pott. & Porc.* v. 79 A vase of the form known as *mei p'ing* with green Imperial dragons in a yellow ground and the Wan Li mark. 1953 S. JENYNS *Ming Pott. & Porc.* iv. 39 Other blue and white pieces.. are the *mei ping* vases decorated with sages. 1968 J. UPDIKE *Couples* ii. 152 An empty but perfect blue vase of *mei ping* form. 1972 *Country Life* 21 Dec. 1715/3 The Mei Ping (a vase designed with an aperture to hold a single flower). 1973 *Ibid.* 2 Aug. 290/1 A Mei-P'ing, literally 'plum blossom vase', a shape which first appeared in the 10th century..small mouth, wide shoulders and tapering body. 1974 *Times* 3 Apr. 1/5 She paid £160,000 for an early Ming blue-and-white *mei p'ing* and cover.

meisie, meisje (mē̆ⁱ·si). *S. Afr.* [Afrikaans *meisie*, f. Du. *meisje*, cogn. w. Eng. MAIDEN *sb.*] Girl; young lady or woman.
1890 *Digger's Doggerel* 27 Romeo Troilus Giddy McSmack Has been granted, secure from all legal attack, The sole right to practise and use osculation On the 'meisjes' and 'vrauws' of the Transvaalsche Nation. 1895 H. A. BRYDEN in *Blackw. Mag.* July 131/1 A little wizened Hottentot has been..making coffee for the *baas* and *meisje.* 1910 *Cape Times* 7 Oct. 10 There is enough to be done before nightfall for an idle meisje to bestir herself without waiting to be called. 1927 W. PLOMER *I speak of Afr.* 47 He was stared at, especially by fat meisjes here and there, flappers with matronly figures. 1934 'N. GILES' *Ridge of White Waters* II. ii. 207 That child of Isaak Prinsloo's grows into a pretty meisie. 1953 J. COLLIN-SMITH *Locusts & Wild Honey* II. iv. 146 The lass loves you; which is more than can be said for me of that wilful 'meisjie' your sister. 1965 *Economist* 17 Apr. 339 Milling about in front as well as behind the open counters are the attractive *meisies*, known as hostesses.

Meissen (mai·sĕn). The name of a town near Dresden used *attrib.* or *absol.* to designate a hard-paste porcelain made there since 1710; Dresden porcelain (*DRESDEN). Also *fig.*
1863 W. CHAFFERS *Marks Pott. & Porc.* 176 *Dresden. Meissen.* The two swords crossed..; used about 1730. 1882 'OUIDA' *Bimbi* 50 A lovely little lady..made of the very finest and fairest Meissen china. 1938 *Burlington Mag.* Dec. p. xvii/2 Pieces of Meissen and other porcelains. 1951 'J. TEY' *Daughter of Time* i. 11 People wouldn't send you a lot of fool nonsense when you were flat on your back, and bossy bits of Meissen wouldn't expect you to read them. 1951 [see *CHANTILLY 1]. 1961 J. WADE *Back to Life* xiv. 195 Her face..had the smooth, pleased prettiness of Meissen. 1964 MRS. L. B. JOHNSON *White House Diary* 5 June (1970) 156, I recognize it as Meissen from Germany. 1968 *Michelin Guide N.Y. City* 52 German stoneware, faience and porcelain (Meissen, Nymphenburg and Hochst). 1970 *Canad. Antiques Collector* May 29/1 Meissen was the first true porcelain made in Europe. It was produced in a factory at Meissen, about 10 miles from Dresden, from about 1710 onwards.

Meissner effect (mai·snəɹ). *Physics.* [named after Fritz Walther *Meissner* (1882–1974), German physicist, who with R. Ochsenfeld published an account of the phenomenon in *Naturwissenschaften* (1933) XXI. 787.] The existence of zero, or very low, magnetic induction in a superconducting material even in the presence of a magnetic field; esp. the (partial or complete) expulsion of magnetic flux when the material becomes superconducting in a magnetic field.
[1935 *Proc. R. Soc.* A. CXLIX. 73 Equation (6) says more than (2), so far as it includes Meissner's effect.] 1935 *Ibid.* CLII. 10 The magnetization curve for decreasing fields is in striking contradiction to the classical electromagnetic description of a Supraconductor, and provides an independent confirmation that B = 0 is characteristic of the ideal supraconducting state (Meissner effect). 1953 C. F. SQUIRE *Low Temperature Physics* vii. 116 The complete Meissner effect (B = 0) is difficult to obtain experimentally. *Ibid.* 117 (*heading*) A permanent magnet floats above the superconducting tin dish because of the Meissner effect. 1968 C. G. KUPER *Introd. Theory Superconductivity* iii. 44 It is rather unusual in physics for purely topological properties of a body to be important, but superconductivity provides an example. In a multiply-connected superconducting body, although the Meissner effect ensures that the magnetic flux will be expelled from the material of the body, flux may be trapped in the holes.

Meistersinger (mai·stəɹzi:ŋəɹ, -s-), *sb. pl.* and *sing.* [G.: cf. MASTER-SINGER.] German lyric poets and musicians in the 14th to 16th centuries organized in guilds and having an elaborate technique; (*sing.*) a member of such a guild.
1845 LONGFELLOW *Poets & Poetry of Europe* (1847) 373/2 These Chambers [of Rhetoric] were to Holland, in the fifteenth century, what the Guilds of the Meistersingers were to Germany. [1854 A. G. HENDERSON tr. *Cousin's Philos. of Kant* i. 5 The poetry of this period is to be found in the songs of the *Minnesängers* and the *Meistersängers.*] 1954 *Grove's Dict. Mus.* (ed. 5) VII. 912/2 The Meistersinger were either preoccupied with the observance of rules or were rarely visited with inspiration. 1968 *Encycl. Brit.* XV. 118/2 The *Meistersinger* were not popular figures, as Wagner's opera *Die Meistersinger* suggests; they were largely ignored by professional men.
fig. 1924 R. CAMPBELL *Flaming Terrapin* iv. 56 Across the night with dismal hum The hurricanes, your meistersingers, come.

mejliss, var. *MAJLIS.

mekometer. Add: **b.** Also **Mekometer.** A device for the accurate measurement of distances in which light elliptically polarized at a microwave frequency is beamed at a reflector at the distance to be measured and the polarization of the reflected light analysed to find the amount by which the distance of the reflector exceeds a whole number of modulation half-wavelengths.
1961 FROOME & BRADSELL in *Jrnl. Sci. Instruments* XXXVIII. 458/1 The paper describes an experimental equipment (known as the 'N.P.L. Microwave Mekometer') using elliptical polarization modulation of a light beam at 9·4 Gc/s (9400 Mc/s) and intended for the accurate measurement of distances of the order of 50 m. 1962 *New Scientist* 26 July 207/2 At the NPL the Mekometer is to be used for the verification of surveyors' precision tapes. 1964 *Trans. Soc. Instrument Technol.* XVI. 31 By September 1960 patents had been filed and in early 1961 the mekometer had become a reality using the Pockels effect at 9·375 G c/s in a device for the accurate measurement of distance. 1970 *Physics Bull.* Aug. 349/2 Prof R G Mason and his collaborators at Imperial College used a new electromagnetic distance measuring instrument, the mekometer, developed by Dr K D Froome..and in the course of nearly two years detected movements of a few centimetres.

mel (mel). *Acoustics.* [f. MEL(ODY *sb.*] A unit of subjective pitch, defined so that the number of mels is proportional to the pitch of a sound, and the pitch of a 1000-hertz note (often, one forty decibels above the listener's threshold of hearing) is 1000 mels.
1937 S. S. STEVENS et al. in *Jrnl. Acoustical Soc. Amer.* VIII. 188/1 The numbers on the pitch scale are related to each other as the subjective magnitudes of the pitches. A pitch of 1000 units (mels) is subjectively twice as high as a pitch of 500 units... The name *mel* was chosen as a name for the subjective pitch unit. 1957 J. L. HUNTER *Acoustics* viii. 249 A tone having an apparent pitch twice as high has a pitch of 2,000 mels..regardless of its frequency. 1962 P. LADEFOGED *Elem. Acoustic Phonetics* vi. 79 Frequency data about speech sounds are often converted into mel units before being presented graphically. 1966 R. L. SURI *Acoustics* I. ii. 21 The unit of pitch has been given the name 'mel' such that the pitch at 1000 cycles is 1000 mels.

melamed (mĕlā·mĕd). Also **melammed.** Pl. **melamdim.** [Heb.] A teacher of elementary Hebrew.
1892 I. ZANGWILL *Childr. Ghetto* II. 16 A *Melammed*, or Hebrew teacher. 1902 H. HAPGOOD *Spirit of Ghetto* i. 19 He is carried by his father to the school and received there by the 'melamed', or teacher. 1968 L. ROSTEN *Joys of Yiddish* 232 Pitying, even derogatory, tales about *melamdim* abound among Jews. In all of them, the *melamed* is not the hero but the goat—hapless, unlucky, unresourceful. 1970 L. M. FEINSILVER *Taste of Yiddish* ii. 233 Where the melamed used the ruler and threatened with his strap, the modern Hebrew teacher takes all he can and then sends the kid to the office. 1972 F. B. MAYNARD *Raisins & Almonds* 163 A world peopled by rabbis, starving *melameds* (teachers), matchmakers, [etc.].

melamine. Now usu. pronounced (-īn). Add: **2.** Also **Melamine.** Melamine resin, or a plastic derived from it.
1940 *Plastics* IV. 162/3 The Westinghouse panels for refrigerators making use of the new plastic were made, initially at least, with a white 'Melamine' coating over phenolic laminated sheet. 1943 *Jrnl. Oil & Colour Chem. Assoc.* XXVI. 187 There has been some tendency to regard melamine and melamine resin as meaning the same thing. 1958 *Spectator* 26 Sept. 406/3 Much of the tough new plastic tableware is made of melamine. 1972 *Daily Tel.* 6 June 13/4 Young couples usually eat in the kitchen, and kitchen/dining tables with white melamine tops are attractive. 1973 'D. HALLIDAY' *Dolly & Starry Bird* ii. 20 Charles.. poured him a noggin..into a yellow Melamine cup.
3. *attrib.* and *Comb.*: *melamine-formaldehyde, -surfaced* adjs.; **melamine resin,** any of the synthetic resins made by the condensation of melamine with an aldehyde, used in making tableware, as coatings for working surfaces, in adhesives, and for treating cloth and paper.
1941 *Industr. & Engin. Chem.* June 771/2 In order that melamine-formaldehyde resins are to be of use to the paint and varnish industry. 1962 J. T. MARSH *Self-Smoothing Fabrics* vi. 81 Melamine-formaldehyde finishes suffer little loss in strength on chlorination but the fabric becomes yellow. 1967 *Times Rev. Industry* June 74/3 No kitchen interior is acceptable to-day without its melamine-formaldehyde laminated surfaces. 1939 *Brit. Plastics* June 28/2 (*heading*) Creaseless fabrics containing melamine resin. 1943 *Jrnl. Oil & Colour Chem. Assoc.* XXVI. 192 The American paint industry is showing a great interest in melamine resins. 1966 *McGraw-Hill Encycl. Sci. & Technol.* XIV. 216/2 Melamine resins are especially valuable as adhesives for laminates of paper or fabrics. 1960 *Times* 16 Mar. (Canberra Suppl.) p. viii/5 Bulkheads and deckheads..will be covered in either melamine-surfaced laminated plastic, or the softer material commonly known as P.V.C. 1971 *Woman's Own* 27 Mar. 53/3 Elegant shelves can be made from white melamine-surfaced Contiplas.

melampyre (me·læmpaiᵊɹ). [f. mod.L. *Melampyrum*, the generic name of cow-wheat: see MELAMPYRIN.] = COW-WHEAT.
1905 E. PHILLPOTTS *Secret Woman* III. v. 244 The melampyre's lemon blossoms and the orange stars of the woody loosestrife mingled close at hand.

Melanesoid (melānī·zoid), *a.* [f. *Melanes(ia* (see MELANESIAN *a.* and *sb.*) + -OID.] Similar in racial type to the Melanesian; resembling a Melanesian.
1932 *Man* Dec. 283 The term Melanesoid was used to indicate the offspring of tribes from Southern China (Yunnan) and Northern Indo-China (Tonkin). 1939 F. WEIDENREICH in *Peking Nat. Hist. Bull.* XIII. 170 The morphological analysis of the three skulls..led to the result that they typify three different racial elements, best to be classified as primitive Mongoloid, Melanesoid and Eskimoid types. 1943 —— *Skull of Sinanthropus Pekinensis* 251 Skull of a female... The female, B, shows the greatest similarity to the Melanesians of today. For this reason I have designated the type as 'Melanesoid'. 1946 [see *ESKIMOID *a.*].

mélange, *sb.* Add: **3.** (See quot. 1935.)
1922 JOYCE *Ulysses* 245 I'll take a *mélange*, Haines said to the waitress... He sank two lumps of sugar deftly longwise through the whipped cream. 1935 H. SIMPSON *Cold Table* 264 Café Mélange. Ingredients: Coffee, milk; water; whipped cream; castor sugar. 1971 *Guardian* 27 Feb. 7/4 The famous Viennese coffee houses where you can sip your *melange* (or a dozen other varieties of coffee).

melanic, *a.* Add: **2.** Esp. in reference to the darker varieties of moths and other animals that have developed in certain industrial areas. (Later examples.) Also as *sb.*, an animal characterized by melanism.
1915 R. C. PUNNETT *Mimicry in Butterflies* viii. 101 In some parts of England the common peppered moth, *Amphidasys betularia*, has been almost entirely supplanted by the darker melanic form *doubledayaria.* 1940 H. B. COTT *Adaptive Coloration in Animals* I. i. 17 The distribution of such melanic varieties, coinciding as it often does with industrialism, is very significant. So is the fact that the melanic forms have not..become established, even though they are present, in rural districts. 1958 H. B. D. KETTLEWELL in *New Scientist* 3 July 298/2 New theories were evoked to account for the rapid spread of species [of moths] changing from light coloration to black—the 'industrial melanics'. 1970 *Nature* 12 Sept. 1155/1 The murk of nineteenth century Manchester fostered the melanic form *carbonaria* of the peppered moth, *Biston betularia* (L.). 1971 *Ibid.* 29 Oct. 586/3 The value of smoke abatement is demonstrated by the almost immediate decline in the frequency of the melanics where smoke control zones are introduced. 1973 *Ibid.* 21/28 Dec. 535/1 Intensive selective predation by birds on moths, which largely accounts for the presence of melanics in both industrial and some non-polluted rural areas. 1975 *Sci. Amer.* Jan. 90/2 The first [problem] is the question of why there is a difference in the proportion of melanics found among populations of various moth species that share the same environment. *Ibid.* 95 (*caption*) Frequency of melanic form is different in various species.

melanization (melănəizēⁱ·ʃən). [f. MELANIZ(E *v.* + -ATION.] The process or result of becoming melanized.
1945 W. M. KROGMAN in R. Linton *Sci. Man in World Crisis* 47 A possible exception may be found in the 'protective' melanization or skin coloring of tropic-dwelling peoples. 1954 *Sci. News* XXXIV. 94 The pink and brown colour of insectorubin is completely obscured by the melanization of the outer cuticle. 1974 *Nature* 10 May 187/1 The epidermal cells..showed progressive granulation and melanisation.

melano (mĭlā·no). [f. Gr. μελαν-, μέλας black: after ALBINO.] An animal distinguished by an abnormal development of black pigment in the epidermis, hair, feathers, etc.; opposed to ALBINO.
1902 *Ann. & Mag. Nat. Hist.* IX. 59 Spotted tiger-cats of the *F. Macrura* group. The small specimen is a melano, but shows indications of the normal spotted condition.

melano-. Add: **melanoblast** (me·lăno-, mĭlæ·noblast) *Zool.* [a. G. *melanoblast* (S. Ehrmann *Das melanotische Pigment* (1896) viii. 20): see -BLAST], a cell that produces melanin; also, a precursor of a melanin-forming cell; **melanocra·tic** *a. Petrol.* [ad. G. *melanokrat* (W. C. Brögger *Eruptivgesteine des Kristianiagebietes* (1898) III. 263), f. Gr. κρατ-εῖν to rule, prevail)], (of a rock) dark-coloured; rich in dark-coloured minerals; **me·lanoderm** (also

Melano-) *sb.* and *a.* [Gr. δέρμ-α skin], (of, pertaining to, or being) a person of a dark-skinned (negroid) race; **melanodermic** *a.*, (*b*) (naturally) dark-skinned, negroid; **melanoge·nesis** *Physiol.* [-GENESIS], the formation of melanin; **melanophore** (me·lăno-, mĭlæ·nofōᵊɹ) *Zool.* [ad. G. *melanophor* (R. Keller 1895, in *Arch. f. Physiol.* LXI. 141): see -PHORE], a cell containing melanin; esp. such a cell in the lower vertebrates which is contractile and confers the ability to change the depth of colour (see quot. 1953); **me·lanosome** *Physiol.* [*-SOME⁴], a particle in the cytoplasm of melanocytes in which melanin is thought to be formed from tyrosine (see quot. 1961).
1902 *Encycl. Medica* XI. 182 Ehrmann.., from his investigations of amphibians and reptiles, concludes that the epithelial colouring matter is elaborated in special connective tissue corpuscles (chromatoblasts or melanoblasts), from material derived from the red blood corpuscles. 1942 G. H. BOURNE *Cytol. & Cell Physiol.* viii. 329 A spread of melanin formation from the melanoblasts to similar but non-pigmented dendritic cells of the white area of the skin. 1953 *Science* 5 June 640/1 The term melanoblast is used by medical investigators for the mature cell elaborating melanin, whereas in biology the term melanoblast refers to an immature pigment cell during its migration from the neural crest. *Ibid.* 640/2 The term melanoblast for the mature pigment-forming cell as originally suggested by Bloch is objectionable. *Ibid.* (*table*) Recommended terminology of pigment cells. Immature melanin-forming cell: Melanoblast. 1968 H. HARRIS *Nucleus & Cytoplasm* i. 11 Enucleate fragments of prospective pigment cells (melanoblasts) from the developing neural crest of urodele embryos also survive for long periods *in vitro.* 1909, 1954 Melanocratic [see *leucocratic* adj. s.v. *LEUCO-]. 1963 D. W. & E. E. HUMPHRIES tr. *Termier's Erosion & Sedimentation* 411 *Peridotite*, an ultrabasic (melanocratic) crystalline rock. 1924 A. C. HADDON *Races of Man* (ed. 2) 13 Among the xanthoderms and melanoderms the irides are almost uniformly dark brown in colour. 1927 H. H. WILDER *Pedigree of Human Race* vi. 345 The Melanoderm Race has developed several moderately high civilizations, especially on the west of the Sahara. 1935 *Jrnl. R. Anthrop. Inst.* LXV. 123 The three main divisions of mankind—Leukoderms (Caucasians), Xanthoderms (Mongolians) and Melanoderms (Western and Eastern Negroes)—did not prove to have a blood-group factor of their own. 1924 A. C. HADDON *Races of Man* (ed. 2) 144 A tropical jungle may have been a refuge for an indigenous melanodermic folk from fairer intruders into the country. 1928 *Funk's Stand. Dict.*, Melanogenesis. 1964 *Oceanogr. & Marine Biol.* II. 409 An interesting feature of melanogenesis in the ophiuroids is that in light-coloured individuals which are regenerating arm-tips the regenerate is dark, suggesting that under conditions of wound-healing and repair the normal inhibitor of melanogenesis..is overcome or absent. 1903 *Proc. Amer. Acad. Arts & Sci.* XXXIX. 261 The conspicuous black bodies of Anolis..well buried in the derma and sending branching processes outward toward the epidermis, correspond to the melanophores described by Keller. 1946 *Nature* 7 Sept. 344/1 The pigmentary hormone of *Dixippus*, the regulator of colour change in this animal, activates the melanophores of frogs..by causing expansion of them. 1953 *Science* 5 June 640/1 Investigators in biology and medicine at the present time are using different terms for the same cell. For example, the term melanophore which has long been used by biologists refers to certain dendritic-shaped cells in the skin of fish, amphibians and reptiles which have 'contractile' properties. The melanin contained in the melanophores may, in response to certain stimuli, disperse into the dendrites or concentrate in the perikaryon, thus accounting for the color change. In human cytology and pathology the term melanophore is a macrophage. *Ibid.* 640/2 (*table*) Recommended terminology of pigment cells. 'Contractile' cell: Melanophore. 1965 LEE & KNOWLES *Animal Hormones* x. 128 A characteristic feature of hormonally controlled melanophores is that colour change takes hours to occur, whereas this is achieved in minutes if there is nervous control. 1961 M. SEIJI et al. in *Jrnl. Investigative Dermatol.* XXXVI. 251/2 Because the melanin granule has been shown by electronmicroscopy to be structurally distinct from mitochondria, to be unique in its localization within the cytoplasm of mammalian melanocytes, and to contain a specialized metabolic pathway that converts tyrosine to melanin, we would like to suggest that during its enzymically active stages it be called a melanosome... The term 'melanin granule' could be reserved for the mature, fully melanized particle which has lost its tyrosinase activity and is no longer confined to the cytoplasm of the melanocyte. 1973 *Nature* 26 Oct. 436/3 The melanocyte's contribution [to pigmentation] begins with the synthesis of melanosomes, and the oxidation of tyrosine to melanin.

melanocyte (me·lăno-, mĭlæ·nosəit). *Zool.* [f. MELANO- + -CYTE.] A mature melanin-forming cell; also, a melanophore.
1890 in BILLINGS *Med. Dict.* II. 128/1. 1935 *Chem. Abstr.* XXIX. 8125 Melanocyte reaction of the preparations of pituitary body and the urine of cancer patient. 1953 *Science* 5 June 640/2 (*table*) Recommended terminology of pigment cells. Mature melanin-forming cell: Melanocyte. 1957 *Times* 4 Oct. 5 There are 500 to 2,000 melanocytes per square millimetre of skin surface, except on the face, forehead and behind the ears, where there are 1,500 to 4,000 per square millimetre. 1971 *New Scientist* 12 Aug. 366/1 It is difficult to understand why MSH is still secreted by the pituitary gland of birds and mammals, when these animals have lost the capacity for varying the colour of their body through the movements of pigment granules in melanocytes.
Hence **melanocyte-stimulating hormone,** a hormone that stimulates melanocytes or melanophores and causes darkening of the skin; abbrev. **MSH.*
1953 LERNER & FITZPATRICK in M. Gordon *Pigment Cell Growth* 329 There are many indications that the pituitary gland produces a melanocyte-stimulating hormone (MSH) Rattner *et al.*). 1954 A. B. LERNER et al. in *Jrnl. Clin. Endocrinol. & Metabolism* XIV. 1465 We have called the pituitary factor which darkens skin the melanocyte-stimulating hormone or MSH, a term first suggested by W. H. Rattner. Names used previously include melanophore hormone, melanophore-dilating principle, intermedin, and others. 1956 *Jrnl. Biol. Chem.* CCXXI. 958 Two distinct melanocyte-stimulating hormones (α- and β-MSH) are present in hog pituitary gland. 1964 [see *INTERMEDIN]. 1971 *New Scientist* 12 Aug. 365/2 Melanocyte-stimulating hormone and corticotrophin..are two of the many polypeptide hormones secreted by the pituitary gland.

melanotekite (me:lănotī·kəit). *Min.* [ad. Sw. *melanotekit* (G. Lindström 1880, in *Öfversigt af K. Vetenskaps-Akad. Förhandl.* XXXVII. VI. 56), f. Gr. μελανο- MELANO- + τήκ-ειν to melt, dissolve: see -ITE¹.] A black to dark grey silicate of lead and iron (see quot. 1962).
1882 *Jrnl. Chem. Soc.* XLII. 291 Melanotekite, so called from its behaviour before the blowpipe (fusion to a black bead),..is a black to blackish-grey mineral, often with a bluish tinge. 1962 *Arkiv för Mineral. och Geol.* III. 143 Kentrolite and melanotekite are isostructural. They form a series with the end-members $Pb_2Mn_2Si_2O_9$ and $Pb_2Fe_2Si_2O_9$... Kentrolite can be defined as those members of this series which contain more than 50 mol per cent $Pb_2Mn_2Si_2O_9$, and the others are called melanotekite. 1968 *Amer. Mineralogist* LIII. 1278 The synthesis of melanotekite ($Pb_2Fe_2Si_2O_9$), kentrolite ($Pb_2Mn_2Si_2O_9$) and of intermediate members of this complete solid solution series has been reported elsewhere (Ito and Frondel, 1967). Using similar procedures, we have synthesized the Sc, Ga and Cr analogues of these compounds.

melanovanadite (me:lănovæ·nădəit). *Min.* [f. MELANO- + VANAD(IUM + -ITE¹.] A black opaque oxide of calcium and vanadium that occurs as bunches of acicular monoclinic crystals.
1921 W. LINDGREN in *Proc. Nat. Acad. Sci.* VII. 249 In 1920 Mr. W. Spencer Hutchinson, Consulting Engineer for the Vanadium Company of America, brought to my attention three specimens of a mineral collected by him at Mina Ragra, Peru. He suspected that it was a new mineral, and this opinion was proved correct by chemical and optical examination. The formula is $2CaO.3V_2O_5.2V_2O_4$ and I wish to propose for it the name of melanovanadite, in allusion to it being practically the only vanadium mineral of a deep black color. 1956 *Science* 1 June 990/3 Along fractures in the ore-bearing sandstone there is commonly a thin massive coating of an undescribed vanadium oxide. On this coating duttonite occurs as crusts and coatings... Associated minerals are melanovanadite and abundant crystals of hexagonal native selenium. 1968 I. KOSTOV *Mineral.* II. 470 Simplotite and melanovanadite have perfect {010} cleavage.

melatonin (melătōᵘ·nin). *Biochem.* [f. Gr. μέλα-ς black + *SERO)TONIN.] An indole derivative, $CH_3O{\cdot}C_8H_5N{\cdot}CH_2{\cdot}CH_2{\cdot}NH{\cdot}CO{\cdot}CH_3$, which is formed in the pineal gland in various mammals (principally from serotonin) and may be concerned with the regulation of certain physiological activities, esp. the reproductive cycle.
1958 A. B. LERNER et al. in *Jrnl. Amer. Chem. Soc.* LXXX. 2587/1 We wish to report isolation from beef pineal glands of the active factor that can lighten skin color and inhibit MSH. It is suggested that this substance be called melatonin. *Ibid.* 2587/2 In preventing darkening of frog skin by MSH, melatonin..was at least..5,000 times as active as serotonin. 1965 *Sci. Amer.* July 60/3 The ability of melatonin to modify gonadal function suggests, but does not prove, that its secretion may have something to do with the timing of the estrus and menstrual cycles. 1968 PASSMORE & ROBSON *Compan. Med. Stud.* I. xxiv. 36/1 Evidence from experiments in rats has led to the discovery that the pineal produces a hormone named melatonin, capable of antagonizing the effects of gonadotrophic hormones, slowing the oestrous cycle and reducing the weight of the ovaries. Production of melatonin falls when a rat is in a constantly lit environment... How far these findings relate to man is uncertain; but secreting tumours of the pineal in children lead to delayed sexual development, as if too much melatonin were being released. 1970 *Nature* 30 May 804/1 Melatonin controls diurnal and seasonal adjustments of activity in many species. 1974 *Sci. Amer.* June 62/3 The pineal..makes a hormone called melatonin that inhibits the activity of sex glands.

Melba (me·lbă). The stage name Nellie *Melba* (adopted from *Melbourne*, Australia, by Helen Mitchell (1861–1931), an Australian operatic soprano): used to designate certain foods, etc., named in her honour, as **peach Melba** (also **melba, pêche à la Melba, pêche Melba** (pēʃ me·lbă)), a confection of ice-cream and peaches flavoured with raspberry sauce, etc.; **Melba sauce** (see quot. 1951); **Melba toast** (or **toast Melba**), thinly-sliced bread toasted to crispness.
1905 [see *COUPE¹]. 1907 G. A. ESCOFFIER *Guide Mod. Cookery* xx. 778 *Pêches Melba.* Poach the peaches in vanilla-flavoured syrup. Dish them in a timbale upon a layer of vanilla ice-cream, and coat them with a raspberry purée. 1907 *Yesterday's Shopping* (1969) 33/2 *Sauces..* Melba (Escoffier's)—1/6. [1909 A. G. MURPHY *Melba* xxiii. 141 Melba's..tour for 1898–99. By this time innumerable soaps and sauces, ribbons and ruffles, had been named after her.] 1909 W. J. LOCKE *Septimus* iii. 40 The spoonful of peach Melba which she was going to put in her mouth. 1923 *Mrs. Beeton's All about Cookery* 418/1 Peach Melba... Halve and peel the peaches... serve them piled around a mould of vanilla ice cream... pour over a rich raspberry syrup. 1925 I. C. B. ALLEN *Mrs. Allen on Cooking* xii. 222 Patti Bread or Melba Toast. 1926 KIPLING *Debits & Credits* 28 Filet béarnaise, Woodcock and Richebourg '74, Pêches Melba, Croûtes Baron. 1928 D. L. SAYERS *Unpleasantness at Bellona Club* xviii. 214 You're as cold as a *pêche Melba.* 1938 *Times* 19 Aug. 7/3 Future generations may want more chickens and more peach melbas and less bread and cheese. 1938 M. L. RITZ *César Ritz* xiv. 204 Toast thin slices of bread once, then cut it through again, and again toast it... The result was Escoffier's justly famous *toast Melba...* During that year [*sc.* 1897] Melba had returned from America very ill... I had heard Escoffier discuss her *régime.* Dry toast figured on it... 'Call it *toast Melba*,' I said. 1951 *Good Housek. Home Encycl.* 552/1 *Melba Sauce*, a bright-red sweet sauce made from fresh raspberries and served with fruit sundaes, peach melba and similar desserts. 1953 E. TAYLOR *Sleeping Beauty* ix. 161 Sundaes, shakes, parfaits, whips, melbas. 1964 A. LAUNAY *Caviare & After* 137 *Melba*, large or small cuts of meat garnished with truffles, mushrooms, [etc.].. and served with a Port-flavoured sauce. *Also* fruit served with vanilla ice-cream and thick raspberry syrup. 1970 SIMON & HOWE *Dict. Gastron.* 292/2 The peach is historically associated with ice-cream because when Dame Nellie Melba, the great Australian singer, gave a party at the Savoy Hotel in London in 1892, Escoffier, who was then the chef at the Savoy, created the famous *Pêche Melba* in her honour. 1972 S. ATTERBURY *Waste Not—Want Not* 142 The Slow Cooking Method is ideal for Melba toast. 1972 *Vogue* June Special 128/3 *Melba sauce.* Blend 8 oz. fresh raspberries and pass through a sieve. Mix in 4 tablespoons icing sugar and add Kirsch to taste.

meld, *v.²* and *sb.* Add: Also in other card games, esp. canasta and rummy. Also as *v. intr.* Hence **me·lding** *vbl. sb.* (Further examples.)
1952 *Times Rev. Year* 1 Jan. p. v/2 Canasta has begun to influence the language, if slightly. It has spread the use of natural..as a noun, and figurative jobs have been found for meld, a combination of three or more cards of the same rank. 1958 'J. WELCOME' *Run for Cover* iv. 83 He melded several more times—small melds which did not look dangerous from a canasta point of view. 1964 A. WYKES *Gambling* vii. 163 In most card games, players aim either to make specific card combinations (or 'melds') as in rummy, or to take tricks, as in whist. *Ibid.* 164 Games in which 'melding' (making specific combinations of cards) is a basic principle. These are the *rummy* games, which include all forms of *poker.*

meld (meld), *v.³* orig. and chiefly *U.S.* [perh. a blend of MELT *v.¹* and WELD *v.*; but cf. E.D.D. *melder* entanglement, mental confusion; *meldered*, mixed, entangled.] *trans.* and *intr.* To merge, blend; to combine, incorporate. Hence as *sb.* and **me·lding** *vbl. sb.*
1939 *New Yorker* 23 Sept. 31 (Advt.), Schenley's exclusive process—melding—which 'marries' the whiskey blend so perfectly that it retains its rich flavor. 1952 *N.Y. Times* 18 Jan. 4/6 (*citing W. S. Churchill's Address to U.S. Congress*) What matters most is not the form of fusion, or melding—a word I learned over here. 1952 *Time* 29 Dec. 37/1 Taylor hopes to remodel the entire museum... He has plans to meld his eleven departments into five. 1959 M. STEEN *Tower* I. iii. 50 Our small..existence couldn't meld with his design for living. 1960 S. PLATH *Colossus* (1967) 14 Then I was seeing A melding of shapes in a hot rain. 1971 *Nat. Geographic* Oct. 560/2 Then clouds and gray sea melded and a steady rain slanted across the dry sides of the island. 1973 *Times Lit. Suppl.* 3 Aug. 911/5 The craft..of melding brisk jollity with real death. 1974 *Spartanburg* (S. Carolina) *Herald* 21 Apr. A6 (Advt.), A lovely cushioned group that offers a compatible meld of Ivy Leaves and light scrollwork. 1975 *New Yorker* 19 May 12/3 Jones' meld of traditional techniques with radical conceptions is in itself a radical conception, and gives the photographs their special, quiet interestingness.

melded (me·ldĕd), *ppl. a.* [Blend of MELT *v.¹* and WELDED *ppl. a.*: cf. prec.] Formed from or using man-made fibres that have an outer sheath which has been melted to bind the fibres together into a fabric.
1969 *Sci. Jrnl.* July 77 (*caption*) New possibilities have now been opened up by side by side or core/sheath bi-component fibres which, when heated enough to melt one component, result in 'melded' (melt welded) fabric. 1970 *Cabinet Maker & Retail Furnisher* 23 Oct. 174/1 The most recent development of durable non-woven fabrics for upholstery is the production of melded fabrics. 1972 *Guardian* 5 Apr. 9 Melded fabrics have arrived from ICI Fibre.

Melean, var. *MELIAN *sb.* and *a.*

mêlée. Add: Also **melée, melee. 2.** [perh. a different word.] Small diamonds less than about a carat in weight.
1911 L. COHEN *Reminisc. Kimberley* 267 (Pettman), On a certain day I had entrusted him with two or three hundred carats of melée—small stones—to sell. 1920 *Daily Tel.* 22 June 1/2 Stones of various weights from 1½ carats downwards, and a quantity of melee. 1962 R. WEBSTER *Gems* I. ii. 39 Farther down the scale are mêlée

which are crystals less than 1 carat in weight. 1972 *Daily Tel.* (Colour Suppl.) 30 June 10/2 Stones under one carat, known as Melée, are divided into fewer categories, but with subdivisions of these main ones according to colour, quality.., and of course size, there are well over 2,000 kinds.

Melian (mī·liăn), *sb.* and *a.* Also **Melean.** [f. *Melos* (Gr. Μῆλος), the name of an island in the Ægean Sea + -IAN.] **a.** *sb.* An inhabitant of Melos. **b.** *adj.* Of or pertaining to the island of Melos.
1550 T. NICOLLS tr. *Thucydides' Hystory* [*Peloponnesian War*] v. xi. sig. Bb3ᵛ The Meliens..refused to be under the obeissaunce of the Athenyans. 1629 T. HOBBES tr. *Thucydides' Peloponnesian Warre* v. 341 Dialogue betweene the Athenians and Melians. 1709 I. LITTLEBURY tr. *Herodotus' Hist.* II. VIII. 300 The Hermionians..were ejected by Hercules and the Melians out of that Country which is now call'd Doris. *Ibid.* 302 The Melians..arriv'd in Gallies of fifty Oars. 1832 J. BELL *Syst. Geogr.* II. 513 Marbles of many varieties are abundant in Greece... Another variety was..the Melian. 1887 L. E. UPCOTT *Introd. Greek Sculpture* viii. 110 The statue of Aphrodite in the Louvre, called the 'Melian Venus' (Venus de Milo), found in 1820 in the island of Melos. 1910 W. JAMES *Mem. & Stud.* (1911) 270 The Meleans say that sooner than be slaves they will appeal to the gods. 1936 *Discovery* Oct. 325/1 One of the finest Egyptian masterpieces in the world, a royal portrait, was actually executed in Melian obsidian. 1946 G. MURRAY *Euripides & his Age* v. 128 The Melians..answer as best they can. 1968 *Encycl. Brit.* XV. 132/1 The Melian earth was employed as a pigment by ancient artists. *Ibid.* 132/2 In 416 B.C. the Athenians attacked the island and compelled the Melians to surrender.

melibiose (melibəi·ōᵘz, -s). *Chem.* [a. G. *melibiose* (Scheibler & Mittelmeier 1889, in *Ber. d. Deut. Chem. Ges.* XXII. 1684), f. G. *meli-tose* MELITOSE: see BI-² and -OSE².] Glucose-6-α-galactoside, $C_{12}H_{22}O_{11}$, a crystalline sugar obtained from raffinose.
1889 *Jrnl. Chem. Soc.* LVI. 953 This disaccharide is, therefore, not lactobiose (lactose), which it closely resembles, but a new carbohydrate, for which the author proposes the name *melibiose.* 1948 [see *DIOSE]. 1970 G. C. WHITING in A. C. Hulme *Biochem. Fruits* I. i. 8 Kliewer..detected trace amounts of maltose and melibiose in the grape. 1975 *Sci. Amer.* Jan. 82/3 The molecule bears a close resemblance to two common plant galactosides: melibiose, a disaccharide (two-sugar compound), and raffinose, a trisaccharide.
Hence **melibi·ase** [a. G. *melibiase* (A. Bau 1895, in *Chemiker-Zeitung* 16 Oct. 1874/1): see *-ASE], an enzyme which brings about the hydrolysis of melibiose.
1899 J. R. GREEN *Soluble Ferments* ix. 136 Barr [*sic*] has extracted the enzyme from low fermentation Froberg yeast but he attributes it only to the final stage in the hydrolysis, the conversion of melibiose into glucose and galactose, and he has named it melibiase in consequence. 1943 H. TAUBER *Enzyme Technol.* i. 28 Ale yeast does not contain melibiase, whereas lager yeast does. 1956 *Nature* 25 Feb. 383/1 (*heading*) Complementary action of melibiase and galactozymase on raffinose fermentation.

melik, var. MALIK.
1920 *Blackw. Mag.* Nov. 669/1 Each hilla, or village, has its sheik, each group of hillas is under a melik.

melioidosis (me:li,oidōᵘ·sis). *Path.* [mod.L., f. Gr. μηλί-ς a distemper of asses, prob. glanders + -OID + -OSIS.] An infectious disease similar to glanders which is caused by the bacterium *Pseudomonas pseudomallei*, is endemic in rodents in certain (chiefly tropical) regions, and is occas. transmitted to man (in whom it is usu. fatal) and to other animals.
1921 STANTON & FLETCHER in *Trans. 4th Congr. Far Eastern Assoc. Trop. Med.* 197 The Greek physicians described under the name 'Melis' a variety of conditions resembling glanders. We propose for this disease of such varied form the name 'Melioidosis'. 1925 —— in *Lancet* 3 Jan. 10/2 Melioidosis..apart from one case in Singapore, has been recognised nowhere except in the towns of Rangoon and Kuala Lumpur. 1939 *Nature* 11 Nov. 801/1 The great plague of Athens in 430 B.C. may, Scott believes, have been melioidosis. 1952 M. E. FLOREY *Clin. Applic. Antibiotics* I. viii. 237 Five West African soldiers employed in Burma were found to be suffering from melioidosis. 1970 JUBB & KENNEDY *Path. Domestic Animals* (ed. 2) I. iii. 162/2 Outbreaks of melioidosis, as well as isolated cases, occur in sheep, goats, and pigs.

meliorant (mī·liŏrănt). [ad. late L. *meliōrantem* pr. pple. of *meliōrāre*: see MELIORATE *v.*] Something that makes better; an improver.
1920 G. SAINTSBURY *Notes on Cellar-Bk.* 102, I am afraid that the 'whets' of our ancestors were rather stimulants to drinking..than meliorants of appreciation.

meliorative, *a.* Delete † *rare* and add: *spec.* in *Linguistics*, giving or acquiring a more favourable meaning or connotation (opp. *pejorative*). (Later examples.)
1902 *Encycl. Brit.* XXXI. 678/1 The so-called meliorative and pejorative developments in word-meaning. 1916 F. SWINNERTON *Chaste Wife* xvii. § 2. 254 Its note had been meliorative rather than optimistic. 1933 G. A. VAN DONGEN *Amelioratives in Eng.* I. 146 A meliorative sense-development can be shown to have taken place in a far greater number of words than is generally supposed. 1954 PEI & GAYNOR *Dict. Ling.* 134 *Meliorative suffix*, a suffix which gives a word a more favorable or flattering connotation. 1967 R. A. WALDRON *Sense & Sense Devel.* vii. 159 *Shrewd* is certainly most often a term of praise nowadays, though it comes from ME *shrewede* 'wicked, vicious' ..and passes through the 'cunning, crafty' stage before reaching its present mainly meliorative sense. 1971 *Archivum Linguisticum* II. 36 Not only is *silly old* — recognizably an example of a 'meliorative-pejorative' adjectival compound in English but, more generally, the example has..been shown to be ambiguous. 1973 *Times Lit. Suppl.* 8 June 647/3 His qualities were British, in the meliorative sense.
b. As *sb.* A word which has had a meliorative sense-development.
1933 G. A. VAN DONGEN *Amelioratives in Eng.* I. 115 The delimitation of the scope of content displayed by those words..may result in the word becoming a meliorative or a pejorative.

melisma. Add: Pl. **melismata, melismas.** Also, in singing, the prolongation of one syllable over a number of notes. (Further examples.)
1938 *Oxf. Compan. Mus.* 557/1 The melisma is a feature of eighteenth-century vocal music, often used merely for display purposes but also descriptively and for emotional expression (Handel, 'Rejoice greatly' and 'Thou shalt break them' in *Messiah*; Bach in such words as 'wept' and 'scourged' in his Passions). 1957 *Times* 22 Oct. 3/1 Some of the tunes were of considerable interest, varying from oriental melismata to a modal melody like an English folksong. 1959 *Listener* 8 Jan. 80/2 The opening cantabile theme..continuously flows into more ornate melismata. *Ibid.* 15 Jan. 145/2 Little arabesques and melismas. 1962 A. HUXLEY *Island* xiv. 242 Long-drawn, almost bird-like melismata on a single vowel sound. 1963 *Times* 27 Dec. 4/7 The melismas with altered vowels ('I saw her yesterday-ee-ay') which have not quite become mannered. 1965 *Observer* 5 Sept. 24/8, I enjoy the graceful vocal melisma of the improvisations. 1971 *Daily Tel.* 25 June 10/4 The melismata of Boulez' 'Pli selon Pli' are wearing well. 1972 E. T. SITHOLE in T. Kochman *Rappin' & Stylin' Out* 70 The black man..appreciates the rhythm of his speech and retains it in his songs by avoiding the melisma (many notes to one syllable of a word), as most of the Negro spirituals illustrate.

melismatic, *a.* (Examples.)
1909 C. H. H. PARRY *J. S. Bach* 99 A recitative for soprano with a beautiful melismatic close. 1936 *Jrnl. Theol. Stud.* XXXVII. 165 Does that mean that these sequences were originally intended to be sung as they stand, and that afterwards the melismatic form was combined with the syllabic? 1971 *Daily Tel.* 6 May 14/6 The English performances..entirely lacked the minute melismatic ornaments..that gave the Greek singers' performances their idiomatic character. 1972 *Times* 29 May 5/4 The sopranos, soaring effortlessly through melismatic phrases like 'music', 'praise' and 'rejoice'.

melkbos (me·lkbǫs). *S. Afr.* Also **melkbosch.** [Afrikaans, f. *melk* milk + *bos* bush.] A deciduous shrub or small tree, *Diplorhynchus condylocarpon*, of the family Apocynaceæ, which has a milky latex; also = *milk-bush* (*d*) s.v. *MILK *sb.* 10 b. Also *attrib.*
1862 'A LADY' *Let.* 23 Apr. in L. G. Ross *Life at Cape* (1963) ix. 100 Here also you find tufts of reed called the 'melkbosch', which are full of a milky fluid, only grateful to goats, for sheep will not touch it. 1898 in T. R. Sim *Forests & Forest Flora Cape Good Hope* (1907) 316 The local name of the plant [*sc. Euphorbia cervicornis*] is Olifant Melkbosch. 1939 'D. RAME' *Wine of Good Hope* I. vii. 92 They camped below melk-bosch trees. 1951 L. G. GREEN *Grow Lovely* xxiv. 199 An even more historic landmark..is the gnarled melkbos tree known as 'the old slave tree'. 1957 *Cape Times* 18 Feb. 2/3 Seedlings from the old Melkbos tree at Mossel Bay should be cultivated. *Ibid.* 12 Sept. 2/4 The Board decided to give a melkbosch tree for an afforestation scheme. 1973 PALMER & PITMAN *Trees S. Afr.* III. 1911 One of the features of the tree [*sc. Diplorhynchus condylocarpon*] is its milky latex which when 'dry' is soft, sticky and rubber-like and which gives it the common names of 'rubber tree' or 'melkbos'.

Melkite (me·lkəit), var. MELCHITE.
1902 *Encycl. Brit.* XXVII. 237/2 The emperor appointed a new bishop of Alexandria, whose adherents the Copts styled Melkites or Imperialists. 1927 B. J. KIDD *Churches Eastern Christendom* xviii. 460 Originally a 'Melkite' meant a Christian who accepted the Emperor's religion. 1933 *Times Lit. Suppl.* 14 Dec. 890/2 They [*sc.* Christians] actually tended to welcome the Moslem conquest as releasing them from the persecution of the Melkite Church. 1947 G. EVERY *Byzantine Patriarchate* xii. 158 In Palestine..most Syrian Christians were Melkites. 1961 N. ZERNOV *Eastern Christendom* ii. 68 The official hierarchy.. was nicknamed *Melkites* (King's men).

melktert (me·lktẹᵊrt). *S. Afr.* [Afrikaans, f. *melk* milk + *tert* tart, pie.] A kind of pie with a cinnamon-flavoured custard filling.
1944 I. D. DU PLESSIS *Cape Malays* 42 Many old Cape dishes such as *melktert*..are still to be found in the Malay home. 1947 *Cape Argus* 22 Feb. 2 Melktert, which the Queen enjoyed at Paarl this week, is one of those old Cape afternoon tea dishes which are as popular to-day as they were two centuries ago. 1949 L. G. GREEN *In Land of Afternoon* iv. 61 She was able to bake a cake and prepare *melktert* and *poffertjies* for Queen Mary. 1958 *Cape Argus* 1/2 Dr. and Mrs. Verwoerd gave them tea, melktert and sandwiches. 1972 *Good Housek. World Cookery* 406/2 (*heading*) Melktert (Dutch Milk Tart).

mell, *v.²* For 'Now *arch.* and *dial.*' substitute 'Now chiefly *arch.* and *dial.*' **1.** (Further examples.)
a 1889 G. M. HOPKINS *Poems* (1967) 185 May Mells blue and snowwhite through them, a fringe and fray Of greenery. 1959 *Encounter* Aug. 60/2, I published an excerpt from *Road* (melling it with parts of *Visions of Neal*).
7. Delete † *Obs.* and add later example.
1941 E. R. EDDISON *Fish Dinner* (1968) vii. 92, I find close harbours of discontentment:..foolish and furious designs. Go, I'll mell me with no flirtations but them as end in bed.

mellah (me·lă). [Etym. unknown.] The Jewish quarter in a Moroccan or Turkish city; cf. GHETTO.
1874 tr. *G. Rohlfs's Adventures Morocco* vi. 105 There is a Jew's quarter (Mälha) or Ghetto, for there is no town in Morocco, and scarcely an oasis in the desert where Jews are not to be found. 1893 S. BONSAL *Morocco* xvii. 315 In the large towns the Jews are obliged to reside in a separate quarter called the Mellah, the gates of which are closed and placed under guard at night. 1925 P. GUEDALLA *Napoleon & Palestine* v. 41 Old men in Turkish *mellahs* muttered the undying *Esperança de Israel.* 1963 *Guardian* 17 July 8/6 The mellahs, as the ghettoes of Morocco are called. 1972 *Country Life* 17 Feb. 414/3 The different quarters of the town: the kasba or quarter of government officials and Europeans; the Medina or Moorish quarter and the Mellah or Jewish town.

Melle, Melli, varr. *MALI *a.* and *sb.*

mellophone (me·lofōᵘn). Chiefly *U.S.* [f. MELLO(W *a.* + *-PHONE.] A type of brass instrument similar to the orchestral horn; an alto or tenor horn.
1927 *Melody Maker* May 491/1 The Mellophone has a distinct tone of its own; mellow, as its name implies. In colour I can only liken it to a cross between the tone of the French horn and trombone. In looks the instrument.. closely resembles a French horn. 1929 *Ibid.* Apr. 369/3 A mellophone is used and scored for..as part of the brass section. 1946 R. BLESH *Shining Trumpets* (1949) vii. 166 The brass section was rounded out by Isidore Barbarin on mellophone (alto horn). 1958 P. GAMMOND *Decca Bk. Jazz* xix. 229 He was mainly featured on an E-flat tenor horn, which produced a mellophone-like sound. 1966 *Crescendo* Sept. 35/2 Roy Fox's *Whispering* (with Nat himself soloing on mellophone).
Hence **mello·phonist,** one who plays a mellophone; **mellopho·nium,** a mellophone.
1962 *Melody Maker* 7 July 9 The Stan Kenton–Tex Ritter Capitol album soon to be released, has mellophoniums, trombones, and rhythm only, plus the hawaiian guitar. *Ibid.* 21 July 5 Two odd instruments currently in London. Above,..the skoonum... Below, Jimmy Deuchar with his new mellophonium. 1975 *New Yorker* 5 May 7/1 (Advt.), Mellophonist Don Elliot's quartet is on hand Wednesday and Thursday.

Mellotron (me·lŏtrǫn). *Mus.* [f. *mellow* + *electronic.*] (See quots.)
1969 *Britannica Bk. of Year* (U.S.) 800/2 *Mellotron*, an electronic musical instrument programmed to imitate the sounds of orchestral instruments. 1970 *Times* 22 Dec. 11 The cold, windswept string tone of the Mellotron, a keyboard instrument which simulates—but not quite—the sound of an orchestra. 1974 *Melody Maker* 20 Apr. 46 The Mellotron, which was first introduced in 1963, cannot be slotted into any known category, but is best described as a series of controlled tape machines manipulated by a single keyboard. Each key relates to a tape on which has been pre-recorded a single note of an orchestral instrument. When the key is depressed the tape is played.

mellow, *a.* Add: **5. b.** *U.S. slang.* Satisfying, attractive, skilful, pleasant.
1942 Z. N. HURSTON in A. Dundes *Mother Wit* (1973) 224/2 If they's white, they's right! If they's yellow, they's mellow! If they's brown, they can stick around. 1944 D. BURLEY in *Ibid.* 207/1 The whole town's copping the mellow jive. 1945 L. SHELLY *Jive Talk Dict.* 15/1 *Mellow* (adj.) superlative. *Ibid.* 29/1 *Mellow fellow*, a satisfactory person. *Mellow mouse*, attractive female. 1960 WENTWORTH & FLEXNER *Dict. Amer. Slang* 336 *Mellow*,.. skillful; sincere, heart-felt; said of a jazz performance. 1970 C. MAJOR *Dict. Afro-Amer. Slang* 81 *Mellow*, gentle, sincere, satisfying; cool. 1973 *To our Returned Prisoners of War* (Office of U.S. Secretary of Defense) 7 *Mellow*, pleasant or satisfied. Just good.
6. b. *U.S. slang.* (See quot.)
1946 MEZZROW & WOLFE *Really Blues* (1957) 376 *Mellow*: feeling good, especially after smoking marihuana.
7. b. *mellow yellow* (*U.S. slang*), banana peel used as an intoxicant; also as *adj.*
1967 *Boston Sunday Herald Mag.* 30 Apr. 28/3 So I offered him—grass, acid, speed, magic—mushrooms, DMT, hash, and mellow yellow. 1968–70 *Current Slang* (Univ. S. Dakota) III–IV. 82 *Mellow yellow*, intoxicated from smoking a banana peel.

mellow (me·lo͞u), *sb.* *U.S. slang.* [perh. *catachr.* f. MELODY *sb.*] (See quots.)
1942 BERREY & VAN DEN BARK *Amer. Thes. Slang* § 568 *Mellows*, spirituals or religious songs. 1958 P. GAMMOND *Decca Bk. Jazz* i. 21 There were sustained 'mellows'—hollers in which a single idea or phrase might be repeated with numerous variations until the singer tired of it or thought of another.

melo (me·lo), colloq. abbrev. of MELODRAMA.
1889 E. DOWSON *Let.* 24 Mar. (1967) 54, I could no more go *twice* than I could to an Adelphi *melo.* *Ibid.* 24 Dec. 121 Even bad melo doesn't cause me to vomit as it did of old. 1920 A. HUXLEY *Let.* 23 Apr. (1969) 183 What one wd call a West End melodrama as opposed to a Lyceum melo. 1971 *Radio Times* 26 Aug., True life was *melo* about the first woman to win the George Cross. (As a stump word, '*melo*' is short for '*melodrama*'.) 1973 *Ibid.* 20 Dec. 18/4 *The Roots of Heaven*..John Huston's melo about elephant conservation.

melodeon, melodion. 1. (Earlier and later examples.)
1847 W. G. HAMMOND *Jrnl.* 30 Sept. in G. F. Whicher *Remembrance Amherst* (1946) 183 Under the lead of Goodale with his melodeon [they] sing 'Sparkling and Bright'. 1850 *Rep. Comm. Patents 1849* (U.S.) 278 Improvement in Melodeons. 1964 *Amer. Folk Music Occasional* I. 39 As a youngster he heard his father's fiddle and listened to his mother's singing of the old songs to the strain of a wheezy melodeon.
3. Delete ? from ? *U.S.* and add examples.
1840 *Boston Transcript* 1 Jan. 3/1 A grand vocal and instrumental concert will be given by Mr. John Bartlett, at the Melodeon, Washington street, on Saturday evening. 1861 E. COWELL *Diary* (1934) 283 Hattie James..was burned to death, lately at the Gayities or Melodeon. 1948–9 *Northwest Ohio Q.* Winter 19 Parker occupied Cleveland's Melodeon again on June 12, 1854, when he delivered 'The Progress of Mankind' to a 'crowded house'.

melodicon (mĭlǫ·dikǫn). [ad. Gr. μελῳδικόν neut. of μελῳδικός pertaining to melody, f. μελῳδία: see MELODY *sb.*] (See quot.) *Psalm melodicon*: see PSALM *sb.* 3.
1876 STAINER & BARRETT *Dict. Mus. Terms* 285/2 *Melodicon*, an instrument made of steel bars in different lengths tuned to the diatonic scale, struck with hammers held in the hand. 1938 *Oxf. Compan. Mus.* 789/1 Grenié's invention, the *Orgue Expressif*, became the parent of a progeny including the..Melodicon.

melodramatics (melodrămæ·tiks), *sb. pl.* [f. MELODRAMATIC *a.*] Melodramatic behaviour, action, or writing.
1915 *Nation* (N.Y.) 11 Feb. 161/2 We do not know when we have witnessed so disgusting a misuse of Federal authority, heightened as it was by the melodramatics with which Dr. Allen took the stand on his own behalf. 1929 J. C. POWYS *Meaning of Culture* 374 No tricky affectations or morbid self-consciousnesses, no melodramatics with regard to art or with regard to one's own originality. 1959 *Economist* 7 Feb. 493/1 The melodramatics of 'massive resistance'. 1959 *Times* 19 Oct. 6/7 Then it [*sc.* a film] falls back defeated, although the melodramatics are exciting enough. 1963 *Guardian* 18 Sept. 20/4 No melodramatics had followed Mr. Macmillan's receipt of the Denning report.

melody, *sb.* **6.** (Later examples.)
1934 S. R. NELSON *All about Jazz* i. 27 In his melody section, Whiteman had a complement of violins, 'cellos, saxophones, trumpets and trombones. *Ibid.* ii. 56 The more attractive section in the orchestra, in which we have the melody instruments. 1955 L. FEATHER *Encycl. Jazz* (1956) 64 The slide trombone was at first considered no less a rhythm than a melody instrument in jazz.

melomaniac. (Further examples.)
1921 W. J. TURNER *Mus. & Life* 18 It seems characteristic of that bogus science rampant among us to call persons fond of music 'melomaniacs'. 1926 C. GRAY in Gray & Heseltine *Carlo Gesualdo* I. 46 This musical education and culture was by no means confined to the male sex... Lucrezia d'Este, afterwards Duchess of Urbino, was a veritable melomaniac. 1973 *Times Lit. Suppl.* 17 Aug. 955/5 A silent musical feast, this one, but a must for all connoisseurs and melomaniacs.

melon¹. Add: **1. c.** *to cut the melon*, to decide a question.
1911 H. QUICK *Yellowstone Nights* xii. 308 The O.M. as usual cuts the melon with a word.
d. Abundant profits to be shared among a number of people. Esp. in phr. *to cut the melon.* Hence **melon-cutting** *vbl. sb.*, in Stock Exchange and Betting slang, the dividing up or sharing of profits.
1908 *Daily Report* 24 Aug. 2/4 The theory that any prospective melon-cutting will be postponed until next year. 1909 *N.Y. Even. Post* (Semi-Weekly ed.) 7 Oct. 2 A purse of $25,000 will be distributed among employees. About 8,000 men will participate in the cutting of the melon. 1911 *Daily Colonist* (Victoria, B.C.) 11 Apr. 14/2 The Suez Canal..has been earning a gross revenue of upward of $20,000,000 a year, forming one of the juiciest melons every year anywhere in the world. 1927 C. A. & M. R. BEARD *Rise Amer. Civilization* II. xx. 203 All went well until The Credit Mobilier in 1868 'cut a melon' in the form of dividends composed of the stocks and bonds of the Union Pacific. 1927 *Sunday Express* 24 July 6/4 As the company distributed some Preference shares only a short while ago we should think it unlikely there will be any further melon-cutting yet. 1928 *Weekly Dispatch* 24 June 6/4 The Union Pacific's portfolio [of outside investments] is one of the biggest potential 'melons' on the American horizon. 1939 *New Statesman* 7 Jan. 7/1 The enemy could practically destroy our commerce and industry... Every nation of the world would have an incentive to have a free cut at the melon. 1941 DICE & EITEMAN *Stock Market* (ed. 2) 462 *Cutting a melon*, the declaration of a large extra dividend. 1942 BERREY & VAN DEN BARK *Amer. Thes. Slang* § 734/6 *Melon cutting*, the division of heavy winnings by a group of bettors. *Ibid.* § 745/7 *Melon cutting*, the division of heavy winnings by a group of gamblers. 1948 *Aurora* (Illinois) *Beacon News* 7 Nov. (Suppl.) 39/2 This year, a record number of your friends and neighbors will split a record 'melon' in our 1948 savings clubs. 1964 P. WYCKOFF *Dict. Stock Market Terms* 163 *Melon*, slang expression referring to the sum total of extraordinary profits waiting to be divided.
e. = *melon pink* below.
1930 MAERZ & PAUL *Dict. Color* 199/1 Melon... *Melon Yellow.* 1975 *Harper's & Queen* June 172/3 Striped swimsuit... Cassis, citron, melon,..chocolate.
4. melon pink, a yellowish-pink colour.
1949 *Dict. Colours for Interior Decoration* (Brit. Colour Council) III. 17/2 *Melon pink*,..a descriptive colour name, from the fruit, used in the textile trade. 1975 *Country Life* 6 Mar. 561/2 Daylilies..provide a show of yellows, melon pinks and apricots.

melongene (me·lǫndʒīn). [ad. F. *mélongène* aubergine.] A West Indian name for the aubergine, *Solanum melongena*, or its fruit; = MELONGENA.
1939 *Nature* 29 July 178/1 The very general use to which the [Trinidad Low Temperature Research] Station could be put was recognized..and..work was..extended to tomatoes, limes,..melongenes, cucurbits of various kinds, and to the assortment of vegetables that can be grown in the tropics. 1952 S. SELVON *Brighter Sun* ii. 37 Lettuce taking good... But ochro and melongene would take good. 1969 G. SIMS *Sand Dollar* iv. 50 We might.. fry some jacks and melongene, that's a kind of egg-plant. 1973 *Express* (Trinidad & Tobago) 26 June 3/6 Outside of the same building melongene is offered at 20 cents a pound.

melon-hole: see MELON³.

melopœic (melopī·ik), *a.* [f. MELOPŒ(IA + -IC.] Of, pertaining to, or characterized by melopœia.
1927 *Contemp. Rev.* July 73 Wagner..turned to the melopœic chant, dramatised but not lyricised, for the expression of the words. 1940 *Burlington Mag.* Oct. 109/2 This melopoeic pattern is the chief beauty of Augustan verse.

melos (me·lǫs, mī·lǫs). *Mus.* [a. Gr. μέλος song, tune.] Song, melody; *spec.* the succession of tones considered apart from rhythm; an uninterrupted flow of melody.
1740 J. GRASSINEAU *Mus. Dict.* 129 *Melos*, is no more than a song or piece of melody. 1811 BUSBY *Dict. Mus.* (ed. 3), *Melos*, a term applied by the ancients to the sweetness of any melody; or to that quality or character by which a melody was rendered agreeable. 1876 STAINER & BARRETT *Dict. Mus. Terms* 286/2 Melos... A succession of musical sounds as opposed to *noises.* A tune. A song. 1887 E. DANNREUTHER tr. *Wagner's On Conducting* 18 He [*sc.* a conductor] found the right tempo whilst persistently fixing the attention of his orchestra upon the *melos* of the symphony. The right comprehension of the *melos* is the sole guide to the right tempo. 1903 A. W. PATTERSON *Schumann* 205 The tasteful *melos* or *arioso* which, throughout, takes the place of recitative. 1947 *Penguin Music Mag.* II. 13 The 'melos' by which he [*sc.* Wagner] means the unifying thread of line that gives a work its form and shape. 1963 *Times* 5 June 15/4 Size and melos and purposeful rhythm are the chief characteristics of Professor Josef Krips's Beethoven readings. 1964 C. HOWELL tr. *Gélineau's Voices & Instruments Christian Worship* i. 20 By means of melos he lets the words take on the colors of the prism of sentiments which illuminates them.

melt, *sb.³* Add: **2.** (Further examples.) Also *fig.*
1962 SIMPSON & RICHARDS *Physical Princ. Junction Transistors* iii. 39 A small single crystal, known as the seed crystal, is inserted in the melt and withdrawn slowly in the vertical direction... A large single crystal having the same orientation as the seed is thus formed. 1966 *New Statesman* 22 Apr. 591/2 Pan American has now chucked a large, expected crystal into the melt by ordering 25 jumbo-jets from Boeing. 1971 *Nature* 8 Jan. 80/3 In the Vale of York, the most densely populated area liable to flooding discussed in the report, the chief culprits are snow melt..and violent thunderstorms. 1971 I. G. GASS et al. *Understanding Earth* i. 17/1 The composition of the olivine formed by crystallization from a silicate melt is of course a function of the availability of the two elements.
5. Special Comb.: **melt spinning,** the process of extruding a substance (esp. a polymer) which has been softened by heat through a spinneret so as to form a fibre; hence (as back-formations) **melt-spin** *v. trans.*, **melt-spun** *ppl. a.*; **melt-water** *Geol.*, water resulting from the melting of ice or snow, esp. that of a glacier.
1950 R. W. MONCRIEFF *Artificial Fibres* xvii. 197 Nylon is melt spun. 1973 *Materials & Technol.* VI. iv. 325 The molten polymer produced is melt-spun directly. 1939 W. W. TRIGGS *Brit. Pat.* 528,455 27 Apr., In melt spinning, the spinneret is generally placed so that the filaments fall vertically from the spinneret. 1963 A. J. HALL *Textile Sci.* ii. 75 With the introduction of nylon an entirely new method of fibre-spinning was established—so-called melt-spinning in which the polymer..is melted in a novel device above the spinneret so that it can by means of a pump be extruded through the multi-holed spinneret into cold air to ensure the immediate solidification of the issuing fluid streams into filaments. 1973 *Materials & Technol.* VI. iv. 297 Melt spinning is used for the majority of thermoplastic man-made fibre-forming materials. 1948 SCHMIDT & MARLIES *Princ. High-Polymer Theory & Pract.* viii. 340 Nylon, polyvinylidene-vinyl chloride (e.g. saran), and Fiberglas are commercially important examples of melt-spun fibers. 1962 *Economist* 20 Jan. 241/1 Part of the industry where there was overlapping of effort, namely the melt-spun fibre field. 1934 *Geogr. Jrnl.* LXXXIII. 79 The amount of melt-water is so great that temporary rivers and lakes form. 1970 *Nature* 24 Oct. 352/1 Because most of the volcano is covered by glaciers, a great deal of melt-water has caused floods. 1973 C. BONINGTON *Next Horizon* x. 144 The face of the Pillar was sheer, clean and dry, but the slabby flanks were dotted with snow patches and running with melt water.

melt, *v.¹* **2. d.** (Later examples.) Also const. *away.*
1934 A. THIRKELL *Wild Strawberries* vi. 118 David.. melted from the room. 1957 C. MACINNES *City of Spades* I. viii. 58, I could see no sign of Hamilton, and hoped he'd melted. 1959 M. DOLINSKY *There is no Silence* iv. 61 What friends?.. They melted with the first headline. 1970 'D. HALLIDAY' *Dolly & Cookie Bird* vii. 72 People had started to melt, walking fast round the side of the house. 1975 P. SOMERVILLE-LARGE *Couch of Earth* iii. 48 Otway came bursting out... The girl melted away immediately.

meltdown (me·ltdaun). [f. vbl. phr. *to melt down* (MELT *v.¹* 8 c).] **a.** The process of melting.
1937 *Ice Cream Trade Jrnl.* Mar. 36/1 The Sod. Alg. ice cream melts down cleanly in the mouth... Due to the clean melt-down..a cooler sensation results in the mouth than with gelatin ice cream. 1965 *New Scientist* 15 Apr. 161/3 Overheated fuel may result in 'meltdown' and general contamination of the reactor system. 1975 *New Yorker* 12 May 98/2 He was worried about loss-of-coolant accidents, core meltdowns, and breaches of containment walls.
b. A melted mass.
1973 *Sci. Amer.* Aug. 114/2 They recycle..bottles into gemlike necklaces, the meltdown fanned by the bowl bellows of ancient Egypt.

meltemi (melte·mi). Also **meltem.** [a. mod. Gr. μελτέμι, Turk. *meltem.*] A wind experienced in the north-eastern Mediterranean; etesian wind.
1921 *Handbk. Macedonia* (Admiralty, Naval Intelligence Div.) ii. 50 At the coast of the Aegean Sea, and at Constantinople, the northerly and north-easterly winds (the 'Etesian' winds of the Greeks and the 'Meltemi' of the Turks) become more definitely the prevailing wind. 1942 *Turkey* (Admiralty, Naval Intelligence Div.) I. v. 196 The west coast also has north-west (*etesiae* or *meltemi*) winds in summer. 1957 L. DURRELL *Bitter Lemons* 73, I will take you to a special place of mine to taste the *meltemi* wind. 1958 *Times* 20 Sept. 9/6 In Aegina the *meltemi* will be blowing, a fresh salty breeze which tempers the noonday heat. 1959 *Chambers's Encycl.* VII. 775/2 The prevailing wind [in Istanbul]..in summer..is the *meltem*, a refreshing breeze. 1967 J. RATHBONE *Diamonds Bid* xx. 171 The gusty dry wind, which the Turks call 'meltemi', sending clouds of dust down the boulevards.

Meltonian, *a.* (Example.)
1901 *Edin. Rev.* Jan. 100 The red evening coat in which fox-hunters dine may be traced to the Meltonian dandies.

melusine (meliūsī·n). Also **Melusine.** [a. Fr. *mélusine*, a kind of felt. Connection w. *Mélusine*, the name of a fairy in French folk-lore, is likely.] A silky, long-haired felt, used for making hats. Also *attrib.*
1908 *Westm. Gaz.* 24 Oct. 19/2 The Parisienne's latest love in millinery is the hat of silky beaver felt that she calls melusine. 1952 C. W. CUNNINGTON *Eng. Women's Clothing* ii. 62 Toques in Melusine beaver. 1959 *Housewife* June 98 My white melusine hat is now a little soiled. 1962 A. SOUTHERN *Millinery* ii. 29 Melusine is a high-grade fur felt which is made in a range of beautiful colours.

mem, var. MA'AM. (Further examples.)
1854 B. P. SHILLABER *Life & Sayings Mrs. Partington* 47 'This is grand weather, mem, for poor people' said Mr. Tigh, the rich neighbor of Mrs. Partington. 1867 *Goodwife at Home* i. 5 Eh! Dear be here, mem, is this you, In sic a byous day? 1876 E. B. RAMSAY *Reminisc. Scottish Life & Character* (ed. 21) iv. 78 Then I canna engadge wi' ye, mem; for 'deed I wadna gie the crack i' the kirkyard for a' the sermon. 1877 G. MACDONALD *Marquis of Lossie* III. ix. 161 But, mem,..I canna lee.

mem: see *MEM-SAHIB.

mem, *v.* [f. the sb.] *trans.* To note or write down as a memorandum.
1915 W. J. LOCKE *Jaffery* v. 61 Once having 'mem-ed' an unpleasant thing in my diary, the matter is over.

member, *sb.* Add: **1.** Also *virile member*: see VIRILE *a.* and *sb.* 3.
b. Delete *Obs.* and add later example.
1966 L. COHEN *Beautiful Losers* (1970) I. 24 This member of mine rigid as a goal post.
4. b. (Later examples.)
1888 J. RUNCIMAN *Chequers* 187 You're a red-hot member! 1891 *Sporting Life* 28 Mar. 3/5 But, warm a member as our hero was, standing in front of a blazing furnace for hours..was too hot even for Jem's *sanguinary* temperament. 1922 JOYCE *Ulysses* 298 Gob, he's a prudent member and no mistake. *Ibid.* 228 Hot members they were all of them, the Geraldines.
d. Usu. *attrib.*, applied to a country, nation, state, etc., belonging to an international organization.

1931 *Times Lit. Suppl.* 28 May 429/4 Common action by a society of States against a member-State. **1959** *Ibid.* 13 Feb. 79/2 The member-nations' extra-European commitments. **1959** A. H. ROBERTSON *European Institutions* iv. 93 The Member States would consult together. **1962** *B.S.I. News* June 11/2 Copies of the national standards.. would go to each member-country. **1971** W. LAQUEUR *Dict. Politics* 361 Departmental ministers of member countries. *Ibid.* 362 Representatives of the member states' Chiefs of Staff. *Ibid.* 525 All member nations have one voice and one vote.

c. *U.S. slang.* A Negro.

1964 L. HAIRSTON in J. H. Clarke *Harlem* 290 Three more, one of 'em a member,..sailed over. **1970** H. E. ROBERTS *Third Ear* 10/1 Member, a fellow black person.

12. member bank *U.S.*, a bank which holds shares in, and has representation on the board of directors of, a Federal Reserve Bank (see also quot. 1930); **member-mug** *slang* and *dial.* (see *E.D.D.*) [f. sense 1 b + MUG *sb.*[1]], a chamber-pot.

1914 *Federal Reserve Act* § 1 The term 'member bank' shall be held to mean any national bank, state bank, or bank or trust company which has become a member of one of the reserve banks created by this Act. **1923** E. A. SALIERS *Accountants' Handbk.* 865 Member banks may rediscount short-time commercial notes with federal reserve banks. **1930** J. M. KEYNES *Treat. Money* I. 9 The typical modern Banking System consists of a Sun, namely the Central Bank, and Planets, which, following American usage, it is convenient to call the Member Banks. **1948** G. CROWTHER *Outl. Money* (ed. 2) ii. 43 The banks other than the Central Bank are usually called 'joint-stock banks' in Great Britain and 'member banks' in the United States (i.e. members of the Federal Reserve System). **1699** B. E. *New Dict. Canting Crew*, *Member-mug*, a Chamber-pot. **1785** GROSE *Dict. Vulgar T.*, *Member Mug*, a chamber pot. **1932** AUDEN *Orators* III. 104 The war-memorials decorate with member-mugs.

member, *v.* Add: **2.** Aphetic form of REMEMBER *v.* Freq. written as **'member**.

1899 KIPLING *Stalky & Co.* 254 'Member the snow all white on his eyebrows, Tertius? **1936** M. MITCHELL *Gone with Wind* lxi. 1009, I gave him to you, once before—'member?—before he was born. **1945** 'O. MALET' *My Bird Sings* II. x. 167 'I remember Papa!' shouted out Amaryllis... 'So do I 'member Papa!' said Acanthus. **1971** *Black World* June 72/2 You member the day I left, Carrie Jean? **1973** *Amer. Speech 1970* XLV. 76 'Member the day I saw you on Broad Street?

memberess (me·mbərės). Also **membress**. [f. MEMBER *sb.* + -ESS[1].] A female member; *spec.* a female member of Parliament. (Not in freq. use.)

1867 J. MACGREGOR *Rob Roy on Baltic* x. 126 It would ..be worth while being an M.P...to see the Chancellor of the Exchequer badgered..by Dr. Emma Blew, Memberess for Honiton. **1876** C. M. YONGE *Three Brides* I. xix. 338 You proved yourself the fittest memberess for the future parliament. **1933** H. BELLOC in *G. K.'s Weekly* 7 Dec. 212/1 Your member, or membress, of Parliament is all for what he or she will call temperance.

‖ **membra disiecta, disjecta,** varr. DISJECTA MEMBRA *Lat. phr.*

1957 N. R. KER *Catal. Manuscripts containing Anglo-Saxon* p. liv, One volume of **29** belonged to Cotton, as well as *membra disiecta* of **22, 83** [etc.]. *Ibid.* p. lxii, The *membra disiecta* have come into existence for various reasons and especially because one part of a manuscript seemed more important than another part or had a different sort of interest. **1963** A. E. ELSEN *Rodin* 175 Vitrines filled with the sculptor's arsenal of membra disjecta. **1970** *Times Lit. Suppl.* 23 July 787/1 As they fed wanly on their ration of membra disjecta, they had scant hope of introduction to the more invigorating study of the specifically human phenomenon of language as a whole.

membrana·ceously, *adv.* [f. MEMBRANACEOUS *a.*] With membranaceous material.

1821 W. P. C. BARTON *Flora N. Amer.* I. 14 Stem erect, ..four-sided, membranaceously winged on the angles, smooth, nearly naked.

membrane. Add: **3. membrane filter,** any of various filters made of cellulosic material and capable of retaining objects as small as bacteria; so *membrane filtration*.

1951 *Jrnl. Amer. Water Works Assoc.* XLIII. 975/1 It appears reasonably certain that membrane filters have approached a degree of refinement that suggests their extended application. **1969** *Methods in Microbiol.* I. vii. 207 Membrane filters, as manufactured by the Millipore Corporation. **1951** *Jrnl. Amer. Water Works Assoc.* XLIII. 945/2 Recent German development in the field of membrane filtration. **1957** G. E. HUTCHINSON *Treat. Limnol.* I. ix. 646 The probability..that seston collected by membrane filtration or centrifugation is very largely detritus.

membranophone (membrēi·nŏfōun). [f. MEMBRANO- + *-PHONE.] A musical instrument which employs a stretched membrane to produce the sound.

1937 *Times Lit. Suppl.* 17 Apr. 288/2 Those [instruments] which employ a stretched membrane (membranophones), such as drums. **1958** *Oxf. Univ. Gaz.* 10 Mar. 772/1 Membranophones, instruments with taut membranes made to vibrate by percussion, friction, or sympathetically. **1960** *Times* 18 Mar. 4/6 But aerophones, idiophones, membranophones..are universal and they are the concern of organology. **1971** *Sci. Amer.* Dec. 90/3 A drum is a membranophone.

‖ **membrillo** (membrī·lʸo). [Sp. *membrillo* quince.] A preserve of quinces.

1920 A. E. W. MASON *Summons* viii. 75, I..made money..by selling dulces and membrilla and almond rock. **1935** M. HUXLEY *Let.* 4 May in A. Huxley *Lett.* (1969) 393 Cherries..have been preserved in all ways possible. Bottled, made into jams, made into queso, like membrillo, under alcohol. **1947** M. LOWRY *Under Volcano* vii. 239 There were big green barrels of jerez, habanero, catalán, parras, zarzamora, malaga, durazno, membrillo, raw alcohol at a peso a litre. **1966** P. V. PRICE *France: Food & Wine Guide* 263 A thick quince jelly, like the Spanish *membrillo*. **1970** *Guardian* 20 Nov. 11/1 *Membrillo*..is the famous Spanish quince cheese.

memento. Add: **6.** ‖ **memento vivere** [L. = 'remember (that you have) to live' (used in conscious opposition to *memento mori*): a reminder of life; a reminder of the pleasure of living.

1928 BLUNDEN *Undertones of War* ii. 17 Sitting in the headquarters dugout with 'La Vie Parisienne' as a *memento vivere*. **1931** A. HUXLEY *Cicadas* 8 Rosy among the funeral black (*Memento Vivere*) a naked girl. **1966** *Punch* 19 Oct. 603/2 A *memento vivere* from an asylum sought, found, but ultimately rejected.

memoirist. (Later examples.)

1889 G. W. CABLE *Strange True Stories Louisiana* ii. 48 Carlo was beginning to swear 'fit to raise the dead', writes the memoirist, at the tardiness of the Norman pair. **1907** *Daily Chron.* 11 Jan. 3/2 In almost every section of the volume he advances, as a memoirist, a moralist,..or a translator.., someone whose name deserves to be rewritten over a faded tomb. **1914** *N.Y. Times* 31 May, These memoirists are as frankly revealing as any that described the daily life of the Grand Monarch's Court. **1970** *Daily Tel.* (Colour Suppl.) 13 Nov. 54/4 Hack memoirists often describe [Greta] Garbo as 'lonely' or 'loveless'.

memomotion (me·momōᵘʃən). [f. *memo* (? repr. L. *memor* mindful, and derived words) after *MICROMOTION.] A term used (usu. *attrib.*) in place of *MICROMOTION when time-lapse photography is used in place of ordinary cinematography.

1950 M. E. MUNDEL *Motion & Time Study* xiii. 253 Memomotion study is the name given to the special form of micromotion study in which the pictures are taken at unusually slow speeds, one frame per second being the most common. **1961** *Engineering* 6 Oct. 449 Both memomotion and micromotion filming have developed into very useful study aids. **1967** G. WILLS in Wills & Yearsley *Handbk. Managem. Technol.* x. 183 With techniques from work study using memo-motion cameras and a variety of other photographic techniques, an accurate picture of behaviour can be deduced. **1970** R. HAMMELL in D. Baker et al. *Physical Design Electronic Syst.* I. xi. 507 A highly precise form of activity sampling, called memomotion study, employs a motion picture camera.

memorandize (memŏræ·ndəiz), *v.* [f. MEMORAND(UM *sb.* + -IZE.] *trans.* and *intr.* To make memoranda (of).

1881 W. WHITMAN *Specimen Days* (1882–3) 178 Now he is sitting on the limb of an old tree..—seems to be looking at me while I memorandize. **1912** G. MALLORY *Boswell* ix. 246 Miss Burney has left us an admirable account of Boswell's deportment when in the act of 'memorandising' Dr. Johnson's conversation.

memorandum, *sb.* **6.** *memorandum-book* (earlier example).

1748 S. RICHARDSON *Clarissa* V. xlii. 301 On which she observes in her memorandum-book as follows.

memorial. A. *adj.* **3. a.** (Later examples.)

1891 S. W. MITCHELL in *Century Mag.* Dec. 287/2 The man thus imprisoned within himself recovers by effort a vast amount of memorial property presumed to have been lost. **1920** *Times Lit. Suppl.* 20 May 320/2 A link of material transmission..which..puts the theory of simple memorial piracy definitely out of court. **1959** F. BOWERS *Textual & Lit. Crit.* iii. 71 A memorial lapse, but not a misreading, must be posited. **1963** *Medium Ævum* XXXII. 23 It is often impossible to be sure whether scribal or memorial transmission has produced a given cluster of variants.

memoried, *a.* **2.** (Further example.)

1951 N. M. GUNN *Well at World's End* i. 9 She had a rosy wrinkled face like some memoried moony fruit.

memorist. 2. (Later example, not U.S.)

1920 H. G. WELLS *Outl. Hist* 115/2 Here we have..the medicine-man, the shrine-keeper, and the memorist, developed, with the development of the community.

memory, *sb.* Add: **1.** In *Psychol.* freq. subcategorized according to its manifestation or the bodily process with which it is believed to be connected. (Further examples.)

1883 F. GALTON *Inquiries into Human Faculty* 106 One favourite expedient was to associate the sight memory with the muscular memory. **1897** tr. *T. Ribot's Psychol. of Emotions* 153 Others..recall the circumstances *plus* the revived condition of feeling. It is these who have the true 'affective memory'. **1899** *Amer. Jrnl. Psychol.* XI. 7 He found that recollection could be mediated..(1) through visual images, (*a*) successive in time or space, or (*b*) grouped..*plus* motor memory. **1906** C. S. SHERRINGTON *Integrative Action Nervous Syst.* ix. 330 The relative haste with which an animal when hungry approaches food.. suggests that conation attaches to the visual reaction by association through memory with affective tone. **1921** B. RUSSELL *Analysis of Mind* ix. 157, I shall attempt the analysis of memory-knowledge..because memory, in some form, is presupposed in almost all other knowledge. **1955** H. E. GARRETT *Gen. Psychol.* x. 381 The phenomena of memory may be classified under the four headings *fixation* or acquisition, *retention*, *recall*, and *recognition*... Each of these four processes is a necessary part of memory. **1961** F. H. GEORGE *Brain as Computer* viii. 280 The hippocampus and hippocampal gyrus are therefore important features in recent memory. **1967** M. B. ARNOLD in Appley & Trumbull *Psychological Stress* 139 How intensely we experience stress may depend on the strength of this affective memory. **1969** T. FREEMAN *Psychopathol. of Psychoses* viii. 134 Sometimes the patient appears to have no capacity for short-term recall and yet at other times this function is quite intact. Similar inconsistencies of performance affect long-term and immediate memory.

c. The capacity of a body or substance for manifesting effects of its previous state, behaviour, or treatment; the effects themselves.

1887, etc. [see *magnetic memory* (*MAGNETIC *a.* and *sb.* A. 5)]. **1935** *Proc. R. Soc.* A. CXLIX. 72 The [magnetic] field..is to be regarded as 'frozen in' and represents a permanent memory of the field which existed when the metal was last cooled below the transition temperature. **1949** *Proc. Internat. Congr. Rheology 1948* i. 5 In the usual form of the theory of viscosity it is assumed that the rate of deformation is so slow compared with the relaxation process, that only slight deviations from the equilibrium state will be found... When the relaxation-time is large and the rate of deformation high..the deviation from the equilibrium state will then show traces of a more distant past. Here we see manifested a kind of 'memory' on the part of the flowing medium. **1950** *Physical Rev.* LXXVIII. 341/2 (*heading*) Memory in simple ferromagnetic domain crystal. **1964** J. M. BLATT *Theory Superconductivity* ix. 332 Since the supercurrent acts in such a direction as to make the total flux approach more closely to an integral number of flux quanta, this initial value of m_0 remains unchanged..and preserves a 'memory' for the initial, external flux. **1971** *Nature* 5 Mar. 28/1 The palaeomagnetic memory of most igneous rocks resides in grains of magnetite and titanomagnetite which are chemically unstable. *Ibid.* 30/2 Titomagnetite retains a memory of its original magnetization after oxidation.

d. The capacity of a body or substance for returning to a previous state when the cause of the transition from that state is removed.

1956 *Chem. Abstr.* L. 4577 The conversion of a naturally occurring twinned quartz crystal to single crystals by torsion often results in the formation of an unstable state, and the crystal reverts to its original twinned state. Natural crystals exhibiting 'memory' usually contain large amounts of impurities. **1961** *Chem. & Industry* 12 Aug. 1261/2 [This] could have accounted for that part of the diameter increases observed when the polymer solution flowed out of a capillary, which could not be attributed to memory of the convergent flow at the entrance to the capillary. (This memory could not account for all the diameter increase because as the length of the capillary was increased, the diameter increase decreased, but did not tend to zero.) **1964** A. S. LODGE *Elastic Liquids* x. 236 'Bouncing putty'..may be said to have a 'memory' of a few seconds in the sense that if a sample is first rapidly elongated and then held at constant length for a few seconds, no recovery occurs on release; whereas, on immediate release following the initial elongation, appreciable recovery (i.e. decrease in length..) occurs.

2. d. A device (usu. part of a computer) in which data or program instructions may be stored and from which they may be retrieved when required. Freq. *attrib.* (see *12).

1946 *N.Y. Times* 15 Feb. 16/4 Numerical values covering a wide range of scientific 'constants' are interjected as and when they are needed. There are four kinds of 'memory' in the Eniac to accomplish this. **1946** *Nature* 20 Apr. 527/2 The units in which addition is carried out provide a 'memory' with a capacity of about twenty numbers. **1948** *Math. Tables & Other Aids to Computation* III. 123 The instructions governing the routine operations that the machine is to perform can be stored in the memory in exactly the same manner in which the numbers on which the machine is to operate are stored. **1959** *Listener* 25 June 1109/1 In a typical ferrite-core memory several thousand tiny ferrite rings..are threaded on to thin copper wires. **1959** *Times* 13 Oct. 11/7 Data are stored in an electronic 'memory' or 'information bank', using digital techniques. **1962** F. I. ORDWAY et al. *Basic Astronautics* iv. 137 When a particle traverses the two scintillators..the pulse height from one of the counters is processed by a channel analyzer. Data are accumulated in the analyzer's magnetic core memory and then read out serially. **1970** O. DOPPING *Computers & Data Processing* x. 134 A secondary memory..cannot exchange data directly with any unit other than primary memory. .. In an external memory..the information contents can be changed not only by programmed operations, but also by manual operations... However, many people use the word external memory for what has been called secondary memory in this book.

12. (sense 1) *memory-bowed, -lit, -masking, -moving, -sweet* adjs.; *memory-cell, hole, -idea, -image, -judgement, -knowledge, lane, -mirror, -process, sketch, -trace, work*; (sense *2 d) *memory device, disc, element, store, unit*; **memory bank,** the memory device of a computer; also *fig.* of the human memory; **memory-belief,** the faith, probably unverifiable, that a person has in the truth of his memories; **memory book** *U.S.*, a blank book in which cuttings from newspapers and the like are pasted for preservation; a scrap-book; **memory cycle** *Computers*, (the time taken by) the process of replacing one unit of data in a memory by another; **memory drug,** a drug supposed to improve the memory; **memory drum,** (*a*) *Psychol.*, a revolving device on which material to be learnt appears; also *transf.*; (*b*) a drum-shaped memory device in a computer; **memory effect,** an effect arising from 'memory' (senses *1 c, d); **memory span** *Psychol.*, the amount of material learnt under controlled conditions which is capable of being recalled; so *memory-span test*; **memory trace** *Psychol.*, a trace hypothetically left in the nervous system by the act of memorizing.

1955 *Astounding Sci. Fiction* Jan. 56 The memory banks of the computers would still contain all data pertaining to the course set for the EDS. **1970** E. TIDYMAN *Shaft* (1971) i. 16 Every face that passed him on the street became a deposit of his memory bank. **1971** J. H. SMITH *Digital Logic* vi. 119 A shift register is a memory bank in which the numbers may be moved. **1972** J. D. BUCHANAN *Professional* v. 63 He ran the last two or three assignments through his memory bank to see whether or not they bore a residual potential for violence. **1921** B. RUSSELL *Analysis of Mind* ix. 159 Everything constituting a memory-belief is happening *now*. *Ibid.*, It is not logically necessary to the existence of a memory-belief that the event remembered should have occurred. **1925** C. D. BROAD *Mind & its Place* v. 233 Memory-beliefs..are not reached by inference. **1948** *Mind* LVII. 17 According to this theory a memory-belief has an 'intrinsic' probability; it carries its evidence, as it were, on its face. **1931** *Publishers' Weekly* 14 Feb. 843/1 Another demand..is that for inexpensive memory books used by grammar school children. **1925** BLUNDEN *Eng. Poems* 28 And to my spirit memory-bowed The world with all its wars and wails Seems turning slow. **1892** VAN LIEW & BEYER tr. *Ziehen's Introd. Physiol. Psychol.* 156 These numerous sensory cells transmit their excitation further to one other ganglion-cell, a memory-cell. **1964** *Ann. N.Y. Acad. Sci.* CXV. 655 Most of the LINC's instructions require from one to four memory-cycle times of eight microseconds each for execution. **1970** O. DOPPING *Computers & Data Processing* x. 136 In a memory cycle, the computer can either erase the contents of a cell and write new information into it, or read the content of a cell and re-write it. **1959** *Times* 9 Oct. 7/4 On a memory device within the machine are stored details of the votes going to each candidate in each constituency in 1955. **1969** J. J. SPARKES *Transistor Switching* v. 118 The commonest memory device nowadays is the ferrite magnetic core. **1961** *Flight* LXXIX. 464/1 The equipment contains 10,000 diodes, 1,500 transistors, 3,500 resistors, 670 capacitors and a memory disc. **1965** M. SPARK *Mandelbaum Gate* v. 129 Are these the memory drugs? **1966** *New Scientist* 4 Aug. 249/3 The new memory drug..is said to increase brain RNA production by 30 to 40 per cent. **1951** E. R. HILGARD in S. S. Stevens *Handbk. Exper. Psychol.* 547 (*caption*) Memory drum... The material to be memorized appears in the small aperture as the drum revolves. **1953** C. E. OSGOOD *Method & Theory Exper. Psychol.* III. xii. 502 Lists of 12 nonsense syllables..are learned in constant order on a memory drum. **1962** [see *fish-finder* (*FISH *sb.*[1] 7)]. **1964** C. DENT *Quantity Surveying by Computer* iii. 22 The memory-drum of a Pegasus computer. **1971** *Jrnl. Gen. Psychol.* LXXXV. 137 Whether the results would be similar to those found with the use of the more..traditional memory drum. **1957** *Jrnl. Chem. Physics* XXVII. 93/2 'Memory' effects..can occur particularly in nucleation processes involving condensed phases. **1971** *Physica* LXII. 393 (*heading*) Memory effects and dynamical correlations in liquid argon and sodium. **1972** *Sci. Amer.* Nov. 40/2 Hysteresis arises because of memory effects in the magnetic materials surrounding the coil. **1958** *Electronic Engin.* XXX. 1/1 Although considerable research on switching and memory elements is still proceeding the basic design of the digital and analogue computer is fairly well established. **1949** 'G. ORWELL' *Nineteen Eighty-Four* I. 40 When one knew that any document was due for destruction, or even when one saw a scrap of waste paper lying about, it was an automatic action to lift the flap of the nearest memory hole and drop it in, whereupon it would be whirled away on a current of warm air to the enormous furnaces which were hidden somewhere in the recesses of the building. **1894** CREIGHTON & TITCHENER tr. *Wundt's Lect. Human & Animal Psychol.* xix. 282 Memory-ideas are aroused by sense-perceptions, and again interrupted by new impressions. **1890** W. JAMES *Princ. Psychol.* I. xv. 620 He thinks it is almost entirely the amount to which the memory-image of the first impression has faded when the second one overtakes it, which makes us feel how wide they are apart. **1895** R. P. HALLECK *Psychol. & Psychic Culture* 106 Memory images are those which most nearly represent existing things. **1904** G. S. FULLERTON *Syst. Metaphysics* iii. 37 He may easily introduce into the memory-image elements..not present..in the original. **1921** B. RUSSELL *Analysis of Mind* 207 A memory-image of a particular occurrence. **1964** M. CRITCHLEY *Developmental Dyslexia* viii. 52 Orton believed that during the normal processes of early visual education, storage of memory-images of letters and words takes place in both hemispheres. **1896** L. T. HOBHOUSE *Theory of Knowl.* I. i. 35, I substitute for the memory-judgment proper..the apprehension of the content of the memory judgment as an idea. **1937** *Mind* XLVI. 211 The only natural interpretation is to take it as questioning not the accuracy of memory-judgments but the worth of clear and distinct perception itself. **1931** *N. & Q.* 5 Sept. 180/1 The position is illustrated and supported by comparison of our knowledge of the physical world with our memory-knowledge. **1943** *Mind* LII. 192 We do possess both knowledge of the past generally and memory knowledge in particular, even though we may be mistaken in particular cases. **1954** DANNETT & RACHEL (*title*) Down memory lane. **1958** *Times* 5 June 16/6 *Liberty Hall*, an unhappily managed trip down memory lane. **1967** A. WILSON *No Laughing Matter* IV. 422 You were all down Memory Lane, no doubt, judging by the laughter. **1970** R. HILL *Clubbable Woman* vii. 203, I don't live down memory lane. What this photograph says to me is not that happiness is gone for ever, but that it's repeatable. **1973** *Ottawa Jrnl.* 5 Feb. 11/3 It was a journey down memory lane, full of anecdotes of the good old days. **1933** W. DE LA MARE *Fleeting* 72 Those eyes that, memory-lit, Now ponder on my own. **1923** BLUNDEN *To Nature* 37 I'm not rejected then, my mind's delight Was not a play of memory-masking fancy. **1938** R. GRAVES *Coll. Poems* 163 Where port in Limerick glasses Glows twice as red reflected In the memory-mirror of the waxed table. **1908** HARDY *Dynasts* III. v. v. 212 Amid no memory-moving urgencies. **1897** C. H. JUDD tr. *Wundt's Outl. Psychol.* 241 The process that arises under such circumstances is a memory-process. **1906** *Daily Chron.* 16 Apr. 3/5 Some clever memory sketches of the Franco-British Exhibition. **1925** R. FRY *Let.* 7 Sept. (1972) II. 581, I managed to do one picture... This is a memory sketch of the composition. **1917** H. J. HUMPSTONE (*title*) Some aspects of the memory span test. **1930** R. S. WOODWORTH *Psychol.* (ed. 8) iii. 76 If the list of numbers to be memorized exceeds the memory span, several readings are necessary before it can be recited. **1951** E. R. HILGARD in S. S. Stevens *Handbk. Exper. Psychol.* 547/2 Related to what has sometimes been called the span of attention or the span of apprehension is the immediate memory span—the number of items that can be learned in one trial when they are presented serially at a controlled rate. **1955** H. E. GARRETT *Gen. Psychol.* x. 386 One of the simplest ways of determining the efficiency of immediate memory (fixation) is to test the memory span. **1969** C. N. COFFER in Talland & Waugh *Path. of Memory* 219, I would anticipate that the first locus of pathology would lie in the memory span, that is, the size of the core of actually retained list members. **1964** C. DENT *Quantity Surveying by Computer* iii. 19 For the computer's main memory store, magnetic core storage is now frequently used. **1938** W. DE LA MARE *Memory* 3 Still memory-sweet its old decoy. **1924** J. RIVIÈRE et al. tr. *Freud's Coll. Papers* I. 63 Both the memory-trace and the affect attached to the idea are there once and for all. **1951** C. T. MORGAN in S. S. Stevens *Handbk. Exper. Psychol.* 781/1 Learning problems in which memory traces had to last some time in order for the animal to make the correct choice. **1953** C. E. OSGOOD *Method & Theory Exper. Psychol.* III. xiii. 588 The gestalt theory..must make two predictions: (1) The modifications within the memory trace for visual forms must be in the direction of reducing the stresses present in the original perception. **1967** HILGARD & ATKINSON *Introd. Psychol.* (ed. 4) xii. 321/2 This particular hypothetical construct means that the memory trace does exist and that we may some day discover its nature... Hydén has proposed the theory that ribonucleic acid (RNA) might well be the complex molecule that serves as a chemical mediator for memory. **1959** *Science* 16 Oct. 957/1 Although tables of probabilities..containing over 300 items were used in the present study, they did not exhaust the capacity of the computer's memory unit. **1966** C. R. TOTTLE *Sci. Engin. Materials* vi. 141 Application of ferroelectrics includes memory units in computers and in capacitors. **1939** F. J. BROWN *Sociol. Childhood* 456 It has..encouraged or even compelled him to do 'memory work'.

Mem-sahib. Add: Also **memsahib** and abbrev. **mem**. Also, one who behaves like a European woman. (Further examples.)

1873 C. M. YONGE *Pillars of House* III. xxxii. 205 She.. heard that 'it was but a few steps, and the Mem Sahib was waiting'. **1884** KIPLING *Plain Tales from Hills* (1888) 236 The half-caste woman that we call the *Mem Sahib*... The *Mem Sahib* looks very old now. **1939** 'N. BLAKE' *Smiler with Knife* i. 19 Snug little billet, this. How does the Mem like country life? **1971** N. BARBER *War of Running Dogs* xi. 124 The mems met at tea-parties to roll bandages. **1971** R. DENTRY *Encounter at Kharmel* iii. 45 'Memsahib!' he exclaimed incredulously. 'Correct,' said Pepper. 'One memsahib, two memsahibs.' She held up two fingers. *Ibid.* 47 Respectfully, mem, you are tired. You come to the rest-house. **1971** R. RUSSELL tr. *Ahmad's Shore & Wave* ii. 23 These nawabs are turning their daughters into mem-sahibs. **1971** *Femina* (Bombay) 30 Apr. 23/1 The shopkeeper is all smiles at having produced something exactly that the *mem-sahib* wanted. **1972** A. CHRISTIE *Elephants can Remember* x. 151 You are speaking of what you might call the memsahib days..when there were Service communities in Malaya. **1972** 'J. ROSS' *Here lies Nancy Frail* ix. 114 The room was a monument to a long-dead British Raj... She had the demeanour of a mem sahib in the making.

menace, *sb.* Add: **f.** *colloq.* Applied to a person.

1936 D. CARNEGIE *How to win Friends and influence People* 63 A few doors down the street lived a 'menace', as they say out in Hollywood—a bigger boy who would pull the little boy off his tricycle and ride it himself. **1942** BERREY & VAN DEN BARK *Amer. Thes. Slang* § 583/16 *Villain*..menace. *Ibid.* § 636/23 *Formidable opponent*.. menace. **1944** J. H. FULLARTON *Troop Target* III. xiii. 95 That B.S.M.'s a bloody menace.

menadione (menădəi·ōᵘn). *Pharm.* orig. *U.S.* [f. ME(THYL + NA(PHTHALENE + *-DIONE.] 2-Methyl-1,4-naphthoquinone, $C_{11}H_8O_2$, a yellow crystalline powder which is a synthetic analogue of vitamin K and is used, often in the form of its sodium bisulphite compound, in treating hæmorrhage due to hypoprothrombinæmia. In the *British Pharmacopœia* called *MENAPHTHONE.

1941 *Jrnl. Amer. Med. Assoc.* 15 Mar. 1054/1 The Council has adopted the term 'menadione'..and has authorized its use as a nonproprietary name to describe the substance 2-methyl-1,4-naphthoquinone. **1942** T. SOLLMANN *Man. Pharmacol.* (ed. 6) 496 Various other naphthoquinone derivatives have the same action, for instance..'menadione', N.N.R. (menaquinone),..which is three or four times more potent than the K vitamin. **1955** J. G. DAVIS *Dict. Dairying* (ed. 2) 566 It has been claimed that feeding menadione..to lactating cows will increase the keeping quality of the milk. **1968** J. H. BURN *Lect. Notes Pharmacol.* (ed. 9) 83 Menaphthone (menadione or methyl-naphthoquinone) is a substance which..increases the production of prothrombin in those deficient. It is commonly given to mothers a day or two before the baby is born.

ménage. Add: **4.** ‖ **ménage à trois** [Fr., lit. 'household of three'], an arrangement or relationship in which three people live together, usually consisting of a husband, his wife, and the lover of one of these. Also *transf.* and *fig.* Cf. *À TROIS. Hence ‖ **ménage à deux,** an arrangement of two people living together.

1891 G. B. SHAW *Quintessence of Ibsenism* 116 An elderly gallant who quite understands how little she cares for her husband, and proposes a *ménage à trois* to her. **1911** —— *Getting Married* 114 Nelson..formed a *menage à trois* with Sir William and Lady Hamilton. **1929** D. H. LAWRENCE *Lovely Lady* (1932) 106 The scheme of a *ménage à deux* with her mother had not succeeded. **1933** *Times Lit. Suppl.* 19 Oct. 713/3 He meets, and marries.. a highly sexed waitress, who inevitably tires of..the mother's dominance of the *ménage à trois*. **1944** H. G. WELLS *'42 to '44* 102 They [*sc.* a Roman Catholic husband and wife]..submit to a domestic *ménage à trois*, with the priest as controlling intervener. **1958** P. KEMP *No Colours or Crest* xii. 249 That curious *ménage à trois*, the Anglo-American honeymoon with Russia. **1959** N. MAILER *Advts. for Myself* (1961) 285 A *ménage-à-trois* was completed—the bohemian and the juvenile delinquent came face-to-face with the Negro, and the hipster was a fact in American life. **1959** *Times* 28 Dec. 3/6 This happy *ménage-à-trois*—the errant wife, the lover and the unsuspecting husband. **1960** *Times* 30 Apr. 9/2 Primarily it [*sc.* a song] is a marriage of voice and verse, but since an instrument is soon added it turns the relationship into a *ménage à trois*. **1973** *Country Life* 14 June 1768/3 Pauline Viardot..the opera singer who with her husband and Turgenev formed a *ménage à trois*.

Menangkabau, var. *MINANGKABAU *a.* and *sb.*

menaphthone (mĕnæ·fþōᵘn). *Pharm.* [f. ME(THYL + NAPHTH(ALENE + *-DI)ONE.] The name of *MENADIONE in the *British Pharmacopœia*.

1943 *Brit. Pharmacopœia 1932* Add. VI. 19 Menaphthone should be kept in a well-closed container, protected from light. **1954** *Thorpe's Dict. Appl. Chem.* (ed. 4) XI. 369 Menaphthone, B.P.,..has outstanding activity, being several times as active as the natural vitamin-K. **1962** J. H. BURN *Drugs, Med. & Man* v. 60 It is fortunately possible to restore the prothrombin in the blood by giving vitamin K, or, as an alternative, a synthetic substance menaphthone, which is allied to vitamin K in composition. **1968** [see *MENADIONE].

menarche (mĕnā·ɹkī). [a. G. *menarche* (E. H. Kisch 1895, in *Berl. klin. Wochenschr.* 30 Sept. 848/1), f. Gr. μήν month + ἀρχή beginning.] The first appearance of menstruation.

1900 in DORLAND *Med. Dict.* **1905** *Index-Catal. Library Surg.-General's Office* X. 670/1 (*heading*) Menarche. *See* Menstruation (Commencement of). **1910** M. E. PAUL tr. *Kisch's Sexual Life of Woman* i. 83 The diseases of the female genital organs at the time of the menarche are very various. **1949** E. B. HURLOCK *Adolescent Development* ii. 41 It may be only a few days or it may be nearly a year after the menarche before the pubescent girl menstruates again. **1949** M. MEAD *Male & Female* III. viii. 178 In Samoa, there is no social stress on menarche. **1966** *Lancet* 31 Dec. 1453/2 No significant association was found between cervical cancer and abstinence from intercourse during or shortly after the menses; median age at menarche; regularity of the menstrual cycle, [etc.].

menazon (me·năzǫn). [f. ME(THYL + *n* (in *AMINO-) + AZ(O- + *thi*)*on*(*ate* (s.v. *THIO-1).] An organophosphorus compound, $(CH_3O)_2PS \cdot S \cdot CH_2 \cdot C_3N_3(NH_2)_2$, which is used as an insecticide for aphides (see quot. 1961).

1961 A. CALDERBANK et al. in *Chem. & Industry* 13 May 630/1 A new class of heterocyclic thiophosphate ester has been prepared which combines outstanding aphicidal properties with low toxicity to other insect species and to mammals... One of the most active members is *S*-(4,6-diamino-1,3,5-triazin-2-ylmethyl)*O*,*O*-dimethyl phosphorothiolothionate.., which has been given the common name 'menazon'. **1964** *Which?* Apr. 114/1 The organo-phosphorus compounds, such as dimethoate, malathion, menazon, are mainly poisonous to sucking insects, such as greenfly. **1972** *Bull. Entomol. Res.* LXII. 177 In order to ensure protection from secondary spread of rosette virus on the sprayed plots, menazon was applied five times at weekly intervals.

mench: see *MENSH.

mend, *v.* Add: **5. a.** *to mend up*: delete (? obs.) and add later examples. Phr. *to mend one's fences*: see *FENCE *sb.* 5 c.

1747 in *Amer. Speech* (1940) XV. 228/2, I went to mamacock & Crossman Lot & mended up fence. **1833** S. SMITH *Life & Writings J. Downing* lvi. 192 They've got their clothes pretty much mended up, and they look quite tidy. **1854** M. L. CHARLESWORTH *Ministering Children* ii. 19 Mamma is going to give me all Edward's old warm stockings, if I mend them up quite neat!

d. *to mend a pen* (earlier example).

1820 KEATS *Lett.* (1958) II. 262, I have been writing with a vile old pen the whole week... The fault is in the Quill: I have mended it.

6. b. Also *to mend up*.

1877 A. SEWELL *Black Beauty* (*c* 1878, ed. 5) xliv. 217 The farrier said he [*sc.* a horse] might mend up enough to sell for a few pounds.

10. c. To recover from, get better *of*, grow out *of*.

1881 J. FOTHERGILL *Kith & Kin* III. ii. 43 He had always trusted that the boy would mend of such outlandish indifference.

mendang (mendæ·ŋ). Also **mendong, mendung**. [Tibetan.] A sacred wall composed of flat stones carved with Buddhas or religious texts. Also *attrib.*

1925 J. A. HAMMERTON *Countries of World* VI. 3946/2 The mendangs—long walls in the middle of the road composed for the most part of inscribed stones. **1939** M. PALLIS *Peaks & Lamas* vi. 65 Each entrance to a village is marked by a *mendong* or *Mani* wall, a low cemented breast-work upon which innumerable flat stones carved with sacred texts in low relief have been laid, the accumulated offerings of local piety. **1952** H. W. TILMAN *Nepal Himalaya* I. iv. 39 These walls or 'mendongs', which are seven or eight feet high, must be passed on the left. **1960** 'S. HARVESTER' *Chinese Hammer* x. 94 A *mendang* wall in the town, the sacred wall. **1972** J. & R. GODDEN *Shiva's Pigeons* II. 88 Mendungs—low stone walls—carved..are built beside mountain tracks and along the trade routes of the north-eastern passes. *Ibid.*, It is reverent to pass a mendung on the right-hand side.

Mende (me·ndi), *sb.* and *a.* Also **Mendi**. **A.** *sb.* **a.** A group of Negro peoples inhabiting Sierra Leone and Liberia; a member of this group. **b.** The language of the Mende. **B.** *adj.* Of or pertaining to the Mende or their language.

1887 *Encycl. Brit.* XXII. 44/2 The following are the more important races [of Sierra Leone] that can be distinctly classified..Mendis, 3088. **1908** T. G. TUCKER *Introd. Nat. Hist. Lang.* 147 *Mende*, of Upper Guinea, including dialects of Mandingo, Bambara, etc. **1911** *Encycl. Brit.* XVI. 542/1 Some of the Krumen are coarse and ugly, and this is the case with the Mende people. **1936** G. GREENE *Journey without Maps* I. iii. 62 He could speak Mende; he was picking up Buzie. **1948** —— *Heart of Matter* I. I. i. 7 His Mende sergeant clicked his heels. **1957** C. MACINNES *City of Spades* 216 Mr. Karl Marx Bo.. planned to send..a tendentious report on the trial to the Mendi newspaper of which he was part-time correspondent. **1970** P. OLIVER *Savannah Syncopators* 112 *Mende*, Sierra Leone tribe situated near the coast. **1971** E. JONES in J. Spencer *Eng. Lang. W. Afr.* 68 The official news bulletin put out daily over the Sierra Leone Broadcasting Service by the Ministry of Information, as well as other important government statements, are broadcast in Krio as well as English, Mende and Temne. **1972** J. L. DILLARD *Black English* III. 130 There will probably never be any reliable evidence as to whether the slaves themselves differentiated between the Vai-Hausa forms and the Mende form in pronunciation. **1974** *Times* 4 May (Suppl.) p. i/6 Typically a [Sierra Leonean] minister will be a Mende or a Temne with a Creole permanent-secretary.

mendelevium (mendĕlī·viŏm, -ĕi·viŏm). *Chem.* [f. the name of Dmitri Ivanovich *Mendeleev* (1834–1907), Russian chemist + -IUM.] An artificially produced transuranic element, the longest-lived isotope of which has a half-life of two months. Atomic number 101; symbol Md (formerly Mv).

1955 A. GHIORSO et al. in *Physical Rev.* XCVIII. 1519/2 We would like to suggest the name mendelevium, symbol Mv, for the new element in recognition of the pioneering role of the great Russian chemist, Dmitri Mendeleev. **1967** *New Scientist* 21 Sept. 598/2 The new mendelevium isotope, with 101 protons and 157 neutrons, falls into the odd-odd class... The long half-life will enable quite large quantities of mendelevium to be made. **1971** C. KELLER *Chem. Transuranium Elements* II. ix. 596 Six isotopes of mendelevium are known with mass numbers 252 and 254–258... They can be produced by the bombardment either of uranium or plutonium with heavy ions, or of einsteinium with α particles. *Ibid.* 597 Since ^{256}Md can only be obtained by nuclear reactions with accelerated ions, it is impossible to produce weighable quantities. Consequently, the chemical investigation of mendelevium is restricted to tracer work.

Mendelian, *a.* Add: **B.** *sb.* One who adheres to or supports Mendel's principles of heredity.

1903 K. PEARSON in *Phil. Trans. R. Soc.* A. CCIII. 57 If we were 'pure Mendelians' we should for the purpose of character classification make $v = w$. **1907** *Nature* 23 May 73/1 It would be regarded as a demonstration of the falsity of the doctrine of gametic purity by everyone who was not a Mendelian. **1925** A. HUXLEY *Let.* 25 Feb. (1969) 242 You Mendelians have made all that..philosophy look.. dubious. **1941** J. S. HUXLEY *Uniqueness of Man* ii. 82 Twenty-five years ago..the field of heredity was still a battle-field. The Mendelians and the Biometricians were disputing for its possession. **1972** *Science* 12 May 623/3 The Mendelians, with their insistence upon large mutations as the agent of evolutionary change, undercut the Darwinian assumption of insensibly graded variation.

Mendelianism (mendī·liăniz'm). *Biol.* [f. MENDELIAN *a.* + -ISM.] = *MENDELISM.

1903 K. PEARSON in *Biometrika* II. 215 Mendelianism fails also for skin-colour in crosses between the black and white races of man. **1906** *Mem. & Proc. Manch. Lit. & Philos. Soc.* L. XI. 11 What is the essential feature of that which is called Mendelism by those who believe in it, and Mendelianism by those who do not? **1947** JACOBS & STERN *Outl. Anthropol.* 309 *Mendelianism*, the theory regarding the process of inheritance of characters, first published by Gregor Mendel in the 1860's and rediscovered about 1900. **1965** J. D. WATSON *Molecular Biol. of Gene* i. 27 All four wrote major works, each showing..how Mendelianism and Darwinism were indeed compatible.

Mendelism (me·ndĕliz'm). *Biol.* (now chiefly hist.). [f. the name *Mendel* (see MENDELIAN *a.*) + -ISM.] The theory that discrete bodies (now called genes) control the inheritance of any particular character and that these are inherited in accordance with certain simple laws first propounded by Mendel; the body of knowledge (forming the basis of the modern science of genetics) founded on the experiments performed and laws propounded by Mendel.

1903 *Science* 20 Mar. 451/2 The breeder wants to preserve the desirable characters or traits and eliminate the undesirable ones, but under the strict interpretation of Mendelism this is difficult. **1905** R. C. PUNNETT (*title*) Mendelism. **1906** L. H. BAILEY *Plant Breeding* (ed. 4) 168 Already so many adjustments have been made of the Mendelian principles that it is becoming difficult to determine what Mendelism is. **1920** *Edin. Rev.* July 135 Mendelism promises to furnish the stock-breeder with better and more certain means of increasing the economic value of his stock. **1927** HALDANE & HUXLEY *Animal Biol.* ii. 62 The comparatively new science called Mendelism. **1942** [see *BIOMETRY 2]. **1966** E. A. CARLSON *Gene* ii. 11 If Weldon had stopped here he might have won the battle and delayed the inevitable victory of Mendelism for more years than it took.

Mendelist (me·ndĕlist). *rare.* [f. as prec. + -IST.] = *MENDELIAN *sb.*

1910 H. WALKER *Lit. Victorian Era* I. iii. 230 There are Mendelists and Mutationists as well as Darwinians. **1951** *Biol. Abstr.* XXV. 657/1 The author opposes what he describes as the 'theories of the Mendelists and the Morganists' that the origin of life on earth is an extraordinarily rare occurrence.

Mendelize (me·ndĕləiz), *v.* Also **mendelize**. [f. as prec.: see -IZE.] *intr.* To behave or be transmitted in accordance with Mendel's laws of inheritance.

1906 L. H. BAILEY *Plant Breeding* (ed. 4) 171 We do not know what plants will Mendelize until we try. **1913** M. HARTOG *Probl. Life & Reproduction* vii. 190 He found the characters thus acquired behaved as unit ('allelomorph') hereditary characters to the normal ones, and Mendelised just as do the unit characters which have arisen spontaneously, or have been transmitted for generations. **1924** E. W. MACBRIDE *Introd. Study Heredity* viii. 212 'Sports'.. which breed true when crossed with their like, but which 'mendelize' when crossed with the type. **1973** *Sci. Amer.* Apr. 8/2 Mendel repeated Nägeli's work with seeds supplied by him and to his disappointment found that they did not Mendelize.

Hence **Me·ndelizing** *ppl. a.*

1918 BABCOCK & CLAUSEN *Genetics Rel. Agric.* 286 Those changes in specific factors which result in the appearance of new Mendelizing characters. **1952** *Amer. Scientist* XL. 89 The differences between subspecies are hereditary and they are based upon mendelizing differences. **1969** *Nature* 14 June 1101/2 Their results..constitute the first thorough genetic analysis of a maternally determined Mendelizing character.

Mendelssohnian (mendĕlsōᵘ·niăn), *a.* [f. the name of the German composer Felix *Mendelssohn*-Bartholdy (1809–47) + -IAN.] Of, pertaining to, or characteristic of Mendelssohn, or his musical productions.

1887 G. B. SHAW *How to become Mus. Critic* (1960) 132 The modern Mendelssohnian 'culture', with all its refinement, its elegance, its reticence, and its 'chastity'. **1900** *Musical Standard* LIX. 283/1 The second movement was rendered with a Mendelssohnian elegance. **1909** H. G. WELLS *Ann Veronica* iii. 60 The organ..was, in its Mendelssohnian way, as glad as ever it could be. **1928** *Observer* 29 Jan. 14/4 A clearly conceived and agreeable piece of music with Mendelssohnian orchestration. **1974** *Daily Tel.* 28 Oct. 11/1 Helen Donath's singing of the consolatory, very Mendelssohnian soprano solo gained in security after a nervous opening.

Also **Mendelsso·hnic** *a.*

1889 G. B. SHAW *London Music 1888–89* (1937) 164 Her Mendelssohnic sense of form in composition. **1890** *Ibid.* 300 This Mendelssohnic curse of speed for speed's sake. **1928** *Observer* 4 Mar. 14/5 Ask yourself what are the real Beethoven touches in the Mendelssohnic finale.

mending, *vbl. sb.* Add: **1. c.** Also *attrib.*

1856 C. M. YONGE *Daisy Chain* I. vii. 63 Ethel had to fetch her mending-basket. **1863** 'G. HAMILTON' *Gala-Days* 41, I find myself in a mending-basket. **1867** A. D. WHITNEY *Summer in L. Goldthwaite's Life* viii. 165 What should we do without our mending-day? **1899** A. NICHOLAS *Idyl of Wabash* 18 Taking his hose from the mending basket and darning them. **1966** *Olney Amsden & Sons Ltd. Price List* 28 Mending cotton, white only. **1974** J. STUBBS *Painted Face* vii. 107 Bessie..opened her mending basket.

2. a. (Also in *sing.*)

1891 *Harper's Mag.* Sept. 579/1 Mrs. Dorset was on the bench in the porch, the basket of mending by her.

mendong, var. *MENDANG.

mendopo, var. *PENDOPO.

mendung, var. *MENDANG.

meneer, var. MYNHEER.

1939 S. CLOETE *Watch for Dawn* i. 5 What are these questions that you torment me with, meneer? **1946** *Cape Times* 16 Nov. 4 Meneer is a strange man. **1948** *Cape Argus* 9 Aug. 9/5 The meneer found those spoons... Mr. Erlank swung back to the constable. 'Did you find spoons there?' he asked. **1970** G. CROUDACE *Scarlet Bikini* vii. 79 I'm listening, *meneer*.

‖ **mengkuang** (meŋkwæ·ŋ). [Malay.] A tree belonging to one of the larger species of *Pandanus*, providing leaves that can be woven into matting, etc. Also *attrib.*

1900 H. N. RIDLEY in *Jrnl. Straits Branch Roy. Asiatic Soc.* XXXIII. 170 *Pandanus atrocarpus* Griff. 'Mengkuang'. The biggest species here, often 40 feet high with very long leaves. **1908** L. WRAY in A. Wright *20th Cent. Impressions Brit. Malaya* 235/3 Of late years a fairly large industry has sprung up in Negri Sambilan in the manufacture of mat hats... The finer are of *Pandan* leaves, and the coarser of *Mengkuang* leaves. **1935** I. H. BURKILL *Dict. Econ. Products Malay Peninsula* II. 1644 Such species [of *Pandanus*] as are used for matting are called by the Malays 'mĕngkuang'. **1947** R. O. WINSTEDT *Malays* 175 Dish covers are sometimes made of this mad plait, or they are made of strips of the white inner sheath of bamboo..stuck over a conical *meng-kuang* lining shaped like Chinese hats. **1954** R. E. HOLTTUM *Plant Life Malaya* ii. 23 The big swamp Pandan called Mengkuang, with trunks sixty feet or more high and leaves twenty feet long, rivals most palms in bulk. **1965** C. SHUTTLEWORTH *Malayan Safari* ii. 29 A patch of earth near the cooking fires was swept clean and laid with broad *mengkuang* leaves. **1972** M. SHEPPARD *Taman Indera* 158 Coarser matting for everyday household use and sacking are made of *mengkuang*. Pandan is frequently dyed and woven into elaborate patterns. *Mengkuang* is more often used in its natural form.

mengkulang (me·ŋkulaŋ). [Malay.] A timber tree belonging to the Malaysian genus *Tarrietia*, esp. *T. simplicifolia*, or its wood.

1940 E. J. H. CORNER *Wayside Trees Malaya* I. 622 The species of Tarrietia are mostly timber-trees known to Malays as Mengkulang. **1956** *Handbk. Hardwoods* (Forest Prod. Res. Lab.) 150 Locally, mengkulang is used for interior construction, flooring and furniture. **1972** *Timber Trades Jrnl.* 13 May 39/2 Mengkulang is not easy [to obtain], presumably because of the competition from the plywood mills.

Ménière (‖ meniẹ̄r). *Path.* The name of Prosper *Ménière* (1799–1862), French physician, used in the possessive to designate a disease of the membranous labyrinth of the ear associated with dizziness, tinnitus, etc., and causing progressive deafness in the ear affected (described by Ménière in 1861).

Accentuation of the name varies. The forms *Menière*, used by the physician himself, and *Ménière*, used by many writers about him and now by his descendants, are both common. The unaccented form *Meniere* is used in some modern English works.

1876 *Edin. Med. Jrnl.* XXI. 716 (*heading*) Case of Menière's disease. **1885** *Encycl. Brit.* XIX. 39/1 A diminution of the power of co-ordinated action, as in Menière's disease. **1907** *Brit. Med. Jrnl.* 11 May 1107/2 Ménière's disease..is a rare disease, while cases exhibiting Ménière's symptoms are not by any means uncommon. **1938** *Proc. R. Soc. Med.* XXXI. 1317 (*heading*) Observations on the pathology of Ménière's syndrome. **1955** W. GADDIS *Recognitions* III. i. 732 If it is Ménière's syndrome, we'll have you up staggering around in no time. **1968** HARRISON & NAFTALIN (*title*) Ménière's disease. **1968** J. L. PULEC (*title*) Meniere's disease. **1971** *Daily Tel.* 30 Jan. 3/2 Grounded by Ménière's syndrome, a malady of the inner ear which was causing increasing deafness, he secretly went into hospital.

meningioma (menindʒiōᵘ·mă). *Path.* Pl. **-omas, -omata.** [mod.L., f. by shortening *mening*(*othel*)*ioma*, f. MENINGO- + *ENDO)-THELIOMA.] A tumour, usu. benign, arising from the meninges (esp. those of the brain).

1922 H. CUSHING in *Brain* XLV. 282 (*heading*) The meningiomas (dual endotheliomas): their source, and favoured seats of origin. *Ibid.* 285 The term *meningothelioma* was first proposed, but it has, on further consideration, been shortened to *meningioma*. This word, consequently, will be used to indicate the entire group of tumours which appear to arise from the pachymeninx. **1961** *Lancet* 16 Sept. 656/2 Of particular interest was the activity of the cells comprising the whorls of a meningioma. **1974** PASSMORE & ROBSON *Compan. Med. Stud.* III. xxxiv. 110/2 About 18 per cent of intracranial tumours are meningiomata.

meningo-. Add: **meni:ngoco·ccal, -co·ccic** *adjs.*, of, pertaining to, involving, or caused by a meningococcus; **meni:ngo-ence·phalocele** [ENCEPHALOCELE], the protrusion of brain substance and meninges through a hole in the skull; the mass so protruded; **meni:ngo-ence:phalomyeli·tis** [*encephalomyelitis* s.v. *ENCEPHALO-], inflammation of the meninges, brain, and spinal cord; **meningo-myelocele** (examples); = *MYELOMENINGOCELE.

1907 *Edin. Med. Jrnl.* LXIV. 227 One case..associated with tubercular meningitis..clearly falls..into line with meningococcal arthritis. **1949** H. W. FLOREY in H. W. Florey et al. *Antibiotics* I. i. 23 Meningococcal carriers were successfully cleared of micro-organisms. **1966** *Jrnl. Amer. Med. Assoc.* 2 May 391 (*heading*) Meningococcal disease, 1965. **1907** *Jrnl. Med. Res.* XVII. 229 Seven different antigonococcic serums and a meningococcic serum. **1950** *Amer. Jrnl. Med.* VIII. 468 (*heading*) Cause of death in meningococcic infection. **1891** F. P. FOSTER *Med. Dict.* III. 2277/1 *Meningo-encephalocele*, a tumour of the head analogous to hydro-myelocele. **1901** *Brit. Med. Jrnl.* 22 June 1542/1 (*heading*) Case of meningo-encephalocele treated by excision of the mass. **1964** S. DUKE-ELDER *Parsons' Dis. Eye* (ed. 14) xxxiii. 529 Clinically they [*sc.* dermoid cysts of the orbit] may be mistaken for meningo-encephaloceles, protrusions of the cerebral contents, which usually occur at the upper and inner angle where there are most sutures between bones. **1900** DORLAND *Med. Dict.* 381/1 Meningo-encephalomyelitis. **1966** WRIGHT & SYMMERS *Systemic Path.* II. xxxiv. 1199 Myalgic Meningoencephalomyelitis. An extensive epidemic of a benign myalgic encephalomyelitis occurred among the staff of the Royal Free Hospital in London, in 1955. **1885** *Trans. Clin. Soc.* XVIII. 340 Protrusion of the membranes together with the spinal cord and its appertaining nerves, meningo-myelocele. **1966** WRIGHT & SYMMERS *Systemic Path.* II. viii. 1234/2 In its least serious form, spina bifida occulta, there is incomplete closure of one or more of the vertebral arches... In a more serious and commoner variety, meningomyelocele, the spinal cord is involved, and portions of its posterior columns may be stretched out in the wall of the subcutaneous cystic swelling.

Mennecy (me·nəsi), *a.* Also **Menecy**. [Name of a town in France, near Paris.] The designation of a soft-paste porcelain made at Mennecy. Also *absol.*

1863 W. CHAFFERS *Marks Pott. & Porc.* 203 Menecy, marked in blue on a soft paste egg cup, of very early manufacture. **1869** C. SCHREIBER *Jrnl.* Aug. (1911) I. 32 Mennecy china box. **1870** *Ibid.* 71 We found a..Mennecy basket and cover with coloured flowers in relief. **1874** *Ibid.* 255 A small piece of Mennecy at Van Houtum's. **1957** *Encycl. Brit.* XVIII. 351/2 At its best, Mennecy porcelain is of unsurpassed quality, mellow in tone and texture and of a warm white colour. **1960** R. G. HAGGAR *Conc. Encycl. Cont. Pott. & Porc.* 314/1 The Mennecy porcelain manufacture originated in the rue de Charonne, Paris..in 1734. **1960** H. HAYWARD *Antique Coll.* 181/2 Mennecy porcelain, soft-paste porcelain..made at Mennecy, Ile-de-France, between 1735 and 1785. **1972** *Country Life* 8 June (Suppl.) 58/1 An ormolu birdcage with Mennecy flowers.

Mennist (me·nist). *U.S.* [irreg. f. *Menno* (see MENNONITE) + -IST.] = MENNONIST, MENNONITE. Also *attrib.*

1771 G. TAYLOR *Voy. N. Amer.* 170 In the City of Philadelphia you see Churchmen, Quakers, Lutherans, Calvinists, Moravians, Catholics, Menists, [etc.]. **1869** *Atlantic Monthly* Oct. 474/1 The Mennists in many outward circumstances very much resemble the Society of Friends. *Ibid.*, In the interior of the Mennist meeting, a Quaker-like plainness prevails.

meno (me·no), *adv. Mus.* [It.] Less; used in musical directions, as *meno mosso*, rather slower, less animated (lit. 'less moved').

1876 STAINER & BARRETT *Dict. Mus. Terms* 286/2 *Men*, *meno*... Less; as *meno forte*, not so loud. **1880** GROVE *Dict. Mus.* II. 311/2 *Meno mosso*, a direction, which, like Più lento, generally occurs in the middle of a movement. *Ibid.* 312/1 Beethoven uses 'Meno mosso e moderato' in the Fugue for strings in B♭, op. 133, and 'Assai meno presto'—'very much less quick'—in the Trio of Symphony No. 7.

menologist (mĕnǫ·lŏdʒist). [f. MENOLOGY: see -IST.] A compiler of a menologium.

1895 F. TUPPER in *PMLA* X. 224 The Menologist (*Cælendcwide*, 56) ushers in Easter thus: 'Aprelis monað on þam oftust cymð [etc.]'. **1948** K. MALONE *Middle Ages* I. 35 For sources the menologist presumably used church calendars and the like... In any case the menologist was no mere clerk, learned in Church Latin only. He was steeped in classical Old English poetry, as his style and choice of words reveal.

menology. Add: In sense 1 b now usu. in form *menologium*. (Earlier and later examples.)

1709 E. ELSTOB tr. *Ælfric's Homily on St. Gregory* App. 26 The Original MS. of the Menologium is in the Cottonian Library. **1735** M. SHELTON tr. *Wotton's Short View G. Hickes's Anc. Northern-Lang.* 20 The first of these is a Poetical Calendar, translated into Latin, wherein it is observable, that this Menology, wrote at the Command of some Anglo-Saxon Bishop, or King, who reigned after the Dissolution of the Heptarchy, makes mention of no Saints except Pope Gregory, and Austin the Monk. **1807** S. TURNER *Hist. Anglo-Saxons* (ed. 2) II. 280 The elegant Menology in the Cotton Library. **1830** B. THORPE tr. *Rask's Gram. Anglo-Saxon Tongue* 138 Another remarkable instance of this [lack of strophic or stanzaic divisions] is the conclusion of the *Menologium Saxonicum*. **1905** E. D. HANSCOM in *Jrnl. Eng. & Germ. Philol.* V. 441 The *Menologium* preserves the popular names of October and November, *Winterfylleð* (184) and *Blotmonað* (195). **1932** CHADWICK & KERSHAW *Growth of Lit.* I. 380 The *Menologium* or metrical calendar at the beginning of MS. C of the *Saxon Chronicle*.

Menominee (mĕnǫ·minī). *U.S.* Also **Menomini, -onee**. [ad. Ojibwa *manōmini*, lit. 'wild-rice people'.] **1.** The name of a tribe of Algonquian Indians first discovered near the mouth of the Menominee River in Michigan and Wisconsin; also *attrib.*; also applied to the language spoken by this people.

1762 T. HUTCHINS *Jrnl.* 26 June in *Mich. Hist. Mag.* (1926) X. 369, I delivered the same Message to the Meynomeneys that I had done to the Sax and Reynard Nations. **1830** [see *HALF-BLOOD 4]. **1835** C. F. HOFFMAN *Winter in West* II. 3 The Mê-nó-mê-nê, or wild-rice-eaters, is a broken band that served with effect against the Sauks and Foxes in the Indian difficulties of 1832. **1920** L. BLOOMFIELD in C. F. Hockett *Leonard Bloomfield Anthol.* (1970) 90 Have been writing down Menominee words and stories. They are a delightful people, of good culture: it must have been an elaborate and beautiful culture 200 years ago. **1922** —— in *Ibid.* 99 These stems are a living (freely formed) derivation in Menomini. **1953** A. HUXLEY *Let.* 17 Aug. (1969) 683 The Menomini are Indians in a reservation in Wisconsin. **1964** E. BACH *Introd. Transformational Gram.* I. 2 To learn Greek or Menomini..it is not enough to read through a compact presentation of paradigms or rules. **1975** *Times* 8 Jan. 6 Troops of the Wisconsin National Guard were called out today to deal with a group of Menominee Indians occupying a large estate... The Menominees have a reservation near by.

2. In full, *Menominee whitefish*. The round whitefish, *Prosopium cylindraceum*, found in lakes of northern North America.

1884 G. B. GOODE *Fisheries U.S.: Nat. Hist. Aquatic Animals* 541 *Coregonus quadrilateralis*. The only name which I have heard applied to this fish is that of 'Menomonee White-fish'. **1902** *Rep. U.S. Comm. Fisheries 1901* 653 Table showing, by States, the products of the fisheries of the Great Lakes in 1899..White-fish (Menominee) fresh ..White-fish (Menominee) salted. **1944** G. L. NUTE *Lake Superior* 186 Menominees, a kind of whitefish, are listed at 14,940 pounds, all caught in American waters.

menopausal (menŏpǭ·zăl), *a.* [f. MENOPAUSE + -AL.] Of, pertaining to, or connected with the menopause. Also *fig.*

1910 *Practitioner* June 787 Permanent cessation of the menses, so long as the ovaries have been conserved, is not associated with so-called 'menopausal' symptoms. **1964** L. MARTIN *Clin. Endocrinol.* (ed. 4) vii. 233 If human menopausal gonadotrophin units (HMG) are used, the average normal result is 30 units and values above this may be found in cases of malignant testicular tumours. **1969** D. WIDGERY in Cockburn & Blackburn *Student Power* 126 It was inevitable that students would eventually come to revolt against the menopausal leadership of NUS and its flaccid policies. **1974** M. SPARK *Abbess of Crewe* III. 83 She is behaving in a most menopausal way, and she claims there is a plot against her to prevent her being elected Abbess.

menopause. Add: (Further examples.) Also *fig.*

1954 W. FAULKNER *Fable* 69 It was winter again now, the long unbroken line from Alps to sea lying almost quiescent in mud's foul menopause. **1962** J. H. BURN *Drugs, Med. & Man* xv. 152 Stilboestrol is also used at the menopause when menstruation ends. **1970** JENNER & SEGAL *Men & Marriage* 144 In the nineteen-fifties a spate of 'popular medicine' articles flooded the press bearing the news that men, too, suffered from the Change of Life and middle-aged men found themselves swamped with sympathy over their newly recognised menopause. **1972** F. WARNER *Lying Figures* III. 19 Having your mental menopause?

menorah (mĕnōᵘ·rā, -ōə·rā). [Heb. *mᵉnōrāh* sabbath or festival candelabrum.] A holy candelabrum having seven branches used in the ancient temple in Jerusalem; also a candelabrum having any number of branches used in modern synagogues.

1888 (*title*) Menorah monthly. **1936** *Jews in Palestine Campaign* (Menorah Club) 41 A Club under the name of Menorah (seven branched candlestick)..was established. **1958** B. MALAMUD *Magic Barrel* (1960) 128 Don't those peaks..look like a Menorah?.. Like a seven-branched candalabrum..? **1966** L. DAVIDSON *Long Way to Shiloh* ii. 31 The great seven-branched lamp, the Menorah, has been the symbol of Judaism for some thousands of years; of..Israel, for the past fifteen or so... Titus took the lamp when he destroyed the Temple in August of 70. **1970** L. M. FEINSILVER *Taste of Yiddish* 247 The Sabbath menorah has seven holders, to accommodate a candle for each day of the week. The Chanuka menorah has nine holders. **1973** *Jewish Chron.* 2 Feb. 16/5 The Mayor of Hackney..was presented with an Israeli-made menorah. **1973** *Country Life* 20 Sept. 784/2 The symbol of the Menorah, the seven-branched candelabrum, was originally placed in the tabernacle in the Sinai Desert and in the First Temple in Jerusalem, and was carried away by Titus. Today the Menorah, framed by two olive branches, is the emblem of the modern state of Israel.

men's: see *MAN *sb.*[1] 23.

mensa. Add: **3.** (With capital initial.) Adopted as the name of an organization of people with above-average intelligence quotients. Also *attrib.*

Members qualify by passing a test equivalent to an I.Q. of 148 or more on the Cattell scale.

1962 *Mensa Register* p. ii, The Mensa Register is published to satisfy members' curiosity as to who the other members are, and to allow such individual contact to be made as is desired. **1968** *Brit. Mensa* Dec. 6/2 Mensa is essentially a social organisation, designed to bring mutually congenial people together. *Ibid.*, Examination of the Bulletin will show thriving Mensa groups in every geographic region. **1971** *Sear* (Manch. Branch Brit. Mensa) Nov., Mensa is a social club with the entry requirement that members have an IQ higher than 98% of the general population. **1972** *Times* 2 Nov. 9/4 All participants have mensa minds that have read Wittgenstein. **1972** W. GARNER *Ditto, Brother Rat!* vi. 46 Mensa is a sort of high IQ élite. Anyone can try their tests. Very few pass.

‖ **mensch** (menʃ). Also **mensh**. [Yiddish, a. G. *mensch* person.] A person of integrity or rectitude; one who is morally just, honest, or honourable.

1953 S. BELLOW *Adventures A. March* 43, I want you to be a *mensch*. **1959** H. PINTER *Birthday Party* III. 55 You'll be a mensch... You'll be a success. **1968** L. ROSTEN *Joys of Yiddish* 234 The key to being 'a real *mensh*' is nothing less than—character: rectitude, dignity, a sense of what is right, responsible, decorous. Many a poor man, many an ignorant man, is a *mensh*. **1970** *New Statesman* 30 Oct. 556/1 Mr Nixon is seen as an essentially decent man,..but not as a *mensch* on the scale of Roosevelt, Eisenhower, Kennedy. **1972** *New Yorker* 24 June 26 What is a *mensch*?.. It means you're a substantial human being.

mensh (menʃ). Also **mench**. Colloq. abbrev. of MENTION *sb.* and *v.*; freq. in phr. *don't mensh*, = *don't mention it* (see MENTION *v.* 1 c).

1937 in PARTRIDGE *Dict. Slang* 517/1. **1955** 'G. CARR' *Corpse at Camp Two* iii. 40 'Pray forgive my tactlessness.' .. 'Don't mench, sir.' **1961** J. DAWSON *Ha-Ha* ii. 31 'I'm sorry to hear it.' 'Thank you, not at all, don't mensh.' **1968** C. WATSON *Flaxborough Crab* iii. 35 'Thank you very much.' 'Don't mensh.' **1973** *Times* 27 Apr. 4/3 One little boy asked if we could 'have a mensh [mention] for Aunty Dora who fell down the stairs and broke her leg'. **1974** F. NOLAN *Oshawa Project* iii. 24 'Thanks, Lucky.' 'Don't mensh, don't mensh,' Luciano said.

Menshevik (me·nʃĭvik), *a.* and *sb.* [a. Russ. *men'shevik*, f. *mén'she*, compar. of *mályĭ* little. The Russ. pl. *men'sheviki* has been used by some English writers.] **A.** *adj.* Of, pertaining to, or characteristic of, the Mensheviks or Menshevism.

1907 [see *BOLSHEVIK B. *adj.*]. **1919** J. REED *Ten Days that shook World* iii. 47 Said the Menshevik *Dien*, 'The Government ought to defend itself and defend us.' *Ibid.* iv. 91 Raising his voice to a shout he [*sc.* Khintchuk] read the Menshevik declaration. *Ibid.* viii. 204 The familiar faces of the Menshevik and Socialist Revolutionary intellectuals. **1920** *Glasgow Herald* 14 Oct. 9 Martov (Zedarbaum), who headed the Menshevik opposition when Lenin broke up the Russian Social Democratic Party in 1903. **1923** E. A. ROSS *Russ. Soviet Republic* 323 The imprisonment of the Menshevik members of the Moscow printers' union. **1971** D. SMITH *Russia of Tsars* viii. 128 The Petrograd Soviet was equally hesitant... Alexander Kerensky, its Menshevik vice-president,..used his influence. **1975** *Times Lit. Suppl.* 4 July 740/5 When accused..of holding menshevik positions, he replies..that revolutionary virginity is not worth preserving at the price of inaction.

B. *sb.* A member of the political group or party forming the smaller part of the Russian Social-Democratic Party after the split with the Bolsheviks in 1903 and denounced as counter-revolutionaries after the 'October' Revolution of 1917. Cf. *BOLSHEVIK *sb.* Also *transf.* and *fig.*

1917 [see *BOLSHEVIK *sb.*]. [**1919** J. REED *Ten Days that shook World* p. xiv, *Russian Social Democratic Labour Party*. Originally Marxian Socialists. At a party congress held in 1903, the party split, on the question of tactics, into two factions—the Majority (Bolshinstvo) and the Minority (Menshinstvo). From this sprang the names 'Bolsheviki' and 'Mensheviki'—'members of the majority' and 'members of the minority'.] **1923** E. A. ROSS *Russ. Soviet Republic* 322 The Mensheviks can get no paper, which is a government monopoly, for pamphlets or leaflets at election time. **1926** *Contemp. Rev.* Sept. 274 He was an outsider—a 'menshevik' (the 'minority' man). **1935** N. MITCHISON *We have been Warned* I. 67 Idlers, parasites, mensheviks, defeatists. **1973** *Listener* 1 Feb. 135/3 The Provisionals..by playing the Bolsheviks to the Officials' Mensheviks—though not in ideology of course—have indeed become the party of the majority.

Menshevism (me·nʃĭviz'm). [a. Russ. *men'shevizm*: see *MENSHEVIK *a.* and *sb.*] The doctrines and practices of the Mensheviks.

1920 *Glasgow Herald* 14 May 9 Communism as it is offered to Trans-Caucasia has assumed the form of Menshevism. **1926** *Contemp. Rev.* Sept. 275 Marx would prove it: but that would be 'Menshevism'. **1928** *Observer* 1 July 9/2 Comrade Trotsky's theory of permanent revolution is a variety of Menshevism.

Menshevist (me·nʃĭvist). [a. Russ. *men'shevist* (now disused) *MENSHEVIK *a.* and *sb.*] A Menshevik; a supporter of Menshevism. Also *attrib.* or as *adj.*

1919 *Times Lit. Suppl.* 14 Aug. 432/3 A pleasing description of the Menshevists. **1926** *Contemp. Rev.* Sept. 274 He was known to return in all intricate cases to his menshevist fallacies. **1931** *Times* 7 Mar. 12/2 The Menshevists had always been opposed to the violent overthrow of the Bolshevists. *Ibid.*, Documentary evidence of the alleged Menshevist propaganda all, strangely enough, got destroyed.

Men's Lib.: see *MAN *sb.*[1] 23.

‖ **mens rea** (menz rī·ǎ). *Law.* [mod.L., lit. 'guilty mind'.] The criminal state of mind accompanying an act which condemns the perpetrator of the act to criminal punishment; criminal intent.

1861 LEIGH & CAVE *Crown Cases Reserved* (1866) 53 The *mens rea* is an essential ingredient in every offence. **1914** D. A. STROUD *Mens Rea* i. 14 There is no real contradiction in describing a mere absence of mind as a *mens rea* or guilty mind. **1955** *Times* 15 July 11/6 If the customer used it to apply to some other dress she would be guilty of fraud. One is not anxious to multiply criminal offences in which there is no *mens rea*. **1959** *Chambers's Encycl.* IV. 241/1 Drunkenness may render the accused incapable of forming the specific intent necessary to commit the crime—when *mens rea* will not be present and the accused must be acquitted. **1965** *Listener* 16 Dec. 1006/2 The compulsory death penalty for murder, which led to a special interpretation of *mens rea*, has been abolished. **1972** *Police Rev.* 10 Nov. 1446/3 The prosecution has failed to adduce sufficient evidence of the presence of *mens rea* (or guilty knowledge).

men's room: see *MAN *sb.*[1] 23.

‖ **mens sana in corpore sano.** Lat. phr. (occurring in Juvenal *Satires* x. 356): a sound mind in a sound body, esp. regarded as the ideal of education. Also *ellipt.*, as *mens sana*.

c **1605** J. HARINGTON or S. DANIEL *Prayse of Private Life* in J. Harington *Lett. & Epigrams* (1930) iii. 329 Desyrynge of God noe more than the Satericke Poet wished, *Mens Sana, in Corpore Sano*. **1629** in *Camden Misc.* (1854) III. 10 To beare *mens sana in corpore sano*, a valient heart in such a bodie..wch hath never bene debilitated. **1749** CHESTERFIELD *Let.* 2 Oct. (1932) IV. 1415 *Mens sana in corpore sano*, is the first and greatest blessing. **1824** J. WIGHT *Mornings at Bow Street* 55 The General angrily declared he was altogether *mens sana in corpore sano*. **1851** GEO. ELIOT *Let.* 11 Sept. (1954) I. 359, I am really strong and healthy and hope to bring a mens sana in corpore sano to London. **1966** 'K. NICHOLSON' *Hook, Line & Sinker* xiii. 159 Cold bath first thing: *mens sana*, and all that. **1967** S. JOHNSON *Gold Drain* ii. 22 [They] accused him of suffering from the effects of a public-school education, from the *mens sana* approach. **1973** *Listener* 23 Aug. 260/1 One lady, supporting *mens sana in corpore sano* against Jo Grimond, wrote that Christ must have kept himself fit.

‖ **mensur** (me·nsūr). [G.] In Germany, a fencing duel between students fought with partially blunted weapons.

1911 L. KNOWLES *Day with Korps-Students in Germany* (ed. 2) 6 My visit was not to admire Heidelberg, but to see my *Korps*-student friends,..for on the morrow there was to be a *Mensur*, or fighting-bout. *Ibid.* 4 Even to Germans, these *Mensuren*, or lighter student-duels,..have their humorous side. **1960** *Times* 17 Sept. 9/7 The *Mensur* or duel..is, however, frowned upon by officialdom. **1967** A. ARENT *Gravedigger's Funeral* xiii. 221 That final accolade of Teutonic narcissism, the face scars or *Mensur* marks of the military caste. **1970** 'M. HEBDEN' *Mask of Violence* (1971) xxv. 219 Anarchical young men of..the older universities, with the *Schmisse*—the duelling scars—on their cheeks, were already shouting the old *Mensur* wish before the commencement of a fight: '*Waffenschein!*' **1972** *House & Garden* May 142/4 Over the river, Hirchgasse where the 'mensor' [*sic*] duel is still fought.

mensuralist. Add: **2.** An advocate of a style of plainsong in which the rhythm depends on using notes of fixed length. Also *attrib.*

1905 *Grammar of Plainsong* I. i. 3 The 14th century brought a further falling off in the chant. At that time.. the harmonists and mensuralists were making their mark even on the Church's song, and destroying its rhythm. **1911** *Catholic Encycl.* XII. 147/1 The mensuralists.., with Decheurens as their principal representative, hold that the notes of plain chant are subject to strict measurement. **1929** *Music Q.* XV. 18 Even the mensuralist.. expressed the opinion that the Gregorian composers probably were not clearly conscious of writing metric arrangements. **1954** *Grove's Dict. Mus.* (ed. 5) VI. 820/1 The Mensuralists, in so far as they agree among themselves, hold that the notes in early medieval manuscripts are not..of approximately equal duration, but that they represent longs and shorts, the longs always being twice the duration of the shorts. **1959** *Times Lit. Suppl.* 13 Feb. 83/4 For more than fifty years many books and articles have been published expounding the 'mensuralist' view with great ingenuity. **1964** E. CARDINE *Is Gregorian Chant Measured Mus.?* 10 The *a priori* mensuralist appears perfectly clearly from the Introduction [of Vollaerts' book] onwards.

men's wear, menswear: see *MAN *sb.*[1] 23.

mental, *a.*[1] and *sb.* Add: **A.** *adj.* **1.** (Further examples.)

1843 MILL *Logic* II. vi. iv. 506 These differences of mental susceptibility..may be consequences of the previous mental history of those individuals. **1874** J. SULLY *Sensation & Intuition* i. 7 To understand fully all the facts of a single mental history. **1879** W. JAMES *Coll. Ess. & Rev.* (1920) 139 Ever some vital factor of our mental life will rebel and refuse to be dragged the same way with the rest. **1883** F. H. BRADLEY *Princ. Logic* 182 If we say, 'I wish S–P were a fact',..the judgment..is concerned with nothing but my mental attitude. **1890** W. JAMES *Princ. Psychol.* I. ii. 48 A most interesting effect of cortical disorder is *mental blindness*. This consists not so much in insensibility to optical impressions, as in *inability to understand them*. *Ibid.* II. xx. 205 Sounds certainly play a far more prominent part in the mental life of the blind than in our own. **1897** C. H. JUDD tr. *Wundt's Outl. Psychol.* 326 The law of mental growth is as little applicable to all contents of psychical experience as any other psychological law of development. **1919** L. M. TERMAN *Measurement of Intelligence* i. 6 The large majority of these belong to the moron grade; that is, their mental development will stop somewhere..between 9 and 12. **1921** B. RUSSELL *Analysis of Mind* i. 25, I should say that images belong only to the mental world. *Ibid.* iii. 68 Since Kant it has been customary to recognize three great divisions of mental phenomena, which are typified by knowledge, desire and feeling. *Ibid.* vi. 109 When you realize that you are glad to meet him, you acquire knowledge of a mental fact. *Ibid.* viii. 141 The sensation, as a mental event, will consist of awareness of the colour. **1932** *Mind* XLI. 137 A man whose whole life had been devoted to thought, and whose intense mental energy came out all the more clearly as his bodily powers declined. **1936** *Discovery* Jan. 22/2 Only in this way is it possible to render intelligible the mental attitude of the Chinese to other peoples. **1949** G. RYLE *Concept of Mind* i. 16, I shall probably be taken to be denying well-known facts about the mental life of human beings. *Ibid.* vi. 163 The argument, then, that mental events are authentic..must be rejected. **1949** R. G. SIMPSON *Fund. Educ. Psychol.* ii. 22 Growth of mental functions is not as easily traced as that of physical abilities, primarily because mental growth is composed of many characteristics. **1958** K. LOVELL *Educ. Psychol. & Children* x. 127 The effects of training would be transferred from one mental function to another of the same type. **1966** *Guardian* 16 May 3/7 The students at Lancaster confess to a mental block about using nuclear weapons in any circumstances. **1968** J. A. SHAFFER *Philos. of Mind* i. 5 The only evidence we have for the existence of these faculties is the existence of the very mental phenomena they were intended to explain. **1969** C. HODDER-WILLIAMS *98·4* iv. 41 Having a mental block about one's past is all right... Security work makes a virtue out of forgetting things. **1970** K. CAMPBELL *Body & Mind* vi. 113 The new view allows, indeed requires, that mental states be causes. *Ibid.* 125 Mental objects such as images and dreams..resist being brought within the fold of Materialism. **1973** A. KENNY *Wittgenstein* viii. 142 When we hear a sentence in a language we know, there are mental events—feelings, images, etc.

b. (*a*) *mental breakdown, deficiency, derangement, disease, disorder, handicap, illness, incapacity, retardation, subnormality*, etc.: general terms indicating temporary or permanent impairments of the mind, due to heredity, birth injury, environment, or accident, which usually need special care; *mental health*, health of mind as distinct from physical health; *mental hygiene*, mental health; measures directed towards the preservation or improvement of mental health;

(*b*) *mental case, defective, incapable, patient*, etc.: persons suffering from some kind of mental impairment; persons under medical care for mental illness;

(*c*) *mental home, hospital, institution, ward*, etc.: places where those with mental disorders are confined or treated;

(*d*) *mental nurse, specialist*, etc.: persons specializing in the treatment or care of those with mental disorders.

(*a*) **1794** H. L. PIOZZI *Brit. Synonymy* II. 6 Phrenzy, madness and distraction are the poetical expressions of what we call mental derangement, or disordered spirits, in elegant conversation. **1833** MILL *Lett.* (1910) I. 38, I..am so far from being in better mental health than yourself, that I need sympathy quite as much. **1839** POE *Fall House of Usher* in *Burton's Gentleman's Mag.* Sept. 150 And now, some days of bitter grief having elapsed, an observable change came over the features of the mental disorder of my friend. **1848** GEO. ELIOT *Let.* c 14 May (1954) I. 261 Study *mental hygiene*—take long doses of 'dolce far niente'. **1856** *Educ. of Imbecile* (Home for Invalid & Imbecile Children, Edinburgh) 11 His appearance and bearing gave evidence of great mental deficiency. **1869** GEO. ELIOT *Let.* 21 Sept. (1956) V. 56, I have such a horror of a mental breakdown. **1902** B. RUSSELL *Let.* in A. Wood *B. Russell* (1957) viii. 79 Since I finished my book, I have devoted myself to what you [*sc.* Beatrice Webb] would call mental hygiene, with good results so far. Beyond reading a mathematical MS. of Whitehead's, I have done no work for the last fortnight, but have spent my whole days out of doors basking in the return of summer. **1904** *Lancet* 27 Aug. 598/2 Scientific views regarding mental disease have..been undergoing great changes. *Ibid.* 17 Sept. 838/1 Those cases of mental incapacity arising from incipient or oncoming insanity. **1907** [see *BREAK-DOWN 1 c]. **1908** *Lancet* 12 Sept. 812/2 It was difficult..to secure effective treatment for early undeveloped cases of mental disorder. **1913** *Act* 3 & 4 *Geo. V* c. 28 (*title*) Mental Deficiency Act. **1914** W. B. DRUMMOND tr. *Binet & Simon's Mentally Defective Children* ii. 16 Some heads of schools..have fixed to almost a year the mental retardation of the child as compared with normal children of the same age. **1921** C. BURT *Mental & Scholastic Tests* ii. 163 The central problem of this memorandum —the line of demarcation for mental deficiency. **1946** C. P. BLACKER (*title*) Neurosis and the mental health services. **1946** *Amer. Jrnl. Psychiatry* CIII. 323/1 At the present time it cannot be said that we *cure* any of the more important and more fixed mental diseases. **1957** S. H. KRAINES *Mental Depressions* xv. 422 The goal of such mental hygiene is to remove and correct those unhealthy attitudes which add to the physically based tension. **1960** *Guardian* 16 Mar. 4/6 The most important change in terminology made by the new Act [*sc.* Mental Health Act, 1959] is that the expressions 'mental deficiency' and 'mental defectives' should be abolished... 'Mental disorder' is introduced as a new term covering all forms of mental ill health. **1962** E. CLEAVER in A. Dundes *Mother Wit* (1973) 10/2 The continued application of these judgments is the cause of an untold amount of mental illness and frustration. **1964** J. TIZARD *Community Services for Mentally Handicapped* I. ii. 18 There is a statutory obligation placed upon the local authorities in Britain to ascertain cases of mental subnormality..and to make provision for them. **1970** 'T. COE' *Wax Apple* (1973) vi. 50, I automatically *had* felt superior. After all, I'd never had a mental breakdown. **1970** L. J. KARMEL *Measurement & Evaluation in School* III. vii. 182 Some states will use 75 or even 80 as the cut-off point for mental retardation. **1971** *Rand Daily Mail* 4 Sept. 2/1 Nothing he had heard in court suggested that Muller has ever had a mental disorder. **1971** *Guardian* 18 Oct. 10/3 Mental illness is an *illness* needing therapeutic treatment... Mental handicap..is an innate condition for which there is no cure. **1972** R. QUILTY *Tenth Session* 58 A lifetime in mental-health nursing. **1973** *Black World* May 7/1 A crucial matter here is the role The Black Church can play in our mental hygiene and group advancement. **1974** *Guardian* 28 Jan. 6/7 The patient is no longer suffering from mental illness and..[should] be discharged.

(*b*) **1899** *Pop. Sci. Monthly* Apr. 747 (*title*) Mental defectives and social welfare. **1904** *Lancet* 17 Sept. 838/1 The Home Secretary..has at last been impressed with the total inadequacy of the provision made for..mental incapables. **1908** *Ibid.* 12 Sept. 813/2 Dr W. C. S. Clapham ..had many suspected mental cases sent to his outpatient department. **1913** E. MEYNELL *Life F. Thompson* 279 Many a time I've asked him to have his bit of lunch with me and the other 'mental'—O yes, she's a mental case, as I may have told you. **1922** W. R. INGE in *Edin. Rev.* July 34 The Eugenics Education Society..actively supported the Act for the compulsory segregation of mental defectives. **1938** *Brit. Jrnl. Psychol.* July 37 The implications of Wiersma's results obtained from mental patients have therefore been shown to hold for normal subjects. **1944** H. G. WELLS '*42 to '44* 182 The highly bred dogs are mostly physical and mental defectives. **1970** 'T. COE' *Wax Apple* (1973) iv. 33 The majority of mental patients who have been hospitalized once will be hospitalized several times more. **1973** 'D. SHANNON' *No Holiday for Crime* (1974) iv. 61 If you're going to say a mental patient, I don't think it was that kind of hospital.

(*c*) **1898** H. MORTEN *Compl. Syst. Nursing* 249 A year's training in a general hospital, and two years' training in a mental hospital, would be the proper scheme to turn out the best mental nurse. **1908** *Lancet* 26 Dec. 1943/1 (*heading*) Mental observation wards in Glasgow Parish Hospital. **1928** *Daily Mail* 25 July 8/3 One in ten of the people who reach the age of 40 in the State of New York will already have passed some time in a mental home or institution. **1932** KIPLING *Limits & Renewals* 151 They pushed him into a Mental Home, And that is like the grave. **1958** 'N. SHUTE' *Rainbow & Rose* v. 182 He had three young children and disliked the thought of bringing them up in the surroundings of a mental home. **1964** *Penguin Bk. Austral. Ballads* 18 The discharged mental-hospital patient..with a certificate to *prove* that he was sane. **1970** 'T. COE' *Wax Apple* (1973) ii. 17 The Midway was a haven for people recently out of mental institutions. **1972** R. BLOCH *Night-World* (1974) viii. 49 The mental wards were like a prison, only worse. **1974** N. FREELING *Dressing of Diamond* 74 We're looking at every affair..linked to mental homes, prison releases..psychopathic anything. **1974** *Guardian* 24 Jan. 15/8 The 'oriental girl'..had been led down from her ledge and taken off to mental hospital.

(*d*) **1898** [see sense 1 b (*c*) above]. **1922** *Lancet* 5 Aug. 290/1 The two duties of mental nurses. **1932** *Daily Tel.* 25 Apr. 9/3 Dr. Caton, the eminent mental specialist,.. has completed his findings of the state of Lt. Massie's mind at the time.

c. *colloq.* Mentally disordered or defective. Also in phr. *to go mental*, to become mentally disordered.

1927 D. L. SAYERS *Unnatural Death* iv. 41, I gather she was a little queer towards the end—a bit mental, I think you people [*sc.* nurses] call it? **1930** J. B. PRIESTLEY *Angel Pavement* ii. 68, I don't care if Mr. Dersingham goes mental, we're going to be lucky. **1958** 'N. SHUTE' *Rainbow & Rose* iv. 151 Everybody goes a bit mental in a war. **1959** S. GIBBONS *Pink Front Door* iv. 48 They had the same round eyes, only his hadn't got that almost mental look. **1973** 'J. PATRICK' *Glasgow Gang Observed* iii. 34 They must be mental... Shit-bags the lot o' them. **1974** N. FREELING *Dressing of Diamond* 65 There's somebody mental floating about.

2. Also, taking place in the mind. (Further examples.)

1676 LOCKE *Jrnl.* 26 Aug. in *Ess. Law Nature* (1954) 278 The thing to be believed was a proper object of faith and not of sense, a mental proposition, viz. that the God of the three children was the true God. **1843** MILL *Logic* I. I. iii. 66 His mental image..of the sun, and his idea of God, are thoughts. **1865** —— *Exam. Hamilton's Philos.* vi. 72 The belief is without understanding, for we form no mental picture of what we believe. *c* **1873** W. JAMES in R. B. Perry *Tht. & Char. W. James* (1935) I. 499 Under each and all of these formulations lurks the same mental act, that of insisting that things do not exist, so to speak, only once, but are in a manner duplicated. **1880** —— *Coll. Ess. & Rev.* (1920) 161 This complex aggregate of afferent feelings..renders absolutely precise and distinctive our mental image of the exact strength of movement to be made. **1882** L. STEPHEN *Sci. of Ethics* vi. 231 And thus the same words may call up the mental images which would be generated in the most and in the least sympathetic witness. **1889** Mental arithmetic [see *GUESSING *vbl. sb.*]. **1921** B. RUSSELL *Analysis of Mind* i. 19 The whole question of the relation of mental occurrences to objects grows very complicated. *Ibid.* vi. 109 'Introspection' is supposed to furnish data for knowledge of our mental processes. **1925** J. RIVIERE et al. tr. *Freud's Coll. Papers* IV. 14 In the psychology which is founded on psycho-analysis we have accustomed ourselves to take as our starting-point the unconscious mental processes. **1932** *Mind* XLI. 37 There are four corresponding 'mental ex-

periences' (παθήματα ἐν τῇ ψυχῇ, 511 D). **1935** C. BURT *Subnormal Mind* i. 41 The reading test and the mental arithmetic test..will yield a fair estimate of the child's educational level. **1944** J. S. HUXLEY *On Living in Revolution* iv. 56 The deliberate 'mental operation' of psychoanalysis or other form of psychotherapy. **1949** G. RYLE *Concept of Mind* 8 The logical categories in terms of which the concepts of mental..operations have been co-ordinated have been wrongly selected. *Ibid.* vii. 200 Stomach-aches ..have physiological attachments which threaten to sully the purity of the brook of mental experiences. *Ibid.* viii. 254 In short, there are no such objects as mental pictures. **1963** R. CARNAP in P. A. Schilpp *Philos. R. Carnap* 8 All mental processes are intimately connected with the brain. **1971** P. A. CHILTON tr. *Piaget & Inhelder's Mental Imagery in Child* p. xiii, Having attempted to analyse some of the characteristics of perceptual development, it was necessary to go on to tackle the question of the evolution of mental images. **1972** W. C. COE *Challenges Personal Adjustment* iv. 76 Mental telepathy is the transfer of thought from one person to another without overt communication.

5. Special collocations: **mental age**, the degree of mental development of a person, expressed as the age at which a similar level is attained by an average person; **mental chemistry** *Psychol.*, J. S. Mill's term for the psychological processes by which complex ideas, sensations, etc., are formed from an aggregate of simple ones; **mental cruelty**, conduct which inflicts suffering on the mind of another person, esp. *U.S.* as constituting grounds for legal separation or divorce; **mental healing**, healing effected solely by the mind of the healer; so *mental healer*; **mental ratio** = *intelligence quotient* (s.v. *INTELLIGENCE *sb.* 8); **mental set**, the set (SET *sb.*[1] 12) or predisposition of the mind which governs reactions to stimuli; **mental test** = **intelligence test*; so *mental tester, testing*; **mental year**, the average mental attainment of each year of growth, used as a unit of measurement of mental development.

1912 *Pedagogical Seminary* XIX. 189 Table I. Mental Ages..showing the chronological and mental age distribution of the two thousand public school children graded by the Binet tests. **1919** L. M. TERMAN *Measurement of Intelligence* iii. 40 A mentally defective child of 9 years may have a 'mental age' of only 4 years, or a young genius of 9 years may have a mental age of 12 or 13 years. **1937** 'M. INNES' *Hamlet, Revenge!* III. iii. 259 His mental age's about eight. **1972** J. AIKEN *Died on Rainy Sunday* 88 Really Ellie *is* a child, she thought; in some ways her mental age can't be more than twelve or thirteen. **1843** MILL *Logic* II. vi. iv. 503 These therefore are cases of mental chemistry: in which it is proper to say that the simple ideas generate, rather than that they compose, the complex ones. **1890** W. JAMES *Princ. Psychol.* II. xx. 202 Discrimination, association, [etc.]..are quite capable of giving us all the space-perceptions we have so far studied, without the aid of any mysterious 'mental chemistry'. **1929** W. McDOUGALL *Mod. Materialism* v. 119 J. S. Mill's use of the expression 'mental chemistry' had suggested some analogy between such mental and such physical syntheses. **1948** R. S. WOODWORTH *Contemp. Schools Psychol.* (ed. 2) iii. 44 This idea of a 'mental chemistry' reappeared from time to time in later psychological theories. **1928** *Harper's Mag.* July 159/1 The Nevada laws..necessitate a minimum of fraud... The judges demand very slight evidence to prove mental cruelty. A woman may obtain her freedom on the ground that her husband has..told her to go to hell once too often. **1936** M. ALLINGHAM *Flowers for Judge* vii. 125 Not a care in the world except Mr. Brande's neglect and mental cruelty to her. **1967** R. S. MASTER *Elem. Psychiatry* xx. 398 The petitioner has to prove..that the spouse can differentiate between right and wrong and understand the nature of the mental cruelty. **1968** *Globe & Mail* (Toronto) 17 Feb. 29/1 The third wife of wealthy playboy John Jacob Astor III was granted a divorce yesterday on grounds of mental cruelty. **1920** in WEBSTER, Mental healer. **1957** O. NASH *You can't get there from Here* 67 The characters in many of our popular songs are fair game for the nearest mental healer. **1970** T. S. SZASZ *Ideology & Insanity* (1973) xiii. 240 The precise activities of the various 'mental healers' are rarely defined by their practitioners. **1888** L. E. WHIPPLE (*title*) Mental healing. **1902** W. JAMES *Varieties Relig. Experience* iv. 121 Their savage and primitive philosophy of mental healing. **1924** GALSWORTHY *White Monkey* I. xii. 100 No need of any of these vitamins, false teeth, mental healing. **1921** C. BURT *Mental & Scholastic Tests* ii. 151 If a child's mental age be divided by his chronological age, the quotient will state what fraction of ability the child actually possesses... This fraction may be termed..the child's 'intelligence quotient' or, more euphoniously.., his 'mental ratio'. **1927** A. HUXLEY *Proper Stud.* 67 A mental ratio is obtained by dividing the child's mental by his chronological age. **1934** *Brit. Jrnl. Psychol.* Jan. 312 Not all candidates with high mental ratios succeed. **1913** E. L. THORNDIKE *Educ. Psychol.* II. xii. 356 In the case of alternative systems of bonds there is then often an inhibition for a time, reducing to zero as the two systems of bonds get organized in connection with two systems of mental sets. **1922** R. S. WOODWORTH *Psychol.* iv. 72 Much used also are 'adjustment' and 'mental set', the idea here being to liken the individual to an adjustable machine which can be set for one or another set of work. **1934** H. C. WARREN *Dict. Psychol.* 164/2 *Mental set*,..the attitude or determination of an individual before receiving an expected stimulus or with reference to a problem or general situation. **1953** *Jrnl. Exper. Psychol.* XLVI. 50 The aspect of directed thinking commonly called 'mental set'. **1963** J. LYONS *Structural Semantics* ii. 35 He [*sc.* the hearer] must be thought of as being in a certain state of 'expectancy' (or 'mental set'), in which he is disposed to hear certain units rather than others. **1890** J. M. CATTELL in *Mind* XV. 373 Psychology cannot attain the certainty and exactness of the physical sciences, unless it rests on a foundation of experiment and measurement. A step in this direction could be made by applying a series of mental tests and measurements to a large number of individuals. **1934** H. C. WARREN *Dict. Psychol.* 165/1 Mental test. **1958** M. ARGYLE *Relig. Behaviour* vi. 58 A child's mental age is the average chronological age of children showing the same degree of intelligence, as measured by mental tests. **1972** L. S. HEARNSHAW in Cox & Dyson *20th-Cent. Mind* I. vii. 234 Francis Galton..attempted.. to devise techniques for measuring intelligence. These techniques were christened 'mental tests' by the American psychologist, J. M. Cattell, in 1890. **1952** *Brit. Jrnl. Psychol.* Feb. 23 When the subjects are in full sympathy.. but the mental tester rarely has such a favourable situation. **1959** *Chambers's Encycl.* XII. 655/1 Mental testing began with the testing of intelligence (i.e. of general intellectual capacity). **1921** C. BURT *Mental & Scholastic Tests* ii. 147 The estimate for the intelligence of every child has been converted into terms of mental years. **1942** F. J. SCHONELL *Backwardness in Basic Subjects* iv. 74 Rene C. was, in arithmetic, almost two mental years ahead of her achievements in reading.

B. *sb.* Restrict †*nonce-uses* to senses in Dict. and add: **c.** *colloq.* A mentally-deranged person; a mental patient.

1913 [see 1 b (*b*) above]. **1937** PARTRIDGE *Dict. Slang* 517/1 *Mental*, a person mentally deranged, mad. **1942** BERREY & VAN DEN BARK *Amer. Thes. Slang* § 529/16 Mental, nut, psycho, psychot, *a psychopathic case*. **1973** F. DE FELITTA *Oktoberfest* (1974) xiv. 162 'What's to prevent him from going?' 'He is a mental.'

mentalism. Restrict *rare* to sense 1 in Dict. and add: **2.** Esp. the theory that physical and physiological phenomena are ultimately only explicable in terms of a creative and interpretative mind.

1917 A. S. PRINGLE-PATTISON *Idea of God* xx. 392 The position is open..to the general objections which have been brought against Monadism and Mentalism. **1932** *Times Lit. Suppl.* 28 Apr. 314/4 Scientists who repudiate mechanism only to involve themselves in thoroughpaced mentalism. **1954** *Essays in Crit.* IV. 235 But MacNeice has tried mentalism and so raged against it..that he nearly knocked his brains out with rhyme. Therefore we must settle for monadism. **1964** *Language* XL. 124 Linguists who conceive of their science as a discipline.. often pride themselves on their freedom from mentalism. But freedom from mentalism is an inherent feature of the taxonomic conception of linguistics. **1972** *Ibid.* XLVIII. 418 Writing grammars..was motivated for Boas by a theoretical concern, a 'mentalism' derived partly from the German tradition of Herder, Humboldt, [etc.]. **1973** *Word* 1970 XXVI. 93 We can arrive at a knowledge of the *signifié* independently of the particular *signifiant* under study, yet still avoid mentalism.

mentalist. Restrict *rare* to sense 1 in Dict. and add: **2.** (Further examples.) Also *attrib.* or as *adj.*

1927 B. RUSSELL *Outl. Philos.* xxvii. 303 Idealists in the technical sense, or mentalists, as Dr. Broad more appropriately calls them. **1933** L. BLOOMFIELD *Lang.* ix. 142 For the mentalist, language is the expression of ideas, feelings, or volition. **1966** *Philos.* XLI. 141 We must give a sense to this mentalist and substance vocabulary. **1972** *Language* XLVIII. 418 A mentalist tradition more recent than the 'Cartesian'.

mentalistic (mentăli·stik), *a.* [f. MENTALIST + -IC.] **1.** Of or pertaining to the processes of the mind or to processes of a similar nature. *rare.*

1962 *Listener* 29 Mar. 568/3 Other practitioners may turn it [*sc.* treatment of insanity] into a purely mentalistic and verbal technique, letting slip all the gains which psycho-analysis has achieved by its attention to physical processes.

2. Of or pertaining to mentalism (sense *2). Hence **mentali·stically** *adv.*

1917 A. S. PRINGLE-PATTISON *Idea of God* x. 191, I feel it to be important..to free the position I am defending from any supposed dependence on the Mentalistic doctrines which have often been used to support it. **1925** C. D. BROAD *Mind & its Place* xiv. 611, I class these two alternatives together under the name of 'Mentalistic Neutralism'. **1927** *Mod. Philology* Nov. 225 This theory.. makes the hitherto otiose mentalistic factors play an actual part in linguistic change. **1950** D. JONES *Phoneme* vii. 23 This is evidence in favour of a mentalistic view of the phoneme. **1965** N. CHOMSKY *Aspects of Theory of Syntax* i. 4 Hence, in the technical sense, linguistic theory is mentalistic, since it is concerned with discovering a mental reality underlying actual behavior. **1967** [see *identity theorist* (*IDENTITY 10 b)]. **1971** D. CRYSTAL *Ling.* 95 Very little of language..can be tested in the same way as material objects.., as mentalistically-inclined linguists have been quick to point out. **1973** *Word* 1970 XXVI. 150 The chapter on Syntax gives a traditional mentalistic statement of the function of the cases and moods.

mentality. Add: **3.** Mental character or disposition; outlook; kind or degree of intelligence.

1895 *Funk's Stand. Dict.*, *Mentality*..cast or habit of mind. **1911** BEERBOHM *Zuleika D.* viii. 132 Your words show up your good heart... Your mentality, too, is bully, as we all predicate. **1922** JOYCE *Ulysses* 132, I speak the tongue of a race the acme of whose mentality is the maxim: time is money. **1926** FOWLER *Mod. Eng. Usage* 348/1 'The mentality of the politician is a constant source of amazement to the engineer.' Twenty years ago, no-one would have written that. **1931** L. WATT *Future of Capitalism* ii. 23 It is useless to pretend that there will be anything but hostility between the partners in industry so long as this mentality persists. **1958** R. F. C. HULL tr. *Jung's Coll. Works* XI. vii. 480 Something of the kind happens as soon as we are confronted with the Eastern mentality.

mentalize, *v.* Add: In nonce-uses: **b.** To stimulate the mind. **c.** To realize mentally or invest with mental qualities. So **me·ntalized** *ppl. a.*, **me·ntalizing** *vbl. sb.*

1807 T. CAMPBELL *Let.* 6 July in W. Beattie *Life & Lett. T. Campbell* (1849) II. iv. 110, I sometimes call on the satirical Gifford..who *mentalises* me for a few moments, but the impression lasts too short. **1837** H. SMITH *Let.* 2 Oct. in A. Mathews *Mem. Charles Mathews* (1839) IV. xviii. 447 While others satisfied themselves with endeavouring to *embody* their originals, he made it his study to *mentalize* them. I am obliged to coin a word, but my meaning is, that while he surpassed all competitors in the mere mimicry of externals, he was *unique* in the subtlety, acuteness, and truth with which he could copy the *mind* of his prototype. **1927** D. H. LAWRENCE *Let.* 28 May (1962) II. 981 One emerges with a body all right— but a different one, perhaps, not so mentalised. **1935** W. DE LA MARE *Early One Morning* II. 404 Seeming stupidity may be due to a highly individualised way of mentalizing things. **1936** *Downside Rev.* LIV. 551 The body continues to produce its own characteristic actions, but along with these it produces a new action—thought. Body becomes mind, or rather mentalised. **1936** J. R. KANTOR *Objective Psychol. Gram.* v. 66 We react to printed words..as though they were words heard in conversation. .. It is precisely the mentalizing procedure that produces the confusion of responses to language activity. **1973** D. RAMSAY *Deadly Discretion* 76 Want to try mentalizing? .. First empty your mind... Then wait for me to send you the message.

mentally, *adv.* **b.** (Further examples.)

1894 *Psychol. Rev.* I. 425 It is..the mission of the mentally defective to give us..object lessons in psychology. **1934** H. S. SULLIVAN in H. S. Perry *Fusion of Psychiatry & Social Sci.* (1964) 9 Disorder of the emotional life of the mentally deranged is too striking to be ignored. **1964** J. TIZARD (*title*) Community services for the mentally handicapped. **1970** L. J. KARMEL *Measurement & Evaluation in School* III. vii. 182 The lowest 2 per cent of the population..could be justifiably described as mentally defective.

menthol. Add: *menthol cigarette*, a cigarette flavoured with menthol.

1952 M. TRIPP *Faith is Windsock* i. 10 He threw a packet of American menthol cigarettes across the hut. **1962** L. DEIGHTON *Ipcress File* xx. 131 She offered me one of those menthol cigarettes that taste like paint remover. **1969** O. HESKY *Sequin Syndicate* xi. 114 She..might have been an advertisement for some brand of menthol cigarette.

mentholated (me·nþǫlēⁱtėd), *ppl. a.* [f. MENTHOL + -ATE[3] + -ED[1].] Treated or impregnated with menthol; containing menthol.

1933 *Tobacco World* 1 May 13 (Advt.), And now say 'hello' to KOOL, B & W's mild mentholated new cigarette. **1962** P. MOYES *Death on Agenda* ix. 149 He was lighting up yet another mentholated cigarette. **1970** G. GREER *Female Eunuch* 259 One vaginal deodorant is..flavoured with peppermint... Others are mentholated.

menthone (me·nþōⁿn). *Chem.* [f. L. *menth-a* mint + -ONE.] 3-Methyl-6-isopropylcyclohexanone, $C_{10}H_{18}O$, an optically active cyclic ketone whose lævorotatory form is a liquid with an odour of peppermint and occurs in American peppermint, geranium, and other oils.

1889 ROSCOE & SCHORLEMMER *Treat. Chem.* III. v. 470 Menthone, $C_{10}H_{18}O$, is formed by heating menthol with sulphuric acid and potassium bichromate. **1929** *Org. Syntheses* IX. 53, *l*-Menthone can be prepared by the oxidation of rhodinol with a chromic-sulphuric acid mixture. **1969** W. TEMPLETON *Introd. Chem. Terpenoids & Steroids* iii. 53 Menthone..is found in oils of peppermint, geranium and pennyroyal, mainly in its (−)-form. **1970** [see *isovaleraldehyde* (*ISO- b)].

menticide (me·ntisəid). [f. L. *menti-, mens* mind + -CIDE 2.] A word coined by J. A. M. Meerloo to designate the undermining or destruction of a person's mind or will by 'psychological intervention and judicial perversion'; also in extended use. Cf. *BRAINWASHING.

1951 J. A. M. MEERLOO in *Amer. Jrnl. Psychiatry* Feb. 595/1 Such an organized system of psychological intervention and judicial perversion, in which a powerful tyrant synthetically injects his own thoughts and words into the minds and mouths of the victims he plans to destroy by mock trial, may well be called menticide. **1957** —— *Mental Seduction & Menticide* Pref., The modern words 'brainwashing', 'thought-control', and 'menticide' serve to provide some indication of the..methods by which man's integrity can be violated. **1973** *Black Panther* 20 Oct. 17/1 No treatise on prisons in 1973 can possibly be complete and up-to-date without dealing with the scientifically modernized versions of these conditions and practices as they are manifested in the forms of menticide and genocide.

mention, *sb.* Add: **2. f.** *Mil.* A commendatory reference made to a person in an official military dispatch (abbrev. of *mention in dispatches*).

1915 A. N. LYONS *Kitchener Chaps* 81 No, sir, there's no particular cop about these 'mentions'; only something for your pals to read. **1921** S. C. JOHNSON *Medal Collector* viii. 180 No more than one leaf, however, may be fixed to the ribbon even though the wearer may have been the subject of several mentions. **1958** M. DICKENS *Man Overboard* iii. 35 Other people came out of the war with Mentions and worthwhile gongs. **1964** T. WHITE tr. *Leullietté's St. Michael* 221 I've got the Médaille Militaire, the Croix de Guerre and seven mentions.

mention, *v.* Add: **1. a.** *to be mentioned in dispatches*: to receive a 'mention' (see *MENTION *sb.* 2 f). Also *transf.*

1915 A. HUXLEY *Let.* June (1969) 73 Poor Bob Gibson is killed. He was as good a soldier as he was a don. He was mentioned in dispatches. **1922** JOYCE *Ulysses* 449, I fought with the colours..and was disabled at Spion Kop and Bloemfontein, was mentioned in dispatches. **1961** PARTRIDGE *Dict. Slang* Suppl. 1182/2 *Mentioned in despatches*. To have one's name appear in a newspaper, a parish magazine, or even on a notice board: jocular. **1975** *Times* 27 Aug. 20/3 In proud and loving memory of Captain Jocelyn Fulke Dalrymple Radice, The Queen's Bays, mentioned in British and French Army dispatches, ..who died of wounds on August 27, 1944.

b. (Later examples.)

1863 G. MACDONALD *D. Elginbrod* III. III. ix. 158 He mentioned to Miss Talbot that he had been his guest that night. **1929** C. K. S. MONCRIEFF tr. *Proust's Cities of Plain* I. II. ii. 268, I mentioned to him that I had thought I heard him come upstairs.

c. (Earlier example.)

1841 LYTTON *Night & Morning* I. II. ii. 264 'I am going to leave your house, ma'am; and I wish to settle any little arrears of rent, &c.' 'Oh! sir! don't mention it,' said the landlady.

mento-. mento-meckelian, also as *sb.*, a small bone formed by the ossified end of Meckel's cartilage.

1892 J. A. THOMSON *Outl. Zool.* 449 The mento-meckelians seem to arise from two lower labial cartilages. **1925** J. S. KINGSLEY *Vertebr. Skeleton* 71 The other cartilage bone is a mentomeckelian, best known in Anura, at the symphysis of the lower jaw.

menton (me·ntǫn). *Anat.* [a. F. *menton* chin:— L. *mentum*.] = *GNATHION.

1937 *Amer. Jrnl. Physical Anthrop.* XXII. 483 Menton. .. By all authors, this is designated as the lowest point on the lower border of the mandible in the median plane. **1954** T. C. WHITE et al. *Orthodontics* ix. 165 The Mandibular Plane passes from the menton (lowest point on the cross-section of the mandibular symphysis)..to the lowest point on the angle of the mandible. **1968** [see *GNATHION]. **1974** *Nature* 8 Mar. 165/1 Six bilateral and four single roentgenographic landmarks were delineated. The single landmarks were: sella; anterior nasal spine; incisal point; menton.

‖ **mentri** (me·ntri). Also 9– **mantri.** [Mal. *mentri*, ad. Skr. *mantri* *MANTRI.] A title used in the Malay states for a minister. Also **mentri besar**, a title for the chief minister of a Malay state.

Mentri was used *spec.* of a Perak territorial chief in the 19th century.

1839 T. J. NEWBOLD *Pol. & Statistical Acct. Straits of Malacca* I. v. 239 The Mantris were the privy councillors to the Panghúlus, two in number. **1880** F. A. SWETTENHAM *Some Acct. Independent Native States Malay Peninsula* i. 7 He had..got the Mĕntri to assist him in farming the revenues of Krian, a Province claimed by the Mĕntri. **1900** W. E. MAXWELL in W. W. Skeat *Malay Magic* iii. 65 The grave was discovered..by workmen employed by the Mĕntri of Perak. **1907** F. A. SWETTENHAM *Brit. Malaya* vi. 118 The Malay chief of the Lărut district was styled the Mantri, and he was one of the four high officers of state. **1955** *Times* 1 Aug. 6/6 They would be the symbol of loyalty of all classes and of all races loyal to Malaya. 'I only hope that in a short time their highnesses will see fit to appoint as their mentris besar (chief ministers) elected members of state councils.' **1958** J. SLIMMING *Temiar Jungle* ii. 27 *Mentri* is a Malay title superimposed on the Temiar system of leadership. **1972** *Straits Times* (Malaysian ed.) 4 May 6/4 Malaysia and Indonesia will be closer, and grow better in education when a common spelling system is used by both governments, Mentri Besar Inche Ja'afar bin Hassan said yesterday.

menu. Add: **2.** (Further *transf.* examples.) Also *menu card* (earlier example).

1881 C. C. HARRISON *Woman's Handiwork* II. 124 A menu card, with gilded edge and lettering. **1922** JOYCE *Ulysses* 161 Maul her a bit. Then the next thing on the menu. **1940** H. G. WELLS *Babes in Darkling Wood* II. iii. 206 All the intelligent ones feel baffled at the menu of these degree courses. **1971** *Computers & Humanities* VI. 104 The musical symbols at the top of the workscope are the 'menu'. **1974** *Some Technical Terms & Slang* (Granada Television), *Menu*, a list of items read out at the start of a magazine programme.

‖ **menus plaisirs** (mənü plęzīr), *sb. pl.* [Fr., lit. 'small pleasures'; pocket-money.] Simple pleasures; small personal expenses or gratifications; fanciful or trifling objects bought with one's pocket-money. Also *attrib.*

1697 VANBRUGH *Relapse* I. ii. 8, I shall see you stand in damnable need of some Auxiliary Guineas, for you[r] *Menu Plaisirs*. **1779** H. WALPOLE *Let.* 14 Jan. (1904) X. 363 My *menus plaisirs*, a few sprinkled visits of charity from a few friends that remained in town. **1814** JANE AUSTEN *Mansf. Park* II. v. 114 He will have a very pretty income ..and as..he will still live at home, it will be all for his *menus plaisirs*. **1883** M. E. BRADDON *Golden Calf* II. xi. 293 Whatever honorarium he received for his work was expended upon his *menus plaisirs*. **1915** F. M. HUEFFER *Good Soldier* III. v. 193 He may have had five hundred pounds a year English, for his menus plaisirs. **1966** 'M. INNES' *Change of Heir* v. 40 'It's in her letter. The bit about..*menus frais*.' 'What's that?' 'The same as *menus plaisirs*. Pocket-money on the scale appropriate to an English gentleman.'

Meo (mī·o). Also 9 **Mewoh.** [Native name.] An Indian people of Rajasthan and the Punjab, whose religion is a blend of Hinduism and Islam; a member of this people. Also as *adj.* Cf. *MEWATI *sb.* and *a.*

1832 J. TOD *Ann. Rajast'han* II. 393 At this period.. the mountainous region, called Mēwat, was inhabited by a daring and ferocious banditti, called Mewohs, who pillaged in gangs. **1855** H. H. WILSON *Gloss. Judicial & Revenue Terms* 339/2 *Meo*, (?) H. A class of cultivators in the province of Dehli. **1873** E. BALFOUR *Cycl. India* (ed. 2) III. s.v., Meo, cultivators in the Delhi province, are a tribe of people inhabiting the low hills about Gurgaon. **1880** P. W. POWLETT in *Rajputana Gazetteer* III. 201 Meos do not marry into their own pál, or clan, but are lax about forming connections with women of other castes, whose children they receive into the Meo community. **1885** G. C. WHITWORTH *Anglo-Indian Dict.* 204/2 *Mewáti*, the name of a branch of the Jádon Rájputs;..they rule in Alwar, and are connected with the Meo tribe of the same country. **1896** W. CROOKE *Tribes & Castes N.-W. Provinces & Oudh* III. 487 Whatever their connection with the Minas may be, the Meos themselves pretend to be of Rājput descent. **1916** R. V. RUSSELL *Tribes & Castes Cent. Provinces India* IV. 235 Abu Fazl writes that the Meos were in his time famous runners, and one thousand of them were employed by Akbar as carriers of post. **1970** D. G. MANDELBAUM *Society in India* II. xxix. 558 This intensified Islamization is shown by the Meos, the dominant jati in some two thousand villages on the borderlands of Rajasthan and Punjab. *Ibid.* 559 The Meos have adopted more of Muslim practices in the last 17 years than they had in the previous 450 years.

Meo, var. *MIAO *sb.* and *a.*

meow (miau), var. MIAOW *int.* and *sb.*

1873 *Young Englishwoman* May 217/2 'Meow!' And an enormous cat came out. **1947** M. LOWRY *Under Volcano* vi. 166 Provocative nocturnal meows. **1956** H. G. DE LISSER *Cup & Lip* vii. 81 'What you really mean is that but for the help of the coolie..you couldn't have had Miss Ludford over here.' 'Meow, meow,' laughed Arthur.

mepacrine (me·păkrin, -īn). *Pharm.* [perh. f. ME(THYL + PA(LUDISM + A)CR(ID)INE.] A tricyclic base, $C_{23}H_{30}N_3OCl$, derived from acridine and usually administered in the form of its dihydrochloride dihydrate salt, which is a yellow crystalline compound formerly widely used in the treatment of malaria and now used mainly as an anthelmintic. Cf. *ATEBRIN, *QUINACRINE.

1943 *Lancet* 29 May 699/1 The only synthetic antimalarial compound of any real value so far in general use is mepacrine. **1956** W. SLIM *Defeat into Victory* xvi. 352 The jaundice yellow of all our complexions as a result of the daily dose of mepacrine. **1962** *Lancet* 29 Dec. 1387/2 A related drug, chloroquine, has proved superior to mepacrine in the suppressive treatment of malaria, rheumatoid arthritis, lupus erythematosus, and light eruptions. **1971** B. W. ALDISS *Soldier Erect* 74 The site where hundreds of BORs had pretended to swallow the morning mepacrine tablet and instead had ground it underfoot, happily risking malaria in the exercise of their own free will.

meperidine (mĕpe·ridīn). *Pharm.* Chiefly *U.S.* [f. ME(THYL + PI)PERIDINE.] = *PETHIDINE.

1947 *Jrnl. Amer. Med. Assoc.* 25 Jan. 243/2 In this country only 1 case of spontaneous addiction to meperidine hydrochloride is reported. **1962** J. H. BURN *Drugs, Med. & Man* x. 104 The disadvantage can however be countered by choosing one of the substitutes for morphine which have been made in the laboratory in the last twenty years. The first of these was meperidine (in Britain called pethidine). **1971** J. J. BURNS in B. N. La Du et al. *Fund. Drug Metabolism* xvii. 343 The rhesus monkey and the dog metabolize..meperidine and antipyrine much more rapidly than does man.

mephenesin (mĕfe·nĕsin). *Pharm.* [f. ME(THYL + PHEN(YL + CR)ES(OL + -IN[1].] A colourless crystalline compound, $CH_3 \cdot C_6H_4O \cdot CH_2CH(OH)CH_2OH$, which is used as a muscle-relaxant in the treatment of spastic, hypertonic, and hyperkinetic conditions, and as a tranquillizer.

1952 *Brit. Pharmaceutical Codex 1949* Suppl. 40 Mephenesin consists of 3-(2-methylphenoxy)propane-1:2-diol. **1955** *Sci. News Let.* 16 July 40/3 In the great majority of 109 cases [of cerebral palsy] treated before dental surgery with mephenesin,..the drug produced feelings of well-being and relaxation, and nervousness diminished or disappeared. **1962** J. H. BURN *Drugs, Med. & Man* xii. 127 Another study was made by other workers to test the claim that the drug mephenesin exerts a specific effect on subjective anxiety and feeling of tension, and also on objective manifestations of this state. **1972** *Neuropharmacology* XI. 733 Amino-oxyacetic acid did not protect rats against the audiogenic seizures that constitute the barbiturate abstinence syndrome whereas mephenesin ..did provide protection.

mephitical, *a.* Delete † and add later examples.

1930 D. L. SAYERS *Strong Poison* viii. 95 In one corner an anthracite stove, glowing red and mephitical, vied with a roaring gas-oven in another corner. **1933** *Punch* 6 Dec. 641/3 In the sphere of things political An atmosphere malign, mephitical, Infects the universe.

meprobamate (meprōu·bămēⁱt). *Pharm.* [f. ME(THYL + PRO(PYL + *car*)*bamate* (s.v. CARBAMIDE).] A colourless crystalline compound, $CH_3CH_2CH_2C(CH_2O \cdot CO \cdot NH_2)_2CH_3$, which is a mild tranquillizer used in the treatment of motion sickness, neuroses, and insomnia.

1955 *Dispensatory U.S.A.* (ed. 25) 1861/1 Meprobamate. Equanil (Wyeth). Miltown (Wallace).—This is 2-methyl-2-*n*-propyl-1,3-propanediol dicarbamate. **1962** A. HUXLEY *Island* xiii. 201 Mass opiates in the form of television, meprobamate, positive thinking and cigarettes. **1965** J. POLLITT *Depression & its Treatment* iv. 48 When tension is more prominent than depression, chlordiazepoxide, meprobamate or even amylobarbitone in average doses will help tide the patient over difficult phases. **1965** *Evening Standard* 13 Sept. 10/3 He had an envelope full of pep pills—amphetamine—in his left coat pocket and an envelope full of Equanils—meprobamate—in his right pocket. **1967** *Martindale's Extra Pharmacopoeia* (ed. 25) 391/2 Suicide attempts with meprobamate are not infrequent... There is a serious addiction risk. **1969** *Punch* 5 Mar. 340/3 The classic example was the test of the tranquilliser meprobamate, sold here under the brand names Miltown and Equanil. It had been one of the most profitable drugs ever marketed—but in the trials, the placebo proved just as effective.

mepyramine (mepi·rămīn). *Pharm.* [f. ME(THYL + PYR(IDINE + AMINE.] A crystalline substituted amine, $C_{17}H_{23}N_3O$, or its maleate, which is an antihistamine drug used in treating allergic conditions.

1949 *Lancet* 9 July 47/2 The histamine antagonist, N(2-dimethylaminoethyl)N(*p*-methoxybenzyl)2-aminopyridine maleate..later known on the continent as 'Neo-antergan' and in this country as 'Anthisan' or mepyramine maleate..was soon recognised as the most powerful and specific of the histamine antagonists. **1964** W. G. SMITH *Allergy & Tissue Metabolism* v. 61 One of these was histamine whose actions could be blocked by mepyramine. **1972** *Nature* 21 Apr. 369/1 Certain responses..are not antagonized by the classical antihistaminic drugs, such as mepyramine.

mer (mɔ̄ɹ). *Chem.* [f. POLY)MER.] The repeating unit of a polymeric molecule.

1936 W. O. CAROTHERS in *Trans. Faraday Soc.* XXXII. 49 Assume that the weight of the mer or unit is 100. **1947** R. L. WAKEMAN *Chem. Commercial Plastics* xx. 554 The word 'mer'..has..been used by Carothers to designate the individual unit of repetition in both addition and condensation polymers (as distinguished from the monomeric substances themselves from which the polymer is produced). **1952** *Jrnl. Polymer Sci.* VIII. 258 The slight variability in composition and chemical structure permitted..for high polymers results from the presence of end groups, occasional branches, variation in orientation of the mers (monomeric units), irregularity in the sequence of different types of mers (in copolymers), and sometimes other irregularities. **1967** MARGERISON & EAST *Introd. Polymer Chem.* i. 3 As soon as more than one mer is involved, a wide diversity of polymer structure and type becomes possible.

-mer (mɔɹ), terminal element repr. Gr. μέρος part, occurring in various chemical terms: orig. in *polymer* and *isomer*, and usu. in words denoting particular kinds of polymer (as *dimer, elastomer*) or isomer (as *epimer*).

meranti (mĕræ·nti). [Malay.] A hardwood timber produced by trees of the genus *Shorea*, belonging to the family Dipterocarpaceæ, and native to Malaya, Sarawak, and Indonesia; also, a tree yielding this wood.

1783 W. MARSDEN *Hist. Sumatra* 130 *Maranti maracooly* and *murbow*, are in much estimation for building. **1839** T. J. NEWBOLD *Pol. & Statistical Acct. Straits of Malacca* I. vii. 442 The trees chiefly in use for purposes of house and ship are the Chingei,..the red and white Meranti for planks, etc. **1900** W. W. SKEAT *Malay Magic* 109 The *mĕranti* is a fine hard-wood forest tree. **1935** I. H. BURKILL *Dict. Econ. Products Malay Peninsula* II. 2002 Red Mĕranti is the commonest Malayan timber. **1955** *Nomencl. Commercial Timbers* (*B.S.I.*) 74 For the light to medium-weight timbers it should be noted that *meranti* is the group name for species of *Shorea* in Malaya, Sarawak, Brunei and Indonesia. **1956** *Archit. Rev.* CXX. 117 All windows and doors are framed in red meranti hardwood. **1956** *Handbk. Hardwoods* (Forest Prod. Res. Lab.) 150 In Malaya this timber is commonly classified for export mainly on the basis of colour and weight as light red meranti (the most important), dark red meranti, yellow meranti and white meranti. **1966** D. FORBES *Heart of Malaya* x. 113 Their box-like wooden houses..are dull and plain in contrast with the charming thatched cottages of red meranti wood, in which the Malays live. **1971** *Timber Trades Jrnl.* 14 Aug. 42/2 Malaysian and Singapore plywoods, of which meranti and mengkulang are the most important, accounted for 11·9% of deliveries of 102 500m³.

merbau (mɔ̄·ɹbǭu). Also **marbow, merabau, murbow.** [Malay.] A hardwood timber obtained from *Intsia bijuga* or *I. palembanica*, of the family Leguminosæ, trees native to Malaysia and Indonesia; also, the tree itself.

1783 [see *MERANTI]. **1836** J. LOW *Diss. Soil & Agric. Penang* v. 202 *Marbow*—this is a high tree, affording large planks. **1911** *Encycl. Brit.* XVII. 472/2 Good hard-wood timber is found in plenty, the best being the *merabau*, *penak, rasok* and *chengal*. **1935** I. H. BURKILL *Dict. Econ. Products Malay Peninsula* II. 1243 Mĕrbau is used for all kinds of high-grade construction on land, for furniture, and for sleepers. **1940** E. J. H. CORNER *Wayside Trees Malaya* I. 396 This small genus [*sc.* *Intsia*] includes the well-known timber-tree *Merbau*. **1958** J. SLIMMING *Temiar Jungle* iv. 60 The children play with spinning tops, made of *Merbau* wood. **1965** C. SHUTTLEWORTH *Malayan Safari* i. 16 There are heavy hardwoods such as merbau, balau and chengal. **1972** *Timber Trades Jrnl.* 13 May 39/2 Of the lesser species merbau and kapur are readily available.

Mercalli (mɔɹkæ·li). *Seismology.* The name of Giuseppe *Mercalli* (1850–1914), Italian geologist, used *attrib.* to designate an arbitrary 12-point scale he devised (in 1897) to express the intensity of an earthquake at any place.

1921 C. DAVISON *Man. Seismol.* iii. 45 The Mercalli scale..is suitable for strong earthquakes and is adopted in Italy. **1927** —— *Founders Seismol.* vi. 106 Mercalli found that the higher degrees of the Rossi-Forel scale were insufficient... His modification of that scale contains the germs of the Mercalli scale now so widely used. **1972** *Times* 28 Nov. 5/2 The tremor (eight on the 12-point Mercalli scale) damaged many buildings in the old part of the city. **1973** *Daily Colonist* (Victoria, B.C.) 8 Apr. 10/5 An earthquake can have only one magnitude but it can have several intensities and these are registered on what is known as the modified Mercalli intensity scale.

mercapto(-) (mɔ̄ɹkæ·pto). *Chem.* [f. MERCAPT(AN + -O.] **a.** As an inseparable formative element in chem. names indicating the presence of the radical —SH, as *mercapto-acetic* adj., *mercaptoethanol*; **mercaptopu·rine** *Pharm.*, any of the monomercapto derivatives of purine, *spec.* 6-mercaptopurine, $C_5H_4N_4S$, which is a yellow crystalline powder and is a cytotoxic agent used chiefly in treating leukæmia in children.

1971 *Nomencl. Org. Chem.* (I.U.P.A.C.) (ed. 2) C. 211 When —SH is not the principal group, the prefix 'mercapto-' is placed before the name of the parent compound to denote an unsubstituted —SH group.

1884 ROSCOE & SCHORLEMMER *Treat. Chem.* III. II. 88 Thioglycollic acid, or mercapto-acetic acid, $CH_2(SH)CO_2H$, is obtained by the addition of chloracetic acid to an aqueous solution of potassium hydrosulphide. **1947** *Biochem. Jrnl.* XLI. 49/2 BAL is also effective in detoxicating this arsenical, whereas a concentration of 2-mercaptoethanol equivalent to twice that of BAL is almost completely inactive. **1965** PEACOCKE & DRYSDALE *Molecular Basis Heredity* vii. 73 By inhibiting nuclear division in sea urchin eggs with mercaptoethanol, Mazia provided evidence that the prophase chromosome contains at least four strands. **1954** *Amer. Jrnl. Med. Sci.* CCXXVIII. 376/2 We have in mercaptopurine a new type of agent which has demonstrated activity against experimental tumors and leukemias in the mouse and against leukemia in man. **1961** *Lancet* 7 Oct. 821/1 Methotrexate..and mercaptopurine..are folic-acid antagonists, which.. interfere with nucleic-acid synthesis by inhibiting folic-acid reductase. **1965** Mercaptopurine [see *immunosuppressant* s.v. *IMMUNO-]. **1973** *Arthritis & Rheumatism* XVI. 139 Thirteen patients with psoriasis and arthritis were treated with 6-mercaptopurine in dosages ranging from 20 to 150 mg/day.

b. As an independent word, usu. *attrib.* and sometimes joined by a hyphen to the second element: the radical —SH, present in mercaptans; sulphydryl.

1896 *Jrnl. Chem. Soc.* LXX. I. 412 Unsymmetrical homologues of thiodiglycollic acid..are prepared by the interaction of the neutral sodium salts of mercapto-acids and those of haloid fatty acids. **1930** *Chem. Rev.* VII. 511 Mercapto acids may be made in the same way as hydroxy acids. **1957** R. H. BARRY in E. Sagarin *Cosmetics* xx. 468 These inventors obtained coverage also in Great Britain for other organic mercapto compounds which function in alkaline-reacting media as depilatories.

mercapturic (mɔ̄ɹkæptiū·ɹik), *a.* *Chem.* [ad. G. *bromphenyl*)*mercaptur*(*säure* bromophenylmercapturic acid (Baumann & Preusse 1879, in *Ber. d. Deut. Chem. Ges.* XII. 807), f. *mercapt*(*an* MERCAPTAN + *ur*(*in* URINE *sb.*[1]: see -IC.] *mercapturic acid*: any of the acids of the formula $RSCH_2CH(NH \cdot CO \cdot CH_3)COOH$ (where R is an aryl radical), some of which are excreted in the urine, probably as detoxication products of aromatic compounds.

1879 *Jrnl. Chem. Soc.* XXXVI. 803 An examination of the urine of a dog, to which 3 to 4 grams of bromobenzene had been daily administered for six months, showed... The urine contains bromophenyl-mercapturic acid. **1884** *Ibid.* XLVI. 1395 A body strongly lævorotatory, which on treatment with acids yields mercapturic acid. **1934** *Times Lit. Suppl.* 9 Aug. 555/2 Naphthalene is detoxicated in the rabbit partly by conjugation with cysteine and subsequent excretion as the mercapturic acid. **1971** H. G. MANDEL in B. N. La Du et al. *Fund. Drug Metabolism* x. 168 Acetylation takes place in the liver..and the resulting mercapturic acid is then excreted in the urine.

Mercator (mɔ̄ɹkēⁱ·tǫ̆ɹ). Also **mercator.** [L., = 'merchant'.] **a.** The name of Gerhardus *Mercator* (= L. equivalent of Gerhard Kremer) (1512–94), Flemish cartographer, used *attrib.* and in the possessive with reference to the orthomorphic cylindrical map projection first used by him in 1568, in which meridians are represented by equidistant straight lines at right angles to the equator and any course that follows a constant compass bearing is represented by a straight line.

1669 Mercator's projection [see PROJECTION *sb.* 7 b]. *a* **1877** KNIGHT *Dict. Mech.* II. 1419/2 *Mercator-chart*, a mode of projection invented by Gerald Mercator, in which the meridians and parallels are straight and parallel lines. **1883** *Encycl. Brit.* XV. 520/2 By 1601 Mercator's projection was in use for all sea charts. **1908** G. R. PUTNAM *Naut. Charts* 9 The Arcano del Mare, 1646, was the first marine atlas in which all the maps were drawn on the mercator projection. **1912** A. R. HINKS *Map Projections* iii. 29 The great distortion in the north and south makes Mercator's projection altogether unsuitable for a land map. **1938** L. M. MILNE-THOMSON *Theoret. Hydrodynamics* v. 138 An illustration of conformal mapping is afforded by an ordinary map on Mercator's projection. **1960** C. ECKART *Hydrodynamics of Oceans & Atmospheres* 280 The vertical co-ordinate, *μ*, on a Mercator chart of the Sphere is defined by $d\phi = \cos\phi\, d\mu$.

b. *absol.* (A map drawn on) Mercator's projection.

1879 *Encycl. Brit.* X. 208/2 The meridian Mercator drawn as described in the last paragraph..may be made to serve the important purpose of enabling one to trace on the ordinary Mercator's chart the track of a great circle joining any two places. **1953** A. H. ROBINSON *Elem. Cartogr.* iii. 43/1 The poles cannot be represented for they are 'at' infinity on the conventional Mercator.

Mercatorial (mɔ̄ɹkătǭ°·riăl), *a.*[2] Also **mercatorial.** [f. prec. + -IAL.] Of, pertaining to, or derived from Mercator or Mercator's projection or chart.

a **1888** P. F. SHORTLAND *Naut. Surveying* (1890) 144 Their Mercatorial meridians will be inclined to each of them respectively about 1'·2. *Ibid.* 346 To reduce a plane sheet to a Mercatorial projection proceed as follows. **1942** *Jrnl. R. Aeronaut. Soc.* XLVI. 21 It is..possible to include these scales in certain navigational devices, so that rhumb lines and mercatorial bearings may be projected if required without the aid of an actual graticule.

merchandise, *v.* Restrict *arch.* to senses in Dict. and add: **3. a.** To put on the market; to promote the sale of (goods, etc.).

1926 *Publishers' Weekly* 22 May 1687/2 When an author suddenly springs into prominence with a best seller.. I would make that best seller work retroactively and I would merchandise all of his preceding books. **1959** I. ROSS *Image Merchants* (1960) xv. 270 A new breakfast food or a new form of aspirin can easily be merchandised. **1970** R. LOWELL *Notebk.* 43 This typing paper..only merchandised in Maine. **1971** *Daily Tel.* 9 June 5/3 Books can be merchandised like other products.

b. *transf.* To advertise (an idea or person); to publicize; to 'put over'.

1973 J. RYDER *Trevayne* (1974) xxxix. 309 Andrew Trevayne could be merchandised with extraordinary effect. He has all the qualifications. **1974** *Radio Times* 20 Jan. 9/3, I was never merchandised in my career as any kind of symbol. **1974** *Saturday* (Charleston, S. Carolina) 20 Apr. 1-A/3 Paul said the effect of trouble will depend on how the 'unfortunate situation' is merchandised.

merchandiser. Delete *Obs.* except *arch.* and add later examples.

1965 F. SARGESON *Mem. Peon* iv. 68 A merchandiser of building material. **1971** *Daily Tel.* 10 June 25 (Advt.), We are seeking to appoint Merchandisers who will be responsible within a Sales Team for maximising the Company's sales in cash and carries and supermarkets. **1973** *Ibid.* 3 Jan. 19 (Advt.), A vacancy exists for a Merchandiser to develop fashion ranges of men's shirts.

merchandising, *vbl. sb.* Delete † and add later examples. Also, the promotion of sales (of goods, etc.) by advertising and publicity; the process of selling goods (see *MERCHANDISE *v.* 3).

1922 S. LEWIS *Babbitt* iv. 39, I guess that'll show Chan Mott..something about modern merchandizing! **1940** *Economist* 11 May 860/1 The distinctive feature of the chain stores' success is the wedding of efficient merchandising with the control of production from the retail end. **1958** *Ibid.* 18 Oct. 200 (Advt.), Merchandising..means following up your advertising. It means getting retailers on your side, and giving them all they need to help you sell. **1968** *Globe & Mail* (Toronto) 17 Feb. B 6 (Advt.), Must be experienced in all areas of supermarket operations and merchandising. **1971** *Daily Tel.* 10 June 25 (Advt.), Candidates should be in their twenties with good education and preferably have experience in selling or merchandising in the Grocery field. **1973** *Times* 10 Mar. 14/6 In the past two decades the merchandizing of Disney's cartoon creations in a plethora of T-shirts, key-rings and assorted trivia may have obscured their creator's rightful claim to be regarded as one of the most original film producers in cinema history.

attrib. (Later examples.)

1926 *Publishers' Weekly* 16 Jan. 171/1 His frank criticism of 'price-cutting for bait' indicates that he has found other merchandising practices of more value in business building. **1930** *Daily Express* 9 Sept. 2/6 A wide assortment of issues rose from 1 to 3 points, with utilities, merchandising stocks, coppers, and specialities prominent.

merchant, *sb.* and *a.* Add: **A.** *sb.* **1. d.** (Further examples.)

1704 S. KNIGHT *Jrnl.* (1825) 56 [The Indians] give the title of merchant to every trader. **1784** J. F. D. SMYTH *Tour U.S.A.* I. 99 Linen-drapers, grocers, stationers, &c. are not known here; they are all comprehended in the single name and occupation of merchant, or store-keeper. **1809** 'D. KNICKERBOCKER' *Hist. N.Y.* II. VII. x. 251 If peradventure some straggling merchant of the east, should stop at his door, with his cart load of tin ware or wooden bowls, [etc.]. **1818** in *Trans. Illinois State Hist. Soc. 1910* 162 Dry goods are geting very cheap, the country is full of them; we have more merchants than any thing else. **1853** *Harper's Mag.* Aug. 425/1 The subject, we take it, is the 'merchant' of a country-store; quite a different variety from the big bugs of the trade in the Great Metropolis. **1871** R. SOMERS *Southern States since War* 129 Few are able at the end of the year to square accounts with 'the merchant'. **1897** J. L. ALLEN *Choir Invisible* i. 5 A heavy roll of home-spun linen, which she was taking to town to her aunt's merchant as barter for queen's-ware pitchers. **1924** *Scots Mag.* Aug. 342 The merchan's in an awfu' ill teen the day, swearin' like a' that. **1961** *Buchan Observer* 6 June 5 (Advt.), For Sale, by Private Bargain, this well-sited General Merchant's Business, at the junction of the main Fraserburgh and Peterhead Roads. **1975** 'E. LATHEN' *By Hook or by Crook* xiii. 129 An ancient panama..part of the summer uniform demanded of city merchants.

f. merchant of death, one who makes a profession of war; *spec.*, a dealer in armaments; a mercenary soldier.

1934 ENGELBRECHT & HANIGHEN *Merchants of Death* xviii. 261 The business of the arms industry is steadily increasing..and governments are everywhere drawing closer the ties which bind them in a virtual partnership with the merchants of death. **1956** C. W. MILLS *Power Elite* viii. 189 Top generals..lived through the general anti-militarist peace of the 'twenties and 'thirties, begging for appropriations, denying the merchants-of-death charges. **1959** E. AMBLER *Passage of Arms* ix. 233 Nowadays..we don't hear the phrase 'merchants of death' very much. **1963** *Guardian* 11 Jan. 8/1 A mist of superstition and loathing of the trade in arms, of the 'Merchants of Death'. **1974** D. SEAMAN *Bomb that could Lip-Read* xi. 98 Inflation would hit a merchant of death as much as any working man.

3. Delete †*Obs.* and add later examples. Now usu. with a qualifying word, as *speed merchant*, denoting one who has an interest in or partiality for the thing specified. *slang.*

1886 *Referee* 17 Oct. 2/3 The success of 'Indiana' mainly depends upon the extravagant humours of the chief low-comedy merchant. **1909** J. R. WARE *Passing Eng.* 175/2 The theatre coming to be called the 'shop', actors dubbed themselves 'merchants', qualified by their line. **1914** *Automotor Jrnl.* 246/2 It may be that when the new road has been built the speed merchant and the road-hog..may pay their money and betake themselves to their favourite seaside haunt at any speed they like. **1919** J. BUCHAN *Mr. Standfast* ix. 176 Some movie-merchant had got a graft with the Government, and troops had been turned out to make a war film. **1923** *Daily Mail* 15 Feb. 6 The chauffeur of a motor-car has a rain-blurred windscreen, and the goggled 'speed-merchant' cannot see so well as usual. **1929** A. CONAN DOYLE *Maracot Deep* 244 Storr, the googlie merchant, had a better showing with four for ninety-six. **1933** D. L. SAYERS *Murder must Advertise* xviii. 316 He was now faced by the merchant with the off-break. The first two balls he treated carefully. **1957** *Railway Mag.* Nov. 752/2 One wonders how many drivers, other than the confirmed speed merchants, will even attempt to run the 8.20 a.m. from Kings Cross from Hitchin to Huntingdon in 24 min. **1963** *Pix* 28 Sept. 63 Being a good weather 'merchant' is one of the finest refinements in surfing. **1970** D. FRANCIS *Rat Race* vi. 83 Here is this bloody bomb merchant running around loose with no one knowing what he'll do next. **1971** *Guardian* 24 Sept. 13 Anthony Tucker on America's leading doom merchant Mister Catastrophe. **1971** G. SIMS *Deadhand* II. iii. 97 Sorry to be such a gloom merchant. But..we're broke, you see.

4. Delete †*Obs.* and add later examples.

1899 KIPLING *Five Nations* (1903) 8 The pot-bellied merchant foreboding no wrong With headlight and side-light he lieth along. **1905** J. MASEFIELD *Mainsail Haul* 65 There's a fat merchant on the coast... We're going out for her.

6. a. *merchant shipper.*

1912 *Pitman's Commercial Encycl. & Dict. Business* II. 822/1 We will take as an example an indent for cotton goods, such as a large firm of merchant shippers would receive. **1919** *Brit. Manufacturer* Nov. 16/2 Orders may.. be distributed by the merchant shipper of this country.

b. merchant prince (earlier and later examples); hence **merchant-princely** *a.*

1843 L. M. CHILD *Lett. from N.Y.* viii. 53, I sometimes ask whether the age of Commerce is better than the age of War? Whether our 'merchant princes' are a great advance upon feudal chieftains? **1961** NEW ENG. BIBLE *Rev.* xviii. 23 Your traders were once the merchant princes of the world. **1967** *N.Y. Times* (Internat. ed.) 11-12 Feb. 4/1 Prince Philip, describing himself as a merchant prince working to increase British exports, told 500 businessmen yesterday that no matter where he traveled he was questioned about Britain's financial woes. **1874** TROLLOPE *Way we live Now* (1875) I. x. 59 He..had twice dined.. amidst all the magnificence of merchant-princely hospitality in Grosvenor Square. **1928** *Daily Express* 20 Dec. 6 Even the final figure, the three million dollar trade, is nothing breath-bereaving nor merchant-princely.

merchantability (mɜ̄ːɹtʃăntăbiˑlĭti). [f. MERCHANTABLE *a.*: see -ITY.] The condition or state of being fit or prepared for market; the ability to be bought or sold.

1939 *Credit Manual of Commercial Laws* (Nat. Assoc. Credit Men) 31 *Merchantability.* Where goods are bought by description from a dealer in goods of that kind..there is an implied warranty that they shall be of a merchantable quality. **1961** *Listener* 21 Sept. 417/1 This condition, unlike the condition of merchantability, may be expressly excluded. **1972** *N.Y. Law Jrnl.* 31 Oct. 17/6 The findings on such issues..constitute a breach of the implied warranty of merchantability. **1973** *Ibid.* 1 Aug. 4/6 The issue is whether the pacemaker was fit for the ordinary purpose for which pacemakers were used (merchantability).

merchant bank (mɜ̄ˑɹtʃănt bæŋk). [MERCHANT *a.* 1 + BANK *sb.*³ 7.] A bank whose main business is the providing of long-term credit and the support and financing of trading enterprises. So **merchant banker**, a member of such a bank (also *pl.*, the bank as a firm); **merchant banking** *vbl. sb.*, the activity undertaken by a merchant bank.

1904 H. T. EASTON *Hist. Banks* xxi. 215 The chairman of the London and Westminster Bank stated some time ago that 'acceptances drawn by mercantile firms abroad is [*sic*] part of the business that ought to be carried on by a merchant or merchant-banker'. **1924** L. LE M. MINTY *Eng. Banking Methods* ii. 51 The Merchant Bankers..are a class of large importing and exporting firms that, in addition to their regular business in trade, carry on an accepting business... Many firms referred to as 'merchant bankers' are merchants no longer. **1926** W. LEAF *Banking* vii. 184 Several of the great merchant banking houses of the Continent found it worth their while to transfer their international offices to London. **1928** *Daily Mail* 25 July 18/1 Dr. David Jochelman... Merchant Banker. **1930** S. E. THOMAS *Princ. Banking* ii. 18 In a sixth class [of banks] are the various concerns in the City which are known as 'merchant bankers' or 'merchant banks'. Most of these function mainly as 'accepting houses'. **1933** B. ELLINGER *This Money Business* ii. 16 Merchant Bankers or Accepting Houses, whose business it is to accept bills for other people and earn a commission by doing so. **1940** T. S. ELIOT *East Coker* iii. 10 They all go into the dark.. The captains, merchant bankers, eminent men of letters. **1959** *Chambers's Encycl.* II. 100/2 Mention may also be made of the London merchant bankers. Their traditional function has been to facilitate the finance of international trade by 'accepting' (guaranteeing) bills of exchange. **1961** M. SPARK *Prime of Miss Jean Brodie* vi. 161 A successful business man who varied in his line of business from canned goods to merchant banking. **1965** J. L. HANSON *Dict. Econ.* 276/1 Many foreign merchant bankers found it convenient to transfer their headquarters to London—hence the number of merchant banks in England with foreign names. **1965** PERRY & RYDER *Thomson's Dict. Banking* (ed. 11) 375/1 Merchant bank. The great merchant banking houses of today have evolved from the merchants who, coming to this country in the early or middle years of the eighteenth century, built up their business by lending their names to bills financing particular transactions and by accepting such bills made them first-class bills freely discountable. **1966** *Times Rev. Industry* Oct. 99/1 When I became chairman of Lazards last year I was able to put down my occupation as 'merchant banker' instead of the less reputable 'company director'... I certainly was not a 'merchant' and not really a 'banker'. **1972** *Accountant* 19 Oct. 500/2 The work of an accountant in the field of merchant banking or taxation are [*sic*] worlds apart from the work in manufacturing industry or in practice.

merchanting, *vbl. sb.* Add: (Further examples.) Also *attrib.*

1957 *Listener* 28 Nov. 876/1 There is London as a merchanting centre... I have in mind the work done by the commodity markets and the large merchanting houses in buying and selling throughout the world. **1963** *Times* 16 Feb. 6/1 The merchanting trade has a considerable processing element, about a third of the 9m. tons of compound feed produced in the United Kingdom being manufactured by country compounders. **1967** *Economist* 28 Oct. 420/2 Lord Cromer, when still Governor of the Bank of England, put a figure of £200 million on the overseas earnings from all of Britain's insurance, merchanting, brokerage and banking services in 1965. **1971** *Timber Trades Jrnl.* 21 Aug. 13/1 The merchanting and retail side of the timber trade. **1973** *Daily Tel.* 31 July 17 (Advt.), More depots have been acquired in the merchanting and retailing section where we now have 87 outlets.

merchanting (mɜ̄ˑɹtʃăntiŋ), *ppl. a.* [f. MERCHANT *v.* + -ING².] Engaged in trade as merchants.

1930 *Observer* 9 Feb. 18/1 The merchanting body itself is in no greatly better case. **1957** *Times* 23 Dec. 11/2 Two main divisions—the producing subsidiary companies with their mills and the merchanting subsidiaries in Manchester and London.

merchant mill. *U.S.* ? *Obs.* [f. MERCHANT *a.* 1.] A mill engaged in the grinding of grain for the purpose of trade.

1774 P. V. FITHIAN *Jrnl.* (1900) 111 Mr. Carter's Merchant Mill begins to run to-day—She is calculated to manufacture 25,000 Bushels of Wheat a Year. **1816** U. BROWN *Jrnl.* in *Maryland Hist. Mag.* (1915) X. 273 He has a Merchant Mill, a Saw Mill,..all on the Waters of Bath. **1837** W. JENKINS *Ohio Gazetteer* 68 The village contains..1 large merchant mill,..with four run of stones. **1851** C. CIST *Sk. Cincinnati in 1851* 194 It proposes to perform in a small compass..the work of a merchant mill. **1877** W. A. JOHNSON *Hist. Anderson County, Kansas* 252 In the spring of 1874 Chris. Bouch..commenced the construction of a first-class merchant mill.

merchant navy (mɜ̄ːɹtʃănt nēⁱˑvi). [MERCHANT *a.* 2 + NAVY¹.] A fleet or number of ships used in trade and not for purposes of war.

In 19th-cent. official sources called 'merchant shipping', 'merchant service', 'mercantile marine', 'mercantile navy', but not 'merchant navy'.

1875 T. FARRER in A. Hurd *Merchant Navy* (1921) I. 80 The actual increase of our Merchant Navy is a most remarkable fact. **1921** A. HURD *Merchant Navy* I. 2 Within a few months of the opening of hostilities, the King, in a message of appreciation of the services of the merchant seamen, referred to 'his Merchant Navy',..and the Prime Minister..described the Merchant Navy as 'the jugular vein of the nation'. **1928** *Times* 14 Feb. 14/6 The King has appointed the Prince of Wales Master of the Merchant Navy and Fishing Fleets. The announcement will appear in the *London Gazette* to-night. **1929** *Ann. Reg. 1928* I. 10 On February 13 the King paid a graceful compliment to the British seafaring population outside the ranks of the Royal Navy by creating the Prince of Wales 'Master of the Merchant Navy and the Fishing Fleets'—a new title for which there was no exact precedent in British history or usage. **1937** *Discovery* June 165/1 Seamen, whether of the Royal Navy, the Merchant Navy, or the Fishing Fleet. **1965** S. J. HARLAND (*title*) The dustless road: a career in the Merchant Navy.

merchant-tailor. Delete *Obs.* etc. and add further examples.

1653 *Rec. Early Hist. Boston* (1886) X. 1, I Robert Keayne, Cittizen and Merchant Taylor of London by freedome,..now dwelling at Boston. **1818** H. B. FEARON *Sk. Amer.* 33 Taylors are numerous: they are denominated, (in conformity with the accustomed vanity of the country), 'Merchant Taylors'. **1889** *Kansas City* (Missouri) *Times & Star* 13 Nov., Call on Ed. Howe, the artistic merchant tailor. **1922** JOYCE *Ulysses* 717 The establishment of George Mesias, merchant tailor and outfitter. **1975** *Times* 5 July 14/7 The following were elected officers of the Merchant Taylors' Company for the year beginning July 14.

mercurate (mɜ̄ˑɹkiŭᵊrēⁱt), *v. Chem.* [f. MERCUR(Y *sb.* + -ATE³.] *trans.* To convert into a mercury derivative, esp. by replacing a hydrogen atom in an aromatic ring by a group containing mercury

1923 *Org. Syntheses* III. 14 Phenol has been mercurated in water solution. **1942** FUSON & SNYDER *Org. Chem.* xxxi. 430 Hydrocarbons are mercurated by treatment at 90–160° for one or more hours. **1967** I. L. FINAR *Org. Chem.* (ed. 5) I. xx. 583 Aromatic compounds containing almost all the common functional groups have been mercurated.

So **meˑrcurated** *ppl. a.*; **mercuraˑtion**, the process of mercurating or of becoming mercurated.

1921 *Jrnl. Amer. Chem. Soc.* XLIII. 621 The 3 mercurated phenols were boiled with aqueous potassium iodide. **1923** *Org. Syntheses* III. 100 p-Tolylmercuric chloride has been prepared..by the direct mercuration of toluene with mercuric acetate. **1965** PHILLIPS & WILLIAMS *Inorg. Chem.* I. xi. 420 Mercuration of aromatic systems occurs very rapidly, presumably through a similar intermediate. **1967** I. L. FINAR *Org. Chem.* (ed. 5) I. xx. 584 The position of the acetoxymercuri-group can be found by treating the mercurated compound with halogen.

Mercurey (‖ merkūre, mɜ̄ˑɹkiŭri). Also **mercurey.** [Fr., name of a vineyard of the Côte Chalonnaise.] A Burgundy wine produced in the Chalonnais district of France.

1907 A. BENNETT *Grim Smile of Five Towns* iii. 225 It was a bottle of mercurey, a wine which has given me many dreadful dawns, but which I have never known how to refuse. **1928** P. M. SHAND *Bk. French Wine* iv. 173 Mercurey is the only real 'name' of the Côte Chalonnaise. *Ibid.*, Mercurey owes its classical name to a former temple of Mercury. **1958** A. L. SIMON *Dict. Wines* 110/1 Most of the *Mercurey* wines are red, and they are also of better quality than the few white wines of that district. **1972** D. MCLACHLAN *No Case for Crown* xiii. 165 The wine was his favourite *Mercurey*, which he ordered himself from France.

mercurial, *sb.* **3.** Add. to def.: Also more widely, any compound that contains mercury. (Further examples.)

1971 *Sci. Amer.* May 18/3 Among other inorganic mercurials, some of the oxides, such as the red oxide used in antifouling paint for ship bottoms may be potentially hazardous. **1971** *Nature* 23 July 222/1 The build-up of alkyl mercurials in the environment and their toxicological properties was the subject of many of the contributions.

mercurochrome (mɜ̄ɹkiūᵊˑrokrōᵘm). *Pharm.* Chiefly *U.S.* [f. MERCUR(Y *sb.* + -O + Gr. χρῶμ-α colour.] A mercury derivative, $C_{20}H_8O_6Br_2HgNa_2$, of fluorescein which forms greenish iridescent scales, dissolves in water to give a red solution, and is used as a weak antiseptic. (Orig. given a wider meaning: see quot. 1919.)

1919 H. H. YOUNG et al. in *Jrnl. Amer. Med. Assoc.* 15 Nov. 1484/1 To the substance obtained by substituting one atom of mercury in the molecule of dibromfluorescein we have given the name 'mercurochrome-220'. [*Note*] The name 'mercurochrome' will be applied generally to all the mercury-bearing dyes that we are investigating, the individual being distinguished by the laboratory number following the name. **1942** W. SIMPSON *One of our Pilots is Safe* iii. 63 They left my face uncovered, painted with scarlet mercurochrome. **1951** R. BRADBURY *Illustr. Man* (1952) 178 He smelled the cutting odours of iodine, raw adhesive, and pink mercurochrome. **1960** O. NASH *Boy is Boy* (1961), It's easy to spot a boy... He smells of licorice, he smells of mice, Of Mercurochrome, and vanilla ice. **1960** E. GURR *Encycl. Microsc. Stains* 1. 262 Mercurochrome is an excellent substitute for eosin, as unlike the latter it is extremely fast to light and preparations stained with it do not fade even when exposed to strong sunlight for a very considerable time. **1961** J. HELLER *Catch-22* (1962) xiv. 100 Give me a minute or two to put some mercurochrome on these cuts.

mercury, *sb.* Add: **11. mercury arc lamp** (also *ellipt.* as *mercury arc*) = **mercury vapour lamp*; **mercury arc rectifier,** a rectifier consisting of one or more graphite or iron anodes and a mercury pool cathode enclosed in an envelope from which the air has been pumped out; **mercury fulminate,** fulminate of mercury; **mercury gilding** (see quot. 1960); **mercury lamp** = **mercury vapour lamp*; **mercury pool,** a mass of liquid mercury, esp. one used as an electrode; **mercury vapour lamp,** a lamp in which light (rich in the ultra-violet) is produced by an electric discharge through mercury vapour, the envelope being often coated with a fluorescent substance so as to produce more visible light (cf. **fluorescent lamp*); **mercury vapour pump,** a pump for producing high vacua which works by entraining molecules of the gas to be evacuated in a jet of mercury vapour; **mercury vapour rectifier** = **mercury arc rectifier*.

1906 *Trans. Amer. Inst. Electr. Engin.* XXIV. 372 The constant current mercury arc rectifier system, as used for operating..mercury arc lamps..is sketched diagrammatically. **1916** F. B. PIDDUCK *Treat. Electr.* ix. 363 The absence of strong lines in the red unfits the mercury arc for certain purposes. **1936** TEAGO & GILL *Mercury Arcs* ii. 13 Steel containers are used for demountable mercury arcs of large output. **1971** *New Scientist* 3 June 564/1 The conventional sources of infrared radiation are hot bodies (the globar or mercury arc lamp). **1906** Mercury arc rectifier [see *mercury arc lamp* above]. **1930** *Engineering* 24 Jan. 109/3 A 650-volt direct-current supply from mercury arc rectifiers. **1966** G. F. ALLEN *Brit. Rail after Beeching* v. 133 In the Eastern Region's multiple-units.. the resultant distortions in the electrical circuitry were aggravated by serious failures of the mercury-arc rectifiers. **1904** *Kynoch Jrnl.* Oct.–Dec. 199 The flame from the mercury fulminate..ignites the charge of explosive contained in..the case. **1973** L. RUSSELL *Everyday Life Colonial Canada* xii. 150 The hammer [of a gun] was reduced to a simple head, designed to strike the top of the copper [percussion] cap. Under this top was a pinch of mercury fulminate, a compound so unstable that a moderate impact causes it to explode. [**1910** R. L. HOBSON *Worcester Porc.* xii. 100 A radical change came over the ware about 1780, when the cheaper process of mercurial gilding came into use... This later gilding has a more metallic and brassy appearance.] **1957** MANKOWITZ & HAGGAR *Conc. Encycl. Eng. Pott. & Porc.* 95/2 Mercury gilding was introduced about 1785. **1960** R. G. HAGGAR *Conc. Encycl. Cont. Pott. & Porc.* 207/1 s.v. *Gilding*, Mercury-gilding came in about 1780, and consisted of a mixture of gold with mercury which could be painted onto the ware, the mercury being driven off as a vapour during the firing process. **1971** *Country Life* 10 June 1419/2 This mercury gilding was harder, longer wearing and more brilliantly lustrous than honey gilding, but tended to display a brassy tinge. **1904** *Trans. Amer. Inst. Electr. Engin.* XXII. 73 Hewitt's investigations..are.. pioneer work in the field of the mercury lamps. **1966** HEWITT & VAUSE *Lamps & Lighting* xviii. 276 Mercury lamps are used for street lighting and to a limited extent in industrial installations. **1907** FRANKLIN & ESTY *Elem. Electr. Engin.* II. ix. 172 The mercury-arc rectifier consists essentially of a highly exhausted glass bulb..with two iron or graphite electrodes..and the mercury-pool electrode. **1956** *Nature* 11 Feb. 267/2 Industrially, the polarograph is finding an ever-increasing application as a continuous service indicator. For such purposes, the mercury-pool anode is unsatisfactory. **1970** J. SHEPHERD et al. *Higher Electr. Engin.* (ed. 2) xxv. 799 The connexions apply equally to 3- and 6-anode mercury-arc rectifiers in which case all the 'cathodes' are common and are in fact the mercury pool. **1904** *Brit. Pat.* 3657 2 The light given out is similar to that which is given out by an ordinary mercury vapour lamp. **1909** *Westm. Gaz.* 6 Mar. 3/2 A 'new' process of sterilising milk by exposing it to the ultra-violet rays of a mercury-vapour lamp. **1943** J. B. PRIESTLEY *Daylight on Saturday* i. 2 The factory inside is..lit with innumerable mercury-vapour lamps that produce a queer greenish-white mistiness of light. **1955** E. B. FORD *Moths* i. 15 Owing to the high power and surface-brightness of a mercury-vapour lamp, such a source may provide a more efficient means of collecting than an ordinary light. **1972** *New Yorker* 26 Aug. 20/3 In the glare of mercury-vapor street lamps, the Bronx presents a

silent, bland, greenish face. **1926** J. H. SMITH tr. *Dunoyer's Vacuum Practice* i. 42 In mercury vapour pumps the size of the orifice through which the vapour is driven is..of secondary importance. **1966** ADAM & EDWARDS tr. *Diels & Jaeckel's Leybold Vacuum Handbk.* 1. vi. 90 Special forms of cold traps with maximum practicable conductivity have been designed for large mercury vapour pumps. **1908** M. SOLOMON *Electr. Lamps* xi. 283 This property of the mercury-vapour arc has led to the invention and development of mercury-vapour rectifiers. **1966** R. G. KLOEFFLER *Electron Tubes* ix. 194 The mercury-vapor rectifier diode uses a hot cathode and mercury vapor under a low pressure.

mercy, *sb.* Add: **8.** For *Obs.* read *Obs.* exc. *arch.* or *Hist.* (Later examples.)

1890 W. P. BAILDON *Sel. Civil Pleas* I. 44 Let them have their seisin thereof, and James is in mercy for the unjust detention. **1895** POLLOCK & MAITLAND *Hist. Eng. Law* II. II. iv. 512 At first the declaration that a man is in the king's or the lord's mercy implies that the king or lord may, if he pleases, take all his goods. **1914** G. F. DEISER *Year Bks. of Richard II: 12 Richard II* 161 The judgment was that the plaintiffs take nothing by their writ, but be in the mercy for their false plaint.

10. *attrib.* passing into *adj.* (orig. *U.S.*): now esp., administered or performed out of mercy or pity in order to terminate or relieve pain or distress (cf. *mercy-stroke* in Dict.), as *mercy flight, mercy killing* (so *mercy killer; mercy-kill* v. nonce-wd.), *mission, murder*, etc.

1927 *Daily Express* 24 Mar. 3/6 The 'mercy bullet'.. contains a chemical which is released on striking the animal. The fluid in the blood will cause temporary unconsciousness. **1969** *Listener* 23 Jan. 111/3, I had representatives in touch with both sides to get the mercy corridor working, to try and get flights in by day as well as by night. **1933** *Meccano Mag.* Mar. 195/2 Never a month passes without a 'mercy flight' being carried out. **1944** *Beaver* Dec. 40/2 The aeroplane..is proving of the greatest value as a means of conveying medical assistance. .. Such 'mercy flights' are increasing annually. **1973** *Guardian* 14 Feb. 4/4 Two American prisoners of war freed by the Communists were making mercy flights home tonight to be with their families. **1957** *Observer* 8 Dec. 14/3 This [melodrama], with its brilliant young scientist strangling his wife and volunteering for deep freeze and being mercy-killed by his boss. **1935** *Amer. Speech* X. 120/2 On the trail of the..mercy killer..the public follows day by day. **1951** J. D. SALINGER *Catcher in Rye* xxi. 194 He feels sorry for it [a child]..sticks this blanket over her face..and makes her suffocate... He was a mercy killer. **1935** A. A. BRILL in *Vital Speeches* 16 Dec. 165 (*title*) Is 'mercy killing' justified? **1952** J. CARY *Castle Corner* Pref. 6 Think of Oates' suicide in the Antarctic, or thousands of 'mercy killings', which are technically murder. **1974** *Times* 6 Feb. 6/1 A Long Island doctor who was accused of a 'mercy killing' of a dying cancer patient, was acquitted. **1972** *Reader's Digest* Mar. 76/1 (*heading*) Their mercy missions reach new heights of skill and ingenuity. **1974** *Times* 23 Apr. 4/6 An Army convoy of 20 four-ton lorries left Lyneham..yesterday on the start of a 3,000-mile mercy mission to Niger. **1930** *Commonweal* (N.Y.) 16 July 293 Mercy murders once more: euthanasia. **1943** S. M. EMERY *Commander of Clouds* 7 He turned the nose of his own kite toward the mercy ship and thrust on more power. **1944** F. CLUNE *Red Heart* 8 On an average, Doctor John Grieve Woods travels over 17,000 miles annually by air on mercy-trips.

merd. Delete † *Obs.* and add later examples, freq. in Fr. form *merde*. Also as *adj.*

1920 T. S. ELIOT *Ara Vos Prec* 11 The goat coughs at night in the field overhead; Rocks, moss, stonecrop, iron, merds. *a* **1930** D. H. LAWRENCE *Phoenix* (1936) 13 Don't you see, in his [*sc.* a rabbit's] very immobility, how the whole world is *merde* to him? **1960** 'A. BURGESS' *Doctor is Sick* xv. 111 Lurching from one delightful street-smell to another—merds, garlic, mutton-fat, urine, food-tins. **1962** *Listener* 26 July 141/1 It's still evasion to ignore the new tragedy of what might be called the *merde* world. **1968** J. UPDIKE *Couples* ii. 167 'Janet, you disgust me,' Harold said. 'How can you unload this *merde* on three people who adore you?' **1970** *New Yorker* 7 Mar. 36 We cannot have that kind of pornographic *merde* in this majestic and high-minded sentence.

b. As *int.* (in Fr. form *merde*). *coarse slang.*

1920 E. E. CUMMINGS *Let.* 14 Oct. (1969) 74, I am *not* self-sufficient do I hear you say? Merde! **1924** D. H. LAWRENCE *Let.* 9 Aug. (1932) 606 A great *merde!* to all latter-day Joan-of-Arcism. **1933** 'G. ORWELL' *Down & Out* iii. 25 '*Merde!*' he used to shout, 'you here again?' **1961** I. JEFFERIES *It wasn't Me!* x. 132 'Oh, well, Merde!' He stuck out his hand and I shook it. 'Merde to you.' **1974** N. FREELING *Dressing of Diamond* 78 He never lost his temper or said merde.

‖ **merdeka** (mɜ̄ɹdēˑkă). [Malay.] Freedom, independence; *spec.* independence for Indonesia or Malaya.

1954 N. EPTON *Islands of Sunbird* iii. 58 South-west of the capital to the distant cocoanut palms and freedom, 'Merdeka', at last! Merdeka... An Indonesian official once told me that Merdeka is still a rousing rallying cry for many Indonesians. **1955** *Times* 16 Aug. 6/7 A banner bearing the word merdeka ('liberty'). **1961** P. KEMP *Alms for Oblivion* ii. 85 The nationalists did not spare their prisoners whom they usually put to death by hewing in pieces to the greater glory of *Merdeka*. **1969** *Guardian* 2 Sept. 10/3 Tunku Abdul Rahman's Alliance Party..the old guard who had 'won' Merdeka. **1972** 'G. BLACK' *Bitter Tea* iii. 36 *Merdeka*..is how the Malays label their national sovereignty. The word means freedom.

mere, *sb.*¹ Add: **2.** Freq. used of Grendel's abode in the Old English poem *Beowulf*.

Beowulf [in Dict.]. **1849** A. D. WACKERBARTH tr. *Beowulf* 53 It is not far,—a Mile from here, Where stands the Monster's sluggish Meer. **1855** B. THORPE tr. *Beowulf* 107/2 Departed home thence the gold-friend of men,.. and on the mere they gaz'd. **1898** G. SAINTSBURY *Short Hist. Eng. Lit.* I. i. 5 Beowulf..soon hears from the King that his adventure is not done, and determines to finish it in the mere itself. **1912** *PMLA* XXVII. 208 (*title*) The haunted mere in *Beowulf*. **1951** *Speculum* XXVI. 33 Grendel's mere has other attributes... The hart pursued by hounds chooses rather to give up its life than to hide its head in the grove surrounding the pool. **1957** *Rev. Eng. Stud.* VIII. 6 All this radiance is eclipsed when under cover of darkness Grendel's mother carries off Æschere to her retreat below the mere.

mere, mear, *sb.*² **1, 1. b.** (Later examples.)

1956 G. E. EVANS *Ask Fellows who cut Hay* xxv. 227 Robert Savage recently used it [*sc.* the word *mere*] to indicate the division between the *yards* on the Common. 'They called them *meres*: they were usually made wide enough so that each man could have a load o' muck taken to his yard.' **1972** *Parl. Debates Commons: Rep. Standing Comm. D: Local Govt. Bill, 18th Sitting* 27 Jan. 1043 A 'mere' is, in fact, a boundary.

mere, *a.*² Add: **5. c.** Esp. in *predic.* use: insignificant, ordinary; foolish, inept.

1732 SWIFT *Let.* 19 Feb. in *Corr.* (1965) IV. 4, I..am as meer a Monk as any in Spain. **1893** BEERBOHM *Lett. to R. Turner* (1964) 64 She is still very mere but not quite so mere—in the strict sense of the word—as she was four weeks ago. **1919** D. ASHFORD *Young Visiters* v. 46 Peaple who have got something funny in their family and who want to be less mere if you can comprehend. **1942** M. DICKENS *One Pair of Feet* ix. 191 Sissons, who didn't count, because she was the Junior and mere. **1955** T. H. PEAR *Eng. Social Differences* viii. 182 In such situations American and Dominion soldiers have always felt less 'mere' than the English.

mere, *v.*² Delete 'exc. *dial.*', restrict *Obs.* to sense 2, and add: **1.** (Further examples.) Also, to record the position of (a boundary) by specifying its relation *to* a visible feature on the ground.

1925 A. S. GREEN *Hist. Irish State to 1014* xiii. 241 On all sides fertile soil was reclaimed for tillage, partitioned, mered, and fenced. **1932** *Instructions to Field Revisers* (Ordnance Survey) App. A. 47 Revision of Boundaries... It often happens that detail is placed on the actual boundary and the original fence entirely obliterated. Before mereing the new detail it is usual to consult the owner or tenant, and not mere on the supposed custom of the district. *Ibid.* 48 Fences to woods..are, as a rule, maintained by the owner or occupier of the wood,..and are, therefore, mered accordingly on Ordnance maps. **1971** *House of Commons Bill 1971–2, No. 2: Local Govt.* Sched. I. III. 192 The boundaries of the new local government areas shall be mered by Ordnance Survey. *Ibid.* 193 Any such boundary defined on the map annexed to any order under..the Local Government Act 1958 by reference to proposed works shall..be mered as if the boundary had not been so defined. **1975** J. B. HARLEY *O.S. Maps* i. 12 The surveyor who completes the map detail also..perambulates and meres administrative boundaries. *Ibid.* iii. 39 Because boundaries are invisible..their precise location in relation to visible ground features is recorded by perambulating the boundary line and 'mereing' it to those features.

‖ **mère** (mɛ̄r). [Fr., mother.] An identifying word appended to a name (usu. a surname) to distinguish her from others of the name.

1862 Q. VICTORIA *Let.* 12 Nov. in R. Fulford *Dearest Mama* (1968) 130 All you say about poor Marie Leiningen Mère interests me. **1881** TROLLOPE *Ayala's Angel* I. xviii. 222, I am told by Tringle mère that I am less acceptable than old Traffick. **1947** M. LOWRY *Under Volcano* i. 23 Taskerson mère had taken a fancy to the French boy. **1967** J. FLEMING *No Bones about It* 98, I have never known Borgan *père* and *mère* attend a point-to-point. **1968** J. HAYTHORNE *None of us cared for Kate* 15 Prentice *mère* has been bombarding the Secretary of State with letters. **1973** *Listener* 1 Nov. 607/3 The execution of Hugo *mère's* lover.

-mere (miᵊɹ), terminal element repr. Gr. μέρος part, occurring with the sense 'part', 'segment' in various biological terms, as *centromere, genomere, hyalomere, metamere.*

Meredithian (meridiˑþiăn), *a.* and *sb.* Also **-ean.** [f. the proper name *Meredith* + -IAN.] **A.** *adj.* Of, pertaining to, or characteristic of George Meredith (1828–1909), English novelist and poet, or his works. **B.** *sb.* An admirer of Meredith.

1889 E. DOWSON *Let.* 7 July (1967) 89 Excuse me for straining after the Meredithian but I have been reading 'Diana' until I am pink and my brain quavers. **1891** *Lit. World* 29 May 510/3 Even the girl-actress catches the infection, and swathes her thought in Meredithean phrase. **1892** *Review of Reviews* Jan. 95/2 Meredithians owe a debt of gratitude to the publishers. **1909** *Daily Chron.* 22 Jan. 3/4 The bookseller, an enthusiastic Meredithian. **1910** *Westm. Gaz.* 9 Apr. 13/2 Your crazy Meredithian will tell you that this is because it is the least worthy of the master's works. **1928** *Daily Tel.* 5 June 12/7, I was never initiated into the inner Meredithian cult, though I have always been a great admirer of his poetry. **1935** J. AGATE *More First Nights* (1937) 123 One or two of my colleagues have dismissed this piece as 'Meredithian', as though that were a term of disparagement. **1936** CHESTERTON *As I was Saying* xvi. 100 In a score of ways, the modern world has followed the Meredithean model for the world. **1968** *Listener* 18 July 91/3 Inevitably, little of the Meredithian style remains.

mereness. Restrict † *Obs.*⁻⁰ to sense in Dict. and add: **2.** The state or quality of being merely something or of being small or insignificant.

1960 *Times Lit. Suppl.* 25 Mar. 193/3 Your reviewer.. and others have made something of the fact that Bevin used to refer to Lord Attlee as the Little Man. Lloyd George's colleagues used exactly the same phrase about him: both Prime Ministers lacked inches. Is it wise to see a value judgment in this statement of physical mereness? **1966** W. V. QUINE in *Jrnl. Philos.* LXIII. 658 The word 'concept', which Russell applied to these nonexistents, connotes mereness.

merengue (mĕreˑŋge). Also **meringue** (mĕræŋg). [f. Haitian Creole *méringue* and Amer. Sp. *merengue* (Webster).] A dance popular in Dominica and Haiti. Also *attrib.*

1936 L. A. HUDSON tr. *Possendorf's Damballa Calls* iv. 48 Dancing a second Merinque with Diane. **1956** M. STEARNS *Story of Jazz* iii. 25 The Haitian *meringues* sometimes sound a little like our ragtime without the force and drive. **1957** *New Yorker* 26 Oct. 166/2 In the cafés and restaurants one hears everything from Vivaldi to Afro-Cuban merengues. **1964** W. G. RAFFÉ *Dict. Dance* 318/2 The *Merengue* folk dance is very popular in the Dominican Republic as a favourite example of Caribbean rhythm. **1965** *Evening Standard* 17 Sept. 8/6 The kid who danced was doing the merengue... The solo merengue kid was dervishing around. **1972** *New Yorker* 21 Oct. 34/2 They danced tangos, merengues..to the strains of the Nagy Hungarian band. **1973** *Sunday Advocate-News* (Barbados) 18 Mar. 14/4 [Haiti] Now cruise ships are calling once more and holiday-makers from the United States, Canada and Europe come by the planeloads to a festive welcome by meringue bands at the airport.

mereology (meriǫˑlŏdʒi). *Logic.* [ad. F. *meréologie*, irreg. f. Gr. μέρος part + -OLOGY.] (See quots. 1946, 1962.)

[**1937** A. TARSKI in J. H. Woodger *Axiomatic Method in Biol.* App. E. 161 The general theory of the concept 'part of' has been developed by S. Leśniewski under the name of *Meréologie*.] **1946** *Mind* LV. 368 The attempt to provide a foundation for mathematics in logic led Lesniewski to formulate a system consisting of 'Protothetic'..'Ontology' ..and 'Mereology' (which is formally similar to the Boolean algebra except for the exclusion of the null class). **1956** J. H. WOODGER tr. *Tarski's Logic, Semantics, Metamath.* ii. 25 The geometry of solids is based upon mereology, in the sense that the relation between part and whole is included in the system of primitive notions of the geometry of solids. **1962** W. & M. KNEALE *Devel. of Logic* vi. 426 An extended system of Boolean algebra in which there are two operations of a non-finitary character, namely that of taking the logical sum or union of all the elements of some specified set..and that of taking the logical product or intersection of all the elements of some specified set... This extended system is closely related to ..'mereology'. In fact the difference is simply that Leśniewski's system excludes the possibility of a null element included in every other element. **1967** H. SKOLIMOWSKI *Polish Analytical Philos.* iv. 106 In mereology the word 'part' is given the meaning it has in everyday usage, i.e. by a 'part' is understood a section of a given object, and such a section as is not identical with the object itself.

merese (mĕrīˑz). [Etym. unknown.] A rib, flange, or collar, on the stem of a glass vessel. Also *attrib.*

1923 H. J. POWELL *Glass-Making in Eng.* iii. 44 *Merese*, a sharp-edged button between bowl and leg of wine glass, or connecting sections of stem or in place of step. **1935** *Burlington Mag.* Oct. 150/1 The two mereses and the foot-stem cast are of kinds which occur repeatedly. **1949** W. A. THORPE *Eng. Glass* (ed. 2) vi. 170 Meantime..the foot merese is moving up the stem, regardless of its use as camouflage and boundary mark. **1960** H. HAYWARD *Antique Coll.* 182/1 *Merese knop*, a knop or protuberance in the stem of a drinking glass or other glass vessel of sharp-edged, flattened form. **1968** O. N. WILKINSON *Old Glass* vi. 100 'Mereses' (flat discs) and knopping appear, more commonly at the top and bottom of stems to disguise welding than as a decorative feature.

meresman. (Earlier and later examples.)

1828 R. GRIFFITH *Let.* 26 Sept. in C. Close *Early Yrs. Ordnance Survey* (1969) 112 Any casualty among the meresmen or expense incurred by calling back the boundary surveyors from distant parts..must be charged to the Ordnance. **1935** *Rep. Progress Ordnance Survey* 1. 6 In 1841..the Survey Act of that year authorized the appointment, by Justices of the Peace in England..and by Sheriffs in Scotland, of 'meresmen' to aid the surveyors in identifying and locating the parish boundaries. *Ibid.*, The legally appointed meresmen often disagreed. **1972** *House of Commons Rep. Standing Comm. D: Local Govt. Bill* 27 Jan. 1043 A 'meresman' is someone whose job it is to ascertain.. boundaries.

merestone. (Later examples.)

1970 H. BRAUN *Parish Churches* xx. 235 Let it [*sc.* the church] rise from the land untroubled by fencing. Its yard can be kept from the plough by mere-stones. **1972** *Parl. Debates Commons: Rep. Standing Comm. D: Local Govt. Bill, 18th Sitting* 27 Jan. 1044 A 'merestone' is a boundary stone.

merge, *v.* Add: **3. b.** Of firms or trading companies: to combine or amalgamate; to combine *with* another.

1926 *Office Appliances* Feb. 49/1 With regard to merger reports..two propositions had been received—one to sell out; the other to merge with another company. **1930** J. B. PRIESTLEY *Angel Pavement* ii. 54 That's the way things are going all the time now,..big combinations—merging away till you don't know where you are. **1965** NEWMAN & LOGAN *Business Policies & Cent. Managem.* (ed. 5) xiv. 353 Poor health or old age makes them [*sc.* large stockholders] willing to merge at an attractive price. **1965** J. F. BRADLEY *Administrative Financial Managem.* xxviii. 577 If Company A and Company B are to merge, Company A might take over Company B.

Hence **merged** *ppl. a.*

1965 NEWMAN & LOGAN *Business Policies & Cent. Managem.* (ed. 5) xiv. 356 The character and the operating policy of the merged firms may differ..sharply.

mergee (mɜɹdʒīˑ). [f. MERGE *v.* + -EE¹.] One who takes part in a merger.

1964 *Punch* 30 Dec. 989/2 Mergees are going to need twice the stuff. **1967** *Time* 24 Nov. 60 Royal Little, retired founder of Textron, Inc., counsels..students on the pitfalls of getting together [in mergers]. These include such dangers as whether the mergee's inventory is all he says it is. **1971** *Atlantic Monthly* Oct. 85 Elsewhere in the hotel—we learn later—the rival corporation, Penta, has chosen a more ingenious method: they tell their mergees that it seems important only that the change be comfortable for everyone.

merger¹. **1. b.** For '*U.S.*' read 'orig. *U.S.*' and add: Also *transf.* (Further examples.)

1915 E. WALLACE *Man who bought London* vi. 54 The Shearman Anti-Trust Law which..drove what is known on the other side of the Atlantic as the 'mergers' out of business. **1928** *Manch. Guardian Weekly* 10 Aug. 104/2 Washington..is shown by the dispatches to suspect something like an Anglo-French alliance or merger or pooling behind the text of the naval compromise. **1930** J. B. PRIESTLEY *Angel Pavement* ii. 55 Along comes a big merger—a bit of syndicate and trust work—and up they go. **1931** CHESTERTON in *London Mercury* Feb. 342 The great moral institutions of modern times, the Straddle, the Wheat Corner, the Merger, and the rest. **1949** I. DEUTSCHER *Stalin* iii. 79 A merger between Mensheviks and Bolsheviks was the main item. **1949** WODEHOUSE *Uncle Dynamite* ii. 24, I have always hoped that you and Sally would eventually form a merger. **1961** *Listener* 16 Nov. 834/3 Theology seems to have made a 'merger' with philosophy for three centuries. **1970** T. LUPTON *Managem. & Social Sci.* (ed. 2) iv. 103 If an organization is being created *de novo*, as the result, say, of a merger. **1971** *Farmer & Stockbreeder* 23 Feb. 18/3 The crunch comes when we look at the most important reason for industrial mergers—marketing. **1974** WODEHOUSE *Aunts aren't Gentlemen* ii. 10, I had once asked her to marry me... In the season I suggested a merger. But apparently I was not the type, and no business resulted.

3. *attrib.* and *Comb.*

1914 W. H. LOUGH *Corporation Finance* (rev. ed.) xiv. 231 The average merger bond should prove a safe investment. **1926** *Office Appliances* Feb. 49/1 With regard to merger reports..two propositions had been received. **1928** *Daily Mail* 3 Aug. 18/2 There is no compulsion on the Preference shareholders to accept an offer from the merger company. *Ibid.* 18/3 The merger shares issued to the shareholders may bring a bigger income than the.. Preference shares. **1964** G. G. FISCH *Organization for Profit* x. 140 The merger-acquisition specialist is the official corporate professional 'wheeler-dealer'. **1967** H. B. MAYNARD *Handbk. Business Administration* ix. xi. 125 The SEC's financial statement requirements in 'merger proxies' are..similar to those of Form S-1. *Ibid.*, In the merger situation, the acquired company will attempt to obtain SEC approval to terminate its reporting requirements.

meri². Add: Now usu. in the spelling *mere*. (Earlier and later examples.) Also, a miniature greenstone version.

1829 J. F. ATKINS in R. McNab *Hist. Rec. N.Z.* (1908) I. 692 The chief..shot him through the head, and with his *maree* (a short stone club, with a sharp edge) he split his skull. **1833** in R. McNab *Old Whaling Days* (1913) 48 An instrument made of a beautiful blue or green marble stone, which they call a Mary. **1874** A. BATHGATE *Colonial Experiences* xviii. 258 The Maories used them [*sc.* stone implements] to kill the Moriories rather than..degrade their own meres. **1905** W. B. *Where White Man Treads* 94 He yearned to take the taiha and mere from his hand, and substitute the axe and spade. **1921** *Outward Bound* June 46/2 The valuable greenstone from which the Maoris fashion all their meres, their tikis, canoe-chisels, brooches and other articles. **1938** R. FINLAYSON *Brown Man's Burden* 10 In the clear space by the flagpole old Tamarua is delivering a speech of welcome, mere in hand. **1942** N. MARSH *Death & Dancing Footman* ix. 158 It hung on the wall there... It came from New Zealand... It's called a mere. **1956** M. DUGGAN *Immanuel's Land* 15 The chipped greenstone *mere*. **1957** *N.Z. Listener* 22 Nov. 4/2 We know what a 'mere'..or a 'hangi' is, but they remain essentially Maori in idea. **1975** D. BAGLEY *Snow Tiger* iv. 53 The greenstone *mere*—the Maori war axe.

meridianal, var. MERIDIONAL *a.*

1900 *Geogr. Jrnl.* XV. 540 The great meridianal systems, *e.g.* the Urals and the Rocky mountains. **1936** *Discovery* Mar. 69/2 The endeavour of the continents themselves to rotate is generally in a different direction from that of the poles of the earth, so setting up meridianal directed forces acting westward.

meridionality. Add to def.: the state of having an alignment or direction along a meridian (i.e. north or south). (Later example.)

1973 *Nature* 30 Mar. 295/1 During the present period of meridional global circulation the specific tendency over the European Arctic and sub-Arctic seas has been towards northerly meridionality and climatic deterioration... The increased northerly airflow has been held responsible.. for the extension of sea ice to the north Icelandic coast.

mering, mearing, *vbl. sb.* Also **mereing.** Restrict *Obs.* to sense 1, delete 'exc. *dial.*' and add: **2.** (Further examples.) (See also quot. 1975.) Also *fig.*

1900 *Century Mag.* Feb. 605 Billy starts up an' lifts the tether to lead the goat to the mearin' of the parishes. **1920** E. POUND *Umbra* 113 At midnight mirk, In secrecies I nurse My served make In heart; nor try My melodies At other's door nor mearing. **1925** W. B. YEATS *Early Poems & Stories* 449 Some world out of sight and misty, that has for its mearing the colours that are beyond all other colours. **1932** *Instructions to Field Revisers* (Ordnance Survey) App. A. 46 When boundaries have been surveyed, the initials of mereings are shown in ink along the surveyed features. *Ibid.* 48 Mereings on O.S. maps show to which field or inclosure the boundary fence..belongs, whether or not the land on both sides is owned or occupied by one and the same person. **1975** J. B. HARLEY *O.S. Maps* iii. 39 The term mereing has also been extended to apply to the written statement indicating the precise relationship of a boundary to the adjacent detail (for example, 4 ft RH = 4 ft from root of hedge)... Where space permits, mereings in this sense are placed parallel to the boundary, and when the type of mereing changes (as from the edge of a curb to the face of a wall) a special symbol.. indicates the point of change.

3. *mering-fence, -hedge, -stone, -wall.*

1846 T. KEIGHTLEY *Notes Bucolics & Georgics of Virgil* 9 On this side is the meering or boundary-hedge of sallows between you and your next neighbour. **1911** J. CAMPBELL (*title*) Mearing stones: leaves from my note-book on tramp in Donegal. **1928** R. A. S. MACALISTER *Archæol. of Ireland* i. 26 A mearing-wall (as boundary-walls are called in Ireland). **1933** *Irish Press* 10 Nov. 1/4 A huge earth bank which serves as a mearing fence between the lands.

meringue¹. Add: **b.** *meringue (à la) Chantilly*: a meringue filled with Chantilly cream; also *fig.*

1901 C. H. SENN *New Cent. Cookery Bk.* xxvii. 561 *Meringues à la Chantilly*... Fill the shells with whipped cream sweetened with vanilla sugar. **1949** *New Statesman* 15 Oct. 422/2 The food in the cafeteria is excellent, but it is in the dining-room that you get the Meringues Chantilly. **1959** *Listener* 2 July 39/1 For the meringues Chantilly take 3 egg whites. **1962** K. O'HARA *Double Cross Purposes* i. 5 An incredible figure wearing fly-leg heels, a meringue-chantilly hat and a pint of perfume. **1970** SIMON & HOWE *Dict. Gastron.* 264/1 (*heading*) Meringue Chantilly.

meringue², var. *MERENGUE.

merino. Add: **1. b.** *fig.* **pure merino** *Austral. slang*, an early immigrant to Australia with no convict origins; a member of a leading family in Australian society; a person of fine breeding or good character. Hence as *attrib. phr.*, first-class; well-bred; excellent.

1827 P. CUNNINGHAM *Two Yrs. New South Wales* II. xxiv. 116 The *legitimates*..such as have *legal* reasons for visiting the colony; and the *illegitimates*, or such as are free from that stigma. The *pure Merinos* are a variety of the latter species, who pride themselves on being of the *purest blood* in the colony. **1827** *Monitor* 13 Jan. 2 It operates greatly to the credit of the Pure Merino Bank. **1936** M. FRANKLIN *All that Swagger* vii. 69 It was too early to regard Delacy as a deserter,..he was a pure merino who would not abscond from a fine young wife. **1941** BAKER *Dict. Austral. Slang* 46 *Merino, pure*, originally, a free settler. Later, members of the alleged 'leading families'. Also (adj.) first-class, superlative in quality. **1954** T. RONAN *Vision Splendid* 113 'Old Mentmore,' he explained, 'is one of your pure merino sportsmen: member of all the big racing clubs down south.' **1966** G. W. TURNER *Eng. Lang. Austral. & N.Z.* i. 9 These sterling characters were also known as *pure merinos* when the sheep industry began to be based on imported merino sheep.

2. (Earlier examples.)

1818 M. EDGEWORTH *Let.* 29 Oct. (1971) 130 Tell me which you prefer the Merino or the Queens cloth... The Merino looks much the best in the piece. **1820** G. KEATS *Let.* in Keats *Lett.* (1958) II. 357 We will send Miss Brawn an india Crape dress or merino shawl or something scarce with you, but cheap with us.

b. (Earlier examples.) Also, a merino shawl.

1845 M. M. NOAH *Gleanings* 16 Thin kid shoes and gloves [are worn by the lady of fashion]; a fine merino over her arm. **1848** MRS. GASKELL *Mary Barton* I. iv. 44 She put on her pretty new blue merino, made tight to her throat. **1849** C. BRONTË *Shirley* II. ii. 46 These fine sunny days began to make me ashamed of my winter merino; so I have furbished up a lighter garment.

4. A variety of potato.

1849 E. EMMONS *Agric. N.Y.* II. 41 Merino Potato... Not highly esteemed for the table. **1853** *Trans. Mich. Agric. Soc.* V. 208 A great many varieties of the potato are cultivated in Europe and America. Some of the more approved kinds are..the Merino [etc.]. **1887** A. W. TOURGÉE *Button's Inn* 178 He picked up the potatoes,.. —delicate white 'Kidneys',..and coarse red 'Merinoes'.

merisis (meˑrisis). *Biol.* [mod.L., f. Gr. μερίς part (cf. MERIS) + *-sis*, Gr. suff. of action.] (See quot. 1962.)

1940 [see *AUXESIS 2]. **1962** R. RUGH *Exper. Embryol.* (ed. 3) 486/2 *Merisis*, growth by cell multiplication (in plants).

meristele (meristīˑl, -stīˑli). *Bot.* [f. Gr. μερίς part + STELE.] A strand of vascular tissue made up of xylem surrounded by phloem. Hence **meristeˑlic** *a.*

1894 S. H. VINES *Student's Text-bk. Bot.* I. 153 When the stem is monostelic, each leaf receives a portion, termed a meristele, of the stele of the stem. **1902** A. G. TANSLEY in *Encycl. Brit.* XXV. 413/1 Such a leaf-bundle contains parts of all the tissues of the stele, and is hence called a meristele. *Ibid.* 414/2 (*caption*) Meristelic arc of collateral bundles, united by conjunctive. **1911** [see *DICTYOSTELE]. **1956** *Nature* 4 Feb. 218/1 There is a reduction in the number of meristeles in the axis and in the leaves [of *Dryopteris aristata* under conditions of starvation]. **1964** A. J. BROOK *Living Plant* xxxv. 441 Each vascular strand or meristele, consists of a central core of xylem surrounded by a ring of phloem.

meristem. (Later examples.) **meristematic** *a.* (later examples).

1908 W. R. FISHER *Schlich's Man. Forestry* (ed. 2) V. I. i. 9 The widening of the cells of the meristem, or primary growing tissue of the yearling shoot, is caused by the water taken from the plasmic contents of the nascent organs. **1954** *Biol. Rev.* XXIX. 62 If the development is delayed, the bud appears as a 'detached meristem' in the sense that it is separated from other meristematic tissues by enlarging vacuolating cells. **1961** F. A. L. CLOWES *Apical Meristems* xi. 110 The primary meristem mantle..is prominent in some monocotyledons. **1967** *Times* 15 Mar. 7/1 The revolutionary method of propagating orchids by 'meristem' cuttings aroused great interest yesterday... This new method consists of cutting a tiny piece from the centre of the stem of an orchid..and placing it in a tube of culture solution. In two months this meristem cutting proliferates, producing as many as 20 tiny plants. **1974** *Nature* 2 Aug. 382/2 All divisions within the apical meristems of the shoots give rise to daughter cells with different fates: some remain meristematic.

merit, *sb.* Add: **5. b.** In Buddhism (and Jainism), the good actions in one of a person's successive states of existence which help determine his fate in the next; esp. in phr. *to acquire merit*.

1832 C. COLEMAN *Mythol. Hindus* xiii. 220 Like the Buddhas, they [*sc.* the Jains] believe that there is a plurality of heavens and hells; that our rewards and punishments in them depend upon our merit or demerit. **1834** *Jrnl. Asiatic Soc. Bengal* III. 383 These attributes appertain to persons subject to mortal births and deaths, of which the series is as little limited as is that scale of cumulative merits to which it expressly refers. **1853** R. S. HARDY *Man. Budhism* ix. 450 Let him who has the opportunity of acquiring merit, by being born when the precepts of Budha are taught, be careful not to let his privileges pass away without improvement. **1863** E. SCHLAGINTWEIT *Buddhism in Tibet* iv. 25 Emancipation takes place either instantaneously, on account of the merit accumulated in previous existences, or by assiduous attention to the various exercises prescribed. **1876** *Encycl. Brit.* IV. 433/1 This seed of existence Buddhism finds in 'Karma', the sum of merit and demerit. **1901** KIPLING *Kim* xiv. 380 'She has acquired merit beyond all others,' said the lama. 'For to set a man upon the way to Freedom is half as great as though she had herself found it.' 'Umm,' said Kim thoughtfully, considering the past. 'It may be that I have acquired merit also... At least she did not treat me like a child.' **1909** W. J. LOCKE *Septimus* xv. 225 I've done little enough in the world..! Give me this chance of —the Buddhists call it, 'acquiring merit'. **1920** —— *House of Baltazar* iv. 43 'A strong man keeps temptation at his elbow in order to defy it.' 'In that way, honourable master, is merit acquired.' **1935** *Chambers's Encycl.* II. 527/2 According to Buddhist belief, when a man dies he is immediately born again, or appears in a new shape; and that shape may, according to his merit or demerit, be any of the innumerable orders of being composing the Buddhist universe—from a clod to a divinity. **1967** D. T. KAUFFMAN *Dict. Relig. Terms* 309/2 Buddhism sees merit as the result of selfless love. **1972** 'E. PETERS' *Death to Landlords!* xi. 162 The easiest way to peace of mind is to give. It is a fairly cheap way to acquire merit. **1974** *Encycl. Brit. Micropædia* VIII. 303/2 Merit making involves *dāna* (giving, such as offering food and robes to monks or donating a temple or monastery); *sila* (the keeping of the moral precepts); and *bhāvanā* (the practice of meditation).

9. merit increase, an increase in pay resulting from **merit rating*; **merit money,** the money awarded in a merit increase; also *attrib.*; **merit rating,** the assessment of an employee's ability to do his job; a measurement of this ability; also *attrib.*; hence *merit rate*.

1952 *Federal Register* (U.S.) 15 Feb. 1466/1 The type of increase (such as 'ten percent increase', 'merit or length of service increase', 'inter-plant inequity'). **1959** J. H. TAYLOR *Personnel Administration* vii. 146 The descending line of authority should have a systematic basis for consideration of merit increases. **1961** R. LIKERT *New Patterns Managem.* ii. 15 Motivational forces..are involved in reactions to rating systems linked to merit increases. **1967** H. B. MAYNARD *Handbk. Business Administration* xi. x. 109 Generally, merit increases are granted for improved performance. **1973** *Daily Tel.* 10 Mar. 2/5 The inquiry found considerable support for general changes in salary levels to be clearly distinguished from individual 'merit' increases. **1947** O. DE R. FOENANDER *Industr. Regulation Austral.* ix. 84 'Merit money', whether purely voluntary, or under pressure from the Court in profitable industries, had, too, expanded. **1963** *Times* 11 Feb. 7/2 On the question of 'merit money' a spokesman said the title was misleading, and though theoretically such payments were at the discretion of supervisory staff they were not always based on merit and were sometimes virtually automatic. **1964** *Daily Tel.* 16 Jan. 23/1 (*heading*) Teachers' merit money urged by Bow Group. **1967** *Economist* 15 Apr. 278/2 The company's previous merit-money system has

proved both unfair and inflationary. **1946** F. H. JOHNSON et al. *Job Evaluation* xiii. 264 Experience has shown that merit rating's principal effect is to control the flow of newer workers as they go up the ladder. **1947** GILBRETH & COOK *Foreman in Manpower Managem.* xiv. 183 Requests for detailed discussion of the foreman's merit-rating sheet with executives will furnish a series of points of view ..which should be invaluable. **1954** *Merit Rating* (Brit. Inst. Managem.) ii. 6 An employee's merit rating history is invaluable for promotion purposes. **1959** *Gloss. Terms Work Study* (B.S.I.) 26 *Merit rating*, the systematic assessment of the behaviour and/or ability of workers in their work... *Merit rate*, the wage increment for a worker's merit. **1967** COULTHARD & SMITH in Wills & Yearsley *Handbk. Managem. Technol.* 206 Regrettably, the merit-rating forms usually place the emphasis on personality traits rather than performance... A more limited group have taken the merit-rating report to the next logical stage and set up a system of annual feedback meetings at which the manager and the subordinate carry out a post-mortem on the past year's work.

meritable, *a.* Delete ?*Obs.* and add later examples.

1927 *Observer* 20 Feb. 13/1, I should still have some meritable pieces left over in case the visitor wished to go to a few matinées. **1928** *Ibid.* 15 Apr. 8 What an odd yet meritable set of people.

meritocracy (meritǫˑkrăsi). [f. MERIT *sb.* + -OCRACY.] Government by persons selected on the basis of merit in a competitive educational system; a society so governed; a ruling or influential class of educated people. Hence **meˑritocrat** *sb.* and *a.*; **meːritocraˑtic** *a.*

1958 M. YOUNG *Rise of Meritocracy* iv. 71 Before the meritocracy was fully established, age-stratification as a substitute for the hereditary order may have been necessary for the sake of social stability. **1958** *Economist* 1 Nov. 407/1 Mr Young's meritocratic Britain, though described with ostensible enthusiasm, is an odious place. **1960** *Guardian* 28 Apr. 10/4 If it [*sc.* Oxbridge] adopts purely intellectual criteria it..is accused..of creating an isolated meritocracy. **1960** *20th Cent.* Apr. 358 They teach their future rulers, oligarchic or meritocrat, how to acquire and wield power. **1961** *Harper's Bazaar* Feb. 56/1 The grammar schools..have given birth to a new class, the meritocracy. **1961** D. JENKINS *Equality & Excellence* vi. 110 One of [Michael] Young's most unlikely prognostications is that the best public schools will be taken over by the meritocrats. **1964** M. ARGYLE *Psychol. & Social Probl.* xi. 146 Selection in Britain is not only producing a meritocracy, but is perpetuating a class system. **1966** *Guardian* 9 July 4/3 The lobby which says the independent schools produce 'insensitive meritocrats'. **1967** *New Scientist* 10 Aug. 308/1 Public subsidy of the future meritocrat elite is efficient, but inequitable. **1969** *Sci. Jrnl.* June 9 Pronouncements on the issue by scientists alone might appear to be the jabberings of a meritocracy. **1973** *Guardian* 21 Apr. 13/8 The monarchy..is, perhaps, our last stronghold..against the relentless advance of the meritocracy. **1973** *New Statesman* 28 Sept. 410/3 On the surface his background was impeccable for a rising Labour politician in the meritocratic Forties and Fifties. **1975** *Times Lit. Suppl.* 2 May 471/2 Keynes..was a meritocrat —not a democrat.

merkin. Add: **b.** Also, an artificial vagina. (Further examples.)

1886 R. F. BURTON tr. *Arabian Nights' Entertainments* X. 239 For the use of men they have the 'merkin', a heart-shaped article of thin skin stuffed with cotton and slit with an artificial vagina. **1962** E. WILSON *Night Thoughts* 203 Said Philip Sidney, buttoning his jerkin 'Allow me, darling: you have dropped your merkin.' **1967** G. LEGMAN *Fake Revolt* 31 A sort of ice-cold dildo-and-merkin combination. **1972** *Variant* I. vi. 54 *Variant* reporters interviewed a French pubic wig maker, the head of one of the world's most important firms making merkens and other 'intimate wigs'.

merle (mɜ̄ɹl), *a.* [f. dial. *mirlet, mirly* speckled.] Of a dog, especially a collie: having blue-grey fur speckled or streaked with black. Also as *sb.*, a dog coloured in this way. Cf. MARLED *ppl. a.*², MIRLED *ppl. a.*

1905 C. H. WHEELER in J. Watson *Dog Bk.* v. 351 The remainder of the litter [of collies] were blue merles. **1936** A. W. MEYER *Dogs* 262 Merle—Bluish-gray with some black, as seen in certain collies. **1948** C. L. B. HUBBARD *Dogs in Brit.* xviii. 193 The eyes [of collies] are medium in size, almond in shape, and coloured brown except in merle dogs when they are 'wall' (a blue-and-white or china shade). *Ibid.* 463 Merle.—The term applied to a blue-grey mixture flecked or streaked with black, uncommon except in working sheepdogs. **1968** J. F. GORDON *Beagle Guide* 240 *Merle*, bluish-grey color marbled with black, seen in working sheepdogs. **1971** F. HAMILTON *World Encycl. Dogs* 67 (*caption*) Blue merles provide an extra challenge for Collie breeders, because of the difficulty of producing correct markings and a good color. Silva Gale from Shiel was not only a lovely merle but certainly lost no points in head.

mermaid. Add: **5. b.** A vigorous climbing rose with single yellow flowers, produced by crossing *Rosa bracteata* (**Macartney rose*) with a yellow tea rose and introduced in 1918 by William Paul & Son.

1918 *Rose Annual* 148 The most sensational Rose among Climbers of the year was undoubtedly that fine Hybrid, Mermaid. The huge golden lemon single flowers charmed everyone who saw it. **1922** T. G. W. HENSLOW *Rose Encycl.* xix. 409/2 Mermaid..Wm Paul & Son 1918: —Flower sulphury yellow,..of great size, single, produced continuously from early summer till late in autumn. **1965** G. S. THOMAS *Climbing Roses* iv. 62 A few old roses..., and some newer varieties like 'Mermaid' and 'New Dawn', show that grace and recurrence of bloom can be combined. **1974** *Times* 26 Oct. 12/4 Plant a fast-growing rose like Mermaid or Albertine to cover the pergola.

mero-¹. Add: **meˑrocrine** (-krəin) *a. Physiol.* [ad. F. *mérocrine* (L. Ranvier 1887, in *Jrnl. de Micrographie* XI. 9), f. Gr. κρίν-ειν to separate], of, pertaining to, or designating a gland in which secretion is unaccompanied by any substantial change in the secreting cells; **merocyˑanine** *Chem.* [so called from having part of the structure of cyanine dyes], any of a class of neutral dyes (many of which are used as sensitizers for photographic emulsions) in which a nitrogen atom (usu. part of a basic heterocycle) is linked to a carbonyl group (usu. part of an acidic heterocycle) by a conjugated chain of carbon atoms; freq. *attrib.* in *merocyanine dye*; **meromiˑctic** *a.* [ad. G. *meromiktisch* (I. Findenegg 1935, in *Internat. Rev. d. ges. Hydrobiol.* XXXII. 377), f. Gr. μικτός mixed], applied to a lake in which, when overturn occurs, water below a certain depth does not take part in it owing to its high density (usu. the result of a high salt concentration); so **meromiˑxis** [Gr. μίξις mixing], the state of being meromictic; **meromyˑosin** *Biochem.*, either of the two fractions obtained from myosin by the action of enzymes, of which the lighter fraction consists of long rod-shaped molecules from the 'tail' of the myosin molecule and the heavier one consists of molecules having a short rod-shaped portion attached to a globular 'head'; **meroplaˑnkton** *Biol.* [back-formation from the adj.], a collective term for aquatic organisms that are meroplanktonic; **meroplanktoˑnic** *a.* [ad. G. *meroplanktonisch* (E. Haeckel *Plankton-Studien* (1890) iii. 25)], passing only part of the life-cycle drifting or swimming weakly in water; **merozoˑite** *Zool.* [ad. F. *mérozoïte* (P. L. Simond 1897, in *Ann. de l'Institut Pasteur* XI. 551): see *-ZOITE], (in many sporozoa, esp. in the orders Coccidia and Hæmosporidia) any of the cells produced by multiple fission (schizogony) of a schizont.

1905 GOULD *New Med. Terms* 358/1 Merocrine. **1928** E. V. COWDRY *Special Cytol.* I. ii. 36 The sweat, or sodoriferous glands, are of distinctly different nature... Ranvier speaks of these glands as 'merocrine'. **1958** *Gray's Anat.* (ed. 32) 1302 Most of the sweat glands are merocrine in nature, i.e. produce their thin watery secretion without demonstrable changes in the gland epithelium. **1965** LEE & KNOWLES *Animal Hormones* v. 91 [In the thyroid gland] the colloid is secreted from the surface of the epithelium into the lumen of the acinus (merocrine secretion). **1937** L. G. S. BOOKER *U.S. Pat.* 2,078,233 27 Apr. 6/1 It has recently been proposed to call the dyes of this new and very large class, merocyanine dyes. **1955** D. GRAHAM in H. A. Lubs *Chem. Synthetic Dyes & Pigments* xii. 676 The λ max. of a strongly polar merocyanine dye shifts to a shorter wavelength with an increase in polarity of the solvent. The λ max. of a weakly polar merocyanine shifts to longer wavelengths. **1956** K. M. HORNSBY *Basic Photogr. Chem.* iii. 34 While these merocyanines are..useful sensitisers, those containing a >C=S grouping..can be converted to more complex dyes. **1973** *Nature* 21–28 Dec. 508/1 This was achieved in a giant axon using a merocyanine dye; in a stained axon a single action potential gave rise to a fluorescence increase which was detectable with a signal-to-noise ratio greater than 10:1. **1937** *Trans. Connecticut Acad. Arts & Sci.* XXXIII. 74 The lake is in fact meromictic, to use Findenegg's (1935) useful term. **1970** *Limnol. & Oceanogr.* XV. 363 (*heading*) Physicochemical limnology and geology of a meromictic pond on the Red Sea shore. **1955** *Mem. Ist. Ital. Idrobiol. de Marchi* VIII. Suppl. 141 (*heading*) Längsee: a history of meromixis. **1970** *Limnol. & Oceanogr.* XV. 363 The meromixis is described of a small (140 × 50 m) coastal pond on the Sinai shore of the Red Sea. **1952** A. G. SZENT-GYÖRGYI in *Federation Proc.* XI. 297/1 The fraction with the lower sedimentation constant (meromyosin-L), has the peculiar solubility of myosin... The fraction with the higher S_{20} (meromyosin-H), is soluble at any KCl concentration at $_pH$ 7 and precipitates at $_pH$ 5·1. **1966** *McGraw-Hill Yearbk. Sci. & Technol.* 258/2 These experiments have shown that the myosin molecules aggregate in an antiparallel, overlapping fashion, the straight L-meromyosin forming the backbone of the filament and the H-meromyosin constituting the cross bridges. **1909** GROOM & BALFOUR tr. *Warming's Oecol. Plants* xxxviii. 161 These terms 'neritic' and 'pelagic' or 'oceanic' plankton approximately correspond to Haeckel's 'meroplankton' [*printed* 'neroplankton'] and 'holoplankton'. **1942** H. U. SVERDRUP et al. *Oceans* xvii. 814 This temporary element, or meroplankton as it is sometimes called, is especially abundant in the neritic waters. **1967** *Oceanogr. & Marine Biol.* V. 241 The results are summarized and brought together in a voluminous thesis on the larvae of Crustacea Decapoda of the meroplankton of the Gulf of Marseilles. **1973** *Nature* 16 Feb. 475/2 There are few comparable estimates of growth efficiencies for meroplankton species. **1893** G. W. FIELD tr. *Hæckel's Planktonic Stud.* in *Rep. U.S. Comm. Fisheries 1889–91* 583 The meroplanktonic organisms..are found swimming in the sea only for a part of their lives, passing the other part vagrant or sessile in the benthos. **1903** *Amer. Naturalist* XXXVII. 516 The meroplanktonic stage..was apparently suppressed even in the earliest species of Fulgur. **1963** J. E. G. RAYMONT *Plankton & Productivity in Oceans* xiv. 371 Even with coastal or neritic plankton it is not the meroplanktonic species which cause the main seasonal fluctuations. **1900** *Jrnl. R. Microsc. Soc.* 336 In the author's nomenclature this process of asexual multiplication is known as schizogony, the mother-cells are schizonts, and the daughter-cells merozoites. **1940** L. H. HYMAN *Invertebrates* I. iii. 144 The growing vegetative parasite is called a trophozoite. When this..undergoes multiple fission directly into agametes, it is called a schizont or agamont, the multiple fission is termed schizogony or agamogony, and the agametes are known as merozoites. **1967** J. H. WILMOTH *Biol. Invertebr.* ii. 40/2 Among the Coccidia, multiple division occurs during both asexual and sexual phases. *Eimeria schubergi* parasitizes the intestinal cells of the centipede, *Lithobius*. Infective sporozoites invade epithelial cells of the host. Schizogony produces many merozoites which are freed to invade new cells.

Meroitic (meroˌiˑtik), *a.* and *sb.* [f. *Mero-ë*, the name of the capital of an ancient Nubian kingdom in north-east Africa, + *-itic*.] **A.** *adj.* Of, belonging to, or pertaining to an ancient Nubian kingdom (see above), its language, or its inhabitants. **B.** *sb.* The language of this kingdom.

1852 tr. *C. R. Lepsius's Discoveries in Egypt* xx. 227 Three royal personages..had built the principal temples in Naga, Ben Naga and in Wadi Temêd, and belonged..to the most shining period of the Meroitic Empire. **1902** E. A. W. BUDGE *Hist. Egypt* VIII. 169 The Meroïtic inscriptions have not as yet been deciphered. **1911** F. L. GRIFFITH *Karanòg* vii. 83 It may be that, while Meroitic was the official language for writing, Nubian was the mother-tongue of Lower Nubia. **1911** A. H. SAYCE in J. Garstang et al. *Meroë* ix. 50 Whether *u*ᶜ was actually the word for 'land' in Meroitic is..uncertain. *Ibid.* 51 The destruction is more complete even than that of the Meroitic temples at Kerma and Kawa. **1934** A. TOYNBEE *Study of Hist.* II. 117 This Meroitic Power lived on, as a politically independent embodiment of the Egyptiac Society, until the third century of the Christian Era. **1955** A. J. ARKELL *Hist. Sudan* vii. 160 The Meroitic hieroglyphic..script seems to have been invented on the basis partly of the Meroitic cursive script, and partly of Egyptian hieroglyphs. **1967** P. L. SHINNIE *Meroe* v. 133 There is also just a hint that Meroitic may be related to the little-known group of languages known as Koman, now spoken in a limited area up the Blue Nile and around Jebel Gule. *Ibid.* vii. 156 The numerous finds of glass vessels imported as containers for oils and unguents as well as the mirrors show that Meroitic ladies, at least of the richer classes, took considerable trouble over their appearance.

meroola, var. *MARULA.

merrily, *adv.* Delete 'Somewhat *arch.*' and add further examples.

1912 G. B. SHAW *John Bull's Other Island* Pref. p. vi/2 He chalks up No Surrender merrily, and puts up one of the famous fights of history. **1926** W. R. INGE *Lay Thoughts* III. i. 185 The process [*sc.* increase of population] went on merrily at first because the new countries produced far more food than they needed for themselves. **1939** JOYCE *Finnegans Wake* III. 615 On the top of the longcar, as merrily we rolled along, we think of him looking at us yet. **1966** *Listener* 17 Feb. 247/1 Frontier wars ..continue as merrily in our own nuclear age as ever they did before 1914. **1972** R. ADAMS *Watership Down* xxxiii. 261 Why not a water-rabbit? I shall float merrily along—. **1974** *Observer* 3 Mar. 34/3 McKenzie merrily complained that the Battleground machine wasn't telling him anything.

merry, *a.* Add: **1. b.** Also *Merrie England*, freq. in ironic or satirical use; so *Merrie Englander*.

1839 G. DANIEL in *Bentley's Misc.* v. 98 (*title*) Merrie England in the olden time. *c* **1882** (*title*) Merrie England its kings & queens..with pictures and stories from English history. **1893** *Ladies' Home Jrnl.* Feb. 13/1 The old Roman custom..was transplanted to merrie England. **1902** HOOD & GERMAN (*title*) Merrie England: a new and original comic opera in two acts. **1912** A. HUXLEY *Let.* 16 June (1969) 43 It [*sc.* a monument] is made of a peculiar grey stone, which looks just like that horrible papier-mâché stone of which ruined castles are made in exhibitions of Merrie England. **1930** E. WAUGH *Labels* 22 The Merrie-Englanders have so eloquently upheld the cause of freedom that a subdued but smouldering resentment is now one of our national characteristics. **1946** J. B. PRIESTLEY *Bright Day* x. 301 The war was over, democracy saved, and here was Merrie England. **1970** H. BRAUN *Parish Churches* xx. 236 The happy breed of the wool-rich yeomen followed in that Merrie England which was later to turn to mourning for Englishmen clubbed to death by neighbours in the Great Rebellion. **1971** B. INGLIS *Poverty & Industr. Revolution* iv. 197 Cobbett.. was far from being a Merrie Englander, in the sense of seeing the past bathed in a romantic glow. **1973** *Radio Times* 13 Dec. 27 (*caption*), The redoubtable Dr Who returns..to fight *The Time Warrior* in not-so-merrie England: 5.10. **1974** *Punch* 23 Oct. 654/1 Mecca had planned to build a Merrie Englande entertainment complex to rival Disneyland.

3. c. *merry hell*: see *HELL *sb.* 10 q.

f. *the more the merrier* (later examples).

1614 JONSON *Barth. Fair* (1631) I. vi. 83, I, and Salomon too, Win, (the more the merrier) Win, we'll leaue Rabby Busy in a Booth. **1811** JANE AUSTEN *Sense & Sens.* II. iii. 38 The more the merrier say I, and I thought it would be more comfortable for them to be together. **1874** TROLLOPE

MERRY-GO-ROUND

Way we live Now (1875) I. xxxiii. 208 The more the merrier. Ruby'll have enough for the two o' you, I'll go bail. **1922** E. O'NEILL *Hairy Ape* (1923) v. 47 De more de merrier when I gits started. **1928** A. HUXLEY *Point Counter Point* xi. 168 The more the merrier was her principle; or if 'merrier' were too strong a word, at least the noisier, the more tumultuously distracting. **1952** 'M. COST' *Hour Awaits* 28 The more the merrier,..for with her cousins also crowding the Hotel, additional 'cover' would be afforded. **1974** 'D. FLETCHER' *Lovable Man* II. 120, I moved over to features. More the merrier. I'm free-lance now.

merry-go-round. Add: **2.** Also **merry-go-rounder**, a cause of astonishment.
1838 DICKENS *O. Twist* II. xxv. 81 Oh, my eye! here's a merry-go-rounder!—Tommy Chitling's in love!
3. Used *attrib.* or as *adj.* of a railway system, whereby a train of coal hoppers runs perpetually on a circular route between consignor and consignee.
1963 *Mod. Railways* Jan. 23 Some of the desirable characteristics of the 'Merry-go-round' railway..are found in the Tyne Dock–Consett iron ore traffic of the N.E.R. **1966** G. F. ALLEN *Brit. Rail after Beeching* viii. 255 Inauguration of 'merry-go-round' coal supply by rail, one of the outstanding concepts of the Beeching era. **1970** *Railway Mag.* Oct. 552/1 Wagons, each capable of carrying up to 32 tons, are permanently coupled into 'Merry-go-round' trains, making 'non-stop' journeys between collieries and generating stations. *Ibid.*, The fully-automated 'Merry-go-round' system works efficiently. *Ibid.* 584/2 Merry-go-round workings in the Knottingley area have seen a variety of motive power recently. **1972** P. LEVY *Spotlight on Trains* 32 British Rail came up with the idea of the 'merry-go-round' train..a long train of coal hoppers running in a circular route between the coal mine and the power station. **1973** *North Berks Herald* 13 Dec. 1/2 Sixteen merry-go-round coal trains have been arriving at Didcot power station each week.

merry-maker. (Earlier example.)
1827 G. GRIFFIN *Holland-Tide* iii. 236 Music..was constantly in high request..among the merry-makers.

merry man, merryman. Add: **1.** *pl.* Also *colloq.* (somewhat jocular): followers, subordinates.
1873 TROLLOPE *Phineas Redux* (1874) I. i. 2 Moderate Liberals had been glad to give Mr. Daubeny and his merry men a chance. **1921** *Daily Colonist* (Victoria, B.C.) 8 Apr. 4/3 The result of their round robin was equally effective;..it took the place of the robber's mask so that no one could tell the leaders from their merrymen. **1932** D. L. SAYERS *Have his Carcase* xvi. 205 It's Umpelty and his merry men. Pass me the field-glasses, Bunter. **1939** JOYCE *Finnegans Wake* I. 48 *Hurleyquinn* the zitherer of the past with his merrymen all. **1960** N. MITFORD *Don't tell Alfred* ii. 24 He will keep Bouche-Bontemps and his merry men in a state of chronic perplexity. **1970** *Punch* 17 June 890/1 The Tories, according to Harold Wilson and his merry men, are dishonest, heartless, reckless and dangerous. **1973** *Times* 24 Nov. 11/4 Tomorrow night Miss Laine, Mr Dankworth and his three merry men will be in concert in Glasgow. **1973** R. PERRY *Ticket to Ride* iv. 61 Abbott and his merry men still weren't on my trail.

Merry Widow. a. The English name of Franz Lehár's operetta *Die Lustige Witwe* (first produced (in German) in Vienna, 1905, and (in English) in London, 1907) used allusively, freq. joc., of an amorous or designing widow.
1907 *Times* 10 June 4/3 The fame of *Die Lustige Witwe* must have preceded the coming of the opera, for the appearance of the composer was greeted with thunders of applause... *The Merry Widow*..is a genuine light opera. .. Perhaps, in the original, Sonia..is a 'merry' widow. **1922** JOYCE *Ulysses* 209 Woos and wins her, a whoreson merry widow. **1942** H. C. BAILEY *Dead Man's Shoes* xxvii. 105 Randolph also found the marriage in the paper that morning... He burst out laughing..'Queen Caroline! The merry widow!' **1961** *Studies in Eng. Lit.* (Houston, Texas) I. IV. 23 (*heading*) *Lady Susan*: Jane Austen's character of the Merry Widow. **1965** J. M. CAIN *Magician's Wife* (1966) iv. 31 She'll be a Merry Widow, that we know for sure, but not with your help.
b. Used *attrib.* and *absol.* to designate a type of ornate, wide-brimmed hat.
1908 *Daily Chron.* 9 July 1/4 The women in the galleries took off their 'Merry Widow' hats, and waved them frantically. **1909** *Ibid.* 21 Jan. 7/3 A huge Merry Widow of the approved Occidental pattern from China. **1922** JOYCE *Ulysses* 554 Under the umbrella appears Mrs. Cunningham in Merry Widow hat and kimono gown. **1956** C. H. B. KITCHIN *Secret River* i. 61 Mrs Ashworth in a Merry Widow hat, in which she thought she had looked ravishing. **1966** *Times* 15 June 15/5 When *The Merry Widow* music echoed across Europe, Lily Elsie wore *the* hat every night at Daly's. It caught on. From the opening night, June 8, 1907, everybody wore the Merry Widow hat.

mersalyl (mɜ̄·ɪsălil). *Pharm.* [f. MER(CURY *sb.* + SAL(IC)YL.] The sodium salt, $C_{13}H_{16}NO_6HgNa$, of *o*-[(3-hydroxymercuri-2-methoxypropyl)carbamoyl]phenoxyacetic acid, which is a powerful diuretic formerly used in the treatment of œdema.
1936 *Brit. Pharmacopœia 1932* Add. I. 46 (*heading*) Mersalyl. **1940** *Jrnl. Amer. Med. Assoc.* 23 Nov. 1786/1 (*heading*) The effect of mersalyl (Salyrgan) on plasma volume. **1958** *Times* 23 Apr. 15/4 He had referred to ammonium chloride as a diuretic normally used in conjunction with mersalyl. **1972** *Nature* 7 Apr. 301/2 Mersalyl, a potent inhibitor of myofibrillar ATPase, was without effect on the contraction induced by calcium.

Mersenne (mɜ̄ɪ-, ‖ mɛrse·n). *Math.* The name of Marin *Mersenne* (1588–1648), French mathematician and musician, used *attrib.* and in the possessive to designate numbers of the form $2^p - 1$ (where *p* is a prime number).
1892 *Messenger of Math.* XXI. 40 The riddle as to how Mersenne's numbers were discovered remains unsolved. **1911** *Encycl. Brit.* XIX. 863/1 Similar difficulties are encountered when we examine Mersenne's numbers, which are those of the form $2^p - 1$, with *p* a prime; the known cases for which a Mersenne number is prime correspond to *p* = 2, 3, 5, 7, 13, 17, 19, 31, 61. **1939** USPENSKY & HEASLET *Elem. Number Theory* iv. 82 Numbers of the form $2^p - 1$ are called Mersenne's numbers because of a statement made concerning them in the preface to his 'Cogitata physico-mathematica', published in 1644. **1966** OGILVY & ANDERSON *Excursions in Number Theory* ii. 22 The Mersenne number $2^{11} - 1$ is composite (it equals 23 × 89). **1974** *Nature* 16 Aug. 610/3 The largest known Mersenne prime, by 1971, is $2^p - 1$ where $p = 19937$.

Mersey (mɜ̄·ɪzi). The name of the river *Mersey*, on which stands the city of Liverpool, applied *attrib.* in **Mersey beat, Mersey sound,** to the kind of popular music associated with 'The Beatles' [*BEATLE].
1963 *Meet the Beatles* 12/2 The Beatles, undoubted monarchs of the Mersey Beat scene. **1965** S. JEPSON *Third Possibility* v. 36 The Mersey sound banged and twanged into the night. *a* **1966** M. ALLINGHAM *Cargo of Eagles* (1968) iv. 60 The adenoidal moan of the Mersey beat: '*I wanna be your rave*...' **1969** C. BOOKER *Neophiliacs* viii. 203 The 'Liverpool phenomenon' and the 'Mersey Sound' were now [*sc.* in 1963] arousing interest far beyond circles normally interested in pop music.

Merseysider (mɜ̄·ɪzisəideɪ). [f. *Mersey side* (SIDE *sb.*[1] 7 a) + -ER[1].] A dweller on a bank of the river Mersey in England, *spec.* within the area of Liverpool. Cf. *CLYDESIDER.
1943 *Bombers over Merseyside* (Liverpool Daily Post) 25 (*caption*) It was spectacles of destruction, such as this, that steeled the Merseysiders' resolution to defeat the Luftwaffe's brazen challenge. **1963** *Times* 17 Apr. 3/6 Everton took root in Birmingham territory, yet..the Birmingham defensive covering was such that there were serious doubts about the Merseysiders' ability to create the vital chink. **1966** 'L. LANE' *ABZ of Scouse* Foreword, A Scouser is a Merseysider who conducts his ordinary, everyday conversations in Scouse. **1971** N. FISHER *Rise at Dawn* vi. 95 This was driven by Lance-Corporal Simmonds ..a taciturn Merseysider. **1973** *Guardian* 4 June 16/4 The peculiar inspirational quality of the city..laughs at the rest of the world who has never known the magic of being ..a Merseysider.

Merthiolate (mɜ̄ɪþəi·ŏlēit). *Pharm.* Also **merthiolate.** [f. *mercurithiosalicylate*, f. MERCURI(CO- + THIO- + SALICYLATE *sb.*] A proprietary name (in the U.S.) for thiomersal, sodium ethyl mercurithiosalicylate.
1928 *Official Gaz.* (U.S. Patent Office) 6 Nov. 14/2 Eli Lilly and Company... *Merthiolate* for medicine or pharmaceutical preparation—viz., sodium mercurithiosalicylate or organic mercury compound solution useful in antisepsis. **1931** *Amer. Jrnl. Hygiene* XIII. 310 Merthiolate is found to have certain unusual properties..which makes it well adapted to tissue antisepsis. **1957** F. & R. LOCKRIDGE *Tangled Cord* (1959) xii. 150 Hilda Graham.. had found merthiolate in a medicine cabinet in the bathroom and poured it on Ferris's foot. **1973** *Nature* 21/28 Dec. 521/2 The supernatant was decanted, a sample removed for protein estimation by the method of Lowry, and stored at 4° C with the addition of 1/10,000 merthiolate as preservative.

Mertonian (mɜ̄ɪtōᵘ·niăn), *sb.* and *a.* [f. the name of *Merton* College (founded by Walter de Merton in 1264) + -IAN.] **A.** *sb.* A member of Merton College, Oxford.
1883 *Fortn. Rev.* XXXIX. 34 Another Mertonian, John Tatham,..was elected Rector of Lincoln College. **1899** B. W. HENDERSON *Merton College* 172 Not a few Mertonians have been appointed to University Chairs. **1954** *Postmaster* (Merton College, Oxf.) Sept. 13 The only other Mertonian to appear in New London was Louis MacNeice. **1961** D. KNOWLES *Eng. Mystical Trad.* iii. 41 In mathematics and kindred sciences the series of great Mertonians at Oxford, Thomas Bradwardine, Richard Swineshead, William Heytesbury and Ralph Strood were the masters of the academic world of their day. **1971** E. GRANT *Physical Sci. in Middle Ages* iv. 25 The Mertonians arrived at a precise definition of uniform acceleration.
B. *adj.* Of or pertaining to Merton College or its members; used *spec.* with reference to a school of mathematics and astronomy that existed there in the 14th century.
1899 B. W. HENDERSON *Merton College* 278 The society ..entertained a large Mertonian company at a dinner in Hall. **1947** G. SARTON *Introd. Hist. Sci.* III. i. 116 Our knowledge of the early Mertonian scientists is very insufficient, because a good part of the Merton library and archives was sold as waste paper about the middle of the sixteenth century. **1959** A. C. CROMBIE in M. Clagett *Critical Probl. Hist. Sci.* 91 In finding expressions for rates of change, they [*sc.* Oxford mathematicians] formulated sophisticated concepts like those of acceleration and instantaneous velocity..and reached important results like the Mertonian Mean Speed Law. **1974** A. J. POMERANS tr. *Clavelin's Nat. Philos. Galileo* ii. 80 This proof.. remained indirect, as did all the Mertonian attempts to prove the mean-speed theorem.

Meru (me·ru), *sb.* and *a.* Also **Mweru.** [f. the name of a town and district in central Kenya.] **A.** *sb.* **a.** A Bantu tribe inhabiting the Meru region of Kenya; also, a member of this tribe. **b.** The language of this people. **B.** *adj.* Of or pertaining to the Meru people or their language.
[**1883** R. N. CUST *Sk. Mod. Lang. Afr.* II. xii. 342 A little to the North of the Ma-Konde dwell the Wa-Mwera, not a large tribe, but..they have a separate language.] **1909** J. H. PATTERSON *In Grip of Nyika* xxviii. 350 They approached, decked out in all the finery of the Meru belles, and each with a broad smile..selected a favourite warrior. **1910** C. W. HOBLEY *Ethnol. of A-Kamba* vi. 156 Mweru is the name of a very large tribe living on the North and N.E. slopes of Kenia and on the Jombeni range. **1919** H. H. JOHNSTON *Compar. Study Bantu & Semi-Bantu Lang.* iii. 112 The Meru dialect..is said to be markedly distinct. It is spoken in the north-east portion of the Kikuyu area. **1942** *Man* XLII. 58/1 To the north of the Mwimbi and Tharaka, near neighbours also of both the Kikuyu and Kamba peoples, are to be found the Meru-speaking peoples, numbering roughly some 150,000. *Ibid.* 59/1 The most significant feature of the Meru tribal organization is the intricate system of age grades which cuts across family and clan loyalties. **1944** W. H. LAUGHTON *Meru* 8 The Meru live in families in small groups of huts built on the numerous hillsides. **1953** J. MIDDLETON in *Ethnogr. Survey Afr.: E. Cent. Afr.* (Internat. Afr. Inst.) v. 40 The Meru have a system of age-sets based on circumcision. **1963** *Times* 25 May 8/7 Fighting broke out when..Somalis surrounded the polling station..intent on preventing Turkana and Meru tribes people from voting.

merula, var. *MARULA.

mesa. Delete ‖ and *South. U.S.*, and add further examples.
1948 C. A. COTTON *Landscape* (ed. 2) x. 139 (*caption*) The Schlern mesa, of dolomite, South Tyrol. **1951** WODEHOUSE *Old Reliable* v. 68 There he was..under the dressing-table, with his fanny sticking up like a mesa in the Mojave desert. **1963** A. LUBBOCK *Austral. Roundabout* 73 The descent on the northern side of the plateau winds down, between table-top mesas and rugged bluffs. **1970** R. J. SMALL *Study of Landforms* iii. 72 To the east of Lyme Regis..very broad valleys are separated in interfluves occasionally surmounted by butte-like hills..and mesa-like plateaus developed in the near horizontal Upper Liassic sandstones. **1974** H. MACINNES *Climb to Lost World* x. 163 Only the proud tops of the neighbouring mesas stood out above the swirling clouds. *Ibid.* xii. 217 We saw a magnificent panorama of mesas: a weird contorted skyline of grotesque sandstone figures towards the Venezuelan part of the summit.
fig. **1963** V. NABOKOV *Gift* iii. 150 Out of the total of five hundred copies printed, four hundred and twenty-nine still lay, dusty and uncut, forming a neat mesa in the distributor's warehouse.
2. *Electronics.* In some transistors and semiconductor diodes, a raised, flat-topped portion of *n*- or *p*-type semiconductor surrounded by an area from which the upper layer has been etched away to expose the underlying *p*- or *n*-type material (respectively). Usu. *attrib.*, as *mesa diode, transistor.*
1958 C. H. KNOWLES in *Electronic Industries* (Philadelphia) Aug. 55 Recent developments at several locations have resulted in a new line of VHF–UHF Transistors... We will refer to this new line of transistors as Mesa Transistors. *Ibid.*, The Mesa Transistor gets its name from its physical configuration... A basic part of its structure is the 'Mesa', (Spanish name for 'table'), which is the active region of the transistor. **1962** SIMPSON & RICHARDS *Physical Princ. Junction Transistors* viii. 169 Following construction of the device the germanium surrounding the electrodes was etched away to form the table-like area or 'mesa'. **1966** *McGraw-Hill Encycl. Sci. & Technol.* VII. 316/1 Silicon mesa diodes are also used for high-speed, low-power applications. **1972** BOYLESTAD & NASHELSKY *Electronic Devices & Circuit Theory* iii. 137 The diffusion technique is employed in the production of mesa and planar transistors, each of which can be of the diffused or epitaxial type. **1973** *Nature* 12 Jan. 92/3 Improved cut-off frequency..and noise performance result in useful performance up to 4 GHz with promise of extending this to at least 8 GHz by reverting to an improved version of the old 'mesa' design.

mesarch (me·zāɪk), *a.* [f. MES(O- + Gr. ἀρχή beginning.] **1.** *Bot.* [a. G. (H. zu Solms-Laubach *Einleitung in die Paläophytologie* (1887) xi. 263).] Of the development of the primary xylem, spreading in two or more directions from the earliest parts formed.
1891 [see *EXARCH *a.*]. **1902** A. G. TANSLEY in *Encycl. Brit.* XXV. 414/2 (*caption*) Typical siphonostele (represented as mesarch) with internal phloem. **1954** *Biol. Rev.* XXIX. 63 Such studies resulted in the recognition of the now well-known centripetal and centrifugal patterns of differentiation, the exarch, endarch, and mesarch types of xylem. **1965** K. ESAU *Plant Anat.* (ed. 2) xv. 378 In the third [type of xylem] the differentiation progresses in two or more directions from the first mature xylem elements. The resulting primary xylem is called mesarch.
2. *Ecology.* (See quot. 1923.)
1923 G. E. NICHOLS in *Ecology* IV. 171 A third type [of succession] is here suggested, *viz.*, mesarch... Mesarch series are those which originate in mesophytic habitats such as are afforded by moist, rich soils, and in which the vegetation likewise becomes progressively more and more advanced. **1960** [see *HYDRARCH *a.*].

mesatipellic (me:sătipe·lik), *a.* *Anthrop.* [f. Gr. μέσ(σ)ατ-ος midmost (superl. of μέσος

middle) + -I- + πέλλ-α wooden bowl (taken in sense PELVIS) + -IC.] Characterized by an index of the pelvic brim between 90 and 95.
1886 W. TURNER *Rep. Crania* in *Rep. Sci. Results Voy. H.M.S. Challenger* XVI. 40 In the males the same index was at or about 91, so that they were mesatipellic. **1924** *Proc. Soc. Antiquaries Scotl.* LVIII. 34 An index of 93·22 is mesatipellic, and considerably higher than that of the average European male. **1966** B. J. ANSON *Morris's Human Anat.* (ed. 12) iv. 282/2 Pelves with an index below 90 are platypellic; from 90 to 95, mesatipellic; and above 95, dolichopellic.

mesaxon (mesæ·ksǫn). *Anat.* [f. MES(ENTERY (see quot. 1955) + *AXON.] In a nerve fibre, a structure composed of a pair of parallel membranes contiguous with the plasma membrane of the Schwann cell, forming in an unmyelinated fibre a channel leading from the outside of the cell to the axon it surrounds and in a myelinated fibre a spiral round the axon that constitutes the myelin sheath.
1955 H. S. GASSER in *Jrnl. Gen. Physiol.* XXXVIII. 713 The membrane sharply visible on the outside of the Schwann tubes may be taken as the starting point. To this membrane..the axons are in continuous attachment. On account of the analogy with the mesentery the attachments have been designated mesaxons. **1961** *Lancet* 16 Sept. 656/1 H. F. de Webster..in a study on the demyelination which occurs in experimental diphtheritic neuritis found that the earliest changes amount to focal fragmentation of the mesaxons. **1970** J. PICK *Autonomic Nervous System* v. 127/1 According to Gasser.., an unmyelinated nerve fiber usually consists of several axons which are placed in the infoldings of their Schwann cell. The plasma membrane of this sheath cell doubles up and together with the intervening channel forms the mesaxon. Upon reaching the axon, the membranes of the mesaxon separate and surround the axolemma.

mescal. Add: Also 8–9 **mascal, mescale, mixcal, muscal(e. 1.** (Further examples.)
1833 B. LUNDY *Jrnl.* 16 Oct. in *Life B. Lundy* (1847) x. 61 They have a kind of whiskey here, called muscal, which is distilled from a plant called Maguey. **1854** in *Southwestern Hist. Q.* (1931–2) XXXV. 235 About one dozen men came over, bringing..muscal liquor... This Liquor has a tast between whiskey and brandy, and considerable intoxicating power. **1926** D. H. LAWRENCE *Plumed Serp.* iv. 80 The fiery white brandy distilled from the maguey: mescal, tequila. **1947** M. LOWRY *Under Volcano* x. 319 There had been until recently several drinks of mescal (why not?—the word did not intimidate him, eh?) waiting for him outside in a lemonade bottle. **1974** *Encycl. Brit. Macropædia* X. 972/2 The alcoholic liquor mescal is distilled from pulque.
2. Substitute for def.: Any of several plants of the genus *Agave* found in Mexico and the southwestern United States that are used as sources of fermented liquor, food, or fibre, esp. the American aloe or maguey, *A. americana*, a stemless plant having long spiny leaves. (Earlier and later examples.)
1743 J. LOCKMAN tr. *Trav. Jesuits* I. 399 On the Mountains grew Mescales, a fruit peculiar to the Country, and is gathered all the year round. **1759** [see *2 b]. **1808** W. SHALER *Jrnl. Voy. between China & Amer.* (1935) 53 They also have a plant called the mixcal. **1848** W. H. EMORY *Notes Mil. Reconn.* 59 This afternoon I found the famous mezcal, (an agave,) about three feet in diameter, broad leaves, armed with teeth like a shark. **1891** *Century Mag.* Mar. 653 Along deserts bristling with spines of the cactus, spanish bayonet, mescal and palo verde. **1914** C. F. SAUNDERS *With Flowers & Trees in Calif.* 139 The mescal buds are capable of making by distillation one of the fieriest intoxicants known. **1951** KEARNEY & PEEBLES *Arizona Flora* II. 192 The names century-plant and mescal are applied to the large, paniculate species, and some of the small species are known as lechuguilla and amole.
b. The cooked root of the mescal as an article of food.
1759 tr. *Venegas's Nat. & Civil Hist. Calif.* I. 44 The mountains and forests yield the mezcal,..the root of which boiled is a principal ingredient of the mexcalli. **1831** J. O. PATTIE *Personal Narr.* 63, I afterwards ascertained that it was a vegetable called by the Spanish mascal (probably maguey). **1844** J. GREGG *Commerce Prairies* I. 290 Those [Apaches] that are found east of the Rio del Norte are generally known as Mezcaleros, on account of an article of food in use among them, called mezcal. *Ibid.*, Mezcal is the baked root of the maguey. **1881** *Amer. Naturalist* XV. 875 The 'mescal' of the Arizona Apaches, that is, the baked head of the *Agave palmeri* and *Agave parryi*. **1951** KEARNEY & PEEBLES *Arizona Flora* II. 192 The name 'mescal' is applied also to the food obtained by roasting the caudex and emerging flower stalk. *Ibid.* 193 Even at the present time a small quantity of mescal is made by the Papagos.
3. a. A small desert cactus, *Lophophora williamsii* (formerly *Anhalonium lewinii*, etc.), found in northern Mexico and southern Texas and having a soft, segmented body a few inches high in the form of a flattened globe. Cf. *PEYOTE.
1885 [see *mescal head* in *4]. **1895** *Therapeutic Gaz.* XI. 579/2 In connection with the physiological action of the mescal, its use by the Indians is of great interest. **1896** *Ibid.* XII. 8/1 The exercises open with a prayer by the leader, who then hands each man four mescals, which he takes and eats. **1911** *Encycl. Relig. & Ethics* IV. 736/1 The Nahuatl *peyotl*..under the incorrect title of *mescal*.. is well known to the whites. **1920** *Sci. Amer.* 14 Feb. 157 The peyote, often popularly miscalled 'mescal' through confusion with the maguey cactus from which a fiery intoxicant is prepared, is a species of small cactus widely used for both medicinal and ceremonial purposes by the Indian tribes of the southwestern U.S. **1937** J. BORG *Cacti* 209 Lophophora Williamsii (*Lem.*) *Coulter*... This is the famous mescal or peyotl of the ancient Mexicans.
b. A preparation of the cactus for ingestion.
1896 *Brit. Med. Jrnl.* 5 Dec. 1625/2 The history of the use of mescal by the Indians of New Mexico is very well known in the United States. **1899** *Jrnl. Physiol.* XXX. 83 'Mescal' never gives rise to merriment, but rather to a condition of ideal content, and produces wakefulness. **1937** J. BORG *Cacti* 209 The dried tops of the plant made into bunches or wreaths used to be sold as mescal.
4. *attrib.* and *Comb.*, as *mescal-eater, intoxication*; *mescal-inspired* adj.; **mescal button,** a dried disc-like top of the cactus *Lophophora williamsii*, eaten or drunk as a decoction for its intoxicating and hallucinogenic effects (cf. *MESCALINE); also, the plant itself, = sense *3 a; **mescal head** = **mescal button.*
1888 *Therapeutic Gaz.* IV. 232/1, I undertook to experiment with a drug..which they had received from Mexico. In its own country the drug is said to be called 'Muscale Buttons', and is used as a narcotic, food, or relish. *Ibid.*, *Anhalonium Lewinii*, nov. spec., Mexico. Local name, 'Muscale Buttons'. **1896** *Brit. Med. Jrnl.* 5 Dec. 1625/2, I took fully 1½ drachm of an extract of which each drachm represented one mescal button. **1913** *Jrnl. Nervous & Mental Dis.* XL. 427 We endeavoured further to extend knowledge of pathological mental states by producing mental conditions nearly allied to generally recognized types of insanity... For this purpose we used the Mexican drug [*printed* drag] pelotte—the mescal button. **1959** O. & M. LEESE *Desert Plants* iv. 41 There are too the Lophophoras or Anhaloniums, known as the Mescal Button or Peyote, which are devoid of spines and look somewhat like a denizen of the sea. **1934** R. HAMER in R. Skelton *Poetry of Thirties* (1964) 88 The mescal-eater's almost heard Omnipotent transcendental word. **1885** *Outing* Oct. 24/2 The old and young squaws..had brought down from the hillsides donkey-loads of mescal-heads. **1933** L. SPIER *Yuman Tribes* 55 The mescal heads were baked in a pit. **1946** in M. Lowry *Let.* 2 Jan. (1967) 61 The mescal-inspired phantasmagoria. or heebie-jeebies, to which Geoffrey has succumbed. **1897** *Lancet* 5 June 1541/2 It cannot be said (from my experience) that the pleasure of mescal intoxication lies in any resultant passive emotional state such as is produced by tea or alcohol, but strictly in enjoyment of the colour visions produced.

mescaline (me·skălīn, -īn). Also **mescalin,** † **mezcaline.** [ad. G. *mezcalin* (now *mescalin*) (A. Heffter 1896, in *Ber. d. Deut. Chem. Ges.* XXIX. 222), f. Sp. *mezcal* MESCAL: see -IN[1], -INE[5].] **a.** The alkaloid 3,4,5-trimethoxyphenethylamine, $(CH_3O)_3 \cdot C_6H_2 \cdot CH_2CH_2NH_2$, which is the chief active principle of mescal buttons, producing effects similar to those of LSD but much less strongly.
1896 *Jrnl. Chem. Soc.* LXX. I. 267 Alkaloïds of *Anhalonium Lewinii*, Hennings (*Lophophora Lewinii*, Rusby).—The dried 'discs' (Scheiben) of this cactus are used by the Indians of N. Mexico for the preparation of an intoxicant, and are brought into commerce under the name of 'Muscale buttons'... They were extracted with 70 per cent. alcohol, and the residue obtained by evaporating this extract was extracted with ether and then with chloroform. The chloroform dissolved an alkaloïd which was named mezcaline. **1899** *Jrnl. Physiol.* XXV. 82 Mezcaline, whilst agreeing precisely with the other alkaloids as regards cardiac and respiratory action, etc., appeared more effective in the production of colour visions. **1900** DORLAND *Med. Dict.* 384/1 Mescalin. **1936** *Mind* XLV. 68 My first experience of a similar kind was during an experiment with mescaline. **1956** A. HUXLEY *Heaven & Hell* i. 11 The soul is transported to its far-off destination by the aid of..mescalin. **1958** M. ARGYLE *Relig. Behaviour* ix. 112 A number of drugs produce quasi-mystical experiences, the best-known being mescaline. **1962** *Q. Rev.* XVI. 136 The outstanding result of mescaline ingestion is visual hallucination, but depersonalisation and time distortion frequently occur. **1966** T. PYNCHON *Crying of Lot 49* i. 17 The experiment he was helping the community hospital run on effects of LSD-25, mescaline, psilocybin, and related drugs on a large sample of suburban housewives. **1974** *Sci. Amer.* June 66/1 Other hallucinogenic agents such as mescaline and amphetamine, on the other hand, are related in structure to noradrenaline.
b. *attrib.*, as *mescaline hallucination, psychosis.*
1913 *Jrnl. Nervous & Mental Dis.* XL. 432 A mescalin hallucination is felt to be much more objective than a mere idea such as a visual image. **1941** *Arch. Neurol. & Psychiatry* (Chicago) XLV. 130 (*heading*) Mescaline hallucinations in artists. **1940** *Jrnl. Mental Sci.* LXXXVI. 36 The feeling of unreality, both as regards the self and the external world, so often found in schizophrenics, is one of the typical features of the mescaline psychosis. **1958** M. ARGYLE *Relig. Behaviour* ix. 114 It is acute schizophrenia which most resembles 'mescaline psychosis', and Andis has described a number of psychotic patients whose reported experiences resembled mescaline psychosis more than the classical type of psychosis.

mescalism (me·skăliz'm). [f. MESCAL + -ISM.] **a.** The practice of taking mescal buttons. **b.** The state of intoxication produced by them.
1902 *Amer. Anthropologist* Oct.–Dec. 789/1 Through mescalism one seems almost to 'attain an objective knowledge of one's own personality'. **1936** *Mind* XLV. 70 Like the starry and bejewelled scenes observed in mescalism, spots of light or their relatives seen in hypnagogic images have also a retro-retinal origin.

mese, var. *MEZE.

‖ **meseta** (mĕse·tă). [Sp., dim. of *mesa* MESA, f. L. *mēnsa* a table.] A plateau; *spec.* the high plateau of central Spain. Also *attrib.*
1904 T. H. HOLDICH *Countries of King's Award* xiii. 366 The moraines streak the surface of the 'mesetas', which here represent the beds of ancient lakes and are disposed more or less in the form of terraces. **1905** *Spectator* 28 Jan. 112/2 The region of the Patagonian Andes..presents a jumble of mountains, rivers, 'mesetas', 'pampas', forests, and glaciers. **1958** FISHER & BOWEN-JONES *Spain* I. ii. 20 By far the largest proportion of Spain is occupied by the Meseta, which, however, is far from being a uniform mass. It is best regarded as comprising two enormous tablelands, or upland basins, which are separated by a diagonally-running fold mountain series. **1962** R. WAY *Geogr. Spain & Portugal* xiii. 257 The peasants..have long journeys to make by mule or ox before they begin their day's work, and then have to endure the vagaries of the meseta climate, particularly strong winds. **1963** *Guardian* 13 July 6/4 The regions of the high *meseta*—the Castiles, Aragon, Leon, and Estremadura. **1969** J. MANDER *Static Society* ii. 81 The landscape resembles..the *meseta* of Spain.

mesh, *sb.* Add: **1. c.** *Electr.* A closed loop of windings or other impedances connected in series.
1881 J. C. MAXWELL *Treat. Electr. & Magnetism* (ed. 2) I. II. vi. 374 If the conducting wires form a simple network and if we suppose that a current circulates round each mesh, then the actual current in the wire which forms a thread of each of two neighbouring meshes will be the difference between the two currents circulating in the two meshes. **1892** S. P. THOMPSON *Dynamo-Electr. Machinery* (ed. 4) xxiv. 709 The three coils may be joined..in a closed mesh joined with the three lines at its corners. **1970** J. SHEPHERD et al. *Higher Electr. Engin.* (ed. 2) ii. 47 Circuits involving multiple meshes may be solved by considering either the meshes (mesh analysis) or the junctions (node analysis).
d. With prefixed numeral, e.g. *50 mesh*, designating a screen with that number of square openings per unit length (e.g. per inch), and applied to materials which will pass through such a screen (but, usually, not through the next finest screen).
1930 *Engineering* 22 Aug. 223/2 The dust cloud which it would encounter would consist..of particles ranging from 60 mesh to beyond 200 mesh. **1932** RILEY & JOHANNSEN *Med. Entomol.* xvii. 264 A 16-mesh screen is ordinarily employed. **1933** W. T. READ *Industr. Chem.* vii. 65 This relationship has been established by the United States Bureau of Standards for 200-mesh screens so that the wire has a diameter of $2 \cdot 1 \times 10^{-3}$ in. and each opening a width of $2 \cdot 9 \times 10^{-3}$ in., thus giving 200 openings per linear inch. **1948** PIERCE & HAENISCH *Quantitative Analysis* (ed. 3) v. 63 The usual sample for analysis should pass a screen of 80–100 mesh or smaller. **1971** *Nature* 25 June 524/2 Samples of powder were first ground to 400 mesh and then briquetted.
3. c. *Building.* A steel network used as reinforcement in concrete.
1904 C. F. MARSH *Reinforced Concrete* ii. 44 The ribs on the Cottançin system are considered as N-girders, of which the joints are absolutely fixed, the mesh forming the tension bracing and the concrete the compression bars. **1936** E. PROBST *Princ. Plain & Reinforced Concrete Construction* ii. 97 Ready-made reinforcements are often used for slabs and also for beams and columns. An example of this is the triangular mesh, made from interwoven round steel bars. **1948** L. J. MURDOCK *Concrete Materials & Pract.* xvii. 251 Expanded-metal (steel) reinforcement..is made by cutting slits in blank steel plate and sheets and then expanding them into diamond-shaped meshes. **1971** B. P. HUGHES *Limit State Theory for Reinforced Concrete* 397 The fabric in sections is indicated by heavy dashed lines—for oblong mesh, long or short dashes according to whether the section is parallel or at right angles to the main wires.
4. b. *In*, or *out of, mesh.* Of gearwheels or their teeth: engaged, or not engaged, with each other. So *into mesh.*
1904 A. B. F. YOUNG *Compl. Motorist* 78 When the top gear is engaged, none of the other gears are in mesh, although they rotate. **1905** R. T. SLOSS *Bk. Automobile* 207 The gears must be thrown into mesh sharply or not at all. **1921** [see *COASTING *vbl. sb.* 4 c]. **1948** *Motor Manual* (ed. 33) vii. 126 If one wheel has 20 teeth and another has 40, the two being in mesh, then the larger one will turn exactly half as fast as the smaller one. *Ibid.* 129 At one time it was common to slide one gear along splines on its shaft..so that its teeth came into mesh with those of its mate. **1972** 'J. & E. BONETT' *No Time to Kill* xi. 143 The gears of his brain, he reflected, were not in mesh. A walk before breakfast might re-engage them.
6. *mesh-bag*; (sense *1 c) *mesh-connected* adj., *-connection.*
1911 *Daily Colonist* (Victoria, B.C.) 25 Apr. 6/4 (Advt.), Solid Gold Mesh Bags, nothing more intrinsically beautiful in the category of Hand Bags. **1920** *Edin. Rev.* Oct. 349 The mesh-bag in which the Mexican hunters carried their arrow heads. **1960** L. HELLMAN *Toys in Attic* I. 27 Gives Anna a large gold mesh bag. **1896** D. C. & J. P. JACKSON *Alternating Currents* viii. 395 In a three-phase machine, if the armature is mesh-connected, the pressure between any two collector rings is equal to the pressure developed in one coil. **1954** E. HUGHES *Fund. Electr. Engin.* vi. 221 (*caption*) Conventional representation of a mesh-connected winding. **1896** D. C. & J. P. JACKSON *Alternating Currents* xiii. 552 The arrangements are either of the star or mesh

connection. **1971** *Gloss. Electrotechnical, Power Terms* (*B.S.I.*) II. i. 7 *Mesh connection*, in a polyphase device or system of devices. The arrangement in which the end of each phase is connected to the beginning of the next in sequence so as to form a ring, each point of connection being connected to a terminal.

mesh, *v.* Add: **3. b.** (Further examples.) Also *trans.*, to cause (gears, esp. those of a motor vehicle) to become engaged; to put into mesh.
1890 *Cent. Dict.*, *Mesh*, to engage (the teeth of wheels or the teeth of a rack and pinion) with each other. **1907** C. WHEELER *Bicycles in Making* 78 Small pinion wheels.. also mesh with what is called a fulcrum pinion. **1913** R. KENNEDY *Bk. Motor Car* II. 194 Then..gear wheels which are revolving have to mesh with gear wheels which are stationary. **1926** J. A. MOYER *Gasoline Automobiles* (ed. 2) vii. 237 The rod *A*..meshes the gear wheel *B* with the flywheel *C*. **1935** M. M. ATWATER *Murder in Midsummer* xx. 192 He meshed the gears and the old car moved slowly away. **1957** *Laboratory Investigations* VI. 562 Racks are mounted on the sides of the blades to mesh with the idling gear in each plate. **1961** L. GRIBBLE *Wantons die Hard* i. 15 He meshed the gears and the car headed out of.. Tyler Place. **1972** H. BUCKMASTER *Walking Trip* 58 Norman..meshed the gears noisily as he watched for an entrance into the traffic.
d. *intr.* (occas. *trans.*). To fit in; combine. Also const. *with*. (Cf. sense 3 b.)
1944 H. G. WELLS *'42 to '44* 65 Such perplexities and failures to mesh are by no means confined to Anglo-Russian relationships. **1951** *Good Housek. Home Encycl.* 291/2 The units are generally designed to 'mesh' together. **1963** *New Society* 7 Nov. 19/1 Many young people are bewildered by school and unable to mesh with it. **1964** I. L. HOROWITZ *New Sociol.* 15 The general theory of action is really a general theory of how the parts mesh to form a whole. **1967** *Listener* 9 Nov. 609/3 What has always meshed best in his verse is precisely the neat machinery of ideas which he shared with half a dozen contemporaries, such as Mr Alvarez and Miss Jennings. **1968** *Economist* 17 Feb. 54/4 The difficulties of meshing management and staff at Holland's Amro bank, four years after that merger. **1971** *Nature* 12 Nov. 61/1 The TXE-4..cannot mesh with the pulse code modulation digital transmission systems which the Post Office is installing.

meshing, *vbl. sb.* Add to def.: a meshed structure; mesh-work.
1907 *Daily Chron.* 25 Sept. 8/4 Splash! go the dredges, small scoops of steel meshing. **1926** *Brit. Weekly* 26 Aug. 430/2, I had a copper frame constructed with a panel of copper meshing to which the letters were fixed. **1968** J. ARNOLD *Shell Bk. Country Crafts* 300 While the Honiton makers worked the pattern first and then 'grounded' it with meshing,..in the East Midlands pattern and ground were made in a single process.

meshuga, meshugga (mĕʃu·gă), *a.* *slang.* Also **mash-, meshuggah, meshuger, mishugge,** etc. [ad. Yiddish *meshuge*, f. Heb. *mĕshuggā'* part. of *shāgag*, to go astray, wander; cf. G. *meschugge* crazy.] Mad, crazy; stupid. The adj. has the form **meshugener, meshugenah,** etc., when it precedes its noun. This form is also used as *sb.*
1892 I. ZANGWILL *Childr. Ghetto* I. 156 She's *meshuggah* —quite mad! **1900** *Atlantic Monthly* LXXXVI. 108/2 'Meschugener,' leered the banker. **1922** JOYCE *Ulysses* 157 Meshuggah. Off his chump. **1930** *Amer. Mercury* Dec. 455/1 Me broad gets caught in a snow-storm [= becomes drugged with cocaine] an' goes meshuga. **1952** V. GOLLANCZ *My Dear Timothy* xii. 110 My father probably murmured to my mother.., 'The boy's *meshuggah* (which means cracked).' **1959** B. KOPS *Hamlet of Stepney Green* II. ii. 44, I don't like saying this, Bessie, but your boy is meshuger. **1961** B. VAWTER *Conscience of Israel* i. 22 The 'son of *nabi*' sent by Elisha to anoint Jehu is called by Jehu's companions 'this madman'—*meshugga*. **1962** J. BALDWIN *Another Country* (1963) III. i. 377 We finally got that *meshugena* of a broken-down movie star in town. **1968** L. ROSTEN *Joys of Yiddish* 237 A crazy man is a *meshuggener*. A crazy woman is a *meshuggeneh*. That *meshuggener*! Has he gone yet? **1971** *Sunday Times* (Johannesburg) 28 Mar. 4/3 Going steady!.. What kind of a meshugenah idea is this? **1972** *Listener* 24 Aug. 243/3 My favourite speaker was known to us all as Meshuggener Moishe. *Ibid.* 244/1 That Moishe, bless him, is he a meshuggener! **1973** *Jewish Chron.* 19 Jan. 24/1 The kids at school call me meshuga. That means crazy.

meshugaas (mĕʃu·gās). *slang.* Also **mishugas.** [ad. Yiddish *meshugaas* f. Heb. *mĕshuggā'* (see prec.).] Madness, craziness; nonsense, foolishness.
1907 I. ZANGWILL *Ghetto Comedies* 59 'Hannah, will you explain to me what this *meshuggas* (madness) is?' cried S. Cohn, lapsing into a non-Anglicism. **1970** L. M. FEINSILVER *Taste of Yiddish* i. 59 Everyone has his own wackiness—often heard as 'We all have our own mishugas'. **1971** *Sunday Times* (Johannesburg) 28 Mar. 4/2 He is so well-acquainted and up-to-date with genes and other anthropological meshugaas.

meshumad, meshummad (mĕʃu·mād). Pl. **-im.** [Yiddish, f. Heb. *mĕshummādh*, lit. one who is destroyed.] An apostate from Judaism.
1892 I. ZANGWILL *Childr. Ghetto* I. 14 The new-fangled Jewish minister..rigged out like the Christian clergyman, has been mistaken for..a *Meshumad*, and pelted with gratuitous vegetables. **1903** R. T. HERFORD *Christianity in Talmud & Midrash* II. ii. 336 'Meshummadim' are those who wilfully transgress some part of the ceremonial law, and thereby proclaim their apostasy. **1938** *Vallentine's Jewish Encycl.* 45 The Heb. words for apostate are *min*,.. *mumar*,..*meshumad* (one who has renounced his religion or, perhaps, one who has become baptized), and *apikoros*. ..The terms *mumar* and *meshumad* are often used indiscriminately... The *Meshumadim*, from early Middle Ages, developed a super-zeal for their new religion. **1962** *Stand. Jewish Encycl.* 1308 *Meshummad* (Heb.): a convert to Christianity; an apostate (used opprobriously). **1968** L. ROSTEN *Joys of Yiddish* 238 Jews distinguish forced converts, or *anusim*, from those who joined another faith of their own volition, *meshumadim*.

mesic (me·sik, mī·sik), *a.*[1] *Ecology.* [f. Gr. μέσ-ος middle + -IC.] Having, or characterized by, a moderate amount of moisture.
1926 [see *HYDRIC *a.*[2]]. **1967** M. E. HALE *Biol. Lichens* vii. 91 The curves of frequency show the relative abundance of lichens in oak-dominated woods and their rarity ..in mesic climax forests. **1974** *Environmental Conservation* I. 60/1 A small tracked vehicle..was employed to establish a series of repetitive passes in a level mesic 'meadow'.

mesic (mī·zik, me·zik), *a.*[2] *Nuclear Physics.* [f. *MES(ON[3] + -IC.] Of, pertaining to, or being a meson; applied *spec.* to a system analogous to an atom in which a meson takes the place of either an orbital electron or the nucleus.
1939 *Physica* VI. 877 'Mesic' charges. **1952** R. E. MARSHAK *Meson Physics* iv. 153 There are also selection rules for radiative and mesic absorption. **1956** S. TOLANSKY *Introd. Atomic Physics* (ed. 4) xix. 328 Since mesons only exist virtually within unexcited nuclei and are only made free to exist alone by collision processes, there are no mesic atoms in normal unexcited materials. **1969** *Sci. Jrnl.* July 44/1 The pion can be used to prepare mesic atoms in which a π^- replaces an electron in the atom. **1972** *Physics Bull.* Mar. 148/2 The meson then cascades down first by Auger transitions and, as it reaches lower orbits, increasingly by radiative transitions, emitting mesic x rays.

mesmerize, *v.* Add: **a.** Also *transf.* and *fig.*, to fascinate, spellbind.
1936 G. B. SHAW *Simpleton* I. 50 *Vashti*... Would you not die for me? *Iddy.* (*mesmerized by her eyes*) Oh DEAR!!! Yes: your eyes make my heart melt. **1940** W. FAULKNER *Hamlet* II. i. 113 She seemed to be momentarily mesmerised by a complete inert soft surprise. **1955** *Economist* 26 Mar. 1068/1 Members seemed to be mesmerised by an uncouth demagogue sitting with his acolytes in the public gallery. **1975** *Times* 24 July 4/7 Sir G. Howe said..Labour ministers..had been mesmerized by their own verbal fantasies.

meso-. Add: Now usu. with pronunc. (mī·zo) or (mī·so). **me:saorti·tis** (also **mesoaortitis**) *Med.*, inflammation of the middle layer of the aorta; **mesaxo·nic** *a. Zool.* [Gr. ἄξων, ἀξον- axis], of the feet of certain ungulate mammals: having the axis in the central toe; **mese·ctoderm** *Embryol.* [ad. G. *mesektoderm* (J. B. Platt 1894, in *Archiv f. mikrosk. Anat.* XLIII. 913)], (*a*) that part of the mesenchyme which is derived from ectoderm rather than from mesoderm; (*b*) (see quot. 1956); **mese·ndoderm, mese·ntoderm** *Embryol.* [ad. and a. G. *mesentoderm* (J. B. Platt 1894, in *Archiv f. mikrosk. Anat.* XLIII. 913)], (*a*) that part of the mesenchyme which is derived from endoderm rather than from mesoderm; (*b*) (see quot. 1957); **Meso-Ame·rica, Mesoamerica** [f. Sp. *Mesoamérica* (P. Kirchhoff in *Acta Americana* (1943) I. 92)], the central region of America, from northern Mexico to Nicaragua, which was civilized in pre-Spanish times; **Meso-Ame·rican** *a.*, of or pertaining to Meso-America; also as *sb.*, an inhabitant of Meso-America; **mesoaortitis**, var. *mesaortitis* above; **me·soconch** (-kǫŋk), **mesoconchic** (-kǫ·ŋkik), **-conchous** (-kǫ·ŋkəs) *adjs. Anthropol.* [Gr. κόγχ-ος eye-socket], having orbits of moderate height in relation to their width, as expressed by the orbital index (see quots.); so **meso·conchy** (-kǫŋki), the property of being mesoconchic; **me·soform** *Physical Chem.* = *mesophase* below; **mesoku·rtic** *a. Statistics* [Gr. κύρτ-ος bulging], applied to (a graph of) a frequency distribution having the same kurtosis as the normal distribution; hence **mesokurto·sis** [*KURTOSIS], the property of being mesokurtic; **me·sopause,** the boundary between the mesosphere and the thermosphere, at an altitude of about 80 km. (50 miles), where the temperature stops decreasing with height and starts to increase; **me·sophase** *Physical Chem.* [a. G. *mesophase* (Zocher & Birstein 1929, in *Zeitschr. f. physikal. Chem.* A. CXLI. 415)], a mesomorphic phase; **mesophi·lic** *a. Biol.* [*-PHILIC], (of an organism, esp. a bacterium) flourishing at moderate temperatures; so **me·sophile,** a mesophilic organism; **meso·sa·lpinx** *Anat.* [SALPINX 2], an upper fold of each of the broad ligaments of the uterus which contains and supports the Fallopian tube; **mesosa·probe, mesosapro·bic** *adjs.* [ad. G. *mesosaprobe* (Kolkwitz & Marsson 1908, in *Ber. Deutsch. Bot. Ges.* XXVI. 507): see *SAPROBE], of running water: partially polluted; **me·soscaph(e** [ad. F. *mésoscaphe* (J. Piccard): cf. *BATHYSCAPH(E], a submersible vessel designed for exploration of the sea at moderate depths; **mesosco·pic** *a. Geol.* [after *micro-, macroscopic*], large enough for examination with the naked eye but small enough for examination as a single entity; **me·sosphere,** the layer of the earth's atmosphere between the stratopause below and the mesopause above; hence **mesosphe·ric** *a.*; **mesotarsal** *a.* (example); **mesothelium,** delete entry and see *MESOTHELIUM; **mesothe·rmal** *a. Petrol.*, of, pertaining to, or designating mineral and ore deposits formed by hydrothermal action at intermediate temperature and pressure.
1909 *Cent. Dict.* Suppl., Mesaortitis. **1910** *Practitioner* Apr. 422 A large proportion of all cases of aneurysm in young and middle-aged men are due to a syphilitic mesaortitis. **1962** *Lancet* 28 Apr. 889/2 The response to the mesoaortitis of syphilis is primarily an endarteritis. **1898** A. S. WOODWARD *Outl. Vertebr. Palæont.* 319 They [*sc.* Perissodactyla] are all digitigrade quadrupeds, with the axis of both feet passing through the digit No. iii (hence mesaxonic). **1933** A. S. ROMER *Vertebr. Paleont.* xvi. 300 In the majority of ungulates..the third toe was the longest, and the axis of symmetry of the foot lies through this digit (mesaxonic). **1974** *Nature* 8 Mar. 174/2 In the development of a mesaxonic foot perissodactyls have reduced the astragalo-cuboid contact. **1894** *Jrnl. R. Microsc. Soc.* 544 Miss J. B. Platt has studied the ontogenetic differentiation of the ectoderm in *Necturus*. The 'mesoderm' in the head is differentiated by the yolk spherules which it contains into two sharply separable layers—mesectoderm and mesendoderm. **1921** *Jrnl. Compar. Neurol.* XXXIII. 4 In 1894 Miss Platt elaborated the idea..introducing the terms mesectoderm and mesentoderm for mesenchyme derived from the ectoderm and endoderm, respectively. **1938** *Nature* 23 Apr. 754/1 The mesectoderm..normally migrates ventrally over the mesentoderm. **1956** C. H. WADDINGTON *Princ. Embryol.* xii. 256 'Mesectoderm'..is also used for the epiblast of a blastoderm before the mesoderm has invaginated and thus become separated from the ectoderm. **1965** L. B. AREY *Developmental Anat.* (ed. 7) ii. 22 Mesenchyme is predominantly derived from the mesoderm,..but some of it comes from the ectoderm and this contribution is often called mesectoderm. **1894** Mesendoderm [see *mesectoderm* above]. **1964** H. W. MANNER *Elem. Compar. Vertebr. Embryol.* v. 50 The morphological result of this invagination [of the blastula] is a cup-like structure, composed of two layers of cells, an outer ectoderm and an inner mesendoderm, so called because it contains the presumptive material for both the mesoderm and endoderm. **1921, 1938** Mesentoderm [see *mesectoderm* above]. **1957** *Dorland's Med. Dict.* (ed. 23) 818/2 *Mesentoderm*, the inner layer of an amphibian gastrula not yet separated into mesoderm and entoderm. **1948** A. L. KROEBER *Anthropol.* (rev. ed.) xviii. 793 We have seen that native Meso-America..consisted of most of what now is Mexico and of Guatemala and that it constituted the North American half of prehistoric Nuclear America. **1952** tr. *P. Kirchhoff's Mesoamerica* in S. Tax *Heritage of Conquest* 23 We include these tribes in Mesoamerica, because of the very considerable number of..Mesoamerican cultural traits. **1967** L. DEUEL *Conquistadors without Swords* xv. 189 The close parallels between the formative cultures of Mesoamerica and the Central Andes became evident. **1974** *Encycl. Brit. Macropædia* XI. 934/2 About half of Mexico; all of Guatemala, British Honduras, and El Salvador; and parts of Honduras and Costa Rica are included in Meso-America. **1948** A. L. KROEBER *Anthropol.* (rev. ed.) xviii. 786 The term 'South Mexican–Central American' would.. be more accurate..but it is cumbersome; and *Meso-American* has been suggested and employed as a convenient coinage that runs no risk of being confused. **1956** R. REDFIELD *Peasant Society & Culture* 74 Those Meso-American peoples. **1965** *Canad. Jrnl. Ling.* Spring 101 Jicaque, a Mesoamerican language. **1967** L. DEUEL *Conquistadors without Swords* xv. 189 Its importance as a prototype of Mesoamerican architecture may well be compared to the step pyramid of Zoser in early dynastic Egypt. **1969** J. MANDER *Static Society* ii. 83 What is true of the Aztecs is true of the Meso-American and Andean civilisations. **1974** *Encycl. Brit. Macropædia* XI. 935/1 The only linguistic groups that played any great part in Meso-American civilization were the Mixtec and Zapotec. *Ibid.*, The Meso-Americans reached stages of development unknown away from those areas. **1909** *Cent. Dict.* Suppl., Mesoconch. **1920** H. H. WILDER *Lab. Man. Anthropometry* i. 67 Chamaeconch..below 76. Mesoconch..76–85. **1960** M. F. A. MONTAGU *Introd. Physical Anthropol.* (ed. 3) 606 Mesoconch..[Orbital index of] 76·0–84·9. **1909** *Cent. Dict.* Suppl., Mesoconchic. **1960** J. COMAS *Man. Physical Anthrop.* vii. 409 Mesoconchic..[an orbital index of] 83·0 to 88·9. **1885** *Jrnl. R. Anthrop. Inst.* XIV. 71 Mesokonchous..[orbital index of] 80·1 to 85·0. **1904** *Biometrika* III. 214 (*table*) Mesoconchy. **1933** *Trans. Faraday Soc.* XXIX. 1008 For a mesoform to appear, it is necessary for these [binding] forces to persist in either one or two dimensions after loosening of the third. **1946** *Thorpe's Dict. Appl. Chem.* (ed. 4) VII. 350/2 Some substances show only one of these mesophases; some show both, in which case the smectic is always the lower-temperature form. The matter is somewhat complicated, however, by the existence of substances which show three, four, and even five mesoforms. **1905** Mesokurtic [see *LEPTOKURTIC *a.*]. **1972** R. B. CAIN *Elem. Statistical Concepts* xix. 159 Any curve which has the same degree of kurtosis as the standard normal curve is said to be mesokurtic. **1905**

K. PEARSON in *Biometrika* IV. 173 The mesokurtosis of the Gaussian curve is not a universal characteristic of frequency distributions. **1943** M. G. KENDALL *Adv. Theory Statistics* I. v. 129 [For the normal distribution] we also have $\beta_2 = 3$, $\gamma_2 = 0$, which accounts for the standard adopted for mesokurtosis. **1950** S. CHAPMAN in *Jrnl. Atmospheric & Terrestrial Physics* I. 121 The..upper boundary..would be the stratopause, and the mesosphere would extend from this level to the mesopause. **1963** *New Scientist* 25 July 169/2 At a height of some fifty miles there is a sudden drop in the temperature of the atmosphere from freezing point to about −80°C. It is here, in the 'mesopause', that the so-called 'noctilucent' clouds occur. **1972** *Nature* 28 Jan. 215/1 A sounding rocket was launched..chiefly to study the ion composition around the cold summer mesopause. **1929** *Brit. Chem. Abstr.* A. 870/1 Mesophases occupy an intermediate position between crystals and purely amorphous substances. **1937** O. B. DARBISHIRE tr. *A. von Buzágh's Colloid Syst.* vii. 135 Mesophases only show a symmetrical arrangement in certain given directions, and not in all three directions of space. **1972** *Physics Bull.* May 279/1 The characteristic anisotropy of the mesophase is now known to result from the strong tendency of the constituent molecules to lie with their long axes parallel. **1928** P. H. FOSTER in C. M. Hilliard *Text-bk. Bacteriol.* viii. 95 Mesophiles..may be further subdivided to distinguish organisms which have as their optimum temperature 37°C.. and those growing best below this, usually between 20° and 30°C. **1969** *New Scientist* 27 Nov. 450/2 Mesophiles (organisms with a maximum temperature for growth of 37°C and a minimum less than 20°C) multiply rapidly. **1897** LEHMANN & NEUMANN *Atlas & Essent. Bacteriol.* 98 Mesophilic bacteria: minimum at 10°–15°, best at 37°, maximum at about 45°. **1964** *New Scientist* 12 Nov. 445/2 The method of the Microalgae Research Institute is to separate its algal strains into high-temperature or 'thermophilic', normal temperature or 'mesophilic', and low-temperature or 'psychrophilic' varieties. **1890** *Syd. Soc. Lex.*, *Mesosalpinx*, the fine fold of peritonaeum which continues the mesovarium to the Fallopian tube in the foetus. **1901** *Gray's Anat.* (ed. 15) 1013 The Fallopian tube is contained in a special fold of the broad ligament, which is attached to the part of the ligament near the ovary, and is known by the name of the mesosalpinx. **1970** L. J. A. DIDIO *Synopsis Anat.* 297/2 The position of the ovarian ligaments divides each broad ligament into an upper mesosalpinx and a lower mesometrium. **1927** *Glasgow Herald* 11 June 4/2 Investigators distinguish a.. meso-saprobe zone which carp, tench, sticklebacks, and the like can endure. **1925** *Bull. Illinois Nat. Hist. Survey* XV. 441 The polluted or mesosaprobic zone represents the next step towards purification of the waters. A variety of higher water-plants may exist..and there may be considerable amounts of dissolved oxygen present. **1970** tr. *J. Schwoerbel's Methods Hydrobiol.* 8 This self cleansing of running water..leads from the poly-saprobic zone through the meso-saprobic zone to the oligo-saprobic zone with only small residues of pollution. **1955** *Britannica Bk. of Year* 490/1 *Mesoscaphe*, an underwater helicopter designed by Auguste Piccard. **1963** *Observer* 13 Jan. 14/4 The mesoscaphe, a cylindrical steel shell sealed at each end, will be lighter than water even when fully loaded with passengers. **1969** *New Scientist* 19 June 626/1 The submersible..is a direct descendant of the *Auguste Piccard*—the 'mesoscaph' that, carrying 40 passengers at a time, gave thousands of visitors..a glimpse of the depths of Lake Geneva. **1973** *Times Lit. Suppl.* 13 July 815/4 Twenty years ago Auguste Piccard..conceived the idea of a mesoscaphe or exploratory vessel that could operate at intermediate depths. In 1964 his son Jacques.. launched the first mesoscaphe. **1957** WEISS & MCINTYRE in *Jrnl. Geol.* LXV. 577/2 The different kinds of mesoscopic structures found in the area. [*Note*] The use of the terms 'microscopic', 'mesoscopic', and 'macroscopic' in this paper accords with the definitions suggested elsewhere by one of the writers (Weiss, 1957 [= 'at press': *published* 1958]), as follows:..2. *Mesoscopic*: This covers fields ranging in size from a single hand specimen to a single continuous exposure (generally, but not always, of small size) in which data can be measured with sufficient accuracy and continuity to allow determination of its over-all structural geometry. **1965** G. J. WILLIAMS *Econ. Geol. N.Z.* vi. 66/1 Grindley (1963) noted that post-metamorphic folding in southern Westland was accompanied by axial-planar cleavage on mesoscopic shear folds. **1969** *Nature* 22 Nov. 827/1 Attention is to be confined to structures on what is called the mesoscopic scale, that is, visible in anything between hand specimen and outcrop. **1950** S. CHAPMAN in *Jrnl. Atmospheric & Terrestrial Physics* I. 121, I propose the name *mesosphere* for the layer between the top of the stratosphere and the major minimum of temperature existing somewhere below 100 km. **1961** *New Scientist* 30 Nov. 568/3 Winds at heights of 80 to 100 km in the Earth's atmosphere—in the so-called mesosphere. **1973** *Physics Bull.* Dec. 727/1 The fifth experiment aboard Concorde concerned emission from the stratosphere and mesosphere arising from the electronic transition between the metastable $^1\Delta_g$ and the ground $^3\Sigma_g^-$ states of molecular oxygen. **1960** *Meteorol. & Geoastrophysical Abstr.* XI. 1535 Mesospheric temperatures. **1972** *Nature* 28 Jan. 194/1 It is, perhaps, of interest to enquire whether the low mesospheric temperatures existing at the time of the flight render the weakly-bound cluster ions more stable. **1897** PARKER & HASWELL *Text-bk. Zool.* II. 366 The ankle-joint of the bird is a meso-tarsal joint. **1922, 1969** Mesothermal [see *hypothermal* adj. c s.v. *HYPO- II].
2. *Chem.* (Often italic.) **a.** [app. first used in *mesotartaric*.] As an inseparable formative element in chem. names denoting the meso isomer (see below), as *meso-erythritol, -inositol*; *mesotartaric* adj. (in Dict.).
1936 *Chem. Abstr.* XXX. 7625 While *meso*-inositol can serve as a part of the bios complex, it has never been isolated from yeast. **1937** F. C. WHITMORE *Org. Chem.* 385 Since the liquid 1,4-dibromide gives *meso*-erythritol it must be the *cis* form of configuration (A). **1968** I. L. FINAR *Org. Chem.* (ed. 4) II. vii. 231 On oxidation, D-erythrose gives *meso*-tartaric, and on reduction gives *meso*-erythritol. **1971** *Biochim. & Biophys. Acta* CCXLI. 204 Pea chloroplasts are also impermeable to the six-carbon cyclic polyhydroxy alcohol, *meso*-inositol.
b. Used as quasi-*adj.* (either as a separate word, usu. *attrib.*, or joined by a hyphen to a following sb.) to denote an isomer which has one or more pairs of enantiomorphic structural units so arranged that the molecule as a whole is optically inactive; as *meso form, isomer.*
1896 *Jrnl. Chem. Soc.* LXX. I. 412 It is impossible to say which is the racemic acid and which the meso-form. **1907** J. B. COHEN *Org. Chem. Adv. Students* I. ii. 94 Ordinary inositol from beans and flesh is a meso compound. **1937** F. C. WHITMORE *Org. Chem.* 53 These are stereoisomers of the same type as the tartaric acids, one being meso and the other racemic. *Ibid.* 371 Usually the meso-form can be separated from the *dl*-form by patient application of fractionation. **1956** *Nature* 11 Feb. 281/1 When it [*sc.* the filtrate] was acidified..the meso-isomer precipitated. **1968** R. O. C. NORMAN *Princ. Org. Synthesis* v. 154 As with other examples of diastereoisomers, the properties of *meso* forms are different from those of the isomeric mirror-image pairs.

mesocephal (meso·se·făl, mī·zo-). [f. MESOCEPHALI *sb. pl.*] One who has a mesocephalic skull.
1900 [see *DOLICHOCEPHAL]. **1901** [see *BRACHYCEPHAL]. **1935** HUXLEY & HADDON *We Europeans* vi. 185 In South Arabia..they [*sc.* the Jews] are preponderantly dolicocephals, and in North Africa dolicocephals or low mesocephals. **1957** C. G. SELIGMAN *Races Afr.* (ed. 3) iv. 78 They are essentially mesocephals, perhaps reaching the lower grades of brachycephaly.

mesocratic, *a.* (In Dict. s.v. MESOCRACY.) Add: **2.** *Petrol.* [after *leuco-, melanocratic.*] Of a rock: intermediate between a leucocratic and a melanocratic rock.
1904 *Amer. Geologist* XXXIV. 134 The main body of the boss is made up of a coarsely crystalline, mesocratic, hornblende gabbro. **1954** [see *leucocratic* adj. s.v. *LEUCO-].

mesolithic, *a.* **2.** (Earlier and later examples.)
1866 *Jrnl. Anthropol. Soc.* IV. p. clxxxiv, The author [*sc.* H. M. Westropp] described in some detail the characteristic forms of the gravel drift, flint implements of Ireland, and polished stone implements. The following terms were proposed by the author to distinguish them—Palæolithic, Mesolithic, Kainolithic. **1932** *Jrnl. R. Anthrop. Inst.* LXII. 257 Until recently the true Mesolithic was unknown in Palestine, although it was the custom among local archaeologists to describe as such the industry containing small unpolished axes or adzes which is found so abundantly on the surface all over the country. **1936, 1970** [see *EPIPALÆOLITHIC *a.*]. **1975** *Guardian* 21 Jan. 6/3 The Mesolithic or Middle Stone Age. *Ibid.* 6/7 Mesolithic man achieved partial domestication of the red deer.

mesomerism (mesǫ·mĕriz'm). *Chem.* [(in sense a) ad. F. *mésomérie* (A. Cornillot 1927, in *Ann. de Chim.* VIII. 267), f. *meso-* MESO- after *tautomérie* TAUTOMERISM.] † **a.** A kind of tautomerism (see quot.). *Obs.*
1928 *Chem. Abstr.* XXII. 2155 The term *mesomerism* is proposed to designate affinitive tautomerism and distinguish it from structural tautomerism (or desmotropism) and activation tautomerism (or tautomerism proper).
b. The property exhibited by certain molecules of having a structure which cannot adequately be represented by a single structural formula but can only be said to be intermediate between two or more graphical structures which differ in the distribution of electrons; resonance.
1934 *Nature* 23 June 947/1 Mesomerism and tautomerism are different concepts. **1961** I. L. FINAR *Org. Chem.* I. ii. 15 Heisenberg (1926), from quantum mechanics, supplied a theoretical background for mesomerism; he called it resonance, and this is the name which is widely used. **1973** B. J. HAZZARD tr. *Organicum* III. iv. 147 The capacity of a substituent for entering into mesomerism with a neighbouring double bond is defined as the mesomeric effect, and its sign is determined..by the polarization of the substituent accompanying the mesomerism.
So **meso·me·ric** *a.*, exhibiting or arising from mesomerism (sense b).
1933 C. K. INGOLD in *Jrnl. Chem. Soc.* 1124 The permanent polarisation associated with the tautomeric effect was originally distinguished by the name 'electronic strain', but this term has not proved convenient and, on account of considerations indicated later, the designation *mesomeric* effect is now substituted. **1956** E. DE B. BARNETT *Mechanism Org. Chem. Reactions* i. 5 Neither the double-bonded formula nor the dipolar formula is a true representation of the molecule, which must be regarded as a mesomeric hybrid to which both structures contribute. **1973** B. J. HAZZARD tr. *Organicum* III. iv. 146 A mesomeric system may involve not only double and triple bonds between the various bond partners but also free electron pairs conjugated with them.

mesomorph (me·somǫɪf, mī·zo-). [f. MESO- + Gr. μορφ-ή form.] **1.** *Anthropometry.* A person with a powerful, compact body-build in which the physical structures developed from the mesodermal layer of the embryo, i.e. the bones, muscles, and connective tissues, predominate: one of W. H. Sheldon's three constitutional types (cf. *ECTOMORPH, *ENDOMORPH 2).
1940 W. H. SHELDON *Varieties Human Physique* iii. 35 Bones, muscles, connective tissue..predominate overwhelmingly in the variants of type 2. We therefore call these variants mesomorphs. **1944** A. HUXLEY *Let.* 19 July (1969) 508 The football team entirely composed of big-boned, large-faced mesomorphs. **1951** AUDEN *Nones* (1952) 56 Behold the manly mesomorph Showing his splendid biceps off. **1971** J. Z. YOUNG *Introd. Study Man* xxxix. 573 In mesomorphs muscle and bone predominate, ..and they are adventurous, aggressive, extroverted, and dominating.
Hence **me·somorphy,** the mesomorphic body-build; the quality of being mesomorphic.
1940 W. H. SHELDON *Varieties Human Physique* i. 5 Mesomorphy means relative predominance of muscle, bone, and connective tissue. **1944** [see *ENDOMORPHY]. **1962** [see *ECTOMORPH]. **1969** DOWNS & BLEIBTREU *Human Variation* vii. 248 A football player, for example, would be high in mesomorphy.
2. *Physical Chem.* = *mesophase* (s.v. *MESO-). *rare.*
1969 [see *MESOMORPHISM].

mesomorphic (mesomǭ·ɪfik, mī·zo-), *a.* [f. MESO- + Gr. μορφ-ή form + -IC.] **1.** *Physical Chem.* [ad. F. *mésomorphe* (G. Friedel 1922, in *Ann. de Physique* XVIII. 273).] Existing in, pertaining to, or designating the state of a liquid crystal, intermediate between the ordered state of matter in crystals and the disordered state in ordinary liquids.
1923 *Chem. Abstr.* XVII. 3267 (*heading*) The mesomorphic states of matter. **1938** [see *liquid crystal* s.v. *LIQUID *a.* 7]. **1940** GLASSTONE *Text-bk. Physical Chem.* vii. 506 Mesomorphic behavior is restricted to substances having long chains. **1962** [see *liquid crystal* s.v. *LIQUID *a.* 7].
2. Characteristic of or resembling a mesomorph; of or pertaining to mesomorphy.
1940 W. H. SHELDON *Varieties Human Physique* i. 5 The mesomorphic physique is normally heavy, hard, and rectangular in outline. **1944** A. HUXLEY *Let.* 19 July (1969) 508 The ectomorph and his endomorphic wife in the museum, looking at the mesomorphic ideal of Greek sculpture. **1959** *Chambers's Encycl.* XI. 335/2 Correlation coefficients as high as 0·8 have been obtained..between the somatotonic component and extreme mesomorphic physique. **1963** *Punch* 25 Dec. 935/3 Mesomorphic McCarthies.

mesomorphism (mesomǭ·ɪfiz'm, mī·zo-). *Physical Chem.* [f. *MESOMORPH(IC *a.* + -ISM.] The state of being mesomorphic.
1933 [see *LYOTROPIC *a.* 2]. **1969** J. S. DAVE et al. in G. H. Brown et al. *Liquid Crystals 2* II. 229 The authors have investigated mixed mesomorphism in the binary mixtures wherein one of the components is a pure smectic mesomorph and the other a non-liquid crystalline Schiff's base.

mesomorphous (mesomǭ·ɪfəs, mī·zo-), *a.* *Physical Chem.* [f. as prec. + -OUS.] = *MESOMORPHIC *a.* 1.
1930 *Engineering* 7 Nov. 575/3 The so-called liquid crystals of Lehmann, which he regards as mesomorphous crystals. **1972** P. A. WINSOR in Brown & Labes *Liquid Crystals 3* I. 66 An idealized phase diagram showing the relationships between the amorphous solution phase and the succession of fused mesomorphous solution phases found in binary amphiphile/water system[s] is shown.

‖ **meson**[2] (mesǫ·n). [a. Sp. *mesón* an inn.] A type of inn in Mexico (see quot. 1861).
1824 J. R. POINSETT *Notes on Mexico* iv. 32 We entered the town..and drove through it to the meson. **1847** G. A. F. RUXTON *Adventures Mexico & Rocky Mts.* viii. 52 The meson was better than usual, being the stopping place of the diligencia to Fresnillo. **1861** E. B. TYLOR *Anahuac* 209 The *meson* of Mexico is a lineal descendant of the Eastern Caravanserai... It consists of two courtyards, one surrounded by stabling and the other by miserable rooms for the travellers, who must cook their food themselves, or go elsewhere for it. **1910** A. SANTLEBEN *Texas Pioneer* 164 Meson means an inn or hostelry. **1955** *Amer. Speech* XXX. 258 *Meson*, an inn.

meson[3] (mī·zǫn, me·zǫn). *Nuclear Physics.* [Alteration of the earlier name *MESOTRON: see *-ON[1].] **a.** Any of a group of unstable subatomic particles (first found in cosmic rays) that are intermediate in mass between an electron and a proton; the name is now commonly restricted to particles that are strongly interacting and have zero or integral spin (cf. *MUON), certain of which occur in atomic nuclei as transmitters of the binding force between the nucleons.
1939 H. J. BHABHA in *Nature* 18 Feb. 276/2 The name 'mesotron' has been suggested by Anderson and Neddermeyer..for the new particle found in cosmic radiation with a mass intermediate between that of the electron and proton. It is felt that the 'tr' in this word is redundant, since it does not belong to the Greek root 'meso' for middle; the 'tr' in neutron and electron belong, of course, to the roots 'neutr' and 'electra'... It would therefore be more logical and also shorter to call the new particle a meson instead of a mesotron. **1942** POLLARD & DAVIDSON *Appl. Nucl. Physics* ii. 13 The positive and negative

mesotrons or mesons should be mentioned here. 1947 *Sci. News* IV. 125 Mesons are the most penetrating component of cosmic rays. 1948 *Times* 10 Mar. 4/7 Other scientists spoke of mesons as the 'cosmic cement' which is believed to hold together the nucleus of the atom. 1963 S. Tolansky *Introd. Atomic Physics* (ed. 5) xxiii. 391 They [sc. muons] were once called mu-mesons but it was recognised that they were not in fact mesons since they had different spins. They are fermions, not bosons, and now the name meson is reserved for bosons only. 1971 *Sci. Amer.* Oct. 42/3 Such high-energy cosmic ray collisions produce a shower of secondary particles, principally kaons (K mesons) and pions (pi mesons), which in turn decay spontaneously into muons. *Ibid.*, The detector, a 2,000-ton array of concrete water tanks, light-collecting tubes and gas-filled cylinders, was designed to record the arrival of muons, or mu mesons.

b. Special Comb.: **meson factory** *colloq.*, an establishment having a high-energy accelerator for producing an intense beam of mesons and the equipment for making experimental use of such a beam.

1966 *Physics Today* Dec. 21/1 Meson factories, which will produce beams of nucleons and mesons thousands of times more intense than those presently available, can be expected to take us to our next level of understanding of nuclear structure... By meson factory I mean a complete nuclear-physics installation based on an accelerator of 500–1000 MeV, which is capable of providing at least 100 microamperes of high quality external beam. All ancillary facilities for performing high precision experiments with the primary and secondary beams are included: targeting, beam transport, data-handling equipment and beam dumps for the residual beams. 1972 *Physics Bull.* Mar. 148/3 With the completion of high intensity proton accelerators, meson factories, around the world.., it will soon be possible to produce muonic atoms and molecules copiously.

mesonic (mī-, mezǫ·nik), *a. Nuclear Physics.* [f. *MESON[3] + -IC.] = *MESIC *a.*[2]

1939 *Nature* 8 July 78/1 The most important problem is the construction of a classical model of the heavy mass by means of a mesonic field in the same manner as the electronic mass is built up from an electromagnetic field. 1953 *Physical Rev.* XCII. 801/2 The use of the μ^- meson as a nuclear probe is possible because the mesonic orbits are much closer to the nucleus than the corresponding electron orbits. 1954 *Ibid.* XCVI. 1145 The effects of vacuum polarization on the energy levels of π and μ mesonic atoms are computed. 1959 *Chambers's Encycl.* XIII. 241/2 The measurement of mesonic mass is very difficult. 1968 M. S. Livingston *Particle Physics* vi. 122 The first estimate of the mass of the muon came from studies of the ionization density of the tracks of mesonic particles in cloud chamber photographs of cosmic rays.

mesonomic (mesonǫ·mik, mī-zo-), *a. Law.* [f. Meso- + Gr. νόμ-ος law + -ic.] In the terminology of A. Kocourek: 'a jural relation which does not directly affect the natural physical freedom of a human being with the support of the law but yet has legal consequences in its evolution'. Opp. *zygnomic *a.*

1927 A. Kocourek *Jural Relations* v. 69 Compared with zygnomic relations, mesonomic relations are inferior in legal potency. A zygnomic relation directly constrains the servus of the relation with the support of the law, but a mesonomic relation either does not constrain the servus directly or lacks the support of the law. 1930 — *Introd. Sci. of Law* iv. 294 Put broadly, the function of mesonomic relations is to initiate or destroy zygnomic relations. *Ibid.*, Mesonomic relations have many different forms and functions... The power to commit a tort is mesonomic. The duty to pay damages for the tort is zygnomic. The power to pay damages for the tort is zygnomic. The power to violate a contract is mesonomic. The duty to pay damages is zygnomic... A duty owed by a subject to the sovereign not to commit a crime is mesonomic. The power of a prosecutor to prosecute a criminal offender is zygnomic. 1964 J. Stone *Legal Syst. & Lawyers' Reasonings* iv. 148 Kocourek's mesonomic-zygnomic distinction is no doubt a valuable one, even if we shrink back at the neologisms. *Ibid.* 149 For citizens' counsellors, as well as for citizens .., it is predominantly in the range of alternative courses left open by privileges and powers, the mesonomic relations, that attention and concern mainly moves.

mesopelagic (mesopīlæ·dʒik, mīzo-), *a. Biol.* [f. Meso- + Pelagic *a.*] Of, pertaining to, or designating the intermediate depths of the sea, *spec.* those between 200 and a thousand metres down.

1951 Allee & Schmidt *Hesse's Ecol. Animal Geogr.* (ed. 2) xiv. 312 The eyes of many mesopelagic animals are especially large. 1957 *Mem. Geol. Soc. Amer.* LXVII. I. 643 Among the typical animals of the mesopelagic zone..are such well-known fishes as *Idiacanthus, Chauliodus, Stomias*, and the lantern fishes. 1964 *Oceanogr. & Marine Biol.* II. 367 The photophores of most mesopelagic animals shine downwards (if the creature swims with its ventral surface downwards), and the colour of the light they emit is approximately that of..transmitted sun- and sky-light. 1970 [see *dysphotic* adj. s.v. *dys-]. 1975 *Nature* 5 June 452/2 The mesopelagic migrants, spending the daytime at depths down to perhaps 1,000 m and rising to within 200 m or so of the surface at night, show an increased respiratory rate not only with increased temperature but also with increased pressure.

Mesopotamia. For 'A country between two rivers' read 'A tract between two rivers' and add later examples.

1886 *Pall Mall Gaz.* 23 June 13/2 Every Oxford man has known and loved the beauties of the walk called Mesopotamia. 1944 G. B. Cressey *Asia's Lands & Peoples* xxvi. 401/2 The land between the rivers, that is, the true 'Mesopotamia', is thus capable of easy irrigation from either side. 1963 A. R. Woolley *Clarendon Guide Oxf.* 130 A path continues between the river and a mill-race and this is known as Mesopotamia.

b. = *Belgravia.

1864 E. Yates *Broken to Harness* (ed. 3) xv. 271 A house in Great Adullam Street, Macpelah Square, in that district of London whilom known as 'Mesopotamia'.

2. As the type of a word which is long, pleasant-sounding, and incomprehensible; used allusively for something which gives irrational or inexplicable comfort or satisfaction to the hearer (see quots.).

1827 Scott *Chron. Canongate* 1st Ser. I. v. 109 She resembled exactly in her criticism the devotee who pitched on the 'sweet word Mesopotamia', as the most edifying note which she could bring away from a sermon. 1870 *Brewer's Dict. Phr. & Fable* 572/1 *The true 'Mesopotamia' ring..i.e.*, something high-sounding and pleasing, but wholly past comprehension. The allusion is to the story of an old woman who told her pastor that she 'found great support in that comfortable word *Mesopotamia*'. 1886 'M. Gray' *Silence of Dean Maitland* III. iii. iv. 94 It was said of the Bishop of Belminster that he could pronounce the mystic word 'Mesopotamia' in such a manner as to affect his auditors to tears; but of the dean it might be averred that his pronunciation of 'Mesopotamia' caused the listeners' hearts to vibrate with every sorrow and every joy they had ever known, all in the brief space of time occupied by the utterance of that affecting word. 1906 F. M. Parsons *Garrick & his Circle* 245 Whitefield possessed the inborn gift of preaching to the nerves, and there is an edifying, though probably fallacious, report of Garrick's having remarked that he could pronounce 'Mesopotamia' in such a way as to move any audience to tears. 1908 G. B. Shaw *Platform & Pulpit* (1962) 47 There are people who will swallow as inspired revelation any sort of stuff that, so to speak, has the word Mesopotamia in it. 1924 L. Parks *What is Modernism?* p. xviii, The reaction to words is a curious psychological—I suspect pathological—phenomenon. The oft-told tale of the good woman who found 'Mesopotamia' a soothing word is one example.

Mesopotamian, *a.* Add: (Further examples.) Also as *sb.*, a native or inhabitant of Mesopotamia (the larger part of which is now Iraq).

1616 T. Coryat *Traveller for English Wits* 42 Your generosities most obliged Countreyman,..the Hierosolymitan - Syrian - Mesopotamian - Armenian - Median.., Thomas Coryate. 1673 A. Woodhead *Guide to Controversies Relig.* (ed. 2) iii. viii. 283 The Assyrians, Persians, and Mesopotamians, are ranged also under a new Patriarch. 1935 Huxley & Haddon *We Europeans* vi. 175 A North Syrian group which received accretions from Mesopotamian, Egyptian, Hittite, Amorite, and perhaps other sources. 1952 Gerth & Martindale tr. *Weber's Anc. Judaism* i. i. 6 The rise to political prominence of Babylonian power at the end of the third millenium and the continuous ascendancy of Babylon..constituted later aspects of Mesopotamian influence. 1954 H. Frankfort *Art & Archit. Anc. Orient* p. xxv, The Mesopotamians congregated in cities. *Ibid.* 6 The Mesopotamian deeply felt the enormity of the presumption that man should offer residence to a deity. 1974 *Encycl. Brit. Micropædia* VI. 820/1 Because religion provided the only total view of existence in ancient Mesopotamian civilization, religious themes, attitudes, and presuppositions were frequently included in quasi-secular writings.

mesorrhine, mesorhine, *a.* Add: **B.** *sb.* A person or skull having a nasal index intermediate between leptorrhine and platyrrhine. Hence **me·sorrhiny**, the state or quality of being mesorrhine.

1904 *Biometrika* Mar.–July 214 Nasal Index. Mesorrhiny. 1953 Beals & Hoijer *Introd. Anthropol.* iv. 98 Mesorrhines, like the Polynesians, who live in tropical regions.

mesoscale (me·soskēil, mī·zo-). *Meteorol.* Also **meso-scale.** [f. Meso- + Scale *sb.*[3]] An intermediate scale, between that of high- and low-pressure systems on the one hand and that of microclimates on the other, on which such phenomena as storms occur. Freq. *attrib.*

1956 T. Fujita *Mesoanalysis of the Illinois Tornado of Apr. 9, 1953* (Univ. of Chicago Dept. Meteorol. Technical Rep. to U.S. Weather Bureau, Contract No. Cwb 8613) 2 A method of analyzing mesoscale charts. 1959 *Modernizing the Aviation Weather Service* (U.S. Weather Bureau) 39 This meso-scale is considered a necessary link between the present synoptic scale and local weather events. 1969 *Britannica Bk. of Year* (U.S.) 518 Basic research continued into cyclonic storms and other aspects of the mesoscale or so-called secondary circulation of the atmosphere, including tropical cyclones (hurricanes and typhoons), mid- and high-latitude lows, and tornadoes. 1970 *Nature* 15 Aug. 646/2 Is there any sense in seeking to understand the effects of, say, major afforestation schemes on climate in the surrounding region without knowing much more about the behaviour of the atmosphere on the mesoscale? 1971 *Sci. News Let.* 30 Jan. 81 Dr. H. A. Panofsky.. pointed to weaknesses in the knowledge of mesoscale meteorology, the meteorology of areas 10 to 20 miles in diameter. 1974 *Times Lit. Suppl.* 13 Sept. 982/3 Given the 3 to 5 kilometre photographic resolution, satellite data are most usefully applied to macroscale phenomena such as the major global wind patterns, or to mesoscale phenomena ranging in scale from 10 to 100 kilometres.

mesosiderite (mesosi·dĕrəit, -səidīə·rəit, mīzo-). *Geol.* [ad. G. *mesosiderit* (G. Rose 1865, in *Amtlicher Bericht 39. Versammlung deutscher Naturforscher und Ärzte, Giessen 1864* 111); see Meso- and Siderite[1].] Any of the stony-iron meteorites in which the silicates are principally present as pyroxene and plagioclase.

1868 *Geol. Mag.* V. 76 Meteorites have long since been arranged under two great divisions, the irons and the stones... In examining a certain number of these masses it has been thought convenient by some to establish a third, or intermediate division, to which the names of *Mesosiderites, Lithosiderites*, or of *Siderolites*, have been given. *Ibid.* 78 For these meteorites M. Rose has proposed a special name, that of *mesosiderites*. 1918 G. T. Prior in *Mineral. Mag.* XVIII. 151 In the Rose–Tschermak–Brezina classification of meteorites the intermediate class of stony-irons, in which iron and stony matter occur in approximately equal amounts and to which Maskelyne gave the name siderolites, is divided into (1) siderolites proper..and (2) lithosiderites, comprising the groups of the mesosiderites, grahamites, and lodranites. Mesosiderites are defined as consisting of iron and crystalline olivine and bronzite... As seen in the following pages, chemical and microscopic examination of typical members of these two groups fails to reveal any real distinction between them based on the amount of felspar they contain. For the combined group, all the members of which contain abundant felspar, it is proposed to retain the earlier name of mesosiderite. 1962 B. Mason *Meteorites* viii. 120 The stony-irons are a minor group of meteorites... They are divided into two major groups according to the nature of the silicate minerals, the pallasites (olivine stony-irons) and mesosiderites (pyroxene-plagioclase stony-irons).

mesosoma. Add to defs.: Also, the central part of the body of certain other invertebrates. (Later examples.)

1932 Borradaile & Potts *Invertebrata* xv. 450 The lung books are found on segments 3–6 of the mesosoma. 1962 D. Nichols *Echinoderms* xiv 174 The lophophore itself is borne by the mesosome, and there is a strong partition between this division of the body and the metasome.

2. *Bacteriology.* (Used in the form **mesosome.**) A cytoplasmic structure in many bacteria which is principally composed of membranes, probably being formed by invagination of the cell membrane, and may be a site of active respiratory activity and the place of attachment of the bacterial genome.

1960 P. C. Fitz-James in *Jrnl. Biophysical & Biochem. Cytol.* VIII. 508/2 This [sc. 'dense body'] is probably an unsuitable name. Certainly they are not always electron dense... Since they appear to be attached to the surface membranes of the cell, one should adopt the suggestion of Robertson..for such membrane-attached cytoplasmic structures and use the term 'mesosomes'. Henceforth in this paper this term will be used. 1968 H. Harris *Nucleus & Cytoplasm* vi. 123 The loss of the ability to form the cell wall appears to be associated with the disappearance from the cytoplasm of the cell of specific membranous structures, known as 'mesosomes', which appear to be essential for the formation of the wall. 1974 *Nature* 1 Feb. 303/1 These properties together with the possession of fimbriae and the absence of mesosomes are commonly shared by Gram-negative, but generally absent in Gram-positive bacteria.

mesothelioma (me:soþīliōu·mă). *Path.* Pl. **-omas, -omata.** [f. next + *-oma.] A tumour of mesothelium; formerly, † a tumour composed of cells derived from the embryonic mesothelium.

1909 J. G. Adami *Princ. Path.* I. 647 Of mesothelial origin: Tumors (mesotheliomas) whose characteristic constituents are cells derived in direct descent from the persistent mesothelium of the embryo. (*a*) Typical.—Adenoma of kidney, testicle...'mesothelioma' of pleuræ, peritoneum, etc. *Ibid.* 746 If a convenient term is required for all this order of tumors, the transitional adenocarcinomas of adrenal, kidney, ovary, and testis, we have, from embryogenetic considerations, suggested the term mesothelioma. 1921 *Lancet* 23 July 173/2 The true nature of these tumours has been the subject of much controversy as is seen by the variety of names given to them—e.g. endothelial cancer,..mesothelioma, and so on. 1966 Wright & Symmers *Systemic Path.* I. i. 6/1 Primary tumours of the pericardium..are very rare: most of those that have been described appear to have arisen from mesothelial cells, and may in consequence be termed 'mesotheliomas'. 1971 *Brit. Med. Bull.* XXVII. 71/2 Wagner.. discovered the first of a large group of pleural and peritoneal tumours—mesotheliomata—apparently related to exposure to crocidolite. 1973 *Nature* 8 June 352/2 Mesotheliomas of the pleura and peritoneum have been linked with the inhalation of asbestos fibres.

mesothelium, delete entry in Dict. (s.v. Meso-) and substitute:

mesothelium (meso-, mīzoþī·liǝm). *Embryol.* and *Histology.* Pl. **-thelia.** [f. Meso- + Epi)thelium.] In a vertebrate embryo, epithelium that forms the surface layer of the mesoderm and lines the body-cavity; in a post-natal organism, the tissue derived from this that forms the lining of the pleuræ, peritoneum, and pericardium (by some writers, esp. pathologists, not regarded as epithelium).

1886 C. S. Minot in *Buck's Handbk. Med. Sci.* III. 176/1 The whole of the mesoderm..does not go through this metamorphosis, but..a part remains closely compacted; but ultimately it is only the single layer of cells immediately bounding the cœlom, and the cells constituting the myotomes.., which remain thus close together. These cells, therefore, have all the characteristics of an epithelium, so that the cœlom is limited by an epithelium of cuboidal cells, for which I have proposed the name *mesothelium*. *Ibid.* 176/2 Derivatives of the mesothelium (epithelial mesoderm). 1945 W. J. Hamilton et al. *Human Embryol.* vi. 80 The mesoderm bounding the coelom [of the embryo] will persist throughout life as an epithelium (so-called mesothelium) which forms the visceral and parietal linings of the peritoneal, pleural and pericardial cavities. Parts of the peritoneal mesothelium become specialized to form the germinal epithelium, the paramesonephric ducts..and the cortex of the suprarenal gland. 1950 A. W. Ham *Histol.* xi. 140/1 Endothelium and mesothelium, structurally, are both typically epithelial... Nevertheless, pathologists usually prefer to consider endothelium and mesothelium as apart from ordinary epithelium because endothelial and mesothelial cells behave differently from ordinary epithelial cells. 1960 F. D. Allan *Essent. Human Embryol.* xvii. 165 The cells of the cortical primordium are derived from the coelomic mesothelium. 1968 Passmore & Robson *Compan. Med. Stud.* I. xvii. 3/2 Endothelia and mesothelia arise only from mesoderm.

Hence **mesothe·lial** *a.*

1892 C. S. Minot *Human Embryol.* (1897) vii. 159 Although the pancreas, ovary, and spinal cord all contain connective tissue, we do not call them mesenchymal, but respectively entodermal, mesothelial, and ectodermal. 1946 B. M. Patten *Human Embryol.* xv. 469 In the case of a part of the gut tract which lies within the body cavity, the corresponding connective-tissue layer will be supplemented by an epithelial layer derived from the mesoderm which lines the coelomic cavities... This mesothelial layer provides a smooth, moist surface which permits the viscera within the body cavity to change shape and position with a minimum amount of friction. 1950 [see above]. 1968 Passmore & Robson *Compan. Med. Stud.* I. xvi. 5/2 (*heading*) Mesothelial or serosal sac membranes.

mesotherm. Add: (Later example.) Hence **mesothe·rmic** *a.*

1903 W. R. Fisher tr. *Schimper's Plant-Geogr.* 236 The most important family of the north temperate zone among the *Polycarpicae*, that of the Ranunculaceæ, is mesothermic and microthermic. 1960 N. Polunin *Introd. Plant Geogr.* x. 201 Examples [of climatic relics] are the mesothermic plants to be found in some boreal areas that have cooled at least since the 'postglacial optimum' when such plants presumably migrated to these areas. *Ibid.* 285 The great world vegetational zones..depend primarily on temperature, and we find it convenient to distinguish between megatherms (plants favouring warm habitats), microtherms (plants favouring cold habitats), and the intermediate mesotherms.

mesothorium (mesoþōə·riǝm, mīzo-). *Chem.* [mod.L. (O. Hahn 1907, in *Ber. d. Deut. Chem. Ges.* XL. 1469): see Meso- and Thorium.] Either of two radioactive nuclides in the thorium decay series: *mesothorium I* (symbol MsThI), the isotope of radium with mass number 228, produced by the alpha decay of thorium 232, or *mesothorium II* (symbol MsThII), the isotope of actinium with mass number 228, produced by the beta decay of mesothorium I.

1907 *Jrnl. Chem. Soc.* XCII. ii. 359 The suggestion is made that thorium is not directly transformed into radiothorium, but into an intermediate product, 'Mesothorium', which then yields radiothorium. 1935 *Jrnl. R. Aeronaut. Soc.* XXXIX. 857 The mesothorium preparation produces a point source of gamma radiation suitable for the inspection of thick metal objects for which X-ray investigation becomes very expensive. 1937 *Nature* 21 Aug. 318/1 *Lemna* [duckweed] accumulates mesothorium I and concentrates it a hundredfold as compared with its content in the water. 1969 R. F. Lang tr. *Henglein's Chem. Technol.* 546 Mesothorium is present in monazite to the extent of 1 mg per t, and serves as a substitute for radium.

mesotron (mī-, me·zotrǫn, me·sotrǫn). *Nuclear Physics.* Now *rare* or *Obs.* [f. Meso- + *-tron.] The name orig. given to the meson (*meson[3]).

1938 Anderson & Neddermeyer in *Nature* 12 Nov. 878/2 The existence of particles intermediate in mass between protons and electrons has been shown in experiments on the cosmic radiation... We should like to suggest..the word 'mesotron' (intermediate particle) as a name for the new particles. It appears quite likely that the appropriateness of this name will not be lost, whatever new facts concerning these particles may be learned in the future. 1939 *Ibid.* 13 May 796/2 The meson or mesotron was first introduced by Yukawa to serve as the connecting link between β-ray disintegration and the forces responsible for nuclear binding. 1942 *Endeavour* Jan. 22/2 Skobeltzyn..has argued against the theory of Heisenberg and Euler that the spontaneous disintegration of mesotrons is the mechanism responsible for the formation of the electronic component observed in the lower layers of the atmosphere. 1949 Koestler *Insight & Outlook* x. 136 The so-called 'elementary particles'—electrons, photons, neutrons, mesotrons, and so forth.

Mespot (me·spǫt). Also **Mess-pot.** Slang abbrev. of Mesopotamia.

1917 *To-Day* 6 Jan. 246/3 *Mesopotamia.*—So you call it the 'Mess-pot', do you? 1920 *Sat. Rev.* 3 July 4/1 The 'Messpot' madness. 1933 J. Buchan *Prince of Captivity* i. iii. 107 What front were you on—the Western, Palestine, Mespot? 1943 C. S. Forester *Ship* viii. 48 Mr. Whipple.. had gone off and joined the army and had done his bit in Mespot.

mesquite, mesquit[2]. Add to forms: **masketo, moscheto, musqueto, mus-, mesquito; musquet, -quit, -kit; muskeete. 1.** (Earlier and later examples.)

1759 tr. *Venegas's Nat. & Civil Hist. Calif.* I. 100 Their most usual device was to hold up in their hands some little tablets of wood made with great labour, for want of iron tools of mesquite. 1806 M. Lewis in *Deb. Congress U.S.* (1852) 9th Congress 2 Sess., App. 1083 A bean that grows in great plenty on a small tree resembling a willow called masketo. 1838 'Texian' *Mexico v. Texas* 70 Even where a tree appears it is sure to be a mesquite (*Mimosa nilotica*). 1838 C. Newell *Hist. Revolution Texas* 147 Live oak,.. black walnut, cypress, and musquit prevail. 1909 'O. Henry' *Roads of Destiny* viii. 129 Near the store, scattered among the mesquite and elms, stood the saddled horses of the customers. 1948 *Reader's Digest* Jan. 70/1 Anywhere in America has its particular hearth perfume—..mesquite floating out of desert chimneys in the Southwest. 1969 T. H. Everett *Living Trees of World* 201/1 The mesquite..[is] a native of the southwestern United States, Mexico, Central America and the West Indies.

b. A thicket of mesquite trees.

1834 A. Pike *Prose Sk. & Poems* 63 We emerged from the broken hills into the mesquito. 1910 W. M. Raine *Bucky O'Connor* 224 The Irishman..kept his party in the mesquit till the headlight of an approaching train was visible. 1945 *New Yorker* 25 Aug. 26 A railroad bull came walking along the tops of the cars and kicked me into the mesquite. 1974 *Sci. Amer.* Apr. 104/1 The particular dung beetles concerned would colonize only pads dropped in open pastures, however, and not those dropped in the dense growth of mesquite, where the cattle sought shelter from the heat of the day.

2. (Earlier examples.)

1823 W. B. Dewees *Lett. from Early Settler Texas* (1852) 35 The musquit grass grows very thick and about three feet high, and looks very much like a blue grass pasture. 1831 M. Holley *Texas* (1833) vi. 69 The pasturage here.. called Muskit grass, (pronounced Muskeet) bears a strong resemblance to the blue grass.

3. (Earlier and later examples.)

1806 M. Lewis in *Deb. Congress U.S.* (1852) 9th Congress 2 Sess., App. 1103 Some small cultivated fields, fenced round with small cedar and moscheto brush. 1831 J. O. Pattie *Personal Narr.* 59 We found the river skirted with very wide bottoms, thick set with the musquito trees. *Ibid.* 83 There is here little timber, beside musqueto wood, which stands thick. 1834 A. Pike *Prose Sk. & Poems* 56 The valley was..full of small hills interspersed with mezquito bushes. 1846 J. A. Quitman *Diary* 13 Aug. in J. F. H. Claiborne *Life & Corr. J. A. Quitman* (1860) I. 239 The steamer purchases musket-wood at $2.50 per cord for dry, $2.25 green. 1867 A. D. Richardson *Beyond Mississippi* xix. 226 Another waiting coach was soon rolling us forward among mesquite groves. 1905 A. Adams *Outlet* 16 The horses had run idle during the winter in a large mesquit pasture. 1957 J. Kerouac *On Road* (1958) 166 Tucson is situated in beautiful mesquite riverbed country.

mess, *sb.* Add: **1. c.** (Earlier and later U.S. examples.)

1697 S. Sewall *Diary* (1878) I. 455 Betty gets her Mother a Mess of English Beans. 1775 B. Romans *Florida* 12 (Th.), He told me that his mother had an inclination to eat fish, and he was to come to get her a mess. 1842 *Knickerbocker* XIX. 557 Sally couldn't hardly bring in the pail, she gave such a mess. 1861 O. W. Norton *Army Lett.* (1903) 26 H. and I got enough [potatoes] for a mess, and some parsnips. 1872 *Rep. Vermont Board Agric.* I. 197, I tested their milk by weighing every mess for a month. 1877 *Ibid.* IV. 54 We took off what cream arose on the night's mess, and churned it. 1883 J. C. Harris *Nights with Uncle Remus* iii. 30 Brer Rabbit, he hop in, he did, en got 'im a mess er greens, en hop out ag'in.

e. *N. Amer.* A quantity or number *of* something. See also quot. 1970.

1830 *Mass. Spy* 23 June (Th.), We saw yesterday a large mess of early potatoes. 1834 C. A. Davis *Lett. J. Downing* 40 With that, he out with his wallet, and unrolled a mess on 'em. 1854 *Knickerbocker* XLI. 502 There was wolves in the Holler—an unaccountable mess of 'em. 1872 *Rep. Vermont Board Agric.* I. 634 They will dress a mess of ore to any required per cent. 1939 L. M. Montgomery *Anne of Ingleside* xxxviii. 300 Tell Susan Baker I'm much obliged for that mess of turnip greens she sent me. 1956 B. Holiday *Lady sings Blues* (1973) xix. 154 Just before I was set to go on for the second set a big mess of gardenias arrived backstage. 1966 W. T. E. Kirkeby *Ain't Misbehavin'* v. 51 This is Fats Waller, the baby who plays that mess of organ over at the Lincoln. 1970 C. Major *Dict. Afro-Amer. Slang* 81 *Mess*, large quantity;..someone ..[who] is remarkable or puzzling.

3. d. *colloq.* Excrement, esp. of an animal; usu. in phr. *to make a mess.*

1903 in *Eng. Dial. Dict.* s.v. Mess, *sb.*[4] 4. 1928 Kipling *Limits & Renewals* (1932) 50 It [sc. a dog]'s made a mess in the corner. 1937 V. Woolf *Years* 245 The pigeons were a nuisance, he thought, making a mess on the steps. 1939 A. Huxley *After Many a Summer* i. x. 138 A lovely stinking little baby who still made messes in its bed. 1940 N. Mitford *Pigeon Pie* ix. 144 Perhaps, she thought, the bird wants to go out... It made a mess on her skirt. 1960 *Woman's Own* 6 Aug. 49/2 It's the dog. It made a mess on the carpet. 1972 *New Yorker* 30 Sept. 44/2 Happy..had helped them unpack by..pulling one of Mrs. Webster's dresses from a hanger and then making a mess on it.

e. *slang.* An objectionable, ineffectual, or stupid person.

1936 M. Mitchell *Gone with Wind* i. vi. 122 'Oh,' thought Scarlett...'To have that mealy-mouthed little mess take up for me!' 1938 E. Bowen *Death of Heart* i. ii. 40 From what you say, her mother was quite a mess. 1965 M. Spark *Mandelbaum Gate* iv. 104 These were lapsed Jews, lapsed Arabs, lapsed citizens, runaway Englishmen, dancing prostitutes, international messes.

4. b. Also, the place where food is served (see quot. 1886 in Dict.). Phr. *to lose the number of one's mess* (earlier and later examples).

1807 in A. Paget *Paget Papers* (1896) II. 314 If we are going against Copenhagen many of us will lose the number of our mess. 1911 C. E. W. Bean *'Dreadnought' of Darling* xxx. 260 That meant a ride out to the blacks' camp and some of them losing the number of their mess. It did not matter who was shot. 1934 G. B. Shaw *Too True to be Good* ii. 76 The conversation in the officers' mess doesnt suit me. *Ibid.* iii. 81 *Sweetie.* Well, of course. Youre in the sergeants' mess. 1969 D. Hay *Man in Hot Seat* vi. 60 At the Aldershot end of the airfield..stood the long blue-carpeted Mess of the Empire Test Pilots' School.

7. (sense 3) *mess-maker*; (sense 4) *mess-bag, -basket, -bill, -chest* (earlier examples), *-cloth, -dinner, -fire, fund, -gear, -hall, -man, -pan, -pork* (earlier examples), *-pot, -table* (earlier and later examples), *-tent* (later example); **mess-boy** *Naut.*, one who waits at table in a mess-room; **mess-jacket** (add definition): a short tailless jacket reaching just below the waist-line (further example); **mess-kit,** (*a*) utensils for cooking or handling food; (*b*) (military) uniform designed to be worn at meals.

1885 *Outing* VII. 55/1 From some dark corner of the messbags, or petacas, he unearthed a handful of dried apples. 1839 *Knickerbocker* XIII. 211 Tell Margaret to prepare the mess-basket. 1916 F. M. Ford *Let.* 29 Nov. (1965) 78 Also I have found a considerable portion of my Mess Bill out of Auction. 1917 W. Owen *Let.* 24 May (1967) 464 There have been a number of Mess Bills, & other cheques drawn lately. 1934 E. Bowen *Cat Jumps* 67 If mess-bills ran up..was John to blame? 1963 *Times* 24 May 14/7 Chits were signed, which sooner or later found their way to the Mess and appeared on one's mess-bill. 1818 'A. Burton' *Adventures J. Newcome* ii. 74 The Mess-boy to the Coppers dragged it [sc. a pudding]. 1955 C. S. Forester *Good Shepherd* 116 A big pot of coffee. And a sandwich. Tell the mess-boy I want one of my specials. 1964 O. E. Middleton in C. K. Stead *N.Z. Short Stories* (1966) 210 The messboy's discreet ropesoles pad patiently to and fro. 1970 *Islander* (Victoria, B.C.) 22 Feb. 5/1 The American trading schooner Nanook arrived. Its messboy had hanged himself for some unknown reason. 1830 J. F. Cooper *Water Witch* III. ix. 258 The smaller booms with the mess-chest and shot-boxes were all that lay between the group..and the depths of the ocean. 1858 T. Vielé *Following Drum* 15 A camp-kettle, mess chest, bundle of canvas, and set of tent poles. 1839 C. F. Briggs *Adventures H. Franco* I. xxiii. 250 Throwing down my knife..I leaped on to the messcloth, and gave him a blow in the eye. 1803 Scott *Let.* 10 July (1932) I. 195, I can proceed no further being alarmed by the Bugle Call not indeed to summon to battle but to the less hazardous task of a Mess Dinner where our Society tho' somewhat noisy is very good humoured. 1825 H. Wilson *Mem.* II. 116 The mess-dinner at Lewes..must more resemble a..private party than a mess-room, as they seldom mustered more than seven or eight persons together at table. 1885 A. Edwardes *Girton Girl* I. xiv. 283 The usual guest-night at mess. Curious how precisely alike all mess dinners are. 1837 W. Irving *Capt. Bonneville* (1895) I. 24 The various mess-fires were surrounded by picturesque groups. 1850 L. H. Garrard *Wah-to-Yah* xii. 180 He..walked to a messfire. 1850 J. J. Hort *Horse Guards* 70 He cannot act otherwise than by occasionally adding to the mess and the band funds. 1876 Voyle & Stevenson *Mil. Dict.* (ed. 5) 254/1 Married men pay one half if they do not regularly attend the mess, but they pay all contributions to the mess fund. 1970 V. Canning *Great Affair* xii. 216 Ex-pilot officer Robinson, cashiered for fiddling the mess funds. 1890 *Cent. Dict.*, Mess-gear. 1918 L. E. Ruggles *Navy Explained* 98 *Mess gear*, the tableware, plates, cups, saucers, food containers and implements used by mess cooks. 1862 G. C. Strong *Cadet Life at West Point* 66 We were as unfortunate in the mess-hall as out of it. 1958 *Listener* 6 Nov. 717/1 Food eaten at the communal mess-hall [in China]. 1905 A. M. Binstead *Mop Fair* i. 10 A tall blonde in a heel-tipping skirt and mess-jacket of blue herring-bone. *a* 1877 Knight *Dict. Mech.* II. 1421/2 *Mess-kit*, that portion of camp equipage consisting of cooking utensils. 1953 J. Masters *Lotus & Wind* vii. 89 His mess-kit was grey and black with silver facings. 1954 W. Faulkner *Fable* (1955) 343 Bottles, old messkits. 1929 D. H. Lawrence *Pansies* 89 It is hard.. to put up with the clever mess-makers. 1850 *Punch* 20 July 33/1 Messman wanted for a Cavalry Regiment. 1920 *Chambers's Jrnl.* May 285/2 Understanding the messman to have come off from the Colon with plenty of oysters. 1813 *Niles' Reg.* III. 295/2 [List of Military Supplies] Mess pans. 1846 R. B. Sage *Scenes Rocky Mts.* xxvii. 227 A large tin mess-pan, and a tin-cup and plate for each of its number. 1861 O. W. Norton *Army Lett.* (1903) 35 New knapsacks, canteens..mess pans and a complete outfit. 1832 *Louisville* (Kentucky) *Public Advertiser* 10 Mar., Mess and Prime Pork in barrels and half barrels, for sale. 1848 *Rep. Comm. Patents 1848* (U.S.) 527 The finest and fattest [hogs] making clear and mess pork. 1857 R. Glisan *Jrnl. Army Life* (1874) viii. 86, I shall be able to say good bye to the messpots of Uncle Sam. 1819 J. A. Quitman *Diary* 15 Nov. in J. F. H. Claiborne *Life & Corr. J. A. Quitman* (1860) I. 42 Mrs. and Miss Griffith, charmed with our mess-table, became our boarders. 1895 M. A. Jackson *Mem. Stonewall Jackson* xi. 191, I took my meals with him and the staff at their mess-table. 1916 in 'Contact' *Airman's Outings* (1917) 266 A great wind..whines past the mess-tent.

mess, *v.* Add: **4.** Examples of *to mess about in boats.* Also const. *around.*

1908 K. Grahame *Wind in Willows* i. 7 There is *nothing*—absolutely nothing—half so much worth doing as simply messing about in boats. 1932 *Amer. Speech* VII. 334 Mess around, to 'kill time'; to interfere; to meddle. 1957 D. Robins *Noble One* iv. 41, I can't see the attraction of messing around with a lot of sick animals. 1962 J. Cannan *All is Discovered* iv. 87 What he enjoyed was messing about in boats and sitting in the sun. 1964 *Evening Post* (Wellington, N.Z.) 4 Jan., A few audible reminders to himself to 'stop messing around' failed to help matters. 1970 H. E. Roberts *Third Ear* 10/1 Mess around, to engage in a great deal of purposeless activity. 1973 *Times* 18 Apr. 3/2 (Advt.), And if you ask the Chancellor of the Exchequer what a company like ours is doing 'messing about in boats' he'll tell you we're the world's largest manufacturer of quality inflatable leisure craft.

b. *to mess with*: to interfere or get involved with; to make a mess of; to trouble or annoy. *U.S. colloq.* or *dial.* Cf. sense 7 in Dict.

1903 *Dialect Notes* II. 299 *Mess with*, to meddle with; also, to make a mess of. 'Don't mess with your food.' 1913 *Ibid.* IV. 5 *Mess with*, to associate with. 'We don't mess with those people.' 1955 S. Whitmore *Solo* 27 And what little lady is going to mess with you. 1955 Shapiro & Hentoff *Hear me talkin' to Ya* 374 The really good musicians are too smart to mess with it. 1956 B. Holiday *Lady sings Blues* (1973) iv. 46 This talk about a big tone messed with Lester for months. 1968 E. Gaines in A. Chapman *New Black Voices* (1972) 101, I never messed with a woman I didn't love. 1971 *Black World* Apr. 66 You hit her with a chair leg. You didn't have no right to mess with that poor girl.

c. *to mess up*: to make a mess or muddle of a situation; to get into trouble; to become mixed up or involved. *U.S. colloq.*

1933 *Amer. Speech* VIII. iii. 29/2 Boy, I ain't a-goin' t' mess up no more from now on. 1938 M. K. Rawlings *Yearling* xxiii. 290 If she's nothin' but one o' them leetle ol' chipperdales, why do he mess up with her? 1956 B. Holiday *Lady sings Blues* (1973) iii. 33 When the time came to take those bills off the table, I was always messing up. 1969 H. Kohl in T. Kochman *Rappin' & Stylin' Out* (1972) 110 He wanted to learn badly, they told me, and was messing up by memorizing the signs in the neighborhood and thinking that's all there was to reading.

5. (Further examples.) Also, to spoil, ruin.

1909 *Dialect Notes* III. 349 The house is all messed up. 1919 G. B. Shaw *Heartbreak House* ii. 85, I get my whole life messed up with people falling in love with me. 1959 I. Fleming *Goldfinger* xiv. 194 Bond only prayed that she hadn't got some private plot involving him or Goldfinger that was going to mess up his own operation. 1966 *Word Study* Dec. 3/2 There's no real point in worrying anyhow. It just messes you up.

6. Change '*vulgar*' to '*colloq.*' and add further examples. Also, to inconvenience or annoy. Also with *around.*

1934 A. P. Herbert *Holy Deadlock* 276 Why *should* our private lives be spied upon—and messed about in a Court like this? 1955 *Times* 20 May 14/6 If industry is once again going to be messed about by Government interference—more nationalization, more controls. 1957 M. Spark *Comforters* v. 110 Her great desire to travel by train was dispersed by the obvious necessities of going to Mass, and of not messing Laurence around any further. 1973 *Time Out* 2–8 Mar. 14/1 My impression is that when the surge of violence was on—at its height, two years ago—quite a few teachers got messed about. It was somewhere beyond extreme rudeness, but short of your actual NAS physical assault.

‖ **messa di voce** (me·sā di vō·tʃe). Pl. **messe di voce.** [It., lit. 'placing of the voice'.] In singing, a gradual crescendo and diminuendo on a long-held note.

1801 in Busby *Dict. Mus.* 1876 Stainer & Barrett *Dict. Mus. Terms* 286/2 *Messa di Voce* (*It.*), the swelling and diminishing of the sound of the voice upon a holding note. 1938 *Oxf. Compan. Mus.* 567/2 [In the eighteenth century] every long note (irrespective of the sense of the words) was expected to bear a *messa di voce*. 1958 *Listener* 14 Aug. 250/3 The tenor's wonderful *messa di voce* at the cadence leading to the reprise. 1964 *Conc. Oxf. Dict. Opera* 259/1 *Messa di voce*.., the art of swelling and diminishing tone on a single note.

message, *sb.*[1] Add: **1. b.** (Further *transf.* examples.) Now esp., the broad meaning (of something); a view expressed in a piece of writing, etc., esp. one communicating a criticism of a social or political matter.

1936 *Time* 17 Feb. 46/2 In *Modern Times*, the 'message' has been underlined rather than, as in the old days, subconsciously implied. 1940 'G. Orwell' *Inside Whale* 156 It will be seen that once again I am speaking of these people as though they were not artists, as though they were merely propagandists putting a 'message' across. 1949 H. Nicolson *Diary* 7 Sept. (1968) 174 T. S. Eliot.. is off to lecture in Germany. He asked me whether they would expect a 'message'. I said the only thing to do was to treat them as ordinary members of cultured society. 1955 *Times* 12 May 13/4 Mr. Kauffmann resembles other talented American novelists in that his real gifts for storytelling and satirical observation are somewhat obscured by too heavy an insistence on the 'message' implicit in the plot. 1969 J. Argenti *Managem. Techniques* ix. 54 If the chairman of the company shows an informed interest in techniques just occasionally, this should be enough for the message to get through to all levels. 1970 *Guardian* 27 Aug. 6/2 David Halliwell is..the black sheep of the student militants... 'Do you approve?' asked a nicely-dressed teenager, identifying the shock with the message. 1975 *Broadcast* 23 June 17/1 The medium may not be the message, but undeniably the characteristics of the medium shape the message.

c. (Earlier example.) Also, a communication from Parliament to members of the royal family, or between the two houses of Parliament themselves.

1566 *House of Commons Jrnl.* 30 Sept. (1742–62) I. 73/1 At which Thirtieth Day of September, 1566,..Mr. Comptroller..with a convenient Number, went up to the Lords with that Message. 1621 *Ibid.* 13 Feb. (1742–62) I. 520/1 Having faithfully discharged themselves in their Message to the King. 1701 *Ibid.* 20 June (1742–62) XIII. 638/1 A Message was sent to the House of Commons, by Mr. Baron Tracy and Mr. Baron Berry. 1818 *Ibid.* 6 June LXXIII. 424/1 Resolved, *Nemine Contradicente*, That a Message be sent from this House to congratulate their Royal Highnesses the Duke and Duchess of Cambridge, on their happy nuptials. 1820 *House of Lords Jrnl.* 15 Aug. LIII. 367/2 Ordered, *Nemine Dissentiente*, That a Message of Condolence be sent from this House to His Royal Highness The Duke of York..and that The Duke of Wellington and The Earl Graham do attend His Royal Highness with the said Message. 1844 Erskine May *Law of Parl.* xvi. 249 A message is the most simple and frequent mode of communication; it is daily resorted to for sending bills from one house to another [etc.]. *Ibid.* xvii. 268 Messages are frequently sent by both houses to members of the royal family, to congratulate them upon their nuptials ..or other auspicious events..or to condole with them on family bereavements. 1950 *Times* 21 Oct. 3/5 Before 1855 ..messages were always carried to the Lords by members of the Commons themselves, and from the Lords to the Commons by Masters in Chancery or judges.

2. c. Phr. *to get the message*: to understand a position stated or implied.

1964 'C. E. Maine' *Never let Up* xv. 149 'Let's go in and have a drink anyway.' 'Now you're getting the message,' he commented. 'Lead the way.' 1967 O. Wynd *Walk Softly, Men Praying* viii. 134 They didn't ask me in for a drink... Richard..realized it would only be postponing the inevitable. He had got the message at last. 1972 D. Lees *Zodiac* 90 They don't seem able to make up their minds whether to warn me off or knock me off but I do get the message loud and clear and..I'm going. 1974 *Times* 9 Jan. 14/8 (*heading*) Will Nato get the message?

4. *message bag*; **message-lad** = *message-boy*; **message-stick,** also used in Norway.

1917 'Contact' *Airman's Outings* vii. 184 The contact patrol buses..by means of message bags dropped over brigade headquarters report progress to the staff. 1948 'N. Shute' *No Highway* xi. 289 Stubbs came back with the message bag. 1836 Dickens *Sk. Boz* (1837) 2nd Ser. 101 The mother had got the boy a message-lad's place in some office. 1860 *Leisure Hour* 3 May 287/1 To this day the people in Norway are called together for the despatch of public business, in a somewhat similar manner. A bud-stick, or message-stick..is painted and stamped with the royal arms.

messageless (me·sėdʒles), *a.* [f. Message *sb.*[1] + -less.] Without a message; having no message to communicate.

1925 *Brit. Weekly* 18 June 268/2 He lost his faith in the Bible..and (to use his own significant words) 'I became a messageless man'. *a* 1930 D. H. Lawrence *Last Poems* (1932) 283 The dark and gleaming beauty of the messageless gods.

‖ **messagerie** (mesaʒri). [Fr.: see Messagery.] Usu. *pl.* The transportation or delivery of goods, messages, or people; a conveyance for these. Also *attrib.* So **messageries maritimes,** the transport of goods, etc., by sea; the name of a shipping-line.

1792 A. Young *Trav. France* i. 11 In ten miles we met not one stage or diligence; only two messageries, and very few chaises. 1878 Lady C. Schreiber *Jrnl.* 29 June (1952) 150 We took our passage in the French Messageries boat. 1907 G. B. Shaw *John Bull's Other Island* p. x, Many Englishmen like Frenchmen better than Englishmen, and never go on board a Peninsular and Oriental steamer when one of the ships of the Messageries Maritimes is available. 1931 E. Waugh *Remote People* 12 French colonial officials, their wives and disorderly children..make up the bulk of a normal Messageries Maritimes passenger list. 1934 A. Christie *Murder on Orient Express* ii. xiv. 169 He saw me on board a French Messagerie boat for Smyrna. 1974 S. Coulter *Chateau* ii. vi. 286 She had come down by the Messageries coach.

Messalina (mesălī·nă). The name of Valeria *Messalina*, third wife of the Roman emperor Claudius, used allusively for a licentious and scheming woman.

1887 *Athenæum* 8 Oct. 467/1 His heroine is a New York Messalina who fastens herself upon a villain of the worst type. 1925 W. J. Locke *Great Pandolfo* xx. 257 I'm not either a Messalina, or one of the *grandes amoureuses*. 1931 G. J. Renier *The English* iv. 85 Messalinas are exceptions. A woman needs only one partner. 1946 Koestler *Thieves in Night* 146 Particularly Gaby, our red-haired Viennese Messalina who, having a year ago left Max for Mendl, has now left Mendl too. 1975 'J. Lymington' *Spider in Bath* ii. 39 'Unfaithful bitch!' he cried. 'Messalina! Medusa! Gorgon!'

messaline (me·sălīn). [Fr., = *Messalina* (see prec.).] A soft, lightweight, and lustrous twilled-silk fabric. Also *attrib.*

1909 *Public Ledger* (Philadelphia) 26 June 7/7 Sale of Pongee & Messaline Dresses (title). *Ibid.*, Beautifully made of extra fine natural pongee silk and messaline. 1920 *T. Eaton & Co. Catal.* Spring & Summer 1/2 This delightful Frock of Messaline Silk. 1921 *Daily Colonist* (Victoria, B.C.) 11 Oct. 5/3 (Advt.), Black Messaline Silk, a beautiful heavy quality for making skirts and dresses. *Ibid.* 23 Oct. 16/4, 36-Inch Black Messaline, in an exceptionally good dye. A well woven silk in which you will realize the best wear. 1968 J. Ironside *Fashion Alphabet* 241 *Messaline*, named after the wife of the Emperor Claudius (Messalina) this is a lustrous..satin-weave silk fabric.

Messapian (mĕsēi·piăn), *sb.* and *a.* Also **Messa·pic.** [f. L. *Messāpi-us* + -an.] **A.** *sb.* **a.** A native or inhabitant of the ancient district of Messapia (now Apulia and Calabria) in southern Italy. **b.** The language of the Messapians. **B.** *adj.* Of or pertaining to the Messapians or their language.

1773 [see *Iapygian *a.* and *sb.*]. 1876 *Encycl. Brit.* IV. 650/2 The inhabitants [of Calabria] were Sallentines and Calabrians or Messapians, both probably of Pre-Hellenic or Pelasgic race. 1880, 1882 [see *Iapygian *a.* and *sb.*]. 1932 W. L. Graff *Lang.* 375 Venetic of Northeastern Italy and the South Italian Messapic. 1932 *Times Lit. Suppl.* 21 July 530/4 The three groups of Apulian vases, Messapian, Peucetian and Daunian. 1948 D. Diringer *Alphabet* ii. viii. 471 The Messapic or Messapian was the earliest European offshoot of the Greek language. 1949 *Oxf. Class. Dict.* 560/2 The Apulian Peucetii..and Daunii ..also spoke Messapic. 1966 M. S. Beeler in Birnbaum & Puhvel *Anc. Indo-European Dial.* 52 The undoubtedly Illyrian Messapic of the southeastern end of the peninsula.

messeigneurs, pl. of Monseigneur.

messenger. Add: **1. e.** Used as the name of a newspaper, periodical, etc.

1834 (*title*) Southern literary messenger. 1886 *Encycl. Brit.* XXI. 109/2 Many excellent literary journals and magazines..among these..the time-honoured *Viestnik Yevropî* ('Messenger of Europe'). 1922 Joyce *Ulysses* 460 *Messenger of the Sacred Heart* and *Evening Telegraph* with Saint Patrick's Day Supplement. 1975 (*title*) Kent Messenger.

f. *Biol.* A molecule or substance that carries genetic information. Freq. *attrib.* (cf. *messenger RNA* in *7).

1961 Brenner, Jacob, & Meselson in *Nature* 13 May 576 The paradox..can be resolved by the hypothesis, put forward by Jacob and Monod [in *J. Mol. Biol.* (in the press)], that..ribosomes are non-specialized structures which receive genetic information from the gene in the form of an unstable intermediate or 'messenger'. 1962 *Listener* 8 Mar. 413/2 'Messenger' molecules of RNA. 1969 A. M. Campbell *Episomes* ix. 116 The genes of one operon are all transcribed onto the same messenger molecule. *Ibid.* 117 The rate of messenger synthesis. 1971 *Nature* 2 July 12/1 Where there are no operons messengers are evidently monocistronic.

4. Also, 'any line sent ahead by which a larger line is run to a dock, buoy or similar use' (G. Bradford *Gloss. Sea Terms* 1927).

5. b. A device which may be sent down a line in order to trip some mechanism attached to it.

1929 *Jrnl. du Conseil* IV. 193 When a messenger is sent down the line it first hits the arm holding the stop. By the blow the stop is forced away, the messenger below is set free. 1959 H. Barnes *Oceanogr. & Marine Biol.* iii. 113 When the required depth has been reached, the bottle is allowed to remain there for 3 to 5 minutes, so that the thermometer may reach equilibrium and a 'messenger' is then sent down the wire. The hook is released and the springs contract closing the bottle. 1963 H. F. P. Herdman in M. N. Hill *Sea* II. vi. 124 When the messenger hits this catch, the lids are immediately closed and the bottle falls away through 180°.

7. *messenger boy* (examples); **messenger cable,** a cable used to support a power cable or other conductor of electricity; a suspension cable or wire; **messenger RNA** *Biol.*, RNA which, after being synthesized in a cell nucleus in accordance with the genetic information carried by a gene ('transcription'), passes out of the nucleus and carries this information to a ribosome, where it determines which particular protein is synthesized there ('translation'); abbrev. *mRNA* (*M 5); **messenger wire** = *messenger cable.*

1876 J. S. Ingram *Centenn. Exposition* 713 The messenger boys were seen everywhere conspicuous in their neat uniforms. 1886 *Pall Mall Gaz.* 26 Aug. 11/1 A Wall-street banker..sent a note by a district messenger boy to the office of his broker. 1922 Joyce *Ulysses* 84 Messenger boys stealing to put on sixpence. 1959 N. Mailer *Advts. for Myself* (1961) 208, I was amateur agent for it, messenger boy, editorial consultant..and I made a hundred mistakes. 1916 *Stand. Rules Amer. Inst. Electr. Engin.* § 778 A messenger wire or cable is a wire or cable running along with and supporting other wires, cables or contact conductors. 1948 *Building, Estimating & Contracting* (Amer. Techn. Soc.) IX. 48 A stranded steel messenger cable is strung over the chord of the trusses. 1961 Brenner, Jacob, & Meselson in *Nature* 13 May 577/1 Model III implies that a special type of RNA molecule, or 'messenger RNA', exists which brings genetic information from genes to non-specialized ribosomes. 1961 Jacob & Monod in *Jrnl. Molecular Biol.* III. 350 A small fraction of RNA, first observed by Volkin & Astrachan (1957) in phage infected *E. coli* and recently found to exist also in normal yeasts..and coli.., does seem to meet all the qualifications listed above. This fraction (which we shall designate 'messenger RNA' or M-RNA) amounts to only about 3% of the total RNA. 1970 Ambrose & Easty *Cell Biol.* iii. 113 Three types of RNA are involved in protein synthesis

in the cytoplasm—messenger, transfer, and ribosomal RNA. 1973 *Sci. Amer.* Aug. 21/1 Some of the RNA (messenger RNA) determines the structure of the proteins (primarily enzymes) that constitute or manufacture all the tissues of the organism. 1898 E. J. Houston *Dict. Electr. Words* (ed. 4) 849/2 Messenger wire of aerial cable. 1916 Messenger wire [see *messenger cable* above]. 1948 *Man. Uniform Traffic Control Devices* (U.S. Pub. Roads Admin.) iii. 117 All overhead cable shall be supported by a suitable aerial messenger wire whenever there is a span of more than 30 feet.

Messenian (mesī·niăn), *sb.* and *a.* [f. L. *Messenius*, Gr. Μεσσήνιος Messenian + -an.] **A.** *sb.* **a.** A native or inhabitant of Messenia, a region in the south-west Peloponnese bordered on the east by Laconia. **b.** The dialect of this region. **B.** *adj.* Of or belonging to Messenia.

1579 North tr. *Plutarch's Lives* 674 The Lacedæmonians brake of from this general peace, and..made warre, in hope to recouer the Messenians contrie. 1600 Holland tr. *Livy's Romane Hist.* xxxvi. 936 But the Messenians.. sent the embassadors away. 1794 T. Taylor tr. *Pausanias' Descr. Greece* I. iv. iv. 346 A disagreement, for the first time, took place between the Messenians and Lacedæmonians. 1826 *Kaleidoscope* 14 Feb. 261/2 Ephemerus, the Messenian, advanced this paradoxical opinion. 1830 W. M. Leake *Trav. Morea* I. 366 Andhrússa..is advantageously situated, overlooking the rich Messenian plain. 1910 C. D. Buck *Introd. Study Gk. Dial.* 10 The Doric Group... 2. *Messenian.* There is scarcely any material until a late period, when the dialect is no longer pure. 1911 *Encycl. Brit.* XXV. 610/1 Under Alcamenes and Theopompus a war broke out between the Spartans and the Messenians, their neighbours on the west. 1958 R. Liddell *Morea* ii. 95 Currants from the rich Messenian vineyards lay everywhere. 1960 A. R. Burn *Lyric Age Greece* ix. 182 A Messenian hero, Aristomenes..gained fame in song. 1968 V. Ehrenberg *From Solon to Socrates* ii. 33 The kings Polydorus and Theopompus, the contemporaries of the First Messenian War. *Ibid.* vii. 275 The place was held by Messenians from Naupactus. 1969 A. Toynbee *Some Probl. Greek Hist.* iii. ii. 164 It is probable that the Laconians and Messenians..were of mixed origin. 1972 A. Bartoněk *Classification West Gk. Dial.* 91 In principle..Messenian produces the impression of representing an archaic form of Laconian. 1974 *Encycl. Brit. Micropædia* VI. 822/3 Many modern historians believe that there were only two early Messenian wars; the first (*c.* 735–*c.* 715) was the Spartan conquest of Messenia; but a Messenian revolt precipitated a second war, in which the Spartans were ultimately successful.

messer[2] (me·səɹ). *colloq.* [f. Mess *v.* + -er[1].] One who makes a mess; a muddler, bungler (see also quot. 1951).

1937 E. Pound *Let.* July (1971) 296 The respectable and the middle generation, illustrious punks and messers, fakes like Shaw, stew like Wells, nickle cash-register Bennett. 1949 F. Sargeson *I Saw in my Dream* xiii. 131 Though of course he's an old messer. 1951 Partridge *Dict. Slang* (ed. 4) 1108/2 *Messer*,..a 'near' prostitute; an amateur not above taking money or a present... A man, or a woman, that does not keep to one lover. 1966 'J. Hackston' *Father clears Out* 69 Where money and accounts were concerned she was the greatest messer living.

Messerschmitt (me·səɹʃmit). Also (erron.) **Messerschmidt.** [f. the name of Willy *Messerschmitt* (b. 1898), German aircraft designer.] Any of several types (esp. fighters) of German military aircraft used in the 1939–45 war. Also as *adj.*

1940 I. Halstead *Wings of Victory* i. ii. 52 The Messerschmitts attacked us from the rear. 1948 A. M. Taylor *Lang. World War II* (rev. ed.) 130 *Messerschmitt*, twin-engined German fighter monoplane, especially adapted to bomber escort work. 1955 E. M. Hull in G. Conklin *Sci.-Fiction Adventures in Dimension* 12 If a flight of Messerschmitts attack us in the next forty minutes, our machine-guns won't be much good. 1957 L. G. S. Payne *Air Dates* 276 Munich, where Messerschmitt 262 jet fighters were being built. 1969 *Listener* 1 May 595/3 A wartime Spitfire chasing about after Messerschmidts over Kent. 1971 L. Deighton *Declarations of War* 155 Above them the Messerschmitts and Spitfires were now only a mile apart.

messor. Restrict *Obs.*[−0] to sense in Dict. and add: **2.** *Ent.* [mod.L. (A. Forel 1890, in *Ann. Soc. Ent. Belg. Bull.* p. lxviii).] A member of the genus of harvesting ants so called. Also *attrib.*

1924 J. A. Thomson *Sci. Old & New* xii. 68 In the case of the Messor ants of the Sahara there are deep and spacious underground galleries, in which food is accumulated for the dry season.

mess-up. *colloq.* [f. *to mess up* (Mess *v.* 5).] = Mess *sb.* 3.

1902 C. J. C. Hyne *Mr. Horrocks, Purser* 111, I should say he feels this mess-up more than any of us. 1920 W. J. Locke *House of Baltazar* xxii. 274 It is ruin to your career and a mess up of your whole life. 1929 *Star* 21 Aug. 12/3, I am afraid there has been a bit of a mess-up.

messy, *a.* Add: **b.** *colloq.* Immoral; unethical.

1924 Lawrence & Skinner *Boy in Bush* xxi. 299, I can't bear to think of Monica messy with Easu. 1928 D. H. Lawrence *Woman who rode Away* 198 He was a perfectly decent boy, and there would never have been anything messy to fear from him. 1960 Wentworth & Flexner *Dict. Amer. Slang* 337/1 *Messy*,..immoral; unethical.

mestang, var. Mustang.

1834 A. Pike *Prose Sk. & Poems* 74 Lewis & Irwin obtained young and unbroken wild horses, (or, as the hunters call them, mestangs). 1837 W. Irving *Capt. Bonneville* II. xix. 28 She was mounted on a mestang or half wild horse.

mestizo. Add: (Further examples.) Also applied to other persons of mixed blood, or to a Central or South American Indian who has adopted European culture.

1878 C. Hallock *Amer. Club List & Sportman's Gloss.* p. vii/2 *Mestizo* (Sp.), a cross between an Indian and a negro. 1909 Webster *Mestizo*... In Spanish America and the Philippines, a person of mixed blood; esp., the offspring of a European or person of European stock and an (East) Indian, Negro, Malay, or other person of dark, non-European stock; often specif., *Phil. I.*, a person of Chinese and native blood. 1930 R. Macaulay *Staying with Relations* vi. § i. 75 They [sc. the Spaniards] go mestizo sooner or later, and are the better for it; a little Indian blood gingers them up. 1941 R. Humphreys *Latin Amer.* 6 In Brazil..half the population is white, but Indians predominate in the interior, mestizos in the north, and the negro element is strong in Bahia. 1959 [see *Cholo, cholo]. 1962 N. Maxwell *Witch-Doctor's Apprentice* v. 52 Cholo means mestizo, half-breed. Seems it's more polite to call a man a half-breed than an Indian. 1967 Webster, *Mestizo*... 2: a completely acculturated Central or So. American Indian. 1969 *Time* 14 July 14/2 *Mestizo*, person of mixed Spanish and Indian blood, as are most Mexican Americans. 1973 *Nat. Geographic* May 642/1 Juan himself is a mestizo, part Indian, part Spanish. 1974 *Encycl. Brit. Micropædia* VI. 824/3 In some countries—*i.e.*, Ecuador—the word has acquired social and cultural connotations; a pure-blooded Indian who has adopted European dress and customs is called a mestizo (or cholo).

b. *mestizo labourer, town.*

1970 L. Grebler et al. *Mexican-Amer. People* xiv. 322 A leisured 'Spanish' *hidalgia* and a mass of *mestizo* and Indian laborers. 1887 L. Oliphant *Episodes* (ed. 4) vi. 118 There was absolutely nothing to see in the sleepy little *mestizo* town.

‖ **mesto** (me·sto), *a. Mus.* [It., f. L. *maestus* sad.] As a direction in music: sad, mournful.

1811 in Busby *Dict. Mus.* (ed. 3). 1880 Grove *Dict. Mus.* II. 315/2 *Mesto*, 'sadly'; a term used three times by Beethoven, in the pianoforte sonatas, op. 10, no. 3, and op. 59, and in the slow movement of Quartet op. 18, no. 7. The slow movement of the first of these is called Largo e mesto, and of the second and third Adagio molto e mesto. It is also used by Chopin in the Mazurkas, op. 33, nos. 1 and 4. 1938 *Oxf. Compan. Mus.* 568/1 *Mesto* (It.), 'mournful', sad. 1959 *Collins Mus. Encycl.* 424/1 *Mesto*.., sad.

mestranol (me·strănǫl). *Pharm.* [f. Me(thyl + *œ)stra(diol + -n- + -ol.] The 3-methyl ether, $C_{21}H_{26}O_2$, of ethinylœstradiol which has actions similar to, but more potent than, those of œstradiol and is used in treating disorders of menstruation, fertility, and pregnancy, and (together with a progestational agent) as a contraceptive.

1962 *Approved Names* (Brit. Pharmacopœia Comm.) 15 Mestranol. 1963 *Federation Proc.* XXII. 481 A 2-year.. test was conducted..to determine the safety of Ortho-Novum, an oral contraceptive preparation containing norethindrone..and mestranol..in a ratio of 1,000 to 6. 1968 *Times* 21 Dec. 13/8 Most of the young quail fed on mestranol became sterile. 1971 *Brit. Med. Bull.* XXVII. 26/1 No difference could be detected between the risks associated with two different types of oestrogen (mestranol and ethynyloestradiol [*sic*]) when used in the same dose.

Mesvinian (mesvi·niăn), *a.* and *sb. Archæol.* [ad. F. *mesvinien*, f. *Mesvin* in Belgium: see -ian.] **A.** *adj.* Belonging to the middle palæolithic period or culture of Belgium. **B.** *sb.* The Mesvinian period.

1911 W. J. Sollas *Anc. Hunters* 109 The Mesvinian, now accepted as the oldest of the Palaeolithic horizons. *Ibid.* 111 The Mesvinian implements are ruder than the Chellean, and the characteristic boucher is lacking. 1922 *Proc. Prehist. Soc. E. Anglia* III. 602 The Mesvinian Industry..might well be the precursor of the Mousterian Industry. 1948 [see *eolithic *a.*].

met.[1], **met,** colloq. abbrev. of Metropolitan *a.* (as in *Metropolitan Railway, Metropolitan Opera House, Metropolitan Police*, etc.). In *pl.*: stocks or shares in the London Metropolitan Railway.

1886 H. Baumann *Londinismen* 107/2 *Mets*, Aktien der unterirdischen (Metropolitan) Eisenbahn. 1896 Farmer & Henley *Slang* IV. 305/2 *Met.* 1. A member of the Metropolitan (or New York) Base-Ball Club. 2. in *pl.* (Stock Exchange), Metropolitan Railway Shares. 3. *The Met* (London), the Metropolitan music-hall. 1908 *Daily Report* 7 Feb. 1/4 Both 'Mets.' and 'Districts' have moved in contrast with the Railway market generally. 1926 [see *metroland]. 1937 'C. McCabe' *Face on Cutting-Room Floor* iii. 18 Then I was suddenly in the crowd of clerks and typists rushing towards King's Cross Met station. 1944 'D. Hume' *Toast to Corpse* ix. 91 You haven't had thirty years in the Mets for nothing, and you've been about a bit. 1946 L. Baker *Out on Limb* viii. 85 She walked like a queen,..sang like head bird at the Met, and had a brilliant career. 1953 *Manch. Guardian Weekly* 1 Oct. 2/4 There is very little Wagner at the Met this year. 1968 'M. Underwood' *Man who killed too Soon* xiv. 120 Inspector Drew's colleagues in other forces, especially those in the Met. 1971 [see *firm *sb.*[1] 2 c]. 1972 P. Marks *Collector's Choice* i. 15 The crowds came to ogle the Van Goghs and the multi-million-dollar Rembrandts—the Met had thirty-seven. 1973 *Radio Times* 26 Feb. 7 It is no longer cheap to travel to and from London on the Met. 1973 'S. Harvester' *Corner of Playground* iii. iv. 202 An operatic mezzo-soprano, who had sung at the Met, La Scala, Covent Garden. 1974 S. Gulliver *Vulcan's Bulletins* 112 Something or other might be found to interest the Mets or the Home Office... Your trading days would be over. 1974 D. Ramsay *No Cause to Kill* ii. 147 The television set blared baseball. The New York Mets were playing the Los Angeles Dodgers.

met.[2], **met,** colloq. abbrev. of Meteorological *a.* or of *Meteorological Office(r)*. Freq. with capital initial.

1940 *War Illustr.* 26 Jan. 19/2 One of the 'Met. men' studying the big thermometer attached to one of the interplane struts of his 'Gladiator'. 1942 T. Rattigan *Flare Path* i. 113 What's the met. report like? 1943 L. Cheshire *Bomber Pilot* iii. 45 The latest 'met' forecast came in... Navigation should be easy. 1943 Hunt & Pringle *Service Slang* 45 *Met.* or *Mets.*, the Meteorological Officer. [R.A.F.] 1943 *R.A.F. Jrnl.* Aug. 32 The Met. Officer has sent his obs. through each hour. 1958 *New Scientist* 15 May 9/2 The met. observations become meaningful only when compared with the observations made by the other parties all over the continent. 1964 *Punch* 7 Oct. 527/1 Metmen ask each other. 1970 *New Scientist* 24 Dec. 569/1 It is almost inevitable..that the Met Office will come in for derision over their latest idea. 1972 K. Campbell *Thunder on Sunday* 9 He picked up the blue weather folder that the Met Officer at Keflavik had given him. 1973 W. M. Duncan *Big Timer* i. 14 The met men promise a hard winter. 1974 L. Lamb *Man in Mist* viii. 50 The Superintendent turned..to ask what the met. report had been.

me·ta[2], **Meta,** abbrevs. of Metaldehyde. *Spec.* a block of metaldehyde used (*a*) as fuel for cooking and heating, (*b*) for killing slugs. Also *attrib.*

Registered as a proprietary term by Lonza Elektrizitätswerke, Switzerland, in 1924 *Trade Marks Jrnl.* 26 Mar. 717 (as a fuel) and in 1938 *Ibid.* 18 May 603 (slug-killer).

1925 E. F. Norton *Fight for Everest 1924* v. 102 Our loads..comprise one 10-lb. tent, two sleeping bags, food and 'meta' (solid spirit). *Ibid.* 706 The meta cooker doing its indifferent best to produce half a pot of warm water. 1938 *Times* 26 Nov. 15/6 Tablets of Meta crushed and used by itself or mixed with bran, may be put about among plants before dark and the victims collected in the morning. 1947 F. Smythe *Again Switzerland* vi. 120 He carries with him a packet of 'meta' fuel. 1952 E. R. Janes *Flower Garden* 119 Metaldehyde (Meta) is a very fine slug killer... Under some conditions in the dead of winter it is hardly possible to apply Meta. 1955 P. Bauer *Kanchenjunga Challenge* iv. iv. 196 We went on travelling light with perhaps the Zdarsky sack and a small meta-tablet cooker.

meta (me·tă), *a. Chem.* Now usu. italicized. [f. Meta- *prefix.*] Characterized by or relating to (substitution at) two carbon atoms separated by one other in a benzene ring; at a position next but one *to* some (specified) substituent in a benzene ring. Also as *adv.*

1876 *Jrnl. Chem. Soc.* XXIX. 237 The author places the sulpho-acids obtained by the action of sulphuric acid on.. iodobenzene in the meta (1:3) series. 1924 E. J. Holmyard *Outlines Org. Chem.* xix. 367 The *para* isomer $C_6H_4X_2$ can yield only *one* trisubstitution product $C_6H_3X_3$, while the *ortho* can give *two* and the *meta* can give *three*. 1938 L. F. Fieser in H. Gilman *Org. Chem.* I. ii. 146 There is..no such enormous difference in reaction rate.. as there is between the ortho-para directing amines and the meta orienting ammonium salts. 1949 [see *orient *v.* 4 a]. 1968 R. O. C. Norman *Princ. Org. Synth.* xi. 357 Position *a*, which is *ortho* to acetamido and *meta* to methyl, is more reactive than position *b*, which is *meta* to acetamido and *ortho* to methyl.

meta-. Add: **1. meta-e·thics,** a name applied to the study of the foundations of ethics, esp. of the nature of ethical statements; hence **meta-e·thical** *a.*; **meta-hi·story,** inquiry into the principles governing historical events; so **meta-histo·ric(al** *adjs.*, transcending history, controlling the course of history; **meta-philo·sophy,** inquiry into the problems ulterior to philosophy or philosophical theories; so **metaphiloso·phical** *a.*; **metasocio·logy** (see quot. 1970); so **metasociolo·gical, meta-sociologi·stic** *adjs.*; **meta-theo·logy,** (*b*) the philosophical study of the nature of religious language or statements; hence **meta-theo·lo·gian, meta-theolo·gical** *a.*

1949 A. J. Ayer *Philos. Ess.* (1954) x. 246 All moral theories..are neutral as regards actual conduct. To speak technically, they belong to the field of meta-ethics, not ethics proper. *Ibid.*, Expounding my meta-ethical theory. 1957 D. M. Mackinnon *Study in Ethical Theory* i. 10 The moral philosopher is..preoccupied with..problems of meta-ethics, with the question of where ethical language belongs on the language-map. *Ibid.* 11 This problem of the relation of duty and good is in some sense meta-ethical. 1966 G. C. Kerner *Revolution in Ethical Theory* 1 The problems of ethical theory are thus..problems..of moral language;..they are 'meta-ethical'. *Ibid.* ii. 70 Meta-ethics is conceived to be a purely theoretical and ethically neutral enterprise. 1973 *Nature* 26 Jan. 249/2 With someone who prefers an incoherent picture of nature, I have no idea how to proceed—just as in meta-ethics, one is powerless to proceed with someone who regards a universe without sentience as possessing greater intrinsic value than one with sentience. 1929 R. HUGHES tr. *Deissmann's New Testament in Light of Mod. Res.* vi. 172 The holy is pre-historic and metahistoric. 1945 G. DIX *Shape of Liturgy* ix. 264 These meta-historical facts of the resurrection and ascension. 1949 *Mind* LVIII. 411 The value of morals as 'meta-historical reason' controlling history and determining the future. 1969 P. A. ROBINSON *Freudian Left* 148 The typical practicing psychoanalyst carefully distinguished the discrete precepts and techniques of his therapeutic science from the ambitious meta-historical adventures in which Freud had indulged. 1957 *Times Lit. Suppl.* 27 Dec. 782/2 Metahistory (which stands in much the same relation to history as metaphysics does to physics). 1964 C. S. LEWIS *Discarded Image* viii. 175 What Virgil puts forward in a mythical form is precisely meta-history. 1942 *Mind* LI. 284 'Why are no philosophical disputes ever settled?' It is with this 'metaphilosophical' problem..that Professor Ducasse's book..is concerned. 1964 *Philos. Rev.* LXXIII. 554 Blakeley..proposes an original and provocative metaphilosophical thesis. 1970 M. LAZEROWITZ in *Metaphilosophy* I. 91 Metaphilosophy is the investigation of the nature of philosophy, with the central aim of arriving at a satisfactory explanation of the absence of uncontested philosophical claims and arguments. 1959 R. BIERSTEDT in L. Gross *Symposium Sociol. Theory* 137 The distinction between methodological (or metasociological) theory on the one hand and substantive (or sociological) theory on the other. 1964 P. MEADOWS in I. L. Horowitz *New Sociol.* 448 Formulations which phrase a meta-sociologistic model, that is, the theme that beyond the teeming and changing varieties of social life and differentiated functions there are social patterns generating and guiding the social work life. 1958 W. STARK *Sociol. of Knowl.* I. iv. 197 A metasociology which would be..a study of man as he appears in all societies, of man *as such*. 1970 G. A. & A. G. THEODORSON *Mod. Dict. Sociol.* 254 *Metasociology*, the branch of sociological theory that is concerned with the methods and logic of sociological inquiry, rather than with propositions, principles, and generalizations about social life. 1967 *Philosophy* XLII. 197 The meta-theologian..claimed that Christian discourse, as it stands, is incoherent. 1969 R. S. HEIMBECK *Theol. & Meaning* i. 20 Since 1955, the quantity of metatheological literature has multiplied many times over. 1957 I. M. CROMBIE in B. Mitchell *Faith & Logic* ii. 77 It is from reading theology, not meta-theology, that one can come to understand how theological statements work. 1959 P. MUNZ *Probl. Relig. Knowl.* 12 The meta-theology which I have put forward neither stands nor falls with any one particular theological opinion which I have expressed or implied. 1967 *Philosophy* XLII. 195 One piece of meta-theology which has won wide acceptance..is that 'God' is not a substance-word.

b. Prefixed to various classificatory words to designate concern with the ulterior or underlying principles peculiar to that classification, as *metacriterion*, *metacriticism*, (hence *metacritical* adj., *-ally* adv.), *metasystem*, *metatheorem*, *metatheory* (so *metatheoretic*, *-ical* adjs.).

1953 C. E. BAZELL *Ling. Form* v. 63 Universality of application is only one meta-criterion for the choice of criteria. 1954 C. F. HOCKETT in *Word* X. 233 Neither any existing version of IA nor any existing version of IP meets all the metacriteria. 1963 *Listener* 3 Jan. 21/1 They [*sc.* the techniques of modern criticism] could make exciting sense (if not in strictly critical terms, then in metacritical ones) of works which would have seemed absurd if taken literally. 1970 A. RODWAY *Truths of Fiction* i. 9 Concentrate, metacritically, on what the text refers to... Study of form is purely critical, of content either critical or metacritical; of what the work leads to, whether in the way of causes or effects or general topics, purely metacritical. 1966 *Philosophy* XLI. 320 The aesthetician..is concerned (among other things) with metacriticism. 1970 A. RODWAY *Truths of Fiction* i. 6 The logical primacy of intrinsic criticism suggests that extrinsic criticism might also be called *metacriticism*. 1956 J. H. WOODGER tr. *Tarski's Logic, Semantics, Metamath.* 116 It is possible to construct a particular science, namely the 'metasystem', in which the given system is subjected to investigation. 1964 P. MEADOWS in I. L. Horowitz *New Sociol.* 452 Metasystem or general systems theory. 1969 *New Scientist* 4 Sept. 461/1 What Professor Beer is asking for is that we approach the problem at a higher level—the level of the 'metasystem'. 1940 W. V. QUINE *Math. Logic* ii. 89 We establish theorems wholesale, by arguments which show that the appropriate sequences *could* be found for each particular case. Such principles, describing general circumstances under which statements are theorems, will be called metatheorems. 1943 *Mind* LII. 267 Closely connected with the distinction between use and mention is that between a theorem and a metatheorem, the latter being, as the name suggests, a theorem *about* theorems, wherein symbols are mentioned and names of symbols used. 1971 G. HUNTER *Metalogic* p. xii, Complete proofs for metatheorems (theorems *about* a system) are..more laborious for natural deduction systems than for axiomatic ones. *Ibid.* I. 11 A theorem about a theorem (also called a metatheorem) is a true statement about the system expressed in the metalanguage. 1965 B. MATES *Elem. Logic* viii. 128 We are now in a position to..give informal proofs of a number of metatheoretic generalizations about the theorems of logic. 1953 *Mind* LXII. 557 The metatheoretical problems of logical calculi, such as independence of axioms, completeness, and decision methods. 1956 E. H. HUTTEN *Lang. Mod. Physics* iii. 81 When we want to explain how scientific theories are constructed ..we must speak *about* them; and this requires a suitable terminology. This meta-theory, or methodology, is as necessary to science as grammar is to ordinary language. 1963 *Language* XXXIX. 208 A metatheory for semantics must also exhibit the relations between semantics and other areas of linguistics. 1974 *Sci. Amer.* May 122/3 He outlines a metatheory in which the universe at every micromicroinstant branches into countless parallel worlds.

4. me·tacneme [Gr. κνήμη tibia; cf. CNEMIAL *a.*], a secondary mesentery which develops in some Zoantherians; so **metacne·mic** *a.*; **metanephri·dium** [a. G. *metanephridium* (B. Hatschek *Lehrbuch der Zoologie* (1889) II. 162): see NEPHRIDIUM], in certain invertebrates, a nephridium with a ciliated opening into the coelom; so **metanephri·dial** *a.*; **metaphase**, delete **metaphasis** and substitute for def.: [a. G. *metaphase* (E. Strasburger 1884, in *Arch. f. mikrosk. Anat.* XXIII. 260)], the stage in mitotic or meiotic nuclear division which follows prophase and precedes anaphase, during which the chromosomes become arranged with their centromeres on the equatorial plate; a dividing nucleus at this stage; (later examples); **metaphlo·em**, a constituent of primary phloem which is formed after the earliest development of the shoot; **metatra·cheal** *a. arch.*, usu. in phrase *metatracheal parenchyma*, describing the structure of wood in which concentric bands of parenchyma independent of the vessels are formed; **metaxy·lem**, a constituent of the primary xylem which is formed after the earliest development of the shoot.

1900 J. E. DUERDEN in *Johns Hopkins Univ. Circular* XIX. 47/2 The first six pairs of mesenteries are found to differ so essentially in their mode of origin and significance from the mesenteries appearing later that I find it convenient to have some word which will include them either as a whole or individually. I therefore propose for them the term 'Protocnemes', and shall refer to the mesenteries subsequently developed as 'Metacnemes'. 1902 *Ann. & Mag. Nat. Hist.* IX. 397 The different fundamental types of metacnemic sequence now known within the Actiniaria and Madreporaria. *Ibid.*, The metacnemes arise as unilateral pairs at one, three, seven, etc. regions within all the six primary exocoeles. 1940 L. H. HYMAN *Invertebrates* I. vii. 589 In most forms [of sea anemone] additional septa called metacnemes arise in pairs. *Ibid.* ii. 37 The nephridial system of the coelomate invertebrates is of the metanephridial type, i.e., the nephridial tubules begin as coelomic openings. 1963 R. P. DALES *Annelids* v. 98 The metanephridial funnels or postnephridial solenocytes lie in the coelomic fluid. 1930 W. R. COE in *Biol. Bull.* LVIII. 208 This type of excretory organ may be designated a metanephridium in order to distinguish it from the more usual type, protonephridium, found in nemerteans. 1967 E. J. W. BARRINGTON *Invertebr. Struct. & Function* xii. 236 The nephridium occurs in two main forms, the protonephridium and the metanephridium. 1924 E. W. MACBRIDE *Study of Heredity* ii. 42 The formation of the equatorial plate and of the spindle is known as the metaphase. 1961 *Lancet* 26 Aug. 489/1 Metaphases in freshly aspirated sternal and iliac crest marrow were analysed. 1962 *Ibid.* 26 May 1098/2 Rich crops of cells in metaphase were obtained twice. 1969 *Times* 20 June 7/3 Shortly before ovulation the oocyte goes through the process of cell division and then starts to divide a second time, a stage known as metaphase 2. 1973 *Nature* 1 June 290/2 Well-spread metaphases were photographed with a 95× fluoride objective. 1902 *Encycl. Brit.* XXV. 413/1 In many cases external protophloem..can be distinguished from metaphloem. 1965 K. ESAU *Plant Anat.* (ed. 2) xii. 292 The sieve elements of the metaphloem are commonly longer and wider than those of the protophloem. 1908 BOODLE & FRITSCH tr. *Solereder's Systematic Anat. Dicotyledons* II. 1143 The wood parenchyma generally forms tangential bands (known as the 'metatracheal' parenchyma in contrast to the 'paratracheal' parenchyma, aggregated round about the vessels). 1933 *Tropical Woods* XXXVI. 9 *Metatracheal parenchyma*, aggregated wood parenchyma forming concentric laminae, mostly independent of the vessels and vascular tracheids. 1970 WILSON & WHITE *Jane's Struct. Wood* (ed. 2) vi. 116 Apotracheal parenchyma may occur as..tangentially arranged sheets of cells..or in more extensive tangential bands... The two latter types are sometimes referred to as metatracheal parenchyma but this term is better avoided. 1902 *Encycl. Brit.* XXV. 415/1 Sometimes..the centre of a bulky root stele has strands of metaxylem..scattered through it. 1965 K. ESAU *Plant Anat.* (ed. 2) xi. 243 The metaxylem, which appears after the protoxylem, is in the process of differentiation while the shoot is elongating.

5. b. Prefixed to the names of rocks or of classes of rock to indicate that they have undergone metamorphism, as *metadiorite*, *metadolerite*, *metagranite*, *metasediment* (hence *metasedimentary* adj.), *metasyenite*, *metavolcanic*; also *metaigneous* adj.

1876 J. D. DANA in *Amer. Jrnl. Sci. & Arts* XI. 121 The rocks are..*Metamorphic doleryte*, *metamorphic diabase*, and *metamorphic melaphyre*... To distinguish these metamorphic rocks from the igneous of the same composition, they are named, on my suggestion, *metadoleryte*, *metadiabase*, and *metamelaphyre*. The examples are part of a long series of rock species which have representatives both among igneous (or intrusive) and metamorphic rocks. Other kinds are *dioryte* and *metadioryte*, *syenyte* and *metasyenyte*, *felsyte* and *metafelsyte*, etc. 1920 A. HOLMES *Nomencl. Petrol.* 154 *Meta-*, a prefix used before the names of igneous rocks to signify that the mineral and chemical composition of the latter have been modified by alteration. 1942 M. P. BILLINGS *Struct. Geol.* xii. 215 Metasediments, metavolcanics, and meta-igneous rocks are metamorphic rocks derived, respectively, from sedimentary, volcanic, and igneous rocks. 1961 J. CHALLINOR *Dict. Geol.* 126/1 *Meta-(rock)*. A metamorphosed rock which was originally of the kind or type included in the name. Thus 'metasediment' or 'metasedimentary rock', 'meta-igneous rock', 'metadolerite', &c. 1973 *Nature* 21 Sept. 120/1 The metasediments occur in a (refolded) syncline among granitic gneisses. *Ibid.* 139/2 The lithology of some of these enclaves strongly suggests that they were originally supracrustal rocks similar to those that occur at Isua, including metasedimentary ironstones.

6. a. metapro·tein, an intermediate product in the hydrolysis of a protein which is soluble in acids and alkalis but insoluble in water.

1909 *Cent. Dict.* Suppl., Metaprotein. 1911 *Encycl. Brit.* XIX. 922/1 The first result of the action of this secretion on protein matter is to render it soluble—a metaprotein or acid albumin (syntonin) being formed. 1949 G. B. BACHMAN *Org. Chem.* xviii. 220 Primary derivatives: proteins, metaproteins, and coagulated proteins.

7. a. Add to def.: *spec.* denoting (partial) dehydration.

b. Delete entry in Dict. and see sense *5 b.

metabio·logy. Also **meta-biology.** [f. META- 1 + BIOLOGY.] A hypothetical or postulated science dealing with phenomena of living organisms beyond the scope of conventional biology, or treating them in a more fundamental way. Chiefly in non-scientific use, freq. with allusion to Shaw.

1921 G. B. SHAW *Back to Methuselah* Pref. p. lxxxv, As the conception of Creative Evolution developed I saw that we were at last within reach of a faith which complied with the first condition of all the religions that have ever taken hold of humanity: namely, that it must be, first and fundamentally, a science of metabiology. 1936 *Scrutiny* Mar. 377 And Keats's genius..is not really illuminated by the procedure of *Keats und Shakespeare* or, except as another of Metabiology's cloudy trophies, exalted. 1945 K. R. POPPER *Open Society* I. v. 72 Plato's idealist historicism ultimately rests..upon a kind of meta-biology of the race of men. 1962 A. HUXLEY *Let.* 1 Mar. (1969) 929 He would radiate a kind of religious enthusiasm—about Dostoevsky and his ideas, about 'metabiology', about Lawrence as 'The Son of Man', the 20th-century Messiah. 1968 *New Scientist* 21 Nov. 415/2 They will be searching for a new integral approach to biology, in which organisms will be described *as a whole*, rather than simply in the terms of the molecules from which the organisms are constructed... Monod offered to these members of a future biological avant-garde the term 'meta-biology'.

So **me:tabiolo·gical** *a.*

1921 G. B. SHAW (*title*) Back to Methuselah: a metabiological pentateuch. 1935 *Theology* XXX. 89 The metabiological reality which Mr. Murry would substitute for Deity. 1960 C. S. LEWIS *Four Loves* v. 125 A theory [of the love of lovers] more likely to be accepted in our own day is what we may call Shavian—Shaw himself might have said 'metabiological'—Romanticism. According to Shavian Romanticism the voice of Eros is the voice of the *élan vital* or Life Force. 1967 *Listener* 3 Aug. 141/2 Belonging to this period was the spirited exchange of letters between Bernard Shaw, Julian Huxley and others which lasted from the beginning of November 1942 to well into March 1943—referred to in the office as the metabiological marathon.

metabiosis (metăbəiō̆u·sis). *Biol.* [f. META- + Gr. βίωσις mode of life, but formed as back-formation from the adj.] A type of symbiosis in which one of the organisms modifies the environment before the second is able to live in it. So **metabio·tic** *a.* [ad. G. *metabiotisch* (C. Garré 1887, in *Correspondenz-Blatt für Schweizer Aerzte* 1 July 390).]

1899 *Knowledge* July 151/2 It [*sc.* the yeast organism] is dependent upon its predecessor for its particular action—that is to say, we have here a condition of metabiosis. *Ibid.* 152/1 This implies nothing more or less than metabiotic relationships between the different kinds of the bacteria concerned. 1966 F. H. MEYER in S. M. Henry *Symbiosis* I. iv. 172 The nitrite bacteria are dependent on ammonia-producing organisms, while the nitrate bacteria are again dependent on the activity of the nitrite bacteria. For such direct living 'one after another', Garré gave [*sic*] the name metabiosis.

metabolic, *a.* Add: **5.** *Biol.* Of unicellular organisms, exhibiting metaboly.

1906 M. HARTOG in *Cambr. Nat. Hist.* I. v. 125 Such movements, permissible by the perfectly flexible but firm pellicle, are termed 'metabolic' or 'euglenoid'. 1926 G. N. CALKINS *Biol. Protozoa* vi. 254 In all cases of amoeboid and metabolic forms the cell symmetry is variable. 1955 *New Biol.* XIX. 116 Change in shape in unicellular organisms is referred to by the somewhat confusing term 'metaboly' and organisms which exhibit it are termed 'metabolic'.

Hence **metabo·lically** *adv.*, in, or as regards, metabolism.

1913 *Jrnl. Amer. Med. Assoc.* 18 Oct. 1465/1 The total metabolically active tissues of the body. 1928 *Biochem. Jrnl.* XXII. 1049 Patients, suffering from various complaints, though presumed to be metabolically sound. 1964 *Oceanogr. & Marine Biol.* II. 178 This classification is important metabolically. 1967 [see *GUANASE].

metabolism. 1. (Further examples.) Add to def.: The rate at which the body functions over all; the sum of the chemical changes undergone in the body by any particular substance.

1951 A. GROLLMAN *Pharmacol. & Therapeutics* iii. 77 Metabolism of alcohol..proceeded at about the same rate irrespective of whether the subjective was at rest or engaged in muscular activity. 1962 A. PIRIE (*title*) Lens metabolism in relation to cataract. 1966 WRIGHT & SYMMERS *Systemic Path.* II. xxxi. 1098/1 Thyroid hormone stimulates metabolism, increases oxygen consumption, and causes a rise in heat production. 1968 PASSMORE & ROBSON *Compan. Med. Studies* I. xxxi. 1/1 The liver holds a key place in the metabolism of the body. 1969

J. H. GREEN *Basic Clin. Physiol.* xvii. 96/1 If the thyroid gland is underactive (hypothyroidism),..there will be a reduction in the body's metabolism. 1971 B. N. LA DU et al. (*title*) Fundamentals of drug metabolism and drug disposition. 1972 *Internat. Jrnl. Biochem.* III. 294 (*heading*) An anti-anabolic role of adenosine 3′, 5′-cyclic monophosphate in the control of liver metabolism.

metabolite. Substitute for def.: **a.** Any substance formed from another by metabolism. **b.** A substance necessary to metabolism or to a particular metabolic process. (Further examples.)

1923 *Jrnl. Physiol.* LVII. 248 The increased combustion of sugar is immediately followed by the physiological oxidation of the accumulated fatty acid metabolites. 1946 P. H. MITCHELL *Textbk. Biochem.* xii. 322 The modern view of typical cases of biological oxidation is that enzymes catalyze the transfer of H_2 from a fuel food (the metabolite) to another substance. 1951 M. ABERCROMBIE et al. *Dict. Biol.* 159 Most metabolites are made by the organism in the course of metabolism; others must be taken in from the environment... Autotrophic organisms need to take in only inorganic metabolites, e.g. water, carbon dioxide, nitrates... Heterotrophic organisms need ..a wide range of organic metabolites from the environment. 1965 LEE & KNOWLES *Animal Hormones* iv. 76 Progesterone appears to be an intermediary metabolite in the formation of adrenal cortical hormones. 1967 [see *FUSIDIC *a.*]. 1967 M. E. HALE *Biol. Lichens* iv. 58 Nitrogen is an essential metabolite for synthesizing proteins in both the alga and fungus. 1970 G. R. TAYLOR *Doomsday Bk.* vi. 130 In the body..DDT tends to convert into similar substances known as DDE and DDD, the three being known..as 'DDT and its metabolites'.

metabolizable (metæ-bǒləizăb'l), *a.* [f. METABOLIZ(E *v.* + -ABLE.] Capable of being metabolized: applied to (*a*) substances that can be utilized by the body, and (*b*) energy that can be made available or produced by metabolic processes.

1905 *Bull. Bureau Animal Industry, U.S. Dept. Agric.* No. 74. 7 The metabolizable, available and utilizable energy of the hay. 1957 G. E. HUTCHINSON *Treat. Limnol.* I. ix. 617 Metabolizable material derived from the mud. 1961 *Ann. Rev. Physiol.* XXIII. 17 The metabolizable energy supplied by a feed. 1968 *New Scientist* 28 Nov. 496/1 They have investigated the possibility of obtaining metabolizable sugars from sawdust on a commercial basis.

Hence **meta:bolizabi·lity**, capability of being metabolized.

1929 MITCHELL & HAMILTON *Biochem. Amino Acids* vii. 360 The experiments..demonstrate the ready metabolizability of histidine, as compared..with..imidazole lactic acid. 1949 *Jrnl. Agric. Res.* LXXVIII. 487 The digestibility and metabolizability of two rations by sheep are reported.

metabolize, *v.* Add: **2.** *intr.* To perform metabolism.

1934 in WEBSTER. 1943 *Bacteriol. Rev.* VII. 139 The animal tissue cell is formed and metabolizes in an environment which is stabilized within narrow limits compared with the wide range of..conditions to which bacteria are subjected. 1971 *Sci. Amer.* Dec. 36/3 The anoxybiotic species does not stop metabolizing as soon as it encounters anaerobic conditions; it continues to metabolize for as long as six days, even though the new metabolic end product is unusual and potentially harmful.

†metabolon (mětæ-bǒlǫn). *Physics. Obs.* [ad. Gr. μεταβόλον, neut. of μεταβόλος changeable, f. μεταβάλλειν to change.] A radioactive atom produced in the process of radioactive disintegration.

1903 RUTHERFORD & SODDY in *Phil. Mag.* V. 586 At each stage [of disintegration] one or more α 'rays' are projected, until the last stages are reached, when the β 'ray' or electron is expelled. It seems advisable to possess a special name for these now numerous atom-fragments, or new atoms, which result from the original atom after the ray has been expelled, and which remain in existence only a limited time... We would..suggest the term *metabolon* for this purpose. 1904 E. RUTHERFORD *Radio-Activity* x. 324 The various metabolons from the radio-elements are distinguished from ordinary matter by their great instability and consequently rapid rate of change.

metaboly (metæ-bǒli). Substitute for entry: *Biol.* [ad. G. *metabolie* (M. Perty *Zur Kenntnis kleinster Lebensformen* (1852) 127), f. Gr. μεταβολή change.] The changes of shape characteristic of certain unicellular organisms; euglenoid movement.

1926 G. N. CALKINS *Biol. Protozoa* vi. 254 Metaboly may still be observed. 1969 F. E. ROUND *Introd. Lower Plants* ii. 13 A characteristic of *Euglena* is the spirally striate pellicle which is pliable and allows the cell to assume a variety of shapes—metaboly—an unfortunate term.

metacentric, *a.* Add: **2.** *Cytology.* [*-CENTRIC 2.] Of a chromosome: having the centromere in or near the centre.

1939 C. D. DARLINGTON in *Jrnl. Genetics* XXXVII. 357 Two sister chromatids would become the concurrent arms of a new metacentric chromosome. 1945, 1946 [see *ACROCENTRIC *a.*]. 1962 *Lancet* 2 June 1158/1 The 46 chromosomes..in the cultured cells of the blood of their patient included an unpaired metacentric chromosome. 1970 *Nature* 5 Dec. 938/2 Somatic mouse cells have forty acrocentric and telocentric and no metacentric chromosomes.

Hence as *sb.*, a metacentric chromosome.

1945 M. J. D. WHITE *Animal Cytol. & Evolution* iv. 56 Whether such a metacentric could pass through an indefinite number of mitoses without being frequently disrupted is open to doubt. 1961 *Lancet* 26 Aug. 463/2 The usual human Y..is normally about the same length as the smallest metacentric..or very slightly longer. 1971 *Nature* 15 Oct. 481/2 The karyotype presented as typical by Kao and Puck was interpreted by them as having three hamster telocentrics replaced by two human metacentrics.

metachromasia (metăkrō̆umē̆i·ziă). *Biol.* Also in anglicized form **metachromasy** (-krō̆u·măsi). [mod.L., f. META- + Gr. χρῶμα, χρωματ- colour: see -IA[1].] The property exhibited by certain biological materials and structures of staining a different colour from that of the stain used; also, the corresponding property of certain stains of changing colour in the presence of certain biological materials and structures.

1903 *Lancet* 18 July 177/1 The cells which contain granules contain also a store of ferment (zymogen), whilst the cells that are destitute of granules exhibit the reaction of mucin (metachromasy). 1956 *Nature* 3 Mar. 428/1 Anaphylactic shock brought out..degranulation and decrease of the metachromasia of the remaining granules in most of the cells. 1960 L. PICKEN *Organization of Cells* x. 481 The amoebae..are found to become metachromatic.. in increasing numbers as aggregation approaches... The appearance of metachromasy seems, however, to precede sensitivity to acrasin. 1964 W. G. SMITH *Allergy & Tissue Metabolism* iii. 33 Both tissue and blood mast cells are characterised by a coarse granular cytoplasm possessing a strong affinity for basic dyes, some of which change colour (exhibit metachromasia) as staining occurs. 1967 *New Scientist* 2 Feb. 275/2 These fragments are then encapsulated by a high polymer (producing the green metachromasia) and enter some kind of 'spore' stage. 1971 E. GURR *Synthetic Dyes* I. ii. 63 The significance of the dimer spectra in terms of theories of metachromasy is discussed.

metachromatic, *a.* Add: **2.** *Biol.* Exhibiting or involving metachromasia.

1897 MUIR & RITCHIE *Man. Bacteriol.* i. 11 It is..very probable that the occurrence of metachromatic granules in a bacterium indicates the onset of degenerative changes. 1902 *Jrnl. R. Microsc. Soc.* 89 (*heading*) Metachromatic granules in sporiferous bacteria. 1925 C. H. BROWNING *Bacteriol.* ii. 29 Sometimes with methylene-blue these beads stain of a different tint from the rest of the bacillus (metachromatic staining). 1957 *New Biol.* XXIV. 52 Structures which change the colour of the stain in this way are said to be metachromatic. 1964 W. G. SMITH *Allergy & Tissue Metabolism* iii. 36 Metachromatic staining of the faded granules began.

Hence **metachroma·tically** *adv.*

1908 *Anatomical Rec.* II. 106 Sections from such material stained in toluidene blue show the mucous secretion metachromatically stained, but no metachromatism is visible in the demilune cells. 1957 *New Biol.* XXIV. 54 Polysaccharides of high molecular weight other than heparin stain metachromatically. 1971 *Nature* 24 Sept. 264/2 The lipid mass was stained metachromatically pale orange-brown with cresyl violet..and thionin.

metachromatism. Add: **b.** *Biol.* = *METACHROMASIA.

1893 *Jrnl. R. Microsc. Soc.* 563 (*heading*) Metachromatism of parasitic sporozoa and carcinoma cells. 1904 [in Dict.]. 1908 [see *METACHROMATICALLY *adv.*]. 1917 C. E. MARSHALL *Microbiol.* (ed. 2) ii. 43 These bodies..are stained violet-red by most of the basic dyes, aniline blue or violet... By reason of this property of metachromatism, they have been called metachromatic granules.

metachrome. Add: **B.** *adj. Dye Chem.* Designating mordant dyes and their mordants that may be applied simultaneously in the same bath, and the method of dyeing by this process.

1901 *Jrnl. Soc. Dyers & Colourists* XVII. 66/1 Metachrome Brown B Paste: This colour is the first of a new series of (metachrome) dyes which have the property of dyeing in the single bath with metallic salts. *Ibid.*, With less than 3 per cent. dyestuff, 3 per cent. metachrome mordant must be used, but with larger amounts of dyestuff an equal amount. 1927 HORSFALL & LAWRIE *Dyeing Textile Fibres* ix. 263 The meta-chrome process as originally introduced..was confined to a comparatively small number of dyestuffs derived from picramic acid. 1963 A. J. HALL *Textile Sci.* iv. 184 The third (metachrome) method enables both dye and bichromate to be applied at the same time and it depends on the fact that no appreciable combination occurs between the bichromate and the wool or the dye under alkaline conditions, but this holds only for a limited number of dyes. 1971 R. L. M. ALLEN *Colour Chem.* iv. 45 In 1900 the Berlin Aniline Company devised the metachrome method whereby selected dyes can be applied simultaneously with a chroming agent.

metachronism. Add: **2.** *Biol.* The coordination of the movement of parts, esp. cilia, into a progressive wave.

1905 *Jrnl. Exper. Zool.* II. 408 It is the principal object of this paper to discuss the cause of metachronism in ciliary action. 1928 J. GRAY *Ciliary Movement* vii. 117 Any particular cilium is slightly in advance of the cilium behind it in the series and slightly behind the one just in front of it. This regular sequence is known as metachronism. 1972 M. S. GARDINER *Biol. Invertebr.* IV. xi. 189/1 In Metazoa, the cilia on adjacent cells may exhibit a similar coordination in metachronism.

So **metachro·nal** *a.*, exhibiting or characteristic of metachronism (sense *2); **metachro·nally** *adv.*

1905 *Jrnl. Exper. Zool.* II. 407 The cause of metachronal action is..to be sought..in the mechanical effect of one cilium on another. *Ibid.*, These swimming plates are arranged in rows and the members of each row, like ordinary cilia, beat metachronally, not synchronally. 1928 J. GRAY *Ciliary Movement* vii. 118 Although the direction of the metachronal wave..differs in different tissues, it is remarkably constant in each particular case. 1940 G. S. CARTER *Gen. Zool. Invertebr.* viii. 359 Metachronal rhythm is not confined to metazoan cilia; it can be well seen in such ciliate protistans as *Opalina*. 1962 D. NICHOLS *Echinoderms* viii. 99 In at least one urchin, *Diadema*, the spines show metachronal rhythm during locomotion, and they move the animal across the ocean floor with considerable speed. 1971 *Nature* 12 Feb. 491/1 The cilia of these last two cell types beat metachronally.

metacommunica·tion. [META- 1.] Communication that takes place with, or underlies, a more obvious form of communication; principles or theories about communication derived from the study of communication. Hence **metacommunica·tional, metacommu·nicative** *adjs.*

1951 RUESCH & BATESON *Communication* vi. 152 He is, also,..making implicit metacommunicative statements about his own position and stock of information. *Ibid.* vii. 203 (*heading*) Communication between two persons and metacommunication. 1963 T. A. SEBEOK in J. A. Fishman *Readings Sociol. of Lang.* (1968) 28 The metacommunicative messages used by rhesus monkeys, enabling them to distinguish between play and nonplay, have received particularly careful attention. 1967 J. A. MEERLOO in L. Thayer *Communication* 54 We cannot, of course, recover man's contemplations about himself from fossil remains, and data from living nonliterate men are lamentably deficient in metacommunicational material. 1967 P. WATZLAWICK et al. *Pragmatics Human Communication* i. 40 When we no longer use communication to communicate but to communicate *about* communication, as we inevitably must in communication research, then we use conceptualizations that are not part of but *about* communication. In analogy to metamathematics this is called metacommunication. 1974 *Publishers Weekly* 29 Apr. 47/3 The author is one of the students of 'metacommunications' or body language—Gregory Bateson and Raymond Birdwhistell are the two best-known names in this field, though their work derives from anthropologists such as Lorenz.

metacone (me·tăkō̆un). *Zool.* [f. META- + CONE *sb.*[1]] An external cusp on the outer back corner of a mammalian upper molar tooth. Hence **metaco·nal** *a.*

1888 H. F. OSBORN in *Amer. Naturalist* XXII. 1072 Proposed terms... Metacone. 1896 *Proc. Zool. Soc.* 570 The dental germ presenting the appearance of a high cone with a large posterior heel (metaconal region) and a slight internal extension. *Ibid.*, The posterior extension representing the metacone. 1933 A. S. ROMER *Vertebr. Paleont.* xii. 248 Inside the metacone [there is often] a smaller one [*sc.* cusp], the metaconule. 1971 P. HERSHKOVITZ in A. A. Dahlberg *Dental Morphol. & Evolution* viii. 103 In the first molar of *Potamogale*, the first indication of the metacone is a wedgelike indentation. *Ibid.* 129 In cercopithecoids, crista V is transverse and meets the buccal and metaconal portion of the plagiocrista.

metaconid (metăkō̆u·nid). *Zool.* [f. *METACON(E + *-ID[5].] A cusp on a mammalian lower molar tooth corresponding to the metacone on an upper molar.

1888 H. F. OSBORN in *Amer. Naturalist* XXII. 1073 Only the paracone and metaconids and hypoconids [of *Amphitherium*] have been observed heretofore. 1904 *Ann. & Mag. Nat. Hist.* XIII. 409 The antero-internal cusp not or scarcely divided into its constituent paraconid and metaconid. 1919 [see *HYPOCONID]. 1933 A. S. ROMER *Vertebr. Paleont.* xii. 248 The lower teeth tend also to square up..so that the four definitive cusps are the protoconid, metaconid, entoconid, and hypoconid. 1968 R. ZANGERL tr. *Peyer's Compar. Odontol.* 186 In the lower jaw..paraconid and metaconid are lingual. 1975 *Nature* 31 July 402/2 Paraconid and metaconid were clearly separated at their bases on the medial face of the tooth.

metacontrast (metăkǫ·ntrɑst). *Psychol.* [ad. G. *metakontrast* (R. Stigler 1910, in *Arch. f. ges. Physiol.* CXXXIV. 386), f. Gr. μετα- META- + *kontrast* CONTRAST *sb.*] A change, esp. a diminution, in the after-effect of a visual stimulus as a result of a second stimulus following shortly afterwards.

1950 *Jrnl. Optical Soc. Amer.* XL. 796/1 An investigation of the effect of variation of the luminance, exposure asynchrony, and spatial separation of the stimuli on the magnitude of metacontrast. 1971 *Jrnl. Gen. Psychol.* LXXXIV. 86 Masking and metacontrast studies do, however, demonstrate that a second stimulus may impair the perception of the first stimulus. 1972 *Science* 13 Oct. 179/3 They suggest that temporally backward and spatially lateral inhibition (metacontrast) occurs when the duration of the flash is long enough.

metaconule (metăkō̆u·niul). *Zool.* [f. *METACON(E + *-ule*.] An intermediate cusp between the hypocone and the metacone of a mammalian upper molar tooth.

1888 H. F. OSBORN in *Amer. Naturalist* XXII. 1074 The Bunodont series are universally characterized by the initial or advanced development of the proto- and metaconules in the upper molars. 1905 *Amer. Geologist* XXXV. 244 The intermediate cusps (protoconule and metaconule) are both well-defined. 1968 R. ZANGERL tr. *Peyer's Compar. Odontol.* 187 Intermediate cusps occurred.., a metaconule between protocone and metacone. 1971 W. D. TURNBULL in A. A. Dahlberg *Dental Morphol. & Evolution* ix. 163 (*caption*) Specimen consists of the protocone and metaconules.

metacryst (me·tăkrist). *Petrol.* [f. META- + *-cryst* after PHENOCRYST.] A large crystal formed in a metamorphic rock by recrystallization.

1913 W. LINDGREN *Mineral Deposits* xi. 158 An individual in another may be briefly called a metasome; if the metasome develops strongly with crystal outlines it may be called a metacryst. 1932 F. F. GROUT *Petrogr. & Petrol.* 365 Metacrysts commonly have abundant inclusions..and in some the inclusions are oriented so as to show that the metacrysts grew by replacement. 1963 D. W. & E. E. HUMPHRIES tr. *Termier's Erosion & Sedimentation* xvii. 339 Dolomite is present in limestones as rhombic 'metacrysts' which can cut across original structures (for example, oolites) and fossils.

metadyne (me·tădəin). *Electr.* [ad. F. *métadyne* (J. M. Pestarini 1930, in *Rev. gén. de l'Électr.* XXVII. 355/1), f. META- + Gr. δύναμις power.] A rotary direct-current generator in which the output voltage can be varied by a small signal applied to a control field perpendicular to the main field and which is used in position- or speed-control systems.

1930 *Sci. Abstr.* B. XXXIII. 376 The 'Métadyne' is a direct-current machine having more than two brush axes per pole-pair. 1945 [see *AMPLIDYNE]. 1951 F. J. TEAGO in P. Kemp *Electr. Engin.* III. 89/1 One of the chief reasons for using metadyne control is that the alternating current may be kept constant. 1970 J. SHEPHERD et al. *Higher Electr. Engin.* (ed. 2) xiv. 496 Metadyne generators are uncompensated or undercompensated cross-field generators.

metagalaxy (metăgæ·lăksi). *Astr.* [f. META- 1 + GALAXY *sb.*] The entire system of galaxies (see quot. 1930); also, a cluster or group of galaxies. So **metagala·ctic** *a.*

1930 H. SHAPLEY *Flights from Chaos* xiii. 141 Corresponding to individual stars, multiples, and star clusters we have galaxies, multiple galaxies, and clusters of galaxies... To designate the system including all of these I propose to use Lundmark's term, the Metagalactic system—or, more briefly, the Metagalaxy. 1951 *Astron. Jrnl.* Apr. 47/2 (*heading*) Differential rotation of the inner metagalaxy. 1957 H. SHAPLEY *Inner Metagalaxy* p. v, The terms 'Metagalaxy' and 'metagalactic' refer to the total recognized assemblage of galaxies. The Metagalaxy includes also whatever there may be in the way of gas, particles, planets, stars, and star clusters in the spaces between the galaxies. It is essentially the measurable material universe. 1965 *Rev. Mod. Physics* XXXVII. 654/1 In principle there may be several condensations of the initial plasma so that there may be other metagalactic systems in the universe. *Ibid.* 663/2 According to Klein there may be other metagalaxies in the universe. 1965 J. D. NORTH *Measure of Universe* ii. 20 His prediction of a high collision rate between the nebulae of a single metagalaxy. *Ibid.* App. 408 The galaxies assigned to the Local Group increased in number and the 'local metagalaxy' took its place with the other known clusters. 1970 *Nature* 12 Dec. 1069/1 According to this scheme, an initial contraction of the metagalaxy (containing equal amounts of matter and antimatter) resulting from its self-gravitation was turned into an expansion by the pressure of radiation produced by annihilation reactions. *Ibid.*, Although the gravitational and radiation fields in this case are metagalactic, the scale of the matter–antimatter separation is determined by the magnetic field and is likely to be much more local.

metagenesis. 1. For 'generation' read 'generations' in both cases.

metageometry. Add: Also **me:tageometri·cian** = METAGEOMETER.

1903 *Science* 16 Jan. 106/2 Our metageometricians tried to derive the basic geometrical principles from pure reason but failed.

metagnomy (metæ·gnǒmi). *Psychics.* [ad. F. *métagnomie* (Boirac, 1917), f. META- + Gr. γνώμη thought.] The acquisition of information by supernormal means; divination. So **me·tagnome**, one who has the power of metagnomy; a medium; **metagno·mic** *a.*, of or pertaining to metagnomy.

1919 W. DE KERLOR tr. *Boirac's Psychol. of Future* xi. 232 Clairvoyance, or 'metagnomy'. *Ibid.* 256 It is especially in the somnambulistic state, natural or provoked, that metagnomic manifestations occur. *Ibid.* 257 The mesmerist or hypnotist..evokes the metagnomic faculty. 1933 T. BESTERMAN tr. *Driesch's Psychical Res.* I. i. 2 The subject of the investigation, the sensitive, the medium, the metagnome, or whatever one likes to call him. 1960 *New Scientist* 28 July 306/2 Prosopopesis, metagnomy, telergy and teleplasty: if these are accepted, what remains of the 'laws' of physics, chemistry, biology and psychology? 1965 *Listener* 29 Apr. 639/3 With the existence of such phenomena as telepathy, metagnomy, precognition, etc., now well established,..there is surely a possibility that some astrologers may be psychically gifted individuals.

metagon (me·tăgǫn). *Biol.* [f. META- + Gr. γόν-ος offspring.] (See quot. 1968.)

1962 GIBSON & BEALE in *Genetical Res.* III. 25 As a provisional hypothesis..we proposed that the cytoplasm of mate-killer paramecia contained, in addition to the visible mu particles, certain other factors, here denoted 'metagons', which are assumed to be formed only in the presence of one or other of the genes M_1 and M_2. 1964 *New Scientist* 6 Aug. 322/2 Particles of RNA called 'metagons'. 1968 R. RIEGER *Gloss. Genetics & Cytogenetics* 285 *Metagon*, presumably, a primary, gene-initiated product in *Paramecium* which is RNA (complementary to the DNA of a specific gene) in nature, conditionally stable, infectious (capable of transmission from one *Paramecium* to another through the cytoplasm and the external medium), and capable of replication under certain conditions.

metake (me·take). [Japanese.] A tall slender Japanese bamboo, *Pseudosasa* (or *Arundinaria*) *japonica*.

1896 A. B. FREEMAN-MITFORD *Bamboo Garden* 69 *Arundinaria japonica* or Métaké... The word Métaké, or, more correctly, Médakè, means in Japanese 'female Bamboo', but there is no scientific reason for using the word 'female' in connection with this species. *Ibid.* 72 The Japanese gardeners consider Ya-daké and Mé-také to be two different plants. 1966 F. A. MCCLURE *Bamboos* 293 *Pseudosasa japonica* (*Arundinaria japonica*; *Sasa japonica*). Metake; Yadake; arrow bamboo. 1971 *Country Life* 18 Nov. 1371/1 The common Metake, *Arundinaria* or *Pseudosasa japonica*..[is] a particularly fine species [of bamboo] with slender canes quite 10 ft. high and glossy green leaves that are glaucous beneath.

metakinesis (metăkəinī·sis). Now *rare*. Pl. -kineses. [mod.L., f. META- + Gr. κίνησις motion.] **1.** *Cytology.* [coined (in Ger.) by W. Flemming in *Zellsubstanz, Kern und Zelltheilung* (1882) xx. 268.] **a.** (See quot. 1968.)

1888 [see *homœotyplical* adj. (*HOMŒO-)]. 1899 *Jrnl. R. Microsc. Soc.* Apr. 168 Karyokinesis in the Root-tips of Allium... Anaphase. (5) After the longitudinal segmentation of the chromosomes, which, as a general rule, does not begin until the chromosomes are in the equatorial plane, the daughter chromosomes are gradually pulled apart... This stage is known as metakinesis. 1903 *Bot. Gaz.* XXXV. 251 Heuser ('84) seems to have been the first to call attention to the double character of the daughter chromosomes in the diaster stage of *Tradescantia virginica*, but he interpreted the separation of the daughter segments during metakinesis of the first mitosis as a transverse division. 1968 R. RIEGER *Gloss. Genetics & Cytogenetics* 286 *Metakinesis*, the separation of the two chromatids of each chromosome and their movement to opposite spindle poles during anaphase of mitosis.

b. [given this sense by F. Wassermann 1926, in *Zeitschr. f. Anat. u. Entwicklungsges.* LXXX. 399.] (See quot. 1968.)

1948 W. ANDREW tr. *E. D. P. de Robertis's Gen. Cytol.* viii. 182 The prometaphase generally begins with the disintegration of the nuclear membrane... When the nuclear membrane has disintegrated, a more fluid zone is noted in the center of the cell in which the chromosomes..begin to be displaced in apparent disorder toward the equator. This mechanism of equatorial arrangement was called metakinesis (Wassermann). 1968 R. RIEGER *Gloss. Genetics & Cytogenetics* 286 *Metakinesis*,..chromosome congression to the spindle equator.

2. A manifestation of consciousness or mental phenomena.

1890 C. L. MORGAN *Animal Life* xii. 467 We call manifestations of energy 'kinetic' manifestations, and we use the term 'kinesis' for physical manifestations of this order. Similarly, we may call concomitant manifestations of the mental or conscious order 'metakinetic', and may use the term 'metakinesis' for all manifestations belonging to this phenomenal order. *Ibid.* 488 When, in man, the metakineses associated with these neural kineses assume the form of hypotheses, theories, interpretations of nature, moral ideals, and religious conceptions, these are..no longer subject to the law of natural selection. 1892 K. PEARSON *Gram. Sci.* ix. 401 This metakinesis does not appear to be more than a metaphysical name for nonconscious life, for there is no sense-impression that we have of such life that we can describe as metakinetic. 1903 L. F. WARD *Pure Sociol.* 156 Morgan's metakinetic energy is therefore the same as my conative energy or form of causation, and the difference between kinesis and metakinesis is the difference between motion produced by physical or ordinary efficient causes and motion produced by psychic or conative causes.

Hence **metakine·tic** *a.*

1890, etc. [see 2 above]. 1925 *Glasgow Herald* 11 July 4 Who can be sure that there is not a psychical or metakinetic side to the mountain and the precious stone, the waterfall and the great sea?

metal, *sb.* Add: **11.** (Earlier and later examples.)

1782 in *Sc. Nat. Dict.* (1965) VI. 259/3 The mettle for the road is not to be got but at the south end of the road. 1815 T. TELFORD *Life T. Telford* (1838) 483 The metal to be of the best blue or red whin. c 1906 P. C. COWAN *Making & Maintenance Roads* 17 The old macadam surface was first carefully levelled up and solidly rolled with any necessary amount of new metal. 1970 *N.Z. Listener* 21 Sept. 14/3 The lush pastures gave way to upland scrub country; the road metal became pumice and then clay merely.

13. b. *metal-detector*; *metal-clattering*, *-cutting*, *-using* adjs. **c.** *metal-bushed*, *-clad*, *-faced*, *-lustred*, *-rimmed*, *-studded* adjs. **d.** similative, as *metal-blue*, *-cold* adjs.

1930 BLUNDEN *Poems* 309 The metal-blue cucumber slices. 1883 *Man. Seamanship for Boys' Training Ships R. Navy* (Admiralty) (1886) 14 *Spindle*..passes through a metal bushed hole in the partners, up through the centre of the barrel. 1926 *Gloss. Terms Electr. Engin.* (*B.S.I.*) 87 *Metal-clad*, a qualifying term applied to apparatus to denote that the conducting parts are entirely enclosed in a metal casing. 1931 *Flight* 22 May 461/1 So gradually..the general theory of the metalclad airship was mathematically and experimentally proved. 1956 *Proc. Inst. Electr. Engin.* CIII. A. 82/1 When closing a metalclad switch I have observed..sparks jumping from one metal part to another. 1932 D. GASCOYNE *Roman Balcony* 12 A metal-clattering cavalcade Advanced Across the beach. 1918 D. H. LAWRENCE *New Poems* 47 Is it all nought? Cold, metal-cold? 1934 WEBSTER, Metal-cutting *adj.* 1961 *Times* 29 Dec. 12/4 Metalcutting machines. 1971 B. ST. J. WILKES *Nautical Archaeol.* vi. 120 Metal detectors are sold quite extensively in the US and Canada to amateur 'prospectors' to aid their hunt for gold. 1975 *Guardian* 1 Oct. 2/1 Chicago police will be using hand-held metal detectors on all crowds who come near President Ford. 1934 *Archit. Rev.* LXXV. 34/1 There are various metal-faced papers on the market. 1967 *Jane's Surface Skimmer Systems 1967–68* 60/2 The float table's conveying surface is of metal-faced plywood construction. 1862 G. M. HOPKINS *Vision of Mermaids* (1929), Others small braids enclustered Of glassy-clear Aeolis, metal-lustred With growths of myriad feelers. 1932 C. ISHERWOOD *Memorial* I. iii. 36 Eric's tall bony figure, with his metal-rimmed glasses and the odd pauses in his speech. 1909 *Q. Rev.* Jan. 148 Motor-cars,..having armoured or metal-studded tires damage the surface. 1928 C. DAWSON *Age of Gods* iii. 50 What we term the neolithic age in Europe was really the first stage in the diffusion of the higher metal-using culture of the Near East. 1964 T. L. KINSEY *Audio-Typing & Electr. Typewriters* vii. 77 A large proportion of the remainder are at work in the engineering and other metal-using industries.

14. metal age *Archæol.*, the period or stage of development in the human race in which copper and bronze were used for making weapons and tools; **metal arc welding**, arc welding in which the melting of a metal electrode provides the joining material; **metal fatigue**, fatigue (sense 1 b, in Dict. and Suppl.) of metal; **metal Mike** *Naut. colloq.* (see quot. 1961); **metal rectifier** *Electr.*, a rectifier in which rectification takes place at the junction of a metal and another solid substance (such as copper oxide or selenium); **metal thread, yarn** = *metallic thread, yarn* (*METALLIC *a.* 1 f); **metal-to-metal** *a.*, used of a contact or connection; **metal-works**, a factory where metal is produced.

1927 PEAKE & FLEURE *Hunters & Artists* vii. 112 The dawn of the Metal Age. 1951 *Proc. Prehist. Soc.* XVII. 1 The beginnings of a Metal Age in the Middle East are known to go back before the earliest written documents. 1963 H. N. SAVORY in Foster & Alcock *Culture & Environment* iii. 33 We can no longer think of a Secondary Neolithic element in south Wales as something introduced only a short while before the dawn of the Metal Age. 1926 *Jrnl. Iron & Steel Inst.* CXIV. 611 Metal arc welding can also be used for this purpose. 1952 FUCHS & BRADLEY *Welding Pract.* II. v. 107 One of the characteristic features of metal arc welding..is the highly localised intensity of heat input. 1968 J. GIACHINO et al. *Welding Technol.* iv. 70 Gas metal-arc welding was considered a high current density, small diameter filler wire process. 1954 *This Week's Listening* 30 Dec. 18 *No Highway* is about metal fatigue, which one of the characters describes as 'a disease of the metal'. 1958 *Economist* 11 Oct. 169/2 The Comet should, by rights, have been in service on the North Atlantic four years ago: the lapse of time is a measure of the work needed to stiffen its skin against metal fatigue that sent two of the original Comets in quick succession to the bottom of the Mediterranean. 1973 P. DICKINSON *Green Gene* ix. 180 It's like metal fatigue. You stand the stresses OK for years, so you think you'll stand them for ever. Then you snap, under no load at all. 1929 *Yachting* June 41/2 While in the act of setting 'Metal Mike' on the new course, we sighted a small ship's boat a point on the starboard bow. 1961 F. H. BURGESS *Dict. Sailing* 145 *Metal Mike*, the 'automatic helmsman'. 1927 *Wireless World* 30 Nov. 733/1 (*heading*) Battery charging rectifier incorporating the new dry 'metal' rectifier. 1971 B. SCHARF *Engin. & its Lang.* xx. 276 Three important types of rectifier are diode rectifiers, mercury arc rectifiers and metal rectifiers. 1959 *Times* 28 Apr. 20/6 A Chinese silk and metal-thread carpet. 1967 E. SHORT *Embroidery & Fabric Collage* ii. 49 Gold and silver threads couched by hand, or synthetic metal threads (Lurex) used on the machine. 1906 Metal-to-metal [see *leather-faced* adj. s.v. *LEATHER *sb.* 5 c]. 1910 *Daily Chron.* 2 Feb. 5/1 The surface where the wheel had been on the axle showing a bright metal-to-metal contact. 1922 *Encycl. Brit.* XXX. 36/2 The head of steel being secured to the liner with a plain metal-to-metal joint by bolts from the head to the crank-case. 1971 *Flying* Apr. 5/1 Aluminium honeycomb panel construction and metal-to-metal bonding. 1908 *Westm. Gaz.* 3 Oct. 10/1 On the hours of work in foundries and metal-works generally the Committee felt itself still imperfectly informed. 1913 J. M. MATTHEWS *Textile Fibres* (ed. 3) i. 12 Bayko metal yarn is a textile product recently introduced.

metalanguage (metălæ·ŋgwědʒ). [See META- 1, *1 b.] A language which supplies terms for the analysis of an 'object' language; a system of propositions about other propositions.

[1935 A. TARSKI in *Studia Philosophica* I. 282 Die Namen der Ausdrücke der ersten Sprache und der zwischen ihnen bestehenden Relationen gehören schon zu

der zweiten Sprache, der sog. Metasprache (welche übrigens die Grundsprache als Fragment enthalten kann).] 1936 *Mind* XLV. 486 The concepts *analytic* and *contradictory* in the language *L*, for instance, cannot be defined in *L*, as Carnap has shown. In order to escape from these restrictions one must build up a new language (a so-called *meta-language*) disposing of more means of expressing thoughts than the former. 1947 H. REICHENBACH *Elem. Symbolic Logic* i. 9 We say that signs of signs constitute a language of a higher level, which we call *metalanguage*. 1948 L. HJELMSLEV in *Studia Linguistica* I. 75 This would mean, in logistic terms, that linguistics is a metalanguage of the first degree, whereas phonetics and semantics are metalanguages of the second degree. 1954 A. J. AYER *Philos. Ess.* 12 Particulars, considered as occasions, would be referred to only at the level of the meta-language; when it was a question not of using the language, but of talking about its use. 1959 *Listener* 1 Oct. 520/1 We can then avoid the clumsy term 'meta-language'—the name for the language in which we talk about the terms of any formalized language. 1960 E. DELAVENAY *Introd. Machine Transl.* vii. 110 Between metalanguage and pure poetry, from the clear and distinct expression of a scientific representation to the synthetic expression of the vibrations of the poet's ego at the centre of his individual universe, there exists a whole vast range of untranslatables. 1962 *Times Lit. Suppl.* 13 Apr. 252/5 Machine translation research may be of value..in that, because one of its basic techniques is the establishment of a meta-language or intermediate language of ideas, it is in principle capable of constructing automatically an abstract. 1964 C. CHERRY in *Endeavour* Jan. 13/2 A meta-language is a formal language-system for describing language or a linguistic source. Strictly, a book of grammar is written in a meta-language—it defines a syntax but is itself not literature. 1973 *Computers & Humanities* VII. 223 Each segment is analyzed as to content and translated into a 'meta-language', which allows unambiguous comparisons among the variant readings.

Hence **me·ta-me·tala:nguage**, a language used in the description of another language which is itself a meta-language; the universal linguistic or symbolic system from which a particular metalanguage derives.

1954 I. M. COPI *Symbolic Logic* App. B. 341 The first of these is the meta-metalanguage's synonym for the name relation in the metalanguage. 1957 N. CHOMSKY *Syntactic Struct.* vi. 54 Linguistic theory will thus be formulated in a metalanguage to the language in which grammars are written—a metametalanguage to any language for which a grammar is constructed. 1963 H. B. CURRY *Found. Math. Logic* ii. 31 In that case we use a third language, L_3, customarily called the metametalanguage. 1967 *Encycl. Philos.* VII. 352/2 A proof of adequacy..requires an inductive argument of the meta-metalanguage.

metalation (metălē̆i·ʃən). *Chem.* [f. METAL *sb.* + -ATION.] The introduction into an organic compound of an atom of a metal in place of one of hydrogen (usu. one attached to an aromatic ring).

1934 GILMAN & YOUNG in *Jrnl. Amer. Chem. Soc.* LVI. 1415/1 The term metalation is proposed for reactions involving replacement of hydrogen by a metal to give a true organometallic compound. *Ibid.*, Metalations were effected by metals, organometallic compounds and salts. 1937 F. C. WHITMORE *Org. Chem.* 724 The aniline sulfate solution contains the strongly positive group $—NH_3$ and consequently gives *m*-substitution. The 'metalation' of benzine apparently offers an exception to this generalization. 1957 E. G. ROCHOW et al. *Chem. Organometallic Compounds* iii. 54 Metalation reactions occur only with the derivatives of the strongly electropositive alkali and alkaline earth metals and, rarely, magnesium. 1968 R. O. C. NORMAN *Princ. Org. Synthesis* xi. 392 Electrophilic metalations by metal salts should not be confused with the metalation of aromatic compounds with metal.

Hence (as a back-formation) **me·talate** *v. trans.*, to bring about metalation in; **me·talated, me·talating** *ppl. adjs.*

1939 *Jrnl. Amer. Chem. Soc.* LXI. 109/2 Phenyl ether was not metalated by triphenylmethylsodium. 1954 *Organic Reactions* VIII. vi. 260, *n*-Butyllithium metalates thiophene in the 2 position. *Ibid.* 261 The usual metalating agents..yield volatile acids which are easily separated from the acids of higher molecular weight obtained from the metalated products themselves. 1966 *McGraw-Hill Encycl. Sci. & Technol.* IX. 402/2 The most commonly used 'metalating agent'..is *n*-butyllithium in ether solution.

metalaw (me·tălǭ). [f. META- 1 + LAW *sb.*[1]] A hypothetical legal code based on the principles underlying existing legal codes and designed to provide a framework of agreement between diverse legal systems (orig. conceived as between terrestrial and possible extra-terrestrial beings); so **metale·gal** *a.*, of questions, etc.: pertaining to the basic principles underlying legal systems or upon which laws are formulated.

1956 *N.Y. Times* 20 Sept. 12/3 Andrew G. Haley..sought to codify some general principles of space law or 'metalaw' as he called it, by analogy to metaphysics. 1956 A. G. HALEY (*title*) The present day developments in space law and the beginnings of metalaw. 1957 *Observer* 20 Oct. 14/4 Still more tricky are the laws which should govern our relations with any extra-terrestrial intelligent beings we may meet. Here, something which Mr. Haley [an American lawyer] calls 'Metalaw' must prevail. 1959 BENN & PETERS *Social Princ. & Democratic State* iii. 58 What criteria must a rule satisfy to be a valid law?.. How are legal rules related to particular decisions..? These questions..involve analysis of the formal structure of a legal system, and of the relation between norms of different levels of generality; they might be termed metalegal questions. 1969 *New Scientist* 2 Jan. 36/3 The legal aspects of space—what is now called 'metalaw'—a subject that has an interest of its own. 1971 G. SCHWARZENBERGER *Internat. Law & Order* iv. 29 Alleged rules of international *jus cogens* which are based on no other evidence than postulates of natural law or other metalegal norms must be ignored on the level of *lex lata*.

metaldehyde. Add: (Later examples.) Cf. *META[2].

1949 *New Biol.* VI. 29 The most usual and one of the most successful methods of killing large numbers is to put out heaps of bran mixed with metaldehyde—the bran attracts the slugs which feed on the mixture, while the metaldehyde causes them to slime so profusely that they remain on the surface of the soil in close proximity to the bait. Here they die partly as the result of the metaldehyde acting as a stomach poison and partly owing to desiccation. 1963 *Which?* Mar. 88/2 A comparatively 'safe' poison is metaldehyde, now the most generally used for killing slugs. 1974 *Country Life* 24 Oct. 1228/1 Work at the Terrington EHF has compared metaldehyde and methiocarb [slug] pellet baits.

metalimnion (metăli·mniǫn). Pl. **-limnia.** [f. META- + Gr. λιμνίον, dim. of λίμνη lake.] The layer of water in a stratified lake which lies beneath the epilimnion and above the hypolimnion and in which the temperature decreases rapidly with depth. So **metalimne·tic** *a.* [cf. Gr. λιμνήτης living in marshes], of or within the metalimnion.

1935 P. S. WELCH *Limnol.* II. iv. 54 The term *thermocline* was first used by Birge in 1897... Since then, the terms *transition zone*, *mesolimnion*, and *metalimnion* have been proposed. 1957 G. E. HUTCHINSON *Treat. Limnol.* I. vii. 428 It is convenient to define the widely used term *metalimnion* to designate the whole of the region in which the temperature gradient is steep. *Ibid.* 464 In general, the smaller lakes..showed very striking metalimnia at depths of between 3 and 10 m. 1957 Metalimnetic [see *EPILIMNION]. 1960 *Limnology & Oceanogr.* V. 216 (*heading*) The cause of a metalimnetic minimum of dissolved oxygen. 1974 *Nature* 8 Feb. 393/2 During the summer, in the metalimnion of Lake Kinneret, the photosynthetic, green sulphur bacterium, *Chlorobium phaeobacteroides* can reach concentrations as high as 10^7 cells ml^{-1}.

metalingual (metăli·ŋgwăl), *a.* [f. META- + LINGUAL *a.*] = *METALINGUISTIC *a.*

1950 *Mind* LIX. 490 Due..to the common confusion of words and things its metalingual character has been overlooked. 1961 *Word* XVII. ii. 128 To these R. Jakobson has recently added three more functions—the poetic, phatic and metalingual. 1964 E. A. NIDA *Toward Sci. Transl.* iii. 45 In this volume the words *metalanguage* and *metalingual* refer to that part of language which is used in speaking about language itself, namely, the terms designating all the various features of language and the way these are used in describing and talking about languages.

metalinguistic (metăliŋgwi·stik), *a.* and *sb.* [f. META- + LINGUISTIC *a.* (cf. *METALANGUAGE).] **A.** *adj.* Of or pertaining to a metalanguage, or to metalinguistics (see B). **B.** *sb. pl.* Trager's term for that branch of linguistics which is concerned with the relation of language to the other elements of a culture (see also quot. 1974). Hence **metali·nguist** *sb.*, **metalingui·stically** *adv.*

1944 *Mind* LIII. 26 It cannot occur at the zero-level (it is a metalinguistic statement). 1944 H. REICHENBACH in P. A. Schilpp *Philos. B. Russell* 53 The use of a metalinguistic vocabulary is not a sufficient criterion for a more advanced state of logical analysis. 1949 G. L. TRAGER in *Studies in Ling.: Occasional Papers* I. 7 The full statement of the..relations between the language and any of the other cultural systems will contain all the 'meanings' of the linguistic forms, and will constitute the metalinguistics of that culture. 1951 TRAGER & SMITH *Outl. Eng. Struct.* 83 The metalinguist can turn it [*sc.* a datum] into a conclusion by clearly identifying out the microlinguistic characteristics of the speech. 1951 *Language* XXVII. III. 212 The discussion of linguistic research techniques is not a linguistics as we have known it, but rather a metalinguistics. 1952 R. M. HARE *Lang. Morals* I. iii. 38 A metalinguistic analysis is tempting. *Ibid.*, Some hypothetical imperatives might be analysed metalinguistically. 1953 C. E. BAZELL *Ling. Form* viii. 98 The study of microcriteria, whether phonetic or semantic, belongs to metalinguistics. 1966 J. J. KATZ *Philos. of Lang.* v. 222 S is metalinguistically true if, and only if, the semantically interpreted underlying phrase marker for its constituent sentence satisfies the condition in the reading for its metalinguistic predicate. 1967 C. L. WRENN *Word & Symbol* 4 The rather programmatic 'science' of metalinguistics recognises the need to study language and culture in intimate relationship. 1972 G. H. FISHER *Public Diplomacy* v. 119 This aspect of comparative linguistics is in its infancy. Anthropologists and linguistic scientists.. call it 'metalinguistics'. 1974 *Encycl. Brit. Micropædia* VI. 827/3 Some linguists use the term metalinguistics in reference to the study of metalanguages, languages or codes used to discuss or describe other languages.

metalled, *ppl. a.* **2.** (Earlier and later examples.)

1825 J. C. LOUDON *Encycl. Agric.* 511 In a road from a highway to a farmery, it may often be advisable to place the metalled road in the middle. 1955 *Times* 25 June 7/6 They live in this far-away valley at the end of a good metalled road that accompanied by power and telephone lines winds over the pass from Malakand. 1969 *Jane's Freight Containers 1968–69* 139/2 It has its own drinking water system, sewage, drains, metalled roads and mains.

metallic, *a.* and *sb.* Add: **A.** *adj.* **1. e.** *metallic circuit* (Telegr.), a circuit composed entirely of metal conductors, as opposed to one in which the return path of the current is through the earth; similarly *metallic return*.

1854 W. F. COOKE *Electr. Telegr.* 25 If, from the copper at one end, a piece of wire, or metallic circuit is carried.. round to the zinc..a current of electricity immediately passes through the wire. 1928 A. WILLIAMS *Telegr. & Telephony* ix. 131 The first telephone companies had a hard fight to interest the public in their services, and to keep down costs—which would have been nearly doubled by a metallic return—only one wire was used for each subscriber. *Ibid.*, The result was that lines with earth-return lines had to be converted into metallic circuits.

f. *metallic thread, yarn*: thread made from metal, or a synthetic material resembling metal.

1904 J. M. MATTHEWS *Textile Fibres* i. 4 Metallic threads are largely imitated by coating linen yarns with a thin film of gold or silver. 1963 A. J. HALL *Textile Sci.* ii. 105 Metallic yarns are made by various methods which include bonding metal. 1968 J. IRONSIDE *Fashion Alphabet* 211 Today, aluminium or plastic has been substituted for the rather heavy metals and 'metallic' yarns are now light, soft and non-tarnishing.

g. *metallic soap*: any of a class of soaps that are salts of carboxylic acids with an alkaline-earth metal or a heavy metal (instead of with an alkali metal as in ordinary soap) and are soluble in organic solvents but not in water, some of which are used in waterproofing materials, finishing textiles, and making anti-oxidants, lubricants, and fungicides.

1918 H. SEYMOUR *Reproduction of Sound* ii. 44 The metallic soaps so long in use [for the moulding and shaping of disc record blanks] have been superseded on the score of efficiency..but the [new] substance is more difficult to handle than the metallic soaps. 1940 A. WOOD *Acoustics* xviii. 504 The recording wax is a circular slab..which is composed of a metallic soap and has a highly polished plane surface. 1952 R. A. PINGREE in H. C. Speel *Textile Chem. & Auxiliaries* xx. 404 Air permeable water-resistant treatments were first obtained by depositing a metallic soap in and upon the fibers of the fabric. 1971 *Materials & Technol.* IV. ii. 70 Metallic soaps of long chain fatty acids will gell lubricating oil fractions and the properties of the grease will be governed mainly by the metallic radical.

h. *metallic arc welding* = *metal arc welding* s.v. *METAL *sb.* 14.

1927 *Jrnl. Iron & Steel Inst.* CXV. 909 The welding processes considered by the author are thermit welding, resistance welding..and metallic arc welding. 1948 F. KOENIGSBERGER in H. W. Baker *Mod. Workshop Technol.* I. ix. 181 Metallic-arc welding is the arc-welding process most frequently used in general engineering.

3. a. (Further examples.)

1936 *Discovery* Dec. 367/2 The marvellous, metallic blue *Morphos* are eminently characteristic of Tropical America. 1968 *Motor* 21 Dec. 66/3 (Advt.), Lotus Elan.. metallic blue with matching interior. 1975 G. V. HIGGINS *City on Hill* xi. 233 A metallic blue Opel.

c. Also, of the taste of tea made in a metal tea-pot.

1909 *Chambers's Jrnl.* Nov. 693/1 The objection to metal is simply that there is a danger of giving the tea what is known as a 'metallic' taste.

metallically, *adv.* Add: **1.** (Later examples.)

1909 *Physical Rev.* XXVIII. 159 The clamp *K* containing the molybdenite is metallically connected with the binding post *H*. 1913 V. B. LEWES *Oil Fuel* vi. 173 A second platinum rod also passes through the cover, being metallically connected with it.

4. In the manner of a metal or metals.

1944 *Physical Rev.* LXVI. 326/1 There are some substances such as the aniline dyes and $KMnO_4$ which reflect metallically in narrow wave-length regions.

metallization. (Further examples, corresponding to *METALLIZE *v.* 1 b.)

1876 *Chem. News* 7 July 9/2 (*heading*) Metallisation of organic bodies to render them fit to receive galvanic deposits. 1934 *Archit. Rev.* LXXV. 33/3 Metal surfaces may now be obtained by a new process called metallization, which enables any dry non-greasy surface to be sprayed with metal. 1952 J. DELMONTE *Plastics Molding* xiv. 430 In vacuum metallization, the art has been successfully developed for coating molded plastic parts. 1967 *Electronics* 6 Mar. 25 Litton engineers haven't decided whether to use a single or two-layer metalization to interconnect the circuits within the wafers.

metallize, *v.* Add: Also **metalize. 1.** (Later example.)

1912 G. B. BARHAM *Devel. Incandescent Electr. Lamp* iv. 37 When the untreated carbon filament, used as a base, is metallized, its weight is reduced by about 8 per cent.

b. To coat or cover with metal.

1911 *Engin. & Mining Jrnl.* XCI. 532/1 It [*sc.* the process] mainly consists of throwing finely distributed liquid metal..against the surface to be metallized. 1929 *Daily News* 16 Jan., Doors, window sashes, wainscotings, panel boards and panelling can be artistically metalized with one metal or another. 1952 J. DELMONTE *Plastics Molding* xiv. 431 Another technique of metallizing plastic parts..

is the chemical deposition of a silver film upon a carefully cleaned plastic surface. **1973** M. I. Kohan *Nylon Plastics* xvii. 585 Small nylon parts can often be vacuum metallized without predrying.

metallized *ppl. a.* (further examples); also **me·tallizer**, (*a*) a machine or plant for metallizing; (*b*) a person or organization involved in metallizing; **me·tallizing** *vbl. sb.*

1905 J. C. Howell in *Electrician* 28 July 590/1 On account of the positive-resistance curve and physical characteristics of these filaments, they have been given the name 'metallised filaments'. **1911** *Engin. & Mining Jrnl.* XCI. 532/1 The metallizing process invented by M. U. Schoop, of Zurich, is..different from any process so far known. **1912** G. B. Barham *Devel. Incandescent Electr. Lamp* iv. 37 It is said that the intense temperature of the metallizing process changes the composition of the carbon filament by removal of the ash residue and by the volatilization of the hydrocarbon contained in the filament. **1952** J. Delmonte *Plastics Molding* xiv. 429 The metallizing of plastics parts has become a popular procedure for some molded parts. **1955** *Sci. News Let.* 21 May 336/3 Game includes five metalized, full-color replicas of regulation traffic signs. **1969** W. R. R. Park *Plastics Film Technol.* viii. 195 As the applications for metallized film continued to grow, more sophisticated continuous metallizers have been evolved. **1969** R. F. Lang tr. *Henglein's Chem. Technol.* 807 Metallizing of fabrics increases their value. **1973** M. I. Kohan *Nylon Plastics* xvii. 585 The composition of these lacquers is proprietary with metallizers. **1974** *Sci. Amer.* Mar. 112/2 A simple measurement of the focal length can be made by standing the mirror on edge, directing a flashlight toward the metallized surface [etc.].

metallo-. Add: **meta·lloenzyme** *Biochem.*, an enzyme which is a metalloprotein; **me:tallogene·tic** *a.*, of or pertaining to metallogeny; hence **me:tallogene·tically** *adv.*; **metalloge·nic** *a.*, (*a*) (of an element) occurring in ores or as the native metal, rather than in rocks; (*b*) = **metallogenetic* adj.; **metallo·geny** *Geol.* [ad. F. *métallogénie* (L. de Launay *Sci. Géol.* (1905) ix. 263): see -GENY], (the study of) the origin of mineral deposits, esp. as related to petrographic and tectonic features; **me:talloorga·nic** *a. Chem.* = *organometallic* adj. s.v. ORGANO-; also (*rare*) **me:tal(l)orga·nic** *a.*; **metallophone**, (*b*) (later examples); **meta·lloprotein** *Biochem.* (see quot. 1964).

1955 *Adv. Protein Chem.* X. 321 The metalloenzyme can be isolated from its matrix, retaining all of its metal complement in the 'natural' state. **1971** Wiseman & Gould *Enzymes* iii. 43 It is mainly this involvement in metalloenzymes and metal-activated enzymes which is responsible for the requirement for these metal ions..in the food of animals. **1972** *Nature* 15 Dec. 417/1 Nitrogenase is a metalloenzyme containing iron and molybdenum both of which are essential for catalytic activity. **1909** *Cent. Dict.* Suppl., Metallogenetic province. **1910** *Q. Jrnl. Geol. Soc.* LXVI. 281 His [*sc.* de Launay's] object is to delineate the various regional types of ores.., the regional types being termed 'metallogenetic provinces'. **1965** G. J. Williams *Econ. Geol. N.Z.* vi. 61/2 The interplay of opinion reflecting metallogenetic fashions current from time to time. **1926** *Mineral. Abstr.* III. 133 Igneous activity was subordinate in Palaeozoic times, but very extensive and of much significance metallogenetically during the late Mesozoic in both Japan and Korea. **1920** H. S. Washington in *Jrnl. Franklin Inst.* CXC. 782 It may be as well to suggest here, and to use henceforward, two terms... We may call the 'rock elements' petrogenic and the 'ore elements' metallogenic. **1959** *Nature* 28 Nov. 1693/1 During the past few decades the problem of construction of metallogenic maps of various countries..has been discussed. **1974** *Encycl. Brit. Micropædia* VI. 828/2 Among the excellent examples of metallogenic provinces in North America are the gold province on the Canadian Shield [etc.]. **1905** *Nature* 13 Apr. 576/1 On the possible *rôle* of slipping in metallogeny. **1908** O. C. Williams tr. *L. de Launay's World's Gold* p. xiv, Those who..are interested in the manner of the concentration of metals in the earth or in what I call their 'Metallogeny'. **1959** *Nature* 28 Nov. 1693/1 Valuable contributions to the metallogeny of various ores were recently made by N. S. Shatsky. **1971** *Mineralium Deposita* VI. 404/1 This note is only intended to link some past ideas of the authors with what may emerge in the future as a key model of metallogeny. **1886** E. F. Smith tr. *V. von Richter's Chem. Carbon Compounds* 141 Most of the metallo-organic compounds can be prepared by the direct action of the metals or their sodium amalgams upon the bromides and iodides of the alkyls. **1946** *Nature* 30 Nov. 791/1 This suggests that some of the polyvalent metal in soil exists as an insoluble metallo-organic complex with some of the organic matter. **1974** *Sci. Amer.* Oct. 75 (*caption*) Metallo-organic complexes, such as titanium diisopropoxide [$Ti(OR)_2$], also fix nitrogen under ambient conditions. **1961** K. P. Wachsmann in A. Baines *Mus. Instruments* i. 31 Whereas the metallophones can be dated—the saron to not much before A.D. 900 and the gender to not later than A.D. 1157—the ancestry of the xylophone is quite obscure. **1961** P. Kemp *Alms for Oblivion* vi. 102 A *gamelan* of musicians invited by Le Mayeur to play us Balinese music... There were metallophones with polished bronze keys of different pitch. **1969** *Listener* 10 July 58/1 One has read a lot about the exotic Balinese gamelan orchestra with its myriad metallophones. **1972** *Where* Sept. 252/2, I was..unprepared, and unfortunately quite inequipped, for the spate of song books now also coping for xylophones, chime bars, guitars, tuned percussion, autoharps, Indian bells, metallophones, wood blocks, glockenspiels and maracas. **1940** *Biochem. Jrnl.* XXXIV. 1163 The metallo-protein compounds present in the red blood corpuscles. **1964** *Ann. Rev. Biochem.* XXXIII. 331 The term metalloprotein is used to designate those types of metal-protein complexes in which the strength of the binding is so great that a metal atom can be considered an integral part of the structure of the protein. **1971** *Nature* 10 Sept. 136/2 The ferredoxins are members of a class of metalloproteins known as iron-sulphur proteins. **1890** Webster, *Metalorganic*... Written also *metallorganic*. **1965** *Polymer Rev.* VIII. p. v, The Editors invited Professor Andrianov to write a monograph on his pioneering work on metalorganic polymers—or 'elemento-organic polymers', as they are called in the Soviet Union. **1974** *Nature* 27 Sept. 307/1 The surfaces were covered with a monolayer of metallorganic soap.

metallographer. (Earlier and later examples.)

1902 *Jrnl. Iron & Steel Inst.* LX. 242 Samples cut from ..test bars have already been..microscopically examined by an eminent metallographer. **1961** *Evening Standard* 17 July 14/5 Metallographer required..for examination of ferrous and non-ferrous materials. **1966** D. G. Brandon *Mod. Techniques Metallogr.* 3 The successful metallographer combines an understanding of the effects of specimen preparation with a knowledge of the optics of image formation.

metallographical (metălogræ·fikăl), *a.* [f. METALLOGRAPHIC *a.* + -AL.] = METALLOGRAPHIC *a.*

1902 *Jrnl. Iron & Steel Inst.* LX. 242 No regard has been paid to the chemical or purely metallographical alterations..connected with the process of heating or annealing any iron material with subsequent slow or sudden cooling. **1950** *Engineering* 10 Nov. 341/2 However carefully determined the mechanical and metallographical factors might be [etc.].

Hence **me:tallogra·phically** *adv.*, by metallographic methods.

1936 L. R. Van Wert *Introd. Physical Metallurgy* ii. 23 Crystallographers distinguish six distinct crystal types. Only three of these are of any interest metallographically. **1967** A. H. Cottrell *Introd. Metall.* xx. 373 Determining the extent of the isothermal transformation metallographically. **1969** J. A. Scott in A. F. Madayag *Metal Fatigue* iii. 92 Microscopic examination of a metallographically prepared sample is the usual method of inspection for inclusions.

metalloid, *sb.* **b.** Add to def.: Also (now the usual sense), an element intermediate in its properties between a typical metal and a typical non-metal. (Further examples.)

Cf. quot. 1836 in Dict. (given corrected and more fully below).

1836 W. T. Brande *Man. Chem.* (ed. 4) v. 318 And then, adverting particularly to the unmetallic substances, he [*sc.* Berzelius] subdivides them into three classes, namely, — 1. Permanently elastic or gaseous bodies (Gazolyta)... 2ndly. Metalloids: sulphur, phosphorus, carbon, boron, and silicon. 3rdly. Salifying substances (Halogenia). **1876** *Encycl. Brit.* V. 476/1 The non-metallic elements are also sometimes termed metalloids, but this appellation.. strictly belongs to certain elements which do not possess the properties of the true metals although they more closely resemble them than the non-metals in many respects. **1894** G. S. Newth *Text-bk. Inorg. Chem.* I. ii. 8 The element arsenic possesses many of the physical properties of a metal, but in its chemical relations it is more allied to the non-metals; such elements as these are often distinguished by the name *metalloids.* **1959** *Nomencl. Inorg. Chem.* (I.U.P.A.C.) 10 The word metalloid should not be used to denote non-metals. **1959** B. Chalmers *Physical Metall.* ii. 72 In these structures, the anion is a 'metalloid', such as sulfur, selenium, tellurium, tin, or antimony, and the cation is a transitional metal, such as chromium, nickel, iron, etc. **1964** E. G. Rochow *Organometallic Chem.* (1965) ii. 10 There also are some elements which look like metals but are brittle and have the electrical properties of semiconductors; these are commonly called the metalloids.

metallurgic, *a.* (Earlier example.)

1778 J. C. Lettsom *Hist. Orig. Med.* vi. 98 Metallurgic chemistry was one of the most remote inventions.

metallurgically (metălǿ·ɪdʒikăli), *adv.* [f. METALLURGIC, -ICAL *adjs.*: see -ICALLY.] From a metallurgical point of view; as regards metallurgy.

1890 in *Cent. Dict.* **1911** F. W. Harbord in Harbord & Hall *Metall. Steel* (ed. 4) I. viii. 199 What is to all intents and purposes metallurgically the Bertrand–Thiel process is at work in the Hoesch Works in Germany. **1968** *Physics Bull.* Dec. 411/2 A metallurgically good type II superconductor shows electrical resistance in the vortex state.

metallurgist. Add: Now commonly with main stress on the second syllable (mĭtæ·lǿɪdʒist).

metallurgy. Also with pronunc. (mĭtæ·lǿɪdʒi). Add to def.: Now understood as including the scientific study of the structure, properties, and behaviour of metals. (Further examples.)

1914 W. Rosenhain *Introd. Study Physical Metall.* i. 1 The scope of Physical Metallurgy..brings it well over the border-land of several sister-sciences—such as chemistry .., physics.., and that branch of knowledge generally known as 'strength of materials'. **1948** R. H. Harrington *Mod. Metall. Alloys* p. ix, The House of Metallurgy has, today, many rooms filled with stacks of measured data concerning heat treatments, physical properties, microstructures, and crystal lattice measurements. **1950** *Sci. News* XV. 138 Modern metallurgy borders also on physics and chemistry, mining and mineral dressing, applied mechanics and physical chemistry. **1967** A. H. Cottrell *Introd. Metall.* i. 1 Metallurgy is now a disciplined applied science based on a clear understanding of the structures and properties of metals and alloys.

metally, *a.* **1.** Delete † *Obs.* and add later *poet.* example.

1898 G. Meredith *Empty Purse* in *Poems* II. 202, I can hear a faint crow Of the cock..As down the new shafting of mines, A cry of the metally gnome.

metalogic. Add: (Further examples.) See also quots. 1936, 1937.

1902 R. R. Marett in H. Sturt *Personal Idealism* v. 232 The no-man's-land of dogmatic 'Metalogic'. **1936** *Mind* XLV. 482 If a contradiction appears at the end of a chain of inferences one at least of two cases must be realised according to a very evident theorem of metalogic: (1) either there is a fallacy in the chain..or (2) the premises are contradictory. *Ibid.* 485 Lukasiewicz, the great Polish logician, has generalised that idea by introducing what he calls *metalogic*, which has to ordinary logic the same relation as metamathematics to mathematics. **1937** A. Smeaton tr. *Carnap's Logical Syntax of Lang.* 9 The Warsaw logicians..have spoken of..*metalogic*... The word 'metalogic' is a suitable designation for the subdomain of syntax which deals with logical sentences in the narrower sense. **1955** A. N. Prior *Formal Logic* 64 The consideration of these deductions from outside may be called 'metalogic', and that is in fact the name now commonly applied to it. **1963** R. Carnap in P. A. Schilpp *Philos. R. Carnap* 54 At that time I defined the term 'metalogic' as the theory of the forms of the expressions of a language. Later I used the term 'syntax' instead.

metalogical, *a.* Add: (Further examples.) Cf. prec.

1951 J. Łukasiewicz *Aristotle's Syllogistic* 103 The.. metalogical principle of traditional logic: 'utraque si praemissa neget, nil inde sequetur'. **1955** A. N. Prior *Formal Logic* 65 'Metalogical' demonstrations ought..to be themselves 'logical' in the sense of being cogent.

Hence **metalo·gically** *adv.*

1946 C. I. Lewis *Analysis of Knowledge* v. 129 Analytic statements in non-logical terminology, like 'All birds are bipeds', will not be metalogically derivable from its postulates. **1955** A. N. Prior *Formal Logic* 132 This rule is metalogically derivable.

metalogi·cian. [f. METALOGIC: see -ICIAN.] One who is versed in metalogic.

1902 R. R. Marett in H. Sturt *Personal Idealism* v. 237 A condition of existence which even the 'metalogician' finds it difficult to conceive. **1964** *Amer. Philos. Q.* I. 236/2 As used by grammarians, metalogicians, etc., they mention a type.

metalogue (me·tălǫg). $rare^{-1}$. [f. META-, on model of *prologue* and *epilogue*.] A speech delivered between the acts or scenes of a play.

1956 Auden & Kallman *Magic Flute* [at end of Act I] 57 Metalogue (To be spoken by the singer taking the role of Sarastro.)

metamathematical (me:tămæþĭmæ·tikăl), *a.* [f. as next, after *mathematics, mathematical*.]

a. (In Dict. s.v. META- 1.) *rare.*

b. Of or pertaining to metamathematics.

1926 F. P. Ramsey in *Mathematical Gaz.* XIII. 188 It is contended that the principles used in the metamathematical proof that the axioms of mathematics do not lead to contradiction, are so obviously true that not even the sceptics can doubt them. **1933** M. Black *Nature of Math.* 150 It is extremely probable that a metamathematical proof of the consistency of the whole of pure mathematics is impossible. **1952** S. C. Kleene *Introd. Metamath.* vi. 140 In either case, the rules of inference must have the character of schemata, i.e. they must employ metamathematical variables, since infinitely many applications have to be provided for. **1971** *Nature* 8 Jan. 104/2 Hilbert's programme for demonstrating the consistency of arithmetic by metamathematical methods resulted in a new outburst of activity which culminated..with Gödel's incompleteness theorem.

Hence **me:tamathema·tically** *adv.*, by means of, or from the point of view of, metamathematics.

1937 *Philos. of Sci.* IV. 329 Gödel has proved metamathematically that there are surely theorems and problems about the natural numbers which cannot be proved or solved by elementary means. **1956** J. H. Woodger tr. *Tarski's Logic, Semantics, Metamath.* 116 Let us now consider the situation metamathematically. **1967** S. C. Kleene *Math. Logic* § 39. 214 Ackermann in 1924–5 thought he had proved metamathematically the consistency of N.

metamathematician (me:tămæþĭmăti·ʃăn). [f. next, after *mathematics, mathematician.*] An expert in metamathematics.

1935 *Mind* XLIV. 394 The metamathematician may use mathematical induction for the recursive definition of simple functions. **1952** S. C. Kleene *Introd. Metamath.* iii. 64 The interpretation motivates the metamathematician in his choice of the particular formal system which he introduces by his definitions.

metamathematics (me:tămæþĭmæ·tiks), *sb. pl.* (const. as *sing.*). [f. META- 1 + MATHEMATICS *sb. pl.*] (In Dict. s.v. META- 1); *spec.*

[after G. *metamathematik* (D. Hilbert 1923, in *Math. Ann.* LXXXVIII. 153)], the field of study concerned with the structure and formal properties of mathematics and similar formal systems.

1890 in *Cent. Dict.* **1926** *Encycl. Brit.* II. 831/1 The science on which Hilbert is now (1926) engaged, which takes for its subject matter the meaningless formulae of mathematics, he calls metamathematics, and believes it to be capable of establishing many important results relating to the multiplicative axiom or axiom of selections, and to the continuum problem. **1940** *Mind* XLIX. 242 The investigation of calculi, in this light, independently of all regard for the 'meaning' of the component complex marks, was christened metamathematics. **1952** S. C. Kleene *Introd. Metamath.* iii. 63 The formal systems which are studied in metamathematics are (usually) so chosen that they serve as models for parts of informal mathematics and logic with which we are already more or less familiar, and from which they arose by formalization. **1956** E. H. Hutten *Lang. Mod. Physics* ii. 35 That we require a meta-mathematics in order to prove the consistency and completeness of mathematics..does not mean that we are confronted with the final break-down of logic and mathematics. **1963** G. T. Kneebone *Math. Logic* 381 Metamathematics began as a study of particular formal systems, and it was initially concerned with questions of consistency and completeness of given sets of axioms. But it now embraces investigations of altogether wider scope, designed to yield fundamental information about entire classes of formal systems of some particular kind. **1966** J. J. Katz *Philos. of Lang.* iii. 27 Metamathematics, as the general theory of the formal structure of the language of mathematics, was closely parallel to Carnap's idea of logical syntax, as the general theory of the formal structure of scientific and factual discourse. No wonder, then, that metamathematics appealed to Carnap as a model for his theory of logical syntax.

meta-metalanguage: see *METALANGUAGE.

metamict (me·tămikt), *a. Min.* [ad. Da. *metamikt* (W. C. Broegger 1893, in *Salmonsens Konversationsleksikon* I. 743/2), f. Gr. μετα- META- + μικτ-ός mixed, blended.] Of a mineral: converted into an amorphous state as a result of the radioactive decay of atoms contained in it. Also applied to the state itself.

1927 *Phil. Mag.* IV. 525 Metamict substances like xenotime and thorite have probably been formed under very high pressure. **1950** *Science* 24 Mar. 312/2 For materials which are completely metamict, it would be possible to determine only a minimum age. **1965** G. J. Williams *Econ. Geol. N.Z.* xiii. 203/2 He suggested the red colouration occurs only in pegmatites sufficiently old for the transformation of the radioactive mineral to the metamict state. **1970** *Nature* 11 Apr. 147/2 Quartz is thus a 'metamict' mineral, like zircon.

Hence **me:tamictiza·tion**, the process of becoming, or state of being, metamict; **me·tamictness**, metamict character.

1952 *Amer. Mineralogist* XXXVII. 142 There has been some speculation as to the reasons for variations in degree of metamictness of different samples of the same mineral. *Ibid.* 154 More powerful sources of radiation..might prove useful in the study of metamictization. **1966** *McGraw-Hill Encycl. Sci. & Technol.* VIII. 296a/1 X-ray methods do not permit a close estimate of the degree of metamictness. **1970** *Science* 30 Jan. 617/1 The U and Th contents of baddeleyite are quite low, which explains the high transparency and the lack of metamictization.

metamorphic, *a.* Add: **2.** Also as *sb.* (usu. *pl.*), a metamorphic rock.

1881 King & Rowney *Old Chapter Geol. Rec.* 49 The rocks of the locality are well-bedded metamorphics. **1923** *Univ. of Toronto Studies, Geol. Ser.* xvi. 7 A few blocks of the same granite were observed along with metamorphics like those seen in the till. **1970** *N.Z. Jrnl. Geol. & Geophysics* XIII. 72 Piroutet..regarded all the metamorphics as pre-Cambrian rocks.

metamorphize, *v.* Delete † *Obs.* and add later U.S. example.

1943 W. Stevens *Let.* 29 Mar. (1967) 444 It would not help to change them to something else, any more than it would help to metamorphize this, that or the other (clouds into foam, living blossoms to blossoms without life, heat to a form of heat).

metamorphose, *v.* Add: **1. b.** *intr.* with *into.*

1927 Haldane & Huxley *Animal Biol.* ix. 180 When the tadpole metamorphoses into the frog, some of its tissues start to dedifferentiate.

metamorphosize (metămǭ·ɪfǿsəiz), *v.* [f. METAMORPHOS(IS + -IZE.] *trans.* and *intr.* = METAMORPHOSE *v.* Hence **metamo·rphosized** *ppl. a.*

1908 *Dict. Nat. Biogr.* V. 1312/1 Dowsing's acquaintance with 'Lating'..led him to metamorphosise Dr. Billingford into a maid recommending her daughter's soul to the Virgin Mary. **1928** *Observer* 29 Jan. 9/3 It may well be found that Unwin and his biographer between them have metamorphosised the study of economic history in this country. **1969** *Daily Tel.* 4 Aug. 15/1 It [*sc.* Shaw's *As Far as Thought Can Reach*] envisages the world in A.D. 30,000, when children of 17 are hatched from eggs to spend a few years on adolescent pursuits—love, mating, music, the dance—before metamorphosising into Ancients. **1972** *Ibid.* 17 May 12/6 The parable of metamorphosised man is beautifully photographed by Walter Lassally.

metanalysis (metănæ·lisis). *Philol.* [f. MET(A- + ANALYSIS.] Reinterpretation of the division between words or syntactic units: as *adder* < OE. *nædre* by analysis in ME. of *a naddre* as *an addre*. Hence **meta·nalyse** *v. trans.*

1914 O. Jespersen *Mod. Eng. Gram.* II. v. 141, I have ventured to coin the word 'metanalysis' for the phenomenon frequent in all languages that words or word-groups are by a new generation analyzed differently from the analysis of a former age. **1940** —— *Ibid.* V. xix. 308 A good many sentences of this type are double-barrelled and present the possibility of a 'metanalysis', by which 'It is good for a man/not to touch a woman' may come to be apprehended 'It is good/for a man not to touch a woman'. **1957** G. V. Smithers *Kyng Alisaunder* II. 138 *ikyr* < OE. *nicor* by a process of metanalysis in which an initial consonant is treated as the final consonant of the preceding word, or a final consonant is attracted into the beginning of the next word. **1962** R. Quirk *Use of English* viii. 127 When the French word *crevice*..was introduced into Middle English, its connexion with 'sea food' caused people to *metanalyse* the final syllable as *-fish* ('crayfish'). **1970** B. M. H. Strang *Hist. English* iv. 250 Assimilations account for the /m/ (earlier /n/) in such words as *comfort*, *noumpire* (later metanalysed as *umpire*). *Ibid.* 268 *For then ones* metanalysed as *for the nonce.* **1972** J. L. Dillard *Black English* iv. 162 *Metanalysis* describes the analysis of words or groups of words into new elements... *a napron* was metanalyzed to *an apron.*

metanoia (metănoi·ă). [Gr. μετάνοια, f. μετανοεῖν to change one's mind, to repent.] Penitence, repentance; reorientation of one's way of life, spiritual conversion.

1873 M. Arnold *Lit. & Dogma* vii. 196 Of 'metanoia', as Jesus used the word, the lamenting one's sins was a small part; the main part was something more active and fruitful, the setting up an immense new inward movement for obtaining the rule of life. And 'metanoia', accordingly, is: *a change of the inner man.* **1881** *Amer. Church Rev.* July 167 What a Metanoia was there, to both Jesus and John! .. And what a Metanoia had come also upon the disciples of John and upon Israel! **1918** *Encycl. Relig. & Ethics* X. 733/2 'Repentance' has an emotional tone; μετάνοια is ethical and intellectual; the former is negative—a turning away from sin; the latter is positive—an enthusiasm for righteousness. **1939** V. A. Demant *Relig. Prospect* ix. 237 If we understand St. Paul's use of the word 'spiritual', not in our misleading sense of 'non-material' but as the nature of a creature turned to God, we see how this *metanoia*, this turning about, brings a restored understanding of the order of human powers and faculties. **1945** A. Huxley *Let.* 10 Apr. (1969) 520 Virgil's *metanoia* was in the nature of a death-bed repentance. **1969** F. de Graeve in J. Kerkhofs *Mod. Mission Dialogue* p. xvi, It must reveal the Church..as the community in which the religious intentionality of all people can blossom into that newness of life that is the real metanoia. **1973** E. Powell *No Easy Answers* xii. 123 To entertain this idea and to be penetrated with it is the change of mind, repentance, *metanoia*, of which the baptist was not the announcer but the forerunner.

metaphone (me·tăfōun). *Phonetics.* [f. META- + PHONE *sb.*[1]] (See quots.)

1930 H. E. Palmer *Princ. Romanization* II. 52 Contrasted with *monophones* we have *metaphones*, which we may define as two or more phones which serve jointly as units of meaning within the limits of a given linguistic community. **1934** Yuen-Ren Chao in M. Joos *Readings in Linguistics* (1958) 38/2 A dynamophone is a metaphone which contains two or more phones differing..in quality. **1966** M. Pei *Gloss. Ling. Terminol.* 161 *Metaphone*, a free allophonic variant chosen in preference to another because regarded as more suitable to the type of speech used (*tomahto*, *eyether*, used in a given situation instead of the more customary pronunciations of *tomato*, *either*). **1972** Hartmann & Stork *Dict. Lang. & Ling.* 140 *Metaphone*, a free variant of a phoneme, e.g. [ai] or [i:] in *neither*.

metaphony. Add: (Later examples.) Hence **metapho·nic** *a.*

1950 *Trans. Philol. Soc.* 1949 23 The name *apophony* may be retained for the quantitative alternations, and.. the qualitative alternations may be distinguished as *metaphony*. **1953** K. Jackson *Lang. & Hist. Early Brit.* 371, -āui- > eu appears to be due to metaphony rather than to palatalisation. **1957** *Archivum Linguisticum* IX. 107 Metaphonic Spanish and Portuguese verbs. **1970** *Ibid.* I. 22 Moreover, in Lena all adjectives of the type *blanco* have three forms: one ending in -u which undergoes metaphony and is used in agreement with masc. nouns. **1973** A. H. Sommerstein *Sound Pattern Anc. Greek* ii. 76 There are three main ablaut rules—Vṛddhi, Vowel Weakening and *e/o* ablaut or metaphony.

metaphorize, *v.* (Earlier and later examples.)

1731 J. Constable *Reflections Accuracy of Style* 100, I wish too, that..the vanity of metaphorizing did not put several upon a greater stretch of thoughts than their wit will bear. **1909** *N.Y. Even. Post* 27 Nov. 5 However agitated or depressed they may be, they must go on metaphorising. **1949** Wellek & Warren *Theory of Lit.* xv. 202 We metaphorize also what we love, what we want to linger over, and contemplate.

metaphyseal, metaphysial: see *METAPHYSIS.

metaphysic, *sb.*[1] **1. a, b.** (Later examples.)

1918 B. Russell in *Monist* XXVIII. 496 In the present lectures, I shall try to set forth..a kind of logical doctrine which seems to..result from the philosophy of mathematics..: a certain kind of logical doctrine, and on the basis of this a certain kind of metaphysic. **1968** J. M. Ziman *Public Knowl.* iii. 38 One further condition may be necessary for consensible knowledge—a common metaphysic. Those who participate in the consensus must already share many beliefs. **1969** *N.Y. Rev. Bks.* 16 Jan. 15/1 The United States, whose metaphysic, according to Lionel Trilling, is always material and practical, has much more closely fulfilled the nineteenth-century Comtean vision of the future. **1972** D. Bell in Cox & Dyson *20th-Cent. Mind* I. vi. 211 On this central point Bosanquet's metaphysic remains impenetrably obscure, sustained by a combination of piety and metaphor.

metaphysical, *a.* Add: **1. b.** Hence *absol.*

1935 E. R. Eddison *Mistress of Mistresses* x. 192 You may have a nose for metaphysicals..but here you cry out upon no trail. I know nothing. Only, I am.

c. Applied to concepts or propositions which, by relying on abstract principles, are not considered verifiable in terms acceptable to some logical positivist or linguistic philosophers. Cf. *METAPHYSICS *sb. pl.* 1 c.

1865 Mill in *Westm. Rev.* XXVII. 344 The mode of thought which M. Comte terms Metaphysical, accounts for phænomena by ascribing them, not to volitions either sublunary or celestial, but to realized abstractions. **1922** tr. *Wittgenstein's Tractatus* 151 *Where* in the world is a metaphysical subject to be noted? **1936** A. J. Ayer *Lang., Truth & Logic* i. 31 We may accordingly define a metaphysical sentence as a sentence which purports to express a genuine proposition, but does, in fact, express neither a tautology nor an empirical hypothesis. **1950** R. Carnap *Logical Found. Probability* ii. 38, I hope that nobody will misinterpret my statement of the objectivity of logical relations as a metaphysical statement of the 'subsistence' of these relations in a Platonic heaven. **1953** G. E. M. Anscombe tr. *Wittgenstein's Philos. Investigations* § 116 What *we* do is to bring words back from their metaphysical to their everyday use.

3. d. *Christian Science.* Applied to that which transcends matter or the physical; *metaphysical healer*, one who heals physical ills through metaphysics; so *metaphysical healing.*

1876 A. B. Alcott *Jrnls.* (1938) 466 A wider acquaintance with idealism in its various phases will be serviceable to these 'Metaphysical Healers' and 'Christian Scientists' as they call their school. **1884** M. B. Eddy in *Christian Sci. Jrnl.* 2 Feb. 2 Metaphysical healing, or Christian Science, is a demand of the times. **1907** 'Mark Twain' *Christian Sci.* I. 70 Mrs. Eddy is president of the Trust's Metaphysical College in Boston, where the student of C. S. healing learns the game by a three weeks' course. **1939** M. Gregory *Psychotherapy* iv. 178 Students, intoxicated with her doctrines, came there for instruction in metaphysical 'obstetrics' and moral science. **1958** T. L. Leishman *Why I am Christian Scientist* viii. 116 Now, as then, signs and wonders are wrought in the metaphysical healing of physical disease. **1972** *Pioneers in Christian Sci.* (Mary Baker Eddy Mus., Longyear Hist. Soc.), Asa Gilbert Eddy..studied with her and began the practice of metaphysical healing.

5. Add to def.: In more recent use, of poetry which expresses emotion within an intellectual context; (further examples); also *ellipt.* as *sb.*, a metaphysical poet. Hence **metaphysica·lity**, the quality of being metaphysical.

1898 G. Saintsbury *Short Hist. Eng. Lit.* vii. ii. 412 Crashaw is perhaps the chief metaphysical, the type of the whole class. **1906** H. J. C. Grierson *First Half 17th Cent.* ix. 374 In Donne's love-poetry there is a real metaphysical strain, while the range of erudition from which he draws his imagery was something altogether new. **1921** G. Saintsbury in *Times Lit. Suppl.* 27 Oct. 698/2 [Dryden] might..so use 'metaphysics' as equivalent to 'second thoughts', things that come *after* the natural first; and, once more, this definition would, I think, fit all the poetry commonly called 'metaphysical'. *Ibid.*, I only hope that in a second edition he will add a few more 'specimens' of the lighter metaphysicality. **1936** *Essays & Stud.* XXI. 137, I use 'metaphysical' not in the Johnsonian sense of 'conceited', but as implying a certain refinement and complexity of thought—a kind of ecstasy of intellectual parturition as we find it in Shakespeare and Donne. **1947** C. Day Lewis *Colloq. Element Eng. Poetry* 21 The poets of the Romantic Movement gave an outlet to new ideas.. beneath the crust of their age, just as the Metaphysicals had done for the new ideas of their time. **1955** R. Graves *Crowning Privilege* ii. 27 Dryden's conversion from metaphysicality was not to the poetic faith of his English predecessors, but to current French theory. **1960** D. Daiches *Crit. Hist. Eng. Lit.* II. xiv. 1114 The insistence that intellect and emotion should work together in poetry and that one should seek to recover the 'unified sensibility' of the metaphysical poets. *Ibid.* 1116 But the poets who began to write in the late 1920's and 1930's saw both Hopkins and Eliot as their masters, as well as the metaphysicals..and the popular singers of the English music hall.

7. in appos. use, as *metaphysical-aesthetical*, *metaphysical-epistemic* adjs.

1963 *Times Lit. Suppl.* 4 Jan. 4/4 Dr. Kramrisch does not disdain any of the resources of jargon, metaphysical-aesthetical, aesthetical-theosophical. **1971** *Hist. of Sci.* X. 111 A consistent and effective metaphysical-epistemic understanding of science.

metaphysician. Add: (Later examples.)

1936 A. J. Ayer *Lang., Truth & Logic* i. 33 The metaphysician fails to see this [*sc.* 'substance' fallacy] because he is misled by a superficial grammatical feature of his language. **1956** J. O. Urmson *Philos. Analysis* vii. 110 But let us now suppose that the technical statements we are brought up against are the technical statements of metaphysicians. **1958** G. J. Warnock *Eng. Philos. since*

1900 ix. 121 Even if..the conclusion could really be drawn that the metaphysician's employment of language was without significance.

2. One who practises metaphysical healing. Cf. *METAPHYSICAL *a.* 3 d.

1881 M. B. Eddy *Sci. & Health* (ed. 3) iii. 151 Metaphysicians can heal the sick, absent from them: space is no obstacle to mind.

metaphysico-. (Further examples.)

1848 D. G. Rossetti *Let.* 20 Aug. (1965) I. 40 Hunt's stanzas..partook of the metaphysico-mysterioso-obscure. **1850** H. L. Mansel *Lett., Lect. & Rev.* (1873) 12 Another metaphysico-grammatical theory. **1905** *Spectator* Suppl. 28 Jan. 119/1 One of those extremely clever and almost painfully 'up-to-date' metaphysico-theological books which America produces in such abundance. **1930** D. B. W. Lewis *Stuffed Owl* p. xx, Verse which..crystallises the metaphysico-theologo-cosmologo-nigology to which the best modern thought..inclines. **1936** L. S. Stebbing in Day Lewis & Stebbing *Imagination & Thinking* 15 The truth of this statement is logically independent of any metaphysico-psychologist's *-ism*. **1955** H. B. Acton *Illusion of Epoch* 3 The criticisms that apply no more to it than they do to other metaphysico-ethical systems.

metaphysics, *sb. pl.* Add: **1. a.** (Later examples.)

1951 G. J. Warnock in Edwards & Pap *Mod. Introd. Philos.* (1973) 781, I do not know..how the term 'metaphysics' really ought to be defined. I suspect..that it is useless to try to divide philosophy into compartments. **1959** P. F. Strawson *Individuals* viii. 247 If metaphysics is the finding of reasons, good, bad or indifferent, for what we believe on instinct, then this has been metaphysics. **1960** C. C. Gillispie *Edge of Objectivity* xi. 496 Comte had to..repudiate not only metaphysics but also ontology. Thus would he deprive science of any and every claim to deal with objective reality or with any truth deeper than consistency or efficacy. **1967** J. W. Yolton *Metaphysical Analysis* (1968) xi. 189 The typical double level of metaphysics: the level of exposition, which is always external to the system being developed, and the level of participation in the system.

b. (Further examples.) Occas. construed as *sing.*

1859 Mill in *Fraser's Mag.* LIX. 489/2 His [*sc.* Austin's] book on the *Province of Jurisprudence* stepped at once into the very highest authority on what may be termed the metaphysics of law. **1958** W. Stark *Sociol. of Knowl.* I. iv. 197 A metasociology which would be, not a metaphysics, in so far as metaphysics is divorced from the empirical, but a study of man as he appears in all societies. **1964** A. W. Gouldner in I. L. Horowitz *New Sociol.* 209 This.. is a metaphysics of the underworld, in which conventional society is seen from the standpoint of a group outside of its own respectable structures.

c. Used by some followers of positivist, linguistic, or logical philosophy: concepts of an abstract or speculative nature which are not verifiable by logical or linguistic methods.

1865 Mill in *Westm. Rev.* XXVII. 347 In repudiating metaphysics, M. Comte did not interdict himself from analysing or criticising any of the abstract conceptions of the mind. **1936** A. J. Ayer *Lang., Truth & Logic* 30 Philosophy, as a genuine branch of knowledge, must be distinguished from metaphysics. **1937** A. Smeaton tr. *Carnap's Logical Syntax of Lang.* 8 The sentences of metaphysics are pseudo-sentences which on logical analysis are proved to be either empty phrases or phrases which violate the rules of syntax. **1956** J. O. Urmson *Philos. Analysis* vii. 106 The view of Wittgenstein that metaphysics was not merely outdated as the old positivism had it, but was a logically impossible enterprise, being excluded by the essential nature of language. **1957** S. Körner in C. A. Mace *Brit. Philos. in Mid-Cent.* 126 This principle [of verification]..has dominated European thought for at least three centuries..and its acceptance does not mean, therefore, the end of metaphysics as had been believed by some logical positivists. **1966** R. Sternfeld *Frege's Logical Theory* ii. iii. 55 Thus, the general problem of existence is simply replaced by arithmetical attribution, plus Frege's philosophic superstructure. And this portion of metaphysics is replaced by arithmetic.

metaphysis. Restrict $rare^{-0}$ to sense in Dict. and add: **2.** *Anat.* [after DIAPHYSIS, EPIPHYSIS.] The region either at one or at both ends of a growing long bone which lies between the diaphysis and the epiphysial cartilage and which is the site of advancing ossification.

1913 Dorland *Med. Dict.* (ed. 7) 560/2 *Metaphysis*, the end of the diaphysis of a long bone where it joins the epiphysis. **1926** *Biochem. Jrnl.* XX. 380 The upper end of the tibia was isolated and slices..containing the epiphyseal cartilage, the metaphysis and a little of the shaft were made. **1947** *Radiology* XLIX. 347/1 In the rat the isotopes that seek bone come down not only in old bone but predominantly in the zone of new bone growth in the metaphysis. **1966** *Lancet* 31 Dec. 1430/1 The proximal metaphysis contained less bone than the distal metaphysis in both control and treated animals.

Hence **metaphyseal, -physial** *adjs.* (metăfi·ziăl, -fisĭ·ăl), of or pertaining to a metaphysis.

1913 Dorland *Med. Dict.* (ed. 7) 560/2 *Metaphyseal*, pertaining to a metaphysis. **1931** D. M. Greig *Clin. Observations Surg. Path. Bone* vii. 190 This slim exostosis has doubtless been one of the earliest manifestations of the metaphysial growth disturbance. **1961** *Lancet* 30 Sept. 744/2 Vertebral osteomyelitis commences in the metaphyseal region of the vertebral body. **1966** *Ibid.* 31 Dec. 1430/1 This appearance resembled the metaphysial sclerosis sometimes seen in vertebræ of patients with secondary hyperparathyroidism.

meta-political, *a.* **1.** Delete † *Obs.* and add later example.

1937 J. M. Murry *Necessity of Pacifism* ii. 38 The deep-seated social urge towards life becomes necessarily anti-political in its manifestation. Perhaps meta-political would be a better word. It points and thrusts *beyond* politics.

metapsychic (metăsəi·kik), *a.* [ad. F. *métapsychique*; cf. *METAPSYCHICS *sb. pl.*] = next.

1905 *Westm. Gaz.* 25 Feb. 10/3 The new President [*sc.* M. Richet]..suggested the term 'metapsychic sciences' as a substitute for 'modern miracle'. **1923** S. De Brath tr. *Richet's 30 Yrs. Psychical Res.* 4 The terms 'supernatural' and 'supernormal' must therefore be rejected along with 'the occult'... I proposed the term Metapsychic.

metapsychical (metăsəi·kikăl), *a.* [f. META- + PSYCHICAL *a.*, after METAPHYSICAL *a.*] That is beyond the sphere of ordinary psychology; pertaining to *METAPSYCHICS.

1905 L. I. Finch tr. J. Maxwell (*title*) Metapsychical phenomena. **1914** A. L. Teixeira de Mattos tr. *Maeterlinck's Unknown Guest* 50 Nevertheless it may be said that these regions quite lately annexed by metapsychical science are as yet hardly explored. **1927** *Glasgow Herald* 19 Apr. 12 The author of this book was the director of the Metapsychical Institute of Paris.

metapsychics (metăsəi·kiks), *sb. pl.* [ad. F. *métapsychique* (C. Richet 1905, in *Proc. Soc. Psychical Research* XIX. 2), f. Pol. *metapsychika* (W. Lutosławski 1902, in *Wykłady Jagiellońskie* II), after METAPHYSICS *sb. pl.*] A name applied to a science or study of certain phenomena which are 'beyond the scheme of orthodox psychology'. Cf. *PARAPSYCHOLOGY.

1905 O. Lodge in L. I. Finch tr. *Maxwell's Metapsychical Phenomena* p. xi, To emphasise the fact that these occurrences are at present beyond the scheme of orthodox psychology..Professor Richet has suggested that they be styled 'meta-psychical phenomena', and that the nascent branch of science..be called for the present 'Metapsychics'. **1922** B. Miall tr. *Maeterlinck's Gt. Secret* xi. 249 The recent researches of Dr. W. Crawford which have made a sensation in the world of metapsychics. **1957** *Encycl. Brit.* XXI. 245/1 A group of investigators..are not prepared to accept the explanation in terms of human survival and..therefore dislike the term spiritualism, preferring to employ some noncommittal term such as metapsychics or parapsychology.

Hence **metapsy·chism**; **metapsy·chist**, a student of metapsychics.

1922 *Glasgow Herald* 27 Oct. 4 Modern spiritualism, under the term Metapsychism, is favourably viewed. **1922** B. Miall tr. *Maeterlinck's Gt. Secret* xi. 216 Our occultists, who are now assuming the name of metapsychists. **1927** *Brit. Jrnl. Psychical Res.* I. vii. 207 It was said that the English spiritualists did not understand the French metapsychists. **1928** *Daily Express* 27 June 6/4 What a palpitating problem for the psychologists and the metapsychists!

metapsychological (metăsəikolǫ·dʒikăl), *a.* [f. next.] Of or pertaining to metapsychology.

1922 J. Strachey tr. *Freud's Group Psychol.* 63 It is much more difficult to give a clear metapsychological representation of the distinction. **1924** W. B. Selbie *Psychol. Relig.* 295 These questions..cannot be settled on psychological or even on metapsychological grounds. **1944** *Scrutiny* XII. 137 The metapsychological stage for the poet is..an attempt..to achieve some sort of compromise between a desire for ultimate belief..and the many confused modern influences which have moulded his sensibility. **1952** E. Weiss in Alexander & Ross *Dynamic Psychiatry* 61 The history of metapsychological concepts, as they developed in the study of psychoanalysis. **1972** M. Schur *Freud: Living & Dying* III. xiv. 375 All these metapsychological explanations can be applied to neurotic anxiety.

metapsychology (metăsəikǫ·lŏdʒi). [f. META- 1 + PSYCHOLOGY.] A name given to speculative inquiry regarding the ultimate nature of the mind and its functions which cannot be studied experimentally.

1909 in *Cent. Dict.* Suppl. **1914** A. A. Brill tr. *Freud's Psychopathol. Everyday Life* 309 We venture to explain in this way the myths of paradise and the fall of man, of God, of good and evil, of immortality and the like—that is, to transform *metaphysics* into *meta-psychology*. **1946** J. H. Masserman *Princ. Dynamic Psychiatry* 285/1 *Metapsychology*, a psychological theory that cannot be verified or disproved by observation or reasoning. **1970** *Jrnl. Gen. Psychol.* LXXXIII. 71 Those who like to conceptualize their data from the point of view of psychoanalytic metapsychology. **1970** H. F. Ellenberger *Discovery of Unconscious* x. 754 He [*sc.* L. Daudet] also wrote nonfiction about daydreams and human personality, notably on the ego and the Self, and he called his own psychological system a metapsychology.

metapsychosis (metăsəikōᵘ·sis). Pl. **-oses.** *rare.* [f. META- + PSYCHOSIS 2.] The supposed psychic action of one mind upon another.

1885 *Proc. Soc. Psychical Res.* III. 422 It would be a grave retardation of science were it assumed that this strange metapsychosis was a medical curiosity alone. **1902** J. M. Baldwin *Dict. Philos. & Psychol.* II. 668/2 The term metapsychosis has been suggested to designate mental stata especially of the percipient under telepathic conditions. **1945** C. Williams *All Hallows' Eve* vii. 112 The terrible metapsychosis gnawed at him.

metarhodopsin (me:tărodǫ·psin). *Biochem.* Also **meta-rhodopsin.** [f. META- + *RHODOPSIN.] Either of two interconvertible intermediates (the orange metarhodopsin I and the yellow metarhodopsin II) that are formed when rhodopsin is bleached by light, being produced from lumirhodopsin and undergoing spontaneous hydrolysis to retinal and opsin.

1950 G. Wald et al. in *Science* 17 Feb. 180/1 If the solution of lumi-rhodopsin is warmed to about −20°C, a further change occurs in darkness. The absorption band shifts another 7–9 mμ toward the blue, with little further change in height or shape... We shall call this second product meta-rhodopsin. **1970** G. S. Brindley *Physiol. Retina* (ed. 2) i. 11 Figure 1.4 shows the effect of pH on the absorption spectrum of metarhodopsin at 3·2°C. Increase of temperature favours the acid form ('metarhodopsin II'). *Ibid.*, It seems that when rhodopsin is bleached by light the alkaline form ('metarhodopsin I') is always formed first, and this passes quickly into the equilibrium mixture appropriate to the pH and temperature. **1975** [see *LUMIRHODOPSIN].

metarule (me·tărūl). [f. META- 1 + RULE *sb.*] A convention or universal rule in a symbolic system, esp. a linguistic system.

1957 *Encycl. Brit.* XIV. 319/1 The inference schemas which make up the calculus are never confused with the meta-rules for their reduction to one another. **1966** J. R. Ross in Reibel & Schane *Mod. Stud. in English: Readings in Transformational Gram.* (1969) 289, I propose that the following metarule, or convention, be added to the theory of grammar. **1970** *Language* XLVI. 28 Then objective guidelines are needed, meta-rules, to decide on the priority of application. **1972** W. Labov in Stockwell & Macaulay *Ling. Change & Generative Theory* 160 We can then argue for an alpha-switching meta-rule which reverses the direction of the nasality constraint.

metascience (me·tăsəiĕns). [f. META- 1 + SCIENCE.] (See quot. 1938.) So **metascienti·fic** *a.*, of or pertaining to metascience; **metascienti·fically** *adv.*

1938 C. W. Morris in *Internat. Encycl. Unified Sci.* I. i. 69 We may introduce the term 'metascience' as a synonym for 'the science of science'. **1953** *Brit. Jrnl. Psychol.* XLIV. 52 [Scientific] problems are periodically discarded ..because they are intrusions from 'meta-science' and depend, therefore, upon emotional bias for working solution. **1954** J. G. Kemeny in P. A. Schilpp *Philos. R. Carnap* (1963) 712 Degree of confirmation..is used meta-scientifically, it is applied to theories. **1965** P. Caws *Philos. of Sci.* iv. 28 The clarification and refinement of concepts..is a metascientific activity. **1971** J. Wiatr in R. Klibansky *Contemp. Philos.* IV. 308 Historical materialism, defined as philosophical meta-scientific reflection of the greatest importance for all the social sciences.

metasequoia (metăsĭkwoi·ă, -sīkoi·ă). [mod. L. (S. Miki 1941, in *Jap. Jrnl. Bot.* XI. 261), f. META- + SEQUOIA.] A deciduous, coniferous tree of the genus so called, belonging to the family Pinaceæ and known only from fossil remains until the single living species, *Metasequoia glyptostroboides*, was discovered in the Szechuan province of China in 1941; also called dawn redwood or water fir. Also *attrib.*

1948 *Science* 6 Feb. 140/2 This living *Metasequoia* is a large tree... It is deciduous, the trees being leafless in the winter months. **1948** *Jrnl. N.Y. Bot. Garden* XLIX. 204 (*caption*) The landlord of the property where the second lot of *Metasequoia* trees was discovered. **1950** *Ecology* XXXI. 262/2 The general region within which lies the home of Metasequoia is a mountainous tableland cut by deep valleys, southeast of the Yangtze River. **1960** *Times* 19 Sept. 5/3 The 'living fossil' tree, *Metasequoia*. **1971** *Country Life* 2 Sept. 553/3 This border has a backing of large shrubs that include bamboos..and is dominated by the metasequoia.

metasomatic, *a.* Add: Hence **metasoma·tically** *adv.*

1921 *Geol. Mag.* LVIII. 553 The great eastern mass of rauhaugite is metasomatically derived from sövite. **1965** *Amer. Mineralogist* L. 1485 Hibschite or hydrogrossular.. has long been known from..metasomatically altered anorthosites and pyroxenites.

metasomatize (metăsōᵘ·mătəiz), *v. Geol.* [f. METASOMAT(ISM + -IZE.] *trans.* To change as a result of metasomatism. So **metaso·matized** *ppl. a.*

1942 *Mineral. Abstr.* VIII. 242 Xenoliths of siltstone or sandstone country-rock are metasomatized to granophyre. **1955** *Mineral. Mag.* XXX. 681 The carbonatite contains abundant fragments of fine-grained phlogopite-rock, which can be shown to represent metasomatized fragments of brecciated gneiss. **1963** Metasomatized [see *JOHANNSENITE].

metastability (metăstăbi·lĭti). [f. next, after *stable, stability.*] The property or state of being metastable.

1901 *Jrnl. Physical Chem.* V. 270 (*heading*) The metastability of the Weston cadmium cell. **1914** W. Rosenhain *Introd. Study Physical Metall.* xiv. 328 The metallic state, in the conditions which prevail at the surface of the earth, is for the majority of metals a state of chemical meta-stability. **1924** *Nature* 13 Dec. 859/2 Atomic theory would suggest that these lines were connected with a faint metastability of certain possible quantum states of the atoms. **1950** *Engineering* 3 Feb. 137/3 This type of undercooling is made possible only by the metastability of iron carbide. **1968** C. G. Kuper *Introd. Theory Superconductivity* i. 4 The latter is a small and very simple device which exploits the metastability of superconducting persistent currents.

metastable (metăstēⁱ·b'l), *a.* [Irreg. f. META- + STABLE *a.*, as tr. of G. *metastabil* (coined in sense 1 by W. Ostwald, in *Lehrb. d. allgemeinen Chem.* (1893) II. I. 517).] **1.** Of a physical system: persisting (in its existing state) when undisturbed or subject to disturbances smaller than some small or infinitesimal amount, but passing to a more stable state when subject to greater disturbances.

1897 *Jrnl. Chem. Soc.* LXXII. II. 309 The author gives to the above form of instability, where the change to solid can only be caused by the presence of the solid itself, the name of 'metastabile' [*sic*]. *Ibid.* 777/1 (Index), Equilibrium, metastable and labile. **1899** J. Walker *Introd. Physical Chem.* xi. 101 A supercooled liquid may be kept for a very long time without any solid appearing, but as soon as the smallest particle of the substance in the more stable solid phase is introduced, the less stable, or, as it has been called, the metastable phase is transformed into it. **1940** Glasstone *Text-bk. Physical Chem.* vi. 459 The term metastable is used to describe a definite equilibrium, which is nevertheless not the most stable equilibrium at the given temperature; a metastable system undergoes spontaneous change on the addition of the stable phase. **1941** J. H. Keenan *Thermodynamics* xxiii. 404 A system is in a state of metastable equilibrium if for all infinitesimal possible variations $\Delta S)_E < 0$, that is, $\Delta E)_S > 0$, while for some finite possible variations $\Delta S)_E > 0$, that is, $\Delta E)_S < 0$. A marble at rest at the bottom of the higher of two depressions in a continuous surface is an example of metastable equilibrium. **1954** *Electronic Engin.* XXVI. 60/1 The 11 points *a, c, e,...t, u* are the stable positions corresponding to the digits 0, 1, 2,...9 and the reset point [of the counting tube], whereas the 10 points *b, d, f,...* are metastable. Suppose the anode current is at the point corresponding to *b* and a small disturbance shifts the beam to the left... The beam will then move farther away..and soon reach the point corresponding to *c* at which it will remain despite any small disturbances. **1968** C. G. Kuper *Introd. Theory Superconductivity* i. 11 Long-lived metastable states cannot be rejected *a priori* as impossible. Diamond at room temperature shows no tendency to graphitize, nor glass to crystallize. But in all cases like these the excess entropy is frozen in by the *immobility* of the atoms at low temperatures.

2. *Physics.* Of an excited state of an atom, nucleus, or other quantum-mechanical system: having an exceptionally long lifetime because the transitions to states of lower energy are forbidden transitions. [Introduced in this sense (in Ger.) by Franck & Knipping 1919, in *Physik. Zeitschr.* XX. 485/2.]

1922 A. D. Udden tr. *Bohr's Theory of Spectra* III. 86 These experiments showed that the impact of electrons could bring helium into a 'metastable' state from which the atom cannot return to its normal state by means of a simple transition. **1942** J. D. Stranathan *'Particles' of Mod. Physics* xi. 455 It seems that certain nuclei can exist for considerable time in an excited state, known as a metastable state. **1961** Powell & Crasemann *Quantum Mech.* xii. 458 The ground state of the orthohelium system ..is therefore stable with respect to optical transitions of the usual kind, and has a correspondingly long lifetime. For this reason, it is frequently referred to as a metastable state. **1973** *Sci. Amer.* Feb. 93/2 The singlet and triplet metastable states have long lifetimes because they have no means of radiating their energy.

metastably (metăstēⁱ·bli), *adv.* [f. prec. + -LY².] In a metastable state.

1938 A. Johannsen *Descr. Petrogr. Ign. Rocks* IV. 401 Clinopyroxenes..may form metastably at lower temperatures. **1963** W. A. Deer et al. *Rock-Forming Min.* IV. 181 The higher temperature forms..can exist metastably below their inversion temperatures. **1973** A. D. Edgar *Exper. Petrol.* iii. 61 For some compositions it may be necessary to avoid certain very reactive components, e.g. α-cristobalite, which is known to persist metastably well below its equilibrium stability field.

Metastasian (metăstēⁱ·siăn, -ziăn), *a.* [f. the name of Pietro *Metastasio* (1698–1782), Italian poet and librettist + -AN.] Of or pertaining to P. Metastasio or characteristic of his works.

1947 A. Einstein *Mus. Romantic Era* ix. 89 Canzonets on Metastasian texts. **1961** *Times* 16 Feb. 16/7 The Metastasian masterpiece, *Il Zio*. **1963** *Listener* 14 Feb. 313/2 Dissatisfied with the Metastasian libretto, he looked to Paris for an alternative. **1964** *Conc. Oxf. Dict. Opera* 259/2 It was against the abuses to which Metastasian opera lent itself—above all the halting dramatic progress, with the plot continually arrested to make way for demonstrations of vocal skill—that Gluck rebelled.

metastasize (metæ·stăsəiz), *v. Path.* [f. METASTAS(IS + -IZE.] *intr.* Of a disease, esp. a tumour: to pass from one part or organ to another; to undergo metastasis (sense 2 a).

1907 *Jrnl. Med. Res.* XVII. 187 As might be expected the tendency to metastasize is much greater in certain tumors than in others. **1947** *Nature* 4 Jan. 15/1 Many prostatic cancers which have metastasized stand still or disappear after the removal of the patients' testicles. **1974** Passmore & Robson *Compan. Med. Stud.* III. xix. 14/2 This tumour invades neighbouring structures, advances along nerves and often recurs after local removal. It may eventually metastasize to distant organs.

metate. (Earlier and later examples.)

1834 in *Southwestern Hist. Q.* (1942) XLV. 330 Mrs. Roark had a Mexican utensil for grinding corn, called a *metate*. It was a large rock which had a place scooped out of the center that would hold a peck of corn. **1847** W. S. Henry *Campaign Sk. War with Mexico* 134 The eldest was on her knees at the medatstone, grinding corn. **1932** E. Wilson *Devil take Hindmost* xviii. 199 In the caves there are only the metates left—the big stones on which they [*sc.* the Indians] ground their meal. **1972** *Sci. Amer.* May 89/3 The *manos*, or stone rollers, and *metates*, or shallow stone troughs, that are used together to grind maize. **1975** *Ibid.* Jan. 100/3 Other fragments showed evidence of splitting, indicating that the grain had been prepared not by pounding but by being rolled back and forth on a stone *metate*.

metathesis. Add: **1. b.** (Further examples.) Also, *quantitative metathesis, metathesis of quantity*, a change of sequence *long vowel + short vowel* to *short vowel + long vowel.*

1891 D. B. Monro *Gram. Homeric Dial.* (ed. 2) 15 *Metathesis.* This term has been employed to explain a number of forms in which a short vowel is lost before a liquid, and the corresponding long vowel follows the two consonants thus brought together: as ἔμ-βλή-την *met*, Mid. βλῆ-το *was struck* (βαλ-, βέλ-ος). *Ibid.* 51 G. Curtius.. made the counter-supposition that..the successive steps might be ὁράοντες, ὁρώοντες and (by metathesis of quantity) ὁρόωντες. **1901** H. Oertel *Lect. Study Lang.* 227 What goes under the name of 'quantitative metathesis' in Greek hardly belongs here. The 'metathesis' is confined to adjacent vowels, and the quantitative increase of the second vowel is probably compensative in its nature, going parallel with a quantitative reduction of the first. **1931** B. F. C. Atkinson *Gr. Lang.* 206 Another curious phenomenon of the metre is the metathesis of quantities or shortening of a long vowel or diphthong before hiatus. **1933** C. D. Buck *Compar. Gram. Gr. & Lat.* 93 Long vowels are shortened before other vowels in various dialects... When the second vowel is short it may be lengthened, resulting in what is known as 'quantitative metathesis'. **1959** A. Campbell *Old Eng. Gram.* 184 By full metathesis a consonant moves from immediately before a vowel to immediately after it, or the reverse. *Ibid.* 185 More usual than full metathesis is reversal of order in consonant groups, so that e.g. *sk > ks, sp > ps.* **1973** A. H. Sommerstein *Sound Pattern Anc. Gr.* 70 Another non-occurring vowel sequence is [ɛ̄ō]. Quantitative Metathesis can be extended without difficulty to modify this sequence by shortening the first vowel (there is no need to lengthen the second, it being already long).

metathesize (metæ·þĭsəiz), *v.* [f. METATHESIS 1 b + -IZE.] **a.** *intr.* To undergo metathesis. **b.** *trans.* To transpose (sounds, etc.) by metathesis; to subject (a word, etc.) to metathesis of some of its constituent sounds. Hence **meta·thesized** *ppl. adj.*

1920 H. C. Wyld *Hist. Mod. Colloq. Eng.* 300 In Received Standard we use many metathesized forms, such as *wright* O.E. *wyrhta*. **1946** E. A. Nida *Morphol.* ix. 180 A preconsonantal *y* metathesizes with the following consonant. **1951** *Traditio* VII. 411 About the second theme, *-ferth*, there can be little room for disagreement. It is obviously a metathesized form of *frith* (peace) which occurs in many Germanic names. **1959** M. Schlauch *Eng. Lang. in Mod. Times* vi. 161 There are instances [in Northern dialects] of metathesised *r*, as in *gert* for *great*. **1968** F. G. Lounsbury in J. A. Fishman *Readings Sociol. of Lang.* (1968) 52 Cayuga has many metathesized pairs [of syllables] in similar alternation, where the metatheses affect the relative order of vowels and the glottal stop. **1971** *Canad. Jrnl. Ling.* Fall 5 The only way to save the hypothesis of sibilant dissimilation would be to posit some rule metathesizing /z/ and /s/, and there appears to be no real evidence for voicing metathesis in modern Greek dialects. **1973** A. H. Sommerstein *Sound Pattern Anc. Greek* ii. 30 Send them first to /dz/; then we will need to metathesize this to [zd].

metatony (metæ·tŏni). *Linguistics.* Also **metatonie.** [ad. F. *métatonie* (F. de Saussure 1894, in *Mémoires de la Société de Linguistique de Paris* VIII. 429), f. META- + TONE *sb.* + -Y³.] In Baltic and Slavonic languages, one of certain kinds of substitution of one distinctive intonation for another in a given syllable; these substitutions collectively. Hence **metato·nic** *a.*

1936 *Trans. Philol. Soc.* 17 The long monophthongs, which have now a rising intonation, are abnormal. Except in loan words, they are due to various obscure causes, known collectively as *metatonie*. *Ibid.* 29 But *metatonie* in the Slavonic period (i.e. *nowocyrkumfleksowa*) may be present in some of these cases. **1949** Entwistle & Morison *Russian & Slavonic Languages* iv. 74 The process by which acutes change to circumflexes and *vice versa* is known as metatony. **1957** C. S. Stang *Slavonic Accentuation* 21 Thus neo-acute does not arise as a result of metatony, if by this term is implied a change of intonation within one and the same stressed syllable. *Ibid.* 23 An assumption based on the behaviour of metatonic circumflex. **1960** W. K. Matthews *Russ. Hist. Gram.* ii. 42 The Russian form *voróna*..illustrates a shift of stress forward in the word, which is known as progressive metatony and is the outcome here of the greater energy of the acute accent as compared with that of the circumflex. *Ibid.* vi. 99 The modern instances of metatony or shift of stress.. have parallels in the fourteenth century. **1965** G. Y. Shevelov *Prehist. of Slavic* xxxiii. 532 He [*sc.* Kozwadowski] called [the] whole phenomenon of their rise and all the changes associated with it metatony... The original idea of metatony as a pitch mutation resulting in the appearance of the two new intonations was a typical product of the Neogrammarian approach. *Ibid.* 533 Metatonic changes are most marked in those Sl[avic] languages or dialects which preserve distinctive (phonemic) pitch. **1973** T. Mathiassen in A. Ziedonis et al. *Baltic Lit. & Linguistics* 165, I am inclined to assume for such cases..only the working of metatony, which involves merely a shift of *intonation* from circumflex to acute.

Metatron (mĕtæ·trǫn). [Etym. uncertain.] In Jewish theology, a supreme angelic being, usually identified with either Michael or Enoch.

1865 C. D. Ginsburg *Kabbalah* 27 The Kabbalistic description of *Metatron* is taken from the Jewish angelology of a much older date than this theosophy... Metatron ..is the Presence Angel..the visible manifestation of the deity. **1904** *Jewish Encycl.* VIII. 519 *Meṭaṭron*,..name of an angel found only in Jewish literature... Meṭaṭron combines various traits derived from different systems of thought. **1914** *Encycl. Relig. & Ethics* VII. 625/1 This mediator is usually known as Meṭaṭron... Meṭaṭron appears in a praiseworthy capacity in his relations with Moses... Michael and Metatron are sometimes interchanged. **1941** G. G. Scholem *Major Trends Jewish Mysticism* ii. 66 The Metatron mysticism which revolves round the person of Enoch..who..was raised..to the rank of first of the angels. **1959** D. D. Runes *Conc. Dict. Judaism* 168/2 *Metatron*, the highest angel, identified with the archangel Michael, or with Enoch who was transformed into a heavenly being; a mystic figure. **1967** D. T. Kaufman *Dict. Relig. Terms* 310 *Metatron*, in Jewish belief, a supreme angelic being.

metatrophic (metătrǭᵘ·fik), *a. Biol.* [ad. G. *metatroph* (A. Fischer *Vorlesungen über Bakterien* (1897) v. 47), f. META-: see TROPHIC *a.*] Needing the presence of organic substances for nutrition.

1900 A. C. Jones tr. *Fischer's Struct. & Functions Bacteria* v. 48 The metatrophic bacteria..cannot live unless they have organic substances at their disposal. **1902** *Encycl. Brit.* XXV. 439/1 Some Angiosperms are persistently heterotrophic, either as saprophytes (metatrophic) or as parasites (paratrophic), and this manner of life involves some kind of symbiosis. **1930** S. Thomas *Bacteriol.* (ed. 2) v. 75 Metatrophic bacteria..include the bacteria causing fermentation, decay, and the decomposition of fatty substances.

Metawileh (metă·wile). Also **Matawila, Metawaileh, Metawala,** etc. [ad. Arab. *matāwila*, pl. of *mutawālī* one who professes to love 'Alī.] Name of a sect of the followers of 'Alī who live in Lebanon and Syria.

1799 W. G. Browne *Travels* 406 Some precaution was necessary against the *Metaweli*, Mohammedans of the sect of Ali, who once formed a powerful and ferocious tribe. **1845** *Encycl. Metrop.* XXV. 395/1 'Alí, whose followers, here called Motewális, or Mutewellis, are persecuted by the Sunnis. *Ibid.*, This word..has been variously spelt Metwali, Motwâli, &c. **1860** D. Urquhart *Lebanon* I. xi. 94 The Metuali race... In religion..are Shiites, in race Arabs. *Ibid.* 96 The Metuali..do not exceed 60,000. **1875** *Encycl. Brit.* III. 177/1 In 1400 it [*sc.* Baalbec] was pillaged by Timur..and afterwards it fell into the hands of the Metaweli, a barbarous predatory tribe. **1888** *Ibid.* XXIII. 711/1 The present town [of Tyre] has arisen since the Metâwila occupied the district in 1766. **1900** G. Bell *Let.* 8 June (1927) I. 119 The Metawaileh (he belonged to that peculiar Muhammadan sect—please note that in the plural the accent is on the second syllable—Metâwaileh). **1909** T. E. Lawrence *Let.* 2 Aug. (1938) 68 Nabatiyeh, the Metawileh (Xtian) Headquarters in the hills. **1917** —— *Let.* 10 July (1938) 231 The Metowala [*sic*] of the Jebel Amr. **1926** —— *Seven Pillars* (1935) v. lviii. 331 On the higher slopes of the hills clustered settlements of Metawala, Shia Mohammedans from Persia generations ago. **1969** H. Z. Hirschberg in A. J. Arberry *Relig. Middle East* II. xviii. 340 A harsh judgement awaits the Mutāwila (the adherents of 'Alī).

Metaxa (metæ·ksă). Also **metaxa, -as.** In full, *Metaxa brandy.* A proprietary name for a dark Greek brandy; a drink or glass of this brandy.

1940 H. J. Grossman *Guide to Wines, Spirits & Beers* xviii. 184 A great deal of brandy is distilled in Greece... A sweet, dark, resinous brandy, which is quite popular and sells in fair quantities, is Metaxa. **1948** S. Ellin *Dreadful Summit* (1958) ix. 91 Dr Cooper said to the bartender, '..Make it two Metaxas.' **1962** K. Royce *Night Seekers* v. 69 Spurling ordered Turkish coffee..and a bottle of Metaxa. **1966** 'A. Hall' *9th Directive* xxiii. 213, I went into the bar and drank a Greek Metaxa brandy. **1967** E. Ambler *Dirty Story* I. iii. 25, I sipped Metaxa. **1974** C. Spencer *How Greeks kidnapped Mrs Nixon* x. 68 Kalvos gave them a bottle of Metaxas brandy. **1975** *Times* 6 Sept. 5/4 Metaxas is the big name in Greek brandy.

metaxenia (metăzī·niă). *Bot.* [f. META- + XENIA.] (See quots.)

1928 W. T. Swingle in *Jrnl. Heredity* XIX. 263/1 In order to be able to refer to this newly discovered action of pollen on the ovarial tissues of the mother plant..I proposed at Phoenix, Arizona, in February, 1926, the term metaxenia to designate the action of the male parent on tissues of the mother plant outside of the embryo and endosperm. The term metaxenia is formed..by analogy with the term xenia which denotes the influence exerted by the male parent on the endosperm of the seed. **1949** Darlington & Mather *Elem. Genetics* 426 Xenia, the effect of more distantly related, as contrasted with more closely related, pollen on the maternal tissue of a fruit... Metaxenia has been used for true xenia. **1973** Proctor & Yeo *Pollination of Flowers* xi. 351 The pollinations which produce larger seeds yield dates with more flesh (i.e. maternal tissue), the pollen having influenced the flesh indirectly ('metaxenia') through its direct genetic effect (xenia) on the seed.

metazoa, *sb. pl.* Add to etym.: (E. Haeckel 1874, in *Jenaische Zeitschr. Naturw.* VIII. 10). (Later examples.)

1940 L. H. Hyman *Invertebrates* I. v. 249 Even the simplest Metazoa are two-layered or diploblastic. **1963** A. Remane in E. C. Dougherty et al. *Lower Metazoa* ii. 26 No biologist regards *Protohydra* as the most primitive metazoön.

metazoan, *a.* and *sb.* (Later examples.)

1940 L. H. Hyman *Invertebrates* I. v. 250 The ancestral metazoan would not have required a digestive sac or mouth. *Ibid.* 252 The three lowest metazoan phyla, the Porifera, the Cnidaria, and the Ctenophora, are commonly stated to have remained at the gastrular level of construction. **1965** B. E. Freeman tr. *Vandel's Biospeleol.* vii. 64 Those Metazoan groups which are almost exclusively marine are represented underground by very few species. **1971** *Nature* 28 May 260/1 Gravity corers are widely used for the collection of the smallest marine metazoans (meiofauna) from subtidal grounds.

metempsychosic (mete:mpsəikōᵘ·sik), *a.* [f. Metempsychos(is + -ic.] Relating to metempsychosis.

1905 W. J. Locke *Morals M. Ordeyne* xviii. 217, I have bemused myself with gnostic and metempsychosic speculations.

metencephalic, *a.* (Example.)

1899 *Proc. Zool. Soc.* 1024 The metencephalic fossa of the Pygopodes.

meteor. Add: **6. d.** meteor bumper *Astronautics*, a structure on the outside of a spacecraft that serves to protect it from the impacts of meteoroids; **meteor trail**, a bright streak of ionized gas formed by a meteor passing through the upper atmosphere, which can provide a reflector for radio communication.

1951 *Jrnl. Brit. Interplanetary Soc.* X. 275 For space station lifetimes of more than one year, it is concluded that either a heavily-armoured hull or a 'meteor bumper' will be required. **1960** *Aeroplane* XCVIII. 680/2 Dorsey described the space-laboratory as a double shell, three compartment, cylinder with convex end domes. The outer shell, made of beryllium, would serve as a 'meteor bumper' and thermal shield. **1962** F. I. Ordway et al. *Basic Astronautics* xi. 450 A double-wall meteor bumper in which the outer wall serves to break up the impacting meteor. **1930** R. H. Baker *Astron.* vi. 246 (*caption*) Meteor trail. **1935** —— *Introd. Astron.* ix. 195 It is estimated that a single observer can see an average of ten meteor trails in the course of an hour on clear, moonless nights. **1958** *Jrnl. Atmospheric & Terrestrial Physics* XII. 329 At an early stage it became apparent that reflections from meteor trails play an important part in v.h.f. ionospheric forward scattering. **1968** *Radio Communication Handbk.* (ed. 4) xii. 17/2 There are times when the more orthodox modes of propagation fail to provide a satisfactory means of communication and one or other of the various forms of propagation by scatter may then offer a useful alternative: these include tropospheric scatter, ionospheric scatter..and meteor-trail scatter.

meteorette (mī:tĭ,ǫre·t). [f. Meteor + -ette.] A small meteor.

1876 *Gentl. Mag.* XVI. 552 Specks flitting like meteorettes over the crests of the billows.

meteoric, *a.* Add: **2. a.** (Later examples.)

1965 G. J. Williams *Econ. Geol. N.Z.* viii. 94/2 Morgan (1924) had difficulty in visualizing how meteoric solutions could descend to the depth at which veins have been found. **1969** C. Ollier *Volcanoes* ii. 17 The water emitted by hot springs is usually of meteoric origin, that is derived from rain, and only a very minor part is likely to be derived from magma.

meteorically, *adv.* Add: **b.** With the suddenness and speed of a meteor.

1916 A. S. Neill *Dominie's Log* xiv. 154 She dons the bridal white, and at once she rises meteorically in the social scale. **1955** *Sci. Amer.* Apr. 38/3 Meanwhile they also spread through the Straits of Mackinac into Lake Michigan and increased meteorically there. **1973** *Current Hist.* May 204/1 Born in 1926, he had joined the Communist party in 1943, fought with the Partisans, and risen meteorically in the Croatian Communist party.

meteorite. Add: (Earlier example.)

1824 *Phil. Mag.* LXIV. 113, I shall..pass in review..the principal new facts..respecting igneous meteors and meteorites, which have been made known..during the year 1823.

Hence **meteori·tical** *a.*, **meteori·tically** *adv.*

1919 Beerbohm *Seven Men* 5 At the end of Term he settled in—or rather, meteoritically into—London. **1939** *Pop. Astron.* XLVII. 328 The Editor ventures the following suggestions relative to three..meteoritical terms. **1946** *Ibid.* LIV. 430 These amendments went into effect at the adjournment of the meeting on September 10; whereupon the name of the Society was officially changed from 'The Society for Research on Meteorites'..to 'The Meteoritical Society'. (It is hardly necessary to add that *meteoritical* is merely the adjective of *meteoritics*, defined as 'the science of meteorites and meteors'.) **1950** *New Mexico Q.* Autumn 270 The Institute of Meteoritics, the first scientific organization, at least in the English-speaking world, devoted to meteoritical research. **1974** *Geotimes* Mar. 8, Aug. 7–9 Meteoritical Society, ann. mtg, Los Angeles.

meteoritics (mī:tĭ,ǫri·tiks), *sb. pl.* (const. as sing.). [f. Meteorit(e + -ics, as ad. Russ. *meteoritika* (Yu. I. Simashko 1889, in *Niva* XX. 82/2).] The scientific study of meteors and meteorites.

1934 in Webster. **1946, 1950** [see *meteoritical *a.*]. **1952** *Jrnl. Brit. Interplanetary Soc.* XI. 243 The astronomer in particular is interested in the composition of meteorites as chemical samples of the universe... The techniques of investigation are, however, not astrophysical but rather mineralogical and metallurgical, and it is hoped that the coining of the new term 'meteoritics' for the subject..may signify an increased co-operation between astronomers and the laboratory investigators. **1960** I. Vidziunas tr. *Krinov's Princ. Meteoritics* i. 29 One of the most important areas of meteoritics is the study of the motion of meteoric bodies in the Earth's atmosphere and the observation of bolides. **1963** *Nature* 5 Jan. 38/1 Meteoritics is a young science. Only since around 1945 has extensive research been undertaken. **1975** *Sci. Amer.* Jan. 29/1 In recent years the discipline of meteoritics has moved beyond the taxonomic stage, and sound geochemical and physical reasoning has been applied in interpreting the masses of data.

Hence **meteori·ticist**, an expert in meteoritics.

1952 *Jrnl. Brit. Interplanetary Soc.* XI. 243 The first problems that meteoriticists (!) must tackle are the location and classification of falls, and a statistical analysis of the information so obtained. **1971** *New Scientist* 3 June 556/1 From studies of the amount of xenon present in various meteorites, meteoriticists had speculated that Pu-244 once existed in achondritic meteorites. **1975** *Sci. Amer.* Jan. 29/2 Some meteoriticists boldly construct multistage scenarios of condensation, agglomeration, accretion, heating, metamorphism and differentiation to explain the accumulated facts.

meteorogram (mī·tĭ,ǫrǒgræm). [f. Meteoro(graph + -gram.] A record furnished by a meteorograph.

1904 *U.S. Dept. Agric. Monthly Weather Rev.* XXXII. 121/2, (1) instrumental errors, (2) errors in exposure of instruments when comparing with standards, (3) errors in reading from meteorograms. **1923** N. Shaw *Forecasting Weather* (ed. 2) v. 144 The trace given by a barometer is called a 'barogram', that by a thermometer a 'thermogram', while a trace of either the direction or force of wind is called an 'anemogram'. When two or more of these traces are all combined in one picture..the whole is called a 'meteorogram'. **1955** W. J. Saucier *Princ. Meteorol. Analysis* xii. 386 (*heading*) 'Meteorograms' of surface weather elements for 27 February through 3 March 1950.

meteorological, *a.* Add: **b.** *Meteorological Office, officer.*

1930 *Daily Express* 6 Oct. 2/4 Mr. M. A. Giblett, M.Sc., meteorological officer, R 101..organised the necessary meteorological services for airship flights to Canada. **1937** *Discovery* Jan. 5/1 For this a Benndorf self-registering electrometer was kindly lent by the Meteorological Office. **1967** *Ships' Code & Decode Bk.* (Meteorol. Office) (ed. 7) 66 Meteorological Officers. **1968** *Measurement Upper Winds by Pilot Balloons* (Meteorol. Office) (ed. 4) 37 The Meteorological Office has produced electric torches for use with pilot balloons. **1975** *Times* 12 Aug. 3/4 The Meteorological Office held a competition..for the best ideas for a new television presentation of weather.

meter, *sb.*³ Add: **1. d.** = *exposure meter.

1920 *Sat. Westm. Gaz.* 22 May 16/2 The golden rule is to expose by meter on the holiday, and leave all else to the return home.

e. *U.S. slang.* (See quots.)

1940 *Music Makers* May 37/2 *Meter*, quarter, twenty-five cents. **1960** Wentworth & Flexner *Dict. Amer. Slang* 337/2 *Meter*, a quarter... From the coin which often is needed to operate a gas meter. Orig. Negro use. Never common. **1970** C. Major *Dict. Afro-Amer. Slang* 81 *Meter*, twenty-five cent coin.

f. = *parking meter.

1960 *Daily Tel.* 31 Mar. 15/1 What promises to be the most important experiment in traffic control starts next Monday, when car parking over the whole of Mayfair becomes subject to meters. **1973** *Weekly News* (Glasgow) 11 Aug. 2/2 After a search, I found a parking place in Glasgow on a Saturday afternoon. I was standing next to the meter..when a man who'd been standing on the opposite pavement came over. **1974** *Times* 21 Jan. 12/8 Since many meters in the Whitehall area have been suspended because of bomb threats, the wardens had a field day.

2. b. Special Combs.: **meter-feeder**, a motorist who illicitly extends his parking time by putting more money in the meter instead of moving away; so **meter-feeding** *vbl. sb.*; **meter maid** orig. *U.S.* (see quot. 1958); **meter-park** *v. trans.*, to park a vehicle at a parking meter; **meter-parking**, the use of parking meters; a place at a parking meter; **meter-reader**, a person responsible for reading gas or electricity meters; **meter-reading**, the reading of a meter or meters; **meter zone**, a limited area where the parking of vehicles is controlled by meters.

1965 *Guardian* 5 Apr. 8/2 The would-be meter feeder is tipped off if a hostile scout appears. **1972** *Ibid.* 8 May 9/6 Scotland Yard..was setting up a special vigilante squad whose job was to keep tabs on meter feeders in Central London. **1966** *Punch* 5 Jan. 15/1 Driving dangerously, you may think, is obviously much more criminal than meter-feeding. But meter-feeding is a deliberately calculated breach of the law, whereas dangerous driving may be accidental. **1969** *Guardian* 28 Mar. 22/4 We realize councils must do something to stop meter feeding. **1957** in *Amer. Speech* (1961) XXXVI. 282 Surveys conducted in cities using 'meter maids' have found that their meter revenue increased. **1958** *Britannica Bk. of Year* 519/2 *Meter maid*,..a woman police official with the task of patrolling metered parking-sites and reporting parking offences. **1968** *Harper's Mag.* Feb. 41 A Meter Maid was soon watching me censoriously. **1970** S. Ellin *Man from Nowhere* xxx. 150 Some meter maid found him when she looked in the car where it was parked uptown. **1970** *Sunday Times* 3 May 28/7 Why do meter maids..never look glamorous at all? **1971** H. C. Rae *Marksman* iii. iii. 212 The Mark 10 was meter-parked close to the side entrance. **1959** *Daily Tel.* 15 Dec. 1/7 In the past month, 81 motorists were convicted of contravening the existing meter parking scheme. **1963** D. B. Hughes *Expendable Man* (1964) iv. 109 There were no meter parkings open near the courthouse. **1963** *Times* 8 Jan. 9/3 In an ideal world, everyone in uniform, from guardsmen to meter readers, would be immaculately turned out and dressed in strict conformity with the rules of their service. **1971** A. Price *Alamut Ambush* x. 123 A duty caller like the meter reader and the postman. **1974** *Times* 18 Apr. 14 Where the occupiers are out when he calls, the meter-reader leaves a card, inviting them to be at home next day. **1957** V. J. Kehoe *Technique Film & Television Make-Up* viii. 96 The middle of the ratio scale on a meter reading. **1959** *Daily Tel.* 15 Dec. 1/7 Larger meter zone by April. **1963** *Sunday Express* 3 Mar. 4/4, I had paid out over £220 in parking fines in the past 21 months because I lived in a 'meter zone'.

meter, *v.* (in Dict. after Meter *sb.*³). Add: Also, to supply through a meter. (Further example.)

1968 *Brit. Med. Bull.* 192/2 It is meaningful to envisage computer power being metered to separate offices.

b. *trans.* To regulate the flow of; to deliver (fluid) in regulated amounts *to*.

1932 *Compression Ignition Engines* xi. 132 Injection Pump.—A mechanical device which meters the liquid fuel in controllable quantities and delivers it at a set pressure to the working cylinders. **1963** *Adv. Space Sci. & Technol.* Suppl. I. 173 In addition to atomizing and mixing the liquids, it meters the flow to the combustion chamber. **1972** *Sci. Amer.* Aug. 21/2 Oil could also be metered to the critical contact region.

2. To measure (the parking-time) of motorists, etc., by means of parking meters; to provide with parking meters. So **me·tered** *ppl. a.*

1957 *New Yorker* 26 Oct. 36/1 You can park scooters almost anywhere without getting a ticket. In a metered area, the usual thing is to park them sidewise, between two parked cars. **1960** *Guardian* 13 Jan. 6/7 When all Mayfair becomes metered..the position of the motorists will be worse than ever. **1961** *Ibid.* 27 June 5/6 A motorist ..may only park his car in a metered parking place. **1962** *Times* 16 Apr. 11/3 Brighton has proposed to meter the sea-front motorists' parking-time. **1963** 'W. Haggard' *High Wire* iii. 24 The comfortable little street..had recently been metered... This was a metered parking ground.

metered (mī·təɹd), *ppl. a.* In the senses of the vb. Cf. prec.

1928 M. Rittenberg *Mail-Order made Easy* xiv. 177 'Metered' mail—where a printed impression is put on the envelope instead of an adhesive stamp. **1957**, etc. [see *meter *v.* 2]. **1968** *Listener* 27 June 851/1 The trouble is that in this country taxi-drivers who work for garages and not for themselves are commonly paid no set wages, but instead a proportion of the metered fare—say, 40 per cent.

metering (mī·təriŋ), *vbl. sb.* [f. Meter *v.* + -ing¹.] Measuring; freq. *attrib.* as *metering-point, pump, station*, etc. Also in other senses of Meter *v.* (in Dict. and Suppl.).

1913 *Wireless World* May 93/2 We must take a point just behind the metering-point..as the beginning of the plant essential to the particular system under consideration. **1935** *Geogr. Jrnl.* LXXXV. 538 At each metering station a sufficient number of discharge measurements is taken. **1935** *Economist* 7 Dec. 1143/1 The development of the hand-operated metering pump affixed to an underground tank at garages..revolutionised the service of distribution. **1959** *Daily Tel.* 24 Dec. 6/2 To improve off-street parking by general metering of the kerbs. **1962** *Rep. Comm. Broadcasting 1960* 264 in *Parl. Papers 1961–2* (Cmnd. 1753) IX. 259 Metering and access-barring apparatus. **1963** *Lancet* 5 Jan. 17/1 In order to do this, a multichannel metering pump was made which consists of laminæ massaging silicone-rubber or plastic tubes. **1973** *Country Life* 20 Sept. 826/1 Full aperture metering gives you a bright clear image of the subject... The FD range consists of fully automatic metering lenses.

metestrus, -um, varr. *metœstrus.

metethereal, var. Metetherial *a.*

1911 T. Flournoy *Spiritism & Psychol.* ii. 57 We participate..in a higher order of phenomena—of a metethereal world, as Myers called it. **1924** T. H. Y. Trotter *Mus. & Mind* I. ii. 28 A metethereal world,..transcendental and spiritual. **1963** *Listener* 24 Jan. 175/3 Esoteric volumes on astral projection, crammed with words like 'metethereal'.

meth (meþ), colloq. abbrev. *Methedrine; also, a Methedrine tablet.

1967 *N.Y. Times* 16 Oct. 53 She was a good kid, if she hadn't been so freaked out on meth. **1968** *Guardian* 21 Nov. 20/3 The pushers passing out bombers, meths, dexes, and blues. **1972** M. J. Bosse *Incident at Naha* i. 38 She snorted Methedrine... The Feeler offered me a snort of Meth, but I never use Speed. **1972** J. Wambaugh *Blue Knight* (1973) v. 74 She's a meth head and an ex-con.

meth, var. *meths.

meth- (meþ), used as comb. form of Methyl before a vowel in a few chemical names, as **me·thacrylic acid**, 1-methylacrylic acid, CH_2:$C(CH_3)COOH$, a colourless compound melting at 15°C, which polymerizes when distilled and which is used in the manufacture of methacrylate resins; so **metha·crylate**, (*a*) a salt of methacrylic acid; (*b*) any of the esters of methacrylic acid, some of which are used in making resins by polymerization (see *methyl methacrylate* s.v. *methyl c); **metho·xide**, (a salt of) the anion CH_3O^-, derived from methanol.

1865 *Jrnl. Chem. Soc.* XVIII. 142 Ethylic methacrylate is readily decomposed by boiling alcoholic potash. *Ibid.*, Silver methacrylate is best obtained pure by precipitating ammonium methacrylate with nitrate of silver. **1936** *Industr. & Engin. Chem.* Mar. 269/1 The monomeric methacrylates cannot be stored or transported with safety. **1962** *Economist* 6 Oct. 74/3 An airdrying paint (based on methacrylates) has been used..for several years. **1964** G. H. Haggis et al. *Introd. Molecular Biol.* iv. 91 (*caption*) A thin section preparation showing T_2 bacteriophage infection of *E. coli* bacteria (osmium tetroxide fixation and methacrylate embedding technique..). **1865** *Jrnl. Chem. Soc.* XVIII. 142 Methacrylic acid is a colourless oil, which..possesses a faint odour, like pyrogallic acid. **1889** G. M'Gowan tr. *Bernthsen's Text-bk. Org. Chem.* 166 Meth-acrylic acid..is found in small quantity in Roman camomile oil, and smells like decaying mushrooms. **1972** *Materials & Technol.* IV. xii. 428 Technical production of methacrylic acid employs the oxidation of isobutene by a mixture of nitric acid and nitrogen dioxide to hydroxyisobutyric acid, which is then dehydrated. **1885** *Jrnl. Chem. Soc.* XLVIII. 1031 Anhydrous sodium methoxide, NaOMe, was prepared by heating the solution [of sodium in methyl alcohol] at 180–200° in a current of dry hydrogen. **1934** *Ibid.* 1204 The agreement between the values for the mobility of the methoxide ion is fair. **1965** *New Scientist* 26 Aug. 513/3 Ordinary methyl alcohol is sufficiently acidic to form salts with alkali metals like sodium and potassium, the 'methoxides', whose general formula is $CH_3O^-M^+$ (where M^+ is any metal ion).

metha: see *metho¹.

methadone (me·þădōᵘn). *Pharm.* orig. *U.S.* Also **-on** (-ǫn). [f. Meth(yl + *a(mino- + d(i-² + -one in 6-dimethylamino-4,4-diphenyl-3-heptanone, the systematic chemical name.] A powerful synthetic analgesic, $(CH_3)_2N{\cdot}CH(CH_3){\cdot}CH_2{\cdot}C(C_6H_5)_2{\cdot}CO{\cdot}CH_2{\cdot}CH_3$, which is similar to morphine in its effects but less sedative and is used (usu. as the hydrochloride) as a substitute drug in the treatment of addiction to morphine or heroin.

1947 *Jrnl. Amer. Med. Assoc.* 23 Aug. 1483 The Council voted to recognise the word Methadon as the generic designation for 6-dimethylamino-4,4-diphenyl-3-heptanone. **1948** *Ann. N.Y. Acad. Sci.* LI. 17 It is not certain whether the Germans succeeded in the complete resolution of *dl*-methadone. **1952** W. T. Salter *Textbk. Pharmacol.* vi. 81/1 The danger of addiction of racemic methadon and levo-methadon is about equal to that of morphine. **1962** K. Orvis *Damned & Destroyed* xxiii. 174 A new drug, methadon, is being used successfully as a reduction treatment... The patient starts regaining strength—moral and physical strength—almost at once. **1967** *Martindale's Extra Pharmacopoeia* (ed. 25) 783/1 Symptoms of morphine withdrawal in addicts may also be mitigated by methadone substitution; addiction to methadone is easier to treat than addiction to morphine. *Ibid.* 800/2 Methadone hydrochloride may be used for the relief of severe pain of all types. **1970** *Sunday Times* 8 Mar. 1/2 She was taking big doses of heroin three times a day, together with the morphine-antagonist, methadone, ..every four hours. **1971** *Guardian* 4 June 3/3 The aim of the methadone programme is to switch the mainliner's addiction from heroin to methadone, an addictive drug.. which allows its addicts to lead a normal life... The methadone programme was begun in New York in 1965 by the Beth Israel Medical Centre. **1972** *Science* 26 May 882/2 Methadone not only reduces the hunger for heroin, it affects respiration, digestion, and sexual behavior; it also, as does any opiate, affects social and psychological behavior.

methamphetamine (meþæmfe·tămīn, -in). *Pharm.* [f. *meth- + *amphetamine.] A methyl derivative, $C_6H_5{\cdot}CH_2{\cdot}CH(CH_3){\cdot}NH(CH_3)$, of amphetamine, used in the form of the hydrochloride, a white crystalline compound, as a stimulant of the central nervous system; its effects resemble those of amphetamine but are more rapid in onset and longer lasting.

1949 *Jrnl. Amer. Med. Assoc.* 22 Jan. 228 New and nonofficial remedies. Methamphetamine hydrochloride... *d*-Desoxyephedrine hydrochloride. **1951** A. Grollman *Pharmacol. & Therapeutics* xi. 212 Methamphetamine hydrochloride..may be considered as the methyl derivative of amphetamine. It is a potent analeptic and, like amphetamine, has been used in the treatment of narcolepsy, postencephalitic parkinsonism, alcoholism, and in certain depressive states..and in the treatment of obesity. **1960** J. J. Lewis *Introd. Pharmacol.* viii. 287 Amphetamine, dexamphetamine and methamphetamine have qualitatively similar properties but vary in potency. Methamphetamine is more potent than dexamphetamine which is in turn more potent than its racemate amphetamine. **1970** C. M. B. Pare in C. W. H. Havard *Current Med. Treatm.* (ed. 3) xv. 479 Following injection of the methamphetamine the doctor directs the interview towards topics which he thinks may be particularly stressful.

methanation (meþănēⁱ·ʃən). [f. Methan(e + -ation.] Conversion (esp. of carbon monoxide and hydrogen) into methane.

1956 M. Greyson in P. H. Emmett *Catalysis* IV. vi. 506 The development of commercial methanation procedures is normally undertaken on the assumption that means exist, or will exist, for the conversion of coal to mixtures of hydrogen and carbon monoxide. **1969** R. F. Lang tr. *Henglein's Chem. Technol.* 425 This reaction is used for the production of especially high-grade fuel gases from water gas ($CO + 3H_2 - 48{\cdot}87$ kcal $= CH_4 + H_2O$) (methanation) ($CO_2 + 4H_2 - 39{\cdot}1$ kcal $= CH_4 + 2H_2O$). **1973** *Nature* 7 Dec. 326/2 It is the catalytic enrichment (a methanation) which is the key to the production of SNG [*sc.* substitute natural gas] and the Gas Board is loath to reveal the full details of its operation.

Hence (as a back-formation) **me·thanate** *v. trans.*, to convert into methane, to subject to methanation.

1963 Dirksen & Linden *Pipeline Gas from Coal by Methanation of Synthesis Gas* (U.S. Inst. Gas Technol. Res. Bull. No. 31) p. iii, This program established that a variety of nickel catalysts were capable of methanating synthesis gas to 900 Btu/SCF pipeline gas. **1974** *Sci. Amer.* Apr. 9/3 Methanating animal wastes from feed-lots in the Middle West is probably the best application of this idea.

methane. Add: Now usu. pronounced (mī·þēⁱn).

methanol (me·þănǫl). *Chem.* [f. Methan(e + -ol.] Methyl alcohol, CH_3OH, a colourless, volatile, poisonous liquid with a pungent odour which is produced mainly by the high-pressure reduction of carbon monoxide or dioxide with hydrogen and is used as an intermediate in the synthesis of formaldehyde, as a solvent, and as a denaturant for ethyl alcohol.

The adoption of *methanol* as the systematic name of methyl alcohol was a consequence of Resolution 15 (on alcohols and phenols) of the report of 1892 of the International Conference on Chemical Nomenclature (see *Nature* (1892) 19 May 58–9).

1894 G. M'Gowan tr. *Bernthsen's Text-bk. Org. Chem.* (ed. 2) 87 Methyl alcohol. [*Note*] 'O[fficial] N[ame]' Methanol. **1932** *Discovery* May 165/2 Besides its important use as a methylating agent, methanol is the raw material for formaldehyde. **1958** *Times Rev. Industry* Dec. 57/2 Refueller..designed for replenishing aircraft with..methanol. **1973** *Daily Tel.* 27 July 2/7 Hawkshaw ..knew that methanol was an anti-freeze substance used on buses and for stripping paint and cleaning paint-brushes.

Hence **methano·lic** *a.*, in or of methanol.

1953 *Jrnl. Polymer Sci.* X. 375 The results of viscosity and molecular weight determination obtained in methanolic solution are presented in Figure 2. **1975** *Nature* 3 Jan. 8/1 Solvolysis of $[\{(C_5H_5)_2Ti\}_2N_2]$ with methanolic HCl yields mainly nitrogen.

methaqualone (meþæ·kwălōᵘn). *Pharm.* [f. *2-methyl-3-o-tolyl-4(3H)-quinazolinone*, the systematic chemical name (with inserted *a*), f. Methyl + Tolyl + *quinazolin(e* (f. Quin(oline + *azol(e* (f. Az(o- + *-ole) + -ine⁵) + -one.] A hypnotic and sedative drug used generally in the form of its hydrochloride, $C_{16}H_{14}N_2O{\cdot}HCl$, a white crystalline powder with a bitter taste.

1961 *Brit. Med. Jrnl.* 21 Jan. 173/1 Methaqualone in a single dose of 150 mg. is a reliable hypnotic. **1969** *Daily Tel.* 17 Dec. 13/1 Methaqualone, a drug used by thousands of people in sleeping tablets and slimming capsules, is to be referred to the Committee on Safety of Drugs. **1973** *N.Y. Times* 10 June 1. 59/1 Nonmedical use quickly spread when users found that methaqualone can produce a tingly, relaxed, uncoordinated drunken euphoria. It also developed a reputation—some say unfounded—as a powerful aphrodisiac. **1973** *Daily Tel.* 11 July 2/8 Mandrax is the trade name for a sleeping tablet otherwise known as methaqualone and has been controlled as a potentially dangerous drug since 1971 following cases of abuse by drug addicts.

Methedrine (me·þĕdrin, -īn). *Pharm.* [f. Meth(yl + *benz)edrine.] A proprietary name for methamphetamine.

1939 *Trade Marks Jrnl.* 11 Jan. 46/2 Methedrine... Pharmaceutical and veterinary preparations. The Wellcome Foundation Ltd.,..London,..manufacturing chemists. **1943** *Brit. Med. Jrnl.* 20 Mar. 348/2 Methedrine has been found to be an effective blood-pressure-raising drug. **1965** J. Pollitt *Depression & its Treatment* iv. 47 Methedrine produces euphoria, mental alertness and pressure of talk. **1967** *Economist* 11 Nov. 628/2 Their deaths were just one instance of the violent crime that is being associated with the increasing use of an amphetamine with the trade name of Methedrine and the nick-name of 'speed'. **1975** *Harpers & Queen* May 126/2 Out come..a half-dozen bottles of Methedrine.

metheglin. (Later examples.)

1789 J. Morse *Amer. Geogr.* 197 The perry..is an agreeable liquor, having something of the harshness of claret wine, joined with the sweetness of metheglin. **1839** J. K. Townsend *Narr. Rocky Mts.* v. 203 Among the rest, was some *metheglen* or diluted alcohol sweetened with honey. **1855** T. C. Haliburton *Nat. & Hum. Nat.* II. ix. 267 All the friends of the new married couple..did nothing for a whole month, but smoke, drink metheglin, [etc.]. **1971** J. Doxat *Drinks & Drinking* 152 *Metheglin*, old name for mead, but probably, from its derivation, from an ancient Welsh word for 'spicey potion'.

methenamine (meþī·n-, mĕþe·nămīn). *Pharm.* [f. Methen(e + Amine.] The name given to *hexamethylenetetramine* (s.v. *hexa-) in the *Pharmacopeia of the U.S.A.*

1926 *Pharmacopeia U.S.* 238 Methenamine. **1926** *Dispensatory U.S.A.* (ed. 21) 695/2 Not only are there a large number of compounds of methenamine upon the market, but the official drug is sold under an almost innumerable number of trade names. **1948** E. Pound *Pisan Cantos* (1949) lxxiv. 17 Methenamine eases the urine. **1951** A. Grollman *Pharmacol. & Therapeutics* xxv. 526 Hexamethylenetetramine ($(CH_2)_6N_4$), official as methenamine (U.S.P.) and hexamine (B.P.)..is of interest from its liberating formaldehyde in the course of its excretion in the urine. **1969** J. H. Thompson in J. A. Bevan *Essent. Pharmacol.* xlix. 511 Methenamine mandelate is a mixture of methenamine and mandelic acid.., two old urinary antiseptic agents which today are rarely given separately.

methicillin (meþisi·lin). *Pharm.* [f. *meth- + *pen)icillin.] A penicillin, $C_{17}H_{19}N_2O_6NaS{\cdot}H_2O$, which is especially useful for its activity against staphylococci which produce penicillinase. Also called *methicillin sodium* or *sodium methicillin*.

1961 *Brit. Med. Jrnl.* 25 Mar. 865/1 During a period of 12 weeks one ward was sprayed at regular intervals daily with 4g. of methicillin. **1963** *Brit. Pharmacopœia* 486 Methicillin Sodium is the monohydrate of sodium 6-(2,6-dimethoxybenzamido)penicillinate. **1968** *New Scientist* 18 Jan. 118/3 There were signs of bacterial infection in the lungs, and the doctors prescribed methicillin. **1970** *Daily Tel.* 23 Oct. 13/3 The second [semi-synthetic penicillin], methicillin,..was active against the resistant staphylococci and was soon accepted as a major life-saving antibiotic.

methionine (meþəi·ǒnīn). *Chem.* [Blend of *meth- + Thion-: see -ine⁵.] A sulphur-containing amino-acid, $CH_3{\cdot}S{\cdot}CH_2{\cdot}CH_2{\cdot}CH(NH_2){\cdot}COOH$, which is probably a constituent of all proteins.

1928 Barger & Coyne in *Biochem. Jrnl.* XXII. 1418 Since the amino-acid has a good title to be regarded as a constituent of protein, a shorter name than γ-methylthiol-α-aminobutyric acid seems desirable, and..we suggest for it the name *methionine*, in allusion to the characteristic grouping. **1956** *Nature* 11 Feb. 280/1 The amino-acid requirement of this organism [*sc.* a trypanosomid flagellate] could be satisfied by methionine alone. **1962** [see *isoleucine]. **1971** J. Z. Young *Introd. Study Man* xxiii. 315 Old people are said to require more methionine than young. **1972** *McGraw-Hill Yearbk. Sci. & Technol.* 123/1 Since methionine is the only metabolic precursor of homocysteine in the human diet, deliberate reduction in the amount of dietary animal protein, which contains more methionine than plant protein, may prove to be effective in preventing arteriosclerosis.

methisazone (meþi·sæ̆zōᵘn). *Pharm.* [f. Meth(yl + Is(atin + *thiosemicarb)azone (cf. the chemical name given in def.).] A fine orange-yellow powder that has prophylactic activity against smallpox; *N*-methylisatin β-thiosemicarbazone, $C_{10}H_{10}N_4OS$.

1964 *Brit. Med. Jrnl.* 5 Sept. 621/1 Methisazone breaks new ground in the rational therapy of virus infections. **1966** *New Scientist* 17 Nov. 336/3 Dr D. J. Bauer and Dr K. Apostolov..had shown methisazone ('Marboran') to be active against certain adenoviruses growing in tissue culture. **1970** Passmore & Robson *Compan. Med. Stud.* II. xx. 36/1 In man, methisazone has been shown to be effective in preventing disease in contacts of patients suffering from smallpox... Methisazone has not been shown to be effective in treating established infections.

metho¹ (me·þo). *Austral.* and *N.Z. colloq.* Also **metha** (*rare*). [f. *meth(ylated spirit* + *-o.] **1.** Methylated spirit. Also *attrib.*

1933 *Bulletin* (Sydney) 18 Oct. 10/2 A metho. drinker—a regular visitor—came into the pharmacy the other day. *Ibid.* 1 Nov. 11/1 Charged in a Melbourne court with having drunk metho., an elderly gent. at first declared [etc.].

1935 K. Tennant *Tiburon* ii. 23 The two metho-drinkers were escorted out firmly. **1938** 'R. Hyde' *Nor Years Condemn* x. 190 They drink a lot of metha, don't they? Great names they've got for it—Phar Lap, and Johnny Gee, and White Light. **1944** W. E. Harney *Taboo* (ed. 3) 33 She is a 'metho' addict and, as a result, is blind and half paralysed. **1945** O. E. Burton *In Prison* 116 A half-crazed sponge for absorbing 'metho'. **1953** A. Upfield *Murder must Wait* xvii. 153 The straight-out metho drinkers and city crooks. **1973** *Sunday Mail* (Brisbane) 18 Mar. 25/5 The oldtimer who'd tried to smuggle it through, by carrying it nonchalantly in his hand, surrendered it sadly. When it comes to metho spotting, Envoy Smith has the nose of a connoisseur.

2. A person who is addicted to drinking methylated spirit. See also *meths.

1933 *Bulletin* (Sydney) 1 Nov. 11/1 A John Hop who has helped to deal with many 'methos.' tells me not a few prefer petrol. **1945** Baker *Austral. Lang.* 166 Addicts of these noxious drinks are known as *meths, methos*, [and] *metho artists.* **1966** G. W. Turner *Eng. Lang. Austral. & N.Z.* vi. 118 Words in -*o* are more specifically Australian, though..*metho* (a drinker of methylated spirits) and *smoko* (with the sense 'morning or afternoon tea'..) are current in New Zealand. **1973** *Brisbane City Mission* Sept. 17/2 The methos... When a man takes to drinking methylated spirits, you know..that he has no thought for betterment of his lot.

Metho² (me·þo). *Austral.* Colloq. abbrev. of Methodist 4.

1941 in Baker *Dict. Austral. Slang* 46. **1961** P. White *Riders in Chariot* viii. 232 Only the civil servants are Roman Catholics here... Arch and me are Methoes, except we don't go; life is too short.

method, *sb.* Add: **I. 2. c.** (Further examples.)

1843 Mill *Logic* I. iii. viii. 450 These two methods may be respectively denominated the Method of Agreement, and the Method of Difference. *Ibid.* 462 This method may be called the Indirect Method of Difference, or the Joint Method of Agreement and Difference. *Ibid.* 465 The Method of Residues is one of the most important among our instruments of discovery. *Ibid.* 470 The method by which these results were obtained, may be termed the Method of Concomitant Variations. **1916** H. W. B. Joseph *Introd. Logic* (ed. 2) xx. 434 So obvious is the difficulty of finding such instances as these canons require, that Mill, having begun by mentioning four methods (of Agreement, of Difference, of Residues, and of Concomitant Variations), adds a fifth, which he calls the Joint Method of Agreement and Difference. **1929** A. N. Whitehead *Process & Reality* 6 When the method of difference fails, factors which are constantly present may yet be observed under the influence of imaginative thought. **1953** I. M. Copi *Introd. Logic* xii. 365 We may begin with an example or two in which the scrupulous use of the Methods results in a more or less conspicuous failure to discover the cause of a given phenomenon. **1965** S. F. Barker *Elem. Logic* vii. 237 When we infer from these data that the hamburgers caused the ptomaine, we are employing what Mill called the method of concomitant variation.

e. *Theatr.* A theory and practice of acting associated with the Russian actor and director Konstantin Stanislavsky (1863–1938), in which the actor seeks the complete illusion of reality by identifying himself as closely as possible with the part he plays. Often *attrib.*, as *method-acting, actor, school*, etc. Also *method-act* vb., *method-acted* ppl. adj. Hence (not in common use) *methodic, methody* adjs.; *methodism, methodist* sbs.

1923 O. M. Sayler *Russian Theatre* 254 'My method, though imperfect,' he [*sc.* Stanislavsky] says, 'I consider psychologically natural.' **1925** H. Carter *New Spirit European Theatre* xviii. 225 The third movement also was a form of protest by theatrical reformers, who were sick of M. Stanislavsky's hair-for-hair actualistic method. *Ibid.* 230 Stanislavsky was unable to change his method of production and style, which were suited to a particular species of bourgeoisie play. **1954** F. M. Whiting *Introd. Theatre* vi. 136 The actor, if sufficiently sensitive to the drives and motives of the character he portrays, will 'instinctively' sense what the basic responses should be. To achieve this goal, Stanislavski devised his method. **1956** K. Tynan in *Harper's Mag.* Mar. 63/2 This is Stanislavsky without Freud, physiological acting without the psychiatric glosses beloved of American 'method' actors. **1957** *Observer* 15 Dec. 10/4 Eli Wallach, the American disciple of the Method.., was admirably solid, but not noticeably more absorbed in his role than some of our own best actors look to be in theirs. **1958** E. Dundy *Dud Avocado* i. vii. 114 The Method..the Stanislavsky Method: working for realism through improvisations and sense memory, and emotional recall. **1958** *Times* 11 Nov. 4/5 It was surely not for nothing that [Jackson] Pollock himself belonged to the country that has promulgated the cult of 'method' acting. **1959** *Times Lit. Suppl.* 27 Mar. 172/1 The whole 'Method' school about which so much pedantic and often absurd controversy has raged are dealt with mordantly. **1959** *Spectator* 29 May 765/1 This was a Method *Don Giovanni*, with real people caught up in a real drama. **1959** *Guardian* 15 Oct. 8/3 What emotional exercises had he done in preparation for Romeo? The real Methodist is not so much different in kind as different in degree. **1960** R. Lewis *Method—or Madness?* i. 4 One hears that Method actors are 'mumblers'. *Ibid.* iv. 75 Another reason that some actors..seem 'Methody', or inexplicably involved, is that their approach is *too* analytical. **1960** *New Left Rev.* Sept.–Oct. 65/1 The method-acted *Tomorrow With Pictures*. **1961** *20th Cent.* Feb. 134 Other factors include the impact of Method-ism in the theatre. **1962** *New Statesman* 19 Jan. 97/1 This long film forfeits seriousness..by being..too thunderously repetitious, too Methodic, strenuous and symbolic. **1963** *Times* 17 Jan. 4/6 Even the Method groups that were flourishing in London a few years ago have now mainly vanished. **1967** *Spectator* 7 July 22/2 His method acting is not merely an attack on their academic ennui, but a mask for his own profound frustrations. **1970** J. Quartermain *Man who walked on Diamonds* xvii. 90 Method-act yourself into a super-sleuth. **1971** J. Willett in A. Bullock *20th Cent.* 243/1 In America..the Group Theater..applied the realistic Stanislavsky 'method' for social ends. **1972** *Daily Colonist* (Victoria, B.C.) 24 Feb. 44/1 A method actor will use his own memory to simulate an emotion for the cameras. **1973** H. McCloy *Change of Heart* xii. 146 If she could feel like that, maybe she would look like that. 'Method' acting. Stanislavsky.

3. e. *Campanology.* (See quot. 1901.)

1668 'Campanista' *Tintinnalogia* 2 Before I Treat of the method and diversity of Peals, I think it not impertinent to speak something of the Properties wherewith a Young Ringer ought to be qualified. **1852** B. Thackrah *Art Change Ringing* 14 These twenty-four changes may be rung in several other methods. **1879** W. Banister *Art & Sci. Change Ringing* (ed. 2) 14 The Plain Bob Method—is applicable to any number of bells, but is properly an even bell method. **1901** H. E. Bulwer *Gloss. Technical Terms Bells* 10 *Method*, any special way in which continuous 'changes' on five or more bells are produced by the regular and orderly movement of all, without repetition of any one 'change'. **1901, 1928** [see *extent *sb.* 7]. **1962** G. E. Evans *Ask Fellows who cut Hay* (ed. 2) xviii. 147 The old ringer answered.., 'No tunes. We allus rang the *method*, the same as we did in the steeples.' **1965** W. G. Wilson *Change Ringing* xviii. 149 We showed how variations in a method could be made in the Plain Bob type of method by means of second place.

II. 5. Phr. *method in one's madness* (also varr. of this expression, normally with allusion to quot. 1602 in Dict.): reason, orderliness, or sense lying behind one's apparent insanity or stupidity.

1843 Poe *Gold-Bug* in *Dollar Newspaper* (Philadelphia) 21 June 1/7 My friend, about whose madness I now saw, or fancied that I saw, certain indications of method. **1850** F. E. Smedley *Frank Fairleigh* xxix. 241 A fear of completely knocking up..induced me to preserve some little method in my madness. **1894** A. Conan Doyle *Mem. Sherlock Holmes* 128, I have usually found that there was method in his madness. **1911** Brereton & Rothwell tr. *Bergson's Laughter* i. 2 The comic spirit has a logic of its own, even in its wildest eccentricities. It has a method in its madness. **1922** Chesterton *Man who knew too Much* 110 He may be mad, but there's method in his madness. There nearly always is method in madness. It's what drives men mad being methodical. **1953** A. Huxley *Let.* 21 July (1969) 680 There is, in the long run (at least I hope so), some kind of method in my madness.

III. 10. **methods engineer**, a person concerned with method study and methods engineering in a business; **methods engineering**, the organization of business methods through method study; **method(s) study**, the use of time-and-motion study and other systems to determine the most efficient methods to use in business activities.

[**1928** R. C. Davis *Princ. Factory Organization & Managem.* vi. 72 The methods manager or industrial engineer is concerned primarily with the efficient application of the human forces of the organization.] **1939** Maynard & Stegemerten *Operation Analysis* i. 5 Time formulas are useful mathematical devices that the methods engineer employs. **1944** —— *Guide to Methods Improvement* ii. 9 A methods engineer secured a position in a textile mill in the South. **1953** J. R. Immer *Materials Handling* iii. 30 This part of materials movement is.. more often the concern of the methods engineer than of the materials-handling engineer. **1960** *News Chron.* 28 July 8/8 (Advt.), Methods Engineer..experience of Time and Methods Study and Ratefixing. **1939** Maynard & Stegemerten *Operation Analysis* i. 1 Before discussing the methods-engineering procedure in detail, it will be advisable to formulate a clear statement of what the term covers... Methods engineering is the industrial science which is chiefly concerned with increasing labor effectiveness. **1944** —— *Guide to Methods Improvement* i. 5 In order to continue to make improvements in methods, a procedure known as 'method engineering' has been developed. **1953** J. R. Immer *Materials Handling* iv. 41 The methods-engineering department has the responsibility for 'planning the manual part of the operation'. **1932** A. H. Mogenson *Common Sense appl. to Motion & Time Study* i. 11 More attention is being given to this phase of methods study at present. **1939** Maynard & Stegemerten *Operation Analysis* i. 2 A methods study always begins with a careful primary analysis of existing conditions. **1955** *Furniture Devel. Council Newslet.* June 3/2 Efficient method study cannot be put into practice without a sound knowledge of the work in hand. **1959** *Gloss. Terms Work Study* (*B.S.I.*) 6 *Method study*, the systematic recording and critical examination of existing and proposed ways of doing work, as a means of developing and applying easier and more effective methods and reducing costs. **1969** J. Argenti *Managem. Techniques* i. 2 One glance round the factory is enough to reveal to an expert whether the manager there has been using Method Study, for example. *Ibid.* 167 The Method Study Officer ..analyses the purpose and function of each step in the process.

Methodee, var. Methody (in Dict. and Suppl.).

‖ **Methodenstreit** (meto·dənstrəit). [G., lit. 'methods struggle'.] Discussion or dispute of opinions concerning the methodology of a field of study.

1958 *Spectator* 7 Feb. 173/2 In history, since the *Methodenstreit* of sixty years ago, the flood of 'scientific' research [etc.]. **1968** *Explorations in Entrepreneurial Hist.* VI. 75, I would like to narrow my comments on this twentieth-century *Methodenstreit* to three questions. **1970** *Computers & Humanities* V. 2 A literature of criticism is just beginning to emerge, but it has not yet created the type of *Methodenstreit* that has apparently debilitated scholarship in economic history.

methodic. B. *sb.* **2.** Add to def.: esp. in the teaching of languages.

1962 P. Strevens *Papers in Lang.* (1965) v. 70 This organizational framework for arranging linguistic terms into a teachable pattern is known in Britain as *Methodics*. **1964** *New Statesman* 6 Mar. 378/1 (Advt.), Applications are invited for a post as Lecturer in the methodics of language teaching. **1964** M. A. K. Halliday et al. *Ling. Sci.* x. 200 Methodics is a framework of organization for practical language teaching, in which pedagogical techniques and linguistic theory cross-fertilize each other.

Methodism. Add: **3.** The doctrines and practice of the 'methodic' school of physicians.

1896 T. C. Allbutt in *Classical Rev.* X. 346/2 A second school was Methodism, which was satisfied to refer all symptoms and all disease to the variations of the 'strictum' and the 'laxum'; that is, to the restriction or laxity of the secretions and other fluids of the body. **1958** D. Guthrie *Hist. Med.* (rev. ed.) v. 67 This theory, known as 'Methodism', was elaborated by Themison (123–43 B.C.), the pupil of Asclepiades.

methodize, *v.* Add: **1. c.** To convert into Methodists.

1846 A. Wiley in *Indiana Mag. Hist.* (1927) XXIII. 428 Some disciples who are hard cases, will be hard to methodize.

3. *Theatr.* = *method-act* vb. (see *method *sb.* 2 e).

1958 *Spectator* 31 Jan. 136/3 The pouting and posturing and methodising. **1960** *Ibid.* 18 Nov. 782/3 John Dexter's production has been much condemned... Perhaps he should have varied the pace, Methodised the delivery.

methodology. Add to def.: Also, the study of the direction and implications of empirical research, or of the suitability of the techniques employed in it; also *attrib.* (Further examples.)

In some contexts weakened to mean little more than 'method'.

1932 W. E. D. Allen *Hist. Georgian People* vii. 91 Alp-Arslan and his commanders were tacticians who evolved a new methodology in mediæval Asiatic war. **1937** T. Parsons *Struct. Social Action* ii. v. 178 (*heading*) The methodology and main analytical scheme. **1944** F. Kaufmann *Methodology of Social Sci.* p. vii, A reconsideration of the problem how the logical analysis of scientific procedure (methodology) is related to deductive logic. **1949** Shils & Finch tr. *Weber's Methodology of Social Sci.* iii. 114 The most significant achievements of specialist methodology use 'ideal-typically' constructed conceptions of the objectives. **1960** *Amer. Speech* XXXV. 212 The authors tell too little about their methodology. **1970** O. Dopping *Computers & Data Processing* xxii. 343 So far, what we have considered is a straight simulation which does not give rise to any methodology difficulties. **1971** *World Archaeol.* III. 210 The main challenge appears to be in establishing connections that bridge the methodologies of different fields of study. **1972** *Jrnl. Social Psychol.* LXXXVII. 127 None of these methodologies have been extended into the realm of applied psychology. **1974** *Nature* 22 Mar. 294/3 The interfacing of the two techniques..is discussed in depth together with the methodology of operating them together.

methodological *a.*, **methodologically** *adv.*, **methodologist**: further examples.

1914 B. Russell *Our Knowl. External World* viii. 238 It is necessary to practise methodological doubt, like Descartes, in order to loosen the hold of mental habits. **1937** T. Parsons *Struct. Social Action* i. i. 23 These [assumptions] may lie in a number of different directions. But the ones to which special attention should now be called are the 'methodological'. **1941** J. C. Ransom *New Crit.* ii. 208 In my mind Dante's beliefs are very bold speculations at which the accusing finger has pointed steadily for a long time now, but substantively are better grounded, and methodologically far more consistent, than Shelley's beliefs. **1942** *Economica* IX. 286 That he [*sc.* the social scientist] systematically starts from the concepts which guide individuals in their actions, is the characteristic feature of that methodological individualism which is closely connected with the subjectivism of the social sciences. **1949** Shils & Finch tr. *Weber's Methodology of Social Sci.* iii. 113 The professional methodologist will take umbrage at many of Meyer's formulations. **1952** K. R. Popper *Open Society* (ed. 2) I. iii. 31, I use the name methodological essentialism to characterize the view..that it is the task of pure knowledge or 'science' to discover and to describe the true nature of things, i.e. their hidden reality or essence. **1962** F. Williams tr. *Sartre's Imagination* p. vi, This truism has some far-reaching methodological implications, however. **1964** Gould & Kolb *Dict. Social Sci.* 425/2 The methodologist examines systematically and logically the aptness of all research tools, varying from basic assumptions to special research techniques, for the scientific purpose. *Ibid.* 471/1 Methodologically the term [*sc.* non-empirical] is used to describe all methods of acquiring belief other than those of positive science. **1965** B. B. Wolman *Scientific Psychol.* 18 One form of this practice is an extreme and radical methodological reductionism. Instead of developing methods and concepts derived from human life, they imitate..physical science. **1969** R. Blackburn in Cockburn & Blackburn *Student Power* 203 What is sometimes called 'methodological individualism' (the bourgeois doctrine that all statements about society can be reduced to statements about individuals) is a device for evading such facts as these. **1972** *Computers & Humanities*

VII. 86 When one kind of analysis is conveniently included in the package but a more methodologically sound analysis is not, inappropriate analysis may result. **1973** *Amer. Speech 1970* XLV. 124 Labov has led something of a methodological revolution.

methods-time. [f. pl. of Method *sb.* + Time *sb.*] Used *attrib.*, esp. in phr. *methods-time measurement* (also with capital initials): see quots. 1948², 1968. Abbrev. *MTM.*

1948 H. B. Maynard et al. *Methods-Time Measurement* p. v, The methods-time measurement procedure eliminates the necessity for judging the performance level at which an operator works while being observed. *Ibid.* ii. 12 Methods-time measurement is a procedure which analyzes any manual operation or method into the basic motions required to perform it and assigns to each motion a predetermined time standard which is determined by the nature of the motion and the conditions under which it is made. *Ibid.* iii. 25 The desirability of having accurate methods-time data available has long been recognized by methods engineers. *Ibid.* xv. 129 A question which was uppermost all during the period when the methods-time standards were being developed. **1952** A. G. Shaw *Purpose & Pract. Motion Study* iii. 60 The Methods-Time Measurement technique analyses a movement cycle into elements. **1959** V. H. Rotroff *Work Measurement* iii. 30 Applying the MTM technique, the analyst summarizes all motions required to perform the job properly. **1968** Johannsen & Robertson *Managem. Gloss.* 87 *Methods Time Measurement*, a predetermined motion time work measurement technique devised by the Methods Engineering Council (USA) which analyses manual operations and methods into basic motions required to perform them... There are nine basic MTM motions.

Methody. Add: Also 8 **-dee.** (Earlier and later examples.)

1753 J. Wesley *Jrnl.* 8 Aug. (1913) IV. 78 As soon as we entered the town [*sc.* Mevagissey] many ran together, crying, 'See, the Methodees are come.' **1794** W. B. Stevens *Jrnl.* 19 Mar. (1965) 140 A mighty good Methody Parson comes every day. **1834** [see *battleship c]. **1907** T. E. Kebbel *Lord Beaconsfield* xvi. 259 Do I not remember many a sturdy villager who was always ready with a gibe at 'the Methodies', as they were called. **1933** *Bulletin* (Sydney) 25 Oct. 20/1 You should have told him Methodies don't dance. **1962** G. Lawton *John Wesley's English* ix. 211 Three other nicknames for 'Methodist'—'Culamite', 'Macabee' and 'Methodee' may..be considered here.

methonium (meþōᵘ·niəm). *Pharm.* [f. *meth- + *-onium.] Any of various polymethylene bistrimethylammonium cations, $[(CH_3)_3N(CH_2)_xN(CH_3)_3]^{2+}$ (where x is an integer), or salts of these ions, some of which are used as ganglionic blocking agents in the treatment of hypertension.

[**1949**: cf. *decamethonium, *hexamethonium* (s.v. *hexa-).] **1950** *Lancet* 26 Aug. 353/1 (*heading*) 'Medical sympathectomy' in hypertension. A clinical study of methonium compounds. **1952** *Pharmacol. Rev.* IV. 220 The name of 'methonium' compounds for the members of the polymethylene bistrimethylammonium series has been approved by the British Pharmacopoeia Commission, and it is under this name, preceded by the appropriate numerical prefix, that they are now known. **1969** *Jrnl. Neurochem.* XVI. 1173 Some members of the methonium series are..taken up by a saturable mechanism and..the uptakes can be described quantitatively by Michaelis-Menten type kinetics.

methotrexate (meþo-, mīþotre·ksēⁱt). *Pharm.* [Origin unknown.] 4-Amino-10-methylfolic acid, $C_{20}H_{22}N_8O_5$, an orange-brown powder which is a folic acid antagonist and is used in the treatment of tumours, esp. cancer and leukæmia.

1955 *Sci. News Let.* 5 Feb. 88/2 The new drug is an anti-folic acid compound very similar to Aminopterin. It is called Methotrexate by its manufacturers, Lederle Laboratories, Pearl River, N.Y. **1961** [see *mercaptopurine* s.v. *mercapto(-)]. **1970** *Times Lit. Suppl.* 26 Mar. 332/4 Burkett's [*read* Burkitt's] tumour..can be treated easily and successfully by dosing with methotrexate. **1971** *Nature* 22 Oct. 518/3 The Food and Drug Administration has decided to allow the controversial drug methotrexate to be used for severe cases of psoriasis. **1973** *Brit. Pharmacopœia* 300/1 Methotrexate.

methoxy(-) (meþǫ·ksi). *Chem.* [f. Methoxy(l (cf. Hydroxy-).] **a.** As an inseparable formative element in chem. names, indicating the presence of a methoxyl group, as *methoxyacetophenone, -benzoic* adj., *-pyridine, -succinic* adj.; **metho·xychlor**, a crystalline compound, $(C_6H_4OCH_3)_2CHCCl_3$, of similar chemical structure to DDT, which is used as an insecticide, esp. for veterinary hygiene.

1942 Fuson & Snyder *Org. Chem.* xxi. 288 A remarkable instance of catalysis has been observed in the formation of *p*-methoxyacetophenone from anisole by treatment with acetic anhydride. **1964** N. G. Clark *Mod. Org. Chem.* xviii. 479 Like the aromatic aldehydes, the ketones are frequently employed as synthetic intermediates, and as ingredients in perfumery. Some simple derivatives possess quite delightful odours, for example, *p*-methylacetophenone (hawthorn), *p*-methoxyacetophenone (heliotrope). **1895** Perkin & Kipping *Org. Chem.* II. xxvi. 397 The methylcresols..are oxidised by chromic acid, yielding the corresponding methoxybenzoic acids, $C_6H_4(OCH_3){\cdot}COOH$. **1949** *Jrnl. Econ. Entomol.* XLII. 845/1 Ellenville line flies exhibited resistance to residues of both DDT and methoxychlor. **1959** H. Martin *Sci. Princ. Crop Protection* (ed. 4) x. 203 When methoxychlor was used instead [of DDT], Carter *et al.* were unable to detect the insecticide in the milk. **1969** K. A. Hassall *World Crop Protection* II. iv. 88 Methoxychlor..shows a lower tendency than DDT to dissolve in animal fats and also possesses a much lower acute oral toxicity to higher animals. It is therefore used on and near livestock. **1973** *Globe & Mail* (Toronto) 8 Sept. 8/3 City golf courses and cemeteries were relatively mosquito-free, having been fogged all along with methoxychlor, a new insecticide that doesn't have the harmful properties of DDT. **1889** G. M'Gowan tr. *Bernthsen's Text-bk. Org. Chem.* 486 Methoxy-pyridine and methyl-pyridone, which result from these two forms by exchange of H..for CH_3, are known. **1893** *Jrnl. Chem. Soc.* LXIII. 229 Inactive methoxysuccinic acid may be resolved into two active acids, of equal and opposite rotatory powers.

b. As an independent word, usu. *attrib.* and sometimes joined by a hyphen to the second element: = Methoxyl.

1900 Perkin & Kipping *Org. Chem.* (rev. ed.) xxxiii. 485 By far the greater number of alkaloids contain one or two, sometimes three or more, methoxy-groups.., united with a benzene nucleus. **1946** *Jrnl. Pharmacol. & Exper. Therap.* LXXXVIII. 413 Substitution of chlorine by a.. methoxy, butoxy or amyloxy group yields ineffective compounds. **1969** *Times* 10 Feb. 10/8 Dr. Shulgin..prepared many derivatives of mescalin, differing from the parent compound in their pattern of methoxy substitution on the phenyl ring. **1971** *Nomencl. Org. Chem.* (I.U.P.A.C.) (ed. 2) C. 154 The following contractions for oxygen-containing radical names are recommended... Methoxy CH_3—O—.

meths (meþs). Also **meth.** Colloq. abbrev. of *methylated spirit(s)*. Also *attrib.*, freq. as **meth(s)-drinker**, one who is addicted to drinking methylated spirits. (See also *metho¹.)

1933 M. Marshall *Tramp-Royal on Toby* 367/1 *Meth*, methylated spirit. **1935** [see *jake *sb.*²]. **1939** 'N. Shute' *What happened to Corbetts* i. 18 A Primus stove.. and paraffin and meths. **1947** [see *hottie, hotty *sb.*]. **1959** *Listener* 21 May 893/3 A lovely copper kettle with a meths-burner underneath. **1961** *Times* 15 Aug. 11/3 The surly greetings of the meth-drinkers. **1966** J. Bingham *Double Agent* x. 149 It sounds like a meth-drinkers' camp. **1968** *Daily Tel.* 12 Nov. 16/2 Young people who work at clearing slag heaps, helping meths-drinkers or cleaning canals to make them navigable. **1970** M. Tripp *Man without Friends* i. 7 I'd seen a meths drinker spewing blood the night before. **1973** D. Jordan *Nile Green* xxxiv. 167, I shook my head like a meths drinker.

Methuselah. Add: **2.** *transf.* A very large wine-bottle (see quots.).

1935 A. L. Simon *Dict. Wine* 172 *Methuselah*, double Jeroboam, holding 8 reputed quarts or 6·40 litres, equal to 225·350 fluid ounces. **1951** *Bohemian Life* (Bohemian Distrib. Co., Los Angeles, U.S.A.) Apr., Out-size bottles include the..Methuselah..8 bottles or 213·30 ounces. **1959** *Gloss. Packaging Terms* (*B.S.I.*) 28 *Methuselah*, a wine bottle—capacity 9 reputed quarts. **1962** [see *Balthazar]. **1972** E. Meigh *Story Glass Bottle* 73 Wine bottlers ransacked the Old Testament for names of kings and captains to name the bottles in which their valuable liquids were confined: thus, in ascending order of magnitude are jeroboams, rehoboams, methuselahs.

methyl. Add: **c. methyl alcohol**, add to def.: = *methanol; **methylcellulose**, any of a range of white, tasteless compounds which are produced by etherifying cellulose with various proportions of methyl chloride or sulphate and are used as thickening, emulsifying, and stabilizing agents, esp. in the food industry, as laxatives, and in adhesives; **methyldo·pa** *Pharm.*, a whitish powder, $C_{10}H_{13}NO_4$, which is used as a hypotensive agent; **methyl ethyl ketone**, a colourless volatile liquid, $CH_3COCH_2CH_3$, which is widely used as a solvent for organic materials; butanone; **methylglyo·xal**, a yellow liquid aldehyde, CH_3COCHO, with a pungent odour, which readily polymerizes and is an intermediate in carbohydrate metabolism; **methyl isobutyl ketone**, a liquid ketone, $(CH_3)_2CHCH_2COCH_3$, which is manufactured by the hydrogenation of mesityl oxide and widely used as a solvent; hexone; **methyl methacrylate**, (*a*) the methyl ester of methacrylic acid, a volatile colourless liquid that readily polymerizes to resinous glass-like materials that are sold under trade-names such as *Lucite, Perspex*, and *Plexiglas*; (*b*) polymerized methyl methacrylate; **methyl orange**, an orange crystalline compound, $(CH_3)_2N{\cdot}C_6H_4{\cdot}N{:}N{\cdot}C_6H_4{\cdot}SO_3Na$, which is the sodium salt of an azo-dye made by coupling diazotized sulphanilic acid with dimethylaniline and is chiefly used (in dilute aqueous solution) as an acid–base indicator in volumetric analysis, giving a pink colour at a pH of 3 and a yellow colour at a pH of 4·4; **(6α-)methylpredni·solone**, a synthetic compound, $C_{22}H_{30}O_5$, which functions as a glucorticoid and is used mainly in treating rheumatoid arthritis and rheumatic fever; **methyl red** [tr. G. *methylrot* (Rupp & Loose 1909, in *Ber. d. Deut. Chem. Ges.* XLI. 3905)], a red crystalline compound, $(CH_3)_2N{\cdot}C_6H_4{\cdot}N{:}N{\cdot}C_6H_4{\cdot}COOH$, that is analogous to methyl orange (anthranilic acid being used in its preparation in place of sulphanilic acid) and is used similarly, the colour of an alcoholic solution being red at a pH of 4·4 and yellow at a pH of 6; **methyl rubber**, an early synthetic rubber made by polymerization of dimethylbutadiene; **methyl salicylate**, a colourless or pale-yellow liquid, $C_6H_4(OH)CO{\cdot}OCH_3$, which is the chief constituent of oil of wintergreen and sweet birch oil and is used as a flavouring material, in perfumery, and in liniments and ointments; **methyltesto·sterone** *Pharm.*, any of the methyl derivatives, $C_{20}H_{30}O_2$, of testosterone; *spec.* 17-methyltestosterone, a white crystalline powder with similar actions and uses to those of testosterone; **me:thylthiou·racil**, any compound which is a methyl and a thio derivative of uracil; *spec.* 6-methyl-2-thiouracil, $C_5H_6N_2OS$, an antithyroid substance used to control thyrotoxicosis.

1921 *Jrnl. Chem. Soc.* CXIX. i. 79 A methyl cellulose of this limiting methoxyl content can be prepared which is representative of the whole of the original cellulose. **1947** Winding & Hasche *Plastics* iii. 119 Water solubility plus the ability to produce viscous solutions has made methyl cellulose useful as a thickening agent in the textile, food, and adhesive industries. Since it is edible, it can be used as a base for salad dressings. **1954** *Jrnl. Pharmacy & Pharmacol.* VI. 731 Methylcellulose eye-drops..are liable to contamination by moulds. **1966** *Punch* 5 Jan. 11/1 In its survey of strawberry jam it [*sc. Which?*] found two specimens in which the composers had succeeded in soaring far above the sordid world of methyl cellulose and lecithin. **1954** *Arch. Biochem. & Biophysics* LI. 456 Two compounds, α-methyl-3,4-dopa (α-MD) and α-methyl-3-hydroxyphenylalanine, respectively, added in a low range of concentrations, accelerated, and at higher concentrations, inhibited, kidney dopa decarboxylase. **1961** *New Zealand Med. Jrnl.* LX. 569/2 Methyl dopa, like other potent hypotensive agents, produces side-effects including fatigue. **1972** *Materials & Technol.* V. xxi. 792 Methyldopa and pargyline cause the replacement of noradrenaline by a weaker transmitter. **1876** *Jrnl. Chem. Soc.* XXIX. 897 By the reduction of methyl-ethyl ketone, pinacone $C_8H_{18}O$ is obtained. **1938** H. P. Starck *Princ. Org. Chem.* vii. 186 Methyl ethyl ketone was formerly used as a solvent in the preparation of cordite from gun cotton and nitroglycerine. **1959** *Times Rev. Industry* Aug. 98/1 Methylethylketone for removing waxy materials. **1898** *Jrnl. Chem. Soc.* LXXIV. i. 224 Methylglyoxal is formed from dihydroxyacetone..when the latter is distilled with dilute sulphuric acid. **1913, 1951** Methylglyoxal [see *glyoxalase]. **1962** A. Pirie *Lens Metabolism Rel. Cataract* 431 Experiments were done in the hope of detecting methylglyoxal which Salem and Crooke (1950) stated was formed from hexosediphosphate by liver. **1888** *Jrnl. Chem. Soc.* LIV. 125 Its ketone..is therefore identical with methyl isobutyl ketone. **1960** *Times Rev. Industry* Dec. 16/3 The chloride is dissolved in water, and then treated with an organic solvent called methylisobutyl ketone (MIBK) which extracts the iron. **1973** H. M. Stanley in E. G. Hancock *Propylene* i. 9 The Shell Oil Company..in 1937–8 commenced the manufacture of acetone derivatives, including diacetone alcohol, mesityl oxide and methylisobutyl ketone. **1933** *Chem. Abstr.* XXVII. 4363 Thermoplastic products are obtained by polymerizing methyl methacrylate..by exposure to light or heat. **1936** *Industr. & Engin. Chem.* Oct. 1161/2 These properties make methyl methacrylate an outstanding plastic. **1964** N. G. Clark *Mod. Org. Chem.* x. 194 Acetone cyanohydrin..is an important intermediate in the manufacture of methyl methacrylate, the material used for making the polymer 'Perspex'. **1967** *Times Rev. Industry* June 74/3 Methyl methacrylate and/or polyester reinforced glass fibre are used for hand basins, baths, complete bath room heat units, heating ducts, roof lights, &c. **1881** G. Lunge in *Chem. News* 16 Dec. 288/1 It is in reality a salt of sulpho-benzene-azo-dimethylamin, for which long name I propose the short and sufficiently clear name 'Methyl-orange'. **1930** Field & Weill *Electro-Plating* 56 The ordinary indicators used by the chemist such as litmus, methyl orange and phenolphthalein. **1969** H. A. Flaschka et al. *Quantitative Analytical Chem.* II. ix. 110 Methyl orange is therefore unsuited to the titration of a weak acid. **1957** *Ann. Rheumatic Dis.* XVI. 298/2 6-Methyl-prednisolone was administered under various conditions to 41 patients with active peripheral rheumatoid arthritis. **1960** *Antibiotic Med.* VII. 704 The subcutaneous injection of depot methyl prednisolone has demonstrated significant permanent relief of refractory anal pruritus in 14 patients. **1973** *Jrnl. Amer. Med. Assoc.* 19 Feb. 896/1 Large doses of methylprednisolone were used intravenously as the sole steroid agent in 44 transplants in 43 renal allografted patients for the first two to three weeks after operation. **1910** *Jrnl. Chem. Soc.* XCVII. 2490 As a means of measuring colorimetrically the concentration of hydrogen ions in a solution, methyl-red would probably be found valuable when such concentration lies between 10^{-4} and 10^{-6}. **1951** Whitby & Hynes *Med. Bacteriol.* (ed. 5) xii. 193 When *Bact. coli* is grown for 5 days at 30° C. in the glucose phosphate medium described above an acid reaction is maintained which is sufficient to give a red colouration on the addition of 0·04 per cent aqueous methyl red as indicator. **1969** Fischer & Peters *Brief Introd. Quantitative Chem. Analysis* viii. 212 At pH 4·2 (and below)..methyl red is present as the red acid form and..it exists as the yellow base form if the pH becomes

6·2 (and above). 1919 *India-Rubber Jrnl.* 16 Aug. 18/2 The dye works laid themselves out for the manufacture of large quantities of methyl-rubber hot polymerisation products. 1936 *Trans. Faraday Soc.* XXXII. 91 The sharpness of the fibre pattern depends on the sample of methyl rubber investigated. 1972 *Materials & Technol.* V. xiv. 456 During the first World War the Germans worked out..the first factory-scale synthetic rubber plant which produced the so-called methyl rubber... The quality of this rubber was poor; it was only useful for the preparation of hard rubber battery boxes for submarines. 1876 *Encycl. Brit.* V. 572/2 Oil of wintergreen..contains methyl salicylate. 1948 *Jrnl. Amer. Med. Assoc.* 28 Feb. 651/2 Methyl salicylate is a potent and fairly rapidly acting poison. 1972 *Materials & Technol.* IV. xii. 448 Methyl salicylate, which is a known anti-rheumaticum, is made by the Fischer esterification of salicylic acid. 1936 *Biochem. Jrnl.* XXX. 292 Various..derivatives methylated at position 17, including methyltestosterone. 1955 W. GADDIS *Recognitions* II. v. 526 Crossing the Atlantic Ocean to get laid. He can't even get it up without a dose of methyltestosterone. 1959 *Brit. Med. Jrnl.* 31 Jan. 259/2 Methyltestosterone was introduced to clinical medicine..in 1939, and since that time must have been used for its androgenic and protein anabolic effects in thousands of cases. 1970 PASSMORE & ROBSON *Compan. Med. Stud.* II. xii. 11/2, 17α-Methyltestosterone is active by oral administration. 1974 *Nature* 13 Dec. 585/2 We have induced reversible infertility, with little effect on libido, in five healthy young men by giving them tablets containing methyltestosterone and ethynyloestradiol. 1944 *Q. Jrnl. Pharmacy & Pharmacol.* XVII. 318 Methylthiouracil produces hyperplasia of the thyroid gland of the rat. 1960 *Lancet* 19 Mar. 653/2 Professor Wayne's references..to methylthiouracil prompt me to record the long-term follow-up of 100 consecutive cases of thyrotoxicoses. 1968 *Listener* 18 July 70/1, I had recently been diagnosed as hyperthyroidic and had with remarkable benefit embarked on a course of methylthiouracil tablets.

methylate, *v.* Substitute for first part of def.: To introduce one or more methyl groups into (a compound or group). Add further examples of this use.

1861 *Jrnl. Chem. Soc.* XIII. 324 These crystals were not analysed, but there can be no doubt that they were the methylated phosphorus-urea of the allyl-series. 1917 *Ibid.* CXI. 848 Experiments have proved the wide scope of the use of formaldehyde for methylating amino-compounds. 1929 *Jrnl. Amer. Chem. Soc.* LI. 2535 We have found methyl bromide to be a satisfactory methylating agent with both malonic ester and acetoacetic ester. 1967 I. L. FINAR *Org. Chem.* (ed. 5) I. xxvi. 683 Veratrole ..may be prepared by methylating catechol with methyl sulphate in alkaline solution.

methylated *ppl. a.* (examples of *ellipt.* use, = methylated spirit).

1912 D. H. LAWRENCE *Let.* 2 Sept. (1962) I. 142 We take rucksacks..with food and methylated, cook our meals by some stream. 1938 MRS. BELLOC LOWNDES *Let.* 29 Sept. (1971) 165, I also got..rather more methylated, rice and matches than usual.

methylene. Add: **methylene chloride,** dichloromethane, CH_2Cl_2, a volatile liquid used as a solvent, esp. in paint-removers and for extractions in pharmacy, and as a refrigerant.

1880 *Jrnl. Chem. Soc.* XXXVIII. 307 Methylene chloride, CH_2Cl_2, is best obtained by cautiously adding hydrochloric acid to a mixture of alcohol, chloroform, and metallic zinc. 1953 O. E. ANDERSON *Refrigeration in Amer.* xii. 193 Other absorption machines were brought out for air-conditioning work. One of these used methylene chloride as the refrigerant. 1957 H. R. SHEPHERD in E. Sagarin *Cosmetics* xxxvi. 803 Ethyl chloride, methylene chloride, and others..among the earliest aerosol propellents, are not presently being used for cosmetic purposes. 1973 *Oxf. Univ. Gaz.* CIII. Suppl. 5. 10 The goddess has now been degreased with lustrations of methylene chloride and acetone.

methylol (me·þilǫl). *Chem.* [f. METHYL + -OL.] = *hydroxymethyl* s.v. *HYDROXY- 3.

1898 *Jrnl. Chem. Soc.* LXXIV. I. 506 Nitrobutylic alcohol..readily condenses with formaldehyde (1 mol.) to form a glycol (2-methylol-2-nitro-1-butanol). 1937 F. C. WHITMORE *Org. Chem.* 521 Formaldehyde condenses with urea to give methylol ureas, $NH_2CONHCH_2OH$ etc. The methylol groups can combine with more urea to form more and more complex molecules until a resin is formed. 1962 J. T. MARSH *Self-Smoothing Fabrics* i. 3 The fundamental basis of the process remains unchanged and depends on the application of methylol compounds within the fibre of the fabric followed by their further condensation.

methysergide (meþisɜ̄·ɹdʒaid). *Pharm.* [f. *METH- + *L)YSERG(IC *a.* + AM)IDE.] 1-Methyl-D-lysergic acid butanolamide, $C_{21}H_{28}N_3O_2$, a serotonin antagonist that is administered, usu. in the form of its maleate, in the prophylaxis of recurrent migraine.

1962 *Amer. Jrnl. Med. Sci.* CCXLIII. 152/1 Fifty-seven patients..were treated with antiserotonin agent, methysergide (UML-491). 1962 *Daily Tel.* 30 July 11/6 Methysergide (Deseril), is being tested as a preventive against chronic forms of migraine. 1968 J. H. BURN *Lect. Notes Pharmacol.* (ed. 9) 30 In some patients methysergide causes dizziness and nausea and produces aching in the legs.

metic. (Further examples.)

1921 *Times Lit. Suppl.* 10 Nov. 731/2 The proportion of the horrific in Poe's poetry..is entirely different; he is a metic in the land of shadows. 1936 *Ibid.* 14 Nov. 918/4 Sir Harry Preston was not Brightonian by birth—but then most of the town's great men have been metics. 1974 *Sci. Amer.* Sept. 95/1 In Athens the free immigrants called *metics*, who were permanent residents rather than passing traders, may have outnumbered the slaves.

meticulosity. (Later example.)

1972 *Country Life* 16 Mar. 657/2 The meticulosity of manners was necessary inside or outside the novels, in the villainy of a manor house character or at the tea-table.

meticulous, *a.* Add: **2.** In present usage: careful, punctilious, scrupulous, precise.

1952 W. D. JACOBS *William Barnes, Linguist* i. 9 They [*sc.* Barnes's linguistic studies] present a possible solution to mongrelized English and an alternate program of greater scale and with more meticulous plan than any previous to Barnes. 1964 *New Statesman* 8 May 710/1 Wilson is far too meticulous a constitutionalist not to appreciate the impropriety—not to mention the impossibility—of an opposition's seeking to govern a country in advance. 1973 *Times* 4 May 2/7 Dr Ramsey said in a meticulous English accent: 'Can I just say [etc.].'

meticulously, *adv.* Add to def.: In present usage: in a careful or punctilious manner. (Cf. *METICULOUS *a.* 2.)

1961 'J. WYNDHAM' *Consider her Ways* 90 He had..a black felt hat meticulously brushed. 1971 *Nature* 4 June 334/3 It is essentially a reference book, meticulously cross-referenced, with a huge bibliography and a good index. 1974 *Physics Bull.* Mar. 104/2 With the aid of a meticulously detailed account of the formal basis of quantum mechanics, Professor Scheibe has provided a precise characterization of the orthodox interpretation of the theory.

meti·culousness. [-NESS.] The quality of being meticulous; meticulosity.

1923 *Sunday at Home* Mar. 382/2 He was measured and cautious in his statements to the point of meticulousness. 1927 *Daily Tel.* 2 Mar. 5/5 Examples..handed down to posterity in paint with a meticulousness that..never fails to charm us. 1928 *Observer* 1 July 15/6 The paragraph of last Sunday did not do justice to the meticulousness of the index to Mr. Shaw's 'Intelligent Woman's Guide to Socialism'. 1937 *Discovery* Jan. 32/2 The pleasing meticulousness of a Chinese drawing.

métier. (Later examples.)

1950 T. S. ELIOT *Cocktail Party* III. 157 You understand your *metier*, Mr. Quilpe—Which is the most that any of us can ask for. 1958 *Listener* 27 Nov. 874/1 It is fashionable to regard technical and physical craft [in the theatre] as adventitious and to have no sense whatever of what the French call *métier*. 1975 *Times* 18 Sept. 36/2 The parallels started to extend from the man to the *métier*.

metis. Add: (Earlier and later examples.) Also applied to other persons of mixed blood and *transf.* to animals and plants.

1816 C. ROBERTSON in *Publ. Hudson's Bay Rec. Soc.* (1939) II. 248 Your European Servants and Metiss are in many places deserting over to the North West Company. 1949 *Amer. Speech* XXIV. 95 A common term to indicate a crossbreed between Persian lamb and other Asiatic species of lamb is *metis*, derived from the French for 'half-breed'. 1966 *Kingston* (Ontario) *Whig-Standard* 29 July 6/3 A settlement of about 1500 Cree Indians and 300 Metis. 1972 *Guardian* 4 July 16/2 The estimated 15,000 to 50,000 *métis* children in South Vietnam—children of American soldiers, either half-black or half-white. 1972 D. BLOODWORTH *Any Number can Play* v. 37 He nodded towards a plump Eurasian... 'Bonjour, les gars,' murmured the métis. 1974 *Sci. Amer.* June 115/1 In California the work has taken mainly the direction of producing new *métis* (crosses within species) by crossing Europe's temperate-climate and Mediterranean vines with a view to obtaining better grapes for the fertile but desertlike Central Valley. 1975 *Time* 3 Feb. 8/3 The oppression and mistreatment of the Indians and Métis..never become anything more than a rhetorical device.

‖ **métissage** (metisāʒ). [Fr.: cf. METIS.] Cross-breeding. Also *fig.*

1895 in *Funk's Stand. Dict.* 1959 *Times* 13 Aug. 9/7 He is, too, on historical grounds, a firm believer in the value of *métissage*—the cross-fertilization of civilizations. 1972 *Guardian* 8 July 9/1 Léopold Senghor, President of Sénégal,..is a great believer in the virtues of *métissage*, or racial mixture.

métisse (metīs). Fem. of METIS.

1895 in *Funk's Stand. Dict.* 1955 G. GREENE *Quiet Amer.* III. i. 196 Across the way a *métisse* with long and lovely legs lay..reading a glossy woman's paper. 1961 E. BRUTON *King Diamond* vi. 87 'She's not African. Not entirely anyway... She's a metisse.'..'I know he wouldn't have gone for an African.' 'He didn't exactly. She's half and half.'

metœstrus (metī·strǝs). *Biol.* Also -um, (*U.S.*) metestrus (-e·strǝs, -ī·strǝs), -um. [f. MET(A- + ŒSTRUS, ŒSTRUM.] The short period following œstrus in many mammals during which sexual activity subsides.

1900 W. HEAPE in *Q. Jrnl. Microsc. Sci.* XLIV. 8 Metœstrum or the Metœstrous Period.—If conception does not take place during œstrus the activity of the generative organs gradually subsides during a definite period, which I have called the metœstrum; and this is followed..by a long period of rest. 1923 *Amer. Jrnl. Anat.* XXXII. 340 Follicular phase [of the sexual cycle]: *a*, proestrus. *b*, oestrus. *c*, metoestrus (in certain species where ovulation is separated from oestrus by an interval). 1966 *McGraw-Hill Encycl. Sci. & Technol.* XI. 476/1 After metestrus a relatively longer period or diestrus occurs before proestrus recurs... All these phases can be demonstrated in the guinea pig and rat. 1973 *Nature* 12 Jan. 129/2 When the drug was administered on other days of the cycle, metoestrus, dioestrus or pro-oestrus, it failed to increase ovulation rate.

Hence **metœ·strous** *a.*

1900 [see above]. 1956 *Nature* 4 Feb. 235/1 These groups were of pro-œstrous, œstrous, metœstrous and diœstrous rats.

metonymic, *a.* (Examples.)

1952 M. MCCARTHY *Groves of Academe* (1953) v. 85 As he had paused in the hall outside Domnar's door..he had not yet..felt the metonymic urge that would prompt him ..to substitute the effect for the cause. 1969 P. ANDERSON in Cockburn & Blackburn *Student Power* 261 'I' is no longer I in the opaque, metonymic *double-entendre* of Freud's patients.

me too (mī tū). Phr. used alone or in various collocations (see quots.) of a person who, or a course of action which, adopts or acquiesces in the views, policies, etc., of someone else, often *spec.* those of one's political or other opponents. Freq. in a derogatory sense. orig. *U.S.* Also (with hyphen) as *vb.* Hence **me-too-er, me-too-ism.**

The colloq. ejaculation 'Me too!' signifying either that the speaker shares another person's view or experience or that the speaker wants the same share as another is getting, is the basis of the modern use.

1745 CHESTERFIELD *Let.* 13 Apr. (1932) II. 596 You must mark out Lord Granville by exterminating without quarter all who belong to him... If you take this resolution,..I empower you to make what use you please of my name as quitting with you; and I say as Will Seymour did, *And me, too, sweet Jesus.* 1851 H. MELVILLE *Moby Dick* I. xxxix. 275 Me too; where's your girls? 1873 L. M. ALCOTT *Work* II. xi. 336 'Me too!' cried little Ruth, and spread her chubby hand above the rest. 1921 H. WILLIAMSON *Beautiful Yrs.* 128 'Where are you going?' 'Out.' 'Where to man?' 'Mr. Norman's.'.. 'Norman's, you said? Right-o. Me, too!' 1922 JOYCE *Ulysses* 459 *Mrs. Mervyn Talboys*... He implored me..to give him a most vicious horsewhipping. *Mrs. Bellingham.* Me too. *Mrs. Yelverton Barry.* Me too. 1924 A. S. NEILL *Dominie's Five* xii. 245 'I'm going to start to-day to learn to read.' 'Me too,' said Donald. 1940 H. L. ICKES *Diary* (1954) III. 312, I think that Willkie overlooked the best chance that he had by being content merely to 'me too' the President on his foreign policies and most of his domestic ones instead of striking out for himself in a bold and positive way. 1949 *Time* 7 Feb. 10/3 We [*sc.* the Republicans] suffered because we tried to me-too the New Deal. 1949 *Sun* (Baltimore) 14 Dec. 2/1 A recommendation that..the Republican party..divest itself of 'me-tooism' and go to the people with a program clearly and unmistakably in opposition to that now offered by our opponents. 1951 *Economist* 10 Nov. 1109/1 He is against any 'Me Too' procedures, by which he means opposition acceptance of Administration policy at home or abroad. 1952 *Manch. Guardian Weekly* 1 May 3 If General Eisenhower gets the Republican nomination he will have to risk the shame of me-tooing the Truman policy. 1958 *Spectator* 1 Aug. 156/3 With the collapse of doctrinaire Socialism, and the adoption instead of a modified me-tooism, the range of controversy is being still further reduced. 1959 *Guardian* 28 Dec. 6/1 Wendell Willkie, Governor Dewey, and General Eisenhower were all..'me-tooers'—men whose chief appeal was to the vague, uncontroversial idealism of the non-partisan. 1960 *New Left Rev.* Jan./Feb. 1/2 The champions of 'me-too' advance into..an 'American' future. 1960 J. R. ACKERLEY *We think the World of You* 82 The almost mad stare with which her starting eyes pierced and searched my own for the answer to the only question in the world: 'Me too?' 1962 *Sunday Times* 19 Aug. 18 'I'm a rugged individualist: I think for myself.' 'Me too.' 'Same here.' 1965 *Economist* 14 Aug. 607/2 The unions have few independent ideas to contribute to this design for a better America; they were 'me tooers' in Roosevelt's New Deal, Truman's Fair Deal and Kennedy's New Frontier. 1967 *Spectator* 13 Oct. 419/2 The usefulness of 'me-tooism' in capturing the floating voter should not be underestimated. 1975 G. V. HIGGINS *City on Hill* i. 18 Sam gets round to saying it on the floor a week later, and he's me-tooing.

Metopirone (metopi·rōun). *Pharm.* Also **metopirone.** [f. 2-*met*hyl-1,2-di(*pyr*id-3-yl)-propan-1-*one*, the chemical name (see *METYRAPONE), with alteration of *y* to *i* and insertion of *o*.] A proprietary name for metyrapone.

1960 *Lancet* 17 Dec. 1332/2 Metopirone is an 11 β-hydroxylase inhibitor which presumably prevents the production of aldosterone by the adrenal gland. 1961 *Trade Marks Jrnl.* 8 Feb. 172/1 Metopirone... Pharmaceutical preparations and substances for human use and for veterinary use. CIBA Limited.., Basle, Switzerland. 1962 *Lancet* 19 May 1041/2 The duration of the action of metopirone was investigated, in order to determine the best dosage schedule in tests of pituitary function. 1973 *Arch. Internat. de Pharmacodynamie et de Thérapie* CCII. 93 Metopirone was found to provoke in normal, but not in adrenalectomized rats, a three-fold increase in blood sugar and a four-fold increase in plasma adrenaline.

metoposcopy. **2.** (Later example.)

c1886 L. HEARN in G. M. Gould *Concerning L. Hearn* (1908) 87 'Are you not a Greek?' I asked, for there was no mistaking the metoposcopy of that head. Yes; he was from Zante.

metovum. Substitute for def.: An ovum in its second stage, e.g. a meroblastic ovum after formation of the food-yolk; also called *deutovum* (DEUTO- 2).

Metrazol (me·trăzǫl). *Pharm.* Also **metrazol.** [f. *penta)me(thylenete)trazol*, its chemical name, f. PENTA- + METHYLENE + TETRA- + AZO- + -OL.] A proprietary name of *LEPTAZOL.

1928 *Official Gaz.* (U.S. Patent Office) 11 Sept. 265/1 E. Bilhuber, Inc... *Metrazol* for medicine for the heart and the vascular system. 1932 *Arch. Internat. de Pharmacodynamie et de Thérapie* XLII. 200 Cardiazol (metrazol) is pentamethylenetetrazol. 1950 *Brit. Jrnl. Psychol.* Dec. 104 Metrazol was used to induce convulsions in another unsuccessful attempt to apply therapy at a physiological level. 1973 *Physiol. & Behaviour* X. 94/2 Convulsive doses of Metrazol..may not actually prevent memory storage but rather modify the conditions under which these processes take place.

metre, *sb.*² Add: **a.** Now one of the base units of the International System of Units, and redefined in terms of the wavelength of a spectral line (at 605·8 nanometres) of an isotope of krypton (see quot. 1970). (Further examples.)

1961 *Nature* 21 Jan. 195/1 The eleventh General Conference of Weights and Measures was held in Paris during October 11–20... One epoch-making scientific decision was taken, namely, to redefine the metre in terms of a natural atomic standard, the wave-length of light, thus deposing the platinum–iridium bar—the International Prototype Metre—from the supremacy it has held in the field of length measurements since 1889. *Ibid.* 196/2 The specification of the conditions of excitation and observation [of the radiation] ensures a reproducibility of the new optical metre to about 1 part in 100 millions. 1970 *Internat. System of Units* (*B.S.I.*) 5 The metre is the length equal to 1 650 763·73 wavelengths in vacuum of the radiation corresponding to the transition between the levels $2p_{10}$ and $5d_5$ of the krypton-86 atom. (11th CGPM (1960), Resolution 6). 1971 *Physics Bull.* July 397/1 The definition of the metre in terms of the wavelength of the orange spectral line of a krypton discharge was internationally accepted in 1960 and brought about a new interest in interferometry for length measurement.

b. **metre-angle** *Ophthalm.* [tr. G. *meterwinkel* (A. Nagel, in Graefe & Saemisch *Handbuch der gesammten Augenheilkunde* (1880) VI. x. 479)], a unit of convergence equal to the angle between the line of sight of either eye and the median line passing between them when the eyes are fixating a point on that line one metre away; **metre-candle** = *LUX *sb.*; **metre-kilogramme-second,** used *attrib.* to designate a system of units in which the basic units of length, mass, and time are respectively the metre, the kilogramme, and the second, and which was taken as the basis of the International System of Units; commonly abbreviated *M.K.S., *m.k.s.*

1886 C. M. CULVER tr. *Landolt's Refraction & Accommodation of Eye* ii. 187 We are indebted to Nagel for the ingenious idea of rendering the mensuration of convergence so simple and..so practical. He calls this unit-angle the 'Meterwinkel' *metre-angle*. 1949 H. C. WESTON *Sight, Light & Efficiency* i. 27 The value of the metre-angle depends on the distance apart of the two eyes, but is commonly equal to about 1·75°. 1964 S. DUKE-ELDER *Parsons' Dis. Eye* (ed. 14) xxviii. 455 With an emmetropic person the amount of convergence, reckoned in metre angles, is the same as the amount of accommodation reckoned in dioptres. 1909 *Cent. Dict.* Suppl., Meter-candle. 1915 R. A. HOUSTON *Treat. Light* xx. 362 It is often necessary to measure in foot-candles or metre-candles the degree of illumination of a surface. 1939 A. W. BARTON *Text Bk. Light* ix. 197 A metre-candle is equal to a lumen per square metre,..so that there are 10,000 metre-candles in a phot. 1970 M. V. KLEIN *Optics* iv. 126 Lux are also called meter-candles. 1940 *Chambers's Techn. Dict.* 544/1 Metre-kilogramme-second (M-K-S) system of units. 1943 LEMON & FERENCE *Analytical Exper. Physics* ii. 38/1 Recently adopted by an international congress as the official system of metric units is the meter-kilogram-second (MKS) system. 1963 *Listener* 24 Jan. 156/1 The metre-kilogram-second system has been preferred by the Institute of Electrical Engineers.

metric, *a.*¹ and *sb.* Add: **A.** *adj.* **2.** (Examples.) Add to def.: *esp.* relating to, involving, or defining distance; *metric geometry* = *metrical geometry* s.v. METRICAL *a.*¹ 2; *metric space* [tr. Ger. *metrischer raum* (F. Hausdorff *Grundzüge der Mengenlehre* (1914) vii. 211)], a set together with a metric defined for all pairs of elements of the set.

1873 *Proc. Lond. Math. Soc.* IV. 387 This metric geometry is due to Prof. Cayley. 1910 VEBLEN & YOUNG *Projective Geom.* i. 12 The difference between projective and the ordinary Euclidean metric geometry. 1916 *Monthly Notices R. Astron. Soc.* LXXVI. 701 The line-element *ds* must be invariant for all transformations, and it entirely characterises the metric properties of the four-dimensional time-space. 1923 J. RICE *Relativity* xiii. 312 Space-time is a metric four-dimensional manifold. 1927 *Bull. Amer. Math. Soc.* XXXIII. 14 The following illustrations convey some notion of the scope of the concept metric space. If the aggregate *P* denotes the linear continuum of all real numbers and $(p, q) = |p - q|$, the resulting space is metric. Similarly euclidean space is also metric. 1963 R. A. ROSENBAUM *Introd. Projective Geom. & Mod. Algebra* i. 12 Note that all the items of this list of non-projective properties involve magnitudes of lengths and angles. Such 'metric' properties are the concern of traditional elementary geometry. The situation may be loosely described by stating that metric geometry treats of more highly restricted properties than those of projective geometry. 1968 E. T. COPSON *Metric Spaces* i. 3 The study of the properties of sets of 'points' in a 'space' whose only geometrical property is the existence of a 'distance' between each pair of 'points' is called metric space topology. *Ibid.* ii. 21 Different choices of metric on a given set *E* give rise to different metric spaces. 1971 *Nature* 5 Nov. 35/1 It has the properties of a metric function in a space, the elements of which are finite non-empty sets.

B. *sb.* **1.** (Further examples.)

1892 *Mod. Lang. Notes* VII. 100 Metrics and æsthetics go hand in hand. 1952 G. SARTON *Hist. Sci.* I. xx. 521 Ancient music included not only music as we understand it but also metrics, poetry. 1970 *Jrnl. Eng. & Gmc. Philol.* LXIX. 81 Rules for syntax and metrics in *Beowulf*.

2. *Math.* and *Physics.* A metric function, i.e. one defining a distance or an abstract quantity analogous to distance (see quot. 1962).

1921 *Proc. R. Soc.* A. XCIX. 104 In the non-Euclidean geometry of Riemann, the metric is defined by certain quantities, $g_{\mu\nu}$, which are identified by Einstein with the potentials of the gravitational field. 1934 C. C. KRIEGER tr. *Sierpiński's Introd. Gen. Topology* vi. 90 In every metric space *M* a metric, which is equivalent to the given one, can be established,..such that the new distances between the elements of *M* are all ≤1. 1956 E. H. HUTTEN *Lang. Mod. Physics* iii. 114 Riemann gave a better representation for the multiplicity of geometries by taking space to be a three-dimensional manifold, and it is the metric given by a mathematical expression referring to distance that decides the type of geometry. 1962 B. H. ARNOLD *Intuitive Concepts Elem. Topology* viii. 138 The function of *d* is a metric in *X* if and only if the following conditions are satisfied for all points *x*, *y*, and *z* of *X*. (1) $d(x, y) \geq 0$. (2) $d(x, y) = 0$ iff $x = y$. (3) $d(x, y) = d(y, x)$. (4) $d(x, y) + d(y, z) \geq d(x, z)$. 1965 J. D. NORTH *Measure of Universe* iv. 63 In 1908 Minkowski introduced into the context of the Special Theory of Relativity the metric now known by his name: $ds^2 = dt^2 - (1/c^2)(dx^2 + dy^2 + dz^2)$. The null geodesics of Minkowski space-time represent the paths followed by light.

3. = METRE *sb.*¹ 1 a.

1933 T. S. ELIOT *Use of Poetry* ii. 38, I do not even believe that the metric of *The Testament of Beauty* is successful. 1941 R. GIRVAN in *Proc. Brit. Acad.* 1940 331 The metric is astonishing and cannot be paralleled in Anglo-Saxon poetry. 1951 T. S. ELIOT *Poetry & Drama* ii. 20 He [*sc.* Yeats] wrote plays in verse..in a metric which..is not really a form of speech quite suitable for anybody except mythical kings and queens.

4. (See quot. 1934.)

1934 H. C. WARREN *Dict. Psychol.* 166/2 *Metric*, a system or standard of measurement in terms of which the conclusions stated hold. 1968 *Language* XLIV. 715 The simplicity metric demands that one choose the unmarked segment as the underlying one, since its choice leads to less complexity in the phonological representation. 1973 A. H. SOMMERSTEIN *Sound Pattern Anc. Greek* iii. 94 This statement does not depend on the acceptance of a feature-counting simplicity metric. P. H. Matthews, in a seminar at Cambridge, has criticized the assumption that such a metric is the appropriate evaluation measure for grammars.

metric, *a.*² Add: *metric ton*, 1000 kilogrammes (2204·6 lb. avoirdupois, or 0·9842 ton); = *TONNE.

1924 *Times Trade & Engin. Suppl.* 29 Nov. 238/3 The output of certain important goods has considerably increased:..sugar to 318,987 metric tons, against 270,279. 1957 G. E. HUTCHINSON *Treat. Limnol.* I. viii. 546 A quantity of salt of the order of 100,000 metric tons per year is transported. 1973 *Guardian* 25 Jan. 7/5 The Government yesterday took a hard line with newsprint manufacturers by allowing increases of only £2 a metric ton (tonne).

b. Having the metric system as the principal system of weights and measures.

1919 [see *NEWTON, 2]. 1960 *Nature* 2 July 30/2 There is no strong feeling in industry or commerce that Britain is being adversely affected..in exports to metric countries by the retention of the Imperial system. 1961 *B.S.I. News* Apr. 16/1 The controversial issue of whether this country should 'go metric'. 1971 *Sci. Amer.* Sept. 76/2 A metric America..would seem to be desirable in terms of our stake in world trade. 1974 *Times* 22 May 6 Virtually every country in the world is metric.

B. *sb.*² Metric measurement; metric weights and measures collectively.

1969 *Times* 21 July p. vii/5 Metric is so much easier to teach and to learn. 1970 *Daily Tel.* 21 May 10/5 Metric was so simple that a toddler could learn it, he said. And the litre was not strange to car-buyers. 1971 *Nature* 13 Aug. 439/1 The deliberate encouragement of the tendency towards metric could quite quickly create a situation in which the metric system was as widely used as the conventional system in the United States. 1973 *Country Life* 27 Dec. 2186/1 Once we get used to metric, things should be simpler for all of us.

-metric (me·trik), a terminal element of adjs. corresponding to sbs. ending in -METER or -METRY.

metrical, *a.*¹ Add: **1.** *spec.* applied to Old or Middle English verse.

1802 J. RITSON (*title*) Ancient Engleish metrical romancëes. 1807 S. TURNER *Hist. Anglo-Saxons* (ed. 2) II. 294 This poem [*sc.* *Beowulf*] is certainly a metrical romance in the Anglo-Saxon language. 1830 B. THORPE tr. *Rask's Gram. Anglo-Saxon Tongue* 150 The Anglo-Saxons..in many M.S.S., carefully separate the verse by metrical points. 1897 *Mod. Lang. Notes* XII. 79 This regularity of arrangement holds only for the half line, the metrical unit. 1923 G. SAINTSBURY *Hist. Eng. Prosody* (ed. 2) I. II. i. 90 The metrical romances present by far the largest section..of earlier fourteenth-century verse-literature. 1930 FRENCH & HALE *M.E. Metrical Romances* p. v, The metrical romances are the first large body of English fiction. 1946 *Trans. Philol. Soc.* 1943–6 59 What neither Sievers nor any other writers..have ever pointed out, is that the 'five types' are language patterns not metrical patterns. 1953 *Speculum* XXVIII. 449 The discovery of..the non-existence of metrical formulas in the poetry of lettered authors. 1963 R. QUIRK in Brown & Foote *Early Eng. & Norse Stud.* 159 Metrical units in variation.

metricalization. [-IZATION.] A making metrical in character.

1924 *Glasgow Herald* 13 Nov. 10 The question of the metricalisation of our coinage.

me·trically, *adv.*² [f. METRICAL *a.*² + -LY².] In or with the metric system.

1969 *Daily Tel.* 5 Sept. 25/2 A builder of iron and steel plants has found his design time reduced by 15 per cent. when working metrically.

metricate (me·trikēit), *v.* [Back-formation from next: see -ATE³.] **a.** *intr.* To change to or adopt the metric system of weights and measures.

1965 [see *METRICATION]. 1968 *Guardian* 27 July 3 It may not be practicable to metricate, except possibly over a far longer period than that suggested. 1970 *Daily Tel.* 14 May 18 Current Admiralty Tide Tables, now beginning to metricate in earnest, propose to use decimetres. 1972 *B.S.I. News* June 17/2 Did it not take..nearly a century for Britain to decide to metricate?

b. *trans.* To convert or adapt to the metric system.

1970 *Times* 28 Oct. 7 The cost of metricating road signs..would eventually have to be considered. 1972 *Daily Tel.* 13 Oct. 19/3 The Royal Navy..started to metricate Admiralty charts in 1970.

Hence **me·tricated** *ppl. a.*, made or sold in accordance with metric measurements; converted to or using the metric system.

1970 *Daily Tel.* 14 May 18 Your metricated cook cannot have, standing alongside the kitchen scales, a set of SI weights... There will be no SI weights. *Ibid.* 1 Dec. 3/2 When clothes are fully metricated, at least two things will still be recognisable in the old shapes, brassieres and shoes. 1971 *Timber Trades Jrnl.* 14 Aug. 54/3 This is the second season of metricated wood and the system is now accepted as normal practice among the merchants. 1972 *Bookseller* 18 Mar. 1642/3 Publishers have already invested hundreds of thousands of pounds in metricated editions of educational works. 1972 *B.S.I. News* June 12/1 The encouragement of the use of 'metricated' products.

metrication (metrikēi·ʃən). [f. METRIC *a.*² + -ATION.] The process of converting to the metric system of weights and measures; the adoption of the metric system.

1965 J. V. DUNWORTH (Director, Nat. Physical Lab.) in *Times* 29 Nov. 11/5 Earlier this year the National Physical Laboratory sought the guidance of the editor [*sc.* E. McIntosh] of the *Concise Oxford Dictionary* on this matter... His reply was as follows:—'Consider the following: Carbon(ize); decimal(ize);..methyl(ate); oxygen(ate). The modern tendency is to use -ize, -ization in forming new words rather than -ate, -ation. It seems to me that either *metrication* or *metricization* could be used.' In the light of these comments the N.P.L. and the Ministry of Technology chose 'metrication' on the grounds of brevity and euphony. The corresponding verb 'metricate' seems very satisfactory. 1966 *New Scientist* 6 Jan. 32/1 Some British scientists who have worked for years with dynes and oersteds, calories and millimetres of mercury, may think they need not share the worries of their fellow citizens about metrication. 1968 *Observer* 21 Apr. 7/1 The Board of Trade has refused to announce a programme for metrication. 1969 *Guardian* 27 Jan. 3/4 The metrication of large sections of British industry will soon be an accomplished fact. 1971 *Daily Colonist* (Victoria, B.C.) 25 Nov. 26/2 Creeping metrication is upon us and the CFL [*sc.* Canadian Football League] has taken official cognizance of the fact.

metrification. Add: **2.** = *METRICATION.

1965 *Observer* 5 Dec. 40/5 If of old measures we're forsakers..How describe it? There are cries For Metrification, Metricize. 1970 *Times* 17 Apr. 6 Now there was a threat of metrification of weights and measures. 1973 *Daily Tel.* 27 July 16 What angers us is the manner in which the Government has tried to persuade us that VAT..was to our benefit, like decimalisation, two-tier postal systems and now metrification. 1974 *Publishers Weekly* 25 Nov. 12/3 [*heading* 1980: Target Date for Metric Conversion in U.S.] Metrification has been called the non-issue of the century.

metrify, *v.* Add: **2.** *intr.* = *METRICATE *v.* a. *rare.*

1968 *Sunday Times* 31 Mar. 10 The Confederation of British Industry hopes that 75 per cent. of Britain's industries will have metrified by 1975.

metrizable (metrǝi·zāb'l), *a. Math.* [f. as next + -ABLE, tr. G. *metrisierbar* (P. Urysohn 1924, in *Math. Ann.* XCII. 275).] Of a topological space: capable of being assigned a metric which makes it a metric space identical to the original space.

1927 *Bull. Amer. Math. Soc.* XXXIII. 25 It is therefore of interest to formulate the conditions that a space be metrizable in terms of continuous functions. 1968 E. T. COPSON *Metric Spaces* ix. 142 General topology is a generalization of the theory of metric spaces since there are topological spaces which are not metrizable.

So **metrizabi·lity**, the property of being metrizable.

1927 *Bull. Amer. Math. Soc.* XXXIII. 23 Axiom 1 is a sufficient condition for metrizability. 1964 W. J. PERVIN *Found. Gen. Topology* x. 158 In the case of separable metric spaces, Urysohn..found necessary and sufficient conditions for metrizability.

metrization (metrǝizēi·ʃən). *Math.* [ad. G. *metrisation* (P. Urysohn 1924, in *Math. Ann.* XCII. 275): see prec. and -ATION.] The process of assigning a metric to a metrizable topological space.

1927 *Bull. Amer. Math. Soc.* XXXIII. 14 The Metrization Problem. The problem is to state in terms of the concepts point, and point of accumulation the conditions that a topological space be metric. 1937 *Ibid.* XLIII. 141 This theorem gives conditions for the metrization of neighbourhood spaces and a comparatively simple method of introducing the metric. 1964 W. J. PERVIN *Found. Gen. Topology* x. 158 The metrization theorem.

metro² (me·tro). *colloq.* [Fr., abbrev. of (*Chemin de Fer*) *Métropolitain* Metropolitan Railway.] The Metropolitan Underground Railway of Paris (usu. in form **métro**). Hence applied to the underground railway in other countries. (Applied to London trains *metro* (me·tro) is an abbreviation of METROPOLITAN *a.* rather than a use in English of F. *métro*.)

1904 A. BENNETT *Jrnl.* (1932) I. 202 Wandering down through the Palais Royal and then taking the Métro. 1919 MENCKEN *Amer. Lang.* 110 In England..a subway is always a *tube*, or the *underground*, or the *Metro*. 1924 S. STORY *Dining in Paris* 12 Business men, clerks, stock-brokers..have lit the eternal cigarette..before going to a restaurant for dinner or taking omnibus or 'metro' to their distant homes. 1927 W. E. COLLINSON *Contemp. Eng.* 66 Recently attempts have been made to put the short form metro [for the Underground railway in London] before the public. 1953 X. FIELDING *Stronghold* iv. ii. 263, I realized why the ibex..smelt so strongly of the Paris *métro*. 1963 *Listener* 7 Mar. 418/1 It was in the way of a roundabout and the new metro [in Rotterdam]. 1966 *Ibid.* 1 Dec. 801/1 The underground system..is called the *métro*, but, typical of bilingual Montreal, is more often referred to as the subway. 1973 *Nat. Geographic* May 658/1 The Metro, Mexico's new subway system, is one of the wonders of the city.

Metro³ (me·tro). *Canad.* [Abbrev. of METROPOLITAN *a.* 2.] The Metropolitan area of Toronto and other Canadian cities. Also *attrib.* or as *adj.*

1957 *Maclean's Mag.* 17 Aug. 3/3 Metro chairman..said a Bloor Street subway would be a mistake; Metro would study others. 1962 *Time* (Canad. ed.) 26 Jan. 10/2 Canada's second experiment with metropolitan government... The Metro [of Winnipeg] did not follow Toronto Metro's example. 1963 J. N. HARRIS *Weird World Wes Beattie* (1964) xi. 138 The threat that the Metro Police would try it if the provincials didn't was sufficient to settle the matter. 1966 *Globe & Mail* (Toronto) 25 June 3/4 [He] faced an indirect challenge during the election campaign from a ginger group of eight Liberal in metro Winnipeg ridings. 1968 *Ibid.* 5 Feb. 1/9 There were 25,000 [drunks] charged in Metro Toronto last year. 1970 *Toronto Daily Star* 24 Sept. 35/2 (Advt.), This outstanding offer is good in Metro Toronto only. *Ibid.* 35/3 Metro-area delegates. 1975 *Globe & Mail* (Toronto) 11 June 3/4 A Toronto criminal lawyer says Metro police have told him some members of the Rastafarian Brethren are walking time bombs as far as violent crime in Metro is concerned.

metroland (me·trolænd). [f. METRO(POLITAN *a.* + LAND *sb.* 3.] The area surrounding a metropolis; *spec.* the district around London served by the (Metropolitan) underground railway. Also, *collect.* the people inhabiting these areas. Hence **me·trolander** *sb.*

1926 R. MACAULAY *Crewe Train* III. v. 295 That house at Great Missenden..will suit them exactly. In metroland, and such nice people all about... They *must* have a car, though; relying entirely on the Met. is too awkward, with so many strikes and so few late trains... After all, it's not London; metro-land can't be London. *Ibid.* vi. 298 Metro-landers have *pieds-à-terre* in London. 1938 J. BETJEMAN *Oxf. Univ. Chest* v. 102 The houses of Metroland and beechy Bucks dot the landscape. 1940 GRAVES & HODGE *Long Week-End* viii. 114 In 1923 the London Underground, wishing to popularize 'Metroland'..published two guide-books. 1951 R. HOGGART *Auden* v. 137 The appropriate isolation for Auden is..isolation in a vast anonymous metroland such as New York. 1963 *Times* 24 May p. iii/3 Under that great general manager Mr. R. H. Selbie, the Metropolitan invented Metroland. 1973 *Radio Times* 26 Feb. 7 The Metro-landers who bought houses took a great pride in their gardens.

Metroliner, metro-liner (me·trolǝi:nǝɹ). [f. METRO(POLITAN *a.* 2 + LINER² 8.] A high-speed inter-city train in the United States.

1969 'O. BLEECK' *Brass Go-Between* xiv. 156 Between Washington and New York they finally got one new high-speed Metroliner running. 1970 *Guardian* 19 Jan. 4/1 The metroliner, a Government-supported experiment in fast inter-city transport..has cut the travelling time from Washington to New York..by nearly ninety minutes. 1972 *Daily Tel.* 23 Nov. 9/2 The official said that the trains at present operating as metro-liners between the three [American] cities were capable of 150 mph. 1973 J. DI MONA *Last Man at Arlington* (1974) 57 The Metroliner to Washington waited in the station.

metrologist. (Later examples.)

1969 *Physics Bull.* Sept. 365/2 A number of speakers considered that would-be metrologists would be better advised to make a broad study in science or engineering at the undergraduate level and specialize in metrology by spending a further year taking a MSc course in this discipline. 1970 *Daily Tel.* 14 May 18 For heaven's sake, exasperated metrologists explain, do try to remember that shop scales do not show the weight of apples; they show the force exercised by the apples in that particular gravitational environment.

metrology. **1. b.** (Later examples.)

1969 *Physics Bull.* Sept. 365/2 A large proportion of conference participants deprecated the term 'measurement science' and the use of the term 'metrology' appeared to have a large measure of support. 1971 *Inside Kenya Today* Mar. 29/2 We might have weights manufactured to International Specifications, i.e. to specification of Legal Metrology.

metronome. Add: **b.** *fig.* Also as *v. intr.*

1959 *Listener* 2 Apr. 600/2 They..listened to the stillness of the white moonlight metronomed to the trot, trot, trot of the horse's hooves. 1962 L. DEIGHTON *Ipcress File* ii. 21 Pin-tables metronoming away the sunny afternoon.

metronomic, *a.* Add: **b.** *fig.* Resembling the action of a metronome.

1959 *Times* 30 May 3/1 The Hungarian..is metronomic as she unwinds the rallies. 1963 *Listener* 3 Jan. 45/2 The metronomic dance music rhythm of our time. 1975 'E. LATHEN' *By Hook or by Crook* xi. 107 Miss Martineau sobbed with metronomic regularity.

metronomically, *adv.* (Later examples.)

1970 R. P. WARREN *Incantations* 49 The disturbance you are so metronomically creating. 1974 *Daily Tel.* 14 Sept. 11/8 Both in the opening of the symphony and in the Adagio Mr Davis adopted and almost metronomically maintained tempi as fast as Beethoven has presumably imagined them.

me:tronomiza·tion. [f. METRONOME + -IZATION.] The determining or indicating of the rate at which music should be played.

1923 A. BETTI in *Music & Lett.* Jan. 3 Can the metronomisation of a piece be absolutely exact?

metronym (mī·tronim). [f. Gr. μητρ-, μήτηρ mother + ὄνομα, Doric ὄνυμα name.] A metronymic name.

1904 *Nature* 5 May Suppl. p. xiii/2 The acceptance of metronyms in the genealogies as proofs of female kinship, while patronyms are rejected.

metronymic, *a.* and *sb.* **a.** *adj.* Add to def.: Also applied to a people or state of society where such a system of naming prevails.

1896 F. H. GIDDINGS *Princ. Sociol.* 158 In a metronymic group all relationships are traced through mothers; paternal relationships are ignored. 1903 L. F. WARD *Pure Sociol.* 339 The metronymic family. 1944 H. P. FAIRCHILD *Dict. Sociol.* 192/2 *Metronymic*, deriving the personal or family name from the mother or other matrilineal relative. 1960 C. WINICK *Dict. Anthropol.* 521/2 *System, metronymic*, tracing kinship exclusively through the mother.

metrop (mĭtrǫ·p). Colloq. abbrev. of METROPOLIS 2.

1888 [see *KILO²]. 1919 WODEHOUSE *My Man Jeeves* 216, I think we've had about enough of the metrop. for the time being. 1925 —— *Carry on, Jeeves!* v. 105 Dear old Rocky made him look like a publicity agent for the old metrop! 1974 'A. GILBERT' *Nice Little Killing* i. 13 'Know the Metrop?' She shook her head. 'You want to be a bit careful.'

metropole. **1.** Delete † *Obs.* and add later examples.

1937 A. HUXLEY *Let.* 3 June (1969) 422 Our nearest railway is Santa Fe, 70 miles away, and our metropole is Denver at 350 miles. 1970 *Financial Times* 23 Mar. 19/1 What Algeria lacked could easily come from the metropole. 1973 *Caribbean Contact* Jan. 10/2 The ill-defined..countries of the Caribbean Sea are becoming states in their own right, and are slowly..severing their traditional ties with their respective European metropoles.

3. A luxury hotel.

1890 W. BOOTH *In Darkest Eng.* II. vi. 209 A superior lodging-house, a sort of poor man's Metropole. 1925 E. SITWELL *Troy Park* 74 That child is the small wicked ghost Of Metropoles and oyster bars.

metropolitan, *a.* and *sb.* Add: **A.** *adj.* **2. a.** (Later examples.)

1930 H. CRANE *Let.* 29 Nov. (1965) 359 In one of the many Doubleday-Doran shops in the metropolitan area. 1936 [see *HINTERLAND]. 1958 *Listener* 11 Dec. 981/2 Black cities; white suburbs—that is how the current trend is often summarized... The pattern of 'metropolitan segregation' (as it has been called) has opened a new, and a frightening, chapter of the 'American Dilemma'. 1961 E. A. POWDRILL *Vocab. Land Planning* v. 89 The 'metropolitan region' is a giant urban regional system, different less in kind than in size from that of any urban region. 1963 *Times* 7 June 3/7 It contains a detailed breakdown of the rates levied by the 83 county boroughs and 28 metropolitan boroughs. 1969 *Daily Tel.* 12 June 23/2 Three areas—Manchester, Birmingham and Liverpool—will become Metropolitan authorities, with the key functions of planning, transportation and development... These three will have below them metropolitan district authorities running education, the personal social services, health and housing. 1971 *Ibid.* 17 Feb. 9 Six new metropolitan county councils are proposed by the Government in its plans for reorganisation of local government published today. 1972 *Times* 12 Feb. 14/8 His Lordship had not failed to observe the practice in some metropolitan courts. 1973 *Times* 12 May 1/1 The 36 metropolitan district councils in England and 37 district councils in Wales created under the Local Government Act 1972, will take over in 12 months' time the statutory powers of the present bodies which they will replace. 1974 *Daily Tel.* 1 Apr. 6/4 The 45 new counties will contain 332 districts, including 36 metropolitan districts covering towns in the six metropolitan counties.

b. Of or pertaining to an underground railway serving a large city; *spec.* of the London Underground Railway, now extended overground to serve an extensive suburban area. Also *ellipt.*

1867 TROLLOPE *Claverings* II. xix. 233 He was very keen at the present moment about Metropolitan railways. c1875 'BRENDA' *Froggy's Little Brother* (new ed.) iii. 33 A Metropolitan train was just in, and a crowd of passengers, as usual, came swarming up the steps into the street. 1883 E. W. HAMILTON *Diary* 13 Nov. (1972) II. 505 The proposal of the Metropolitan Railway to run an underground line from the back of the India Office to Knightsbridge, and thence across the Park to the Marble Arch and up the Edgeware Road. 1909 CHESTERTON *Tremendous Trifles* 244 A Metropolitan station, where I took a train home. 1934 H. G. WELLS *Exper. Autobiogr.* II. ix. 817 Moscow also is making an imitative tube system... It will be the least stable 'Metropolitan' in the world. *Ibid.* 819 The constructors of the new Metropolitan. 1959 [see *DISTRICT *sb.* 3 g]. 1974 M. BIRMINGHAM *You can help Me* ii. 38 The Metropolitan train from Euston Square..[to] Aldgate East.

c. Of a type of early English pottery found in or near London: (see quots.).

1891 J. E. & E. HODGKIN *Examples Early Eng. Pott.* 6 Of a less decorative character than most of the slip-decorated pieces is the ware which we have classed..as *Metropolitan Slip*, the pieces in this group having been mainly found in or near London. *Ibid.* 9 Metropolitan Slip... A Jar of elegant shape, recently dug up..near Bishopsgate Street. 1903 R. L. HOBSON *Catal. Eng. Pott. Brit. Mus.* 108 Examples of Metropolitan Slip ware, made of red clay, with ornament in white slip and a transparent yellowish lead glaze. 1924 RACKHAM & READ *Eng. Pott.* iii. 28 Another type of ware, showing the same technical methods as the Staffordshire slipwares, has been given the name of 'Metropolitan', because it has usually been found in or near London. *Ibid.*, The earliest date on a piece of Metropolitan ware is on a jug.., inscribed..1638. 1957 MANKOWITZ & HAGGAR *Conc. Encycl. Eng. Pott. & Porc.* 149/1 *Metropolitan slipware*, the name given to a class of red earthenware decorated with white trailed slip... Examples dated from 1638 to 1659 are recorded. *Ibid.* Plate 66 (caption) Metropolitan jug. 1967 *Times* 14 Mar. 21/7 (Advt.), A Metropolitan ware silver-mounted jug.

3. Freq. with reference to France: of or pertaining to the home country (as distinct from colonial territories). (Further examples.)

1910 *Encycl. Brit.* X. 795/2 The organization of the 'metropolitan troops' [in France] by regiments. 1943 H. NICOLSON *Diary* 4 Feb. (1967) 278 In Metropolitan France de Gaulle is the great symbol. 1958 *Optima* Mar. 22/1 Were peace to be restored, metropolitan France would have, within the following few years, to devote 2½ per cent. of her national revenue to raising the Algerian standard of living. 1959 B. & R. NORTH tr. *Duverger's Pol. Parties* (ed. 2) II. ii. 330 The practice of alliances..made it possible for the Centre parties to gain 61% of the seats in metropolitan France with 51·4% of the votes. 1972 *Sci. Amer.* Apr. 19/3 Back in colonial times the metropolitan countries certainly maintained peace in and among their colonies.

metroscope². (Earlier example.)

1845 in C. Cist *Cincinnati Misc.* 270 A very ingenious instrument, called a Metroscope, which has been lately invented for the purpose of taking the measure of the human head so as to furnish an exact fit of hats.

metrostyle (me·trǒstǝil). [f. Gr. μέτρον measure + STYLE *sb.*] A device for regulating the speed of a mechanical piano. (Now disused.)

1904 E. NEWLANDSMITH *Temple of Art* 152 A totally wrong rendering of the work..is obviated in the Pianola by an apparatus called the 'metrostyle'. By means of following with the metrostyle pointer a certain line drawn on the paper music-roll [etc.]. 1907 *World* 16 July 140/2 The Model 'K' is a pianola equipped with the Metrostyle. 1909 H. G. WELLS *Tono-Bungay* I. ii. 76 There was a different grand piano with a painted lid and a metrostyle pianola.

Hence **me·trostyle** *v. trans.* and *intr.*, to regulate the speed of (a mechanical piano roll) by a metrostyle; to employ a metrostyle.

1908 G. KOBBÉ *Pianolist* ii. 31 Grieg—here are a couple of rolls from his 'Peer Gynt' suite metrostyled by himself. 1920 E. NEWMAN *Piano-Player* 147 The roll [of the piano-player] should be metrostyled by some artist who knows the work thoroughly. *Ibid.* 148 Careful metrostyling would no doubt do away with the necessity for most of the time signs.

‖ **metteur en scène** (mętǒr aṅ sę̄n). [Fr., lit. 'one who puts on the stage'.] A producer of a play; a director of a film.

1911 *Proc. Musical Assoc.* Mar. 94 The producer or '*metteur en scène*' of a play draws up a plan of the whole action in every detail. 1921 CONRAD *Let.* 23 Oct. in G. Jean-Aubry *J. Conrad: Life & Lett.* (1927) II. 262 One of our most clever producers (*metteurs en scène*). 1930 *Times Lit. Suppl.* 17 Apr. 333/2 Mr Jacques Arnavon is a great *metteur-en-scène*. 1963 *Movie* Feb. 36/2 To treat *The Barber of Stamford Hill* as a tentative work is to predict that Wrede could become a notable *metteur en scene*. 1968 *Times Lit. Suppl.* 26 Sept. 1079/1 For Clair's *écriture*, as he readily admits, is primarily a matter of words on paper; from then on he is literally a *metteur en scène*. 1968 L. DURRELL *Tunc* iv. 196 But she is being directed and rehearsed by the *metteur-en-scène*. 1974 *Times* 4 Jan. 8/7 Murnau's greatness as *metteur-en-scène* is unimpaired by time.

metump (line). *N. Amer.* = TUMP-LINE.

1754 in *Coll. New Hampsh. Hist. Soc.* (1824) I. 279 The deponent sold the said Indians two shirts,..and there was next to their skin tied a number of small metump lines, not such as are usually made for tying packs. 1963 W. S. AVIS et al. *Dict. Canad. Eng., Intermediate Dict.* 561/1 *Metump*, a broad strap or headband that is passed around the forehead and attached to a load carried on the back.

‖ **mettwurst** (me·tvŭɹst). [G.] A type of smoked German sausage.

1895 *Army & Navy Co-op. Soc. Price List* 104/2 Mettwurst. 1911 [see *BLUTWURST]. 1966 W. S. RAMSON *Austral. Eng.* 161 Some unrecorded borrowings from German may be in local use in parts of South Australia. Price noted that..*mettwurst*,..'sausage',..had 'penetrated to the English colonists'. 1971 *Sunday Times* (Colour Suppl.) 27 June 50/1 Mettwurst can be found in the form of a small sausage ('ends') or a horseshoe-shaped 'Westphalian' ring. Very heavily smoked; lightly spiced, no garlic. Mettwurst is eaten cold, sliced, but is also very good poached, sliced fairly thickly and served with boiled potatoes and cabbage.

metyrapone (meti·-, metǝi·ə·răpōun). *Pharm.* [f. 2-*met*hyl-1,2-di(*pyr*id-3-yl)*p*ropan-1-*one*, the chemical name (f. METHYL + DI-² + PYRIDYL + PROPAN(E + *-ONE), with insertion of *a*.] A whitish crystalline compound, $C_5H_4N{\cdot}CO{\cdot}C(CH_3)_2{\cdot}C_5H_4N$, which inhibits the synthesis of cortisone and hydrocortisone and is used for testing the function of the anterior pituitary.

1962 *Lancet* 8 Dec. 1199/1 This paper reports the prevention of massive adrenal necrosis due to D.M.B.A. by the concurrent administration of metyrapone. 1970 PASSMORE & ROBSON *Compan. Med. Stud.* II. vi. 14/2 The metyrapone test depends upon the fact that the feedback control mechanism between the adrenal cortex and the hypothalamo-pituitary axis is mediated by the circulating concentration of unbound cortisol. 1973 *Clin. Pharmacol. & Therapeutics* XIV. 455 After 1 year of treatment urinary steroids and their responses to oral metyrapone were unchanged.

‖ **meunière** (mönyę̄r), *a.* and *adv. Cookery.* [a. F. (*à la*) *meunière*, lit. 'in the manner of a miller's wife'.] Cooked in or served with hot butter (see quots.).

Usu. following the sb. qualified, as *trout (à la) meunière*, etc.

1846 A. SOYER *Gastronomic Regenerator* 102 Brill à la Meûnière. 1895 G. A. SALA *Thorough Good Cook* 155 Sole à la Meunière. 1903 C. H. SENN *Pract. Gastron.* 74 (Truite) à la Meunière (Trout, Meunière Style).—Braised trout served with burnt butter, breadcrumbs, and chopped parsley. 1958 J. GROSSINGER *Art Jewish Cooking* (1962) 34 Baked stuffed brook trout meunière. 1959 *Good Food Guide* 145 Good filleted haddock meunière. *Ibid.* 161 Rainbow trout meunière. 1974 *Times* 1 June 13/3 We..opted for a well-known Mexican delicacy, *crialladas*... Served meunière, they tasted delicious.

Meursault (mörso). [Name of a commune in the department of Côte d'Or, France.] A white wine of Burgundy, produced near Beaune.

1833 C. REDDING *Hist. Mod. Wines* v. 99 Between Volnay and Meursault the vineyard of Santenot is situated; it..produces a celebrated white wine, called Meursault. 1907 *Yesterday's Shopping* (1969) 97/2 White Burgundy..Meursault. 1928 E. WAUGH *Decline & Fall* II. vi. 198 He dined at Basso's..off bouillabaisse and Meursault. 1963 *Sunday Times* (Colour Suppl.) 27 Oct. 51/1 Just before serving, poach stuffed trout in fish stock and Meursault. 1967 A. LICHINE *Encycl. Wines* 354/1 Most Meursault is white and the wines are soft, round and feminine in texture, with a bouquet that eludes description, bordering now on the scent of violets, now on the aroma of almonds.

mevalonic (mevălǫ·nik), *a. Chem.* [f. ME(THYL + VAL(ERIC *a.* + LACT)ON(E + -IC.] *mevalonic acid*: 3,5-dihydroxy-3-methylvaleric acid, $C_6H_{12}O_4$, a crystalline compound which is a growth factor for some lactobacilli and is a precursor of cholesterol in animals and of carotenoids in plants.

1957 D. E. WOLF et al. in *Jrnl. Amer. Chem. Soc.* LXXIX. 1486/1 The name 'mevalonic acid' is now being used to designate β,δ-dihydroxy-β-methylvaleric acid. 1962 H. A. KREBS in A. Pirie *Lens Metabolism Rel. Cataract* 351 The reaction sequence is inhibited by cholesterol itself and some of its immediate precursors by stopping an early stage, probably that giving rise to mevalonic acid. 1973 *Plant Physiol.* LI. 110/1 Mevalonic acid-2-^{14}C was readily incorporated into the free, esterified, and glycosidic sterol fractions of tobacco..seedlings.

Hence **meva·lonate**, the anion of mevalonic acid.

1959 A. WHITE et al. *Princ. Biochem.* (ed. 2) xix. 488 Mevalonic kinase..catalyzes the formation of phosphomevalonic acid and ADP from mevalonate and ATP. 1971 HUNTER & ROSE in Rose & Harrison *Yeasts* II. vi. 243 Brewer's solubles were shown to contain a compound which could replace acetate in sterol biosynthesis, and this compound was shown to be mevalonate.

Mewari (mewā·ri), *sb.* and *a.* [f. *Mewar*, a former native state of India, also known as Udaipur, now part of the state of Rajasthan.] **A.** *sb.* The language spoken in Mewar. **B.** *adj.* Of or pertaining to Mewar or its inhabitants.

1888 KIPLING *From Sea to Sea* (1899) I. 48 The Thakur..spoke shrilly in Mewari. The Englishman replied in English-Urdu. *Ibid.* 51 Mewari [Mewarri, 1888 newspaper publ.]..is a heathenish dialect, something like Multani to listen to. 1925 A. HUXLEY *Let.* 21 Dec. (1969) 262 Bikaner [is]..inhabited..by a community which counts more millionaires to the thousand than any other population in India, perhaps in the world. These are the Mewari merchants, the Jews of India. 1958 [see *MEWATI *sb.* and *a.*].

Mewati (mewā·ti), *sb.* and *a.* Also **Mewatti.** [Native name (see def.).] **A.** *sb.* **1.** An Indian people native to Mewat, a region south of Delhi and now part of Rajasthan; a member of this people, *spec.* one professing Islam. Cf. *MEO. **2.** The language of this people, a dialect of Rajasthani. **B.** *adj.* Of or pertaining to this people or their language.

1788 J. RENNELL *Mem. Map Hindoostan* p. xlix, Of the state of the internal government of Hindoostan, a judgment may be formed, by the punishment inflicted on the Mewatti, or the Banditti tribe, which inhabit the hilly tract, within 25 miles of Delhi. 1824 J. MALCOLM *Mem. Cent. India* (ed. 2) II. xiv. 174 The Mewatties, a well-known Mahomedan tribe in Hindustan, have long resorted to Central India. 1832 J. TOD *Ann. Rajast'han* II. 393 Girdhur..with a small but select band hunted the Mewatti leader down, and..slew him in single combat. 1855 H. H. WILSON *Gloss. Judicial & Revenue Terms* 340/2 *Mewáti*,..a tribe of Rajputs inhabiting the province of *Mewat*, now known as *Mucheri*, and formerly notorious for their turbulent and predatory character. 1880 P. W. POWLETT *Rajputana Gazetteer* III. 170 These repeated expeditions against the Mewáttis did not render them quiet. *Ibid.* 171 Rewári is referred to as being in the hands of a Mewátti chief. *Ibid.* 174 Humayun..conciliated them..by causing his minister Bairám Khan to marry a younger daughter of the same Mewátti. 1896 W. CROOKE *Tribes & Castes N.-W. Provinces & Oudh* III. 493 Common proverbs..mean that, in dealing with a Mewâti, you had better kick or abuse him before you do business with him. 1901 [see *JAIPUR]. 1908 G. A. GRIERSON *Ling. Survey India* IX. II. 44 Mêwâtî is, properly speaking, the language of Mewat, the country of the Mêôs, but it covers a larger tract than this. It is the language of the whole of the State of Alwar, of which only a portion is Mewat. *Ibid.* 45, I am not acquainted with any literary work in the Mêwâtî dialect. 1908 *Imperial Gazetteer India* XVII. 313 The Muhammadan Meos call themselves Mewātīs. 1914 H. A. ROSE *Gloss. Tribes & Castes Punjab & N.-W. Frontier Province* III. 80 In the Muhammadan historians the Meos appear to be unknown by that name, but the Mewátis were notorious throughout the Muhammadan period. 1957 [see *JAIPUR]. 1958 B. N. PRASAD in V. K. Narasimhan et al. *Lang. India* 90 The New or Modern Indo-Aryan languages... Rājasthānī (Mārwarī, Mewārī, Mālawi, Jaipuri and Mewatī). 1974 *Encycl. Brit. Micropædia* VIII. 395/1 Rajasthani's 20 dialects are classified into four main groups: the northeastern Mewati [etc.].

Mewoh, var. *MEO.

mews. Add: **2. c.** A mews converted into accommodation for people; freq. *attrib.*

1805 [in Dict., sense 2 a]. 1848 THACKERAY *Vanity Fair* xxxiv. 305 Come down with me to Tom Corduroy's, in Castle Street Mews, and I'll show you such a bull-terrier. 1885 *List of Subscribers, Brighton* (South of Eng. Telephone Co.) 4 Nye, Sons & Silverthorne—Regency-mews. 1932 *Times Lit. Suppl.* 10 Nov. 841/1 Elizabeth..lived in the inevitable mews flat. 1954 T. S. ELIOT *Confid. Clerk* I. 9 And the flat in the mews? How soon will that be ready for him? 1958 *Observer* 18 May 16/4 Those saucy, disreputable mews dwellings. 1964 D. FRANCIS *Nerve* ii. 21 The big converted mews garage which served her as sitting-room, bedroom and rehearsal room. 1974 M. BABSON *Stalking Lamb* xvii. 127 He intended to make the mews house his operational headquarters.

Mex (meks), *a.* and *sb.* **a.** *U.S.* Colloq. abbrev. of MEXICAN *a.* and *sb.*

In quot. 1913, 'Mexican money'.

1854 G. D. BREWERTON in *Harper's Mag.* Apr. 584/2, I thought it proper to consult with one of the Quartermaster's agents..which resulted in my receiving the information that the 'United States Hotel' upon the 'Plaza' provided 'chicken fixins and corn doins'—or, if a 'stranger' wanted 'Mex livin', *frijoles* and *tortillas* to boot—in better style than any other establishment in Santa Fé. 1906 *McClure's Mag.* June 121/1 'Where'd you get the coat?' I asked the Mex. 1913 *Ibid.* Mar. 119 Is that gold or Mex, dear? 1929 J. PARKER *Old Army* 246 He was evidently not one of those who looked with pleasure upon 'Mex' officers. 1934 J. M. CAIN *Postman always rings Twice* 97 Ensenada is all Mex. 1970 G. JACKSON *Let.* 22 Mar. in *Soledad Brother* (1971) 187 An Italian..killed a Mexican in Folsom because the Mex suddenly started telling everyone not to trust someone. 1974 *Times* 21 Feb. 11/6 John Wayne's *The Alamo* ('better Tex than Mex') is a Goldwater Western.

b. *U.S. Forces' slang.* Foreign currency, esp. that of the Philippine Islands.

1898 *Amer. Soldier* (Manila, Philippine Islands) I. x. 8/1 We will send a set..to any address in America for 25 cents (mex). 1907 *Army & Navy Life* (U.S.) June 679 M stands for money, which is mostly in 'Mex'. 1926 ANDERSON & STALLINGS *What Price Glory?* in *Three Amer. Plays* II. 67, I wouldn't take a hundred dollars Mex. for that. 1941 T. CRUSE *Apache Days & After* 287 The rent was $100 a month 'Mex', which translated as $50 in gold.

Mexican, *a.* and *sb.* Add: **A.** *adj.* **b.** *Mexican blanket, cotton, eagle, flycatcher, saddle, trader, wagon*; **Mexican fruit fly,** a central American insect pest, *Anastrepha ludens*; **Mexican hairless (dog),** a small dog of the breed so called, lacking hair except for tufts on the head and tail; **Mexican hog,** the peccary; **Mexican orange (-blossom, -flower)** = *CHOISYA; **Mexican overdrive** *U.S. slang* (see quots.); **Mexican poppy** (earlier and later examples); **Mexican thistle** (examples); **Mexican War,** the war of 1846–8 between the U.S.A. and Mexico in which the allegiance of Texas was the most important issue.

1834 A. PIKE *Prose Sk. & Poems* 74 We gave him a red and gaudy Mexican blanket. 1894 *Harper's Mag.* Jan. 299/1 He had parted with Pedro for forty dollars, a striped Mexican blanket, and a pair of spurs. 1827 *Western Monthly Rev.* I. 82 The kinds of cotton which are chiefly cultivated are Louisiana, green seed, or Tennessee, and recently Mexican. 1834 R. BAIRD *View of Valley of Mississippi* xxiv. 304 Cotton is the chief staple. Three kinds are cultivated,—sea island, Mexican, and green seed. 1835 in *Southwestern Hist. Q.* (1925) XXVIII. 190 The most common birds and fowls found [in Texas] are the Mexican eagle, the hawk, [etc.]. 1836 M. HOLLEY *Texas* v. 100 The Mexican eagle, which is among the smallest of the aquiline tribe. 1870 *Amer. Naturalist* III. 473 A solitary Mexican Fly Catcher..gave a specimen of the summer group of migrants. 1924 *Monthly Bull. Calif. Dept. Agric.* XIII. 55 The Federal Horticultural Board placed a quarantine against the Mexican fruit fly in Mexico. 1947 *Jrnl. Econ. Entomol.* XL. 483/1 Neither four applications of a DDT spray nor three applications of DDT dust gave significant reductions in populations of the Mexican fruit fly, *Anastrepha ludens*. 1972 SWAN & PAPP *Common Insects N. Amer.* 627 The Mexican Fruit Fly, *Anastrepha ludens*, occurs in Central America, Mexico, and southern Texas, attacks citrus and mangoes chiefly; it is conspicuous—considerably larger than a house fly, brightly colored with attractive wing pattern. 1899 R. B. LEE *Hist. & Descr. Mod. Dogs Gt. Brit. & Ireland (Non-Sporting Division)* (new ed.) xix. 410 A peculiar looking object, and one which requires a great stretch of the imagination to call handsome, is the hairless, or, rather, the crested dog... He is mostly known as the Mexican hairless dog. 1948 C. L. B. HUBBARD *Dogs in Brit.* 296 The Mexican Hairless Dog..is recognised by the American Kennel Club and classified as a Toy Dog. 1970 *New Yorker* 28 Feb. 31/1 Larry Wolf..totally bald..plays a Mexican hairless. 1971 F. HAMILTON *World Encycl. Dogs* 537 Although not recognized by the British or the American Kennel Clubs, the Xoloizcuintli (pronounced Shollosquintly) or Mexican Hairless Dog, is one of the oldest breeds, now almost extinct. 1821 T. NUTTALL *Jrnl. Trav. Arkansa* ix. 216 The *Sus tajassu* or Mexican hog is not uncommon some distance higher up Red river. 1836 M. HOLLEY *Texas* v. 95 The Pecari or Mexican hog is even yet occasionally met with on the frontiers, in considerable gangs. 1899 G. JEKYLL *Wood & Garden* vi. 71 On the southern sides of the same gateway are two large bushes of the Mexican orange-flower (*Choisya ternata*), loaded with its orange-like blooms. 1951 *Good Housek. Home Encycl.* 115/2 Mexican Orange, a bright-leaved, white-flowering shrub. 1971 H. L. V. FLETCHER *Pop. Flowering Shrubs* viii. 176 *Choisya ternata*, an evergreen, is called the Mexican Orange Blossom. It has white scented flowers all spring and summer. It is easy in some gardens but of doubtful hardiness. 1975 *Times* 5 July 10/5 (*caption*) The Mexican orange..and *Senecio laxifolius*..both respond well to hand trimming. 1961 *Amer. Speech* XXXVI. 272 Mexican overdrive... Coasting down hill with gears disengaged. 1971 M. TAK *Truck Talk* 105 *Mexican overdrive*, the 'gear' a trucker uses when, in going downhill, he throws the transmission out of gear and lets the truck coast down in neutral. 1848 W. H. EMORY *Notes Mil. Reconn.* 13 We find in the bottoms Mexican poppy. 1936 F. CLUNE *Roaming round Darling* xxi. 211 We crossed a score of creeks in ten miles, fringed by tobacco-bush with privetlike leaves, and Mexican poppy simulating a Scotch thistle. 1965 *Austral. Encycl.* VII. 188/2 Most widespread and troublesome of introductions among the Papaveraceae is the Mexican poppy (*Argemone mexicana* and its variety *ochroleuca*), which was naturalized about Sydney more than a century ago. 1846 in *Calif. Hist. Soc. Q.* (1942) XXI. 203 Many of them [*sc.* Indians] had Mexican saddles, cartridge boxes, and different parts of the Mexican dress. 1848 *Knickerbocker* XXXI. 328 Their strong, gaunt horses were equipped with rusty Spanish bits, and rude Mexican saddles. 1865 *Atlantic Monthly* Jan. 59/2 A Mexican saddle,—out of which you can scarcely fall. 1910 J. HART *Vigilante Girl* 345 She galloped on in her high-peaked Mexican saddle. 1837 J. MACFADYEN *Flora Jamaica* I. 20 *Argemone Mexicana*. Mexican or Gamboge Thistle. 1906 F. BLERSCH *Handbk. Agric. S. Afr.* 144 Mexican poppy or yellow poppy, usually called Mexican thistle at the cape. 1826 H. G. Rogers in H. C. Dale *Ashley-Smith Explor.* (1918) 208 He is what they term here [*sc.* in the Los Angeles area] a Mexican trader. 1862 M. D. Colt *Went to Kansas* 37 So here may be seen the huge Mexican wagon, stubborn mule, swarthy driver. 1846 *Dollar Newspaper* (Philadelphia) 27 May 3/1 (*heading*) The Mexican War. 1881 *Harper's Mag.* Jan. 258/2 The Mexican War..the Abolitionists declared..was waged to obtain new territory for the extension of slavery. 1931 E. O'Neill *Mourning becomes Electra* (1932) I. 19 He went to the Mexican War and come out a major.

c. *Comb.*, as **Mexican-American,** of or pertaining to Mexican settlers or their descendants in the U.S.A.; also as *sb.*

1953 *Jrnl. Social Issues* IX. I. 26 Another fortunate situation has been the fact that our Mexican-American membership has been the most insistent and aggressive in the fight against the illegal alien from Mexico, popularly called the wetback. 1964 S. M. Miller in I. L. Horowitz *New Sociol.* 293 This urban poor is composed of many strands:..Puerto Ricans and Mexican-Americans. 1972 *Jrnl. Social Psychol.* LXXXVII. 3 Mexican-Americans comprise one of the largest minority groups in the United States. 1973 D. Barnes *See Woman* (1974) p. i, To the north, Hollywood's own unique night life seethed... In the east, the Mexican-American community slept. 1973 *Black Panther* 21 July 14/1 The Mexican-American workers in the canneries.

B. *sb.* **2.** (Earlier examples.)

1827 J. F. Cooper *Prairie* I. v. 149 A foal that is worth thirty of the brightest Mexicans that bear the face of the King of Spain. 1836 *Knickerbocker* VIII. 580 The lad could not change the Mexican which I gave him. 1845 J. J. Hooper *Some Adventures Simon Suggs* 76 There's an old friend of mine..that's got three or four hamper baskets-full o' Mexicans.

3. A variety of sheep.

1878 I. L. Bird *Lady's Life Rocky Mts.* (1879) x. 173 The flocks are made up mostly of pure and graded Mexicans. 1887 *Scribner's Mag.* II. 511/1 The season comes for the shearing of Southdowns or rough-fleeced Mexicans.

Mexicanize, *v.* Add: Also, to subject to the influence or domination of Mexicans. (Further examples.)

1844 J. Gregg *Commerce Prairies* II. 119 To this great ball, however, no Americans were invited, with the exception of a Mexicanized denizen or two. 1872 'Mark Twain' *Roughing It* 178, I had never seen such wild, free, magnificent horsemanship..as those picturesquely clad Mexicans, Californians, and Mexicanized Americans displayed. 1900 *19th Ann. Rep. U.S. Bureau Amer. Ethnol. 1897–98* p. xvi, These Indians, now practically Mexicanized. 1904 *Baltimore American* 15 June 6 With the prospective passing of President Diaz, of Mexico, all the world will hope that his country will not revert to that condition which once led to the invention of the word Mexicanized. 1910 *N.Y. Even. Post* 13 Oct. 8 Some object to describing the Roosevelt plan as one to Mexicanize our government. But that is precisely what it is. 1941 R. Humphreys *Latin Amer.* 23 To Mexicanize the Indian and to make the Mexican master in his own country. 1973 *Nature* 13 July 66/3 Mexico recently passed two laws, one to regulate transfer of technology into the country in an attempt to make sure that the imported technology is suited to Mexico's needs, and the other to limit foreign investment, with the long-term goal of Mexicanizing industry.

So **Me:xicaniza·tion.**

1878 *Detroit Free Press* 2 June 4/1 'Mexicanization'—Taking evidence to see whether election protests were forged, and, if so, who forged them, who connived at the forgery, or were aware of it. 1890 *Congress. Rec.* 5 June 5655/1 Gentlemen, do you know what a single silver standard means in this country? It means Mexicanization. 1938 *Newsweek* 28 Mar. 20/3 Cárdenas' swift action caught the British and American Governments off guard. Following developments in anxious silence, they feared the 'Mexicanization' campaign might next strike their mining interests. 1963 *Times* 6 July 18/5 After many months of negotiation..one of the major obstacles to the company's Mexicanisation proposals has now been cleared.

Meyerbeerian (məiˌəɪbīə·riăn), *a.* [f. the name of the German operatic composer Giacomo *Meyerbeer* (1791–1864) + -ian.] Of, resembling, or characteristic of the style or work of Meyerbeer.

1890 G. B. Shaw *London Music 1888–89* (1937) 359 Mr Goossens seemed to me to be imperfectly in sympathy with the electrical Meyerbeerian atmosphere. 1955 E. Dent in H. van Thal *Fanfare for E. Newman* 103 Balthazar is a typically Meyerbeerian character, and his music is more like Meyerbeer than Donizetti.

Hence **Meyerbee·riad,** something resembling the style or work of Meyerbeer; **Meyerbee·rianism** *sb.*; **Mey·erbeerish** *a.*

1947 A. Einstein *Mus. Romantic Era* xvii. 308 Serov..wrote a Biblical drama *Judith* and a grand opera *Rogneda*—both of which might be characterized as merely gaudy Meyerbeeriads. 1955 F. Toye in H. van Thal *Fanfare for E. Newman* 163 *Don Carlos* is permeated with Meyerbeerianism from beginning to end. 1962 *John o' London's* 8 Mar. 233/3 A certain amount of Meyerbeerish trafficking with 'grand opera'.

meyerhofferite (məiˌəɪhǫ·fĕrəit). *Min.* [f. the name of Wilhelm *Meyerhoff* (1864–1906), German chemist + -ite[1].] A colourless to white hydrated calcium borate that is found chiefly as an alteration product of inyoite.

1914 [see *inyoite]. 1951 C. Palache et al. *Dana's Syst. Min.* (ed. 7) II. 357 The meyerhofferite occurs as pseudomorphs after inyoite with a fibrous internal structure, as small transparent colorless crystals on the surface of these pseudomorphs, and as masses of interlaced glassy crystals or fibrous aggregates embedded in clay. 1967 [see *inderborite].

‖ **mézair** (mezę̄r'). [Fr., f. It. *mezzaria* middle gait.] (See quot. 1960.)

1754 R. Berenger tr. *Bourgelat's New Syst. Horsemanship* xvii. 115 The Mezair is higher than the Action of *Terre-a-Terre*, and lower than that of *Curvets*: we may therefore conclude, that the *Terre-a-Terre* is the Foundation of the Mezair, as well as of *Curvets*. 1928 *Daily Express* 22 June 11/3 There is the 'piaffe' in which the horses keep time without advancing, and the 'mezair', that sets them mincing forward, balancing on their hind legs. 1956 L. Mins tr. *Seunig's Horsemanship* (1958) iii. 313 The *mézair* is performed by having the horse fall back on its haunches after a gallopade. 1960 A. Podhajsky *Spanish Riding School Vienna* 36 The Mézair is a series of Levades following upon each other at short intervals, after each of which the fore-legs always touch the ground for an instant, the hind-legs following in a jump and then the Levade is repeated, so that a small increase of space forward occurs.

mezcaline, obs. var. *mescaline.

‖ **meze** (me·ze). Also **mese, mezée, mezze, mezzeh.** [Turk. *meze* snack, appetizer.] A type of hors-d'œuvre served esp. with an aperitif in Greece and the Near East. Also *attrib.*

1926 *Manch. Guardian Weekly* 19 Feb. 151/2 It is taken as a habit with the dried fish and *mézés* that accompany a summer beer. 1950 E. David *Bk. Mediterranean Food* 146 Tarama is the name given to the dried eggs of grey mullet pressed and sold out of a barrel—a favourite mézé in Greece and Turkey. 1955 *Times* 16 July 1/5 It is customary, throughout Greece and the Near East, to serve Mezedes with your aperitif. 1957 L. Durrell *Bitter Lemons* 25 We..shared a stirrup-cup and a *meze*. 1958 R. Liddell *Morea* II. ii. 58 The bit of cheese or the meat ball, brought to me with a slice of tomato by way of *mézé*, when I drank my ouzo. 1966 J. Aldridge *Statesman's Game* ix. 67 He drained his martini and said 'Louise! Bring in that mezée trolley will you?' 1968 J. Cleary *Season of Doubt* xii. 216 Lucille..passed around a tray of *mezzeh*, the traditional Lebanese hors-d'oeuvre. 1974 *Times* 16 Feb. 15/2 *Mezes*—a distinctive form of eating throughout the Middle East, where the concept of hors d'oeuvres is Lucullan.

mezz (mez). *slang.* [f. the name of *Mezz* Mezzrow (1899–1972), a jazz clarinettist and drug addict.] Marijuana; **me·zzroll,** a marijuana cigarette.

1938 *Amer. Speech* XIII. 188/1 *Mezz*, marijuana. 1946 Mezzrow & Wolfe *Really Blues* (1957) 215 New words came into being..*mezzroll*, to describe the kind of fat, well-packed and clean cigarette I need to roll. 1960 *Time* 25 Jan. 87/2 In U.S. slang marijuana is called..mezz. 1972 *Sunday Sun* (Brisbane) 2 July 14/2 Detectives from the CIB Drug Squad in Brisbane are becoming quite familiar now with words like..mezz, Mary Jane.

‖ **mezzadria** (medzădrī·ă). Also **mezzeria.** [It.] A system of land tenure in Italy whereby the farmer pays a proportion (orig. half) of the produce to the landowner as rent, the landowner usually supplying the stock, seeds, etc. Also *attrib.*, as *mezzadria district, system*, etc. So **me·zzadro** (pl. **-i**), the tenant farmer of this system.

1875 *Encycl. Brit.* I. 415/1 A system..in certain provinces of Italy..called mezzeria..the halving, that is, of the produce of the soil between landowner and landholder. 1909 King & Okey *Italy Today* (new ed.) viii. 171 In the *mezzadria* districts there are comparatively few agricultural labourers and therefore less pauperism. *Ibid.*, Occasionally,..the rent is proportioned to the crop, and thus the system slides imperceptibly into *mezzadria*. 1928 L. B. Register tr. *Calisse's Hist. Italian Law* III. xx. 727 In the years following..the Germanic invasions land was not as yet capable of returning an adequate compensation for the labor required for cultivation... It was preferable to fix the return as a proportion of the harvest, a half for example, as in the 'mezzeria'. 1949 L. Olschki *Genius of Italy* (1950) i. 10 *Mezzadria* imposes hard penalties on the peasant. 1962 *Listener* 30 Aug. 309/2 He has always worked on the share-cropping system known as the *mezzadria*. 1964 *Economist* 4 July 55/2 The *mezzadri* strike, vote communist.., or leave for the cities. 1967 C. Seton-Watson *Italy from Liberalism to Fascism* viii. 303 The *bracciante* resented the *mezzadro*'s reliance upon his own family and his reluctance to employ paid labour. 1970 I. Origo *Images & Shadows* ix. 212 Our land was worked on the system which had been almost universal in Tuscany for nearly six centuries, the *mezzadria*,..the tenant—called *mezzadro*,..contributed the..labour. 1973 *Daily Tel.* 23 Jan. 17/3 Mrs Doxat has plans for her vineyards, which are run under the *mezzadria* system, whereby the Italian tenant works them and takes 52 per cent. of the profits, while all capital investment is the owner's responsibility.

mezzani (medzā·ni). [It., pl. of *mezzano* middle, medium.] A type of medium-sized macaroni.

1895 'M. Ronald' *Century Cook Bk.* II. vii. 225 The macaroni called 'Mezzani', which is a name designating size, not quality, is the preferable kind for macaroni dishes made with cheese. 1958 *Catal. County Stores, Taunton* June 17 *Naples Macaroni*. Long—Mezzani, Tagliatelle, Linguine. 1964 M. Waldo *Art Spaghetti Cookery* (1965) 9 There are larger types [of pasta] such as *mezzani*, so large that merely one, stuffed with a filling, makes a portion.

mezzanine. Add: **1. d.** In a theatre or cinema: see quot. 1961. *U.S.*

1927 Sexton & Betts *Amer. Theatres of Today* 3/2 If..the site is unusually small, or if, due to its location or to the high cost of land, the maximum number of seats are required, balconies and mezzanines are necessitated. *Ibid.* 4/2 Seats lost by reducing the length of the main balcony are obtained in a mezzanine balcony. 1933 *Radio City News* 1 May 1/3 It is now possible for patrons to reserve seats in the first smoking mezzanine for any performance of the week. 1957 *New Yorker* 29 June 22/1, I was in a movie house, fairly plush, in a sort of mezzanine, or balcony. 1961 Bowman & Ball *Theatre Lang.* 219 *Mezzanine*, a seating area just above the orchestra, or the forward part of such an area; the first balcony.

mezza voce, *adv.* Add: Also as *sb.* and *adj.*

1877 G. B. Shaw *How to become Mus. Critic* (1960) 29 The critics will fall into raptures over his exquisite management of the *mezza voce*. 1927 *Sunday Times* 6 Mar. 7/3 Though he can achieve an extraordinary *mezza voce* at times, his voice is really far too big for a small hall. 1954 C. K. Scott *Fund. Singing* iii. 163 This is the principle of *mezza voce* which gives perhaps the most appealing sound that can be made. 1955 *Times* 29 June 3/1 He does not suffer from the common fault of Italian tenors of yelling continuously; his *mezza voce* is pleasing. 1958 *Times* 10 Nov. 14/3 She..impressed upon her excellent accompanist..a similar *mezza voce* treatment. 1973 *Guardian* 12 Mar. 8/3 A clear, *mezza voce* climax.

mezze(h, varr. *meze.

mezzeria: see *mezzadria.

mezzo, *a.* Add: *mezzo forte*, also as *sb.*

1955 *Times* 30 May 3/5 It seems a pity not to take advantage of it in place of the unblinking mezzo-forte adopted by Mr. Jones. 1967 *Times* 23 Nov. 8/8 She scarcely ever spoke at less than a shout and needs to find the occasional mezzo forte.

mezzo-brow, *sb.* and *a. colloq.* Now *rare.* [f. It. *mezzo* middle + Brow *sb.*[1]] = *middle-brow* (see *middle *a.* 6).

1925 N. Playfair *Story Lyric Theatre* i. 6, I am not a 'high-brow', but what I believe is now called in America a 'mezzo-brow', an Uncompromising Mezzo-brow! And if you imagine that to be a 'mezzo-brow' means that one has no positive opinions..I give you the lie. 1925 *Times Lit. Suppl.* 12 Nov. 751/2 (*heading*) A mezzo-brow manager: the story of the Lyric Theatre, Hammersmith. 1935 *Punch* 19 June 734/2 He deplores..the red rash of rural villadom. But, resolutely mezzo-brow, he has a good word for Blackpool beach and the roadside Lido. 1940 [see *book *sb.* 13]. 1945 Mencken *Amer. Lang.* Suppl. I. 325 The search for a term to designate persons neither *high-brows* nor *low-brows* has led to the suggestion of *mizzen-brow* and *mezzo-brow*..but they have not caught on.

Mezzofanti (medzofæ·nti). The name of Giuseppe *Mezzofanti* (1774–1849), an Italian cardinal who was master of more than fifty languages, used to denote a person of exceptional linguistic ability. Hence **mezzofa·ntic** *a.*

1872 G. M. Hopkins *Let.* 3 Dec. (1956) 238 We have a half-English half-Italian young sucking Mezzofanti among us who could have written in seven languages to my knowledge. 1904 A. Vambéry *Story my Struggles* I. i. 88 The high-flown announcements of my mezzofantic perfections remained without the slightest result. 1939 Joyce *Finnegans Wake* II. 260 Long Livius Lane, mid Mezzofanti Mall. 1972 *Times Lit. Suppl.* 30 June 752/5 Misspellings are legion and mezzofantic in scope.

‖ **Mezzogiorno** (medzoˌdʒiǭ·rno). [It.] The southern part of Italy, including Sicily and Sardinia.

1932 Joyce *Let.* 13 May (1966) III. 245 You seem to be cruising in the mezzogiorno. 1952 P. H. Bonner *SPQR* (1953) vi. 54 The cities of the Mezzogiorno, like Naples and Salerno and Bari. 1960 E. & R. Chevallier *Mezzogiorno* 7 Magic is by no means a thing of the past in the Mezzogiorno. There are still witches in the land, and the people still fear the 'evil eye'. 1963 'W. Haggard' *High Wire* viii. 92 He had..a loving little wife in Naples, but he hadn't seen her for a year. That was a long time to leave a woman in the mezzogiorno. 1972 A. Thorne *Rome & S. Italy* ii. 45 The houses of the Mezzogiorno have far more in common architecturally with those of Byzantium and the islands of the Aegean. 1973 A. B. Mountjoy *Mezzogiorno* i. 10 An abundance of children characterized the villages and towns of the Mezzogiorno.

Mganda, var. *Muganda.

‖ **mganga** (mˌga·ŋgă). Also **m'ganga, nganga.** Pl. **mgangas, waganga.** [Swahili *mganga*, pl. *waganga* medicine man.] In Tanzania and other parts of East Africa, the name given to a native doctor or witch-doctor. (In quots. 1864, 'an object used in magic rites, a charm'.)

1864 J. A. Grant *Walk Across Afr.* p. xvi, M'ganga or Onganga; a general term for a charm, or for a man who divines events. *Ibid.* iii. 39 The only superstitious observance we noticed was in a field at the foot of a tree; a grass model of a hut was erected for the rain-god..and called, as usual, a 'M'ganga'. 1895 H. H. Johnston *River Congo* (ed. 4) xvi. 273 Somebody is suspected of having caused the death by supernatural means, and the horrid old *nganga* or 'medicine man' who holds the inquest..is called upon to detect the guilty person. 1930 *Discovery* Aug. 265/1 The mganga (as the native calls his medicine-man) is the surgeon, physician, neurological expert, herbalist, toxicologist, and veterinarian. 1947 *E. Afr. Ann. 1946–7* 55/1 The 'mgangas', or general practitioners..prescribe or apply remedies, and in some cases even set fractures and do simple operations. 1965 J. Listowel *Making of Tanganyika* iv. 35 For thousands of years, the relationship between chiefs and witch doctors, in Swahili called *waganga*, had been a close one. 1971 N. Q. King *Christian & Muslim in Afr.* 67 The *mganga*, wrongly but regularly translated as 'witch-doctor', is a healer and 'putter-together', with knowledge of herbal and spirit powers.

Mgr. *R.C.Ch.* Abbrev. of Monsignor, -nore. Also **Msgr.**

1853 J. R. Logan in *Jrnl. Indian Archipelago* VII. 53, I have received valuable assistance..from Mgr. Pallegoix, Mgr. Le Fevre and several other learned missionaries. 1858 N. P. S. Wiseman *Let.* in M. F. Roskell *Mem. F. K. Amherst* (1903) v. 217 Mgr Searle joins me in these feelings. 1897 Addis & Arnold *Cath. Dict.* (ed. 5) 425/1 Mgr. Pap-Szilagyi has made a methodical compendium of these documents. 1922 Joyce *Ulysses* 312 The rt rev. Mgr M'Manus. 1959 E. Waugh *Life R. Knox* III. ii. 244 'Father Knox', a name prominent in letters since 1912 gave place to the less familiar 'Mgr Knox'. 1965 W. Mitchell tr. *Huyghe's Relig. Orders Mod. World* p. v, Contents... What do we mean by Religious? Mgr. Gerard Huyghe... The call to holiness in the Church Mgr. Charue.

mi, abbrev. of Minor *a.* 7.

1791 [see Minor *a.* 7]. 1867 J. A. Symonds *Let.* Mar. (1967) I. 703 Our tutor lists will need to be filled up next term... I claim B mi as my own. 1932 Wodehouse *Louder & Funnier* 12 Faber *mi* got hold of the manuscript and refused to give it up, and Faber *ma*..hit him over the head. 1963 *Times* 5 June 14/3 Frost mi..was only allowed to shake his head.

miacid (məi·ăsid). *Palæont.* [f. mod.L. family name *Miacidæ*, f. the generic name *Miacis* (E. D. Cope 1872, in *Proc. Amer. Philos. Soc.* XII. 470): see -id[3].] A small, carnivorous mammal of the family Miacidæ, known from North American fossil remains of the Palæocene and Eocene epochs. Also as *adj.*

1966 R. & D. Morris *Men & Pandas* viii. 191 About thirty million years ago the ancestors of all the modern carnivores appeared on the scene. These were the miacids and they were small creatures rather like present-day civets. 1972 T. A. Vaughan *Mammalogy* xii. 194/2 Miacids were small and perhaps mostly arboreal carnivores. 1973 R. F. Ewer *Carnivores* v. 225 It is impossible to believe that any carnivore could have evolved a dentition of miacid type without some corresponding behavioural adaptations. 1973 *Nature* 14 Dec. 391/1 The earliest carnivora are the miacids *Protictis* and *Ictidopappus*.

‖ **miai** (mi·əi). [Jap., f. *mi* seeing + *ai* mutually.] The first step in a Japanese arranged marriage whereby the prospective partners meet briefly in company with their families to decide if they are mutually acceptable.

1890 B. H. Chamberlain *Things Japanese* 221 The middleman arranges for what is termed the *mi-ai*, literally, the 'mutual seeing'—a meeting at which the lovers (if persons unknown to each other may be so styled) are allowed to see, sometimes even to speak to each other. 1902 L. Hearn *Kottō* x. 90 To arrange for the *miai* ('see-meeting') tomorrow. 1966 P. S. Buck *People of Japan* (1968) v. 62, I have a young friend who was married several months ago, who saw his wife for the first time at the miai seven weeks before the wedding.

Miao (miɑu·), *sb.* and *a.* Also **Meo.** [Native name.] **A.** *sb.* **a.** A member of the Miao, a mountain-dwelling people of China and Indo-China; the Miao people. **b.** The language spoken by the Miao. **B.** *adj.* Of or pertaining to the Miao people or their language.

1917 S. Couling *Encycl. Sinica* 4/2 There have been frequent revolts of the Miao, the last great one being during the T'ai P'ing rebellion. The Miao women wear wonderfully embroidered clothes and short, white kilted skirts. 1937 E. Snow *Red Star over China* v. iii. 195 The Reds had a method. They had already safely passed through the tribal districts of the Miao and the Shan peoples, aborigines of Kweichow and Yunnan, and had won their friendship. 1939 L. H. Gray *Foundations of Lang.* 390 Miao, Yao, Khāmtī, the extinct Āhom, etc. 1948 D. Diringer *Alphabet* 418 Miao spoken on the borders of northern Siam. 1948 R. A. D. Forrest *Chinese Lang.* i. 33 Miao..has indeed close relationship with Chinese; but it is a relationship of influence and not of kinship. 1955 *Times* 9 May 8/4 The Meos, or Miaos as they are more commonly known in China, are more recent emigrants from Kweichow and Yunnan provinces. 1963 *Times* 24 Apr. 10/3 According to Prince Souvanna Phouma, the Pathet Lao has also demanded the withdrawal of trained Meo tribesmen, and it is not clear whether the rightist troops mentioned are Meo, or, for that matter, whether the latter have descended from their hills. 1970 *Daily Tel.* 4 Mar. 16/4 The Meo general and his predominantly Meo hill-tribe force..form the Government's only credible shield against the North Vietnamese. 1972 *Nat. Geographic* Feb. 275 Soon Meo harvesters will scrape off the sticky sap, boil it, and sell the crude narcotic for at least $25 a pound. 1972 *Times* 4 July 16/1 Some of the Meo have succeeded in preserving themselves and their cultural identity by not taking sides in the war. 1974 R. Butler *Buffalo Hook* xii. 105 They're an odd bunch, the Hainanese... The island's aboriginal people are called, of all things, Miao.

Miaotse (miɑu·tsɪ). Also **Miao-chia, Meaou-tse, Miautsz', Miaotsze,** etc. = prec., sense A.

1836 J. F. Davis *Chinese* I. vii. 287 The Chinese law prohibits all marriages between subjects and foreigners, and even forbids any alliances between the unsubdued mountaineers, called Meaou-tse, in the interior of the empire, and its own people in the neighbouring plains. 1848 S. W. Williams *Middle Kingdom* I. iii. 147 The unsubdued Miautsz'..occupy the north-east portion of the [Western] province, in the mountain fastnesses between it and Kweichau. 1883 *Encycl. Brit.* XVI. 223/2 In figure the Miautse, both men and women, are shorter and darker-complexioned than the Chinese. 1911 *Ibid.* XVIII. 354/2 The emperor K'ien-lung..attacked the Miaotsze, who suffered a crushing defeat, and were compelled to purchase peace by swearing allegiance to their conquerors. 1917 S. Couling *Encycl. Sinica* 4/2 *Miao-chia*,..or *Miao-tzu*..is the name given by Chinese to tribes calling themselves Mhong, whose head-quarters are in Kueichou, but who are also found in south Yünnan, Ssûch'uan and Hunan. 1932 W. L. Graff *Lang.* 421 The Miaotse or Mautzy group, probably related to Tai.

miaow, *int.* Add: Often used to imply that the person addressed is a 'cat' (see Cat *sb.*[1] 2 a).

1937 M. Irwin *Stranger Prince* 321 'Congratulations, sweetheart! I did not know you had secured him...' 'Miaow!' 1948 'E. Crispin' *Love lies Bleeding* xii. 134 'I sometimes wonder if she has any *deep* emotions at all.' 'Miaouw,' said Fen gently. Elspeth grinned. 'All right, I *am* being catty.' 1962 E. Bradford *Touchstone* xi. 91 'He's always..where the bar is.' She leaned forward. 'Miaouw!' 1965 L. Brain *It's Free Country* xv. 146 'P'raps she just likes men,' said Howard. 'Men with money,' said Susan. 'Miaow!' 1967 'S. Woods' *And shame Devil* 89 'He probably has some money of his own. Otherwise, why should she have married him?' 'Miaow,' said Antony.

miaow, *v. intr.* (Further example.)

1975 J. Symons *Three Pipe Problem* xvii. 174 It miaowed in faint protest, arched its back.

miarolitic (mīˌăroli·tik), *a. Petrol.* [ad. Ger. *miarolitisch* (H. Rosenbusch *Mikrosk. Physiogr.* (ed. 2, 1887) II. 39), f. It. dial. *miarolo*, name of a kind of granite containing cavities + *-lit* -lite: see -ic.] Characterized by irregular cavities into which well-formed crystals project; also applied to the cavities themselves.

1895 A. Harker *Petrol.* ii. 31 Vacant spaces are apt to occur, into which project the sharp angles of well-formed crystals. This miarolitic or drusy structure is more or less marked in some granites (e.g. the Mourne Mts in Ireland). 1931 A. Johannsen *Descr. Petrogr. Igneous Rocks* I. iii. 36 Miarolitic cavities are generally irregular and angular in form, and are seldom more than a few inches in diameter. 1970 *Nature* 28 Nov. 850/2 The nepheline gabbro..is a subangular block coated by a thin crust of MnO_2... The texture is ophitic and miarolitic.

miasmal, *a.* (Later example.)

1919 T. S. Eliot *Hippopotamus* in *Poems*, The True Church remains below Wrapt in the old miasmal mist.

miasmatic, *a.* (Later and *fig.* examples.)

1938 M. Brinig *May Flavin* iv. 396 An incomparable California sun pushing the miasmatic mists back into the sea. 1968 'Han Suyin' *Birdless Summer* I. vi. 132 Pao and his benevolent brothers bemoaned the miasmatic disorder in which they were involved.

miasmic, *a.* Add: (Later *fig.* example.) Hence **mia·smically** *adv.*

1938 E. Waugh *Scoop* I. v. 81 Suddenly, miasmically, in the fiery wilderness, there came an apparition. 1961 B. Fergusson *Watery Maze* iii. 80 The Pantellaria project and another miasmic one were abandoned.

mic, mic. (məik). Colloq. abbrev. of Microphone. (Cf. *mike *sb.*[5])

1961 A. Berkman *Singers' Gloss. Show Business* 58 Microphone: (Abbr. mike or mic). 1973 *Sci. Amer.* Apr. 2/1 (Advt.), Eight input controls for complete mic/line mixing. 1974 *Some Technical Terms & Slang* (Granada Television), *Mic.*, microphone.

mica. Add: **2.** *water mica*, a trade name for clear, colourless mica.

1905 *Jrnl. Franklin Inst.* Sept. 200 The clear kind is known to the trade as 'water mica'.

3. *mica-packed, -scaled, -topped* adjs.; **mica flap,** the flap of a mica valve; **mica valve,** a device consisting of a flap of mica hinged at the top which is used to allow air to flow in one direction only.

1906 J. W. Thomas *Ventilation, Heating & Lighting of Dwellings* ii. 27 There are many chimney ventilators on the market, some having mica flaps..to shut against a down current. 1934 E. L. Joselin *Ventilation* iii. 44 Outlets with mica flaps are frequently fitted into chimney breasts near the ceiling. 1964 J. S. Scott *Dict. Building* 207 Mica-flap valve. 1909 *Westm. Gaz.* 9 Mar. 4/3 A new three-point sparking-plug..which has no asbestos or mica-packed joints. *a* 1963 S. Plath *Crossing Water* (1971) 44, I shone, mica-scaled, and unfolded To pour myself out like a fluid. 1958 A. Wilson *Middle Age of Mrs Eliot* III. 381 David banged his fist on the mica-topped table in front of him so that its little contemporary steel-tube legs rattled. 1880 S. S. Hellyer *Plumber & Sanitary Houses* (ed. 2) xxii. 262 When no convenient place can be found for leaving the mouth of the induct-pipe open to the atmosphere, a mica-valve can be fixed over it. 1909 G. B. Shaw in *Trans. Medico-Legal Soc.* VI. 217 One day on the Health Committee a question came up with regard to a mica valve not being in order.

micaceously (məikē̆i·ʃəsli), *adv.* [f. Micaceous *a.* + -ly[2].] Like mica.

1933 H. G. Wells *Bulpington of Blup* i. 18 It had walls and pinnacles of a creamy sort of rock that glittered micaceously.

Micarta (məikā·ɪtă). Also **micarta.** [perh. f. Mica + It. *carta* paper.] A laminated electrical insulating material, originally consisting of paper, mica, and enamel: now a proprietary term for one composed of layers of paper or fabrics bound by a resin and used esp. in the form of sheets and tubes.

1912 *Sci. Abstr.* B. XV. 230 The coil is wrapped with several layers of a foil termed 'Micarta Folium', consisting of paper, mica, and enamel; the whole coil is then..compressed. 1914 *Official Gaz.* (U.S. Patent Office) 10 Feb. 585/2 Westinghouse Electric & Manufacturing Co... *Micarta*..Electrical insulating sheets and tubes. Claims use since Apr. 4, 1912. 1921 *Raw Material* IV. 449/2 The method of manufacturing *micarta*..consists of impregnating the base material with a solution of the binder, evaporating the solvent and then fusing together laminations of impregnated base material by the action of heat under pressure. 1923 *Trade Marks Jrnl.* 3 Jan. 42 Micarta... Electrical insulating sheets and tubes... Westinghouse Electric and Manufacturing Company. 1938 *Jrnl. R. Aeronaut. Soc.* XLII. 9 For those engineers dealing with wooden or micarta airscrews or metal airscrews of less usual form. 1966 *Sci. & Technical Aerospace Rep.* IV. 667/1 The ablation rate of contoured Micarta (laminated plastics) specimens immersed in solid propellant exhaust gases was measured under closely controlled conditions.

Micawber (mikǭ·bər). The name of Wilkins *Micawber*, a character in Dickens's novel 'David Copperfield', applied *gen.* to a feckless optimist with a habit of 'waiting for something to turn up'. Also *attrib.* and *Comb.*, as **Micawber-like** *a.*; **Micaw·ber** *v. intr.*, to behave like Micawber; **Micawberish** (mikǭ·bəriʃ) *a.*, characteristic of a Micawber, irresponsible; **Micaw·berishly** *adv.*; so **Micaw·berism; Micaw·berite** = Micawber.

1852 Geo. Eliot *Let.* 2 June (1954) II. 31 No good news yet, but I have a Micawber-faith that something will turn up. 1880 J. Hollingshead *Plain Eng.* 2 Undeceived by the Micawberism of one class, or the dazzling brilliancy of the other. 1882 W. D. Hay *Brighter Britain!* I. vi. 143 A Micawber-like roll of his voice. 1920 *Glasgow Herald* 19 June 6 He was in a state of what may be described as 'Micawberish embarrassment'. 1927 *Observer* 22 May 8 Nancy's father, a sort of Micawberish Costigan. 1939 *Times Lit. Suppl.* 18 Mar. 162/3 We find a grand collection of literary people who have become household words, like..Micawber. 1948 E. Gowers *Plain Words* 48 The present inclination of the official is in the opposite direction. He is a Johnsonian rather than a Micawberite. 1949 I. Deutscher *Stalin* iii. 70 Lenin treated such Socialists contemptuously as the romantic Micawbers of revolution. 1950 *Mind* LIX. 84 An opportunistic, capricious, or Micawberish policy. 1961 *Times* 5 June 15/3 The British Government can hardly be blamed for being somewhat Micawber-like in their approach to this problem. 1963 *Punch* 24 July 126/1 Every author must spend half his life Micawbering, waiting for something to turn up. 1969 *Times* 2 May (Suppl.) p. viii/3 History has been on the side of the Micawbers. 1971 *Guardian* 16 July 10/2 Mr Wilson's social contract..will..look remarkably like the 1964 or 1966 vintage, with a dash of Micawberism added. 1972 *Ibid.* 11 Oct. 12/2 The Green Paper's hope that growth will provide comes dangerously near the Micawber tradition. 1972 'J. & E. Bonett' *No time to Kill* x. 138 Idle men who take a job in the sunshine that leads nowhere...today's Micawbers, aimlessly awaiting what will never turn up. 1973 *Nation Rev.* (Melbourne) 31 Aug. 1463/3 Despite portraying a Micawberlike optimism, Plorn died destitute. 1974 *Times* 7 Jan. 5/5 The Micawberish disasters that threatened them so long as the father—that sad barrister without briefs—was still alive. 1974 *Times* 4 Mar. 15/3 The Liberal Party..has always Micawberishly hoped that some rich sympathizer would turn up.

micell, var. next.

micella. Add: Also **micell(e.** [ad. G. *micell* (C. Nägeli *Mikrophysik* (1877) 424.] Substitute for def.: **a.** Each of the minute ordered aggregates of macromolecules from which the microfibrils of many natural and artificial fibrous materials are made up. (Earlier and later examples.)

1881 *Encycl. Brit.* XII. 12/1 Nägeli concluded that these structures were made up of crystalline doubly refracting particles or micellæ, each consisting of numerous atoms and impermeable by water. 1937 O. B. Darbishire tr. *A. von Buzágh's Colloid Syst.* vii. 133 The above structural picture of the cellulose micelle is in agreement with a number of physical and physico-chemical properties of cellulose. Chief among them is the well-known great stability of the cellulose micelle, as shown, for instance, by the insolubility of cellulose in water. 1946 *Nature* 10 Aug. 199/2 The main component of starch..is considered to be a network of primary valency chains, which are linked in crystalline micells in which water is bound in the lattice. 1959 *Chambers's Encycl.* III. 222/1 Later evidence suggests rather that the micelles are areas in which the molecular chains are regularly arranged and crystalline; these merge into amorphous areas with possibly some of the long chains extending from one micelle to another. 1965 Micella [see *crystallite 4].

b. An ultramicroscopic aggregate in a colloid consisting of some tens or hundreds of ions or molecules.

1901 *Jrnl. Chem. Soc.* LXXX. II. 231 A 'micelle' is used to denote the smallest quantity of a colloid which possesses all the physical properties of the colloid and is formed by the association of molecules of large size. 1926 H. S. Hatfield tr. *Freundlich's Colloid & Capillary Chem.* 371 For the micella of the gold sol we must take into account the fact that foreign substances enter into its structure, which largely determine its chemical properties. 1927 H. S. van Klooster tr. *Kruyt's Colloids* vi. 102 The micell is the particle plus the entire double layer. 1949 Alexander & Johnson *Colloid Sci.* I. ii. 31 It is generally agreed that the physical properties of soap solutions, such as surface activity, conductivity, osmotic coefficients, solubilization of organic compounds, the Krafft phenomenon, etc., are due..to the occurrence of micelles. 1969 *New Scientist* 21 Aug. 379/1 The granules correspond to what are generally referred to as casein micelles, being formed by the denaturation and aggregation of the milk proteins during the manufacture of cheese. 1971 *Nature* 9 July 118/1 The bulk of DDT carried in contaminated water is probably in an organic environment, dissolved in suspended liquid fats, in soap and in detergent micelles.

micelle, now the usual form of prec.

micellization (mise:ləizē̆i·ʃən). *Chem.* [f. Micell(a + -ization.] The formation of micelles.

1966 *Jrnl. Amer. Chem. Soc.* LXXXVIII. 247/1 DPI [*sc.* dodecylpyridinium iodide] was chosen because a large spectral change accompanies its micellization and thus provides a self-indicator of the micellization process. 1972 *Nature* 3 Mar. 32/2 The results further suggest that at concentrations of ganglioside above 30% the formation of membranous structures would be inhibited by lipid micellization.

Michael. Add: **4.** = *Mickey Finn. *U.S. slang.*

1942 Berrey & Van den Bark *Amer. Thes. Slang* § 509/9 Opiate; 'knockout drops'... *Michael*. 1957 B. Buckingham *Boiled Alive* xxiv. 178 He only pretended to trust me and just slipped me a Michael in my drink. I passed out in the car a few minutes after leaving the bar.

5. Slang phr. *to take the Michael* (*out of*) = *to take the micky* (*out of*) (see *micky[1] 6).

1959 H. Pinter *Birthday Party* I. 9 They won't come. Someone's taking the Michael... It's a false alarm. 1962 *Spectator* 23 Feb. 242/2 Like many satirists Mayakovsky takes the michael out of both sides. 1966 L. Davidson *Long Way to Shiloh* xi. 157 Jesus, did we take the Michael! We used to chat 'em up, these old bats out looking for prospects.

Michaelangelesque, var. Michelangelesque *a.*

1845 R. Ford *Hand-bk. for Travellers Spain* II. viii. 635/2 Two grand subjects in *chiaro oscuro* on a gilded ground..are quite Michael-Angelesque. 1856 M. D. Wyatt in O. Jones *Gram. Ornament* xix. 4 Primaticcio, a master whose style of drawing was founded upon the Michael-Angelesque system of proportion. 1864 [see Michelangelesque *a.*]. 1874 E. Eastlake tr. *Kugler's Handbk. Painting: Italian Schools* (ed. 4) I. iv. i. 251 In addition to his larger Michael-Angelesque peculiarities *Luca* may be known by the squareness of his forms in joints and extremities. 1886 [see Michelangelesque *a.*]. 1934 R. Campbell *Broken Record* iv. 84 He looked very Mosaic in the Michaelangelesque sense of the word. 1935 *Burlington Mag.* June 278 Some Michaelangelesque drawings by Maurice Delacre.

miche, *v.* **2, 2. b.** (Further examples of the spelling *mitch*, which is now usual.)

1867 W. F. Rock *Jim an' Nell* 6 Wan vomoon Hur mitched vro' schule. 1888 'Q' *Troy Town* xi. 117 Turn your back, an' they'd be mitchin' in a brace o' shakes. 1907 J. M. Synge *Playboy of Western World* II. 42 You're pot-boy in this place, and I'll not have you mitch off from us now. 1933 Davies & Thomson tr. *O'Sullivan's 20 Yrs. A-Growing* i. 6 What would you say for us to go mitching? 1939 Dylan Thomas *Map of Love* 14 When I whistled with mitching boys through a reservoir park. *a* 1953 —— *Quite Early One Morning* (1954) 84 He cribbed, mitched, spilt ink, rattled his desk and garbled his lessons with the worst of them. 1960 A. Clarke *Later Poems* (1961) 84, I mitched from miracles. 1968 *TV Times* 28 Sept.–4 Oct. 69/1, I used to mitch a lot from school because I simply dreaded it.

Michelangelesque, *a.* (Later examples.)

1903 [see *contrapposto]. 1932 R. Fry *Let.* 10 Mar. (1972) II. 666 The cubico-Michelangelesque affair may best express the spirit of your house. 1956 K. Clark *Nude* iv. 134 The same Michelangelesque motive is used. 1963 A. E. Elsen *Rodin* 55 The Michelangelesque motive of the right elbow crossed over to the left thigh..pleased the artist. 1970 T. Hilton *Pre-Raphaelites* viii. 196 In such Michelangelesque subjects as *The Wheel of Fortune*, Burne-Jones's relations with the Italian masters were always to be uneasy.

Michelin (mi·tʃəlin, ‖miʃelæ̃). The name of André *Michelin* (1853–1931) and Édouard *Michelin* (1859–1940), French manufacturers of motor vehicles, used to designate motor-vehicle tyres produced by the company they founded; freq. with allusion to the symbols included in its advertising, or to the gastronomic and touring guides produced by the company.

1902 C. L. Freeston in A. C. Harmsworth et al. *Motors* 226 The tyre which was most favourably known abroad, i.e. the Michelin. *Ibid.*, The Michelin tyre is made in various sizes. 1921 W. J. Locke *Mountebank* iv. 45 The avocations that had led him to know the Inns of France with the accuracy of a Michelin guide. 1933 A. G. Macdonell *England, their England* vii. 99 A blazer of purple-and-yellow stripes..surmounted by a purple-and-yellow cap that made him somehow reminiscent of the Michelin twins. 1934 H. Miller *Tropic of Cancer* 18 A caricature of a man... Thyroid eyes, Michelin lips. 1951 G. Greene *End of Affair* III. vii. 145, I can't stand twenty-four hours of maps and Michelin guides. 1954 *Observer* 18 July 10/3 The Michelin man (whose name is..Bibendum) is..evocative of such delights as foreign travel, luscious food, the best maps in the world. 1958 E. Dundy *Dud Avocado* I. i. 11 Baedekers and Michelins and museum catalogues..discarded as too boring and corny. 1959 *Motor Manual* (ed. 36) v. 130 The unusual Michelin X, in which three layers of steel cords are built into the cover between the casing and the tread. 1962 L. Deighton *Ipcress File* 221 Joigny a Michelin-starred town a hundred kilometres south of Paris. 1968 A. Diment *Gt. Spy Race* viii. 125 The old charlatan has got a Michelin star to his name. 1971 C. Bonington *Annapurna South Face* App. B. 244 They [*sc.* down jackets] had the advantage of being quite close-fitting and did not make one feel like the Michelin tyre man. 1972 *Guardian* 20 Oct. 13/2 The spirit of the French Michelin man is quietly pacing the British byways. Michelin will bring out their first guide to British hotels and restaurants in March 1974. 1975 R. Hill *Jackdaw* 147 He opened his Michelin..and studied the list of hotels.

Michelsberg (mi·χĕlzbęrk). [Place-name in Baden, Germany.] A Neolithic Belgo-Germanic culture illustrated by remains (esp. pottery) found at Michelsberg. Also *attrib.* Cf. *Cortaillod.

1929 Peake & Fleure *Way of Sea* i. 12 The Michelsberg culture occurs at many sites near the Rhine between Lake Constance and Cologne. *Ibid.* ii. 21 The pottery..was introduced from France into Switzerland..about 2200 B.C., and here this culture developed into that known as Michelsberg, with its tulip-shaped vases, about a century later. 1954 S. Piggott *Neolithic Cultures* ii. 32 A similar mingling of agricultural and hunter-fisher strains is probably perceptible in the Michelsberg culture in Belgium and the Rhineland. 1970 Bray & Trump *Dict. Archaeol.* 146/2 *Michelsberg*, a Neolithic culture of Belgium, north France, the Rhineland and parts of Switzerland... There are many regional sub-groups. The Belgian one has leaf-shaped arrows, antler combs, flint mines, and enclosures similar in construction to causewayed camps, and may have links with the Windmill Hill culture.

Michelson (məi·kĕlsə̆n). *Physics.* The name of A. A. *Michelson* (1852–1931), German-born U.S. physicist, used *attrib.* and in the possessive to designate (*a*) the Michelson-Morley experiment (see next); (*b*) the type of interferometer that was used in this experiment (invented earlier by Michelson).

1902 *Encycl. Brit.* XXX. 250/1 Lodge..studied the behavior of a beam of light which was divided into two portions by a semi-transparent mirror as in Michelson's interferometer. 1956 E. H. Hutten *Lang. Mod. Physics* iii. 99 The theory of *special* relativity rests mainly on the Michelson experiment. 1966 Michelson interferometer [see *Michelson-Morley]. 1965 M. Garbuny *Optical Physics* vi. 308 That such a wavetrain moving through space with a finite length has actually a physical significance..can be demonstrated by the Michelson experiment. 1972 C. Møller *Theory of Relativity* (ed. 2) i. 28 Michelson's experiment was only the first of a long series of attempts to determine the motion of the earth relative to the ether.

Michelson-Morley (məi:kĕlsə̆n mǭə·ɪli). *Physics.* The names of A. A. *Michelson* (see prec.) and E. W. *Morley* (1838–1923), U.S. chemist and physicist, used *attrib.* to designate an experiment first performed by them in 1887 in which a beam of light is divided into two parts which are made to travel over paths at right angles to one another before being reunited, the behaviour of the resulting interference fringes (e.g. when the whole apparatus is rotated through 90 degrees) showing that the speed of light is the same in both directions, in contrast to what would be expected if the earth were in motion through an 'ether'; so *Michelson-Morley apparatus*.

1913 O. Lodge *Continuity* 20 Many forms of statement of the famous Michelson-Morley experiment are misleading. 1925 I. A. Richards *Princ. Lit. Crit.* x. 71 No new facts nor any new hypotheses—no Michelson-Morley experiment, nor any widened purview—led up to the separate value theory of art. 1965 D. Bohm *Special Theory of Relativity* xxii. 107 The Michelson-Morley experiment may be regarded as an excellent confirmation of the Lorentz contraction. 1966 *McGraw-Hill Encycl. Sci. & Technol.* VII. 505/1 The Michelson-Morley apparatus..consists of a horizontal Michelson interferometer with its two arms at right angles.

michenerite (mi·tʃĕnĕrəit). *Min.* [f. the name of C. E. *Michener*, 20th-cent. Canadian mineralogist + -ite[1].] A palladium bismuthide, $PdBi_2$, or perhaps a telluride and bismuthide of palladium and platinum (see quot. 1963), which is found as greyish-white isometric crystals.

1958 Hawley & Berry in *Canad. Mineralogist* VI. 200 Two palladium bismuthides from nickeliferous ores of the Frood Mine, Sudbury, Ontario, detected and described by C. E. Michener some years ago, are re-described and named michenerite and froodite. 1963 *Mineral. Abstr.* XVI. 283/1 Michenerite associated with the moncheite and kotulskite is shown to have a composition approximating to $Pd_{0.75}Pt_{0.25}BiTe$ rather than $PdBi_2$. 1972 *Ibid.* XXIII. 316/1 Moncheite and michenerite occur as disseminations between sulphides and rock-forming silicates and as idiomorphic grains.

Michigan (mi·ʃigăn). **a.** The name of the State on the Great Lakes of North America, used *attrib.* to denote a thing or type found in, peculiar to, or characteristic of Michigan.

1835 C. Bradley in *Ohio Archaeol. & Hist. Soc. Publ.* (1966) XV. 257 My conquered enemy was the massassagua, the Michigan rattlesnake. 1838 *State of Indiana Delineated* (J. H. Colton & Co.) 26 A rail road is located from Madison to Indianapolis, and the great Michigan road through the state commences here. 1855 'Q. K. P. Doesticks' *Doesticks, what he Says* xii. 97 One was afflicted with the measles, and the other had the Michigan itch. 1857 *Trans. Illinois Agric. Soc.* III. 496 For this purpose [*sc.* subsoiling] the Michigan double or subsoil plow is used. 1884 G. B. Goode *Fisheries U.S.: Nat. Hist. Aquatic Animals* 505 The Michigan Grayling..is at present most interesting to the angler. 1900 *Atlantic Monthly* LXXXV. 102/2 Heavy horses or oxen draw a brace of huge wheels for hauling. (This is the Michigan buggy.) 1922 H. Titus *Timber* 12 Didn't Michigan Pine build th' corn belt? 1932 *Evening Sun* (Baltimore) 9 Dec. 31/5 *Michigan roll*, a bankroll..of stage money with a genuine banknote..outside. 1955 *Publ. Amer. Dial. Soc.* XXIV. 116 If the *bundle* consists largely of small bills, it is called a *michigan bankroll*, or a *mish*. 1962 E. Lucia *Klondike Kate* ix. 195 It proved to be a Michigan bank roll, for the only ten[-dollar bill] was on the outside, all the others being ones. 1971 M. Tak *Truck Talk* 105 *Michigan rig*, a double-bottom rig out of Michigan, where such combinations are legal.

b. *U.S.* A card game similar to Newmarket.

1944 A. H. Morehead *Pocket Bk. Games* 80 Michigan is a simple game, yet rewards careful attention. 1946 'L. Ford' *Honolulu Story* 176 'I want to play hearts. On the floor.' 'Michigan,' Mary said. 'Michigan it is.' 1952 E. Kempson et al. *Hoyle Up-to-Date* 215 (*heading*) Stops family. Newmarket (Boodle, Newmarket, Chicago, Saratoga, Michigan, Stops). 1972 R. Harbin *Waddingtons Family Card Games* 110 In the USA, Newmarket is known as Michigan, or Chicago, or Boodle, or Stops; and they play it slightly differently. But I think that our name, and our rules, are better. 1974 W. B. Gibson *Hoyle's Mod. Encycl. Card Games* 161 *Michigan*, a modern game of the 'stops' type, played with a standard fifty-two-card pack, with cards running in ascending value. *Ibid.* 164 *New Market*, the English counterpart of *Michigan*.., the only difference being that the extra hand, or 'widow', is never taken up for play but is 'dead' from start to finish.

Michigander (miʃigæ·ndəɪ). [f. prec.] A native or inhabitant of the State of Michigan.

1848 A. Lincoln *Coll. Works* (1953) I. 509, I mean the military tail you Democrats are now engaged in dove-tailing on to the great Michigander. 1865 *Harper's Mag.* May 814/1 One of the Michiganders uses the goose-quill to some purpose. 1897 *Outing* XXX. 293/1 (*heading*) Michi-gander lawmakers. 1952 *Manch. Guardian Weekly* 8 May 4/2 Jackson would like to have been known thereafter as 'The Republican City' and several boosters and patriotic Michiganders did their best to make that name take.

Michiganian (miʃigē̆i·niăn). [f. *Michigan + -ian.] = prec.

1813 *Niles' Reg.* V. 185 The Michiganians. 1835 [see Wolverene, -ine 3]. 1837 H. Martineau *Society in Amer.* I. I. 65 The Michiganians were in the singular position of having a state government in full operation, while they were excluded from the Union.

miching, *vbl. sb.* and *ppl. a.* Add: Also **mitching.**

Examples s.v. *miche *v.* 2, 2 b.

Michler (mi·χləɪ, mi·k-). The name of Wilhelm *Michler* (d. 1889), German chemist, used in the possessive to denote certain compounds first studied by him, as **Michler's hydrol,** 4,4'-bis(dimethylamino)benzhydrol, $[C_6H_4N(CH_3)_2]_2CHOH$, a crystalline compound which is made by the condensation of dimethylaniline and formaldehyde or by reduction of Michler's ketone and is an important intermediate for the synthesis of triphenylmethane dyes; **Michler's ketone,** the ketone $[C_6H_4N(CH_3)_2]_2CO$ corresponding to Michler's hydrol, which is made by treating phosgene or carbon tetrachloride with dimethylaniline.

1910 *Jrnl. Chem. Soc.* XCVIII. I. 451 Tetramethyldiaminobenzhydrylphosphinous acid.., prepared from Michler's hydrol, occurs in colourless crystals. 1971 R. L. M. Allen *Colour Chem.* viii. 110 Michler's hydrol..condenses with bases such as dimethylaniline to give triphenylmethanes which by oxidation and treatment with acids yield rosanilines. 1897 *Jrnl. Chem. Soc.* LXXII. I. 157 The author obtained crystalline leuco-bases of dyes by the condensation of Michler's ketone with various phenols. 1949 P. W. Vittum tr. *Fierz-David & Blangey's Fund. Processes Dye Chem.* i. 140 Michler's ketone and its homologs are important starting materials for the preparation of valuable triphenylmethane dyes such as Victoria blue B and Victoria pure blue BO.

Michurinism (mitʃūə·riniz'm). Also **michurinism.** [f. the name *Michurin* + -ism.] Belief in or advocacy of the views of the Russian horticulturalist I. V. Michurin (1855–1935): = *Lysenkoism.

1949 J. S. Huxley *Soviet Genetics & World Sci.* i. 23 Michurinism is..an essentially non-scientific or pre-scientific doctrine. 1955 *Bull. Atomic Sci.* June 210/1 Genetics seems to be the field of 'natural' science which is most abused with political and other special interests, in their attempts to fabricate theoretical bases for their practices, as in the case of Hitler's racist obsessions and Stalin's Michurinism. 1974 *Encycl. Brit. Micropædia* VI. 862/2 Michurin's theories of hybridization, labelled michurinism, which accepted completely the inheritance of acquired characteristics, were adopted as the official science of genetics by the Soviet regime.

Hence **Michu·rinist,** one who believes in or advocates Michurinism; also as *adj.*

1949 *Ann. Reg. 1948* 210 The Michurinists, who regarded environment as the chief determinant of heredity. 1949 C. Zirkle *Death of a Science in Russ.* vii. 174 The most advanced, materialistic, biological science is the Michurinist trend in the Soviet agrobiology. 1957 C. Hunt *Guide to Communist Jargon* xxviii. 100 In 1948 it was the turn of the biologists, and a number of them were deprived of their posts for supporting Weissmann-Morganist as opposed to Michurinist genetics.

Mick[1]. **a.** For 1882 in first quot. in Dict. read 1872, and add earlier and later examples. Also sometimes applied derogatorily. Also *attrib.* or as *adj.*

1856 *Butte Record* (Oroville, Calif.) 20 Sept. 3/3 One of the 'bucks' jerked something from his belt..and made for a Mick. 1894 P. L. Ford *Hon. Peter Stirling* lvii. 369 Fortunately it's a Mick regiment, so we needn't worry over who was killed. 1913 J. London *Valley of Moon* 27 They's been too much drink, an' you know what the Micks are for a rough house. 1932 E. Wilson *Devil take Hindmost* vii. 38 The Communists..have..recruited..a considerable number of seedy unemployed—niggers, micks. 1943 [see *Geordie 2]. 1961 *Spectator* 12 Aug. 254 Labels such as Wop, Polack and Mick. 1970 M. Kenyon *100,000 Welcomes* iv. 32 Where's Ireland, huh? Who needs Micks?

b. A Roman Catholic. Also **Mickey.** Freq. derogatory. Also *attrib.* or as *adj.*

1924 P. Marks *Plastic Age* 201, I suppose you refer to..my one mick friend, although he isn't Irish. 1948 P. White *Aunt's Story* 258 He says that Mother is wrong to send a girl to a convent with a lot of micks. 1956 'N. Shute' *Beyond Black Stump* ii. 57 Stanley and Phyllis went to Church of England schools..but all the rest of us are Micks. 1960 *Times* 28 Jan. 15/5 Religious rivalries from Salvationists to Plymouth Brethren—united in hatred of the 'Mickeys' or Catholics. 1971 *Guardian* 27 May 13/7 Curiously, in the circumstances of the Australian fondness for the 'o' suffix, a Roman Catholic is apparently never a Catho, but remains, in lower-level Protto usage, a Mick. 1973 *Times* 31 July 12/7 On this theory the Council of Ireland can be presented as a second inter-Parliamentary tier—'a place where the Micks and Prods can get together occasionally' as one Assembly member put it—built on a structure that already exists.

c. Examples of various *slang* (usu. derogatory) extended uses.

1928 F. Scott Fitzgerald in *Sat. Even. Post* 28 Apr. 4/2 In sordid poverty, below the bluff two hundred feet away, lived the 'micks'—they had merely inherited the name, for they were now largely of Scandinavian descent. 1937 in Partridge *Dict. Underworld* (1949) 437/2 *Mick*, a road mechanic. 1941 J. Smiley *Hash House Lingo* 37 *Mick*, Englishman. 1958 L. A. G. Strong *Light above Lake* 94 He's a dismal ould mick of a God. 1974 *Amer. Speech 1971* XLVI. 81 *Mexican*: greaser, spick, wetback, mick, halfbreed.

d. *mad mick*: a pick. *slang.*

1924 *Truth* (Sydney) 27 Apr. 6 *Mad mick*, a pick. 1935 A. J. Pollock *Underworld Speaks* 74/1 *Mad mick*, a pick (prison) 1973 F. Huelin *Keep Moving* 78 Well, I won't buy drinks f'r any bloody ganger, just f'r a chance to swing a mad mick.

mick[2] (mik). *Austral. slang.* [Origin unknown.] The head, or sometimes the reverse, of a penny (see also quot. 1919).

1919 W. H. Downing *Digger Dial.* 33 *Mick*, (1) the Queen's head on a coin... (2) a queen in a pack of cards. 1938 J. Robertson *With Cameliers in Palestine* xx. 198 'A pair of Micks', which means that the offerings [in the game of two-up] are not accepted. 1941 Baker *Dict. Austral. Slang* 46 *Mick*, the 'head' of a penny. 1953 T. A. G. Hungerford *Riverslake* 126 'Ten bob he tails 'em!' he intoned,.. 'I got ten bob to say he tails 'em—ten bob the micks!'

mick[3] (mik). *slang.* Also **mickey, micky.** [var. of Mike *sb.*[3] reinterpreted as a proper name.] *to do a mick*, etc., to go away, to clear off (see *mike *sb.*[3]).

1937 Partridge *Dict. Slang* 519/1 *Mick, do a* [equated with *do a mike*]. 1959 I. & P. Opie *Lore & Lang. Schoolch.* x. 192 *Sending away*,..do a mickey. 1961 S. Chaplin *Day of Sardine* xi. 225, I laid the ring on the notepaper and did a mickey as soon as I heard the front doorbell go.

mick[4] (mik). *slang.* [Origin unknown.] A seaman's hammock.

1929 *Papers Mich. Acad. Sci., Arts & Lett.* X. 308/1 *Mick*, an abridgement of 'face like a scrubbed hammock'. 1946 J. Irving *Royal Navalese* 115 *Mick*, hammock. 1961 Partridge *Dict. Slang* Suppl. 1183/1 *Mick*,.. a seaman's hammock.

mick[5] (mik). *dial.* Also **mickey, micky.** [Origin unknown.] A pigeon.

1940 *N. & Q.* 3 Aug. 79/1 Mick was the usual word for a pigeon, especially the domesticated kind [in Cheshire]. 1965 *Jrnl. Lancs. Dial. Soc.* Jan. 7 Woodpigeon..Mick, Micky: Southport, Liverpool. 1966 F. Shaw et al. *Lern Yerself Scouse* 23 *De mickeys are lettin on de roof*, the pigeons are alighting on the roof. 1966 'L. Lane' *ABZ of Scouse* 68 *Mickey-snatcher*, a person who steals municipal pigeons.

mickery (mi·kĕri). *Austral.* Also **mickerie.** [Origin unknown.] **a.** A type of well (see quot. 1934). **b.** Marshy ground; = Soak *sb.* 2 d. Also *attrib.*

1934 *Bulletin* (Sydney) 6 June 20/2 A mickery was a timbered well-shaft sunk into the sandy bed of a creek; it was worked by means of a pole placed across a forked stick, the pole having a bucket attached to one end and a weight to the other. 1935 H. H. Finlayson *Red Centre* x. 98 The Wonkonguroos, themselves an offshoot of the Aruntas, occupying until recently what is known as the mickerie country, a desolation to the north-west of the Yalliyandas. 1945 Baker *Austral. Lang.* iii. 58 *Mickery country*, country which holds moisture after rain or which is of swampy 'soak' type.

Mickey (mi·ki). *U.S. slang.* [f. *Mickey Mouse.] A type of radar-assisted bombsight. Also *attrib.*

1944 *News* (San Francisco) 28 Nov. 7/3 'Mickey', a sensational radar device which 'sees' through darkness, clouds and artificial smoke. 1944 *Time* 11 Dec. 24 Mickey was a British invention, freely given to the U.S. U.S. scientists later developed an improved model. It was first used by U.S. bombers just over a year ago; today every fleet of heavy bombers over Europe presumably is or will be equipped with it, on day or night raids. 1945 *News* (San Francisco) 15 Aug. 10 (Advt.), Mickey radar sends out super-high frequency radio waves which bounce back from solid targets and are picked up by the radar receiver and transformed into a radar picture on a screen like the one in a home television receiver. 1954 Berrey & Van den Bark *Amer. Thes. Slang* § 819 *Mickey*, the improved American airborne radar. 1955 M. Reifer *Dict. New Words* 131/2 *Mickey*.., a radar-type bomb-sight that permits air navigation in zero visibility and pinpoint strategic bombing despite overcast or darkness. Named for the cartoon figure, Mickey Mouse, created by Walt Disney.

Mickey, var. *Mick[1] b, *Mickey Finn, *Mickey Mouse.

mickey: see *mick[3], *mick[5], *micky[1].

Mickey Finn (mi:kiˌfi·n). *slang* (orig. *U.S.*). Also **Mickey Flynn** and ellipt. **Mickey.** [Origin uncertain.] A strong alcoholic drink; a drink adulterated with a narcotic or purgative substance, usu. administered to a person with the deliberate intention of stupefying or greatly discomforting him; the adulterant itself; also *transf.* and *fig.* (Cf. **knock-out drops*.) Hence **mickey-fi·nn** *v. trans.*, to adulterate with a Mickey Finn; to stupefy with a Mickey Finn; also *ellipt.*

1928 M. C. Sharpe *Chicago May* xii. 99, I got a bottle of brandy... He was lit up..but I shot a few more Mickey Finns (double drinks) into him. 1931 *Amer. Mercury* Mar. 316/1 But he never slipped an obstreperous customer the croton oil Mickey Finn of the modern night club. 1934 *Cosmopolitan* Apr. 131/1 She had been mickey-finned. 1934 J. O'Hara *Appointment in Samarra* (1935) iii. 86 The cheap bastard... I'd like to give him a Mickey Finn. 1935 N. Ersine *Underworld & Prison Slang* 53 *Mickey Finn*, a drink spiked with knockout drops. 1936 *Amer. Speech* XI. 124/1 *Mickey Flynn* or *Mickey Finn*, a knockout dose (often cigar ashes in a carbonated drink) administered to an addict or a sucker. 1938 A. J. Liebling *Back where I came From* 88 Mickeys..act so drastically that one may kill a drunk with a weak heart. 1951 Wodehouse *Old Reliable* v. 71 She had been about to suggest that the butler might slip into Adela's bedtime Ovaltine what is known as a knockout drop or Mickey Finn. 1957 P. Frank *Seven Days to Never* iv. 118 This was no ordinary hangover... He had been expertly mickey-finned. 1960 *Guardian* 17 June 7/4 The reader cannot be sure when Mr. Colquhoun is about to slip him a Bible..and when a Mickey Finn. 1962 E. Lucia *Klondike Kate* iv. 107 However, the dames seldom rolled the miners or slipped them a Mickey. 1966 R. W. Taylor *Doomsday Square* (1967) xiii. 144, I Mickey-Finned him with my little pressure syringe. Let him sleep. 1971 'C. Franklin' *Home Secretary Affair* vii. 99 Two men..had mickey-finned his drink. 1972 M. Pugh *Murmur of Mutiny* xix. 144 I'd a couple or so in the sergeants' mess... They mustaf been mickied. 1973 D. Bagley *Tightrope Men* xii. 93 Meyrick was probably knocked out by a Mickey Finn in his nightly Ovaltine.

Mickey Mouse (mi·ki mɑus). **1.** The name of a mouse-like character in a series of animated cartoons designed by the American cartoonist Walter Elias ('Walt') Disney (1901–66), used with direct or indirect reference to the character in these cartoons. Also *ellipt.* **Mickey,** a Mickey Mouse cartoon. Also *attrib.* or as *adj.*, characteristic of, resembling, or featuring Mickey Mouse. Frequently (esp. in *U.S.*) designating something small, insignificant, or worthless.

1934 R. Stout *Fer-de-Lance* v. 66 Everybody..would be..laughing at us instead of going to a movie to see Micky Mouse. 1935 S. Lewis *It can't happen Here* ix. 84 Dr. Goebbels..is privily known throughout Germany as 'Wotan's Mickey Mouse'. 1936 N. Coward *To-night at 8.30* II. 49 We'll miss the Mickey. 1936 'G. Orwell' *Let.* 26 Aug. in *Coll. Ess.* (1968) I. 228 You have moved too much away from the ordinary world into a sort of Mickey Mouse universe where things and people don't have to obey the rules of space and time. 1937 *Cinema Arts* July 24 The light which the electrician is aiming at Mr. Gable's head is the smallest 'spot' used in the studio, and is therefore known as a Mickey Mouse. 1939 *War Illustr.* 18 Nov. 310/3 Special 'Mickey Mouse' gas-masks in various colours and having separate eye-pieces and a little nose are being made for small children who are repelled by the ordinary ones. 1941 *San Francisco Examiner* 2 Mar. Pict. Rev. 7/1 George Graham, a timid, middle-aged mickey mouse who was afraid of crowds, people, anything. 1942 N. Streatfeild *I ordered Table for Six* 20 That little bit of the room at the top must have been a nursery once, there's some Mickie Mouses on that wall-paper. 1951 *San Francisco Examiner* 3 Feb. 1/3 (*heading*) 40th Division. Troops Lack Equipment. Woes of 'Mickey Mouse Army'. 1957 T. Gunn *Sense of Movement* 54 Bent keys, Italian grammars, Mickey Mouse caps. 1958 *Amer. Speech* XXXIII. 226 A 'Mickey Mouse course' means a *snap* course. 1962 *New Statesman* 19 Jan. 96/2 A Mickey Mouse clock ticks beside the marital bed. 1963 *Amer. Speech* XXXVIII. 176 [Kansas University slang] An assignment which is regarded as foolish and a waste of time is a *Mickey Mouse*. 1966 Mrs. L. B. Johnson *White House Diary* 4 June (1970) 387 It [*sc.* the graduation ceremony] is such a mass production affair—'Mickey Mouse', as she calls it—not quite the sophisticated thing to do anymore. 1968 *Sunday Times* 20 Oct. 8/4 Cunningham referred to 'Micky Mouse' (unnecessary and childish) operations. 1969 C. Davidson in Cockburn & Blackburn *Student Power* 329 The forms given us for our self-government are of the Mickey Mouse, sand-box variety. 1972 J. Mosedale *Football* ii. 25 One reason for the AFL's reputation as a Mickey Mouse league is that it gave new life to NFL rejects. 1973 R. Ludlum *Matlock Paper* xix. 161 The average purchase of chips was $200 to $300. Hardly 'Micky Mouse'. 1974 *Globe & Mail* (Toronto) 13 Nov. 1/2 The titles kept the press and broadcast media from thinking 'it was such a Mickey Mouse operation'. 1975 *Amer. Speech 1972* XLVII. 152 You getting flew coy with the gray boys, Mickey Mouse? *Ibid.* 153 *Mickey mouse n*, white; black who 'puts on a white face'.

2. *Mus.* Inferior dance-band music, similar to that played as a background to a cartoon film; usu. *attrib.*

1938 *Down Beat* Apr. 25/2 A strictly 'mickey mouse' band is still box office. 1945 *Ibid.* 1 Sept. 15/1 Strictly mickey mouse is what they [*sc.* night clubs] advocate, and many of the truly good musicians are forced to play the tripe in order to get work. 1954 *San Francisco Chron.* 17 June 17 (*heading*) Glenn Miller's music was all Mickey Mouse. 1958 *Amer. Speech* XXXIII. 225 A mickey or Mickey Mouse band is not merely a 'pop tune' band,..but the kind of pop band that sounds as if it is playing background for an animated cartoon. 1966 *Crescendo* May 4/2 The jazzman's struggle for self-expression in a world of commercial Mickey Mouse bands. 1970 C. Major *Dict. Afro-Amer. Slang* 81 *Micky Mouse* (*music*), popular, commercialized music.

3. *Air Force slang.* A name applied to an electrical device which releases bombs from aircraft.

1941 *N.Y. Times* 6 Apr. 30/5 This war is producing a new batch of army slang... Many of the words are peculiar to certain units, and very few are yet in general use... An R.A.F. pilot calls his cockpit the 'pulpit'... The instrument releasing the bombs, an electrical distributor, is called a 'Mickey Mouse'. 1943 Hunt & Pringle *Service Slang* 46 *Mickey Mouse*. The bomb-dropping mechanism on some types of bomber aircraft is so called because it strongly resembles the intricate machinery portrayed in Walt Disney's cartoons. 1943 L. Cheshire *Bomber Pilot* i. 8 If I knew how to use 'Mickey Mouse', to open and shut bomb doors, or use a bomb-sight.

Hence **Mickey-mousing** (see quots.).

1957 Manvell & Huntley *Technique Film Music* ii. 38 'Mickey-Mousing' in film musical terminology..means the exact, calculated dove-tailing of music and action. 1960 *Times* 1 Dec. 7/3 Odds and ends of Albeniz..treated to what film composers would call 'mickey-mousing', with every detail of the music mirrored with stupefying literalness in the dancing. 1973 *New Society* 2 Aug. 288/1 What is known in the film-music trade as 'Mickey-Mousing'—the action of the film gets a simultaneous musical parody. When Tarzan swings down on his rope, a harp plays a descending glissando.

mickle, muckle, *a.*, *sb.* and *adv.* Add: **A.** *adj.* **1. d.** (Further example.)

1887 W. S. GILBERT *Ruddigore* II. 44 His gallantries were mickle.

3. muckle-mouthed *a.*, = *mickle-mouthed*.

1951 J. D. SALINGER *Catcher in Rye* xi. 93 She was sort of muckle-mouthed... When she was talking and got excited,..her mouth sort of went in about fifty directions.

micky[1]. Add: Also **mickey**. **1.** (Further examples.) Also **mick**.

1934 *Bulletin* (Sydney) 1 Aug. 46/3, I lifted nearly two hundred Poolpee micks on my way back and took 'em home with me. 1958 *Amer. Speech* XXXIII. 167 *Mickey*, *maverick*, a wild young bull. 1966 BAKER *Austral. Lang.* (ed. 2) iii. 63 *Micky* or *mick*, an unbranded steer, perhaps from the Aboriginal *micky*, quick.

2. (Earlier example.)

1858 J. D. LOVETT in *Harvard Mag.* July 267 While Mickey there stands, A-wringin' his hands, And Biddy is wipin' her eyes on her slave.

3. *Austral.* A honeyeater, *Myzantha melanocephala*.

1911 J. A. LEACH *Austral. Bird Bk.* 173 Noisy Miner, Garrulous Honeyeater, Snake-Bird, Cherry-eater, Soldier, Micky, Squeaker, *Myzantha garrula*. 1931 N. W. CAYLEY *What Bird is That?* 81 Noisy Miner *Myzantha melanocephala*... Also called.. Soldier-bird, Micky, and Squeaker. In small parties, frequenting open forest country and partly cleared lands. 1971 *Courier-Mail* (Brisbane) 24 July 12/8 Mickeys, or Soldier Birds, or Noisy Miners, are great little fighters and battle the hawks, crows and goannas quite fearlessly.

4. Chiefly *Canad. colloq.* A flask of liquor.

1914 JACKSON & HELLYER *Vocab. Criminal Slang* 58 *Micky*, current amongst bottle drinkers. A corruption of Michael..a flask of liquor. 1926 J. BLACK *You can't Win* vi. 66 A four-bit micky, a fifty-cent bottle of alcohol. 1935 A. J. POLLOCK *Underworld Speaks* 76/1 *Mickey*, half pint of bootleg whiskey. 1950 H. SUTTON *Footloose in Canada* 5 An American pint holds 16 ounces, a Canadian 'mickey', 12 ounces of rye, or 13 ounces of Scotch. 1971 *Islander* (Victoria, B.C.) 21 Mar. 16/2 Mark, the public nuisance that he was, bought his liquor in mickeys that he hid in the woodpile around the cannery. 1972 *Regional Lang. Stud.—Newfoundland* IV. 11 *Flask* is used in the Atlantic provinces only. In central and western Canada this would be a *micky*.

5. *slang.* The penis. *rare.*

1922 JOYCE *Ulysses* 765 Ill put on my best shift and drawers let him have a good eyeful out of that to make his micky stand for him.

6. Colloq. phr. *to take the micky* (*out of*) (someone): to act in a satirical, disrespectful, or teasing manner (towards). Cf. *MIKE *sb.*[6] Hence **micky-take** *v.* and *sb.*, **micky-taking** *ppl. a.* and *vbl. sb.*

The more usual spelling of this sense is 'mickey'.

1952 'J. HENRY' *Who lie in Gaol* iv. 66 She's a terror. I expect she'll try and take the mickey out of you all right. Don't you stand for nothin'. 1954 A. HECKSTALL-SMITH *Eighteen Months* xi. 136 For a while everyone tried to take 'the Micky' out of Bobbie, but they soon gave up trying after the first foolish failure. Then their ridicule turned to admiration. 1956 A. WILSON *Anglo-Saxon Att.* II. i. 206 'You're not going to take the mickey out of me, Vin Salad, I'll spatter you,' he cried. 1957 L. P. HARTLEY *Hireling* 134 He had no great regard for Constance, except in so far as she sometimes took the mickey out of Hughie. 1958 *Observer* 28 Dec. 3/1 'Tonight' is not only a tough and irreverent programme, but glib and smart and anxious to take the mickey. 1959 R. STOREY *Touch it Light* in *Plays of Year* XVIII. 419 You don't stop at nothing, do you? As long as you can take the mickey. Mr. Funnyman, that's you. At any price. 1959 *Oxford Mail* 13 June 6/3 This modern dress, mickey-taking version of Aristophanes. 1959 *Vogue* Dec. 101 They'd think you were a nutter and laugh and mickey-take. 1960 E. W. HILDICK *Jim Starling & Colonel* ix. 76 The servers must have thought that no boy would dare to take the mickey in such circumstances. 1962 *Times* 20 July 10/6 The micky-taking ways of their own brand of insolent wise-cracking. 1965 *Listener* 25 Mar. 451/2 Those 'fruits' don't like people trying to, how do you say, take the mickey. 1967 *Spectator* 3 Nov. 535/2 One looks forward after reading this brilliant exercise in mickey-taking to Miss Tracy's next novel. 1968 J. LOCK *Lady Policeman* ix. 85 He keeps up a barrage of mickey taking. 1968 *Listener* 7 Nov. 622/3 He parried Kenneth Allsop's micky-take. 1971 B. W. ALDISS *Soldier Erect* 101 Geordie looked anxiously at me, in case I thought he was taking the micky too hard. 1973 'J. PATRICK' *Glasgow Gang Observed* xiv. 127 Big Sheila indulged in the same sort of 'mickey-taking'.

micky[2] (mi·ki). *N.Z.* Also **micky-mick**, **miki-miki**, etc. Representing a dialectal variant (within the Maori language) of *MINGIMINGI.

1898 MORRIS *Austral Eng.* 294/2 Mingi..in south New Zealand..is often called Micky. 1907 'G. B. LANCASTER' *Tracks we Tread* Gloss., *Mic-a-mic*, scrub. 1933 *Press* (Christchurch, N.Z.) 25 Nov. 15/7 *Scrub*..The best known components in Canterbury are Manuka, Wild Irishman Miki-miki (of which the shepherds make walking sticks), and Mountain pine. 1944 *Mod. Jun. Dict.* (Whitcombe & Tombs) 258 *Micky-mick*, a corrupt form of mingi-mingi, the Maori name of a class of shrubs, one of which makes a good walking-stick. 1949 F. SARGESON *I saw in My Dream* xiv. 229 I'll make a garland for her hair. Out of micky-mick. 1963 B. PEARSON *Coal Flat* vii. 122 Willows and the trees they called..mickeymick, mockamock, whiteywood, birch.

micky, var. *MICK[3], *MICK[5].

Micmac (mi·kmæk), *sb.* and *a.* [Native name, lit. 'allies'.] **A.** *sb.* **a.** An Indian people of the Maritime Provinces and Newfoundland in Canada; a member of this people. **b.** The Algonquian language of this people. **B.** *adj.* Of, pertaining to, or designating the Micmacs or their language.

1830 W. S. MOORSOM *Lett. from Nova Scotia* 110 The tribe to which the Indians of Nova Scotia belong is called the Micmac. *Ibid.* 112, I am not aware that any one Indian claims authority over the whole Micmac tribe. 1877 L. H. MORGAN *Anc. Society* II. vi. 174 They affiliate more closely with the Micmacs than with the New England Indians south of the Kennebeck. 1891 [see *INDIAN *sb.* 2 b]. 1911 A. L. PRINGLE *Home of Evangeline* iv. 139 He taught her the Micmac, and proud she was in his absence to be catechist to the Indians. 1933 [see *CREE *sb.* and *a.*]. 1964 P. K. BOCK in J. A. Fishman *Readings Sociol. of Lang.* (1968) 217 A recurrent situation on the Micmac Indian Reserve studied by the author in 1961 will be used to illustrate how the method of structural description..may be applied. 1969 *Canad. Jrnl. Ling.* XV. 78 Micmac morphophonemics. 1969 *Observer* (Colour Suppl.) 18 May 25/1 The French..offered the Micmac Indians a bounty for every scalp they took from the Beothuk of Newfoundland. 1972 G. V. HIGGINS *Friends of Eddie Coyle* vi. 35, I get five of these Micmacs come in, real Indians, for a change.

Micoquian (mĭkōu·kiăn), *a.* and *sb.* Also **Micoquean**. *Archæol.* [ad. F. *Micoquien* (O. Hauser *La Micoque* (1916) 55), f. La *Micoque* (see below).] **A.** *adj.* Of or pertaining to a late Acheulian culture (earlier called *Moustérien supérieur*) of southern France and England, represented by the remains found at La Micoque, near Les Eyzies (Dordogne) in S.W. France. **B.** *sb.* The Micoquian culture.

1937 GARROD & BATE *Stone Age Mt. Carmel* I. 120 By far the most instructive site for comparison..is La Micoque, type-station of the Tayacian and the Micoquian. *Ibid.*, Well-made Micoquian hand-axes. 1945 *Proc. Prehist. Soc.* XI. 21 The Summertown terrace gravels were deposited contemporaneously with the Upper Acheulian or Micoquian culture stage. 1946 F. E. ZEUNER *Dating Past* vi. 173 From the *base* of the Younger Loess I, a Micoquian handaxe was recovered. *Ibid.* 195 A new stage of the Acheulian cannot be recognized clearly before the advent of the Micoquian. 1957 *Encycl. Brit.* II. 238/2 The Micoquian, or Final Acheulean, is characterized by elongated hand axes. *Ibid.* 539/2 Some pieces belonging to the Micoquian..industry were excavated in 1937 in the terraces of the stream of Cubuksuyu at Etiyokusu near Ankara.

micrergate (məikrə·ɪgĕit). *Zool.* [f. Gr. μικρ-ός small + ἐργάτης worker.] A small worker ant.

1907 W. M. WHEELER in *Bull. Amer. Mus. Nat. Hist.* XXIII. 56 The micrergate, or dwarf worker, is a worker of unusually small stature. It appears as a normal or constant form in the first brood of all colonies that are founded by isolated females. 1915 H. ST. J. K. DONISTHORPE *Brit. Ants* 41 The micrergate is a worker of unusually small size.

micrite (mi·krəit). *Geol.* [f. MICR(OCRYSTALLINE *a.* + -ITE[1].] Microcrystalline calcite present as an interstitial constituent or matrix material in some kinds of limestone; a limestone consisting chiefly of this.

1959 R. L. FOLK in *Bull. Amer. Assoc. Petroleum Geologists* XLIII. 17 Finally, it was decided to use a composite word, the first part of which refers to the allochem composition... Whether the rock is type I or type II is shown by the second part of the name '-sparite' for those with sparry calcite cement and '-micrite' (pronounced 'mick-rite') for those with microcrystalline ooze matrix... Type III limestones, almost entirely ooze, are designated simply as 'micrite' without any allochem prefix. 1970 K. C. JACKSON *Textbk. Lithology* vi. 401 The proportion of micrite in limestone varies from a minor interstitial filling between allochems to the entire rock. 1973 *Nature* 2 Mar. 41/2 The patch reef, 13 m long and over 2 m high, tongues out to the west into 2 m of thinly bedded calcitic micrites of quiet water origin.

Hence **micri·tic** *a.*, containing a high proportion of micrite.

1962 LEIGHTON & PENDEXTER in *Mem. Amer. Assoc. Petroleum Geologists* I. 37/2 Photographs *A* and *B*.. illustrate limestones composed of 90 per cent, or more, micrite. Two subtypes of these micritic limestones are recognized. 1973 *Nature* 20 July 145/2 The deep water Alpine Triassic is represented by cherty, micritic, radiolarian limestones with thin-shelled pelecypods.

micro. **1.** Substitute for def.: Abbrev. of MICROLEPIDOPTERA *sb. pl.*, used to refer to a moth belonging to any of several families whose members are mostly smaller than those of interest to collectors, the Macrolepidoptera. Add examples.

1869 *Rep. Comm. Agric. 1868* (U.S. Dept. Agric.) 313 The presence of 'micros' is indicated by discolored lines... It may be ascertained whether the 'micro' is at home by holding the leaf up to the light. 1907 R. SOUTH *Moths Brit. Isles* 1st Ser. 6 Possibly, when this new order of things is more generally understood the so-called 'Micros' will receive their proper share of attention. 1972 L. E. CHADWICK tr. *Linsenmaier's Insects of World* 172/2 As for the 'micros', they are neither a uniform grouping nor properly to be separated from the rest of the moths.

2. *Fashion.* Short for *micro-skirt* (*MICRO- 1 c).

1968 *N.Y. Times* 22 Jan. 36 Hemlines go to all lengths. In extremes, there are micros, which barely cover the buttocks; minis, maxis and the nineteen-thirties length. 1969 D. CLARK *Death after Evensong* ii. 40 She's some bird... Legs just a bit skinny for a micro, but still a good shape.

micro-. Add: In some words, esp. in *Med.*, the pronunc. (mi·kro) also occurs. **1.** For 'object' read 'entity' and add: *microabscess*, *-aneurysm*, *-biota*, *-chromosome*, *-constituent*, *-crater*, *-earthquake*, *-environment* (hence *-environmental* adj.), *-event*, *-explosion* (hence *-explosive* adj.), *-fossil*, *-fracture* (so *-fracturing* vbl. sb.), *-fungus* (later examples), *-graver*, *-instability*, *-lens*, *-metazoan*, *-metazoon*, *-particle*, *-phenocryst*, *-plankton* (hence *-planktonic* adj.), *-population*, *-pore* (further examples, also of *-porous* adj.) (hence also *-porosity*), *-powder*, *-quantity*, *-state* [STATE *sb.* 30], *-system*, *-tektite*, *-vegetation*, *-zone*; **mi·croatoll**, a circular growth of coral a few metres in diameter and with a central depression, such as is found in intertidal areas in warm seas and on the flats inside a coral reef; **mi·crobeam**, a very narrow beam of radiation; **mi·croblade** *Archæol.*, a flake chipped from a prepared flint core; **mi·crobody** *Cytology*, one of the round or egg-shaped microscopic particles occurring in cytoplasm which are surrounded by a membrane and contain oxidases; **mi·crocapsule**, a minute capsule used to contain, and render temporarily inactive, drugs, dyes, etc.; **mi·crocirculation** *Physiol.*, circulation of the blood in the smallest blood vessels; **mi·crocolony** *Biol.*, a group of animals or plants, esp. bacteria, found in a microhabitat; a very small group of cells in culture; **mi·crocontinent** *Geol.*, an oceanic, often submarine, plateau that is thought to be an isolated fragment of continental material; so **mi:crocontine·ntal** *a.*; **mi·crocrack**, a microscopic crack; hence **mi·crocracked** *ppl. a.*, **mi·crocracking** *vbl. sb.*; **mi·croculture** *Biol.*, (a) culture (of tissue, micro-organisms, etc.) on a small scale; **mi·croele:ctrode**, an electrode with a very fine tip, such as one suitable for investigating the electrical properties of individual cells; **microfi·bril** *Biol.*, a small fibril in living tissue that is visible only under the electron microscope, esp. one of a group that together make up a fibril (such as a cellulose fibril in the wall of a plant cell); hence **microfi·brillar** *a.*; **microfila·ria** *Zool.*, the minute larval form of a filaria; **mi:croha·bitat** *Ecol.*, a habitat which is of small or limited extent and which differs in character from some surrounding more extensive habitat; **mi·crologic**, micro-electronic logic (*LOGIC *sb.* 3*); freq. *attrib.*; **mi·cromodule** *Electronics*, a miniaturized module, consisting of a stack of interconnected microelements; **micromuta·tion** *Biol.*, (an instance of) mutation that has a superficially small or trivial effect on the phenotype; **mi·cro-operation** *Computers*, a simple operation carried out in response to a microinstruction (see also sense *2 a); **mi·cro-order** *Computers*, a microinstruction produced by a microprogram; **mi·croplate** *Geol.*, a relatively small lithospheric plate; **mi·cro-process**, a process that occurs on a minute scale; **micropro·cessor** *Computers*, a processor small enough to be accommodated on a single chip, or just a few chips, and capable of serving as the central processing unit of a computer of comparable size; **mi·cropulsa:tion**, a small oscillation in the strength of the earth's magnetic field; †**microray** = *MICROWAVE (*obs.*); **microsphere**, (*c*) any minute spherical object; *spec.* one of those obtained by the cooling of a solution of proteinoid; **mi:crostratifica·tion**, (a) small-scale horizontal stratification of the water of a pond or lake; **mi·croswitch**, a switch which can be operated rapidly by a small movement; **microtri·chium** *Ent.* [Gr. θρίξ, τριχ- hair], usu. in pl. **microtri·chia**, minute hair-like structures found on the wings of certain insects, esp. those of the order Diptera; **microtu·bule** *Biol.*, any of the small, relatively rigid tubules, typically about 25 nanometres in diameter, that are present in the cytoplasm of many plant and animal cells and are thought to have a structural function and to be involved with cell motility; so **microtu·bular** *a.*; **microvi·llus** *Biol.* (*pl.* **-vi·lli**), one of a number of minute projections from the surface of some cells; any process similar to a villus but smaller; **mi·crowire**, very fine glass-coated wire; **mi·croworld**, a realm or world (WORLD *sb.* 13) very restricted in its dimensions or variety.

1946 *Nature* 5 Oct. 487/2 Thrombo-angiitis, phlebitis and lymphangiitis with micro-abscess formation can be found on study of sections of these structures. 1962 *Lancet* 1 Dec. 1134 The kidneys showed the changes of acute and chronic pyelonephritis, with microabscesses and gross fibrosis. 1948 PARSONS & DUKE-ELDER *Dis. Eye* (ed. 11) xvii. 371 These droplets..tend to obscure the lumen of the vessel, forming a small micro-aneurysm. 1962 H. HEATH in A. Pirie *Lens Metabolism Rel. Cataract* 366 The fact that microaneurisms [*sic*] develop only in some cases of long-standing diabetes would seem to indicate a derangement in some slow metabolic process. 1933 P. H. KUENEN in *Snellius-Expedition* V. II. iii. 64 Conditions unfavourable to coral growth above low tide level certainly favour the formation of the micro-atolls. 1963 D. W. & E. E. HUMPHRIES tr. *Termier's Erosion & Sedimentation* xiii. 286 Pools are frequently observed on the surface of the reef-flats... These pools contain micro-atolls which have overhanging margins, indicating a reaction against choking muds. 1967 *Oceanogr. & Marine Biol.* V. 488 Sometimes this calcareous worm produces small reefs fringing the piers, or even micro-atolls. 1950 *Engineering* 5 May 498/3 The Cambridge work on micro-beams was most promising. It was nearing the stage when an investigator could look down a microscope, see a particle which he had etched, and direct X-rays on to that one particle, thus obtaining its diffraction pattern and identify it. 1962 *Times* 17 May 14/4 Investigate the effects of irradiation with a microbeam of ultraviolet light. 1972 *Science* 3 Nov. 461/2 Sea urchin sperm flagella are irradiated near their base with a laser microbeam. 1968 *New Scientist* 1 Aug. 223/1 Professor James B. Lackey.. has been looking at..esturine [*sic*] habitats, concentrating on the microbiota, especially the growth and decline of algae and protozoa. 1974 *Nature* 8 Feb. 361/2 Only a selection of the diverse Bungle Bungle microbiota [from Western Australia] is presented here. 1969 *Britannica Bk. of Year* (U.S.) 101 Late in the Upper Paleolithic and during the ensuing Mesolithic, it became the fashion to make smaller and smaller bladelettes. Commonly inserted as 'side blades' into lateral grooves in antler and bone projectile points, such 'microblades' lacerated the flesh of wounded game animals and thus promoted free bleeding and rapid death. 1972 *Sci. Amer.* Jan. 51/1 Working near Telegraph Creek in northwestern British Columbia in 1969 and 1970, Jason W. Smith and his colleagues from the University of Calgary unearthed several hundred delicate obsidian 'microblades', flakes seldom more than an inch long or a fifth of an inch wide, along with eight of the 'cores' that yield such blades. 1954 J. RHODIN *Correlation Ultrastruct. Organization & Function in Tubule Cells Mouse Kidney* ii. 21 The number, size and shape of the microbodies varies from cell to cell with a mean of about 10 in each cell. 1973 *Plant Physiol.* LI. 905/2 Electron micrographs frequently show association of microbodies with the rough endoplasmic reticulum. 1961 *New Scientist* 20 Apr. 115/3 The National Cash Register Company first developed microcapsules of a dye in a colourless form, which were spread on paper. 1967 *Britannica Bk. of Year* (U.S.) 803/2 *Microcapsule*, a tiny capsule containing a liquid or solid substance (as a chemical or medicine) that is released when the capsule is broken, melted, or dissolved. 1969 *New Scientist* 3 Apr. 18 Microcapsules—tiny bags made of synthetic membrane containing enzymes or other active materials—show promise as a means of treating kidney failure. 1972 *Chem. Processing* Apr. 21/2 Wass, Pritchard have developed a dry system for the application of fragrance or flavour microcapsules during the printing process. 1905 E. B. WILSON in *Jrnl. Exper. Zool.* II. 375 Especially large or small chromosomes may be designated as 'macrochromosomes' or 'microchromosomes', irrespective of their behavior. 1958 C. P. SWANSON *Cytol. & Cytogenetics* xiii. 459 Apart from accessory chromosomes, which are generally..smaller than the usual chromosomes, microchromosomes are found throughout the plant and animal kingdom. 1969 BROWN & BERTKE *Textbk. Cytol.* 755/1 *Microchromosome*, very small metaphase chromosome of a karyotype possessing only two sizes of chromosomes, very small ones and quite large ones; the latter are called macrochromosomes. 1959 *Angiology* X. 241/1 Pressure differences in the vessels of the microcirculation are small. 1969 *Sci. Jrnl.* Apr. 20/2 To increase an impaired microcirculation patients are commonly treated with histamine which is a strong vasodilator and rapidly relaxes the walls of capillaries. 1959 *New Biol.* XXX. 53 The fact that a large number of nests are collected each year may be attributed to birds successfully breeding in the inaccessible parts of the caves. These micro-colonies form reservoirs from which the caves are restocked annually. 1968 *Brit. Med. Bull.* XXIV. 246/1 It is often found that cells grown in culture after irradiation show a spread of colony size;..there are often micro-colonies notably smaller than those growing from unirradiated controls. 1968 L. E. CASIDA in Gray & Parkinson *Ecol. Soil Bacteria* 100 Many soil bacteria grow as microcolonies within the larger pores of soil aggregates. 1973 R. G. KRUEGER *Introd. Microbiol.* iii. 31/2 Microcolonies are formed by some photosynthetic bacteria that produce buds from the ends of filamentous outgrowths of the cell. 1901 TAYLOR & WHITE *U.S. Pat.668,269* 19 Feb., When treated with higher heats..the steel of the tools show under the microscope a distinctly larger grained structure in many cases interspersed with austenite, a micro-constituent of steel discovered by Osmond. 1930 *Times* 29 Mar. 17/1 The metallurgy of steel castings, and the discovery of about 25 micro-constituents of steel. 1966 D. G. BRANDON *Mod. Techniques Metallogr.* i. 1 Micro-constituents were successfully identified. 1965 HEEZEN & THARP in *Phil. Trans. R. Soc.* A. CCLVIII. 98 We may, of course, in referring to these aseismic ridges as micro-continents, infer that they are either (1) fragments of former continents, or (2) nuclei of growing continents. 1966 R. W. FAIRBRIDGE *Encycl. Oceanogr.* 373/2 Madagascar is clearly a microcontinent. *Ibid.* 874/1 In the mid-ocean there have been identified..(3) true mid-ocean plateaus, to which Heezen has applied the genetic connotation 'microcontinent'. 1970 *Nature* 13 June 1044/2 We suggest that the East Canaries block is a microcontinent or sialic continental fragment detached from the African margin. 1966 R. W. FAIRBRIDGE *Encycl. Oceanogr.* 964/2 Submarine ridges and rises in other oceans..often have foundations of 'microcontinental' blocks. 1974 *Nature* 14 Mar. 204/2 This suggests that the Rockall–Faeroe Plateau as a whole may form a single microcontinental fragment. 1950 *Welding Res.* Suppl. Sept. 473-s/2 The microcracks have produced..highly localized strain. 1960 W. D. BIGGS *Brittle Fracture of Steel* v. 154 Ingot irons and spectrographically pure irons do contain quantities of iron oxide in the form of inclusions in which micro-cracks may be nucleated. 1972 *Science* 2 June 1015/1 This can be attributed to the absence of water in the lunar rocks combined with the effects of porosity and microcracks. 1970 *Times* 22 May 27 At low thicknesses the deposits of chrome thrown by the new method are microporous, while thicker coatings are microcracked. 1956 *Welding Res.* Suppl. Feb. 78-s/3 The pearlite area..showed no evidence of microcracking. 1972 *Sci. Amer.* Nov. 92/2 Concrete is an inhomogeneous mixture of materials and is subject to local microcracking. 1965 *Jrnl. Appl. Physics* XXXVI. 3701/1 From the crater fields observed at low power density irradiation it can be concluded that the formation of microcraters is an early stage in the development of laser-induced radiation damage of metal surfaces. 1970 *Daily Tel.* 7 Jan. 3 (*caption*) A grain of lunar soil highly magnified to reveal a microcrater. 1892 *Phil. Trans. R. Soc.* B. CLXXXIII. 132 (*caption*) Glass micro-culture chamber in use with gas generators. 1973 *Nature* 5 Oct. 263/2 We analysed microcultures containing small numbers of cells (10^4 to 10^5) and the antibody response could be followed for up to 40 d. 1967 *N.Y. Times* 23 May 29 Before a volcano erupts it generates micro-earthquakes, which we are able to detect on our seismographs. 1974 *Nature* 24 May 308/1 [They] used sonobuoys in an attempt ..to see if, notwithstanding the lack of large earthquakes, they could detect microearthquake (magnitude less than 4·0) or earthquake swarm activity. 1917 *Amer. Jrnl. Physiol.* XLIV. 521 The electrodes were of copper or platinum, the muscle resting on an indifferent plate and subject above to the light contact of an active needle, microscopically sharpened—the micro-electrode. 1946 *Nature* 20 July 97/2 As soon as the results of Granit's micro-electrode experiments on the retinæ of animals were published, it was clear that a method was wanted for obtaining similar information with regard to the colour vision of man. 1955 C. R. N. STROUTS et al. *Analytical Chem.* II. xviii. 576 A microelectrode usually consists of a platinum wire about 0·5 mm. in diameter, sealed into the side of a glass tube. 1966 T. PYNCHON *Crying of Lot 49* iii. 55 Something tidal began to reach feelers in past eyes and eardrums, perhaps to arouse fractions of brain current your most gossamer microelectrode is yet too gross for finding. 1954 *New Biol.* XVII. 7 Dr. Edney's account of the water relations of woodlice is a study in comparative physiology, within a single order, linked closely to the differences in environments—and indeed micro-environments—of the species concerned. 1962 F. I. ORDWAY et al. *Basic Astronautics* xiii. 515 The apparent simplicity of the concept of a microenvironment within the spaceship leaves one serious gap. 1973 *Nature* 2 Mar. 20/2 Commitment may not occur until stem cells enter the micro-environment of the thymus or bursa (or bursa equivalent). 1971 *World Archaeol.* III. 128 A village..in a different micro-environmental zone. 1974 *Nature* 27 Sept. 317/2 Oscillations of microenvironmental pO_2 in the medium overlaying attached cultures of 10^7 cells per 50 mm Petri plate were measured amperometrically with calibrated Pt/Ir micro-oxygen cathodes polarised between 0·6 and 0·8 V. 1949 A. PAP *Elem. Analytic Philos.* vii. 137 The belief in micro-events involving micro-objects which are the hidden causes of observable events. 1957 C. DAY LEWIS *Poet's Way of Knowledge* 14 During the last half-century, physical scientists have moved towards the study of the micro-event. 1959 K. R. POPPER *Logic Sci. Discovery* viii. 196 Hypothetical and not directly observable 'micro events'. 1968 *Sci. Jrnl.* Nov. 44/2 The rate of current growth must be increased and the plasma contained in a special non-cylindrical shape—the so-called plasma focus. However, the process itself will be so brief as to resemble a strong explosion. If the dimensions of the system are not too large this can be manifested as a 'micro-explosion' which will be a good source of both fast neutrons and hard x-rays. 1971 *Jrnl. Physics D* IV. 1941 Joule heating immediately underneath an operating site raises the local temperature of the metal above its boiling point... The resulting micro-explosion..gives rise to a crater with a surrounding lip of electrode debris. 1974 *Sci. Amer.* June 24/1 Calculations were undertaken..to try to find out what would happen when tiny deuterium-tritium pellets were imploded to thermonuclear conditions by intense beams of laser light. It was also proposed that the fusion microexplosions could be applied to the generation of power. 1971 *Jrnl. Physics D* IV. 1945 The micro-explosive disruption of the layer following Joule heating of the underlying metal causes collapse of the high trapped-ion field and the quenching of local emission. 1938 *Nature* 19 Nov. 902/1 It may be observed ..that the micelles themselves are aggregated into 'microfibrils', separated by spaces rather larger than the usual intermicellar spaces. 1962 J. T. MARSH *Self-Smoothing Fabrics* xx. 351 It is also probable that with cotton, there is some binding or entanglement of the microfibrils and even of the growth layers. 1970 T. S. & C. R. LEESON *Histol.* (ed. 2) ii. 37/1 In many cells cytoplasmic microfibrils are present, probably consisting of elongated protein molecules. 1971 *Sci. Amer.* June 44/3 Using the electron microscope to inspect the formation of ligament and tendon during the very first stages, we found no sign of the amorphous component in the embryonic elastic fibers. At this time the fibers appeared to consist only of the 110-angstrom microfibrils. 1956 *Nature* 18 Feb. 319/1 Sikorski and Woods pointed out the difficulty of reconciling the macrofibrillar type of structure observed in the follicle of a wool fibre with the occurrence of extended microfibrillar sheets in disintegrated fibres. 1972 *Canad. Jrnl. Bot.* L. 479 The microfibrillar material transports sucrose in pulses at about 400 cm h^{-1}. 1878 *Lancet* 23 Mar. 440/2 Micro-filariæ in chylous urine. 1946 *Nature* 21 Dec. 913/1 The disease [*sc.* equine dermal filariasis] is associated with the presence of microfilariæ in the skin lesions. 1966 *New Scientist* 1 Sept. 482/3 Adult worms spawn progeny called microfilariae. They are ingested by, and develop in, several varieties of mosquito. 1924 *Bull. Amer. Assoc. Petroleum Geologists* VIII. 539 (*heading*) The value of micro-fossils in petroleum exploration. 1961 J. CHALLINOR *Dict. Geol.* 127/2 There is no definite size limit for the category of microfossil, but one recent worker..has taken 2 mm. 1969 *Times* 11 Jan. 15/8 It is..possible that the rocks may have been contaminated by chemical fossils and microfossils after they were laid down. 1939 *Amer. Mineralogist* XXIV. 73 The streaks, patches, and veinlets of coarsely crystalline kaolinite and halloysite are sedimentary in origin and appear to be related to re-crystallization along the micro-fractures in the clay. 1973 *Jrnl. Biomech.* VI. 5/2 Preliminary results from our study of human materials have demonstrated evidence of healing or healed microfractures in the subchondral bone taken at autopsy from patients with early signs of degenerative joint disease. 1969 *Nature* 27 Sept. 1306/1 All the rock samples are characterized by small surface pits lined with glass, areas of spattered glass and whitish markings which are small areas of microfracturing. 1920 Microfungus [see *ENCROACH *sb.*]. 1971 N. E. HICKIN *Wood Preservation* 66 Micro-fungi... An imprecise term of no taxonomic significance, usually used to include those fungi whose classification is in doubt as no sexual process has been observed, the *fungi imperfecti*, and including also the smaller *Ascomycetes* with small sporophores. 1926 *Guide Antiquities Stone Age Dept. Brit. & Mediaeval Antiquities Brit. Mus.* (ed. 3) 90 The early sites with the angle-graver or true graver, and the later with the micro-graver as the typical implement. 1939 V. G. CHILDE *Dawn European Civilization* (ed. 3) i. 6 Pigmy flints or microliths, ingeniously worked into regular geometrical shapes..or into micrograver. 1933 *Ecol. Monogr.* III. 169 Niche—the microhabitat or ultimate division of the habitat including recognition of its modifying factors. 1970 *Watsonia* VIII. 93 Like several other British ferns, it shows a partiality for railway platforms, which often supply a very special moist and calcareous microhabitat. 1971 *Sci. Amer.* Sept. 105/3 All habitats—from the expanses of sea ice and open water to the microhabitats of tidal flats, leeward waters and protected valleys—are utilized [by Eskimos]. 1962 W. B. THOMPSON *Introd. Plasma Physics* viii. 235 Since the wavelength is so small, this instability is almost independent of the macroscopic geometry, and represents one of the 'micro-instabilities'. 1971 *Science Year 1972* 222 Fusion researchers have been trying to find magnetic-bottle configurations that will be able to reduce the microinstabilities to acceptable proportions. 1964 S. DUKE-ELDER *Parsons' Dis. Eye* (ed. 14) vii. 82 Micro-lenses resting on the cornea are easier to fit and to wear. 1971 *Time* 19 July 48 Walon Green..used microlenses and extreme slow motion to get awesome footage of mayflies living out their brief lives. 1960 *Electronics* 25 Nov. 107/1 (*heading*) Micrologic computer. 1965 *New Scientist* 25 Mar. 769/1 Marconi..has a pilot 'micro-logic' circuit manufacturing plant at its laboratories. 1971 J. H. SMITH *Digital Logic* i. 4 The Integrated Circuit. This is often referred to as micro logic. 1972 *Sci. Amer.* Sept. 2/1 (Advt.), This revolutionary small computer lets your people use the languages which best express the problems you need to solve. Its variable micrologic processor operates efficiently with each language by instantly restructuring itself as it shifts from one program or program segment to the next. 1964 *Oceanogr. & Marine Biol.* II. 382 This remarkable interstitial fauna..consisting of protozoans and micrometazoans.. has been eagerly studied since the twenties by Kiel zoologists. 1969 *New Scientist* 3 Apr. 34/2 The dozen or so micrometazoa with which they are working. 1959 *Times* 19 Mar. 10/5 New miniature military radios, radar controls and other products of a revolutionary programme of micro-module electronics. 1965 *Listener* 1 July 6/1 I.B.M. already has 80 per cent. of the world computer market. With its micro-modules it looks like keeping it. 1940 R. GOLDSCHMIDT *Material Basis of Evolution* 324 Among these evolutionary steps there are many of a type which preclude an evolution by slow accumulation of micromutations. 1972 *Science* 12 May 623/3 In the 1930's population geneticists recognized micromutation as the raw material of evolutionary change. 1953 Micro-operation [see *MICROPROGRAM]. 1970 O. DOPPING *Computers & Data Processing* xii. 197 A program step, i.e. the execution of an instruction in a computer, consists of a number of micro program steps, or micro operations. 1953 Micro-order [see *MICROPROGRAM]. 1972 D. LEWIN *Theory & Design Digital Computers* iv. 71 The main function of the control unit is to decode the order digits of an instruction word, thereby generating the necessary sequence of control waveforms (micro-orders) to allow the instruction to be executed. 1969 *Encycl. Sci. Suppl.* (*Grolier*) 78 The search continues for facts about the similarities and differences between 'organized elements', microfossils, and microparticles formed from amino acid polymers. 1926 G. W. TYRRELL *Princ. Petrol.* v. 86 The terms *microporphyritic* and *microphenocrysts* may be used when the texture can only be made out with the use of a microscope. 1946 *Jrnl. Geol.* LIV. 27/2 Microphenocrysts of augite are set in a microcrystalline to cryptocrystalline and hyaline groundmass. 1973 *Nature* 9 Feb. 374/1 It occurs as inclusions in both the other minerals, and forms skeletal microphenocrysts about 0·05 mm across. 1903 *Jrnl. R. Microsc. Soc.* 638 Spinelli publishes a first contribution to the marine flora of Sicily... The microplankton of the Sicilian coast has not been included. 1969 BENNISON & WRIGHT *Geol. Hist. Brit. Isles* xiv. 319 Marine microplankton has also been found [in the Wealden area]. 1971 *Nature* 30 Apr. 562/1 Cita and Blow compared the microplanktonic succession in these stratotypes with that of the tropical zonation scheme. 1972 *Ibid.* 4 Feb. 253/2 From the point of view of plate tectonics the Mediterranean is a complicated part of the world, being a region of numerous small interacting plates (microplates)

rather than a part of one large one. 1967 *Oceanogr. & Marine Biol.* V. 543 There is a highly detailed zoning of ecological conditions and micro-populations in the supra- and the mediolittoral zone. 1956 *Nature* 17 Mar. 502/1 The same type of micro-pore structure as occurs in the coking coals. 1967 M. CHANDLER *Ceramics in Mod. World* iv. 133 Translucent alumina ceramics that contain no micropores. 1939 *Chem. Abstr.* XXXIII. 4099 With increasing percentages this reduction continues until with 5% of Cu_2O there is no change in vol. However, this is in part only apparent, since Cu_2O causes microporosity. 1961 A. TAYLOR *X-ray Metallogr.* iii. 32 For the more subtle flaws and fine detail, such as hairline cracks or microporosity, the much more sensitive photographic method must be employed. 1963 J. OSBORNE *Dental Mech.* (ed. 5) xiv. 325 A shrinkage develops causing a microporosity throughout the entire casting. 1955 *Sci. News Let.* 29 Jan. 67/1 Oxygen..diffuses through a microporous filter disk made of porcelain and having 800,000,000 holes to the square inch. 1970 *Times* 22 May 27 At low thicknesses the deposits of chrome thrown by the new method are microporous. 1953 *Electronic Engin.* XXV. 233/1 Micropowder magnets can be made—after the powder has been produced—in a great variety of shapes. 1965 *New Scientist* 6 May 368/1 This metal could be either in very finely divided state (micropowder) or else as a sintered iron oxide compound. 1936 *Mind* XLV. 275 Jordan's attempt to account for it in terms of an intensification of acausal micro-processes is not supported by the facts. 1962 *Times* 13 Apr. 19/7 A higher resolution..is still more attractive on account of the prospects it affords of exploring those structural details and micro-processes upon which the functioning of living matter depends. 1970 *IEEE Trans. Computers* XIX. 710 LX-1 is an integrated circuit prototype of a microprocessor which is being used as a design vehicle to study the problems associated with the design and implementation of a similar computer constructed with large-scale integrated circuits. 1974 *Computer* July 22/2 There are at least four major classifications of systems that can be designed using microprocessors: calculators, controllers, data processors, and general-purpose computers. Controllers and calculators are the most likely candidates for single-chip CPUs. 1974 *Ibid.* Aug. 34/1 The present system uses a microprocessor and a small number of other custom LSI devices to control the spark ignition timing and EGR valve position based on a number of input engine variables. 1975 *Sci. Amer.* May 34/2 In 1971..the Intel Corporation, which had undertaken to develop a calculator chip, chose to design it as a more versatile programmable, single-chip microprocessor. 1949 *Proc. Japan Acad.* XXV. IX. 24 A remarkable micropulsation of dH/dt at the sudden commencement [of the magnetic storm] frequently took places [*sic*] during the summer and equinox, while it is weak during the winter. 1960 *Jrnl. Geophysical Res.* LXV. 1843/1 Micropulsations belong to the family of disturbances that have been related to the arrival of solar terrestrial particles. 1968 *McGraw-Hill Yearbk. Sci. & Technol.* 241/1 The instability also generates geomagnetic micropulsations and..very low frequency radio emissions. (Micropulsations are small quasi-sinusoidal oscillations of the Earth's field having periods ranging from about 1/5 sec to several hundred seconds.) 1946 F. SCHNEIDER *Qualitative Organic Microanalysis* vi. 164 The method can be readily adapted to the use of microquantities by the use of a reflux tube. 1971 *Nature* 7 May 11/2 The microquantities of compounds or isotopes which have now become detectable can serve as tracers to distinguish between, say, older and more recent material on the surface. 1931 *Electrician* 3 Apr. 509/2 The great advance made..has opened up the range of wave-lengths between 10 cm. and 1 m. for practical use... These short wavelengths have been designated 'micro-rays'. 1934 *Discovery* Sept. 242/2 The micro-rays used have the great advantage of freedom from disturbance by other wireless transmissions. 1937 *Jrnl. R. Aeronaut. Soc.* XLI. 47 The micro-ray equipments used on the commercial Lympne–St. Inglevert and the experimental Escalles–St. Margaret's [radio] links. 1960 *Science* 22 July 204/2 A tendency to yield microspheres having diameters in a bacterial range is illustrated. *Ibid.*, Another property of the microspheres is the tendency to shrink in sodium chloride solution hypertonic to that in which they are produced. 1965 R. S. YOUNG in S. W. Fox *Orig. Prebiological Syst.* IV. 349 It is difficult to imagine that this phenomenon of microsphere formation, so easily demonstrable in the laboratory, did not occur in nature if there were a suitable accumulation of building blocks. Whether this had anything to do with the origin of the cell is purely speculative at this stage, but it is certainly suggestive. *Ibid.* 354 The microsphere is a remarkably stable structure. 1974 *Nature* 10 May 177/1 About 100,000 microspheres 25 μm in diameter labelled with ^{46}Sc were injected into the left ventricle over 30 s. 1962 *Economist* 20 Jan. 204/1 The micro-states left behind by France—Niger, Chad, Dahomey, Togo. 1970 *Britannica Bk. of Year* (U.S.) 463 Other events included..proposals that a special UN membership category be created for 'microstates'. 1974 *Austral. Outlook* XXVIII. I. 24 No clear and universally acceptable definition of a microstate has yet emerged. 1937 *Science* 26 Feb. 224/1 (*heading*) Microstratification of the waters of inland lakes in summer. 1956 *Nature* 17 Mar. 520/1 With a small amount of plankton in such a large body of water [as Lake Tanganyika], it is possible that its distribution is uneven due to currents and microstratifications. 1957 G. E. HUTCHINSON *Treat. Limnol.* I. vi. 396 In very many lakes..elaborate microstratification occurred throughout the hypolimnion. 1958 *Jrnl. Exper. Analysis of Behaviour* I. 173 The *S* read from loose printed pages; every time he stuttered, *E* pressed a microswitch which activated an Esterline-Angus recorder. 1961 *New Scientist* 13 Apr. 16/1 Carbon dioxide is used for inflation, its flow being controlled..for patients with very slight residual power, by an electrical system using microswitches. 1967 E. CHAMBERS *Photolitho-Offset* ix. 128 The exposed film enters the processor at the film feeding station, the film activating a microswitch that starts a metering pump. 1940 *Ann. Rev. Biochem.* IX. 609 Since this is exactly the field served by the widely used Warburg manometric apparatus, and since the sensitivity of the diver method is about 1500 times as great as that apparatus, many possibilities of application to the microsystem exist. 1969 *New Scientist* 27 Feb. 452 One early form of life could have been a thermodynamically open, self-assembled, proteinaceous microsystem, capable of propagating its own kind through the use of preformed polyamino acids. 1967 *Nature* 22 Apr. 374/1 It is concluded that the glassy objects [in deep-sea sediments] discussed in this report are microtektites, and that they constitute a portion of the Australasian strewn field which extends from Thailand to Tasmania. 1973 *Ibid.* 16 Feb. 431/2 Microtektites (diameters < 1 mm) are also associated with the two most recent strewn fields. 1934 C. H. CURRAN *Families & Genera N. Amer. Diptera* 488 Microtrichia—The smaller abundant hairs of the wing. When these are present the wing is said to be villous. 1957 RICHARDS & DAVIES *Imms's Gen. Textbk. Entomol.* (ed. 9) I. 13 Microtrichia..are minute hair-like structures, found, for example, on the wings of the Mecoptera and certain Diptera. They resemble very small covering hairs. 1963 D. B. SLAUTTERBACH in *Jrnl. Cell Biol.* XVIII. 384/2 It may be generally recognized that cells possess a 'microtubular system'. 1971 *Sci. Amer.* Aug. 52/3 At the base of every eukaryotic flagellum and cilium is a distinct microtubular structure: the basal body. 1963 D. B. SLAUTTERBACH in *Jrnl. Cell Biol.* XVIII. 367 Small cytoplasmic tubules are present in the interstitial cells and cnidoblasts of hydra. They are referred to here as 'microtubules'. 1974 *McGraw-Hill Yearbk. Sci. & Technol.* 124/2 Microtubules are hollow cylinders about 250 A..in diameter which are present in a wide variety of cellular structures, including the mitotic spindle, cilia and flagella, and neural axons... They probably provide an internal 'cytoskeleton' that produces or maintains asymmetric cell structure, and they appear to be involved in the production of certain types of cell motility. 1956 *Nature* 4 Feb. 221/2 It seems quite plausible..that the stream of nutrient salts and metabolites excreted by the cuticle serves as the nutrient medium of the epiphyllic micro-vegetation. 1958 *Blumea* IX. 206 Barnacles were collected in order to get an impression of the algal microvegetation..growing on them. 1953 BORYSKO & BANG in *Bull. Johns Hopkins Hosp.* XCII. 259 The cell surface in contact with the allantoic fluid was consistently characterized by the presence of variable numbers of small projections..which we have named 'microvilli' to distinguish them from other types of surface projections (cilia, pseudopods, brush borders, blebs, etc.). These microvilli appeared as club-shaped projections extending into the allantoic sac. They were approximately 0·1 micron in width, ranging from 0·1 to 1·0 micron in length. 1968 *New Scientist* 28 Nov. 513/1 The skin of many tapeworms possesses a number of very small, finger-like projections. They are known as microvilli, from their resemblance to the larger, but similar, projections called villi found in the small intestine of most types of vertebrate. 1969 *Nature* 11 Oct. 116/2 (*caption*) This normal baby hamster kidney (BHK) cell is magnified about 3,300 times. The tendril-like processes are microvilli which appear at particular phases of the cell cycle, noticeably when the cell rounds up before division. 1963 *Times* 21 May 20/2 World scientists are taking their turn in a growing queue for Britain's newest export—glass-sheathed copper microwire 50 times thinner than a human hair. 1970 *New Scientist* 30 Apr. 228/3, 100 grammes of microwire, about 100 km long, would cost £1000. 1955 O. KLEIN in W. Pauli *Niels Bohr* 117 Einstein..may have felt that on the side of the quantum physicists the importance of the general relativity claim in the search for the laws of the microworld was usually underestimated. 1967 *New Scientist* 18 May 383/2 The pavilion [at London Zoo] is an assemblage of micro-worlds in which the jumpers can jump, the climbers climb, the burrowers burrow and the nocturnal sleepers sleep. 1973 C. BONINGTON *Next Horizon* ii. 46 In the micro-world of an expedition the pettiest details, like an unwashed pan or an irritating mannerism, are blown up out of all proportion. 1957 G. E. HUTCHINSON *Treat. Limnol.* I. ix. 609 By covering lake mud in the laboratory with oxygenated water, he was able to show that a microzone of oxygen-poor water was rapidly produced above the mud. 1964 *Oceanogr. & Marine Biol.* II. 137 Cores often exhibit a well marked horizontal stratification into the microzones of Perfiljew.

b. Other terms in which *micro-* indicates reduced size or scale, but not of what is denoted by a following sb., as *micro-distribution*; **mi·crobreccia** *Geol.* (see quot. 1972); **mi·crocamera**, a camera used in photomicrography; **mi:crocosmopo·litan**, one who regards himself as a citizen of every part of a particular society; a citizen of a defined and limited world; **mi·cro-evolu:tion**, evolutionary change within a species or smaller group of plants or animals, taking a relatively short time; hence **mi:cro-evolu·tionary** *a.*; **mi·croglossary**, a glossary or dictionary of terms in a particular subject; **microplasti·city**, plastic flow which occurs in small areas of a material at stresses below the elastic limit of the bulk material; so **micropla·stic** *a.*; **mi·crorelief** *Physical Geogr.*, small-scale relief (RELIEF[3] 3 b); **mi:crostimula·tion** *Physiol.*, stimulation applied to a very small area; **mi·crotheory**, a theory about one particular aspect or part of some subject or phenomenon.

1948 *Jrnl. Geol.* LVI. 149 (in facing Table 3), Siltstone (gritty). a. siltstone proper. b. Micro-conglomerate or micro-breccia. 1951 E. B. KNOPF tr. *B. Sander's Contrib. Study of Depositional Fabric* 28 The fragments in this deformation breccia are calcitized crinkled dolomite and various sedimentary microbreccias. *Ibid.* 29 Some fragments in the microbreccia are themselves fragments of breccia. 1963 D. W. & E. E. HUMPHRIES tr. *Termier's Erosion & Sedimentation* xiii. 294 The true reefs contain several facies:..breccias and microbreccias formed of fragments of calcareous organisms; oyster limestones; [etc.]. 1972 *Gloss. Geol.* (Amer. Geol. Inst.) 449/2 *Microbreccia*, (*a*) A poorly sorted sandstone containing relatively large and sharply angular particles of sand set in a very fine silty or clayey matrix; e.g. a graywacke. It is somewhat less micaceous than a siltstone. (*b*) A breccia within fragments of a coarser breccia (Sander, 1951, p. 28). 1973 *Nature* 23 Mar. 252/2 A feldspathic microbreccia from the Descartes region of the lunar highlands contains an unusual assemblage of pyroxene fragments. 1928 *Daily Express* 21 June 12 Modern science has at its disposal, 'doctors, chemists, biologists, ultra-violet lamps, micro-cameras, and spectroscopes'. 1958 *Newnes Compl. Amat. Photogr.* XXV. 219 Apart from the micro-camera, a plate camera is perhaps the next best. 1961 A. TAYLOR *X-ray Metallogr.* vi. 163 Equally good resolution may be obtained with quite small cameras provided the focus of the X-ray tube is restricted and the geometry of the camera and the specimen is suitably modified, as, for example, in certain microcameras. 1938 S. BECKETT *Murphy* xi. 240 It was as though the microcosmopolitans had locked him out. 1966 *Punch* 12 Jan. 50/1 Liberal Jewish New Yorkers..are cosmopolitans, but New York is their cosmos. They are really microcosmopolitans. 1964 *Oceanogr. & Marine Biol.* II. 111 The study of.. micro-distribution with reference to mixing processes in the sea. 1971 *Nature* 25 June 524/2 So far it has not been possible to study the micro-distribution of lead in materials, except by using electron microprobes or rather crude microchemical techniques which are only really valid if the lead concentration is high. 1940 R. GOLDSCHMIDT *Material Basis of Evolution* 199 Microevolution by means of micromutation leads only to diversification within the species. 1963 DAVIS & HEYWOOD *Princ. Angiosperm Taxon.* ii. 72 They [*sc.* biosystematists] have gone a long way in elucidating the processes of evolution at and below the species level (micro-evolution). *Ibid.* xii. 405 The significance of ecological modifications in a micro-evolutionary context has probably been underestimated. 1974 T. HEYERDAHL *Fatu-Hiva* ii. 86, I was on constant look-out for animals to save in bottles and tubes for the study of trans-oceanic migration and micro-evolution. 1955 LOCKE & BOOTH *Machine Translation Lang.* 11 The main deficiency at the present time is the absence of adequate field dictionaries for various technical fields and for all relevant languages. Preparation of these *microglossaries*..is important, though tedious work. 1956 *Nature* 7 Jan. 1/1 To solve the problem of word-order, the grammatical structure of the language must be investigated in detail and micro-glossaries—stem-ending dictionaries in the specified subject—must be compiled and coded. 1960 E. DELAVENAY *Introd. Machine Transl.* vi. 91 Research since 1949 has led to the provisional conclusion that in scientific texts non-grammatical polysemantic nouns and verbs do not present any great difficulty within the limits of the restricted vocabulary of any given science or technical subject. Thus special restricted dictionaries—microglossaries—should be constituted. 1960 *Mech. Engin.* July 71/3 Future investigations should be extended to a variety of materials to investigate further the validity of microplastic-strain hysteresis energy as a criterion for fatigue fracture. 1966 *Jrnl. Strain Analysis* I. 415/2 It is..a fact that plastic flow occurs in some parts of the cross-section even if the material is within the nominal elastic limit. This plastic deformation, which is of the same order of magnitude as the elastic strains, will be termed as [*sic*] micro-plastic throughout the paper. *Ibid.*, It is believed that micro-plasticity constitutes an important source of failure in the case of metal fatigue. 1972 T. IMURA in G. Thomas *Electron Microsc. & Struct. Materials* 129 The microplasticity can be attributed solely to the motion of edge dislocations. 1932 FULLER & CONARD tr. *Braun-Blanquet's Plant Sociol.* xi. 272 Causes of injury are..direct effect upon soil formation by stirring of the fine earth and changing the microrelief (hummocks and paths). 1938 *Geogr. Jrnl.* XCII. 271 [Articles] deal with glaciation, with melting and formation of micro-relief, and with run-off and thawing of glaciers. 1968 R. W. FAIRBRIDGE *Encycl. Geomorphol.* 795 Relief refers to relative height, while microrelief refers to small-scale differences in relief. Microrelief consists of any minor undulations on the surface of the land, usually of a scale which would not show on a normal topographic map. 1946 *Nature* 28 Dec. 947/1 Two point sources of light of about the same brightness were presented to an observer by means of the micro-stimulation apparatus. 1972 *Jap. Jrnl. Pharmacol.* XXII. 635 (*heading*) Re-examination of a centrally-induced cough in cats using a micro-stimulation technique. 1956 W. H. WHYTE *Organization Man* (1957) 28 Perhaps an overall theory of behavior is..a will-o'-the-wisp. If so, our efforts may still be rewarded by the salvage of microtheories about limited areas. 1971 *Jrnl. Gen. Psychol.* LXXXIV. 155 Almost all of them have been extremely cautious about extending their data to explanatory microtheories of the perceptual phenomena.

c. *Fashion.* Denoting extreme shortness of a woman's garment, as in *micromini*, *-shift*, *-skirt* (hence *-skirted* adj.).

1967 *Word Study* Dec. 3/2 *Mini*..has undergone further compounding in the formation *micromini* denoting an exceptionally short miniskirt. 1971 'V. X. SCOTT' *Surrogate Wife* 40 She wore micro-minis to show lots of leg. 1967 *Punch* 1 Mar. 308/2 The latest collection included a transparent micro-shift worn over an emaciated G-string. 1966 *Courier-Mail* (Brisbane) 29 Aug. 11 Mini-skirts, Micro-skirts, Maxi-skirts—the fashion battle rages on in London. 1967 *Punch* 21 June 893/2 Micro-skirts are designed to keep elders in their place, for no woman over forty can wear them without looking idiotic. 1969 *Guardian* 27 May 2/4 Women students of Pretoria University turned up..in ankle-length and calf-length dresses, in a campaign against 'micro-skirts'. 1973 J. DI MONA *Last Man at Arlington* (1974) viii. 73 A young blonde.. in a microskirt. 1967 *Times Educ. Suppl.* 13 Oct. 763 (*caption*) Mortar-boarded candidate..for student office at Boston's North-eastern University and his mortar-boarded micro-skirted assistant.

2. a. (*Micro-* now often implies simply a smallness of scale of the subject, rather than any use of microscopy, and so this use passes into 6. Cf. *MACRO- 1 e.) *microanatomy* (hence

-anatomist), *-chemistry* (in Dict.: hence also *-chemist*), *-cinematography* (so *-cinematographic* adj.), *-dissection* (hence *-dissect* vb.), *-ecology*, *-injection*, *-metallurgy*, *-operation* (see also sense *1) (so *-operative* adj.), *-palæontology* (hence *-palæontologic*, *-logical* adjs., *-palæontologist*), *-physiology* (hence *-physiologist*), *-sociology* (hence *-sociological* adj.); **mi:cromachi·ning**, the process or technique of shaping objects on a very small scale by non-mechanical means; **mi:cropinocyto·sis** *Biol.*, a submicroscopic form of pinocytosis in which material is taken into a cell as a result of the invagination and pinching off of the cell surface; hence **mi:cropinocyto·tic** *a.*

1964 G. H. HAGGIS et al. *Introd. Molecular Biol.* i. 1 The concentration of anatomists and microanatomists on finer and finer detail may be illustrated by four centuries of work on the structure of the vascular system and the blood. 1935 *Times Lit. Suppl.* 24 Oct. 662 What he has done is to present in a lively—almost exciting—way the more significant facts of human physiology and micro-anatomy as far as they have yet been discovered. 1946 *Nature* 26 Oct. 578/1 His researches have been mainly in the fields of micro-anatomy, embryology and the more physiological side of zoology. 1962 *Science Survey* III. 252 Her main interests have been in the micro-anatomy of the retina and its relationship to visual function in different animals. 1940 *Ann. Rev. Biochem.* IX. 610 Little used by biochemists or microchemists. 1971 *Nature* 7 May 11/1 There is also methane..in the indigenous rocks on the surface of the Moon. Is this..the remnant of the primordial gas from which the Moon was formed? That, no doubt, is a question that will preoccupy lunar microchemists for a long time to come. 1940 *Chem. Abstr.* XXXIV. 289 (*heading*) Micro tearing machine for the photomicrographic and microcinematographic examination of materials. 1962 *Lancet* 1 Dec. 1172/1 The misshapen lymphocytes may reasonably be regarded as dying cells; this view is supported by the microcinematographic studies of human blood-cells by Bessis. 1952 *Chem. Abstr.* XLVI. 13336/3 Microcinematography. 1971 *Nature* 10 Dec. 352/1 We are continuing to use single smooth muscle cells in dynamic studies of contraction recorded with microcinematography. 1973 *Ibid.* 2 Mar. 52/1 Cytoplasm was micro-dissected from salivary glands of larvae. 1915 *Science* 19 Feb. 291/1 The cells were isolated and studied by means of microdissection and vital staining in a hanging drop of the insect body fluid. 1972 *Sci. Amer.* June 95/1 A small bundle of fibers from a single receptor is separated by microdissection from the main trunk of the optic nerve and placed on an electrode. 1963 *New Society* 3 Oct. 26/1 A pioneering work in the micro-ecology of housing estates. 1969 *Nature* 29 Nov. 846/1 The microecology of the blowfly's gut has not been well explored. 1921 *Science* 28 Oct. 411/2 The microdissection and microinjection of marine ova and of animal and plant cells. 1970 *Nature* 22 Aug. 857/2 For micro-injection we used a hydraulic system of mineral oil with the micropipette tip filled with tracer solution. 1960 K. R. SHOULDERS in *Proc. Western Joint Computer Conf.* 251/2 The most highly resolved construction process that we have any control over is what we call electron beam activated micromachining. *Ibid.* 256/1 Electron-beam micromachining is the combination of certain methods of deposition, resist production, and etching. 1970 *New Scientist* 16 Apr. 101/1 Micromachining has received much publicity in the past, and forms the mainstay of Laser Associates' range of systems. 1899 *Proc. R. Soc.* LXV. 85 (*heading*) Experiments in micro-metallurgy:—Effects of strain. 1958 *Engineering* 11 Apr. 458/1 The accomplishments in micrometallurgy at the laboratory soon led to the determination of some of the properties of the new metal. 1913 *Jrnl. R. Microsc. Soc.* 207 (*heading*) Apparatus for micro-operations. 1968 *Sci. Jrnl.* Nov. 54/1 Micro-operations such as removing parts of the embryo or grafting tissues from one embryo to another. 1922 *Anatomical Rec.* XXIV. 18 Instruments for micro-operative work. 1941 *Bull. Amer. Assoc. Petroleum Geologists* XXV. 1208 The progressive multiplication of the micropaleontologic groups studied is considered. *Ibid.* 1219 These micropaleontologic objects he called 'microzoa'. 1972 *Biol. Abstr.* LIII. 3865/2 (*heading*) Lithologic and micropaleontologic study of the Lias..of Mayorca. 1929 *Jrnl. Paleont.* III. 229 (*heading*) Micropaleontological activities. 1973 *Nature* 13 July 74/2 Theyer presented palaeomagnetic and micropalaeontological data which seemed to invalidate the timing of what was thought to be one of palaeontology's most reliable datum planes. 1928 *Jrnl. Paleont.* II. 159 This is no reflection upon the micropaleontologist. 1972 *Daily Tel.* 31 Aug. 21 (Advt.), The palynologist should preferably have experience in the Mesozoic/Tertiary, the micropalaeontologist preferably on Mesozoic/Tertiary Foraminifera. 1883 A. H. FOORD (*title*) Contributions to the micropalæontology of the Cambro-Silurian rocks of Canada. Part 1. 1928 *Jrnl. Paleont.* II. 158 Micropaleontology should, in no way, radically depart from the general field and fundamental principles of paleontology. It is merely a study of the smaller fossils whose characteristic features are studied by the aid of a microscope. 1957 *New Biol.* XXIV. 11 Micro-palaeontology, which is mainly the study of fossil foraminifera, has become an important and highly specialized profession and practically the whole of current research on these animals is devoted to this aspect of the matter. 1974 *Sci. Amer.* May 136/3 Two clever micro-physiologists have measured the power involved, and they conclude that one gliding algal filament..used about 7,300 molecules of ATP per second to move. 1954 *New Biol.* XVI. 37 We cannot expect to close the evolutionary gap preceding the establishment of cellular organisms until we know far more of intracellular histology and micro-physiology. 1956 *Jrnl. Biophysical & Biochem. Cytol.* II. Suppl. 107 The particles, possibly coated with protein, become attached to the plasma membrane. Minute invaginations of the membrane, with the adherent granules, develop. The invaginated membrane is pinched off, resulting in the formation of an intracellular vacuole. Possibly a type of micropinocytosis takes place. 1973 *Nature* 2 Mar. 57/1 Electron microscopic studies..suggest that some synaptic vesicles form by a process of micropinocytosis at the presynaptic terminal membrane. *Ibid.* 57/2 A micropinocytotic origin of synaptic vesicles remains uncertain. 1942 *Jrnl. Legal & Pol. Sociol.* I. 55 We will forego here any prolonged analysis of this 'micro-sociological aspect', and will concentrate all our attention on the functional relationship between democracy and types of particular groups and all-inclusive societies. 1944 *Man* XLIV. 21/1 The most valuable contribution of the social anthropologist may well still lie in this micro-sociological field. 1972 P. SHERIFF tr. *Rocher's Gen. Introd. Sociol.* I. i. 4 The microsociological plane of different types of social links. 1941 *Jrnl. Philos.* XXXVIII. 486 The problems of mass, community, and communion arise in 'microsociology'. 1944 *Man* XLIV. 21/1 Much of the anthropologist's work has lain hitherto in what may be called micro-sociology—the study of small groups or of small units in larger groups. 1959 G. D. MITCHELL *Sociol.* i. 23 These two traditions may be said to be the forerunners of the modern tendency for sociology to bifurcate into what may be called macro- and micro-sociology. 1966 P. A. SOROKIN *Sociol. Theories Today* IV. xiv. 470 Micro-sociology studies the simplest manifestation of social reality—sociability, that is, 'the multifarious ways of being bound by a whole and in a whole'. 1974 *Times Lit. Suppl.* 22 Nov. 1304/1 Micro-sociology, indeed any serious sociology of an analytic kind, has to begin with a study of the most elementary social relations.

b. *micro-texture*.

1965 G. J. WILLIAMS *Econ. Geol. N.Z.* x. 157/2 The hardness and toughness of greenstone depends partly on its mineral composition, these qualities resulting from a micro-texture of very closely felted and interwoven minute fibres. 1971 *Good Motoring* Sept. 9/2 A road surface which gives grip at low vehicle speeds must incorporate roadstone of a harsh rather than a polished micro-texture.

c. *micro-slide* (examples).

1909 in *Cent. Dict.* Suppl. 1951 *Electronic Engin.* XXIII. 8/1 The combination of a television camera with a microscope makes it possible for a large group of students to watch simultaneously the events taking place on a microslide. 1971 *McGraw-Hill Yearbk. Sci. & Technol.* 359/2 A rapid detection method for the major components of cannabis..involves the use of microslides for thin-layer chromatography.

3. microgy·ria, abnormal smallness of the gyri of the brain; **microma·stia** [Gr. μαστ-ός breast], the condition in a post-pubertal woman of having an abnormally small breast; **microma·zia** [Gr. μαζ-ός breast] = prec.; **microphthalmia**: also **microphtha·lmos**, abnormal smallness of the eye; **micropsia**: also **mi·cropsy**.

1840 BILLINGS *Med. Dict.* II. 154/1 Microgyria. 1905 *Brit. Med. Jrnl.* 28 Oct. 1100/1 In this case of microgyria the right side of the brain was less than two-thirds the size of the left. 1920 *Brain* XLIII. 26 Microgyria, which has hitherto been usually known under the name of hemiatrophy of the brain or arrested development of the nervous system, has been comparatively rarely described, and merits further study. 1961 *Lancet* 2 Sept. 513/2 Microgyria was apparent in the posterior part of both lobes. 1918 DEAVER & MCFARLAND *Breast* iii. 45 These terms are applied to conditions in which there is congenital total absence of one or both of the mammary glands, or in which one or both glands are undeveloped or rudimentary in development. To the latter condition the term 'micromastia' might more correctly be applied. 1953 *New Statesman* 5 Sept. 254/3 About 4,000,000 young American women suffer in some degree from micromastia (immature breasts). 1965 L. B. AREY *Developmental Anat.* (ed. 7) xxiii. 452 Retention of the prepubertal condition (micromastia). 1890 *Syd. Soc. Lex.*, Micromazia. 1894 W. R. WILLIAMS *Monogr. Dis. Breast* iii. 33 When the defect is less complete than in the above cases, we get a very small imperfectly-developed gland, like the normal male breast, or smaller—micromazia. 1928 F. Z. SNOOP *From Monotremes to Madonna* 22 The breast may be absent, amazia; or very small, micromazia. [1845 *Dublin Jrnl. Med. Sci.* XXVII. 29 Microphthalmus is the term applied by continental writers to that peculiar condition of the eye, when there appears to be an arrest of development of this organ at some particular period of its growth, without either atrophy or disease.] 1850 *Boston Med. & Surg. Jrnl.* XLII. 421 (*heading*) Microphthalmos, complicated with congenital cataract in both eyes. 1934 *Arch. Ophthalm.* XI. 516 It is justifiable to consider the two forms of microphthalmos as one genetic entity with the morphologic distinction between microphthalmos with external, or orbital, cysts and microphthalmos with intraocular cysts. 1971 *Amer. Jrnl. Ophthalm.* LXXI. 1128/1 In colobomatous microphthalmos, the pathogenesis of the small eye is related to faulty closure of the embryonic fissure. 1890 W. JAMES *Princ. Psychol.* II. xix. 93 In consequence of this so-called micropsy, Aubert relates that he saw a man apparently no larger than a photograph.

4. microae·rophil(e = *microaërophilous* (s.v. MICRO- 4); also as *sb.*, a microaerophile organism; **mi:croaerophi·lic** = *microaërophilous* (s.v. MICRO- 4); **microce·llular**, containing or characterized by minute cells; **micro·phagous** *Zool.* [-PHAGOUS], feeding on minute particles; **microva·scular**, of or pertaining to the smallest blood-vessels.

1903 *Science* 6 Mar. 371/1 The microaerophiles will grow luxuriantly under normal conditions under diminished oxygen pressure. 1909 *Cent. Dict.* Suppl., Micro-aërophile *a.* 1957 G. E. HUTCHINSON *Treat. Limnol.* I. xiii. 757 It would seem that the limnetic purple bacteria are microaerophil rather than anaerobic. 1970 PASSMORE & ROBSON *Compan. Med. Stud.* II. xviii. 46/2 These microaerophiles often prefer a concentration of CO_2 that is higher than normal, e.g. 5–10 per cent. 1903 *Science* 6 Mar. 371 *Clostridium Pasteurianum*..is an anaerobe, but it also grows in symbiosis with aerobic forms; it is, therefore, microaerophilic. 1971 *Nature* 9 July 132/2 *Campylobacter*, a genus of microaerophilic bacteria including the organism formerly called *Vibrio fetus*. 1909 WEBSTER, Microcellular. 1958 *Punch* 8 Oct. 476/3 Bottines have micro-cellular rubber soles and are very neat and light. 1965 *Biol. Abstr.* XLVI. 5347/1 (*heading*) Complex treatment of the microcellular bronchial carcinoma by means of the nitrogen mustard and X-rays. 1971 D. G. JONES in C. M. Blow *Rubber Technol. & Manuf.* x. 397 Microcellular soling containing minute discrete air cells and having a specific gravity as low as 0·3 is also used extensively. 1923 J. W. FOLSOM *Entomol.* (ed. 3) xiii. 373 According to the nature of their food, most insects may be classified as follows:..*microphagous* ([feeding] on micro-organisms, as bacteria, yeasts, etc.). 1950 Microphagous [see *CHEMOTAXIS]. 1963 R. P. DALES *Annelids* iii. 76 The evolution of the Polychaeta is thus mirrored by the evolution of feeding methods; a transition in one direction towards burrowing, tube building, sedentary life and microphagous feeding; and in the other to a more active scavenging or predatory existence. 1959 *Ann. N.Y. Acad. Sci.* LXXXII. 236 Can anything be learned about the mechanism that damages blood vessels in diabetes mellitus by studying the dynamic morphology of the superficial microvascular system in man? 1967 *Sci. News Let.* 9 Sept. 262 Precise formations of the microvascular system and other spaces in organs of dead animals are revealed in detail when this liquid silicone compound is injected.

5. a. *microbar* [*BAR *sb.*[6] 1, 2], *-calorie*, *-curie*, *-henry* [*HENRY[3]], *-inch*, *-mho*, *-poise*, *-rad*, *-second*, *-watt*; **mi·crodegree**, one millionth of a degree centigrade (kelvin). Also duplicated to denote division by a million million (corresponding to the single prefix *PICO-), as in *micromicrocurie*, *-farad*.

1918 Microbar [see *BAR *sb.*[6] 2]. 1963 JERRARD & MCNEILL *Dict. Sci. Units* 22 Acoustic pressure of the order of one dyne cm^{-2} is now described by the microbar. 1971 *Nature* 1 Jan. 15/3 The figures..indicate that a pressure wave moving at 330 m s^{-1} generates a displacement of the Earth's surface of 10 to 15 nanometres per microbar with periods between 20 and 100 s. 1969 *Sci. Jrnl.* Aug. 42 Heat flow through the ocean crust is on average 1·2 microcalories/cm^2/s except in active regions such as mid-ocean ridges. 1911 *Sci. Amer.* 29 Apr. 429 The 'optimum' or most favorable dose of radiation, which developed the mold in four days, was found to be ½ microcurie per cubic centimeter of air. 1958 *Immunology* I. 29 The status of the grafts was assessed at monthly intervals after grafting by administering 5 microcuries of ^{131}I intraperitoneally. 1970 PASSMORE & ROBSON *Compan. Med. Stud.* II. vi. 8/2 When used for diagnostic purposes, radioiodine is usually given by mouth in a dose of 5–15 microcuries. 1957 *Science Progress* XLV. 415 Another..task would be the study of lattice specific heats at microdegree temperatures. 1971 *Physics Bull.* Dec. 713/1 Recent work, such as the specific heat measurement by Ahlers (1971) is characterized by temperature resolution as small as a microdegree. 1909 *Cent. Dict.* Suppl., Microhenry. 1911 *Physical Rev.* XXXII. 612 (*heading*) Inductance in microhenrys. 1964 R. F. FICCHI *Electr. Interference* iv. 39 Experience has shown that as the L exceeds 0·025 microhenries in magnitude, the effectiveness of the bond diminishes rapidly. 1941 OBERG & JONES *Machinery's Handbk.* (ed. 11) 1776 Assume that the quantities *a*, *b*, *c*, etc., equal the various profile measurements in micro-inches. 1962 *B.S.I. News* June 24/2 Another firm..even specified surface finished [? *read* finishes] in micro-inches on their drawings. 1970 *Sci. Amer.* Mar. 142/3 The rotor is driven magnetically on a superb ball-bearing axis. (The tolerances are specified to five or 10 microinches.) 1919 *Electric Jrnl.* XVI. 322 Micromhos per mile of each conductor. 1940 *Bell System Technical Publ. Monograph B-1268* 7 One interesting commercial tube in which one stage of secondary electron multiplication is added to an ordinary tetrode, has a transconductance of 14,000 micromhos. 1962 SIMPSON & RICHARDS *Physical Princ. Junction Transistors* viii. 175 Present values of its mutual conductance are small (~100 micromhos). 1961 *Daily Tel.* 10 Oct. 18/3 In the past 23 days, the average amount of radioactivity in every kilogram of air has been 3·42 micro-microcuries. 1909 *Cent. Dict.* Suppl., Micro-microfarad. 1921 *Physical Rev.* XVIII. 143 This plate had coatings so small that its normal capacity was only about 0·67 micro-micro-farads. 1962 CORSON & LORRAIN *Introd. Electromagn. Fields* ii. 60 A sphere one meter in radius has a capacitance of about 100 micro-microfarads. 1941 *Ann. Reg. 1940* 353 For hydrogen vapour it [*sc.* the viscosity] is *c.* 10 micropoise at 14·5° K. 1963 JERRARD & MCNEILL *Dict. Sci. Units* 107 The viscosity of gases is frequently given in micropoises (μP); (air 181 μP at 20° C). 1969 *Times* 2 Sept. 10/4 On some of the islands of the atoll..the intensity of fall-out radiation range[s] from three to seven microrads an hour. 1906 A. E. KENNELLY *Wireless Telegr.* ix. 99 If we call the one millionth part of a second one microsecond for convenience of description, then one complete wave would pass off in 1/2·5 microsecond. 1942 J. D. STRANATHAN *'Particles' of Mod. Physics* xii. 481 These particles are radioactive with a half life of a few microseconds. 1966 T. PYNCHON *Crying of Lot 49* v. 114 'Nearly three weeks it takes him,' marvelled the efficiency expert, 'to decide. You know how long it would've taken the IBM 7094? Twelve micro-seconds.' 1909 *Cent. Dict.* Suppl., Microwatt. 1914 R. STANLEY *Text-bk. Wireless Telegr.* xvii. 261 Duddell carried out experiments to find the minimum power required to produce audible signals in a telephone receiver at different frequencies, and found that, while 430 microwatts were required at 300 frequency, only 7·7 microwatts were required at 900 frequency. 1970 *Physics Bull.* Sept. 403/2 Sound power of the order of a microwatt from a dripping tap.

6. Substitute for 'Prefixed..as': Prefixed to the names of instruments and techniques with the sense 'specially designed for dealing with or measuring small effects or small quantities of material'. (In the names of techniques this use passes into 2 a.) *microammeter, -balance, -burette* (U.S. *-buret*), *-calorimeter* (hence *-calorimetric* adj., *-calorimetry*), *-densitometer* (hence *-densitometric* adj., *-densitometry*), *-determination, -electrophoresis* (hence *-electrophoretic* adj., *-electrophoretically* adv.), *-estimation, -gasometer* (hence *-gasometric* adj., *-gasometrically* adv., *-gasometry*), *-gravimetric* adj., *-Kjeldahl* (used *attrib.* or *absol.*: cf. *KJELDAHL), *-manometer* (hence *-manometric* adj., *-manometrically* adv.), *-method, -photometer* (hence *-photometric* adj., *-photometry*), *-pipette, -respirometer* [ad. G. *mikrorespirometer* (T. Thunberg 1904, in *Zentralbl. f. Physiol.* 3 Dec. 553)] (hence *-respirometric* adj., *-respirometry*), *-spectrograph* (hence *-spectrographic* adj., *-spectrography*), *-spectrophotometer* (so *-spectrophotometric* adj., *-spectrophotometrically* adv., *-spectrophotometry*), *-syringe, -technique*; **mi·croburner**, a small Bunsen burner for giving a single small flame; **microdiffu·sion** *Chem.*, diffusion of the vapour of a substance in an open container into an adjacent container in which there is a second substance, by which the first may be detected; usu. *attrib.*; **mi·crofilter**, a filter for separating out small quantities of material or very fine particles; **mi:cro-incinera·tion**, a process by which tissue sections are heated to a high temperature so as to remove organic matter and facilitate chemical analysis *in situ* of the inorganic constituents left behind; **mi·croneedle**, a very fine needle used in micromanipulation; **mi·croprobe** = *MICROANALYSER; also as *v. trans.*, to analyse with a microanalyser.

1930 *Telegraph & Telephone Jrnl.* Dec. 47/2 The deflection on the microammeter is proportional to the speed at which the tongue of the standard relay is oscillating, and a direct speed reading is given on the microammeter scale in terms of words per minute or cycles per second. 1964 R. F. FICCHI *Electr. Interference* vii. 127 A useful indicating device for locating leaks can easily be constructed using a tuned circuit consisting of a coil and variable condenser together with a 0-100 microammeter and crystal diode detector (1N34). 1903 *Jrnl. Chem. Soc.* LXXXIV. II. 571 A micro-balance with torsional control is described, having a sensitiveness of 0·0380 mg. per scale division. 1966 *McGraw-Hill Encycl. Sci. & Technol.* II. 74/1 Microbalances, used to weigh masses of a fraction of a gram, may be of the beam or the torsion type. 1926 *Chem. Abstr.* XX. 2543 A microburet for measuring minute drops is described. 1946 *Nature* 19 Oct. 556/2 The apparatus is completed by a microburette mounted on a movable arm. 1964 Micro buret [see *BURETTE 2]. 1911 *Jrnl. Infectious Dis.* VIII. 351 Natural gas for the microburner may be improved by causing it to pass through alcohol or benzine. 1938 *Jrnl. Laboratory & Clin. Med.* XXIV. 310 Two cubic centimeter samples of unhemolyzed serum are digested in pyrex test tubes..with 1·5 c.c. of concentrated sulfuric acid... The digestion may be carried out..rapidly by the direct flame of a microburner. 1962 *B.S.I. News* July 28 Specifies the following items of apparatus for use in micro-chemical analysis: (i) Crucible holder... (ii) Micro-burner jet for use with coal gas, giving a non-luminous flame up to about 5 cm in height. 1911 *Jrnl. Physiol.* XLIII. 261 (*heading*) A new form of differential micro-calorimeter, for the estimation of heat production in physiological, bacteriological, or ferment actions. 1959 DAWSON & LONG *Chem. of Nucl. Power* i. 9 A microcalorimeter has been developed at Harwell for the rapid estimation of small amounts of polonium. 1924 *Chem. Abstr.* XVIII. 2723 (*heading*) Utilization in biology of the microcalorimetric method. 1971 *Nature* 22 Oct. 560/2 Microcalorimetric studies of sperm whale ferri-MbCN have demonstrated a change in heat capacity (C_p) in the range 35°±18°C. 1924 *Chem. Abstr.* XVIII. 4483 Microcalorimetry. 1973 *Nature* 16 Feb. 473/1 For research purposes, microcalorimetry provides highly informative data regarding microbial metabolism. 1935 *Discovery* Nov. 324 Micro-densitometer for analysing sound track. 1966 *Aviation Week & Space Technol.* 5 Dec. 99/3 The CRT system will be used..as a flying spot scanner and will scan through a precision optical chain providing a high-resolution microdensitometer. 1959 *Listener* 12 Mar. 451/2 Micro-densitometric tracing of the profile of the crater. 1973 H. L. SNYDER in L. M. Bibermann *Perception of Displayed Information* iii. 89 The X-Y luminance patterns of any given object or area of a scene can then be determined by scanning microdensitometric measurements for a film transparency input to the television system. 1957 A. ENGSTRÖM in V. E. Cosslett et al. *X-Ray Microsc. & Microradiogr.* 32 Microdensitometry of the fine-grained photographic emulsion presents the same type of problems as does direct microspectrography of biological material. 1973 *Nature* 17 Aug. 413/1 More has been achieved from a digitized array of optical densities of the image obtained by microdensitometry. 1925 *Analyst* L. 302 (*heading*) Micro-determination of methoxyl. 1947 CHERONIS & ENTRIKIN *Semimicro Qualitative Org. Analysis* ii. 72 The Alber specific-gravity pipettes that are commercially available may be used for the microdetermination of densities. 1967 *Oceanogr. & Marine Biol.* V. 173 A colorimetric method has recently been described for microdetermination of lipids. 1939 E. J. CONWAY *Micro-Diffusion Analysis & Volumetric Error* i. 4 (*heading*) Scale and accuracy of the micro-diffusion methods described. *Ibid.* ii. 7 (heading) A standard micro-diffusion apparatus or unit. 1956 *Nature* 31 Mar. 623/2 The asparaginase activity of the extract was estimated from the ammonia formed (by the Conway microdiffusion method). 1959 *Times* 14 Oct. 14/5 Ultracentrifuge and micro-electrophoresis equipment. 1971 *Nature* 22 Oct. 567/1 Cell electrophoresis was performed with a thin walled cylindrical cell in a particle micro-electrophoresis apparatus. 1961 *Lancet* 16 Sept. 656/2 Microelectrophoretic analyses of the ribosenucleic acid formed in the neurone and in the glia. 1973 *Neuropharmacology* XII. 77 A technique has been described for stereotaxically performing microelectrophoretic studies on single brain cells in awake, non-paralyzed cats. 1963 *Federation Proc.* XXII. 625/2 Responses of neurones to microelectrophoretically applied Acetylcholine (ACh)..have been obtained. 1973 *Jrnl. Pharmacy & Pharmacol.* XXV. 309 Responses of single cortical neurons to microelectrophoretically applied noradrenaline at pH 3·1 and 5·0 and to hydrogen ions were compared in the halothane-anaesthetized cat. 1922 *Analyst* XLVII. 80 (*heading*) Micro-estimation of nitrogen and its biological applications. 1972 *Analytical Chem.* XLIV. 1879 (*heading*) Simultaneous microestimation of choline and acetylcholine by gas chromatography. 1911 *Jrnl. Chem. Soc.* C. II. 225 (*heading*) A micro-filter for the treatment of small quantities of precipitate. 1958 *Ann. Rep. Chief Inspector of Factories on Industr. Health 1957* 23 in *Parl. Papers 1958–9* (Cmnd. 558) XIII. 183 He wore a micro-filter respirator. 1972 *Physics Bull.* Aug. 455/1 The difference is due to the use of a microfilter in an aircraft fuelling system. 1951 *Amer. Jrnl. Clin. Path.* XXI. 1153 (*heading*) A practical microgasometer for estimation of carbon dioxide. 1972 *Analytical Biochem.* XLV. 112 The magnetic diver microgasometer is operated at a constant pressure so that the enclosed gas bubble can contract or expand freely. 1956 *Nature* 28 Jan. 185/2 In 1937, Linderstrøm-Lang introduced a microgasometric method based on the principle of the Cartesian diver. 1967 *Internat. Jrnl. Neuropharmacol.* VI. 266 The electron microscopic–cytochemical technique has been applied directly to unfixed isolated neurons following microgasometric analysis. 1968 *Progress Brain Res.* XXIX. 41 Ultracytochemistry was then applied to individual neurons, which had been analyzed microgasometrically. 1957 G. E. HUTCHINSON *Treat. Limnol.* I. xvii. 879 A technique involving characteristically ingenious microgasometry..has been used by Krogh and Lange. 1972 *Analytical Biochem.* XLV. 115 The dimensions of the ampullas most frequently used for magnetic microgasometry. 1931 *Industr. & Engin. Chem. (Analytical Ed.)* III. 345 An apparatus is described for micrograviometric analyses, such as sulfate, halide, and phosphate determinations. 1966 *McGraw-Hill Encycl. Sci. & Technol.* II. 199/2 Microgravimetric analyses, in which weighings are made to ±0·002 mg on a Kuhlman beam balance. 1924 *Analyst* XLIX. 52 (*heading*) Micro-incineration applicable to histochemical investigation. 1969 BROWN & BERTKE *Textbk. Cytol.* iii. 23/1 Micro-incineration involves the ashing of tissue sections at a temperature of about 600°C. This method can provide information regarding general distribution of certain minerals. 1923 *Jrnl. Amer. Chem. Soc.* XLVI. 2069 A new micro-Kjeldahl method..has been devised. 1946 *Nature* 30 Nov. 791/1 The extract was then filtered and its nitrogen content determined by micro-Kjeldahl. 1973 *Analytical Biochem.* LIII. 36 We have developed two simple modifications which minimize, if not eliminate, the excessive foaming encountered during micro Kjeldahl digestion of biological materials. 1897 *Jrnl. Physical Chem.* I. 596 The author's micromanometer (l.c., 1895), giving measurements accurate to 0·0033 mm water or 0·00024 mm mercury was used. 1949 O. G. SUTTON *Sci. of Flight* 201 The most sensitive micromanometer in general use, the Chattock-Fry gauge, is simply a glorified U-tube. 1972 *Physics Bull.* Aug. 491/2 The new vacuum micromanometer has been designed for applications requiring sensitive differential pressure measurements at high vacuum levels. 1937 *Discovery* July 224/2 We have new micromanometric techniques which can be applied directly to small-celled animals or plants in life. 1956 *Jrnl. Laboratory & Clin. Med.* XLVII. 642 The development of the micromanometric methods described herein was undertaken because of the need for a precise but rapid method for detecting sudden alterations in acid-base balance. 1973 *Jrnl. Neurochem.* XX. 1029 The rate of oxygen consumption has been measured micromanometrically in fresh mouse neuroblastoma cells. 1920 *Jrnl. Soc. Chem. Industry* 15 Mar. 206A/1 A micro-method in which 1–2 c.c. of urine and 1·5–3 c.c. of alkali need only be used. 1946 *Nature* 10 Aug. 199/2 We have developed a new micromethod for X-ray diffraction investigation of biological objects. 1967 *Oceanogr. & Marine Biol.* V. 171 A micromethod for determination of protein in extremely small quantities by the quenching of dye fluorescence has also been described. 1974 *Nature* 3 May 37/2 Studies of lymphocyte transformation *in vitro* were carried out by our micromethod. 1921 *Science* 28 Oct. 411/2 The method of making the glass micro-needles and pipettes. 1940 C. SHERRINGTON *Man on his Nature* iv. 116 The protein coat of the fertilized egg-cell..can be cut by the 'microneedle' without loss of its rigidity. 1971 *Nature* 2 July 28/2 The microneedles are made from 1 mm 'Pyrex' or borosilicate glass rod drawn out to form a thin (~0·3 mm) shaft, ~50 mm long. 1899 J. HARTMANN in *Astrophysical Jrnl.* X. 325 The apparatus..may be designated as a microphotometer, since it is a combination of microscope and photometer. 1947 *Jrnl. Brit. Interplanetary Soc.* VI. 162 Meteor light-curves, which must be examined photographically, with the help of a recording microphotometer. 1971 *Tsitologiya* XIII. 1530 The scanning and integrating microphotometer permits to define the quantity of substance, area and the linear dimension of micro-objects. 1952 *Chromosoma* V. 341 The microphotometric evaluation of cytochemical color reactions, such as the Feulgen-reaction on desoxyribose nucleic acid (DNA), has in recent years attained increasing importance. 1960 *Jrnl. Histochem. & Cytochem.* VIII. 4/1 It is of importance in some microphotometric studies of cells and tissues to determine the protein content of a single cell, nucleus, or nucleolus. 1937 *Monthly Notices R. Astr. Soc.* XCVIII. 113 A slit ·025 mm. wide and < 0·4 mm. high was used in all the microphotometry. 1973 *Acta Cytol.* XVII. 94 It is the objective of this paper to examine the feasibility of using automated scanning microphotometry with computer analysis of the cell images for the purpose of differentiating nonmalignant from malignant cells in pleural fluid. 1918 *Biol. Bull.* XXXIV. 134 The capillary attraction in the lumen of a micropipette..is quite sufficient for the purpose. 1922 *Anatomical Rec.* XXIV. 2 With the micropipette..one can..inject substances into..a cell. 1955 *Sci. Amer.* Aug. 98/2 Deposit five thousandths of a milliliter of serum on the ruled strip with a calibrated micropipette. 1968 *Times* 14 Nov. 8/7 A few of the cells, typically about three, are sucked up into a micropipette and injected through the slit in the first embryo. 1960 *Oxf. Univ. Gaz.* 19 Feb. 743/2 X-ray Fluorescent Micro-probe. A study is being made of the possible archaeological applications of this very new technique. 1969 *Awake!* 22 Oct. 19/2 A painting that was supposedly done by a sixteenth-century painter was exposed as a forgery by this laser device, which is called a microprobe. 1973 *Nature* 12 Jan. 87/1 Six olivines (three from Venezuela and three from Ghana) microprobed at the Geophysical Laboratory in Washington showed..an extremely constant Ni content. 1974 *McGraw-Hill Yearbk. Sci. & Technol.* 291/2 The ion microprobe mass spectrometer uses primary ions of argon or oxygen..that are focused to a 1–2-μm spot before the sample is bombarded. 1905 *Jrnl. Chem. Soc.* LXXXVIII. II. 44 By means of an apparatus termed the 'micro-respirometer'..the respiratory exchanges in small objects like nerves can actually be measured. 1946 *Nature* 27 July 126/2 A Cartesian diver micro-respirometer. 1965 B. E. FREEMAN tr. *Vandel's Biospeleol.* xx. 342 The measurement of respiration is carried out in respiratory chambers or micro-respirometers. 1905 *Jrnl. Chem. Soc.* LXXXVIII. II. 44 Micro-respirometric investigations. 1970 *Acta Soc. Bot. Poloniae* XXXIX. 497 The microrespirometric technique has been used for many research works on leaves and leaf segments. 1960 E. J. BOELL in Sasser & Jenkins *Nematology* viii. 109 (*heading*) The Cartesian diver technique in microrespirometry and enzyme assay. 1973 *Soil Biol. & Biochem.* V. 271 A new method has been developed for microrespirometry utilizing gas chromatography. 1934 *Photogr. Jrnl.* LXXIV. 518/2 The application of the microspectrograph to the identification of organic compounds. 1953 *Experientia* IX. 422/2 The transmission curves obtained with the microspectrograph were transformed into extinction curves. 1950 T. CASPERSSON *Cell Growth & Cell Function* iii. 61 There are two groups of cell substances that are more easily studied by microspectrographic procedures than any others. These are the proteins..and the polynucleotides. 1947 *Acta Path. & Microbiol. Scand.* XXIV. 417 (*heading*) Ultraviolet microspectrography as an aid in the study of the nucleotide content of bacteria. 1957 Microspectrography [see *microdensitometry* above]. 1951 *Rev. Sci. Instruments* XXII. 866/2 The electronic part of the microspectrophotometer consists of the two photocells, one above the other. 1966 *McGraw-Hill Encycl. Sci. & Technol.* II. 595/1 Such microspectrophotometers, capable of carrying out spectral analyses within the dimensions of a single cell, will play an increasingly powerful role in furthering our knowledge of cell biology. 1950 T. CASPERSSON *Cell Growth & Cell Function* ii. 45 This is a further reason for the use of high-aperture lenses in microspectrophotometric work. 1970 *Nature* 17 Oct. 255/2 Microspectrophotometric analyses of the salic particles in the visible and near ultraviolet show a continuous absorption spectrum increasing towards the ultraviolet. 1951 *Exper. Cell Res.* II. 301 In order to obtain significant data on the content of a certain absorbing substance of a whole cell the only as yet available way is to work microspectrophotometrically with photographic procedures. 1971 J. M. PAULUS *Platelet Kinetics* 330 Ploidy measurements of megakaryocytes have usually been obtained by determining microspectrophotometrically the total relative extinction of megakaryocyte nuclei. 1935 *Chem. Abstr.* XXIX. 1117 (*heading*) Ultraviolet absorption spectrum of sea-urchin eggs. Technic of microspectrophotometry. 1972 *Amer. Jrnl. Bot.* LIX. 829/1 The two-wavelength method of microspectrophotometry..was employed in the Feulgen-DNA studies. 1958 *Listener* 9 Oct. 563/1 The bacterial virus is really a tiny syringe—a microsyringe. 1973 *Internat. Jrnl. Peptide & Protein Res.* V. 208/1 All dilutions and sampling by volume were performed with a microsyringe..that delivered volumes accurate to $10^{-3} \pm 5 \times 10^{-4}$ ml. 1892 *Jrnl. R. Microsc. Soc.* 555 (*heading*) Zimmermann's botanical microtechnique. 1956 *Nature* 28 Jan. 153/1 The general climate of micro-chemistry has changed greatly in the interval. In 1949 the applications of micro-techniques were few and not very widely used, but at the present day they are ubiquitous. 1964 *Oceanogr. & Marine Biol.* II. 111 The micro-techniques described here are primarily those for small or micro-amounts of sea water.

7. Prefixed to a sb. (or used *attrib.* without a hyphen) to indicate that the object designated has been reduced in size by the use of microphotography, or is used in connection with such an object, as *micro book, edition, -record, -recording, -reproduction, -text*; **mi·croreader, -viewer**, an apparatus that produces from microfilm or microprint an image enlarged sufficiently to be readable.

1970 *New Scientist* 31 Dec. 601/1 The micro book..opens up the prospect of vast reductions in the world's consumption of paper. 1971 *Brit. Printer* Jan. 80/1 An inventor, who has been working for two years to develop a practicable 'micro book', has come up with a solution. *Ibid.*, Microfilming was not the simplest or most economical way of producing a micro book. 1970 *New Scientist* 31 Dec. 601/2 Publishers of technical journals..could put out in micro editions some of the..specialised research material that cannot be included in their normal issues. 1971 *Brit. Printer* Jan. 80/2 PVC paper also has the advantage, for micro editions, of being durable, difficult to tear and waterproof. 1949 M. C. KEELEYSIDE in *Summary Proc. 3rd Ann. Conf. Amer. Theol. Library Assoc.* 11 The reading machines are new on the market. I am proud to say that our library has the first microreader that any library in the world ever possessed. 1970 *Publishers' Weekly* 8 June 152/1 Obviously it is not possible for people to read microfilmed material without some magnification

device, hence the microreader which you can see in most libraries. 1948 *Sci. News* VII. 90 The Airgraph scheme..made one advantage of such micro-records obvious to the general public. 1957 J. BURKETT *Microrecording in Libraries* i. 8 Microrecording was first put on a business-like footing in 1928 when Eastman Kodak introduced its Recordak Division. *Ibid.* ii. 13 An opaque microrecord is ..a positive print made from a photographic negative. 1971 *Brit. Printer* Jan. 80/1 Davies started on the project mainly as a result of an inquiry from the American Council for Library Resources, which was interested in obtaining a small, cheap and portable method of retrieving micro-records. 1938 G. VAN ITERSON in *Trans. 14th Conf. Fédération Internat. de Documentation* I. 149 (*heading*) The preparation and reading of micro-reproductions of treatises. 1958 *Engineering* 4 Apr. 443/3 To this [*sc.* the decrease in library accommodation] there is one reasonable solution and that is micro-reproduction. 1944 F. RIDER *Scholar & Future of Research Library* II. i. 99 Why might we not combine the micro-texts of our books, and the catalog cards for these same books, in one single entity? *Ibid.* 100 A fair amount of micro-text can be put on the back of a standard-size catalog card. 1958 *Times Rev. Industry* Mar. 61/3 The United States has been a fertile ground for the spread of microtexts. 1973 *Computers & Humanities* VII. 163 Anyone who uses the subject part of an ordinary library catalog, whether in card, book, or microtext form, would benefit from this explanation. 1972 M. J. BOSSE *Incident at Naha* iii 153, I feel I'm turning into a library. I'll start dreaming I'm a card catalog or a microviewer.

8. *micro* is now freely prefixed to sbs., often resulting in trivial or nonce words; from being used as an independent word without a hyphen it passes into a quasi-*adj.* with the meanings: **a.** Microscopic, minute; small-scale, small.

The examples are arranged in chronological order.

1922 *Encycl. Brit.* XXX. 34/2 Micro-investigation of glued joints proved the value of carefully preparing the timber and glue. 1926 R. W. LAWSON tr. *Hevesy & Paneth's Man. Radioactivity* i. 2 Very small quantities of two different gases..gradually accumulate within the tube, and can be detected by the methods of micro-gas analysis. 1931 *Boys' Mag.* XLV. 157/1 Any good crystal detector will do. One with fine or micro adjustment is to be preferred. 1935 *Discovery* Nov. 320/1 Whenever it finds a patch of..dry, dusty ground, there it arrives... Such patches, for the purposes of that *Acrotylus*, are deserts. They are, in fact, micro-deserts. 1946 KOESTLER *Thieves in Night* 279 Pen-holders of olivewood with a tiny inlaid lens through which one could see a micro-panorama of Jerusalem. 1958 *Spectator* 15 Aug. 236/2 Micro, corneal and contact lenses. 1967 G. WILLS in Wills & Yearsley *Handbk. Managem. Technol.* x. 176 At the macro level, marketing management became concerned with understanding its social environment; at the micro level, the concept of the marketing mix postulated a co-ordinative and integrative activity for product distribution and communication. 1969 *Jane's Freight Containers 1968–69* 552/2 Ten models of these 1-ton micro-fork lift trucks are available for loading and unloading containers. 1970 *Globe & Mail* (Toronto) 26 Sept. 9/5 More attention will be given to the ubiquitous 'micro' problems in non-sugar sectors of the [Cuban] economy. 1973 *Physics Bull.* Oct. 626/1 Finally, going lower than micro, we might mention the HP-45, Hewlett-Packard's new pocket 'scientific' calculator, said to be the most powerful of its size and price. 1974 *Sci. Amer.* July 134/2 A computer patiently runs the long repetitive scans looking for tiny needles in microhaystacks.

b. *Chem.* Of or pertaining to microanalysis. Cf. *micromethod* in *6, *MICRO-SCALE.

1931 J. W. BROWN in C. A. Mitchell *Recent Adv. Analytical Chem.* II. xv. 306 The work of Emich inspired Fritz Pregl..to attempt to carry out organic determinations on a micro scale. 1937 *Ann. Rev. Biochem.* VI. 85 While many important studies have been made on reducing sugars in recent years, most of these have not involved the use of methods which are strictly micro. 1946 BELCHER & WILSON *Qualitative Inorg. Microanalysis* i. 1 Micro methods handle solid samples over the range of 0·1 to 1 mg., and volumes of solution ranging from 0·02 to 0·2 ml. 1955 C. R. N. STROUTS et al. *Analytical Chem.* I. xiv. 314 The economy of time afforded by many micro procedures favours their adoption even when the amount of sample available is sufficient for macroanalysis. 1971 *Nature* 19 Mar. 194/2 Any laboratory where the analysis of metals is practised at macro, micro or trace levels. 1974 *Encycl. Brit. Macropædia* IV. 79/2 Samples..can be classified as macro (10^{-1} g), semi-micro (10^{-1}–10^{-2} g), micro (10^{-3}–10^{-4} g), sub-micro (10^{-4}–10^{-6} g), nanogram (10^{-9} g), or picogram (10^{-12} g).

microabscess, -aerophil(e, -aerophilic, -ammeter: see *MICRO- 1, 4, 4, 6.

microanalyser (məikro,æ·nălaizəɹ). Also (chiefly *U.S.*) **-analyzer.** [f. MICRO- 6 + ANALYSER, -ZER.] An instrument in which a beam of radiation (usu. electrons) is focused on to a minute area of a sample and the resulting secondary radiation (usu. X-ray fluorescence) is analysed to yield chemical information about the area.

1944 HILLIER & BAKER in *Jrnl. Appl. Physics* XV. 665/1 Figure 1 is a photograph of the first experimental model of an electron microanalyzer. 1959 *Electronic Engin.* XXXI. 680/1 In the Hinxton Hall microanalyser the specimen surface is scanned at slow speed by an electron beam focused to a diameter of less than one micron. 1960 *Archaeometry* III. 36 During the past eighteen months the Laboratory has been engaged upon the construction of an X-ray microanalyser for archaeological applications. 1966 D. G. BRANDON *Mod. Techniques Metallogr.* 173 (*caption*) Secondary-ion emission microanalyser. 1973 *Histochem. Jrnl.* V. 176 EMMA-4 combines the full facilities of a conventional transmission electron microscope and an electron probe X-ray microanalyser.

microanalysis (məikro,ænæ·lĭsis). [f. MICRO- 2 a + ANALYSIS.] The analysis for chemical information of very small samples, or very small areas of an object; now *spec.* the quantitative analysis of samples weighing only a few milligrammes (contrasted with semimicroanalysis and ultramicroanalysis). Cf. *MICRO- 8 b.

1856, 1904 [in Dict. s.v. MICRO- 2 a]. 1937 *Discovery* Aug. 227/2 The addition..of the delicate weapon of micro-analysis to the analyst's armoury. 1966 D. G. BRANDON *Mod. Techniques Metallogr.* iii. 172 In all the methods of microanalysis discussed so far, a point-by-point determination of the x-ray emission or absorption spectrum is related to the chemical composition in order to find the spatial distribution of the components.

Hence **microa·nalyst**, one who performs or is skilled in microanalysis; **mi:croanaly·tical** *a.*, of or using microanalysis.

1924 E. FYLEMAN tr. *Pregl's Quantitative Org. Microanalysis* i. 4 He has worked out a microanalytical method for the determination of glycerine. *Ibid.* ii. 13 A watch-maker's lens..should always be in the pocket of the microanalyst. 1938 *Ann. Reg. 1937* 354 Paneth and Glückauf showed, by microanalytical methods, that when neutrons are produced by bombarding beryllium with gamma-rays the beryllium..breaks down into helium. 1964 N. G. CLARK *Mod. Org. Chem.* xxiv. 496 He later returned to Graz, where he established an international reputation as a micro-analyst.

microanatomist, -anatomy, -aneurysm, -atoll: see *MICRO- 2a, 2a, 1, 1.

microbalance, -bar: see *MICRO- 6, 5a.

microbarom (məikrobæ·rǫm). *Meteorol.* [f. MICRO- + BAROM(ETER.] A minute oscillation of atmospheric pressure with a period of the order of 5 seconds.

1939 BENIOFF & GUTENBERG in *Bull. Amer. Meteorol. Soc.* XX. 424/1 The waves..exhibit no relation to microseisms nor to the barometric conditions in this region. No hypothesis as to their origin has yet been suggested. We designate them microbaroms. 1953 *Jrnl. Acoustical Soc. Amer.* XXV. 796/1 Microbaroms apparently originate in ocean storms associated with low pressure areas. 1967 *Times* 21 Nov. 3/5 The detection of microbaroms depends very much on the availability of sensitive equipment.

microbe. Add: Also *fig.*

1890 in J. R. Ware *Passing Eng.* (1909) 175/2 The abdication by the Radical party of its proper functions has an unfortunate tendency to foster..the microbe of sectionalism.

microbeam: see *MICRO- 1.

microbial, *a.* Add: Also **micro·bially** *adv.*, by or with microbes.

1971 *Flying* Apr. 88 (Advt.), Microbially induced corrosion of fuel tanks. 1974 *Nature* 27 Sept. 316/2 Plants in the three microbially contaminated pots produced proteid roots.

microbiology. Add: Also **microbiolo·gic** *a.* = MICROBIOLOGICAL *a.*; **microbiolo·gically** *adv.*, by microbiological methods.

1909 *Westm. Gaz.* 9 Dec. 2/1 He is..reminded of the doctrine of the etiology of infectious diseases before the advent of the microbiologic epoch. 1942 *Industr. & Engin. Chem. (Analytical Ed.)* 15 Aug. 667/1 The extracted residue of dried skim milk was microbiologically inert, while the residue of alfalfa leaf meal exerted a little stimulatory action. 1956 *Nature* 4 Feb. 221/1 The relation between seven different tree-supports..and two epiphytes was examined microbiologically. 1966 *McGraw-Hill Encycl. Sci. & Technol.* IX. 620/2 Penicillin can also be chemically or microbiologically degraded and then built up to form new penicillins. 1969 WILSON & MIZER *Microbiol. in Nursing Pract.* i. 11/2 Since Leeuwenhoek's first glimpse of bacteria and protozoa, microbiologic investigations have resulted in the accumulation of a vast body of knowledge..that has lightened immeasurably the burden of human disease.

microbiota: see *MICRO- 1.

microbism (mai·krobiz'm). *Med.* [f. MICROB(E + -ISM.] Infection with microbes.

1904 *Lancet* 18 June 1724/1 The various explanations of 'return' cases were considered, including..the possibility of a relapse of the original disease, of latent microbism, or of missed cases. 1963 *Biol. Abstr.* XLI. 553/2 (*heading*) Investigations on the role of staphylococci in cutaneous microbism.

microbody, -burette: see *MICRO- 1, 6.

microburin (məikrobiūə·rin) *Archæol.* [f. MICRO- 1 + BURIN.] (See quot. 1970.)

1932 *Antiquity* VI. 364 A feature of the industry is the presence of the burin of upper Palæolithic type... The microburin is not mentioned either as occurring or as absent. 1955 [see *IBERO-]. 1957 V. G. CHILDE *Dawn European Civilization* (ed. 6) i. 6 At Parpalló in Eastern Spain even microburins occur from the Solutrian layers upward. 1958 *Man* Apr. 57/2 The ends are then shaped either by twisting off, or by retouching, and even by the microburin technique of notching and then snapping... The microburins, the 'anti-microburins' and the obliquely snapped waste..were all backed as well. 1970 BRAY & TRUMP *Dict. Archaeol.* 147/1 *Microburin*, a by-product of the manufacture of microliths. A blade is notched, and then snapped off... One piece becomes a microlithic tool, while the residue (the 'microburin') still shows traces of the original notch and fracture.

microburner: see *MICRO- 6.

microbus (məi·krobʊs). [f. MICRO- 1 + BUS *sb.*[2]] A small vehicle designed to carry passengers in seats fitted as in a bus.

1959 *Camb. Rev.* 7 Mar. 428/2 (Advt.), Micro-buses available on continent, fitted camping facilities. Make up your own party. 1960 [see *JAMBOREE]. 1962 E. KIMBROUGH *Pleasure by Busload* ii. 17 'This is a Volkswagen Microbus,' Sophy answered... No one had told me it looked as big as a trailer and that the span from ground to floorboard was a span well over three feet. 1968 *New Scientist* 19 Sept. 596/1 Caltech relied on a less streamlined red and white 1958 Volkswagen microbus. 1971 *Telegraph* (Brisbane) 14 Sept. 22/4 Mr. Ralph Nader claimed the famous Beetle was loaded with safety hazards... He claimed the VW microbus was even worse.

microcalorie, -calorimeter (etc.), **-camera, -capsule:** see *MICRO- 5a, 6, 1b, 1.

microcard (məi·krokāɹd). Also **Microcard.** [f. *MICRO- 7 + CARD *sb.*[2]] An opaque card bearing microphotographs of a number of pages of a book, periodical, etc. (A proprietary name in the U.S.)

1944 F. RIDER *Scholar & Future of Research Library* II. i. 99 Why might we not combine the micro-texts of our books, and the catalog cards for these same books, in one single entity?.. I called this new concept..a 'micro-card'. 1950 *Official Gaz.* (U.S. Patent Office) 24 Jan. 907/1 The Micro Library Incorporated, La Crosse, Wis.... *Microcard* for publications in card form, reproduced by printing, photographing and otherwise. Claims use since November 1947. 1950, 1953 [see *MICROFICHE]. 1958 *Engineering* 31 Jan. 155/1 Micro-card can also be used on the reverse side for such purposes as abstracts in normal type and catalogue notations. 1958 *Times Rev. Industry* Mar. 61/1 For recording such material as technical periodicals or newspapers, the microcard system is often the better method. 1964 M. MCLUHAN *Understanding Media* (1967) iii. 49 Today with microfilm and micro-cards..the printed word assumes again much of the handicraft character of a manuscript. 1973 *Computers & Humanities* VII. 168 Few agencies even assemble material on microfiches or micro-cards.

So **mi·crocard** *v.*, to reproduce on microcards; **mi·crocarded** *ppl. a.*

1944 F. RIDER *Scholar & Future of Research Library* II. vii. 189 What types of material, and..what titles..ought, therefore, to be micro-carded first. c 1960 *Microcards* (Microcard Foundation), Reports coming in from men in the field are collated and the completed reports are Microcarded... Literally shelves of Microcarded books and periodicals can be stored in one file cabinet.

microcellular, -chemist, -chromosome, -cinematography (and **-graphic**): see *MICRO- 4, 2a, 1, 2a.

microcircuit (məi·krosəɹkit). *Electronics.* Also **micro-circuit.** [f. MICRO- 1 + CIRCUIT *sb.*] An integrated circuit or other minute circuit.

1959 *Electronics* 4 Sept. 49/3 New circuit fabrication techniques such as are used in microcircuits. 1960 *IRE Trans. Military Electronics* IV. 461/1 (*caption*) Detailed sketch of a silicon integrated microcircuit. 1965 *Economist* 25 Sept. 1228/1 A micro-circuit is no bigger than a pin-head, replacing transistor circuits as big as a large envelope knitted out of wires. 1970 *Daily Tel.* 8 May 21 Not only do they have the advantage of smaller size: microcircuits are also more reliable and faster—a vital factor for computer use. 1971 *Physics Bull.* Jan. 45/1 The chips measure ⅛ in[2] and contain 1434 microcircuit elements each, integrated into 128 memory and 46 support circuits.

So **mi·crocircuitry**, microcircuits collectively; the branch of electronics concerned with microcircuits.

1959 *Electronics* 4 Sept. 44/1 The microcircuitry program is concerned with the development of a circuit fabrication technique based on the combination of elementary materials rather than on the assembly of individual components. 1960 *IRE Trans. Military Electronics* IV. 459/2 Integrated microcircuitry. 1970 *Daily Tel.* 21 Sept. 19 Recent developments in microcircuitry..have made it possible to reduce the circuitry of an amplifier to the size of an ordinary jacket button. 1974 *Country Life* 5 Dec. 1763/1 By..a brilliant piece of microcircuitry, the oscillations are translated into a constant, one second pulse.

microcirculation: see *MICRO- 1.

microclimate (məi·krokləimĕt). Chiefly *Ecol.* and *Meteorol.* [f. MICRO- 1 + CLIMATE *sb.*] The climate of a very small or restricted area, or of the immediate surroundings of any individual or object of interest, esp. as it differs from the climate generally.

1925 *Rev. Appl. Mycol.* IV. 471 Temperature, absolute

and relative humidity, aeration, insolation, &c., which prevail at various levels..are comprised by him [*sc.* L. F. Roussakov] under the term 'microclimate'. 1934 [see *MICROCLIMATOLOGY]. 1951 R. GEIGER in T. F. Malone *Compendium Meteorol.* 994/1 We shall first consider microclimate as the special climate prevailing in a layer of air about two meters in height adjacent to the surface of the ground. In this layer friction between the air and the earth's surface plays a decisive role. 1958 *Engineering* 14 Mar. 352/1 A basic item influencing the design of an air-conditioning system is the 'micro-climate'—the climate existing immediately around the structure to be air conditioned. 1967 T. J. CHANDLER *Air around Us* x. 116/1 Climates can differ from one side to the other of a hedge or garden wall, or from the base of a bush to its leafy crown. Such conditions are usually described as microclimates. 1972 *Physics Bull.* June 342/1 Cool air is introduced into the microclimate between the pilot's clothing and his skin. 1973 *Times* 17 Nov. 12 Provided the compost is kept always moist a moist 'microclimate' is created around the plants.

Hence **microclima·tic** *a.*, **-clima·tically** *adv.*

1929 *Conf. Empire Meteorologists, Agric. Section* II. 131 The actual, *i.e.* the microclimatic, conditions often bear only a very remote relation to the general conditions in the habitat. 1952 *Archit. Rev.* CXII. 323 The total site area (shaded) is over 45,000 acres but more than half of this will be devoted to forestry—for microclimatic control, for experimental purposes, and as capital investment. 1972 *Nature* 10 Mar. 64/1 The immigration and establishment of local small stands of trees in microclimatically favourable habitats. 1974 *Ibid.* 15 Mar. 261/3 The alternative explanation..relates the higher soil moisture beneath vegetation to the modified microclimatic conditions in such situations.

mi:croclimato·logy. *Meteorol.* [f. MICRO- 2 a + CLIMATOLOGY.] The study of microclimates.

1934 L. A. RAMDAS in *Current Sci.* II. 445/2 The aims of 'micro-climatology' are (1) to investigate the physical laws underlying the deviations of 'micro-climate' from 'macro-climate'..and (2) to apply the theoretical results to practical ends. 1944 V. CONRAD *Methods in Climatol.* i. 14 The layering of the air is of interest..as one of the most important features from which 'Microclimatology', and especially 'Orographic Microclimatology', makes its start. 1972 *Last Whole Earth Catalog* (Portola Inst.) 83/1 This appears to be the definitive text on microclimatology: the climatic conditions within six feet or so of the earth's surface. 1974 *Daily Colonist* (Victoria, B.C.) 28 Aug. 6/4 Principles of microclimatology and its relationship to plant growth will be discussed.

Hence **mi:croclimatolo·gical** *a.*, **-climato·logist.**

1937 *Geography* XXII. 87 (*heading*) Reality in climate: the results of recent microclimatological studies, with special references to temperatures. 1951 R. GEIGER in T. F. Malone *Compendium Meteorol.* 993/2 Location considerations in plant growth are always microclimatological in nature; whether a plant is located in an open field or at the north side of a large boulder..is a decisive factor with respect to the conditions affecting its growth. *Ibid.*, The biology of bees can be understood only if one knows the microclimate of the beehive. It is the problem of the microclimatologist to investigate these small-scale climates.

microcolony, -constituent, -continent: see *MICRO- 1.

microcopy (məi·krokǫpi), *sb.* [f. *MICRO- 7 + COPY *sb.*] A copy of the text of a book, periodical, etc., that has been reduced in size by the use of microphotography; *in microcopy*, in the form of a microcopy or microcopies. Also as *v. trans.*, to make such a copy of; also *absol.*; **mi·crocopying** *vbl. sb.*

1934 WEBSTER, Microcopy *n.* 1936 *Microphotogr. for Libraries* (Amer. Library Assoc.) 3 We microcopy because, by reducing the size of letters, we can reduce tremendously the cost of making copies. *Ibid.* 4 Whenever we contemplate reproducing something in microcopy for a number of users, we should give especial attention to materials cost. *Ibid.* 13 The production of microcopies is relatively easy and inexpensive. 1942 *Punch* 14 Jan. 4/2 By means of photographic 'micro-copying', miniature negatives of all kinds of documents can be taken on continuous lengths of..films. 1944 F. RIDER *Scholar & Future of Research Library* II. ii. 115 All that we have to do is to take two copies of the book that we are proposing to micro-copy. *Ibid.* vi. 170 The library can usually supply a borrower at once with a duplicate micro-copy. 1950 [see *flat film* (*FLAT *a.* 15)]. 1962 *B.S.I. News* Feb. 39 Specifies size ranges for transparent bases for microcopies—roll or strip microfilm, microfiches and unexposed rolls of sensitized film. 1972 *Computers & Humanities* VII. 46 The National Library will acquire most periodicals required in order to back up these services, as well as other references available in microcopy.

microco·smically, *adv.* [f. MICROCOSMIC *a.*: see -LY[2].] In relation to the microcosm.

1881 MAX MÜLLER tr. *Kant's Critique Pure Reason* I. ii. 363, I might call the two former [ideas], in a narrower sense, cosmical concepts (macrocosmically or microcosmically) and the remaining two transcendent concepts of nature. 1939 [see *MACROCOSMICALLY *adv.*]. 1942 *Mind* LI. 235 A microcosmic being exists microcosmically under divine reference in constitutive *communitas* with a complement that is not absolutely other; and under self-reference in indefinitely continuing commerce of co-operation and opposition with an imaginative other.

microcosmopolitan, -crack (etc.), **-crater:** see *MICRO- 1 b, 1, 1.

micro-crystal. Add: Also **microcrystal.** (Earlier and later examples.)

1886 *Jrnl. R. Microsc. Soc.* VI. 725 (*heading*) Preparing microcrystals. 1964 L. MARTIN *Clin. Endocrinol.* (ed. 4) viii. 267 Alternatively, long-acting injections of micro-crystals of œstradiol monobenzoate (Ovocyclin I.M.) may be given in doses of 10 mg.

microcrystalline, *a.* Add: *microcrystalline wax*, a mixture of hydrocarbons of higher molecular weight than those in paraffin waxes and with a melting point of up to 90°C which is obtained from the residual lubricating fraction of crude oil and is used in making waxed paper, adhesives, and polishes.

1943 *Jrnl. Amer. Pharmaceutical Assoc.* (*Sci. Ed.*) XXXII. 111/2 A series of 'Petro-waxes', one of which melts at about 160°F. It is a white microcrystalline wax that is available in quantity. It was thought that due to the microcrystalline structure of this wax, it might make stable ointments containing as much as 50% water. 1944 H. BENNETT *Commercial Waxes* i. 60 The designations of microcrystalline and amorphous waxes are now being used synonymously although the former is a more accurate designation. 1957 VAN DER HAVE & VERVER *Petroleum* xi. 339 As its name implies, 'micro-crystalline wax' has a very fine crystal structure; it is a flexible and often somewhat sticky product; it has a high molecular weight (550–600) and a melting point within the same range as ceresin. 1972 *Oxf. Univ. Gaz.* CII. Suppl. No. 3. 10 They [*sc.* metal objects] are boiled to remove chlorides and after drying in an infra-red cupboard are impregnated with microcrystalline wax.

Hence **mi:crocrystalli·nity**, the property or state of being microcrystalline.

1946 *Nature* 28 Dec. 930/1 Microcrystallinity and a capacity to yield strong and pliable fibres are properties by no means confined to the linear poly-amides. 1967 *Encycl. Polymer Sci. & Technol.* VII. 207 It is possible to distinguish microcrystallinity, detectable by x-ray examination, from macrocrystallinity, visible through a polarizing microscope.

microculture, -curie: see *MICRO- 1, 5a.

microcyclic (məikrosəi·klik), *a.* *Bot.* [f. *MICRO- 1 b + CYCLIC *a.*] Of a plant rust: having a short life cycle.

1926 [see *MACROCYCLIC *a.* 1]. 1950 E. A. BESSEY *Morphol. & Taxon. Fungi* xii. 396 These last two are properly speaking microcyclic rusts. 1970 J. WEBSTER *Introd. Fungi* 376 Some rusts have only telia, with or without pycnia, and are said to be microcyclic.

microcyte. Add: (Now freq. with pronunc. (mi·kro-).) Hence **microcytic** (-si·tik) *a.*, typical or characteristic of a microcyte; characterized by microcytes; **microcytosis** (-səitǝʊ·sis) = MICROCYTHÆMIA.

1890 BILLINGS *Med. Dict.* II. 154/1 *Microcytosis*, production of microcytes. 1925 *Amer. Jrnl. Med. Sci.* CXXX. 684 Measurements..showed the majority of the cells to be microcytic. 1932 [see *MACROCYTE]. 1938 H. DOWNEY *Handbk. Hematol.* III. xxxi. 2286 One cannot fail to notice that the average size of the red cells is less than normal; there is a distinct microcytosis. 1962 *Lancet* 12 May 1004/2 Clinical and hæmatological examinations..showed anæmia with anisocytosis, microcytosis, macrocytosis, and reticulocytosis. 1966 *Ibid.* 24 Dec. 1398/2 Prasad described a syndrome of microcytic anæmia.

microdegree, -densitometer (etc.), **-determination, -diffusion:** see *MICRO- 5a, 6, 6, 6.

microdiorite (məikrodəi·ǒrəit). *Petrol.* [f. MICRO- + DIORITE.] (See quots. 1920, 1961.)

1920 A. HOLMES *Nomencl. Petrol.* 157 *Micro-*, a prefix commonly added—..(2) to the names of phanerocrystalline rocks to indicate a microcrystalline rock or groundmass of corresponding mineral composition and texture; e.g., *microgranite, microdiorite, microsyenite*, etc. 1961 J. CHALLINOR *Dict. Geol.* 127/1 *Microdiorite*, the medium-grained equivalent of a diorite, usually porphyritic (diorite porphyrite, or porphyrite). 1963 D. W. & E. E. HUMPHRIES tr. *Termier's Erosion & Sedimentation* iv. 94 (*caption*) The eroded surface of a microdiorite dike. 1965 G. J. WILLIAMS *Econ. Geol. N.Z.* x. 150/1 Positive nickel reactions are also produced by many of the less basic rocks from the 'mineral belt' such as the gabbros, leucogabbros and quartz microdiorites.

microdissect(ion), -distribution: see *MICRO- 2a, 1b.

microdot (məi·krodǫt), *sb.* [f. MICRO- (here merely emphasizing the smallness implied by *dot*) + DOT *sb.*[1]] **1.** A photograph, esp. of printed or written matter, reduced to about the size of a dot. Freq. *attrib.*

1946 J. E. HOOVER in *Reader's Digest* May 50/2 It was incredibly ingenious and effective, this micro-dot gadget. It perfectly counterfeited a typewritten or printed dot. The young Balkan agent, for example, had four telegraph blanks in his pocket, carrying Lilliputian spy orders that looked like periods; 11 micro-dots on the four papers. 1961 *Daily Mail* 8 Feb. 10/1 There was also a piece of glass with three microdots between the pieces of glass. These little dots are tiny pieces of film... By enlarging the dots again you can see what they contain. One microdot can contain a great deal of writing. 1964 M. GOWING *Britain & Atomic Energy* ix. 246 The communication of the message was in the best spy thriller traditions; it was by a microdot concealed in the hollow handle of a doorkey. 1965 I. FLEMING *Man with Golden Gun* viii. 108 The minor tools of espionage—codes, microdot developers, cyanide. 1968 *Listener* 1 Aug. 150/1 A house stacked with high-frequency transmitters, microdot readers.

2. A tiny capsule or tablet of LSD.

1971 *Oxford Times* 10 Dec. 1/5 Produced 2½ microdot tablets in foil which were later found to contain LSD. 1972 *Daily Colonist* (Victoria, B.C.) 5 Jan. 1/1 Called a microdot, the deceptively small, purple pill packs superconcentrated doses of the hallucinatory drug LSD. 1973 R. BUSBY *Pattern of Violence* v. 81 Gelatine micro-dots of the hallucinogenic drug LSD.

microdot (məi·krodǫt), *v.* [f. prec. sb.] *trans.* To make a microdot or microdots of. So **mi·crodotting** *vbl. sb.*

1957 *Time* 28 Oct. 22 Some messages were recorded on a film which could be softened and rolled into a ball; others were microdotted, i.e., whole pages of printing were reduced on film to pinhead size. 1961 *Daily Tel.* 22 Mar. 23/5 The wireless transmitter and equipment for micro-dotting at the house. 1963 J. JOESTEN *They call it Intelligence* I. vi. 61 Colonel Abel..did a brisk business in micro-dotting. 1969 D. LAMBERT *Angels in Snow* xix. 251 He decided to micro-dot the information and despatch it to Washington.

microearthquake, -ecology: see *MICRO- 1, 2a.

mi:cro-econo·mics, *sb. pl.* (usu. const. as *sing.*). Also **microeconomics.** [f. MICRO- 2 a + ECONOMIC *sb.* 2 c.] That branch of the science of economics which deals with the individual (firm, product, consumer, etc.) rather than the aggregate.

1948 K. E. BOULDING *Economic Analysis* (rev. ed.) xiii. 259 There are two main branches of modern economic analysis, to which the names 'microeconomics' and 'macroeconomics' may conveniently be given. Microeconomics is the study of particular firms, particular households, individual prices, wages, incomes; individual industries, particular commodities. 1949 L. H. HANEY *Hist. Econ. Thought* (ed. 4) xxxvii. 734 The preceding chapters have mostly dealt with 'micro-economics'—economics concerned with the determination of *particular* prices or values. Now comes a chapter which deals with 'macro-economics', or national *aggregates* in the shape of total national income as affected by total spending or not-spending. 1960 A. CAIRNCROSS *Introd. Econ.* (ed. 3) ii. 23 When he is talking about the decisions made in an individual business the economist is in the realm of micro-economics. 1965 *McGraw-Hill Dict. Mod. Econ.* 324 Micro-economics deals with the division of total output among industries, products, and firms and the allocation of resources among competing uses. 1968 *Economist* 11 May 66/2 If the Government can provide the micro-economics that industry wants, then industry will willingly co-operate to help to provide the macro-economics that the Government needs. 1974 *Sci. Amer.* Jan. 27/2 A social decision, one not based entirely on microeconomics must be made on the value of the land to the society as a whole.

So **mi:cro-econo·mic** *a.*, **mi:cro-eco·nomist.**

1949 L. H. HANEY *Hist. Econ. Thought* (ed. 4) xxxiv. 692 The 'over' theories are mostly micro-economic. The 'under' theories all tend toward macro-economics, and are concerned with totals and averages, not margins. 1958 HENDERSON & QUANDT *Microecon. Theory* i. 3 Microeconomic theories are sufficiently flexible to permit many variations in their underlying assumptions. 1965 H. I. ANSOFF *Corporate Strategy* (1968) i. 16 The so-called microeconomic theory of the firm..sheds..little light on decision-making processes in a real-world firm. 1968 *Economist* 13 Jan. 46/1 Mr. Merton Peck..is mainly a 'micro-economist', interested in marketing questions and industrial problems. 1973 *Spectator* 3 Mar. 281/1 The most distinguished body of micro-economic thinkers, gathered together in one place, in the world. 1974 *Times Lit. Suppl.* 12 Apr. 398/1 Those conventional micro-economists who presume that their rather scholastic abstractions provide adequate criteria for decisions in nationalized industries.

microelectrode: see *MICRO- 1.

microelectronics (məi:kro,ĭlektrǫ·niks, -elektrǫ·niks), *sb. pl.* (usu. const. as *sing.*). Also **micro-electronics.** [f. MICRO- 2 a + *ELECTRONICS.] **a.** The branch of technology concerned with the design, manufacture, and use of microcircuits. **b.** Microelectronic devices or circuits.

1960 K. AMIS *New Maps of Hell* (1961) i. 19 Even if the problem of fitting all that machinery into a container on the human scale would require the development of a kind of micro-electronics that for the time being, one would imagine, is at a rudimentary stage [etc.]. 1967 *New Scientist* 11 May 342/2 Complete radio and TV sets have been built experimentally with 90 per cent microelectronics. *Ibid.* 342/3 There are..practical difficulties in packaging microelectronics in order to make the best use of their inherently high reliability. 1968 *Brit. Med. Bull.* XXIV. 192/2 Micro-electronics is having a profound effect on the application of computer power. 1970 *Daily Tel.* 8 May 21 Microelectronics, developed for American space and military hardware, is a way of miniaturising large circuits so that up to 100 transistors and associated components take up no more space than a pinhead. 1974 *Nature* 8 Feb. 338/2 Every week seems to bring a new idea for making use of ion implants in microelectronics.

So **mi:croelectro·nic** *a.*

1960 *Proc. Western Joint Computer Conf.* 251/1 Microelectronic data processing systems are analyzed. 1967

New Scientist 11 May 342/1 Many ground and airborne radar systems, missiles and communications receivers for all three services are being developed in microelectronic form. 1973 *Sci. Amer.* Apr. 65/1 The size of microelectronic circuits has decreased to the point where their surface dimensions are measured in microns..and their thickness in angstroms.

micro-electrophoresis (etc.): see *MICRO- 6.

microelement (məi·kro,elĭmĕnt). [f. MICRO- + ELEMENT *sb.*] **1.** *Plant Physiol.* = *MICRONUTRIENT.

1936 *Chem. Abstr.* XXX. 1496 Yields of oats..were greatest with the doubled concn. of nutrients with a special addn. of microelements. 1950 CURTIS & CLARK *Introd. Plant Physiol.* xiii. 367 We prefer the terms trace or micro elements. 1966 F. M. IRVINE tr. *Lundegårdh's Plant Physiol.* vi. 251 Manganese takes up an intermediate position between a macro- and a microelement.

2. *Electronics.* A thin, flat, miniaturized circuit made with standardized length and width for assembly into a micromodule.

1959 *Proc. IRE* XLVII. 897 (*caption*) Micro-module, exploded view, before stacking and interconnecting. As many micro-elements as desired may be assembled to form the module... A 0·01-inch space is allowed between micro-elements. 1963 S. F. DANKO et al. in E. Keonjian *Microelectronics* iii. 142 The 0·016-in.-diameter riser wires are soldered to appropriate metallized notches in the stack of microelements.

microencapsulation (məi:kro,enkæpsiulēi·ʃən). [f. MICRO- 2 a + *ENCAPSULATION.] The process of enclosing substances in microcapsules. Hence **microenca·psulate** *v. trans.*, **-enca·psulated** *ppl. a.*, **-enca·psulating** *vbl. sb.*

1961 *New Scientist* 20 Apr. 115/2 Pre-packed flavour and aromas, contained in microscopic plastic capsules, might be added to beverages, cokes and the like—this is one of the applications envisaged for the techniques of 'microencapsulation' being developed in the United States. 1967 *Britannica Bk. of Year* (U.S.) 803/2 *Microencapsulate, vb.*, to enclose a small amount of a substance in a microcapsule (*microencapsulated* aspirin); *microencapsulation*. 1969 *New Scientist* 3 Apr. 18/1 These results led us to suggest the possibility of using microencapsulated enzymes for experimental enzyme replacement. 1970 BAKAN & ANDERSON in L. Lachman et al. *Theory & Pract. Industr. Pharmacy* xiii. 402/2 The process is capable of microencapsulating liquids and solids. *Ibid.*, By vertical stacking of the microencapsulating units, production rates of 50 pounds per hour have been achieved. 1970 *Daily Tel.* 2 Oct. 5/1 The different synthetic aromas are contained in small labels employing the technique of micro-encapsulation. 1971 *New Scientist* 15 July 122/3 The technique has the advantage over other attempts that are being made to microencapsulate enzymes for therapy that liposomes are constructed of 'natural' materials. 1971 *Reader's Digest* (U.S.) Oct. 206/1 (Advt.), Through a special microencapsulation process, hundreds and hundreds of tiny granules of pure aspirin are concentrated in each Bayer Timed-Release tablet.

microenvironment(al), -estimation, -event, -evolution(ary), -explosion, -explosive: see *MICRO- 1, 6, 1, 1 b, 1, 1.

microfauna (məi·krofǭnă). *Biol.* [f. MICRO- 1 + FAUNA.] A fauna made up of minute animals, or one found in a microhabitat.

1902 *Geogr. Jrnl.* X. 323 The average amount of Plankton (micro-fauna) contained in them [*sc.* mountain lakes] is only the tenth..part of that contained in the standing waters of..lowlands. 1910 H. F. OSBORN *Age of Mammals* iv. 254 Of the microfauna we first observe among the castroids that the genus *Chalicomys* replaces the *Steneofiber* of the Oligocene. 1924 *Bull. Amer. Assoc. Petroleum Geologists* VIII. 549 All of the evidence at hand is studied and weighed in the light of intensive study of the microfaunas of the Gulf Coast region. 1947 *Antiquity* XXI. 190 Both diggers and sieve-men became very skilful at descrying even the smallest specimen... The sieve-men would pick out tiny bones of microfauna with amazing verve. 1957 G. E. HUTCHINSON *Treat. Limnol.* I. xv. 817 This rise [*sc.* in soluble copper] may contribute to the disappearance of certain members of the microfauna and microflora from the lake. 1975 *Times* 27 May 14/7 The upper part of those river-borne sediments [at Hoxne] contains a rich microfauna of small mammals, fishes and amphibia.

Hence **microfau·nal** *a.*

1935 *Bull. Geol. Soc. Amer.* XLVI. 498 The microfaunal evidence in the Wheeler Canyon section is not quite conclusive. 1964 *Oceanogr. & Marine Biol.* II. 385 The species minimum at about 5‰ salinity..has also been found in more recent investigations on..various microfaunal groups. 1972 *Times* 11 Dec. 1/8 The microfaunal remains have shown that fishes, birds and crabs played almost as important a part in the diet of the population as that of cattle. 1973 *Nature* 7 Dec. 347/1 Experiments introducing modern lichens into the normal preparation of microfaunal samples using hydrogen peroxide failed to mineralise the plant structures.

microfibril(lar): see *MICRO- 1.

microfiche (məi·krofīʃ). Pl. **microfiche, -fiches.** [f. *MICRO- 7 + F. *fiche* slip of paper, index-card.] A flat piece of film, usually the size of a standard catalogue card, containing microphotographs of the pages of a book, periodical, etc. Also shortened to **fiche.**

1950 *Rev. Documentation* XVII. 216/1 Perhaps the French decision of reserving the word microcard for the opaque microcopy and using microfiche for the transparency may show us the way [to avoid confusion in terminology]. 1953 *Library Sci. Abstr.* 193 The translucent microfiche is considered by some people to be better than the opaque microcard. 1959 *Times* 2 Apr. 11/7 Research work..can be economically published on *microfiche*... The cost of a *fiche* is very low, about 2s. 6d., and many research reports would not need more than two or three *fiche* to cover them. 1967 J. R. U. PAGE in De Reuck & Knight *Communication in Sci.* 151 Master microfiche are made for NASA and the Space Documentation Service. 1969 *Daily Tel.* 19 Mar. 14/8 The fiches will allow universities and other institutions to have vast reference library resources in small spaces. 1970 *Library Assoc. Rec.* Mar. 97/1 The National Aeronautics and Space Administration..makes available microfiches of the documents. *Ibid.* 97/2 The bibliographic information, together with the document and microfiche copy, is sent to NASA. 1970 *Publishers' Weekly* 8 June 152/1 A typical microfiche has 72 printed pages in an area 4 by 6 inches. 1972 J. POYER *Chinese Agenda* (1973) xiv. 210 He tossed a slim creased manila envelope..to Gillon. It felt empty... 'Microfiche,' Liu explained. 1974 *Bookseller* 8 June 2600/2 Colour printing on microfiche is now being done.

microfilaria: see *MICRO- 1.

microfilm (məi·krofilm), *sb.* Also with hyphen. [f. *MICRO- 7 + FILM *sb.*] **1.** (A length of) photographic film containing microphotographs of the pages of a book, periodical, etc.

1935 *Library Jrnl.* LX. 145/1 The only way to find out how well the film, made by M. Dagron, has preserved its original qualities during its sixty-four years, was to consider it like one of our modern micro-films and try to print enlargements from it. 1936 *Science* 1 May 403/1 Bibliofilm Service copied upon 35 mm film material in the library [*sc.* U.S. Department of Agriculture library], substituting microfilm for loan of the books and journals. 1936 *Dialect Notes* VI. 528 Documents in microfilm form ..will cost approximately 1 cent a page. 1939 *Nature* 11 Mar. 392/2 The so-called micro-film..offers one means for the photographic reproduction of bulky reports. 1940 A. HUXLEY *Let.* 14 Oct. (1969) 461 All the big libraries make these micro-films now. 1948 *Times* 17 Feb. 7/4 A brochure describing the micro-film reproduction of *The Times* from its first issue in 1785 to the present day. 1957 *Listener* 24 Oct. 646/2, I should not be able to go to California or Russia. I should just have to go home and work with microfilms. 1961 L. MUMFORD *City in Hist.* xvii. 546 What is visible and real in this world is only what has been transferred to paper or has been even further etherialized on a microfilm or a tape recorder. 1972 *English Studies* LIII. 533 He reads Eckhart in Latin at the public library from microfilm.

2. Special Comb.: **microfilm reader**, a small projector used to produce a readable image from microfilm; **microfilm viewer** (see quot.).

1950 *Amer. Documentation* I. 139/2 (*caption*) Griscombe portable microfilm reader. 1962 A. GÜNTHER *Microphotogr. in Library* (Unesco) 8 A system whereby..positive microfilm strips copied from the negative master are kept near the microfilm reader offers the best advantages. 1936 *Science* 1 May 403/2 Microfilm viewer—a small monocular optical device for reading 35 mm microfilms a line at a time, suitable for inspecting film or for use while travelling.

microfilm (məi·krofilm), *v.* [f. the sb.] *trans.* To record on microfilm. Hence **mi·crofilming** *vbl. sb.*

1940 *N. & Q.* 12 Oct. 253/2 Steps are being taken to have parish registers microfilmed... Micro-filming is a very cheap as well as speedy process. 1955 *Times* 28 July 9/4 It is intended to continue the microfilming..as and when funds permit. 1955 H. VAN THAL *Fanfare for E. Newman* v. 63 The Germans, who have always held Bizet in high respect,..micro-filmed the manuscript. 1972 M. J. BOSSE *Incident at Naha* ii. 76 Any microfilming service would do the job at ten cents a page. 1973 F. DE FELITTA *Oktoberfest* (1974) xiv. 164 There are a number of sources of wartime documentation which Yad Vashem has not yet microfilmed.

microfilter: see *MICRO- 6.

microflora (məi·kroflōᵊrā). *Biol.* [f. MICRO- 1 + FLORA.] A flora made up of minute plants, or one found in a microhabitat.

[1905 *Ann. Rep. Board of Regents Smithsonian Inst. 1904* 351 Has any competent hand celebrated the mikroflora of the highest ridges, those tiny, vivid forget-me-nots and gentians.] 1932 FULLER & CONARD tr. *Braun-Blanquet's Plant Sociol.* viii. 235 The microflora of the soil..is composed of countless bacteria, fungi, and algae. 1969 *New Yorker* 12 Apr. 85/1 The exobiologists have insisted that..the astronauts' microflora—bacteria and other organisms—have all been typed and catalogued. 1974 *Nature* 8 Feb. 361/2 This previously unknown microflora of blue-green and green or red algal affinities is from ..the Bungle Bungle dolomite which outcrops in the Osmond Range of Western Australia.

microform (məi·krofǭɹm). [f. *MICRO- 7 + FORM *sb.*] Microphotographic form; a microphotographic reproduction on film or paper of a manuscript, book, etc., requiring magnification to produce a readable image.

1960 *N. & Q.* Jan. 2/1 The genealogical and heraldic material listed as being already in 'microform' includes the Huddersfield Parish Registers, 1606–1812. 1962 A. GÜNTHER *Microphotogr. in Library* (Unesco) 14 The first supplementary note [to a catalogue entry] should indicate the microform. 1969 *R. & E. Coordinator* (Res. & Engin. Council Graphic Arts Industry) Apr. 4/2 Libraries make microform reproductions of everything from rare books to technical publications. 1971 *Amer. N. & Q.* X. 11/2 A selection of the major works from this catalogue is being offered in microform by the Erasmus Press. 1974 *Reprographics Q.* VII. 85/1 The cost/benefit of issuing parts lists and maintenance manuals on microforms.

microfossil, -fracture, -fracturing: see *MICRO- 1.

microgametophyte (məikrogæmī·tofəit). *Bot.* [f. MICRO- + *gametophyte* (s.v. GAMETE).] A gametophyte that develops from a microspore; a male gametophyte.

1907 *Amer. Naturalist* XLI. 360 In our material of Agathis the protoplasm is unfortunately very much shrunken..but this fortunately does not interfere with the understanding of the general conditions present in the microgametophyte. 1938 G. M. SMITH *Cryptogamic Bot.* II. v. 132 Microgametophytes of other genera also have but one vegetative cell. 1959 FOSTER & GIFFORD *Compar. Morphol. Vascular Plants* viii. 161 Early stages in formation of the microgametophytes [in *Selaginella*] begin while the microspores are still within the microsporangium.

microgasometer, -metric (etc.): see *MICRO- 6.

microgenic (məikrodʒe·nik), *a.* Chiefly *Journalists' colloq.* [f. MICRO(PHONE + *-GENIC b.] Of a voice: that sounds well when transmitted by microphone; well suited to broadcasting.

1931 S. GOLDWYN in *Sat. Rev.* 14 Feb. 220/1 An actor may be 'photogenic'..but that is not enough. He must also be 'microgenic', that is to say, he must have a voice that is suitable to the microphone. 1944 *Evening Standard* 5 Dec. 6/4 It seems strange that words like flak..should escape censure by the austere..while such bitter eloquence is directed against..microgenic (good for broadcasting).

microglia (məikro-, mikrǫgləi·ă). *Anat.* [f. MICRO- 1 + NEURO)GLIA.] Neuroglial cells derived from mesoderm and functioning as macrophages (scavengers) in and about the central nervous system (now regarded as components of the reticulo-endothelial system); a tissue composed of such cells. Usu. const. as *pl.* So **microgli·al** *a.*

1924 *Jrnl. Nervous & Mental Dis.* LIX. 346 In 1918 del Rio-Hortega succeeded..in impregnating clearly and specifically a homogeneous group of cells... These cells he called microglia. 1929 *Amer. Jrnl. Path.* V. 452 The microglial reaction around a sterile puncture wound of the cerebrum. 1932 *Times Lit. Suppl.* 10 Nov. 843/3 The rod-cells..are in fact generally held to be a variety of microglia. 1962 *Gray's Anat.* (ed. 33) 56 The astrocytes and oligodendroglia are ectodermal in origin..whilst the microglia is of mesodermal origin. 1966 W. ANDREW *Microfabric of Man* x. 147 (*caption*) Microglia in gray matter of spinal cord. Cell bodies of these cells are often angular or elongated. *Ibid.* 148/1 The microglia..are capable of remarkable transformations. 1973 PALAY & CHAN-PALAY *Cerebellar Cortex* vi. 178/2 Often microglia laden with such inclusions can be found to have wandered even into the molecular layer. *Ibid.*, The processes of a microglial cell can even invade the confines of the myelin sheath.

microglossary: see *MICRO- 1 b.

micrograph. Add: **3.** An enlarged image of an object (as seen through a microscope) obtained either by hand drawing or (now more usu.) photographically.

1904 *Electrochem. Ind.* Mar. 88/2 This embodies..the determination of oxygen content, the preparation of micrographs, and..the planimetric measurements of enlarged micrographs, with calculation of the percentage of oxygen. 1916 *Trans. Amer. Inst. Mining & Metall. Engin.* LI. 832 A microscopic examination was made of the threaded end of each of the 108 tensile bars. The micrographs shown at the end of this report reproduce the average structure of the top and bottom tensile bars. *Ibid.* 833 All micrographs are magnified 100 diameters. 1941 R. M. ALLEN *Photomicrogr.* i. 1 The word *micrograph* used alone has had to be broadened in meaning to include pictures of minute objects either drawn by hand or produced through photographic processes. 1953 *Jrnl. Appl. Physics* XXIV. 616/1 The simplest method of obtaining x-ray micrographs is to place the specimen in contact with a photographic film of the maximum resolution type and expose it to the radiation from a normal x-ray tube. All the enlargement is obtained photographically. 1960 D. F. LAWSON *Technique Photomicrogr.* ii. 4 Drawings made from images projected from a microscope are referred to as micrographs. 1968 H. HARRIS *Nucleus & Cytoplasm* iii. 59 Some electron micrographs of isolated 'polysomes' also appeared to show the ribosomes connected by some form of strand.

micrographically (məikrogræ·fikăli), *adv.* [f. MICROGRAPHIC *a.* + -AL + -LY[2].] By means of micrography or micrographs.

1898 *Jrnl. Iron & Steel Inst.* LIV. 191 The presence of thin decarbonised outer layers..may be frequently recognized micrographically. 1908 *Analyst* XXXIII. 289 All further changes (hardening) which take place inside the B constituents cannot be followed micrographically, because B appears to be a uniform substance by reflected light.

1971 (title) The compact edition of the Oxford English Dictionary: complete text reproduced micrographically. 1974 Nature 19 Aug. 709/2 Grain growth [in metals] has been studied micrographically for many years.

micrography. Add: **1. b.** The technique of producing micrographs (sense *3), or of studying objects by means of micrographs.
1908 Analyst XXXIII. 289 (heading) The micrography of cement. 1953 Jrnl. Appl. Physics XXIV. 603/1 The experiments so far made support the theoretical indications that x-ray shadow micrography is a practicable technique at least up to the resolution of the optical microscope. 1973 Jrnl. Bacteriol. CXIV. 413 (heading) Electron micrography of bud formation in Metschnikowia krissii.

micrograver, -gravimetric: see *MICRO- 1, 6.

microgroove (məi·krogrūv). [f. MICRO- 1 + GROOVE sb.] A very narrow groove on a gramophone record; a record having such grooves. Freq. attrib.
1948 Amer. Speech XXIII. 252 CBS announces a new-type record—microgroove. 1948 Electronic Engin. XX. 333/1 The groove size, from which the name 'Microgroove' has been derived, is only 0·0027 in. to 0·003 in. 1949 Wireless World Apr. 146/1 What is a 'microgroove' record? It is a name given to the 7-, 10- or 12-in pressings in high-grade vinylite, with a rotational speed of 33⅓ r.p.m... As it was found necessary to reduce the groove width to about one-third the size of normal record grooves, the name 'Microgroove' was coined originally by Columbia Records, Inc., for their 33⅓ r.p.m. records, but it would now appear to be used in America as a generic term for all such fine-pitch records. 1951 SACKVILLE-WEST & SHAWE-TAYLOR Record Guide 715 They increased the playing time of each side by..a much narrower groove—cut known as 'microgroove'. Ibid. 717 Even a 7-inch 33⅓ microgroove would apparently be too long—or too expensive—to meet this case. 1957 Times 13 Dec. 18/2 The 78 r.p.m. record began to lose favour. On the other hand, the demand for microgroove records continued strongly and sales of both 45 r.p.m. and 33⅓ r.p.m. records increased. 1958 Times 13 Sept. 9/1 This old Weingartner set has now reappeared..skilfully transferred to LP microgroove by Pathé-Marconi's engineers. 1972 Jazz & Blues Nov. 35/2 Some microgroove reissues contain edited versions of these titles.

microgyria, -habitat: see *MICRO- 3, 1.

microhardness (məikrohā·ɹdnės). [f. *MICRO- 1 b + HARDNESS.] The hardness of a very small area of a sample, as measured by an indenter.
1921 Trans. Amer. Soc. Mech. Engin. XLII. 1109 If we let e be the microhardness. 1934 Amer. Mineralogist XIX. 163 We have undertaken to obtain microhardness values for each of the nine minerals from talc to corundum. 1954 A. R. BAILEY Text-bk. Metall. xiii. 457 Micro-hardness testers make it possible to obtain comparative hardness figures for the different constituents of alloys,..and may aid in the identification of constituents and inclusions. 1966 D. G. BRANDON Mod. Techniques Metallogr. i. 5 In soft materials it is almost impossible to eliminate surface deformation by repeated mechanical polishing and etching... A detectable increase in microhardness can be observed up to 5 µm below the polished surface.

microhenry, -inch, -incineration, -injection, -instability: see *MICRO- 5 a, 5 a, 6, 2 a, 1.

microinstruction (məi·kro,instrʌkʃən). Computers. [f. MICRO- 1 + INSTRUCTION.] One of a sequence of instructions produced by a computer in response to some more comprehensive instruction; spec. one that corresponds to one of the smallest, most elementary operations that can occur in the computer and is produced in accordance with a microprogram.
1959 E. M. GRABBE et al. Handbk. Automation, Computation, & Control II. ii. 254 Many microinstructions can be constructed from properly sequenced microinstructions. 1964 T. W. McRAE Impact of Computers on Accounting i. 24 This problem has been solved by programmers devising..autocodes so that a single macroinstruction written down by the programmer generates the required set of microinstructions during translation by the computer from the autocode to the absolute machine code. 1966 P. B. JORDAIN Condensed Computer Encycl. 318 Each microinstruction is coded as several bits, one bit for each functional unit or data path in the computer hardware. The one-for-one correspondence between bit positions and functional units allows for very simple interpretation of each microinstruction.

micro-Kjeldahl, -lens: see *MICRO- 6, 1.

Micro-lepidoptera, sb. pl. (Later example.) Also microlepidopterous a. (Later examples.)
1860 Rep. Brit. Assoc. Adv. Sci. ii. 122 (heading) On some peculiar forms amongst the Micro-Lepidopterous larvæ. 1972 L. E. CHADWICK tr. Linsenmaier's Insects of World 226/1 A number of families have been included under the collective concept of the Microlepidoptera, but today the tendency is to consider them as separate groups. Ibid. 226/2 The microlepidopterous families are not isolated in their phylogeny.

microlinguistics (məikroliŋgwi·stiks), sb. pl. (const. as sing.). [f. MICRO- 2 a + LINGUISTIC sb. b.] (See quots. 1949, 1972.) So microlinguistic a.
1949 G. L. TRAGER in Studies in Ling.: Occasional Papers i. 2 The whole of the field concerned with language..we shall call Macrolinguistics. The three subdivisions we shall call Prelinguistics, Microlinguistics, Metalinguistics. Ibid. 4 Microlinguistics..deals with the analysis of language systems. 1953 Internat. Jrnl. Amer. Ling. XIX. ii. Suppl. 38 Smith presented the conceptual scheme developed at the Foreign Service Institute which divides linguistics into three main compartments:..'microlinguistics' deals with the analysis of linguistic systems. Ibid. 29 He gave an example of the sort of methodological error which Smith's separation of microlinguistic and metalinguistic levels would avoid. 1955 S. & M. CHASE Power of Words x. 101 Microlinguistics..takes a long time to teach a unit as large as the sentence. 1963 Amer. Speech XXXVIII. 238 Since distinctive articulatory features form part of a system of contrastive correlations, any analysis based on this structure is microlinguistic; but not extralinguistic. 1967 D. STEIBLE Conc. Handbk. Ling. 78 Microlinguistic meaning. The term refers to the identity or difference in meaning which results when a part of a larger structure is replaced by a different part. 1972 HARTMANN & STORK Dict. Lang. & Ling. 141/2 Microlinguistics, those aspects of linguistic studies which are concerned with the direct analysis of linguistic material, e.g. phonology, grammar, lexicology.

microlite. 2. Add: Cf. CRYSTALLITE 2 in Dict. and Suppl. (Further examples.)
1926 [see *CRYSTALLITE 2]. 1954 H. WILLIAMS et al. Petrogr. ii. 13 Extremely minute incipient crystals..are called microlites, provided they are birefringent; if they are even smaller, spherical, rod- and hair-like isotropic forms, they are called crystallites.

microlith. Add: **2.** Archæol. A small stone tool with a sharpened edge used with a haft, characteristic of Mesolithic cultures.
1908 H. G. O. KENDALL in Man VIII. 103 Palæolithic Microliths... By microliths I mean tiny flakes or other pieces of flint which have been trimmed or used by man at some part of the edge. 1927 PEAKE & FLEURE Hunters & Artists vii. 96 This [Capso-Tardenoisian] industry is characterized by the presence of very small flints of geometric shapes, chiefly of trapezoid, rhomboid, and triangular forms; these are commonly known as microliths. 1932 J. G. D. CLARK Mesolithic Age in Brit. p. xx, By a 'microlith' we understand a narrow flake blunted on one or both edges by steep secondary chipping, but devoid of secondary work on either face. 1960 New Scientist 11 Aug. 418/3 Finds included 52 microliths (tiny points which must have been hafted as arrow-heads and other weapons). 1971 World Archaeol. III. 157 Often termed 'microliths' by Old World archaeologists, these tools fall into three formal categories.

microlithic, a.¹ Add: **b.** Archæol. Of or pertaining to microliths (*MICROLITH 2); characterized by the use of microliths.
1923 A. L. KROEBER Anthropol. xiv. 407 Terminal Capsian..was a local phase..with the microlithic flint industry especially conspicuous... In Africa, the development of the extreme microlithic forms..has been most clearly traced. 1937 GARROD & BATE Stone Age Mt. Carmel I. i. viii. 114 We there found a layer containing an abundant microlithic industry, without pottery, superimposed on eroded breccia with fluids of Levalloiso-Mousterian type. 1947 J. G. C. HAWKES Prehist. Brit. (rev. ed.) 19 The reductions of the size of flints to a 'microlithic' scale. 1972 World Archaeol. III. 157 Numbers of fragmentary flakes with backed (or 'microlithic') retouch occur in many levels of the site.

micrologic: see *MICRO- 2 a.

micrologist. (Later example.)
1960 [see *MICRURGY].

micrology. 2. (Examples.)
1906 M. F. GUYER (title) Animal micrology. 1907 Nature 18 Apr. 583/1 The term 'micrology' has not received any general acceptance on this side of the Atlantic. 1914 (title) Journal of micrology.

micromachining: see *MICRO- 2 a.

micromanipulation (məi:kromănipiulē·ʃən). [f. MICRO- 2 a + MANIPULATION.] The performance of extremely delicate operations (such as the isolation of a single yeast cell from a culture) under the microscope, usu. with the aid of a micromanipulator; an operation so performed.
1921 Science 28 Oct. 411/2 (heading) A simple apparatus for micro-manipulation under the highest magnifications of the microscope. 1931 Chem. Abstr. XXV. 1409 (heading) Micromanipulations on latex in dark fields. 1949 New Biol. VII. 72 Although the chromosomes of most cells are very small objects,..in a few favourable cases it has been possible to study their physical properties by micromanipulation. 1971 Nature 2 July 33/1 The direct manipulation of chromosomes of living human cells in vitro is a potential means of obtaining transplantable genetic material.

So micromani·pulator, an instrument which is used in conjunction with an optical microscope to perform micromanipulations and which allows a microneedle, micropipette, etc., to be moved with great control through the field of view.
1921 Science 28 Oct. 413/1 There are two models of the micro-manipulator, a simple and a more elaborate form. 1949 A. G. SANDERS in H. W. Florey et al. Antibiotics II. xvi. 678 Spores from a conidium can be picked off singly with the point of a needle. Some people can do this with the needle held in the hand, others prefer to use a micromanipulator. 1968 Sci. Jrnl. Nov. 18 (caption) Integrated circuit chip is shown above being placed in its case with aid of a micromanipulator. 1973 Nature 4 Feb. 263 Zeiss micromanipulators were used to position micropipettes.

micromanometer (etc.), **micromastia, -mazia:** see *MICRO- 6, 3, 3.

micromesh (məi·kromeʃ). [f. *MICRO- 1 + MESH sb.] Material (esp. nylon) consisting of a very fine mesh. Freq. attrib.
1959 Manch. Guardian 27 July 4/4 Plaza..have..a cheap..micromesh. 1960 Harper's Bazaar Apr. 98 Seamfree in sheer micro-mesh. 1962 Which? Apr. 113/2 Two of the leading brands of seamfree nylon stockings. 1963 R. A. R. HIGHAM Handbk. Papermaking i. 6 Empty the screen on its micro-mesh wire circle and gently wash with distilled water. 1974 Times 26 Apr. 7/7 Initially only in micromesh, the tights..are in three sizes.

micrometallurgy, -metazoan, -metazoon: see *MICRO- 2 a, 1, 1.

micrometeor (məikromī·tīǫɹ). [f. MICRO- 1 + METEOR.] = *MICROMETEOROID.
1957 Seattle Times 10 Nov. 16/5 Density of micrometeors and meteoric dust, and the like. 1973 Times 16 May 1/1 The micrometeor shield was designed to protect the outer skin of the space station from minor damage but also played an important role in controlling Skylab's temperature.

Hence micrometeor·ic a.
1958 Times 30 Aug. 6/1 It also carried recording instruments to register impacts of micro-meteoric particles.

micrometeorite (məikromī·tīǫrəit). [f. MICRO- 1 + METEORITE.] A micrometeoroid; spec. one that has entered the earth's atmosphere (cf. the distinction between METEORITE and METEOROID sb.).
1949 F. L. WHIPPLE in Science 28 Oct. 438/1 The term micrometeorite is here defined as an extraterrestrial body that is sufficiently small to enter the earth's atmosphere without being damaged by encounter with the atmosphere. 1956 Spaceflight I. 27/1 The density of meteorites and micro-meteorites (interstellar dust) in space will also be of interest to the designers of manned research vehicles. 1967 Technology Week 23 Jan. 47/1 Unlike a terrestrial observatory, it will have to have airtight quarters, well-protected against radical temperature change, hard radiation and micrometeorite bombardment. 1971 I. G. GASS et al. Understanding Earth iii. 115/1 Many granules are so small that they do not offer sufficient air resistance to become incandescent; they sink to the ground as micrometeorites. 1973 McGraw-Hill Yearbk. Sci. & Technol. 271/2 Bodies less than about 0·1 mm—too small to be called meteorites, since their slow spin permits them to radiate the tremendous heat energy due to friction with the atmosphere before it can cause vaporization. These smaller bodies are usually called micrometeorites, or while still in space, micrometeoroids.

Hence micrometeoriti·c a.
1960 H. E. NEWELL in J. A. Ratcliffe Physics Upper Atmosphere iii. 123 Micrometeoritic material may account for a small portion of the E-region ionization. 1974 Nature 20/27 Dec. 669/2 Micrometeoritic craters.

micrometeoroid (məikromī·tīǫroid). [f. MICRO- 1 + METEOROID sb.] A solid particle in space, of extraterrestrial origin, which is small enough to survive entry into the earth's atmosphere. Cf. prec.
1954 Mineral. Abstr. XII. 242 There may be 'micro-meteorites' (? micrometeoroids..) too small to suffer ablation and falling as cosmic dust. 1962 Daily Tel. 26 Aug. 1/4 Another earth satellite, Explorer 13, was put into orbit to-day. It will measure dust-like particles in space known as micrometeoroids. 1968 Analog Sept. 30/2 The suit will be made to protect the astronaut..from tiny particles known as micrometeoroids. 1973 [see prec.]. 1974 Nature 6 Sept. 17/1 It sent back more than 300 pictures of Jupiter, as well as measurements of magnetic fields, energetic particles,..and micrometeoroids.

Hence micrometeoroi·dal a.
1972 Science 1 June 979/1 The scale of hypervelocity impact craters extends from a few microns down to submillimeter and submicron micrometeoroidal pits on rock surfaces.

micrometeorology (məi:kromītīǫrǫ·lŏdʒi). [f. MICRO- 2 a + METEOROLOGY.] The study of the meteorological characteristics of a small area; the study of small-scale meteorological phenomena.
1930 Flight 18 Apr. 442 The progress of European meteorology in the past fifteen years..is due to the introduction of micrometeorology,..for detailed observations from an ever-increasing number of weather stations. 1953 O. G. SUTTON (title) Micrometeorology: a study of physical processes in the lowest layers of the earth's atmosphere. 1968 New Scientist 5 Dec. 555/1 By incorporating fluctuations of temperature and humidity, one can arrive at similar expressions for the transport of heat and water vapour. The determination of these three vertical transports is one of the central problems in micrometeorology. 1974 Physics Bull. Feb. 66/2 The heat balance of plants and animals and the micrometeorology of crops.

Hence mi:crometeorolo·gical a., -meteoro·logist.

1942 Univ. Chicago Inst. Meteorol. Misc. Rep. No. 3. 1 The study of the exact nature of these more or less localized thunderstorms suggests the use of a micrometeorological network of stations. Such a network exists over the Muskingum Watershed in..Ohio where the United States Soil Conservation Service..has maintained a dense observational network. The stations are no more than 8 miles apart. 1953 O. G. SUTTON Micrometeorol. p. vii, I have attempted to meet the needs of meteorologists..who require detailed information about physical processes in the regions of the atmosphere where life is most abundant. It is my hope that such an account will help to increase the number of micrometeorologists. 1960 Times 20 Aug. 2/2 The work will involve physical and micrometeorological problems of measurement. 1968 New Scientist 5 Dec. 554/2 While the micrometeorologist's most productive tools are certain statistical concepts of fluid mechanics and aerodynamics, much experimental work needs to be done before these concepts can be applied to the atmosphere.

micrometer, var. *MICROMETRE.

micromethod: see *MICRO- 6.

micrometre (məi·kromītəɹ). Also (U.S.) -meter. [f. MICRO- 5 a + METRE sb.²] A millionth of a metre: = MICRON, MIKRON (in Dict. and Suppl.).
The word has only recently become common (cf. quot. 1968), and now forms part of the International System of Units.
1880 Jrnl. R. Microsc. Soc. III. 327 The same reasoning..leads us, however, to recommend the adoption of 'micrometre' instead of 'micromillimetre', which would secure the uniformity desired, besides being a more convenient word. 1968 KAYE & LABY Tables Physical & Chem. Constants (ed. 13) 2, µ is widely employed as an abbreviation of µm (10^{-6} m or micrometre) and is then called micron. 1968 Nature 16 Nov. 652/1 By resolution No. 7 the conference [sc. the thirteenth General Conference of Weights and Measures] decided to proscribe further use of the name 'micron', with the symbol µ attributed to this name, for the millionth part of the metre. The symbol µ is now the recognized prefix for the decimal sub-multiple 10^{-6} and the appropriate name for the millionth part of the metre is 'micrometre', with symbol µm. 1971 I. G. GASS et al. Understanding Earth xiv. 336/1 (caption) The amplitude..used for determining the magnitude is half the displacement between the two arrows—in this case 5 micrometres. 1973 Physics Bull. Nov. 662/3 They contain solid particles..with dimensions of about a tenth of a micrometre, a mass of 10^{-15} g each. 1974 Sci. Amer. July 45 (Advt.), Neutron radiography can resolve a few micrometers.

micromho, -microcurie, -microfarad, -mini: see *MICRO- 5 a, 5 a, 5 a, 1 c.

microminiature (məikromi·nitiŭɹ, -mi·niătiŭɹ), a. [f. *MICRO- 1 b + MINIATURE a.] Much reduced in size, as a result of microminiaturization; even smaller than a size regarded as miniature.
1958 Electr. Manufacturing Aug. 94/1 Intensive efforts at the Diamond Ordnance Fuze Laboratories..in the field of microminiature packaging have recently resulted in extremely small and compact assemblies. 1963 Listener 16 May 832/3 Microminiature integrator, for rocket guidance system. The integrator..is 100 times smaller than the conventionally built equipment it replaces. 1967 Electronics 6 Mar. 37 (Advt.), These microminiature relays are direct descendants of our military, aero/space designs. 1971 J. H. SMITH Digital Logic i. 5 Thin film circuits are often associated with microminiature components.

mi:crominiaturiza·tion. Also with hyphen. [f. *MICRO- 1 b + *MINIATURIZATION.] Extreme miniaturization; spec. the development or use of techniques for making electronic components and devices of greatly reduced size (smaller than those produced by 'miniaturization').
1955 Proc. IRE XLIII. 1897/2 The most attractive features [of opto-electronic devices] seem to be the possible microminiaturization (which, in the present state of the art, is the ultimate in miniaturization) resulting in small weight and extremely low power consumption, [etc.]. 1959 Listener 28 May 930/2 The most startling recent development [in transistors] is the technology of what is called micro-miniaturization. 1963 Daily Tel. 22 May 21 A stage of 'micro-miniaturisation' has been reached which makes many of the marvel-working components almost invisible. 1965 Times Rev. Industry Winter 12/1 Micro-miniaturization is now accepted as starting at a packing density exceeding 50 parts per cubic inch. 1971 Nature 10 Dec. 315/3 The new on-board computer and guidance system makes great use of micro-miniaturization. 1973 Sci. Amer. Dec. 24/2 Its nervous system is a miracle of microminiaturization, and some of its independently evolved behavior patterns are not unlike our own.

So micromi·niaturize v. trans., to produce in a very much smaller version; micromi·niaturized ppl. a.
1959 J. E. SENSI in Horsey & Shergalis Proc. Symposium Microminiaturization of Electronic Assemblies i. v. 54 To realize fully the military and civilian potential of microminiaturized devices, optimum means of producing these devices in large quantities are necessary. 1959 Electronics 11 Dec. 51/2 Much of the knowledge gained in designing electronic circuits can be used directly, with only minor changes, in microminiaturizing circuits. 1963 Listener 16 May 832/3 Already..the sort of computer that only yesterday took up a whole room can be 'microminiaturized' so that it will go into a small suitcase. 1967 N.Y. Times 20 May 49 The transistorized lock..could be microminiaturized and adapted to any number of combinations. 1972 Lebende Sprachen XVII. 133/1 Plessey has evolved a small microminiaturized computer.

micromodule: see *MICRO- 1.

micromotion (məi·kromōuʃən). Also with hyphen. [f. MICRO- 1 + MOTION sb.] A small bodily movement made during the performance of some task, esp. when recorded cinematographically for purposes of work study. Usu. attrib., esp. designating this method of work study.
1913 F. A. TALBOT Pract. Cinematogr. xiv. 174 (heading) Micro-motion study: How increased workshop efficiency is obtainable with moving pictures. 1913 Technical World XIX. 189/1 'This micro-motion study furnishes a means for the transference of skill from man to machine,' the general manager further stated. 1947 J. J. GILLESPIE Dynamic Motion & Time Study 2 Motion study..with its motion cameras, therbligs, micromotion clocks..has become a complex, unwieldy technique. 1948 Sci. News VII. 107 For the so-called 'micro-motion' analysis of repetition movements a cine-record is necessary. Instead, small lamps can be attached to the moving parts of the operator's body, their light being recorded over one cycle of the operation by working in a dimly lit room while the camera shutter is left open. 1961 [see *MEMOMOTION].

micromutation: see *MICRO- 1.

micron, mikron. Add: The spelling mikron was never common and is now Obs. (Earlier and later examples.)
In the International System of Units the word has been replaced by *MICROMETRE.
[1880 Procès-Verbaux des Séances du Comité Internat. des Poids et Mesures 1879 45 Le Comité international des Poids et Mesures adopte, pour un usage officiel, le système suivant des signes abréviatifs pour les poids et mesures métriques... (table) Mesures de longueur... Micron... µ.] 1885 Jrnl. R. Microsc. Soc. V. 140 Proposal..(2) to use micron in place of micromillimetre. 1966 Electronics 14 Nov. 25 The detector..is believed to be the first photovoltaic detector of infrared radiation in the 8- to 14-micron range. 1968 [see *MICROMETRE].

microneedle: see *MICRO- 6.

Micronesian, a. and sb. Add: Also (rare) Mikronesian. (Earlier and later examples.)
1847 J. C. PRICHARD Res. Physical Hist. Mankind (ed. 3) V. 157 A certain difference has been noted between the Micronesians and the Polynesians in general. Ibid., Micronesia, or the Micronesian Archipelago. 1877 Jrnl. Anthrop. Inst. VI. 392 The language of Yap..is the only Micronesian language in which the numerals are formed in the manner in question. 1884 W. TURNER Rep. Crania in Rep. Sci. Results Voy. H.M.S. Challenger X. 82 The islands of the Mikronesian group. 1890 D. G. BRINTON Races & Peoples 235 Some ethnographers would make the Polynesians and Micronesians a different race from the Malays. 1945 Language XXI. 214 In the vocabularies of the Micronesian languages..evidence of earlier use of English has remained in the form of loan-words. 1974 Times 21 Jan. 12/6 Some Micronesian families connected by marriage with Japanese see themselves in a strong compradore position. Ibid. 12/8 The High Commissioner..expects that a Micronesian will succeed him.

micronize (məi·krŏnəiz), v. [f. MICRON + -IZE, or perh. a back-formation from Micronizer (proprietary name in U.S.).] trans. To break up into very fine particles. So mi·cronized ppl. a., mi·cronizing vbl. sb. Also microniza·tion.
1940 Jrnl. Econ. Ent. XXXIII. 481/1 Good kill obtained with micronized cubé indicates that fine grinding may increase the toxicity. 1941 Ibid. XXXIV. 560/2 Micronization (fine grinding) of derris and cubé[4] increased kill over that obtained by the regular product. [Note] [4]Prepared by the Micronizing Processing Co., Moorstown N.J. 1952 M. E. FLOREY Chem. Applic. Antibiotics I. iii. 111 A ball mill for the production of a very finely divided ('micronized') mixture of potassium penicillin and glucose. 1958 Vogue July 5 This creamy-smooth makeup contains micronised powder, 2½ times finer than ordinary face powder. 1968 W. A. GRAY Packing of Solid Particles vii. 105 If, however, the particles are reduced in size to a few microns (micronized) and suitable pressure is applied, then a strong compact can be formed. 1968 Materials & Technol. I. viii. 263 Micronizing is used not only on sulphur but also for insecticides, fungicides, pharmaceutical preparations, grinding and polishing powders, iron oxide, etc. 1972 Jrnl. Econ. Entomol. LXV. 1446/1 Colorimetric analysis indicated that some decomposition took place during preparation and micronization of the dust.

micronutrient (məi:kroniū·triĕnt). Biol. [f. MICRO- + NUTRIENT sb.] Any of the chemical elements which are required by plants (or, less commonly, animals) in trace amounts for normal growth and development.
1943 Ann. Rev. Biochem. XII. 525 List of essential micronutrients.—The essentiality of boron, manganese, and zinc for higher plants is no longer open to dispute. 1953 J. RAMSBOTTOM Mushrooms & Toadstools viii. 80 Manganese, zinc, boron, copper, molybdenum, have been proved definitely to be micro-nutrients for certain fungi. 1970 [see *MACRONUTRIENT]. 1974 Nature 8 Feb. 392/1 Since 1957, selenium has been recognized as an essential micronutrient for animals.

micro-opaque (məikro,opēi·k). [f. *MICRO- 7 + OPAQUE a. (sb.).] A type of microform produced on card or paper instead of film. Also attrib. or as adj.
1952 Aslib Proc. IV. 154 Micro-opaques have made far greater psychological appeal to the user than microfilm. 1956 Amer. Documentation VII. 169/1 Writers in several countries have concerned themselves with the general theme of micro-opaques. 1962 A. GÜNTHER Microphotogr. in Library (Unesco) 7 Micro-opaque cards have another advantage in that they do not need a protective envelope and may be readily filed. 1963 Amer. N. & Q. June 155/1 Original public records are being published in a micro-opaque series of 3 × 5-inch cards issued by the Public Record Office of Great Britain.

micro-operation: see *MICRO- 1, 2 a.

micro-operative, -order: see *MICRO- 2 a, 1.

micro-oven (məi·kro,ʌv'n). [f. *MICRO(WAVE + OVEN sb.] A microwave oven. Hence micro-o·vening vbl. sb., cooking in such an oven.
1962 Punch 21 Nov. 738/2 A cooked meal that is quick-frozen and then re-heated in a matter of seconds in a micro-oven. 1965 Economist 22 May 941/3 Two micro-ovens could do the same work in half the time, using less space. 1971 Daily Tel. 26 July 15/2 Thus their formula for the Big Chance: cheap to use some traditional greasy cafes, manufacture centrally some basic dishes for instant micro-ovening and bounce into branded home cooking.

micropædia (məikropī·diă). [f. MICRO- + Gr. παιδεία education.] A section of the 15th edition of Encyclopædia Britannica (published in 1974) in which information is presented in a condensed form. (Cf. *MACROPÆDIA, *PROPÆDIA.) Hence micropæ·dic a.
1974 Daily Tel. 12 Jan. 3/5 Consisting of 30 volumes, it will consist of what are described as a propaedia, a micropaedia and a macropaedia. 1974 Times Lit. Suppl. 11 Oct. 1120/4 About 180 writers are given biographies proper, the rest have to make do with a Micropaedic note.

micropalæontology (etc.), **-particle, -phagous, -phenocryst:** see *MICRO- 2 a, 1, 4, 1.

microphone. 2. Add to def.: Now applied to any instrument designed to convert sound waves impinging upon it into corresponding variations in voltage or current, which may then be amplified or transmitted for reconversion into sound (as in broadcasting and the telephone) or recorded; esp. one made as an independent unit (colloq. abbrev. *MIKE sb.⁵). (Further examples.)
This function is that of all telephone 'transmitters'. Hughes's instrument was simply a particularly sensitive one by the standards then current (hence the name: see quot. 1889), and modifications of it were for a time almost universally used as telephone transmitters, so that microphone became synonymous with transmitter and acquired its present more general meaning.
1889 PREECE & MAIER Telephone iv. 37 Hughes' Microphone... The microphone is nothing but a telephonic transmitter, but it owes its name..to its power to convert vibrations of feeble intensity into undulatory currents, which, passing through a receiving telephone, produce sonorous vibrations of much greater intensity than those of the original source. 1891 F. C. ALLSOP Telephones ii. 20 Prof. Hughes's microphone..forms the basis on which all the modern carbon transmitters are constructed. 1923 W. S. CHURCHILL World Crisis 1915 291 Already the microphone or hydrophone for detecting the beat of a submarine propeller in the distance had been discovered. 1923 E. W. MARCHANT Radio Telegr. vi. 76 Suppose..that the emission of waves from the transmitting aerial is controlled by means of a microphone, such as is employed in the ordinary telephone transmitter, the stream of waves given out by the antenna will be varied in accordance with the fluctuations in the current passing through the microphone. 1929 Morning Post 24 May 12/7 The engineer..in film-direction..has the last word as to whether the actor is speaking the line effectively for the microphone. 1935 H. C. BRYSON Gramophone Record iii. 59 Three main kinds of microphone are in common use: the carbon microphone, the condenser microphone, and the moving coil microphone. 1962 A. NISBETT Technique Sound Studio ii. 42 As an example of the sort of set-up which can be adopted in this type of studio is one where no less than six microphones were used for a quarter-hour playlet. 1970 M. L. GAYFORD in T. L. Squires Telecommunications Pocket Bk. iv. 36 The standard modern telephone sets now in production in most countries represent a considerable improvement over earlier sets... Improved designs of microphone and receiver give a generally better frequency response and transient response.

3. Special Comb.: **microphone boom,** a boom (*BOOM sb.² 1 d) with a microphone at the end.
1931 L. COWAN Recording Sound for Motion Pict. 377 Microphone boom, crane-like device for supporting and manipulating microphone. 1954 Time 12 July 47/3 Joan Diener, instead of being forced to stand near a microphone boom in order to be heard, was able to move at will in a TV studio by means of a tiny concealed microphone transmitter.

microphoned (məi·krŏfōund), a. [f. MICROPHONE + -ED².] **1.** Picked up and transmitted by a microphone.
1927 Daily Express 21 Sept. 11/5, I heard in Archie de Bear's room at the Vaudeville..a loud speaker carrying from the stage what was only a microphoned reproduction of a gramophone record made by the Revellers.

2. Containing or furnished with a microphone.
1933 News Chron. 13 June 1 At one end is a high microphoned platform for the President. 1934 Punch 11 Apr. 416/1 Put a commentator in a microphoned observation-car on each long-distance train. 1960 Guardian 20 July 5/7, I lived in a room in Moscow University which every-one..said was microphoned.

microphoneme (məi:krofōu·nīm). Linguistics. [f. MICRO- 1 + PHONEME.] (See quot. 1935.) Hence microphone·mic a.
1935 W. F. TWADDELL in Lang. Monogr. XVI. v. 39 A term of an ordered class of minimum phonological differences among forms is a microphoneme. 1936 Language XII. 56 If one restricted one's attention to the series kill: till: pill: hill, one could arrive at alternative microphonemic classes. 1953 Trans. Philol. Soc. 84 All that we have..are..certain microphonemes and greater or lesser sets of microphonemes.

microphonic (məikrofǫ·nik), a. and sb. [f. MICROPHON(E + -IC.] **A.** adj. **1.** (In Dict. s.v. MICROPHONE.)

2. a. Characterized by or pertaining to the production of variations in electrical potential in response to sound waves or vibrations.
1879 Telegraphic Jrnl. VII. 132/1 The communication of the current with the vibrating plate..is effected by means of two small springs which are lightly pressed by the membrane..which act as a weak microphonic contact. 1919 R. STANLEY Text-bk. Wireless Telegr. (new ed.) II. vi. 116 With some valve designs there is a microphonic effect in L.F. amplifiers so that the slightest jar given to the apparatus, such as tapping it with the finger or even walking near it, is strongly magnified. 1931 Jrnl. Physiol. LXXI. p. xxix, I conclude that the effect is due to some kind of microphonic action by which vibrations produce changes in the potential between different points in the inner ear. 1940 A. WOOD Acoustics xvii. 478 The microphonic response of the cochlea of a cat to a pure tone of 1000 cycles/sec. is amplified and analysed with a wave analyser. 1951 Electronic Engin. XXIII. 429/2 An attempt has been made..to determine the absolute microphonic performance of a valve by relating its electrical output to the frequency and intensity of a sound field in which it is situated. 1970 Nature 11 July 184/2 The microphonic response of the cochlea.

b. Of an electrical signal: generated in response to sound waves or vibrations.
1929 Proc. IRE XVII. 1621 Microphonic output is caused by relative motions between the various elements of the tube. 1930 ZWORYKIN & WILSON Photocells ix. 114 Mechanical vibration of the elements produce what are generally called microphonic noises. 1947 Jrnl. Appl. Physics XVIII. 242/2 A simple experiment was set up to check Eq. (7) by mechanically driving a tube and measuring the resulting 'microphonic' output signal.

B. sb. **1.** pl. (In Dict. s.v. MICROPHONE.) rare.

2. a. A microphonic signal generated in the cochlea.
1938 STEVENS & DAVIS Hearing xii. 319 When the voltage of the cochlear microphonic is plotted against the logarithm of the sound-intensity..the function appears as a sigmoid curve. 1962 Laryngoscope LXXII. 432 Is the reduction in amplitude of the cochlear microphonics due to an impairment of sound transmission or to a more direct effect on the organ of Corti? 1974 Nature 10 May 162/2 The 'poor' cochlear microphonic obtained from BALB/c mice after priming is indicative of hair cell damage.

b. Electronics. An undesired signal or modulation produced (e.g. in a valve) by mechanical vibration. Usu. pl.
1929 Proc. IRE XVII. 1622 The simplest test for microphonics consists of an audio-frequency amplifier of fairly high gain, the tube under test being used in the first stage. 1947 Jrnl. Appl. Physics XVIII. 245/2 It may well be advantageous to operate the input stage of a high gain automatic volume controlled amplifier with fixed bias to reduce microphonics. 1960 IRE Trans. Microwave Theory & Techniques VIII. 372/1 Waveguide under a high acoustical field can definitely contribute to microphonics via the mechanism of phase modulation. 1965 GEWARTOWSKI & WATSON Princ. Electron Tubes v. 152 Vibration of grid wires in grid-controlled tubes is a principal source of 'microphonics'.

microphonism (məikrǫ·fŏniz'm). Electronics. [f. MICROPHON(E + -ISM.] = *MICROPHONY 2.
1947 Jrnl. Appl. Physics XVIII. 239/1 This suggests a method of design..by which the microphonism of a tube may be reduced since..most of the microphonic difficulties encountered in tubes are caused by the mechanical movement of the grid. 1950 Proc. IRE XXXVIII. 529/1 The effects of microphonism are most usually encountered in high-gain amplifiers.

microphony. Restrict rare⁻⁰ to sense in Dict. and add: **2.** Electronics. [f. MICROPHON(E + -Y³.] The generation of microphonics in electrical apparatus.
1934 F. J. CAMM Everyman's Wireless Bk. v. 91 (table) Microphony: This may result in ringing, booming, or other sounds. 1946 Electronic Engin. XVIII. 336/2 The ruggedness which is required to give long service will normally ensure freedom from microphony. 1951 Ibid. 431/2 The design of a valve with low microphony is a long and tedious process. 1963 Mullard Technical Communications VII. 227/1 Microphony is occasionally experienced in a tape recorder when high gain a.f. valves are used in the first stages.

microphoto (məi:krofōu·to), colloq. abbrev. of MICROPHOTOGRAPH 2.
1949 Jrnl. Geol. LVII. 370/1 The phenomenon of corrosion between crystals (Perrin and Roubault, 1939, p. 160, pl. 5, microphotos 7 and 8). 1972 K. MASTERS Spray Drying xiv. 518 Each microcapsule contains numerous minute droplets of perfume oil. A microphoto of an encapsulated fragrance is shown in an article by Barreto.

microphotograph. 1. (Further examples.)
1867 Pop. Sci. Rev. VI. 54 'Microphotograph' is a very long name, recently introduced, to denote a very small object; it refers to the minute photographic reductions of portraits or views so often shown as curiosities under the microscope. 'Photomicrograph', on the contrary, is a name given to the photographic enlargement of a microscopic object. 1878 Jrnl. R. Microsc. Soc. I. 300 Microphotograph.—Mr. Langenheim..has photographed the Lord's Prayer on the ten-thousandth of a square inch. 1940 A. HUXLEY Let. 14 Oct. (1969) 461, I would like to have..micro-photographs suitable for reading by means of a reading machine (all the big libraries make these micro-films now) of Part III of this book. 1957 R.A.E. News Nov. 8/2 The National Coal Board publicity office tried to excuse..an advertisement that described a picture, about four inches by six, as a 'microphotograph'. In the O.E.D. this word certainly denotes both a small photograph of a large-or-small object and a large-or-small photograph of a small object. However, the ambiguity proved such a nuisance that after international discussion between various bodies, the words 'microphotograph' and 'photomicrograph' respectively were agreed on. 1962 A. GÜNTHER Microphotogr. in Library (Unesco) 5 Microphotographs are photographs which are reduced to a minute or even microscopic size. 1966 M. R. D. FOOT SOE in France viii. 181 A microphotograph in the false bottom of a matchbox. 1969 Proc. R. Microsc. Soc. IV. 142 Sixth Report of the Nomenclature Committee. The Committee recommends the following names and definitions:... Microphotograph. A very small photograph, intended to be viewed with a microscope.

2. (Further examples.)
1927 HALDANE & HUXLEY Animal Biol. ii. 51 Fig. 18. Micro-photograph (× 150) of a section through the ovary of a mammal (cat). 1944 R. SOUTH Caterpillars Brit. Butterflies 5 Of the eggs and chrysalids, the former [illustrations] are from microphotographs by A. E. Tonge and drawings by Horace Knight. 1968 Punch 21 Aug. 271/3 There are blackboard diagrams, and micro-photographs of sperm and ova. 1974 Nature 22 Feb. 511/3 Species of pine and hardwood trees, grasses and cultivated plants, for example, are represented in the atlas of 136 microphotographs.

microphotometer (etc.), **microphthalmos:** see *MICRO- 6, 3.

microphysics (məi·krofiziks), sb. pl. (const. as sing.). Also micro-physics. [f. MICRO- 2 a + PHYSICS.] The part of physics that is concerned with bodies and phenomena on a microscopic or smaller scale, esp. with molecules, atoms, and sub-atomic particles.
1885 [in Dict. s.v. MICRO- 2 a]. 1956 E. H. HUTTEN Lang. Mod. Physics v. 175 These rules work well enough when we describe macro-physical phenomena; but their insufficiency is revealed in micro-physics. 1962 N. R. HANSON in Quanta & Reality v. 86 Von Neumann advanced a 'proof' that it would be impossible to reinstate a classical determinism in microphysics; it was logically pointless to look for 'hidden variables' underlying quantum processes. 1971 Nature 24 Dec. 433/2 Research in Britain concentrates on the fundamental microphysics of fogs.

Hence microphy·sical a., of or pertaining to microphysics.
1902 [see macrophysical s.v. *MACRO- 1 e]. 1936 Discovery Mar. 96/2 Microphysical inquiries into which chemistry is now being absorbed. 1956 E. H. HUTTEN Lang. Mod. Physics v. 179 Two micro-physical events are connected in a different way from two macro-physical events. 1962 S. TOULMIN in Quanta & Reality 16 In the macroscopic world, all statistical statements..report the overall aggregates or averages of large numbers of individual events..; but the microphysical statements of quantum mechanics were statistical in a more absolute sense. 1973 Nature 14 Dec. 378/3 According to Audretsch the effect of expansion is significant for particle creation when the age of the Universe is comparable to the microphysical time scale—when the Universe is only about 10^{-23} s old!

microphysiologist, -physiology, -pinocytosis, -pipette, -plankton(ic): see *MICRO- 2 a, 2 a, 2 a, 6, 1.

microplastic(ity), -plate, -poise, -population, -porosity, -powder: see *MICRO- 1 b, 1, 5 a, 1, 1, 1.

microprint (məi·kroprint). [f. *MICRO- 7 + PRINT sb.] **a.** A photographic print of text reduced by microphotography. **b.** Printed matter so reduced. Hence mi·croprinting vbl. sb., the production of microprint.
1933 Library Jrnl. LVIII. 913/1 With these photographic prints on sensitized paper there is no more question of fire hazard and the necessity of projection is likewise eliminated... We have, however, to overcome one difficulty, namely to make the 'micro prints' easily legible. 1951 Amer. Documentation II. 151/1 Microprint is delivered in labelled, cloth slip-cases ready for shelving. 1961 T. LANDAU Encycl. Librarianship (ed. 2) 121/1 Microfilming and the production of microprint are camera methods. 1962 A. GÜNTHER Microphotogr. in Library (Unesco) 5 Micro-writing and micro-printing were known centuries ago. 1970 Brit. Printer Jan. 80/1 Trial runs have satisfied him that micro printing to a 10th of normal size..can be done successfully on a standard offset-litho press. 1970 New Scientist 31 Dec. 601/2 People of all ages have found that they can read microprint without strain.

microprism (məi·kropriz'm), a. Photogr. [f. MICRO- + PRISM.] Applied to an area of the focusing screen of some reflex cameras which is covered with a grid of tiny prisms and which splits up the image when the subject is not in focus; also applied to such a focusing system.
1966 H. KEPPLER Asahi Pentax Way 319 In using the central microprism grid for focusing, you will find that it snaps images in and out of focus with medium focal length lenses more efficiently than with either wide-angle or extremely long lenses. 1968 Newnes Compl. Amat. Photogr. (ed. 3) iv. 67 A new alternative [to a rangefinder] is the use of a 'microprism' zone, which gives a dotted effect when the image is out of focus. 1970 Amat. Photographer 11 Mar. 4 (Advt.), A very advanced camera offering a 6–1 zoom ratio and microprism focusing. 1973 Country Life 20 Sept. 826/1 (Advt.), Microprism focusing means speedy and accurate focusing.

microprobe, -process: see *MICRO- 6, 1.

microprogram (məi·kroprōugræm), sb. Computers. [f. MICRO- 1 b + PROGRAM, PROGRAMME sb.] A program that causes any machine instruction to be transformed into a sequence of microinstructions.
1953 WILKES & STRINGER in Proc. Cambr. Philos. Soc. XLIX. 230 The operation called for by a single machine order can be broken down into a sequence of more elementary operations... These elementary operations will be referred to as micro-operations. Basic machine operations, such as addition, subtraction, multiplication, etc., are thought of as being made up of a micro-programme of micro-operations, each micro-operation being called for by a micro-order. 1962 HUSKEY & KORN Computer Handbk. XVI. 32 Once a given set of instructions has been designed into the machine by the wiring of a microprogram plugboard, the machine may be used as a stored-program computer in the normal fashion. 1969 [see *MICROPROGRAMMED ppl. a.]. 1971 New Scientist 7 Jan. 27/1 This memory can hold up to 1024 words of microprogram which can be changed easily and inexpensively and tailored exactly to suit individual requirements.

microprogramming (məikroprōu·græmiŋ), vbl. sb. Computers. [f. MICRO- 2 a + *PROGRAMMING vbl. sb.] The technique of making machine instructions generate sequences of microinstructions in accordance with a microprogram, rather than initiate the desired operations directly, so that by changing the microprogram the set of possible machine instructions can be varied.
1953 WILKES & STRINGER in Proc. Cambr. Philos. Soc. XLIX. 230 It is also necessary that provision should be made for conditional micro-orders which play a role in micro-programming similar to that played by conditional orders in ordinary programming. 1962 HUSKEY & KORN Computer Handbk. XVI. 32 The concept of microprogramming suggests that the machine be designed and constructed in such a way that the individual basic command steps, such as clear the accumulator or left shift one place, are accessible for alteration. 1967 P. A. STARK Digital Computer Programming xix. 360 Microprogramming is a technique for letting the programmer manufacture complex instructions..from small portions of instructions. 1971 New Scientist 17 June 685/2 ICL is understood to be taking its own steps along the microprogramming path in its next range of computers.

So micropro·grammed ppl. a., employing microprogramming with a number of different microprograms; (as a back-formation) micropro·gram v. trans.; micropro·grammer, one who writes microprograms; a specialist in microprogramming.
1956 Jrnl. Assoc. Computing Machinery III. 79 One of the basic requirements of a microprogrammed computer is a means of ensuring the microprogrammer ready control of the subcommands with which he is working. It is necessary that the microprogram be physically easy to set up on the machine. 1957 Ibid. IV. 161 The arithmetic floating point operations of addition, subtraction and multiplication can be micro-programmed with about 70 micro-orders. 1969 P. B. JORDAIN Condensed Computer Encycl. 319 It is cheaper to establish a microprogram memory which can hold many microprograms and thus allow for a large set of computer instructions which can be interpreted and implemented on simple and inexpensive computer hardware. In such computers, sometimes called microprogrammed computers, each computer instruction operation code defines the beginning of the corresponding microprogram stored in the microprogram memory. 1970 S. S. HUSSON Microprogramming i. 13 The user programmer tells the system or device what to do by placing instructions in the high-speed main storage. The microprogrammer tells the system or device how to do it by controlling which storage and logic elements are used and how they are used for each operation. Thus the machine instruction which the programmer considered to be the lowest level of communication with the system can now be viewed as a closed subroutine broken down into a sequence of more elementary functions called microinstructions. Each microinstruction is designed to specify the control gates that are opened at a particular point during the machine cycle. 1975 Sci. Amer. May 36/3 Some multichip machines have the advantage that they can be 'microprogrammed' by the user.

microprojection (məi:kroprŏdʒe·kʃən). [f. MICRO- + PROJECTION sb.] The process of projecting an enlarged image of a microscopic specimen.
1904 Jrnl. R. Microsc. Soc. 582 The firm..manufacture an optical bench and appliances for..micro-projection, and optical lantern projection. 1932 Ibid. LII. 134 The microprojection of a subject for reconstruction work is best accomplished by using an apparatus affixed in the vertical position, projecting the image upon a horizontal table for ease in drawing. 1972 Jap. Jrnl. Pharmacol. XXII. 636 The reactive points were identified with the aid of microprojection apparatus to make the composite map from the serial sections obtained from individual animals.

Hence microproje·ctor, an apparatus for microprojection.
1932 Jrnl. R. Microsc. Soc. LII. 136 A method devised whereby a standard microprojector is altered to allow use in the vertical position. 1948 Times 15 June 7/5 He introduced into his lectures the latest methods of teaching, first by lantern slides, secondly by microprojector, and later by cine-films. 1957 B. M. ROVINSKY et al. in V. E. Cosslett et al. X-Ray Microsc. & Microradiogr. 277 The high resolution of the microprojector with its great sensitivity in distinguishing densities of microscopic areas makes it possible to use the device for examination of a wide range of objects.

micropsy: see *MICRO- 3.

micropublishing (məi·kropʌ:bliʃiŋ), vbl. sb. [f. *MICRO- 7 + PUBLISHING vbl. sb.] The publication of copies of books, periodicals, etc., in microform. Hence (as a back-formation) mi·cropublish v. trans.; also mi·cropublished ppl. a., micropu·blisher.
1969 R. & E. Coordinator (Res. & Engin. Council Graphic Arts Industry) Apr. 4/1 According to a recent booklet..micropublishing sales are running at an estimated $25 million a year. 1971 Publishers' Weekly 23 Aug. 41/1 A new company, Congressional Information Service, which collects, indexes and micro-publishes some 450,000 pages per year of government documents. 1973 Ibid. 19 Mar. 52/2 Many micropublished products are compilations of public domain materials. Ibid. 52/3 The micropublisher is particularly affected by this unofficial copyright. 1974 Bookseller 8 June 2600/1 The Oxford English Dictionary in two volumes with a magnifying glass is a significant half way stage to micro publishing.

micropulsation, -quantity, -rad: see *MICRO- 1, 1, 5 a.

microradiography (məi:krorēidiǫ·grăfi). [ad. F. microradiographie (P. Goby 1913, in Compt. Rend. CLVI. 686): see MICRO- 2 a and RADIOGRAPHY.] Radiography of the fine structure of an object.
1913 P. GOBY in Jrnl. R. Microsc. Soc. 373 The new method of employment of X-rays, which I have the honour of introducing to you under the name of Micro-radiography, aims at rendering easily visible the internal structure of microscopic objects. 1944 Jrnl. Appl. Physics XV. 44/1 The early work on microradiography was confined chiefly to light alloys and made use of low voltage, long wave-length radiation. 1966 D. G. BRANDON Mod. Techniques Metallogr. 76 The two major advantages offered by microradiography as opposed to ordinary optical metallography are, firstly, the sensitivity to chemical composition and, secondly, the ability of the technique to integrate over a large volume of material. 1973 Calcified Tissue Res. XI. 176 This work investigated the effect of fluoride on the mineralization of rat dentine using tetracycline-labelling and microradiography.

Hence microra·diogram, the original image obtained on a sensitive plate or film in microradiography; microra·diograph, a photographic enlargement of a microradiogram; mi:croradiogra·phic a., of or obtained by microradiography.
1913 Jrnl. R. Microsc. Soc. 375 Plate XVI is a reproduction of a micro-radiogram, enlarged nineteen times, of the anterior and posterior limbs of the three-toed lizard. 1944 Jrnl. Appl. Physics XV. 43 Investigations have been made of the various factors considered to influence the sensitivity of the microradiographic method for the examination of alloys. Ibid. 52 (caption) Microradiographs of cartridge brass with Co radiation at 20 kv at 75 ×. 1957 A. ENGSTRÖM in V. E. Cosslett et al. X-Ray Microsc. & Microradiogr. 32 In order to take care of the inhomogeneous distribution of the absorbing material, various types of scanning photometry of the microradiogram are now being used. 1966 D. G. BRANDON Mod. Techniques Metallogr. 76 The detail seen in a microradiograph is projected from a slab of metal up to 0·1 mm in thickness. 1969 HOBDELL & BOYDE in Möllenstedt & Gaukler Internat. Symposium X-Ray Optics & Microanalysis 611 The interpretation of the microradiographic image was found to be in good correlation with the interpretation of the scanning electron microscopic image of bone surfaces.

micro-ray, -reader, -record(ing), -relief, -reproduction, -respirometer (etc.): see *MICRO- 1, 7, 7, 1b, 7, 6.

micro-scale (məi·kroskēil). Also micro scale, microscale. [f. MICRO- 1 + SCALE sb.³] A small or microscopic scale; spec. in Chem., the scale of microanalysis.
Micro is freq. apprehended as an adj. qualifying scale (cf. *MICRO- 8).
1931 [see *MICRO- 8 b]. 1946 BELCHER & WILSON Qualitative Inorg. Microanalysis i. 3 This does not arise on the micro scale, because of the larger amounts of ethanol used to ensure complete precipitation. 1956 Nature 25 Feb. 378/2 Schindler's procedure, which was adapted to microscale working, was followed. 2–5 mgm. of vitamin B_{12} is dissolved in 500 µl. of 10 per cent aqueous ammonium chloride solution. 1961 New Scientist 16 Mar. 670/3 These materials are, however, crude on a microscale—they are porous and they contain impurities. 1964 N. G. CLARK Mod. Org. Chem. xxiv. 496 It became imperative to develop techniques demanding smaller samples. This was achieved in 1911 by Pregl, using 1–5 mg of material (micro-scale). 1966 D. G. BRANDON Mod. Techniques Metallogr. 103 It is now possible to perform a very large number of metallurgical operations in [sic] a microscale inside the electron microscope. 1968 [see *MACRO-SCALE].

microscope, sb. Add: **1. c.** An instrument analogous to an optical microscope in function but employing radiation other than visible light (e.g. electrons or X-rays). (Cf. electron microscope s.v. *ELECTRON² 2 b.)
1927 G. L. CLARK Appl. X-Rays i. 5 The ultraviolet microscope..discloses a fine structure which appears perfectly homogeneous under visible light rays. 1939 Electronics & Television & Short-Wave World XII. 637/3 The electron source..might be made completely independent from the microscope. 1957 DUNCOMB & COSSLETT in V. E. Cosslett et al. X-ray Microsc. & Microradiogr. 374 The purpose of the microscope is to form a picture of a surface by its X-ray emission and to analyze the elements in a selected volume of about one cubic micron in the surface by the characteristic lines emitted. 1964 G. H. HAGGIS et al. Introd. Molecular Biol. 346 Higher voltages on the microscope increase resolution, for a given section thickness, but reduce contrast. 1966 McGraw-Hill Encycl. Sci. & Technol. VIII. 371/2 In the electron, proton, x-ray, and β-ray microscopes, the image is usually recorded on a fluorescent screen or is photographed.

microscope, v. Add: So mi·croscoping vbl. sb., examination by microscope.
1868 G. H. LEWES Jrnl. 10 Jan. in Geo. Eliot Lett. (1956) IV. 416 With him I spent the greater part of the time at Bonn, discussing microscoping etc. 1919 S. PAGET Sir V. Horsley II. i. 143 He..spoke his mind against that sort of pathology which hardly gets beyond the microscoping and exhibiting of diseased organs.

microsecond: see *MICRO- 5 a.

microsegment (məi·krosegmĕnt). Linguistics. [f. MICRO- 1 + SEGMENT sb.] A unit of sound enclosed between two open junctures.
1958 C. F. HOCKETT Course in Mod. Ling. vi. 60 Apart from stresses, a microsegment consists of segmental phonemes—that is, vowels and consonants. 1963 Amer. Speech XXXVIII. 55 Miss Sivertsen describes the phonological system within which she operates; phonemes are described in a framework of macro- and microsegments which, in turn, are divided into syllables. 1971 Language XLVII. 739 Each section of the description is..worked out in detail..; the main headings are phonemes, syllables, micro-segments, meso-segments, macro-segments, and mega-segments.

microseism. Substitute for def.: A minor earthquake; in mod. use, any imperceptible disturbance of the earth's crust which is capable of being registered on a sensitive seismometer but which is not caused by an earthquake (see quots.). (Further examples.)
1903 [see *MACROSEISM]. 1924 Bull. Seismol. Soc. Amer. XIV. 28 They [sc. earth-tremors] are now generally known as microseisms. 1959 Observer 6 Sept. 4/8 Another advance is the linking of 'microseisms'—minute earth tremors which constantly vibrate the recording pens of seismographs—with storms at sea. 1965 A. HOLMES Princ. Physical Geol. (rev. ed.) xxv. 916 As a background to the P, S and L waves and their many associates there are small irregular earth tremors and quiverings going on all the time... These microseisms..set a limit to the degree of magnification that can usefully be employed, since they only confuse the earthquake record if they are made too big. Some of the more conspicuous microseisms are caused by distant traffic, others by the pounding of breakers on rocky coasts, while others have been traced to changes of atmospheric pressure and especially to hurricanes and typhoons. But after all such regional increases of microseismic activity have been accounted for, there still remains a world-wide background of chaotic seismic 'noise'. Don and Florence Leet have suggested that these microseisms are caused by the strained condition of the crust, which 'hums' or 'sings' like a highly strained piece of steel. 1971 Nature 12 Feb. 452/3 Microseisms are of practical importance in predicting the onset of a storm in areas where meteorological observations are scarce.

microseismic, a. Add: (See also quot. 1972.) (Further examples.)
1972 Gloss. Geol. (Amer. Geol. Inst.) 452/2 Microseismic data, earthquake measurement or observation by instrumental means, as opposed to macroseismic observations. The term is not to be confused with the connotation of the term microseism. 1973 Sci. Amer. Apr. 31/1 It was a great thrill when we scanned along a seismogram made at Jamestown of an explosion on Novaya Zemlya and there—Eureka!—at the travel time predicted for a wave reflected seven times was an unmistakable tiny pulse nestling in the valley of microseismic background noise.

microshift, -skirt(ed): see *MICRO- 1 c.

microsleep (məi·kroslīp). [f. MICRO- 1 + SLEEP sb.] A transitory state of sleep, esp. in a person deprived of his normal sleep; a period or occasion of such sleep.
1945 W. T. LIBERSON in Digest Neurol. & Psychiatry XIII. 106 Another feature which we described is 'micro-sleep', a paroxysmal sleep of 1–10 seconds. 1959 Jrnl. Amer. Med. Assoc. 5 Sept. 14/1 Sleep deprivation brings with it an increasing burden of drowsiness..and brief lapses of awareness or 'microsleeps'. 1969 Sunday Times (Colour Suppl.) 16 Feb. 44/4 Tiny seizures of sleep known as microsleeps and lasting two or three seconds have occurred in people deprived of sleep. 1970 Sci. Jrnl. May 14/3 The subjects developed serious anomalies in brain wave pattern accompanied by 'microsleep'—a compulsive tendency to drop off unless kept constantly active. 1974 Times 3 Dec. 3/1 A driver can be asleep for short periods without knowing it and without anyone else being aware of it... A neurologist..described the state as 'micro-sleep'. .. They were in fact asleep for periods lasting from five to 10 seconds.

microsmatic (məi:krǫzmæ·tik), a. Zool. [f. MICR(O- + Gr. ὀσμ-ή smell + -ATIC.] Having poorly developed olfactory organs.
1890 W. TURNER in Jrnl. Anat. XXV. 106, I propose..to arrange the Mammalia in relation to the development of the olfactory apparatus into three groups:—(a) Macrosmatic... (b) Microsmatic, where the olfactory apparatus is relatively feeble... (c) Anosmatic. 1891 Proc. Zool. Soc. 582 The olfactory bulbs and ethmo-turbinals are present, but only moderately well-developed in this animal [sc. the platypus]; it therefore belongs to the group named by Sir W. Turner..Microsmatic. 1962 Science Survey III. 260 In the microsmatic bats and in man only a very small area [of the nasal labyrinth] is olfactory. 1971 Nature 6 Aug. 396/2 Olfactory communication is now known to be important in determining the interaction between the sexes in a microsmatic higher primate, namely, the rhesus monkey.

microsocial (məikrosōu·ʃăl), a. [f. MICRO- + SOCIAL a. and sb.] Of or pertaining to a small society or community.
1909 MRS. C. WEEKES tr. Goll's Criminal Types in Shakespeare iii. 134 Crimes committed by her..are looked upon as typical..of micro-social family assertion, arising out of her love for the little group. 1969 A. COCKBURN in Cockburn & Blackburn Student Power 10 Within this framework the personnel officer can bring to bear his knowledge of micro-social dynamics.

microsociological, -sociology: see *MICRO- 2 a.

microsomal (məikrŏsōu·măl), a. Biol. [f. MICROSOM(E + -AL.] Of or pertaining to microsomes.
1897 Jrnl. R. Microsc. Soc. 22 The fibrils around the centrosomes have an exquisite microsomal structure. The microsomes occur in groups on the fibrils at equal distances from the centrosome. 1955 Federation Proc. XIV. 262/1 It appears that all the microsomal RNA is in the small particles. 1970 [see *MICROSOME b]. 1971 Nature 9 July 84/1 Perhaps these ribosomes also make microsomal membrane proteins.

microsome. Add: **b.** Any of the particles which constitute the lightest fraction obtained by ultracentrifugation of cell contents under specific conditions and which are believed to be formed from fragmented endoplasmic reticulum and attached ribosomes; also (esp. formerly), a ribosome in an intact cell.
1943 A. CLAUDE in Science 21 May 453/2 In order to differentiate the small particles from the other, already identified elements of the cell, it may be convenient in the future to refer to this new component under a descriptive name which would be specific. For this purpose the term microsome appears to be the most appropriate. The term microsome..was applied originally by Hanstein (1880) to any granules, as seen in living protoplasm. The use of the word was progressively narrowed down, being retained as a general term to designate any small granules of undefined nature. Under these conditions, it seems proper to suggest that the term microsome..should be restricted to designate the small particles exclusively. 1955 Federation Proc. XIV. 262/1 It was found that the microsomes are morphologically identical with the vesicular and tubular elements of the endoplasmic reticula (ER) of intact cells. 1957 C. P. SWANSON Cytol. & Cytogenetics (1958) ii. 32 (caption) The structure of the cytoplasm of rat liver cells as revealed by electron microscopy, showing the endoplasmic reticulum..and the attached particles (microsomes) which are between 100 and 200 A in diameter. 1960 New Biol. XXXI. 30 The microsomes themselves, which are obtained after the cell has been broken down, are vesicles made of two components: a membrane, which is rich in proteins and lipids, and very small granules... These small particles contain as much as 40–50 per cent RNA and..most of the cellular RNA is present in them. 1968 L. L. LANGLEY Cell Function (ed. 2) i. 9 The endoplasmic reticulum is seen..to form a series of small canals

through the cytoplasm... Closely associated with the membranes which line the canals are tiny granules termed microsomes. Because they contain such a high concentration of RNA, they are often referred to as ribosomes. 1970 Ambrose & Easty *Cell Biol.* v. 164 Microsomes are in fact small spherical vesicles formed from disrupted endoplasmic reticulum. The microsomal fraction of homogenized cells may also contain ribosomes.

microspecies (məi·krospī:ʃīz, -spī:sīz). *Taxonomy.* [f. Micro- + Species *sb.*] A species differing only in minor characters from others of its group, often one of limited geographical range forming part of an aggregate species.

1916 J. P. Lotsy *Evolution by Means of Hybridization* i. 22 They have subsequently been called: mikrospecies, Jordanian species, subspecies, small species or elementary species indiscriminately. 1922 G. C. Druce *Flora Zetlandica* in *Bot. Exchange Club* VI. Suppl. 45 Twenty-two micro species of Hieracia are found in the Faroes. 1946 *Nature* 19 Oct. 535/2 The genus *Psalliota*..consists of 'species' or 'microspecies' extremely difficult to separate one from another. 1953 E. Mayr et al. *Methods & Princ. Syst. Zool.* ii. 39 Although conventionally referred to as races, reproductively isolated chromosomal populations are more logically designated (micro)species. 1963 Davis & Heywood *Princ. Angiosperm Taxon.* xiii. 425 In large critical groups of species the difficulty of deciding to which species various subspecies should be subordinated encourages the taxonomist (often against his inclinations) to maintain numerous microspecies.

microspectrograph (etc.), **-spectrophotometer** (etc.), **-sphere:** see *micro- 6, 6, 1.

microspore. **2.** Substitute for def.: The smaller of the two kinds of spores in heterosporous cryptogams; also, the homologous structure in seed plants (i.e. the immature pollen grain). (Add further examples.)

1964 E. J. H. Corner *Life of Plants* x. 177 The male spores [in *Selaginella*, etc.], called microspores because of their small size, are produced in large numbers in microsporangia. 1965 K. Esau *Plant Anat.* (ed. 2) xviii. 540 A microspore develops into the male gametophyte, the pollen grain.

Hence **mi:crosporoge·nesis**, the development of microspores; **mi:crosporogene·tic** *a.*

1946 *Nature* 12 Oct. 520/1 In the microsporogenesis of buckwheat autotetraploids during the first metaphase of meiosis only very rarely were exclusively quadrivalents observed. 1961 *Developmental Biol.* III. 241 Microsporogenesis in the higher plants offers an opportunity for analyzing various stages of differentiation within the essentially closed system afforded by the microsporocyte. 1973 *Nature* 6 July 35/2 The cold treatment slowed down the microsporogenesis considerably, enabling the microsporogenetic stages to be clearly recognized.

microsporidian (məikrospori·diăn), *sb.* and *a.* *Zool.* [f. mod.L. name of order *Microsporidia* (G. Balbiani 1883, in *Jrnl. de Micrographie* VII. 349), f. Micro- 1 + Spore + Gr. dim. suff. -ιδιον.] **A.** *sb.* A protozoan parasite affecting arthropods or fishes, belonging to the order Microsporidia. **B.** *adj.* Of or pertaining to a parasite of this kind.

1910 *Encycl. Brit.* IX. 386/1 The genus *Myxocystis*..has been shown..to be a true Microsporidian. 1912 E. A. Minchin *Introd. Study Protozoa* xvi. 411 The existence of the polar capsule in the Microsporidian spore was discovered by Thélohan. 1930 R. R. Kudo in Hegner & Andrews *Probl. & Methods of Res. in Protozool.* xxxiii. 327 The spore is perhaps the most important stage of a microsporidian from the taxonomic standpoint. 1964 T. C. Cheng *Biol. Animal Parasites* v. 141/1 The infective form of microsporidians is the spore, which is typically covered by a keratinous membrane. 1972 Swan & Papp *Common Insects N. Amer.* 291 Silkworms suffer from fungus, virus, and microsporidian (protozoan) infections.

microsporidiosis (məikrosporidiō·sis). [f. *microsporidi(an *sb.* and *a.* + -osis.] = *nosema.

1911 Fantham & Porter in *Proc. Zool. Soc.* 626 Microsporidiosis (due to *Nosema apis*) had probably been introduced from the Continent into British apiaries. 1937 W. Herrod-Hempsall *Bee-keeping* II. 1425 Nosema Disease (Microsporidiosis).

microstate, -stimulation, -stratification: see *micro- 1, 1 b, 1.

microstrip (məi·krostrip). *Electr.* [f. *micro(wave + Strip *sb.*²] A transmission line for microwaves that consists of dielectric material with a metallic film forming a conducting strip along one face and a metallic coating that serves as an earth covering the opposite face; also *collect.* (without *a*), as a material.

1952 Grieg & Engelmann in *Proc. IRE* XL. 1644 (*heading*) Microstrip—a new transmission technique for the kilomegacycle range. *Ibid.* 1645/1 Because of the ease of manufacture and the apparent similarity to conventional wiring, the generic name of microstrip has been given to this transmission system. 1957 *Jrnl. Appl. Physics* XXVIII. 299/1 The 'microstrip' is a strip transmission line which consists of a thin metallic strip pasted on a dielectric layer which is in turn a coating on a ground plane. *Ibid.*, Sections of microstrip are useful as microwave filters. 1967 *Electronics* 6 Mar. 251/2 The microstrip is formed on the alumina board by depositing a silver circuit pattern and ground plane. 1971 *New Scientist* 1 July 26/1 A traditional waveguide must have dimensions comparable to the wavelength of the microwave signal it is handling—a few centimetres... No such physical restrictions apply to microstrip, which has the added advantage of a constructional technique compatible with modern microelectronics. *Ibid.* 26/2 The one [earth] plate can form the second conductor for all the microstrips used in the receiver.

mi·crostructure. Also micro-structure. [f. Micro- + Structure *sb.*] Structure on a microscopic or very small scale; = *fine structure 2.

1885, 1898 [in Dict. s.v. Micro- 2 b]. 1956 *Nature* 25 Feb. 380/2 The inhomogeneous microstructure of the muscular tissue. 1959 B. Wall tr. *Teilhard de Chardin's Phenomenon of Man* I. i. 50 We owe our knowledge of the macro-structure and micro-structure of the universe far more to increasingly accurate measurements than to direct observations. 1960 E. H. Gombrich *Art & Illusion* viii. 274 The clue of texture..is basically also a clue of regularity and one which proves so reliable because the microstructure of things is least affected by accidents. 1962 F. I. Ordway et al. *Basic Astronautics* iv. 153 Some hope to learn from gravitational research more about the microstructure of the building blocks of nature, the subatomic particle. 1973 *Sci. Amer.* Feb. 65/2 It is now customary to apply the term 'microstructure' to oceanic physical processes on the scale of a few centimeters or less.

Hence **microstru·ctural** *a.*, of or pertaining to the microstructure; **microstru·cturally** *adv.*

1893 A. Geikie *Text-bk. Geol.* (ed. 3) 596 *Macro-structural* metamorphism, having the external structure (morphology) changed, as where an amorphous condition becomes schistose; *micro-structural*, having the internal structure (histology) wholly changed, with or without a macro-structural alteration. 1937 O. B. Darbishire tr. *A. von Buzágh's Colloid Syst.* iv. 45 With increasing crystal size the microstructural composition asymptotically approaches the stoichiometric composition. *Ibid.*, The solution of this paradox is found in a 'microstructurally correct computation'. 1966 C. R. Tottle *Sci. Engin. Materials* x. 224 Steels are susceptible to microstructural changes at the temperatures required for stress relief, and for this reason relief is often avoided.

microsurgery (məikrosɜ̄·ɹdʒəri). [f. Micro- 2 a + Surgery.] Manipulation (as by injection, dissection, etc.) of individual cells with the aid of microscopy; surgery of such delicacy as to necessitate microscopy.

1927 *Protoplasma* II. 203 Micro-surgery of plant cells shows that the impression of fluidity which one gets from mere microscopic observation is illusory. 1928 E. V. Cowdry *Special Cytol.* i. 7 With the apparatus of Chambers it is feasible to isolate a cell, to take it up with a pipette, to incise part of it and to inoculate into its cytoplasm a small amount of fluid. This microsurgery has been applied to the investigation of many problems. 1941 *Bot. Rev.* VII. 355 Harder (1926), by microsurgery, isolated the penultimate cell of a growing hypha of *Coprinus sterquilinus*. 1960 *McGraw-Hill Encycl. Sci. & Technol.* IV. 571/2 As gastrulation and embryo formation proceed, microsurgery becomes more and more the instrument of choice for analysis. The defect experiment tests the effect of removal of one part or rudiment. 1971 *Nature* 2 July 28/1 The techniques of microsurgery on single cells have been applied to a variety of cell types. 1973 *Daily Colonist* (Victoria, B.C.) 10 June 28/2 Microsurgery has been used in brain operations for the last five years, beginning in Europe.

So **microsu·rgical** *a.*

1963 *Science* 5 July 46/3 (*heading*) Ruby laser as a microsurgical instrument. 1968 H. Harris *Nucleus & Cytoplasm* iv. 81 Nuclei which were divested of virtually all their cytoplasm by microsurgical procedures none the less retained their ability to produce specific puffing patterns. 1974 *Nature* 29 Mar. 450/1 Microsurgical removal of the root cap of *Zea* has been shown to prevent the response of the roots to gravity.

microswitch, -syringe, -system, -technique, -tektite: see *micro- 1, 6, 1, 6, 1.

microtelephone (məikrote·lĭfōᵘn). *Teleph.* [f. Micro- + Telephone *sb.*] **1.** Any of various modifications of the Bell telephone transmitter which were supposed to render it more sensitive.

1879 J. Ochorowicz in *Nature* 27 Mar. 482/2 The microtelephone is regulated once for all, and transmits the feeblest word with a truly perfect precision. 1879 T. A. Edison *U.S. Pat.* 222,390 9 Dec. 311/1 This invention I term the 'micro-telephone' in consequence of the same responding to minute vibrations.

2. [f. Micro(phone.] = *handset.

1895 A. R. Bennett *Telephone Syst. Europe* xvii. 296 Usually the transmitter and receiver are attached to the same handle, in 'micro-telephone' form. 1930 [see *handset].

microtext, -texture, -theory: see *micro- 7, 2 b, 1 b.

microtine (məi·krotĭn, -əin), *sb.* and *a.* *Zool.* [f. mod.L. subfamily name *Microtinæ* (f. generic name *Microtus* (F. von P. von Schrank *Fauna Boica* (1798) I. 66), f. Micr(o- + Gr. ὠτ-, οὖς ear): see -ina².] **A.** *sb.* A member of the rodent subfamily Microtinæ, which includes voles and lemmings. **B.** *adj.* Of or pertaining to this subfamily.

1926 M. A. C. Hinton *Monogr. Voles & Lemmings* I. 16 Some of the most characteristic features of the Microtine skull are shown in longitudinal vertical sections. 1936 *Proc. Prehist. Soc.* II. 53 In the case of the microtine mammals—another rapidly evolving group—detailed work along these lines has been done. 1948 A. L. Rand *Mammals E. Rockies* 165 In some areas at least in Canada the scarcity of microtines seems to cause a scarcity of certain furs. 1956 *Nature* 10 Mar. 445/2 A recognizable description of an Indian mole-vole..of the genus *Ellobius*, a microtine. 1972 T. A. Vaughan *Mammalogy* xv. 190 (*caption*) Population densities of several species of Microtines.

mi·crotome, *v.* [f. the sb.] *trans.* To cut in sections with a microtome. So **mi·crotomed** *ppl. a.*

1893 *Brit. Med. Jrnl.* 23 Sept. 685/2 The divided roots were microtomed and examined in serial sections. 1898 *Phil. Trans. R. Soc.* B. CXC. 95 The following nerves of muscles were microtomed for detection of some fibres. 1972 *Physics Bull.* Nov. 668/1 Evidence for twisting fibrils has also been obtained from transmission electron micrographs of microtomed specimens.

microtone (məi·krotōᵘn). *Mus.* [f. Micro- 1 + Tone *sb.*] An interval smaller than a semitone.

1920 *Outward Bound* Oct. 77/1 The ancient Greeks also recognised and used these microtones. The Greeks found twenty-four in the octave and the Indians usually recognise twenty-two. 1927 *Observer* 23 Oct. 14 But these microtones (seventy-two in the octave) give quarters of the untempered chromatic semitone. 1946 *Mod. Music* Spring 113 The Oriental sound of Hovhaness' music is realized with occidental instruments not by preparing the instruments, nor by employing microtones, but solely through the character of the melody. 1957 Manvell & Huntley *Technique Film Music* iii. 170 Volume was controlled by varying the exposure;..portamento by a rapid series of micro-tones. 1970 W. Apel *Harvard Dict. Mus.* (ed. 2) 527/1 Long a structural feature of Asian music, the use of microtones in Western music, although far from new, has been..far less extensive.

Hence **microto·nal** *a.*, of or pertaining to a microtone or microtones; employing or producing microtones; **microtona·lity**; **microto·nally** *adv.*

1942 *Scrutiny* XI. 6 The quarter-tone or *sruti* is the microtonal interval between notes of the 'scale'. 1946 R. Blesh *Shining Trumpets* (1949) v. 106 The microtonally flatted fifth also occurs. *Ibid.* x. 234 The blues scale..gives all the opportunity needed for microtonality. 1946 *Mod. Music* Spring 113 The microtonality of Alois Haba and others presents an Oriental characteristic not found in any of the music discussed above. 1959 *Chambers's Encycl.* IX. 627/2 The microtonality of the Czech Alois Haba (1893–), who cut the scale into ½- and ⅓-tones. 1967 *Times* 29 May 6/4 Glissandos on the microtonal harp were pure magic. 1974 *Country Life* 24 Jan. 136/1 Mozart..as an infant..could remember microtonal differences of tuning in instruments heard weeks previously.

microtopography (məikrotǫpǫ·grăfi). [f. Micro- 2 a + Topography.] The surface features of a material, or of the earth or other body, on a small or microscopic scale.

1956 *Nature* 24 Mar. 564/1 Prof. Tolansky selected as typical subjects for examination some of his own studies on the optical testing of surface finish, of microtopography in general, the examination of micro-hardness and directional hardness [etc.]. 1962 F. I. Ordway et al. *Basic Astronautics* iii. 65 Photometric observations..suggest that the microtopography [of the moon] is rough, probably being pitted. 1969 *Ecology* L. 740 Lichen cover and microtopography were positively associated.

Hence **mi:crotopogra·phic, -topogra·phical** *adjs.*

1958 *New Biol.* XXVI. 40 Even in apparently flat and uniform grassland, the distribution of the buttercup species is usually found to be associated with microtopographical features which determine slight differences in the rate of drainage and the speed with which the soil dries out during drought periods. 1960 S. Tolansky *Surface Microtopogr.* vi. 71 Its [sc. diamond's] unique hardness makes it an ideal material for microtopographic studies. 1971 *Nature* 22 Jan. 248/1, I have had occasion recently to re-read Goethe's *Theory of Colours* (1810) and have found that he described the nature of the microtopographical interference pattern given by a crystal surface.

microtrichium: see *micro- 1.

microtron (məi·krotrǫn). [f. *micro(wave + *-tron.] A variant of the cyclotron for accelerating electrons by passing them repeatedly through a cavity in which they are accelerated by microwaves, the amount of the acceleration on each passage being such as to allow for the increase in their time of revolution that results from their relativistic increase in mass.

1946 *Rev. Sci. Instruments* XVII. 9/1 The microtron, suggested by J. S. Schwinger and by V. Veksler, is an electron cyclotron in which the particles slip one cycle of phase on each circuit through a constant magnetic field. 1958 *Times* 21 June 4/7 It is called the 'microtron' because the electrical pulses used in acceleration are got from a pulsed microwave oscillator. 1974 *Nature* 1 Feb. 250/3 A prototype continuous microtron (a cyclic accelerator in which the electrons are accelerated by microwaves) has been constructed in Obninsk... Until now, it has proved possible to operate microtron accelerators in a pulsed mode only, with a relatively low repetition frequency.

microtubular, -tubule: see *micro- 1.

microunit (məi·kro-, mi·kroyū:nit). [f. Micro- 5 a + Unit *sb.*] A millionth part of a unit, esp. of an international unit (as of insulin).

1900 Dorland *Med. Dict.* 392/1 *Micro-unit*, a unit of small measurements. 1911 Stedman *Med. Dict.* 533/1 *Microunit*, the millionth of an ordinary unit, such as a meter, gram, ohm, etc. 1959 *Nature* 21 Nov. 1649/2 Two to three micro-units of human insulin were detectable in the experiment depicted. 1962 *Lancet* 13 Jan. 73/2 This difference is highly significant statistically..whether the figures are considered at the level of net glucose uptake or after conversion into microunits of insulin per millilitre.

microvascular, -vegetation, -viewer, -villus, -watt: see *micro- 4, 1, 7, 1, 5 a.

microwave (məi·krowēⁱv). [f. Micro- 1 + Wave *sb.*] **1.** An electromagnetic wave with a wavelength between about one millimetre and 30 centimetres (corresponding to a frequency between 300 gigahertz and one gigahertz); one whose wavelength is such that it is convenient to use hollow waveguides for its transmission.

The precise figures taken as extremes of wavelength or frequency are arbitrary and vary somewhat with different writers; formerly waves much longer than the present-day maximum were described as microwaves.

1931 *Telegraph & Telephone Jrnl.* XVII. 179/1 When ..trials..with wavelengths as low as 18 cm. were made known, there was undisguised surprise..that the problem of the micro-wave had been solved so soon. 1935 *Proc. IRE* XXIII. 1503 (*heading*) Diffraction of microwaves over the curve of the Earth. *Ibid.*, Radio waves below ten meters in length, at the present time termed 'microwaves', may be received at moderate distances over the horizon from the transmitter. 1940 *Amat. Radio Handbk.* (ed. 2) xvi. 232/1 Wavelengths below 1 metre are referred to as 'micro-waves' or 'centimetre waves', and demand special treatment. 1942 J. C. Slater *Microwave Transmission* 1 The range of waves called microwaves: wave lengths perhaps in the range from a half meter to a few centimeters. 1965 *New Scientist* 10 June 724/3 Microwaves, or waves of even shorter wavelength, must be used because it is only when such short wavelengths are employed that a highly directional beam can be produced. 1968 *Maclean's Mag.* Dec. 19/1 Signals from the satellite will be received by ground stations and relayed by microwave or cable to home receivers. 1973 *Sci. Amer.* Sept. 74/2 Electromagnetic radiation at microwave frequencies is now widely used for cooking purposes. It may also be practical to use microwaves to 'cook' agricultural pests in the soil: insects, weeds, fungi and so on.

2. a. *attrib.*

1933 *Electrician* 6 Jan. 3/1 We decided..to concentrate our efforts on the generation and efficient radiation of what may be termed a medium wavelength on the microwave scale—that is, a wavelength of the order of half-a-metre, i.e., 600 000 k.c. 1946 *Proc. IRE* XXXIV. 775/1 The extent of the microwave spectrum has been variously defined. This paper will concern itself with the range of frequencies between approximately 2000 and 30,000 megacycles per second. 1946 *Jrnl. R. Aeronaut. Soc.* L. 956/1 During the war airborne microwave radar was used to provide in the aircraft a map or picture showing the main features, e.g., towns and rivers, of the country below. 1947 *Electronic Engin.* XIX. 17 Ultra-short and microwave radio links. 1952 *Sci. News* XXV. 30 From a study of such rotational spectra of asymmetric molecules in the microwave region the distance between the atoms can be found very accurately. 1957 *B.B.C. Handbk.* 59 The vision signals from remote outside broadcast points are carried back to the main television network by BBC microwave or VHF radio links. 1959 *Daily Tel.* 18 May 12/6 The sky platform would hover in a fixed position. Helicopter type rotary wings would be turned by microwave energy beamed from stations on earth. 1964 *Ann. Reg. 1963* 163 The microwave telecommunications network linking Ankara, Karachi and Teheran. 1967 D. Wilson in Wills & Yearsley *Handbk. Managem. Technol.* iii. 44 Data transmission is concerned with the transmission of data over private or G.P.O. lines or by using microwave techniques. 1971 *Brit. Printer* Jan. 67/2 Microwave dryers are currently at an advanced stage of development. 1972 *Sci. Amer.* Feb. 13/1 Although the microwave region of the electromagnetic spectrum is not precisely defined, we use the term to describe radiation of wavelengths ranging from 30 centimeters to three millimeters.

b. Special Comb.: **microwave oven,** an oven in which food is cooked by passing microwaves through it, the resulting generation of heat inside the food making rapid and uniform cooking possible.

1965 *Economist* 22 May 941/2 (*heading*) Microwave ovens. A meal a minute. 1968 *New Scientist* 24 Oct. 175/1 The microwave oven..is now coming into quite wide use in homes in the United States. 1973 *Daily Tel.* 26 Apr. 11/8 Microwave ovens which are still in the experimental stage, would take eight minutes to produce a standard loaf, instead of 30.

microweld (məi·kroweld), *v.* [f. Micro- 1 b + Weld *v.*] *trans.* To join by a very small weld. So **mi·crowelded** *ppl. a.*, **mi·crowelding** *vbl. sb.*

1962 *Flight Internat.* LXXXII. 1027/1 According to Aerojet-General Corporation, a micro-welding device developed by the Corporation's Astrionics Division 'will make a reliable weld in the time it takes you to dot an "i"... But the dot of your "i" will be 30 times bigger than the weld.' 1963 E. Keonjian *Microelectronics* 380 Micro-welded connections. 1968 *Sci. Jrnl.* Nov. 17 (*caption*) The microwelding machine below is used to bond leads to the circuit. 1970 *New Scientist* 13 Aug. 338/2 An army of girls some 30 000 strong is employed..in micro-welding hair-sized connections from each silicon chip. 1975 *Country Life* 3 Apr. 838/1 The use of microwelding—an engineering skill put to very effective use as in a bracelet that seems to be a web of fine interlaced wires.

microwire, -world, -zone: see *micro- 1.

micrurgy (məi·krɜ̄ɹdʒi). [f. Micr(o- + *-urgy* after Metallurgy.] The performance of delicate manipulations under the microscope, esp. on biological material such as individual cells. So **micru·rgical** *a.*

1927 *Protoplasma* II. 189 (*heading*) The structural organization of plant protoplasm in the light of micrurgy. *Ibid.* 191 The extensive micrurgical researches that have been made on animal cells. 1935 *Industr. & Engin. Chem.* (*Analytical Ed.*) VII. 218/1 The word 'micrurgy' was coined by Peterfi..to describe biological microdissection and later expanded by Titus and Gray..to include dissection and examination under the microscope of non-biological material. Since the term, by its pure Greek derivation, means 'operations on a small scale or work with minute quantities', it can logically be used as a general term to include microchemistry, microanalysis, and chemical microscopy... It is here so used, and is suggested for general adoption. 1960 *New Scientist* 20 Oct. 1049/1 With the new technique of micrurgy the gap between the living cell and the micrologist had been bridged. 1973 *Nature* 2 Mar. 47/2 For micrurgical manipulation, two- or eight-celled eggs from superovulated BALB/c mice were deposited into a suspension of somatic cells and virus in an 'egg-well'.

mid, *a.*, *sb.*¹, and *adv.* Add: **A.** *adj.* **1. c.** (Further examples.)

1898 G. B. Shaw *Perfect Wagnerite* 34 A..melodic bogey to mid-century ears. 1906 *Dialect Notes* III. 146 The mid-term examinations will begin the last of January. 1923 D. H. Lawrence *Birds, Beasts & Flowers* 103 A disgusting bat At mid-morning. 1928 T. S. Eliot *For Lancelot Andrewes* 76 Arnold turned from mid-century Radicalism. 1938 *Encycl. Brit. Bk. of Year* 700/1 There are 1,200 firms in Britain now operating such amenities as a ten-minute mid-morning break. 1951 I. Shaw *Troubled Air* xv. 248 The mid-afternoon coffee was put up in containers. 1952 C. P. Blacker *Eugenics* 259 At about mid-pregnancy, identical twins differ more in size than do fraternal. 1953 *Manch. Guardian Weekly* 1 Oct. 1 The Administration party loses seats at the 'mid-term' elections. 1958 P. Shore in N. Mackenzie *Conviction* 23 All over Britain a new mid-twentieth-century society is coming vigorously to life. 1959 P. Townsend *Died o' Wednesday* xii. 216 Chance customers who dropped in for mid-morning coffee, light lunches and afternoon tea. 1961 *Lancet* 5 Aug. 280/1 The gradual change in uterine activity begins as early in pregnancy as midterm. 1968 A. Diment *Gt. Spy Race* iv. 53 The obscure type faces which these mid-century tailors use to advertise their dens. 1972 J. McClure *Caterpillar Cop* ix. 144 The last lesson before mid-morning break. 1973 *Guardian* 19 Apr. 14/2 The people who fear it [sc. the Watergate scandal] will hurt them are Republican candidates for next year's mid-term elections.

d. *mid-flight, -race, -sentence, -stride*; **mid-Atlantic,** (*a*) the middle of the Atlantic Ocean; (*b*) something that has both British and American characteristics, or is designed to appeal to both the British and the Americans; also *attrib.* or as *adj.*; **mid-band** *a. Electronics*, of or pertaining to the middle of a band of frequencies; **mid-calf,** (*b*) *attrib.* or as *adj.*, describing a garment that reaches half-way down the calf of the leg; **mid-cycle** *a.* and *sb. Physiol.*, (occurring at) the middle of the menstrual cycle; **mid-main** *poet.*, mid-ocean; **mid-square** *a. Math.*, describing a method of generating a pseudorandom sequence of digits by squaring an arbitrary large number, taking the middle digits of the result (usu. half the total number of digits, a zero being added at the left if necessary to make the total even), and using these as the first digits of the series and as the number to be squared to provide the next digits, and so on indefinitely; **mid-town, midtown** chiefly *U.S.*, the middle of a town; a central area in a town or city; freq. *attrib.*

1892 'Mark Twain' *Amer. Claimant* 65 Two shipments would meet and part in mid-Atlantic. 1897 *Proc. Zool. Soc.* 351 *Monachus*, the Seal of the Mid-Atlantic. 1940 G. Greene *19 Stories* (1947) 184 Central heating gave it the stuffy smell of mid-Atlantic. 1957 M. Swan *Brit. Guiana* i. 25 He was drowned..when the ship in which he was returning to England from the United States was wrecked in mid-Atlantic. 1958 *Spectator* 7 Feb. 164/1, I crossed swords with the redoubtable Douglas Fairbanks Jr. on whether or not his television films could justly be called 'British'..; my argument was that as they aim at the American—or 'mid-Atlantic'—market, they do not deserve the name. 1962 *Radio Times* 22 Nov. 41/2 A spell in Hollywood where he was sent by the studio to which he was under contract to acquire a 'mid-Atlantic' accent. 1970 *Observer* 8 Feb. 30/6 There's another sameness too, a mid-Atlantic look. 1956 *Nature* 25 Feb. 392/2 At a mid-band wave-length of 3·2 cm., an input-voltage standing-wave ratio of 0·9 or better can be achieved over a 10 per cent band-width. 1962 Simpson & Richards *Physical Princ. Junction Transistors* xiii. 300 The mid-band voltage gain. 1967 *Harper's Bazaar* Sept. 45 The mid-calf length for day. 1969 *Guardian* 30 July 7/1 Courrèges..does in fact show one or two midcalf dresses for late day. 1974 *Country Life* 24 Jan. 181/1 Slim, mid-calf dresses in pin stripes. 1951 C. K. Weichert *Anat. Chordates* ix. 399 This is the condition at the time of ovulation, and the endometrium is said to be of the mid-cycle type. 1952 E. S. Taylor *Man. Gynecol.* iii. 19 At mid-cycle a ripened ovum is extruded from the ovary. 1965 J. H. Burn *Lect. Notes Pharmacol.* (ed. 8) 93 The function of the progestogen is to prevent 'break-through' bleeding in the mid-cycle. 1974 Passmore & Robson *Compan. Med. Stud.* III. xxviii. 33/2 These pills relieve dysmenorrhoea, premenstrual tension, midcycle pain or bleeding, and regulate previously irregular periods. 1896 Midflight [in Dict., sense A. 1]. 1932 W. Faulkner *Light in August* v. 98 It did not vanish in midflight. 1948 R. Graves *Coll. Poems* 219 That sea-birds of all sorts that flock About the Bass, repeatedly Collide in mid-flight. 1969 *Listener* 6 Feb. 163/1 (*caption*) The Lunar Excursion Module (centre) after mid-flight reassembly of the spacecraft has taken place, to allow two astronauts to enter the module from the main unit. 1862 G. M. Hopkins *Vision of Mermaids* (1929), Mermaids..ring the knells Of seamen whelm'd in chasms of the mid-main. 1959 *Times* 5 Oct. 4/7 It is easy to criticize Eldon for his wasteful short bursts in midrace. 1901 H. G. Wells *First Men in Moon* xiii. 10? I looked up, and stopped in mid-sentence. 1967 *Coast to Coast 1965–66* 30 She paused in mid-sentence. 1951 *Appl. Math. Ser. Nat. Bureau of Standards* (U.S.) XII. 33 (*heading*) The mid-square method of generating digits. 1968 P. A. P. Moran *Introd. Probability Theory* i. 46 The midsquare method has, however, been shown to be unsatisfactory. 1932 W. Faulkner *Light in August* xix. 437 The one stopped in the act of crouching from the leap, the other in midstride of running. 1957 T. Hughes *Hawk in Rain* 51 And his foot hung like Statuary in mid-stride. 1934 Webster, Mid-town (s.v. *mid-*). 1952 S. Kauffmann *Philanderer* (1953) ix. 157 They went to the roof garden of a midtown hotel. 1959 J. Cary *Captive & Free* 41 The mid-town terraces which can and have so easily become slum tenements. 1963 *Listener* 31 Jan. 202/2 In the very middle of mid-town, just off (and even just on) Broadway, the whole street is sometimes used as an open-air loading bay and temporary warehouse. 1974 *Times* 19 Jan. 10/1 New York is not yet on a three-day week, but mid-town restaurants are doing badly.

f. *mid-clavicular, -sternal, -ventral* adjs.; *mid-Victorian* (earlier and later examples); hence as *sb.*; also † *mid-Vic* sb. and adj. in the same senses; *mid-Victorianism.*

1902 H. J. Stiles in D. J. Cunningham *Text-bk. Anat.* 1184 In a well-proportioned subject, the mid-clavicular line, if prolonged downwards, will be found to be continuous with the vertical Poupart line. 1961 *Lancet* 2 Sept. 573/2 A hyperactive precordium with the maximum apical impulse in the left fifth intercostal space outside the midclavicular line was noted. 1902 H. J. Stiles in D. J. Cunningham *Text-bk. Anat.* 1184 The vertical lines are the mid-sternal, the lateral sternal, [etc.]. 1963 *Lancet* 19 Jan. 111/1 She had a severe midsternal pain which continued overnight, preventing sleep. 1904 *Amer. Naturalist* Feb. 123 The median vein lies along the mid-ventral line of the swollen abdomen. 1925 J. T. Jenkins *Fishes Brit. Isles* 315 Its colour was almost black dorsally..with an irregular mid-ventral streak. 1933 E. Hamilton *Halcyon Era* 6 The papers, over which the poor Midvics yawned in ill-concealed boredom, dealt almost exclusively with Court news. *Ibid.* 32 The studied Midvic pose. *Ibid.* 39 No one was ever allowed to take anything off in Midvic days. 1901 F. H. Burnett *Making of Marchioness* II. xiv. 280 'She was so respectable?' 'She was even a little Mid-Victorian, dear Mary.' 1927 Mid-Victorian [see *Edwardian *a.* 3]. 1965 N. St. John-Stevas in Bagehot's *Coll. Works* I. 79 Bagehot was never in the least infected by the vulgar *No Popery* of the mid-Victorians. 1969 H. Perkin *Key Profession* iii. 81 The mid-Victorian reform which reduced the power of the 'Heads of Houses'. 1971 *Daily Mail* 12 Feb. 13 The artist..had an eccentric taste for mid-Victorianism. 1927 W. E. Collinson *Contemp. Eng.* 62 The peg-top trousers of mid-Victorianism.

2. a. *spec.* (*a*) *Phonetics.* Of a vowel-sound: produced with the tongue or some part of it in a middle position between high and low. Freq. *Comb.*

1876 [see High *a.* 4 b]. 1908 H. Sweet *Sounds of Eng.* 25 If the tongue stops exactly half-way, we obtain the normal 'mid' position, as in the first elements of *ei* and *ou*, which are mid-front and mid-back respectively. 1927 J. J. Hogan *Eng. Lang. in Ireland* 60 He notices the representation of M.E. ē slack by a mid-vowel: 'They pronounce the words tea, sea, please, as if they were written tay, say, plays; instead of tee, see, pleese.' 1935 *Harvard Stud. Philol. & Lit.* XVII. 44 A diphthong whose first element was at first a mid-front vowel, and later..front-slack, mid-back-tense, or possibly 'neutral'. 1961 R. B. Long *Sentence & its Parts* xix. 415 For mid-central /ə/ the tip of the tongue is characteristically pulled back and elevated slightly. 1965 *Language* XLI. 346 Ngbaka shows only four sequences: high-mid, mid-high, mid-low, and low-mid. 1965 [see *down-glide]. 1965 [see *high tone (*high *a.* 22 a)].

(*b*) Of a colour: occupying a middle position in a range of shades.

1916 *Daily Colonist* (Victoria, B.C.) 1 July 12/3 (Advt.), The colors include white, cream, .. mid-brown, lemon, [etc.]. *Ibid.* 23 July 6/1 Gowns..in navy, black, mid-blue, [etc.]. 1929 *Radio Times* 8 Nov. 439/2 (Advt.), This stylish coat... In bottle green, burgundy, dk. brown, mid-brown, navy and black. 1937 *Discovery* Oct. 325/2 The complementary..of mid-yellow is violet-blue. 1971 *Vogue* 15 Oct. 8 Dress..in violet, fir green and mid grey.

b. *mid-position, -section.*

1888 J. Rose *Mod. Machine-Shop Practice* II. xxxvii. 379/1 As the eccentric is in mid-position (*e* being equidistant from B and D), the valve will be in mid-position. 1896 *Rules governing Printing of Specifications* (U.S. Govt. Printing Office) 53 Mid-position. 1953 L. T. C. Rolt *Railway Adventure* iv. 106 We succeeded after some difficulty in doing this,..locking the valve in mid-position by screwing up the spindle gland nuts dead tight. 1957 R. W. Zandvoort *Handbk. Eng. Gram.* vi. 251 Mid-position of an adverb is apt to entail a brief pause between the adverb and the object. 1971 *Engineering* Apr. 47/1 When the sensed dimension is such that the core B is in its mid-position the primary flux is [etc.]. 1961 Webster, *Midsection*, a section midway or about midway between the extremes; midriff. 1969 *Publ. Amer. Dial. Soc.* LI. 5 *Midsection*, representation of the kayak's profile if intersected by a plane perpendicular to the keel-line at the mid-point of the craft. 1972 *Sci. Amer.* Sept. 136/3 A four-foot structure with a cathode at one end and a collector at the other and a large electromagnet surrounding the midsection. 1973 *Ibid.* July 24/2 Actually all the horns called conical incorporate a certain amount of cylindrical tubing in their midsection. 1974 *Plain Dealer* (Cleveland, Ohio) 26 Oct. 4D/1 He is not throwing any punches. Instead, he is permitting Williams to pound him in the midsection.

d. **mid-body** *Cytology* [tr. G. *zwischenkörper* (W. Flemming 1891, in *Arch. f. mikrosk. Anat.* XXXVII. 690)] (see quots.); **mid-brow** *sb.* and *a.* = **middle-brow* (*middle A. 6); **midcrop,** a crop harvested between the main crops; **Midcult, midcult** orig. *U.S.*, middle-brow culture; also *attrib.*; **Mid-East, mid-East, Mideast** = *Middle East; also *attrib.*; **mid-European** *a.* = *Middle-European* adj. (*middle A. 6); also as *sb.*; **mid-fi,** sound-reproduction equipment of a slightly lower quality than *hi-fi; also *attrib.* or as *adj.*; **midsa·gittal** *a. Anat.* = Median *a.*¹; **mid-shot** *Cinemat.* and *T.V.* (see quot. 1953); **mid-west, mid-West** = *Middle West* (*middle A. 6); **mid-western, mid-Western** *a.* = *Middle Western* (*middle A. 6); **mid-westerner, mid-Westerner,** an inhabitant of the Middle West; (also in form *Midwest*, etc.); **mid-wing** *a. Aeronaut.* having the main wings placed approximately halfway between the top and bottom of the fuselage.

1896 E. B. Wilson *Cell* 338 *Mid-body* ('Zwischenkörper'), a body or group of granules, probably comparable with the cell-plate in plants, formed in the equatorial region of the spindle during the anaphases of mitosis. 1969 Brown & Bertke *Textbk. Cytol.* xix. 417/1 The portions of the continuous [spindle] fibers where cytokinesis is to occur, become thickened to form the mid-body, 'stem body' of Belar, or 'Flemming body' which has been seen in cytokinesis of other animals, including *Hydra*. 1928 *Sunday Express* 1 July 12/6 Delighting the low-brow, the mid-brow, and the high-brow with equal facility. 1959 *Manch. Guardian* 10 Aug. 3/2 The dramatic policy is good average mid-brow. 1964 *Punch* 19 Feb. 268/1 When the Midbrows are first showing signs of moving in. 1966 R. A. Downie tr. *O. Del Buono's Bond Affair* 158 It is superficial, banal, *midbrow* wherever psychological analysis is attempted. 1957 *Times* 28 Dec. 10/1 Only the small West African midcrops standing in the way of an absolute Brazilian control over international cocoa offerings until the gathering of the new West African main crops in the autumn. 1973 *Times* 13 July 21/8 Henry Stephens & Sons (London) reported that whatever happened, even at best, the 1973–74 crop would be a late one and the light and mid-crops would not be sufficient to meet internal demand for local industries. 1960 D. Macdonald in *Partisan Rev.* XXVII. 592 This intermediate form—let us call it Midcult—has the essential qualities of Masscult—the formula, the built-in reaction, the lack of any standard except popularity—but it decently covers them with a cultural fig-leaf... Midcult has it both ways: it pretends to respect the standards of High Culture while in fact it waters them down and vulgarizes them. 1962 *Listener* 22 Nov. 863/1 There may seem to be phases in broadcasting history when British broadcasting..has favoured what Dwight Macdonald has called 'midcult', something which is neither for the few nor for the many. 1966 *Ibid.* 27 Jan. 142/2 Even Wilson he feels (though he admires him), has had to sacrifice to 'Midcult' values. *c* 1944 (*newspaper-title*) Mid-East Mail. 1969 *Daily Tel.* 5 Feb. 24 (*heading*) Nixon decision on Mid-East 'in few days'. 1971 R. Thomas *Backup Men* xix. 168 A juke-box blared out some Mideast music. 1972 *Newsweek* 31 July 28/1 There seemed to be some hope for a break in the Mideast diplomatic logjam. 1974 *Publishers Weekly* 7 Jan. 25 (Advt.), 3 months ago this was to be a major study of the Middle East Conflict. Today it is the first major study of the Mideast including the October War. 1960 J. Stroud *Shorn Lamb* iv. 41 Mad Mid-Europeans camping with their chattels in the office porch. 1961 *Guardian* 1 Apr. 4/7 The hall mark of mid-European free dance. 1974 'J. Le Carré' *Tinker, Tailor* xi. 89 Toby Esterhase's faithful mid-European echo. 1970 J. Earl *Tuners & Amplifiers* 7 There is still a margin between what the audiophiles term true hi-fi and general 'domestic quality', and here an entirely new and highly popular range of equipment is emerging. Some call this 'mid-fi' equipment,..but it is noteworthy that such equipment is already rising above the basic 'domestic quality' and entering into the hi-fi fringes. 1971 *Hi-Fi Sound* Feb. 64/3 Today's hi-fi will be tomorrow's mid-fi. 1975 *Gramophone* Jan. 1389 (Advt.), Some of those 'mid-fi' stereo systems of a few years back have a good record player and good speakers. 1957 R. T. Woodburne *Essent. Human Anat.* i. 3/2 The median plane is a vertical plane through the body reaching the surface at the mid-line in front and behind. This plane is also known as the midsagittal plane of the body. 1967 G. M. Wyburn et al. *Conc. Anat.* viii. 203/2 Cut the eyeball in a midsagittal plane. 1968 Chomsky & Halle *Sound Pattern Eng.* 302 Consonantal sounds are produced with a radical obstruction in the midsagittal region of the vocal tract. 1953 K. Reisz *Technique Film Editing* 280 *Mid-shot*, shot taken with the camera nearer to the object than for a long shot but not so near as for a close-up; in relation to the human subject, a shot of the human figure approximately from the waist upwards. 1926 E. Ferber *Show Boat* v. 80 To the farmers and villagers of the Midwest..the show boat meant music, romance, gaiety. 1948, 1968 Midwest, Mid-West [see *Far West]. 1889 Farmer *Americanisms* 365/2 *Mid-Western States*, W. Virginia, Kentucky, Tennessee, Missouri, Kansas, and Arkansas. 1906 (*title*) The Midwestern (Des Moines, Iowa). 1923 *Collier's Mag.* 25 Aug. 24/3 One of the economic causes of Mid-Western discontent is the feeling that the Mid-West is the object of discrimination. 1936 *Mind* XLV. 218 After leaving Harvard, Boodin seemed for many years to be lost in the obscurity of a small mid-western college. 1972 M. J. Bosse *Incident at Naha* ii. 93 You can never wholly shed a *Midwestern* background. 1927 *Scribner's Mag.* Oct. 480/2 These midwesterners are alike unto Americans in other rural areas. 1969 I. Kemp *Brit. G.I. in Vietnam* iii. 47 A sun-tanned, crew-cut, athletic looking mid-Westerner from Green Bay, Wisconsin. 1971 *Guardian* 10 July 9/1 Stolid and conservative Mid-westerners..are..concerned about the plight of the Indians, because the Indians are part of Middle West life. 1934 *Flight* 15 Feb. 156/1 The machine is a mid-wing cantilever monoplane with the wing in three sections. 1942 *R.A.F. Jrnl.* 16 May 16 A twin-engine mid-wing monoplane of creditable modern design.

C. *adv.* **b.** *Comb.*

1876 G. M. Hopkins *Wreck of Deutschland* xxxiv, in *Poems* (1967) 62 Mid-numberèd he in three of the thunder-throne! 1960 *Farmer & Stockbreeder* 23 Feb. 69/3 The Colman-Fella tedder..can be operated in conjunction with a mid-mounted mower. *Ibid.* 8 Mar. 75/2 There is much to be said for mid-mounting of tools that need accurate steerage.

mid air. (Examples in aeronautical contexts.)

1928 R. Fry *Let.* 25 May (1972) II. 627 Reports of aeroplanes that catch fire and grill the passengers in mid-air. 1958 'P. Bryant' *Two Hours to Doom* 58 Mid-air re-fuelling. 1970 *Guardian* 21 Aug. 5/1 If something is not done soon about these near misses, there is bound to be a mid-air collision.

Midas. Add: **1. b.** Esp. in phr. (*the*) *Midas touch.*

[1879 C. M. Yonge *Magnum Bonum* III. xxxix. 899 He would talk of the touch of Midas.] 1883 *Authors & Publishers* (G. P. Putnam's Sons) 12 'From the authors he seized brains and from the public gold.' Certainly a most desirable result, and the picture of our publisher, in the guise of a prestidigitateur, exercising an infallible King-Midas touch on the material submitted to him, is a very fascinating one. 1938 M. Kennedy (*title*) The Midas touch. 1960 *Times* 18 July 3/4 Picasso, with his Midas touch, has at first try made the lino-cut a more dignified medium. 1970 *Observer* 29 Nov. 10/5 [Art] objects which start out as sincere personal statements are turned, by the fatal Midas touch of capitalism, into gold.

2. b. midas-fly, mydas fly, a large fly of the family Mydaidæ.

1895 J. H. Comstock *Man. Study Insects* 461 The Midas-flies rival the robber-flies in size, and quite closely resemble them in appearance. 1972 Swan & Papp *Common Insects N. Amer.* 609 Mydas flies..are elongated, moderate to very large in size, resemble wasps and.. robber flies.

mid-course. Add: **1. b.** In contexts of interplanetary flights. Also *attrib.*

1959 *IRE Trans. Military Electronics* III. 150/1 It will become apparent that no interplanetary mission even with the crudest requirements would have a reasonable probability of success unless some midcourse or terminal guidance is carried out. *Ibid.* 159/1 The midcourse corrections may also remove residual parallel components. 1964 *Times* 1 Aug. 6 The small rocket engine which can manoeuvre the space craft slightly in mid-course. 1969 *Times* 16 July 5/8 MCC, midcourse correction.

midder (mi·dəɹ). *Med. slang.* [f. Mid(wife *sb.* or Mid(wifery + *-er⁶.] Midwifery; a midwifery case, childbirth. Also *attrib.*

1909 A. N. Lyons *Sixpenny Pieces* iii. 23 It was no good waiting breakfast for Fatty..because Fatty's 'call' was a 'midder'. 1931 'F. Iles' *Malice Aforethought* iv. 78 Dr. Bickleigh was wishing that a call would reach him..., a midder-case even. 1937 A. J. Cronin *Citadel* II. vii. 153 We can't go! I've got a positive conviction I'm having a midder case next Sunday evening! 1948 M. Allingham *More Work for Undertaker* xxiv. 279 You've got a midder, you say, doctor? 1965 M. Polland *Thicker than Water* (1967) iii. 29 Although he..did his medicine in Edinburgh, he came here to the Rotunda for his midder.

middle, *a.* and *sb.* Add: **A.** *adj.* **2. b.** Of a colour = *mid *a.* 2 a (*b*).

1869 *Bradshaw's Railway Manual* XXI. 460 (Advt.), Brunswick Green dark middle, and pale. 1926–7 *Army & Navy Stores Catal.* 299/1 Washable water paint... Colours ..Light Stone—Middle Stone—Dark Stone. 1950 J. Cannan *Murder Included* vii. 158 A wool frock of a dull middle-blue.

6. Middle Academy, name given to the mainly sceptic school of philosophy developed in the third century B.C. by Arcesilaus (316/15–242/1 B.C.) when he was head of the Academy founded by Plato; **Middle America,** (*a*) a geographical region comprising central America, Mexico, and the Antilles; (*b*) the 'silent' majority of Americans, regarded as a homogeneous group; hence **Middle American** *a.* and *sb.*; **middle article** = Middle B. 12; **middle-brow, middlebrow,** (*a*) *sb.*, a person of average or moderate cultural interests; (*b*) *adj.*, claiming to be or regarded as only moderately intellectual; **middle common room,** a common room for graduate students; also graduate students collectively; **middle distance,** (*b*) *Athletics*, a distance for a race longer than a sprint but shorter than a long-distance race, esp. one of 440 yards, 880 yards, or a mile (or corresponding metric distances); also (with hyphen) *attrib.*; **middle distillate,** a petroleum fraction that comes off at intermediate temperatures (about 180° to 340°C) in fractional distillation, from which is obtained paraffin, diesel oil, and heating oil; **middle eight** *colloq.*, the eight bars in the middle of a conventionally structured popular tune, often of a different character from the other parts of the tune; the B section in a tune of the form A, A, B, A; the 'release'; **Middle-European** *a.*, of, pertaining to, or characteristic of central Europe or its people; cf. *Mittel-European *a.* and *sb.*; **middle game,** the part of a game of chess between the opening and the end-game; **middle ground,** (*c*) a place half-way between extremes; an area of moderation or compromise; also *attrib.*; **middle guard** [Guard *sb.* 3 b] *Cricket*, the position occupied by a batsman so that his bat defends the middle stump; **middle income,** an average income; also (with hyphen) *attrib.*; **Middle Kingdom** (later examples); (*b*) in ancient Egypt, the Eleventh and Twelfth Dynasties, which ruled from the 22nd to the 18th century B.C.; **middle lamella** *Bot.* (see quots.); **middle leg** *slang*, the penis; **middle-length** *attrib.*, (of a story, etc.) of medium length; **middle management** orig. *U.S.* (see quot. 1957); also (with hyphen) *attrib.*; **middle-middle,** (*a*) the middle or centre; (*b*) a member of the middle-middle-class; **middle-middle-class,** the class of society midway between the 'upper' and the 'lower' class; also *pl.* in the same sense; **middle name** orig. *U.S.*, (*a*) a name between one's first Christian name and one's surname; (*b*) *fig.*, the outstanding characteristic of a person; **middle-off, -on** *Cricket* = Mid-off, -on; **middle passage** (earlier and later examples); (see also quot. 1949); **middle period,** the middle phase (of a culture, artist's work, etc.); also *attrib.*; **middle rail,** (*b*) the 'live' central rail of an electric railway; **middle-range** *attrib.*, designating a thing or things that occur in the middle of a range of items; **middle-rank,** a body of things or persons of intermediate status or value; also *attrib.* or as *adj.*, of neither high nor low rank or value; hence *middle-ranking* adj. (cf. *high-ranking *a.*); **middle rib,** in beef: one of the ribs between the fore ribs and the chuck ribs; **middle-road** *attrib.*, = *middle-of-the-road; **middle school,** (*b*) the middle forms in a grammar or independent school (see quot. 1960); (*c*) a separate post-primary school within the educational system of a state for children aged between about nine and thirteen years; also *attrib.*; **Middle States** (examples); **middle-tone** = Half-tone *sb.* 2; **middle-water** *attrib.*, applied to fishing, or to ships engaged in fishing, at a medium distance from land; **middle weight,** substitute for def.: a man of average weight; *spec.* (in various sports) used to designate an intermediate weight class; esp. in professional boxing, a boxer whose weight is not more than 11 stone 6 lbs.; also *attrib.*; (add earlier and later examples); **Middle West,** the north central states of the U.S.A., as distinct from the West or Far West (see quot. 1949); so **Middle Western** *a.*; **Middle White,** a Yorkshire breed of pig; **middle wicket** (earlier examples).

1659 T. Stanley tr. *Sextus Empiricus's Pyrr. Hyp.* in *Hist. Philos.* iv. 33 Arcesilaus, Institutor and President of the middle Academy, seems to me to participate so much of the Pyrrhonian reasons, as that his Institution and ours is almost the same. 1744 W. Guthrie *Morals of Cicero* p. xiii, We are now arrived at the middle Academy, the Founder of which was Arcesilas. 1845 G. H. Lewes *Biogr. Hist. Philos.* II. viii. iv. 165 The Middle Academy and the New Academy we thus unite in one; although the ancients drew a distinction between them, it is difficult for moderns to do. 1899 M. M. Patrick *Sextus Empiricus & Gk. Scepticism* iv. 77 Sextus himself claims a close relation

between the Middle Academy and Pyrrhonism. 1970 *Oxf. Class. Dict.* (ed. 2) 95/1 The term 'Middle Academy' may also derive from Antiochus [of Ascalon]. 1898 *Pop. Sci. Monthly* Nov. 1 (*title*) Was Middle America peopled from Asia? 1952 S. Tax (*title*) Heritage of conquest: the ethnology of Middle America. 1957 *Social & Econ. Stud.* (Kingston, Jamaica) VI. iii. 380 (*title*) Haciendas and plantations in Middle America and the Antilles. 1966 West & Augelli *Middle Amer.* i. 1 Middle America is an arbitrary geographic expression which refers to a mosaic of people, places, and cultures. Mexico, Central America, and the West Indies, the area which the term usually defines, share a general focus on the Gulf of Mexico and the Caribbean Sea and an intermediate location between North and South America. 1968 *Sunday Times* 29 Sept. 8 What is seriously wrong with Mr Nixon's new Middle America is that it is virtually all white. 1971 *Guardian* 5 Apr. 4/1 Mr Agnew has continued to reflect the prejudices and confusions of Middle America. 1972 P. Dickinson *Lizard in Cup* x. 159 We've got to show that we can build, but..Middle America will like that even less than the bombs. 1973 *Tucson* (Arizona) *Daily Citizen* 22 Aug. 28/1 The Braunlichs will also tell you that, sad as it is, middle America is leery of things it gets for free. 1926 F. F. Blom *Tribes & Temples* I. i. 4 There are maps of most of the Middle American countries, and the greater number of them..are remarkably inaccurate. 1969 *Collier's Encycl. Year Bk.* 3 Her [*sc.* Mrs. Richard Nixon's] looks and taste are classic Middle American, even as her husband's are. 1970 *Time* 5 Jan. 9 Who precisely are the Middle Americans?.. They make up the core of the group that Richard Nixon now invokes as the 'forgotten Americans' or 'the Great Silent Majority'. 1971 *Guardian* 5 Jan. 10/6 If the chips were to come down now, the final word would undoubtedly rest with the middle Americans. 1975 *Atlantic Monthly* Jan. 28/1 The phrase 'Middle American' was first used..by Joseph Kraft in the spring of 1968, when he was writing about the municipal workers in New York City... He specifically referred to their ethnic character: Irish policemen, Italian sanitation workers, Jewish schoolteachers, and so on. *a* 1894 C. H. Pearson in W. Stebbing *Charles Henry Pearson* (1900) viii. 90 T. C. Sandars.. created the so-called middle article—the essay on social topics. 1966 *Listener* 27 Oct. 621/3 Those 'light' middle articles which used to be a feature of the highbrow weeklies. 1925 *Punch* 23 Dec. 673/3 The B.B.C. claim to have discovered a new type, the 'middlebrow'. It consists of people who are hoping that some day they will get used to the stuff they ought to like. 1928 *Observer* 17 June 26 The standard of 'middle-brow' music and plays is always rather low. 1934 C. Lambert *Music Ho!* iv. 247 Hindemith is the journalist of modern music, the supreme middlebrow of our times. 1958 Middle-brow [see *glossy *a.* a]. 1972 L. Alcock *By South Cadbury* i. 22 The onerous and unrewarding task of Secretary was filled by Geoffrey Ashe, writer of distinguished middle-brow books on the problems of the historical Arthur. 1958 *Times* 28 Oct. 12/5 Lincoln College is to be the first Oxford college to establish a special common room for postgraduate students of the college. It will be known as the Middle Common Room. 1969 *Rep. Comm. on Relations with Junior Members Univ. Oxf.* 48 One shall be a president of a Middle Common Room elected by the conference of M.C.R. Presidents. 1971 *Guardian* 21 Dec. 1/6 The ability of the junior and middle common rooms to play their essential rôle in a collegiate university. 1891 *Harper's Young People* 7 Apr. 384/2 Among middle-distance men, as among sprinters, there are various types of runners. 1901 *Encycl. Sport* I. 56/1 It is fairer to describe a Quarter Mile as one of the middle distances. *Ibid.*, The sprinter must use different tactics to the middle-distance runner. 1929 G. M. Butler *Mod. Athletics* v. 74 Speed..should be the middle-distance runner's main objective. 1960 Middle distance [see *front-runner* s.v. *front *sb.* 14]. 1956 *Nature* 10 Mar. 460/1 Prof. Morton's research work has been concerned with the constitution of petroleum, [and] with methods of separation of hydrocarbons, particularly in the middle-distillate boiling range. 1973 *People's Jrnl.* (Inverness & Northern Counties ed.) 15 Dec. 1/1 The nature of North-Sea crude oil is more suited to what we call 'middle distillates'—diesel oil, oil fuel for heating and sulphur by-products. 1966 *Melody Maker* 16 Apr. 8 Doesn't sound as though there's a middle eight. It's good, though not as good as some of their previous records. 1968 *Listener* 1 Feb. 157/3 Popular song has long been confined to an appalling eight-bar monotony. An eight-bar section, repeated; a 'middle eight' and the first eight again. 1939 M. Allingham *Mr. Campion & Others* 86 She's frightfully susceptible. It's her Middle-European blood. 1949 E. Coxhead *Wind in West* iv. 115 That's enough gloom, turn on the Middle-European gaiety. 1972 C. Drummond *Death at Bar* vi. 159 Dee had liked middle-European victuals. 1894 J. Mason *Princ. Chess* 184 No true knowledge of it [*sc.* the opening] is possible independently of a fair knowledge of..the middle game and end. 1958 *Times Lit. Suppl.* 14 Nov. 664/4 The main portion of the work.. is concerned with middle-game tactics. 1959 *Listener* 5 Mar. 434/1 Coming to middle game, it is instructive to see how a great player defends himself when he is in trouble. 1961 A. Smith *East-Enders* vi. 102 The middleground is missing... There's no common place inside the East End for everyone as there used to be. 1972 *Language* XLVIII. 278 In the case of terms like *possible* and *impossible*, there is no middle ground; but that is simply the nature of terms which permit no qualification. 1972 *Guardian* 11 Oct. 12/2 The Laureate has to be a middle-ground man, and there are very few of them around. 1973 *Ibid.* 24 Mar. 22/6 Ultimately as the Alliance leaders know, the organisation of middleground politics is of secondary importance. 1871 'Thomsonby' *Cricketers in Council* 25 Hold your bat up straight, on the popping crease, and ask for 'middle' guard. 1941 *Economist* 12 Apr. 475/1 Direct taxation on what have come to be known as the 'middle incomes'. 1958 B. A. Smith in N. Mackenzie *Conviction* 59 The major beneficiaries of these changes were the middle-income groups. 1971 M. McCarthy *Birds of America* 54 The subspecies they belonged to—white, middle-income intelligentsia. 1848 S. W. Williams (*title*) The Middle Kingdom: a survey..of the Chinese Empire and its inhabitants. 1890 F. L. Griffith *Antiquities Tell el Yahûdîyeh* 39/1 The earliest dateable antiquities from Tell el Yahûdîyeh are of the middle kingdom. 1906 J. H. Breasted *Hist. Egypt* III. viii. 156 The stable organization, which enabled her [*sc.* Egypt] about 2000 B.C. to enter upon her second great period of productive development, the Middle Kingdom. 1928 C. Dawson *Age of Gods* viii. 173 The Middle Minoan period corresponds to the Middle Kingdom. 1969 V. G. Kiernan *Lords of Human Kind* v. 150 That China was the Middle Kingdom, the one truly civilized realm, was..an axiom to its inhabitants. 1971 J. R. Harris *Legacy of Egypt* (ed. 2) 3 Pieces of Middle Kingdom jewellery were reproduced at Byblos in the second millennium B.C. 1925 Eames & MacDaniels *Introd. Plant Anat.* ii. 25 When a pronounced secondary wall is present the primary wall is commonly called the middle lamella. 1947 *Ibid.* (ed. 2) ii. 28 The intercellular layer is the middle layer of this group of three or five layers and the term middle lamella should be restricted to this. 1965 K. Esau *Plant Anat.* (ed. 2) iii. 34 On the basis of development and structure three parts are commonly recognized in plant cell walls: the intercellular substance or middle lamella, the primary wall, and the secondary wall. 1922 Joyce *Ulysses* 443 Are you going far, queer fellow? How's your middle leg? 1935 Dylan Thomas *Let.* Feb. (1966) 151 Men should be two tooled and a poet's middle leg is his pencil. 1928 *Scholartis Press Catal.* July, A volume of five middle-length (*not* short) stories by Norah Hoult. 1946 'G. Orwell' *Shooting Elephant* (1950) 168 The usual middle-length review. 1957 Clark & Gottfried *Dict. Business & Finance* 228/2 *Middle management*, in general, the group or class of junior executives and senior supervisory personnel in the direct line of authority and communications between the top levels of management and the first line supervisory personnel. 1966 S. Phipps *God on Monday* vi. 76 The opposite swing of the same pendulum may be seen in the groups of middle-management-type bungalows that are appearing round the fringes of English villages. 1966 *Punch* 28 Sept. 485/1 The productivity agreement says a good deal for the efficiency of the group, at least at middle-management level. 1975 *Harper's & Queen* May 101 The Saudis need to swell their middle management by 200,000. 1914 E. M. Forster *Maurice* (1971) xiii. 202 The clientele of Messrs Hill and Hall was drawn from the middle-middle classes. 1926 D. H. Lawrence *David* iii. 21 And only from the middle-middle of all the worlds, where God stirs amid His waters, can strength come to us. 1934 H. Read *Art & Industry* iv. 126/1 He..is a middle-middle-class man with a nice little house in the suburbs. 1936 'G. Orwell' *Keep Aspidistra Flying* iii. 54 The derelict spinsters of the middle-middle classes. 1955 T. H. Pear *Eng. Social Differences* iii. 90 When 'middle-middles' become 'upper-middles' they..drop middle-class euphemisms. 1973 *Listener* 22 Feb. 249/3 Their milieux range from lower-middle to middle-middle class. 1835 *Harvardiana* II. 23 [He] then asks their middle names. 1919 Wodehouse *Damsel in Distress* ii. 31 My dad ran a Bide a Wee Home for flowers, and I used to know them all by their middle names. *Ibid.* xvi. 203 Everyone told me your middle name was Nero. 1920 Ade *Hand-Made Fables* 92, I take it that Mixer is your middle name. 1926 A. Christie *Murder R. Ackroyd* xi. 144 'Modesty is certainly not his middle name.' 'I wish you wouldn't be so horribly American, James.' 1932 *N. & Q.* 3 Sept. 177/2 There is also a proverbial saying, 'Money is his middle name', 'Art is her middle name', meaning one's forte. 1972 P. Cleife *Slick & Dead* xvii. 134 If I had a dollar for every time I've said that, my middle name would be Rothschild. 1972 J. Porter *Meddler & her Murder* i. 11 Tact was far from being the Hon. Con's middle name. 1851 W. Clark in W. Bolland *Cricket Notes* 137 The middle off, cover point, long slip, and long stop should all save one run. 1843 'Wykehamist' *Pract. Hints Cricket* (*caption*) Short leg or middle on. 1887 F. Gale *Game of Cricket* 139 Middle-on and middle-off were.. about equal distance from either wicket, standing back some fifteen yards from the centre between the wickets, opposite each other. 1788 T. Clarkson *Essay on Slavery* (ed. 2) III. iii. 98 The captain of a ship, then on the middle passage, had lost a considerable number of his slaves by death. 1949 C. Lloyd *Navy & Slave Trade* i. i. 5 The Round Trip..was commonly divided into three 'passages'. On the outward passage the cargo consisted of textiles, hardware, alcohol and antiquated firearms. These were traded on the coast for slaves, who were shipped to America and the West Indies on the notorious Middle Passage. The principal cargoes taken on there for the homeward passage were sugar, tobacco and rum. 1969 *Listener* 22 May 713/3 Even the unsqueamish stomachs of the 18th century were turned by accounts of the Middle Passage. 1873 C. M. Yonge *Pillars of House* IV. xxxvi. 50 Here's the dining room... This is the middle period, the Stewart style part. 1894 G. B. Shaw *Music in London 1890–94* (1932) III. 157 Those features of the middle period Beethovenism of which we all have to speak so very seriously. 1930 W. S. Maugham *Cakes & Ale* i. 17 The novels of his middle period reflected..the strain. 1951 T. S. Eliot *Poetry & Drama* ii. 20 His [*sc.* Yeats's] middle-period *Plays for Dancers*. 1905 Goodchild & Tweney *Technol. & Sci. Dict.* 401/2 *Middle rail*, a heavy conductor in the form of a rail carried on insulating supports, which is laid between the running rails of an electric railway to supply current to the motors. 1964 G. F. Arnold in D. Abercrombie et al. *Daniel Jones* 17 The middle-range percentages..are provided by comparisons of phonemes. 1967 M. Argyle *Psychol. Interpersonal Behaviour* ix. 161 In the matter of price, for example, some Ss [*sc.* salesmen] show the middle-range item first, others show the most expensive. 1961 *Times* 28 Dec. 11/4 That solid middle-rank of literature. 1969 *Daily Tel.* 10 Oct. 3 Sterling, too, is a middle-rank currency these days—apparently safe from heavy selling pressure but not exactly in demand by speculators. 1972 *Guardian* 13 Nov. 2/3 Middle-rank American and Vietnamese officials in Saigon. 1959 *Encounter* Aug. 18/1 Instructions transmitted..from on high through a number of middle-ranking personages down to floor polishers. 1747, 1844 Middle rib [see *chuck *sb.*⁴ 2]. 1963 A. L. Simon *Guide Good Food & Wines* 406/1 The *Middle Ribs* and *Chuck Ribs*, sometimes called *Wing Ribs*, are both uneconomical and ungainly as joints owing to the larger proportion of bone to meat. 1958 A. Wilson *Middle Age of Mrs Eliot* III. 343 A nice middle-road historian's position, he thought to himself with comforting irony. 1971 *Guardian* 14 May 1/3 With no obviously dominating candidate to step into Nasser's shoes, they chose the safe middleroad candidate—Sadat. 1914 'I. Hay' *Lighter Side School Life* viii. 224 The occasion of his first attendance at a meeting of the Middle School Debating Society. 1933 Middle school [see *Aryan *sb.* 2]. 1960 *Where?* III. 15 *Middle school*, usually the third and fourth forms of a school (a grammar school expression). 1962 E. J. King *World Perspectives in Educ.* III. vii. 142 The new Danish two-year programme established for all twelve-year-olds in 1959..was..a nationwide formalization of what used to happen in many urban middle schools. 1971 *Guardian* 20 Oct. 1/8 Surrey County Council..passed a scheme providing for middle schools followed by 12–18 comprehensives. 1973 *New Society* 10 May 294/1 In 1970, the Department of Education and Science introduced a new category called 'middle schools' into its yearly *Statistics*... What is a middle school? The official version is..a school which straddles the traditional primary-to-secondary transfer age of eleven. 1784 G. Washington *Diary* 4 Oct. (1925) II. 326 The middle States with the Country immediately back of them. 1848 J. F. Cooper *Oak Openings* I. xiii. 197 Who ever heard of the 'tribe' of New England, or..of the 'tribe of the Middle States? 1912 M. Nicholson *Hoosier Chron.* 59 There had been an infusion of population from New England and the Middle States. 1909 Webster, Middle tone. 1961 M. Levy *Studio Dict. Art Terms* 57 *Half-tone*, the tone value in a painting which is halfway between the dark and the light. Sometimes called Middle Tone. 1962 Middle-water [see *distant-water* s.v. *distant *a.* 8]. 1967 *Times Rev. Industry* May 28/2 The near and middle-water fleet consists of boats between 80 and 140 feet long which sail mainly from Grimsby, Fleetwood, Lowestoft, Aberdeen, Milford Haven and North Shields for fishing grounds around the Faroes, in the North Sea and to the west of the British Isles. 1889 Middle weight [see Feather-weight 3]. 1909 *Westm. Gaz.* 9 Feb. 12/1 Some fine wrestling has been seen, more especially in the middle-weight class. 1947 E. Gruhn *Text Bk. Wrestling* (ed. 4) 84 *Middle-weight*, up to 174 lb. (79 Kilos). 1953 D. Murray *Weight Lifting* iii. 56 The body-weight classes used in standard international lifting competition are as follows..165½ pounds—Middleweight. 1972 F. Butler *Hist. Boxing in Brit.* xix. 132 The middleweight division was started in England in 1786... The middleweights can claim more superb champions than any section outside the heavyweight. 1898 M. H. Catherwood (*title*) Heroes of the Middle West. 1909 Middle West [see *dripped *ppl. a.*]. 1917 *Nation* (N.Y.) 17 May 589/2 The personal tour of the Secretary of the Treasury through the Middle West, to speak at public meetings, is a wise arrangement. 1949 *Oxf. Jun. Encycl.* III. 452/2 The Middle West region is oddly named, because the states of Ohio, Indiana, Illinois, Michigan, Wisconsin, Minnesota, Iowa, and Missouri, which make it up, are really neither middle nor west. 1909 'O. Henry' *Options* 310 I'm only a little Middle-Western girl. 1916 A. Huxley *Let.* 30 June (1969) 103 Mere text-book papers, such as would be set by a Middle Western College out your way. 1967 Middle-western [see *fraternity 7]. 1893 L. M. Douglas *Man. Pork Trade* p. xiv (*caption*) Small and Middle White Yorkshire Pigs. 1912 Middle White [see *fatstock* s.v. *fat *a.* 14]. 1953 J. Jobson *Househ. & Country Crafts* vi. 65 Almost every county in England has produced its own breed of pig, and we have amongst others Large and Middle White. 1786 Middle wick't [see *bat *sb.*¹ 3 c]. 1816 W. Lambert *Cricketers' Guide* (ed. 6) iii. 42 Middle Wicket Off. The man should stand on the off side, not far from the Bowler's wicket, and about 23 yards from the Striker's wicket. *Ibid.* 44 (*heading*) To cover the Point and Middle Wicket.

B. *sb.* **1.** *to knock* (a person) *into the middle of next week*: see Week *sb.* 6 d.

b. *U.S.* A strip of unplanted ground between rows of cotton, corn, etc. Usu. *pl.*

1829 L. Covington *Diary* 28 May in *Documentary Hist. Amer. Industr. Society* (1910) I. 238 Two Ploughs breaking middles in Popular tree cut. 1907 T. F. Hunt *Forage & Fiber Crops in Amer.* 352 The field is made up into alternate beds and middles or into 'back' furrows and 'dead' furrows. 1946 *Democrat* 11 Apr. 1/5 Two and three year old kudzu stands that have not covered the middles will be greatly helped if the middles are broken out.

2. b. Delete 'Now *rare* or *Obs.*'

c. Slang phr. *in the middle*: in a difficult, dangerous, or untenable position; in trouble. Also *slang* (orig. *U.S.*).

1930 *Amer. Mercury* Dec. 457/1 What's the idea? Trying to put me in the middle with the law? 1943 R. Chandler *Lady in Lake* (1944) xxxiv. 179 The other guy could have knocked him out to put him in the middle. 1954 'N. Blake' *Whisper in Gloom* II. xvi. 217, I still don't like it. How d'ya know he's not leaving us in the middle? 1972 J. Burmeister *Running Scared* x. 131, I was the man in the middle. If your note giving my location should go astray..I could quietly starve to death.

3. b. The part of a side of bacon which is left when the fore-end and the gammon are removed.

1892 P. L. Simmonds *Dict. Trade Products* (rev. ed.) Suppl. 473/2 *Middles*,..a name for sides of bacon and pork, there are long and short middles. 1917 G. J. Nicholls *Bacon & Hams* 70 These middles are cured in dry salt. 1923 R. E. Davies *Pigs & Bacon Curing* 29 The side may be cut into three parts, comprising the fore-end, the middle, and the gammon with corner.

13. *Cricket.* = *middle guard* (see *A. 6).

1866 'Capt. Crawley' *Cricket* 22 The batsman should, after asking the umpire for middle, and taking his block at a bat's length from the stumps, stand..in the position shewn. 1904 F. C. Holland *Cricket* 1 What guard is to be chosen? Some cricketers take centre, some the leg stump, and many middle and leg. 1960 I. Peebles *Bowler's Turn* 187 He had batted on middle and off and shown a readiness to hook.

14. *colloq.* A middle-class person.

1955 T. H. Pear *Eng. Social Differences* 101 Wealthy 'middles' are now admitted to some formerly exclusive hunts. **1967** *Listener* 21 Dec. 802/1 If a man spoke rather loudly..keeping his vowels open, then he was an Upper. If he attempted all this and just failed, then he was a Middle.

middle, *v.* Add: **9.** *Cricket.* To strike (a ball) with the middle of the bat; also with the bowler or the stroke as object.

1894 J. H. Pinkleton *Ashes crown Year* xi. 112 Hutton was in grand form. He middled Lindwall with confidence. **1955** F. Peebles *On Ashes* iii. 29 The batsman started by showing every sign of good form..middling his strokes with ominous regularity. **1955** Miller & Whitington *Cricket Typhoon* ii. 189 May began to middle the ball.

middle age, *sb.* Add: **2.** (Earlier examples.)

a. *sing.* **1621** Donne *Sermon I Tim.* i. 15 (1661) 192 It is a perplex't question in the School, (and truly the Balance in those of the middle age, very even) whether if Adam had not sinned, the son of God had come into the world, and taken our nature and our flesh upon him. **1624** Wotton *Elem. Archit.* sig. ¶ 4 After the reuiuing and re-polishing of good Literature, (which the combustions and tumults of the middle Age had vnciuillized).

b. *pl.* **1616** Spelman *De non temerandis Ecclesiis* (ed. 2) App. 194 Thus the eldest and newest Expositors are wholly for mee, many also (& of the best of them) of the middle ages. **1699** M. Lister *Journey to Paris* 108 It would have been some satisfaction to have seen by the Pictures, what the middle Ages, at least, had thought of them [sc. animals].

3. Also *attrib.* in sense 1, as *middle-age bulge, spread* (cf. next, sense b).

1937 'John o' London's' 29 Jan. 742/1 (Advt.), Join the happy throng who have learnt to control the 'middle-age spread' by wearing the..supporting belt. **1969** *Times* 5 May 9/1 The butcher wants his beer before it has developed a middle-age spread. **1972** J. Porter *Meddler & her Murder* iii. 48 She was fighting a losing battle against the middle-age bulge.

middle-aged, *a.* Add: **1. a.** (Earlier and later examples.) Also *transf.* and *fig.*

1608 Topsell *Serpents* 73 The elder looke to the family, placing in due order that hony which is gathered and wrought by the middle-aged Bees. **1611** Coryat *Crudities* 251 He was a middle-aged man, as about forty yeares old. **1918** W. Owen *Let.* 22 July (1967) 566, I have on unused boots with me, but I left a delicate middle-aged pair in the Kitchen Cupboard. **1937** [see *eclair]. **1940** [see *auto *sb.*³]. **1950** T. S. Eliot *Cocktail Party* I. ii. 57 Only since this morning I have met myself as a middle-aged man. **1960** M. Spark *Ballad of Peckham Rye* vii. 150 The chief barmaid had a tiny nose and a big chin; she was a middle-aged woman of twenty-five. **1974** *Broadcast* 28 Oct. 20/2 Young men go to bed with young ladies, middle-aged men (presumably) with middle-aged ladies, and old men with very young ladies.

b. *spec.* **middle-aged spread,** paunchiness in a middle-aged person; also *transf.* and *fig.* Cf. *middle age *sb.* 3.

1931 H. G. Wells *Work, Wealth & Happiness of Mankind* (1932) xv. 768 Impermanence is the lot of all encyclopædias, and though the Britannica..shows now these marks of advanced maturity, of 'middle-aged spread', there is no reason for supposing that the spirit of Diderot is dead. **1942** D. Powell *Time to be Born* (1943) ii. 43 Erase that middle-aged spread. **1957** J. Braine *Room at Top* i. 7, I hadn't then begun to acquire a middle-aged spread. **1962** *Listener* 20 Sept. 430/1 That impish sense of the ridiculous..which will always stop 'Tonight' from acquiring the pompous middle-aged spread that so often overtakes programmes.

2. (Earlier and later examples.)

1611 T. James *Treat. Corruption of Scripture* Advt. to Christian Reader, sig. *2 The open or secret wrongs done vnto Fathers, auncient, middle-aged, or moderne writers, by the Papists. **1846** Dickens *Pictures from Italy* 5 The first chapter of a Middle Aged novel.

middle-ageing, *ppl. a.* Delete '*nonce-wd.*' and add further examples. Also **middle-aging.**

1916 E. Pound *Lustra* 104 With middle-ageing care I had laid out just the right books. **1956** D. M. Davin *Sullen Bell* II. iv. 130 The job was getting him. If not the job..his own middle-ageing self. **1969** D. Lambert *Angels in Snow* v. 74 A successful middle-ageing woman. **1973** *Publishers' Weekly* 8 Jan. 62/1 They get acquainted with a pair of middle-aging Jewish bachelor doctors.

middle-ager. orig. *U.S.* [f. Middle age *sb.* + -er¹.] A middle-aged person.

1949 *Labor & Nation* Jan.–Feb. 7 Not every youngster or middle-ager..is a double dealer. **1956** *N.Y. Times Mag.* 29 July 5 (*heading*) America's unknown middle-agers. **1962** *Punch* 12 Sept. 392/3 A tough competent middleager.

middle class. Add: **a.** (Earlier examples.)

1766 Queen Caroline Matilda of Denmark *Let.* 25 Dec. in *Mem. Unfortunate Queen* (1776) 21 There is no such thing here as a middle class of people living in affluence and independence. **1792** A. Young *Trav. France* I. xxii. 549 Knowledge, intelligence, information, learning, and wisdom ought to govern nations; and these are all found to reside most in the middle classes of mankind; weakened by the habits and prejudices of *the great*, and stifled by the ignorance of the vulgar.

c. *middle-class morality.*

1926 F. M. Ford *Man could Stand Up* I. iii. 49 What *should* keep them apart?... Middle Class Morality? **1966** *Punch* 13 July 78/2 The tyranny of middle-class morality must be conquered, especially in the field of the homosexuality laws. **1968** A. MacLeod *Dam* vi. 62 A concession to your middle class morality. **1975** *Times* 15 Jan. 15/5 It was refreshing to see the virtues of middle class morality applauded.

Hence **middle-classdom, -classism,** the middle class as a whole; their characteristics, interests, or position; **middle-classness** (further examples); **middle-classy** *a.*, suggestive of the middle class.

1930 *Observer* 14 Sept. 7 The secret of Denmark's somewhat stuffy middle-classdom. **1963** *Times* 2 May 18/2 Incipient middle-classdom; outward conformity. **1930** *Working Men's College Jrnl.* Apr. 77 Mr. Lupton..did not think the question before the House was one of aristocracy *v.* middle-classism. **1923** A. Huxley *Antic Hay* ix. 142 The dreadful middle-classness of her Art and Craftiness. **1970** *Guardian* 2 Mar. 9/2 There is nothing so guaranteed to preserve English middle classness..as being surrounded..by foreign parts. **1926** *Glasgow Herald* 21 May 8 Highly respectable, middle-classy railway clerks.

Middle East. [f. Middle *a.* + East *sb.*] States lying between the Near and Far East, esp. Egypt and Iran and the countries between them. Also *attrib.*

The name *Middle East* has been used with considerable freedom: see esp. quot. 1958. Cf. *Far East, *Near East.

1902 A. T. Mahan *Retrospect & Prospect* 237 The Middle East, if I may adopt a term which I have not seen, will some day need its Malta, as well as its Gibraltar. **1903** V. Chirol *Middle Eastern Question* i. 5 'The Middle East', that is to say..those regions of Asia which extend to the borders of India or command the approaches to India. **1923** [illegible] The interests of Great Britain and Russia in the Middle East..are in reality irreconcilable. **1925** A. Toynbee *Survey Internat. Affairs 1920–23* I. i. 3 The affairs of Turkey and the other countries of the Middle East. **1944** J. S. Huxley *On Living in Revolution* [illegible] Organizations like the Middle East Supply Council. **1958** Ld. Vansittart *Mist Procession* vi. 82 [illegible] 'floppy modernism' which lumps everything from the Mediterranean to Bengal as Middle East... Persia, Baluchistan, Afghanistan, India were the Middle East. **1974** *Encycl. Brit. Micropædia* VI. 871/2 Middle East..has come to be applied to the lands around the southern and eastern shores of the Mediterranean Sea, extending from Morocco to the Arabian Peninsula and Iran and sometimes beyond.

So **Middle Ea·stern** *a.*, of or pertaining to the Middle East.

1903 V. Chirol (*title*) The Middle Eastern question; or, some political problems of Indian defence. **1909** A. Hamilton *Probl. Middle East* p. xi, No study of Middle Eastern politics can avoid encroaching upon those of the Near East and of the Far East. **1925** A. Toynbee *Survey Internat. Affairs 1920–23* I. 21 The most important permanent results of this meeting of the Supreme Council in the Near and Middle Eastern field were [etc.]. **1966** *Listener* 27 Oct. 600/1 Their children look at the world from a middle-eastern not a European standpoint. **1973** A. Mann *Tiara* iii. 28 What was the rôle of the dark girl? Her first name was Middle Eastern, not Italian.

middleman. Add: **4. c.** *Mountaineering.* The middle climber of a team. Also *attrib.*, as **middleman('s) loop, noose,** etc., a knot used by a climber to tie himself on to the middle of a rope.

1892 C. T. Dent et al. *Mountaineering* iv. 102 The 'fisherman's bend', used as a middleman noose for instance, has only 65 per cent of the strength of the rope. **1909** G. C. E. Benson *Brit. Mountaineering* ii. 35 The Middleman Loop is the only one that should be used for this purpose. **1951** E. Cornish *One Green Bottle* vi. 160 Each child was tied on with a middleman's noose. **1968** R. Franklin *Dict. Knots* [illegible] also called the *Englishman's Loop*, (in America) the *Fisherman's or Angler's Loop*... It is a useful loop knot tied in the bight by one of at least four different methods. Once much used for the middleman on a rope in climbing, but now superseded. **1971** J. Lovelock *Climbing* iii. 43 The middle man can tie on to the rope in a number of ways.

5. a. (Examples.)

1870 O. Logan *Before Footlights* 248, I give it up, Brudder Bones, as the middle man at the minstrels always does the end man's conundrums. **1880** [see *interlocutor¹ c]. **1930** C. Wittke *Tambo & Bones* iv. 136 Lively repartee between the endmen and the middleman.

c. *N. Amer.* One who paddles or rows in the middle of a canoe or boat.

1761 A. Henry *Trav. & Adventures Canada* (1901) ii. 14 They engage..the middle-men at one hundred and fifty livres and the end-men at three hundred livres, each. **1801** A. Mackenzie *Voy. from Montreal* p. xxviii, The canoe men are of two descriptions, foremen and steersmen, and middlemen. **1839** J. K. Townsend *Narr. Rocky Mts.* xv. 355 The middle-men ply their oars; the guides brace themselves against the gunwale of the boat, placing their paddles edgewise down her sides. **1968** [see *bowsman].

middleness (mi·d'lnės). *rare.* [f. Middle *a.* + -ness.] The fact or quality of being middle, average, or middle-class.

1929 D. H. Lawrence *Pansies* 120 Their middleness is only an unreality separating two realities. **1963** *Times* 25 May 9/5 Some few fortunates have special attributes in highest degree, the majority have varying degrees of middleness, and the tail is sadly lacking in endowments.

middle-of-the-road. Phr., often used *attrib.* or quasi-*adj.*, pertaining to or designating a person who, or a course of action, etc., which, is moderate or unadventurous, tending to avoid extremes; orig. *spec.* in *U.S.* with reference to the views of the Populist party. Hence **middle-of-the-roader.**

[**1777** P. Thicknesse *Year's Journey* I. vi. 43 It is necessary..to keep in the *middle* of the road, so as not to be too suddenly surprised. **1892** *Rocky Mountain News* (Denver, Colorado) 17 July 1 Side tracks are rough, and they're hard to walk, keep in the middle of the road.] **1894** *Iowa State Register* (Des Moines) 5 Sept. 8/5, I am a middle-of-the-road man, but I don't propose to lie down across it so no one can get over me. Nothing grows in the middle of the road. **1896** *N.Y. Tribune* 21 July 9/2 If the Bryan faction predominates, the Middle-of-the-Roaders will bolt the Democratic ticket. **1896** *Congress. Rec.* 20 Dec. [illegible] The only honest Populist is the 'middle-of-the-road' Populist. **1927** *Amer. Speech* II. 443/1 The 'middle of the road' is the secured path followed by compromising politicians who desire to promote their own or their party's fortunes. **1950** *Ann. Reg. 1949* 331 The Japanese press had said that..efforts to establish a 'middle-of-the-road' democracy had failed. **1951** [see *functionalism 1]. **1959** *Economist* 10 Jan. 101/1 He has been ultra-conservative, demagogically radical, or firmly middle-of-the-road. **1971** D. E. Westlake *I gave at the Office* (1972) 151 They are neither right-wing nor left-wing but middle-of-the-roaders. *Ibid.*, There are no revolutionaries on this island, no extreme left-wing, not right-wing, and most certainly not middle-of-the-road revolutionaries. **1973** S. Harvester *Corner of Playground* ii. i. 80 [They] wanted him, the old middle-of-the-road liberal democrat, to be their first president.

middlescent (mid'le·sent), *a.* and *sb.* [f. Middle *a.* + -escent after Adolescent *sb.* and *a.*] **A.** *adj.* Of, pertaining to, or taking place in middle age. **B.** *sb.* A middle-aged person. Hence **middle·scence,** the period of middle age.

Not current in the U.K.

1965 *N.Y. Times* 5 Dec. iv. 10/3 Middlescence, that in-between stage between youth and premature grave in which the parent undergoes alarming physical and emotional change. *Ibid.* 10/5 The middlescent parent is a highly emotional creature of quixotic moods. **1969** *Woman's Day* (Sydney) 12 June 40/4 Although most middlescent romantic folly doesn't get off the planning or fantasy stage, there's a lot to be said for being in love, perhaps especially at 40. **1969** *Britannica Bk. of Year* (U.S.) 802/1 *Middlescent, n.* middle-aged individual, esp. one who has just turned 40; *middlescence, n.* **1973** *Daily Colonist* (Victoria, B.C.) 19 Sept. 29/4 Middlescents don't know how to deal with middlescence..it might find itself 'drained of crucial managerial resources', because the middlescent is a troubled man.

middle town. 1. The centre of a town.

1835 S. Rodman *Diary* (1927) 324/1, I went to the middle town [of Shirley Village, Massachusetts] on two first days and heard good sermons.

2. Usu. **Middletown.** A typical middle-class community. orig. *U.S.* Hence **Middletowner,** the average middle-class person.

1929 R. S. & H. M. Lynd (*title*) Middletown: a study in contemporary American culture. **1954** Koestler *Invis. Writing* v. 65 Another factor..is the uniform dreariness of the average Russian town, and its lack of architectural character... This uniformity reminds one of the middle-towns of America. **1974** *Times* 11 Feb. 4/7 'Middletown' is bored nearly to tears by the television coverage of the [election] campaign. *Ibid.*, I have watched the coverage after pounding the streets of 'Middletown', and what I have heard from political leaders has had little to do with what is bothering the many 'Middletowners' I have met. *Ibid.* 28 Feb. 18/7 Middletowners..are not political animals... There [illegible] Middletown suggests to me that no party leader has tried to find out what bothers its people.

Middleveld (mi·d'lfelt, -velt). *S. Afr.* Also **-veldt.** [Partial tr. Afrikaans *Middelveld*, lit. 'intermediate region'.] A region in East and West Transvaal lower than the Highveld but higher than the Lowveld, between 3,000 and 4,000 feet above sea-level. Also *transf.*

1877 C. Warren *Diary* 17 July in *On Veldt in Seventies* (1902) 225 We arrived at a settlement station in the Middle Veldt, forty miles from Worcester, 2800 feet above it. **1878** A. Aylward *Transvaal To-Day* 25 The nearer and more recently settled district called the 'Middleveld' of the Orange Free State. **1929** [see *lowveld]. **1931** *Discovery* Aug. 259/2 There are two cotton breeding stations in the Transvaal; one at Rustenberg, where problems connected with the improvement of cottons for middle-veld areas are dealt with. **1939** tr. *E. N. Marais's My Friends the Baboons* v. 55 It was in the middle-veld between the Palala and the Magalakwen. **1955** J. H. Wellington *S. Afr.* I. i. iii. 72 These two regions..have important points of dissimilarity, the chief of which are the greater altitude and humidity of the highveld and the greater areas of pre-Karoo surfaces in the middleveld. **1972** *Farmer's Weekly* (S. Afr.) 21 Apr. 11 Normal dryland conditions in the Rhodesian middleveld.

middling, *sb.* Add: **1. b.** A person who or a thing which is mediocre or second-rate (cf. Middling *a.* 3 b); freq. in dial. phr. *among the middlings*, of a mediocre class; also, in a moderate condition of health.

1877 *Sunday Mag.* 182 'How are you getting on, Dick?'..'Well, only among the middlings, Sir.' **1885** R. Holland *Gloss. County of Chester* 226, I said to his employer, 'What sort of a man is your team-man?' The answer was, 'Well! he's just about among the middlings;' so I did not engage him. **1931** R. Campbell *Georgiad* iii. 62 They're all members of the self-same school, And drilled..to enforce on all The standards of the middling and the small. **1964** R. Church *Voyage Home* ii. 28 Whenever I asked after his permanently ailing wife, he beamed with benevolence and replied: 'Oh, amongst the middlings, you know, amongst the middlings.'

3. c. (Earlier and later examples.)

1743 W. Ellis *Mod. Husbandman* IV. iii. 63 Its second, or Middling, or that Meal commonly made Use of by Farmers for spending it in their Families. **1786** G. Washington *Diary* 13 Sept. (1925) III. 116 My Corn being out, or nearly so, I was obliged to have middlings and ship stuff mixed for bread. **1960** *Farmer & Stockbreeder* 12 Jan. 107/1 A balanced ration for weaner piglets is 40 parts middlings, 25 parts barley meal, [etc.]. **1969** G. E. Evans *Farm & Village* vi. 72 The miller..used to charge us..for grinding the corn. But we got the offal as well as the flour, and the best middlings we fed to the pigs.

d. Of minerals.

1869 *Amer. Jrnl. Sci.* XCVII. 9 The amount of heavy lubricating oil was largely increased, and the 'middlings' correspondingly diminished. **1909** Webster, *Middling*,.. *pl.* The second quality of ore obtained by washing. **1969** G. J. Williams *Econ. Geol. N.Z.* ix. 135/2 Low-grade magnetic concentrates and a high proportion of middlings.

4. *U.S.* (Earlier examples.) Also in *sing.*

1777 *Calendar Virginia State Papers* (1875) I. 288 Bakin in hams, middlings, shoulders, &c. **1831** J. M. Peck *Guide for Emigrants* 153 The hog..is cut up into hams, shoulders, and middlings or broadsides. **1834** D. Crockett *Narr. Life* xi. 79, I got also a large middling of bacon, and killed a fine deer. **1848** *Rep. Comm. Patents 1847* (U.S.) 527 [illegible] middlings. **1857** 'Porte Crayon' *Virginia Illustr.* i. 31 Fried middling and hot coffee were then served and we all partook. **1904** E. Glasgow *Deliverance* 51 She had to fry the middling in the kitchen, and mother complains so of the smell.

middlingness. (Further examples.)

1929 D. H. Lawrence *Pansies* 120 Nothing that transcends the bourgeois middlingness. **1957** *Essays in Crit.* VII. 216 The umbrella of 'moderation'—the ideal of the new slogans—means too frequently that middlingness, lack of real poetic talent and ambition has found yet another cosy corner for itself in our watered-down, democratic culture.

middy¹. [For etym. read: [f. Mid *sb.*¹ + -y⁴.] Add: (Earlier example.)

1818 'A. Burton' *Adventures J. Newcome* 145 A Middy rudely said, He'd sell them [sc. prisoners] for five pounds a head.

2. In full, **middy blouse.** A kind of loose blouse, often extending below the waistline, similar to that worn by midshipmen. Also *attrib.*

1911 *Daily Colonist* (Victoria, B.C.) 13 Apr. 24/3 (Advt.), Children's Middy Dress, in white duck. Square neck and short sleeves. **1913** T. Eaton & Co. *Catal.* Spring & Summer 51/1 Plain Galatea 'middy'. [illegible] **1917** D. Canfield *Understood Betsy* (1922) iii. 37 The necktie of her middy blouse fell forward. **1929** Wodehouse *Mr. Mulliner Speaking* [illegible] **1957** [illegible] **1963** [illegible] **1965** Mrs. L. B. Johnson *White House Diary* 6 Aug. (1970) 307 Carol Channing in her white velvet bell-bottom trousers trimmed in red and a middy blouse top.

middy² (mi·di). *Austral. slang.* [f. Mid *sb.*³ + -y⁴.] A measure of beer or other liquor (see quots.); the glass containing it.

1945 Baker *Austral. Lang.* iii. 76 The *middy*, a beer glass containing nine ounces, is a measure used only in N.S.W. hotels. **1956** S. Hope *Diggers' Paradise* 230 A middyglass contains ten liquid ounces. **1957** 'N. Culotta' *They're a Weird Mob* (1958) ii. 25 Those big glasses are called schooners and those small ones are called middies. **1968** *Southerly* XXVIII. 38 He ordered two more middies. **1970** *Observer* (Colour Suppl.) 15 Feb. 39/2 Getting 'full' on frosty 'middies' of beer is still a serious manly business. **1974** K. Cook *Bloodhouse* 79 'Middy of rum, Mick,' said the youth... Ten ounces of rum sold over the bar cost four dollars.

midear (midiə·ɪ), var. *my dear* (see My *poss. adj.* 2 b, Dear *sb.*²), used as a form of address.

1959 J. Braine *Vodi* xiv. 191 Anything you want, midear. **1967** J. Aiken *Ribs of Death* xxv. 176 Do-ee think it *could* be true, midear?

mid feather. 1. (Further examples.)

1905 Goodchild & Tweney *Technol. & Sci. Dict.* 402/1 *Midfeather*, (*Paper Manufac.*) a partition fixed in the 'breaker' to promote circulation of the pulp. **1930** *Engineering* 22 Aug. 223/1 Numerous tests showed that, provided the midfeather was inserted at the fan end of the duct, completely uniform distribution of the dust was obtained. **1940** *Chambers's Techn. Dict.* 547/1 *Mid-feather* (*Join.*), a cross-tongue.

mid-field. Add: Also **midfield.** (Further examples.) Also as quasi-*adv.*

1897 *Badminton Mag.* IV. 422 They [sc. rooks] quickly shift their position to safer quarters in mid-field. **1909** *Westm. Gaz.* 24 Nov. 12/2 C. Lyle shot well, but was very slow in midfield. **1956** *Times* 16 Apr. 13/1 Haynes, in spite of the presence of Evans in mid-field, showed himself one of the very few inside forwards in the British Isles with a spark of true genius. **1960** *Times* 30 Nov. 3/6 Certainly they had no penetrative midfield player. **1968** *Listener* 7 Nov. 625/3, I was aware..that Gilzean would be playing mid-field: I knew my Spurs. **1969** [see *link *sb.*² 3 h]. **1973** *Liverpool Echo* (Football ed.) 17 Mar. 18/1 With Hughes pushing them forward with some strong play in mid-field, Liverpool hit their post. **1974** *Times* 23 Feb. 14/8 Even eight, nine and ten-year-olds these days are taught by games masters in terms of..'midfield provider', 'sweeper' and the rest. **1974** *Guardian* 19 Aug. 16/2 The midfield men preferred the easy option of overlapping fullbacks to making their own breaks. *Ibid.* 16/3 The painstaking midfield play of Kendall and Campbell.

midge. 4. (Earlier example.)

1865 C. M. Yonge *Clever Woman* I. ii. 52 One of the midges, or diminutive flies used at Avonmouth, came to the door.

midget. 2. For 1859 read 1848.

4. A small vehicle, aircraft, etc.

1930 *Daily Express* 8 Sept. 3/7 The midgets will function as destroyers, and the heavier tanks will discharge the duties of land battleships. **1933** M. Arlen *Man's Mortality* ii. 32 He climbed into the midget [sc. aircraft], and called out to the cadet: 'Altitude?'

5. *attrib.* **a.** In sense 'weak, puny', as *midget effort*. **b.** 'small, small-scale, tiny', as *midget submarine, warship*, etc.; **midget golf,** a form of miniature golf, usually played indoors; so *midget-golf course*, etc.

1908 *Daily Chron.* 7 Aug. 4/4 The spiritual intelligences of the dead..are making midget efforts to comprehend and explain the circumambient infinite. **1922** Joyce *Ulysses* 600 Marcella, the midget queen. **1930** *Daily Express* 6 Nov. 3/7 Sydney's midget golf boom. **1930** *Daily Tel.* 1 Dec. 19/6 (Advt.), An 18-hole midget golf course complete. **1933** *Pop. Sci. Monthly* Mar. 22/1 (*caption*) Midget submarine ready for its first test. **1934** *Discovery* Nov. 324/1 The combined radio-gramophone and electric clock..three midget valves. **1942** *Hutchinson's Pict. Hist. War* 18 Mar.–9 June 20 (*caption*) One of the Royal Navy's midget war ships, a motor gun-boat. [illegible] Japanese midget submarines are believed to exist. **1966** T. Pynchon *Crying of Lot 49* [illegible] [illegible] Nov. 6/3 The National Federation of Midget Automobile Manufacturers and Dealers' Associations revealed..that a total of 78,055 new midget automobiles were registered in October. *Ibid.*, Demand for midget commercial vehicles would increase.

Midi (mi·di), *sb.* [Fr.] The south of France. Also *attrib.*

1883 H. James *Little Tour in France* (1885) xvi. 110, I was at [illegible] **1936** C. Connolly *Rock Pool* [illegible] **1961** [illegible] **1967** A. Wilson *No Laughing Matter* [illegible] **1972** *Guardian* 22 Feb. 4/4 For electoral purposes the Midi is usually divided into the Midi-Pyrénées and the Midi-Méditerranée or Languedoc-Roussillon.

midi-, comb. f. of Mid *a.*, Middle *a.*, in imitation of *maxi- and *mini-, denoting garments longer than mini- but shorter than maxi-, normally extending to mid-calf. Hence **midi** *sb.*, such a garment. So *midi-length*.

1967 *Word Study* Dec. 3/2 In contrast to the miniskirt a skirt whose hemline comes as low as midcalf has been called a *midiskirt*. **1968** *Chicago Tribune* 9 July [illegible] A swaggering midi-coat that gives a new proportion to the trouser costume. **1969** *Sun* 13 Mar. 8/2 The newsy midi hem-length (note midi, *not* maxi). **1969** *Internat. Herald Tribune* (Paris) 6 Nov. 6/2 Almost every designer is nervously groping for the right length; it's touch and go whether the average woman will go for the maxis and midis. **1970** *Daily Tel.* 14 Jan. 15/3 Hardy Amies proved yesterday..that the midi length only works if it is waisted or broken by a hip-belt. His spring collection was awash with midis. *Ibid.* 15/4 Midi-dresses with big ballooning sleeves, hip sashes, polo necks. **1971** *Petticoat* 17 July 23 Baggage and General midi bumper boots, £4. **1974** *Country Life* 14 Feb. 334/2 A slim-cut, midi-length trench [coat].

‖ **midinette** (mid*i*net). [Fr. Perh. orig. a portmanteau word f. *midi* mid-day + *dînette* light dinner: cf. 1922 *Larousse* s.v., les *midinettes* sont celles qui se contentent d'une *dînette à midi*.] A milliner's female assistant, esp. in Paris.

1909 *Westm. Gaz.* 7 Aug. 15/1 The Parisian..is tired of the absurd hat. The midinette and those of her kind have made it impossible. **1919** Beerbohm *Seven Men* 11 A midinette who..murdered, or was about to murder, a mannequin. **1925** *Brit. Weekly* 19 Feb. 498/2 She was in the same class with the midinettes of Paris, with whom travelling Englishmen spend a passing hour and quickly forget. **1932** Sellar & Yeatman *And now All This* vi. 60 In the South, the beautiful *Midi*, one does *not* perceive the beautiful Race of the *Midinettes* (*hélas*). **1958** J. MacGregor *Glamour in Your Lens* 84 She'll..be prepared to transform herself on the spot into..a French midinette. **1961** R. Searle *Which Way did he Go?* 38 He firmly believes that the Boulevards abound with Midinettes who are dazzled with admiration for him. **1975** R. Cobb *Paris & its Provinces* 3 Some story in which a *midinette* meets a *Duc*.

Midland, *sb.* and *a.* Add: **A.** *sb.* **b.** Used *ellipt.* for the names of companies or organizations, as Midland Bank, Midland Railway.

1869 *Bradshaw's Railway Manual* XXI. 86 The question was..referred to the arbitration of Captain Galton, who decided that the Midland might work the local line. **1959** *Chambers's Encycl.* XI. 459/2 When the Midland and the Glasgow and South Western decided to throw in their lot together, approval was not given, and the only tangible result..was the adoption by the Scottish company of the Midland lake livery. **1972** C. Drummond *Death at Bar* i. 14 His banking account, with the Midland, is divided into business and private.

c. The central area of the United States (see quot. 1896), esp. regarded as a dialectal area of American English.

1896 *Dialect Notes* I. iv. 438 Midland: a belt separating the North from the South and extending from the Atlantic City and the Mississippi (including Long Island, New York City and the adjoining counties, New Jersey, Del., all but the northern strip of Penn., the upper prong of West Virginia, southern Ohio, middle Ind., middle Ill., and St. Louis county, Mo.). **1937** *Amer. Speech* XII. 316/1 For many years *Midland West*..has reigned in solitary grandeur. **1963** R. I. McDavid *Mencken's Amer. Lang.* 405 The 'General American' area is really made up of two major dialects: one, Inland Northern..; the other, Midland, based on the speech of Pennsylvania. **1972** H. Kurath *Stud. Area Ling.* 44 The transition area between the North and the Midland reflects partly the complicated history of the settlement.

B. *adj.* **1. b.** (Earlier and later examples.) Also *pl.*

1756 A. Butler *Lives Saints* I. 45 St. Cedd..first preached to the Midland English. **1849** D. Rock *Church of our Fathers* I. i. v. 351 [illegible] **1922** Joyce *Ulysses* [illegible] The Link line railway laid..between the cattle park, Liffey junction, and terminus of Midland Great Western Railway..in proximity to the terminal stations or Dublin branches of Great Central Railway, Midland Railway of England [etc.]. **1942** *Short Guide Gt. Brit.* (U.S. War Dept.) 4 The great 'midland' manufacturing cities of Birmingham, Sheffield, and Coventry. **1954** Darby & Terrett (*title*) The Domesday geography of Midland England. **1972** [illegible] **1974** D. Woodhouse *Mama Doll* xiii. 180 The few words he did speak were spoken in a Midland accent.

c. Of or pertaining to the Midland of the United States or the regional type of American English spoken there.

1896 *Dialect Notes* I. ii. 57 But the differences in the different sections of the country are not so great that we can properly speak of a New England dialect, a southern dialect, a Midland dialect. **1900** B. B. Smyth *Plants & Flowers Kansas* 45 This is the Midland adder-tongue, as named by Prof. Knerr of Atchison. **1944** H. Kurath in *Language* XX. 151, I hope to be able to show before long, on the basis of the Atlas materials, that we must recognize a large Central or Midland speech area in addition to the Southern, the Northern, and the [illegible]. **1949** —— *Word Geogr. Eastern U.S.* p. v, There is an extensive Midland speech area that lies between the traditionally recognized 'Northern' and 'Southern' areas. **1959** E. Tunis *Indians* vi. 86 Midland Indians were farmers first and, like them all, hunted buffalo.

Midlander. (Further examples.) Also one who lives in the Midland of the United States.

1889 G. B. Shaw *London Music 1888–89* (1937) 243, I was at Leicester, delivering to the midlanders an impassioned appeal. **1914** Belloc *Green Overcoat* v. 191 'I know you would!' said the big Midlander. **1932** *Times Lit. Suppl.* 14 Jan. 19/2 Himself a Midlander, he is very insistent on the special virtues which Shakespeare inherited from his Midland birth. **1972** H. Kurath *Stud. Area Ling.* 53 The proportion of 'Northerners' and 'Midlanders' in these two states as a whole differs little. **1972** *Country Life* 14 Dec. 1638/1 It is always with mild surprise that you recall that Dr. Johnson was a Midlander.., that he was born not within the sound of Bow Bells but of the bells of Lichfield Cathedral.

mid-line (mi·dləin). [Mid *a.*] **1.** *Zool.* A median line; also, the median plane or plane of bilateral symmetry.

1868 [in Dict. s.v. Mid *a.* 2 d]. **1927** Haldane & Huxley *Animal Biol.* xii. 285 (*caption*) The lateral nerve-trunks have united in the mid-line. **1957** R. H. Smythe *Conformation of Dog* 61 Behind the base of the neck in the midline of the back. **1959** Southwood & Leston *Land & Water Bugs Brit. Isles* 31 The males of this genus have a patch of wax-producing glands beneath the abdomen on each side of the mid-line. **1961** *Lancet* 26 Aug. 443/1 (*caption*) Coronal section of brain of patient who survived for 6 weeks. Note that the only abnormal appearances are slight dilatation of the ventricles and a tear (arrow) on the underside of the corpus callosum to the right of the midline. **1974** *Nature* 8 Mar. 165/1 The roentgenographic landmarks were joined to form triangles on both sides of the midline.

2. The middle of a line of poetry.

1882 G. M. Hopkins *Lett. to R. Bridges* (1955) 164 The word is *more* and is a midline rhyme to *score*.

midlittoral (midli·tŏrăl), *a.* and *sb.* *Ecol.* [f. Mid *a.* + Littoral *a.* and *sb.*] **A.** *adj.* Designating that zone on the sea-shore which is both covered and uncovered by the neap tides. **B.** *sb.* The zone so delimited.

1948 *Austral. Jrnl. Sci. Res.* B. I. 196 (*heading*) The mid- and upper-littoral. **1949** T. A. & A. Stephenson in *Jrnl. Ecol.* XXXVII. 298 The most difficult zone to rename is the Balanoid. This is the middle zone on the

shore... On the whole we favour the name *midlittoral zone* as being in keeping with the other terms proposed. *Ibid.*, The word *midlittoral* is less common, though it has been used by Dakin *et al.* (1948) in their recent account of the New South Wales Coast, even if not as an exact equivalent of our Balanoid zone. **1967** *Oceanogr. & Marine Biol.* V. 469 The lowest levels of the midlittoral. **1970** R. C. Newell *Biol. Intertidal Animals* i. 2 The second zone recognised by Stephenson and Stephenson (1949) is the midlittoral zone, characterized by barnacles and limpets. **1974** *Encycl. Brit. Macropædia* IV. 804/1 The midlittoral of sand and mud shores is usually inhabited by several types of polychaete worms, the most typical being lugworms.

midmost, *a.* and *adv.* Add: **A.** *adj.* **1. b.** *absol.* (passing into *sb.*) (Further *poet.* examples.)

1912 E. Pound tr. *Cavalcanti's Sonnets & Ballate* 19 As though I'd won unto his heart's mid-most. **1915** —— *Cathay* 7 And within, the mistress, in the midmost of her youth. **1955** —— *Classic Anthol.* II. 142, I have held him in love so long From heart's mid-most be it song Not to be lost.

midnight, *sb.* Add: **4. a.** *midnight mass* (earlier and later examples).

1665–7 J. Lauder *Jrnls.* in *Publ. Scottish Hist. Soc.* (1900) XXXVI. 118 The rest of our Scotsmen ware so curious as to go hear Midnight Masses. **1848** J. H. Newman *Let.* 25 Dec. (1962) XII. 382 The midnight mass was a high one. **1866** G. M. Hopkins *Lett. to R. Bridges* (1955) 16 It is not much more than an hour to Xmas day: I am sitting up for the midnight mass. **1960** A. Christie *Adventure of Christmas Pudding* 33 Who's going to brave the snow and go to midnight mass?

5. † **midnight banquet** = *midnight feast*; **midnight black,** an intense black colour; also *attrib.*; **midnight blue,** a very dark shade of blue; also *attrib.* or as *adj.*; **midnight feast,** a feast at midnight; *spec.* a children's secret feast in a school dormitory or the like; **midnight matinée,** a special theatrical performance presented at midnight; **midnight-tide** *poet.*, midnight.

1896 C. M. Yonge in C. Coleridge *C. M. Yonge* (1903) ii. 114 She invited the other young ladies to a midnight banquet..in their night-caps and dressing gowns. **1925** J. Gregory *Bab of Backwoods* xvi. 201 Monte made out vaguely the slim form in white, a whitish blur against a midnight-black curtain. **1922** H. Crane *Let.* 2 Mar. (1965) 80 Pastel tinted flowers on a background of midnight black. **1916** *Daily Colonist* (Victoria, B.C.) 8 July 2/1 (Advt.), The same style [of shoe] in jungle brown, midnight blue and bronze. **1935** R. Hichens *Afterglow* 86 A silken collar and a midnight blue tie. **1937** *Times* 10 July 15/4 Dressed in midnight-blue crêpe-de-chine. **1940** R. Chandler *Farewell, my Lovely* xxxviii. 289 His dinner clothes were midnight blue. **1972** *Guardian* 5 Dec. 11/2 Toga dress..in..midnight blue. **1938** A. Christie *Appointment with Death* i. vii. 52 This is rather fun... Rather like the midnight feasts we used to have at school. **1964** C. Hodder-Williams *Main Experiment* i. vi. 57 'When will you be back from London?' 'Late tonight.' 'We'll have a midnight feast, then.' **1972** *Listener* 2 Nov. 614/1 Midnight feasts, practical jokes and all the fun of the dormitory. **1952** Granville *Dict. Theatr. Terms* 118 Midnight matinée. **1963** *Daily Express* 25 Sept. 1/4, I don't honestly see why the Stationery Office have to put on a midnight matinee for the Denning Report, even if it has got a U Certificate. **1918** W. de la Mare *Motley* 4 At cold of midnight-tide.

mid-ocean. [Mid *a.*] **a.** The middle of an ocean.

1697, 1881 [in Dict. s.v. Mid *a.* 1 d]. **1905** *Smart Set* Sept. 116/2 It was as surely and as irrevocably lost as though it had sunk into a thousand fathoms of midocean. **1930** S. Spender *20 Poems* 3 Like Icarus mid-ocean-drowned. **1938** *Times Lit. Suppl.* 3 Sept. 572/4 His repairs in mid-ocean to his small auxiliary engine.

b. *attrib.*, esp. in **mid-ocean ridge,** one of the mountainous ridges, several kilometres high and over a thousand wide, that rise abruptly from the abyssal plain in the middle of each ocean basin and form a connected worldwide system, marked along much of its length by a central rift that overlies earthquake foci.

1956 *Geophysical Monogr.* I. 75/1 Mid-ocean islands. **1961** Ewing & Landisman in M. Sears *Oceanogr.* i. 5 The remaining 18% [of the earth's surface]..contains parts of the mid-ocean ridge system. **1963** G. L. Pickard *Descriptive Physical Oceanogr.* ii. 10 The Mid-ocean Ridge is probably the most extensive single feature of the earth's topography. **1968** M. J. Keen *Introd. Marine Geol.* iii. 46 The abyssal plains may be cut..by mid-ocean canyons. **1968** *Times* 8 Oct. 7/7 Present-day oceans are marked by underwater mountain ranges, called mid-ocean ridges, along which molten material is injected upwards from the earth's mantle and spreads out on either side of the ridge. **1971** J. R. Heirtzler in A. E. Maxwell *Sea* IV. i. iii. 99 The Red Sea..is bisected by the mid-ocean ridge which passes south through the Gulf of Aden and the Indian Ocean.

So **mid-ocea·nic** *a.*

1961 Ewing & Landisman in M. Sears *Oceanogr.* i. 9 On every crossing the ridge has been found where it was expected, with its median rift accurately following the mid-oceanic earthquake belt. **1966** R. W. Fairbridge *Encycl. Oceanogr.* 506 The Mid-Atlantic Ridge is that portion of the mid-oceanic ridge system which lies within the limits of the Atlantic Ocean. **1970** L. R. A. Capurro *Oceanogr. for Practicing Engineers* vii. 139 The sea bottom can be divided into three major morphologic divisions: continental margin, ocean-basin floor, and midoceanic ridge.

mid-off. In etym. for 'Mid *a.* 6' read 'Mid *a.* 1 d'. Add earlier and later examples.

1865 J. Lillywhite *Cricketer's Compan.* 25 Mr. Lyttelton from a fine hit, was cleverly had by Mr. Voules at deep mid off. **1867** J. Lawrence *Handbk. Cricket Ireland 1866–67* 106 Harrow drive or Mid off. **1955** *Times* 9 May 15/2 He was dropped off him at mid-off when he had got to 29. **1959** [see *deepish *a.*].

mid-on. (Earlier and later examples.)

1870 *Times* 20 July 10/3 At 21 Smith was caught at mid-on. **1873** J. Pycroft *Cricket Field* (ed. 6) xii. 289 Some bowlers prefer to dispense with long-leg... In case of his removal it would be well to place him at mid-on. **1955** *Times* 12 May 4/5 Twenty-five minutes afterwards Watson lost patience with himself and mis-hit Anderson to mid-on and then Lowson, chancing his arm in the same sort of way, was well caught also at mid-on. **1959** [see *check *v.*¹ 16 d]. **1974** *Observer* 9 June 24/7 He hit one excellent four..and a single, and then holed out to mid-on.

mid-ra·nge. [Range *sb.*¹] **1.** *Statistics.* The arithmetic mean of the largest and the smallest values of a group, esp. a sample.

1949 *Ann. Math. Statistics* XX. 257 The significance tests investigated..are based on the quantity [(sample midrange) − (hypothetical mean)]/(sample range). **1951** Dixon & Massey *Introd. Statistical Analysis* xvi. 238 It is probably not advisable to use the midrange for samples of more than five observations since its efficiency drops below that of the median beyond this point. **1973** M. H. Belz *Statistical Methods for Process Industries* ii. 22 There are four commonly employed measures of sample location, the sample mean, the sample median, the sample mode and the sample mid-range.

2. a. The middle part of the range of audible frequencies. Freq. *attrib.*

1955 E. T. Canby *Home Mus. Syst.* (ed. 2) x. 160 The newest wrinkles in the speaker department are the tiny super tweeter, for crispness in extremely high tones, and the large mid-range tweeter, to take over the middle 'intelligence' highs. **1960** C. Brown *Introd. Hi-Fi* iv. 89 A third speaker to handle the mid-range can be introduced if required. **1971** *Hi-Fi Sound* Feb. 68/1 The radiogram has speaker units which are..surprisingly sensitive, particularly in the mid-range. **1975** *Gramophone* Jan. 1425/1 The tweeter maintains the smooth and wide dispersion of energy as it takes over from the bass and mid-range unit.

b. A loudspeaker designed to reproduce mid-range signals with fidelity while being relatively unresponsive to those of low or high frequency.

1955 E. T. Canby *Home Mus. Syst.* (ed. 2) x. 160 A cheap and effective alternative is an ordinary medium-sized cone speaker used as a mid-range. **1975** *Hi-Fi Answers* Feb. 68/1 It was specifically developed as a good mid-range for the new Leak 2000 series of speakers.

midrib. Add: **2.** (Later examples.)

1901 *Chambers's Jrnl.* May 301/2 The leaves are thoroughly dried. In testing their dryness the mid-rib or vein should not be overlooked. **1965** K. Esau *Plant Anat.* (ed. 2) xvi. 434 In some large grasses the median part of the blade is thickened into a midrib. **1974** *Country Life* 7 Mar. 479/2 Gracillimus has narrow leaves with a silver midrib.

b. *Archæol.* A structure similar to the principal rib or vein of a leaf.

1929 V. G. Childe *Danube in Prehist.* 126 Kite-shaped daggers with rivet-holes and a rudimentary midrib. **1938** *Proc. Prehist. Soc.* IV. 283 Three flanged axes were found with a tanged spearhead with short blade and pronounced mid-rib.

midriff. Add: **1. b.** *Fashion.* (*a*) The mid-portion of the torso; (*b*) that part of a woman's garment which covers the midriff, *esp.* if cut separately from the upper bodice; (*c*) *U.S.* a woman's garment which leaves the midriff uncovered.

1941 *Fashion Digest* Fall 74 Playsuit with wide-midriff and full-pleated skirt. **1948** *Harper's Bazaar* Jan. 99 Inside the elasticized bodice is a lightly boned midriff, an uplift brassiere. **1953** P. Binder *Muffs & Morals* i. 31 The midriff mode..came in just before the Second World War... It shocked nobody. **1964** *McCall's Sewing* 12/1 Wide sashes, contrasting belts, set-in midriffs will give a lift to the waistline. **1966** L. Halliday *Fashion Makers* ii. 41 A modern girl will cheerfully bare her midriff on the beach. **1970** *Daily Tel.* 23 Feb. 13 A real Slim Jane? Then only you, one prays, will dare to wear the new bare-midriff styles. **1972** E. Larsen *Creative Dressmaker* viii. 73 Cover the midriff with a piece of lace.

mid-season. 2. (Further examples.)

1921 Wodehouse *Indiscretions of Archie* xix. 210, I was just putting old Bill through it..with a view to getting him into mid-season form. **1946** —— *Joy in Morning* v. 35 Though crushed to earth..he rises again—not absolutely in midseason form, perhaps, but perkier than you would expect. **1969** *Burpee Catal.* 81/2 Sweet Corn... A midseason corn. *Ibid.* 123/2 Delicious [straw]berries in midseason.

midshipman. Add: **3. midshipman's hitch** (see quot. 1886); **midshipman's nuts** (earlier examples); **midshipman's roll** (see quot. 1857).

1808 D. Lever *Young Sea Officer's Sheet Anchor* 9 To make a Midshipman's Hitch. **1886** *Encycl. Brit.* XXI. 591/2 *Midshipman's Hitch.*—Take two round turns inside the bight, the same as a half-hitch repeated; stop up the end; or let another half-hitch be taken or held by hand. Used for hooking a tackle for a temporary purpose. **1828** *Night Watch* II. viii. 50 'You shall have a fistful of midshipman's nuts to crack for your supper.'.. He gave me some broken biscuits. **1846** H. Melville *Typee* vi. 38, I took a double handful of those small, broken, flinty bits of biscuit which generally go by the name of 'midshipmen's nuts'. **1826** W. N. Glascock *Naval Sketch-Bk.* (ed. 2) I. 8 Get your hammock slung... 'Cause,..none o' your 'midshipman's rolls', you know! **1857** H. E. Davenport *Rovings on Land & Sea* 280 If any of the hammocks are lashed in a slovenly manner, or merely bundled up in what is called a 'midshipman's roll', the owner is..made to secure it in a more ship-shape manner.

midstream. Add: **2.** *Med.* Used, usu. *attrib.*, to designate any portion of urine passed by an individual other than that first passed or last passed in an act of urination.

1958 *New Engl. Jrnl. Med.* CCLIX. 764 (*heading*) A comparison of bacterial counts of the urine obtained by needle aspiration of the bladder, catheterization and midstream-voided methods. **1962** *Lancet* 15 Dec. 1246/1 The urine was examined microscopically for pus cells... In men and children midstream specimens were used. **1972** *Ibid.* 2 Sept. 452/2 Specimens of midstream urine were collected in sterile universal bottles.

midsummer. Add: **3. midsummer madness** (further examples).

1914 C. Mackenzie *Sinister St.* II. iii. v. 594 The freshmen..celebrated the beauty of the season with a good deal of midsummer madness. **1921** Galsworthy *To Let* II. ix. 200 Come, be reasonable, Fleur! It's midsummer madness! **1922** Joyce *Ulysses* 482 This is midsummer madness, some ghastly joke again. **1970** M. Peters in *Midsummer Variations* 203 Midsummer madness they all said That summer day when we were wed.

midway. Add: **A.** *sb.* **3.** *U.S.* Freq. with capital initial. At an exhibition, fair, or the like: a central avenue along which the chief exhibits or amusements are placed; hence, any cheap place of amusement; *slang*, a hall. Also *attrib.*

The use originated in the inclusion of the 'Midway Plaisance' of Chicago in the grounds of the exposition held there in 1893.

1893 *Outing* Dec. 208/1 The waiters were the hardiest set on the Midway. **1901** *World's Work* Aug. 1097/2 Nowadays we frankly admit that the Midway is the strongest magnet of a big fair. **1901** *Everybody's Mag.* Oct. 424 Can I arrange with you for placing a first-class Midway on your grounds? *Ibid.* 427/1 A mile and a half of kerwhango and clash—That is the Midway show. **1927** K. Nicholson *Barker* 149 *Midway Confab*, news column in the *Billboard*, the weekly trade paper for outdoor showfolk. **1932** *Evening Sun* (Baltimore) 9 Dec. 31/5 *Midway*, a hall. **1949** *Sat. Even. Post* 17 Sept. 25/1, I have worked in..big-time railroad shows whose midways had the glitter of a handful of diamonds. **1956** H. Gold *Man who was not with It* (1965) i. 3 There he is on the midway, Grack the Frenchie, talking for his Countstore.

C. *adv.* **1.** Also *U.S.* const. *of*.

1812 J. J. Henry *Accurate Acct. Heroes Campaign against Quebec* 192 About midway of the horn [of the moose],..there is a broad flat part. **1896** [in Dict.]. **1903** *Nation* (N.Y.) 17 Sept. 234 He died midway of his 70th year. **1927** *Sat. Even. Post* 24 Dec. 44/3 She stopped midway of her sentence.

midwife, *sb.* Add: **5. midwife toad,** a European toad, *Alytes obstetricans*, the male of which cares for the eggs until they hatch.

1901 H. Gadow in *Cambr. Nat. Hist.* VIII. vi. 158 *A[lytes] obstetricans*, the 'Midwife-toad', has the general appearance of a smooth toad... The pairing and the peculiar mode of taking care of the eggs by the male..has given it the specific name *obstetricans*, the midwife. **1934** *Times Lit. Suppl.* 16 June p. iv/2 A number of interesting toads have been received lately at the London Zoo, including a collection from Germany of 20 midwife toads with their eggs. **1954** G. Durrell *Three Singles to Adventure* viii. 180 The midwife toad of Europe, instead of leaving its eggs in the nearest water to hatch unattended, hands them over to the male, who winds them round his hind legs and carries them about until they hatch. **1971** Koestler *Case of Midwife Toad* i. 14 Kammerer's undoing was a grotesque amphibian creature: the midwife toad, *Alytes obstetricans*.

midwife, *v.* Add: **2. b.** (This sense not *rare*.) (Further examples.)

1959 *Times Lit. Suppl.* 4 Dec. 709/3 It also midwifed the late editor of *Commentary* himself, Elliot E. Cohen. **1971** B. Malamud *Tenants* 83 Lesser, dreaming of new light in his book, beheld in his dark thoughts Bill Spear, potential executioner, requesting him to midwife his bloody fable. **1972** *Sunday Tel.* 30 Apr. 7/7 As befitted the man who midwived 'That Was The Week That Was', Sir Hugh also got the best laugh at the Colonels' expense.

midwifely, *a.* Delete † *Obs.* and add to def.: also, characteristic of a midwife.

1936 'R. West' *Thinking Reed* i. 15 'When a woman is very tired,' he said, with a return of midwifely sententiousness, 'she does not know what is the matter with her.'

mid-year. Add: **3.** The middle of the year. Also *attrib.*

1901 *U.S. Dept. Agric. Yearbk.* 154 To teachers the series of meetings is a series of mid-year institutes. **1909** *Cent. Dict.* Suppl. s.v., Rents due at the midyear. **1932**

Irish Press 10 Mar. 5/3 This corresponded to an annual rate of 14·6 per thousand of the estimated mid-year population for the year 1931. **1973** E. Taylor *Serpent under It* (1974) v. 82 This may set us back months. Instead of getting our degrees at midyears, we'll probably have to wait now until next June.

Miehle (mī·lə). The name of Robert *Miehle*, 19th-century American printer, used *attrib.* or *absol.* to designate a flat-bed, cylinder printing press invented by him in 1884, or later developments of this machine.

1887 *Inland Printer* Oct. 35/1 (*heading*) The Miehle two-revolution press. **1894** *Amer. Dict. Printing & Bookmaking* 378/1 *Miehle Press*, a new press, lately devised by Robert Miehle, a pressman of Chicago. The bed is carried in harmony with the movement of the cylinder while printing, then it is gradually slowed up, carried smoothly over the centre, and returned for the back movement... Great speed is claimed for this press. **1915** *Southward's Mod. Printing* (ed. 3) II. viii. 101 As a standard machine of the very highest class, the Miehle is widely in use. *Ibid.* 102 (*caption*) Miehle Two-Revolution Press. *Ibid.* 104 Miehle presses are also used in tandem, coupling up two or more machines for the production of colour work. **1963** *Printing & Mind of Man: Catal. Display of Printing Mechanisms & Printed Materials Brit. Mus. & Earls Court* (Earls Court Section) 80/2 The Miehle machine, in several models, continues to be marketed and sold all over the world. The cylinder revolves continuously, is raised after the impression in order to clear the forme during its second revolution, and is then brought down on impression again by the action of a cam and eccentrics. **1966** Berry & Poole *Ann. Printing* 258/2 Cassell's were the first printers in this country to import Miehles... For many years the Miehle was in a class by itself for the new three-colour letterpress. **1973** J. Moran *Printing Presses* xi. 159 The Miehle, which had great impressional strength, accurate register and durability..kept its lead for some eighty years.

mielie (mī·li). *S. Afr.* [Afrikaans.] = Mealie. Also *attrib.* and *Comb.* as *mielie cob, land*; **mieliepap,** mealie meal porridge (cf. quots. s.v. *mealie).

1804 J. T. van der Kemp in *Trans.* [London] *Missionary Soc.* (ed. 2) I. 438 There is another kind of corn, which they call *bona*, and is known in the Colony by the name of *meelis*. **1818** G. Barker *Diary* 7 Dec. (MS.), Began to clear the beans & mielies in the garden. **1835** A. Smith *Diary* 28 Nov. (1940) II. 285 The Hottentots..had no pumpkins, melons, mielies nor calabases. **1926** O. Schreiner *From Man to Man* i. 78 Near the mielie lands. *Ibid.* iii. 117 The maids were lighting the evening fires with mielie cobs. **1952** *Cape Times* 11 Jan. 3/4 The nearby Vergelegen Estate with its acres of corn and mielies. **1953** P. Abrahams *Return to Goli* vi. 200 It was the kind of food I had not eaten for fourteen years: a thick *mielie pap*. **1959** *Cape Times* 26 Oct. 9/4 The mieliepap..is the staple food of millions in Africa. **1971** *Cape Herald* (Bonus) 15 May 5/1 (Advt.), With mieliepap you need never fear that some of the goodness has been lost in making. **1974** *Eastern Province Herald* 23 July 13 He is wise to discard his bowler hat when crossing a field of mielies.

mien (mīn), *sb.*² [Chinese, lit. = wheat flour.] Wheat flour noodles. (Cf. *chow mein.)

1934 in Webster. **1936** P. Fleming *News from Tartary* v. ix. 235 At noon they woke me for a meal of mien and boiled vegetables. **1950** R. Alley *Leaves from Sandan Notebk.* 46 They gave him a meal of mien. **1956** B. Y. Chao *How to cook & eat in Chinese* xix. 218 Wheat is eaten more than rice as the staple food... When it is eaten wet, it is in the form of *mien*, or unraised noodles. **1971** N. Froud *Far Eastern Cooking for Pleasure* 82 *Mien*, Chinese noodle or wun tun paste. This dough is used for Chinese noodles, wun tun and various other patties and dumplings.

mierkat. = Meerkat 2.

1901 O. Schreiner *Thoughts on S. Afr.* (1923) 19 There was a possibility that the red African mier-kat might ultimately creep back into its hole in the red African earth. **1949** *Cape Argus Mag.* 3 Sept. 1/9 To most people it is just a 'mierkat', but in reality it is a rodent, a squirrel, and quite different from the insect-eating, slender-tailed mierkat that is to be found in the same vicinity.

miersite (məi·əɪzəit). *Min.* [f. the name of Sir Henry Alexander *Miers* (1858–1942), English mineralogist + -ite¹.] An iodide of silver and copper, (Ag, Cu)I (with the ratio of silver to copper approximately 4:1), which is found as yellow isometric crystals at Broken Hill, New South Wales, Australia.

1898 L. J. Spencer in *Nature* 14 Apr. 574/2 The new mineral has been named in honour of Mr. H. A. Miers, F.R.S., Professor of Mineralogy at Oxford, who first correctly determined the crystalline form of marshite, a mineral so closely resembling miersite in appearance that the two species are only distinguished by chemical tests. **1901** [see *isomorphously *adv.*]. **1922** *Mineral. Abstr.* I. 305 Silver iodide and copper iodide when fused together show..specks of birefringent iodyrite in a base of isotropic miersite. **1951** C. Palache et al. *Dana's Syst. Min.* (ed. 7) II. 20 Both miersite and iodyrite are obtained by metathical [*read* metathetical] reaction in water solution at ordinary temperature.

Miesian (mī·ziăn), *a.* and *sb.* [Name of the German-American architect, L. *Mies* van der Rohe (1886–1969) + -ian.] **A.** *adj.* Of, pertaining to, or characteristic of the style of architecture of Mies van der Rohe. **B.** *sb.* A devotee or follower of Mies's style.

1956 *Archit. Rev.* CXX. 238/2 The job has some obviously Miesian passages. **1958** H. R. Hitchcock *Archit. 19th & 20th Cent.* III. xxiii. 390 The 'Miesian' is today almost a sub-school of the new architecture. **1960** A. Drexler *Ludwig Mies van der Rohe* 32 The measure of Mies's authority is this: it no longer seems possible to rebel against the Miesian discipline except in Miesian terms. **1963** tr. *G. Hatje's Encycl. Mod. Archit.* 194/2 Inside the enclosure there was the usual Miesian spatial continuity. **1965** M. Bradbury *Stepping Westward* 10 The new crematorium..was designed by a devout Miesian. **1969** *Archit. Rev.* CXLVI. 313 A long, low, not quite Miesian exercise in structural and not so structural steel and glass. **1973** *Current Affairs Bull.* (Sydney) Aug. 8/2 The shell-shapes.. are of Miesian derivation.

‖ **mietjie** (mī·tʃi, mī·kʸi). Also **meitjie, michi.** [Afrikaans.] = *Klaas's cuckoo* (*Klaas).

1853 *Edin. New Philos. Jrnl.* LV. 82 The pretty notes of the *michi* and *diedrick* further enliven the growing day. **1936** E. L. Gill *First Guide S. Afr. Birds* 108 The usual call of Klaas's Cuckoo is quite well represented by its Afrikaans name 'mietjie'. **1939** [see *diederik, diedrik]. **1970** McLachlan & Liversidge *Roberts's Birds S. Afr.* (ed. 3) 242 The Afrikaans name [of Klaas's cuckoo] is onomatopoeic, 'Meitjie' and gives the call pretty closely.

miff, *v.* Add: **1.** (Earlier and later examples of *transf.* use.)

1883 *N. & Q.* 6 Oct. 267/2 A curious word came under my notice of late with regard to a flower losing its strength and beauty. I was speaking to a Surrey gardener about some fading plants, and he remarked that they were 'miffing off'. **1907** R. Farrer *My Rock-garden* v. 71 Here it flowers once, then it miffs off without any apparent reason. **1960** F. C. Stern *Chalk Garden* vii. 73 We have sometimes been successful with it [*sc.* a peony], but it miffs off for no apparent reason.

2. (Earlier and later examples.)

1811 C. Mathews *Let.* Dec. in A. Mathews *Mem. Charles Mathews* (1838) II. 177 You give me much gratification by your explanation of the word that miffed me. **1889** *Kansas City* (Missouri) *Times & Star* 27 Nov., Dr. G. W. Fitzpatrick..is badly miffed because he wasn't appointed surveyor of the port. **1904** E. Robins *Magnetic North* i. 252 Don't get mifft, Colonel. **1907** *N.Y. Even. Post* (Semi-Weekly ed.) 2 Sept. 4 He is a little miffed to find that there are other lawyers in the Cabinet whose advice the President prefers to his own. **1957** *New Yorker* 26 Oct. 82/2 The feminine contingent..was..more than a little miffed, to learn that London..had got married. **1972** *Times Lit. Suppl.* 18 Feb. 173/3 Understandably, he is more than a little miffed by all this. **1973** *Daily Tel.* (Colour Suppl.) 23 Feb. 54/2 He told us a slightly improper story. The girls were not shocked but were rather miffed at his thinking they would not be.

miffish (mi·fiʃ), *a.* [f. Miff *sb.* + -ish¹.] = Miffy *a.* So **mi·ffishly** *adv.*

1957 *New Statesman* 2 Nov. 572/2 For all his shyness and laconicism, he had a green heart—miffish and virginal perhaps, but seldom sentimental. **1968** E. Hyams *Gardener's Bedside Bk.* 35 Failure with this lily was more common than success for so many years that gardeners still have the impression that it is very difficult and miffish. **1968** 'B. Mather' *Springers* i. 5 He went on miffishly: 'Did I ask for the job?' **1973** —— *Snowline* xv. 180 There was a lot of banter..directed at the earnest ones, who received it in silent and miffish dignity.

miffy, *a.* Add: Of a plant (further examples).

1871 S. Hibberd *Amateur's Flower Garden* 146 *M[yosotis] dissitiflora*..is a most valuable species for early flowers, but 'miffy', and therefore needing perpetual renewal. **1907** R. Farrer *My Rock-garden* i. 19 A sound perennial in one place, soil and climate, but a miffy delicate untrustworthy creature half a mile away. **1934** *Gardeners' Chronicle* 19 May 330/3 A few plants invariably become 'miffy', from no apparent reason, and die out.

Also **mi·ffily** *adv.*, in a miffy manner.

1958 'A. Bridge' *Portuguese Escape* 104 'All right, all right,' Melplash said, rather miffily. 'There's no rush.'

might, *sb.* Add: **7.** *dial.* A considerable quantity or amount.

1834 W. A. Caruthers *Kentuckian in N.Y.* I. 28 I'm 'bliged to do a might of business in Baltimore afore I can go on. **1878** J. H. Beadle *Western Wilds* ii. 29 It took a might of time. *Ibid.* 43 It was a might o' comfort, though, to see 'em 'fore they died. **1903** in *Eng. Dial. Dict.* (Yks., Suffolk). **1955** J. Masters *Coromandel!* i. 20 A sleeveless leather jerkin that..hid a might of queer things.

8. A possibility as distinct from a certainty.

1901 M. Franklin *My Brilliant Career* xvi. 133 'We might have both been drowned,' he said sternly. 'Mights don't fly,' I returned. **1922** Joyce *Ulysses* 344 There was just a might that he might be out. **1961** C. H. D. Todd *Pop. Whippet* 57 You don't want any *might* about it—you need to know.

might-be. Add: **B.** *adj.* That might be; (remotely) possible.

1934 R. Campbell *Broken Record* iv. 102 This was during the Douanier rage, when every painting-railway-porter was a might-be genius. **1941** L. MacNeice *Plant & Phantom* 26 A curfew..imposed Upon their might-be-wanderings; their might-be-applications For resurrection. **1956** Auden *Old Man's Road*, With might-be maps of might-have-been campaigns.

might-have-been. Add: (Further examples.) Also *attrib.*

1931 R. Campbell *Georgiad* iii. 63 That moonlit people of the might-have-been. **1955** A. L. Rowse *Expansion Eliz. Eng.* 90 The might-have-beens of history are not a very profitable subject. **1956** [see prec.]. **1970** *Guardian* 11 Aug. 11/8 One of history's might-have-been mysteries. **1974** R. Thomas *Porkchoppers* xxvi. 230 The name that symbolized the might-have-been world of Donald Cubbin. **1974** A. Goddard *Vienna Pursuit* III. 174 Overwhelmed with horror for the might-have-been, she paused.

mighty. Add: **B.** *adv.* (Later examples.)

1931 W. G. McAdoo *Crowded Yrs.* ii. 23 That seemed to me to be mighty good pay. **1958** *Times* 16 Oct. 17/1 They left it till mighty near no-side before they got their noses thankfully in front.

migma (mi·gmă). *Geol.* [a. Gr. μίγμα mixture, f. μ(ε)ιγνύναι to mix.] A mixture of solid and molten rock.

1943 *Proc. Geol. Assoc.* LIV. 73 The migma, if the amount of its liquid portion became great enough, would flow and could intrude itself into its surroundings in typical eruptive or intrusive fashion. **1952** T. F. W. Barth *Theoret. Petrol.* IV. xiii. 364 In the root parts of the folded mountains..a migma is formed that corresponds to the magma of the higher levels. **1974** *Encycl. Brit. Micropædia* VI. 880/1 Between these extremes is a concept of a pore fluid, or migma, generated by the differential melting or partial fusion of the root portions of mountains.

migmatite (mi·gmătəit). *Petrol.* [ad. Sw. *migmatit* (J. J. Sederholm 1907, in *Bull. Comm. Géol. Finlande* V. xxiii. 88), f. Gr. μίγμα (see prec.): see -ite¹.] A rock composed of a metamorphic host rock with streaks or veins of a granitic rock.

1907 J. J. Sederholm in *Bull. Comm. Géol. Finlande* V. xxiii. 110 For the gneisses here in question, characteristic of which are two elements of different genetic value,.. the author proposes the name of *migmatites*. **1942** *Proc. Geol. Assoc.* LIII. 77 *Pseudodioritic migmatite* is a field-term used to name a group of fairly homogeneous foliated hornblende-rich rocks having a dioritic aspect. **1965** G. J. Williams *Econ. Geol. N.Z.* xiii. 203/2 The monazites and xenotine [*read* xenotime] in migmatites of the Charleston-Fox River area, do not seem to have produced a red colour in the adjoining felspars. **1974** *Encycl. Brit. Micropædia* VI. 880/1 Many migmatites probably represent the partial fusion of the metamorphic host during extreme metamorphism.

Hence **migma·tic, migmati·tic** *adjs.*, composed of migmatite.

1926 *Mineral. Abstr.* III. 84 Three types of pegmatites are distinguished: I, normal pegmatites; II, contact-pegmatites..; and III, migmatic pegmatites. **1942** *Proc. Geol. Assoc.* LIII. 67 The region around Bettyhill on the north coast of Sutherland furnishes a splendid display of migmatitic rocks produced by the injection of granitic and pegmatitic material into the different lithological types found in the Moine series. **1968** K. R. Mehnert *Migmatites* i. 5 It is inappropriate to burden petrographic investigations of metamorphic or migmatic rocks from the beginning with the question about the possible parent rock. **1969** Bennison & Wright *Geol. Hist. Brit. Isles* iii. 53 The Moinian rocks are generally metasediments showing little metamorphic segregation, although in the central migmatitic core..they are gneissose.

migmatization (migmătəizē·ʃən). *Petrol.* [f. *migmat(ite + -ization.] The process by which a migmatite is formed.

1932 E. G. Woods *Baltic Region* vii. 67 A great deal of migmatization having taken place has masked very effectively the relationship of these rocks to one another. **1958** *Geol. Mag.* XCV. 383 The infrastructure, although undergoing migmatization, retains the trends of the older orogenic belt from which it is largely formed. **1971** *Scottish Jrnl. Geol.* VII. 323 The frequent spatial coincidence between areas of migmatization and centres of high-grade regional metamorphism has long been recognized.

Hence **mi·gmatized** *ppl. a.*, converted into a migmatite.

1958 *Geol. Mag.* XCV. 383 During the Caledonian orogeny, both the old crystalline rocks and a low structural level of the Caledonian geosynclinal rocks themselves became migmatized and rose as diapirs. **1971** *Scottish Jrnl. Geol.* VII. 305 In central and eastern Sutherland the Moine rocks are extensively migmatized.

migniardise. Delete † *Obs.* and add: Now only **mignardise.** Also, a fancy cake or similar delicacy, usu. served at the end of a meal. (Later examples.)

1931 Belloc in *One Hundred & One Ballades* 16 Elderly women full of Mignardises—The latest fruit of Adam's ancient sin. **1935** W. Stevens *Let.* 19 Dec. (1967) 303 Poulenc is a beautiful instance of *mignardise* in music. **1974** S. Coulter *Château* i. xvii. 132 Their digestive systems were dealing with Brioche de foie gras, ..ananas voilé, mignardises and fruit.

2. (See quot. 1950.) Also *attrib.*

1872 *Young Englishwoman* Nov. 606/2 A band ornamented with hem-stitch and a mignardise braid. **1873** *Ibid.* June 302/1 For this larger rosette commence with the centre star by joining the ends of a length of mignardise. **1882** Caulfeild & Saward *Dict. Needlework* 111/2 Mignardise is used almost entirely to form narrow edgings for underlinen and children's dresses. *Ibid.* 112/1 Scalloped edging.—Formed of two rows of Mignardise braid. **1950** 'Mercury' *Dict. Textile Terms* 345 *Mignardise*, a variety of crochet formed by inserting narrow fine ribbons into the design as the heavy part of the pattern that would otherwise be formed of crochet.

MIGNONETTE

mignonette. Add: **1. d.** A perfume derived from the flowers of the mignonette.

1897 *Sears Roebuck Catal.* 19/2 Perfumes..Crab Apple.. Mignonette..Sweet Pea. **1913** T. Eaton & Co. *Catal.* Spring & Summer 177/1 Perfumes..Jasmin Mignonette Opoponax. **1972** *Guardian* 22 Aug. 9/4 Jacksons have revived these flower perfumes..the shop plan to reintroduce other fragrances including wallflower, mignonette, and honeysuckle.

migod (mig͡ɒ·d), *int.* ¶ Representation of a colloq. pronunc. of *my God!* (My *poss. adj.* 3).

1953 M. Teague *Joyful Condemned* xx. 282 Migod, what they done to Trix! **1968** M. Richler *Cocksure* ii. 16 Migod, Mortimer thought. **1970** J. Cleary *Helga's Web* v. 78 Oh migod, isn't it terrible!

migraine. Delete ‖ and add further examples. Also *attrib.* and *fig.*

1892 [see *day *sb.* 19]. **1937** *Tablet* 23 Oct. 553/2 We feel quite anxious for this young man who has fierce migraine-like black-outs. **1961** R. Graves *More Poems* 5 Love is a universal migraine, A bright stain on the vision Blotting out reason. **1971** *Brit. Med. Bull.* XXVII. 33/2 The common concept of migraine is of a syndrome in which severe unilateral periodic headache is accompanied by nausea or vomiting and preceded by a warning which is usually visual. **1972** 'D. Shannon' *Ringer* (1972) ix. 149 He's not at all well. A migraine headache. **1975** J. Symons *Three Pipe Problem* x. 72 Taking a pill for a mild migraine attack.

‖ migraineur (migrɛnœ̄r). [Fr. (cf. Migraine.] One who suffers from migraine.

1977 *Times* 25 Jan. 6/6 The migrainous subject (or migraineur). **1971** *Daily Tel.* 21 Feb. 11/3 The general and specific measures recommended for the prevention of migraine..the alleviation of the acute attack will prove a disappointment to the inveterate *migraineur* in search of relief but reflect the unpredictable and individual response to treatment. **1977** *Listener* 15 Apr. 480/3 Classical migraine is (pace G. B. Shaw) as uncommon among migraineurs as 'normal' vision among the rest of us.

migrainous, *a.* Add: Also, subject to attacks of migraine. (Further examples.)

1971 [see prec.]. **1973** *Tucson* (Arizona) *Daily Citizen* 22 Aug. 32/1 Migrainous women seldom put on weight, and they rarely develop any serious illness. **1974** *Radio Times* 6 June 66/1 Thousands of migrainous readers who read your feature 'The biggest aspirin splitter in the world'.

migrant. Add: **B.** *sb.* **c.** *Bot.* A plant whose distribution has changed or extended.

1905 F. E. Clements *Res. Methods Ecol.* iv. 219 *Migrant*, a plant that is migrating or invading. **1960** N. Polunin *Introd. Plant Geogr.* vi. 165 These recent migrants were aided by natural means—wind, water or animals—or by Man, through intentional or accidental importation.

migrate, *v.* Add: **1.** (Later examples, in *Chem.* (cf. *migration a.).)

1899 J. Walker *Introd. Physical Chem.* xx. 210 Had no silver ions migrated from the anode, the rise in concentration would have been 32–2. **1931** J. C. Ware *Analytical Chem.* i. ii. 11 Ions move or migrate independently in a solution and at different rates. **1938** E. S. Wallis in H. Gilman *Org. Chem.* I. viii. 724 The ease with which different groups migrate within the molecule is not wholly a property of the group itself, but is dependent to a varying extent on the molecule as a whole. **1967** J. N. Richards et al. *Elem. Org. Chem.* xi. 194 A carbonium ion is produced as an intermediate, and a methyl group migrates from an adjacent carbon to the positive center.

migration. **a.** Add: (Further examples.) *spec.* in *Chem.*, the (non-random) movement from one place to another of an atom or group, e.g. within a molecule as part of a rearrangement of its structure, or towards an electrode during electrolysis.

1879 *Encycl. Brit.* VIII. 108/2 For fused electrolytes a W-shaped tube..is sufficient; with solutions..the separation is more difficult, owing to the 'migration of the ions' and other causes. **1894** tr. *E. Goblet d'Alviella's Migration of Symbols* 82 Is it not the Winged Circle, whose migrations I trace in another chapter? **1898** *Jrnl. Chem. Soc.* LXXIII. i. 456 One of the following initial changes may occur. (1) Migration of the OH group. (2) Migration of the OSO_3H group as a whole. (3) Migration of a hydrogen atom of the ring, in the meta-position with regard to the side group. **1929** *Times* 13 Nov. 13/1 A serious obstacle to the work of archæologists, historians and others..is the migration of manuscripts. **1938** A. L. Raymond in H. Gilman *Org. Chem.* II. xvii. 148 The migration of the benzoyl group from position three to six in monoacetoneglucose. **1962** D. H. Calam in A. Pirie *Lens Metabolism Rel. Cataract* 439 Although the differences in migration of a group of mono-amino, mono-carboxylic acids..are small, they are well separated by chromatography.

c. *Nat. Hist.* Also, of plants, change or extension of distribution.

1905 F. E. Clements *Res. Methods Ecol.* iv. 216 Migration results when spores, seeds, fruits, offshoots, or plants are moved out of their home. **1932** Fuller & Conard tr. *Braun-Blanquet's Plant Sociol.* xiii. 306 The first step in the development of vegetation is 'migration'. **1966** E. Palmer *Plains of Camdeboo* xvi. 270 White and Sloane wondering at the reasons for plant migrations. **1973** Polunin & Smythies *Flowers of S.-W. Europe* ii. 86 The sierra ..has acted as a refuge for a number of montane species, which had in all probability previously undergone migrations and recessions culminating in the last ice age.

migrationist. Add: **2.** One who emphasizes the importance of migration in the distribution of species.

1918 L. Huxley *Life J. D. Hooker* II. xxxii. 98 Darwin was a migrationist; Forbes and others pushed the extension theory to excess.

‖ mihrab (miyrā·b). Also **9 mehhra·b, mehrab, mehreb, mihrâb.** [Arab. *miḥrāb* praying-place.] **1.** A niche, chamber, or slab in a mosque, indicating the direction of Mecca. Cf. Kiblah.

1816 H. M. Williams tr. *Alb Bey's Travels* II. xxi. 217 In the wall at the end of the nave is the mehreb or niche where the Iman places himself to direct the prayer. **1836** E. W. Lane *Acct. Manners & Customs Mod. Egyptians* I. iii. 94 In the centre of its exterior wall is the *mehhra·b* (or niche). **1839** J. Pardoe *Beauties Bosphorus* 81 The *mihrab*, or niche at the eastern extremity of the edifice. **1845** R. Ford *Hand-bk. for Travellers Spain* I. 376/2 The exquisite niche, the *Mihrab* or sanctuary, in which the Koran was deposited. **1883** E. O'Donovan *Merv* xx. 242 A large deep recess, furnished with a *mihrab*, or devotional station, with Bass 8 Fath. **1898** O. Pickthall Pattison's *Fact & Fancy* 38. The Sayyid took his station at the mihrab. **1930** E. Waugh *Labels* 141 Agia Sophia..whose whole architectural rectitude has been fatally disturbed by the reorientation of the *mihrab*. **1931** *N. & Q.* 17 Oct. 271/1 The charm of the mosque with its *mihrab*—a niche..and its arcades—which carry nothing. **1950** *Chambers's Encycl.* VII. 763/2 The focus on the side nearest to Mecca was much deeper than the others and formed a sanctuary with a central niche (*mihrab*) in its back wall. **1974** *Observer* (Colour Suppl.) 24 Nov. 43/2 The focal point of a mosque is the *mihrab*, a niche in the wall nearest to Mecca to which the congregation turns in prayer.

2. A niche motif on an Oriental prayer rug, resembling the shape of a mihrab in a mosque.

1900 J. K. Mumford *Oriental Rugs* x. 172 *Niche* or *Mihrab*..the name applied to the pointed design at one end of a prayer rug. **1931** A. U. Dilley *Oriental Rugs & Carpets* viii. 201 Above this field is either one large prayer niche (called a mihrab), shaped like a tent, or a row of smaller prayer niches. **1963** tr. *J. Schlosser's European & Oriental Rugs & Carpets* 38 The mihrab, at first curved, later became steep and pointed... Scrolling stems with large flowers, or small branches..fill the border, the spandrels, and the cross panel which is usually above the mihrab. **1972** P. L. Phillips in *Formenton's Oriental Rugs & Carpets* 23 The mihrab, the niche in the centre of the rug, is often separated by two columns.

Mikado. Add: **2.** Mikado pheasant, a pheasant native to the island of Formosa, *Syrmaticus mikado*, first described in 1906 from specimens in the Mikado's collection in Tokyo.

[**1906** W. R. Ogilvie-Grant in *Bull. Brit. Ornith. Club* XVI. 123 Among the Mikado's collection of live animals and birds, at Tokio, there are said to be a pair of Pheasants from Formosa belonging to an undescribed species.] **1922** C. W. Beebe *Monogr. Pheasants* III. 200 In appearance,..the Mikado Pheasant resembles the tragopans and impeyans, being heavy bodied and rather thick necked. *Ibid.* 201 Several healthy hybrids with the Elliot pheasant..strongly resemble the female Mikado. **1965** P. Wayre *Wind in Reeds* xv. 211 The beautiful Mikado Pheasant, of which the male is a deep bluish purple with red wattles and a purple and white barred tail,..found only on..Taiwan. *Ibid.* 212 There the Mikado inhabits the bamboo and juniper thickets..above five thousand feet. **1972** *Shooting Times & Country Mag.* 1 July 19/1 Next on the list for rehabilitation in Formosa (Taiwan to most, these days) is the Mikado pheasant. The Trust is breeding this bird and hopes soon to have enough for a further transplant.

mikan (mī·kăn). [Jap.] A Satsuma orange.

[**1922** T. Tanaka in *Jrnl. Heredity* XIII. 243/2 The leading orange grown in Japan is a kind of mandarin, *Unshû Mikan*, called the Satsuma orange in the United States.] **1947** J. Bertram *Shadow of War* 193 Mikans—the sweet, thin-skinned, tangerine-oranges. **1972** *Nat. Geographic* CXLI. 672/2 Obasan offered up sliced raw fish,..and finally the Japanese orange, the mikan, for dessert at the Japanese..... A. Brosnowski *Take One Ambassador* v. 56 Drink cans, *mikan* peel..used chopsticks, everywhere.

mike, *sb.*[3] Substitute for def.: A rest; a period of idleness; a waste of time; esp. in the phr. *to do* or *have a mike* = to be idle, escape from or evade work, go away. Add further examples. Cf. *mick[2].

1899 E. Whiteing *No. 5 John St.* xxiv. 238 It was pleasant to..share the tobacco and biscuit, and make sure of a good 'mike' on this side of a life to come. **1925** Fraser & Gibbons *Soldier & Sailor Words* 155 *Mike, to* (to do a), to make off. To avoid duty. **1940** M. Marples *Public School Slang* 119 To *do a mike* (St Bees, 1915 +), to break bounds. **1955** 'N. Shute' *Requiem for Wren* iii. 83 That's a good mike for you, but you'll have plenty to do later on. **1958** *Times* 26 Sept. 19/1 The day of the cheerful veteran forward, gratefully relying upon opportunities for a mild 'mike', may be coming to an end.

Mike, *sb.*[4] Add: Phr. *for the love of Mike*, see *love *sb.*[1] 7 a.

mike (maik), *sb.*[5] Colloq. abbrev. of Microphone 1; also *attrib.* Cf. *mic.

1927 *Melody Maker* June 579/3, I think it is more that he plays too loudly than that he is too near the 'mike'. **1928** Ade *Let.* 7 July (1973) 134 Open the act with a fake microphone all set and adjusted for broadcasting. You come out and talk into the 'mike' announcing the name of a fake station in the town..and say you have a very interesting program ahead and then you can read it into the mike.. **1937** *Daily Herald* 16 Feb. 19/6 He is unlikely to be affected with 'mike' fright, because, in his line, he has found visible audiences in far more truculent mood than will be his unseen Midland Regional listeners. **1939** *Evening News* 7 Nov. 4/5 To follow the players about, the 'mike' is moved across the floor on a long arm called a 'mike boom', and its operator is a 'mike slinger'. **1943** J. B. Priestley *Daylight on Saturday* xi. 68 He delighted in entertainment, liked to make his little speech at the mike. **1958** B. Holliday *Lady sings Blues* (1973) iii. 38, I got to the mike somehow and grabbed it. **1962** *Listener* 12 Apr. 656/3 Robert Kee avoided..entangling his mike-cable in the mob. **1971** D. E. Westlake *I gave at the Office* (1972) 188 'I am a soldier,' he said, 'not a baseball player. No interview.' Frankly, I think he had mike fright.

mike (maik), *sb.*[6] In slang phr. *to take the mike out of* = to take the micky out of. Cf. *micky[1] 6.

1935 G. Ingram *Cockney Cavalcade* i. 14 He wouldn't let Pancake 'take the mike' out of him. **1935** T. E. Lawrence *Mint* (1955) vi. 17 But, mate, you let the fight down, when he takes the mike out of you every time. **1940** *N. & Q.* 1 June 382/1 'Taking the mike out of' anyone means pulling his leg, having a game with him. **1956** J. Cannan *People to be Found* i. 14 They won't 'all take the mike out of him.' **1973** 'B. Mather' *Snowline* vi. 75 Watch it... The Swami don't dig taking the mike out of the gods.

mike (maik), *sb.*[7] *slang.* [Abbrev. of *microgram.*] A microgram, *spec.* of lysergic acid diethylamide (LSD).

1970 N. Saunders *Alternative London* xxii. 168 Lysergic Acid Diethylamide is the most common hallucinogen—and by far the most powerful, in that you only need a few millionths of a gram (micro-grams, 'mikes') to trip for eight hours. **1970** R. Platt *Pushbutton Butterfly* (1971) 43 Janet Sanders could be on acid, dropping the usual LSD tab of two hundred and fifty mikes. That's the standard trip..a penny a microgram. **1973** J. Wood *Nerds Ha'e* 12 Five They wanted me to tell where I got the mikes... The acid, see?

mike, *v.*[1] Add to def.: to avoid work; go away, escape; also with *off.* (Further examples.) Hence **mi·king** *vbl. sb.*

1894 A. Morrison *Tales of Mean Streets* 47 'I ain't settled with you yet, my gal,' he added to Lizer; 'mikin' about at 'ome an' 'idle money.' **1899** J. Flynt *Tramping with Tramps* iv. 83 'They had no trouble..' **1912** [....]. **1920** [....] **1938** '....' A. Morrison *Child of Jago* xxvi. 251 He..'d make a far day's work for a fiar day's pay. *Ibid.* 84 Molden was simply 'miking' on a grand scale. *Ibid.* 85 A policeman is paid to work his beat, not 'mike' with or without permission. **1959** N. Loftus *Heaven in your Hand* 145 There was nobody to send. Both my young b—s have miked off. **1974** P. Levi in J. Burnett *Useful Toil* iii. 336 [He would] spy on us as we worked, and then..thunder at any one he thought was miking.

mike (maik), *v.*[2] *colloq.* [f. *mike *sb.*[5]] *trans.* To place a microphone in (a place) or near (a person) for recording purposes; freq. = *bug *v.*[2] 2. So **miked** *ppl. a.*

1961 M. Procter *Body to Spare* xvi. 124 He was put in a cell with Cosy, and the cell was 'miked'. **1965** D. Tosh *Diplomatic Cover* v. 82 The Russian..turned a switch on the bigger recorder... Christ, they've miked us! **1968** '.... Tension Line' 19, I take it this Dean is the American Counsellor whose bungalow you're trying to mike? **1969** A. Hall' *Striker Portfolio* xvi. 196 My one task for the day was to find out if the room was miked because I didn't want them to hear my movements. **1969** *Rolling Stone* (.... Collectors'...) 14 Jan. 1969 *Rolling Stone* ... May 12/3 What came out was a beautifully mixed and miked package of spiritual soul. **1972** *Jazz & Blues* Oct. 30/1 Mezz is too closely miked for one to be able to follow the soprano clearly at all times. **1974** *Listener* 4 Apr. 437/3 [The] film..owes a great deal to..unobtrusive miking techniques..and the quiet manoeuvring of the crew.

miker (mai·kəɹ). *dial.* and *slang.* [f. Mike *v.*[1] + -er[1].] = Micher *sb.*

1863 J. D. Robertson *Gloss. Words County of Gloucester* 94 *Miker* is used for a truant. **1928** *Daily Tel.* 9 Oct. 10/5 It is reported that the casual ward of Edmonton Workhouse was known far and wide as the 'Mikers' Paradise'. **1931** C. Williams *Three Plays* 29 You always saw and sneaked your profit out Like a kerchief-miker.

Mikimoto (mikimōᵘ·to). The name of Kokichi *Mikimoto* (1858–1954), Japanese pearl farmer, used *attrib.* of pearls cultured by means of a technique which he perfected.

1956 R. Eunson *Pearl King* ii. 28 At Toba Bay Mikimoto pearls are harvested by the crop and sacked up like wheat. **1959** R. Kirkbride *Tamiko* iii. 20 Diamonds and great clusters of Mikimoto pearls gleamed in the candlelight. **1969** J. Bennett *Dragon* i. 6 A string of good cultured Mikimoto pearls around her neck.

mikva (mi·kvā). Also **mikve(h, mikwe(h.** Pl. **mikvaoth.** [Heb. *miqwāh*, lit. collection, mass, esp. of water; pool of water.] A bath in which certain Jewish ritual purifications must be performed; the action of taking such a bath. Also *attrib.*

1843 De Sola & Raphall tr. *18 Treat. from Mishna* 356 Treatise Mikvaoth. (Contains laws that relate to diving baths for the cleansing of persons.) **1904** *Jewish Encycl.* VIII. 588 *Mikweh*... Because of the use made of this word in connection with ritual purification.., it has become the term commonly used to designate the ritual bath... The mikweh must contain sufficient water to cover entirely the body of a man of average size. **1962** B. Abrahams tr. *Life Glückel of Hameln* v. 108 Every Jewish community had its *mikveh*—ritual communal bath. **1966** *New Statesman* 6 May 648/2 All women about to marry must endure an interview with a woman in the rabbinate who issues a ticket to the ritual bath (*mikva*). **1968** L. Rosten *Joys of Yiddish* 242 Today, only very religious Jewish women observe the *mikva* custom—or attend a bathhouse for *mikvah*. **1970** L. Mc Friedhaven *Taste of Yiddish* 249 *Mikve* immersion is also part of Orthodox conversion ritual. *Ibid.*, The average American Jewish couple would be surprised to learn..that the wife should then visit the *Mikve* before union. **1974** *Observer* (Colour Suppl.) 10 Nov. 37/3 By ancient Jewish teaching..a woman becomes virtually unclean by the act of menstruation and she must abstain from sexual relations during it and for seven days after it is finished. Before she recommences relations with her husband she should immerse herself in the *mikva* (ritual bath). Brides prior to marriage should also be purified in the *mikva* and also mothers after childbirth. *Ibid.*, Reform Jews do not use the mikva.

mil. Add: **1.** (Later example.)

1973 *Sci. Amer.* Feb. 66/3 Away from coastal areas.. the salinity of the ocean varies from 32 to 37 grams of dissolved solid per kilogram of seawater, expressed as 32 to 37 ‰. (The symbol ‰ is read 'per mil'.)

2. (Later examples.) Also *attrib.*

1883 *Encycl. Brit.* XVI. 734/1 Another proposal starts from the present pound as unit. It is to be divided into 10 florins (as.), which would contain 100 mils (or farthings reduced 4 per cent.). A new coin, 10 mils (2s. 4d.) would probably have to be introduced. **1920** *Rep. R. Comm. Decimal Coinage* 13 in *Parl. Papers* (Cmd. 628) XIII. 467 The pound and mil provides no exact equivalent of the penny. Of the nearest equivalents 4 mils is 4 per cent. less, 5 mils is 20 per cent. more than the penny. **1920** *Glasgow Herald* 10 Apr. 4 If there were any demonstrable superiority in this 'pound mil' system it might be worth while to face all.. delays. **1960** *News Chron.* 4 May 3/4 The day when we.. pay 100 mils for a packet of cigarettes has drawn a little closer. **1963** *Rep. Comm. Inquiry Decimal Currency* 9, xiv, in *Parl. Papers 1962–3* (Cmnd. 2145) XI. 195 We sometimes refer to 'mil' systems..as three-place decimal systems. **1967** *Guardian* 26 Apr. 2/6 The committee..was considering an amendment to introduce the pound-florin-mil system.

3. (Examples.)

1896 F. Bedell *Princ. Transformer* xv. 306 For conductors larger than 50,000 circular mils, flat copper ribbons are used. **1962** A. Nisbett *Technique Sound Studio* 253 *Coarse-groove*, The groove normally used for 78 rpm recordings... Width 6 mils, depth 2–3 mils. **1973** *Sci. Amer.* July 43/1 Graphite fibers, produced by the carbonization of rayon or acrylic fibers, average about a third of a mil in diameter.

5. The name of a coin whose value is a thousandth part of the unit of currency, in Cyprus (and formerly in Palestine and Egypt) and Hong Kong.

1902 *Encycl. Brit.* XXXI. 292/1 *Hong Kong*,...the denominations are..the cent and the mill in bronze. **1908** G. B. Rawlings *Coins* xi. 157 Queen Victoria's issues for Hong Kong consist of silver dollars, half-dollars..and mils. *Ibid.*, The mil follows the Chinese fashion and is pierced in the centre. On the one side it has a crown, V.R., and the date, and Hong Kong One Mil. **1929** *Whitaker's Almanack* 447 Palestine..Silver—100, 50 mils, 5 mils. **1931** *Ibid.* 21 Commercial *Coins Mod. World* 4 Palestine Mandated Territory... Currency in Mils on a decimal basis, similar to that in use in Egypt. *Ibid.* 106 Egypt... Currency unit... Piastre of 10 Milliemes or Mils. **1956** *Whitaker's Almanack* 783 Cyprus... 1,000 mils=£1 Sterling. **1968** C. Narbeth *Coin Collectors' Encycl.* 65 Cyprus... A new currency was introduced in 1955 of 100 mils (equals 2s.). **1970** C. Wood *Terrible Hours* v. 56 'How much?' 'Forty-five mils!' **1970** G. Lyall *Judas Country* iii. 20, I served 'em a compulsory breakfast I got two mils.

6. A unit of angular measure equal to 1/1600 of a right angle, which is approximately the angle subtended by one metre at a distance of 1000 metres.

1907 O. M. Lissak *Ordnance & Gunnery* xiii. 507 The horizontal deflection scale..is graduated, in sights for field artillery, to thousandths of the range. These graduations are called mils. **1920** Carter & Arnold *Field Artillery Instruction* vi. 219 The angle of site scale and the deflection scales on all instruments are graduated in mils. **1931** F. C. Finch *Topogr. Maps* 138 The vertical lines of the sketching screen, being one inch apart and twenty inches from the eye, determine an angle of fifty mils. **1941** *Amer. Math. Monthly* XLVIII. 88 Most American mobile artillery units as well as many heavy railway mounts have the scales on their sights, azimuth circles, and quadrants graduated in mils. **1955** C. R. Wylie *Plane Trigonometry* i. 21 Show that on any circle a central angle of 1 mil intercepts an arc whose length is approximately 1/1,000 of the radius of the circle. **1970** *Daily Tel.* 1 June 15/7 The Services no longer measure bearings in 360 degrees but in 6,400 mils.

mil, mill, var. *mill *sb.*[6] and *sb.*[7]

Milan[1]. Add: **2.** (See quot. 1968.) Also *attrib.*

1895 *Montgomery Ward Catal.* 131/3 Misses' or children's, Union Milan and fancy braid. **1948** K. Hardy *Costume Design* x. 209 *Milan*, finer version of leghorn. **1950** 'Mercury' *Dict. Textile Terms* 345/2 *Milan hat*, a hat of fine straw, originally manufactured in the province of Milan, Italy. **1968** J. Ironside *Fashion Alphabet* 253 *Milan*, a fine, closely woven straw from Milan, Italy, used for expensive hats.

Milanese, *a.* and *sb.* Add: **A.** *adj.* (Earlier examples.)

1617 J. Chamberlain *Let.* 21 June (1939) II. 82, I met with a Milanese gentleman of some qualitie. **1753** M. W. Montagu *Let.* 10 Oct. (1967) III. 39 A Milanese lady being now professor of Mathematics in the University of Bologna.

2. Of a warp knit fabric made on a Milanese loom usually from silk or rayon yarns; of a garment made of this fabric. Also *ellipt.* as *sb.*

1897 *Sears, Roebuck Catal.* 231/1 The New Four-Button Pure Silk Glove... Guaranteed all pure Milanese silk. *Ibid.*, Black Milanese Silk Mitts, the softest, finest and most durable of all silks. **1916** *Daily Colonist* (Victoria, B.C.) 19 July 14/5 (Advt.), Nothing could be more appropriate for wear at the big Fete Saturday than one of these Suits, of Milanese Silk. **1922** *Tatler* 10 Aug. Advts. p. 6, Ladies' exceedingly dainty Cami-Knickers, made of Milanese. **1933** *Daily Mail* 5 July Advts. p. m, Pure Silk Milanese Vest. **1937** *Glasgow Herald* 10 Apr. 10 What your supplies and your Milanese may be, you alone may care. But one takes of one's hat to the 'shimmering crepe de soie'. **1945** M. D. Potter *Fiber to Fabric* 238 Milanese has a distinctive diagonal cross effect. Originally made from silk, this..... **1968** J. Ironside *Fashion Alphabet* 241 *Milanese*, a warp-knit fabric made on a Milanese loom, used mainly for women's underwear, very fine and lustrous.

B. *sb.* **3.** The Milanese dialect.

1643 J. Howell *Instructions Forreine Trav.* xi. 138 There is in Italy.. the Milanese, the Parmasan, the Piemontese, and others, which have each several Dialects and Idioms of Speech. **1818** [see *Genoese *sb.*]. **1880** *Encycl. Brit.* XII. 496/2 It may be added that the Milanese..retains the ... as a compound or reduplication in the manner of the *mi-mi*. **1966** T. G. Griffith *Migliorini's Italian Lang.* xi. 405 A Milanese, for example ... the Milanese dialect and dialect against Giceriani.

mild, *sb.*[2] Eng. and U.S. dial. var. of Mile *sb.*[1]

1701 in *Essex Inst. Hist. Coll.* (1900) XXXVI. 83 To run the line of division from Ipswich riverline running..six mields. **1725** in *Early Rec. Lancaster, Mass.* (1884) 231 We traueled to Groton 12 milds... We marcht up the riuer aboutt 8 milds. **1777** E. Lincoln *Papers* 9 July in *Papers* (1904) 15 They were Engaged in Carring gun bots..over land about one mild into Lake George. **1883** 23 *Knickerbocker* VIII. 352, I expect we are a mild and a half from the city. **1886** F. T. Elworthy *West Somerset Word-Bk.* 476, I count 'tis up vower mild yer-vrom. **1903** *Dialect Notes* II. 321, I haven't walked a mild in a year. **1927** *Amer. Speech* III. 10 Most natives [sc. Ozarkers] use *mile* for both singular and plural, but some of the old-timers use the plural form *milds.*

mild, *a.* Add: **5. b.** (Further examples of *absol.* use.) Also *attrib.* phr. *mild-and-bitter*, a mixture of mild and bitter ale or beer.

1894 A. Morrison *Martin Hewitt* ii. 69 'Had his glass o' beer, has he?'... 'Has two glasses of mild a-day... Never puts on flesh.' **1933** D. L. Sayers *Hangman's Holiday* 157 Half of mild-and-bitter, please. **1964** Dylan Thomas *Let.* 11 Sept. (1966) 267 If t's time for the Black Lion But there is only the mild. **1931** Ibid. *Times* 10 Nov. ... E. Hyams *Sylvester* xxiv. 22 The chaps in the local, drinking a pint of mild and bitter. **1957** J. Braine *Room at Top* ii. 22 We used to..live on onions and cheese washed down with mild-and-bitter. **1963** *Times* 25 May 9/7 'But t'brig isn't t'world', a swinging shop-worker says over his glass of mild. **1974** W. Haggard *Kinsmen* vii. 70 He went to the bar. Mysteriously four evident locals were already inside and drinking mild.

mildewed, *ppl. a.* (Later *fig.* examples.)

1925 in J. Manchon *Le Slang.* 1930 K. Campbell *Adamastor* 56 Worse than death 'The palsied soul, the mildewed brain'. **1969** J. K. & P. Opie *Lore & Lang. of Schoolch.* ix. 161 Juvenile repugnance continues to be expressed by the old standbys:..mardy, mildewed, mingy, mincy-making, (etc.).

mildly, *adv.* Add: Esp. in *colloq.* phr. *to put it mildly*, to express an idea without exaggeration; freq. ironical, with an implication of understatement.

1939 Joyce *Finnegans Wake* 439 What I'm wondering to myselfwhose for there's a strong tendency, to put it mildly, by making use of the medium. **1949** E. E. Cummings *Let.* 9 July (1969) 191 Thank you much more than kindly for a ... (putting it very mildly) luxurious gift. **1958** *Spectator* 22 Aug. 242/1 This, to put it mildly, was improbable. **1973** *Listener* 6 Apr. 448/3 Suggestions that conditions (on Mars) may periodically change, causing caps to grow and retreat or so, are—to put it mildly—highly speculative as yet.

mile, *sb.*[1] Add: **1. d.** (Further examples.)

1888 *Punch* 12 Dec. 281/1 The fellows generally bag his mugs, and make him play the 'Mikado' which is miles better. **1919** 'C. Dane' *Legend* 64 He had heard nothing... He was miles away. **1922** Joyce *Ulysses* 639 Bloom..picked it up..meaning to return it to him... whose thoughts were miles away from his hat at the time. **1932** E. M. Delafield *Thank Heaven Fasting* i. iv. 65, I should have thought he'd be miles better than no one. **1943** J. B. Priestley *Daylight on Saturday* vii. 38 Not that the boys drew back..but..not one of them came within miles of being..Mister Right. **1951** E. Paul *Springtime in Paris* xv. 287 Pierre Vautier, smiling, gesturing and talking a mile a minute to an attractive brunette. **1961** M. Kelly *Spoilt Kill* iii. 166 Freddy was miles out in alleging Corinna's coldness. **1961** Partridge *Dict. Slang* Suppl. 1184/1 *Miles away, be*, to be either day-dreaming or lost in thought. **1964** C. Willock *Enormous Zoo* viii. 133 Ken Beaton's original estimate of the elephant, buffalo and hippo population had been miles out. **1969** A. La Bern *Nice Class of People* ii. 11 Ann Corrie had to repeat the sentence... He was miles away. **1970** J. Porter *Rather Common Sort of Crime* i. 17 She *was* on the scrounge... You could spot it a mile off. **1970** *Globe Mag.* (Toronto) 26 Sept. 9/3 The meeting..is going well. Stephen has made some miles by talking about the proposed shutdown of the Garden City Paper Mills. **1973** L. Snelling *Heresy* I. vii. 52 The French technicians are all ..very Lefty... He can keep them from getting within a mile of the set. **1974** N. Freeling *Dressing of Diamond* 176 This hasn't been done cold-bloodedly for money... Makes it all miles easier. **1974** D. Gray *Dead Give Away* ii. 28 'Aunt Milly's miles out,' said Marion. 'As usual.'

5. b. *mile-consuming* ppl. adj., *-deep* (earlier example), *-high* (examples), *-wide.*

1932 W. Faulkner *Light in August* i. 8 He drove on, the wagon beginning to fall into its slow and mileconsuming clatter. **1888** J. R. Lowell *Fitz Adam's Story* in *Heartsease & Rue* iv. 151 Mile-deep the glaciers brooded here, they say. **1963** J. Lusby in B. James *Austral. Short Stories* 229 A black speck raced towards me along the rim of a mile-high blood-red cliff of cumulus. **1968** *Listener* 27 June 841/2 The ozone-drenched, mile-high city of Denver. **1866** J. G. Whittier *Snow-Bound* 22 Where Salisbury's level marshes spread Mile-wide as flies the laden bee. **1903** Kipling *Five Nations* 56, I heard the mile-wide mutterings of unimagined rivers.

c. mile-a-minute *attrib.*, travelling at a rate of a mile a minute; covering a mile a minute; **mile-eater** (cf. *eat *v.* 18 g) *colloq.*, a fast driver or traveller; **mile-heat,** a racing heat of one mile; **mile-post** (earlier and later examples); also *fig.*; hence **mile-posted** *ppl. a.*

1957 *Railway Mag.* Nov. 752/2 In view of the Eastern, North Eastern and Scottish changes, a new table of their mile-a-minute runs for the winter appears here. **1961** *Christian Science Monitor* 9 Jan., The daredevil courage needed to flash down a hard-packed track at better than a mile-a-minute speed. **1908** *Westm. Gaz.* 20 Aug. 12/1 These mile-eaters go early to bed and prefer their steam-horses to our live ones. **1957** S. Moss *In Track of Speed* i. 9 The driver himself must possess those faculties which go to make the expert mile-eater. **1802** N. Macon *Let.* 10 Sept. in J. Steele *Papers* (1924) I. 315 He says there are no regular Mile heats at that turf. **1868** H. Woodruff *Trotting Horse* vi. 70 You can tell by the way he finishes..whether he will be likely to stand the mile-heat out and to repeat it. **1768** in *Maryland Hist. Mag.* (1907) II. 317 As we returned (besides the Mile Posts) we erected Marks on the Tops of all the High Ridges. **1909** *Westm. Gaz.* 26 Mar. 12/2 Oxford paddled up to the mile-post. **1926** Ade *Let.* 29 Aug. (1973) 107 The day in 1894 when the editor put me in charge of a department was an important mile post. **1941** W. Temple *Citizen & Churchman* i. 2 It is in reality a mile-post marking the distance away from it which thought has travelled. **1954** R. D. Burnell *Oxf. & Cambr. Boat Race* vi. 116 Cambridge led at once... At the Mile Post they led by seven seconds. **1896** 'Mark Twain' in *Harper's Mag.* Jan. 294/2 The road was mile-posted with English fortresses, so to speak.

mileage. Add: **1. c.** A rate per mile charged for the use of railway vehicles carrying goods or passengers over another company's line.

1837 *Penny Mag.* Suppl. 31 Mar. 115/1 Mileage on the whole mail. **1863** *Great Western Mag.* Aug. 74 The Clearing House..will debit each company on the journey with its proper amount of 'mileage'. **1873** *Cassell's Mag.* VIII. 400/1 As..the Caledonian Company will receive the fares of the passengers, they will be required to pay for the use of the carriages conveying them; in other words the.. Company will be charged 'mileage'. **1926** Huebner & Johnson *Railroad Freight Service* vii. 129 The prevailing rate of mileage paid by one railway to another for the use of foreign cars was three quarters of a cent per car per mile.

2. b. (Further examples.)

1945 N. L. McClung *Stream runs Fast* p. xi, There was good mileage in me yet. **1962** L. Deighton *Ipcress File* xxv. 158 In the Café Budapest..'kezet csokolom' (kiss your hand) had given good mileage with the younger waitresses. **1970** W. V. Quine *Philos. of Logic* v. 71 We can still get a little additional mileage by superimposing the virtual theory. **1971** *Rolling Stone* 24 June 36/1 Sam's got mileage that any PR firm might envy: front-page space in both San Francisco dailies,.. and wire service coverage. **1971** A. Price *Alamut Ambush* ix. 114 The newspapers had got very fair mileage out of the bomb explosion. **1973** *Times* 5 Feb. 19/5 When..profits began to tumble, the..senior management of the time imagined that there was no more milage in the product. **1973** *Sci. Amer.* Apr. 103/3 Mathematical notions whose foundations have been matters of continuous debate have often yielded the most mileage. **1974** T. Allbeury *Snowball* iii. 14 They'd enjoy stirring up the Canadians, and the French-Canadians would..get a lot of political mileage.

c. (Earlier and later examples.)

1869 *Bradshaw's Railway Manual* XXI. 367 Advances are made by the Italian Government, to be made good afterwards out of the mileage subventions promised... The Italian and Pontifical Governments have given mileage guarantees. **1884** J. B. Pope *Railway Rates* 22 They [sc. Railway Companies] shall be allowed to vary or graduate the mileage charges according to distance. **1908** *Westm. Gaz.* 14 Nov. 15/2 An instrument, made by one of the largest clock-makers in the world,..combines a time-piece, mileage recorder, speed indicator [etc.]. **1928** C. E. R. Sherrington *Econ. Rail Transport Gt. Brit.* II. vii. 98 With the coming of the Parliamentary trains,..the equal mileage basis became..stabilized. **1969** *Observer* (Colour Suppl.) 23 Mar. 31/3 The White Car is running on Shell *with* the mileage ingredient. **1970** *Motoring Which?* Oct. 151/1 Check the mileage recorder—see how long the road test has been. **1971** 'J. Ashford' *Bent Copper* iv. 34 He left and went out to..his car, a battered Ford, which he ran on a mileage allowance. **1973** *Radio Times* 20–27 Dec.

19/1 (Advt.), Explore Canada for a week with Avis for £37·50, with no mileage surcharge. **1975** *New Yorker* 26 May 60 (Advt.), In mileage tests conducted by the Environmental Protection Agency, Seville got 13 miles per gallon in the city test.

mile-castle. Add: Also **milecastle.** (Later examples.)

1935 *Antiquity* IX. 92 The recent excavations at High House Turf-Wall milecastle. **1936** *Nature* 25 July 156/2 A milecastle and three quarters of a mile of the wall itself were presented to the National Trust. **1963** E. S. Wood *Collins Field Guide Archaeol.* II. ii. 177 The milecastles [of Hadrian's Wall] were about 75 by 60 feet, the turrets about 20 feet square.

mileometer, var. *milometer.

miler[2]. Add: (Earlier and later examples.)

1889 E. Sampson *Tales of Fancy* 31, I..was in private trials one of the fastest 'milers' of my time. **1955** *Times* 31 Aug. 3/5 To-night B. S. Hewson, potentially perhaps the greatest miler in the world, is to defend the trophy against a good field. **1965** *Illustr. London News* 4 Sept. 14/4 (*heading*) The first African miler to break four minutes. **1971** L. Koppett *N.Y. Times Guide Spectator Sports* viii. 159 The miler who excited the world's track fans in the 1920s was Paavo Nurmi, a Finn.

2. *colloq.* A walk or journey of a specified number of miles.

Properly the second element of a compound.

1856 Dickens *Let.* 14 Nov. (1938) II. 811, I went out this morning for a 12-miler.

‖ miles gloriosus (mī·lēiz *or* mai·līz glōᵊriōᵘ·səs). Pl. **milites gloriosi.** [f. L. *miles* soldier + *gloriosus* boastful, conceited.] The name of a comedy by Plautus (*c* 250–184 B.C.), used allusively to designate a braggart soldier. Also *attrib.*

1917 K. M. Westaway *Orig. Element in Plautus* ii. 28 Other plays of Plautus contain *milites gloriosi* of smaller fame. **1936** P. Fleming *News from Tartary* vii. iv. 343 One..was a glib Turki from Turfan, the shoddiest type of the *miles gloriosus*. **1950** A. Bonjour *Digressions in Beowulf* 18 In spite of Beowulf's biting allusion to Grendel's security, we should not take this as an entirely idle vaunt of some *miles gloriosus*. **1962** G. K. Hunter *John Lyly* iv. 238 The version of *miles gloriosus* habits found here is without the menace that accompanies its adult presentation, in Pyrgopolynices (*Miles Gloriosus*) or Thraso (*Eunuchus*). **1964** *Rev. Eng. Stud.* XV. 385 A typical figure in Gleig's tales in particular was the Peninsular miles gloriosus. **1969** E. Segal tr. *Plautus' Three Comedies* Introd. 8 The *miles gloriosus* is by no means a Plautine invention, although the boastful officer is one of the Roman comedian's favorite characters.

Milesian, *a.*[1] and *sb.*[1] (Earlier and later examples.)

1550 T. Nicolls tr. *Thucydides* iv. sig. Siiii verso, They had in their compaignie the succours of Milesyans, of Andryens and of Caristians. **1600** P. Holland tr. *Livy* xliii. 1159 The Milesians for their part said, That hitherto they had done nothing. **1602** I. B. in E. Beaumont *Salm. & Hermaphr.* Pref. Verses, Or wanton Nymphs in watry bowres haue woue, With fine Mylesian threds, the verse he sings. **1607** [see *Calabrian *a.* and *sb.*]. *a* **1635** Randolph *Hey for Honesty* (1651) iv. iii. 472 You told her, The Milesians were valiant in the daies of yore. **1961** L. Mumford *City in Hist.* vii. 191 Since it would be erroneous to call this Hippodamian planning, I shall follow Roland Martin and call it Milesian, after Miletus, the chief point of origin. **1968** *Encycl. Brit.* XV. 444/1 The Milesians..were already rebuilding their city on a new grid plan to the type invented..by the Milesian Hippodamus.

Milesian, *a.*[2] and *sb.*[2] (Earlier and later examples.)

1596 Spenser *State Irel.* (1633) 31 All which are in truth fables, and very Milesian lyes, as the later proverbe is: for never was there such a King of Spaine, called Milesius. **1705** M. Kennedy (*title*) A Chronological Genealogical and Historical Dissertation of the Royal Family of the Stuarts, beginning with Milesius the stock of those they call the Milesian Irish, and of the old Scotish Race. **1839** Carlyle *Chartism* iv. 28 The wild Milesian features ..salute you on all highways and byways. *Ibid.*, The English coachman..lashes the Milesian with his whip. **1910** D. Hyde in R. M. Dorson *Peasant Customs* (1968) II. 704 Some of the Scotch stories may have been bequeathed to the Gaelic language by those races who were displaced by the Milesian Conquest in the fifth century. **1921** *Edin. Rev.* Jan. 167 And lastly the Milesians. The Milesian being a literary and honorific cognomen of the Firbolgs' conquerors, the Gaels. **1971** *It* 2–16 June 24/1 The first groups of wholly human invaders to reach these shores.. were called the Milesians or 'Sons of Miledh'.

milestone, *sb.* Add: **3.** *Naut. slang.* (See quots.)

1946 J. Irving *Royal Navalese* 116 *Milestones*, the heavy, green seas which break inboard in bad weather. **1962** Granville *Dict. Sailors' Slang* 77/1 Milestones refer to the homeward trip, for, like milestones on country roads they seem to make the journey longer and harder, and one's progress slower in consequence.

milestone, *v.* [f. the sb.] *trans.* Used *fig.*, to mark (stages) as if by milestones.

1902 J. H. M. Abbott *Tommy Cornstalk* 157 And the road was mile-stoned by the parched hides and whitened bones of horses, mules, and oxen. *a* **1910** 'Mark Twain' *Autobiogr.* (1924) I. 299 You could look back over that speech and you'd find it dimly milestoned along with those commas. **1922** *Chambers's Jrnl.* Dec. 861/1 The Overland is mile-stoned with our bones. **1973** J. Cleary *Ransom* iii. 98 Malone's life was milestoned by friends he had never made. **1973** J. Wainwright *High-Class Kill* 149 The book..will make passing reference to these things—as a means, perhaps, of milestoning his climb to the rank of chief constable.

milice. Delete † *Obs.* and add: In revived use, *spec.* a force employed by the occupied French state of 1940–44 to repress internal dissent. (In quot. 1945, by substitution for *milicien.)

1945 H. Nicolson *Let.* 11 Mar. (1967) 440 The *milices* had beaten him up. **1958** *Listener* 21 Aug. 277/2 The Cagoulards, who in the event became the Kernel of the Vichy Milice. **1968** A. Diment *Bang Bang Birds* vi. 98 Their disposal was left to the Gestapo and the Milice. **1974** T. Allbeury *Snowball* vi. 32 An officer..captured in Perpignan by the Milice in 1944 and handed over to the Gestapo.

‖ miliciano (milipiā·no). [Sp.] A militiaman of the irregular Republican force formed during the Spanish Civil War. Also **miliciana,** a militiawoman of this force.

1938 J. Riesenfeld *Dancer in Madrid* viii. 159 A woman *miliciana* was being helped up by two *milicianos*. **1957** P. Kemp *Mine were of Trouble* ii. 23 Women, too, enlisted in the militias... They were also employed as jailers to guard female political prisoners, several of whom have told me that they suffered much worse treatment from the *milicianas* than from the men. *Ibid.* 27 The only opposition came from the *milicianos*, who fought with courage but without discipline or military training. **1962** D. A. Puzzo *Spain & Great Powers* iii. 69 Against German and Italian air power and matériel, the *milicianos* could not hope to advance. **1965** C. D. Eby *Siege of Alcázar* (1966) iii. 65 At the Plaza de Padilla they were stopped by some *milicianos*.

‖ milicien (milisiæ̃n). [Fr.] A member of the Milice (see *milice above).

1945 *Tomorrow* (N.Y.) Feb. 11/1 These fellow-soldiers.. would..comb a building for Germans or the dreaded Miliciens. **1961** P. de Vomécourt *Who lived to see Day* ix. 107 Almost to a man the *miliciens* were thugs on the make. ..Many of them were convicted criminals. **1966** M. R. D. Foot *SOE in France* v. 120 Miliciens were Frenchmen who lived and worked in their home towns and villages. **1967** *Listener* 16 Nov. 640/3 He is liquidated on suspicion of being one of the *miliciens*.

milieu. Delete ‖ and add earlier and later examples.

1854 Geo. Eliot *Let.* 6 Apr. (1954) II. 149, I could no more live out of my *milieu*, than the haddocks I daresay you are often having for dinner. **1955** *Times* 19 May 13/5 Its [sc. a book's] understanding of the poet's *milieu*. **1958** H. A. Wilmer *Social Psychiatry in Action* i. 21 The crucial point, however, is that a milieu was created that permitted recovery, rather than driving patients deeper into insanity. **1975** J. Rossiter *Golden Virgin* i. 11 Whitehall, a *milieu* in which you could look revoltingly nude without a bowler hat.

2. *Comb.*, as **milieu therapy** *Psychol.*, a form of group psychotherapy which relies on the social environment evolved by the staff and patients in the treatment unit.

1940 *Amer. Jrnl. Orthopsychiatry* X. 905 In environmental (manipulative, external, reality, milieu) therapy, it is assumed that the child's difficulty is in the social situation... Since the difficulties in the child are resultants of the difficulties in the environment, a milieu therapy is the truly rational therapy. **1961** R. Kee *Refugee World* 1 Their [sc. the refugees'] removal from a filthy, over-crowded hut to shelter considered fit for human beings is disguised as 'milieu therapy'. **1963** *New Society* 5 Sept. 17/3 Community services (or milieu therapy) for young deviants. **1964** M. Argyle *Psychol. & Social Probl.* v. 71 *Milieu therapy* and *therapeutic community* treatment consists of a residential institution run on more relaxed and permissive lines than is usual. **1972** G. Sereny *Case of Mary Bell* iv. 214 There are four of these experimental units in Britain..operated loosely on Aichhorn's 'Milieu Therapy'. They are designed to provide for persistently 'asocial' children..a secure environment.

miling, *vbl. sb.*[2] [f. Mil(e *sb.*[1] + -ing[1].] The action of running a mile (as an athletic event).

1913 S. A. Mussabini *Compl. Athletic Trainer* 73 This is miling of the best sort, disdaining the waiting tactics which so many adopt. **1955** R. Bannister *First Four Minutes* ii. 21 It is this controlled tension about to break down that gives miling its great excitement for the spectators. **1963** *Times* 27 May 5/6 Miling..received a needed fillip..when A. J. Harris won the Surrey race in 4 min. 2·4 sec.

‖ militaire (militẹ̄r). [Fr.] A soldier. (Cf. Militare *a.*)

1746 G. Townshend *Let.* 1 Oct. in J. H. Jesse *George Selwyn* (1843) I. 114 They look upon the *militaires* with abhorrence. **1827** Disraeli *Viv. Grey* III. v. vi. 101 He was a starch *militaire*, with a blue frock coat buttoned up to his chin. **1847** Thackeray *Van. Fair* (1848) xxxvi. 328 That young woman..forgot her charge in the society of this *militaire*. **1938** *Times Lit. Suppl.* 11 June 417/2 'Strathmore' can never be matched by an age that has learned to see *les militaires* with the cold eye of the elderly Ouida.

militancy. (Further examples.)

1912 in E. Pankhurst *My own Story* (1914) iii. iii. 258 The leaders..have so often warned the Government that unless the vote were granted to women in response to the mild militancy of the past, a fiercer spirit of revolt would be awakened. **1913** L. A. Harker *Ffolliots of Redmarley* xii. 156 Eloquent forgot her militancy. **1975** D. Ramsay *Descent into Dark* ii. 56 Militancy was her bag, and..she looked like someone who spent a lot of time at the barricades.

militant, *a.* and *sb.* Add: **A.** *adj.* **1. c.** Applied to or adopted as a designation by those who seek political or industrial change by employing or advocating the use of direct action, demonstrations, etc.; freq. applied to union leaders who hold out for high wage settlements, refuse to take part in discussions, etc.

1907 M. McMillan in 'B. Villiers' *Case for Women's Suffrage* 114 *Why* did the militant Suffragette ever come to the door of the House of Commons? **1914** E. Pankhurst *My own Story* i. iii. 37 That visit was one of the contributory causes that led to the foundation of our militant suffrage organisation, the Women's Social and Political Union. **1930** *Daily Express* 6 Oct. 11/6 Mr. Maxton leads a group of I.L.P. members who have brought a militant policy with them. **1960** *Economist* 8 Oct. 120/3 Mr Ted Hill's boilermakers..are incensed at the 'more militant than thou' attitude which Mr Greene is thus able to assume. **1969** *Rep. Comm. on Relations with Junior Members Univ. Oxf.* 158 Militant students believe that..they have a special position and function. **1975** *Times* 10 Apr. 17/3 The militant left are a menace to the welfare of Europe.

B. *sb.* **c.** A person who is a militant in the senses above.

1909 *Englishwoman* Apr. 323 That bias has been greatly intensified amongst almost all classes of suffragists by the tactics of the militants. **1914** E. Pankhurst *My own Story* I, (*heading*) The making of a Militant. **1939** *Theology* XXXIX. 437 The lives of certain of the militants reveal that Jocism stands for strength through holiness and self-sacrifice. **1968** *Daily Tel.* 12 Nov. 25/1 Ultra-left militants in the Electrical Trades Union are planning another demonstration today. **1969** *New Yorker* 17 May 114/2 A mysterious black militant. **1973** *Black World* Sept. 96/2 The young militants look down upon Anna's poetry. **1975** *Daily Mirror* 29 Apr. 4 They defeated a bid by union militants for an increase which would have broken the Social Contract. **1975** *Times Lit. Suppl.* 9 May 512/5 Intransigent NUJ [sc. National Union of Journalists] militants who now say they will discuss no press charter with the proprietors.

militaria (militē³·riă). [f. Militar(y *a.* and *sb.* + -ia[2].] Military articles of historical interest.

1964 *Exchange & Mart* 26 Mar. 12/4 Wanted: militaria, early model soldiers, uniforms, head-dress, accoutrements, paintings, by keen collector. **1970** *Times* 23 Dec. 10 During his last years he spent much of his time in trying to gather together writings, militaria and other possessions left by the Field-Marshal. **1973** *Inverness Courier* 31 July 8/4 A wider range of collectors items than ever before will be on show, and..there will be specialist stands featuring fine antique glass, weapons, militaria, old postcards, and early newspapers. **1974** *Country Life* 25 Apr. 1031/2 Sale of: Sporting Guns..Hand Guns and Nazi Militaria.

militarist. Add: **B.** *adj.* = Militaristic *a.*

1934 R. Campbell *Broken Record* i. 19 This form of historical instruction is no doubt at the back of the modern militarist-political unrest. **1944** J. S. Huxley *On Living in Revolution* iii. 33 Shall it be peaceful, co-operative, democratic, or shall it be militarist, totalitarian, brutal?

militarize, *v.* (Later examples.)

1922 P. N. Miliukov *Russia* 205 The climax was reached when the Bolsheviks decided to militarize labor. **1972** *Daily Tel.* 11 Jan. 9/8 Will the generals militarise the Government? **1973** *Listener* 14 June 796/3 During the Cultural Revolution..Chinese politics were again militarised.

Hence **mi·litarized** *ppl. a.*

1922 *Edin. Rev.* July 28 Appalling slaughter and suffering, patiently endured by a militarized people.., has disillusioned the Germans.

military, *a.* and *sb.* Add: **A.** *adj.* **3. b. military academy,** a place of training in the military art (cf. Academy 5); **military age,** the age at which one becomes liable for military service; **military attaché,** an army officer serving with an embassy; also, one attached as an observer to a foreign army; hence *military attachéship*; **military band** [Band *sb.*[3] 4], a band attached to a military unit; **military braid** (see quots.); **military brush** = *military hairbrush*; **military college** (cf. College *sb.* 4 e) = *military academy*; **Military Cross** (abbrev. *M.C.), a decoration instituted in 1915 and awarded to officers for gallantry in the face of the enemy; **military hairbrush,** a hairbrush without a handle; **military honours:** see Honour, honor *sb.* 5 and 5 d; **military hospital** (earlier and later examples); **military law** (earlier example); **Military Medal** (abbrev. *M.M.), a decoration of similar distinction to the Military Cross which was instituted in 1916 for 'other ranks'; **military-minded** *a.* [Minded *ppl. a.* 5], having a mind of a military character; **military orchid, orchis,** a European orchid, *Orchis militaris*, with pinkish-grey, helmet-shaped flowers, now very rare in Britain; also called soldier orchid; **military police,** the body of soldiers responsible for police duty in the armed forces; hence **military policeman; Military Secretary,** an army staff officer who acts as personal and confidential secretary to the Commander-in-Chief or certain other specified officers (see quot. 1876); hence *Military Secretaryship*; **military service** (cf. Service[1] 12), now, service in the armed forces (further examples); **military two-step,** in old-time dancing, a variation of the two-step.

1776 *Jrnls. Continental Congress U.S.* (1906) VI. 860 Resolved, That the Board of War be directed to prepare a plan for establishing a..Military Academy. **1802** [see Royal *a.* 6]. **1805** J. Orrok *Let.* 7 Aug. (1927) 79 He is.. a Lieut. although not in orders, which he cannot be untill he has been a few months at a..Military Academy. **1934** *Amer. Speech* IX. 313/1 The United States Military Academy. **1974** *Hartsville* (S. Carolina) *Messenger* 22 Apr. 2-A/7 He made the decision to earn an appointment to a Military Academy. **1920** Webster, Military age. **1934** G. B. Shaw *Too True to be Good* iii. 88 Had I been of military age I should have been a conscientious objector. **1941** *Manch. Guardian Weekly* 26 Sept. 194/2 It would not be fair to raise the military age..without a thorough comb-out of the younger men. **1857** *Foreign Office List* X. 18 Military Attaché to the Embassy, Lt. Col. E. L. Claremont, C.B. **1877** H. Ponsonby *Let.* 18 Nov. in A. Ponsonby *Henry Ponsonby* (1942) 167 We have dozens of Military Attachés with the Armies in the field. **1961** J. Masters *Road past Mandalay* vi. 76 It was Persia, and the Military Attaché was doubtful. **1882** Military attachéship [see Attachéship]. **1775** *Westm. Mag.* May 231/1 Three military bands, composed of fifes, drums, cymbals, etc. **1836** Military band [in Dict., sense A. 1]. **1912** G. Miller (*title*) The military band. **1964** A. Sexton *Sel. Poems* 7 While a military band plays a Strauss waltz. **1950** 'Mercury' *Dict. Textile Terms* 346 *Military braid*, a broad braid such as is worn on the tunics of soldiers. **1966** *Olney Amsden & Sons Ltd. Price List* 39 Rayon Military Braid in fashion shades 66/6 gross yards. **1968** J. Ironside *Fashion Alphabet* 76 *Military braid*, a flat braid with a diagonal weave. **1926** *Daily Colonist* (Victoria, B.C.) 6 Jan. 2/1 (Advt.), Gentleman's French Ivory Military Brushes. Concave back, a fine quality bristle. **1969** *H. Knowles-Brown Ltd.* (*Hampstead*) *Christmas Catal.*, Ivory backed military brushes from; per pair £19 0 0. **1809** G. L. Wardle *Charges against Duke of York* 334 Mr. Froome came to town to settle some old accounts of mine as treasurer to the Royal Military College. **1837** W. Dyott *Diary* 17 Jan. (1907) II. 246 Young Palmer, a candidate for the army at the Military College at Sandhurst. **1915** *London Gaz.* 1 Jan. 7/1 Royal Warrant instituting a new decoration 'The Military Cross'. **1917** W. Owen *Let.* 9 Apr. (1967) 451, I think Capt. Green..will get a Military Cross, which he has long deserved—for 2½ years active service. **1969** S. Mays *Fall out Officers* viii. 52 Wearing among his campaign ribbons that of the Military Cross. **1894** *Country Gentlemen's Catal.* 148 Two Ivory Military Hair Brushes, in Solid Leather Case, 57/-. **1972** J. Porter *Meddler & her Murder* i. 7 A couple of quick passes with her silver-backed military hair brushes. **1778** *Crit. Rev.* Sept. 189 An account of military honours paid to crowned heads and to other persons. **1853** Military honours [see Honour, honor *sb.* 5 d]. **1901** Military honours [in Dict., sense A. 1]. **1975** *Times* 2 Aug. 2/2 The UVF, a Protestant paramilitary group, said they would receive full military honours. **1777** *Jrnls. Continental Congress U.S.* (1907) VII. 162 An Inspector General of the Army..[shall] visit the Military hospitals..to examine the medicines and instruments. **1789** P. Thicknesse *Year's Journey* (ed. 3) I. iii. 25 The *Silver Lion*..is..preferable to Dessein's, as the drains from the Military Hospital run under the latter. **1860** F. Nightingale *Notes on Nursing* iii. 23 The ordinary run of military hospitals. **1969** S. Mays *Fall out Officers* iii. 25, I was not taken to the Cambridge Military Hospital, Aldershot, as I had thought. **1737** *London Mag.* Aug. 492/2 'Tis certain the military law may be made much stricter and more severe than the common Law can be made. **1916** *London Gaz.* 5 Apr. 3647/1 Royal Warrant instituting a new medal entitled 'The Military Medal'... We do..institute and create a silver medal to be awarded to non-commissioned officers and men for individual or associated acts of bravery on the recommendation of a Commander-in-Chief in the field. **1917** A. G. Empey *Over Top* 300 *Military Medal*, a piece of junk issued to Tommy who has done something that is not exactly brave but is still not cowardly. **1922** *Encycl. Brit.* XXXI. 892/1 *The Military Medal*.—Instituted in March 1916 for award to non-commissioned officers and men of the army for individual or associated acts of bravery in the field. **1957** *Ibid.* XVI. 638/1 Other awards of lesser degree in this class [awarded for gallantry] are: Distinguished Service Order, instituted in 1886 and the Military Cross (1914); only officers are eligible for the award... Also the Military Medal (1916); only 'other ranks' are eligible for the award. **1910** W. James *Mem. & Stud.* (1911) xi. 288 Commonwealths fit only for contempt, and liable to invite attack whenever a centre of crystallization for military-minded enterprise gets formed anywhere in their neighbourhood. **1939** *Ann. Reg. 1938* 266 The Portuguese are rapidly becoming a military-minded people. **1934** M. J. Godfery *Monogr. & Iconogr. Native Brit. Orchidaceæ* 168 *Orchis militaris* L. Soldier Orchid, Military Orchid. **1948** J. Brooke *Military Orchid* i. 17 The Military Orchid had taken on a kind of legendary quality, its image seemed fringed with the mysterious and exciting appurtenances of soldiering, its name was like a distant bugle-call, thrilling and rather sad. **1969** J. E. Lousley *Wild Flowers of Chalk & Limestone* (ed. 2) vii. 90 The largest..was 14 inches (35 cm.) tall with no less than 26 flowers. This must be about the finest Military Orchid seen in England. *Ibid.* 91 It is the resemblance of the hood to an ancient helmet which has led to the plant being called the Soldier or Military Orchid. **1812** W. Withering Jr. *Withering's Brit. Plants* (ed. 5) II. 29 (*heading*) Narrow-lipped Military Orchis. *O. militaris.* **1884** W. Miller *Dict. Eng. Names Plants* 99/2 Orchis,..Military. *Orchis militaris.* **1950** G. Brenan *Face of Spain* vii. 156 Under an olive tree I picked a specimen of that rare plant, the Military Orchis, which I knew from the plate in Bentham's Flora but had never found before. **1827** J. J. Snodgrass *Narr. Burmese War* v. 59 In every village there appeared a small party of military police. **1933** J. Buchan *Prince of Captivity* i. ii. 66 The military police arrived in quest of him. **1974** J. Wainwright *Hard Hit* 52 One evening session had seen..seven full-weight smash-ups start—and end with the arrival of the Military Police. **1973** J. Stranger *Walk Lonely Road* v. 44 My own father was a policeman. His father was a military policeman. I suppose it's like the Services. It runs in families. **1812** J. Orrok *Let.* 26 May (1927) 130 To Colonel Torrens, Military Secretary to His Royal Highness the Commander-in-Chief. **1853** J. H. Stocqueler *Mil. Encycl.* 179/2 All military correspondence with the commander-in-chief should be sent through the military secretary. **1876** Voyle & Stevenson *Mil. Dict.* (ed. 3) 255/2 *Military Secretary*, an officer attached to the staff of the commander-in-chief, to governors of provinces, and to an officer commanding an army in the field. His duties to some extent are confidential, and he relieves the officer under whom he is serving of a great deal of personal correspondence. **1947** R. G. Jessel *G, A, & Q* iii. 9 At the end of the course, a report on each student is sent to the Military Secretary at the War Office. **1778** E. Draper *Let.* 12 Feb. in *N. & Q.* (1944) 29 July 51/1 Report says, that Dick Sulivan is coming home, in that case, your Brother, I suppose, Steps into the Military Secretaryship. **1863** *Act* 26 & 27 *Vict.* c. 65 § 17 Her Majesty may direct the Lieutenants of Counties..to call out the Volunteer Corps..for actual Military Service. **1909** G. B. Shaw *Press Cuttings* 25 What women need is the right to military service. **1922** *Encycl. Brit.* XXXII. 39/2 Mr. Lloyd George persuaded the Supreme Council to accept the principle that all the enemy Powers should be obliged to abolish compulsory military service. **1911** *Ball Room* Mar. 4/1 (Advt.), Finnigan's,..Manchester... Inventor of the original Military Two Step. **1949** V. Silvester *Old Time Dancing* 49 The Military Two-step is often danced with varying distances separating the partners at different times. **1950** M. Gwynne *Old Time & Sequence Dancing* 78 Military Two Step... A championship dance. Commence as in the Veleta. **1966** *Listener* 24 Nov. 783/3 The military two-step ended with the ripping of Freddie's entire sleeve.

c. *Comb.* with other *adjs.*, as *military-industrial, -political, -scientific, -technological* adjs.

1961 D. Eisenhower in *N.Y. Times* 18 Jan. 22/4 In the councils of Government, we must guard against the acquisition of unwarranted influence, whether sought or unsought, by the military-industrial complex. **1973** *Times* 13 Jan. 19/5 President Eisenhower, who warned the country of the dangers of a military-industrial complex, must be twitching in his grave. **1965** H. Kahn *On Escalation* 291 The military-political outcome of a war. **1960** *Encounter* Oct. 9 Military-scientific research. **1962** *Times* 26 Feb. (Canada Suppl.) p. ii/2 The 'military-technological complex' eating into American business and politics alike.

B. *sb.* **1.** (Later N. Amer. examples.) Also with sing. verb.

1968 *Globe & Mail* (Toronto) 3 Feb. 11/5 The military use special film to photograph hidden features of the ground below. **1968** Mrs. L. B. Johnson *White House Diary* 8 Oct. (1970) 718 I've come to have a lot more understanding of what the military puts up with and especially the wives. **1970** *New Yorker* 3 Oct. 44/3 Other branches of the military..were not impressed.

2. Delete † and add later example.

1962 *Listener* 1 Nov. 723/3 A few splendid portraits of the Austrian court nobility and of the high militaries.

militate, *v.* Add: **1. c.** To display industrial or political intransigence; to act in the manner of a militant (sense *c).

1951 E. Paul *Springtime in Paris* xvi. 321 Busse knew all too well what happened to French Communists who showed disloyalty, or even who failed to 'militate'. **1969** *N.Y. Rev. Bks.* 30 Jan. 4/3 Simone Weil going to work in a factory and eventually starving herself to death in order to share the diet of the people of occupied France was answering the same 'call' as..Silone militating in the underground, in clandestinity.

militia. Add: **4.** In later use, *spec.* as part of the British armed forces assembled in 1939.

1939 *War Illustr.* 16 Dec. 427/3 We have taken, besides the Militia classes which have been called up, over 85,000 voluntary recruits since the war began.

militiaman. Add: *spec.* One called up in 1939 as part of the armed services at the outbreak of the war.

1939 *War Illustr.* 14 Oct. 129 (*caption*) Regular soldiers ..and the Militiamen—all have made an excellent impression at home and in France.

Also **mili·tiawoman,** a woman in a militia force; = *miliciana.

1936 *New Statesman* 21 Nov. 802/1 The C.N.T., since it contains a great many Murcians.., includes the most alarming of the faces met with in Barcelona and also most of the young militia-women. **1938** 'G. Orwell' *Homage to Catalonia* iv. 47 There were three militiawomen there who did the cooking.

Milium[2] (mi·liəm). The proprietary name of a type of insulating fabric. Also **milium.**

1950 *Rayon & Synthetic Textiles* June 95/1 Milium has been chosen as the trademark name to designate the new 'warmth without weight' fabric development of Deering, Milliken & Company. **1951** *Official Gaz.* (U.S. Patent Office) 4 Sept. 39/1 The Vadium Corp., Wilmington, Del. Milium. For Textile Fabrics of Cotton, Rayon, Nylon, and Mixtures Thereof Having Heat Reflective and/or Heat Retentive Properties. Claims use since Apr. 20, 1950. **1954** Potter & Corbman *Fiber to Fabric* (ed. 2) vi. 110 A wool outer fabric with a milium satin lining is approximately equal in warmth to an untreated satin lining plus an 8½-ounce wool interlining with the same outer fabric. **1956** *N.Y. Times* 20 Mar. 30/6 Milium is a metal-impregnated lining material valued for its lightness and insulating quality. **1968** J. Ironside *Fashion Alphabet* 241 Milium insulates against cold and allows the body to retain heat, while in summer it protects from extreme heat. **1972** *Guardian* 30 June 7 This superb new Milium Ironing Board Cover..is scorch resistant.

milk, *sb.* Add: **1. c.** *mother's milk* (further examples); also as a slang name for various liquors (see quots.).

c **1821** 'W. T. Moncrieff' *Tom & Jerry* (1828) iii. iii. 67 *Log.* What, my lily! here, take a drop of mother's milk. (Gives black child gin out of measure he has received from Landlord.) **1846** *Swell's Night Guide* 125/2 *Mother's milk*, rum boose, good liquor. *c* **1863** T. Taylor in M. R. Booth *Eng. Plays of 19th Cent.* (1969) II. 156 Brandy do a man harm! It's mother's milk. **1922** G. M. Trevelyan *Brit. Hist. 19th Cent.* xxiii. 363 Britons had sucked in fear of Napoleonic conquest with their mother's milk. **1966** 'L. Lane' *ABZ of Scouse* 71 *Mother's milk*, Guinness, a popular brand of stout. **1972** *Guardian* 22 Aug. 4/6 A six-month-old-baby..is being kept alive by mother's milk supplied by volunteers.

g. *ellipt.* = Milkman 1. *colloq.*

1895 W. P. Ridge *Minor Dialogues* 79, I know *all* the comic songs..and I sing 'em whilst I'm a doing up the front steps; and the milk, *he* says he reckons it'll end in me going on the stage. **1933** A. Thirkell *High Rising* ii. 36 The London tradesmen..called her Miss, until a fateful day when the Milk, so she told Laura, had called her Miss once too often. **1967** 'A. Gilbert' *Visitor* x. 174 She hadn't informed the postman and anyone can put out a note for the Milk. **1975** B. Meyrick *Behind the Light* xv. 202 The disappearance of George the Milk's horse.

h. Milk-white colour. Cf. sense 11 in Dict.

1899 Swinburne *Rosamund* i. i. 2 White I know from red, and dark from bright, And milk from white in hawthorn-flowers.

2. b. *pure milk*: something of the purest or finest quality.

1931 *Daily Express* 15 Oct. 2/4 Men like Mr. Runciman, who hitherto represented the purest milk of the Cobdenite gospel. **1955** *Times* 6 July 11/3 Broadcasting..probably remains the most effective way, within the compass of an election campaign, of distributing the pure milk of party doctrine. **1975** 'W. Haggard' *Scorpion's Tail* vii. 103 For the pure milk of doctrine she cared not a damn. She saw communism as a convenient front.

c. *milk of human kindness* (further examples); *spilt milk*: see also Spilt *ppl. a.* 2 b; *to bring* (a person) *to his milk* (U.S.): to bring (him) to his senses; to compel (him) to acquiesce or submit; *to come* (or *go*) *home with the milk*: to arrive home at the time when the milkman calls, *i.e.* early in the morning.

1775 Sheridan *Rivals* iii. iv. 57 The thunder of your words has soured the milk of human kindness in my breast! **1839** Dickens *Nickleby* xxxviii. 377 What's come of my milk of human kindness? It turns into curds and whey when I look at him. **1857** J. G. Holland *Bay-Path* 209 There ain't anything that'll bring you to your milk half so quick as a good double-and-twisted thrashin. **1857** S. A. Hammett *Sam Slick in Texas* iv. 22 When you cum to bring 'em down to thar milk, they'll turn out greener than Buffalo Bayou in September. **1917** Wodehouse *Man with Two Left Feet* 238 You talk of a man 'going home with the milk' when you mean that he sneaks in in the small hours of the morning. **1923** W. J. Locke *Moordius & Co.* ii. 17 The family has nothing to do with the way the governess spends her evenings..except if she comes home with the milk after her evening out. **1947** W. S. Maugham *Creatures of Circumstances* 100 Every party's got to come to an end, and next day it doesn't matter much if you went home with the milk or if you left while the fun was in full swing. **1956** G. Durrell *My Family & other Animals* xviii. 237 Overflowing with the milk of human kindness, the family had invited everyone they could think of, including people they cordially disliked.

3. c. *the milk in the coconut*: a puzzling fact or circumstance; a crux. *colloq.* (orig. *U.S.*).

1840 *Spirit of Times* 21 Mar. 25/2 All of 'vich'..fully accounts..for the milk in the cocoa-nut. **1853** *Knickerbocker* XLII. 50 The milk in the cocoa nut was accounted for. **1898** Mrs. Lynn Linton *Let* in G. S. Layard *Mrs. Lynn Linton* (1901) xxiv. 362 The Koran is very interesting—but oh, the milk in the cocoanut! It is so queerly disjointed and non-sequential, far more so than the Epistles, and they have their full share of that milk in the cocoanut. **1972** L. Meynell *Death by Arrangement* i. 17 'Nobody can really be christened Waveney: it's a river...' 'In East Anglia. Hence, as they say, the milk in the coconut. Rolffe's father..called his eleven children after East Anglian rivers.'

4. *milk of magnesia*: a proprietary name for a white suspension of magnesium hydroxide in water, taken as an antacid.

1880 *Trade Marks Jrnl.* 3 Mar. 95 Milk of Magnesia... Charles Henry Phillips,..New York, United States of America; manufacturing chemist.... Preparations of magnesia for medical purposes, especially hydrate of magnesia, and also proprietary medicines. **1924** H. Crane *Let.* 30 Nov. (1965) 194, I had..taken a great deal of Alkalithia and milk of magnesia. **1961** J. Heller *Catch-22* (1962) xxvi. 282 Aarfy had a date..with a Red Cross girl..whose father owned an important milk-of-magnesia plant.

9. a. (*a*) *milk-loaf*; (*b*) *milk-bloom, bottle, -car, carton, -churn, -ejection, -gland, -jug* (earlier example), *lorry, -pitcher, saucepan, -shop* (earlier example), *stand, tanker, truck, wagon*; *milk-boy* (earlier and later examples).

1855 TENNYSON *Maud* xxi. 70 The slender acacia would not shake One long milk-bloom on the tree. **1905** G. F. M'CLEARY *Infant Mortality & Infants Milk Depôts* viii. 129 (*caption*) Packing the milk bottles in ice before sending them to the city. **1957** M. SUMMERTON *Sunset Hour* v. 67, I could hear..milk bottles being handled in and out of crates. **1959** I. & P. OPIE *Lore & Lang. Schoolch.* i. 9 Cigarette cards..are being replaced in flicking games by milk-bottle tops. **1972** C. FREMLIN *Appointment with Yesterday* xii. 93 'Everything goes down the waste-disposal!' But not dead matches. Or milk-bottle tops. **1972** J. MOSEDALE *Football* x. 139 He stepped on a broken milk bottle, severing all the tendons in his foot. **1847** THACKERAY *Vanity Fair* (1848) vii. 59 The groom..did not care to descend to ring the bell; and so prayed a passing milk-boy to perform that office for him. **1865** A. & E. KEARY *Little Wanderlin* 125 After the milk-boy came the vegetable women. **1964** F. WARNER *Early Poems* 13 A milk-boy, whistling down the wind. **1890** *Railways of Amer.* 146 The different kinds of cars which are now used.. Mail car, Milk car, Oil-car [etc.]. **1916** JOYCE *Portrait of Artist* ii. 69 Often they drove out in the milkcar. **1964** 'E. LATHEN' *Accounting for Murder* (1965) vi. 46 He mounted the stairs..startling two typists who were precariously balancing milk cartons. **1967** R. LOWELL *Near Ocean* 24 Milk cartons, kidney heaped to spoil, Two plates sheathed with silver foil. **1931** A. UTTLEY *Country Child* iii. 48 He was backed into the loading place, waiting for the milk-churns. *Ibid.* xxi. 278 The milk-churns were rattled and banged across the railway line. **1963** *Times* 16 Feb. 5/5 An estimated 120,000 milk churns, valued at £5 each and believed to have been taken by householders to store water, are missing from dairies. **1967** *Ibid.* 26 Sept. 1/1 He inadvertently rolled a milk churn under an oncoming express during his first day as an apprentice at Hatfield station. **1950** *N.Z. Jrnl. Agric.* June 540/3 The problems related to the process of 'milk ejection' in the cow. **1927** HALDANE & HUXLEY *Animal Biol.* xiii. 320 The saucer-shaped depression into which the milk-glands open. **1832** F. TROLLOPE *Dom. Manners Amer.* I. xii. 170 An intimation accompanied the milk-jug, that the milk must be fresh. **1910** *Practitioner* June 801 Milk-loaf, scones. **1939** G. HOUSEHOLD *Rogue Male* 110, I saw..a couple of milk lorries bobbing about..to collect the cans set out on wooden platforms by the road. **1971** W. J. BURLEY *Guilt Edged* i. 5 The milk lorry on its way back to the factory after morning collections from farms. **1855** *Harvard Mag.* I. 420 We were..a good deal incommoded by the diminutive size of the milk-pitchers. **1907** *Yesterday's Shopping* 213/2 Milk saucepan with earthenware lining. **1975** J. SYMONS *Three Pipe Problem* xvii. 163 He burnt the milk saucepan dry. **1847** DICKENS *Dombey* (1848) xxi. 207 A neighbouring milk-shop. **1950** *N.Z. Jrnl. Agric.* Apr. 378/3 Milk stands erected as part of the releaser room.. have given satisfactory results. **1965** in P. Jennings *Living Village* (1968) 66 The milk..is collected by a milk tanker. **1972** *Guardian* 16 Oct. 10/2 Until the rains came..villagers were collecting water in buckets from milk tankers in High Furness. **1910** *Daily Chron.* 22 Apr. 1/3 The..express ..ran into a milk truck and a guard's van. **1947** E. HODGINS *Mr. Blandings builds his Dream House* vii. 94 Her husband got run over by the milk truck. **1973** R. L. SIMON *Big Fix* (1974) xx. 170 A trio of milk trucks from a dairy. **1883** *Wheelman* Apr. 28/1 A superannuated 'bus-driver, with a conveyance strongly resembling a milk-wagon. **1960** T. HUGHES *Lupercal* 46 Light and birdsong come Walloping up roads with the milk wagon.

b. *milk producer, strainer*; *milk-marketing, -producing* vbl. sbs. and adjs.

1933 *Statutory Rules & Orders* No. 789. 21 This scheme may be cited as the Milk Marketing Scheme, 1933, and applies to England and Wales. **1936** Milk Marketing [see *GRADE *sb.* 5 c]. **1968** *Listener* 4 July 15/2 In the new milk marketing case this truth is emphasised... Under the Milk Marketing Scheme the milk producers sell their milk to the Milk Marketing Board. **1870** *Rep. Comm. Agric. 1869* (U.S. Dept. Agric.) 449 The annual meeting of the Milk Producers' Association of Massachusetts and New Hampshire. **1950** *N.Z. Jrnl. Agric.* Feb. 163/3 It is the responsibility of the milk producer to ensure..that only milk of the highest quality leaves the farm. **1946** *Nature* 12 Oct. 523/1 The recording movement may..progress to greater service to the milk-producing industry. **1975** *Country Life* 12 June 1590/1 Different cows have different milk-producing capabilities. **1686** S. SEWALL *Letter-Bk.* (1886) I. 33 Five Duz. of milk strainers of the smaller sort. **1872** W. S. JONES *Let.* 20 Mar. in G. N. Jones *Florida Plantation Rec.* (1927) 199 The milk strainer is also in bad repair.

d. *milk-blue, -dim, -green, -pale* adjs.

1917 D. H. LAWRENCE *Look! We have come Through!* 77 The milk-blue, morning lake. **1945** W. DE LA MARE *Burning-Glass* 41 Pulsing beneath the silken skin The milk-blue blood rills out and in. **1955** E. POUND *Section: Rock-Drill* xc. 65 Moon's barge over milk-blue water. **1926** H. READ *Coll. Poems* 55 Oh, turn your milk dim eyes To outer things! **1912** D. H. LAWRENCE *Let.* 2 June (1962) I. 130 The pale, milk-green river. **1895** W. B. YEATS *Poems* 33 And at his cry there came no milk-pale face Under a crown of thorns and dark with blood. **1910** W. DE LA MARE *Three Mulla-Mulgars* xviii. 248 At each thorn-tip, as the flame licks near, wells out and gathers a milk-pale globe of poison.

10. milk bank, a bank (*BANK *sb.*[3] 7 f) of human milk; **milk-bar,** a place where drinks made from milk (and often also other refreshments) are sold (see BAR *sb.*[1] 28); **milk-blooded** *a.*, cowardly, spiritless; **milk chicken,** a chicken that has been fed on milk and ground oats; **milk chocolate,** † (*a*) a beverage made from chocolate and milk; (*b*) eating chocolate (CHOCOLATE 2) made with milk; (*c*) a brown colour; also *attrib.*; **milk-coffee,** coffee made with milk; white coffee; also *attrib.*, of a light brown colour; **milk factor,** a factor causing disposition towards mammary cancer which is transmitted to offspring in milk; **milk-fish,** (*a*) substitute '*Chanos chanos*' for '*Chanos salmoneus*'; (examples); (*b*) *Austral.* = TREPANG; **milk-float,** (*a*) a float (FLOAT *sb.* 14) for the conveyance of milk; (*b*) a small electrically-driven milk-cart; **milk-flour,** a preparation of desiccated milk; **milk kitchen,** a special kitchen at a maternity hospital or the like, where babies' feeds are prepared; **milk line** *Embryol.* [tr. G. *milchlinie* (O. Schultze 1892, in *Anat. Anzeiger* VII. 266)], (the line occupied by) a ridge of thickened ectoderm that appears on either side of mammalian embryos, extending from the front to the rear limb buds, on which the mammary glands later form in females; **milk-name,** the name given to a Chinese child at a ceremony held one month after birth: it is later superseded by more formal names but continues to be used in particular situations (see quot. 1911); **milk powder,** a preparation of desiccated milk; **milk-ranch** *U.S.*, a ranch producing milk; **milk ridge** *Embryol.* [tr. G. *milchleiste* (O. Schultze 1893, in *Verhandl. d. physik.-med. Ges. zu Würzburg* XXVI. 173)] = **milk line*; **milk-room** orig. *U.S.*, a room in a house or dairy in which milk is kept; **milk-round,** (*a*) = **milk-route*; (*b*) *transf.*, a regular trip or tour in which one calls at several places; spec. *R.A.F. slang* (see quot. 1945); hence *milk-roundsman*; **milk-route** orig. *U.S.*, a route on which milk is regularly collected from farmers or delivered to customers; **milk-run** = **milk-round*; **milk-shake** orig. *U.S.*, a beverage composed of milk, flavouring, etc., mixed by shaking or agitation; **milk stout,** formerly, a kind of sweet stout made with lactose; also *attrib.* and *fig.*; **milk-toast** *U.S.*, toast which is softened in milk; **milk-train,** (*a*) a railway train chiefly transporting milk, usu. very early in the morning; (*b*) *R.A.F. slang* (see quot. 1943); **milk-tube,** (*b*) a milking tube; **milk-walk** (later examples); also, a dairy business; **milk-wine,** a beverage obtained from fermented milk; **milk-woman** (later example).

1948 *Archit. Rev.* CIV. 21 (*caption*) Mothers' milk bank. **1972** *Guardian* 22 Aug. 4/6 If there should be a shortage of mother's milk, the hospital will get supplies from the National Milk Bank. **1935** *Forres Gaz.* 20 Nov. 1/2 The milk bar, or place where milk drinks are sold, is a popular institution all over Australia, and plans are on foot for installing..them in Britain. **1938** E. WAUGH *Scoop* II. iv. 200 Legend..told and retold over the milk-bars of Fleet Street. **1957** J. BRAINE *Room at Top* xxx. 254 A milk bar near the railway station. **1971** J. PHILIPS *Escape a Killer* (1972) I. v. 69 There's a milk bar in the village. **1847** E. BRONTË *Wuthering Heights* I. xi. 259, I wish you joy of the milk-blooded coward. **1910** *Blackw. Mag.* Feb. 183/2 The sooner we give up all this milk-blooded, blue-spectacled, pacificist talk the better. **1902** *Encycl. Brit.* XXXI. 882/2 Chickens fattened quite young..and known as *petits poussins* or 'milk chickens'. **1723** J. NOTT *Cook's & Confectioner's Dict.* sig. I8 (*heading*) To make Milk Chocolate. **1752** M. W. MONTAGU *Let.* 16 Feb. (1967) III. 5 As soon as I am risen, I constantly take 3 cups of milk coffee, and two hours after that a large cup of milk chocolate. **1904** 'SAKI' *Reginald* 101 They all sat down to play progressive halma, with milk-chocolate for prizes. **1910** *Encycl. Brit.* X. 614/2 Milk powder..is largely employed in the preparation of so-called milk chocolates. **1926** C. BEATON *Diary* 15 Apr. in *Wandering Yrs.* (1961) iv. 80, I..bought some bars of milk chocolate. **1955** *Radio Times* 22 Apr. 21/3, 14 milk chocolate caramels. **1958** S. HYLAND *Who goes Hang?* xiv. 63 A large table splendidly covered with milk-chocolate-coloured leather. **1969** *Vogue* 15 Mar. 81/1 An edging of milk chocolate suede. **1974** 'E. LATHEN' *Sweet & Low* xii. 124 The creamy satisfaction of milk chocolate. **1695** J. LIGHTBODY *Every Man his own Gauger* 62 If you would make Milk Coffee, you must, to every Pint of Water, put a quart of Milk. **1752** [see *milk chocolate* above]. **1972** H. OSBORNE *Pay-Day* II. iv. 42 The girl at the desk..was a milk-coffee negress. **1939** J. J. BITTNER in *Public Health Rep.* (U.S. Public Health Service) LIV. 1115 The breast cancer observations ..may be explained by a theory..assuming that three 'factors' are needed... These factors are: (A) A 'breast cancer-producing influence' transferred through the milk of high-cancer stock females to their progeny. This has been designated as the 'milk factor' in the tables. **1943** C. G. GESCHICKTER *Dis. Breast* xxxv. 800 This so-called milk-factor or milk-influence has been extracted from the mammary glands of lactating cancer-susceptible mice by Bittner and shown to increase the incidence of mammary cancer whether injected in, or fed to young mice. **1966** WRIGHT & SYMMERS *Systemic Path.* I. xxviii. 990/1 There is no evidence that a milk factor plays any part in the occurrence of carcinoma of the breast in women. There is no way by which such a factor can be demonstrated. **1880** *Proc. Linn. Soc. New South Wales* V. 128 Another species [of Trepang] is the 'milk fish', or 'cotton fish', so called from its power of emitting a white viscid fluid.., which clings to any object like shreds of cotton. **1905** D. S. JORDAN *Guide to Study of Fishes* II. iii. 44 The *Chanidæ*, or milkfishes, constitute another small archaic type, found in the tropical Pacific. They are large, brilliantly silvery, toothless fishes. *Ibid.* 45 The single living species is the *Awa*, or milkfish, *Chanos chanos*, largely used as food in Hawaii. **1962** K. F. LAGLER et al. *Ichthyol.* vi. 203 The irregular movements of the Asiatic milkfish (*Chanos*) into and out of fresh water make it amphidromous. **1971** *Daily Colonist* (Victoria, B.C.) 30 Mar. 5/4 Taiwan and Indonesia produce much needed sea protein in traditional milkfish 'farm ponds'. **1974** *Nat. Geographic* Dec. 788/2 Across southern Asia, from the Philippines to India, commercial aquaculturists have begun to raise milkfish, a food species that subsists on plant life. **1887** Milk float [in Dict., sense 9 a]. **1935** N. COLLINS *Three Friends* viii. 143 A horse attached to a milk float wore a hat made of newspaper. **1951** *Engineering* 20 July 95/3 Pedestrian-controlled vehicles (such as hand-operated electric milk floats). **1974** M. BABSON *Stalking Lamb* xi. 73 An electric milk float trundling down the street. **1902** *Chambers's Jrnl.* 22 Feb. 191/1 The milk-flour is completely soluble in water. **1965** *Nursing Times* 5 Feb. 181/1 In some hospitals the labour wards and the milk kitchens were each centralized. **1893** *Jrnl. R. Microsc. Soc.* 304 Prof. O. Schultze finds in the embryos of pig, rabbit, mole, fox, and cat, that the first rudiment of the mammary glands is seen as a linear epithelial thickening on each side of the body... This 'milk-line' stretches from the anterior to the posterior limb-rudiment. **1946** B. M. PATTEN *Human Embryol.* ix. 241 Some animals (for example, the sow and the bitch) develop a series of nipples spread over nearly the entire length of the milk line. *Ibid.*, Not infrequently supernumerary nipples may occur at other levels along the course of the milk line. **1960** F. D. ALLAN *Essent. Human Embryol.* xix. 179 The primordia of the [mammary] glands form thickened strips of ectoderm located bilaterally from axilla to groin which are called the milk ridges or lines. **1836** J. F. DAVIS *Chinese* I. vii. 288 The birth of a son is of course an occasion of great rejoicing; the family or surname is first given, and then the 'milk name', which is generally some diminutive of endearment. **1911** J. D. BALL *Chinese at Home* vii. 75 The milk name.. clings to him or her through life, being used by parents, relatives, and most intimate friends, as well as by superiors. **1931** C. L'E. EWEN *Hist. Surnames* 8 The Chinese receive a number of names..the 'milk-name' when a month old. **1975** O. SELA *Bengali Inheritance* iii. 23 His milk name had been Chan Yan-Wo, and at school he had changed it to Richard.., the first step in accepting Western ways. **1834** *India Jrnl. Med. Sci.* I. 1. 32/2 *Milk Powder*... Specimens of an article he has advertised under the designation of Pulverised Milk. **1910** *Encycl. Brit.* X. 614/2 Milk powder is manufactured under various patents. *Ibid.*, Milk powder made from skim-milk keeps well for considerable periods. **1972** D. BLOODWORTH *Any Number can Play* xii. 96 A big American air force general with a complexion like milk-powder. **1856** *Calif. Pathfinder* (San Francisco) 13 Nov. 2/4 The milk ranch that burned down beyond the Mission yesterday morning. **1909** BAILEY & MILLER *Text-bk. Embryol.* xvi. 449 In embryos of six to seven mm., or even less, a thickening of the epidermis occurs in a narrow zone along the ventro-lateral surface of the body (Strahl). In embryos of 15 mm. this thickening, known as the milk ridge, extends from the upper extremity to the inguinal region. **1960** Milk ridge [see *milk line* above]. **1836** *Knickerbocker* VIII. 706 In the rear, is quite a city of additions, in the shape of bed-rooms, bath-rooms, milk-rooms, buttery [etc.]. **1970** *Cape Times* 28 Oct. 21/1 (Advt.), Lean-to, barn for animals, dairy and milk room, 4 calf pens. **1900** *Oxford Times* 13 Jan. (Advt.), Wanted, a single man to serve a milk-round. **1927** R. B. FORRESTER *Fluid Milk Market Eng. & Wales* 96 Retail Delivery. Milk Rounds..no close or detailed survey of actual roundsman systems has ever been made in this country. **1945** PARTRIDGE *Dict. R.A.F. Slang* 39 *Milkround*, a run made fairly regularly by a Squadron or a Force, if it returns to its station or base in the early morning. **1952** E. F. DAVIES *Illyrian Venture* x. 191 We did a 'milk-round' of all the jails in Vienna, picking up and setting down prisoners at every stop. **1958** *Times* 9 Aug. 7/7 Strange though the urgent masochism of the milk round may seem, that is how the great majority of Americans still see us, from the windows of a coach. **1970** *Times* 17 Nov. 19/8 Like other business organisations, we make what is known as the annual milkround, going to every university at the recruiting time. **1972** *Guardian* 8 Feb. 13/7 Fund-raising must be..centralized, instead of the monthly 'milk rounds' by volunteers. **1972** *Accountant* 14 Sept. 327/1 'Farmer's Wife' branded goods —cream, yogurt, butter, eggs, potatoes, bread, margarine, bacon, sausages and poultry—constitute an increasingly important part of the milk division's turnover and profit. Sales are helping to maintain regular milk rounds when other industries find rising costs of such personal service a constant headache. **1940** F. KITCHEN *Brother to Ox* xiii. 202, I want to say what a pleasant job it is being a milk-roundsman. **1874** *Rep. Comm. Agric. 1873* (U.S. Dept. Agric.) 246 The most economical method of managing the delivery of milk at the factory is by establishing milk routes. **1897** 'MARK TWAIN' *Following Equator* xliv. 464 The vested rights..are frequently the subject of sale or mortgage. Just like a milk-route. **1959** N. MAILER *Advts. for Myself* (1961) 372 The milk companies..are saved most of the costs of local distribution by delivering the orange juice on their milk route. **1943** K. TENNANT *Ride on Stranger* (1968) ix. 103 Shannon did not know anyone who wanted half a milk-run. **1944** J. H. FULLARTON *Troop Target* iii. 24 Isobel married that joker with a milk-run out Henderson way. **1944** T. H. WISDOM *Triumph over Tunisia* vi. 54 It was General Doolittle who organised the 'milk-run' Fortress raids on the ports of Tunis and Bizerta. **1964** *Observer* (Colour Suppl.) 11 Oct. 17/2 Similar risks must be taken by transport aircraft pilots, flying their daily 'milk runs' to supply jungle-bound positions along the 1,000-mile frontier [of Borneo]. **1969** *Daily Tel.* 11 Oct. 11/5 Another way of island hopping down to Grenada..is to catch the early morning 'milk-run' plane from Antigua, which calls in at Dominica, St. Lucia,

Martinique and Barbados, collecting and unloading passengers, mail and newspapers as it goes. **1972** *Guardian* 30 Dec. 13 Woe betide any who suddenly discovers he has to go to Brussels the next morning. The businessmen's milkrun is always booked days ahead. **1889** *Harper's Bazaar* 4 May 330/3 You needs some milk shake..an' I got some nice new w'iskey to putt in. **1911** H. S. HARRISON *Queed* vii. 85 You ain't feelin good, are you, Doc? You're lookin' white as a milk-shake. **1937** *Daily Herald* 20 Feb. 11/3 (*caption*) Mrs.—..sampling a milk shake after she had opened a milk bar in Tottenham Court-road yesterday. **1952** 'J. TEY' *Singing Sands* xii. 196, I had a coupla bananas and a milk shake in Leicester Square. **1953** E. TAYLOR *Sleeping Beauty* xiii. 200 She saw herself translated to the Corner House, but to the same sundaes and parfaits and milk-shakes. **1968** *Blues Unlimited* Sept. 10 Joe worked in the kitchens at the Cafe six nights a week, pouring cokes and milk shakes. **1942** *R.A.F. Jrnl.* 13 June 32 'There you are, gentlemen,' boomed a rich, milk-stout voice. **1959** M. GILBERT *Blood & Judgement* i. 12 A lady..[was] addressing herself to a glass of milk stout. **1965** S. M. TRITTON *Tritton's Guide to Better Wine & Beer Making for Beginners* 133 Milk Stout... Pour hot ..water over the patent malt and stir in the flaked barley. .. Boil the hops in 2 pints of water... Dissolve sugar and lactose..and add to bulk. Follow by the yeast and ferment to completion. **1974** G. MANN *Home Wine & Beer Making* 99/1 Sweet, or milk, stout dark and sweet, but still with an underlying bitter twang. **1855** J. R. BESTE *Wabash* II. 260 Large platters of milk toast. This delicacy is made of slices of toast, buttered and sprinkled with pepper and salt, and laid in a dish of warm milk, which serves as a sauce to the rest. **1903** K. D. WIGGIN *Rebecca* xxiii. 258 She's just asked me for some milk-toast. **1853** *Knickerbocker* XLII. 532 The '*milk-train*' still had the right of way. **1897** [see *HIGHBALL 2]. **1930** WODEHOUSE *Very Good, Jeeves!* ix. 251 Her intention was..to..leave by the next train, even if that train was a milk-train, stopping at every station. **1943** HUNT & PRINGLE *Service Slang* 46 *The Milk Train*, appropriate name for the modern 'Dawn Patrol' on early morning reconnaissance flights. **1955** *Railway Mag.* May 359/1 A daily milk train which was worked by the L.N.E.R. **1877** *Rep. Vermont Dairymen's Assoc.* VIII. 106 The milk must be drawn by means of a catheter, or milk tube. **1851** H. MAYHEW *London Labour* I. 435/2 My father had a milk-walk. *c* **1864** BROUGH & HALLIDAY in M. R. Booth *Eng. Plays of 19th Cent.* (1973) IV. 240, I have a horse and cart, Miss Penelope, and a first-rate milk walk. **1905** G. B. SHAW in *Grand Mag.* Feb. 111 He..had..bought an agent's business as a doctor buys a practice or a dairyman a milkwalk. **1917** WODEHOUSE *Man with Two Left Feet* 247 He was.. owner of a milk-walk in the most fashionable part of Battersea. **1911** M. I. NEWBIGIN *Mod. Geogr.* vii. 189 A milk-wine or koumiss, produced by the fermentation of milk, is the characteristic drink. **1879** F. R. STOCKTON *Rudder Grange* v. 56 She had spent the night in a wooden rocking-chair at the milk-woman's.

b. milk-bush, (*c*) (example); (*d*) in South Africa and Australia, a name used for several shrubby plants, often succulent, which have a milky latex, esp. various species of *Euphorbia* (cf. sense *a*); **milk-parsley** (later example); **milk-plant,** (*b*) = **milk-bush* (*d*); **milk thistle** (further examples); **milk-wood,** (*c*) add: or other Australian trees of the genus *Alstonia* (further examples); (*d*) and (*e*) (examples).

1818 C. I. LATROBE *Jrnl. Visit S. Afr. 1815–16* 133 The milk-bush (*ficus*), a tree not unlike a Portugal laurel. **1861** J. A. GRANT *Jrnl.* 27 May in *Walk across Afr.* (1864) v. 79 After we had entered the first milk-bush enclosure, there were several cleanly-swept windings. **1883** 'R. IRON' *Story Afr. Farm* I. i. 3 The milk-bushes with their long, finger-like leaves..were touched by a weird and an almost oppressive beauty as they lay in the white light. **1907** *Nature* 17 Jan. 288/1 The common milkbush of the karroo and karroid regions of the interior, viz. Euphorbia mauritanica. **1926** [see *CAUSTIC *a.* 1 e]. **1965** *Austral. Encycl.* VI. 84/1 Milkbush, a popular name for several shrubs or small trees with a milky sap, especially *Wrightia saligna* in the family Apocynaceæ. **1966** E. PALMER *Plains of Camdeboo* xvi. 259 The milk bush grows here, the *Euphorbia mauritanica* of botanists, with its long, smooth, fleshy, yellow-green stems. **1974** *Country Life* 3 Oct. 923/1 The double-brooding swallowtail..—from milk-parsley-eating caterpillar..to Britain's largest butterfly. **1965** *Austral Encycl.* VI. 84/1 Milk plant is sometimes applied to members of the genus *Euphorbia*. **1787** W. WITHERING *Brit. Plants* (ed. 2) II. 875 *Carduus marianus*... Leaves with a net-work of white veins... Milk Thistle. Ladies Thistle. **1880** BRITTEN & HOLLAND *Dict. Eng. Plant-Names* 335 Milk Thistle... *Sonchus oleraceus*, L., in allusion to its milky juice. **1883** W. ROBINSON *Eng. Flower Garden* 270/2 If a few plants are raised in the garden and planted out in rough and somewhat bare places or banks, &c., the Milk Thistle will soon establish itself permanently. **1960** *Oxf. Bk. Wild Flowers* 34/1 Corn Sowthistle or Field Milk-thistle (*Sonchus arvensis*). A perennial with a hollow stem and milky juice. *Ibid.* 150/2 Milk Thistle (*Silybum marianum*). This rather rare annual or biennial Thistle has large, solitary, often drooping, purple flower-heads. **1862** L. PAPPE *Silva Capensis* (ed. 2) 24 *Sideroxylon Inerme Lin.* (Milk-wood; Melkhout)... Wood whitish, very hard, close, and durable. **1889** J. H. MAIDEN *Useful Native Plants Austral.* 570 *Melaleuca leucadendron*... Called 'Milkwood' in the Northern Territory. **1907** T. R. SIM *Forests & Forest Flora Cape Good Hope* 252 *Sideroxylon inerme.* (White Milkwood;..). *Ibid.* 254 *Mimusops obovata.* (Red Milkwood;..). **1908** E. J. BANFIELD *Confessions of Beachcomber* I. i. 37 On Timana are gigantic milkwood trees (*Alstonia scholaris*) which need great flying buttresses to support their immense height. **1917** [see *JAKKALSBESSIE]. **1928** D. COTTRELL *Singing Gold* III. i. 191 A thin ribbon of smoke showed against great milkwood trees. **1932** [see *JAKKALSBESSIE]. **1946** L. G. GREEN *So Few are Free* (1948) x. 137 Vaillant noted a large milkwood tree growing out of a rocky crevice. **1973** PALMER & PITMAN *Trees S. Afr.* III. 1737 The wooden rails of the little railway on the Durban Bluff in the early days..were hewn out of milkwood trees.

milk, *sb.*[2] Colloq. abbrev. of MILKSOP.

1881 *Punch* 10 Sept. 110/2 Patriotic? Well, them as talks Muggins like that to our gurls must be milks. **1923** in J. MANCHON *Le Slang.*

milk, *v.* Add: **1. f.** *colloq.* To put milk into or on to.

1877 *Trans. Devon Soc. Adv. Sci.* IX. 134 Have you *milked* your tea? **1969** J. WAINWRIGHT *Big Tickle* 52 She milked and sugared both mugs of tea.

4. e. *Theatr. slang.* (See quots.).

1939 HIXSON & COLODNY *Word Ways* xvi. 142 To overplay an audience for applause is called *milking the audience.* **1942** BERREY & VAN DEN BARK *Amer. Thes. Slang* § 593/24 *Milk a scene* or *the audience*, to try to get more laughs or applause out of a part than it deserves. **1962** *Times* 15 May 13/3 Too many of the other acts, however, have no idea how to..milk a laugh. **1971** M. BABSON *Cover-up Story* ix. 107 They milked the applause for all it was worth, then Bart held up his hand again.

milking, *vbl. sb.* Add: **4.** *milking machine* (earlier and later examples), *-shed, -yoke*; **milking bail** = BAIL *sb.*[3] 5; **milking-parlour,** a shed specially equipped for milking cows; **milking shorthorn,** a type of shorthorn developed specially for producing milk.

1890 W. H. S. ROBERTS *Hist. Oamaru* 54 A stockyard and milking-bail had also been erected. **1850** *New England Farmer* II. 282 Gutta percha patent milking machines are in use on Long Island. **1945** 'G. ORWELL' *Animal Farm* v. 37 A windmill, which could..supply the farm with electrical power..and would also run..an electric milking machine. **1974** 'E. LATHEN' *Sweet & Low* xiii. 131 Shots of ..the dairies,..complete with milking machines. **1946** *Agric. Overseas Rep.* I. 8 The milking parlour arrangement is only at the experimental stage. **1952** *Blackw. Mag.* Feb. 97/1 A thoroughly sound, commonsense opinion on..a milking-parlour. **1972** *Country Life* 2 Nov. Suppl. 3 The Farm buildings include 2 milking parlours, cubicles for 120, stock yards. *a* **1930** 'H. STONE' in Murdoch & Drake-Brockman *Austral. Short Stories* (1951) 125 It was a silent milking shed that evening. **1937** *Discovery* July 214/2 The rear parts serve as tool and store sheds, the front as byre and milking-shed. **1970** *Kenya Farmer* Feb. 36/1 'Oh Gawd!' said Fred the farmer as he came stomping up the verandah steps, his boots looking more as if he had walked from Timboroa than from the milking sheds two hundred yards away. **1910** *Encycl. Brit.* V. 539/2 The non-pedigree milking Shorthorn of the north of England is an excellent cow. **1957** Milking Shorthorn [see *dairy shorthorn* s.v. *DAIRY *sb.* 4]. **1910** J. MASEFIELD *Ballads & Poems* 42 Sleepy men bear milking-yokes Slowly towards the cattle-byre.

milko (mi·lko), *sb.* and *int.* Also **milk-o, milk-oh.** [f. MILK *sb.* + O *int.*] **A.** *sb.* A milkman. *slang.*

1911 *Answers* 11 Feb. 362/3 Milk-O! on the Make... Many of the milk-o! fraternity boast [of]..their..stealings. **1933** *Bulletin* (Sydney) 31 May 12/1 It is difficult to imagine a man looking like Il Duce while filling a billycan, but I have seen our milko perform the feat. *Ibid.* 5 July 20/3 One of Sydney's milk-ohs. **1958** *Daily Mail* 25 Oct. 5/2 The milko..has become the *milk salesman.* **1968** D. IRELAND *Chantic Bird* i. 6, I remember the milk tap I turned on in old Bay Road, the milko chased me all the way home.

B. *int.* A shout made by a milkman to indicate that milk is available.

1916 'TAFFRAIL' *Pincher Martin* xii. 232 'E wus drivin' one o' these 'ere milk-carts an' shoutin' 'Milk-o!' artside th' 'ouses. **1967** L. DEIGHTON *London Dossier* 132 The milkman shouting 'milko' as he leaves milk on a doorstep.

milkstone. Add: **2. b.** A hard deposit formed in or on dairy equipment by precipitation from milk.

1949 *N. Z. Jrnl. Agric.* Nov. 487/2 Milkstone is the casein of milk which has become attached to metal in the form of encrustations. **1963** *New Scientist* 14 Nov. 387 The alloy resisted the build-up of 'milkstone' better than polished stainless steel.

milk-warm, *a.* (Later examples.)

1894 KIPLING *Seven Seas* (1896) 36 Milk-warm wi' breath o' spice an' bloom. **1918** A. HUXLEY *Defeat of Youth* 34 At your mouth, white and milk-warm sphinx. **1922** —— *Mortal Coils* 177 Shelley had been drowned in this milk-warm sea.

milkweed. Add: **1.** (Earlier and later examples.)

1598 FLORIO *Worlde of Wordes* 199/2 Lattaria, herbe Tithimale, spurge or milk-weede. **1955** G. GRIGSON *Englishman's Flora* 227 Sun Spurge... Milk-weed, Ess[ex], Herts, E[ast] Ang[lia]. *Ibid.* 392 Sow Thistle... Milkweed, Som[erset].

2. (Earlier and later examples.)

1814 J. BIGELOW *Florula Bostoniensis* 62 *Asclepias Syriaca*, Common Silk weed or Milk weed,..is used as a substitute for feathers, fur, cotton, &c. **1854** THOREAU *Walden* 252 One very calm October after noon, for such days especially they settle on to the lakes, like the milkweed down, having looked in vain over the pond for a loon, suddenly one, sailing out from the shore.., set up his wild laugh. **1923** W. CATHER *Lost Lady* 17 The silvery milkweed was just coming on. **1933** M. DE LA ROCHE *Master of Jalna* vii. 69 A milkweed pod having burst, its hoard was released and the silvery particles,..swam delicately on the light breeze. **1957** L. EISELEY *Immense Journey* 69 There dassed before my eyes the million airy troopers of the milkweed pod. **1968** *Times* 2 Oct. 12/5 Butterflies reared on the milkweed species Asclepias curassavica.

3. In names of various North American insects feeding on milkweed: **milkweed beetle,** a brightly coloured beetle, *Tetraopes tetraophthalmus*; **milkweed bug,** either of two species of bugs of the family Lygæidæ, *Oncopeltus fasciatus*, the large milkweed bug, or *Lygæus kalmi*, the small milkweed bug; **milkweed butterfly** = MONARCH *sb.*[1] 3 (in Dict. and Suppl.).

1842 T. W. HARRIS *Insects Injurious to Vegetation* 455/2 Milk-weed beetle. **1954** BORROR & DELONG *Introd. Study Insects* xxii. 398 *T[etraopes] tetraophthalmus* Forster is a common species feeding on milkweed, and is often called the red milkweed beetle. **1905** V. L. KELLOGG *Amer. Insects* x. 211 The milkweed-bug, *Oncopeltus fasciatus*, about ⅝ inch long, orange above with most of head and prothorax except the margins black, and a broad black band across the middle of the fore wings and large black blotch on their tips, is a common showy bug on various species of milkweed. **1970** *Nature* 3 Jan. 82/1 We have treated eggs of the milkweed bug, *Oncopeltus fasciatus*, with a juvenile hormone analogue. **1972** SWAN & PAPP *Common Insects N. Amer.* xii. 125 In the Small Milkweed Bug, *L[ygæus] kalmi*, the black on the pronotum is separated from the black on the front of the wings. **1974** *Nature* 12 Apr. 556/3 Experiments in this same general field carried out on the milkweed bug *Oncopeltus.* **1889** *Insect Life* I. 221 The following is a brief account of a migratory movement of enormous numbers of the common so-called Milk-weed Butterfly. **1906** R. SOUTH *Butterflies Brit. Isles* 107 The actual number of specimens of the Milkweed, or, as it is sometimes called, Monarch butterfly, seen or caught in England..does not much exceed thirty. **1972** *Country Life* 6 Apr. 846/1 The milkweed or monarch butterfly is a denizen of North America.

milky, *a.* Add: **3. b.** (Earlier examples.)

1765 G. WASHINGTON *Diaries* (1925) I. 210 Note, the [mulberry] Stocks were very Milkey. **1768** *Ibid.* 282 Some [wheat] whose straw and head was green but the grain of full size and Milky. **1789** J. MORSE *Amer. Geogr.* 52 About the time that it begins to turn from its milky state and to ripen, they run their canoes into the midst of it.

4. Also (*slang*), cowardly.

1936 J. CURTIS *Gilt Kid* ii. 18 They just talk that way to make you turn milky. **1938** G. GREENE *Brighton Rock* II. i. 62 I'm not milky... I just don't want another killing. **1954** 'N. BLAKE' *Whisper in Gloom* xvi. 217 Look at da kid. He's not milky. **1969** H. CARVIC *Miss Seeton draws Line* ix. 171 'Getting milky?' scoffed Doris.

b. Of a noise, song, etc.: soft. *poet.*

1924 E. SITWELL *Sleeping Beauty* x. 38 Goats gold as wheat With a kind white milky bleat. **1925** —— *Troy Park* 76 A white bird sang a milky song Of easy heaven and feathered rest.

5. *milky-coloured, -sapped, -toothed* adjs.

1906 W. B. YEATS *Poems 1899–1905* 268 Hold up your hands to him, that you may pluck That milky-coloured neck out of the noose. **1923** D. H. LAWRENCE *Birds, Beasts & Flowers* 19 Folded upon itself..And milky-sapped, sap that curdles milk and makes *ricotta.* **1905** E. F. BENSON *Image in Sand* i. 2 Brown-faced, milky-toothed Arabs were there.

milky (mi·lki), *sb. slang.* Also **milkie.** [f. MILK *sb.* + -Y[6], -IE.] A familiar or nursery name for: (*a*) a milkman or milk-boy; (*b*) milk.

1886 H. BAUMANN *Londinismen* 108/1 *Milky*, milkman. **1922** JOYCE *Ulysses* 363 Go home to nicey bread and milky and say night prayers with the kiddies. **1923** 'R. CROMPTON' *William Again* xii. 203 ''Ello, kids!' said the milk-boy... ''Ello, Milky!' **1946** P. H. SIMPSON *If you'd care to Know* 134 Box in which the 'milkie' places the bottled milk. **1966** 'L. LANE' *ABZ of Scouse* 68 She keeps ther milkie on ther doorstep fer ars (hours). **1975** *Evening News* 21 Apr. 4/4 He appeared his normal easy-going self and all he said to me was, 'Hullo milkie.'

milky way. Add: **2. a.** (Later examples.) Also *attrib.*

1851 H. MELVILLE *Moby Dick* I. xl. 291 Leaving a milky-way wake of creamy foam. **1916** D. H. LAWRENCE *Amores* 93 Pleiads of people are Deployed around me, and I see The street's long outstretched Milky Way.

mill, *sb.*[1] Add: **1. b.** *to go* (*pass*, etc.) *through the mill* (further examples).

1818 SCOTT *Heart Midl.* in *Tales my Landlord* 2nd Ser. III. iv. 96 Frank here won't hear of our putting her through the mill. **1837** *Knickerbocker* IX. 356, I had been 'through the mill' of a pre-concerted, artificial revival. **1840** R. H. DANA *Two Yrs. before Mast* 50 I've been through the mill. **1868** H. WOODRUFF *Trotting Horse* vi. 76 It was thought that they would be ruined for service if they were 'put through the mill'. **1887** *Contemp. Rev.* Jan. 10 Certain persons who have gone through the mill of what is known as our 'higher education'. **1903** G. GISSING *Private Papers H. Ryecroft* 138 His hardships were never excessive; they did not affect his health or touch his spirits; probably he is in every way a better man for having..'gone through the mill'. **1904** J. C. LINCOLN *Cap'n Eri* ii. 29 Jerry's the only one of us three that's been through the mill. **1940** H. READ *Annals of Innocence* II. i. 75 A boy who is destined to be a teacher, a doctor, a technician or a scientist, must go through the mill and acquire the necessary qualifications. **1959** I. & P. OPIE *Lore & Lang. Schoolch.* x. 200 *Running the Gauntlet.* Although well known by this name, the ordeal is also termed ..'Through the Mill'. **1965** *Listener* 1 July 21/1, I am a collector—and one who has gone through the mill. I started..in the basement with bus tickets.

4. (Further example.)

1919 *Brit. Manufacturer* Nov. 26/2 In the linen industry a 'mill' means the works where flax is spun into yarns, while a 'factory' means the place of the further evolution of the yarns being woven into cloth.

6. (Earlier example.)

1835 DICKENS *Sk. Boz* (1836) 1st Ser. I. 334 The mill's a d—d sight better than the Sessions.

b. *transf.* A prison or guard-house.

1851 H. MAYHEW *London Labour* I. 352/2 A few weeks after I was grabbed for this, and got a month at the mill... When I came out of prison, I went to Epsom races. **1853** WHYTE-MELVILLE *Digby Grand* I. ix. 229 The latter worthy..gave a policeman such a licking the other night, that he was within an ace of getting 'a month at the mill'. **1889** H. H. MCCONNELL *Five Years a Cavalryman* 194 Very few, indeed, are they who during their term of service can say: 'They never had *me* in the mill.' **1916** E. C. GARRETT *Army Ballads & Other Verses* 21 And they put me in 'the mill'. **1928** L. H. NASON *Sergeant Eadie* 78 Why, put 'em in the mill! **1951** J. JONES *From Here to Eternity* IV. xlii. 636 'You were here when one of the old ones was in the mill, weren't you, Jack?' 'Two,' Malloy said. 'Both of them during my first stretch.' **1960** WENTWORTH & FLEXNER *Dict. Amer. Slang* 339/1 *Mill*.., a prison; a guardhouse.

6*. *U.S. slang.* A typewriter.

1913 *Writer's Bulletin* Oct. 103/2 After I got a good idea I would hustle to my 'mill' and pound out some copy. **1922** N. A. CRAWFORD *Weavers With Words* 22 And sometimes..I'll start to say, 'Jim, got a good cigarette?' and turn toward his battered old 'mill'. **1932** C. D. MACDOUGALL *College Course in Reporting* 498 *Mill*, typewriter. **1948** MENCKEN *Amer. Lang.* Suppl. II. 717 Writers' cramp was cured..on the advent of the *mill*, *i.e.*, the typewriter.

7. (Earlier example.)

1819 T. MOORE *Tom Crib's Memorial to Congress* 36 We who're of the fancy-lay, As dead hands at a mill as they.

b. *U.S.* A circling movement of cattle. (Cf. MILL *v.*[1] 12.)

1897 E. HOUGH *Story of Cowboy* 146 By shouts and blows he did all he could to break the 'mill' and get the cattle headed properly. **1903** A. ADAMS *Log of Cowboy* iv. 27 We soon had a mill going which kept them [*sc.* cattle] busy and rested our horses. **1942** E. E. DALE *Cow Country* 55 Those behind them would follow and a 'mill' would be established in which the animals would swim around and around in a circle until they drowned unless it were quickly broken up and the leaders again headed for the opposite shore.

8*. *slang.* The engine of an aircraft or a 'hot rod' racing car.

1918 *Atlantic* Sept. 414 Motor is 'moulin'—to start it, one 'turns the mill'. **1923** G. H. MCKNIGHT *Eng. Words & their Backgrounds* 56 *Tail* and *joystick* and *mill* (French *moulin*) were names for different parts of the airship. **1937** E. C. PARSONS *Great Adventure* vi. 60 To nurse one of the grunting old mills up to that height,..and keep it running for an hour, was in itself quite a stunt. **1948** MENCKEN *Amer. Lang.* Suppl. II. 724 There are others [*sc.* new terms] that remain the private property of the men working in automobile plants and of those who sell or repair cars. A few specimens; Bald-head. A worn tire... Mill. An engine [etc.]. **1954** R. F. & B. W. YATES *Sport & Racing Cars* ii. 24 The additional motor 'moxie' provided by a reground camshaft is truly amazing, and all of this without running too much risk of a cranky 'mill' at low idling speeds. **1954** *Amer. Speech* XXIX. 100 *Mill*,..any engine. *Ibid.* 97 *Full mill*,..an engine with all necessary speed racing accessories. **1975** B. GARFIELD *Hopscotch* xv. 152 This was an old car but it must have had a souped-up mill.

9. *mill-bag, -brook, -lot, -room* (later example), *-yard*; *mill-cut, -like* adjs.

1832 J. P. KENNEDY *Swallow Barn* I. xv. 155 With the large canvass mill-bags spread out for saddles. **1851** R. GLISAN *Jrnl. Army Life* (1874) vi. 58, I..endeavored to throw [it] in a mill-bag style over my shoulder. **1636** *Official Rec. Springfield, Mass.* (1898–9) I. 159 The lotts.. are ordered to lye adjoining to Mill Brooke. **1864** T. L. NICHOLS *40 Yrs. Amer. Life* I. ii. 20 Grist-mills which ground our corn, and saw-mills which supplied our timber, were upon a mill brook. **1866** G. M. HOPKINS *Poems* (1967) 171 The streams are full And millbrook-slips with pretty pace Gallop along the meadow grass. **1925** *Glasgow Herald* 2 Apr. 9 To import into this country a sufficient number of mill-cut houses to supply the shortage. **1854** *Poultry Chron.* I. 148 The 'mill-like motion of the gizard'. **1918** MRS. E. LIDDELL in J. Gott *Lett.* 66 Here is a man who, alike in the mill-like grinding of life in Leeds..and amid the urgent claims of a diocese, always found time to love and to remember. **1746** *Boston News-Let.* 16 Nov., Seven Acres..to be laid out to the Right of the 30 Acre Mill-Lot, granted to Thomas Richardson. **1833** B. SILLIMAN *Man. Sugar Cane* 45 The length of the mill-room A is 64 feet. **1824** *New Hampsh. Hist. Soc. Coll.* I. 246 A saw mill torn down and twelve thousand of boards in the mill-yard carried away. **1936** *Discovery* Sept. 288/1, I saw the mill cat crossing the wall from our garden to the mill yard. **1955** J. R. R. TOLKIEN *Return of Ring* 296 The low wall of the mill-yard.

10. mill band, an endless belt for the wheels of mill machinery; **mill finish,** of paper, not subjected to any extra processing after being made; **mill log** *U.S.*, a log cut at a saw mill; **mill-power,** water-power for driving a mill; also, a unit for measuring this (see quots. 1903, 1911); **mill privilege, right** *U.S.*, the privilege or right of using water for driving a mill; **mill-run,** (*d*) applied to timber sawn to the usual specifications; (see also quot. 1957); also *transf.*, of average or mediocre quality, 'run-of-the-mill'; **mill-saw** (earlier example); **mill-scale,** read: *Metallurgy*, a deposit of iron oxide formed on iron or steel during hot working; **mill-seat** (earlier example); **mill site** *U.S.* = *mill-seat*; **mill-stream** (later examples); **mill tail** (see quot. 1835); also *attrib.* and *fig.*; **mill town, village,** a town or village characterized by the presence of mills; **mill-work** (earlier example).

1858 P. L. SIMMONDS *Dict. Trade Products* 247/2 *Mill-band maker*, a manufacturer of bands for machine shops, and for driving wheels. **1869** *Bradshaw's Railway Manual* XXI. App. 103 Manufacturers of..Engine Hose, Fire Buckets, Mill Bands, &c. **1957** J. BRAINE *Room at Top* ix. 86 The smells of East Warley tugging at me for attention..—malt, burning millband, frying fish. **1907** CROSS & BEVAN *Text-bk. Paper-Making* (ed. 3) x. 270 In hand-made paper the 'mill-finish' is obtained by pressing the sheets of paper one against another. **1952** E. J. LABARRE *Dict. Paper* (ed. 2) 163/2 Mill finish is synonymous with machine finish, and merely indicates that the paper has received its finish on the paper-machine. **1795** T. B. HAZARD *Diary* (1930) 171/2, I helpt brother Robert flote mill logs to mill. **1849** D. NASON *Jrnl.* 99, I asked the guide if there were any mill-logs among it. **1833** *Chambers's Edinb. Jrnl.* II. 167/3 This stream..at some after time may be turned to account as a mill power. **1903** *Trans. Amer. Soc. Mech. Engin.* XXIV. 983 Wherever water-power is sold it is customary to use the turbines as meters. .. From tables and curves made up from..tests of.. turbines,..data are obtained from which to compute the actual discharge. This is referred to a given head and thence reduced to mill-powers, the values of which vary with the locality. **1911** *Encycl. Brit.* XIV. 92/2 A mill-power is defined as 38 cub. ft. of water per sec. during 16 hours per day on a fall of 20 ft. This gives about 60 h.p. effective. **1734** *New Hampsh. Probate Rec.* (1914) II. 508, I also give unto my son..the one half of my mill Priviledge on the southerly side of ye River at Lole-End. **1892** *Rep. Vermont Board Agric.* XII. 134 Many mill privileges with excellent water power are afforded. **1794** *Mass. Hist. Soc. Coll.* III. 147 The principal object of the original settlers being lumber, more attention was paid to mill-rights than to the soil. **1847** W. I. PAULDING *Antipathies* III. iii, in J. K. & W. I. Paulding *Amer. Comedies* 262 There's a man at Jack O'Lantern's that owns land and mill-rights **1881** *Chicago Times* 1 June, The supply of choice mill-run lumber was generally quite limited. **1928** FOY & HARLOW *Clowning through Life* 299 He thought himself far too good for the ordinary mill run of melodramas which prevailed at that house. **1957** *N.Z. Timber Jrnl.* Nov. 59/2 *Mill run.* Usually implies all saleable output of timber from a saw-mill. **1856** 'MARK TWAIN' *Let.* 14 Nov. in *Adventures T. J. Snodgrass* (1928) 25 Everybody was a bobbin up and down like a mill saw. **1880** *Encycl. Brit.* XIII. 357/1 During rolling this film [of oxide] becomes somewhat thick and peels off, forming 'mill-scale'. **1902** BREARLEY & IBBOTSON *Analysis of Steel-Works Materials* VII. 229 (*heading*) Mill scale. **1940** SIMONS & GREGORY *Steel Manuf.* xvi. 109 The charge consists of steel scrap and grey phosphoric ..to which are added millscale..or iron ore, and lime or limestone. **1968** T. H. ROGERS *Marine Corrosion* vi. 78 Steel with mill-scale either in or on the surface, when exposed to sea water, will pit very severely wherever any couple between scale and steel occurs. **1770** G. WASHINGTON *Diaries* (1925) I. 365 Mr. Ballendine and myself leveled Doeg Run in order to fix on a Mill Seat. **1831** J. M. PECK *Guide for Emigrants* 196 There are but few good mill sites in the State. **1896** C. H. SHINN *Story of the Mine* 81 Water claims and mill sites were taken up almost as soon as work had fairly begun on the Comstock. **1956** H. EVANS *Mountain Dog* 107 Earth and stones had been bulldozed to the water's edge to form a millsite. **1815** D. DRAKE *Nat. View Cincinnati* i. 58 In summer and autumn, it [*sc.* Licking River] is a moderate mill-stream. **1840** *Knickerbocker* XVI. 22 A wooden bridge which crossed a mill-stream. **1939** JOYCE *Finnegans Wake* 175 But the Mountstill frowns on the Millstream while their Madsons leap his Bier. **1975** J. B. HARLEY *O.S. Maps* iii. 44 In Ordnance Survey usage the term 'mill race' is given to the water leading to a mill, and 'mill stream' to the water leaving it. **1835** J. ABBOTT *Expos. Princ. Hydraulic Engine* 126 *Mill tail*, the water which has passed through the wheel race; or is below the mill. **1922** BLUNDEN *Shepherd* 64 No water ever ran so blithe As that same mill-tail stream, I'd say. **1925** —— *Eng. Poems* 24 Master-fish by bridges In freshened milltails leaping. **1951** E. PAUL *Springtime in Paris* ix. 171 If the priests were right, and Busse, as a card-carrying Communist, was doomed to the foulest mill tails of hell, [etc.]. **1847** D. P. THOMPSON *Locke Amsden* x. 199 [The paper] came into town all damp from the press of Mill-Town Emporium. **1902** S. E. WHITE *Blazed Trail* xxi. 155 He arrived out of breath in a typical little mill town. **1925** T. DREISER *Amer. Trag.* (1926) I. II. vi. 195 He decided to remain—later sitting down to dinner with a small group of milltown store and factory employees. **1944** *Reader's Digest* Dec. 16/1, I used to live, years ago, in a mill town in the Deep South. **1863** A. D. WHITNEY *Faith Gartney's Girlhood* xxiii. 218 It needs just such a man [as minister] among mill-villages like these, he says. **1770** G. WASHINGTON *Diaries* (1925) I. 381 Ball and his People went about 12 oclock to Framing the Mill Work.

mill, *sb.*[5] Add: **a.** (Earlier and later examples.)

1786 in *Amer. Museum* (1789) II. 182 Mills, the lowest money of account, of which one thousand shall be equal to the federal dollar, or money unit. **1974** *News & Reporter* (Chester, S. Carolina) 24 Apr. 1-A/8 Board Chairman J. F. (Buddy) Martin told him that in 1970 the county had raised from one mill to two the amount of money that was available for the hospital from the county.

Also *attrib.*, as *mill tax.*

1848 *Indiana Hist. Soc. Publ.* (1895–1903) III. 514 The former will pay on a mill tax $200. **1853** in *Trans. Mich. Agric. Soc.* (1856) VII. 293 A mill tax is annually levied to purchase books for these libraries. **1903** *Scribner's Monthly* Oct. 486 They support the Universities by a direct mill tax levied upon the assessed valuation of the State.

mill (mil), *sb.*[6] Also **mil.** Colloq. abbrev. of MILLIMETRE (esp. in *Photogr.*, designating a size of film).

1960 E. MORGAN *You're a Long Time Dead* 386 Sandy, I'll be getting pictures of you in that outfit, don't worry, as good as anyone can take—*Best Man*, What, on 35 mill? **1971** *Guardian* 25 Oct. 8/3 'Is it videotape or 35 mill?' she asks. **1974** S. GULLIVER *Vulcan Bulletins* 26 'What do you want?' 'Eighty-one mil. mortar bombs.'

mill., mill (mil), *sb.*[7] Also **mil.** Colloq. abbrev. of MILLION.

1955 R. J. SCHWARTZ *Compl. Dict. Abbrev.* 112/3 *Mill.*, million. **1975** *New Yorker* 20 Jan. 29/1 Thanks a mil for your letter. **1975** D. LOWDEN *Bellman & True* xi. 55 'How much will he take then?' 'Anything we can give him, up to two mill.'

mill, *v.*[1] Add: **2. e.** *Soap manufacture.*

1902 *Chambers's Jrnl.* Apr. 204/1 When quite hard, this fine soap is milled, or cut into very small shreds, after which it is pressed in moulds into fancy shapes. **1967** *Everyman's Encycl.* XI. 271/2 The chips are transferred to a mixer where the dyes and perfumes are added, which are then milled to make the soap plastic and homogeneous.

f. *slang.* To send to the treadmill; to send to prison (cf. MILL *sb.*[1] 6 in Dict. and Suppl.).

1838 DICKENS *O. Twist* II. xxv. 83, I shouldn't have been milled if it hadn't been for her advice. But..what's six weeks of it?

12. Also *transf.* (of persons, vehicles, etc.) and *fig.*

1910 W. M. RAINE *Bucky O'Connor* 227, I expect you were able to make out, even if I did get the letters to milling around wrong. **1911** H. QUICK *Yellowstone Nights* v. 127 The main thing the matter was that failure o' his a-millin' through his mental facilities. **1919** L. F. CODY *Memories Buffalo Bill* 302 Indians and soldiers milled, the Indians fighting with their knives, the soldiers with their guns. **1927** H. E. FOSDICK *Pilgrimage to Palestine* 262 We look down upon the throng milling around the Chapel of the Sepulcher. **1935** *Punch* 29 May 648/2 The sergeants are milling round like madmen with last-minute instructions. **1957** J. KEROUAC *On Road* (1958) v. 33 First we milled with all the cowboy-dudded tourists..at bars. **1968** R. M. PATTERSON *Finlay's River* III. 164 To follow their wanderings in detail would be pointless. They milled around like that for the next two days, obsessed with this ridge-climbing idea.

c. *fig.* To turn *over* in one's mind.

1905 *Smart Set* Oct. 17/1 No,..I ain't buyin' no dishes. I was just kind o' millin' things over to myself. **1923** R. D. PAINE *Comrades of Rolling Ocean* xvii. 298 Judson, on guard in the cabin, was milling this problem over. **1958** 'A. GILBERT' *Death against Clock* viii. 111 Barney's milled it over and over..and we can't think of any reason. **1964** M. GOWING *Britain & Atomic Energy* ix. 250 Nor did they [*sc.* the American engineers] want to spend much time in milling over alternative approaches to problems for which they had chosen..their own solution.

millable (mi·lăb'l), *a.* [f. MILL *v.*[1] or *sb.*- + -ABLE.] **a.** Suitable for milling (sense 1).

1905 A. COCKAYNE in *13th Rep. Dept. Agric.* (N.Z.) Appendix x. 402 Most of the Phormium [in flax swamps] was short, but there were considerable quantities of good millable fibre... All the millable flax is being rapidly cut down, and no doubt these swamps will in time become good agricultural land. **1955** *Times* 2 May 23/5 As far as millable grades were concerned, the wheat market showed little feature during the past week. **1959** *Cape Argus* 15 Aug. 1/5 The fire..destroyed about 2,000 tons of millable cane before it could be brought under control. **1963** *Times* 23 Feb. 5/1 The millable oat championship went to a Blenda sample from Mr. W. Sharp, of Banff, Aberdeen.

b. Suitable for cutting in a sawmill.

1924 *Times Trade & Engin. Suppl.* 29 Nov. 250/2 It is proposed..to open up 500,000 acres of farming land and 1,000,000,000 feet of millable bush. **1969** *Northern Territory News* (Darwin) *Focus '69* 83/1 There is a considerable stand of timber again reaching 'millable' stage. **1972** *Country Life* 16 Mar. 653/3 Some sort of grading [of timber]..based on veneer, planking and millable butts in the case of hardwoods.

‖ **mille-feuille** (milfœy). Pl. **mille-feuilles.** [Fr., lit. 'a thousand leaves'.] A rich pastry consisting of thin layers of puff pastry filled with jam, cream, etc. Also *attrib.* and *fig.*

1895 G. A. SALA *Thorough Good Cook* xiv. 426 Mille Feuilles (Italian Pyramid). **1902** S. BEATY-POWNALL *'Queen' Cookery Bks.* XI. vi. 169 *Millefeuilles, Gâteau* (also known as *Gâteau Milfras*). **1918** A. BENNETT *Roll-Call* I. viii. 169, I should like a strawberry ice, and a lemon-squash, and a millefeuille cake. **1945** N. MITFORD *Pursuit of Love* xviii. 146 Soon Davy was falling upon *éclairs* and *mille feuilles* with all the abandon of a schoolboy. **1967** *Listener* 21 Dec. 802/2 When strangers meet, and nature calls, our society splits into a *mille-feuille* of social strata: each one of us clinging to our own euphemism. **1971** COOMBES & WAKELIN *Good Housek. Advanced Cooking is Fun* xv. 218 *Mille-feuilles*... Spread stiffly whipped cream onto the centre of the puff paste layer.

millefleurs. Add: Also **millefleur. a.** (Earlier example.)

a **1850** J. ATKINSON in A. Davis *Package & Print* (1967) plate 8 (Advt.), Essence of Millefleur, Bouquet, Marechalle.

b. (Freq. hyphenated or written as two words.) A pattern of flowers and leaves used for tapestry, porcelain, etc. Usu. *attrib.*

1908 J. F. BLACKER *Chats on Oriental China* x. 123 The colour is not nearly as brilliant as in the '*mille fleurs*' class, though the same wavy porcelain is to be noted. **1933** *Burlington Mag.* July 35/1 The style of the tapestry..the particular type of *mille fleurs* pattern. **1933** *New Statesman* 14 Oct. 444/1 The two mille fleur boots, entirely encrusted with sharp microscopic forget-me-nots. **1938** *Burlington Mag.* June p. xxix/1 The fifteenth century French armorial tapestry with its typical *mille-fleurs* decoration of varied flowers and leafwork on the coveted red ground. **1960** H. HAYWARD *Antique Coll.* 183/2 *Mille fleurs*, decoration occurring on Chinese porcelain with panels of growing plants reserved on a flower-covered ground. **1970** *Centennial Acquisitions Mus. Fine Arts Boston* 57 (*caption*) The relatively small size of this *mille-fleurs*..tapestry suggests that it may be part of a larger hanging.

millegrain (mi·lĭgrēin). Also **milligrain.** [f. L., *mille* thousand + GRAIN *sb.*[1] 12.] (See quot. 1951.) Also *attrib.*

1948 W. A. JACKSON *Jewellery Repairing* iv. 49 If the setting be a millegraine one and the tiny beaded edge be worn down..a new edge can sometimes be brought up by the burnisher. **1951** M. FLOWER *Victorian Jewellery* 253 *Millegrain*, a kind of setting in which the metal gripping to stone is decorated with a line of tiny grains or beads. **1960** H. HAYWARD *Antique Coll.* 183 *Millegrain setting*, a means of setting a gemstone whereby the stone is held in a mount ornamented with a band of very small beads of metal. **1970** E. BRUTON *Diamonds* xx. 349 In the nineteenth century many small diamonds were set in a form known as milligrain, in which the metal used to hold the diamonds was crenelated by drawing a hardened steel ring round the metal.

milleme, var. *MILLIEME.

‖ **Mille Miglia** (mī·le mī·lya). [It., lit. 'a thousand miles'.] The name of a sports-car race run over approx. one thousand miles of roads in Italy from 1927 to 1957; also *fig.*

1933 B. LYNDON *Combat: Motor Racing Hist.* viii. 133 The entry list suggested that there would be a merciless fight between the Bugatti team and the Alfa-Romeos, which had just won the Mille Miglia. **1957** S. MOSS *In Track of Speed* ii. 27 The Italians had made a real Roman holiday of the Mille Miglia and..had come to regard themselves as almost unbeatable. **1963** P. DRACKETT *Motor Rallying* v. 75 The galloping Gaul's remarkable record included first in the Alpine and German and second in the Mille Miglia. **1966** W. COURT *Power & Glory* III. xviii. 166/1 The Mille Miglia, run in traditional form on 23 occasions between 1927 and 1957, was a.. throw back to the days that had ended with Paris–Madrid. **1969** J. LEASOR *They don't make them like that any More* i. 1 The pinheaded nutters in the Minis..roared about as though they were overtaking each other in the Mille Miglia. **1970** M. O'BRINE *Crambo* xvii. 77 He took a four-berth cabin..that..had cockroaches doing a Mille Miglia around the wash-basin.

millenarianism. (Earlier and later examples.)

1856 [see *EBIONITISM]. **1894** J. H. BLUNT *Dict. Sects* (ed. 2) 329 Millenarianism is to a certain degree not unorthodox, but..Millenarians who uphold the doctrine of a sensual or Judaic Millennium are unequivocally condemned by theologians. **1937** A. REESE *Approaching Advent Christ* 304 Millennarianism has need to pray frequently to be saved from its friends. **1957** *Oxf. Dict. Chr. Ch.* 900 In the early Church, Millenarianism was upheld principally among the Gnostics and Montanists, but also upheld by more orthodox writers... Although primitive Millenarianism lingered on..down to the end of the 4th cent., it received its death-blow from Origen. **1961** B. R. WILSON *Sects & Society* II. ix. 189 Mrs Eddy..was unrelenting in her attacks on spiritualism, mesmerism .. and millenarianism.

millenarism. Delete †*Obs.* and add later examples.

1894 F. C. CONYBEARE *Apology of Apollonius* 358/1 Millenarism of Gospel. **1957** P. WORSLEY *Trumpet shall Sound* 232 Where millenarism survives in countries with popular secular political organizations, it is generally escapist and quietist. **1962** N. COHN in S. L. Thrupp *Millennial Dreams in Action* II. 32 The oldest form of millenarism of which much is known is the messianic hope of the Jews. *Ibid.* 35 Marxists have sometimes tried to interpret the millenarism of the Spirituals..as a protest by poor peasants. **1971** *Catholic Dict. Theol.* III. 280 This decision amounted to an assessment that a mitigated Millenarism..was an error in faith, even if it could not yet be called a heresy.

millenarist. Add: (Later examples.) Also *attrib.* or as *adj.*

1894 F. C. CONYBEARE *Apology of Apollonius* 287 They began to compromise with the world..and they laid up their old millenarist system of faith and morality. **1957** P. WORSLEY *Trumpet shall Sound* vii. 137 These Papuans are typical of all millenarists. **1962** M. ELIADE in S. L. Thrupp *Millennial Dreams in Action* III. 143 The millenarist movements became savagely anti-Christian. **1967** R. A. MARKUS in *Cambr. Hist. Later Greek & Early Medieval Philos.* 408 He had..abandoned all traces of millenarist thinking. **1971** *Catholic Dict. Theol.* III. 279 Justin ..has been cited..as a Millenarist. *Ibid.*, This is frankly Millenarist. *Ibid.* 280 In the Middle Ages Joachim of Flora revived Millenarist ideas.

millenary, *a.* and *sb.* Add: **B.** *sb.* **1. b.** A thousandth anniversary or its celebration; a millennium.

1897 F. HARRISON (*title*) Millenary of King Alfred. **1955** *Times* 18 Aug. 9/6 The few simple features which moulded the town in the past govern its shape still, so that, for all that it is new, it is recognizably Kassel. (The town celebrated its millenary in 1913). **1974** E. LEMARCHAND *Buried in Past* vi. 114 The borough plans to celebrate its alleged Millenary in August. I say alleged, because its claim to have received a charter from King Edgar in the year 973 has been discredited for some time. **1975** *Church Times* 14 Feb. 18/3 St. John's, Little Missenden, celebrates its millenary.

mille·nnialism. [-ISM.] Belief in the coming or the present existence of the millennium (MILLENNIUM 2).

1906 *Pall Mall Gaz.* 4 Jan. 1 In a spirit of fatuous millennialism the constabulary force..was being reduced. **1937** A. REESE *Approaching Advent Christ* 305 The indictment..fails to take notice of the panics that Post-millennialism provoked in the Middle Ages. **1945** D. H. KROMMINGA *Millennium in Church* 6 My own view I would designate as Covenantal Millenialism. **1967** J. WILSON in B. R. Wilson *Patterns Sectarianism* x. 353 The upsurge of post-adventual millennialism induced some to make comparison of their own country with the location of the forthcoming Kingdom of God.

millennialist. Add: (Examples.) Also as *adj.*

1903 D. D. RUTLEDGE *Christ, Anti-Christ & Millennium* xi. 291 One Post-Millennialist misunderstood this view. *Ibid.* 301 The reply of the Post-Millennialists is to the effect that Revelation xx. is all symbolical and..cannot be relied on. **1932** A. STEWART *Christianity, Communism, Adventism* 5 'Jesus will soon be here' said one, an ardent pre-millennialist. *Ibid.* xi. 111 Post-millennialists..labor with abounding hope. **1967** R. ROBERTSON in B. R. Wilson *Patterns Sectarianism* ii. 72 A more strictly millennialist belief that 'the end of all things is at hand'.

miller[1]. Add: **3*.** *Austral.* = *floury miller* (*FLOURY *a.* d).

1896 *Rec. Austral. Museum* II. 107 The same kind of *Cicada* is known by different names in different localities, such as 'Miller', 'Mealy-back', etc. **1941** BAKER *Dict. Austral. Slang* 46 *Miller*, a nickname for a cicada.

Miller[2] (mi·ləɪ). *Cryst.* The name of W. H. *Miller* (1801–80), English scientist, used *attrib.* with reference to the method for specifying the positions of planes in crystals that he used in his *Treatise on Crystallography* (1839), esp. in *Miller index* (INDEX *sb.* 9 c).

1890 G. H. WILLIAMS *Elem. Crystallogr.* ii. 31 To change the signs of any Miller symbol is to change the plane to its parallel and therefore equivalent plane on the opposite side of the crystal. **1900** MOSES & PARSONS *Elem. Mineral.* I. i. 10 The Miller Indices may be obtained from Weiss's parameters by first dividing each by the common multiple of their numerators and taking the reciprocal of the result. **1940** GLASSTONE *Text-bk. Physical Chem.* v. 336 The fact is represented by a bar over the Miller index, e.g., (1$\bar{1}$1) for a face which has intercepts *a*, -*b* and *c*, on the axes *OX*, *OY* and *OZ* respectively. **1966** C. R. TOTTLE *Sci. Engin. Materials* iii. 55 When a plane is to be described, the notation used is that of Miller indices. The plane is extended to cut the *x*, *y*, and *z* axes, and the intercepts in steps of the unit cell dimensions *a*, *b*, *c* are written down. The reciprocals of these intercepts are next obtained, and fractions cleared to give the smallest integers again. *Ibid.* 59 There are six such planes in the cell, all having the last index in the Miller notation equal to 0, since they are all parallel to the hexagonal axis.

Miller effect. *Electronics.* [named after John Milton *Miller* (1882–1962), U.S. physicist.] The effect whereby capacitance (esp. inter-electrode capacitance) in the output of a valve or transistor increases its input impedance.

1931 in S. R. ROGET *Dict. Electr. Terms* (ed. 2) 209/2. **1934** J. H. REYNER *Television* x. 130 In radio practice these bypass condensers approximate to 100 or 300 μμF. and reduce the Miller effect considerably. **1967** *Electronics* 6 Mar. 130/1 In applying this principle, called the Miller effect, to a transistor, the equivalent base-to-emitter capacitance..becomes the rated value of the capacitor multiplied by the voltage gain of the transistor. **1971** J. H. SMITH *Digital Logic* iv. 74 This feedback effect of a capacitor from the output of an amplifier to its input, is known as the Miller effect.

Millerian (milīə·riăn), *a. Cryst.* [f. the surname *Miller* + -IAN.] Of or pertaining to W. H. Miller or his system of specifying the positions of crystal faces (cf. *MILLER[2]).

1896 C. J. WOODWARD *Crystallogr. for Beginners* iii. 24 In the Millerian system..fractions of the parameters are taken, and the denominators of these fractions form the indices. **1944** C. PALACHE et al. *Dana's Syst. Min.* (ed. 7) I. 25 When the lattice is rhombohedral, Millerian three-index symbols are also given. **1964** HARTSHORNE & STUART *Pract. Optical Crystallogr.* i. 9 If *hkl* are the Millerian indices, the intercepts along the axes are *a*/*h*, *b*/*k*, *c*/*l*.

Millerism. (Earlier and later examples.)

1843 *Niles' Reg.* 240/3 Millerism. Father Miller lately visited Rochester, New York, and devoted one whole week to dealing out exhortations. *a* **1852** F. M. WHITCHER *Widow Bedott Papers* (1883) xii. 44 When Millerism was makin' such a noise,..the Wiggletown folks raly thought ther was something in it. **1961** B. R. WILSON *Sects & Society* III. xii. 239 There was little organisation..of the scattered converts from Millerism and Campbellism in Britain in the 1860s. **1962** G. SHEPPERSON in S. L. Thrupp *Millennial Dreams in Action* II. 51 He emphasizes how the reaction against Millerism..'speeded the adoption of a fervent postmillennialism'.

millet[1]. **2.** (Further examples.)

1889 J. H. MAIDEN *Useful Native Plants Austral.* 97 *Panicum decompositum*,..'Australian millet', 'Umbrella grass'... One of the most valuable of the Darling Downs (Queensland) grasses. **1896** *Australasian* 14 Mar. 488/5 One of the very best of the grasses found in the hot regions of Central Australia is the Australian millet, *Panicum decompositum.*

‖ **millet**[2] (mi·let). [Turk. *millet* nation, group of co-religionists, f. Arab. *milla* religion.] A division of the subjects of the Ottoman Empire according to allegiance to a religious leader (see quot. 1902). Often used *spec.* of non-Muslims.

[**1861** G. FINLAY *Hist. Greek Revolution* I. i. 9 The Greek language was the language of the Church and the law which ruled the whole assemblage of nations called by the Othoman administration, *Roum meleti*, or Roman nation.] **1900** 'ODYSSEUS' *Turkey in Europe* vii. 296 The Turk divides the population of the Ottoman Empire into Millets, or religious communities. **1902** *Encycl. Brit.* XXX. 395/1 All Moslems..are included in the *millet*..of Islam. The Rûm, or Roman (*i.e.*, Greek) *millet* comprises all those who acknowledge the authority of the Œcumenical Patriarch,..the Bulgar *millet* comprises the Bulgarians who accept the rule of the exarchate; the other *millets* are the Katolik (Catholics), Ermeni (Gregorian Armenians), Musevi (Jews) and Prodesdan (Protestants). **1933** *Times Lit. Suppl.* 14 Dec. 890/2 The system of organization in *millets*..which existed under the Sultans goes back to the early days of Islam. **1942** *Turkey* (Geogr. Handbk. Ser. B. R. 507, Admiralty, Naval Intelligence Div.) I. viii. 290 Each *millet* had complete autonomy, under the authority and leadership of an ecclesiastical functionary. **1971** A. MANGO *Discovering Turkey* i. 69 The Ottoman Empire..was a multi-national state,..the criterion of differentiation among its subjects was religion and not nationality. *Millet*..which in modern Turkish means 'nation', was until recently used to describe a religious community. **1973** *Times Lit. Suppl.* 29 June 754/2 They could coexist in the same territory under the same loose domination, as many different *millets* did in the Ottoman Empire.

millet-seed. Add: **b.** *Petrol.* Used *attrib.* to designate (grains of) sand of almost spherical shape as a result of abrasion produced by the wind.

1935 T. EASTWOOD *Northern Eng.* (*Brit. Regional Geol.*) 7 Thick red sandstones (Penrith Sandstone)..betray their origin as desert sand by the presence of 'millet-seed' grains. **1971** I. G. GASS et al. *Understanding Earth* xiii. 169/2 In desert sands..the quartz grains are described as 'millet-seed' sands, a reference to the almost perfect, spherical shape.

mill-horse. c. (Earlier example.)

1859 MILL *On Liberty* v. 203 The official body are under the constant temptation of sinking into indolent routine, or, if they now and then desert that mill-horse round, of rushing into some half-examined crudity which has struck the fancy of some leading member of the corps.

milli-. Add: *millibarn* [*BARN *sb.* 1 d], *-calorie* (or *-calory*), *-curie, -gal* [*GAL[2]], *-henry* [*HENRY[3]], *-joule, -kayser, -kelvin, -lambert, -micron, -mol(e* [*MOLE *sb.*[7]] (hence *-molar* adj.), *-poise* [*POISE *sb.*[2]], *-rad, -radian, -rem, -roentgen, -watt*; **mi·llidegree,** a thousandth of a degree centigrade (kelvin); **mi:lliequi·valent,** one thousandth of a gramme-equivalent; **milliosmol, -mole** (mili,ǫ·zmǫl, -mōul) [f. blend of OSMOTIC *a.* and *MOLE *sb.*[7]], an amount of any osmotically effective ion in solution equal to a milligrame divided by the atomic weight of the ion; hence milliosmo·lar *a.*

Also combined with *micro-* to denote division by a thousand million (corresponding to the single prefix *NANO-), as in *millimicroampere, -mole, -second.*

1955 *Physical Rev.* XCVII. 88 The cross section..has been calculated to be approximately 35 millibarns. **1958** O. R. FRISCH *Nucl. Handbk.* I. 12 Nuclear cross sections are usually expressed in barns or millibarns. **1909** *Cent. Dict.* Suppl., Millicalory. **1937** *Geogr. Jrnl.* LXXXIX. 543 The cooling power of the atmosphere at 2.0 p.m...averages 6·7 millicalories per square centimetre per second. **1953** *Brit. Jrnl. Psychol.* Nov. 280 The heat intensity required to produce pain was measured with a radiometer.. the threshold being expressed in millicalories per second per square centimetre. **1910** E. RUTHERFORD in *Nature* 6 Oct. 430/2 This matter was left for the consideration of the standards committee; the latter suggested that the name Curie should be used as a new unit to express the quantity or mass of radium emanation in equilibrium with one gram of radium (element). For example, the amount of emanation in equilibrium with one milligram of radium would be called 1/1000 Curie or one millicurie. **1947** *Sci. News* IV. 126 One millicurie of radium expels thirty million alpha particles per second. **1963** Millicurie [see *CURIE 1]. **1951** *Jrnl. Chem. Physics* XIX. 1161/1 The Curie temperatures are of the order of tens of millidegrees. **1974** *Physics Bull.* Mar. 93/2 The lighter isotope, [3]He, had until recently shown no anomalous behaviour even when cooled to temperatures of a few millidegrees. **1929** C. J. ENGELDER *Textbk. Elem. Quantitative Analysis* vii. 139

MILLIAMMETER

There is a gram-milliequivalent weight in 1 cc. of a normal solution. **1946** *Nature* 19 Oct. 556/2 The mean value for agouti mice is 121 milliequivalents per litre, and for black 124·5. **1965** *Math. in Biol. & Med.* (Med. Res. Council) i. 37 The computer prints out in milli-equivalents per litre all the major constituents of the plasma, cells and alveolar gases. **1914** Milligal [see *GAL²]. **1934** *Geogr. Jrnl.* LXXXIII. 446 The Bouguer anomalies are small on the coast and decrease steadily westward to about −150 milligals in Shansi. **1969** *New Scientist* 10 July 88/2 The gal is..a unit of one cm per second per second and a milligal thus approximately one millionth of the normal gravity of the Earth. **1909** *Cent. Dict.* Suppl., Millihenry. **1922** Glazebrook *Dict. Appl. Physics* II. 421/1 All the coils are used in series for the higher range (9 to 105 millihenries), but only portions of each in series for the lower range (0·7 to 12 millihenries). **1950** *Engineering* 7 Apr. 398/3 In the rectifier positive lead, a 10-millihenry air-core reactor is connected. **1972** *Physics Bull.* Apr. 205/1 Methane–air can be reliably ignited by a spark discharge of only about one millijoule. **1968** *Chem. Abstr.* LXVIII. 4308/2 The Lamb shifts of the 4*S* and 3*S* levels of the II.. are, resp., 60 ± 4 millikaysers and 138·5 ± 1·6 millikaysers. **1970** G. K. Woodgate *Elem. Atomic Struct.* i. 2 The term values T_i are written as positive numbers in units of cm^{-1}, recently re-named the Kayser (K). The new name is more commonly found when the sub-unit milli-kayser is used; 1 mK = 10^{-3} cm^{-1}. **1972** *Physics Bull.* Feb. 85/3 Philips shows very thin metal foils between 2 and 10 μm thick being used to solve heat exchange problems at very low temperatures, a few millikelvin. **1918** Webster Add., Milli-lambert. **1920** E. N. Harvey *Nature Animal Light* iii. 64 The brightness of a surface is measured in lamberts or millilamberts... A millilambert is 1/1000 lambert. **1970** *Nature* 24 Jan. 347/2 Each pattern subtended 7° at the eye, and was projected..at an average screen illuminance of 3 millilamberts. **1956** Millimicro-ampere [see *COSMOTRON]. **1960** *Cambr. Rev.* 8 Oct. 21/2 The techniques developed for handling and measuring the amounts of steroids of only a few millimicromoles likely to be present in experimental samples have made steroid chromatography a leading branch of microanalysis. **1904** C. Hering *Ready Reference Tables* I. 31, 1 milli-micron (spectroscopy) or micro-millimeter (microscopy) = 10 Angstroem units. **1966** C. R. & T. S. Leeson *Histol.* ii. 22/2 One millimicron is one thousandth of a micron, i.e., 10 Å. **1956** *Proc. CERN Symposium* II. 69 (*heading*) Recent advances in millimicrosecond counting techniques. **1964** F. L. Westwater *Electronic Computers* viii. 127 Switching speeds of a few millimicroseconds appear possible. **1934** Webster, Millimolar. **1941** *Jrnl. Physiol.* C. 61 The value of η represents the total indiffusible substance (as millimols), and of ε the difference between the total negative and positive charges on the indiffusible molecules (expressed as milliequivalents or valencies multiplied by the millimolar concentration). **1972** *Nature* 10 Mar. 57/3 Using lower than customary magnesium concentrations (in the region of millimolar), Allet finds that lithium and caesium chloride produce similar results. **1904** C. Hering *Ready Reference Tables* I. 60, 1 millimol = 0·001 mol or gram molecule. **1934** *Biochem. Jrnl.* XXVIII. 285 If we consider the titration of 1 ml. of *N*/10 acid delivered for example from a 1 ml. simple pyrex pipette..the coefficient of variation in millimols of KOH is 25 × 10^{-6}. **1937** Pierce & Haenisch *Quantitative Analysis* vi. 83 The choice of unit, millimole or mole, milliequivalent or equivalent, depends upon the unit of volume chosen. **1954** [see *ISOHYDRIC *a.* b]. **1970** R. W. McGilvery *Biochem.* xxv. 614 It requires 2·3 millimoles of H^+ to react with the hemoglobin in a liter of blood before the pH can change by 0·1 unit. **1939** J. L. Gamble *Chem. Anat., Physiol. & Path. Extracellular Fluid* Notes to chart 2 The term milliosmol is used to distinguish ionic from molecular concentration. The osmolal value of a solution of Na+Cl, for instance, is double the molal value. **1942** *Ibid.* (4th printing), The milliosmolar and milliequivalence values for the univalent ions are obviously identical. The chemical equivalence of the divalent ions is twice their milliosmolar value. The term milliosmolar is used instead of millimolar to make clear the additive osmotic effect of individual ions. **1959** *Pediatric Clinics N. Amer.* VI. 272 The nephritic patient who is unable to concentrate urine ..will need a water allowance..of approximately 3 ml. per milliosmol of total urinary solute. **1963** *Lancet* 12 Jan. 77/1 The last twelve patients admitted to this hospital with diabetic acidosis have had a calculated serum osmolarity greater than 300 milliosmoles per litre. **1972** *Science* 19 May 815/3 Severely disabled chicks..had plasma osmolality values (in milliosmoles per kilogram) of 335 ± 10 as compared with 309 ± 7 for controls. **1934** *Jrnl. Amer. Chem. Soc.* LVI. 998/2 We find the freezing point of our maximum density water to be 3·82°, and the viscosity at 20°, 12·6 millipoises. **1954** *Brit. Jrnl. Radiol.* XXVII. 247/1 The value is 30 millirad per week. **1973** *Times* 31 July 5/4 He said a total dose of 0·1 of a millirad —one-third of the normal background radioactive level— had been recorded during the 21 hours following the blast. **1956** *Spaceflight* I. 28/1 Despite errors caused by ionospheric refraction of the signal, the use of data from a number of stations will, it is claimed, enable the satellite's position to be established to within a fraction of a milliradian. **1971** *Sci. Amer.* June 65/1 A similar analysis of the collision kinematics indicates that the resolution in angle should be a fraction of a milliradian, which is about three minutes of arc. **1954** *Brit. Jrnl. Radiol.* XXVII. 246/1 Basic permissible weekly doses for the critical organs. Whole body exposure...0·3 millirems per week in blood-forming organs, the gonads and the eyes. **1971** *Sci. Amer.* Aug. 115/2 The U.S. Atomic Energy Commission requires that personnel working with radiation materials receive not more than 100 millirems of radiation per 40-hour week. **1955** *Sci. News Let.* 20 Aug. 116/1 A milliroentgen is one-thousandth the unit of quantity of X-rays. **1963** B. Fozard *Instrumentation Nucl. Reactors* i. 3 It is..often necessary to measure dose rate; a convenient practical unit for this quantity is the milliroentgen per hour (mr/h). **1929** *Papers Inst. Post Office Electr. Engin.* CXXIX. 22 It is..sent to line at the power of 1 milli-watt. **1956** *Nature* 25 Feb. 392/1 This feature of existing methods [for the measurement of power] is particularly evident at low power-levels, of the order of 1 milliwatt, at wave-lengths of 3 cm. and less.

milliammeter (miliæ·mītəɹ). [f. Milli- + Ammeter or Milliam(père + -meter.] An instrument for measuring currents of the order of milliamperes; = *milliampère meter*.

1902 G. B. Massey in H. R. Bigelow *Internat. Syst. Electro-Therapeutics* (ed. 2) B. 150 (*caption*) Weston milliammeter, arranged specially for medical work. **1946** *Nature* 28 Dec. 943/1 The received signal was recorded continuously on a recording milliammeter. **1966** *McGraw-Hill Encycl. Sci. & Technol.* I. 323/2 A rectifier-type ac milliammeter consisting of a small copper oxide or germanium bridge rectifier feeding a conventional dc milliammeter has low losses, good overload capacity, and adequate accuracy.

milliamp (mi·li,æmp). Colloq. abbrev. of Milliampère.

1923 *Radio Times* 28 Sept. 19/2 Another [voice], 'Five milli amps. No! Sorry! I thought you were Cardiff.' **1927** *Sunday Express* 17 Apr. 8/3 (Advt.), Every milliamp of electricity is made by the dynamo. **1957** P. Lafitte *Person in Psychol.* ii. 12 In the sciences, the fact is strictly the instrument reading..and the report of twenty milliamps..is in a sense an interpretation. **1970** D. F. Shaw *Introd. Electronics* (ed. 2) xiii. 323 Equation 13.13 (*b*) becomes $V_a = 400 - 40I_a$ where V_a is given in volts and I_a in milliamps.

milliampère. Add: Now usually written without an accent and stressed on the penultimate syllable. (Earlier and later examples.)

1885 *Jrnl. Soc. Telegr. Engin.* XIV. 465 It [*sc.* the current] was 2·632 milliampères. **1922** F. W. Aston *Isotopes* 48 The bulb is arranged to take from 0·5 to 1 milliampere at potentials ranging from 20,000 to 50,000 volts. **1956** [see *COSMOTRON].

Hence **millia·mperage**, current expressed in, or of the order of, milliamperes.

1909 *Cent. Dict.* Suppl., Milliamperage. **1937** *Discovery* Feb. 54/1 The meter then reads to the corrected milliamperage of the total body surface-area. **1961** *Med. X-Ray Protection up to 3 Million Volts* (U.S. Nat. Bureau Standards Handbk. 76) 18 A beam monitoring device fixed in the useful beam is recommended to indicate any error due to incorrect filter, milliamperage, or kilovoltage.

Millian (mi·liăn), *a.* and *sb.* [f. the name of John Stuart *Mill* (1806–73), English philosopher + -ian.] **A.** *adj.* Of or pertaining to Mill or his philosophical or political theories. **B.** *sb.* A follower of Mill or his theories. (See also *Millite *a.* and *sb.*)

1859 J. A. Symonds *Let.* Apr. (1967) I. 183 He has a truly *Millian* contempt for public opinion. **1950** E. Nagel in Mill *Philos. Scientific Method* p. xlvii, There are doubtless no perfect Millians alive today. **1958** *Victorian Studies* I. 253 Later Millians,..such as Bain, Croom Robertson and Sully, were eclipsed by the bold but philosophically shallow eloquence of men like Huxley and Tyndall. **1972** *Times Lit. Suppl.* 14 Apr. 420/5 Practical criticism operated on the John Stuart Millian assumption that out of this bedlam the truth would ultimately emerge.

milliardaire (mi:liăɹdēə·ɹ). Now *rare*. [f. Milliard after Millionaire.] A person possessing a 'milliard of money'.

1924 *New Internat. Year Bk. for Year 1923* 270/2 A South-American milliardaire who can have all he wants. **1926** C. H. Herford *Mind of Post-War Germany* vii. 34 The hero, son of a milliardaire, is engaged in manufacturing a gas more powerful than all known fuels. **1927** *Spectator* 23 Apr. 720 Including..a respectable proportion of the Royal Families and aristocracy of most European countries, besides a sufficiency of 'milliardaires' from both Americas.

millibar (mi·libāɹ). *Meteorol.* [f. Milli- + *Bar *sb.*⁹] The usual unit of barometric pressure, equal to one thousandth of a bar (*Bar *sb.*⁹ 2), i.e. 1000 dynes per sq. cm.

1910, etc. [see *Bar *sb.*⁹ 2]. **1914** *Q. Jrnl. R. Meteorol. Soc.* XL. 187 The megadyne per square centimetre has been adopted and called the 'bar'... The bar is less convenient for printing and conversation than the millibar, its 1/1000th part, and the latter has therefore been generally adopted; it is equivalent to the pressure produced by about three-hundredths of an inch of mercury or by three-quarters of a millimetre. **1924** *Glasgow Herald* 23 Dec. 5 So far as I can judge from the synoptic chart for December 9, 7 a.m.,..the pressure at New Pitsligo at 7 a.m. seems to have been 1016 millibars, and by noon to have risen to about 1012 millibars, equivalent to 30·12 inches. **1942** V. C. Finch et al. *Elem. Meteorol.* v. 104, 1/10 inch on a barometer or barograph is equivalent to about 3 millibars. Either inches or millibars may be used in numbering the isobars on a weather map. **1963** G. M. B. Dobson *Exploring Atmosphere* i. 3 A scale at the side of the frontispiece gives the average pressure at different heights in millibars (mb), where 1000 mb is approximately the average pressure at sea level.

millieme (mī·lyēm). Also milleme, millime. [F. *millième* thousandth.] A unit of currency, equal to a thousandth of the main unit, orig. in Egypt, now in the United Arab Republic, the Sudan, and Tunisia.

1902 *Encycl. Brit.* XXVII. 700/1 The unit is the Egyptian pound, which is divided into 100 piastres, the piastre being again divided in 10 milliemes. *Ibid.*, Pieces of 20, 10, 5, 2, and 1 piastres in silver; 5, 2, and 1 milliemes in nickel. **1919** W. H. Downing *Digger Dial.* 34 *Milleme*, a small Egyptian coin. **1942** C. Barrett *On Wallaby* vi. 122 His son went round with a tambourine collecting stray milliemes. **1956** E. E. Evans-Pritchard *Nuer Relig.* iii. 88 A Nuer refused to accept a milleme coin ..because a boy of his family, when playing with a milleme, got it into his ear and died. **1972** *Whitaker's Almanack 1973* 987 Tunisian Dinar of 1,000 Millimes. *Ibid.*, Sudanese Pound of 100 Piastres or 1,000 Milliemes. **1973** *Country Life* 20 Sept. 805/3 [Tunisia] One morning I was offered a necklace for '800 millimes and packet fags'.

milligrain, var. *MILLEGRAIN.

millilitre. Add: Also (*U.S.*) -liter. (Later examples.)

1896 C. R. Honiball *Engin. Arithmetic & Mensuration* xli. 308 Millilitre = ·001 Litres. **1935** C. J. Smith *Intermediate Physics* (ed. 2) ix. 172 The litre and the millilitre (ml.) are now more frequently chosen as the unit of volume, burettes, flasks, etc. being marked in millilitres and fractions thereof. **1960** *Science* 22 July 204/2, 1 billion of these units result from treating 15 milligrams of the proteinoid with 2·5 millilitres of hot water and allowing the clear solution to cool. **1969** *Guardian* 28 Feb. 4/4 Every patient prescribed liquid medicine by his doctor will receive a plastic spoon which will hold five millilitres of medicine; and all prescriptions will be in five millilitre units. **1974** *Daily Tel.* 10 Oct. 2/4 [He] had 120 milligrammes of alcohol in 100 millilitres of blood, 40 over the legal limit.

millimetric (milime·trik), *a.* [f. Millimetr(e + -ic.] **a.** *fig.* Minute.

1909 *Milton Memorial Lect.* 194 Those millimetric distinctions by which human character declines or ascends. **1937** E. Pound *Let.* 30 Nov. (1971) 300 As I haven't yet a projector, the small but not millimetric photos would save time. **1965** *Economist* 28 Aug. 772/2 Participation in the congress and the administrative bureaucracy is to be shared with 'millimetric' equality.

b. Of the order of a millimetre in length; employing or characterized by electromagnetic waves of this wavelength.

1962 *Newnes Conc. Encycl. Electr. Engin.* 883/2 The use of waveguides of practicable size is restricted to waves in the microwave region (i.e. centimetric and millimetric wavelengths). **1969** A. L. Cullen in F. A. Benson *Millimetre & Submillimetre Waves* i. 5 For many purposes.. microwave radar was found to possess advantages over millimetric radar. **1973** *Physics Bull.* Feb. 99/2 The properties measured include..the absorption of millimetric microwaves. *Ibid.* Nov. 651/2 These direct measurements initially suggested that the millimetric background temperature was much *more* than 2·7 K.

mill-in: see *-IN *suffix*³.

milliner, *v.* [f. Milliner 2.] To make up articles of women's clothing, esp. hats. Also *fig.*

1885 G. B. Shaw *Cashel Byron's Profession* (1886) iii. 34 We will go to Paris, and be millinered there. **1895** —— *Our Theatres in Nineties* (1932) I. 205 The displays of fashionable life..are now millinered and tailored ..by the artists and tradesmen who equip the real fashionable world. **1907** in C. W. Cunnington *Eng. Women's Clothing* (1952) ii. 79 A modified *cloche* 'millinered' in a light rough-surfaced cloth.

millinering, *vbl. sb.* (Earlier example.)

1857 Geo. Eliot *Scenes Clerical Life* (1858) II. 99 It was hard for Mrs. Raynor to have to work at millinering— a woman well brought up.

millinery. Add: **3.** *millinery wire*.

1912 *Every Woman's Encycl.* VII. 4447/2 Get some dark green millinery wire covered. **1932** D. C. Minter *Mod. Needlecraft* 174/1 *Lace Trimmings*..are always mounted on thin millinery wire. **1966** *Olney Amsden & Sons Ltd. Price List* 29 Millinery wires..Satin wire..7/- dozen.

milling, *vbl. sb.* Add: **1.** (Further example.)

1877 *Amer. Miller* Mar. 39/1 The process of milling in the United States is carried on under two different systems, namely, low milling and high, or grits, milling.

3*. The action of Mill *v.*¹ 12 a.

1874 J. McCoy *Hist. Sk. Cattle Trade* 101 Drovers consider that the cattle do themselves great injury by running round in a circle, which is termed in cow-boy parlance, 'milling'. **1924** *Scribner's Mag.* Dec. 607 Jack..stood outside the door and watched the milling of the excited, hysterical women. **1943** C. H. Ward-Jackson *Piece of Cake* 41 *Milling, milling around*, flying at high speed in and out across one another's path; or flying in a defensive circle, with the nose of one aircraft a few yards from the tail of another.

4. milling machine, (*a*) *Engin.*, a machine in which a work-piece fixed to a carriage is subjected to the action of rotating cutters; (*b*) a machine for fulling cloth.

1876 [in Dict.] **1888** *Encycl. Brit.* XXIV. 661/2 The cloth to be fulled is well saturated with hot soap and water ..and rubbed between rollers in the milling-machine while so heated and soaped. **1953** L. E. Doyle *Metal Machining* xii. 266 A milling machine must hold and rotate a cutter and have means to hold a workpiece and move it uniformly in at least one direction. **1962** W. J. Onions *Wool* xi. 245 In the operation of milling..fabrics are thickened and shrunk to a desired width..The operation is usually carried out in the rotary milling machine.

milling, *ppl. a.* **3.** (Further examples.)

1919 L. F. Cody *Memories Buffalo Bill* xiii. 289 Here the buffalo thundered along in their milling herd, while Will and the assembled cowboys circled them. **1931** H. F.

Pringle *Theodore Roosevelt* I. viii. 99 The milling crowds of the cities.

million. Add: **1. a.** (*a*) (Further example.) Also in phr. *thanks a million*: see *THANK *sb.*

1934 J. B. Priestley *Eng. Journey* ix. 316 This [ship-yard]..had been a spectacular failure in which over a million of money had been lost.

b. (Further examples.) Also in phr. (chiefly *U.S.*) (*like*) *a million dollars*: excellent, splendid, magnificent (usu. prec. by *to feel* or *to look*).

1924 Wodehouse *Carry on, Jeeves!* iii. 55 It was one of those topping mornings, and I had just climbed out from under the cold shower feeling like a million dollars. **1933** E. E. Cummings *Eimi* 199 Something which might be port ..'Looks like a million dollars'—and which tastes like awfully watered vino rosso à l'américain. **1933** *Amer. Let.* 13 Nov. (1973) 176 Those [farm buildings] that have been repaired..and painted look like a million dollars compared with most of the nearby so-called improvements. **1947** *Time* 27 Mar. 4/3 You'll go home feeling like a million dollars, rested and refreshed as never before! **1956** D. Gascoyne *Night Thoughts* 35 You'd look a million dollars if you went. [illegible] **1973** *Times* 23 Apr. 5/6 A formula to double millions makes of the word look [illegible]

2. c. *a million to one*: a million chances to one; hence, an expression indicating very low probability. Freq. *attrib.*

1787 Sterne *Tr. Shandy* IV. ix. 73 Calculate it fairly.. and it will turn out a million to one, that..the forceps should have the ill luck just to fall upon..that one part. [illegible]

d. *pl.* In full, *millions fish*; = *Guppy*¹.

[illegible] **1908** *Science* 18 Dec. [illegible] 'Millions' are among the small native natural enemies of mosquitoes. **1924** *Glasgow Herald* 2 June 11/4 In Barbados..the natives are persuaded to keep in their rain barrels the minute fish known as 'millions'... Hence the absence of mosquitoes. **1966** D. W. Tucker *Ichthyol.* [illegible] Guppy, Millions Fish, Venezuela, Barbados, Trinidad, parts of northern Brazil and Guiana.

e. (*one*) *in a million*: (a person, thing, etc.) that is very rare, unusual, or valuable.

1900 Conrad *Lord Jim* viii. 99 The occasion was obscure, insignificant—what you will; a lost youngster, one in a million—but then he was one of us. **1906** E. Nesbit *Railway Children* xiii. 288 Take care of your Mother.. She's a woman in a million. **1925** Wodehouse *Carry on, Jeeves!* ii. 50 You know, Jeeves, you're by way of being rather a topper.. One in a million, by Jove! **1931** *Times Lit. Suppl.* 5 Feb. 96/3 Emily was, of course, a chimp..in a million.

f. *gone a million*: (having) completely lost, done for, in a hopeless state. *Austral.* and *N.Z. colloq.*

1913 C. J. Dennis *Battle of Wazzir* in A. H. Chisholm *Making of Sentimental Bloke* (1963) 132 Fer young Bill was gone a million, we'd never guessed the game. **1922** A. Wright *Colt from Country* 142 What hope would you have when that came out? You'd be gone a million. **1939** K. S. Prichard *Haxby's Circus* xvi. 187 If it weren't for you, I'd be gone a million on the minx. **1941** *Coast to Coast 1941* No doubt about it, Sim was right. If they drop their bundles they're gone a million. **1958** *N.Z. Listener* 19 Sept. 6/4 We scraped in in that game, about two nil.. ..they would have gone up [illegible]

3. (*b*) million-dollar *a.*, worth or costing a million dollars; also (*transf.* and *fig.*) expensive-looking; magnificent, splendid; very attractive; (*c*) *million-dollared* (also) in quot. = 'a person who has a million dollars'), *-footed*, *-pointed*, *-voiced*, adjs.; millionheiress, an heiress to a 'million of money'; millionseller, a gramophone record, book, etc., of which a million copies have been sold.

1892 Milliondollar [in Dict.]. **1921** *Discovery* Feb. 48/1 The water hyacinth..is a beautiful aquatic plant..but its spread in St. John's River and the enormous sums spent in attempting its suppression have earned it the name of the 'million-dollar weed'. **1932** *Amer. Speech* VII. 250 Big Crosby plaintively croons that he has 'Found a Million Dollar Baby in the Five and Ten Cent Store'. **1959** *Listener* 2 May 332/1 A very ugly, million-dollar kidnapping case in California. [illegible] **1967** *Listener* [illegible] I've got this million-dollar leg wound that will take me out of combat. **1972** *News & Observer* (Raleigh, N. Carolina) 30 Dec. 4/2 We don't hear much [nowadays] about..million dollar rains, and undertakers. **1857** J. G. Whittier *Barefoot Boy* in *Poetical Works* II. 230 Let the million-dollared ride! Barefoot, trudging at his side, Thou hast more than he can buy. **1865** W. Whitman *Drum-Taps* 61 When million-footed Manhattan, unpent, descends to its pavements. **1885** W. B. Yeats *Island of Statues* II. iii, in *Dublin Univ. Rev.* July 137/1 Though I be Far fleeter than the million-footed sea. **1919** W. De Morgan *Old Madhouse* 457 His mind took kindly to the interruption of this young man's nuptials with an American millionheiress of startling beauty. **1949** Berrey & Van den Bark *Amer. Thes. Slang* § 417 *Millionheiress*, a woman who is a millionaire. **1923** Million-pointed [see *million-voiced* below]. **1969** N. Cohn *AWopBopaLooBop* (1970) iii. 27 He [sc. Elvis Presley] has racked up twenty worldwide million sellers. **1971** *Sunday Times* [illegible] Frank Sinatra's recording of Love & of course the near-millionseller Story Untold. **1973** *Jazz & Blues* Sept. 11/2 Fats Domino cut 18 million-sellers in this town. **1894** 'Mark Twain' *Those Twins* 331 The two sat unconscious of the million-voiced music of the mosquitos. **1923** Belloc *Sonnets & Verse* 149 To-night in million-voicèd London I Was lonely as the million-pointed sky.

millionaire. Add: **1. c.** *Millionaires' Row*: a street containing the residences of very rich people.

1950 'J. Guthrie' *Is this what I Wanted?* iii. 56 Charles drove off past the park, the Broad Walk and Millionaire's Row. **1954** 'N. Blake' *Whisper in Gloom* iii. 43 [The Bentley]..swept through the doorway of 'Millionaires' Row'. **1964** B. Wynne *Spies Within* iv. 37 What about this millionaire's Row? **1972** *Guardian* 19 Dec. 11/1 My Gozan friend had ragged, empty pockets.. but he lived on Millionaires' Row.

2. Used *attrib.* of a town with more than a million population.

1950 C. B. Fawcett in Salamon & Kuchar *Mélanges de Géogr. offerts à V. Švambera* 52 A 'millionaire-city' is to be understood as a conurbation which contains at least one million inhabitants. **1958** D. L. Linton in *Geography* XLIII. 258 The emergence of millionaire cities and even Millionaire cities in tropical countries is now widespread and continuing. **1961** *Land Use in Urban Environment* (Dept. of Liverpool, Dept. Civic Design) 11 The great concentration of population, housing, offices and factories in the millionaire cities.

millionai·reship. [-SHIP.] The position or state of a millionaire.

1901 *Chambers's Jrnl.* Apr. 217/2 The flour industry gave the late Mr. Charles A. Pillsbury the means of millionaireship. **1930** W. O. Stapledon *Last & First Men* 78 For the individual, the goal imposed by his religious teaching was conspicuous advance in aeronautical prowess, legal sexual freedom, and millionaireship.

millionairess. Delete *jocular* and add further examples.

1887 *Atlantic Monthly* LX. 222 Stuffs which none but an empress or a millionairess would dare to look at. **1906** L. Bell *Carolina Lee* 25 You'll have to go on being a millionairess, whether you will or no. **1958** [see *REFRAGE]. **1957** *Nature* 17 Dec. 577/1 Mrs Mary Lasker, a New York Socialite's *Frankensteiner* of World (ed. 2) 98 Bernard Shaw [illegible]... **1972** *Times* 13/4 We could be millionairesses if we were concerned only with money.

Millipore (mi·lipōəɹ). Also millipore. [f. Milli- + Pore *sb.*¹] Designating membrane filters made by the Millipore Filter Corporation of Watertown, Mass., or by a foreign subsidiary of this company.

1956 *Jrnl. Bacteriol.* LXXI. 499 Direct staining procedures for microorganisms collected on the 'millipore filter'. **1964** F. I. Ordway et al. *Basic Astronautics* iv. 131 Air moves through the opening into the 2-cm diameter chamber at the rear of which is located a millipore filter in a removable housing. **1967** *Oceanogr. & Marine Biol.* V. 171 Filtration through fine Millipore filters removes more of the finely dispersed material. **1969** *Methods in Microbiol.* I. 201 The standard Millipore filters.

millisecond (mi·lisekənd). [f. Milli- + Second *sb.*²] One thousandth of a second.

1924 Glazebrook *Dict. Appl. Physics* III. 411/1 The arrangement of the coils is astatic and the time constant is a few milliseconds. **1939** *Papers Inst. Post Office Elect. Engin.* CXXIX. 11 It would be necessary to measure time intervals to an accuracy of less than about 1 millisecond. **1966** E. Delavenay *Introd. Machine Transl.* vi. 78 The lexical memory of the machine should provide random access to any word in not more than 10 milliseconds (ms). **1973** C. Egleton *Seven Days to Killing* i. 16 The car disintegrated, and a millisecond later the blast wave shattered the windscreen in the cab of the truck.

millisite (mi·lisəit). *Min.* [f. the name of F. T. *Millis* (see quot. 1930) + -ite¹.] A light grey or white hydrated basic phosphate of sodium, potassium, calcium, and aluminium, $(Na,K)CaAl_6(PO_4)_4(OH)_9.3H_2O$.

1930 Larsen & Shannon in *Amer. Mineralogist* XV. 309 The name millisite is proposed for the species after F. T. Millis who sent the original specimens to the U.S. National Museum. **1951** *Ibid.* XXVII. 294 Millisite is invariably associated with wardite as alternating layers in spherules or crusts. It is light gray to white in color, and normally is present as layers of fine fibers normal to the layering. **1965** *Enci. Sci.* Soc. *Amer.* XLI. 281 Crandallite and millisite occur in the lateritic profiles derived from apatite and barrandite and the weathering, in the presence of phosphate, of the carbonate and volcanic rocks [on Christmas Island].

Millite (mi·ləit), *a.* and *sb.* [f. the name *Mill* (see *Millian *a.* and *sb.*) + -ite² 1 a.] = *Millian *a.* and *sb.*

1917 J. Morley *Recoll.* I. iv. 52 It was, in fact, the pure milk of the Millite word. *Ibid.* vi.87 He [*sc.* Bagehot]..was not in the least an orthodox Millite. **1949** G. B. Shaw *Sixteen Self-Sketches* x. 56 Both societies were strongly Millite.

millivolt. Add: (Examples.)

In quot. 1861 *volt* is being used for what is now called an ohm.

1861 *Electrician* 9 Nov. 3/2 The application of the French system of notation in the table, is evidently faulty, the thousandth part of a volt being one millivolt (*sic*), and not one kilovolt, as stated. **1885** *Jrnl. Soc. Telegr. Engin.* XIV. 196 Differences of potential greater than such millivolts as are concerned in thermo-electricity. **1971** *Jrnl. Gen. Psychol.* LXXXIV. 11 A sensitivity of about 60 to 200 millivolts per millivolt was used for the average response.

Hence **millivo·ltmeter**, an instrument for measuring voltages of the order of millivolts.

1907 H. N. Norris *Introd. Study Electr. Engin.* xii. 346 The instrument..usually known as a milli-voltmeter. **1971** *Nature* 15 Jan. 167/2 Screen leads connected the coils to a millivoltmeter arranged to read from − 50 to + 50.

millocracy. Delete *nonce-wd.* and add further examples.

1860 G. S. Haight in Geo. Eliot *Let.* VI. 286 A small boarding school for girls..drawn largely from the millocracy, rich and profoundly ignorant. **1970** R. Blake *Conservative Party* i. 21 Why should not the landed aristocracy join hands with the socially dispossessed..against the northern 'millocracy'?

millpond. b. Add:

The use is anticipated in, and perhaps originally suggested by, the following passage:—

1813 'H. Bull-Us' *Diverting Hist. John Bull & Bro. Jonathan* (ed. 2) i. 5 He put himself in a boat, and paddled over the mill-pond to some new lands. *Ibid.* ii. 12 The tenants began to carry their grain to different parts of the great mill-pond.

Mills (milz). The name of Sir William *Mills* (1856–1932) used *attrib.*, as *Mills bomb, grenade*, etc., to designate a type of hand grenade, serrated on the outside to form shrapnel on explosion, invented by him.

1916 *War Illustr.* 30 Sept. 162/1 (*caption*) 'Lobbing' a Mills bomb over into the enemy lines. **1917** A. Ainslie *Hand Grenades* 8 Grenade hand no. 5 Mark I. Mills hand grenade. **1917** A. G. Empey *From Fire Step* xii. 74 The standard bomb used in the British Army is the 'Mills'. It is about the shape and size of a large lemon. **1923** Kipling *Irish Guards in Gt. War* I. 75 The Mills bomb..was not born till the autumn of 1915. **1935** C. Day Lewis *Time to Dance* 38 But the aeronauts, knowing iron the coinage here, had brought Mills bombs and revolvers. **1942** R. Gorman *Mod. Weapons of War* vii. 111 The segmented jacket of a Mills bomb or hand-grenade is not unlike that of a pomegranate's ridged outer skin. **1967** J. Wainwright *Pride of Pigs* 17 Any damn fool can work a Mills hand grenade..slip your finger through the split-ring, jerk out the pin and throw. *Ibid.* He smuggled a Mills bomb home in..his kitbag as a memento of El Alamein. **1974** *Daily Tel.* 6 May 1/6 A fully-primed Mills bomb which had been used as a mantelpiece ornament for 56 years was handed to police at Nottingham yesterday.

Mill's Methods (milz me·þədz). *Logic.* The Methods of Agreement, of Difference, of Joint Agreement and Difference, of Residues, and of Concomitant Variations which form the five canons of inductive inquiry proposed by J. S. Mill (1806–73) for discovering, and establishing the validity of, causal relations between phenomena. Cf. *Method *sb.* 2 c.

1896 J. Welton *Man. Logic* II. v. v. 142 (*caption*) As Mill's Methods have obtained general currency they demand some examination. **1930** W. E. Johnson *Logic* II. x. 215, I hold..that Mill's methods can and should be exhibited as strictly formal. **1954** R. Blanché *Ind. Philos.* (1944) 197/2 Mill's methods, inductive methods formulated by John Stuart Mill for the discovery of causal relations between phenomena. **1953** S. E. Toulmin *Philos. of Sci.* ii. 9 The accumulation of confirming instances, Mill's Methods and the probability-calculus: such things form the staple of most expositions. **1966** P. Caws *Philos. of Sci.* xxxiii. 251 Mill's methods..are an elegant recipe for detecting the constant conjunctions of which Hume speaks. **1973** M. C. Beardsley *Primer of Logic* xii. 423 (*heading*) Mill's methods for discovering causes: agreement and difference.

Milltown: see *Miltown.

millwrighting, *vbl. sb.* (Further example.) Also as *ppl. a.*

1821 A. Constable *Let.* 25 June in J. Constable *Correspondence* (1962) I. 66 Your millwrighting jobs coming in [so] that I cannot keep my money in my pocket. **1949** K. S. Woods *Rural Crafts Eng.* ii. iv. 76 Where local smiths have lost the art the bills are sent away to millwrighting firms.

millyum (mi·lyəm), representing a colloq. pronunciation of *million*. *rare.*

1940 E. Pound *Cantos* liv. 40 Jobs for two millyum men. *Ibid.* lvi. 58 Thus saved several millyum lives of those chinamen. **1955** —— *Section: Rock-Drill* lxxxix. 53 And in '34 presumably will be 2 millyum.

milo (məi·lo). Also millo, milo maize. [ad. Sotho *maili*.] One of a group of drought-resistant varieties of the grass *Sorghum vulgare*,

introduced from Africa to suitable regions elsewhere. Also *attrib.*

1882 *Rep. Comm. Agric. Georgia 1881–82* 23 My attention was some time since called to the claims of 'Ivory wheat' and 'Millo Maize' to a place in our long list of profitable food crops. **1887** *Florida Dispatch* 10 Jan. 34/3 The head of yellow millo maize is formed in sections of smaller heads, lying very close and compact. *Ibid.* 14 Feb. 165/1 Mr. Jones recommends the substitution of Kaffir corn, Milo Maize, etc., for a part of the corn crop. **1930** *U.S. Dept. Agric. Farmers' Bull.* No. 1147. 3 Milo has long since passed the experimental stage as a farm crop in the southwestern United States... Milo made its first appearance in this country soon after 1880. **1968** T. Capote *In Cold Blood* (1966) i. 7 One of these barns.. housed a dark, pungent hill of milo grain. **1970** M. Doggett *Sorghum* iii. 88 The milos of the U.S.A. stem from one introduction. **1974** *Islander* (Victoria, B.C.) 12 Mar. 11/1 Milomaize is a white corn, something like wheat. We ground it in an old coffee grinder for bread, and boiled it for porridge. **1973** *Houston* (Texas) *Chron.* 21 Oct. 26 Cotton, wheat, milo, corn and soybean production can be increased tremendously.

milometer (məilǫ·mītəɹ). Also mileometer. [f. Mile *sb.*¹ + -ometer.] An instrument which is fitted to a vehicle to record the distance in miles travelled by it.

1953 A. Smith *Blind White Fish in Persia* 218 Every time another 1,000 miles was registered by the milometer, those in front would cheer a little. **1963** 'B. Graeme' *Almost without Murder* xv. 170 We hit a fold road..where the milometer ticked off the miles with boring rapidity. **1968** *Punch* 31 July 172/1 One army jeep being driven aimlessly around by Georg in order that its milometer should record that monthly mileage total considered by higher authority appropriate for army vehicles. **1974** *Times* 15 Oct. 14/1 Guilty..of..offering for sale cars whose milometers had been altered to show a lower reading. **1975** G. Lyall *Judas Country* xxi. 153 From the milometer I'd guess she was only just run in.

milord. Add to def.: *spec.* an Englishman travelling in Europe in aristocratic style. (Earlier and later examples.) Hence **milo·rdi-ness**, milo·rdism.

[**1598** J. Chamberlain *Let.* 17 Sept. (1939) I. 45 Yet me thinks still I am out of my element when I am among Lords, and I am of Rabelais minde that they looke but *comme un milord d'Angleterre*.] **1758** M. W. Montagu *Let.* May (1967) III. 149 He heaps of having done his duty in waiting on the two Milords. **1822** L. Simond *Trav. Switzerland* I. 157 Accustomed to the *Milords Anglais* of former times. **1930** A. Huxley *Leda* 44 They behaved like English aristocrats in a French novel.. I tried to imitate their milordliness. **1932** R. Cobb *Reactions to French Revolution* [illegible] A tall young man, very shy and nervous, very English, but trying to hide these insular virtues behind the assumption of lofty milordism. **1945** E. Waugh *Brideshead Revisited* I. viii. 185 It's not much as though he lived like a Milord. **1954** [see *Edwardian *a.* 2]. **1961** A. Wilson *Old Men at Zoo* i. 9 The Zoo authorities had been very indulgent to a number of 'milord' whims that were perhaps more in keeping with an aesthetic undergraduate. **1969** *Listener* 6 Feb. 180/3 The great man scampered about, huffing and puffing, playing the magnificent milord.

Milori (milō·ri). Also Milory. The name of the 19th-cent. French colour-maker A. *Milori* used *attrib.* in Milori blue, a particularly pure variety of Prussian blue (Prussian *a.* 2); Milori green = *chrome green* (Chrome 3).

1885 J. G. Taylor *Field's Chromatography*, *Modernized* 194 *Milory green*, syn. Chrome Green. **1890** B. W. Warhurst *Colour Dict.* 47 *Milori*, full to dark green; also Chinese blue. **1924** F. W. Weber *Artists' Pigments* 2 Prussian Blue..was put on the market under quite an array of names...Turnbull's Blue, Paris Blue,..Milori Blue, Chinese Blue [etc.]. **1952** R. Mayer *Artist's Handbk.* 37 There is a great difference in clarity and beauty of colour as well as in permanence between the common varieties of Prussian blue and the very best, well-washed grades, such as the pure Chinese or Milori blues. **1963** *New Yorker* 8 June 20 Wool carpets only come in these colors: ..Milori Green. **1967** [see *iron-blue *a.*].

milpa (mi·lpă). [Mexican Sp.] In Central America and Mexico, a small cultivated field, usually of corn or maize; also, designating a method of cultivation practised in tropical regions (see quot. 1936).

1844 J. Gregg *Commerce Prairies* I. 150 The *labores* and *milpas* (cultivated fields) are often..without any enclosure. **1869** J. R. Browne *Adventures Apache Country* 164 Our horses were closely picketed in the milpas, or corn-fields, down by the river. **1934** A. Huxley *Beyond Mexique Bay* 116 The peasant may need only two or three acres for his milpa. **1936** *Nature* 26 Dec. 1090/1 The cultivation of maize..is on the milpa system; that is, a plot, after being burned off, is cultivated for two years, when it is allowed to revert to forest conditions, taking about eight to ten years to become completely reestablished and ready for burning off again. **1956** R. Redfield *Peasant Society & Culture* iv. 118 To the Maya Indian labor on the milpa is dignified by its connections with religion and many virtue. **1964** *Sci. Amer.* Nov. 101/1 In Central America the Mayas,..who depended greatly on the milpa system of agriculture, were forced to abandon their cities and move north into Mexico. **1974** *Environmental Conservation* I. 17/1 This is the so-called 'slash-and-burn' or 'milpa' system, as practised in the humid tropics.

Milquetoast (mi·lktōst). *orig. U.S.* Also with small initial. [f. the name of Caspar *Milquetoast*, a cartoon character created by H. T. Webster in 1924.] A timid or unforthcoming person. Also *attrib.* or as *adj.*

1938 M. Fishback *Safe Conduct* vi. 70 Don't be a Milquetoast either, and be afraid to add it [*sc.* the bill] up. **1939** C. Morley *Kitty Foyle* xxx. 305 What is it makes a man with brains so milquetoast when he gets away from the blackboard? **1961** M. Beadle *These Ruins are Inhabited* iii. 34 American men are Milquetoasts. **1972** L. O'Donnell *Phone Calls* vi. 73 You couldn't expect Norah to respect a Milquetoast. **1973** *Observer* (Colour Suppl.) 4 Nov. 23/4 Any of those milquetoast settings that Britten gave Wilfred Owen's poems in the 'War Requiem'. **1974** *Ottawa Citizen* 3 Sept. 71/1 I'm wondering if a judge would consider this a perfectly good but milquetoast name to something no one will forget, like—Jake Sexchamp.

milreis. Add to def.: replaced by new currency in Portugal in 1911 (see quot.); and in Brazil in 1942 (see *cruzeiro).

1913 *Statesman's Yearbk.* 1155 The Decree of the Provisional Government of May 22, 1911, established a new monetary system... The unit is the gold *escudo*, of 100 centavos, which is equivalent to the 1-milreis gold piece.

Milton (mi·ltən). **1.** Name of the English poet *John Milton* (1608–74) used in the phr. *mute inglorious Milton* (see quot. 1751) to symbolize the idea of native ability frustrated by lack of opportunity.

1751 Gray *Elegy* 8 Some mute inglorious Milton here may rest. **1883** *Authors & Publishers* (G. P. Putnam's Sons) 61 We do not believe that our American prairies conceal any Charlotte Brontës to whom the opportunity for expression and fame has been denied, or that a careful search through American villages would develop any 'mute, inglorious Miltons' rusting away their undeveloped lives. **1922** H. L. Mencken *Prejudices* (1923) 3rd Ser. iii. 89 A genuine artist..would have thoughts and feelings of his own, and the impulse to give them objective form would be irresistible... There are no mute inglorious Miltons, save in the hallucinations of poets. The one sound test of a Milton is that he functions as a Milton. **1933** J. Buchan *Prince of Captivity* I. iv. 125 We've got to see that our Miltons don't remain mute and inglorious, but above all that our Hampdens are not left to rot on a village green. **1948** T. S. Eliot *Notes Def. Culture* vi. 102 The Equality of Opportunity dogma..derives emotional reinforcement from the belief in the mute inglorious Milton. This myth assumes that a great deal of first-rate ability—not merely ability, but genius—is being wasted for lack of education.

2. [Perh. a different word.] *slang*. An oyster.

1841 Thackeray *Professor* ii, in *Bentley's Misc.* Sept. 285 Mrs. Grampus herself operated with the oyster-knife, and served the morsels to the customers. **1845** 'Bon Gaultier' *Bk. Ballads* 35 Fill me once more the foaming pewter up! Another board of oysters, ladye mine! To-night Montgomery himself shall sup, These mute inglorious Miltons are divine.

miltonia (miltō·niă). [mod.L. (J. Lindley 1837, in *Bot. Reg.* XXIII. 1976), f. the name of Charles William Wentworth Fitzwilliam, Viscount *Milton*, later 3rd Earl Fitzwilliam (1786–1857), English politician and horticulturist + -ia¹.] An epiphytic orchid of the tropical, South American genus so called, belonging to the family Orchidaceæ and bearing brilliantly coloured flowers.

1845 J. Lindley *Sertum Orchidaceum* plate 21 It [*sc. Miltonia candida*] differs in the structure of its column and labellum..from the original Miltonia. **1890** W. Watson *Orchids* iii. 315 Miltonias are easily propagated. **1930** T. W. Briscoe *Orchids for Amateurs* viii. 120, I have found that a small portion of partly decayed oak or beech leaves has had very beneficial results on all the Miltonias. **1965** *Times* 6 Feb. 12/3 One exhibit of orchids contains many beautiful cypripediums and miltonias.

Miltonian, *a.* Add: As *sb.*, an admirer or imitator of Milton.

1842 Dickens *Let.* 19 Oct. (1974) III. 352, I have been going, every day, to write to you about the Miltonians. **1907** *Illustr. London News* 4 May 672/1 Nearly all Englishmen are either Shakespearians or Miltonians... Each represents something in the make-up of England. **1947** A. J. A. Waldock *Paradise Lost & its Critics* i. 9 Grierson..observed, supervised and corrected; but not yet had England produced a Miltonian. **1960** *Times* 22 Sept. 15/4 The most dyed-in-the-wool Miltonian.

Miltonic, *a.* (and *sb.*). Add: **2.** As *sb. pl.*, verses, or style, typical of Milton.

1928 O. Barfield *Poetic Diction* x. 177 No one would have dreamed of employing the stale Miltonics, which lay at the bottom of so much eighteenth-century 'poetic diction', in prose, however imaginative. **1944** F. R. Leavis in *Scrutiny* XII. 202 Johnson's disapproval of Gray's Pindarick sublimities goes with his disapproval of Miltonics.

Miltonically, *adv.* (Earlier example.)

1853 De Quincey *Autobiogr. Sk.* I. i. 2, I here record the fact of my brothers and sisters..and Miltonically I include myself.

Miltonism. Add: (Later example.) Also Milto·nicism.

1936 F. R. Leavis *Revaluation* ii. 46 A common, limply pompous Miltonicism. **1938** E. Pound *Let.* 8 May (1971) 316, I do, however, prefer your 'supreme Hippocrates' [line 137]...Miltonism tho' it may be.

Miltonize, *v.* Add: Hence **Mi·ltonizing** *ppl. a.* and *vbl. sb.*

1936 F. R. Leavis *Revaluation* iv. 116 The meditative-Miltonizing poetic modes. **1944** —— in *Scrutiny* XII. 189 The author of *The Vanity of Human Wishes* has, as critic, no weakness..for the Miltonizing habit of his age. **1953** H. House *Coleridge* iii. 65 This Miltonising is not a mere matter of poetical echoes..it is part of a conscious political act. **1958** *Essays & Stud.* v. 69 These larger Miltonizing politico-social pieces like *Religious Musings*.

Miltown (mi·ltaun). *Pharm.* Also (*erron.*) Milltown. A proprietary name for meprobamate; a tablet of this drug.

1954 F. M. Berger in *Jrnl. Pharmacol. & Exper. Therap.* CXII. 413 2-Methyl-2-*n*-propyl-1,3-propanediol carbamate was unique in that it possessed a muscle relaxant and sedative action of an unusual kind. This compound has been named Miltown. **1956** A. Huxley *Let.* 20 Oct. (1969) 810 The conference on meprobamate was quite interesting and I made some pleasant acquaintances —Dr. (F. M.) Berger, the inventor of Miltown, [etc.]. **1957** *Trade Marks Jrnl.* 27 Feb. 191/2 Miltown... Pharmaceutical preparations in pill, tablet and powder form, or in liquid form for intravenous injection, all for use in the treatment of tension, as muscular relaxants and as anti-convulsants. Carter Products, Inc... New York. **1960** C. Fitz Gibbon *When Kissing had to Stop* vii. 91 All myths are to a greater or lesser extent tranquillisers in one way or another. But a government can't function on Milltown. **1964** M. McLuhan *Understanding Media* (1967) vii. 77 Would it not seem suddenly to be a conspiracy to make the artist a frill, a fribble, or a Milltown? **1969** [see *meprobamate].

miltsiekte (miltsi·ktĕ). *S. Afr.* Also meltziekte, miltsiek. [Afrikaans, f. *milt* spleen + *siekte* sickness.] = Anthrax 2.

1835 A. Smith *Diary* 1 Feb. (1939) I. 241 Bloedsiekte, or anthrax. It is also known in Afrikaans as miltsiekte. **1877** *Queenstown Free Press* 1 Dec. (Pettman), The oldest and most experienced of Kurveyors confess themselves flabbergasted by meltziekte. **1947** H. C. Bosman *Mafeking Road* xi. 63 After making their purchases they whiled away the time in discussing politics and the mealie-crops and the miltsiekte. **1959** *Cape Times* 12 Mar. 2/2 Miltsiek has broken out in certain areas in the Bushmanland. **1972** L. G. Green *When the Journey's Over* (1973) v. 51 You had to guard your oxen against redwater, *meltsiekte* and lung sickness.

miltz (milts). *Cookery*. [ad. G. *milz*.] The spleen. (Cf. Milt *sb.*)

1909 *Daily Chron.* 25 Feb. 3/5 No more cooking of chleb borsch schave and stuffed miltz. I shall have a cook of my own. **1951** L. W. Leonard *Jewish Cookery* xiv. 189 The butcher will be glad to make the incision in the side of a beef miltz as large as you desire. **1966** *Bloom's Jewish Restaurant Menu, Entree*..Stuffed Miltz.

mim, *a.* Add non-*dial.* and U.S. examples.

1891 R. T. Cooke *Huckleberries* 96 She was a mim, soft-spoken woman, but guileful and gliding as a snake. **1959** 'J. Ross' *Boy in Grey Overcoat* viii. 93 That mim curl of grey hair across the back of her head. **1960** Auden *Homage to Clio* 56 Before you catch it for your mim look and gnostic chirrup.

mimamsa (mīmā·msa). Also mimansa. [Skr. *mīmāṃsā* profound thought, consideration, investigation, f. *man* to think, consider.] The name of one of the six systems of orthodox Hindu philosophy, called more fully the *Pūrva-*, or 'earlier', *mimamsa*, which was founded by Jaimini and concerns itself with the interpretation of Vedic ritual and text.

The term also occurs in the name of a closely-related school, the *Uttara*, or 'later', *mimamsa*, often called *Vedānta*, which deals with the nature of Brahma.

1788 *Asiatick Researches* I. 352 Of the Philosophical Schools it will be sufficient here to remark, that..the two Mimánsà's, of which the second is often distinguished by the name of Védánta, [seem analogous] to the Platonick. **1811** W. Ward *Acct. Writings, Relig. & Manners Hindoos* I. ii. 370 Like the schools of philosophy among the Greeks, these several systems have each originated with a single and a different head, or founder..Jüyüminee of the Meemangsa. **1841** M. Elphinstone *Hist. India* I. ii. v. 215 The prior Mímánsá, which teaches the art of reasoning with the express view of aiding the interpretation of the Védas, is, so far, only a school of criticism. **1861** H. H. Wilson *Ess. & Lect. Relig. Hindus* I. 12 Jaimini, by Siva's orders, composed the Mímánsá, which is heretical, in as far as it inculates works in preference to faith. **1915** *Encycl. Relig. & Ethics* VIII. 648/2 The Mīmāṃsā teaches that the relation of word and meaning is not dependent on general agreement, but that the meaning is naturally inherent in the word. **1964** *Language* XL. 113 He did not go any further with his study of the *mīmāṃsā* system. **1971** *Illustr. Weekly India* 11 Apr. 35/2 The Indian philosophical systems such as Buddhism, Nyaya, Vaisheshika, Yoga and Mimamsa.

mimbar (mi·mbaɹ). Also 9 mambar, monbar, 9– minbar (the best form). [Arab. *minbar* pulpit.] The pulpit in a mosque. Also *attrib.*

1816 *Travels of Ali Bey* II. vi. 84 El Monbar, or The Tribune of the Priest of Fridays, is on one side of the Makam Ibrahim, at fourteen feet distance, and in front of the northern angle of the Kaaba. **1836** E. W. Lane *Acct. Manners & Customs Mod. Egyptians* I. iii. 94 To the right of this [*sc.* the mihrab] is the *mim'bar* (or pulpit). **1839** T. J. Newbold *Pol. & Statistical Acct. Straits of Malacca* I. v. 249 The Khatib..recites the Khatbeh..in

the mosque, from the three steps of the mimbar, a species of rostrum. **1855** R. F. Burton *Pilgrimage* II. xvii. 141 The Mambar, or pulpit, was the invention of a Medinah man of the Beni Najjar. **1875** *Encycl. Brit.* II. 446/2 Near this was a pulpit (*mimbar*). **1885** T. P. Hughes *Dict. Islam* 349 *Minbar*, generally pronounced *mimbar*... The pulpit in a mosque from which the khuṭbah (or sermon) is recited. It consists of three steps, and is sometimes a moveable wooden structure, and sometimes a fixture of brick or stone built against the wall. **1932** *Times Lit. Suppl.* 10 Nov. 840/1 The charming panels in the side walls of the *mimbar* steps at the Ravali Masjid. **1967** V. Pritchard *Eng. Medieval Graffiti* 35/1 There is a whole panel of swastika-peltae, in a simplified form, carved on a wooden mimbar in Kairwan, North Africa.

mime, *sb.* Add: **4. b.** (The art of) gesture, movement, etc. (as distinct from words) used to express emotion and dramatic action or character; dumb show; = Pantomime *sb.* 4.

1932 I. Mawer *Art of Mime* ii. i. 125 The aim of mime is not a performance of certain physical exercises which can be welded into some kind of whole, nor is it merely 'gesture'—gesture is one branch only. **1953** *Ballet Ann.* VII. 22 There is always a great misunderstanding of the word *mime*... In ballet it means the formal gesture language used in the narration of the classics. **1967** *Listener* 13 Apr. 503/3 Some sort of research is required..to find out the best way of using mime on television.

6. Comb. *mime-ballet*, *-drama*, *-writer*.

1955 *Times* 11 May 7/6 Two mime-ballets by Rocca and Dallapiccola respectively. **1931** A. Nicoll *Masks, Mimes & Miracles* i. 78 The Oscan mime became one of the most popular divertisements there. **1968** J. Winearls *Mod. Dance* (ed. 2) vii. 145 There have been Masques, Dance Plays, Mime Dramas and every combination of the fundamental expressions of movement and voice. **1957** N. Frye *Anat. Crit.* iv. 285 Classical mime-writers like Herodas.

mime, *v.* Add: **1. c.** (Later examples.)

1915 M. E. Perugini *Art of Ballet* xiii. 115 The two well-known dancers..mutely mimed the actions and emotions of the leading characters. **1959** W. Golding *Free Fall* xiv. 250 The maker they mimed for you in your Victorian slum was the old male maker, totem of the conquering Hebrews.

d. Of a singer, to present a pre-recorded song by mime, usu. on television.

1965 G. Melly *Owning-Up* xi. 131 A weekly [T.V.] programme featuring the new releases and illustrating them visually by..the artists miming to their own records. **1966** *Crescendo* Jan. 8/1 He seems content to mime 'Tears' to a gaggle of unbelieving teen-agers on *Top Of The Pops*. **1966** *Listener* 11 Aug. 204/3 Since the singers and dancers are so expert in miming to their own recordings, it is possible to eliminate all microphones from the stage.

Also **mimed** *ppl. a.*

1910 *Daily Chron.* 9 Apr. 7/5 The marvellous power of facial expression to convey an emotion..is brought home ..by the intense interest one feels in these 'mimed' plays. **1965** *Listener* 2 Dec. 908/3 Mimed opera... It soon became painfully obvious that sound was out of synchronization with vision.

mimeo (mi·mio), *sb.* [Abbrev. of Mimeograph *sb.*] A copy of a document, newspaper, etc., reproduced by means of a mimeograph (see Mimeograph *sb.* in Dict. and Suppl.). Also *attrib.*, as *mimeo mag, newspaper, stencil*. Hence **mimeo**, **mimeo-stencil** *vbs. trans.*, to reproduce by means of a mimeograph; **mimeoed** *ppl. a.*

1943 J. S. Huxley *TVA* 139 TVA, Mimeo, 1941. **1967** Karch & Buber *Offset Processes* vi. 243 Within a few minutes a phototube will scan and transfer all details to a mimeo-stencil or offset plate. **1967** *Maclean's Mag.* Jan. 56 Why Sandra Peredo's patronizing attitude toward the little magazines? She states, for instance, that the mimeo mags are interested in dirty words and anti-Establishment statements as their main themes. **1969** *Harper's Mag.* May 82 We learn later that the statement was mimeoed and a few reporters were walking off with it. **1972** M. Joos in *Language* XLVIII. 260 Having mimeo-stenciled a number of text pages without the footnoting for which I was leaving room on each stencil, I was now putting the stencils through my typewriter a second time. **1973** *Times* 17 Apr. (Liberia Suppl.) p. ii/5 These tribes have their own newspapers called mimeo-newspapers.

mimeograph, *sb.* Substitute for def.: A duplicating machine for producing copies from a stencil; also *attrib.* (Add later examples.)

From 1903 to 1948 a proprietary name.

1914 [see *duplicating *vbl. sb.*]. **1935** M. M. Atwater *Murder in Midsummer* xviii. 160 Miss Marsh ran off on the mimeograph sheets of instructions all neatly tabulated. **1952** S. Kauffmann *Philanderer* (1953) v. 84 The mimeograph ink of the mail-room. **1963** F. C. Archer *Gen. Office Pract.* (ed. 2) xvii. 233 Mimeograph paper comes in a variety of colors to contrast or to harmonize with the ink. **1967** Karch & Buber *Offset Processes* iv. 72 Letter shops, once using mimeograph-type equipment, added the offset process..to expand..their production.

mimeograph *v. trans.* (later examples); **mimeographed** *ppl. a.* (later examples).

1912 J. C. Dana *Mod. Amer. Library Econ.* v. 6 Mimeographed copies of tests for books and printed forms for book notes are to be distributed when needed. **1937** M. L. Hanley *Word Index to Joyce's Ulysses* p. xii, Mimeographing: Dan Marlow. **1946** *R.A.F. Jrnl.* May 152 The *Bulletin*..was an unpretentious affair of a few mimeographed sheets. **1949** Shurr & Yocom *Mod. Dance* 3 Would you consider mimeographing this material? **1972** *Jrnl. Social Psychol.* LXXXVII. 23 Choices were to be indicated on a mimeographed answer form.

mimesis. Add: **1.** (Earlier and later examples.) Also *transf.*

1550 R. Sherry *Treat. Schemes & Tropes* sig. E 3 Mimisis, that is a folowing eyther of the wordes or manoures whereby we expresse not onlye the wordes of the person, but also the gesture. **1593** H. Peacham *Garden of Eloquence* (1954) 138 Mimesis is an imitation of speech whereby the Orator counterfaiteth not onely what one said, but also his utterance, pronunciation, and gesture. **1962** S. E. Finer *Man on Horseback* xii. 240 This is to reckon without *mimesis*. Among unstable states, particularly those with a rage for innovation, military intervention has proved to be highly contagious. **1962** *Listener* 13 Dec. 1006/1 Bunyan stands with Malory and Trollope as a master of perfect naturalness in the mimesis of ordinary conversation. **1965** M. Bradbury *Stepping Westward* vi. 292 Walker had never before heard in anyone's speech vocabulary—*mimesis, epistemology, mythopoeic*. **1965** M. Cohen in M. Black *Philos. in Amer.* 115 If Le Corbusier has encouraged the doctrine of *mimesis* in designing the chapel at Ronchamp, he has rejected it firmly at the Villa Savoie.

2. For '= Mimicry 3' read '= Mimicry 2'.

3. *Sociol.* The deliberate imitation of the behaviour of one group of people by another as a factor in social change.

1934 A. J. Toynbee *Study of Hist.* III. 245 The problem of bringing the uncreative rank and file of a growing society into line with the creative pioneers..cannot be solved in practice, on the social scale, without also bringing into play the faculty of sheer mimesis—one of the less exalted faculties of Human Nature which has more in it of drill than inspiration. **1962** *Listener* 8 Feb. 257/3 There is..an important distinction between *mimesis*, which is imitation of those above you, and solidarity, which is imitation..of those with whom you find yourself in a common situation. **1965** P. Laslett *World we have Lost* ix. 213 Only if imitation, mimesis, is taken to constitute 'solidity' can the phrase the solid middle class be made to apply to any substantial part of the population.

mimetism. (Examples.)

1918 B. Miall tr. *Hamon's Lessons of World-War* ix. 271 Men hide their guns..under canopies of leaves, in order to conceal them from the view of aeroplanes. Man adopts the mimetism of Nature. **1970** *Times Lit. Suppl.* 18 Dec. 1480/5 The later history of the Inca dynasty is a depressing one of mimetism and osmosis.

mimi, var. *maimai.

mimic, *v.* Add: **4. b.** *Med.* Of a drug: to produce an effect very similar to (that of some other cause).

1971 *Nature* 12 Feb. 497/1, *d*-Amphetamine closely mimicked the excitatory and inhibitory effects of *l*-NA. **1974** *Ibid.* 31 May 473/1 Colchicine..has been reported to mimic the effects of denervation on mammalian skeletal muscles.

mi·micable, *a.* [f. Mimic *v.* + -able.] Capable of being mimicked.

1955 J. L. Austin *How to do Things with Words* (1962) viii. 96 The phatic act, however, like the phonetic, is essentially mimicable.

mimicry. Add: **2.** Also used of similar resemblances in plants. (Later examples.)

1931 R. N. Chapman *Animal Ecol.* viii. 190 It seems likely that many controversies over protective coloration, mimicry, and resemblance might find the solution if they were investigated from the viewpoint of their contributions to the maintenance of the population of the species. **1951** *Dict. Gardening* (R. Hort. Soc.) III. 1304/1 There is often such resemblance between plants which themselves possess no special protective apparatus and those that do as to suggest that 'mimicry' occurs among them in the same way as it does among insects. **1966** E. Palmer *Plains of Camdeboo* xvi. 268 Some plants are barely recognizable as plants at all, and these are the mimicry plants which are some of the most famous plants in the world. **1968** R. D. Martin tr. *Wickler's Mimicry in Plants & Animals* x. 100 Some insects resembling ants live within the society of ants and devour their hosts. This latter case would seem to be a case of aggressive mimicry. **1975** *Nature* 17 Jan. 191/1 Mimicry is a phenomenon of evolutionary convergence or parallelism by which an edible mimic species gains some measure of protection from predators by virtue of its close resemblance to a model species which is unpalatable (that is, distasteful or dangerous).

miming, *vbl. sb.* (See under Mime *v.*) (Later examples in sense *1 d of the verb.)

1965 *Melody Maker* 3 Apr. 16 Miming is slowly disappearing from the pop scene. **1965** *Listener* 2 Dec. 908/3 The first programme of the series was made almost unwatchable by the process of miming.

mimosa. Add: **2*.** A yellow colour resembling that of the mimosa. Also *attrib.*

1909 *Cent. Dict.* Suppl., *Mimosa*, same as thiazol yellow. **1928** A. Christie *Mystery of Blue Train* vii. 61 The little mimosa suit of crêpe de chine. **1966** *Harper's Bazaar* Sept. 41 Evening dress... In mimosa, toast or turquoise. **1971** R. Busby *Deadlock* xiii. 201 The immaculate coffee-coloured tussore suit..with a fresh mimosa shirt.

3. *mimosa-bush* (earlier example); *mimosa-yellow* adj.

1856 F. Fleming *S. Afr.* xii. 264 A broad valley, covered with rich pasturage and dotted with '*mimosa*' bushes, stretched out over several acres. **1910** *Encycl. Brit.* VIII. 747/1 The following list includes the principal coal-tar colours... mimosa yellow.

mimosine (mimǝᵘ·sīn). *Chem.* Also † mimosin. [ad. G. *mimosin* (J. Renz, at the suggestion of K. Suessenguth, 1936 in *Zeitschr. f. physiol. Chem.* CCXLIV. 154), f. Mimos(a + -ine⁵.] An amino-acid, $C_5H_4O_2N{\cdot}CH_2CH(NH_2)COOH$, found in the tree *Leucæna glauca* and in *Mimosa pudica*, the sensitive plant.

1937 *Brit. Chem. Abstr.* A. iii. 50/2 The sap from the tubular cells of young shoots and leaf stalks of *Mimosa pudica*, L., and *Leucaena glauca*, Benth., yield mimosin. **1949** T. A. Henry *Plant Alkaloids* (ed. 4) 5 Kostermanns has also investigated mimosine with results indicating that it is a derivative of 3:4-dihydroxypyridine. **1971** *Toxicon* IX. 241 When animals ingest..mimosine they suffer from growth retardation, alopecia, skin irritation and cataract formation.

mimp, *sb.* and *a.* **A.** *sb.* Substitute for quots.

1786 J. Burgoyne *Heiress* iii. ii. 54, I am preparing the cast of the lips for the ensuing winter—thus—It is to be call'd the Paphian mimp. **1822** M. Edgeworth *Let.* 16 Jan. (1971) 321 The famous learned Mrs. Somerville..no *set* smile or prim look—no *mimps* with her mouth.

mimsey, *a.* Add: Also mimsy. (Further examples.)

Lewis Carroll's *mimsy*, which may be an invented word, has influenced all subsequent uses.

1855 'L. Carroll' *Rectory Umbrella & Mischmasch* (1932) 139 All mimsy were the borogoves. *Ibid.* 140 *Mimsy*, (whence *mimserable* and *miserable*). 'Unhappy.' **1911** C. Mackenzie *Passionate Elopement* xxi. 186 Four shillings and sixpence, ma'am, for a little mimsy book not so thick as the magick history of Jack the Giant Killer. **1920** D. H. Lawrence *Touch & Go* 6 Good plays? You might as well say mimsy bomtittle plays, you'd be saying as much. **1933** W. de la Mare *Lord Fish* 171 Treading mimsey as a cat. **1934** *Times Educ. Suppl.* 24 Mar. p. iv/2 A people unimaginative enough to accept a mimsy and scrannel 'P.R.' in place of the organ music, the soul-uplifting harmony of 'Proportional Representation'. **1936** *Punch* 10 June 650/1 'It's the glamour of it,' sighed Josephine. 'Whenever I smell a programme I go quite mimsey—honestly I do.' **1937** 'N. Blake' *There's Trouble Brewing* i. 24 An affected mimsy sort of voice that she reserved presumably for cultural pronouncements: Nigel preferred her normal, unmitigated boom. **1956** J. Cannan *People to be Found* vii. 91 With horror they had seen the lawns of the Botanic Gardens torn up and replaced by a mimsy pseudo-Elizabethan rose-garden. **1963** *Times* 8 Feb. 14/3 Moreover his interpolated variation in the first act, danced to the normally unused *andante* of the *pas de trois* and consisting largely of slow *pirouettes en attitude*, looked as mimsy as the borogroves [*sic*], and could not be regarded as successful.

mimulus. Add: **1.** (Later example.)

1794 *Curtis's Bot. Mag.* VIII. 283 Mimmulus is a classical word for the Pedicularis, or Lousewort.

2. Substitute for def.: [Linnæus 1741, in *Acta Soc. Reg. Scient. Upsaliensis* 82.] An annual or perennial herb of the large genus so called, belonging to the family Scrophulariaceæ and widely distributed in America, Asia, and Africa, esp. *Mimulus luteus*, the monkey-flower, a yellow-flowered perennial, native to western North America but naturalized elsewhere. (Earlier and later examples.)

1768 P. Miller *Gardeners Dict.* (ed. 8) s.v. Mimulus. Upright Mimulus with oblong linear leaves. **1794** *Curtis's Bot. Mag.* VIII. 283 Linnæus first gave to it [*sc. Mimulus ringens*] the name of *Mimulus*. **1900** L. H. Bailey *Cycl. Amer. Hort.* II. 1018/1 There is nothing difficult in the culture of Mimulus. Some of the finest plants have been self-sown on a rubbish heap. **1963** *Times* 25 Apr. 14/6, I think of a turn in a wood and a solitary exquisite butterfly-orchis; of a wide green hillside and a patch of mimulus beside a grooved trickle of water. **1974** A. Scott-James *Sissinghurst* xiii. 143 The mimulus continues as the centre-point.

min (min), *sb.*³ Shortened form of Minute *sb.*¹

1890 in *Cent. Dict.* **1892** *Field* 14 May 735/3 He.. 'clocked' 2 min 2 sec all the way. **1959** P. Bull *I know Face* i. 18 Miss Pereira..told me to meet her in the ABC up the street in ten mins.

Min (min), *a.* and *sb.*⁴ **A.** *adj.* Of or pertaining to the Min district or dialect in Fukien province, S.E. China. **B.** *sb.*⁴ The dialect of this district.

1902 *Encycl. Brit.* XXVII. 27/1 The Min forts at the entrance of the Foochow river. **1910** *Ibid.* VI. 202/1 The French fleet attacked and destroyed..the forts which were built to guard the entrance to the Min river. **1959** *Chambers's Encycl.* III. 488/2 The Min dialects of Fukien, with well-marked varieties in Foochow and Amoy. **1964** M. A. K. Halliday et al. in J. A. Fishman *Readings Sociol. of Lang.* (1968) 146 There are six major dialects in modern China: Mandarin, Cantonese, Wu, North Min, South Min and Hakka. **1975** *Language* LI. 258 Cháozhōu, a Mǐn dialect spoken on the southeast coast of mainland China.

minable, var. Mineable *a.* (in Dict. and Suppl.).

Minæan (minī·ăn), *sb.* and *a.* Also **Minean.** [f. L. *Minæus*, f. Arab. *Ma'īn*, + -AN.] **A.** *sb.* **a.** A native or inhabitant of an ancient kingdom of southern Arabia. **b.** The Semitic language of the Minæans. **B.** *adj.* Of or pertaining to the Minæans, their kingdom, or their language. So **Mina·ic** *sb.* and *a.*

1601 Holland tr. *Pliny's Nat. Hist.* xii. xiv. 366 There is another tract by it selfe confronting this countrye [*sc.* Arabia], wherein the Minæans doe inhabite. **1833** A. Crichton *Hist. Arabia* I. iv. 148 The Minæans were rich in palm-groves, flocks, and fertile fields. **1844** C. Forster *Hist. Geogr. Arabia* I. ii. 133 We find Carman reg., the Minæan metropolis..peopled by Pliny with the Charmæi. **1886** *Encycl. Brit.* XXI. 654/1 The other [dialect], which expresses the causative by *sa*.., is the Minaic. To this latter branch belong the numerous South Arabic inscriptions recently found in the north of the Ḥijáz, near Hejr where the Minæans must have had a commercial settlement. *Ibid.*, The singular manner in which districts containing Sabæan inscriptions and those containing Minaic alternate with one another. **1902** in Gottheil & Jastrow *Semitic Study Ser.* I. (Advt.), Selected Sabaean and Minaean inscriptions. **1934** L. H. Gray *Introd. Semitic Compar. Ling.* i. 6 South Arabic is represented only by inscriptions (Minaean,..Qatabānian,..). **1936** F. Stark *Southern Gates Arabia* 5 The Arabian empires rose and fell—Minean, Sabæan, Katabanian. **1951** W. F. Albright in H. H. Rowley *Old Testament & Mod. Study* ii. 32 The difference between Sabaean and Minaean, the former of which has an *h*-causative whereas the latter has the *s*-causative. **1974** *Encycl. Brit. Macropædia* I. 1044/1 Four main states are known to have established themselves in South Arabia—Ma'īn (of the Minaeans), Saba' (of the Sabaeans), Qatabān (Qitbān) and Ḥaḍramawt. *Ibid.* XIX. 1084/1 Yemen..was successively the centre of the Minaean (13th to 7th centuries BC), Sabaean.., and Himyaritic..civilizations.

Minamata disease (minămā·tă). *Path.* [Named after *Minamata*, the name of a town in Kumamoto prefecture, Japan, where it was first recognized.] A disease, caused by ingestion of alkyl mercury compounds, which is characterized by impairment of cerebral functions such as speech, sight, and muscular coordination and which is usually permanent and sometimes fatal.

1957 *Acta Path. Jap.* VII. 605 The so-called Minamata-disease may be recognized as a toxic encephalopathia. The toxic substances contained in fish and shell-fish have not been yet found. **1971** *Sci. Amer.* May 15 The Minamata disease, as it came to be called, produced progressive weakening of the muscles, loss of vision, impairment of other cerebral functions, eventual paralysis and in some cases coma and death. **1973** *Biol. Conservation* V. 143/2 Minamata disease first appeared in cats, which died from eating fish scraps from the village kitchens. **1975** *Times Lit. Suppl.* 19 Sept. 1042 (*caption*) Demonstrators for the victims of 'Minamata Disease', a defect of the central nervous system caused by eating fish poisoned by mercury in industrial waste.

Minangkabau (mi:næŋkăbau·), *a.* and *sb.* Also **Manikabowe, Menangkabau, Menangkabo,** 8 **Menangcabow(e),** 9 **Menangkabow, Minangkabauer.** Pl. **Minangkabau, -baus. A.** *adj.* Of or pertaining to Minangkabau, a Malay territory in the highlands of Sumatra. **B.** *sb.* A native or inhabitant of Minangkabau.

1783 W. Marsden *Hist. Sumatra* 279 Between the Menangcabow people, those of Rou.., and the Achenese, wars used to be perpetual. **1808** *Asiatick Researches* X. iii. 165 The Menangkábow race..seem at an early period to have ruled the whole island of Sumatra. **1821** J. Leyden tr. *Malay Annals* ii. 37 All the Menangcabows were surprised at his appearance. **1839** T. J. Newbold *Pol. & Statistical Acct. Straits of Malacca* I. v. 200 The inhabitants [of Naning]..are always styled 'Manikábowes', or settlers from Menángkábowe, in Sumatra. *Ibid.* II. xiv. 219 The information..was derived..from various Menangkabowe chiefs of Rumbowe and Naning. **1911** J. Frazer *Golden Bough: Magic Art* (ed. 3) I. iii. 58 The Minangkabauers of Sumatra..will call in the help of a wizard. *Ibid.* 193 The Minangkabau people of Sumatra. **1947** R. O. Winstedt *Malays* viii. 154 Some of the sayings of the Minangkabau tribes of Negri Sembilan rise to poetry. **1968** *Encycl. Brit.* XV. 144/1 The Menangkabau have much in common with the coastal Malay, but differ radically in social organization. **1972** M. Sheppard *Taman Indera* 65 The *Randai*, the Minangkabau dance drama from West Sumatra.

minasragrite (minăsrā·grəit). *Min.* [f. *Minasragra*, name of its locality near Cerro de Pasco, Peru + -ITE¹.] An acidic hydrated vanadyl sulphate, $(VO)_2H_2(SO_4)_3 \cdot 15H_2O$, found as a blue efflorescence of monoclinic crystals on patronite.

1915 W. T. Schaller in *Jrnl. Washington Acad. Sci.* V. 7 The following very brief notes of four new minerals are given in order to secure priority... Minasragrite is a blue hydrous vanadium sulphate from Minasraga Peru. **1934** *Amer. Mineralogist* XIX. 198 Specimens of patronite in the Harvard mineral collection..showed in places a vivid blue incrustation which proved to be minasragrite. **1968** I. Kostov *Mineral.* II. ix. 517 Minasragrite, found with patronite from the well known vanadium deposit of Minasragra..in Peru, is monoclinic.

minatory, *a.* and *sb.* Add: Hence **mi·natoriness,** threateningness. *rare.*

1961 J. N. Findlay *Values & Intentions* viii. 344 Acknowledging a God capable of such universal minatoriness.

‖ **minaudière** (minodiɛr). [Fr. fem. adj., lit. affected, coquettish.] †**1.** An affected or coquettish woman. *Obs.*

1716 M. W. Montagu *Let.* 21 Nov. (1965) I. 282 The Saxon Ladys..[illegible] **1818** Lady Morgan *Fl. Macarthy* III. ii. 93 She struck me to be a mere *minaudière*!

2. A small case for a woman's cosmetics, jewellery, etc.

1940 *Shopping News* (Springfield, Mass.) 15 May, [illegible] **1957** A. Adburgham in *Punch* 18 Dec. [illegible] **1967** *N.Y. Times* 22 June 43, [illegible] **1969** *New Yorker* 8 Mar. 27 (Advt.), [illegible]

mince, *sb.* Add: **3.** *Rhyming slang.* = *MINCE-PIE 4. (Usu. in *pl.*)

1937 in Partridge *Dict. Slang* 532/1. **1958** F. Norman *Bang to Rights* 149 [illegible] **1960** *News Chron.* 16 Feb. [illegible] **1962** R. Cook *Crust on its Uppers* (1964) [illegible]

mince, *v.* Add: **5. a.** Also with *out.*

1862 Mrs. H. Wood *Channings* II. v. 73 'You—are—very—kind—to—take—up—Arthur Channing's cause!' they mince out. [illegible] 255 'Ah—"[illegible]"'—said Sir John, mincing out his pun as though he loved it.

mincemeat. Add: **2.** Also, to beat decisively or easily in a contest.

1876 *Coursing Calendar* 193 Maniac made mincemeat of Smoker, who was so stiff that he could scarcely raise a gallop. **1955** *Times* [illegible] mincemeat of all their opponents in the Grand Eights.

mincemeaty, [f. Mincemeat + -y¹.] Suggestive of mincemeat.

1870 P. Brooks in A. V. G. Allen *Phillips Brooks* (1900) 246 Huxley's new Lay Sermons..is like..most books for the people that popularize science. It is patronizing and mincemeaty.

mince-pie. Add: **4.** *Rhyming slang.* An eye. (Usu. in *pl.*)

1857 'Ducange Anglicus' *Vulgar Tongue* 13 *Mince pies*, eyes. [illegible] as I closed [illegible] *Jerusalem's* [illegible] like a pump. [illegible] thinks he'll [illegible] **1922** Joyce [illegible] kid. **1928** M. [illegible] pies, eyes.

‖ **Mincha** (mi·nxā). Also **Minchah, Minha(h).** [Heb. *minḥāh*, lit. gift, offering.] (See quot. 1962.)

[**1706** I. Abendana [illegible] The Captives [illegible] at Jerusalem [illegible] Offerings..[illegible] corruptly for [illegible] ANDER *Hebrew* [illegible] Minchah... [illegible] as offering; [illegible] the evening. [illegible] they go to the [illegible] noon Service. [illegible] Knowest thou [illegible] Minchah. [illegible] pronounced [illegible] tions. **1962** [illegible] term for the [illegible] times the term [illegible] but it acquired [illegible] *Misnah.* The [illegible] **1973** *Synagogue* [illegible] during *Mincha* [illegible]

Mincing Lane (mi·nsiŋ lēⁱn). Used *absol.* and *attrib.* with reference to an auction-room for tea and other commodities which was originally situated in the London street of this name.

1913 *Times* 13 Oct. (Finance & Commerce) 18/1 The Mincing-lane Tea and Rubber Share Brokers' Association (Limited). **1914** in *Conc. Oxf. Dict.* **1923** *Westm. Gaz.* 31 Jan., You must go to Mincing-lane for an excuse for the rise [in the price of tea]. **1959** *Chambers's Encycl.* XI. 235/2 Business in rubber, tea, coffee, cocoa, sugar, spices and condiments was carried on at the offices of Mincing Lane until the destruction of the Commercial Sale Room by bombing in 1941. *Ibid.* XIII. 490/2 Requirements [of tea] are mainly obtained by purchases in the Mincing [illegible]

mind, *sb.*¹ Add: **11. g.** *to pay no mind*, *not to pay any mind*: to pay no heed or attention (to someone or something); not to care or worry. *U.S. colloq.* and *dial.*

[illegible]

to Teachers xi. 114 If he wants to get ideas on any subject, he sits down to work at something else, his best results coming through his mind-wanderings. **1923** U. L. Silberrad *Lett. J. Armiter* xiii. 264 But—I am tired! Footweary as well as mind-weary. **1890** W. James *Princ. Psychol.* I. vi. 154 Somewhere, then, there is a transformation... The question is, Where—in the nerve world or in the mind-world? **1951** W. de la Mare *Winged Chariot* 44 The world without; the mind-world in our head.

b. mind-bender, a person who, or thing which, influences or alters the mind; *spec.* a psychedelic drug; so **mind-bending** *a.*; **mind-**

blower *slang*, something that blows one's mind (see *BLOW *v.*¹ 24 j); so **mind-blowing** *a.*; (as a back-formation) **mind-blow** *v.*; **mind-blown** *a*; **mind-body** *Philos.* and *Psychol.*, usu. *attrib.*, a term used in relation to the question of whether a distinction can be made between mental and physiological events; **mind-boggling** *a.*, that causes the mind to boggle or be overwhelmed; **mind-changer,** (*a*) a person who changes his mind; (*b*) a psychedelic drug; **mind-curist** = *mind-curer* (*b*); **mind-dust,** in materialist evolution hypotheses, the particles of 'mind' or mental substance of which mind-stuff is composed; also *attrib.*; **mind-expanding** *a.* = *PSYCHEDELIC *a.*; so **mind-expander; mind-healing** (earlier examples); **mind-read** *v. trans.*, to discern what is passing in the mind of (another person); **mind-set,** habits of mind formed by previous events or earlier environment which affect a person's attitude (cf. *mental set* s.v. *MENTAL *a.*¹ 5); **mind-sight** (further examples); **mind-stuff** (further examples); **mind-transference,** telepathy.

1963 J. Kennaway (*title*) The mind benders. *Ibid.* xxv. 158 Oonagh has said that there were instincts in man laid too deep for the most skilful mind-bender to probe. **1966** *New Scientist* 15 Dec. 639/3 *The mind-bender*, a persuasive person who, for subversive political purposes or financial gain (often both), bends others to his will. Increasingly, he employs drugs, particularly of the hallucinogen group. **1967** *Sci. News Let.* 22 July 80 STP is a new, untested drug, resembling both amphetamine pep pills and the active ingredient in mescaline, the cactus-derived mind-bender. **1970** K. Platt *Pushbutton Butterfly* (1971) vii. 70 LSD, marijuana—any of the mind-benders. **1973** B. Turner *Hot-Foot* xvi. 122, I felt..as if I were hearing a soapy disc-jockey play the groovy zip of some discordant group of mind-benders. **1965** *Economist* 25 Sept. 1215/3 The Socialist Labour League, furious exegetes of the gospel according to Trotsky, with their mind-bending vocabulary full of 'Pabloism' and that mythical entity the 'rank-and-file'. **1966** *New Scientist* 21 Apr. 151/1 Already 'mind-bending' gases for military purposes are said to be at an advanced stage of development. **1970** *Sunday Times* 25 Jan. 29 The theoretical mathematics of the situation [*sc.* the mining of metals] are positively mind-bending. **1971** *New Scientist* 27 May 531/1 Heroin and other mind-bending agents. **1973** *Times* 25 May 4/3 Viscount Weymouth this afternoon unveiled the mind-bending paintings and sculpture-painting with which he has encrusted the walls of his stately home inches deep, acres wide and of fathomless significance. **1970** *Listener* 22 Oct. 540 It can mind-blow a long-haired GI to know he'll have to live straighter to survive in Sweden than in the Army or in America. **1968** *New Scientist* 27 June 703/1 Two chemicals with almost identical structures can have very different psychedelic properties: one might be a real mind-blower and the other as ineffective as a sugar lump. **1969** *Gandalf's Garden* vi. 11/1 *Mindblower*, an experience or idea which changes one's thought-pattern, enlivening the mind and emotions. **1967** *Jazz Monthly* Dec. 16/3 While the music lasted little of this was evident; the spectacular mind-blowing ferocity of it all simply carried the group through. **1968** *Times* 4 May 21 The poet celebrates the mindblowing effects of LSD and laments at the same time his lost childhood. **1973** *Guardian* 8 May 18/1 For the hopeful voyeur of sheer obscenity in modern urban life, Glasgow is hard to beat. 'Mind blowing', was the concise description of a former civil servant. **1974** H. McCloy *Sleepwalker* ii. 16 A mind-blowing mustard yellow for the woodwork and on the walls a psychedelic splash of magenta and orchid and lime. **1969** B. Patten *Notes to Hurrying Man* 23 Bloated We lie dreamless, mind-blown in its ruins. **1907** *Mind* XVI. 620 The essential significance of the mind-body relationship. **1920** S. Alexander *Space, Time & Deity* II. 355 The acts of its mind-body would take the place of our organic or motor sensa. **1925** I. A. Richards *Princ. Lit. Crit.* 84 The Mind-Body problem is strictly speaking no problem. **1963** A. Kaplan in P. A. Schilpp *Philos. R. Carnap* 841 The situation is like that of the attempts to soften the mind-body dualism by introducing subtle interactions. **1970** H. Feigl in C. V. Borst *Mind-Brain Identity Theory* i. 35 The crucial..puzzle of the mind-body problem, at least since Descartes, has consisted in the challenge to render an adequate account of the relation of..mental facts (intentions, thoughts, volitions, desires, etc.) to the corresponding neurophysiological processes. **1964** *Punch* 19 Feb. 257/1 A lot of mind-boggling statistics. **1973** C. Bonington *Next Horizon* x. 146 A monstrous bergschrund, a huge, mind-boggling chasm about fifteen feet across. **1931** *Punch* 4 Nov. 494/2 Things and opinions change so quickly in these days that no one is going to crow over a graceful mind-changer. **1958** A. Huxley in *Sat. Even. Post* 18 Oct. 110/3 Within a few years there will probably be dozens of powerful but—physiologically and socially speaking—very inexpensive mind-changers on the market. **1965** Mind-changer [see *HALLUCINOGENIC *a.*]. **1894** 'Mark Twain' *Lett.* (1920) 316 A patient had actually been killed by a mind-curist. **1890** W. James *Princ. Psychol.* I. vi. 146 Evolutionary psychology demands a mind-dust. *Ibid.* xiii. 492 It is the mind-dust theory, with all its difficulties in a particularly uncompromising form. **1970** *Times* 26 Mar. 7 Hallucinogenic..agents such as L.S.D., mescaline and other so-called mind-expanders. **1973** *Black World* Apr. 16/1 The play is truly a mind-expander. **1963** *News-Call Bulletin* (San Francisco) 29 May 1/6 Professors Richard Alpert and Timothy Leary..started several years ago to experiment with 'psychedelic' or 'mind-expanding' drugs. **1972** D. Bloodworth *Any Number can Play* xv. 135 This dim, suffocating chamber..was..decorated with..mind-expanding daubs in fluorescent paint. **1883, 1891** Mind-healing [see *CHRISTIAN SCIENCE]. **1968** H. Waugh *Con Game* viii. 81 These psychiatrists are too damned smart. All the time I was talking to him I had the feeling he was mind-reading me. **1972** J. Quartermain *Rock of Diamond* xxiv. 156 If I mind-read you, you're looking for a fall guy to do your dirty work. **1934** *Agric. Hist.* June 86 The 'oats' motif can be traced throughout Johnson's life. As a schoolboy the young Sam was given oatmeal porridge for breakfast, and if he was like many children, this may have given him a mind-set for life. **1964** *Spectator* 6 Mar. 303/2 It was thereafter always a wonderful fight, although the mindset one brought to it made it impossible to recognise what was happening. **1971** *Amer. Benedictine Rev.* Dec. 424 The mind-set and aspirations of a vibrant hellenistic culture. **1930** Blunden *Leigh Hunt* ii. 22 The grace which the mind-sight of those merry young scholars awoke in him. **1935** —— *Edward Gibbon* 15 He felt that there was in the world actual need of a history such as I have mentioned hovering in the mindsight of his age. **1947** S. Spender *Poems of Dedication* 27 She drinks his acres of light Which..Beyond mind-sight and eye-sight Reach a womb where his rays Penetrate her night. **1930** A. D. Lovejoy *Revolt Against Dualism* viii. 272 Mind-stuff is not supposed to be the same kind of thing as either data or the awareness of them. **1937** E. Upward in C. Day Lewis *Mind in Chains* 42 A poet's images or a novelist's characters are not created out of pure mind-stuff, but are suggested to him by the world in which he lives. **1886** *Science* 17 Dec. 559/1 Mr. Hodgson is now engaged..in some experiments on the subject of mind transference, or the occasional communication of mental impressions independently of ordinary perceptions. **1897** 'Mark Twain' *Following Equator* 317 Here was a clear case of mental telegraphy; of mind-transference.

mind, *v.* Add: **5. a.** Esp. in colloq. phr. *don't mind me*: take no notice of me; do not worry about me; do as you please. Often ironical.

1867 Trollope *Phineas Finn* (1869) I. viii. 69 'It's the meanest trade going... I don't know whether you are in Parliament, Mr. Finn.' 'Yes, I am; but do not mind me.' **1901** G. B. Shaw *Devil's Disciple* II. 39, I never care much for my tea. Please dont mind me. **1911** —— *Doctor's Dilemma* III. 70 B.B. Well, what is our friend Dubedat? A vicious and ignorant young man with a talent for drawing. *Louis.* Thank you. Dont mind me. **1926** M. A. von Arnim *Introd. to Sally* ii. 16 You go a'ead, sir, when she come back, and don't mind me. **1967** S. Knight *Window on Shanghai* III. xx. 86 Ah, how philosophic I wax (and wane?) Don't mind me. **1973** L. Koenig *Little Girl* (1974) iii. 34 'Don't mind me,' she said, dragging the chair back..and filling its place with the table... 'But the table belongs here.'

c. *mind you* (later examples).

1959 *Listener* 22 Jan. 154/2 The Japanese—who were scattered over a very large part of Asia at that time, mind you—would have all fought it out to the death. **1969** *Ibid.* 3 Apr. 467/2 And so to the book's title. It's not, mind you, 'why we are in Vietnam', but a question.

8. a. *do you mind?*: also used to mean 'do you mind not doing that?', i.e. 'please do not do that'.

1961 J. I. M. Stewart *Man who won Pools* xvii. 167 He was trying to put a hand on Phil's arm. '*Do* you mind?' Phil had never got it out more arrogant. Moore fell back.

d. *I don't mind if I do*: a colloq. phr. of acceptance or agreement, used esp. in accepting the offer of a drink.

c **1847** J. S. Coyne in M. R. Booth *Eng. Plays of 19th Cent.* (1973) IV. 193 You *are* a regular brick, and I don't mind if I do take some of your pickles. **1849** C. Brontë *Shirley* I. vii [*sc.* viii]. 184 'Take another glass,' urged Moore. Mr. Sykes didn't mind if he did. **1870** D. J. Kirwan *Palace & Hovel* v. 65 Tell ye me 'istry, is it? Vell. I don't mind if I do. *Ibid.* 69 'I'll give you a drink, me oul wiper.'..'Well, Billy, I don't mind if I do.' **1926** C. Beaton *Diary* in *Wandering Yrs.* (1961) 143 Everyone 'talked common'... 'I don't mind if I do, oo-er!' **1932** A. Christie *Peril at End House* viii. 101 'Come and have a drink,' I said... 'I don't mind if I do.' **1946** T. Kavanagh *Tommy Handley in Holidayland* 5 'Bitter, sir?' interrupted the Colonel [Chinstrap]..., 'I don't mind if I do.' **1967** J. Porter *Dover & Unkindest Cut* xi. 122 'Another cup of tea, Mr Dover?' 'I don't mind if I do,' said Dover, passing his cup.

10. a. *to mind one's step*: see *STEP *sb.* 6.

d. *to mind out*: to look out, be careful; freq. *imp. colloq.* and *dial.*

1886 R. Holland *Gloss. County of Chester* 227 *Mind out*,..to be on one's guard. **1890** *Dialect Notes* I. 65 Mind out what you are doing. **1892** *Ibid.* 233 That is the word with the bark on it; you better mind out. **1894** W. Raymond *Love & Quiet Life* xii. 136 If I don't min' out, woone o' these days..he'l vall off. **1938** J. Stuart *Beyond Dark Hills* iii. 59 John's got a bad boy. He'll go to the pen if he don't mind out—that boy will. **1946** *Amer. Speech* XXI. 56 English children whizzing around on bicycles..will warn each other to keep out of the way by shouting 'Mind out!'.

11. *spec.* To look after (a baby or child), esp. in the absence of its parents; to look after (a shop or store) (also *transf.*).

1839 G. C. Lewis *Gloss. Herefordshire* 67, I ha left Bill at home to mind the children. **1876** C. C. Robinson *Gloss. Mid-Yorks.* 84/1 Minding the bairns and the house. **1899** Kipling *Stalky & Co.* 63 Arrah, Patsy, mind the baby; just ye mind the child awhile! **1902** *Dialect Notes* II. 239 Mind the baby while I'm gone. **1953** A. Upfield *Murder must Wait* x. 93 He accused me of neglecting the baby and said he'd..let his..secretary mind it. **1957** J. Blish *Fallen Star* vi. 75 The cabin door opened and the Commodore came out... 'Who's minding the store?' I asked him. 'Hanchett. We're on autopilot and he's watching the instruments.' **1963** J. H. Harris *Weird World Wes Beattie* (1964) iii. 32 If you will just mind the shop, I'll be on my way. **1970** G. F. Newman *Sir, You Bastard* ii. 70 The CID room was quiet. DC Jones was minding the shop. **1971** *Where* Nov. 343/2 A certificate is issued specifying the number of children and the hours and days when they can be minded. **1973** P. O'Donnell *Silver Mistress* vi. 109 'Who's running your section at the moment?' 'I am. With limited authority... I'm just minding the shop.'

12. *intr.* To matter, be important. (Not in standard use.)

1915 F. H. Lawrence *Let.* 16 Mar. in T. E. Lawrence *Home Lett.* (1954) 680, I had a room without a door or a window..but that did not mind. *Ibid.* 682 Bullets really don't mind much. **1961** *Listener* 5 Oct. 500/1 (*child's composition*) You migth see a earwig gust coming out of [an apple] but that dosenet mind gust pick it of.

minded, *ppl. a.* Add: **4. c.** With prefixed sb. forming adjs. with the sense 'interested in or enthusiastic about (the thing specified)'.

1928 [see *air-minded* adj. s.v. *AIR *sb.*¹ B. III. 1]. **1932** *Daily Express* 27 June 11/3 Get travel-minded. **1933** *Times Lit. Suppl.* 30 Mar. 205/3 Today we are an inland-dwelling folk, car-minded and mechanical. **1949** Helicopter-minded [see *'COPTER, COPTER]. **1956** *Planning* 23 July 153 PEP is intensely research-minded, and would rather not reach any conclusion than plump for one which cannot survive searching criticism. **1966** C. Mackenzie *Paper Lives* viii. 109 You *must* get computer-minded and appreciate that with the development of automation more and more people will be out of employment. **1972** *Guardian* 18 Aug. 9/2 There is also a bowling green for the sports-minded customer.

Mindel (mi·ndĕl). *Geol.* The name of a tributary of the Danube in Bavaria, W. Germany, adopted by A. Penck (in Penck & Brückner *Die Alpen im Eiszeitalter* (1909) I. I. 110) and used *attrib.* to designate the second (antepenultimate) Pleistocene glaciation in the Alps, and in conjunction with *Riss to designate the following interglacial period. Also *absol.*

1910 [see *GÜNZ]. **1939** [see *INTERGLACIAL *a.*]. **1957** G. E. Hutchinson *Treat. Limnol.* I. i. 6 Relatively early in the Pleistocene, probably at the time of the Mindel glaciation, the Caspian was occupied by a high-level lake. **1972** *Sci. Amer.* Mar. 60/1 Deposits laid down in the subsequent interglacial period, the Mindel-Riss, contain a few fossil remains of a true cave bear. A bear skull preserved in Mindel-Riss sediments at Swanscombe in England shows the domed forehead that is characteristic of the species.

minder. Add: **3. b.** *spec.* (*a*) a machine-minder; (*b*) one who minds (see *MIND *v.* 11) a baby or child; a baby-sitter.

(*a*) **1835** [in Dict.]. **1888** *Encycl. Brit.* XXIII. 709/2 If he is a machinist, he may superintend or be a 'minder' or he may be a layer-on or a taker-off of the sheets. **1938** *Amer. Speech* XIII. 271 The man who runs the presses is a *pressman* in America, but a *machinist* or a *machine minder* or simply a *minder* in England. **1970** *Financial Times* 13 Apr. 4/5 Among minders, the proportion on litho presses is forecast to go up from 20 per cent. in 1967 to 28 per cent. in 1972.

(*b*) **1863** [see *baby-minder* s.v. *BABY *sb.* B. 1 b]. **1874** [in Dict.]. **1941** [see *child-minder* s.v. *CHILD *sb.* 22]. **1957** *Times* 25 Nov. 11/4 Of course we will not worry about the children, we assure the kind minder as we blow farewell kisses. **1971** *Daily Tel.* 4 Nov. 9/3 Mothers, forced to work to make ends meet, ship their toddlers to unregistered minders, paid to keep an eye on them in 'cramped rooms, with no toys or stimulus'.

c. *slang.* A person employed to protect a criminal; a thief's assistant.

1924 E. Wallace *Room 13* xii. 61 Glancing down into the street, he distinguished one of the 'minders' his father had put there for his protection. **1928** —— *Flying Squad* xvi. 144 Whizzers..had 'minders', whose business it was to kick and disable the poor souls who found themselves robbed and attempted to recover their own. **1960** *Observer* 25 Dec. 7/6 A climbing team..was most often three-handed. Driver, minder, and climber... The minder stays at the foot of the pipe or ladder. His job is to safeguard the climber's rear and collect any gear he may sling down. **1968** C. Drummond *Death & Leaping Ladies* vi. 160 At school he was a juvenile fence and money-lender, with a couple of tough, simple-minded older boys as his 'minders'. **1973** E. McGirr *Bardel's Murder* ii. 35 Comes of a whole family of wrong 'uns... A high class 'minder' around the big gambling set.

mindlessly, *adv.* (Further examples.)

1963 *Times* 11 June 3/7 Dumitrescu worked away mindlessly like a machine at the head of the group. *a* **1966** M. Allingham *Cargo of Eagles* (1968) xv. 170 Temper and hysteria had fought mindlessly for the upper hand. **1972** *Publishers' Weekly* 21 Aug. 71/3 The result is mindlessly entertaining.

mine, *sb.* Add: **3.** In naval warfare. such a receptacle used on the high seas as well as at the entrance to a harbour. Also, a receptacle containing explosive placed in or on the ground as a weapon of war. Freq. with qualifying word indicating the kind or use: for *acoustic, land, magnetic,* etc., *mine*, see under the first elements.

1880 *Encycl. Brit.* XI. 309/2 The arrangements for exploding submarine mines or fixed torpedoes, where hostile vessels are passing close to them, are somewhat of the same character. **1889** J. T. Bucknill *Submarine Mines* xx. 219 Just as the mines themselves form a grand obstruction to the passage of large vessels. so smaller obstructions can. impede. the passage of small craft whose aims may be to attack the mines. **1890** [see *land-mine* s.v. *LAND *sb.*¹ 12]. **1915** W. Owen *Let.* 8 Jan. (1967)

313, I am not exposed to chances of mine or torpedo. **1942** *War Illustr.* 16 Oct. 238/1 A man can walk over a mine easily enough, but a tank or lorry'll send them up in a proper earthquake. **1968** Visct. Montgomery *Hist. Warfare* xviii. 416 In 1855, in the Baltic, Russia made the first serious use of floating mines.

5. a. *mine-dust, -head, -mouth, -slime, -stamp* (Stamp *sb.*³ 9), *-worker*; (sense 3) *mine-crater, -warfare.*

1917 'Contact' *Airman's Outings* 269 Two huge mine-craters sentinel it, left and right. **1886** J. Barrowman *Gloss. Scottish Mining Terms* 45 *Mine dust*, the riddlings of calcined ironstone. **1923** *Daily Mail* 16 Mar. 9 The present intention is to send labour squads..to successive mine-heads. **1886** J. Barrowman *Gloss. Scottish Mining Terms* 45 *Mine mouth*, the point where a mine leaves the surface of the ground. **1921** *Chambers's Jrnl.* Apr. 262/1 A savage land of rocks and lakes and mine-slime and active and derelict mine-workings. **1926** J. Masefield *Odtaa* xiv. 233 It is a kind of a mine-stamp, or engine of some sort. **1953** D. Lessing *Five* III. 128 The mine-stamps thudded day and night, coming loud or soft, according to the direction of the wind. **1910** *Blackw. Mag.* June 895/2 The immense development of mine warfare. **1957** *Encycl. Brit.* XV. 531/2 A self-contained continental power..has little to lose and everything to gain by unrestricted mine warfare. **1974** *Saturday* (Charleston, S. Carolina) 20 Apr. 3-A/2 Like wood, the fiberglass hull is non-magnetic and is considered good for mine warfare. **1901** *Edin. Rev.* Apr. 496 An agreement..that the mine-workers should receive an advance of 10 per cent. in their rates of wages. **1928** *Britain's Industr. Future* (Liberal Industr. Inquiry) IV. 266 The pensioning of older mine-workers. **1975** *Times* 18 Feb. 2/1 Mineworkers' leaders in the traditionally militant areas of Scotland and Yorkshire yesterday voted to back the £140m pay deal negotiated by their national union leaders with the National Coal Board last week.

6. mine-boy *S. Afr.*, a native African who works in the mines; **mine-car** (see quots.); **mine-detector,** an instrument which, by its reaction to metal, indicates the presence of mines; **mine-dump** *S. Afr.*, a pile of refuse material on the surface of a mine, esp. a gold-mine (cf. Dump *sb.*⁴ 1); **mine-field** (later examples); also, an area of land in which mines have been laid; also *fig.*; **mine-hunt** *v. intr.*, to hunt or sweep for mines; **mine-hunter** = *mine-sweeper*; **mine-hunting** = *mine-sweeping*; **mine-layer,** a ship or aeroplane equipped to lay mines; **mine-laying,** the operation of laying explosive mines; also *attrib.* or as *adj.*; **mine-sinker,** a device for keeping a mine submerged; **mine-sowing** = *mine-laying*; **mine-sweeper** (later examples); **mine-sweeping** (later examples); **mine-thrower** [tr. G. *minenwerfer*], a trench-mortar; **mine-tipple** *N. Amer.* [Tipple *sb.*³], a tip (Tip *sb.*⁵ 3) at a mine.

1945 P. Abrahams *Song of City* 73 The tom-tom beat of the Maraba..danced away the seething bitterness that is attendant with repression... On the morrow the house-boy would be a good..houseboy!.. And the mine-boy. **1953** P. Lanham *Blanket Boy's Moon* I. v. 51 One Monare, a mine-boy, friend of Ntoane here, was watching. **1954** P. Abrahams *Tell Freedom* III. iii. 108 Fights often flare up between mine-boys and house-boys. **1956** F. S. Atkinson in D. L. Linton *Sheffield* 270 Diesel locomotives with large mine-cars are being used extensively underground, particularly in Yorkshire. **1967** *Gloss. Mining Terms* (*B.S.I.*) x. 11 *Mine car*, a large tub (usually spring mounted and over 60 ft.³ capacity) used primarily for mineral haulage underground. **1943** *Hutchinson's Pict. Hist. War* 12 May—3 Aug. 125 (*caption*) The soldier in the foreground is sweeping the ground with a mine detector and wearing earphones as he listens intently for the buzz indicating the presence of a mine. **1945** *Finito! Po Valley Campaign* 41 The glass-topped Topf mines that fooled the mine-detectors. **1955** E. S. Gardner *Case of Glamorous Ghost* (1960) x. 121, I used what is known as a mine detector..an electronic device so designed that when it is moved over the surface of the ground it will give a peculiar squeal when it is moved over a metallic object. **1926** S. G. Millin *S. Africans* iii. 77 They are the mine-dumps, the refuse of stamp-mill and cyanide tank. **1956** V. Jenkins *Lions Rampant* v. 76 Dominating the landscape..are the huge yellow-white mine dumps which mark the gold mines. **1971** *Gloss. Soil Sci. Terms* (Soil Sci. Soc. Amer.) 11/1 *Mine dumps*, areas covered with overburden and other waste materials from ore and coal mines, quarries, and smelters, and usually with little or no vegetative cover. **1889** J. T. Bucknill *Submarine Mines* xiv. 168 Secrecy is essential... Any artifice which ingenuity can suggest should be undertaken... False reports concerning the mine fields should be spread. **1917** A. Conan Doyle *His Last Bow* viii. 298 It would brighten my declining years to see a German cruiser navigating the Solent according to the minefield plans which I have furnished. **1938** *Encycl. Brit. Bk. of Year* 432/2 Other anti-tank agencies being given consideration are mine-fields and 'catch trenches' (*i.e.* tank traps). **1942** *Times* 8 June 4/6 The 'Cauldron', an area..between..the upper and lower gaps which Rommel forced through our minefield. **1957** *Encycl. Brit.* XV. 532/2 By the end of 1914 Britain had established extensive mine-fields in the English Channel. **1963** *Times* 16 Feb. 4/6 This bold venture into the mine-fields of satire is one to be supported and preserved. **1968** Visct. Montgomery *Hist. Warfare* xxi. 512 Wavell once compared the tactics of war in the desert to war at sea; minefields were laid in the desert very much as they were in the sea. **1973** *Times* 30 Nov. 19/1 He is..operating..in a political minefield, requiring more circumspection than he has always managed. **1915** Kipling *Fringes of Fleet* 74 He's mine-hunting, I expect, just now. **1964** *Navy News* Nov. 5/5 After refitting and converting..H.M.S. *Iveston* commissioned on October 16th as a Coastal Minehunter. **1974** *Times* 18 Apr. 1/4 It is the diver's job to locate the unexploded bomb on the canal floor after it has been picked up by the mine-hunter's sonar equipment. **1964** *Navy News* Nov. 5/5 She will join the First Minehunting Squadron in December and will be based at Port Edgar in the Firth of Forth. **1974** *Observer* 17 Mar. 8/4 Mine-hunting ships—the Navy's new name for minesweepers. **1974** *Saturday* (Charleston, S. Carolina) 20 Apr. 3-A/2 The mine-hunting vessels include two with wood hulls and one with an experimental fiberglass hull... All three mine hunting vessels have sonar to detect mines and other objects like bombs. **1909** *Q. Rev.* Oct. 575 Six second-class cruisers of the Naval Defence Act have been converted into mine-layers. **1923** W. S. Churchill *World Crisis 1915* 260 The mine-layer *Nousret* had on March 18 thirty-six mines ready for laying. **1939** *War Illustr.* 16 Dec. 435 Raid by British warplanes..on the German seaplane base at Borkum. They had been looking for the mine-layers—and they had found them. **1944** *Aeronautics* July 32/3 Suitable for duties as a torpedo bomber, a dive bomber, or a mine layer, the Fairey Barracuda was first reported in action on the 3rd April 1944. **1911** *Q. Rev.* Oct. 466 The money..is now expended upon a large and increasing fleet of mine-laying and mine-sweeping ships. **1921** *Flight* 19 May 348/2 The American naval authorities are reported to have been carrying out in Chesapeake Bay experiments in mine-laying from the air. **1928** C. F. S. Gamble *Story N. Sea Air Station* xviii. 310 The U.C. boats were of a type designed both for mine-laying and torpedo work. **1939** *Flight* 30 Nov. 429/2 The mine-laying machines cannot behave as the enemy reconnaissance machines have been behaving of late, namely arriving at a great height and hurrying away at the first hint of defence activity. **1955** *Times* 24 May 8/4 The risk of more serious incidents, such as Egyptian minelaying and Israel retaliation, should also lessen. **1958** P. Kemp *No Colours or Crest* iii. 37 The fast mine-laying cruiser *Manxman*. **1889** C. Sleeman *Torpedoes* (ed. 2) vi. 119 The only practicable method of testing the efficiency of mine anchors or sinkers is to moor a buoyant mine to an anchor. **1926** *Spectator* 21 Aug. 271/1 He..produced 50,000 minesinkers at a very low cost. **1940** Bartlett & Williams *War of 1939* II. iii. 104 The Germans..had turned to indiscriminate mine-sowing. **1940** *War Illustr.* 5 Jan. 568/3 Particular attention being paid to the favourite areas of the mine-sowing seaplanes in the Thames Estuary. **1914** *Illustr. London News* 22 Aug. 286/3 Mine-sweepers..precede the fleet to sea and clear its path of hostile mines. **1940** Bartlett & Williams *War of 1939* II. iii. 106 Mine damage at this time was not confined to merchant vessels, the British minesweepers *Mastiff* and *Aragonite* being sunk with casualties. *Ibid.* 107 The British Admiralty were quick..to appeal for two hundred additional drifters to act as minesweepers. **1972** *Daily Tel.* 21 Jan. 2/8 More mine-sweepers and helicopters will be used from this summer to improve coastal fishery protection. *Ibid.*, Helicopters..will make spotter sweeps to help the minesweepers to guard the six-mile limit fishing areas against unauthorised vessels. **1915** *Chambers's Jrnl.* May 294/2 The operation of taking up mines is known as 'mine-sweeping'. **1958** *Times* 11 Nov. 9/1 Two coastal mine-sweeping flotillas..are already under N.A.T.O. command. **1973** *Listener* 26 Apr. 535/1 Mine-sweeping operations off Haiphong. **1915** *Illustr. London News* 13 Feb. 204/2 The Germans..had actually provided..themselves with mortars of this description, the so-called *minen-werfer*—mine-throwers. **1923** *Daily Mail* 17 Jan. 7 They captured 7 Frenchmen, 15 local policemen, 3 minethrowers, and a machine gun. **1930** J. Dos Passos *42nd Parallel* I. 111 In the middle of squirrels and minetipples. **1974** *Beautiful Brit. Columbia* Spring 27/2 Coal is still mined there in large quantities. The highway passes underneath a portion of the mine tipple.

mine, *v.* Add: **9.** *U.S.* (See quot. 1937.)

1937 *Amer. Speech* XII. 105 They *mine* the soil; that is, they use up fertility without restoring it. **1972** *New Yorker* 25 Nov. 42/3 When they moved here, the land had been farmed out—'mined', the local phrase was—and the one undiscouraged crop was the wild strawberries.

mineable, *a.* Add mod. examples of the form **minable.**

1971 *Sci. Amer.* May 17/1 There are minable cinnabar deposits in many regions around the world, and man was attracted to its use as early as prehistoric times. **1972** *Nature* 14 Apr. 332/1 And is it likely that economically minable lead ores are equally distributed between North America and the rest of the world?

‖ **minenwerfer** (mī·nənvɛrfəɪ, mī·nənwɔ̄ɪfəɪ). *Mil.* [G., f. *minen* (pl.) mines + *werfer*, f. *werfen* to throw.] A German trench mortar. Cf. *mine-thrower*, *MINNIE².

Also called *minnenwerfer* from an expansion of Minnie², the British soldiers' nickname for this gun.

1915 A. D. Gillespie *Lett. from Flanders* (1916) 80 About Christmas time the Germans fetched up a thing called a *Minen Werfer*, a kind of trench mortar which throws 600 lb. of gun cotton. **1915** [see *mine-thrower*]. **1916** F. M. Ford *Let.* Aug. (1965) 68 The Germans seem to have nothing but minenwerfers. **1917** A. G. Empey *Over Top* 300 *Minnenwerfer*, a high-power trench mortar shell of the Germans, which makes no noise coming through the air... Tommy nicknames them 'Minnies'. **1923** Kipling *Irish Guards in Gt. War* I. 101 He retaliated..along the line of seven-inch minenwerfers.

miner¹. Add: **6. miner's right** (earlier and later examples); also *N.Z.*

1855 in *Occasional Papers Univ. Sydney Austral. Lang. Res. Centre* (1966) No. 9. 15 It shall be lawful for the Governor..to cause documents to be issued each of which shall be called 'The Miner's Right' and shall be granted to any person applying for the same upon payment of a fee of one pound. **1858** in *Ibid.* 16 It is not generally known, ..that..any one interested in the workings at the gold-fields should hold a Miner's Right which can be had at the Treasury. **1863** *Rules & Regulations Otago Gold Fields* 7 Every person residing on a Gold Field and engaged in mining for gold, shall take out a Miner's Right; such Miner's Right to be carried on the person, and produced for inspection when demanded. **1868** V. Pyke *Province of Otago* 41 The only qualification is the possession of a 'Miners' Right'. **1950** *N.Z. Jrnl. Agric.* Feb. 189/2 Exhibits [in an Arrowtown museum, Central Otago, include] ..a miner's right. **1959** Baker *Drum* 127 *Miner's right*, a licence to dig for gold granted to a miner, orig. in the 1850s.

b. *miner's nystagmus.*

1879 [see Nystagmus 2]. **1959** *Chambers's Encycl.* X. 152/2 Owing to the inadequate illumination in mines a large number of miners get a condition known as 'miner's nystagmus'. **1962** H. C. Weston *Sight, Light & Work* (ed. 2) ix. 262 The occupational disease known as *miner's nystagmus* is so named on account of the ocular movements commonly associated with it.

mineragraphy (minĕræ·grăfi). *Min.* [f. Minera(l *sb.* + -graphy.] = *mineralography.*

1924 *Amer. Mineralogist* IX. 177 Mineragraphy, the study of minerals in polished section with the metallographic microscope for the purpose of determining their identity and paragenesis, has generally been carried on under vertical illumination. **1965** G. J. Williams *Econ. Geol. N.Z.* ix. 136/2 (*heading*) Mineragraphy of titanomagnetite.

So **minera·grapher,** one who practises mineragraphy; **mineragra·phic** *a.*

1931 *Amer. Mineralogist* XVI. 209 In the course of some recent mineragraphic work on the manganiferous iron ores of the Cuyana Range..it was necessary to make quick identification of the various oxides of iron and manganese. **1953** *Austral. Jrnl. Chem.* VI. 443 The most comprehensive mineragraphic report..records the presence of more than 30 mineral species. **1964** *Weekly News* (Auckland) 10 June 8/5 The mineragrapher determines the intimacy of the admixture of minerals composing the ore. **1965** G. J. Williams *Econ. Geol. N.Z.* ix. 136/2 Titanium being an embarrassment in any metallurgical process, attention has been given to mineragraphic studies with a view to locating this element.

mineral, *sb.* Add: **4. d.** = Mineral water. (Usu. in *pl.*)

1885 *List of Subscribers* (United Telephone Co.) p. xv, We have had a run upon minerals, and are nearly out. *Ibid.*, We are out of minerals. Kindly send us..one gross of seltzer, one gross of soda. **1922** Joyce *Ulysses* 597 They might hit upon some drinkables in the shape of a milk and soda or a mineral. **1927** *Glasgow Herald* 15 Apr. 11 There will be..supper with ale and minerals at Osborne's Hotel.

6. mineral dressing, treatment of ore so as to remove gangue and concentrate the valuable constituents; so *mineral dresser*; **mineral rod** (later examples).

1895 *Funk's Stand. Dict.*, *Mineral dresser*, a machine for trimming or dressing mineralogical specimens. **1957** *Sci. News* XLVI. 35 This upgrading [of ores] is the work done by the mineral dresser. **1939** A. M. Gaudin *Princ. Mineral Dressing* i. 1 Mineral dressing is commonly regarded as the processing of raw minerals to yield marketable products and waste by means that do not destroy the physical and chemical identity of the minerals. **1957** *Sci. News* XLVI. 37 The methods of mineral dressing most in use before World War I were limited to gravitation in pulsing or streaming currents of water.., the use of magnets on ferro-magnetic ores, and such hydro-metallic processes as the cyanidation of gold and the leaching of copper. **1974** *Encycl. Brit. Macropædia* XI. 1063/2 Modern processes in mineral dressing have become increasingly directed toward using fundamental principles governed by laws of physics, chemistry, and electricity. **1849** T. L. Clingman in C. Lanman *Lett. from Alleghany Mts.* 187 Travelling about the country under the guidance of *mineral rods* or dreams in search of mines. **1902** A. D. McFaul *Ike Glidden* xxi. 184 They've sent for Squire Blunt to come up here in the morning, with a mineral rod, to assist them.

mineral, *a.* Add: **3. b.** Esp. in *mineral spring.* Also *attrib.*

1783 S. Tenney *Let.* 1 Sept. in *Mem. Amer. Acad. Arts & Sci.* (1793) II. 43, I mentioned some mineral springs in the vicinity of this place. **1797** [in Dict.]. **1843** W. W. Mather et al. *Geol. N.Y.* IV. 308 In this district the only mineral springs of interest are the salines, the sulphur springs, [etc.]. **1911** *Daily Colonist* (Victoria, B.C.) 22 Apr. 1/4 Craig..was probably fatally shot by a burglar late tonight in his room at a mineral springs resort. **1972** *Gloss. Geol.* (Amer. Geol. Inst.) 456/1 *Mineral spring*, a spring whose water contains enough mineral matter to give it a definite taste, in comparison to ordinary drinking water, esp. if the taste is unpleasant or if the water is regarded as having therapeutic value.

5. mineral soil, any soil in which the organic constituents are small in proportion to the inorganic ones; **mineral teeth,** artificial teeth.

1924 F. E. Bear *Soil Managem.* iv. 26 The tendency of mineral soils to be similar in chemical composition, irrespective of their source of origin, is very nicely shown in the following table. **1960** Teuscher & Adler *Soil & its Fertility* i. 10 The fundamental distinction should be made between mineral soils and organic soils, depending upon the proportion of organic matter which they contain... In general, a soil is designated as organic when it contains 20 per cent or more of organic matter. **1851** C. Cist *Sk. Cincinnati in 1851* 220 Mineral Teeth. One factory. **1885** *List of Subscribers, Classified* (United Telephone Co.) (ed. 6) 230 Manufacturers of Mineral Teeth and every Dental Requisite.

b. mineral brown (see quot. 1930); **mineral grey** (earlier and later examples); **mineral**

violet = *manganese violet*; **mineral white** (earlier example).

1869 T. W. Salter *Field's Chromatogr.* (new ed.) xvii. 342 Under the names of Euchrome and Mineral Brown, they [*sc.* Cappah browns] have been introduced into commerce for civil and marine painting. **1930** Maerz & Paul *Dict. Color* 167/1 *Mineral brown*, this name, or its synonym, *Metallic Brown*, is occasionally found given to specific colors in paints. It has..long been used to refer to any native earth colored by iron oxide, etc. **1869** T. W. Salter *Field's Chromatogr.* (new ed.) xix. 375 Mineral gray..is obtainable from the lapis lazuli, after the blue and ash have been worked out. **1958** M. L. Wolf *Dict. Painting* 179 A substance known as gangue (vein-stone) is often offered as mineral gray, but it is not a successful substitute. **1913** B. Brown *Painter's Palette* iv. 21 Taking these paints..and making them..account for themselves in the matter of value, we find that their values fall on the value-scale thus:..O.—Burnt Sienna. Rose Madder. Mineral Violet. **1934** H. Hiler *Notes Technique Painting* ii. 132 Mineral violet (manganous metaphosphate). **1958** M. L. Wolf *Dict. Painting* 170 Manganese pigments... The violets are particularly varied, some of the shades including mineral-, permanent-, and Nuernberger violet. **1875** E. Spon *Workshop Receipts* 93/2 White Pigments ..Mineral white.—Precipitated carbonate of lead.

mineralization. Add: **1.** (Further examples.)

1898 *Q. Jrnl. Geol. Soc.* LIV. 81 The silicification, or more generally the mineralization, of these conglomerates is, in my opinion, the result of secondary processes of infiltration and crystallization. **1918** *Mining Mag.* XIX. 196/2 The supposed essential similarities between the conglomerates and ordinary lode deposits and quartz veins seem rather to have been based on ignorance of the various factors identical to formation of fissure deposits than on clear conceptions as to how the conglomerates received their mineralization. **1935** *Amer. Jrnl. Sci.* XXX. 115 'Mineralization' was shown to be a process of chemical action in which the mineralizer, HCl, was regenerated after performing transport and combination of the Fe_2O_3 with the MgO. **1971** *Nature* 24 Dec. 460/2 We intend to use this method to monitor the growth and dissolution of crystals in gels, especially in connexion with mineralization and demineralization in biological systems.

2. (Further examples.)

1970 *Daily Tel.* 24 Oct. 13/3 Two of the three drills encountered insignificant mineralisation. **1971** *Nature* 6 Aug. 393/2 In regions of mercury mineralization, however, up to 100 μg/kg has been reported.

mineralize, *v.* **2.** (Further examples.)

1911 *Encycl. Brit.* XVIII. 255/1 Shales, sandstones and igneous rocks may be silicified and mineralized under suitable conditions. **1973** *Nature* 7 Dec. 347/1 Experiments introducing modern lichens into the normal preparation of microfaunal samples using hydrogen peroxide failed to mineralise the plant structures.

mineralized, *ppl. a.* Add: **2.** Also of methylated spirit (see quot. 1906), and of carbon electrodes. (Further examples.)

1906 *Act 6 Edw. VII* c.20 § 4 (1) The expression 'mineralized methylated spirits' means..spirits which, in addition to being methylated..have mixed with or dissolved in them..mineral naphtha. **1915** K. Tornberg tr. *Rasch's Electric Arc Phenomena* vii. 149 The radiation from the luminous arc flame of mineralized carbons, thus, is not always sufficient to compensate for any considerable deficiency occasioned by artificial lengthening of the arc. **1967** *Martindale's Extra Pharmacopoeia* (ed. 25) 86/2 Mineralised methylated spirits is the only variety that may be sold by retail in Great Britain for general use.

mineralizer. Add: **1. b.** *Petrol.* A volatile substance dissolved in a magma which aids the formation of minerals by altering the properties of the magma but is not necessarily present in the final mineral; also, a substance which promotes the artificial synthesis of a mineral.

1909 A. Harker *Nat. Hist. Igneous Rocks* xii. 290 Mineralisers are, before all, powerful fluxes. One of their most important offices is that of reducing the viscosity of a magma. **1921** *Jrnl. Geol.* XXIX. 205 The presence of mineralizers has lowered the temperature of freezing below the inversion range of these minerals. **1943** R. D. George *Minerals & Rocks* xx. 434 When a rock is poured out of a volcano the mineralizers have a chance to escape before they have done their work. **1958** J. H. de Boer in Everett & Stone *Struct. & Properties Porous Materials* 290 We had a mixture of alumina and silica—I don't remember if it was in the right proportion—but heating at a relatively low temperature..mullite was already formed with lithium as a mineralizer. **1974** L. N. Kogarko in H. Spenser *Alkaline Rocks* vi. iv. 480/2 Mineralizers may not enter into the composition of minerals but exert catalytic action on the process of their growth.

mineralocorticoid (mi:nĕrălokǭ·ɹtikoid). *Biochem.* [f. Mineral *sb.* + -o + *corticoid.] Any of the steroid hormones produced in the adrenal cortex which are esp. concerned with maintaining the salt balance in the body; any analogous synthetic compound.

1950 H. Selye *Physiol. & Path. of Exposure to Stress* 646 Mineralo-corticoids predispose to the formation of lung edema. **1964** L. Martin *Clin. Endocrinol.* (ed. 4) v. 165 The mineralocorticoids promote the retention of salt and water and the excretion of potassium by the renal tubules. **1965** Lee & Knowles *Animal Hormones* iv. 70 The division of adrenal cortical hormones into gluco-corticoids and mineralocorticoids is a convenient one. **1970** Passmore & Robson *Compan. Med. Stud.* II. vi. 12/2 Nowadays, the mineralocorticoid of choice in the treatment of defective secretion of endogenous aldosterone is the synthetic agent 9α-fluorocortisol.

mineralogic, *a.* For † *Obs.* read: Now chiefly *U.S.* (Add later examples.)

1952 G. Sarton *Hist. Sci.* I. xxi. 560 A mineralogic analysis. **1972** *Science* 3 Nov. 497/1 Approximately one-third of the book is devoted to mineralogic and petrologic attributes of sandstones.

mineralography (minĕrălǫ·grăfi). *Min.* [f. Mineral *sb.* + -o + -graphy.] The study of the physical and chemical microstructure of minerals, *spec.* of polished sections using the reflecting microscope.

1916 J. Murdoch *Microsc. Determination Opaque Minerals* p. iii, This new method [*sc.* using the reflecting microscope]..opens an entire new field of geologic science that promises to do for the ores what petrography has done for the rocks; for this the name of 'Mineralography' is proposed. **1921** H. L. Alling in *Jrnl. Geol.* XXIX. 194 The rock-forming minerals can be studied..as the end products of crystallization of melts—geophysical chemistry, or as here proposed—mineralography. **1937** *Amer. Mineralogist* XXII. 492 Attention is restricted to instrumental and manipulative procedure with the opaque minerals; no consideration is given..to those more interpretative and philosophical phases of mineralography, such as textures.. deformation, alteration etc. **1942** *Chem. Abstr.* XXXVI. 1267 (*heading*) The imprint method in mineralography.

Hence **mineralogra·phic** *a.*, of or using mineralography.

1916 J. Murdoch *Microsc. Determination Opaque Minerals* p. v, Mineralographic methods show that..the sulphide minerals tend to be microscopically mingled and intergrown in most intimate fashion. **1937** *Amer. Mineralogist* XXII. 492 Many requests..have come for description or discussion of mineralographic technique as employed at Harvard. **1971** *Daily Tel.* 29 Apr. 25 (Advt.), Mineralographic and related analyses on ores and metallurgical products.

mineralogy. Add: **2.** (A description of) the mineralogical features of a region or a specimen.

1798 R. Jameson (*title*) An outline of the mineralogy of the Shetland Islands, and of the Island of Arran. **1878** *Mineral. Mag.* II. 106 (*heading*) The geognosy and mineralogy of Scotland. **1933** P. G. H. Boswell *On Mineral. Sedimentary Rocks* p. v, Having devoted much time to a digest of the literature of the mineralogy of sediments, I published a review of the work. **1952** *Jrnl. Geol.* LX. 107 These processes..have produced significant changes in the mineralogy and textural relationships of the plutonic acidic rocks. **1971** I. G. Gass et al. *Understanding Earth* i. 27/2 Chemical weathering is very effective in producing a mineralogy which differs greatly from that of the igneous and metamorphic rocks.

mineraloid (mi·nĕrăloid). *Min.* [a. G. *mineraloid* (J. Niedźwiedzki 1909, in *Centralbl. f. Min., Geol. u. Paläont.* 662), f. *mineral* Mineral *sb.*: see -oid.] A substance that might be regarded as a mineral but is amorphous rather than crystalline.

1913 A. F. Rogers in *Proc. Amer. Philos. Soc.* LII. 608 The question of names for colloidal or amorphous minerals arises... Niedzwiedzki has proposed the term *mineraloid* for the natural amorphous substances. **1917** —— in *Jrnl. Geol.* XXV. 526 Lechateliérite is a glass and may be considered along with other natural glasses as a mineraloid. *Ibid.* 540 The hydrocarbons may be included under Niedzwiedski's term *mineraloid*. As Niedzwiedski used this term for all naturally occurring amorphous substances, this changes somewhat the original definition of mineraloid. Such substances as opal, cliachite, limonite, collophane, halloysite, etc., are definite enough to be called minerals even though they are amorphous. The term 'mineraloid' seems appropriate for the less definite mineral-like substances. **1941** C. S. Hurlbut *Dana's Man. Min.* (ed. 15) iii. 94 There are a number of mineral substances whose analyses do not yield definite chemical formulas and further show no signs of crystallinity. They have been called gel minerals or mineraloids. **1944** A. Holmes *Princ. Physical Geol.* iv. 37 Only a few non-crystalline substances are regarded, by common usage, as minerals, and these are generally distinguished as mineraloids. **1951** [see *jordisite]. **1972** G. S. Fay *Rockhound's Manual* iii. 37 Mineraloids are substances which look like and are often grouped with minerals, but mineraloids have an amorphous, or noncrystalline, structure as is evident when viewed under high-power microscopes. Opal is a mineraloid.

minestra (mine·stră). [It.] An Italian vegetable soup; = next.

1750 [see Macaroni 1]. **1871** *Monthly Packet* Sept. 362 One of the waiters here came up to me, announcing that 'la minestra' was ready. I had ordered no soup. **1907** J. Webster *Jerry Junior* xii. 184 They supped on *minestra* and *fritto misto*. **1935** M. Morphy *Recipes of all Nations* 127 This is the traditional 'minestra' served at Easter and Christmas in the province of Parma. **1941** 'M. Home' *Place of Little Birds* vi. 89 Donati..himself saw to the making of the *minestra*. **1948** E. Pound *Pisan Cantos* (1949) lxxviii. 65 'No, there is nothing to pay for that bread.' 'Nor for the minestra.' **1972** P. Eve *European's Cook Bk.* 130 *Minestra and Minestrone.* The basic difference between these two soups is that the former is all vegetable whereas the Minestrone is made with meat stock.

minestrone (ministrǝ̄u·ni). [It.] A thick soup containing vegetables and rice or pasta.

1891 E. Duret *Pract. Househ. Cookery* i. 63 The minestrone is a real Italian national soup. It is composed of a mixture of vegetables and rice. **1903** [see *fritto misto]. **1922** *Blackw. Mag.* Feb. 143/2 There were slices of melon, olives and tunny and a minestrone. **1936** Auden & Isherwood *Ascent of F6* II. iii. 95 Just as you like. What about soup? Minestrone, I think? **1959** *Sunday Times* 29 Mar. 21/6 There are..dozens of versions of Minestrone, which is a really solid soup thick with vegetables and cheese and rice or pasta and intended, with bread and wine, to constitute the entire midday meal of hungry working people. **1972** N. Froud *World Bk. Soups* 26 A really good Italian minestrone should have a foundation of salt pork or gammon and white haricot beans.

minette¹. Add: **2.** A low-grade oolitic iron ore found mainly in Luxemburg and Lorraine. Freq. *attrib.*

1902 *Encycl. Brit.* XXIX 584/2 Of these [ores] the chief is the lean but very cheap 'Minette' ore of the enormous deposits of Luxemburg and Lorraine. **1919** J. M. Keynes *Econ. Consequences Peace* iv. 90 The German Delegation made strong efforts to secure the inclusion of a provision by which coal and coke to be furnished by them to France should be given in exchange for *minette* from Lorraine. **1940** *Economist* 13 July 40/1 A new process of steelmaking ..was..applied by Continental steelmakers to the phosphoric pig-iron produced from their minette ores. **1966** P. T. Flawn *Mineral Resources* vi. 139 The famous minette iron ores of Lorraine were exploited first on the outcrop in Germany and then followed down dip into France.

Ming (miŋ), *sb.*² [Chinese, lit. 'bright, clear'.] The name of a dynasty which ruled in China from 1368 to 1644; a ruler belonging to this dynasty. Also *attrib.*

[**1671** J. Ogilby tr. *Montanus' Atlas Chinensis* 485 Thus ended a Prince, perhaps the greatest in the World..: Together with him, the Name of the Empire, viz. Taiming, that is, Of great Brightness, after it had continu'd two hundred sixty six Years..was utterly extinguish'd. **1676** *China & France* 19 The antient Kings of China were of the Family of Min, which signifies Light.] **1795** W. Winterbotham *Hist., Geogr. & Philos. View Chinese Empire* i. 5 The whole of their emperors, abstracting from those who are said to have reigned in the fabulous times, are comprehended in twenty-two dynasties... 21. Ming,..1368. 22. Tsing, 1645. **1836** J. F. Davis *Chinese* I. v. 185 The first Emperor of the Ming dynasty, which expelled the Mongols in 1366, had been servant to a monastery of bonzes, or priests of Budh. **1854** *Hist. China* v. 56 Houng-nan, or the first Ming, died in 1398, after a reign of thirty years. **1940** E. Pound *Cantos* lx. 91 And Japan kept peace even all through the great Ming rebellion. **1967** D. Bloodworth *Chinese Looking Glass* vii. 68 When in the 17th century the eunuchs betrayed their Ming ruler, and the Forbidden City of Peking fell to rebels, the Emperor wrote a valedictory message on his yellow robe and strangled himself with his silken girdle.

b. Used *attrib.* and *absol.* of the porcelain of the Ming period (of which the finest examples are extremely rare and valuable).

1892 J. D. Ball *Things Chinese* 309 A street hawker may be seen..ladling iced syrup out of Ming bowls, and there is hardly a butcher's shop without a large Ming jar. **1898** W. G. Gulland *Chinese Porc.* I. 3 We must also remember that a large amount of Ming porcelain must have been destroyed at the end of the Ming dynasty. **1907** E. Wharton *Fruit of Tree* II. xiii. 209, I want to show you a set of Ming I picked up the other day. **1936** R. Lehmann *Weather in Streets* III. iv. 318 Just because she hasn't even the guts to put her own stockings on—she's to be treated like a Ming vase. **1939** T. S. Eliot *Old Possum's Pract. Cats* 24 Down from the library came a loud *ping* From a vase which was commonly said to be Ming. **1970** *Oxf. Compan. Art* 234/2 The standard Ming porcelain body was refined and white, capable of thin potting when necessary, and covered with a fairly even, clear glaze.

c. Used *attrib.* to denote colours characteristic of Ming porcelain, as *Ming blue, green, yellow.*

1926 *Textile Mercury* 19 June 566/3 The Textile Colour Card Association's..new greens are Locarno, elfin, ming,.. and Paradise greens. **1931** B. Rackham in R. L. Hobson et al. *Chinese Ceramics* ii. 141 The blue..is quite different from the many tones of Ming blue,..and whatever may be the attractive qualities of the Ming blues in isolation, there can be no doubt that this later blue and white porcelain is far more effective. **1935** J. P. Marquand (*title*) Ming yellow. **1966** *Country Life* 30 June 1761/2 Shift-dresses.. in white, Ming-green, navy or camel. **1969** *Vogue* 1 Mar. 99 White wool tabard, edged with waves of Ming blue.

minge (mindʒ), *sb. dial. or slang.* [Origin obscure.] The female pudendum; hence, by extension, women regarded collectively as a means of sexual gratification.

1903 *Eng. Dial. Dict.* IV. 118/2 *Minge*, the female pudendum. **1925** Fraser & Gibbons *Soldier & Sailor Words* 156 *Minge*, female society (similar to Binge—*q.v.*), *e.g.*, 'His failing is Binge and yours Minge.' **1936** J. Curtis *Gilt Kid* viii. 80 I'm going to give you a kick in the minge if you don't shut up. **1974** *New Direction* IV. iv. 19/2 They've all..scented and talced their minges.

‖ **mingei** (miŋge·i). Also **Mingei.** [Jap. f. *min* people + *gei* arts.] Japanese folk-art; traditional local Japanese handicraft. Also *attrib.*

1960 B. Leach *Potter in Japan* viii. 183 Sen cha (green tea) taste, different from 'Matt cha' and certainly not 'Mingei'. **1967** H. H. Sanders *World of Jap. Ceramics* 177 The brown clay of Ryūmon-ji, when decorated with white slip..was considered..to be one of the distinctive *mingei* products of Japan. **1969** *Sat. Rev.* (U.S.) 13 Sept. 90/1 *Mingei*, or folk art, restaurants..with décor drawn from

Japan's rich folk tradition, hand-made plates and bowls by master potters.

mingily (mi·ndʒili), *adv.* [f. *mingy *a.*: see -ly².] Meanly, stingily.

1958 *Listener* 6 Nov. 722/2 The most mingily ungenerous gathers backed by elastic.

mingimingi (mi·ŋimiŋi). Also **mingi.** [Maori.] An evergreen shrub, *Cyathodes acerosa* (or *C. fasciculata*) belonging to the family Epacridaceæ, native to New Zealand, Victoria, and Tasmania, and bearing tiny, green flowers and red or white berries. Cf. *micky².

1889 T. Kirk *Forest Flora N.Z.* 213 The wood of the mingi is of a light-brown colour. 1906 T. F. Cheeseman *Man. N.Z. Flora* 411 *C*[*yathodes*] *acerosa*... Abundant from the North Cape southwards. Sea-level to 2500 ft. Mingimingi. 1929 W. Martin *N.Z. Nature Bk.* II. viii. 126 The mingi-mingi..is a rigid, pungent-leaved shrub with either white or red berries. 1963 *Weekly News* (Auckland) 10 July 37/3 The small shrub called mingimingi or black teatree is as good as manuka as a source of heat. 1966 G. W. Turner *Eng. Lang. Austral. & N.Z.* viii. 168 Another shrub *mingi* is said by Morris to have the form *micky* in the South Island. This is likely, as the South Island dialects of Maori have *k* for North Island *ng*.

Mingrelian (miŋgrī·liăn, min-), *sb.* and *a.* Also 7–8 **Mengrelian.** [f. *Mingrelia* (see below) + -an.] **A.** *sb.* **a.** A member of the people inhabiting Mingrelia, an area of the Kutais region of the Caucasus. **b.** The language of this people. **B.** *adj.* Of or pertaining to this people.

1639 [see *Circassian *sb.* 1]. 1690 Locke *Hum. Und.* i. iii. 18 The Mengrelians, a People professing Christianity. *a* 1791 [see Georgian *a.*² 1]. 1876 *Encycl. Brit.* V. 257/2 The Mingrelians..extend from the Zenesquali on the east to the Ingur and the Black Sea on the west, while the lower course of the Rion may be considered as constituting their limit on the south. 1883 *Ibid.* XVI. 437/1 The Mingrelians (still almost exclusively confined to the Mingrelian territory, and numbering 197,000) are closely akin to the Georgians. 1921 [see *Kirghiz *sb.* and *a.*]. 1939 L. H. Gray *Foundations of Lang.* xii. 375 South Caucasian.. consists of four languages: Georgian.., Mingrelian, Laz, and Svanian. 1959 *Chambers's Encycl.* VIII. 356/1 The chief languages are:..Mingrelian, Lazic, Svanetian. *Ibid.* XI. 430/1 The Georgians, Mingrelians and Armenians of the Caucasus. 1973 *Observer* 15 Apr. 39/1 The Beria Papers by Alan Williams... Brash English journalist and rip-roaring Russian defector, with linguistic help from beautiful Mingrelian.

mingy (mi·ndʒi), *a. colloq.* Also † **mingee.** [Perh. f. M(ean *a.*¹ + St)ingy *a.*, or a blend of Mangy *a.* and Stingy *a.*] Mean, stingy, niggardly; disappointingly small.

1911 J. W. Horsley *I Remember* xi. 254 'Mingee' for greedy. 1912 R. Brooke *Let.* May (1968) 382, I called you a mingy and coprologous Oxford poetaster. 1918 W. Owen *Let.* 19 Aug. (1967) 569, I rushed off a note in time for this evening's post, which may seem very mingy. 1926 C. Beaton *Diary* in *Wandering Yrs.* (1961) vii. 148 A mingy little tray he had picked up from heaven-knows-where. 1930 E. V. Lucas *Down Sky* 223 It's dear, but we are not going to be mingy. 1940 [see *clip *sb.*¹ 2 e]. 1972 *Guardian* 30 Aug. 9/5 The opening for filling steam irons with distilled water is usually mingy, and the thing overflows.

Comb. 1959 *Times* 28 Dec. 3/1 Both..were determined ..not to let the mingy-minded weather spoil the jubilee match. 1966 'L. Lane' *ABZ of Scouse* 68 *Mingy-arsed bastard*, a miserly person.

Hence as *sb.*, a mean person. *rare.*

1939 M. Egan *To Love & Cherish* ii. 48 Don't be a mingy, father; they only cost a shilling.

Minha(h, varr. *Mincha.

mini (mi·ni), *sb.* **1.** Abbrev. of *minicar, minicab* (see *mini- b); *spec.* the proprietary name of a small car made by British Leyland (formerly the British Motor Corporation). Also *attrib.* and *Comb.*

1961 *Economist* 24 June 1327/2 Taxi-men and mini-men have tested their vocabularies in London this week: the mini-men are confident of profit. 1961 *Engineering* 17 Nov. 658 The Mini's astonishing success is due purely and simply to good engineering. 1962 *Listener* 18 Oct. 634/2 The designer of the Morris Minor and the 'minis'. 1963 *Times* 19 Apr. 17/1 The company also announces the appointment of Mr. A. A. Issigonis, designer of the 'mini' range and the technical director, to the corporation's board. 1964 *Times* 11 Feb. 11/7 At present a young man who passes a test in a mini is legally entitled to drive an eight-wheeler weighing 24 tons at 60 m.p.h. on a motorway. 1970 G. F. Newman *Sir, You Bastard* iii. 99 Sneed squeezed his mini on to the drive where four other cars were parked. 1971 *Times* 6 Aug. 7/6 Feeding the fantasies of mini-drivers, convincing them..that anything the film's crack stunt-drivers could do they could do.

2. Abbrev. of *mini-skirt* (see *mini- b).

1966 *Guardian* 27 July 6/4 The new thing about the Scherrer mini is that it flares. 1967 *Punch* 4 Jan. 1/1 The lengths of female laid bare by minis. 1968 *Listener* 12 Dec. 790/3 One after another, Arab states are banning the mini. 1971 B. Malamud *Tenants* 42 She wore..a plain white mini with purple tights.

mini (mi·ni), *a.* [Abbrev. of Miniature *a.* Cf. next.] Very small, tiny.

1963 *Daily Tel.* 17 Dec. 13 A 'mini' census covering one householder in ten will be taken by the Government in 1966. 1966 *Ibid.* 14 Nov. 10/3 M. Redlus insists: 'My minis will be the most mini in Europe but they'll be decent.' 1966 *Punch* 7 Dec. 857/1 Leg make-up..gives sitting-down confidence to the wearer of the miniest skirt. 1966 *Sunday Express* 18 Dec. 23/5 Of course 25,000 cases is pretty mini. 1967 *Word Study* Dec. 3/2 There's nothing mini about their wages. 1969 *Daily Tel.* 25 Sept. 21/1 Girls prance on the longest legs, in the mini-est skirts and the kinkiest boots.

mini-, *pref.*, combining form of Miniature *a.* (reinforced by the first letters of Minimum *a.*), used to designate things that are very small of their kind.

A prefix much in vogue from the 1960s. Only a selection from the virtually unlimited number of combs. is illustrated here. The examples are arranged in alphabetical order of the combs. for convenience of reference.

1966 *Daily Tel.* 24 Oct. 11/1 The demand for a prototype female briefcase has been underlined by a strong season of mini-bag fashions—little swinging double-sided dog lead bags that made the absence of a briefcase for women all the more apparent. 1962 *Punch* 12 Dec. 878/1 A cycle firm is bringing out a mini-bike. 1970 *Time* 2 Nov. 56 Half-size (or even smaller) motor-cycles... Recession or no, minibikes seem to be all over, but nowhere are they more visible than in Los Angeles. 1960 *House & Garden* Oct. 135 (Advt.), Hygena think of details—..like built-in bread bins,..Minibins, refrigerators. 1968 *Economist* 10 Feb. 46/3 Capital investment shows no signs (yet) of re-energising—and it shouldn't, despite the mini-boom. 1968 *Observer* 22 Dec. 21/7 A mini-bottle of Gala's nail polish (Little Gems 4s. 4d. a bottle). 1970 *Jrnl. Gen. Psychol.* LXXXIII. 155 Either *S* or *E* could activate the green light by depressing the button on a minibox situated in front of each of them. 1961 'R. M. Dashwood' *Provincial Daughter* 55 Squalid piles of dust, marbles and minibricks. 1966 *Times Rev. Industry* Sept. 63/3 With the 'mini-budget' having withdrawn another £500m. from internal demand. 1971 *Daily Colonist* (Victoria, B.C.) 24 Nov. 1/7 The vessels were described as 'mini-bulkers', small ships..for ferrying cargoes. 1936 *Miniature Camera Mag.* Dec. 4/2 It is perhaps to be expected that all sorts and conditions of industries and businesses should have sprung up around the successful Minicamera. 1964 *Punch* 21 Oct. 592/1 Mini-holidays, mini-cameras, mini-tellies. 1971 *Author* LXXXII. 108 Outside such writer-populated districts as Hampstead or Chelsea, even minor authors may be mini-celebrities, and so news. 1968 *Economist* 27 Apr. 79/2 America..might well take chief responsibility for producing the Olivetti mini-computers as that is where chief demand lies. 1973 *Business Week* 8 Dec. 69/1 Today, a $2,000 minicomputer is more powerful, more reliable, and easier to use than the big, $100,000 machines of a decade ago. 1963 *Aeroplane* 24 Jan. 25/2 A one-man autogyro with an estimated selling price of £1000 was demonstrated recently in South Africa... This 'mini-copter' is said to be in production already. 1971 *Time* 19 Apr. 60 But, like a stabbing pain that passes quickly, the mini-crisis was a warning that the dollar faces more trouble. 1967 *Spectator* 15 Sept. 300/2 Now a new type of winter holiday is offered by several lines—the 'mini' cruise. 1969 *Daily Tel.* 13 Feb. 27/6 Results in the 'mini-elections' in India's four northern States..show the once all-powerful Congress Party to be humbled also in the Punjab and Bihar. 1969 *Times* 27 Jan. 10/8 It will have a 30-acre 'mini-farm' for practical training on its doorstep. 1973 C. Bonington *Next Horizon* xix. 259 There was a long pause and then we saw the green miniflare which was the signal to follow. 1954 *Sat. Rev.* (U.S.) 24 Apr. 60 The world famous Philips Minigroove 33⅓ Long Playing Records. 1968 *New Yorker* 16 Mar. 43 Armed with three 7·62-mm. machine guns, called miniguns, which could fire a hundred rounds per second. 1969 I. Kemp *Brit. G.I. in Vietnam* vi. 117 The deep sustained roar of the dragon-ships' mini-guns. 1963 *Daily Tel.* 12 Nov. 17/3 A proposal that 'mini-houses' should be built to cater for people in lower income groups. 1963 *Ibid.* 6 June 15/5 (*heading*) Mini-jet as Paris 'ferry' to air show. 1967 *Word Study* Dec. 3/1 The sphere of clothing is..well represented with ..*mini-jupe*, [etc.]. 1973 *Publishers Weekly* 18 June 35/1 Underage hopefuls in mini-jupes. 1969 *Sunday Times* 6 Apr. 53 Of course, describing these as mini-kilts makes the purist splutter in his porridge. 1974 *People's Jrnl.* (Inverness & Northern Counties ed.) 24 Aug. 3/5 Last Friday, during the great float-out, the pair donned mini-kilt outfits and became hostesses to the many guests attending the ceremony. 1963 *Daily Tel.* 13 Apr. 15/4 The mini-lifeboats..will be normally manned by a crew of two. 1963 *Aeroplane* 21 Mar. 9/2 The third D.H. 125 'mini-liner' is shortly to go into service with Bristol Siddeley Engines. 1973 *Courier & Advertiser* (Dundee) 7 Aug. 7/2 The Fred Olsen mini-liner Basel is due over the weekend with 500 tons of paper from Norway. 1972 *N.Y. Times* 4 June 4/8 The six-mile Crazylegs minimarathon. 1963 *Times* 29 May 15/2 Most of these countries have tiny internal markets—'mini-markets' Mr. Gates calls them. 1966 J. Porter *Sour Cream* iv. 45, I must have been out of my mini-mind to let myself be manoeuvred into this ridiculous, and dangerous, situation. 1965 *Guardian* 25 Aug. 16/8 Irresponsible action by a few mini-minded strikers. 1959 *Motor Show Catal.* 497 Mini-Minor Saloon de Luxe. 1973 *Guardian* 8 May 2 (*heading*) Mini-Nuclear arms seen as 'wishful-thinking'. 1954 *Life* 29 Nov. 83 The new 'mini-pig'..is ideal for medical research, which is what he was bred for. 1970 *Daily Tel.* 29 May 5/1 The new mini pigs..at about 140 pounds are four to five times lighter than the farm variety. 1970 *New Scientist* 29 Jan. 187/1 The minipill was developed for one reason alone: because it was believed to provide safe contraception. 1967 *Sunday Times* 14 May 12/7 Expo's own minirail system passes on a slender viaduct. 1967 *Word Study* Dec. 3/2 A financial analyst reports the good news that 1967 has experienced only a mini-recession. 1970 *Times* 17 Mar. 27 Another aspect of the current apprehension is linked to the 'mini-recession' in the American motor industry. 1967 *Courier-Mail* (Brisbane) 14 Nov. 18 The collection will include mini-shifts which can double as pant tops teamed with matching slacks or shorts, or can be co-ordinated with bikinis for beach wear. 1969 *Daily Tel.* (Colour Suppl.) 24 Jan. 17/2 A girl in a mini-shift. 1971 *Time* 1 Feb. 32 In Paris, minishorts are an every-night, run-of-the-disco affair. 1973 *Times* 27 Nov. 16/2 The arrival of the mini-short in preference to the old baggy maxi-pants. 1967 *Time* 17 Mar. 36 For added balance, ski bobbers wear mini-skis fitted with braking crampons on both feet. 1974 *Maclean's Mag.* Jan. 25/3 On our GLM mini skis, we were led out to the beginners' hill, which could not have been more than 100 feet away over almost perfectly level ground. 1966 *Economist* 31 Dec. 1385/1 The 'mini-states' must, for the sake of their own reputations and the UN's, accept a system of weighted voting. 1968 *N.Y. Times* 26 Jan. 70 South Africa's economic predominance radiates from here to the three ministates of Botswana, Lesotho and Swaziland. 1973 *Nation* (Barbados) 16 Dec. 4/2 Without the eventual federation of the unit territories or a confederation of independent ministates, the West Indies have no future. 1959 *Chambers's 20th Cent. Dict.* Suppl., Mini-sub(marine). 1966 *New Scientist* 22 Sept. 655/2 The mini-sub *Alvin* which was used for recovering the H-bomb lost off the Spanish coast. 1963 *Flight Internat.* LXXXIII. 25/1 There has been a great deal of unofficial talk about 'mini-submarines' carrying one or two Polaris apiece, and also of a larger vessel carrying eight missiles. 1973 *Reader's Digest* Apr. 58/2 To complement the divers, mini-submarines less than 20 feet long are used by the airman for underwater surveying. 1968 *Saturday Night* (Toronto) Sept. 36 Elaine Bedard, wearing a flared pink leather minisuit, awaited him in his Mercedes-Benz. 1971 'V. X. Scott' *Surrogate Wife* 254 Under the coat she wore a cream-coloured minisuit. 1967 *Times* 18 May 23 Whisky is to be exported from Scotland by pipeline and 'minitanker', the shipping firm of Christian Salvesen and Co., of Leith, announced yesterday. 1971 *New Scientist* 2 Sept. 520 Petrol is supplied free, either by the client filling up at the company's garage or from a mini-tanker which regularly visits the special parking places. 1967 *Economist* 29 Apr. 484/3 BMC has come off even worse, working short time on truck manufacturing and likely to drop its ill-fated mini-tractor. 1966 *Punch* 14 Sept. 380/3 Hardy Amies's mini-trousers Are the latest passion-rousers. 1960 *Guardian* 30 Dec. 12/1 The first of the new vehicles to be used for road patrols will be Austin Minivans. 1968 H. C. Rae *Few Small Bones* ii. viii. 142 Small personal items were taken by mini-van to the new flat. 1966 *Economist* 10 Dec. 1112/2 The day may eventually come when the big powers will stand back and permit a nuclear mini-war between smaller countries.

b. Special combinations: **minicab** [Cab *sb.*³], a small taxi-cab; **minicam,** a miniature camera; so as *v. trans.*; **minicar,** (*a*) a small motor car (cf. *mini *sb.* 1); (*b*) a child's toy model of a motor car; **minicell** *Biol.*, a miniature cell, without nuclear material, produced by the division of individuals of a particular strain of the bacterium *Escherichia coli*; **minicoat,** a short coat, one not reaching to the knee; **mini-dress,** a dress with a mini-skirt; **Mini-Moke, mini-moke,** a small motor vehicle resembling a jeep; cf. *Moke⁴; **mini-ness** = Tininess; **Minipiano,** (a proprietary name of) a small piano; also **minipiano; Miniprinter,** a proprietary name of a type of small machine for printing tickets; **mini-skirt, miniskirt,** a very short skirt; hence **miniskirted** *a.*, wearing or having a mini-skirt.

1960 *Economist* 12 Nov. 711/3 Current regulations regarding London taxis would not allow the introduction of what Mr. Dennis Vosper, speaking for the Home Office, called 'minicabs'. 1961 *Daily Tel.* 6 June 19/6 London's taxi war, between regular taximen and minicab operators, took a new turn yesterday. 1965 *Spectator* 12 Mar. 322/3, I travelled by mini-cab from Baker Street to Kensington. 1973 *Times* 29 Nov. 4/7 A mini-cab operator..was sentenced to four years. 1937 *Amer. Speech* XII. 236/2 A professor at the University of Wisconsin minicammed his students during an examination. 1939 Webster Add., *Minicam*, short for *miniature camera.* 1940 Graves & Hodge *Long Week-End* xxv. 432 Their photographs, largely contributed by 'minicam' amateurs. 1948 C. Day Lewis *Otterbury Incident* iii. 28 Penknives, Minicars, balls ..the sort of oddments you keep in your pocket. 1949 *Light Car* Dec. 599/2 Three-wheelers... The 122 c.c. Bond Minicar. 1963 *Spectator* 1 Nov. 558 To say Britain came late into the minicar race is to miss the point. 1967 H. I. Adler et al. in *Proc. Nat. Acad. Sci.* LVII. 321 A newly isolated strain of *Escherichia coli* K12 regularly produces a large number of unusually small anucleate cells during the logarithmic phase of growth. These small cells do not divide... In this report we communicate information regarding some of the basic properties of these minicells. 1971 *Nature* 3 Sept. 11/1 Although minicells, small enucleate *Escherichia coli*, have been used occasionally by molecular biologists, it seems safe to say that they became available too late in the game. 1966 *Guardian* 8 Sept. 7/3 Topcoats were mini-coats—a little more than loose jackets. 1967 *Britannica Bk. of Year* (U.S.) 338 Rabbit had been dyed in new and heady shades—orange, mauve, navy, shocking pink, bright green—to fashion double-breasted minicoats and pea jackets. 1965 *Christian Sci. Monitor* 20 Nov. 30 The fashion pages of British papers sport mini-dresses. 1966 *Daily Tel.* 17 Oct. 10/6 The skinny model girl loped down the runway at Marlborough House in her horizontally-striped mini-dress. 1962 *Ibid.* 13 Feb. 15/4 In the BMC vehicle, named Mini-Moke, the drive is confined to the front wheels. 1972 D. Francis *Smokescreen* i. 10 No one had bothered to put the canvas over the Minimoke. 1974 H. MacInnes *Climb to Lost World* iii. 46 Jonathan drove a mini-moke with Greek lettering emblazoned upon it. 1964 *Punch* 21 Oct. 592/1 Car itself victim of over-exposure, flaunting its mini-ness. 1967 *New Scientist* 13 Apr. 94/1 The current preoccupation with 'mini-ness' has now extended into the realm of..microbiology. 1934 *Trade Marks Jrnl.* 4 Apr.

434/2 Minipiano... Pianos. Brasted Bros. Ltd.,..London,..piano manufacturers; and C.A.V. Lundholm Aktiebolag (a Joint Stock Company organised under the laws of Sweden),..Sweden; merchants. 1943 H. W. Van Loon *Lives* xviii. 558 He..went over to the little minipiano which had been wished on Frits and now stood at the foot of the stairs. 1947 A. H. Howe *Sci. Piano Tuning* (rev. ed.) xxi. 96 Tuning the 73 note minipiano is quite different from the conventional mode. 1949 *Electronic Engin.* XXI. 461/2 It is intermediate in size between a minipiano and a small upright piano. 1949 *Railway Gaz.* 6 May 510/2 The equipment exhibited by Westinghouse Garrard Ticket Machines Limited at..Olympia, includes... The Westinghouse mini-printer,..with four or six printing units. 1958 *Times* 11 Feb. 15/3 The company also supplies rapid and mini printers, as used in booking offices of the London Transport Executive. 1965 *Vogue* Aug. 53 Mini skirt,..snakeskin belt. 1965 *Economist* 20 Nov. 862/1 The Fashion House Group of London dumbfounded the..audience of American buyers quite as much by the sight of the British aggressively selling as by their mini-skirts and kooky outfits. 1970 G. F. Newman *Sir, You Bastard* viii. 243 There were more miniskirts around three o'clock on a warm May afternoon. 1971 B. Malamud *Tenants* 54 Had Mary Kettlesmith described his acrobatics with her miniskirt? 1966 *Listener* 30 June 955/1 The mini-skirted or levied young. 1972 F. Warner *Maquettes* 19 Pretty, miniskirted, and attractive young lady.

miniature, *sb.* and *a.* Add: **A.** *sb.* **5. d.** *Chess* [tr. G. *miniatur*]. A problem involving few men, *spec.* one in which not more than seven pieces are used; a game of relatively few moves.

1903 *Brit. Chess Mag.* 91 It shows the composer's various styles better to give No. 76, which is a four-move 'miniature'. 1907 S. S. Blackburne *Terms & Themes Chess Probl.* 29 Very light-weight problems are known as 'miniatures'. 1970 A. Sunnucks *Encycl. Chess* 309 There is no fixed number of moves which determine whether a game can be classified as a miniature, but the term is generally used to describe a game of under 20 moves.

e. Something that is much smaller than the size normal for things of its class; *spec.* (*a*) a very small bottle of spirits; (*b*) a miniature camera.

1939 *Sun* (Baltimore) 26 June 18/2 The sale of so-called 'miniatures' in Baltimore is fostering juvenile drinking. 1954 A. Lee *Round Many a Bend* vii. 68 Sunday was also the day on which we sold most 'miniatures'. (*footnote*) This is the name for the small bottles of spirits in the trade. 1955 [see *definition 5 c]. 1958 *Spectator* 1 Aug. 167/3 The miniatures are obviously going to be the fashionable gimmick. The Pye pocket portable measures only 4 × 7 × 1½ inches. 1962 'H. Howard' *Double Finesse* vi. 67 Didn't I see you knock back a miniature of whisky? 1971 C. Bonington *Annapurna South Face* x. 123 We had plenty of whisky—over four hundred miniatures and seventy-two full bottles.

f. A short piece of music.

1958 *Listener* 18 Dec. 1051/1 This has no connexion whatsoever with the writing of miniatures or the use of short lyric forms. 1962 *Ibid.* 15 Mar. 489/2 Schumann was a master of epigram, and the epigrammatic *Einfall* was most eminently suited to his vocal and instrumental miniatures.

7. *miniature painting* (earlier example).

1765 T. H. Croker et al. *Compl. Dict. Arts & Sci.* II. s.v. *Marbling*, The consistence of the solution should be nearly that of strong gum-water, used in miniature painting.

B. *adj.* (Further examples.) Also, designed on a small scale; much smaller than the normal size; tiny; *spec.* of a camera (see quot. 1943); hence applied to photography, films, etc., involving the use of such a camera.

1869 S. R. Hole *Bk. about Roses* x. 151 The Miniature or Pompon Provence,..the 'baby Roses' and the 'pony Roses' of our childhood. 1887 (*title*) Payne's Miniature Scores. No. 1. Mozart Quartet in G major. 1893 G. B. Shaw *Widowers' Houses* iii. 73 With photographic portrait of Blanche on miniature easel on the top. 1911 *Daily Colonist* (Victoria, B.C.) 9 Apr. 1/5 The Department of militia has issued notice of a competition at local miniature ranges to be fired between the 11th and 19th of April. 1913 A. G. Fulton *Notes on Rifle Shooting* 7 Miniature shooting teaches almost all that is necessary to make a man a good shot with the Service rifle. 1914 *Physical Rev.* Apr. 255 From them a miniature universe could be constructed exactly similar in every respect to the present universe. 1917 *Autocar Handbk.* (ed. 7) i. 13 (*heading*) The miniature car. 1921 *Sci. Amer.* 25 June 514/2 (*caption*) Miniature camera with magazine mounted for use.. and the miniature projector. 1941 J. du Mont (*title*) 200 miniature games of chess. 1943 C. Duncan *Man. Miniature Camera* (ed. 2) iii. 12 When the Royal Photographic Society found it necessary to define the term 'miniature camera' they ruled that it was one designed to make negatives 'not larger than six square inches'. 1943 *Gloss. Terms Electr. Engin.* (*B.S.I.*) 144 Miniature film radiography. 1953 J. du Mont (*title*) Chess: more miniature games. 1956 A. L. Sowerby *Dict. Photogr.* (ed. 18) 466 The Leica, though by no means the first camera to make small pictures, was undoubtedly the camera which popularised miniature photography. 1958 *Spectator* 1 Aug. 167/2 The new miniature transistor radios. 1959 *Chambers's Encycl.* VII. 246/1 Increased use of mass miniature radiography is resulting in many suspicious cases being discovered which cannot be definitely diagnosed. *Ibid.* XIV. 778/2 The technical quality of the present-day miniature film approximates closely to that of full-sized film. 1970 *Jrnl. Gen. Psychol.* LXXXIII. 4 The more sophisticated paradigms simulating the therapy-interview situation..and miniature systems of conditioning in the mass.

b. Applied to a dog of a breed or variety smaller than average; also as *sb.*

1902 *Daily Tel.* 13 Feb. 6/4 [Cruft's Dog Show at Royal Agricultural Hall]. In one of the annexes of the hall are shown the Griffons, the Maltese and other miniatures. 1903 R. B. Lee *Hist. & Descr. Mod. Dogs Gt. Brit. & Ireland: Terriers* (ed. 3) xv. 401 Little dogs of these colours and toy white English terriers will not have any kind of classification, unless special arrangements are made for grouping them as a section of their own, called 'smooth-coated terriers (miniature) other than black and tan'. 1904 H. Compton *20th Cent. Dog* I. 301 The miniature black and tan terrier—to give it its new Kennel Club title—is more familiarly known..by its original one of the 'Toy Terrier'. 1912 *Encycl. Brit.* VII. 375/2 *Non-Sporting* [Dogs].—Bulldog, bulldog (miniature)..black and tan terrier (miniature). 1924 [see *bench *v.* 3 c]. 1945 C. L. B. Hubbard *Observer's Bk. Dogs* 53 A miniature Dachshund should be a sturdy little sportsman. 1948 —— *Dogs in Brit.* iii. xx. 319 The Miniature Poodles in Britain are mostly descended from imported French dogs although many are 'bred down' from small standard Poodles. 1959 *Observer* 1 Feb. 12 The breed standard describes miniature poodles as active and intelligent... Height at shoulder should be under 15 in.

c. Applied to a version of golf played on a miniature course.

1915 F. M. Hueffer *Good Soldier* i. i. 10 Sitting together..in front of the club house, let us say, at Homburg ..watching the miniature golf. 1930 *Glasgow Herald* 25 Sept. 17/5 Miniature Golf Course in Glasgow. What is claimed to be a real golf course in miniature is being laid out. 1930 *Daily Express* 6 Oct. 8/3 The Government's experts have been investigating the subject because of the demand the miniature golf establishments are creating for materials. 1966 J. Ball *Cool Cottontail* (1967) xi. 118 He and his date played miniature golf, had dinner, and saw a movie.

miniature, *v.* **2.** (Earlier example.)

a 1706 Evelyn *Diary* an. 1686 (1955) IV. 531 His booke of Birds, Fish: flowers, shells &c drawn & miniatured to the life.

miniaturization (mi:nit-,mi:niătiŭraizēi·ʃən). [f. Miniature *a.* + -ization.] The process of miniaturizing; an instance of this.

1947 *Technical News Bull., U.S. Bureau of Standards* Jan. 8/1 With the end of the war and the declassification of the principle of the VT fuse, industry is manifesting interest in the program of miniaturization of electronic devices. 1950 *Engineering* 21 Apr. 447/1 The trend towards 'miniaturisation', i.e. the production of smaller components and greater compactness of layout, is continuing. 1960 Koestler *Lotus & Robot* ii. viii. 211 That patient biding of one's time to avenge an assumed insult by a subtle twist—a miniaturization of the mediaeval vendetta. 1967 M. Chandler *Ceramics in Mod. World* iv. 134 If miniaturization was to proceed, clearly much higher permittivities were needed.

miniaturize (mi·nit-, mi·niătiŭraiz), *v.* [f. as prec. + -ize.] *trans.* To produce in a smaller version; to render small.

1946 *Jrnl. R. Aeronaut. Soc.* L. 945/1 The weight added to the aircraft by the installation is about 120 lb., but.. this weight is based on a war-time design which could be miniaturised. 1950 *Ibid.* LIV. 281/2 There has been a general move to miniaturise radio components with consequent saving in bulk and weight. 1957 *Times* 28 Aug. (Radio and TV Suppl.) p. xv/2 The printed circuit gives the designer added ability to miniaturize his equipment. 1957 *Spaceflight* I. 77/2 A single satellite would be quite unable to carry all the instruments which scientists want to send aloft, even when they are miniaturized. 1963 A. Smith *Throw out Two Hands* xiii. 135 The camp had been miniaturised. Micro-huts had been built, with midget cooking-pots smouldering over dwarf fires. 1972 *Lebende Sprachen* XVII. 133/1 Nowadays micro-film can be miniaturized to the point where the library on the desk can be of virtually any size required.

Hence **mi·niaturized** *ppl. a.*

1951 *Electronic Engin.* XXIII. 478/2 Miniaturized components generally are becoming more and more readily available. 1967 M. Chandler *Ceramics in Mod. World* iv. 135 A whole new technology for the production of miniaturized capacitors. 1971 *N.Y. Times Bk. Rev.* 19 Dec. 7/4 The pages fall open easily,..four of the original appearing in each page of the miniaturized volumes.

minibus. Delete ? *Obs.* and add earlier and later examples. (The modern use is influenced by the prefix *mini-.)

1845 *Scotsman* 15 Feb. 3/5 (Advt.), Important sale of horses, harness, and carriages... One excellent 12-inside *omnibus*, nearly as good as new... One handsome *minibus*, in good order. 1958 *Oxford Mail* 9 June 1/1 A Morris Mini-bus which could possibly save rural services threatened with extinction through high running costs went on exhibition today... It carries twelve people, including the driver, and is designed to operate without a conductor. 1960 *Guardian* 19 Nov. 5/6 The company arrived by mini-bus. 1962 [see *commercial *sb.* 2]. 1965 M. Morse *Unattached* v. 180 Transport difficulties..were largely solved when the worker secured the use of a mini-bus. 1972 *Daily Tel.* 19 Feb. 11/3 The company minibus (or coach if there are enough of you interested) is available for excursions.

minification (mi:nifikēi·ʃən). [f. Minify *v.* + -ation, after Magnification.] Diminution or reduction in size, appearance, importance, etc.

1904 *Jrnl. R. Microsc. Soc.* June 281 A magnifying power which exactly balances the ten-fold minification before spoken of. 1961 L. Mumford *City in Hist.* viii. 242 These are symptoms of the end: magnifications of demoralized power, minifications of life. 1974 *New Scientist* 3 Oct. 61/1 The illustrations are generous in number and in size, though, talking of size, in many cases an indication of the minification would have been instructive.

minim, *a.* and *sb.* Add: **B.** *sb.* **9.** A very small Roman bronze (or occas. silver) coin, usu. produced locally. Pl. **minimi, minims.**

1896 W. C. Hazlitt *Coin Collector* ix. 247 *Minim*, a term which, for want of a better one, has been assigned to a class of bronze money of Roman type, probably of the fourth or fifth century B.C., which may have been of local or provincial origin, and is of unusually small module. 1935 *Discovery* July 196/2 A hoard of 800 minims, small coins in some cases not more than 5 mm. in diameter, buried beneath the floor of the stage. 1962 C. R. Josset *Money in Brit.* i. 7 An attempt to overcome the shortage was made by a local production of crude bronze coins which are known as 'minimi' because of their smallness. 1971 *Daily Tel.* 13 July 9/1 Two rare British silver minims, or small coins, of the first century A.D. have been discovered during excavations taking place on a Roman occupation site in Chapel Street, Chichester.

minimal, *a.* Add: **b.** spec. *Linguistics.* (*a*) Distinguished only by a single feature; usu. applied to a pair of similar forms; (*b*) other uses (see quots.).

(*a*) 1939 *Amer. Speech* XIV. 122 Words can be distinguished by the minimal opposition of vowel nasality and [n]. 1942 C. F. Hockett in *Language* XVIII. 7 The term 'contrastive pair', meaning any pair between which there are differences in a context of similarity, any pair usable for the listing of features, is used here instead of the traditional term 'minimal pair'. 1950 D. Jones *Phoneme* vi. 15 When a distinction between two sequences occurring in a language is such that any lesser degree of distinction would be inadequate for clearly differentiating words in that language, the distinction is termed a 'minimal one'... Minimal distinctions are very commonly effected by the addition or subtraction of a phoneme. 1955 C. F. Hockett *Man. Phonol.* vi. 212 Before analysis is complete, one cannot be certain that a given pair is 'minimal' in a strict phonologic sense—that there is but a single difference, at the level of ultimate phonologic constituents. 1961 H. A. Gleason *Introd. Descr. Ling.* (ed. 2) i. 16 In calling *bill* and *pill* a minimal pair we assume that they differ by only one phoneme. 1964 R. A. Hall *Introd. Ling.* 81 Minimal pairs are not essential to show that two sounds do not belong to the same phoneme. 1971 *Archivum Linguisticum* II. 48 In the days of 'classical' phonemics much play was made of the 'minimal pair' in order to establish throughout a language such lexical differences as those between *pin, bin, tin, din*, [etc.].

(*b*) 1930 H. E. Palmer *Princ. Romanization* 52 We may designate by the term *monophone* any phone of the first or second degrees of abstraction of which the concrete members are so similar in point of production and of acoustic effect even when observed by a competent observer, that it may be regarded as a minimal unit of pronunciation (i.e. practically insusceptible of sub-division). 1941 G. L. Trager in L. Spier et al. *Lang., Culture & Personality: Ess. in Memory of E. Sapir* 133 Intensity of tone is manifested as relative height of pitch: maximal intensity is *high* tone, minimal intensity is *low* tone, medial intensity is *middle* tone. 1949 J. R. Firth in *Trans. Philol. Soc. 1948* 142 The weak, neutral, or 'minimal' vowel. 1955 Quirk & Wrenn *Old Eng. Gram.* 129 Taking the consonants of a word as its minimal root. 1962 B. M. H. Strang *Mod. Eng. Struct.* vi. 84 They [*sc.* central nouns] can follow directly in minimal constructions (i.e. be head-word to) a closed system of words we shall call *determiners.* 1964 R. A. Hall *Introd. Ling.* 15 Morphemes ..are the minimal units which carry meaning. 1971 D. Crystal *Ling.* 187 The grammar..*could* be analysed in terms of identifying a set of minimal units.

c. In *Art*, used of a form of painting and sculpture which is characterized by the use of simple or primary forms, structures, etc., often geometric and massive. Hence **mi·nimalism, mi·nimalist** (also *attrib.* or as *adj.*), **minima·lity.**

[1958 *Listener* 23 Oct. 647/2 Creating huge, minimal forms with the palette knife in a few colours.] 1965 R. Wollheim in *Arts Mag.* Jan. 26 In a historic passage Mallarmé describes the terror, the sense of sterility, that the poet experiences when he..confronts the sheet of paper..and no words come to him... Why could not Mallarmé, after an interval of time, have simply got up from his chair and produced the blank sheet of paper *as* the poem which he sat down to write?.. Such a gesture ..would provide us with an extreme instance of what I call minimal art. 1967 C. Greenberg in E. Lucie-Smith *Movements in Art since 1945* (1969) i. 13 The working out of this problem, whose solution seems to have arrived in the form of what is called Primary Structures, ABC, or Minimal Art. The Minimalists appear to have realized ..that the far-out in itself has to be the far-out as an end in itself. 1969 *Britannica Bk. of Year* (U.S.) 800/2 *Minimal sculpture*, sculpture in the idiom of minimal art. 1969 *Time* 3 Jan. 42 Minimal forms still massively demand their unrewarding space, but they are countered by weirdly eccentric shapes that are frankly frivolous, at least unpredictable. 1969 E. Lucie-Smith *Movements in Art since 1945* viii. 240 His sculptures have been described as 'minimal art', or as examples of the 'single-unit Gestalt'. *Ibid.* 242 Judd's colleague, Robert Morris, defends minimality in equally emphatic terms. 1969 *Manch. Guardian Weekly* 1 May 20 Call it minimalism if you like, but the real point is that the thing insists on being just itself and nothing more. *Ibid.*, Tony Smith, usually taken as the original minimalist sculptor..is well represented by large sculptures. 1971 *Rolling Stone* 24 June 36/5 It remained only for minimal sculpture to come along, with its emphasis on the self-contained object (sometimes just a log, rock, or mound of dirt). 1973 *Phaidon Dict. 20th*

Cent. Art 254/1 The immediate predecessors of the Minimalists were Ad Reinhardt and Josef Albers, who brought to their canvases the 'exclusive, negative, absolute, and timeless' quality so desired by the Minimal artists. 1973 *Times* 30 June 12/4 There will be works of American, Continental and British artists, abstract expressionism, Pop, kinetic, minimal and conceptual art. 1973 *Times Lit. Suppl.* 9 Nov. 1363/4 The pop, minimalist, new realist, and other artists who have succeeded the abstract impressionists.

minimalist (mi·nimălist). [f. Minimal *a.* + -ist, or (in sense 1) ad. F. *minimaliste*, tr. Russ. *men'shevik* *Menshevik *a.* and *sb.*] **1.** (Also with capital initial.) = *Menshevik *sb.*; more widely, a person who advocated small or moderate reforms or policies. Also *attrib.* or as *adj.*

1907 I. Zangwill *Ghetto Comedies* 408 'Ah, you're a Maximalist,' said the beadle. 'No, I am only a Minimalist. I merely want the minimum—that we save our own lives.' 1917 [see *Bolshevik *sb.*]. 1917 *Times* 23 June 7/1 At the 'All Russia' Congress of the Workmen's and Soldiers' Delegates the 'Minimalist Socialists' have defined their programme. 1918 E. P. Stebbing *From Czar to Bolshevik* iii. 25 The Social Democrats consisted chiefly of Bolsheviks with a smaller Menshevik group. The Social Revolutionaries were subdivided into Maximalists and Minimalists. 1922 *Blackw. Mag.* June 820/2 The delegation represented not only Communists, but also Minimalists and the converted intelligentsia. 1954 *Ann. Reg. 1953* 153 There were the 'maximalists' (Germany and the Netherlands),..and the 'minimalists' (France) who wanted the E.P.C. to be little more than a system of inter-governmental association and co-operation. 1972 *Times* 19 Oct. 1/6 A minimalist summit, dealing only with well tried issues of economic integration.

2. See *Minimal *a.* c.

minimality (minimæ·lĭti). [f. Minimal *a.* + -ity.] **1.** *Linguistics.* The quality or character of being minimal.

1953 C. E. Bazell *Ling. Form* i. 10 The criterion of minimality is fulfilled in either case. 1963 J. Lyons *Structural Semantics* ii. 29 What can be done, however, is to apply the criterion of minimality after the establishment of the several meaning-relations between sentences in context. 1969 *Word* XXV. 235 Because of the minimality requirement in the definition of the morpheme, the second class is not equivalent to a class of bound morphemes.

2. See *Minimal *a.* c.

minimally (mi·nimăli), *adv.* [f. Minimal *a.* + -ly².] To a minimal extent or degree.

1935 H. F. Twaddell in *Lang. Monogr.* XVI. 42 In American English, the forms *beet: bit: bait: bet: bat* are minimally phonologically different. 1936 *Amer. Speech* XI. 298 They generally cite only minimally different pairs. 1951 R. Firth *Elem. Social Organiz.* i. 35 Even minimally, their orientations are affected by its presence. 1971 *Sci. Amer.* Sept. 40/5 In order to meet not only the food requirements but also a minimally reasonable quality of life, the contributions that can be made by the use of energy in various forms are essential. 1973 *Nature* 6 July 40/2 *In vitro* methods for the short-term culture of erythrocytic forms of *Plasmodium knowlesi* and for isolating free merozoite preparations minimally contaminated with host cells. 1973 *Publishers Weekly* 13 Aug. 52/1 Such a glaring omission will surely put off some serious movie buffs. Probably, however, sales will be damaged only minimally.

minimax (mi·nimæks), *sb.* and *a.* [f. Mini(mum *sb.* and *a.* + Max(imum.] The smallest of a set of maxima; usu. *attrib.* (passing into *adj.*: see b), *spec.* designating (*a*) a strategy that minimizes the greatest loss or risk to which a participant in a game or other situation of conflict will be liable; (*b*) the theorem of game theory that states that, for a finite, zero-sum game with two players, the smallest maximum loss that a player can make himself liable to by a suitable choice of strategy is equal to the greatest minimum gain that he can guarantee himself. Cf. *Maximin *sb.* and *a.*

[1928 J. von Neumann in *Mathematische Annalen* C. 307 (*heading*) Beweis des Satzes Max Min = Min Max. 1944 —— & Morgenstern *Theory of Games* xvii. 154 A slightly more general form of this Min-Max problem arises in another question of mathematical economics.] 1941 Courant & Robbins *What is Math.?* vii. 345 (*heading*) Minimax points and topology. 1947 A. Wald in *Econometrica* XV. 282 We shall refer to an admissible decision function η for which (1·6) takes its minimum value as a minimax solution of the problem. *Ibid.* 283 An element a_0 in *A* is said to be a minimax strategy of player 1 if $\text{Inf}_b K(a, b)$ takes its maximum value with respect to *a* for $a = a_0$. 1949 *Ibid.* XVII. 230 Often Nature's strategy is completely unknown. In that case Wald suggests that the statistician play a minimax strategy: that is, the statistician should select that decision procedure which minimizes the maximum risk. 1951 *Ann. Math. Statistics* XXII. 466 The minimax risk is ¼(1−α). 1957 Luce & Raiffa *Games & Decisions* i. 2 Although Borel gave a clear statement of an important class of game theoretic problems and introduced the concepts of pure and mixed strategies, von Neumann points out that he did not obtain one crucial result—the minimax theorem—without which no theory of games can be said to exist. 1958 *Engineering* 21 Mar. 369/2 Player B, on the other hand, will determine the largest value in each column, and the smallest of these, the 'minimax'. 1961 L. Weiss *Statistical Decision Theory* v. 89 The minimax criterion has been criticized as being too conservative. 1962 *Sci. Amer.* Dec. 114/2 Both firms were guided by the principle of the minimax, choosing the best of the worst outcomes. 1968 *Brit. Med. Bull.* XXIV. 232/2 This would be the Bayesian Strategy of decision theory.., but other strategies, such as minimax or restricted Bayes, could also be used. 1970 [see *Maximin *sb.* and *a.*]. 1972 *Computer Jrnl.* XV. 277 (*heading*) On minimax solutions of linear equations.

b. Used predicatively as adj.: equal to, or resulting in, a minimax value.

1952 *Ann. Math. Statistics* XXIII. 587 For such weight functions the supremum of the risk over all θ ∈ Ω is infinite for every decision function, so that every decision function is minimax. 1965 Wozencraft & Jacobs *Princ. Communication Engin.* ii. 118 The decision rule which has the smallest maximum probability of error is called the minimax decision rule. Which of the eight rules is minimax? 1970 R. B. Ash *Basic Probability Theory* viii. 250 Find a minimax test, that is, a test that minimizes max (α, β). It is immediate from the definition of admissibility that an admissible test with constant risk (i.e., α = β) is minimax.

minimax (mi·nimæks), *v.* [f. prec. sb.] **a.** *trans.* To make equal to a minimax value. **b.** *intr.* To adopt or employ a minimax strategy. So **mi·nimaxing** *vbl. sb.* and *ppl. a.*

1964 Gould & Kolb *Dict. Social Sci.* 573/2 The goal may be assumed to take the form of maximizing (or, in game theory, minimaxing) the expected value, over some time interval, of a utility function. 1968 *Listener* 4 Apr. 438/1 Just like that 50–50 combination of heads and tails, the minimaxing peasant might pick a 40–60 combination of wheat and millet. 1973 *Nature* 23 Feb. 507/1 The fundamental mechanism underlying all this work [on machines designed to play games] has been a cycle of processes: lookahead, evaluation and minimaxing.

minimization. (Further examples.)

1940 *Jrnl. Obstetr. & Gynaecol. Brit. Empire* XLVII. 236 As regards anaesthesia, the minimization of shock is the main indication. 1969 *Nature* 29 Nov. 845/1 The arrangement of molecules..points to a minimization of the electric quadrupole–quadrupole interaction energy of the crystal. 1973 *Ibid.* 27 Apr. 556/1 Some potential energy calculations on β-pleated sheets have been done by Chothia.., who finds that a right-handed twist in the direction of the polypeptide chain, which corresponds to dihedral angles within one zone of the conformational map, makes for minimization of energy. 1974 *Clin. Pharmacol. & Therapeutics* XV. 444/2 The new method, called *minimization* for brevity, is compared here with randomization in regard to its vulnerability to experimenter bias and chance skewing.

minimize, *v.* Add: **3.** *intr.* To attain a minimum value.

1973 *Physics Bull.* Dec. 725/3 One of the unexpected discoveries made using hvem is that for a particular incident electron energy the diffracted intensity may minimize rather than maximize. 1975 *Ibid.* Jan. 13 Other parameters used to measure the intensity of the solar cycle tend to minimize at this time.

minimum, *sb.* and *a.* Add: **B.** *adj.* **b.** Special collocations: *minimum rate, wage*; **minimum free form** (see quot. 1926); **minimum vocabulary** (see quot. 1944).

1926 L. Bloomfield in C. F. Hockett *Leonard Bloomfield Anthol.* (1970) 130 A minimum free form is a *word*.. which may be uttered alone (with meaning) but cannot be analyzed into parts that may (all of them) be uttered alone (with meaning). 1958 C. F. Hockett *Course in Mod. Ling.* xix. 171 Minimum free forms and lexemes also do not meet this requirement. 1810 Minimum rate [in Dict.]. 1866 *Leisure Hour* 8 Dec. 783/2 All the workmen insist on is a *minimum* rate of wages. 1909 *Act* 9 *Edw. VII* c. 22 § 4 Trade Boards shall, subject to the provisions of this section, fix minimum rates of wages for timework for their trades. 1974 Minimum rate [see *minimum wage* below]. 1944 B. Russell in P. A. Schilpp *Philos. B. Russell* 14, I mean by a 'minimum vocabulary' one in which no word can be defined in terms of the others. All definitions are theoretically superfluous, and therefore the whole of any science can be expressed by means of a minimum vocabulary for that science. 1947 *Mind* LVI. 358 Every minimum vocabulary adequate to describing the world of ordinary experience must contain at least one universal-name. 1860 *Trades' Societies & Strikes: Rep. Comm. appointed by Nat. Assoc. for Promotion of Social Sci.* 299 And so the minimum wage which the Institute stipulates for..is constantly reduced. 1908 *New Age* 18 July 223/2 A serious objection to the fixing of a minimum wage in England for employed is that it would be most difficult to make provision for the large number of incompetent, inferior, and slow workers. 1940 *Economist* 6 July 4/2 The establishment, by law or by bargaining, of a decent minimum wage in all trades is an indispensable foundation. 1974 *Encycl. Brit. Micropædia* VI. 916/1 Minimum-wage legislation or machinery for fixing minimum rates now exists in most of the nations of Latin America, Africa, and Asia, as well as in more industrialized countries.

c. In combs. used *attrib.*, as **minimum-cost,** involving the smallest possible cost; **minimum-iron,** of a garment: requiring only a small amount of ironing; **minimum-security,** of a prison: having a minimum amount of restrictions on prisoners.

1962 A. Battersby *Guide to Stock Control* vi. 52 The best period may be found by a minimum-cost procedure .., that is, by minimizing the total costs of ordering and stockholding. 1959 *Harrods News* Summer 11 Little girl's sundress of minimum-iron cotton. 1963 A. J. Hall *Textile Sci.* v. 267 Very large amounts of cotton and viscose rayon materials are finished with what is now known as..'minimum-iron' finishes. 1965 G. Jackson *Let.* 18 Apr. in *Soledad Brother* (1971) 73, I can also obtain a parole faster there or a transfer to some minimum security camp. 1970 *Globe & Mail* (Toronto) 25 Sept. 1/6 A 26-year old escaper from the William Head minimum security prison near Victoria.

mining, *vbl. sb.* Add: **3.** *mining camp* (examples), *captain, company, engineer, -man, population, recorder, town* (earlier and later examples); **mining geology,** geology as applied to mining.

1902 'Mark Twain' in *Harper's Monthly Mag.* Feb. 431/2 Don't you put on any exclusiveness in a mining-camp. 1966 'E. Lathen' *Death shall Overcome* i. 9 Wall Street is power. The talk..closes mining camps in the Chibougamou. 1853 *Harper's Mag.* Mar. 442/2 We are accompanied by Captain John Cox, the mining captain. 1859 L. Sawyer *Diary* 23 Sept. in *Way Sk.* (1926) 116 The ..river mining companies which have not already proved failures. 1872 *Vermont Board Agric. Rep.* I. 629 Captain Thomas Pollard..had formerly been mining engineer. 1897 'Mark Twain' *Following Equator* 687 The mining engineers from America. 1941 R. Peele (*title*) Mining engineers' handbook. 1906 J. Park *Textbk. Mining Geol.* i. 1 Economic or Mining Geology, which bears more directly on mining, and the development of the mining industry. 1874 R. W. Raymond *Statistics of Mines* 499 He talked over the scheme with many railroad and mining-men. 1930 J. Dos Passos *42nd Parallel* i. 128 The bars.. were full of ranchers and miningmen. 1854 A. Delano *Life on Plains* xxvii. 382 There is arable land enough..to supply the whole mining population with vegetables, fruit and grain. 1876 *White Pine News* (Hamilton, Nevada) 22 July 3/1 An election took place on Treasure Hill on Thursday for Mining Recorder. 1968 Mining recorder [see *file *v.*² 1 c]. 1829 Dr. Willard in R. G. Thwaites *Early Western Trav.* (1905) XVIII. 359 The mining towns are mostly dependent on their supplies from abroad. 1856 *Hutchings Mag.* July 33/2 [We had] to.. make a 'pilgrim's progress' to the nearest mining town. 1944 N. W. Ross *Westward the Women* 133 She appeared in the mining town of Murray.

mining, *ppl. a.* Add: **2.** **mining bee,** a solitary bee of the family Andrenidæ, including many British and American species which nest in tunnels in the ground, sometimes grouped in colonies.

1893 L. N. Badenoch *Romance Insect World* iii. 72 (*caption*) Profile view of nest of a Mining Bee (Andrena vicina). 1912 Sanderson & Jackson *Elem. Entomol.* xvii. 268 None of the short-tongued bees live in colonies, and many of them make their nests in the ground, which has given them the name of 'mining bees'. 1974 *Country Life* 21 Feb. 351/1 The spoil-heaps excavated by mining-bees (*Andrena armata*) when they make their nests.

miniscule, erron. var. Minuscule *a.*

Now with stress on first syllable.

1898 J. Southward *Mod. Printing* I. xxii. 139 Each of the text letters already named has its own lower case or 'miniscule' letters. 1948 *N.Y. Times* 12 Dec. vii. 5 Now once again these miniscule land areas have faded from our interests. 1955 *Ibid.* 10 Apr. x. 27 Upland meadows are carpeted with miniscule wild flowers. 1961 *Economist* 16 Dec. 1118/1 Many 'gardens' would be miniscule affairs. 1967 [see *integrated *ppl. a.* b]. 1970 *Daily Tel.* 24 Apr. 1/3 If these conditions were fulfilled the risk from the pill was 'miniscule'. 1973 *Orcadian* 2 Aug. 4/4 The most interesting feature of this miniscule nation..is the strength of its national culture.

minister, *sb.* Add: **3. c.** **Minister of State,** a government minister, now usu. regarded in the U.K. as holding a rank below that of a head of department; **Minister of the Crown,** a minister or the head of a department in the U.K. government (see also quot. 1946); **Minister without Portfolio,** a government minister who has Cabinet status but is not in charge of a specific Department of State.

1696 Minister of State [in Dict. sense 3 a]. 1735 Bolingbroke *Diss. upon Parties* (ed. 2) p. xxv, But This will not become a *Matter of State*, though you are a *Minister of State*. 1864 Salisbury in *Q. Rev.* CXVI. 253 Ministers of State are case-hardened by practice. 1950 W. S. Churchill in *Hansard Commons* 31 Oct. 16, I like to see this reverence and respect for the past and all we owe to those who have gone before, and to see Ministers of State shake themselves clear from the obsession into which they fall from time to time. 1957 *Act* 5 & 6 *Eliz. II* c. 20 § 13 'Minister of State' means a member of Her Majesty's Government in the United Kingdom..who neither has charge of any public department nor holds any other of the offices specified in the Second Schedule to this Act. 1958 A. Chanda *Indian Administration* iii. i. 64 The appointment of Ministers of State to take charge of independent portfolios was an innovation [in India]... In France, a Minister of State ranked higher than a Minister and was usually entrusted with some special functions. In the UK, a Minister of State was, however, a minister of the second rank, functioning as the principal aide to a Cabinet Minister. 1963 Harvey & Bather *Brit. Constitution* xv. 251 When the work is particularly heavy or involved, or when it entails frequent visits abroad, Ministers of State, who act as deputy ministers, may be appointed. 1970 J. Harvey *How Brit. is Governed* xii. 147 In departments where the work is particularly heavy, the present-day practice is to appoint Ministers of State who virtually act on behalf of the minister. 1776 J. Hatsell *Coll. Cases Privilege Parl.* v. 196 The increase of their consequence in the state, and their influence in the management of public affairs, rendered them more an object of the attention of the Ministers of the Crown. 1844 Erskine May *Law of Parl.* xvii. 262 Another form of communication from the Crown to either house of Parliament, is in the nature of a verbal message, delivered, by command, by a minister of the Crown to the house of which he is a member. 1848 Disraeli in *Hansard Commons* 20 June 961 Surely, the people of this country are not accustomed to wait to express their opinion, till it may chance to be elicited by some captious expression of a Minister of the Crown. 1892 W. R. Anson *Law & Custom of Constitution* II. i. 10 The present dependence of the Ministers of the Crown, for their existence as a Ministry, upon the maintenance of a majority in the House of Commons. 1937 *Act* 1 *Edw. VIII* & 1 *Geo. VI* c. 38 § 3 If and so long as any Minister of the Crown to whom this section applies is a member of the Cabinet. 1946 *Act* 9 & 10 *Geo. VI* c. 31 § 8 (2) 'Minister of the Crown' means the holder of an office in His Majesty's Government in the United Kingdom, and includes the Treasury, the Admiralty, the Board of Trade, the Army Council, and the Air Council. 1956 Abraham & Hawtrey *Parl. Dict.* 113 Ministers of the Crown... In its widest sense..it means any member of the Government, of whatever rank (this does not, of course, include parliamentary private secretaries, who have no official status at all). 1975 *Listener* 7 Aug. 183/1 A speech by a Minister of the Crown. 1915 *Hansard Commons* LXXII. p. iv, Minister without Portfolio—Rt. Hon. the Marquess of Lansdowne, K.G. 1921 H. H. Asquith in *Ibid.* 23 June 1630, I was the first Prime Minister in modern times during the last half-century or more, to have in his cabinet a Minister Without Portfolio. 1954 Ld. Templewood *Nine Troubled Yrs.* x. 136 The Law Officers at once intervened to say that a Minister with a special Portfolio could not be a Minister without Portfolio. 1955 *Times* 30 June 8/5 M. Nguyen Huu Chau, Minister without portfolio in the Government of southern Viet Nam. *Ibid.* 18 July 8/3 The Bishop of Derby will open a debate on mining subsidence in the House of Lords on Wednesday, when the Government's views will be put forward by Lord Munster, Minister without Portfolio.

7. = *horned pout* (*horned *a.* 2 b). (Earlier and later examples.)

1839 D. H. Storer in Storer & Peabody *Rep. Fishes, Reptiles & Birds Mass.* 102 The Horned Pout..is known in the interior of the state by the vulgar names of 'Horn pout', and 'Minister'. 1849 Thoreau *Week Concord Riv.* 34 The Horned Pout..[is] sometimes called Minister. 1884 G. B. Goode *Fisheries U.S.: Nat. Hist. Aquatic Animals* 628 The common 'Horned Pout', 'Bullhead', 'Bull-pout', or 'Minister' of the Northern and Eastern States is the most generally abundant and familiar representative of this family [*sc.* Siluridæ].

ministeriable (ministiə·riăb'l), *a. rare.* [f. Ministeri(al *a.* + -able.] = *ministrable *a.*

1923 J. A. Spender *Life H. Campbell-Bannerman* II. 127 His test of complete co-operation was that Lord Rosebery should be definitely within the circle of ministeriable ex-Ministers.

ministerial, *a.* **4.** (Further examples.)

1879 Gladstone *Gleanings* I. viii. 229 Ministerial responsibility comes between the Monarch and every public trial and necessity, like armour between the flesh and the spear. 1969 *Sci. Jrnl.* Sept. 6 As an example of apparent ministerial unconcern he refers to recent exchanges in Parliament on this very problem. 1973 H. Trevelyan *Diplomatic Channels* i. 16 There are the 'grey eminences' in the corridors of power, who shun the light, who learn when to feed those dangerous animals, ministers, pacing up and down their party cages, who acquire an unrivalled knowledge of how government actually works beneath the misleading surface of ministerial responsibility. 1974 *Times* 19 Sept. 1/4 In a ministerial broadcast on all channels last night Mr Wilson said [etc.].

ministering, *ppl. a.* Add to def.: esp. in phr. *ministering angel*; *spec.* a kind-hearted person, usu. a woman, who helps and comforts people in distress; freq. a nurse.

1602 Shakes. *Ham.* v. i. 263 A Ministring Angell shall my Sister be. 1808 [in Dict.]. 1912 'Saki' *Chron. Clovis* 223 Martin Stoner rose heavily to his feet and followed his ministering angel along a passage..into a large room lit with a cheerfully blazing fire. 1922 Joyce *Ulysses* 349 A sterling good daughter was Gerty just like a second mother in the house, a ministering angel too with a little heart worth its weight in gold. 1924 A. Huxley *Little Mexican* 140 Emmy was good, was kind, a ministering angel. 1931 M. D. George *England in Transition* iv. 92 Nurses paid by the parish—creatures compared with whom Mrs. Gamp would have been almost a ministering angel. 1956 F. F. Darling *Pelican in Wilderness* i. 10 Such detached thought as I could muster was of feeling sorry for the air hostess. She remained throughout the gentle ministering angel, imperturbable as she picked her way through the shambles. 1974 W. Foley *Child in Forest* ii. 188 She came in her nurse's uniform, and I thought 'ministering angel' a very apt description.

ministrable (mi·nistrăb'l), *a.* and *sb.* [a. F. *ministrable* in the same sense: cf. -able.] **A.** *adj.* Fit or likely to become a minister; = *cabinetable *a.* **B.** *sb.* One who is likely or hopes to become a minister.

In quot. 1968 a use of the Fr. word.

1921 *Contemp. Rev.* Mar. 289 The anti-British feeling [in France]..oozes out of declarations by Ministers and Ministrable politicians. 1927 *Blackw. Mag.* Feb. 277/2 Those that are ministrables have done the best for themselves. 1968 *Economist* 9 Mar. 53/2 The *ministrables* were more inclined to vote for the government than against it.

Ministry. Add: **5. b.** The name given to certain departments of the British government.

1916 *Whitaker's Almanack* 223/2 Munitions, Ministry of, Minister, Rt. Hon. D. Lloyd George, M.P. 1942 *R.A.F. Jrnl.* 18 Apr. 24 The Ministry of Food announces that the milk ration for children will be increased. 1963 *Listener* 28 Feb. 368/1 It [*sc.* the Treasury] has also done hitherto the work of a ministry of economic affairs. *Ibid.* 7 Mar. 422/1 The Ministries of the three Services are to be reorganized under a unified Ministry of Defence. 1968 *Times* 16 Dec. 7/1 An attempt at a Ministry takeover and a threat to a much valued independence.

minitrack (mi·nitræk). Also **Minitrack.** [f. *mini*mum-weight *track*ing.] A system for tracking satellites in which a very light-weight oscillator is fixed in the satellite and its position is calculated from the phase difference between radio signals received by each of a spaced pair of fixed aerials on the ground. Usu. *attrib.*

1956 *Spaceflight* I. 27/2 A system of radio triangulation, called Minitrack, using phase comparison techniques, will establish the satellite's position and orbital path. *Ibid.* 28/1 Minitrack will enable a limited amount of research information to be telemetered to earth. 1958 *Observer* 2 Feb. 1/3 The minitrack stations have highly sensitive directional antennae which can 'focus' on the satellite. They also collect on tape recorders all the information it transmits. 1961 *Daily Tel.* 18 Jan. 20/4 One of two remote-controlled rotating aerials at the minitrack radio tracking station at Winkfield, Berks, which is now nearly ready to go into operation. 1966 *McGraw-Hill Encycl. Sci. & Technol.* VIII. 504/1 A two-axis Minitrack system was constructed so as to measure the north-south and east-west angular positions simultaneously. 1974 G. Perry in H. Miles *Artificial Satellite Observing* vii. 145 The NASA Minitrack system provides directional observations with an accuracy of about 100 seconds of arc.

mink. Add: **1.** Also, a garment made of this fur.

1951 F. Loesser in Swerling & Burrows *Guys & Dolls* (1960) ii. i. 44 (*song-title*) Take back your mink. *Ibid.* 45 Take back your mink Those worn out pelts And go shorten the slevees [*sic*] For somebody else. 1969 S. Ellin in *Ellery Queen's Grand Slam* (1971) 19 Two cars, a new mink whenever you feel like it. 1972 J. Gill *Tenant* iii. ii. 87 On their tenth wedding anniversary he had given her a silver mink.

2. Substitute for def.: A small, semi-aquatic, stoat-like mammal belonging to one of several species of the genus *Mustela*, esp. the American mink, *M. vison*, which is farmed for its dark brown fur. (Later examples.)

1914 W. T. Hornaday *Wild Life Conservation* iv. 146 In farming communities, the Mink, Weasel, Skunk, Raccoon, and even the Opossum all become so destructive to poultry as to constitute pests. 1965 D. Morris *Mammals* 282 American Mink are farmed extensively for their lustrous, rich-brown pelts, and these animals have been deliberately introduced into other parts of the world because of their economic importance... In Britain, specimens which have escaped from fur farms have set up breeding colonies that are causing some concern.

b. A dark brown colour.

1955 *Punch* 16 Mar. 349/2 Colours are black, white, mink, and blue. 1961 *Guardian* 16 Jan. 4/7 There is a new colour available in the range [of blankets]—'mink'. 1971 *Vogue* 15 Sept. 30 This exciting trouser suit..[is] available in Mink/Black and Mulberry/Dark Navy.

4. *mink coat, farm, farmer, farming, oil, ranch, ranching, skin* (earlier example), *stole.*

1928 A. Christie *Mystery of Blue Train* x. 71 Perfectly dressed in a long mink coat and a little hat of Chinese lacquer red. 1958 *Daily Mail* 21 Mar. 5/2 A mink coat for his wife? 'Good gracious, I don't think so.' 1975 'M. Duke' *Death of Holy Murderer* 7 An inside pocket of her mink coat. 1965 in P. Jennings *Living Village* (1968) 246 On the mink farm..mating took place early in March. 1967 *Guardian* 26 Aug. 3/6 There is a mink farm near by, and..householders have been protesting..about the smell. 1961 A. Wilson *Old Men at Zoo* i. 53 A drunken failed mink farmer from Essex. 1916 *Yukon Territory* (Canada Dept. Interior) 188 The following practical hints on mink farming have been recently published. 1965 *Harper's Bazaar* June 46 To take care of cuticles, some extravagant materials are used—the latest being mink oil. 1974 *Times Herald Record* (Middletown, N.Y.) 22 Apr. 12/2 What about *mink oil*? It is close to the human lipid and helps to control the moisture balance and moisture retention. 1948 A. L. Rand *Mammals E. Rockies* 92 In Alberta the number of mink ranches has increased steadily from 35 in 1929 to 773 in 1943. 1916 *Yukon Territory* (Canada Dept. Interior) 188 Mink ranching will become an important industry. 1678 *Rec. Court of New Castle on Delaware* (1904) 349, 22 mincq skins great and smal. 1969 'E. Lathen' *Murder to Go* (1970) 230 Mrs. Chester Brewster..in a mink stole. 1974 D. Ramsay *No Cause to Kill* ii. 111 A mink stole or a diamond bracelet.

b. *fig.* in *attrib.* and *Comb.* uses: opulent, sumptuous, wealthy.

1960 *Spectator* 4 Nov. 681 A mink-stoled Republican residential suburb of Chicago. 1966 *Radio Times* 22 Sept. 22/1 Kenton gives the mink touch to half-price electric heating.

minke (mi·nkə). [f. the name *Meincke*; see quot. 1971.] In full, *minke whale.* A small whalebone whale, *Balænoptera acutorostrata*; also called the piked whale or lesser rorqual.

1939 *Geogr. Jrnl.* XCIII. 190 Minke and killer whales were cruising quietly in all parts of the bay [*sc.* the Bay of Whales in the Antarctic]. 1970 *Islander* (Victoria, B.C.) 19 July 13/1 We watched seven killer whales kill and eat a minke whale. 1971 F. D. Ommanney *Lost Leviathan* ii. 39 It [*sc.* the piked whale or lesser rorqual] is also known to the Norwegians as the Minke whale after a whaling gunner named Meincke who accidentally shot one in mistake for a Blue and thus achieved a rather dubious immortality. 1973 *Sci. Amer.* Aug. 43/1 The danger exists that intensive hunting of sperm and minke whales could eventually reduce both species to the level of severe endangerment.

minky. Also **minky-winky.** *colloq.* Quasi-childish nonce-words for Milk *sb. rare*⁻¹.

1930 D. H. Lawrence *Nettles* 8 Eat your pap, little man, like a man! Drink its minky-winky, then, like a man! *Ibid.* 9 Drop of whiskey in its minky?

minne-drinking (mi·nə,dri:ŋkiŋ), *vbl. sb.* [f. G. *minne* love + Drinking *vbl. sb.*] Originally, a heathen practice among Germanic tribes at grand sacrifices and banquets, in honour of the gods or in memory of the absent or deceased. Later, a similar practice said to survive in certain localities in Germany.

1880 J. S. Stallybrass tr. *Grimm's Teutonic Mythol.* I. iii. 62 Minne-drinking, even as a religious rite, apparently exists to this day in some parts of Germany.

‖ **minnelied** (mi·nəlīt). Pl. -lieder (lī:dəɹ). [G., f. *minne* love + *lied* song.] A love-song written by a minnesinger, or in the style of the minnesingers.

1876 Stainer & Barrett *Dict. Mus. Terms* 292/2 Minnesingers..devoted their talents to the production of love songs (Minnelieder). 1881 *Encycl. Brit.* XII. 90/1 The first lyrical writer of Holland was John I...who practised the *minnelied* with success.

minnenwerfer: see *Minenwerfer.

Minnepoesy (mi·nəpōu:ėsi). *arch. rare.* [f. G. *minne* love + Poesy *sb.*] = next.

1845 Longfellow *Poets & Poetry of Europe* 182/2 In the fresh and youthful Minnepoesy, all art has acquired the appearance of nature.

Minnepoetry (mi·nə,pōu:ėtri). *arch. rare.* [f. G. *minne* love + Poetry.] The poetry of the Minnesingers.

1887 *Amer. Jrnl. Philol.* Dec. 454 The classical representative of Minnepoetry, Walther von der Vogelweide.

Minnesong (mi·nəsǫŋ). [ad. G. *minnesang*, f. *minne* love + *sang* Song *sb.*; cf. Minnesinger.] One of the songs of the Minnesingers. Also *collect.*

1845 Longfellow *Poets & Poetry of Europe* 182/2 This is the reason that all the Minnesongs..seem still to resemble each other. 1907 F. C. Nicholson *Old German Love Songs* p. iii, English works on the subject of the German Minnesong are..scanty in number. 1915 K. Breul *Cambr. Songs* 36 Several..of the 'Cambridge Songs' may be considered as direct forerunners of the early Minnesong.

Minnesota (minĭsōu·tă). [See next.] In phr. *Minnesota Multiphasic Personality Inventory*: a personality test made up of over 500 items, the responses to which are graded for various personality traits in accordance with criteria calculated from groups of normal subjects and subjects with clinically diagnosed psychiatric disorders. (Cf. *M.M.P.I.)

1946 *Jrnl. Appl. Psychol.* XXX. 517 The present paper presents preliminary data on the use of the Minnesota Multiphasic Personality Inventory (MMPI) with respect to differential diagnosis, with secondary findings upon the subject of overall identification of 'abnormals' from people in general. 1952 L. W. Ferguson *Personality Measurement* viii. 235 (*caption*) Classification of items in the Minnesota Multiphasic Personality Inventory. 1973 *Jrnl. Genetic Psychol.* CXXII. 65 The tasks presented to the subjects consisted of Block Counting..and the Depression scale from the Minnesota Multiphasic Personality Inventory (MMPI).

Minnesotan (minĭsōu·tăn). [f. the name of *Minnesota* (see below) + -an.] A native or inhabitant of Minnesota, a State in the north-central United States.

1888 D. D. Field *Speeches, Arguments & Misc. Papers* (1890) III. 369 Nebraskan, Kansan, Arkansan, Minnesotan, are the true designations of the citizens of those flourishing States. 1939 F. Scott Fitzgerald *Let.* Mar. (1964) 53 He's a Minnesotan and seems to me an altogether admirable fellow. 1965 Mrs. L. B. Johnson *White House Diary* 20 Jan. (1970) 229 Next we visited the Shoreham, which was full of Minnesotans.

Minnie², **minnie** (mi·ni). *Military slang.* Also **minny.** [abbrev. G. *minenwerfer* trench-mortar.] A German trench-mortar, or the bomb discharged by it. Also *attrib.* Hence as *v. trans.*, to attack with such a trench-mortar.

1917 A. G. Empey *From Fire Step* 36 A German 'Minnie' (trench mortar) had exploded in the next traverse. 1927 E. Thompson *These Men, thy Friends* 116 A minnie had been established in the enemy line. 1930 Blunden *De Bello Germanico* iv. 46 He might have 'minnied' or gunned us out in a few light-hearted rounds.

Ibid. vi. 73 The German minnie-man knew how to upset our domestic programme. **1933** —— & Norman *We'll shift our Ground* 14 M. M. for bombing a minny-crew out. **1950** G. Wilson *Brave Company* iii. 40 That bloody moaning Minnie... It's a hell of a weapon.

minnie, minny (mi·ni), *v. Sc.* and *north. dial.* [f. Minnie.] *trans.* To mother; to act as a mother towards (a lamb); to find (a lamb) its mother; also *refl.*, of a lamb: to find (itself) a mother.

1772 in *Sc. Nat. Dict.* (1965) VI. 283/1 Four or six lambs broke off from the flock of old sheep..and run [*sic*] to the ewes, and minnied or mothered themselves by sucking. **1825** Jamieson *Suppl.*, s.v., It is given as a proof of the accuracy of a shepherd's acquaintance with his flock.. that, after the lambs have been separated from the ewes, he can *minnie ilka lamb*. **1861** F. O. Morris *Rec. Animal Sagacity & Character* 117 There was not a single ewe.. which did not minny her lambs—that is, assume the character of mother towards the offspring from which she had been separated.

minnow. Add: **3.** minnow-twisting *vbl. sb.*, erratic movement or behaviour, resembling that of a minnow.

1935 L. MacNeice *Poems* 52 The minnow-twistings of the latinist who alone Nibbles and darts through the shallows of the lexicon.

minnowed (mi·nǝud), *a. poet.* [f. Minnow + -ed².] Containing or abounding in minnows.

1889 W. B. Yeats *Wanderings of Oisin* 71 Hour by hour He ruffles with his bill the minnowed streams.

Minoan (minǝu·ăn, mǝin-), *a.* and *sb.* [f. L. *Mīnōs* (Gr. Μίνως), the name of a legendary king of Crete + -an.] **A.** *adj.* Of or pertaining to ancient Crete, *spec.* to the Bronze Age civilization extending from the early part of the third to the end of the second millennium B.C., or to its people, culture, or language; also, to this civilization (or aspects of it) discovered elsewhere in the Aegean area. **B.** *sb.* **a.** An inhabitant of Minoan Crete or other parts of the Minoan world. **b.** The language or scripts associated with the Minoan civilization. Hence **Mino:aniza·tion**; **Mino·anized**, **Mino·anizing** *adjs.*

The precise chronology of the successive phases and of the collapse of Minoan civilization is highly controversial.

1894 A. J. Evans in *Jrnl. Hellenic Stud.* XIV. 367 At a time when 'Minoan' Crete and Mycenaean Greece had..evolved independent systems of writing. **1902** *Nature* 20 Nov. 58/1 The dominions of the proud Minoan thalassocrats disappeared. *Ibid.*, The hieroglyphic of their tutelary deity may have been used by the Minoans as a sort of heraldic device. **1904** *Ibid.* 15 Sept. 482/2 It was this faience that the Minoan potters imitated. **1921** *Spectator* 5 Mar. 293/1 A hanging lamp..with a yellow and orange spiral..and with a sort of Minoan design on it in orange, terra-cotta, and black. **1931** *Times Lit. Suppl.* 16 Apr. 308/1 He gives proof of his belief that the language of the Minoan script is Greek. **1939** J. D. S. Pendlebury *Archaeol. Crete* iv. 230 The Minoanization of the Mainland. *Ibid.*, Though superficially Minoanized, the mainland still kept a good deal of its native culture and taste. **1950** G. E. Daniel *100 Yrs. Archaeol.* vi. 195 [illegible] in Egyptian and Minoan. *Ibid.*, It seems likely that the language of the Minoans was Anatolian. **1960** H. L. Lorimer *Homer & Monuments* i. 38 The strongly Minoanizing regions of LH II. **1966** *Jrnl. Hellenic Stud.* LXXXVI. 1 Documents in the script known as 'Minoan Linear B' were unearthed at Knossos in Crete over fifty years ago. **1966** C. H. Gordon (*title*) Evidence for the Minoan language. **1970** Bray & Trump *Dict. Archaeol.* 149/1 'Minoan' is strictly a cultural term. Surviving bones show that the Bronze Age Cretans were a racially mixed group. **1970** *Oxf. Classical Dict.* (ed. 2) 1060/2 A Minoan colony of pre-eruption days [at Thera] is being excavated.

Min of Ag (min ǝv æg), *colloq.* abbrev. of *Ministry of Agriculture* (, *Fisheries and Food*).

1946 M. Dickens *Happy Prisoner* vi. 106 D'you know what the Min. of Ag. want Fred to do... They want him to plough up the hill field. **1962** *Observer* 6 May 21/3 A man at the Board of Trade will talk about 'those Min of Ag people'. **1972** *Guardian* 22 Mar. 13/1 'The Archers' began ..[as] a sort of broadcast Min of Ag handout.

Minol (mǝi·nǝl). Also **minol**. [Prob. f. Min(e *sb.* + -ol.] A mixture of ammonium nitrate, T.N.T., and aluminium used as the explosive in depth charges.

1946 *Battle of Atlantic* (H.M.S.O.) 57 About this time [*sc.* January 1943] the shattering power of our naval depth-charge was increased by the use of a new explosive called 'Minol'. This would crack the pressure hull of a U-boat at 25 feet. **1947** Crowther & Whiddington *Science at War* iv. 161 From January 1943, all Mark VII depth charges were filled with Minol. **1972** *Materials & Technol.* IV. xix. 725 The ammonals..and the minols—ammonium nitrate 40%, TNT 40% and aluminium 20% —were used for commercial and military applications, but are liable to evolve gases owing to the interaction of aluminium with moist ammonium nitrate.

minor, *a.* and *sb.* Add: **A.** *adj.* **2.** *minor poem, public school, road*; **minor league** (chiefly *N. Amer.*), the lower associations of teams in baseball, etc. (opp. *major *a.* 1 e); also *attrib.* and *fig.*; hence **minor leaguer**; **minor loyalty**: adherence to an institution, church, trade union, or the like, which is subordinate to loyalty to one's country or its government; **minor piece**: in *Chess* (see quot. 1847); **minor suit**: in *Bridge*, diamonds or clubs; **minor tactics**: the tactics or handling of bodies of troops in the immediate face or expected presence of the enemy.

1820 Keats *Let.* 2 Jan. (1958) II. 26 It is my intention to wait a few years before I publish any minor poems. **1820** J. S. Bingham tr. *E. Del Rio's Incomparable Game of Chess* i. 26 Two minor Pieces, for a Rook and two Pawns, may be considered an equal contract. **1847** H. Staunton *Chess-Player's Handbk.* i. iii. 24 The Bishop and Knight, in contradistinction to the Queen and Rook, are called the minor Pieces. [illegible] **1906** *Cincinnati Enquirer* [illegible] minor leaguers. [illegible] **1928** H. Rowan-Robinson *Some Aspects Mechanization* 3 The study of the minor tactics of petrol-driven forces. **1932** *Times* 28 Mar. 20/1 After six years in the minor leagues he has become catcher for the Washington Senators. **1936** 'J. Tey' *Shilling for Candles* vii. 81 The Grammar schools..turned out a very fine type of boy. Better often..than came from the minor public schools. [illegible]

6. d. (Earlier example.)

1820 J. Severn in Keats *Let.* (1958) II. 342 Here I must change to a Minor Key—Miss C fainted..I was very ill.. Keats ascended his bed.

B. *sb.* **9.** (Earlier and later examples.)

1821 P. Egan *Real Life in London* I. vi. 92 Mr Gloss'em, who is a *shining character* in the theatrical world, at least among the minors of the metropolis. **1931** *Times Lit. Suppl.* 1 Jan. 1/2 The vast galleries of the patented theatres and the cramped benches of the 'minors' were thronged with a new audience.

10. In American universities and colleges, a subject or course of study to which less attention is given, or for which fewer credits are given, than for a major. Also, this subject seen as a qualification. (See also quot. 1969.)

1890 in T. W. Goodspeed *Hist. Univ. Chicago* (1916) 142 The plan of majors and minors..has been arranged in order to meet this difficulty. **1909** Webster, *Minor*, a subject of study, pursued by a candidate for a higher degree, less time being devoted to it than to his major subject. **1919** *Univ. Texas Bull.* No. 1925. 105 The student will note that it is possible to arrange his minor..[so] as to take in effect two majors. **1926** [see *elective B]. **1930** *Bull. Ada* (Okla.) *Even. News* 2 July 6/2 Alfred Pryor-Smith..will graduate at East Central in July, with a major in history and a minor in sociology. **1969** *Amer. Heritage Dict.*, *Minor*..One who studies a minor: *a chemistry minor*.

11. In *Bridge*, = *minor suit*; also, a card in a minor suit.

1927 [see *major *a.* 1 d]. **1958** *Listener* 4 Dec. 965/1 [illegible] both minors. **1960** *Times* 16 Nov. 17/4 A two-suiter in the minors only, with Diamonds as long as, or longer than the Clubs.

minor (mǝi·nǝɹ), *v.* Chiefly *N. Amer.* [f. Minor *a.* and *sb.*] *intr.* Of a university student: to take, or qualify *in*, a minor (see prec., sense *B. 10).

1934 in Webster. **1967** *Oxf. Mag.* 10 Feb. 205/1 [Canada] They intend to major in life sciences and minor in Phys. Ed.

minorate, *v.* Delete † *Obs.* and add later examples.

1789 J. Morse *Amer. Geogr.* 212 Their design is by quantity to depreciate the value of their bills; and lands mortgaged for public bills will be redeemed in these minorated bills. **1922** Joyce *Ulysses* 387 Assuefaction minorates atrocities (as Tully saith of his darling Stoics).

minority. Add: **2.** (Later example of *in minority*.)

1920 *Act 10 & 11 Geo. V* c. 64 § 2 A husband of full age,.. whose wife is in minority, shall be her curator during her minority.

3. b. A small group of people separated from the rest of the community by a difference in race, religion, language, etc. Also in quasi-*adj.* use.

1921 H. W. V. Temperley et al. *Hist. Peace Conf. Paris* V. ii. 112 These treaties provide for the protection of racial, linguistic, or religious minorities included within the boundaries of the specified States. **1930** *Economist* 29 Nov. 1001/1 The Nazis and the Stahlhelm have been conducting reprisals against the Polish minority in Germany. **1937** *Times* 24 Feb. 13/3 The concessions made by the Prague Government to sections of the German minority.. are disconcerting to any plans..for a large-scale minorities campaign. **1945** *Ann. Reg. 1944* 26 Some broad general declaration by the United Nations deprecating the ill-treatment by a state of its minorities. **1945** L. Wirth in R. Linton *Sci. of Man* 347 The existence of a minority in a society implies the existence of a corresponding dominant group enjoying higher social status and greater privileges. **1970** *Washington Post* 30 Sept. B. 14/4 A hard-hat worker who loses his job to a minority worker. **1973** *Black Panther* 3 Mar. 13/3 In civil rights, aid to minority business enterprises is stressed, rather than to the minority poor. **1975** *Atlantic Monthly* Jan. 29/1 By 'minority' today, we mean a disadvantaged group of citizens. *Ibid.*, Today, by a 'minority' we mean not the privileged at the top, but the underprivileged at the bottom.

5. minority carrier *Electronics*, a charge carrier of the kind carrying the smaller proportion of the electric current in a semiconducting material (cf. *majority carrier* s.v. *Majority 7); **minority group**, a group forming a minority (sense *3 b); **minority language**, a language spoken by a minority group if different from that of the majority; **minority man**, one who is in a minority or tries to secure recognition of the claims of minorities; **minority member**, a member appointed to a board, committee, or the like to represent a minority; **minority movement**, a movement to secure justice or proper representation for minorities; **minority report** (examples); **minority rights**, rights granted to minorities to act as a safeguard of their interests and help prevent discrimination against them by the majority.

1951 Minority carrier [see *majority carrier* s.v. *Majority 7]. **1962** Simpson & Richards *Physical Princ. Junction Transistors* iv. 60 Another variation in charge in the vicinity of the junction arises from the injection or extraction of minority carriers under conditions of forward or very small reverse bias respectively. **1969** J. J. Sparkes *Transistor Switching* i. 3 The current from one junction to the other is carried by minority carriers. **1942** Locke & Stern *When Peoples Meet* i. iv. 125 Such.. inclusion of minority groups is in marked and often ironic contrast to the more normal 'divide and rule' policy..of dominant groups. **1964** M. Argyle *Psychol. & Social Probl.* vii. 94 Several studies show that prejudice increases with the density of the minority group in an area. **1971** R. Bendix in A. Bullock *20th Cent.* 357/1 The citizenship of racial minorities remains an unresolved problem. Members of minority groups are denied rights which are formally theirs. **1939** L. H. Gray *Foundations of Lang.* v. 118 Minority-languages may thus be not merely linguistic in interest, but may also constitute political problems of all degrees of importance. **1927** *Observer* 1 May 17/1 It was a curious moment..to choose for legislation calculated..to revive the power of 'minority men' and direct actionists in Britain. **1874** *Porcupine* 31 Jan. 693/2 The city of London has already conceded a minority member. **1927** *Daily Tel.* 6 Sept. 7/3 He did not agree with the Minority Movement. **1833** *Reg. Deb. Congress U.S.* 2 Mar. 1927 A new set of majority and minority reports are to be launched upon the public. **1940** B. de Voto (*title*) Minority report. **1958** *Everyman's Encycl.* XII. 532/2 With Sidney Webb..she [*sc.* Beatrice Webb] issued the minority report which initiated the Socialist agitation for the reform of the old Poor Law. **1924** R. W. Seton-Watson *New Slovakia* vi. 104 Such international opinion as regards the 'Minority rights' provided for by the Peace Treaties, as a moral obligation assumed by all members of the League of Nations. **1955** I. L. Claude *National Minorities* xii. 166 Whenever the issue must be squarely faced, the U.S. takes the lead..in opposing the concept of special minority rights. **1972** 'E. Lathen' *The Longer the Thread* viii. 74 She was a Puerto Rican—yet she had never served on a commission about minority rights.

b. quasi-*adj.* Of, for, composed of, or appealing to a minority (of people); freq. with the suggestion of 'serious, intellectual, highbrow' (as opposed to *Mass *sb.*² 10).

1930 F. R. Leavis (*title*) Mass civilization and minority culture. **1932** Q. D. Leavis *Fiction & Reading Public* ii. iv. § 3. 199 Enough attention has perhaps been given to the effects of minority values. **1941** *Partisan Rev.* Mar.–Apr. 111 Leisured people who have been brought up in a minority culture. **1944** L. MacNeice *Christopher Columbus* 9, I am in favour of occasional special [radio] programmes for small minority audiences. **1958** R. Williams *Culture & Society* iii. iii. 235 Those extreme forms of the idea of a 'minority culture'. **1959** *20th Cent.* Nov. 324 Serious minority programmes [on television] at peak hours. *Ibid.* 333 A place remains for it [*sc.* sound radio].. in the evenings for special minority interests and for music. **1960** *Housewife* Apr. 10/2 The Editor..considers very few subjects indeed too 'minority' or too apparently trivial to be given a sensible airing. **1971** *Guardian* 22 Feb. 8/1 Even from the BBC..a complete diet of minority-interest programmes..would not be..acceptable. **1972** P. Black *Biggest Aspidistra* i. iv. 42 The heat was taken off the minority drama.

Minotaur. (Later examples of allusive use.)

1939 Spender & Rees tr. *Büchner's Danton's Death* i. iv. 42 The people is a Minotaur that must be fed with corpses every week if it is not to eat the Committee alive. **1945** Auden *Sea & Mirror* ii. 51 Home to your promiscuous pastures where the minotaur of authority is just a roly-poly ruminant and nothing is at stake. **1950** T. S. Eliot *Cocktail Party* iii. 165 We talk of darkness, labyrinths, Minotaur terrors. **1964** A. W. Gouldner in I. L. Horowitz *New Sociol.* 196 The lair of this minotaur [*sc.* Max Weber]..is still regarded by many sociologists as a holy place.

Minox (mǝi·nǫks). The proprietary name of a type of miniature camera.

1952 *Brit. Jrnl. Photogr.* 5 Sept. 433/2 Several miniature cameras have already appeared on the market. One notable example is the Minox. This uses 16-mm. film.. and the negatives produced give moderate size enlarged prints of good quality. **1962** L. Deighton *Ipcress File* ii. 20 Chico..would have been making time with the Minox on [illegible] camera. **1966** I. Bingham *Double Agent* v. 72 Golchenko [illegible] rolls of Minox film. **1966** *Trade Marks Jrnl.* 6 Apr. 463/1 Minox... Photographic apparatus and instruments and parts and fittings there for. Minox Gesellschaft mit beschränkter Haftung (a Limited Liability Company organised under the laws of the German Federal Republic), Giessen-Heuchelheim, Germany; manufacturers. **1969** K. Benton *24th Level* vi. 116 [He] took out his Minox..and took photographs.

minstrel, *sb.* Add: **2.** *spec.* One of the Old English period.

1767 Percy *Ess. Anc. Eng. Minstrels* 9 The privileges and honours which were so lavishly bestowed upon the northern scalds, were not wholly withheld from the Anglo-Saxon Minstrels. **1840** T. Wright *Ess. Lit. & Learning under Anglo-Saxons* 3 All literary genius centred in one person, the minstrel, who equally composed and sang. **1892** J. Earle *Deeds of Beowulf* 136 Rieger.. understands that the minstrel did not merely narrate, but improvised. **1928** W. W. Lawrence *Beowulf & Epic Trad.* 46 The lines at the beginning [of *Widsith*] introducing the minstrel, and those at the end glorifying his profession. **1951** D. Whitelock *Audience of Beowulf* 77 The poet, if he had not wished, was not forced to make the minstrel sing of the Creation. **1966** G. Stanley *Continuations & Beginnings* 129 Sometimes a minstrel working in the oral-formulaic tradition coined a phrase, for every phrase must have been new before it grew old.

4. (Earlier examples.)

1843 G. C. D. Odell *Ann. N.Y. Stage* (1928) IV. 668 The Ethiopian Serenaders, or Boston Minstrels. **1846** *Illustr. London News* 24 Jan. 61/2 The Ethiopian serenaders. A party of American minstrels..commenced..a series of concerts.

4*. *slang.* (See quots.)

1967 M. M. Glatt et al. *Drug Scene* 125 Minstrel (*black and white*), Durophet. **1971** E. E. Landy *Underground Dict.* 133 Minstrel, 12.5 mg. capsule of an amphetamine and a sedative.

5. *attrib.* (Further examples, esp. in sense 4.)

1865 *Chicago Tribune* 10 Apr. 1 Buckley and Budd's Minstrel house is blazed. **1870** O. Logan *Before Footlights* 414 A clever actor..who wrote a burlesque..for a minstrel show. **1885** W. B. Yeats *Island of Statues* ii. iii, in *Dublin Univ. Rev.* July 139/2 He who hath the halcyon's wing ... **1907** A. Einstein *Mus. Romantic Era* xvii. 331 Of his [*sc.* S. Foster's] songs, *Oh! Susanna* was one of the oldest [1848]..and was most widely circulated by the 'minstrel shows'. **1949** *Radio Times* 15 July 18/3 A black-faced minstrel show with boxes, tambourines, corner men, stump speeches. **1975** *Listener* 3 Apr. 454/3 At the time of the Civil War, the minstrel show became less complex in its treatment of Negro life.

mint, *sb.*¹ Add: **6. mint condition** = *mint-state*; **mint par, parity (of exchange)**, the ratio between the gold equivalent of the currency units of two countries; the rate of currency exchange between two countries based on this ratio; **mint-state**, also applied to books and other objects in pristine condition.

1902 *Connoisseur* May 69/1 Nothing is more marked in present day stamp collecting than the insistence..upon what is expressively termed a 'mint' condition in unused specimens. **1923** *Punch* 7 Feb. 130 Here we have Holbein's portrait of the first earl... His *chef-d'œuvre*, in mint condition. **1955** L. Murdoch *Flight from Enchanter* i. 9 The books were chaotic, but in mint condition, since reading was not a popular activity. **1975** 'D. Rutherford' *Mystery Tour* iii. 39, I did have half a dozen vintage Rolls-Royces..but..it was becoming more and more expensive to keep the cars in mint condition. **1882** Mint par [see Par *sb.*¹ 2 a]. **1891** G. Clare *Money-Market Primer* 74 A Mint Par can only be established between countries that employ the same standard of value. **1928** W. F. Spalding *Dict. World's Currencies & Foreign Exchanges* 134/2 The mint par of exchange is the rate of exchange at which the standard coin of one country is convertible into that of another country according to the terms of their respective mint laws. **1965** Perry & Ryder *Thomson's Dict. Banking* (ed. 11) 376/2 The Mint Par between two countries never varies unless one of them alters its coinage regulations. **1928** L. D. Edie *Money, Bank Credit & Prices* iii. 52 A so-called mint parity existed between all gold standard countries. **1940** G. Crowther *Outl. Money* ix. 317 The 'mint parity' being $4.86⅔ = £1, whenever the exchange rate..fell..it became cheaper to buy gold from the Bank of England..and..sell it to the Federal Reserve Bank for dollars. **1965** Seldon & Pennance *Everyman's Dict. Econ.* 285 *Mint Parity of Exchange*, the exchange rate between two currencies both of which are legally convertible at fixed rates into..gold.... The mint parity then expresses the ratio between the two legal rates. **1931** *Times Lit. Suppl.* 16 Apr. 305/3 Copies of 'Waverley'..are excessively rare in mint state.

mint, *sb.*² Add: **1. c.** = Peppermint 2 b. Also a sweet, or chocolate, flavoured with or containing mint. Also *adj.*

[illegible] **1964** *Times* 21 Nov. 2/3 [illegible]

3. *mint-green* adj.; **mint cake** (later examples); **mint jelly**, mint-flavoured jelly, usu. eaten with roast lamb; **mint julep** (earlier and later examples); **mint rock** (example); **mint-sling** *U.S.*, a drink containing some alcoholic beverage flavoured with mint; **mint vinegar**, mint-flavoured vinegar.

1958 E. Newby *Short Walk in Hindu Kush* xiv. 170 [illegible] **1804** *Balance* 13 Mar. 86 (Th.) Three Mint Slings. **1846** Soyer *Gastr. Regenerator* [illegible] **1935** *Daily Express* [illegible] **1952** E. Acton *Mod. Cookery* (ed. 2) v. 137 Green mint vinegar..The mint itself..will keep well in vinegar, though the colour will not be very good. **1957** E. Craig *Collins Family Cookery* 893 Mint Vinegar.. [illegible] then pour in mild vinegar to overflowing.

mint, *a. ellipt.* for *in mint condition* (see *mint *sb.*¹ 6).

1902 *Connoisseur* Jan. p. xiv, A hitherto unknown stamp..unused (mint). **1928** *Humphris'* (Norwich) *Catal.* No. 149. 13/1 'Mint' signifies as New. **1950** J. Carter *A.B.C. for Bk.-Collectors* 59 A cloth-bound, boarded or a wrappered book may be called unopened, uncut, mint. **1976** *Ibid.* 120 Dust-jacket defective, otherwise mint. **1977** P. Oliver *Screening Blues* 3 Other conditions reflect the popularity of singer or song, and some [records], grey on one side and 'mint' on the other, betray hard service in a juke-box. **1972** *Drew & Muir* (Takeley, Hertfordshire) *Catal.* No. 35. 16/1 This book is one of the prettiest small publications of the period..Superb copy in original 'hard currency'. **1840** J. P. Kennedy *Quodlibet* 106 Tumblers by the dozen..Mint Drops, and the Lord knows what.

mint-drop. **2.** (Earlier examples.)

1835 Hawthorne *Passages from Note-Bks.* in *Atlantic Monthly* (1866) Jan. 3/2 The bar-keeper had one Ten-dollar [illegible] **1837** *Congress. Globe* 29 Sept. App. 339/1 [The money flowed to Mobile] by the aid of the far-famed Specie circular, in 'mint drops' and 'hard currency'. **1840** J. P. Kennedy *Quodlibet* 106 Tumblers by the dozen..Mint Drops, and the Lord knows what.

Mintech (mi·ntek). Also **Mintec, Min. Tech.** Colloq. abbrevs. of *Ministry of Technology*.

1967 *Guardian* 26 Aug. 6/6 A Min Tech review of the whole problem. **1967** *Spectator* 15 Dec. 750/3 The gritty 'technological' phrasemaking of you-know-who and his Mintech. **1970** *Guardian* 25 Feb. 10/3 Mintech exists... They really do give this little name..to the great big Ministry of Technology. **1970** *Punch* 17 June 920/1 Every available taxpayer's penny is needed for wasting at Min-Tech. **1971** *New Scientist* 17 June 673/1 The civilian aspects of the defence research establishments—so laboriously cultivated under Mintech auspices.

minto (mi·ntǝ). Also **Mintoe, mintoe.** [f. Mint *sb.*²] The proprietary name of a type of boiled sweet flavoured with peppermint.

1935 *Trade Marks Jrnl.* 16 Oct. 1298/2 Nuttall's Mintoes... Boiled sugar sweetmeats flavoured with mint. William Nuttall Limited,..Doncaster..; confectioners. **1949** Dylan Thomas *Let.* 13 Oct. (1966) 328 Yesterday I broke a tooth on a minto. **1953** —— *Under Milk Wood* (1954) 2 Coughing like nannygoats, sucking mintoes, fortywinking hallelujah. **1957** *Sheat's Compl. Confectioner* (ed. 13) 63 Mintoes. Boil the sugar, glucose and water..kneed in the oil of peppermint and.. spin out and cut. *Ibid.* 64 (caption) Kiss Cutting on 'Mintoe' Machine. **1963** *Listener* 21 Mar. 501/2, I wanted to keep on sucking mintoes for ever.

Minton (mi·ntǝn) (1766–1836), used *attrib.* to designate the pottery made at Stoke-on-Trent, Staffs., from 1793 onwards, by him and his successors. Also *ellipt.*, = Minton ware.

1857 J. Marryat *Hist. Pott. & Porc.* (ed. 2) xii. 302 Staffordshire is now the site of the great manufactures. From among these may be mentioned Minton, Copeland, Ridgway. **1863** W. Chaffers *Marks Pott. & Porc.* 125 Minton's, established in 1791 by Mr. Thomas Minton.... The name, indented on the ware, is generally adopted both for china and earthenware. **1888** [see Tile *sb.*¹ 1 c]. **1926** S. T. Warner *Lolly Willowes* i. 11 An amateur of china, who had dowered all his nieces..with Minton. **1960** K. Collier *House called Memory* iii. 44 The best silver tea-service and the Minton tea-set. **1964** 'R. Raine' *Wreath for America* iii. 23 A big silver tray bearing two Minton cups and saucers. **1972** R. Player *Oh! Where are Bloody Mary's Earrings* vii. 182 A scraping of boots on the Minton tile floor.

minuet, *v.* Add examples of *fig.* use.

1881 G. M. Hopkins *Let. to R. Bridges* (1955) 125 The magic nib has..minuetted..into the syllables of your name. **1972** *Newsweek* 17 July 21/3 MacGregor might have to minuet with White House aides for a Presidential audience.

‖ **minuetto** (miniue·tǝ). [It.] = Minuet.

1724 *Short Explication Foreign Words in Musick Bks.* 46 *Minuetto*, a Minuet. **1883** *Grove Dict. Mus.* II. 342/2 'Minuetto' is often used as..Trio..or Air belonging thereunto. **1888** [see Minuet 1]. **1923** [see *Grape *sb.*³ 1]. **1931** Blom & Westrup *Everyman's Dict. Mus.* (ed. 5) 431/2 Minuetto (It.), 'little minuet', which is the correct It. name for the Minuet.

minus. Add: **2.** *spec.* as part of an examiner's mark, as in *a*— (read as *alpha minus*).

1924 A. Huxley *Brave New World* iv. 75 He..called to a lounging couple of Delta-Minus attendants to come and push his machine out. **1958** [see *Beta 3]. **1962** M. Drabble *Summer Bird-Cage* i. 13, I..got on the train, where I read.. *Tender is the Night* (beta minus).

b. *minus quantity* (later example); also *transf.*, insignificant.

1916 R. Fry *Let.* 14 Aug. (1972) II. 401, I fear my recommendation would generally prove a minus quantity. **1922** Wodehouse *Clicking of Cuthbert* iii. 221 He might be a pretty minus quantity in a drawing-room or at a dance, but in a bunker or out in the open with a cleek, I assure you..you'd be surprised. **1966** F. Sinclair *Most Unnatural Murder* xv. 174 Cherub-faced Phyl's a minus quantity.

c. Also, absent.

1853 Dickens *Down with Tide* in *Househ. Words* 5 Feb. 483/2 Being, when called upon to answer the question what Waterloo described as 'Minus', or, as I humbly conceived it, not to be found. **1858** A. Mayhew *Paved with Gold* ii. xiv. 342 If we ain't minus in less than no time, we're blowed upon.

e. Followed by the name of a colour to designate the complementary colour, i.e. that of white light from which the specified colour has been removed; so *minus colour*.

1901 Cadett & Shepherd *Orthochromatic & Three-Colour Photogr.* 14 [In the Sanger Shepherd process] the prints for the minus green or pink, and minus blue or yellow positives are printed (superimposed) together. **1901** [*Chambers's Jrnl.* 4 May 366/1 The prefix 'minus' attached to a primary colour..[implies] that this particular colour is cut out of the spectrum of white light, and that the negatively-named compound is a blend of the hues remaining, 'White minus red', 'white minus green', and 'white minus blue' would be the complete expressions; they are ordinarily termed complementary colours. **1936** *Discovery* Jan. 2/1 It [*sc.* Monastral Fast Blue BS] is a true 'minus-red' pigment for three-colour printing. **1939** J. H. Coote *Making Colour Prints* 17 The three colours which are used for 'subtraction' are described as 'minus' colours. **1970** D. L. MacAdam *Sources of Color Sci.* 130 Red, green, and blue, being the colors in positive synthesis, minus red, minus green, and minus blue, or cyan blue, bright crimson and yellow are the printing colors.

minuscule, *a.* Add: **2.** (Further examples.)

Now usu. stressed on first syllable. See also *miniscule.

1963 *Ann. Reg. 1962* 17 Such minuscule militants as the boot and floor polish manufacturers. **1969** *Listener* 30 Jan. 155/3 One is impressed inevitably by the intensity and concentration of human effort on a project whose rewards, however satisfying, are bound to be minuscule. **1972** *Time* 17 Apr. 24/1 The Gallup organization, in all of its national soundings, has shown McGovern running between a minuscule 3% and 6% when pitted against his rivals for the nomination.

minute, *sb.*¹ Add: **1. b.** (Further examples.) Also in phr. *up to the minute*, completely modern.

1795 tr. *C. P. Moritz's Trav.* 93 Composing a sermon.. should not thus have been put off to the last minute. **1913** *Vanity Fair* (N.Y.) Dec. 9/2 Look at your Christmas shopping in this light... Don't put it off until the last minute.

1920 F. M. Ford *Let.* 27 July (1965) 117 Does 'for a minute' = 'at present'? or that you wouldn't think for a minute of knowing our establishment? **1937** A. J. Cronin *Citadel* ix. 75 Doctor! I think you'd be interested in our new indexometer. It has a multiplicity of uses, is absolutely up to the minute..and the price is only two guineas. **1955** R. Macaulay *Last Lett. to Friend* (1962) 208 Having the two children made it fun; they loved every minute of it. **1956** E. S. Gardner *D.A. takes Chance* iii. 24 A very attractive young woman, vivacious, up to the minute, a thoroughly modern young woman. **1958** *Spectator* 22 Aug. 249/1 These ought to be worth a trial to give ballet that shake-up it badly needs. Not merely for Art's sake, but, any minute now, for the sake of the box office. **1972** *Daily Tel.* 26 Aug. 20/4 A good atmosphere helps me to do my best, and while at the minute it doesn't feel like an Olympic Games, I think I can psych myself up when the time comes.

d. A distance expressed in the number of minutes needed for it to be traversed (on foot, etc.).

1886 *Taunt's New Map Thames* (ed. 5) Advts. 45 Hotel ..Adjoining the River, 3 minutes from Railway Station. **1907** *Daily Chron.* 18 Sept. 3/7 (Advt.), St. Pancras Station..is within a few minutes of the City. **1922** Joyce *Ulysses* 224 Can you send them by tram? Now?..—Certainly, sir. Is it in the city? —O, yes... Ten minutes. **1931** R. Campbell *Georgiad* i. 17 Up-to-date methods: breezy situation: And only twenty minutes from the station. **1934** G. B. Shaw *On Rocks* i. 203 The Isle of Cats... Down the river, Sir Arthur. Twenty minutes from your door by underground. **1962** J. G. Bennett *Witness* vii. 86, I went to his apartment, a few minutes from where we lived.

6. *minute of dissent*, a minute recording a person's disagreement with something.

1886 Kipling *Departmental Ditties* (ed. 2) 23 No longer Brown reverses Smith's appeals, Or Jones records his Minute of Dissent. **1930** *Times* 15 Mar. 7/1 All the members have signed the report, but Lord Ebbisham did so subject to a 'minute of dissent' which is attached to the main report.

7. minute-gun (earlier and later examples); **minute steak** (see quot. 1934); **minute-to-minute** *attrib.*, from one minute to the next.

1728 G. Carleton *Mem. Eng. Officer* 205 The first Guns that were fir'd from Gorge's battery were the Minute-Guns for his Funeral. **1936** H. Nicolson *Diary* 28 Jan. (1966) 241 The King's funeral. I stay in..all morning and do not hear more than the minute-guns firing dolefully in the distance. **1970** *Brewer's Dict. Phr. & Fable* (rev. ed.) 714/2 *A Minute gun*, a signal of distress at sea, or a gun fired at the death of some distinguished person. **1934** Webster, *Minute steak*, a small thin steak that can be quickly cooked. **1959** *Good Food Guide* 204 The grills..range from 6/6 (minute steak) to 9/6 (mixed grill, including vegetables). **1966** *Listener* 27 Jan. 134/3 The minute steak is the gastronomic symbol of the age. **1948** 'G. Orwell' in *Adelphi* XXIV. 249/2 One ought, apparently, to live in a continuous present, a minute-to-minute cancellation of memory. **1968** G. M. B. Dobson *Exploring Atmosphere* (ed. 2) v. 105 In practice certain precautions have to be taken to allow for the minute-to-minute changes in the general electric field.

minute, *v.* Add: **2. b.** To inform (someone) about a matter by means of a minute or memorandum.

1918 G. S. Gordon *Let.* 13 Dec. (1943) 87 Milford has minuted me about the Oxford Trivium. **1952** *Punch* 10 Sept. 353 He had minuted General Ismay. **1964** M. Gowing *Britain & Atomic Energy* v. 174 Lord Cherwell was still minuting Mr Churchill that the British diffusion method was much superior. **1974** 'J. Le Carré' *Tinker Tailor* xxvi. 221 In no case should I phone him or minute him; even the internal lines were taboo.

Minute-man. Restrict *Hist.* to sense in Dict. and add: **1. b.** In extended uses, a 'watchdog' or political activist; a member of an organization devoted to specific political issues. Also *gen.* (with small initial), a vigilant, observant, or enterprising person. (Now usu. without hyphen or as one word.)

1923 *Minute Men of the Constitution Roster* (inside front cover), The Minute Men of the Constitution is a non-partisan association, organized to obtain delegates from Illinois to the Republican and Democratic State and National Conventions... The above is a movement for good government. **1930** P. R. Leach *That Man Dawes* x. 190 Dawes called..a number of close friends and organized 'The Minute Men of the Constitution'. The purpose of the Minute Men..was to function as an open-eyed body of men throughout the state to watch elections, prevent vote frauds at the polls and counteract unfair propaganda. The society was not secret, had no passwords or meeting places and was non-political. **1952** *Manch. Guardian Weekly* 23 Oct. 3/1 An appeal went out for minute-men to tap the moneybags in their own localities. **1961** *Guardian* 25 Nov. 7/2 The Minute Men and Birchites. **1961** Webster, *Minuteman*, one who resembles a Revolutionary minuteman esp. in qualities of vigilance and readiness to take prompt action. **1964** D. Bell *Radical Right* i. 4 Three elements [in the 1960s] conjoined to attract public attention to the radical right... The John Birch Society. .. Seminars of anti-Communist 'schools'... And, third, there was the disclosure of the existence of extreme fanatic groups, such as the Minutemen, who organized 'guerilla-warfare seminars', complete with rifles and mortars, in preparation for the day when patriots would have to take to the hills to organize resistance against a Communist-run America. **1964** *Economist* 19 Dec. 1350/2 The Minutemen, a mysterious, probably negligible, vigilante group, has drawn the gloomily righteous conclusion that America has seen its last free election. **1965** T. H. White *Making of President 1964* iii. 89 Across the sky of politics there began to float new names like the John Birch Society, the Minutemen, the National Indignation Convention, [etc.]. **1970** *Peace News* 17 Apr. 8/1 It would be easy for the casual observer to link the Panthers with..the Minutemen and American Nazis.

2. (With capital initial.) A type of intercontinental ballistic missile (see quot. 1971).

1961 *Daily Tel.* 2 Feb. 1/4 The American Air Force achieved a spectacular advance today when it launched the first Minuteman missile. **1969** *Guardian* 6 Dec. 2/2 They claim that the three five-megaton warheads fitted to these missiles are unnecessarily large and accurate for use against any target but the hardened US Minuteman missile sites. **1971** E. Luttwak *Dict. Mod. War* 130/1 The Minuteman series are three-stage missiles with solid-propellant motors designed as simplified and low-cost weapons for launching from 'hard' silo sites. They need very brief pre-launch preparation.

minuter. **1.** Delete † *Obs.* and add later example.

1911 *Pitman's How to take Minutes* 14 The minuter must take sufficient notes of the proceedings.

minuterie (miniū·tǝri). [Fr., = clockwork, timing mechanism, minuterie, f. *minute* Minute *sb.*¹] **a.** A light switch incorporating a timing mechanism that automatically turns it off a short time after it has been (manually) turned on. **b.** An electric light controlled by such a switch.

1955 W. Gaddis *Recognitions* i. ii. 72 Crémer opened the door, and the light of the minuterie threw his flat shadow across the sill. **1958** E. Dundy *Dud Avocado* i. iii. 52 Racing from minuterie to minuterie to keep the stairs in a blaze of 40-watt bulbs. **1962** A. Williams *Long Run South* v. i. 160 He went up in the lift and pressed the minuterie. **1963** 'D. Rutherford' *Creeping Flesh* iii. 153 The stairs were illuminated with a tasteful glow. No penny-wise *minuterie* here. **1971** M. McCarthy *Birds of America* 302 On the second landing he found it..in the weak light of the *minuterie*.

minuting, *vbl. sb.* (See under Minute *v.*) Add earlier example.

1856 Dickens *Dorrit* (1857) ii. viii. 387 The work of form-filling, corresponding, minuting, memorandum-making.

minx. **2. b.** Delete † *Obs.* and add later examples.

Sense 2 a is still the more usual sense.

1939 Joyce *Finnegans Wake* 80 What are you doing your dirty minx and his big treeblock way up your path? *Ibid.* 496 There wasn't an Archimandrite of Dane's Island and the townlands nor a minx from the Isle of Woman..would come next or nigh him. **1941** J. Smiley *Hash House Lingo* 38 *Minx*, prostitute.

minxish (mi·ŋksiʃ), *a.* [f. Minx + -ish¹.] Having the character of a minx; like a minx. Hence **mi·nxishly** *adv.* Also (*rare*) **minxy** (mi·ŋksi) *a.*; (*dial.*) **mi·nxin** *a.*

1870 *Porcupine* 12 Feb. 443/3 Through a door, left slightly ajar, he..sees another minxish 'Girl of the Period' waiting the return of her companion. **1883** M. R. Lahee *Acquitted though Guilty* 55 Hoo wur a honest dacent woman: noan-like the, the minxin slut. **1919** W. De Morgan *Old Madhouse* xxv. 387, I do *not* believe that Lucy..is half as minxish as she made out. **1927** *Daily Express* 27 Apr. 3/2 The mannequin wore it minxishly, for it was a frock for a minx. **1935** *Times Lit. Suppl.* 3 Oct. 613/3 A weak, attractive, minxish child. **1939** C. Morley *Kitty Foyle* 91 We got some right minxy columbines round here.

miny, *a.* Add: Also **miney**. **1.** (Later example.)

1907 *World* 16 July 113/1 What do you say to mines? No, there's nothing miney about me.

minyan (mi·nyān), *sb.*¹ Pl. **minyanim**. [Heb. *minyān*, lit. 'count, reckoning'.] The quorum of ten males over thirteen years of age required for formal Jewish worship.

1753 *Jewish Ritual* 53 This..is executed by the Priest in the Presence of ten Jews, which Number of Jewish Men, they call in Hebrew, A Minyon. **1891** M. Friedländer *Jewish Relig.* 472 Women are disqualified for forming the quorum (*minyan*) required for public worship. **1893** I. Zangwill *Childr. Ghetto* (ed. 3) i. xii. 111 There was never lack of *Minyan*—the congregational quorum of ten. **1960** L. P. Gartner *Jewish Immigrant* vii. 215 In the early 1860's [in Leeds], there were barely the ten Jews required for a minyan. **1962** B. Abrahams tr. *Life Glückel of Hameln* iv. 78 A company of more than two *minyanim*. **1973** *Jewish Chron.* 2 Feb. 10/1 After a life of almost 33 years the congregation's list of members had of late diminished to the point where a minyan had become very difficult to secure.

Minyan (mi·niăn), *a.* and *sb.*² [f. L. *Minyae*, Gr. Μινύαι Minyans + -an.] **A.** *adj.* Also 6 **Mynian**, 9 **Minyean** (mini,ī·ăn) [Gr. Μινύειος, adj. f. Μινύαι.] **a.** Of or pertaining to the Minyans. **b.** Designating a type of very smooth grey pottery first found at Orchomenus and originally attributed to the Minyans. **B.** *sb.* A member of a possibly historical ancient people said to have inhabited parts of central Greece (chiefly Orchomenus in Boeotia and Iolchus in Thessaly), with whom the legends concerning Jason and the voyage of Argo are associated. Also with *pl.* **Minyae, Minyai**.

1598 Chapman tr. *Homer's Seaven Bookes of Iliades* 35 Those who in Aspledon dwelt, and Mynian Orchomen. **1709** I. Littlebury tr. *Herodotus' Hist.* I. iv. 416 Their Answer was, that they were Minyans, Grandsons of those Heroes who sail'd in the *Argos*. *Ibid.*, When the Lacedemonians heard they were of Minyan Extraction, they sent another Messenger. **1867** Max Müller *Chips* II. xvi. 67 The Minyans..reigned chiefly in Iolkos, in Southern Thessaly. **1881** H. Schliemann in *Jrnl. Hellenic Stud.* II. 152 A wall of unwrought stones.. which Professor Sayce holds to be the ancient Minyean city wall. **1894** tr. *A. Holm's Hist. Greece* I. vii. 65 The oldest accounts of the Minyae are to be found in Homer where the Boeotian Orchomenus is mentioned as a Minyan city. **1912** Wace & Thompson *Prehist. Thessaly* 21 Minyan Ware. This class of pottery was first found in any quantity by Schliemann at Orchomenos. **1925** V. G. Childe *Dawn European Civilization* v. 78 It is only in Central Greece that Minyan ware occurs in such quantities... True Minyan is wheel-made and has a silver-grey colour to the reduction of the iron oxides in the clay. **1928** C. Dawson *Age of Gods* viii. 190 The Greek mainland was conquered by a new people from the north, the bearers of the so-called Minyan culture. **1929** F. W. Hasluck *Christianity & Islam under Sultans* II. xxv. 366 The role of the magician-philosopher-engineer Plato..proves to be similar to that of the Minyans in Boeotia. **1958** L. Cottrell *Anvil of Civilisation* x. 138 These invaders are characterised by a certain type of plain-grey pottery called Minyan. **1964** E. Vermeule *Greece in Bronze Age* iii. 72 The tribal name Minyan must be strenuously distinguished from the archaeological label 'Minyan' for the invaders who bring the Middle Helladic Age to Greece. We have no idea at all what these newcomers called themselves. **1969** A. Toynbee *Some Probl. Greek Hist.* i. ii. 16 The Minyai were a maritime commercial people who came by sea. *Ibid.*, We shall find independent legendary evidence of a Cretan origin for..the Minyan settlements. **1970** Bray & Trump *Dict. Archaeol.* 149/2 *Minyan Ware*, a grey or yellow wheel-made ware of high quality... It was ancestral to Mycenaean pottery, and may represent a movement of new peoples into the Aegean area.

Min Yuen (min yiŭe·n). [ad. Chinese *min yuan* people's movement.] Name of an underground Communist supply organization formed during the Malayan Emergency, 1948–1960. Also *attrib.*

1951 *Communist Banditry in Malaya* (Federation of Malaya, Dept. Information) 9 During the year [*sc.* 1949] the main incidents were carried out by small gangs operating either as killer squads or as part of the Min Yuen Organization. It was the duty of this latter Organisation to arrange for the collection of food supplies and money from the local population, and for the dissemination of Communist propaganda. **1954** V. Bartlett *Rep. from Malaya* iii. 41 There is a branch of the Communist Party known as Min Yuen (literally 'the Helpers'), which collects food for the terrorists. **1966** E. O'Ballance *Malaya* iv. 92 Supporting the insurgent army was the clandestine Min Yuen, an underground organisation that provided money, food, intelligence and communications. *Ibid.*, Min Yuen is a contraction of Min Chung Yuen Thong, which has been variously translated as 'popular mass movement' or 'people's revolutionary movement'. **1970** T. Lilley *Projects Section* ii. 20 The Min Yuen—the Communist Liberation Army's supply organisation. **1971** N. Barber *War of Running Dogs* ii. 32 The Min Yuen (which means 'Masses Movement'),..consisted of ostensibly normal, innocent citizens who would in fact back up the army.

minyulite (minyū·lǝit). *Min.* [f. the name of *Minyulo* Well, Dandaragan, W. Australia, near which the mineral was first found + -ite¹.] A hydrated basic phosphate of potassium and aluminium occurring as colourless to white needles and fibres similar to wavellite.

1933 Simpson & LeMesurier in *Jrnl. R. Soc. Western Austral.* XIX. 13 In examining the Minyulo outcrop there was seen to be a fibrous mineral, closely resembling wavellite, which occurred in small quantities in a hard phosphatic ironstone carrying partly altered apatite nodules. .. The mineral has been proved in the laboratory to be a well defined new species for which the authors suggest the name Minyulite. **1961** *Mineral. Abstr.* XV. 210/2 Minyulite..occurs on Visean shales and phthanites in Belgium as fine silky fibers, ranging 60 to 300 μ, often grouped together in radiating rosettes (1 to 3 mm).

miocene, *a.* (Earlier example.)

1831 [see *eocene *a.* 1].

miogeoclinal (mǝi:odʒī,oklǝi·nāl), *a. Geol.* [abbrev. of next.] = *miogeosynclinal *a.* So **mioge·ocline** = *miogeosyncline.

1971 *Nature* 2 July 41/1 Several models..can be invoked to explain the eventual orogenic deformation of miogeoclinal continental margins. *Ibid.* 42/1 Fossil arcs.. must face away from the miogeocline in the activation model but towards the miogeocline in the collision model. **1972** [see *miogeosynclinal *a.*]. **1972** *Sci. Amer.* Mar. 33/1 The present Atlantic marine deposits closely resemble the ancient miogeoclinal foldbelts of the Paleozoic era and earlier.

miogeosynclinal (mǝi:odʒī,osinklǝi·nāl), *a. Geol.* [ad. G. *miogeosynklinal* (H. Stille *Einführung in den Bau Amerikas* (1940) i. 15),

f. Gr. μείων less: see Geosynclinal *a.* and *sb.*] Of or pertaining to a miogeosyncline. So **miogeosy·ncline**, a geosyncline in which the process of sedimentation appears to have been accompanied by little or no volcanism; *esp.* one situated between a larger, volcanic geosyncline (a eugeosyncline) and an area of the crust that has achieved stability (a craton).

1942 *Bull. Geol. Soc. Amer.* LIII. 1642 In contrast to the Magog eugeosyncline, the Champlain belt contains dominant carbonates of shallow-water origin, unaffected by subsequent volcanism; it is a miogeosyncline. *Ibid.* 1643 The eastern, miogeosynclinal belt has the well-described sequence of southeastern Idaho. **1951** *Mem. Geol. Soc. Amer.* XLVIII. 4 As volcanic rocks are practically absent in the orthogeosynclines that adjoin the North American early Paleozoic craton, they are thus miogeosynclines. **1969** Bennison & Wright *Geol. Hist. Brit. Isles* iii. 58 No vulcanicity is known and the succession is probably indicative of deposition on an unstable shelf at the margin of a miogeosyncline. **1971** *Nature* 24 Sept. 252/1 The West Andean geosyncline was divided into a 'eugeosynclinal' volcanic (coastal range) zone and a 'miogeosynclinal' sedimentary (longitudinal valley) zone during the Jurassic. **1972** *Sci. Amer.* Mar. 30/2 When one examines the structure of ancient folded mountains, one finds that the classic geosyncline is divided into a couplet: two adjacent and parallel structures consisting of a eugeosyncline (true geosyncline) and a miogeosyncline (lesser geosyncline), often shortened to eugeocline and miogeocline.

Miohippus (mǝiohi·pǝs). [mod.L. (O. C. Marsh 1874, in *Amer. Jrnl. Sci.* 3rd Ser. VII. 249), f. Mio(cene *a.* + Gr. ἵππος horse.] A small fossil horse of the genus once so called, known from North American remains of the Oligocene period, and now included in the genus *Anchitherium*.

1877 T. H. Huxley in *Pop. Sci. Monthly* Jan. 294 Next [in the equine series] comes the *Miohippus*, which corresponds pretty nearly with what I spoke of as the *Anchitherium* of Europe. **1933** A. S. Romer *Vertebr. Paleont.* xvii. 325 In *Mesohippus*..and the somewhat larger *Miohippus* we find the beginning of a series of functionally three-toed horses. **1958** C. L. & M. A. Fenton *Fossil Bk.* xxxii. 419 *Mesohippus* was followed by *Miohippus*, a larger and somewhat more horselike beast.

miombo (miǫ·mbǝ). Also **miomba**. [Swahili.] A tree of the tropical African genus *Brachystegia*, belonging to the family Leguminosæ; woodland composed mainly of these trees. Also *attrib.*

1864 J. A. Grant *Walk across Afr.* vi. 88 The wands from the Miombo, a kind of banyan, afford the natives the fibre which they attach to their wool. **1910** *Encycl. Brit.* XI. 772/2 The silk-cotton tree (*Bombax ceiba*), miomba, tamarisk, copal tree..are frequent [in German East Africa]. **1934** *Jrnl. Ecol.* XXII. 220 A strip of 'Miombo' through which a narrow, fairly shallow ravine..was running. **1947** *E. Afr. Ann.* 1946–7 95/2 The considerable forests of Unyamwezi, largely miombo, provide much honey. **1959** R. W. J. Keay in *Vegetation Map Afr.* Explanatory Notes 9/1 In these woodlands, commonly known by the vernacular name 'myombo', several species of *Brachystegia*..are dominant. **1948** *Cape Argus* 3 Dec. 5/4 Surrounded by uninhabited miombo bush. **1966** C. A. W. Guggisberg *S.O.S. Rhino* iii. 53 Black rhinoceroses frequent savannah country, mopane and miombo forests, bushveld and dry thornbush. **1970** A. T. Semple *Grassland Improvement* i. 14 African farmers have found that their ground-water supplies improve when they extend their settlements in the 'miombo' (*Brachystegia* woodland).

miosis, also var. Myosis (in Dict. and Suppl.).

miotic, var. Myotic *a.* and *sb.* (in Dict. and Suppl.).

mipafox (mi·păfǫks). [f. bis(*mono-isopropylamino*)*fluorophosphine oxide*, its chemical name.] An organic phosphorus compound, $[(CH_3)_2CHNH]_2POF$, which is used as an insecticide.

1953 *Recommended Common Names for Pesticides* (*B.S.I.*) II. 5 Mipafox: fluorobisisopropylaminophosphine oxide. **1956** *Nature* 28 Jan. 186/1 It has recently been found that 8×10^{-7} *M* mipafox..completely inhibits the non-specific serum cholinesterase. **1971** *Brit. Jrnl. Pharmacol.* XLI. 21 The ileum was then incubated with mipafox (10 μg/ml) for 75 min.

Mipolam (mi·pǫlæm). Also **mipolam**. A proprietary name for plastics composed of polyvinyl chloride which are used for chemically resistant piping and containers.

1936 *India-Rubber Jrnl.* 4 Nov. 19/2 Mipolam is a purely thermoplastic mass, and behaves accordingly in electrical tests. **1939** H. R. Simonds *Industr. Plastics* (1940) xiii. 324 A new polyvinyl chloride plastic in Germany known as Mipolam is resistant to acids and to many other chemicals and is being used for beer, alcohol and other prime containers, and also in the form of piping and pumping fixtures in chemical plants. **1943** *Jrnl. R. Aeronaut. Soc.* XLVII. 140 The group of polymerization resins includes, among others, the plastic of German production—mipolam—remarkable for its high fire resistance, toughness and elasticity. **1951** *Trade Marks Jrnl.* 24 Oct. 980/1 Mipolam... Mouldable plastics for use in manufactures. Dynamit-Actien-Gesellschaft Vormals Alfred Nobel & Co... Troisdorf, Cologne, Germany; manufacturers. **1969** R. F. Lang tr. *Henglein's Chem. Technol.* 215 Mipolam pipes can be enlarged thermally, beaded, bent and cemented with solvents.

Mir (miǝɹ), *sb.*¹ Also **meer**. [a. Hindi and Pers. *mīr*, ad. Arab. *amīr* leader, commander: see Ameer, Emir.] = Ameer, Emir.

1625 Purchas *Pilgrimes* i. iii. xii. 282 The Mir of Aden sent a Boat and a Messenger aboord. **1787** C. Hamilton *Hist. Relation Rohilla Afgans N. Provinces Hindostan* 41 Omdat al Moolk..was at this period Meer Buchshy or Paymaster-General of the Empire. **1792** R. Heron tr. *Niebuhr's Trav. Arabia* II. xxiii. vii. 155 Mr Kniphausen had agreed with Mir Naser, Prince of Bender Rigk,..that the Dutch should..be allowed to seat their factory there. **1800** F. Gladwin tr. *Ayeen Akbery* i. ii. 231 (*heading*) The office of Meer Behry, or Admiralty. **1840** J. B. Fraser *Trav. Koordistan* i. iii. 80 The jealousy of the Meer extends only to strangers travelling in the country without apparent business. **1873** E. Balfour *Cycl. India* (ed. 2) III. 284/1 Mir Jaffir, in 1702, was appointed dewan of Bengal... He was of a poor brahmin family..but was.. converted to mahomedanism. **1885** T. P. Hughes *Dict. Islam* 350/1 *Mīr*. A title of respect used for the descendants of celebrated Muhammadan saints. More generally used for Saiyids, or descendants of Fāṭimah, the Prophet's daughter. **1928** *Blackw. Mag.* May 709/1 To get the savour of the Mir's table talk, you must hear a little more about its setting. **1973** C. Bonington *Next Horizon* xviii. 252 On my first morning I attended the Court of the Mir of Hunza... At ten o'clock, the Mir, an absolute monarch with complete control over the internal affairs of his 20,000 people, walks from his palace to the Durbah. *Ibid.* 253 The Mir sits on a small rostrum, and his Court, in strict order of precedence, squat on carpets in two lines on either side of him. **1973** *Observer* (Colour Suppl.) 30 Sept. 51/1 In some circumstances,..the Mir (ruler) of Hunza could, from personal knowledge of his State's history, verify ages.

‖ **mir** (mīǝɹ), *sb.*² [Russ.] A village community in pre-revolutionary Russia. Also *attrib.*

1877 D. M. Wallace *Russia* I. viii. 179 The Mir is the most peculiar of Russian institutions. **1878** E. C. Grenville-Murray *Russians To-Day* 21 The Mir system may be summed up in a few words; it has simply caused the peasant to exchange the domination of his old master for the more grinding tyranny of many masters. **1905** J. H. Rose *Devel. European Nations* xi. 294 The ownership of the soil of Russia by the Mirs, the communes of her myriad villages. **1916** C. E. Bechhofer *Russ. at Cross-Roads* 71 No period is known in Russian history when the Mir did not exist. **1925** *Contemp. Rev.* Jan. 60 They pointed out that the land-holding peasant..did not cease to be a member of the *mir*. **1967** *Listener* 2 Nov. 558/3 The break-up of the *mir* and the..transition from communal to hereditary tenure. **1975** *Times* 8 Jan. 15/7 The democratic and civic traditions of Russia, from Kievian Rus to the *mirs* and the Zaporozhean Republic.

Mir (miǝɹ), *sb.*³ [ad. *Mirabad*, the name of a town in the Sarawan district, S.W. of Arak, Iran.] A rare and fine quality Saraband rug woven in Mirabad. Also *attrib.*

1900 J. K. Mumford *Oriental Rugs* xi. 197 In the Sarawan district the *tereh* Mir, so called from the village where it is said to have originated, is the almost universal design. *Ibid.*, Artisans in other localities have copied the Mir Saraband, changing the borders or coloration to suit their fancy. **1931** A. U. Dilley *Oriental Rugs & Carpets* iv. 125 All the fine Sarabands, known as Mir from the town of that name which an earthquake destroyed, were 'Farmayashti', or special-order weavings. **1953** A. C. Edwards *Persian Carpet* ix. 144/2 *Miri* no doubt refers to the Mir carpets of the early nineteenth century which were the prototypes of the present-day Serabands. **1972** P. L. Phillips tr. *Formenton's Oriental Rugs & Carpets* 69 The botch design came from the Seraband region and..was used at the beginning in the Mir carpets. **1973** *Times* 2 Apr. 6/5 (Advt.), A superb Mir carpet in soft pastel dyes.

mirabelle. Add: **2.** An alcoholic spirit distilled from mirabelles, *spec.* those grown in Alsace, France.

1940 H. J. Grossman *Guide to Wines, Spirits & Beers* xviii. 185 The brandy obtained from plums is known by several names, depending on the country of origin:.. Quetsch or Mirabelle when it comes from Alsace in France. **1963** *Times* 24 Apr. 11/7 'Try that,' he said, proffering a spoon dipped into the last vat of limpid 85° Mirabelle. **1967** N. Freeling *Strike Out* 67 Bernhard's last bottle of mirabelle.

‖ **mirabile dictu** (mirā·bili di·ktu), *Lat. phr.* [L. *mirabile*, neut. of *mirabilis* wonderful + *dictu*, supine of *dicĕre* to say: cf. Virgil *Georgics* ii. 30 quin et caudicibus sectis (mirabile dictu) truditur e sicco radix oleagina ligno.] Wonderful to relate.

1831 *Athenæum* 12 Mar. 173/2 An unassuming young female relative, whom she gives in marriage to the son of her (*mirabile dictu!*) honest attorney. **1837** J. F. Cooper *Recoll. Europe* I. 318 The late king was the Miller, and, *mirabile dictu*, the Archbishop of Paris did not disdain to play the part of the Curé. **1841** Dickens *Let.* 21 July (1969) II. 334 A man is coming here at 8, to settle (mirabile dictu) the matter of Mrs. Macrone's book. **1861** S. B. Birch *Constipated Bowels* ii. 15 The tradesman..is permitted to go so far as to usurp the functions of the physician, and is even encouraged (*mirabile dictu!*) to prescribe himself on the strength of the pharmacopœia. **1943** L. Marchal *Vichy* xiii. 195 The Marshal was anxious to save Otto Abetz, who, *mirabile dictu*, was considered in certain Vichy circles as a 'friend of France'.

miracidium (mǝiǝ:rāsi·diǝm). *Zool.* Pl. **miracidia**. [mod.L., f. Gr. μειρακίδιον, dim. of μειράκιον boy, stripling.] The ciliated, first larval form of a digenetic trematode. Also *attrib.* So **miraci·dial** *a.*

1898 A. Sedgwick *Student's Text-bk. Zool.* I. v. 229 The miracidium generally becomes a sporocyst, rarely a redia. **1904** C. W. Stiles *Illustr. Key to Trematode Parasites of Man* 18 Miracidium ciliated, develops after eggs leave the host. **1952** J. Clegg *Freshwater Life* viii. 127 The miracidium stage [of *Bucephalus*] is passed in the freshwater Mussels *Anodonta* and *Unio*. **1962** J. D. Smyth *Introd. Animal Parasitol.* xxxii. 388 The miracidial immobilisation test is based on the immobilisation of miracidia in immune serum. **1963** *New Scientist* 29 Aug. 441/3 Those [schistosome] eggs which reach the outside world hatch in fresh water and the microscopic miracidia which emerge swim around. **1971** *Oxf. Bk. Invertebr.* 24/1 When the pasture is damp, the first larva [of the liver fluke], the miracidium, hatches and this swims by its cilia in the film of water on the pasture until it encounters a snail. **1974** *Nature* 22 Mar. 361/2 A better understanding of the intricacies of how miracidia and cercariae find and penetrate hosts.

miracle, *sb.* Add: **2. a.** Also with defining word prefixed designating a remarkable development in some specified area.

1959 M. Crosland tr. *Rovan's Germany* 78 The fact that millions of new arrivals were housed and employed.. is the most miraculous aspect of the 'German miracle'. **1963** *Listener* 21 Feb. 321/2 This contract production is an important factor in the Polish miracle. **1973** A. Price *October Men* ix. 128, I don't mind him being part of the Italian economic miracle.

5. *miracle drug, -worker* (further examples); also *miracle-bred* adj.; **miracle man**, delete † and add further examples; **miracle rice**, a modern hybrid rice seed that yields more than the traditional varieties.

1928 W. B. Yeats *Tower* 62 Even the grey-leaved olive tree Miracle-bred out of the living stone. **1953** J. Ramsbottom *Mushrooms & Toadstools* xxiii. 279 Penicillin was really on the stage at last; it dwarfed other performers into insignificance. It was hailed as a 'miracle drug'. **1962** *Lancet* 2 June 1137/1 A miracle drug that will charm away any and every form of infection. **1970** P. Moyes *Who saw her Die?* xvi. 209, I got hold of the streptomycin..and the miracle drug worked. She was cured. **1914** F. L. Packard (*title*) The miracle man. **1926** A. Conan Doyle *Hist. Spiritualism* I. iii. 54 He [*sc.* Andrew Jackson Davis] was a miracle man, the inspired, learned, uneducated apostle of the new dispensation. **1951** L. MacNeice tr. *Goethe's Faust* 185 In priestly robes and wreath a miracle-man Will now fulfil what he in hopes began. **1959** I. & P. Opie *Lore & Lang. Schoolch.* x. 179 They are willing to acclaim anyone who habitually comes top of the class without apparent effort, naming him Genius, The Brains, Miracle Man. **1969** *Americana Ann.* 408 The introduction of U.S.-supplied 'miracle rice' notwithstanding, the country [*sc.* Laos] was still unable to produce enough rice for its own people. **1972** *Times* 8 May 14/1 Dr Norman Borlaug, the discoverer of the 'miracle' rice and wheat strains..was awarded the Nobel Peace Prize in 1970. **1923** Kipling *Independence* 29 He was something of a magician, if not a miracle-worker. **1970** G. Greer *Female Eunuch* 277 She twice..took a little pill... She thinks they are little miracle workers.

miraculous, *a.* Add: **2. a.** (See quot. 1965.) *dial.*

1879 F. M. Fetherston *Oops & Doons T. Goorkrodger* 27, I say when a man's drunk, *he's miraculus and mad*. **1925** L. P. Smith *Words & Idioms* 142 *Miraculous* has changed its meaning to 'very drunk'. **1965** *Sc. Nat. Dict.* VI. 286/1 *Miraculous*, in a stupefied or incapable condition, esp. from drink, very intoxicated.

mirador. Add: (Later examples.) Also *fig.*

1950 G. Brenan *Face of Spain* iv. 80 We went into the house. One wing, that of the *mirador* or tower, had been set aside for our books and furniture. **1955** J. Thomas *No Banners* xxxi. 310 The *miradors* (watch-towers) around Buchenwald were manned by *Wehrmacht* troops. **1971** *Homes & Gardens* Aug. 36/1 Spanish women were generally kept in the background, and watched the world from behind their wooden *miradors*. **1971** *Nat. Geographic* Oct. 547/1, I spent several weeks last spring in this brilliant mirador of capitalist colonialism on the edge of China.

mirage. Add: (Now usu. with pronunc. (mi·rāʒ).) **c.** A wave-like appearance of warmed air visible just above ground level. Also *attrib.*

1913 A. G. Fulton *Notes on Rifle Shooting* 18 When mirage can be seen, it provides the best means of estimating allowances for gentle but tricky winds. **1958** J. A. Barlow *Elem. Rifle Shooting* (ed. 5) iv. 50 The effect of wind on such air disturbance is readily remarked, since the result of a cross wind is to give the impression of the mirage flowing either slowly or swiftly in the same direction as the wind, just like a clear stream of water over a pebbly bed. **1962** *Amer. Speech* XXXVII. 270 *Mirage puddle*, a heat-caused illusion which makes the road appear wet in the distance.

2. *attrib.* and *Comb.*, as *mirage dream, -water*; *mirage-bright, -lifted, -making, -reflected* adjs.

1924 E. Sitwell *Sleeping Beauty* xvii. 65 Mirage-bright It lies, that dusty gold. **1908** *Daily Chron.* 3 Aug. 3/2 It was like one of those mirage dreams which lure the gambler to his doom. **1874** J. G. Whittier *Sea Dream* in *Atlantic Monthly* Aug. 160 And watched the mirage-lifted wall Of coast, across the dreamy bay. **1902** J. H. M. Abbott

Tommy Cornstalk i. 13 And the wide rolling downs quivered and danced with the same beautiful mirage-making islands of kopjes. **1923** R. Graves *Whipperginny* 9 Mirage-reflected drink At the clear pool's brink. **1918** E. Sitwell *Clown's Houses* 11 Then, mirage-waters as they flow, Or dream-perfumes, they fade and go.

miraged (mirā·ʒd), *a.* [f. Mirage + -ed^{2}.] Seen in a mirage; of the nature of a mirage.
1920 *Blackw. Mag.* June 817/2 A dim outline of miraged date palms. **1925** *Chambers's Jrnl.* Aug. 486/2 Framed anew, in mystic space—Miraged dream past all believing—Looms the cradle of a race.

Miranda (miræ·ndă). *U.S. Law.* [f. the name Ernesto A. *Miranda* (see quot. 1966).] The name given to a set of rules specified by the Supreme Court in the U.S. whereby law enforcement officers are required to apprise a person suspected of a crime of his rights to counsel and his privileges against self-incrimination prior to his being interrogated. Also *attrib.*
[**1966** *Washington Post* 14 June A7/1 Here are excerpts from yesterday's Supreme Court opinion on the admissibility of statements obtained in questioning of a person in police custody, and on the Fifth Amendment protections such a person is to be afforded against self-incrimination. This opinion..encompasses four cases—Ernesto A. Miranda v. the State of Arizona, [etc.].] **1967** *Time* 3 Mar. 49/1 The Supreme Court's famous *Miranda* decision.. wrought vast changes in police procedure. *Ibid.* 49/2 *Miranda* does not seem to have helped Miranda and his mates. **1970** W. Wager *Sledgehammer* xxii. 173 I've got quite a few motions—the complete kit. I'll start with *Miranda* when the trial begins. **1972** J. Mills *Report to Commissioner* 270 Do you know what a Miranda warning is? **1973** *N.Y. Law Jrnl.* 2 Aug. 13/5 At this point Miller read to Dixon his four required warnings under Miranda.

miration (məiᵊrēⁱ·ʃən). *U.S. regional colloq.* [Abbrev. of Admiration.] An expression of admiration, wonder, or surprise; a fuss, to-do. So **mira·te** *v. intr.*, to feel or express surprise or astonishment.
1893 in H. Wentworth *Amer. Dial. Dict.* (1944) 390/2. **1903** *Dialect Notes* II. 321 *Mirate*, to wonder at; to admire. **1909** *Ibid.* III. 349 *Miration*, an expression of admiration. 'He made a great miration over the baby.' **1926** in J. F. Dobie *Rainbow in Morning* (1965) 82 He may have made *a great miration* over the place, but it's my private opinion that he has something up his sleeve. **1935** Z. N. Hurston *Mules & Men* (1970) i. vi. 125 Aw, man, you done seen Tookie and her walk too much to be makin' all dat miration over it. **1946** *Publ. Amer. Dial. Soc.* vi. 20 *Miration*, exaggerated and pretentious wonderment, a carrying-on. 'They made a great miration about my killing that squirrel.' **1950** *Ibid.* xiv. 46 *Mirate*, to make a miration over; to express surprise, admiration over a person or thing. **1950** A. Lomax *Mr. Jelly Roll* (1952) i. 4 This lady displayed me in saloons, setting me on the bar and..making mirations.

Mirdita (mə̄ɹditā). Also **Mirditë** (mə̄ɹditə). The name of a region on the river Drin in Albania used *attrib.* and *absol.* to designate the tribal people living there. Also anglicized as **Mirdite** (mə̄·ɹdəit).
1861 G. Finlay *Hist. Greek Revolution* I. i. ii. 43 The Mirdites are considered the most warlike of the Christians. **1920** *Q. Rev.* Jan. 65 The late Mirdite Prince, Prenk Bib Doda. **1939** A. Toynbee *Study of Hist.* IV. 367 The present Mirdite territory would have been a natural site for the plantation of a Mardaite settlement by a seventh-century Roman statesman. **1954** M. Hasluck *Unwritten Law in Albania* xv. 154 There had been attacks on the Montenegrins by the Mirditë, which had followed on plenary assemblies of the Mirditë tribes. **1959** *Chambers's Encycl.* I. 220/1 The Mirdita region south of the Black Drin gorge.

mire, *sb.*^{1} Add: **2. d.** Dung. *rare.*
1922 Joyce *Ulysses* 649 Bloom..with Stephen passed through the gap of the chains..and, stepping over a strand of mire, went across towards Gardiner street lower.
3. *mire-bestrowed, mire-smirched* adjs.
1908 Hardy *Dynasts* III. vi. viii. 281 Cavalry in the cornfields mire-bestrowed. **1960** S. Plath *Colossus* 12 Common barnyard sows, Mire-smirched, blowzy.

mire, *v.*^{1} **II.** *intr.* **3.** Delete † before 'Also' and add further examples.
1835 H. Evans *Jrnl.* in *Mississippi Valley Hist. Rev.* (1927) XIV. 195 In crossing some of these creeks some of our horses and pack mules mired down. **1840** W. Sewall *Diary* (1930) 221 The roads being soft..I mired down. **1941** *Amer. Speech* XVI. 184 *Mire down*, to stick in mud.
b. To defecate. *rare.*
1922 Joyce *Ulysses* 48 An archway where dogs have mired. *Ibid.* 649 Slowly..he [*sc.* the horse] mired.

mired, *ppl. a.* **2.** (Later examples.)
1891 'Q' *Noughts & Crosses* 207 My mired boots played havoc with the neatly sanded floor. **1897** T. C. De Leon *Jealous God* v, in *Novelette Trilogy* 127 Lifting tenderly the mired, limp and senseless form of a shriveled old woman, struck down by them [*sc.* 'bus horses].

mirepoix (mirpwa). *Cookery.* [f. the name of the Duc de *Mirepoix* (1699–1757), French diplomat and general.] A mixture of diced vegetables used for flavouring or served as a vegetable dish.
1877 E. S. Dallas *Kettner's Bk. of Table* 65 Set the steak to boil,..add to it a Mirepoix of red wine, and let it simmer... As for taste it is perilous to attempt to improve upon a good Mirepoix. *Ibid.* 304 Mirepoix..is..the convenient name for the faggot of vegetables that flavours a stew or sauce. **1877** *Cassell's Dict. Cookery* 425/2 *Mirepoix*, a flavouring for made dishes. **1906** *Mrs. Beeton's Bk. Househ. Managem.* lxi. 1648 A mirepoix is the foundation for flavouring sauces, braised meats, and a number of thick soups. **1936** Lucas & Hume *Au Petit Cordon Bleu* 173 *Mirepoix*, carrot, onion, celery, turnip, and French beans cut into very small, even dice. **1960** *Guardian* 15 June 16/4 A mirepoix..is..another interesting way of serving early vegetables..cut into dice.

mirex (məiᵊ·reks). orig. *U.S.* [etym. unknown.] An organochlorine insecticide active esp. against ants.
1962 *Bull. Entomol. Soc. Amer.* VIII. 89/1 The Committee on Insecticide Terminology announces the following proposed common names... Mirex. Dodecachloro-octahydro-1,3,4,-methano-2H-cyclo-buta(cd)pentalene. **1963** *Jrnl. Econ. Entomol.* LVI. 296/1 The newly developed mirex..(formerly known as GC 1283) is an analog of Kepone and has exhibited excellent ant poison characteristics with low toxicity to higher animals. **1972** *Nature* 18 Feb. 353/1 'Mirex', the chief weapon used against the fire ant, has been criticized because of its persistence in the environment and its possible toxicity to animals and fish.

mirid (məiᵊ·rid, mi·rid), *sb.* and *a.* [f. mod.L. family name *Miridæ*, f. the generic name *Miris* (J. C. Fabricius *Entomologia Systematica* (1794) IV. 183), f. L. *mīrus* wonderful, extraordinary: see -id^{3}.] **A.** *sb.* A leaf bug of the family Miridæ, formerly called Capsidæ (see *capsid *a.* and *sb.*^{1}), which includes a large number of insects that live on the sap of plants, often causing damage to the plants affected. **B.** *adj.* Of or pertaining to an insect of this kind.
1941 *Bull. Illinois Nat. Hist. Survey* XXII. 2/1 The eggs of most mirids hatch early in the season. *Ibid.* 3/2 A majority of mirid species produce only one generation per year. **1957** Richards & Davies *Imms's Textbk. Entomol.* (ed. 9) iii. 461 Other Mirids do considerable damage to cultivated plants. **1962** Metcalf & Flint *Destructive & Useful Insects* (ed. 4) vi. 225 Anyone who sits down in a grassy, weedy spot in early summer..can scarcely fail to make the acquaintance of some of the hundreds of kinds of mirids that crawl about over the vegetation and feed on its sap. **1972** Swan & Papp *Common Insects N. Amer.* xii. 117 The mirids comprise a fairly large family commonly called plant bugs or leaf bugs.

mirifically (məiri·fikăli), *adv.* [f. Mirifical *a.* + -ly^{2}.] So as to excite wonder or admiration; wonderfully, superbly.
1922 W. J. Locke *Tale of Triona* v. 49 Into the lounge filled with mirifically vestured fellow-creatures.

mirliton. Delete ? *Obs.* and substitute for def.: A toy instrument resembling a kazoo. Add further examples.
1865 M. Eyre *Lady's Walks S. of France* xx. 217 In returning home the pilgrims unite in bands, singing;.. while others, provided with *mirlitons*, play the wildest accompaniments. **1894** G. Du Maurier *Trilby* I. ii. 155 Taffy and Jeannot and Little Billie made the necessary music on their *mirlitons*. **1938** [see *kazoo]. **1970** P. Oliver *Savannah Syncopators* 109 *Kazoo*, submarine-shaped tube mirliton, played in blues bands.

mirnyong, var. *mirrnyong.

miro^{1}. Substitute for def.: A large, evergreen tree, *Podocarpus ferrugineus*, of the family Taxaceæ, native to New Zealand, or the timber produced by it. Also *attrib.* (Earlier and later examples.)
1832 G. Bennett in *London Med. Gaz.* X. 793/1 The Miro tree..is named Miro by the natives of New Zealand. **1905** W. B. *Where White Man Treads* 19 When the red miro berries were ripe..the Maori smiled. *Ibid.*, On the ranges of the interior there are certain waterless districts, where the miro is abundant. **1926** H. Guthrie-Smith *Tutira* (ed. 2) 99 Black pine, miro. **1966** *Encycl. N.Z.* II. 568/2 Miro was known and cherished by the Maoris because of its fleshy covered fruit, about an inch long, on which the native pigeon feeds... Miro is found in lowland forests throughout North, South, and Stewart Islands. **1971** *Daily Tel.* 23 Dec. 3/7 [On Pitcairn Island] Henry may give Warren..a block of scarce miro wood for carving.

miro (miᵊ·ro), *sb.*^{2} Also **miro-miro**, 9 **mirro mirro.** [Maori, adopted as a generic name by R. P. Lesson in *Traité d'Ornithologie* (1831) 389.] Either of two New Zealand flycatchers of the genus *Petroica*, the black-and-white tomtit, *P. macrocephala*, or the greyish-brown New Zealand robin, *P.* (formerly *Miro*) *australis.*
1843 J. E. Gray in E. Dieffenbach *Trav. N.Z.* II. 191 *Miro Forsterorum*..Turdus minutus..Mirro mirro of the natives of Queen Charlotte's Sound. **1848** R. Taylor *Leaf from Nat. Hist. N.Z.* 9/1 *Miromiro*, small land bird, very tame; can be caught by the hand. **1879** *Trans. N.Z. Inst.* XII. 119 The miro-miro is the little Petroica toitoi, which runs up and down trees peering for minute insects in the cavities in the bark. **1930** W. R. B. Oliver *N.Z. Birds* 458 *Genus Petroica*..Distinguished from Miro by its smaller size. *Ibid.* 460 White-breasted Tit. Miromiro. **1966** R. A. Falla et al. *Field Guide to Birds N.Z.* 200 Tomtit... Other names: Miromiro (North Island).

‖ **mirrnyong** (mə̄·ɹnyɒŋ). *Austral.* Also **mirnyong.** [Native word.] A mound of shells, ashes, and other debris accumulated in a place used for cooking by Australian Aborigines; an Aboriginal kitchen-midden.
1878 R. B. Smyth *Aborigines Victoria* I. 238 (*heading*) Mirrnyongs, shell-mounds, and stone-shelters. *Ibid.* 239 The sites for *Mirrn-yong* heaps appear to have been chosen generally in localities near water. **1888** R. M. Johnston *Syst. Acct. Geol. Tasmania* 337 (Morris), With the exception of their rude inconspicuous flints, and the accumulated remains of their feasts in the 'mirnyongs', or native shell-mounds, along our coasts,..we have no other visible evidence of their former existence. **1896** A. H. Keane *Ethnol.* v. 94 Australia, numerous *mirrnyongs* (ash-heaps, shell-mounds, &c.) mainly confined to the eastern and southern regions. **1964** *Mod. Encycl. Austral. & N.Z.* 666/1 *Mirrnyong Heaps or Kitchen Middens*, aboriginal cooking areas which have grown through thousands of years of use. Consist of ashes, shells and other debris; some have an area of 5,000 sq ft and are 10 ft high. Mainly found in coastal and Murray Valley areas. **1965** *Austral. Encycl.* I. 35/1 Continued use of the same cooking-place builds up enormous deposits of ashes, such as those in Riverina and Murray River districts of New South Wales, Victoria and South Australia, where the ovens or Mirrnyong heaps are up to 125 feet long, 50 feet wide, and 12 feet thick.

mirr-n'yong, var. *murrnong.

mirror, *sb.* Add: **7. a.** *mirror-gazer, -hall, -light, -scroll, -stand, -trick; mirror-topped* adj.
1937 G. Barker *Calamiterror* 9 The mirror-gazer self-betrayed. **1923** Blunden *To Nature* 10 From the mirror-lights on the dressing table. **1970** R. Lowell *Notebk.* 207 Your wall-mirror in a mat of plateglass sapphire, mirror-scroll and claspleaves, holds our faces. **1817** J. Constable *Let.* 10 July in *Corr.* (1964) II. 228, I am glad I have not made a purchase of the mirror stand (called a Canterbury). **1960** H. Hayward *Antique Coll.* 187/1 *Mirror-stand*, an adjustable mirror mounted on a shaft and tripod base, resembling a pole-screen; popular at the end of the 18th cent. **1949** D. Smith *I capture Castle* iii. xiv. 260 A mirror-topped table. **1950** M. Allingham *Mr. Campion & Others* xiii. 270 He remembered Geoffrey's face at the other end of the mirror-topped table. **1940** W. Faulkner *Hamlet* iv. i. 247 Something to be repudiated with contempt, like a mirror trick.
b. *mirror-eye, finish, -sheen; mirror-dark, -flat, -polished, -resembling, -scaled* adjs.
a **1955** W. Stevens *Opus Posthumous* (1957) 51 Your gowns..came shining as things come That enter day from night, came mirror-dark. **1923** D. H. Lawrence *Birds, Beasts & Flowers* 98 The red-gold mirror-eye [of a fish] stares and dies. **1951** Koestler *Age of Longing* ii. viii. 306 She felt herself reflected in their watchful mirror-eyes, and was forced to see herself as they saw her. **1897** *Sears, Roebuck Catal.* 112/3 Heavy nickel plated and polished to a mirror finish. **1926–7** *Army & Navy Stores Catal.* 199/3 Polishing wool..gives to silver, electro-plate, gold, etc., that beautiful 'mirror' finish of newly manufactured articles. **1971** *Engineering* Apr. 118/2 A second tool produces the final truly 'mirror' finish. **1923** D. H. Lawrence *Birds, Beasts & Flowers* 98 This red-gold, water-precious mirror-flat bright eye. **1936** J. Steinbeck *In Dubious Battle* iv. 29 One line of worn and mirror-polished rails extended ahead. **1937** *Discovery* Feb. 57/1 Pickle mirror-polished silver sheet in sulphuric acid. **1927** W. B. Yeats *October Blast* 9 All those things whereof Man makes a superhuman Mirror-resembling dream. **1934** —— *King of Gt. Clock Tower* 40 The mirror scaled serpent is multiplicity. **1960** S. Plath *Colossus* 22 River lapsing Black beneath bland mirror-sheen.
c. mirror drum, a scanning device, used in early television transmitters and receivers, which consists of a rotating drum with its curved surface covered with a number of equally spaced plane mirrors, there being as many mirrors as there are scanning lines in the picture; **mirror embroidery** = **mirror-work*; **mirror fugue** *Mus.*, a fugue that can be played in a reversed or inverted manner, as if read in a mirror placed at the end of or underneath the music; **mirror-glass** (later examples); also *attrib.*; **mirror image,** something that resembles an image in a mirror in having left and right interchanged or its constituent parts arranged in reverse order (but being otherwise identical); also *transf.* and *fig.*; **mirror machine** *Nuclear Physics*, a linear device in which plasma is confined by means of magnetic mirrors; **mirror nucleus** *Nuclear Physics*, a nuclide having as many neutrons as another nuclide (of the same atomic number) has protons, and as many protons as the other has neutrons; also (more correctly) **mirror nuclide**; **mirror-painting** (see quot. 1960); also, the process of such painting; **mirror-picture** = **mirror-painting*; also, a

picture as seen in a mirror; **mirror-plate** (later example); also, a type of metal plate used for fixing two things together; **mirror room,** a room with mirrors set into the walls; **mirror scale,** a scale provided with an adjacent mirror so that parallax errors may be avoided when taking readings; **mirror-script** = *mirror-writing*; **mirror-wall,** a wall entirely covered with a mirror; **mirror-work,** small rounds of mirror appliquéd on fabric; **mirror-writer** (later example); hence (as a back-formation) **mirror-write** *v.*; **mirror-writing** (earlier and later examples).
1927 *Wireless World* 20 Apr. 480/2 As the mirror drum revolves, these seven beams trace seven lines at once on the screen, and then pass over another adjacent track of seven lines until the entire screen has been covered. **1935** M. G. Scroggie *Television* iii. 22 The scanner which has been used for the last few years to transmit the B.B.C. programmes by the Baird low-definition system, and also in a large proportion of the receivers, is the mirror drum. **1968** *Brit. Med. Bull.* XXIV. 261/2 The National Biomedical Research Corporation is developing a mirror-drum scanner, of a type pioneered at Harwell. **1967** E. Short *Embroidery & Fabric Collage* i. 17 (*caption*) Indian 'Shisha' or mirror embroidery. **1931** D. F. Tovey *Compan. to 'Art of Fugue'* 61 The original edition [of Bach's 'Art of Fugue']..should not have printed the mirror-fugues in succession instead of in mirror-reflection. **1962** *Listener* 27 Dec. 1109/2 The fifth fugue is again for strings only, as are the rectus versions of the 'mirror' fugues XII and XIII [of Bach]. **1973** *Times* 23 Apr. 16/2 A concert-goer who can recognize a mirror-fugue merely by listening to it has no need of assistance. **1876** J. S. Ingram *Centenn. Exposition* ix. 287 Inside was an oblong square, formed of mirror-glass, which reached to the top of the case. **1934** *Heal & Son Catal.: Better Furnit.* 8 Dressing-table,..pink mirror-glass top. **1953** *Glass for Glazing* (*B.S.I.*) 18 (*heading*) Mirror glass. **1885, 1929** Mirror image [see *enantiomorph]. **1937** 'G. Orwell' *Road to Wigan Pier* xii. 244 Fascism..is a sort of mirror-image..of a plausible travesty of Socialism. **1949** Mirror image [see *asymmetric *a.* b]. **1961** *Lancet* 26 Aug. 447/2 The hemispheres are not exact replicas but mirror images of each other. **1962** W. Nowottny *Lang. Poets Use* vi. 141 The formal correspondence between the lines..makes 'time is setting with me' the mirror-image of 'The wan moon is setting ayont the white wave'. **1964** *Language* XL. 247 The phonological disintegration characteristic of the aphasic's linguistic regressions is a mirror-image of the child's acquisition of its sound pattern. **1966** *New Statesman* 27 May 775/3 The Black Muslim creed is the mirror image of the white racialist one. **1972** J. McClure *Caterpillar Cop* v. 68 An element of variety had been introduced by building the bungalows in pairs and making one the mirror image of the other. **1954** R. F. Post *16 Lect. Controlled Thermonucl. Reactions* (Univ. Calif. Radiation Laboratory, UCRL-4231, 2 Feb.) p. vi, A general principle involved in the mirror machine's conception was the establishing as an initial condition that the plasma should be created by injection and trapping of a space-charge neutralized energetic ion beam into an otherwise evacuated chamber. **1958** *New Statesman* 6 Sept. 266/3 In the mirror machines the molecules of heavy hydrogen are violently injected into a chamber and go spiralling along until magnetic forces at the ends of the chamber reflect them and send them spiralling back again—from one magnetic mirror to another. **1969** *New Scientist* 25 Sept. 639/1 If mirror machines are feasible, then a fusion reactor based on the system would be relatively easy to build. **1947** H. A. Bethe *Elem. Nucl. Theory* ii. 7 If the binding energies of a pair of nuclei which differ only in the interchange of neutrons and protons are compared, a difference in binding energy which increases with the charge of the nuclei is found. Examples of such 'mirror' nuclei are: $^{1}H^{3}$ $^{2}He^{3}$; $^{3}Li^{7}$ $^{4}Be^{7}$;.. $^{14}Si^{29}$ $^{15}P^{29}$. **1962** H. D. Bush *Atomic & Nucl. Physics* vi. 127 Evidence for the equality of *n–n* and *p–p* forces is provided by certain positron emitters where the parent and product of the decay are mirror nuclei. **1955** Richtmyer & Kennard *Introd. Mod. Physics* (ed. 5) x. 510 Among the many unstable nuclides..a set of particular importance are the so-called 'mirror nuclides'. **1960** H. Hayward *Antique Coll.* 187/1 *Mirror-painting*, a type of glass picture in which the glass was first coated at the back with an amalgam of tin and mercury to make it into a mirror. The parts to be painted were then scraped away and painted in as required. **1970** G. Savage *Dict. Antiques* 275/1 Most surviving mirror-paintings are Chinese and belong to the 18th century... Mirror-painting, being on the back of a sheet of glass, meant working in reverse. **1939** *Burlington Mag.* May p. xv/1 A pair of decorative Chinese mirror-pictures in Chippendale frames. **1959** E. Pulgram *Introd. Spectogr. of Speech* v. 46 Oscillograms of repetitive waves whose half-cycles are not mirror pictures of one another. **1964** *Amer. N. & Q.* Jan. 72/1 *Mirror pictures*..this kind of repeated or reflected picture of a picture of a picture, ad infinitum. **1940** *Chambers's Techn. Dict.* 550/1 *Mirror plate*,..(1) Plate glass for silvering.—(2) A fixing device in the form of a small metal plate, one end being screwed..to the object..and the other fixed to the base. **1966** A. W. Lewis *Gloss. Woodworking Terms* 56 *Mirror plate*, small metal plate screwed to the backs of frames, etc., so that they can be fixed to a wall. **1926** A. Huxley *Jesting Pilate* i. 70 These mirror rooms at Amber. **1901** M. W. Travers *Exper. Study Gases* vi. 56 *The mirror scale* (Jolly).—In reading barometers, manometers, etc., it is usual to employ a glass scale ruled in millimetres. The scale is etched on the surface of a strip of glass about 5 mm. in thickness, which is then silvered on the second surface. **1961** M. D. Armitage *Basic Princ. Electronics & Telecommunications* xi. 290 (*caption*) Use of mirror scale. **1890** W. James *Princ. Psychol.* I. x. 399 The subjects, e.g., often write backwards, or they transpose letters, or they write mirror-script. **1964** *Listener* 20 Aug. 264/2 A prism, or perhaps tent-shaped room, some eighty feet high, whose two inclined faces are all mirror; hidden in the ridge are two film cameras... So that film image as well as the constantly moving crowd are repeated *ad infinitum* in the mirror-wall, as if it were the inside of a kaleidoscope. **1969** *Guardian* 1 July 9/2 Mirror-work is a traditional Indian craft which looks enchanting..densely applied to a gipsy-type waistcoat or belt. **1973–4** *Oxfam Catal.* 12 Typical Gujarat embroidery and mirrorwork covers this bag. **1960** I. Bennett *Delinquent & Neurotic Children* ix. 367 Left-handed, tends to mirror-write, poor wrist coordination. *Ibid.* 370 Mirror-writer, below average in every subject. **1776** G. Campbell *Philos. Rhetoric* I. ii. iii. 420 If the analogy of the language must be preserved in composition, to what kind of reception are the following entitled..homedialect, bellysense, and mirrourwriting? **1924** R. M. Ogden tr. *Koffka's Growth of Mind* v. § 6. 293 Certain children can read mirror-writing at first just as well as they can ordinary writing. **1969** 'E. McBain' *Shotgun* iii. 35 A blotter was the only other thing on the desk. Carella automatically checked it for any mirror writing that might have been left on it. **1970** D. Bowden tr. *Luria's Traumatic Aphasia* xiii. 332 (*caption*) Writing disturbance in visual agraphia; 'mirror writing'.

mirth, *sb.* Add: **5.** *mirth-maker* (later example), *-provoker*; *mirth-lit* adj.
1849 C. Brontë *Shirley* II. iv. 100 Lifting up her mirth-lit face to the gallery. **1969** *Daily Tel.* 15 Feb. 14/6 Kenneth Horne..was acknowledged as one of radio's top mirth-makers. **1895** W. Archer *Theatr. 'World' 1894* lii. 341 The pun, as a 'mirth-provoker', is dead.

mirthful, *a.* Add: **1.** (Later example.)
1940 W. Faulkner *Hamlet* iv. i. 279 His constant expression of incorrigible and mirthful disbelief had left him now.
2. (Later example.)
1877 *Athenæum* 13 Oct. 475/2 The piece..is one of the most mirthful and original that has, during late years, been seen on the stage.

mirthquake (mə̄·ɹþkwēⁱk). *colloq.* Also **mirthquaker.** [f. Mirth *sb.* + Quake *sb.*, after Earthquake.] An entertainment that excites convulsive mirth.
1928 *Daily Express* 24 Apr. 4/2, I found Prince George ..among the first to see Harold Lloyd's new 'mirthquake'. **1938** *N.Y. Times* 13 Jan. 19/1 Robert Benchley's new mirthquake *After 1903—What?* Always in hot water—that's Benchley, America's Mogul of Mirth. **1939** *Amer. Speech* XIV. 4 A 'screamario' is the scenario for a comedy, which proves to be a 'mirthquake'. **1942** Berrey & Van den Bark *Amer. Thes. Slang* § 281/4 Something humorous; joke... Mirthquake, mirthquaker. **1965** Leitner & Lanen *Dict. French & Eng. Slang* 90/1 *Mirthquake*,.. film, pièce, etc., très amusant.

Mirzapur (mə̄·ɹzăpūᵊɹ). Also **Mirzapore.** The name of a town in the state of Uttar Pradesh in Northern India used *attrib.* and *absol.* to designate a type of carpet manufactured there.
c **1882** *Cardinal & Harford's Price List Oriental Carpets & Rugs* 10 Indian carpets from Masulipatam, Mirzapore, etc... The Mirzapore can be manufactured in less time than any other. **1900** J. K. Mumford *Oriental Rugs* (1901) xiv. 264 The designs of the old Mirzapur carpets..showed a pronounced Hindu character. **1967** *Times* 23 Feb. 24/6 (Advt.), Really superb plain off-white 'Mirzapur' carpets with 1 in. virgin wool pile.

mis-, *prefix*^{3}: see Miso- (in Dict. and Suppl.).

mis (miz), colloq. abbrev. of Miserable *a.* Cf. *miz, mizz.
1886 in H. Baumann *Londinismen.* **1939** N. Monsarrat *This is Schoolroom* ii. ix. 205 'I'm mis,' she volunteered immediately, in a muted babyish voice. **1952** 'C. Brand' *London Particular* x. 121 Rosie was ackcherly utterly *mis.* about..poor, darling Thomas. **1974** *Observer* 27 Oct. 5/5, I wouldn't care to guess what proportion of the population has some sexual problem that makes them mis.

mis', var. Miss *sb.*^{2} 5 (in Dict. and Suppl.).

misact, *v.* Delete † *Obs.* and add later example.
1925 *Contemp. Rev.* Nov. 624 The wonderful scene in Pimen's cell..was completely 'mis-acted', if I may use a non-existing word.

misadventurer (misădve·ntiŭrəɹ). *rare.* [f. Misadventure *sb.* + -er^{1}.] One who meets with or suffers misadventures; an unfortunate person.
1886 Hardy *Mayor Casterbr.* II. xviii. 250 His mood was no longer that of the rebellious, ironical, reckless misadventurer.

misadve·rtence. *rare.* [Mis-^{1} 4.] Carelessness, thoughtlessness, absent-mindedness.
1870 Tennyson *Holy Grail* 43 Once by misadvertence Merlin sat In his own chair, and so was lost.

misali·gnment. Also misalinement. [Mis-^{1} 4.] Bad or imperfect alignment.
1924 J. F. Hobart *Tulley's Handbk.* (ed. 7) II. xv. 582 Misalignment refers either to that of the turbine and generator, or that of the rotating and stationary elements. **1937** *Marconi Rev.* May–Aug. 27 The tracking does not allow phase misalignment greater than ±5 degrees. **1947** *Brit. Jrnl. Psychol.* Dec. 57 It is easy to present a misalinement to an operator, and then screen his eyes just before he makes his corrective movement. **1963** C. R. Cowell et al. *Inlays, Crowns & Bridges* xii. 130 Difficulty is rarely encountered with cases of gross misalignment [of the teeth]. **1974** *Sci. Amer.* May 54/3 There is often a misalignment of the eyes too.

So **misali·gned** *a.*
1948 *Aircraft Power Plants* (Northrop Aeronaut. Inst.) 216 The crank extension shaft may be misaligned. **1954** S. Duke-Elder *Parsons' Dis. Eye* (ed. 12) xxix. 483 In concomitant squint, as opposed to paralytic squint, although the eyes are misaligned, they retain their abnormal relation to each other in all movements. **1962** A. Nisbett *Technique Sound Studio* iv. 81 (*caption*) Misaligned recording or reproducing head. **1973** *New Scientist* 22 Nov. 544 Directional roof aerials pointed accurately at Wrotham for optimum BBC reception are, likely as not, hopelessly misaligned for signals from Croydon.

misalloca·tion. [Mis-^{1} 4.] Failure to allocate in an efficient or correct way what is to be assigned or distributed.
1950 *Jrnl. Pol. Econ.* Apr. 118 The crop-share lease has not resulted in the gross misallocation of land that would have occurred. **1961** *Ann. Reg. 1960* 474 There can be no doubt that misallocation and waste continued in 1960. **1965** H. Kahn *On Escalation* viii. 152 There has been a startling misallocation of official emphasis. **1971** *Nature* 9 July 82/2 Mondale's argument that the shuttle is a far grosser misallocation of priorities than the SST..may carry great weight.

misandry, misarchist: see *miso-.

misappliance (misăplai·ăns). *rare*$^{-1}$. [f. Mis-^{1} 4 + Appliance.] Placing (of oneself) amiss.
1903 H. James *Ambassadors* iii. 58 He scarce knew where to sit for fear of a misappliance.

misappropriation. (Further examples.)
1838 W. H. Prescott *Hist. Reign Ferdinand & Isabella* III. ii. xxv. 490 He made a strict inquisition into the funds of the military orders, in which there had been much waste and misappropriation. **1922** Joyce *Ulysses* 718 Forgery, embezzlement, misappropriation of public money. **1952** M. A. Elliott *Crime in Mod. Society* iii. 59 Misappropriation of funds of private character is seldom punished unless there is a glaring embezzlement.

misarticula·tion. [Mis-^{1} 4.] Inability to articulate correctly.
1959 *Jrnl. Speech & Hearing Res.* II. 244 (*heading*) A phonetic study of misarticulation of /r/. **1972** *Language* XLVIII. 492 It is important that the study of misarticulation be integrated into this dependency relationship, for the mutual benefit of speech therapy and linguistic theory.

misascri·ption. Also with hyphen. [Mis-^{1} 4.] False ascription.
1923 E. K. Chambers *Elizabethan Stage* II. xii. 30 He light-heartedly accuses my friend Mr. Pollard, me, and others of perpetuating an old mis-ascription. **1946** *Mod. Lang. Notes* LXI. 61 The initial and continued misascription of the fourteen line poem, 'The Muses, fairest light in no darke time', to John Cleveland. **1971** A. Kirk-Greene in J. Spencer *Eng. Lang. W. Afr.* 137 These lexical items [in West African English] consist of creation in two ways: deliberate coining and misascription.

misassimila·tion. *rare.* [Mis-^{1} 4.] Incomplete or unsuccessful assimilation.
1934 E. Sitwell *Aspects Mod. Poetry* 69 Here, too, is another misassimilation of Hopkins by the same writer.

misattribu·tion. [Mis-^{1} 4.] Attribution, usu. of a work of art, literature, etc., to the wrong person. Hence **misattri·butor,** one who makes a misattribution.
1873 [see Mis-^{1} 4]. **1927** *Sunday at Home* Jan. 216/1 The hurried journalist is a great misquoter and a great misattributor too. **1963** *Times* 4 June 14/4 Highmore, even until recent years, has suffered from misattributions, either of his best work to others, Hogarth in particular, or of the inferior work of others to him. **1973** *Times* 3 Nov. 15/6 Misattributions of this kind should be avoided. **1975** *Sotheby & Co. Catal.* 28–29 July 4/2 Books may not be returned nor will their sale be set aside..for possible misattribution of authorship where the authorities are in disagreement.

misbelead, *v.* Delete † *Obs.* and add *rare.*
1909 Galsworthy *Strife* ii. ii. 251 *Thomas.* I haf ears to my head... *Jago.* Your ears have misbeled you then.

misbou·nd. [f. Mis-^{1} 2 + Bound *ppl. a.*^{2} 6.] Of a book: badly or wrongly bound.
1802 D. Wordsworth *Jrnl.* 5 Feb. (1941) I. 107 The Chaucer not only misbound but a leaf or two wanting. **1889** [see Mis-^{1} 2]. **1952** J. Carter *ABC for Bk.-Collectors* 120 When a leaf or leaves, or an entire gathering, has been wrongly folded or misplaced by the binder, it is called misbound. Provided that nothing is missing, and that the amount of matter misbound is not too great or its misplacing too glaring, collectors commonly take a more charitable view of the result than readers.

miscarriage. Add: **2. a.** (Further examples.)
1784–5 Burns *Poem on Pastoral Poetry* in *Poems & Songs* (1968) I. 191 Scarce ane has tried the Shepherd-sang But wi' miscarriage? **1880** 'Mark Twain' *Lett. to Publishers* (1967) 123 Chatto waits..without asking a solitary question about the book, and then pitches into *me* about the miscarriage.
3. (Later examples.)
1967 M. M. Bookmiller et al. *Textbk. Obstetr.* (ed. 5) xxii. 333/2 The term early abortion refers to expulsion of the fetus up to 12 weeks' gestation. When expelled from 12 to 28 weeks it is called a late abortion and from 28 to 36 weeks it is said to be premature termination of pregnancy. Many laymen still associate the word abortion with illegal

interference and prefer to call a spontaneous termination of pregnancy a miscarriage. The nurse should be guided accordingly. **1970** *Sci. Jrnl.* June 75/2 Most hospitals regard the loss of a pregnancy before 20 weeks as a miscarriage. **1971** Rugh & Shettles *From Conception to Birth* (1972) x. 157 There are generally signs of an impending miscarriage (another term for a spontaneous abortion). **1974** Passmore & Robson *Compan. Med. Stud.* III. xlii. 1/2 Abortion may be defined as the termination of pregnancy before the fetus is viable. In Britain this is considered to be before the 28th week of pregnancy. However, fetuses expelled before the 28th week occasionally survive, and to allow for this possibility the 20th week is sometimes used as the watershed between viability and non-viability... In many North American centres abortion is defined as the expulsion of a fetus weighing less than 500 g. A precise definition is required for medicolegal purposes and in Britain abortion means the termination of pregnancy before the 28th week. Miscarriage is synonymous with abortion and is often more acceptable to the lay public.

miscast, *sb.* Restrict † *Obs.* to sense 1 and add: **2.** *Theatr.* An actor or actress who is miscast (see *miscast *v.* 4); an instance of miscasting.
1907 G. B. Shaw *Lett. to G. Barker* (1956) 81 Barker was capable of anything in the way of casting..Miss Sterling McKinlay was..a very obvious miscast. **1908** *Ibid.* 142 It was not her fault; it was simply a miscast.

miscast, *v.* Add: **4.** *Theatr.* In passive, of an actor: to be cast in an unsuitable rôle; of a play: to have unsuitable actors performing in it; also *fig.*
1927 *Observer* 10 July 15/1 We say that so-and-so was good or adequate or miscast or unhappy in his part. We do not appraise the acting as we appraise the play. *Ibid.* 21 Aug. 9/4 'The Climax'..was brought to England seventeen years ago... It was then admittedly miscast. **1944** M. J. MacManus *Eamon de Valera* xiii. 288 Poor Mr. Thomas..was sadly miscast on the diplomatic stage. **1957** *Observer* 8 Sept. 10/4 As Tigre, M. Barrault is mournfully miscast. **1972** *Daily Tel.* 9 Feb. 13/6 This was partly because John Neville was miscast in the main role.

miscasting, *vbl. sb.* Add: **3.** *Theatr.* The allotting to an actor of a part which does not suit him.
1926 *Spectator* 25 Sept. 470/2 In a long list of characters there is not one case of miscasting. **1927** *Observer* 26 Sept. 13/4 Those instances of what I thought was miscasting have made me long to try my hand at casting a play. **1946** *Weekly Rev.* 20 June 156/1 The extraordinary mis-casting of the Elder who was made up to look like an emaciated crook suffering from T.B. **1971** *Daily Tel.* 24 Nov. 10/4 Act Four [of 'Swan Lake'] suffered..from the miscasting of Miss Mason.

misca·talogued, *ppl. a.* [Mis-^{1} 2.] Erroneously or inaccurately entered in a catalogue.
1963 [see *information retrieval* s.v. *information 8]. **1974** *Times* 10 Jan. 4/6 Low-season sales are always well scrutinized by dealers looking for miscatalogued bargains.

miscegenation. (Further examples, not *U.S.*)
1927 M. M. Bennett *Christison* ii. 29 'Miscegenation' being official jargon for what Governor Bourke called 'detaining black women by force'. **1971** *Sunday Times* 20 June 29/6 [He] must inaugurate 'creative miscegenation' by marrying a Chinese girl.
miscegen (further examples); also **miscege·nic** *a.*; **misce·geny,** miscegenation.
1865 S. S. Cox *Eight Yrs. Congress* 354 A very sprightly suffragan of the miscegen stamp. *Ibid.*, The result would be an average miscegen and a superior patriot. **1941** 'R. West' *Black Lamb* I. 527 It was a fusion, lovely but miscegenic, of the Byzantine and the baroque styles. **1935** *Punch* 14 Aug. 176/1 Since miscegeny is not a bad British trouble, *Shanghai* is a film that is more likely to interest America than ourselves.

mischancing, *vbl. sb.* For *rare*$^{-0}$ read *rare* and add example.
1929 W. Faulkner *Sartoris* v. 357 A period of history which had seen brothers and husband slain in the same useless mischancing of human affairs.

mischief, *sb.* Add: **9. c.** (Further example.) Also phr. *like the mischief.*
1876 T. E. Brown *Doctor* 36 And them givin' sheet Like the mischief. **1895** *Century Mag.* June 279/2 And there's kindnesses and kindnesses, Mr. Ludovic. There's some that cost like the mischief. **1907** J. M. Synge *Lett. to Molly* (1971) 123, I am coughing away like the mischief today. **1922** Joyce *Ulysses* 354 She wished..they'd take the..twins and their baby home to the mischief out of that. **1942** Berrey & Van den Bark *Amer. Thes. Slang* § 20/5 Indefinite eminence in degree. (Preceded by.. 'like')... The (very) mischief.
10. mischief night, an evening, orig. 30 April, now 4 November, on which children indulge in mischievous pranks; also *attrib.*
1865 W. S. Banks *List Provincial Words Wakefield* 47 *Mischief neet.* Boys, thirty years ago, used to go about damaging property, believing the law allowed them, on this night. Happily the practice is over at Wakefield, and the time forgotten. **1871** *N. & Q.* 17 June 525/1 The eve of May Day was formerly known as 'Mischief Night' throughout South Lancashire, and prior to the epoch of the 'new policeman', many were the strange pranks, rude practical jokes, and mortifying degradations committed. **1959** I. & P. Opie *Lore & Lang. Schoolch.* xii. 255 In the nineteenth century April the thirtieth..was..the traditional Mischief Night... How Mischief Night has..come to be transferred to the other end of the year is one of the mysteries of the folklore calendar. *Ibid.* 279 '20 Boys in Mischief Night Raid at Ayton.' **1969** —— *Children's Games* ii. 68 They even celebrated Mischief Night on 4 November as do their northern contemporaries. **1972** 'J. Ripley' *My Word you should have seen Us* 119 It was 'Mischief Night'—the evening before 'Bonfire Night'—and an annual happening peculiar to the northern provinces.

mischievous. ¶. Later examples of dial., vulgar, and jocular uses of *mischevious, mischievi(e)ous.* Also *mischeivious.*
1571–1747 examples in Dict. s.v. sense 3 *β.*
1847 C. M. Yonge *Scenes & Characters* xv. 194 You thought mischievous was meant in Hannah's sense, when she complains of master Reginald being very mische-vious. **1861** —— *Young Step-Mother* xxxi. 476 For shame, to be so *mischievieous*; such a great boy as you. **1913** Mrs. P. Campbell *Let.* 5 Feb. in *B. Shaw & Mrs. Campbell* (1952) 81 Some mischeivious personal experience. **1952** F. Swinnerton in *Bks. of Month* Nov.–Dec. 31/1 Wells,.. friendly with everybody, mischevious, quick-thinking, nonsensically inventive.

misch-masch, var. Mish-mash *sb.*

mischmetal(l (mi·ʃmet'l). Also **misch metal.** [ad. G. *mischmetall*, f. *misch-en* to mix + *metall* Metal *sb.*] A mixture of lanthanons containing about 50 per cent cerium which is obtained usu. by electrolysis of the fused chlorides from monazite and is used in lighter flints.
1923 B. S. Hopkins *Chem. Rarer Elements* x. 168 The most important of these [cerium alloys] is the alloy called misch metal, mixed metal, commercial cerium, or simply 'cerium'. **1924** J. W. Mellor *Comprehensive Treat. Inorg. & Theoret. Chem.* V. xxxviii. 608 A. Sieverts and G. M. Goldegg studied the action of hydrogen and nitrogen on the mischmetals. **1954** H. E. Kremers in C. A. Hampel *Rare Metals Handbk.* xvi. 343 The most common ferrous alloy of the rare earths is the common lighter flint, which contains about 30 per cent iron, the balance being misch metal. **1966** Phillips & Williams *Inorg. Chem.* II. xix. 18 A particularly interesting lanthanide alloy is 'mischmetall', with 50 per cent Ce, 25 per cent La, and the rest mostly Nd and Pr. This alloy is added to partly purified iron when it preferentially removes elements such as C, O, S, N, and P into the slag as lanthanide compounds.

mischsprache (mi·ʃ,ʃpraχə). [G., 'mixed language'.] = *mixed language* (*mixed *ppl. a.* 11).
[**1885** M. Grünbaum in *Sammlung gemeinverständlicher wissenschaftlicher Vorträge* XX. 613 (*title*) Mischsprachen und Sprachmischungen.] **1930** J. T. Hatfield et al. *Curme Vol. Ling. Stud.* 12 The question as to the justification of calling this tongue German and whether it is a *Mischsprache.* **1963** *English Studies* XLIV. 9 There are cases where..a scribe *half*-transforms his original, producing a sort of *Mischsprache.*

misch(t)y, dial. var. Mischief *sb.*
1890 S. S. Buckman *John Darke's Sojourn in Cotteswolds* xvi. 150 Er's harmless enow when he comes yereby, er 'oodn't do none mischy. **1895** Hardy *Jude* i. i. 9 Just now he's a-scaring of birds for Farmer Troutham. It keeps un out of mischty. **1896** —— *Under Greenw. Tree* ii. v. 117 Bless ye, my sonnies! 'tisn't the pa'son's move at all. That gentleman over there..is at the root of the mischty [mischief, 1872, 1876, 1891 eds.].

miscible, *a.* Add: Usu. *spec.* of a liquid: capable of forming a true solution *with* another liquid. (Further examples.)
1960 Hamill & Williams *Princ. Physical Chem.* ix. 244 While phenol and water at this temperature and pressure are only partly miscible the addition of acetone increases their mutual solubility. **1964** G. I. Brown *Introd. Physical Chem.* xxiii. 252 Ether will dissolve a little water (about 1·2 per cent at room temperature) to form a homogeneous solution, and water will also dissolve a little ether (about 6·5 per cent at room temperature) to form a similar solution. Within these limits ether and water are completely miscible. As their mutual solubilities are limited, however, ether and water are only partially miscible.

misco·de, *v.* [Mis-^{1} 1, 3.] *trans.* To code incorrectly. So **misco·ding** *vbl. sb.*
1965 *Math. in Biol. & Med.* (Med. Res. Council) ii. 54 Gross mistakes in the date are probably rare, and are due only to mispunching and/or miscoding. **1970** *Sci. News Let.* 23 May 510 The mutant DNA miscoded a single amino acid in the sequence of structural protein in the membrane. As a result..the entire membrane was defective.

misconduct, *v.* **2.** (Further examples.)
1901 *Daily Colonist* (Victoria, B.C.) 11 Oct. 2/1 The Jacksons..induced girls to misconduct themselves with the belief that it was a necessary part of their religious devotions. **1922** Joyce *Ulysses* 318 Belle in her bloomers misconducting herself. *Ibid.* 457 He made improper overtures to me to misconduct myself at half past four p.m. on the following Thursday.

misconstrue, *v.* **1. a.** (Further *absol.* example.)
1842 C. Lever *Jack Hinton* (1843) xlviii. 308 It is so easy, when people have no peculiar reasons to vindicate another—to misconstrue—perhaps condemn.

miscontent, *v.* Delete † *Obs.* and add later example.
1920 M. Hewlett *Light Heart* v. 33 'I have had words come by me,' she said, 'that you are beguiling my Thordis. That miscontents me.'

miscri·ticize, *v. rare.* [Mis-^{1} 1.] To criticize adversely or wrongly.
1877 Swinburne *Lett.* (1960) III. 275 Attacked and miscriticized in the Saturday Review and the Academy.

mis-cue, *sb.* Add: Now usu. **miscue. b.** Hence in other sports, and *transf.* or *fig.*, an error resulting in a failure of some sort.
1883 in *Amer. Speech* (1965) XL. 130 When I escort a lady I..have a..Havana And puff it all the time And should she make a small miscue And knock it from my mouth, [etc.]. **1920** Ade *Hand-Made Fables* 26 It suggested that there had been a Miscue at the Christening. **1942** Berrey & Van den Bark *Amer. Thes. Slang* § 170/2 Error; mistake; blunder... Miscue. **1958** J. A. Barlow *Elem. Rifle Shooting* (ed. 5) v. 56 The idea..is to be able to put sufficient force behind the bolt to eliminate any chance of a miscue—in other words, a failure to close the bolt. **1970** *Washington Post* 30 Sept. D1/7 One Oriole scored, another took third on the miscue and John Oates' sacrifice fly made it 2–2.
miscue *v.* (further examples); also *transf.* and *fig.* So **miscue·d** *ppl. a.*; **miscue·ing** *vbl. sb.*
1929 *Times* 1 Nov. 7/4 At the end of the last of these breaks he missed a difficult *massé* cannon, through partly miscueing. **1941** G. Heyer *Envious Casaca* xii. 219 Mathilda shied away from the thought, miscued, and straightened herself. **1941** J. Smiley *Hash House Lingo* 38 *Miscue*, make an error. **1955** I. Peebles *On Ashes* x. 100 Bailey tried to hook Johnston and miscued so that the ball bounced off his pads. **1962** *Times* 3 Jan. 3/7 His miscued strokes always went up to score disconcertingly. **1962** *Sunday Times* 25 Feb. 22/7 Miss Truman was now mis-cueing. **1970** *Sunday Tel.* 1 Nov. 32/7 Southgate went nearest to a goal when Neale cut in from the left but Walker miscued badly.

misdemeanour, *sb.*^{1} **2.** All distinctions between a felony and a misdemeanour were abolished by the Criminal Law Act of 1967.

misdescriptive, *a.* (Further example.)
1938 R. G. Collingwood *Princ. Art* v. 80 Calling it by that misdescriptive name, we patronizingly license the child to go on with it.

misdi·agnose, *v.* [Mis-^{1} 1.] *trans.* To diagnose wrongly, make a wrong diagnosis of (a condition); also, to diagnose wrongly the condition of (an individual).
1928 *Daily Express* 6 Oct. 8/6 Internal troubles which have been misdiagnosed are the most common. **1949** G. Ryle *Concept of Mind* iv. 107 The..distresses of which such feelings are diagnosed, or mis-diagnosed, as signs are not themselves feelings. **1963** *Lancet* 12 Jan. 95/1 In spite of general recognition for nearly half a century that the two conditions simulate each other, nobody can say how often torsion is misdiagnosed as 'orchitis'. **1968** *Globe & Mail* (Toronto) 5 Feb. 13/6 Many children with minimal brain dysfunction are still being misdiagnosed. **1972** *Village Voice* (N.Y.) 1 June 36/3 His primary theme, however, is that far too many of the kids being drugged have been misdiagnosed.

misdiagno·sis. [Mis-^{1} 4.] A wrong diagnosis.
1949 G. Ryle *Concept of Mind* iv. 105 Such misdiagnoses are more common in children than in grown-ups. **1965** J. Pollitt *Depression & its Treatment* iii. 43 Migraine, angina pectoris, and prolapsed intervertebral disc are the commonest misdiagnoses in this field. **1971** *Daily Colonist* (Victoria, B.C.) 28 Feb. 28/5 This is a depressingly high rate of misdiagnosis.

misdi·al, *v.* [Mis-^{1} 1.] *intr.* To dial (usu. by mistake) a number other than that required on a telephone.
1964 W. Markfield *To Early Grave* (1965) xii. 244 You *could* have been all along misdialing. **1967** J. Gardner *Madrigal* i. 6 He misdialled, then got it right... The signal brut-brutted at the far end of the line. **1970** *Guardian* 27 Apr. 10/2 The Post Office say the numbers are similar to his, and people misdial.

misdirected, *ppl. a.* Add: Hence **misdire·ctedness.**
1965 A. Farrer in J. Gibb *Light on C. S. Lewis* 38 The primary function of mental pain, says Lewis, is to force our misdirectedness on our attention.

misdirection. Add: **1. b.** Of the action of a conjurer, thief, etc.: distraction, guidance (of a person's attention) away from (something).
1943 A. Christie *Moving Finger* x. 124 A conjuring trick... You've got to make people look at the wrong thing and in the wrong place—Misdirection, they call it. **1949** *Amer. Speech* XXIV. 40 When the performer directs your attention, by word, glance, or gesture, away from a secret function, he calls the process *misdirection.* **1955** *Publ. Amer. Dial. Soc.* xxiv. 16 The thieves can apply misdirection (verbal, kinesic, tactile) to take his mind off his wallet—if only for a few seconds—and he will be astonished to find that he has been robbed. **1971** P. O'Donnell *Impossible Virgin* iv. 71 There had been callers at the house... None of them had been genuine... It was all part of a misdirection play.

misdistribution. [Mis-^{1} 4.] Wrong or faulty distribution. So **misdistri·bute** *v.*
1914 G. B. Shaw *Misalliance* Pref. p. xxxvi, Obstructing the way of the proper organization of childhood, as of everything else, lies our ridiculous misdistribution of the national income. **1920** S. Alexander *Space, Time & Deity* II. iii. ix. 280 Evil is misdistribution, and vice is a feature of character which wills such misdistribution. **1958** *Archit. Rev.* CXXIV. 338/1 The misdistribution of people has inevitably brought problems, chiefly that of lack of housing. **1968** *Economist* 27 July 39/1 The schools are to have another $33 million, similarly misdistributed.

misdivision. Add: **b.** *spec.* in *Cytology.* The abnormal transverse (instead of longitudinal) division of a centromere at meiosis or mitosis.
1939 C. D. Darlington in *Jrnl. Genetics* XXXVII. 341 (*heading*) Misdivision and the genetics of the centromere. *Ibid.* 346 Misdivision in..107 univalents consisted of the centromere dividing into two halves, as it does normally; but to these halves the wrong chromatids were attached or no chromatids at all. *Ibid.*, The simplest misdivision is that where two short arms are carried to one pole, two long arms to the other. **1972** tr. *J. Sybenga's Gen. Cytogenetics* v. 223 It happens occasionally that the centromeric region breaks up (mis-division), which results in two half chromosomes (the arms) each with a part of the centromere. *Ibid.*, After centromere mis-division, functional telocentric chromosomes may arise.

misdraw, *v.* Add: **1.** (Later example.)
1885 *Proc. Soc. Psychical Res.* III. 427 There were also 40 diagrams..all misdrawn.
Hence **misdraw·n** *ppl. a.*, badly or wrongly drawn (up).
1867 Bagehot *Eng. Constitution* 268 The practical arguments and the legal disquisitions in America are often like those of trustees carrying out a misdrawn will.

‖ **mise au point** (mīzopwæṅ). [Fr.] A focusing or clarification of an obscure subject or problem.
1946 *Word* II. 113 A general *mise au point* of the linguistic side of semantic problems is thus overdue. **1949** *Archivum Linguisticum* I. 126 The Geneva School retorted with an important *mise au point.* **1971** *Ibid.* II. 64 For a recent *mise au point*, see P. H. Matthews,..in *New Horizons in Linguistics.*

‖ **mise-en-page** (mīzaṅpāʒ). [ad. Fr. *mise-en-pages* imposition.] The design of printed pages, including the layout of text and illustrations; also, the composition of pictures.
1926 R. Fry in J. M. Cameron *Victorian Photographs* 13 Here the artist has been able to control everything, the *mise-en-page*, the disposition of the drapery, and the illumination. **1930** *Times Lit. Suppl.* 3 July 556/1 More intent on realizing the volumes than on the mise-en-page. **1938** *Burlington Mag.* Sept. p. xiii/1 There is hardly a single plate..which can be considered a really satisfactory *mise-en-page.* **1939** *Ibid.* Apr. 177/1 No difference as far as the mise-en-page is concerned between the portrait said to be by Bellini and a Turkish work. **1957** *Listener* 5 Dec. 947/2 The wrapper is deplorable..and the *mise-en-page* rather poor..nevertheless a book to be thankful for. **1959** *Times* 26 Mar. 3/4 Neither the tonal harmony..nor the *mise-en-page* is always quite strong enough in these latter pictures. **1963** *Ibid.* 18 Apr. 7/6 Of special interest are the publications in which students are given the opportunity of gaining practical experience in the design of the book and the *mise-en-page.* **1968** *Listener* 25 Jan. 106/3 Television cannot emphasise in the manner of a newspaper's *mise-en-page.* **1972** *Guardian* 19 June 11/2 Aubrey Beardsley..contributed..the art of *mise en page* ..the art of making dramatic use of big blank spaces.

‖ **mise-en-scène** (mīzaṅsęn). [Fr.] **a.** The staging of a play; the scenery and properties of a stage production; the stage setting.
1833 W. C. Macready *Diary* 14 Dec. (1912) I. 85 Saw the play, *Coriolanus*, in so disgraceful a state that it was useless to bestow a word upon the *mise en scène.* **1840** A. Bunn *Stage* II. xi. 298 More attention was paid to the *mise en scène* than to the acting. **1891** 'L. Malet' *Wages of Sin* I. iii. ii. 131 Only look at the walls of our exhibitions, look at the *mise en scène* of our theatres! **1911** G. B. Shaw *Blanco Posnet* Pref. 340 The *mise-en-scène* of a play is as much a part of it as the words spoken on the stage. **1951** *Oxf. Compan. Theatre* 715/1 Perhaps Antoine..never realized the force of some of his *mises-en-scène.* **1961** K. Tynan *Curtains* iii. 391 In the newer *mises en scène* one got a feeling of dehydrated Luhtishness. **1974** *Listener* 17 Jan. 92/3 The newspaper office *mise-en-scène*..gave it [*sc.* a radio play] early vitality.
b. *transf.* and *fig.* The setting, surroundings, or background of an event or action.
1872 E. Braddon *Life in India* i. 8 Novelists..sometimes select India as the *mise en scène* of their tales. **1894** [see *impressionize *v.*]. **1901** *Q. Rev.* CXCIII. 314 She [*sc.* Queen Victoria] was unrivalled in her sense of the proper *mise en scène* of a formal ceremonial. *a* **1916** H. James *Ivory Tower* (1917) 270, I manage to treat myself to some happy..mise-en-scène or exploitation of my memory of (say) California. **1924** Earl of Birkenhead *Amer. Revisited* vi. 165 She would have dictated peace, I should imagine, at Buckingham Palace; for the *réclame* of the *mise-en-scène* would, on the whole, have been greater than that of Versailles. **1940** Wodehouse *Eggs, Beans & Crumpets* 12 You have simply got to get your atmosphere right... Chance your arm with the *mise en scène*, and before you can say 'what ho', you've made some bloomer. **1972** J. Hodgson *Uses of Drama* xvi. 183 In certain dramatized situations 'people' are actors, and the rush of social events, mise-en-scene.

mise·mphasis. [Mis-^{1} 4.] Incorrect emphasis.
1927 *Sat. Rev. Lit.* 12 Mar 651/1 The historic church has too frequently specialized in morbid and bogey-ish suggestions until the natural trend of a faith that is faith has been diverted to misemphasis on guilt and introspective self-analysis under the obsession of fear. **1962** R. B. Fuller *Epic Poem on Industrialization* 3 Full of fanciful misemphasis. **1965** H. Kahn *On Escalation* viii. 152 This example of past misemphasis is now widely known and is often cited by analysts.

misère. Add: **2.** = Misery.
1897 E. Dowson *Let.* *c* 10 June (1967) 385 Writing to you has somewhat cheered me, but all my misères will return in a moment. **1903** [see *God-forsaken *ppl. a.*]. **1964** *Punch* 2 Sept. 359/3 Recommended to those who still enjoy being stunned by the grandeurs of empire with the *misères* guaranteed absent.

miserere. 4. (Earlier *attrib.* example and later *transf.* example.)
1848 B. Webb *Sk. Continental Ecclesiol.* ii. 39 Four beautiful stalls, with miserere-seats and canopies. **1959** *Punch* 30 Dec. 675/2, I..selected a train with the longest carriages I could find... To get the full vista, I sat on the miserere at the front.

misericord, *sb.* **1.** For † *Obs.* read *Obs.* except *arch.* and add later example.
1922 Joyce *Ulysses* 380 They had had ado each with other in the house of misericord where this learning knight lay.

misery. Add: **2. c.** *misery me!*, an interjection expressing self-pity, distress, or general wretchedness.
1888 W. S. Gilbert *Yeomen of Guard* ii. 48 Misery me, lackadaydee! He sipped no sup and he craved no crumb, As he sighed for the love of a ladye! **1968** N. Marsh *Clutch of Constables* i. 15 'O misery, misery, misery me,' she wrote with enormous relish.
6. *to put* (a person or animal) *out of* (his) *misery*: see Put *v.*^{1} 48 d.
8. *misery-line, -threshold.*
1902 W. James *Varieties Relig. Experience* vi. 135 The sanguine and healthy-minded live habitually on the sunny side of their misery-line. *Ibid.* v. 135 We might speak of a 'pain-threshold', a 'fear-threshold', a 'misery-threshold'.

misestimate, *sb.* (Earlier example.)
1843 Mill *Logic* II. v. iv. 398 A positive mis-estimate of evidence actually had.

misfe·lt, *ppl. a.* [Mis-^{1} 2.] Felt incorrectly or imperfectly.
1935 L. MacNeice *Poems* 42 We whose senses give us things misfelt and misheard.

misfield, *v.* Add: (Earlier and later examples.)
1870 *Times* 10 Aug. 5/5 Both of these [*sc.* hits for four] ..were unaccountably misfielded by Smith, who rarely makes mistakes. **1954** J. B. G. Thomas *On Tour* vi. 66 Dixon..looked at the South African wing for a moment and misfielded. **1974** *Times* 8 Jan. 7/8 The new beak picks up the pill and chucks at McCallum who misfields it.
Also as *sb.*, failure to gather the ball properly.
1909 *Daily Chron.* 25 Feb. 8/4 Guy's scored after a misfield of a high kick by Batchelor.

misfire, *sb.* Add: (Earlier example.)
1839 Ure *Dict. Arts* 478 A mis-fire is hardly ever experienced with the fire-arms made at the Royal manufactory.
b. The state or action of misfiring.
1966 B. C. Macdonald *Car Doctor* iv. 45 If there is a regular or rhythmical type of misfire, together with a blackish exhaust, then the mixture is too rich. **1973** J. B. Edwards in Springer & Patterson *Engine Emissions* ii. 47 A significant level of misfire may exist in a poorly maintained and/or adjusted engine.

misfire, *v.* Add: **b.** Of an internal combustion engine: to fail to explode the charge, or to explode it at the wrong instant. So **misfi·ring** *vbl. sb.*
1905 *Motor Cycle* 6 Mar. 218/3 If an engine is back firing, that is, giving explosions in the silencer, it is also misfiring. Misfiring may result from a defect in the ignition system. **1928** Montagu & Bourdon *Cars & Motor-Cycles* III. 1244 An engine is said to misfire when the gas in one or more of the cylinders does not ignite. **1966** B. C. Macdonald *Car Doctor* ix. 80 Other causes of misfiring are: incorrect valve timing, weak or broken valve springs, sticking valves, too much carbon in engine, [etc.]. **1974** Haynes & Ward *Audi 100 Owners Workshop Manual* iv. 89/2 If the engine misfires regularly, run it at a fast idling speed.
c. *fig.* To fail, to be misdirected, to make a mistake. So **misfi·red** *ppl. a.*
1942 Berrey & Van den Bark *Amer. Thes. Slang* § 170/6. **1949** Koestler *Promise & Fulfilment* xii. 132 The other after-effects of the misfired operation were in the same strain. **1974** 'J. Graham' *Bloody Passage* viii. 120 'What a wonderful idea,' he said, and then the thing misfired slightly.

misfit, *sb.* Add: *transf.* and *fig.* *spec.* a person unsuited to his environment, work, etc.
1903 [in Dict.]. **1936** *Discovery* Sept. 280/1 The selection and training of personnel to eliminate as far as possible the misfit and (what is far more prevalent) the partial misfit who just stands the test of results but has really missed his vocation. **1939** T. S. Eliot *Family Reunion* i. ii. 55 The very moment when you are wholly conscious Of being a misfit, of being superfluous. **1959** *Times Lit. Suppl.* 30 Jan. 57/1 He is a determined individualist, wears Afrika Korps uniform while serving in the British Army and is something of a misfit. **1975** *Times* 20 Aug. 4/8 The police..said young misfits were taking as their victims other car drivers.
attrib. **1910** *Encycl. Brit.* XIV. 223/2 The advantage of this combination is that..it..lessens the danger of making 'misfit' pig iron, *i.e.* that which, because it is not accurately suited to the process for which it is intended, offers us the dilemma [etc.]. **1961** A. Miller *Misfits* xi. 117 Nothin' but misfit horses, that's all they are, honey.
2. *Physical Geogr.* A stream which, if its average flow in the past was at present-day levels, would be expected to have eroded a larger or a smaller valley than it has done. Usu. *attrib.* or as *adj.*
1910 Lake & Rastall *Text-bk. Geol.* iii. 47 In this case the lower part of valley A is left dry, or with an insignificant stream only, which appears to be too small to have eroded the valley in which it flows. Such a stream is called a misfit. **1932** *Jrnl. Geol.* XL. 486 (*heading*) Misfit streams. **1964** *Prof. Papers U.S. Geol. Survey* No. 452-A. 6/1 Streams recognized as misfit are so usually underfit that the two names are frequently interchanged. **1968** R. W. Fairbridge *Encycl. Geomorphol.* 706/2 The most commonly recognized case of a misfit stream is the underfit river.

misgotten, *ppl. a.* **1.** (Later example.)
1903 W. S. Jackson *Nine Points of Law* vi. 171 Here was he with a trunk-load of misgotten gold wandering haphazard..over two countries.

mish^{2} (miʃ). Colloq. abbrev. of Missionary *sb.*
1939 J. Cary *Mr. Johnson* 164 That's wot's wrong with some of these teetotal mishes. **1946** C. S. Archer *China Servant* vii. 105 You've no idea what a bunch of mishs will do, when sex rears its head.

mishellene (mishe·līn). [ad. Gr. μισέλλην, f. μισ(ο)-, f. μισεῖν to hate + Ἕλλην Hellene, after Philhellene *a.* and *sb.*] One who dislikes or is opposed to Greece or the Greeks. Hence **mishelle·nic** *a.*
1958 R. Liddell *Morea* i. 16 Those two learned, grumpy, and extremely mishellenic travellers, Edward Dodwell and Sir William Gell. *Ibid.* 23 To the grumpy Mishellene, it is all merely boring and tiresome. **1972** W. St. Clair *That Greece might still be Free* xxxi. 350 Henry Lytton Bulwer..became violently pro-Turkish in the Greek–Turkish questions later in the century. David Urquhart..also became a noted mishellene.

mishit (mi·shit), *sb.* [Mis-^{1} 4.] In cricket, tennis, etc., a faulty or bad hit.
1882 *Australians in Eng.* 25 He made two mishits which fell harmless. **1898** [in Dict. s.v. Mis-^{1} 4]. **1928** *Daily Tel.* 11 May 18/1 Lyon has never played a better innings... I did not notice that he made even a mis-hit. **1963** *Times* 11 June 4/6 Bear's bold effort ended in a mishit to leg slip.
So **mishi·t** (also erron. **miss-hit**) *v. trans.*, to hit (a ball) faultily. Also **mishi·tting** *vbl. sb.*
1904 P. F. Warner *How we recovered Ashes* ix. 189 Gregory..mis-hit Braund and Rhodes caught him easily at backward point. **1930** *Times* 17 Mar. 4/4 Three minutes from the end Craig made his only mistake of the game, miss-hitting a bumping ball. **1963** *Times* 6 May 3/5 Some flashes of brilliance,..together with a good deal of mis-hitting,..suggests that the final next Sunday may provide an attractive game. *Ibid.* 13 June 3/2 Booth batted sensibly for an hour to claim 33 out of 47 before mishitting a towering catch to backward point. **1974** *Country Life* 21 Feb. 360/1, I wondered whether he had mishit his tee shot.

mish-mash, *sb.* Add: (Further examples.) Also **misch-masch, mish-mosh, mish-mush.**
c **1855** 'L. Carroll' (*title*) Mischmasch. *a* **1922** T. S. Eliot *Waste Land Drafts* (1971) 27 From such chaotic misch-masch potpourri What are we to expect but poetry? **1957** *Time* 2 Sept. 34/2 Paul Gregory's *Crescendo*, a mish-mash of American music, with Ethel Merman, Rex Harrison, Louis Armstrong, &c. **1959** N. Mailer *Advts. for Myself* (1961) 335 You call this a report?.. It is nothing but a mish-mosh. **1962** *John o' London's* 18 Jan. 68/2 How did the cast make out amid the mish-mush? **1964** G. Marx *Let.* 24 Feb. (1967) 305, I suggest you learn how to pronounce 'mishmash'. It is not pronounced 'mash',..but rather as though it were spelled *mosh.* **1965** P. Ziegfeld *Ziegfelds' Girl* i. 7 How the elegant and impeccably British Mr. Maugham had come to write this mishmash nobody could figure out. **1975** *Listener* 30 Oct. 589/2 The original *Panorama* had consisted of a mish-mash of disconnected and frequently frivolous items.

Mishnaic, *a.* (Later examples.)
1953 *Jrnl. Theol. Stud.* IV. 7 St. Mark is an apocalyptic, St. Matthew is a mishnaic writer. **1973** *Jewish Chron.* 2 Feb. 22/4 The Amoraim ('interpreters'), the post-Mishnaic teachers in the great schools of Palestine..and Babylon.

mis-hook (mishu·k), *v. Cricket.* [f. Mis-^{1} 1 + Hook *v.* 8 c.] *trans.* To hook (a ball) faultily.
1955 *Times* 19 July 12/1 The new ball was losing its shine when he mishooked a bouncer gently to mid-wicket. **1963** *Times* 12 June 4/4 Bear completely mishooked a ball and was caught at 19.
So **mi·shook** *sb.*, a faulty hook.
1961 *Times* 16 May 4/1 Minney..was caught off a mis-hook.

misidentifica·tion. [Mis-[1] 4.] Erroneous identification.
1902 *Encycl. Brit.* XXV. 468/1 The chief defects in practice were (1) frequent failure to identify, (2) liability to mis-identification. 1946 F. E. Zeuner *Dating Past* III. v. 132 The implications of such misidentifications..are obvious. 1968 *Listener* 27 June 828/3 Much more serious ..is..the kind of terrible events which were foreshadowed in the novel *Fail Safe*, whereby one has accidental mis-identification leading to some international disaster. 1973 *Nature* 13 July 74/3 They criticize Theyer's micro-palaeontological work by giving chapter and verse on a series of supposed misidentifications and miscorrelations.
So **miside·ntify** *v.*, to identify erroneously.
1895 L. Stejneger in *Ann. Report U.S. Nat. Museum 1893* 449 The specimens so recorded which I have had an opportunity to examine have either been misidentified, or else the locality was very doubtful. 1924 *Geogr. Jrnl.* LXIV. 457 A marks the position of Camp IV. (26,700 feet) and G the point reached by Mallory, Norton, and Somervell in 1922, which was probably misidentified and placed too high in the theodolite measure of 1922, made in very difficult circumstances. 1951 *Essays in Crit.* I. 332 He [*sc.* Iago] pretends ignorance, he pretends helpfulness, he intentionally misidentifies. 1952 *Mind* LXI. 422 He misidentifies the test he is using. 1971 *Nature* 19 Mar. 149/3 This name has been widely used for specimens of the genus in the Mediterranean and northern Europe. Templeman regards them all as having been misidentified.

Misima, var. *Massim.

misinform, *v.* (Later example.)
1974 *Physics Bull.* Jan. 3/2 All the above is well known. ..Those attempting to misinform the World Scientific Community are fully aware of my real circumstances.

misinfo·rmative, *a.* [Mis-[1] 6.] That gives wrong information. So **misinfo·rmatory** *a.*
1912 *Times Lit. Suppl.* 15 Feb. 64/3 To modify these few misinformative parts of a work which otherwise is.. valuable. 1927 *Observer* 3 Apr. 25/3 A so-called Informatory Double that does not contain top card strength is better termed 'misinformatory'.

miski·ck, *v.* [Mis-[1] 1.] *intr.* In various sports, to fail to kick the ball properly.
1901 [in Dict. s.v. Mis-[1] 1]. 1963 *Times* 21 Feb. 3/6 Lancaster presented Harris with an open goal, but the Watford winger completely miskicked. 1973 *Times* 24 Apr. 8/8 He miskicked and mishandled and could take pleasure only from his two successive penalty shots.
So **miski·ck** *sb.*
1973 *Times* 1 Jan. 17/6 His miskick only emphasized his mistake.

Miskito (miski·to), *a.* and *sb.* Also **Misskito,** Mosquito, Musquito. [f. *Misquito*, a section of the eastern coast of Nicaragua.] **A.** *adj.* Of or pertaining to an American Indian people living on the Atlantic coast of Nicaragua and Honduras. **B.** *sb.* **a.** A member of this people. **b.** The language of this people.
1789 O. Equiano *Interesting Narr. Life O. Equiano* II. xi. 172 He had a mind for..cultivating a plantation at Jamaica and the Musquito Shore... I found with the Doctor four Musquito Indians... One of them was the Musquito king's son. 1830 *Honduras Almanack* 11 The canoe, a paddle and a harpoon, are the Mosquito man's whole wealth. 1907 F. W. Hodge *Handbk. Amer. Indians* I. 948/2 *Mosquito Indians*, a tribe named from its habitat on Mosquito lagoon. 1911 *Encycl. Brit.* XVIII. 902/2 The Mosquito Coast is so called from its principal inhabitants, the Misskito Indians, whose name was corrupted into Mosquito by European settlers and has been entirely superseded by that form except in the native dialects. 1932 *U.S. Bureau Amer. Ethnol. Bull.* No. 106. 10 Ethnographical Survey of the Miskito and Suma Indians of Honduras and Nicaragua. 1959 *Chambers's Encycl.* I. 343/2 The Lenca, Xicaque,..and coastal Misquito of the Atlantic pocket of Nicaragua-Honduras missed the stream of Maya-Nahua influence. 1964 *Amer. Speech* XXXIX. 47 The Mosquitos are Caribs. 1964 E. A. Nida *Toward Sci. Transl.* ix. 194 In Miskito, a language of Nicaragua and Honduras, one encounters the following transliterations. 1966 C. F. & F. M. Voegelin *Map N. Amer. Indian Lang.* (*caption*) Misumalpan Family 1. Miskito.

misknowing, *vbl. sb.* **2.** (Later example.)
1892 F. S. Ellis *Lexical Concordance Shelley* p. viii, Well would it be for the world if no more was known of any poet's life, except through his works, than is known of Shakespeare's; how greatly should we then be delivered from misknowing!

misla·belling, *vbl. sb.* [Mis-[1] 3.] Incorrect labelling.
1952 M. Gardner *In Name of Science* i. 5 The government charged that displays of the book [*sc. Look Younger, Live Longer* by G. Hauser] next to jars of blackstrap constituted, because of the book's sensational claims, a 'mislabeling' of the product. 1960 *Times* 20 Sept. (Pure Food Suppl.) p. i/5 There still remains..the possibility of ..mislabelling in retail shops. 1964 *Economist* 6 June 1098/2 Some American firms..have been charged with mislabelling.

misleadingness. (Earlier and later examples.)
a 1866 J. Grote *Treat. Moral Ideals* (1876) 379 An element of deceptiveness and misleadingness. 1957 *Essays in Crit.* VII. 342 The misleadingness of any implication that I was scraping the barrel for evidence. 1975 *Nature* 10 Jan. 79/2 One example of the misleadingness of demonstration must suffice.

misline (mislai·n), *v.* [Mis-[1] 1.] *trans.* To print with lines omitted or arranged in the wrong order. So **mislinea·tion,** the result of a mistake of this kind.
1922 Joyce *Let.* 20 Mar. (1966) III. 62 Besides misprints I see..that it is in part mislined. 1964 F. Bowers *Bibliogr. & Textual Crit.* vi. iv. 182 It may be that some mislineation at the beginning of *Othello* was caused by the trouble he [*sc.* the compositor] experienced in adjusting text about the ornamental initial letter.

mislocate, *v.* Delete *rare*[−1] and add later examples.
1954 *Word* X. 239 In diagnosing the conflict between mentalism and mechanism he [*sc.* Bloomfield] mislocates the issue. 1959 P. F. Strawson *Individuals* II. viii. 244 Such an account runs the risk of mislocating the problem altogether. 1971 *New Scientist* 18 Mar. 611/1 The flash is both harder to see and it is mislocated. 1973 *Nature* 7 Sept. 12/2 Nuclear explosions fired on Amchitka are regularly mislocated by tens of kilometres.

mislocation. (Later examples.)
1874 L. Bacon *Genesis New England Churches* p. x, Every careless mislocation of words in the structure of a sentence. 1974 *Nature* 29 Nov. 369/2 All the main features are fairly well reproduced although there are some small mislocations.

mismatch, *sb.* Add to def.: A discrepancy; lack of correspondence; also, an unequal or unfair sporting contest.
1954 F. C. Avis *Boxing Reference Dict.* 71 *Mismatch*, a contest between two boxers of very different standards of ability. 1958 *Optima* Mar. 40/2 The mismatch between arisings and normal disposal methods. 1959 *New Scientist* 9 Apr. 802/3 There is a bad mismatch which will run right through the crystal. 1961 *Times* 21 Nov. 3/3 There had been suggestions that this [*sc.* a boxing match] had been a mismatch. 1971 *Daily Tel.* 28 Apr. 5/3 There is a definite mis-match between what universities are producing and what industry is wanting. 1973 *Computers & Humanities* VII. 139 The computer..compares the corresponding words..until a mismatch occurs.

mismatching, *vbl. sb.* (Later examples.)
1971 *Jrnl. Gen. Psychol.* LXXXV. 188 The interference occurs when the pictorial and linguistic symbolic representations of the color concepts are combined together with mismatching. 1974 *Nature* 7 June 566/2 This difference in thermal stability would indicate about 5% mismatching between the Mu MTV DNA and the tumour RNA.

mismate (mismēi·t), *v. rare.* [Mis-[1] 1: back-formation from Mismated *pa. pple.* and *ppl. a.*] *intr.* and *refl.* To mate or match (oneself) unsuitably.
1891 Hardy *Group of Noble Dames* I. iii. 113 No syllable would have been breathed of how I mismated myself for love of you! 1946 *Mod. Lang. Notes* Feb. 73 This is Milton's longest account of this Biblical episode, and it is clear than here the sons of God are pious men who mismate with the daughters of him 'who slew his brother'.

mismated, *pa. pple.* and *ppl. a.* (Later examples.)
1914 F. L. Packard *Miracle Man* i. 10 Trousers..torn at the ankles where they flapped around miss-mated socks and shoes. 1931 *Weekend Rev.* 16 May 743/2 It presents him [*sc.* Bulwer-Lytton]..as the child of his mismated parents. 1952 W. Styron *Lie down in Darkness* v. 194 At that age he was clear-headed enough to understand that he was not alone in a world of mismated passions: others betrayed and were betrayed, and got tired of loving.

mismating (mismēi·tiŋ), *vbl. sb.* [Mis-[1] 3.] Unsuitable mating or matching; *spec.* wrong assorting (of printing types).
1900 T. L. De Vinne *Pract. Typogr.: Treat. Type-Making* vi. 236 This difficulty tempts founders to make one set of small capitals serve for two or more distinct faces. An inexpert can seldom detect the mismating. 1934 *Punch* 10 Jan. 55/2 There is enough cruelty, halfwittedness, epilepsy, moral depravity and the mismating and obstetrical bungling that sometimes prelude these horrors. 1968 *Punch* 7 Feb. 205/3 Mis-matings are most common in such surroundings.

Misnagid, var. *Mitnagged.

‖ **miso** (mi·so). [Jap.] A paste, made from soya beans and barley or rice malt, used by the Japanese in preparing various foods.
1727 J. G. Scheuchzer tr. *Kæmpfer's Hist. Japan* I. I. ix. 121 Of the Meal of these Beans is made what they call *Midsu*, a mealy Pap, which they dress their Victuals withal. 1905 *Chambers's Jrnl.* 25 Mar. 270/2 Soya beans.. from which miso, soya and tofu are made. 1930 *Economist* 4 Jan. 24/1 The higher Japanese price-level is accounted for largely by such 'sheltered' goods as red beans, miso, dried bonito, [etc.]. 1966 P. S. Buck *People of Japan* (1968) xiv. 167 Chicken with *miso* paste and raw vegetables. 1970 J. Kirkup *Japan behind Fan* 104 A red lacquer bowl of rich *miso* soy (bean paste) soup.

miso-. Add: **misandry,** the hatred of males; **misarchist,** one who hates or opposes government in any form; **misogela·stic** *a. nonce-wd.* [Gr. γελαστ-ός laughable (see Agelast; cf. Agelastic *a.* and *sb.*)], hating laughter; **misosophy** (later example); so **misoso·phical** *a.*
1946 *Scrutiny* XIII. 249 In the absence of feminine precedents, she [*sc.* Beatrice] could do no better than what she very sensibly does do: follow masculine example, and answer to their affected misogyny with the affectation of misandry. 1960 B. Kaye *Upper Nankin St.* xii. 232 Such women are common in..Kwangtung Province, where there is a tradition of misandry. 1898 L. F. Ward *Outl. Sociol.* x. 228 These misarchists see the beneficent influences of natural law in the industrial world interfered with. 1877 G. Meredith *Ess. Comedy* in *New Q. Mag.* VIII. 2 It is but one step from being agelastic to misogelastic. 1937 *Philos.* XII. 332 The disposition to be convinced of ill-founded or unfounded doctrines, or unconvinced of well-founded ones, is a 'misosophical' disposition. *Ibid.* 319 A fraternity of persons of kindred credulities could only constitute a school of 'misosophy'.

misogyne. Delete *rare*[−1] and add further examples. Also **misogyn.**
1877 C. Reade *Woman-Hater* I. v. 113 Misogyn consented, but sighed. 1919 Beerbohm *Seven Men* 186 'Tis a goodly jest! The confirm'd misogyn a ladies' man!

misorder, *v.* **1.** (Later example.)
1909 *Daily Chron.* 24 Aug. 4/3 [He] charged the Admiralty with having so misordered the Navy as to expose the nation to the gravest jeopardy every hour.

misorienta·tion. [Mis-[1] 4.] Variation in orientation.
1952 *Acta Crystallogr.* V. 162 When aluminium is cold-rolled, the original grains break up into smaller particles. X-ray micro-beam back reflexion photographs permit the determination of the mean particle size and misorientation. 1966 D. G. Brandon *Mod. Techniques Metallogr.* ii. 82 If two neighbouring regions differ slightly in orientation,..and if the misorientation is extensive, characteristic radiation will only image one region at a time. 1973 *Nature* 3 Aug. 276/2 The large more misorientated subgrains (misorientation normally >5°) could be directly related to the optical subgrains.
So **miso·rient, -o·rientate** *vbs.* [Mis-[1] 1], to orient differently or variably; also, to orient badly. (Chiefly as pa. pples.)
1951 *New Yorker* 22 Sept. 94 The result of misorienting the Secretariat and using glass so exuberantly is to create a building that functionally is often windowless on all sides. 1953 *Acta Crystallogr.* VI. 167/1 A large number of grains in which there is widely misoriented material will be irradiated. *Ibid.* 177/1 This mosaic structure consists of particles slightly misorientated with respect to one another. 1967 *Sunday Times* 9 Apr. 52 We have the uneasy suspicion that our communal sense of values has become misorientated; and so it has. 1970 *New Scientist* 8 Oct. 65/1 Adjacent grains are misoriented with respect to one another. 1973 Misorientated [see above].

mispercei·ve, *v.* [Mis-[1] 1.] *trans.* To perceive wrongly or incorrectly; to mistake.
1924 J. E. McTaggart in J. H. Muirhead *Contemp. Brit. Philos.* (1st Ser.) 265 Consequently H will be misperceived by G as existing in time. 1953 *Mind* LXII. 207 The bedpost was not being misperceived when I dreamed that I saw the Eiffel Tower. 1973 E. Bullins *Theme is Blackness* 174 No, of course I don't think you're the bad guy but you're misperceiving me wrongly. 1974 *Sci. Amer.* Jan. 84 One might succeed in correcting the eyes shown here so that they are perceived as gazing downward and leftward, but at that very moment the mouth is uncorrected and expresses sorrow rather than pleasure. Conversely, one might correct the mouth and misperceive the eyes.

misperce·ption. [Mis-[1] 4.] The action of misperceiving or condition of being misperceived.
1722, 1893 [see Mis-[1] 4]. 1949 H. C. Weston *Sight, Light & Efficiency* iii. 108 One very familiar example of this mis-perception occurs in reading when we see a group of letters as the word they ought to represent, even though the printed word is mis-spelt.

misprint, *v.* Add: **b.** *intr.* Of deer: to leave foot-prints in a pattern different from the usual one.
1909 W. A. & F. Baillie-Grohman in Edward, Duke of York *Master of Game* 262 A hind..misprints, that is sometimes the hind foot will be placed beside the fore foot, sometimes inside or in front of it. 1957 F. J. T. Page *Field Guide Brit. Deer* 71 Misprint. To step irregularly: failure to register [*i.e.* to place the hind feet in the slots made by the fore feet].

mispronou·ncer. [Mis-[1] 5.] One who pronounces words incorrectly.
1885 *Educ. Times* 1 June 207/1 Warnings more adapted to American mispronouncers than to English.

misquo·te, *sb.* [Mis-[1] 4.] An incorrect quotation, a misquotation.
1855 J. A. Symonds *Let.* Sept. (1967) I. 61 How very kind of M. de Condolle it was to lend Mdlle the three missquotes. *a* 1953 E. O'Neill *Touch of Poet* (1957) I. 30 Disdainfully, emphasizing his misquote of the line from Byron. 1968 C. M. Vines *Little Nut-Brown Man* ix. 149 'Preferred not to' had a different nuance from 'was not well enough to'..it seemed a curious misquote. 1974 *Daily Tel.* 21 Oct. 16 Mr Heath made little attempt to correct the misquotes and inaccurate statements made by various Labour politicians.

misrecollection. (Later example.)
1922 O. Jespersen *Lang.* iii. 70 He speaks..of other linguistic changes as well. These he refers to the following causes...(1) Mishearing and misunderstanding; (2) misrecollection; [etc.].

misre·gister, *sb. Printing.* [Mis-[1] 4.] The incorrect positioning of printed matter in relation to other printed matter on the same sheet, esp. of two or more colours in relation to each other.
1931 *National Lithographer* Feb. 27/2 It was determined to study first the factors causing misregister. 1949 R. F. Reed *What Lithographer should know about Paper* (Lithographic Technical Foundation, N.Y., Technical Bull. No. 8) 50 While misregister in the ordinary sense never occurs until two or more colors have been printed, it is obvious that improper printing of the first color may be.. the real cause of register trouble. 1963 C. W. Latham *Advanced Pressmanship* xvii. 130 When the first color plate does not register with another color plate it may be termed internal misregister or misfit. 1966 R. R. Coupe *Sci. of Printing Technol.* ii. 47 The need to avoid misregister in multicolour work is usually regarded as the major reason for air-conditioning, but there are many others. 1968 *Gloss. Terms Offset Lithogr. Printing (B.S.I.)* 30 *Mis-register*, the appearance of the printed image out of its correct position. 1970 R. F. Reed *What Printer should know about Paper* (Graphic Arts Technical Foundation, Pittsburgh) iii. 33 Distortion due to wavy-edged paper is probably the most common and serious cause of misregister on sheet-fed offset presses.
Hence **misre·gistered** *ppl. a.*
1963 C. W. Latham *Advanced Pressmanship* xvii. 130 If the plate is at fault, the same misregistered image will show the same error sheet after sheet.

misre·gister, *v.* [Mis-[1] 1.] *trans.* To form from elements that are not properly aligned or positioned.
1969 G. L. Hansen *Introd. Solid-State Television Syst.* xi. 265 It is imperative that the optical alignment of the light paths be precise. If they are not, the reproduced color image will be misregistered. That is, each image will not be located at exactly the same point on the camera tube..and a blurred presentation will result. 1969 P. B. Jordain *Condensed Computer Encycl.* 278 If a line is misregistered irregularly, that is, if some characters are high, others low, etc., then line registration is bypassed in favor of character registration.

mi:sregistra·tion. [Mis-[1] 4.] Faulty or imperfect registration (i.e. alignment or positioning) of images, *spec.* of the three fields that compose a colour television picture.
1942 H. C. McKay *Photographic Negative* IV. x. 649 The neutralized areas will print out gray, and the lines of misregistration will print dark. 1952 *Electronics* Nov. 216/2 Color edging includes color fringing, misregistration, etc. 1960 *Electronic Engin.* XXXII. 71 Misregistration is the least favourable feature of colour television as at present conceived. 1969 P. B. Jordain *Condensed Computer Encycl.* 87 Character misregistration takes the form of character high, character low, or character skew. 1971 [see *Facture 4]. 1971 H. E. Ennes *Television Broadcasting* x. 469 This beam pulling or dynamic misregistration results in colored edges.

misruling, *vbl. sb.* (Later example.)
1927 *Daily Tel.* 25 Oct. 8/5 If in refereeing I give a flagrant legal mis-ruling..the Rugby Union can send for me and say what they think about me and my mis-ruling.

miss, *sb.*[1] Add: **7. a.** An unsuccessful gramophone record. Opp. *hit *sb.* 4.
1965 *Listener* 9 Sept. 391/1 Persons invited to give their verdict..are not being asked to say whether the songs are good or bad but merely whether they will be 'hits' or 'misses'. 1966 *Melody Maker* 16 July 20 Dusty's new single may be one of her misses.
b. Examples of transf. uses of *to give a miss*; also transf. use of the billiards phrase *to give the miss in baulk*.
1907 Westbrook & Wodehouse *Not George Washington* II. xxi. 228 And James..is actually giving this the miss in baulk! 1919 B. Ruck *Disturbing Charm* I. ii. 10 The Professor chose (as he often did) to give lunch a miss. 1923 Wodehouse *Inimitable Jeeves* iii. 31 Anyway, it never even occurred to me for a moment to give her the miss-in-baulk. 1927 A. Huxley *Let.* 17 May (1969) 286 The result of this will be that we must, alas, give Paris a miss. 1930 *Morning Post* 16 July 8/3 The leek is..among the..vegetables that are too often given a miss. 1950 J. Cannan *Murder Included* vii. 183 I'm afraid I've given church a miss this morning. 1973 Boyd & Parkes *Dark Number* ix. 91, I think the CID would be happier if you gave the whole place a miss.

Miss, *sb.*[2] Add: **2. b.** *Miss Nancy* (further and *attrib.* examples); so *to talk Miss Nancy*, to speak politely. Hence *Miss-Nancyfied, -Nancyish* adjs., effeminate; *Miss-Nancyism* (earlier example).
1848 A. Brontë *Tenant of Wildfell Hall* I. iii. 53 You will treat him like a girl—you'll spoil his spirit, and make a mere Miss Nancy of him. 1855 'Q. K. P. Doesticks' *Doesticks, what he Says* 298, I could overlook the boarding-school-ism of the Miss Nancyish 'Journal'. 1863 'G. Hamilton' *Gala-Days* 117 A man's hair is shag... Ceasing to be shag, it does not become beauty, but foppishness, effeminacy, Miss Nancyism. 1870 A. W. Drayson *Young Dragoon* viii. 61 Officers and men must be thorough soldiers—not 'Miss Nancy' sort of fellows. 1874 *Southern Mag.* XIV. 353 Poh! 'Miss-Nancyfied' men! 1916 W. Riley *Netherleigh* xv. 152 Talkin' Miss Nancy as if 'e was a dancin' master. 1928 'Brent of Bin Bin' *Up Country* ii. 8 He actually carried sleeping attire about with him, and a tooth-brush, Miss Nancy habits derided by the men.
d. *Miss Milligan*, a kind of patience played with two packs of cards.
1899 M. W. Jones *Games of Patience* 5th Ser. x. 27 Miss Milligan Patience. 1914 C. Mackenzie *Sinister St.* II. III. xiv. 782 She used to sit playing 'Miss Milligan'..and said ..that she had really enjoyed Patience for the first time. 1934 H. G. Wells *Exper. Autobiogr.* I. i. 29, I have played a spread-out patience called Miss Milligan for the past fifteen years. 1938 C. Morgan *Flashing Stream* II. ii. 205 Karen, make it four and bridge. Oh, you don't play. Five and poker. (*No answer.*) Hell, I'll play Miss Milligan. 1975 J. Symons *Three Pipe Problem* xviii. 200 She played all sorts of patience games from simple single-pack patiences like Miss Milligan and the elegant Windmill to complicated double-pack games like French Blockade and Triple Line.
e. A young woman, *Miss America, Miss England, Miss Europe, Miss World*, etc., chosen for beauty, personality, etc., to represent a country, region, etc.; also *transf.*
[1905 R. H. Davis (*title*) Miss Civilisation.] 1922 *N.Y. Times* 5 Sept. 19/6 Miss Margaret Gorman of Washington, winner of the 1921 contest, will be known as 'Miss America'. 1927 *Maclean's Mag.* 1 June 40 'Miss Toronto' wearing a stylish Aberley of the attractive 'Bird of Gladness' design in which she won the cup at the 1926 Beauty Contest at Sunnyside. 1929 *Daily Tel.* 8 Feb. 11/4 'Miss Europe' was chosen to-night..from among the seventeen girls who had been selected as the most beautiful women of their respective countries. 1935 M. Campbell *My 30 Yrs. Speed* ix. 202 Sir Henry Segrave was killed on Lake Windermere during attempts on the water-speed record with *Miss England II*. 1953 S. Spewack *Under Sycamore Tree* II. i. 35 Attention, everybody. We now bring you the results of the beauty contest..to pick Miss Human Ant of nineteen fifty-three. 1958 *Listener* 23 Oct. 662/1 Sport, travelogues, space-rockets, Miss World.. succeed each other rapidly and effortlessly. 1962 E. Cleaver in A. Dundes *Mother Wit* (1973) 14/2 A..blue eyed 'white' girl is..proclaimed as..Miss Universe. 1968 *Radio Times* 28 Nov. 70/3 The first Miss World in 1951 measured 37-23-36. 1972 G. Bromley *In Absence of Body* iii. 27 Poised at a desk on a low dais—as though she might have been Miss Great Britain..was a ravishing receptionist. 1974 *Times* 8 Mar. 3/4 (*heading*) Miss World stripped of title.
f. *Miss Ann(e, Annie* (see quots.).
1926 C. Van Vechten *Nigger Heaven* 286 *Miss Annie*, a white girl. 1942 Z. N. Hurston in A. Dundes *Mother Wit* (1973) 224/2 Miss Anne used to worry me so bad to go with me. 1965 [see *Charley, Charlie 7]. 1966 *Publ. Amer. Dial. Soc.* 1964 XLII. 45 *The Man* and *Miss Ann* refer more specifically to the boss and the fair, young white lady of the plantation... Both..are used ironically. 1970 C. Major *Dict. Afro-Amer. Slang* 81 *Miss Ann*, a white woman—carry-over from Southern terminology, but now used with a good-natured sneer or with outright maliciousness.
g. *Miss Willmott's ghost*, a large sea holly, *Eryngium giganteum*, so called in allusion to Ellen Ann Willmott (1860–1934), English horticulturist, who was responsible for the introduction of many plants.
1956 A. M. Coats *Flowers & their Histories* 89 The biennial sea-holly..has a spectral look in the twilight which might well justify its name of Miss Willmott's Ghost. It is said that when visiting gardens, Ellen Willmott used surreptitiously to drop a few seeds of this plant here and there, to surprise the owners in due course. 1963 *Oxf. Bk. Garden Flowers* 154/1 The biennial species, *E*[*ryngium*] *giganteum*, becomes quite white and desiccated after it has flowered, and is often called 'Miss Willmott's Ghost', after that great gardener Miss Ellen Willmott. 1974 R. L. Fox *Variations on Garden* 127 It [*sc. Eryngium giganteum*] is also called Miss Willmott's Ghost, Miss Willmott being a former plantswoman of the home counties with a tongue, and tastes, as sharp as a thistle's spine.
h. Occas. uses, as *Miss Lonelyhearts* (see *lonely *a.* 6); *Miss Right*, a woman who would be a perfect wife; *Miss White*, a lavatory.
1922 Joyce *Ulysses* 347 When she wanted to go where you know she said she wanted to run and pay a visit to the Miss White. *Ibid.* 640 He would one day take unto himself a wife when Miss Right came on the scene.
3. d. A pert girl.
1818 Keats *Lett.* (1958) II. 13 She is a downright Miss without one set off—we hated her. 1864 C. M. Yonge *Trial* I. vi. 100, I came down upon little Miss at last for her treatment of the doctor. 1937 M. Allingham *Dancers in Mourning* xi. 158 A sulky little miss if ever I saw one.
e. *Miss sahib*, in India, the daughter of a mem-sahib, a European girl.
1888 Kipling *Soldiers Three* (1889) 8 *Bund karo* all the Miss Sahib's *asbab* an' lookslippy! 1892 —— & Balestier *Naulakha* xx. 236 'Has the miss sahib any orders,' asked Dhunpat Rai. 1971 R. Dentry *Encounter at Kharmel* iii. 46 Oh, memsahib... Is the miss-sahib unwell? 1973 'B. Mather' *Snowline* ix. 105, I saw the sahib... He..passed close to a group of goras and dirty miss-sahibs, who called out to him.
f. A female schoolteacher; an English governess in France.
1924 A. D. Sedgwick *Little French Girl* I. vi. 51 The 'Misses' of her childhood. 1951 R. Senhouse tr. *Colette's Chéri* 21 No 'Fräulein', no 'Miss' was ever to be seen at Chéri's side. 1968 L. Berg *Risinghill* 16 Girls are caned as well as boys in Islington. 'Miss said no one should come in the class during the dinnertimes.' *Ibid.* 227 That's not a bad thing for a child to copy—to think 'Sir's mod!' or 'Miss is mod!' 1973 *Guardian* 20 Mar. 17/3, I would like to subject some of the 'misses' and some of the 'sirs' to the indignities and fears that they have heaped upon my kids.
4. (Examples of the use designating sizes and styles of clothing.)
1880 in *Amer. Mail Order Fashions* (1961) 20 A Misses' bathing costume. The pattern..is in 6 sizes for misses from 10 to 15 years of age. 1892–3 *T. Eaton & Co. Catal.* Fall & Winter 11/1 In misses' and small women's coats, we are still unexcelled. 1930 E. Wallace *Lady of Ascot* viii. 67 She catered for what they call in America the 'Miss', and had as her principal clients thousands of working girls, who, through the Carawood stores, were able to dress fashionably. 1951 *Vogue* Feb. 94/1 We pass through the Baby Linen on our way to the Misses. 1954 M. Corey *McCall's Compl. Bk. Dressmaking* 50 In junior sizes, as in misses' and women's sizes, the size that you take in a dress pattern is also the right size for your coat or your suit. 1970 *Vogue Sewing Bk.* II. 108 The Misses' figure is considered the statistically 'average' figure. 1973 *Philadelphia Inquirer* 7 Oct. 9 (Advt.), Misses' nationally famous Separates. Coordinated sets. *Ibid.*, Every winter coat for misses, juniors, women reduced Monday only.
5. (Further examples.) Also used conventionally of a married woman in public life.
1790 N. Webster in *Gazette U.S.* 17 Nov. (Th.), The use of Miss for Mistress in this country is a gross impropriety. 1819 *Mass. Spy* 12 May (Th.), I concluded he had resolved to marry Miss Spruce, but found upon inquiry that his name was Spruce, and Miss Spruce was his wife. 1838 Dickens *Nickleby* (1839) xxv. 246 The company ..fell to, immediately: Miss Petowker blushing very much when anybody was looking, and eating very much when anybody was *not* looking. 1854 —— *Hard T.* in *Househ. Words* 12 Aug. 598/2 Miss Josephine Sleary..was then announced... 'Here 'th Jothphine hath been and got married to E. W. B. Childerth, and shee hath got a boy.. three yearth old.' 1873 M. Holley *My Opinions* 166 Miss Aster would give up her bedroom to me, or mebby she would make Mr. Aster sleep with one of the boys, and have me sleep with her. 1878 R. T. Cooke *Happy Dodd* x. 99 Mis' Potter sent that. 1888 L. D. Powles *Land of Pink Pearl* 154 No married woman, not even excepting the Governor's wife, is ever accorded the title of 'Mrs.' but all ladies, married or single, are called 'Miss' or 'Missey' indiscriminately. 1936 Mencken *Amer. Lang.* (ed. 4) 124 The vulgar American misuse of..*Mis'* (pro. *miz*) for *Mrs*... was so widespread by 1790 that..Webster denounced it as 'a gross impropriety'... It survives unscathed in the speech of the common people. 1937 N. Marsh *Vintage Murder* i. 6 What about Miss Dacres? Or should I say Mrs. Meyer? I never know with married stars. 1974 *Daily Tel.* 3 Oct. 10/6 Miss Blyton seems to have indulged in a few affairs before marrying her devoted surgeon second husband. 1975 *Times* 3 Apr. 14/4 Miss [Eileen] Fowler attributes her success in her unusual field partly to the fact that she started her working life as an actress... 'My husband was a bit overweight when we married,' she said. *Ibid.* 4 Apr. 1/1 Miss Ure..was found collapsed by her husband, Mr. Robert Shaw, the actor.

miss (mis), *sb.*[4] Colloq. abbrev. of Miscarriage 3.
1897 W. S. Maugham *Liza of Lambeth* x. 167, I've 'ad twelve, ter sy nothin' of two stills an' one miss. 1951 J. Cannan *And All I Learned* v. 70, I heard of a girl who'd had eleven misses. 1959 'J. Ross' *Boy in Grey Overcoat* x. 125, I didn't care what happened either to me or the child. I hoped I would have a miss. 1971 'D. Shannon' *Murder with Love* (1972) viii. 138 She had a miss, that time, lost the baby.

miss, *v.*[1] Add: **5. c.** [ellipt. use of 5 a.] *intr.* Of a motor vehicle or an engine: to fail to explode the mixture in a cylinder. (In quot. 1904 *transf.*) Phr. *to miss on all* (or *four*, etc.) *cylinders*: see *cylinder *sb.* 6.
1904 *Peel City Guardian* 14 May 3/2 Hargreaves was 'missing' very badly. 1917, 1932 [see *cylinder *sb.* 6]. 1953 A. Smith *Blind White Fish in Persia* x. 199 The departure from the Consulate was unceremonious, for the truck was missing badly, stalled several times and finally pulled us out through the gate. 1973 D. MacKenzie *Postscript to Dead Let.* 8 The motor started missing a few miles back..then it died completely.
d. *he* (or *she*, etc.) *never misses* (or *has not missed*, etc.) *a trick*: he, etc., never fails to seize an opportunity, advantage, etc. *colloq.*
1962 *Oxford Times* 28 Dec. 15/2 Peter Butterworth and Joe Black are pantomime professionals who never miss a trick. 1965 *Harper's Bazaar* Feb. 66/1 Fenwicks..never misses a trick when it comes to picking up a new accessory idea. 1967 'E. Lathen' *Murder against Grain* iv. 32 You have to hand it to them. Those boys haven't missed a trick. 1967 O. Norton *Now lying Dead* vi. 109 He never missed a trick. 1973 S. B. Jackman *Guns covered with Flowers* viii. 131 Clever chap... Doesn't miss a trick.
9. c. In various colloq. phrases, as *to miss the boat*; *to miss the bus*: see *bus *sb.*[2] 1 b.
1929 F. C. Bowen *Sea Slang* 91 *Miss the boat*, to be late for anything. 1930 *Times Educ. Suppl.* 8 Mar. 106/3 Boys of average ability..often had to neglect other pursuits.. for fear that they might 'miss the boat'. 1930 *Aberdeen Press & Jrnl.* 3 Sept. 4/5 As a medium for a dull debut, 'A Devil's Disciple' by Bernard Shaw.., to use an Americanism, missed the boat by twenty years. 1931 *Time & Tide* 29 Aug. 1001 There are ten men in the Cabinet... There are three more who, by strange irony of circumstance, have missed the train. 1934 T. E. Lawrence *Let.* 18 Aug. (1938) 875, I fear I have missed the boat, for lately a viking ship came from you: so I place you in Norway. 1939 H. Nicolson *Let.* 18 July (1966) 406 But Anthony..is in fact missing every boat with exquisite elegance. 1973 *Times* 24 Mar. 2/4 Some firms were missing the boat because their managements were not prepared to be adventurous.

22. b. To fail to menstruate at the normal time, to miss a period.
1947 C. Willingham *End as Man* 9 A beautiful but wicked girl of a good Port George family missed one month. Then she missed another month. She went to a doctor and found out the truth. 1961 G. Greene *Burnt-Out Case* vi. i. 184, I think I have a baby on the way.. I've missed twice. 1971 'P. Hobson' *Three Graces* i. 8, I think I'm pregnant. This is the second time I've missed.
25. *miss out* (*on*) ——. To fail (esp. to achieve something); to make a mistake (over something); to omit; to be omitted.
1929 D. Scarborough *Can't Get Red Bird* xxvii. 405, I feel sorry for a poor sucker that misses out on any one of 'em. 1934 *Hound & Horn* VII. 393 They have a way of missing out on emotional experience, either through timidity and caution or through heroic renunciation. 1938 R. Franken *Gold Pennies* xix. 224 Morton lumbered behind her, fearful that he might be missing out on something. 1942 Berrey & Van den Bark *Amer. Thes. Slang* § 262/2 *Fail*,.. miss out (on). 1944 D. Runyon *Runyon à la Carte* 100 He will lay them according to how he figures their word..if Brandy Bottle misses out. 1952 G. W. Brace *Spire* (1953) xii. 105 It was Flanders who was planning the reception... But he missed out on Wilfred Stearns. 1959 *Listener* 15 Jan. 115/2 We had to by-pass Tippaburra and miss out on that Christmas Eve spree. 1960 S. H. Courtier *Gently dust Corpse* iii. 32 They..had missed out when prosperity hit the Mallee. 1960 I. Cross *Backward Sex* 116 You and I are going to miss out, y' know. 1961 J. Wade *Back to Life* ix. 120 Sorry I missed out on that report. 1963 A. Lubbock *Austral. Roundabout* 48 We didn't want the kids to miss out... They don't often get the chance to have a bit of fun. 1965 M. Morse *Unattached* i. 38 It's a terrible feeling missing out. 1969 *New Yorker* 12 Apr. 56/2 The motivation derives from the desire not to miss out on any information that could be essential later. 1972 *National Observer* (U.S.) 27 May 21/5 (Advt.), Don't 'miss out' on any of the fresh, new kind of reporting that makes The Observer the national newspaper for the *business of living*.

missable (mi·săb'l), *a.* [f. Miss *v.*[1] + -able.] That can be or is likely to be missed.
1924 *Glasgow Herald* 15 June 11/6 Of course it was not a record, but he holed out everything missable. 1955 [see *holeable *a.*]. 1959 *Times* 11 Sept. 5/1 [He] holed all the missable putts.

missalist. (Later example.)
1909 *Daily Chron.* 9 Oct. 4/4 The three brothers Maris might be re-incarnations of the Van Eyck brothers, or the de Limburg missalists.

missed, *ppl. a.* Add: **c.** Med. *missed abortion*: the retention of a fœtus in the womb for a period after it has died; also, the fœtus itself; *missed labour*: the retention of a fœtus in the womb beyond the normal period of pregnancy.
[1847 H. Oldham in *Guy's Hosp. Rep.* V. 109 A female carries a child in the womb to the full period of gestation; but the process of labour is literally missed, and lactation follows on completion of gestation. *Ibid.*, Cases resembling this, in its principal feature of labour being missed, have been recorded.] 1864 *Med. Times & Gaz.* 22 Oct. 449/2 Dr. Greenhalgh did not consider that Dr. Williams' cases could be placed under the head of missed labour... Dr. Oldham said that he had used the term as the most appropriate he could find. It was a case in which the time of natural labour passed by without any pains, and the child was not expelled. 1878 J. M. Duncan in *Ibid.* 28 Dec. 730/1 In a case of missed abortion..the important element of suspicion as to the real conditions may not have come into the mind either of the patient or her physician. *Ibid.*, This is a case in which you have..slight protraction of pregnancy, and then the condition of missed labour. 1936 W. Shaw *Text-bk. Gynæcol.* xii. 262 In missed abortion the signs of pregnancy disappear. 1971 E. S. Taylor *Beck's Obstetr. Pract.* (ed. 9) xxix. 436/1 Missed labor, unlike missed abortion, is extremely rare. Pregnancy continues to term in the normal manner, but near the expected date of confinement labor starts and then ceases after a time. 1972 W. Barr *Clin. Gynaecol.* xi. 148 There is a risk that, if a missed abortion is left in situ for over four weeks, defibrination of the blood can occur.
d. *missed approach* [*approach *sb.* 13], in *Aeronaut.*, an approach that is discontinued for any reason; esp. (with hyphen) *attrib.*
1951 *Gloss. Aeronaut. Terms* (*B.S.I.*) III. 20 *Missed-approach altitude*, the minimum height at which a final approach should be discontinued if it cannot be completed. *Ibid.* 21 *Missed-approach procedure*, procedure to be followed when an aircraft cannot complete final approach. 1971 *Flying* Apr. 42/3 Sometimes, the missed-approach procedure directs you to an NDB. 1973 *Black Panther* 13 Oct. 14/2 The Aero Commander pilot requested of Midway Tower a 'missed approach' (that he be allowed to go round again and make a second landing attempt).

mi·ssense, *a. Biol.* [Mis-[1] 4.] Causing or involving the insertion of a different amino-acid at a particular point in a polypeptide or protein molecule from that which is usual.
1961 Levinthal & Davison in *Ann. Rev. Biochem.* XXX. 651 An alteration in any base pair by a mutagen raises the possibility of a class of so-called nonsense mutations, as well as a class of missense mutations... A mis sense mutation in this context is one which causes a substitution of one amino acid for another at a particular position in the protein. 1974 *Nature* 2 Aug. 412/1 The missense mutation *trpA36* results in a Gly (GGA)→Arg (AGA) amino acid substitution at position 211 of the tryptophan synthetase A protein. *Ibid.* 413/1 The anticodons of the suppressor tRNAs are complementary to the missense codons suppressed by these tRNAs.

missfire. Add: (Further example.) Also *attrib.*
1914 W. Owen *Let.* 28 Aug. (1967) 281 After your deplorable miss-fire fashion. 1933 *Times Lit. Suppl.* 2 Mar. 148/4 Narrating the missfire of a tired New York business man who hoped his wife and daughter would fly with him.

missie, var. Missy (in Dict. and Suppl.).

missile, *a.* and *sb.* Add: **B.** *sb.* **1. b.** *Mil.* A destructive projectile that during part or all of its course is self-propelling and directed by remote control or automatically.
1945 (27 Aug.), etc. [see *guided *a.* b]. 1945 *Sci. Amer.* Nov. 283/3 Communications systems between air and ground making possible the most intricate maneuvers.. by..pilotless missiles. 1946 *Aeroplane Spotter* 1 June 128/1 A rocket-propelled missile carries its oxygen internally, whereas a jet-propelled missile takes in its oxygen from the atmosphere. 1954, etc. [see *ballistic *a.* d]. 1956 *Newsweek* 7 May 19/1 The 'Nike B' is designed as an antimissile missile as well as an improved version of the anti-aircraft 'Nike' weapons. 1959 *Times Lit. Suppl.* 13 Feb. 79/2 Some of the smaller European members of N.A.T.O. have given notice that they will not allow American intermediate-range missiles to be stationed in their territory in time of peace. 1964 *Ann. Reg.* 1963 221 Another factor may have been a realization that the arms race was becoming ruinously expensive, and..a partial test ban would prevent the United States from outpacing the U.S.S.R. in research on an anti-missile missile. 1973 *Sci. Amer.* Nov. 27/2 Under the SALT I agreement the U.S.S.R. is allowed about 25 percent more offensive missiles than the U.S.
c. *attrib.* and *Comb.*, as *missile base, carrier, gap* [*gap *sb.*[1] 6 a], *silo, site, submarine*; *missile-armed, -firing, -launching* adjs.
1959 *Daily Tel.* 20 May 10/2 The nuclear-powered, missile-armed submarine offers itself as the costly major strategic weapon of the future. 1969 *New Scientist* 28 Aug. 421/2 The Royal Navy has only three fast patrol boats in service and none of these is missile-armed. 1956 W. A. Heflin *U.S. Air Force Dict.* 329/1 Missile base. 1958 *Listener* 3 July 7/2 The Soviet Union's aim is to attract Iceland out of the Nato alliance because she is fearful of the American missile bases there. 1957 *Economist* 21 Dec. 1025/2 The wheels of the missile-carriers rumble towards Europe from east and west. 1958 *Listener* 20 Feb. 303 The missile-firing nuclear-powered submarine will..prove the most potent form of nuclear attack in the future. 1959 *Economist* 13 June 1020/2 The Air Force gets an additional $170 million to help close the 'missile gap'. 1962 *Listener* 19 Apr. 697/1 The passages on the 'missile gap' are a little dated, since Mr Kennedy has now told us that it scarcely ever existed. 1958 *New Statesman* 2 May 600/1 A missile-launching site. 1965 H. Kahn *On Escalation* vii. 141 The Soviets could increase pressure on us by simultaneously stationing missile-launching submarines or ships off our coasts, by sabotaging communications, [etc.]. 1967 *Economist* 28 Oct. 371/1 Their programme includes planes and medium-range missile silos as well as three Polaris-type submarines. 1974 *Times* 24 Oct. 6/8 Under the 1972 agreements the Soviet Union was allowed to have 1,618 operational land-based missile silos. 1949 *Aviation Week* 21 Feb. 11/3 (*heading*) Defense chiefs ask guided missile site. 1962 *Listener* 29 Mar. 553/1 On-site inspections of airfields and missile-sites. 1959 *New Statesman* 3 Jan. 6/3 A similar sum, according to one estimate, will be needed for the 40 Polaris missile-submarines which the Navy says it requires. 1974 L. Deighton *Spy Story* xii. 119 They'll have one missile submarine close enough to fire.

mi·ssileman. Also missile man, missile-man. [f. Missile *a.* and *sb.* + Man *sb.*[1]] One who is engaged in the construction, design, flying, or operation of a missile (sense *B. 1 b).
1951 [see *bird *sb.* 4 d]. 1961 *Amer. Speech* XXXVI. 234 The verb *destruct* is used by missile men to describe the push-button blowing up of a missile. 1962 *Daily Tel.* 7 Feb. 1/6 (*heading*) 300 missile men to lose jobs. 1963 *Ann. Reg.* 1962 35 The rapid development of the missile-men's craft. 1964 *Economist* 29 Feb. 803/3 The plight of the redundant missile-men.

missilery (mi·sail[ə]ri). *N. Amer.* Also **missilry.** [f. Missile *sb.* + -ery.] Missiles collectively; a collection of missiles.
1880 *Harper's Mag.* Sept. 506/1 There were in her mainmast eighteen large grape, and sixteen musket-balls, besides smaller missilery in profusion. 1957 *Britannica Bk. of Year* 512/1 In military affairs, the prominence of the guided missile as a weapon was indicated by such terms as *missilry*, a collective term for such missiles. 1959 *Life* (Internat. ed.) 13 Apr. 20/1 We do have a mass of evidence to indicate that the Soviets have gone all-out in missilery. 1973 *Daily Colonist* (Victoria, B.C.) 27 Apr. 4/7 The [transistor] devices became important foundation stones of military aviation, missilery, and space exploration, where it is vital to save space and weight.

missing, *ppl. a.* Add: **1.** (Further examples.) spec. *the missing*, soldiers (sailors, etc.) neither present after an action nor known to have been killed or wounded; so *missing, presumed dead* (in quot., *fig.*). In wider use: (*to be*) *among the missing*: to be absent, to absent oneself (*U.S. colloq.*).
1855 T. C. Haliburton *Nat. & Hum. Nat.* I. i. 10 If a person inquires if you are at home, the servant is directed to say, No, if you don't want to be seen, and choose to be among the missing. 1859 in Bartlett *Dict. Amer.* (ed. 2) 273 There comes old David for my militia fine. I don't want to see him, and think I will be among the missing. 1917 'Contact' *Airman's Outings* p. xii, Once eleven of our machines were posted as 'missing' in the space of two days. 1918 W. Owen *Let.* 8 Oct. (1967) 581 Must now write to hosts of parents of Missing, etc. 1962 *Listener* 11 Oct. 585/1 His [*sc.* Schönberg's] music seemed dead-alive on more than one occasion... Ernest Newman..reported him missing, presumed dead, just because he did not seem to have made it in time.
4. **missing link,** (*b*) (earlier and later examples); **missing person,** a person whose whereabouts are unknown (and who is being sought); also *attrib.* of an organization, etc., recording information about such persons.
1862 G. du Maurier *Let.* Oct. in *Young G. du Maurier* (1951) 178, I..said that if he would take the trouble to make a post mortem on the Irish roughs I intend to kill next Sunday in the Park, he might convince himself that the 'missing link' had been found. [1876 tr. *E.H.P.A. Haeckel's Hist. Creation* II. xxii. 293 Although the preceding ancestral stage is already so nearly akin to genuine Men that we scarcely require to assume an intermediate connecting stage, still we can look upon the speechless Primæval Men (Alali) as this intermediate link.] *a* 1930 D. H. Lawrence *Phoenix II* (1968) 569 One woman.. wrote to me out of the blue: 'You, who are a mixture of the missing-link and the chimpanzee, etc.'—and told me my name stank. 1936 A. O. Lovejoy *Great Chain of Being* viii. 235 By 1760 the triumphs of the missing-link hunters were being celebrated in verse. 1966 R. & D. Morris *Men & Apes* v. 126 Albertus [Magnus] made the first attempt to bridge the gap between man and the rest of the animal world by means of a kind of 'missing link' in the shape of the pygmy and the ape. 1876 Geo. Eliot *Dan. Der.* III. III. xx. 39 There were safer means than advertising: men might be set to work whose business it was to find missing persons. 1943 R. Chandler *Lady in Lake* (1944) ii. 15 It will mean going to the Missing Persons Bureau. 1967 R. Rendell *Wolf to Slaughter* ii. 20 They didn't want to add to their Missing Persons list if they could help it. 1970 *Guardian* 7 Jan. 18/2 A 'missing person' poster for station notice boards will be issued soon. 1975 'E. Lathen' *By Hook or by Crook* xv. 147 The.. kids had been missing persons all through the war.

missiology (misiǫ·lŏdʒi). [f. Missi(on *sb.* + -ology.] The study of the methods, purpose, etc., of religious missions.
1937 *Tablet* 23 Oct. 545/1 (*heading*) The science of missiology. First principles of mission work. *Ibid.*, Missiology is a new word for a new science. 1961 A. V. Seumois in G. H. Anderson *Theol. Christian Mission* II. 133 Practical missiology, scientifically theological, covers the study of mission spirituality and mission methodology. 1971 *Sunday Times* (Johannesburg) 28 May 30/5 (Advt.), Senior Lecturer in Ecclesiastical History and Missiology. 1975 *Church Times* 11 July 2/5 Miss Myrtle Langley is to be head of the new department of missiology..at Trinity College, Bristol.
So **missiolo·gical** *a.*, of or pertaining to missiology; **missio·logist,** one who concerns himself with missiology.
1951 *Theology* LIV. 349 The aim of missions..is.., as a distinguished Roman Missiologist has said, to 'found' the Church'. 1957 *Scottish Jrnl. Theol.* X. 300 One meets here the biblical theologian, the scientist in religions, the missionary and 'missiologist'. 1961 A. V. Seumois in G. H. Anderson *Theol. Christian Mission* II. 123 The writing which marked the launching of the modern missiological movement. 1970 J. Power *Mission Theol. Today* p. ix, There is a lacuna that so far has not been filled either by missionaries themselves or by missiologists. 1971 N. Q. King *Christian & Muslim in Afr.* 90 Leaving aside such theological and missiological considerations, a large number of African Christians ask us to compare some of the Christians of the older generation.

mission, *sb.* Add: **2. b.** orig. *U.S.* A military operation or project; esp. the dispatch of an aircraft or spacecraft on an operational flight; also *transf.*
1929 E. W. Dichman *This Aviation Business* v. 107 Night, heavy long-distance, slow, and large all describe a certain type of airplane designed to accomplish a particular mission. 1939 *Aircraft Yearbk.* iii. 64 Many missions were flown, day and night, by participating bombardment and attack units. 1944 *Amer. N. & Q.* Apr. 15/2 *Mission*, ordered operation against the enemy, such as dropping bombs, strafing ground troops and ships, dropping parachute troops, flying diversions (missions intended to draw the enemy away from the main objective), taking photographs, etc. 1962 J. Glenn in *Into Orbit* 43 The clock is pre-set on the ground according to a timing for retro-fire which we have computed before the mission. 1968 Mrs. L. B. Johnson *White House Diary* 1 July (1970) 694 Lynda heard from Chuck... He had returned from a mission and..had only time to read two or three letters, write her.., stuff a few more into his pocket, and leave on another mission! 1969 *Times* 23 May 1/2 Twice during the critical hours before the separation of the lunar module.. from the command service module..the mission's future looked doubtful. 1971 *Daily Tel.* 19 July 7 Apollo 15, America's eighth manned flight for the Moon, is due to be launched on a 12-day mission..from Cape Kennedy a week today.
10. *mission-chapel* (earlier U.S. example), *farm, -house* (earlier U.S. examples), *land, station, style*; also *mission-bred, -trained, -ward* adjs.
1909 *Times Lit. Suppl.* 7 Jan. 3/2 He makes capital fun of the mission-bred Kaffir's misuse of book-learning. 1871 *Scribner's Monthly* I. 497 His church was only a mission chapel, supported by a richer society of the same denomination. *a* 1861 T. Winthrop *John Brent* (1883) ii. 13 He had found his early way to California, bought a mission farm, and established himself as a ranchero. 1794 C. I. Latrobe tr. *Loskiel's Hist. Mission among Indians N. Amer.* II. xii. 166 The mission-house on the Mahony.. was..burnt. 1824 W. H. Keating *Narr. Exped. St. Peter's River* (1825) 150 At the time we passed at the Carey mission-house, this gentleman was absent on business. 1851 *Whig Almanac 1852* 18/2 The Commissioners are required to report to the Secretary of the Interior the tenure by which the Mission lands are held. 1828 I. McCoy in *Kansas Hist. Q.* (1936) V. 243 Here we intersected a waggon road leading from the settlements on Missouri River to Harmony Mission Station. 1844 J. McDonogh *Papers* (1898) 78 One of these young men..is now at the mission station at Settra Kroo, Liberia, keeping a school for the native youth. 1876 W. Booth in H. Begbie *Life W. Booth* (1920) I. xxv. 417 What is a Mission Station?.. It is not a building..; it is not even a society, but a band of people united together to mission,..to christianize an entire town or neighbourhood. 1884 Lady Martin *Our Maoris* i. 15 The Bishop came a day or two later, as he had to visit a mission station on the way. 1971 *Scope* (S. Afr.) 19 Mar. 31/1 The maternal grandfather.. came to take up the same mission station. 1909 Wells & Hooper *Mod. Cabinet Work* 257 In America there has been a similar movement, known as 'The Mission Style', which is more or less a revival of Gothic and Jacobean forms applied to modern work. 1930 J. Dos Passos *42nd Parallel* I. 122 I've got several magnificent mission style bungalows. 1948 A. L. Kroeber *Anthropol.* (rev. ed.) xii. 484 Since the American occupation, the buildings and ruins of the Spanish period have stood out as landmarks and have set the model for a type of architecture: the Mission style, which in essentials is nothing but Spanish Moorish architecture. 1972 M. Mead *Blackberry Winter* x. 117 The apartment Luther had found [in 1923] was small enough for the furniture we had: a strong round folding table,..Luther's mission-style desk, [etc.]. 1965 *Listener* 27 May 766/1 The men at the top in Uganda and Tanganyika were both mission-trained teachers. 1925 T. Dreiser *Amer. Trag.* (1926) I. I. i. 8 The missionward march was taken up.

b. **mission control** *collect.*, a group or organization responsible for directing a spacecraft and its crew; **mission furniture** *U.S.*, a plain, solid style of furniture said to have been modelled originally on the furniture of the Spanish missions in North America; **mission oak** *U.S.*, mission furniture made of oak; **mission stiff** *U.S. slang*, (*a*) a missionary; (*b*) one who frequents missions, esp. a tramp who is religious or who pretends to be religious so as to get free food and lodging.
1964 J. L. Nayler *Dict. Astronautics* 165 *Mission Control Center.* The Center..is due to be operational in 1964 for *Gemini* rendezvous flights. 1969 *Times* 16 July 5/8 Key abbreviations used by mission control and the astronauts will be [etc.]. 1973 *Guardian* 21 May 2/7 After the first switch-over, mission control commanded the computer to move back to the primary coolant circuit. 1900 *Harper's Bazaar* 28 Apr. 388/1 She stumbled upon an artistic small shop filled to overflowing with what the salesman called Mission furniture. 1910 *Daily Chron.* 24 Jan. 3/5, I have often wondered why the modest designs of the mission furniture are so attractive. 1967 *Boston Sunday Herald* 9 Apr. (Show Guide) 15/1 The turn-of-the-century Mission furniture,..coming into popularity with the recent art nouveau revival. 1927 U. Sinclair *Oil!* 223 Inside was furniture of a style called 'mission oak'. 1973 *Washington Post* 13 Jan. F.1/8 (Advt.), Big mission oak library table. 1904 'No. 1500' *Life in Sing Sing* 256/2 *Mission stiff*, missionary; a convert. 1931 'D. Stiff' *Milk & Honey Route* v. 58 You may hang on to the good life for a time, while your erstwhile companions in sin dub you a 'mission stiff'. 1948 Mencken *Amer. Lang.* Suppl. II. 676 At the bottom of the pile are the poor wretches..who.. gravitate dismally toward the big cities, to become beggars and mission-stiffs.

missional (mi·ʃənăl), *a. rare.* [f. Mission *sb.* + -al.] Relating to or connected with a religious mission; missionary.
1907 W. G. Holmes *Age of Justinian & Theodora* II. 687 Several prelates, whose missional activities brought over whole districts and even nationalities to their creed.

missionarism (mi·ʃənăriz'm). *rare.* [f. Missionary *a.* and *sb.* + -ism.] = Missionizing *vbl. sb.*
1890 H. S. Holland in S. Paget *Henry Scott Holland* (1921) II. iv. 197 If I believe anything at all, I believe, with it, all that missionarism involves.

missionary, *a.* and *sb.* Add: **A.** *adj.* **1.** (Further examples.)
1841 Geo. Eliot *Let.* 21 June (1954) I. 98 We yesterday heard him preach his last Missionary sermon. 1849 [see *au fait *advb. phr.*]. 1854 C. M. Yonge *Castle Builders* v. 69 The chimney-piece ornamented with missionary boxes and cards for shilling and penny subscriptions. 1854 in C. C. Richards *Village Life Amer.* (1912) 47 If we wanted to take shares in the missionary ship, *Morning Star*, we could buy them at 10 cents apiece. 1866 J. C. Patteson *Let.* 1 Jan. in C. M. Yonge *Life J. C. Patteson* (1874) II. x. 160, I value much these memorials of the first Missionary Bishop of the Church of England. 1873 C. M. Yonge *Pillars of House* II. xviii. 150 Mrs. Shapcote sent out invitations to a missionary tea in honour of him [*sc.* a missionary]. 1932 T. S. Eliot *Sweeney Agonistes* 23 I'll convert you! Into a stew. A nice little, white little, missionary stew. 1933 P. A. Eaddy *Hull Down* v. 108 On some missionary hooker where they'll want to dish up two prayer meetings in the one week to all hands. 1942 A. P. Jephcott *Girls growing Up* iii. 58 We redeemed ourselves..by working hard for a missionary box. 1971 J. Manton *Sister Dora* xvi. 272 Selwyn of Lichfield..had been the first missionary bishop in Melanesia.
b. *missionary position*: the position for sexual intercourse in which the woman lies underneath the man and facing him.
1969 *Daily Tel.* (Colour Suppl.) 10 Jan. 7 In six States [in the U.S.] a woman may still be awarded a divorce if her husband makes love to her in any other than the missionary position. 1971 *Vogue* Nov. 60/2 The face-to-face 'missionary position' (so called because it is virtually unknown in primitive races) is actually said to have been invented by Roman courtesans to hinder conception. 1971 'V. X. Scott' *Surrogate Wife* 54 His wife would allow only one position—male-on-top—called the Missionary position by some and the Mamma-Papa position by others.
B. *sb.* †**4.** *N.Z.* A Christian. *Obs.*
1834 E. Markham *N.Z. or Recollections of It* (MS.) 32 They are all Missionaries as they call the Christians. 1840 *N.Z. Jrnl.* 5 Dec. 292/2 The natives..call themselves missionaries, having embraced Christianity. 1841 *Ibid.* 10 Apr. 87/2, I asked whether it was true, that he had given up all fighting intrigues, and become a missionary. 1854 R. E. Malone *Three Years' Cruise Australasian Colonies* iii. 22 Mihaneri (missionaries—the universal name, in New Zealand, for Protestant Christians).
5. The sweet-brier, *Rosa eglanteria*. *N.Z.*
1881 F. Larkworthy *N.Z. Revisited* 30 The sweetbriar, which here [*sc.* at Tarawera] goes by the name of the 'Missionary', blocking the roads and vacant spaces. [1898 W. P. Reeves *Long White Cloud* i. 17 The sweetbriar.. covers whole hillsides, to the ruin of pasture. Introduced, innocently enough, by the missionaries, it goes by their name in some districts.] 1912 B. E. Baughan *Brown Bread from Colonial Oven* iii. 48 'Missionary', in the North Island is frequently an alternative spelling for 'sweet-brier', which is a pest. 1921 H. Guthrie-Smith *Tutira* xxvii. 274 Sweet-briar..'Missionary' as it is still called, has been spread abroad by the horse.

mi·ssionary, *v.* [f. the sb.] **a.** *intr.* To act as a missionary. **b.** *trans.* To act as a missionary towards (someone).
1862 *Independent* (N.Y.) 24 Apr. 3/1 He [*sc.* the Rev. S. H. Tyng] was always fond of missionarying. *c* 1876 J. Albery *Man in Possession* I, in *Dramatic Works* (1939) II. 105, I know I *was* [dreadfully wicked], but I'm not now, Teddy. I've been missionaried, and preached to. 1884 'Mark Twain' *Huck. Finn* xix. 183 Preachin's my line, too, and workin' camp-meetin's, and missionaryin' around. 1893 K. D. Wiggin *Polly Oliver's Problem* (1894) vii. 87 Boys hate to be missionaried, and I'm sure I don't blame them.

missionate, *v.* (Examples.)
1828 *Richmond* (Virginia) *Enquirer* 19 Aug. 4/1 (Th.), [Mr. Weed] was next heard of in the southern tier of counties, missionating for the administration. 1896 *Home Missionary* Oct. 303 To make professional visits, or to 'missionate' to the farmer, will not serve the purpose. 1966 *Publ. Mod. Lang. Assoc.* LXXXI. II. 7/2 Their fellow linguists who are interested in..missionating to convert the National Council of Teachers of English. *Ibid.* 8/1 Our missionaries should at least know what they are talking about before they set out to missionate.

missionee (miʃənī·). *rare.* [f. Mission *v.* + -ee[1].] One who is susceptible to the arguments of an emissary or a missionary.
1951 'J. Tey' *Daughter of Time* v. 70 George could obviously be talked into anything. He was the born missionee.

mi·ssioning, *vbl. sb.* [f. Mission *v.* + -ing[1].] The conducting of a religious mission.
1886 in H. Baumann *Londinismen*. 1961 B. R. Wilson *Sects & Society* III. xiv. 293 Costly missioning in pagan lands.

missionist (mi·ʃənist). [-ist.] One who does mission work.
1909 M. B. Saunders *Litany Lane* I. iii. 34 These were wood-carvers, church artists, metal-workers, window designers, architects, carpenters and missionists.

missis, missus. Add: **1.** (Earlier and later examples.)
1833 Dickens *Let.* c 10 Dec. (1965) I. 34 Hint this delicately to your *Missus*. 1836 —— *Let.* c 20 July (1965) I. 155 My Missis furthermore desires me to say [etc.]. 1934 T. S. Eliot *Rock* ii. 65 Lor-love-a-duck, it's the missus! 1946 K. Tennant *Lost Haven* (1947) I. 23, I wouldn't let any missus of mine..go gallivanting with another chap. 1975 *Daily Mirror* 29 Apr. 25 If you fancy taking the missus for a day out, you take her virtually free.
2. (Earlier and later examples.) *spec.* used by N. American Negroes and in *India* and *S. Africa* of a white employer, and loosely of any (esp. a white) woman.
1790 J. B. Moreton *Manners & Customs West India Islands* 154 Then missess fum me wid long switch. 1835 Dickens *Sk. Boz* (1837) 2nd Ser. 87 The servant..has utterly disregarded 'Missis's' ringing. 1852 Mrs. Stowe *Uncle Tom's Cabin* II. xxxiv. 203 'Missis,' said Tom, after a while, 'I can see that, some how you're quite 'bove me in everything; but there's one thing Missis might learn, even from poor Tom.' 1901 M. Franklin *My Brilliant Career* xiii. 107 I'll tell the missus on you as sure as eggs. 1924 E. Lewis *Harp* II. vii. 98 Does the missis hear the young masters in the stable? 1940 in H. Wentworth *Amer. Dial. Dict.* (1944) 392/2 Mighty fine for young missus. 1942 P. Abrahams *Dark Testament* II. iii. 113 As soon as she saw it was a white person she ran back into the house. 'Ma! Ma! There's a missus at the door!' 1943 'B. Knight' *Covenant* (1944) II. x. 132 He smiled at the Ayah and said, 'I will speak to the *Missis* and perhaps she will be willing to go.' 1950 L. Bennett et al. *Anancy Stories & Dial. Verse* 48 Eee-Hee Missis, is me same one Sidung yah all de time. 1952 P. Abrahams *Path of Thunder* III. i. 190, I work for old missus when I was a child. 1971 *Weekend World* (Johannesburg) 9 May 2/2 Langton told her to cook. 'When I refused, he slapped my face hard, accusing me of making myself a "missus".'

Missisauga (misisǭ·gă). Also **Messasague, Messisauger, Missasago, Missasauga, Missis-(s)a(u)ga(h), Mississagua,** etc. [Ojibwa, lit. 'people of the Mississagi River' (in Ontario).] **1.** A tribe of Algonquian Indians; a member of this tribe. Also *attrib.* or as *adj.*
1703 tr. *Lahontan's New Voy. N.-Amer.* I. xxv. 230 A list of the savage nations of Canada... The Missisagues. 1749 G. Clinton *Let.* 3 June in E. B. O'Callaghan *Docs. rel. Colonial Hist. New-York* (1855) VI. 486 To meet the Missisaque Indians at Oswego. 1772 in *14th Rep. R. Comm. Hist. Manuscripts* App. X. 85 in *Parl. Papers 1895* (C. 7883) LIX. 1 The Chippawaes and Mississagaes are by far the most numerous and powerful nation with whom we have any connection in North America. 1798 B. S. Barton *New Views Origin of Tribes & Nations N. Amer.* (ed. 2) App. 4 The Messaugers, or Messasagues. The language of these Indians is, undoubtedly, very nearly allied to that of the Chippewas. 1831 A. S. Withers *Chron. Border Warfare* 299 Their force consisted of four thousand warriors, and was led on by a Missasago chief. 1838 A. Jameson *Winter Stud. & Summer Rambles Canada* I. 16 One solitary wigwam.., the dwelling of a few Missassagua Indians. *Ibid.* 296 The scene of bloody conflicts between the Hurons and the Missassaguas. 1888 *Jrnl. Amer. Folklore* I. 151 (*heading*) Notes on the history, customs, and beliefs of the Mississagua Indians. *Ibid.* 152 These are the most advanced in civilization of the Mississaguas. 1948 *Southwestern Jrnl. Anthropol.* Spring 100 These people, particularly the Missasaugas, seem to have occupied the southern end of the Park a century ago. 1960 D. Jenness *Indians of Canada* (ed. 5) iv. 40 The cultivation of maize had spread..to some adjacent Algonkian tribes,..the Missasauga on the north shore of lake Huron. *Ibid.* xvii. 282 Many Missisauga moved into the old territory of the Hurons between lakes Huron and Erie.
2. = Massasauga.
1843 W. Oliver *Eight Months Illinois* 150 The inhabitants recognize two kinds of rattlesnakes, to wit, the wood- and the prairie-rattlesnake, or missisauga, of which the latter is much the smaller and less dangerous. 1961 *Listener* 16 Nov. 826/1 A mississaga rattler I once knifed.

Mississippian (misisi·piăn), *sb.* and *a.* [f. *Mississippi* (see below) + -an.] **A.** *sb.* **1.** A native or inhabitant of Mississippi, a state on the Gulf of Mexico.
1775 J. Adair *Hist. Amer. Indians* 93 'The ugly yellow French,' (as they [*sc.* Indians] term the Mississippians). 1867 *Harper's Mag.* June 1/1 Two of us New Englanders, and a Mississippian. 1948 *Daily Ardmoreite* (Ardmore, Okla.) 4 May 1/6, I recognize our Negroes, as do all good white Mississippians, as a part of our citizenry. 1973 A. Dundes *Mother Wit* 37 Mississippian David L. Cohn.
2. *Geol.* The Mississippian period or system.
1910 *Encycl. Brit.* V. 310/2 It became the practice to distinguish a 'productive' [Upper]..and an 'unproductive', barren..Lower Carboniferous; these two groups correspond in North America to the 'Carboniferous' and 'Sub-Carboniferous' respectively, or, as they are now sometimes styled, the 'Pennsylvanian' and 'Mississippian'. 1969 Bennison & Wright *Geol. Hist. Brit. Isles* ix. 184 The Carboniferous System is traditionally divided into the Lower and Upper Carboniferous in Britain and western Europe. The two systems in North America, the Mississippian and the Pennsylvanian, correspond broadly to these divisions.
B. *adj.* **1.** Of, pertaining to, or peculiar to Mississippi.
1835 J. H. Ingraham *South-West* II. 79 Of every variety of gaited animals..the Mississippian pacer is the most desirable. 1963 *Economist* 10 Aug. 509/2 A man who, in Mississippian terms, is a relative moderate on the race issue.
2. *Geol.* [Named after the Mississippi River, on the bluffs of which in Iowa and Missouri the system is exposed.] Of, pertaining to, or designating a period and system of the Palæozoic Era in North America that succeeded the Devonian and preceded the Pennsylvanian, and corresponds more or less to the Lower Carboniferous in Europe.
[1870 A. Winchell *Sk. Creation* xii. 136 The Mountain limestone, or Lower Carboniferous mass, which I have proposed to designate the Mississippi Group, because so extensively developed in the valley of the Mississippi River.] 1891 H. S. Williams in *Bull. U.S. Geol. Survey* No. 80. 135 As these formations are bound together by a common general fauna and constitute a conspicuous feature in the geology of this region, it is proposed to call them the Mississippian series. 1933 R. C. Moore *Historical Geol.* xvii. 257 The consensus of judgment among American geologists increasingly supports the view that the Mississippian and Pennsylvanian deposits should be reckoned as independent geologic systems rather than as subordinate divisions (series) of a so-called system that combines them. *Ibid.* 260 Red shale and sandstone.. with a maximum thickness of about 3,000 feet form the upper part of the Mississippian system in much of the Appalachian region. 1967 *Oceanogr. & Marine Biol.* V. 131 Evaporite sediments of Mississippian age have caused similar uplift in nearby Nova Scotia.

Misskito, var. *Miskito *a.* and *sb.*

MISS-MARK

mi·ss-mark. rare. [f. Miss v.¹ Cf. Mark sb.¹ 7 e.] A person who misses the mark, or who fails in a purpose.
1908 Hardy Dynasts III. vii. ix. 520 So, as it is, a miss-mark they will dub me.

Missouri (mis-, mizūə·ri). U.S. [The name of a river and a state in the U.S.] **1.** A member of an American Indian people of the Sioux family, first encountered by Europeans near the Missouri River; also, the language of this people.
1703 tr. Lahontan's New Voy. N.-Amer. I. 130 We..arriv'd on the 18th at the first Village of the Missouris. 1807 P. Gass Jrnl. 26 Six of them were made chiefs, three Otos and three Missouris. 1933 L. Bloomfield Lang. iv. 72 The Siouan family includes many languages, such as..Missouri, Winnebago, [etc.]. 1947 St. Louis (Missouri) Globe-Democrat 16 Mar., The Missouris were a comparatively insignificant tribe.
2. Colloq. phr. to be (or come) from Missouri: to be very sceptical; to believe nothing until it is demonstrated. (Originally I come from Missouri. You have got to show me.)
1900 Missouri State Tribune (Jefferson City) 13 Dec. 4/1 Ex-Lieut.-Gov. Chas. P. Johnson thinks he knows the origin of the extensively-used expression: 'I'm from Missouri; you'll have to show me'; at least he can recall its use twenty years ago in Colorado. 1901 Columbia Missouri Statesman 13 Dec. 1/3 You gentlemen are from Kentucky, Texas, Tennessee and Arkansas and seem to trust each other, but 'I'm from Missouri and you must show me'. 1912 C. McCarthy Wisconsin Idea 291 In the words of the current slang phrase, every Wisconsin legislator 'comes from Missouri' and you have to 'show him'. 1931 Amer. Speech VI. 205 I'm from Missouri, I don't believe that; you'll have to show me, or prove it to me. 1963 J. Mitford Amer. Way of Death iii. iv. 132 If you suggest..that Destiny led him there, he will give you an I'm-from-Missouri look.
3. attrib. and Comb., as **Missouri antelope** = Pronghorn sb.; **Missouri Compromise** Hist., an arrangement made in 1820 which provided that Missouri should be admitted to the Union as a slave state, but that slavery should not be allowed in any new state lying north of 36° 30′; also attrib.; **Missouri Indian** = sense 1 above; **Missouri mule,** a mule bred in Missouri; **Missouri question** Hist., the question of the conditions under which Missouri should be admitted into the Union, and the connected problems regarding slavery; **Missouri skylark,** a variety of pipit, Anthus spraguei.
1806 Lewis & Clark in Deb. Congress U.S. (1852) 9th Congress 2 Sess. App. 1046 The Missouri antelope, (called Cabri' by the inhabitants of the Illinois). 1820 in T. H. Benton Exam. Dred Scott Case (1857) 102 The line is..nominated..by its popular descriptive appellation of 'the Missouri Compromise Line'. 1847 J. K. Polk Diary 16 Jan. (1910) II. 335 The line of the Missouri Compromise, viz., 36° 30′. 1943 E. B. White One Man's Meal 16 The Missouri Compromise had temporarily settled the slavery question. 1949 D. S. Freeman in B. A. Botkin Treas. S. Folklore p. x, Wirt..did not issue his life of Henry until almost the time of the Missouri Compromise. 1765 R. Rogers Conc. Acct. N. Amer. 194 The inhabitants on this river are called the Missouri Indians. 1817 J. Bradbury Trav. Interior Amer. 41 It is customary amongst the Missouri Indians to register every exploit in war, by making a notch for each on the handle of their tomahawks. 1923 Nation (N.Y.) 17 Oct. 432 Then there is the Missouri mule. He it was who won the war. 1972 Listener 21 Dec. 858/2 Not for nothing did the idiom 'as stubborn as a Missouri mule' come into the language. 1819 J. Adams Let. 21 Dec. in T. Jefferson Writings (1903) XV. 236 The Missouri question, I hope, will follow the other waves under the ship, and do no harm. 1884 J. G. Blaine 20 Yrs. Congress I. 15 The 'Missouri question'..formally appeared in Congress in the month of December, 1818. 1858 S. F. Baird in Rep. Explor. Route to Pacific (U.S. War Dept.) IX. 234 Neocorys Spraguei, Sclater. Missouri Skylark. 1940 E. T. Seton Trail of Artist-Naturalist 299 The strictly prairie birds were gone—of the Missouri skylark, for instance, I saw not one.

Missourian (mis-, mizūə·riăn), sb. and a. [f. prec. + -an.] **A.** sb. A native or inhabitant of the State of Missouri. **B.** adj. Of or belonging to, native or peculiar to, Missouri.
1820 Deb. Congress U.S. (1855) 26 Jan. 945, I cannot believe that I, or any other man or men, are better capable of governing Missourians than they are of governing themselves. 1862 Harper's Mag. Sept. 450/2 Teamsters, many of whom were young Missourians embarked for the first time upon a prairie trip. 1885 'Mark Twain' in Century Mag. Dec. 201 The Masons gave us a Missouri country breakfast, in Missourian abundance. 1944 B. A. Botkin Treas. Amer. Folklore ii. 318 One of the harshest of these [sc. slang names] is Puke, for a Missourian. 1948 Mencken Amer. Lang. Suppl. II. vii. 173 D. S. Crumb..unearthed a great deal more that was specially Missourian, e.g...buckshot land, poor clay soil. 1957 Encycl. Brit. XIII. 258/2 About 1,700 armed Missourians invaded Kansas and stuffed the ballot boxes. 1973 R. Rosenblum Mushroom Cave (1974) 83 The junior diplomat, a glad-handing Missourian.

missourite (mis-, mizūə·rait). Petrogr. [f. *Missour(i + -ite¹ (see quot. 1896).] A grey, granular, igneous rock composed mainly of pyroxene, leucite, and sometimes olivine.
1896 Weed & Pirsson in Amer. Jrnl. Sci. II. 323 This rock is a new type, and it fills a place which has hitherto been vacant in all systems of rock classification... We have therefore called it missourite from the Missouri River, the most prominent and best known geographical object in the region where it occurs. 1927 Ibid. XIV. 179 These granitoid augite-leucite rocks will be called missourite, although Pirsson's original Montana missourite contained less leucite and about 15 per cent of olivine. Ibid. 180 The presence of olivine in missourite need not be regarded as essential. 1963 Mineral. Abstr. XVI. 388/1 The first missourites recorded from Russia are in an outcrop of alkaline gabbroic rocks near the Lomachan River.

mi·ss-out. [f. vbl. phr. to miss out: see *Miss v.¹ 25.] **a.** pl. In Gambling, loaded dice. **b.** In Craps, a losing throw: see *Craps b; also, the action of losing the right to throw.
1928 J. O'Connor Broadway Racketeers xiv. 157 The game keeper has all sorts of crooked ones, those known as 'Shapes', others called 'Miss-Outs'. 1936 Detective Fiction Weekly 21 Mar. 139/1 If the dice are cut as 'passers', the percentage in favor of the shooter is 'mild'; but cut for a banking or fading advantage (called 'miss-outs' or 'missing dice') the advantage is stronger. 1942 W. Faulkner Go down, Moses 114 The spots on them miss-out dice. 1961 in Partridge Dict. Underworld (1968) 846/1 Misses... (2) Crooked dice that are gaffed to make more miss-outs than passes. 1964 A. Wykes Gambling v. 109 He loses if he throws a Two, Three, or Twelve (a 'crap' or 'miss-out').

mis-speak, v. Add: **3. b.** refl. To fail to convey the meaning one intends by one's words.
1890 in Cent. Dict. 1894 Congress. Rec. 19 Jan. 1051/1, I simply wanted to bring that matter out plainly... I believe he misspoke himself. 1973 Harper's Mag. June 38/2 'The President,' Ziegler said, 'misspoke himself.' He explained that the President had noted his error in reviewing the transcript of the press conference. 1975 G. V. Higgins City on Hill ii. 53 If I gave that impression, I misspoke myself.

mi·ss-stays. [f. as Misstay v.] Of a ship: the act or fact of failing to go about.
1878 D. Kemp Man. Yacht & Boat Sailing 245 A 'miss-stays' may be the consequence.

misstay, v. Add: (Earlier examples.) Also miss-stay.
1829 G. Griffin Collegians (ed. 2) I. xii. 241 Ahoy! ahoy! Have an oar out in the bow, or she'll miss-stay in the swell. 1849 N. Kingsley Diary (1914) 88 We misstayed in but 11 feet of water, but the bottom is very muddy and not dangerous.

mis-step, sb. (Earlier and later examples.)
a 1800 Spirit of Farmer's Museum (1801) 205 The Squire..can sit on the sessions, and fine poor girls for natural missteps. 1837 Yale Lit. Mag. III. 8 (Th.), Forgetting the round door block, he made a mis-step. 1931 F. L. Allen Only Yesterday 101 The publishers of the confession magazines..concentrated on the description of what they euphemistically called 'missteps'. 1934 D. Sargent Thomas More iii. 62 As in all diplomatic conferences each side spends a great deal of time waiting for the other side to make a misstep. 1949 Sat. Even. Post 1 Oct. 20/3 Russians..turn sick with fear if they make the slightest little misstep. 1963 J. Walsh Shroud (1964) vi. 47 His headlong disregard for bodily care and his courting of the injuries or death that lay only a mis-step away. 1974 Publishers Weekly 4 Feb. 64/2 Henry Keller..picks up a hitchhiker... For awhile, he shares her sexual favours... Becky is killed accidentally. The police don't care much; neither does Henry's wife, who forgives his misstep. 1974 Sci. Amer. Oct. 87/2 A misstep would not necessarily lead to a stumble or a fall.

mis-step, v. Delete † Obs. and add later U.S. example.
1869 S. Bowles Our New West v. 102 Mules don't misstep, and even the top-heavy pack jacks..carried their burden and themselves unharmed to the top.

missus: see Missis in Dict. and Suppl.

missy, sb. (Later examples.)
1919 Wodehouse Damsel in Distress vii. 107 Those—them—over there are Ayrshires, missy. 1922 Joyce Ulysses 733 Little chits of missies they have now singing. 1924 R. Macaulay Orphan Island xv. 202 'You're very smart and proud, missie,' her uncle told her. 1967 P. Roth When she was Good iii. 274 Either you calm down with that bossy little voice, missy, or you get out.

mist, sb.¹ Add: **1.** (Example of a techn. definition.)
1972 Meteorol. Gloss. (Meteorol. Office) (ed. 5) 182 Mist, a state of atmospheric obscurity produced by suspended microscopic water droplets or wet hygroscopic particles. The term is used for synoptic purposes when..the associated visibility is equal to or exceeds 1 km; the corresponding relative humidity is greater than about 95 per cent.
e. A colour suggestive of a mist.
1926 Daily Colonist (Victoria, B.C.) 21 July 16/4 (Advt.), A 4-ply worsted wool in shades of pink, mist, [etc.]. 1927 Daily Express 12 Mar. 3/5 Mist, a subdued mauve, suggesting the atmospheric effects of sunset. 1937 Discovery July 217/2 Our silk stockings are..described as..sun-tan, sandalwood, mist. 1963 New Yorker 29 June 57 Black, mist, rust, or olive.
5. mist-belt, -light, -magic, -mote, -pavilion, -plash, -sheet, -thread, -veil; mist-circled, -cold, -coloured, -dimmed, -green, -pale, -shrouded (later examples), -tracked, -veiled, -wild, -wreathed, -wrought adjs.; **mist-blower,** a device for spraying insecticide into the tops of trees; so mist-blowing vbl. sb.; **mist-net,** a net made of very fine threads, used to trap birds etc. for ringing or examination and subsequent release; also as v. intr., to trap in a mist-net; hence **mist-netter,** one who uses a mist-net; **mist-pond** = *Dew-pond; **mist propagation,** a method of rooting plant cuttings in which high humidity is maintained in a greenhouse by an automatic system of watering with fine spray at regular intervals; **mist propagator,** an installation for this type of cultivation.
1906 Rep. Brit. Assoc. Adv. Sci. 1905 594 Passing either east or west of this 'mist-belt' the rainfall rapidly diminishes. 1946 Potts & Friend in Bull. Connecticut Agric. Exper. Station No. 501. 48 The development of a mist blower which will apply thoroughly a small quantity of a concentrated insecticide. 1969 Nature 9 Aug. 558/2 Only the mistblower, which is mounted on a tractor, seems feasible for large-scale applications. 1960 Farmer & Stockbreeder 12 Jan. 97/2 Mistblowing of fruit trees is a practical proposition. 1935 W. Empson Poems 27 Starlit, mistcircled, one whole pearl embrowned. 1889 W. B. Yeats Wanderings of Oisin iii. 34 Came now the sliding of tears and sweeping of mist-cold hair. 1890 Cent. Dict., Mist-colored. 1929 W. Faulkner Sartoris 224 That 'ere mist-colored stallion. 1880 'Mark Twain' Tramp Abroad 398 Along their mist-dimmed heights [i.e., of the Alps]. 1961 A. Sillitoe Key to Door iv. xxvi. 389 Green fields rolling up to..Catstone Wood, a mist-green spearblade of sky above. 1930 Blunden Poems 40 The wolfish shadows in the eerie places Sprawl in the mist-light. 1921 R. Graves Pier-Glass 12 Cold fog-drawn Lily, pale mist-magic Rose. 1923 H. Crane Let. 15 Apr. (1965) 132 The eerie speed of the shutter..catching even the transition of the mist-mote into the cloud. 1956 Brit. Birds L. 450 Mist nets, a traditional Japanese method of catching birds, were introduced to British ringers at a meeting of the Bird Observatories Committee in January, and by the early autumn of 1956 it is probable that over a hundred nets were in use. 1961 New Scientist 23 Mar. 728/1 The watchers gather with note-books, binoculars, mist nets, Heligoland traps and boxes of rings. 1971 Daily Tel. (Colour Suppl.) 18 June 7/1 Birds are..snared in a 'mist net' (a long net erected on poles, which is of so fine a mesh that the birds cannot see it, and so fly into it, where they are entangled). 1972 Science 1 Sept. 806/3 Bats are mist-netted near cattle. 1973 Country Life 1 Feb. 263/3 Helpers..ring large numbers of passerines, which they trap in mist-nets. 1960 Brit. Birds LIII. 526 The capture of rare and difficult forms..is now within the province of every mist-netter. 1849 C. Brontë Shirley III. vi. 142 He would rather have appointed tryste with a phantom abbess, or mist-pale nun. 1925 C. Day Lewis Beechen Vigil 26 Now from blue mist-pavilion You may see King Silence go Royally through the forest. 1916 Blunden Harbingers 33 So heavily drives the rain, and lashes The open pool into white mist-plashes. 1893 Dartnell & Goddard Gloss. Words Wiltshire 104 Mist-pond, a pond on the downs, not fed by any spring, but kept up by mist, dew, and rain... More commonly called dew-ponds. 1931 N. & Q. 22 Aug. 141/2 High up on the hills, in various parts of the country, are to be found ponds. Some call them dew ponds, but a more correct title is mist pond. [1941 Amer. Nurseryman 1 May 5 (title) Propagation under mist.] 1953 Ibid. 1 Aug. 63/2 Results of some of the work we have done at Koster Nursery appear to indicate the equal value of constant mist propagation in the open. 1961 Amat. Gardening Suppl. 28 Oct. 34/2 A technique known as mist propagation... The main feature of this technique is the automatic provision of a fine mist spray. 1969 New Scientist 10 Apr. 70 (caption) After five weeks' mist propagation, two-leaf cuttings of Iceberg [sc. a rose] are well-rooted young plants. 1972 Country Life 1 June 1419/3 Mist Propagation equipment. Maximises health and growth of plants. 1971 'J. Fraser' Death in Pheasant's Eye xxxi. 187 He'd have sufficient cash to buy a proper heating and ventilating system... Aye, and perhaps a mist propagator! 1917 D. H. Lawrence Look! We have come Through! 47 A thick mist-sheet lies over the broken wheat. 1919 V. Woolf Night & Day v. 62 Lonely mist-shrouded voyagings. 1957 Manvell & Huntley Technique Film Music iii. 164 Calm scenes of mist-shrouded lakes and shots of dew-spangled vegetation. 1888 W. B. Yeats Phantom Ship in Wanderings of Oisin (1889) 87 Hang the mist-threads for a little while Like cobwebs in the air. 1867 M. Arnold Heine's Grave in New Poems 204 And mist-track'd stream of the wide. 1928 Blunden Retreat 14 Or mist-veil brushed thee, fine as yet was wove For moonmaid's clothing. 1908 Daily Chron. 14 Nov. 4/4 Down the damp roadway move long lines of mist-veiled traffic. 1936 L. B. Lyon Bright Feather Fading 48 Mist-wild you melt now, gossamer fawn. 1849 M. Arnold Resignation in Strayed Reveller 123 Make, whistling, towards his mist-wreath'd flock. 1909 E. Pound Personae 43 Slender as mist-wrought maids and hamadryads.

mistake, sb. Add: **1. d.** An instance of a woman's becoming pregnant unintentionally; an unplanned baby.
1957 New Yorker 12 Jan. 30/3 Owing to a 'mistake', Bernadette was probably 'caught'. She was beginning to 'show'. 1959 Times 2 Mar. 5/3 We all know the baby is a 'mistake',..but surely it is a mistake which is understandable. 1963 in Sc. Nat. Dict. (1965) VI. 303/1 The peer lassie was pitten awa frae hame for makin a mistak.
2. b. Also in mistake for.
1906 Galsworthy Man of Property i. ii. 39 Old Jolyon ..gave the driver a sovereign in mistake for a shilling. 1923 World's Work May 563/1, I remember looking at him and..expecting that I had been arrested in mistake for him.
c. Also make no mistake (about) (something): have no doubt about it.

1885 W. S. Gilbert Mikado ii. 27 Ah, pray make no mistake, We are not shy; We're very wide awake. 1911 G. B. Shaw Shewing-up of Blanco Posnet 390 It wont make any difference to us: make no mistake about that. 1962 Listener 27 Sept. 463/2 But the present terms do confront us with this choice: make no mistake about it. 1963 Ibid. 21 Feb. 341/1 Make no mistake about Mr Bennet: we are meant to disapprove thoroughly of his detachment. 1974 Times 22 Mar. 11/7 Make no mistake. We had a major work of television last night.
3. mistake-free a.
1969 F. I. Dretske Seeing & Knowing ii. 63 What we might call a mistake-free way of seeing D.

mistake, v. Add: **4. c.** to mistake one's man: to judge incorrectly, or underestimate, the capabilities, character, etc., of the person with whom one has to deal.
1794 Mass. Spy 16 Apr. (Th.), If he supposes I am to be frightened by his pompous accusations, he has much mistaken his man. 1841 Congress. Globe 18 June 75/3 Mr. G. said that he was not to be coughed or cried down; gentlemen mistook their man if they supposed he was to be affected by the machinery of the political party.

mister, sb.² Add: **1. e.** Mister Big, Mister Fixit: see *Mr. 2 e; Mister Charlie: see *Charley, Charlie 7.
2. b. Colloq. shortening of Mister Mate (Mate sb.² 4 a). Freq. as vocative.
1909 F. H. Shaw Daughter of Storm xx. 177 'All right, sir,' said the second mate to Steadman... 'West by north,' said Steadman... 'I'll go and turn in, mister.' a 1966 C. S. Forester Hornblower & Crisis (1967) v. 40 The mate was marking up the traverse board. 'What's the course, Mister?' asked Hornblower. 1972 Listener 6 Jan. 18/3 The Captain..addressed the Mate as William, except when he thought he was getting uppish, when he called him 'Mister'.

mistify: see Mystify v.¹, v.²

misting, vbl. sb. Add: **2.** misting-up, the act of obscuring as with mist, the process of becoming thus obscured; also concr.
1964 B. Gaston Drifting Death ii. 22, I rinsed out the face mask against misting-up. 1966 T. Wisdom High-Performance Driving x. 108 A film of dirt or misting up on the inside [of the windscreen] can make the fog seem twice as thick as it is in reality. 1969 Listener 6 Feb. 161/1 There were no major hitches during the flight, and only a few minor ones, such as the misting-up of windows.

mistletoe. Add: **1. b.** (Earlier U.S. examples.)
1806 Lewis & Clark in Deb. Congress U.S. (1852) 9th Congress 2 Sess. App. 1142 Mistletoe, thistle, wild hemp, bulrush. 1819 E. Evans Pedestrious Tour 318 In this..country [sc. Louisiana] grows the celebrated plant called mistletoe. 1838 J. Hall Notes on Western States ii. 28 The mistletoe is seen hanging from the branches of the trees throughout the whole course of the Ohio.
3. mistletoe bird Austral., a small black, white, and crimson bird, Dicæum hirundinaceum, which feeds on nectar, pollen, and berries; **mistletoe cactus**, a tropical American epiphytic cactus of the genus Rhipsalis, esp. R. cassytha and other species bearing white fruits resembling those of mistletoe.
1908 E. J. Banfield Confessions of Beachcomber i. iii. 96 Flower-pecker or Mistletoe Bird. 1944 A. Russell Bush Ways iii. 19 Already there is a mistletoe-bird, with crimson breast, swaying itself on a mistletoe twig. 1965 Austral. Encycl. VI. 104/2 Mistletoe-bird, a small arboreal bird (Dicaeum hirundinaceum), the only representative in Australia of a group found throughout southern Asia..and belonging to a family known as flower-peckers. [1850 J. Macfadyen Flora Jamaica II. 182 Rhipsalis...Pseudo-parasitic plants, growing on trees, leafless, with small flowers, and with berries white, resembling those of the mistletoe.] 1889 W. Watson Cactus Culture for Amateurs 227 They [sc. the flowers]..are succeeded by white berries, exactly like those of the Mistletoe, whence the name Mistletoe Cactus, by which this species [sc. Rhipsalis cassytha] is known. 1967 Elbert & Hyams House Plants xi. 102 R[hipsalis] cassytha, the Mistletoe Cactus, is a hanging mass of succulent branches dripping with white berries.

mistress, sb. Add: **13, 14 a.** Also W. Indies.
1957 F. A. Collymore Notes for Gloss. Barbadian Dial. (ed. 2) 57 The archaic nominative of address has survived in Barbados, and may be heard any day on the lips of any servant or huckster addressing the mistress of the household, as, I want some more butter, mistress. Mistress, you want any useful limes? 1966 Evening Standard 1 Feb. 8/4, I would be very glad to get out of this hard country [sc. Jamaica], mistress. 1966 Guardian 14 Dec. 8/4 We go..to see Mistress Gladys Walker... Here [sc. in Barbados], Mrs is often spoken out, in full.

mistressly, a. Add: **3.** Characteristic of a man's mistress. nonce-wd.
1939 A. Huxley After many a Summer i. xiii. 180 Flirting with him all through dinner, so that you got the old man hopping jealous of him. That was masterly. Or should one say mistressly?

mistrust, v. **3. b.** (Further U.S. examples.)
1840 C. F. Hoffman Greyslaer i. x. 109, I mistrust that your Injun friend there..didn't help you much..in finding out old Josie. 1861 O. W. Holmes Elsie Venner vii. 73, I mistrusted he didn't mean to come. 1867 Harper's Mag. July 147/1 They have left the Atlantic coast, given up by physicians as in the last stage of consumption—a fact that would never be mistrusted from their present robust condition. 1898 A. Nicholas Idyl of Wabash 188 Before early apples were ripe I mistrusted what was keeping him. 1909 Dialect Notes III. 349, I mistrusted he was at the bottom of it. Not common.

mistrusting, vbl. sb. (Later example.)
1921 Spectator 30 Apr. 556/2 Class bitterness and the mistrusting of the employer by the employed.

mistune, v. (Further examples, in Radio.) Hence **mistu·ning** vbl. sb.
1914 R. Stanley Text-bk. Wireless Telegr. 134 In the Telefunken transmitter the circuits are slightly mistuned, the aerial circuit having a free wave length about 2 per cent. higher than that of the primary circuit, and this mistuning is increased with the closeness of the coupling. 1970 J. Earl Tuners & Amplifiers iii. 73 It is possible to mistune a stereo transmission and collect a terrific amount of sideband noise. Ibid., The quality of f.m. is significantly impaired by even slight mistuning.

misty, a.¹ Add: **3.** misty-eyed adj., that brings tears to the eyes; having tears in one's eyes.
1956 W. H. Whyte Organization Man (1957) iii. xiii. 156 He can grow as misty-eyed as the next man at the banquet honoring the Grand Old Man. 1974 M. Cecil Heroines in Love vi. 151 Misty-eyed emotion and passionate declarations of love.

mis-ty·ping, vbl. sb. [Mis-¹ 3.] A bad or false typing; a typing error.
1936 F. M. Ford Let. 2 July (1965) 252 The 4 being a mistyping for ', I having just changed my machine.

misuse, sb. **2.** Delete † Obs. and add later poet. example.
1881 Swinburne Mary Stuart i. i. 10 The Catholics naked here to all misuse Fall off in numbered force, in means and power.

misuser¹. (Later example.)
1927 Manch. Guardian Weekly Oct. 315/2 An exercise in most delicate raillery at the expense of all the misusers of the English language.

misvocaliza·tion. [Mis-¹ 4.] The insertion of incorrect vowel-signs in forms of writing consisting mainly or entirely of consonants.
1932 Times Lit. Suppl. 14 Jan. 20/2 Surely kābhōdh 'glory'..is a misvocalization of kābhēdh 'liver'. 1942 Jrnl. Theol. Stud. XLIII. 155 An act. verb is required by the context on Gunkel's restoration of the text; the misvocalization is due to the unusual form.

‖ **mit** (mit). colloq. or jocular. [G., with.] With (esp. with apparent ellipsis of 'me' or 'us').
1885 W. James Let. 19 Feb. (1920) I. 241, I..suppose Mrs. Godkin will come mit. 1922 Joyce Ulysses 505 Will some pleashe pershon not now impediment so catastrophics mit agitation of firstclass table-numpkin? 1959 D. Barton Loving Cup 237 Why not come along mit.

Mitanni (mitæ·ni). Name of the people and language of Mitanni, a Hurrian kingdom centred on the Habur and Upper Euphrates which flourished in the fifteenth and early fourteenth centuries B.C. Also attrib. or as adj. So **Mita·nnian** sb., an inhabitant of Mitanni; the language of Mitanni; **Mita·nnian, Mita·nnite** (rare) adjs., of or pertaining to Mitanni, its people, or its language.
1897 A. H. Sayce in Proc. Soc. Biblical Archæol. XIX. 285 Tesup was the Mitannian Air-god. 1900 —— Ibid. XXII. 176 The Mitannian language was highly agglutinative. Ibid. 182 The verb in Mitannian has hardly been differentiated from the noun. 1907 —— Archæol. Cuneiform Inscr. vi. 169 On the west..the Mitannians found themselves confronted by another northern population, the Hittites. 1910 Encycl. Brit. XIII. 539/1 Whether the Mitanni..were racially kin to the Hatti, cannot be determined at present. 1911 Ibid. XVIII. 182/2 From cuneiform sources we know the names of six other Mitanni rulers. Ibid., The language of the Mitanni state..was neither Aryan nor Semitic. Ibid. 183/1 The Hittite King's interference restored the Mitannite state as a protectorate. 1933 L. Bloomfield Lang. iv. 65 Extinct languages of an older time..Mitanni, east of Mesopotamia, from around 1400 B.C. 1939 L. H. Gray Foundations of Lang. 380 Northern Mesopotamia was the home of the Subaraean group,..divided into Mitannian and Khurrian (or Kharrian), which were very similar, if not identical. 1948 D. Diringer Alphabet 90 Hurrian..differs but very little from the language of the Mitanni. 1948 G. R. Driver Semitic Writing iii. 131 The proto-Elamites in the East and the Hittites and the Mitanni in the north devised their own systems of pictographic and cuneiform writing. 1952 Trans. Philol. Soc. 117 Chance..has also apparently left us without a ā-stem among the Indoeuropean borrowings into certain languages of the ancient Near East—Mitanni, Kassite, Lycian and the like. 1955 T. Burrow Sanskrit Lang. i. 28 The author..was a Mitannian called Kikkuli. 1970 [see *Elamite sb. and a.]. 1973 A. Malamat in D. J. Wiseman Peoples Old Testament Times vi. 149 The Aramaeans were always strongly influenced by the specific local environment, in Mesopotamia by the remnants of the Mitanni culture and by the Assyrians. 1973 H. A. Hoffner in Ibid. ix. 223 It was these Mitannians who were responsible for the introduction into the Near East of scientific techniques for the breeding and training of chariot horses. Ibid. 224 The Hittite conqueror caught his Mitannian foes off guard and routed all opposition.

mitch: see *Miche v.

Mitchell (mi·tʃĕl). The name of Sir Thomas Livingstone Mitchell (1792–1855), Scottish-born explorer of Australia, used attrib. in **Mitchell grass** to designate an Australian fodder grass of the genus Astrebla.
1883 F. M. Bailey Synopsis Queensland Flora 660 Used for food by the natives. The most valuable fodder-grass of the colony. True Mitchell-grass. 1902 Encycl. Brit. XXXII. 108/2 The 'Mitchell grasses' (Astrebla pectinata) and its varieties, viz., the Wheat (triticoides), the weeping (elymoides), and the curly (curvifolia),..have the most extraordinary vitality. 1909 Chambers's Jrnl. Dec. 809/2 Mitchell grass is said to be able to survive a rainless period extending over three years. 1927 M. M. Bennett Christison v. 55 Curly Mitchell grass shimmered gold and silver. 1934 Bulletin (Sydney) 27 June 22/1 Bull Mitchell grass and its seed is easily the best feed we have. 1936 I. L. Idriess Cattle King xix. 178 Thudding through the Mitchell grass to the crackling of sticks and flying gravel, the mob galloped on. 1948 V. Palmer Golconda xix. 153 League upon league of crinkly Mitchell grass. 1965 Austral. Encycl. IV. 365/1 Four species of Astrebla (Mitchell grasses) are known. Because the dry leaves remain attached to the plant and support stock in times of drought, they have become famous as fodder plants.

mitching, vbl. sb. and ppl. a.: see *Miche v.

mite¹. Add: **3.** mite-borne adj.; so **mite-borne typhus**, scrub typhus; also called **mite typhus.**
1939 Brit. Encycl. Med. Pract. XII. 347 The classical form of the disease is the Japanese river fever which bears the same relation to mite-borne typhus fevers as does Rocky Mountain fever to the tick-borne. 1974 Passmore & Robson Compan. Med. Stud. III. xii. 73/2 The rash spares the face in louse-borne typhus but not in the mite-borne disease. 1921 Indian Med. Gaz. LVI. 370/1 (table) Mite typhus. 1939 Brit. Encycl. Med. Pract. XII. 348 Scrub or rural form of tropical typhus..has been found to be mite-typhus conveyed by T[rombicula] deliensis. 1959 C. Ogburn Marauders (1960) viii. 265 He had all of the three worst scourges of the organization, in combination: mite typhus, amoebic dysentery and malaria.

mite². Add: **4.** Used adverbially.
1939 L. M. Montgomery Anne of Ingleside i. 9 You needn't be a mite afraid to sleep in that bed. I aired the sheets to-day. 1955 H. Croome Mountain & Molehill ii. 29 We were a mite surprised to see so many German names on your prospectus. 1958 Spectator 22 Aug. 241/2 This, to me, is a mite depressing. 1972 J. Porter Meddler & her Murder viii. 107 'There was no need to go to all that expense, dear,' said Miss Jones, a mite huffily. 1972 J. Wainwright Night is Time to Die 66 If..the farmer buys pigs ..and wishes to move them..he must obtain such a licence... This..may seem a mite bureaucratic. 1974 Wodehouse Aunts aren't Gentlemen iii. 20 Last night..it may be that I became a mite polluted, but that rarely happens.
6. mite society, a 19th-century society whose object was to collect funds for some charitable purpose by small contributions (see sense 1 c).
1822 Missionary Herald (Boston, Mass.) XVIII. 21 Female Mite So[ciety] for Cher[okee] and Choc[taw] missions [gave $]25. 1823 Baptist Mag. IV. 133, I have also assisted in the organization of two Female Mite Societies. 1872 Newton Kansan 26 Sept. 3/2 The Mite Society will hold a ten cent sociable in the school room this Thursday evening. 1878 Harper's Mag. Jan. 203/1 By means of 'mite' societies..sufficient money was raised to inclose it [sc. the grave-yard]. 1883 C. F. Wilder Sister Ridnour's Sacrifice 262 We call upon certain poor, we attend the 'Dorcas', the socials, the festivals, and mite societies.

mitella. 2. Substitute for def.: [Adopted as a generic name by J. P. de Tournefort in Institutiones Rei Herbariæ (ed. 3, 1719) I. 241.] A perennial herb of the genus so called, belonging to the family Saxifragaceæ, native to North America and north-east Asia, and bearing racemes of small flowers; usually called Mitre-wort. (Later example.)
1882 Harper's Mag. Nov. 853/2 Why should the starry blossom of the fringed mitella seek the snow-flake as its model?

mithan. Add to forms: mithong, mithun. (Later examples.)
1921 Blackw. Mag. Feb. 258/2 There was no rice, no water, no fences, no herds of mithun. 1923 Ibid. Feb. 186/2 The mithong were once more stalled beneath the houses. 1937 Jrnl. R. Anthrop. Inst. 15 Over the neck of the bound and trembling Mithan.

Mithraistic (miþrei,i·stik), a. [f. Mithraist + -ic.] = Mithraic a.
1900 Open Court May 290 Mithraistic Cameos. Showing Mithras born from the rocks between the Dioscuri, surrounded by Mithraistic symbols, among them the cup and bread of the Eucharist. 1920 Glasgow Herald 1 May 4 Hymn-writing..ran too much to gloomy terrorism and Mithraistic images of wounds and blood.

mithril (mi·þril). [Invented word.] Name given by J. R. R. Tolkien to a mythical precious metal.
1954 J. R. R. Tolkien Fellowship of Ring ii. iv. 331 Here alone in the world was found Moria-silver, or true-silver as some have called it: mithril is the Elvish name...

The beauty of mithril did not tarnish or grow dim. Ibid., Bilbo had a corslet of mithril-rings. 1955 —— Return of King v. x. 165 The coat of mithril-mail that Frodo had worn wrapped in his tattered garments. Ibid. vi. ix. 308 On her finger was Nenya, the ring wrought of mithril, that bore a single white stone flickering like a frosty star.

miticide (mai·tisaid). [f. Mit(e¹ + -i- + -cide.] Any substance used to kill mites.
1946 Travis & Morton Use of Insect Repellants & Miticides (U.S. Dept. Agric., Bureau Entomol. & Plant Quarantine, E-698) 4 Benzyl benzoate is very effective as a miticide. 1955 Sci. News Let. 18 June 392/2 Each of the six dusts is a mixture of insecticides, miticides and fungicides, and prevents damage by aphids, leaf-hoppers, spider mites, and by the plant diseases, mildew and blackspot. 1972 Daily Colonist (Victoria, B.C.) 30 July 47/5 Spray thoroughly now with a good miticide such as kelthane or malathion.

mitigate, v. **4. a.** Delete ? Obs. and add later example.
1965 W. Mitchell tr. Huyghe's Relig. Orders Mod. World i. 6 The mendicant Orders..while in part adopting the monastic and canonical forms of organization..plainly mitigated them to enable their members to go out and preach..in the highways and by-ways.

mitigated, ppl. a. Add to def.: spec. designating or pertaining to a religious order less austere than other orders.
1671 [in Dict.]. 1694 Earl of Perth Let. 17 Sept. (1845) 44 They are called rich Clarisses, because the poor Clarisses are of a far more rigid order..; these are far more mitigated, and they gave us an entertainment of musick. 1888 [see *Carmelite sb. and a. 1 b]. 1948 W. S. Maugham Catalina ix. 43 Since it was a convent of the mitigated order they enjoyed a good deal of freedom.

Mitin (mai·tin). A proprietary name for certain mothproofing agents, spec. Mitin F.F., a substituted urea, $Cl_2C_6H_3 \cdot NH \cdot CO \cdot NH \cdot C_6H_3(Cl)O \cdot C_6H_3(Cl)SO_3Na$, which is used for treating woollen goods such as carpets.
1938 Trade Marks Jrnl. 17 Aug. 994/1 Mitin Mothproof... Moth repelling and destroying preparations. J. R. Geigy.., Basle,..Switzerland; manufacturers. 1945 Chem. Abstr. XXXIX. 764 A new attack based on the insecticidal activity of p,p′-dihalodiphenyl sulfones, sulfoxides and sulfides, and resulting in the introduction of the H_2O-sol. dye Mitin F.F. showed the fundamental toxicity of this type of compd. 1958 House & Garden Apr. 7/3 (Advt.), Stockwell [carpet]... All-woollen pile. Mitin-processed; guaranteed mothproof for life. 1969 Chem. Abstr. 18 Aug. 206/2 For the protection of wool, mohair, camel's hair, or alpaca, mixts. of 1 part dieldrin,..Mitin FF, naphthalene..or alkylyl-2-thiazolyl sulfide and 0·33–12 parts epoxide, polyamide, and (or) acrylic resins are used.

mitla (mi·tla). [Native name.] An unidentified animal said to inhabit the forests on the borders of Bolivia and Brazil.
a 1925 P. H. Fawcett Exploration Fawcett (1953) xiv. 173 In the forests [of Bolivia] were various beasts still unfamiliar to zoologists, such as the mitla, which I have seen twice, a black dog-like cat about the size of a foxhound. 1965 Sun 28 Sept. 3/1 He will spend two months looking for a mitla..in the forests of Rio Abuna.

Mitnagged (mitnā·gĕd). Also **Misnagid.** Pl. **Mitnaggedim.** [ad. Heb. miṯnăggēd opponent.] The name given by the Chasidim to their religious opponents; hence any Jew who is not a Chasid.
1904 Jewish Encycl. VIII. 623/1 Mitnaggedim, (lit. 'opponents'), title applied by the Hasidim to their opponents, i.e., to the Orthodox Jews of the Slavonic countries who have not become adherents of Hasidim... 'Mitnagged' now means..simply a non-Hasid. 1907 I. Zangwill Ghetto Comedies 409 'Me join the Misnagdim!' cried the cobbler in horror. 'Never will I join with those who deny the Master-of-the-Name.' 1936 M. Goldstein Thus Religion Grows iii. 269 The Hasidim as well as the Mitnaggedim, both were confronted by the challenge of a new heaven and a new earth. 1964 S. Bellow Herzog 139 She..seemed to be seeing the Old World—her father the famous misnagid.

mitochondrion (maitokɒ·ndriɒn). Biol. Pl. **-chondria.** [a. G. mitochondrion (C. Benda 1898, in Arch. f. Anat. u. Physiol. (Physiol. Abth.) 397), f. Gr. μίτος thread + χονδρίον, dim. of χόνδρος granule or lump (of salt).] An organelle that is present (usu. in great numbers) in the cytoplasm of all cells with a true nucleus and primarily functions to store and release energy by the reactions of the Krebs cycle (see *Krebs).
1901 Jrnl. R. Microsc. Soc. 14 Meves believes that this term [sc. Nebenkern] should lapse, and himself employs for the separate granules Benda's term mitochondria, and for the Nebenkern which may be formed by their union, the term mitochondrial corpuscle. 1911, 1920 [see *Chondriosome]. 1949 Amer. Jrnl. Physiol. CLVII. 136 The residue..which contained only mitochondria and microsomes was transferred to a vial. 1962 Science Survey III. 169 Each mitochondrion is bounded by an outer limiting membrane, then there is a narrow space and then another membrane bounding an inner chamber. 1964 A. L. Lehninger Mitochondrion p. vii, It is now some fifteen years since the mitochondria were first recognized to be the 'power plants' of aerobic cells. 1970 New Scientist 24 Sept. 626/1 The amount of DNA in a chick liver cell mitochondrion..is only about 0·5 per cent of that in a bacterium. 1971 [see *Krebs].
So **mitocho·ndrial** a., of or pertaining to a mitochondrion or to mitochondria.
1901 [see *Mitochondrion]. 1920 L. Doncaster Introd. Study Cytol. ii. 23 Some investigators maintain..that every mitochondrial body arises only from a pre-existing one. 1962 Lancet 19 May 1056/1 Increased permeability of the mitochondrial membrane leads to the escape of enzymes and cofactors. 1970 Sci. Amer. Feb. 100/1 Heavy mitochondrial damage was followed by an almost total absence of dehydrogenase activity.

mitogenetic (maitodʒĕne·tik), a. Biol. [f. as next + -genetic.] Mitogenic; applied spec. to a type of radiation supposed by some to be emitted by dividing cells and to stimulate mitosis in other tissues.
1927 Biol. Abstr. I. 1042/2 The blood through the admission of air sends out mitogenetic currents... The mitogenetic oxidation reaction is cared for through oxyhemoglobin. 1928 Ibid. II. 406/2 Experiments..lead the author [sc. A. Gurwitsch] to believe in the physical nature of what he terms mitogenetic rays. 1930 Jrnl. Marine Biol. Assoc. XVII. 65 Various authors have framed the hypothesis of a mitogenetic radiation emanating from cells in the act of division. Gurwitsch first of all pointed out the mitogenetic influence of cells of embryonic tissues. 1952 G. H. Bourne Cytol. & Cell Physiol. (ed. 2) ii. 94 The school of Gurwitsch has..claimed that mitogenetic radiation is closely associated with many other forms of cellular activity.

mitogenic (maitodʒe·nik), a. Biol. [f. Mito(sis + *-genic.] Inducing or stimulating mitosis. So **mi·togen,** a substance or agent which has a mitogenic effect.
1962 Hereditas XLVIII. 619 (heading) On the effect of mitogenic plant extracts (phyto-hemagglutinin) on human white blood cells cultivated in vitro. 1963 Proc. 9th Congr. European Soc. Haematol. II. i. 78 (heading) Factors stimulating cell division in cultured leucocytes. Chemical, serological and immunological studies on mitogens. 1970 Nature 26 Sept. 1351/1 When certain mitogenic agents are added to lymphocytes cultured in serum, the cells are activated to transform and divide. 1971 Ibid. 31 Dec. 508/2 Pokeweed mitogen is one of the plant lectins which..can be used to stimulate lymphocytes in vitro.

mitomycin (maitomai·sin). Biochem. [f. mito- (perh. representing Gr. μίτος thread or Mitosis, *Mitochondrion, etc.: the allusion is not explained by the (Japanese) authors of the name) + *-mycin.] An antibiotic active against some bacteria and tumour cells that is produced by the soil bacterium Streptomyces cæspitosus; also, any of the three (or more) slightly different molecular species (as those designated mitomycin A, B, and C) into which preparations of this antibiotic can be resolved.
1956 T. Hata et al. in Jrnl. Antibiotics (Tokyo) A. IX. 141 (heading) Mitomycin, a new antibiotic from Streptomyces. Ibid. 145 Mitomycins have a high antibacterial activity against gram positive and gram negative bacteria. 1958 Antibiotics & Chemotherapy VIII. S[treptomyces] caespitosus, which produces mainly mitomycin A and B in some cultural conditions, produces mitomycin C exclusively under other conditions. 1968 [see *Induction 9 d].

mitosis. Add: Pl. **mitoses** (-ōu·sīz). [First formed in Ger. by W. Flemming in Zellsubstanz, Kern und Zelltheilung (1882) xxiv. 376.] Substitute for def. of senses a and b: The process of nuclear division by which a cell nucleus gives rise to two daughter nuclei identical with the parent nucleus; an instance of this; commonly also used to refer to the whole process of mitotic cell division, i.e. division of the cytoplasm as well as the nucleus; also, a cell or nucleus undergoing this. (Earlier and later examples.)
1887 Jrnl. R. Microsc. Soc. 163 (heading) Showing mitosis in brain of tadpole. 1918 Surg., Gynecol. & Obstetr. XVIII. 205/2 These cells are large and mostly polyhedric. ..Mitoses are found, though rarely. 1937 C. D. Darlington Recent Adv. Cytol. (ed. 2) i. 22 Nuclei divide by the characteristic process of mitosis in the course of which the whole nucleus, apart from the nucleoli, resolves itself into longitudinally split threads, the chromosomes. 1962 D. G. Cogan in A. Pirie Lens Metabolism Rel. Cataract 292 Although showing few mitoses normally, the lens epithelium may be activated into lively mitotic activity when stimulated by heat. 1970 Ambrose & Easty Cell Biol. i. 20 The actual process of cell division, however, known as mitosis, is remarkably similar for all cell types.

† **mitosome** (mai·tosōum). Cytology. Obs. Also 9 **mitosoma.** [a. G. mitosoma (G. Platner 1889, in Arch. f. mikrosk. Anat. XXXIII. 199), f. Gr. μίτο-ς thread + σῶμα body.] (See quots.)
1891 Jrnl. R. Microsc. Soc. 461 The portion of the mitosoma which becomes attached to the nucleus becomes chromatic and wanders to the anterior end of the spermatozoon. 1895 G. W. Field in Jrnl. Morphol. XI. 237 In view of the fact that the term 'Nebenkern' has come to be applied to many sorts of intracellular structures..it seems best, since the history of the middle piece or 'Nebenkern' is now better understood, that this noncommittal term..should be replaced by some term which gives a hint as to the nature of this body. Since the middle piece or 'Nebenkern' of the echinoderm spermatozoön is formed from the mitotic spindle, the term 'mitosome', introduced by Platner, has been adopted and will be used to designate the middle piece, = Nebenkern = corpuscle accessoire of other writers. 1920 L. Doncaster Introd. Study Cytol. vii. 95 In the spermatocyte, as the cell enlarges, the mitochondrial bodies increase in size... In the young spermatid they unite to form a fairly compact mass near the nucleus at the side of the cell at which the tail will grow out. [Note] This mitochondrial mass ('mitosome') constitutes the 'Nebenkern' of some authors, but as the word has been used to designate the remains of the division-spindle.., the 'idiozome'..and other cell-structures, it is now dropping out of use. 1925 E. B. Wilson Cell (ed. 3) iv. 366 One of these [elements of doubtful nature] is the spindle-remnant, sometimes called the 'mitosome'... This body was believed by some earlier observers to play an important part in the sperm-formation, and was confused with the nebenkern or with the acroblast; but later studies seem to show that it disappears without taking any definite part in the sperm-formation. 1934 L. W. Sharp Introd. Cytol. (ed. 3) xiv. 218 In insects generally the chondriosomes form a single more or less compact body, the nebenkern (the mitosome of Gatenby).

mitotically, adv. (Examples.)
1896 Jrnl. R. Microsc. Soc. 494 The cells of the inner theca layer begin to divide mitotically. 1946 [see *Amitotic a.]. 1973 Nature 30 Mar. 299/3 The thymus is usually thought of as a staging house through which pass stem cells and their mitotically amplified products.

mitral, a. **2. b.** (Further examples.)
1872 Half-Yearly Abstr. Med. Sci. LV. 103 (heading) The physical signs of mitral stenosis. 1926 Daily Colonist (Victoria, B.C.) 14 Jan. 4/6 In 1921 the patient, a young girl, came up to the London Hospital suffering from mitral stenosis, or a gradual closing of one of the four valves of the heart. 1966 Wright & Symmers Systemic Path. I. xxiv. 50/1 In mitral stenosis, two further complications may arise which still further burden the already overstressed heart:..arterial fibrillation, and..increasing incompetence of the valve with progressively greater regurgitation of blood into the left atrium during ventricular systole.

mitre, sb.¹ Add: **2. f.** A medieval type of woman's headwear resembling a bishop's mitre. Also attrib.
1877 Encycl. Brit. VI. 469/2 Some of the more popular of these strange varieties of head-gear have been distinguished as the 'horned', the 'mitre', [etc.]. 1906 H. Druitt Man. Costume vi. 258 The next development shows the cauls curving outwards and upwards, and terminating above the head in a pair of horns. This form is called the horned, lunar, mitre or heart shaped head-dress according to the shape which it assumes. 1960 Cunnington & Beard Dict. Eng. Costume 136/2 Mitre head-dress.

‖ **Mitsein** (mi·tzain). Philos. [ad. G. mit with + sein being.] A term used by Heidegger to express the concept of man's being in its relationship with others.
1955 J. Macquarrie Existentialist Theol. iv. 89 Thus 'being-in-the-world' implies 'being-with-others' (Mitsein). 1957 H. Barnes tr. Sartre's Being & Nothingness III. iii. 413 The very existence of this grammatical form [sc. the word we] necessarily refers us to a real experience of the Mitsein. 1963 Philos. XXXVIII. 275 The whole world of human Mitsein is regarded by M. Bastide as his province. 1966 A. Manser Sartre vi. 97 The essence of relations between consciousnesses is not Mitsein (being together), it is conflict.

mitsumata (mi:tsuma·tā). [Japanese.] A deciduous shrub, Edgeworthia papyrifera, bearing clusters of fragrant yellow flowers, belonging to the family Thymelæaceæ, and native to China, although widely cultivated in Japan, where its bast fibre is used in the manufacture of paper. Also attrib.
1889 J. Rein Industries of Japan iii. 402 Mitzu-mata (Edgeworthia) paper has also a distinctly marked yellow colour. 1891 B. H. Chamberlain Things Japanese (ed. 2) 334 Several plants and trees contribute their bark to the manufacture of Japanese paper... The one most easily recognised by the unlearned is the Edgeworthia papyrifera, which has the peculiarity that its branches always divide into three at every articulation, whence the Japanese name of mitsu-mata, or 'the three forks'. 1936 D. Hunter Papermaking Pilgrimage Japan, Korea & China iii. 45 The so-called 'vellum'—that smooth, long-fibred, natural-toned paper sometimes used in the printing of fine books and etchings..is made largely from the bark of the mitsumata shrub. 1947 —— Papermaking (ed. 2) ii. 57 The origin of mitsumata (Edgeworthia papyrifera) as a papermaking material is uncertain, but there is a record stating that in the year 1597 a papermaking family was granted the privilege of gathering mitsumata bark in a certain locality of Japan. 1974 G. Usher Dict. Plants used by Man 223/2 E[dgeworthia] papyrifera... Cultivated for the bark fibres which are used, particularly in Japan to make a hand-made paper (Nepal Paper, Mitsumata Paper).

mitt. Add: **2. b.** U.S. A protective glove worn in baseball by the catcher or first baseman.
1902 Sears Catal. 326 Boys' Canvas Mitt, made of canvas throughout; a good, cheap mitt for boys; well stuffed. 1949 Nat. Geogr. Mag. June 738/1 On the ball field the Indians waved their mitts.

3. For U.S. slang read slang (orig. U.S.). Add earlier and later examples.
1896 Ade Artie xiii. 116, I thought them was gloves you had on. Gee, is them your mits? 1914 Joyce Dubliners 74 He was also handy with the mits. 1940 R. Chandler Farewell, my Lovely ii. 12 'Freeze the mitts on the bar.' The barman and I put our hands on the bar. 1959 I. & P. Opie Lore & Lang. Schoolch. x. 197 The commonest challenge is 'Put up your mitts'.
c. the glad mitt: a warm or friendly reception; = glad hand (*Glad a. 4 e); the frozen (or icy) mitt: an unfriendly reception; rejection; the 'cold shoulder'. slang.
1904 'No. 1500' Life in Sing Sing 255/2 Glad mitt, warm welcome. 1907 J. London Road 187 The erstwhile hospitable farmers met us with the icy mit. 1925 Fraser & Gibbons Soldier & Sailor Words 156 He tried to make up to me but I gave him the frozen mit. 1937 M. Sharp Nutmeg Tree ix. 111, I expected any number of black eyes, Julia darling, but not the frozen mitt. 1960 A. Prior in Pick of Today's Short Stories XI. 179 She'd have taken it and then handed me the frozen mitt.
d. attrib. and Comb., as **mitt camp** U.S. slang, a palmist's or fortune-teller's establishment, tent, etc.; **mitt joint** U.S. slang, (a) (see quot. 1914); (b) = *mitt camp; **mitt-reader** U.S. slang, a palmist; a fortune-teller.
1942 Berrey & Van den Bark Amer. Thes. Slang § 466/5 Mitt camp or joint, a fortune-telling establishment. Ibid. § 626/7 Mitt camp or joint, a fortune teller's tent or booth. 1956 H. Gold Man who was not with It (1965) i. 4 She would..take the tickets to Palmistry Pauline's mitt-camp. 1914 Jackson & Hellyer Vocab. Criminal Slang 59 A 'mitt joint' is a gambling house where victims are 'steered' for fleecing by means of deceptively 'sure thing' hands. 1923 C. R. Cooper Under Big Top 60, I have seen a couple halt before a 'mitt joint' where a greasy Mexican or Syrian or anything else but a gypsy stands. 1942 Mitt joint [see mitt camp above]. 1928 Amer. Speech III. 414 Mitt reader, a palmist, or fortune teller. 1956 H. Gold Man who was not with It (1965) xiv. 120 How do you know? You a mitt reader like your mother.

‖ **Mittagessen** (mi·tāge:sən). Also **Mittagsessen,** and with lower-case initial. [Ger.] In Germany: a midday meal; lunch.
1880 Geo. Eliot Let. 11 July (1956) VII. 303 There is a magnificent drive to Baden which can be reached in seven hours (including time for the Mittagsessen and the rest). 1899 R. Fry Let. Oct. (1972) I. 174 The Galleries shut at the absurd hour of 3..in order that the officials may have some absurd meal, a mittagsessen or something. 1941 M. Treadgold We couldn't leave Dinah xv. 238 The heavy mittagessen upon which Germany nourishes young and old. 1972 Sat. Rev. (U.S.) 25 Mar. 72/1 The Mittagessen (lunch) always begins with soup.

Mittel-Europa (mi·t'l,yūərōu·pă). Also **Mittel Europa, Mitteleuropa.** [Ger.] Central Europe. Also attrib.
1918 M. J. Davoren tr. Gettlich's German Grip on Russia 24 It is in this combination of a community of the interests of Budapest with those of the financial circles of Berlin that the idea of the new politico-commercial consortium, known to-day by the name 'Mitteleuropa', arose. 1918 Z. N. Preev Russian Riddle iii. 36 It would be worth her while to abandon..her grandiose plans of Mittel-Europa. 1931 Times Lit. Suppl. 9 Apr. 281/3 The glorification of the Mittel-Europa ideal which Frederick the Great failed to realize. 1950 A. Christie Murder is Announced v. 48 Very temperamental you'll find her. Mittel Europa refugee of some kind. 1956 Auden & Kallman Magic Flute (1957) 116 We'll never lack friends back in Mittel-Europa. 1971 C. Fick Danziger Transcript (1973) 29 He had a slight Mittel-Europa accent, but at Claridge's, who doesn't? 1975 J. Symons Three Pipe Problem xviii. 176 Would the totally unknown actress be able to resist the charm of Mitteleuropa?

Mittel-European (mi·t'l,yūərōp ī·ăn), a. and sb. [ad. G. mittel-europäisch, f. mittel middle.] **A.** adj. = *Middle-European adj.
[1937 Wyndham Lewis Blasting & Bombardiering xii. 189 With his dramatic miteuropean accent he gave this suggestion such a rich Austrian welcome as no suggestion ever had before or since.] 1930 C. Beaton My Royal Past xiii. 143 The destined lot of mittel-European royalty. 1957 P. Kemp Mine were of Trouble x. 187 The sinister but intriguing personality of a Mittel-European barman known as Otto. 1959 J. Drummond Black Unicorn xv. 110 The fringe of the fashionable area, where the big shops give way to little mittel-European establishments. 1974 Times 31 Dec. 7/3 The pretence of Mittel-European dilettantism.
B. sb. A native or inhabitant of central Europe.
1950 A. Christie Murder is Announced xi. 118, I expect Mitzi, our Mittel European, would love that. 1958 Times 24 July 5/3 A mittel-European exploding with conversation like a fire-cracker.

‖ **Mittelschmerz** (mi·t'lʃmĕrts). Gynæcology. Also **mittelschmerz.** [G., lit. 'middle pain'.] Pain in the lower abdomen regularly experienced by some women between menstrual periods, perhaps related to the occurrence of ovulation.
1895 Lancet 28 Dec. 1625/1 Dr. J. Halliday Croom read a paper on So-called Mittelschmerz, sometimes called a form of dysmenorrhœa. 1942 Mazer & Israel Diagn. & Treatm. Menstrual Disorders ix. 135 The pain undoubtedly emanates from the ovaries, since neither hysterectomy nor resection of the presacral nerve..eliminates mittelschmerz, but bilateral oophorectomy does. 1971 Vogue Nov. 60/2 Have intercourse as near ovulation as possible. This can be detected..by the pain that some women experience (Mittelschmerz). 1971 I. A. McDonald Method Obstetr. & Gynaecol. xiii. 300 Some females regularly have a spotting of blood on day 14 of the menstrual cycle... It may be associated with lower abdominal pain (Mittelschmerz).

mitten, sb. Add: **1. d.** (c) Handcuffs.
1880 G. Webster in Sc. Nat. Dict. (1965) VI. 306/1 My lad was made fast an' a pair o' mittens clappit on wi' little mair adee. 1937 'D. Hume' Halfway to Horror 3 'Mittens' are handcuffs; a padlock is a 'monkey'.
4. mitten crab (see quots.).
1934 Times 1 Feb. 17/5 If we want an English equivalent for the German Wollhandkrabbe it might be called the mitten crab. 1934 Nature 9 June 856/1 The pincer claws are clothed with long soft hair, and a writer in the Times has suggested 'mitten crab' as an appropriate name for it... The mitten crab must have been introduced into German rivers before 1912... In 1923 the species was found to be established in the..Elbe and was determined as Eriocheir sinensis. 1959 Chambers's Encycl. IX. 455/1 Mitten crab, a popular name applied to a greenish grapsoid crab, Eriocheir sinensis, with conspicuous brown tufts of long silky hairs on the pincers of the male.

Mitty² (mi·ti). Also **Walter Mitty.** [f. the name of Walter Mitty, hero of James Thurber's short story The Secret Life of Walter Mitty (in New Yorker (1939) 18 Mar.).] A person who indulges in day-dreams; one who imagines a more adventurous or enjoyable life for himself than he actually leads; the characteristics of such a person. Freq. attrib. or quasi-adj. Hence **Mittye·sque, Mi·ttyish, Mitty-like** adjs.
1950 B. Schulberg Disenchanted (1951) xvii. 313 I've had daydreams of how I'd come back. Walter Mitty stuff about arriving in style. 1953 Sunday Times 14 June 8/4 The Mitty me, I notice, will risk his life for a trifle, but never gets his hands dirty. 1958 Times Lit. Suppl. 16 May 274/1 Greave takes refuge from the horrid realities of life in Mittyesque fantasies, pretending he is a high-powered American salesman. 1960 Harper's Bazaar Apr. 125/1 The average motoring man is..a visionary, a Walter Mitty locked in a private world of fantasy... Vintage cars seldom fail to spark off the Mitty in a man. 1960 Sunday Express 12 June 14/5 Women..live in a dream world of their own imagining—a Walter Mitty-ish 'Other Life'. 1961 John o' London's 28 Sept. 363/3 I'm in the delirious position of being able to indulge my Mitty-like obsession. 1968 S. Brittan Left or Right v. 106 The whole Walter Mitty idea of a private line from Downing Street to the White House. 1972 Guardian 11 July 10/5 Both men are Mittyesque failures. 1974 N. Freeling Dressing of Diamond 33 Bernard was doing his hospitality act... This was no Mitty performance.

‖ **mitzvah** (mi·tsvā). Judaism. Also **misva, mitsva(h), mitzwa.** Pl. **misvot, mitswoth, mitzvot, mitzwoth.** [Heb. miṣwāh commandment.] A precept; something which should be done; hence a good deed done as a religious duty, without expectation of earthly reward. Cf. *Barmitzvah.
1650 E. Chilmead tr. Leon Modena's Hist. Rites of Jews i. i. 2 Precepts of the Written Law..they call..Mizuoth de Draita, that is to say, Præcepta Legis, Precepts of the Law. The second sort are..Mizuoth de Rabbanan,..Precepts of the Wise men. 1753 Jewish Ritual 33 All the eight Cakes together, for the Service of both Nights, are call'd Mitzwoth, i.e. the Statutes. 1831 Ascamot, or Laws & Regulations Jewish Congregation vii. 52 If these three Misvot as well as the Misva of saying the Zemirot, shall have been given..it shall not be given to any other. 1893 I. Zangwill Childr. Ghetto (ed. 3) i. iii. 41 Mitzvah is a 'portmanteau-word'. It means a commandment and a good deed, the two conceptions being regarded as interchangeable. 1932 L. Golding Magnolia St. i. ii. 33 To break some of the less severe laws is almost accounted a mitzvah, a pious duty. 1959 W. L. Gresham Houdini xxi. 154 To Harry a good deed was something to be done quietly: 'So when you do a mitzvah you don't take along a brass band.' 1973 Jewish Chron. 2 Feb. 40/2 'Neither,' he adds scornfully, 'do we give out mitzvot according to money or position or influence.' 1973 Synagogue Light Sept. 11/1 There is great apprehension as to the meaning and message of this act of removing the shoe, which forms an essential part in the right and mitzvah of Chalitzah.

mivvy (mi·vi). slang. [Origin uncertain; perh. a corruption of Marvel sb. The three senses may not be connected.] **1.** A marble.
1856 N. & Q. 5 Apr. 283/2 Mivvies, marbles. 1917 H. H. Richardson Fortunes Richard Mahony i. iv. 33 You were always a good one at striking a bargain, my boy! What about: 'Four mivvies for an alley!'
2. A contemptuous term for a woman; the landlady of a lodging-house.
1881 Punch 10 Sept. 110/1 Lor' bless yer, they don't know the ropes, these old mivvies don't, more than a mug. 1887 Punch 10 Sept. 111/1 And talk about stodge! Jest you arsk the old mivvey as caters for me at the crib where I lodge. 1892 E. J. Milliken 'Arry Ballads 13/2 Barearmed old mivvies you meet spread out pink in a theatre stall. 1923 J. Manchon Le Slang 195 Mivvy, femme.
3. A person who is adept at something; a 'marvel'.
1906 E. Pugh Spoilers xv. 162 He's a mivvy at makin' things easy. 1959 'O. Mills' Stairway to Murder xi. 121 He's a mivvy with anything like that.

Miwok (mī·wɒk, mai·wɒk). [Native name.] A Penutian Indian people of California; a member of this tribe; also, the language spoken by the tribe. Also attrib. or as adj.
1877 S. Powers in Contrib. N. Amer. Ethnol. III. 346 By much the largest nation in California, both in population and in extent of territory, is the Miwok, whose ancient dominion extended from the snow-line of the Sierra Nevada to the San Joaquin River, and from the Cosumnes to the Fiesno. 1916 Univ. California Publ. Amer. Archaeol. & Ethnol. XII. iv. 141 With the Miwok the moiety has no subdivisions. Ibid. 142 That totemic symptoms of one sort or another are present in the Miwok organization cannot be denied. 1949 Los Angeles Times 10 Apr. ii. 5/2 The name Yosemite was given to this valley by the Miwoks. 1964 Gould & Kolb Dict. Social Sci. 436/1 A Miwok couple about to marry endogamously. 1966 C. F. & F. M. Voegelin Map N. Amer. Indian Lang. (caption) Miwok-Costanoan Family i. Sierra Miwok. 2. Coast-Lake Miwok. 1972 Language XLVIII. 847, 6b is given..by Sedlak for ..Miwok, Mongol (Dagor),..and Siona. 1973 Black Panther 28 Apr. 10/1 He had come to participate in the Tuolumne Acorn Festival, an annual traditional ceremony of his people, the Miwok Indians.

mix, sb.² Add: **1. a.** (Further examples.) Also, a number of ingredients mixed together, or intended for mixing; spec. the prepared ingredients of a cake, etc., sold ready for cooking; more generally, the proportion of different constituents that make up a product, plan, policy, etc.; a combination of various components into an integrated whole.
1912 L. & M. Greenbaum Pract. Dentistry xxvii. 457 It is best to subject the mass to the least amount of stirring conducive to a homogeneous mix. 1922 Mojonnier & Troy Technical Control of Dairy Products xiii. 276 Ice cream made from mix No. 9 will feel about 12·00 per cent warmer to the tongue than ice cream made from mix No. 1. 1938, etc. [see cake-mix s.v. *Cake sb. 9]. 1945 B. Macdonald Egg & I (1946) 66, I could use automatic biscuit mix for the crust. 1959 Life (Internat. ed.) 13 Apr. 20/2 The new term is 'mix'. There is, says the Administration, an adequate 'mix' of forces to maintain our deterrent. 1962 Listener 29 Mar. 579/2 When the egg mix is light and foamy add it to the sauce. 1962 Times 26 Apr. 9/5 (headline) Canada may ban food mix sales. 1962 L. Deighton Ipcress File xviii. 110 A bottle of Scotch, gin, some assorted mixes. 1964 M. McLuhan Understanding Media (1967) v. 60 Oral societies are made up of people differentiated..by their unique emotional mixes. Ibid. 64 Chaplin ..hit upon the wondrous media mix of ballet and film. 1966 [see *Jersey¹ 1]. 1967 Daily Tel. 17 May 17/7 Without sufficient statistics on the question of socially mixed entry to public schools, the committee believes it impossible to induce artificially a 'social mix'. 1967 Times Rev. Industry Aug. 16/2 Retailing strategy will probably go on trying to maximize interest in stores by an optimum mix of heavily advertised manufacturer-branded lines and an increasing proportion of housebrand lines. 1970 Daily Tel. 9 Apr. 36/5 The mix of dancing, cabaret, drinking and gaming created an undesirable temptation to young people. 1971 Guardian 9 June 13/1 Alcoholic mixes like bitter lemon. 1971 M. McCarthy Birds of America 31 First a cake made with a mix and then a real one. 1972 Publishers Weekly 6 Mar. 25/1 In publishing, I think you have to have a good mix, and not try to reach just one segment of the population, but to cover all segments. 1972 Guardian 5 Aug. 11/8 By expelling the Russians he [sc. Sadat] has drastically altered the strategic mix.
c. Add to def.: spec. (a) a state of confusion; a mess or muddle; (b) a fight. (Add earlier and later examples.)
1841 S. Bamford Passages in Life of Radical I. xv. 94, I had expected being conducted to London alone, and certainly was not prepared for a mix-up with these men. 1913 Chums 25 Jan. 361/2 Then for a full minute the two engaged in a 'mix-up'. 1913 Collier's 1 Feb. 27 (caption) A bad mix-up. 1919 [see *Gov]. 1923 Wodehouse Inimitable Jeeves viii. 85 When the driver started making a fuss, there was a bit of a mix-up. 1932 J. C. Powys Glastonbury Romance (1933) II. xxiii. 737 God! What a mix-up it all is. 1971 D. E. Westlake I gave at the Office (1972) 139 As soon as the mix-up was brought to light I was freed.
2. Cinemat. and Broadcasting. The action or result of mixing two pictures or two sounds (see *Mix v. 7 a).
1922 [see *Fade sb.¹ 2]. 1932 Wireless World 16 Mar. 276/1 A sound-mix, which corresponds to a picture-mix (when the picture dissolves into another picture) is done by starting with one microphone set at its gain figure and the second microphone set at zero gain. Ibid., For a smooth mix, the two knobs must be turned as nearly as possible at the same speed. 1960 [see *Fade sb.¹ 2]. 1961 Listener 2 Nov. 716/1 If phrases overlap musically the visual counterpart is a mix; and..a mix between two different angles on the same performer is usually upsetting to the viewer. 1962 A. Nisbett Technique Sound Studio ix. 152 It is the way that fades and mixes are carried out.. that distinguishes the polished, finished recording from the one that sounds amateur.

mix, v. Add: **1. g.** Colloq. phr. to mix one's drinks: to drink various kinds of alcoholic liquor in succession; spec. to become intoxicated by drinking both wine and liquor made from grain. Also ellipt. and absol.
1898 J. D. Brayshaw Slum Silhouettes 238 'E was gettin' a bit beargered—not that 'e'd 'ad so much, but 'e would keep mixin'; first one thing an' then annuver. 1933 J. B. Priestley Wonder Hero vii. 269 Had a thick night last night... Mixed 'em a bit. Always a mistake—mixing

'em. **1950** G. GREENE *Third Man* ii. 13 If you stayed around in a hotel lounge, sooner or later..one mixed one's drinks. **1961** J. B. PRIESTLEY *Saturn over Water* iii. 35, I mixed my drinks too much last night—I feel better now.

h. Slang phr. *to mix it*: to fight or quarrel; to start fighting; freq. const. *with*; also, to cause trouble. So *to mix it up*: to fight vigorously.

1900 A. CONAN DOYLE *Green Flag* 165 This round must decide it. 'Mix it oop, lad; mix it oop!' the iron-men whooped. **1905** C. H. DAY *Actress & Clerk* xv. 149 As Hard Knox would have himself described in the technique of the ring, the men 'mixed it'. **1906** H. GREEN *At Actors' Boarding House* 359 They're goin' to mix it up. The little un'll win out, see if she don't. My eye! dames is allus fightin'. **1918** E. M. ROBERTS *Flying Fighter* 91 He grew angry and we mixed it. I gave him a black eye. **1919** WODEHOUSE *Coming of Bill* (1920) II. xiv. 239, I thought I could stay around and part 'em if they got to mixing it. **1941** E. C. SHEPHERD *Mil. Aeroplane* 14 Many a German bomber in daylight raids over England has accepted failure rather than 'mix it' with the British fighters which came to dispute with it. **1941** *Time* 8 Dec. 22/1 The tanks ..face the approaching column... Then they begin to mix it up. **1945** *Aeronautics* Feb. 50/3 Pilots took full advantage of every opportunity of 'mixing it' with the Hun. **1950** P. TEMPEST *Lag's Lexicon* 136 *Mix it, to*, to put one man against another. To make trouble. **1958** F. NORMAN *Bang to Rights* 28 The screw who's giving evidence against you starts telling a load of bleeding lies and mixing it for you. **1973** D. LEES *Rape of Quiet Town* vi. 92 These lads don't want to fight for nothing. If they can get away without mixing it they will.

4. a. *to mix in with*: delete † and add later examples. Also *to mix in*: to start or join in a fight (*slang*). Cf. 4 c in Dict. and *1 h in Suppl.

1870 J. P. SMITH *Widow Goldsmith's Daughter* vi. 69 Of course they couldn't expect to mix in with the rich children. **1895** M. HALSTEAD *100 Bear Stories* 117 Elk killing didn't seem half so great an achievement as it had before the bear had mixed in with the proceedings. **1912** R. A. WASON *Friar Tuck* xxi. 158 'Well, what if he did shoot,' sez Slim, 'we wouldn't have to mix in, would we?' **1971** WODEHOUSE *Much Obliged, Jeeves* ix. 89 If you see any more gnats headed in her direction, hold their coats and wish them luck, but restrain the impulse to mix in.

d. To be sociable.

1816 JANE AUSTEN *Emma* III. vii. 113 Mr. and Mrs. Elton..showed no unwillingness to mix, and be..agreeable. **1905** *Dialect Notes* III. 88 He doesn't mix much. **1940** A. CHRISTIE *Sad Cypress* I. vi. 83 Rather a funny crowd of people, but I don't mix much. You told me once that I wasn't a good mixer. **1965** M. SPARK *Mandelbaum Gate* I. 7 He was in no great hurry for the flat, preferring hotel life where one need not mix.

7. *Cinemat.* and *Broadcasting*. **a.** *trans.* To combine (two pictures or sounds) temporarily by fading one out as the other is faded in; freq. *intr.*, to pass from one picture *to* another in this way.

1922 L. C. MACBEAN *Kinematogr. Studio Technique* ix. 82 On occasions..it is necessary..to fade or mix titles into a scene to which they relate. **1929** *Radio Times* 8 Nov. 389/1 The D.C. Panel..which allows a producer.. to mix and fade speech, music, and sound-effects. **1953** K. REISZ *Technique Film Editing* I. i. 25 From the scene in the present, Griffith simply mixed to the earlier scene and then mixed back again. **1961** G. MILLERSON *Technique Television Production* xvi. 305 Mixing opposite directions of movement can sometimes arouse feelings of expansion or impact. **1962** *Listener* 6 Dec. 983/1 Mr Cooper uses..the language and grammar of film-making in his radio plays. He will cut sharply from scene to scene; or he will mix from one scene into the next.

b. *trans.* To combine (two or more sound signals) into one, either linearly, by adding together a fraction of each in a mixer (sense *2 b), or non-linearly, by causing one signal to modulate a second in a mixer (sense *2 c).

1928 [see *BALANCE *v.* 4]. **1931** C. DREHER in L. Cowan *Recording Sound for Motion Pict.* xxiv. 345 He also mixes the output of the microphones when several are used. **1958** W. F. LOVERING *Radio Communication* viii. 172 Suppressor-grid control may be used to 'mix' two signals to produce a modulated output. **1962** J. H. & P. J. REYNER *Radio Communication* ix. 364 A pentode valve was used as an anode bend detector.., while a small triode, assembled round the same cathode, generated the local oscillation, and the two were mixed by using a common bias resistor in the cathode circuit. **1974** *Encycl. Brit. Macropædia* XII. 549/1 When several microphones are used on the set, their outputs are often mixed and reproduced on a single film or tape.

mixed, *ppl. a.* Add: **3.** (Earlier and later examples of *mixed motives*.)

1818 M. EDGEWORTH *Let.* 13 Oct. (1971) 114 There were mixed motives I grant... Lord Byron was distressed for money. To be sure he could have had other fortunes but then there was vanity. **1939** L. MACNEICE *Autumn Jrnl.* iii. 18 None of our hearts are pure, we always have mixed motives.

6. b. *Phonetics.* Of a vowel sound = *CENTRAL *a.* 1 d.

1867 [see *BACK *a.* 1 c]. **1890** H. SWEET *Primer Spoken Eng.* 4 In the vowels we distinguish three horizontal positions, or degrees of retraction of the tongue: back, mixed, front. **1918** D. JONES *Outl. Eng. Phonetics* 17 An example of a mixed vowel is the vowel in *bird*. **1966** M. PEI *Gloss. Ling. Terminol.* 164 *Mixed vowel*, Migliorini's term for middle vowel.

11. mixed bag, a heterogeneous collection of people, objects, items, etc.; **mixed bathing,** simultaneous bathing in the same place by people of both sexes; **mixed bed,** a flower bed containing an assortment of plants, arranged in irregular groups; **mixed blessing,** a blessing (BLESSING *vbl. sb.* 4) that has unpleasant elements in it; **mixed blood,** (*a*) descent from two or more races; (*b*) a person of mixed descent; **mixed border,** in a garden, a long bed containing a mixture of hardy herbaceous plants and shrubs, hardy and half-hardy plants, bulbs, etc.; **mixed breed,** a cross-breed; **mixed-celled** *a. Path.*, involving or containing cells of more than one kind; **mixed company,** (*a*) company comprising both men and women; (*b*) company comprising people of different classes or characters; **mixed crystal** *Physical Chem.*, a homogeneous crystal formed of more than one crystalline substance; **mixed economy,** an economic system containing both private and state enterprise; **mixed farming,** farming which combines the raising of livestock and the cultivation of arable crops; so *mixed farm*; **mixed flow,** flow (in a turbine or the like) that consists of two or more types (usu. radial and axial) in succession; usu. *attrib.* (with hyphen); **mixed grill,** a dish consisting of several different grilled or fried items of food; also *fig.*; **mixed language,** a language made up of a mixture of elements from two or more languages; a creolized language; cf. *MISCHSPRACHE; **mixed-manned** *a.*, pertaining to or designating a military force comprising people of more than one nationality; so *mixed manning*; **mixed marriage** (earlier and later examples); (*b*) a marriage between persons of different races; also *transf.* and *fig.*; **mixed media,** (*a*) = **mixed technique*; (*b*) an entertainment, work of art, etc., which combines various media; = *MULTI-MEDIA *a.*; also *attrib.* or as *adj.*; **mixed-pressure** *a. Engin.*, applied to a steam turbine powered by both high- and low-pressure steam; **mixed technique** (see quots.).

1936 C. C. R. MURPHY (*title*) A mixed bag. **1943** K. TENNANT *Ride on Stranger* (1968) v. 47 This mixed bag began, not ill-naturedly, to re-arrange itself in the seating. **1973** A. BEHREND *Samarai Affair* i. 13 Representatives of the press, a mixed bag in age, but not in sex. **1901** *Graphic* 31 Aug. 268/3 The case against mixed bathing has passed into the academic or empty stage. **1930** *New Statesman* 27 Dec. 356/1 He..could obtain any sum he pleased for writing on any subject he pleased, from the League of Nations to the ethics of mixed bathing. **1964** M. LASKI in S. Nowell-Smith *Edwardian England* iv. 167 Bexhill where mixed bathing has just [*sc.* circa 1901] been introduced, though the custom will quickly spread. **1871** W. ROBINSON *Hardy Flowers* iii. 13 A mixed bed, carefully arranged as to the height, and tastefully as to the quality and disposition of the contents. In this kind of bed I should have no band or circle whatever, but simply a careful following out of the mixed principle. **1873** *Young Englishwoman* May 238/1 We hope this year to see a great improvement in the bedding system—mixed beds introduced in the place of those of one sort of plant, and of one colour. **1933** *Discovery* Oct. 309/2 The introduction of European influences may prove a mixed blessing. **1960** *News Chron.* 21 Sept. 6/3 In theory it was a Good Thing.. In practice it turned out a mixed blessing. **1973** *Guardian* 16 June 11/3 Mr Duggan regards Mr Bloom's twin tub machine..as a very mixed blessing; a machine..which.. uses the same lot of suds time and time again. **1817** S. BROWN *Western Gazetteer* 244 About one half of the Cherokee nation are of mixed blood by intermarriages with the white people. **1858** THOREAU in *Atlantic Monthly* Aug. 306/2 The two mixed bloods..went off up the river. **1935** HUXLEY & HADDON *We Europeans* i. 23 If a Scottish or Irish clan is of 'mixed blood', what likelihood is there of purity of descent among the millions that make up the population of any great modern nation? **1960** *Press* (Vancouver) Dec. 13 A new and dominant element, the mixed bloods, descended from French and Scottish fathers and Indian mothers. **1963** A. LUBBOCK *Austral. Roundabout* 125 The largest..group of Aborigines are those, both full and mixed bloods, who have been completely detribalized... There is a fourth group of mixed-blood Aborigines. **1973** *Guardian* 20 June 11/3 The left-overs of the Korean war—the mixed-blood children fathered and then deserted by GIs... Today only 5 per cent of children placed by Harry Holt's [adoption] agency are mixed-bloods. **1868** D. THOMSON *Handy Bk. Flower Garden* xii. 326 A mixed border of hardy and half-hardy plants.. would be effective anywhere. **1871** W. ROBINSON *Hardy Flowers* i. 5 The mixed border is capable of infinite variation as to plan as well as to variety of subjects. The most interesting variety is that composed of hardy herbaceous plants, bulbs, and alpine plants. **1899** G. JEKYLL *Wood & Garden* xvi. 200, I have a rather large..mixed border of hardy flowers. **1903** W. ROBINSON *Alpine Flowers* (ed. 3) I. 34 The mixed-border system rightly done enables us to cultivate..many of the more vigorous alpine plants as edgings. **1957** C. LLOYD *Mixed Border* i. 11 The mixed border stands mid-way between two extremes: the shrubbery on the one hand and the herbaceous border on the other. **1970** P. COATS *Flowers in Hist.* 39 'Mixed border' is the fashionable phrase, today, for a border of herbaceous plants with a backing of shrubs. **1974** A. SCOTT-JAMES *Sissinghurst* xiii. 137 At its [*sc.* a wall's] foot is a mixed border..planted with roses, shrubs, herbaceous flowers and irises. **1775** in *South Carolina Hist. & Geneal. Mag.* (1916) XVII. 99 Breakfasting with his mixed breed daughters. **1789** P. THICKNESSE *Year's Journey* (ed. 3) II. xlv. 107 If a male or female of this species [*sc.* Orang Outang] were to cohabit with an European of the contrary sex, they would..produce a mixed breed. **1838** H. COLMAN *1st Rep. Agric. Mass.* (Mass. Agric. Survey) 53, I have had some of the full-blood and some of the mixed breed. **1908** *Practitioner* Feb. 235 Leucocythaemia..may be qualified by such descriptive titles as mixed-celled leucocythaemia..or lymphocytic leucocythaemia. **1964** S. DUKE-ELDER *Parsons' Dis. Eye* (ed. 14) xxv. 369 The cells [of a melanoma of the choroid] are usually spindle-shaped; they may be cylindrical or palisade-like, arranged in columns or around blood vessels, or even endothelial in appearance; most tumours are mixed-celled. **1816** JANE AUSTEN *Emma* II. xvii. 329 Walk half-a-mile to another man's house, for the sake of being in mixed company till bed-time. **1820** HAZLITT in *London Mag.* Sept. 253 The conversation of authors..is better than any other. That of mixed company becomes utterly intolerable. **1901** G. B. SHAW *Three Plays for Puritans* 305 As far as my social experience goes (and I have kept very mixed company) there is no class in English society in which a good deal of Drinkwater pronunciation does not pass unchallenged save by the expert phonetician. **1973** A. BROINOWSKI *Take One Ambassador* ix. 131 Nance Donnelly..objected ..to jokes about sex in mixed company. It was alright, a bit of dirt between men... But not with ladies present. **1892** *Jrnl. Chem. Soc.* LXII. I. 265 (*heading*) Solubility of mixed crystals, especially of two isomorphous substances. **1916** R. H. RASTALL *Agric. Geol.* i. 4 Many of the most important rock-forming minerals are not pure chemical compounds; they are to be regarded rather as mixtures of various compounds possessing the property of isomorphism; in other words they are mixed crystals. **1952** T. L. TIPPELL tr. *Guinier's X-Ray Crystallogr. Technol.* viii. 207 There are substances which can never be classified; these are mixed crystals (solid solutions) whose unit cell size varies continuously with their composition. **1963** C. R. BERRY et al. in J. J. Gilman *Art & Sci. Growing Crystals* xii. 229 Mixed crystals of AgCl and AgBr..have been grown. **1938** *Encycl. Brit. Bk. of Year* 171/2 Under this mixed economy there is a large and developed system of trade unionism among producers, and another..system of co-operative societies among consumers. **1949** E. ESTORICK *Stafford Cripps* xix. 362 Cripps' role as the 'master planner of Britain's mixed economy' now had its international complement. **1973** *Guardian* 1 June 12/2 Neither Lonrho's shareholders nor Labour's fundamentalists provide an argument for abandoning the mixed economy. **1892** W. E. SWANTON *Notes on N.Z.* ii. 89 The farm upon which, as in England, both sheep and cattle are carried and also crops are grown..is what is generally called a 'mixed farm'. **1917** C. S. ORWIN *Determination of Farming Costs* ii. 13 The analytical method is probably more useful in connexion with the highly-developed mixed farms of this country. **1942** *E. African Ann. 1941–2* 128/2 Friends who own a mixed farm ask us over. The first signs of their successful management are the calves... This is also a most successful pyrethrum farm. **1973** *Country Life* 15 Mar. 713/2 In the southern region of England..the average price of mixed farms of 10–49 acres with vacant possession..rose to £915 an acre. **1872** *Trans. Illinois Dept. Agric.* IX. 66 The majority of farmers, fruit-growers or others, generally succeed best by what is called mixed farming. **1908** KIPLING *Lett. to Family* vi. 52 Providence..did not intend everlasting wheat in this section [in Canada]... Are you interested in mixed farming? **1913** W. K. HARRIS *Outback in Austral.* xxiv. 169 The district is an ideal one for 'mixed farming' (wheat and sheep). **1959** A. MCLINTOCK *Descr. Atlas N.Z.* p. xiv, Typical of the [Canterbury] plains is the patchwork of fields, indicative of a system of mixed farming, with supplementary fodder crops being necessary for stock. **1889** Mixed-flow [see *AXIAL *a.* 4]. **1958** *Engineering* 21 Mar. 376/1 The runner vane shape is of particular importance, especially in mixed-flow machines, where the vanes are three-dimensional surfaces. At present it is possible only in the case of axial-flow machines to define runner vane shapes in a systematic way. **1969** *Trans. Inst. Engin. & Shipbuilders Scotl.* CXII. 221 (*heading*) Mixed flow pumps and fans. **1913** W. PETT-RIDGE (*title*) Mixed grill. **1922** A. HUXLEY *Mortal Coils* 201 'Two mixed grills,' I said..to the waiter. **1959** *Good Food Guide* 205 The best dishes are such things as sauté river trout, mixed grill and cold turkey. **1973** 'D. HALLIDAY' *Dolly & Starry Bird* xvii. 258 The whip..was pointing to me. I was wondering, if I got marked like a mixed grill, whether Charles would still love me. **1973** *Times Lit. Suppl.* 21 Dec. 1555/3 The audience will be a mixed-grill of faculty, students, alumni, businessmen, and perhaps a few who have simply wandered in from the rain. **1888** H. SWEET *Hist. Eng. Sounds* 55 A very intimate mixture of two languages is always a prelude to the complete extinction of the weaker one, and this is why few..of these thoroughly mixed languages become permanently fixed. **1922** O. JESPERSEN *Lang.* xi. 224 These [pidgin] languages are not 'mixed languages' in the proper sense of that term. **1932** W. L. GRAFF *Lang.* II. x. 390 In several instances English used by people of an entirely different linguistic type results in a mixed or creolized language, the best known example being the so-called *pidgin English* of the Far East. **1972** R. ANTTILA *Introd. Historical & Compar. Ling.* xiii. 171 The various kinds of borrowing, that is, vocabulary, adstratum phonetics, and syntax, have led to the notion of a *mixed language*. **1963** *Economist* 16 Mar. 980/3 The mixed-manned, mixed-money nuclear force that Mr Merchant is..trying to create. **1964** *Ann. Reg. 1963* 28 On June 4 Admiral C. V. Ricketts..arrived in London to sell the mixed-manned fleet to a reluctant Mr Thorneycroft. **1963** *Times* 11 June 13/6, I fail to understand the fuss about mixed manning for the proposed Nato nuclear surface fleet. **1698–9** Mixed marriage [see WORLD *sb.* 4 b]. **1829** K. H. DIGBY *Broad Stone of Honour: Godefridus* xviii. 212 They are the last to admit those monstrous and impious plans..which divide the children of mixed marriages, by training some to receive as truth, what others are to protest against as error. **1961** C. MCCULLERS *Clock without Hands* ix. 178 The conditions of the Negro in the North are appalling—mixed marriages, nowhere to live and lay his head. **1962** *Sunday Times* 8 July 32/5 David Franklin's libretto reinterprets the Orpheus legend in terms of a mixed marriage between a white composer and

a coloured cabaret-star turned opera-singer. **1966** *New Statesman* 15 Apr. 548/3 In this mixed marriage, she brings her talent from the world of working-girl and art student, weds it to the dandy showmanship that goes with a certain kind of aristocracy. **1970** J. BROWN *Un-melting Pot* vii. 105 Even mixed marriages between Jamaicans and Barbadians. **1972** *Listener* 21 Dec. 854/3 They understand very well the power of the Catholic Church, and the processes by which it is maintained: clerical control of the educational system, the mixed-marriage laws and the preservation of high fertility rates. **1962** R. G. HAGGAR *Dict. Art Terms* 214/2 *Mixed media*, painting with water colors and paste, Indian ink, oil color, crayon, or some similar combination. **1970** BURTON & LANE *New Directions* iii. 66 The most extensive development of mixed media work are the Theatre Folk Ballads which Charles Parker has developed. **1972** G. F. BROMMER *Drawing* iv. 49 Paul Klee experimented freely with mixed media, as did his contemporaries in the German Bauhaus. **1972** *Guardian* 24 June 9/2 The newer, mixed-media work using a lot of sound, space, smell, and participation. **1973** *Sunday Times* 28 Oct. 37/1 (Advt.), Don't miss Contemporary Dance Theatre in Robert Cohan's *Stages*! A mixed media dance production with Bob Downes open music. **1909** *Engineering* 5 Feb. 198/1 A turbine of what is called a mixed-pressure type, having high-pressure stages, in which the live steam may..keep the turbine running during the periods of insufficient supply of exhaust steam. **1929** T. M. NAYLOR *Steam Turbines* i. 9 Mixed pressure turbines use both high-pressure and exhaust or low-pressure steam. Often the exhaust steam supplied is intermittent, and so to obtain constant power from the turbine, high-pressure steam is admitted and it is controlled by a governor to enable the turbine to work at constant power. **1971** B. SCHARF *Engin. & its Lang.* xv. 209 We may..distinguish between straight condensing turbines, pass-out turbines, back-pressure turbines, exhaust turbines and mixed-pressure turbines. **1935** E. NEUHAUS tr. *Doerner's Materials of Artist* v. 240 Painting with tempera into wet resin-oil color (mixed technique). This technique is better suited to a deliberate, stylistic type of painting. **1969** R. MAYER *Dict. Art Terms & Techniques* 246/1 *Mixed technique*, in painting, the technique of combining tempera colors with paints of an oleoresinous medium... Mixed technique first became known in the U.S. and Britain in 1934.

b. In Combs. used *attrib.*

1908 J. M. SULLIVAN *Criminal Slang* 16 *Mixed-ale oration*, a cheap political harangue containing bad English grammar. *Mixed-ale philosopher*, a drunken know it all. **1948** *Penguin Music Mag.* June 49 The best members of these choirs..incorporated in..a secular mixed-voice choir. **1956** *Railway Mag.* Nov. 729/2 Many Drummond mixed-traffic 4-4-0 tender engines were tried on this route. **1963** *Guardian* 28 Jan. 7/4 Everyone asks Mr Morris..how mixed-ability classes affect the number of grammar school entrances. **1963** *Daily Tel.* 25 June 1/1 President Kennedy and Dr. Adenauer agreed in Bonn to-night that the proposed mixed-crew NATO nuclear force was 'a good instrument for serving all members of the alliance in combining their defence efforts'. **1964** *Economist* 17 Oct. 228/2 A.. contribution to the mixed-fleet command. **1969** *Jane's Freight Containers 1968–69* 444 A racing Mini being loaded into a mixed-traffic, British United Airways, VC 10. **1971** *Guardian* 17 Sept. 1/5 A delegation of six mixed-race Rhodesians. **1974** *Times* 21 May 3/1 Some teachers found that their brighter pupils started slacking when they were put in mixed-ability classes.

c. mixed up, mixed-up, involved, embroiled, intermingled; (mentally) confused, unbalanced, neurotic; hence **mixed-up-ness,** confusion. Cf. MIX *v.* 6.

1862 QUEEN VICTORIA *Let.* 15 Jan. in R. Fulford *Dearest Mama* (1968) 41, I only want your advice—not to get you further mixed up. **1884** 'MARK TWAIN' *Huck. Finn* xlii. 432 Aunt Sally she was one of the mixed-upest looking persons I ever see. **1888** 'R. BOLDREWOOD' *Robbery under Arms* II. viii. 144 How were any police..to keep the run of a few men..among such a mixed-up mob? **1903** H. JAMES *Ambassadors* VI. xv. 206 How comes Chad so mixed up, anyway? **1927** H. T. LOWE-PORTER tr. *T. Mann's Magic Mountain* I. iv. 123 'A fine mixed-up state of affairs,' said Hans Castorp. **1937** *New Yorker* 23 Jan. 12/1 M. Dali, standing alone and thinking dark, mixed-up thoughts. **1939** D. PARKER *Here Lies* 61 I've been all sort of mixed up to-day... Everything so strange. **1945** G. ENDORE *Methinks the Lady* (1947) xi. 264 Such confusion, such mental mixed-up-ness, occurs at times in all of us. **1955**, etc. [see *CRAZY *a.* 4 f]. **1966** J. BINGHAM *Double Agent* x. 157 Poor damned old mixed-up queer. **1967** A. LASKI *Seven Other Years* xi. 154 It wasn't even as if she had made some kind of distinction between God and Christ and seen One as protecting her and the Other being Them: just a general mixed-upness. **1973** C. BONINGTON *Next Horizon* xi. 157 Layton was like a big, slightly mixed-up puppy, in need of love and care.

mixer. Add: **1. b.** One who mixes drinks; a bartender. orig. *U.S.*

1858 LONGFELLOW *Catawba Wine* in *Atlantic Monthly* Jan. 271 To the sewers and sinks With all such drinks, And after them tumble the mixer. **1919** T. K. HOLMES *Man from Tall Timber* viii. 93 He..drank several insidious concoctions of the hotel's most famous 'mixer'. **1934** M. ALLINGHAM *Death of Ghost* xxiii. 262 There was too much gin in the cocktails, he decided, and reflected that the fault was a common one among unprofessional mixers. **1939** C. ISHERWOOD *Goodbye to Berlin* 21 Bobby is a mixer at a west-end bar.

c. *Cinemat.* and *Broadcasting*. One who operates the mixer (sense *2 b) during sound recording and is responsible for balancing different signal sources and producing transitions from one source to another.

1929 F. GREEN *Film finds its Tongue* xiii. 186 A 'mixing booth' or monitor room had been built about 15 feet up... It had glass sides through which the 'mixer' could view everything that was going on. **1948** L. LEVY *Music for Movies* iv. 35 Just before writing this I had been sitting at the control desk in a mixer's cabinet during the rehearsals for a new picture. **1957** MANVELL & HUNTLEY *Technique Film Music* iv. 183 Regular consultation with the Music Mixer is necessary to ensure that the complete intention of the composer is..being registered on the sound-track.

2. *spec.* A domestic electrical appliance for mixing foods, ingredients for cookery, etc. See also **cake-mixer*, **electric mixer*.

1957 M. GAIR *Sapphires on Wednesday* xi. 134 It was a kitchen,..American style, complete with refrigerator, dishwasher, mixer, and garbage disposal unit. **1960** *Harper's Bazaar* Oct. 98/2 You must become accustomed to your oven, to your mixer, to your omelette pan. **1974** *Trafford Catal.* Spring–Summer 781/1 Three machines in one—a mixer, a liquidiser and a mincer.

b. *Cinemat.*, *Broadcasting*, etc. A device designed to receive two or more separate signals, from microphones or other sources, and combine them (usu. in variable proportions) in a single output.

1929 F. GREEN *Film finds its Tongue* xi. 171 When more than one microphone was used, each of them had an amplifying dial for monitoring. A panel containing more than one such dial was called a 'Mixer'. **1935** NILSON & HORNUNG *Pract. Radio Communication* viii. 356 The program fed into the mixer does not always come directly from a microphone but may be fed in from a phonograph pick-up. **1962** A. NISBETT *Technique Sound Studio* ix. 152 Where no mixer is available, it may be that linking music, etc., can be cut in by editing the tape; but I would regard a simple mixer—with two faders—as the very minimum for creative programme work. **1968** C. N. G. MATTHEWS *Tape Recording* xi. 109 To do full justice to any group of instrumentalists you need more than one microphone and also a mixer.

c. *Electronics.* A valve or circuit that produces an output signal containing frequencies equal to the sum and the difference of the frequencies of two input signals. Freq. *attrib.*, as *mixer tube*, *valve*, etc.

1936 *Proc. IRE* XXIV. 208 (*caption*) Typical mixer circuit using 6L7 tube. *Ibid.* 210 The conversion conductance of a mixer tube. **1938** F. E. TERMAN *Fund. Radio* ix. 249 The heterodyne detector.., commonly referred to as first detector, converter, or mixer, is required to develop a difference frequency..by combining the incoming radio wave with a local oscillation differing in frequency by the desired amount. **1952** E. ARMITAGE *Wireless Fund.* xviii. 324 The frequency-changing stage of a superhet consists of two parts—(1) a beat oscillator generating a signal of constant amplitude, and (2) a mixer valve into which are fed, on to different grids, both the aerial signal and the beat oscillator signal. **1968** *Radio Communication Handbk.* (ed. 4) 48/2 Any non-linear circuit element will act as a mixer, that is to say if frequencies f_1, f_2 are combined in the element, frequencies f_1, f_2, f_1+f_2 and f_1-f_2 will be present in the output.

3. A person in respect of his capacity for mixing with others; a sociable person; esp. *good mixer*, one who mixes readily with others. orig. *U.S.*

1896 ADE *Artie* xii. 105 I'm a good mixer and I've kind o' got next to the live ones. **1904** W. H. SMITH *Promoters* xx. 287 He was a most excellent 'mixer', told a story well [etc.]. **1917** WODEHOUSE *Man with Two Left Feet* 91 Some men are shy and some men are mixers. **1925** W. S. MAUGHAM in *Good Housekeeping* Sept. 15/2 He was a good mixer, and in three days he knew everyone on the ship. **1930** R. MACAULAY *Staying with Relations* xix. 281 Good plumbers, but poor mixers; that's what we are. **1937** [see *CEREBROTONIC *a.* and *sb.*]. **1955** *Times* 18 Aug. 4/6 A man who is a ready mixer and with a natural flair for salesmanship may make £1,000 a year. **1967** N. FREELING *Strike Out* 148 You needed to..get on well with all the murderers? A 'good mixer'—a 'good team man'. **1975** *Listener* 24 July 126/3 He is neither so astute a tactician nor so gifted a mixer with his own men.

4. A social gathering for making people acquainted with each other. *U.S. colloq.*

1916 *Dialect Notes* IV. 277 A very successful mixer was given on Charter day. **1948** *Downers Grove* (Illinois) *Reporter* 21 Oct. 1/8 The Trojan Fathers Fall Mixer will take place Tuesday, Oct. 26 at the high school auditorium.

5. A drink with which an alcoholic liquor is diluted, as soda water, etc.; an alcoholic drink used in cocktails. Also *attrib.* orig. *U.S.*

1938 D. BAKER *Young Man with Horn* III. i. 147 They worked all possible combinations of bootleg gin and mixers, orange juice, lemon juice, grapefruit juice..and.. root beer. **1948** *Sun* (Baltimore) 1 Jan. 15/1 In the case of the Tom Collins, it's largely a matter of which mixer should be used, plain soda or ginger ale. **1961** *Encounter* XVI. v. 81 Using Coca-cola as what the Americans call 'a mixer'. **1961** *Guardian* 21 Nov. 13/1 Vodka has become, like gin, an international 'mixer',..used to make martinis or to drink with tonic water. **1965** *Economist* 11 Dec. 1251/1 It [*sc.* Beechams] has just introduced a number of 'mixer' drinks to compete with Schweppes. **1968** *Daily Tel.* (Colour Suppl.) 13 Dec. 41/4 Port was traditionally enjoyed here as a mixer—mulled, say, in the way Dr Johnson liked it, with spices, orange, and cloves. **1975** T. TEAL tr. *Sjöwall & Wahlöö's Cop Killer* vii. 69 'Do you drink?' 'Yes... But..not lukewarm vodka with no mixer.'

6. A trouble-maker. *slang.*

1938 PARTRIDGE *Dict. Slang* (ed. 2) 1014/1 He's a reg'lar mixer! **1964** J. BURKE *Hard Day's Night* i. 21 He's a king mixer. *Ibid.* iv. 84 'The old mixer,' growled Paul. 'Come on—we'll have to put him right.' *Ibid.* v. 116 'Look,' said Paul slowly and deliberately, 'he's a mixer and a trouble-maker!' **1966** A. E. LINDOP *I start Counting* xviii. 226, I knew what a mixer she was, and I knew she was not capable of keeping a secret.

7. *attrib.*, as **mixer tap,** a tap through which both hot and cold water can be drawn in various proportions; **mixer valve,** a valve by which the proportions of the fluids or gases in a mixture are regulated.

1936 *Archit. Rev.* LXXX. 325/2 Mixer taps or hand showers are worth the extra cost of £2 or £3 they entail. **1972** *House & Garden* Dec.–Jan. 85/3 Mixer taps by Trufords..gold-plated. **1904** *Electr. World & Engin.* 2 Jan. 22/1 This magnet controls a compressed air valve, and this compressed air valve in turn controls a large mixer valve.

mixing, *vbl. sb.* Add: **c. mixing valve,** a valve in which separate supplies of hot and cold water are mixed together; = **mixer tap*.

1902 R. STURGIS *Dict. Archit.* III. 977 General Morin called attention to the necessity of the 'mixing valve'. **1951** *Good Housek. Home Encycl.* 281/2 *Mixing valves*: These are used mainly for showers and baths. **1973** J. WAINWRIGHT *Pride of Pigs* 59 A corner shower with a mixing valve that really worked.

Mixmaster (mi·ksmāstəɹ). The proprietary name of a type of electrical food-mixer. Freq. *transf.* and *fig.* (see quots.).

1931 *Good Housekeeping* (N.Y.) Oct. 223/3 (Advt.), See what you get *now* in Mixmaster. A food mixer that beats everything—a juice extractor—an automatic salad oil dripper. **1935** *Official Gaz.* (U.S. Patent Office) 11 June 278/2 Chicago Flexible Shaft Co. *Mixmaster* for Food Mixers, Fruit Juice Extractor, [etc.] filed Nov. 21, 1934. **1946** *Britannica Bk. of Year* (U.S.) 832/2 *Mixmaster*, army bomber propelled by two rear counter-revolving propellers. Made by Douglas Aircraft Co., Inc. **1951** R. MALKIN *Boxcars in Sky* xiii. 156 In the business, they refer jocularly to the helicopter by a number of descriptive names: 'eggbeater', 'windmill', and 'mixmaster', to mention a few. **1963** *Economist* 11 May 537/2 A 'mixmaster' force of ships with American missiles and international crews. **1967** *Punch* 22 Mar. 410/1 On all those important matters our opinions are divided, and so therefore are our teachers; the whole situation is confused in the extreme. The mix-master whirrs, the beaters rotate, but unfortunately the recipe is missing. **1971** M. TAK *Truck Talk* 107 *Mixmaster-special*, any transmission with two shift levers; so named because, in running through the gears, the trucker's hands are constantly in motion.

mixo-. Add: **mixoha·line** *a.* [Gr. ἅλιν-ος of salt], brackish; **mixonephridium, mixonephrium,** in certain annelids, an organ in which the nephridium is combined with the cœlomoduct; **mixosco·pia, mixo·scopy** *Psychol.* [ad. G. *mixoskopie* (A. Moll *Die Conträre Sexualempfindung* (1891) v. 136), f. Gr. σκοπεῖν to look at], (see quot. 1939); hence **mixosco·pic** *a.*

1959 *Archivio di Oceanogr. e Limnol.* XI. Suppl. 243 At the final session of the Venice Symposium a revised classification of marine waters according to salinity was adopted and recommended for universal application... The term 'brackish', as a classificatory term, was avoided because of its ambiguous meaning and the term 'mixohaline' was proposed to indicate diluted sea water. **1969** G. VEVERS tr. *Friedrich's Marine Biol.* vii. 419 Waters with intermediate salinities are known as brackish or mixohaline. **1946** E. S. GOODRICH in *Q. Jrnl. Microsc. Sci.* LXXXVI. 119 When, as in the majority of Polychaeta, the coelomostome is so completely fused to the inner end of the nephridium as to form an apparently simple large-funnelled organ..it may be called a mixonephridium... This new name is proposed to draw attention to the difference between these two kinds of organ (the metanephridium, and the compound nephromixial organ). **1963** Mixonephridium [see *CŒLOMO-]. **1972** M. S. GARDINER *Biol. Invertebr.* xiii. 505/2 In a mixonephridium, the nephrostome has presumably become occluded and lost. **1958** G. A. KERKUT *Borradaile & Potts's Invertebrata* (ed. 3) viii. 287 Mixonephrium. The coelomoduct is so closely associated with the nephridium that they form an apparently simple funnelled organ. **1939** G. R. SCOTT *Encycl. Sex* 200/2 *Mixoscopia* or *mixoscopy*, the securing of sexual orgasm or excitation as a result of seeing human beings or animals engaged in copulation. **1940** HINSIE & SHATZKY *Psychiatric Dict.* 350/2 *Mixoscopia*, a form of sexual perversion, deriving pleasure from watching the act of coition between the desired one and another person. **1903** *Alienist & Neurologist* May 167 Mixoscopic. **1905** H. ELLIS *Stud. Psychol. Sex* IV. 188 Founded on the sense of vision also we find a phenomenon, bordering on the abnormal, which is by Moll named mixoscopy. This means the sexual pleasure derived from the spectacle of other persons engaged in natural or perverse sexual actions.

mixolimnion (miksoli·mniŏn). Pl. **-limnia.** [f. MIXO- + *-limnion*, after *EPILIMNION, *HYPOLIMNION.] The upper, freely circulating layer of a meromictic lake. Cf. *MONIMOLIMNION.

1937 G. E. HUTCHINSON in *Trans. Connecticut Acad. Arts & Sci.* XXXIII. 74 A meromictic lake may be regarded as consisting of an upper region capable of complete mixture at the time of the overturn, and here called the mixolimnion, and a lower region not undergoing mixture at the overturn and termed by Findenegg the monimolimnion. **1968** *Limnol. & Oceanogr.* XIII. 273 Both lakes are permanently stratified. The chemocline, separating the mixolimnion..from the monimolimnion.., lies at 18 m in Green Lake and at 24 m in Round Lake.

mixologist (miksǫ·lŏdʒist). *U.S. slang.* [f. MIX *sb.*² or *v.* + -OLOGIST.] One who is skilled in the mixing of drinks; = *MIXER 1 b. Hence **mixo·logy.**

1856 *Knickerbocker* XLVII. 615 Who ever heard of a man's..calling the barkeeper a mixologist of tipicular fixins? **1870** W. F. RAE *Westward by Rail* xv. 201 The most delicate fancy drinks are compounded by skilful mixologists in a style that captivates the public. **1908** W. G. DAVENPORT *Butte & Montana* 45 Brandy and cigarettes were furnished by an expert mixologist from the Thornton Hotel. **1952** *John o' London's* 25 Jan. 82/4 There [*sc.* at Miami], it seems, a mixer of drinks at a bar is referred to as a mixologist. The art of mixing a cocktail is consequently known as mixology.

mixoploid (mi·ksoploid), *a.* (and *sb.*). *Biol.* [f. MIXO- + *-PLOID.] Containing cells which are of differing ploidy or, more widely, have differing numbers of chromosomes. Also as *sb.*, a mixoploid individual.

1931 B. NĚMEC in *Rep. Proc. 5th Internat. Bot. Congr. 1930* 233 Many plants contain under normal conditions both diploid and polyploid cells. It is easy to get experimentally plants containing a varying number of polyploid cells. The author designates such plants mixoploid. *Ibid.*, Mixoploid organs tend to eliminate the polyploid cells when they form a minority, especially when they are highly polyploid. **1939** *Hereditas* XXV. 111 As a result of the agar treatment entire plants occasionally become mixoploids. **1944** *Jrnl. Heredity* XXXV. 359/2 Mixoploid tissue resulted from colchicine treatment. *Ibid.* 361/1 Individual anthers were mixoploid. **1951** *Nature* 2 June 891/1 In many species the diploid tissue grows at the expense of the tetraploid, and such a mixoploid reverts rapidly to the diploid condition. **1967** *Current Sci.* XXXVI. 307 (*heading*) Preferential elimination of diploid cells in a colchicine-induced mixoploid tissue. **1971** J. HAMERTON *Human Cytogenetics* I. vii. 224 Autosomal mosaics are less common, and about 1% of Down's syndrome patients are estimated to be mixoploids.

So **mi·xoploidy,** the property or state of being mixoploid.

1931 B. NĚMEC in *Rep. Proc. 5th Internat. Bot. Congr. 1930* 233 (*heading*) Mixoploidy and the cellular theory. **1958** C. P. SWANSON *Cytol. & Cytogenetics* ix. 300 This gradual change, if caught before the completion of the entire process of doubling, will give both tissues of mixed ploidy and cells with an uneven degree of polyteny in the individual chromosomes. Mickey (1946, 1947) has also shown this type of mixoploidy to be true for the spermatogonia in the testes of the grasshopper, Romalea. **1963** LEVAN & MÜNTZING in *Portugaliae Acta Biol.* A. VII. 10 The present writers would suggest that the term mixoploidy should include all cases in which cell populations contain more than one chromosome number irrespective of whether the numbers are euploid or aneuploid. Mixoploidy, thus, covers all cases of mosaicism and chimeric constitution in which the heterogeneity between different elements involves differences in chromosome number. Mixoploidy may originate by all kinds of mitotic irregularities, by cellular and nuclear fusions, or even by amitotic processes, provided they give rise to viable products. **1971** J. HAMERTON *Human Cytogenetics* I. vii. 225 *In vitro* studies on mixoploidy should include the use of single cell clones from a mixoploid culture so that the only genetic difference is confined to a single chromosome.

mixotrophic (miksotrōu·fik), *a. Biol.* [a. G. *mixotroph* (W. Pfeffer *Pflanzenphysiologie* (ed. 2, 1897) I. vii. 349), f. MIXO- + Gr. τροφικός nursing.] Living by a mixture of autotrophic and heterotrophic nutrition; pertaining to nutrition of this kind.

1900 A. J. EWART tr. *Pfeffer's Physiol. Plants* I. vii. 364 By others [*sc.* plants] a portion only of the organic food is drawn from the external world, the rest being supplied by the imperfectly developed chlorophyll-apparatus; such may be termed 'mixotrophic' plants. *Ibid.*, All stages of transition between pure autotrophism and heterotrophism are exhibited among obligate or facultative mixotrophic plants. **1940** L. H. HYMAN *Invertebrates* I. iii. 58 Nutrition [of Protozoa] is holophytic, saprozoic, holozoic, or mixotrophic. **1965** BELL & COOMBE tr. *Strasburger's Textbk. Bot.* 5 Such mixotrophic organisms [*sc.* unicellular Protista] (which obtain their energy partly directly as autotrophs, and partly indirectly as heterotrophs) are regarded as the starting-point of two great developmental series which have led on the one side to the 'typical' plants and on the other to the 'typical' animals.

Mixtec (mī·ʃtek). Also **Mixteca, Mixteco.** [Sp., f. native name.] A people of central America; a member of this people; also, their language. Also *attrib.* So **Mixte·can.**

1850 R. G. LATHAM *Nat. Hist. Varieties Man* 409 South of Mexico we have several languages of a small..area... *Mixteca*, spoken in Oaxaca [etc.]. **1928** D. H. LAWRENCE *Woman who rode Away* 267 His mother was a Mixtec Indian woman. **1934** A. HUXLEY *Beyond Mexique Bay* 288 In the waste land..stood a church..and, beside it, the Mixtec ruins which it had been built to exorcise and sanctify. **1948** D. DIRINGER *Alphabet* I. vii. 123 The Zapotecs and Mixtecs..in ancient times probably played the part of cultural intermediaries between the Maya Old Empire of the East, and the Toltec 'Empire' of the West. **1952** E. FISCHER-JØRGENSEN in E. P. Hamp et al. *Readings in Linguistics II* (1966) 307 An exception of a different kind is formed by languages of the Mixteco-type. In Mixteco the minimal utterance is cvcv or cvv, containing two syllabic bases. **1965** *Canad. Jrnl. Ling.* Spring 152 We made the first attempt with the sixteenth century Mixtec dictionary. *Ibid.* 153 We began on the Mixtec, first listing all the affix elements we knew. **1968** CHOMSKY & HALLE *Sound Pattern Eng.* 377 In languages such as Mixtecan or Chinese, each vowel in the word may have its own distinctive prosodic features. **1972** *Language* XLVIII. 847 Otomi nasal vowels, Mixteco,..and Zoque. **1975** *Times* 24 May 10/1, I went to Mexico with only the most cursory knowledge of..the Toltecs, the Olmecs, the Mixtecs, the Zapotecs. *Ibid.* 10/5 Friezes made up of exquisitely curved geometric patterns..were Mixtec work.

mixture. Add: **3. a.** *the mixture as before*: medicine to be taken in a similar dose as on a previous occasion (as a set phrase often found on medicine bottles); freq. *transf.* and *fig.*, something that has already been encountered, used, etc.

1834 DICKENS *Let.* 3 Sept. (1965) I. 40, I have taken a wine-glass full of 'the mixture as before' twice a day, varying the amusement with an occasional pill. **1920** WODEHOUSE *Jill the Reckless* (1921) viii. 120 There he sat, surrounded by happy, laughing young men, each grasping a glass of the good old mixture-as-before. **1959** *Listener* 26 Feb. 363/1 What special or new responsibilities do these developments place upon industry? Some will say that it is the mixture as before, but in larger and more frequent doses. **1973** R. HILL *Ruling Passion* II. iv. 113 After ten minutes, all Pascoe had was the mixture as before.

d. In an internal-combustion engine, the mixture of vaporized or gaseous fuel with air that enters the combustion chamber to form the explosive charge.

[**1848** *Chambers's Edin. Jrnl.* 6 May 303/1 What is the moving power? The answer will be heard with surprise: the successive explosions of a mixture of gas and air in the boxes at the root of the wings, by which means they will be made to flap about twelve times a minute! The balloon ..is a mere reservoir for gas. The explosion is to be effected in the four boxes by the electric spark.] **1894** B. DONKIN *Text-Bk. Gas, Oil & Air Engines* I. 6 Sometimes an auxiliary pump is used for compressing the mixture. **1914** W. D. NEWTON *War* iii. 19 Brun shut off mixture, and, slowing down, he swung from the motor-cycle. **1943** A. P. FRAAS *Aircraft Power Plants* vi. 107 The power loss resulting from mixtures 10 or 20 per cent leaner than that for best power is not large as compared with the reduction in fuel consumption. **1968** R. H. BACON *Car* iii. 23 Modern carburettors have..cold starting devices to give very rich mixtures when starting the engine from cold. **1973** J. LEASOR *Host of Extras* i. 21 The engine..was under no stress at all, with tiny valves that let it take such delicate breaths of mixture, it never grew fussed like engines of lesser breeds.

e. Petrol to which has been added a small proportion of oil, used as a combined fuel and lubricant in some two-stroke engines; = *PETROIL.

[**1927**: see *PETROIL.] **1952** *Cyclemotor Manual* ('Motor Cycling') iii. 20 Use of oil of too thick a grade may result in the mixture being too heavy to pass through the carburetter jet in sufficient quantity. **1960** J. QUEENBOROUGH *Garage & Service Station Handbk.* xv. 263 Avery-Hardoll Ltd., market the *Petroiler* which can deliver a choice of two mixtures; there are tanks for petrol (13½ gallons) and two grades of oil (1½ gallons each). **1967** P. E. IRVING *Two-Stroke Power Units* vii. 131 As the proportion of oil is for convenience fixed at one figure, usually..six per cent, it may on occasion be necessary to add a little more oil or dilute the mixture with more petrol to obtain the proportion recommended for any particular engine. **1972** J. STEVENS *Scooter* iv. 99 The first scooters had engines calling for a 6% oil content in the mixture.

mix-up: see MIX *sb.*² c in Dict. and Suppl.

mixy (mi·ksi), *a.* [f. MIX *v.* + -Y¹.] **a.** Adapted for mixing. **b.** *colloq.* Sociable.

1929 R. BRIDGES *Testament of Beauty* ii. 41 Nor that the unwholesomeness of mixy pollen was by the flowers contrived for their own benefit. **1942** BERREY & VAN DEN BARK *Amer. Thes. Slang* § 363/14 Sociable,..mixy. **1968** P. G. HOLLOWELL *Lorry Driver* vii. 181 Lorry drivers aren't so mixy nowadays since there are so many of them.

miz (miz). [Shortening of MISTRESS *sb.*] **1.** Prefixed as a title to the name of a married or unmarried woman, = 'Mrs.' or 'Miss'. *southern U.S.*

1907 'O. HENRY' *Heart of West* i. 5, I ain't reflectin' none on Miz Yeager—she's the finest little lady between the Rio Grande and next Christmas. **1913** H. KEPHART *Our Southern Highlanders* xiii. 290 A married woman is not addressed as Missis by the mountaineers, but as Mistress when they speak formally, and as Mis' or Miz' for a contraction. **1937** M. MITCHELL *Gone with Wind* xlv. 800 'Don't you question Miz Wilkes' word,' said Archie. **1952** V. WILKINS *King Reluctant* I. iii. 46, I kin' 'spicion wat Miz Fell gwine ter say about dis-yere chile, Miz Virgie! **1975** E. BERCKMAN *Indecent Exposure* v. 53 Miz Tor is such a wonderful lady.

2. Repr. pronunc. of *Ms².

1972 *Village Voice* (N.Y.) 1 June 28/3 Cavett addressed her as Mrs. Morgan and asked her if she would rather be called a miz and she said she didn't care. **1974** J. PHILIPS *Power Killers* (1975) III. iv. 183 Nice to have you back, ma'am... Or should I say 'Miz'? **1975** P. G. WINSLOW *Death of Angel* i. 48 'Smoky Angel, I believe he is called, Mrs Jones.' 'Miz,' she corrected him.

miz, mizz, colloq. abbrevs. MISERY, MISERABLE *a.* Cf. *MIS.

1918 *Chambers's Jrnl.* Mar. 156/2 He won't get any peace now we've seen him. We'll make his life a mizz. **1954** J. B. PRIESTLEY *Magicians* ii. 27 Don't look so miz. Are you hating it? **1968** *Times* 24 Feb. 21/3 We feel a teeny bit miz—reality does depress.

mizen, mizzen. 3. *mizen-staysail* (examples).

1757 in J. S. MCLENNAN *Louisbourg* (1918) 209 Fore stay sail, Main and Mizen stay sail all blown away. **1964** R. E. LLOYD in *Roving Commissions 1963* 228 On again next day with, to our joy, a mizzen staysail breeze.

Mizo (mī·zo), *sb.* and *a.* Pl. **Mizo, Mizos.** [Native name, lit. 'highlander', f. *mi* person + *zo* hill.] **A.** *sb.* A native or inhabitant of the territory of Mizoram in north-eastern India, formerly the Lushai Hills District or the Mizo Hills District. **B.** *adj.* Of or pertaining to these people. Cf. *LUSHAI *a.* and *sb.*

1832 *Asiatic Res.* XVII. 375 *Krisong*..is esteemed as being the more martial and decided character, and his influence..with the *Mizhus*..is consequently greater. *Ibid.*, *Ruding*, a Chief of the *Mizhū* tribe. *Ibid.* 400 The *Reiga* tribe are on the western side of the great river, beyond the *Pasi* and *Mizong* tribes. **1912** J. SHAKESPEAR *Lushei Kuki Clans* v. 110 Some Mizo (natives of these Hills) who were passing through the village also heard the song of those who knew magic. **1954** *Current Indian Statutes* II. 107 The district is largely inhabited by tribes who are collectively known as 'Mizos'—'Lusei' being one of the tribes. **1968** CHAPMAN & CLARK *Mizo Miracle* 85 The pioneer missionaries Lorrain and Savidge toured the district, learning the Mizo language and reducing it to writing. **1969** S. BARKATAKI *Tribes of Assam* 82 Mizo is a generic term which includes several sub-tribes..Lusei, Ralte, Hmar, Pawi (Poi). *Ibid.* 83 The *lingua franca* of the Mizos which has become popularly known as the Mizo language or *Dulhian*, is the dialect of the Lusei clan. *Ibid.* 128 One can hardly imagine the difficulty with which a Mizo or Zemi Naga has to eke a livelihood out of his little patch of land. **1975** *Times* 17 May 6/1 'Hostile' Mizos have been up in arms for more than a decade. *Ibid.*, The Mizo National Front has joined hands with the underground movement of the Nagas.

Mizpah (mi·zpā). [ad. Heb. *Miṣpah* place-name in ancient Palestine (Gen. xxxi. 49).] An expression or token of association ('The Lord watch between me and thee'), esp. used *attrib.* to designate an ornament with 'Mizpah' inscribed upon it, as given by a lover.

[**1887** *Mizpah* 21 Feb. 7 A deeper, more mysteriously beautiful meaning than appears at first sight, is comprehended in the word 'Mizpah'—light within light, circle within circle, revelation within revelation. *Ibid.* 19 It is our duty to examine from the Tower of Mizpah this wonderful Advent of Divinity.] **1898** *T. Eaton & Co. Catal.* Spring & Summer 133/5, 10k Mizpah ring. **1907** B. M. CROKER *Company's Servant* xv. 152 Was she wearing his presents? the sapphire brooch..the 'Mizpah' bangle. **1909** P. WEBLING *Story of Virginia Perfect* i. 11 The thick wedding-ring, with the heavy 'Mizpah' ring squeezed on her finger over it. **1940** H. G. WELLS *Babes in Darkling Wood* I. iv. 118 Mizpah, as they say inside the engagement rings. *Ibid.*, 'Anyhow, Gemini, Mizpah!'..'Mizpah,' he responded. Like most English people they thought that was a pledge between two young lovers. **1961** M. K. ASHBY *Joseph Ashby* xv. 202 The pedlars..used to bring the silver 'Mizpah' brooches. **1970** I. ORIGO *Images & Shadows* ii. 69 He gave me the gold Victorian locket—engraved with the word Mizpah, 'God watch between us two'—which he had given to his fiancée.

Mizrach (mi·zrāχ). [ad. mod.Heb. *mizrāḥ*, f. Heb. *mizrāḥ* east, f. *zāraḥ* to rise.] The east, the direction of Jerusalem; a sacred picture facing east; the Judaic practice of turning towards Jerusalem in prayer. Hence **Mizra·chi,** a religiously traditionalist Zionist organization, amalgamated with Hapoel Hamizrachi in 1957.

1892 I. ZANGWILL *Childr. Ghetto* I. 46 A crudely-coloured *Mizrach* on the east wall, to indicate the direction towards which the Jew should pray. **1911** *Zionist* Oct. 107/2 The Misrachi profess to see in such work a danger to orthodox Judaism of which they claim to be the bulwark in Zionism. *Ibid.* Nov. 123/2 The Misrachi Federation has issued a manifesto..appealing for loyalty and labour for the Zionist movement. **1922** JOYCE *Ulysses* 689 His gaze turned in the direction of Mizrach, the east. **1934** M. M. KAPLAN *Judaism as Civilization* III. xiv. 174 The *Mizrahi* organization which is orthodox in its constituency, is devoid of any systematic philosophy of Judaism. **1956** *Ann. Reg. 1955* 285 The Mizrachi (Religious) Zionists of the world. **1973** *Jewish Chron.* 19 Jan. 2/1 (*heading*) Mizrachi in stormy debates. *Ibid.*, A stormy debate as a resolution about diaspora leaders..marked the closing session here.. of the 22nd Mizrachi–Hapoel Hamizrachi conference.

mizzle, *sb.*¹ (Later examples.)

a **1963** S. PLATH *Crossing Water* (1971) 34 This mizzle fits me like a sad jacket. **1975** P. G. WINSLOW *Death of Angel* x. 204 The neighbours would not be likely to air themselves in their gardens in a steady mizzle.

mizzle, *sb.*² *rare.* [f. MIZZLE *v.*²] Phr. *to do a mizzle*: to depart suddenly.

1923 J. MANCHON *Le Slang* 195 *To do a mizzle*, se trotter, se barrer.

mizzle, *v.*² Add: (Further examples.) Naut. phr. *to mizzle one's dick*: to miss one's passage.

1842 BARHAM *Lay St. Cuthbert* in *Ingol. Leg.* 2nd Ser. 229 Come, mizzle!—be off with you!—go! **1849** DICKENS *Dav. Copp.* (1850) xxii. 236 Now you may mizzle, Jemmy (as we say at Court). **1925** J. MASEFIELD *Sard Harker* III. 134 He had broken his word..and missed his passage; 'mizzled his dick', as Pompey Hopkins called it. **1970** 'R. LLEWELLYN' *But we didn't get Fox* iii. 47 There was a girl with him... He fell behind the table, and she mizzled.

mizzle (mi·z'l), *v.*⁵ [Perh. f. MOAN *v.* + GRIZZLE *v.*²; cf. MISERY.] *intr.* To complain, whimper; used also of fretful children.

a **1935** T. E. LAWRENCE *Mint* (1950) I. xiv. 49 The question took a self-pitiful turn, and I mizzled gently in the white-walled silence. **1945** BAKER *Austral. Lang.* vi. 134 To complain: *to mizzle* and *to whinge* (whence come the nouns *mizzler* and *whinger*, and the verbal nouns *mizzling* and *whingeing*).

mi·zzler². *slang.* [f. prec. + -ER¹.] One who complains.

1945 [see prec.].

Mlimo (m,lī·mo). Also **Umlimo** (umlī·mo). [Bantu; see quots.] The name given by the Matabele and other East African tribes to their god. Also, one of the prophets or priests of this cult.

[**1833** S. KAY *Trav. Caffraria* I. ix. 236 A few indeed there were who seemed to have some confused notion of invisible powers, whom they designated *Mooreemo* and *Booreemo*, and of whom they were taught by their sorcerers to stand in constant dread.] **1861** E. CASALIS *Basutos* II. xiii. 248 Every being, to whom the natives render adoration, is called *Molimo*, the signification of which shows that it is by no means of heathen origin. It is evidently composed of the prefix *mo*, which belongs to almost all those words representing intelligent beings, and of the root *holimo—above, in the sky*. *Moholimo*, or the abbreviation Molimo, therefore, signifies, *He who is in the sky*. **1896** F. C. SELOUS *Sunshine & Storm Rhodesia* xxvi. 236 The Umlimos or prophets..exist among all the tribes in Mashunoland. **1928** P. NEILSON *Matabele at Home* ii. 38 Before the occupation by the white people of this country, it had become customary for many of the Matabele to go to the Umlimo's cave..at regular intervals, and there to make obeisance by presenting the Umlimo with cattle and beer. **1955** E. C. TABLER *Far Interior* iv. 112 These magicians were diviners and hereditary priests of the cult of *Mwari* (Sintabele, *Mlimo*), a deity that was believed to have made the world and therefore controlled the rain... Lobengula, who seems to have had no faith in the *Mlimo*, nevertheless refrained from wiping out this nest of tricky parasites. **1965** L. H. GANN *Hist. S. Rhodesia* iv. 128 The Matabele, having settled north of the Limpopo, transferred their allegiance from *Nkulunkulu*, the Zulu high god, to *Mlimo*, the Karanga deity. *Ibid.* 129 *Mlimo* originally appears to have been regarded as some remote *deus absconditus*, the Ancient of Days, who took no notice of tribal affairs... But when the Matabele fell on evil days the god gained increasing political importance, and his cult..provided an effective machinery of revolt.

Mlle. [See MADEMOISELLE 1.] (Examples.)

1792 F. BURNEY *Jrnl.* May (1972) I. 155 Mlle Planta had told her I was going to be married!..why what I had said..a week a go, should not have reached this Mlle nor her informant, is marvellous. **1820** M. WILMOT *Let.* 3 Mar. (1935) 53 Poor nurse had her share—Mlle hers—and Nanny hers. **1896** C. K. SHORTER *Charlotte Brontë & her Circle* iv. 109 Madame Héger..has been gibbeted for all time in the characters of Mlle. Zoraïde Reuter and Madame Beck. **1975** B. GARFIELD *Hopscotch* i. 13 The German folded and then it was *Mlle* Stein's turn.

mm, m'm ('m). Also mm-m, (*rare*) mn. [Imit.] Used to express a hesitating or inarticulate utterance of interrogation, assent, reflection, or satisfaction on the part of a speaker. Cf. UM *int.*

1922 JOYCE *Ulysses* 56 A sleepy soft grunt answered:—Mn. **1924** *Dialect Notes* V. 273 *M*-: —, — -m. **1966** G. N. LEECH *Eng. in Advertising* xxii. 197 By tomorrow they'll [*sc.* sandwiches] lose not a single 'M'm' of flavoursome moisture! **1967** *Boston Herald* 8 May 24/4 (*caption*) Mm... Another dud, but not bad for a dull movie. **1967** *Listener* 21 Sept. 367/2 Restraint is unnatural, mm? **1968** [see *CHASE *v.*¹ 7 c]. **1974** D. FRANCIS *Knock Down* ii. 21 'You'd thought of that, had you?'..'Mm,' I said.

Mme. [See MADAME 1.] (Examples.)

1806 *Young Ladies' Assistant in Writing French Lett.* i. 6 In hand-written letters, *Monsieur*, *Madame*, *Mademoiselle*, before family names, are abbreviated thus: *Mr. Mde. Mlle.*; in printed ones, we find *M.* for *Monsieur*, *Mme.* for *Madame*, *Melle.* for Mademoiselle. **1860** D. G. ROSSETTI *Let.* 9 June (1965) I. 368 My best address would be: Chez Mme Houston (as above): the said Mme is English and very obliging. **1896** C. K. SHORTER *Charlotte Brontë & her Circle* iv. 102 Mme. Héger was an accomplished spy. **1975** *Listener* 7 Aug. 190/1 Mme Laure Bernardini and her companion, Mme Thérèse, are very old.

mneme (nī·mi). *Psychol.* and *Physiol.* [a. G. *mneme* (R. Semon *Die Mneme als erhaltende Prinzip im Wechsel des organischen Geschehens* (1904)), f. Gr. μνήμη memory.] The capacity which a living substance or organism possesses for retaining after-effects of experience or stimulation undergone by itself or its progenitors.

1913 M. HARTOG *Probl. Life & Reproduction* 275 The mnemic possibilities of an organism may be termed, collectively, its 'mneme'. **1921** L. SIMON tr. *Semon's Mneme* 12 The capacity for such after-effect of stimulation constitutes what I have called the Mneme. **1928** J. T. MACCURDY *Common Princ. Psychol. & Physiol.* ii. 15 The mneme and memory are thus reduced, fundamentally, to physico-chemical phenomena. **1966** E. ENG tr. *Strauss's Phenomenological Psychol.* I. iii. 61 *Mneme* must not be limited to organic substances.

mnemic (nī·mik), *a.* [f. as prec. + -IC.] Pertaining to, of the nature of, or involving mneme. Hence **mne·mically** *adv.*, **mne·micness,** the state or quality of being mnemic.

1908 *Daily Chron.* 3 Sept. 5/7 Alleging the existence of a mnemic factor in the life of plants. **1913** [see prec.]. **1921** B. RUSSELL *Analysis of Mind* iv. 78 Following a suggestion derived from Semon..we will give the name of 'mnemic phenomena' to those responses. **1921** L. SIMON tr. *Semon's Mneme* 11 Instead of speaking of a factor of memory, a factor of habit, or a factor of heredity, and attempting to identify one with another, I have preferred to consider these as manifestations of a common principle, which I shall call the mnemic principle. This mnemic property may be regarded from a purely physiological point of view, inasmuch as it is traced back to the effect of stimuli applied to the irritable organic substance. **1925** C. D. BROAD *Mind & its Place* viii. 377 Experiences which are owned in senses (2) or (3) may be said to be 'mnemically owned'. **1941** *Mind* L. 417 The only perceptible difference between conscious and non-conscious behaviour is *mnemicness*. **1943** A. M. FARRER *Finite & Infinite* xvii. 201 That already depends on the mnemic content, which disposes us to attention in one direction rather than another, because certain elements of memory are pressing closest to the gate. **1963** O. L. ZANGWILL tr. *Luria's Restoration of Function after Brain Injury* vi. 213 The patient retained the content of this particular thought, and..his difficulties were dynamic in nature rather than mnemic. **1968** S. BOGOCH *Biochem. Memory* vi. 194 That is, the glycoproteins of the nervous system represent the mnemic substances in which experiential information s encoded.

mnemon (nī·mǫn). *Psychol.* [f. Gr. μνήμ-η memory + *-ON¹.] A unit of memory (see quots. 1965, 1966).

The coiner of the term appears to be Cherkin (quot. 1966), whose forthcoming paper is mentioned by Young in 1965. Cherkin's paper was communicated to the editor of the *Proceedings* on 19 November 1965.

1965 *New Scientist* 23 Dec. 861 In the author's [*sc.* J. Z. Young's] view, memory is localized in small combinations of brain cells, which he calls 'mnemons'. **1966** A. CHERKIN in *Proc. Nat. Acad. Sci.* LV. 88 The proposed unit is defined as the minimum physical change in the nervous system that encodes one memory... The name proposed for the unit is the 'mnemon' (mneme = memory; -on = suffix denoting a fundamental particle). **1971** J. Z. YOUNG *Introd. Study Man* xix. 252 The time to begin accumulating such units of memory (mnemons) would be as soon as they are ready.

mnemonist. Add: Also, a professional entertainer who practises recollection.

1969 *Observer* 26 Jan. 28/7 Eventually he became a mnemonist or professional 'memory man'. His powers of recall were not so far from total.

mnemotechnic, *a.* and *sb.* Add: **B.** *sb.* Also as *sing. rare.*

1922 JOYCE *Ulysses* 503 See, you have forgotten. Exercise your mnemotechnic.

mnemotechnist (nīmote·knist). [f. MNEMOTECHNY + -IST.] = MNEMONIST.

1891 *Chambers's Encycl.* VII. 240/2 The mnemotechnist who has a succession of things to be remembered.. compels himself to detect some association..between each of them and one of the 'hieroglyphs' which are to serve as memorial links.

mo (mōu), *sb.*² Also mo'. Colloq. or slang abbrev. MOMENT *sb.* Chiefly in ellipt. phr. *half a mo*: wait for half a moment, *i.e.*, for a short time.

1896 in J. R. Ware *Passing Eng.* (1909) 9/2 In half a mo' —half a mo' Your pluck and perseverance you can show. **1903** [see *'ARF]. **1905** H. G. WELLS *Kipps* II. v. 234 Chitterlow hesitated. 'Half a mo', my boy,' he said. **1929** N. C. JAMES *Sleeveless Errand* 176 Well, wait a mo, while I get my tata on. **1934** [see *'ARF]. **1938** AUDEN & ISHERWOOD *On Frontier* III. i. 98 Wait a mo. Gimme a torch. **1972** J. WILSON *Hide & Seek* vi. 110 Hang on... Hang on a mo. Look, you can't pin nothing on me.

mo, *sb.*³ Austral. and N.Z. slang abbrev. MOUSTACHE, MUSTACHE *sb.*

1936 M. FRANKLIN *All that Swagger* xli. 383 Darcy was a man. He had a 'mo'. **1946** D. STIVENS *Courtship Uncle Henry* 18 'I'll warm your pants for you..' he'd warn me, the ends of his long black mo shaking. **1947** 'A. P. GASKELL' *Big Game* 17 Never mind Henry,..we'll soon shave Hitler's mo off.

mo' (mǭ), *a.* (*sb.*) and *adv. U.S.* Also **mo.** An abbrev. (chiefly found in written Black English) of MORE *a.* (*sb.*) and *adv.*, esp. in phr. *no mo'*.

1902 J. D. CORROTHERS *Black Cat Club* i. 23 Read dat piece o' yo's once mo' an' let me die a-listenin' to it! *Ibid.* ii. 40 Black cats is dead bad luck. Dey's hoodooed me mo' den once. **1944** C. HIMES *Black on Black* (1973) 196 Pour me some mo' of that licker. **1953** S. A. BROWN in A. Dundes *Mother Wit* (1973) 41/2 Dere ain't no mo scufflin'. **1962** N. E. WHITTEN in *Ibid.* 402/1 White folks don' put much stock in roots and the like no mo'. **1969** S. SANCHEZ in S. Henderson *Understanding New Black Poetry* (1973) III. 272 White people Ain't rt bout nothing No mo. **1973** *Black World* Sept. 35 Me an' my baby's Got two mo' ways, Two mo' ways to do de Charleston!

-mo (mo), *suffix.* The final syllable of terms derived from the abl. sing. masc. of L. ordinal numerals which are used to denote book sizes by the number of leaves into which a sheet of the paper on which the book is printed has been folded, e.g. *duodecimo*, *sextodecimo*, etc., and, by analogy, *vicesimo-quarto*, *tricesimo-secundo*, etc., which may be pronounced or written as *12mo*, *16mo*, *24mo*, *32mo*, etc., or *twelvemo*, *sixteenmo*, *twenty-fourmo*, *thirty-twomo*, etc.

c **1716** T. RAWLINSON *Let.* in T. Hearne *Remarks & Coll.* (1901) V. 178, I..would willingly know something of y^r Sylloge Epistolar., whither MSS. & unpublish'd, or a 12mo of A^o 1640. **1742** in *N. & Q.* (1855) 2 June 419 History of the Adventures of Joseph Andrews, &c., 12mo., in 2 vols., No. 1500, with alterations. **1776** in T. Harmer *Observations Divers Passages Scripture* (ed. 2) I. 484 (Advt.), The New Testament in Greek,..2 vol. 12mo. **1801** *Schedule of Presswork Prices* in E. Howe *London Compositor* (1947) iii. 98 Twelves: Pot, such as ladies and christian ladies table part, 6mo. 35 Pica ems wide, 26 long. **1810** *Scale of Prices for Compositors' Work* Art. 6, in *Ibid.* vi. 176 English and larger type, not less than 7s... English 12mo. to be paid not less than 10s. 6d. **1841** W. SAVAGE *Dict. Art of Printing* 798 A sheet of paper folded into twenty-four leaves, forty-eight pages, is termed twenty-fourmo. **1894** *Amer. Dict. Printing & Bookmaking* 548/1 *Trigesimo-secundo*, the bibliographical term for thirty-twomo; written shortly 32mo. **1927** R. B. MCKERROW *Introd. Bibliogr.* I. iv. 34 Duodecimo, sextodecimo, vicesimo-quarto, and tricesimo-secundo, but often called 'twelvemo', 'sixteenmo', 'twenty-fourmo', and 'thirty-twomo'. *Ibid.* II. ii. 167 Both in a 16mo and a 32mo the watermark is, however, often absent. *Ibid.* 170 (*heading*) Twelve-mo by cutting. **1949** F. BOWERS *Princ. Bibliogr. Descr.* v. 193 Sexagesimo-quarto—64° *or* 64mo. **1952** J. CARTER *ABC for Bk.-Collectors* 89 The principal sizes, with their common abbreviations, are: Folio... Quarto... Octavo... Duodecimo (12mo, pronounced twelvemo). Sextodecimo (16mo, pronounced sixteenmo). Vicesimo-quarto (24mo, pronounced twentyfourmo). Tricesimo-secundo (32mo, pronounced thirtytwomo). **1973** *Collins's Authors & Printers Dict.* (ed. 11) 406/2 *Sixty-fourmo*,..a book based on 64 leaves, 128 pages, to the basic sheet..abbr. 64mo (no point).

moa. Add: **b.** *attrib.*, as *moa bone*.

1875 J. VON HAAST in *Trans. N.Z. Inst.* VII. 89 The older occupants, probably the Moa-hunters, who were inhabitants, or at least frequent visitors at the Moa Bone Point Cave. **1957** J. FRAME *Owls do Cry* xiv. 61 A moa bone.

moa-hunter, Moa-hunter (mōu·ă hʊ·ntəɹ). *N.Z.* [f. MOA + HUNTER.] The name given to early Maori inhabitants of New Zealand. Also *attrib.* Hence **moa-hunting** *ppl. a.*

1870 J. VON HAAST in *Proc. Zool. Soc.* 53, I have been so fortunate as to find a large Moa-hunters' encampment, with their cooking-places and kitchen-middens. **1872** —— in *Trans. N.Z. Inst.* IV. 78 Proceeding now to an examination of the traces left by the moa-hunting population. *Ibid.* 80, I discovered a moa-hunter encampment of considerable extent. **1873** A. TROLLOPE *Austral. & N.Z.* II. xxiii. 379 From these fractures Dr. Haast draws the conclusion that there were, before the Maoris, a race of moa-hunters. **1874** A. BATHGATE *Colonial Experiences* xvii. 241 The moa was hunted and used as food by man... Were these moa-hunters the ancestors of the Maories, or some more ancient race? **1892** W. L. BULLER in *Trans. N.Z. Inst.* XXV. 92 Long after the moa-hunters had disappeared. **1950** R. DUFF *Moa-Hunter Period Maori Culture* 7 The Moa-hunter phase of Maori culture, as defined and isolated here, is in my opinion clearly distinct from pre-European Maori culture, although it is probably ancestral to it. **1962** *Antiquity* XXXVI. 169 The existence of a widespread Moa-hunter culture in both [North and South] islands inferred by Duff has indeed been demonstrated by a number of excavations. **1974** *Nat. Geographic* Aug. 196 The men who hunted *Dinornis*..were called by later Polynesians *tangata whenua*... But the name by which they are commonly known in English is the more appropriate one: moa-hunters.

Moal, Moallakat. see *MU'ALLAQAT *sb. pl.*

moan, *sb.* Add: **1. c.** A grievance, a grumble; an 'airing' of complaints. orig. *Services' slang.*

1911 'GUNS' & 'THEELUKER' *Middle Watch Musings* 12 'Guard and Steerage 'ammicks, Sir!' I wake up with a groan; Why can't I sleep till 7 a.m.? Once more I had a moan. **1927** *Daily Express* 5 Oct. 3/4 The midshipmen fling their moanful forms into chairs, and one says:—'Come on, you chaps, let's have a moan!' **1942** 'DUGGIE' in Forbes & Allen *Ten Fighter Boys* 20 We all had a moan to the C.O. about it, and he in turn was in full agreement. **1974** *Times* 6 Apr. 14/8 It's the one moan I have about international rugby. There ought..to be referees from neutral countries.

moan, *v.* Add: **3. c.** (See quot. 1925.) orig. *Services' slang.*

a **1922** T. S. ELIOT *Waste Land Drafts* (1971) 57 line 42 So this injurious race was sullen, and kicked; Complained too of the ship... So the crew moaned. **1925** FRASER & GIBBONS *Soldier & Sailor Words* 156 *To moan*, to complain, to grumble, to be a pessimist. (Navy—equivalent to the Army 'grouse'.) **1948** *Landfall* II. 112 He felt through his pockets for a cigarette, found a butt and lit it. Why moan?

moaner (mōu·nəɹ). *colloq.* [f. MOAN *v.* + -ER¹.] One who moans; a complainer, a murmurer; a pessimist.

In quots. 1927, 1932 'a singer of blues songs'.

1927 [see *HONKY-TONK 1]. **1929** *Papers Mich. Acad. Sci., Arts & Lett.* X. 308/2 *Moaner*, a pessimist. **1932** *Amer. Speech* VII. 247 Clara Smith evidently deems it a mark of distinction to be known as 'The World's Greatest Moaner'.

1942 Berrey & Van den Bark *Amer. Thes. Slang* § 406/1 Gloomy or irritable person; pessimist..complainer; fault-finder..moaner. 1952 *Landfall* VI. 202 Anyone who questions too often is a 'moaner', yet in New Zealand the moaner is common. 1959 I. & P. Opie *Lore & Lang. Schoolch.* x. 187 Croydon boys have twenty names for a cry-baby:..*moaner* [etc.]. 1969 I. Kemp *Brit. G.I. in Vietnam* v. 95 Burmeister..once said to me, 'Limey, you and Goad are the two biggest moaners in my squad.' 1969 F. Sargeson *Joy of Worm* iii. 87 But my boy, I did not intend to write you a letter to make myself out what our fellow countrymen call a 'moaner'.

moaning, *ppl. a.* Add: *spec.* **moaning minnie** (also with capital initials). **a.** Either of two German types of mortar (*minenwerfer* or *nebelwerfer*); also, a shell from one of these mortars. **b.** An air-raid siren. **c.** = *Moaner.

1941 R. Greenwood *Mr. Bunting at War* xiv. 191 'One up now,' said Chris, listening to the drone of an engine. 'Hope Moaning Minnie doesn't sound, and bring mother downstairs.' 1944 *Hutchinson's Pict. Hist. War* 12 Apr.–26 Sept. 376 (*caption*) When the Germans beat a hasty retreat from Cagny among the material they abandoned was this multiple mortar, or 'Moaning Minnie'. 1950 [see *Minnie]. 1960 *Sunday Times* 31 Jan. 33/4 Another said that she '..just didn't believe these moaning minnies'. 1972 *N.Z. News* 26 Apr. 6/5 I don't want to give the impression of being a moaning Minnie but may I ..make a special plea to the railmen to..get back to work.

moanism (mōu·niz'm). *rare.* [f. Moan *v.* + -ism.] The practice of lamenting; emotionalism.

1916 R. Frost *Let.* 24 May (1964) 34 Moanism and swounding.

mob, *sb.*¹ Add: **1.** Also in *Social Psychol.*

1897 E. A. Ross in *Pop. Sci. Monthly* July 390 (*heading*) The mob mind. *Ibid.*, Great mental instability marks the true mob. 1931 E. S. Bogardus *Fund. Social Psychol.* (ed. 2) xiv. 315 A mob is a crowd in a very high state of suggestibility. 1940 E. A. Strecker *Beyond Clin. Frontiers* iv. 60 Lynching mobs, certain strike mobs, etc. furnish examples of positive reality-evasion action en masse. 1959 Gill & Brenman *Hypnosis* ii. 103 Such people are sometimes said to be hypnotized. They show regressive phenomena similar to those revealed by a mob. 1973 G. R. Leslie et al. *Order & Change: Introd. Sociol.* xi. 389 Although scientific interest in mobs and riots as mechanisms of social change is relatively new, the phenomena themselves have existed throughout history.

4. (Further Austral. examples.) Also in New Zealand use, without disparaging implication, a crowd, a group, a gang of workmen.

1834 in R. McNab *Old Whaling Days* (1913) App. C. 424 A mob of natives came running into the hut where we stopped. 1845 *N.Z. Company Rep.* XIX. 70 The Pah is small, and occupied by a few Natives, the crops, as I understood, being the property of several 'mobs' in different parts of the Sound. 1852 *Austral. & N.Z. Gaz.* 10 Jan. 11/2 In Major Hornbrook's words 'the *Stedfast's* mob is a much jollier mob than that of the *Duke of Bronte*'. 1860 G. Duppa in S. S. Crawford *Sheep & Sheepmen Canterbury* (1949) v. 48 Commence shearing with a strong mob of shearers. 1863 F. E. Maning *Old N.Z.* iv. 65 It was 'our mob' coming to the rescue. 1907 W. H. Koebel *Return of Joe* 257 [He] had but a few hours ago formed one of their 'mob', and [was] the most skilful bushwhacker in the district. 1941 *Coast to Coast* 214 The mob around the bar was thinning down, with chaps grabbing their bundles and going off home. 1944 *Living off Land* viii. 155 You may meet the Abo. He may be only a poor specimen on the outskirts of a township, or he may be one of the wild mob in the furthest nor' west. 1960 S. Ashton-Warner *Incense to Idols* 23, I know one girl from another, course you'll be at the..mob. 1968 K. Weatherly *Roo Shooter* 37 Hunter pushed his way through the mob, every one of whom he knew,..and sat down at the table.

b. (Later Austral. examples.) Also quasi-*adv.*

1934 A. Russell *Tramp-Royal in Wild Austral.* xiii. 97 There'll be mobs of water on the track, we'll get mobs of beef at the runs, the stages'll be mobs shorter, an' there'll be mobs better camping grounds..and of course we'll be able to take it mobs easier. 1942 C. Barrett *On the Wallaby* iii. 41 Even an offer of..mobs of tucker..failed to gain me a guide.

c. (Further examples.)

1838 T. Walker *Month in Bush* 8, I beheld a level plain,..with..'mobs' of cattle scattered over the surface. 1843 J. Cotton *Let.* July in Billis & Kenyon *Pastures New* (1930) xiii. 227, I inspected a mob (as it is termed here) of the cattle. 1850 *Household Words* 6 Apr. 45/1, I was going down to Sydney with a mob of horses. 1853 R. Gough *Let.* 24 Sept. in J. Deans *Pioneers of Canterbury* (1937) 255, I should like to put all the calves in one mob. 1875 *Trans. & Proc. N.Z. Inst.* VII. 130 For about 400 birds of this large size to have been roasted in so small a compass in one mob would be a physical impossibility. 1933 *Bulletin* (Sydney) 5 July 21/3 The dog was turning the mob to move over the fallen animal. 1936 I. L. Idriess *Cattle King* vii. 63 A squatter was overlanding a big mob of stock. 1940 *Geogr. Jrnl.* XCV. 242 There is now only one firm remaining which has a mob of mules. 1964 *Sunday Mail Mag.* (Brisbane) 27 Sept. 3/5 Behind him..were the ragged outriders of the mob... I knew he didn't stand a chance of clearing those terrible hooves. 1968 K. Weatherly *Roo Shooter* 35 A small mob of wild pigs, mostly white with black spots. 1972 P. Newton *Sheep Thief* ii. 20 The two men had taken out a mob of ewes.

e. *Mil. slang.* A battalion, a regiment; a military unit.

1916 J. N. Hall (*title*) Kitchener's mob. 1916 *Anzac Book* 32 Yes; some d-d gobblers thought they would catch our mob nappin' but missed the bus. 1925 Fraser & Gibbons *Soldier & Sailor Words* 156 *Mob*, any collection or body of troops. A very old Army term. 1948 Partridge *Dict. Forces' Slang* 120 *Mob*, unit, not necessarily in a derogatory sense. 'What mob are you from, chum?' 1972 M. Pugh *Murmur of Mutiny* iv. 34 You must have heard of Sharjah and the Trucial Oman Scouts. This mob is modelled on them.

5. a. (Earlier example.) *swell mob*: see also Swell *a.*

1839 H. Brandon *Poverty, Mendicity & Crime* 164/1 *Mob*—companions. Working with mobs. Robbing with companions.

b. *U.S.* A more or less permanent association or gang of violent criminals. *The Mob*, a supposed permanent gang controlling much of organized crime in the U.S. and elsewhere; cf. *Mafia. Also *attrib.* and *Comb.*, amongst gangs, on behalf of a mob or 'The Mob'.

1927 *Amer. Speech* II. 385/1 Any kind of a gang was known as a push, a word credited to Australia, but I think it is a sister of the mob of the city underworld. 1930 Wodehouse *Very Good, Jeeves!* xi. 300 By the time he had come to the surface, a sort of mob-warfare was going on at the other side of the field. 1932 Turkus & Feder *Murder, Inc.* xv. 445 One more hood..moved in to the extent that he had his picture taken in a friendly pose with a candidate for the vice-presidency of the United States. 1968 P. Oliver *Screening Blues* iv. 134 Within a year control of the New York numbers racket passed into the dominion of a mob enforcer. 1969 *Guardian* 24 Jan. 7/6 The Mob from its Chicago headquarters runs the subcontinent. 1973 'A. Thackeray' *One Way Ticket* 23 Better watch out... It could be the Mafia, the Mob, or whatever they call it these days.

6. a. *mob action, behaviour, -condemnation, control, -culture, -emotion, -fever, -hysteria, -indignation, -madness, mania, mind, -movement, -orator* (earlier and later examples), *-oratory, -psychology, -reaction, scene, -sensation, -storm, -sycophancy, -tide, violence* (examples), *-will, -worship*; **b.** *mob-inspiring* adj.

1972 Turner & Killian *Collective Behavior* (ed. 2) iii. 49/2 Mob action is frequently nothing more than culturally sanctioned punishment carried out by unauthorized persons without 'due process'. 1973 G. R. Leslie et al. *Order & Change: Introd. Sociol.* xi. 389 The distinction between mob behavior and riot behavior has to do with the degree to which the hostility is focused upon a single object or class of objects. When the hostility is so directed and is concentrated on destruction of particular scapegoats, the term 'mob' is used. 1929 D. H. Lawrence *Pornogr. & Obscenity* 5 When it comes to the so-called obscene words, the first reaction is almost sure to be mob-reaction, mob-indignation, mob-condemnation. 1971 D. E. Westlake *I Gave at the Office* (1972) 85 The gym is pretty much the universal of mob control. 1968 L. Durrell *Tunc* v. 269 A surrogate mob-culture. 1928 D. H. Lawrence *Woman who rode Away* 231 A steam of hot mob-emotions! 1929 Galsworthy in *Story-Teller* May 478 (?) ... mob-emotion... 1884 *Harper's Mag.* (?) ... mob-fever... 1968 Lawrence... mob-hysteria... 1888 J. Burroughs (?) ... mob-mind... 1882 H. Spencer (?) ... mob-movement... 1891 'Mark Twain' *Prince & Pauper* (?) ... mob-orator... 1891 ... mob-oratory... 1909 ... mob-psychology... 1969 M. B. Arnold in T. Mischel *Human Action* 192 Emotions, whether mob-violence or passive love-ins, seem to have been chosen by many young people as sole guide of their actions. 1924 J. Masefield *Sard Harker* 51 Never had I thought that my fellow-citizens of Las Palomas would try to impose the mob-will upon the individual. 1893 E. Dowson *Let.* c 27 Aug. (1967) 288 'One View of the Question', a beautiful piece of satire on English mob-worship.

7. mob-man = Mobsman.

1747 in *New Jersey Archives* (1883) 1st Ser. VII. 428 He discoursing with several of the mobbers,..has heard them [say]..that the King himself was unable to quell mobs in England. 1835 *Maryland Hist. Mag.* IX. 160 You must have large companies of worthies marching to and fro, and a mob man, as such, cannot be seen.

mob, *v.*² Add: **1. b.** Esp. in bird behaviour, to engage in *mobbing (*vbl. sb.*² 1 c).

1927 E. M. Nicholson *How Birds Live* iii. 87 An owl appears and is surrounded by a clamorous crowd of small birds which proceed to mob it. 1936 *Brit. Birds* XXX. 28 Usually when seen it was sitting in a tree, and it was much 'mobbed' by Rooks. 1938 T. H. White *Sword in Stone* xvii. 250 They [sc. rooks] have got the courage to mob their enemies. I should think it takes some courage to mob those beasts. It takes a pack of you. 1965 P. Wayre *Wind in Reeds* x. 62 The falcon, sometimes.. accompanied by the tiercel, would fly out and mob us.

mobbing, *vbl. sb.*² Add: **1. b.** *Sc. Law.* (See quot. 1959.)

1800 D. Hume *Commentaries Law of Scotl.* IV. xvi. 228 A multitude may be convened for a criminal purpose without being guilty of mobbing. 1889 [in Dict.]. 1898 J. Chisholm *Green's Encycl. Law of Scotl.* 370 In the law of Scotland mobbing, or 'the Tumultuous Convocation of the Lieges', includes the several degrees and stages of disorder which are known in the law of England under the names of Riot, Rout, and Unlawful Assembly. 1959 *Chambers's Encycl.* XII. 327/2 Mobbing is where a number of persons combine together for an illegal purpose... The crime is not constituted unless there is a threatening action taken in an effort to obtain a definite end and this distinguishes it from rioting and breach of the peace. 1970 *Observer* 4 Feb. 29/4 We almost got ye for mobbin' and riotin'.

c. Esp. in bird behaviour, a type of display in which a group of small birds engages to drive off a predator, or a similar kind of display exhibited by one or two birds, in which they fly close to the object of their apparent aggression. Also *attrib.*

1919 F. Finn *Bird Behaviour* ix. 275 The mobbing of Hawks and Owls is no doubt often dictated by revenge. 1927 E. M. Nicholson *How Birds Live* vii. 93 Akin to fighting and play, but not identifiable with either, is the mobbing habit. 1936 *Brit. Birds* XXX. 27 The 'mobbings' of nesting Rooks by other members of the colony..are usual in origin. 1949 *Ibid.* XLII. 64 This behaviour is certainly in no way analogous to the so-called 'mobbing' of predatory birds. 1962 *Behaviour* IV. 288 (*title*) The motivational organisation controlling the mobbing calls of the Blackbird.

mobbish, *a.* (Later example.)

1920 *Q. Rev.* July 166 This mobbish or, as it may be termed, 'synnomic' character of primitive mentality is well known.

mo·bbishness. [-ness.] Tendency to mobbism; the practice of acting in groups.

1920 *Q. Rev.* July 166 The savage enjoys no privacy, but is always in some sort of a crowd,..experiencing therefore all those peculiar mental effects which mobbishness brings in its train. 1949 W. Deeping *Kitty* xxiv. 320 You would still hope for your super-scientist..who, by pressing a button in his laboratory, could efface all mobs and mobbishness.

mobbism. (Earlier U.S. example.)

1794 *Mass. Spy* 16 Apr. 3/2 A few days since, we experienced a scene of the most unlicensed mobbism.

mob-cap. (Earlier and *transf.* examples.)

1795 T. Wilkinson *Wandering Patentee* II. 132 On she came in a frock and a little mob-cap, and sung the song. 1971 *Daily Tel.* 19 Jan. 11/2 There's a whole range of Victoriana too, including variable mob-cap lamp-shades in lace and embroidered cotton.

mobese (mǫbī·z). [f. Mob *sb.*¹ + -ese.] The cant of American professional criminals.

1931 *People* (Austral.) 19 Oct. 13/1 A few over felt..that Dewey should be 'hit'—which is mobese for gang murder. 1965 I. Fleming *Man with Golden Gun* xiii. 173 In mobese, he 'gave him the works'.

mo·b-handed, *a. colloq.* [f. Mob *sb.*¹ + Handed *a.* 1 b.] In considerable numbers, constituting a large body.

1934 M. Allingham *Cheapjack* xix. 254 His companions had been 'mob-handed', that is to say, working in a big group. 1966 A. Prior *Operators* vi. 64 ... had returned home penniless to find the police mob-handed. 1970 *Sunday Tel.* 9 Aug. 27/7 The evergreen X class [of yachts] turned out mob-handed at 60 strong.

mo-bike (mōu·baik), colloq. abbrev. of *motor bicycle*.

1925 *Punch* 22 Apr. 433/1 Just think of going out on a mo-bike in top-hats and tail-coats! 1971 *Ceylon Observer* (Mag. ed.) 19 Sept. 7/3 A Japanese mo-bike (over 180 c.c. of course!).

mobile (mōu·bail), *sb.*³ [Subst. use of Mobile *a.*] **1. a.** (Also with pronunc. (mōu·bīl).) A form of decoration consisting usu. of abstract designs in metal, plastic, etc., contrived (as by suspension) so as to be mobile. Cf. *Stabile *sb.*

[1936 P. Nash in *Archit. Rev.* LXXX. 208/3 Shadows bring us to Calder, who is..so far as I know, the original inventor of mobile sculpture and so, also, of the '*objet*

mobile'. The mobile object is not now confined to Calder's invention; both Max Ernst and Duchamp have made various pieces of this nature.] 1949 *Ibid.* CVI. 117 Alexander Calder's work on the 'stabile' is not as well known in England as is his work on the now well established 'mobile'. In fact Calder has always done 'still' sculpture, and the term stabile, given to it by Hans Arp, appears to be some months older than the name mobile, which was invented by Marcel Duchamp. 1952 *Granta* 29 Nov. 8/1 We find it agreeable..to hang pastel-tinted antlers on the wall near a mobile. 1957 *Times* 18 Nov. 11/1 Mobiles at Heal's include a life-size black cat, and a cut-out set of small figures costs from 4s. and can be set up by children of about 10 years without help. 1958 E. Dundy *Dud Avocado* I. ix. 156 He picked up a wire coat-hanger and some string and a couple of paint-brushes and a shoe, and started making them into a Mobile. 1971 D. D. Boyden *Introd. Mus.* (ed. 2) 524 Someone has aptly noted that chance music is rather like the mobiles of Calder. 1972 *Sci. Amer.* Mar. 76/1 They were shown an arrangement of three colored geometrical objects in a 'mobile'.

b. *transf.* and *fig.*, esp. in *Mus.* (see quot. 1967).

1961 *Punch* 11 Jan. 116/3 Admirers of the hard, glittering mobiles constructed..by Stan Kenton may be pleased by *Standards in Silhouette*... A big, big band, making big, big sounds, in what was once a daringly experimental manner. 1967 *Listener* 2 Feb. 176/3 The crystallization of these new formal principles was the 'mobile'. It connotes a dynamic arrangement of musical thoughts in which several patterns are possible, depending on the decision of the interpreter... A 'mobile' is made up of finite (musical) thoughts of fairly conventional dimensions... As units they remain constant; but the arrangement of their sequence varies, subject to certain pre-compositional order. 1970 'J. Morris' *Candywine Devel.* xvi. 184 Five electric guitars and a mobile of drums were backstopping a sleek, oxblood brown singer. 1971 E. Borroff *Mus. in Europe & U.S.* xxvii. 656 Other works include..*Symphonies* (1955), for fifteen soloists; *Mobile* for two pianos (1958).

2. Short for: (*a*) mobile canteen; (*b*) (*Austral.* and *N.Z.*) mobile barrier; (*c*) mobile police.

1940 *New Statesman* 9 Nov. 466/1 Go up and have a cup of tea at the mobile. 1969 *Australian* 24 May 34/4 Fifth.. over this trip and from behind the mobile here last week. 1971 W. J. Burley *Guilt Edged* iv. 62 Control to all mobiles: keep look out for red Mini-Cooper saloon. 1974 J. Gardner *Corner Men* ii. 14 Put out a call. There must be some mobiles around. We need them here.

Mobile, *sb.*⁴: see *Mobilian *sb.*

mobile, *a.* Now usu with pronunc. (mōu·bail) in the U.K. **1. a.** Delete † *Obs.* and add later examples.

1927 *U.S. Daily* 22 Nov. 2/3 Mobile stations must be established in such a way as to comply as concerns frequencies and types of waves, with the general provision of Article 5. 1935 *Discovery* May 151/2 The treatment of patients far removed from the well-equipped hospitals, by means of mobile units complete with laboratories. 1937 *Archit. Rev.* LXXXI. 19 (*caption*) A mobile sculpture by Alexander Calder. 1938 *Encycl. Brit. Bk. of Year* 122/1 Mobile recording vans with a new method of editing records made it possible to broadcast composite sound records of events only a few minutes after their occurrence. 1940 *Economist* 5 Oct. 422/2 Cooking facilities have been provided by the L.C.C. Some mobile kitchens are at work. 1940 *New Statesman* 5 Oct. 321/2 Feeding centres and mobile canteens. 1955 *Radio Times* 22 Apr. 11/3 Franklin Engelmann with a BBC mobile recording unit visits Dudley to meet local people. 1959 *Economist* 14 Mar. 991/1 Mobile shops have more than doubled in number since the war, and are still taking extra trade on to the roads at the rate of perhaps fifteen a week. 1960 *Library Assoc. Rec.* Aug. 262/2 *Mobile library*, a vehicle devised, equipped and operated to provide, as far as reasonably practicable, a service comparable to a part-time branch library. 1961 L. D. Stamp *Gloss. Geogr. Terms* 258/1 *Mobile or footloose industries*, broadly those generally called 'light'. 1962 *Economist* 17 Nov. 665/3 The mobile lounges will not only provide transport..between the terminal and the parked aircraft but will also serve as waiting rooms. 1971 *Guardian* 10 June 1/1 A 22-bed mobile hospital..was among the supplies which left for Calcutta. 1974 *Camping & Caravanning* Sept. 15/1 Fees 30p per unit per night, C.C.Y. 5p. Mobile shop. Walking distance of beach.

f. *Sociol.* Of a person: able to move into different social levels, or a different environment or field of employment. Of a society: not rigidly stratified, in which upward or downward movement between social levels can take place, and also movement between fields of employment, etc., within the same social level.

1927 P. A. Sorokin *Social Mobility* II. vii. 138 Such a type of social stratification may be styled open, plastic, penetrable, or mobile. *Ibid.* v. xvii. 427 Unskilled labor is more mobile than skilled labor. 1940 K. Mannheim *Man & Society* II. vi. 93 The significance of the mobile elements in social and cultural life. 1945 C. W. Mills in *Jrnl. Econ. Hist.* V. Suppl. v. 39 For laissez faire, the pattern of success might involve a larger proportion of upwardly mobile persons. 1959 V. Packard *Status Seekers* (1960) xviii. 256 Many socially declining or downward-mobile people turn to alcohol or drugs for support.

g. *Philol.* = *Movable *a.* 7 b.

1955 T. Burrow *Sanskrit Lang.* iii. 80 The so-called mobile *s*. Indo-European *s* when it formed the first member of an initial consonant group, was an unstable sound, and liable to disappear under conditions which it has not been possible accurately to define. 1965 G. Y. Shevelov *Prehist. of Slavic* 230 The mobile consonants, i.e. consonants sometimes used, sometimes dropped on word boundaries... *n*-mobile..*s*-mobile... Problem of *k*-mobile.

h. Special collocations: as, **mobile barrier** *Austral.* and *N.Z.*, in *Trotting*, a foldable barrier designed to facilitate a flying start; **mobile home**, a large caravan permanently parked and used as a residence.

1965 *Weekly News* (Auckland) 8 Dec. 59/1 The controversy that is developing over the use of the mobile barrier in trotting. 1954 *N.Y. Herald Tribune Bk. Rev.* 13 June 12 Books pertaining to trailer houses—or mobile homes—published within the last five years. 1961 *Daily Tel.* 22 Nov. 17/4 Mobile homes for 1,200 families could be provided on 300 vacant sites in London. 1969 *Eugene* (Oregon) *Register-Guard* 3 Dec. 5D/1 Mobile homes—the things people used to call trailer houses. 1973 *People's Jrnl.* (Inverness & Northern Counties ed.) 1 Dec. 22/5 *Mobile Home*, Coats Caravans. 32 ft. × 9 ft. 6 in. Astral Mobile Home. Double End Bedroom, Bathroom, Kitchen and Lounge. Solid Fuel Fire. £895.

2. b. (Earlier example.)

1853 Mill in *Edin. Rev.* XCVIII. 432 They [sc. the Athenians] were not fickle, but (a very different quality, vulgarly confounded with it) mobile; keenly susceptible.. to the feeling and impression of the moment.

3. Also of police.

1938 F. D. Sharpe *Sharpe of Flying Squad* i. 11 The Flying Squad has about twenty cars, and they are very different cars to those of the Mobile Police. 1955 *Radio Times* 22 Apr. 13/3 The Police. A series of eight talks. 7. Beats and Mobile Patrols. 1967 N. Lucas *C.I.D.* vi. 78 The new mobile patrol toured the streets with the hidden detectives scanning..the crowds.

-mobile. Used freely in the 20th century as the second element in combinations: **a.** Portable, or travelling under its own power. **b.** In occas. uses of immobile objects or structures, usu. having a function pertaining to, or being an imitation of, an automobile or other form of transport. (Examples, as *bloodmobile*, *bookmobile*, *clubmobile*, are entered under the first element in this Supplement.)

Mobilian (mobi·liăn), *sb.* Also **Mobile** (mo-bī·l). [? f. the town of *Mobile* in Alabama + -ian.] A lingua franca or trade language used formerly in south-eastern North America (see quot. 1907). Also as *adj.*

1840 G. Bancroft *Hist. U.S.* III. xxii. 249 The whole country south-east, south, and west of the Cherokees.. was in the possession of one great family of nations, of which the language was named by the French the Mobilian, and is described by Gallatin as the Muskhogee-Choctà. 1907 F. W. Hodge *Handbk. Amer. Indians* I. 916/1 The so-called Mobilian trade language was a corrupted Choctaw jargon used for the purposes of intertribal communication among all the tribes from Florida to Louisiana, extending northward on the Mississippi to about the junction of the Ohio. It was also known as the Chickasaw trade language. 1928 W. A. Read *Indian Place-Names Louisiana* 6 The identity of *Manchac* with the Mobilian *imashaka*, 'rear entrance', is rendered more plausible by my discovery of the form *Mashake*. 1937 *Amer. Speech* XII. 212 Besides the Shawnee words one Algonquian term has slipped in through the medium of the Mobilian trade language. 1947 P. S. Martin et al. *Indians before Columbus* 68 In the Southeast a Choctaw jargon called 'Mobilian' was spoken from Florida to Louisiana and up the Mississippi River as far north as the Ohio. 1964 *Amer. Speech* XXXIX. 16 He [sc. Charles P. G. Scott] attempted to push *O.K.* back to the Mobile trade language, current along the Gulf of Mexico in the eighteenth century.

mobiliary, *a.* Add: **2.** *mobiliary art* = *art mobilier* (*Art *sb.* VI. c).

1927 [see *Home *sb.*¹ 14]. 1960 *Times Lit. Suppl.* 2 Sept. 565/4 'Mobiliary' art in the form of sculpture, carving and engraving on bone, antler, ivory and stone.

mobility¹. Add: **1. c.** *spec.* in *Sociol.* The possibility of movement between different social levels that exists in a society (*vertical mobility*); also the possibility of movement to different fields of employment or interest, or to new areas, within the same social level (*horizontal mobility*).

1900 *Amer. Jrnl. Sociol.* VI. 377 (*caption*) Mobility of type. 1927 P. A. Sorokin *Social Mobility* II. vii. 136 The intensiveness of the vertical mobility may be measured in the same way in the field of the political and occupational stratifications. *Ibid.* 160 Horizontal mobility, in spite of the great importance of the problem, is not an object of this study. 1938 T. H. Marshall *Class Conflict* 111 The use of mobility as an excuse for inequality is usually associated with a measure of self-deception. 1956 C. W. Mills *Power Elite* xv. 349 Only if the criteria of the top positions were meritorious..could we smuggle merit into such statistics of mobility. 1965 B. B. Wolman *Handbk. Clin. Psychol.* xxxiv. 979 While mobility has been frequently related to schizophrenia, R. Freedman..failed to find positive correlation between high mobility and high hospital admission rates of schizophrenics. 1972 S. Cotgrove *Sci. of Society* (rev. ed.) vii. 230 Restricted mobility facilitates the protection of privileges by a stratum and the development and persistence of a unitary culture.

f. *Chem.* and *Physics.* [tr. G. *beweglichkeit* (given this specific sense by F. Kohlrausch 1876, in *Nachrichten von d. K. Ges. d. Wissensch. und d. G.-A.-Universität zu Göttingen* 17 May 220).] The degree to which a charge carrier undergoes movement in a definite direction in response to an electric field, now usu. expressed as the average speed (in cm. per second) in a field of one volt per cm. divided by the net number of charges on the carrier.

1895 C. S. Palmer tr. *Nernst's Theoret. Chem.* II. vii. 315 The term mobility (Beweglichkeit) or velocity of transport will mean..the velocity with which 1 g.-ion will be transported under the influence of a pull of 1, *e.g.* 1 kilogram weight. *Ibid.* 316 If we denote the mobility of the positive and negative ions by U and V, then their respective velocities will be in the same ratio as their mobilities. *Ibid.* 317 The conductivity of a solution of a binary electrolyte is greater in accordance as it contains more free ions, and according as these have a greater mobility. 1912 *Jrnl. Chem. Soc.* CI. II. 1276 The following figures..are the results of the chief researches on the mobility of the hydrogen ion... All values are expressed in terms of the reciprocal ohm. 1924 J. R. Partington in H. S. Taylor *Treat. Physical Chem.* I. xi. 539 The equivalent conductance at infinite dilution is the sum of the mobilities of anion and kation at a given temperature. 1946 [see *Carrier 1 k (ii)]. 1950 W. J. Moore *Physical Chem.* xv. 434 With two exceptions, the ionic mobilities in aqueous solutions do not differ as to order of magnitude, being all around 6×10^{-4} cm^2 $sec.^{-1}$ $volt^{-1}$. The exceptions are the hydrogen and hydroxyl ions with the abnormally high mobilities of $36{\cdot}2 \times 10^{-4}$ and $20{\cdot}5 \times 10^{-4}$. 1963 B. Fozard *Instrumentation Nucl. Reactors* ii. 15 In a uniform electric field the electrons and ions may be regarded as acquiring a uniform drift velocity which is superimposed upon their kinetic motion and directed towards the electrodes [of the ionisation chamber]... The term mobility may be used to indicate the ease with which ions may be caused to drift.

mobilization. Add: **1.** (Later examples; also examples in *Surg.*: cf. *Mobilize *v.* 1 b.)

1890 Law & Jewell tr. *Gruber's Text-bk. Dis. Ear* xvii. 479 (*heading*) Mobilisation of the stapes. 1901 Rose & Carless *Man. Surg.* (ed. 4) xvi. 400 A most valuable adjuvant in the treatment of fractures is massage..whilst in some cases early mobilization is also desirable. 1930 *Morning Post* 7 Aug. 11/6 The credit mobilisation in London by the Australian banks to meet Governmental commitments..was approved. 1967 *Economist* 14 Oct. 120/1 A growing number of Labour MPs are sponsoring the case for government 'mobilisation' of the £3,200 million odd of foreign shares held by private British investors. 1967 S. R. Mawson *Dis. Ear* (ed. 2) xxi. 515 Relief of conductive deafness due to stapedial ankylosis by mobilization of stapes.

b. *Sociol.* The organizing of some hitherto unused form of social energy to bring about changes within a society.

1953 K. W. Deutsch *Nationalism & Social Communication* vi. 104 The processes of mobilization and assimilation may be illustrated rather strikingly in the case of Finland. 1964 G. Germani in I. L. Horowitz *New Sociol.* 395 We understand by mobilization the 'excess'..of group participation in relation to the level defined by the old society as 'normal'. 1968 A. Etzioni *Active Society* xv. 393 Whatever the form of mobilization, whether it be direct or indirect, the process entails a shift of control and/or a shift of the usage of assets. 1972 Turner & Killian *Collective Behavior* (ed. 2) iv. 62/2 The crowding together of Negro Americans in the black ghettoes is a type of mobilization that has contributed to urban insurrections.

mobilize, *v.* Add: **1. b.** *Surg.* To restore mobility to (an ankylosed bone); to free or detach so as to render more accessible.

1894 O. Dodd tr. *Politzer's Text-bk. Dis. Ear* 318 The stapes is mobilized by means of a single or double hook inserted between the crura. 1914 Rose & Carless *Man. Surg.* (ed. 9) xxxv. 1076 If a stone is lodged behind the second piece of the duodenum, it may be possible to manipulate it up, and make it accessible above the intestine; but otherwise the duodenum must be mobilized by dividing the peritoneum on its outer edge. 1953 *N.Y. State Jrnl. Med.* LIII. 2653/1 The stapes was mobilized, and the hearing improved on the operating table. 1967 G. M. Wyburn et al. *Conc. Anat.* iv. 121/2 Next the temporal lobes [of the brain] should be mobilised from the floor of the middle cranial fossa.

c. *Sociol.* To bring into circulation (hitherto unused social assets or energies).

1953 K. W. Deutsch *Nationalism & Social Communication* vi. 100 Within any geographical setting and any population, economic, social, and technological developments mobilize individuals for relatively more intensive communication. *Ibid.*, Population mobilized for mass communication. 1968 A. Etzioni *Active Society* xv. 409 Two disparate organizations seem to mobilize a collectivity more effectively than one. 1968 G. Myrdal in W. Ewald *Environment & Change* iv. xvi. 261 Their [sc. the lower strata of American society's] low rate of participation in elections, when they are not mobilized and exploited by the political machines.

2. Also *fig.*

1871 L. W. M. Lockhart *Fair to See* III. xxxiii. 100 A hundred times he had paraded the line of arguments he meant to employ, and the reserve which, in case of their failure, he held in readiness, and, so to speak, mobilised. 1951 D. B. Truman *Governmental Process* ix. 271 The political party in the United States most commonly is a device for mobilizing votes.

mobilized *ppl. a.*, **mobilizing** *vbl. sb.*: further examples.

1913 in W. S. Churchill *World Crisis 1911–14* (1923) viii. 190 A very large staff would be employed at all the mobilizing centres to report upon the whole workings of the mobilization. 1953 K. W. Deutsch *Nationalism & Social Communication* vi. 101 The rate of growth of the mobilized population..and the changes in its sociological level could all be calculated. 1968 A. Etzioni *Active* Society xv. 408 Strains and conflicts among the organizational arms of any mobilizing collectivity should be 'routine' and expected.

Möbius (mǒ·biŏs). Also **Moebius.** The name of August Ferdinand *Möbius* (1790–1868), German mathematician, used, chiefly in *Möbius band, strip*, to designate a surface having only one side and one edge, formed by twisting one end of a rectangular strip through 180 degrees and joining it to the other end.

1904 E. R. Hedrick tr. *Goursat's Course in Math. Analysis* I. 546 Möbius' strip. 1941 Courant & Robbins *What is Math.?* v. 260 If the Moebius strip is cut along this [center] line..we find that it remains in one piece. 1950 Möbius band [see *Klein bottle]. 1960 F. Land *Lang. Math.* xi. 171 We have in the Moebius strip the strange object which consists of one continuous surface bounded by one continuous curve. 1965 Möbius strip [see *Klein bottle]. 1970 *New Yorker* 28 Nov. 58/3 Barbara, taking a pair of scissors..turned the clipping on fragrant Christmas gifts into a Möbius strip. 1974 *Times* 10 Sept. 14/8 If the curtain at Covent Garden three weeks hence goes up on yet another ring, circle, hoop, saucer, dish or Möbius band I shall be heard to blow my nose in an ominous falling figure..which..betokens Doom.

mobocracy. 2. (Later examples.)

1921 G. B. Shaw *Back to Methuselah* Pref. p. lxii, Bastiat had proved convincingly that Nature had arranged Economic Harmonies which would settle social questions far better than theocracies or aristocracies or mobocracies. 1949 F. L. Wright (*title*) Genius and the mobocracy.

mobsman. Add: **2.** (Earlier and later examples.) Also, = next.

1846 *Swell's Night Guide* p. iii, The tophic blacklegs and swell mobsman, who can pluck a pigeon with the sang froid of a ripened friendship. 1935 A. J. Pollock *Underworld Speaks* 77/1 *Mobsman*, a gangster, a thug; ruffian; gunman. 1974 *Daily Tel.* (Colour Suppl.) 29 Nov. 86/4 It is doubtful if the Victorian Londoner needed warning, for the artful mobsmen, toolers and dippers, together with their stickman accomplices, were everywhere among the crowds.

mobster (mǫ·bstəɹ). *slang* (orig. *U.S.*). [f. Mob *sb.*¹ + -ster.] A member of a group of criminals; cf. *Gangster. Also *attrib.* and *transf.*

1917 *Lincoln* (Nebraska) *Evening News* 11 July 4 Many mobsters have left the city, it is asserted, and leaders of the mob are going to be hard to find. 1940 *New Yorker* 13 July 17/1 A mob nickname he got from the mobsters. 1947 E. Hyams *William Medium* x. 199 South African diamond mobsters. 1947 J. Mulgan *Report on Experience* x. 125, I never lived in Chicago, but have a wide vicarious acquaintance from films and paperbacks of mobster-rule and gang-law. 1957 *Observer* 3 Nov. 19/3 Mr. [Marc] Lawrence, renowned for his portraits of flinty Hollywood mobsters. 1962 D. Warner *Death of Bogey* I. ii. 10 A scurrying horde of spivs and pimps, fiddlers and tweedlers, tearaways, mobsters. 1964 D. Varaday *Gara-Yaka* xviii. 159 The dead mobsters were mangy, disease-ridden outcasts of the dog world. 1972 D. E. Westlake *Cops & Robbers* (1973) xvi. 251, I was afraid to think about Vigano and his mobsters.

mocamp (mōu·kæmp). [f. Mo(tor *sb.* + Camp *sb.*²] (See quot. 1970.)

1967 *Times* 1 Nov. (Suppl.) p. ii/1 If you're touring Turkey, stay at BP Mocamps. Here's a welcome for campers and caravanners. 1970 *Britannica Bk. of Year* (U.S.) 798/3 *Mocamp*, a camp providing tourists with a protected area for tents and trailers and offering various conveniences and services. 1972 *Maclean's Mag.* June 54/3 In Turkey camping facilities called Mocamps are found all over the country.

moccasin. Add: **1. b.** A type of shoe for informal wear, resembling those worn by American Indians.

1895 *Montgomery Ward Catal.* 513/3 Infants' Moccasins, made from soft dongola stock, with silk lace and tassel. 1944 H. McCloy *Panic* ii. 13 The loose sweater, kilted skirt, and flat-heeled moccasins she had put on so hastily. 1970 B. Knox *Children of Mist* iii. 53 He wore a blue sports shirt with brown shoes, his feet were in light tan moccasins.

c. *Austral.* and *N.Z.* (See quots.)

1929 H. B. Smith *Sheep & Wool Industry Austral. & N.Z.* (ed. 3) x. 75 The shearers..arrayed in their working clothes, with bowyangs..and moccasins on (a kind of shoe made out of wool-pack, after the style of the foot-covering of the Red Indian). 1965 *N.Z. Listener* 26 Feb. 15/2 *Moccasin*, the shearer's home-made footwear, usually made of sacking or felt.

2. moccasin flower (earlier and later examples).

1680 in J. Ray *Hist. Plant.* (1688) II. 1926/1 Helleborine flore rotundo luteo, purpureis venis striato. The Mockasine flower. 1954 C. J. Hylander *Macmillan Wild Flower Bk.* 65 The Lady's-slippers or Moccasin-flowers are easily recognizable by the inflated sac which forms the lip of the flower. 1970 R. T. Northen *Home Orchid Growing* (ed. 3) 197/2 Cypripedium. These are the moccasin-flowers or ladyslippers of our woods and moist meadows.

3. For '*Ancistrodon*' substitute '*Aghistrodon*'. (Earlier and later examples.)

1784 J. Filson *Discovery Kentucke* 27 The horned and the mockason snakes. 1784 J. Smyth *Tour U.S.A.* I. vii. 54 The most noxious, virulent, and deleterious of the species, the rattle, moccason, and horn-snakes. 1788 [see *Copper-belly]. 1965 R. & D. Morris *Men & Snakes* iv. 80 As recently as 1943, it was claimed that moccasin venom had been used with success in the treatment of rheumatoid arthritis.

4. *attrib.*, as **moccasin telegram** or **telegraph** *N. Amer.*, a means for the rapid or surreptitious transmission of information, orig. by an Indian runner; = **bush telegraph*.

1908 A. C. Laut *Conquest Gt. Northwest* II. 35 Word of the white woman ran before the advancing traders by 'moccasin telegram'. 1909 A. D. Cameron *New North* 349 And now, apprised by moccasin telegraph, we are all on the *qui vive* to catch sight of a floating bride. 1927 *Sat. Even. Post* 23 July 3/3 That agency known to white men as the Moccasin Telegraph, by which odd bits of news are flashed from one isolated native camp to another. 1969 *Islander* (Victoria, B.C.) 15 June 5/1 Word of the new constable was relayed to Simon by moccasin telegraph.

moccasined, *a.* **1.** (Earlier and later examples.)

1829 J. F. Cooper *Borderers* III. iii. 89 The two chiefs left the piazza in the noiseless manner of the moccasined foot. 1951 I. Shaw *Troubled Air* viii. 138 Moving a moccasined foot gently back and forth. 1968 M. Woodhouse *Rock Baby* xvii. 176 She started back towards the camp, her moccasined feet leaving damp footprints.

mocha¹. Add: **4.** A type of English pottery, made from the late eighteenth to the early twentieth century, with white or cream body decorated with coloured bands on to which moss- or fern-like patterns have been applied. Freq. *attrib.*

1837 S. Shaw *Chem. of Compounds used in Manuf. Porc.* I. v. 346 The readiness with which they combine with earths..renders them very useful in the Mocha and dipped ware. *Ibid.* II. i. 410 The peculiar kind of clay, in this neighbourhood,..veins of which are still kept open for supplying the same, fine in grain, and dark in colour, for the mocha dip. 1953 *N. Teulon-Porter Coll. Mocha Pott.* (Stoke-on-Trent Museum & Art Gallery Comm.) 1 There are a few collectors in the field and soon Mocha will vie for attention with lustre or transfer printed pottery. 1961 L. G. G. Ramsey *Connoisseur New Guide Antique Eng. Pott., Porc. & Glass* 67 Mocha ware, so named because of its resemblance to the quartz mocha stone, was in demand for kitchen jugs and mugs and large cups and saucers after its invention about 1780. 1968 *Canad. Antiques Collector* June 17/2 What is Mocha Ware? Sometimes referred to as 'Leeds Ware' or 'banded creamware' it is a creamware decorated with seaweed or tree silhouettes. This was made from 1787 up to 1903.

mocha². Add: **1. b.** Used *attrib.* of cakes, puddings, etc., flavoured with coffee, or coffee and chocolate.

1892 A. B. Marshall *Larger Cookery Bk.* xii. 475 *Moka Cake*... Prepare a Genoise paste mixture.., bake for an hour and a quarter, then turn out, and, when cold, mask over with Coffee glace. 1908 J. Kirkland *Mod. Baker* III. lxxxi. 390 A butter cream flavoured with coffee and vanilla..will be useful for flavouring and decorating the popular cakes known as *Mocha fancies*. 1963 V. Nabokov *Gift* v. 305 Old Stupishin, whose spoon was working its way through a wedge of mocha cake. 1972 M. J. Bosse *Incident at Naha* ii. 102 His skin was like rich mocha chocolate. 1975 *New Yorker* 3 Feb. 26/2, I sold schnecken, cookies, apple turnovers, and mocha tarts.

c. A shade or tint of the colour of mocha coffee; a dark brown colour. Also *mocha brown*.

1895 *Montgomery Ward Catal.* 274/3 Men's Fedora Hats ..Colors: Blue black, mocha brown and slate. *Ibid.* 275/1 Fedora Hats... Colors: Black, dark brown, gray, tan and mocha. 1919 T. S. Eliot *Sweeney among Nightingales* in *Poems*, The silent man in mocha brown Sprawls at the window sill and gapes. 1971 *Vogue* 15 Sept. 15 Eye shadow in earthy wine, rose, turquoise and mocha. 1974 *Harrods Christmas Catal.* 15 Leather clutch bag... Black, mocha brown, navy, rust, or terracotta.

3. (See quot. 1968.) Also *attrib.*

1895 *Montgomery Ward Catal.* 290/2 Men's dressed mocha kid gloves. 1922 [see *Boulton *a.*]. 1938 E. Bowen *Death of Heart* II. iv. 235 The stitching on her brown mocha gloves. 1968 J. Ironside *Fashion Alphabet* 238 *Mocha*, a fine, soft, hard-wearing leather made from sheepskin from Arabia, Africa and Persia. The chief outlet was Mocha in Arabia, hence the name.

‖ **mochi** (mǫ·tʃi). Also 7 musho. [Jap.] A cake made from pounded, glutinous rice.

In some quots. *mochi* has the honorific prefix *o-*.

1616 R. Cocks *Diary* 10 Feb. (1883) I. 109 Shezque Dono..came to the English howse and brought a present of *mushos*, wyne, and redish. 1880 I. L. Bird *Unbeaten Tracks Japan* I. 235 *Mochi*, a small round cake of unbaked rice dough, though insipid, is not unpalatable, and is much in favour. 1891 A. M. Bacon *Jap. Girls & Women* i. 5 Cakes of mochi, or rice paste. 1936 K. Tezuka *Jap. Food* 49 *Mochi* (glutinous rice boiled and pounded) is a special New Year food. 1960 B. Leach *Potter in Japan* ii. 62 We were taken indoors and fed on what was described to me as good country food, which included O Mochi. These are dumplings made of pounded steamed rice of a particularly glutinous variety. *Ibid.* 239 *Mochi*, steamed and dried cakes of glutinous rice. 1970 J. Kirkup *Japan behind Fan* 106 We would sip green tea and beer and nibble peanuts and seaweed biscuits and *o-mochi*, or sweet bean cakes.

Mochica (motʃī·kă), *a.* and *sb.* Also **Mochican, Moche, 9 Moxa.** [Sp., f. an Indian word; cf. *Moche*, the name of an archæological site in the valley of the same name in the coastal region of northern Peru.] **A.** *adj.* Of or pertaining to the Mochica, a pre-Inca people living on the Peruvian coast, or their modern descendants, or the language spoken by them. **B.** *sb.* **a.** The name of this people or a member of it. **b.** The language of the Mochica.

1853 F. L. Hawks tr. *Rivero & Von Tschudi's Peruvian Antiquities* v. 97 The Moxa language has strictly no declension. 1871 *Jrnl. R. Geogr. Soc.* XLI. 283 Bishop Luis Geronimo de Orè..is the only source from which we get specimens of the Puquina and Mochica languages. *Ibid.* 323 We have..the Lord's Prayer in the *Mochica*, a dialect spoken in the valleys of Runahuanac and Huarco. *Ibid.* 326 Tribes along the Peruvian coast..the *Chimus* (the *Yuncas* of Carrera); the *Mochicas* (the *Chinchas* of Garcilasso). 1877 E. G. Squier *Peru* viii. 128 You should have seen my Moche children forty years ago. 1927 *Geogr. Rev.* XVII. 42 The inhabitants of Chan-Chan spoke a language known as Yunga (or Yunca) or Mochica of which today vestiges survive in the vicinity of Eten. 1948 [see *Bichrome *a.* and *sb.*²]. 1953 A. C. Kinsey et al. *Sexual Behavior Human Female* vii. 231 The Mochican pottery of ancient Peru..depicts practically every petting and coital technique. 1959 [see *Huaco]. 1961 J. B. Priestley *Saturn over Water* v. 57 A monochrome ceramic, of the Mochica Culture, representing a warrior in ambush. 1965 A. Emmerich *Sweat of Sun & Tears of Moon* ii. 15 The Mochica also continued the use of elaborate gold headdresses by important personages. 1967 R. Peniston-Bird tr. *Gallo's Gold of Peru* 15 The Mochicans of the Peruvian coast..having as their centre the pyramids of the sun and moon at Moche developed the science of metallurgy before most of the Peruvian peoples. 1971 L. A. Boger *Dict. World Pott. & Porc.* 229/2 The decoration on Mochica pottery which is painted or modeled or a combination of both, gives a vivid picture of the life and customs of the people.

mochree, var. *Machree.

mock, *sb.*¹ Add: **1. b.** *to put the* (or *a*) *mock*(*s*) *on* (someone): see quot. 1943. *Austral. slang.*

1911 E. Dyson *Benno* 33 It's up t'me t'put a mock on that tripester. 1938 X. Herbert *Capricornia* xxxii. 482 'He put the mocks on me,' roared Norman... 'What's he saying, dear?' 'He..reckons I told the police on him.' 1943 Baker *Dict. Austral. Slang* (ed. 3) 51 *Mocks on, put the*, to upset someone's plans, to spoil a person's calculations. Also, 'put the mock on'. 1965 W. Grout *My Country's 'Keeper* xx. 206, I hope I am not 'putting the mock' on Norm because my feelings are the same as the rest of the Australian Test players: When O'Neill is a doubtful Test starter the job always looks grimmer.

mock, *a.* Add: **1. c.** (Earlier and later examples of *mock modesty*.)

1749 J. Cleland *Mem. Woman Pleasure* II. 12 The mask of mock-modesty was compleatly taken off. 1880 Swinburne *Heptalogia* 90 Thank my stars I'm as free from mock-modesty, friend, As from vulgar fatuity. 1962 I. S. Black *High Bright Sun* i. 8 She had..none of the island girls' self-consciousness, none of their mock modesty.

d. Designating an examination set by a school to give pupils practice for a specified public examination. Also *ellipt.* as *sb.*

1960 *Guardian* 22 June 6/4 A prefect enters... 'It was a long time ago that we did Mock.' 1960 *Where?* III. 15 '*Mock*' GCE, an internal examination..run by some schools as a rehearsal for the normal GCE examinations. 1964 C. Dale *Other People* iii. 71 June..had done Tennyson for mock GCE. 1967 *Guardian* 2 May 6/5 Some interviewers..asked if students had done 'mock' 'A' levels at school. 1969 C. Fremlin *Possession* xvii. 138 How could she ever get through her Mocks next term?

2. mock auctioneer, the auctioneer at a mock auction; **mock croc**: see *Croc² 1 b.

1959 *Daily Tel.* 13 Mar. 23/4 The requirement that a mock auctioneer should display his name and address was still valid though it was not enforced by the police. 1959 *Listener* 9 July 72/1 The mock-auctioneer in back street or fairground.

3. a. (Further examples.)

1931 *Times Lit. Suppl.* 18 June 479/1 The sumptuous mock-Tudor mansion. 1933 L. Bloomfield *Lang.* xxiii. 421 Mock-learned words, like *scrumptious, rambunctious, absquatulate*. 1936 *Discovery* Oct. 321/2 The short 'Brutus' curls of regency mock-classical beauties. 1949 Koestler *Insight & Outlook* vii. 105 His facial expression and whole attitude must be mock-aggressive. 1951 W. de la Mare *Winged Chariot* 58 Mock-solemn creatures, with our jackdaw airs. 1952 S. Kauffmann *Philanderer* (1953) iii. 38 Russell inquired, in customary mock-religious tones, about the state of the Street [sc. Wall Street] and cotton futures. 1958 *Spectator* 8 Aug. 193/1 The viewer who is sitting proudly in mock-antique splendour. 1968 *Listener* 18 July 92/3 A baron who mock-diffidently invites him to dinner. 1969 *Ibid.* 9 Jan. 43/3 We all went off to the pub: mock Tudor, phoney like the rest of us. 1975 J. Howlett *Christmas Spy* II. ii. 42 Her mock-Jacobean entrance hall.

mock, *v.* Add: **4. c.** *to mock up*: to make a mock-up of (see *Mock-up); also, to counterfeit, simulate, imitate; to contrive or improvise; freq. **mocked-up** *ppl. a.*

1911 *Encycl. Brit.* XXIV. 971/2 The shapes and sizes of the armour plates are sometimes obtained by the 'mocking up' process, in which the surface of the armour is represented in three dimensions. 1914 in W. S. Churchill *World Crisis 1911–14* (1923) 527 It is necessary to construct without delay a dummy fleet... They are then to be mocked up to represent particular battleships of the 1st and 2nd Battle Squadrons. *Ibid.* 528 The utmost secrecy must be observed, and special measures taken to banish all foreigners from the districts where the mocking-up [of the battleships] is being done. 1950 *Jrnl. R. Aeronaut. Soc.* LIV. 305/1 The first type should be used..provided that

certain sections, as for example the engine installation, are mocked up accurately, if necessary as a separate mock-up. 1952 *Archit. Rev.* CXII. 55 It consisted of fabrics and prototype furniture by Terence Conran, arranged in a room cunningly mocked-up with a couple of venetian blinds and a bamboo ceiling. 1955 A. H. N. Green-Armytage *Portrait St. Luke* vii. 121 Mocked-up discourses in the biography of a man whose trade it was to deliver discourses of his own. 1959 *Observer* 5 April 18/4 The fuddy-duddy diplomat whose mocked-up vicissitudes make the story. 1961 *Listener* 12 Oct. 576/3 Denis Mitchell mocked up a couple of glimpses of America. 1967 *Ibid.* 2 Feb. 175/3 Not a very good play, perhaps, with some mocked-up dialogue and sex brought in as a perfunctory afterthought.

mockage. 1. (Later examples.)

a 1916 A. R. Macewen *Hist. Church in Scotl.* (1918) II. xxvii. 176 In their mockage they termed every thing that repugned to their corrupt affections 'devout imagination'. 1922 E. A. Parry *What Judge Thought* viii. 133 It is interesting to remember that in Lewis Carroll, an ironist of a different type from Maule, we have another example of a deeply scientific mathematician revelling in the expression of ludicrous antiphrasis and quaint ridicule and mockage of commonplace humanity.

mocker¹. Add: **1. c.** *transf.*

1611 Bible *Prov.* xx. 1 Wine *is* a mocker, strong drink is raging. 1972 N. Marsh *Tied up in Tinsel* vii. 177 'He was a wine-bibber,' Nigel shouted. 'Wine is a mocker.'

d. Slang phr. *to put the mocker*(*s*) *on*: to thwart or bring bad luck to (a person, enterprise, etc.), to deride, mock, denigrate. Also *to have* (*got*) *the mockers on, to give* (one) *the mockers*, etc. Cf. *Mock *sb.*¹ 1 b.

1923 C. Drew *Rogues & Ruses* 115 They'll ave to race without me to-morrow. I've got a mocker hung on me. 1949 L. Glassop *Lucky Palmer* vii. 62 It's that sheila... She's put the mocker on us. 1970 'B. Mather' *Break in Line* v. 61 'He isn't bad at all.' 'Then what did you put the mockers on him for?' 1970 J. Porter *Dover strikes Again* ii. 36 This investigation had got the mockers on it from the start. 1974 'J. Ross' *Burning of Billy Toober* xi. 102, I tailed him... Not to do anything. Just to let him see he was being tailed. To give him the mockers.

mocker² (mǫ·kəɹ). *Austral.* and *N.Z. slang.* Also **mokker.** [Origin obscure.] Clothes; a dress. So **mo·ckered-up** *a.*, dressed up.

1938 Partridge *Dict. Slang* (ed. 2) 1014/2 *Mockered up*, dressed in one's best. 1945 Baker *Austral. Lang.* vi. 119 *All laired up* and its synonym *all mockered up* may also be noted. 1953 —— *Australia Speaks* iv. 106 *Mocker*, clothes in general. 1959 G. Slatter *Gun in my Hand* 51 Gives us a hand sometimes on the mixer or labourin about. Gets into his old mocker and gets stuck in. 1965 M. Shadbolt *Among Cinders* xxv. 250 She was mockered up to the nines.

mocker-nut. (Earlier and later examples.)

1814 F. Pursh *Flora Americæ Septentrionalis* II. 638 *Juglans tomentosa*... This is known under the name of Mocker Nut, White-heart Hickory or Common Hickory. 1832 D. J. Browne *Sylva Amer.* 187 In the part of New Jersey which lies on the river Hudson, this species is known by the name of Mockernut Hickory. 1926 E. M. Roberts *Time of Man* 364, I see you dance under the mockernut tree. 1947 Collingwood & Brush *Knowing your Trees* 152/1 Attaining maturity at 250 to 300 years, mockernut hickory sometimes reaches a height of ninety or a hundred feet. 1969 T. H. Everett *Living Trees of World* 98/2 The mockernut..forms a handsome specimen that grows up to 90 feet high with a trunk up to 3 feet in diameter... Its nuts have thick, hard shells that contain a very small amount of sweet meat.

mock-up (mǫ·kʊp). [f. *Mock *v.* 4 c.] **a.** An experimental model (often full-sized) of a projected aircraft, ship, apparatus, etc., used esp. for study, testing, practice, or display. Also *attrib.*

1920 *Flight* 19 Feb. 218/1 If the dimensions of such parts are difficult to determine on paper, the use of 'mock-ups' should be resorted to for this purpose. 1933 *Jrnl. R. Aeronaut. Soc.* XXXVII. 759 A mock-up of the eventual design was built to full scale with every instrument, lever, and fitting installed. 1944 U. Sinclair *Presidential Agent* (1945) IV. xv. 292 At the moment the Germans had the fastest fighter [plane], but Robbie had a new one in the 'mock-up' stage that was going to knock them all cold. 1944 [see *Console *sb.* 3 b]. 1951 R. Bradbury *Illustr. Man* (1952) 106 It's only a mockup... When they plan a rocket they build a full-scale model first, of aluminium. 1966 T. Pynchon *Crying of Lot 49* iv. 84 She was gazing at a mockup of a space capsule. 1968 *Daily Tel.* (Colour Suppl.) 8 Nov. 23 (*caption*) This instrument panel mock-up..resembles as closely as possible the eventual airliner. 1971 *Sunday Australian* 8 Aug. 3/3 A 747 training mock-up at Sydney Airport.

b. *transf.* and *fig.* A plan, model, conception; an imitation.

1954 G. Smith *Flaw in Crystal* xvi. 163 These coloured sheets..were mock-ups for a new children's magazine. 1957 J. F. Horner *Summary of Scientology* 79 Because they do not immediately alter to fit his mock-up, he is constantly failing. 1959 J. Cary *Captive & Free* xiii. 68 'What I want to know,' she said, 'is when we're going to see the mock-up of the new front page.' 1959 *Guardian* 4 Dec. 13/5 The thing is written in a half-bantering Anglicised mock-up of the Spanish idiom. 1961 L. Mumford *City in Hist.* Note to plate 38, 'All under one roof' may prove just a mock-up for the terminal form of the anti-city.

mocky (mǫ·ki). *U.S. slang.* Also **mockey, mockie.** [Origin uncertain; perh. f. Yiddish *makeh* a boil, sore.] A Jew. Also *attrib.* or as *adj.*

1931 D. Runyon in *Collier's* 10 Jan. 10/3, I consider this..disrespectful, like calling Jewish people mockies, or Heebs, or geese. 1937 E. H. Sutherland *Professional Thief* i. 12, I was over on the East Side and there saw this Jew who was pointed out to me as one of their best mocky cannons. 1943 I. Wolfert *Tucker's People* xxix. 481 Love thy neighbor if he's not..a mockie or a slicked-up greaseball from the Argentine. 1955 *Publ. Amer. Dial. Soc.* xxiv. 90 Jewish organizations are referred to as *mocky mobs* or *mocky jew mobs*. 'Mocky is not a Jew. It's a Sixth Avenue Jew.'

mocock (mokǫ·k). *N. Amer.* Also **makak, makuk, mocuck, mohcock, mokuk, muccuck.** [American Indian.] (See quot. 1827.)

1779 J. Long *Jrnl.* (1904) II. 155 We were reduced to a few fish and some wild rice, or *menomon* (which are kept in *muccucks*, or bark boxes). 1827 T. L. McKenney *Sk. Tour to Lakes* 194 A mocock is a little receptacle of a basket form, and oval, though without a handle, made of birch bark, with a top sewed on with *wattap* (the fine roots of the red cedar, split,) the smaller ones are ornamented with porcupines' quills, died red, yellow, and green. 1839 C. M. Kirkland *New Home* xx. 138 The Indians bring in immense quantities [of whortle-berries] slung in panniers or mococks of bark on the sides of their wild-looking ponies. 1859 P. Kane *Wanderings among Indians N. Amer.* 32 My companion was cooking some fish in a moh-cock, Indian fashion (for we had lost our kettle). 1905 *N.Y. Even. Post* 6 May, An old squaw stopped to offer a small mocock, a birch-bark box, holding perhaps a pound of maple sugar. 1931 G. L. Nute *Voyageur* 80 In the spring maple sugar was also bought by the *makuk* (a birch-bark vessel) from the squaws. 1959 E. Tunis *Indians* iii. 53/1 By far the commonest birch-bark container was the mocuck, with a square bottom larger than its round top, that served as box or basket at need.

mod (mǫd), *sb.*² Colloq. abbrev. of Modification. Also occas. as abbrev. of Modify *v.*

1943 C. H. Ward-Jackson *Piece of Cake* 43 Has this Wimpey got the new escape gear mod? 1958 'N. Shute' *Rainbow & Rose* i. 17 Captain Pascoe had it modded, special. 1967 *Autocar* 5 Oct. 24/1 This Healey had all the works racing mods which brought the engine power up to 210 b.h.p. 1967 *New Scientist* 14 Dec. 654/2 If the 'mods' are minor, the production line could absorb them without much interruption. 1974 *Publishers' Weekly* 12 Aug. 54/2 A rising generation of behavioral psychologists..is fanning out into our society to do its 'behavior mod' thing... These lab-trained mod squads have begun infiltrating schools, American family life, [etc.].

mod (mǫd), *sb.*³ and *a.* Also with capital initial. [Abbrev. of Modern *a.* and *sb.* or Modernist.] **A.** *sb.* A teenager who is characterized by his sophistication and tidiness; freq. contrasted with *Rocker¹. Also *attrib.* **B.** *adj.* Modern, sophisticated, stylish, esp. in dress.

1960 *New Left Rev.* Sept.–Oct. 4/2 Teds and Mods, Beatniks and Ravers. 1963 *Guardian* 13 May 18/1 Fights between the 'mods' and the 'rockers'. 1964 *Observer* 24 May 12/2 Mods and Rockers have co-existed comparatively well for a year or so—the Mods, neatly dressed and on scooters, the Rockers in studded leather jackets and on motor-bikes. 1964 *Punch* 3 June 815/1 (*heading*) Modgirl. 1965 *Granta* Summer 10 Student activists..have taken more account of the mod image cultivated by some universities. 1966 *Punch* 5 Oct. 505/2 He is naturally anxious that your furnishings should include what the enlightened store now calls 'Mod Gear'. 1968 J. Ironside *Fashion Alphabet* 22 The Mods, both girls and boys, were very clean and neat and both wore close cut hair. *Ibid.* 194 The 'Mod' hair-cut, as opposed to long-haired 'Rockers', is short, neat and cut close to the head. 1970 G. Jackson *Let.* 4 Apr. in *Soledad Brother* (1971) 207 This running dog ..was transmitting the credo of the slave to our youth, the mod version of the old house nigger. 1972 *Daily Colonist* (Victoria, B.C.) 25 Feb. 4/2 The operation by the Metropolitan Police..of a 'mod squad'..whose members stroll the streets in 'mod' dress and unshorn hair. 1973 E. Bullins *Theme is Blackness* 167 Everybody in our integrated circle of mod people is with it, man. We're the Now Crowd. 1975 *Islander* (Victoria, B.C.) 3 Aug. 4/3 Jerry, a mod young priest from San Diego.

mod (mǫd; *also read as* 'modulo'), *prep. Math.* Also **mod.** (with point). Abbrev. of *Modulo *prep.*

The notation $b \equiv c$ (mod. a) ('b is congruent to c modulo a') was introduced by Gauss (*Disquisitiones Arithmeticae* (1801) I. 2).

1854 *Cambr. & Dublin Math. Jrnl.* IX. 85 Each of these quantities must be congruent to zero, that is $A_0 \equiv 0$, $A_1 \equiv 0$, ... $A_{p-1} \equiv 0$ (mod. p). 1860 *Rep. Brit. Assoc. Adv. Sci. 1859* I. 230 The congruence $\phi(x) \equiv 0$, mod P, is said to be solved, when all the integral values of x are assigned which make the left hand number of the congruence divisible by P. 1949 Uspensky & Heaslet *Elem. Number Theory* vi. 128 Two congruences with the same moduli can be added or subtracted, member by member, like equalities. In other words, from two congruences $A \equiv a$ (mod m), $B \equiv b$ (mod m) it follows that $A \pm B \equiv a \pm b$ (mod m). 1949 W. Ledermann *Introd. Theory Finite Groups* i. 17, $ax \equiv bx$ (mod. m) implies that $a \equiv b$ (mod. m). 1971 D. Gorenstein in Powell & Higman *Finite Simple Groups* ii. 67, q is an odd prime power congruent to -1 (mod 4).

modacrylic (mǫdăkri·lik). [f. Mod(ified *ppl. a.* + *Acrylic *sb.*] A type of synthetic fibre consisting of molecules with between 35 and 85 per cent by mass of $—CH_2CH(CN)—$ units (derived from acrylonitrile), which is used for children's nightwear, dresses, and household textiles.

1959 *Federal Register* (U.S.) 10 Feb. 981/2 The following generic names for manufactured fibers..are hereby established... Modacrylic. 1964 *Which?* Aug. 253/1 Fabrics made from modacrylic are easy to wash in warm water, and require little or no ironing. 1973 *Materials & Technol.* VI. vii. 495 Materials such as Saran and modacrylics do not burn but shrink away from the flame. 1974 *Encycl. Brit. Micropædia* VI. 958/2 Wigs made of modacrylics have had good acceptance.

modal, *a.* and *sb.* **A. 4.** Add to def.: Esp. in various collocations, as *modal logic*, that branch of logic which is concerned with the study of modal propositions (see also quots.).

1932 Lewis & Langford *Symbolic Logic* vi. 153 (*heading*) Consistency and the modal functions. 1943 *Mind* LII. 265 The ideal textbook in mathematical logic would include extensive discussion of the intensional and modal logics. 1957 *Jrnl. Symbolic Logic* XXII. 176 (*title*) New foundations for Lewis modal systems. 1957 A. N. Prior *Time & Modality* 133 There are some modal logicians who feel that statements containing sequences of modal operators like *MM, MML*,..are one and all 'meaningless'. 1962 W. & M. Kneale *Devel. of Logic* x. 613 We turn to consider the possibility of using quantifiers to operate across modal signs. 1968 Hughes & Cresswell *Introd. Modal Logic* p. xi, Modal logic can be described briefly as the logic of necessity and possibility, of 'must be' and 'may be'. *Ibid.* ii. 25 Because of the non-truth-functionality of modal operators..the initial account does not lead to any obvious formal definition of validity for modal formulae. 1970 J. N. Findlay tr. *Husserl's Logical Investigations* I. 32 Modal distinctions play a central part in phenomenological theory, and Husserl is now contributing importantly to their own phenomenology.

5. a. *Esp.* in phr. *modal auxiliary*.

1933 E. H. Grout *Stand. Eng.* v. 122 The modal auxiliaries *may, might, can, could, must, ought*,..give a cast to the whole sentence in which they are. 1961 R. B. Long *Sentence & its Parts* vi. 138 A category of modal auxiliaries is often set up for modern English, to include various verbs expressing ideas of possibility, constraint, and desire.

6. a. *Statistics.* Of or pertaining to a mode (sense *7 c); occurring most frequently in a sample or population.

1897 *Proc. R. Soc.* LXII. 175 Probable error of modal frequency y_0. 1900 K. Pearson *Gram. Sci.* (ed. 2) 383 The average value of the character is very frequently taken as determining the type instead of the modal value. 1938 A. E. Waugh *Elem. Statistical Method* iv. 46 In the first place it is necessary to locate the modal class. By this we mean the class which contains the most items. 1954 M. Beresford *Lost Villages* ix. 288 In the receipts of 1377 we have only the constables' names to add flesh and blood to the averages, modal ranges and medians of statistical calculations. 1968 *Listener* 25 July 101/1 The administrators we saw..had averaged only 2·8 years in all their completed jobs in the class; in fact, the modal (most frequently occurring) period in completed jobs was two years. 1973 *Jrnl. Genetic Psychol.* CXXII. 248 The modal age of the youngsters was 13.

b. Representative, typical; *modal personality*, an imaginary personality in which each component trait or characteristic is present to an extent equal to the modal value of a particular society or group or, more widely, which is taken as in some way representative of it.

1944 C. Du Bois *People of Alor* I. i. 3 Modal personality, then, is the product of the interplay of fundamental, physiologically and neurologically determined tendencies and experiences common to all human beings acted upon by the cultural milieu. *Ibid.* 5 On such a base line data will show central tendencies that constitute the modal personality for any particular culture. 1948 K. Davis *Human Society* xv. 427 The modal divorce now occurs in the third year of marriage. 1949 R. K. Merton *Social Theory* i. 57 The characteristic (modal) pattern for handling a standardized problem. 1954 Inkeles & Levinson in G. Lindzey *Handbk. Social Psychol.* II. xxvi. 980/2 In our opinion, 'national character' ought to be equated with modal personality structure. 1956 W. H. Whyte *Organization Man* (1957) 281 What might be called the modal man, however, is a twenty-five-to-thirty-five-year-old white-collar organization man. 1968 *McGraw-Hill Yearbk. Sci. & Technol.* 50/1 Examples of these typical motor patterns are the deep cooing and bowing of the domestic pigeon, the butting of a billy goat, [etc.]... These motor patterns have been given a wide variety of names, such as modal action patterns..and instinctive movements. 1968 J. O. Ellefson in E. Norbeck et al. *Study of Personality* ix. 142 Field primatologists report impressions of the existence of modal personalities, often designated as temperament, that characterize species. 1970 E. McGinnies *Social Behavior* iii. 70 Linton (1945) has conceived of national character as a modal personality structure, or configuration that appears with considerable frequency in a society. *Ibid.* 71 Identification of such modal behavior configurations makes it possible to arrive at certain generalizations about any given society and to describe more succinctly the differences between one society and another.

7. *Petrol.* Of or pertaining to the mode (sense *5 b) of a rock; as indicated by a mode.

1902 W. Cross et al. in *Jrnl. Geol.* X. 609 A Modal Variety..may be defined as a rock having a mode with a slightly different development of the quite subordinate component minerals. 1938 *Nature* 17 Sept. 495/2 The second volume includes all the rocks with more than 5 per cent of modal quartz. 1962 A. E. J. & C. G. Engel in A. E. J. Engel et al. *Petrologic Stud.* 48 The thin sections used for modal analyses of two-pyroxene amphibolites were slightly thicker than is conventional, to accentuate the color difference.

B. *sb.* **2.** *Gram.* A modal verb (see A. *adj.* 5).

1959 *Rep. 10th Ann. Round Table Meeting Ling. & Lang. Stud.* (Georgetown Univ. Inst. Lang.) vi. 112 Parallel rules apply to most of the modals and conjugators. 1965 N. Chomsky *Aspects of Theory of Syntax* ii. 63 *May* is a verbal auxiliary..and..a Modal. 1971 J. Anderson in A. J. Aitken et al. *Edin. Stud. Eng. & Scots* 69 These phenomena are often well documented in grammars purporting to give an account of the modals.

modality. Add: **1. b.** In diplomacy, politics, etc.: a procedure or method; a means for the attainment of a desired end.

1957 G. F. Kennan in *Listener* 28 Nov. 868/2 The modalities of German unification must flow from the will of the German people, expressed in free elections. 1960 *Guardian* 23 Aug. 7/6 He did hear nine members of the Council praise his statesmanship and the procedures ('modalities' is the new and foolish word) he had adopted. 1970 *New Yorker* 17 Oct. 162/2 The new word that is constantly being heard here is 'modalities'. Everyone involved in the peace talks agrees that the military modalities of a cease-fire are more easily negotiated than the political modalities.

2. b. (Later examples.)

1949 Hutten & Reichenbach tr. *H. Reichenbach's Theory of Probability* x. § 80. 404 Like probabilities, the modalities must be regarded as properties not of individual propositions but of propositional sequences. 1951 G. H. von Wright *Ess. Modal Logic* 3 Related to the problems of mixed modalities are the problems of superimposed or higher order modalities.

4. *Psychol.* **a.** (See quot. 1909.)

1895 *Amer. Jrnl. Psychol.* VII. 84 Sinn, sense, sensibility, modality. 1909 *Cent. Dict. Suppl.*, *Modality*, in Psychol.: (*a*) the nature or character of sensation or stimulus as determined by the sense-department to which it belongs or appeals: a term proposed by Helmholtz, to avoid a confusing use of *quality*.. Hence (*b*) the sense-department itself; as, the sensations of different *modalities*. 1925 G. B. Phelan (*title*) Feeling experience and its modalities. 1951 G. Humphrey *Thinking* ii. 57 Sensory presentations of various modalities—auditory, kinaesthetic, and so on. 1971 tr. *H. von Helmholtz's Sel. Writings* xiv. 369 The most fundamental [distinction] is that among sensations which belong to different senses, such as the differences among blue, warm, sweet, and high-pitched. In an earlier work I referred to these differences in the *modality* of the sensations. 1972 D. R. Kenshalo in Kling & Riggs *Woodworth & Schlosberg's Exper. Psychol.* (ed. 3) v. 119/1 We insist that each primary sensory modality has its own nerve pathway, the tactile, pain, and temperature senses fail to qualify as different modalities because their nerves are intermingled.

b. A term used to denote qualitatively different attributes or traits of personality.

1946 R. B. Cattell in *Brit. Jrnl. Psychol.* May 159 Three classes or 'modalities' of traits: (1) Dynamic traits, e.g. dispositions, sentiments, neurotic symptoms, ergs; (2) Temperament traits, e.g. general emotionality, surgency, preservation, hyperthyroidism, personal tempo; (3) Abilities or cognitive traits, e.g. native general intelligence, acquired perceptual and executive skills. 1962 E. R. Hilgard *Introd. Psychol.* (ed. 3) xvi. 452/2 Guilford.. writes of seven 'modalities of traits', indicating that the kind of trait we see depends upon the direction from which we view personality. 1964 L. J. Bischof *Interpreting Personality Theories* xiv. 594 Formally, Cattell proceeds to divide traits into three *modalities*: temperament, dynamics, and ability. 1972 *Jrnl. Social Psychol.* LXXXVII. 52 The conditioning events taking place during this phase leave an indelible imprint on the psychological modalities referred to as personality.

modalize, *v.* Add: (Later examples.) Hence **modali·zable** *a.*, **modaliza·tion**, **mo·dalized** *ppl. a.*

1955 A. N. Prior *Formal Logic* iii. i. 202 'Modalization' here plays a part. *Ibid.*, Provided that no propositional variable occurring in *β* occurs in *α* unless it is 'modalized', i.e. is either immediately preceded by '*M*' or '*L*' or occurs as part of a propositional formula which as a whole is preceded by '*M*' or '*L*'. *Ibid.* 209 A similar hexagon.. could be constructed for modalized conjunctive and alternative forms. 1957 —— *Time & Modality* 137 The only modalizable forms are single variables with or without a preceding sequence of N's. 1963 A. Kenny *Action, Emotion & Will* ix. 189 Descriptions of formal objects can be formed trivially simply by modalizing the relevant verbs. 1973 J. J. Zeman *Modal Logic* v. 80 Systems containing only a finite number of non-equivalent modalities will.. be called 'systems of complete modalization'. *Ibid.*, A formula is completely modalized if the prefixing of modal operators to it does not change its 'modal quality'.

mod. con. (mǫd kǫn). Also **mod. cons.** Colloq. abbrev. of *modern convenience*(*s*) (see *MODERN *a.* 3 a). Also *transf.*

1934 *Punch* 24 Jan. 86/2 An advertisement..describing just such a house as we wanted. Just the right number of rooms, 'five minutes from the station, h. & c. in all bedrooms, all mod. cons.' 1952 A. Hocking *Best Laid Plans* ix. 135 Four bedrooms, two sitting-rooms, k. and b., as the house agents say; every mod. con. 1963 *Times* 13 Feb. 11/4 Such a sophisticated 'mod. con.', the equivalent of a hot bath to splash in, is, generally speaking, beyond the ambition of the normal duck. 1966 'H. MacDiarmid' *Company I've Kept* xiii. 188 We had no 'mod cons', and were getting too old to put up with really primitive conditions. 1972 G. Durrell *Catch me a Colobus* v. 99 It's a modest little place..but it's got all mod. con. and that sort of thing.

moddam, moddom, moddum: see *MODOM.

mode, *sb.* **I. 2.** Delete † *Obs.* and add later examples. Also freq. *attrib.*

1933 L. Bloomfield *Lang.* xvi. 273 In English..the unreal appears only in clauses introduced by *if* or *though*, or in combination with the phrasal mode-forms (*he would help us*, unreal of *he will help us*). 1946 H. Hoijer et al. *Ling. Struct. Native Amer.* 97 The forms of the verb [in Algonquian] fall into five *orders*. Each order consists of one or more *modes*, each with a full set of forms. 1946 C. Morris *Signs, Lang. & Behavior* v. 125 A third possibility ..may be called the *mode-use classification*. 1961 R. B. Long *Sentence & its Parts* 495 Five modes are recognized here: common (or 'indicative'), subjunctive, infinitival, gerundial, participial. 1965 *Canad. Jrnl. Ling.* Spring 175 We have now considered all the verb prefixes except those for mode-aspect. The Eyak mode-aspect system is relatively well-proportioned and clear-cut.

b. *Linguistics.* (See quots.)

1954 K. L. Pike *Lang. in Rel. Human Behavior* i. iii. 35/2 On any level of focus each..emic unit..is divided structurally into three specific kinds of complex overlapping components which I shall call *modes*. 1967 W. A. Cook *On Tagmemes & Transforms* i. 9 Every linguistic sign is defined by its meaning, form and distribution... These are included in Pike's three modes: the manifestation mode, the feature mode, and the distribution mode. *Ibid.* 10 The tagmeme can be fully defined, parallel to the phoneme and morpheme, with its own peculiar feature, manifestation, and distribution modes.

3. b. (Later examples.) Also in wider use (see quots.).

1937 A. Smeaton tr. *Carnap's Logical Syntax of Lang.* iv. § 68. 247 Some of the known examples of intensional sentences belong to the autonomous mode of speech. 1941 O. Helmer tr. *Tarski's Introd. Logic* viii. 175 The proof of Theorem 1—like any other indirect mode of inference—can be brought under the schema sketched above. 1946 C. I. Lewis *Analysis of Knowl.* i. iii. 39 It is desirable to recognize two further modes also, which will here be called, respectively, comprehension and signification. 1965 J. O. Urmson *Philos. Analysis* 37 'Xs are logical constructions' is in the material mode of speech. 1966 W. V. Quine *Ways of Paradox* xiii. 156 Whatever may be said about necessity may be said also, with easy and obvious adjustments, about the other modes. 1970 A. E. Blumberg tr. *Stegmüller's Main Currents Contemp. Germ., Brit. & Amer. Philos.* vi. 234 Two modes may exclude each other (necessity and impossibility). *Ibid.* vii. 308 Put in the formal mode his thesis simply states that all thing-statements can be translated into a sense-data language (a phenomenalistic language).

4. c. *Physics.* Any of the distinct kinds or patterns of vibration that an oscillatory system can sustain.

1867 J. Tyndall *Sound* v. 188 When we make the same passage [from a fundamental tone to the first overtone] in a stopped pipe, we obtain a note a fifth above the octave. No intermediate modes of vibration are..possible. 1877 Rayleigh *Theory of Sound* I. vi. 141 When a string vibrates in its gravest normal mode, the excursion is at any moment proportional to sin *πx/l*. *Ibid.*, The production of 'harmonics' by lightly touching the string at the points of aliquot division is a well-known resource of the violinist. All component modes are excluded which have not a node at the point touched. 1911 *Encycl. Brit.* XXV. 454/1 In fig. 34 the stationary wave systems of the first four modes are represented. 1949 H. E. Penrose *Princ. & Pract. Radar* 626 Energy may be propagated in a wave-guide in a doubly infinite series of modes analogous, to a small extent, to a singly infinite series of modes represented by a fundamental note and its harmonics. The modes are distinguished by the patterns of the lines of force traversing the fields. 1950 Stephens & Bate *Wave Motion & Sound* 386 In the theory of the specific heats of a solid the thermal vibrations are supposed to result from many simultaneous modes, whose phases have a random distribution. 1962 *Newnes Conc. Encycl. Electr. Engin.* 883/2 The various higher-order modes travel [along a waveguide] with different velocities. 1969 L. Allen *Essent. Lasers* ii. 13 The large number of modes at infrared or optical frequencies which are present in any cavity of reasonable size poses problems. This is because of the need for a high level of spontaneous emission to ensure that sufficient photons go into any one particular mode, to maintain the rate of stimulated emission.

5. b. *Petrol.* The quantitative mineral (as distinct from chemical) composition of a rock sample. Cf. *NORM 2.

1902 W. Cross et al. in *Jrnl. Geol.* X. 604 We introduce two terms..as substitutes for the cumbrous and oft-repeated expressions, standard mineral composition (that calculated from the rock analysis) and actual mineral composition. For the first we propose the word norm, and for the second the word mode. 1932 A. Johannsen *Descr. Petrogr. Igneous Rocks* II. 111 (*heading*) Table 63. Modes of sodaclase-granites. 1962 H. R. Cornwall in A. E. J. Engel et al. *Petrologic Stud.* 361 In ash flows in cooling unit 3 of the Bullfrog Hills caldera..lithologic differences between the lower and upper parts indicate quite certainly that more than one flow is present. The differences are shown by variations in the mode. 1974 *Nature* 16 Aug. 562/2 The Clare Castle gneisses are coarse grained and weakly banded... Modes vary within the range: garnet 20–30%, sillimanite 5–20%, plagioclase 25–30%, potash feldspar 10–20% and quartz 10–40%.

II. 7. c. *Statistics.* The value or range of values of a variate for which there is a maximum number of instances in a given population.

1895 K. Pearson in *Phil. Trans. R. Soc.* A. CLXXXVI. 345, I have found it convenient to use the term *mode* for the abscissa corresponding to the ordinate of maximum frequency. Thus the 'mean', the 'mode', and the 'median' have all distinct characters. 1906 R. H. Lock *Rec. Progress in Study of Variations* 89 When dealing with a symmetrical curve the position of the mode is identical with that of the median. 1947 O. L. Davies *Statistical Methods in Res. & Production* iii. 27 In most industrial applications, however, distributions with more than one mode (multimodal) are, or should be, rare. The presence of two or more modes usually means that the sample is not homogeneous, i.e. that two or more distinct distributions have been combined. 1948 L. D. & A. Crow *Educ. Psychol.* xx. 393 The score in a given set of data that appears most frequently is called the mode. 1973 *Jrnl. Genetic Psychol.* CXXIII. 87 The Kuhlman–Anderson scores for the one group of Ss from a fifth-grade class yielded a mean of 93·91, a range of 46, a median of 93, and a mode of 92.

d. [Shortened form of F. *gris mode* fashion grey.] The name given to a variety of shades of grey (see quot. 1930).

1895 *Montgomery Ward Catal.* 125/3 Kid gloves... Colors: Black, brown, tan, mode, slate. 1930 Maerz & Paul *Dict. Color* 167/1 Mode..was a term used in the nineteenth century to indicate a *class* of colors..usually on the pale order, running from neutral grays to strongly tinted greys of all hues... The old pattern books contain hundreds of samples of different colors, of every conceivable hue, all called 'Mode'. 1957 M. B. Picken *Fashion Dict.* 224/2 Mode..pale, bluish-gray color, sometimes drab.

14. mode-locking *Physics*, a technique by which the phase of each mode of oscillation in a laser is 'locked' to those of the two adjacent modes (so that a fixed phase relationship arises between all the modes), resulting in the emission at intervals of about a nanosecond of short trains of extremely short pulses whose duration is of the order of picoseconds; so **mode-locked** *a.*, applied to a laser in which this technique is employed and to the resulting pulses; (as a back-formation) **mode-lock** *v. trans.*, to subject (a laser) to mode-locking.

1966 *Appl. Physics Lett.* VIII. 182/1 The YAG:Nd laser was mode locked over a frequency width of order 12·6 Gc/sec (42 × 300 Mc/sec). 1971 *Physics Bull.* Dec. 718/2 Flashlamp pumped dye lasers have been successfully modelocked to produce high power pulses of transform-limited durations of 2–3 ps. 1965 *IEEE Jrnl. Quantum Electronics* I. 16/2 The light pulses from a mode-locked laser were observed with a magnetically focussed photomultiplier tube. 1972 *Sci. Amer.* June 24/1 As the energy in the laser cavity is built up and then decays, a string of such mode-locked pulses emerges from the partially transmitting mirror in the front of the cavity. 1966 *IEEE Jrnl. Quantum Electronics* I. 16/1 (*heading*) Effect of mode-locking. 1967 *Science* 23 June 1558/3 In the first experimental demonstration of 'mode-locking', helium-neon and argon lasers were used.

model, *sb.* Add: **I. 2. d.** *Dentistry.* A positive copy of the teeth or oral cavity, which is cast in metal, plaster, etc., from an impression (sense *2 e) and which may be used to construct dental appliances.

1839 C. A. Harris *Dental Art* xxi. 348 The obtaining of a model of the alveolar ridge, or ridges, when one for each jaw is required, though apparently very easy, is nevertheless often attended with some difficulty. 1857 *Brit. Jrnl. Dental Sci.* I. 579/1 Mr. Saunders thought that the plan of bending down the front part of the model could be fatal to a correct impression. 1927 F. A. Peeso *Crown & Bridge-Work* vii. 140 When the plaster for the impression has been tinted, the impression and model are easily distinguished by the difference in coloring. 1938 *Dental Rec.* LVIII. 14, I think, from a study of the original models, that there had probably been pyorrhoea for some years. 1940 [see *IMPRESSION sb. 2 e]. 1973 D. H. Roberts *Fixed Bridge Prostheses* v. 66 Only one model can be poured from each impression.

e. A simplified or idealized description or conception of a particular system, situation, or process (often in mathematical terms: so *mathematical model*) that is put forward as a basis for calculations, predictions, or further investigation.

Cf. sense 1 c.

1913 N. Bohr in *Phil. Mag.* XXVI. 1 To explain the results of experiments on scattering of *α* rays by matter Prof. Rutherford has given a theory of the structure of atoms. According to this theory, the atoms consist of a positively charged nucleus surrounded by a system of electrons [etc.]... Great interest is to be attributed to this atom-model. 1923 [see *Bohr]. 1938 R. W. Lawson tr. *Hevesy & Paneth's Man. Radioactivity* (ed. 2) viii. 89 These and other fundamental facts were responsible for the introduction of the model of the atom already described. In spite of these facts, however, it was later necessary to replace the atomic model by conceptions of a less concrete nature..in order to be able to interpret more complicated spectra. 1939 H. Levy *Mod. Sci.* xxx. 515 Models of the universe have been erected that enable us with varying degrees of definiteness to picture these earlier stages. 1940 *Econ. Jrnl.* L. 91 Previous models of the Trade Cycle..have thus mostly been based on the assumption of statically stable situations, where equilibrium would persist if once reached. 1949 *Econometrica* XVII. 193 The Mathematical Model discussed here..is a generalization of the Leontief Inter-Industry Model. 1958 *Listener* 11 Dec. 972/1, I want to discuss the cosmological theories which are generally classed as the evolutionary models of the universe. 1969 J. Argenti *Managem. Techniques* 170 Any set of mathematical equations, linking together a complexity of factors and used to study the effects of change, is a Model. 1969 *Sci. Jrnl.* Dec. 27/1 Mathematical models of the global atmosphere..can now be constructed with the aid of large, fast computers to handle the enormous quantities of data and the complex equations that represent the movements and heat balance of the atmosphere. 1970 *Nature* 21 Nov. 719/2 The equivalence of the corpuscular and wave models in the theory of optics. 1971 *Daily Tel.* (Colour Suppl.) 3 Dec.

24/3 Forrester has designed a model of the world system to try to discover the long term effects of pollution and overpopulation. The model, processed through a computer, predicts a variety of different futures. 1972 G. H. A. Cole in Cox & Dyson *20th-Cent. Mind* I. viii. 250 A system of physical concepts and quantities that describes the main features of a situation is known as a model. 1973 *Sci. Amer.* Dec. 117/1 There are many other varieties of speech error. All of them must be accounted for in a model of speech production.

f. *spec.* in *Mathematical Logic.* A set of entities that satisfies all the formulas of a given formal or axiomatic system.

1940 W. V. Quine *Math. Logic* vi. 271 The fact that such classes constitute a model of the traditional real number system was pointed out by Dedekind. 1948 *Jrnl. Symbolic Logic* XIII. 16 (*heading*) Models of logical systems. 1952 S. C. Kleene *Introd. Metamath.* ii. 25 When the objects of the system are known only through the relationships of the system, the system is abstract... Then any further specification of what the objects are gives a representation (or model) of the abstract system, i.e. a system of objects which satisfy the relationships of the abstract system and have some further status as well. These objects are not necessarily more concrete, as they may be chosen from some other abstract system (or even from the same one under a reinterpretation of the relationships). 1963 W. V. Quine *Set Theory* vi. 135 We have provided a model of arithmetic in set theory when we have provided a way of so reinterpreting arithmetical notations in set-theoretic terms as to carry the truths of arithmetic into truths of set theory. 1974 *Encycl. Brit. Macropædia* XI. 639/2 By Gödel's completeness theorem of 1930, if a formal system based on the first-order functional calculus *F* is consistent, there is a model in which the objects are the natural numbers.

II. 7. d. An article of apparel of a particular design; a specified type or design of clothing; freq. with defining word prefixed.

1880 *Queen* 12 June (Advt.), Messrs Jay import from the first houses in Paris. Models of every style. 1906 *Bazaar, Exchange & Mart* Suppl. 3 Oct. 1308/1 Great bargains in ladies' wearing apparel, new and equal to new. Paris models. 1912 *Tatler* 23 Oct. 105 The forthcoming models are more than usually extravagant and..the latest creations all seem to have been designed 'regardless of cost'. 1933 N. Coward *Design for Living* I. 22 A silly pride made me show off to you, parade my attraction for you, like a mannequin. New spring model, with a few extra flounces! 1958 [see *CASUALNESS]. 1975 *Times* 29 July 8/5 The Valentino collection is untypically small... Strikes..have dogged the production of the models.

e. A motor vehicle of a particular design; a vehicle produced in a specified year; also *transf.*, one of a series of varying designs of the same type of object; also *fig. Model T*, an early type of car produced by the American Ford Company; also allusively, of a person or thing that is outmoded, mass-produced, etc. Also *attrib.* or as *adj.*

1900 *Automobile Topics* 22 Dec. 366/1 For sale. Two-passenger Winton, 1900 model, in first-class condition. 1901 *Ibid.* 21 Sept. 848 Type No. 2. Model 'C'—12 HP. Double Cylinder Gasoline Engine. 1909 *Automobile* 7 Jan. 9/1 Henry Ford made a name for himself which will cling for all time when he handed out a replica of a full-fledged automobile of the four-cylinder type at the price of a runabout—nay, at the price of the cheapest runabouts. Model T of the Ford line is in the same class, in that it is all automobile and no price. 1909 *Westm. Gaz.* 2 Nov. 5/1 The engine of the 20-h.p. model..is of the monobloc order. 1930 *Ibid.* 4 Jan. 5/1 Mr. Huff pays a visit to Europe..to inspect the new models at Olympia. 1912 V. W. Pagé *Mod. Gasoline Automobile* xii. 618 The Ford car is one of the most popular of moderate-priced automobiles and over 100,000 of the Model 'T' are now on the road. *Ibid.* 619 (*caption*) Outlining the distinctive control system of Ford Model 'T' automobile. 1927 *Motor Cycling* 7 Dec. 102 (*heading*) Road tests of 1928 models. 1930 L. Mumford *City Devel.* (1946) 63 One might call this the model T dilemma. Mass-production..suffers..from rigidity. 1930 H. Crane *Let.* 29 Dec. (1965) 360 The middle west business man, approved panic model of 1931. 1932 A. Huxley *Brave New World* ii. 27 Twenty-three years after Our Ford's first T-Model was put on the market. 1942 E. Paul *Narrow St.* ii. 17 Mary drew from somewhere inside her waist a dog-eared American passport of a model no longer in vogue. 1945 *Amer. Speech* XX. 148/1 *Model T*, non-contemporary. 1947 *Reader's Digest* Jan. 119/1 Such simple demands as wages, hours and working conditions are strictly Model T. 1955 W. Gaddis *Recognitions* iii. ii. 752 The minute you get used to the goddam thing some bastard puts out a new model. 1963 R. Wolff *I, Keturah* (1964) i. vii. 47 The Model T chugged up the hill. 1966 *Economist* 26 Mar. 1251/2 A return to the one-off, custom-built job, is playing right back into Britain's hands and away from the undeveloped nations churning out marine model-Ts. 1968 *Listener* 23 May 670/1 There are still 405-line-only models, which cannot be converted, on sale. 1970 *Globe & Mail* (Toronto) 26 Sept. B 1/5 During the past model year, the trend was accelerated by the..fact that the car companies had concentrated in their Canadian plants production of the models that happened to be most successful in the marketplace.

10. b. *Biol.* An animal or plant to which another bears a mimetic resemblance.

1877 *Encycl. Brit.* VI. 127/1 Probably this beetle shared in the immunity from attack accorded to its model. 1907 *Nature* 31 Oct. 673/2 An insect thus resembled by another is spoken of as its 'model', the imitating insect is called a 'mimic'. 1930 R. A. Fisher *Genetical Theory Nat. Selection* 148 The resemblance which is favourable to the mimic will be for the same reason disadvantageous to the model. 1968 R. D. Martin tr. *Wickler's Mimicry in Plants & Animals* i. 16 (*caption*) Leaf beetles..serve as models for roaches.., which are palatable and resemble their models so closely that they are also avoided by predators. *Ibid.* iv. 43 This weed [*sc.* rye-weed]..is less demanding and tougher than its model, the wheat plant.

11. b. Substitute for def.: A person, freq. a woman, who is employed to display clothes by wearing them, or to appear in displays of other goods. (Add further examples.)

a 1911 D. G. Phillips *Susan Lenox* (1917) II. i. 7 She was dressed in the sleek tight-fitting trying-on robe of the professional model. 1958 *Woman's Own* 5 Feb. 16/2 The first lesson every model learns is to stand and walk correctly. 1959 *Guardian* 26 Oct. 7/7 If the men were only going to become part-time models, they would need to go on doing one or two lessons a fortnight. 1962, 1970 [see *fashion-model* s.v. *FASHION sb. 13]. 1971 B. Patten *Irrelevant Song* 55 Their beauty more awkward than even the topmost models.

c. A euphemism for 'prostitute'.

1963 [see *company director* s.v. *COMPANY sb. 10]. 1968 J. Lock *Lady Policeman* ii. 19 There had been an increase of newsagents' notice-board ads for 'Models'. 1970 G. Greer *Female Eunuch* 195 Working as hostesses in high-class clubs, as 'models' or simply walking the streets.

13. For 'see 14' read 'see 15'.

b. *Sc. colloq.* A model lodging-house (see sense 15 in Dict.).

1899 'J. Flynt' *Tramping with Tramps* ii. 233 The price..is threepence a night, and this is the common price all over Great Britain, except in the so-called 'Models', where a penny more is charged simply for the very deceitful name. 1927 [see *flop-house* s.v. *FLOP sb. 5]. 1935 MacArthur & Long *No Mean City* xix. 282 'A model' in Gallowgate—one of those buildings which are ironically termed 'Working Men's Hotels'.

15. a. (Further examples.)

1849 J. S. Buckingham *National Evils* 25 My thoughts were thus..directed to..the desirability of forming at least one Model Town. 1857 Geo. Eliot *Scenes Clerical Life* (1858) I. 109 What a hobby farming is with Lord Watling!.. It is really a model farm. 1898 E. Howard *To-Morrow* iv. 41 Another site for a model city could be purchased. 1909 *Chambers's Jrnl.* Feb. 87/1 It [*sc.* Kinlochleven] is built on the most approved lines—a veritable model town. 1967 *Boston Globe* 28 May 14/1 The new Congress..almost stopped the Model Cities program. 1970 G. E. Evans *Where Beards wag All* xi. 117 A horseman applied for a job with a farmer who had a few years before built himself a *model* farm with the most up-to-date farm buildings and all the latest equipment.

b. *model-maker* (earlier and later examples); also *model-building, -making.* Also freq. attrib. in sense 2 a, as *model aeroplane, aircraft, boat, engine, railway, soldier, train, yacht.*

1920 Model aeroplane [see *model-making* below]. 1973 *Times* 27 July (Suppl.) p. iv/4 Thousands of different hobbies and specialisations: toy soldiers, model aeroplanes, cigarette cards. 1951 *Catal. of Exhibits, South Bank Exhib., Festival of Britain* 126/2 Model aircraft. 1974 *Times* 8 Feb. 15/5 Pursuits such as whippet racing and model aircraft flying. 1912 W. Owen *Let.* 2 July (1967) 147 Bournemouth Cliffs..and especially the Model-Boat-Canal were..familiar to me. 1974 *Country Life* 3–10 Jan. 56/4 Model boats wanted... Cased or uncased. 1957 B. F. Skinner in Saporta & Bastian *Psycholinguistics* (1961) 235/2 Model-building has a special status in the field of verbal behavior. 1972 *Computers & Humanities* VII. 79 His study is one of the richest we have for its methodological innovation, its model-building, and its attempt to treat the political culture of the time both as a system and as a block of evidence for larger concerns. 1906 E. Nesbit *Railway Children* i. 3 Peter had a birthday—his tenth. Among his other presents he had a model engine. 1598 Florio *Worlde of Wordes* 106/2 *Disegnante*, a map or modle maker. 1960 *Nature* 26 Dec. 928/2 He obtained a job as a model-maker to a firm of instrument manufacturers. 1965 *Math. in Biol. & Med.* (Med. Res. Council) iv. 132 Friendly model-makers from the physical sciences are tempted to construct theories of 'how the brain works' on the basis of a few isolated and easily mathematized facts. 1920 *Glasgow Herald* 29 Oct. 9 The Prince of Wales ..accepted from him a model aeroplane with which he won first prize in the junior section of a model-making competition. 1946 *Nature* 14 Sept. 361/2 The two tanks and propeller-testing tunnel were in operation, and it was possible to see every stage in the process of model-making and testing. 1963 L. Loevinger in H. W. Baade *Jurimetrics* 32 The basic elements that these procedures have in common are explication, model-making (or operational organization or programming), and the production of a testable conclusion. 1974 *Country Life* 12 Dec. 1860/1 The exhibition is divided into several sections each dealing with a different aspect of model-making. 1909 (*title of periodical*) Model railways and locomotives. 1972 *Times* 7 Aug. 2/4 A branch from an elm tree fell on to a model railway at Blenheim Palace. 1938 *Daily Herald* 21 Dec. 6/1 The collecting of model soldiers is a nursery pastime elevated into the dignity of an adult occupation by the word 'research'. 1973 *Country Life* 17 May 1385/1 Model soldiers have gone far beyond the realm of child's play and toy soldiers. 1969 D. E. Westlake *Up your Banners* (1970) xxxviii. 274 [He] never had a model train set when he was young. 1903 A. Bennett *Truth about Author* ii. 16 He sailed model yachts for us on the foulest canal in Europe. 1967 M. Waddell *Otley Pursued* viii. 64 A solitary soul wore gumboots and carried a large white model yacht.

c. model agency, an agency that supplies models (sense 11 or *11 b); **model girl** = sense *11 b; also *attrib.*; **model school,** (*a*) a school intended to be a model in organization, teaching methods, etc.; (*b*) a school where models (sense *11 b) are trained; **model theory,** the theory of models (sense *2 e or, esp., *2 f), dealing with their construction, the conditions of their validity, etc.; so *model-theoretic, -theoretical* adjs., *model-theoretically* adv.

1945 *Glamour* Nov. 166/2 First, you have an interview with one of the leading model agencies, such as Conover or Powers. *Ibid.* 260/2 The model agency rarely selects girls for a given assignment. Invariably, the photographer calls for the girls he wants. 1950 J. D. MacDonald *Brass Cupcake* (1955) iv. 41, I went down to New York City... A model agency took me on. 1956 S. Bellow *Seize the Day* (1957) i. 27, I got hold of the artist and he gave me the number of the model agency. 1972 A. MacVicar *Golden Venus Affair* vi. 62 He got me work with this model agency. 1973 *Guardian* 25 May 5/5, I mentioned reports that a model agency apparently puts men in touch with call girls. 1962 *John o' London's* 4 Jan. 20/1 Witches who are model girls gone macabre. 1973 A. Price *October Men* vii. 92 A woman in a big hat, slender like a model-girl. 1974 R. Harris *Double Snare* iii. 19 Her figure..must once have been willowy in the model girl fashion. 1854 Dickens *Hard T.* i. iii. 15 To think of these vagabonds.. attracting the young rabble from a model school. 1935 *Discovery* Nov. 342/1 The prime cause of the excavation was the announcement that a new model school was to be built on a field..which..covered part of the site of Camulodunum. 1966 A. Prior *Operators* iv. 40 Robin..said he knew some people in modelling who could help her... He had not been able to afford the so-called Model School. 1957 *Bull. Amer. Math. Soc.* LXIII. 289 (*heading*) Model-theoretic and decidability theorems concerning generalized products. 1963 A. Robinson *Introd. Model Theory* p. vi, The model-theoretic approach to set theory. 1958 *Notices Amer. Math. Soc.* V. 673 (*heading*) Some model-theoretical results concerning weak second-order logic. 1973 J. J. Zeman *Modal Logic* p. vi, The systems are studied model-theoretically. 1960 *McGraw-Hill Encycl. Sci. & Technol.* VIII. 525/2 On the basis of model theory, a small, readily modified model can be built and tested at low relative cost and the results applied to the full-scale device. 1967 S. C. Kleene *Math. Logic* § 53. 117 In the predicate calculus, proof theory has the advantage over model theory. 1969 W. A. J. Luxemburg (*title*) Applications of model theory to algebra, analysis and probability.

model, *v.* Add: **2. b.** [after *MODEL *sb.* 2 e.] To devise a (usu. mathematical) model of (a phenomenon, system, etc.).

1965 C. H. Springer et al. *Adv. Methods & Models* ii. 57 We 'modeled' a business process with the aid of a ready-made algebraic model. 1971 *Nature* 18 June 425/1 The first attempts to model the urban system were made by traffic engineers. 1974 *Physics Bull.* Feb. 84/3 The UKAEA has modelled the diffusion of particles in a fluid acted on by buoyancy, winds, currents and turbulence. 1972 *Sci. Amer.* May 97/3 Our hope is that the maps will.. eventually be of aid to meteorologists who are modeling the present circulation in the atmosphere. 1974 *McGraw-Hill Yearbk. Sci. & Technol.* 250/2 All models [of land use] seek to allocate land-absorbing activities on some type of spatial network of subareas. The size of the subareas will, of course, vary depending on the total area being modeled.

8. *trans.* and *intr.* To act as a model (MODEL *sb.* 11 and *11 b); to display (clothes) as a model.

1915 W. B. Yeats *Reveries* (1916) 153 A pretty gentle-looking girl was modelling in the middle of the room. 1927 *Cleveland Press* 4 Mar., Vivian..will model Saturday in the shoe section of the Bailey Co. 1931 *Durant* (Okla.) *Daily Democrat* 29 Oct. 3/2 See them [*sc.* coats] modeled during style promenade tomorrow. 1948 'J. Tey' *Franchise Affair* xxii. 260 A natural blonde with the clothes and figure of a girl who has 'modelled' clothes. 1957 M. Summerton *Sunset Hour* iv. 61 Lolly..had modelled bikinis and gossamer underwear before blasé camera crews. 1969 *Guardian* 30 June 7/1, I was watching this Negro modelling sleepwear.

modelling, *vbl. sb.* Add: **3. c.** The action of *MODEL *v.* 8; the work of a fashion-model. Also *attrib.*

1949 *Chicago Tribune* 17 Feb. 10/3, I never thought of modeling as a career. 1959 'J. Christopher' *Scent of White Poppies* iv. 55, I did do some modelling there. 1963 G. Marx *Let.* 11 Apr. (1967) 63 She is abandoning the modeling school and plans to embark upon..settlement work. 1964 [see *fashion-modelling* vbl. sb. s.v. *FASHION sb. 13 a]. 1973 *Sun* 25 May 3 She started her modelling career when a fashion photographer spotted her on a beach near Rimini.

d. The devising or use of abstract or mathematical models (*MODEL *sb.* 2 e).

1965 C. H. Springer et al. *Adv. Methods & Models* i. 4 The best way to untangle the confusion which many people have about mathematical modeling as a method for solving important business problems is to untangle the whole idea of model building as a way of thinking about the world we live in. 1971 J. Howlett in B. de Ferranti *Living with Computer* ii. 13 The computer has..made it possible to apply this method—mathematical modelling followed by numerical solution of the resulting equations —to problems that would be quite intractable without its aid. 1974 *Nature* 2 Aug. 450/3 It also gives the reader some feel for the vast problems involved in any worthwhile quantitative modelling of climatic change.

‖ **modelletto** (mǫdĕle·to). Pl. **modelletti.** [It., dim. of *MODELLO.] = *MODELLO.

1937 *Burlington Mag.* Mar. 133/1 The other collectors ..added the bozzetti and modelletti to their bronzes and sculptures. 1938 *Ibid.* Oct. 141/2 Tiepolo has developed his theme in masterly fashion, perhaps as a 'modelletto' for an altar-piece. 1962 R. G. Haggar *Dict. Art Terms* 215/1 *Modello*... Another name is *modelletto*.

‖ **modello** (mǫ̆de·lo). Pl. **modelli, modellos.** [It., see MODEL *sb.*] A sketch, often executed

in detail, for a larger painting which is prepared for a patron's approval; also, a small model for a larger sculpture. Cf. *BOZZETTO, *MAQUETTE.

1937 *Burlington Mag.* Oct. 188/1 We find the more finished *modello* replaced by the quick colour sketch. 1959 *Times Lit. Suppl.* 20 Mar. 154/5 The relation of genuine *modelli* to 'engraver's copies' is left in a state of confusion. 1964 *Listener* 12 Nov. 767/2 Vasari, on his visit to Titian's house in 1566, noticed many of these modellos propped against the wall. 1967 W. Gaunt *Compan. Painting* 63 The more detailed *modello*..gave the full conception of the painting. *Ibid.* 67 The *modelli* of the Italians. 1972 *Country Life* 6 Apr. 857/1 It is possible that the picture is a modello for a larger work, intended for a church or chapel. 1974 *Daily Tel.* 29 July 8/6 The recently acquired picture ..is the modello for the painting of 'The Senators of Florence swearing allegiance to the Grand Duke Ferdinand II'.

modelly (mǫ·dĕli), *a.* [f. MODEL *sb.* + -Y[1].] Resembling a model (*MODEL *sb.* 11 b); having the characteristics of a fashion-model.

1961 I. Jefferies *It wasn't Me!* vi. 70 Ghislaine was a bit too modelly to be true. 1965 *Observer* (Colour Suppl.) 5 Sept. 6/2 They wanted actresses because they thought models were too modelly.

modem (mōu·dem). [f. MO(DULATOR + *DEM(ODULATOR.] A combined modulator and demodulator (such as is used in connecting a computer to a telephone line) for converting outgoing signals from one form to another and converting incoming signals back again.

1961 G. L. Evans et al. in *Convention Rec. 5th National Symposium Global Communications* 100 A recent study of wireline data communications..has underlined the need for advancement in the field of data modulators-demodulators (modems). 1963 *Daily Tel.* 16 Dec. 15/2 The direct current (DC) signals which flow through a computer cannot be transmitted over telephone lines, so it is necessary to convert them... The black box which does this conversion at either end of the line is called a modem. 1971 *New Scientist* 7 Jan. 18/1 It is around 3.30 pm in the communications control room of Time Sharing Ltd, in London's West End. The modems are silent... Power has been off for half an hour. 1975 *Daily Colonist* (Victoria, B.C.) 13 May 27/1 They communicated over regular telephone lines, using teletypewriters connected to the lines by electronic devices known as modems.

Modena. Add: **2.** In full, *Modena pigeon.* A pigeon of the variety so called, distinguished by its stocky build and red legs.

1879 L. Wright *Pract. Pigeon Keeper* xix. 220 (*heading*) Modena flying pigeons. 1936 W. A. Dalley in *Pigeons of Today* 105 When once here the Modena..became an exhibition or fancy pigeon pure and simple. 1965 W. M. Levi *Encycl. Pigeon Breeds* 247 The Modena is one of the most popular breeds in the United States. 1969 H. H. Shrives *Fancy Pigeons* 81 Modena pigeons are divided in Gazzi or Pied, Schietti or Self-coloured, and Magnani or Harlequin.

modenature. Delete *Obs. rare*[−1] and add later examples.

1953 *Archit. Rev.* CXIV. 91 The two outstanding features of the doctrine of Auguste Perret are the insistence on modenature—the profiling and management of projecting features—and the philosophy of the concrete frame. 1959 P. Collins *Concrete* ix. 197 The only modifications to this structural composition [in Greek architecture] consisted of surface modulations intended to make the forms more pleasing to the eye (namely *modénature* or profiling), and optical corrections.

Modenese (mǫdĭnī·z), *a.* and *sb.* [f. MODENA + -ESE.] **A.** *adj.* Of or pertaining to Modena or its inhabitants. **B.** *sb.* A native or inhabitant of Modena.

1813 J. C. Eustace *Tour through Italy* I. vi. 131 The important 'Bucket'..was carried off from a well in one of the streets of Bologna, by a party of Modenese troops. 1839 K. H. Digby *Mores Catholici* ix. iii. 80 Peace was then made between all the Modenese,..more than twenty thousand of the Modenese went to Reggio and Parma; and those two cities made peace with each other. 1936 G. F.-H. & J. Berkeley *Italy in Making* II. xvi. 245 At Fivizzano, the Modenese soldiers fired on the crowd. 1957 *Encycl. Brit.* XV. 633/2 The wooden bucket captured by the Modenese from the Bolognese in the affray at Zappolino.

moderate, *a.* and *sb.* Add: **A.** *adj.* **2. b.** (Later examples.)

1924 J. T. Gwynn *Indian Politics* iii. 18 The Moderate or Co-operating party is to-day so unpopular that it takes some strength of mind to remain a Co-operator. 1954 B. & R. North tr. *Duverger's Pol. Parties* i. i. 46 In the nineteenth century parties were based upon the caucus and weak articulation; today most Conservative, Moderate, and 'Liberal' parties in Europe still display these two essential characteristics. 1973 *Perthshire Advertiser* 17 Feb. 1/1 Another member of the Moderate-Independent Association of Perth town councillors, Councillor Henry Giulianotti, has resigned from the association. 1973 T. K. Derry *Hist. Mod. Norway* x. 325 In the summer of 1934 Hjort had failed to put through a scheme for linking the Agrarian Party with the Moderate Liberals.

3. b. (Later examples.)

1949 J. D. B. Wilson *Southern Highlands* 175 Further east are two ribs beyond a shallow gully which give defined climbs..of moderate standard. 1956 A. J. J. Moulam *Tryfan & Glyder Fach* 85 Ordinary route. About 200 feet. Moderate. 1966 M. Woodhouse *Tree Frog* xv. 122 If the climb had reached any level of difficulty higher than Moderate, which is the Climbers' Club's polite way of labelling a gumshoe doddle, we'd have died. 1971 N. Tennent *Islands of Scotl.* i. 32 Arran rock offers little choice between easy to moderate scrambles and hard, strenuous routes. *Ibid.* iv. 78 *Broad Buttress*, 450 ft., Moderate.

c. Of prices, charges, etc.: not excessive, reasonable, low.

1904 *Punch* 6 Apr. p. ii (Advt.), Hotel..standing high in its own beautiful park... Moderate tariff or inclusive terms. 1923–4 *Guide to Oxf.* 19 (Advt.), Norfolk Hotel,.. Central Position. Moderate Terms. 1971 *Bibliotheck* VI. 57 The New Aldis..at the moderate price of £4.50 and with more than 1600 additional entries..will be especially welcome. 1973 *Michelin: France* 40 Good meals at moderate prices.

4. (Later examples.)

1925 V. Woolf *Common Reader* 134 Six moderate-sized volumes. 1959 *Guardian* 9 July 5/3 The play..has a moderate-sized cast.

B. *sb.* **a.** (Later examples.)

1920 H. V. Lovett *Hist. Indian Nationalist Movement* iii. 69 The Moderates were pushed out of a hall and assailed with stones and mud. 1924 J. T. Gwynn *Indian Politics* iii. 18 It used to be..the fashion to decry the Moderates and Co-operators as if they were a party of weak men and time-servers. 1969 *Listener* 28 Aug. 268/3 'Moderates' (who include men with a very militant record) are at present containing the extremists and hot-heads. 1975 *Times* 6 Jan. 2/3 The moderates should exercise their overwhelming strength in the trade union movement.

moderate, *v.* Add: **1. d.** *Nuclear Sci.* To slow down (a neutron); also, to provide (a reactor) with a moderator.

1956 H. Seligman in A. Pryce-Jones *New Outl. Mod. Knowl.* 158 There are very few substances which can be used for this slowing-down process. There is carbon in pure form, or water which, instead of the normal hydrogen, has the 'heavy' hydrogen atom in its molecule. These substances are called moderators, as they are moderating the neutrons. 1958 O. R. Frisch *Nucl. Handbk.* x. 4 The type of thermal neutron spectrum obtained from a reactor depends upon the conditions under which the neutrons are moderated. 1959 *Listener* 19 Nov. 873/1 They [*sc.* the reactors]..are cooled by carbon dioxide and moderated with graphite. 1966 *New Scientist* 24 Feb. 484/3 In a fast-reactor system, so called because the neutrons causing fission are not slowed down or 'moderated' but react at high energies, it is possible in theory to consume all the uranium. 1973 *Nature* 2 Feb. 317/1 In the reactors cooled and moderated by water, the reactor vessel must also be at this pressure.

3. a. (Later examples.)

1968 *N.Y. Times* 26 June 1/5 Being shown live at the time was a panel discussion on the 'underground press' moderated by Steven V. Roberts, a reporter for The New York Times. 1975 *Listener* 9 Jan. 39/1 The closed circuit at the University of Kent at Canterbury was used recently to televise a meeting whose potential participants promised to be too numerous for any of the available halls... There were two 'teams' moderated by the chaplain.

b. (Further U.S. examples.)

1766 T. Clap *Ann. Yale-Coll.* 15 Mr. Andrew moderated at the Commencements. 1778 E. Stiles *Lit. Diary* (1901) II. 311 The first Commencements were private. Rector Pierson moderated and gave Degrees till his Death.

moderated, *ppl. a.* Add: **3.** *Nuclear Sci.* Of a reactor: provided with a moderator. Of a neutron: slowed down by a moderator.

1945 [see *GRAPHITE b]. 1950 F. Gaynor *Encycl. Atomic Energy* 114 In a moderated reactor there remain more free neutrons to sustain and propagate the fission chain reaction of U[235]. 1962 *Newnes Conc. Encycl. Nucl. Energy* 244/2 Most of the fissions [in a thermal reactor] are produced by these moderated neutrons.

moderation. Add: **1. e.** *Nuclear Sci.* The action or process of slowing down neutrons by the use of a moderator.

1945 H. D. Smyth *Gen. Acct. Devel. Atomic Energy Mil. Purposes* ii. 20 The process of slowing down or moderation is simply one of elastic collisions between high-speed particles and particles practically at rest. 1958 O. R. Frisch *Nucl. Handbk.* v. 17 The moderation of fast neutrons to thermal velocities occurs by transfer of energy on elastic collision. 1969 *New Scientist* 25 Sept. 639/2 The three essential functions of a blanket are neutron moderation, tritium breeding, and heat-transfer.

moderationism (mǫdərēi·ʃəniz'm). [f. MODERATION 2 + -ISM.] A policy or doctrine of being moderate or acting with moderation.

1960 [see *CENTRISM].

moderatism. (Later examples.)

1923 G. M. Trevelyan *Manin & Venetian Revolution* iii. 49 The event resounded through Italy and Europe. It shook the somewhat self-complacent 'moderatism' of Giobertian waiters on opportunity. 1970 *Guardian* 1 Oct. 19/4 Mr. Crouch's apologia for 'moderatism'..is as unexciting in print as it was ineffective in the worst university crises.

moderator. Add: **4.** *spec.* A chairman of a television discussion (also in extended use). *N. Amer.*

1952 in A. Rothe *Current Biogr.* 433/2 Since 1946 he [*sc.* Senator Blair Moody] has also been moderator of the radio and television program *Meet Your Congress.* 1972 *Village Voice* (N.Y.) 1 June 28/3 She kept qualifying her aggression with apologies and withdrawals and cancellations..and maintaining the good will of the moderator. 1972 *Evening Telegram* (St. John's, Newfoundland) 24 June 1/2 Ron Pumphrey, moderator of an open-line program on St. John's radio station VOCM. 1973 *Guardian* 17 Oct. 12/4 The present moderators..were picked by London broadcasting to do a specific job, that of answering the phone and referring the questions to a guest expert. 1974 R. Thomas *Porkchoppers* xxv. 212 The program's moderator..was..the syndicated political columnist who specialized in political muckraking.

7. d. *Nuclear Sci.* A substance that slows down neutrons passing through it; *spec.* one used in a reactor to reduce the speed of fast neutrons so that they cause fission more readily.

1945 H. D. Smyth *Gen. Acct. Devel. Atomic Energy Mil. Purposes* ii. 20 The light elements are most effective as 'moderators', i.e., slowing-down agents, for neutrons. 1958 W. K. Mansfield *Elem. Nucl. Physics* iv. 38 If the U[235] content exceeds about 50 per cent it is possible to achieve a chain reaction without the use of a moderator. 1961 G. R. Choppin *Exper. Nucl. Chem.* viii. 115 Since it is desirable to increase the probability of capture by slowing the neutrons down, a moderator containing a low Z element with a low capture cross section surrounds the source. 1962 *Newnes Conc. Encycl. Nucl. Energy* 509/1 The most common moderators are: light water (H_2O), heavy water (D_2O), graphite, beryllium, beryllia and organic liquids. 1966 C. R. Tottle *Sci. Engin. Materials* x. 236 In nuclear reactors..a moderator, or slowing-down material, is used to decrease the energy of fast neutrons produced in fission until they are more readily captured in the fissile material at lower energies. 1969 Bennison & Wright *Geol. Hist. Brit. Isles* i. 18 Since carbon is a good moderator of neutrons, carbonaceous rocks are liable to give a spurious indication of porosity.

moderatorial (mǫdĕrĕtōə·riăl), *a.* [f. MODERATOR + -IAL.] Of, pertaining to, or characteristic of a moderator or chairman.

1867 'T. Lackland' *Homespun* ii. 155 This moderatorial edict is echoed up in the bell-tower. 1926 *Scots Observer* 13 Nov. 4/4 There was a moderatorial flavour in the eloquence of the evening. 1968 *Guardian* 19 Mar. 16/1 In his moderatorial address to the council last night, the Rev. Edward Rogers, a former president of the Methodist Church, said the ecumenical movement had reached the point where it presented a challenge. 1975 *Church Times* 27 June 14/4 The most excitingly growing church I found during my Moderatorial year was such a union.

modern, *a.* and *sb.* Add: **A.** *adj.* **2. a.** spec. (*the*) *modern Babylon*: London; *modern Greats*: at Oxford University, the school of Philosophy, Politics, and Economics.

1835 J. M. Wilson *Tales of Borders* I. 356, I proceeded to London... Months passed away, and I was still a wanderer upon the streets of the modern Babylon. 1847 Disraeli *Tancred* III. v. v. 77 London is a modern Babylon. 1850 Dickens *Dav. Copp.* xxxvi. 374 Bidding adieu to the modern Babylon. 1904 A. Bennett *Great Man* viii. 79 The human tide which beats for ever on the shores of modern Babylon. [1909 *Wanted! A New School at Oxf.* 4 It is a plan for, as it were, a modern-side Greats, based on Philosophy, but also..containing an admixture of certain other subjects.] 1922 Joyce *Ulysses* 611 The sights of the great metropolis, the spectacle of our modern Babylon. 1925 *Times* 15 July 19/3 The examiners in the Final Honour School of Philosophy, Politics, and Economics ('Modern Greats') issued the following class list. 1935 N. Mitchison *We have been Warned* ii. 146 He was.. thinking what a rotten school Modern Greats was. 1971 D. Scott *A. D. Lindsay* iii. 50 A new School of Politics, Philosophy and Economics—'Modern Greats'.

c. (Examples.)

1699 M. Lister *Journey to Paris* 108 Another Book overwritten in a small Modern Greek Hand, about 150 years ago. 1748 Smollett *R. Random* I. xxx. 275, I asserted that the modern Greek was as different from that spoke and written by the ancients, as the English used now from the old Saxon spoke in the time of Hengist. 1841 Borrow *Zincali* I. ii. i. 235 The number of Persian, Sclavonian, and modern Greek words with which it [*sc.* the language of the 'Gitános of Estremadura'] is chequered. 1900 Clarke & Murray *Dent's School Gram. Mod. French* p. v, An attempt has been made to make this a grammar of modern French. 1927 S. Jónsson *Primer Mod. Icelandic* p. v, The first suggestion that I should write a text book of modern Icelandic was made to me in 1917. 1971 B. S. J. Isserlin *Hebrew Word-Bk.* 1 Grammatically modern Hebrew differs less from Biblical Hebrew than Old English does from Modern English. 1971 N. Fisher *Rise at Dawn* x. 169 He was in Corfu and he speaks good modern Greek.

d. (Earlier and later examples.) Cf. quots. 162. and 1706 under sense 2 a in Dict.

1821 H. C. Robinson *Diary* 3 Oct. (1967) 70 Sara Coleridge..has taught herself modern languages, and is said to have great talent. 1932 Auden *Orators* i. i. 19 The really disgusted—the teacher of modern languages. 1961 R. B. Long *Sentence & its Parts* xvii. 377 In *modern-language teaching* what is modern is the languages; in *modern language teaching* what is modern..is the teaching of languages.

e. *modern school*, also, a secondary modern school.

1944 *Ann. Reg.* 1943 62 Three main types of secondary schools—grammar, modern, and technical. 1957 *Listener* 13 June 951/1 The 1944 [Education] Act had brought the modern school into being. 1975 *Times* 29 Aug. 10/4 When university entrance is considered there is a 47 per cent advantage of the combined grammar and modern schools over the comprehensives.

g. *Typogr.* Used to designate a group of type-faces developed in the late eighteenth and early nineteenth centuries, distinguished by flat serifs, increased contrast between the thick and thin parts of the letters, and an effect of greater precision and vertical emphasis in use. Also *modern-cut, -face*(*d*) adjs.

1808 C. Stower *Printer's Gram.* Specimens of Printing Types, (*heading*) Specimens of modern-cut printing types, from the founderies of Messrs. Fry and Steele, and Messrs. Caslon and Catherwood. 1819 R. Austin in A. F. Johnson *Type Designs* (1934) iii. 97 The modern or new fashioned faced printing-type at present in use was introduced by the French, about 20 years ago. 1874 G. Simpson in Geo. Eliot *Lett.* (1956) VI. 44 Tinted paper ought never to be used with modern faced type. It suits ancient face only. 1894 *Amer. Dict. Printing & Bookmaking* 379/2 *Modern faces*, these are those kinds of Romans which have been cut since the beginning of the century. 1902 T. L. De Vinne *Pract. Typogr.: Treat. Title-Pages* 234 Of modern-cut types we have many varieties. 1926 S. Morison *Type Designs* 31 The main distinguishing characteristic of the modern-face is that the serif is thinner, longer and more refined than in the old-face. The difference as between the stem and the hair-line is more marked, and the general note of 'modern' is that of extreme precision and a certain perpendicularity. 1934 A. F. Johnson *Type Designs* iii. 73 During the eighteenth century the design of our roman types underwent a radical change, resulting in the style which we know as modern face, the type of the nineteenth century and still the type used in our newspapers and most of our books. 1972 P. Gaskell *New Introd. Bibliogr.* 210 By the second decade of the nineteenth century English printers were using modern face almost exclusively.

h. Of a movement in art and architecture, or the works produced by such a movement: characterized by a departure from or a repudiation of accepted or traditional styles and values. Cf. *ABSTRACT A. 4 d.

[1849 *Art Jrnl.* XI. 69/3 Between this society and one begun some years ago for the encouragement of modern Art and native artists, there should be no rivalry.] 1895 R. Muther *Hist. Mod. Painting* I. 10 Because this distinction between the eclectic and the personal, the derived and the independent, has not yet been carried out with sufficient strictness..it has hitherto..been found so difficult to discover the distinctive *style* of modern art. 1927 C. Bell *Landmarks 19th-Cent. Painting* 5 Géricault and then Delacroix were the new influences in France; in England the innovator was Constable. From these points of departure you can trace the whole glorious history of modern art. 1929 H. R. Hitchcock *Mod. Archit.* xvii. 201 There is..little to compare with the unconsciously 'modern' work of those architects who continued the English tradition. 1938 O. Lancaster *Pillar to Post* 74 When, shortly after the War, the Modern Movement.. was first brought to public notice it led to a natural and healthy reaction against the excessive ornament..of the previous generation. 1958 S. W. Cheney *Story Mod. Art* (rev. ed.) p. v, I have accepted here the broadest traditional usage of the term 'modern art' as covering the course of creative invention since 1800. 1972 P. M. Bardi *Archit.* xix. 117/1 The flight of refugees from the Nazis.. scattered the pioneers of the Modern movement across western Europe and America. 1973 *Times* 19 June 14/4 *A Child of Six Could Do It*, [an exhibition of] cartoons about modern art at the Tate.

i. In full, *modern first* (*edition*). A bookseller's term for the first edition of a book published after about 1900. Also *absol.* or as *sb.*

1922 M. Sadleir *Excursions in Victorian Bibliogr.* 7 The dapper expert in ingenious moderns with his prefaces, his cancel-titles, [etc.]. 1952 J. Carter *ABC for Bk.-Collectors* 122 *Modern firsts*, a category widely employed but.. impossible to define with any precision, since its use among antiquarian booksellers is, and probably always will be, quite unstandardised. At present (1951) it commonly extends as far back as 1900, and will often include books published before that date if their author's hey-day was after it. 1968 *Bertram Rota Ltd. Catal.* No. 158 (verso front cover), We are always pleased to receive offers of modern first editions and private press books in fine state. 1973 *Directory of Dealers Secondhand & Antiquarian Bks. Brit. Isles 1973–75* 240 E. H. Bucknall..modern first editions. 1975 *Bibliognost* Aug. 5 The chance that someone will find a simple first edition of a modern author is slim, because [John F.] Fleming's firsts are very special... Each would be priced well above the average cost of 'moderns'.

3. a. In spec. phrases: *modern convenience*, an amenity, device, fitting, etc., such as is usual in a modern house; freq. *pl.*; cf. *MOD. CON.; *modern dance*, a free expressive style of dancing distinct from classical ballet (see quots.); hence *modern dancer, dancing* vbl. sb.; *modern jazz*, jazz of a type which originated during and after the war of 1939–45.

1859 [see *dish-lift* s.v. *DISH *sb.* 10]. 1912 E. L. Urlin *Dancing Anc. & Mod.* p. xv, Modern dances..[are] derived from some primary human instinct, such as Worship, Mimicry, Love, or War... Modern dancing begins where..the art survives solely on account of the pleasure it gives to the performer, or to the spectator. 1926 *Times* 6 May 1/6 (Advt.), Superior accommodation in lady's quiet house..all modern conveniences. 1933 J. Martin *Mod. Dance* 2 There are as many methods and systems of modern dancing as there are dancers. *Ibid.* 3 By the modern dance we..imply by a method of negation those types of dancing which are neither classic nor romantic. 1937 E. St. V. Millay *Conversation at Midnight* i. 15 Peace and Quiet poured down the sink, In exchange for a houseful of 'modern conveniences'. 1955 D. Gillespie in Shapiro & Hentoff *Hear Me Talkin' to Ya* xix. 300 No one man or group of men started modern jazz. 1957 G. B. L. Wilson *Dict. Ballet* 188 *Modern dance*, the term used to designate a variety of styles which are not founded on the *danse d'école* (i.e. the Classical ballet). *Ibid.* 189 Modern Dance claims to make much use of 'natural movements' and it is also a reflexion of a state of mind. 1961 *Metronome* Apr. 12 By the 1950's 'modern' jazz, as the more advanced developments were termed, had to free itself both from esoteric tendencies within jazz itself and from over-dependence on Western European classical traditions. 1968 J. Winearls *Mod. Dance* (ed. 2) 9 The title 'Modern Dance' distinguishes those kinds which have been invented, developed, or adapted from various sources during the past half-century and which are clearly marked by an expressive style quite different from that of other forms such as National, Folk, Musical Comedy or Ballet. *Ibid.*, Modern Dancers consider that Ballet cannot deal satisfactorily with all possible dance-subjects.

5. *modern-built* (earlier example), *-day, -minded, -style* adjs.

1811 Jane Austen *Sense & Sens.* III. vi. 111 Cleveland was a spacious, modern-built house. 1909 *Westm. Gaz.* 5 July 2/3 No one can fail to be impressed by the seriousness of modern-day cricket. 1965 *Times Lit. Suppl.* 25 Nov. 1075/4 Philip Bummidge is a modern-day Hamlet. 1907 *Dublin Rev.* July 191 The author is at times betrayed into making Neri almost impossibly modern-minded. 1970 *Guardian* 15 Jan. 12 Nigeria needs to prove that it is stable, modern-minded and representative. 1883 'Mark Twain' *Life on Mississippi* 427 Inviting modern-style pleasure resorts. 1927 *Melody Maker* Aug. 781/2 The voice again is heard together with trumpet and saxophones alternating in a modern-style extemporisation. 1975 *Guardian* 20 Jan. 8/3 Modern-style choreography.

B. *sb.* **1. b.** Delete † *Obs.* and add later example.

1975 *Country Life* 2 Jan. 37/2 In the visual arts the Walker Art Centre houses a world-famous collection of moderns.

moderner. Delete † *Obs. rare* and add later examples.

1926 *Brit. Weekly* 29 July 356/1 With a moderner's contempt of ailments. 1971 *Ceylon Daily News* 17 Sept. 4/6 It is psychologically impossible for a moderner to live in the same world that his predecessors lived.

modernism. Add: **3.** *Theol.* A tendency or movement towards modifying traditional beliefs and doctrines in accordance with the findings of modern criticism and research, esp. a movement of this kind in the Roman Catholic Church at the beginning of the twentieth century.

1901 G. Tyrrell *Let.* 18 Aug. in M. D. Petre *Autobiogr. & Life Geo. Tyrrell* (1912) II. ii. 52 He [*sc.* Kegan Paul] simply rants against Modernism, and glories in what ought to be our shame. 1903 —— *Let.* 19 Nov. (1920) vii. 132, I hope that the Catholicism in which we may eventually unite will be one in which all that is good and true in Modernism will be saved and sanctified. 1907 tr. *Pius X's Encyclical Let. Doctrines Modernists* 15 If we..seek to know how the believer, according to Modernism, is marked off from the Philosopher, it must be observed [etc.]. 1913 A. Fawkes *Stud. Modernism* 373 The name Modernism was given to the present phase of the liberalising movement in the Church of Rome by the *Civiltà Cattolica.* 1915 *Hastings' Encycl. Relig. & Ethics* VIII. 763/1 Modernism is the name given by the papal encyclical which condemned it to a complex of movements within the Roman Communion, all alike inspired by a desire to bring the tradition of Christian belief and practice into closer relation with the intellectual habits and social aspirations of our own time. 1923 *Edin. Rev.* Jan. 62 Between English Modernism and the now discredited Roman Modernism there is a deep cleavage. 1927 H. D. A. Major *Eng. Modernism* 18 In the Roman Church Modernism is opposed to Mediævalism; in the English Church Modernism, as in Holland, is opposed to Traditionalism; in America Modernism is opposed to Fundamentalism. 1955 *Times* 17 Aug. 9/5 The mistake of modernism is to try to force a supernatural revelation into the mould of natural science and scholarship. 1972 L. F. Barmann *Baron F. von Hügel & Modernist Crisis in Eng.* p. ix, In the sixty years following that conflict much has been written about modernism, though mostly from partisan standpoints.

4. The methods, style, or attitude of modern artists (*MODERN *a.* 2 h); *spec.* a style of painting in which the artist deliberately breaks away from classical and traditional methods of expression; hence, a similar style or movement in architecture, literature, music, etc.

1929 H. R. Hitchcock *Mod. Archit.* xvii. 205 A city [*sc.* New York], whose 'modernism' consists in copying the poorest French models of the New Tradition. 1934 R. Blomfield *Modernismus* viii. 145, I have already called attention to the disastrous effect of Modernism on architecture, painting, and sculpture... I find the same insidious and repulsive influence at work in a good deal of contemporary music. 1946 R. Wellek in W. S. Knickerbocker *20th Cent. Eng.* 69 The fine arts themselves and the art of literature reacted against realism and naturalism in the direction of symbolism and other 'modernism'. 1955 *Times* 1 June 8/7 Professor A. E. Richardson..foresaw a revival of the Hellenic influence in art and the decline of 'modernism'. 1961 *Listener* 23 Nov. 848/1 The American modernism introduced by Mr. T. S. Eliot, following Mr. Ezra Pound. 1971 *New Society* 25 Mar. 496/3 'Vision in Motion'..a brilliant manual, or compendium, of modernism. 1973 *Observer* 5 Aug. 28/3 Sassoon became..increasingly embittered at his neglect by the new pundits of 'modernism'. 1973 *Black World* Sept. 19/1 During the Twenties, Jorge de Lima became an important member of the literary movement known as Modernism. 1974 *New Yorker* 25 Feb. 122/1 Modernism, we are told, is passé; the Harvard English Department lists a course, 'American Modernism', that treats of 'American writing from 1900 to 1930'.

‖ **modernismus** (mǫdəɹni·zmŭs). [G.] = *MODERNISM 4.

1934 R. Blomfield *Modernismus* p. v, Since the war, Modernism, or 'Modernismus', as it should be called on the German precedent, has invaded this country like an epidemic. 1934 C. Lambert *Music Ho!* iv. 248 Listening to his [*sc.* Hindemith's] firmly wrought works we seem to see ourselves in a block of hygienic and efficient workman's flats built in the best modernismus manner. 1948 Wyndham Lewis *Let.* 18 Oct. (1963) 464 'The Waste Land', which bit of modernismus appeared a few years before. 1958 *Listener* 17 July 102/1 A city that was to be the last word in *modernismus*, a technician's 'New Jerusalem'. 1970 *Daily Tel.* (Colour Suppl.) 9 Oct. 65/2 A bale of soft material, patterned in that modernismus which was so big in the Thirties, and now survives mainly in British Rail.

modernist. Add: **4.** *Theol.* One who inclines to, supports, or advocates theological modernism. Also *attrib.* or as *adj.*

1907 tr. *Pius X's Encyclical Let. Doctrines Modernists* 6 It is one of the cleverest devices of the Modernists (as they are commonly and rightly called) to present their doctrines without order and systematic arrangement. 1918 M. D. Petre *Modernism* v. 101 These words are not written in a spirit of hostility to the Catholic modernist position. 1920 W. Sanday *Divine Overruling* 67, I do not disclaim the name of Modernist. The name describes justly what I aim at being. I aim at thinking the thoughts and speaking the language of my own day, and yet at the same time keeping all that is essential in the religion of the past. 1923 *Edin. Rev.* Jan. 62 Roman Modernists took Newman's doctrine of development. 1931 J. S. Huxley *What dare I Think?* vii. 228 Accounts of God which are as modernist as could be desired. 1970 *Times* 19 Aug. 10/8 The appointment of another avowed Modernist with a strong sense of the Church and a flair for knowing men, proved to be the eirenic answer. 1972 L. F. Barmann *Baron F. von Hügel & Modernist Crisis in Eng.* p. ix, Modernists have been concerned to defend themselves and their aims, and anti-modernists have been equally concerned to justify the conduct of the Roman authorities and the positions of Roman theologians.

5. An artist, architect, writer, etc., whose work is characterized by modernism (see *MODERNISM 4). Also *attrib.* or as *adj.*

1927 F. J. Mather *Hist. Mod. Painting* 372 Modernist pictures are becoming discreet, almost cautiously monotonous in colour. *Ibid.* 373 In comparison the Modernists have attained nothing of the coherence or authority of a school. 1934 R. Blomfield *Modernismus* vii. 125 The Modernist seems to glory in his own obscurity. *Ibid.* 132 In these Modernist works there is no brushwork worth looking at. 1935 H. G. Wells *Things to Come* xii. 99 One sits on a chair of modernist form. (All furniture is metallic.) 1937 S. W. Cheney *World Hist. Art* (1938) xxviii. 862 This wing of modernists—with Cézanne as their prime exhibit—brought in the familiar studio talk about *form* as the indispensable creative force in painting. 1961 *Listener* 23 Nov. 863/1 Rosenberg, unlike the modernists, does not go in for phanopoeia or free association. 1963 *Ibid.* 21 Mar. 518/1 The writers were, in a sense, traitors in the modernist camp.

6. One who plays, appreciates, or supports modern jazz (see *MODERN *a.* 3 a). Also *attrib.* or as *adj.*

1955 J. Prentice in A. J. McCarthy *Jazzbook 1955* 105 As I understand that the traditional field will be covered in another article I will confine myself to the modernists. 1958 V. Bellerby in P. Gammond *Duke Ellington* ii. 143 Such extensions and developments only lead to a destruction of its [*sc.* Jazz's] real and lasting value, as has unquestionably happened with some of the more involved modernist experiments. 1962 *Sunday Times* (Colour Suppl.) 10 June 3 The strange division between the extrovert 'traditionalists'..and the very different *avant garde* 'modernists' (whose idol is the revolutionary Charlie Parker).

modernistic (mǫdəɹni·stik), *a.* [f. MODERNIST + -IC.] Of, pertaining to, or suggestive of modernism or modernists; having affinity to or sympathy with what is modern. Hence as *sb. pl.*, examples of modernistic art, music, etc.

1909 *Daily Chron.* 12 June 1/3 'L'Unione' is denounced as reeking with modernistic and kindred ideas opposed to the principles and dogmas of the Roman Church. 1924 *Public Opinion* 17 Oct. 383/1 The New Testament in relation to its own time is essentially and boldly modernistic. 1927 *Sunday Times* 13 Feb. 20/4 The audience liked its florid style as a change from more modernistic music. 1935 *Archit. Rev.* LXXVII. 82 The beautiful new dance-hall, with its modernistic sofas and lalique panels. 1958 P. Gammond *Decca Bk. Jazz* xv. 190 George Shearing will be remembered for his very accomplished playing in the 1940s, not for his more recent vapid modernistics. 1960 *Times* 29 Sept. 16/7 There is much to admire in this farrago of 'modernistics'. 1973 A. Behrend *Samarai Affair* iii. 30 A modernistic portrait in oils of the latest pilot boat.

‖ **modernus** (mǫdə̄·ɹnŭs). Pl. **moderni.** [L.] A modern person; someone who is characterized by, or notable for, his modernity.

1953 W. R. Trask tr. *Curtius's European Lit.* ii. xiii. 490 He is a *modernus* and considers that the ancients had loaded down their poetic narratives with a superfluity of similes, rhetorical figures, and digressions... These *moderni* of 1175..either consciously break with the theory of *imitatio* or restrict it very considerably. 1962 M. McLuhan *Gutenberg Galaxy* 183 Francis Bacon, PR voice for the *moderni*, had both his feet in the Middle Ages.

QUENCY MODULATION), or (*b*) the method by which the modulation is applied (as in *grid modulation).

1919 *Proc. IRE* VII. 193 (*heading*) Modulation. **1921** J. SCOTT-TAGGART *Thermionic Tubes* xiii. 355 Two general methods of modulation are used at present; either the amplitude of the continuous waves is varied by the microphone, or the wave-length is altered... Sometimes both wave-length and amplitude modulation occur at the same time. **1924** W. JAMES *Wireless Valve Transmitters* ix. 200 The strength of the note received is dependent.. on the degree of modulation—that is, on the extent to which the amplitude of the oscillations vary. **1931** *Proc. IRE* XIX. 2146 Amplitude modulation by means of vacuum tubes can be effected in two different ways: as plate modulation or grid modulation. **1932** LADNER & STONER *Short Wave Wireless Communication* v. 73 Rectification is essential at the receiver, for the purpose of extracting the modulation. **1943** F. E. TERMAN *Radio Engineers' Handbk.* VII. 581 In phase modulation, intelligence is transmitted by varying the phase of the transmitted wave. **1949** H. E. PENROSE *Princ. & Pract. Radar* xvi. 310 The klystron and the reflex klystron depend for their action primarily upon velocity modulation of the electron stream. **1952** [see *MODULATE *v.* 5 a]. **1953** W. A. EDSON *Vacuum-Tube Oscillators* xvi. 386 The presence or absence of an output signal, in conjunction with an appropriate code, permits the communication of information. Such keying represents the simplest possible form of modulation, and is applicable to all kinds of oscillators. **1962** A. NISBETT *Technique Sound Studio* vi. 107 The equipment is lined up on a standard 1,000 c/s tone sent from the studio. The standard used is 1 milliwatt in 600 ohms..and is equivalent to 40% modulation at the transmitter. **1968** *Radio Communication Handbk.* (ed. 4) ix. 5/2 Choke modulation employs a choke as the coupling impedance between the modulator and the r.f. stage. **1972** *Sci. Amer.* Sept. 101/2 The most widely used processes of modulation are amplitude modulation (AM), frequency modulation (FM), pulse amplitude modulation (PAM) and pulse code modulation (PCM).

b. *transf.* The action or result of varying the magnitude, degree, etc., of something.

1964 *Science* 15 May 816 Modulation of transfer RNA species can provide a workable model of an operator-less operon. **1970** R. W. MCGILVERY *Biochem.* xxiii. 543 (*heading*) Regulation [of metabolism] by modulation of enzyme activity. **1971** *Physics Bull.* July 388/1 An acoustic wave produces a periodic modulation of the density of the medium. **1973** *Sci. Amer.* Aug. 50/2 The conduction between them can be controlled by the modulation of the charge in a channel between them.

c. *attrib.*, as **modulation envelope**, the envelope of an amplitude-modulated carrier wave; **modulation factor** = *modulation index*; **modulation frequency**, the frequency of a wave used to modulate another wave; **modulation index**, a coefficient representing the degree of modulation of a carrier wave; *spec.* the ratio of the difference between the maximum and minimum frequencies of a frequency-modulated carrier to the frequency of the modulating signal.

1930 *Proc. IRE* XVIII. 2161 If this leakage is slower than the rate at which the modulation envelope decreases, then the condenser voltage cannot follow the modulation envelope. **1950** P. PARKER *Electronics* x. 301 In the radio frequency stages of a receiver, distortion is important only in so far as it makes the modulation envelope of the signal voltage different from the wave-form of the modulating sound. **1970** J. EARL *Tuners & Amplifiers* ii. 47 To the centre-tap..is fed the mono and subcarrier stereo components, and the action of the switching transistors..is such that a 'modulation envelope' is formed, one side carrying the left-channel information and the other side the right-channel information. **1939** *Amat. Radio Handbk.* vi. 93/1 When using a continuous pure tone (sine wave) for modulating..the percentage modulation can be obtained by the Heising formula:—If I_O = R.M.S. value of unmodulated aerial current..I_M = ditto when modulated. m = modulation factor. Then $I_M = I_O\sqrt{(1+m^2/2)}$. **1930** *Proc. IRE* XVIII. 2162 The rate of decrease of the modulation envelope depends upon the modulation frequency *f* and the degree of modulation *m* of the signal. **1962** SIMPSON & RICHARDS *Physical Princ. Junction Transistors* xviii. 453 The carrier and modulation frequencies are applied to one or both of the input electrodes. **1930** B. VAN DER POL in *Proc. IRE* XVIII. 1200 The whole practical problem of the amount of disturbance arising from frequency modulation depends therefore upon the value of the ratio $m = \Delta\omega/p$ of the absolute frequency deviation to the imposed audio frequency *p*. Owing to the importance of this parameter *m*..it may be found useful to designate it by a special name for which we suggest the expression 'frequency modulation index'. **1931** H. RODER in *Ibid.* XIX. 2151 B. van der Pol has introduced the expression 'modulation index' for m_f. We shall use this term for both m_p and m_f [where m_p represents the degree of phase modulation in a wave $i = A_o \sin(\omega_o t + m_p \sin \mu t)$]. **1974** *Encycl. Brit. Macropædia* XVIII. 91/2 The larger the modulation index (and hence the wider the bandwidth required for transmission), the more effectively FM performs.

8. *Biol.* Reversible variation in the activity or form of a cell in response to a changing environment.

1939 P. WEISS *Princ. Devel.* I. 94 This physiological, strictly nonprogressive fluctuation of a cell in response to its environmental conditions may be called modulation. It provides for a certain latitude within which a cell can comply adequately with certain variable functional demands of the developed body. **1964** N. T. SPRATT *Introd. Cell Differentiation* ii. 20 We..cannot accurately draw a line between differentiations and modulations. **1970** AMBROSE & EASTY *Cell Biol.* xiii. 441 Hormones are known to affect the synthetic function and the size of certain organs. This is an example of what Weiss has called 'modulation'.

modulator. Add: **1. b.** *spec.* A device that produces modulation of a wave (*MODULATION 7). Also *transf.*, a regulator, a controlling mechanism.

1919 *Proc. IRE* VII. 193 The curvature of the characteristic..makes possible its employment as a modulator and detector. **1930** *Discovery* Dec. 398/2 The output from the subscriber's telephone is amplified in the transmitting voice frequency amplifier and passes to the low frequency modulator. **1952** [see *MODULATE *v.* 5 a]. **1964** M. BROTHERTON *Masers & Lasers* xv. 182 If we feed into a modulator a voice frequency of 256 cps along with a carrier frequency of 50,000 cps, the modulator reacts by producing two new frequencies—one the difference frequency at 49,744 cps and the other the sum frequency at 50,256 cps. **1970** R. W. MCGILVERY *Biochem.* xxiii. 547 The deoxycytidine phosphates are not modulators of any of the reductions. **1970** *Sci. Amer.* Nov. 120/1 The use of the laser for mass communication..awaits the development of a practical modulator: an apparatus for impressing multiple signals on the light beam. **1972** *Jazz & Blues* Nov. 21/1, I..checked out this new modulator which a lot of people..are now using. You put in your voice and you get another sound on it. **1974** *Nature* 17 May 250/1 Such a test might assess the role of putrescine as a potential modulator of growth in normal and neoplastic tissue.

module, *sb.* Add: **4. c.** A length chosen as a basis for the dimensions of parts of a building, items of furniture, etc., to facilitate their co-ordination, so that all lengths are an integral multiple of it; *spec.* one of 4 inches (101·6 millimetres). Also *attrib.*

1936 BEMIS & BURCHARD *Evolving House* III. iv. 64 A dimension of 4″ for the module..is selected because it is the nominal greatest common divisor of the wood-frame house, which represents the bulk of American housing. **1945** *Archit. Record* Jan. 103/1 The architect can best realize the advantage of the coordination of masonry and metal window dimensions by doing preliminary building layouts on the familiar cross section paper of the module system, each grid line representing 4 inches. **1946** *Industr. Standardization* XVII. 269/1 'Module' furniture is designed on a coordinated 6-inch scale so that all pieces are interchangeable. **1949** *Architects' Jrnl.* 20 Oct. 432/1 The planning grid on which the Hertfordshire County Council structure is based was an 8 ft. 3 in. module. **1955** *Sci. News Let.* 1 Jan. 13/1 Houses of the future may all be built using a four-inch cube called a module as the structural 'atom'. **1966** L. B. ANDERSON in G. Kepes *Module, Symmetry, Proportion* 117 Now the idea of the module is again asserted, with emphasis on its ability to encompass growth and change.

d. One of a series of production units or component parts that are standardized to facilitate assembly or replacement and are usu. prefabricated as self-contained structures.

1955 *Sci. Amer.* Aug. 30 (*caption*) Assembled module consists of a stack of wafers coated with opaque plastic. Vertical wires through notches in the wafers provide electrical connections between the parts. **1959** HORSEY & SHERGAUS *Proc. Symposium Microminiaturization of Electronic Assemblies 1958* I. iv. 44 The electronic 'module'..is an individually fabricated subassembly that may be replaced *in toto* when repair becomes necessary. **1964** R. F. FICCHI *Electr. Interference* iii. 23 For circuits and components that can be grouped together, e.g., modules (a group of components mounted on a nonconducting board and wired together), and chassis drawers (groups of modules mounted in a rack), the following wiring rules should be followed. **1969** W. V. TIPPING *Introd. Mech. Assembly* ix. 240 The modules are bolted together to form a full machine ready for final troubleshooting. **1970** J. EARL *Tuners & Amplifiers* vi. 141 The vast majority of tuner-amplifiers are now transistored, the designs being based on printed circuit boards or 'modules'. **1970** *New Yorker* 3 Oct. 28/1 In prefabrication, flat pieces of a house are built in a plant and assembled on a site. Plumbing, electrical wiring, and heating are then installed by conventional methods. Modules are larger, three-dimensional units, which are completely finished in the factory and then bolted together at the site in a much shorter time. Modular housing gives you more control in the plant. **1971** [see *MODULAR *a.* 1 b]. **1971** *Real Estate Rev.* Fall 48/2 Our housing needs in the next nine years must be met with factory-built modules, assembled on site.

e. *Astronautics.* A separable section of a spacecraft that can operate as an independent unit.

1961 *New Scientist* 4 May 241/3 To deal with its dual function, the *Apollo* craft will have three separate sections, or modules: first, a command centre module..; secondly, a propulsion module..; and finally, a so-called 'mission' module. **1963** *Ann. Reg. 1962* 400 A capsule was to be fired from the earth into orbit round the moon, when a special part of it, christened the lunar excursion module, would detach itself. **1969** *Listener* 24 July 123/1 The ITV audience was being guided..towards the first climax of the night, the landing of the module on the Moon's surface. **1970** *Sci. Jrnl.* Aug. 35/2 As additional modules are placed in orbit and docked with the first module, some could be devoted to specialized activities.

f. One of a number of distinct, well-defined units from which a computer program may be built up or into which any complex process or activity is analysed (usu. for computer simulation), each of which is complete in itself but bears a definite relationship to the other units.

1963 L. SCHULTZ *Digital Processing* xv. 340 Ideally, the total program system could be segmented into completely independent parts (called modules) that exhibit interdependence only through a central communication pool. **1964** FISHER & SWINDLE *Computer Programming Syst.*

iv. 220 Because of such varying input/output requirements among programs and among machine configurations, it is now possible with many of the new IOCS packages to specify which modules of an IOCS [*sc.* an input/output control system, a set of routines for reading input and writing output data] are necessary for the most efficient processing of input/output requirements. **1965** *Economist* 17 July 272/2 This network is split up into sections or 'modules' that can be used in block form whenever a similar network is required. Shell used the example of the construction of chemical plant which the company put up around the world. The 'modules' in this instance are the particular items of plant, such as pressure vessels, which are standard. **1968** F. F. MARTIN *Computer Modeling* ix. 193 Modules are of two types: system or auxiliary. System modules simulate a specific function in system (or operation) logic... For example, a detection routine in a radar model is a system module that simulates the detection function of the radar... Auxiliary modules are nonsystem modules and are not necessarily unique to any given model. For example, a random statistical variates generator routine is an auxiliary module that may be applied in any stochastic model to generate random numbers. **1971** B. DE FERRANTI *Living with Computer* ix. 82 The computer programmer breaks his problem down into modules and gives the modules names so that they can be handled. **1973** R. M. ARMSTRONG *Modular Programming in COBOL* v. 60 The components of an operational control information system may be delineated as follows. Logistics: 1. Raw materials. 2. Production. 3. Salable product. Physical assets: 1. Property and equipment. 2. Capital projects... Manpower: 1. Payroll. 2. Benefits. 3. Personnel administration. We might describe these components as applicational modules at the systems level. *Ibid.* 63 In applying the above [characteristics] to a module in a computer system, a more precise definition is necessary. 1. The functions of input and output are well defined. 2. The module has a single entry point and a single exit point. 3. It exits to a standard return point in the module from which it was executed.

g. A unit or period of training or education. Cf. *MODULAR *a.* 1 c.

1966 *Economist* 3 Dec. 1005/1 Eventually the sort of retraining envisaged could fit in with the notion..of periodic training 'modules', whereby skilled men would take repeated periods off productive work to renew their perhaps rusty skills and learn new ones. **1968** *Daily Tel.* 12 Nov. 19/5 A training manual for a module course on 'Repairs and Restoration' has been produced.

5. b. *Math.* [ad. G. *modul* (R. Dedekind in P. G. L. Dirichlet *Vorlesungen über Zahlentheorie* (ed. 2, 1871) Suppl. x. 442).] Orig., a set that is a subset of a ring and is closed under addition and subtraction; now usu. defined as a commutative additive group whose elements may be multiplied by those of a ring (usu. a ring with an identity element), the product being in the group and the multiplication obeying the associative and distributive laws; *left, right module* (see quot. 1970[1]). Formerly also **modul.**

1927 *Amer. Math. Monthly* XXXIV. 64 The class concept was introduced by Dedekind as follows. A set *M* of elements of *I*(*a*) which is closed under subtraction, and hence under addition and subtraction, is called a module. If *M* be such that, if *β* is any element of *M*, and *γ* is any element of *I*(*a*), then *βγ* is an element of *M*, the module *M* is called an ideal. **1937** A. A. ALBERT *Mod. Higher Algebra* i. 9 An additive abelian group is frequently called a modul. *Ibid.* xi. 252 A set 𝔐 of elements of a ring 𝔄 is called a modul of 𝔄 if *a* − *b* is in 𝔐 for every *a* and *b* of 𝔐... Thus a modul is simply an additive subgroup of the additive abelian group 𝔄. **1948** O. ORE *Number Theory* vii. 159 The integers 0, ±1, ±2,... form a modul, but the natural numbers 1, 2, 3,... do not. *Ibid.* 161 A ring is a modul that is closed under multiplication. **1970** HARTLEY & HAWKES *Rings, Modules & Linear Algebra* v. 70 A module over a ring *R* (or *R*-module) is an Abelian group *M* (almost always written additively) together with a map $(r,m) \to rm$ from $R \times M$ to *M* satisfying the conditions $r(m_1+m_2) = rm_1+rm_2$, $(r_1+r_2)m = r_1m+r_2m$, $(r_1r_2)m = r_1(r_2m)$, $1m = m$, for all $r, r_1, r_2 \in R$ and all $m, m_1, m_2 \in M$. It would be more accurate to call what we have just defined a left *R*-module. There is a similar definition of a right *R*-module in which the elements of *R* are written on the right. *Ibid.* 69 A module is a construct of great versatility. It turns up in many seemingly unlikely guises and has the knack of manifesting some of the quintessential features of a wide variety of mathematical structures. **1970** *Nature* 19 Dec. 1234/2 A vector space is built up linearly by means of 'scalar' multipliers from a number field. The more general concept of a module replaces the field by an arbitrary ring (with unity) related to an Abelian group so that a 'product' is defined satisfying the usual distributive and associative laws.

8. *Engin.* The pitch diameter of a gear wheel in millimetres (or inches) divided by the number of teeth.

1909 in WEBSTER. **1912** G. T. WHITE *Toothed Gearing* ii. 17 Module = Dmm/N = ..25·4/diametral pitch = circular pitch" × 8·085. **1964** MORRISON & CROSSLAND *Introd. Mech. Machines* ii. 142 One set of standard proportions is that the addendum should be equal to the module *m* (or the reciprocal of the diametral pitch) and the dedendum.. greater than this by an amount of one-twentieth of the circular pitch or 0·157*m*.

modulo (mǫ·diŭlo), *prep. Math.* [abl. of L. *modulus*.] With respect to a modulus of. Also *attrib.*, = modular.

1897 *Bull. Amer. Math. Soc.* III. 381 Congruences irreducible modulo *p* (*p* = prime). **1903** J. BOWDEN *Elem. Theory Integers* x. 242 The statement of congruence is written *α* ≡ *β* (mod. *μ*), which is read '*α* is congruent with *β*, modulo *μ*'. **1939** L. E. DICKSON *Mod. Elem. Theory Numbers* i. 8 Any positive integer *n* is congruent modulo 9 to the sum *s* of its digits. **1966** OGILVY & ANDERSON *Excursions in Number Theory* iv. 43 Since 8, 15, 22, 29, etc., and also −6, −13, etc. are all congruent to 1 (mod 7), they are members of the same residue class modulo 7... In arithmetic modulo 7, 8 is equivalent to 1 in many senses. *Ibid.* 44 In number theory one must often work with very large numbers. If these can be reduced to equivalent smaller numbers, much time-consuming labor can be avoided. Herein lies one of the great contributions of modulo arithmetic. **1971** J. H. SMITH *Digital Logic* vi. 115 Modulo-7 counters count normally up to six (110). The next count must reset the counter to zero. **1973** *Nature* 20 Apr. 541/3 A sufficiently dense integer sequence is well distributed in most arithmetic progressions modulo most (small enough) primes.

modulor (mǫ·diŭlǫɹ). [Fr.] A scale with initial dimensions of 7 feet 5 inches (226 cm.) and 3 feet 9 inches (114·3 cm.), devised by the French architect Charles-Edouard Jeanneret, called Le Corbusier (1887–1965), based on the proportions of a human figure six feet tall, and intended to be used for the co-ordinated design of buildings, their contents, and their surroundings. Also *attrib.*

1948 M. GHYKA in *Archit. Rev.* Feb. 39 Le Corbusier's Modulor, and the concept of the Golden Mean... The Modulor is a linear scale in the form of a tape. Its object is to provide a means by which practical realization may be given to the need for efficiency and harmony in the dimensioning of objects and their containers, in architectural design and in town-planning... The Modulor differs from such a [conventional] scale in that the markings on it are related to each other by the principle of the Golden Mean. **1954** DE FRANCIA & BOSTOCK tr. *Le Corbusier's Modulor* 55 The necessities of language demanded that the golden rule should be given a name. Of several possible words, the MODULOR was chosen... The 'Modulor' is a measuring tool based on the human body and on mathematics. **1962** *Listener* 27 Sept. 472/1 The modulor which he [*sc.* Le Corbusier] invented..seemed to have served this purpose. It..forced him to develop the shell from the inside in several stages, starting with the modulor as the smallest unit. *Ibid.* 472/2 His faithful students soon learned to assemble their modulor units to very much the same buildings they had designed before. **1964** J. SUMMERSON *Classical Lang. Archit.* vi. 46 It was not till the early years of World War II that le Corbusier created the system..which he has called the 'Modulor'. **1972** *Times* 30 Nov. (Books Suppl.) p. ii/5 Furneaux Jordan quotes the story of how Corbusier's six foot 'modulor' was chosen because he had heard that the 'good-looking man' in English detective novels..is always six feet tall.

modulus. Add: **2. d.** (Examples.) [Introduced in this sense by Gauss in *Disquisitiones Arithmeticae* (1801) I. 1.]

1845 *Encycl. Metrop.* I. 642/2 Numbers of the same form with respect to any modulus, are all those which can be represented by the same formula. Thus, 13, 17, 21, &c. are all of the form 4*n* + 1; and 19, 25, 31, &c. of the form 6*n* + 1; 4 and 6 being the moduli. **1888** C. SMITH *Treat. Algebra* xxviii. 487 If two numbers *a* and *b* leave the same remainder when divided by a third number *c*, they are said to be congruent with respect to the modulus *c*. **1949** [see *MOD *prep.*]. **1967** J. E. SHOCKLEY *Introd. Number Theory* iii. 48 Corollary 6.2 furnishes the most efficient method for solving several systems of congruences when the same sets of moduli appear in each system.

modus. Add: **5. b.** *modus operandi* (later examples).

1935 G. POOLE *Haulage & Winding* xiii. 324 Assuming that an overwind has taken place,..the *modus operandi* is as follows:—The engine-driver at once closes his throttle and immediately he has to call an assistant to go to the overwinding device; [etc.]. **1947** *Sci. News* V. 14 The *modus operandi* of insulin was, in Sakel's original theory, upon the nerve cells and the hormones which, he believed, excited their activity especially in the 'vegetative centres'. **1949** [see *HOW *sb.*[3] 2]. **1955** *Bull. Atomic Sci.* Feb. 58/3 Their *modus operandi* against limited aggression would be simple. **1972** *Mod. Law Rev.* XXXV. 28 The modus operandi of a small claims judge should approximate more closely to that of a police detective.

d. modus ponens (*Logic*), 'mood that affirms'; the rule that from *if p then q* together with *p*, *q* may be inferred; an argument of this form. In full **modus ponendo ponens.**

a **1856** W. HAMILTON *Lect. Metaphysics & Logic* (1860) I. 344 We can always easily convert an hypothetical syllogism of one form into another,—the *modus ponens* into the *modus tollens*. **1870** W. S. JEVONS *Elem. Lessons Logic* xix. 161 The argument is said to be of the *modus ponens*, or mood which posits or affirms. **1916** H. W. B. JOSEPH *Introd. Logic* (ed. 2) xv. 335 The argument is said to be in the *modus ponens*. **1957** P. SUPPES *Introd. Logic* ii. 32 The Latin name for the Law of Detachment is *modus ponendo ponens*. **1972** *Computer Jrnl.* XV. 230/2 Applying the rule of modus ponens on the above instance.

e. modus tollens (*Logic*), 'mood that denies'; the rule that from *if p then q* together with *not-q*, *not-p* may be inferred; an argument of this form. In full **modus tollendo tollens.**

a **1856** [see *modus ponens*]. **1881** MAX MÜLLER tr. *Kant's Critique Pure Reason* II. II. i. 678 The *modus tollens* of reasoning, from consequences to their grounds, is not only perfectly strict, but also extremely easy. **1916** H. W. B. JOSEPH *Introd. Logic* (ed. 2) xv. 337 The *modus tollens* is of the form [etc.]. **1940** *Mind* XLIX. 208 An inference in the *modus tollendo tollens*..yields the contrary of the original contrary hypothesis. **1965** E. J. LEMMON *Beginning* peculiar idioms and expressions, coloured considerably by Kathakali. 1950 K. Ambrose *Classical Dance & Costumes India* ii. Mohiniattam...was prevalent in Malabar right up to the late 'twenties, and it is very largely composed of a fusion of the Tanjore temple-dance and the *Kathakali* dance style of Malabar. 1953 F. Bowers *Dance in India* 69 Mohini Attam was performed exclusively by women and was largely used by prostitutes to attract patrons. Because of its immoral associations, Mohini Attam fell into great disfavor and declined. 1967 Singha & Massey *Indian Dances* xii. 115 Mohini came to be synonymous with the essence of feminine beauty and allurement, and Mohini Attam is a dance which displays just such qualities. It is a solo dance, reserved exclusively for women... Mohini Attam is for the most part nritta although abhinaya is by no means absent. 1969 *Cultural News from India* Nov. 13 Kanak Rele...has done deep research into South Indian dance... Her repertoire consists of three major items: Mohini-attam; Kansa-natakam and pure Kathakali. 1971 *Femina* (Bombay) 2 Apr. 11/1 She was also one of the very first women in our time to learn Kathakali and Mohini Attam.

moho¹. Delete 'extinct' in definition. See also Notornis and *takahe.

Moho³ (mōᵘ·ho). *Geol.* Also **moho.** [Abbrev. of *Mohorovičić.*] = *Mohorovičić discontinuity.

1956 *Adv. Geophysics* III. 118 The boundary...is now called the Mohorovičić discontinuity (vulgarly 'The Moho'). 1959 *Daily Tel.* 21 Apr. 13/1 The bore must itself be several miles deep to penetrate to the mysterious 'moho' and find out what the bulk of the world is really made of. 1960 *New Scientist* 19 May 1278/3 It is...generally accepted that both the density and the seismic wave velocities change at the Moho. 1972 *Nature* 15 Dec. 383/3 The graben gives clear Moho refractions with normal upper mantle velocities.

mohonono (mohonōᵘ·no). *S. Afr.* [ad. Lozi *muHonono.*] An evergreen tree with grey-green leaves, *Terminalia sericea*, of the family Combretaceæ, native to southern Africa; also called Transvaal silver leaf and vaalboom.

1864 T. Baines *Explor. S.-W. Afr.* xx. 441 Passing through a thick mopane and mohonono forest, [we] outspanned about 7 p.m. in a more open flat. 1878 K. Johnston *Africa* xxiv. 427 The silvery mohonono...is in form like the cedar of Lebanon. 1949 K. L. Simms *Sun-Drenched Veld* iii. 33 There is...the cedar-spread of the mohonono.

mohoohoo (mŏhū·hū). Also **mahoohoo, mohohoo, mohohu, mohuhu, monooho(o), muchocho.** [Sechuana.] The white rhinoceros, *Ceratotherium simus*, found in central Africa and Zululand.

1835 A. Smith *Diary* 11 July (1940) II. 107 The Baquana say that two kinds [of rhinoceros] are only in this country. The black, which they call *muchli* and the white *mohohoo*. 1842 R. Moffat *Missionary Labours & Scenes S. Afr.* xxvii. 461 The mohohu, the largest species [of rhinoceros], has been known even to kill the elephant, by thrusting the horn into its ribs. 1849 A. Smith *Illustr. Zool. S. Afr.: Mammalia* plate 19 Localities abounding in grass are therefore the haunts of the Mohoohoo. 1850 R. G. Cumming *Five Yrs. Hunter's Life S. Afr.* xii. 77/2 With considerable difficulty we separated the horn of the muchocho from the skin by means of a long sharp knife. 1856 C. J. Andersson *Lake Ngami* xxx. 387 The common white rhinoceros...called Monoohoo by the Bechuanas. 1866 *Chambers's Encycl.* VIII. 236/2 The White R[hinoceros]...or Muchuco, or Monooho, is the largest of the well-ascertained African species. 1875 W. H. Drummond *Large Game & Nat. Hist. S. & S.-E. Afr.* 84 *Rhinoceros simus*, the mohohu of the Bechuanas. 1879 E. P. Wright *Animal Life* 141 It [sc. the white rhinoceros] is called by the Bechuanans, Mahoohoo, and is considered by them to be one of the original animals of their country. 1884 J. Mackenzie *Day-Dawn in Dark Places* 189 It is said the flesh of the mohuhu, or white rhinoceros is very good. 1894 R. Lydekker *Royal Nat. Hist.* II. 480 The individuals exhibiting this form being known to the Bechuanas by the name of mohohu.

Mohorovičić discontinuity (mohorōᵘ·vitʃitʃ). *Geol.* Also **Mohorovicic discontinuity.** [f. the name of A. *Mohorovičić* (1857–1936), Yugoslav seismologist.] The discontinuity between the earth's crust and the mantle which is believed to exist at a depth of about 10–12 kilometres under the ocean beds and 40–50 kilometres under the continents.

1936 J. B. Macelwane in Macelwane & Sohon *Introd. Theoret. Seismol.* I. viii. 204 We shall speak of this major boundary above which the various surficial layers lie as the Mohorovičić discontinuity, after A. Mohorovičić who discovered it. 1956 [see *Moho³]. 1958 [see *mantle *sb.* 10*]. 1971 I. G. Gass et al. *Understanding Earth* iii. 58/2 On the basis of seismic velocities, the mantle is usually divided into three regions—from the Mohorovicic discontinuity down to 400 km, from 400 to 1000 km, and from 1000 km to the core boundary.

Mohs (mōᵘz). The name of Friedrich *Mohs* (see Mohsine), used *attrib.* and in the possessive (chiefly in *Mohs('s) scale*) with reference to a scale of hardness he devised in which ten reference minerals that include very soft and very hard ones are assigned values of one to ten in order of increasing hardness.

The ten minerals of the scale are: 1, talc; 2, gypsum; 3, calcite; 4, fluor-spar; 5, apatite; 6, orthoclase; 7, quartz; 8, topaz; 9, corundum; 10, diamond.

1879 *Mineral. Mag.* II. 265 Its density is 1·0025; hardness between 1 and 2 of Mohs's scale. 1897 *Amer. Jrnl. Sci.* CLIV. 409 We shall describe only a preliminary series of tests with the minerals of the Mohs scale. 1951 D. Tabor *Hardness of Metals* i. 2 The Mohs hardness scale has been widely used by mineralogists and lapidaries. 1962 R. Webster *Gems* I. iv. 74 The hardness of beryl is 7¾ on Mohs's scale. 1974 *Sci. Amer.* Aug. 64 Each material in the Mohs scale, up to Mohs 9 (corundum), is about 1·2 times as hard as the preceding material. Thus the scale is almost logarithmic from Mohs 1 through Mohs 9, a range that includes all but a few substances.

¶ Erron. written as *Moh* (or *moh*).

1903 J. E. Marr *Agric. Geol.* ii. 28 Moh's scale of hardness. 1934 J. A. Perry *Chem. Engineers' Handbk.* xvi. 1558 The hardness of a material as measured by the moh scale is not always a criterion of its resistance to crushing. 1969 B. R. Schlenker *Introd. Materials Sci.* iv. 69 A Moh's hardness determination set of minerals.

mohua, mohur, var. Mahwa (in Dict. and Suppl.).

mohuhu, var. *mohoohoo.

Moi (mōᵘ·i), *a.* and *sb.* [Native name.] **A.** *adj.* Of or pertaining to a people of Indo-Australoid origin who were among the original inhabitants of Vietnam and are now found in the Southern mountain region (see quot. 1959). **B.** *sb.* **a.** The name of this people. **b.** A member of this people. **c.** The name of their language. Cf. *Montagnard 1 b.

1845 *Encycl. Metrop.* XVI. 780/2 The southern branches of those mountains, and the long ridge which divides Anam from Kambója are inhabited by the Múóng and Moï, or Ké-moïs tribes, whose language also differs from that of the lowlanders. 1853 J. R. Logan in *Jrnl. Indian Archipelago* VII. 36 The Moi or Ka-moi...on the opposite side of the Mekong occupy the broad expansion of the Anam chain towards Kamboja. 1911 *Encycl. Brit.* XIV. 491/1 The population of French Indo-China falls into five chief divisions—the Annamese...the Khmers...the Chams...the Thais...and the autochthonous tribes classed by the other inhabitants as Mois or Khas ('savages'). 1928 H. Hervey *Trav. French Indo-China* 248 Here the Moi, aboriginal tribes of ancient Chiampa, mingle with the Laotians and the Annamites. 1957 J. Oliver tr. *Riesen's Jungle Mission* i. 14 With her to interpret, I could understand and express the most varied ideas, increasing my French-Moï vocabulary. 1959 E. Brockett tr. *Bertrand's Jungle People* 11 The rather vague term 'Moi Plateau' is applied to the area bounded to the north by the River Mékong, to the west by the Cambodian jungle and to the east by the mountain range of Annam. *Ibid.* 20 On the floor...a square sod of earth placed on a tin slab formed the traditional Moï hearth. 1961 D. Lancaster *Emancipation French Indochina* 4 The Mois...are a handsome, bronze-skinned people akin to the Dyaks of Borneo and the Bataks of Sumatra. 1965 B. Newman *Background to Viet-Nam* vii. 73 Clever French philologists had adapted the Moïs languages to the Latin alphabet, so that reading and writing could be taught. 1968 *Listener* 13 June 760/3 Hill tribes...whom...the Vietnamese call *Moi*...practise a primitive agriculture, live in thatched long-houses...and are much addicted to gongs, rice wine and animal sacrifice.

moiety. Add: **2. c.** Chiefly *Biochem.* and *Pharm.* A group of atoms forming part of a molecule.

1935 Dorland & Miller *Med. Dict.* (ed. 17) 842/1 *Carbohydrate moiety*, the non-nitrogenous residue of the amino acids resulting from deamination. 1945 *Jrnl. Biol. Chem.* CLIX. 311 The lactone moiety [of pantothenic acid] can replace pantothenic acid for growth of the above organisms. 1954 A. White et al. *Princ. Biochem.* xii. 265 The other penicillins have the same type of structure but have different side chains replacing the benzyl ($C_6H_5CH_2$—) moiety. 1962 *Lancet* 29 Dec. 1381/1 A.I.C. has been shown to be a key intermediate...in the biosynthesis of the purine moiety of inosinic acid. 1970 Passmore & Robson *Compan. Med. Stud.* II. xxxii. 9/2 Species differences may vary strikingly when the compound under examination is not itself toxic but becomes metabolized to release the toxic moiety. 1974 *Nature* 13 Dec. 586/2 Its molecular structure (containing both an indole and a phenylethylamine moiety) suggests the possibility of an interaction with brain monoamines.

4. Also a type of social division found in varying forms amongst tribes in other parts of the world, though rarely in Africa. Also *attrib.*

1914 W. H. R. Rivers *Kinship & Social Organisation* iii. 72 The dual system in which there are only two social groups or moieties. 1934 R. H. Lowie *Introd. Cultural Anthropol.* xiv. 261 The moiety system can not explain why the Murngin and Miwok permit only one kind of cross-cousin to be married, when both belong to the proper moiety. 1936 R. Linton *Study of Man* xii. 207 Where both moieties and clans occur, the former are ordinarily more limited in their functions and of less social importance, possibly because the larger size of the moiety makes the establishment of well-defined attitudes...more difficult. 1937 R. H. Lowie *Hist. Ethnol. Theory* xi. 182 Some Melanesians...have totemic moieties. 1938 G. A. Reichard in F. Boas *Gen. Anthropol.* ix. 431 One of the Toda moieties considers itself superior to the other and has definite and important religious functions. 1944 B. Malinowski *Sci. Theory of Culture* vi. 162 The family, an extended kinship group, a clan, or a moiety, constitute one type. 1949 M. Mead *Male & Female* 422 Three different forms of marriage, and several types of age-grading and moiety systems, resulted in a rich and complex form of social organization. 1952 A. R. Radcliffe-Brown *Struct. & Function Primitive Society* vi. 118 Such moiety totemism...is found in a number of different varieties in Australia, and still other varieties are found in Melanesia and in North America.

Moine (moin). *Geol.* [f. *the Moine* (Gael. *A' Mhòine*), name of an area east of Loch Eriboll in northern Sutherland where rocks of the series are exposed.] Used *attrib.* and as *the Moine(s)* to denote a highly folded series of metamorphic rocks, chiefly granulites and schists, in NW. Scotland and western Ireland which are thought to have been deposited in Pre-Cambrian (possibly Torridonian) times and metamorphosed later.

1888 *Q. Jrnl. Geol. Soc.* XLIV. 425 The various stages in the production of the Moine schists. *Ibid.* 438 Other intrusive igneous rocks pierce the micaceous flagstones of the Moine series. 1931 Gregory & Barrett *Gen. Stratigr.* i. 27 The lower part corresponds to the British Lewisian and Moine. 1934 *Geol. Mag.* LXXI. 303 The view that the Moines are Torridonian metamorphosed in post-Cambrian time is associated with the name of Peach. 1953 *Q. Jrnl. Geol. Soc.* CVIII. 100 The problems of the age of the Moine Series, and of the relations of the series with the other Highland formations, have been a source of controversy for more than a century and the metamorphism of the series has been assigned to periods ranging from early Pre-Cambrian to Lower Palæozoic. 1961 *Science Progress* XLIX. 715 Younger Dalradian rocks...overlie the Moines and separate the Scottish and Irish outcrops. The Moines rest on a Lewisian basement and cannot accordingly be older than 1600 million years. *Ibid.* 716 The Moine Series is the last great formation of sedimentary origin in the British Isles whose stratigraphical relations remain to be established. 1969 Bennison & Wright *Geol. Hist. Brit. Isles* iii. 53 The unmetamorphosed equivalent of the lower part of the Moine may be the Greywacke Group of the Torridonian of Islay and Oronsay.

Moinian (moi·niăn), *a.* *Geol.* [f. prec. + -ian.] Of, pertaining to, or designating the Moine series of rocks (see prec.). Also *absol.*, the Moine series.

1938 A. K. Wells *Outl. Hist. Geol.* vi. 50 There is nothing inherently impossible in the suggestion that the Moinian rocks may be metamorphosed Torridonian. *Ibid.*, It is as well to remind ourselves that the great overthrusts separate an area where the succession is Lower Palæozoic rocks, on Torridonian, on Lewisian, from one...in which the succession is: Dalradian, on Moinian, on Lewisian. 1954 *Q. Jrnl. Geol. Soc.* CX. 46 Many bodies of Lewisian gneiss are marked on the Geological Survey maps of the Moine area of Ross-shire. Preliminary studies...suggest that some at least of these bodies resemble the so-called inliers of Scardroy and Fannich in being of Moinian age. 1969 Bennison & Wright *Geol. Hist. Brit. Isles* iii. 46 The Lewisian also forms the basement to the Moinian Series, a dominantly arenaceous sedimentary sequence but everywhere separated from the Torridonian by the Moine thrust. 1971 *Geol. Mag.* CVIII. 193 Moinian and Dalradian metamorphic rocks occur in the north-east extension of the Ox Mountains [in Eire]. The petrology and migmatization of the Moinian are discussed and Dalradian rocks are described for the first time from this area.

moiré, *a.* and *sb.* Add: **A.** *adj.* (Further examples.) Also applied to other materials (as paper, linoleum) having an appearance resembling watered silk.

1873 *Young Englishwoman* Apr. 199/2 The bottom is covered with moiré-paper. 1947 D. Hunter *Papermaking* (ed. 2) 528, 1806... The commencement of the use of moiré papers, embossed by heated cylinders, used in bookbindings. 1972 J. K. P. Edwards *Floors* v. 69 Linoleum can be manufactured to give many different patterns and a wide variety of types, including plain, jaspé, moiré, marble and others.

b. Applied to the wavy or geometrical pattern of light and dark stripes ('fringes') that is observed when one pattern of lines, dots, etc., is visually superimposed on another similar pattern, or on an identical one that is slightly out of alignment with the first; chiefly in *moiré fringe, pattern.*

1940 *Chambers's Techn. Dict.* 553/1 *Moiré effect* (Photog.), a 'watered-silk' pattern...arising from interference between two line-screens; a defect for which occasional uses are found. 1950 *Proc. R. Soc.* A. CCI. 189 With the [diffraction] gratings made in the manner described above the moiré pattern consisted of a very complicated wavy pattern which repeated itself with great uniformity every quarter of an inch, this being the pitch of the lead-screw. 1953 H. A. Chinn *Television Broadcasting* ii. 70 The problem of a moiré pattern caused by 'beating' of the scanning lines with the lines of the target mesh does exist. 1956 J. Guild *Interference Syst. Crossed Diffraction Gratings* p. v, Though attention was called to them by Lord Rayleigh as long ago as 1874 the fringes produced by crossed diffraction gratings, now usually termed moiré fringes, have somehow escaped mention in the textbooks of physical optics. 1963 *Sci. Amer.* May 54/3 The only general requirement for a moiré pattern is that the interacting figures have some sort of solid and open regions. The solid regions can be lines (straight, curved or wiggly), dots or any other geometric form... In the typical moiré pattern the moiré effect materializes when two sets of straight lines are superposed so that they intersect at a small angle. *Ibid.* 63/1 Moiré patterns sometimes plague the printer whenever he is obliged to print two or more halftone impressions one atop the other,

which he must do in making multicolored reproductions. 1971 J. H. Smith *Digital Logic* i. 3 Automatic programmed machines are...made to an extremely high accuracy, e.g. 5000 steps (digits) per inch are easily realised using moiré fringe devices. 1973 *Sci. Amer.* Oct. 120/2 Two sets of concentric circles that overlap generate moiré patterns in the form of radial lines. Concentric circles that overlap a grid of comparable spacing generate ellipses, parabolas or hyperbolas, depending on the angle of inclination between the plane of the concentric circles and the plane of the grid.

B. *sb.* **1.** (Further examples.) Add to def.: A moiré pattern or effect.

1953 H. A. Chinn *Television Broadcasting* i. 27 The most objectionable moiré is probably that produced by the blanking pulses because they usually represent the greatest possible black-to-white contrast ratio. 1963 *Sci. Amer.* May 55 (*caption*) Moiré composed of beats is produced from nonintersecting parallel lines when the spacing of one set differs from that of another. 1967 E. Chambers *Photolitho-Offset* xi. 160 The Agfa-Gevaert Diffusing Diaphragm is also a most useful asset in eliminating undesirable moiré and avoiding the need for excessive retouching. 1967 V. Strauss *Printing Industry* iv. 194/2 Moiré can also be introduced in presswork by defective overprinting.

moissanite (moi·s-, mwa·sănəit). *Min.* [f. the name of H. *Moissan* (1852–1907), French chemist: see -ite¹.] A green (sometimes black or bluish) silicon carbide, SiC, with metallic lustre, which is known only from the meteoric iron of Canyon Diablo, Arizona, and is made artificially as *carborundum.

1905 G. F. Kunz in *Amer. Jrnl. Sci.* XIX. 397 As this is the first instance in which this compound [sc. carborundum] has been proved to occur in nature, and therefore, as a mineral, is entitled to a distinct mineralogical name, it would seem that the name of Professor Moissan himself should be associated with it. I would, therefore, propose for it the name of Moissanite. 1968 I. Kostov *Mineral.* 105 Artificial moissanite is called carborundum.

moist, *a.* and *sb.* Add: **8.** **moist-lipped, -nosed, -skinned, -tinged** adjs.

1962 I. Murdoch *Unofficial Rose* xxix. 284 She was, as he looked down at her, tense, moist-lipped. 1934 Dylan Thomas *Let.* 9 May (1966) 120, I would have introduced a...paragraph all about her nasty little moist-nosed muse. 1957 J. S. Huxley *Relig. without Revelation* (rev. ed.) ix. 216 The reptiles replaced the moist-skinned amphibians. 1943 D. Gascoyne *Poems 1937–42* 55 Obscurely still beneath a moist-tinged blank Sky like the inside of a deaf mute's mouth.

moist, *v.* Add: **3.** *intr.* To rain slightly, to drizzle. *U.S. rare.*

1916 H. L. Wilson *Somewhere in Red Gap* iii. 117 It was moisting when we started, and pretty soon it clouded up. *Ibid.* 118 It wasn't moisting any more—it was raining for fair.

moistly, *adv.* Delete † *Obs.* and add later examples.

1905 H. G. Wells *Kipps* ii. iii. 197 The bull really came at them...opened a mouth below his moistly glistening nose, and booed. 1906 B. von Hutten *What became of Pam* 78 March had moistly melted into April. 1927 H. V. Morton *In Search of England* x. 180 Three of those prim, sallow, enthusiastic, middle-aged lovers of England ...were...regarding moistly the bare rooms in which Brewster and the 'Pilgrim Fathers' were imprisoned. 1939 L. M. Montgomery *Anne of Ingleside* vii. 50 Walter hated to be kissed moistly on the forehead. 1957 R. A. Heinlein *Door into Summer* (1960) i. 14 He shook hands moistly, sat me down,...and attempted to take my bag.

moisture, *sb.* Add: **4.** **moisture-seal**; **moisture-bearing, -holding, -loving** (later examples), **-proof** (hence **-proofness**) adjs.; **moisture-proofing** vbl. sb.; **moisture content,** the proportional amount of moisture in any substance; **moisture cream,** a cosmetic cream which keeps the skin moist; a type of face-cream; cf. *moisturizer; **moisture lotion,** a liquid preparation for moisturizing the skin; **moisture meter,** an instrument for indicating the moisture content of a substance (commonly by measuring its electrical resistivity).

1922 W. G. Kendrew *Climates of Continents* 290 The rainfall increases steadily towards the east and south. An irregularity is caused by the Appalachians,...within range of the moisture-bearing winds from both the Gulf and the Atlantic. 1923 H. M. Bunbury *Destructive Distillation of Wood* iii. 20 The moisture content of the limbs or branches is also greater than that of the trunk. 1936 *Burlington Mag.* Nov. 229/2 It was necessary to bring the panels to the agreed moisture-content before these repairs were put in hand. 1952 L. M. Thompson *Soils & Soil Fertility* x. 159 Increasing the moisture content of soils increases the uptake of phosphorus. 1971 *Materials & Technol.* II. 601 The moisture content of coke may...be high because of its high porosity. 1957 *New Yorker* 2 Nov. 61/2 (Advt.), Tussy Moisture Cream and Lotion...won't let your skin dry out. 1961 *Which?* Mar. 58/1 Basically, like other types of face cream moisture creams are emulsions of oils, fats or waxes, with water. 1952 A. G. L. Hellyer *Sanders' Encycl. Gardening* (ed. 22) 407 Culture of Pears: Soil, well drained and well supplied with humus to improve moisture-holding capacity. 1959 *Times* 7 Dec. (Agric. Suppl.) p. viii/5 The moisture-holding properties of old clay wastings. 1957 Moisture lotion [see *moisture cream* above]. 1974 *Country Life* 28 Nov. 1694/3 Their own products include moisture lotion...eye shadow and lipsticks. 1911 E. M. Clowes *On Wallaby* x. 272 A thick grove of moisture-loving palms. 1949 V. G. Childe *Prehist. Communities Brit. Isles* (ed. 3) viii. 135 The land molluscs of this period are...no longer the moisture-loving woodland species found in the neolithic camps. 1974 *Country Life* 28 Feb. 443/2 The moisture-loving hydrangeas. 1952 *Sci. Amer.* Sept. 61 (*caption*) Moisture meter...senses moisture in textile yarn by electrical conductivity and controls the speed of the drying drum by feedback. 1970 *Harrods Catal.* May 20/2 Moisture meter indicates condition of soil. ...47/–. 1904 *Electr. Rev.* (N.Y.) XLV. 422/1 These transformers are mounted in moistureproof iron cases and may be operated either dry or with oil. 1929 *U.S. Patent* 1,737,187 4/1 By the term 'moistureproof'...we mean the ability to resist the diffusion of water vapor to an extent at least as great as...that displayed by ordinary waxed papers...and functioning to resist the penetration of water vapor therethrough. 1960 *Farmer & Stockbreeder* 19 Jan. (Suppl.) 27/1 The birds being vacuum packed in moisture-proof bags. 1975 *Sci. Amer.* July 123/1 The cell that supports the mirror is also made of moistureproof plywood. 1929 *U.S. Pat.* 1,737,187 4/1 A sheet or film of regenerated cellulose combined with a moisture-proofing composition. 1971 J. F. Hanlon *Handbk. Package Engin.* x. 8 For moisture protection a layer of asphalt between two layers of paper is generally used. It is the least expensive moistureproofing, but in cold weather it becomes stiff. 1929 *U.S. Pat.* 1,737,187 4/1 The product...is one which has all of the desired properties of moistureproofness, flexibility,...and lack of odor. 1962 *Punch* 22 Aug. 256/1 We have improved on the moisture-proofness of Cellophane. 1960 Moisture-seal [see *crisper b].

moisturize (moi·stiŭrəiz, -stʃər-), *v.* [f. Moisture *sb.* + -ize.] To render moist, used esp. of a cosmetic cream applied to the skin. Hence **moi·sturized** *ppl. a.*; **moi·sturizing** *vbl. sb.* and *ppl. a.*

1945 Mencken *Amer. Lang. Suppl.* I. 402 Verbs made of common nouns:..*to moistureize.* 1958 *Inside the ACD* (Amer. College Dict.) X. ii. 1/1 *Moisturize*: Television advertisers have been giving this word hard use. Creams and lotions now *moisturize* the skin and scalp. 1958 *Times* 22 Sept. 13/5 Intended for all types of skin, it [sc. a cream] is said to have balancing qualities which include moisturizing and mild stimulating action. 1958 *Vogue* July 5 This make-up...is moisturized to help guard your skin against the sun's chief danger, dryness. Wear it over...exquisite moisturised foundation. 1959 *News Chron.* 14 July 6/3 A very light type of fluid foundation...with a moisturising content. 1963 *Word Study* Dec. 6/2 The word *to moisturize*, currently fashionable in cosmetic and beauty-care advertising and attacked by some linguists as superfluous and barbaric. 1965 M. Frayn *Tin Men* i. 5 Tropical plants...perpetually watered by invisible built-in moisturising systems. 1966 *Harper's Bazaar* Sept. 73/2, I cover my face, lips too, with Veiled Radiance—it's essential that I use a moisturised base. 1971 *Homes & Gardens* Sept. 99/2 So much for the face, but the rest of the body will appreciate a little bit of moisturising as well. 1973 *Country Life* 26 Apr. 1194/4 The creamy moisturised lipsticks are available in pinks, browns, red and tawny tones. 1974 *She* Jan. 5 Skin Tone...moisturises and conditions your skin.

moisturizer (moi·stiŭrəizəɹ, -stʃər-). [f. *moisturize *v.* + -er¹.] A preparation that renders or keeps the skin moist; a cosmetic cream.

1957 *Daily Mail* 17 Oct. 10/3 If the face then feels uncomfortably taut use a skin moisturiser before applying a foundation. 1966 J. S. Cox *Illustr. Dict. Hairdressing* 99/2 Moisturizer permanent wave. 1967 *Times* 17 Jan. 13/1 A moisturiser under the foundation is essential if the skin is dry. 1972 *Guardian* 7 Nov. 11/3 The Vichy preparations include complete ranges of cleansers, toners, moisturisers, night creams.

moit² (moit). *dial.* and *Austral.* [var. Mote *sb.*¹] A particle of wood, stick, or some other substance caught in the wool of a sheep. Hence **moi·ting** *vbl. sb.* (see quot. 1862), **moi·ty** *a.*

1862 C. C. Robinson *Dial. Leeds* 359 *Moiting*, a process in the manufacture of cloth, by which the wool, subsequent to being scoured...and preparatory to its passing through the 'willey', is cleansed from 'moits' or shivs. 1878 *Yorkshireman* 17 Aug. 97/2 I've a splendid lot [of wool] in just now—...not moity, and free from burr. 1878 'R. Boldrewood' *Ups & Downs* viii. 83 The 'heavy and moity' parcels were not touched by the cautious operators at any price. 1903 *Eng. Dial. Dict.* IV. 144/2 T'sliver's full o' moits an' as rough as a bear's back. 1945 Baker *Austral. Lang.* iii. 67 Many terms used by Australian wooltraders come from old English dialect, including...*moits*, short pieces of stick and scrub, principally found in neck wool. 1959 —— *Drum* 127 *Moity* (of wool), carrying vegetable matter other than burr. 1965 J. S. Gunn *Terminol. Shearing Industry* ii. 4 *Moit(s)*, pieces of stick and rubbish matted in wool, especially the neck wool.

Mojave, var. *Mohave.

mojo¹ (mōᵘ·dʒo). *local U.S.* [Prob. of Afr. orig.: cf. Gullah *moco* witchcraft, magic, Fula *moco'o* medicine man.] Magic, the art of casting spells; a charm or amulet used in such spells. Also *attrib.* and as *vb.*

1926 N. N. Puckett *Folk Beliefs Southern Negro* i. 19 The term *mojo* is often used by the Mississippi Negroes to mean 'charms, amulets, or tricks', as 'to work mojo' on a person or 'to carry a mojo'. 1930 R. Bass in A. Dundes *Mother Wit* (1973) 382/2 There are a few signs that are more or less common to all mojo-workers. *Ibid.* 384/2 [Of the deepest South] Their mojo is not fakery. It is not trickery. It is magic... Mojo is making its last stand. It has retreated to the swamp-lands. 1934 B. A. Botkin in W. T. Couch *Culture in South* xxvi. 585 Fragments of hoodoo and conjuration, whose spells,...mojos,...goofer bags are the special province of the Negro 'root doctor' or 'hoodoo man'. 1962 N. E. Whitten in A. Dundes *Mother Wit* (1973) 406/1 Local names for amulets are 'mojo', 'monjo', 'lucky hand'...and 'jomo'. 1966 *Crescendo* Aug. 3/2 With his weather mojo working overtime he got four hot sunny days. 1970 R. Welburn in A. Chapman *New Black Voices* (1972) 356 It is overdue time To mojo the demons... Now it is time for mojo. 1971 *Black World* Apr. 81 A Mojo...a Mojo workin for you. 1973 C. Himes *Black on Black* 18 Slick (in a whining voice): You got the best go and the mojo.

mojo² (mōᵘ·dʒo). *U.S.* [Orig. unknown: see quot. 1955.] An addict's name for any narcotic drug, esp. morphine.

1935 A. J. Pollock *Underworld Speaks* 77/1 *Mojo*, any of the poisonous habit forming narcotics (dope). 1955 *Amer. Speech* XXX. 87 *Mojo* (probably from Sp. *mojar*, 'to celebrate by drinking')...a euphemism for morphine. 1963 R. I. McDavid *Mencken's Amer. Lang.* 725 An addict well supplied is *on the mojo* and is said to be in high. 1971 E. E. Landy *Underground Dict.* 134 *Mojo*,...morphine.

moke². Add: **2.** (Examples.)

1855 D. G. Rossetti *Let.* 25 Nov. (1965) I. 282 He has an irreconcilable grudge against a poor moke of a fellow called Archer Gurney. 1873 J. A. Mair *Handbk. Proverbs* 459 *Moke*, an old person, disrespectfully spoken to. 1915 *Dialect Notes* IV. 199 Terms of disparagement...*moke* about the same meaning and usage as *mutt*, or *boob*.

3. (Earlier and later examples.)

1879 in L. A. Smith *Music of Waters* (1888) 356 For many a long time he tried a Derby to win, But I was the moke to carry him in. 1888 'R. Boldrewood' *Robbery under Arms* I. viii. 105, I am regular shook on this old moke. 1901 M. Franklin *My Brilliant Career* xxvi. 220 My old moke flung a shoe and went dead lame. 1928 'Brent of Bin Bin' *Up Country* xiii. 210 If I lend you my mokes, the bridles are mine if you win. 1943 'W. Hatfield' *I find Austral.* 53 He's an old quiet moke these days.

4. (Examples.)

1856 C. White *Oh, Hush!* 9 Rose, don't you interfere, I'll show dis moke a sight. 1882 G. W. Peck *Peck's Sunshine* 53 They want to hear old fashioned negro melodies, and yet these mokes will tackle Italian opera. 1901 *Cosmopolitan* Sept. 637/2 Across the stage have paraded long processions of 'musical mokes', 'knock-abouts' and 'monologists'. 1945 Mencken *Amer. Lang. Suppl.* I. 635 *Moke* was thrown into competition with *coon* in 1899 by the success of 'Smokey Mokes', a popular song by Holzmann and Lind.

Moke⁴ (mōᵘk). = *Mini-Moke* (see *mini- b).

1963 *Daily Tel.* 8 Jan. 11/1 BMC believes the twin-engine Moke could be useful to farmers and for military and industrial use. 1967 *Guardian* 26 June 5/2 The Moke, a risky, insolent-looking little vehicle...has the minute quality of a 'Dinky' car. 1969 [see *fade *sb.*¹ 1]. 1974 A. Williams *Gentleman Traitor* ix. 135 Keys to Thackeray Mansions and the Moke.

mok-e-mok, var. *moki-moki.

moki¹. Substitute for def.: Either of two New Zealand marine fishes, *Latridopsis ciliaris*, which is blue-grey and white, or the red moki, *Chironemus spectabilis*, which is reddish-brown, with dark brown bars on its sides; also, formerly, the blue cod, *Parapercis colias*. (Further examples.)

1845 E. J. Wakefield *Adventure in N.Z.* I. iv. 93 The moki is also a well-flavoured fish, weighing 10 lbs. or 12 lbs. 1851 E. Ward *Jrnl.* 14 Feb. (1951) 129 A Maori brought us a fish today which he called 'Moko'—it was a very fair fish. 1960 Doogue & Moreland *N.Z. Sea Anglers' Guide* 235 Red Moki... Belongs to the same family [sc. the Chironemidæ] as tarakihi and porae, but the stouter body form and colour distinguish it immediately. 1968 *N.Z. Listener* 15 Mar. 6/1 I've left a moki and a feed of crayfish in the outside safe. 1971 *N.Z. News* 3 Feb. 3/2 A member of the Whangarei Underwater Club...shot a six-pound red moki.

moki². Add: (Earlier and later examples.) Also **mogey, moggie.**

1835 J. A. Wilson *Jrnl.* 25 Aug. in *Missionary Life & Work N.Z.* (1889) ii. 25 The moki only carries one person at a time. You sit as on horseback. 1860 W. C. *Let.* 16 Apr. in J. H. Beattie *Pioneers explore Otago* (1947) vii. 45 He started in the dead of winter, on a moggie by himself... He crossed and re-crossed the lake several times. 1884 R. C. Reid *Rambles on Golden Coast N.Z.* 53 It had been used by the diggers as the keel of their 'mogey'. 1963 *Evening Post* (Wellington, N.Z.) 9 Dec., He made the journey from Kingston to Queenstown on a mokihi.

moki, var. *moki-moki.

‖ **mokihana** (mōkihā·nā). Also **mokehana.** [Hawaiian.] A Hawaiian tree, *Pelea anisata*, of the family Rutaceæ.

1888 W. Hillebrand *Flora Hawaiian Islands* 64 *P[elea] anisata*... Nat[ive] name: 'Mokehana'.—All parts of the tree, but particularly the capsules, when bruised, emit a strong spicy odor of anise. 1915 W. A. Bryan *Nat. Hist. Hawaii* xv. 221 The most highly scented of all are the seed pods of the mokihana used in making leis. 1954 *Ellery Queen's Mystery Mag.* Oct. 45/1 She...hung around her neck a *lei* of *mokihana* and *maile.*

‖ **moki-moki** (mǫ·kimǫki). *N.Z.* Also **moke-mok, moki, moki-mok, mokky.** [Maori.] = *makomako¹; Moko-moko 2.

1928 *Daily Express* 6 June 3 What is the moki-moki? 1935 E. Hodgkinson in A. E. Currie *Cent. Treas. Otago Verse* (1949) 33 The forest rings with tuis' and mokis' chiming. 1940 N. Marsh *Surfeit of Lampreys* (1941) xiv. 218 English birdsong there was pierced by the colder and deeper notes of bell-birds and mok-e-moks. *a* 1948 L. G. D. Acland *Early Canterbury Runs* (1951) 361 Bell-bird.—Mako-mako, or Moki-mok... The name is still sometimes used; but the bird is now generally called a 'Mokky'.

moksha (mǫ·kʃā). *Hinduism* and *Jainism.* Also 8 **moksh,** 9— **moksa.** [Skr. *moksha* liberation, emancipation, f. *muc* to loose, release.] The final liberation of the soul when it is exempted from further transmigration; the bliss attained by this liberation. Also called *mukti.

1785 C. Wilkins tr. *Bhăgvăt-Gēētā* xvi. 115 The divine destiny is for *Mōksh*, or eternal absorption in the divine nature. 1828 H. H. Wilson in *Asiatick Res.* XVI. 104 An important consequence of this doctrine is the denial of *Moksha*, in its more generally received sense, that of absorption into the universal spirit. 1831 V. Kennedy *Res. Anc. & Hindu Mythol.* vi. 185 The terms employed for final beatitude, that is identification with the supreme Being...are *kaivalyam, moksha,* or *jivanmukti.* 1840 H. H. Wilson *Two Lect. Relig. Pract. Hindus* ii. 64 All the philosophical schools propose for their object the ascertainment of those means by which the wanderings of the soul may be arrested,...its emancipation from bodily imprisonment and degradation be effected for ever. This is what is termed *Moksha*, or *Mukti.* 1875 M. Williams *Indian Wisdom* vii. 131 There are three 'gems' which together effect the soul's *Moksha*...right intuition...right knowledge...right conduct. 1915 *Encycl. Relig. & Ethics* VIII. 772/1 *Mokṣa* as the supreme aspiration runs through that best known portion of the *Mahābhārata* epic called the *Bhagavad-Gītā.* 1933 J. Baillie *And Life Everlasting* (1934) v. 118 In the concurrent tendency to regard *nirvana*—or *moksha*...as a state not of utter annihilation but of (unconscious) blissful union with the divine, we may surely find a foreshadowing...of a positive doctrine of eternal life. 1962 A. Huxley *Island* ix. 136 Murugan calls it dope... We, on the contrary, give the stuff good names—the *moksha*-medicine, the reality-revealer, the truth-and-beauty pill. 1971 *Illustr. Weekly India* 11 Apr. 9/2 They are the Tirthankaras, the Finders of the ford by which to cross the Ocean of Rebirth, or the Guides that show the way to *moksha.*

mokuk, var. *mocock.

mol: see *mole *sb.*⁷

mola² (mōᵘ·lă). [Native name.] A square of brightly coloured, appliquéd cloth worn as a blouse by Cuna Indian women of the San Blas Islands, Panama. Also *attrib.*

1941 *Nat. Geogr. Mag.* Feb. 217 With the skirt is worn a short-sleeved waist called a *mola*. The mola is of true Indian manufacture. 1964 I. Salem tr. *M. & H. Larsen's Forests Panama* viii. 99 These *molas*...are perhaps the only really interesting things produced by the Cuna. Layers of different-coloured materials are cut into various shapes and superimposed one on another to produce a very colourful raised design. *Ibid.*, Some of these *molas* still depict the rather naïve, stylized animals portrayed on pre-Columbian pottery. 1966 J. M. Kelly *Cuna* i. 26 The 'Mola' tops had been introduced by the Spaniards who didn't like the women walking around with their breasts showing. 1972 *Islander* (Victoria, B.C.) 23 July 7/3 The women are known for their 'molas'. A mola is a square of five layers of material, and each piece is cut to show the colors of the different layers underneath. 1973 M. Mackintosh *King & Two Queens* v. 79 'These are original Cuna Indian molas,' said Rodger, 'and the most modern things in this upper hall.' 1974 *Daily Colonist* (Victoria, B.C.) 10 July 18/5 Molas are worn by women and girls of the Cuna tribe. The rectangular panels, approximately 16 by 24 inches, are worn in matched sets on the front and back of the blouse atop wrap-around print skirts.

molal (mōᵘ·lăl), *a.* *Physical Chem.* [f. *mole *sb.*⁷ + -al.] **a.** = *molar *a.*³ a, *molecular *a.* 1 a.

1908 Goodwin & Mailey in *Physical Rev.* XXVI. 49 In the fifth column we have computed the values of the fluidity *F* divided by the corresponding molecular volume *ϕ* of the liquid; this we have denoted by *f* and called the molal fluidity. 1923 Lewis & Randall *Thermodynamics* iii. 22 If *w* is the molal mass of a substance (also known as molal weight or molecular weight), a mol of the substance in question is defined as *w* grams. 1972 *Science* 2 June 1011/2 Carbon dioxide...has a molecular weight *M* of 44 and a partial molal volume of 34·8 ml/mole.

b. Of a solution: containing one mole, or a specified number of moles, of solute per kilogramme of solvent or (now *rare*) per litre of solution. Of a concentration: expressed in terms of these quantities. Cf. molar *a.*³ b.

1923 Lewis & Randall *Thermodynamics* iv. 33 When we speak of a molal or tenth molal aqueous solution..., we shall refer, not to the concentration, but to the number of mols of solute in 1000 grams of water, which we may call the molality. 1924 J. C. W. Frazer in H. S. Taylor *Treat. Physical Chem.* I. vii. 236 Concentrations expressed in terms of 1000 cc. of solution and using the above methods of expressing the amounts of substances are known as molal, formal or normal concentrations. 1930 W. T. Hall *Textbk. Quantitative Analysis* iii. 26 A solution is said to be molal in hydrogen ions if it contains 1·008 g of H^+ per liter. 1938 Lundell & Hoffman *Outl. Methods Chem. Analysis* xxxv. 127 One author will regard a solution containing one mole of solute per liter of solution as molal, whereas another calls such a solution molar. 1965 Phillips & Williams *Inorg. Chem.* I. vii. 250 The activity coefficient is then around 10^{-14} for molal solutions of these ions. 1970 [see molality below].

Hence **mola·lity,** the molal concentration of a solution.

1923 [see sense b above]. 1931 [see *molarity]. 1963 W. J. Moore *Physical Chem.* (ed. 4) v. 117 One advantage of molality is that it is easy to prepare a solution of given molality by accurate weighing procedures. 1970 *Man. Symbols & Terminol. Physicochem. Quantities & Units* (I.U.P.A.C.) 12 A solution having a molality equal to 0·1 mol kg^{-1} is sometimes called a 0·1 molal solution. 1973 [see *molarity].

Molale (molā·li). Also **Molele.** [Native name.] A Penutian Indian people of Oregon; a member of this people; also, their language. Also *attrib.*

1846 H. Hale in *U.S. Exploring Expedition 1838–42: Ethnogr. & Philol.* 561 The Wailatpu family [of languages]... Wailatpu... Moléle. 1848 *Trans. Amer. Ethnol. Soc.* II. i. 15 These Indians [sc. the Wailatpu] include two tribes, the Cayuse...: and the Molele, west of the Cayuse...in the mountainous territory about Mounts Hood and Vancouver...; probably extinct. *Ibid.* 77 Wailatpu,...*Language*...Molele. 1891 *7th Ann. Rep. U.S. Bureau Amer. Ethnol. 1885–86* 127 Hale established this family and placed under it the Cailloux or Cayuse or Willetpoos, and the Molele... The Moláles were a mountain tribe and occupied a belt of mountain country south of the Columbia River. *Ibid.* 128 There are 31 Molále now on the Grande Ronde Reservation, Oregon. 1965 *Canad. Jrnl. Ling.* Spring 124 Inland, no...Molale, or Cayuse were found. 1966 C. F. & F. M. Voegelin *Map N. Amer. Indian Lang.* (*caption*) Molale language isolate. 1972 *Language* XLVIII. 380 A Molale myth.

molar, *a.*² Add: **b.** *Psychol.* (See quot. 1932.)

1925 C. D. Broad *Mind & its Place* xiv. 616, I will lump together all such changes under the name of 'molar behaviour', as contrasted with 'molecular behaviour'. 1932 E. C. Tolman *Purposive Behavior* I. i. 7 On the one hand, he [sc. J. B. Watson] has defined behavior in terms of its strict physical and physiological details, i.e., in terms of receptor-process, conductor-process, and effector-process per se. We shall designate this as the molecular definition of behavior. And on the other hand, he has come to recognize...that behavior...is more than and different from the sum of its physiological parts. Behavior...has descriptive and defining properties of its own. And we shall designate this latter as the molar definition of behavior. 1946 C. Morris *Signs, Lang. & Behavior* ii. 55 Behavioristics naturally attempts to supplement its description of behavior in molar or macroscopic terms. 1971 *Jrnl. Gen. Psychol.* LXXXIV. 157 Attention has been concentrated on those papers which do not support the notion that single cell behavior is reflected in molar behavior.

molar (mōᵘ·lăɹ), *a.*³ *Physical Chem.* [f. *mol(e *sb.*⁷ + -ar¹.] **a.** Of or pertaining to one mole of a substance; = *molal *a.* a, *molecular *a.* 1 a.

1902 A. Findlay tr. *Ostwald's Princ. Inorg. Chem.* vi. 89 The ratio of the weight of a given gas to that of an equal volume of the normal gas under the same conditions, is called its molecular weight or its molar weight. Since the former name has been derived from certain hypothetical notions regarding the constitution of the gases, notions which are not essential to the actual facts, we shall give preference to the name molar weight although, at present, the other is still the one most used. 1906 [see *gramme-molecular* adj. s.v. *gramme b]. 1946 J. R. Partington *Gen. & Inorg. Chem.* i. 7 This is called the molar volume V_m, and Avogadro's hypothesis shows that it is the same for all substances. 1962 P. J. & B. Durrant *Introd. Adv. Inorg. Chem.* xi. 303 If both sides of the equation are multiplied by the molecular weight *M* and divided by the density *ρ*, $(k-1)M/(k+2)\rho = \frac{4}{3}\pi N\alpha = P_M$ where *N* is Avogadro's number and P_M is known as the molar polarisation. *Ibid.* 307 A quantity *R* known as the molar refraction. 1967 *Oceanogr. & Marine Biol.* V. 154 Some normal calcite and a small amount of dolomite with a molar ratio of $CaCO_3$:$MgCO_3$ of 1·1:1 were also present.

b. Of a solution: containing one mole, or a specified number of moles, of solute per litre of solution. Of a concentration: expressed in terms of these quantities. Cf. *molal *a.* b.

1927 C. J. Engelder *Textbk. Elem. Qualitative Analysis* i. 4 Thus 34·468 grams of HCl or 58·46 grams of NaCl, dissolved and diluted to one liter, are molar solutions of these compounds. 1946 *Nature* 26 Oct. 585/1 The molar concentration required to produce a given effect is approximately one third that of the preceding member [of a homologous series]. 1970 *Man. Symbols & Terminol. Physicochem. Quantities & Units* (I.U.P.A.C.) 12 A solution with a concentration of 0·1 mol dm^{-3} is often called a 0·1 molar solution. 1971 G. D. Christian *Analytical Chem.* i. 9 A one-molar solution of silver nitrate and a one-molar solution of sodium chloride will react on an equal-volume basis.

Hence **mola·rity,** the molar concentration of a solution.

1931 J. C. Ware *Analytical Chem.* i. i. 3 Where the solubility is very small, the molarity and the molality of the solution will be approximately the same. 1951 M. A. Paul *Princ. Chem. Thermodynamics* vii. 388 The molarity of a given solution evidently varies in general with temperature. 1970 *Man. Symbols & Terminol. Physiochem. Quantities & Units* (I.U.P.A.C.) 12 Concentration is sometimes called 'molarity' but this name is both unnecessary and liable to cause confusion with molality and is therefore not recommended. 1973 Block & Holliday *Mod. Physical Chem.* ix. 203 In general, molarities cannot easily be related either to mole fraction or to molality, unless the changes in volume on dissolution are either negligible or can be allowed for.

molarization (mōᵘlărəizēⁱ·ʃən). *Zool.* [f. Molar *a.*¹ and *sb.* + -ization.] The assumption, during the course of evolution, of the characteristics of a molar tooth by a non-molar tooth (usu. a premolar).

1937 *Proc. Zool. Soc.* B. CVII. 104 The molarization of the premolars is a case of convergent evolution. 1953 J. S. Huxley *Evolution in Action* ii. 56 The so-called molarization of the pre-molars.—The gradual conversion of the front grinders, which have milk-teeth predecessors, so as to resemble the true molars, which develop later and have no 'milk' fore-runners. 1972 *Nature* 24 Nov. 236/1 A morphological complex characterizes the earliest hominids. This complex...includes...lowering of the molar cusps, molarization of the premolars, [etc.].

So **mo·larized** *ppl. a.*, showing the consequences of molarization.

1971 A. A. Dahlberg *Dental Morphol. & Evolution* viii. 99 The more distant from the canine, the more molarized or cuspidate the tooth. 1973 *Nature* 14 Sept. 106/1 In other cretaceous eutherians P⁴ is always three-rooted and partly molarized.

molasses. Add: **1. b.** *fig.*

1925 T. Dreiser *American Tragedy* (1926) I. ii. xvii. 127 'You're the cutest thing here,' whispered Clyde, hugging her fondly. 'Gee, but you can pour on the molasses, kid, when you want to,' she called out loud. 1972 *N.Y. Times Book Review* 26 Nov. 1/1 The mournful molasses of his [sc. Eisenhower's] prose.

3. *molasses barrel, cake, candy, cask, cookie, gingerbread, hogshead, jug, scone, taffy, tierce.*

1846 D. Corcoran *Pickings* 29 Isn't that cotton bale dancing a quadrille with the molasses barrel? 1836 W. G. Simms *Mellichampe* II. xxiv. 192 The negro broke his molasses-cake evenly between himself and the soldier. 1863 B. Taylor *H. Thurston* i. 19 The distribution of wedges of molasses-cake. 1869 *Harper's Mag.* Oct. 753/1 With that penny he bought a 'bolivar'—as the huge molasses-cake was called in those days. 1809 'D. Knickerbocker' *Hist. N.Y.* II. vii. iii. 195 Each...he patted on the head...and gave him a penny to buy molasses candy. 1850 N. Kingsley *Diary* (1914) 103 [We] are privately enjoying ourselves over a dish of molasses candy. 1945 *This Week Mag.* 15 Dec. 2/2 Cookies, molasses candy, popcorn balls and cider were brought out. 1834 H. J. Nott *Novelettes* I. 79 He was in a molasses cask. 1851 A. O. Hall *Manhattaner* 5 It was a modest commercial plain...with...molasses casks, and corn sacks. 1887 I. Alden *Little Fishers* xxi. 373 Dough-nuts, molasses cookies, and soft gingerbread. 1974 'D. Shannon' *Crime File* (1975) iv. 66 She asked for my recipe for molasses cookies. 1832 L. M. Child *Amer. Frugal Housewife* 70 A very good way to make molasses gingerbread. 1864 T. L. Nichols *40 Yrs. Amer. Life* I. 36 The spectators...ate molasses-gingerbread. 1849 Thoreau *Week Concord River* 198 An 'untameable fly' buzzed at my elbow with the same non-chalance as on a molasses hogshead. 1863 'G. Hamilton' *Gala-Days* 76 He...came back with a molasses-hogshead. 1839 *Southern Lit. Messenger* V. 65/2 Behind the bar were...a molasses jug, a bottle of vinegar, and...decanters. 1906 F. Lynde *Quickening* 111 She went ...to fill the molasses jug. 1927 M. De La Roche *Jalna* xxii. 275, I am bringing you a molasses scone to stay you, Mamma. 1928 S. V. Benét *John Brown's Body* 113 An awful molasses-taffy voice Behind them yelled 'Halt!' 1946 Partridge & Bettmann *As We Were* 13 Molasses taffy was cooked in a kettle on top of the kitchen stove and when ready for pulling was served out in gobs. 1910 *Chambers's Jrnl.* Feb. 88/2 A few rusty molasses-tanks. 1851 H. Melville *Moby Dick* I. xxii. 168 Have an eye to the molasses tierce, Mr. Stubb.

Hence **mola·ssed** *a.* = Molassied *a.*

1941 *Nature* 3 May 532/2 Nehring & Schramm also tested the effect...of molassed bran plus urea [on sheep]. 1960 *Farmer & Stockbreeder* 16 Feb. 128/3 Dried molassed beet pulp can replace an equal weight of oats in rations for store cattle and calves.

moldavite (mǫ·ldăvəit). [ad. G. *moldawit*, f. *Moldau*, G. name of the Vltava River in western Czechoslovakia: see -ite¹.] A tektite from the tektite field of Czechoslovakia; formerly, the substance (considered a local variety of obsidian) of which such tektites consist.

1896 A. H. Chester *Dict. Names Minerals* 179 *Moldawite*, an obs. syn. of obsidian. 1909 [see *billitonite]. 1915 *Bull. Geol. Soc. Amer.* XXVI. 281 The moldavites occur here and there, but never in any manner indicative of their origin. 1935 *Times* 28 Jan. 15/4 'Tektites' from Bohemia and Moravia have for more than 150 years been cut as gem stones under the names 'obsidian', 'water chrysolite' and 'moldavite'. 1964 *New Scientist* 16 Jan. 160/2 The moldavites from Bohemia and Moravia, about 14·8 million years old (Miocene). 1972 *Science* 11 Aug. 519/2 In 1970 and 1971, two noteworthy finds of double moldavites were made in the fields of southern Bohemia.

Moldo-Wallachian (mǫ:ldo,wǫlēⁱ·kiăn), *a.* Also **Moldavo-Wallachian.** [f. Mold(avian *a.* and *sb.* + -o + Walachian, Wallachian *a.*] Of or pertaining to both Moldavia and Wallachia, principalities of Rumania united in 1859. Also as *sb.*

1848 [see *anarchial *a.*]. 1869 *Bradshaw's Railway Manual* XXI. 358 A promise that the whole of the Moldo-Wallachian network should be under one supervision. 1929 W. G. East *Union Moldavia & Wallachia* iii. 54 The alleged desire of the Moldo-Wallachians to form one national state. *Ibid.* 58 Napoleon was secretly planning a war with Austria during his advocacy of Moldo-Wallachian union. 1971 P. Clinca tr. *Giurescu's Making of Romanian Unitary State* 117 The twin Moldavo-Wallachian town of Focsani lying on the former border.

mole, *sb.*² Add: **6.** (Earlier examples.)

1881 H. W. Nesfield *Chequered Career* vii. 75 We...met men in rough flannels and dirty soil-stained moles. 1888 G. O. Preshaw *Banking under Difficulties* xxvi. 163 These moles are 12s. a pair.

6*. A shade of grey. Also as *adj.*

1908 *Westm. Gaz.* 29 Aug. 13/2 Mole has always been recognised...as a shade universally harmonious. *Ibid.* 26 Sept. 13/2 How charming with a mole suit is a mole hat, massed with roses! 1914 [see *beaver¹ 2 c]. 1923 [see *brick *sb.*¹ 4 c]. 1971 *Vogue* 15 Sept. 49 Rich autumn colours like wine, mole and teak.

8. **mole ditch** = **mole drain*; so *mole-ditching* vbl. sb.; **mole drain,** a drain made by a mole-plough; **mole-drainer** = Mole-plough.

1868 *Rep. Comm. Agric. 1867* (U.S. Dept. Agric.) 232 *Under draining* wet, heavy places, with 'mole ditches', or blind drains. 1868 *Rep. Iowa Agric. Soc. 1867* 154 Underdraining by mole-ditching has been tried. 1939 Webster Add., Mole drain. 1950 *N.Z. Jrnl. Agric.* June 534/1 A number of standard mole drains was drawn with the 3 in. plug at depths of 18 to 20 in. 1957 E. Blunden *Poems of Many Years* 280 Waters drawn together From gully and moledrain by contour, chance and weather. 1859 *Trans. Ill. Agric. Soc.* III. 361 Dragging the mole drainer all over our lands.

b. **mole snake,** a non-venomous colubrid snake, *Pseudaspis cana*, native to Southern and E. Africa, and feeding on rats and mice.

1893 J. Noble *Illustr. Official Handbk. Cape & S. Afr.* 84 The Colubrinæ include the large and abundant 'black' or 'mole' snake. 1911 *East London* (Cape Province) *Dispatch* 1 Sept. 7 When alarmed the mole snake is very pugnacious. 1931 *Discovery* Mar. 74/2 The South African mole snake is a non-poisonous species. It constricts its prey and swallows it whole. 1951 R. Campbell *Light on Dark Horse* ii. 36 A mole-snake has not only an apparent immunity [to bites from other snakes] but also the power to strangle other snakes. 1966 E. Palmer *Plains of Camdeboo* xiii. 220 Sometimes we see a long, graceful, chinless snake in the veld, with a dark muscular body shining as if carefully polished. This is the mole snake and is harmless to man, killing its prey by constriction. 1970 [see *boomslang].

mole (mōᵘl), *sb.*⁷ *Physical Chem.* Also **mol** (formerly as an alternative spelling, now usu. as an abbrev.). [a. G. *mol* (W. Ostwald *Grundlinien d. anorg. Chem.* (1900) viii. 163), f. *mol-ekül* Molecule.] That amount of any particular substance having a mass in grammes numerically the same as its molecular or atomic weight; now defined equivalently in the International System of Units as the quantity of specified elementary entities (molecules, ions, electrons, or the like) that in number equals the number of atoms in 0·012 kilogramme of carbon 12.

1902 A. Findlay tr. *Ostwald's Princ. Inorg. Chem.* viii. 156 When one gram-molecule or one mole (the molar or molecular weight of a substance expressed in grams) of any substance is dissolved in a litre or 1000 gm. of water, the solution produced freezes at −1·850°. 1923 Lewis & Randall *Thermodynamics* iii. 22 The mol is not defined unless the chemical formula is established by universal usage or is definitely stated. 1954 *Physiol. Rev.* XXXIV. 342 Estimations of the molar concentration (mols per kilogram of water or osmols if the salts are known...). 1959 N. Feather *Introd. Physics Mass, Length & Time* xv. 282 A mass equal to 'the molecular weight in grammes', which we define as 1 mole of a substance, contains the same number of molecules (Avogadro's constant), whatever the substance. 1963 W. J. Moore *Physical Chem.* (ed. 4) ix. 326 One mole of electrons would be one faraday of electrical charge. 1970 [see *molality]. 1971 G. D. Christian *Analytical Chem.* i. 8 Each mole of silver ion will react with one mole of chloride ion. 1972 *Physics Bull.* Jan. 40/1 The 14th General conference of Weights and Measures (CGPM) met in Paris on 4–7 October 1971... Amongst the main decisions taken by the conference were the final adoption of the 'mole' (mol) as an SI base unit. 1973 Block & Holliday *Mod. Physical Chem.* xi. 256 In the chemical reaction occurring in the Daniell cell...two moles of electrons are required to convert one mole of copper (II) to one mole of zinc (II).

2. *Comb.* **mole fraction,** the ratio of the number of moles of a component in a solution to the total number of moles of all components present.

1923 Lewis & Randall *Thermodynamics* xxii. 261 The mol fraction of bromine in a solution containing 160 grams bromine and 154 grams carbon tetrachloride is ½ if we are considering the formula Br_2. 1973 A. W. Adamson *Textbk. Physical Chem.* ix. 345 The partial pressure of a component becomes proportional to its mole fraction in the limit of zero concentration.

‖ **mole** (mōᵘ·li), *sb.*⁸ [Mexican Sp., ad. Nahuatl *mulli, molli* sauce, stew.] A highly spiced sauce made chiefly from chilli and chocolate and served with various meats.

1932 H. W. Bentley *Dict. Sp. Terms in Eng.* 169 *Mole* ..., a sauce used in Mexican cookery in connection with the serving of meats. 1948 *Sat. Even. Post* 2 Oct. 52/3 Señora Gonzalez does her stuff on such fabulous and sustaining dishes as chicken mole—boiled chicken bathed in a sauce of exotic Mexican spices, cinnamon, chili, mashed-up peanuts and even a dash of chocolate [etc.]. 1957 'B. Buckingham' *Boiled Alive* xviii. 121 Turkey swimming in *mole*, a hot sauce made of chilli and chocolate, stuffed sweet peppers and mounds of pink-brown beans. 1966 *Punch* 9 Mar. 364/2 We were...sated with rich turkey *mole.*

mole, *v.*² Add: **2.** *to mole out*: also, to elicit, bring to light.

1924 W. M. Raine *Troubled Waters* vii. 70 Tait would mole out quite enough evidence against him without any additional data supplied by indiscretion. 1932 L. C. Douglas *Forgive us our Trespasses* (1937) xiii. 251 Maybe I'll ask you to mole out further data.

b. *intr.* To behave in the manner of a mole.

1856 W. G. Simms *Eutaw* xii. 129 How he snaked, and moled, and cooned,...we need not undertake to narrate.

molecular, *a.* Add: **1. a.** *spec.* applied to numerous physical quantities that involve the molecular weight of the substance concerned in their calculation (for most of which *molar* is a more appropriate designation; see *molar *a.*³ a).

1867 C. L. Bloxam *Chem.* 515 One equivalent of each of these hydrocarbons in the state of vapour occupies four volumes. [*Indexed, p.* 660, *as*] Molecular volumes of olefines. 1880 *Jrnl. Chem. Soc.* XXXVIII. 294 For every substance, therefore, (A − 1)/*d* is a constant... If this constant be multiplied by the molecular weight P of any body, then P.(A − 1)/*d*, referred to chemically comparable quantities, is the molecular refractive index, called in the rest of this paper the molecular refraction of the body. 1886 *Ibid.* L. 294 The author [sc. W. Ostwald] in continuing his researches on 'Electrical Conductivity'...has examined about 130 different inorganic and organic acids in different states of dilution (gram-molecules in *n*-litres). He...expresses his results in terms of molecular conductivity. 1921 Hood & Carpenter *Text-bk. Pract. Chem.* vi. x. 489 The molecular conductivity is that of a solution containing 1 gram-molecule of the substance, as above [sc. when placed between electrodes of indefinite size and 1 cm. apart]. 1938 H. A. Perkins *Course of Physics* xlvii. 624 The number of gram molecules per liter, or moles per liter, is called molecular concentration. 1959 N. Feather *Introd. Physics Mass, Length & Time* xv. 283 Equation (126) is now the equation of state for a mass *m* of an ideal gas of molecular weight, or, strictly, of molar mass, *M*. 1966 Gucker & Seifert *Physical Chem.* (1967) xiii. 346 One molecular weight of the salt contains *N* ions of *each* element in the molecular volume *V*, equal to the molecular weight divided by the density.

b. Applied to the name of a science to denote a branch of it that deals with phenomena at the molecular level.

1950, etc. [see *molecular biology]. 1960 *N.Y. Times* 3 Jan. 1F/2 The chief executive officer...disclosed that Westinghouse was actively engaged in 'molecular electronics'. 1961 R. D. Baker *Essent. Path.* i. 6 With the electron microscope we are entering an era of molecular pathology. 1963 *Adv. Computers* IV. 141 'Integrated' circuits fabricated using 'molecular electronics' techniques. 1963 *Listener* 17 Jan. 121/2 It seemed to me that soon what Russian biology will mean will not be Lysenko, but some first-class molecular genetics. 1966 *McGraw-Hill Yearbk. Sci. & Technol.* 43/1 This view of the gene has changed considerably in molecular genetics, since the gene is now recognized as the information necessary to assemble a protein molecule. 1970 *Sci. News Yearbk.* 225 Molecular astronomy is a science only a few years old. Its purpose is to determine what chemical molecules can be found in interstellar space and, if possible, how they got there and what their being there means for theories of galactic and stellar evolution and of cosmology. 1973 *Sci. Amer.* Apr. 52/3 That was more or less the picture of the interstellar medium in 1968, the year marking the birth of molecular astronomy as we now know it. 1974 *Nature* 23 Aug. 685/1 Now it is the turn of immunochemistry, or as some would have it, molecular immunology.

† **2.** *Biol.* Of tissue: consisting, or assumed to consist, of molecules (sense *3); finely granular. *Obs.* (but the use survives in *molecular layer* (see *5)).

1826, 1851 [see *molecule 3]. 1856 *Quain's Elem. Anat.* (ed. 6) III. 26 The intermediate portion, *f*, has a longitudinally-striped appearance, and is formed by the fine fibres of Müller..., which intervene between the two parts of the granular layer, together with a homogeneous molecular uniting material.

3. *Philos.* Designating a proposition, sentence, etc., consisting of simpler propositions, sentences, etc. connected by a conjunction. Also *ellipt.* as *sb.*

1892 *Mind* I. 237 The molecular proposition—which cannot be expressed as a synthesis of more elementary propositions—involves a single (absolute or relative) predication and a number of interconnected individual subjects. *Ibid.* 239 The synthesis of moleculars having the same subject is represented by a synthesis of the predications. 1914 B. Russell *Our Knowl. External World* ii. 54 'Molecular' propositions are such as contain conjunctions—*if, or, and, unless*, etc.—and such words are the marks of a molecular proposition. 1933 [see *atomic *a.* 2a]. 1937 A. Smeaton tr. *Carnap's Logical Syntax of Lang.* iii. § 28. 88 $\mathfrak{S}_1$ is called a *molecular sentence* when $\mathfrak{S}_1$ is either an atomic sentence itself, or is formed from one or more such by means of symbols of negation and junction (and brackets). 1965 B. Mates *Elem. Logic* iii. 44 A formula that is not an atomic formula is called *general* if it begins with a universal or existential quantifier; otherwise it is called *molecular.*

4. *Psychol.* Concerned with or pertaining to an elementary unit of behaviour such as a physiological response. Cf. *molar *a.*² b.

1925, 1932 [see *molar *a.*² b]. 1970 F. A. Logan in W. S. Sahakian *Psychol. of Learning* xvi. 307 The molar response 'bar depression' includes a number of responses distinguishable at a more molecular level, such as depression with the right paw, left paw, etc. 1973 *Jrnl. Genetic Psychol.* CXXIII. 99 The continuous-recording technique did...yield a more molecular analysis in that incidence of behaviors could be subdivided into frequency and duration.

5. Special collocations: **molecular fossil,** a molecular species found in ancient rock that is regarded as evidence of the early development of life; **molecular layer** *Anat.*, (*a*) either of the two plexiform layers of the retina; (*b*) the outermost layer of the cortex of the cerebellum and cerebrum, containing a mass of nerve fibres with many synapses but relatively few cells; **molecular orbital,** an orbital in a molecule; **molecular sieve,** a substance containing in its crystal structure pores and channels of molecular dimensions that permit the entry of certain small molecules but are impervious to others; *esp.* a zeolite used as a selective adsorbent.

1965 M. Calvin in *Proc. R. Soc.* A. CCLXXXVIII. 443 We are going to spend much of our time tracing organic evolution back in so far as we can trace it back in terms of 'molecular fossils' from the earliest well recognized fossils of morphological form. 1969 *Sci. Jrnl.* Apr. 36 The exciting studies of Calvin, who has attempted to prove the existence of what he calls 'molecular fossils', have not, however, firmly established the biogenic nature of these substances. 1971 *Nature* 30 July 325/2 The search for geological evidence about the origin and evolution of life on Earth has led to detailed searches for morphological and molecular fossils in Early Precambrian sedimentary rocks. [1856 Molecular layer: cf. sense 2 above]. 1867 *Quain's Elem. Anat.* (ed. 7) III. 726 The retina...exhibits a series of dissimilar strata, together with structures not confined to one stratum. (1st) Externally is the columnar layer; (2nd), in the middle is the granular layer, comprising the external nuclear, the internuclear, the internal nuclear, and the molecular layers; and (3rd) internally is the nervous layer. 1874 A. E. J. Barker tr. *Frey's Histol. & Histochem. of Man* 593 Externally—that is, directed towards the surface of the cerebellum—these large ganglion cells send off several (generally two) characteristic protoplasm processes through the so-called 'molecular layer of Hess'. 1955 P. D. Trevor-Roper *Ophthalm.* i. 19 The outer molecular layer is a reticular layer in which the end-knobs of the axons of the rod and cone cells arborise with the dendrites of the bipolar cells. *Ibid.* 20 The inner molecular layer consists of the arborisations of the axes of the bipolar cells with the dendrites of the ganglion cells. 1968 Passmore & Robson *Compan. Med. Stud.* I. xxiv. 46/1 In the older parts of the [cerebral] cortex, such as the rhinencephalon, only three layers of cells may be distinguished: the superficial molecular layer of fibres, the intermediate granular layer and a deep layer of pyramidal cells. 1932 R. S. Mulliken in *Physical Rev.* XLI. 50 The method followed here will be to describe unshared electrons always in terms of atomic orbitals but to use molecular orbitals for shared electrons. 1965 Phillips & Williams *Inorg. Chem.* I. iii. 67 The problem will be introduced by a summary of the two important wave-mechanical approximation methods, the L.C.A.O. (linear combination of atomic orbitals) molecular-orbital method and the valence-bond method. 1926 *Colloid Symp. Monogr.* IV. 11 Collander's work on the sieve structure of semipermeable membranes of copper ferrocyanide has shown that the interstices between the micelles or aggregates are about 4 Ångstrom units in diameter. I would suggest a better molecular sieve, namely such crystals as dehydrated zeolites, etc. 1949 *Discussions Faraday Soc.* VII. 135 By cation interchange and by burning out interstitial ammonium ions a diversity of modified molecular-sieve sorbents can be produced. *Ibid.*, Some natural crystalline zeolites fall into three classes of molecular-sieve sorbent each capable of separating mixtures by selective occlusion. 1965 Phillips & Williams *Inorg. Chem.* I. xiv. 544 The molecular-sieve zeolites, sodium and calcium aluminosilicates, are used to fractionate noble gases and low molecular weight molecules generally. 1966 *McGraw-Hill Encycl. Sci. & Technol.* VIII. 547/1 Molecular sieves are capable of drying gases and liquids to extremely low residual water concentrations.

molecular biology. [f. Molecular *a.* + Biology.] Biology at the molecular level, esp. that branch of biology which is concerned with the formation, organization, and activity of macromolecules essential to life (i.e. nucleic acids, proteins, etc.).

1950 *Harvey Lectures* XLVI. 3 The name 'molecular biology' seems to be passing now into fairly common use. .. It implies...an approach from the viewpoint of the so-called basic sciences with the leading idea of searching below the large-scale manifestations of classical biology for the corresponding molecular plan. It is concerned particularly with the forms of biological molecules, and with the evolution, exploitation and ramification of those forms in the ascent to higher and higher levels of organization. 1963 *Listener* 17 Jan. 121/2 The newest and most refined field of genetics, which deals with definite chemical substances and is known as molecular biology. 1964 G. H. Haggis et al. *Introd. Molecular Biol.* p. x, We may say that molecular biology is primarily concerned with the structure of proteins, nucleic acids and other large biological molecules, and with the detailed structure of myofilaments, chromosomes, ribosomes, membranes and other cell components. But the study of structure cannot be divorced from the study of function. 1970 *Nature* 4 July 13/1 The synthesis of genes *de novo* has been an aim of molecular biology since its very inception. 1974 *Daily Tel.* 3 Sept. 6 Molecular biology and genetic engineering have reached the stage where results of research could plunge the world into far deeper problems than those raised by the atom bomb.

molecularity. Add: **2.** *Chem.* The number of reacting molecules involved in a (real or postulated) single step of a chemical reaction.

1939 FOWLER & GUGGENHEIM *Statist. Thermodynamics* xii. 500 In general there is no simple connection between the molecularity of the mechanism..and the order of the reaction. 1950 K. J. LAIDLER *Chem. Kinetics* iii. 56 Whereas the order of a reaction is deduced directly from the experimental results, the molecularity can be determined only on the basis of additional arguments about which there is sometimes some uncertainty. 1970 PRETTRE & CLAUDEL *Elem. Chem. Kinetics* v. 64 If we accept that the reaction $I_2 + H_2 \rightarrow 2HI$ is a simple one, this means that it involves a molecule of iodine and a molecule of hydrogen and that consequently its molecularity is two.

molecularly, *adv.* (Further examples.) Also, on a molecular scale.

1927 L. B. LOEB *Kinetic Theory Gases* vii. 273 The surfaces are not completely molecularly rough. 1969 *Physics Bull.* June 227/1 Muscovite mica..gives molecularly smooth surfaces on cleavage.

molecule. Add: **1. c.** (Earlier example.)

1867 C. L. BLOXAM *Chem.* 515 One equivalent of each of these hydrocarbons in the state of vapour occupies four volumes. [*Note*] Or one molecule occupies two volumes (H = 1 vol.).

† **3.** *Biol.* A minute but functional particle of animal tissue that is invisible or barely visible under the light microscope. *Obs.*

1826 KIRBY & SPENCE *Introd. Entomol.* IV. xxxvii. 3 Comparative anatomists have considered the nervous system of animals as formed upon three primary types, which may be called the molecular, the ganglionic, and cerebro-spinal. The first is where invisible nervous molecules are dispersed in a gelatinous body, the existence of which has only been ascertained by the nervous irritability of such bodies, [etc.]. 1841 T. R. JONES *Gen. Outl. Animal Kingdom* 6 In animals belonging to this division, no nervous filaments or masses have been discovered... The contractile molecules of their bodies are not yet aggregated into muscular fibre. 1851 W. WILKINSON *Outl. Physiol.* 9 Molecular and granular matter consists of particles that vary in size from immeasureable minuteness to 1·10,000th of an inch in diameter; and these particles are called molecules or granules, according to the appearance they present when examined with a magnifying power of 300 diameters. *Ibid.*, Molecules are merely indistinct granules; but under a higher magnifying power molecules become granules, and new molecules appear.

moler[2] (mōᵘ·ləɹ). Also **Moler.** [a. Da. *moler*, f. northern Da. dial. *mo* loose chalky soil (= Norw., Sw. *mo* sandy heath) + *ler* loam, clay (= Norw., Sw. *ler* clay).] (See quot. 1923.) (A proprietary name.)

1910 *Chem. Abstr.* IV. 2363 'Moler' is an argillaceous diatomaceous earth found in a limited region in the fiords of Denmark. 1923 A. B. SEARLE *Sands & Crushed Rocks* I. iii. 127 Moler is a sandy material somewhat resembling kieselguhr or diatomaceous earth.., but it is less refractory and contains a considerable proportion of clay and volcanic ash, for which reason it is self-binding and can be made into bricks without any other bond. 1932 A. G. GEESON *Gen. Building Repairs* (ed. 2) I. i. 48 Plaster, terra cotta, moler and breeze slabs have the advantage that they can be sawn to size. 1936 *Times* 9 Nov. 20/1 (Advt.), An extensive deposit of Diatomite of a high quality, known as 'Moler', which is suitable for many of the above purposes, exists on the Island of Für in Denmark. 1948 *Archit. Rev.* CIV. 57 The remaining blocks will be finished in a 4½ in. external concrete-brick cladding with an internal 4 in. skin of molar [*sic*] blocks with a 2 in. cavity. 1963 *Trade Marks Jrnl.* 6 June 766/1 Moler... Insulators and insulating materials. Aktieselskabet Skarrehage Molervaerk.., Nykbøing Mors, Denmark; manufacturers; and Refractulation Limited,.. London,.. merchants.

Moler, var. *MALER *sb.* and *a.*

moleskin. Add: **4.** *moleskin trousers* (earlier example); **moleskin squatter** *Austral.* and *N.Z.* (see quot. 1941).

1941 BAKER *N.Z. Slang* v. 40 Among other terms we have derived from farm life and sheep stations are *moleskin squatter*, 'a working man who has come to own a small sheep run.' *a* 1948 L. G. D. ACLAND *Early Canterbury Runs* (1951) 387 *Moleskin squatter*... A correspondent writes: 'The last time I heard it was when the Government cut up Cheviot in 1893... Since then *m.ss.* have become too numerous to attract attention.' 1873 TROLLOPE *Harry Heathcote* (1874) i. 3 The young man.. had on a flannel shirt, a pair of moleskin trousers, and an old straw hat.

molestive (mole·stiv), *a.* [f. MOLEST *v.* + -IVE.] Tending to annoy; troublesome, interfering.

1905 *North Amer. Rev.* Nov. 657, I suppose that the stranger always finds the patriotism of a country molestive. 1929 W. DEEPING *Roper's Row* vi. 58 If the young Prossers had persecuted the son, Mrs. Prosser had been equally molestive to the mother.

moley (mōᵘ·li), *sb. slang.* [Origin obscure.] (See quot. 1950.)

1950 J. D. CARR *Below Suspicion* xi. 132 'Use the moleys when you catch 'im.' The moley was an ordinary potato, its surface jagged with the edges of safety-razor blades. They ground it into your face, twisted it, and... 1959 *Spectator* 6 Mar. 314/1, I suppose if I go on criticising him I shall end up by having the boys with the moleys call on me one dark night.

‖ **molinete** (mǫline·te). Also **molinet.** [a. Sp. *molinete*, lit. (toy) windmill, little mill.] In *Bullfighting*, a decorative pass made by a matador (see quot. 1959[2]).

1932 [see *FAROL]. 1959 V. J. KEHOE *Aficionado!* 19 Manolo then did a *derechazo*,..ending the series with a *molinete* on his knees. *Ibid.* 208 The *molinete*. Performed with either hand, this pass is started by the matador citing the toro as for a cambiado, then spinning out of the pass in the opposite direction of the toro's charge. 1971 J. LEIBOLD *This is Bullfight* xvii. 202 The molinete can be executed with either the right or left hand... Timing, grace and elegance of execution are of prime importance in accomplishing the molinete.

moll, *sb.* Add: **1. c. moll heron** (later example).

1939 F. THOMPSON *Lark Rise* v. 97 All legs and wings, like a moll-heron.

2. (Further examples.) Also, a girl, woman; a girl-friend or sweetheart, esp. of a criminal; a female pickpocket or thief. See also **gun moll. slang.*

1753 J. POULTER *Discoveries* (ed. 2) 34 To nap the Slangs from the Cull or Moll; that is,..to take the Things from the Man or Woman. 1840 H. D. MILES *Dick Turpin* xxi. 250 You might ha' knowed his moll, a spicy, swellish sort of a bit o' muslin. 1872 G. P. BURNHAM *Mem. U.S. Secret Service* 190 Doctor Blake and his 'moll' visited the town of Toms River. 1891 J. BARON *Blegburn Dickshonary* 44 'Aw'm gooin' to meet mi Moll to-neet' is a varra common sayin' wi' factory lads: some o' th' better soort say 'woman' i' th' place o' Moll, but nooan so mony. 1923 [see *DAME 2 c]. 1931 *Amer. Speech* VII. 111 *Moll.* 1. A gangster's sweetheart or mistress. 2. Any girl whether associated with the underworld or not. 1934 DYLAN THOMAS *18 Poems* 22 In this our age the gunman and his moll, Two one-dimensioned ghosts, love on a reel. 1946 K. TENNANT *Lost Haven* (1968) i. 20 He went off with that bloody moll whose name I wouldn't speak. 1955 *Publ. Amer. Dial. Soc.* xxiv. 99 A woman pickpocket is..a *moll.* 1962 N. MARSH *Hand in Glove* iii. 90, I can see you're in a fever lest slick Ben and his moll should get back..before you make your getaway. 1975 C. FREMLIN *Long Shadow* xxvi. 190 The Psychopath's Moll. I'm doing it again, thought Imogen..saving him from the consequences of his follies.

b. *attrib.* and *Comb.*, as **moll-shop** *slang*, a brothel; **moll-tooler** *slang*, a female pickpocket.

1923 J. MANCHON *Le Slang* 196 Moll-shop. 1957 M. K. JOSEPH *I'll soldier no More* (1958) 181 Pretty faces.. peered shyly into the street. 'Looks like a moll-shop,' said Connolly. 1859 HOTTEN *Dict. Slang* 63 *Moll tooler*, a female pickpocket.

moll, *v.* Add: (Further examples.) So **molled** *ppl. a.*, associating with or accompanied by a woman.

1882 *Sydney Slang Dict.* 6/1, I see yer the other night when yer was Molled up and too proud to speak. 1935 *Flynn's* 19 Jan. 87/2 With each man molled, and his moll posing as his wife, they would not attract the suspicion which would be directed against a mob of men living together.

moll-buzzer. Add: (Further examples.) See also quots.

1900 'FLYNT' & WALTON *Powers that Prey* 225 Her gift for mathematics made it clear that 'moll-buzzing' was much more remunerative than sleeping in cellars and peddling Park Row literature. 1903 Moll-buzzing [see *GRAFT *sb.*[4]]. 1910 *N.Y. Even. Post* 25 Aug., To have the country cousin clutch his arm and enquire whether that rough-looking customer coming out of a Chatham Square saloon is a dip, a yegg, a stall, a moll-buzzer, a Fagin, or a gun. 1912 A. BERKMAN *Prison Mem. Anarchist* II. xxii. 278 The 'gun'..gathers messages for their 'moll buzzers' [*f. n.* women thieves]. 1912 *Collier's* 23 Nov. 12/2 When he came out he began as a 'moll buzzer', which, you understand, means a pickpocket who specializes on the fair sex; he buzzes the 'molls'. 1936 *Evening News* 9 Dec. 8/5 Buzzers are male pickpockets who specialise in opening women's handbags; moll buzzers are the females of the species. 1955 *Publ. Amer. Dial. Soc.* xxiv. 169 The *moll buzzer* (male or female) is not strictly speaking a pickpocket. ..But the thief takes money from the purses or handbags ..carried by women. *Ibid.*, There's a reason for moll buzzing.

mollifyingly (mǫ·lifəi,iŋli), *adv.* [f. MOLLIFYING *ppl. a.* + -LY[2].] In a mollifying manner.

1920 E. O'NEILL *Beyond the Horizon* I. ii. 44 Mayo (*mollifyingly*). Of course I will, boy, and be glad to. 1928 'S. S. VAN DINE' *Greene Murder Case* ii. 25 'You understand, of course,' he added mollifyingly, 'that I shall not interfere with your activities in any way.' *a* 1953 E. O'NEILL *Long Day's Journey* (1956) 30 Tyrone (*impressed —mollifyingly*). I know you may have thought it was for the best.

molluscicide (mǫlʊ·skisəid). Also **molluscacide.** [*molluscicide* f. MOLLUSC + -I- + -CIDE; *molluscacide* f. MOLLUSCA *sb. pl.* + -CIDE.] Any substance used to kill molluscs.

1947 DORLAND & MILLER *Med. Dict.* (ed. 21) Molluscacide. 1948 *Jrnl. Parasitol.* XXXIV. Suppl. 33 (*heading*) Results of screening tests on chemicals as molluscicides. 1949 *Ibid.* XXV. 475 (*heading*) Influence of some potential molluscacides on the oxygen consumption of *Australorbis glabratus.* 1965 *New Scientist* 22 Apr. 229/2 No efforts had previously been made to wipe out the snails with molluscicides. 1972 *Country Life* 2 Mar. 527/3 A new molluscicide spray is now being used..to eradicate the pest. 1974 *Nature* 22 Mar. 363/1 The widely used molluscicide sodium pentachlorophenate is a repellant to snails.

Hence **mollusci·cidal** *a.*

1950 *Jrnl. Parasitol.* XXXVI. 152 The molluscacidal activity of a series of 42 organic compounds has been studied. 1950 *Biol. Abstr.* XXIV. 3249/2 In this series of tests alpha-chloro-esters and amides of aliphatic monobasic acids showed a more marked molluscicidal activity than the unchlorinated derivatives. 1968 *New Scientist* 4 Jan. 24/3 Chlorinated hydrocarbons and organo-phosphates were found to have no molluscicidal activity.

molluscous, *a.* **3.** (Earlier example.)

1837 *London Med. Gaz.* XIX. 860/2 The structure of the molluscous tubercles.

molluscum. 1. Substitute for def.: Any of various disorders characterized by soft rounded tumours or nodules on the skin, *spec.* (and orig.) molluscum contagiosum. Freq. in mod.L. collocations, as **molluscum contagiosum,** a viral disorder characterized by small, smooth, pinkish nodules with a central depression, that are painless, yield a milky fluid when squeezed, and usu. occur in groups; **molluscum sebaceum,** (*a*) molluscum contagiosum (? *Obs.*); (*b*) = *kerato-acanthoma* s.v. *KERATO-. (Add earlier and additional examples.)

This use is thought to derive from the adjectival use in quot. 1793.

[1793 C. F. LUDWIG in W. G. Tilesius *Historia Pathologica Singularis Cutis Turpitudinis* 6/1 Verum enim vero Rheinhardi visu foedum corpus tectum est verrucis mollibus sive molluscis et madidis sive myrmeciis.] 1813 T. BATEMAN *Pract. Synopsis Cutaneous Dis.* 268 Molluscum. This form of tubercular disease is noticed rather as a singularity, which occasionally occurs, and of which a few instances are recorded, than as an object of medical treatment. 1817 —— *Delineations of Cutaneous Dis.* Explan. of Plate LXI, Molluscum contagiosum. This singular eruption had not been noticed by Dr. Willan, and was unknown to myself till after the publication of two editions of my Synopsis. 1837 *London Med. Gaz.* XIX. 860/1 (*heading*) A few remarks on molluscum; with two cases of molluscum non-contagiosum. *Ibid.* 860/2 The non-contagious molluscum is characterized by tumors of various sizes, some of them as large as a hen's egg. 1868 *Jrnl. Cutaneous Med.* I. 53 (*heading*) On molluscum sebaceum. 1870 *Med.-Chirurgical Trans.* XXXV. 225 The definition of Plenck could hardly be applied to any disease but Molluscum Contagiosum; the case recorded by Tilesius appears certainly to have been one of Molluscum Fibrosum. *Ibid.* 230 The patient was covered with the tumours of Molluscum, or, as he [*sc.* Virchow] prefers to term it, Fibroma Molluscum. 1932 R. L. SUTTON *Introd. Dermatol.* xv. 280 Molluscum contagiosum. Synonyms.—Molluscum sebaceum; Molluscum epitheliale; [etc.]. 1936 MACCORMAC & SCARFF in *Brit. Jrnl. Dermatol.* XLVIII. 625 The microscopic architecture bears a resemblance to molluscum contagiosum, and it is possible that some of the recorded cases of giant molluscum contagiosum are examples of this tumour. We suggest the name 'molluscum sebaceum' as a convenient label. 1950 Molluscum sebaceum [see *kerato-acanthoma* s.v. *KERATO-]. 1960 J. MARSHALL *Dis. Skin* xxvi. 684 Molluscum sebaceum presents as a solitary lesion usually on the centre face, sometimes..elsewhere. 1961 D. M. PILLSBURY et al. *Man. Cutaneous Med.* ix. 205 Molluscum is of no more than cosmetic importance, except when the conjunctivae are involved. 1967 H. MONTGOMERY *Dermatopathology* II. xxxiv. 1036/2 Cutaneous tags or papillomas have been given many different names in the older literature, including acrochordon, fibroma molluscum, molluscum fibrosum, and soft warts. 1970 PASSMORE & ROBSON *Compan. Med. Stud.* II. xviii. 116/2 Cowpox may develop after contact with infected animals. .. In contrast, molluscum contagiosum is contracted only from human cases.

b. A soft nodule characteristic of molluscum.

1841 *Edin. Med. & Surg. Jrnl.* LVI. 216 In the 2d and 3d Figures are represented the free and attached surfaces of a tubercle, consisting of three mollusca, each with its proper aperture. 1890 J. L. MILTON *On Path. & Treatm. Dis. Skin* (ed. 3) ix. 449 A boy, twenty-two months old, was brought to St. John's Hospital with several molluscum spots on the left side of the face and neck... His mother, who had suckled him, had now a molluscum on her breast. 1960 J. MARSHALL *Dis. Skin* xv. 336 When they appear the papules [of molluscum contagiosum] are 1 to 2 mm. in diameter and the fully developed lesions are usually 5 to 10 mm.; but much larger 'giant mollusca' are sometimes seen.

c. *attrib.*, as *molluscum tumour*; **molluscum body,** † **corpuscle,** one of the characteristic ovoid bodies that are found in the core of the nodules of molluscum contagiosum and are thought to be degenerate epidermal cells; also, a cytoplasmic inclusion in a cell that is in the early stages of degeneration.

1892 *Edin. Med. Jrnl.* XXXVIII. 283 The so-called molluscum bodies..are not independent animal organisms. 1937 E. H. MOLESWORTH *Introd. Dermatol.* vii. 228 The so-called molluscum bodies are degenerate and deformed cells contained in the core of the lesion. They are not the infective agents. 1966 WRIGHT & SYMMERS *Systemic Path.* II. xxxix. 1571/2 The epidermal cells that are infected by the virus [of molluscum contagiosum] undergo premature keratinization and form the so-called molluscum bodies, which contain the virus inclusions. 1974 PASSMORE & ROBSON *Compan. Med. Stud.* III. II. xxxi. 58/1 Cytoplasmic inclusions (molluscum bodies) form in the cells of the stratum spinosum. 1886 C. H. FAGGE

Princ. & Pract. Med. II. 688 The white material seems to be made up almost entirely of characteristic oval transparent bodies..without a nucleus... These have been described as molluscum corpuscles. *Ibid.*, A molluscum tumour resembles an ordinary sebaceous cyst or steatoma, but the contents are white instead of yellow.

Mollweide (mǫ·lvəidə). The name of Karl B. *Mollweide* (d. 1825), German mathematician and astronomer, used in the possessive, *attrib.*, and *absol.* to designate a homalographic map projection in which the surface of the globe is represented by an ellipse, with lines of latitude represented by the major axis and straight lines parallel to it (spaced more closely towards the poles) and meridians represented by the minor axis and equally spaced elliptical curves.

1901 C. F. CLOSE *Map Projections* II. 22 (*heading*) Homalographic equal-area (Mollweide's). Sometimes called Babinet's. 1912 A. R. HINKS *Map Projections* vii. 73 For the whole world on a single sheet we have Mollweide, which is useful for distribution diagrams, but can scarcely be called a map. 1937 F. DEBENHAM *Exercises in Cartography* x. 118 Draw an interrupted Mollweide projection to show continental areas with the least distortion. *Ibid.*, Make an approximate graticule on the Oblique Mollweide suitable for showing the British Empire Seas. 1947 H. D. THOMPSON *Fundamentals Earth Sci.* v. 52 Mollweide's Homalographic equal-area projection (1805) shows the entire surface of the earth inside an ellipse, the major, or equatorial axis of which is twice as long as the minor axis. 1960 F. LAND *Lang. Math.* vi. 70 Mollweide's map..is an equal-area map, preserving area relations at the expense of shapes. 1971 R. W. PURTON *Let's look at Maps & Mapmaking* 14 Mollweide's projection can be drawn with any line of longitude as its centre and can be used to draw attention to one particular part of the world.

molly, *sb.*[1] Add: **4. Molly dancer** *dial.* (see quots.).

1959 I. & P. OPIE *Lore & Lang. Schoolch.* xii. 261 In north Manchester, gangs of boys calling themselves 'Molly Dancers' wear old clothes, mostly women's, and make-up their faces, and go around singing..'Spare a copper for the Molly dancers'. 1971 *Jrnl. Lancs. Dial. Soc.* xx. 8 *Molly Dancers*, performers of a traditional folk play, of which a number survived in North Staffordshire into the 1930's. I assume the phrase to be a popular etymology for *Morris Dancers.*

molly (mǫ·li), *v.* [f. MOLLY[1] or MOLLY-CODDLE *v.*] **1.** *intr.* (See quot.) *dial.*

1884 R. LAWSON *Upton-on-Severn Words* 23 *Molly*, to do woman's work indoors, being a man. ''E were a good un to molly for 'isself, were old Joe.'

2. *trans.* = MOLLY-CODDLE *v.*

1907 M. C. HARRIS *Tents of Wickedness* II. ii. 138 Paul hasn't been mollied, and I hope he's a nice fellow.

molly, var. MALLEE[1].

molly-coddler (mǫ·likǫːdləɹ). [f. MOLLY-CODDLE *v.* + -ER[1].] One who molly-coddles.

1863 A. J. MUNBY *Diary* 31 Aug. (1972) 172 It keeps up a wholesome protest against the Mollycoddlers, to see a whole countryful of stout lasses devoted to field-labour only. 1933 J. CARY *Amer. Visitor* 213 Minnie was a molly coddler, in Bewsher's phrase, and liked to take temperatures.

mollycot. (Earlier example.)

1826 *Blackw. Mag.* XX. 846/2 What in vulgar English is called a Molly-cot.

molly-dook (mǫ·lidūk), *a. Austral. slang.* [f. MOLLY[1] (or perh. MAULEY) + DUKE *sb.* 7.] Left-handed. So **molly-dooker, -hander,** a left-handed person; **molly-duked** *a.*, left-handed.

[1926 'J. V. MARSHALL' *Timely Tips to New Australians*, *Mauldy*, left-handed.] 1934 *Bulletin* (Sydney) 21 Mar. 11/3 Hence the trade is taboo to the molly-hander. 1941 BAKER *Dict. Austral. Slang* 47 *Mollydooker*, a left-handed person. Whence, 'mollydook' (adj.): left-handed. 1969 *Southerly* XXIX. 8 It could be being written by someone else with the same absurdly decorous aim, someone molly-duked, atheist, over-educated.

mollyhawk. (Earlier and later examples.)

1880 L. V. BRIGGS *Jrnl.* 19 Oct. in *Around Cape Horn on Bark 'Amy Turner'* (1926) 105, I caught and skinned two mollyhawks, or mollymoke, as some sailors call them. 1917 [see *John Down* (*JOHN 4)]. 1923 D. H. LAWRENCE *Kangaroo* viii. 169 A big albatross swung slowly down the surf: albatross or mollyhawk, with wide, waving wings. 1927 M. M. BENNETT *Christison of Lammermoor* i. 16 The mollyhawks and albatrosses planed and circled, keeping up with the ship with scarcely a movement of their great wings. 1969 *Landfall* XXIII. 103 In front of his house a mollyhawk was patiently taking up and dropping a pipi to break it on the hard sand by the water.

Molly Maguire. Add: **a.** (Further examples.) Now *Hist.*

1908 L. H. DAWSON *Nicknames & Pseudonyms* 202 *Molly Maguires*, young men, usually dressed in women's clothes, who formed an Irish association to resist distraint for rent in 1843. Their name was derived from Connor.. Maguire, 2nd Baron of Enniskillen, who in 1641 took a prominent part in the Catholic conspiracy. 1909 *Westm. Gaz.* 19 Feb. 2/3 There were Molly Maguires in Ireland.. long before the time of Mr. William O'Brien. 1971 E. NORMAN *Hist Mod. Ireland* iii. 57 The secret societies formed to stimulate local violence on behalf of individual tenant claims;..societies which adopted the awful names of 'Whiteboys', 'Rockites', 'Levellers', 'Whitefeet', 'Molly Maguires', 'Ribbonmen', and so on.

b. (Further examples.)

1886 P. STAPLETON *Major's Christmas* 86 There come among us some men that belonged to a band of murderers that called themselves Molly Maguires. 1970 *Guardian* 14 May 10/1 The Molly Maguires were a group of rebel miners who fought a brave and reckless losing battle with employers who stopped at nothing to destroy them.

mollymawk, mollymauk: now more usual spellings of MALLEMUCK. Cf. *MOLLYHAWK.

1913 H. K. SWANN *Dict. Eng. & Folk-Names Brit. Birds* 153 *Mallemuck*, an old Dutch-mariner's name for the Fulmar. Now corrupted into 'Molly-mawk', and applied to various other species such as the Black-browed Albatross. 1933 *Geogr. Jrnl.* LXXXI. 217 The mollymauk is a wild sea-bird which inhabits the regions of Cape Horn. 1959 D. A. BANNERMAN *Birds Brit. Isles* VIII. 193 Dr Cushman Murphy describes this mollymauk [*sc.* the black-browed albatross, *Diomedea melanophrys*] as 'the commonest albatross in the southern hemisphere, the most sociable and the most fearless of man while at sea'. 1972 M. F. SOPER *N.Z. Birds* 181 Albatrosses and mollymawks are tube-nosed birds..that, except for a few species inhabiting the North Pacific, are confined to the Southern Hemisphere. In New Zealand waters the term 'mollymawk' is conveniently applied to the smaller forms, all of which have black backs, and the term 'albatross' to the two 'great' albatrosses, the Wandering and the Royal, the adults of which have white backs.

molossic, *a.* (Earlier example.)

1864 H. BUSHNELL *Work & Play* i. 34 You distinguish ..the solemn, religious spondee, the swift trochaic run of eagerness or fear, the heavy molossic tread of grief or sorrow.

Molotov (mǫ·lǫtǫf). Also **Molotoff,** and with lower-case initial. The name of Vyacheslav Mikhailovich *Molotov* (1890–), U.S.S.R. Minister for Foreign Affairs from 1939 to 1949, used *attrib.* in **Molotov bread-basket,** a container carrying high explosive and scattering incendiary bombs; **Molotov cocktail,** a makeshift incendiary hand-grenade, consisting of a breakable container filled with inflammable liquid, and a means of ignition; also *ellipt.* as *Molotov.*

1940 *Illustr. London News* 6 Apr. 446/2 The 'Molotov Breadbasket'..appeared to consist of two types. 1940 *Hutchinson's Pict. Hist. War* 7 Aug.–1 Oct. 91 Used with success in the Finnish war, the so-called 'Molotov cocktails' are considered an effective weapon against armoured divisions and have been adopted by the Home Guard. 1940 *Flight* 26 Sept. 244/1 A Molotoff bread basket dropped some 50 incendiary bombs in a S.E. district of London. 1942 J. T. GORMAN *Mod. Weapons of War* ix. 147 Stationary tanks can be disabled by dropping grenades or Molotov cocktails through the ventilating openings. 1943 T. W. LAWSON *Thirty Seconds over Tokyo* 32 A 500-pound incendiary, something like the old Russian Molotov Breadbasket. 1951 J. CARSWELL in A. Somerville *Autobiogr. Working Man* p. xiv, Feargus O'Connor was trying to popularize the ginger-beer bottle filled with inflammable mixture (a version of the Molotov Cocktail to which Englishmen seem to turn when ammunition is short). 1969 *Oz* Apr. 5/1 Issue number three carried instructions on how to make a molotov with diagram and said where in the university to strike with same. 1970 R. JOHNSTON *Black Camels* vi. 90 Sulaiman was on the last bottle. Half fill it with petrol, drop in one end of the wick, plug the neck of the bottle and, hey presto, a Molotov Cocktail... Select a target, light the wick, throw the bottle. Bottle hits target, bottle smashes, petrol spills out, flame from wick ignites petrol. 1972 R. K. SMITH *Ransom* IV. 168 Thirty cars zooming out of the night loaded with molotovs.

† **molrowing** (mǫ·lrau,iŋ). *Obs. slang.* Also moll-rowing. [Perh. f. MOLL *sb.* + ROW *v.*[3] + -ING[1].] **a.** (See quot. 1860.) So **mo·lrower,** a wencher or whoremonger. **b.** Caterwauling, row, noise.

1860 HOTTEN *Dict. Slang* (ed. 2) 174 *Molrowing*, 'out on the spree', in company with so-called 'gay women'. In allusion to the amatory serenadings of the London cats. 1892 E. J. MILLIKEN *'Arry Ballads* 42 A pootily piped 'Blooming Lavender',..as meller as blackbirds in June, Beats 'Andel's molrowings a buster. 1894 G. A. SALA *Things I have Seen* II. xiii. 121 The scene wound up with a great concert of practical cats on the roof, whose diabolical mollrowings still ring in my ears. 1896 FARMER & HENLEY *Slang* IV. 332/1 *Molrower*, a whoremonger.

molten, *ppl. a.* Add: **4.** *Comb.*, as *molten-blue, -crystal, -golden* adjs.

1862 G. M. HOPKINS *Poems* (1930) 132 As though some sapphire molten-blue Were vein'd and streak'd with dusk-deep lazuli. 1950 D. GASCOYNE *Vagrant* 27 Moulting molten-crystal plumes of birds of paradise. 1849 POE *Bells* in *Sartain's Union Mag.* Nov. 304/1 From the molten-golden notes,..What a liquid ditty floats.

‖ **molto** (mǫ·lto), *a.* and *adv.* [It.] (See quot. 1801.)

1801 T. BUSBY *Dict. Mus.*, *Molto*, very or much. A word used in conjunction with some other, by way of augmentation; as *Molto Allegro*, very quick, *Molto Adagio*, very slow. 1889 G. B. SHAW *London Music 1888–89* (1937) 148 The *molto sostenuto* indicated by Gounod, which is the characteristic effect of the movement, was quite lost. 1928 [see *MARTELLATO *a.*] 1959 *Times* 15 Jan. 12/6 Mr. Buckley did not play them *molto dramatico* in octaves as Rachmaninov used to do. 1970 *Times* 23 Feb. 12/3 That striking but rarely heard effect, *molto pianissimo* from a large choral body.

moly[2] (mǫ·li), colloq. abbrev. of MOLYBDENUM. Also used for *molybdenum disulphide* (a lubricant).

1961 in WEBSTER. 1963 R. F. WEBB *Motorists' Dict.* 155 The [lubricating] action of 'moly' has been likened to a pack of cards resting on a table. 1970 *Financial Times* 13 Apr. 27/8 Not that Amax is remaining dormant on the moly front. The bringing in of the Henderson mine..will nigh on double production capacity.

molybdenian (mǫlibdī·niăn), *a. Min.* [f. MOLYBDEN(UM + *-IAN 2.] Of a mineral: having a (small) proportion of a constituent element replaced by molybdenum.

1930 W. T. SCHALLER in *Amer. Mineralogist* XV. 571 Molybdenum—molybdenian. 1953 *Amer. Mineralogist* XXXVIII. 904 Prior to oxidation the rock contained..a few grains of molybdenian scheelite.

molybdenum. Add: Now usu. pronounced (mǫ̆li·bdĕnǫ̆m). **b. molybdenum blue,** a complex oxide or mixture of oxides of pentavalent or hexavalent molybdenum with a strong blue colour that is produced, usu. as a colloidal solution, when an acidic solution of a molybdate is reduced, and is used in chemical analysis and occas. as a dye; also, the colour of this substance.

1901 *Jrnl. Chem. Soc.* LXXX. II. 163 Molybdenum blue does not appear to contain the dioxide, MoO_2,..but the semipentoxide, Mo_2O_5. 1920 F. A. MASON tr. *G. von Georgievics's Text-bk. Dye Chem.* 542 Molybdenum blue.. was used experimentally for a certain time on silk and cotton materials. 1951 *Amer. Mineralogist* XXXVI. 610 Ilsemannite is soluble in water, first producing a greenish blue solution which later deepens to a typical molybdenum blue. 1952 KIRK & OTHMER *Encycl. Chem. Technol.* IX. 203 The present importance of molybdenum blue rests on the fact that several colorimetric methods depend upon it: (1) the determination of trace quantities of phosphorus or phosphates by the preferential reduction of phosphomolybdic acid; and (2) a number of other methods employing phosphomolybdic acid as a reagent (for phenols, tyrosine, tryptophan, etc.) for reducing compounds. 1965 D. ABBOTT *Inorg. Chem.* xii. 642 Molybdenum 'blue' is an oxide formed when an acidified molybdate solution is greatly reduced. It has variable composition, usually about 67–68 per cent Mo.

molybdian (mǫ̆li·bdiăn), *a. Min.* [alteration of *MOLYBDENIAN *a.*] = *MOLYBDENIAN *a.*

1951 C. PALACHE et al. *Dana's Syst. Min.* (ed. 7) II. 1078 Seyrigite... A molybdian scheelite, with MoO_3 24·01 per cent.

molybdophyllite (mǫ:libdofi·ləit). *Min.* [ad. G. *molybdophyllit* (G. Flink 1901, in *Bull. Geol. Inst. Univ. Upsala* V. I. 91), f. Gr. μόλυβδο-ς lead + φύλλ-ον leaf: see -ITE[1].] A hydrous silicate of lead and magnesium.

1901 *Jrnl. Chem. Soc.* LXXX. II. 664 Molybdophyllite —This new mineral is of rare occurrence with hausmannite in granular limestone or dolomite at Långbanshyttan, in Wermland, Sweden. 1968 I. KOSTOV *Mineral.* 327 Perfect cleavage in one direction is displayed by molybdophyllite.

molysite (mǫ·lisəit). *Min.* [said by Dana, the coiner of the name, to be f. Gr. μόλυσις stain, which was app. taken (in error for μόλυσμα) as the sb. corresp. to μολύνειν to stain, perh. by confusion with μώλυσις scalding (f. μωλύ(ν)ειν to scald, parboil): see -ITE[1].

The statement in quot. 1935 as to the origin of the name is incorrect. In *Atti d. R. Accad. d. Sci. fis. e matem.* (1874) VI. IX. 43, Scacchi attributes the name *molysite* to Dana (in *Syst. Min.* (ed. 5)), and renders it in It. as *molisite* so as to make it conform to It. orthography (as he explains in *Lo Spettatore del Vesuviano* (1887)).]

Natural ferric chloride, $FeCl_3$, formed as a sublimation product on lavas surrounding fumaroles and occurring (before being hydrated by the air) as a yellow to red film or incrustation.

1868 DANA *Syst. Min.* (ed. 5) 118 Molysite. 1935 J. W. MELLOR *Comprehensive Treat. Inorg. & Theoret. Chem.* XIV. lxvi. 41 A. Scacchi observed in recent eruptions a yellowish or brownish-red incrustation associated with the lavas of fumaroles or steam holes. The coloration was attributed to the presence of ferric chloride, $FeCl_3$, and A. Scacchi called the mineral molisite, which J. D. Dana altered to molysite—from μόλυσις, a stain. 1968 I. KOSTOV *Mineral.* II. iii. 195 Molysite $FeCl_3$.

mom (mǫm). Also **Mom.** Colloq. abbrev. (chiefly *U.S.*) of MAMMA[1], MAMA; *spec.* the typical American matriarchal mother. Also *attrib.* and *Comb.*, as *mom cult, culture; mom-bashing, -like*, etc.; *spec.* **mom-and-pop** *U.S.*, used *attrib.* to denote a small shop or store, etc., of a type often run by a married couple.

1894 *Dialect Notes* I. 332 *Mam, mom, mæ*, for mamma or mother. 1911 R. W. CHAMBERS *Common Law* v. 156 City-wearied fathers of youngsters who called their parents 'pop' and 'mom'. 1950 *Brit. Jrnl. Psychol.* June 230 The 'mom' culture in which American adolescents are often said to live. 1951 Mom-and-Pop [see *HOLE *sb.* 7 b]. 1957 *Economist* 28 Dec. 1121/2 Tin Pan Alley, Fundamentalism, and the Mom cult had it coming to them. 1958 *Manch. Guardian* 7 June 4/6 Three leading serious comedies are studies in parent-hating or mom-bashing. 1961 J. HELLER *Catch-22* (1962) i. 9 The hot dog, the Brooklyn Dodgers. Mom's apple pie. That's what everyone's fighting for. 1961 J. MCCABE *Mr. Laurel & Mr. Hardy* (1962) v. 102 The troubles of the boys in their continuing endeavour to get away from their domineering, Mom-like wives. 1962 *Listener* 20 Dec. 1053/1 These 'mom and pop' stores and the somewhat larger ones certainly have no love for the supermarkets. 1966 *Economist* 9 July p. xxi/3 'Mom-and-pop' dealers in smaller communities and suburban fringes. 1968 *Globe & Mail* (Toronto) 17 Feb. 46/8 (Advt.), A family kitchen with mom in mind boasts a.. built-in range and oven. 1972 *New Yorker* 21 Oct. 30/1 The mom-and-pop grocery store was the New York State headquarters of the McGovern campaign. 1975 *Ibid.* 13 Jan. 36/2 'Of course we will, Mom,' I said, and I patted her hand and laid it aside.

mombin (mǫ̆mbī·n). [Amer. Sp. *mombin*, f. Caribbean native name.] A West Indian tree of the genus *Spondias*, esp. *S. lutea*, of the family Anacardiaceæ, or its yellow or purplish fruit, resembling a small plum; = HOG-PLUM a.

1837 J. MACFADYEN *Flora of Jamaica* I. 228 The fruit [of *Spondias graveolens*] is called Mombin by the French colonists. 1920 W. POPENOE *Man. Tropical & Subtropical Fruits* iv. 156 The red mombin is a small tree, often spreading in habit. *Ibid.* 158 The red mombin is abundant in Mexico and Central America. *Ibid.* 159 The Yellow Mombin..is generally considered inferior in quality to the red mombin. 1972 A. F. SIMMONS *Growing Unusual Fruit* 201 Mombin fruit... Spondias species. *Ibid.* 202 Yellow mombin or hog plum (*Spondias mombin*). The egg-shaped yellow fruits are about 1 in long and sub-acid to sweet.

mome[2]. (Later example.)

1923 E. SITWELL *Bucolic Comedies* 17 An old dull mome With a head like a pome.

mome (mōᵘm), *a.* A factitious word introduced by 'Lewis Carroll' (see quot. 1855).

Also occurs in *Through the Looking-Glass* (1871) i. 21.

1855 'L. CARROLL' *Rectory Umbrella & Mischmasch* (1932) 139 All mimsy were the borogoves; And the mome raths outgrabe. *Ibid.* 140 *Mome* (hence *solemome, solemone* and *solemn*), 'grave'. 1960 M. GARDNER *Annotated Alice* 195/1 'Mome' has a number of obsolete meanings such as mother, a blockhead,..none of which, judging from Humpty Dumpty's interpretation, Carroll had in mind. 1970 R. D. SUTHERLAND *Lang. & Lewis Carroll* vii. 149 Humpty Dumpty is reporting the generally accepted meanings... The information he imparts is 'as sensible as a dictionary'... He admits some difficulty with *mome.*

moment, *sb.* Add: **1. c.** *not for a moment*: emphatically not; *of the moment*: of importance at the time in question; esp. *man of the moment*; *never a dull moment*: a catch-phrase designating constant variety; *to have one's* (or *its*) *moments*: to be impressive, etc., on occasions; *to live for* (or *in*) *the moment*: to live without concern for the future; (*at this*) *moment in time*: now, the present instant.

1871 BROWNING *Pr. Hohenst.* 28 Well, that's my mission, so I serve the world, Figure as man o' the moment. 1889 J. K. JEROME *Three Men in Boat* ix. 85 There is never a dull moment in the boat while girls are towing it. 1926 I. MACKAY *Blencarrow* xxiii. 207 With the proper sort of company the situation might have had its moments. 1927 *Sat. Even. Post* 9 Apr. 119 Even a mailman has his moments. 1932 *Weekend Rev.* 14 May 617/1 There is never a dull moment. 1935 E. CARR *Jrnl.* in *Hundreds & Thousands* (1966) 185 The three dogs lie on the bed,.. trusting me to attend to their wants, living for the moment. 1935 *Discovery* Aug. 220/1 Land utilisation is the problem of the moment. *Ibid.* 221/1 This is the need dictated by world economic conditions *of the moment.* 1936 *Punch* 12 Aug. 170/2, I don't suggest for a moment that these are finished ideas. They are no more than artists' roughs. 1950 L. KAUFMAN *Jubel's Children* xx. 223 Jubel says there's never a dull moment. 1963 W. H. MISSILDINE *Your Inner Child of Past* xii. 127 Because they 'live in the moment', they actually tend to be blindly unaware of the feelings of others. 1970 O. NORTON *Dead on Prediction* vii. 137 That's something I *didn't* know... You have your moments! 1972 G. BROMLEY *In Absence of Body* iii. 30 What can we actually do to help at this moment in time? 1972 'T. COE' *Don't lie to Me* (1974) xxii. 169 Hargerson had his moments; happily this was one of them. 1972 R. ADAMS *Watership Down* xx. 126 'No one hurt?' 'Oh, several have been hurt, one way and another.' 'Never a dull moment, really,' said Bigwig. 1973 *Guardian* 12 Mar. 11/2 The usual stuff about meaningful confrontations taking place..at this moment in time. 1974 C. EGLETON *October Plot* i. 11 There were five similar [flak] towers..but at this moment in time, they were only of passing interest. 1974 V. BROME *Day of Destruction* xv. 153 Here's our man of the moment... Hail the conquering hero comes.

d. *moment of truth*: the time of the final sword-thrust in a bull-fight (Sp. *el momento de la verdad*); *transf.*, a crisis or turning-point; a testing situation.

1932 E. HEMINGWAY *Death in Afternoon* vii. 68 The whole end of the bullfight was the final sword thrust, the actual encounter between the man and the animal, what the Spanish call the moment of truth. 1949 *New Statesman* 15 Jan. 61/1 A good detective story should be like a good bull-fight... The author plunges the unexpected explanation into him like a sword—the moment of truth, as the Spaniards call it. 1954 H. CASTEEL *Running of Bulls* v. 106 The Spanish call the kill the hora de la verdad (the hour (moment) of truth)... this is still the climactic second to which the whole fight has been dedicated. 1956 I. BROMIGE *Enchanted Garden* III. iii. 140 This, thought Fiona, was the moment of truth. 1957 'D. RUTHERFORD' *Long Echo* viii. 150 This was his Moment of Truth, but there was no arena packed with spectators to give him vicarious courage. 1959 *Listener* 6 Aug. 215/2 Now comes 'the moment of truth'... The matador causes the bull to stand square..and plunges his sword between its shoulder-blades. 1965 H. I. ANSOFF *Corporate Strategy* (1968) i. 16 We shall have..very little [to say]..about the utility function applied at the ultimate 'moment of truth' in selecting the preferred alternative. 1972 H. MACINNES *Message From Malaga* i. 9 You've become a self-centred bastard, he told himself... He blamed this moment of truth on the combination of guitar, scent of flowers, night sky.

2. d. *Geol.* [a. F. *moment* (proposed in this sense by E. Renevier 1882, in *Congrès géol. internat.: Compt. Rend. de la 2me Session, 1881* 540).] A period of geological time corresponding to a stratigraphical zone (as defined by its fossil content).

1933 W. J. ARKELL *Jurassic Syst. Gt. Brit.* i. 21 Strictly speaking,..'moment' should take priority over secule, and Diener adopts it, with the variations time-moment and zone-moment. 1958 *Bull. Geol. Soc. Amer.* LXIX. 113/2 At the Congress in Bologna in 1881..the only concrete suggestion came from the Swiss delegation (Renevier, 1882), which formally proposed 'moment' for the time equivalent of a zone. *Ibid.*, The term 'moment' has priority, as was..recognized at an early date by Diener (1919). 1969 BENNISON & WRIGHT *Geol. Hist. Brit. Isles* ii. 23 A zone is defined as strata deposited during an interval of time (known as a secule or moment, though these terms are not widely used) throughout which a particular faunal or floral assemblage existed.

8. c. *Physics.* The distance between the two poles of a simple bar magnet, or the two charges of an electric dipole, multiplied by the strength of either of them (called more fully the *magnetic* and the *electric moment*, respectively); more widely, the maximum torque that any given system of poles, or charges, can experience in a uniform magnetic, or electric, field (respectively) of unit strength; more fully called (*magnetic* or *electric*) *dipole moment.* Similarly the moment of order 2 (*quadrupole moment*) is a certain tensor of rank 2, whose components are quadratic functions of the spatial co-ordinates of the poles or charges; so *octupole moment* (moment of order 3), etc. Cf. *MULTIPOLE *sb.*

1865 *Phil. Mag.* XXIX. 441 The product of the strength of the poles into the length between them is called the magnetic moment of the magnet. 1884 J. T. SPRAGUE *Electricity* (ed. 2) iii. 110 The magnetic moment of any uniformly magnetized substance is proportional to its volume. 1892 O. LODGE *Mod. Views Electricity* 445 A circuit conveying a current exactly imitates a magnet of definite moment, the equivalent moment being..μnAC where A is the mean area of the coil, *n* the number of turns of wire, C the current, and μ a constant characteristic of the medium inside the coil. 1903 S. J. BARNETT *Elements of Electromagnetic Theory* xi. 282 The quantity $M = mL = T/H \sin \theta$ is called the magnetic moment of the magnet (analogous to the electric moment of an electret). 1916, etc. [see *MULTIPOLE *sb.*]. 1920 *Chem. Abstr.* XIV. 3014 Assuming the mol. to be a dipole it is possible to obtain the distance between the 2 charges from the moment of inertia and from this in turn to calc. the elec. moment by application of the quantum theory. 1934 *Physical Rev.* XLV. 761/1 We derived the magnetic moment of the deuton in the same way as the magnetic moment of the proton. 1942 J. D. STRANATHAN *'Particles' of Mod. Physics* x. 401 The magnetic moment of a neutron must be regarded as a fundamental property of the particle. 1953 *Sci. News* XXX. 7 The magnetization [of a piece of iron] is the magnetic moment per unit volume, or the pole strength per unit cross-sectional area. 1958 CONDON & ODISHAW *Handbk. Physics* IV. ii. 19/2 A quadrupole term is related to the second-order moments of the charge distribution... (Various slightly different definitions of the quadrupole moment are to be found in the literature.) *Ibid.* 20/1 The spherical-harmonic addition theorem.. permits the writing of $V_k(\mathbf{R})$ in the form

$$V_k(\mathbf{R}) = \frac{1}{R^{k+1}} \sum_{m=-k}^{+k} Q_{km} F_{km}(\Theta) e^{-im\Phi}$$

where the 2^k-pole moment is fully described by the $2k+1$ quantities,

$$Q_{km} = \frac{4\pi}{2k+1} \int \rho(\mathbf{r}) F_{km}(\theta) e^{im\phi} dv.$$

1962 CORSON & LORRAIN *Introd. Electromagn. Fields & Waves* v. 210 We set $\mathbf{m} = I\mathbf{S}$ as the dipole moment of the current loop, where **S** is a vector whose magnitude is equal to the area of the loop. It is perpendicular to the loop. *Ibid.*, It is possible to generalize the concept of magnetic dipole moment to any distribution of current in space... For a current distribution **J** in a volume τ, $\mathbf{m} = ..\frac{1}{2}\int \mathbf{r} \times \mathbf{J} d\tau$. 1970 G. K. WOODGATE *Elem. Atomic Struct.* ix. 177 In order to be able to discuss departures of the nuclear charge distribution from spherical symmetry we should like to attribute to the nucleus electric multipole moments. *Ibid.* 180 The quadrupole moment *Q* is positive if the nuclear charge distribution is elongated along the direction of **I** [*sc.* the nuclear spin] (prolate) and negative if the distribution is flattened (oblate).

d. *Statistics.* Each of a series of quantities (*first, second*, etc., *moment*) that express the average or expected value of the first, second, etc., powers of the deviation of each component of a frequency distribution from some given value, usu. the mean (giving a *central moment*) or zero.

1893 K. PEARSON in *Nature* 26 Oct. 615/2 Now the centre of gravity of the observation curve is found at once, also its area and its first four moments by easy calculation. 1925 R. A. FISHER *Statistical Methods for Research Workers* iii. 70 The standard error of the variance is $\sqrt{[(\mu_4 - \mu_2^2)/N]}$, where N is the number of samples, and μ_2 and μ_4 are the second and fourth moments of the theoretical distribution. 1938 A. E. WAUGH *Elem. Statistical Method* vi. 114 The formula for the first moment about the mean involves $\sum x$ and..it reduces to zero, since $\sum x = 0$. We can thus say some things about the moments of all curves in advance: (1) $v_0 = 1$, (2) $v_1 = 0$, (3) $v_2 = \sigma^2$. *Ibid.*, In any symmetrical curve the odd moments, being based on the sums of odd powers of deviations, will equal zero. 1959 G. & R. C. JAMES *Math. Dict.* 257/2 $\mu_r = \int_{-\infty}^{\infty} (x-a)^r f(x) dx$ is the *r*th moment of *x* around the point *a*, where *x* is a random variable with frequency function $f(x)$. 1961 D. V. HUNTSBERGER *Elem. Statistical Inference* v. 102 The first moment about the origin is the mean μ of the theoretical distribution... The variance σ^2 of a random variable *Y* is defined as the second moment about the mean, the average value of $(Y - \mu)^2$.

10. *moment-to-moment* attrib. phr., of an immediate requirement, necessity, etc.

1934 T. S. ELIOT *Elizabethan Essays* 8 Contemporary literature, like contemporary politics, is confused by the moment-to-moment struggle for existence. 1957 A. MILLER *Coll. Plays* (1958) Introd. 23 It wanted..a kind of moment-to-moment wildness in addition to its organic wholeness.

momentaneity (mōᵘmĕntĕnī·iti, -ĕi·Iti). [f. MOMENTANE(OUS *a.* + -ITY, after SPONTANEITY.] Transitory character; momentariness.

1921 D. H. LAWRENCE *Sea & Sardinia* viii. 314, I felt my sound Sardinian soul melting off me, I felt myself evaporating into the real Italian uncertainty and momentaneity. 1923 —— *Kangaroo* vii. 157 All her high moments would have this Bacchic, weapon-like momentaneity.

momentaneous, *a.* Restrict † *Obs.* to sense 3 and add: **1.** (Later examples.) **momentaneously** *adv.* (later example).

1921 H. POUTSMA *Characters of Eng. Vb. & Expanded Form* i. 2 The actions expressed by verbs..may be.. momentaneous, i.e. covering only one moment, or comprised between two closely contiguous moments, so that the beginning and the end practically synchronize. *Ibid.* 5 [Actions expressed by verbs may be] momentaneously iterative, which may be graphically represented by a succession of dots. 1921 E. SAPIR *Language* v. 108 'Cry out every now and then' or 'cry in fits and starts' is momentaneous-iterative. 1923 D. H. LAWRENCE *Kangaroo* vii. 163 Do I want this curious transparent blood of the antipodes, with its momentaneous feelings? 1926 H. POUTSMA *Gram. Late Mod. Eng.* II. § 2. 545 When distinctly verbal in nature, the attributive past participle mostly has a momentaneous or resultative aspect.

momentarily, *adv.* Add: **1.** (Later examples.)

1928 'S. S. VAN DINE' *Greene Murder Case* iii. 39 The situation probably drove everything momentarily from their minds except the two victims of the shooting. 1975 *Times* 8 Sept. 12/3 One MP..was 'growled down'... He wondered momentarily if he had wandered into the Lords by mistake.

3. (Later example.)

1899 W. J. LOCKE *White Dove* (1900) iii. 43 Sylvester.., having done all that was momentarily possible, was at last able to reflect.

4. At any moment. *N. Amer.*

1928 *Sun* (Baltimore) 13 Aug. 1/2 Arrests were expected momentarily as police continued their investigation of the fantastic murder mill. 1944 *Musical America* 25 Mar. 7/4 An addition to the family..is expected momentarily. 1970 *New Yorker* 3 Oct. 30/3 Miss Loren had been delayed in traffic but would arrive momentarily. 1970 *Daily Tel.* 22 Oct. 18 The [Canadian] captain came on the intercom. 'Please fasten your seat belts,' he said. 'We shall be taking off momentarily.' 1972 *New York* 8 May 6/2 We are expecting our first child momentarily. 1975 *Publishers Weekly* 10 Feb. 52/1 There is one woman who imagines she is in a motel and that her deceased husband will momentarily arrive to pick her up.

momento (mŏme·nto), var. MEMENTO 3.

1871 GEO. ELIOT *Jrnl.* 19 Mar. in J. W. Cross *George Eliot's Life* (1885) III. xvi. 129 My present fear is that I have too much matter—too many *momenti.* 1951 DYLAN THOMAS *Let.* 12 Apr. (1966) 356 And the London frowsty Casino, a momento of which I enclose.

momentum. Add: **6.** Special Comb.: **momentum space** *Physics*, a three-dimensional space in which each particle of a physical system is represented by a point whose three Cartesian co-ordinates are numerically equal to the components of its momentum in the directions of the three co-ordinate axes.

1948 W. HUME-ROTHERY *Electrons, Atoms, Metals & Alloys* xx. 143 The momentum space is divided into little cells of side h/L, and volume h^3/L^3. 1970 I. E. MCCARTHY *Nuclear Reactions* I. v. 111 We consider the matrix element in momentum space. 1970 G. K. WOODGATE *Elem. Atomic Struct.* vi. 98 Under this assumption the electrons are moving freely, with no forces acting on them, and their translational momenta can be taken to be directed isotropically in momentum space.

momie-cloth: see MUMMY-CLOTH 2. Also *ellipt.*

1881 C. C. HARRISON *Woman's Handiwork* I. 46 (*heading*) Raw silk momie cloth, for draperies,..fifty inches wide, at $3.50... Cotton momie cloth..in all the new shades for $1.10,.. Basket momie cloth, in woollen, costs $3. 1895 *Montgomery Ward Catal.* 4/2 Blade Crepon.. similar to Momie or Armure. 1960 CUNNINGTON & BEARD *Dict. Eng. Costume* 264/2 *Momie cloth*,..of cotton or silk warp and woollen weft, resembling a fine crepe; usually black and used for mourning.

momism[2] (mǫ·miz'm). *U.S.* Also **Momism.** [f. *MOM + -ISM.] Excessive attachment to, or domination by, the mother.

1942 P. WYLIE *Generation of Vipers* xi. 185 Hitherto.. man has shown a considerable qui vive to the dangers which arise from momism and freely perceived that his 'old wives' were often vixens, dragons, and xanthippes. *Ibid.* 197 The mealy look of men today is the result of momism and so is the pinched and baffled fury in the eyes of womankind. 1947 *Sat. Rev.* (U.S.) 18 Jan. 19/2 Philip Wylie's latest diatribe about momism is welcome since it frankly implies that men have relinquished their duties as fathers. 1958 *Times Lit. Suppl.* 17 Oct. 595/1 The great *condottiere* of the lodes was a victim of 'momism'. His mother, it seems, made marriage impossible. 1962 R. E. FITCH *Odyssey of Self-Centred Self* iv. 129 We reject the Father God, and turn to Mother Nature. This is metaphysical Momism. 1968 'E. V. CUNNINGHAM' *Cynthia* (1969) ix. 117 You never made a real pass at me. You suffer from reverse Momism.

momma (mǫ·mă). Add: Also **Momma.** (Earlier and later examples.) Also *attrib.* and *Comb.* Cf. *MOM.

1884 *Century Mag.* Apr. 859/1 This incident..illustrates the position of the 'momma' or 'mammy' in a Southern family in the olden time. 1929 *Amer. Speech* IV. 476 A final note may be added on the word *momma* 'negro nurse'. .. The term is associated with the negroes..but is not an example of negro dialect, but rather of American English. 1939 L. M. MONTGOMERY *Anne of Ingleside* viii. 58 You're nothing but a sweet little girl. Momma's Pet! 1947 AUDEN *Age of Anxiety* (1948) v. 116 I've made their magic but their Momma Earth is His stone still. 1951 'J. WYNDHAM' *Day of Triffids* ii. 43 How'd that be, Momma? 1960 G. W. TARGET *Teachers* 223 A real nice Yiddisher Momma. 1961 *John o' London's* 9 Nov. 519/2 Mommy is the all-American Momma-figure. 1970 G. GREER *Female Eunuch* 158 Baby-talk, even to the extent of calling..the wife 'momma'. 1972 M. J. BOSSE *Incident at Naha* iii. 128 A real live trollop..the last of the red-hot mommas. 1972 *Times* 29 Sept. 20/6 How well we have seen the result of this [price-] battle in the position of the small retailer, affectionately known as the 'Momma and Poppa shop'. 1973 *Black World* Apr. 57 My momma always set more places than there was folks to eat.

momme. Add: Also with pronunc. (mǫ·me). Also 8 **mome, momi.** (Earlier and later examples.)

1727 J. G. SCHEUCHZER tr. *Kæmpfer's Hist. Japan* I. IV. vii. 349 The yearly value of the *Cobanj*..is from 55 to 59 *Mome*, or *Maas* in silver. *Ibid.* viii. 362 The highest value of the *Cobang*, as current in the country, is of sixty *Momi*, or *Maas*, of silver. 1902 L. HEARN *Kottō* vii. 58, I have already been able to put by about one hundred *mommé* of silver. 1965 *Economist* 25 Dec. 1432/1 We buy gold by the —tael in Shanghai; momme in Tokyo; tola in Karachi; hong ping tael in Macao; and kilogram in Kabul. 1974 S. MARCUS *Minding the Store* (1975) ix. 184 This select 5% are not..our best customers, but..our most exacting ones. They know..the difference between a silk scarf made of twelve-momme weight and one made of sixteen-momme.

Mommy, mommy[2] (mǫ·mi). Chiefly *U.S.* Var. MAMMY and MUMMY *sb.*[2] Cf. *MOM, MOMMA.

1902 in *Dialect Notes* II. 239. 1974 HAWKEY & BINGHAM *Wild Card* vi. 59 Mommy'll bring you a glass of water. 1975 *Listener* 18 Sept. 386/2 'You and Mommy are always with people,' she complains.

‖ **mompei, mompe** (mǫ·mpe). [Jap.] Baggy working trousers worn in Japan.

1947 J. MORRIS *Phoenix Cup* II. 24 Most of the women wear *mompei*, rather ugly baggy trousers normally worn only when working in the fields. 1959 R. K. BEARDSLEY et al. *Village Japan* v. 101 For field work, women wear a surplice-cut blouse of dark cotton work cloth tucked into roomy, gathered-ankle field trousers (*mompei*) of similar material. 1960 B. LEACH *Potter in Japan* v. 109 To keep moderately clean, I wear a pair of Hamada's 'mompei' (baggy overall trousers) and rubber boots. 1965 W. SWAAN *Jap. Lantern* xix. 224 The bending figures in their *mompe* or baggy trousers. 1972 *Nat. Geographic* May 672/2 Groups of women in *mompe*, the traditional baggy work trousers worn in rural areas, shifted the seaweed by hand.

momser, momza, momzer, varr. *MAMZER.

Mon (mōᵘn), *sb.*[2] and *a.* Also **Moan, Mun, Mwun.** [Native name.] **A.** *sb.* A people of Indo-Chinese origin, also called *TALAING, now inhabiting eastern Burma and western Thailand but having their ancient capital at Pegu in southern Burma; a member of this people. Also, their language. **B.** *adj.* Of or pertaining to the Mon. Cf. *MON-KHMER.

1798 F. BUCHANAN in *Asiatick Researches* V. 235 To the kingdom, the natives of which call themselves Moan, we have given the name of Pegu, a corruption of the vulgar appellation of its capital city Bagoo. 1828 J. CRAWFURD *Jrnl. Embassy to Siam & Cochin China* xv. 448 Of the Mon, or Pegu, race, although the Siamese be in possession of no part of their territory, there are a considerable number in Siam. 1854 F. MASON in *Jrnl. Amer. Oriental Soc.* IV. 284 *Moan*, the name by which the Talaings now call themselves. 1873 *Jrnl. Asiatic Soc. Bengal* XLII. 34 The people of Pegu..call themselves Mun, Mwun, or Mon... The Burmese, since the conquest of Pegu..in 1757–58, had strongly discouraged the use of the Mun language. 1877 [see *ANNAMITE *a.* and *sb.*]. 1881 C. J. F. S. FORBES *Compar. Gram. Languages of Further India* iv. 49 In three savage and more primitive dialects we find further traces of affinity with the Mon. 1907 V. C. S. O'CONNOR *Mandalay* IV. ii. 400 Under Binya Dala a strenuous effort was made by the Môn people to recover their ancient ascendency in Burma, and restore the glories of Pegu. 1939 L. H. GRAY *Foundations of Lang.* xii. 393 *Môn-Khmêr* has six sub-divisions: Central (Môn or Talaing, Khmêr or Cambodian, Bahuar, etc.). *Ibid.* 394 The earliest Môn inscription is of 1084 (?). 1948 D. DIRINGER *Alphabet* vii. 410 The early Mon character was not only the ancestor of the modern Mon script, but also of the Burmese and some Shan characters. 1962 *Listener* 25 Oct. 646/2 U Nu proposed to..create separate states for the Mon and Arakanese peoples. 1971 *National Geographic* Mar. 356 Masterpieces of lacquer ware... The craft may have originated in China and spread across Southeast Asia to the Mons, who brought the technique to the capital. 1972 W. B. LOCKWOOD *Panorama Indo-Europ. Lang.* 228 The group [*sc.* Mon-Khmer] takes its name from two of its members: Mon, the vernacular of the coastal districts round the Gulf of Martaban between Rangoon and Moulmein, and Khmer or Cambodian, the state language of Cambodia.

mon[3] (mʊn). Colloq. abbrev. of MONEY *sb.* See also *MUN.

1895 W. C. GORE in *Inlander* Nov. 64 *Mon*,..money. *a* 1911 D. G. PHILLIPS *Susan Lenox* (1917) II. v. 119 In little old New York..you've got to have the mon. or you get..the swift, hard, kick. 1958 A. WILSON *Middle Age of Mrs. Eliot* II. 122 *I'm* only going because she's got lots of lovely 'mon' and might leave it to me. 1963 'J. BELL' *Flat Tyre in Fulham* iii. 30 All that good 'mon' down the drain.

monact, *a.* and *sb.* (Examples.)

1885 tr. F. E. Schulze in *Rep. Sci. Results Voy. H.M.S. Challenger, Zool.* XXI. 29 It seems to me, however, that those spicules, called by Carter 'Clavulæ'..are really monacts. *Ibid.* 37 The derived nature of a monact spicule is in many cases determinable.

monad. Add: **2.** (Later examples.) Also *attrib.*

a 1914 C. S. PEIRCE *Coll. Papers* (1931) I. 146, I therefore divide all objects into monads, dyads, and triads; and the first step in the present inquiry is to ascertain what are the conceptions of the pure monad, free from all dyadic and triadic admixtures. *Ibid.* ii. 149 Now in order to convert that psychological or logical conception into a metaphysical one, we must think of a metaphysical monad as a pure nature, or quality, in itself without parts or features, and without embodiment. 1929 R. BRIDGES *Test. Beauty* i. 19 It was no flaw In Leibnitz to endow his monad-atoms with Mind. 1937 [see *ACTIO IN DISTANS]. 1965 *New Statesman* 18 June 942/2, I had been very impressed, wandering around housing estates, at the growth of what..seemed 'monad' politics... People didn't connect, except through mass media, but found images difficult to accept as reality.

monadic, *a.* Add: **1. b.** *Philos.* Of a proposition, fact, function, etc., or the predicate contained therein, when the predicate is non-relational and applies to only one subject term.

1897 C. S. PEIRCE in *Monist* VII. 167 A non-relative name with a substantive verb, as '—is a man',..has one blank; it is a *monad*, or *monadic relative.* 1921 W. E. JOHNSON *Logic* I. 203 The number of substantival references are respectively one, two, three and four, and the corresponding adjectives or propositions may be called monadic, diadic, triadic and tetradic. 1939 *Mind* XLVIII. 486 It appears sometimes to be assumed that the elementary statement must be monadic, *i.e.*, must have the form of a one-termed predicate, *ϕx.* 1946 C. MORRIS *Signs, Language & Behavior* iii. 78 'Black' is in this sense *monadic*... 'Deer' and 'black' are both monadic... But it would often be added that 'deer' designates an object and 'black' a quality of an object. 1956 J. O. URMSON *Philos. Analysis* ii. 17 A fact in which a particular has some absolutely simple and determinate characteristic was known as a monadic fact. 1963 W. W. LAMBERT in S. Koch *Psychol.* VI. 177 Most studies in general experimental psychology are carried out in terms of a monadic rather than a dyadic character. 1965 HUGHES & LONDEY *Elem. of Formal Logic* xxxix. 274 Such a schema might contain only monadic predicate variables (e.g. '$fx \supset gy$').

monadically, *adv.* (Later examples.)

1934 *Proc. Brit. Acad.* XIX. 71 Thus, Carnap holds, if we insist that protocol propositions must refer to content, then each protocol-language can be used only *monadically.* 1950 W. V. QUINE *Methods of Logic* (1952) § 23. 130 A capital letter occurs monadically, i.e., with variables attached singly.

monadist (mǫ·nădist). Also **Monadist.** [f. MONAD + -IST.] A follower of monadism; Leibniz himself.

1855 J. M. D. MEIKLEJOHN tr. *Kant's Critique Pure Reason* 274/2 Against the assertion of the infinite subdivisibility of matter..objections have been alleged by the Monadists. 1900 B. RUSSELL *Crit. Exposition Leibniz* § 78. 136 A Monadist must, with Leibniz, maintain the mutual independence of substances. 1932 *Mind* XLI. 518 Leibniz ..the great Monadist.

monadistic, *a.* Delete *rare* and add later examples.

1903 B. RUSSELL *Princ. Math.* IV. xxvi. 221 Given, say, the proposition *aRb*, where *R* is some relation, the monadistic view will analyse this into two propositions.. ar_1 and br_2. 1917 A. S. PRINGLE-PATTISON *Idea of God* ix. 179 On this monadistic theory, the organic vesture of the spirit and its environmental conditions are both resolved into innumerable quasi-spiritual centres. 1937 *Mind* XLVI. 275 He rejects a pantheist version of the Absolute in favour of a monadistic one. 1941 D. H. PARKER *Experience & Substance* xvii. 348 The merits of a monadistic as opposed to a holistic interpretation of the 'wisdom of the body'.

monadnock (mǫ̆næ·dnǫk). *Geomorphol.* [The name of a mountain in New Hampshire, U.S.A., having this character.] A hill or mountain of erosion-resistant rock rising above a peneplain.

The place-name appears in Melville's *Moby Dick* (1851), 'his great, Monadnock hump'.

1893 W. M. DAVIS in *Nat. Geogr. Mag.* V. 70 The continuity of the plateau-like uplands [of New England] is interrupted in two ways; isolated mountains rise above it, and branching valleys sink below it. Mount Monadnock is a typical example of the former, with its bold summit more than a thousand feet above the surrounding plateau. .. It is simply an unconsumed remnant of the greater mass of unknown dimensions and form, from which the old lowland was carved... In my teaching, Monadnock has come to be recognized as an example of a distinct group of forms, and its name is used as having a generic value. A long paragraph of explanation is packed away when describing some other mountain as a 'monadnock' of greater or less height. 1900 J. E. MARR *Sci. Study Scenery* 147 The peneplain will be cut up by denudation, which may give rise to new hills, carved out of the plain, and marked at first by the possession of flat tops; to these Davis gives the name 'monadnocks'. 1935 *Geogr. Jrnl.* LXXXVI. 268 Only here and there a monadnock stands out distinctly, as for instance the Umanak Rifkol, which forms a landmark 900 feet high. 1947 AUDEN *Age of Anxiety* (1948) v. 108 O stiffly stand, a staid monadnock, On her peneplain. 1968 R. W. FAIRBRIDGE *Encycl. Geomorphol.* 709/1 Monadnocks may take the form of hills, ridges or ranges.

monadological (mǫ:nădolǫ·dʒikăl), *a.* [ad. G. *monadologisch*, F. *monadologique*: see -AL.] Of or pertaining to monadology. So **monadolo·gically** *adv.*

1895 A. C. ARMSTRONG tr. *Falckenberg's Hist. Mod. Philos.* vii. 278 Leibnitz's..monadological and harmonistic principles. 1897 C. H. JUDD tr. *Wundt's Outl. Psychol.* 313 Matter is thought of as made up of similar [mental] atoms of a lower order (monistic, or monado-logical spiritualism). 1937 *Mind* XLVI. 503 Must we think, monadologically, of as many object worlds as there are knowing minds, though these many object worlds may have common features, in the sense of numerically diverse instances of the same universals? 1938 *Jrnl. Theol. Stud.* XXXIX. 427 It must be a matter for consideration how far it is of any profit to be familiar with (e.g.) Leibniz's monadological religious metaphysic without his doctrine of the judgement. 1943 M. FARBER *Found. Phenomenology* xvi. 533 The objective world is 'innate' in the monadological world. 1957 R. MANHEIM tr. *Cassirer's Philos. Symbolic Forms* III. II. iv. 170 'Monadological' time is the πρότερον τῇ φύσει, and..we can only thence arrive at mathematical-physical time.

Mona Lisa (mōᵘ·nă lī·ză). [It.] The name of a portrait painted by Leonardo da Vinci (1452–1519), used allusively and attrib. of an enigmatic smile or expression such as that of the woman in this painting. Also *fig.* See also *GIOCONDA.

1923 D. H. LAWRENCE *Birds, Beasts & Flowers* 162 She, smiling with goaty munch-mouth, Mona Lisa, arranges it so. 1930 J. COLLIER *His Monkey Wife* xiv. 197 She eyed Emily with what had been called, in her last year at school, her Mona Lisa smile. 1934 T. S. ELIOT *Elizabethan Essays* 60 It [*sc. Hamlet*] is the 'Mona Lisa' of literature. 1950 A. WILSON *Such Darling Dodos* 59 She twisted her face into what she felt to be a more than usually Mona Lisa smile. 1967 J. CREASEY *Famine* xii. 113 She stood in front of him with that half-smile, a Mona Lisa kind of inscrutability. 1975 G. LYALL *Judas Country* xxx. 215 Mitzi was looking at me with a little mousey Mona Lisa smile.

Mona marble (mōᵘ·nă). [f. *Mona*, Roman name for Anglesey + MARBLE *sb.*] A serpentine limestone from the metamorphic beds of the island of Anglesey, off the N. coast of Wales.

1816 *Repository of Arts* (Ackermann) Oct. 243 (*heading*) Mona marble chimney-piece. *Ibid.* 244/2 The beautifully variegated tints of the mona marble. **1911** *Encycl. Brit.* XVII. 677/1 Mona marble is an ophicalcite from the metamorphic series of the Isle of Anglesey. **1971** *Country Life* 6 May 1089/1 'A chaste and elegant structure of Mona Marble', inscribed in Latin and Welsh.

‖ **mon ami** (mon ami). Also fem. **mon amie.** [Fr.] 'My friend', as a term of address.
1786 W. Cowper *Let.* 31 Aug. (1904) III. 93 Adieu, *mon ami*, yours faithfully, W.C. **1856** *Newsp. & Gen. Reader's Pocket Compan.* II. § 1564 All in good time *mon ami*—all in good time. *Ibid.* § 1788 Throughout the journal she is designated emphatically as *mon amie.* **1896** C. M. Yonge *Release* II. 95 'Are you speaking of Beaudésert, *mon amie?*' said M. de Nidemerle. **1911** 'I. Hay' *Safety Match* xv. 236 Flirtation is a crooked business, and you are straight, *mon amie.* **1966** A. Christie *Third Girl* xxi. 216 'I was quite sure she *wouldn't* walk out on me.' 'Ah yes. And then, *mon ami*, she did.' **1973** 'A. Hall' *Tango Briefing* vi. 72 The desert is different from other places, *mon ami.*

monamine. Substitute for def.: = *MONOAMINE. Add further examples.
1943 Sumner & Somers *Chem. & Methods of Enzymes* xv. 269 It is probable that all of these enzymes are one and the same... The most suitable name for the enzyme would be 'monamine oxidase'. **1969** *Times* 21 Feb. 10/7 The patients were receiving anti-depressant drugs known as monamine oxidase inhibitors. **1973** *Sci. Amer.* Sept. 121/1 It was also discovered that iproniazid acted to inhibit the enzyme monamine oxidase. (Monamines, such as serotonin, norepinephrine and dopamine, are believed to act as chemical transmitters between nerve cells in the brain.)

monarch, *sb.*[1] Add: **3.** In full, *monarch butterfly.* Substitute '*Danaus*' for '*Danais*' in def. (Earlier and later examples.)
1890 Hyatt & Arms *Insecta* 186 The monarch or milk-weed butterfly..is abundant during July and August wherever the milk-weed grows. **1945** E. B. Ford *Butterflies* i. 23 The Milkweed, Monarch, or Black-veined Brown Butterfly, *Danaus plexippus*... The first specimen [in Britain] was caught by Mr. J. Stafford at Neath, South Wales, on September 6th, 1876. **1974** A. Dillard *Pilgrim at Tinker Creek* xiv. 252 Each monarch butterfly had a brittle black body and deep orange wings limned and looped in black bands. A monarch at rest looks like a fleck of tiger. **1975** *Country Life* 20 Feb. 450/1 The magnificent black-veined brown or monarch butterfly, that rare migrant to Britain.

monarch, var. *MONIKER.

monarcho- (mǫnā·ɹko), comb. f. Monarchic *a.*, esp. in *monarcho-fascist*, in Communist phraseology, of a fascist form of government with a king as titular head of state, such as that established in Greece after the war of 1939–45.
1947 *Times* 10 May 3/5 Marcos speaks of the 'Monarcho-Fascist régime of the British occupation'... He refers further to the 'Greek Monarcho-Fascist forces' who collaborated with the occupation forces. **1950** R. Leeper *When Greek meets Greek* xii. 195 Our critics on the Left had got so much into the habit of denouncing us as reactionaries (I myself was generally dubbed a monarcho-fascist) that perhaps they had come to believe it. **1966** *Economist* 27 Aug. 809 Aeschylus expounds what would sound today like 'monarcho-fascist' theories. **1971** W. H. McNeill in A. Bullock *20th Century* 51/2 A new guerilla movement..claimed to represent the democratic will of the Greek people against the 'monarcho-fascist' regime.

monarda. Read: (monā·ɹdă). Also 8 **monardus.** [mod.L. (Linnæus *Hortus Cliffortianus* (1737) 11), f. the name of Nicolas Monardes (1493–1588), Spanish physician and botanist, one of the earliest writers on American plants.] An aromatic herb of the North American genus so called, belonging to the family Labiatæ, esp. *Monarda didyma*, which has bright red flowers and is also called bergamot, bee balm, or Oswego tea.
1712 [in Dict.]. **1789** W. Aiton *Hortus Kewensis* I. 36 Scarlet Monarda or Oswego-tea. **1865** [in Dict.]. **1931** M. Grieve *Modern Herbal* I. 95/1 Another American swamp plant...*M[onarda] didyma*, the Scarlet Monarda, is said to yield an oil of similar composition [to that of thymol]. **1970** C. Lloyd *Well-Tempered Garden* i. 45 For plants of an intricately branching habit like *Aster acris*, *Salvia superba*, or monardas, peasticks are ideal [support]. **1974** A. Scott-James *Sissinghurst* xiv. 149 A charming lilac-pink monarda called Beauty of Cobham.

-monas [a. Gr. μονάς: see Monad], an ending of some mod.L. names of genera of simple organisms (chiefly bacteria, protozoa, and algæ): cf. *CHLAMYDOMONAS, *LEPTOMONAS.

Monastral (mǫnæ·străl). The proprietary name of two groups of pigments having exceptional fastness: (*a*) blue and green pigments that are derivatives of phthalocyanine and are widely used esp. in paints, printing inks, and plastics; *spec.* copper phthalocyanine (Monastral Fast Blue B, or BS); (*b*) red and violet pigments that are derivatives of quinacridone.
1936 *Trade Marks Jrnl.* 2 Jan. 2/2 Monastral... All goods included in Class 1, but not including chemical photographic developers and not including any goods of a like kind to chemical photographic developers. British Dyestuffs Corporation Limited, Imperial Chemical House,..London,..manufacturers. **1936** *Discovery* Jan. 2/1 A new blue pigment, which combines fastness to light, heat, acids and alkalis with tinctorial strength, has recently been introduced by Imperial Chemical Industries, Ltd. Known as Monastral Fast Blue BS. **1952** K. Venkataraman *Chem. Synthetic Dyes* II. xxxvii. 1133 There are only three [phthalocyanine derivatives] which are extensively used for coloring printing inks, paints, varnishes, enamels, plastics, and other materials. These are metal-free phthalocyanine (Monastral Fast Blue G, ICI), copper phthalocyanine (Monastral Fast Blue B) and highly chlorinated copper phthalocyanine (Monastral Fast Green G). **1958** *Times* 4 Dec. 3/2 Some series of squares [of mosaic] were made to order in a precise shade of colour indicated by Mr. Piper himself, a rich monastral blue and a delicate pale yellow being examples. **1961** H. F. Payne *Org. Coating Technol.* II. xx. 904 At present a family of three distinctly different pigment colors is commercially available, namely: Monastral Red B: a dark, transparent, blue-shade red toner; Monastral Red Y: a yellow-shade red toner, lighter and more opaque in mass tone than red B; Monastral Violet R: a dark, transparent, red-shade violet toner. **1970** *Sunday Times* (Colour Suppl.) 22 Nov. 69/4 Monastral Blue, an incredibly powerful synthetic pigment of fairly recent origin..is sold under..different names: Windsor Blue, Phthalocyanine Blue.

monaural, *a.* Add: (Further examples.)
1935 *Brit. Jrnl. Psychol.* XXV. 270 Differences in the time of arrival of sounds from a single source at the two ears, or differences in intensity,..are far smaller than the monaural or binaural thresholds for such time or intensity differences. **1955** *Sci. News Let.* 24 Sept. 204/3 Splitting sounds so that one ear hears it first is known as 'monaural delay'. Its use by the Air Force and airlines promises better and clearer communications in the air. **1972** L. A. Jeffress in J. V. Tobias *Found. Mod. Auditory Theory* II. ix. 360 The case where the noise is binaural and in phase at the ears and the signal is monaural..is an interesting one.
2. = *MONOPHONIC *a.* 2.
1931 *Jrnl. Acoustical Soc. Amer.* Apr. 483 The fundamental difficulty is that we are at present restricted to monaural recording. **1948** *Electronics* Aug. 88/2 A person listening to sound through a binaural system has the illusion that the sound originates in the room rather than in the [head]phones. The effect is striking to anyone used to hearing monaural sound from a headset. **1951** *Wireless World* Mar. 85/1 Current radio broadcast and recording techniques are almost exclusively monaural, using a single microphone and transmission channel. **1959** *Sci. News* LIII. 57 A discerning listener could never be deceived into believing that even the best monaural reproduction was an original performance. **1964** P. J. Guy *Disc Recording* x. 148 Two loudspeakers are better than one, even for monaural reproduction. **1968** *Times* 29 Nov. p. iv/7 In July, 1967, E.M.I. ceased to produce new classical recordings in the monaural form. **1970** [see *MONOPHONIC *a.* 2]. **1971** M. McCarthy *Birds of America* 212 Their gramophone and scratched monaural-78 records. **1974** *Encycl. Brit. Macropædia* XVII. 53/2 Stereophonic systems..produce a far greater sense of realism than single-channel (or monaural) systems.

monaurally (mǫnǭ·răli), *adv.* [f. prec. + -LY[2].] **a.** With or to one ear.
1933 *Jrnl. Acoustical Soc. Amer.* IV. 288 The observer is placed in a sound field which is substantially that of a plane progressive wave, facing the source and listening monaurally. **1948** *Psychol. Bull.* XLV. 194 He observed that a sound heard binaurally was always louder than the same sound heard monaurally. **1972** N. I. Durlach in J. V. Tobias *Found. Mod. Auditory Theory* II. x. 385 The tone is presented monaurally.
b. In a monaural (monophonic) manner; = *MONOPHONICALLY *adv.* b.
1958 *Manch. Guardian* 30 June 5/3 Dvořák's Fourth Symphony is first played through one speaker, then monaurally through both, and finally stereophonically through both. **1964** P. J. Guy *Disc Recording* x. 147 It was the original aim of the record manufacturers to make stereophonic records playable monaurally with an ordinary pickup.

monbar, var. *MIMBAR.

Monbazillac (‖ moṅbaziyak, mǫnbæ·zilæk). Also **Mont Bazillac, Montbazillac.** [Name of a village south of Bergerac in the Dordogne department of S.W. France.] A sweet, white dessert wine, similar to Sauternes.
[**1833** C. Redding *Mod. Wines* v. 158 The best wines are produced on the hills, upon the left bank of the Dordogne, in the communes of St. Laurent and Monbazillac.] *Ibid.* 404/2 (*index of wines*) Mont Basillac, 158. **1926** P. M. Shand *Bk. Wine* v. 66 Montbazillac..is a label still met, which prepares one for a rich, sweet white wine that is made from the Sémillon grape. **1935** C. W. Berry *In Search of Wine* viii. 186 We then had a bottle of 1918 Monbazillac—a trifle sweet for me. **1955** J. Thomas *No Banners* v. 39 A bottle of Montbazillac stood on the little table between their chairs. **1967** A. Lichine *Encycl. Wines* 131/2 With one exception—sweet, white Monbazillac—the wines of Bergerac are simply not in the same class [as those of Bordeaux]. **1975** *Times* 17 May 11 Montbazillac is made with grapes that have been subjected to the action of *botrytis cinerea* or 'noble rot'.

‖ **mon cher** (moṅ ʃɛ̄r). [Fr.] 'My dear fellow', as a term of address.
1680 Rochester *Poems* 24 But now Moncher, dear Pug, says she, adieu. **1888** 'Rita' *Vivienne* I. ii. 23 Good bye, *mon cher*, and don't overwork yourself. **1920** F. M. Ford *Let.* 30 May (1965) 100 You see, mon cher, there are plenty of other people..with the temperament of poets. **1975** A. Christie *Poirot's Last Case* xv. 162 So far, *mon cher*, this X has operated with so much ability that he has defeated me, Hercule Poirot!

monchiquite (mǫntʃī·kwəit). *Petrogr.* [a. G. *monchiquit* (Hunter & Rosenbusch 1890, in *Tschermak's mineral. u. petrograph. Mittheilungen* XI. 447), f. Serra de *Monchique*, name of a mountain range in southern Portugal where such a rock is found: see -ITE[1].] A lamprophyre containing small phenocrysts of olivine and augite, and usu. also biotite or an amphibole, in a glassy ground-mass consisting of or containing analcite.
1891 *Ann. Rep. Arkansas Geol. Survey 1890* II. iv. 110 The Brazilian dikes..form an olivine-free class of the rock that Rosenbusch calls monchiquite. **1896** *Jrnl. Geol.* IV. 679 (*heading*) On the monchiquites or analcite group of igneous rocks. **1961** [see *CAMPTONITE]. **1965** [see *FERROAN *a.*].

Mond (mǫnd). The name of Ludwig *Mond* (1839–1909), German-born British chemist, used *attrib.* to designate processes devised by him and the plant and products of these processes, esp. (*a*) a method of manufacturing producer gas by passing air and an excess of steam into a gas producer so that most of the nitrogen of the coal is converted into ammonia and can be recovered; (*b*) a process for obtaining high-purity nickel by decomposing by heat the volatile nickel carbonyl, which is obtained from crude nickel oxide by reduction with hydrogen followed by reaction with carbon monoxide.
1887 *Encycl. Brit.* XXII. 244/1 The Mond process, of all the many sulphur-recovery processes yet introduced, is the best. **1903** *Jrnl. Franklin Inst.* CLV. 390 The success of the Mond nickel process, as carried on in England, has been seriously hindered by the discovery that the process is dangerous to the health of the men employed. **1907** E. L. Rhead *Metallurgy* (new ed.) vi. 76 Mond gas is generated by passing into the producer air containing the maximum amount of water vapour that the heat produced is capable of decomposing. **1936** Bone & Himus *Coal* xxiv. 424 The trades using Mond Gas are diverse, including..manufacture of motor cars and component parts, foundries.., chemical works, [etc.]. **1946** J. R. Partington *Gen. & Inorg. Chem.* xvii. 485 Mond gas is formed with a large excess of steam which keeps the temperature low. **1947** G. W. Himus *Elements Fuel Technol.* xv. 324 A number of Mond plants were installed and found to be economic. **1969** H. T. Evans tr. *Hägg's Gen. & Inorg. Chem.* xxxiv. 727 In the so-called Mond process a very fine powder is desired, and the reduction is done with water gas at 350–400°C.

‖ **mondaine** (moṅdɛ̄n), *sb.* (and *a.*). [F. *mondaine*, f. *mondain* worldly, ad. L. *mundānus*, f. *mundus* world: cf. Mundane *a.* (*sb.*)] A woman belonging to fashionable society. Also **mondain** (moṅdæ̃n), a man belonging to fashionable society.
1888 Mrs. H. Ward *R. Elsmere* II. iv. xxvi. 285 He was in a mood to be impatient with a *mondaine's* languid inquiries into clerical work. **1889** E. Dowson *Let.* 24 Mar. (1967) 55 Here you have the refinement of the mondaine with the independence of the cocotte. **1902** [see *ETIQUETTE 4]. **1906** B. von Hutten *What became of Pam* xii. 192 He was a dandy and a *mondain*, who..had never done anything noteworthy. **1908** *Westm. Gaz.* 3 Oct. 13/2 The bold attempt of the milliners to coax mondaines to the smaller hat will be watched with interest. **1912** *Tatler* 23 Oct. p. vi (Advt.), Not only have the requirements of the smart mondaine been carefully considered but [etc.]. **1920** A. Huxley *Let.* 4 May (1969) 185 Cocteau is a man of fabulous cleverness, but not serious—He is just a mondain. **1920** *Q. Rev.* July 38 The provocative elegance of the Flavian *mondaines*. **1924** *Public Opinion* 7 Nov. 458/2 She was in point of fact a most accomplished mondaine. **1970** G. Greer *Female Eunuch* 266 Flatly contemptuous words like *kept-woman* and *call-girl* have taken over..from..*courtesan, mondaine.*
b. *adj.* Attached to things of the world; worldly.
1889 E. Dowson *Let.* c 11 Jan. (1967) 23, I will free myself from the intolerable *corvée* of the *mondain* dinner. **1896** C. M. Yonge *Release* II. viii. 144 Ah! you are *mondaine*, you are Protestant, madame. **1927** H. Nicolson *Some People* 88 When I got to bed I realised that by 'worldly' he had meant 'mondaine'. **1929** L. Rea *Six Mrs. Greenes* II. 58 Lavinia, mondaine, vivid, with a delicate certainty of touch that enabled her [etc.]. **1962** *Listener* 24 May 920/3 His pleasure in the company of children and simple people—as opposed to *mondain* people.

mondanity, var. Mundanity.
1911 Mrs. H. Ward *Case of Richard Meynell* xii. 242 With her grey hair, and her plain widow's dress, she threw her sister's charming mondanity into bright relief.

Monday. Add: **3. Monday Club,** a right-wing Conservative club (cf. Club *sb.* 13) that originally held its meetings on Mondays; so **Monday-clubber,** a member of this club; **Monday-morning** *attrib.*, suggestive of lethargy

or disinclination after a busy or eventful weekend; *Monday pop* (earlier example).
1962 *Evening Standard* 12 Feb. 6 (*caption to cartoon*), Lord Salisbury has become patron of the 'Monday Club', a new group of Young Conservatives formed to 'keep the Conservative Party Conservative'. **1966** *Observer* 16 Oct. 12/5 To dismiss them simply as acts of political vengeance or as another example of 'African savagery' is Monday Club mumbo-jumbo. **1970** G. K. Roberts *Political Parties & Pressure Groups in Britain* [illegible] Recent examples in Britain are: the Monday Club, which tries to maintain what it chooses to regard as [illegible]. **1972** P. Copping *Story of Monday Club* 5 Rebelling against Macmillan's 'Winds of Change' speech a tiny group of young conservative dissentients formed the Monday Club on 1st January 1961. **1973** *Times* 20 June [illegible] The executive of the Monday Club, the right-wing Conservative pressure group, decided last night..to expel the Essex branch of the club for its associations with the National Front. **1973** *Guardian* 27 June 1/3 Mr Enoch Powell..banned reporters from a Monday Club meeting last night. **1972** *Times* 21 July 12/1 His successful conservative opponent, the Monday Clubber, Harold Soref. **1921** D. H. Lawrence *Sea & Sardinia* 261 The rather inert, narrow, Monday-morning market. **1938** R. G. Collingwood *Princ. Art* v. 95, I get back to my book with that Monday-morning feeling. **1862** Monday pop [see Pop *sb.*[4]].

mondial, *a.* Restrict † *Obs.* to sense in Dict. and add: **2.** [ad. mod. F. *mondial.*] Pertaining to, affecting, or involving the whole world; world-wide, universal. Hence **mondialization.**
[illegible]

‖ **mon Dieu** (moṅ djø). [Fr.] = *My God!* (cf. God 7).
1768 Sterne *Sent. Journey* I. 57 The poor monk blush'd..'*Mon Dieu!*' said he. **1849** Thackeray *Pendennis* I. viii. 80 'You may come and live down there, *mon Dieu!* for ever' (said the Major, with a dreary sigh). [illegible]

Mondism (mǫ·ndɪz'm). [f. the name of Alfred Mond (1868–1930), British politician.] The views of A. Mond, especially his advocacy of co-ordination and co-operation as bases of industrial enterprise.
1927 F. W. Ford *Let.* 5 Jan. (1965) 82 What is Ministry! Everything one has fought against one's life. Northcliffism, ..Mondism, Ballidism! **1928** *Daily Tel.* 20 July 10 After further disorder Mr. Cook was ordered out of the room. He attacked 'Mondism', which he declared was the infant of Fascism and should be killed at birth. **1962** T. L. Johnston *Collective Bargaining in Sweden* vii. 169 This labour peace [illegible] the Swedish attempt at Mondism.

‖ **mondo** (mǫ·ndo). [Jap., f. *mon* asking + *do* answering.] An instructional technique of Zen Buddhism consisting of rapid dialogue of questions and answers between master and pupil.
1927 D. T. Suzuki *Essays in Zen Buddhism* I. vi. 256 The following 'mondo' or dialogue (literally questioning and answering) will give us a glimpse into the ways of Zen. **1960** Koestler *Lotus & Robot* x. 233 In its mondos and koans, Japanese ambiguity reaches its metaphysical peak. **1961** *Times Lit. Suppl.* 17 Feb. 8/3 Perhaps the most significant part of its [*sc.* a book's] contribution consists in its argument..that the zen *mondo* are full of direct, coherent, systematic thinking [illegible].

Mondrian (mǫ·ndriăn), *a.* [f. the name of P. C. ('Piet') Mondrian, *Mondriaan* (1872–1944), Dutch painter.] Resembling the geometrical abstract style of Mondrian.
1958 *Listener* 27 Nov. 888/1 A Mondrian scheme of simple primary colors. **1963** *Times* 30 Apr. 15/5 The Mondriaan-like luminous coloured panels.

Monégasque (monegask), *sb.* and *a.* Also **Monegasque.** [Fr.] **A.** *sb.* A native or inhabitant of Monaco. **B.** *adj.* Belonging to or characteristic of Monaco or its inhabitants.
1882 T. H. Pickering *Monaco* i. 11 That which may be, and is, done in Riviera towns, is impossible in the Monégasque metropolis on the Mediterranean. *Ibid.* vii. 96 Three harvests of four different products make up a total, according to Monégasque arithmetic, of twelve crops. *Ibid.* iii. 59 The Monégasques enjoy the privilege of being governed—and well governed—without being called upon to contribute in any way towards the expenses. [illegible] **1962** *Listener* 7 May 867/1 The population [of Monaco] is 23,000, but only 3,000 of them are real Monegasques. **1966** G. Greene *Comedians* III. i. 64, I am uncertain whether she was French—perhaps she was even Monégasque. [illegible] **1968** *Guardian* 5 Jan. 6/6 Installed as manager is Georges Penna, a Monegasque of Italian parentage. **1973** *Daily Tel.* 19 Mar. 3/1 In 1956 there was great rejoicing when Prince Rainier banned them from Monaco after Lady Docker tore up a paper Monégasque flag.

monekeer, monekur, varr. *MONIKER.

Monel (mǭ·nĕl). Also **monel.** [(Altered form of) the name of Ambrose *Monell* (d. 1921), president of the International Nickel Company when that firm introduced the alloys.] Used as a proprietary name (usu. *attrib.*) to denote alloys composed of about 68 per cent nickel and 30 per cent copper with small amounts of other elements, which have a high tensile strength and good resistance to corrosion, particularly towards sea-water.
1906 *Trade Marks Jrnl.* 17 Feb. 286 Monel Metal... Metal alloy. International Nickel Company,..Manhattan, New York, U.S.A. **1928** *Chambers's Jrnl.* Mar. 207/1 These discs are made of monel metal. [illegible] **1970** *Motor Boat & Yachting* 16 Oct. 49/2 The original stainless fastenings have been superseded by monel fastenings. **1970** W. J. Patton *Mod. Manufacturing Processes* [illegible] **1977** T. P. McMahon *Issue of Bishop's Blood* (1972) xiii. 193, I was shuddering at what Meredith was going to charge. He's got a thing about always using Monel metal on what he calls 'furnishings'.

monellin (mǫne·lɪn). *Chem.* [f. the name of the *Monell* Chemical Senses Center, in Philadelphia, U.S.A., where it was first isolated + -IN[1].] A protein with a sweet taste isolated from the berries of the tropical plant *Dioscoreophyllum cumminsii.*
1972 Morris & Cagan in *Biochim. & Biophys. Acta* CCLXI. 114 We propose the name 'monellin' for the sweet material we have isolated. **1973** *Daily Colonist* (Victoria, B.C.) 24 Feb. 25/6 [Scientists] had isolated a new, low-calorie sweetener said to be 3,000 times more intense by weight than sugar. Named 'Monellin' the product is derived from a wild red berry that grows abundantly in tropical West Africa. **1973** *Nature* 27 July 196/3 Both thaumatin and monellin exhibit sweetness to a prodigious degree.

moneme (mǫ·nīm). *Linguistics.* [a. F. *monème*, f. Mon(o- + *-EME.] In the terminology of certain linguists, the smallest meaningful unit of language: = *MORPHEME.
[illegible]

monenergist, monenergistic, *a.*: the more correct forms of Monergist, Monergistic *a.* (see s.v. Monergism).
1915 *Encycl. Relig. & Ethics* VIII. 821/2 The Monenergistic or monothelete controversy seems at first glance to be a mere sequel to the Monophysite conflict. *Ibid.* 822/1 The Monenergists were possessed with the idea that the redemptive activity of the God-man emanated wholly and solely from His divine nature.

monetarism (mǫ·n-, mɐ·nĭtăriz'm). [f. Monetary *a.* + -ISM.] The economic doctrine or theory of a monetarist or of monetarists.
1969 *Newsweek* 6 Jan. 48 The combination of Stansian horse-and-buggy finance with Friedmanian go-go monetarism. **1970** *Times* 12 Nov. 17 The lecture was a full-blooded onslaught on 'the new monetarism', the doctrines of the Chicago school of economists led by the celebrated Professor Milton Friedman. **1971** *Times* 2 Aug. 13/4 Professor Paul Samuelson..describes monetarism as a disease and defines it as a 'pathological belief..that only the rate of growth of the money supply can affect significantly the state of the economy and its level of unemployment'. **1971** *Sunday Times* 8 Aug. 4 Monetarism, the belief that the state of the economy can be decisively manipulated through regulating the flow of money, became an accepted cult in the White House after Dr Burns had left it to assume the chairmanship of the Federal Reserve Board. **1975** *Times* 18 Aug. 14/5 The Treasury['s]..distrust of 'monetarism' arises from a belief that it has only one equation, that which links the money supply and the money national income.

monetarist (mǫ·n-, mɐ·nĭtărist), *a.* and *sb.* [f. Monetary *a.* + -IST.] **A.** *adj.* Of a monetary character or having a monetary basis.
1914 Geddes & Thomson *Sex* x. 239 This order of things—avowedly mechanical, industrial, monetarist at best,..seems to many of us..[illegible] form of industrial civilisation. **1971** *Times* 2 Aug. 13/4 The father of the current monetarist school of economic thought. **1972** *Times* 6 Dec. 22/8 The distinctive monetarist assertion is that inflation is attributable solely to increase in the money supply. **1973** *Times* 16 Aug. 15/2 It is not..true—even in terms of the monetarist theories on which Mr Powell relies—that the Government deficit uniquely determines the money supply. **1974** *Financial Times* 21 Oct. 16/8 The basic argument of the sadly mislabelled 'monetarist' school is that attempts by monetary and fiscal stimulation to lower the level of unemployment below a sustainable minimum, determined by underlying forces, will not bring about a lasting drop in the level of unemployment.
B. *sb.* One who places emphasis on monetary matters, *spec.* one who advocates tighter control of the money supply as an important remedy for inflation.
1963 *Economist* 27 Apr. 299/2 To control inflation by curtailment, as prescribed by the 'monetarists'. **1965** *Ibid.* 25 Sept. p. xxiv/1 The main battle was between the 'monetarists' (who..believe that screwing down the money supply is always the answer to inflations of any sort) and the 'structuralists'. **1971** *Daily Tel.* 29 Jan. 12/2 It is argued by monetarists that lower rates of interest would contribute to inflation. **1971** *Times* 2 Aug. 13/4 In a highly controversial article in today's *Washington Post* influential monetarists in the United States are compared to water diviners.
Hence **monetari·stic** *a.*
1972 *Publishers' Weekly* 6 Mar. 60/1 Nixon's two-and-a-half-year dalliance with laissez-faire and the monetaristic theories of Milton Friedman.

monetary, *a.* **2.** (Later examples.)
1936 J. M. Keynes *Gen. Theory Employment* xv. 203 A monetary policy which strikes public opinion as being experimental in character or easily liable to change may fail in its objective. **1944**, etc. [see *International Monetary Fund* s.v. *INTERNATIONAL *a.* 2]. **1951** R. Firth *Elements of Social Organization* iv. 142 Even where monetary rewards for labour are largely current, he has noted that work may be undertaken for other than money symbols. **1961** *Ann. Reg. 1960* 475 The measures ranged from the traditional restraints to a new and previously untried monetary instrument. **1973** *Times* 6 July 21/7 The 'Smithsonian' agreement, which President Nixon characterized..as 'the most significant monetary agreement in the history of the world'.

monetite (mǫ·nĕtəit). *Min.* [f. *Moneta*, name of a small island near Puerto Rico: see -ITE[1].] A hydrogen phosphate of calcium, $CaH(PO_4)$, occurring as translucent, pale yellowish white crystals.
1882 C. U. Shepard in *Amer. Jrnl. Sci.* CXXIII. 400 The Monetite mineral, which we call monetite from its locality, is accompanied by two other species. **1968** I. Kostov *Mineral.* 458 [illegible]

money, *sb.* Add: **3. c.** Wages, salary; one's pay.
1887 Parish & Shaw *Dict. Kentish Dial.* 103 He's getting good money, I reckon. **1916** G. B. Shaw *Pygmalion* II. 143 His proper trade's a navvy; and he works at it sometimes..and earns good money at it. **1920** E. O'Neill *Beyond the Horizon* II. i. 85 If that's the case, you can go to the devil... You'll get your money tomorrow when I get back from town. **1963** H. Garner in R. Weaver *Canad. Short Stories* (1968) 2nd Ser. 45 Nobody really liked working for Malloy-Harrison, but the money was better than most places.
5. big money: see *BIG *a.* B. 2; **dirty money:** see *DIRTY *a.* 6 b.
6. a. (Further examples.)
1611 Bible *1 Tim.* vi. 10 The loue of money is the root of all euill. c **1645** [see Fool *sb.*[1] 1 d]. **1661** A. Behn *Rover* II. iii. 1 43 Money speaks sense in a Language all Nations understand. **1748** B. Franklin *Advice to Young Tradesman* in *Writings* (1905) II. 371 Remember that *time* is money. **1845** C. Lever *Let.* in L. Stevenson *Dr. Quicksilver* (1939) ix. 149 You have paid your money, and you may take your choice. **1846** *Punch* 3 Jan. 17 (*caption*) You pays your money, and you takes your choice. **1853** T. T. Lynch *Lect. Self Improvement* v. 113 Money is power—power for bread and power for tinsel. **1898** G. B. Shaw *Arms & Man* II. iii. 57 A twenty leva bill! Sergius gave me that, out of pure swagger. A fool and his money are soon parted. **1903** *Sat. Even. Post* 5 Sept. 12/1 When

money talks it often merely remarks 'Good-by'. **1905** 'O. Henry' in *N.Y. World Mag.* 12 Nov. 8/1 Money talks. But you may think that the conversation of a little old ten-dollar bill in New York would be nothing more than a whisper. **1927** E. O'Neill *Marco Millions* I. ii. 36 He'll have time enough for that, but with us time is money. *Ibid.* iii. 44 Money isn't everything, not always. **1930** G. B. Means *Strange Death of President Harding* iv. 72 One can do nothing—be nothing, without money, not even in the White House. Money is power. **1952** W. G. Hardy *Unfulfilled* 199 Money isn't everything. **1956** A. Wilson *Anglo-Saxon Attitudes* II. ii. 277 Yeah, he's on the Market... You know the sort of stuff. Money talks and so on. **1965** *Times* 14 July 8/4 In the Government today were known supporters of C.N.D. Why? Because the Prime Minister put them there. Why? Because he shared their views? Because it was expedient to do so? Because he was practising some duplicity? Why? 'You pays your money and you takes your choice.'
b. *for my money* (earlier and later examples); also, in my opinion; *to have a run for one's money*: see Run *sb.*[1] 1 d.
1549 [see Go *v.* 24 b]. **1566** E. H. tr. *Erasmus's Diuersoria* sig. B2ᵛ This behauiour doth well beseme Frenchmen peraduenture, how be it the fashions of Duche lande shall go for my monye when all is done. **1589** [see Go *v.* 24 b]. **1840** *Spirit of Times* 21 Nov. 447/3 Give me the gall, I say, that *has an eye for dirt*, for she is the gall for my money. **1932** D. L. Sayers *Have his Carcase* xi. 138 Peter's the man for my money. He won't see a hardworking man lose a job for want of a good news story. **1943** *N.Y. Times* 9 May II. 5/5 Glenn was, and for my money is still tops. **1954** J. Symons *Narrowing Circle* xiv. 59 For my money, Marian was responsible..for the trouble we'd had. **1969** D. Clark *Death after Evensong* iv. 97, I wouldn't mind not finding who did Parseloe in. For my money he deserved it.
c. *to cost money*: see *COST *v.* 1 d.
e. *money burns* (*a hole*) (*in*) *one's pocket* (or † *purse*) (and similar phrases): one is impatient to spend one's money. Cf. Burn *v.*[1] 16 in Dict. and Suppl.
1529 More *Dyaloge of Ymagys* II. x. f.lxi, Hauyng a lytell wanton money whyche hym thought brennyd out the botom of hys purs. **1702** Farquhar *Inconstant* v. iii. 77 My time lyes heavy on my hands, and my Money burns in my Pocket. **1875** S. Smiles *Thrift* viii. 125 A man who has more money about him than he requires..is tempted to spend it... It is apt to 'burn a hole in his pocket'. **1943** M. Lasswell *Suds in your Eye* xiv. 103 Her money was burning a hole in her pocket. **1958** L. Durrell *Balthazar* xiii. 227 I've scraped a dowry together over the years... The money burns my pocket. **1972** A. S. Neill *Neill! Neill! Orange Peel!* II. 238 Today, I don't see the young..saving money. It burns a hole in their pockets.
f. *your money or your life*: a formula used by highwaymen, etc., in threatening to kill a person if he does not hand over money.
1841 F. A. Burney *Jrnl.* 23 Feb. (1926) 321 Mr. Dixon attempted expostulation, upon which the Highwayman drew out a Pistol,..exclaiming, with an oath, 'Your money or your life!' **1848** J. A. Froude *Let.* 16 May in W. H. Dunn *Froude* (1961) I. 118 Nothing will open rich John Bull's understanding but a hand at his throat and 'Your money or your life'. **1864** J. Payn *Lost Sir Massingberd* II. xiii. 212 A pistol, was protruded into the carriage. 'Your money or your life!..,' said a rough voice.
g. *in the money*: among the prize-winners in a competition, show, or the like; amply or sufficiently supplied with money; rich.
1902 'D. Dix' *Fables of Elite* 48 It is True that when the Spurt is over I am generally in the Money. **1928** *Morning Post* 20 Oct. 6/1 One of them is to-day a full champion, the other three all winners, and 'in the money', as the fanciers say, whenever shown. **1945** G. Casey in *Coast to Coast 1944* 3 Shift her round like you was doin' yesterday and you'll be in the money. **1946** L. Bromfield *World we live In* 325 Being in the money at the moment, I said that of course I'd lend her any reasonable amount. **1969** T. Parker *Twisting Lane* 200 She said we could stay there rent free until I was in the money again.
h. *money for jam, money for old rope* (and similar phrases): a profitable return for little or no trouble; a very easy job; someone or something easy to profit from, beat, etc.
1919 *Athenæum* 8 Aug. 727/2 The great use of jam in the Army..originated a number of phrases, such as 'money for jam' (money for nothing). **1927** T. E. Lawrence *Let.* 22 Sept. (1938) 540 Recently I made nearly ten pounds out of reviewing eleven books. Money for jam, as the airman says. **1936** J. Curtis *Gilt Kid* xiii. 134 He would spin her a fanny about the marriage laws, tie the poor kid up. It ought to be money for old rope. **1942** E. Waugh *Put out More Flags* 150 At the moment there were no mortars and he was given instead a light and easily manageable counterfeit of wood which was slung on the back of his haversack, relieving him of a rifle. At present it was money for old rope. **1958** 'A. Gilbert' *Death against Clock* iv. 47 If he saw the wallet it must have seemed money for jam. **1966** 'L. Lane' *ABZ of Scouse* 70 *Money fer owd rope*, something for nothing. Other similar phrases are *money fer dirt*; *money fer raggety kecks*; *money fer jam*, etc., etc. **1973** A. Hunter *Gently French* iii. 24 Wasn't no risk, it was money for jam. **1974** N. Bentley *Inside Information* v. 52 I'll advance you another two hundred. Christ, that's money for old rope.
i. *not everybody's money*: not to everybody's liking.
1923 J. Manchon *Le Slang* 196 *You ain't everybody's money*, vous ne pouvez pas plaire à tout le monde. **1933** 'G. Orwell' *Let.* c 10 Dec. in *Coll. Ess.* (1968) I. 128 As to the actual writing in *Ulysses*, it isn't everybody's money, but personally I think it is superb in places.
j. *to put one's money on*: to bet on (a horse, etc.); also *fig.*, to favour or depend on; to expect the success of.
1931 T. R. G. Lyell *Slang, Phrase & Idiom* 528 If you've got any sense, you'll put your money on that horse I told you of. **1963** *Listener* 21 Feb. 341/3 She does not put all her money on love. **1969** *Ibid.* 8 May 636/1 A century hence, prophesied one critic, it would be only 'the careless glance of curiosity, or the student's all-ranging eye' that would turn upon the Little Nells and Paul Dombeys; he put his money instead on Dickens's humour.
k. *to put* (or *get*) *one's money where one's mouth is*: to produce, bet, or pay out money to support one's statements or opinions. orig. *N. Amer.*
1942 Z. N. Hurston in A. Dundes *Mother Wit* (1973) 224/1 'Put your money where your mouth is!' he challenged. **1951** *Amer. Speech* XXVI. 99/1 *Get your money where your mouth is*, a phrase [in poker] which means, 'put up or shut up'. **1970** *Globe & Mail* (Toronto) 26 Sept. 7/3 Eventually it got to the point when he suggested that maybe I was the guy who should take it on. Sort of put your money where your mouth is. **1975** A. Price *Our Man in Camelot* v. 95 The squadron betting book the barman keeps..for guys who are ready to put their money where their mouth is.
l. *to have money to burn*: see *BURN *v.*[1] 8 d.
7. a. *money-economy, -fear, -flow, -hunger, -lust, -purse, -rent, -sense, -slave, -supply, -system, -till.*
1942 L. B. Namier *Conflicts* 50 It is not easy to translate into exact figures this barter business, which is..contrasted with the money-economy and transactions of the Western Powers. **1962** H. R. Loyn *Anglo-Saxon Eng.* iv. 159 The earliest law-codes give evidence..of the importance of a money-economy. **1927** Money-fear [see *money-lust* below]. **1953** C. F. Hockett in Saporta & Bastian *Psycholinguistics* (1961) 64/1 Money-flow (at least in one direction) is *income.* **1891** Stevenson & Osbourne *Wrecker* (1892) vii. 120 Far from the money-hunger of the West. **1965** *Times Lit. Suppl.* 25 Nov. 1047 (Advt.), A young manhood of scheming and money-hunger in Chicago. **1927** D. H. Lawrence *Let.* 18 Dec. (1962) II. 1027 They must first overthrow in themselves the money-fear and money-lust. **1930** E. Pound *XXX Cantos* xiv. 61 The perverts, who have set money-lust Before the pleasures of the senses. c **1821** J. W. Masters in *Eng. Dial. Dict.* (1903) IV. 149/2 He brought our Jack a leather cap An' Sal a money-puss. **1878** B. F. Taylor *Between Gates* 273 We stood under fig-trees hung with money-purses filled with seeds. **1966** A. R. Scammell *My Newfoundland* 33, I kept mine [*sc.* a dollar bill] for weeks in my little money-purse (we never called them purses). **1792** A. Young *Trav. France* I. iv. 340 Much the greater part of the lands of France are not let at a money-rent, but at one-half or one-third produce. **1848** Mill *Pol. Econ.* I. II. viii. 366 An attempt to introduce..a system of money rents and capitalist farmers. **1865** D. G. Rossetti *Let.* 14 Sept. (1965) II. 571 This might have been..executed..more profitably in a money-sense than what I did do. **1963** *Times Lit. Suppl.* 25 Jan. 62/3 Their hair-dressing or their money-sense. **1929** D. H. Lawrence *Pansies* 116 He can't help being a slave, a wage-slave, A money-slave. **1878** F. A. Walker *Money* iv. 76 (*heading*) The importance of the money supply. **1975** *Times* 18 June 29/7 For this year a 15 per cent rise in the money supply and a 12½ per cent rise in gnp..would be the right interim targets. **1929** D. H. Lawrence *Pansies* 109 Why don't we do something about the money system? **1857** *Quinland* I. II. 289 If the stars were extinguished, it would not disturb him, unless his money-till were upset. **1937** C. M. Arensberg *Irish Countryman* 175 His or her remittances would eventually reach his money-till.
b. *money-grabber, -loser, -raiser, -spender; money-changing, -earning, -grabbing, -losing, -raising, -spending, -sucking* vbl. sbs. and ppl. adjs.; *money-conscious, -directed* adjs.; **c.** *money-mad* (further examples); *money-minded* adj.
1938 G. Greene *Nineteen Stories* (1947) 76 Mr. Calloway sat on his usual seat staring out over the money-changing booths at the United States. **1933** Money-conscious [see *CONSCIOUS *a.* 12]. **1963** *Times* 1 May 15/4 The parties being money-conscious to the highest degree. **1970** T. Hilton *Pre-Raphaelites* vi. 171 The destructiveness of capitalist society, its callous and money-directed disregard of culture. **1912** J. London *Let.* 7 Sept. (1966) 364 My long stuff is pretty good at money-earning. **1903** *Eng. Dial. Dict.* s.v., He's a regular money grabber. **1933** *Times Lit. Suppl.* 27 Apr. 283/3 A money-grabber, notorious in a money-grabbing age. **1920** D. H. Lawrence *Touch & Go* 9 We say it is a mere material struggle, a money-grabbing affair. **1928** *Weekly Dispatch* 6 May 15 About 30 [musical comedies] prove to be money-losers. **1963** E. Humphreys *Gift* II. iv. 239 If a reputation for being a money-loser had reached as far as Barrot's ears, there was very little hope of him ever making a film again. **1870** J. K. Medbery *Men & Myst. Wall St.* 200 It is the greatest money-making and money-losing spot on the globe. **1960** *Farmer & Stockbreeder* 16 Feb. 87/2 All your life you have heard that farming is a money-losing proposition. **1929** D. H. Lawrence *Pansies* 79 Fear of my money-mad fellow-men. **1965** G. Jackson *Let.* 16 Mar. in *Soledad Brother* (1971) 69 The shocks and strains of this money-mad society are enough to ruin the purest of minds. **1974** A. Ross *Bradford Business* 169 Money-mad property developers. **1588** W. Kempe *Educ. Children* sig. C3 One of these money minded parents. **1957** *Times Lit. Suppl.* 20 Dec. 769/4 The insistence of the more money-minded directors on its premature exploration. **1909** *Westm. Gaz.* 11 Aug. 1/3 This remarkable man began his career as a money-raiser fifteen years ago with an £8,000 collection. **1955** W. Dean in H. Van Thal *Fanfare for E. Newman* v. 61 *Don Giovanni* in Trianon's version and Nicolai's *Merry Wives of Windsor* had to be put on as money-raisers. **1960** *Farmer & Stockbreeder* 29 Mar. 105/3 The huge success of money-raising efforts. **1920** *Edin. Rev.* July 163 Meleager was always a money spender rather than a money maker. **1900** J. London *Let.* 1 Mar. (1966) 97 The habit of money spending. **1911** H. Granville Barker *Madras House* III. 91 The Middle Class Women of England form one of the greatest Money Spending Machines the world has ever seen. **1921** Galsworthy *To Let* II. v. 162 A lot of slow-fly money-sucking officials.
8. money-back *a.*, designating a system, agreement, etc., whereby a customer will be refunded the money he pays, if he is not satisfied with the goods or service provided; **money-belt** orig. *U.S.*, a belt designed for carrying money; **money bug** *U.S. slang*, a person having great wealth or financial power; **money centre** *U.S.*, a place of pre-eminent importance in the financial affairs of a region or country; *spec.* New York; **money crop** *U.S.*, a crop that is grown mainly for selling and not for the grower's consumption; = *cash-crop*; **money-gold** *rare*, gold coin; **money illusion** (orig. *U.S.*), the illusion that money has a fixed value in terms of its purchasing power; **money king** *U.S.*, a magnate in finance; a person of great wealth; **money-man,** delete † and add later examples; **money market** (earlier and later examples); (*b*) a place in which the financial activity of a region is centred; **money-order** (further examples); **money-player,** (*a*) *U.S.*, a type of gambler (see quot. 1935); (*b*) a professional, as opposed to an amateur; **money-power,** (*a*) the power to coin money, regulate its use, etc.; (*b*) the power exercised by money or by wealthy people, firms, etc.; **money-shark** orig. *U.S.*, an avaricious money-dealer; **money shop,** a shop where money can be obtained; *spec.* an establishment which performs more conveniently many of the functions of a bank, and specializes in arranging loans; **money-spinner,** (*b*) (earlier and later examples); also, a person who, or thing which, makes a lot of money; something that is very profitable; hence **money-spinning** *vbl. sb.* and *ppl. a.*
1922 *Weekly Dispatch* 12 Nov. 4 (Advt.), All our business is conducted on the 'Moneyback' principle; that is to say, if you are not perfectly satisfied with your purchase return it to us within seven days and we will refund your money in full by return of post. **1955** *Radio Times* 22 Apr. 2/3 Sisco paints are sold on a 'money back' guarantee. **1972** *Farm & Country* 19 Dec. 13/3 We are certain that you will be satisfied and offer a money back guarantee. **1846** *St. Louis* (Missouri) *Reveille* 9 Sept. 3/2 The stock consists, in part, of Shirts, Collars,..Money Belts. **1923** *Outward Bound* Mar. 408/2 Among cowboys..one might..leave one's money-belt full of gold and notes beside the fire. **1958** P. Kemp *No Colours or Crest* viii. 174, I had a hundred gold sovereigns in my money belt. **1898** *People* 20 Mar. 4/4 The happiness or the misery of 3 millions of people wholly dependent on the whims and caprices of, say, half a dozen 'money bugs', as they are called in the States. **1922** *Public Opinion* 11 Aug. 132/2 The profiteering class, money bugs as the Americans call them. **1838** D. D. Barnard *Speeches & Rep.* 36 Composed of twenty-six local sovereignties, of all which New-York is the money centre, as London is the money centre of half the world. **1900** *Congress. Rec.* 17 Feb. 1897/2 Gilt-edged paper can be placed in the money-centers at a small per cent. **1881** *Harper's Mag.* Oct. 723/1 Cotton is the money crop. **1904** T. Watson *Bethany* 5 They never failed to make it their object to produce on the farm the necessary supplies, tobacco or cotton being merely the surplus crop, the money crop. **1974** S. Marcus *Minding the Store* (1975) xii. 243 Until about 1932, Texas had essentially an agriculturally based economy. By 1905 cotton was the money crop. **1841** *N.Z. Jrnl.* No. 32. 92 Natives talk about money-gold. **1842** *Lett. Settlers in Wellington, Nelson & New Plymouth* (1843) 137 We can get them [*sc.* pigs] from the natives for blankets, or for 'money gold' as they call it, which we call sovereigns. **1925** J. Gregory *Bab of Backwoods* xxiii. 283 Gold that had..been dull bits of ore dug from rocky hillsides; that men had taken and made into money-gold. **1928** I. Fisher *Money Illusion* (1929) i. 4 The 'Money Illusion'..., the failure to perceive that the dollar, or any other unit of money, expands or shrinks in value. **1975** *Times* 30 June 12/8 Money illusion is, as in Germany, dying..as people spend indexed wages rather than save. **1838** D. D. Barnard *Speeches & Rep.* 106 To see him [*sc.* the President] sit as a great money king over the nation. **1900** *Congress. Rec.* 7 Feb. 1610/1 Where ought control of the currency to rest?.. At present the banks and the money kings wield this power. **1928** Money man [see *DATE *v.* 2 d]. **1958** Money-man [see *CACKLE *sb.* 3 a]. **1973** *Times* 2 Feb. 14/7 New York money men have been known to quake in the knowledge that they have on deposit $1,000m of overnight money. **1791** A. Hamilton *Establishment of Mint* in *Wks.* (1810) I. 291 In Holland, the greatest *money* market of Europe, gold was to silver..as 1 to 14·88. **1816** Scott *Antiquary* II. vii. 190 In the present state of the money-market. **1883** *Century Mag.* Sept. 691/2 Wall Street..is the money market of the whole country. **1964** *Financial Times* 12 Mar. 10/2 South Africa has developed a money market which, in relation to national income, handles a larger volume of funds than the traditional London discount market. *a* **1861** A. H. Clough *Mari Magno* in *Poems* (1871) 320 The money-order had been cashed. **1972** *Post Office Guide* 590 Inland money orders are issued by and payable at all money order offices. **1935** A. J. Pollock *Underworld Speaks* 77/1 *Money player*, the tougher the game, this particular

gambler excels on account of having lots of nerve. **1944** *Gen* 11 Mar. 30 It is one of the die-hard notions that no money-player is fit to lead an England team. **1829** *H. R. Doc. 21st U.S. Congress 1 Sess.* No. 6. 12 The application of the money power of the Government to regulate the unequal action. **1831** T. H. Benton in *Reg. Deb. Congress U.S.* 2 Feb. 50/2 The money power of the bank is both direct and indirect. **1840** J. S. Mill in *Edin. Rev.* LXXII. 11 The additions to the 'money-power' of the higher ranks, consist of the riches of the *novi homines* who are continually aggregated to that class from among the merchants and manufacturers. **1926** R. H. Tawney *Religion & Rise of Capitalism* ii. 89 When he [*sc.* Luther] looks at German social life, he finds it ridden by a conscienceless money-power, which incidentally ministers..to the avarice and corruption of Rome. **1959** *Ann. Reg. 1958* 90 The United Party was pictured as being..dominated by jingoes and in the hands of the money-power. **1844** *Congress. Globe* 28th Congress 2 Sess. App. 37/2 Banks..managed..by a set of irresponsible money sharks. **1972** G. F. Newman *You Nice Bastard* 347 Moneyshark, unlicenced money-lender (operating at especially high rates of interest). **1816** Scott *Let.* 23 July (1932) I. 502 You had better be looking out & inquiring after some money-shop, as we shall have enough of bills. **1972** *Guardian* 9 June 11 The new 'money shop' branches which are sprouting up in the High Streets... Money shops are not really banks at all... 'Loan shop'..is still the best shorthand description. **1859** G. A. Sala *Twice round Clock* 69 The clown, the dunderheaded moneyspinner who votes that books are 'rubbish'. c **1880** A. W. Pinero *Money Spinner* (1890) 22 Have you forgotten my father's [gaming] house...Have you forgotten what they called me then, because of my never-failing good fortune—because of my *luck*. They called me the *Money Spinner*! **1952** W. Granville *Dict. Theatr. Terms* 119 *Money-spinner*, a successful play or artiste. **1954** G. Smith *Flaw in Crystal* 87 He found he had a pretty knack of writing so that ordinary men and women could understand and feel something of the beauty he saw around him. And he turned that touch into a money-spinner. **1958** A. Wilson *Middle Age of Mrs Eliot* II. 237 If he publishes anything it'll have every chance of being a money spinner. **1970** *Times* 23 Jan. 25/5 (Advt.), Their products range from fertilizers and basic heavy chemicals to..complex petro-chemicals..big money-spinners for Britain's export. **1936** J. Buchan *Island of Sheep* vi. 105 He's a stockbroker—a one-man firm which he founded himself. His interests? Not financial exclusively—indeed, he professes to despise the whole money-spinning business. **1973** *Courier & Advertiser* (Dundee) 1 Mar. 2/2 The money spinning lager boom already accounts for some 12 per cent. of the beer drunk in Britain.

money-grubber. (Earlier and later examples.)
1840 *Bentley's Miscellany* VIII. 449 The money-grubber, pent up in a close city. **1932** A. Huxley *Lett.* (1969) 361 It's terrifying what can be done by people who have a monopoly of means of propaganda—private money-grubbers.

money-maker. Add: **1. b.** (Later U.S. examples.)
a **1734** J. Comer in *Rhode Island Hist. Soc. Coll.* (1893) VIII. 65 This day came up ye case of ye money makers to trial. **1777** in *New Hampsh. Hist. Soc. Coll.* (1863) VII. 93 To pay Col. Sam'l Folson nineteen pounds eleven shillings for..apprehending Money makers. **1778** in *Ibid.* 149 Jnº Mac Glauglin, a money maker.
2. b. (Earlier and later U.S. examples.)
1899 Ade *Doc' Horne* 22, I expect to have an interest in the Neapolitan Dental Parlors, where I'm working now. It's a sure money-maker. **1940** E. Fergusson *Our Southwest* xix. 349 He has instead snapped up his dances as money-makers.

mong (mɒŋ), *sb.*[2] Austral. slang abbrev. of Mongrel *sb.*
1933 *Bulletin* (Sydney) 16 Aug. 39/1 And the mong fired him! **1934** *Ibid.* 30 May 20/3 The silliest dogs for miles around live in Warrnambool; these mongs crouch down when a car approaches and spring at the front wheels with snappy barks. **1956** G. Casey in *Coast to Coast 1955–6* 80 Reckon Charlie Spend's having trouble with that mong of his. **1966** [see *HEEL *v.*[1] 4]. **1967** [see *BLUDGE *v.*]. **1967** Partridge *Dict. Slang* Suppl., *Mong*, a mongrel dog... **2.** Hence, pejorative for any dog, even of the best pedigree. **1970** J. Cleary *Helga's Web* v. 77, I can't *lose.* I backed a mong last night that had only three legs..and it finished up beating the bunny home.

mongan (mǫ·ŋgăn). *Austral.* [Native name.] A species of ring-tailed opossum, *Pseudocheirus herbertensis*, found in the rain forest of north-eastern Queensland.
1890 C. Lumholtz *Among Cannibals* 173 (Morris), Jimmy..had..found mongan.., a new and very pretty mammal, whose habitat is exclusively the highest tops of the scrubs in the Coast Mountains. **1970** W. D. L. Ride *Guide to Native Mammals of Australia* 76 Mongan, Herbert River Ringtail, *Pseudocheirus herbertensis*... Very dark brown with white under-surface; tip of tail white.

monger (mɒ·ŋgəɹ), *v.* [f. Monger[1].] *trans.* To deal or traffic in.
1928 *Observer* 5 Feb. 15/1 Both American and British opinion is laughing out of court those who monger their scares about the United States Navy. **1962** *Times Lit. Suppl.* 10 Aug. 570/3, I laughed at her and told her, that she cannot monger any money out of my pocket.

Mongo[1] (mǫ·ŋgo). Also **Lomongo.** [Native name.] A Bantu people living in Zaïre (formerly the Belgian Congo); a member of this people; the language of this people. Also *attrib.*
1906 Visct. Mountmorres *Congo Independent State* 45 East of the Launoa..the Mongo, perhaps on account of their contact with the Arabised tribes..appear to have advanced to a much higher state of civilisation than in other parts. **1910** *Encycl. Brit.* VI. 925/1 Farther north and largely occupying the valley of the Ruki are the Mongo, a large forest tribe. **1919** A. Werner *Introd. Sketch Bantu Lang.* 314 *Congo*,..the languages included under this heading are..*Lolo* (Mongo, Lunkundu)—on the Equator within the great northern bend of the Congo. **1928** E. A. & L. Ruskin *Dict. Lomongo Lang.* Introd., Lomongo belongs to the Bantu group of languages and is spoken by at least 300,000 people, dwelling in Middle Belgian Congo, north and south of the equator. **1946** *Nature* 6 July 14/2 Important monographs were produced on the Mongo, Basongo, and Bakongo. **1961** G. Greene *Burnt-Out Case* vi. i. 178 She remembered..her father teaching her the Mongo for 'bread' and 'coffee' and 'jam'. They were still the only three words in Mongo that she knew. **1961** *Listener* 16 Nov. 831/1 If the Mongo tribes had been contained within a single province..the prospect of maintaining an independent Congo as a single state would hardly have been worth considering.

mongo[2] (mǫ·ŋgo). [Mongolian.] A monetary unit of Mongolia, one hundred of which are equal to one tugrik.
1935 *Statesman's Year Bk.* 774 The *Tukhrik*, a silver coin equivalent to half an American gold dollar, and divided into 100 *Mongo*. **1968** C. Narbeth *Coin Collector's Encycl.* 145 The Mongolian People's Republic..first minted coins in copper and silver of Mung and Tugrik denominations. **1969** *Whitaker's Almanack 1970* 984 Mongolia (Outer), Tugrik of 100 Mongo. **1972** *Statesman's Year-Bk.* 1165, 100 *möngö* = 1 *tugrik*.

Mongol, *sb.* and *a.* Add: **A.** *sb.* **2.** (Also with lower-case initial.) A person afflicted with mongolism.
[**1866**: see *MONGOLIAN *a.* and *sb.* A. 3.] **1896** *Brit. Med. Jrnl.* 12 Sept. 617/1 There is probably in the Mongol a more profound and general condition of intrauterine failure of development than in the cretin. **1913** *Jrnl. Nerv. & Mental Dis.* XL. 338 Mentally, mongols are as a rule quiet, good tempered, and easily amused. **1926** *Lancet* 23 Jan. 190/2 The resemblance between the Mongol and the cretin has naturally produced an endocrine theory. **1932** [see B. 2 below]. **1960** *Guardian* 29 Apr. 6/5 The number of adult mongols was likely to increase fairly steadily. **1970** Passmore & Robson *Compan. Med. Stud.* II. xxxi. 17/1 A variety of cytogenetic aberrations have been found amongst the 2–5 per cent of mongols who are neither simple trisomics nor trisomic/normal mosaics.
B. *adj.* **2.** (Also with lower-case initial.) Pertaining to or affected with mongolism; = Mongolian *a.* and *sb.* A. 3 (in Dict. and Suppl.).
1896 *Brit. Med. Jrnl.* 12 Sept. 617/1 There are at present in the Royal Albert Asylum 19 Mongol idiots in a population of 560. **1932** Shrubsall & Williams *Mental Deficiency Pract.* xiii. 206 In mongols, laxity of the joints is very characteristic, and mongol imbeciles tend in sitting naturally, to take up the crossed position of the Buddha in contemplation. **1965** [see *DOWN'S SYNDROME]. **1970** Passmore & Robson *Compan. Med. Stud.* II. xxxi. 18/1 The greatly increased risk of having mongol offspring is the reason why every effort should be made to discover translocation carriers so that appropriate genetic advice may be given. **1974** G. F. Newman *Price* ii. 68 He made contributions to the maintenance of his ex-wife and her mongol son.

Mongolian, *a.* and *sb.* Add: **A.** *adj.* **1.** Also *Mongolian lamb.*
1963 *Sunday Express* 27 Jan. 9/5 A Mongolian lamb hat. **1970** *Vogue* Jan. 22/3 A helmet of white Mongolian lamb.
b. *Mongolian hot-pot* (see quot. 1971).
1967 E. Hunt *Danger Game* viii. 152 He ate oysters at Lo Fan Shan and Mongolian Hotpot. **1968** *Times* 6 Jan. 20/5 The New Hong Kong..are serving a speciality called 'bin lo'..a cross between Mongolian hotpot and Japanese *sukiyaki.* **1971** *Good Food Guide* 421 Mongolian hot-pot (a kind of Mandarin fondue bourguignonne, for which you do your own cooking at the table).
2. b. *Mongolian fold* = *EPICANTHUS; so *Mongolian eye*, one with an epicanthus.
1913 [see *EPICANTHUS]. **1926** H. H. Wilder *Pedigree of Human Race* v. 298 It follows that it is not the Mongolian eye that is racial. **1927** Peake & Fleure *Priests & Kings* xii. 192 The head is..relatively low and rounded with, in many individuals, flattened features and the additional or 'Mongolian' fold of the eyelid. **1935** Huxley & Haddon *We Europeans* iv. 117 A fold of skin, the 'epicanthic fold', covers the inner angle of the eye. This gives the effect known as the 'Mongolian eye'. **1964** *New Statesman* 10 Apr. 559/1 A quick and painless operation will..remove the young lady's Mongolian fold.
3. (Earlier and later examples.) Also with lower-case initial.
1866 J. L. H. Down in *Clin. Lect. & Rep.* (London Hospital) III. 260, I have for some time had my attention directed to the possibility of making a classification of the feeble-minded, by arranging them around various ethnic standards... The great Mongolian family has numerous representatives, and it is to this division, I wish..to call special attention. A very large number of congenital idiots are typical Mongols. *Ibid.* 261 The Mongolian type of idiocy occurs in more than ten per cent. of the cases which are presented to me. **1890** *Jrnl. Mental Sci.* XXXVI. 189 Mongolian imbeciles. **1965** [see *DOWN'S SYNDROME].
4. *Mongolian pheasant*, a variety of pheasant, *Phasianus colchicus mongolicus*, native to southern Russia and Mongolia, introduced to other countries early in the twentieth century; also called the Kirghiz pheasant; also *ellipt.*; *Mongolian spot*, a bluish or brownish spot found, usu. singly, in the sacral region of nearly all new-born babies of Oriental races (and occas. in other races), and which usu. disappears in two or three years.
1903 W. Rothschild in *Field* 20 June 1033/3 The bird known in America as the Mongolian pheasant is not the *Phasianus mongolicus* of Brandt, but the *Phasianus colchicus, var. mongolicus* of Pallas, otherwise our common Chinese ring-necked pheasant (*Phasianus torquatus*) Linné. *Ibid.* 1033/6, I have had a number of the true Mongolian pheasant (*Phasianus mongolicus*) alive at Tring for two years. **1909** J. G. Millais *Nat. Hist. Brit. Game Birds* 104 If the pure and cross-bred Mongolians have a fault it is a tendency to stray more than other pheasants. **1921** *Daily Colonist* (Victoria, B.C.) 23 Oct. 18/1 This boy [*sc.* an old cock pheasant] looked like a Mongolian, seemed too dark for an ordinary ringneck. **1963** D. A. Bannerman *Birds Brit. Isles* XII. 338 The Kirghiz pheasant *Phasianus mongolicus* Brandt. Commonly (but incorrectly) named the Mongolian pheasant, is distinguished by the broad white-ring around the neck interrupted in front, having the mantle, chest and breast bronzy orange-red... Introduced into England for Lord Rothschild in 1900. **1965** *Observer* (Colour Suppl.) 10 Oct. 33/2 There's a Mongolian there..a fine ring neck... That old bird he always roosts in that apple tree there. **1907** *Arch. Pediatrics* XXIV. 428 His findings led him..to a special study of the morphology of the 'Mongolian' spot. **1966** Wright & Symmers *Systemic Path.* II. xxxix. 1507/1 The blue naevus..is analogous to the 'Mongolian spot', but may occur in any race and on any part of the body. **1969** *Beaver* Spring 49/1 A baby's pale Mongolian spots are patted lovingly.

Mongolianize (mǫŋgōᵘ·liănəiz), *v.* [f. Mongolian *a.* + -IZE.] *trans.* To render Mongolian in character or quality.
1923 *Contemp. Rev.* Mar. 299 The Bolshevik, striving originally to make Russia Western, has succeeded merely in Mongolianising her.

mongolism (mǫ·ŋgŏliz'm). *Med.* Also **Mongolism.** [f. Mongol *sb.* and *a.* + -ISM. Cf. Mongolian *a.* and *sb.* A. 3 (in Dict. and Suppl.).] A relatively common congenital form of mental deficiency which is associated with a low expectation of life, is always accompanied by a chromosomal abnormality (usually trisomy for chromosome 21), and is marked by numerous signs, including short stature, short thick hands and feet, a large tongue, a flat face with features somewhat similar to those of Mongolians, and a friendly and cheerful disposition; Mongolian idiocy.
1900 *Lancet* 6 Jan. 23/1 (*heading*) The differential diagnosis of Mongolism and cretinism in infancy. **1924** *Psyche* V. 1 It is..well known that a certain type of mental and physical backwardness has long been described, under the name of 'Mongolism'. **1960** *News Chron.* 1 July 5/4 From a medical dictionary he learned that mongolism was a type of congenital idiocy. **1961** R. D. Baker *Essent. Path.* xxii. 615 Cases of..spina bifida and meningocele, mongolism, and cerebral palsy occur at the rate of 1 or 2 of each of these conditions per 1000 births. **1961** [see *DOWN'S SYNDROME]. **1971** *Brit. Med. Bull.* XXVII. 47/1 The epidemicity of mongolism has also been studied.

Mongoloid, *a.* and *sb.* Add: **A.** *adj.* **1.** (Later examples.)
1934 Webster & Westley *World Civilization* ii. 22 These three races are generally called Negroid, Mongoloid, and Caucasian. *Ibid.* 23 The Malays, Polynesians, and American Indians are included among Mongoloid peoples. **1939** C. S. Coon *Races of Europe* i. 2 From a branch of this hyperborean group there evolved, in northern Asia, the ancestral strain of the entire specialized mongoloid family. **1967** E. A. Hoebel in Rapport & Wright *Anthropol.* 70 The most outstanding Mongoloid physical trait is the slant eye, more elegantly known to anthropologists as the internal epicanthic fold.
b. *Mongoloid eye, fold* = *Mongolian eye*, *Mongolian fold.*
1931 E. A. Hooton *Up from Ape* v. 420 The Mongoloid eye characteristically fills the orbit and protrudes slightly. *Ibid.* 421 A common modification..is the inner or internal epicanthic fold, often referred to as the 'Mongoloid fold'. **1960** J. Comas *Man. Physical Anthropol.* v. 271 Generally, the eyelid opening is horizontal, but in many groups of Asia and America the so-called Mongoloid fold is found.
2. (Further examples.) Also with lower-case initial. Add to def.: characteristic of or resembling a mongol or mongolism.
1949 S. S. Sarason *Psychol. Probl. Mental Deficiency* viii. 210 Ages of the mother and father at the birth of the Mongoloid child are relatively advanced. **1961** *Lancet* 13 May 1028/2 (*heading*) 21-Trisomy/normal mosaicism in an intelligent child with some mongoloid characters. **1962** [see B. 2 below]. **1965** Rosen & Gregory *Abnormal Psychol.* xxi. 462/1 She was mongoloid in appearance, good natured, cheerful and cooperative. **1973** *Times Lit. Suppl.* 19 Oct. 1269/5 The bereaved mother peopling her flat with 'friends' via spiritualist contact with a mongoloid daughter.
B. *sb.* **1.** (Later examples.)
1964 R. T. Anderson in M. F. A. Montagu *Concept of Race* 62 The second [interpretation] holds that the Lapps represent a remnant of an ur-race that was the ancestor of both modern Mongoloids and modern Caucasoids. **1971** *National Geographic* Mar. 323/1 Like the majorities, they are virtually all Mongoloids, and often hard to tell apart.

MONGOOSE

2. (Also with lower-case initial.) = *Mongol *sb.* and *a.* A. 2.
1949 *Amer. Jrnl. Mental Deficiency* LIV. 204 (*heading*) Case report: reproduction in a mongoloid. 1962 *Amer. Jrnl. Human Genetics* XIV. 125 Trisomy of a short acrocentric chromosome in the human complement has come to be recognized as such a common etiologic factor in mongolism..that the discovery of a forty-seven chromosome mongoloid is no longer worthy of note. Only a few cases have been described in which the mongoloid patient had forty-six chromosomes. 1974 E. Tidyman *Dummy* vi. 87 Its..inmates..were mongoloids and schizoids, the emotionally disturbed and the mentally incompetent.

mongoose. Add: **1.** The proper plural form is *mongooses*, but *mongoose*, *mongeese*, and other variants are occasionally used.
1896 R. Lydekker *Hand-bk. Carnivora* 244 (*heading*) The mungooses. Genus Herpestes. 1927 L. Gask *All about Animals* 158 Mongooses..are well known in Africa and in the warmer parts of Asia. 1927 G. Jennison *Nat. Hist.: Animals* 82 Mongoose saved the West Indian sugar plantations from total destruction by rats. 1960 G. Durrell *Zoo in my Luggage* 109 Young pigmy mongooses flatly refuse to feed from bottle or from fountain pen filler. 1971 F. Collymore in J. Figueroa *Caribbean Voices* I. 81 The mongoose..is a troublesome sort of creature... For nobody seems to know with any degree of certainty which to choose of his plural forms—mongooses, mongeese, mongoose or mongooze.

mo·ngrelizing, *vbl. sb.* [-ing[1].] The action of the verb Mongrelize.
1922 *Edin. Rev.* July 36 Unchecked mongrelising destroys the symmetry of a national type.

monic (mǫ·nik), *a.* *Math.* [f. Mon(o- + -ic.] Of a polynomial: having the coefficient of the term of highest degree equal to one.
1937 A. A. Albert *Mod. Higher Algebra* ii. 24 The coefficient a_n is called the leading coefficient of *f* and we shall call *f* a monic polynomial when $a_n = 1$. [*Note*] This notion occurs so frequently in algebra that the author believes it would be wise to adopt the above term. 1963 Beaumont & Pierce *Algebraic Found. Math.* ix. 328 Thus, for any nonzero polynomial $g(x)$, there is a unique monic polynomial $f(x)$ which is a multiple of $g(x)$ by a nonzero element of *F*.

monica, monick(er, varr. *moniker.

monies, irreg. pl. of Money *sb.*

moniker (mǫ·nikəɹ). *slang.* Also monarch, monekeer, monica, monick(er, monniker, etc. [Origin unknown.] A name, a nick-name. Also (*rare*) as *v. trans.*, to apply a name to (a person).
1851 Mayhew *Lond. Labour* I. 218/2 'What is your "monekeer" (name)?'—Perhaps it turns out that one is 'White-headed Bob', and the other 'Plymouth Ned'. 1851 [see Ream *a.*]. 1879 *Macm. Mag.* Oct. 502/1 While at the station they asked me what my monarch (name) was. 1882 *Sydney Slang Dict.* 6/1 Monniker, a person's name (often assumed). *Ibid.* 10/1 The padding ken of Sally Hicks, who's got a new moniker, which is Lushing Loo. 1885 M. Davitt *Prison Diary* I. xvi. 151 Many of them having a fresh 'monicker' (name) each conviction. 1895 *Times* 11 Nov. 3/4 The van is alright; I have had the 'monnick' (slang word for name) taken off it. 1903 'J. Flynt' *Rise Ruderick Clowd* i. 46 In the—Ward 'monakers', or nicknames, mean more than they do in uptown communities. 1907 J. London *Road* (1914) 169 His 'monica' was Skysail Jack. 1926 J. Black *You can't Win* ix. 111 Whether it was his appearance or his careful manner of speech that got him his monoger, 'The Sanctimonious Kid', I never knew. 1936 J. Curtis *Gilt Kid* v. 50 Eileen, or something, that was the sister's monick. 1936 M. Franklin *All that Swagger* lii. 488 'I thought it was mimosa.' 'That's its trade moniker; it's home name is wattle.' 1938 *Amer. Speech* XIII. 195, I monikered him (gave him his moniker, or tramp name). 1944 *Sun* (Baltimore) 7 Aug. 2/7 He came by the name of Omar N. (the N being for Nelson) through his father's searching for a 'distinctive' moniker for his offspring. 1959 *Times Lit. Suppl.* 13 Feb. 84/1 Henry Handel Richardson herself..was able to hide behind the male signature on her books (her maiden name wedded to two favourite family monikers). 1960 *Melody Maker* 31 Dec. 6/3 Oliver Cool is the pseudonym of singer Larry Ellis, who found he won more success when he adopted the 'far-out' monniker. 1964 *Punch* 2 Dec. 826/1 Mildly bitter moniker given to Swiss-based moneymen. 1968 L. Morton *Long Wake* i. 18 The documents came to Jack Roberts for my 'Monica'. 1972 *Village Voice* (N.Y.) 1 June 78/3 Slim and his brother helped bail the pilot out. The Airplane Kid moniker followed.

monilia (mǫni·liă). *Bot.* and *Med.* Also Monilia. Pl. monilia, -iæ, -ias. [mod.L., f. L. *monile*, necklace, in allusion to the chains of spores.] A fungus belonging to the genus *Monilia*, which now (properly) consists only of imperfect forms of *Sclerotinia* and *Neurospora*, several of which cause plant diseases, but formerly included some species (now assigned to *Candida*) which are important human pathogens. Also *attrib.*
1751 J. Hill *Gen. Nat. Hist.* II. 69 Monilia is a genus of Fungi, consisting of a pedicle supporting a number of naked seeds, arranged in series like the beads of a necklace. The Monilia all produce distinct male and female flowers..[Micheli] has..divided this genus into two, under the names of Botrytis and Aspergillus; but the differences, this division are founded upon, are rather specific than generical; we have therefore arranged them all together under one genus. 1916 *Jrnl. Trop. Med. & Hygiene* XIX. 89/1 White, creamy, roundish colonies, characteristic of a monilia, developed within twenty-four hours. 1920 *Lancet* 17 Apr. 851/2 The 20°C. incubator in which the collection of monilias was kept. 1932 *Jrnl. Infectious Dis.* L. 75 Various types of moniliae. 1950 *Jrnl. Amer. Med. Assoc.* 13 May 215/2 A pyoderma in which Monilia may be present merely as saprophytes. 1968 J. H. Burn *Lect. Notes Pharmacol.* (ed. 9) 112 In extreme cases there may be a fatal septicaemia due to monilia. 1975 C. A. Thorold *Diseases of Cocoa* iii. 27 Monilia disease is confined to the north-west of South America and the adjacent part of Central America.
So **moni·lial** *a.*, of, caused by, or pertaining to a monilia or moniliæ; **moni·lioid** *a.*, resembling a fungus of this type.
1928 *Mycologia* XX. 127 (*title*) The monilioid species of *Sclerotinia*. 1947 *Jrnl. Amer. Med. Assoc.* 20 Sept. 192/2 (*heading*) Iodine treatment in monilial infections. 1951 *Ibid.* 27 Jan. 207/1 (*heading*) Monilial infections complicating the therapeutic use of antibiotics. *Ibid.* 208/1 A severe monilial esophagitis. 1973 *Nature* 28 Sept. 214/1 Specimens of clinically uninflamed gingiva were cultured with the addition of nystatin..to prevent monilial overgrowth. 1974 Passmore & Robson *Compan. Med. Studies* III. xiii. 13/1 Trichomonal and monilial infections should be excluded by the examination of smears of vaginal exudate [in the treatment of vulvovaginitis].

moniliasis (mǫniləi·ăsis). *Path.* Pl. -ases. [f. *monil(ia + *-iasis.] Infection with or a disease caused by a fungus of the genus *Candida*.
1920 *Lancet* 24 Apr. 898/2 It is probable therefore that the so-called tea-taster's cough is a moniliasis. 1925 *Jrnl. Obstetr. & Gynæcol.* XXXII. 69 (*heading*) Vaginal monilias and vaginal moniliases. 1947 *Jrnl. Amer. Med. Assoc.* 27 Dec. 1186/2 The moniliases form a group of infections. 1958 *New Biol.* XXVII. 62 Moniliasis, which is a disease or complex of diseases with diverse manifestations caused by several yeast-like fungi of which *Candida albicans* is the most important, has a world-wide distribution... The most familiar aspect of moniliasis is thrush in infants. 1964 Winner & Hurley *Candida Albicans* p. v, We refer to these diseases as candidosis, this being the term, which ..we have preferred to 'candidiasis', 'candidasis' and 'moniliasis'. 'Moniliasis' was the correct term when the causative organism was known as Monilia, but its continued use today leads to confusion. 1974 Passmore & Robson *Compan. Med. Studies* III. xix. 9/2 (*heading*) Candidiasis (thrush, moniliasis).

monimolimnion (mǫ:nimoli·mniǫn). Pl. -limnia. [a. G. *monimolimnion* (I. Findenegg 1935, in *Internat. Rev. d. ges. Hydrobiol. u. Hydrographie* XXXII. 377), f. Gr. μόνιμο-ς stable + *-limnion*, after *epilimnion, *hypolimnion.] The lower, denser, non-circulating layer of a meromictic lake. Cf. *mixolimnion.
1937 [see *mixolimnion]. 1957 G. E. Hutchinson *Treat. Limnol.* I. vii. 464 These lakes have never fully circulated in the ordinary way, since they developed large monimolimnia. 1968 [see *mixolimnion].

monish (mʌ·niʃ), *sb.* [See Money *sb.* 4 ¶.] (Examples.)
1846 *Swell's Night Guide* 125/2 *Monish, tip us the*, give me the money. 1848 *Sinks of London laid Open* 116/1 *Monish, tip us the*, give me the money.

monism. Add: **1. d.** In various uses indicating a theory or doctrine of a single force, source, or system from which all particular instances devolve.
1890 W. James *Princ. Psychol.* I. x. 366 The contrast between the Monism thus reached and our own psychological point of view can be exhibited schematically. 1903 B. Russell *Princ. Math.* i. iv. 43 The subject is important, since the issues between monism and monadism, between idealism and empiricism..all depend, in whole or in part, upon the theory we adopt in regard to the present question. *Ibid.* 44 Numerical identity and diversity are the source of unity and plurality; and thus the admission of many terms destroys monism. 1949 P. A. Schilpp *Albert Einstein* xxiii. 591 The question is still open whether it is possible thereby to arrive at a real monism; for the conversion of Joule into *mkg* does not prove that the two forms of energy are in essence the same. 1964 Gould & Kolb *Dict. Soc. Sci.* 440/1 Monism is the notion..that in any society there must be one supreme power or authority and this must be lodged in the state. 1970 G. A. & A. G. Theodorson *Mod. Dict. Sociol.* 93 *Cultural monism*, a doctrine that advocates the assimilation of ethnic minorities into the dominant culture of a society to attain cultural uniformity. 1971 R. F. C. Hull tr. *Jung's Psychol. Types* in *Coll. Wks.* VI. viii. 318 The idea-oriented attitude must tend towards monism.

monist. Add: (Later examples.)
1934 F. W. Coker *Recent Political Thought* xviii. 504 The pluralist charges the monists with the error of regarding all non-political associations created by the state as dependent for their continued existence upon the will of the State. 1960 L. Lipson *Great Issues of Politics* (ed. 2) vi. 153 The monist, too, has his problems. Desiring to provide society with an integrating focus, he must designate what this should be. 1964 Gould & Kolb *Dict. Soc. Sci.* 439/2 In international law the monists are those who believe that international and municipal law are both parts of a single system of law. *Ibid.* 563/2 Politically, public opinion is regarded by the monists as simply the manifested general will.
Also *attrib.* or as *adj.*, of or pertaining to the doctrine of monism.
1931 *Times Lit. Suppl.* 5 Mar. 164/2 Ramanuja fought against the monist view with as intense spiritual passion as any Christian critic could have shown. 1960 L. Lipson *Great Issues of Politics* (ed. 2) vi. 152 Both philosophies stand unequivocally against something: the pluralist, against organized unity; the monist, against anarchic diversity.

monistic, *a.* (Further examples.)
1890 W. James *Princ. Psychol.* I. x. 367 For *us* it is an unpardonable logical sin,..to change the terms without warning... For monistic idealism, this is the very enfranchisement of philosophy. 1903 B. Russell *Princ. Math.* iv. xxvi. 221 The monistic view..regards the relation as a property of the whole composed of *a* and *b*, and as thus equivalent to a proposition which we may denote by (*ab*)r. 1926 J. C. Smuts *Holism & Evolution* v. 108 The fundamental concept of Holism will bring us nearer to that unitary or monistic conception of the universe which is the immanent ideal of all scientific and philosophic explanation. 1957 S. Aurobindo in Radhakrishnan & Moore *Sourcebk. in Indian Philos.* xvi. 586 In the monistic view the individual soul is one with the Supreme. 1960 L. Lipson *Great Issues of Politics* (ed. 2) xiv. 417 There are classic instances of a monistic state combined with a controllable government. *Ibid.*, A monistic state like Britain, therefore, has its powers both centralized and integrated. 1964 Gould & Kolb *Dict. Soc. Sci.* 563/2 The divisions..can be roughly classified as being between advocates of a monistic, organic view and holders of a pluralistic, numerical theory of public opinion. 1971 J. Z. Young *Introd. Study Man* x. 124 Others [committed by] their addiction to a monistic science, try to find a system that avoids the separation of mind and matter. 1971 R. F. C. Hull tr. *Jung's Psychol. Types* in *Coll. Wks.* VI. viii. 301 Intellectualism is always monistic.

monitor, *sb.* Add: **3. b.** One who is appointed to listen to and report on radio broadcasts, esp. from a foreign country.
1939 *Times* 2 Nov. 5/3 The material, after it is taken down and translated by the monitors, is summarized by an editorial staff. 1945 *Times* 26 May 2/4 In another room monitors were listening to Morse signals. 1974 *Daily Tel.* 16 July 1/2 Radio monitors heard what purported to be a 20-second reading by the Archbishop.
c. *Broadcasting.* A device for indicating or ascertaining the technical quality of a transmission without disturbing the transmission itself; esp. (also *monitor screen, tube*) a television screen for displaying the picture from a particular camera or that being transmitted. Freq. *attrib.*
1931 A. Nadell *Projecting Sound Pictures* ii. 36 The projectionist has only to listen to the sound in his monitor and adjust the guides until the disturbance disappears. 1934 *Proc. IRE* XXII. 450 The monitors are designed for continuous operation, giving a continuous indication of the radio transmitter's frequency, and enabling the radio station operator to tell at a glance whether the transmitter is within the assigned limits. 1944 R. E. Lee *Television* 155 The video engineer clamps down one of the 'pots' on his control panel, draining the light from the monitor screen. 1958 M. Dickens *Man Overboard* x. 155 He caught sight of himself on the small monitor screen at one side of the cameras. *Ibid.* 160, I saw myself on the monitor. 1959 *Listener* 13 Aug. 238/1 In their monitors the searchers could see the boxes televised by the camera 200 feet down beneath the surface of the lake. 1961 G. Millerson *Technique Telev. Production* i. 16 Above each camera's control unit is a monitor-tube on which its picture is displayed. 1961 K. R. Sturley *Sound & Television Broadcasting* v. 338 Equipment known as a 'transmitter executive monitor' may be installed at a station to keep a watch on distortion due to overload and noise such as hum. 1969 *Times* 22 July (Moon Suppl.) p. i/1 There's a great deal of contrast in it and currently it's upside down on our monitor but we can make out a fair amount of detail. 1971 H. E. Ennes *Television Broadcasting* i. 32 On the monitor panel are located adjustments for the kinescope focus and brightness.., oscilloscope (waveform monitor) focus and brightness, and a switch to accommodate monitoring either the line waveform..or the field waveform..on the oscilloscope tube.
d. Any instrument or device for monitoring some process or quantity; *spec.* one for detecting or measuring radioactivity.
1948 *Nucl. Sci. Abstr.* I. 282 The different types of radiation distribution..and detecting devices, such as air monitors..are briefly explained. 1958 O. R. Frisch *Nucl. Handbk.* ii. 9 A special monitor..has been designed for the use of persons leaving a laboratory; this counts α- or β-activity on both sides of the two hands. 1961 *Rev. Sci. Instruments* XXXII. 1098 (*heading*) Transmission current monitor for high energy electron beams. 1962 *IRE Trans. Audio* X. 132/1 The audio-radar monitor recorder is designed to continuously monitor and record 16 audio channels and one radar display on a single reel of magnetic tape. 1966 P. J. O'Higgins *Basic Instrumentation* x. 289 A monitor is an instrument that is used to measure continuously or at intervals a quantity or condition that must be kept within a prescribed limit. An example is the oil gauge in an automobile. 1973 *Biomed. Engin.* VIII. 255/3 There are situations in which a portable monitor can with advantage be replaced by a permanently-installed instrument which performs a 'watch-dog' function and gives warning if the radiation level rises above a pre-set value (or alternatively warns if any radiation is present at times when none should exist). 1973 *Daily Tel.* 10 Aug. 3 (Advt.), A foetal heart monitor..enables medical staff to tune in to the heart-beat of an unborn baby. 1974 *Rev. Sci. Instruments* XLV. viii. p. xi (Advt.), Continuous liquid helium level monitor.
5. Substitute for first part of def.: A large

lizard of the family Varanidæ, found in Africa, Asia, and Australia. (Later examples.)
1958 [see *agamid]. 1965 R. McKie *Company of Animals* i. 3 Not twenty paces away was a six-foot Malayan lizard, a monitor, grey against the shingle.
9. *attrib.* (see also *3 c), as *monitor man, room, speaker*; **monitor lizard** = sense 5.
1869 P. Gillmore tr. *Figuier's Reptiles & Birds* (1870) 13 In the Crocodiles and Monitor Lizards..a mutilated part is not renewed. 1908 E. J. Banfield *Confessions of Beachcomber* i. iii. 99 There is..no iguana (rather, monitor lizard), though a fair variety of other reptiles. 1947 I. L. Idriess *Isles of Despair* xiv. 92 The great carved head of Kodal the crocodile, of Karum the monitor lizard. 1966 D. Forbes *Heart of Malaya* viii. 93 The only creature we saw was a six-foot long monitor lizard slithering across the road. 1929 *Photoplay* (Chicago) Apr. 31/2 *Monitor man*, the person who operates the volume control on talking picture production, modulating sounds as they come through the microphone so as to get a more even and natural tone. 1931 *Star* 8 May 12/1 The 'monitor room' where the sound experts listen to the conversation was moved 50 feet away. 1960 O. Skilbeck *ABC of Film & TV* 86 *Monitor speaker*, loudspeaker used as a check in dubbing, or as an alternative to headphones by recordists. 1975 *Independent Broadcasting* Aug. 11/1 The technical area with two mixing desks..and two monitor speakers.

monitor, *v.* Restrict *nonce-word* to sense in Dict. and add: **2.** To check or regulate the technical quality of (a radio transmission, television signal, etc.).
1924 [implied in *monitoring *vbl. sb.*]. 1929 G. Allighan *Romance of Talkies* 64 During the recording, the mixer operator monitors the record through the light valves, thereby assuring himself that no record is lost. 1930 *Bell System Techn. Jrnl.* IX. 273 The..control desk provides facilities by which the attendant can monitor the incoming voice currents and the outgoing radio signal. 1934 *Electronics* May 139/2 Methods of monitoring program transmission both visually and aurally. 1958 *Times Rev. Industry* Aug. 9/3 The output from the analyser can be monitored with headphones. 1962 A. Nisbett *Technique Sound Studio* 258 The signal is monitored by the limiter and any wave which exceeds a certain volume causes a corresponding increase in feedback, which reduces the signal.
b. To listen to and report on (radio broadcasts, esp. from a foreign country); also, to eavesdrop on (a telephone conversation).
1939 *Times* 2 Nov. 5/3 About 150 news bulletins alone, in many different languages, are monitored each day. 1948 *Amer. Speech* XXIII. 219 From secret radios, constructed and hidden by British prisoners, the B.B.C.'s broadcasts were monitored. 1955 F. Swinnerton *Summer Intrigue* xi. 85 A telephonist had illegitimately monitored a conversation. 1958 'N. Shute' *Rainbow & Rose* i. 7 We've been monitoring their frequency. *Ibid.* 22 The constable sat down to monitor the conversations. 1971 R. Dentry *Encounter at Kharmel* v. 92 If I were running Ziauddin's operation, I'd be monitoring every newscast in Asia. 1972 D. Haston *In High Places* vi. 74 We sat for three weeks in Kleine Scheidegg, monitoring weather forecasts twice a day.
c. In more general use: to observe, supervise, or keep under review; to measure or test at intervals, esp. for the purpose of regulation or control.
1944 *Times* 20 Mar. 5/7 American and British control officers work at the same desks throughout the 24 hours, 'monitoring' the aircraft as they travel across the 3,000 miles of ocean. 1947 C. Goodman *Sci. & Engin. Nuclear Power* I. 255 We must monitor the cooling water to be sure that it contains no dangerous activities. 1957 *Listener* 3 Oct. 501/2 The Royal Observer Corps have been given the vital job of monitoring the extent of fall-out all over the country. 1958 *Observer* 9 Nov. 17/5 The phonocardiograph is useful in monitoring the [unborn] child's heart when the mother is sleeping or is over-anxious. 1959 *Daily Tel.* 14 Apr. 1/6 The committee was set up in November, 1956, to initiate and monitor a programme of research design. 1961 *New Scientist* 21 Dec. 744/1 The principle is now used..to monitor (while in service) the thickness of boiler shell and pipe walls that are subject to corrosion. 1966 *Lancet* 24 Dec. 1387/1 Pulse-rate and blood-pressure were closely monitored. 1967 S. Woodgate in Wills & Yearsley *Handbk. Management Technol.* 94 These work schedules give supervision and management the ability to monitor progress and to detect immediately if delays to schedules are occurring. 1972 *Sci. Amer.* Aug. 110/3 The radius of the mirror was monitored frequently during the grinding operation by a center-of-curvature test. 1974 *Nature* 17 May 204/2 Thirty-five inspectors monitor the 2,000-odd works registered under the Alkali Act. 1974 *Physics Bull.* Nov. 533/1 Traces of metals in air and water as low as 0·01 parts per million can be monitored with the latest instrument.
So **mo·nitored** *ppl. a.*; **mo·nitoring** *vbl. sb.* and *ppl. a.*
1924 *Proc. IRE* XII. 561 A third requirement relates to monitoring facilities. 1929 G. Allighan *Romance of Talkies* 64 Acoustic treatment of the walls of the monitoring room secures the reverberation characteristic of the theatre. 1933 *Electronics* June 160 A monitoring panel and a two-way communication system so that the operator may keep in constant communication with the pilot of a landing aircraft. 1937 *Discovery* Nov. 330/2 At the side of the studio there is a 'monitoring' television receiving set, so that the floor directors and the announcers can see the outgoing programme. 1939 *Times* 2 Nov. 5/3 Monitoring consists of listening to and recording information contained in programmes broadcast by other countries. *Ibid.*, The monitoring service [of the B.B.C.] deals with some 250,000 words every 24 hours. 1945 *Times* 7 Apr. 4/7 The Japanese Government intended to continue its efforts..according to a monitored report from the United States Office of War Information. 1949 *Atomic Energy & Life Sci.* (U.S. Atomic Energy Comm.) 61 Automatic monitoring equipment, measuring radioactivity of water ..would stop flow if radiation were above safe limit. 1961 *Lancet* 22 July 187/2 Continuous monitoring of the atrial pressure would provide early warning of a dangerous rise. 1962 *B.S.I. News* Mar. 24/2 Two types of plug are dealt with in B.S. 196: two-pole and two-pole and pilot contact, the latter being chiefly required for use with electrical interlocks or with 'monitored earth' circuits. 1971 *Hi-Fi Sound* Feb. 66/1 Systems of this standard are, or can be, of 'monitoring' class—the sort of equipment *Hi-Fi Sound* would use for examining the qualities of new recordings or evaluating new cartridges or other components. 1972 *Accountant* 17 Aug. 214/1 A period of..regulated and monitored practical training.

monk, *sb.* Add: **4.** Also *spec.* in *Printing*, a blotch or area where the ink is excessive.
1683 [in Dict.]. 1771 P. Luckombe *Hist. & Art of Printing* 500 *Monk*, when the Press-man has not distributed his balls and the ink lies in blotches, it is called a Monk. 1892 A. Powell *Southward's Pract. Printing* (ed. 4) xliv. 409 Be very careful not to let the roller 'run into the ink'.., or the roller will be 'smothered'... If not perceived, and the roller is applied to the type, there will be great blotches of ink in different places, perhaps filling it up and causing 'monks', or black patches, in the impression. 1915 *Southward's Mod. Printing* (ed. 3) II. i. 4 This is called rolling the forme... If inefficiently performed there will be too much ink on the impression, or even blotches, technically, 'monks'. 1938 *Amer. Speech* XIII. 272 A dark spot on a page is a *monk*. 1956 *Collins's Authors' & Printers' Dict.* (ed. 10) 265/2 *Monk*,..a patch of letterpress with too much ink.
5. b. monk bond *Building* (see quot. 1936); **monk's bench** = *monk's table*; **monk's cloth** (later examples); **monk('s) shoe** (see quot. 1969); also *ellipt.* as *monk*; **monk's table**, a convertible wooden seat, the back of which is hinged to swing over and rest horizontally on the arms, thus forming a table.
1936 *Archit. Rev.* LXXIX. 241/3 Monk bond, which is a better ordered arrangement than Yorkshire or Flying bond, is popular in the North of Europe. Two stretchers are followed by one header in every course, the headers being so disposed that verticality of their axial lines is little apparent, and a striking result is obtained of diagonal lines of stretchers, which look like a series of corbels or cantilevers embedded in the wall. 1952 *Ibid.* CXII. 87 A double monk bond was adopted, the whole of the façade being set out to a large scale to ensure that the pattern over them was an even one. 1925 Brodhurst & Layton *Gloss. Eng. Furnit.* 110 Sometimes settles were treated in a similar manner and called Monks' Benches. 1950 S. Howard *Our Furnit. through Ages* viii. 162 A table-settle ..was of Cromwellian inspiration and was in consequence most uninspiring; it is sometimes called a 'Monk's Bench'. 1950 '*Mercury*' *Dict. Textile Terms* 352/2 *Monks' cloth*, a heavy all-cotton fabric woven in a basket weave from coarse and rough yarns. 1972 P. Marks *Collector's Choice* i. 26 The yards of monk's cloth and fawn-colored carpet lining his galley. 1938 L. MacNeice *I crossed Minch* vii. 92 Ginger-coloured monk suede shoes. 1953 J. Korn *Boot & Shoe Production* xv. 142 The Monk Shoe... Like the Cromwell, this was formerly a quarter-over shoe. 1968 *Times* 6 Dec. 10/6 The best-selling shoes are..boots and monks (almost boots). 1969 R. T. Wilcox *Dict. Costume* (1970) 217/1 *Monk's shoe*, a low shoe of soft but heavy leather with heel and plain toe and a strap passing over the instep and buckled at the side. 1907 *Yesterday's Shopping* (1969) 130/4 Bamboo Monks' Table... This can be used either as a seat or table, the top being made so that it will slide off to form back of seat. 1972 *Country Life* 23 Mar. 723/3 The so-called monk's table, serving as seat and table..was developed during the 17th century... In shape it was a low-backed box-chair with a solid round or rectangular back, wider than the seat, and with horizontal arms.

monk (mʌŋk), *sb.*[2] Colloq. abbrev. of Monkey *sb.*
1843 *Spirit of Times* 1 Apr. 54/1 P. Fowler..is one and identical with the celebrated jockey of Miss Foote, in stable parlance, ycleped Monk... The word Monk in this case..is in fact an abbreviation of a noun substantive by which a certain animal is recognised, the countenance of which is said strongly to resemble that of the very jockey in question. 1901 Greenough & Kittredge *Words & their Ways* (1902) vi. 61 Slang is fond of clipped words: as, *monk* for *monkey*. 1903 A. M. Binstead *Pitcher in Paradise* xii. 285 He swore at the monkey freely, and the monk retaliated in cordial but vehement dumbshow with every bitter curse that was known to the forest. 1958 G. Barker *Two Plays* 70 That's the one. His little monk's dead.

monkey, *sb.* Add: **1. c.** (Further examples.) Also *transf.* and *fig.*
1890 Kipling *Let.* in C. E. Carrington *Rudyard Kipling* (1955) v. 157 The blandishments of the people to whom a new writer-man is as a new purple monkey on a yellow stick. 1926 W. de la Mare *Connoisseur* 317, I went bobbing over its boulders and chasms like a jack-in-a-box or a monkey-on-a-stick. 1927 W. E. Collinson *Contemp. Eng.* 8 The monkey on a stick, often used to indicate someone very fidgety and restless. 1966 G. Butler *Nameless Coffin* xii. 192 A 'monkey' is a man who has never been in prison..but is the known associate of criminals. He is a monkey on a stick and it is usually only a matter of time before his stick breaks and he's down there with his friends.
d. *three* (*wise*) *monkeys*: a conventional sculptured group of three monkeys, one with its paws over its mouth ('speak no evil'), one with its paws over its eyes ('see no evil'), and one with its paws over its ears ('hear no evil'); hence used allusively.
1926 *Army & Navy Stores Catal.* 197/3 The three wise monkeys. 'Speak no evil, see no evil, hear no evil.' Per group of three Monkeys -/4. 1969 B. Knox *Tallyman* v. 89 The three wise monkeys would be non-starters against those people. 1969 G. Mitchell *Dance to your Daddy* x. 119 Perhaps there are three other wise monkeys in this house besides yourselves. 1970 A. Draper *Swansong for Rare Bird* viii. 66, I know the score. I'm like the three monkeys. 1974 J. Stubbs *Painted Face* vii. 108 He was like those three monkeys... Neither heard, saw nor spoke evil.
2. b. (Further examples.) Also *spec.* in various *slang* uses (see quots.).
1912 A. H. Lewis *Apaches of N.Y.* xi. 225 'Did youse lobsters hear me handin' it to th' monkeys?' he asked... 'That chink, Low Foo, snakes two of me shirts. I sends him five, an' he on'y sends back three. So I caves in his block wit' a flatiron.' 1914 Jackson & Hellyer *Vocab. Criminal Slang* 59 *Monkey*, a man, used in the mildly indifferent sense of a stranger... Sometimes used to signify a 'boob'. 1928 *Amer. Mercury* Aug. 399/2 Several chorus girls from a Broadway show entered the room against the vigorous protests of the head waiter. One of the monkies, as they are called, sang and did an infinitely torrid dance. 1941 J. Smiley *Hash House Lingo* 38 *Monkey*, dish washer. 1942 Berrey & Van den Bark *Amer. Thes. Slang* § 451/2 Plebeian,..*monkey*. *Ibid.* § 460/5 *Monkey*,.. a Chinese. *Ibid.* § 511/3 (*Bridge*) *monkey*, a bridge builder. *Ibid.* § 576/22 Orchestra leader,..*monkey*. 1966 [see sense 1 c above]. 1967 G. Jackson *Let.* 2 Nov. in *Soledad Brother* (1971) 137, I am the object of the severest ridicule (coon, monkey, shoe..). 1970 P. Laurie *Scotland Yard* 291 *Monkey, a*, an unpleasant person.
c. Slang phr. *to make a monkey* (*out*) *of*: to make a fool of (someone); to deceive, dupe; to ridicule. orig. *U.S.*
1900 Ade *Fables in Slang* 164 His friends would stand and watch him make Monkeys of these anaemic Amateurs. 1900 F. P. Dunne *Mr. Dooley's Philos.* 192 'Willum Waldorf Asthor has busted th' laws iv hospitality, an' made a monkey iv a lile subjick iv th' queen,' he says. 1931 M. Allingham *Police at Funeral* viii. 101, I don't want to put up any idea that isn't useful, and if I'm making a monkey of myself you mention it. 1932 Kaufman & Ryskind *Of Thee I Sing* ii. i. in *Famous Plays* (1933) 668 The man who ought to love me Tried to make a monkey of me. 1934 J. O'Hara *Appt. in Samarra* (1935) vii. 211 So then you turn around and pay him back by.. making a monkey out of him right in his own spot. 1938 E. Bowen *Death of Heart* i. v. 97 You make a monkey of me. 1952 [see *agit-prop]. 1973 'M. Innes' *Appleby's Answer* xxi. 180 The plain fact was that Bulkington had.. made a monkey of her. It was all very mortifying.
d. orig. *U.S.* A modern dance. Hence as *v. intr.*
1964 [see *discothèque]. 1965 [see *frug]. 1969 N. Cohn *AWopBopaLooBop* (1970) ix. 85 Dance-crazes bossed pop right up until the Beatles broke. There was the Hully Gully, the Madison,..the Monkey. 1969 *New Yorker* 11 Oct. 50/3 Gamely they tried to Frug (or was it Monkey?) to the plangent anthems of a younger generation. 1975 *Time* (Canadian ed.) 25 Aug. 49/1 The frug, the boogaloo, the monkey and similar 'hang-loose' mating rituals.
10*. *Austral.* and *N.Z.* (See quots.).
1933 *Press* (Christchurch, N.Z.) 4 Nov. 15/7 *Monkey*, a handle made by putting a strap between two dees on a saddle and rolling it round itself. It is to hold on to when riding a bucking horse. 1945 Baker *Austral. Lang.* iii. 71 A *monkey* or *monkey-strap*, a looped strap on the offside of a saddle pommel, used by inferior 'rough-riders'.
12. b. Slang phr. *cold enough to freeze the balls off a brass monkey*: extremely cold. Also in similar and allusive phrases.
[1835 F. Chamier *Unfortunate Man* I. iv. 117 He was told to be silent, in a tone of voice which set me shaking like a monkey in frosty weather.] 1928 *Amer. Speech* IV. 123 Cold enough to freeze the tail off a brass monkey. 1937 Partridge *Dict. Slang* 528/2 Cold enough to freeze the balls off a brass monkey. 1972 *Evening Telegram* (St. John's, Newfoundland) 5 Aug. 3/1 Here's Smallwood still putting up a brass-monkey face right to the bitter end. 1973 *Guardian* 20 July 9 (*headline*) Brass monkey weather. *Ibid.*, You ought to buy yourself a brass monkey and then you will know what a freeze is.
c. *I'll be* (or *I am*) *a monkey's uncle*: a colloquial expression of surprise.
1926 Maines & Grant *Wise-Crack Dict.* 5/2 *Be a monkey's uncle*, be surprised. 1961 Partridge *Dict. Slang* Suppl. 1142/2 I'll be a monkey's uncle!
d. Slang phr. *not to give* (or *care*) *a monkey's* (*fuck*, etc.): not to care at all; to be completely indifferent or unconcerned.
Even without the last bracketed word a fairly coarse expression.
1960 G. W. Target *Teachers* 100 The Old Man's door opened and the pair of them came out, Stillwell not seeming to give a monkey's, but too casual, and poor Jimmy Taylor with his hands clenched before him like the broken forelegs of a ginned rabbit. 1961 Partridge *Dict. Slang* Suppl. 1188/1 *Monkey's f*ck, not to care a*, not to care a rap: low (esp. Naval). 1968 M. Woodhouse *Rock Baby* xii. 116, I don't give a monkey's knee if he was with the Resistance or the Mafia. 1970 *Observer* 10 May 33/5 Tony Martin has booked himself a vasectomy... 'I was brought up a Catholic,' he said, 'but I don't give a monkey's; you've got to be practical.' 1972 J. Brown *Chancer* iii. 44, I don't give a monkey's if you drop down dead. 1975 J. Wainwright *Square Dance* 26 'Not,' snarled Sugden, 'that I give a solitary monkey's toss what you wear.'
13. (Earlier examples.)
1833 B. Webster *Golden Farmer* ii. ii. 40 The Golden Farmer,..ven his monkey's up, vould go through me like a flash of lightning through a gooseberry bush. c1852 J. R. Planché *Good Woman in Wood* ii. ii. 27 I'm short in

stature—that I don't deny, But put my monkey up, I'm six feet high!
b. *to have a* (or *the*) *monkey on one's back*: (*a*) to be angry or enraged (*Obs.*); (*b*) orig. *U.S.*, to be a drug-addict (see also quot. 1942); hence *monkey* = addiction to, or habitual use of, drugs. *slang.*
1860 Hotten *Dict. Slang* (ed. 2) 174 A man is said to have his monkey up, or the monkey on his back, when he is 'riled', or out of temper. 1942 Berrey & Van den Bark *Amer. Thes. Slang.* § 509/28 *Have a Chinaman* or *monkey on one's back*, to manifest withdrawal distress. 1949 N. Algren *Man with Golden Arm* 60 He *wants* to carry the monkey, he's punishin' hisself..'n don't even know it. *Ibid.*, Then I got forty grains 'n went up to the room 'n went from monkey to nothin' in twenny-eight days 'n that's nine-ten years ago 'n the monkey's dead. 1960 C. L. Cooper *Scene* xii. 168 We're just little kids with the monkey and a couple bucks. Well, *our* monkey weighs just as much as yours, and it hurts just as much, and it takes the same stuff to get it off as it does yours! 1970 E. R. Johnson *God Keepers* (1971) vi. 61 An addict's greatest worry would not be his, since Vito would feed his monkey. *Ibid.* 68 Having a monkey on your back.. always worked out logically to be the first purpose in a junkie's life. 1972 H. C. Rae *Shooting Gallery* iii. 204, I didn't have the monkey then. That's how I got started.
14. Delete *Betting* and add further examples.
1922 E. Wallace *Flying Fifty-Five* x. 57, I lost a monkey on the last race, and a monkey on the Coventry. That's a thousand, isn' it? 1972 *Sunday Sun* (Brisbane) 9 Jan. 18/4 Lady from pub A told lady from pub B that the boss had given her a monkey as a bonus. Lady from pub B interpreted monkey as the slang term for $500 and immediately attacked her boss for his stinginess in only handing out $100. 1973 *Times* 9 Jan. 4/8 It looks like you are going to be roped into that theft from the pub but it will be all right. It will cost you a monkey (£500).
16. *monkey fur, -god, -mischief, -people, -skin*; *monkey-fashion* adv. (earlier example).
1868 A. J. Munby *Diary* 7 Sept. (1972) 245 She sprang or climbed, monkeyfashion. 1920 in C. W. Cunnington *Eng. Women's Clothing* (1952) v. 160 Afternoon frock.. with..fringe of monkey fur. 1960 *Sunday Express* 23 Oct. 14/5 The latest fur coat for cult-clothes addicts is in monkey fur. 1883, 1886 Monkey-god [see *Hanuman 1]. 1962 *Listener* 22 Feb. 350/2 The monkey-gods of Stupidity, Mediocrity, and Servility to be always defied. 1778 P. Thicknesse *Year's Journey* (ed. 2) II. xlv. 103 All [negroes] have a degree of monkey cunning, and even monkey mischief. 1886 C. M. Yonge *Chantry House* I. xv. 142 He made no secret of his contempt for the insufferable dulness of the country, enlivening it by various acts of monkey-mischief. 1895 Kipling *2nd Jungle Bk.* 140 Ere Mor the Peacock flutters, ere the Monkey-People cry. 1910 W. de la Mare *Three Mulla-Mulgars* viii. 105 Mishcha and Môha..wouldn't touch monkey-skin. 1962 E. E. Evans-Pritchard *Ess. Soc. Anthropol.* v. 112 His head was covered with a monkey-skin cap.
17. a. **monkey bag** (see quot.); **monkey band** = *monkey orchestra*; **monkey bridge** *Naut.* (see quots.); **monkey business** orig. *U.S.*, foolish, trifling, or deceitful conduct (cf. Monkey *v.* 2); **monkey-chaser** *U.S. slang*, (*a*) a Negro from the West Indies or other tropical regions; (*b*) (see quot. 1952); **monkey coat** = *monkey-jacket*; **monkey-doodle business** *U.S. slang* = *monkey business*; **monkey-eating eagle**, also **monkey eagle**, the largest eagle found in the Philippine Islands, *Pithecophaga jefferyi*; **monkey forecastle** *Naut.* (see quot. 1948); **monkey gland**, a gland or testicle from a monkey, grafted on to a man as a possible means of rejuvenation; also *attrib.*; **monkey-house**, also *transf.* and *fig.*; **monkey-hurdler** *U.S. slang* (see quots.); **monkey island** *Naut. slang*, a small bridge above the pilot-house; **monkey-jacket** (earlier example; see also quot. 1968); **monkey-man**, (*a*) *U.S. slang*, a weak and servile husband; (*b*) a man resembling a monkey; **monkey meat** *U.S. Army slang*, tinned meat; *spec.* tinned beef-and-potato hash; **monkey orchestra**, a group of Meissen or other porcelain figures representing monkeys playing musical instruments; **monkey** (or **monkey's, monkeys'**) **parade** *slang* (see quots.); also **monkey-parading** *vbl. sb.* and *ppl. a.*; **monkey poop** *Naut. slang* (see quot. 1929); **monkey pox**, a virus disease of monkeys (and of human beings), similar to smallpox; **monkey's fist**, a thick knot made at the end of a rope to give it weight when it is thrown; **monkey shaft** *Austral.* and *N.Z. colloq.*, a small trial shaft; **monkey-shines** (earlier and later examples); also *sing.*; **monkey's island** *Naut. slang* = *monkey island*; **monkey strap** = sense 10* above; **monkey suit**, † (*a*) a type of child's suit (see quot. *a* 1901); (*b*) *slang* (orig. *U.S.*), a uniform; a formal dress suit, evening dress; **monkey's wedding** *S. Afr. colloq.*, a situation of alternating or simultaneous sunshine and rain; **monkey-trap** *colloq.* or *dial.*, something decorative worn by women to make themselves attractive to men; **monkey trial**, the trial in 1925 of a Tennessee school-teacher, J. T. Scopes, for teaching evolutionary theories; also *transf.* and *attrib.*; **monkey trick**, a mischievous, foolish, or underhand trick or act; an antic; usu. *pl.*; **monkey-wrench**, also *fig.*, esp. in colloq. phr. *to throw* (or *hurl*) *a monkey-wrench into the machinery*, etc.: to act as an obstruction or hindrance; to 'throw a spanner into the works'; hence as *v. trans.*, to turn with a monkey-wrench.
1847 H. Melville *Omoo* xxxiv. 134 A small leather wallet—a 'monkey bag' (so called by sailors)—usually worn as a purse about the neck. 1934 W. B. Honey *Dresden China* iv. 118 Some of the best known of all Meissen figures are the *Monkey Band*.., an assemblage of more than twenty separate figures with a conductor, which is said to have been modelled in ridicule of Count Brühl's orchestra. 1927 G. Bradford *Gloss. Sea Terms* 115/2 *Monkey bridge*, usually above the pilot or chart house where the standard compass is commonly set. Sometimes the fore and aft bridge on a sailing ship. 1967 C. Jokstad *Captain & Sea* 100 We placed the body on the monkey bridge, which is the very after end or stern of the vessel. 1883 G. W. Peck *Peck's Bad Boy* 109 There must be no monkey business going on. 1902 H. Hapgood *Spirit of Ghetto* vii. 209 The 'monkey business' of learning had ruined the child. 1934 C. Day Lewis *Hope for Poetry* vi. 29 A pandemonium of slogans, national anthems, headlines..manifestos, monkey business. 1972 'H. Carmichael' *Naked to Grave* vi. 79 Because I've seen her talking with one of the neighbours isn't to say there was any monkey-business between them. 1926 C. Van Vechten *Nigger Heaven* 286 *Monkey-chaser*, a Negro from the British West Indies. 1952 *New Yorker* 30 Aug. 15/3 Monkey chasers are gin and ice, with a little sugar and a trace of water. That term 'monkey chaser' comes from Georgia. 1859 *Times* 27 Jan. 9/4 Jones was at the time dressed not as he now is, but in what they call a 'monkey coat'. 1928 *Sat. Even. Post* 12 May 41/2 They ought not to go abroad with such monkey-doodle business. 1909 *Westm. Gaz.* 6 Sept. 5/3 The general plumage of the Monkey Eagle is a rich brown above and creamy white in the under parts. 1909 R. C. McGregor *Man. Philippine Birds* I. 226 (*heading*) *Pithecophaga jefferyi* Grant. Monkey-eating eagle. 1966 R. & D. Morris *Men & Apes* vi. 204 The heavy-billed, shaggy-crested *Pithecophaga* eagle of the Philippines is popularly called the monkey-eating eagle. 1971 J. E. DuPont *Philippine Birds* 47 (*heading*) Monkey-eating Eagle. 1873 *Rep. R. Comm. Unseaworthy Ships* 808/1 in *Parl. Papers* (C. 853) XXXVI. 315 Has a half poop and monkey forecastle. 1948 R. de Kerchove *Internat. Maritime Dict.* 473/1 *Monkey forecastle*, a short low forecastle open on the after side and used solely for anchor gear (windlass, and so on). 1924 G. B. Shaw *St. Joan* p. xix, Fortified..against old age by ..weekly doses of monkey gland. *Ibid.* p. xx, Which is the healthier mind? the saintly mind or the monkey gland mind? 1929 *Encycl. Brit.* XI. 747/2 The grafting into men of testicles from apes (the so-called 'monkey glands') has been practised by Voronoff and others with resulting rejuvenation. 1971 R. Dentry *Encounter at Kharmel* iv. 72 Stop talking like Noel Coward after a shot of monkey glands! 1914 'I. Hay' *Knight on Wheels* (ed. 2) xxiii. 221 The rooms..were described by the agent and Timothy as 'a lovely little bachelor suite' and 'a self-contained monkey-house' respectively. 1923 A. Huxley *Antic Hay* xvii. 246 The world would soon be a..bear-garden..and a monkey-house. 1924 J. Buchan *Three Hostages* ix. 131 It's up to the few sahibs like him in that damned monkey-house at Westminster to make a row about it. 1966 J. B. Priestley *Salt is Leaving* iv. 49 'Why do you have this window thing—here along the wall?' 'So I can see what's happening in my club... Come in later one night—..when it's all lit up down there—and you're looking through a window at a monkey house.' 1936 *Amer. Mercury* May p. x/2 Monkey hurdler, organist. 1951 W. Morum *Gabriel* II. ix. 250 Nelson's a monkey hurdler... He plays one of those Wurlitzer organs at the talkies. 1912 'Aurora' *Jock Scott* xix. 243, I was on Monkey Island (a pet name for the upper bridge) for hours. 1963 P. J. Abraham *Last Hours* xi. 134 Up on the monkey island he had realized there would be no power for the lights. 1830 N. Ames *Mariner's Sk.* 187 My wardrobe consisted of a 'monkey' jacket, bought in Gravesend, [etc.]. 1968 J. Ironside *Fashion Alphabet* 37 *Monkey-jacket*. This was originally a warm, heavy, thick, jacket worn by sailors... But more usually now it refers to a shorter waist-length jacket similar to a bell-boy, middy or mess jacket. 1924 J. E. & L. Martin (*song-title*) Monkey man. 1927 S. A. Brown in C. Cullen *Caroling Dusk* 132 Had a stovepipe blonde in Macon Yaller gal in Marylan In Richmond had a choklit brown Called me huh monkey man—Huh big fool monkey man. 1928 *Collier's* 5 May 36/3 A money man is a contemptuous thing among Negroes, friendless, distrusted, and generally despised. But a monkey man! Well, there's nothing lower than a monkey man. 1935 E. R. Burroughs *Pirates of Venus* iii. 54 Venus was inhabited! But by what?.. Were they a species of monkey-man? 1960 P. Oliver *Blues fell this Morning* 117 The man who is proud of his sexual prowess, real or imaginary, takes pleasure in ridiculing the 'monkey men': the West Indians and other Negroes that he despises whom he compares with the apes. 1966 *Southern Folklore Q.* XXX. 230 The Monkey Man is the servile mate who without protest turns over his hard-earned money to a woman... The husband who stayed and put up with it all became a Monkey Man and a standing joke. 1968 P. Oliver *Screening Blues* vi. 258 The 'creeping man' and the 'monkey man' are objects of contempt in the blues, but they are the lovers who succeed when the singer fails. 1918 in Cowing & Cooper *Dear Folks at Home* (1919) x. 87 He was crying pitifully, and we gave him cigarettes and some of our monkey meat and hardtack. 1919 *Red Cross Mag.* Feb. 37/1 When you hear the soldier's side of the story, his mess in camp and on the line consists of 'monkey meat' (canned beef-and-potato hash). 1929 *Papers Mich. Acad. Sci. & Arts* X. 309 *Monkey meat*, canned beef and potato-hash. 1906 R. L. Hobson *Porcelain* xiii. 122 A Frenchman named Acier.. worked at Meissen from 1764 till he was pensioned off in 1799, and modelled the celebrated Cries of Paris..and the Monkey Orchestra. 1960 R. G. Haggar *Conc. Encycl. Continental Pott. & Porc.* 22/2 Monkey orchestra: a series of twenty-one porcelain figures of monkeys playing instruments with a conductor, modelled by Kändler at Meissen in 1747. 1910 H. G. Wells *New Machiavelli* (1911) i. iii. 66 These twilight parades of young people, youngsters chiefly of the lower middle-class, are one of the odd social developments of the great suburban growths—unkindly critics..call them, I believe, Monkeys' Parades. 1914 E. Pugh *Cockney at Home* 116 If you don't know what a monkey parade is ask Anderson here.. 'It's a place where the elite of the beau monde of Suburbia meet nightly, for purposes of flirtation.' 1934 J. B. Priestley *Eng. Journey* iv. 183 Elderly citizens have been protesting against this practice of promenading on Sunday nights [in Bradford]. They have always been disgusted by the sight of young people monkey-parading in this fashion... They could easily do it in a much more civilised fashion than this of monkey-parading. *Ibid.* 185 A Sabbatarian town of this kind, which could offer its young folk nothing on Sunday night but a choice between monkey-parading and dubious pubs. 1929 F. C. Bowen *Sea Slang* 92 *Monkey poop* [misprinted *poor*], the half deck of a flush decked ship. 1959 P. von Magnus et al. in *Acta Path. & Microbiol. Scand.* XLVI. 156 The isolation of a virus from the diseased animals will also be described as well as some studies on the properties of the agent which in this paper will be referred to as monkey pox virus. 1960 *Ann. N.Y. Acad. Sci.* LXXXV. 957 The etiological agent of monkey pox has been isolated from naturally infected animals by using both embryonated chicken eggs and tissue cultures. 1971 *Daily Colonist* (Victoria, B.C.) 25 Aug. 5/1 Monkeypox, a disease very like smallpox, has been discovered for the first time in human beings in Africa. What caused the outbreak is a mystery. 1927 G. Bradford *Gloss. Sea Terms* 115/2 *Monkey's fist*, a complicated knot with weight enclosed, used at the end of a heaving line. 1961 K. Vonnegut *Sirens of Titan* (1967) i. 21 The hard, ball-like knot known as a *monkey's fist*. 1974 *Islander* (Victoria, B.C.) 29 Sept. 10/2 They were so high above us that we had to crane back our necks to look at them, waiting for the lethal lead-filled monkey's fists at the end of their lines to come whistling down. 1880 G. Sutherland *Tales of Goldfields* 69 They began to think they might be already too deep for it, and a small 'monkey'-shaft was therefore driven upwards from the end of the tunnel. c1832 T. D. Rice *Jim Crow* iii, I cut so many munky shines, I dance de gallopade. 1847 'H. Franco' *Trippings of Tom Pepper* I. vi. 43 Let me catch him cutting up any monkey shines in this house, and I'll bea[n] him! 1932 R. Frost *Let.* 13 Dec. (1964) 231 In a way it was a monkey-shine and he needn't have minded poetry's having a little the best of it for once. 1945 Monkey-shines [see *cut *v.* 59 o]. 1973 'H. Howard' *Highway to Murder* xi. 141 Why all the monkeyshines to get rid of Lucy? He'd been divorced before and he could be divorced again. 1917 'Taffrail' *Sub* iv. 110 He kept watch on the 'monkey's island'. 1919 W. Lang *Sea-Lawyer's Log* iii. 33 Now this 'ere is called the forebridge, and that little platform above it is the Compass Platform. Monkey's island, they calls it. 1929 W. Smyth *Bonzer Jones* i. 13 Put a monkey strap on [the buck-jumper for riding]! 1886 C. M. Yonge *Chantry House* I. i. 9 Mrs. Gooch had only to thrust her hand into the little pocket of his monkey suit to convict him on the spot. a1901 — *Autobiogr.* in C. Coleridge *C. M. Yonge* (1903) ii. 66 'Monkey suits', with jacket and waistcoat all in one, and trousers fastened over. 1920 R. Lardner in *Sat. Even. Post* 27 Nov. 42/4, I and the Mrs. and Kate was the only ones there in evening clothes. The others had attended these functions before and knew that they wouldn't be enough suckers on hand to make any difference whether you wore a monkey suit or rompers. 1928 *Daily Mail* 7 May 6/4 *Monkey suit*, flying suit. 1950 Dylan Thomas *Let.* 10 Jan. (1966) 343, I..demothed my monkey-suit and borrowed some proper shoes. 1974 A. Fowles *Pastime* v. 46 He could..hire one of those monkey-suits from Moss Bros. 1949 *Cape Times* 29 Nov. 16/3 The Peninsula had a 'monkey's wedding' rainfall yesterday with the sun shining at intervals and rain falling intermittently. 1961 *Ibid.* 8 Aug. 7 The weather was not exactly 'monkey's wedding' weather last week, sunshine alternating with a drizzle. 1931 W. de la Mare *Seven Short Stories* 85, I take you for an honest man's daughter with not a ha'penny to spare on fal-lals and monkey-traps. 1925 *Evening Sun* (Baltimore) 26 June 17/3 Washington is getting the reflex of increasing European interest in the so-called monkey trial. 1963 *Times* 30 Apr. 11/6 The student teachers, Miss Dori Doss and Miss Martha Powell, are apparently prepared for a second 'monkey trial' if necessary, though they realize that their determination could cost them their careers. 1969 D. F. Horrobin *Science is God* ii. 15 It is associated more with the summer heat of the monkey-trial country of Tennessee rather than with the coolness of the dreaming spires of Oxford. 1653 W. Denton *Let.* 13 Mar. in M. M. Verney *Memoirs* (1894) III. vi. 203 Bringe it [*sc.* a colt] with you in your coach, and then she will teach it all her Monkey tricks. 1742 Richardson *Pamela* (ed. 3) III. xxxiv. 408 Hold him fast, and play over all thy monkey tricks with him, with all my heart. 1809, 1864 Monkey trick [in Dict., sense 16]. 1910 G. K. Chesterton *G. B. Shaw* 152 Shaw treats vengeance as something too small for man—a monkey trick he ought to have outlived. 1951 Monkey trick [see *dot *v.*[1] 6]. 1958 C. Achebe *Things fall Apart* iii. xxv. 184 He had warned Obierika that if he and his men played any monkey tricks they would be shot. 1904 W. N. Harben *Georgians* 267 He..dug down in the road whar his pipe j'ined the main, till he got to it, an' then he monkey-wrenched it off. 1920 *Everybody's Mag.* May 36/3 Don't throw a monkey-wrench into the machinery! 1931 *Daily Express* 16 Oct. 1/2 Mr. Lloyd George hurled a monkey wrench last night into the creaking and decrepit machinery of Liberalism. 1966 D. Varaday *Gara-Yaka's Domain* xi. 126 Just as I was about to squeeze the trigger an ebb wind threw the monkey wrench into the works. The rising ill-wind struck the back of our necks, carrying our scent down to the elephant. 1971 *Black Scholar* Apr.–May 30/2 It is the black inmate who throws the monkey wrench into the works. 1975 *Jewish Chron.* 16 May 2/3 Mr Eban has thrown a monkey-wrench into the Israeli information campaign in the United States.

b. **monkey fiddle**, a West Indian tree or shrub, *Pedilanthus tithymaloides* or *P. angustifolius*, of the family Euphorbiaceæ; **monkey nut** (later examples); **monkey-pod** (**tree**) = *guango; **monkey-puzzle**, substitute for def.: a large, evergreen tree, *Araucaria araucana*, native to Chile and belonging to the family Pinaceæ, whose leaves are densely arranged to cover the whorled branches; also *fig.* (later examples); **monkey-puzzler** = *monkey-puzzle*; **monkey-rope** *S. Afr.*, one of several climbing plants, esp. *Secamone alpinii*, of the family Asclepiadaceæ.
1913 *Publ. Field Columbian Museum Bot. Ser.* II. 360 The peculiar silicious ridges of the stems and branches [of *Pedilanthus bahamensis*] produce a high squeak when they are rubbed together—children play at fiddling with them, hence the local name 'Monkey-fiddle'. 1954 *Caribbean Quarterly* III. i. 6 The duppy fiddle (more often called monkey fiddle) makes a grotesque squeaking when two sticks of it are rubbed together. 1896 H. L. Tangye *New S. Afr.* iv. 297 He brings a whole heap of sweet potatoes.. and monkey nuts, the latter being sometimes known as ground, or pea-nuts. 1916 [see *Arachis]. 1950 T. S. Eliot *Cocktail Party* iii. 142 *Alex.* Three of us have been out on a tour of inspection Of local conditions. *Julia.* What about? Monkey nuts? 1888 W. Hillebrand *Flora Hawaiian Islands* 115 *Pithecolobium Samang*, the Samang or Monkey-pod tree, enjoys great favour as a shade tree. 1937 D. & H. Teilhet *Feather Cloak Murders* ix. 136 They drove down the scented street, overshadowed by the great leafy monkey pod trees which rustled in the night breeze. 1969 T. H. Everett *Living Trees of World* 203/1 The rain tree, monkey pod or saman.., an evergreen or deciduous native of the West Indies and Central America now planted in many other tropical regions... has a wide-spreading, dome-shaped crown and grows up to 100 feet in height. 1923 Dallimore & Jackson *Handbk. Coniferæ* ii. 160 The common name of *Monkey Puzzle* is said to have originated in Cornwall... On one occasion when the owner of a plant was exhibiting it to friends a remark was passed to the effect that 'it would puzzle a monkey to climb that tree', and the owner forthwith adopted the name 'monkey puzzle'. 1938 *Times Lit. Suppl.* 14 May 342/2 The multitudinous little Careys of the future will have to be trained in mental agility on this veritable monkey-puzzle of a pedigree. 1940 [see *Chile]. 1969 T. H. Everett *Living Trees of World* 24/2 The dark green leaves of the monkey puzzle are broad, stiff, leathery and sharp-pointed. They overlap each other and clothe the curious, rather brittle snakelike branches closely. 1906 J. Galsworthy *Man of Property* iii. iii. 298 In the shade of a monkey-puzzler or in the lee of some india-rubber plant. 1936 E. Waugh *Mr. Loveday's Little Outing* 9 The weather..had suddenly blackened into a squall. There had been a scuttle for cover;..a table-cloth lofted to the boughs of the monkey-puzzler, fluttering in the rain. 1838 W. H. Harvey *Genera S. Afr. Plants* 221 S[*ecamone*] *Thunbergii*... known by the colonial name of 'Babianstouw' or 'Monkey-rope', is a voluble [*sic*] climbing shrub, not uncommon in our woods. 1849 [in Dict. s.v. sense 17 a]. 1876 H. Brooks *Natal* 125 The festooned stems of evergreen twiners hang down as 'monkey ropes'. 1907 T. R. Sim *Forests & Forest Flora Cape of Good Hope* 177 Both [species of *Vitis*] form 'Monkey-ropes', which, split up, are much used by the natives for tying down the thatch on hut-roofs. 1972 *Stand. Encycl. Southern Africa* VII. 515/1 Monkey-rope.. (*Secamone alpinii*.) Plant of the family *Asclepiadaceæ*, a scrambler on bushes and trees. When it grows in a forest its old stems form the well-known 'monkey-ropes' hanging down from the trees. *Ibid.* 515/2 Other climbers, also called monkey- or bush-rope, are *Rhoicissus capensis* (monkey grapes), *Cynachum obtusifolium*, and *Dalbergia* spp. ('doringtou'). The runners of most of these plants are particularly strong and are therefore used by baboons and monkeys for climbing.

monkey, *v.* Add: **2.** (Earlier and later examples.) Also, to fool or mess *about* or *around*; to waste time, or spend time aimlessly; to tamper *with*. So **mo·nkeying** *vbl. sb.* orig. *U.S.*
1881 I. M. Rittenhouse *Jrnl.* in *Maud* (1939) ii. 39 What with talking, running back and forth and general monkey-ing Clara slipped and fell. 1884 E. W. Nye *Baled Hay* 38 The young coyote may come and monkey o'er his grave. 1884 *Canon City* (Colo.) *Mercury* 22 Aug. 4/1 This reminds us of a sign in a Michigan planing mill, 'Dont Munkey with the Buz Saw.' 1916 *Dialect Notes* IV. 277 'What did you do after supper?' 'We just monkeyed around.' 1932 E. Waugh *Black Mischief* v. 168 I'm not going to have any monkeying about with my men. You'll lame the whole army in a day if you try to make 'em wear boots. 1939 G. B. Shaw *Geneva* iii. 98 A frontier is a frontier; and there must be no monkeying with it. 1955 *Times* 27 June 2/7 Any attempt to 'monkey about' with the powers or composition of the Upper House would destroy the balance of the constitution. 1957 *Times* 23 Sept. 11/1 Any departure from tradition form or colour would be as serious as monkeying about with the colour of an old school tie. 1972 J. W. Thompson in W. King *Black Short Story Anthol.* 257 Muhdear [*sc.* Mother] monkeyed with my collar again. And for what must have been the twentieth time, she smoothed my tie.

monkey-face. Add: **1.** (Further examples.) Also, a person with a monkey-like or funny face.
1936 C. Sandburg *People, Yes* lvi. 130 Where did ja cop dat monkeyface. *Ibid.* 131 Gwan monkeyface peddle yer papers. 1946 E. O'Neill *Iceman Cometh* (1947) i. 17 Hello, leedle Rocky! Leedle monkey-face! I did not recognise you. *Ibid.*, Laugh, leedle monkey-faces! Laugh like fools, leedle stupid peoples! 1959 I. & P. Opie *Lore & Lang. Schoolch.* ix. 171 A funny-faced person may also be called..'Monkey Face'.
b. A grimace. *West Indies colloq.*
1943 L. Bennett *Jamaican Humour in Dialect* 3 For wen she laugh it like she Dah meck up monkey face. 1946 E. N. Burke *Stories told by Uncle Newton* I. 31 The boys make monkey faces at me and I return the compliment with interest. 1952 S. Selvon *Brighter Sun* iv. 58 'Pewee, come an' go to school!' but Pewee made them a 'monkey face' and ran into the shop.
3. The monkey-faced owl. *U.S. colloq.* or *dial.*
1955 *Amer. Speech* XXX. 180 The barn owl is very generally called *monkey-faced owl*, a designation often shortened to *monkey face*.

monkey-faced, *a.* (Earlier and later examples.) spec. *U.S.*, *monkey-faced owl*, the barn-owl, *Tyto alba*.
1803 *Lett. Miss Riversdale* II. 284, I shall be so much my own enemy, as to take up with any monkey-faced battered rake. 1915 J. Buchan *Thirty-Nine Steps* i. 12, I went home, dressed, dined at the Café Royal, and turned into a music-hall. It was a silly show, all capering women and monkey-faced men. 1917 *Dialect Notes* IV. 426 *Hibou paille*, the American barn owl (Aluco pratincolo): also called *monkey-faced owl*. 1955 [see *monkey-face 3]. 1961 O. L. Austin *Birds of World* 154/1 Barn owls differ from the typical owls in their heart-shaped facial disk, which is so simian in aspect that the birds are often called 'monkey-faced owls'.

Mon-Khmer (mōu-n,k'mēəɹ). [f. *Mon *sb.*[2] and *a.* + *Khmer *sb.* and *a.*] The designation of a group of Indo-Chinese languages, of which the most important are Mon and Khmer, spoken in south-east Asia and considered by some philologists to belong to the Austroasian family. Also *attrib.*, with reference to the peoples who speak these languages.
1887 *Trans. Philol. Soc.* 1885–86 428 The..Indo-Chinese division and its two branches Mōn-Khmer, and Taï-Shan. *Ibid.* 435 The numerals and many words belong to the Mōn-Khmer family. 1904 G. A. Grierson *Linguistic Survey of India* II. 1 The Mōn-Khmēr languages are monosyllabic. 1932 [see *Khmer *sb.* and *a.*]. 1933 [see *Austric *a.*]. 1936 [see *Austro-[2]]. 1964 S. Wavell *Naga King's Daughter* ii. 36 The Semelairs, whose unique language is Mon-Khmer, had lived on the lake as long as they could remember.

Monmouth. 1. (Later examples.)
1777 *Maryland Jrnl.* 22 July (Th.), [He had on a] Monmouth cap, and old coarse shoes. 1938 A. L. Maycock *Nicholas Ferrar* viii. 204 The three masters in their black gowns and Monmouth caps followed by the boys, two and two.

monnick, monniker, varr. *moniker.

‖ **mono** (mǫ·no), *sb.*[1] [Sp.] A boiler-suit, workman's overalls (see also quot. 1937).
1937 F. Borkenau *Spanish Cockpit* ii. 123 The streets and cafés are full of militia, all of them dressed in their *monos*, the new dark-blue uniforms. 1965 C. D. Eby *Siege of Alcázar* (1966) i. 28 Many of them wore their regular overalls—loose-fitting blue boiler-suits called *monos*.

mono (mǫ·no). [The prefix Mono- abstracted from compounds in which it occurs.] **1.** Colloq. abbrev. of *monophonic *a.* 2; also as *sb.*[2], monophonic recording or reproduction.
1959 *Manch. Guardian* 4 Aug. 3/2 For the benefit of non-stereo collectors, most of these records have appeared also in mono. 1961 *Times* 1 July 11/5 The mono recording is fair enough. 1963 *Which?* Jan. 3/1 When a mono orchestral record..is reproduced, the sound from any one instrument generally appears to come from much the same direction as all the other instruments. 1971 M. McCarthy *Birds of America* 24 'Did it *have* to be stereo?' Peter groaned, homesick for their old mono set. 1972 K. Bonfiglioli *Don't point that Thing at Me* ii. 16 The Flying Scotsman whooped stereophonically... I competed in mono. 1974 *Listener* 14 Feb. 219/2 A radio play with a surplus of stereo trickery is not likely to be well received by listeners on mono sets.
2. Colloq. abbrev. of *mononucleosis.
1964 *Amer. Speech* XXXIX. 120 The general student preoccupation with internal disorders and other forms of illness, especially around examination time (cf. the frequent references among students to *mono* for 'mononucleosis'). 1970 *New Yorker* 9 May 45/3 After three sleepless, cold nights..he came down with something resembling mono. 1972 F. Knebel *Dark Horse* (1973) vi. 88, I had a chance to meet the Yugoslav ambassador last summer, but then I came down with mono.

mono: see *monosabio.

mono-. Add: **1.** **monoalphabe·tic** *a.*, denoting a cipher in which each letter corresponds to one letter of the normal alphabet; **mono·anthropism**, belief in the unity or indivisibility of mankind; **mo·noblast** *Biol.* [a. G. *monoblast* (O. Naegeli *Blutkrankheiten und Blutdiagnostik* (ed. 3, 1923) 143): see -blast], a cell that develops into a monocyte; **mo·nocable**, an aerial ropeway in which a single endless rope is used both to support the loads and to move them; usu. *attrib.*; **monocau·sal** *a.*, in terms of a sole cause; **monocistro·nic** *a.* *Genetics*, containing as much genetic information as is carried by a single cistron; **monoclo·nal** *Biol.* and *Med.*, forming a single clone; derived asexually from a single individual or cell; **monoco·lour** = *monochrome *sb.* 2 c; **monocoloured** *a.*: delete *rare*[−1] and substitute *rare*; (later *fig.* example); **monoconsona·ntal** *a.*, containing a single consonant; **mo·nocrystal**, a single crystal; so **monocry·stalline** *a.*, consisting of or constituting a monocrystal; **monode·ntate** *a.* *Chem.* [L. *dentātus*: see Dentate *a.*], (of a ligand) having only one point of attachment to a central atom; **monodiale·ctal** *a.*, speaking only one dialect; hence **monodiale·ctally** *adv.*; **mo·nodiet**, a diet confined to one type of food; **monodime·nsional** *a.*, existing in or having only one dimension; linear; **monodispe·rse** *a.* *Physical Chem.*, (of a dispersed phase) present as particles all of the same mass; (of a sol) containing such a phase; **monoenerge·tic** *a.* *Physics*, (consisting of particles) of the same energy; emitting radiation all of one energy; **mo·nofil, mo·nofilament**, a single strand of man-made fibre; freq. *attrib.*; **monofu·nctional** *a.* *Chem.*, having or corresponding to a single functional group per molecule; **mo·nogerm** *a.* *Agric.*, (of a sugar-beet variety or seed produced by it) having or consisting of only one seed in each seed ball; (cf. *multigerm* s.v. *multi- 3); **monogla·cial** *a.*, applied to the theory that there was only one ice age; so **monogla·cialism**, the monoglacial theory; **monogla·cialist** *sb.* and *a.*, (of or pertaining to) a supporter of this theory; **mo·nolog** *Linguistics* (see quots.); **monomi·ctic** *a.* [Gr. μικτός mixed], applied to a lake in which there is only one overturn each year; **monominera·lic** *a.* *Petrol.*, composed of a single mineral; **mo·nopulse**, used *attrib.* and *absol.* to designate (the mode of operation of) radar in which the direction (and usu. also the range) of a target is precisely determined from a single echo pulse that is detected using two or more adjacent aerials; **mo·no-sentence** *nonce-wd.*, a brief or insufficient sentence; **mo·no-sound** *rare*[−1], an abrupt single sound; **mono-speci·fic** *a.* *Biol.*, affecting or containing only one species; **monosta·tic** *a.*, (of radar) having a single aerial as both transmitter and receiver; **monosymptomatic** *a.*, also in *Psychol.* and more general use; **monothetic** *a.*, (*b*) *Taxonomy*, (of a classification) having groups formed on the basis of a single characteristic, or a series of single characteristics; hence **monothe·tically** *adv.*; **mo·notower**, used *attrib.* and *absol.* to designate a crane whose jib is mounted on a single tower; **mono-unsa·turated** *a.* *Chem.*, (of a compound) saturated except for one multiple bond (usu. between carbon atoms) at which addition can normally occur; **monoxe·nic** *a.* *Parasitology* [Gr. ξένος stranger], applied to a culture, or the cultivation, of a parasitic organism in the presence of a single other species; **monoxe·nous** *a.* *Parasitology* [a. G. *monoxen* (A. de Bary 1867, in *Bot. Zeitung* XXV. 264/1), f. Gr. ξένος stranger], of a parasitic plant or animal: restricted to a single host species during its entire life-cycle.
1927 *Daily Express* 24 Nov. 13 The most simple code is the 'monoalphabetic', in which a letter is always represented by the same sign, letter, or numeral. 1950 A. Huxley *Themes & Variations* 259 The best antidote to nationalistic idolatry is a monotheism with its corollary (since God's fatherhood implies men's brotherhood) of monoanthropism. 1925 *Contrib. Embryol. Carnegie Inst. Washington* XVI. 241 Naegeli..derives the monocyte from the myeloblast of the bone-marrow..and terms this young cell the 'monoblast'. 1938 H. Downey *Handbk. Hematol.* I. v. 386 Ehrich has taken..exception to Bloom's conclusion that monoblasts other than lymphocytes and myeloblasts do not exist. 1968 Passmore & Robson *Compan. Med. Stud.* I. xxvi. 12/2 These cells [*sc.* monocytes] are produced mainly in the spleen and lymph nodes. The parent cell, the monoblast, is like the myeloblast and lymphoblast, a cell of about 15 μm in diameter with a large nucleus, containing 2–3 nucleoli, filling much of the cell cytoplasm. 1926 H. Blyth *Mod. Telpherage & Ropeways* xiii. 111 The mono-cable system has been developed very extensively in this country, and although formerly considered as being suitable only for light duty and small gradients, has now proved to be quite suitable for capacities up to 150 tons per hour and gradients of 1 in 2½. 1958

I. C. F. STATHAM *Coal Mining Pract.* II. vii. 509 The Mono-Cable is the oldest and simplest, being first used in Dantzig in 1644. **1964** *Economist* 14 Nov. 745/2 Using [for bridges] a mono-cable instead of dual cable suspension. **1973** *Gloss. Terms Materials Handling (B.S.I.)* vii. 8 *Mono-cable trestle*, a structure supporting a mono-cable ropeway between the ropeway stations. **1957** C. HUNT *Guide to Communist Jargon* xxiv. 86 Lenin's doctrine illustrates the weakness of monocausal explanations of highly complex phenomena. **1972** R. PLANT in Cox & Dyson *20th-Cent. Mind* I. v. 161 Weber conceded that an economic interpretation of history and social life was possible, but he was not willing to accept any monocausal view in its entirety. *Ibid.* 165 Against the Marxist monocausal view of history therefore, Weber poses a far more complex picture. **1965** *Proc. Nat. Acad. Sci.* LIV. 1193 For these purposes, one would prefer a monocistronic message containing information for only one protein. **1972** *Nature* 21 Jan. 131/2 Each segment is transcribed into a single messenger RNA and at least eight of these ten molecules are translated as monocistronic messengers. **1914** W. E. AGAR in *Phil. Trans. R. Soc.* B. CCV. 422 When a population is known to have been descended asexually from a single common ancestor, it consists of a single clone and may be described as monoclonal. *Ibid.* 442 The monoclonal population was descended from 70 original ancestors, themselves all descended parthenogenetically from a single known female. **1973** *Sci. Amer.* Aug. 44/1 Are the plaque cells transformed cells, in which the genetic material has somehow become changed from that of the normal cell population? If so, the plaque cells could be expected to be monoclonal: a homogeneous population derived from a single cell, rather than a mixed population of cells as in normal human tissue. **1963** *Listener* 31 Jan. 220/2 The idea of using full orchestra was avoided... To paint a picture which should have universality of expression..the monocolour of strings seemed the only possible medium. **1955** *Times* 25 June 5/6 The parties of the extreme left and those of the right favour a 'mono-coloured' Government..since such a one-party Government..would be more or less permanently at the mercy of right or left. **1948** D. DIRINGER *Alphabet* x. 161 The consonantal principle and the selected symbols to represent mono-consonantal words were used by the Egyptians at the beginning of the third millennium B.C. **1965** *Canad. Jrnl. Linguistics* Spring 154 In other cases we find monoconsonantal roots in one language cognate with biconsonantal forms in the other. **1934** *Rev. Sci. Instr.* V. 402/2 Monocrystals with dimensions of the order of 10 cm were readily produced. **1946** *Nature* 10 Aug. 199/2 The X-ray density must be related to monocrystals, and if indeed the starch grain were a monocrystal the numbers found might be considered as a proof. **1973** *Ibid.* 16 Nov. 124/1 Of future plans perhaps the most striking was V. Braginsky's..idea of using a massive monocrystal instead of an aluminium bar with a resultant very high quality factor to look for the sort of signals astronomers expect from nearby galaxies. **1934** *Rev. Sci. Instr.* V. 402/2 The test showed the main body of the growth to be monocrystalline with the seed. **1946** *Nature* 24 Aug. 261/1 An extensive study was made by R. Jacquersson on the effect of torsion on a monocrystalline rod. **1968** C. G. KUPER *Introd. Theory Superconductivity* vi. 97 Pippard.. made measurements with a number of monocrystalline cylindrical specimens of tin. **1949** KIRK & OTHMER *Encycl. Chem. Technol.* IV. 382 Monofunctional ligand groups..are sometimes known as unidentate or monodentate groups. **1974** *Jrnl. Inorg. & Nuclear Chem.* XXXVI. 1221/1 The behaviour of ethylenediamine as a monodentate ligand has also been reported in a few other cobalt..and chromium..complexes. **1972** J. L. DILLARD *Black English* iii. 85 Reading and writing almost presuppose a certain degree of mastery of Standard English, since nothing was written for the monodialectal speaker of Plantation Creole. **1964** *Archivum Linguisticum* XVI. 73 Martinet..[distinguishes] two senses of the term *dialect*: dialect$_1$, a variety of a language..used monodialectally within a whole language area,..and *dialect$_2$*, a local dialect within a larger dialect$_1$ area. **1920** *Chambers's Jrnl.* 1 May 349/2 No matter what the advocate of a mixed-food diet may say to the contrary, there is no disputing the fact that the nearer one gets to the mono-diet the better the health will be. **1953** 'I. DEVI' *Forever Young Forever Healthy* v. 61 The mono-diet..is the kind of diet where you eat only *one* type of food. It makes no difference whether it is only milk,..or anything else. **1952** *New Biol.* XII. 100 Besides such two-dimensional layers of more or less loosely fixed molecules there certainly must exist some analogous linear phenomena along the edges of crystals, and this would make monodimensional strings or chains of adsorbed particles. **1959** *Brno Studies in English* I. 31 The monodimensional character of spoken utterances..as opposed to the..polydimensional character of written utterances. **1925** G. BARGER tr. *H. Freundlich's Elements Colloid Chem.* 132 The preparation of colloidal solutions containing particles of uniform size—so-called unidisperse (monodisperse) sols—is mostly very difficult. **1940** GLASSTONE *Text-bk. Physical Chem.* xiv. 1232 By means of the ultracentrifuge it has been found that many protein solutions are monodisperse. **1971** *Europ. Jrnl. Biochem.* XXIII. 575 The binding of human serum albumin by monodisperse latex particles was studied. **1950** GLASSTONE *Sourcebk. Atomic Energy* xi. 310/1 Another source of monoenergetic neutrons is the $Li^7(p,n)Be^7$ reaction, the energy of the incident protons being carefully controlled. **1954** *Sci. News* XXXIII. 41 If, therefore, we bombard a nucleus of a particular element with a monoenergetic beam of particles, the nuclei which are excited as a result of the collisions taking place will have a wide range of excitation energies. **1971** *Nature* 3 Dec. 261/1 This factor can be experimentally determined through the use of mono-energetic gamma-ray sources. **1916** CROSS & BEVAN *Text-bk. Paper-Making* (ed. 4) i. 10 The [cellulose] solution has been extensively applied to the production of artificial threads of high lustre, so-called 'artificial silk', and various grades of Monofil (*Crin, Crinole*). **1940** *Times* 21 Mar. 5/5 The other two factories will convert the raw material into intermediate products —monofil at the works of I.C.I. [etc.]. **1972** *Trout & Salmon* June 58/2 Nylon monofil slips more easily through the rod-rings. **1949** *Mod. Plastics* Nov. 69/1 Nylon monofilaments are well established as a material for bristles. **1960** *Aeroplane* XCVIII. 43/3 The chafer covering the bead is of monofilament nylon, as in the case of all other Firestone tubeless tyres, to prevent capillary action of the air and so assist in the maintenance of pressure. **1960** M. SHARCOTT *Place of Many Winds* vii. 115 These lures..are attached to nylon monofilament lines in varying lengths. **1973** *Sci. Amer.* Dec. 134/1 New forms of nylon rope without stretch (a braided sheath around a monofilament core) are used for the long, dark climb. **1946** *Jrnl. Amer. Chem. Soc.* LXVIII. 360/2 The monofunctional trimethylchlorosilane acts as a growth terminating agent or 'chain-stopper'. **1949** [see *monodentate* adj. above]. **1964** *Brit. Med. Bull.* XX. 91/1 The carcinogenic action of a series of monofunctional ethyleneimines. **1971** *Nature* 10 Dec. 324/1 When the number of bonds was anything less than twelve, the bifunctional reagent was able to react with only a monofunctional stoichiometry. **1974** *Analytical Chem.* XLVI. 344 (*heading*) Potentiometric titration of monofunctional bases in ion exchanger–aqueous solution medium. **1950** *Proc. Amer. Soc. Sugar Beet Technologists* VI. 156 Monogerm sugar beet varieties have not been developed in either Europe or America. Failure to develop such varieties is due to difficulties in detecting monogerm plants. **1971** *Farmers Weekly* 19 Mar. 38/3 Modern monogerm varieties mean there is less worry about bolting than before. **1972** *Nature* 21 Jan. 136/2 Larks still destroy young seedlings, spaced scientifically and grown from monogerm seeds. **1914** W. B. WRIGHT *Quaternary Ice Age* vi. 125 We are bound to take our stand on the comparatively simple monoglacial hypothesis until we can prove at least one interglacial period. **1961** L. D. STAMP *Gloss. Geogr. Terms* 323/1 The monoglacial theory considers the Pleistocene ice-sheet to have expanded and contracted in only one general movement without any substantial 'interglacial periods' of recession followed by advance. **1957** J. K. CHARLESWORTH *Quaternary Era* II. xxxvi. 920 Upholders of monoglacialism are becoming fewer, and discussion is turning more and more upon the number of the epochs rather than upon the question of their existence. **1959** F. E. ZEUNER *Pleistocene Period* ii. 54 (*heading*) Monoglacialism. **1914** W. B. WRIGHT *Quaternary Ice Age* vi. 125 The Cyprina Clays of Denmark.. were referred by Geikie..to one of the interglacial periods, but..it was very justly held by the monoglacialists that they might easily be of preglacial age. **1946** *Antiquity* XX. 10 Controversy raged with considerable acrimony between the monoglacialist and polyglacialist exponents of Pleistocene geology. **1968** R. W. FAIRBRIDGE *Encycl. Geomorphol.* 458/2 For a long time, a considerable controversy persisted between monoglacialists and polyglacialists, eventually to be conclusively won by the latter. *Ibid.* 913/1 For a long time, there was a minority belief in only one glacial epoch, the so-called monoglacialist school, and a few representatives still survive. **1929** H. E. PALMER *Rep. Res. Activities 1928–9 Inst. Res. Eng. Teaching* (Tokyo) 9 Each of these white spots is a word, that which is contained in a ring is not a *word* but a *word-family*... I propose to call the spots *monologs* and the rings *monologemes* (the termination *on* suggests the individual unit, and the termination *eme* suggests the group unit..). There are times however when..instead of the cumbrous 'monologons or monologemes' we may use the convenient common form *monologs*. **1947** H. BONGERS *Hist. & Princ. Vocab. Control* I. iii. 56 The vocabulary of a language consists of:..Units neither more nor less than single words written without a break... These have been conveniently termed '*monologs*'. **1956** *Proc. Nat. Acad. Sci.* XLII. 84 For the three categories of Forel's classification [*sc.* polar, temperate, and tropical] the terms cold monomictic, dimictic, and warm monomictic have been used by one of us [*sc.* Hutchinson]..in a forthcoming extensive work on limnology. **1957** G. E. HUTCHINSON *Treat. Limnol.* I. vii. 438 It is..proposed to use the term dimictic for any lake circulating twice a year; warm monomictic will be employed as a substitute for tropical, implying winter circulation above 4°C; and cold monomictic as a substitute for polar, implying summer circulation below 4°C. **1968** R. W. FAIRBRIDGE *Encycl. Geomorphol.* 603/2 The monomictic lake waters [of Lake Atitlán, in Guatemala] appear to be fully circulating in the dry season, but Deevey (1957) found that in August the temperature at the surface is 24°C versus 20°C at the bottom. **1917** *Jrnl. Geol.* XXV. 243 The inquiry gives considerable support to the belief that the monomineralic rocks..are generated by the process of collection of crystals under the action of gravity. **1956** *Mineral. Mag.* XXXI. 120 Specimens (24) to (28) inclusive are practically monomineralic rock specimens from which small, platey single crystals may be obtained. **1970** *Nature* 17 Oct. 258/2 The extreme composition range of the lunar sprays can be most readily explained by solar flares melting minute, essentially monomineralic particles on the lunar surface. **1955** *IRE Convention Record* VIII. 132/2 A third method for generating both positive and negative errors simultaneously..is known as monopulse radar, since all information on range and angle errors is generated from each single pulse. This very satisfactory name was proposed by Bell Telephone Laboratories engineers. **1965** D. K. BARTON in R. S. Berkowitz *Mod. Radar* vii. 603 (*heading*) Operation of amplitude-comparison monopulse. *Ibid.* 598 In return for the added complexity in the receiving and antenna system, the monopulse radar offers reduction in tracking errors caused by target scintillation, an improved data rate.., and efficiency in overcoming thermal noise errors on distant targets. **1966** *McGraw-Hill Encycl. Sci. & Technol.* VIII. 577/2 An additional advantage of monopulse tracking as compared to conical scan is that no mechanical action is required in monopulse. **1970** *Sci. Jrnl.* May 46/1 Modern radars, such as those of the chirp or monopulse types, need compression filters to convert long shallow pulses into tall brief ones and expansion filters to do the opposite. **1817** KEATS *Let.* Mar. (1931) I. 13 Your kindness affects me so sensibly that I can merely put down a few mono-sentences. **1875** *Gentl. Mag.* XV. 445 His laugh—with an abrupt, short, monosound— more like a short gasp or snort than a laugh. **1951** WHITBY & HYNES *Med. Bacteriol.* (ed. 5) xvii. 293 One such mono-specific serum thus appropriately prepared will only agglutinate *Br[ucella] melitensis*. **1964** A. L. THOMSON *New Dict. Birds* 309/2 Flocks may be monospecific or composed of several species. **1970** *Nature* 22 Aug. 779 (*caption*) *Digenea subarticulata*..belongs to a genus formerly thought to be monospecific. **1974** *Ibid.* 1 Nov. 78/2 The reactivity of the DeV reagent was abolished by previous incubation with a monospecific rabbit antihuman IgG serum. **1957** R. WATSON-WATT *Three Steps to Victory* xxii. 128 We had already erected a direction-finding hut so close to the others as to show that we were ..confidently counting on wholly monostatic working. That is to say we would obtain all the information required from a sector of our front by observations made at a single site accommodating both transmitter and receiver. **1973** *Nature* 9 Mar. 109/2 Some work has been reported in which a steerable monostatic radar has been pointed obliquely in several directions successively, so as to measure more than one velocity component. **1937** G. W. ALLPORT *Personality* (1938) p. ix, Cultural determinism is one of the monosymptomatic approaches; it has a blind spot for the internal balancing factors and structural tenacity within personality. **1965** Monosymptomatic [see *conversion hysteria*]. **1972** H. J. EYSENCK *Psychol. is about People* iii. 111 Her patients were all..suffering from serious and complex phobic anxieties; there were no monosymptomatic cases. **1962** P. H. A. SNEATH in Ainsworth & Sneath *Microbial Classification* 291 Special classifications are usually based on single characters, or on a series of single characters... Beckner..calls such groups monotypic, because the defining set of features is unique, but since monotypic has other meanings, monothetic is a better term. **1965** *Math. in Biol. & Med.* (Med. Res. Council) III. 82 (*caption*) Group Y is monothetic, and is subdivided monothetically. **1969** E. MAYR *Princ. Systematic Zool.* iv. 83 When the definition of the logicians— 'individuals sharing common characters'—was replaced by 'members of a group having descended from a common ancestor', a monothetic characterization of a taxon was no longer necessary. **1938** *Engineer* 3 June 615/1 Messrs. Butters Brothers are showing..a 'Monotower' electric crane. **1953** *Dock & Harbour Authority* XXXIV. 88/2 The design of the mast for the monotower crane. **1963** *Engineering* 19 Apr. 532 During erection as a monotower up to 90 ft, the crane grows from the bottom. **1971** B. SCHARF *Engin. & Its Lang.* xvi. 230 Monotower cranes..consist of a latticed tower surmounted by a derrick crane. **1972** J. R. ILLINGWORTH *Movement & Distribution Concrete* vi. 84 With monotowers, full circle slewing is obtained, but static base operation limits coverage to the area swept out by a particular jib length. **1939** *Thorpe's Dict. Appl. Chem.* III. 241 (*table*) Fully saturated triglycerides... Mono-unsaturated-di-saturated glycerides. **1950** *Analytical Chem.* XXII. 1261/1 Extinction coefficients..are reported for seventeen pure cis and trans monounsaturated and saturated acids, esters, and alcohols. **1965** E. D. WILSON et al. *Princ. Nutrition* (ed. 2) v. 35 The quantity of the monounsaturated fatty acids in the average American diet matches that of the saturated ones, about 40 per cent. **1953** E. C. DOUGHERTY in *Parasitology* XLII. 259/2 Such an expression as 'two-membered culture' becomes 'monoxenic culture'. **1969** *Oikos* XX. 287 (*heading*) The monoxenic cultivation of some rhabditid nematodes. *Ibid.* 288/1 Some nematode species were studied under controlled conditions on monoxenic agar cultures. **1940** B. G. CHITWOOD in J. R. Christie *Introd. Nematology* II. iv. 243/2 Parasites of animals may also be classified according to the number of hosts necessary for completion of the life cycle. Species in which there is a single host are termed monoxenous. **1962** W. P. ROGERS *Nature of Parasitism* x. 219 A specific parasite (sometimes called species-specific or monoxenous) can live on or in one species of host only. **1973** N. D. LEVINE *Protozoan Parasites* (ed. 2) i. 5/1 A monoxenous parasite has only one type of host—the definitive host. *Ibid.* 5/2 The term monoxenous parasite is used by some authors for a parasite which is restricted to a single host species.

2. Also used to indicate monomeric compounds (e.g. *mononucleotide* below). **mono·ethano·lamine**, 2-aminoethanol $H_2NCH_2CH_2OH$, a viscous high-boiling liquid used in making detergents; **monogly·ceride**, any compound consisting of glycerol esterified at only one of its hydroxyl groups; **mononu·cleotide** [ad. G. *mononucleotid* (Levene & Mandel 1908, in *Ber. d. Deut. Chem. Ges.* XLI. 1906)], any of the compounds whose molecules are formed from a single molecule each of phosphoric acid, a sugar, and a heterocyclic base, and are the units of which nucleic acids are composed; a monomeric nucleotide.

1929 *Chem. Abstr.* XXIII. 3232 If 352 parts of ethylene oxide is added to 3400 parts of 25% aq. NH_3 at 10°, monoethanolamine is obtained. **1959** *Times Rev. Industry* Jan. 67/1 The CO_2 removed by scrubbing with monoethanolamine is used as a purge gas. **1972** *Materials & Technol.* IV. xv. 516 Since monoethanolamine reacts more readily with ethylene oxide than does ammonia, excess of the latter and careful control of the ethylene oxide concentration is essential for appreciable amounts of the mono-alcohol to be formed. **1860** H. DEBUS in *Q. Jrnl. Chem. Soc.* XII. 243 (*heading*) Monoglycerides. **1926** G. D. ELSDON *Edible Oils & Fats* iv. 17 The β monoglycerides have been prepared by using dichlorhydrin. **1954** E. W. ECKEY *Veg. Fats & Oils* iv. 143 An important practical application of alcoholysis of fats is the preparation of monoglycerides which are used in shortenings to improve their cake-making properties and as emulsifiers... Monoglyceride mixtures are prepared also as a first step in the manufacture of oil-modified alkyd resins, and in the preparation of certain detergents. **1972** *Lipids* VII. 433/1 Palmitic, stearic and oleic acids were the predominant fatty acids present in the monoglycerides. **1908** *Jrnl. Chem. Soc.* XCIV. I. 587 The composition of the thyminglucophosphoric acid is almost identical with that of heminucleic acid.., and may be considered to be a mononucleotide. **1956** *Nature* 11 Feb. 271/1 Preparations or fractions of ribonucleic acid were hydrolysed in *N* potassium hydroxide to mononucleotides. **1968** H. HARRIS *Nucleus & Cytoplasm* iii. 55 The mononucleotides released from this material by alkali hydrolysis are contaminated with other compounds. **1974** *Nature* 1 Nov. 75/2 The data show total ^{32}P c.p.m. in the RNA and in the triphosphate termini as mononucleotides and pppNps released by KOH hydrolysis of that RNA.

monoamine (mǫno,ǣi·mīn). *Biochem.* [f. MONO- 2 + AMINE.] **a.** Any compound having a single (primary, secondary, or tertiary) amine group in its molecule, *spec.* one which is a neurohormone; = MONAMINE (in Dict. and Suppl.).

1951 I. L. FINAR *Org. Chem.* I. xiii. 260 Diamines may be prepared by methods similar to those for the monoamines, using alkylene halides instead of alkyl halides. **1965** J. POLLITT *Depression & its Treatment* iv. 53 Monoamines are neurohormones regarded as essential for normal activity of the brain. **1972** *Nature* 31 Mar. 225/2 There is much evidence pointing to changes in monoamine metabolism in mental illnesses characterized by alterations in mood.

b. *Comb.* **monoamine oxidase**, the enzyme or group of enzymes that brings about the oxidation (and consequent inactivity) of monoamines in the body; **monoamine oxidase inhibitor**, any of a class of drugs which inhibit the functioning of monoamine oxidase (so allowing accumulation of serotonin and noradrenaline in the brain), some of which are used as antidepressants.

[**1943**: cf. *MONAMINE.] **1951** *Proc. Soc. Exper. Biol. & Med.* LXXVIII. 157/1 Despite numerous investigations of its enzymatic function, monoamine oxidase has not been significantly purified. **1962** *Lancet* 29 Dec. 1343/1 There is a strong impression that iproniazid is clinically more effective in relieving depression than other monoamine-oxidase inhibitors. **1965** J. POLLITT *Depression & its Treatment* iv. 47 Methedrine and other amphetamines are incompatible with monoamine oxidase inhibitors. **1973** *Nature* 6 Apr. 417/1 Roulston, studying the action of chlordimeform and related compounds on the southern cattle tick, suggested that inhibition of monoamine oxidase (MAO) could be involved.

monoao, var. *MANOAO.

monobloc (mǫ·noblǫk), *a.* (*sb.*) [Fr. (1906 in Robert); cf. MONO- 1 and BLOC(K *sb.*] Made as, or contained within, a single casting. Also as *sb.*, a monobloc engine.

1909 *Westm. Gaz.* 2 Nov. 5/1 The engine of the 20-h.p. model..is of the monobloc order. **1929** *Radio Times* 8 Nov. 448 (Advt.), Batteries beyond present standards... Above is the LDG, 2v. 60 a.h... Below: the monobloc 10v. DMHG with special elements. **1930** *Engineering* 26 Sept. 401/3 The cylinders form a monobloc casting. **1944** *Jrnl. R. Aeronaut. Soc.* XLVIII. 365 In liquid-cooled engines, separate steel cylinders were generally abandoned in favour of monobloc light alloy castings. *Ibid.*, At the outbreak of the present war the monoblocs in use were still not entirely satisfactory. **1963** *Engineering* 2 Aug. 140/3 Of monobloc construction, the new pump is available in cast iron. **1972** *Shooting Times & Country Mag.* 24 June 14/2, I must confess to being disappointed that our top-flight English gunmakers..do not offer an alternative. I visualise a self-opening boxlock, with monobloc barrels, [etc.].

monocaryon, -caryotic, varr. *MONOKARYON, -KARYOTIC *a.*

monochromasy (mǫnokrōu·măsi). *Ophthalm.* Also -chromacy. [f. MONO- + Gr. χρῶμα colour: see -Y³.] = *MONOCHROMATISM b.

1900 *Stud. Yale Psychol. Lab.* VIII. 15 Still another form of color-vision is found in monochromasy. All the visible objects are seen as shades of one color. What this color is, it has as yet been impossible to say. **1922** W. PEDDIE *Colour Vision* ii. 16 The most extreme condition is that of total colour blindness in which every colour in the spectrum can be matched with grey. It is not inconceivable that a case of this kind may be one of true monochromasy, the one colour extending throughout the whole spectrum. **1957** *Jrnl. Optical Soc. Amer.* XLVII. 338/2 (*heading*) 'Blue mono-cone monochromacy': a new color vision defect. **1971** *Jrnl. Physiol.* CCXII. 211 (*heading*) Colour vision in blue-cone 'monochromacy'.

monochromat (mǫnokrōu·mĕt), **-chromate** (-krōumeit), *sb.* *Ophthalm.* [ad. L. *monochrōmat-us*, Gr. μονοχρώματ-ος: see -ATE¹.] A person who suffers from monochromatism (sense *b).

1902 J. M. BALDWIN *Dict. Philos. & Psychol.* II. 793/2 König affirmed that the vision of the faint-light 'monochromates' was in quality blue. **1915** J. H. PARSONS *Introd. Study Colour Vision* II. i. 159 Differing from all these groups there are people who apparently see all parts of the spectrum of one hue... These are the total colour-blind or monochromats. **1956** *Brit. Jrnl. Ophthalm.* XL. 463 This female 'cone monochromate' would thus have to be a compound heterozygote for protanopia and deuteranopia. **1971** *Nature* 9 Apr. 395/1 The recovery of the rod sensitivity after a long full bleach can be followed for a wider than normal range when an area of the monochromat's retina which has few cones is tested.

monochromate (mǫnokrōu·mēit), *v.* [Back-formation from *MONOCHROMATOR.] *trans.* = *MONOCHROMATIZE *v.* So **mo·nochromated, -chromating** *ppl. adjs.*

1955 G. E. BACON *Neutron Diffraction* iv. 80 For any position *XY* of the monochromating crystal there will be some particular wavelength which is at the right glancing angle θ to the crystal to undergo Bragg reflection when it is incident along *PB*. **1955** *Rev. Sci. Instruments* XXVI. 564 A bent crystal monochromator..was constructed to monochromate the specimen-reflected x-ray beam. *Ibid.*, X-ray diffraction laboratories are making increasing use of single crystal monochromated radiation for various types of investigations. **1961** J. THEWLIS et al. *Encycl. Dict. Physics* IV. 827/2 Normally, the monochromating crystal is used in conjunction with a neutron spectrometer. **1971** *Nature* 16 Apr. 436/1 The aim of our experiments was..to check that there was no large disparity between the observed and calculated flux of monochromated synchrotron radiation.

monochromatic, *a.* Add: **1.** Also applied to other radiation that is (nominally) of a single wavelength or energy throughout. (Further examples.)

1923 GLAZEBROOK *Dict. Appl. Physics* IV. 22/2 The crystal is pulverised and the powder compressed into a rod, placed in the axis of a cylindrical photographic film and subjected to 'monochromatic' X-rays. **1955** C. R. N. STROUTS et al. *Analytical Chem.* II. xxvi. 942 It is desirable to use essentially monochromatic radiation for most analytical applications of X-ray diffraction. **1967** R. CASTAING et al. in A. Septier *Focusing Charged Particles* II. iv. 289 An incident monochromatic electron beam.

3. *Ophthalm.* Exhibiting or affected with monochromatism (sense *b); completely colour-blind.

1902 J. M. BALDWIN *Dict. Philos. & Psychol.* II. 793/2 The sensations of those individuals whom they named 'monochromatic'..were dogmatically affirmed to be vision under the form of red or blue or green, it was uncertain which. **1956** *Brit. Jrnl. Ophthalm.* XL. 470 His vision is monochromatic when examined in the colour apparatus. **1974** *Sci. Amer.* Dec. 25/3 The ophthalmologist testified that the witness was actually monochromatic, which meant he could perceive no colors at all.

monochromatically *adv.*, also *fig.*; also **monochromati·city**, the condition of being monochromatic; the extent to which any radiation is concentrated at one wavelength or frequency.

1960 S. TOLANSKY *Surface Microtopography* i. 5 For two-beam work the Wratten 77A didymium glass filter gives excellent monochromaticity. **1963** G. TROUP *Masers & Lasers* (ed. 2) vii. 113 The advantages of maser optical sources are: the extreme monochromaticity, resulting in huge spectral energy-densities; [etc.]. **1972** *Daily Tel.* 31 May 10/3 Janacek's Sonata '1905'..was massively though rather monochromatically delivered. **1972** *Nature* 22 Dec. 483/1 The intensity, coherence and monochromaticity of laser light have made it an ideal radiation source for selective alteration of cell organelles.

monochromatism (mǫnokrōu·mătiz'm). [f. MONOCHROMAT(IC *a.* + -ISM.] **a.** (In Dict. s.v. MONOCHROMATIC *a.*)

b. *Ophthalm.* Complete colour-blindness, in which all colours appear as shades of one colour, probably grey.

1934 in WEBSTER. **1946** W. D. WRIGHT *Res. Normal & Defective Colour Vision* xxiv. 297 In monochromatism, all qualitative differences in stimuli disappear, so that colour matching is reduced to brightness matching. **1956** *Brit. Jrnl. Ophthalm.* XL. 462 There exist rare cases of monochromatism in which the characteristic symptoms accompanying typical total colour blindness are lacking. **1966** *Arch. Internal Med.* CXVIII. 491/2 This report describes a family with atypical congenital monochromatism.

monochromatize (mǫnokrōu·mătəiz), *v.* [f. MONOCHROMAT(IC *a.* + -IZE.] *trans.* To make monochromatic. So **monochro·matized, monochro·matizing** *ppl. adjs.*

1951 *Jrnl. de Physique* XII. 268/1 It is..necessary to 'monochromatize' the beam. *Ibid.* 268/2 The monochromatized neutrons..had de Broglie wavelengths 1·06 Å and hence velocities approximately three times thermal. **1955** G. E. BACON *Neutron Diffraction* iv. 102 Two methods of making diffraction measurements have.. been described using a neutron beam direct from the pile, without first monochromatizing. **1962** *Rev. Sci. Instruments* XXXIII. 875/1 (*heading*) Preparation of bent crystals for monochromatizing x rays. **1974** *Nature* 19 Apr. 671/2 Both structure analyses were based on diffractometer data taken with monochromatised Cu radiation. **1974** *Physics Bull.* Oct. 443/2 The three essential components of a neutron scattering spectrometer: monochromatizing crystal, material sample and crystal for analysis of the scattered beam.

Also **mo:nochromatiza·tion**, the process of making monochromatic.

1955 C. B. WALKER tr. *Guinier & Fournet's Small-Angle Scattering X-Rays* iii. 104 It might be thought that the double monochromatization and the large distance from X-ray tube to film would require a prohibitive increase in the exposure time. **1955** *Rev. Sci. Instruments* XXVI. 564 The usual method of monochromatization consists of inserting (at the required Bragg angle) a single crystal between the x-ray source and the specimen. **1967** J. LECIEJEWICZ tr. *Auleytner's X-Ray Methods in Study of Defects in Single Crystals* ii. 76 The only disadvantage of the method is the very long exposure time caused by the low intensity due to precise monochromatization of the beam.

monochromator (mǫ·nŏkrŏmēitəɹ, mǫnokrōu·mēitəɹ). *Physics.* [f. MONOCHROMAT(IC *a.* + -OR.] Any device used to select radiation of a single wavelength or energy (or, in practice, a very narrow range of wavelengths or energies).

1909 *Chem. Abstr.* III. 2074 The monochromator is a spectrometer arrangement of the Pellin and Broca or Hilger type with constant deviation of 90°, in which the eyepiece can be replaced by a second slit. **1924** E. C. C. BALY *Spectroscopy* (ed. 3) I. iv. 109 Any fixed arm spectrometer can be converted into a monochromator by replacing the eyepiece by a slit placed in the focus of the telescope lens or mirror. **1955** C. B. WALKER tr. *Guinier & Fournet's Small-Angle Scattering X-Rays* iii. 100 Bent crystals were originally introduced in X-ray spectroscopy, serving as analyzers capable of producing intense spectra; they can serve equally well as monochromators. **1969** *Nature* 11 Oct. 146/1 Changes in diffraction geometry, and more especially the use of focusing monochromators, should yield a considerably higher neutron flux, permitting the use of smaller crystals.

monochrome, *sb.* and *a.* Add: **A.** *sb.* **1. b.** *transf.*, esp. in *Photogr.* (see quots.).

1940 *Chambers's Techn. Dict.* 555/2 *Monochrome*, a photographic print in one colour of varying brightness. **1968** *Gloss. Terms Offset Lithogr. Printing (B.S.I.)* 8 *Monochrome*, an original or print in a single colour. **1973** D. A. SPENCER *Focal Dict. Photographic Technologies* 394 *Monochrome*, an image or picture in a single colour which is usually (but need not be) black.

2. b. *Cinemat.* and *Television.* Reproduction in black and white.

1918 H. CROY *How Motion Pictures are Made* 294 Relieving the monotony of the monochrome. **1957** *Encycl. Brit.* XV. 859/2 Monochrome.—For black-and-white photography the principal problem is to make a reasonable compromise between film grain and film speed. **1961** G. MILLERSON *Technique Television Production* 128 Colour media can employ both tonal and colour separation, to distinguish between planes. Colour separation may mean nothing when transformed into monochrome.

c. *fig.*

1962 *Listener* 10 May 828/3 A dramatic exaggeration, associated with Verdi and Puccini, inevitably crept into the gentle monochrome.

3. A paint or glaze of a single colour.

1906 S. W. BUSHELL *Chinese Art* II. viii. 40 Some of the most brilliant monochromes of the time are plain washes of one of the enamel colours used in polychrome decoration. **1933** *Burlington Mag.* Nov. 203/1 Monochromes, which may be divided into celadons and semi-celadons, brown-glazed, white-glazed, and black-glazed wares.

B. *adj.* (Further examples.) Also *fig.*

1906 S. W. BUSHELL *Chinese Art* II. viii. 35 All the new monochrome glazes introduced under his rule. **1957** *Encycl. Brit.* II. 540/1 At Mersin, the burnished and incised monochrome pottery of the Neolithic period was almost entirely replaced by the painted wares. **1959** [see *BICHROME *a.* and *sb.*²]. **1970** *Daily Tel.* 13 Jan. 16 One must have balanced communities; there is nothing worse for society than a monochrome area. **1972** C. STEPHENSON *Merrily on High* xii. 181 As a curate in Cowley St. John I worked in a completely monochrome district as far as the church was concerned.

b. *Photogr.* and *Television.* Reproducing all colours as shades of grey; 'black and white'; pertaining to such reproduction.

1918 H. CROY *How Motion Pictures are Made* 290 A strip of Priezma negative differs but slightly from a monochrome film. *Ibid.* 294 Monochrome effect on the screen is varied successfully by a different process. **1958** *Newnes Compl. Amat. Photogr.* 144 The monochrome photographer who turns to colour is rather apt to think his old principles still apply. **1961** G. MILLERSON *Technique Telev. Production* 128 A further variation can arise through distinctions between techniques when making colour and monochrome motion pictures. *Ibid.* Although colourful design may delight those in the studio, it has no significance to the viewer in monochrome television. **1972** *Sci. Amer.* Feb. 117/1 Intended largely for paramedical students, nurses and technicians, it presents both in sharp monochrome and strong color photographs, all blood red and fat yellow, a review of the general structure of the body.

monochronic, *a.* Delete *rare*⁻⁰ and add examples of the sense: Relating to a single period of time.

1905 F. E. CLEMENTS *Research Methods in Ecology* 319 *Monochronic*, arising but once. **1967** *Philos.* XLII. 140 Geography was at one time in danger of running into a dead monochronic alley by limiting itself to the visible present.

mo·nocle, *v.* [f. the sb.] *trans.* To provide with a monocle. So **mo·nocled** *a.*

1922 M. B. HOUSTON *Witch Man* xii. 154 Major Coberton monocled his eye. **1926** L. P. GREENE *Major—Diamond Buyer* 18 The monocled one coughed deprecatingly. **1940** *Horizon* Mar. 185 The monocled idiot who made good on the fields of Mons and Le Cateau. **1965** *Listener* 30 Dec. 1063/1 Colonel Creighton, monocled, and seated bolt upright in his rickshaw as though it were a Daimler. **1974** *Times* 26 Apr. 18/6 It is ironical that youth and the new guard should be represented by a 63 year old monocled cavalry officer.

monoclinal, *a.* and *sb.* Substitute for defs.: **A.** *adj.* **a.** Consisting of strata that slope in the same direction and at the same angle throughout.

In quot. 1876 in Dict. the word appears to have been used with a slightly different meaning, perh. because 'direction' was interpreted as referring to directions in a horizontal plane rather than a vertical one.

1843 W. B. & H. D. ROGERS in *Rep. 1st, 2nd & 3rd Meetings Assoc. Amer. Geologists & Naturalists* 485 While the phrases, *anticlinal dip*, and *synclinal dip*, sufficiently express the directions of the beds, due to concave and convex flexures, we propose the term *monoclinal*, to signify a sameness in the direction of the dip, and shall term a mountain or valley, in which such sameness prevails, a *monoclinal* mountain, or *monoclinal* valley. **1858** [in Dict.]. **1875** J. W. POWELL *Explor. Colorado River* xi. 160 Monoclinal valleys..run in the direction of the strike between the axes of the fold—one side of the valley formed of the summits of the beds, the other composed of the cut edges of the formation. **1930** *Compt. Rend. 15th Internat. Geol. Congress 1929* II. vi. 361, I shall consider only four major types of blocks included between bounding faults: 1. Tilted or monoclinal blocks; [etc.]. **1942** O. D. VON ENGELN *Geomorphology* xv. 327 Monoclinal ridges, alternating with monoclinal valleys, appear where a series of resistant and weak beds all dip in the same direction. **1957** G. E. HUTCHINSON *Treat. Limnol.* I. i. 111 The channel containing Lenore Lake lies in a zone of fractured basalt on a monoclinal slope.

b. Applied to a fold consisting of a single slope connecting strata that are parallel and more or less horizontal but at different levels.

1877 [in Dict.]. **1880** C. E. DUTTON *Rep. Geol. High Plateaus Utah* ii. 25 The great structural features of the High Plateaus are the faults and monoclinal flexures. **1940** E. S. HILLS *Outl. Struct. Geol.* v. 111 An individual normal fault may pass either laterally or vertically into a monoclinal flexure. **1969** J. BUNDRED *Basic Geol. for Engineers* iv. 115 A monoclinal fold..has basically a single undulation.

B. *sb.* **a.** A monoclinal fold.

1880 C. E. DUTTON *Rep. Geol. High Plateaus Utah* ii. 26 So close is the homology that we are justified in calling a monoclinal in some of its aspects a modified fault. The only difference for structural purposes is that in the case of a typical fault of the simplest form the shearing is along one plane, while in the monoclinal the shearing lies between two planes. **1886** [in Dict.]. **1914** J. PARK *Text-bk. Geol.* ix. 140 In sharply bent monoclinals, the strata in the middle limb are generally drawn out, compressed, or deformed.

b. A monoclinal set of strata; = *HOMOCLINE.

1916 *Bull. Geol. Soc. Amer.* XXVII. 92, I welcome the suggestion of the term homocline. The need of a term for this structure has long been felt in the United States Geological Survey, where confusion has been avoided in editing by using *monoclinal* for the homocline as distinct from *monocline* or monoclinal fold.

monoclinally, *adv.* (further examples).

1880 C. E. DUTTON *Rep. Geol. High Plateaus Utah* ii. 31 At East Fork Cañon the thrown beds..are turned up monoclinally. **1968** R. W. FAIRBRIDGE *Encycl. Geomorphol.* 483/2 At coastlines the same ancient cyclic landsurface may be monoclinally warped down to pass beneath marine sequences of Jurassic or Cretaceous rocks.

monocline. Add: (Further examples.)

1916 [see *MONOCLINAL *a.* and *sb.* B. b]. **1940** E. S. HILLS *Outl. Struct. Geol.* iv. 70 A monocline is a local steepening of an otherwise uniformly dipping or horizontal series, and is composed of an anticlinal bend above, followed by a synclinal bend at a lower level... In oil-field geology, however, the term is used for formations in which the dip is more or less uniform in one direction. **1942** M. P. BILLINGS *Struct. Geol.* iii. 41 In plateau areas.. the strata may locally assume a steeper dip... Such a fold is a monocline. The beds in a monocline may dip at angles ranging from a few degrees to ninety degrees. **1962** READ & WATSON *Introd. Geol.* I. viii. 487 In this way monoclines may be formed and may grade into zones of normal faulting.

b. = *HOMOCLINE.

1912 E. H. C. CRAIG *Oil-Finding* v. 74 Oil may be obtained from monoclines, often in great quantity. In such cases the more gentle the dip, the better. **1925** A. B. THOMPSON *Oil-Field Exploration* I. v. 186 The term 'monocline' is here retained in its customary oil-field phraseology as representing a series of uniformly dipping beds, irrespective of the relationship they may bear to the general structure of the district. **1951** E. N. TIRATSOO *Petroleum Geol.* iv. 79 The monocline can be considered as virtually one limb of an anticline up which the migrating fluids pass, leaving well-marked layers of dynamically separated gas, oil and water.

monocoque (mǫ·nokǫk). [a. F. *monocoque*, f. *mono-* MONO- + *coque* egg-shell.] **a.** *Aeronaut.* A fuselage or other structure having an outer covering in the form of a rigid skin or shell designed to bear all or most of the stresses that arise; now esp. one without longerons or stringers (see quot. 1948). Also *transf.* Usu. *attrib.*

1914 *Sphere* 3 Oct. p. ii/1 The very speedy monocoques of the French. **1919** A. W. JUDGE *Elem. Princ. Aeroplane Design* vi. 98 The Monocoque Type of Body. This type of body..consists of a single shell, conforming with the outside shape of the body, and..so constructed that it can withstand all the stresses which it is called upon to bear, without the necessity for longerons or cross bracing members. **1935** C. G. BURGE *Compl. Bk. Aviation* 595/1 The 'monocoque' is becoming increasingly popular for both military and commercial aeroplanes. **1938** E. W. C. WILKINS *Aeroplane Design* vii. 171 The monocoque fuselage..consists of a thin shell..built around a number of transverse rings or formers... These formers are placed at intervals of about 2 ft. or more in the case of a wooden monocoque, but are very much closer together in the case of the metal monocoque. **1948** C. E. CHAPEL *Aircraft Basic Sci.* ii. 84/2 Monocoques may be divided into three classes (monocoque, semimonocoque, and reinforced shell), and different portions of the fuselage may belong to any of these classes. The true monocoque has as its only reinforcement vertical bulkheads formed of structural members... All stresses are carried by the shell or skin. **1951** *Engineering* 2 Nov. 573/1 A wing of the thin-walled monocoque type. **1969** K. MUNSON *Pioneer Aircraft 1903–14* 123/2 Late in 1911 the ideas of the Swedish engineer Ruchonnet, for a monocoque fuselage shell of moulded plywood, were applied by Armand Deperdussin's designer, Louis Béchereau. **1970** *New Scientist* 12 Nov. 329/2 No longer would an airship need to have a metal fabric-covered skeleton. Instead it could have a 'monocoque' structure—the strength would be in a stiff lightweight outer shell. **1971** *Daily Colonist* (Victoria, B.C.) 6 June 25/6 Launched four years ago, she is of monocoque design—one shell, no ribs, no frame, smooth inside and out.

b. In a motor vehicle, an underframe and body built as a single rigid structure (or in racing cars as a number of box-like sections) throughout which the stresses are distributed. Usu. *attrib.*

1956 MOLLOY & LANCHESTER *Automobile Engineer's Ref. Bk.* XVIII. 2 All-metal bodies fall into two main categories, firstly those which are used in conjunction with a separate chassis frame and secondly which are of the type known as monocoque or unit construction. The former naturally require only body structure panels, whereas the latter, in addition, require many pressings to form a unit (underframe or subframe), which although ultimately part of the body, is [a] substitute for the chassis frame and has provision made for all necessary attachment points for the various mechanical units. **1961** COSTIN & PHIPPS *Racing & Sports Car Chassis Design* v. 43 All these, plus the single transverse torsion bar and the telescopic dampers, pick up on the rear bulkhead of the monocoque. **1963** *Engineering* 1 Feb. 181/1 GT cars featuring monocoque construction of laminated wooden box sections. **1968** *Daily Tel.* (Colour Suppl.) 15 Nov. 29/2 Despite the terrible force of the impact, the integrity of the cockpit area was preserved by the immensely strong monocoque chassis of his car. **1969** 'D. RUTHERFORD' *Gilt-Edged Cockpit* i. 15 With the whole of his body encased in the monocoque shell and no part of his face visible he was an anonymous figure... Already other cars were flashing past. **1973** TERRY & BAKER *Racing Car Design & Devel.* vi. 129 Although the basic space-frame is a comparatively light structure, it requires a separate body and fuel containers, all of which are integral parts of a monocoque, so this tends to nullify any hoped-for weight advantage.

monocular, *a.* Add: **2. b.** Wearing a monocle. *rare.*

1903 G. B. SHAW *Man & Superman* p. xxiii, These college passmen, these well groomed monocular Algys and Bobbies.

c. quasi-*sb.* A glass, esp. a field-glass, in the use of which only one eye is employed in viewing an object.

1936 *British Birds* XXX. 39 It was only when I turned my monocular on the bird in the water that I realized it was a Red-legged Partridge. **1959** *Chambers's Encycl.* IX. 385/2 The modern microscope usually has interchangeable bodies, making it possible to use the same stand with monocular or binocular. **1971** D. WHILLANS in C. Bonington *Annapurna South Face* v. 58 There was a monocular in the tent which I focused on the dark, moving shape.

fig. (Later example.)

1923 *Public Opinion* 8 June 542/3 The bane of cancer research has been its monocular myopia.

monocularly, *adv.* (Earlier and later examples.)

1880 W. JAMES *Coll. Ess. & Rev.* (1920) 177 If under these circumstances the object thus monocularly seen were translocated outwardly, we should have a complete verification. **1972** *Sci. Amer.* Aug. 86/3 An insect disguised as a leaf may be invisible monocularly but stand out in a different depth plane from real leaves when it is viewed stereoscopically.

monoculture (mǫ·nokɒltiŭɹ). [irreg. f. MONO- + CULTURE *sb.* Cf. F. *monoculture* (1842 in Robert).] **a.** The cultivation or exploitation of a single crop, to the exclusion of others that are possible. Also *fig.*

1915 C. R. ENOCK *Tropics* xxxiii. 373 The decline of sugar 'monoculture' may have proved a blessing in disguise. Cotton and many food-stuffs are now produced. **1925** E. F. ROW tr. *Demangeon's Brit. Empire* 134 This plantation system, this exploiting to the uttermost of a single valuable product, involves the dangers of all monoculture. **1948** *Times* 17 June 5 The name [*sc.* groundnut scheme] is too suggestive of monoculture and dustbowls. **1961** J. RUSSELL tr. *Lévi-Strauss's World on Wane* iv. 39 Humanity has taken to monoculture, once and for all, and is preparing to produce civilization in bulk, as if it were sugar-beet. **1962** *Economist* 7 July 60/1 A 'monoculture' economy dependent on only one resource for most of its income. **1970** J. ARDAGH *New France* vi. 206 The monoculture of cheap wine was dangerous for the region's economy. **1974** *Country Life* 28 Mar. 710/1 The replacement of broad-leaved woodland with conifer and increasing monoculture.

b. An area in which monoculture is practised or a single kind of (higher) animal maintained.

1951 *New Biol.* X. 56 Nature abhors a monoculture, and the most carefully tended orchards soon become ecological associations with distinctive flora and fauna. **1970** *New Scientist* 21 May 372/1 The transformation of habitats from virgin forests to cultivated woods, vast crop monocultures, industrial areas and big cities is the most radical change of nature that has struck Europe since the last glaciation. **1974** *Environmental Conservation* I. 17/2 Huge areas are turned into monocultures. **1974** *Nature* 24 May 307/3 One is towards the establishment of monocultures of eland, wildebeest, gazelle or kob.

c. An area in which all the inhabitants share a common culture or way of life.

1968 *Listener* 5 Sept. 298/1 Los Angeles's least endearing characteristic: the tendency to fragment into self-contained, specialised areas—social monocultures. Functional monocultures, too: in Los Angeles you tend to go to a particular place to do a particular thing.

Hence **monocu·ltural** *a.*; **monocu·lturist**, one who practises or advocates monoculture.

1928 E. R. JOHNSON et al. *Princ. Transportation* xxi. 243 The South from earliest colonial days has been agricultural and primarily monocultural. **1964** GOULD & KOLB *Dict. Soc. Sci.* 583/1 The 'Deep South' region of U.S.A. exhibits specialized cultural, political, and economic (monocultural) features. **1965** *New Scientist* 23 July 199/1 Monocultural crop planting on totally cleared areas. **1968** W. E. LAMBERT et al. in J. A. Fishman *Readings Sociol. of Lang.* (1968) 479 It is of special interest to note a basically similar pattern appearing in all three of the American settings, two bicultural and one monocultural. **1973** *Country Life* 30 Aug. 586/2 The monoculturists' ideas of good yields being likely to cause any self-respecting rotational farmer to blush with shame.

monocyclic, *a.* Add: **2. a.** (Examples.)

1882 S. H. VINES tr. *Sachs' Text-bk. Bot.* 601 When the members of a series (calyx, corolla, etc.) are in one whorl, the series is said to be monocyclic. **1895** —— *Students' Text-bk. Bot.* II. III. 501 The simplest case is that in which each series of floral organs..occupies a single whorl, or is monocyclic.

4. *Chem.* Having a molecular structure with a single ring.

1910 *Jrnl. Chem. Soc.* XCVII. 1616 From this glycol we could obtain a monocyclic terpene. **1939** *Nature* 26 Aug. 349/1 The detailed consideration of monocyclic compounds is then taken up. **1967** M. E. HALE *Biol. Lichens* viii. 105 Free monocyclic phenolic substances.

5. *Geol.* Having undergone a single cycle of erosion or (quot. 1969) of mountain-building.

1952 *Jrnl. Geol.* LX. 188/1 Four general types of landscapes, suggested by Bryan's (1943, pp. 472–473) classification of soils: (1) simple, monocyclic landscapes which can be fitted to the idealized cycle; (2) complex, multicyclic landscapes which require historical interpretation; (3) compound landscapes..; (4) exhumed landscapes. **1954** W. D. THORNBURY *Princ. Geomorphol.* ii. 23 Monocyclic landscapes are less common than multicyclic and are in general restricted to such newly created land surfaces as a recently uplifted portion of the ocean floor. **1968** R. W. FAIRBRIDGE *Encycl. Geomorphol.* 286/2 Monocyclic landscapes are rarely found except in very recently formed land surfaces. **1969** BENNISON & WRIGHT *Geol. Hist. Brit. Isles* iii. 36 Within both the Caledonian and Lewisian belts there is evidence of several periods of folding and metamorphism and neither represents a simple monocyclic unit.

monocyte (mǫ·nŏsəit). *Biol.* [ad. G. *monozyt* (Pappenheim & Ferrata 1910, in *Folia Haematologica* X. I. 81): see MONO- and -CYTE.] A kind of large leucocyte which has a single oval or indented nucleus and no coarse granulation in the cytoplasm, constitutes 3–8 per cent of the total leucocytes in human blood, and is the circulating form of the macrophage.

1913 STEDMAN *Med. Dict.* (ed. 2) 556/1 Monocyte. **1927** A. PINEY *Rec. Adv. Hæmatol.* ii. 16 The large hyaline leucocyte or monocyte is not easy to place with certainty into either the granular or the non-granular group. **1970** T. S. & C. R. LEESON *Histology* (ed. 2) viii. 140/1 Monocytes migrate readily through vessel walls and develop into phagocytic cells which cannot be distinguished from macrophages already present within the connective tissues. They are effective in combating tubercle bacilli. **1974** PASSMORE & ROBSON *Compan. Med. Stud.* III. xxi. 5/1 Absence of monocytes or basophils from the blood film or reduction in their number does not have clinical significance.

Hence **monocy·tic** *a.*; **monocyto·sis** [-OSIS], an abnormal increase in the number of monocytes in the blood.

1914 STEDMAN *Med. Dict.* (ed. 3) 576/1 Monocytosis. **1934** WEBSTER, Monocytic. **1935** WHITBY & BRITTON *Disorders Blood* i. 18 The origin of the monocytic cells has been and still is, the subject of much controversy. **1938** H. DOWNEY *Handbk. Hematol.* I. v. 389 Monocytes may appear in large numbers in the peripheral blood in the monocytic leucemias. *Ibid.* II. xv. 1067 In chronic malaria, the irritation of the reticulo-endothelium often leads to a monocytosis in the peripheral blood. **1972** *Nature* 4 Feb. 275/1 Cells of the monocytic series. **1974** PASSMORE & ROBSON *Compan. Med. Stud.* III. xxi. 5/1 Monocytosis is seen in the course of infectious mononucleosis and in some bacterial infections, notably tuberculosis.

monodist. 2. (Later examples.)

1916 STANFORD & FORSYTH *Hist. Mus.* vii. 148 There is no sudden revolution in the method of handling the musical material such as the monodists engineered in 1600. **1974** *Country Life* 14 Mar. 590/3 In the early-17th century ..the Italian Monodists..sought to find a new expressive force by abandoning the use of the contrapuntal device.

monodrama. (Later examples.) Now esp. an opera for one singer.

1954 *Grove's Dict. Mus.* (ed. 5) V. 666/2 The title part of Berlioz's 'Lélio'.., in which a single actor appears, may also be called a 'monodrama'. *Ibid.*, A work like Schoenberg's 'Erwartung', in which a single acting singer appears, is also in the nature of 'monodrama'. **1966** *Listener* 19 May 736/3 *Erwartung*,..a monodrama or 'prose opera' (Stravinsky)... This fragmentary half-hour composition..is scored for soprano (no supporting cast) and large orchestra. **1974** *New Yorker* 29 Apr. 129/1 An extravagant monodrama for singer, pianist, and orchestra.

monodramatic, *a.* (Earlier example.)

1801 *Monthly Mirror* Aug. 136 A German..has lately appeared on this stage, in an Italian mono-dramatic interlude, called Il Cazolaro.

monodromy (mǫ̆nǫ·drŏmi). *Math.* [f. as MONODROMIC *a.* + -Y³.] (See quot. 1909.)

1897 B. RUSSELL *Essay on Foundations of Geometry* i. 24 As regards independence of rotation in rigid bodies (Monodromy). If $(n-1)$ points of a body remain fixed, so that every other point can only describe a certain curve, then that curve is closed. **1903** *Nature* 19 Feb. 382/2 It is pointed out that in the non-Pythagorean geometrics devised by Hilbert, Helmholtz's axiom of monodromy is not verified, inasmuch as it is possible by rotation through four right angles, to bring the points of a line into positions which they do not occupy before the rotation. **1909** *Cent. Dict.* Suppl., *Monodromy*, (*a*) the characteristic property that, if the argument returns by any path to its original value, the function also returns to its original value. (*b*) The property that the curves described by a revolution or rotation through four right angles are closed. **1949** G. & R. C. JAMES *Math. Dict.* 237/1 *Monodromy theorem.* The theorem states that, if the function $f(z)$ of the complex variable z is analytic at the point z_0 and can be continued analytically along every curve issuing from z_0 in a finite simply connected domain D, then $f(z)$ is a function-element of an analytic function which is single valued in D; in other words, analytic continuation around any closed curve in D leads to the original function element.

monœcism. (Further example.)

1966 AINSWORTH & SUSSMAN *Fungi* II. xx. 661 Monoecism and dioecism..are defined on a physiological basis according to whether an organism contributes one or two nuclei to the sexual process.

monœcy (mǫnī·si). *Biol.* [f. MONŒC(ISM + -Y³.] = MONŒCISM.

1949 DARLINGTON & MATHER *Elem. Genetics* xii. 242 Dioecy ensures outbreeding. Monoecy only favours it. **1970** [see *DIŒCY].

monogamic, *a.* Add: Hence **monoga·mically** *adv.*, in a monogamic manner.

1911 G. B. SHAW *Getting Married* 140 In our population there are about a million monogamically superfluous women.

monogamize, *v.* Restrict † *Obs.* to sense in Dict. and add: **2.** *trans.* To make monogamous in character.

1911 H. S. HARRISON *Queed* xxiii. 297 Since, however, the church and the law allowed him but one [wife], he must more drastically monogamize his heart.

monogamous, *a.* Add: Hence **mono·gamousness.**

1946 KOESTLER *Thieves in Night* II. iv. 170 There remains the tyranny of monogamousness.

monogenean (mǫnodʒe·niăn), *sb.* and *a.* [f. mod.L. name of order *Monogenea* (J. V. Carus in W. C. H. Peters et al. *Handbuch der Zoologie* (1863) II. 477), f. MONO- + Gr. γενεά race, generation: see -AN.] **A.** *sb.* A trematode parasite of the order Monogenea, which passes its entire life cycle in one host, esp. an ectoparasitic fluke of fishes; a monogenetic trematode. **B.** *adj.* Of or pertaining to this group of parasites.

1960 *Parasitology* L. 51 (*title*) The attachment of the monogenean *Discotyle sagittata* Leuckart. **1963** J. LLEWELLYN in *Adv. Parasitol.* I. 287 (*title*) Larvae and larval development of monogeneans. *Ibid.*, The two studies were based on samples of monogenean larvae. **1971** *Nature* 23 July 225/1 The targets of the larvae of monogenean flukes are mostly fish of much greater size. **1972** D. A. ERASMUS *Biol. Trematodes* vii. 166 Within their hosts, monogeneans generally occur on the fins, skin, gills, gill chamber and buccal cavity.

monogenesis. Add: **3.** *Linguistics.* The theory that all languages have one common origin.

[**1905** A. TROMBETTI *L'Unità d' Origine del Linguaggio* 56 Noi dunque consideriamo la monogenesi del linguaggio per lo meno come un argomento assai forte in favore della monogenesi dell' uomo.] **1936** *Science & Society* I. 23 At various times scholars have raised the question of monogenesis as opposed to polygenesis of human speech: whether it had a single origin at a given time and place, whence it spread over the rest of the earth, or whether it was independently evolved by different branches of the human family. **1949** *Archivum Linguisticum* I. 76 The issue of monogenesis *versus* polygenesis is likewise evaded, although a critical appraisal of Trombetti's arguments might have been useful. **1949** M. PEI *Story of Lang.* (1952) vii. 357 A few daring linguists, like the Italian Trombetti, have strenuously asserted the thesis of the 'monogenesis', or single common origin, of all the world's tongues. **1966** H. LANDAR *Lang. & Culture* xx. 153 His guess implicates a monogenesis of the world's languages, much as the Italian duck-hunter Trombetti supposed, but a monogenesis which occurred as recently as 30,000 years ago.

monogenic, *a.* Add: **6.** *Biol.* [cf. *GENIC *a.*] Involving or controlled by a single gene.

1939 *Jrnl. Genetics* XXXVIII. 420 These results indicate that the asynaptic condition in *N. sylvestris* is determined by a difference which behaves as a monogenic recessive. **1960** *Phytopathology* L. 766 (*heading*) Dominance of avirulence and monogenic control of virulence in race hybrids of *Ustilago avenae*. **1971** *Brit. Med. Bull.* XXVII. 40/2 Conventional measurements of blood pressure are too variable to distinguish between monogenic and polygenic models of heredity.

Hence **monoge·nically** *adv.*

1970 *Plant Disease Rep.* LVI. 696 A true resistance mechanism which is monogenically controlled.

monoger, var. *MONIKER.

monogram. Add: **5.** (Later example.)

1919 T. WRIGHT *Romance of Lace Pillow* xvi. 223 Monogram Lace, in the form of medallions—the sacred monogram I.H.S. from an antique pattern being most in demand—has been made for Mr. George Smith at Olney during the last thirty years, and sent to the Continent for ecclesiastical purposes.

Hence **mo·nogram** *v. trans.*, to decorate with a monogram; **monogrammed** *a.* (later examples); **mo·nogrammist**, one who applies a monogram.

1939 'A. BRIDGE' *Four-Part Setting* vi. 64 I took those new pillow-cases round to be monogrammed. **1939** *Burlington Mag.* Jan. 40/2 No 'monogrammist' was too petty..to be scrutinised with the same minute and untiring precision. **1951** *Good Housek. Home Encycl.* 139/2 Monogrammed handkerchiefs must be ironed on the wrong side. **1971** *Daily Tel.* 27 Jan. 11 (*caption*) Red wool suit,.. monogrammed in red with Balmain initials. **1974** *Ibid.* 30 Sept. 12 Young Conservatives have their shirts monogrammed 'Y.C.' **1974** *Ashmolean Mus. Rep. Visitors 1972–3* 38 A 19th-century woodcut by the monogrammist *H.Ch.*, *Rothenburg ob der Tauber, Rathaus*. **1975** 'E. LATHEN' *By Hook or by Crook* xvii. 160 Lois..instructed Cartier to monogram a demitasse set.

monographic, *a.* **4.** (Later examples.)

1959 C. SINGER *Short Hist. Sci. Ideas* vii. 274 The earliest modern scientific works of a monographic character, the great books of Belon, Rondelet, Vesalius, Gesner, are exclusively biological. **1970** *Daily Tel.* 12 Sept. 8/5 There is Waterhouse's monograph on Gainsborough..and some exhibition catalogues that are monographic in their range. **1974** *Nature* 18 Jan. 165/3 The book as a whole usefully fills an important gap in the monographic literature.

monographically, *adv.* (Later example.)

1964 R. H. GERHARD in D. Abercrombie et al. *Daniel Jones* 281 The idea of representing monographically the vowels of *go* and *gay*.

monohull (mǫ·nohɒl). [f. MONO- + HULL *sb.*²] A boat with a single hull. Also *attrib.* or as *adj.* Cf. *MULTIHULL.

1967 *Motor Boating* Oct. 135 This year the race was split into two divisions, caused by the catamaran show of superiority: one for monohulls, the other for cats. **1968** *Observer* 9 June 23/5, I would like to see two separate [single-handed Atlantic crossing] races in future,..— one for monohulls and one for multihulls. **1969** *Sunday Times* 2 Feb. 50 The design of the monohull can be improved—there is constant experiment with materials. **1971** *New Scientist* 4 Feb. 265/2 The superiority of the Planesail concept over the conventional monohull or multihull arrangement had already been amply demonstrated. **1972** *Observer* 11 June 18/7, I am confident that Three Cheers, and certainly a multihull, will win—with the radical monohull Strongbow our only serious competitor. **1973** 'A. YORK' *Captivator* iii. 44 She's all but thirty feet long..and being a cat, she's fourteen feet wide. That means she has about the same accommodation as a monohull.

monohybrid (mǫnohəi·brid), *sb.* and *a.* *Biol.* [f. MONO- + HYBRID *sb.* and *a.*] **A.** *sb.* A hybrid that is heterozygous with respect to a single gene. **B.** *adj.* Of, pertaining to, or characteristic of such a hybrid, or a cross resulting in such a hybrid. So **monohy·bridism** (*rare*).

1904 *Rep. Evolution Comm. R. Soc.* No. 2. 126 Unresolved characters in ordinary mono-hybrids. **1907** R. C. PUNNETT *Mendelism* (ed. 2) 34 The cases..concern only a single pair of differentiating characters, that is to say, are cases of monohybridism. **1910** *Amer. Naturalist* XLIV. 67 The crosses have all given a simple mono-hybrid ratio. **1911** FARMER & DARBISHIRE tr. *H. de Vries's Mutation Theory* II. 585 We have given the name of monohybrids to those mongrels whose parents differ from one another in a single elementary character only. **1931** A. F. SHULL *Heredity* (ed. 2) vii. 52 Offspring produced by two parents which differ in only one inherited feature are known as monohybrids. **1964** M. CRITCHLEY *Developmental Dyslexia* x. 64 The author concluded that developmental dyslexia follows a monohybrid autosomal dominant mode of inheritance. **1967** A. MÜNTZING *Genetics: Basic & Applied* v. 46/2 Heterozygotes for one gene pair, monohybrids,.. display monohybrid segregation in the ratio of 3:1.

monohydric, *a.* Add: **b.** Of an alcohol or phenol: containing a single hydroxyl group.

1880 *Jrnl. Chem. Soc.* XXXVIII. 28 (*heading*) Sulphates of mono- and poly-hydric alcohols. **1926** J. READ *Text-bk. Org. Chem.* xxiv. 541 Ordinary phenol, or carbolic acid,.. is a monohydric phenol. **1937** *Jrnl. R. Aeronaut. Soc.* XLI. 737 Fuels containing monohydric alcohols are rendered harmless to magnesium and its alloys by saturating such fuels with complex alkali fluorides soluble in them. **1967** I. L. FINAR *Org. Chem.* (ed. 5) I. vi. 144 The monohydric alcohols form an homologous series with the general formula $C_nH_{2n+2}O$.

monoideal (mǫnoəidī·ăl), *a.* *rare*⁻¹. [f. MONO- + IDEAL *a.* 3 b.] Expressing or conveying only one idea.

1922 JOYCE *Ulysses* 667 The infinite possibilities hitherto unexploited of the modern art of advertisement if condensed in triliteral monoideal symbols.

monoïdeism. Add: Also **monoideism.** Hence **monoidei·stic** *a.*, pertaining to or characterized by monoideism.

1890 tr. *T. Ribot's Psychol. of Attention* iii. 96 This higher form of ecstasy may at times reach the state of complete, absolute monoideism, that is, the state of perfect unity of consciousness. **1902** W. JAMES *Var. Relig. Exper.* 525 Philosophy, with its passion for unity, and mysticism with its monoideistic bent, both 'pass to the limit' and identify the something with a unique God who is the all-inclusive soul of the world. **1904** G. S. HALL *Adolescence* II. 50 The history of philosophy shows that the mono-ideistic thinkers..were the victims of an environment or an age itself overwrought, one-sided and extreme. **1966** D. W. ABSE *Hysteria* viii. 116 It was essentially a narrowing of the attention, or a 'monoideism'.. that ushered in the hypnotic trance.

monokaryon (mǫnokæ·riǫn). *Bot.* Also **monocaryon.** [f. MONO- after *DIKARYON.] A mononuclear cell, spore, or mycelium in higher fungi that have a dikaryotic stage of the life cycle. Hence **monokaryo·tic** *a.*, containing or characteristic of a monokaryon.

1935 E. A. BESSEY *Text-bk. Mycol.* xii. 323 The secondary mycelium (dicaryon phase) is of far greater importance in the life cycle, the primary mycelium (monocaryon phase) occupying but a short period of the life of the fungus. **1936** *Amer. Jrnl. Bot.* XXIII. 310/1 Throughout this paper the terms *monocaryon mycelium* and *dicaryon mycelium* will be employed instead of the older terms *haploid* and *diploid*. **1941** *Bot. Rev.* VII. 412 These give rise either to dikaryotic diploid oidia or to monokaryotic haploid oidia. **1947** F. A. & F. T. WOLF *Fungi* II. xiv. 334 The complete cycle from the monocaryotic to the dicaryotic condition. **1968** R. RIEGER et al. *Gloss. Genetics & Cytogenetics* 299 *Monokaryon*, a uninucleate cell, a spore, or a tissue consisting of such cells, as opposed to a dikaryon. **1972** E. MOORE-LANDECKER *Fund. Fungi* iv. 91 The monokaryon is at first nonseptate. *Ibid.* 92 When the monokaryotic hyphae fuse, the nuclei of one hypha flow into the other.

monokini (mǫnokī·ni). [f. MONO- after *BIKINI b, with reference to BI- *pref.*²] A one-piece beach garment, usu. one equivalent to the lower half of a bikini and worn by women.

1964 *Daily Mirror* (Brisbane) 6 July 6 Monokinis are selling in Paris like iceboxes in Alaska. **1964** *Time* 7 Aug. 36/3 Sunbathing.., she in a bikini.., he in a monokini. **1965** *New Yorker* 10 Apr. 92 He works wearing only a jet monokini, a trained canary perched on his shoulder. **1966** B. E. WALLACE *Murder in Touraine* xi. 96 She was standing there near naked in her brief monokini. **1966** *N.Y. Times* 30 May 20 Hand-loomed Orlon yarn 'monokini' has top and bottom joined by a band. **1968** B. FOSTER *Changing Eng. Lang.* iii. 121 In France in the year 1964.. there was much talk of a topless bathing suit, immediately known as the *monokini*. **1971** *Observer* (Colour Suppl.) 9 May (recto front cover), This monokini is not a tattered pair of cut-down jeans but the latest smart, expensive one-piece swimsuit made of disguised suede. **1974** *Guardian* 4 July 2/7 The monokini is no longer the rage only at St-Tropez. French girls usually go topless in Paris open-air swimming pools.

monolayer (mǫ·nŏlēiəɹ). [f. MONO- + LAYER *sb.*] **a.** A layer or film one molecule thick.

1933 *Biochem. Jrnl.* XXVII. 1581 The structure of proteins in bulk solution is different from that which exists when the protein is extended as a monolayer in a film. **1950** W. J. MOORE *Physical Chem.* xvi. 496 The ordinary three-dimensional states of aggregation have many counterparts in the two-dimensional world of monolayers. **1972** R. G. KAZMANN *Mod. Hydrol.* (ed. 2) iii. 62 Ideally this would form a film one molecule thick on the water surface—a monolayer. **1973** *Nature* 10 Aug. 328/2 In these severe conditions of lack of water (a few monolayers of water adsorbed on soil particles) some growth and metabolic activity of bacteria..does occur.

b. *Biol.* and *Med.* A culture consisting of a layer of cells one cell thick.

1952 *Proc. Nat. Acad. Sci.* XXXVIII. 748 The virus of Western Equine Encephalomyelitis, adapted to chicken embryo, will produce plaques when it is grown on a monolayer of cells obtainable from chicken embryos. **1961** *Nature* 22 Apr. 370/2 The vaccina virus was grown in monolayer cultures of monkey kidney tissue. **1973** *Ibid.* 6 Apr. 399/2 The division of most 'normal' cells in culture is inhibited after they have grown to a confluent monolayer.

monolingual (mǫnoli·ŋgwăl), *a.* and *sb.* [f. MONO- + LINGUAL *a.* 4 b.] **A.** *adj.* Speaking only one language; written in a single language.

1953 *Use of Vernacular Lang. in Educ.* (Unesco) ii. 50 It cannot be denied that the business of government is easier in a monolingual than in a multilingual nation. However, it does not follow that legislation or school policy requiring the use of the official language at all times will give the same results as actual monolingualism. **1954** L. MACNEICE *Autumn Sequel* 54 At times she is monolingual, monotone, At others mistress of the Tower of Babel. **1968** *Amer. N. & Q.* Mar. 109/1 As a monolingual dictionary and a ready reference work..the Larousse has no parallel in any other language. **1972** *Language* XLVIII 910 The English translations do not, I think, allow the monolingual speaker of English to get any sense of how a speaker of Japanese feels when he is addressed with *-san*.

B. *sb.* One who speaks only one language.

1956 E. HAUGEN *Bilingualism in Americas* iv. 77 Any

kind of sampling test which adequately measures monolingual skill can be used to make comparisons between monolinguals and bilinguals speaking the same language. 1971 G. ANSRE in J. Spencer *Eng. Lang. W. Afr.* 147 Essentially,..in cases of both monolinguals and bilinguals incorporating loans, the results seem similar. 1972 H. KURATH *Studies Area Linguistics* vii. 112 A few of them have even been adopted by English monolinguals of Pennsylvania.

Hence **monoli·ngualism**, the ability to speak only one language; **monoli·nguist** = *MONOLINGUAL *sb.*; **monolingui·stic** *a.* = *MONOLINGUAL *a.*

1928 *Observer* 8 Apr. 14/3 There is no one living..speaking only Manx and no English. Ten years before that the monolinguists had dwindled to under half a dozen. 1942 L. B. NAMIER *Conflicts* 3 Union..in monolinguistic national States became in the nineteenth century the political aim of the educated, and in time of the semi-educated, classes in Europe. 1960 J. LODWICK *Asparagus Trench* 66 All the monolinguists who..would opt for Indian tea in Pekin. 1968 Y. MALKIEL *Essays on Linguistic Themes* i. 6 It is..held that bilingualism and even trilingualism are more widely disseminated the world over than is strict monolingualism. 1972 H. KURATH *Studies in Area Linguistics* 124 Monolingualism prevails in large parts of the world.

monolith, *sb.* and *a.* Add: **A.** *sb.* **2.** *transf.* and *fig.* A person or thing resembling a monolith; *esp.* (after Russ. *monolit*; cf. *monolitnost'* monolithic unity of the party) a political or social structure presenting an indivisible or unbroken unity.

1934 H. NICOLSON *Curzon: Last Phase* xi. 323 M. Stamboliisky, the peasant Prime Minister of Bulgaria, was also granted an audience... Curzon was attracted towards this solid, somewhat helpless, monolith. He always felt at his ease with entirely self-made people. 1940 AUDEN *Another Time* 117 The monolith Of State. 1953 *Manch. Guardian* 6 Apr. 4/1 The 'monolith' of Soviet power is stirring. 1957 *Economist* 7 Sept. 766/2 A growing diversity in the economic scene is liable to lead to cracks in the political monolith. 1959 P. H. JOHNSON *Unspeakable Skipton* xxiii. 203 If the Commissioners of Inland Revenue ever caught up with her, she would undoubtedly bring tears to their eyes, reducing them from monoliths to simple, sentimental men with mothers of their own. 1962 J. WAIN *Strike Father Dead* 115 She was a woman of few words, as thick round the middle as an oak, with strong limbs and a big head and shoulders. She wasn't talkative, but..you just wouldn't expect to get a flow of words out of a monolith like that. 1966 *Listener* 12 May 700/3 His [*sc.* Bruckner's] symphonies are towering monoliths. 1975 *Times* 4 Jan. 13/3 State run monoliths whose inspiration would owe more to the opportunism of Benito Mussolini than to idealism.

B. *adj.* **2.** *fig.* = *MONOLITHIC *a.* 4.

1922 E. BLUNDEN *Shepherd* 53 Between great monolith trees.

monolithic, *a.* Add: **1. b.** *Electronics.* Of a solid-state circuit: composed of active and passive components formed in a single chip (or thin film: see quot. 1967).

1963 E. KEONJIAN *Microelectronics* i. 8 (*heading*) Monolithic circuits. *Ibid.* 9 A monolithic piece of material is treated in such a way as to possess an electronic circuit function. 1965 *New Scientist* 20 May 510/2 The multi-chip assembly has some advantages over the monolithic circuit—for example, the various chips can be tested before assembly. 1967 *Encycl. Dict. Physics* Suppl. II. 173/1 *Monolithic circuit*, an integrated circuit which uses either thin film or silicon chip techniques for both active and passive devices but not a mixture of these techniques. Since true monolithic thin film circuits are not commercially available the term is more usually applied to silicon chip construction, where it implies, in particular, that any circuit capacitors do not use evaporated dielectrics. However, monolithic silicon chip circuits may use a deposited pattern of connectors between circuit elements. 1970 [see *INTEGRATED *ppl. a.* b]. 1971 *Physics Bull.* Jan. 45/1 The new computer has a main memory constructed entirely of monolithic circuits and uses silicon memory chips instead of magnetic cores for storage.

4. *transf.* and *fig.* Resembling a monolith, having one or more of the qualities of a monolith; great, massive; immovable, unwavering, unemotional; unified, homogeneous, unchallenged. *Esp.* applied to organizations, parties, governments, etc., which are autocratic or monopolistic (freq. in derogatory use).

1920 D. H. LAWRENCE *England, my England* (1922) 87 Maurice had a curious monolithic way of sitting in a chair, erect and distant. 1923 A. HUXLEY *Antic Hay* xiii. 192 His appearance is monolithic and grim. 1937 *Nation* (N.Y.) 10 July 32/1 The monolithic corporation talks loudly to the workers. 1942 *New Republic* 9 Nov. 598/2 The monolithic power structure of the totalitarian state. 1945 A. L. ROWSE *West-Country Stories* 26 The Fourth Symphony of Sibelius, the most monolithic of them all. 1948 J. TOWSTER *Polit. Power in U.S.S.R.* p. ix, The peculiarities of the monolithic Communist Party. 1952 *Economist* 22 Mar. 702/1 Herr Grotewohl and his monolithic Socialist Unity Party are democratic. 1953 *Times* 11 Apr. 7/2 Wherever there is monolithic rule the autocrat is bound to repress all views that are not his own. 1959 *Listener* 5 Mar. 423/1 There were many contradictory elements in her, but they and she were static, immovable, almost monolithic. 1971 *Nature* 7 May 2/2 But would not the merging of the research councils create too monolithic a central sponsor? 1974 E. AMBLER *Dr. Frigo* II. 124 We were never a monolithic party.

monoli·thically, *adv.* [f. MONOLITHIC *a.*; see -ICALLY.] In a monolithic form or manner (see senses 3 and *4 of MONOLITHIC *a.*).

1932 KIPLING *Limits & Renewals* 80, I was monolithically military. 1952 SPARKES & SMITH *Concrete Roads* xvii. 398 Integral kerbs are those which are constructed monolithically with the slab. 1960 *Times* 17 Feb. 5/6 Insulating concrete can be cast monolithically. 1960 *Sunday Times* 22 May 17/5 Greedy for praise and monolithically uncharitable. 1962 *Friend* 14 Sept. 1116/1 A row of intent figures... Sometimes the families, the lovers and the photographers got mixed up with these still figures. But they monolithically stood on, for three full hours. 1966 *New Statesman* 27 May 762/3 The church [in Spain] can no longer be regarded as an institution monolithically in support of the *status quo*, opposed to change at all costs. 1975 *Times* 11 July 15/4 The contention that the Liberal Party is monolithically moderate.

monolithism (mǫ·nǒlipiz'm). [f. MONOLITH *sb.* + -ISM.] The quality or fact of being monolithic (see *MONOLITHIC *a.* 4); the principle of totalitarian rule.

1927 *Nation* 21 Dec. 703 The tactics..constituted a danger to the 'monolithism' of the party. 1948 J. TOWSTER *Polit. Power in U.S.S.R.* II. vi. 132 Cutting across both the monopoly and monolithism power principles, this act was..condemned as 'open counter-revolution'. 1957 T. KILMARTIN tr. *Aron's Opium of Intellectuals* 129 The Bolsheviks have always had two slogans, one which insisted on monolithism, the other which encouraged conflicts of ideas and tendencies to sustain the vigour of the Party.

monologue, *sb.* Add: **2. c.** Used of Old English verse.

1902 W. W. LAWRENCE in *Jrnl. Eng. & Gmc. Philol.* IV. 462 Ebert expressed the opinion that the Seafarer should be interpreted as a monologue from beginning to end. 1905 E. RICKERT in *Mod. Philol.* II. 372 They are specimens of the *giedd* or short monologue arising from a dramatic situation, such as occurs frequently in *Beowulf*. 1935 A. C. BARTLETT *Larger Rhet. Patterns Anglo-Saxon Poetry* 106 Monologue and duologue, direct and indirect discourse, all are undramatic. 1943 B. F. HUPPÉ in *Jrnl. Eng. & Gmc. Philol.* XLII. 529 The basic outline [of the *Wanderer*] has already been set, with the two contrasting and complementary pagan monologues, framed and bound together by the expository Christian introduction, conclusion and 'bridge passage'.

monomaniac. b. *adj.* (Later example.)

1849 GEO. ELIOT *Let.* 24 Oct. (1954) I. 318 How does she manage to endure her life with that poor monomaniac husband.

monomaniacal, *a.* Add: (Later example.)

1971 *Eng. Studies* LII. 249 He is, in short, a soulless creature set apart from the rest of humanity by his monomaniacal devotion to pure avarice.

Hence **monomani·acally** *adv.*

1856 DICKENS *Dorrit* (1857) I. xxi. 186 Young Sparkler hovering about the rooms, monomaniacally seeking any sufficiently ineligible young lady. 1972 *New York* 3 Apr. 65 Her voice adheres monomaniacally to the same pitch.

Monomark (mǫ·nomăɹk). Also **monomark**. [f. MONO- + MARK *sb.*¹] A combination of letters and/or figures used as an identification mark for goods or personal property. Also *transf.* Hence **mo·nomark** *v. trans.*, to apply a monomark to; **mo·nomarking** *vbl. sb.*

1925 *Westm. Gaz.* 10 July, The Monomark system was explained by its inventor, Mr. William Morris, at a luncheon yesterday in London. A monomark, said Mr. Morris, is the shortest officially recognised postal name and address. 1925 *Glasgow Herald* 10 July 11 The idea..is to set up an international system whereby firms may be be granted a 'monomark', consisting of a combination of symbols preceded by B.C.M. (British Commercial Monomark). *Ibid.*, Any individual, child or adult, who cares to pay a modest fee will be entitled to be 'monomarked'. *Ibid.*, The eternal enigma of the identity of Smith—in England alone there are 530,000 bearers of the name—could be kept from business intercourse by the 'monomarking' of each number of the clan. 1926 *Ibid.* 27 Feb. 10 The persistency of the Scot's mental monomarks applied also to the vocal expression of them. 1928 *Daily Mail* 14 Aug. 9/7 Bench yesterday held that a monomark on a dog's collar did not fulfil the requirements of the law in respect of the address of the animal's owner. 1975 *Times* 18 Sept. 2/1 (Advt.), A Telex at your disposal for £25 p.a... British Monomarks (Est. 1926).

monomer (mǫ·nǒməɹ). *Chem.* [f. MONO- + *-MER.] Any compound from which a polymer might be formed by the combining together of its molecules (sometimes with the molecules of another compound).

1914 *Chem. Abstr.* VIII. 1037 If released from this combination, it [*sc.* chromic acid] gives at once a polychromic acid, which goes back to the monomer only in the presence of strong bases. 1942 *Electronic Engin.* XIV. 668/3 Impurities in the monomer styrene can have a very serious effect in the electrical properties of the finished polystyrene. 1957 *Technology* July 176/2 Will these catalysts polymerize monomers other than ethylene? 1974 *Guardian* 31 Jan. 6/2 Plastics manufacturers are now mounting health hazard investigations of vinyl chloride monomer.

Hence **monome·ric** *a.*, existing in the form of a monomer.

1929 [see *DIMERIC *a.* 2]. 1944 *Electronic Engin.* XVI. 348/3 This 'frost'..was caused by volatile impurities and residual monomeric styrene. 1969 *Nature* 13 Dec. 1060/2 The process by which the monomeric constituents of DNA are produced from their ribose analogues is now recognized to constitute a biochemical control mechanism of the greatest importance.

monomolecular (mǫnomǒle·kiŭlăɹ), *a.* [f. MONO- 1 + MOLECULAR *a.*] **a.** Composed of or pertaining to a single molecule or single molecules.

1877 *Chem. News* 23 Mar. 124/1 (*heading*) On the monomolecular unit of volume for gases and vapours. 1915 *Jrnl. Amer. Chem. Soc.* XXXVII. 191 The anhydride of betaine, $(CH_3)_3NCH_2CO_2$, has the structure of a salt, but no one seems to have determined whether this is monomolecular or dimolecular. 1974 *Biochim. & Biophys. Acta* CCCXLVI. 79 (*heading*) A monomolecular electron transfer chain.

b. In chemical kinetics: having or pertaining to an order or a molecularity of one. Cf. *UNIMOLECULAR *a.*

1899 *Jrnl. Franklin Inst.* CXLVII. 469 Another classical case of the monomolecular reaction is the decomposition of arsine. 1935 J. N. FRIEND *Text-bk. Physical Chem.* II. iii. 56 Radioactive transformations..provide the only known examples of true monomolecular reactions. *Ibid.* 81 As the solvent and catalyst molecules remain virtually constant in amount, many reactions that really are polymolecular are found to conform to the requirements of the monomolecular law. 1943 SUMNER & SOMERS *Chem. & Methods of Enzymes* i. 13 In most instances enzyme reactions classed as monomolecular are only approximately so. 1973 *Biol. Abstr.* LV. 2244/1 The monomolecular reaction equation gives a good fit to disease progress curve in the field within the primary infection period.

c. Of a film or layer: being one molecule in thickness.

1917 *Jrnl. Amer. Chem. Soc.* XXXIX. 1904 On cooling these surfaces to liquid air temperature the surfaces become covered with a monomolecular layer. 1931 *Jrnl. Physical Chem.* XXXV. 859 If a mono-molecular layer of gas is adsorbed at 300°C while the volume is kept constant the pressure increase will be about 8 mm. 1941 *Ann. Reg. 1940* 353 The remarkable properties of monomolecular films of certain organic liquids spread over water, mercury, or gallium. 1965 B. E. FREEMAN tr. *Vandel's Biospeleology* xviii. 317 It is the monomolecular waxy layer which makes the integument of terrestrial arthropods impermeable to water.

monomorphemic (mǫnomǭɹfī·mik), *a. Linguistics.* [f. MONO- + *MORPHEMIC *a.*] Consisting of a single morpheme.

1936 *English Studies* XVIII. 160 Alongside his [*sc.* Trnka's] five 'monomorphemic' or simple phonemes..he is forced to set up a like series of 'dimorphemic' combinations. 1942 [see *BIMORPHEMIC *a.*]. 1953 C. E. BAZELL *Linguistic Form* i. 5 Neither *John start-* nor *start-ed* can be replaced by a monomorphemic constituent. 1971 B. MAFENI in J. Spencer *Eng. Lang. W. Afr.* 110 By far the greater majority of Nigerian Pidgin words are invariable in form and are usually monomorphemic—that is, not divisible into smaller meaningful units. 1973 R. HARRIS *Synonymy & Linguistic Analysis* iii. 66 English *older* consists of two morphemes,..while *sister* is monomorphemic, since neither *sist-er* nor any other division gives units which can plausibly be identified as having 'reasonable similarity of meaning' with phonemically comparable units in other utterances.

monomorphic, *a.* Add: (Further examples.) Now *spec.* in *Genetics*, identical as regards genotype, a particular chromosome, or a part of a chromosome.

1894 W. BATESON *Materials for Study of Variation* 37 The..case in which the whole community, grouped according to the degrees in which they display a given character, forms one Curve of Error, may conveniently be called monomorphic in respect of that character. 1957 *Cold Spring Harbor Symp. Quantitative Biol.* XXII. 397/2 A highly polymorphic population..will be able..to leave progeny of some sort in a wider range of environment than could a monomorphic one. 1960 *Proc. Nat. Acad. Sci.* XLVI. 41 The population fails to reach the level of adaptedness which it would have if it were monomorphic for a genotype with an adaptive value equal to that of the A_1A_2 heterozygotes. Suppose, then, that a mutation produces an allele A_3 such that the fitness of the homozygote A_3A_3 is equal to that of the heterozygote A_1A_2. Natural selection is expected to lead to gradual elimination of A_1 and A_2, and to establishment of a population monomorphic and homozygous for A_3A_3. 1971 *Nature* 17 Sept. 190/2 *D. simulans* is one of the three widespread species of *Drosophila* considered to be monomorphic with regard to chromosome structure. 1973 *Ibid.* 31 Aug. 574/2 Of the thirty-one loci examined eight are monomorphic in all eleven populations. 1974 *Ibid.* 11 Oct. 514/1 *T. vivax* is generally considered monomorphic, although evidence of morphological variation between strains has been reviewed recently.

monomorphism. Add: (Further examples: cf. *MONOMORPHIC *a.*)

1957 *Cold Spring Harbor Symp. Quantitative Biol.* XXII. 399/1 In nature..it is unlikely that..a reversion to monomorphism will often occur. 1960 *Proc. Nat. Acad. Sci.* XLVI. 45 The replacement of the balanced polymorphisms present in the parental races by chromosomal monomorphisms certainly proves that acquisition in evolution of a balanced polymorphic condition need not be irreversible. 1973 *Nature* 31 Aug. 575/1 Most loci tend to monomorphism in reproductively as well as geographically isolated populations.

2. *Math.* A one-to-one homomorphism.

1956 C. CHEVALLEY *Fund. Concepts Algebra* i. 11 A homomorphism which is injective is called a monomorphism; a homomorphism which is surjective is called an epimorphism. 1969 F. M. HALL *Introd. Abstr. Algebra* II. ii. 39 In a monomorphism different objects always have different images. 1971 E. C. DADE in Powell & Higman *Finite Simple Groups* viii. 252 The diagonal map $\sigma \to \sigma \times \sigma$ is a natural monomorphism of G into $G \times G$.

Monomoy (mǫ·nomoi). The name of a peninsula (Monomoy Point) in Mass., U.S.A., used *attrib.* or *absol.* to denote a type of surfboat (see quot. 1966).

1908 *Ann. Rep. U.S. Life-Saving Service 1907* 65 The keeper..returned to the station..to..attempt to reach the steamer in the *Monomoy* surfboat. 1966 *Random House Dict.*, *Monomoy surfboat*, a double-ended surfboat having rather full lines with high carrying capacity and seaworthiness, used by the U.S. Coast Guard. 1967 *Proc. U.S. Merchant Marine Council* Jan. 19/2 The transfer was accomplished by the *Rockaway's* boat, a 26-foot Monomoy, with a crew of 12 men.

Monongahela. (Earlier examples.)

1805 *Mississippi Messenger* (Natchez) 1 July 3/1 *From Pittsburgh*..a Quantity of best Monongahela Whiskey for sale by the barrel. 1831 J. J. AUDUBON *Ornith. Biogr.* I. 504 The women..soon found my flask filled with *monongahela* (that is, reader, strong whisky). 1834 W. G. SIMMS *Guy Rivers* I. 76 Having cleared his throat with the contents of a tumbler of Monongahela which seemed to stand permanently full by his side. 1847 W. I. PAULDING *Madmen All* in J. K. & W. I. Paulding *Amer. Comedies* 192 May I never taste Monongahela again!

mononuclear, *a.* (*sb.*) **b.** *sb.* (Examples.) Add to def.: *spec.* a monocyte.

1908 *Jrnl. Infectious Dis.* V. 175 In the experiments recorded subsequently..+ + [indicates] destruction of the large mononuclears with degeneration of the lymphocytes. 1928 *Amer. Jrnl. Physiol.* LXXXV. 490 There are many forms of large mononuclears which are hard to distinguish from lymphocytes by the ordinary Wright stain. 1956 *Nature* 21 Jan. 138/2 The total leucocyte count was about double.., the mononuclears being increased by about 10,000 and the polymorphonuclears by about 6,500. 1974 PARK & GOOD *Princ. Mod. Immunobiol.* xiii. 171 Other chemotactic substances..derive from the complement components and these influence the mononuclears, as well as polymorphonuclears.

mononucleate (mǫnoniū·klīĕt), *a. Biol.* [f. MONO- + NUCLEATE *a.*] = MONONUCLEAR *a.*

1957 *New Biol.* XXIV. 30 In the primitive type, *Iridia* studied by le Calvez, there is no visible difference between the sexual and the asexual individuals until they begin to reproduce. Both are mono-nucleate. 1971 *New Scientist* 8 July 90/1, I shall be concerned with the mononucleate hybrid cells that are produced when the nuclei of two cells (and occasionally three) fuse together.

mononucleated, *a.* Add: (Further examples.) *spec.* (See quot. 1930.)

1930 W. S. THOMPSON *Population Problems* xvi. 279 The modern city..always has one nucleus which is far more important than any of the others and indeed as a rule it has a 'downtown' area in which are congregated most of the important offices, stores, hotels, amusements, and public buildings... Why have our modern large cities assumed this mononucleated form? 1938 L. MUMFORD *Culture of Cities* vii. 489 Under this mode of design, the planner proposes to replace the 'mono-nucleated city', as Professor Warren Thompson has called it, with a new type of 'poly-nucleated city', in which a cluster of communities, adequately spaced and bounded, shall do duty for the badly organized mass city. 1971 *Nature* 10 Sept. 105/1 Attached cells..were mostly mononucleated, though sometimes binucleated.

mononucleosis (mǫ:noniūkli,ōᵘ·sis). *Med.* [f. MONONUCLE(AR *a.* (*sb.*) + -OSIS.] An abnormally high proportion of mononuclear leucocytes (monocytes or lymphocytes), or of monocytes alone, in the blood; esp. = *infectious mononucleosis* (see *INFECTIOUS *a.* 2 c).

1920, etc. [see *infectious mononucleosis* s.v. *INFECTIOUS *a.* 2 c]. 1928 *Amer. Jrnl. Physiol.* LXXXV. 497 The spleen is an important factor in the production of an emotional relative mononucleosis. 1950 *Biol. Abstr.* XXIV. 1776/1 (*heading*) Exceptional hematological reactions in chronic benzene poisoning. Neutrophilic granulocytosis, lymphocytosis, and mononucleosis. 1968 *New Scientist* 25 Jan. 203/2 Suspecting that the presence of the EB virus and the girl's glandular fever were connected, the Henles..started testing blood samples from other patients who had contracted mononucleosis. 1972 T. P. MCMAHON *Issue of Bishop's Blood* (1973) vii. 88 I'm not too sure that mononucleosis isn't a modern-day version of the megrims of our Victorian ancestors. 1974 J. HELLER *Something Happened* 333 Even people who've been out awhile with hepatitis or mononucleosis have a hard time making their way back. They lack pep.

monooho(o), varr. *MOHOOHOO.

monophasic, *a.* Add to def.: Applied to (a record of) a nerve impulse that is of the same sign throughout, and to experimental arrangements devised to give such records. (Earlier and later examples.)

1883 *Phil. Trans. R. Soc.* CLXXIII. 35 (*table*) Character of variation. Monophasic, −39·1. Diphasic, +11·7, −21·2. 1915 W. M. BAYLISS *Princ. Gen. Physiol.* xii. 380 If electrode B is on a spot [of the nerve] which has been killed, the wave of negativity..disappears in the killed area, so that the only electrical effect seen is that due to the becoming negative of spot A and consists, therefore, of only the first half..of the diphasic response; it is..'monophasic'. 1932 *Jrnl. Cellular & Compar. Physiol.* I. 177 Local asphyxia at one of the leading-off electrodes renders a nerve potential record monophasic even more effectively than crushing does. 1950 *Jrnl. Gen. Physiol.* XXXIII. 654 The problem of obtaining a monophasic lead, by which is meant one in which no membrane changes at the distal electrode are recorded, is one of great importance in view of the complexity of the potentials to be analyzed. Crushing or heating was so unsatisfactory that keeping the distal lead on intact nerve was a preferable arrangement. 1967 L. A. GEDDES in C. C. Brown *Methods in Psychophysiol.* xv. 441 As a result of the intermittent activity of a variety of cells and organs, short duration, asymmetrical (with respect to the time axis) or completely monophasic pulses are presented to the reproducing apparatus.

Hence **monopha·sically** *adv.*, by a technique yielding a monophasic record; **monophasi·city**, the property of being monophasic.

1928 *Amer. Jrnl. Physiol.* LXXXVI. 187 The size of response was measured by the action current led off monophasically..from the peroneal nerve. 1932 *Jrnl. Cellular & Compar. Physiol.* I. 178 Gerard argues that monophasicity results because the asphyxiated region of the nerve is completely depolarized, while a killed region, if supplied with oxygen, may partially repolarize to give a diphasic record. 1941 S. H. BARTLEY *Vision* xii. 286 The monophasically recorded optic-nerve pattern produced by a flash of light on the retina, is a complex series of oscillations. 1950 *Jrnl. Gen. Physiol.* XXXIII. 654 By the time at which the equilibrium period was completed the fibers were fully blocked. Monophasicity continued without further difficulty to the end of the experiment. 1972 *Science* 2 June 1044/1 The excitability of primary afferent terminals was tested by stimulating the fibers through a low-resistance, glass microelectrode.., and the conducted antidromic volley was recorded monophasically.

monophonematic (mǫ:nofōᵘnīmæ·tik), *a. Linguistics.* [f. MONO- + *PHONEMATIC *a.*] = *MONOPHONEMIC *a.*

1940 *Language* XVI. 249 The argument for the monophonematic character of Dutch *ei*,.. (and the same applies to the English sounds in *bite*..) can be shown not to hold if the total structure of the syllable and the behavior of the..diphthongs are examined. 1952 A. COHEN *Phonemes of Eng.* ii. 24 The problem of how to decide whether we have to do with one or more phonemes (monophonematic or polyphonematic interpretation). 1961 *Brno Studies in English* III. 68 It is..in this medium section of the concerned sound trace that this most essential quality of the monophonematic diphthong must obtain the most obvious prominence.

Hence **monophonema·tically** *adv.*, **monophonemati·city**.

1953 C. E. BAZELL *Linguistic Form* iv. 47 The absence of possibility of being on a morph-boundary..may be a criterion of monophonematicity. 1961 *Brno Studies in English* III. 68 What matters..is..the direction and the zonal extent of the monophonematically evaluated 'gliding diphthong'. 1961 F. W. HOUSEHOLDER in Saporta & Bastian *Psycho-linguistics* 18/1 The chief metaphysical bones hassled over..concern such points as 'biuniqueness', 'monophonematicity' [etc.].

monophonemic (mǫnofonī·mik), *a. Linguistics.* [f. MONO- + *PHONEMIC *a.*] Of or pertaining to a single phoneme; consisting of a single phoneme.

1936 *Amer. Speech* XI. 206 Over against the monophonemic words cited above are to be set the pleiophonemic words made up of two or more phonemes each. 1955 *Archivum Linguisticum* VII. 154 A distinction is drawn between monophonemic short-vowel and polyphonemic long-vowel diphthongs. 1957 S. POTTER *Mod. Linguistics* ii. 42 The simplest instances of the commutation test are provided by such monophonemic forms as occur in most languages.

monophonic, *a.* Add: (Examples.)

1920 E. WALKER in F. S. Marvin *Recent Devel. Europ. Thought* 286 If we exclude some monophonic conceptions that have still their value for us, it [*sc.* music] is barely five hundred years old. 1942 *Scrutiny* XI. 1. 7 The monophonic work of Léonin and Perotin. 1972 *Harper's Dict. Mus.* 206/1 Monophonic vocal music has existed since ancient times.

2. Of sound broadcasts, gramophone records, etc.: involving only one channel, so that there is only one output signal and all the sound appears to the listener to come from a single source; = *MONAURAL *a.* 2. Opp. *STEREOPHONIC *a.*

1958 *Newsweek* 13 Oct. 102/2 Bending an ear to a single-track (monophonic, to hi-fi devotees) record is 'like listening to a concert through a crack in the door'. 1959 *Proc. Inst. Electr. Engin.* CVI. B. Suppl. No. 14. 257/1 The monophonic reproduction on two loudspeakers appears to issue from a very small source situated between them but, when operating in a stereophonic manner, the ambient studio sound appears to fill the whole space between the loudspeakers with positional accent on the individual sound sources. 1960 *New Scientist* 4 Feb. 262/2 Such a stereo system may not be 'compatible' with ordinary monophonic receivers. 1962 A. NISBETT *Technique Sound Studio* 240 In a monophonic recording..noise at a natural level will generally sound excessive. 1968 *Consumer Reports* Sept. 502/1 Only stereo discs are reviewed, since record companies have all but stopped producing classical monophonic discs. 1968 *Times* 29 Nov. (Sound of Leisure Suppl.) p. vi/4 The sound was still monophonic: that is, it still came from a single track or radio signal and could never give us any sensation of listening to real musicians spread naturally across an audio stage. 1970 *Jrnl. Gen. Psychol.* LXXXIII. 256 The pure tones..were ..recorded on magnetic tapes by a single-track monophonic Tandberg tape recorder. Subsequently, these magnetic tapes were played back to *S* through a pair of Sharpe monaural liquid-filled headphones. 1974 *Nature* 13 Dec. 535/1 Monophonic reproduction gave no explicit directional information, even when reproduced from more than one loudspeaker.

monophonically (mǫnofǫ·nikăli), *adv.* [f. prec. + -AL + -LY².] **a.** *Mus.* As regards monophony.

1942 *Scrutiny* XI. 12 It is well that we should understand how in 'progressing' music lost much, polyphonically as well as monophonically.

b. In a monophonic (monaural) manner; as regards monophonic reproduction; = *MONAURALLY *adv.* b.

1959 *N.Y. Times* 8 Feb. 14X/3 The disks are available stereophonically and monophonically. 1959 *Proc. Inst. Electr. Engin.* CVI. B. Suppl. No. 14. 207/1 If coincident directional microphones are employed and we add left A to right B..the resultant may well be unsuitable monophonically. 1963 *Which?* Jan. 8/2 Both these could play stereo records, without conversion, though the sound was reproduced monophonically. 1974 *Encycl. Brit. Macropædia* XVII. 53/2 A record that can be played stereophonically over stereophonic equipment or monophonically over monophonic equipment is called a compatible recording.

monophony. Add: (Examples.)

1964 W. LOVELOCK *Student's Dict. Mus.* 56/2 *Monophony*, music which consists simply of a melodic line without any form of accompaniment. 1968 *Encycl. Brit.* XV. 745/2 *Monody*,..a musical term that in England is often used to describe music for a single melodic line, though in the U.S. monophony is preferred for this meaning.

2. Monophonic recording or reproduction.

1959 *Proc. Inst. Electr. Engin.* CVI. B. Suppl. No. 14. 201/1 The best microphone positions for monophony and stereophony frequently differ considerably and are often quite incompatible. 1961 *Atlantic Monthly* Jan. 146/3, I am sure their transients, meaning the quick high overtones of fiddles and trumpets, are as hard to handle in monophony as in stereo.

Monophoto (mǫnofōᵘ·to). [f. MONO(TYPE *sb.* and *a.* + PHOTO.] The proprietary name of a photo-composing machine which uses a perforated paper tape produced by a keyboard unit to control filmsetting by a second unit which replaces the type-caster of a Monotype machine. Also *attrib.*

1954 [see *FILMSET *v.*]. 1958 *Times Lit. Suppl.* 18 Apr. 209/3 How important the correction factor is may be seen from the history of the Monophoto..which is likely to become one of the more successful photo-composing machines. 1967 COX & GROSE *Organiz. Bibliogr. Rec. by Computer* II. 27 (*caption*) These [*sc.* trial galley proofs] were produced on a Monophoto composing machine. 1973 S. JENNETT *Making of Books* (ed. 5) v. 85 The mechanical Monophoto may be called a first-generation photo-composing machine.

monophthongize, *v.* Add: (Further examples.) Also *intr.*

1914 H. C. WYLD *Short Hist. Eng.* vi. 107 In W.S. the *ea* which resulted from *æ* preceded by a front consonant was monophthongized in L.O.E. itself, and became *e*. 1927 J. J. HOGAN *Eng. Lang. in Ireland* 69 M.E. *ai*, *ei*. These were levelled under *ai* in M.E. This was monophthongized, and converged with M.E. *ā*. 1965 W. S. ALLEN *Vox Latina* ii. 62 Although French has monophthongized to *o*, the *au* diphthong must have survived long enough to cause the change of *c* to *ch* in..*chose* from *causam*. 1968 —— *Vox Graeca* ii. 71 The [ę] of Boeotian was the result of monophthongizing the diphthong [ai]. 1970 B. M. H. STRANG *Hist. Eng.* ii. 111 The diphthong, in either form, began to monophthongise.

monophyletic, *a.* Add: (Later examples.) Hence **mono·phyly** (see quot. 1961).

1927 N. P. WILLIAMS *Ideas of Fall* viii. 516 The 'monophyletic' theory of the origin of humanity (that is, the view which regards the whole of mankind as descended from a single pair of ancestors). 1945 G. G. SIMPSON in *Bull. Amer. Mus. Nat. Hist.* LXXXV. 17/1 It is not useful to set up a classification in which groups with different names cannot be distinguished morphologically, but this does happen if theoretical monophyly is too strictly demanded. 1946 *Nature* 16 Nov. 719/1 This clear distinction between the genera at so deep a level in the Palæocene is not suggestive of a monophyletic origin for the family. 1953 E. MAYR et al. *Methods & Princ. Systematic Zool.* iii. 41 One of the objects of taxonomists in the post-Darwinian period was to construct a classification of animals composed of monophyletic groups. 1961 G. G. SIMPSON *Princ. Animal Taxonomy* iv. 124 Monophyly is the derivation of a taxon through one or more lineages (temporal successions of ancestral-descendant populations) from one immediately ancestral taxon of the same or lower rank. 1963 DAVIS & HEYWOOD *Princ. Angiosperm Taxonomy* ii. 46 It is a widely expressed principle that taxonomic groups (at least above the species level) should be monophyletic. 1969 E. MAYR *Princ. Systematic Zool.* iv. 75 The issue of monophyly has been confounded by various confusions.

monopitch (mǫ·nopitʃ), *sb.* and *a.* [f. MONO- + PITCH *sb.*²] **A.** *sb.* Uniformity of pitch in speaking or singing.

1939 in WEBSTER Add. 1942 A. T. WEAVER *Speech Forms & Princ.* x. 243 Monopitch, frequently miscalled monotone, results first, from unemotional speaking and, second, from a dull or inactive mind which fails to see distinctions in the meanings of the language which is being spoken. 1964 CRYSTAL & QUIRK *Systems Prosodic & Paralinguistic Features in Eng.* ii. 16 These deficiencies also mar his discussion of..pitch (including 'monopitch', step up and down, and slide).

B. *adj.* Of a roof: consisting of a single uniformly sloping surface. Of a building: having such a roof.

1941 *Archit. Rev.* LXXXIX. 47 The mono-pitch roof was employed for simplicity and to enable all the rain-water to be collected at the front of the house owing to the limited fall available for the drains. 1961 *Guardian* 25 Mar. 5/5 A simple and comely monopitch building. 1971 *Daily Tel.* 14 Aug. 7/3 The crematorium's group of mono-pitch grey slate roofs appear to form a large pyramid.

monoplane (mǫ·noplēⁱn). [f. MONO- + PLANE *sb.*³] An aeroplane or glider having only one 'plane' or main supporting surface on either side of the fuselage (so called because in the earliest monoplanes the wing on each side was part of a single structure extending across the fuselage); † the wing itself. Also *attrib.* or as *adj.*

1907 *Sci. Amer.* 16 Nov. 358/1 (*heading*) The latest French aeroplanes and their records. The Pelterie monoplane. *Ibid.* 358/3 One end of the monoplane (which is made in two halves) was broken. *Ibid.*, The Pelterie aeroplane resembles the monoplane machine with which M. Bleriot experimented unsuccessfully last spring. 1907 *Nature* 5 Dec. 106/2 Another aëroplane which is also attracting considerable attention at Paris is the 'monoplane' of M. Robert Esnault Pelterie. This, unlike most recent types, has only a single transverse supporting surface. 1908 *Times* 20 May 7/6 A monoplane aeroplane..made some successful evolutions yesterday. 1908 *Sci. Amer.* 18 July 44/1 The rear half of the monoplane, at its outer ends, has movable planes for correcting the transverse stability. 1910 *Blackw. Mag.* July 4/1 The aeroplane, whether monoplane, biplane, or other. *a* 1918 J. T. B. MCCUDDEN *Five Yrs. in R. Flying Corps* (1919) 63 Two Taube monoplanes came over St. Omer. 1922 *Glasgow Herald* 8 Aug. 8 A Swiss pilot on a motorless monoplane succeeded in making a flight lasting 45 seconds. 1935 C. G. BURGE *Compl. Bk. Aviation* 123/1 Gliders are usually of monoplane type, though biplane constructions have been used. 1960 C. H. GIBBS-SMITH *Aeroplane* x. 54 The year 1907 saw the crystallisation of the two basic forms of early aeroplane—the pusher biplane and the tractor monoplane. 1969 K. MUNSON *Pioneer Aircraft 1903–14* 98/1 Levasseur's first full-size aeroplane, a bird-like monoplane tested in 1903, was an utter failure. 1973 J. DI MONA *Last Man at Arlington* (1974) xi. 84 A single engine monoplane thundering toward an ammunition ship.

monoploid (mǫ·nǫploid), *a.* (and *sb.*). *Biol.* [a. G. *monoploid* (O. F. I. Langlet 1927, in *Svensk Bot. Tidskr.* XXI. 1): see MONO- and *-PLOID.] = *HAPLOID *a.* (and *sb.*) (see quot. 1968). Hence **mo·noploidy**, the state or condition of being monoploid.

1928 *Jrnl. Genetics* XIX. 138, I do not think it necessary to replace haploid with monoploid as Langlet (1927) proposes. 1943 L. W. SHARP *Fund. Cytol.* xiv. 211 In the sporophytes of plants true monoploidy occurs very rarely. 1944 *Genetics* XXIX. 232, 16 monosomes and five trisomes were found in the 11 aberrant offspring of a monoploid pollinated by a diploid. 1952 SRB & OWEN *Gen. Genetics* xi. 215 Doubling the chromosomes of a monoploid can give rise to diploid individuals homozygous for all the gene pairs in the organism. 1955 G. F. SPRAGUE *Corn & Corn Improvement* iv. 131 Monoploid or haploid plants, which have one set of chromosomes, arise spontaneously by the parthenogenetic development of unfertilized eggs. Much less frequent are the androgenetic monoploids. *Ibid.*, Genes for a high incidence of monoploidy. 1968 R. RIEGER *Gloss. Genetics & Cytogenetics* 299 *Monoploid*, of cells or individuals (monoploids) having one chromosome set with the basic number of chromosomes, i.e., the lowest haploid number of chromosomes in a polyploid series.

monopode, *sb.* and *a.* Add: (Later examples of spelling *monopod.*) Also *transf.*

a 1963 C. S. LEWIS *Poems* (1964) 43 Ran till the sunrise shone upon the bouncing Monopods at their heels. 1971 C. BONINGTON *Annapurna South Face* App. F. 286 Each member of the team was also issued with a special clamp to fit on the head of an ice axe with a ball-and-socket attachment for the camera, which made a very useful and compact monopod. 1972 QUICK & LABAU *Handbk. Film Production* xiii. 100 A 'monopod' is a single fixed leg terminating in a screw base that matches the bottom of the camera.

monopodium. Add: **2.** (See quot. 1970.) Also, the support for an early 19th-century table, sideboard, etc., comprising an animal's head with a single foot. (Cf. MONOPODE *sb.* 1.)

1807 T. HOPE *Househ. Furnit.* 36 Little round monopodium or stand, of which the top..is capable of being raised or lowered at pleasure. 1948 M. JOURDAIN *Regency Furnit. 1795–1820* (ed. 2) 105 Console table supported by carved gilt lion monopodia. 1959 G. SAVAGE *Antique Collector's Handbk.* 195 The monopodium, or single leg surmounted by a human or animal head and torso, is peculiarly Regency. 1970 —— *Dict. Antiques* 275/2 *Monopodium*..is applied to certain small tables of the Regency and Empire period which have a three-sided support usually terminating in brass claws at the bottom corners. 1971 *Country Life* 22 July 225/3 From about 1815 the marble top of a console table might be supported at each front corner by an eagle monopodium with boldly outcurving breast and widespread wings.

monopolar (mǫnopōᵘ·lăɹ), *a.* [f. MONO- + POLAR *a.*] **a.** Of, having, or using a single electrode; applied esp. to techniques and apparatus (*a*) for passing electric currents through the body using two electrodes of different sizes or natures, one being usually much larger than the other, and (*b*) for measuring electric potentials in the body, where one electrode is inserted in the region to be studied and the other acts as a reference electrode.

1906 *Practitioner* Dec. 772 The patient, if the monopolar bath had to be administered, grasped a metal bar, suspended from the ceiling, and in connection with the battery swung over the bath, the other electrode remaining in the water. 1936 H. H. U. CROSS *Electr. in Therapeutics* xiii. 296 In the spark-gap system the patient may be connected to the solenoid in monopolar or bipolar fashion. 1956 S. LICHT *Some Clin. Applic. Electroneurophysiol.* i. 8 Chauveau..introduced the monopolar method of stimulation into physiology. 1962 BLAKE & TROTT *Periodontology* xiii. 135 Monopolar diathermy of this type is particularly suitable for minor oral surgery. 1973 M. SCHWARTZ *Physiol. Psychol.* i. 20 In monopolar recordings, one electrode is placed on some relatively neutral or 'indifferent' location, such as the ear, for the EEG, while the other is placed on the scalp.

b. Of or pertaining to a *MONOPOLE³.

1971 I. G. GASS et al. *Understanding Earth* xvi. 239/2 The dipolar rather than monopolar nature of the Earth's magnetic field.

monopole¹ (mǫ·nǒpōᵘl). Restrict † *Obs.* to senses in Dict. and add: **4.** [Fr., monopoly.] (See quot. 1967.)

1883 *Daily News* 31 Aug. 6/2 The familiar pop of the champagne cork being very rarely heard. Perhaps those learned authorities, Dr. Oliver and Dr. Myrtle..have 'shut down' on 'monopole' and 'extra dry'. 1885 *Christie's Catal. Wines* 14 Apr. 9 Three Dozens of Champagne, Heidsieck's Dry Monopole, 1874. 1886 *St. Stephen's Rev.* 13 Mar. 12/1 The pint of dry Monopole, with which he sustained exhausted nature. 1915 T. BURKE *Nights in Town* 399 He shouted for a half-of-bitter with the solemnity of one who commands that two bottles of dry Monopole be put on the ice. 1967 A. LICHINE *Encycl. Wines* 356 *Monopole.* This designation, frequently found on wine labels, literally means 'an exclusive'. Blank Monopole would be the proprietary brand of the wine shipper whose name would occupy the blank.

monopole³ (mǫ·nǒpōᵘl). [f. MONO- + POLE *sb.*²] **1.** A single electric charge or, esp., a single magnetic pole, having a spherically symmetric field. Freq. *attrib.* or as *adj.*

1937 J. W. T. SPINKS tr. *G. Herzberg's Atomic Spectra & Atomic Struct.* i. 63 The system gives an external electric field, which..falls off more rapidly with increasing distance than that of the dipole, which itself falls off more rapidly than that of the monopole. 1950 D. HALLIDAY *Introd. Nuclear Physics* ii. 59 A single charge at the origin is a simple monopole, with no higher moments. 1951 *Proc. Cambr. Philos. Soc.* XLVII. 196 Dirac has suggested that the quantization of electric charge could be explained by the existence of magnetic monopoles. *Ibid.* 206 The main difference of behaviour between electric particles and monopoles lies in the greater ionizing power per centimetre of the monopole, except near the end of the path. 1958 CONDON & ODISHAW *Handbk. Physics* VII. iv. 59/1 The first term of Eq. (4.6) corresponds to a monopole or simple charge, the second to the electric-dipole moment, the third to the electric-quadrupole moment, the fourth to the electric-octupole moment, etc. A similar multipole expansion is possible with magnetic interactions..leading to magnetic-dipole terms, etc. 1962 CORSON & LORRAIN *Introd. Electromagn. Fields* ii. 54 The first term is merely the potential which we would have at P if the whole charge were concentrated at the origin. It is called the monopole term and is zero only if the net charge is zero. 1967 W. R. HINDMARSH *Atomic Spectra* v. 49 We must extend the concept of the nucleus by attributing to it magnetic and electric moments (in addition to the electric monopole or nuclear charge) and a finite volume. 1969 *Physical Rev.* CLXXXIV. 1393 (*heading*) Search for magnetic monopoles in deep ocean deposits. *Ibid.* 1401/2 This work would indicate that there is less than one monopole/4000 m³ of the earth. 1973 *Nature* 20 Apr. 527/1 The scheme..ignores all other atomic monopoles (including those on the methyl carbons and hydrogens which closely approach the negative charge) as well as all higher moments. 1974 *Ibid.* 3 May 15/1 Searches for quarks, magnetic monopoles and intermediate vector bosons have all so far proved negative. 1975 *Physical Rev. Lett.* XXXV. 489/2 We conclude that we have detected a magnetic monopole of strength $g = 137e$.

2. *Radio.* An aerial consisting of a single conducting rod with the electrical connection made at one end, the length of which is usually about a quarter of the wavelength to be transmitted or received.

1950 *Proc. IRE* XXXVIII. 1040 (*heading*) Measured directivity induced by a conducting cylinder of arbitrary length and spacing parallel to a monopole antenna. 1966 L. V. BLAKE *Antennas* iv. 176 When the quarter-wave vertical antenna (monopole) is fed at its base with the other side of the feed line connected to ground, its radiation resistance and input impedance are just half the values for the half-wave dipole in free space. 1974 *Nature* 5 Apr. 493/1 The transmitter antenna was a vertical monopole lifted to 1,000 to 1,500 m by balloon.

monopolistic, *a.* Add: (Further examples.) Spec. *monopolistic competition*, a type of imperfect competition characterized by monopolistic trading.

1903 R. T. ELY *Studies in Evolution of Industrial Soc.* iv. 196 Private favoritism monopolies are businesses not naturally monopolistic. 1933 E. CHAMBERLIN *Theory of Monopolistic Competition* iv. 69 Monopolistic competition..concerns itself not only with the problem of an *individual* equilibrium (the ordinary theory of monopoly), but also with that of a *group* equilibrium (the adjustment of economic forces within a group of competing monopolists, ordinarily regarded merely as a group of competitors). 1937 [see *IMPERFECT *a.* 11]. 1938 E. W. SHANAHAN *Mod. Econ. Organisation* x. 235 The movement towards combination..has been successful in establishing monopolistic enterprises. 1942 W. B. TAYLOR *Financial Policies of Business Enterprise* xxxi. 750 The warp and woof of business enterprise are interwoven with ever-changing threads of competition and monopoly; and when both are found co-existing in a given enterprise, that business is said to be carrying on its functions under conditions of monopolistic competition. 1962 *Daily Tel.* 31 Jan. 10/2 The Monopolies Commission can report on monopolistic practices referred to it. 1964 GOULD & KOLB *Dict. Social Sci.* 441/1 Monopolistic competition denotes a condition of partial market control due to the exclusive possession of the trade in some commodity on the part of a seller and limited by the fact that one commodity may be substituted for another with varying degrees of ease. 1969 *Times* 27 Jan. 10/7 It is starting to match monopolistic selling power against monopolistic buying power in the grain market. 1969 D. C. HAGUE *Managerial Econ.* II. iv. 87 The first of these [kinds of imperfect competition] can be dealt with quite briefly. It is known as monopolistic competition, though the originator of the term, the American economist Professor E. H. Chamberlin intended this to be used in a slightly different sense.

Hence **monopoli·stically** *adv.*

1923 *Glasgow Herald* 30 July 6 We need no longer cry out to obstructive departmental control and monopolistically inclined private enterprise—'A plague on both your houses.' 1957 K. A. WITTFOGEL *Oriental Despotism* ii. 45 They [*sc.* industries] were for the most part either directly managed or monopolistically controlled by the hydraulic governments. 1966 *Economist* 26 Feb. 774/1 Rises in dividends are determined individually, not monopolistically.

monopolizable (mǫ̆nǫ·pǒlaizăb'l), *a.* [f. MONOPOLIZE *v.* + -ABLE.] That can be monopolized.

1898 W. J. LOCKE *Idols* xxii. I'm not a monopolisable woman.

monopolizer. Add: Also **monopoliser.**

1779 W. SMITH in *15th Rep. R. Comm. Hist. Manuscripts* App. VI. 429 in *Parl. Papers 1897* (C. 8551) LI. 1 The common people sick of the War, and think they are made a prey of by commissaries, quarter-masters, forage-masters, monopolisers, schemers, &c., &c.

monopoly. **1.** Add to def.: In U.K. law, a situation in which one supplier or producer controls over one third of the market.

1948 *Act* 11 & 12 *Geo. VI* c. 66 Prelim. Note, This Act sets up a permanent Commission to investigate conditions in industry and trade which tend towards a monopoly. 1964 *Times Rev. Industry* Mar. 16/1 The Act also laid down the definition of a monopoly that is still operative: a situation in which one company or group of companies acting together control one third of an industry. 1970 *Daily Tel.* 15 July 17/5 By law, over a third of the market is defined as monopoly.

6*. (With capital initial.) The proprietary name of a game (invented by Mr. Charles Darrow) in which the players use imitation money to engage in simulated financial dealings. Also *attrib.* **Monopoly money**, money that is not 'real', valueless currency.

1935 *N.Y. Times* 24 Nov. x. 6/2 Heading all other 'board games'..is the season's craze, 'Monopoly', the game of real estate. 1938 C. BEATON *Diary* 1 Aug. in *Wandering Years* (1961) 344 The game of 'Monopoly' wasn't very enjoyable this morning. 1939 G. GREENE *Lawless Roads* Prol. 5 A game called 'Monopoly' played with a picture-board and dice and little counters. *Ibid.*, There were 'Monopoly' parties. 1954 *Trade Marks Jrnl.* 23 June 634/2 *Monopoly*..Board games John Waddington Limited..Leeds. 1960 *Guardian* 9 Dec. 9/7 My knowledge of the world of big business..I gained entirely from playing Monopoly. 1972 K. BONFIGLIOLI *Don't point that Thing at Me* xiii. 110 Martland's word was as good as his bond, but his bond was mere Monopoly money. 1972 *Times* 8 Apr. 1/3 Private tenants in rented flats are banding together..against property speculators whom they accuse of ruthlessly playing Monopoly with their homes. 1974 *Times* 27 Apr. 10/7, I think they are playing with Monopoly money because some people say things like 'I have lost £10,000'.

7. *monopoly licence, problem, profits, -trend*; **monopoly capitalism**, a capitalist system typified by trade monopolies in the hands of a few people; **monopoly value**, the value of anything the supply of which is controlled by the holders of a monopoly; *spec.* applied to licensed premises.

1937 C. DAY LEWIS *Mind in Chains* 13 The quality of intellectual production is inevitably debased under monopoly-capitalism. 1943 H. READ *Politics of Unpolitical* i. 6 It [*sc.* the oligarchy of trade unions] is now openly merging itself with the ascendant oligarchy of monopoly capitalism, to form what James Burnham has called 'the managerial class'. 1955 —— *Grass Roots of Art* (rev. ed.) vii. 139, I feel fairly sure that that barrack-room will have more amenities under monopoly capitalism than in the totalitarian State. 1964 GOULD & KOLB *Dict. Social Sci.* 442/2 Monopoly capitalism in Marxist terminology is the stage of capitalism in which control over means of production and distribution has become highly concentrated in the hands of a small group of capitalists. 1905 J. J. COCKSHOTT *Licensing Act 1904* 3 The State, in conferring upon me a monopoly license, had also granted me a commercial asset of great value. 1955 *Times* 15 June 5/4 The monopoly problem is discussed usually in terms of a single seller or a small group of sellers acting in concert, who are able to restrict supplies. 1931 *Economist* 26 Dec. 1222/1 The 'ring', producing a sectionalised product, will be able to exaggerate the monopoly profits already exacted from the helpless consumer. 1964 GOULD & KOLB *Dict. Social Sci.* 543/1 Profits include..permanent monopoly profits, more properly termed monopoly rents. 1944 *Horizon* Jan. 67 What is valid..is no more than the good old perception of monopoly-trends in finance and industrial organization. 1883 F. A. WALKER *Polit. Econ.* III. i. 90 Here is a high degree of value..where yet no labor has been... This is an instance of what may be called 'monopoly-value', or as some prefer to call it, scarcity-value. 1904 *Hansard Commons* 20 Apr. 735 We think that when any new licence is granted the monopoly value should go to the public. 1905 J. J. COCKSHOTT *Licensing Act 1904* 6 The monopoly value will be secured to the public instead of being 'handed over to wealthy brewery companies for the mere asking'. 1934 DUMMEIER & HEFLEBOWER *Econ. with Applications to Agric.* viii. 166 Monopoly value refers to exchange value for any commodity of which the supply is sufficiently under one control that the exchange value or price which prevails for it is different..than it would be under ordinary competitive conditions. 1936 JATHAR & BERI *Introd. Econ.* ix. 170 In all cases of monopoly value, the monopolist must carefully weigh..the nature of the demand, and..the expenses of production per unit. 1953 B. SPILLER *Innkeeping* x. 190 On the grant of a new licence the justices must exact the payment of a *monopoly value*, i.e. the difference between the value of the premises licensed and unlicensed.

Monoprix (mǫ·noprī). [Fr., lit. 'one price'.] One of a chain of multiple stores (in France) in which a cheap class of goods is sold (orig. all at the same price).

1937 W. FORTESCUE *Sunset House* ix. 164 Bulky goods hastily thrust into bags at the Monoprix. 1965 V. CANNING *Whip Hand* v. 48, I bought six pairs of pants at a Monoprix. 1966 G. GREENE *Comedians* I. i. 23 Cynicism is cheap—you can buy it at any Monoprix store. 1973 D. ORGILL *Jasius Pursuit* i. 8 She had seen *Mademoiselle* in the Monoprix down in Cannes.

monopropellant (mǫnoprǒpe·lănt), *sb.* and *a.* [f. MONO- + PROPELLANT *a.* and *sb.*] **A.** *sb.* A substance which can be used as a rocket fuel without needing an additional oxidizing agent.

1949 G. P. SUTTON *Rocket Propulsion Elements* v. 115 Monopropellants are stable at ordinary atmospheric conditions, but decompose and yield hot combustion gases when heated and pressurized. 1958 C. C. ADAMS et al. *Space Flight* iii. 79 A monopropellant is generally unstable and delivers energy through its own decomposition, which is generally induced by a catalytic agent such as potassium permanganate (with hydrogen peroxide). 1972 *Materials & Technol.* IV. xix. 741 Hydrogen peroxide was employed as a monopropellant to provide power for the turbine-driven pumps [in the V2 rocket].

B. *adj.* Employing a monopropellant.

1949 G. P. SUTTON *Rocket Propulsion Elements* v. 115 The feed system of monopropellant units is usually simple, because only one liquid needs to be supplied. 1954 K. W. GATLAND *Devel. Guided Missiles* (ed. 2) i. 23 Rockets can be classified into two main types: (*a*) Bi-propellant, in which the fuel and oxygen supplies are..injected separately into the combustion chamber, and (*b*) mono-propellant, in which the fuel and oxygen are combined in a single substance. 1967 *Technology Week* 23 Jan. 28/2 Anhydrous hydrazine will fuel the liquid monopropellant engine.

monops (mǫ·nǫps), *a.* and *sb.* [f. Gr. μόν-ος MONO- + ὤψ eye.] † **A.** *adj.* One-eyed. *Obs. rare*⁻⁰. **B.** *sb.* A one-eyed individual. *rare*⁻¹.

1857 R. G. MAYNE *Expos. Lex. Med. Sci.* (1860) 716/2 *Monops*, having but one eye: one-eyed. 1876 G. MEREDITH *Beauch. Career* III. ix. 155 He would have been a Nelson of politics, if he had been a monops, with an excuse for not seeing. 1900 DORLAND *Med. Dict.* 396/1 *Monops*, a fetus having but a single eye. [And in later med. dicts.]

monopsony (mǫ̆nǫ·psǒni). *Econ.* [f. MON(O- + Gr. ὀψωνεῖν to buy provisions + -Y³.] A condition in which there is only one buyer for the product of a large number of sellers; cf. MONOPOLY. Also in extended use (see quot. 1971). Hence **mono·psonist**; **monopsoni·stic** *a.*

1933 J. ROBINSON *Econ. Imperfect Competition* xviii. 219 When the market changes..to a single buying agency..this may be described as monopsony buying. *Ibid.* 224 Just as we have price discrimination for a monopolist, so we may have price discrimination for a monopsonist. *Ibid.* xxvi. 296 Monopsonistic exploitation can also arise where..the supply of labour to each firm is less than perfectly elastic. 1941 R. TRIFFIN *Monopolistic Competition & General Equilibrium Theory* iii. 113 Other types of interrelationships, intermediate between monopsony and pure competition, can be distinguished. 1948 *Commerce & Industry* (Pretoria) Mar. 280 The supplying of sawlogs on a contract term of 20 years..may place the sawmiller in a monopsonistic position in a particular forest region. *Ibid.* Bargaining power will be fairly evenly divided between the two parties. The dangers of monopsony..will thus be limited. 1955 *Times* 15 June 5/4 The term 'monopsonist' for a monopoly buyer was coined in the thirties—but has got little foothold in popular discussion. 1964 *Economist* 1 Feb. 434/1 Bargaining between a monopsony buyer and a suppliers' ring. 1968 *Internat. Encycl. Social Sci.* X. 464 Analogous to a seller's monopoly power, a buyer is said to have 'monopsony' power when he can significantly affect the price of what he buys by varying the quantity bought. 1969 D. C. HAGUE *Managerial Econ.* II. iv. 98 When there is a single buyer in any market, he is often described as a *monopsonist*. 1971 M. COOPER in *Royal Soc. Health Jrnl.* XCI. 220/2 The main risks of the B.M.A. scheme however are not of this type. Rather they lie in breaking up the State monopsony. *Ibid.* 221/2 A monopsony is a consumer so large that it can exert pressure on price merely by the threat of withdrawing its custom. 1973 *Times Lit. Suppl.* 9 Mar. 255/2 The granting of monopsonistic concessions to wholesale merchants.

monoptic, *a.* Delete † *Obs.*⁻⁰ and add to def.: Pertaining to or involving vision with one eye. (Later examples.)

1960 KOLERS & ROSNER in *Amer. Jrnl. Psychol.* LXXIII. 4 One may distinguish at least three modes of viewing: monocular, binocular, and a third which we shall call 'dichoptic'. [*Note*] Consistency of usage would require that the first two be called 'monoptic' and 'dioptic' respectively. 1968 *Perception & Psychophysics* III. 237/2 Under the monoptic condition, both disk and ring were presented to the same eye, half the time to the left eye and half the time to the right eye. 1971 *Jrnl. Gen. Psychol.* LXXXIV. 163 Dichoptic masking was nearly as strong as under monoptic and binocular conditions.

Hence **mono·ptically** *adv.*, with one eye.

1965 *Jrnl. Exper. Psychol.* LXIX. 199/1 The extent of masking by pattern was slightly less dichoptically than monoptically. 1972 *Nature* 22 Dec. 480/1 The observer perceives a pattern which approximates very closely a square-wave distribution, just as if the two components were superimposed physically and viewed monoptically.

mono-rail. Also **monorail**. Substitute for def.: A single rail astride which or suspended from which railway vehicles or the like run; also, a vehicle that travels in this way. Also *attrib.* (Add further examples.) Hence **mo·no-railer**; **mo·norailing** *vbl. sb.*

1902 *Encycl. Brit.* XXXII. 143/2 In the Langen monorail the car is hung from a single overhead rail; a line on this system is worked between Barmen and Elberfeld, a distance of about 9 miles. 1949 *Sun* (Baltimore) 3 Oct. 2/7 These chutes will pull the guns out of the rear clam shell doors of the C-82 along monorails attached to the roof of the fuselage. 1958 *Daily Mail* 17 Jan. 9/4 A trip on the world's first monorail. 1960 *Farmer & Stockbreeder* 16 Feb. (Suppl.) 35/1 One is an overhead monorail down the centre feeding passage to carry a food container from which the rations may be semi-automatically fed into the troughs. 1962 R. B. FULLER *Epic Poem on Industrialization* 44 Their 1929 failure With world monopoly 'Monorailing' Was only caused by the fact that Their first trial section of single rail Was not long enough. 1963 *Punch* 7 Aug. 194/2 The monorailer slid to a halt. 1964 *Daily Tel.* 7 Jan. 15/1 A free monorail system may be built in North Buckinghamshire. 1971 *Ibid.* 16 Apr. 6/7 A French demonstration hovertrain has been running for 18 months on an 11-mile stretch of elevated monorail outside Orléans. 1973 R. KERROD *First Look at Railways Today* 45 It is called a monorail, for it moves along a single rail.

b. **monorail camera**, a technical camera having as part of its stand a bar or monorail which allows considerable adjustment of the apparatus and may support additional components.

1958 D. CHARLES *Commercial & Industr. Photogr.* i. 10 Monorail camera..a recent, mainly metal, engineer-built construction, with a full array of 'movements'. 1971 P. D. JAMES *Shroud for Nightingale* iii. 52 The Yard photographer manœuvred his tripod and camera—a new Cambo monorail.

monorheme (mǫ·norīm), *sb. Linguistics.* Also **monorrheme.** [f. MONO- + *RHEME.] A single element or word with a phrasal meaning. So **monorrhe·mic** *a.*

1937 J. ORR tr. *Iordan's Introd. Romance Linguistics* iv. 379 Bally proceeds to view the sentence in its genetic and psychological aspects, its development from the 'monorheme' to the 'dirheme' by a process of integration or 'condensation' of two originally independent units. 1939 L. H. GRAY *Foundations of Lang.* viii. 229 Another type of monorrhemic verbal sentence is seen in verbs expressing meteorological conditions, such as Latin *pluit*. *Ibid.* 230 A verb is the sole part of speech which can form a complete sentence, i.e., which can contain both a subject and a predicate in a single monorrheme. 1950 *Archivum Linguisticum* II. 186 The nature of the 'monorhemes' varies between a wish or request, a questioning attitude,..and..a mere report of something perceived. 1969 tr. *Akhmanova & Mikhael'an's Theory of Syntax in Mod. Linguistics* iii. 55 Phrases are called 'monorhemes' when they consist of a single element, and 'dirhemes' when the number of elements is two. 1970 *Language* XLVI. 923 In these apparent monorhemes a hic-et-nunc adverb has been deleted.

‖ **monosabio** (mǫnosā·bio). [Sp., f. *mono* monkey + *sabio* wise, trained.] A picador's assistant in the bullring. Also *ellipt.* as **mono.**

1897 *Encycl. Sport* I. 158/2 *Monos sabios*, boys and men employed in the service of the bull fight. 1924 E. HEMINGWAY *In our Time* xii. 22 The horse's entrails hung down in a blue bunch..the *monos* whacking him on the back of his legs with the rods. 1927 —— *Men without Women* 38 The monos were leading a horse out by the bridle toward the bull. 1928 *Daily Tel.* 17 July 9/3 In another bull-ring at Vista Allegre..the matador Albageno and a 'monosabio' were dangerously wounded by bulls yesterday. 1932 E. HEMINGWAY *Death in Afternoon* vi. 58 You will see..the picadors arriving on the horses they have ridden in from town, these horses having been ridden from the bull ring by the red-bloused monos or bullring servants. *Ibid.* xvi. 187 [The] San Sebastian monosabios should be, by rights, policemen... They carry..puntillas, broad-headed knives. 1959 V. J. KEHOE *Aficionado!* 37 The *monosabios*..handle the picadores' horses and aid them in getting into the correct position for the charge of the toro. *Ibid.* 53 The areneros and monosabios generally wear a red shirt, and dark blue pants. 1973 J. M. WHITE *Garden Game* 61 An image from the *corrida de toros*..the *monosabios* smoothing the cruelly trampled area.

monosaccharide (mǫnosæ·kăraid). *Chem.* Also † -saccharid. [f. MONO- + SACCHARIDE.] Any sugar which cannot be hydrolysed to give simpler sugars.

1896 W. D. HALLIBURTON *Essent. Chem. Physiol.* (ed. 2) i. 10 (*table*) Monosaccharides or glucoses. 1902 *Encycl. Brit.* XXXI. 723/2 The simplest carbohydrates constitute the group of monosaccharids, of which dextrose..is the most important. 1936 C. S. GIBSON *Essent. Princ. Org. Chem.* xv. 445 The production of a ketohexose from the preliminary aldohexose may be due to the usual interconversion of monosaccharides in the presence of alkalis in aqueous solution. 1968 *Punch* 11 Sept. 356/3 'Oh, ah,' said Walter Gabriel, 'especially considering that, at the same time, she'm also producing monosaccharides, trisaccharides, and a whole perishing range of they oligosaccharides.' 1972 *Sci. Amer.* Oct. 71/1 Lactose is a disaccharide composed of the monosaccharides glucose and galactose.

monosaccharose (mǫnosæ·kărōᵘz, -s). *Chem.* [f. MONO- + SACCHAROSE.] = *MONOSACCHARIDE.

1902 J. B. COHEN *Theoret. Org. Chem.* xx. 276 The monosaccharoses possess strong reducing properties. 1919 S. B. SCHRYVER *Introd. Study Biol. Chem.* v. 199 All the native monosaccharoses are optically active. 1951 I. L. FINAR *Org. Chem.* I. xviii. 357 Monosaccharides (monosaccharoses).

monoscope (mǫ·nǒskōᵘp). *Television.* [f. MONO- + -SCOPE.] (See quot. 1953.)

1938 C. E. BURNETT in *RCA Rev.* II. 414 'Monoscope' is the name which has been given to a developmental type of tube designed to produce a video signal of a test picture or pattern enclosed in the tube. Since the picture must be enclosed in the tube, the Monoscope is not suitable for developing a video signal which represents action. 1953 AMOS & BIRKINSHAW *Television Engin.* I. iv. 79 A monoscope is a camera tube containing a target on which a pattern or photograph is printed and which, by scanning the target, generates a picture signal corresponding to the printed image. 1955 G. M. GLASFORD *Fund. Television Engin.* xv. 480 A useful device for generating a video test signal is the monoscope.

monose (mǫ·nōᵘz, -s). *Chem.* [f. MON(O- + -OSE².] = *MONOSACCHARIDE; (see also quot. 1948).

1892, 1903 [see *BIOSE]. 1935 TIPSON & STILLER in Harrow & Sherwin *Textbk. Biochem.* ii. 77 (*heading*) The action of alkalis on monoses. 1948 [see *DIOSE].

monosemantic (mǫnosimæ·ntik), *a.* [f. MONO- + SEMANTIC *a.*] Of a word or phrase, having only one meaning. Also **monosemante·mic** *a.*

1957 S. POTTER *Mod. Linguistics* vii. 147 The simplest words are manifestly those which symbolize single things or concepts, like proper names. These may be described as *monosemantemic*, and therefore unambiguous. 1960 E. DELAVENAY *Introd. Machine Translation* vi. 97 The vocabulary is classified into monosemantic and polysemantic grammatical words. 1966 *English Studies* XLVII. 56 Such case-forms are not strictly monosemantic, but can vary in their function according to the context.

monosemic, *a.* Add: **b.** = *MONOSEMANTIC *a.*

1969 *Computers & Humanities* III. 251 The verses contain a very high proportion of monosemic words and polysemic words used in their primary meaning. 1971 O. A. WOJTASIEWICZ tr. *Pelc's Studies in Functional Logical Semiotics of Natural Lang.* 23 The English word 'bay' has more usages... 'Orchid', on the other hand, is a monosemic word.

monosemy (mǫ·nosīmi). [f. MONO- + SEM(ANTIC *a.* + -Y³.] Of a word or phrase, the quality of having only one meaning. Cf. *POLYSEMY.

1951 S. ULLMANN *Princ. Semantics* ii. 107 Multiple meaning—which includes a number of non-lexical elements—comprises everything running counter to 'monosemy' in the language system. 1969 S. HOOK *Lang. & Philosophy* III. 255 Now on to the comments concerning polysemy and monosemy. 1972 HARTMANN & STORK *Dict. Lang. & Linguistics* 144/1 *Monosemy*, when a word or phrase has only one semantic meaning, e.g. *ball-point pen*, as opposed to polysemy or multiple meaning.

monosexual (mǫnose·ksiuăl), *a.* [f. MONO- + SEXUAL *a.*] Of one sex, or with the attribute of being either male or female; pertaining to one sex exclusively. (See also quot. 1970.)

1964 *Oceanogr. & Marine Biol.* II. 209 Peripheral cells of the adult thallus undergo meiosis and give rise to attached haploid leafy appendages which are monosexual and produce anisogamous male and female gametes respectively. **1966** *Listener* 23 June 918/3 A film whose monosexual character is emphasized by a soundtrack which contains only three words..spoken by a female voice. **1969** *Sunday Times* 31 Aug. 47/7 The £1.8 million belonging to well over half the university's monosexual seats of collegiate learning. **1970** *Diogenes* LXXII. 61 Mono-sexual, in need of no deep or lasting relation with a partner of the opposite sex. **1971** *Guardian* 2 June 9/4 The Edwardian situation in which half the bachelor population were always of doing monosexual Empire-building.

So **mo:nosexua·lity**, the property or state of being monosexual.

1910 A. A. Brill tr. *Freud's Three Contributions to Sexual Theory* 7 The conception which we gather from this long known anatomical fact is the original predisposition to bisexuality, which in the course of development has changed to monosexuality, leaving slight remnants of the stunted sex. **1958** B. Magee *Go West Young Man* ii. 43 Yale has female graduate students, but no female teachers or undergraduates. This monosexuality is Yale's biggest fault. **1970** *Diogenes* LXXII. 61 In Oceans times this monosexuality took the form of imperialistic sex.

monosign (mǫ·nosain). [f. Mono- + Sign *sb.*] A 'sign' or word used with only one meaning at a time: opp. *plurisign. Hence **mono·signa·tion**, **monosi·gnative** *a.*

1940 *Kenyon Rev.* II. 266 The atomic ingredient of literal language is the monosign (called in logic the 'term'); the atomic ingredient of poetic language tends to be the plurisign. *Ibid.*, I am not inquiring whether the ideal of monosignation is ever perfectly realized. *Ibid.* 267 A logician..requires monosignative clarity. **1949** Wellek & Warren *Theory of Lit.* xv. 190 Instead of aiming at a system of abstractions consistently expressed by a system of monosigns, poetry organizes a unique, unrepeatable pattern of words, each an object as well as a sign. **1954** P. Wheelwright *Burning Fountain* 378 The plurisign is..distinguished from the monosign, or steno-term. **1965** *Encycl. Poetry* 760/1 The depth symbol..defies any adequate analysis into monosignative components.

monosigni·ficant, *a.* [f. Mono- + Significant *a.*] Having only one meaning. Also **monosigni·ficance**.

1940 *Kenyon Rev.* II. 503 *Monosignificance*. Mr. Wheelwright uses..in two ways, as formal logical abstraction..and as a useful temporarily consistent selection. *Ibid.* 505 If..words are not statically and isolatedly monosignificant in the prose of science and instruction, what better sense can be made of..the realm of poetry—connotation—plurisignification. **1947** H. Jacob *Planned Auxiliary Lang.* 20 A high degree of precision..can be attained by assigning to each element one invariable meaning, *i.e.*, each element is monosignificant. *Ibid.*, No language..has so far achieved absolute monosignificance.. The principle of monosignificance is..understood to apply to the smallest but complete unit of thought, the sentence.

mono-ski (mǫ·noski). [f. Mono- + Ski *sb.*] A ski on which a person can stand with both feet. Hence as *v. intr.*, to use mono-skis; **mono-skier**, one who uses mono-skis.

1963 *Austrian Information* (U.S.) 31 Jan. 4/2 Monoskis are at present being displayed in Viennese sporting goods stores. They are single skis on which beginners or below-average skiers can stand with both feet. **1963** *Water Skier* Spring 12/2 Carry out series of connected wraps across wakes on mono ski (feet on front and rear bindings). **1966** *Ibid.* Sept.–Oct. 15/2 Anne and Pat are confident monoskiers. **1968** A. Diment *Gt. Spy Race* viii. 129 It looked pretty amateur from the beach. Round here if you can't mono-ski you're not worth watching. **1971** D. Halliday *Dolly & Doctor Bird* xiv. 201 Dolly's launch passed with a man and a girl on monoskis.

mono-sodium. (Later examples.)

1929 etc. [see *glutamate]. **1954** A. White et al. *Princ. Biochem.* xx. 588 Guanine gout has been reported in this species [sc. the pig]; this is..analogous to human gout, in which monosodium urate may accumulate in the cartilages. **1975** *Listener* 14 Aug. 205/1 Cargoes of factory food, bound for the supermarket shelves.. 'The Taste of Monosodium Glutamate.'

monosome (mǫ·nǒsōᵘm). *Cytology.* [f. Mono- + *-some⁴.] **1.** A chromosome in a diploid chromosome complement which lacks its homologous partner; a diploid individual having such a chromosome in its complement.

1921 [see *hexasome s.v. *hexa-]. **1944** *Genetics* XXIX. 231, 16 monosomes and five trisomes were found in the 21 aberrant offspring. **1970** *Caryologia* XXIII. 359 In this paper, we report on the rates of transmission of disomes (21 II), monosomes (20 II + 1 I), double monosomes (19 II + 2 I) and plants carrying isochromosomes and telocentric chromosomes in two aneuploid series of wheat.

2. A single ribosome attached to a molecule of messenger RNA.

1964 Haselkorn & Fried in *Proc. Nat. Acad. Sci.* LI. 308 In a previous communication we presented evidence indicating that the addition of RNA from turnip yellow mosaic virus (TYMV) to purified *E. coli* ribosomes results in the formation of a complex containing one 70s ribosome and one molecule of RNA. We shall call such complexes monosomes. **1971** *Nature* 10 Sept. 107/2 Most ribosomes appeared as 'free' monosomes, while in the reticular cells most of them were attached to membranes in typical 'polysome' patterns. **1972** *Ibid.* 31 Mar. 237/2 Fig. 1 shows a typical distribution of monosomes and polysomes isolated from non-infected and infected Proctor barley leaves.

monosomic (mǫnosōᵘ·mik), *a.* (*sb.*) *Cytology.* [f. *monosom(e + -ic.] Having or being a diploid chromosome complement in which one (or occas. more than one) chromosome lacks its homologous partner. Hence as *sb.*, a monosomic individual.

1926 *Univ. Calif. Publ. Bot.* XI. 61 This paper is devoted to a second form called 'fluted', which has been found on cytological examination to be a monosomic (2n – 1). *Ibid.* 67 Monosomic forms may be used advantageously for a determination of linkage groups. **1932** *Genetics* XVII. 689 Comparatively few investigations have been concerned with a study of monosomics (2n – 1 types) and their mode of inheritance. **1957** C. P. Swanson *Cytol. & Cytogenetics* (1958) vi. 190 Probably the best known monosomic type is the haplo-IV Drosophila. **1961** *Lancet* 23 Sept. 723/2 The normal karyotypes of the monosomic twin are the X, and the XO-female can reach maturity in reasonably good health. **1973** *Hereditas* LXX. 132/1 A number of monosomic lines of Chinese Spring were used as the female parents. **1974** *Nature* 19 Apr. 714/1 Monosomics and trisomics have seldom been developed in the same species.

So **mo:no-somy**, the character or condition of having a monosomic chromosome complement.

1948 *Nature* 5 June 873/1 Genetical experience shows that, like polyploidy, monosomy and polysomy in a wide range of plants and animals can characterize whole individuals. **1961** *Lancet* 16 Sept. 625/1 Both cases show monosomy and trisomy involving the X chromosome or an autosome of similar size and morphology. **1972** *Science* 5 May 518/3 Not so well documented..are the occasional observations of monosomy, trisomy and other aneuploidies.

monostable (mǫ·nostē¹b'l), *a.* *Electronics.* [f. Mono- + Stable *a.*] Stable in one position or state only: characteristic of a device with this property.

1952 [see *astable *a.*]. **1966** *McGraw-Hill Encycl. Sci. & Technol.* VIII. 651/2 The monostable multivibrator has frequent application in electronic circuits where timing and gating operations are required. **1974** Lowe & Evans *Electronic Princ. & Circuits* xiii. 346 Monostable and bi-stable circuits require a triggering source in order for switching to take place.

monosyllabicity (mǫnosilăbi·siti). [f. Monosyllab(le *sb.* + -icity.] = Monosyllabism.

1961 N. C. Bodman in *Language* XXVII. 205 There has been no change..in the monosyllabicity of the Chinese morpheme from the most ancient times that we know of to the present. **1956** J. Whatmough *Language* iii. 42 The ideal seems to be..somewhere between the extremes of monosyllabication and polysynthesis. **1957** *Archivum Linguisticum* IX. 73 Grappin might have mentioned..the monosyllabicity of ..lot. **1968** *Language* XLIV. 131 The simultaneous trend toward polysyllabic creations in pseudo-scientific language and monosyllabicity (*the mods, the frug*).

monosystemic (mǫnosiste·mik), *a.* *Linguistics.* [f. Mono- + Systemic *a.* 2.] Based on a single system of language analysis.

1959 *Trans. Philol. Soc.* 1957 127 Such studies I should describe as paradigmatic and monosystemic in principle. **1961** *Archivum Linguisticum* III. 130 The structural method still tends to concern itself with units evolved by a monosystemic analysis at the word level. **1968** J. Berry in J. A. Fishman *Readings Sociol. of Lang.* (1968) 739 The initial consonant in the Kikuyu word for 'book'..in terms of a monosystemic analysis, is the same phoneme contrasting with other (original) b-sounds..but relatively rare. **1969** R. H. Robins *Gen. Linguistics* iv. 158 In this sense phonemic analysis is monosystemic. **1967** F. P. Dinneen *Introd. Gen. Linguistics* ii. 309 The American structuralist approach [is] 'monosystemic', whereas Firth's approach is designed to be 'polysystemic'.

monotechnic (mǫnote·knik), *a.* [f. Mono- + Technic *a.*: after Polytechnic *a.* and *sb.*] Dealing with or providing instruction in a single technical subject; also designating an educational institution providing such instruction. Also *ellipt.* as *sb.*, an institution of this kind.

1904 *Jrnl. Inst. Electr. Engin.* 25 Feb. 458 Many of the American schools of engineering are practically monotechnic institutes in contradistinction to the polytechnics here. **1904** G. S. Hall *Adolescence* I. 170 Thousands of our youth of late have been diverted from secondary schools to the monotechnic or trade classes. **1931** *Times Educ. Suppl.* 15 Aug. 313/4 The development of the American university..now includes within its walls monotechnic colleges and polytechnics, while one forms a monotechnic. **1962** *Technology* Dec. 276 Furniture. Monotechnic... The local College of Further Education is one of only two centres offering a complete course in furniture manufacture. **1967** *Times Rev. Industry* July 82/1 Monotechnic education, all the rage 20 years ago, is out. *Ibid.* 82/3 A comparatively small monotechnic cannot keep pace with all these developments.

monothematic (mǫnoþĭmæ·tik), *a.* [f. Mono- + Thematic *a.*] Having a single dominant theme or element, *spec.* in *Mus.* So **mono-the·matism**.

1886 F. Praeger tr. *Naumann's Hist. Mus.* II. 1188 The employment of *leit-motiv*..is nothing but a return to monothematism. **1907** R. Boughton *Bach* 15 Monothematic music in polyphonic style was bound to result in a fugue. **1934** C. Lambert *Music Ho!* v. 326 The work is entirely monothematic, and..should be regarded as a continuation of the formal principles first evolved by Liszt. **1958** *Times* 13 Sept. 10/4 The basically monothematic nature of the opening movement is unusual in a symphony. **1961** *Times* 1 Nov. 16/3 Edwin Muir, a dour, almost mono-thematic poet. **1962** H. R. Lowe *Anglo-Saxon Eng.* iii. 125 These names are often..dithematic such as Wulfhold, Wulfhere, Leofric, and the like rather than monothematic.

monotocardian (mǫnǫtokā·diăn), [f. mod.L. name of suborder *Monotocardia*, f. Mon(o- + Gk. ὠτό-ς gen. of οὖς auricle + καρδία heart.] A prosobranch mollusc having only one auricle and one row of gill leaflets, belonging to the suborder Pectinibranchia, formerly called Monotocardia. Also as *adj.* So **monoto-ca·rdiac** *a.*

1901 *Q. Jrnl. Microsc. Sci.* XLIV. 258 It is fairly easy to derive the Monotocardian type of nervous system, radula, gill, and reproductive system from the corresponding organs of existing Diotocardia. *Ibid.* 259 In the adult Monotocardian a single kidney alone is present. *Ibid.* 261 The monotocardian kidney. **1902** *Rep. Brit. Assoc. Adv. Sci.* 630 Recent research..has resulted in proof that the dextral torsion which leads to the monotocardiac condition, does not affect all organs lying primitively to the left of the rectum. **1906** G. A. Kennet *Boreal Mollusca* Heteropoda..are pelagic monotocardians. **1972** *Oxf. Bb. Invertebrates* 56 Some archaeogastropods show the primitive features of paired gills within the mantle cavity... But others, the 'monotocardians',..have lost the second gill.

monotocous, *a.* (Later examples.)

1936 *Nature* 12 Sept. (Suppl.) 453/2 Conceptual thought could only have arisen in a monotocous mammalian type with an arboreal ancestry. **1937** A. Huxley *Ends & Means* xiv. 264 After achieving a stable inner environment, placental and, in some cases, monotocous birth, highly developed eyes and ears, ..and co-ordinated nervous system, all but one [sc. mammal] proceeded to specialize. **1968** *Economist* 15 June 49/2 The rest of us, monotocous individuals that we are, certainly ought to be fascinated by all the bizarre phenomena involved in multiple births.

monotone, *a.* and *sb.* Add: **A.** *adj.* **2.** *Math.* [ad. G. *monoton* (C. Neumann *Ueber die nach Kreis-, Kugel- und Cylinderfunctionen fortschreitenden Entwickelungen* (1881) iii. 26).] = *monotonic *a.* 2, Monotonous *a.* 3.

1905 J. Pierpont *Lect. Theory of Functions of Real Variables* I. ii. 68 A sequence $A = a_1, a_2, \ldots$ whose elements satisfy the relations $a_n \leq a_{n+1}$, $n = 1, 2, \ldots$ is called an increasing sequence... If, on the other hand, $a_n \geq a_{n+1}$, $n = 1, 2, \ldots$ *A* is said to be a monotone decreasing sequence. **1937** *Duke Math. Jrnl.* III. 489 We assume that $\phi(x)$ is completely monotone in $0 \leq x < \infty$. **1966** [see *monotonicity]. **1968** F. A. P. Moran *Introd. Probability Theory* vi. 186 We say that a sequence, A_n, is monotone if we always have $A_1 \subset A_2 \subset A_3 \subset \ldots$, or $A_1 \supset A_2 \supset A_3 \supset \ldots$ [Etc.] **1970** R. B. Ash *Basic Probability Theory* viii. 255 ϕ is a monotone nondecreasing power function.

Hence **mo-notonely** *adv.* = *monotonically *adv.* 2.

1948 Bochner & Martin *Several Complex Variables* vii. 133 If a sequence of (real) harmonic functions which are collectively bounded from above are monotonely decreasing, then either they converge uniformly to a harmonic function proper, or they diverge 'uniformly' to $-\infty$.

monotonic, *a.* Add: **2.** Of a function or quantity: varying in such a way that it either never increases or never decreases. Of a sequence: consisting of terms that vary in this way. = *monotone *a.* and *sb.* A. 2, Monotonous *a.* 3.

1901 *Ann. Math.* II. 116 It follows that *f*(*t*) is a monotonic function that actually decreases in parts of the interval $0 \leq t \leq 1$. **1908** T. J. I'A. Bromwich *Introd. Theory Infinite Series* i. 4 A sequence in which $a_{n+1} \geq a_n$ for all values of *n* is called an increasing sequence; and similarly if $a_{n+1} \leq a_n$ for all values of *n*, the sequence is called decreasing. Both increasing and decreasing sequences are included in the term monotonic sequences. **1923** *Proc. Cambr. Philos. Soc.* XXI. 633 *Z* shall be monotonic considered as a function of *ρ*. **1957** G. E. Hutchinson *Treat. Limnol.* I. v. 251 The velocity at any point in a fluid could be expressed as a continuous monotonic or regularly oscillating function of time. **1968** E. T. Copson *Metric Spaces* i. 16 Every bounded monotonic sequence of real numbers is convergent. **1970** *Nature* 3 Oct. 36/1 The rate of uptake of calcium by the isolated reticulum..also displayed monotonic dependence on the concentration of free Ca^{++} in the medium. **1973** *Jrnl. Genetic Psychol.* CXXII. 327 Young-adult responses were..even more clearly a monotonic function of actual figure age than were preschool responses.

monotonically, *adv.* Add: **2.** In the manner of a monotonic function, i.e. either without ever increasing or without ever decreasing.

1904 E. R. Hedrick tr. *Goursat's Course in Math.*

Analysis I. iv. 148 A function *f*(*x*) is said to increase monotonically in a given interval (*a*, *b*) if for any two values *x*′, *x*″ in that interval *f*(*x*″) ≥ *f*(*x*′) whenever *x*″ > *x*′. **1934** *Amer. Math. Monthly* XLI. 421 Equation (13) defines *Y* as a single-valued, continuous, monotonically decreasing function of *X*. **1968** *Jrnl. Geol.* LXVI. 164/2 The rate of solution should increase monotonically as the fresh solvent is brought up at ever increasing rate. **1974** *Nature* 2 Aug. 451/3 These provided the dramatic highlight of the programme, which declined monotonically in interest for its final hour.

monotonicity (mǫnǫtǫni·sĭti). [f. Monotonic *a.* + -ity.] The property or state of being monotonic. **a.** Of a function or variable (see *monotonic *a.* 2).

1948 *Duke Math. Jrnl.* XV. 313 (*heading*) The complete monotonicity of certain functions derived from completely monotone functions. **1961** [see *entropy 2 b]. **1972** *Computer Jrnl.* XV. 218/2 We now derive a monotonicity relation which will be used in Section 5 to prove convergence.

b. Of sounds (see Monotonic *a.*).

1971 *Nature* 23 Apr. 524/2 Speech remained unaffected except for slight slowing and slurring of words and the presence of some monotonicity.

monotonous, *a.* **3.** (Further example.)

1935 T. Chaundy *Differential Calculus* iii. 51 Over the whole domain we may call the function 'monotonous'.

monotonously, *adv.* Add: **2.** = *monotonically *adv.* 2.

1938 *Physical Rev.* LIV. 948/1 *F*(*a*) is a monotonously increasing function of *a*. **1970** *Internat. Jrnl. Quantum Chem.* IV. 611 E decreases monotonously from 0 to *R* = 0.94. **1974** *Nature* 27 May 243/2 The torque might have been expected to decrease monotonously as the polymer dissolved.

monotropic, *a.* Add: **3.** *Physical Chem.* Exhibiting monotropy.

1897 W. D. Bancroft *Phase Rule* iii. 32 It is probable that the substances classified as 'monotropic' by Lehmann ..are analogous to phosphorus. **1924** A. E. Hill in H. S. Taylor *Treat. Physical Chem.* I. ix. 390 In other cases it is not easy to determine whether a given form is monotropic or enantiotropic. **1955** J. Zernike *Chem. Phase Theory* ii. 29 Cubic and hexagonal white phosphorus..are monotropic with respect to the violet form. **1964** G. I. Brown *Introd. Physical Chem.* xx. 217 In monotropic substances there is no definite transition temperature, and one form of the substance is more stable than the other.

monotropism (mǫnǫ·trǫpiz'm). *Physical Chem.* [f. *monotrop(ic *a.* 3 + -ism.] = next.

1924 A. E. Hill in H. S. Taylor *Treat. Physical Chem.* I. ix. 390 An example of monotropism is found in the case of phosphorus. **1950** W. J. Moore *Physical Chem.* v. 109 (*heading*) Enantiotropism and monotropism.

monotropy (mǫnǫ·trǒpi). *Physical Chem.* [ad. G. *monotropie* (O. Lehmann *Molekularphysik* (1888) I. 119), f. Gr. μονο- Mono- + τροπή turning.] The existence of two polymorphs of a substance, one of which is stable and the other metastable over the whole range of their existence, so that conversion of the latter into the former can occur but not *vice versa*.

1902 A. Findlay tr. *Ostwald's Princ. Inorg. Chem.* 255 (*heading*) Enantiotropy and monotropy. **1932** J. N. Friend *Text-bk. Physical Chem.* I. xi. 181 It may..so happen that, in certain cases of presumed monotropy, the transition point lies very much below the melting point and has not been realised experimentally for this reason. **1970** Chakravarti & Mitra *Textbk. Physical Chem.* I. vi. 232 Examples of monotropy are white and violet phosphorus, diamond and graphite, silica, iodine monochloride [and] α- and β-benzophenone.

monotype, *sb.* and *a.* **A.** *sb.* **2.** Delete def. and substitute: In graphic art, a print taken from oil-colour or printer's ink painted on a sheet of glass or metal, the process being such that prints are produced singly. Also, the method of producing such a print. Also *attrib.* (Later examples.)

1926 L. Binyon *Engraved Designs of W. Blake* ii. 13 Out of such experiments may have come the idea of the 'monotype' which Blake was to use to such good purpose. **1954** *Oxford Mag.* 21 Jan., Anyone who does not know what Monotypes are can spend a profitable twenty minutes at his exhibition at Black Hall in deducing the nature of Mr. Baynes's technique. **1955** P. Heron *Changing Forms of Art* xi. 174 No English artist living excels him in this medium, especially where it involves the process known as 'monotype'. **1962** *Listener* 18 Oct. 624/2 An exhibition of monotypes and off-set drawings completed by Robert Colquhoun just before he died. **1970** *Oxf. Compan. Art* 736/1 G. B. Castiglione made monotypes on etching plates in the 17th c., William Blake's monotypes were apparently done on pieces of card.

3. Substitute for def.: (With capital initial.) The proprietary name of a composing machine consisting of two units, a keyboard which produces the perforated paper tape used to control the caster, which produces type in individual characters. (Earlier and later examples.)

1893 *Official Directory World's Columbian Exposition* (Chicago) 459/1 Lanston Monotype Machine Co., Washington, D.C. Monotype Machine. **1932** A. Esdaile *Student's Man. Bibliogr.* iii. 58 The monotype method..involves two machines. **1959** [see *Intertype]. **1965** J. Moran *Composition of Reading Matter* vi. 65 The Monotype machine consists of two units—the keyboard and the caster. By operation of the keyboard a paper ribbon is perforated by means of compressed air. The ribbon is fed into a caster which carries a matrix-case. This moves to different positions in accordance with the perforated ribbon and molten metal is pumped into the appropriate matrix. Each type is cast, ejected singly and assembled in a channel until a line is completed. **1973** S. Jennett *Making of Books* (ed. 5) iv. 79 The type produced by the Monotype is of excellent quality, and in use is indistinguishable from founders' type.

monovalent, *a.* Add: **2.** *Med.* **a.** Containing or being an antigen from a single strain of a micro-organism.

1930 H. Child tr. *Besredka's Immunity in Infectious Dis.* iii. 44 The same conclusion..that the streptococcus is unique, because the monovalent serum acts against all the human streptococci. **1960** *Jrnl. Amer. Med. Assoc.* 19 Mar. 1230/1 The influenza pandemic which occurred in the United States in 1957–1958 provided an excellent opportunity to determine the efficacy of monovalent vaccines. **1968** *Amer. Jrnl. Epidemiol.* LXXXVIII. 149/1 CCA units of B/Maryland/1/59 antigen. **1971** *Gen. Sci.* XIII. 871 The monovalent influenzal antigen used was the strain A0/Swine (Shope) vaccine treated with deoxycholate to eliminate possible toxic effects.

b. Of an antigen or antibody: having only one site at which it can become attached to antibody or antigen, respectively.

1951 *Nat.* ... The Divalent... ..carried out with antibodies specific for two haptenic groups of dissimilar nature... It was essential for the experiments involving hapten-antibody combination *in vivo* that the haptens be molecularly dispersed, since aggregation of hapten *in vivo* would result in the hapten no longer being monovalent. **1966** *Jrnl. Exper. Med.* CXXIII. 229 Hapten-specific delayed hypersensitivity was only transiently suppressed by intravenous injections of monovalent conjugates of arsanilic acid and *N*-acetyltyrosine. **1970** J. T. Barrett *Textbk. Immunol.* vi. 127/2 Monovalent (incomplete) Rh antibodies can be responsible for erythroblastosis. **1972** *McGraw-Hill Yearbk. Sci. & Technol.* 142/2 The agglutinin used in this experiment was a monovalent rather than a divalent agglutinin, that is, a molecule which binds to the cell surface with its one remaing active site but which cannot cause agglutination because it lacks a second active site.

3. *Cytology.* = *univalent *a.*

1906 *Rep. Brit. Assoc. Adv. Sci. 1905* 570 Whether each half of the chromosome is to be regarded as a monovalent chromosome is doubtful.

monovular (mǫnǫ-, mǫnōᵘ·viulăı), *a.* *Biol.* [f. Mon(o- + Ovular *a.*] = *monozygotic *a.*

1936 *Yale Jrnl. Biol. & Med.* VIII. 589 We think of human monovular twins as having identically the same inheritance and as being so similar in height, weight, build, appearance, and mentality as to confuse relatives and friends. **1962** *Biol. Abstr.* XL. 1726/1 While neonatal polycythemia in monovular twins is obviously due to intrauterine transfer of blood.., the postnatal polycythemia in single newborn infants remains unexplained. **1966** *Arch. Surg.* XCII. 865/1 The occurrence of Hashimoto's thyroiditis in a mother and her three daughters, two of whom are monovular twins.

monozygotic (mǫnozəigǫ·tik), *a.* (*sb.*) *Biol.* [f. Mono- + Zygot(e + -ic.] Of twins (also *rarely* of triplets, etc.): derived from a single ovum (and therefore identical). Also as *sb.*, such an individual.

1917 H. H. Newman *Biol. of Twins* 3 Such twins, quadruplets, or larger sets of offspring are known as monozygotic. **1930**, etc. [see *dizygotic *a.*] **1957** *Ann. N.Y. Acad. Sci.* LXIV. 759 Because the twins were believed to be probable monozygotics, excess skin was taken from the less extensively burned twin and cross-transplanted to her sister. **1965** *Listener* 24 June 930/2 We have learnt, too, the significance of a natural freak—the monozygotic or identical twin. **1968** [see monozygosity below]. **1970** *Sci. Jrnl.* June 30/2 Such monozygotic twins ..form only a small proportion of all twins.

Hence **monozygo·sity**, the property of being monozygotic.

1928 *Biol. Abstr.* II. 685/2 (*heading*) Value of dactyloscopic and dermatologic methods in diagnosing monozygosity in twins. **1952** *New Biol.* XII. 28 The only method at present available for diagnosing monozygosity involves comparison of as many morphological and physiological characters as possible. **1968** *Obstetr. & Gynecol. Survey* XXIII. 940 Two-thirds of monozygotic twins have a single placenta with a single chorion, and the diagnosis of monozygosity can be definitely established in this two-thirds.

Monrovia (mǫnrōᵘ·viă). The name of the capital city of Liberia used *attrib.* with reference to the political grouping of African states resulting from a conference which took place there in May 1961.

1962 *Ann. Reg. 1961* 304 The Monrovia Powers, which included the heirs of *Afrique noire d'expression française*,.. were more disposed to co-operate with Europe. **1962** *Listener* 18 Oct. 594/2 It [sc. Nigeria] is one of the leading countries of the so-called Monrovia group of African states who have sought to maintain a realistic course on international problems. **1963** Haile Selassie in *Jrnl. Mod. African Studies* Sept. 285 The commentators of 1963 speak, in discussing Africa, of the Monrovia States, the Brazzaville Group, the Casablanca Powers... Let us put an end to these terms. **1965** H. Lewis in C. Legum *Africa* (rev. ed.) 217 It [sc. Nigeria] played a major role in helping to eliminate the friction between the Casablanca and Monrovia groups, and in establishing the Organization of African Unity. **1971** W. Laqueur *Dict. Politics* 313 Tubman..sponsored the Monrovia Conference of 1961.

mons. Add: Used for *mons veneris*, usu. in phr. **mons area**, the sensitive area of the *mons veneris*.

1965 Masters & Johnson in J. Money *Sex Research* iv. 59 The concept of manipulating only the clitoris rather than the entire mons area is grossly in error. **1968** T. Lidz *Person* xiv. 427 Stimulation of the mons area or perineal region in general can rapidly revive the arousal. **1969** 'J' *Sensuous Woman* (1970) iii. 32 It takes longer to come to orgasm with mons manipulation, but it is just as satiating an experience as direct clitoral shaft massage.

Monsieur. 4. (Later example.)

1846 H. Melville *Typee* (rev. ed.) Sequel 304 Jimmy had a talk about the 'Wee-Wees', as the people of the Nukukeva call the Monsieurs.

monsoon. Add: **4. monsoon forest** [tr. G. *monsunwälder* (A. F. W. Schimper *Pflanzengeographie* (1898) II. iii. 281)], a deciduous forest found in regions of heavy seasonal rainfall.

1903 W. R. Fisher tr. *Schimper's Plant-Geography* III. iii. 260 The monsoon forest is more or less leafless during the dry season,..is tropophilous in character, usually less lofty than the rain-forest, rich in woody lianes, rich in herbaceous but poor in woody epiphytes. **1926** Tansley & Chipp *Study of Vegetation* xv. 293 Monsoon forest occurs in regions where the rainfall exceeds 180 cm..but where there is a prolonged dry season during which the forest is more or less leafless. **1960** N. Polunin *Introd. Plant Geogr.* xiv. 439 Widespread though variable are the monsoon forests or similar 'seasonal' forests, developed in regions enjoying abundant rainfall during the wet season, but having this alternating with a pronounced drought lasting from four to six months or sometimes longer.

monsoonal *a.* (further examples); hence also **monsoo·nish** *a.*, suggestive or characteristic of a monsoon.

1886 Kipling *Plain Tales from Hills* (1888) 118 Not a mere hill-shower but a good, tepid, monsoonish downpour. **1913** J. Murray *Ocean* iv. 83 The monsoonal deflection of the tradewinds, for some distance out to sea, out of their course to westward. **1936** *Geogr. Jrnl.* LXXXVIII. 297 The arrival of the deep mass of monsoonal air. **1961** *Aeroplane* CI. 504/1 Hundreds of these birds had made their way to Australia by soaring in front of 'monsoonal fronts'. **1972** *Kent Life* July 82/2 Under the equatorial sun and monsoonal rains, two crops of rice are easily obtained. **1975** *Bangladesh Observer* 22 July 5/3 Two friends from Dacca and I were enjoying a monsoonish evening in Delhi. It had rained not torrentially but heavily enough to flood some areas and bring the temperature down.

monster, *sb.* and *a.* Add: **A.** *sb.* **5.** (Later example.)

1966 K. Martin *Father Figures* i. 20 Buying from the all-purposes shop bullseyes and, for a penny, 'monsters', which were big bottles of fizzy lemonade.

6. a. *monster-browed*, *-headed* adjs.; **b.** *monster-killing*, *-queller* (later example), *-quelling*, *-slayer*, *-slaying*; **c.** *monster-spouted* adj.; **d.** *monster-cloud*, *-machine*, *-mask*.

1929 Blunden *Near & Far* 49 Though the full cloud Frowns monster-browed. **1934** —— *Mind's Eye* 16 A sky of freakish monster-clouds. **1869** 'Mark Twain' *Innoc. Abr.* xx. 147 The more immediate scenery consisted of fields and farm-houses outside the car, and a monster-headed dwarf and a moustached woman inside it. These latter were not show-people. **1966** *English Studies* XLVII. 141 Digressions..thrown in by the poet.. as the fair means by which an experienced teller of long stories overcomes the monotony of a series of monster-killings. **1970** G. Jackson *Let.* 4 Apr. in *Soledad Brother* (1971) 211 He was giving to us all of the life force and activity that the monster-machine had left to him. **1969** L. J. Chiaramonte in Halpert & Story *Christmas Mumming in Newfoundland* 89 They are great favourites with little children, who beg their parents for money to buy a new monster mask. **1959** A. G. Brodeur *Art of Beowulf* 81 Outside the climate of mutual love.., Beowulf would be little more than the monster-queller and marvelous swimmer of folk-tale. **1948** K. Malone in *English Studies* XXIX. 167 Beowulf's mention of sea-monsters..takes us back to the swimming match with Breca, one detail of which is precisely this monster-quelling on the part of the hero. **1948** L. Spitzer *Linguistics & Literary Hist.* iii. 98 She must ask Hippolyte to resemble his father, the monster-slayer, by slaying the monster that *she* has become. *Ibid.* 95 We learn that Hippolyte has not, to his regret, equaled his father in his feats of monster-slaying. **1942** Blunden *Romantic Poetry & Fine Arts* 17 Monster-spouted fountain.

B. *adj.* **1.** (Earlier and later examples.)

1837 (*title of play*) The elements—earth, air, fire water; or, The monster ballroom of 1837 (R.A.M. 15/5/1837). **1952** 'R. Crompton' *William & Tramp* ii. 83 Their jaws never ceased to move rhythmically around a couple of Monster Humbugs. **1961** *Guardian* 27 Sept. 8/3 Many older housewives..find great satisfaction in a monster weekly 'bake'.

‖ **monstre** (moṅstr), *sb.*² [Fr.] In phr. **monstre sacré** (lit. sacred monster), a striking and

eccentric public figure; a false idol, esp. in the world of entertainment.

[**1940** J. Cocteau (*title*) Les monstres sacrés.] **1959** *Manch. Guardian* 17 July 4/3 Unlike Mlle Sagan she seems vulnerable, and capable of development, not a born 'monstre sacré'. **1965** *Listener* 10 June 870/1 We must forgive Mailer because he is both exceptional *and* typical—the paradigm who is also a *monstre sacré*. **1969** *Ibid.* 9 Jan. 62/1 Compare her with the late Edith Piaf, a *monstre sacré* of a different order. **1969** *Times* 26 Nov. 9/4 That most fabulous of *monstres sacrés*, Sarah Bernhardt. **1975** *Times* 30 Oct. 8/5 Half saint, half satyr, wholly *monstre sacré*, the face [of Bertrand Russell] looks out upon us from the photographs.

monstrous, *a.* **8. b.** Delete 'now *rare* or *Obs.*' and add 'now mainly *U.S.*' (Examples.)

1848 W. T. Thompson *Major Jones's Sketches of Travel* iii. 23 Augusty's a monstrous pretty city; but it ain't the place it used to was, not by a grate sight. It seems like it was rottin off at both eends, and ain't growin much in the middle. **1884** 'Mark Twain' *Huck. Finn* 134 Behind a monstrous long raft. **1907** *Dialect Notes* III. 233 'Yes'm, I'se *monstrous* hungry.' **1942** Berrey & Van den Bark *Amer. Thes. Slang* § 20/14 *Very; exceedingly*,..mighty, monstrous.

montage (mǫ·ntā̃ʒ). [Fr., f. *monter* to Mount.] **1.** *Cinemat.* and *Television.* **a.** The selection and arrangement of separate cinematographic shots as a consecutive whole; the blending (by superimposition) of separate shots to form a single picture; the sequence or picture resulting from such a process.

1929 I. Montagu tr. *Pudovkin's On Film Technique* 179 It is important to gain a clear conception of the activities embraced here by the word *Editing*. The word used by Pudovkin, the German and French word, is *montage*. Its only possible English equivalent is *editing*. **1930** *Observer* 5 Oct. 20/4 Montage, or constructive cutting..is simply the method of building up a film from broken and isolated strips of celluloid. **1936** *World Film News* July 48/3, I recall the fate of *montage* when it was adopted in England. **1939** H. G. Wells *Holy Terror* IV. ii. 398 Using the cinema in the schools to display the rapid development of the new life of mankind..with a montage in which the Beethovenised head of Rud, the ruler and guide, was displayed in such a manner as to evoke and sustain a world-wide enthusiasm. **1959** *Punch* 10 June 787/2 Very good photography..and montage, or editing..in the jazz-club scene behind the credit titles. **1961** *Listener* 10 Aug. 217/1 The programme opened with a *montage* of newspapers cascading from the presses. **1974** *Some Technical Terms & Slang* (Granada Television), *Montage*, a sequence of film images cut together, usually to music.

b. *attrib.* passing into *adj.*

1948 A. Huxley *Ape & Essence* (1949) 125 A series of montage shots exhibits, in twenty seconds, the slow, hour-long advance of Loola and Dr. Poole. **1949** *Here & Now* (N.Z.) Oct. 29/3 It doesn't matter if it's [sc. a film is] documentary, expressionist, montage, poetic, or just plain cinematic. **1961** K. Reisz *Technique Film-Editing* (ed. 9) II. vi. 112 The term *montage sequence*..refers to the quick impressionistic sequence of disconnected images, usually linked by dissolves, superimpositions or wipes, and used to convey passages of time, changes of place, or any other scenes of transition. **1962** *Listener* 4 Oct. 536/2 An edgy.. musical score, perfectly geared to the montage shots of computers, drawing-boards, teleprinters. *Ibid.* 29 Nov. 909/2 The rows of cells and their inhabitants are arranged like a montage sequence from a Russian silent film. **1963** *Movie* Jan. 5/3 Montage sequence of individual crows attacking each child.

2. The act or process of producing a composite picture by combining several different pictures or pictorial elements so that they blend with or into one another; a picture so produced.

1938 A. Cooper *Making a Poster* 34 Used as 'montage' in conjunction with blue skies, or green fields, photographs of human beings appear singularly unattractive. **1938** *Times* 26 Feb. 10/3 The colours, photographs, and lettering which in the ordinary poster are brought into an effect of unity by skilful 'montage'. **1941** A. Huxley *Let.* 27 May (1969) 467 Use either the portrait of Father Joseph..or else a montage made of the various illustrations in the book. **1958** *Observer* 18 May 16/5 They see, upon the pink and gold jacket [of a book], a *montage* of representative faces of the period. **1972** *Islander* (Victoria, B.C.) 16 July 10/2 This art [sc. découpage] is not, however, to be confused with either collage or montage, both of which are used as pictures rather than as a decoration for furniture.

3. *transf.* and *fig.* (The process of making) a mixture, blend, or medley of various elements; a pastiche; a sequence, miscellany.

1934 C. Lambert *Music Ho!* v. 329 The montage or pastiche of the neo-classicists. **1941** B. Schulberg *What Makes Sammy Run?* iii. 45 It [sc. a night club] was a montage of hot music, drunken laughter, loud wisecracks and hostesses like lollypops in red, green and yellow wrappers. *Ibid.* ix. 172 It was flashing through my mind like a montage nightmare. **1961** *Listener* 23 Nov. 863/1 It [sc. a poem] is certainly not modernist, if by modernist one thinks of a play of images, a montage in free verse. **1963** T. A. Sebeok in J. A. Fishman *Readings in Sociol. of Lang.* (1968) 19 When describing speech phenomena, many linguists,..continue to employ a mixed nomenclature, drawing for their technical vocabulary now on articulatory, now on acoustic, and sometimes on perceptual phonetics; the resulting montage may show seams, but the total picture makes sense. **1973** *Guardian* 21 Apr. 10/2 She has used a montage of sound, from the throbbing ..[of] your own heart beat to various street sounds, even a street musician.

Montagnais (mo:ṅtanyeˑ), *sb.* and *a.* Also Montagnois, Mountaine(e)r, Mountainier. [a. F. *montagnais* mountaineer; hence, member of a mountain tribe.] **A.** *sb.* An Algonquian Indian people of eastern Canada; a member of this people; also, their language. **B.** *adj.* Of or pertaining to this people.

1625 Purchas *Pilgrimes* IV. viii. vi. 1607 They were three Nations when they went to war; the *Estechemins*, *Algoumequins*, and *Mountainers*. **1703** tr. *Lahontan's New Voy. N.-America* I. 207 He deals with the other Savage Nations, namely, the *Montagnois*, and the *Papipanachois* in Arms and Ammunition. **1792** G. Cartwright *Jrnl.* I. Explanation of frontispiece, His jacket.., sash, and rackets are Mountaineer. **1800** *Mass. Hist. Soc. Coll.* VI. 16 The ensuing vocabulary I transcribed *viva voce* from Gabriel, a young Mountaineer Indian. *Ibid.* 17 There is evidently a great resemblance between the Skoffie and Mountaineer. **1863** H. Y. Hind *Explor. Labrador Peninsula* II. xxvii. 101 The Nasquapees, like their friends and allies the Montagnais, hate the Esquimaux. **1894** *11th Ann. Rep. U.S. Bureau Amer. Ethnol. 1889–90* 267 The Indians of the Ungava district are locally known as Naskopie, a term of reproach applied to them by the mountaineers (the Montagnais of the early Jesuit missionaries). **1916** *Trans. R. Soc. Canada* X. I. 314 It..is possibly a transition between the Ojibwa and Montagnais snowshoe. **1933** [see *Cree *sb.* and *a.*]. **1934** D. Jenness *Indians of Canada* (ed. 2) xviii. 271 The Montagnais country was a well-wooded area abounding in moose. **1941** *Beaver* Mar. 27 The Montagnais shoe is the commonest form used throughout the Labrador Peninsula. **1966** C. F. & F. M. Voegelin *Map of N. Amer. Indian Lang.* (*caption*) Algonquian Family. 1. Cree—Montagnais—Naskapi. **1973** Howat & Taylor *Dict. World Hist.* 1013/1 *Montagnais Indians*, group of Canadian Algonquian tribes, discovered (1603) at the mouth of the Saguenay by Champlain, who enlisted them in an expedition (1609) against the Iroquois.

Montagnard. Add: **1. b.** An aboriginal people living in the highlands of South Vietnam; a member of this people; also *attrib.* or as *adj.* Cf. *Moi *a.* and *sb.*

1962 *Courier-Mail* (Brisbane) 1 Aug. 2/6 Montagnard scouts report that they have seen elephants carrying artillery pieces. *Ibid.*, Whether a montagnard could distinguish between a mortar and a field piece [etc.]. **1966** *Listener* 29 Dec. 949/2 Next..the mountains and the high plateaux where the Montagnard tribesman lives. **1968** *Ibid.* 13 June 760/3 The Highlands take up half the physical area of South Vietnam but they contain less than one tenth of its people. Moreover most of these are not ethnic Vietnamese but hill tribes of South Seas descent whom the French called Montagnards and the Vietnamese call *Moi*, meaning savages. **1972** *Times* 31 May 5/4 (*caption*) A Montagnard woman comforts her husband as he lies in Kontum hospital. **1972** *Guardian* 3 June 11/1 The name Montagnard was first given to the hill tribes of Vietnam by the French colonisers—it means simply 'the mountain people'. **1974** *Nature* 15 Mar. 186/3 The Montagnards (the highlanders) conduct much of their agriculture in cleared areas of forests, and so their fields were often sprayed during attacks on the inland forests.

‖ **montagne russe** (moṅtanʸ rüs). *Canad.* [F. *montagnes russes* switchback.] A switchback, scenic railway, roller coaster.

1838 *United Service Jrnl.* July 302 Wherever there is a hill or declivity, a '*montagne Russe*' is speedily prepared. **1885** [see *cul-de-sac 2]. **1947** *Horizon* July 32 Political excitement..'ran high', but..it seems like a switchback, a *montagne russe* compared with Mount Everest.

Montagu¹ (mǫ·ntāgiu). The name of George *Montagu* (1751–1815), British naturalist, used *attrib.* or in the possessive to designate animals first described by him, as **Montagu's blenny**, a small marine fish, *Coryphoblennius galerita*; **Montagu's harrier**, *Circus pygargus*, the smallest British harrier; also *ellipt.*; **Montagu's sea-snail**, a small marine fish, *Liparis montagui*; **Montagu shell**, the shell of a small marine bivalve of the genus *Montacuta* which lives as a commensal with an echinoderm; **Montagu's sucker, sucking-fish** = *Montagu's sea-snail*; **Montagu's Venus**, the shell of the marine bivalve *Venerupis pullastra*.

1836 W. Yarrell *Hist. Brit. Fishes* I. 219 (*heading*) Montagu's Blenny. **1969** A. Wheeler *Fishes Brit. Is. & N.-W. Europe* 439/1 Montagu's blenny has a very restricted distribution in the British Isles, being found only on rocky shores on the south-western coasts. **1843** W. Yarrell *Hist. Brit. Birds* I. 101 It will..be an advantage, as well as a gratification, to call this bird in future Montagu's Harrier and *Circus Montagui*. **1880** *Encycl. Brit.* XI. 492/1 This was called by him the Ash-coloured Falcon, but it now generally bears his name, and is known as Montagu's Harrier. **1930** J. S. Huxley *Bird Watching* ii. 27, I could take a map and mark down just where I saw my first dartford warbler, my first smew, Montagu's harrier, and so on. **1954** A. W. P. Robertson *Bird Pageant* ii. 38 While hunting, the montagu usually flies into the wind. **1971** *Country Life* 20 May 1245/3 The old, wild, undrained Fens..where the Montagu's harrier still nests in the reed-beds. **1835** L. Jenyns *Man. Brit. Vertebr. Anim.* 473 Montagu's Sea-Snail... Discovered by Montagu, at Milton, on the south coast of Devon. **1925** J. T. Jenkins *Fishes Brit. Is.* 106 The illustration shows eggs of Montagu's Sea Snail. **1969** A. Wheeler *Fishes Brit. Is. & N.-W. Europe* 505/1 Montagu's sea snail is a more littoral species, found on the shore up to mid-tide level. **1901** E. Step *Shell Life* 101 The rusty Montagu-shell (*M[ontacuta] ferruginosa*) is more oblong, greyish white in hue, but this is hidden by a rusty-looking incrustation. **1961** P. Street *Shell Life on Seashore* viii. 122 The Montagu shells attach themselves near the mouth of the urchin so that they can extract food particles. **1836** W. Yarrell *Hist. Brit. Fishes* II. 277 (*heading*) Montagu's Sucking-fish. *Ibid.* 278 Montagu's Sucker, in the adult state, is from two inches and a half to three inches long. **1925** J. T. Jenkins *Fishes Brit. Is.* 106 Montagu's Sucker is common from the north of Norway to the Channel. **1819** W. Turton *Conchol. Dict. Brit. Is.* 243 *Venus Montacuti.* Montagu's Venus.

Montagu² (mǫ·ntāgiu). *Naut.* [The name of Rear-Admiral Victor Alexander *Montagu* (1841–1915): see quot. 1974.] Used to designate a rig used for small boats in the Royal Navy. Hence **Montagu whaler**, a whaler carrying this rig.

1925 A. Moore *Last Days Mast & Sail* vii. 228 The Montagu whalers recently introduced into the Navy carry a standing lug. **1939** A. S. Bennett *June of Rochester* i. 12 *Hecla* was only a converted Montagu whaler with barely sitting headroom in the tiny cabin. **1948** R. de Kerchove *Internat. Maritime Dict.* 473/2 *Montagu rig*, a two-mast fore-and-aft service rig for ship's boats in the British Navy. The sail plan includes a stay foresail, standing lug mainsail, and jib-headed mizen. A centerboard is fitted to boats with Montagu rig. **1974** W. E. May in *Maritime Monogr. & Rep.* (Nat. Maritime Museum) No. 15, 18 It was not until the early years of the present century that the Montagu whaler appeared. This boat, partially following the proposals of Rear-Admiral Victor Alexander Montagu, had a fuller body aft and was rigged with a standing-lug mainsail, triangular foresail and small triangular mizen.

Montague (mǫ·ntāgiu). *Hairdressing.* Also **Montague curl**. [Origin unknown.] A flat curl, several of which were used to dress the front of the hair, often forming a fringe.

1881 *Mme. Demorest's 'What to Wear' 1881–82* Autumn & Winter 68/1 The flat Montague curls and *crève-cœur* locks are being affected. **1882** *Mme. Demorest's 'What to Wear' 1882* Spring & Summer 80/1 Montagues continue, as formerly, to please the fancy. They may be flattened or combed up; the latter is preferred. **1905** A. B. Moler *Man. Barbering* 86 Artificial curls... Montagues are made by fastening a strand of hair to a style of hairpin for this purpose. *Ibid.* 90 Montagues, or short curls, for the front of the face,..are usually dressed on a wig block. **1960** Cunnington & Beard *Dict. Eng. Costume* 138/1 *Montague Curls*,..an evening dress coiffure, the front hair arranged in a crescent-shaped fringe of curls gummed to the forehead. **1966** J. S. Cox *Illustr. Dict. Hairdressing* 99 *Montague*,..a pin curl mounted on a hairpin.

Montaignesque (moṅtēⁱne·sk), *a.* [See -esque.] Characteristic of or relating to the French philosopher and essayist, Michel de *Montaigne* (1533–92), or his writings. Also **Montai·gnian** (-iăn), **Montai·gnish** (-iʃ) *adjs.* So **Montai·gnism**, the theories or ideas of Montaigne.

1831 Disraeli *Young Duke* III. v. vi. 175 In the Lords, I admire the Duke... There is a gruff, husky sort of a downright Montaignish *naïveté* about him. **1917** J. Strahan *Andrew Bruce Davidson* iv. 65 One cannot but be struck by the curiously Montaignesque style. **1959** *Jrnl. Philos.* LVI. 544 Hume..has also affinities to the Montaignian sceptical tradition. **1964** *Philos.* XXXIX. 205 He..found it necessary to refute Montaignism. **1965** *Amer. Philos. Q.* II. 19/1 The same scepticism, with the same Montaignian arguments.

Montana (mǫntā·nă, -tæ·nă). The name of a state in the north-west of the U.S.A. used *attrib.* in **Montana tree**, a type of saddle or saddle-tree.

1891 *Harper's Mag.* June 7/2 Their saddle is what is known as the Montana tree.

Montanian (mǫntā·niăn, -tæ·n-), *a.* Also **Montanan.** [f. prec. + -ian or -an.] A native or inhabitant of Montana.

Montanan is now the more usual form.

1870 *Montana Pick & Plow* (Bozeman) 29 July 2/2 That desire to know, so prominently a characteristic of our countrymen, is perhaps enhanced among Montanians, by reason of their comparative isolation. **1894** *Harper's Mag.* Aug. 482/1 (*caption*) The unterrified Montanians. **1932** *Frontier* (Missoula, Montana) Mar. 206/2 The better Montanan you are..the better American. **1972** *Fairbanks* (Alaska) *Daily News-Miner* 3 Nov. 2/4 Montanans struggled for years with the question of whether to have a convention.

montan wax (mǫ·ntæ̃n wæks). Also **Montan wax.** [f. L. *montān-us* of or belonging to a mountain + Wax *sb.*¹] A hard, brittle substance that is extracted from lignite by means of benzene or other solvents, consists mainly of higher fatty acids and their esters, and is used in making polishes and as an electrical insulator.

1908 *Chem. Abstr.* II. 2875 Montan wax is obtained by extracting brown-coal by benzene and purifying the product by repeated distillation with superheated steam. **1920** *Times* 4 Feb. 11/3 Another by-product recovered from this particular lignite..is Montan wax... This mineral is very valuable on account of its high melting point, and for this reason it is unequalled in the manufacture of high-grade boot polishes, gramophone records, and other articles. **1936** Bone & Himus *Coal* v. 55 The raw coals contain..anything up to about 7·5 per cent. of 'Montan wax' (m.p. usually 78° to 88°). **1947** *Times* 23 June 2/7 The shortage of mined coal, together with the value of montan wax, has given added value to lignite. **1969** R. F. Lang tr. *Henglein's Chem. Technol.* 411 Montan wax is obtained from strongly bituminous, dry coal by extraction with benzene or benzene-alcohol.

‖ **montaría** (mǫntarī·ă). [Pg.] A dugout canoe used in the Amazon region.

1933 P. Fleming *Brazilian Adventure* I. xvi. 132 The other two boats belonged to smaller and faster types. One was a *montaria*, also clinker-built and actually I think a more finished and substantial bit of work than the true montaría of the Amazon. **1936** *Discovery* Dec. 382/2 Whole families, with their *montarias*, the hire of which was included in the daily wage, were recruited near the collecting-grounds.

Mont Bazillac, Montbazillac, varr. *Monbazillac.

‖ **mont de piété** (moṅ də pieté). [Fr.] = *Mount of piety* (see Mount *sb.*¹ 5 b).

1854 Thackeray *Newcomes* I. xxviii. 265, I saw his grandeur when I went lately to Strasbourg, on my last pilgrimage to the Mont de Piété. **1890** W. Booth *In Darkest England* II. vi. 215 There would be no difficulty.. of instituting a private Mont de Piété. **1923** Wodehouse *Inimitable Jeeves* iv. 43 The idea of—er—pledging the pearls at the local Mont de Piété was, you will readily understand, repugnant to us. **1967** S. Pakenham *Sixty Miles from England* ii. 31 The Mont de Piété was full of gold earrings, pawned by fishwives, for the price of fish was low. **1969** [see *hock *sb.*⁷ b].

monte¹. Add: (Earlier example.)

1824 J. R. Poinsett *Notes on Mexico* (1825) iii. 37 We found a numerous assembly of men gambling deeply, at a game called *monte*.

b. *attrib.* and *Comb.*, as *monte-banker*, *blanket*, *card*, *game*, *operator*, *table*, etc.; **monte-bank**, a monte table; also used as the name of the game itself.

a **1861** T. Winthrop *Life in Open Air* (1863) 128 A background of mustangs, monte-banks, and lynch-law. **1939** T. King *21 Games to play for Money* 27 To take its [sc. Faro's] place, Monte Bank has come into being. **1855** F. S. Marryat *Mts. & Molehills* xiv. 267, I was soon asleep, notwithstanding..the clinkings of the monté-bankers, and the noise of the crowd below. **1898** H. S. Canfield *Maid of Frontier* 78 His long and angular shadow fell across the monte blanket spread flat upon the ground. **1897** *Sears, Roebuck Catal.* 356/3 'Spanish Monte Cards', 48 cards in pack, assortment of black and colors. **1899** T. W. Hall *Tales* 276 Judge Leander Quinn was lured away from a monte game with a couple of buck Indians. **1961** J. Scarne *Compl. Guide to Gambling* xix. 520 Countless Monte operators plied their trade on the steamboats of the Ohio and Mississippi..in the 1850s. **1873** J. H. Beadle *Undevel. West* iv. 92 We take our stand near the monte table, where a considerable crowd gathers. **1889** K. Munroe *Golden Days* ii. 15 This influx of gold caused monte-tables, and other gambling layouts, to spring up.

2. (See *monty.)

Monte³ (mǫ·nti). Colloq. shortening of *Monte Carlo. Also = *Monte Carlo rally.

1928 P. de Ketchiva *Confessions of Croupier* xvi. 194 After her extraordinary luck at Monte,..she had lost all her winnings the following night at the Cannes Casino. **1956** R. Baxter in M. Couper *Rallying to Monte Carlo* p. v, 'Doing the Monte' is like being a dog owner. Once you start, it's difficult to stop. **1968** *Autocar* 25 Jan. 35/3 Driving genius helps, but Montes are finally won by masterly planning.

Monte Carlo (mǫ:ntĭ kā·ılo). [Name of a resort in Monaco famous for its gambling casino.] **1. a.** Used *attrib.* to designate methods of estimating the solution to numerical problems that involve the random (or pseudo-random) sampling of numbers with some chosen frequency distribution.

1949 *Math. Tables & Other Aids to Computation* III. 546 This method of solution of problems in mathematical physics by sampling techniques based on random walk models constitutes what is known as the 'Monte Carlo' method. The method as well as the name for it were apparently first suggested by John von Neumann and S. M. Ulam. **1950** *Nucleonics* May 27 (*heading*) Random sampling (Monte Carlo) techniques in neutron attenuation problems. **1955** *Sci. Amer.* May 90/3 The accuracy of a Monte Carlo approximation improves only as the square of the number of trials. **1961** *New Scientist* 16 Mar. 678/3 This method of imitating the apparently random fluctuation of events in the real world which it is unnecessary or impossible to study in detail, is the crux of the Monte Carlo simulation technique. **1964** *Guardian* 19 June 6/6 (Advt.), Familiarity with Monte Carlo methods and computer programming would be an advantage. **1970** O. Dopping *Computers & Data Processing* xx. 328 An important field of application of random numbers is in the Monte Carlo methods. They are based on the repetition of a computation in a great number of cases, where one or more parameters are allowed to vary in a random fashion from one computation to the next. *Ibid.*, Monte Carlo methods can also be used for computing determined results. For instance, we can, in principle, calculate an approximation of $\pi/4$ by investigating which fraction of a number of points, chosen at random in a square, are inside the inscribed circle. **1972** *Computers & Humanities* VII. 39 Monte Carlo methods of simulation were applied to a population model for Paleolithic human populations.

b. *absol.*

1951 *Nat. Bureau of Standards Appl. Math. Ser.* (U.S.) XII. iii. 6/1 In some instances..a judicious combination of Monte Carlo applied to the physical model with analysis of the equations is indicated. **1962** D. Slayton in *Into Orbit* 23 A branch of probability mathematics which the scientists call 'Monte Carlo'—because it boils down, really to nothing more than a complex way of figuring the odds. **1969** *Nature* 15 Mar. 999/2 Monte Carlo involves, essentially, a sequence of operations on a sequence of random numbers (u_i), where, ideally, the (u_i) are independent and identically distributed. **1969** J. Argenti *Managem. Techniques* 172 Monte Carlo is a form of simulation..and both often require the use of a computer. It is possible, however, to use Monte Carlo with pencil and paper.

c. *Monte Carlo fallacy*: the fallacy that the probability of any particular outcome to one of a series of repeated but independent events of chance is inversely dependent upon the previous outcomes (so that, e.g., a succession of failures is thought to increase the probability of success on the next occasion).

1957 *Sci. Amer.* Nov. 136/2 In our studies of the manifestations of subjective probability in gambling we have given particular attention to the Monte Carlo fallacy; the well-nigh unanimous belief that after a run of successes a failure is inevitable, and *vice versa*. **1973** A. J. Ayer *Central Questions of Philos.* viii. 164 The notorious Monte Carlo fallacy consists in assuming that because the odds against an even number's coming up ten times in succession are over a thousand to one, there is equally little chance of the tenth spin yielding an even number, if the previous nine have done so.

2. *Monte Carlo rally*: an annual international car rally, first held in 1911, of which the final stages take place in Monte Carlo (also shortened to *Monte Carlo*).

1950 R. Lowry *Monte Carlo Rally* ix. 76 There are Rallies ideally suited to such tastes but the Monte Carlo is not one of them. **1973** *Guardian* 26 Jan. 13/1 Irate Monte Carlo Rally drivers..complained about speed traps set up by the French police... On the Monte Carlo, two reports of excessive speeding mean automatic disqualification.

‖ **monte di pietà** (mǫ·nte di pyeta). Also **monte de pietà**. Pl. **monti di pietà**. [It.] = *Mount of piety* (see Mount *sb.*¹ 5 b).

1654 J. Howell *Parthenopoeia* Epistle to Reader sig. A1ᵛ *Monte de pietà*, an Hospital of 60000. Duckets of yeerly Revenue. **1805** P. Beckford *Familiar Lett. from Italy* I. 231 The *Monte di Pietà* was established first at Florence, in the year 1496, to restrain the usury of the Jews. **1883** *Athenæum* 18 Aug. 218/1 The..186 first-class pictures from the Monte di Pietà, where they had been pawned ages ago by destitute aristocratic families. **1973** *Listener* 18 Jan. 81/2 These Christian pawnshops, still in existence in Italy today, were known as *Monti di Pieta* or 'funds of charity'. When they were first founded, in Umbrian towns of the mid-15th century, they charged low interest rates.

Montefiascone (mǫntefiaskōᵘ·ne). Also **Monte Fiascone.** The name of a town in Latium, central Italy, used (freq. *attrib.*) to designate the sweet white wine made there.

1673 J. Ray *Observations Journey Low-Countries* 363 Heer [sc. in Rome] is great variety of Wines..: as Greco, Lagrime of Naples..Monte Pulciano, di Monte fiascone. **1803** C. Wilmot *Let.* 1 Jan. in *Irish Peer* (1920) 136 The Montefiascone wine is celebrated; we found it excellent. **1833** C. Redding *Hist. Mod. Wines* ix. 240 Of the home growths..another is the Monte Fiascone, of a fine aroma, and intoxicating. It is grown near the Lake Bolsena. **1966** H. Johnson *Wine* 136 Fugger went no further, but ended his days drinking Montefiascone, and no doubt boring his boon-companions to death with the story of Est Est Est. **1967** A. Lichine *Encycl. Wines* 356 *Montefiascone*, white wine grown in the foothills of the Volsini Mountains, Italy.

Monteith² (mǫ̆ntī·þ). A type of coloured cotton handkerchief with a white design (see quots.).

1882 Caulfeild & Saward *Dict. Needlework* 350 *Monteiths*, a description of Cotton Handkerchiefs, which are dyed of one uniform colour, but have a pattern of white spots occurring at regular distances... These goods are known by the name of the manufacturers, at Glasgow. **1957** M. B. Picken *Fashion Dict.* 226/1 *Monteith*.., cotton handkerchief, having colored background with white design, made by discharging dye. **1958** A. & N. L. Clow in C. Singer et al. *Hist. Technol.* IV. viii. 249, 5000 looms were employed in 1796 in the neighbourhood of Glasgow making pullicates for the Turkey-red dyers. These became known throughout Europe as monteiths, after Henry Monteith, who acquired Dalmarnock from Macintosh.

monte-jus (moṅt,ʒüs). Also **-juice**. [Fr., f. *monter* to raise + *jus* juice, liquid.] In the making of sugar, an apparatus for raising the level of the liquid by means of air or steam pressure.

1872 P. Soames *Manuf. Sugar* 133 The cane-juice runs along the gutter to the monte-jus c, where it is elevated into the clarifiers, d. **1921** *Dict. Occup. Terms* (1927) § 449 *Montejuice attendant*, *montejuice boy*: stands by valves of montejus apparatus. **1924** A. J. Wallis-Tayler *Sugar Machinery* viii. 147 This monte-jus should be sunk in the ground, and enclosed with brickwork or masonry in a sort of well. **1950** N. Deerr *Hist. Sugar* II. xxxiii. 578 The montejus was introduced into the industry in 1819 by Dubrunfaut.

Montelian (mǫntī·liăn), *a.* *Archæol.* [f. the name of the Swedish archæologist, Oscar *Montelius* (1843–1921) + -an or -i)an.] Applied to a system of classification and nomenclature devised by Montelius.

1937 G. E. Daniel in *Antiquity* XI. 185 The implications of the Montelian theory of the dolmen are fivefold. **1938** —— in *Ibid.* XII. 302 These modifications are not enough to make the Montelian classification a really workable typology. *Ibid.* 303, I use the word dolmen throughout here in the Montelian sense of a small single chamber. **1941** *Proc. Prehistoric Soc.* VII. 21 Archaeologists who neglect the B-Dolmens of northern Europe are destroying ..any case they might make for the defence of the Montelian sequence. **1950** G. E. Daniel *Hundred Yrs. Archaeol.* xii. 235 The Montelian periods were rendered difficult to use because of their very close linkage in typology and name with the various classes of collective tombs.

Montélimar (mǫnte·limāı). The name of a town in the department of Drôme in S.E. France used *attrib.* and *absol.* to designate a type of nougat orig. made there.

1908 J. Kirkland *Mod. Baker* IV. 16 *Nougat Montélimar* is another kind of mixture in which almonds are used. **1923** R. Whymper *Manuf. Confectionery* 139 Montélimar nougat for chocolate coating should not be as hard as that which is sold in blocks. **1968** W. J. Fance *International Confectioner* 569/1 Nougat Montélimar which is very well known, originally came from the East. **1972** 'J. Melville' *Ironwood* iii. 50, I wanted to cheer him up. I had some bags of montelimar and marchpane.

Montenegrin(e, *a.* and *sb.* Add: Also **Montenegran.** **A.** *adj.* (Further examples.)

1939 W. B. Yeats *Statesman's Holiday* in *Last Poems & Plays* (1940) 77 Here's a Montenegrin lute, And its old sole string Makes me sweet music And I delight to sing. **1942** *R.A.F. Jrnl.* 13 June 24 The Montenegrin guerrillas attacked the hotel. **1958** P. Kemp *No Colours or Crest* x. 220 The German and Albanian forces..had started an offensive against the Montenegrin Partisans. **1970** *Internat. & Compar. Law Q.* XIX. 334 The Montenegran Code of Property. **1972** *Times* 20 Oct. 9/7 Montenegran officials and a guard of honour awaited the Queen's arrival.

Montepulciano (mǫntepultʃyā·no). The name of a town in southern Tuscany, Italy, used to designate the red wine made there.

1673 [see *Montefiascone]. **1702** Lady Verney *Let.* 25 Aug. in M. M. Verney *Verney Lett.* (1930) I. vii. 114 Mr. Cheret..has sent you a hamper of Montepunchana [*sic*]... A dozen bottles. **1817** T. Jefferson *Let.* in A. Lichine *Encycl. Wines* (1967) i. 5 There is still another wine to be named to you, which is the wine of Florence called Montepulciano. **1958** A. L. Simon *Dict. Wines* 112 *Montepulciano*, one of the better red table wines of Tuscany. **1966** C. Ray *Wines of Italy* xiv. 135 *Marsicano*..lighter in colour and in alcohol than Montepulciano and fresh to the taste when drunk young.

‖ **montera** (mǫntēᵊ·ra). [Sp.: see Montero.] The black hat worn by a bullfighter.

1838 *Quarterly Rev.* LXII. 416 He [sc. the matador] then throws his cap, 'montera', on the ground with peculiar action. **1932** E. Hemingway *Death in Afternoon* vi. 62 As the matadors come in front of the president's box they bow low and remove their black hats or monteras. **1967** McCormick & Mascareñas *Compl. Aficionado* iv. 130 With a single traje de luces costing the equivalent of $225, a good montera costing $30, and all the other equipment in that ratio, to launch a novillero requires capital such as Ayala does not have. **1969** A. Arent *Laying on of Hands* (1971) xi. 125 He yanked off the sombrero and replaced it with the *montera*, the black bullfighter's hat.

Monterey. Monterey cypress = *macrocarpa (earlier and later examples); **Monterey pine** = *insignis (earlier and later examples).

1873 J. Lester *Atlantic to Pacific* 92 Beneath the shade of a Monterey Cypress, is the sarcophagus which holds the cherished dust of Starr King. **1969** T. H. Everett *Living Trees of World* 35/2 The Monterey cypress.., restricted to a very small area on the coast of central California,.. sometimes is as much as 90 feet tall. **1874** J. S. Hittell *Resources of Calif.* (ed. 6) xi. 358 The Monterey pine (*Pinus insignis*) is extensively cultivated as an ornamental tree. **1935** *Discovery* Mar. 82/1 The Monterey Pine (*Pinus radiata*) is the tree that has proved most successful in Australia as a forestry product. **1957** [see *insignis]. **1971** *Countryman* LXXVI. iv. 184 The collared dove has.. one stronghold in the Monterey pines on the west side of Porthcressa. **1974** Hawkey & Bingham *Wild Card* xxi. 168 A patio overlooked a large garden planted with mature Monterey pines.

Montessori (mǫntesǭᵊ·ri). The name of the Italian physician and educationalist, Dr. Maria *Montessori* (1870–1952), used to designate her educational system or ideas for the individual development of young children through free but guided play with apparatus designed specially to encourage sense perception and mental interest; also *ellipt.* Hence **Montesso·rian** *a.* and *sb.*, **Montesso·rianism**.

1912 A. E. George tr. M. Montessori (*title*) The Montessori method. **1917** A. S. Neill *Dominie Dismissed* vii.

MONTEVERDIAN

96 'But I thought that was your line..Montessori and all that kind of thing!' 'I don't know what Montessorianism is,' I said; 'I have forgotten everything I ever read about Froebel and Pestalozzi.' **1929** J. van Druten *After All* II. ii. 114 Even if you do succeed in turning our kids' home into a sort of..Montessori school of confidence. **1934** Webster Montessorian, adj. & n. **1949** G. B. Shaw *Let. in To Young Actress* (1960) 188 Montessori is all right while the children are under six, but no use afterwards. **1958** *Times Lit. Suppl.* 31 Jan. 59/4 From the first three of these it is possible to distil the essence of the Montessorian doctrine. **1958** *Times* 20 Mar. 13/4 The question is why all teachers are not Montessorians. *Ibid.*, Montessorianism is sometimes presented more like a 'religious cult than an educational method. **1972** E. Lathen' *The Longer the Thread* viii. 85 I was thinking of something much more ambitious—with preschool training, possibly Montessori teachers.

Monteverdian (mɒntɪvɛ·rdiăn), *a.* [See -an.] Of, pertaining to, or resembling the Italian composer Claudio *Monteverdi* (1567–1643) or his music.

1947 M. F. Bukofzer *Music in Baroque Era* (1948) iii. 98 Schein adopted here the airy dialogue of the Monteverdian continuo canzonetta. **1952** K. Dale tr. *Redlich's Claudio Monteverdi* iv. 160 The realization of the Monteverdian Basso Continuo has at all times been a rock upon which many editors have foundered. **1959** *Times* 21 Mar. 16/2 The Monteverdian recitative should be compulsory training for all second-year operatic students. **1969** D. Arnold *Monteverdi* vii. 138 Whereas no one could find anything Monteverdian in the 'Gombert Mass', the vespers bear the imprint of his style.

Montezuma's revenge (mɒntɪzū·măz rɪve·ndʒ). *slang.* [f. the name of *Montezuma* II (1466–1520), Aztec ruler at the time of the Spanish conquest of Mexico + Revenge *sb.*] Diarrhœa suffered by visitors to Mexico.

1962 *Western Folklore* XXI. 28 The North American in Mexico has coined a number of names for the inevitable dysentery and diarrhea: 'Mexican two-step', 'Mexican fox-trot', 'Mexican toothache', and, less directly if more colorfully, 'Montezuma's revenge', the 'Curse of Montezuma' and the 'Aztec hop'. **1969** *Daily Tel.* 11 Apr. 18/6 Prevent gippy tummy. Also known as Delhi belly, Rangoon runs, Tokyo trots, Montezuma revenge. **1970** *New Scientist* 8 Jan. 47/1 An intestinal attack known as gyppy tummy..in the Middle East; ..Montezuma's revenge in Mexico. **1970** *Times* 8 May 9/3 England's World Cup football squad suffered their first casualty in Mexico on Wednesday, when 20-year-old Brian Kidd was struck down by what is known as 'Montezuma's Revenge'—a stomach complaint.

montgomeryite (mɒntgʌ·mərɪ,əɪt). *Min.* [f. the name of Arthur *Montgomery* (b. 1909), U.S. geologist + -ite¹.] A hydrated basic phosphate of calcium and aluminium, $Ca_4Al_5(PO_4)_6(OH)_5.11H_2O$, which is found as green monoclinic crystals at Fairfield, Utah, U.S.A.

1940 E. S. Larsen in *Amer. Mineralogist* XXV. 321 The name montgomeryite is proposed for this mineral after Mr. Arthur Montgomery of New York City, who..collected the material. **1964** *Ibid.* XLIX. 1119 This apparently represents the first reported occurrence of montgomeryite from a pegmatite.

month¹. Add: **6. a.** *month-brother, -end; month-old* adj.; month clock, a clock which goes for a month between windings.

a 1889 G. M. Hopkins *Poems* (1967) 37 A sister, born for each strong month-brother, Spring's lilies three and three. **1799** *Times* 1 June 4/4 Verkey carpets, a month clock, a quantity of old china. **1848** F. J. Britten *Watch & Clockmaker's Handbk.* 268 Month Clocks have an intermediate wheel and pinion between the great and centre wheels. **1962** E. Bruton *Dict. Clocks & Watches* plate 7 *(caption)* French month clock with skeleton frame. **1929** *Times* 30 Oct. 14/2 The corporations would..withdraw funds from the call money market to meet their month-end requirements. **1969** R. Bissell *Let. Jan.* (1968) 155 On Thursday..I go..to our filthy academy in the Fens..(by a month-old engagement) with a pack of women. **1940** W. Faulkner *Hamlet* iv. 292 The month-old corn-silk beard which concealed most of his abraded face. **1960** *Farmer & Stockbreeder* 1 Mar. 150/3 Day-old pullets..chicks as hatched..month-old pullets.

monthly, *a.* and *sb.* Add: **A.** *adj.* **4.** monthly boat, ship, a ship on a long voyage on which the crew is paid by the month; monthly nurse (earlier examples); monthly rose (earlier and later examples).

1898 *Nautical Mag.* Dec. 827 She passed the ship *Libertas*, evidently a good monthly boat, jogging along the same way under close-reef topsails. **1798** E. Wynne *Diary* 9 Mar. (1952) 294 These ladies say I shall very soon be brought to bed. I am exceedingly well, had the monthly nurse in the house. **1840** Dickens *Sk. Young Couples* 21 If one of a friend's children die, the formal couple are..punctual in sending to the house..if a friend's family be increased the monthly nurse is not more attentive than they. **1864** Evelyn *Kalendarium Hortense* in *Sylva* 73 September..Flowers in Prime, or yet lasting..Muske Rose, and Monethly Rose. **1955** G. S. Thomas *Old Shrub Roses* ix. 60 There was one ancient Rose which under favourable conditions of high culture and special pruning often produced a second crop of flowers in the autumn. This Rose was known in England as the Monthly Rose. **1858** E. I. Barra *Tale Two Oceans* 50 She was what the sailors called a good old monthly ship. **1903** H. Holmes *Life & Adventures on Oceans* 34 'I like a good monthly ship,'

remarks a seaman standing by. 'In these racehorses a fellow's dead horse is hardly worked out before you think of getting into port again.'

B. *sb.* **1.** Now *colloq.* (Further examples.)

1922 Joyce *Ulysses* 361 That squinty one is delicate. Near her monthlies, I expect, makes them feel ticklish. **1960** C. MacInnes *Mr. Love & Justice* 30 'Don't go,' she said. 'I'm sorry—I'm wrought up. I always am a bit just after my monthlies.' **1963** *Observer* 29 Sept. 31/5 One mother..said she had told him most of them [*sc.* the facts of life]—'even about girls and the monthlies'.

2. (Earlier and later examples.)

1833 *Knickerbocker* I. 185 We have articles on Political Economy in the monthlies, the weeklies, and the dailies. **1849** C. Brontë *Let.* 15 Nov. in C. Shorter *Brontës: Life & Lett.* (1908) II. 83, I feel that the flat for which I wait does not depend on newspapers..The monthlies and quarterlies will pronounce it. **1973** *Nation Rev.* (Melbourne) 31 Aug. 1464/1 The emphasis in the early days of the new monthlies and quarterlies was on strong opinion.

Montian (mɒ·ntiăn), *a. Geol.* [ad. F. *montien* (G. J. G. Dewalque *Prodrome d'une Descr. géol. de la Belg.* (1868) x. 185), f. *Mons* (L. *mons, mont-* mountain), name of a town in SW. Belgium: see -ian.] Of, pertaining to, or designating a stage of the Palæocene series that lies above the Danian (and like it is not represented in Britain) and below the Landenian. Also *absol.*

[1882 A. Geikie *Text-bk. Geol.* 848 The most ancient Tertiary deposit of the west of Europe appears to be the limestone of Mons (Système Montien).] **1896** J. W. Judd *Student's Lyell* xv. 220 Paleocene Beds ('Montian')..—In the coarse limestone of Mons in Belgium and in the Marls of Meudon in the Paris basin we have strata which are perhaps older than any in the British Islands. **1924** J. Park *Textbk. Geol.* xxi. 416 The concretionary limestone of the upper division, representing the Montian sub-stage of the Cretaceous system, has yielded..the very characteristic Danian cephalopod, *Nautilus danicus.* **1925** *Geol. Mag.* LXIII. 146 The Montian (Calcaire pisolitique) is now only found in a few isolated patches and seems to have been deposited in certain areas rather than over the whole region. **1955** G. G. Woodford tr. *Gignoux's Stratigr. Geol.* ix. 473 Far from being localized in narrow gulfs, the Montian sea was widespread in a large gulf bounded on the north by the Pays de Bray and Chateau-Thierry and in the south reaching Mantes..and Vertus. **1960** (see *Danian* a.). **1969** Benison & Wright *Geol. Hist. Brit. Isles* xv. 336 The earliest Tertiary beds found on the continent in Belgium and Denmark which together form the Palæocene (though some authors consider the Danian late Cretaceous), were never deposited in the British area.

Montilla (mɒnti·lyă). [Name of a town in Southern Spain.] A dry sherry-type wine made in the vicinity of Montilla.

1793 D. McBride *Gen. Instructions for Choice of Wines* i. 39 Montilla is a light white wine, the chief beverage at Seville. **1846** R. Ford *Gatherings from Spain* xiv. 155 The name of Amontillado..has been given to it from its resemblance in dryness and quality to the wines of Montilla, near Cordova. **1851** C. Redding *Hist. Mod. Wines* (ed. 3) xii. 208 At Montilla, near Cordova, they have a very fine-flavoured dry wine, called Montilla, which is generally consumed in that place. **1876** H. Vizetelly *Facts about Sherry* viii. 86 With these wines we tasted some choice Montillas of a fabulous age. **1935** Schoonmaker & Marvel *Complete Wine Bk.* 96 Montilla..is, in every way, an admirable wine. **1961** *Spectator* 21 Apr. 582 Montilla is fairly rare in this country, and is not, in the very strictest sense of the word, really a 'sherry' at all for it doesn't come from the delimited area around Jerez, but from near Cordoba. **1967** A. Lichine *Encycl. Wines* 355/1 In Spain, Montillas are usually sold from the barrel and in their first year.

montmorillonite. Add: Also, any of the montmorillonite group, or these minerals collectively (see quots. 1954, 1966²). (Earlier and later examples.)

1854, 1934 [see *Delanovite*]. **1935** *Amer. Mineralogist* XX. 482 Montmorillonite, beidellite, and nontronite..all these minerals form a continuous completely miscible series. **1949** P. F. Kerr & Camra *Chem. Constitution & Properties Engin. Materials* xiii. 383 The main constituent is a mineral of the montmorillonite group, either montmorillonite or beidellite. **1954** H. Williams et al. *Petrogr.* xvii. 327 By substitution of magnesium or iron for aluminum, especially in montmorillonite, the composition of clay minerals may vary considerably from the formulas given. Beidellite, for example, is essentially a magnesian montmorillonite, and nontronite is one rich in iron. **1962** *Amer. Mineralogist* XLVII. 145 Beidellites and montmorillonites should be divided at the composition at which the lattice charges from octahedral and tetrahedral substitution equal one another. *Ibid.*, The term beidellite should be used for the aluminium-rich members of the montmorillonite-beidellite series of minerals. **1966** *McGraw-Hill Encycl. Sci. & Technol.* VIII. 587/2 Montmorillonite clays have wide commercial use. The high colloidal, plastic, and binding properties make them especially in demand for bonding molding sands and for oil-well drilling muds. They are widely used to decolorize oils and as a source of petroleum cracking catalysts. **1966** W. A. Deer et al. *Introd. Rock-Forming Min.* III. 264 The name montmorillonite..was originally applied to a clay mineral with composition similar to that of pyrophyllite except for the presence of excess water, $Al_2Si_4O_{10}(OH)_2.nH_2O$. Chemical variation of this basic formula yields a group of clay minerals which are related by a common structure and by similarity of chemical and physical properties, and are therefore classed as the

'montmorillonite group'. According to present usage one member of this group is itself called montmorillonite and has the formula $(Na)_{0.33}(Al_{1.67}Mg_{0.33})Si_4O_{10}(OH)_2.nH_2O$. An alternative term once used for this type of clay is 'smectite', and this has now been revived to describe the group as a whole (Mackenzie, 1957a), which contains the following principal members: montmorillonite, beidellite, nontronite, saponite, hectorite and sauconite. **1971** *Nature* 15 Jan. 157/2 Montmorillonite, a silicate mineral with typical expanding layers, consists of mica-like sheets that are normally separated by one or two layers of water molecules.

b. *attrib.* montmorillonite group, the group of clay minerals (of which montmorillonite is the type) otherwise known as montmorillonoids.

1938 *Mineral. Mag.* XXV. 153 The montmorillonite group is shown to have three end-members, montmorillonite, $Al_2Si_4O_{11}$, nontronite, $Fe_2Si_4O_{11}$, magnesium-beidellite, $Mg_3Si_4O_{11}$. **1940** *Amer. Mineralogist* XXXI. 423 The minerals of the montmorillonite group are characterized by a very wide range in the ions which can proxy Si or Al within the crystal lattice. **1949**, etc. [see above].

montmorillonitic (mɒ:ntmɒrɪlǒni·tɪk), *a. Min.* [f. prec. + -ic.] Resembling or containing montmorillonite.

1943 *Proc. Soil Sci. Soc. Amer.* VII. 134 *(heading)* The adsorption of phosphate by kaolinitic and montmorillonitic clays. **1958** *Nature* 6 Dec. 1596/1 The montmorillonitic clay..occurs at all levels except the Boulder Clay of the Tees. **1970** *New Scientist* 15 Jan. 98/1 In Mexico..the remarkable natural montmorillonitic clay itself provides the mud.

montmorillonoid (mɒntmɒrɪ·lǒnɔɪd). *Min.* [f. as prec. + -oid.] Any of a group of clay minerals that comprises montmorillonite together with some chemically related minerals that likewise undergo a reversible expansion on absorbing water and have a characteristic structure in which water molecules are situated between sheets composed of two layers of silicon atoms sandwiching one nominally of aluminium, the differences between the minerals consisting in the extent and nature of the replacement of silicon and aluminium by other elements. Also *attrib.* or as *adj.*

1951 *Clay Minerals Bull.* I. 195 To avoid this confusion and also the lengthy term 'minerals of the montmorillonoid group'..the term 'montmorillonoid' was suggested [by D. M. C. MacEwan] for this group of minerals. **1951** D. M. C. MacEwan in G. W. Brindley *X-Ray Identification & Crystal Struct. Clay Minerals* iv. 86 *(heading)* The montmorillonite minerals (montmorillonoids). **1953** *Jrnl. Soil Sci.* IV. 233 It is fairly simple to detect by X-ray diffraction the presence of montmorillonoids in clays using glycerol..or ethylene glycol. **1955** *Trans. Faraday Soc.* LI. 582 Considerable attention has been paid in recent years to the retention of organic molecules by clays of the montmorillonoid group. **1956** *Mineral. Mag.* XXXI. IV. 67 The name montmorillonoid is understood as covering a continuous isomorphous series extending from montmorillonite, with ideal formula $Al_2Si_4O_{10}(OH)_2$, to beidellite, with ideal formula $(R,H)_2O.Al_2(Si_3Al)O_{10}(OH)_2$. **1966** *Canad. Jrnl. Soil Sci.* XLVI. 233 Montmorillonoid minerals are sometimes more abundant in the lower horizons of soils in which clay illuviation has been a..process. **1971** *Clay Minerals* IX. 107 Results of..analyses confirm the general montmorillonoid character of the mineral.

Montpellier (mɒnpɛ·lye). The name of a town in the department of Hérault in Southern France used *attrib.*, as **Montpellier butter**, a green-coloured butter flavoured with herbs and anchovies; **Montpellier green** = *verdigris green* [Verdigris 2]; **Montpellier yellow** = **Cassel yellow* [*Cassel]; also called *mineral yellow* [Mineral *a.* 5 b].

1889 Montpellier butter [see *green butter* (*green *a.* 12)]. **1961** E. David *French Provincial Cooking* 117 Montpellier butter is usually served with salmon and the chef's way of doing it is to have a fine piece of middle cut of cold salmon with the skin removed, and the fish thickly spread with the green butter. **1930** Maerz & Paul *Dict. Color* 185/2 An old synonym for Verdigris is 'Montpellier Green', from the name of the French city, which was a center for the color trade in the Middle Ages and during the Renaissance. **1969** R. Mayer *Dict. Art Terms & Techniques* 420/1 Other names for verdigris are Montpellier green, viride aeris..and eruginate. **1835** G. Field *Chromatogr.* ix. 77 Patent Yellow, Turner's Yellow, or Montpellier Yellow..has an excellent body, and works well in oil and water, but is soon injured both by the sun's light and impure air. **1969** R. Mayer *Dict. Art Terms & Techniques* 406/2 Turner's yellow..Among its many names are..Montpellier yellow, [etc.].

Montrachet (mɒntraʃe, mɒnraʃe). Name of a wine-growing district of the Côte d'Or, France, used to designate the white wine produced there.

1833 C. Redding *Hist. Mod. Wines* v. 99 The situation to the south-west of Meursault, where it joins Puligny, is noted for the delicious white wine called Mont-Rachet, of exquisite perfume, and deemed the most perfect white wine of Burgundy, and even of France. **1839** Ure *Dict. Arts* 1300 The insulated part towards the top furnishes the wine called *Chevalier Montrachet*, which is less esteemed..than the delicious wine grown on the middle

height, called *true Montrachet*. Beneath this district, and in the surrounding plains, the vines afford a far inferior article, called *bastard Montrachet*. **1841** *Fraser's Mag.* June 533 A bottle of Montrachet, similarly poured out to us, was very fair. **1873** H. Vizetelly *Wines of World* ii. 24 The distinguishing characteristics of Montrachet are *finesse*, fulness, absolute homogeneity, and softness, combined with a powerful yet delicate flavour and immense richness. **1903** A. Conan Doyle *Adventures Gerard* vi. 218 But there are times when Montrachet is better than the wine of Bordeaux. **1938** W. S. Maugham *Summing Up* 310 A dozen oysters and a pint of Montrachet. **1967** A. Mendelsohn *Dict. Drink* 228 Dumas has said that Le Montrachet should be drunk bare-headed, on one's knees.

Montrealer (mɒntrɪǭ·lǝɹ). [f. *Montreal* + -er¹.] A native or inhabitant of the city of Montreal in Canada.

1888 [see *Seigneur*]. **1900** *Westm. Gaz.* 1 Apr. 9/3 Dr. Macnamara..is always claimed as a Montrealer, because he was born in this city while his father..was quartered there. **1963** *Guardian* 13 Mar. 13/5 Most Montrealers find London in winter..narrow-freezing. **1967** *Times* 28 Feb. (Canada Suppl.) 3/ Two out of three Montrealers speak French.

montroydite (mɒntrɔi·dəit). *Min.* [f. the name of *Montroyd* Sharpe (see quot. 1903) + -ite¹.] Mercuric oxide, HgO, which is found as deep red orthorhombic crystals.

1903 A. J. Moses in *Amer. Jrnl. Sci.* XVI. 262 The name Montroydite is suggested in honor of Mr. Montroyd Sharpe, one of the owners of the mines at Terlingua (in Brewster County, [Texas]. **1934** *Amer. Mineralogist* XVII. 541 The mineral [*sc.* mosesite] was found in association with eglestonite, terlinguaite, montroydite and kleinite in the quicksilver mining district [of Texas]. **1958** *Prof. Papers U.S. Geol. Survey* No. 122 Montroydite occurs as small crusts of growths and as needlelike crystals along the cleavage planes of calcite. **1968** I. Kostov *Mineralogy* ii. iv. 237 Montroydite has a very high and very strong birefringence.

montuno (mɒntū·nǝ). [Amer.-Sp. *montuno* native to mountains, wild, rustic, untamed.] An improvised passage in a rumba.

1951 *Sat. Rev. Lit.* 24 Nov. 75 His moderns grow logically from an amalgam of influences that includes Tatum, Wilson, Hines, Gillespie, and the Cuban *montuno* masters. **1956** M. Stearns *Story of Jazz* (1957) xiv. 248 The *montuno*, or *ad lib* passage, ..is often based on just one, or sometimes two, chords. **1957** J. Kerouac *On Road* (1958) iv. v. 287 The piano montunos showered down on us from the speaker. **1958** E. Borneman in P. Gammond *Decca Bk. Jazz* xxi. 270 Whereas the themes of the bones were usually lilting tunes of obviously Spanish descent, the montunos were unmistakably African.

monty (mɒ·ntɪ). *Austral.* and *N.Z. colloq.* Also **monte.** [Prob. the same word as Monte¹.] A certainty: used esp. of a horse considered certain to win a race.

1894 H. Lawson *Martin Farrell* in *Coll. Verse* (1967) I. 269 Chaps, I've got a vote for Hughie—but it ain't no monty yet. **1901** *Bulletin Reciter* (Sydney) 119 It's the biggest bloomin' monte that 'ere come our way. **1908** E. S. Sorenson *Squatter's Ward* xii. 122 It's a monty the little upset won't let out a yell just as I was gettin' clear. **1910** O'Brien & Stephens *Material for Dict. Austral. Slang* 290–91 (typescript), *Monte*, a take-down game of cards—a sure thing. **1930** K. S. Prichard *Haxby's Circus* iii. 42 She 's the chance of a life-time... The biggest bloomin' monty ever started on a racecourse. **1933** *Bulletin* (Sydney) 29 Nov. 10/1 'Give them an inch, and they'll take an ell,' said I. 'It's a monty that after this they'll get worse.' **1945** G. Stacpoole in *Bulletin* Dec. 1946 *Coast to Coast* 1945 27 He's a monty! We always were lucky. **1947** D. M. Davin *For Rest of our Lives* 397 From Christchurch this chap Sloan, for a monty, with that particular brand of New Zealand English. **1970** *N.Z. Listener* 12 Oct. 14/ Old Jerry wouldn't think of looking for me over this way. That's for a monty. To keep going's the right course.

monument, *sb.* Add: **4. c.** Applied, like mod.F. *monument*, to outstanding survivals of an early literature.

1852 E. A. Andrews *Copious Latin-Eng. Lex.* App. A. 1659 *(title)* Specimens of the oldest monuments of the Latin language. **1897** W. P. Ker *Epic & Romance* ii. 183 *Beowulf* is, at any rate, the specimen by which the Teutonic epic poetry must be judged. It is the largest monument extant. **1949** G. K. Anderson *Lit. Anglo-Saxons* iii. 63 Unquestionably the most important monument of Old English epic literature..is the poem *Beowulf.*

7. **Monument City** = **Monumental City.**

1856 *Life Illustrated* 31 May 33/4 Baltimore is the 'Monument City', from the great battle monument, and several others of note, within its limits. **1906** *Springfield (Mass.) Weekly Republ.* 8 Mar. 4 Baltimore has been known for years as the 'Monument City', and some of these monuments are in reality works of art.

Monumental City. *U.S.* The city of Baltimore, Md. Cf. *Monument City* above.

1827 *National Gaz.* (Philadelphia) 20 Nov. 213 The brave sons of Cincinnatus at the festal board in the 'monumental city'. **1853** F. W. Thomas *John Randolph* 178 In 'the monumental city' I read law. **1882** *Harper's Mag.* June 20/2 No women..equal the fair daughters of the Monumental City. **1904** *Baltimore American* 5 Dec. 14 Detectives..of the Baltimore force, have been working there during the past week, and it is probable that they will return to the Monumental City today. **1949** *Business Week* 24 Sept. 101 'The Monumental City' is the nation's second largest seaport.

monumentalism (mɒnɪume·ntăliz'm). [f. Monumental *a.* + -ism.] A monumental style; building on a grand scale.

1909 *Encycl. Brit.* XXVIII. 451/2 The plain lessons of the Crimean War were unheeded, and monumentalism became the ideal of coast defences. **1970** H. Braun *Parish Churches* xiv. 174 A tower..supplies its church with the element of monumentalism, the dimension of which is height.

monumentality. (Earlier and later examples.)

1888 G. B. Shaw *Let.* 9 Feb. (1965) 184 If I had..physical monumentality, I should bring London up to the mark. **1934** J. J. Sweeney *Plastic Redirections in 20th Cent. Painting* 26 A genuinely new 'archaism'—an extreme simplicity and monumentality of form, with no necessary naturalistic association. **1957** *Archit. Rev.* LXXXII. 49/1 The proposed King George V Memorial in Old Palace Yard..epitomizes all that is implied by the hideous word 'Monumental'..and the chosen site admits of very little monumentality. **1965** *Times Lit. Suppl.* 4 Jan. 2/3 The sculptor Ivan Schadr..who among many other monumentalities has been responsible for the large moveable pink rose. **1969** R. Banham in Jencks & Baird *Meaning in Archit.* 108 According to Baird, the Thinkbelt's avoidance of showy monumentality..will lead to practically every fashionable evil in the book. **1974** *Impressionism* (R. Acad.) 37/1 The strong contrasts between simple planes of colour, and the vigorous forms of tree and house, give the painting a great monumentality.

monumentally, *adv.* Add: **3.** (Further example.)

1956 *Sci. Amer.* Oct. 118/2 The authors deserve admiration for their persistence and energy in carrying out a monumentally tedious series of experiments.

b. In a monumental way (cf. Monumental *a.* 4.)

1928 E. Pound *Let.* c Apr. (1971) 134 Frazer has of course done the whole job monumentally, but good god how slowly, in how many volumes.

‖ **mon vieux** (mɔ̃n vyø). [Fr.] An affectionate form of address: old friend, old man.

1913 Baroness Orczy *Eldorado* i. vii. 55 If you are not satisfied, *mon vieux*..you can send in your resignation when you like. **1927** F. M. Ford *Let.* 5 Jan. (1965) 83 Well, God bless you, mon vieux! **1929** C. Mackenzie *Sylvia & Michael* i. 12 It has been a thoroughly third-class adventure, *mon vieux*. **1924** E. Hemingway *In our Time* i. 9 I'm drunk, I tell you, mon vieux. **1959** L. Durrell in *Spirit of Place* (1969) 360 Wait till you see the menu—it is pure trigonometry, *mon vieux*.

monzonite. Add: **2.** *Petrogr.* [a. F. *monzonite* (de Lapparent 1864, in *Ann. des Mines* VI. 259).] A granular igneous rock with a composition intermediate between syenite and diorite, *spec.* one containing approximately equal amounts of orthoclase and plagioclase.

1895 A. Harker *Petrol.* iii. 44 A special type of augite-syenite is presented by the Triassic intrusions of Monzoni in the southern Tirol (monzonite of authors). **1914** *Jrnl. Geol.* XXXIII. 506 The term 'monzonite' properly will apply only to the rocks with a mixture of andesine and orthoclase. **1958** E. W. Heinrich *Microscopic Petrogr.* iii. 53 Monzonites occur as stocks, laccoliths, dikes, sills, and small plutons. **1974** *Encycl. Brit. Macropædia* IX. 222/2 In general appearance and in field relations monzonites resemble granitic rocks; they differ from granites by reason of their lower quartz content.

Hence monzoni·tic *a.*, (composed) of monzonite or monzonites.

1923 F. F. Grout *Petrog. & Petrol.* 85 The plagioclase of monzonites crystallized before orthoclase, and the resulting mixture of euhedral and anhedral crystals is the typical 'monzonitic' texture. **1936** A. Johannsen *Descr. Petrogr. Igneous Rocks* III. 96 The relations between the orthoclase rocks, the plagioclase rocks, and the monzonitic series. **1959** W. C. Luth in H. Sørensen *Alkaline Rocks* viii. 514/1 The greatest effects, in terms of producing alkaline magmas would be expected of dioritic, monzonitic and gabbroic magmas.

moo (mū), *sb.*² Shortened f. *moola.

1946 F. Fay *How to be Poor* i. 3 In the vernacular of those in the know, moo. (The word 'moola' [sometimes abbreviated 'moo'], meaning 'money'). *Ibid.* xi. 45 What Are We Gonna Do, To Get Rid of All Our 'Moo'? Catchy, isn't it? ..Later on I shall give and teach you—for free—..the lyrics and music for this our battle cry of freedom, using my well-known 'Dum-De-De-Um-Dum' method. After you have mastered 'What Are We Gonna Do, To Get Rid of All Our "Moo"?', I shall expect to hear you full-throatedly singing this chant of Truth from the housetops. **1949** W. R. Burnett *Asphalt Jungle* (1950) vi. 56 After all, as far as she knew, he was still loaded with what the boys vulgarly referred to as moo! **1975** D. Bloodworth *Clients of Omega* iii. 84 Most of my clients do work for moo... But local stuff I pay.

mooch, *sb.* Add: **3.** *slang.* = *moocher 4.

1914 Jackson & Hellyer *Vocab. Criminal Slang* 60 *Mooch*,..current amongst beggars. A mendicant, an alms solicitor. **1953** W. Burroughs *Junkie* (1972) xiv. 146 Cash was a junk mooch on wheels. He made it difficult to refuse.

mooch, *v.* Add: **5.** (Examples.)

1857 'Ducange Anglicus' *Vulgar Tongue* 13 *Mouch, v.*, to go about sponging on your friends. **1975** D. Ramsay *Descent into Dark* ii. 69, I mooch these days because I've given it [*sc.* smoking] up.

6. *trans.* To beg, cadge, scrounge.

1899 'J. Flynt' *Tramping with Tramps* iv. 395 *Mooch*, to beg. **1914** *Sat. Even. Post* 4 Apr. 10/2 And State Street, Chicago, 60. He sure mooched that stem. No nickels. Dimes, buddie... He can beg coin. **1931** 'D. Stiff' *Milk & Honey Route* iii. 33 Loan your safety razor to a bum and he will mooch you for half your blades. **1933** G. Orwell' *Down & Out* xxxiv. 254 Every night I mooched tanners for my kip off of the students. **1953** W. Burroughs *Junkie* (1972) xiv. 147 He would mooch junk off Old Ike, who couldn't turn down anyone. **1964** G. Kersh in D. Knight 100 *Yrs. Sci. Fiction* (1969) 228, I came across a student, mooching drinks, an educated man with no place to sleep. **1972** D. Morell *First Blood* i. iii. 17 First thing I know, a bunch of your friends will show up, mooching food, maybe stealing, maybe pushing drugs.

mocha (mū·tjă). Also moochi(e), muchi, mutsha, mutshi, umutsha. [Bantu.] A short skirt worn as a loin-cloth by the aboriginal inhabitants of eastern South Africa before the introduction of European dress.

1878 H. A. Roche *On Trek in Transvaal* xv. 325 A wee little Kafir boy with nothing upon him but his 'moocha', or tails, driven by a bevy of calves. **1886** R. Haggard *King Solomon's Mines* x. 153 He slipped off the 'moocha' or girdle round his middle, and stood naked before us. **1898** *Chambers's Jrnl.* 18 Jan. 93/1 His only dress consisted of a monkey skin *muchi*, or apron, and in his hand he carried a rifle. **1909** N. Paul *Child in Midst* 139 Presently he appeared, a tall thin native, with no clothing except a cow-hide mutshi, with flaps around his waist. **1923** *Chambers's Jrnl.* Apr. 222/1 The native kilt or *moocha* composed of strips of raw hide. **1936** *Blackw. Mag.* Mar. 304/1 All the bravery of the Zulu war-dress was there:..the leopard-skin karosses, the embroidered *moochas*, the white goat-hide garters. **1936** Williams & May *I am Black* xii. 181 There came before his eyes a boy of fourteen years wearing only a *moocha* of cords polished with pig's fat. **1948** O. Walker *Kaffirs are Lively* 50 The menfolk stride along bare-legged, in swinging *umutshas* of monkey-tails. **1967** E. M. Slatter *My Leaves are Green* i. He wore only the remains of a monkey-skin tails. **1970** *Rand Daily Mail* 29 Oct. 12 (caption) Beaded 'muchi'.

moocher. Add: **4.** *slang.* A beggar, a cadger; one who begs or scrounges. Cf. *mooch *sb.* 3, *v.* 6.

1857 'Ducange Anglicus' *Vulgar Tongue* 13 *Moucher*, beggar. **1899** 'J. Flynt' *Tramping with Tramps* iv. 395 *Moocher*, a beggar. **1901** *Daily Chron.* [..] This word is the generic term for tramps in England. **1925** *Amer. Speech* I. 152/1 'Moocher' is a gem of a word, with its expressive agent-noun 'moocher' in the name for a person who sponges on another. **1931** 'D. Stiff' *Milk & Honey Route* ix. 46 A moocher may do all the things a panhandler does, only he draws smaller dividends because he lacks the salesman's poise. **1933** 'G. Orwell' *Down & Out* xxxii. 236 Some of the cant words now used in London..A moocher—one who begs outright, without pretence of selling anything. **1969** J. Gores *Dead Skip* ii. 6 Every bum, moocher.. **1975** P. Owen *Damned & Destroyed* xi. 75 You moocher, you—don't you respect a lady's natural curiosity? Be nice to me. After all, I'm paying for this party.

moochi(e), varr. *moocha.

moochin (mū·tʃɪn). *Anglo-Welsh colloq.* Also mwchin. Anglicized form of Welsh *mochyn* pig (cf. Ir. *muc*), applied to a person, esp. a child, as a term of reproach or opprobrium. Cf. Pig *sb.*¹ 5.

1940 Dylan Thomas *Portrait of Artist as Young Dog* 58 He would have to get a smack, if I'm on a sledge,' he said, 'pull me, Patricia, pull me like an Eskimo.' 'Up you get, you moochin, or I'll take you home.' **1954** ——*Let.* 30 July (1966) 280 He said, 'Look, what they did, the moochin.' **1953** —— *Under Milk Wood* (1954) 12 Saw him in the bushes..Playing moochins, Willy Wee.

mood, *sb.*¹ Add: **3. e.** *attrib.* and as quasi-*adj.* That is intended to suggest, induce, recall, or depict a particular mood or frame of mind.

1898 G. B. Shaw *Perfect Wagnerite* 124 Beethoven had shown how those inarticulate mood-poems which surge through us when our passions are roused..can be written down in music as symphonies. **1927** *Observer* 14 Aug. 6 On the last side of 16 is Joseph Speaight's expressive little mood-picture for quartet, 'The Lonely Shepherd'. **1940** A. Ogden *Fourteen Radio Plays* 257 *Mood music*, music as background to build mood or scene. **1946** R. Hull in A. L. Bacharach *Brit. Music* xvii. 219 Yet the rare powers of argument are to the fore..but the variations themselves admit of mood-contrasts which keep the music free from any danger of dry impressions. **1948** L. Levy *Music for Movies* x. 111 The sticker, more expressive phrase, 'mood music', which has already imported from the United States. **1955** D. Keene *Who was Wilma Lathrop?* ii. 19 'Would you prefer mood music?'..Wilma stacked the spindle with records. **1957** Manvell & Huntley *Technique Film Music* i. 22 To help them in this task, volumes of classified 'mood' music began to appear as early as 1913, when the Sam Fox Moving Picture Music Volumes by J. S. Zamecnik were published. **1960** *Times* 28 Sept. 15/4 The *Moon of the Caribbees* hardly more than a mood piece. **1962** *Listener* 27 Dec. 1102/1 They seem to me to be mood-poetry in a pejorative sense. **1969** J. Burmeister *Hot & Copper Sky* ii. 39 'What did you sing?'..'Oldies, "Body and Soul". "Smoke Gets In Your Eyes". "Blue Moon". Strictly mood.' **1970** *Daily Tel.* 20 May 16/8 It says much for Mr Wesker's delicate mood-piece that it nevertheless made an effect. **1972** H. Kemelman *Monday the Rabbi Took Off* xiv. 99 The organ had been playing mood music, a series of mournful cadenzas in a minor key.

f. *Comb.* **mood swing** *Psychol.*, an abrupt change of mood without apparent cause which is associated with some forms of mental instability.

1942 *Amer. Jrnl. Orthopsychiatry* IV. 725 Present methods of examining candidates for the air service are good so far as they go since they inquire into the presence of such symptoms as enuresis, somnambulism, mood swings, or phobias. **1943** J. W. Macfarlane in R. Barker et al. *Child Behavior & Devel.* xviii. 315 At year five, four clusters were apparent. The first included quarrelsomeness, mood swings, negativism, and competitiveness. **1965** J. Pollitt *Depression & its Treatment* ii. 5 Manic-depressive type..Commoner among those with a family history of mood swings or similar illness. **1970** P. Lauchlin *Ego & Defenses* xiii. 201 An interesting..application of the principles of Inversion is encountered in the area of mood and mood swings.

mood, *sb.*² Add: **2. c.** *attrib.*

1928 H. Poutsma *Gram. Late Mod. Eng.* (ed. 2) I. i. 31 Some of these verbs..are also used as modal auxiliaries, i.e. as substitutes for actually existing mood-forms. **1931** M. Callaway *Temporal Subjunctive in Old Eng.* 19 The use of the Subjunctive of Antecedent Action, in dependent clauses introduced by particles meaning 'after'..is to be accounted for..by the nature of the main clause...This theory is substantially identical with that advocated by Oskar Erdmann..and by Ernst Bernhardt..hereafter referred to by me as the Erdmann-Bernhardt theory of mood-syntax. **1962** F. Behre *Contrib. Eng. Syntax* 59 Certain uses of the indicative that have not earlier been duly considered by writers on mood-syntax. **1966** —— in *English Studies* XLVI. 99 In both the Old English and Middle English examples the mood-form is indeterminate.

mooded (mū·dĕd), *a. rare.* [f. Mood *sb.*¹ + -ed².] Formed to convey or reflect different moods.

1866 G. M. Hopkins *Jrnls. & Papers* (1959) 166 Fine, with gracefully mooded clouding.

moodle (mū·d'l), *v.* [Origin unknown.] *intr.* To dawdle aimlessly, to idle time away. Const. *about, on.*

1893 G. B. Shaw *Music in London 1890–94* (1932) II. 16 the literary man..hardly able to believe that the conductor can be serious in keeping the band moodling on for forty-five mortal minutes before the singers get to business. **1921** —— *Back to Methuselah* v. 233 That old one who saw you out of your shell has gone off to moodle about doing nothing. **1928** —— *Intelligent Woman's Guide Socialism* lxix. 328 Napoleon often moodled about for a week at a time doing nothing but play with his children or read trash or waste his time helplessly.

moody (mū·di), *sb. slang.* [f. the adj.] A persuasive and insidious approach adopted in order to gain confidence; any flattery or lies intended to suggest compliance. Hence, nonsense, twaddle, 'rubbish', esp. in phr. *(a lot of) old moody*. Also as *vb.*, to bluff or deceive by means of flattery, etc.

1934 P. Allingham *Cheapjack* xii. 148 'Moody' meant a bit of humbug. It is a very common word among the grafters. They would say, 'We'd better moody the landlady a bit,' when they meant that they would pay her a few compliments; and when they did not believe something you were saying they would tell you to 'cut out the moody'. **1958** F. Norman *Bang to Rights* i. 37 'I don't think that you are as bad as you are painted, I would like you if you would to tell me a little about yourself.' This geezer [*sic*] is known as the old moody. **1960** *Times Chron.* 26 Feb. 6 You trying to give me the old moody? **1962** R. Cook *Crust on its Uppers* i. 21 Trying to moody through to the royal enclosure on the knock. **1966** A. Prior *Operators* x. 142 Emmie was always giving me a lot of old moody about young ladies being stood up. **1972** R. Busby *Frightener* xv. 159 The same moody you'd heard a thousand times before.

mooey (mū·i), *sb. slang.* Also moey, mooe. [f. Romany *mooi* mouth, face; cf. Mow *sb.*²] A mouth; a face.

1859 Hotten *Dict. Slang* 64 Mooe. **1896** Farmer & Henley *Slang* IV. 328/1 Moey. **1955** P. Willmott *Against Law* iii. 148 All nylons and high-heeled shoes and paint an inch thick on their mooeys.

Moog (mǝʊg). Also **moog.** [f. the name of R. A. *Moog*, an American engineer, its inventor.] In full, **Moog synthesizer.** An electronic musical instrument (see quots. 1969 and 1971).

The word Moog is a registered proprietary term.

1969 *Punch* 5 Feb. 193/3 A moog synthesiser..is an electronic musical instrument capable of generating all kinds of weird sounds. **1970** *Melody Maker* 31 Oct. 9/4 'Switched-On Bach'..gave a lot of people the idea to throw Moog records on the market very quickly. **1971** *Playboy* Feb. 156/2 Says drummer Graeme Edge: 'We've got a Moog synthesizer that gives us complete control of any sound frequency that the human ear can hear, and we plan to use it to make tapes for the Mellotron. We'll use the two combined and the Mellotron will eventually be a playback machine for the Moog.' **1971** *Guardian* 5 Mar. 9/4 A Moog synthesizer..is a computer that can provide any thing from electronic drum beats against a background of flashing psychedelic lights, to bird song..complete with country scenes. **1971** *Melody Maker* 13 Nov. 39/1 The Moog is basically a collection of electrical modules, some of which you would find in any good recording studio. **1973** *Observer* 21 Jan. 9/3 Jesus Christ Superstar: The music from the show given the full studio treatment (including moog) by an Australian line-up.

1974 *Official Gaz.: Trademarks* (U.S. Patent Office) 5 Feb. TM 21/1 Moog Music Inc., Williamsville, N.Y. Filed Dec. 1, 1972 Moog. For Electronic Musical Instruments..First use Jan. 1, 1954.

mooktee, var. *mukti.

moola (mū·lă). *slang* (orig. *U.S.*). Also **moolah.** [Origin unknown.] Money.

1939 J. O'Hara *Pal Joey* (1940) 97, I never saw the day wherein no matter how much moola I had I could not use some more. **1941** *Penguin New Writing* 10 119 And the thing about a girl is, she likes to see you with a bit of moola in your pocket. **1944** *Bulletin* (Sydney) [..] There was..a good deal of moolah. **1948** Busby & Holtham *Main Line Kill* v. 44 Expensive though. You bloods must spend a lot of moola on that stuff. **1970** *New Yorker* 5 Sept. 32 *(caption)* Gentlemen, nothing stands in the way of a final accord except that management wants profit maximization and the union wants higher wages. **1973** J. Wainwright *Highsoft Class Kill* 154 What did he do for a living?..Where did the moolah come from? **1975** J. Symons *Three Pipe Problem* xiv. 113 Then the only thing to be settled is the lolly, the moolah.

mooley (mū·li). *U.S.* [Var. of Moiley, Muley *sb.*] A hornless cow; a cow. Also mooley cow.

1838 (see Muley *sb.* 1). **1840** *Picayune* (New Orleans) 10 Sept. 2/6 Brought to the Pound..a red and white mooley cow. **1869** 'T. Lackland' *Homespun* ii. 113 They are all so fond of patient 'mooley' too. **1877** *Rep. Vermont Dairymen's Assoc.* VIII. 95 He should know that his cows beef might come from a polled cow or mooley. **1902** A. D. McFaul *Ike Glidden* ii. 5 I couldn't help laughing at the sight of your Mooley shining it for the woods with her tail up and the bell going. **1956** 'Brahms' & 'Simon' *Trottie True* iv. 55 'Here' some sounds just like an old mooley cow,' she giggled, blissfully unaware that she herself had sounded exactly like a young foghorn.

moomba (mu·mbă). *Austral.* [Aboriginal word.] An annual carnival or pre-Lent festival held in Melbourne.

1963 *Times* 26 Feb. 9/1 A procession of floats to mark the beginning of Melbourne's curious Moomba festival. **1969** *Australian Encycl.* IV. 55/1 Melbourne followed in March 1954 with a fortnight of carnival that was termed Moomba. **1969** *Australian* 31 Mar. 11/1, I recall the early Moombas very well: parts of them, at least.

Moomin (mū·min). In the children's tales of the Finnish writer Tove Jansson (b. 1914), one of an imaginary race of small, shy, fat, hibernating creatures that live in the forests of Finland. Also *attrib.* and *Comb.* So **Moomin-like** *a.*, resembling or characteristic of a Moomin.

1951 E. Portch tr. *Jansson's Finn Family Moomintroll* i. 14 One spring morning at four o'clock the first cuckoo arrived in the Valley of the Moomins. *Ibid.* 22 Moomintroll..threw his eggshell into the waste paper basket, for he was (sometimes) a well brought up Moomin. *Ibid.* vii. 162 Now you can all have a wish—the Moomin family first! **1957** H. Allingham tr. T. Jansson *Moomin* x On the Moomins themselves I find myself uncharacteristically reticent. Their appeal is so personal and so intricate that I feel chatter about them is like gossip in public about friends... When Moomin fans discover each other they..are likely to sit down trustingly together. **1965** G. Mitchell *Pageant of Murder* iii. 29 Kitty made an unseemly noise and drew her dressing-gown more closely about her... **1967** B. W. Alderson tr. *Hürlimann's Three Centuries Children's Bks. in Europe* vi. 81 Nonsense..is perhaps found at its best in the Moomin books of this author. The Moomins..composed of shy round and-soft creatures. **1971** K. Hart tr. *Jansson's Moominvalley in Nov.* xvii. 151 I'd like to observe half-a-minute's silence to show our appreciation of the Moomin family.

moon, *sb.* Add: **1. e.** See also Cheese *sb.*¹ 2 a.

3. b. *to be (or jump) over the moon*: to be very happy or delighted.

[*c* **1765** in *Oxf. Dict. Nursery Rhymes* (1951) 205 High diddle, diddle, The Cat and the Fiddle, The Cow jump'd over the Moon.] **1833** S. Smith *Life & Writings* i. *Downing* 18, I could have jumped over the moon. **1840** T. C. Haliburton *Clockmaker* 3rd Ser. vii. 109 Ready to jump over the moon for delight. **1936** W. Kennedy *Together & Apart* iii. 189 She didn't know she had done it, jumped over the moon. **1944** N. Coward *Middle East Diary* 116 The Captain..is absolutely over the moon with pleasure at having this command. **1972** J. Brown *Chancer* iv. 53 She'd be over the moon. **1962** *Listener* 16 Sept. 431/1 To turn on the telly and watch..You're over the moon, and you want to see what you yourself would choose to see is little short of asking for the moon. **1973** *Economist* 2 Oct. 41/2 To hope for equity or justice in any such scheme is to cry for the moon.

c. Add to def.: Esp. in phr. *to ask (cry, wish) for the moon*: to ask or wish for the unattainable.

[**1550** N. Udall *Answer to Commoners of Devonshire & Cornwall* in N. Pocock *Troubles connected with Prayer Bk. of 1549* (1884) 178 They will cry to have a piece of the moon.] **1852, 1860** [in Dict.]. **1928** E. O. Winstedt tr. *Cicero's Lett. to Atticus* II. x. 293 You are asking for the moon. **1939** J. B. Priestley *Angel Pavement* vi. 350 'You'd better not ask too much..' **1946** V. S. Pritchett *Living Novel* 34 They are crying for the moon; they do not cry for the moon. **1966** *Listener* 16 Sept. 431/1 To hope for equity or justice in any such scheme is to cry for the moon.

5. b. *slang.* The buttocks. (Used in *sing.* and *pl.*)

1756 *Life & Mem. E. T. Bates* iii. 31 But his Moon shall never be covered by war or Buck..'till they put down the Ready—and no Brummagums. **1922** Joyce *Ulysses* 82 Or their skirt behind, placket unhooked. Glimpses of the moon. **1938** S. Beckett *Murphy* 220 Placing her hands upon her moons, plump and plain.

c. A moon-shaped mark or area; *spec.* a small area of greater translucency observable by transmitted light in some early porcelains such as Chelsea.

1855 *Poultry Chron.* III. 57 Breast. The best spangled and clearest from tipping with white at the end of the moon. **1865** *Chambers's Encycl.* VII. 535/2 An artificial fly..adorned with two moons from a peacock's tail. **1909** *Chambers's Jrnl.* Sept. 586/2 Chelsea [porcelain]..may be distinguished by..greasy-looking discs known as 'moons' in the paste. **1935** *Amer. Speech* X. 159/1 *Moons*, crescent-shaped nicks which agates [*sc.* marbles] receive from hard hits. The belief was that soaking an agate in vinegar would take out the moons. **1948** W. B. Honey *Old Eng. Porc.* (new ed.) ii. 53 The exceptionally beautiful porcelain of the red-anchor period has a very soft paste of fine grain, often showing by transmitted light the round spots of higher translucency known to collectors as 'moons'. **1959** G. Savage *Antique Collector's Handbk.* 79 From 1750 to 1753 a body noted for its 'moons', or patches of greater translucency, was employed.

8. b. *U.S. colloq.* = Moonshine 4; *spec.* whisky.

1928 *Collier's* 29 Dec. 8/2 Wherever you go in Colorado a bootlegger..says to you 'Mister, this is good stuff. It's Leadville "moon".' *Ibid.* 8/3 The art of producing sugar 'moon' and aging it in charred casks. **1950** *Sat. Even. Post* 27 May 20/1, I would buy a couple of pints of moon.

10. b. (Further examples.) Freq. with *pl.* as *moon.*

1928 E. Wallace *Gunner* vi. 50 Gunner's got three moon for bein' a suspected. **1936** J. Curtis *Gilt Kid* iv. 39, I was doing nine moon for screwing. **1953** K. Tennant *Joyful Condemned* xviii. 167, I got a twelve moon.

15. a. *moon-base, -dew, -dust, -folk, -land, -lander, -landing, -landscape, -maid, -mist, -mountain, -pull, -rising, -rock, -setting, -shadow, -sky, -stuff, -track*; **c.** *moon-bathed, -brightened, -chained, -chequered, -crazed, -dappled, -dazzled, -drawn, -drenched, -flecked, -flooded, -haunted, -horned, -locked, -mad, -misted, -parched, -rich, -shadowed, -warmed, -washed* adjs.; **d.** *moon-blue, -cold, -coloured, -cool, -grey, -pale, -white, -yellow* adjs.

1961 *Economist* 18 Nov. 676/2 Commercial services..between New York and Moonbase Alpha in one day. **1969** *Guardian* 15 July 6/6 One of the things we shall be doing in those early lunar flights is to find a good place for a moon base. **1909** *Daily Chron.* 4 Sept. 7/4 He was looking up at the moon-bathed cliff with a rapt intense look. **1939** Spender & Gili tr. *Lorca's Poems* 51 To the sound of cold tamboures And moon-bathed cithars. **1918** D. H. Lawrence *New Poems* 33 A moon-blue moth goes flittering by. **1922** Joyce *Ulysses* 438 Her moonblue satin slipper. **1911** Moon brightened [see *image crowded* adj. (*image *sb.* 8)]. **1946** Dylan Thomas *Deaths & Entrances* 63 The moon-chained and water-wound Metropolis of fishes. **1910** W. de la Mare *Three Mulla-Mulgars* vii. 95 The deep shadows of the moon-chequered forest. **1933** —— *Fleeting* 119 The autumnal night Hung starry and radiant..O'er moon-cold hills. **1944** E. Sitwell *Green Song* 1 Till the fire of that sun The heart and the moon-cold bone are one. **1922** W. de la Mare *Down-adown-Derry* 20 A beaming of moon-coloured light. **1930** R. Lehmann *Note in Music* 98 The grass..foamed with the moon-coloured profusion of the flower which she called milkwort or Star-of-Bethlehem. **1969** G. Macbeth *War Quartet* 29 Staled Of heat, now moon-cool. **1921** W. B. Yeats *Four Plays for Dancers* 72 The moon-crazed heron Would be but fishes' diet soon. **1910** W. de la Mare *Three Mulla-Mulgars* xix. 252 The moon-dappled forest through which they had so heavily ascended. **1940** C. Day Lewis *Poems in Wartime* 10 All over the countryside Moon-dazzled men are peering out for invaders. **1927** Joyce *Simples* in *Pomes Penyeach*, A moondew stars her hanging hair. **1844** Emerson *Essays* 2nd Ser. viii. 243 In countless upward-striving waves The moon-drawn tide-wave strives. **1922** Joyce *Ulysses* 48 A tide westering, moondrawn, in her wake. **1929** C. Mackenzie *Gallipoli Memories* xi. 189 The grillos shrilling far and wide across that moon-drenched island. **1959** *Britannica Bk. of Year* (U.S.) 753/1 *Moon dust*, meteoric particles on the surface of the moon. **1969** *Guardian* 30 July 1/7 The first experiments of exposing germ-free mice to moon dust. **1913** W. de la Mare *Peacock Pie* 98 Those wallowing, moon-flecked waves. **1945** C. Mann in Murdoch & Drake-Brockman *Austral. Short Stories* (1951) 262 Nights... moon-flooded, starlit. **1901** H. G. Wells *First Men in Moon* xviii. 218 We have shown these moon folk violence. **1927** J. Joyce *She weeps over Rahoon* in *Pomes Penyeach*, His sad heart has lain Under the moon-grey nettles. **1934** L. B. Lyon *White Hare* 34 The moon-grey barn. **1930** Blunden *Summer's Fancy* 42 Whispering their vows below moon-haunted trees. **1894** O. Wilde *Sphinx*, And did you talk with Thoth, and did you hear the moon-horned Io weep? **1960** T. Hughes *Lupercal* 60 And has shorn Summarily the moon-horned river From my bed. **1922** J. A. Dunn *Man Trap* xv. 207 Death cold the nights in this dead land that looks like moonland. **1970** *Sci. Jrnl.* May 7/1 Neil Armstrong climbing down those few fragile steps of his moonlander's ladder to the craggy lunar surface. **1962** *Daily Tel.* 5 July 26/4 A small Moon-landing capsule. **1969** *Guardian* 16 July 17/4 An American moon landing does not constitute an unchallenged lead over Russia. **1954** Koestler *Invisible Writing* iii. 43 Certain scenes..have the rugged and hostile greatness of a moon-landscape. **1943** C. Day Lewis *Word over All* 42 The earth is buoyed in Moon-locked oblivion. **1869**

W. Barnes *Early England & Saxon-English* 103/2 Moonmad. **1920** Blunden *Waggoner* 31 To hear the moon-mad gypsy rave. **1928** —— *Retreat* 14 Or mist-veil brushed thee, fine as yet was wove For moonmaid's clothing. **1910** W. de la Mare *Three Mulla-Mulgars* i. 4 Mutta-matutta's father died from sleeping in the moonmist. **1913** D. H. Lawrence *Love Poems* 13 The moon-mist is over the village. **1936** L. B. Lyon *Bright Feather Fading* 17 Yet parishes moon-misted, Yet Avon, Severn, Humber His name remember. *a* **1821** Keats *Sonnet: To Nile* in *Compl. Poetical Wks.* (1907) 305 Son of the old moon-mountains African! **1927** F. B. Young *Portrait of Clare* 129 An orchard..in which moon-pale apples lay where they had fallen. **1914** C. Mackenzie *Sinister Street* II. iii. vii. 637 Most people..would obtain the finest thrill that Oxford could offer from the sudden sight of..the moon-parched High Street in frost. **1922** D. H. Lawrence *Fantasia of Unconscious* xv. 279 Her [woman's] deep positivity is in the downward flow, the moon-pull. **1925** C. Day Lewis *Beechen Vigil* 30 In a moon-rich garden. **1927** *Observer* 11 Sept. 8 A..short interval between successive moonrisings being favourable for late harvest work. **1969** *Guardian* 15 July 7/2 On July 24, if all goes well, 50 kilogrammes of moonrock will arrive on earth. **1971** *New Scientist* 24 June 766/1 The exchange in Moscow recently of a minute quantity of Apollo moonrock for an even more minute quantity of Luna 16 moonrock. **1905** E. F. Benson *Image in Sand* xviii. 290 The sublimities of the sun-rises and moon-settings were gone from her. **1925** E., O., & S. Sitwell *Poor Young People* 28 The moon-shadows Palely pass and re-pass. **1958** J. Betjeman *Coll. Poems* 270 Near your curious mausoleum Moonshadowed on the grass. **1934** Blunden *Choice or Chance* 55 To silence too that speaks angelic tongues From moon-skies and the sun's November gleam. **1929** E. H. Visiak *Medusa* (1963) xviii. 226 They are too apt already to be taken with such moonstuff. **1969** *New Scientist* 18 Dec. 607/1 We should not go 'up there'..merely to bring back a few pounds of 'moonstuff'. **1858** Lytton *What will he do with It?* (1859) III. vi. x. 196 They were to his eyes the moon track in the ocean of history. **1899** Kipling *Five Nations* (1903) 9 The moon-track a-quiver bewilders our eyes. **1959** R. Graves *Coll. Poems* 316 A moon-warmed world of discontinuance. **1902** W. D. Howells *Literature & Life* 62 A few moon-washed stars pierce the vast vault with their keen points. **1967** *Coast to Coast 1965–66* 131 Vicky sat staring into the mysterious moon-washed night. **1885** W. B. Yeats *Island of Statues* i. ii, in *Dublin Univ. Rev.* May 82/2 Some say that all who touch the island lone Are changed for ever into moon-white stone. **1916** Blunden *Harbingers* 5 Through the moonwhite sea. **1957** C. Day Lewis *Pegasus* 12 O moon-white brow and milky way of flesh! *a* **1963** S. Plath *Crossing Water* (1971) 38 The moth-face of her husband, moonwhite and ill, Circles her like a prey. **1961** R. Graves *More Poems* 29 And in my dreams went chasing here and there A fugitive beacon—your moon-yellow hair.

16. moon-bounce, the use of the moon as a reflector of radio waves (chiefly in the UHF band) aimed at it from one ground station and detected by another; **moonbound** *a.* (orig. *U.S.*), directed or moving towards the moon; **moonbug** *colloq.*, a lunar module; **moon buggy** = **moon-car*; **moon-cake,** a round Chinese cake eaten during mid-Autumn festivities; **moon-car,** a vehicle designed for use on the moon; **mooncraft,** a spacecraft for travel to the moon; also = **moon-car*; **moon crawler** = **moon-car*; **moon-down** *N. Amer.*, the time when the moon goes down or sets; **moon-face,** a moon-shaped face; *spec.* a rounded swollen face characteristic of individuals with excessive amounts of glucocorticoids in the body, e.g. in cases of Cushing's syndrome; **moon-flight,** a flight to the moon; **moonglow,** glowing moonlight; **moonmark** (see quots.); **moon-milk** = *Milk of the moon* (see Milk *sb.* 6); **moon-penny** (examples); **moonport** [after *airport], (*a*) a place from which rockets are launched to the moon; (*b*) a landing-place for rockets on the moon; **moon probe** *Astronautics*, an exploratory space flight made towards the moon by an unmanned vehicle; also, the vehicle itself; **moon race,** competition between nations to be the first to effect a landing on the moon; **moon-rocket,** (*a*) a ride at a fun-fair consisting of rocket-shaped cars travelling round an inclined circular track; (*b*) a rocket designed for flight to the moon; **moon-shaft,** a ray of moonlight; **moon-ship,** a spaceship designed to travel to or towards the moon; **moon-shoot** = **moon-shot*; **moon-shot** (orig. *U.S.*), the launching of a spacecraft to or towards the moon; **moon-suit,** a suit designed to be worn on the moon; **moon-walk,** the action of walking on the moon; so as *v. intr.*, to walk on the moon; **moonwalker,** one who walks on the moon (see also quot. 1950); so **moon-walking** *vbl. sb.*

1960 *Aeroplane* XCVIII. 320/1 The Moon-bounce technique was a form of long-distance radio communication which could be carried out on wavelengths which were not susceptible to ionospheric or other terrestrial disturbances. **1968** *Radio Communication Handbk.* (ed. 4) xii. 20/2 Typical ground station equipment for moonbounce requires a transmitter output power exceeding 100 watts, net aerial gains of the order of 15–20 db, and receivers of 500 c/s bandwidth. **1975** *Nature* 4 Sept. 36/1 On July 7, 1974 while using a Moon Bounce technique on 1,296 MHz I observed the appearance of strange, delayed echoes. **1965** *NASA Facts* III. i. 4 The moonbound Apollo's space navigation system includes two relatively conventional units. **1968** *Guardian* 23 Dec. 1 The world last night saw..life on board the moonbound capsule. **1963** *Britannica Bk. of Year* (U.S.) 856/1 *Moon bug*, the LEM [*i.e.* lunar excursion module]. **1969** *Daily Tel.* (Colour Suppl.) 10 Jan. 21 *(caption)* Parent spacecraft (command and service modules) separates leaving lunar module ('moonbug') attached to rocket stage. **1971** *Guardian* 23 July 11 The astronauts will have a powered moon buggy called Rover. **1938** Chiang Yee *Silent Traveller in London* 38 At this time [*sc.* mid-September] all the sweet-shops produce a great variety of seasonable cakes, which we call 'Moon-Cake', and those especially which are made in Canton in South China are very famous. **1952** D. Yen Hung Feng *Joy Chinese Cooking* xii. 211 Moon cakes are seldom made at home, for the genuine type are difficult to make and require a long list of hard-to-prepare ingredients as well as special wooden forms and cooking implements. **1966** 'Han Suyin' *Mortal Flower* viii. 174 In all the markets, the mooncakes..parade their roundness in all the foodshops, light, feathery, moon-white. **1971** *Nat. Geographic* Oct. 556 *(caption)* Golden yolks,..salted and dried in the sun..are steamed with rice or cooked in Cantonese moon cakes, served at festival time. **1965** *Sci. World* 28 Oct. 8 Would a sticky coating of dust block the window of an astronaut's 'moon car'? **1973** *Daily Tel.* 13 Mar. 19/4 The Lunokhod 2 robot mooncar has..resumed studies of Moon soil and rock and taking photographs. **1962** F. I. Ordway et al. *Basic Astronautics* v. 216 Evidence derived from stationary and mobile mooncraft will doubtless support the belief that any manned bases or outposts on the Moon will have to be underground. **1963** *Times* 31 May 19/7 Two gigantic crawler transport vehicles for use in the Apollo mooncraft launching. **1971** *Daily Tel.* 3 Aug. 1/6 Falcon, the Apollo 15 mooncraft, blasted off yesterday from the lunar surface. **1970** *Guardian* 18 Nov. 1/2 Russia is likely to try to bring its moon crawler Lunokod-I back to earth. **1797** B. Hawkins *Let.* 23 Dec. in *Georgia Hist. Soc. Coll.* (1916) IX. 276 Last evening, just before moon down, his camp had been fired on. **1863** W. B. Cheadle *Jrnl. Trip across Canada* (1931) 125 Camp at moondown in wood close to our lake. **1938** C. H. Matschat *Suwanee River* 68 Five hours till moondown. **1855** Moonface [in Dict., sense 15]. **1944** J. Hoffman *Female Endocrinology* xxxv. 574 The facial adiposity may materially alter the normal features and is responsible for the 'moon face' and 'pig-eyed' expression of the classical case. **1958** *Spectator* 11 July 60/3 Swelling of the face—'moon-face' after cortisone. **1959** J. Fleming *Miss Bones* xii. 136 Her large moon-face remained quite expressionless. **1963** *Daily Tel.* 28 Oct. 12 The Soviet Union, he says, is planning no manned moon-flights. **1975** S. Johnson *Urbane Guerilla* v. 182 The famous moonflight of Apollo 11. **1926** *Blackw. Mag.* Oct. 530/2 She could no longer see the moving shape presently lost among the vague masses of moonglow and darkling shadow which veiled the moor. **1934** W. Hudson et al. *(song-title)* Moonglow. **1965** M. West *Ambassador* viii. 181, I..woke..to a room full of moonglow. **1879** G. M. Hopkins *Poems* (1967) 80 Only I'll Have an eye to the sakes of him, quaint moonmarks, to his pelted plumage under Wings. **1879** —— *Lett. to R. Bridges* (1955) 83 By *moonmarks* I mean crescent shaped markings on the quill-feathers, either in the colouring of the feather or made by the overlapping of one on another. **1885** *Encycl. Dict.*, Moon-milk. **1965** *New Scientist* 3 June 666/2 These crystals form a colloid in water to give the substance that, in British caves, is called moonmilk. **1973** P. O'Donnell *Silver Mistress* ix. 161 They reached a huge stalactite chamber. Around the perimeter shimmered white patches of the calcite deposit called moonmilk. **1866** Lindley & Moore *Treas. Bot.* II. 754/1 Moon-penny, *Chrysanthemum leucanthemum*. **1940** F. Kitchen *Brother to Ox* ix. 152 The grass-reaper..cuts through life, sweeping down the slender moon-pennies and toppling them over. **1963** *New Scientist* 18 Apr. 138/3 NASA has acquired 87,000 acres just north of Cape Canaveral, Florida, as the site for its 'Moonport'. **1965** *New Statesman* 16 July 74/1 To have two Moonports is extravagance enough. **1967** *Britannica Bk. of Year* (U.S.) 803/3 *Moonport*, a place on the surface of the earth equipped for sending spacecraft to the moon. **1958** *Oxford Mail* 16 Aug. 1/8 Engineers are making final checks on the multiple rocket which is expected to be launched tomorrow on America's first 'Moon probe'. **1972** *Guardian* 22 Feb. 2/4 The unmanned Soviet moon probe Luna 20 made a soft landing on the moon's surface last night. **1963** *Ibid.* 17 July 1/5 It is now intended to have the whole question of the 'moon race' brought up in the International Astronomical Union. **1971** S. Cavell *World Viewed* 63 We lash ourselves to these ideas with burning coils of containment, massive retaliation, moon races, yellow perils. **1946** G. Tyrwhitt-Drake *English Circus & Fair Ground* xvii. 202 There were many..rides, such as Airways, Autodrome Aeroplanes, and Moon Rocket. **1953** Pohl & Kornbluth *Space Merchants* (1955) i. 5 The slim V-2s and stubby Moon rockets of the past. **1954** 'R. Crompton' *William & Moon Rocket* i. 20 A fair ain't a fair these days without a Moon Rocket. **1896** E. Phillpotts *Down Dartmoor Way* 201 They turned intu the black wudes all laced wi' mune-shafts. **1930** E. Pound *XXX Cantos* xxi. 98 Yellow wing, pale in the moon shaft. **1949** Moonship [see *gravity 5]. **1963** *Wall St. Jrnl.* 19 Aug. 20 North American Aviation awarded $934.4 million job... Order is for Apollo Moonship. **1970** *Universe* 30 Jan. 9/5 The horror of some future manned moonship being marooned in space is something we don't like to think about. **1958** *Oxford Mail* 14 Aug. 1/9 The United States first 'moon-shoot' has about a one-in-ten chance of success. **1958** *Washington Post* 18 Aug. A. 8/4 Yesterday's moon shot blew up 50,000 feet and 77 seconds after the launching at Cape Canaveral. **1969** *New Scientist* 9 Jan. 56/3 Much heartened by the success of the three-man Moon shot, the National Aeronautics and Space Administration is unveiling ambitious plans for the mid-1970s. **1971** R. Nesbitt *Evolution & Existence of God* 10 If the Universe were in fact a chaos, Apollo moonshots would be inconceivable. **1953** Pohl & Kornbluth *Space Merchants* (1955) xii. 131 Moon suits rented '50 Years Without a Blowout'. **1966** Webster Add., Moonwalk, Moonwalker. **1969** *Daily Tel.* 16 July 22/2 While on the moon, the astronauts will..only moon-walk for about 3½ to four hours. **1969** *Observer* 20 July 1/2 Hints that the 'moonwalk' will also be brought forward were strengthened when the astronauts' physician, Dr Charles Berry, said that he did not now expect the two men to go to sleep on the moon. **1971** *New Scientist* 11 Feb. 286/2 Physical activity exacerbated this condition and contributed to the unusually high heart-rates seen on the moonwalk. **1973** J. Wainwright *Touch of Malice* 31 It was a little like a moon-walk. A slow deliberate progression from point 'A' to point 'B'. **1950** Webster Add., *Moonwalking*, sleepwalking outdoors in bright moonlight.—*Moonwalker.* **1969** *Times* 3 June (Suppl.) p. iii/1 The two moon-walkers will be in the lunar module's upper, or ascent stage. **1969** *Punch* 31 Aug. 336/3 Machines with metal boots are being constructed for moon-walking.

Moon (mūn), *sb.*² The name of Dr. William *Moon* (1818–94), of Brighton, used *attrib.*, *absol.*, or in the possessive to designate the embossed type which he invented to enable blind people to read.

1859 *3rd Rep. Soc. for Supplying Books in Moon's Type* 7 The plan of teaching the Blind..to read by Books embossed in Moon's type has been tested. **1859** M. Fison *Darkness & Light* 19 Moon's System of Embossed Books..Moon's Alphabet consists of the common letters simplified, and therefore is easily learnt... By Moon's method of Stereotyping, the letters are thrown up with such sharpness and prominence, as to be felt even by fingers become dull through age. **1910** *Encycl. Brit.* IV. 63/1 A society was instituted in 1847 by Dr. W. Moon for stereotyping and embossing the Scriptures and other books in 'Moon' type. **1959** *Listener* 14 May 868/3 Books in Braille and Moon. **1973** *Times* 28 Feb. (Victoria Centre, Nottingham, Suppl.) p. ii/1 (Advt.), [Blind people] may prefer to develop a particular handcraft as a hobby, or to concentrate on mobility, braille, Moon type reading, typing. **1973** *Times Lit. Suppl.* 7 Dec. 1511/4 (Advt.), Books in Braille and Moon that are specially prepared for the Blind of all ages.

moon, *v.* Add: **1. c.** *slang.* To expose one's buttocks to (someone). Cf. *moon *sb.* 5 b.

1968–70 *Current Slang* (Univ. S. Dakota) III–IV. 84 *Moon*,..to display one's bare buttocks as a taunt. **1974** *News & Reporter* (Chester, S. Carolina) 24 Apr. 3-A/1 The convention..was to last through Saturday afternoon, and we imagine that it did although we did hear a wild rumor that a cut-up named Fannie might pull her own version of 'the streak' during the Sunday breakfast..just to get the Sunday session off to a good start if the session started getting dull. But, Fannie has assured us that she didn't 'moon' anybody.

moo·n-blind, *sb.* [Cf. the adj.] Moon-blindness.

1668 [see Moon-blind *a.* 1]. **1877** R. I. Dodge *Plains Gt. West* v. 67 There is said to be [a] plains malady, which, however, I cannot vouch for. It is called 'moon-blind'. The idea is that the full rays of the moon affect the eyes of a man sleeping exposed to them, so that he cannot see at night.

moon-calf. Add: **3.** An animal imagined to inhabit the moon.

1901 H. G. Wells *First Men in Moon* xi. 116 We saw..the mooncalf's shining sides... First of all impressions was its enormous size; the girth of its body was some fourscore feet, its length perhaps two hundred. *Ibid.* 120 We came upon another drove of mooncalves bellowing up a ravine. **1955** *Times* 11 Aug. 7/4 It will be nice if, when we make our landfall on the moon, we find the mooncalves prodding their craters with a noiseless drill.

moondoggle (mū·ndɒg'l). *slang* (chiefly *U.S.*). [Blend of Moon *sb.* and *boondoggle *sb.*] Lunar exploration regarded as a 'boondoggle' (see *boondoggle *sb.* b). Hence moo·ndoggling *vbl. sb.*

1962 *Economist* 1 Sept. 780/1 The..taunts of 'moondoggle'..when Mr Kennedy first outlined his plans for exploring the moon. **1963** *Punch* 1 May 620/2 The vast expense of the crash programme to put a man on the moon is..open to the charge of being..economic moondoggling. **1967** *Britannica Bk. of Year* (U.S.) 803/2 *Moondoggle*, useless exploration of the moon that is wasteful of time and money.

mooner. Add: **3.** One who moons (sense *1 c).

1974 [see *mooning *vbl. sb.* 3].

moon-eyed, *ppl. a.* Add: **1. a.** (Later U.S. examples.)

1889 'C. E. Craddock' *Despot of Broomsedge Cove* vi. 105 Do ye know ennything 'bout'n a horse's eyes? I be sort'n 'feard he's moon-eyed, or suthin'. **1905** A. Adams *Outlet* 21 Tolleston took the only blind horse in the herd.. At the time of his purchase, no one could see anything in the eyeball which would indicate he was moon-eyed.

b. (Later examples.)

1886 R. E. G. Cole *Gloss. Words S.-W. Lincs.* 93 Old Jane, his first wife, was moon-eyed. When folks are moon-eyed, they have to gleg at you (look askance) out of the corner of the eye. **1912** *Dialect Notes* III. 583 *Moon-eyed*, half-blind.

4. Drunk; intoxicated. *U.S. slang.*

1737 *Pennsylvania Gaz.* 6–13 Jan. 2/1 He sees two Moons, Merry, Middling, Moon-Ey'd, Muddled, [etc.]. **1940** *Amer. Speech* XV. 447/2 Sid gits moon-eyed every Saturday night.

moon-fish. Substitute for def.: A name used for several pale-coloured marine fishes having thin, moon-shaped bodies, esp. the OPAH, *Lampris guttatus*, a sunfish, *Mola mola*, or a North American fish of the genera *Selene* or *Vomer*.

1873 [see *ANGEL-FISH]. 1878 *Proc. U.S. Nat. Museum* I. 376 *Argyriosus vomer*.—Moon-fish. *Ibid.*, *Selena argentea*.—Moon-fish... Not common. *Ibid.*, *Vomer setipinnis*.—Moon-fish; Sun-fish. 1896 JORDAN & EVERMANN *Check-List Fishes N. & Mid. Amer.* 350 *Lampris luna*... Mariposa; Opah;..Gudlax; Moonfish. 1959 A. HARDY *Fish & Fisheries* II. iv. 75 Another wanderer from warmer waters, coming right into the North Sea and occasionally being taken off the Yorkshire coast, is the most brilliantly coloured of all our fishes, the opah or moon-fish, *Lampris guttatus* (which has in turn also been called the sun-fish). 1963 P. H. GREENWOOD *Norman's Hist. Fishes* (ed. 2) xviii. 329 The order [sc. Lampridiformes] includes the large oceanic Opah or Moon-fish (*Lampris*). 1965 A. J. MCCLANE *Standard Fishing Encycl.* 499/1 The moonfish (*Vomer*) have a head profile that is only moderately high. *Ibid.* 499/2 Both the lookdowns and the moonfish grow to about 10–12 inches.

moon-flower. 2. Substitute for def.: A tropical climbing plant, *Ipomœa alba*, of the family Convolvulaceæ, which bears fragrant, white, trumpet-shaped flowers opening at night; also, other closely-related plants of the genera *Ipomœa* and *Quamoclit*. (Later examples.)

1939 A. H. WOOD *Grow them Indoors* 102 Morning glory, moon flower, dawn flower—there are numerous common names for different members of this group of woody climbers. 1963 W. BLUNT *Of Flowers & Village* 110 My cobaeas and moonflowers have both already germinated. 1969 G. MACBETH *War Quartet* 28 Split tins..Leafed into fronds of moon-flowers.

3. A name used in Africa for *Datura arborea* or *D. suaveolens*, shrubs native to Central America and belonging to the family Solanaceæ, bearing fragrant, white, trumpet-shaped flowers; also called moon-lily or angel's trumpet.

1913 C. PETTMAN *Africanderisms* 325 Moon lily or Moon flower, *Datura Knightii*, which has long, pendulous, strongly scented white flowers. 1973 PALMER & PITMAN *Trees S. Afr.* III. 1976 It [sc. the potato family] yields foods..flowers..shrubs—such as the moonflower, *Datura arborea* L., and *Cestrum* species.

mooniness. (Later examples.)

1929 W. DEEPING *Roper's Row* xiv. 149 The other girls at Fogson and George's observed her 'mooniness'. 1929 A. HUXLEY *Do what you Will* 55 The human countenance ..of pallid mooniness.

mooning, *vbl. sb.* Add: 3. The action of exposing one's buttocks (*MOON *v.* 1 c).

1974 *Guardian* 22 Mar. 17/8 Streaking..seems to be the mainly male equivalent of the mainly female practice that cropped up in campuses across the United States in the late fifties and early sixties. This was known as 'mooning'. .. Mooning consisted..of exposing the bottom in the general direction of whoever the mooner wanted to impress, protest to, or affront.

moonlet. Add: (Further examples.) Also, an artificial satellite.

1895 [see *impact crater]. 1955 *Sci. News Let.* 13 Aug. 107/2 The man-made moonlets will circle the earth many times every day, appearing to rise in the west and set in the east, reversing other sky phenomena. 1965 J. BLISH *Mission to Heart Stars* ii. 27 Since Phobos always kept the same face turned towards Mars, there would be no need to bother stumbling around on the dark side of the moonlet. 1971 *Sci. Amer.* Jan. 47/1 Was the heat provided by the impact of meteorites or of a larger celestial body, or by the collision of 'moonlets' to form the present moon?

moonlight, *sb.* Add: 1. c. The colour of the light of the moon, as a shade in fabrics.

1922 *Daily Mail* 18 Dec. 1 (Advt.), In the following colours: Turquoise, Mastic, Moonlight, Silver, [etc.]. 1927 *Daily Tel.* 26 Apr. 13 (Advt.), Following shades: Sapphire, Moonlight, Bois de Rose, [etc.].

d. *moonlight and roses*: used allusively of a situation, atmosphere, etc., characterized by sentimentality or romance.

1925 BLACK & MORET (*song-title*) Moonlight and roses. 1942 *Amer. Speech* XVII. 58 The moonlight-and-roses kind of Civil War play. 1959 *Encounter* Apr. 79/1 In *Look Homeward Angel* we have the romanticised figure of Laura James, the moonlight-and-roses woven about her.

4. Ellipt. for *moonlight flit*. *colloq.*

1958 G. BELLAIRS *Corpse at Carnival* ix. 131 His bags has gone... He's packed up... Done a moonlight as likely as not. 1971 R. PARKES *Line of Fire* xvi. 149 It's no good him trying to find 'em... Done a moonlight, they did.

6. b. Add to def.: Also, the action of leaving rented accommodation without paying the rent; a stealthy, usu. nocturnal, departure. (Add further examples.)

1821 J. GALT *Ann. Parish* xxxi. 263 He was fain to make a moonlight flitting, leaving his wife for a time to manage his affairs. 1855 [see FLIT *sb.* a]. 1892 STEVENSON & OSBOURNE *Wrecker* v. 79 In the excellent Scots' phrase, I made a moonlight flitting, a thing never dignified. 1924 M. ARGO in *Sc. Nat. Dict.* (1965) VI. 364/1 Takkin' a meenlichty flittin', are ye? 1969 *Listener* 27 Mar. 424/3 Very often we were living somewhere and couldn't pay the rent, so we had to indulge in what was known as a moonlight flit with what furniture and goods were available.

c. **moonlight lustre,** a lustre glaze with a marbled effect used on porcelain in the early 19th century, *spec.* by the Wedgwood factory.

1924 H. BARNARD *Chats on Wedgwood Ware* ix. 236 Lustre ware was not made until the beginning of the nineteenth century, when some very beautiful effects were produced, including the one which has since been called 'Moonlight'. 1966 G. A. GODDEN *Illustr. Encycl. Brit. Pott. & Porc.* p. xxiv, Messrs Wedgwood introduced a decorative type of marbled pink or purple 'gold' lustre called 'Moonlight Lustre'. 1970 *Times* 7 Oct. 10/6 This type of service was first made around 1820 and only the early examples, like that sold yesterday, are splashed in a pink glaze known as moonlight lustre.

moonlight, *v.* Add: 2. To do a 'moonlight flit'. *dial.* and *colloq.*

1903 in *Eng. Dial. Dict.* 1971 A. NIXON *Attack on Vienna* vii. 71 He moonlit out of his luxury flat..and moved into a boarding-house.

3. To do paid work, usu. at night, in addition to one's regular employment. *colloq.* (orig. *U.S.*).

1957 [implied in *MOONLIGHTING *vbl. sb.* 3]. 1960 *Economist* 12 Nov. 657/1 The firemen and police have long been supplementing their pay..by 'moonlighting', that is, by taking outside paid work. 1965 *Daily Tel.* (Colour Suppl.) 7 May 19/1 Some manage to 'moonlight' with two or three jobs—though jobs are harder to find. 1970 P. CARLON *Death by Demonstration* xvi. 175 You think I moonlight? Believe me, one job's enough. 1974 *Times Lit. Suppl.* 15 Feb. 157/4 He..—naturally for one who moonlights as the *Financial Times*'s gardening correspondent when not otherwise engaged as a Fellow of Magdalen—never misses a turn on botanical or horticultural matters.

moonlighter. Add: 2. One who does a 'moonlight flit'. *dial.* and *colloq.*

1903 in *Eng. Dial. Dict.* 1964 *Sunday Mail* (Brisbane) 15 Nov. 12 Brisbane flat owners..estimate that moonlighters—tenants who slip away overnight without paying the rent—are costing them £100,000 a year.

3. One who 'moonlights' (*MOONLIGHT *v.* 3). *colloq.*

1957 *Times* 12 Nov. (Canada Suppl.) p. ix/4 'Moonlighters' take a second job after hours; married women take advantage of modern kitchens to do the same. 1964 *Observer* 9 Aug. 9/2 Even in America, an estimated four million workers are 'moonlighters'—that is, they take on a part-time job on top of their full-time one. 1973 C. EGLETON *Seven Days to Killing* v. 61, I employ a lot of moonlighters, blokes who take a second job at nights.

moonlighting, *vbl. sb.* Add: 3. The act or practice of *MOONLIGHT *v.* 3. *colloq.* (orig. *U.S.*).

1957 *Reporter* (N.Y.) 8 Aug. 11/3 He takes two or three hours off and then..departs for a second job... The practice is known as 'moonlighting'. 1961 *Economist* 16 Dec. 1145/2 Several attempts have been made to ban moonlighting on the ground that it robs the unemployed of jobs. 1972 *Times* 8 Jan. 21/2 What about moonlighting? This is not the distilling of illicit liquor but the taking of a second job to keep body and soul together, or to finance one's own personal brand of extravagance.

moon-man. Restrict † *Obs.* to senses 1 and 2 in Dict., and add: 3. (Further examples.) Also *fig.*

1958 *Spectator* 5 Sept. 305 An electronic sideshow starred Gygan, the Man from the Moon, a seven-foot creature of steel with a sixty-inch chest to keep all the transistors in. He walked spastic fashion with that curious whirring noise which all Moonmen seem to affect. 1962 A. WESKER *Chips with Everything* I. iii. 19 All right you creepy crawly nignogs, moon men that's what you are, moon men.

4. An astronaut who travels to the moon.

1965 *N.Y. Jrnl.-American* 21 Feb. 11/1 The spot where America's Moonmen are most likely to land. 1968 *Daily Tel.* 30 Dec. 1/3 (*headline*) Moonmen greeted by wives. 1970 *Observer* 19 Apr. 8/4 They were virile men, these astronauts. They were patriots, but they were moonmen.

moonquake (mū·nkwēɪk). *Astr.* [f. MOON *sb.* after EARTHQUAKE.] A tremor of the moon's surface. (In quot. 1940 a *poet. nonce-use*.)

1940 R. GRAVES *No More Ghosts* 41 The pallid sky heaved with a moon-quake. 1953 *Jrnl. Brit. Interplanetary Soc.* XII. 66 The work which such an expedition could do is then described: amongst other things, artificial 'moonquakes' would be set off by explosives, so that seismic records could be made. 1961 *Flight* LXXIX. 427/2 Other scientific instruments aboard the Surveyor will include a sensitive seismometer to record moonquakes or meteorite impacts. 1968 *Observer* 29 Dec. 17/3 The astronauts will place on the surface an instrument to register moonquakes. 1970 *Nature* 5 Dec. 906/1 It is now clear..that tidal strain is causing moonquakes when the Moon is at its nearest to the Earth. 1973 *Telegraph* (Brisbane) 13 Mar. 14/1 A seismologist has said that for some mysterious reason almost all moonquakes occur only in two separate belts on the moon.

moonraker. 2. (Earlier example.)

c 1829 D. JERROLD in M. R. Booth *Eng. Plays of 19th Cent.* (1969) I. 173 Now she makes more sail..mounts her royals, moon-rakers and sky-scrapers.

moonrise. Add: (Further examples.) Also (*U.S.*), the time at which the moon rises.

1877 in BARTLETT *Dict. Amer.* (ed. 4). 1884 'MARK TWAIN' *Huck. Finn.* viii. 63 When it was good and dark, I slid out from shore about moonrise. 1913 D. H. LAWRENCE *Love Poems & Others* 26 (*title*) Red moon-rise. 1926 E. M. ROBERTS *Time of Man* (1927) x. 379 I'll be gone at moonrise. 1931 E. O'NEILL *Mourning becomes Electra* I. i. 191 You dasn't stay there till moonrise at ten o'clock. 1975 *Sci. Amer.* Feb. 70/2 A lunar day is 24·8 hours in length, the interval between successive moonrises.

moonscape (mū·nskēɪp). [f. MOON *sb.* + SCAPE *sb.*[3]] The surface or landscape of the moon, or a scene resembling this. Also *fig.*

1926 *World's Work* May 5 It embodies all the scientific facts known to-day, and the 'moonscape' in the foreground is an actual scene as viewed through the telescope. 1960 *20th Cent.* June 509, I..took..off..on a mighty leap across the moonscape. 1961 I. FLEMING *Thunderball* xvi. 173 Now the pale moonscape changed. 1969 *Sun* 22 July 1/3 Armstrong, the first to walk, looked at the desolate moonscape and told the world: 'It has a beauty of its own ..it's very pretty up here.' 1971 D. LEES *Rainbow Conspiracy* vi. 89 The deserted moonscape terrain of Pike's Scar. 1974 V. CANNING *Mask of Memory* ii. 23 His thin, drawn face a moonscape of chalky white.

moonshine. 4. Add to def.: In the U.S., illicitly distilled liquor, esp. whisky.

1875 E. KING *Southern States N. Amer.* 478 Producing from his pocket a flask of 'moonshine' whiskey, [he] invited us to drink. *Ibid.* 479 Would we have some more 'moonshine'? No? 1886 *Century Mag.* XXXI. 432/1 He had a moonshine apparatus over on Sweetwater. 1901 [in Dict.]. 1957 M. MCCARTHY *Memories Catholic Girlhood* vii. 157 It did not smell like the whisky my grandfather drank. It was moonshine, they said; corn whisky. 1960 A. LOMAX *Folk Songs N. Amer.* 257 I'll eat when I'm hungry and drink when I'm dry, If moonshine don't kill me, I'll live till I die. 1973 'B. MATHER' *Snowline* i. 16 All liquor is either bootleg—smuggled—or moonshine, distilled from anything and everything.

moonshine (mū·nʃəin), *v.* [f. the sb.] 1. *trans.* To cheat or deceive by means of 'moonshine' (MOONSHINE 2 a). *rare*⁻¹.

1824 E. WEETON *Let.* 6 July in *Jrnl. of Governess* (1969) II. 298, I visited the Diorama... I was too much gratified to think I was moon-shined out of my money.

2. *intr.* To make liquor, esp. whisky, illicitly. *U.S. colloq.*

1883 [implied in *MOONSHINING *vbl. sb.*]. 1902 W. N. HARBEN *Abner Daniel* 209 We moonshined it together two year, though he never knowed my chief hidin'-place. 1910 'O. HENRY' *Whirligigs* xv. 173 He acknowledged no occupation save that of a squirrel hunter, but he 'moonshined' occasionally by way of diversion. 1949 *Times-Picayune Mag.* (New Orleans) 13 Feb. 6/2 Even a small operator can make fair living moonshining.

moonshiner. Delete *U.S.* and substitute for def.: a. A smuggler. b. *U.S.* A distiller of 'moonshine' (MOONSHINE 4). (Add further examples.)

1877 *N.Y. Even. Post* 16 June, Nelson County, Kentucky, is the home of the Moonshiner; that is, the manufacturer of illicit whiskey. 1927 [see *FEUDIST 3]. 1970 N. ARMSTRONG et al. *First on Moon* iv. 77 Some seven hundred people had to be moved, including quite a few moonshiners. 1973 *Guardian* 12 Oct. 1/8 Senator Sam Ervin..has made a long-playing record of his homespun Southern philosophy. The record features stories about moonshiners. 1973 *Globe & Mail* (Toronto) 9 Nov. 31/4 That same gallon, he says, costs the moonshiner only $2 to $3 to manufacture, and—if he isn't making it for his own use—he usually sells it for around $30.

moonshining, *vbl. sb.* (Earlier and later examples.)

1883 'S. BONNER' *Dialect Tales* 183 You can't stop moonshinin' 's long's there's an honest man in Old Hickory's State. 1946 G. WILSON *Fidelity Folks* 54 He stoutly maintained that a regulated open saloon was to be preferred to the evils of moonshining and boot-legging.

moonshiny, *a.* 3. (Earlier example.)

1857 H. MELVILLE *Confidence Man* xli. 312 And moonshiny as it in theory may be, yet a very practical philosophy.

moon-up (mū·nʌp). *U.S. dial.* [f. MOON *sb.* + UP *adv.* Cf. SUN-UP.] Moonrise.

1907 S. E. WHITE *Arizona Nights* I. v. 90 'Didn't git in till moon-up last night,' he growled. 1941 *Sat. Even. Post* 1 Mar. 84/2 They'd come between first dark and moon-up.

moonward(s, *adv.* (Further examples.)

1879 G. MACDONALD *Hist. Photogen & Nycteris* xiv, in *Graphic* Christmas No. 8/2 The fountain kept rushing moonward. 1900 O. WILDE *Let.* 21 Apr. (1962) 823 He really bayed for boots, as a dog moonwards. 1918 C. W. BEEBE *Jungle Peace* (1919) xi. 270 Here and there in the jungle on each side, where a tree had fallen, or a flue of clear space led moonwards, the effect was of cold electric light seen through trees in city parks. 1959 C. MACKENZIE *Lunatic Republic* i. 21 Every sound without was so muted that the explosion which started the rocket moonwards sounded no more than the noise of one of the old railway trains of my youth.

moor, *sb.*[1] Add: 1. b. *spec.* (Usu. with capital initial.) Dartmoor Prison (cf. *DARTMOOR b).

1869 F. HENDERSON *Six Yrs. in Prisons Eng.* xix. 228 'How long were you at the Moor, Dick?' 'Three years.' 1924 E. WALLACE *Room 13* vii. 69, I was on the 'moor' with him. 1939 [see *ISLAND *sb.* 1 d]. 1958 F. NORMAN *Bang to Rights* I. 22 I'm doing a bleeding neves. I'll be going down the Moor soon that will be the third poxy

time. 1962 D. WARNER *Death of Bogey* IV. iii. 146 A stretch in the Ville or on the Moor. 1967 C. DRUMMOND *Death at Furlong Post* xvi. 192 There's nothing like the Moor, T.B. more or less guaranteed after fifteen years. 1968 *Guardian* 3 Jan. 5/3 One more stretch for GBH, then down to the Moor.

5. *moor-owner*.

1915 R. LANKESTER *Diversions of Naturalist* v. 47 One moor-owner was able to boast that he had on several occasions killed over 500 head of grouse in a single day. 1971 *Country Life* 12 Aug. 396/1 For the moor-owner grouse soon became more profitable than sheep.

moor, *v.*[1] Add: 2. c. With *up*: to secure a seaplane; of a seaplane, to be made secure.

1936 [see *GRABBIT (BOAT) HOOK]. 1942 *Times* 3 Sept. 5/7 Night was slowly mastering daylight as we..landed, to moor up near some naval patrol boats.

Moorcroft (mūᵊ·ɹkrɒft). The name of William *Moorcroft* (1872–1945) used *attrib.* and *ellipt.* to designate pottery produced by him at his workshop in Cobridge, north Staffordshire, and noted for its powdered blue effects and flambé glazes.

1910 *Pottery Gaz.* May 551/2, I cannot now do more than mention the most interesting developments in glaze effects and wonderful colour schemes revealed in the latest examples of 'Moorcroft Faïence'. 1918 *Pottery & Glass Record* Nov. p. i (Advt.), Moorcroft ware, hand-made decorative objects..W. Moorcroft, Ltd., Potters, Burslem... Works adjoining Cobridge Station. 1923 *Pottery Gaz. & Glass Trade Rev.* 2 Apr. 660/1 Baron Hayashi, the Japanese Ambassador, on purchasing two flambé vases, expressed the opinion that the Moorcroft vases were so like early Chinese work that had Mr. Moorcroft's name not been on them experts would have found it difficult to distinguish the difference. 1939 *Times* 8 Mar. 12/4 Moorcroft pottery has for some years been recognized abroad as standing in a class by itself among the modern products of ceramic art. 1946 *Pottery & Glass* June 33/1 Several American museums possess representative collections of Moorcroft ware. 1956 C. G. E. BUNT *Brit. Potters & Pottery Today* 41 The rich and generous decoration is..all hand work. To this is largely due the distinctiveness, the individuality of Moorcroft Pottery. 1972 F. MACCARTHY *All Things Bright & Beautiful* 85 Blue Moorcroft Pottery and Cornish Ware were endlessly in favour [in the 1920s]. 1974 M. KELLY *That Girl in Alley* iii. 40 The Moorcroft vases, the Lalique bowl, the silk runner on top of the piano. 1975 *Times* 17 Jan. 3 Sotheby's Belgravia sale of studio ceramics..underlined the tremendous new collector interest in the art potters of about 1880–1950... Minor Moorcroft was selling like hot cakes.

Moore (mūᵊɹ, mōᵊɹ). Designating an almanac, the first edition of which, compiled by Francis *Moore* (1657–c 1715), was issued in 1700 under the title of *Vox Stellarum*, and which was later known as *Old Moore's Almanac*. Also *ellipt.* Hence (nonce-wd.) *Old Moore's Almanac* v. absol., to engage in predictions of the future.

1817 KEATS *Let.* 10 Sept. (1931) I. 39, I know not whether you prefer..Cinderella and her glass slipper to Moor's Almanack. a 1821 —— *Cap & Bells* lvi, in *Compl. Poetical Works* (1907) 476 Look in the Almanack—*Moore* never lies. 1855 GEO. ELIOT in *Westm. Rev.* Oct. 437 Let him..rival Moore's Almanack in the prediction of political events. 1873 C. M. YONGE *Pillars of House* IV. xl. 170 The mysteries of Moore's Almanac on the wall. 1896 G. B. SHAW *Our Theatres in Nineties* (1932) II. 124 As devoutly as any superstitious washerwoman ever read Zadkiel or Old Moore. 1916 D. H. LAWRENCE *Let.* 24 May (1962) I. 452 But I am afraid I bore you Old-Moore's Almanacking. 1936 *Discovery* Dec. 389/1 He published a sort of Old Moore's Almanac and told people's fortunes, and generally practised the acts of the astrologer. 1975 *Broadcast* 15 Sept. 16/3 Throw the net wide and don't always toss back the little 'uns are old Moore's words of wisdom.

mooreed, var. *MURID.

mooreite (mūᵊ·rəit, mōᵊ·rəit). *Min.* [f. the name of Gideon *Moore* (1842–1895), U.S. chemist + -ITE[1].] A hydrated basic sulphate of magnesium, manganese, and zinc, $(Mg,Mn,Zn)_8SO_4(OH)_{14}.4H_2O$, found as colourless monoclinic crystals.

1929 BAUER & BERMAN in *Amer. Mineralogist* XIV. 165 The cavities in the pyrochroite are lined, and some of the crevices are filled, with glassy white tabular crystals which, upon investigation, proved to be a new mineral, for which the name mooreite is here proposed, in honor of Dr. Gideon E. Moore, an early investigator of Sterling Hill and Franklin minerals. 1951 *Mineral. Abstr.* XI. 384 The action of salt water on the minerals of this vein in the August Victoria mine at Hüls, Westphalia, has produced the following secondary minerals:..mooreite, dickite, sericite, and blödite. 1969 *Amer. Mineralogist* LIV. 973 Mooreite..is monoclinic although the space group is not $P2_1/m$ as previously reported but rather $P2_1/a$.

moor-grass. 3. For the generic name '*Sesleria*' substitute '*Molinia*'. (Earlier and later examples.)

1749 W. ELLIS *Compl. Syst. Sheep* 321 This moor-grass ..they call *Rosa-Solis*, as it is distinguished by Shepherds from other Grasses, who know it by its three-square Leaf, Rapier-like. 1950 W. H. PEARSALL *Mountains & Moorlands* ix. 179 The moor-grass may be found growing on oxygen-poor peats with a moderately high salt content and low acidity. 1972 *Country Life* 13 Jan. 104/1 There was a telling use of..a cultivated form of the extremely common purple moor grass.

mooring, *vbl. sb.* Add: 4. *mooring boat, bridle* (examples), *chain* (later examples), *hook*; **mooring mast,** a strong upright structure to which an airship is moored; **mooring-out,** used *attrib.* to denote a site at which an airship may be moored.

1957 D. G. O. BAILLIE *Sea Affair* xii. 227 The Arab mooring-boat crews were ready at the buoys to cast off our lines. 1967 S. WATERS *Indentures Indorsed* xxvi. 153 The two ships were almost alongside each other with only a small mooring boat containing five men between them. a 1865 SMYTH *Sailor's Word-Bk.* (1867) 484 *Mooring-bridle*, the fasts attached to moorings, one taken into each hawse-hole, or bridle-port. 1919 *Gloss. Aeronaut. Terms* (R. Aeronaut. Soc.) 58 *Mooring bridle*, a rigging passing between two points from a fitting in the length of which the mooring rope is attached. 1948 R. DE KERCHOVE *Internat. Maritime Dict.* 474/2 *Mooring bridle*, the chains or fasts attached to permanent moorings and taken into the hawse holes. 1812 *Dramatic Censor 1811* 223 The monopoly of the East-India Company, Lord Gwydir's mooring-chain monopoly, should be all abolished. 1888 KIPLING *Departmental Ditties* (1890) 68 Through the mooring-chains The wide-eyed corpse rolled free. 1907 J. MASEFIELD *Tarpaulin Muster* viii. 97, I..cast the painter around the mooring-hook. 1919 *Gloss. Aeronaut. Terms* (R. Aeronaut. Soc.) 59 *Mooring mast*, a mast to the head of which an airship may be moored. 1929 *Times* 2 Nov. 9/4, R 101..was walked over to the mast at dawn..in 20 minutes, was at once coupled up to the mooring-mast cable and then waited patiently..for the actual mooring. 1972 *Daily Tel.* 26 June 5/1 Europa was in the air for the first time since its mooring mast snapped and the airship careered for threequarters of a mile. 1934 J. A. SINCLAIR *Airships in Peace & War* 190 The mooring-out station was more of a war measure. 1971 *Country Life* 6 May 1087/1 On flying from our base in Anglesey, or our mooring-out ground at Malahide, [etc.].

Moorish, *a.*[2] Add: 1. Also freq. used with reference to the style of furniture and architecture, popular in England in the nineteenth century, characteristic of that made by the Moors in Spain (8th–15th c.) and in Northern Africa.

1903 A. BENNETT *Leonora* iii. 80 Bessie..set a Moorish stool before the Chesterfield. 1912 —— *Matador* 49 They were having tea at a little Moorish table in..the lounge. 1965 J. ARONSON *Encycl. Furnit.* (1966) 410/1 The X-type chair was common.., a light Moorish type of repeated slats with inlay. 1969 M. ALLINGHAM *Case-Book* 101 Towards the [sea] port..the architecture veers towards Victorian Moorish. 1972 E. JOY *Furniture* vii. 188 Of completely different character was the imported..Moorish furniture which first made Liberty's reputation when they began in business in 1875 as an 'Oriental Warehouse'. 1975 R. BUTLER *Where all Girls are Sweeter* iv. 30 'Four Winds' was all Moorish, Bauschule and Odeon but it stood back in at least five acres.

moorlog. Delete '† *Obs.*' and add later examples.

1933 *Antiquity* VII. 38 In the summer of 1932 the skipper of the steam drifter *Colinda* procured another bone harpoon of Maglemose type in a sample of 'moorlog' (peat) brought up by the trawl. 1939 G. CLARK *Archæol. & Society* ii. 20 Over many parts of the North Sea bed.. fishermen have found the remains of a great freshwater fen in the form of lumps of 'moor log'.

moose[1]. Add: Also used *collect.*

1817 S. R. BROWN *Western Gazetteer* 202 Cabree and moose are plentiful. 1849 J. PRITTS *Mirror Border Life* 60 Their food was principally the entrails of moose, deer, bears. 1957 *Listener* 14 Nov. 778/2 Large wooden corrals into which their moose were driven. 1964 *Globe & Mail* (Toronto) 12 Dec. 8/7 A few point to the slow resurgence of moose, particularly in the southwest of the province.

b. *moose-hair, -hide* (earlier and later examples), *-meat* (earlier and later examples), *-skin* (further examples); **moose-bird** *N. Amer.*, substitute for def.: a name used for several birds, esp. the Canada jay, *Perisoreus canadensis* (further examples); **moose-calling** (earlier and later examples); **moose elm,** the slippery elm, *Ulmus rubra* (earlier examples); **moose fly,** one of several North American horse flies, esp. one of the genus *Chrysops*; **moose maple,** either of two small maples of eastern North America, *Acer spicatum* or *A. pennsylvanicum*; **moose shanks** (see quot. 1887); **moose tick,** a North American tick, *Dermacentor albipictus*, which infests moose and other animals; **moose warden** *U.S.*, a person employed to protect moose; **moose-wood** (earlier and later examples); **moose-yard** (earlier and later examples).

1832 W. D. WILLIAMSON *Hist. State of Maine* I. 150 The moose-bird..feeds on the berries of the moose brush. 1890 S. M. ST. MAUR *Impressions of Tenderfoot* 273 My friend of the woods, the moose-bird... I found his real name was the great northern shrike, *Lanius borealis*. 1941 E. T. SETON *Trail of Artist-Naturalist* 270 Presently the wailing of the moose-bird fell on my anxious ear. 1964 *Atlantic Advocate* Oct. 67/1 This time of year there were always a pair of moosebirds on the maple. 1838 *United Service Jrnl.* July 305 In moose-calling, success mainly depends on the judicious selection of a station. 1956 W. R. BIRD *Off-Trail in Nova Scotia* vi. 167 The September moon is just right for moose calling. 1810 F. A. MICHAUX *Hist. Arbres Forestiers de l'Amérique Septentrionale* I. 39 Red elm,..Slippery elm,..Moose elm,..dans le haut de l'Etat de New York. 1832 D. J. BROWNE *Sylva Amer.* 311 This species of elm..bears the name of Red Elm, Slippery Elm and Moose Elm. 1834 J. J. AUDUBON *Ornith. Biogr.* II. 437 The musquitoes and moose flies did their best to render us uncomfortable. 1913 *Chambers's Jrnl.* Nov. 764/1 The lordly moose finds..relief from the torturing moose-fly in the cool depths of the lakes. 1953 A. R. M. LOWER *Unconventional Voyages* 43 Further south, where they are never so numerous, they [sc. bulldog flies] are called horse-flies, or moose-flies. 1820 D. W. HARMON *Jrnl. Voy. & Trav. Interior N. Amer.* 317 From two points in this bow,..two strips of leather..are suspended, at the ends of which, tassels, composed of moose hair, are fixed. 1956 D. LEECHMAN *Native Tribes Canada* 43 Some used moose hair for embroidery. There is a tuft of pale, stiffish hair about seven inches long, between the shoulders of the moose. 1708 J. OLDMIXON *Brit. Empire in Amer.* I. 395 Every Man has commonly two Wives, whom they.. make 'em do all Slavery; as draw Sledds,..and dress Moose Hides. 1955 R. P. HOBSON *Nothing too Good for Cowboy* vii. 67, I sat up, pulled my moosehide coat over my sweater, [etc.]. 1966 *Globe & Mail* (*Toronto*) 18 May 6/2 When he is racing [on snowshoes], he wears moosehide moccasins. 1839 C. F. HOFFMAN *Wild Scenes Forest & Prairie* 38 We would come to a sort of plateau of swampy land, overgrown with moose maple. 1904 S. E. WHITE *Blazed Trail Stories* vi. 104 He..was holding aside the screen of moose-maples. 1952 J. JENNINGS *Strange Brigade* II. iii. 128 To our left lay the woodsy thickets..beech and chokecherry and moose maple. 1797 C. CHABOILLEZ *Jrnl.* in B. C. Payette *Northwest* (1964) 163 The People returned, they brought 1 Bear skin..8 pieces Mouse Meat half Dryed. 1930 L. MUNDAY *Mounty's Wife* iii. 48 We camped that night with some Indians [and] huge plates of moose meat were served, as good as any beef, and far better than most other meats. 1961 H. C. DODGE *My Childhood Canad. Wilderness* i. 27 We ate them fried, roasted, boiled, but we never grew tired of moose meat. 1969 in Halpert & Story *Christmas Mumming in Newfoundland* 84 This late supper might include a meal of 'bottled moose-meat', warmed-over soup..'pork-buns', [etc.]. 1887 *Harper's Mag.* Feb. 458/2 'Moose shanks' are made by peeling the skin from the hind legs of the animal. The smaller end is then sewn up to form the toe; and thus a moose-hide stocking is formed. 1952 E. BUCKLER *Mountain & Valley* 100 The Rothesays down the road.. wore moose shanks with the hair still on them. 1664 *Essex County, Mass. Probate Rec.* (1916) I. 458 Moose skin sute. 1775 in *Mass. Hist. Soc. Coll.* (1822) 2nd Ser. IX. 162 Donations..1 pair moose-skin breeches. 1957 C. HARRIS *Cariboo Trail* 174 The mooseskin-trousered [one] looked at him darkly. 1971 D. C. BROWN *Yukon Trophy Trails* x. 120 Lonny Johnny, a full-blood Loucheux, was building himself a moose-skin boat... Lonny used the lateral roots of a spruce tree for ribs and two large bull moose hides for the canvas. 1868 *Amer. Naturalist* II. 559 The Moose Tick... When the cow arrived in New York, her sides and back were almost covered with adult ticks. 1853 *Maine Acts & Resolves* 24 The governor shall..appoint one county moose warden for each of the counties. 1883 G. B. GOODE *Rev. Fishery Industries U.S.* 83 He had a canoe of birch, and on it he had burnt..the title 'Moose Warden', and he said he would take care of all the moose that came within the reach of his rifle. 1778 J. CARVER *Trav. N.-Amer.* 507 The Moose Wood grows about four feet high, and is very full of branches. 1933 D. G. CAMERON *Twigs from Oak* 129 The under-growth consisted of..moosewood, dogwood. 1969 T. H. EVERETT *Living Trees of World* 222/2 Conspicuous in winter because of its white-striped bark, the moosewood (*A[cer] pennsylvanicum*) inhabits open woods throughout eastern North America. 1800 C. D. ROUSO D'ERES *Mem.* 117 The animals are overtaken in their retreats (for they herd together..) which is called the Moose yard, formed by them in trampling down the snow. 1946 W. R. BIRD *Sunrise for Peter* 209 Nathan plunged to join him and almost fell into a well-trodden moose-yard.

c. *transf.* and *fig.*

1925 *Dalhousie Rev.* V. 321 There is much meandering in the neighbouring woods; and..it is known as the game of 'moose'. 1941 *Amer. Speech* XVI. 187 *Moose*, a sister, notably a kid-sister, who informs her mother when her brother is playing stick-ball, shooting dice, etc. 1948 *Sat. Even. Post* 16 Oct. 130/3 Schaeffer was a moose of a man. 1963 *Amer. Speech* XXXVIII. 173 An unattractive female date:..Ugliness..ranges from such general terms as beast..to the more specific bear, cow, goose, moose, roach .., squirrel, and wet fish. 1968 *Word Study* Dec. 3/2 *Moose* is sometimes used for any large and strong person, but it is mostly widely applied to large and strong persons of Scandinavian descent.

moose[3] (mūs). *U.S. Forces' slang.* [ad. Jap. *musume* daughter, girl.] A young Japanese or Korean woman; *esp.*, the wife or mistress of a serviceman stationed in Japan or Korea.

1953 in Partridge *Dict. Slang* (1961) Suppl. 1188/2 An eligible female of Japan or Korea is known as a 'moose',..from the Japanese word 'musume'—girl. 1954 *Amer. Speech* XXIX. 302 To spell the word *mus* might be a good transcription from the Japanese but..it is the spelling *moose* that is most encountered in semiformal Army poop sheets; in signs urging Americans..to meet the best mooses in Kyoto, [etc.]. 1964 *Ibid.* XXXIX. 236 The word *moose* refers, not without some disdain, to Korean wives of Americans, often soldiers.

MOOSE (mūs). [See quot. 1968.] A contrivance for the protection of an astronaut working in outer space.

1968 *New Acronyms & Initialisms* (Gale Research Co.) 144 MOOSE, Man (or Manual) orbital operations safety equipment (Space life raft) (NASA). 1969 *Scottish Sunday*

Express 14 Dec. 12/8 American engineers have designed a space 'lifeboat'..known as MOOSE, from the initial letters of Manned Orbital Operations Safety Equipment.

moosh[1] (muʃ), var. *MUSH *sb.*[1] 3 d.

1914 JACKSON & HELLYER *Vocab. Criminal Slang* 60 *Moosh, moush*,..the human face... Also the mouth... Example: 'He's got a harp moosh,' i.e., Irish. 1953 K. TENNANT *Joyful Condemned* x. 88, I don't usually go round pushing my moosh into anyone's business. 1966 'L. LANE' *ABZ of Scouse* 71 'Ey you wit' ther maggerty moosh. 1972 K. BONFIGLIOLI *Don't point that Thing at Me* iv. 34 You ought to see his moosh, where I hit him, it's a treat, honest.

moosh[2] (muʃ), var. *MUSH *sb.*[4] Also (*Austral.*), prison food.

1919 *Athenæum* 8 Aug. 727/2 When a man was 'run in' the guardroom he was in 'clink' or in 'moosh'. 1945 BAKER *Austral. Lang.* vii. 141 Jail food is *moosh*.

moosh[3] (muʃ), var. *MUSH *sb.*[5] Used esp. as a term of address.

1943 *Police Jrnl.* Mar. 69 *Moosh*, a person, an individual. 1950 P. TEMPEST *Lag's Lexicon* 137 'Moosh' is used more as a greeting: 'Hullo, Moosh.' 1961 J. STROUD *Touch & Go* xvii. 183 Waiter!.. Look, moosh, this is cold. 1966 *New Statesman* 23 Dec. 934/3 My old woman's gone to Paris with a black moosh. 1972 J. BROWN *Chancer* i. 12 Look, moosh, you'll strip off or I'll take them off you.

moot, *sb.*[1] Add: 5. Delete 'Now in use only at Gray's Inn' and substitute 'Revived in the Inns of Court after falling into disuse, and introduced into universities where law is studied'.

1926 E. WEEKLEY *Words Anc. & Mod.* 70 The practice of holding at the Inns of Court moots at which law students gain experience by arguing an hypothetical case. 1962 E. MITCHELL *Business Man's Lawyer* 441/2 *Moot*, a gathering of lawyers or law students, to argue—semi-formally—interesting but academic points of law. 1973 *Univ. of Leicester Prospectus* 1974–75 facing p. 33 (*caption*) A 'moot' in the Department of Law. The Department has won the national 'Observer' mooting competition twice in the last three years.

mootah, mooter (mū·tă). *U.S. slang.* Also **moota, mootie, moota, muta,** and other varr. [Origin unknown.] Marijuana.

1933 C. DE LENOIR *Hundredth Man* xiii. 220 This drug.. looked like chopped hay, or dried clover... In short, a 'muggles', 'weed', or 'mootie', cannabis indica. 1938 *Detective Fiction Weekly* 8 Jan. 47/2 He found plenty of marijuana... As always, it came in cigarettes, called 'reefers', 'muggles', 'moocahs', 'mus', 'grifos', 'mootas', or sometimes, playfully, 'Mary Warners'. 1943 *Time* 19 July 54/2 Marijuana may be called muggles, mooter, Mary Warner, Mary Jane, Indian hay, loco weed, love weed, bambalacha, mohasky, mu, moocah, grass, tea or blue sage. 1946 MEZZROW & WOLFE *Really Blues* (1957) 52 All the time he was in the asylum he kept waiting for a big train to pull in with a carload of muta just for him. 1950 H. E. GOLDIN *Dict. Amer. Underworld Lingo* 285/1 *Marijuana*... mota; muggles; stink-weed. 1956 'E. MCBAIN' *Cop Hater* (1958) x. 92 One of the guys was on mootah. So he got a little high. 1971 E. E. LANDY *Underground Dict.* 128 *Marijuana*,..moocah, mota, mutha.

mooti, var. *MUTI.

mootness (mū·tnės). *U.S. Law.* [f. MOOT *a.* + -NESS.] Of a legal case or question: the fact or condition of being hypothetical.

1946 *Univ. Pennsylvania Law Rev.* XCIV. 127 No distinction will be made between cases where only a single issue is alleged to be moot and those in which the entire case stands or falls on a determination of mootness or its absence. 1951 WOLFSON & KURLAND *Robertson & Kirkham's Jurisdiction of Supreme Court of U.S.* xxxvi. 505 The Court may request counsel to discuss an issue as to mootness. 1955 *Univ. Pennsylvania Law Rev.* CIII. 773 Since a court is deprived of jurisdiction when a case becomes moot, the fact of mootness can be raised at any time during the judicial proceeding and, once proved, will prevent decision of the case on the merits. 1960 *U.S. Supreme Court Reports* 362 U.S. § 577 If..George Parker's five-year quest for justice must end ignominiously in the limbo of mootness, surely something is badly askew in our system of criminal justice. 1970 *Harvard Law Rev.* LXXXIII. 1674 This Note will explore a situation, labeled a mootness question, that has become a recurrent problem for the Supreme Court: where events subsequent to the judgment of the trial court have so affected the relations between the parties that the two conditions for justiciability relevant on appeal—adverse interest and effective remedy—have arguably been compromised. 1974 *Ibid.* LXXXVIII. 373 (*heading*) The mootness doctrine in the Supreme Court. *Ibid.* 376 Mootness questions arise only once a court has determined, usually implicitly, that a litigant has standing to bring the action.

mop, *sb.*[2] Add: 1. d. *Mrs. Mop(p*: see *MRS. 2 c.

4. a. *mop-haired* adj., *-headed* (further examples); c. **mop-board**: for *U.S.* read 'orig. *U.S.*' and add earlier example; **mop-nail** (earlier example).

1854 Mop-board [see *baseboard* (*BASE *sb.*[1] 20)]. 1924 L. ECKENSTEIN *Tutankh-aten* v. 46 The..mop-haired Syrian almost daily presented himself at the gate. 1900 M. THORN in W. D. Drury *Bk. Gardening* xi. 360 Trees that are 'mop-headed', or top-heavy, should be supported by stakes. 1926 S. T. WARNER *Lolly Willowes* II. 85 She looked at the large mop-headed blossoms [sc. chrysanthemums]. 1934 *Times Educ. Suppl.* 3 Feb. p. iv/4 The autumn-flowering, mop-headed group of *P[rimula] capitata*. 1966 *Gloss. Landscape Work* (*B.S.I.*) IV. 15 *Mop-headed*, having naturally or induced a compact, rounded head, small in relation to the height of stem. 1972 *Guardian* 26 Jan. 9/6 Straight Jane Ltd..are now offering a Twist-n-Wring mop..with a head of the traditional type that gave mop-headed hair-dos their name. 1974 *Sunday Express* 21 Apr. 6/8 His mop-headed adventures are told in an oddly convoluted style and some will find them decidedly gooey. 1841 S. BAMFORD *Passages in Life of Radical* I. xxxvii. 216 Some had been grinding scythes, others..screw-drivers, rusty swords, pikels, and mop-nails.

mop (mɒp), *sb.*[7] and *int.* *U.S. slang.* [Echoic; cf. *BOP *sb.*[2]] (See quots.)

1944 in R. S. Gold *Jazz Lexicon* (1964) 208 (*tune-title*) Mop mop. 1945 L. SHELLY *Jive Talk Dict.* 15 *Mop*, the finale. 1947 *Britannica Bk. of Year* 840/2 *Mop*, slang, a word connoting surprise, anguish; a mild sort of 'hubba, hubba, hubba' (1944). 1952 B. ULANOV *Hist. Jazz in Amer.* (1958) xxv. 349 Such words as 'mop!' an exclamation of wide currency in the early forties which accurately described a musical device (the final beat in a cadence of triplets, usually bringing the release of a jazz composition to an end). 1959 *Village Voice* (N.Y.) 28 Oct. 13, I wait a while, eyes closed, and I look, mop! I'm in the bathtub, all alone. 1970 C. MAJOR *Dict. Afro-Amer. Slang* 82 *Mop*, the last beat at the end of a jazz number with a cadence of triplets.

mop, *v.*[2] Add: 1. (Later example.)

1927 *Motor Cycling* 7 Dec. 104/2 With a successful chromium plate the finished surface is so hard that it cannot be buffed or mopped.

2. b. To wipe (perspiration, tears, etc.) *from* the face or brow.

1872 R. W. BUCHANAN *Saint Abe & his Seven Wives* i. 4 And mopping from his brow the sweat, The boy glanced round with teeth still set. 1907 *Smart Set* Apr. 18/2 She.. mopped the hot tears from her face.

3. **mop up.** Also *fig.* Various slang uses: To drink greedily (further examples); also with *it*; also, to eat greedily; to get hold of (profits, etc.) (further example); to make an end of, slaughter (further examples); also *Mil.*, to complete the occupation of (a district, etc.) by capturing or killing enemy troops left there (cf. quot. 1901 in Dict.); also *absol.* and *transf.*

1898 T. TROUBRIDGE in W. A. Morgan *'House' on Sport* 393 The birds..come over in one's and two's, and are 'mopped up'. 1902 WODEHOUSE *Pothunters* i. 17, I hope your first man mops you up. 1915 A. HUXLEY *Let.* Dec. (1969) 88 It is *up to you*..to roll in the Texan subscriptions, while our agents..are..to mop up other corners of the dark continent. 1917 P. GIBBS *Battles of Somme* 295 Reserve battalions..came up behind to 'mop-up' the captured ground. 1920 M. A. MÜGGE *War Diary of Square Peg* 221 The second wave going over the top; it 'mops up', 'cleans up' the enemy's dug-outs. 1921 WODEHOUSE *Indiscretions of Archie* vii. 64 'Seacliff always had a—a tendency—a—a weakness—it was a family failing—' 'Mopping it up, do you mean? Shifting it?' 1936 —— *Laughing Gas* iii. 37, I find you here, mopping up the stuff like a vacuum cleaner. 1938 *Sun* (Baltimore) 31 Oct. 4/1 Three fires..were reported to be under control with the fire-fighters 'mopping up' today. 1940 *Times* (Weekly ed.) 10 Jan. 7/4 While mopping up the ground captured, the Zouaves took five machine-guns, 2,000 rounds of ammunition, and 12 prisoners. 1942 *R.A.F. Jrnl.* 30 May 33 The enemy was still fighting behind us but they would be mopped up in time. 1962 A. NISBETT *Technique Sound Studio* 239 Membrane and vibrating panel absorbers.. readily remove sound energy from the air at their resonant frequency and this is then mopped up within the absorber by various forms of damping. 1970 *Toronto Daily Star* 24 Sept. 1/7 King's troops were pursuing the fleeing rebels, mopping up. 1971 *E. Afr. Standard* (Nairobi) 13 Apr. 1/1 Troops were reported to be mopping up remnants of the..insurgents. 1973 *Times* 30 July 18/2 He also had to mop up the ensuing threatening chaos when prices rose much further than had been predicted. 1973 *Press & Jrnl.* (Aberdeen) 7 Aug. 16/8 A perfect crossfield ball to Hay allowed the inside left to dribble to the by-line and cut it back, but Watson was on hand to mop up the danger. 1973 *Times* 27 Dec. 11/1 The Bank of England 'mopped up' a large surplus by selling Treasury bills to both houses and banks. 1974 *Times* 5 Apr. 16/4 While mopping up a Knickerbocker Glory or two we devised a new form of the old game.

mop-up *sb.*, also, an act of 'mopping-up' (further examples).

1917 P. GIBBS *Battles of Somme* 296 The honour of the new attack was given to the 'mop-up' battalions behind. 1921 *Daily Colonist* (Victoria, B.C.) 2 Oct. 1/5 Men.. known among the strikers as 'mop-up gangs' have committed depredations, beaten up a few loyal employees, and threatened and intimidated other employees. 1944 *Sun* (Baltimore) 29 Apr. 2 (*caption*) Sergt. Charles H. Wolverton..prepares to throw a hand grenade into a Japanese dugout during the mop-up of the Empress Augusta Bay area after 18 days of bitter fighting.

mopane (mopā·ni). *S. Afr.* Also **mapani, mopané, mopani.** [Bantu *mo-pane*.] 1. A tree, *Colophospermum mopane*, of the family Leguminosæ, native to areas of low rainfall in southern Africa, and distinguished by rough, flaking, grey or brown bark, racemes of small, pale green flowers, and leaves formed of a pair of triangular leaflets joined at the base, which fold together during periods of intense heat. Also *attrib.*

1857 D. LIVINGSTONE *Missionary Trav. & Res. S. Afr.* 91 In some parts there are forests of mimosæ and mopane. *Ibid.* 162 Numbers of 'Baobab' and 'Mopané' trees abound all over this hard smooth surface [of calcareous tufa]. 1864 T. BAINES *Explor. S.-W. Afr.* xv. 422 The palms and open grassy plain had given place to thickets of mimosa.. and mopane. 1871 J. MACKENZIE *Ten Years North of Orange River* viii. 140 We at length found in a large mopane forest a well-beaten path. 1881 E. E. FREWER tr. *Holub's Seven Yrs. in S. Afr.* I. 48 The mapani-tree, with its oleaginous leaves and porous brittle wood. 1893 F. C. SELOUS *Trav. & Adventure S.-E. Afr.* 160 The fork of a mopani tree. 1906 *Chambers's Jrnl.* Mar. 213/1 Viewed across the hazy expanse of baobab and mopane bush. 1910 J. BUCHAN *Prester John* iv. 87 The mopani trees with their dull green wearied me. 1932 *Discovery* July 224/1 The mopané is a resinous tree, making good fuel. 1947 J. STEVENSON-HAMILTON *Wild Life S. Afr.* iv. 36 The Mopani belt in its extension along the Imbabate River. 1958 L. VAN DER POST *Lost World of Kalahari* vi. 110 We came down the side of the dunes on to a level plain covered with Mopani trees. 1973 PALMER & PITMAN *Trees S. Afr.* II. 842 The mopane is the dominant tree of large parts of northern South and South West Africa where great stretches of hot, arid country..are covered with it. *Ibid.* 843 Farmers in mopane veld look upon the tree as providing valuable fodder. 1975 M. HARTMANN *Game for Vultures* iii. 44 They moved..away into the black mopani bush.

2. **mopane bee, beetle, fly, worm** (see quots.).

1953 S. H. SKAIFE *African Insect Life* xxi. 357 The stingless bees, or Mopani bees, as they are often called, that are found in Africa belong to the genus *Trigona*. 1972 L. VAN DER POST *Story like the Wind* vi. 161 At about ten in the morning..all the billions of mopani beetles, hidden behind the butterfly leaves of the trees, began to sing their Messiah to the day. 1975 M. HARTMANN *Game for Vultures* i. 2 Mopani flies, little black squelches, were skitting around his eyes and ears. 1966 D. VARADAY *Gara-Yaka's Domain* xvii. 182 No less a find than crawling, fat, three-inch-long worms! Mopani worms—Masonja—as the Bantu call these brightly coloured insects, are regarded by them as a great delicacy. 1973 PALMER & PITMAN *Trees S. Afr.* II. 843 Large, dark, spotted caterpillars, the larvae of the moth *Gonimbrasia belina*—5 to 8 cm long and fat as a finger—feed upon the mopane leaves and are used by Africans and Bushmen as food... Mopane worms are much valued by rural Africans and are often their richest source of protein.

mope, *sb.* 1. Delete † and add later examples.

1932 J. T. FARRELL *Young Lonigan* vi. 259 They had a lot of pep, and weren't a bunch of mopes. 1962 E. O'BRIEN *Lonely Girl* i. 8 It happens to country mopes like you, as soon as you dance with a fellow.

mope, *v.* Add: 4. To confine or shut *up* (in a place).

1863 A. D. WHITNEY *Faith Gartney's Girlhood* xvi. 140 The child shouldn't be moped up here, all winter!

moped (mōᵘ·ped), *sb.* [Sw. (1952), f. *tramp-cykel med motor och pedaler*, pedal cycle with engine and pedals; cf. also Ger. *moped*.] A motorized pedal cycle. Also *attrib.*

1956 I. DUNLOP *Going to Britain* (ed. 2) 21 You must have a licence to drive an autocycle ('moped'). 1957 *Times* 19 Nov. 11/3 The accompanying recommendation that the minimum age limit for riding mopeds should be reduced from sixteen to fifteen will be less readily accepted. 1960 F. FARR *Mo-Peds & Scooters* i. 14 The name 'mo-ped' originated in Sweden, was popularized in Germany and is now generally accepted here. The machines are internationally defined..as cycles fitted with engines of under 50 cubic centimetres capacity and retaining the normal characteristics of bicycles. 1971 *Guardian* 4 June 22/4 Cut the minimum age for moped riders from 16 to 14. 1973 *Times* 18 May 21/7 As the rider of a single-seater moped, I am required by law to possess insurance against claims from mythical passengers. 1974 *Guardian* 25 Mar. 21/1, I bought an electric moped. It..has a range of about 20 miles.

mophy, var. *MOFFIE.

mopiness (mōᵘ·pinės). [f. MOPY *a.* + -NESS.] Mopy state or condition.

1927 *Smallholder* 26 Mar. 115/1 Ordinary Diarrhoea [in chicks]. Symptoms. Looseness of the bowels... There may or may not be mopiness but invariably there is inactivity.

mopoke, morepork. Add: 1. (N.Z. examples.)

1849 W. T. POWER *Sketches in N.Z.* ix. 74 Among the commonest birds which frequent the forest is a small owl, generally known..by the denomination of 'More pork'. 1866 M. A. BARKER *Let.* in *Station Life N.Z.* (1870) xiv. 100 The last cry of a very pretty little owl, called from its distinctly uttered cries, the 'more-pork'. 1874 T. H. POTTS *On Recent Changes Fauna N.Z.* 8 The name.. morepork is well-known throughout the country. Australian settlers distinguish a podargus by a similar name. 1882 —— *Out in Open* 118 *Athene Novæ Zelandiæ*, more-pork, ruru rarupeho of the Maoris. 1894 C. W. RICHMOND *Let.* 22 Mar. in *Richmond–Atkinson Papers* (1960) II. x. 597 Once..they had nothing to eat but a tui and a more-pork. 1905 J. M. THOMSON *Bush Boys of N.Z.* iv. 59 The More-pork is the small New Zealand owl, so-called from its peculiar guttural cry. 1963 *Evening Post* (Wellington, N.Z.) 26 Oct., In the..action shots which are skilfully interwoven with Maori legends and traditions about each bird, are the morepork..fantail, bellbird, tui.

b. (Earlier and later examples.)

1845 R. HOWITT *Impressions Australia Felix* 233 'A more-pork kind of fellow' is a man of cut-and-dry phrases;

a person remarkable for nothing new in common conversation. 1946 *Coast to Coast 1945* 132 He was such a helpless sort of a poor mopoke.

2. (Later examples.)

1916, 1934 [see *BOOBOOK].

mo·pper-up. *Mil.* [f. MOP *v.*[2] 3 + -ER[1].] A soldier who 'mops up' an enemy area (see *MOP *v.*[2] 3). Also *fig.* (perhaps in these uses owing as much to MOP *v.*[2] 3 in Dict.).

1917 *London Gaz.* 26 Nov. 12330/1 Pte. Dancox was one of a party of about ten men detailed as moppers-up. 1919 W. H. DOWNING *Digger Dial.* 34 *Mopper-up*, (1) One of a party of men who follow the leading waves of an attack in order to clear the enemy from the ground behind the assaulting troops; (2) a drunkard. 1920 G. K. ROSE *2/4th Oxf. & Bucks. Lt. Infty.* x. 122 The Berks came afterwards as 'moppers-up'. 1923 KIPLING *Irish Guards in Great War* I. 230 The 'mopper-up' who dealt with the débris of attacks. 1928 *Mod. Lang. Rev.* Apr. 136 His plan of attack is described on p. 174 of his *Philosophy of Grammar*; his 'moppers up' follow in the *Modern English Grammar*. 1936 *Daily Tel.* 15 Aug. 15/5 How..fascinating to know that your goods are coming to the London market by the Mopper Up.

moppet[2]. 1. (Later examples.)

1973 *Houston* (Texas) *Chron. Mag. People, Places, Pleasures* 14 Oct. 6/3 Jane Withers..made a long-delayed transition from movie moppet 'heavy' to middle-aged TV pitchlady. 1975 *Time Out* 10 Jan. 45/2 Watching the well-mannered moppets and well-groomed mums roll up in taxis and Bentleys to the Theatre Royal, Haymarket, certainly helped to set the scene.

moppie (mɒ·pi). *S. Afr.* [Afrikaans, ad. Du. *mopje* ditty.] A street-song of the Cape Malays.

1949 *Cape Times* 10 Jan. 2/6 He hoped that the Cape 'moppies' or comic songs and the *liedjies* would become a regular feature of future carnivals. 1953 DU PLESSIS & LUCKHOFF *Malay Quarter* iii. 48 *Moppies* are little songs (often of doubtful content) sung in order to challenge, deride, or irritate the listener, or merely as foolery. When singing a moppie, the singer often includes a person's name, and if the person referred to cannot respond in similar vein, he is laughed at by all present. 1974 *S. Afr. Panorama* June 20 The 'moppie' is a comical song. The soloist uses gestures of the hands and other movements to illustrate it as he goes.

mopping, *vbl. sb.*[2] Add: Also **mopping-up,** the action of the verb *mop up*, in various senses (also *fig.*). Also *attrib.*

1909 *Westm. Gaz.* 3 Sept. 9/4 Persistent 'mopping up' of London stocks [of silver] naturally restricts supplies for Continental and other requirements. 1918 E. S. FARROW *Dict. Mil. Terms* 394 *Mopping up trenches*, the crushing of hostile units which continue the resistances at certain parts of the trench, and the searching of the trenches and bombproofs with a view to making sure that none of the enemy are left in them. 1925 FRASER & GIBBONS *Soldier & Sailor Words* 158 *Mopping-up*, the term for the work allotted to special parties of men appointed to follow close in the track of advancing 'waves' of troops, in order to explore and clear the enemy lines and dug-outs of men remaining behind... The Mopping-up method was first adopted at the Battle of Arras in February, 1917. 1937 *Sun* (Baltimore) 12 July 5/1 Six flood-weary West Virginia communities began mopping-up operations.. today after two surging creeks flooded homes and business houses. 1940 *Economist* 20 Apr. 717/2 We must not be diverted in Narvik, which is now..a mopping-up operation. 1940 *Ibid.* 2 Nov. 552/1 In the last war, the Excess Profits Duty proved itself an excellent 'mopping-up' tax, against an inflationary background. 1940 'GUN BUSTER' *Return via Dunkirk* II. iv. 120 Mopping-up parties of Tommies were investigating the farmhouses. 1959 *Listener* 19 Mar. 521/1 Shakespeare is naturally drawn upon very little, since he had already been the subject of 'mopping up' in every volume of this kind. 1962 W. NOWOTTNY *Lang. Poets Use* i. 24 A mopping-up of successive problems in one area after another until the whole domain of poetic language is occupied and systematized. 1967 G. F. FIENNES *I Tried to Run a Railway* i. 4 We rounded up a few volunteers to do a mopping up operation on the permanent way. 1973 I. BUTLER *Eldest Brother* xix. 339 Mopping-up operations in Gujerat.

mopstick. Add: 1. c. = COCKALORUM 3. *colloq.*

1969 I. & P. OPIE *Children's Games* viii. 257 In Warwick, 'Mollie, Mollie Mopstick, all off! all off!' In Nuneaton: Mopstick, mopstick, bear our weight, Two, four, six, eight, ten. *Ibid.* 260 'Jack upon the Mopstick' (Warwickshire, 1892),..'Johnny on the Mopstick' (Worcester, c. 1930),..'Mopstick' (Kettering, c. 1915).

3. *slang.* a. A stupid man. b. *U.S.* (See quot. 1915.)

1886 H. BAUMANN *Londinismen* 111/2. 1915 *World* (N.Y.) Sunday Suppl. 9 May 14/3 *Mopstick*, one who loafs around a cheap saloon or barrel house and cleans up the place for drinks.

mopsy. 2. Delete ? *Obs.* and add later examples. Also *attrib.* or as *adj.* (Sense passing into that of MOP *sb.*[4] in some examples.)

1916 A. BENNETT *These Twain* xviii. 406, I always knew that girl was a mopsy slut. 1958 J. CAREW *Wild Coast* iii. 44 He don't have juice in his back to fill up a mopsy with delight. 1958 *Times Lit. Suppl.* 14 Mar. 140/5 Poor Swann's pain and frustration..are a simpler matter, Odette de Crécy being the most commonplace of lying mopsies and a born torturer of the sensitive.

mor (mōᵊɹ). *Soil Sci.* [Da., lit. 'humus' (adopted in this specific sense by P. E. Müller 1879, in *Tidsskrift for Skovbrug* III. 7): cf. MOOR *sb.*[1]] Humus which forms a discrete layer on top of the soil with little or no mineral soil mixed with it, which is characteristic of coniferous forests and is generally strongly acid in reaction. Cf. *MULL *sb.*[10]

1931, 1952 [see *MULL *sb.*[10]]. 1971 *Nature* 10 Sept. 133/1 Formation of the mor type of forest litter..has been attributed to a number of factors. 1974 [see *MULL *sb.*[10]].

mora[1]. Add: 3. (Earlier example.)

1832 [see TIME *sb.* 10].

b. In linguistic analysis, the minimal unit of duration of a speech-sound. Also *attrib.*

1933 L. BLOOMFIELD *Language* vii. 110 In dealing with matters of quantity, it is often convenient to set up an arbitrary unit of relative duration, the *mora*. Thus, if we say that a short vowel lasts one mora, we may describe the long vowels of the same language as lasting, say, one and one-half morae or two morae. 1941 G. L. TRAGER in L. Spier et al. *Lang. Culture & Personality* 136 In many cases it will be found that an element smaller than the phonetic syllable functions as the accentual or prosodic unit; this unit may be called, following current practice, the mora... The term *mora*..is useful in avoiding confusion, even if it should turn out to mean merely phonemic syllable. 1964 E. PALMER tr. *Martinet's Elem. General Linguistics* iii. 80 Each of the segments characterized by one of the successive punctual tones is called a mora. 1964 K. L. PIKE in D. Abercrombie et al. *Daniel Jones* 425 A train of alternating stresses..clashes with a different wave train keyed into a mora count beginning from the last suffixual syllable. 1968 W. S. ALLEN *Vox Graeca* ii. 81 For diphthongs in Greek cannot strictly be distinguished as 'short' and 'long'; for accentual purposes they all have the same value of 2 'morae' (time-units), as for a long simple vowel.

mora[4]. (Earlier and later examples.)

1825 C. WATERTON *Wanderings S. Amer.* i. 5 Heedless, and bankrupt, in all curiosity must he be, who can journey on without stopping to take a view of the towering mora. 1918 C. W. BEEBE *Jungle Peace* (1919) vii. 140 A house with roof of pale pink like a giant mora in full bloom. 1949 *Caribbean Quarterly* I. III. 42 Three forest areas..could be ..logged so as to yield about 3 million cubic feet of saleable timber per year, mainly Greenheart but also Mora and other hardwoods. 1956 *Handbk. of Hardwoods* (Forest Prod. Res. Lab.) 154 Mora is a dominant tree of mixed swamp forests and is found in the Guianas, Trinidad and eastern Venezuela. 1958 J. CAREW *Black Midas* vi. 109 Santos stopped under a mora tree. 1971 *Advocate-News* (Barbados) 17 Sept. Guyana Suppl. p. ii/2 Boys at work in pouring rain felling giant mora and wallaba trees.

mora[5]. (Later examples.)

1907 B. M. CROKER *Company's Servant* xviii. 195 Sitting ..on a little 'morah' at her feet. 1971 *Illustr. Weekly India* 4 Apr. 22/3, I resorted to circumlocutions..which invariably produced a *mora* or a chair in the little Chandni Chowk shops.

moraine. Add: b. In rock-gardening, a bed, often raised, constructed of rubble covered with fine chippings, in an attempt to produce suitable conditions for alpine plants. Also *attrib.*

1907 R. J. FARRER *My Rock-Garden* i. 14 There..is my toy-garden, my baby moraine, the particular pet joy of my heart. 1914 H. H. THOMAS *Rock Gardening* 65 The moraine may be described as a bed or border of varying size, preferably situated on a slight slope, and consisting of broken stone, with which, however, a certain proportion of light, sifted soil is mixed. *Ibid.* 72 Favourite Moraine Plants. 1931 *Times Lit. Suppl.* 16 Apr. 297/2 This [sc. bad drainage] is..the cause of most failures and disappointments in moraine gardening. 1961 W. INGWERSEN in W. E. Shewell-Cooper et al. *Alpine & Rock Gardening* vi. 43 If one moraine is to be divided into limestone and lime-free compartments, then the limestone area must be at the bottom.

moral, *sb.* Add: 9. Now chiefly *Austral.* (Further examples.) Also quasi-*adj.*, morally certain.

1916 C. J. DENNIS *Moods of Ginger Mick* 40 'E 'as struck it for a moral. Ginger's found 'is game at last. 1933 *Bulletin* (Sydney) 18 Jan. 20 Had been winning all over the countryside and was a moral for the big jumping event. 1934 [see *ear-biter*]. 1946 R. FRANKLYN in *Coast to Coast 1945* 221, I think we orta go home, Darky. We're a moral to get nabbed. 1957 'N. CULOTTA' *They're a Weird Mob* (1958) v. 72 As I was sayin' this bloke says 'e's a moral. 1972 I. HAMILTON *Thrill Machine* xxvii. 124 Eric said this bloke had a new film... It's a moral to be a load of bull but you never know. 1975 L. RYAN *Shearers* 79 Andy Burns rang last night..reckons Dallas is a moral in the first.

moral, *a.* Add: 1. a. (Further examples.)

1949 M. FORTES *Social Structure* 60 Its form derives from a paradigm..sanctioned by..moral values. 1951 R. FIRTH *Elem. Social Organiz.* vi. 200 This can be illustrated by considering the moral values attached to human personality in certain social situations. 1964 S. M. WILLHELM in I. L. Horowitz *New Sociol.* 184 The scientific ideology simply places the scientist in a moral vacuum. *Ibid.* 186 Recognizing his historical obligation to impose the very moral accountability which certain scientists seek to avoid, the intellectual insists upon his right to direct the affairs of a recalcitrant scientist.

c. (Further examples.)

1817 COLERIDGE *Biogr. Lit.* I. x. 213 My essays contributed to introduce the practice of placing the questions and events of the day in a moral point of view. 1883 W. JAMES *Let.* 23 Jan. in R. B. Perry *Tht. & Char. W. James* (1935) I. 389 Although from a moral point of view your sympathy commands my warmest thanks, from the intellectual point of view, it seems, first, to suppose that I am a bachelor [etc.]. 1951 C. DAY LEWIS *Poet's Task* 19 As an aesthetic judgement this is so bizarre that one can only take it for a moral judgement. 1972 *Jrnl. Social Psychol.* LXXXVI. 158 The mean moral judgment quotient for girls was slightly higher than that for boys.

2. c. *moral psychology*: psychology concerned with the psychological effect of rules of conduct, esp. the sense of virtue and vice, upon behaviour. Hence *moral-psychological* adj.

1859 J. MARTINEAU in *National Rev.* IX. 504 Does an author, who has so distinguished himself in Logical psychology,..doubt that there is also a Moral psychology? a 1865 J. GROTE *Treat. Moral Ideals* (1876) xiii. 297 Moral psychology endeavours to co-ordinate individual experiences. 1867 MILL *Exam. Hamilton's Philos.* (ed. 3) xxvi. 586 This vital truth in moral psychology, that we can improve our character if we will. 1946 *Mind* LV. 189 His interest in moral psychology and the empirical facts of the moral life remained. 1963 J. WIESENFARTH *Henry James* ii. 51 This is precisely the decision that climaxes her coming to grips with the moral-psychological problem in the novel. 1972 J. RAWLS *Theory of Justice* § 29. 181 The parties must consider the general facts of moral psychology.

3. d. *moral tutor*: a university tutor appointed to have a particular concern for the moral well-being of a student. Hence *moral pupil*.

1932 *Oxf. Univ. Handbk.* I. 130 In some colleges an undergraduate is assigned for all his time to a 'moral tutor', who is often not the tutor to whom he is going for his reading, but one who undertakes to keep in touch with him during his career and to help and advise him generally in his life. 1963 J. I. M. STEWART *Last Tresilians* xiv. 106 That's Leech, the Pro-Provost. He's my moral tutor. 1966 *Rep. Comm. Inquiry Univ. Oxf.* II. 383 Plus £1 per moral pupil other than those for whom a fellow is tutorially responsible. *Ibid.*, Paid to 16 moral tutors. 1969 V. DE S. PINTO *City that Shone* xi. 249 Jackie would continue to be my 'moral tutor', but..my 'English tutor' would be Mr Percy Simpson.

4. Also *moral code, norm, order, rule, system*.

1690 LOCKE *Hum. Und.* I. iii. 15, I think it will be hard to instance any one moral rule. 1927 *Amer. Jrnl. Sociol.* XXXII. 736 The same forces which co-operate to create the characteristic social organization and the accepted moral order of a given society or social group determine at the same time..the character of the individuals who compose that society. 1931 H. S. WALPOLE *Judith Paris* II. ii. 240 A rather blowzy red-cheeked lady, who..had a warm heart but an uncertain moral code. 1951 R. FIRTH *Elem. Social Organiz.* vi. 185 The effective standard of judgement..has appeared to be the recognition of offences against a moral code of behaviour. *Ibid.* 187 The moral rules to be found in different types of society. *Ibid.* 202 Divergence from the Western moral norms is seen much more widely in infanticide. *Ibid.* 213 A moral system.. includes the idea of an elaborate interlocking set of judgements. 1957 P. WORSLEY *Trumpet shall Sound* 250 The role of ritual obscenity as providing an occasion for the statement and reinforcement of moral norms is well known from many societies. 1958 R. C. ANGELL *Free Society & Moral Crisis* ii. 16 The most basic—though not the most tangible—aspect of the moral order is the set of common values that motivates the members of a society. 1970 G. A. & A. G. THEODORSON *Mod. Dict. Sociol.* 264 The various moral systems of the world may include many of the same moral ideas.

7. a. (Further examples.) *moral suasion*: see SUASION 1 b.

1841 THOREAU *Jrnl.* 19 Feb. in *Writings* (1906) VII. 217 It is a moral force as well as he. 1913 *Act* 3 & 4 *Geo. V* c. 28 § 1 Moral imbeciles; that is to say, persons who from an early age display some permanent mental defect coupled with strong vicious or criminal propensities. 1927 *Act* 17 & 18 *Geo. V* c. 33 § 1 Moral defectives, that is to say, persons in whose case there exists mental defectiveness coupled with strongly vicious or criminal propensities and who require care, supervision and control for the protection of others. 1951 R. FIRTH *Elem. Social Organiz.* vi. 213 It is in the capacity to generate and adapt moral force that man derives one of the most potent springs to social action. 1952 R. M. HARE *Lang. Morals* I. iv. 71 If, when we did as we were told, the total effects of our so doing.. were always such as we would not have chosen..then we should..work out our own salvation or become moral defectives. 1967 D. J. WEST *Young Offender* v. 106 A paper ..seeking to prove the moral imbecility of habitual criminals. 1968 *Listener* 26 Sept. 408/1 'Moral insanity' was superseded by 'moral imbecility'; this in turn gave way to 'psychopathic personality' (which had developed out of 'constitutional psychopathic inferiority').

c. *moral support*: support or help the effect of which is psychological rather than physical.

1885 [in Dict.]. 1921 W. S. MAUGHAM *Circle* I. 19 You were rather scared... I thought I'd come and give you a little moral support. 1926 F. M. FORD *Man could stand Up* I. ii. 37 He needed her moral support! 1968 'M. CARROLL' *Dead Trouble* iii. 44 Where were you? Couldn't you have given me some moral support? 1973 J. BURROWS *Like an Evening Gone* i. 7 She had it all worked out. All she wanted from her husband was a little moral support.

d. Similarly *moral fibre*; esp. in phr. *lack of moral fibre*, used *euphem.* for lack of courage. Cf. *LMF* s.v. *L 7.

1884 *Boy's Own Paper* 11 Oct. 18/2 Fanshawe was a boy in whom bad instincts had been nourished by his training, and who, from constant lack of moral fibre, had gradually deteriorated. 1942 [see *GROUNDED *ppl. a.*[1] 8]. 1960 *Times* 11 Jan. 13/6 Fail to register for the cut-price course in foundry management and clearly you are lacking in moral fibre. **1970** L. Deighton *Events Last Flight R.A.F. Bomber* xvii. 263 The R.A.F. authorities..would stamp the words 'lack of moral fibre' across the man's documents ..and send him away..an officially recognized coward. **1973** K. Royce *Spider Underground* iii. 52 She was tough all right, of high moral fibre.

e. *moral turpitude*: conduct considered depraved; an instance of such conduct.

1879 [see Turpitude 1 a]. **1931** J. P. Clark *Deportation of Aliens from U.S. to Europe* v. 165 An alien convicted of stealing $15 in Massachusetts was held subject to deportation because of conviction of a crime involving moral turpitude.

f. *Moral Rearmament*: a religious group founded by the American evangelist Frank Buchman (1878–1961); the beliefs or practices of this group; = *Buchmanism. Abbrev. **M.R.A.* (q.v.). Also *transf.* and *fig.* So *Moral Rearmer.*

1938 *Times* 2 Nov. 6/4 Moral rearmament, which is the true basis of national fitness, is an individual responsibility. **1940** Earl of Athlone et al. *Moral Re-Armament* 3 Moral Re-Armament stands for a change of heart, for that new spirit which must animate all human relationships. **1940** N. Mitford *Pigeon Pie* iii. 55 Luke had been guided to ask a hundred people to dinner..to talk about Moral Rearmament. *Ibid.* iv. 80 If the British people had gone all out for moral rearmament and real appeasement, things need never have reached this pass. **1951** V. A. Demant *What is happening to Us?* iii. 22 The Emperor Augustus tried to save the civilization of Ancient Rome by reviving the old cults..and the poet Horace..wrote his lovely Odes to help in this older version of 'Moral Re-armament'. **1956** A. Huxley *Adonis & Alphabet* 118 For *their* campaigns, the Moral Rearmers have a whole arsenal of Christian *Oklahomas* and ethical *Cats on Hot Tin Roofs*. **1961** [see *Buchmanism]. **1969** *Guardian* 20 June 10/4 Senator Ralph Vibert, an advocate and a Moral Re-Armer. **1973** *Ibid.* 11 June 11/6 We were made to confess all our weaknesses... It was like Moral Re-Armament.

8. b. *moral welfare*: see quot. 1965.

1927 *8th Ann. Rep. Archbishops' Advisory Board Preventive & Rescue Work* 9 The League of Nations Report on the Traffic in Women and Children..has awakened an interest in moral welfare work that is bound to help in arousing the public conscience against prostitution. **1944** A. Thirkell *Headmistress* x. 211 Doing wonders with her Moral Welfare Committee. **1965** Hall & Howes *Church in Social Work* 2 'Moral welfare' is a term which..appears to have come into general use during the inter-war period to designate the social work in relation to sex, marriage and the family.

morale. Delete ‖ and add: **3.** *attrib.* and *Comb.* uses of sense 2, as **morale-booster,** an event, occurrence, or saying which raises one's spirits; also *morale-boosting, -building, -raising* vbl. sbs. and adjs.

1943 J. B. Priestley *Daylight on Saturday* xxx. 234 Grandiose and impracticable schemes of morale-building. **1946** *Nature* 30 Nov. 777/2 Being more formally educational and only indirectly a morale-raising agent, instruction in it could be given not only by officers, but also [etc.]. **1956** C. Cockburn *In Time of Trouble* xvi. 205 The quick-firers of..controversy who..for obvious morale-building purposes prophesied that the current crisis..was the final crisis of American capitalism. **1959** *Punch* 2 Sept. 98/3 To find so many and such diverse people coming out into the open on your side is a superb morale-booster. **1960** *Harper's Bazaar* July 21/2 A beauty salon.. has become a morale-boosting relaxant centre. **1961** *Sunday Express* 12 Mar. 10/3 Another morale-boosting idea for the troops. **1961** *Guardian* 3 May 8/3 The Festival ..was a national morale-booster. **1970** C. Kersh *Aggravations M. Ashe* xi. 158 With that morale-boosting thought to end her day, my mother went to bed. **1970** 'B. Mather' *Break in Line* vi. 71 Even though it was a slip of the tongue..it was quite a morale booster. And Lord knows I needed one. **1971** A. Price *Alamut Ambush* xiii. 155 It was immensely morale-raising to see Mary floor him.

morali·stically, *adv.* [f. Moralistic *a.*: see -ically.] In a moralistic manner; by way of moral judgement.

1890 W. James *Princ. Psychol.* II. xxi. 317 The perfect object of belief would be a God or 'Soul of the World', represented both optimistically and moralistically. **1965** *New Statesman* 30 Apr. 677/2 The historical development of psychoanalysis out of medicine, from which it inherited criteria of normality and health which are much more moralistically loaded when applied to psychological than to somatic therapy. **1972** E. Lucie-Smith *Eroticism in Western Art* iii. 66 *The Golden Age* (otherwise, and more moralistically, dubbed *The Corruption of Men Before the Deluge*).

morality. Add: **8.** *attrib.*, as **morality squad,** in Canada, a municipal or provincial police unit dealing with infractions of legislation concerning prostitution, pornography, etc. So *morality officer,* etc.

1963 J. N. Harris *Weird World Wes Beattie* (1964) x. 119 He called an acquaintance on the morality squad of the Ontario Provincial Police in order to get the low-down on the obscene film racket. **1967** *Canad. News Facts* 4 Dec. 183, 18 arrested on gambling charges... Eight of the men were arrested in Montreal by the Montreal morality squad. **1970** *Toronto Daily Star* 24 Sept. 7/5 Man appears to need some censorship to protect himself from his baser tastes. I trust the morality squad will continue its good work. **1976** *Globe & Mail* (Toronto) 17 Jan. 12/4 Pasquale Zappia..died of a heart attack Thursday while being questioned by an Ottawa police morality officer outside a restaurant.

moralizable (mǫrălai·zăb'l), *a.* [f. Moralize *v.* + -able.] That can be rendered moral or expressed in terms of morality; amenable to moralizing.

1916 F. von Hügel *German Soul* ii. 64 They are not really moral, nor even moralisable. **1961** J. N. Findlay *Values & Intentions* ix. 399 Religion can be seen thus to be intrinsically moralizable.

morally, *adv.* **4.** (Later examples of *morally certain*.)

1869 *Bradshaw's Railway Manual* XXI. 174 It is morally certain that there will be enough net income at the end of next half-year to pay the full year's interest. **1966** 'R. Standish' *Widow Hack* ix. 97 Morally certain that I had not been seen, I returned to my car.

‖ **moran** (mǫ·răn). Also Il-moran. [Masai.] The warrior group of the Masai tribe which comprises the younger unmarried males. Also a member of this group.

1885 J. Thomson *Through Masai Land* (ed. 3) x. 425 A certain rite, better known in Africa than in Europe, was performed; and..he was an El-moran—a warrior. **1902** H. H. Johnston *Uganda Protectorate* II. xix. 822 In former days, before the Masai warriors, called 'El Mórran', started on an expedition, they would fortify their courage with a war medicine, which was said to be the bark of *Acacia verrugosa*. **1905** A. C. Hollis *Masai* 298 Their hair is allowed to grow, and as soon as it has grown long enough to plait, they are called Il-muran (warriors). **1947** *E. African Ann. 1946–7* 43/1 The 'moran' are the warrior age-grade... In the past the 'moran' constituted a free military organization within the tribe. **1960** *Times* 8 Nov. 13/2 One Kikuyu was killed and a score were injured by Masai *moran*. **1975** *Times* 9 Apr. 12/8 The *moran* system of organized adolescence.

morassic (moræ·sik), *a.* [f. Morass + -ic.] Of, pertaining to, or characteristic of a morass; morassy.

1897 C. MacMillan in *Minnesota Bot. Stud.* (1894–8) I. 995 All of these morassic formations characterize quiet shores.

Morasthite (mǫ·răsþəit). Also **Morascite, Morashite, Morashtite, Morastite.** [f. *Moresheth* in Gath, name of the home town of Micah.] Epithet of Micah, the Judæan prophet of the 8th century B.C.

a **1390** Wyclif *Micah* i. 1 The word of the Lord..is maad to Mycheas Morastites. **1535** Coverdale *Jer.* xxvi, Micheas the Morascite, which was a prophet vnder Ezechias kinge of Iuda, spake. *Ibid. Micah* i, This is the worde of the Lorde, that came vnto Micheas the Morastite. **1611** Bible *Jer.* xxvi. 18 Micah the Morashite prophecied in the dayes of Hezekiah king of Iudah. *Ibid. Micah* i. 1 The word of the Lord..came to Micah the Morasthite in the dayes of Iotham, Ahaz, and Hezekiah. **1839** *Penny Cycl.* XV. 172/1 Micah..is called in the title to his prophecy the Morasthite. **1845** J. Kitto *Cyclopedia Biblical Lit.* II. 334/1 It is stated that Micah the Morasthite foretold the destruction of Jerusalem in the reign of Hezekiah. **1904** *Jewish Encycl.* VIII. 533/1 Micah..was a Morasthite; that is to say, a native of Moreshethgath. **1944** G. B. Shaw *Everybody's Political What's What?* xxvi. 231 The noble God of Micah the Morasthite. **1963** Grant & Rowley *Hastings's Dict. Bible* 674 *Morashtite*,—a gentilic used in RV to designate the prophet Micah,..(AV *Morasthite*..); probably derived from Morsheth-gath.

moratorial (mǫrătōə·riăl), *a.* [f. Moratori(um + -al.] Pertaining to or payable in respect of a moratorium.

1914 *Economist* 7 Nov. 833/1 On the debts due to them and covered by the moratorium, they are entitled to a moratorial interest of 5 per cent.

moratorium. Delete ‖ and add: Pl. **-ia, -iums.**

1932 *Sun* (Baltimore) 7 Sept. 8/2 There is little evidence that past Farm Board 'moratoriums' on sales of its commodity holdings, or its quota schemes to sift holdings into the markets in small lots, have buoyed the market for these commodities. **1956** [see sense *2]. **1972** *Times* 27 Jan. 14/4 Those who wanted to agree to moratoria to assist businesses in financial difficulties. **1973** *Nation Rev.* (Melbourne) 31 Aug. 1434/1 Will there be resolutions to the federal or sitdown moratoria in Collins street?

2. A postponement, an agreed delay, a deliberate temporary suspension (of some activity, etc.).

1932 *Sun* (Baltimore) 5 Sept. 1/2 The moratorium on picketing ordered by Milo Reno, of Des Moines, national leader of the strike, seemed virtually achieved when only on two highways near Sioux City were farmers turning back live stock and produce trucks. **1934** W. B. Wolfe *Nervous Breakdown* i. 2 In a nervous breakdown the whole personality declares a moratorium of normal activities, and both body and soul join in a cry for help. **1944** *Word Study* May 7/2 A moratorium on 'of course' sentences. **1956** *Jrnl. Amer. Psychoanal. Assoc.* Jan. 66 Societies offer..more or less sanctioned intermediary periods between childhood and adulthood, institutionalized *psycho-social moratoria*. **1957** *Observer* 25 Aug. 8/2 Their new proposal that nuclear tests should be suspended not for ten months but for two years... Under their plan, the launching of the first year's moratorium would depend on an East-West agreement to a partial disarmament plan covering several fields. **1961** I. Murdoch *Severed Head* xiii. 113 For me they [*sc.* words] constituted rather a kind of moratorium, a momentary neutral zone where I could ..absolutely rest. **1969** *Guardian* 13 Oct. 3/2 Moratorium-day on Vietnam..has so captured the attention that Wednesday is expected to see the most wide-spread political protest in America's history. **1970** *Daily Tel.* 29 Apr. 13/4 Could we not now have a moratorium on Dylan Thomas records? **1972** *Nature* 17 Mar. 94/1 A provision calling for a five-year moratorium on the killing of all ocean mammals.

Moravian, *sb.*[2] and *a.*[2] Add: **A.** *sb.* **1.** (Earlier and later examples.)

1555 [see *Lusatian *sb.* and *a.*]. **1611** Coryat *Crudities* sig. B1 A Noble Scholler..George Haunschildt of Furstenfeldt, a Morauian. **1847** [see Zechian].

B. *adj.* **2.** (Earlier example.)

1739 W. Stephens *Jrnl.* 22 Aug. in *Colonial Rec. Georgia* (1906) IV. 393 An heavy Complaint being exhibited against the Moravian Brethren.

morbid, *a.* **2.** (Further examples.)

1902 W. James *Var. Relig. Exper.* ii. 46 The athletic attitude tends ever to break down, and it inevitably does break down even in the most stalwart when..morbid fears invade the mind. *Ibid.* vi. 163 It seems to me that we are bound to say that morbid-mindedness ranges over the wider scale of experience. **1907** —— *Mem. & Stud.* (1911) 248 Dr. Janet has discussed five cases of morbid impulse. **1918** Kipling *Land & Sea Tales* (1923) 113 My Sub's a morbid-minded young animal. **1922** Joyce *Ulysses* 413 Those who are not so intimately acquainted with the minutiae of the municipal abattoir as this morbid-minded esthete. **1935** *Discovery* Nov. 346/2 The 'misfits' and perverts of morbid psychology.

morbidezza. Add: **2.** *transf.* (esp. *Mus.*): delicacy, softness, sensibility, smoothness; sometimes (degenerating into) unwholesomeness, effeminacy, sickliness.

1890 E. Dowson *Let.* c 31 Mar. (1967) 146 As for Browning!.. The subtility, the tact of omission, the Morbidezza! **1895** *Funk's Stand. Dict.*, *Morbidezza*,.. *Mus.* Extreme delicacy: a direction to the performer. **1910** G. B. Shaw *Brieux* 3 The artistic morbidezza of Byron and Victor Hugo. **1934** *Times Lit. Suppl.* 3 May 324/2 In spite of a collective morbidezza on the part of Pansy's lovers, the characterization is clean cut. **1947** C. Gray *Contingencies* 118 In spite of its *morbidezza* and romantic sentimentality it nevertheless retains the purity of line which is the hall-mark of all classic art. **1960** C. S. Lewis *Studies in Words* vi. 160 But in Mrs Radcliffe *sensibility* perhaps implies a more universal *morbidezza*. **1974** *Times* 13 May 14/7 'E lucevan le stelle' and 'O dolci mani' had true *morbidezza*.

morbidity. Add: **3.** *attrib.*, as *morbidity rate, statistics.*

1939 Kurtz & Edgerton *Statistical Dict. Terms & Symbols* 108 *Morbidity rate*, the ratio of the number of cases of a particular disease in a year to the midyear population, usually expressed per 100,000. **1964** Gould & Kolb *Dict. Social Sci.* 445/1 Frequently suitable criteria are provided by the effects of the illness upon everyday life. This is the only possible approach to aggregating morbidity rates from all conditions. **1971** *Brit. Med. Bull.* XXVII. 11/2 The respiratory morbidity-rates were far in excess of those found in English areas with comparable levels of pollution. **1964** Gould & Kolb *Dict. Social Sci.* 445/1 The main uses of morbidity statistics relate to the study of particular conditions, public health work, the economic aspects of illness, the cost of medical care or social security systems.

morbous, *a.* Add later *literary* example.

1922 Joyce *Ulysses* 412 Organisms in which morbous germs have taken up their residence.

morceau. Add: Pl. **morceaux.**

1767 J. Beattie *Let.* 8 Jan. (1820) I. 39, I dare say Metastasio despises those little *morceaux* of sing-song. **1839** Geo. Eliot *Let.* 22 Nov. (1954) I. 34 What I could wish to have added to many of my favourite morceaux is an indication of less satisfaction in terrene objects. **1889** J. K. Jerome *Three Men in Boat* viii. 119 We played *morceaux* from the old German masters. **1960** *Times* 29 July 13/2 Surely it would be easy to intersperse congruent *morceaux* among lyric poems.

b. morceau de musée (də müze): a museum piece; **morceau de salon** (də salon): a well-known tune elaborated by variations, etc.

1964 *English Studies* XLV. 237 Leaving behind the artifact, the morceau de musée, 'Medallion'. **1886** F. Hume *Mystery of Hansom Cab* (1887) viii. 41 When the gentlemen entered the drawing-room a young lady was engaged in playing one of those detestable pieces of music called *Morceau de Salon*, in which an unoffending air is taken and variations embroidered on it till it becomes a perfect agony to distinguish the tune and the perpetual rattle of quavers and demi-semi-quavers. **1906** A. Bennett *Whom God hath Joined* v. 185 A *morceau de salon* of Chaminade's stood on the open piano... Not in Lawrence's house would such examples of the brilliant secondrate have been found conspicuous. **1947** A. Einstein *Mus. Romantic Era* xv. 209 From his [*sc.* Liszt's] pen there are no 'Scenes from Childhood', though there is a series—unfortunately too extensive—of *morceaux de salon*.

morcellated, *ppl. a.* Add: Also *fig.*

1953 E. L. Mascall *Corpus Christi* iii. 57 It is perfectly possible for the parish communion to result in a far more morcellated and fragmented conception of the Church and of the liturgy than can be produced in any number of 'private masses'. **1964** *Welsh Hist. Rev.* II. 52 The centrifugal tendencies inherent in Welsh tribal life and custom did more than anything else to perpetuate the severely-morcellated political system.

morcellation. Add: (Further examples.) Also *fig.*

1968 J. Howkins *Shaw's Textbk. Operative Gynæcology* (ed. 3) xvi. 306/2 The expert at vaginal hysterectomy can

convert the difficult operation into a relatively easy one.. by removing the obstructing myoma by morcellation and vaginal myomectomy. **1973** *Amer. Jrnl. Obstetr. & Gynecol.* CXVI. 258/2 Morcellation was, and is, done when indicated by the size and extent of uterine tumors and when malignancy does not exist. **1975** E. L. Mascall in *Critique 'Eucharistic Agreement'* v. 70 Merely to state the question in this way..presupposes an artificial and perverse separation between Christ and his members, a morcellation or fragmentation of the body of Christ.

morcellement. Add: **b.** *Surg.* = Morcellation (in Dict. and Suppl.).

1903 *Philadelphia Med. Jrnl.* 31 Jan. 199/1 Porcellini suggested morcellement, the danger of which consists in the subsequent necrosis of the stump of the cervix. **1919** F. E. Leavitt *Operations of Obstetrics* xi. 239 When there are tumors to deal with, like the outgrowths in the region of the buttocks and neck, their removal by morcellement must be undertaken. **1954** W. Shaw *Textbk. Operative Gynæcology* vi. 63/1 By processes of morcellement and splitting of the uterus in the midline it is possible to remove relatively large myomata by the vaginal route.

mordant, *a.* Add: **4. b.** Of a dye: becoming fixed on the fibre as a result of forming an insoluble compound with a mordant.

1902 *Encycl. Brit.* XXVII. 559/2 Employed by themselves, Mordant Colours are usually of little or no value as dyestuffs, because..either they are not attracted by the fibre..or they only yield a more or less fugitive stain. Their importance and value as dyestuffs are due to the fact that they act like weak acids and have the property of combining with metallic oxides to form insoluble compounds termed 'lakes', which vary in colour according to the metallic oxide or salt employed. **1917** Fort & Lloyd *Chem. Dyestuffs* xiii. 112 Acid mordant dyes may be first dyed on wool like acid dyes and then after-chromed. **1940** *Thorpe's Dict. Appl. Chem.* (ed. 4) IV. 127/2 Mordant dyes rank amongst the oldest dyes used by mankind for colouring purposes. **1963** [see *after-chrome *a.*]. **1965** E. Gurr *Rational Use of Dyes in Biol.* i. 115 Since sun yellow does not contain a hydroxyl group it cannot be classified as a mordant dye.

‖ **mordida** (mordi·dă). [Central-Amer. and Mex. Sp.] A bribe; an illegal exaction (in Mexico, etc.).

1940 *Life* 2 Dec. 102 In Mexico, and throughout Latin America, it is next to impossible to make headway with petty officials without constant applications of the *mordida*, which literally translated means 'the bite', or bribe. **1946** M. Lowry *Let.* 15 June (1967) 101 The defection was the original defection of my failing to pay the 'mordida'. **1967** *Punch* 18 Jan. 102/1 Just go see my friend, Sr. G., at *Gobernacion*... He's such a good friend of mine that you'll not even have to pay a *mordida*.

Mordvin (mǭ·ɹdvin). Also **Mordv, Mordvian, Mordvine, Mordvinian.** [Russ.] **a.** A member of a Finnish people inhabiting the region of the middle Volga. **b.** The Finno-Ugric language of this people. Also *attrib.* or as *adj.* So **Mo·rdva,** this people collectively.

1736 tr. *P. J. von Strahlenberg's Historico-Geogr. Descr. N. & E. Europe & Asia* xiii. 412 Morduini, Are Pagans, in Russia, who live under the Government of Nischneygorod. **1800** W. Tooke *Hist. Russia* I. i. 19 *The Mordvines*, called by themselves Moræ, and formerly by the Russians Meren. **1802** F. W. Blagdon tr. *Pallas's Trav. S. Provinces Russian Empire* I. 34 The Mordvines..are more solicitous to preserve the forests in the countries they inhabit. **1879** *Encycl. Brit.* VIII. 700/1 Finnic or Ugrian represented by..Mordvinian. *Ibid.* IX. 219/2 The Volga Finns include..the Mordvinian, divided into small communities on both banks of the Volga. **1883** *Ibid.* XVI. 45/2 Even now part of the Mordvinians (of Finnish origin) call themselves Meschers. *Ibid.* 813/2 Mordvinians, more correctly Mordva, or Mordvs, are a people numbering about one million. **1898** R. Brown *Semitic Influence in Hellenic Mythol.* iii. iii. 89 Chaïtan (= Arabic Shaitan, Heb. Satan) appears in the Mordvinian Pantheon. **1925** P. Radin tr. *Vendryès's Lang.* iii. 118 In Chermish and Mordvian, the -*t* forms both the plural of nouns and the third person plural of verbs. **1931** K. Breshkovskaia *Hidden Springs of Russian Revolution* xxiv. 287, I talked also to the Cheremisses and the Mordva. **1933** [see *Cheremis(s)]. **1939** L. H. Gray *Found. Lang.* xii. 369 The languages of the Uralic family are..*Finnish* group: Finnish proper.., Lapp.., Cheremiss, and Mordvin. **1942** K. W. Deutsch in J. A. Fishman *Readings Sociol. of Lang.* (1968) 601 Of the 13 remaining nationalities, nine have formed administrative units on a national, linguistic basis with various degrees of political self-government within the European part of the U.S.S.R.: the Bashkirs, Chuvashs, Cheremiss, Mordvins, [etc.]. **1944** B. H. Sumner *Survey Russian Hist.* i. 23 The Mordva..centred around the junction of the Volga and the Kama. **1960** E. R. Goodman *Soviet Design for World State* ix. 275 Examining any one of the languages of the Soviet Union, e.g., Mordvinian, one is shocked by the discovery that it swarms with Russian words, and that only the suffixes are Mordvinian. **1971** P. Longworth *Cossacks* ii. 55 There was a ruling class of semi-nomadic Tatars and the primitive Ostyaks, Voguls and Mordvins paid them tribute. **1972** W. B. Lockwood *Panorama of Indo-European Lang.* 152 The chief [languages] are Estonian.., Mordvin (1 m[illion], Mordva ASSR).

more, *a.* (*sb.*) and *adv.* Add: **A.** *adj.* **1. h.** (Later examples of (*the*) *more fool* followed by a pronoun.)

1917 D. H. Lawrence *Phoenix II* (1968) 69 'What have you done now?—lost more money?' 'Three thousand marks.' She was silent in deep wrath. 'More fool you!' she said. **1971** E. Fenwick *Impeccable People* iv. 25, I had to go and feel sorry for this Iversen piece, more fool me.

2. a. Phr. *more* (something) *than one cares* (or *likes*) *to think about.*

1894 Somerville & 'Ross' *Real Charlotte* I. xi. 152 He happened to have lost more money at the Galway races than he cared to think about. **1935** E. Glasgow *Let.* 12 July (1958) 188 For more years than I like to think of (a silly figure of speech, for I am not afraid of the years) I have battered my nerves against human cruelty.

3. a. Phrs. *more days, more dollars; more ships than parish churches.*

1843 J. F. Cooper *Ned Myers* II. ii. 40, I knew there were more ships than parish churches. **1898** A. J. Boyd *Shellback* xv. 258 The captain was acting on the principle of 'More days, more dollars'. **1946** W. McFee *In First Watch* i. 28 There were more ships than parish churches. **1962** Granville *Dict. Sailors' Slang* 78/1 *More days: more dollars!* American Merchant seamen's phrase of the Second World War, meaning that the more days they were at sea the more 'danger money' they would receive. The term was used sarcastically by British Naval-men whose pay compared unfavourably with the Americans'.

B. *absol.* and quasi-*sb.* **4. h.** Phr. (*there's*) *more where it* (or *that, this,* etc.) *came from*: eat or use freely, do not be sparing of (a food, commodity, etc.).

1810 J. Porter *Scottish Chiefs* II. v. 92, I..told him not to spare it, it was a chilly night, and I should get more where it came from. **1854** Mayne Reid *Young Voyageurs* xix. 250 He only looks after his lost root with an air of chagrin, and then, reflecting that there is 'plenty more where it came from', kicks up his heels, and once more plunges to the bottom. **1954** R. Bissell *High Water* xxii. 270 Plenty more where this came from.

C. *adv.* **1. c.** Esp. in phr. *I* (or *you,* etc.) *couldn't be more* ——: I, etc., am extremely ——.

1939 N. Marsh *Overture to Death* xvii. 188 'I thought,' flashed Dinah, 'that nowadays the C.I.D. was almost a gentleman's job.' 'Oh, no!' said Alleyn. 'You couldn't be more mistaken.' **1951** —— *Opening Night* vi. 139 'I couldn't be more sorry,' Percival said weakly. **1954** 'N. Blake' *Whisper in Gloom* i. iii. 49 This party couldn't be more heaven, don't you think? **1971** P. D. James *Shroud for Nightingale* i. 12 She could hardly have been more wrong.

2. *more or less*: also used *attrib.* (usu. with hyphens).

1963 *Language* XXXIX. 460 As far as the bees are concerned, it is clear that their behavior resembles that of an analog computer.., that is, a control machine of the more-or-less type, and not at all of a digital computer, of a yes-or-no type. **1964** M. A. K. Halliday et al. *Linguistic Sci.* v. 124 But this, like synonymy, is a 'more or less' not a 'yes or no' relation. **1975** *Country Life* 13 Feb. 386/2 Every more-or-less sound horse was purchased for the Expeditionary Force by Civilian Re-mount Purchasing Officers.

4. a. Add to def.: The phrase *any more* (freq. written as *anymore*) is also used in affirmative as well as negative contexts in the sense 'now, now-a-days, at the present time; from now on'. *dial.* (chiefly *U.S.*).

1898 *Eng. Dial. Dict.* I. 63/1 A servant being instructed how to act, will answer 'I will do it any more'. **1903** *McClure's Mag.* Dec. 215/1 There's just only this one any more. **1920** D. H. Lawrence *Women in Love* (1921) xiii. 159 'Quite absurd,' he said. 'Suffering bores me, any more.' **1938** J. Stuart *Beyond Dark Hills* vi. 120 They tell me this Macey plant only hires the best of men any more. *Ibid.* x. 319 You know, Jesse, any more I don't worry a great lot. **1973** *Capital Times* (Madison, Wis.) 14 Mar. 2/1 Any more, the difference between a white collar worker and a blue collar worker is simply a matter of shirt preference.

5. a. more'n, mor'n *colloq.* (orig. *U.S.*) = more than; also *ellipt.* = no more than.

1848 J. R. Lowell *Biglow Papers* 1st Ser. 23, I felt a thing go thru my leg,—'t wuz nothin' more'n a skeeter! **1885** W. L. Alden *Adv. Jimmy Brown* 142 They say that Squire Meredith and Deacon Willets are mornhalf eaten up by mosquitoes. **1886** *Harper's Mag.* July 323/2 But he (mor'n you and I with all of our might) Could not here always remain. **1903** K. D. Wiggin *Rebecca of Sunnybrook Farm* 8 We've only just started on it,..it's more'n two hours. **1952** E. Wilson *Equations of Love* 196, I know a lot more'n I did. **1973** E. Berckman *Victorian Album* 47, I don't know any more'n I've told you.

c. *Comb.* with following *adj.* or *sb.*

1923 D. H. Lawrence *Birds, Beasts, & Flowers* 99 Cats, and the Neapolitans, Sulphur sun-beasts, Thirst for fish as for more-than-water. *Ibid.* 141 Why do you arch your naked-set eye with a more-than-comprehensible arrogance? **1924** H. E. Palmer *Gram. Spoken Eng.* p. xxvii, The words *trees, towns, boys,* form an association-group through having the..meanings 'more-than-oneness' in common. **1938** R. Graves *Coll. Poems* 150 Nor could a compromise be found Between the giver's thoughtlessness And his own more-than-thanklessness. **1950** *Scott. Jrnl. Theol.* III. 367 Nyberg says it is clear from the beginning that the Servant is a more-than-individual (*överindividuell*) figure, notwithstanding that the description moves throughout on the individual plane. **1961** *Times* 24 Oct. 13/1 The Official Image of Picasso..may well be a more-than-life-size..figure.

‖ **more** (mōə·re). [L., abl. of *mos* manner, custom: see *mores.] In phrases: chiefly with *poss. adjs.*, as **more suo,** in his own fashion, **more meo,** in my own fashion; and with national *adjs.*, as **more hispanico, turcesco,** in accordance with Spanish (Turkish) customs. Also **more majorum,** in traditional manner.

1600 Holland tr. *Livy's Romane Hist.* xxvi. 585 Hee might proceed in the suite at his owne good pleasure, *more maiorum*. i. [according to the auncient manner used by their forefathers]. **1612** T. Shelton tr. *Cervantes's Hist. Don Quixote* iii. vi. 176 Hee spoke to his Lord with his Cap in his hand, his head bowed, and his body bended (*more Turcesco*). **1713** Addison in *Guardian* 3 July 2/1, I have..been drawn in to tattle of my self, *more Majorum*, almost the length of a whole *Guardian*. **1810** Southey *Let.* 27 Sept. (1856) II. 203 The preface, *more meo*, is short and explicit. *c* **1842** H. James *Let.* in R. B. Perry *Tht. & Char. W. James* (1935) I. 43 Let him try, and above all let him forgive *more suo* all my botherings. **1865** Mill *Exam. Hamilton's Philos.* xviii. 358 In a later part of his Lectures ..he, *more suo*, forgets this definition. *a* **1937** J. L. Stocks *Reason & Intuition* (1939) xiv. 212 What Mill, *more suo*, offers us in the *Logic* is a compromise or middle position. **1943** G. Brenan *Spanish Labyrinth* vii. 154 The workmen ..shot the offending mayor and, *more hispanico*, cut off his head and those of the police who had been killed in the fight and paraded them around the town.

Morea, Moria, varr. *Muria *sb.* and *a.*

moreish, *a.* (Later examples of form *morish*.)

1905 *Dialect Notes* III. 63 That tastes morish. **1971** *Times* 3 Dec. 12/4 Home-made, thin ginger biscuits are very 'morish' they slip down so easily you cannot stop munching them.

Morellian (more·liăn), *a.*[2] [f. the name of Giovanni *Morelli* (1816–91), Italian patriot and art critic, + -an.] Of, pertaining to, or characteristic of the critical method of Morelli, which introduced a new, systematic approach to art criticism, insisting esp. on the study of detail as a guide in the attribution of paintings. So **Morellian** *sb.*, a follower of Morelli's critical method; **More·llianism,** the technique of art criticism employed by Morelli.

1890 *Academy* 3 May 306/3 It would no more be possible ..for the art-historian to go back to a pre-Morellian state of things in criticism. **1916** B. Berenson *Study & Criticism of Italian Art* 3rd Ser. 34 Morellianism, surgical, pitiless, iconoclastic even as it seemed, was yet inspired by the Romantic ideal of genius. *Ibid.* 35 We Morellians fought for authenticity with the uncompromising zeal of Legitimists. **1934** *Times Lit. Suppl.* 21 June p. xv/1 The persistence of Morellianism is naturally more apparent in the strictly historical literature of art than in its criticism. **1942** *Burlington Mag.* Oct. 241/1 In this country there are small conventicles, in which foregather members of the strictest sect of the Morellians, where hard-shell elders impart to neophytes the narrow precepts of the old Morellian creed. **1968** *Encycl. Brit.* XV. 835/2 The so-called 'Morellian method' was followed in this and his later *Critical Studies*.

more'n: see *more *adv.* 5 a.

‖ **morena**[2] (morē·nă). Also **marena.** [Native name.] The title given to a chief in Lesotho; hence a respectful form of address.

1835 A. Smith *Diary* 29 Aug. (1940) II. 186 They walked boldly in though they saw the Matabeli, saluted all the white people with '*Moron*' and then went towards the Matabeli fire. **1861** E. Casalis *Basutos* ii. xii. 214 The Basutos give to the princes who govern them the title of *Morena*... *Morena*..signifies, He who watches over the public safety and welfare. **1912** J. C. Macgregor tr. *D. F. Ellenberger's Hist. Basuto* 292 One is agreeably surprised to see..so much politeness, and so many rules of etiquette ..faithfully observed... The chiefs often addressed their subjects as *marena* ('chiefs'), *benghali* ('my masters'). **1923** E. A. T. Dutton *Basuto of Basutoland* vii. 67 You meet a man on the road; he may simply say, 'Lumela, Morena' (*Good morning, Chief*), and to this the reply is a drawn-out 'Eh!' **1953** P. Lanham *Blanket Boy's Moon* i. ii. 17 The Chief is highly respected by his subjects, who call him, *Morena*. **1974** *Sunday Times* (S. Afr.) 1 Sept. 3 If it is known, call him by his name and surname and add 'mister' (Morena), particularly if you are in his homeland.

morencite (mōə·re·nsəit). *Min.* [f. *Morenc-i*, the name of a village in Arizona, U.S.A. + -ite[1].] A variety of nontronite with part of the ferric iron replaced by magnesium, $(Fe, Mg)_2Si_4O_{10}(OH)_2$, which occurs as brownish yellow seams in lime shale.

1904 Lindgren & Hillebrand in *Amer. Jrnl. Sci.* CLXVIII. 457 Considering that the mineral is not a mixture, but optically well individualized, we have..thought it best to designate it by the name morencite, derived from the locality in which it was found. **1928** *Ibid.* CCVIII. 12 There is no good reason for separating müllerite, morencite, and hoeferite from nontronite on account of their lower water content. **1935** *Amer. Mineralogist* XX. 475 Nontronites including morencite, pinguite, faratsihite, stilpnochloran and chloropal are like montmorillonite in structure.

morenosite (more·nosəit, mōərenōu·zəit). *Min.* [ad. Sp. *morenosita* (D. A. Casares 1851, in *Revista Minera* II. 176), f. the name of *Moreno*, 19th-cent. Spaniard: see -ite[1].] A hydrated nickel sulphate, $NiSO_4.7(\text{or } 6{\cdot}5)H_2O$, which occurs as green orthorhombic crystals and is formed by the oxidation of nickeliferous sulphides.

1868 J. D. Dana *Syst. Min.* (ed. 5) 648 (*heading*) Morenosite. **1901** M. F. Heddle *Mineral. Scot.* I. 24 [Millerite

occurs] with rosettes of brilliant green Morenosite in Mesitine Spar. **1949** *Amer. Mineralogist* XXXIV. 193 On standing in the open in dry air, crystals of morenosite generally dehydrate rapidly to the tetragonal hexahydrate, retgersite. **1962** *Mineral. Abstr.* XV. 458/2 The Valtournache morenosite with 6·5 H_2O has the same structure as $NiSO_4.7H_2O$.

Morenu (morē·nu). Also **morenu.** [Heb.] A title of honour conferred on rabbis and Talmudic scholars.

1650 E. Chilmead tr. *L. Modena's Hist. Rites of Jews* ii. iii. 69 In Germany, and Italy, they are to be honoured.. with the Title..of..Morenu,..that is to say, Master... These men, that is to say, the Cacham, Rab, or Morenu, decide all Controversies concerning the Things that are either Lawful or Prohibited, and all other Differences. **1733** tr. *Picart's Ceremonies & Relig. Customs* I. ii. iii. 46 The Cacham Rau, or Morenu, determines all Manner of Debates..and passes Judgement upon all Religious Concerns. **1905** *Jewish Encycl.* IX. 15/2 *Morenu*, (lit. 'our teacher'). Term used since the middle of the fourteenth century as a title for rabbis and Talmudists. *Ibid.* 16/1 It can not be determined definitely why a special title was applied to rabbis, or why the term 'morenu' was chosen. **1962** B. Abrahams tr. *Life of Glückel of Hameln* ii. 18 Rabbi David Hanau, the great scholar who bore the distinguished title 'Morenu'.

mores (mōə·rīz). Normally const. as *sb. pl.* [a. L. *mores* pl. of *mos* manner, custom.] **1.** Those acquired customs and moral assumptions which give cohesion to a community or social group, the contravention or rejection of which produces a reaction of shock and outrage. Also *attrib.*

1907 W. G. Sumner *Folkways* 30 The mores are the folkways, including the philosophical and ethical generalizations as to societal welfare which are suggested by them, and inherent in them as they grow. *Ibid.* 37 The Latin word 'mores' seems to be..convenient..as a name for the folkways with the connotation of right and truth in respect to welfare, embodied in them. **1927** J. Dowd *Negro in Amer. Life* xxi. 154 The adjustment of the South to her new educational problem is only one of the many adjustments incident to her passing from the mores of slavery to the mores of freedom; and any sociologist knows that it is impossible to change the mores of a people suddenly. **1948** *Mind* LVII. 511 It frequently happens that any individual variations from cultural traits cause great indignation. When this is so the cultural traits in question are called *mores-traits*. **1958** J. K. Galbraith *Affluent Soc.* xviii. 200 Television and the violent mores of Hollywood and Madison Avenue must contend with the intellectual discipline of the school. **1958** *Spectator* 1 Aug. 167/1 It is surely a mistake for a nation of independent character and traditions to ape the different and often alien *mores* of another. **1967** *Ibid.* 30 June 765/1 The law, in the shape of the jury, is the custodian of our *mores*. **1970** G. Greer *Female Eunuch* 171 Perhaps the pop revolution..has had a far-reaching effect on sexual *mores*. **1975** *Verbatim* Feb. 4/2 The fact of the matter is that mores, politics, and sentiment change.

2. *Ecology.* The habits, behaviour, etc., of a group of animals of the same kind; also, occasionally, the group itself. Cf. *niche *sb.* 3 c. Also *attrib.*

1911 V. E. Shelford in *Biol. Bull.* XXI. 146 Ecological succession is based upon physiology, habits, behavior, mode of life, and the like, which I have proposed to call mores (opposed to the term form). *Ibid.* 147 It is of course recognized that within rather uncertain limits the mores of a morphological species remain, in a general way, the same throughout its geographic range. **1913** —— *Animal Communities Temperate Amer.* ii. 33 Ecology..considers physiological life histories primarily in nature, for this reason the central problem of ecology is the *mores* problem. **1954** A. M. Woodbury *Princ. Gen. Ecol.* xi. 197 Ecologically an individual may be classified under two or more headings in which the unit is a mores. For example, an insect may pass through an egg stage, a larval stage, a pupal stage, and an adult stage, each of which would represent a mores, an ecological unit in a community. **1962** H. Hanson *Dict. Ecol.* 229 *Mores*, the general behavioural attributes of motile organisms, or groups of animals possessing particular ecological characteristics. **1973** P. A. Colinvaux *Introd. Ecol.* viii. 117 The fish had different physiologies, habits, behavior, and modes of life, a collection of parameters which he [*sc.* Shelford] called the mores of the animals... His word mores, which has an awkward and alien sound, disappeared, but the idea he was seeking to express by it later found an outlet in Elton's 'niche' and became a central ecological concept.

moresca: see Moresco *sb.* 3 (in Dict. and Suppl.).

moresco, *sb.* **3.** For def. read: An Italian dance to which the English morris dance is related. Add further examples of *moresca*. Cf. *morisca, moriska.

1933 E. K. Chambers *Eng. Folk-Play* 150 A *Moresca* or *Morisco* first appears in the fifteenth century used as *intermedii* in the courtly *ludi* of Italy, Burgundy, and France. **1947** A. Einstein *Mus. Romantic Era* xvii. 293 Even the separate regions..began to make themselves heard: in Italy around 1535, for example, the *Canzon villanesca alla napoletana*, or the *Moresca* in the south of the peninsula. **1959** *Collins Music Encycl.* 437/1 *Moresca*, ..a dance of the 15th and 16th cent., most often a sword dance in which a fight between Christians and Mohammedans was represented.

Moreton Bay (mǭ·ɹtən bēⁱ). Also **Morton Bay** (only in sense b). The name of a bay in Queensland, Australia (orig. *Morton*). **a.** Used *attrib.* to designate several trees native to the surrounding region, esp. **Moreton Bay ash (-tree),** *Eucalyptus tessellaris*, of the family Myrtaceæ; **Moreton Bay chestnut,** an evergreen tree, *Castanospermum australe*, of the family Leguminosæ, bearing racemes of yellow flowers; **Moreton Bay fig (-tree),** *Ficus macrophylla* or *F. platypoda*, of the family Moraceæ; **Moreton Bay pine,** the hoop pine, *Araucaria cunninghamii*, of the family Pinaceæ.

1847 L. Leichhardt *Overland Expedition* 75 (Morris), The Moreton Bay Ash (a species of *Eucalyptus*)..was here also very plentiful. **1908** E. J. Banfield *Confessions of Beachcomber* i. i. 15 Moreton Bay ash and other eucalypts. **1944** F. Clune *Red Heart* 64 Here he saw a Moreton Bay ash tree blazed with the letter L, made on Leichhardt's 1846 expedition. **1884** A. Nilson *Timber Trees N.S.W.* 40 C[*astanospermum*] *australe*—Moreton Bay Chestnut; Bean Tree.—A beautiful tree, attaining a height of 130 feet and a diameter of 5 feet. **1908** E. J. Banfield *Confessions of Beachcomber* ii. i. 259 One of the chief vegetable foods of the blacks is the fruit of 'tindaburra' (Moreton Bay chestnut—*Castanospermum australe*). **1947** J. C. Rich *Materials & Methods of Sculpture* x. 286 *Blackbean* is an Australian wood, also referred to as Moreton Bay Chestnut. It is hard and heavy and resembles teakwood. The color is light to dark brown with lighter colored streaks. **1969** T. H. Everett *Living Trees of World* 193/2 Very handsome lumber, resembling walnut and esteemed for cabinet work and furniture, is obtained from the Moreton Bay chestnut. **1883** *Encycl. Brit.* XX. 173/2 The Moreton Bay fig-tree has immense wall-like abutments. **1890** 'R. Boldrewood' *Miner's Right* xliv. 380 (Morris), The..venerable church with its alleys of araucaria and Moreton Bay fig-trees. **1917** 'H. H. Richardson' *Fortunes R. Mahony* iv. ix. 362 Mahony turned in and drove past exotic firs, Moreton Bay figtrees and araucarias. **1933** D. G. Stead *Tree Bk.* xxii. 89 A species with large glossy leaves is the well-known Moreton Bay Fig, which..is so hardy. **1943** K. Tennant *Ride on Stranger* (1968) viii. 86 She was particularly fond of the Domain... Here were great avenues of Moreton Bay figs; great bulks cold as stone with loose-wrinkled grey skin for bark and roots. **1965** *Austral. Encycl.* IV. 58/2 The Moreton Bay fig (*F*[*icus*] *macrophylla*) is extensively planted as a shade and ornamental tree, but in its wild state is not very common. *Ibid.* 59/1 The small-leaved Moreton Bay fig (*F. platypoda*) is the most widely distributed of all the Australian species. **1838** J. C. Loudon *Arboretum et Fruticetum Britannicum* IV. 2444 The Moreton Bay pine is found, as the name imports, on the shores of Moreton Bay: it has also..a range of 900 miles..on the eastern coast of New South Wales. **1884** [see *hoop-pine* (*hoop *sb.*[1] 13 b)]. **1911** E. M. Clowes *On Wallaby* ix. 251 Queensland! the home of the red pine and the kauripine, the red cedar, the Moreton Bay pine, and black-bean; and nine men to guard the interests of all this wealth! **1923** Dallimore & Jackson *Handbk. Coniferæ* 155 Moreton Bay Pine... A tree 150 ft. or occasionally 200 ft. high, with a girth up to 12 ft.

b. *slang.* (See quots.)

1953 Baker *Australia Speaks* v. 134 *Morton Bay* (*fig*), any witness who lays an information, anyone who unwarrantably attends to or meddles in the affairs of others; by rhyme on *gig*,..and which may be a contraction of *fizgig*, an informer. **1966** —— *Austral. Lang.* (ed. 2) xii. 267 *Moreton Bay*, a trickster's victim or *gay*.

Morgan[3] (mǭ·ɹgən). *U.S.* The surname of Justin Morgan (1747–98), American teacher, used *attrib.* or *absol.* to designate a breed of light, thickset horse developed in New England from the progeny of a stallion owned by him.

1843 *Knickerbocker* XXI. 331 The stage generally being able to work its own way, drawn by horses of the Morgan breed. **1849** *New England Farmer* I. 314 There has never been a stock of horses in New England which has proved so generally useful as the Morgan stock of the original Morgan horse, raised by Justin Morgan, of West Springfield, Mass., in 1793. **1869** C. L. Brace *New West* xiv. 187 Each coach well made and comfortable, with six horses, evidently picked Morgans. **1876** *Rep. Vermont Board Agric.* III. 172 You can teach a Morgan colt anything. **1884** J. Hay *Bread-Winners* v. 77 If you don't want to talk, a train of Morgan horses couldn't make you. **1906** W. Churchill *Coniston* 22 The tough little Morgans of that time..have all but disappeared. **1913** J. London *Valley of Moon* xxi. 516 They call her Ramona—some Spanish name: sired by Morellita out of genuine Morgan stock. **1923** R. Frost *Selected Poems* 4 A little Morgan had one forefoot on the wall, The other curled at his breast. **1934** *Dict. Amer. Biogr.* XIII. 183/1 Before the middle of the nineteenth century Morgan horses had become a distinct type or breed, famed throughout the country for their attractive appearance and their endurance, docility, and utility as driving, riding, cavalry, stage, and general-purpose horses. **1973** *Washington Post* 13 Jan. F1/7 (Advt.), Must sell this w[ee]kend.—Dark bay Morgan gelding, 15·2 h, good temperament, excel. hunter prospect.

morgan[4] (mǭ·ɹgən). *Genetics.* [named after Thomas Hunt *Morgan* (see next).] A unit of the relative distance on a chromosome between two linked genes, defined in terms of the frequency of crossing-over between them so that the distance in morgans between two genes is equal to this frequency when they are close enough together for the effect of multiple crossing-over to be negligible. (Sometimes used for one hundredth of this unit.)

1919 J. B. S. Haldane in *Jrnl. Genetics* VIII. 305 Let *x* be the distance between the loci of two factors, *y* their cross-over value, and let the unit of distance be chosen so that when *y* is sufficiently small *x* becomes equal to *y*... The value of *y* for a given value of *x* is the sum of the probabilities of all odd numbers of cross-overs. $\therefore y = ..\frac{1}{2}(1 - e^{-2x})$... It is suggested that the unit of distance in a chromosome as defined above be termed a 'morgan', on the analogy of the ohm, volt, etc. Morgan's unit of distance is therefore a centimorgan. *Ibid.*, The cross-over values ·412 and ·236 correspond, according to curve (*c*),.. to distances of ·549 and ·261 morgans respectively. **1965** J. A. Serra *Mod. Genetics* I. vii. 260 Map distances between gene loci are the cross-over values expressed as percentages of recombination for small intervals. Thus, 1% of recombination is often called one cross-over unit..or a centimorgan... One morgan is equal to 100 centimorgans. [*Note*] Some authors call a cross-over unit one morgan. **1971** Levitan & Montagu *Textbk. Human Genetics* ix. 376 Each morgan is divisible into 100 centimorgans, each centimorgan corresponding to the map distance between two loci showing 1 per cent crossing over between them. For close loci the crossover percentage can be translated directly into centimorgans, but this is not possible for more distant loci, especially those more than 30 centimorgans apart. **1971** *Nature* 24 Dec. 475/1 Pedigree studies by Lawler and her co-workers established a close linkage (0·03 morgans) between *Rh*, the rhesus set of blood group loci, and one of the loci, El_1, that can lead to elliptocytosis of the red cell.

Morganism (mǭ·ɹgăniz'm). *Biol.* [f. the name *Morgan* + -ism.] Mendelian genetics, incorporating a theory of the gene that came to be generally accepted, as propounded by Thomas Hunt Morgan (1866–1945), U.S. geneticist and zoologist. Hence **Mo·rganist** *a.*

1934 H. J. Muller in *Pamyati V.I. Lenina* (Akad. Nauk SSSR) 572 The great bulk of the facts of real significance, subsequent to 1911, and practically all after 1913, were found by the younger workers quite independently of any guidance from him, in experiments which they had planned... Their results and interpretations were, however, later accepted by Morgan and presented chiefly by him to the scientific and lay public, so that these developments have sometimes been referred to, especially in circles farthest removed from contact with the original work, as 'Morganism'. **1950** A. Huxley *Themes & Variations* 167 The canonization of Lysenko and the anathema pronounced on 'reactionary Morganism'. **1966** E. A. Carlson *Gene* xi. 95 The article [published by H. J. Muller in 1934]..dissociates 'Morganism' from the gene concept developed at the Columbia laboratory. *Ibid.*, The controversy..eventually erupted into a clear cut 'Lysenkoist' camp and a formal genetics camp, disparagingly referred to as 'Morganist'.

morganite (mǭ·ɹgănəit). *Min.* [f. the name of John Pierpont *Morgan* (1837–1913), U.S. financier + -ite[1].] A pink lithian variety of beryl which is prized as a gem.

1911 *Amer. Jrnl. Sci.* CLXXXI. 81 Dr. George Kunz described some new and remarkable gems which had been cut from a rose-colored beryl found in Madagascar. He proposed the name morganite for them in honor of Mr. John Pierpont Morgan of New York City. **1955** Brown & Dey *India's Mineral Wealth* (ed. 3) xv. 598 The lovely gem known as morganite is a pink or rose beryl which owes its colour to the presence of small amounts of lithium. **1959** [see *heliodor]. **1968** *Rocks & Minerals* XLIII. 651/2 Morganite is a pink variety of transparent beryl. A few samples of this rare type have been found at Poona but not associated with the emerald workings.

morganize, *v.* (Examples.)

1832 *Amer. Rail Road Jrnl.* I. 165/3 Let him be *Morganized*, and his work suppressed by burning. **1923** J. Manchon *Le Slang* 197 *Morganize*, v. (amér.), faire disparaître un témoin gênant.

Hence **morganiza·tion,** the fact or process of getting rid of or destroying by secret methods.

1919 J. L. Garvin *Econ. Foundations of Peace* 502 When there were fears about the 'Morganisation' of British vessels, he said that it would be sounder for Britain to nationalise her entire shipping.

morgue[2]. Add: **2. a.** *slang.* In a newspaper office, the collection of material assembled for the future obituaries of persons still living.

1903 E. L. Shuman *Practical Journalism* 103 This can be done with the aid of the 'morgue' or cabinet of biographical and obituary materials that is maintained in every wide-awake newspaper office. **1925** B. Benefield *Chicken-Wagon Family* 94, I have written a column of assorted obituaries after having dug up enough material in the 'Transcript's' 'morgue' for six columns. **1951** A. C. Clarke *Sands of Mars* xi. 136 In a score of newspaper offices, the copy culled from the Morgue began to be set up in type. **1964** *Listener* 5 Nov. 732/2 *The Raker* begins in a newspaper obituary department, usually known as the Morgue. **1971** L. M. Harrod *Librarians' Glossary* (ed. 3) 433 *Morgue*, a collection of obituary notices of famous living people kept up to date in newspaper offices. **1975** J. Symons *Three Pipe Problem* xviii. 187 The *Banner* morgue was unusually comprehensive, and it contained some interesting material about Haynes.

b. Hence a 'library' of cuttings, photographs, and information in a newspaper office, film studio, etc., often including sense *2 a. Also *attrib.*

1918 H. Croy *How Motion Pictures are Made* viii. 200 To make sure that the details are true, a studio has filed away in its morgue photographs of the life it wishes to depict. **1923** G. C. Bastian *Editing Day's News* 10 The person in charge of the newspaper files, clippings, and pictures [is] called the 'morgue librarian'. **1927** *Amer. Speech* II. 238/2 A 'morgue' is a necessary department of a newspaper... Here photographs and engravings of prominent

men, obituaries prepared in advance, pictures of buildings, ships, and even animals, and reference clippings on every conceivable subject are filed away ready for instant use. **1937** *Nat. Geogr. Mag.* Feb. 148/2 Beside his [*sc.* Jack London's] California laurel desk, with its oil burning.. lamp..is preserved his neat 'morgue', as writers call their reference library. **1942** Berrey & Van den Bark *Amer. Thes. Slang* § 606/2 Parts of Studio... *morgue*, the film filing room. **1962** *Listener* 12 July 57/1 Cartoonists in the United States rely for day-to-day material on newspaper morgues and wire-photographs. **1972** *Times* 16 May (Wall Street Suppl.) p. vii/2 Over to the *New York Times* for background material from their morgue: yes, the library is officially called that.

3. *Comb.* **morgue-like,** resembling a morgue (sense 1); **morgue-man,** the man in charge of a morgue (sense *2).

1913 *Cassell's Magazine of Fiction & Pop. Lit.* June 50/1 The long deserted tables, shrouded in black covers, would have seemed morgue-like to a casual observer. **1947** G. Greene *Nineteen Stories* 28 Her large morgue-like mouth was full of blackened teeth. **1912** *Outlook* (N.Y.) 14 Sept. 84/1 In some newspaper offices the morgue man and his assistants..work inside of steel cages. **1935** T. E. Lawrence *Let.* 4 Feb. (1938) 851 Let us now pass to the epitaph. Yes, Hogarth did the morgue-men a first sketch of me in 1920, and they are right to overhaul their stocks.

morillo (mǫrī·lyo). *Bullfighting.* Also **morrillo, murillo.** [Sp. *morrillo* fleshy part of the neck of an animal.] The muscle at the back of the bull's neck, one of the targets for the lances of the bullfighters.

1932 E. Hemingway *Death in Afternoon* xvi. 180 The morillo, or hump of muscle that rises from the back of the bull's neck to his shoulders. **1959** V. J. Kehoe *Aficionado!* i. 34 They [*sc.* the picadors] must place a *pica* (lance) into the rear part of the toro's long hump of neck muscle, known as the *morrillo* or *cerviguillo*. **1967** [see *divisa]. **1968** S. Oag *In Presence of Death* 261 They [*sc.* the picadors] should give it a minimum of three varas to test the bull's courage, weaken its powerful tossing muscle, the murillo, and lower its head. **1971** J. Leibold *This is Bullfight* ii. 31 The throwing muscle at the back of the neck (called the *morrillo*) is remarkably developed [in the fighting bull].

Moriori (mǫriōə·ri). [Native word.] An early Polynesian people of New Zealand, now extinct, who preceded the Maori; associated *spec.* with the Chatham Islands, where most evidence of their culture is found; also, a member of this people.

1865 A. S. Atkinson *Jrnl.* 9 Jan. in *Richmond–Atkinson Papers* (1960) II. iii. 145, I had a long talk with a genuine Moriori..or real aboriginal of the Chatham islands & of some of the southern parts of New Zealand. **1874** A. Bathgate *Colonial Experiences* xviii. 258 The Moriori population, which, at the time of the discovery of the islands, in 1791, was estimated at over 2,000, must at one time have been very large. **1882** T. H. Potts *Out in Open* 160 The Moriori was robust in figure, tall of stature, not darker in colour, perhaps, than many a Maori, but of a dull dusky hue. **1905** W. B. *Where White Man Treads* 27 The ancient Moriori mother scooped a shallow hole in the earth, and warmed it with embers, and lined it with dry grass and well-scorched 'kohu kohu' (moss), to nestle her new-born child in. **1949** P. Buck *Coming of Maori* (1950) i. ii. 14 The Moriori became extinct as a separate people, but their blood is present in some mixed Maori families. **1950** R. Duff *Moa-Hunter Period of Maori Culture* 15 Moriori, like Maori, is a recent term probably coined to distinguish its bearers from Europeans, and was probably never used before the eighteen hundreds. *Ibid.* 20 The Maori invasion of 1835 [of the Chatham Islands]..marked the end of the culture which we know as Moriori. **1962** *Listener* 19 Apr. 676/2 The Moriori, now extinct, were a group of Polynesians who settled in the Chathams somewhere after the twelfth century. *Ibid.*, The Morioris carved birds on the front of their houses.

morisca, moriska, varr. *moresca (see Moresco *sb.* 3 in Dict. and Suppl.).

1935 *Discovery* Sept. 265/2 Miss Violet Alford's comparative study of the relation of the Morris to other seasonal dances of Europe and to the Morisca. **1938** C. Sachs *World Hist. Dance* vii. 301 The carved figures with which Erasmus Grasser decorated the Ratssaal in Munich in 1480... These so-called *maruschka* or *moriska* dancers are..beautiful. *Ibid.* 335 In Europe..the name *morisca* covers many choral dances in double file formation. **1950** L. Armstrong *Dances of Spain* II. 13 Moriscas came into being as the Moors were driven southwards.

morish, var. Moreish *a.*

morituri te salutant (mǫritūə·ri tēⁱ sælu·tænt), *phr.* [L., lit. those about to die salute you (see Suetonius *Claudius* xxi. 6).] **a.** The words addressed to the emperor by the gladiators of ancient Rome on entering the arena, in anticipation of their death. **b.** This address used allusively by anyone facing danger or difficulty. Also in third person sing., **moriturus te salutat,** and in first person sing., **moriturus te saluto.**

[**1606** Holland tr. *Suetonius* 165 When he [*sc.* Claudius] was about to let out the water of the mere: Ficinus he exhibited in it a navall fight before: And as they who were to fight this battaile, cryed out to him, Ave Imperator, &c. i. All haile O Emperour; They salute thee and wish thy life who are ready to dye: and he againe made answere Avete vos.] **1704** B. Kennett *Antiquities of Rome* (ed. 3) ii. iii. 269 The Naumachiæ of Claudius which he presented on the Fucine Lake before he drain'd it, deserve to be particularly mention'd, not more for the greatness of the Show, than for the Behaviour of the Emperour: who when the Combatants pass'd before him with so melancholy a Greeting as, Ave imperator, morituri te salutant, return'd in Answer, Avete vos. **1869** S. R. Hole *Bk. about Roses* i. 6 They may say as they enter the arena, with the gladiators of old to the emperor..*Morituri te salutant*. **1894** —— *More Memories* xix. 248 If the slightest ailment affects him, it is a case of *moriturus te salutat*. **1910** *Encycl. Brit.* XII. 65/1 The admirable picture of Gérome which bears the title, 'Ave, Caesar, morituri te salutant.' **1930** E. H. Young *Miss Mole* vi. 59 I'm speaking, as it were, from my deathbed. *Moriturus te saluto!* Tomorrow ..I'm moving on. **1951** N. Marsh *Opening Night* xi. 254 Moriturus, to coin as Miss G. would say, a phrase, te saluto, Caesar. **1967** M. Grant *Gladiators* iii. 73 When the combatants arrived opposite the emperor's platform, they extended their right hands towards him and cried 'Hail, emperor, greetings from men about to die!' (*Ave, imperator, morituri te salutant!*) **1972** A. Roudybush *Sybaritic Death* (1974) xx. 166 'Ave Petersen,' Per read. 'Moriturus te salutat... I killed Clare.'

Morlach (mǭ·ɹlæk), *sb.* and *a.* Also (*pl.*) **Morla·cchi, Morla·cchian, Morla·cco, Morlack, Morlak.** [f. *mavrovlachi*, f. Gr. μαῦρο-ς black + Βλάχος Vlach: cf. Vlach.] **A.** *sb.* A member of a Vlach people centred on the eastern Adriatic port of Ragusa (mod. Dubrovnik) and, from the twelfth to the fifteenth centuries, in parts of maritime Croatia and northern Dalmatia, forming the country known eventually as Morlacchia, being later incorporated with Slavic peoples. **B.** *adj.* Of, pertaining to, or characteristic of Morlacchia or its people.

1778 tr. *A. Fortis's Travels into Dalmatia* 56 The Morlacchi, in general, have little notion of domestick œconomy... Yet the Morlack is a great œconomist in the use of his wearing apparel. *Ibid.* 65 It rarely happens..that a Morlacco carries off a girl against her will... The Morlack women keep themselves somewhat neat till they get a husband. *Ibid.* 83 Simplicity, and want of order..form.. the principal character of the Morlacchian poetry. **1778** *Critical Review* Mar. 378 The Morlacchians chiefly live on milk. **1793** *Universal Mag.* Oct. 268/1 The people called Morlacks, or Morlacchi, inhabit Morlacchia, which is among the inland mountains of Dalmatia. *Ibid.* 271/2 A Morlack village. **1837** *Mirror* 9 Dec. 383/1 At the death of a Morlach..the family weep and howl... The Morlach women go every holiday to renew their lamentations. **1849** A. A. Paton *Highlands & Islands of Adriatic* II. iii. 32 The Morlack principle is to allow the man to grow as the beast of the forest. *Ibid.* 33 The Morlack is the best soldier and the worst citizen in the Austrian empire. **1877** [see *Illyric* adj. s.v. *Illyrian *a.* and *sb.*]. **1881** E. A. Freeman *Sk. Subject Lands of Venice* 184 Are we to believe that the Morlacchi used the turban as their head-dress before the Ottoman came? **1904** L. Villari *Republic of Ragusa* xii. 322 The Morlachs in the Venetian service made raids into Turkish territory. **1911** *Encycl. Brit.* VII. 773/2 The name of *Morlachs, Morlaks* or *Morlacks* commonly bestowed by English writers on the Dalmatian Slavs..is an abbreviated form of *Mavrovlachi*, meaning either 'Black Vlachs', or, less probably, 'Sea Vlachs'... In northern Dalmatia the Slavs of the interior are still called *Morlacchi*; in the south this name expresses contempt. **1920** M. E. Durham *Twenty Years of Balkan Tangle* i. 12 They are known as Morlachs,..and historically are in all probability descendants of the pre-Slav native population. **1922** D. H. Low tr. *Kraljević's Ballads* p. xii, In 1775 a translation by Werthes of the Morlacchian section was published at Berne. **1957** *Encycl. Brit.* XXIII. 230/1 In the 14th century the Mavrovlachi or Morlachs extended themselves towards the Croatian borders, and a large part of maritime Croatia and northern Dalmatia began to be known as Morlacchia. *Ibid.* 230/2 The Morlachs have now become Slavonized. **1968** *Ibid.* XXIII. 93/2 There were also colonies of the Morlachs in the interior of the ancient Serbia.

Mormon. Add: **1.** (Earlier and later examples.)

1830 [see *Campbellite]. **1837** J. M. Peck *Gaz. Illinois* (ed. 2) i. 74 There are..a few Mormons..scattered through the state. **1922** *Lit. Digest* 10 June 40/1 Because a few of the inmates of the Hollywood film colony have 'married from time to time', they are less to be censured as Mormons than pitied as morons. **1928** R. Campbell *Wayzgoose* i. 26 Muses Nine, Those strapping girls whose love, to say the least, Would make a rabid Mormon of a priest. **1963** K. S. Latourette *Christianity in Revolutionary Age* V. ii. 14 In the Rocky Mountain area and on the west coast the Mormons were prominent.

2. (Later examples.)

1913 J. London *Valley of Moon* 430 He had been provost marshal when the Mormon trouble flared up. **1920** *Encycl. Relig. & Ethics* XI. 83/1 The first principle is set forth by the chief Mormon theologian, Parley P. Pratt. **1942** W. Stegner *Mormon Country* 65 He had sprung clear across the Great Basin deserts to locate the Mormon Station. **1948** [see *Mormon cricket*, below]. **1966** *Encycl. N.Z.* III. 64/2 The many similarities between Maori traditions and Mormon teachings. **1968** *Encycl. Brit.* XIII. 795/2 Space, time, matter and the universe in Mormon theology are limitless by nature. **1975** *Times* 20 Sept. 2/5 The fires at..the high school and the Mormon church [in Kirkcaldy], caused little damage.

3. Special Combs.: **Mormon battalion** (now *Hist.*), a company of soldiers from Mormon communities in Iowa enlisted for service in the Mexican war; **Mormon Bible,** the Book of Mormon; **Mormon Church,** the name by which the Church of Jesus Christ of the Latter-Day Saints is commonly known; **Mormon City,** Salt Lake City, Utah; **Mormon cricket,** a long-horn grasshopper, *Anabrus simplex*; **Mormon fly,** a hesperiid butterfly, *Atrytone hobomok*; **Mormon State,** in the U.S., a State in which Mormons are predominant, usu. applied to Utah; **Mormon trail,** the trail followed by Mormon migrants to Utah in 1847; **Mormon war,** disorders arising between Mormon communities and their neighbours, and *spec.* the fighting between Utah Mormons and the federal troops in 1857–8; **Mormon weed** (earlier and later examples); for Latin name substitute *Abutilon theophrasti.*

1848 *Santa Fé* (New Mexico) *Republican* 27 June 1/4 Capt. Hunt of the Mormon battalion, has left the Ciudad de Los Angelos for the Salt Lake. **1929** L. H. Creer *Utah & Nation* iii. 32 While crossing the Sierras, he met the returning members of the Mormon Battalion. **1838** Whittier in W. P. & F. J. Garrison *Life W. L. Garrison* (1885) II. 221 A discussion of the merits of animal magnetism, or of the Mormon Bible, would have been quite as appropriate. **1845** *Southern Lit. Messenger* XI. 476/1 These facts were withheld from the world for the nine years which intervened from the first publication of the Mormon Bible. **1882** J. W. Buel *Metrop. Life Unveiled* 349 These three men have subscribed to an oath which will be found in all Mormon bibles. **1838** *Niles' Reg.* 13 Oct. 103/3 We, and all the Mormon church,..entertain the same feelings and fears toward the Indians. **1948** *Newsweek* 12 Jan. 38/2 Once a basic tenet of the Mormon Church, the doctrine of plural marriage was formally abandoned in 1890. **1957** W. Mulder *Homeward to Zion* xi. 289 In 1887 the Edmunds-Tucker Act..disincorporated the Mormon Church itself. **1878** J. H. Beadle *Western Wilds* 39 We got to the Mormon City all beat out. **1942** Berrey & Van den Bark *Amer. Thes. Slang* § 46/1 *Mormon City*,..Salt Lake City, Utah. **1896** J. B. Smith *Economic Entomology* iv. 97 At the base of the Rocky Mountains, extending up into the foothills..the 'Mormon cricket', *Anabrus simplex*, occasionally multiplies so greatly that it migrates to the plains below, destroying everything in its path. **1934** *Sun* (Baltimore) 1 May 13/1 The devastating 'mormon' crickets, which came into prominence in 1848, when they threatened the first colonists in Utah with starvation, have appeared in a half dozen spots of Southern Idaho to carry on the destruction they started last year. Black creatures two inches long when full grown, they eat everything in sight. **1948** *U.S. Dept. Agric. Yearbk.* 274/2 The early Mormon settlers knew no satisfactory way of fighting the Mormon cricket. **1959** E. Tunis *Indians* 109/1 Grasshoppers and big mormon crickets were driven into trenches from which they could be gathered in baskets. **1972** Swan & Papp *Common Insects N. Amer.* 75 Mormon cricket: *Anabrus simplex*... Dark brown, black, or green; stout-bodied, wings absent. **1975** *Islander* (Victoria, B.C.) 26 Jan. 15/3 Densest pheasant populations occur in areas offering bonanza supplies of grasshoppers, Mormon crickets or some other..insect. **1845** N. F. Moore *Diary* 26 Aug. (1946) 46 Our boat and everything about it is now covered with Mormon flies as they are called, being, I am told, peculiar to this neighborhood. **1847** C. Lanman *Summer in Wilderness* v. 34 They are called the Mormon fly and I was told were found on these rapids alone. [**1909** *Cent. Dict.* Suppl. 828/3 *Mormon*... In *entom.*, an American hesperiid butterfly, *Atrytone hobomok*, which occurs from eastern Canada to the Mississippi valley.] **1882** C. M. Chase *Editor's Run in N. Mexico* 109 Axtell..went so far as to lay plans..to get the territory admitted into the Union as a Mormon state. **1893** L. Wagner *More about Names* 35 Utah, otherwise The Mormon State, is called by the Mormons themselves Deseret. **1920** *Encycl. Relig. & Ethics* XI. 89/1 The economic salvation of the Mormon State arose..from the money spent in Utah by the Forty-Niners. **1948** Mencken *Amer. Lang.* Suppl. II. 640 Utah calls itself the *Beehive State*,.. but the designation *Mormon State* is far more popular, and seems likely to stick. **1850** *Ann. Iowa* (1910) 3rd Ser. IX. 453 Here we struck the old Mormon trail. **1868** 'Mark Twain' in *Galaxy* May 631 The post route..changed partly to the old Mormon trail. **1949** *Kansas Hist. Q.* Feb. 39 The War Department decided..to improve the Mormon trail from Omaha as far as New Fort Kearney. **1833** in *D.A.* s.v., The Mormon War. **1834** *Knickerbocker* Aug. 159/2 *The Mormon War*... The war between the citizens of Jackson county, Missouri, and the disciples of the Book of Mormon. **1846** *Quincy* (Illinois) *Whig* 3 Feb. 2/4 He was a Major in the last '*Mormon War*'. **1857** *San Francisco Plaindealer* 10 Dec. 2/2 The Mormon war will furnish abundant material for Buncombe speeches. **1942** C. C. Rister *Land Hunger* 8 Upon the outbreak of the 'Mormon War', Dave and his older brother Jack, rode away..to join the army. **1872** *Trans. Ill. Dept. Agric.* IX. p. ix, Mr. James H. McConnell..has..prosecuted experiments with the Indian Mallow (*Abutilon Avacennae*) variously known as 'stump weed', 'velvet leaf', 'butter print', 'Mormon weed', etc. **1907** A. B. Lyons *Plant Names* (ed. 2) 8 Abutilon... American Hemp, Indian Hemp, Mormon-weed, Pie-marker, Pie-print, Sheep-weed.

Mormondom (further examples); **Mo·rmonish** *a.*, pertaining to or characteristic of Mormons or their beliefs; **Mo·rmonry** *rare* = Mormondom.

1846 *Quincy* (Illinois) *Whig* 19 Mar. 3/3 It would be desirable by all quiet citizens that they should be able to discover a complete evacuation of everything Mormonish in Nauvoo, before they hazard purchases in that place. **1852** *Oregon Statesman* (Oregon City) 6 Jan. 2/6 Your course looks as Mormonish as anything I ever saw. **1860** *Harper's Mag.* Nov. 857/2 We may be a trifle more fastidious than they are in Mormondom, but we have a notion that all the 'bad words' ought to be skipped when the ladies are around. **1873** Mormondom [see *Dixie[1] 2]. **1920** *Encycl. Relig. & Ethics* XI. 86/2 That Joseph Smith

..was the real master of Mormondom is borne out by.. accounts of him in the height of his power. **1930** *Amer. Mercury* Jan. 6/1 The next step was an accident, one of a series that has displayed God's providence to Mormonry. **1957** W. Mulder *Homeward to Zion* xi. 289 The 'iniquitous' Cullom bill..stirred Mormondom to its center.

Mormoness (mǭɹmǒnė·s). [-ess[1].] A female Mormon.

a **1861** T. Winthrop *John Brent* (1862) ix. 99 Selecting, perhaps, a Mormoness to kidnap to-night. **1862** C. F. Browne *A. Ward his Book* (1865) 76 'Yes,' hollered a lot of femaile Mormonesses, ceasin me by the cote tales & swingin me round very rapid. **1906** *Out West* May 305 (*title*) The Writer Lady and the Mormoness.

Mormonism. Add: (Further examples.) Also *fig.*

1831 *Niles' Reg.* 16 July 353/1 Mormonism..a new religion. **1855** A. L. Phillipps *Mahometanism* vi. 244 Socialism and Mormonism and infidelity taking the place of religion and social order. **1913** E. F. Benson *Thorley Weir* i. 33 'They can't all be serenading me.' 'I cannot imagine why not. A Mormonism of serenading young men is not illegal.' **1929** L. H. Creer *Utah & Nation* i. 3 Herein was also established the principle of modern revelation, one of the basic principles of Mormonism. **1958** Mulder & Mortensen *Among Mormons* p. v, The learned journals have discovered in Mormonism a ripe field for scholarship.

Mormonist. (Further example.)

1842 Dickens *Amer. Notes* I. v. 181, I should like to try the experiment on a Mormonist or two to begin with.

Mormonite, *sb.* and *a.* Add: **a.** *sb.* (Earlier and later examples.)

1831 *Columbian Reporter* (Taunton, Mass.) 24 Aug. 1/5 The Mormonites. We learn..that this infatuated people are again in motion. **1947** B. De Voto *Across Wide Missouri* 185 The 'Mormonites'..had been driven into the unsettled lands of Clay County. **1958** Mulder & Mortensen *Among Mormons* p. vi, Once the new church is founded the religious press is full of letters..about the Mormonites and their New Jerusalem.

b. *adj.* (Earlier examples.) Now *rare.*

1835 C. Bradley *Jrnl.* 24 June in *Ohio Archaeol. & Hist. Q.* (1906) XV. 269 We have, too, a company of Mormanite missionaries aboard. **1859** Mill *Liberty* iv. 164 The article of the Mormonite doctrine which is the chief provocative..is its sanction of polygamy.

mor'n: see *more *adv.* 5 a.

Mornay (mǭ·ɹneⁱ). Also **mornay.** [perh. f. the name of Philippe de *Mornay* (died 1623), a French Huguenot writer.] *Mornay sauce* (also in Fr. form *sauce mornay*): a rich white sauce flavoured with cheese. Also *ellipt.*, a dish served with Mornay sauce.

1906 A. Filippini *International Cook Bk.* 623 Fillet of sole, mornay. **1925** Leyel & Hartley *Gentle Art of Cookery* 20 Sauce mornay, a cheese sauce. **1939** A. L. Simon *Conc. Encycl. Gastron.* I. 39/1 Mornay, sauce. Heat the required amount of *Béchamel*... When ready to serve, add as much Parmesan cheese as desired. **1948** L. Marion *Be your Own Chef* xii. 230 Remove the vegetable marrow from the oven. Throw away the water..and pour the Mornay Sauce over the pieces of vegetable. **1963** R. Carrier *Great Dishes of World* 62 Mornay sauce—for fish, vegetables, poultry, poached eggs, noodle and macaroni mixtures. **1965** *House & Garden* Dec. 84/2 *Mornay.* Just cheese, but the basis of some of the most wonderful recipes there are. **1974** I. Murdoch *Sacred & Profane Love Machine* 311 Well, *oeufs* somehow? Mornay?

morning, *sb.* Add: **1. e.** *Ellipt.* for **morning paper. colloq.*

1961 [see *evening *sb.*[1] 2 f]. **1966** *New Statesman* 7 Oct. 503/3 Thomson already controls the *Scotsman*, three provincial mornings and nine evenings. **1970** K. Giles *Death in Church* v. 121 Did you see the mornings?..The act of killing the clergy seems to enrage English journalists.

2. b. For 'in vulgar or off-hand speech' read 'in informal speech'.

1911 G. B. Shaw *Blanco Posnet* 390 Morning, Elder. (*Passing on*). Morning Strapper. (*Passing on*). Morning, Miss Evans. **1968** G. Butler *Coffin Following* vi. 120 'Morning,' said the ticket collector.

3. a. *morning, noon, and night* (earlier examples).

1808 E. Weeton Let. 1 Apr. in *Jrnl. of Governess* (1969) I. 80 Your praises..have operated almost like a dose of salts, and have worked me morning, noon, and night for these two days. **1864** Trollope *Can you forgive Her?* I. xx. 160 Your sitting here all alone, morning, noon, and night, won't bring him back.

c. *mornings* (later examples).

1936 *S.P.E. Tract* xlv. 192 A peculiar use of the plural form is illustrated in 'The operatic tenor Campanini was engaged to sing mornings'. **1938** T. Wilder *Our Town* I. 11 Quarter of nine mornings, noontimes, and three o'clock afternoon's [*sic*], the hull town can hear the yelling and screaming from those schoolyards.

g. (*the*) *morning after* (*the night before*): a morning on which the effects of the previous night's drinking or revelry are unpleasantly felt; also *attrib.* and *absol.*, these effects or a person suffering from them; also in extended use for any unpleasant aftermath of pleasure. Also *morning after*, applied to a contraceptive pill that is effective when taken some time after sexual intercourse.

1884 *Punch* 31 May 264/1 His method of inoculation for hydrophobia seems uncommonly like the old 'morning-after' remedy, when the chippy one who could 'strike matches on his tongue' was recommended to take 'a hair of the dog that bit him'. **1909** E. Nesbit *Daphne in Fitzroy St.* xvi. 256 'I'm all right,' said the girl... 'That's what we all of us says, when it comes to be the morning after,' said Mrs. Delarue. **1922** Joyce *Ulysses* 231 It was blue o'clock the morning after the night before. **1927** Kipling *Limits & Renewals* (1932) 157 There wasn't much doubt what Jimmy had been up to. He was altogether 'the morning after'. **1942** T. Bailey *Pink Camellia* v. 28 She sat up and looked at herself in the mirror. 'The morning after! And my face looks like it.' **1946** 'Brahms' & 'Simon' *Trottie True* vi. 152 'The Duke's got a morning-after,' she said crisply. **1947** *Sat. Rev.* (U.S.) 12 Apr. 20/1 The nation relaxes, and only the morning-after headaches and the morning-after quarterbacks remain. **1957** A. C. Clarke *Deep Range* iv. 43 He was not pleased to find that the entire stock of 'morning-after' pills had already been consumed. *a* **1963** S. Plath *Crossing Water* (1971) 33 Musky as a lovebed the morning after. **1967** V. C. Clinton-Baddeley *Death's Bright Dart* 169 One of those aristocratic men in the advertisements whose morning-after needs are ministered to by a smiling and provident butler. **1967** *N.Y. Times* 24 Aug. 35 Dr. Chang and Dr. Pincus had been working recently on a new pill, known popularly as the 'morning-after' pill. It affects the egg after ovulation. **1969** *Daily Tel.* 3 July 23/3 Trials of a 'morning after' pill had been promising. **1973** *Express* (Trinidad & Tobago) 17 Mar. 7/1 The Food and Drug Administration will approve the use of diethylstilbestrol or DES as a 'morning after' contraceptive pill for women in emergency situations such as rape. **1973** L. Meynell *Thirteen Trumpeters* xiv. 215 'I felt *awful*. I didn't care whether I lived or died..' 'A classic case of the morning-after-the-night-before feeling.'

7. *morning beam, gold, light* (further example), *performance* (earlier example).

1928 E. Blunden *Retreat* 33 With friends as nimble as morning-beams, Who sped with him to this playground. **1890** Kipling *Barrack-Room Ballads* (1892) 163 He hewed the living rock with sweat and tears, And reared a God against the morning-gold. **1930** R. Campbell *Adamastor* 86 With food and drink, at morning-light, The children met me at the water-side. **1827** T. J. Dibdin *Reminisc.* I. 251 A large play bill from Dublin, announcing morning performances, on account of a partial insurrection or rebellion.

¶ (Further examples.)

1916 D. H. Lawrence *Amores* 125 It seemed that I and the morning world Were pressed cup-shape to take this reiver Bird. **1917** —— *Look! We have come Through!* 77 See, glittering on the milk-blue, morning lake They are laying the golden racing-track of the sun.

b. *morning coat, jacket, suit* (earlier example).

1867 *Harper's Mag.* Aug. 362/1 He got himself a new morning suit for shop use. **1912** 'C. F. Benton' *Fairs & Fetes* 118 They should sell brushes and combs..and morning jackets. **1933** *Week-end Review* 21 Jan. 69/2 Black morning coat and waistcoat, grey striped trousers, grey tie and pearl pin. **1969** R. Godden *In this House of Brede* xix. 395 Japanese gentlemen in morning coats, grey waistcoats, striped trousers.

8. For 'Only *poet.*' read 'Usu. *poet.*' and add: *morning-blue, -bright* (further examples), *-cold, -fair, -gathered, -grey* adjs.

1920 D. H. Lawrence *Lost Girl* xvi. 366 The lovely translucent pale irises, tiny and morning-blue, they lasted only a few hours. **1922** E. Blunden *Shepherd* 48 Cheeks all morning-bright. **1923** D. H. Lawrence *Birds, Beasts & Flowers* 76 My heart which like a lark at heaven's gate singing, hovers morning-bright to Thee. **1945** J. Betjeman *New Bats in Old Belfries* 43 There splashed about our ankles as we waded Those intersecting wavelets morning-cold. **1938** W. de la Mare *Memory* 34 Eyes blue as speedwell, tranquil, morning-fair. **1857** *Leisure Hour* 2 July 421/1 You cannot now get out of hearing of..'strawberries, morning gathered'. **1903** *Farmer & Stockbreeder* 8 June 950/3 The vegetables mostly arrive in freight trains from distant States. They are well packed, and despatched with regularity, but nothing is absolutely fresh or 'morning gathered', as London greengrocers love to shout. **1961** *Daily Mail* 15 Mar. 8/8 'Morning gathered?' (Which morning?) **1943** C. Day Lewis *Word over All* 15 Children look down upon the morning-grey Tissue of mist that veils a valley's lap.

9. **morning caller,** one who pays a formal morning call; **morning coffee,** coffee taken at mid-morning or (a less common use) at breakfast; **morning girl,** a non-resident maid-servant employed during the morning only; **morning line,** a list of probable betting odds established by the bookmaker prior to a sporting event; also *attrib.*; **morning paper,** a newspaper published so as to be on sale during the morning; **morning prayer,** (*a*) (later examples); (*c*) *Mil. slang* (see quot. 1965); **morning room** (earlier example); **morning tea,** tea taken either before rising or at mid-morning; **morning visit,** (*a*) a visit made in the morning; (*b*) a formal 'afternoon' visit (cf. *morning call*); also *morning visiting, visitor.*

1848 Geo. Eliot *Let.* in J. W. Cross *George Eliot's Life* (1885) I. 184 The bliss of having a very high attic in a romantic Continental town..far away from morning-callers, dinners, and decencies. **1863** Mrs. Gaskell *Dark Night's Work* viii. 128 She allowed Fletcher to..usher him into the library just like any common visitor, any morning-caller. **1941** G. Heyer *Faro's Daughter* i. 1 While the butler went to convey this message to the morning-caller, her ladyship tidied her ruffled person. **1855** Geo. Eliot in *Fraser's Mag.* June 705/1 The dark little bedroom, and the arm-chair where he took his morning coffee as he read. **1917** Conrad *Shadow-Line* iv. 131 Presently Ransome brought me the cup of morning coffee. **1959** T. S. Eliot *Elder Statesman* ii. 43 When I asked about morning coffee She said 'I'm not the one for elevens's.' **1973** A. Christie *Postern of Fate* iii. xv. 231, I was just going to bring your morning coffee up. **1921** *Dict. Occup. Terms* (1927) § 900 Daily servant,..morning girl; a non-resident general servant. **1935** A. J. Pollock *Underworld Speaks* 77/2 *Morning line*, the betting odds quoted in poolrooms on horse races. **1941** *Sun* (Baltimore) 26 Dec. 11/5 The old Louisiana Jockey Club abandoned the sport following a controversy over the..ruling eliminating the 'morning line' bookmaking. **1964** A. Wykes *Gambling* viii. 192 The day starts with the numbers of the runners in the first race posted in order, with the 'morning line' odds in lights opposite them. The morning line is the odds established by the racing officials for each horse (on the basis of its past performance) on the morning of a particular racing day. **1968** *Wall St. Jrnl.* 31 Jan. 1/1 Mr. Lewin has been track handicapper at Atlantic City, making up the 'morning line' (early odds) and picking likely winners for the official track program. **1727–41** Morning paper [see Paper *sb.* 8]. **1768** J. Wedgwood *Let.* 13 June (1965) 64 That Morning Paper which is mostly taken in by People of Fashion. **1862** Morning paper [in Dict., sense 7]. **1968** R. B. Heath *Newspapers* i. 13 National daily morning papers. **1857** Trollope *Barchester Towers* III. xv. 264 Eleanor was dressed a full hour before the time fixed in the Ullathorne household for morning prayers. **1916** Joyce *Portrait of Artist* iii. 130 Death..is a blessed moment for him who has walked in the right path..attending to his morning and evening prayers. **1965** B. Sweet-Escott *Baker St. Irreg.* viii. 250 'Morning prayers' in 12th Army headquarters, the daily briefing of the staff, began with a statement of temperature and humidity. **1816** Jane Austen *Emma* II. xiv. 260 That room was the very shape and size of the morning-room at Maple Grove. **1727** Swift *Furnit. Woman's Mind* in *Poetical Wks.* (1967) 329 Can at her Morning Tea run o'er The Scandal of the Day before. **1923** Wodehouse *Inimitable Jeeves* xii. 135 Sometimes when Jeeves has brought in my morning tea and shoved it on the table beside my bed, he drifts silently from the room. **1940** Morning tea [see *evening paper*]. **1948** H. V. Morton *In Search of South Africa* iv. 123, I saw groups of storm-bound people sitting at their morning tea. **1963** A. Lubbock *Austral. Roundabout* 6 The meals called 'tea'. There are three... First you have 'morning tea', which consists of cups of tea and biscuits, or savouries, at eleven o'clock. **1736** Morning visit [in Dict., sense 7]. **1787** A. Young *Jrnl.* 28 June in *Trav. France* (1792) I. 25 We return in time to dress for dinner, at half after twelve or one: then adjourn to the drawing-room.. as the first thing done, by every person who arrives, is to pay a morning visit to each party already in the place. **1828** Morning visit [in Dict., sense 7]. **1864** Trollope *Can you forgive Her?* I. ii. 9 The old lady walked into the drawing-room one morning at eleven o'clock..these morning visits were made almost every day. **1852** C. M. Yonge *Two Guardians* xi. 195 The afternoons were less certain to be agreeable... Marian had become liable..[to] being carried forth..on a morning visiting expedition. **1777** P. Thicknesse *Year's Journey through France* II. 234 The French never give..any refreshment..to their morning or evening visitors. **1831** M. Edgeworth *Let.* 6 Jan. (1971) 469 Scarcely had I taken up my pen and breathed when other morning visitors entered.

morning-glory. 1. Substitute for def.: A twining herb of the genus *Ipomœa* or closely related genera of the family Convolvulaceæ, esp. *Ipomœa* (or *Pharbitis*) *purpurea*, which is native to tropical America and bears large trumpet-shaped flowers; also *attrib.* and *fig.* (Earlier and later examples.)

1814 F. Pursh *Flora Americæ Septentrionalis* I. 146 *Ipomoea Nil*... Flowers beautiful pale blue, only open early in the morning, from which it has been called Morning-glory. **1934** E. Blunden *Choice or Chance* 11 And with wonder saw The morning-glory about the bucket twined. **1947** R. P. T. Coffin *Yankee Coast* 59 The owners..have to make guitars and banjos do. And accordions, phonographs with morning-glory horns, and, latterly, radios. **1963** W. Blunt *Of Flowers & Village* 176 I've not been very successful with morning glories. **1964** C. Willock *Enormous Zoo* v. 89 We found the DC..in a morning glory-covered bungalow up on the Sudan border. **1966** *Times* 22 Mar. 5/1 The Pharmaceutical Society is carrying out laboratory investigations to determine whether any Morning Glory seeds sold in this country could be dangerous. **1968** Peterson & McKenny *Field Guide to Wildflowers Northeastern & Northcentral N. Amer.* 338 Common Morning-glory. Alien. *Ipomoea purpurea*... A twining vine; the garden morning-glory with broad heart-shaped leaves. Flowers purple, pink, blue, or white. **1970** R. Lowell *Notebk.* 101 The jellied green creepers and morning-glories of the saurian sunset.

3. *U.S. slang.* Something which fails to maintain its early achievement; *esp.* in sporting contexts.

1904 *Outing* XLV. 170/1 Didn't I tell you he was nothing but a morning glory—why that dog couldn't beat a ferry boat. **1920** *Harvey's Weekly* 15 May 15/2 You have ..sent one more morning glory statesman to limbo. **1935** A. J. Pollock *Underworld Speaks* 77/2 *Morning glory*, race horse that works in fast time in the morning but fails to race well. **1938** *Amer. Speech* XIII. 30 Clubs which bloom luxuriantly early in the season, but which are expected to wilt ignominiously as the competition becomes hotter, are known as *Morning Glories.*

morningless, *a.* (Later example.)

1930 O. Gogarty *Wild Apples* 15 Those resinous timbers immuned from decay By thunders ancestral and morningless storms.

Moroccan, *a.* and *sb.* Add: **A.** *adj.* (Earlier examples.)

a **1706** Evelyn *Diary* an. 1682 (1955) IV. 285 We had at this time in Lond together The Russian, Morrocan, & Indian Ambassador. **1869** 'Mark Twain' *Innoc. Abr.* vii. 69 The Spaniards built these watch towers on the hills to enable them to keep a sharper look out on the Moroccan speculators.

b. In the names of various preparations of cannabis.

1970 E. Goode *Marijuana Smokers* i. 18 Another recipe book..has exotic Oriental dishes, such as 'Bhang Sherbet', 'Moroccan Majoon' [etc.]. **1972** J. Brown *Chancer* iii. 38 People talk about the effects of cannabis as if there was just one kind... You smoke Moroccan Gold.. you'll just feel, like, a little drunk. But you smoke Egyptian Black, that will stone you out of your head. **1972** *Times* 14 Dec. 2/3 The charges involved 'speed'..and other drugs, referred to in the hearing as..'Moroccan black'.

morocco, *sb.* Add: **II. 3. b.** *morocco-bound*, also used *fig.*; *morocco-covered* adj.

1873 'Mark Twain' & Warner *Gilded Age* II. ix. 97 A very official chair behind a long green morocco-covered table. **1886** A. Hornblow tr. *Normand's Splashes from Parisian Ink-Pot* 83 Plunged in a big green morocco-covered fauteuil, he began to scan over the 'dailies'. **1896** *Sat. Rev.* 12 Dec. 623/2 'Little Eyolf'..has been promoted into a full-blown fashionable theatrical speculation, with a 'Morocco Bound' syndicate in the background. **1958** *Listener* 2 Oct. 520/2 Sentences like 'I have still to see that lofty homeland of the potato and the llama'..belong to a solid, morocco-bound tradition.

moron[2] (mōᵊ·rǫn). orig. *U.S.* [f. L. *mōrus*, Gr. μωρός stupid.] **a.** One of the highest class of feeble-minded; an adult person having a mental age of between eight and twelve. Also *attrib.*

The term was first adopted and given this meaning by the American Association for the Study of the Feeble-minded in 1910.

1910 H. H. Goddard *Let.* 29 Apr. in *Jrnl. Psycho-Asthenics* Sept.–Dec. 65 The other (suggestion) is to call them [*sc.* feeble-minded children] by the Greek word 'moron'. It is defined as one who is lacking in intelligence, one who is deficient in judgement or sense. **1912** —— *Kallikak Family* 54 The type of feeblemindedness of which we are speaking is the one to which Deborah belongs, that is, to the high grade, or moron. **1919** H. Woodrow *Brightness & Dullness in Children* iii. 45 The term *moron*..is a new term... The desirability of this new term arose from the fact that the term feeble-minded, which is used in England to designate only the highest class of mental defectives, had long been used in America to include all three classes. **1937** C. L. Burt *Subnormal Mind* (rev. ed.) ii. 102 All through their years of training, most of them [*sc.* social workers and teachers] had never, to their certain knowledge, set eyes on a moron or an imbecile. **1948** J. W. Bridges *Psychol.* xxiv. 352 The moron has an I.Q. of 50 to 70..and the idiot below 20. This distinction is made on the basis of intelligence test results and is consequently not very reliable. **1968** W. B. Stephens tr. *Inhelder's Diagnosis of Reasoning in Mentally Retarded* ii. 72 The normal child of seven, the moron of twelve, or even the imbecile assimilates an enormous number of varied experiences which are..qualitatively irreducible from one level to the other.

b. *colloq.* A stupid or slow-witted person; a fool. Also *attrib.*

1922 W. R. Inge in *Edin. Rev.* July 48 It is possible that while we are governed by high-grade 'morons' there will be no practical recognition of the dangers which threaten us. **1922** H. Titus *Timber* iii. 37 So this backwoods moron, even, knew something about his affairs that John Taylor did not know. **1922** L. Stoddard *Revolt against Civilization* iv. 129 It is a mutilated, deformed, moron humanity which glowers or drivels at us through expressionist pictures. **1928** Hecht & MacArthur *Front Page* ii. 91 Another one of your moron blunders. **1930** R. Macaulay *Staying with Relations* ii. 22 'You mutt,' said young Mrs. Rickaby..'Catherine will think we're all morons.' **1959** *Punch* 10 June 780/2 It was so obvious it might have occurred to anyone but a complete moron. **1968** M. Woodhouse *Rock Baby* xxiii. 219 What we have here is a poxed-up, paralytic bunch of morons... They're technically incompetent..and they're pretty useless in almost every other field. **1975** G. Moffat *Miss Pink* xii. 169 He'd have had to be a moron not to have guessed that the killer was really after himself.

So **moro·nic** *a.*, **moro·nically** *adv.*

1926 *N.Y. Herald-Trib.* 28 Nov. vii. 4/2 A facetious account of a young girl's discovery that her brother had died of a venereal disease which has the romping moronic quality of a tabloid front page. **1931** *Churchman* 14 Feb. 15 The philosophers go home considerably enheartened when they find Socrates moronically low in his I.Q. score. **1957** R. A. Heinlein *Door into Summer* (1960) ii. 25 It was a moronically simple idea: don't repair, replace. **1957** *Times Lit. Suppl.* 18 Oct. 621/3 The psychopathic juvenile delinquent, moronically searching for 'the truth'. **1959** *Punch* 13 May 641/2 They interrupt Macbeth, they cackle at M. Hulot, they share their moronic comments with their escorts. **1975** D. Bloodworth *Clients of Omega* xxi. 204 Provoking desperate people into believing that they can only bring about unity among men by knocking their moronic heads together.

morone. (Earlier example.)

1808 *Examiner* 6 Mar. 157/2 Coquelicot and morone are on the decline.

moror, var. *maror.

morosity. Delete 'now *rare*' and add later examples.

1923 E. Bowen *Encounters* 73 Aunt Willoughby's baffled, bearded morosity. **1965** C. H. Sisson *Christopher Homm* xxi. 162 Even to Dad Christopher's courtship was of some use, as an exercise for morosity.

morosophist. Delete † and add later example. *rare.*

1870 K. H. Digby *Halcyon Hours* 255 Morosophists who love to boast Are those of course who scorn the most This holy maid.

morph[3] (mǭɹf). *U.S. colloq.* and *slang* shortening of Morphine *sb.* (see also quot. 1912).

1912 *Collier's* 21 Sept. 20/1 White 'dope fiends,' known in the vernacular of the police as 'hops', 'cokes', or 'morphs'. **1914** Jackson & Hellyer *Vocab. Criminal Slang* 60 *M*, or *Morph*,..used by morphine fiends. Sulphate of morphia. **1926** J. Black *You can't Win* xii. 160 About a spoonful of water and some of their meager store of 'morph' were put in the tin box. **1956** H. Gold *Man who was not with It* (1965) viii. 65 No morph, no! I had really kicked that one, and would do my own traveling from now on.

morph[4] (mǭɹf). *Linguistics.* [f. *morph(eme).] **a.** = *allomorph[2]. **b.** A phoneme or series of phonemes forming a variant or a number of variants of a morpheme.

1947 C. F. Hockett in *Language* XXIII. 322 Recurrent partials not composed of smaller ones (we might call them *morphs*)... By definition, a morph has the same phonemic shape in all its occurrences... Morphs are not always composed of continuous uninterrupted stretches of phonemes, but they are always composed of phonemes. Every utterance is composed entirely of morphs. **1948** P. K. Benedict in *Jrnl. Amer. Oriental Soc.* LXVIII. 185 This nomenclature has obvious advantages from a systematic point of view, and might well be extended to morphophonemics; the 'morph' as the descriptive unit, and the 'allomorph' as the member of a morpheme. **1954** C. E. Bazell in E. P. Hamp et al. *Readings in Linguistics II* (1966) 274 The term *morpheme*..was used both for the narrower modern concept and for that of the morpheme-alternant or (in American usage) the morph. **1955** *Trans. Philol. Soc.* 1954 69 Morphemes. Auxiliary classification here would classify (morpheme) order: to morpheme-alternants ('morphs')... Synthetic classification of morphs is implied in the very procedure of defining their morphemic values. For it is on account of their identification in different substitution-classes (selections), that morphs are assigned their different morphemic values, or, as we say, considered to 'represent' different 'morphemes'. **1960** *Amer. Speech* XXXV. 218 He [*sc.* Hoenigswald] believes sound change to be due to nothing but complete borrowing of morphs between only two morphs..., and sounds..which, according to him, are not at all like the Saporta..regards the process..the Gleason *Introd.* *Transformational Gram.*... **1964** E. Bach *Introd. Transformational Gram.*... **1965** *Language* XLI. 23 The markers of case are almost always portmanteau morphs that are also involved in expressing other categories..of the noun phrase. **1968** J. Lyons *Introd. Theoretical Linguistics* v. 184 The word *bigger* is analysable into two morphs... Each morph *represents* (or is the exponent of) a particular morpheme... Morphemes may be represented directly by phonological (or orthographical) segments with a particular 'shape' (that is, by morphs). **1972** *Hebrew-English Dialect Questionnaire* 1 Some of the principal morphs found in polysyllables are..included in the questionnaire.

c. *attrib.* and *Comb.*

1947 *Language* XXIII. 337 Intonational morphs..are found spread through long stretches of forms...such as *yes*, *I know*, *no*, *maybe*, [etc.]. **1954** *Int. Jrnl. Amer. Linguistics* Form..the morph.... **1962** H. M. Hoenigswald in Householder & Saporta *Problems in Lexicography* 105 There can be no objection..to arrangement by morphs, that is by the phonemic shape of the allomorph of a given morpheme. **1964** E. Bach *Introd. Transformational Gram.* vi. 130 The P rules will operate to select the proper morph shapes and allomorphs. **1966** *Amer. Speech* XL. 297 Orrm's use of double consonants, which is found only in morph-final position or in morph-medial position before another morph-initial obstruents if they are preceded by a vowel or nasal consonant.

morph[5] (mǭɹf). [f. Gr. μορφή form.] A variant form of an animal produced by genetic differences, or (in later use) one of several forms exhibited by an animal in the course of its life cycle. Hence **mo·rphic** *a.*; **mo·rphism** = Polymorphism 2.

1955 J. S. Huxley in *Heredity* IX. 2, I propose to introduce the term morphism and its derivatives, morphic and morph. *Ibid.* 3, I restrict the term morphism to genetic polymorphism..in which (usually sharply distinct) genetic variants or morphs coexist in temporary or permanent balance within a single interbreeding population in a single spatial region, and in such frequencies that the rarer cannot be due solely to mutation, or to the spread of a rare by selection... **1957** *Ibid.* 8 The cuckoo shows a general tendency to morphism: for it shows a plumage dimorphism, barred red and unbarred grey. Among adults, the red morph is confined to a minority, all females. **1966** —— in *Proc. R. Soc.* B. CLXIV. 215 A morphic balance-mechanism must be strong enough to prevent the break-up of the morphism through the disappearance of any morph under adverse selection in periods exceptionally unfavourable to it. **1963** E. Mayr *Animal Species & Evolution* vii. 157 Natural selection may modify the dominance of such morphs. **1968** R. D. Martin tr. *Wickler's Mimicry in Plants & Animals* i. 14 A leaf-beetle occurring as two separate morphs is shown. *Ibid.* vi. 61 Both sexes [of *Ityraea nigrocincta*] have two morphs, a green form and a yellow form. **1970** *Nature* 26 Sept. 1368/1 In the autumn the parthenogenetic viviparous morphs of the aphid give birth to sympods which develop into males and oviparous females. **1974** *Ibid.* 22 Feb. 572/2 The frequencies of colour and banding morphs of the dead animals were compared with those prevailing in the population. **1975** *Jrnl. Zool.* CLXXVII. 336 From considerations of morphism, modality and similarity, a descriptive classification of nine categories is advanced to encompass all situations involving pattern polymorphism and sexual dimorphism in butterflies.

-morph (mǭɹf), terminal element (repr. Gr. μορφή form) of sbs. with the sense 'thing having a (certain) form or character', as Isomorph, Polymorph, *allelomorph, *endomorph.

morphallaxis (mǭɹfălæ·ksis). *Zool.* [f. Gr. μορφή form + ἄλλαξις exchange.] Regeneration by the transformation of existing bodily material.

1901 T. H. Morgan *Regeneration* i. 23 At present there are known two general ways in which regeneration may take place... In order to distinguish broadly these two modes I propose to call those cases of regeneration in which a proliferation of material precedes the development of the new part, 'epimorphosis'. The other mode, in which a part is transformed directly into a new organism, or part of an organism without proliferation at the cut-surfaces, 'morphallaxis'. **1924** G. R. de Beer *Growth* viii. 55 In regeneration in some cases (where morphallaxis occurs) it [*sc.* material needed for growth] is derived from the other tissues of the organism which thereby undergo reduction. **1962** C. H. Waddington *Princ. Embryol.* xiv. 304 Animals such as coelenterates, flatworms and oligochaetes in which morphallaxis occurs to a considerable extent. **1965** Kiortsis & Moraitou in Kiortsis & Trampusch *Regeneration in Animals* 259 If the anteriorly facing cut end of the nerve cord [in the polychaete worm *Spirographis spallanzanii*] is deflected, there is a morphallaxis on one or both sides.

morpheme (mǭ·ɹfīm). *Linguistics.* [F. *morphème*, f. Gr. μορφή form, after *phonème* Phoneme.] **a.** An element such as an affix, preposition, conjunction, or stress pattern considered in respect of its functional relations in a linguistic system (now little used by linguists). **b.** The smallest meaningful morphological unit of language, one that cannot be analysed into smaller forms.

1896 R. J. Lloyd in *Neueren Sprachen* III. 615 The phonetic elements of a given word are its *phonemes*. But its significant elements, be they root, suffix, prefix, inflection or aught else, are *morphemes*. **1925** P. Radin tr. *Vendryès' Lang.* II. ii. 74 By semantemes we understand the linguistic elements which express the ideas of the concepts..and by morphemes we understand those elements which express the connexions between the ideas... The morpheme is generally a phonetic element..indicating the grammatical relations between the ideas of the sentence. **1926** L. Bloomfield in *Language* II. 155 A morpheme is a recurrent (meaningful) form which cannot in turn be analyzed into smaller recurrent (meaningful) forms. Hence any unanalyzable word or formative is a morpheme. **1931** *Amer. Jrnl. Philol.* LII. 25 The morpheme of the dependent element indicates merely an unspecialized dependence (gender, number, person, case). *Ibid.* 79 Nouns are semantemes susceptible of morphemes of case. **1933** L. Bloomfield *Language* x. 161 A linguistic form which bears no partial phonetic-semantic resemblance to any other form, is a *simple* form or *morpheme*. Thus, *bird, play, dance, cran-, -y, -ing* are morphemes. Morphemes may show partial phonetic resemblances. **1935** G. K. Zipf *Psycho-Biol. of Lang.* i. 15 One may view the morpheme as a..word..[*sc.* *ungratefulness*] is a sequence of five units which we shall term morphemes..i.e. *un-tru-th-ful-ness*. **1948** *Word* Apr. 20 If it is a bound form, the element must..in order to be a morpheme—be given..with..(of) the morphemic segment or 'morph'...[The earlier distinction of morpheme and semanteme is obsolete except in non-structuralist circles.] **1959** *Archivum Linguisticum* XI. 93 In the *Prolegomena*, morphemes ('inflectional elements') not, however, to be identified with the morphemes either in the Prague or American (Bloomfieldian) sense are regarded as forming a special part of the content form of language. **1962** E. F. Haden et al. *Resonance-Theory for Linguistics* iii. 40 The word is here defined..as a free form comprising one or more morphemes, occurring in a given position between silences or other words. **1967** *General Linguistics* VII. 65 Maslov discusses the shifts in meaning and utilization which the term 'morpheme' has undergone since its coinage by Baudouin de Courtenay in the 1880's. **1973** *Sci. Amer.* Feb. 57/1 Every language has a stock of several thousand morphemes: the bearers of the basic semantic and grammatical content. An expression such as 'can openers' contains four morphemes: 'can', 'open', 'er' and 's'.

c. *attrib.* and *Comb.*, as *morpheme-class, -combination, -configuration, -count, -sequence, -structure, -theory, -unit, -word*; morpheme-*based, -initial, -like, -medial* adjs.; **morpheme alternant** = *allomorph.

1942 *Language* XVIII. 271 We divide each expression in

the given language into the smallest sequences of phonemes which have what we consider the same meaning when they occur in other expressions... The resultant minimum parts we call not morphemes, but *morpheme allernants*. **1947** *Ibid.* XXIII. 402 Different phonemic shapes of a given base appearing before different suffixes.. are morpheme alternants of the same morpheme. **1952**, **1955** [see *morph[4]]. **1963** J. Lyons *Structural Semantics* ii. 11 A word-based grammar seems to be more satisfactory than a morpheme-based grammar for the description of languages of the 'inflecting' type. **1947** *Language* XXIII. 81 Morphemes are assigned to morpheme-classes on the basis of the environments in which they occur. Each environment determines one and only one morpheme-class. **1963** Erwin & Miller in J. A. Fishman *Readings Sociol. of Lang.* (1968) 76 Morpheme classes can be divided into two groups, lexical and function classes. **1955** C. E. Bazell *Linguistic Form* vii. 81 But here it is not simply a matter of sorting out incommutable morphemes, but also a case of including commutable morpheme-combinations. **1935** G. K. Zipf *Psycho-Biol. of Lang.* (1936) iv. 136 If a morpheme-count reveals that accent tends to settle on morphemes of the greatest average interval..we shall have reached the very heart of the dynamics of accent. *Ibid.* 154 The crystallization of phonemes in a morpheme-configuration. **1964** *Language* XL. 575 Morpheme-final *i* is replaced by a if it is preceded by a vowel or a velar and is followed by a vowel. **1962** H. M. Hoenigswald in Householder & Saporta *Problems in Lexicography* 107 There are, in the view of many linguists, morphemes and morpheme-like entities with so-called grammatical meanings. **1965** *Canad. Jrnl. Linguistics* Fall 36 If paralanguage is a dual system, with phoneme and morpheme-like units, ..it would be difficult to imagine the primes of a more rudimentary system exceeding in number the primes of a more complex one, language. **1951** Z. S. Harris *Methods in Structural Linguistics* v. 129 English has morpheme-medial clusters like /rtr/ (*partridge*) but never like /trt/. **1963** *Language* XXIX. 86 The morphophonemics of a language is then a set of rules for translating morpheme-sequences to phoneme-sequences. **1955** *Publ. Amer. Dial. Soc.* xxiv. 20 All agrain spoken in America are obviously patterned on English, though their specialized morpheme structure usually renders them partially or totally unintelligible to those who speak only the language of the dominant culture. **1972** *Language* XLVIII. 372 The symmetrical vowel-harmony rule [in Turkish] is accompanied by an accidental gap in the inventory of suffixes, and a systematic morpheme-structure constraint on polysyllabic roots. **1971** D. Crystal *Linguistics* iv. 197 Proponents of the morpheme theory have been..convinced of its validity..as a general explanation for grammatical phenomena. **1942** *Language* XVIII. 171 The term *morpheme-unit*..refers..only in phrases with *I*, and, which never occurs with *I*, are put into one morpheme unit... A morpheme unit is..a group of one or more alternants which have the same meaning and complementary distribution. **1933** L. Bloomfield *Language* xiii. 209 *Morpheme-words*, consisting of a single (free) morpheme: *man, boy, cut, run, red, big.*

Hence morpheme-mic *a.*, morphe-mically *adv.*; also (*rare*) morphemati-c *a.*, morphema-tically *adv.*, morphe-miciza-tion.

1930 J. R. Firth *Speech* vi. 56 Dutchmen, Danes, Swedes, and Englishmen share many similar morphemic and phonetic habits. **1933** *English Studies* XV. 87 The alternating pairs of phonemes..evidently form special units which, being conditioned by morphematic circumstances, are called *morphonemes*. **1935** G. K. Zipf *Psycho-Biol. of Lang.* (1936) ii. 35 One may..devise a morphemic alphabet to label all the different morphemes (e.g. prefixes, roots, suffixes, and endings) of a language. **1947** *Language* XXIII. 322 We must have..a set of principles on the basis of which to group, or refuse to identify, different stretches of speech as morphemically the same. **1951** C. E. Bazell *Linguistic Form* vi. 63 Only semantic prejudices could lead one to suppose that there was a closer contextual relation with the morphemically distinct *behind*. **1957** S. Potter *Modern Linguistics* i. 16 Plural forms of [some] substantives..which happen to end in a dental (*nuts*, *lords*..). Here the inflexion *-s* or *-es* is felt by the speaker to be morphemic. **1959** R. Quirk in Quirk & Smith *Teaching of English* i. 34 Many of the curious and amusing forms in *Finnegans Wake*..require us to make morphemic associations in unfamiliar environments. **1964** D. Ward in D. Abercrombie et al. *Daniel Jones* 390 Such a system of orthography..would strip the various relationships between expressions which are grammatically and morphemically (or lexically) different, but actually phonemically identical..the distinctions between morphematic modifications. **1964** E. A. Nida *Toward Sci. Transl.* v. 102 Similarly, we must be able to..the morphemic variations. *Ibid.* 116 The phonemicization of *latter/ladder* (to those dialects that clearly have a voiced intervocalic stop would not be immediately relevant and so on. **1971** *Archivum Linguisticum* II. 49 The 'ambiguity' of morphemic -*s* equally characterizes -*ing*, to name only one further function. *Ibid.* 59 A morphemically variable form.

morphemics (mǭɹfī·miks), *sb. pl. Linguistics.* [f. *morphem(e): see -ic 2.] The study and description of language in terms of morphemes.

1947 *Language* XXIII. 321 Morphemics and tactics are both necessarily involved in grammar. **1951** Trager & Smith *Outl. Eng. Structure* ii. 53 The analysis we are now going to do, the *morphemics* of the language, deals with the recurring patterned partials in utterances. **1960** T. B. W. Reid *Historical Philol. & Ling. Sci.* 9 Others maintain that..phonemics and morphemics alone constitute linguistics, whereas phonetics and semantics are independent disciplines which do not interest them. **1961** D. L. Bolinger *Generality, Gradience & All-or-None* iii. 35 There is obligatory lengthening in too many places for us to insist that English morphemics be studied in total disregard of it. **1966** M. Critchley *Developmental Dyslexia* iv. 14 The characters have to cope not only with the phonetic properties of the sound, but also with the problem of meaning. In other words, both phonemics and morphemics must be satisfied. **1966** *Amer. Speech* XLI. 116 If the first and last words in *Betty is in her beddy* are phonetically identical, then for the morphemics, one would have to..set up an allomorph [bet] for the morpheme (bed).

So morphe·micist, one who studies morphemics.

1949 Kirby & Woolf *Philologica: Malone Anniv. Stud.* 312 In analyzing Joyce's work, the morphemicist would be obliged to construct diagrams in layers, piled one on top of the other.

morphically (mǭ·ɹfikăli), *adv.* [f. Morphic *a.* + -al + -ly[2].] In relation to or as regards shape or form; morphologically.

1894 *Proc. Boston Soc. Nat. Hist.* XXVI. 66 This is morphically a free cell.

morphism: see *morph[5].

-morphism (mǭ·ɹfiz'm), terminal element [f. Gr. μορφή form + -ism] of sbs. with the sense 'condition or property of having a (certain) form or character', and in *Math.* 'a transformation or correspondence of a (certain) kind': e.g. Heteromorphism, Isomorphism (in Dict. and Suppl.).

morpho (mǭ·ɹfo). [mod.L. (J. C. Fabricius 1807, in *Magazin für Insectenkunde* VI. 280), f. Gr. Μορφώ, an epithet of Aphrodite.] A large South American butterfly of the genus so called, often with shiny blue wings. Also *attrib.*

1853 A. R. Wallace *Travels on Amazon* 14 Among them..the splendid blue butterflies..the large and splendid large metallic-blue butterflies which are always first noticed by travellers in South America. **1863** H. W. Bates *Naturalist on River Amazons* I. 103 The splendid metallic blue Morphos..are generally confined to the shady alleys of the forest. **1925** *Glasgow Herald* 1 Aug. 4/2 In the parapet of a flat staircase in one of the Oxford colleges there is a piece of Labradorite with a fine shimmering blue colour that recalls the tail of a peacock and the wing of a Morpho butterfly. **1936** *Ibid.* 12 June 4 A morpho may have a million and a half scales on its four great wings. **1969** B. Bishop *Compl. Poems* 170 Oil..like bits of mirror..so, more blue than that: Like tatters of the Morpho butterfly. **1973** *Sci. Amer.* Dec. 67/2 Only here [*sc.* in the tropical rain forest] are found such insects as the brilliant blue morpho butterfly of tropical America.

morphodite (mǭ·ɹfodait). Also **morphodyke, morphodyte.** Colloq. shortening and mispronunciation of Hermaphrodite *sb.*

1896 *Dialect Notes* I. 421 *Morphodite*,..for *hermaphrodite*. **1951** T. Capote *Grass Harp* (1952) i. 9 One of the stories he spread was that Verena was a morphodyke, never stopped going around. **1956** H. Gold *Man who was not with It* (1965) xiii. 192 Brother, morphodyte, but I suppose morphodykes are likely to take it personally when people are around.

morphogen (mǭ·ɹfodʒĕn). *Biol.* [f. Gr. μορφή form + -o + -gen.] A substance that is or might be capable of bringing about or determining morphogenesis.

1952 A. M. Turing in *Phil. Trans. R. Soc.* B. CCXXXVII. 38 The systems actually to be considered consist therefore of masses of tissues which are not growing, but within which certain substances are reacting chemically, and through which they are diffusing. These substances will be called morphogens, the word being intended to convey the idea of a form producer. It is not intended to have any very exact meaning. **1969** *Math. in Biol. & Med.* (Med. Res. Council) vi. 247 Turing considers the behaviour of two chemical substances, A and B, which he calls 'morphogens'. The significance of the name is simply that it is supposed that a high concentration of one or other of these substances can act as an 'inducer' of further differentiation. **1972** *Nature* 25 Oct. 366/1 Many theories for the mechanism of signalling in embryos have been produced; the oldest, and simplest, is that there is a concentration gradient of a 'morphogen' from the animal to vegetal poles, and that cells can measure their ambient concentration and then behave according to their position in the gradient.

morphogenesis. Add: (Examples.)

1899 *Jrnl. R. Microsc. Soc.* 469 Factors in morphogenesis. **1936** P. D. F. Murray *Bones* i. 16 These experiments evidently exclude all functional factors as important agents in primary skeletal morphogenesis. **1951** *Amer. Nat.* LXXXV. 218 Isolated terminal growing points.. afford excellent material for the investigation of the relation between nutrition and morphogenesis. **1974** *Sci. Amer.* Dec. 44/3 They [*sc.* hydras] have a strong capacity for regenerating lost parts, so that one can study morphogenesis in the adult form as well as in developing embryos. **1974** *Nature* 18 Aug. 301/2 Isolated terminal growing points..

b. *Geomorphol.* The formation of landscapes or land forms.

1958 *Geol. Mag.* XCV. 126 Tricart and Cailleux..claim that a 'sufficiently dense cover of vegetation' interferes very significantly with morphogenesis. **1966** J. A. Mabbutt in G. H. Dury *Essays in Geomorphol.* 102 The hilltoot zone is..very sensitive to climatically-induced changes of morphogenesis. **1968** R. W. Fairbridge *Encycl. Geomorphol.* 130/1 An allied form of morphogenesis..in the relatively stable plains, aggrades and deranges drainage in areas of continental glaciation.

morphogenetic *a.* (further examples, in *Biol.* and *Geomorphol.*); so **morphogene·tically** *adv.*, as regards morphogenesis.

1905 *Jrnl. Exper. Zool.* II. 155 It seemed to me scarcely possible that all of these strikingly different kinds of oöplasm, each with its own peculiar developmental history and destiny, were nevertheless morphogenetically identical. **1933** J. S. Huxley *Probl. Relative Growth* ix. 172 Each such region would be characterized both by the possession of its own growth-gradient and by its own morphogenetic field. **1950** *Ann. Assoc. Amer. Geographers* XL. 214 Recently Budel (1944, 1948) has suggested the recognition of 'formkreisen' or 'morphogenetic regions'. *Ibid.*, 215 Within an area in which the climate is essentially homogeneous, the geomorphic effects of climate may be assumed to be similar. These areas are herein called morphogenetic regions. **1954** W. D. Thornbury *Princ. Geomorphol.* iii. 63 The concept of a morphogenetic region is that under a certain set of climatic conditions particular geomorphic processes will predominate and hence will give to the landscape of the region characteristics that will set it off from those of other areas developed under different climatic conditions. **1957** [see *field *sb.* 17 c]. **1965** *Science* 29 Oct. 627/3 (*caption*) Possible relation of collagen distribution to morphogenetically active and inactive areas. **1966** Saunders & Fallon in M. Locke *Major Probl. Developmental Biol.* 289 Death of cells is a normal component of most morphogenetic movements such as foldings, detachments, and the confluence of anlagen. **1969** *Geogr. Annaler* XLI. 68 (*heading*) Morphogenetic aspects of the agrarian landscape of Öland.

morphogenic, *a.* (Examples.)

1904 *Science* 2 Dec. 749/2 The specific morphogenic factors are connected in some way with specific forms of protoplasm. **1958** C. P. Raven *Morphogenesis: Analysis of Molluscan Development* v. 186 The notion of morphogenic substances which should be taken too strictly, as if there were a rectilinear relation between a single substance and the differentiation of a certain organ. **1965** D. Wexler in Maibach & Rovee *Epidermal Wound Healing* iv. 104 It is thought that in the normal epidermis the morphogenic force is the pressure of mitosis.

morphogeny. Delete *Biol.* and add further examples.

1912 *Ann. Assoc. Amer. Geographers* II. 85 Some British observers have recognized levels of erosion or stranding at several different altitudes, on which they base a more complicated scheme of morphogeny. **1939** *Jrnl. Ecol.* XVII. 374 Of fundamental importance in morphogeny was..to be overlooked in adaptation is the correlation of organs and parts, based upon competition within the plant for food. **1950** *Ann. Assoc. Amer. Geographers* XL. 121 Many of the ideas here expressed are explicitly stated in the lectures of Professor Kirk Bryan, whose emphasis on climatic morphogeny has led to the formulation here set forth.

morphographemic (mǭɹfogrăfī·mik), *a. Linguistics.* [f. Gr. μορφή form + *graphemic *a.*] Of or pertaining to the written form of words; *spec.* *morphographemic rule*, a rule of writing in transformational grammar (see quot. 1972).

1964 O. Thomas *Transformational Gram.* iii. 59 If we were interested in how the final sentence was pronounced, rather than in how it is written, we would have a set of morphophonemic rules (i.e., rules of pronunciation) rather than *morphographemic rules* (i.e., rules of writing). **1968** J. Lyons *Introd. Theoretical Linguistics* vi. 263 For a written language, the morphographemic rules would convert the strings. **1972** H. W. Hoenigswald *Linguist Stud.* 255 The immediate output of the lexicon is correct orthography, subject to minor morphographemic adjustments. **1972** R. A. Palmatier *Gloss. Eng. Transformational Gram.* 98 *Morphographemic rules*..a rule, of the final section of the grammar, which converts the morphemes of a terminal string into their graphic representations as written words; a spelling rule.

morpholexical (mǭɹfolĕ·ksikăl), *a.* [f. Gr. μορφή form + Lexical *a.*] Of or pertaining to lexical form (cf. also *morphophonemic *a.*).

1939 L. Bloomfield in *Travaux du Cercle Ling. de Prague* VIII. 105 It is necessary to distinguish these morpholexical alternations, i.e., those which are associated with particular words..and phonetic alternations. **1949** *Archivum Linguisticum* I. 147 The morpholexical variations are called sandhi..internal and external. **1950** *Language* XXI. 256 Besides zero and actual element, pairs of forms as -kamy and -kamy, with the meaning identical but both ('water'), are reserved for consideration as morpholexically conditioned. **1951** C. S. Harris *Structural Linguistics* xiii. 197 It [*sc.* the procedure]..covers regular and irregular phonological alternation, simple morpholexical variation, suppletion, reduplication, and morphophonemic variation. **1962** Crevarria-Aguilar & Penzl in Householder & Saporta *Problems in Lexicography* iv. 243 The Pashto Academy..has attempted..to include both morphological and sometimes morpholexical standards for writers.

morpholine (mǭ·ɹfŏlīn). *Chem.* [ad. G. *morpholin* (L. Knorr 1889, in *Ber. d. Deut. Chem. Ges.* XXII. 2083), f. *morphin* Morphine *sb.* with insertion of -ol.] A cyclic amine,

$$CH_2CH_2NHCH_2CH_2O,$$

which is a hygroscopic oil used as a solvent for resins and dyestuffs and whose fatty-acid salts are emulsifying soaps used in floor polishes.

1889 *Jrnl. Chem. Soc.* LVI. 1218 Morpholine (tetrahydroparoxazine) may be regarded as the 'inner anhydride' of dihydroxyethylamine. **1935** *Industr. & Engin. Chem.* Aug. 871/1 Of the derivatives of morpholine, one of the most extensively investigated in the medical field has been morpholine ethanol. **1955** *Sci. Amer.* Aug. 76/3 Dilute solutions of morpholine neither attract nor repel salmon but were detectable by them in extremely low concentrations. **1972** *Materials & Technol.* IV. xii. 516 Diethanolamine has also appreciable value as the source of the cyclic ether, morpholine, an important solvent and intermediate in the manufacture of medicinals.

morphologic, *a.* (Further examples.)

1936 *Bull. Amer. Assoc. Petroleum Geologists* XX. 859 The peculiar morphologic phenomenon of deep-sea trenches..is confined to the peri-Pacific-Malayan region. **1960** J. Kantor *Objective Psychol. of Grammar* xi. 158 Yielding the structural categories recognized by such as English. **1944** *Federation Proc.* III. 231/1 An..account of the morphologic changes in muscles following denervation. **1964** A. Juilland *Structural Relations* iii. 51 Morphologic representation has been explored especially by American structuralists. **1966** A. E. J. Morris in Gugel's *Tectonics* ii. 6 The two kinds of morphologic distinguish rock types..involves great risks of error. **1967** *Word* XXIII. 280 All morphologic composites are based on the same syntagmatic pattern 'determinant determinatum'. **1975** *Language* LI. 142 This fact the special forms the specific morphologic..infinitives conjugate with the verb (*have*).

morphological, *a.* Add: **5.** Of or pertaining to morphology (sense *4).

This sense can be regarded as including also the senses in Dict.

1895 N. Story-Maskelyne *Crystallography* i. 3 By reference to this system of planes, it is possible to establish the morphological relationship of all crystals of the same substance, whatever faces they may exhibit. **1918** *Amer. Jrnl. Med. Sci.* CLV. 70 The cardiac signs and symptoms produced as a result of hyperthyroidism are not due to morphological alterations in the heart muscle. **1972** D. Herbert *Urban Geogr.* iii. 66 A clear indication of past functional patterns are often be obtained from the morphological evidence which is left in the urban landscape. **1974** D. Mehr *Crystalline Solids* xii. 374 Morphological studies are..applicable only to substances which form well developed crystals.

morphologist. Delete *Biol.* and add further examples.

1935 [see *clinographic *a.*]. **1972** D. Herbert *Urban Geogr.* iii. 70 To incorporate the physical-functional fabric of the urban into the evolution of urban form has long been advocated by urban morphologists. **1974** *Nature* 13 Sept. 100/1 Although glial cells can be classified by morphologists, physiologists know very little about their function.

morphology. Add: **I. 2. b.** *Appos.*, as *morphology–syntax. Comb.*, as *morphology-based* adj.

1965 H. A. Gleason *Linguistics & Eng. Gram.* vi. 118 'Preposition' is available and needed in the morphology-based system. **1945–9** *Acta Linguistica* V. 125 A problem in morphology–syntax, the..**1961** R. Garvin *On Linguistic Method* 157 Morphemic analysis [is]..primarily syntactic for those languages to which the morphology-syntax division applies.

II. 4. Shape, form, external structure or arrangement, esp. as an object of study or classification.

1890 C. P. Mitchell *Philos. Tumour-Dis.* ii. 51 Facts so varied as those comprised in the morphology of tumours. **1894** *Jrnl. Physiol.* XVII. 81 (*heading*) The morphology and distribution of the wandering cells of Mammalia. **1899** N. Story-Maskelyne (*title*) Crystallography: a treatise on the morphology of crystals. **1921** *Geogr. Rev.* XII. 555 (*heading*) Morphology of the Altai Mountains. **1924** A. E. H. Tutton *Nat. Hist. Crystals* vii. 60 The 'elements' of a crystal, together with a list of the forms observed, and a table of the interfacial angles, define the morphology of the crystal. **1935** *Rep. Brit. Assoc. Adv. Sci.* 373 (*heading*) A comparative study of the morphology of the North Downs and the Chiltern Hills. **1950** *Jrnl. Gen. Physiol.* XXXIII. 657 Two separate lines of approach were required: a study of the functional behavior of the nerve fibers, and a study of their morphology. **1954** H. Brainerd *Lost Villages* 23 In the study of village morphology..an opportunity of seeing a medieval village plan without any of the accretions of later building. **1964** R. C. Evans *Crystal Chem.* (ed. 2) xi. 184 Before the discovery of X-ray diffraction, crystals were studied principally from the standpoint of their morphology. **1966** J. B. Sheppard in *Terrestrial Photogrammetry*..of the nerve fibers in a particular direction... **1969** *Sci. Amer.*..a study... **1972** *Nature* 10 Nov. 372/2 The morphologies found—spheroids, rods, single or paired, filaments, segmented or nonsegmented, 'colonial' structures..—have all been described from prokaryotic groups such as the blue-green algae.

morphomaniac (mǭɹfomē·niæk). [Irreg. f. Morphia + Maniac *sb.*] = Morphinomaniac.

1906 G. K. Chesterton *Dickens* ii. 41 These great popular leaders..became drunkards; they become dema-gogues; they become morphomaniacs. **1921** L. J. Vance *Destroying Angel* xii. 185 Some of them..morphine..?

morphometric (mǭɹfome·trik), *a.* [f. Gr. μορφή form + -o + *-metric.] Of or pertaining to morphometry or morphometrics; morphometrical.

1931 L. Wolfson in *Biometrika* XXII. 325 A special advantage of taking measurements on the rounded bones of the skull is that we thus get some idea of the size and shape of relatively small regions..giving us a better appreciation of local asymmetries than from the use of anthropometric measurements can. **1963** *Bull.*..to distinguish the two types of measurements as ethnometric and morphometric. **1932** *Nature* ..in morphometric analyses of the skull. **1934** *Geogr. Jrnl.* LXXXIV. 474 Other questions, such as the morphology of arctic regions, morphometric methods, classification and variation of climate. **1956** *Bull. Geol. Soc. Amer.* LXVII. 1117 Detailed morphometric analysis of basins in five sample areas in the equilibrium stage show distinctive, though small, differences in hypsometric integrals and curve forms. **1968** *Mineral. Abstr.* XIX. 240/1 Cumulative curves for the morphometric parameter of the grains composing a sample of sand are figured. **1970** *Evolution* XXIV. 220/1 Morphometric data from 21 strains and 123 mandibles of 2 and 3 year old male white-tailed deer from Michigan were used in this study. **1972** A. Young *Slopes* i. 3 Morphometric analysis takes the river basin as the fundamental unit; this is a consequence of its origin in hydrological studies, in particular in the attempt to predict flood discharges from the landform characteristics of river catchments. **1974** *Nature* 24 May 374/1 A three-fold enrichment judged by morphometric analysis of electron micrographs of synaptic junctions..was obtained.

morphometrical, *a.* (Further examples.)

1937 *Trans. Connecticut Acad. Arts & Sci.* XXXIII. 67 The lake is primarily eutrophic, secondarily or morphometrically oligotrophic. **1957** G. E. Hutchinson *Treat. Limnol.* I. ii. 170 The number of lakes of maximum depth of over 400 m. will, however, probably be increased when the large lakes of southern South America become known morphometrically. **1960** *Jrnl. Animal Ecol.* XXIX. 310 The parents were most probably very similar morphometrically to the fledglings.

morphometrics (mǭɹfome·triks), *sb. pl.* [f. as prec.: see -ic 2.] Morphometry (esp. of living forms); also, morphometric features or properties.

1960 *Jrnl. Animal Ecol.* XXIX. 309 The shape of an animal, of which morphometrics are a numerical expression, is dependent upon genetically transmitted characters and the modifying influence of the environment on them. *Ibid.* 310 No completely reliable evidence was obtained for either the morphometrics or the behaviour of the parents of two of the populations studied. **1971** Blackith & Reyment *Multivariate Morphometrics* vii. 63 A similar problem in morphometrics is to measure the degree of similarity of two forms. **1974** *Nature* 22 Mar. 303/3 The chapters on multivariate morphometrics and ordination are particularly welcome.

morphometrical *a.* (examples).

1935 *Geol. Mag.* LXXII. 185 Its [*sc.* the clinographic curve's] utility in delicate morphometrical determinations seems to be undoubted. **1935** *Biometrika* XXVII. 463 (*heading*) Morphometrical indices. **1963** *Anti-Locust Bull.* XVI. (*heading on title-p.*) Morphometrical studies on phases of the Desert Locust (*Schistocerca gregaria* Forskål).

morphometry. Add: Esp. in *Geomorphol.* (Further examples.)

1957 G. E. Hutchinson *Treat. Limnol.* I. ii. 168 (*heading*) Morphometry of lakes over 400 m. deep. **1958** *Geogr. Jrnl.* CXXIV. 370 (*heading*) Aspects of the morphometry of a 'polycyclic' drainage basin. **1968** *Geological Abstr.* Sept. 33 Earlier work showing the dominance of rock type in pebble morphometry is strongly supported. **1963** E. R. Weibel (*title*) Morphometry of the human lung. **1970** R. J. Small *Study of Landforms* i. 4 Quantification is applied..to landscape forms, giving rise to the branch of modern geomorphology known as 'morphometry'.

morphon. Add: **b.** *Linguistics.* A term in stratificational grammar for *morphophoneme. Also *attrib.* Hence **morpho·nic** *a.*

1964 S. M. Lamb in *Rep. 15th Ann. Round Table Meeting Ling. & Lang. Stud.* (Georgetown Univ. Inst. Lang.) 115 The morphons are related to phonemes by the morphophonemic realizations. **1966** *Language* XLI. 100, I shall use Lamb's convenient neologism *morphon*. The syntactic component of a correct generative grammar..yields sentences as strings of morphons. **1966** S. M. Lamb *Outl. Stratificational Gram.* ii. 19 For the terms in the alphabet, alternation pattern, the iconic, the morphonic, and the phonemic. **1967** C. F. Hockett *Language, Mathematics, & Linguistics* iii. 95 Relative to the terminal subalphabet T'(G), a morphon string is a simple (linear) string. **1968** *Amer. Speech*..emphasized..that stratificational grammar insists that the morphophonemes, or 'morphons' in their terms, are unanalyzable symbols. **1968** *Language* XLIV. 595 Rules of grammar are expressed..because the objects on the higher stratum, called 'morphons', are distinct from the objects on the lower stratum, which are called phonemes. **1970** G. Sampson *Stratificational Gram.* ii. 33 The large number of neutralisations represented by almost every morphon. *Ibid.* ii. 45 In a complete description of English, there would be two further strata, each with its lexotactics, below the morphon level... I am not in a position..to give the morphonic realizations of the morphon.

morphoneme (mǭ·ɹfonīm). *Linguistics.* [f. Gr. μορφή form + *phon(eme).] = *morphophoneme. So **morphone·mic** *a.*

1933 *English Studies* XV. 87 The alternating pairs (sometimes groups) of phonemes like *f/v*, *t/θ* evidently form special units which, being conditioned by morphematic circumstances, are called *morphonemes*. **1934** *Language* X. 335 The five Russian vowel morphonemes are very variable... The consonant morphonemes are: /p, b, t, d.../. *Ibid.* 340 The interchange here is morphonemic, and does not properly concern us in a purely phonemic study. **1942** *Ibid.* XVIII. 399 Summarizing briefly Trager's article ('The Phonemes of Russian', *Language* 1934, No. 4) and generally valuing its positive outlook, Reformatsky deplores his terminological unclearness connected with [his] use of the word 'morphoneme'. **1966** J. Vachek *Linguistic School of Prague* iv. 81 The generativists after establishing the morphophonemic make of strings proceed, to 'rewrite' the morphonemes directly as the phonemic implementations of these strings.

morphonology (mǭɹfonǫ·lŏdʒi). *Linguistics.* [f. Gr. μορφή form + Ph(onology).] A term used by the Prague school of linguists for the study of the phonology of morphemes. So **morphonolo·gic, morphonolo·gical** *adjs.*

1929 *Travaux du Cercle Linguistique de Prague* I. 85 La grammaire doit comprendre encore un chapitre particulier, qui étudie l'utilisation morphologique des différences phonologiques, et qui peut être appelée la 'morphophonologie', ou, en abrégeant, la 'morphonologie'. **1933** *English Studies* XV. 87 That part of linguistics which examines the phonological structure of morphemes, is called *morphonology*. **1935** *Amer. Speech* X. 225 Another branch of phonology, established by the Prague linguists, and called morphonology, ..examines the phonological structure of morphemes. **1936** *Ibid.* XI. 114 Morphological phonology or 'morphology'..investigates the morphological use of phonemes or groups of phonemes, or more precisely, for phonological means. **1949** H. Spang-Hanssen in *Travaux du Cercle Ling. de Copenhague* V. 69 Morphonology or morphonemic monographs have on direct bearing on this subject either, since they also presuppose that the phoneme inventory has already been established. **1950** *Lingua* II. 241 This morphology, according to Trubetzkoy, has a threefold task: 1, the investigation of the phonological (i.e. phonematic) structure of the morphemes, 2. the investigation of the combinative sound-changes.., 3. the investigation of the series of sound-changes..which have a morphological function. **1966** J. Vachek *Linguistic School of Prague* iv. 81 The other recent trend in linguistic thinking that has substantially contributed to the revival of interest in morphonology is the stratificational approach by Sydney M. Lamb and his colleagues. *Ibid.*, The first language to be characterized from the morphonological point of view was..Russian. **1971** tr. *Akhmanova's Phonology, Morphonology, Morphology* ii. 89 Professor Vachek speaks of the present state of morphonological research... The 'generativist and transformationist' morphonology is, perhaps, best described as 'codificational morphonology'.

morphophoneme (mǭɹfofōᵘ·nīm). *Linguistics.* [f. Gr. μορφή form + *phoneme.] One of the variant phonemes which belong to the same morpheme.

1934 M. Swadesh in *Language* X. 129 Morphologically distinct phonemes are called morpho-phonemes. A morpho-phoneme is one of a class of like phonemes considered as components of actual morphemes which behave alike morphologically, i.e., have a like place in the same mutation series. **1939** L. Bloomfield in *Travaux du Cercle Linguistique de Prague* VIII. 106 The morphological structure..a morpheme is..these symbols..(of) the morphophonemes. **1947** C. F. Hockett in *Language* XXIII. 323 Morphophonemic statements may involve morphophonemes—that is, the symbols used for phonemes, plus special symbols, with special definitions as to phonemic value under varying circumstances—or they may not. **1957** N. Chomsky *Syntactic Structures* iv. 33 The elements that figure in the rules can be classified into a finite set of levels (e.g. phonemes and morphemes; or, perhaps, phonemes, morphophonemes, and morphemes). **1964** E. Bach *Introd. Transformational Gram.* ii. 25 We distinguish the final morphophonemes in *laugh* and *wife* on the basis that in one case..we have a distinct morphophoneme. **1968** *Language* XLIV. 596 The major thesis of this paper is that the most efficient description of the morphophonemics of a language..requires morphophonemes and, moreover, that these morphophonemes do in fact constitute psychologically real units. **1972** *Ibid.* XLVIII. 372 In Turkish, the target is satisfied at a level preceding the application of the first law of vowel harmony, which maps the morphophonemes *I* and *U* onto their respective phonetic reflexes.

Hence **morphophonemi·c** (*rare*), **morphophone·mic** *adjs.*, **morphophone·mically** *adv.*

1939 [see *morpholexical *a.*]. **1944** *Amer. Speech* XIX. 136 The 'intimacy' of the [r] in Hungarian and Finnish is not phonemic but morphophonemic. **1951** Z. S. Harris *Methods in Structural Linguistics* xiv. 236 The morphophonemic symbols cover the various ways..where /z/ represents phonemic zero before /z/.. **1957** *Publ. Amer. Dial. Soc.* xxviii. 58 Morphophonemically, the *be*-form..stands..as an auxiliary, regardless of the verb. **1958** D. W. Reed in *Dialect Studies* 1958 47/2 Evidence for the diphthongal character of Gothic ai..is largely morphophonemic (e.g. the alternation *triu/triwa*). **1964** D. Ward in D. Abercrombie et al. *Daniel Jones* 390 There are some departures from the morphophonemic principle in Russian, including such specific cases as, for example, *neuzdrí* 'nostrils' (cp. *nos* 'nose'). **1965** *Canad. Jrnl. Linguistics* Spring 169 Most other stems [in Eyak] are morphophonemically invariable. **1966** *Language* XLII. 617 There is no special reason for setting up new allomorphs like *n*'s. *Ibid.* 633 Subphonemic...the morphophonemic solution can be regarded as one that was..categorial properties. **1970** *Bible Translator* XXI. 23 There is reason to believe that most function morphemes should be read at sight rather than 'sounded out', and this is especially true of those having morphophonemic variants. **1971** tr. *Akhmanova's Phonology, Morphonology, Morphology* ii. 72 It is quite impossible to formulate any morphophonemic rules to explain the would-be substitution of [-a] for [-i] or [:y].

morphophonemics (mǭɹfofonī·miks), *sb. pl.* [f. prec. + -ICS.] The study and description of the phonemic aspects of the constitution of morphemes.

1939 L. BLOOMFIELD in *Travaux du Cercle Linguistique de Prague* VIII. 105 The present paper describes..the *internal sandhi* or *morphophonemics* of the language. **1953** [see *morpheme-sequence]. **1957** S. POTTER *Mod. Linguistics* vii. 146 As we proceed from morphophonemics, and morphology proper, to syntax, we observe a rise in the scale of semantic values. **1965** B. M. H. STRANG *Metaphors & Models* 15 The study of how a language encodes its different meanings is called *morphophonemics*. **1968** *Amer. Speech* XLIII. 233 The author includes a description of Jamaican Creole..morphophonemics. **1972** *Language* XLVIII. 805 This negative is realized from the sentence modality without going through the morphophonemics of pre-verbal location. **1973** *Computers & Humanities* VII. 218 How does one handle the unbelievably complex morphophonemics in choosing headwords?

morphophonics (mǭɹfofǫ-niks), *sb. pl.* [f. Gr. μορφ-ή form + PHONICS *sb. pl.*] = *MORPHOPHONEMICS *sb. pl.* (see also quot. 1962). Also **mo·rphophone,** a unit representing the class of phonemes occurring in dialectally different pronunciations of morphemes; **morphopho·nic** *a.*

1962 E. F. HADEN et al. *Resonance-Theory for Linguistics* iii. 30 Such sets of allophones, as expressions of morphological units, are *morphophonic expressions*. *Ibid.* 40 Morphophonics..is the intermediate area between phonology and morphology... Morphophonics is the process of transposing phonemic entities from the context of phonology into the context of morphology. **1967** H. L. SMITH in *Language* XLIII. I. 306 A three-stratum, twenty-seven level model for linguistic analysis indicates structural units—the *morphophones*—between phoneme and morpheme. Morphophones are seen as the basic units of the morphemes, and are themselves composed of dialectally different phonemic variants, which are non-contrasting in the same lexical items. *Ibid.* 311 From one point of view, the morphophone can be seen as a sort of 'holding company' or 'super-family' of different phonemes which are non-contrasting in the same words. But from another point of view, the morphophone is a unit of the language as a whole, and as a unit furnishes the basis of what might be thought of as 'higher order' or 'higher level' contrasts. *Ibid.* 323 Substitutions and interchanges are aspects of the morphophonic system. *Ibid.* 341 The donor language's morphophonics have been subject to a pretty thorough Anglicization. **1972** G. L. TRAGER *Lang. & Languages* iii. 49 The diacritic part of morphology, which examines the items of which morphemes are composed, may be called *morphophonics* (more commonly the term is *morphophonemics*, but there are good theoretical reasons for omitting the *-em-*). *Ibid.* 56 Morphemes with more than one allomorph are made up of or composed of morphophones, and it follows that morphemes with only one allomorph are also composed of morphophones... The morphophonic alternations that occur can then be considered as relations between (simple) morphophones.

morphophonology (mǭɹfofǒnǫ·lǒdʒi). *Linguistics.* [f. Gr. μορφ-ή form + PHONOLOGY.] = *MORPHONOLOGY. So **morphophonolo·gical** *a.*

1934 *Language* X. 128 Morpho-phonology includes, in addition to the study of the phonemic structure of morphemes, the study of interchange between phonemes as a morphologic process. **1938** *Trans. Philol. Soc.* 106 The morpheme in the sense of Skalička and of the Prague Linguistic Circle is a morphophonological unit, and the corresponding phonomorphological unit is the phonemic function. **1953** K. JACKSON *Lang. & Hist. in Early Britain* II. 548 The development of IE. *p* cannot be connected in any way with lenition, since it does not follow the special morphophonological rules for its occurrence. **1970** G. C. LEPSCHY *Survey Structural Linguistics* iii. 56 [One of the] main tasks of synchronic phonology [is]..to describe the morphological utilization of phonological differences (morphophonology or morphonology). **1973** *Archivum Linguisticum* IV. 47 No account is being taken here of regular, if often quite subtle morphophonological alternations, as between..viyis and vayas. *Ibid.* 50 The lengthening of the vowel of -ha in the example..belongs rather to the (morpho)phonology of the language.

morphosyntax (mǭɹfosi·ntæks). *Linguistics.* [f. Gr. μορφ-ή form + SYNTAX 2.] A branch of linguistic study which combines the study of morphology and syntax.

1961 F. W. HOUSEHOLDER in Saporta & Bastian *Psycholinguistics* 23/2 So also in morpho-syntax it should be possible to agree on definitions for terms like 'noun phrase', 'verb phrase', 'sentence', etc. **1967** *Word* XXIII. 472 In morpho-syntax the sequential chain model should perhaps not be completely disregarded. **1968** Y. MALKIEL *Essays on Linguistic Themes* i. 3 A few scholars have gone so far as to consolidate all of morphology and syntax into the single domain of 'morphosyntax', which forms the hard, inalienable kernel of linguistics. **1969** tr. *Akhmanova & Mikael'an's Theory Syntax Mod. Linguistics* ii. 26 He [*sc.* Brøndal] not only remained faithful to his conception of 'morphology' and syntax, but also developed it by introducing the notion of 'morphosyntax' to cover such phenomena as the functioning of words in sentences, the semantic modification of words in context, [etc.]. **1972** *Archivum Linguisticum* III. 89 As a final instance of the general validity of a systematic approach to diachronic morpho-syntax, let us examine a well-documented change within the verbal system of Romance.

Hence **morphosynta·ctic** *a.*, **morphosynta·c·tically** *adv.*

1959 F. W. HOUSEHOLDER in *Word* XV. 232 It does not appear that languages have a semantic structure which is separate and distinct from the morpho-syntactic one. **1962** H. C. CONKLIN in Householder & Saporta *Problems in Lexicography* 119 While an 'appeal' to meaning does not improve grammatical analysis, neither does an intuitive appeal to morphosyntactic form yield the most appropriate analysis of meaning and reference. *Ibid.* 121 The compounds *firewater* and *silverfish*..are endocentric morphosyntactically. **1968** *Word* XXIV. 55 The United States dialect refers to any variety of English set off from other varieties of English, phonetically, morphosyntactically, or lexically. **1969** tr. *Akhmanova & Mikael'an's Theory Syntax Mod. Linguistics* v. 119 The *words*..had to have canonical phonemic shapes, to 'behave' morphophonemically and morphosyntactically. **1970** *Language* XLVI. 332 The proposal does not spell out how one gets from the morphosyntactic surface structure to the phonemics.

morphotactics (mǭɹfotæ·ktiks), *sb. pl. Linguistics.* [f. Gr. μορφ-ή form + TACTICS.] The study of the sequence of morphemes in a language (*MORPHEME b). Hence **morphota·c·tic** *a.*; **morphota·ctically** *adv.*

1958 A. A. HILL *Introd. Linguistic Structures* vi. 68 The term 'phonotactics', now widely used, as well as other terms in -'tactics' to indicate sequences of items such as 'morphotactics' and 'logotactics', I owe to an unpublished lecture by Robert P. Stockwell delivered before the Linguistic Institute held at the Georgetown University Institute of Languages and Linguistics in 1954. *Ibid.* viii. 120 With most prebases, it is a morphotactic characteristic which identifies them. *Ibid.*, These are the 'grammatical endings' of familiar terminology. They can be defined morphotactically as morphemes which stand frequently before terminal junctures. *Ibid.* xxi. 414 Morphotactics gives us the partial predictability of morphemes and allomorphs. **1962** E. F. HADEN et al. *Resonance-Theory for Linguistics* iii. 35 Whereas in morphemics our task is to identify the form and the property of a grammatically functioning unit, in morphotactics we focus our attention on such a unit occurring in a sequence of morphemes... The morphotactic identification will be made with the same criteria for determining Form and Property as in morphemics. *Ibid.* 40 Morphotactics complements morphemics with morphology in a manner exactly parallel to phonotactics and phonemics within phonology. **1966** S. M. LAMB *Outl. Stratif. Gram.* ii. 23 When a lexonic ordering is of elements belonging to different morphotactic classes, e.g. verb stem and verb suffix, it is up to the morphotactics to determine their relative order. **1967** D. G. HAYS *Introd. Computational Linguistics* xi. 184 The morphotactic rules come to light as restrictions on sequence are found... The establishment of the morphotactics can be exceedingly simple or as complex as the development of a full syntax. **1970** G. SAMPSON *Stratificational Gram.* vi. 59 The morphotactics generates sequences of 'words' (M-Word), separated by the downward-determined element 'word-boundary'. *Ibid.* 62 The preference value of this derivation is six, one from the semotactic trace, four from the lexotactic, and one from the morphotactic.

morphotectonics (mǭɹfotektǫ·niks), *sb. pl.* [f. Gr. μορφ-ή form + -O + TECTONICS.] The branch of geomorphology concerned with the form and structure of the larger features of the earth's surface (as continents, mountain ranges, river basins); also, the morphotectonic character or features of a region.

1956 E. S. HILLS in *Jrnl. Geol. Soc. Austral.* III. 1 (*heading*) A contribution to the morphotectonics of Australia. *Ibid.*, A model of this size does in fact afford a unique basis for the study of morphotectonics, since no other medium can present a synoptic view of the topography of an area as large as [a] continent. **1961** —— in *Q. Jrnl. Geol. Soc.* CXVII. 79 What I have called morphotectonics implies, as well as structural geomorphology, something broader and applicable to regional study on a large scale, using topography as the primary criterion, but of course not neglecting all that may be known as to geology and geophysics. **1968** R. W. FAIRBRIDGE *Encycl. Geomorphol.* 734/1 The concept of 'Gondwanaland' in the southern hemisphere..evolved in the late nineteenth century..on the basis of extraordinarily similar morphotectonics, as much as from common features of stratigraphy, paleontology and paleoclimate. *Ibid.* 734/2 Up till now only one attempt has been made at a detailed world survey of morphotectonics.

Hence **morphotecto·nic** *a.*, of or pertaining to or seen from the point of view of morphotectonics.

1961 *Q. Jrnl. Geol. Soc.* CXVII. 84 A purely morphotectonic map for Egypt..shows, firstly, that it is not possible to conceive of the region as an elliptical upwarp. **1967** *Geogr. Abstr.* A. 280 The Provence foldings south-east of the Durance and the Verdon confluent give a good summary of the morphotectonic evolution of the Provence Range. **1973** *Nature* 2 Mar. 43/2 The morphotectonic features shown in Figs. 1 and 3 show a close agreement between structure and topography.

morphotropism (mǭɹfǫ·trǒpiz'm). *Cryst.* [f. as next: see -ISM.] = next.

1905 *Amer. Chem. Jrnl.* XXXIV. 104 The chapter on Morphotropism deals with the dependence of the crystal structure on the chemical constitution of the body. **1959** tr. *W. F. de Jong's Gen. Crystallogr.* 184 If a change of structure that occurs by replacement becomes so great that we no longer want to use the term 'isomorphism', we speak of morphotropism. **1966** *McGraw-Hill Encycl. Sci. & Technol.* VIII. 599/1 Organic compounds also show morphotropism.

morphotropy (mǭɹfǫ·trǒpi). *Cryst.* [ad. G. *morphotropie* (P. Groth 1870, in *Ann. d. Physik und Chem.* CXLI. 39), f. Gr. μορφ-ή form + τροπή turning.] (The study of) the progressive change in crystal structure brought about by replacing one of the species of atom or radical in a crystal by other species.

1900 *Rep. Brit. Assoc. Adv. Sci. 1900* 167 Morphotropy and isomorphism have a common cause, and..this is more likely to be discovered by the crystallographic study of substances showing morphotropic relationships than from the examination merely of materials likely to exhibit isomorphism. **1924** W. S. STILES tr. *F. Rinne's Crystals & Fine Struct. Matter* ix. 105 P. v. Groth has founded the science of morphotropy. By comparing crystalline forms, he succeeded in showing how the replacement of H by (OH), NO_2, NH_2, CH_3, Cl, in benzene (C_6H_6) manifests itself in changes of corresponding angles, and ultimately the symmetry. **1939** R. C. EVANS *Introd. Crystal Chem.* iii. 60 The structural importance of the quantitative lattice theory..lies in the light which it throws on..the problems of polymorphism and morphotropy.

So **morphotro·pic** *a.*, characteristic of or exhibiting morphotropy.

1899 *Mineral. Mag.* XII. 66 The three minerals may be members of a morphotropic series, in which the vertical crystallographic axis increases in length with an increase in the amount of lead. **1937** W. L. BRAGG *Atomic Struct. Minerals* ix. 150 The members of the chondrodite series are interesting because of the morphotropic relationships which exist between them. **1959** tr. *W. F. de Jong's Gen. Crystallogr.* 184 A morphotropic range of compounds is one in which the replacing atoms differ gradually in one property—for example, in size or polarizability. **1973** A. I. KITAIGORODSKY *Molec. Crystals & Molecules* i. 18 Morphotropic changes associated with a loss of symmetry are accompanied by an increase in the packing density.

morpion. Restrict † *Obs.* to sense in Dict. and add: **2.** (mǭ·ɹpiǫn) *transf.* Applied scornfully to a person. *rare*⁻¹.

1954 S. BECKETT *Waiting for Godot* II. 48 Moron!..Vermin!..Abortion!..Morpion!

morrillo, var. *MORILLO.

Morris (mǫ·ris). The name of William *Morris* (1834–96), poet and craftsman, used *attrib.* of styles of furniture, wallpaper, etc., designed by him or made in his factory or at the Kelmscott Press, *spec.* as **Morris chair,** a type of easy chair with open padded arms and an adjustable back. Also **Morris-papered** *a.*, papered with Morris wallpaper.

1880 LADY JEBB *Let.* 6 Mar. in A. Adburgham *Shops & Shopping* (1964) xvi. 173, I wish you could see Mrs Sellar's drawing-room..; Morris papers on the walls, Burne-Jones' photographs. **1900** *T. Eaton & Co. Catal.* Spring & Summer 235/3 Morris chair, golden oak or mahogany finish. **1912** T. E. LAWRENCE *Let.* 20 May in *Home Lett.* (1954) 209 The piece of Morris tapestry I have. **1912** E. WORDSWORTH *Glimpses of Past* xii. 140 Those were the days of green serge gowns, and Morris papers, depicting miniature orange trees and pomegranates, daisies, and sunflowers. **1920** J. GALSWORTHY *In Chancery* III. iii. 257 On a wall, not yet Morris-papered, was a print of the Queen. **1925** G. B. SHAW *Let.* 24 Feb. in *To a Young Actress* (1960) 83 Morris made all sorts of chairs: there is no particular Morris chair that I know of. **1935** N. MITCHISON *We have been Warned* I. 107 The Morris curtains would be drawn. **1936** J. DOS PASSOS *Big Money* 248 He let himself drop into the morrischair. **1937** V. WOOLF *Years* 54 There she was among the Morris wall-papers and the cabinets. **1938** J. CARY *Castle Corner* 274 A Bible in Morris binding. *Ibid.* 303 I'm quite sure our Morris sugar sifter is far greater than Milan. **1958** *Listener* 28 Aug. 317/1 The Morris dining-room. **1961** A. MILLER *Misfits* iii. 31 The complete assortment of furniture, from the Morris chair to the studio couch. **1967** O. LANCASTER *Eye to Future* v. 123 A charming cottage, stone-walled and Morris-papered. **1972** D. MARLOWE *Do you remember England?* i. 20 The warm bottle of Malvern water by the bedside, Morris wallpaper, and a train-journey biography.

Hence **Mo·rrisan, Morri·sian** *a.*, of, pertaining to, or characteristic of William Morris or his work; also **Mo·rrisite,** a follower or adherent of the principles, etc., of William Morris.

1897 A. BEARDSLEY *Let.* 26 Dec. (1971) 413 On the art side I suggest that it [*sc.* a projected new periodical] should attack *untiringly and unflinchingly* the Burne-Jones and Morrisian medieval business. **1918** A. BENNETT *Roll-Call* I. ix. 187 A quiet, inexpensive blue dress, embroidered at the neck in the Morrisian manner. **1936** *Archit. Rev.* LXXX. 179/3 Morris's passionate campaign is famous enough, but there were true Morrisites before Morris, men such as Pugin and Semper. **1936** W. R. TITTERTON *G. K. Chesterton* I. i. 23 The S.D.F.ers had always been Marxian in theory, though revolutionary Morrisites in sentiment. **1949** G. B. SHAW *Sixteen Self Sketches* x. 57 A lady in the esthetic dress momentarily fashionable in Morrisan cliques just then. **1970** *New Society* 31 Dec. 1151/2 It seemed plausibly argued that plastic would demean all our sensations and our contact with nature... This kind of approach had wood in mind—more Morrisite than luddite, perhaps. **1972** *Listener* 6 Apr. 458/2 Ernie's father, killed just before the Armistice, had been a Morrisian socialist and preached an end to deference.

Morrison (mǫ·risən). Name of Herbert S. *Morrison*, Secretary of State for Home Affairs and Home Security (1940–5), used *attrib.* in **Morrison hour** (see quot. 1963); **Morrison shelter,** a transportable indoor steel table-shaped air-raid shelter.

1945 S. SPENDER *Citizens in War* 21 In addition to the Anderson shelters, Morrison shelters and Surface shelters were built at the time of the Blitz. **1948** *Hansard Commons* 18 Mar. 2268 Why should Anderson shelters rust more than Morrison shelters? **1952** *Oxf. Jun. Encycl.* X. 89/1 Morrison shelters, also made of steel and shaped like a rectangular dinner-table, could be set up inside a house to give protection if the house collapsed. **1963** T. & P. MORRIS *Pentonville* iv. 84 Some men complain that though at the time of joining they are told of the so-called 'Morrison Hour' they are not told about compulsory overtime. *Ibid.*, The 'Morrison Hour' is one extra hour worked at the end of the day from Monday to Friday and is compulsory. It dates from 1943... Herbert Morrison was Home Secretary at the time.

morrowless (mǫ·roᵘlės), *a. rare.* [f. MORROW *sb.* + -LESS.] Not subject to time; without end.

1871 W. D. HOWELLS *Suburban Sk.* 163 Sometimes this choice company sits on the curbing that goes around the terrace..and then I envy every soul in it, so tranquil it seems, so cool, so careless, so morrowless. **1954** J. R. R. TOLKIEN *Fellowship of Ring* xi. 205 Through halls of iron and darkling door, And woods of nightshade morrowless.

Morse, *sb.*³ Add: (Earlier and later examples.)

1858 in G. B. Prescott *Hist. Electr. Telegraph* (1860) xiv. 191 We work with Morse key and detector. **1916** [see *DASH *sb.*¹ 7 f]. **1931** F. L. ALLEN *Only Yesterday* 13 A mechanically inclined boy has a wireless set, with which, if he knows the Morse code, he may listen to messages from ships at sea. **1972** V. PITT *First Look At Signals* 38 The most famous distress signal is S.O.S... It may be tapped out in Morse Code.

morse *v.*, also *trans.*

1920 *Punch* 31 Mar. 245/1 It can be used for Morsing instructions about breakfast to the cook. **1944** *Jrnl. R. Aeronaut. Soc.* XLVIII. 534 Sudden load changes of 18 kw. (29 h.p. at 85 per cent. efficiency) could be thrown on and off the engine by violently 'morsing' the switch on the load panel without upsetting stability.

Morse (mǭɹs), *sb.*⁴ *Engin.* [prob. f. the name of the *Morse* Twist Drill Co., now Morse Cutting Tools, of New Bedford, Mass., U.S.A.] *Morse taper*: a taper on a shank or socket that is one of a standard series having specified dimensions and angles; also *ellipt.*

The word *Morse* is a registered proprietary term in the U.S.

1894 *Machinery* (N.Y.) Nov. p. ii (Advt.), The Standard Tool Company, Cleveland, Ohio. Manufacturers of..Taps, Milling Cutters, Morse Taper Reamers. **1896** R. GRIMSHAW *Shop Kinks* 96 'Morse' and 'American' Tapers. There are still shops which have twist-drills and sockets of the old 'American' taper, and sometimes it is bothersome to work them in with the 'Morse', as the American is 9–16 inch to the foot, and the Morse is ⅝ inch. **1906** *Official Gaz.* (U.S. Patent Office) 28 Aug. 2956/2 Morse Twist Drill & Machine Company, New Bedford, Mass. Filed May 29, 1905. Used ten years. *Morse* [for] Drills, Reamers, Taps, ..and Milling Cutters. **1930** *Engineering* 25 Apr. 539/2 The centres, both of the workhead and tailstock..are 2½ in. in diameter. They have shanks with a No. 6 Morse taper. **1964** S. CRAWFORD *Basic Engin. Processes* v. 114 The barrel is hollow, at one end a morse-taper bore locates the dead centre, and also provides the location for drill and reamer shanks. **1971** B. SCHARF *Engin. & its Lang.* x. 87 Morse tapers which are numbered from 0 to 7 make it possible to interchange drills, sockets, etc. as required.

mort, *sb.*¹ **4.** (Later example.)

1888 H. JAMES *Let.* 31 July (1920) I. 138 You have become a beautiful myth—a kind of unnatural uncomfortable unburied *mort*.

mortadella (mǭɹtăde·la). Pl. **mortadelle.** Also **mortadel, mortadello.** [It. dim., irreg. f. L. *murtātum* (sausage) seasoned with myrtle berries.] A large spiced pork sausage; Bologna sausage.

1613 [see *CERVELAT 1]. **1732** LD. ESSEX *Let.* 26 Aug. in R. B. Peake *Mem. Colman Family* (1841) I. ii. 30 If you would let your steward buy for me a good Parmesan cheese and some Mortadellos. **1846** R. FORD *Gatherings from Spain* xi. 129 The savoury piquant *embuchados*, which are akin to the *mortadelle* of Bologna, only less hard. **1921** D. H. LAWRENCE *Sea & Sardinia* iii. 118 There is mortadella, the enormous Bologna sausage, thick as a church pillar. **1934** F. SCOTT FITZGERALD *Tender is Night* I. xiii. 76 They ate sandwiches of *mortadel* sausage. **1950** E. HEMINGWAY *Across River & into Trees* xxii. 162 People at home think *mortadella* is a sausage. **1962** G. BUTLER *Coffin in Oxford* xi. 151 A plate of mortadella. **1972** M. HOWELLS *Pop. Ital. Cookery* 13 *Mortadella.* There are many varieties of this sausage, in different sizes and shapes. **1973** E. MCGIRR *Bardel's Murder* ii. 54 If you'd seen the knife he was using to cut up the mortadella sausage, guv'nor, you'd have gone peacefully away.

mortal, *a.* Add: **2. e.** *mortal mind*: according to Christian Scientists, the source in man of all delusion and error, creating the illusion of bodily sensations, pain, and illness.

1875 M. B. EDDY *Science & Health* vii. 341 Faith is all that ever made a drug remedy the ailments of a man. Mortal mind is belief, the immortal is understanding, the latter is Spirit, the former personal sense; we must learn to hold the immortal and mortal mind or belief separate. **1881** —— *Ibid.* (ed. 3) I. ii. 108 Mortal mind, and not muscles, nerves, bones, etc., determines the condition of its own body. *Ibid.* 114 Mortal mind is the remote cause of all suffering and sickness. **1903** 'MARK TWAIN' in *North Amer. Rev.* Apr. 508 In Christian Science terminology, 'claims' are errors of mortal mind, fictions of the imagination. **1910** *Encycl. Relig. & Ethics* III. 577/2 In Christian Science this lying material sense, or sense of evil, is termed 'mortal mind'. **1970** F. S. MEAD *Handbk. of Denominations in U.S.* (ed. 5) 71/1 Certain terms are important in the exposition of Christian Science... Mortal mind is 'the flesh opposed to Spirit'.

mortality. Add: **5.** *mortality rate, statistics.*

1909 *Daily Chron.* 27 Jan. 3/3 Dr. Millsom points out that..the want of air, ventilation, and light are dangerous or injurious to the health of the residents, and that the mortality rate from all diseases (especially consumption) is very high. **1966** *Lancet* 24 Dec. 1371/1 In an attempt to reduce the mortality-rate, clinicians used ever-increasing amounts of antitetanic serum. **1970** PASSMORE & ROBSON *Compan. Med. Stud.* II. xxxiv. 10/1 The death certificate is both a legal and a statistical document. Its form has been agreed internationally to provide for worldwide comparisons of mortality statistics.

mortar, *sb.*¹ Add: **5.** (sense 3) *mortarman.*

1952 G. E. THORNTON *Hand-bk. Weapon Training* x. 109 The auxiliaries are placed in position by the mortarman's assistant, who works backwards in co-operation with the mortarman. **1965** M. W. BROWNE *New Face of War* iii. 19 They lack optical sighting devices, but good mortarmen don't need gadgets like that. **1970** *Globe & Mail* (Toronto) 25 Sept. 9/8 U.S. mortarmen mistakenly fired 36 rounds into an element of U.S. infantrymen.

mortar, *v.*² Add: Hence **mo·rtaring** *vbl. sb.*

1954 M. BERESFORD *Lost Villages* ii. 73 Packing and mortaring had been carefully executed, and both faces of the wall were regular and straight.

mortar, *v.*³ [f. MORTAR *sb.*¹ 3.] *trans.* To direct mortar fire upon; to hit with mortar shells.

1951 in *Conc. Oxf. Dict.* (ed. 4). **1967** C. CONNELL *World's Greatest Sieges* xxix. 239 The Germans followed the rear-guard down to the water's edge..mortaring men and boats indiscriminately. **1974** *Times* 18 Mar. 6/6 In the towns, he said, the Arab garrisons had mortared and bombed Kurdish quarters.

Hence **mo·rtaring** *vbl. sb.*

[**1920** G. M. CHURCHER in *R. Artillery War Commem. Bk.* II. 43/2 The infantry were more than friends... Without their help trench-mortaring..would not have been the fun it was.] **1971** B. W. ALDISS *Soldier Erect* 229 All that shelling, mortaring and machine-gunning hadn't put a single bunker out of action, although it had spread the jungle about the place. **1972** L. LAMB *Picture Frame* ii. 21 Mortaring started just as he reached the Bois Mesnil lane.

mortarium (mǭɹtā·riǒm). Pl. **mortaria.** [L.] A Roman mortar (see quots.).

1842 W. SMITH *Dict. Gk. & Rom. Antiquities* 622/1 *Mortarium*, also called Pila and Pilum. **1863** W. CHAFFERS *Marks & Monograms on Pott. & Porc.* 18, Fig. 24 is a mortarium; 10¾ inches diameter, of light brown ware, unglazed... Mortaria are sometimes found of the red lustrous ware, called Samian. **1937** *Discovery* Apr. 98/1 They have unearthed..bits of a mortarium. **1938** [see *find-spot* s.v. *FIND *sb.* 4].

mortician (mǭɹti·ʃən). *U.S.* [f. L. *mors, mort-* death (cf. MORT *sb.*¹) + -ICIAN.] An undertaker; one who arranges funerals.

1895 *Columbus* (Ohio) *Dispatch* 14 Aug. (Advt.), We, Mank & Webb, are the only Morticians in the city who do not belong to the Funeral Director's Protective Association. **1915** *Literary Digest* 16 Jan. 130/3 The word 'mortician' is a recent innovation due to a need felt by undertakers for a word more in keeping with, and descriptive of, their calling. **1927** *Glasgow Herald* 1 Oct. 10/7 As the jury troop out of their boxes every tread of their heels will mean another call for the mortician. **1930** *Punch* 18 June 696 He passed me on to a mortician... His prices ..were huge, but then he was mortician to all the best movie-stars' husbands. **1948** *Christian Cent.* 4 Feb. 146/1 Give the morticians credit—their cosmetic arts are highly skilled. **1951** M. MCLUHAN *Mech. Bride* (1967) 14/2 This sentiment also fits glovelike over the mortician chapel. *Ibid.* 43/2 It is therefore ironic that the present *Life* feature..should have so mortician-like an air. *Ibid.* 62/1 Our refusal to face death at all in the mortician parlor. **1953** 'M. INNES' *Christmas at Candleshoe* i. 10 He is a bald pale person from Buffalo, where he carries on the profession of mortician. **1958** *Oxf. Mag.* 30 Oct. 49/2 Well—hush!—there it is, a thing of beauty in a mortician's parlour; embalmed but—see there!—quite conscious. **1972** *Lebende Sprachen* XVII. 34/1 US *mortician*..undertaker.

Mortlake. Restrict '*Obs.* exc. *Hist.*' to sense in Dict. and add: **2.** (Written **mortlake, mort-lake** (with hyphen).) = **ox-bow lake.*

1902 LD. AVEBURY *Scenery Eng.* ix. 303 The loop often remains as a dead river-channel or 'Mortlake'. Such loop-lakes are known in America by the special name of 'Oxbows'. **1937** [see *CUT-OFF *sb.* 2]. **1957** G. E. HUTCHINSON *Treat. Limnol.* I. i. 123 The term *mortlake*, presumably derived from the place name, has been used in some English works as the equivalent of *bras mort* or *Altwasser*. **1962** READ & WATSON *Introd. Geol.* iv. 172 When the loop of the meander becomes large it is liable to be cut-off across its neck, leaving an abandoned separated portion which may remain as an oxbow-lake or mort-lake. **1968** *Geogr. Abstr.* B. 112 The fish fauna of various rivers, peat hags and mortlakes was examined.

Morton's Fork (mǭ·ɹtənz fǭɹk). [f. the name of John *Morton* (c 1420–1500), Archbishop of Canterbury and minister of Henry VII + FORK *sb.* 1.] John Morton's method of levying forced loans by arguing that those who were obviously rich could afford to pay, and those who lived frugally must have savings. Cf. quot. 1622 *s.v.* CROTCH 7. Also *transf.*

1889 J. GAIRDNER *Henry Seventh* x. 151 One article suggested an argument familiarly known as 'Morton's fork'. **1930** SELLAR & YEATMAN *1066 and all That* xxix. 49 Henry VII was a miser and very good at statecraft; he invented some extremely clever policies such as the one called Morton's Fork. **1964** *New Statesman* 28 Feb. 318/1 The political consequence of this is that he is now—publicly at least—caught on Morton's Fork. **1965** *Listener* 16 Sept. 429/2, I don't think Mr Smith can altogether pass unscratched between the twin prongs of that Morton's fork. **1973** *Times Lit. Suppl.* 17 Aug. 957/5 A fine Morton's fork technique this—those who went were conspirators and those who stayed away were cowards. **1974** *Cosmopolitan* May 132/1 'Morton's Fork,' she laughed wryly. 'Young and torpid or old and energetic.'

‖**morucho** (moru·tʃo). [Sp.] A half-breed bull reared in Spain; originally the name of a thoroughbred race of cattle in the Salamanca area.

1932 E. HEMINGWAY *Death in Afternoon* xii. 128 Half-bred bulls or bulls in which there is a little fighting blood, called moruchos in Spanish, are often very brave while calves. **1957** A. MACNAB *Bulls of Iberia* i. 8 There do exist animals of mixed race, called *moruchos*, half fighting blood and half tame blood of the Salamanca area. **1959** V. J. KEHOE *Aficionado!* iv. 67/2 Another variety of Spanish toro is the half-breed *morucho*. This type, which extends from the Salamanca area down to the Guadarrama range section, does not have animals of the fine form, perseverance of charge and nobility as do the pure breeds.

morula, var. *MARULA.

mor-yah, var. *MOYA *int.*

mosaic, *a.*¹ and *sb.* Add: **A.** *adj.* **1. c.** Resembling the colours or patterns of mosaic work.

c **1890** tr. *T. de Dillmont's Encycl. Needlework* 133 *Mosaic stitch*.., the first row consists of one short and one long stitch, alternately; the second, of short stitches only, set between the long stitches of the first row; the third row is a repetition of the first, and so on. **1900** E. JACKSON *Hist. Hand-Made Lace* 127 The Lace resembling Duchesse made in Venice in the present day is called Mosaic lace, on account of small sprigs being used to build up the pattern as the pieces of stone and glass are used in Mosaic work. **1934** M. THOMAS *Dict. Embroidery Stitches* 151 *Mosaic filling*, a drawn fabric stitch. **1960** H. HAYWARD *Antique Coll.* 191/1 *Mosaic binding*, book-binding with polychrome decoration. **1961** J. CARTER *ABC for Bk.-Collectors* (ed. 3) 136 *Mosaic bindings*, leather bindings decorated with contrasting colours, whether inlaid, onlaid or painted.

5. *Embryol.* Of, pertaining to, or characterized by that mode of development in which regions in an embryo are predetermined by the corresponding regions in that embryo at an earlier stage of development.

1893 *Jrnl. Morphol.* VIII. 579 (*heading*) Amphioxus, and the mosaic theory of development. **1904** *Jrnl. Exper. Zoöl.* I. 2 Similar views were more or less clearly expressed by Van Beneden, Flemming, Platner and others prior to the definite formulation of the mosaic-theory of development by Roux in 1888. *Ibid.* Such 'mosaic eggs' as those of mollusks or ctenophores. **1933** J. H. WOODGER tr. *L. von Bertalanffy's Mod. Theories Development* x. 143 After injury we obtain partial embryos from mosaic eggs and whole embryos from the regulative ones. **1963** J. COHEN *Living Embryos* 11 In the Annelid worms and the Molluscs, almost all of the interactions between parts of the egg occur before cleavage commences, and in consequence, if the cells are separated at the 4-cell or 8-cell stage they have already been determined in their fates and can only produce parts of animals. This kind of development is called determinate or mosaic, and is also shown by many primitive Chordates. **1970** AMBROSE & EASTY *Cell Biol.* xiii. 422 Eggs of this kind, in which the cytoplasm is clearly divided into different regions required for the development of specific regions of the embryo, are known as mosaic eggs.

6. *mosaic disease* [tr. G. *mosaikkrankheit* (A. Mayer 1886, in *Die Landwirtschaftlichen Versuchsstationen* XXXII. 453)], a virus disease affecting plants, characterized by a mottled pattern of discoloration on the leaves; also *absol.*; cf. *tobacco mosaic* (*TOBACCO 3).

1894 *Jrnl. Mycol.* VII. 382 The first symptom [of tobacco mosaic] is a geographic or mosaic coloring of the leaf surface, light and dark green... The name 'mosaic disease' was given by Dr. Mayer. **1925** *Contemp. Rev.* Dec. 753 Cuba..is finding soil depletion and mosaic disease increasingly serious matters. **1930** *Discovery* June 196/1 The first virus disease to be discovered was the 'mosaic' of tobacco, which manifests itself in a bright mottling and spotting of the foliage. **1940** *Sun* (Baltimore) 29 Jan. 5/6 First reports of a new disease spreading into Pennsylvania peach orchards have located the dread 'mosaic' in the Spring Grove section of York county. **1970** LIEBSCHER & KOEHLER tr. *Fröhlich & Rodenwald's Pests & Diseases of Tropical Crops* 39 A number of virus diseases, such as mosaic disease, infectious chlorosis, and 'heart rot', manifest themselves by retarded growth, discoloration, and leaf curl.

7. *Biol.* Having or composed of cells of two genetically different types.

1902 W. BATESON et al. *Rep. Evolution Comm. R. Soc.* I. 23 Among the large number of capsules examined, there were some of the mosaic type, in which part of the capsule was prickly and the remainder smooth. *Ibid.* 127 'Mosaic' fruits in *Datura*, where..the otherwise pure extracted recessives (thornless) showed exceptionally a thorny patch... Such a phenomenon may be taken as indicating that the germ-cells may also have been mosaic. **1926**, etc. [see *CHIMERA, CHIMÆRA 3 d]. **1968** [see *gynandromorph* s.v. *GYNANDRO-]. **1969** *New Scientist* 16 Jan. 133/1 Cytogeneticists have found human mosaic individuals. **1974** S. L. ROBBINS *Pathologic Basis Dis.* vi. 178/1 Nondisjunction after zygote formation yields a mosaic individual who has more than one chromosome count in his body cells.

8. Photogr. *mosaic screen*, a screen containing a pattern of small filters of each of the primary colours which was placed in front of the emulsion for both exposure and viewing in some methods of colour photography; so *mosaic process.*

1908 *Brit. Jrnl. Photogr.* Suppl. 3 Jan. 13 To make the Krayn mosaic screen..line-screens are again cemented together and form a block. **1935** *Discovery* July 188/2 The well-known 'screen' or 'mosaic' processes, in which the photograph is recorded through a regular or irregular pattern of coloured rulings or grains. **1942** C. B. NEBLETTE *Photography* (ed. 4) xxxii. 786 The irregular mosaic screens are made from a mixture of small colored particles. **1957** R. W. G. HUNT *Reprod. Colour* iii. 30 In photography, the mosaic processes have had a long and distinguished career.

9. Chiefly *Aerial Photogr.* Applied to a composite photograph made up of a number of separate photographs of overlapping areas.

1920 H. E. IVES *Airplane Photography* xxvi. 316 (*heading*) Arranging prints for a mosaic map. **1930** *Air Ann. British Empire* 207 A photographic survey was made for a railway company, the mosaic strip..being produced on a scale of 130 inches to the mile. **1934** *Discovery* June 15/1 The chief photographic contribution..is an aerial mosaic map of an area of about 200 square miles. **1972** *Sci. Amer.* Mar. 10/2 In the past six months the first complete aerial-mosaic map of Manhattan Island has been assembled, and photographic prints are being made on a scale that brings out a wealth of interesting detail.

10. *Ecol.* Applied to an area in which plant associations occur in an alternating pattern.

1930 G. E. DU RIETZ in *Svensk Bot. Tidskr.* XXIV. 496 Phytocoenose-complexes are vegetational units consisting of phytocoenoses with little or no relationship to each other but more or less regularly alternating. They are of several kinds. Good examples for [*sic*] mosaic complexes are furnished by the..Scandinavian bogs, maritime rocks, etc. **1970** P. OLIVER *Savannah Syncopators* 41 As one moves north from the rain forest and into the tropical woodlands and savannah mosaic regions, the trees become fewer and smaller.

11. *Cryst.* Applied to (the structure of) crystals made up of small blocks of perfect lattices set at very slight angles to one another.

1934 W. P. DAVEY *Study Cryst. Struct. & Applic.* xii. 363 It will be of interest..to examine the various ways in which crystals may be grown in the hope of finding mechanisms of crystal growth which will lead easily to the mosaic type rather than the perfect type of structure. **1938** W. A. WOOSTER *Text-bk. Crystal Physics* ii. 62 The mosaic crystal was imagined to be built up of a number of small blocks of perfect crystal, of not more than some 500 A.U. side, arranged nearly parallel to each other. **1964** R. C. EVANS *Introd. Crystal. Chem.* (ed. 2) ix. 206 As normally prepared, a crystal has a pronounced mosaic structure. **1970** R. A. LAUDISE *Growth Single Crystals* i. 17 There is a continuous series of states of order between mosaic structures and structures showing conventional low-angle grain boundaries.

12. Biol. *mosaic evolution* (see quots.).

1963 E. MAYR *Animal Species & Evolution* xix. 598 There is not a steady and harmonious change of all parts of the 'type', as envisioned by the school of idealistic morphology, but rather a 'mosaic evolution'. Every evolutionary type is a mosaic of primitive and advanced characters, of general and specialized features. **1971** J. Z. YOUNG *Introd. Study Man* xxxiii. 457 The human character appears in some features (the gait) before others (size of brain); this is the phenomenon called mosaic evolution.

B. *sb.* **3. b.** *Biol.* An individual (commonly an animal) composed of cells of two genetically different types. Cf. *CHIMERA, CHIMÆRA 3 d.

1902 W. BATESON et al. *Rep. Evolution Comm. R. Soc.* I. 23 These mosaics occurred as rarities both on prickly individuals and on smooth ones still more rarely. **1946** R. R. GATES *Human Genetics* I. ix. 281 Case I. 1 was an albino and it is said that two of her children were albino mosaics, the girl having half her hair white and half black, with one blue eye and one black. **1949** DARLINGTON & MATHER *Elem. Genetics* v. 112 In animals the effects of somatic mutation are slightly different [from those in plants]... The changed cells give flakes and sectors instead of layers and the product is known as a mosaic instead of a chimaera. **1968** *New Scientist* 14 Nov. 383/2 Chimaeras, or mosaics—animals containing cells from two sets of parents—have been made experimentally before. **1974** S. L. ROBBINS *Pathologic Basis Dis.* vi. 187/2 Approximately 2 per cent of 'mongoloids' are mosaics (trisomy 21/normal).

c. *Photogr.* = *mosaic screen* (in A. 8 above).

1911 A. WATKINS *Photography* (ed. 5) xii. 227 The Thames colour plate. This is a regular mosaic, formed by three printings (each being dyed) on bichromated colloid; the pattern originating with a 200-line half-tone screen. **1957** R. W. G. HUNT *Reprod. Colour* iii. 30 The Autochrome plate, which consisted of a random mosaic of red, green and blue starch grains with the interstices filled with carbon black, came on the market in 1907. **1973** D. A. SPENCER *Focal Dict. Photogr. Technol.* 395 After exposure through this mosaic the emulsion is reversal processed to a positive transparency.

d. Chiefly *Aerial Photogr.* A mosaic map or photograph (see A. 9 above).

1920 H. E. IVES *Airplane Photography* xxvi. 316 For a mosaic of any size an accurate outline map must be drawn on the surface of which the prints are to be attached. **1920** *Flight* XII. 187/2 He then showed the similarities and differences between a photographic mosaic and a map, and outlined the various difficulties that had to be contended with. **1940** *War Illustr.* 2 Feb. 45 The map-like pictures are afterwards assembled in a big mosaic..which forms a complete plan of the area photographed. **1971** R. DENTRY *Encounter at Kharmel* vii. 113 The Yanks updated their U2 mosaic of the area. **1972** *Nature* 18 Feb. 391/2 Electron micrographs of the inner plexiform layer at a magnification of 25,000 times were used to construct mosaics covering the area from the ganglion cell layer to the vitreal margin.

e. An array of many small photo-emissive metal plates, each of which temporarily stores a charge dependent on the amount of light falling on it, that forms the target plate in some television camera tubes (e.g. the iconoscope); also, an array of piezo-electric transducers in a detector of ultrasound.

1928 *Discovery* Nov. 337/1 Carey's idea was to replace the mosaic of the retina by a mosaic of a large number of minute selenium cells..and, further, to replace the nerve fibres by separately insulated electric wires carrying an electric current from a battery, and to use this device to vary the light given by a number of very minute electric lamps..so placed that each lamp would correspond in position to each of the selenium cells. **1933** *Proc. Wireless Section Inst. Electr. Engin.* VIII. 220/2 The charge acquired by each element of the mosaic is released by the cathode-ray beam once in each repetition of the picture. **1953** AMOS & BIRKINSHAW *Television Engineering* I. iv. 54 The mosaic must be of very fine construction with a number of individual cells to each element otherwise the cells show up in the reproduced image as a grain. **1961** G. N. PATCHETT *Television Servicing* III. vii. 197 The mosaic is composed of antimony islands which are made photosensitive with caesium. **1966** *McGraw-Hill Encycl. Sci. & Technol.* XIII. 464/2 The mosaic is electrically coupled to the signal plate by electrostatic capacitance between the two. **1969** J. S. WOOD tr. *L. D. Rozenberg's Sources of High-Intensity Ultrasound* II. II. iv. 215 The construction of ultrasonic receivers in which a kind of 'mosaic' was used, consisting of several piezoelectric cylinders mounted on a large-diameter diaphragm.

mosaicism (mozē·isiz'm). *Biol.* [f. MOSAIC *a.*¹ and *sb.* + -ISM.] The quality or state of being mosaic (*MOSAIC *a.*¹ 7).

1926 *Jrnl. Genetics* XVI. 230 No attempt to represent analytically the genetical behaviour of irregular mosaicism in plants has been altogether satisfactory. **1932** *Genetics* XVII. 39 The facts secured from the different breeding tests demonstrate that the mosaicism is due to an unstable translocation. **1947** *Nature* 11 Jan. 66/2 Chemical treatment of female embryos [of *Drosophila melanogaster*] leads to high frequency of mosaicism. **1961** [see *MONGOLOID *a.* and *sb.* A. 2]. **1963** *Lancet* 13 May 1028/2 The term mosaicism, in cytogenetic usage, describes a condition in which a substantial minority of cells differ from the majority in their chromosomal content. **1971** *Nature* 11 June 388/2 We have recently had the opportunity to study the rare association of mixed gonadal dysgenesis in a female child and X/XY/XYY mosaicism in karyotypes from blood and fibroblast cultures.

Mosan (mōᵘ·zăn), *a.* [Fr., of or pertaining to the Meuse river, f. L. *Mosa* Meuse.] Pertaining to the style of decorative art developed in the Meuse valley in the 11th to 13th centuries.

[**1882** C. DE LINAS *L'Art et l'Industrie d'autrefois dans les Régions de la Meuse Belge* i. 12 Le temps a pu faire surgir un art..ni rhênan, ni flamand, ni français, bien qu'il se rattache à tous les trois... Cet art,..il faudrait aujourd'hui lui donner un nom: pourquoi pas *art mosan*?] **1910** *Archaeologia* LXII. 22 The appearance of these medallions is of unusual brilliancy and richness, even for the productions of the Mosan school of enamelling. **1933** *Burlington Mag.* Apr. 198/1 The term Mosan art was introduced forty years ago by Charles de Linas, who maintained that the art of the middle, that is Belgian, section of the Meuse country was original. **1957** G. ZARNECKI *Eng. Romanesque Lead Sculpture* 12 It has been generally accepted that this manuscript was influenced by Mosan art. The blending of Winchester and Mosan features found in it was probably not confined to this manuscript alone. **1972** *Times* 6 Dec. 1/6 (*caption*) The exquisitely delicate little statue..is characteristic of the Mosan style of the eleventh and twelfth centuries.

Mosan (mōᵘ·săn), *sb.* [f. *mōs*, lit. 'four' in various N. Amer. Indian languages + -AN.] The name of a North American language group comprising Chemakuan, Salish, and Wakashan.

1929 *Encycl. Brit.* V. 139/1 The following reductions of linguistic stocks which have been proposed may be looked upon as either probable or very possible:..5, *Mosan*, consisting of Salish, Chimakuan and Wakashan. **1955** H. A. GLEASON *Introd. Descriptive Linguistics* xxiv. 368 The Pacific Northwest from the Columbia River Valley to southern Alaska is occupied by a number of languages... Most of the languages fall into three families which are related in one phylum known as *Mosan*. **1965** *Canad. Jrnl Linguistics* Spring 78 It was suggested by Sapir, in..1929, that Kutenai may be distinctly related to both Algonkian and Mosan. *Ibid.* 98 Algonkian–Wakashan has perhaps been more commonly known as Mosan-Algonkian. **1968** *Encycl. Brit.* I. 728/2 Tsimshian and Maidu have probably been considerably influenced by contact with Mosan.

mosasaur. (Later examples.)

1966 *McGraw-Hill Encycl. Sci & Technol.* VII. 474/1 Suborder Lacertilia... The group..includes modern lizards..and extinct marine mosasaurs (Cretaceous). **1968** A. S. ROMER *Procession of Life* xi. 195 The mosasaurs..had a long, flexible trunk and a highly developed tail, and hence were able to use body and tail undulation for forward progression in somewhat fish-like fashion. **1971** *Nature* 15 Jan. 172/1 In a great many aquatic and marine reptiles, past and present, ichthyosaurs..mosasaurs..the external nares are just in front of the eyes.

moscato (mǫskā·to). Also **moscatto.** [It.: see MUSCAT.] A range of sweet Italian dessert wines; also, the grape from which the wine is produced.

1903 *Consular Reports* (U.S. Bur. Foreign Commerce) July 336 For each liter of sparkling wine, as Margarita, moscatto, moselle, etc., 1 peso (36.1 cents). **1925** A. HUXLEY *Those Barren Leaves* IV. vi. 314 We'll have a bottle of the sweet *moscato*. **1961** *Guardian* 21 Nov. 14/2 (Advt.), *Moscato.* Semi-sweet, fragrant, sparkling..9/0. **1972** *Times* 25 Mar. 10/6 The moscato grape from which Asti Spumante is made.

Moscow (mǫ·sko, *U.S.* mǫ·skau). **1.** The name of the capital of the U.S.S.R. used allusively to describe the government, political influence, ideology, etc., of the U.S.S.R. So **Mo·scowism.**

1938 *Times Lit. Suppl.* 19 Mar. 182/1 'What Are We To Do?' is less a work of new thinking than of the application and adaptation of Moscowism. **1940** *Manch. Guardian Weekly* 17 May 388 'This book on Poland,' says Cardinal Hinsley in his foreword, 'shows that Nazism and Moscowism agree in two broad lines—Godlessness and ruthlessness.' **1957** C. HUNT *Guide to Communist Jargon* xxiii. 82 Moscow-appointed officials of a non-Russian Soviet republic. **1967** H. ARENDT *Origins of Totalit.* (new ed.) xi. 371 The European Communist parties became branches of a Moscow-directed Bolshevik movement. **1971** P. O'DONNELL *Impossible Virgin* ii. 46 He wants a list of our local agents in Prague. Moscow will pay him the earth for that. **1972** T. LILLEY *K Section* viii. 57 A Moscow-trained and dedicated Communist. **1974** *Times* 27 Aug. 4/8 Mr Florakis leads the Moscow-orientated rump of the [Greek Communist] party.

2. *Moscow mule*, a cocktail containing vodka.

1963 *House & Garden* Mar. 114/2 A Moscow Mule (vodka, ginger beer, ice and half a lime). **1965** O. A. MENDELSOHN *Dict. Drink* 229 *Moscow mule*, cocktail of vodka, ginger beer and lime. The concoction is reported to be unknown in Russia. **1969** [see *MARGARITA 2].

Moses. Add: **1. c.** (Earlier examples.)

1840 in *Amer. Speech* (1965) XL. 130 Oh! Moses how they opened dare eyes. **1842** S. LOVER *Handy Andy* xiv. 123 O'Grady, casting a look of mingled rage and contempt on the fisherman, merely uttered the ejaculation, 'Oh Moses!' and threw himself back in his chair.

d. A derogatory appellation for a Jew.

1844 *Punch* 6 July 19/1 (*heading*) A hint for Moses...Passing by the eminent establishment of E. Aaron and Co.,..I was naturally attracted..to examine the beauteous objects of gent.'s attire, which are exhibited in their unrivalled plate-glass windows. **1858** *Ibid.* 10 July 12/2 The Jews are no more a disqualified caste, And *moses* will henceforth in Parliament sit.

2. b. (Later examples.)

1812 *Boston Gaz.* 26 Oct. Suppl. (Advt.) (Th.), On Saturday was picked up, on Dorchester Flats, a small Moses boat. **1970** *Brewer's Dict. Phr. & Fable* 731/2 *Moses boat*, a type of boat made..by a famous boat-builder, Moses Lowell, in the 18th century.

3. Moses basket, a basket used for carrying babies.

1945 N. MITFORD *Pursuit of Love* x. 84 She..returned carrying a Moses basket full of wails. **1948** M. DICKENS *Joy & Josephine* I. i. 13 She was trying to make out whether the young officer were the father of the baby in the Moses basket. **1949** E. TAYLOR *Wreath of Roses* vii. 100 Carrying the baby between them in a Moses basket. **1959** 'A. STAFFORD' *Custody of Anne* I. 45 She shook the moses basket so that the child stirred and whimpered. **1974** 'J. MELVILLE' *Nun's Castle* i. 22 A display of nursery furniture, including an elegant Moses basket of natural straw lined with soft cotton.

mosesite (mōᵘ·zĭzəit). *Min.* [f. the name of Alfred J. *Moses* (1859–1920), U.S. mineralogist + -ITE¹.] A hydrated nitrogen compound of mercury which is found as yellow isometric crystals and has the formula $Hg_2NX.H_2O$, where X represents chiefly chloride and sulphate.

1910 F. A. CANFIELD et al. in *Amer. Jrnl. Sci.* XXX. 202 We propose the name *mosesite*, which name, besides perpetuating the high attainments of Professor Moses in the science of mineralogy, is particularly appropriate as it links his name with a group of minerals which he first definitely put on record. **1953** *Amer. Mineralogist* XXXVIII. 1225 Mosesite from El Doctor, Mexico, contains AsO_4 and very little MoO_4, but that from Terlingua, Texas, has only Cl and SO_4 as anions. *Ibid.* 1234 Mosesite may be easily synthesized by treating HgCl with dilute ammonium hydroxide. **1959** *Prof. Papers U.S. Geol. Surv.* No. 312. 70/1 The kleinite and mosesite..furnish direct evidence that they were formed at temperatures well above those to be expected during weathering.

mosey, *v.* **1** and **2.** (Earlier and later examples.)

1829 *Virginia Lit. Museum* 30 Dec. 459 *Mosey*, to move off. **1836** *Public Ledger* (Philadelphia) 2 Dec. (Th.), You'r not going to smoke me. So mosey off. **1870** 'MARK TWAIN' *Curious Republic Gondour* (1919) 57, I hain't got time to be palavering along here—got to..mosey along. **1874** —— in *Atlantic Monthly* Nov. 592/2 De way dey made dem sojers mosey roun'! **1901** 'H. MCHUGH' *John Henry* 42 Abie had moseyed away. **1902** W. N. HARBEN *Abner Daniel* 59 Now I must mosey on down-stairs. **1918** *Punch* 27 Mar. 206/1 Her funnel's caked with Cape Horn ice and blistered in the sun, She's moseyed round above a bit, and, poor old ship, she's done. **1961** C. MCCULLERS *Clock without Hands* v. 111 While Mr Malone made up the Coca-Cola, Jester moseyed over to the scales and stood on them. **1966** T. PYNCHON *Crying of Lot 49* v. 110 She was moseying along a street..when she collided with a gang of guided tourists. **1973** *Daily Tel.* (Colour Suppl.) 16 Mar. 22/3 Map in hand he moseys round the alleys and yards of Southwark. **1974** *Black World* June 70/2, I had moseyed over to see what they were yakking about. **1974** D. RAMSAY *No Cause to Kill* II. 134, I thought I'd mosey on over to the liquor store.

mosh, dial. var. MUSH *sb.*¹ and *v.*²

1848 A. B. EVANS *Leicestershire Words* 58, I thought that she would have moshed her children then and there. **1914** D. H. LAWRENCE *Widowing of Mrs. Holroyd* III. 74 One of my sons..was shot till 'is shoulder was all of a mosh.

‖ **moshav** (mōᵘ·ʃāv). Also **moshaev, moshavah.** Pl. **moshavim.** [ad. mod. Heb. *mōshābh* dwelling, colony.] In Israel: a group of agricultural smallholdings worked co-operatively. So **moshav-ovdim, moshav-shitufi** (see quot. 1950).

1931 [see *KIBBUTZ]. **1934** *Cook's Traveller's Handbk. to Palestine, Syria & Iraq* (ed. 6) 48 The *Moshav*, in which small holdings of land are allotted to families on hereditary lease by the Jewish National Fund... The co-operative principle is applied in the purchase and sale of produce and agricultural requirements. **1944** H. F. INFIELD *Co-operative Living in Palestine* (1946) iii. 46 The Moshav is intermediate between individualism and the Kvutza. Here the settler lives in his own house, owns a piece of land and some cattle. *Ibid.* xi. 122 From 1936 to 1939, new Kvutzot were 26, new Moshve-Ovdim only 9. The ration of Kvutzot to Moshve established has doubled. **1950** G. MIKES *Milk & Honey* 136 Some halfway solutions have been attempted. First the *Moshav Ovdim*, in which each family has its own unit of land. The purchase of supplies and the marketing of agricultural products is on a cooperative basis. These..are smallholders' settlements in which every farmer shoulders his own risks... Then there is the *Moshav Shitufi*, in which the women have to work four or five hours a day for the community, but otherwise the family is a separate unit. No personal risk, no private enterprise but more private life. **1959** *Economist* 16 May 640/1 The *moshavim* (the farming settlements where the families rent their own land but co-operate over marketing, tractor stations, seed buying, etc.). **1969** P. WORSLEY in Ionescu & Gellner *Populism* 231 Since family-farming units and communitarian village life are perfectly compatible, forms resembling the *moshav*, or the Saskatchewan Group Farm, have been the favoured models. **1972** C. RAPHAEL *Feast of History* i. 13/1 A *moshav* (small-holders' settlement) in Galilee. **1973** *Times Lit. Suppl.* 18 May 544/4 A moshav is a village devoted to agricultural pursuits and run on cooperative lines. It is not as austere as some of the kibbutzim. **1973** *Jewish Chron.* 6 July 3/2 The first families moved into the second moshav..occupying 30 of the 40 houses built.

Moshi, Mosi, varr. *MOSSI.

‖ **moskonfyt** (mǫ·skǫnfə̄i:t). *S. Afr.* [Afrikaans, f. *mos* must + *konfyt* jam.] A thick syrup prepared from grapes.

1891 *Pall Mall Gaz.* Extra No. 58, Oct. 13/2 It was *mos-komfyt*, I found, made with grape-juice. **1905** *Agric. Jrnl. Cape of Good Hope* 483 (Pettman), The first idea in planting vines is to provide *mos confyt*, a kind of grape sugar syrup, which is given as part of their rations to the coloured labourers. **1931** T. J. HAARHOFF *Vergil in Experience S. Afr.* i. 3 '*Moskonfyt*' is used, as it was by the Romans, partly as a syrup and partly to break down the acidity of wines. **1953** *Cape Times* 10 Mar. 2/4 A consignment of 600 110-gallon drums of *moskonfyt*, which will be processed into wine in Britain. **1975** *Daily Dispatch* (East London, Cape Province) 20 Sept. 5 A grape for raisins and also for the production of sweet wines, moskonfyt and even jam.

Moslem, Muslim, *sb.* and *a.* Add: The predominant form is now *Muslim.*

A. *sb.* (Further examples.) *spec.* = **Black Muslim.*

1955 *Times* 3 May 9/4 Lahore witnessed the astonishing sight of hundreds of Sikhs of all people being fêted by Muslims. **1961** [see *Black Muslim* s.v. *BLACK *a.* 19]. **1969** *Sunday Standard* (Bombay) 3 Aug. 1/8 The promised equal opportunities to Buddhists and Muslims in Ladakh had not materialised so far. **1972** J. MILLS *Report to Commissioner* 120 They [*sc.* Black Panthers] aren't like the Muslims. **1973** *Jewish Chron.* 9 Feb. 11/2 The basic credo of the Muslims is a hatred of Whites.

B. *adj.* (Further examples.)

1968 *E. African Law Jrnl.* IV. 17 The establishment of special courts to deal exclusively with Muslim law is peculiar in East Africa to Kenya and Zanzibar. **1970** *Times* 3 Apr. (Arab League Suppl.) p. iii/1 The Indians are the only large non-Muslim group, besides Christian Arabs and Europeans.

b. Moslem (or **Muslim**) **League,** in full **All India Moslem League,** a Muslim political organization founded in India in 1906 whose demands in 1940 for an independent Muslim state led to the establishment of Pakistan; so **Moslem Leaguer.**

1907 *Times* 2 Jan. 3/1 Resolved [on 30 Dec. 1906], that this meeting, composed of Musulmans from all parts of India, assembled at Dacca, decides that a political association be formed, styled the All India Moslem League. **1913** *Times* 31 Oct. 7/4 The All-India Moslem League..Mr. Ameer Ali has tendered his resignation of the presidentship of the London All-India Moslem League..and his Highness the Aga Khan..will retire from the headship of the League in India. **1955** *Times* 7 July 9/2 The Muslim League did not win a majority, but remains the largest single party. **1971** *Hindustan Times Weekly* (New Delhi) 4 Apr. 7/1 Besides the Congress (N)..it includes such disparate elements as..the Muslim League. **1972** *Guardian* 5 Feb. 11/5 Muslim Leaguers are helpless now.

Mosleyite (mōᵘ·zli,əit). [f. the name of Sir Oswald Ernald *Mosley* (born 1896) + -ITE¹.] A follower or supporter of Sir Oswald Mosley or his views; a British fascist. Also *attrib.* or as *adj.*

1932 AUDEN *Orators* III. 104 The Simonites, the Mosleyites and the I.L.P. **1944** H. G. WELLS *'42 to '44* 33 The Mosleyites running like the wind down Regent Street from the Jewish prizefighters. **1960** *Economist* 8 Oct. 141/2 The lunatic Mosleyite fringe. **1961** *Ibid.* 17 June 1266/1 The alliance of politicians, civil servants and 'sound' outside minds which generally rules England did not go Mosleyite. **1973** *Guardian* 10 Apr. 1/2 Mr Jonathan Guiness said: 'I am supposed to say that..it is a question between the Powellites and the Mosleyites..because I am Sir Oswald Mosley's son-in-law.'

mosquito. Add: **2. a.** *mosquito-borne, -breeding, -carried, -thin* adjs.

1961 M. HYNES *Med. Bacteriol.* (ed. 7) xxv. 392 The Venezuelan virus is also mosquito-borne, but the reservoir is uncertain. **1926** M. LEINSTER *Dew on Leaf* 208 Mosquito-breeding ponds. **1909** *Westm. Gaz.* 11 Feb. 2/3 The City of Rio de Janeiro, as regards mosquito-carried diseases, has ..undergone a complete transformation. **1933** R. CAMPBELL *Flowering Reeds* 26 When to their roost the sacred ibis file, Mosquito-thin against the fading West.

b. mosquito-boot, a boot worn to protect the foot from mosquitos; **mosquito coil,** a spiral made from a dried paste of pyrethrum powder with a combustion-supporting substance, which when ignited burns slowly and produces a smoke that inhibits mosquitoes from biting; **mosquito door,** a door designed to exclude mosquitoes from a house; **mosquito-dope** *N. Amer.*, insect repellent; **mosquito hawk** *N. Amer.*, (*a*) the night-hawk, *Chordeiles minor* (earlier and later examples); (*b*) (examples); **mosquito-net** (earlier and later examples); **mosquito room,** a room from which mosquitoes are excluded; **Mosquito State** *U.S.*, a nickname for New Jersey; **mosquito trousers,** trousers designed to protect the legs from mosquitoes; **mosquito wire,** wire mesh used to exclude mosquitoes.

1774 E. LONG *Hist. Jamaica* III. 884 In those places where they [*sc.* mosquitoes] are very numerous..the usual guard for the legs is the muskeeto-boot, or a kind of half-trouzer, made of linen, tied above the knee, and reaching to the shoes. **1945** *Tee Emm* (Air Ministry) V. 36 Mosquito-boots, which are issued to airmen..are meant to be worn in the evenings. This is a malaria precaution. **1956** E. AMBLER *Night-Comers* i. 1 The mosquito boots and the sweat-stained hats. **1963** *Pyrethrum Post* VII. 22/2 Mosquito coils, as their name implies, are used to prevent adult mosquitoes from biting, particularly during the hours of sleep and are therefore designed to burn for about 8 hours. **1971** *Inside Kenya Today* Mar. 46/2 The prospects for a mosquito coils market [*printed* marker] in the Far East are very promising, and if this market is fully exploited, pyrethrum powder will be a more important product of the pyrethrum industry that it is now. **1972** D. SALE *Love Bite* ii. 31 There are some mosquito coils there, too. Light them at dusk..and you won't be troubled. **1967** P. M. HUBBARD *Custom of Country* (1969) vi. 68 A servant who had been sitting in the veranda got up and held open the mosquito door. **1903** S. E. WHITE *Forest* 383 Pack one, or absolute necessities for hard trip. .. Knife; mosquito-dope; compass; match box. **1948** *Chicago Tribune* 11 July 7/1 New mosquito dopes are being tested by fishermen now that the insects are buzzing. **1709** J. LAWSON *New Voyage to Carolina* 144 East-India Bats or Musqueto Hawks, are the Bigness of a Cuckoo, and much of the same colour. **1737** J. BRICKELL *Nat. Hist. N. Carolina* 163 The Muskeetoe-Hawks, are Insects, so called, from their continually hunting after Muskeetoes, and killing and eating them. **1832** J. PICKERING *Inquiries of Emigrant* 59 Thousands of long large flies, similar to the English dragon fly, but a little smaller, are flying about the fields; they are called musquito hawks, on account of their killing and living on those insects. **1851** *Southern Planter* (Richmond, Va.) July 195/2 Unlike the..mosquito hawk of the South [the joint worm] does not make its appearance suddenly. **1866** W. R. KING *Sportsman & Naturalist in Canada* 108 The somewhat rare swallow-tail or musquito-hawk,..soaring in pursuit of insects. **1938** G. CASH *I like Brit. Columbia* 187 A mosquito hawk zoomed belatedly overhead. **1967** E. B. NICKERSON *Kayaks to Arctic* ix. 77 We learned..that in my father's childhood they [*sc.* dragon-flies] had been called 'mosquito hawks'. **1745** *London Mag.* Nov. 551/2 To scorn Umbrellas and Musketto-Nets, as Jamaican and Carolinean Effeminacies. **1867** M. E. HERBERT *Cradle Lands* i. 3 One comfort only was wanting in this hotel, and that was mosquito-nets. **1969** I. KEMP *Brit. G.I. in Vietnam* viii. 160 The unexpected comforts of a camp bed, an air mattress, and a mosquito net. **1974** P. DICKINSON *Poison Oracle* iv. 111 He had packed the medicines, clothes, mosquito net, compass. **1949** M. MEAD *Male & Female* viii. 171 We asked Tomi..to..stop up the crevices between the mosquito wire and the..floor of our mosquito-room. **1923** *World Almanac* (N.Y.) 422 Nicknames of the States... N.J. 'Jersey Blues', 'Garden', 'New Spain', 'Mosquito'. **1930** in G. E. Shankle *Amer. Nicknames* (1937) 376 We get the scornful title, the *Mosquito State*, because we seem to have our share of these industrious and bloodthirsty insects. **1792** SCOTT *Let.* 10 Sept. (1932) I. 22 Wading through the mosses upon this errand, accoutred with the long gun, a jacket, mosquito trowsers, and a rough cap. **1828** MRS. B. HALL *Let.* 27 Apr. in *Aristocratic Journey* (1931) 261 We got her a pair of mosquito trowsers. **1949** Mosquito wire [see **mosquito room*].

c. mosquito boat, (*a*) (see quot. 1914); (*b*) a motor-torpedo boat; **mosquito craft** (examples).

1914 *Dialect Notes* IV. 150 *Mosquito boat*, small river craft carrying light guns. **1940** *Topeka* (Kan.) *State Jrnl.* 29 Apr. 2 (*headline*) Trial run for 'mosquito boat'. **1850** H. MELVILLE *White Jacket* II. iii. 12 All the musquito craft of Abba Thule, king of the Pelews. **1944** in R. W. Zandvoort et al. *Wartime Eng.* (1957) 158 You always drive a boat; but do not refer to any of them as mosquito craft.

Mosquito, var. *MISKITO *a.* and *sb.*

moss, *sb.*¹ Add: **3. b.** (Later examples.)

1853, 1866 [see ROLLING STONE 1]. **1914** G. B. SHAW *Misalliance* p. lxxxiv, We keep repeating the silly proverb that a rolling stone gathers no moss, as if moss were a desirable parasite. **1936** A. CHRISTIE *ABC Murders* xxxiv. 241 You have been the rolling stone—and you have gathered very little moss. You were bitterly jealous of your brother's wealth.

5. d. Ellipt. for **moss green.*

1897 *Sears Roebuck Catal.* 333/1 Crepe Tissue Paper..in the following colors..Apple, Moss, Grass, Nile and Sea Green. **1923** *Daily Mail* 26 May 1 (Advt.), Superb colours, including: Rose Pink,..Moss, Tomato. **1971** *Vogue* Dec. 148/2 Jersey..colours..rosewood, oak, moss.

e. Hair. *slang.*

1942 in BERREY & VAN DEN BARK *Amer. Thes. Slang* § 121/43. **1966** J. S. COX *Illustr. Dict. Hairdressing* 99/2 *Moss.* (1) Head hair. Slang. (2) Pubic hair. Vulgarism.

6. a. *moss-land.*

1887 T. DARLINGTON *Folk-Speech S. Cheshire* 264 *Moss-land*, boggy land. **1934** *Geogr. Jrnl.* LXXXIII. 425 The surviving mosslands of the Lancashire plain. **1971** *Daily Tel.* 3 July 9/2 Greater provision for visitors is planned for the Wildfowl Trust's projected reserve at Martin Mere, on the mosslands east of Southport.

7. b. *moss-bearded, -cankered, -dappled, -hung, -mantled, -shadowed* adjs.

1851 H. MELVILLE *Moby Dick* III. ii. 26 Venerable moss-bearded Daniel Boone. **1959** D. DAVIE *Forests of Lithuania* iv. 37 A moss-bearded oak that bears The weight of full five hundred years. **1904** KIPLING *Traffics & Discoveries* 304 The moss-cankered oak and beech. **1945** J. BETJEMAN *New Bats in Old Belfries* 6 The view from my bedroom of moss-dappled path. **1869** 'MARK TWAIN' *Innoc. Abr.* 421 Look down upon these moss-hung ruins. **1923** R. GRAVES *Feather Bed* 11 By falls of scree, moss-mantled slippery rock. **1931** C. DAY LEWIS *From Feathers to Iron* 19 The virgin spring moss-shadowed near the shore.

d. moss-agate (earlier and later examples); **moss-bag** *Canad.* (see quot. 1865); **moss crêpe** (see quots.); **moss green,** a green colour resembling that of moss; **moss horn** = **mossy horn*; **moss opal,** a variety of opal containing dendritic markings like those of moss-agate; **moss-peat,** peat formed from mosses, esp. those of the genus *Sphagnum.*

1798 G. MITCHELL tr. *D. L. G. Karsten's Description Minerals in Leskean Mus.* I. 98 Milk white Moss Agate, traversed by brownish particles; from Ceylon. **1837** J. D. DANA *Syst. Min.* VI. 343 *Dendrachates*..was our moss agate. **1904** L. J. SPENCER tr. *M. Bauer's Precious Stones* II. 507 Moss-agate is characterised by the presence of green enclosures, such as are found in many specimens of rock crystal. These enclosures..have the general effect of a piece of moss; hence the term moss-agate. **1943** R. D. GEORGE *Minerals & Rocks* xix. 414 Moss agate is similar to agate, but the coloring is in arborescent or moss-like forms. **1947** J. C. RICH *Materials & Methods of Sculpture* viii. 236 Moss agate is a variety of agate whose decorative appearance militates against its use as a sculptural medium. **1962** R. WEBSTER *Gems* I. x. 178 In England and America the term mocha stone is used synonymously with moss agate, but in European countries the green moss agate alone is known as moss agate while the black and red coloured dendritic agate is termed mocha stone. **1865** MILTON & CHEADLE *N.W. Passage by Land* vi. 85 The 'moss-bag' or Indian cradle. This is a board with two side flaps of cloth, which lace together up the centre. The child is laid on its back on the board, packed with soft moss, and laced firmly down, with its arms to its side, and only its head at liberty. **1960** J. J. ROWLANDS *Spindrift* 120 A baby whimpered in its moss bag laced to a back-board. **1963** *Beaver* Summer 37/2 The 15-day-old baby sleeping in a moss bag. **1955** E. OSTICK *Draper's Encycl.* iii. 84 Moss crêpe is a dress fabric which relies for its characteristic effect on a special yarn called a moss crêpe yarn. **1966** E. PALMER *Plains of Camdeboo* ii. 20 The people of Pearston district in dark suits and moss crêpe demurely assembling for church on a Sunday. **1884** Moss green [see sense 7 c in Dict.]. **1938** R. FIELD *All this &*

Heaven Too (1939) xxiii. 302 She..selected her best dress, the moss-green moire with bands of russet. **1972** H. C. Rae *Shooting Gallery* iii. 210 Tight broadcloth pants in moss green. **1944** H. Wentworth *Amer. Dial. Dict.* 398/1 *Mosshorn, n.*, old cattle; an old cowboy. **1948** H. L. Mencken *Amer. Lang.* Suppl. II. xi. 742 *Mosshorn*, an old steer; also, an old cowboy. **1968** R. F. Adams *Western Words* (rev. ed.) 201/1 *Mossy horn*,..also called *moss horn*. **1904** L. J. Spencer tr. *M. Bauer's Precious Stones* II. 386 Milk-opal sometimes exhibits black arborescent markings, or dendrites so-called, similar to those in certain varieties of chalcedony. Opal of this kind is known as moss-opal. It is cut so as to bring the markings..near the surface. **1966** J. Sinkankas *Mineralogy* II. 447 If containing dendritic or mossy inclusions, it is called moss opal. **1875** S. F. Peckham in *Amer. Cycl.* XIII. 217/2 Moss-peat is oftenest fibrous, and when dried forms elastic masses. **1955** *Times* 21 May 10/6 Bracken..is admirable as a mulch, having all the good qualities of moss-peat, plus a high nutrition value. **1974** *Country Life* 2 May 1076/1 Moss peat can be bought from all suppliers of horticultural sundries, in bags or bales.

moss, *v.* **1.** Delete † and add later example.

1939 G. Greene *Lawless Roads* iv. 119 Would the Cupid's bow just moss a little more as the flowers dropped?

moss-back. Restrict *U.S.* to senses 1 and 2 a in Dict. and add: **2. b.** In *Canad.* and occas. *U.K.* use. (Earlier and later examples.)

1878 C. Hallock *Amer. Club List & Sportsman's Gloss.* p. viii, *Mossback*, a settler; a homesteader; a pioneer farmer. (Western.) **1884** *Nor'Wester* (Calgary) 3 June 4/1 The local mossback refers to the recent picking up of stones on Stephen Avenue. **1924** J. Buchan *Three Hostages* i. 10, I was becoming such a mossback that I had almost stopped reading the papers. **1935** E. Pound *Let.* 28 Jan. (1971) 266 With De Vechii at Ministry of Education there wd. be more chance of *action* than with some aesthetic mossback, sentimentalizing over Dela Crusca. **1937** J. D. Carr *To wake the Dead* xi. 162 Oh, Chris, you are an old mossback! **1941** J. Smiley *Hash House Lingo* 38 *Moss back*, conservative eater. **1959** W. A. Leising *Arctic Wings* 97, I listened to old mossbacks, prospectors, and trappers. **1973** 'Trevanian' *Loo Sanction* (1974) 212 The moss-backs of the National Gallery had pulled off quite a coup in securing the Marini Horse for a one-day exhibition.

moss-backed, mossy-backed, *a.* (Earlier and later examples.)

1876 *Congress Rec.* 13 Jan. 411/1 [In the cotton states] those too cowardly to fight..were known as 'mossy-backed rangers' during the war. **1881** *Harper's Mag.* Sept. 640/2 A thorough-bred, mossy-backed mountaineer. **1889** 'Mark Twain' *Let.* (1917) II. 520 Still mouthing empty reverence for those moss-backed frauds. **1963** *Economist* 7 Sept. 826/1 Bankers..found themselves looking hopelessly moss-backed. **1975** G. V. Higgins *City on Hill* vi. 148 He's mossbacked and close to a fascist, but he's perfectly sincere.

Mössbauer (mö·sbauəɹ). *Physics.* The name of Rudolf L. *Mössbauer* (b. 1929), W. German physicist, used *attrib.* with reference to an effect he reported in 1958, in which gamma rays emitted by an atomic nucleus bound in a crystal exhibit very little Doppler line-broadening (owing to the absence of translational motion of the nucleus) and have almost all the energy released in the transition between nuclear states (owing to the entire mass of the crystal being involved in the recoil of the emitting nucleus, which is consequently very small), so that it is possible to obtain resonance absorption in another similar nucleus with a high degree of precision and hence derive information about the energies and widths of nuclear energy levels; so *Mössbauer effect, spectroscopy*, etc.

1960 *Physical Rev. Lett.* IV. 28 (*heading*) Observations on the Mössbauer effect in Fe^{57}. **1960** *Ibid.* V. 364/2 The Mössbauer spectrum was observed with an Fe absorber, enriched to 75% Fe^{57} and of equivalent thickness 2·2 mg/cm². **1962** *Listener* 6 Dec. 964/2 Physicists all over the world scramble to examine the 'Mössbauer effect' to find out by how much gravity makes a clock run slow. **1965** Phillips & Williams *Inorg. Chem.* I. v. 146 The Mössbauer shift can then be correlated with the degree of ionic bonding. Thus the shift of the γ-ray energy from the value found in metallic Sn increases from SnI_4 through $SnBr_4$, $SnCl_4$, and SnO_2 to SnF_4. **1966** *Jrnl. Physics & Chem. Solids* XXVII. 85 The Mössbauer effect was used to measure the line shifts and electric quadrupole splittings of the 14·4 keV level of Fe^{57} in absorbers at 81°K and 300°K of the marcasite structure compounds FeS_2, $FeSe_2$, $FeTe_2$, and $FeSb_2$. **1970** *Physics Bull.* June 247/1 The Mössbauer nuclear resonance is the narrowest spectral line so far available. **1971** Greenwood & Gibb *Mössbauer Spectrosc.* i. 1 The Mössbauer effect has been detected in a total of 88 γ-ray transitions in 72 isotopes of 42 different elements. **1971** *Sci. Amer.* Oct. 86/2 Mössbauer-effect spectroscopy is now used extensively for studying chemical bonds, the architecture of molecules and the distribution of electronic charge around atoms. **1973** G. M. Bancroft *Mössbauer Spectrosc.* iii. 47 In comparison to many spectroscopic experiments, the basic Mössbauer equipment is rather simple and inexpensive. *Ibid.*, A Mössbauer spectrum consists of a plot of the number of gamma ray photons transmitted through an absorber as a function of the instantaneous relative velocity of the source with respect to an absorber.

Mossi (mǫ·si). Also **Moshi, Mosi.** Pl. **Mossi, Mossis.** The name of a negroid tribe living in Upper Volta in West Africa. Also *attrib.*

1858 H. Barth *Trav. N. & Cent. Afr.* IV. 551 The strongest among these pagan kingdoms..is that of the Mósi, although the country is split into a number of small principalities..paying only some slight homage to the ruler of the principality of Wóghodogó. **1911** F. W. H. Migeod *Lang. W. Afr.* I. i. 38 The powerful nation of the Moshis and other kindred tribes evidently prevented the extension of the Nta in a northerly direction. **1930** C. G. Seligman *Races of Afr.* iii. 63 The Mossi are agriculturalists, with millet as the staple crop. **1959** *Chambers's Encycl.* XIV. 197/2 Upper Volta..was originally created ..in order to stimulate the economic development of the densely peopled Mossi country lying north..of the British Gold Coast. **1964** E. A. Nida *Toward Sci. Transl.* iii. 54 The Mossis of Haute Volta in West Africa speak of most emotional states in terms of the heart.

mossie[1] (mǫ·si). *S. Afr.* [Afrikaans, f. Du. *mosje* dim. of *mos* sparrow.] The Cape sparrow, *Passer melanurus*.

1884 R. B. Sharpe *Layard's Birds S. Afr.* (rev. ed.) 479 The 'Mossie', like its cousin the English bird, is essentially a 'cit'. **1936** E. L. Gill *First Guide S. Afr. Birds* 21 The cock Mossie, with his black head, white eye-stripe and cinnamon back, is a handsome and unmistakable bird. **1959** *Cape Argus* 22 Aug. 5/6 The mossies in her neighbourhood never seem to stray more than a few blocks away. **1974** *Stand. Encycl. S. Afr.* X. 209/2 The Cape Sparrow (*Passer melanurus*) has the same size, and much the same habits as the European sparrow, but the male has a different and more handsome plumage, with a black head and throat and a white curved spot on the side. Mossies, as they are called locally, are common birds in South Africa, especially on farms, occurring in flocks during the non-breeding season.

mossie[2] (mǫ·si, mǫ·zi). Also **mozzie, mozzy.** Slang abbrev. of Mosquito.

1941 Baker *Dict. Austral. Slang* 47 *Mossie, mozzie*, a mosquito. **1954** R. S. Porteous in *Coast to Coast 1953–4* 117 Blast these mossies. There's more inside this net than out of it. **1964** M. Dickson *World Elsewhere* v. 182 Sleeping bags, 'mossie' nets and other equipment. **1971** B. W. Aldiss *Soldier Erect* 167 When we were off guard, we kipped on one of the station platforms outside the RTO office, under our mozzy-nets. **1973** *Shooting Times & Country Mag.* 7 July 14/2 If it *has* chosen unwisely, then the newly-hatched mossies rise triumphantly from the surface only to hit their heads on the caterpillars' safety net and fall back into the liquid.

mossite (mǫ·səit). *Min.* [ad. G. *mossit* (W. C. Brögger 1897, in *Skrifter udgivne af Videnskabsselskabet i Christiania: Mat.-nat. Kl.* VII. 4), f. *Moss*, the name of the city in Norway near which it was found: see -ite[1].] **a.** A mineral, $(Fe,Mn)(Ta,Nb)_2O_6$, similar to tapiolite but containing a high proportion of niobium. **b.** The end member of the series obtained by substituting niobium for tantalum in tapiolite.

1898 *Jrnl. Chem. Soc.* LXXIV. II. 387 The new mineral, mossite, occurs with yttrotantalite and columbite in a pegmatite vein at Berg in the parish of Råde, near Moss, in Norway. **1917** *Mineral. Mag.* XVIII. 120 The relationship which strüverite and ilmenorutile bear to rutile, on the one hand, and to tapiolite and mossite, on the other hand, may be graphically illustrated. **1955** *Amer. Mineralogist* XL. 445 These features lead one to question the existence of mossite and of a complete tetragonal mossite–tapiolite series. **1968** I. Kostov *Mineralogy* II. iv. 247 Tapiolite and mossite have a rutile type of structure in which two $[(Ta, Nb)O_6]$ octahedra alternate with one $[FeO_6]$ octahedron along the tetragonal axis.

mosso (mǫ·so), *a.* [It., pa. pple. of *muovere* to move.] Of musical movement: rapid; animated.

1876 Stainer & Barrett *Dict. Mus. Terms* 297/1 *Mosso* (It.), moved, as *più mosso*, more moved or faster; *meno mosso*, less fast. **1952** H. Hunter *Gram. Mus.* 87 *Italian terms relating to speed*... Più mosso, Più tosto, Stretto, [etc.]. **1962** *Listener* 26 Apr. 749/3 The intense urgency of the 'Libera me', especially of the *più mosso* middle section with the threatening syncopations of the brass.

mossy, *a.* Add: **5. b.** Extremely conservative or reactionary; old-fashioned, out of date; old. *U.S. slang.*

1904 *Collier's* 20 Feb. 1 Arthur Lynch's release has the approval of all England except a few peculiarly mossy old Tories. **1932** *Amer. Speech* VII. 402 That's a mossy hat he wears. **1942** Berrey & Van den Bark *Amer. Thes. Slang* § 116/6 Old; aged,..*mossy*. *Ibid.* § 233/12 Old-fashioned, ..*mossy*.

6. mossy crêpe = **moss crêpe*; **mossy horn** *U.S.*, an old steer; also, an old cowboy.

1945 M. D. Potter *Fiber to Fabric* xii. 239 *Mossy crepe*, ..fabric with texture giving fine moss effect. **1885** C. A. Siringo *Texas Cow Boy* viii. 75 They were all old mossy horn fellows from seven to twenty-seven years old. **1944** R. F. Adams *Western Words* 101/2 *Mossy-horn*, a Texas longhorn steer, six or more years old, whose horns have become wrinkled and scaly... The term sometimes is slangily applied to an old cowman. **1955** Foster-Harris *Old West* viii. 227 Little wrinkles would begin to grow up from the bases of the horns, and the older the steer got, the more the wrinkles would show, giving rise to the term 'mossy horns', meaning old-timers. **1973** R. D. Symons *Where Wagon Led* I. vi. 93 It only takes one ole' mossy-horn to take fright at his own shadder to start the whole lot off.

most, *a.* (*sb.*) and *adv.* Add: **A.** *adj.* **II.** *absol.* (quasi-*sb.*) **4. b.** *slang.* The best, (something that is) extremely good, most exciting; also as quasi-*adv.*

1953 *Time* 17 Aug. 3/1 It's the most! **1954** *Time* 1 Feb. 14/1 Last week the general and even the Pentagon conceded that the bop campaign was the most, to say the least. **1958** *Sunday News* (N.Y.) 11 May 2 Of her husband, she said, 'Adam and I dig each other the most. We have a perfect understanding. It couldn't be better.' **1958** *News Chron.* 22 May 4/5 When a man is found to be attractive—he's the most. **1959** *Punch* 16 Sept. 167/2 It has in fact been voted 'The Sleeve With the Most'..by the editors of the *Beat Disc Review and Pop Chronicle*. **1960** *Guardian* 17 Mar. 4/1 American women get the *most*! **1968** *Listener* 7 Mar. 314/3, I would infinitely prefer to listen to the Kenny Everett programme—'the show that's the most with your tea and toast', as that masterly DJ himself puts it. **1969** H. Waugh *Young Prey* (1970) 6 'So—you like the ride?'..'No kidding, it was the most!'

5. e. *the most of*: the greatest part of, the majority of.

1790 in *Eng. Dial. Dict.* (1903) IV. 170/2. **1843** T. Wilson *Pitman's Pay* 27 The myest of which was left below. **1862** M. D. Colt *Went to Kansas* 192, I..have taken leave of the most of my friends. **1879** Stevenson *Travels with Donkey* iii. 96 Without doubt, the most of mankind grossly overeat themselves. **1931** H. Belloc *Hist. Eng.* IV. i. 127 Much of the most of English wealth in 1536 came from tillage. **1963** C. D. Simak *They walked like Men* ix. 49 'You say all my neighbors have lost their leases or sold out?' 'The most of them.'

B. *adv.* **4.** For '*Obs.* exc. *dial.*' read 'Now *dial.* and *U.S.*'

1775 in *Essex Inst. Hist. Coll.* (1877) XIII. 198 It is so long since I saw, or heard direct from you that I most forgit you. **1800** *Farmers' Register* (Greensburgh, Pa.) 8 Nov. (Th.), And though he squeez'd me most to death, I could not help it, no, not I. **1803** *Port Folio* (Philad.) III. 97 (Th.), You know how it most makes you blind, in winter, to look on the snow. **1825** J. Neal *Bro. Jonathan* I. 107 Most off the handle, some o' the tribe I guess. **1883** 'Mark Twain' *Life Miss.* iii. 26 Then they both got at it.., swelling round and round each other, and punching their fists most into each other's faces. **1901** Merwin & Webster *Calumet 'K'* i. 6, I 'most met my death climbing up just now. *a* **1911** D. G. Phillips *Susan Lenox* (1917) II. iv. 87 The street girls..they 'most starve. **1935** in Z. N. Hurston *Mules & Men* (1970) I. i. 26 He moved by jerks, and he had most no tail.

b. With *all, any, every*, etc.

1770 Washington *Diary* 25 Aug. (1925) I. 395 As the Tassels of most all the Corn..was entirely dry. **1775** J. Andrews *Let.* 11 Apr. in *Mass. Hist. Soc. Proc.* (1866) VIII. 403, I think it exceeds most every thing of the kind. **1834** C. A. Davis *Lett. J. Downing* 35 Most all these southern folks are good fellows. **1854** J. E. Cooke *Virginia Comedians* I. viii. 50 'Never argues with women!' adds Kate, 'as if he was not arguing with me all the time 'most!' **1855** [in Dict.]. **1865** 'Mark Twain' *Celebr. Jumping Frog* (1867) 10 He was lucky,..he most always come out winner. **1894** [see *helgramite]. **1897** Kipling *Captains Courageous* i. 5 She's 'most always sick on the ocean. **1904** *Sun* (N.Y.) 16 Aug. 5 Most everybody in the Twelfth Ward was there. **1926** *Publishers' Weekly* 10 July 117 Most any bookbinder will be glad to tell you all about du Pont Fabrikoid. **1956** *Rev. Sci. Instrum.* XXVII. 961/2 If K > 1, which is most always the case, then [etc.]. **1964** W. B. Huie *Hiroshima Pilot* ii. 15 By 1923 most every surviving Confederate veteran had 'been with Pickett'. **1968** *Wall Street Jrnl.* 28 Feb. 16/1 In most any organization, the man who succeeds a reformer faces the task of consolidating the reforms while smoothing feathers the reformer has ruffled. **1968** *Amer. N. & Q.* Mar. 108/1 A book which touches most all of the high spots for the jet set.

mostest (mōu·stėst), *dial.* and *joc.* var. Most *a.* (*sb.*) and *adv.* Esp. *the mostest*, the greatest amount or degree (of something).

Regarded by many users as a double or strengthened superlative.

1885 *Indianapolis Jrnl.* 15 Nov. 10/4 We set around the kitchen fire an' has the mostest fun. **1887** Parish & Shaw *Dict. Kentish Dial.* 104 *Mostest*, farthest; greatest distance. 'The mostest that he's bin from home is 'bout eighteen miles.' **1892** E. Terry *Let.* 4 July in *Ellen Terry & Bernard Shaw* (1931) 9 It's mostest kind to write to me so about my young friend. **1905** *Memphis* (Tenn.) *Commercial Appeal* 14 May II. 4/6 It is one of the favorite arguments of his detractors that General Nathan Bedford Forrest, the noted Confederate cavalry leader, was an illiterate man... One [story]..gives Gen. Forrest's answer to the question, 'How do you manage to win your battles?' 'Git thar fust with the mostest men,' is the reputed reply. **1909** M. Diver *Candles in Wind* xxxv. 360 'I'm her friend, Paul—just as much as you are...' 'No, I'm the mostest. She said so.' **1935** *Word Study* Oct. 2/2 One of their favorite games was called 'The Ten Mostest'. Each player would choose 30 words: 10 that he or she considered the most beautiful words, 10 funniest words, and 10 most descriptive words. **1940** C. Boothe *Europe in Spring* vii. 172 Norway of all countries was the kind of country which, if you 'got there fust with mostest men', you couldn't be got out of in a day or a year, perhaps ever. **1949** O. Nash *Versus* 68 K is for Keeler..The fustest and mostest To hit where they ain't. **1958** *Daily Herald* 25 Mar. 5/7 Here's the hostess with the mostest... Her guests all agreed Sophia was pretty good..well, pretty, anyway. **1959** *Times* 14 Sept. 3/5 A great ambition to be there firstest with the mostest. **1960** *Time & Tide* 24 Dec. 1599/1 Eddie Byrnes..he's the mostest act on scope. **1971** *Engineering* Apr. 57 But we make the mostest. **1973** A. Hunter *Gently French* vii. 64, I reckon you admire the mostest in anything.

most-favoured-nation: see Favoured *ppl. a.*[1] in Dict. and Suppl.

mot, var. Motte.

mot[4], **mott.** = Mort *sb.*[4]

1773 in Partridge *Dict. Underworld* (1949) 451/1 The first time I saw the flaming mot, Was at the sign of the Porter Pot. *a* **1790** H. J. Potter *New Dict. Cant & Flash* (1795) 42 *Morts*, or *motts*, lewd women, whores, shoplifters, &c. **1796, 1812** [see Mort *sb.*[4] b]. **1821** P. Egan *Real Life in London* I. xii. 223 The Hon. Tom Dashall.. was in close conversation with his *mott*. **1837, 1851** [see Mort *sb.*[4] a]. **1846** *Swell's Night Guide* 60, I don't think as how he ever doss'd with a mott or could fake a shiver if he had it for nix. **1866** E. Sellon *Let.* 4 Mar. in 'Pisanus Fraxi' *Index Librorum Prohibitorum* (1877) 394, I! who had expected some swell mot or other, soon found myself seated beside the most beautiful young lady I ever beheld. **1922** Joyce *Ulysses* 247 He saw the image of Marie Kendall, charming soubrette... One of them mots that do be in the packets of fags. **1969** J. Blackburn *Bury him Darkly* 10 'Look at them two mots, Fergus.' Dan pointed at two mini-skirted girls.

mota, var. *mootah, mooter.

‖ **motard** (motār). [Fr.] A French motor-cycle policeman.

1963 'B. Abro' *July 14 Assassination* vi. 43 Two motards were by the door... They..kicked their machines into life. **1966** 'D. Rutherford' *Black Leather Murders* xi. 223 They were stopped by a motor-cycle patrolman... The *motard* raised a white gloved hand to his equally white helmet in salute. **1972** D. Lees *Zodiac* 136 Two blue-shirted, crash-helmeted *motards* sitting on their bikes.

‖ **mot d'ordre** (modǫ̈rdr). [Fr. (Mot[2]), lit. 'word of command'.] A political slogan or watchword; an expression of policy; an oral directive or plan.

1855 *Newsp. & Gen. Reader's Compan.* I. § 159 'A *mot d'ordre* was given by the Legitimist leaders of the fusion'; that is to say, A watchword, &c. **1878** L. W. M. Lockhart *Mine is Thine* I. vii. 144 Another kind of success was to be procured by occasional fits of recalcitrancy against the *mots d'ordre* of the party. **1884** E. W. Hamilton *Diary* 18 Nov. (1972) II. 735 The Tory Party held their meeting at the Carlton this morning; and *Pax* was the *mot d'ordre*, much to the discomfiture of the hot-headed ones. **1905** B'ness Orczy *Scarlet Pimpernel* xii. 119 Evidently expecting to meet their chief—and perhaps to get a fresh *mot d'ordre* from him. **1930** *Observer* 13 Apr. 17/5 The official *mot d'ordre* is that Mr. Gandhi's salt campaign must be treated with a sense of humour. **1947** *Ann. Reg. 1946* 217 The sustained Soviet offensive..against Great Britain and the United States had the appearance of conformity with a *mot d'ordre*.

mote, *sb.*[1] **3. b.** Add to def.: or in cotton.

1842 in J. S. Bassett *Southern Plantation Overseer* (1925) 166 The gin dos not doo good work it draws too many motes through. **1902** W. Hannan *Textile Fibres* 102 The fragmental portions of cotton seeds carry a tuft of attached fibres on the outer membrane; this is termed a bearded *mote*, and is regarded as an imperfection or impurity.

5. *mote-laden, -like* adjs.; **mote-knife,** a knife in a carding machine for removing motes from textile fibres.

1896 W. S. Taggart *Cotton Spinning* I. 132 The cotton is no sooner taken from the feed than it is carried past one or two bars C and D with sharp edges, known as mote knives. **1951** S. Spender *World within World* iii. 123 The characters of his novels radiate round him under a glowing cloud of dirty varnish, not unlike the mote-laden lighting of Fräulein Thurau's apartment. **1956** H. Gold *Man who was not with It* (1965) xii. 102 The mote-laden rays striking guide-rope and stand. *a* **1889** G. M. Hopkins *Poems* (1967) 194 Swayed about Mote-like in thy mighty glow. **1929** W. Faulkner *Sartoris* III. 168 It oozed chaff and a sifting dust, motelike in the sun.

mote, *sb.*[2] Add: **1.** (Later examples.)

1884 G. T. Clark *Mediæval Military Archit. in England* I. ii. 27 Many of these mounds under the name of motes (motæ) retained their timber defences to the twelfth and thirteenth centuries. **1898** D. Christison *Early Fortifications Scotl.* I. i. 3 The motes of which I have to speak are..fortresses..and consist essentially of conical flat-topped mounds, which were defended by palisades. **1919** *Proc. Soc. Antiquaries Scotl.* LIII. 43 Alexander Elphinstone in 1507 took up his residence upon the old deserted Norman mote.

3. *Comb.*, as **mote-castle** = *motte-castle* (*Motte *sb.*[2] 2).

1919 *Proc. Soc. Antiquaries Scotl.* LIII. 42 Such a fortress is well represented in the mote-castles of Dinan, Hastings and Rennes, in the Bayeux tapestry.

motel (mōu·te·l). [f. Mot(or *sb.* and *a.* + Hot)el *sb.*] A hotel catering primarily for motorists, *spec.* one comprising self-contained accommodation with adjacent parking space. Also *attrib.* and *Comb.*

1925 *Hotel Monthly* Mar. 37/2 The Milestone Interstate Corporation..proposes to build and operate a chain of motor hotels between San Diego and Seattle, the hotels to have the name 'Motel'. **1931** *Sociology & Social Research* XV. 372 The new kind of cottage hotel (e.g., 'Motel'), luxuriously furnished and served like any other hotel, but placed directly on the highway. **1947** A. Huxley *Let.* 14 Nov. (1969) 575 From the Blue Bird Motel in Little Rock, Ark, where we have been held up by torrential rains, I take this opportunity of telling you how grateful we all are. **1950** *Archit. Rev.* CVIII. 394 The motel is a new kind of hotel; like an hotel it has an attached restaurant and various services that earlier and more primitive sorts of roadside hostelries omit. **1955** *Times* 28 June 5/7 Although the 'motel' is still regarded as a novelty in the United Kingdom, an establishment of that kind was opened (and is still operating) at Boroughbridge..some time before the war. **1958** *Times* 24 Nov. (Canada Suppl.) p. viii/6 He has almost no contact with..the motel employees. *Ibid.* p. viii/7 The motel business is a field for individual..enterprise. **1962** *Listener* 25 Oct. 692/3 The vast motel-resort..now in the process of erection outside Seoul. **1970** *Homes & Gardens* June 150/1 In a motel the car is meant to be beside your room, but is not invariably. **1970** *New Yorker* 10 Oct. 141/1 A motelkeeper..was serving our breakfast. **1974** *Washington Post* 11 May E23/1 The oldest motel in America is alive and well..in San Luis Obispo, Calif. *Ibid.*, The establishment..opened Dec. 12, 1925, as the Motel Inn.

Hence **mote·l** *v. intr.*, to stay at a motel; **mote·l,land** (*nonce-wd.*), the realm of motels; **mote·lier,** the owner of a motel; **mote·lled** *a.*, provided with motels.

1959 *Guardian* 6 Nov. 8/3 He, the over-fastidious European, is dragged by her through the appalling delights of motelland. **1959** S. H. Courtier *Death in Dream Time* i. 5 A *motelier's* aspirations. **1963** *Economist* 17 Aug. 595/1 The conventional hotel trade from whose custom, as American moteliers..have found out, the growth must be drawn. **1965** *Punch* 1 Dec. 798/1, I drove through twenty-seven states with a freedom and foot-loose mobility unknown in a nation so thinly motelled as Britain. **1968** G. De Fraga *Murder at Cookout* xii. 56 They're motelling in Canberra for a night or two. **1972** *Sydney Morning Herald* 26 Aug. 37/3 Tasmanian motelier Four Seasons Motels Ltd increased profit by 14 per cent.

moteless, *a.* **2.** (Further examples.)

1876 *City-Road Mag.* 141/1 When the air was rendered moteless. **1882** *American* 19 Aug. 298/2 In this moteless air were placed test-tubes containing various organic infusions. **1932** [see **beam-blindness*].

moth, *sb.*[1] Add: **1. f.** A prostitute. *slang.*

1896 in Farmer & Henley *Slang* IV. 360/1. **1935** A. J. Pollock *Underworld Speaks* 78/1 *Moth*, a female of easy virtue.

3. *moth-face, -flit, -hour, -light, -repellent, -trap*; also *moth-coloured, -hung, -resistant, -soft, -white* adjs.; *moth-like* adj. and adv. (further examples); *moth-hunting* vbl. sb.; **moth-borer,** the larval forms of various moths which damage plants by boring into stems, etc., esp. the sugar-cane borer, *Diatræa saccharalis*; **moth-miller,** substitute for def.: one of several moths with white-flecked, powdery-looking wings (earlier example); **moth spot** = *moth patch*.

1900 *Nature* 21 June 182/2 A considerable number of the eggs of the moth borer (which are laid in patches on the leaves of the sugar-cane) are attacked by parasites. **1931** *Discovery* Sept. 300/1 The attacks of the giant moth-borer..and many other pests began in comparatively recent times [in the West Indies]. **1957** *Encycl. Brit.* XXI. 525/2 The giant moth borer (*Castnia licoides*) is particularly destructive in Trinidad and the Guianas. **1974** *Encycl. Brit. Macropædia* XVII. 771/1 The moth borer, *Diatraea saccharalis*, which is widely distributed throughout cane-growing areas, is capable of causing extensive damage when out of control. **1931** V. Woolf *Waves* 254 Jinny's yellow scarf is moth-coloured in this light. *a* **1963** S. Plath *Crossing Water* (1971) 38 The moth-face of her husband, moonwhite and ill, Circles her. **1921** W. de la Mare *Veil* 68 Soundless the moth-flit, crisp the death-watch tick. **1890** W. B. Yeats *Countess Kathleen* (1892) 108 At the moth-hour of eve. **1909** E. Pound *Personae* 23 The moth-hour of our day is upon us Holding the dawn. **1933** W. de la Mare *Fleeting* 161 The secret scent of the moth-hung flower. **1910** —— *Three Mulla-Mulgars* 166 Andy's eyes was never made for moth-hunting. **1930** E. Blunden *Poems* 326 The fete grows late and dark, The King in moth-light traverses the park. **1934** T. S. Eliot *Rock* ii. 84 Moon light and star light, owl and moth light. **1804–8** Blake *Milton* I. xi, in *Compl. Writings* (1972) 492 Her moth-like elegance shone over the Assembly. **1885** W. B. Yeats in *Dublin Univ. Rev.* July 136/1 Gently shook their moth-like wings. **1904** W. H. Hudson *Green Mansions* xiii. 174 She, too, moth-like, had vanished from my side. **1934** S. Spender *Poems* (ed. 2) 36 The mothlike lips at dusk. **1865** *Harper's Weekly* 25 Nov. 747/3 If there is *one* little hole in your linen or paper, some industrious moth-miller is pretty sure to discover it. **1942** G. G. Denny *Fabrics* (ed. 5) II. 119 Dry cleaners and laundries may apply moth-repellent treatment. **1958** S. Hyland *Who goes Hang?* xli. 200 Paradichlorbenzene was sold for the first time in the shops as moth repellant in 1932. **1964** N. G. Clark *Mod. Organic Chem.* xix. 390 It [*sc.* naphthalene] has been employed in the manufacture of fire-lighters, and as a moth-repellent and insecticide. **1970** *Which?* Sept. 265/2 The carpet is moth-resistant. **1876** G. M. Hopkins *Wreck of Deutschland* xxvi, in *Poems* (1967) 60 Or night, still higher, With belled fire and the moth-soft Milky Way. **1958** L. Durrell *Balthazar* i. 17 Brightly lit up in the moth-soft darkness of the Aegean night. **1922** F. Courtenay *Physical Beauty* 26 The so-called 'moth spots', brown spots or patches which appear after middle life, are due to this tan pigment. **1890** *Cent. Dict.*, Moth-trap. **1928** Metcalf & Flint *Destructive & Useful Insects* ix. 263 Mechanical devices, such as..fly traps, moth traps, maggot traps,..have been used successfully for catching and killing a variety of insects. **1970** *Daily Tel.* 19 Oct. 11/8 A moth new to Britain, Plusia accentifera, has been caught in Mr Terry Dillon's moth-trap at Halsted. **1920** D. H. Lawrence *Lost Girl* xi. 275 Irises rearing purple and moth-white.

moth, *sb.*[2] [f. Hind. *moth* moth bean.] In full, *moth bean*. An Indian bean, *Phaseolus aconitifolius*, of the family Leguminosæ, cultivated in tropical and sub-tropical countries as food for both man and livestock.

[**1832** W. Roxburgh et al. *Flora Indica* III. 299 *P*[*haseolus*] *aconitifolius*... *Hind.* Moot. This plant I have reared from seed sent me by Dr. Hunter from the province of Oude where it is much cultivated..and used for feeding cattle.] **1840** *Penny Cycl.* XVIII. 58/1 *P*[*haseolus*] *aconitifolius*, Moth of the natives, is cultivated in the north-western provinces, and used for feeding cattle. **1884** tr. *A. de Candolle's Origin of Cultivated Plants* v. 345 Moth, or Aconite-leaved Kidney Bean... An annual species grown in India as fodder, and of which the seeds are eatable, though but little valued. The Hindustani name is *mout*, among the Sikhs *moth*. **1886** A. H. Church *Food-Grains of India* 152 The moth-bean..is found from the Himálaya to Ceylon... It is not esteemed as a food for man, for although it is rich in nutrients, it is generally thought to possess heating properties. **1923** H. C. Thompson *Veg. Crops* xxiv. 334 The term 'bean' as used in the United States includes..several species of oriental beans, including adsuki,..moth and rice beans belonging to the genus Phaseolus. **1972** G. A. C. Herklots *Vegetables in South-east Asia* 244 (*heading*) Moth bean... A slender annual with decumbent branches growing to 18 inches or a foot in length.

mo·thball, *sb.* Also **moth ball, moth-ball.** [Ball *sb.*[1]] **1.** A ball of naphthalene (sometimes mixed with other substances) used among stored fabrics to repel moths. Also *fig.*

1906 'O. Henry' in *Munsey's Mag.* Dec. 288/1 Me and Solly..prepared to shake off our moth-balls and wing our way against the arc-lights of the joyous and tuneful East. **1911** G. S. Porter *Harvester* xiii. 282 Wouldn't she like me to wear her things better than to have them lying in moth balls? **1916** M. Lees-Dods *Ideal Home* x. 161 When furs are kept by a lady in her own house, a plentiful supply of pieces of camphor, or moth-ball, should be wrapped in tissue paper and distributed amongst the folds. **1932** *Conc. Househ. Encycl.* 808 The round balls sold as moth balls consist of naphthalene, a product of the distillation of coal tar. **1958** S. Hyland *Who goes Hang?* xli. 199 Synthetic camphor wasn't used for moth-balls until 1914! **1964** *Horlicks Home Bk.* 178 In the past naphthalene (usually in the shape of moth balls) was used. **1974** 'J. Melville' *Nun's Castle* i. 10 She had conceived the idea of staging afresh the forgotten Victorian tournament.. wearing the original clothes, most of which were still packed away in mothballs.

2. orig. *U.S.* Used *attrib.* to designate armaments or installations stored away, held in reserve, or disused. Similarly *in mothballs*, laid up, put out of use or action for a long time.

1946 *Amer. N. & Q.* Apr. 7/2 '*Mothball fleet*': inactive U.S. Navy ships to be preserved for long periods by newly-developed techniques. **1946** *Newsweek* 13 May 22/2 The Navy plans to reduce yard personnel to peace-time proportions by early fall. The moth-ball laying-up program will be complete by that time. **1947** *N.Y. Times* 26 Jan. E. 6/5 The moth ball fleet consists of United States warships now undergoing preservative treatment which will enable them to be put back in use rapidly if needed. **1948** *Sun* (Baltimore) 16 Jan. 1/1 The 45,000-ton New Jersey and Wisconsin had been put in 'moth-balls' January 1. **1950** *Hansard Commons* 22 Mar. 2079 We have a 'moth-ball fleet', as it is somewhat vulgarly termed in some countries, reasonably widely dispersed.. round the United Kingdom. **1958** *Oxford Mail* 12 Aug. 1/2 The United States is to sell for cash 47 liberty ships kept in 'moth balls' since the last war. **1967** *Boston Sunday Herald* 30 Apr. I. 7/1 A..port improvement paid for by the MPA has remained in mothballs.., because no container ships have been scheduled to or from Boston. **1970** *Globe & Mail* (Toronto) 26 Sept. 6/3 Seventy-four of these fighter-bombers are going directly into mothballs as they roll off the production line at Canadair Ltd. **1972** *Guardian* 14 Jan. 1/5 Orange dust found in crater Shorty could mean..that Apollo hardware put in mothballs..could be brought out for another final journey. **1974** *Evening Standard* 15 July 11/2 (*heading*) Air fleets may be put in mothballs.

3. *transf.* and *fig.*

1943 Baker *Dict. Austral. Slang* (ed. 3) 51 *Moth balls*, tracer bullets. **1945** L. Shelly *Jive Talk Dict.* 29 *Moth ball*, an aggravating person. **1951** *Football Record* (Melbourne) 8 Sept. 5 And now for eight teams it's mothballs until 1952. **1966** *Listener* 28 Apr. 629/1 The work..was certainly interesting to hear even if it now has to be put back in moth balls. **1971** M. Tak *Truck Talk* 109 *Mothballs*, balls of ether used by truckers to start cold diesel engines. **1974** *Guardian* 31 Jan. 1/5 The distinguished BBC cheerfully took the most improbable creations [*sc.* films] out of mothballs.

mo·thball, *v.* [f. prec.] *trans.* To place in mothballs, in the senses of the sb. Hence **mo·thballed** *ppl. a.*

1943 R. Malkin *Marriage, Morals & War* xi. 159 His civvies, all dusted off, were mothballed, and shipped back home. **1949** *Jane's Fighting Ships 1949–50* 357 *Princeton*, *Tarawa* and *Antietam* are also being 'mothballed'. **1957** P. Frank *Seven Days to Never* ii. 80 We've got two thousand Fifty-Twos and Forty-Sevens mothballed, lined up on every desert in Arizona. **1959** *Times* 18 May 5/3 The best service a retired general could perform 'is to turn in his tongue along with his suit and mothball his opinions'. **1959** *Time* (Atlantic ed.) 24 Aug. 52/3 Both plants..will have to be mothballed unless they can be adapted to produce other chemicals. **1965** *New Scientist* 2 Dec. 642/1 The capability now coming on stream cannot be mothballed. We must use it or see its value erode. **1966** *Aviation Week & Space Technology* 5 Dec. 22/3 Essential consideration to the proposal is that the separately orbited satellite modules could be mothballed in space—fully equipped with all experiments, however—until ready for use. **1973** *Wall St. Jrnl.* 28 Nov. 1/2 Mobil Oil plans to 'mothball' its aging East Chicago, Ind., petroleum refinery. **1974** *Evening Standard* 15 July 11/2 There is no doubt that the Trident One may be mothballed.

mothed, *ppl. a.* (Later example.)

1930 W. de la Mare *Desert Islands* 39 Stuffed, glass-eyed, a little mothed and dusty.

mother, *sb.*[1] Add: **1. a.** Also used colloquially by a husband addressing or referring to his wife.

1855 Dickens *Dorrit* (1857) I. ii. 13 Mother (my usual name for Mrs. Meagles) began to cry so, that it was necessary to take her out. 'What's the matter, Mother?' said I..'you are frightening Pet.'..'Yes, I know that, Father,' says Mother. **1898** J. D. Brayshaw *Slum Silhouettes* 156 'Sit yer down, mother,' said Joe, taking his seat at the head of the table. **1932** A. Christie *Peril at End House* v. 68 Mother and I..feel it's only neighbourly to do what we can. **1970** P. Carlon *Souvenir* ii. 35 Don't you *loathe* the way old folks call each other Mother and Dad?

g. (*just*) *like mother makes* (or *used to make*) (*it*): having the qualities of home cooking; exactly to one's taste; also *fig.*

1919 Wodehouse *Damsel in Distress* i. 18 There's a new musical comedy at the Regal. Opened last night, and seems to be just like mother makes. **1927** W. E. Collinson *Contemp. Eng.* 52 The notice outside some eating-houses, beef-steak pie like mother makes it! **1963** Wodehouse *Stiff Upper Lip, Jeeves* v. 39 Its façade, its spreading grounds..and what not were all just like Mother makes. **1975** D. Clark *Premedicated Murder* iv. 68 Just like my old mother used to make. A bit of candied peel in a bun can't be beat.

h. Used as an exclamation of surprise, dismay, etc.; freq. *my mother!*

1869 'Mark Twain' *Innoc. Abr.* 52 Twenty-five cigars, at 100 reis, 2500 reis! Oh, my sainted mother! **1909** *Sat. Even. Post* 22 May 6/3 'Gee, what a peach of an idea!' 'Oh, mother!' **1959** N. Mailer *Advts. for Myself* (1961) 93 He roared with laughter now. 'Oh, my mother.' **1972** C. Achebe *Girls at War* 107 'Plane!' screamed his boy from the kitchen. 'My mother!' screamed Gladys.

i. *mothers and fathers*: a game in which children act out the roles of mother and father.

1903 G. R. Sims *Living London* XXXIII. 271/1 Sometimes..they [*sc.* the boys] will join the girls in a mimic domestic drama of 'Mothers and Fathers'. **1969** [see *infant-school]. **1972** J. Wilson *Hide & Seek* vii. 130 Shall we play mothers and fathers with our dolls?

j. Ellipt. for **mother-fucker*. *U.S. slang.*

[**1948** Manone & Vandervoort *Trumpet on Wing* 70 'I'll be a motheree if I'll wear any damn bedpan intern's suit,' I screamed.] **1955** S. Whitmore *Solo* iii. 42 Jaeger said,..'He's..so weak now, he can't blow note one.' 'Hell, this mother never could,' Alfred laughed. **1959** N. Mailer *Advts. for Myself* (1961) 358 Old K, he's nothing but a mother. **1967** *Melody Maker* 14 Jan. 8 *Mother*, term of abuse. As in 'You're a mother, baby!' **1972** *Sunday Times* 7 May 10/6 'Man we must just get out of here before those mothers get us all..,' he shouted at me. **1973** *New Yorker* 17 Feb. 62/2 Out the parachute, out the radio, change the c.g., and the mother will go. **1975** *N.Y. Times* 8 Sept. 33/2 'You mothers! I ain't been out five minutes and I just got *outta* the pen this morning!' Her name is Judy, and although she is white, she talks black jive.

2. d. *spec.* in *Mother Russia*.

1966 J. Bingham *Double Agent* v. 71, I love Russia... Great Mother Russia. **1972** P. Ruell *Red Christmas* xv. 153 Came as quite a shock to them when they realised we weren't doing it all for Mother Russia. **1973** D. Bagley *Tightrope Men* xxiv. 164 'See that tower over there?'..'A Russian observation tower. That's Mother Russia.'

3. e. Colloq. phr. *to be mother*: to serve out portions of food or drink; *spec.* to pour out tea.

[**1926** G. B. Shaw *Glimpse of Reality* in *Translations & Tomfooleries* 184 Let us get to work at the supper. You shall be the mother of the family and give us our portions, Giulietta.] **1958** 'J. Brogan' *Cummings Report* ii. 17 We'll go and have tea, and you be Mother. **1967** J. Porter *Dover & Unkindest Cut* iv. 41 MacGregor, hearing the tea cups rattling outside..opened the door again. 'Shall I be mother, sir?' **1974** J. Mitchell *Death & Bright Water* xx. 243 'Shall I be mother?' Callan nodded, and Blythe's strong fingers popped the cork, the champagne foamed into the glasses.

f. The female owner of a pet, esp. of a dog. *colloq.*

1924 Galsworthy *White Monkey* I. vi. 39 Ting was.. trying to climb a railing whereon was..a black cat... 'Give him to me, Ellen. Come with Mother, darling!' **1940** N. Mitford *Pigeon Pie* ix. 139 Many mothers of dogs had fetched their little ones home.

4. b. *Mother Bunch* [f. the name of a noted ale-wife of late Elizabethan times]: (*a*) *Obs. slang*, water; (*b*) a stout or untidy old woman; *Mother Hubbard* (further examples); also, a kind of loose-fitting garment (chiefly *U.S.*).

1600 Dekker *Shoemaker's Holiday* sig. H2ᵛ Am I sure that Paules steeple is a handfull higher than London stone? or that the pissing conduit leakes nothing but pure mother Bunch? **1847** C. Brontë *Jane Eyre* II. iii. 81 You talked of going..to visit the gipsy camp;..one of the old Mother Bunches is in the servants' hall at this moment. **1861** G. J. Whyte-Melville *Market Harborough* viii. 94, I *have* seen mammas whom the fairest of Eve's daughters might be proud to resemble; but it is sometimes hard upon the young Phœbe to have..at her side the shapeless Mother Bunch, into the *fac-simile* of which she must eventually grow. **1878** F. M. A. Roe *Army Lett. from Officer's Wife* (1909) 186, I made a Mother Hubbard apron of white paper-cambric. **1882** *Wide Awake* (Boston, Mass.) Jan. 18/2 A little girl in a sort of Mother Hubbard cap..is starting off on a journey. **1884** *Nat. Police Gaz.* (U.S.) 30 Aug. 12/3 Mattoon's Chief of Police issued an order that any woman appearing on the streets wearing a Mother Hubbard dress would be arrested and lodged in jail. **1911** *Daily Colonist* (Victoria, B.C.) 19 Apr. 24/4 Overall Aprons, of navy print with white dots, made in Mother Hubbard style. **1919** W. S. Maugham *Moon & Sixpence* xlix. 212 Tiaré Johnson..was dressed usually in a pink Mother Hubbard. **1939** J. Steinbeck *Grapes of Wrath* viii. 99 Ma..wore a loose Mother Hubbard of grey cloth... The dress came down to her ankles. **1943** H. W. Krieger *Island Peoples of Western Pacific* 40 The native dress of the women [of the Gilbert Islands], a grass skirt extending from the waist to the knee, has been for the most part replaced by the 'Mother Hubbard', which is made of imported cotton print cloth. **1964** *Guardian* 28 Dec. 6/4 She no more looks like a Mother Bunch than sounds like one.. a fairly plump but elegant, well-dressed woman. **1966** 'R. Standish' *Widow Hack* viii. 87, I can persuade you to swim wearing a Mother Hubbard over your swimsuit. Then I'll be able to concentrate.

II. 10*. A disc with grooves that is made from the plating of an electrotyped master matrix and is used to make a stamper for gramophone records.

1918 H. Seymour *Reproduction of Sound* 182 The obverse impressions of the original matrix are called 'mothers' in the trade, in view of their office in reproducing matrices from the 'master'. **1935** H. C. Bryson *Gramophone Record* vi. 134 The mother, usually about ·03 inch thick, is then stripped from the master by inserting a blunt knife carefully between them and prising them apart. **1952** Godfrey & Amos *Sound Recording & Reproduction* v. 139 A second negative copy known as the 'stamper' or 'working matrix' is obtained from the mother. **1968** *Jazz Monthly* Feb. 4/1 John Steiner..owns the rights to what remains of the Paramount company, including numerous masters and mothers, so it is likely that the actual recording quality will be a great deal better than that on most past Paramount-derived reissues.

13. c. *spec.* **mother-house,** the founding house of a religious order.

1661 *Manifest publisht to their Brethren by General Chapter of Catholick English Clergy* 3 Dr. Leyburn does calumniate us, as being Enemies to our Mother-house, the Colledge of Doway. *a* **1773** A. Butler *Lives Saints* (1779) III. 243 When in 1504 the abbey of Mount Cassino joined this Congregation, it took the name of this mother-house. **1840** [see *branch house* s.v. *branch *sb.* 13]. **1932** C. P. Curran in F. J. Sheed *Irish Way* 269 In this spirit she worked for ten years in the Mother-house and novitiate. **1956** K. Hume *Nun's Story* i. 12 The Order was established at the end of the eighteenth century. How many postulants had passed through this mother house could never be guessed.

14. a. *mother-instinct, -sentiment*; *mother-centred, -dominated* adjs.

1956 Firth & Djamour in R. Firth *Two Studies of Kinship in London* ii. 41 Some United States sociologists ..have suggested the term 'mother-centred families' for households in which the mother has the dominant role. **1965** *Observer* 4 Apr. 26/4 Pirandello..was a mother-centred man. **1963** *Times* 23 Apr. 16/4 The mother-dominated hero. **1881** 'Mark Twain' *Prince & Pauper* 114 Her sharp mother-instinct seemed to detect it. **1920** T. P. Nunn *Education* xi. 131 A thorough-going misogynist could make out a case for applying the adjectives 'mechanical', 'blind', 'unintelligent', even to human mother-instinct. **1920** Mother-sentiment [see *father *sb.* 11 b].

15. a. mother's blessing (see quot. 1861); **mother's boy,** a boy or man who resembles or is dominated by, or excessively attached to, his mother; a sissy; **mother's darling** = **mother's boy*; **Mother's** (or **Mothers'**) **Day** orig. *U.S.*, a day on which mothers are particularly honoured: in the U.S., the second Sunday in May; in Britain = *Mothering Sunday*; also *attrib.*; **mother's help,** a person who helps a mother, mainly by looking after children; **mother's milk** (see Milk *sb.* 1 c in Dict. and Suppl.); **mother's pet** = Pet *sb.*[1] 2 a; also, the youngest child of a family; **mother's ruin** *slang*, gin (cf. Ruin *sb.* 10); **Mothers' Union,** an organization for mothers to meet together regularly.

1861 Mayhew *London Labour* (1862) Extra vol. 245/2 My husband..can't do nothink but give the babies a dose of 'Mother's Blessing' (that's laudanum, sir, or some sich stuff) to sleep 'em when they's squally. **1880** F. Stevenson *Let.* July in J. Pope-Hennessy *R. L. Stevenson* (1974) vii. 142 Louis is, as I know, a mother's boy..and I am sure he looks like you. **1924** D. H. Lawrence *Phoenix II* (1968) 619 Oh, women, beware the mother's boy! **1945** 'L. Lewis' *Birthday Murder* (1951) ii. 26 Stan's happy as he is being supported by his mother. He's a mother's boy. **1973** W. J. Burley *Death in Salubrious Place* ii. 42 Mother's boy—that's his trouble, but it takes all sorts. **1857** Lytton *What will he do with It?* (1859) I. I. i. 7 He looked like a mother's darling—perhaps he was one. **1922** Joyce *Ulysses* 33 That knockkneed mother's darling. **1936** 'J. Tey' *Shilling for Candles* iv. 41 Mother's darlings had those eyes; so, sometimes, had womanizers. **1908** *Congr. Rec.* 9 May 5971/1 *Resolved*, That Sunday, May 10, 1908, be recognized as Mothers' Day. **1926** A. Huxley *Jesting Pilate* IV. 264 In the First Methodist Church..they were going to distribute 'Mother's Day Flowers to all Worshippers'. (On Mother's Day you must wear a red carnation if your mother is alive, a white one if she is dead.) **1958** *Listener* 27 Nov. 874/1 As uniquely and inimitably American as John Foster Dulles or Mothers' Day. **1959** I. & P. Opie *Lore & Lang. Schoolch.* xii. 242 In 1956 the majority of High Street shops [in Britain] were displaying 'Mother's Day' gifts in their windows. **1962** *Listener* 12 Apr. 628/1 Last Mother's Day—a retail selling device imported from the United States—the flower-shops were as busy as ice-cream vans during a heat wave. **1881** *Instr. Census Clerks* (1885) 30 Mother's Help. **1908** A. S. M. Hutchinson *Once aboard Lugger* I. vii. 41 She is not exactly my friend; she is my—my employer. I'm a mother's-help. **1961** *Evening Standard* 14 July 25/5 (Advt.), A Mother's Help..for happy family. **1824** J. Mactaggart *Scottish Gallovidian Encycl.* 348 *Mithers-pet*, the youngest child of a family; the mother's greatest favourite; the *Tony Lumpkin* of the house. **1830** A. Picken *Dominie's Legacy* I. 104 He was..as raw looking, overgrown, gawky a youth, as any mother's pet of a student. **1937** Partridge *Dict. Slang* 535/1 Mother's ruin. **1955** P. Jones *Birthday Honours* I. 10, I have been to a party, darling... What would you like? 'Mother's Ruin'? **1970** *New Scientist* 23 Apr. 165/2 Gin, as shown by the old temperance demonstration of dropping earthworms into adjacent glasses of water and mother's ruin, can certainly eliminate unwanted planarians. **1888** Mrs. G. Sumner *To Mothers of Higher Classes* vi. 55 The 'Mothers' Union', now started in the Winchester Diocese, and in other Dioceses, is a very simple plan. **1972** L. Lamb *Picture Frame* xiv. 123, I shall have to run a mothers' union or something.

16. a. mother-and-baby home, an establishment serving as a maternity home for unmarried mothers, usu. with pre- and post-natal services; **mother-bomb** (see quot. 1971); **mother-cell** *Biol.*, a cell which later undergoes cell division and gives rise to daughter-cells; *spec.* a cell which later undergoes meiotic division; **mother-child** *a.*, of or pertaining to a mother and her child; **mother-city** (further example); (*b*) a city regarded as serving as a mother to someone; **mother complex,** a complex (see *complex 3) about one's mother; **mother-cult,** the worship of a mother-goddess; **mother-daughter** *a.*, of or pertaining to a mother and her daughter; **mother-feryer** *U.S. slang* = **mother-fucker*; **mother figure,** a person or thing endowed with some of the attributes of a mother; **mother fixation,** a fixation (see *fixation 3 b) on one's mother; **mother-fucker** *coarse slang* (orig. and chiefly *U.S.*), a base, despicable person; someone or something that is very unpleasant; hence **mother-fucking** *ppl. a.*, despicable, base; unpleasant; = Bloody *a.* 10; **mother-grabbing** *ppl. a.* (*U.S. slang*) = **mother-fucking* ppl. adj.; **mother image, imago,** the mental or realized image of an idealized or archetypal mother; **mother-in** (or **and**)**-babe,** used *attrib.* to designate a wooden bobbin, the hollow shank of which contains another smaller bobbin; **mother-lode** *Min.*, the principal vein of ore; also *fig.*; **mother-loving** *ppl. a.*, (*a*) that loves one's mother; (*b*) = **mother-fucking* ppl. adj.; freq. used as a vague intensive; **mother mould** *Sculpture*, a rigid mould which holds casting material; **mother plane** orig. *U.S.*, an aircraft which launches or controls another aircraft; **mother-raper** *U.S. slang* = **mother-fucker*; **mother-raping** *ppl. a.* (*U.S. slang*) = **mother-fucking* ppl. adj.; **mother ship,** substitute for def.: (*a*) a ship or airship escorting or having charge of a number of other, usu. smaller, craft; also *transf.*; (*b*) an aircraft or rocket from which another aircraft or rocket is launched or controlled; = **mother plane*; **mothers' meeting** (earlier and later examples); also *fig.*; **mother-son** *a.*, of or pertaining to a mother and her son; **mother substitute, surrogate,** a person or thing that takes the place of the mother; **mother-symbol,** that which is symbolic of the mother or of motherhood; **mother tincture,** in Homœopathy, a pure undiluted tincture of a drug; **mother-to-be,** an expectant mother.

1965 Hall & Howes *Church in Social Work* v. 90 The county council was running its own mother and baby home. **1972** *Guardian* 14 July 11/2 Miss McM will not consider going into a mother and baby home and is against adoption. **1971** *New Scientist* 21 Jan. 135/2 Shrapnel grenades..are dropped individually, or in clusters from canisters ('mother-bombs'). **1845** Mother-cell [in Dict., sense 13 d]. **1875** Mother-cell [see *Ascomycetes]. **1920** W. E. Agar *Cytology* vii. 212 The diagrams start with the pollen mother-cell in the male and the embryo-sac mother-cell in the female—in each case the last cell generation of the diploid phase. **1932** C. D. Darlington *Rec. Adv. Cytol.* i. 5 In a 'mother-cell' two nuclear divisions follow one another rapidly while the chromosomes only divide once. **1959** W. Andrew *Textbk. Compar. Histol.* xii. 486 Earlier workers believed this cell to be a mother cell for the successive groups of

spermatogonia. **1937** H. Read *Art & Society* v. 191 At first these instincts are concentrated on the mother, and any being that threatens to interfere with the mother–child bond incurs the child's enmity. **1963** Auden *Dyer's Hand* 440 The mother–child relationship..stands for the kind of love that is unaffected by time. **1908** *Westm. Gaz.* 4 Aug. 5/1 Capetown..is in the truest sense the μητρόπολις of South Africa, the 'mother-city' from which the rest have sprung. **1935** L. MacNeice *Poems* 20 See Belfast... This was my mother-city, these my paps... I cannot be Anyone else than what this land engendered me. **1919** M. K. Bradby *Psycho-Analysis* v. 59 If sexual fixation takes place at the third stage, the 'mother-complex' will create an obstacle to a man's happiness in married life. **1924** *Spectator* 16 Aug. 229/1, I got a mother-complex. **1936** C. Day Lewis *Friendly Tree* I. vi. 87 He sucks a pipe constantly. The mother-complex. Infantilism. **1948** *Yearbk. Psychoanal.* IV. 172 (*title*) The mother complex in literature. **1960** R. F. C. Hull tr. *Jung's Structure & Dynamics of Psyche* in *Coll. Wks.* (1966) VIII. v. 369 Analysis shows an infantile longing for the mother, a so-called mother complex. **1909** *Westm. Gaz.* 2 Feb. 5/1 From the trend of recent writings in Hindu literature it is suggested that the Mother cult has been revived. **1969** C. Fremlin *Possession* ii. 17 The closeness of the mother–daughter relationship. **1974** —— *By Horror Haunted* 138 An ordinary, typical mother–daughter misunderstanding. **1946** Mezzrow & Wolfe *Really Blues* 4 A motherferyer that would cut your throat for looking. **1953** *New Biol.* XIV. 29 In birds, for instance, we have the astonishing phenomenon called imprinting, by which the sign stimulus of the IRM to follow a mother-figure differentiates. **1957** *Economist* 7 Sept. 821/2 The commonest illusion..is that the United Nations is a miraculous mother-figure which will give suck and shelter to all comers. **1970** *Daily Tel.* 22 Sept. 14 Mrs Meir is both a superb politician and a Jewish mother-figure which the young State [of Israel] may well need. **1971** *Ibid.* 18 Jan. 10/7 The association also says there should be a 'mother figure' in each nursing school to whom students can turn for advice. **1921** *Internat. Jrnl. Psychoanal.* II. 55 Jesus is no longer satisfied to make Joseph his ideal (a hard task for a boy with a strong Mother-fixation of love). **1954** *Scott. Jrnl. Theol.* VII. 393 Rationalism has handed the problem [of worship] over to the psychologist to explain in terms of repressions, mother-fixations, infantile-regressions and the like. **1969** C. Allen *Textbk. Psychosexual Disorders* (ed. 2) xvii. 375 Sailors usually carry..the mother's photograph... There is strong mother-fixation in their choice of prostitutes. **1956** *Amer. Speech* XXXI. 111 This linguistic vacuum is being filled by a new obscenity symbol, *mother-fucker*. **1960** J. Baldwin in *Partisan Rev.* Spring 292 You've got to fight with the elevator boy because the motherf*****'s *white*! **1970** R. D. Abrahams *Positively Black* ii. 45 I'm one motherfucker that don't mind dying. **1971** B. W. Aldiss *Soldier Erect* 142 Jock..looked up into my face. 'The bastards, the fucking mean scab-devouring mankey-minded shower of mother-fuckers!' **1971** J. Mandelkau *Buttons* xiii. 149 These mother-fuckers had been whining they had no food to sell the people—and this place was stacked! **1973** *Black Panther* 21 July 16/1 We will kill any motherfucker that stands in the way of our freedom. **1959** N. Mailer *Advts. for Myself* (1961) 351 They could smash some mother-f——ing Reds. **1968** *Rat* 13–16 May 10/2 The police were interminably long in getting into Math. Good mother-fucking barricades. **1969** P. Roth *Portnoy's Complaint* 106 You muff-diving, mother-fucking son of a bitch! **1974** S. Ellin *Stronghold* 23 'You motherfucking black clown,' Harvey says without heat, 'nothing is changed.' **1959** M. Russ *Half Moon Haven* (1961) I. 60 There isn't one item on this mother-grabbing planet that I 'like'. **1941** L. MacNeice *Poetry of W. B. Yeats* vii. 138 It would be tempting to regard Cathleen ni Houlihan, the Poor Old Woman, as a mother image and so to refer much of Irish nationalism to a mother-fixation. **1968** C. Rycroft *Crit. Dict. Psychoanal.* 93 Conceptions of the mother existing in the infant's mind formed by splitting of the mother image. **1973** J. Singer *Boundaries of Soul* iv. 91 The Mother image appeared under strange circumstances to my analysand Margaret. **1916** B. M. Hinkle tr. *Jung's Psychol. of Unconscious* v. 250 That amount of libido which unconsciously is fastened to the mother-imago. **1956** R. F. C. Hull tr. *Jung's Symbols of Transformation* in *Coll. Wks.* V. II. v. 222 The water and tree symbolism.. likewise refer to the libido that is unconsciously attached to the mother-imago. **1919** T. Wright *Romance Lace Pillow* xiii. 126 *Mother-in-Babe Bobbins*, in the hollowed shank of which a tiny wooden bobbin rattles. **1928** G. Whiting *Tools & Toys of Stitching* 220 The Cow-and-Calf and Mother-and-Babe bobbins—they are a perfect, never-ending joy and a masterpiece of the Midlands! **1969** E. H. Pinto *Treen* xxi. 311 Collectively, they are known as church window bobbins, but those with smaller bobbins inside the windows are described as mother-in-babe types. *c* **1882** J. H. Beadle *Western Wilds* xxxiv. 561 What miners call a 'mother lode' is often like a tree in its upward development: below is the main trunk, above the branches diverge. **1927** B. A. McKelvie *Black Canyon* p. vi, They started up-stream in search of the mother-lode. **1960** *Encounter* XIV. III. 74 The pages of the T.L.S. were the very mother-lode of academic inanity. **1965** G. J. Williams *Econ. Geol. N.Z.* v. 58/1 The mother-lode that had been envisaged as the source of the gold in the Blue-spur 'cement'. **1972** *Times Lit. Suppl.* 25 Feb. 219/1 It is mother lode, with rich ore; but it lacks the refining that the author intended to give it. **1964** O. E. Middleton in C. K. Stead *N.Z. Short Stories* (1966) 198 We'd all be drawing the dole like every other mother-loving beach-comber. **1969** 'J. Morris' *Fever Grass* i. 6 Get her out of that mother-lovin' joint an' into the cab. **1898** C. R. Ashbee tr. *Cellini's Treat. Goldsmithing & Sculpture* 116 Put them into the cavities..in the mould... Or 'mother mould' as the sculptors would call it. **1947** J. C. Rich *Materials & Methods of Sculpture* v. 100 A heavily bodied plaster mix can be applied over the agar impression to form a mother mold or casing. **1969** R. Mayer *Dict. Art Terms & Techniques* 254/1 *Mother mold*, an outer case or container for a negative mold made of gelatin, rubber, or another weak, flexible substance. The mother mold is made of a rigid material. **1936** *Sun* (Baltimore) 6 July 9/1 Progress on the pick-a-back airplane, a combination in which a 'mother' plane will carry on its back a smaller long range seaplane for 'launching' at high altitude is more secret. **1945** *Time* 19 Nov. 52/2 Everything it sees is projected by radio on a screen in the mother plane. **1962** *Daily Tel.* 10 Sept. 18/4 The 'mother' plane would carry pick-a-back a plane which would be launched at the fringe of space. **1966** C. Himes *Heat's On* iii. 30 Some mother-raper is shooting at me with water-melon seeds. *Ibid.* ii. 22 The dirty mother-raping white nigger! **1969** Mother-raping [see *cock-sucking* ppl. adj. s.v. *cock *sb.*[1] 23]. **1902** H. C. Fyfe *Submarine Warfare* v. 108 A wireless message..has been sent to the commander of the 'mother ship'. **1909** *Q. Rev.* Oct. 575 Depôt ships for destroyers, mother-ships for submarines, and oil-supply vessels. **1922** *Encycl. Brit.* XXX. 17/2 Scouts were flown off lighters at sea against airships, and off the decks of battleships and 'mother' ships. **1926** H. T. Wilkins *Marvels Mod. Mech.* 215 An engineer, aboard the airship, opened the telescopic apparatus which left the aeroplane swaying in space some 60 feet below the mother ship. **1938** *Flight* 8 Sept. 197/1 They [*sc.* engineers] designed for catapult launch from a mother ship;..they used diesel engines. **1946** in *Amer. Speech* (1947) XXII. 230/2 The Navy's drones will be sent into the cloud by one mother ship. **1962** J. Tunstall *Fishermen* ii. 46 The fleet of catching vessels transfers its fish..to a mother ship, which processes and freezes the fish at sea. **1967** *Times Rev. Industry* Apr. 48/3 Trailers, operating from the central plant and acting as 'mother ships' to the delivery vans, can replace the depots. **1969** *Observer* 20 July 7/3 After it has docked with the mother ship the astronauts will spend four hours going over every inch of the LEM. **1973** *Sci. Amer.* Nov. 23/1 The small 'killer' satellites in a 'mother ship' equipped with central guidance and detection devices. On command the mother ship would have oriented itself and determined when, at what rate and in what direction to launch its subsatellites. **1865** C. M. Yonge *Clever Woman of the Family* II. xxx. 312 The mothers' meetings for the soldiers' wives. **1925** Fraser & Gibbons *Soldier & Sailor Words* 159 *Mother's meeting*, an occasional name among bluejackets for the captain's address to a ship's company. **1946** D. Hamson *We fell among Greeks* xviii. 195, I noticed one particular squad which was openly idling... 'Why do you stop work and hold a mother's meeting when I go away?' **1927** B. Malinowski *Sex & Repression in Savage Society* II. iii. 100 Not one single case of mother–son incest could be found. **1949** M. Mead *Male & Female* xvi. 326 A mother–son combination is classified as bad for the son. **1943** J. S. Huxley *Evolutionary Ethics* ii. 16 The absence in the infant's life of a mother or effective mother-substitute during the crucial period from about one to three years old. **1965** F. Sargeson *Memoirs of Peon* vi. 173 Two young sparrow-legged ruffians..engaged in selling my mother-substitute a large trolley-load of empty bottles. **1959** *Science* 21 Aug. 422/3 We took the calculated risk of constructing and using inanimate mother surrogates rather than real mothers. **1969** E. Stotland *Psychol. of Hope* viii. 126 Cloth-covered objects..appear to satisfy a need for contact and closeness; the babies cling to these mother surrogates while they do not cling to those made of uncovered wire. **1956** R. F. C. Hull tr. *Jung's Symbols of Transformation* in *Coll. Wks.* V. 301 At this stage the mother-symbol..points towards the unconscious as the creative matrix of the future. **1902** *Encycl. Brit.* XXIX. 312/2 The pure tinctures are denominated 'mother tinctures'. **1906** W. De Morgan *Joseph Vance* xvi. 149 She makes some concession to my feelings on the subject of High Dilutions, and (at great risk to myself, she says) allows me to have Mother-Tinctures. **1960** C. Dale *Spring of Love* i. 26 She would take a magazine..and re-read..Sister Jane's advice to mothers-to-be. **1973** A. Morice *Death & Dutiful Daughter* vii. 68 Has our little mother-to-be surfaced yet?

b. Mother of Commonwealths *U.S.*, Virginia; **Mother of Parliaments,** (*a*) England; (*b*) the British Parliament; **Mother of Presidents** *U.S.*, (*a*) Virginia; (*b*) Ohio; **Mother of States** *U.S.*, (*a*) Connecticut; (*b*) Virginia; **mother of thousands,** (*a*) for '*Linaria cymbalaria*' substitute '*Cymbalaria muralis*'; (earlier and later examples); (*c*) (examples); (*e*) = *helxine.

1879 *Congress. Rec.* 10 Jan. 413/2 To pour out the vials of his impotent wrath upon the 'Mother of Commonwealths'. **1865** J. Bright in *Birmingham Daily Post* 19 Jan. 5/1 We may be proud of this, that England is the ancient country of Parliaments... England is the mother of Parliaments. **1910** *Encycl. Brit.* VII. 15/1 The early date at which the principle of self-government was established in England, the steady growth of the principle, the absence of civil dissension, and the preservation in the midst of change of so much of the old organization, have given its constitution a great influence over the ideas of politicians in other countries. This fact is expressed in the proverbial phrase—'England is the mother of parliaments'. **1918** *Daily Mirror* 12 Nov. 6/2 Never has the Mother of Parliaments seen such a scene of enthusiasm as when Mr. Lloyd George read out the armistice terms yesterday. **1926** Fowler *Mod. Eng. Usage* 548/1 Mother of Parliaments (British Parliament). **1974** *Times* 24 Aug. 2/4 *France Soir*..went on to explain why in the country of the 'Mother of Parliaments' social tension has grown. **1827** A. Sherwood *Gaz. Georgia* 98 James Monroe..was born in Va., the mother of Presidents. **1897** *Chicago Record* 8 Mar. 4/1 Ohio may claim to take rank with Virginia as a 'mother of presidents'. **1904** *N.Y. Tribune* 12 June 8 Virginia concluded not to indorse any candidate. The 'Mother of Presidents' is a trifle particular. **1948** *Chicago Daily News* 21 Apr. 1/5 Ohio is the mother of Presidents, and Taft is one of her sons. **1834** W. A. Caruthers *Kentuckian in N.Y.* ii. 195 Virginia has been the mother of states. **1838** *Yale Lit. Mag.* III. 86 To thee, Mother of States! to thee, good old Connecticut, do our praises most belong. **1855** *Southern Lit. Messenger* XXI. 675/1 Virginia..[was] hailed as 'the Mother of States'. [**1731** P. Miller *Gardeners Dict.* s.v. *Linaria*, The first of these Plants [*sc.* common yellow toad-flax] grows in great Plenty upon the Sides of dry Banks in most Parts of England and is seldom cultivated in Gardens, for it is a very troublesome Plant to keep within Bounds, the roots being very apt to spread under-ground, and rise at a great Distance from the Mother Plant, whereby it greatly injures whatever Plants stand near it.] **1855** A. Pratt *Flowering Plants & Ferns Gt. Brit.* IV. 126 This plant [*sc. Linaria cymbalaria*] is familiarly known to many persons by the name of Mother of Thousands. **1910** T. W. Sanders *Window & Indoor Gardening* xi. 111 Ivy-leaved Pelargoniums..make delightful basket plants... So, too, [do] Saxifraga sarmentosa (Mother-o'-Thousands or Wandering Jew). **1952** A. R. Clapham et al. *Flora Brit. Is.* 713 *H*[*elxine*] *soleirolii* Req. Mind-your-own-business, Mother of thousands... Naturalized on walls and damp banks. **1958** *N. & Q.* Sept. 411/2 Have your readers ever heard of a plant called 'Mother of Thousands' or 'Wandering Sailor'? *Ibid.* Oct. 452/1 Mother of thousands... I believe this plant is *Saxifraga sarmentosa*. *Ibid.* Nov. 488/2 'Mother of thousands'... The name is given to the Ivy-leaved Toad-flax which is also known as 'Mother of Millions'. **1961** *Countryman* LVIII. III. 547 Mother of-thousands (ivy-leaved toadflax). **1971** K. G. Messenger *Flora of Rutland* 69/1 *S*[*oleirolia*] *soleirolii*... Mother-of-Thousands... Pavements, Uppingham, 1961... Intrusive alien. **1975** *Times* 22 Nov. 10/5 The elegant climber..the mother of thousands..produces small plantlets on thread-like runners.

mother, *v.*[1] Add: **5.** (Further examples.) Also, to pick out from a flock the mother of (a particular lamb). Also *intr.*, *transf.*, and const. *up*.

1889 *Zealandia* I. 30 He pretended to give me every opportunity at 'mothering' (as it is called) my missing lambs, but for fifteen ewes I could find but four. **1890** *Cornh. Mag.* Oct. 386, I was shepherding for Gasgarth, and his missus said to me, 'Jem, mother that 'un,' and I went reight intill middle o' t' flock and browt out t'mother on it. **1950** *N.Z. Jrnl. Agric.* Sept. 197/3 All lambs can be properly mothered up before nightfall. *Ibid.* 202/1 Do not move the ewes until all the lambs have mothered. **1957** *New Biol.* XXII. 101 Some lambs failed to 'mother' after treatment. **1961** R. M. Patterson *Buffalo Head* iv. 134 At Bull Creek we stopped for a little while to give the calves a chance to 'mother-up'; very soon every cow had her calf. **1962** [see *hay-bale* (*hay *sb.*[1] 4)]. **1972** *Country Life* 30 Mar. 812/3 Both lambs and ewes are marked with identical letters to facilitate shepherding, particularly the mothering up, at night, which is very important with winter lambing.

mothercraft (mʌ·ðəɹkrɑft). [f. Mother *sb.* + Craft *sb.*] The 'craft' or business of a mother; knowledge of, and skill in, the care of children. Also *attrib.*

1914 *Times* 25 Apr. 10/3 The School of Mothercraft. **1917** B. Vaughan *Menace Empty Cradle* 44 How much better it would be to teach them mothercraft. **1922** *Daily Mail* 28 Oct. 7 A Mothercraft question set by the Association of Infant Welfare. **1945** [see *Karitane]. **1969** *Sydney Morning Herald* 7 June 4/2 A man wearing a white stocking mask with eye-slits stuck a pistol into the ribs of a mothercraft nurse in a $14,200 payroll robbery at the Royal Hospital for Women, Paddington, yesterday. **1970** *Globe & Mail* (Toronto) 28 Sept. 7/6 As the mother of a child in the Mothercraft nursery may I be permitted to clear the distorted view some of your..readers have.

motherhood. Add: **1. c.** *transf.* Used *attrib.* of an issue, report, etc.: protective, withholding the worst aspects. *N. Amer.*

1973 *Ottawa Jrnl.* 4 Apr. 6/2 They think public participation is such a motherhood issue in politics these days that it must be paid homage to at all cost. **1974** *Evening Herald* (Rock Hill, S. Carolina) 18 Apr. 4/4 At the meeting, officials from Kaiser, Reynolds, Southwire, Alcoa and others agreed that a 'motherhood' package concealing the dangers of aluminium wiring was the best way to deal with the hearings of the Consumer Product Safety Commission in Los Angeles. **1974** *Ottawa Citizen* 14 Nov. 37/2 Opposition spokesmen attacked him for presenting a 'motherhood' report which told nothing of the real situation of the peacekeeping forces, only that Canadian troops are loved overseas.

mothering, *vbl. sb.*[2]: see Mother *v.*[2]

mothering, *ppl. a.* Add: **2.** *U.S. slang.* = **mother-fucking* ppl. adj.

1968 L. W. Robinson *Assassin* (1969) ii. 8 We are lost in the motherin' fog. **1970** 'J. Morris' *Candywine Development* xxiii. 253 What I don't see is how those motherin' hijackers expect to get away with it. **1975** *New Yorker* 23 June 40/2 I'm out there cutting that mothering grass all day!

mother-in-law. 4. mother-in-law's tongue, a tropical African herbaceous plant, *Sansevieria trifasciata* or its variety *laurentii*, of the family Liliaceæ, distinguished by a rosette of long, pointed leaves with white or yellow markings, and often cultivated as a house plant.

1958 *Times* 29 Nov. 9/5 Probably the most long-suffering house plant is the 'Mother-in-law's Tongue'. **1961** *Amateur Gardening* 11 Nov. 7/1 The sansevieria, because of its sword-like shape has received the unsympathetic name of Mother-in-Law's Tongue. **1970** 'D. Halliday' *Dolly & Cookie Bird* ii. 15 The place was modern and airy and full of..mother-in-law's tongue in long boxes. **1974** *Times* 5 Oct. 12/1 The real tough ones include aspidistras, ..mother in law's tongue *Sansevieria laurentii* and the rubber plant *Ficus elastica*.

motherless, *a.*[1] Add: Also as an intensive quasi-*adv.*, esp. in phr. *stone motherless broke*. *Austral. colloq.*

1898 *Bulletin* (Sydney) 17 Dec. Red Page, To these are prefixed the adjectives *motherless* and *dead*, thus *dead motherless broke*. 1916 A. Wright *Under Cloud* 35 I'm stone motherless meself an' I could do with that hundred Booth is offerin'. 1925 E. S. Sorenson *Murty Brown* 135 That leaves me stony, motherless broke again. 1945 G. Casey *Downhill is Easier* i. 20 I'd only known Reg when we were both stone motherless broke. 1972 A. Macdonald *Ukulele Player under Red Lamp* 147 Happy, full as a boot, and stone motherless broke.

mother of pearl. Add: **4. b. mother-of-pearl cloud** *Meteorol.* [tr. Norw. *perlemorsky* (H. Mohn 1893, in *Forh. i Vidensk.-Selsk. i Christiania* i)], a kind of iridescent cloud occurring above the tropopause which is sometimes seen in high latitudes after sunset; cf. *noctilucent cloud* s.v. *NOCTI-.

1932 *Q. Jrnl. R. Meteorol. Soc.* LVIII. 307 The remarkable stratosphere clouds which the late Professor Mohn called mother-of-pearl clouds again appeared over Scandinavia several times during the months of January and February this year. 1957 Ludlam & Scorer *Cloud Study* 69 Mother of pearl clouds are rare. They have been seen most often in Norway (but occasionally also in Scotland, Iceland and Alaska), and then only in the months between November and March. 1972 R. Scorer *Clouds of World* v. 74 Mother of pearl clouds are at altitudes between 19 and 30 km. where they may be brightly illuminated by the sun after sunset at the ground, and after the sun has ceased to illuminate the troposphere where most of the blue colour of the sky originates.

mother's brother. *Anthropol.* The maternal uncle, usu. with reference to his importance in kinship affairs and social customs. Also *attrib.*

1871 L. H. Morgan in *Smithsonian Contrib. Knowl.* No. 218 12 My father's brothers and my mother's brothers, in English, are generalized into one class, and the term *uncle* is employed... The relationships to *Ego* of the two classes of persons are equal in their degree of nearness, but not of the same kind; wherefore, the Roman method is preferable, which employed *patruus* to express the former, and *avunculus* to indicate the latter. *Ibid.* 13 In Seneca-Iroquois, for example, my father's brother is my father... My mother's brother, on the contrary, is my uncle. 1934 R. H. Lowie *Cultural Anthropol.* xiii. 251 A preponderant position of the mother's brother may occur even without matrilocal residence and is termed 'avunculate'. 1949 E. E. Evans-Pritchard in M. Fortes *Social Structure* 95 It is worse with a mother's brother's daughter than with a father's brother's daughter. 1951 R. Firth *Elem. Social Organization* ii. 48 Reference has been made..to the importance of the mother's brother relationship in Oceanic communities. 1972 J. S. La Fontaine *Interpretation of Ritual* 166 While all non-agnatic kin ceremonially recognize a girl's maturity, the primary ritual figure, to the exclusion of her father or senior agnates, is her mother's brother.

moth-proof (mǫ·þ,prūf), *a.* Also **moth proof, mothproof.** [f. Moth *sb.* + Proof *a.*] Resistant to damage by moths.

1893 *Ladies' Home Jrnl.* May 13/4 Moth proof bag. 1907 *Yesterday's Shopping* (1969) 275/2 Moth proof Cotton underlay to go between wire bottom of bed and mattress. 1916 *Independent* (N.Y.) 15 May 248/2 Colorfast, reversible moth-proof rugs. 1922 *Sci. Amer.* 22 Oct. 260/1 Making wool moth-proof. 1924 *Ibid.* July 39/4 (*heading*) Making the closet moth-proof. 1931 *Daily Express* 21 Sept. 7/5 Hygienic, dustproof, mothproof and washable. 1959 *Times* 12 Jan. 11/3 Courlene..is..rot and mothproof, washable and quick drying. 1960 *Which?* Jan. 18/2 Wool can be made mothproof by the application of certain chemicals, and mothproof wool is available from some manufacturers and is particularly desirable for carpets.

Hence as *v. trans.*, to render moth-proof; so **moth-proofing** *vbl. sb.*; **moth-proofer**, a substance for moth-proofing clothes or textile fibres.

1925 *Chem. Abstr.* XIX. 184 (*title*) Mothproofing fibrous materials. *Ibid.* 993 (*title*) Moth-proofing fabrics. 1932 *Discovery* Oct. 321/2 Science has provided sufficient mothproofing substances and means for combating this pest [sc. the moth]. 1938 *Chatelaine* May 79/1 Her lovely clothes at Columbia Studios are mothproofed with Larvex. 1958 *Spectator* 4 July 15/3 Expert mending, dying to pattern, mothproofing. 1960 *Which?* June 130/1 For woollens that have not been mothproofed in manufacture, the consumer can use *household mothproofers*—liquids which impregnate the material and kill the larvae—or moth repellents. 1961 *Times* 13 Dec. 21/7 Air fresheners..mothproofers.

mothy, *a.* Add to def.: Characterized by the presence of moths; suggestive of a moth, resembling the movements of the wings of a moth.

1898 Hardy *Wessex Poems* 155 At mothy curfew-tide. 1917 —— *Moments of Vision* 255 Some nocturnal blackness, mothy and warm. 1949 E. Bowen *Heat of Day* vi. 113 Your mothy way of blinking and laziness about keeping your eyelids open.

motif. Add: **1. b.** (Earlier and later examples.) Also in *Folklore*, a recurrent character, event, situation, or theme.

1857 Kingsley *Two Years Ago* I. vi. 161, I owe you a debt, sir, for having furnished me with one of the most striking 'motifs' I ever had. 1877 *International Rev.* (N.Y.) Jan. 73 Having endeavored to discover the *motif* and purpose of George Eliot's work. 1884 A. Lang in R.M. Dorson *Peasant Customs* (1968) I. 312 Turning from the Samoyeds and the Epirotes to Africa, we find the *motif* (escape of brother and sister) in a Kaffir tale. 1902 W. W. Lawrence in *Jrnl. Eng. & Gmc. Philol.* IV. 476 Repetition of motifs and consequent disturbance of logical arrangement is characteristic of Anglo-Saxon poetry. 1935 A. C. Bartlett *Larger Rhet. Patterns Anglo-Saxon Poetry* 80 The conception of the hereafter clothed in definite form is, in Germanic poetry, found only in Anglo-Saxon, and..it is a favorite motif in poems not bound by Bible text. 1941 G. G. Scholem *Major Trends in Jewish Mysticism* ii. 67 Some of the oldest mythical motifs are to be found..in an extremely interesting..text. 1968 P. Oliver *Screening Blues* ii. 63 The parody sermons, as Stith Thompson shows, are one of the more popular of folk-tale motifs and they abound in Negro folk-lore. 1971 *English Studies* LII. 501 In addition to the motif of longing, the themes of former happiness and of present estrangement from friends also unify the poem.

4. *attrib.* and *Comb.*, as *motif classification, framework, research, -work; motif-less, -like* adjs.; **motif-index**, an index of standard folklore motifs.

1966 K. Luomala in M. Jacobs *Anthropologist looks at Myth* 164 Kirtley's *Motif-Index*..follows Stith Thompson's system for motif-classification. 1971 R. L. Welsch tr. *Krohn's Folklore Methodology* iii. 31 The fact that it occurs more or less incidentally in other complexes is due to the influence of the motif framework. 1932 S. Thompson in *Indiana Univ. Stud.* No. 96 (*heading*) Motif-index of folk literature. 1958 T. P. Coffin *Analytical Index to Jrnl. Amer. Folklore* 239 It was deemed advisable not to attempt a complete Thompson-style motif-index of the Journal. 1972 *Jrnl. Amer. Folklore* LXXXV. 278 One would hope this *Index* would be shelved in every library beside the Thompson Type and Motif indexes. 1973 A. Dundes *Mother Wit* 114 These tools include various tale type indices and motif indices... By using tale type and motif indices, Dorson is able to demonstrate the European provenience of the greater portion of the tales. 1970 *Daily Tel.* 27 June 7 This work, while not..unmotivated, is entirely motif-less—introductory clinking of pebbles, toneless blowing into instruments built for lusher uses. 1964 *English Studies* XLV. 446 Two terms which characterize Christina's..reputation through their motif-like occurrence, *caprice* and *banality*, are a source of insight and confusion. 1938 P. S. Watson tr. *Nygren's Agape & Eros: Part 2* I. 23 The answer can only be supplied by motif-research, that is, by an enquiry which detects the motif behind the formed expression. 1948 L. Spitzer *Linguistics & Lit. Hist.* v. 193 This..movement is incarnated in..the rhythm—as in the verbal motif-work.

motile, *a.* Add: **2.** *Psychol.* Of or relating to responses that involve motor imagery. Cf. *MOTOR *a.* 3. Also as *sb.*, one who responds to perceptions more readily in terms of motor or kinæsthetic imagery than in terms of auditory or visual.

1886 *Mind* XI. 415 This division of men into visuals, audiles, motiles and indifferents..is of great interest and importance. 1889 *Proc. Soc. Psychical Res.* V. 536 A 'motile' reading the above question will dimly imagine the movements of his own larynx in uttering the word *elephant*. 1890 W. James *Princ. Psychol.* II. xviii. 61 Persons who belong to this type (*les moteurs*, in French, *motiles*, as Mr. Galton proposes to call them in English) make use, in memory, reasoning, and all their intellectual operations, of images derived from movement. *Ibid.* 62 Professor Stricker of Vienna..seems to have the motile form of imagination developed in unusual strength. 1892 —— *Text-bk. Psychol.* xix. 308 All blind persons must belong to the 'tactile' and 'motile' types of the French authors. 1909 [see *AUDILE *a.*].

Motilon (mōutilōu·n). Also **Motilone.** Pl. **Motilon, Motilones.** [Sp., of Amer. Ind. origin.] A Cariban Indian people of Colombia; a member of this people; also, their language. Also *attrib.*

1913 P. J. Eder *Colombia* xii. 186 The white settlers have at times been driven away by the onslaughts of the Motilones Indians, who dwell in the forests of the Eastern Cordillera to the east. 1937 T. Rourke *Tyrant of Andes* 26 The Motilones are no myth. They..have been sending the six-foot arrows flying from the bush into the American ..geologists working for the oil companies. 1949 L. S. & E. Judson *Let's go to Colombia* v. 213 The Motilone Indians. Of fierce Carib origin the Motilones were never subdued by the Spaniards. 1950 J. A. Mason in J. H. Steward *Handbk. S. Amer. Indians* VI. 232 The Cariban affinities of Motilón..have always been accepted. 1957 *Encycl. Brit.* XXI. 80/1 They [sc. the Carib] reached the Andes on the west and small groups entered Colombia (Motilon, Opon and Carare Indians). 1964 E. A. Nida *Toward Sci. Transl.* iii. 37 The Motilones in Colombia have borrowed the Spanish word *purisima* 'most pure' from the phrase *Maria purisima* 'Mary, most pure', but have given it the meaning of 'the devil'. 1965 Moser & Taylor *Cocaine Eaters* iv. ii. 151 The arrow is the most treasured object a Motilon possesses. *Ibid.* iii. 157 We had been intrigued by the question of death amongst the Motilon. *Ibid.* 179 The Motilon chiefs. 1968 G. Anthony *Colombia: Land of Tomorrow* xiii. 160 It was left as a souvenir of a raid by the Motilone Indians... The Motilones, who live along the mountainous border between Colombia and Venezuela, are still amongst the fiercest and least-known of all Colombia's Indians.

motion, *sb.* Add: **3. b.** *to go through the motions of* (examples). Also *fig.*, to do something only perfunctorily; to pretend. Also with ellipsis of 'of' and following phr. So *going-through-the-motions*.

1816 Scott *Old Mortality* in *Tales my Landlord* 1st Ser. II. x. 255 She pressed her handkerchief to her face, sobbed with great vehemence, and either wept, or managed, as Halliday might have said, to go through the motions wonderfully well. 1850 'M. Tensas' *Odd Leaves Life Louisiana Swamp Doctor* 88 Cuss you! I was only going through the motions; the rifle ain't loaded! 1856 Dickens *Dorrit* (1857) I. xxxvi. 316 Mr. Dorrit urbanely went through the motions of playing a game at skittles with the Collegian who was the next oldest inhabitant to himself. 1900 'Mark Twain' *Man that corrupted Hadleyburg* 298 A Government cannot satisfy all these public opinions; it can only go through the motions of trying. This Government does that. It goes through the motions, and they do not succeed. 1933 Auden *Poems* (ed. 2) 58 Go through the motions of exploring the familiar. 1944 H. Croome *You've gone Astray* xvi. 166 It was pleasant, undoubtedly pleasant. But both of them felt..that in some fundamental sense they were merely going through the motions. 1958 *Spectator* 22 Aug. 247/1 Her gaoler is called her secretary and in public goes through the motions of secretaryhood. 1963 *Times Lit. Suppl.* 11 Jan. 27/4 A mere exasperated going-through-the-motions. 1973 W. J. Burley *Death in Salubrious Place* i. 24 It's really a diplomatic mission... It would be better if we at least go through the motions of an investigation.

4. (Later examples with reference to the agitated condition of water.)

1864 Dickens *Mrs. Lirriper's Leg.* i, in *All Year Round* Extra Christmas No., 1 Dec. 7/1 Not much motion on the whole, though me with a swimming in the head. 1971 E. R. Seary *Place Names Avalon Penin. Newfoundland* v. 90 Petty Harbour Point, now Motion Head..derived its latest name..from The Motion, the heavy cross sea caused by the irregular and broken ground.

15. motion photography, the photographing of moving subjects, later known as *cinematography*; **motion picture**, a 'moving picture'; a cinema film; also *attrib.*; **motion sickness**, nausea induced by motion; travel sickness; **motion study**, the study of the movements necessary for the most effective and least tiring method of performing a task; also *attrib.*

1912 F. A. Talbot *Moving Pictures* xxviii. 312 Motion photography as an educational force. 1913 *Chambers's Jrnl.* Sept. 688/1 In the early days the long exposure necessary rendered motion-photography well-nigh impossible. [1891 *Leisure Hour* Aug. 711/1 A highly composite mechanism which is to be known as the 'kinetograph', or motion-picture.] 1896 *Boston Even. Transcript* 7 Nov. 16/4 The cinematographe motion picture programme will be augmented daily. 1913 F. A. Talbot *Pract. Cinematogr.* 129 Plate, The first motion-pictures of an opening flower, taken at the Marey Institute. The complete opening of a convolvulus is shown in fifteen pictures. 1915 Wodehouse *Something Fresh* v. 117 The maiden in distress..was merely earning the salary paid her by some motion-picture firm. 1917 H. Garland in *Ade Lett.* (1973) 69 What a change is indicated in a few words. The motor car, the telephone,..and the Motion Picture Theater! 1923 R. D. Paine *Comr. Rolling Ocean* vii. 112 There were a few stores, a church,..but not a solitary motion-picture theater. 1929 M. R. Werner *Bryan* 264 Between 1916 and 1919 Bryan was engaged in negotiations for a motion picture on the curse of drink, of which he was to be the star. 1934 *Discovery* June 171/1 Fifty thousand feet of standard motion-picture film will be exposed. 1952 *Manch. Guardian Weekly* 5 June 7/2 The nine judges of the United States Supreme Court handed down a unanimous decision this week that exerts an historic restraint on the censorship of motion pictures. 1972 D. Shipman *Great Movie Stars: Internat. Yrs.* p. v, The coming of Sound to the motion-picture medium killed it as a means of almost effortless international communication. 1973 L. Snelling *Heresy* i. i. 8 This was a visual generation and the motion-picture its natural medium. 1942 *Sci. News Let.* 12 Dec. 378/1 Air sickness can be prevented or cured if you know how. So can seasickness and the other kinds of motion sickness that attack men lurching over battlefields in tanks and jeeps. 1953 A. Smith *Blind White Fish in Persia* x. 189 We had been asked to investigate the effectiveness of Kwells pills against the motion sickness occasionally experienced in riding camels. 1971 *Jrnl. Gen. Psychol.* LXXXV. 87 (*heading*) A psychophysiological test for motion-sickness susceptibility. 1973 *Times* 1 Aug. 6/1 Still suffering from the motion sickness which has put their mission behind schedule, the three Skylab astronauts today started to tidy up the orbiting laboratory. 1911 F. B. Gilbreth (*title*) Motion study. *Ibid.* p. v, The phrase 'Motion Study' explains itself. The aim of motion study is to find and perpetuate the scheme of perfection... Standardizing the trades is the world's most important work to-day, and motion study is the first factor in that work. 1918 C. S. Myers *Present-Day Applications of Psychol.* 17 Yet in regard to industry, the need for good method and the need for systematic training are only just beginning to be realised, as is instanced in the following illustrations of what is now technically known as 'motion-study'. 1933 M. S. Viteles *Industrial Psychol.* xx. 436 In the practical application of motion study techniques the emphasis is upon the ease of work, upon *human* and not primarily upon *economic* efficiency. 1960 M. Spark *Ballad of Peckham Rye* ii. 15 Motion study did marvels in the factory.

motion, *v.* Add: **5.** *trans.* To impart motion to.

1929 Bridges *Test. Beauty* i. 24 Wisdom..choosing to be call'd Athena daughter of Zeus Motion'd the marble to her living grace.

motional, *a.* Add: (Further examples.) In *Electr.* applied *spec.* to an impedance that arises out of the motion of a conductor in a magnetic field.

1892 O. Heaviside *Electr. Papers* I. 446 In general **h** = V**Dv** × 4π, where **D** is the displacement of the field, **v** the velocity, **h** the induced magnetic force, and V is as in equation (3). [*Note*] The force [so defined]..I now term the motional magnetic force, and its companion (21*a*) below, the motional electric force. 1912 Kennelly & Pierce in *Proc. Amer. Acad. Sci.* XLVIII. 113 The difference obtained by subtracting the damped values from the corresponding free values may be called the motional values of resistance, inductance, etc.; since such differences are due to the motion of the diaphragm. *Ibid.* 144 This means that the impedance of the receiver has become increased, through the vibration of the diaphragm, by a motional impedance. 1957 W. Fraser *Telecommun.* xvi. 510 The movement of the speech coil in the air gap induces a back e.m.f. in the coil so that the loudspeaker has a motional impedance under dynamic conditions. 1969 F. A. Bovey *Nuclear Magn. Resonance Spectrosc.* i. 14 In liquids of ordinary viscosity..the local variations in magnetic field strength have become so short-lived that motional averaging is complete.

motionally (mōu·ʃənăli), *adv.* [f. Motional *a.* + -ly².] As regards motion.

1974 *Nature* 7 June 593/3 As the average pore radius approaches 100 Å, the amount of motionally restricted water in the Vycor glass is reduced to below 10% (v/v).

mo·tionlessly, *adv.* [f. Motionless *a.* + -ly².] Without motion, stilly; in a motionless manner.

1887 W. James in *Pop. Sci. Monthly* June 170 It is the very same instinct which leads a boy playing 'I spy' to hold his very breath when the seeker is near, and which makes the beast of prey..motionlessly lie in wait for his victim.

motivate, *v.* Add to def.: To provide with a stimulus to some kind of action; to direct (a person's energy or behaviour) towards certain goals. (Further examples.)

1924 W. B. Selbie *Psychol. Relig.* iv. 89 Freud has shown very clearly that forgetting is not the unconscious process that we think, but that it is generally motivated. 1949 *Here & Now* (N.Z.) Oct. 29/2 The vast majority of pictures are sexually motivated. 1953 'M. Innes' *Christmas at Candleshoe* v. 53 When at fifteen he was eventually packed off to apprenticeship in the city, the action was motivated only by the plain fact that there was nothing else to do with him. 1964 M. Argyle *Psychol. & Social Probl.* xii. 152 A special feature of the teacher's job is the need for him to motivate the pupils—there is no wage incentive for school work, and the competitive marks system is only a partial solution. 1966 J. Partridge *Middle School* vii. 139 A Secondary Modern in which boys are poorly motivated towards their school work at the best of times. 1971 E. Dichter (*title*) Motivating human behavior. 1974 *Daily Tel.* 2 May 22 (Advt.), Can you motivate salesmen? Not an easy task, because these people are supposed to be self-starters.

2. To serve as or provide a motive for; to justify.

1970 *Nature* 4 Apr. 44/1 The publisher motivates the slim size of these volumes by claiming it makes them more likely to be read. 1973 *Physics Bull.* Apr. 234/3 The demand for a relativistically acceptable version of momentum conservation is used to motivate the introduction of relativistic concepts of dynamics.

Hence **mo·tivated, mo·tivating** *ppl. adjs.*

1909 *Proc. Mod. Lang. Assoc.* XXIV. 175 The motivating reason is of another sort. 1928 C. J. Warden *Outl. Compar. Psychol.* v. 126 Punishment was introduced as a motivating stimulus. 1959 B. Wootton *Social Sci. & Social Path.* viii. 245 Their actions appear motiveless or strangely motivated. 1963 *Economist* 23 Nov. 767/2 Some form of sales promotion (the motivated-sell). 1968 R. West *Sketches from Vietnam* iii. 75 The Americans concede with regret that the Vietcong tend to be more highly motivated than the government supporters. 1973 *Jrnl. Genetic Psychol.* CXXII. 198 The majority of subjects..could be regarded as highly motivated to enter university.

motivation. Add: **1. b.** *Psychol.* and *Sociol.* The conscious or unconscious stimulus for action towards a desired goal provided by psychological or social factors; that which gives purpose or direction to behaviour.

1904 *Psychol. Rev. Monogr. Suppl.* V. ii. 11 (*heading*) The egoistic and the social in motivation. 1922 R. S. Woodworth *Psychol.* viii. 137 The instincts are extraordinarily important in the study of motivation. 1932 W. McDougall *Energies of Men* xvi. 245 It is in relation to the problems of motivation that the postulation of subconscious activities is most urgently required. 1937 T. Parsons *Struct. of Social Action* III. xvi. 637 Weber's work necessitated the ideal-typical formulation of systems of ideas which are relevant to concrete motivation. 1950 Lucas & Britt *Advertising Psychol. & Res.* ii. v. 107 Lists of human wants and related advertising appeals are included in this chapter mainly for the purpose of illustrating the theory of motivation. 1954 R. F. C. Hull tr. *Jung's Coll. Works* XVII. vii. 193 The real motivations are sought and real discoveries are made. 1966 *Rep. Comm. Inquiry Univ. Oxf.* I. 69 Weight can properly be given to evidence about their motivation,..the nature of their schooling, or their non-academic interests and activities. 1972 *Sci. Amer.* Feb. 48/1 There are other important motivations for attempting to synthesize speech. One is learning how speech is produced in the human vocal tract. 1973 B. O'Connell *Aspects of Learning* vi. 103 Motivation may be considered as either intrinsic or extrinsic; intrinsic motives include those of exploration and curiosity, and extrinsic those of status and social approval.

2. Manner or means of movement.

1946 M. Peake *Titus Groan* 407 There could be no mistaking that nimble, yet shuffling and edgeways-on—that horribly deliberate motivation that was neither walking nor running. 1948 M. H. Nicolson *Voyages to Moon* v. 195 Swift's flying chariot remains unique in the history of literature..both for its vast size..and for the plausibility of its principle of motivation.

3. *attrib.* and *Comb.*, esp. in connection with market research and advertising, as *motivation analyst, department, psychology, study*; **motivation research**, research, usu. with the object of increasing a product's sales, undertaken to discover the underlying psychological and social motives that influence people; so **motivation researcher**; cf. *MOTIVATIONAL *a.* b.

1958 A. Huxley *Let.* 15 Feb. (1969) 846 American advertising techniques as perfected by the Motivation Analysts. 1957 V. Packard *Hidden Persuaders* iii. 29 The McCann-Erickson advertising agency in New York had five psychologists manning a special motivation department. 1961 *Daily Tel.* 14 Mar. 17/2 The 'motivation psychology' that allows a driver to be placid and content at one moment and at the next makes him rush forward 'like a raging lunatic'. *Ibid.*, He described 'motivation psychology' as the study of what makes the individual road user tick. 1953 *Business Week* 5 Sept. 40/2 The publication of a..book called 'An Introductory Bibliography of Motivation Research'. 1960 *Times Rev. Industry* July p. ii/2 Motivation research and ergonomics are but two examples of new terms... The one seeks the reasons for decisions and actions; the other is devoted to the study of man and his working environment. 1967 M. Argyle *Psychol. Interpersonal Behaviour* ix. 153 A number of special methods are used in motivation research interviews to get at genuine feelings. 1968 M. Richler *Cocksure* xxii. 139 The Motivation Research boys, the pollsters, covered America for us, and came back with 20,000 completed forms. 1954 G. H. Smith *Motivation Res.* i. ii. 23 In order to do profitable work at this level, the motivation researcher should be familiar with the major concepts of psychiatry and abnormal psychology. 1967 G. Steiner *Lang. & Silence* 45 'Motivation researchers', those gravediggers of literate speech, tell us that the perfect advertisement should neither contain words of more than two syllables nor sentences with dependent clauses. 1954 G. H. Smith *Motivation Res.* v. xvii. 206 Perhaps the best single person to guide a motivation study is a social psychologist.

motivational (mōutivēi·ʃənăl), *a.* [f. prec. + -al.] Of or pertaining to motivation.

1931 L. T. Troland in *Psychologies of 1930* 460 Narrowly and popularly conceived, motivational psychology would be concerned primarily with 'motives', but broadly and scientifically considered, it deals with all of the determinative functions and dynamics of mind. 1940 *Q. Rev. Biol.* Dec. 403 These conventional categories proved satisfactory only so long as each individual was looked upon as a discrete action-centre without intimate motivational connections with others of his species. 1943 *Mind* LII. 137 As to its motivational aspects, one can at least imagine that the cross connections just mentioned make it relatively easy to tap at least one of the just active hungers. 1951 G. Humphrey *Thinking* vi. 183 The mechanism of thinking was at least in the earlier work still associational though this was supplemented by motivational factors. 1958 M. Argyle *Relig. Behaviour* xii. 144 This is because the [social learning] theory shows how religion is passed on but does not show the motivational forces maintaining it. 1967 L. B. Archer in Wills & Yearsley *Handbk. Managem. Technol.* 131 Even in the most technical situations, motivational and aesthetic considerations have a material bearing upon performance. 1970 *Nature* 26 Sept. 1369/2 Investigations of animal learning usually include motivational components which activate and direct behaviour to relevant goals. 1974 *Ibid.* 15 Mar. 195/2 Problems of such motivational 'conflicts' are of great interest in finding out how behaviour in general is organised.

b. Special collocations: **motivational analyst, consultant, selling; motivational research** = *motivation research*; **motivational researcher** = *motivation researcher*.

1957 V. Packard *Hidden Persuaders* i. 8 These motivational analysts..are adding depth to the selling of ideas and products. 1959 *Times Rev. Industry* Nov. 4/2 The copywriter has been joined by the art director, and round them have formed phalanxes of executives and research departments and motivational consultants. 1953 *Reporter* (N.Y.) 13 Oct. 27/3 Motivational research means just what it seems to—exploration into..the real reasons why a consumer buys..a product or service. 1957 *Observer* 22 Sept. 11/6 Last week he [sc. Dr. Ernest Dichter] described to an audience of advertising people what motivational research—M.R. to its friends—is, and what it does. 'It takes the social sciences out of the seminar,' he says, 'and into the supermarket.' 1960 *Sunday Express* 12 June 14/5 'You are not typical,' say the Motivational Research boys. 1960 E. Dichter *Strategy of Desire* (1961) 15 Often the assignment given our organization for conducting motivational research was a very sober and concrete one. 1971 —— *Motivating Human Behavior* 3 Some people begin to question the manipulative nature of motivational research. 1964 M. McLuhan *Understanding Media* xxii. 223 These tales from the Vienna Woods, dreamed up by motivational researchers. 1963 *Horizon* (U.S.) July 36/2 The *beau idéal* of all motivational selling is the use of language to conceal the absence of thought.

Hence **motiva·tionally** *adv.*, as regards or in terms of motivation.

1952 T. Parsons *Social System* i. 4 Only in so far as his relation to the situation is in this sense motivationally relevant will it be treated in this work as action. 1967 M. Argyle *Psychol. Interpersonal Behaviour* iii. 52 Synchronization is necessary along a number of different dimensions for smooth and motivationally satisfying interaction to take place.

motivative (mōu·tivētiv), *a.* [f. Motivat(e *v.* + -ive.] Serving to motivate; providing a reason for, or a stimulus to, action or thought.

1949 *Mind* LVIII. 37 It might be said that..commands..are unreliable from the logical point of view, because their real function is motivative. 1953 R. Carnap in P. A. Schilpp *Philos. R. Carnap* 45 Metaphysical theses..often have other meaning components, e.g. emotive or motivative ones.

motivator (mōu·tivēitəɹ). [f. Motivat(e *v.* + -or 2 c.] Something or someone that initiates, or is a stimulus to, action or behaviour.

1943 *Mind* LII. 124 The hypothesis that the prime motivators, such as hunger, thirst, pain-avoidance, and sex, are the driving power, the source of energy behind the whole infinitely complex pattern of socio-psychological behaviour. 1956 A. Huxley *Adonis & Alphabet* 41 Meaningless pseudo-knowledge has at all times been one of the principal motivators of individual and collective action. 1958 *Listener* 18 Sept. 414/2 Sex is clearly vitally associated with survival and is a primary motivator. 1959 Halas & Manvell *Technique Film Animation* 17 Television is now the main motivator of the increased output of animation. 1972 M. Argyle *Social Psychol. of Work* ix. 241 Herzberg and his colleagues concluded that 'motivators' (achievement, etc.) mainly affect satisfaction.

motive, *sb.* Add: **7.** Also in extended use. (Further examples.)

1865 F. B. Palliser *Hist. Lace* xxxv. 425 The French, by adopting what is technically termed eight 'motives', produce their lace of a finer make and more complex pattern. 1868 C. L. Eastlake *Hints Household Taste* ix. 201 Some of the table-china is also very good in what may be called the motive of its design. 1893 *Jrnl. Amer. Folklore* VI. 255 The motive that Brer Wolf seeks protection from his pursuers with Brer Rabbit..is akin to the episode in the Roman de Renart. 1928 *Daily Express* 21 Apr. 6/2 These motives are mounted in various designs on a background of satin. 1930 S. W. Cheney *New World Archit.* i. 20 The primary 'motive' [of New York] is the repeated pier-line, the chief relieving factor the occasional terrace. 1931 A. U. Dilley *Oriental Rugs & Carpets* plate 16 (*caption*) Vestibule of the Madrassa (college) Mader-i-Shah (Mother of Shah) Sultan Hussein (1700), Ispahan. Compare ceiling motive with rug medallion.

b. For '= Motivo, Motif 3' read '= Motivo, Motif 1 c'.

8. motive-hunter, -pattern.

1905 *Spectator* 28 Jan. 141/1 He is..a motive-hunter, seeking on every side for little justifications for his pride. 1941 Koestler *Scum of Earth* 232 No attempt to discriminate, to discover political motive-patterns.

motive, *a.* Add: **2.** *motive power* (later *attrib.* example).

1936 *Discovery* Nov. 380/1 Motive-power accessories—like electric batteries—are not required.

motivelessly *adv.* (Further example.)

1970 G. Greer *Female Eunuch* 37 When men began to grow their hair in our generation they were not acting motivelessly.

motivelessness. (Further example.)

1933 L. Strachey *Characters & Commentaries* iv. ix. 315 This triumphant invention of the motivelessness of Iago has been dwelt upon by innumerable commentators.

motivic (mōu·tivik), *a.* *Mus.* [f. Motiv(e *sb.* + -ic.] Of or pertaining to a musical motive or motives. Also *Comb.* So **moti·vically** *adv.*

1947 A. Einstein *Mus. Romantic Era* vii. 67 The thematic connection of the introductory *Andante* with the *Allegro ma non troppo*, a motivic 'safeguard' that one will not as yet find so clearly stressed in the Classics. *Ibid.* xi. 144 He [sc. Liszt] also feels doubly strongly the need for grappling the parts of the whole together motivically. 1957 E. T. Cone in N. Frye *Sound & Poetry* i. 4 The motivic shapes..of the music. 1959 *Listener* 16 July 114/1 These processes are not entirely explained by either motivic-contrapuntal or expansive-polyphonic methods. 1961 *Times* 13 Oct. 18/6 Webern's instrumentation of the Ricercare is not..so cannibalistic as a single hearing might lead one to suppose. A letter which he wrote..shows that his principal aim..was with its motivic coherence. 1964 *Listener* 13 Aug. 250/2 *Erwartung* is composed without proper themes, without continuity of motivic work, [etc.]. 1970 *Daily Tel.* 19 Nov. 14/5 A wealth of further motivic events in these two slowish movements.

‖ **motiviert** (motivi·rt), *a.* Also **motivirt.** [G., pa. pple. of *motivieren* to motivate.] Motivated. Also **Motivie·rung**, motivation.

1812 H. C. Robinson *Diary* 20 Aug. in *On Bks. & their Writers* (1938) I. 108 He [sc. Coleridge] thinks the character of Faust himself not motiviert. 1866 Geo. Eliot *Let.* 10 Apr. (1956) IV. 237, I feel sure that there are deficiencies..in what the Germans call the *Motivirung* seen from a legal point of view. 1890 G. B. Shaw *Let.* 17 Aug. (1965) 255, I also opined that the fatalistic apathy of Christ..was inadequately motivirt.

‖ **mot juste** (mo ʒüst). [Fr., lit. 'exact word'.] The precisely appropriate expression.

1912 *Nation* (N.Y.) 14 Mar. 264 Here and throughout we have conspicuously the *mot juste*, not one too many and each where it will tell. 1915 E. Pound in Joyce *Lett.* (1966) II. 364 My head is a squeezed rag, so don't expect le mot juste in this letter. 1916 [see *DIM *a.* 4 b]. 1932 Wodehouse *Louder & Funnier* 8 Most of these essays were written..at a time when the wolf at the door left little leisure for careful thought and the patient search for the *mot juste*. 1955 E. Pound *Section: Rock-Drill* lxxxv. 17 Get the mot juste before action. 1966 J. Dos Passos

Best Times (1968) ii. 44 For years I'd been reading Flaubert: letters, short stories... I caught his obsession for the *mot juste*.

‖ **moto** (mōu·to). *Mus.* [It.] In regard to musical tempo: motion, pace; *spec.* in **moto perpetuo**, a rapid instrumental composition consisting mainly of notes of equal value.

1740 J. Grassineau *Mus. Dict.* 145 *Moto*,..is a term that has many significations in music; sometimes it means only a motion or passage from one note to another... Sometimes it regards the quickness and slowness of such motion, as a brisk, slow, lively or languid motion. 1801 Busby *Dict. Mus.*, *Moto contrario* (Ital.), an expression applied to that progression of the different harmonic parts of a composition by which they move in opposite directions. 1884 F. Niecks *Conc. Dict. Mus. Terms* 177 *Moto* (It.), (1) Motion... (2) Movement... *Moto contrario* (It.), contrary motion. *Moto misto* (It.), mixed motion. *Moto obliquo* (It.), oblique motion. *Moto perpetuo* (It.), perpetual motion. 1901 G. B. Shaw *London Music in 1888–89* (1937) 389 A light-fingered and humorous *moto perpetuo* which might have come straight out of a Mendelssohn concerto. 1969 *Times* 30 Oct. 13/4 A brief moto perpetuo (with splendidly clean, disciplined figuration) by way of conclusion.

moto-. Add: **b. moto-cross**, cross-country motor-cycle racing.

1951 *Motor Cycle Sport in Pictures* 9 (*caption*) 1950 Moto-Cross victors. Left to right, B. W. Hall (B.S.A.), G. J. Draper (B.S.A.) and H. H. Lines (Ariel), three of the British team of 12 which won the Moto-Cross des Nations in Sweden. 1960 *Motor Cycling Sports Yearbk.* 81 (*heading*) Scrambles and Moto-cross. *Ibid.*, A dark cloud which has passed over the moto-cross world. 1968 K. Bird *Smash Glass Image* vii. 85 His machine kicked over the ground as if he were taking part in moto-cross.

motoneurone (mōutoniūə·ron). *Biol.* Also **motoneuron.** [f. Moto-: see Neuron in Dict. and Suppl.] = *motor neurone* s.v. *MOTOR *sb.* and *a.* B. 2.

1908 *Nature* 6 Feb. 333/2 The inhibition of the motoneurones is..followed by a superactivity in them. 1943 *Jrnl. Neurophysiol.* VI. 311 Few if any extensor motoneurons could have discharged in the earlier reflex. 1964 J. Z. Young *Model of Brain* iv. 50 In a mammal the largest motoneurons of the spinal cord may have a dendrite spread of 1 m. 1973 *Nature* 3 Aug. 254/1 The authors rule out the possibility that this occurs *via* the neural feedback loop involving the sensory fibres from the muscle to the motoneurone.

Motopia (mǫtōu·piă). [f. Mot(or *sb.* + Ut)opia.] A name for an urban environment designed to meet the needs of a pedestrian society by strict limitation of the use of the motor car. So **Moto·pian** *a.*

1959 *Engineering* 18 Sept. 206 (*heading*) Motopia: where traffic and pedestrians never meet. 1961 G. A. Jellicoe *Motopia* 7 Motopia..arises..from the realization that our present physical conditions are being thrown into chaos by the advent of one car per family and even one per person. *Ibid.* 15 The English preference..fixed the shape and alignment of the Motopian rectangle. 1968 C. A. Doxiadis *Between Dystopia & Utopia* 20 Such efforts as the *Motopia* of 1961, all of whose cars were to circulate on the tops of the roofs of the buildings.

motor, *sb.* and *a.* Add: **A.** *sb.* **1. a.** Restrict † *Obs.* to senses (*a*), (*c*), and (*d*), and add later example of (*b*).

1952 G. Sarton *Hist. Sci.* I. xix. 498 God exists, for it is the necessary principle and end of everything, the first motor.

2. c. A person in whom motor representations of perceptions predominate over auditory or visual ones.

1890 W. James *Princ. Psychol.* II. xviii. 62 The young savage was a *motor*. 1902 —— *Var. Relig. Exper.* xiv. 347 The shrew-type is defined as possessing an 'active unimpassioned temperament'. In other words, shrews are the 'motors', rather than the 'sensories'. 1929 J. Adams *Everyman's Psychol.* vii. 155 Men fall into two classes of sensories and motors... Motors have to receive sense impressions, and sensories have to send out messages leading to actions. *Ibid.* 156 The motor is often on the edge of action, eager to go.

3. b. (Further examples.)

1912 E. M. Forster *Let.* 25 Dec. in *Hill of Devi* (1953) 17 The Luards..got the Maharajah of Indore's motor. 1929 M. de la Roche *Whiteoaks* xxviii. 381 They were going by motor to the lakes.

5. *motor-ambulance, -hearse, -landau, -lorry, -plough, -sledge, toboggan, -tractor, -truck, vehicle, -wagon*; also designating sea vessels, aircraft, and other devices driven or powered by a motor, as *motor-cannon, -craft, -cruiser, -launch, -liner, meter, -mower, -ship, -sloop, torpedo-boat, -vessel, yacht.*

1915 Motor-ambulance [see *JOY-RIDE *sb.*]. 1917 W. Owen *Let.* 25 Mar. (1967) 447, I went a joy ride on a Motor Ambulance, a Daimler. 1931 *Times Lit. Suppl.* 18 June 478/4 The occasional British practice of sending on billeting-parties in motor-ambulances. 1889 F. R. Stockton *Great War Syndicate* 79 She carried one motor-cannon of large size. 1939 *War Illustr.* 9 Dec. 390 It was to a similar type of French engine that the first successful 'motor-cannon', or shell gun, firing between the arms of the 'V', was fitted. 1905 *Daily Chron.* 17 June 9/1 The latest regulation that no small boat should 'lock' with a motor-craft appears to have caused considerable surprise. 1921 *Daily Colonist* (Victoria, B.C.) 9 Oct. 23/2 The motor cruiser Speejacks was reported safe last night. 1936 *Discovery* Sept. 272/2 Charting the coast from a small motor-cruiser. 1975 *Times* 19 Sept. 4 (Advt.), A boat test on the fast 33' motor cruiser Cleopatra. 1924 P. Creswick *Beaten Path* xxiii. 180, I see there are motor-hearses now. 1973 'R. MacLeod' *Burial in Portugal* i. 23 An elderly motor hearse with black paintwork and tarnished chrome..carried a polished oak coffin. 1916 'R. Dehan' *Earth to Earth* 73 The motor-landau in waiting for him at Covertsham Station, Deershire, was destined to receive a second passenger. 1912 W. Owen *Let.* 31 Aug. (1967) 158, I was asked to accompany them & Vicar on a motor-launch-trip. 1935 *Discovery* Dec. 372/2 A little undecked motor-launch for inshore work. 1975 'D. Rutherford' *Mystery Tour* iii. 40 The trip by motor launch through the canals. 1913 *Chambers's Jrnl.* Jan. 31/1 The motor-liner above mentioned and her two sister-ships were built. 1931 *Evening Standard* 16 Jan. 9/1 (*caption*) The new motor-liner Warwick Castle leaving Belfast for her final trials. 1902 Kipling *Traffics & Discoveries* (1904) 391 There's a heavy load of grist just in from Lamber's Wood. Eleven miles it came in an hour and a half in our new motor-lorry. 1926 *Daily Chron.* 13 May 2/6 Three men were each sentenced to three months hard labour for inciting a crowd of people to set fire to a railway motor lorry in the city. 1934 *Discovery* Apr. 94/1 In most portions of the globe explorers under the new order have the valuable aids of aeroplane and motor-lorry. 1903 L. C. Reed *Amer. Meter Practice* iii. 37 These ten watts are the input into the motor meter, and the efficiency of this motor meter will give the ratio of the torque exerted on the armature. 1971 *Gloss. Electrotechnical, Power Terms* (*B.S.I.*) i. iv. 12 *Motor meter*, meter incorporating a motor. 1923 *Punch* 15 Aug. 165 (*caption*) The motor-mower takes charge. 1959 *Oxf. Univ. Gaz.* 1 Dec. 345/2 One new motor-mower was acquired in replacement of an old one now worn out and for which spare parts are no longer available. 1972 'H. Carmichael' *Naked to Grave* vi. 74 Through the window he saw a motor mower, cans of paint, fertilizer in paper sacks. 1901 *Chambers's Jrnl.* Nov. 764/2 Great things are anticipated of a new method of land culture by means of a motor-plough. 1916 *Daily Colonist* (Victoria, B.C.) 25 July 11/3 The motorship City of Portland..has been given her trial trip. 1929 *Times* 2 Nov. 10/2 The Tuscan Star..is the first motorship ordered by the Blue Star Line. 1955 *Times* 3 May 15/6 The construction of two motor ships of about 1,300 tons deadweight, designed to operate in our short sea services. 1910 *Chambers's Jrnl.* Aug. 546/2 The 'final dash' [to the Pole]..will be made with the help of motor-sledges. 1936 *Discovery* July 215/1 Captain Scott took motor-sledges to the Antarctic in 1910. 1931 *Times Lit. Suppl.* 15 Jan. 33/3 M. Knut Stubbendorff..chartered the motor-sloop Isbjörn. 1948 *New Deal for Saskatchewan Fisheries* 20 Motor toboggans..manoeuver in almost any kind of terrain. 1940 Motor torpedo-boat [see *E-BOAT]. 1928 *Punch* 29 Feb. 248/3 The world is still a stage; the ruling factors Are petrol and pink legs and motor-tractors. 1934 *Discovery* June 143/2 Although aircraft and motor tractor have been used they are in no sense in competition with the dog-sledge. 1916 *Daily Colonist* (Victoria, B.C.) 2 July 15/5 The iron-shod war horse of former days has evolved into the padded wheel motor car, motor truck, and motor cycle of 1916. 1937 *Discovery* Nov. 359/1 The sandy wastes of the Gobi are being cut to ribbons by motor-truck tyres. 1890 J. W. Quinn *U.S. Pat.* 431,993, I..have invented certain new and useful Improvements in Motor-Vehicles. 1937 *Discovery* July 194/2 The noise from motor vehicles. 1959 P. Bull *I know Face* ix. 148 He was referring to his motor-vehicle. 1931 *Daily Tel.* 21 May 7/5 Pioneers of motor-vessels to the East. 1969 *Jane's Freight Containers 1968–69* 394/1 This service is carried out by motor-vessels of 600–1,500 tons. 1909 *Chambers's Jrnl.* June 343/2 Large warehouses have their veritable fleets of motor-wagons. 1929 J. Buchan *Courts of Morning* iii. i. 320 The raiders..helped themselves..loading the loot into light motor-wagons. 1905 *Outing* May 218/2 A 90 foot motor yacht was launched late last season. 1973 'I. Drummond' *Jaws of Watchdog* i. 31 A fast, low-profile motor yacht five miles out to sea.

b. Of, pertaining to, or designed for use in motors, motor vehicles, or motoring, as *motor accident, age, bonnet, cap, -chassis, club, -coat, -dealer, -driver, fitter, garage, -goggles, -hooter, horn, -house, insurance, lantern, licence, -mania, -mask, mechanic, oil, race, -racing, -ride, -road, run, salesman, scarf, show, -smash, -spectacles, -tour, track, traffic, transport, -travel, -travelling, trip, trouble, tyre, -veil.*

1910 'Saki' *Reginald in Russia* 17 What I had mistaken for a motor accident was evidently a case of savage assault and murder. 1972 M. Yorke *Silent Witness* vi. 123 Roy's father had been killed in a motor accident. 1937 *Amer. Speech* XII. 317/1 The influence of..the motor age. 1963 *Daily Tel.* 28 Nov. 16/2 The root of the trouble is that our cities were never built for the motor age. 1910 *Blackw. Mag.* Nov. 604/2 My wife..had been patiently sitting in the hall wearing a new motor-bonnet. 1937 J. Laver *Taste & Fashion* x. 151 The latest trimming for these motor bonnets. 1906 *Daily Colonist* (Victoria, B.C.) 5 Jan. 3/4 (Advt.), Motor Caps. 50c, 75c, $1.00 reduced to 40c. 1907 Motor cap [see *EAR-FLAP]. 1915 *Pearson's Mag.* Jan. 25 Great lines of these old motor-chassis, mounted with a serviceable lorry body, are to be met with on all the roads of France. 1934 *Amer. Speech* IX. 111/1 Such motor clubs as the *A.A.A.*, the *A.M.A.*, and the *Automobile Club of Southern California*. 1909 W. J. Locke *Septimus* iv. 57 Shrugging himself into his motor-coat, which the chauffeur brought him. 1910 *Blackw. Mag.* Apr. 480/1 A long leather motor-coat. 1937 *Discovery* May 164/1 Motor dealers in South America..order what they need by air mail. 1902 *Pall Mall Mag.* XXVIII. 410/1 Should motor drivers be subject to an examination as to proficiency? 1968 J. H. Burn *Lect. Notes Pharmacol.* (ed. 9) 7 It is good for a motor-driver driving through the night. 1961 *Evening Standard* 26 July 22/6 Skilled..motor fitter required. 1973 *Times* 16 Oct. 6/8 Christopher Smart, aged 25, a motor fitter. 1974 P. Wright *Lang. Brit. Industry* ix. 77 He's a mechanic, but he'd rather be called a motor fitter. 1902 Motor garage [see *GARAGE *sb.* 1]. 1922 W. J. Locke *Tale of Triona* xxiv. 273 An ex-officer..who has just set up a motor garage. 1914 Beerbohm *Seven Men* (1919) 127 He did not wear motor-goggles. 1922 Joyce *Ulysses* 483 In motor jerkin, green motorgoggles on his brow. 1911 Beerbohm *Zuleika D.* xix. 291 All along the soaked towing-path lay strewn the horns, the rattles, the motor-hooters, that the youths had flung aside. 1909 *Chambers's Jrnl.* June 401/1 The predominant sounds..were not those of wheels or whistles or motor-horns. 1931 M. Allingham *Look to Lady* x. 105 The squawk of a motor-horn. 1970 V. Canning *Great Affair* xii. 211 A chorus of motor horns began to blow in anger at some road block. 1902 D. Salomons in A. C. Harmsworth et al. *Motors* vi. 84 A well-built motor-house should cost nothing in the upkeep. 1971 *Daily Tel.* (Colour Suppl.) 18 June 33/4 There is the medium-sized house with its own tennis court and 'motor house', as garages were first called. 1955 *Times* 3 May 16/1 Motor insurance, which constitutes a substantial portion of our Accident business. 1975 *Times* 24 Sept. 3/2 The estimated cost of motor insurance premiums in 1975 was £580m. 1914 Chesterton *Flying Inn* xxi. 248 Humphrey had hung one of the motor lanterns on to a branch. 1907 *Yesterday's Shopping* (1969) 413/2 The motor licence case. 1957 *Railway Mag.* June 392/1 The beginnings of a reaction from the worst of motor-mania are bringing the uninitiated back on to the narrow gauge. 1966 *Punch* 16 Feb. 232 The ingenuity, energy and wealth that mechanised intellectual Peter Pans devote to indulgence of their motor-mania. 1916 *War Illustr.* V. 451/3 Our goggles..are shaped like a motor-mask. 1930 *Daily Express* 6 Oct. 7/4 James Horn, aged thirty-one, a motor mechanic. 1961 *Evening Standard* 25 July 15/1 (Advt.), Motor Mechanics required. 1939 G. B. Shaw *Geneva* ii. 45 Motor oil is a sanction when you withhold it. Castor oil is a sanction when you administer it. 1900 *Racing Rules & Regulations* (Motor Car Club) 11 In all Motor Races held under the Rules of the Motor Car Club it must be stated on all Entry Forms..that they are so held. 1936 J. B. Priestley *They walk in City* iii. 41 An aeroplane parade, a motor race, a parade of cadets. 1905 *Official Programme Internat. Tourist Trophy* 4 A new departure in the history of motor racing. 1956 *Nature* 11 Feb. 246/2 The establishment of a motor-racing circuit on public roads in the Peak District National Park. 1974 J. Gardner *Corner Men* ii. 19 Most of the books were..concerned with motor cars or motor racing. 1914 Chesterton *Flying Inn* xx. 241 You owe me a motor-ride, you know. 1919 G. B. Shaw *Inca of Perusalem* Prologue in *Heartbreak House* 234 Dresses, hats, furs, gloves, motor rides: one bill after another. 1909 *Westm. Gaz.* 27 Aug. 2/2 Of more dramatic interest is the second part of the Bill, with its proposal for the creation of motor-roads. 1926 T. E. Lawrence *Seven Pillars* (1935) vii. lxxxii. 457 Gilman and Dowsett..had spent months in Wadi Itm, building, like engineers, a motor road through the gorge. 1949 *Archit. Rev.* CV. 118/1 The average speed of 30 m.p.h. here allowed for the surface plan is high, but, even allowing this figure, the advantage of elevated motor-roads is obvious. 1975 *Listener* 14 Aug. 199/1 A new motor road roaring past their windows. 1936 Joyce *Let.* 18 Nov. (1966) III. 393 Perhaps you could make your suggested motor run to the school at Zug? 1936 J. B. Priestley *They walk in City* i. 17 Mysterious and vaguely amorous gentlemen friends, usually commercial travellers or motor salesmen, loom and soar and suddenly vanish. 1973 A. Behrend *Samarai Affair* vi. 65 That Bromborough motor salesman chap separated from his wife. 1899 in A. Adburgham *Shops & Shopping* (1964) xxii. 261 Motor scarves. 1911 *Daily Colonist* (Victoria, B.C.) 28 Apr. 20/6 Waterproof Motor Scarf of good size..each 75¢. 1905 S. A. Barnett *Let.* 2 Dec. in H. Barnett *Canon Barnett* (1918) II. 194 We have been to the Motor Show, when vast Olympia was crammed by a fashionable crowd. 1965 A. Christie *At Bertram's Hotel* i. 13 He could talk racing shop, cricket..give Motor Show information. 1972 *Nature* 31 Mar. 193/1 It is difficult to believe that international motor shows are instrumental in selling cars. 1916 W. Owen *Let.* 3 July (1967) 398, I was so happy to learn that your motor-smash came to no harm. 1931 R. Lehmann *Letter to Sister* 22 Things happen to you... You have been in a motor smash. 1911 D. H. Lawrence *White Peacock* ii. iii. 262 She disengaged her arms to take off his disfiguring motor-spectacles. 1909 *London Mag.* Sept. 18/2 Excursions through the sky..are to become as common as motor-tours. 1915 J. Buchan *39 Steps* iv. 90 That's the end of my Scotch motor tour. 1928 *Manch. Guardian Weekly* 19 Oct. 301/2 Motor traffic and the urbanisation of large rural stretches are producing a hideous outbreak of advertisement hoardings, petrol pumps..and gaunt new motor-tracks. 1933 G. Robey *Looking Back on Life* xxviii. 306 Lord Dewar..made the famous epigram about there being only two classes of pedestrians in these days of reckless motor traffic—the quick, and the dead. 1926 *Brit. Gaz.* 12 May 2/4 At Liverpool there have been a few cases of interference with motor transport, for which heavy sentences have been inflicted. 1936 *Discovery* June 189/1 An ever-extending network of good macadamised roads is increasingly introducing the use of motor transport. 1923 A. E. Housman *Let.* 18 Aug. (1971) 215 Blazing hot all the time while motor-travel could temper it [sc. the weather]. 1909 *Chambers's Jrnl.* June 342/2 Putting the accidents on railways and on the roads together, motor-travelling included. 1934 Joyce *Let.* 25 Apr. (1966) III. 304, I was swept off..to make my first motor trip. 1950 M. Laski in *Contact* May–June 26/1 It is, of course, possible to take an extended motor-trip abroad without ever coming into contact with the Art of Navigation. 1911 *Chambers's Jrnl.* Jan. 55/1 Cleaver and Latham dropped into the Mediterranean..because of motor trouble. 1907 *Nature* 14 Feb. 383/1 Chemical composition of some motor-tyre rubbers. 1937 *Discovery* June 186/1 The wayside was strewn with motor tyres and broken down cars. 1970 Y. Carter *Mr. Campion's Falcon* vii. 59 Packing-cases, old motor tyres, drums of paint. 1907 *Yesterday's Shopping* (1969) 320D/3 Motor veil. In Crêpe de Chine. 1926 W. de la Mare *Connoisseur* 70 She was a rare one for the fashions: scarves and motor-veils, and that kind of thing.

c. Instrumental, as *motor-assisted, -driven, -dusted, -infested, -mad, -paced* adjs.

1954 *Highway Code* 32 To motor cyclists and riders of motor-assisted pedal cycles..*you must not* carry more than one passenger on a two-wheeled machine. 1904 *Motor Cycle* 6 June 551 A motor-driven roundabout. 1937 *Discovery* Dec. 387/1 The motor-driven fan. 1967 Karch & Buber *Offset Processes* v. 168 Motor-driven lenses, or hit-or-miss methods involving tricky out-of-focus photographic and lighting techniques are used. 1974 H. R. F. Keating *Bats fly up for Inspector Ghote* iii. 31 A powerful motor-driven dhow. 1909 M. B. Saunders *Litany Lane* iii. xviii. 236 His eyes were fixed on the white, motor-dusted hedges. 1909 *Q. Rev.* Jan. 143 This singularly congested and motor-infested country. 1931 T. H. Pear *Voice & Personality* 43 Today's world of motor-infested 'beauty-spots'. 1922 W. J. Locke *Tale of Triona* xxvi. 292 England ran motor-mad that summer. 1938 *Encycl. Brit. Bk. of Year* 100/1 Mills also set up a new record of 2 mins. 1¼ secs. for the mile, standing start, unpaced; while Albert Marquet did the distance in 41⅘ secs. from a flying start and motor paced. 1973 *Observer* 16 Sept. 26/6 Instead of winning a motor-paced race by 10 laps, he would come from behind in the final lap to take it on the line.

6. motor-bandit *Obs.*, a thief who uses a motor car; **motor-bicycle** *v. intr.*, to travel on a motor bicycle (*rare*); so *motor-bicycling* vbl. sb., *motor-bicyclist*; **motor-bike** = *MOTOR-CYCLE *sb.*; hence as *v. intr.*, to travel on a motor-bike; so *motor-biking* vbl. sb.; **motor-bus**, a motor-driven omnibus; hence as *v. intr.*, to travel by motor-bus; **motor camp** orig. *U.S.*, a place where a motorist may park his car and set up camp; **motor caravan**, (*a*) a caravan designed to be towed by a motor car; (*b*) (see quot. 1964); **motor-coach**, (*a*) a single-decker motor-bus; (*b*) (see quot. 1940); **motor-coaster**, (*a*) a motorized vessel employed in sailing along the coast; (*b*) a type of big dipper at a fun-fair; **motor cop** *U.S. colloq.* = *motor-cycle cop*; **motor court** *U.S.*, (*a*) (see quot. 1936); (*b*) a motel; **motor-drive**, (*a*) a drive or journey in a motor-car; (*b*) driving power provided by a motor or engine; **motordrome** *Obs.*, a course for motor-racing; **motor generator** *Electr.*, an apparatus consisting of an electric motor and a generator with their armature shafts mechanically coupled which may be used to change the voltage, frequency, or number of phases of a supply; **motor glider**, an aircraft constructed like a glider but having an engine (now used in training glider pilots); **motor-home** *N. Amer.*, a very large vehicle equipped as a self-contained home; **motor hotel**, a hotel designed particularly for use by motorists; a motel; **motor inn** = *motor hotel*; **motor lodge**, a motel; **motor mate** *Obs.*, one who attends to the motor of an airship; **motor park**, (*a*) *U.S.* = *motor court* (*a*); (*b*) a car-park; **motor-sailer**, a boat equipped with both sails and a motor; **motor-school**, a school where the driving of motor vehicles is taught; **motor-scooter**, a two- or three-wheeled vehicle resembling a child's scooter, propelled by a small engine; hence **motor-scooterist**, one who rides a motor-scooter.

1913 *Punch* 19 Feb. 133/3 On top of all this Motor Bandit business comes the news that two men have been charged..with breaking into a bakery and stealing a sponge-cake, value one penny. 1935 *Ibid.* 6 Mar. 253/3 Returning from a jumble sale a man was held up by a motor-bandit. 1960 J. Betjeman *Summoned by Bells* v. 49 He motor-bicycled his life away, Looking for orchids in the Wytham Woods. 1912 W. Owen *Let.* 16 Aug. (1967) 155 Captain Wigan has had a motor-bicycling accident! 1908 R. Brooke *Let.* 4 Aug. (1968) 136 Abercrombie..is a Metrical Motor Bicyclist, a mumbly Wump, but often splendid. 1903 *Work* 18 Apr. 176/3 (Advt.), Motor Bike, 1½-h-p. 1911 W. Elmhirst *Freshman's Diary* (1969) 20 He went over to Malvern yesterday on his motor bike. 1927 H. G. Wells *Short Stories* 627, I remember my wild rush on my motor-bike to London. 1944 F. Clune *Red Heart* 33 He motor-biked to Alice Springs. 1958 J. Betjeman *Coll. Poems* 224 Along the village street the sunset strikes On young men tuning up their motor-bikes. 1972 M. Gee *In my Father's Den* 27 That motor-bike gang. 1913 G. B. Shaw *Let.* 21 Mar. in *B. Shaw & Mrs. Campbell* (1952) 100 It looked like a superb stroke of motorbiking. 1901 H. G. Wells in *Strand Mag.* Dec. 623/2 'You really think such a thing is possible?'.. 'As possible,' said Gibberne, and glanced at something that went throbbing by the window, 'as a motor-bus.' 1914 A. D. Godley *Let.* 10 Jan. in *Reliquiae* (1926) I. 292 What is this that roareth thus? Can it be a Motor Bus? 1915 D. O. Barnett *Lett.* 39 We came on from our last stopping place, whither we motor-bused, in a car. 1972 *Daily Tel.* 24 Mar. 15/1 Trolleys were a merger of motorbus and tram. 1925 *Sat. Even. Post* 10 Oct. 98/1 The average motor camp is too well known to need any description. 1926 *Daily Colonist* (Victoria, B.C.) 4 July 1/4 Motor Camp is Winning Favor. 1970 *N.Z. Listener* 21 Dec. 8/3 Don't join a joker in a motor camp cabin for the night... Try yelling for help in a motor camp in the holiday season, mate. 1930 *Daily Express* 1 Aug. 9/2 The most modern form of holiday transport—the motor-caravan, or trailer caravan. 1959 *Motor* 7 Oct. 246 Motor Caravans as practical vehicles. 1964 *Which?* Apr. 35/1 A motor caravan is basically an ordinary van that has been fitted with windows, beds and cupboards, a table, cooker and sink. 1923 Motor coach [see *COACH *sb.* 1 e]. 1926 *Times* 6 May 3/5 Intimidation of owners and drivers of motor-cars and motor-coaches plying for public hire is reported to-day in several localities. 1940 *Chambers's Techn. Dict.* 559/2 *Motor-coach*, a passenger coach, equipped with its own motors, for use on electrified railways; it is commonly used in conjunction with trailer coaches to make up a multiple-unit train. 1958 J. Betjeman *Coll. Poems* 153 Then the mystery tour By motor-coach inland this afternoon. 1972 *Modern Railways* Sept. 334 BOB [sc. Berner Oberland Bahn] trains are motorcoach-hauled and not fitted for push-pull operation. 1973 J. Porter *It's Murder with Dover* v. 42 All the cars and motor coaches..streaming through the village. 1928 *Daily Express* 5 Dec. 11/4 It is feared that the London motor-coaster, Wander (82 tons), has been lost, with a crew of five. 1974 *Amer. Speech 1971* XLVI. 84 Fast-moving amusement rides on elevated rails: toboggan, high rides,..shoot the chutes, motor coasters. 1915 *Policeman's Monthly* Oct. 14/1 The motor cops were brought in on the 'carpet'. 1918 R. Wagner *Film Folk* 23 If one goes home some afternoon and finds an ambulance or a motor cop outside the door, he instinctively looks for the camera. 1936 *Pop. Mechanics* LXVI. 674/1 The motor court is a recent development of the motor age. It's not a tourist park, it's not an auto camp... It's a collection of miniature homes clustered around a central service and administration building. 1962 I. Fleming *Spy who loved Me* i. 15 'Motel' isn't a good word any longer. It has become smart to use 'Motor Court' or 'Ranch Cabins'. 1906 Beerbohm *Around Theatres* (1924) II. 221 The details of the motor-drive are quite plausible. 1936 *Discovery* Apr. 113/2 The compressor, or vacuum pump, has an independent motor drive. 1971 *Amateur Photographer* 13 Jan. 51/1 The exposure system is powered by the motor drive batteries but a bridge circuit makes it independent of their precise voltage as they run down. 1908 *Westm. Gaz.* 4 Aug. 4/3 Mr. Locke-King has spent a large fortune in building this wonderful motordrome. 1887 *Electrician* 2 Dec. 74/1 We are enabled to reproduce photographs and sectional drawings of the Paris and Scott motor generator which recently underwent some tests at the Newcastle Exhibition. 1907 Parshall & Hobart *Electr. Railway Engin.* ix. 351 The locomotive is equipped with a motor-generator set comprising a single-phase motor directly connected to a continuous-current generator. 1930 *Engineering* 28 Feb. 278/2 The electrical load is consequently divided between the steam-driven sets and motor-generator sets in accordance with the demand for heating steam. 1966 *McGraw-Hill Encycl. Sci. & Technol.* VIII. 617/1 Motor-generator sets are used for a variety of purposes, such as providing a precisely regulated dc current for a welding application, a high frequency ac power for an induction-heating application, or a continuously and rapidly adjustable dc voltage to the armature of a dc motor employed in a position control system. 1923 *Jrnl. R. Aeronaut. Soc.* XXVII. 531 One of the things that instantly occurred to most people..was that the motor glider heralded the dawn of a new era in commercial aeronautics. 1969 *Sailplane & Gliding* XX. 491/2 The flying training will be based on glider and motor glider experience. 1971 D. Piggott *Gliding* (ed. 3) i. 8 The two-seater motor glider has proved an unqualified success for all stages of training. 1966 *Economist* 15 Jan. 181/1 There is the 'motor home', especially built on a lorry or bus chassis, an elaborate 'land yacht'. 1970 *Globe & Mail* (Toronto) 25 Sept. 38/9 (Advt.), Wanted Motor home, used, good condition. 1973 *Sci. Amer.* Apr. 1/1 (Advt.), These air springs are the only ones of their kind on motorhomes. 1974 *Trailer Life* Nov. 92 Writers, editors and publishers, who formerly had only the word trailer to contend with when referring to recreational vehicles now had to wrestle with trailers, pickup campers, chassis-mounts and the few then-existing motorhomes. 1965 *Punch* 1 Dec. 798 They call themselves motor lodges, motor courts, motor hotels, even tourtels and autotels, but motel is the word that blisters the night sky of the American suburbs. 1967 *Autocar* 5 Oct. 44/1 (*caption*) New motor hotels—not motels—specially designed for motorists, are being opened by the Trust House group. 1967 *Time* 21 Apr. 25, 46 restaurants and six motor inns. 1974 *Country Life* 21 Mar. 693/2 This privately owned Motor Inn situated on the water's edge. 1965 Motor lodge [see *motor hotel*]. 1975 *Globe & Mail* (Toronto) 24 May 10/2 Judge Dunlap dismissed two charges of procuring, one of attempting to procure and a charge of keeping a common bawdy house in a motor lodge here. 1928 C. F. S. Gamble *Story N. Sea Air Station* xxii. 411 One of the motor-mates of the amidships gondola raised the black curtain. 1939 *New Yorker* 14 Oct. 72 The usual phrase for cabin colonies in California is 'auto court'; 'motor park', perhaps copied from 'trail park'. 1965 W. Soyinka *Road* i. 1 The motor-park layabouts are sprawled on the floor and on benches. 1972 C. Achebe *Girls at War* 60 Everybody, even a motor-park tout, knows what school fees are for. 1934 *Yachting* (N.Y.) Jan. 39 A consideration of the motor-sailer. 1937 [see *FIFTY-FIFTY *adv.* and *a.*]. 1971 J. R. L. Anderson *Reckoning in Ice* vi. 111 His boat was what is called a motor-sailer, which generally means..more engine than sail. 1974 *Country Life* 24 Jan. 153/1 We have pure sailing craft, motor sailers and motor yachts, excellently crewed..from £15 to £100 per person per day. 1909 *Chambers's Jrnl.* June 342/1 The..gentleman..is now getting his country grooms trained at a motor-school. 1919 Motor-scooter [see *BULLER *sb.*¹]. 1938 *Times* 28 Feb. 12/2 He then tears down the course on a motor-scooter. 1956 *Railway Mag.* Nov. 722/2 Goods agents..have been provided with motor scooters, to enable them to cover ground more quickly. 1957 J. I. M. Stewart *Use of Riches* 58 The age of the motor-scooter had brought little of modernity to Castelarbia as yet. 1960 *News Chron.* 22 Mar. 7/8, I saw White running behind one of the motor-scooterists. 1965 J. Creasey *Toff & Spider* ii. 22 The motor-scooterist had turned into the Square and was slowing down.

B. *adj.* **2.** (Further examples.) *motor cortex*, the part of the cerebral cortex which has a motor function; *motor neurone*, a neurone having a motor function; *motor root*, the anterior or ventral root of spinal or certain cranial nerves, containing axons of motor neurones; *motor unit*, a neuroanatomical unit which comprises a single motor neurone and the muscle fibres on which it acts.

1840 W. J. E. Wilson *Anatomist's Vade Mecum* viii. 387 There are thirty-one pairs of spinal nerves, each arising by two roots, an anterior or motor root, and a posterior or sensitive root. 1878 *Jrnl. Nerv. & Mental Dis.* V. 766 The largest nerves of the human body (sciatics) arise precisely from that point of the lumbar cord where we find the largest, so-called motor, cells. 1890 W. James *Princ. Psychol.* I. ii. 61 The motor cortex might be sensitive as well as motor. 1898 *Jrnl. Compar. Neurol.* VII. 185 The neuraxis of the motor neuron loses its medullary sheath before piercing the sarcolemma. 1925 Liddell & Sherrington in *Proc. R. Soc.* B. XCVII. 516 The experimental results suggest that a reflex maintains maximal response of the individual 'motor-unit' by a degree of central excitation which is commonly 'supramaximal'. 1926 J. S. Huxley *Essays Pop. Sci.* 286 The rest of the [spinal] cord, however, and in particular the motor areas and motor roots, show no increase in size, showing that the number of muscle-fibres to be supplied does not stimulate the growth of the supplying cells. 1942 O. Larsell *Anat. Nervous Syst.* xxiv. 326 The upper motor neuron fibers of the pyramidal tract terminate in synaptic relation to large motor cells in the anterior column of the gray matter of the spinal cord. 1970 M. Hollander tr. *Monnier's Functions Nervous Syst.* II. xxvi. 572 (*heading*) Functions of the motor cortex and pyramidal system. 1970 J. G. Chusid *Correlative Neuroanat. & Functional Neurol.* (ed. 14) ii. 76/2 The motor unit is made up of the anterior horn cell of the spinal cord and the muscle group it innervates. 1973 *Gray's Anat.* (ed. 35) 1002/2 The motor nucleus of the trigeminal nerve gives rise to the fibres of the motor root. 1974 Passmore & Robson *Compan. Med. Studies* III. xxxiv. 53/2 Motor neurone disease may present with any combination of upper and lower motor neurone signs.

3. Substitute for def.: Of, pertaining to, or involving muscular movement; based upon or received through the movement of parts of the body. (Further examples.)

1880 W. James *Coll. Ess. & Rev.* (1920) 163 Wundt says that were our motor feelings of an afferent nature, 'it ought to be expected that they would increase and diminish with the amount of inner or outer work actually effected in contraction.' 1890 —— *Princ. Psychol.* II. xix. 131 The facts shade off into the phenomena of motor automatism, trance, etc. 1899 A. G. Whyte tr. *Binet's Psychol. of Reasoning* 24 When we think of the ball, this idea must comprise the images of these muscular sensations, as it comprises the images of the sensations of sight and touch. Such is the motor image. 1925 C. Fox *Educational Psychol.* 228 Walking is not bringing into use unconscious motor-images. 1936 *Amer. Speech* XI. 88/2 Speech responses are so complex that they can hardly be considered identical with motor reflexes. 1942 A. T. Poffenberger *Princ. Appl. Psychol.* vi. 105 The fineness of motor control or coördination is..an essential factor in many forms of adjustment. 1951 A. D. Woodruff *Psychol. of Teaching* (ed. 3) xvi. 270 Motor behavior that is guided by perceptual cues, such as playing the piano. 1951 C. I. Hovland in S. S. Stevens *Handbk. Exper. Psychol.* xvii. 627/2 An important change with practice in motor learning is change in muscular tension. 1958 M. Argyle *Relig. Behaviour* v. 56 Evidence drawn from actual studies of conversion shows that people converted at public meetings are more easily hypnotized, display more motor automatisms and can therefore be classified to some extent as hysterics. 1962 *Canadian Jrnl. Linguistics* VII. 65 The assumption that the ability to speak is simply a motor-skill which can be measured by tests of imitation and reading aloud. 1969 Miller & McNeill in Lindzey & Aronson *Handbk. Social Psychol.* (ed. 2) III. xxvi. 687 At the other [extreme] would be the so-called 'motor theory' of speech, which holds that a listener can recognize speech only by imitating (at least covertly) the motor movements that would produce it. 1971 *Jrnl. Gen. Psychol.* LXXXV. 208 In the events intervening between stimulus input and motor output, there is ordinarily first the operation of cognition.

motor, *v.* Add: (Further examples.) Also (*trans.*), to traverse (a distance) in a motor vehicle.

1919 Wodehouse *Damsel in Distress* viii. 103, I motored down with a boy I know. 1928 'S. S. Van Dine' *Greene Murder Case* xii. 141 Vance and Ada and I motored the few blocks to 18, Broad Street. 1932 J. Buchan *Gap in Curtain* i. 31 Mayot had motored to Cirencester to meet a friend. 1972 *Daily Tel.* (Colour Suppl.) 27 Oct. 86/3 From Sidmouth I walked (though most people motor) to Salcombe Regis. 1972 *Listener* 21 Dec. 849/1 If we were opening a show with a Boxing Day matinee, we should have finished rehearsing on Friday night and then have motored to our respective homes.

b. *intr.* To travel in a motor-boat; to use the engine in a sailing-boat.

1968 'D. Halliday' *Dolly & Singing Bird* xiii. 147 'You don't deny you were motoring?' Johnson kept his voice reasonable. 'We used the engine..to South Rona and back. At Portree, we changed over to sail.' 1971 P. M. Hubbard *High Tide* xiii. 134 There was no wind to sail with. I could motor down river and hope to pick up something of a sailing breeze outside.

motorable (mōu·tərăb'l), *a.* [f. Motor *v.* + -able.] Of a road or district: suitable for motor vehicles; capable of being travelled over in a motor vehicle.

1920 A. L. Bagley *Holiday Rambles N. Wales* 178 This is certainly not a motorable road. 1953 X. Fielding *Stronghold* 93 The only motorable highway in the whole of Sphakia province. 1972 P. M. Hubbard *Whisper in Glen*

iii. 25 Roads which were a little alarming but just about motorable. 1974 *Country Life* 24 Jan. 144/2 We..thought of a certain lake. Ten hours by pony from the nearest motorable road.

Motorail (mǭᵘ·torēⁱl). [Blend of Motor *sb.* and Rail *sb.*²] A service whereby cars, with their drivers and passengers, are transported by railway. Also *attrib.*
1968 *Daily Tel.* 6 Jan. 10/6 This year British Rail will have nearly 100 more Motorail trains, giving a total carrying space for 90,000 cars. 1971 *Country Life* 25 Nov. 1463/3 British Rail unveiled their plans for 1972 Motorail... The 27 routes to be operated will have a combined capacity for 120,000 cars and over half a million passengers. 1974 *Ibid.* 18 Apr. 965/4 Excellent train service to Penzance..and Motorail links throughout Britain.

Motorama (mǭᵘtǫrā·mā). [Blend of Motor *sb.* 3 b and *-or)ama* after Panorama.] An exhibition of motor vehicles.
1950 *Richmond* (Va.) *News Leader* 18 Jan. 38/1 General Motors opens its auto show here. The display which General Motors calls 'Mid-century Motorama' brings..38 of GM's 1950 model cars. 1954 *Amer. Speech* XXIX. 157 *Motorama*, an automobile exhibit. 1974 *John O'Groat Jrnl.* (Wick) 6 Sept. 2/2 *Motorama*, Wick & District Round Table present Motorama Spectacular at Ackergill on Sunday, 15th September at 2 p.m.

mo·tor-boat, *sb.* A motor-driven boat or launch.
1902 *New Liberal Rev.* Apr. 440 The paraffin motor..is ..impossible in anything but an open motor-boat. 1913 E. F. Benson *Thorley Weir* i. 11 If I must go on the river, give me a motor-boat. 1973 A. Behrend *Samarai Affair* ii. 24 He would..step overside into the boarding punt, a small-sized motor-boat. 1973 *Fisheries Fact Sheet* (Environment Canada Fisheries & Marine Service) No. 1. 2/2 Individual fishermen fishing near their homes from small row-boats or motor-boats.

mo·tor-boating, *vbl. sb.* Also without hyphen and as two words. [f. prec. + -ing¹.] **1.** Travel in a motor-boat.
1918 *Chambers's Jrnl.* Aug. 541/1 The water is never dangerously rough, and provides the finest possible field for motor-boating. 1961 *Guardian* 8 Feb. 10/2 The sports of motor-boating and water-skiing.
2. *Electronics.* Oscillation in an amplifier that is of such a low frequency that individual cycles may be heard, giving a characteristic sound, and caused by feedback from the output to the input of earlier stages, often through a common voltage supply.
1930 J. H. Reyner *Testing Radio Sets* vi. 84 Back-coupling with a mains unit usually takes the form known as motor-boating. The oscillation set up is of a low frequency and gives a continuous 'pop, pop, pop'. 1943 F. E. Terman *Radio Engineers' Handbk.* v. 409 A power supply having low internal impedance at low frequencies is also helpful in eliminating motorboating. 1945 *Electronic Engin.* XVII. 429 Regeneration may lead to low-frequency oscillation (about 5 to 10 c/s) known as 'motor boating'. 1965 G. J. King *Radio & Audio Servicing Handbk.* iii. 78 Decoupling prevents signals which could occur in an element common to two or more stages from getting back into the input circuits and causing oscillation or motorboating.
So **mo·tor-boat** *v. intr.*, (*a*) to travel in a motor-boat; (*b*) (of an amplifier) to exhibit motor-boating.
1922 *Contemp. Rev.* Mar. 409 The scenery through which he tramped or motor-boated. 1932 F. E. Terman *Radio Engin.* vii. 277 Amplifiers receiving their plate voltage from a common dry battery have a tendency to motor-boat when the batteries are near the end of their useful life, particularly when the amplification is high. 1970 *Nature* 28 Nov. 797/2 Two men motor-boating on the loch collided with a large object.

motorcade (mǭᵘ·tǫɹkēⁱd). orig. *U.S.* [f. Motor *sb.* + *-cade.] A procession of motor vehicles.
1913 *Arizona Republican* (Phoenix) 5 June 2/4 The motorcade can make its music self supporting and donate large and salubrious gobs of melody to the natives at all points along the line. *Ibid.* 4 July 4/2 This 'motorcade' came from a suggestion thrown out by the sporting editor of the Republican. It was immediately accepted by several local automobile owners, whereupon, the sporting editor [*sc.* Lyle Abbott] became the busiest man in Phoenix and hammered away at the 'motorcade' a term which, by the way, he had invented sometime before in order that newspapers might keep pace with the developments of vehicular transportation. 1924 *N. & Q.* 19 Apr. 288/1 A parade of motor cars..was termed in the local [Florida] papers a 'motorcade'. 1928 in *Amer. Speech* (1930) VI. 155 The North Dakota farmers' motorcade started to-day on its long drive to the Republican Convention at Kansas City. 1933 *Sun* (Baltimore) 19 Aug. 3/1 The action was taken after a telegram from the Governor was read to workers who came into Mahanoy City in a motorcade. 1936 [see *-cade]. 1942 S. L. A. Marshall *Armies on Wheels* vi. 99 A motor-cade in which the cars become bogged down. 1953 *Times* 13 July 3/5 The word 'motorcade'..is..no novelty. Some years before the war it was employed—and perhaps coined—by the Salvation Army in advertisements of a procession of cars which bore the Army's principal officers on a tour of the main towns of the country. 1969 *Wall St. Jrnl.* 28 Jan. 1/6 He rides in Presidential motorcades. 1972 *Guardian* 25 Oct. 2 The old-style motorcade..is hard to beat as a campaigning technique. 1974 *Northern Times* (Golspie, Sutherland) 23 Aug. 7/6 Mr. George Reid, M.P. for Clackmannan and East Stirling, accompanied by Mr. Sutherland, will be in a motorcade covering Helmsdale, Brora, Golspie, Dornoch and Bonar-Bridge.

motor car. Add: (Further examples.) Also *attrib.* Hence **mo·tor-carist,** a motorist; **mo·tor-carring,** travel in a motor car.
1899 *Motor-Car World* I. 37/2 Many of the disabilities under which motorcarists suffer in England will be removed. 1901 'Saki' *Let.* 17 Aug. in *Square Egg* (1924) 61 Travelling with Aunt Tom is more exciting than motorcarring. 1903 *Westm. Gaz.* 12 Jan. 10/1 The champion motor-carist of the House of Commons. 1909 *Daily Graphic* 26 July 7/4 Numbers came from all the neighbouring towns, while motor-car parties came from considerable distances. 1931 H. G. Wells *Work, Wealth & Happiness of Mankind* (1932) 21 The motor-car 'bandit'. 1970 'D. Halliday' *Dolly & Cookie Bird* i. 4 George is a motor-car salesman. 1972 *Daily Tel.* (Colour Suppl.) 20 Oct. 52/1 The Rolls-Royce approach to making motor cars (they are never just 'cars' at Crewe). 1974 *Times* 4 Dec. 17/2 The motor-car industry..has been the most important single engine of economic development in industrial countries for the past fifty years.

mo·tor-cycle, *sb.* [Motor *sb.* 5.] A form of bicycle, usually larger and heavier than an ordinary bicycle, having a small (usu. petrol) motor by which it is propelled. Also *attrib.*, as **motor-cycle cop** (*U.S. colloq.*), **policeman,** a policeman who rides a motor-cycle.
1896 [see Motor *sb.* 5]. 1902 *Motor Cycling* 12 Feb. 24/1 In a year or two motor cycles will be as plentiful as the ordinary cycle is to-day. 1919, 1928 [see *combination 9* b]. 1927 K. Eubank *Horse & Buggy Days* 33 There's always some fool motor-cycle cop coming along to interfere. 1928 'S. S. Van Dine' *Greene Murder Case* xxv. 288 On North Broadway we were forced to the kerb by a motor-cycle policeman. 1934 T. S. Eliot *Rock* i. 21 Every son would have his motor cycle. 1958 *New Statesman* 22 Feb. 220/1 A motor-cycle policeman seized his shoulder and flung him along the pavement. 1972 C. Weston *Poor, Poor Ophelia* (1973) vii. 39 All those sex-mad motorcycle toughies. 1972 R. Adams *Watership Down* xix. 111 Few places are far from human noise—cars, buses, motor-cycles, tractors, lorries. 1974 *Times* 18 Mar. 7/1 Five people were killed here in a running battle between motorcycle gangs.
Hence **mo·tor-cycle** *v. intr.*, to ride a motor-cycle; **mo·tor-cycling** *vbl. sb.*; **mo·tor-cyclist,** one who uses a motor-cycle.
1902 *Motor Cycling* 12 Feb. 24/1 Now that motor cycling has come to stay..it may not be out of place to say where improvements may be made that will be beneficial to..all those who motor-cycle. *Ibid.* 19 Feb. 32 Motor cyclists will not require palatial club premises. 1912 G. W. E. Russell *Afterthoughts* xx. 184 Motorists and cyclists, and those strange hybrids the motor-cyclists, seem to disregard weather. 1925 E. F. Norton *Fight for Everest, 1924* v. 103, I wore a fur-lined leather motor-cycling helmet. 1956 G. N. Patterson *God's Fool* ii. 17 Pulling on his motor-cycling gloves savagely, he stalked out. 1972 'J. & E. Bonett' *No Time to Kill* i. 5 The chef's elder daughter..motor-cycled out for a free lunch in the hotel kitchen. 1973 J. Wainwright *Pride of Pigs* 92 The motor-cyclist was pinned under his machine.

motordom (mǭᵘ·tǫɹdəm). [f. Motor *sb.* + -dom.] The realm or world of motors; motor vehicles, the people who use them, or those who deal in them, considered collectively.
1900 *Captain* III. 225/1 In the world of motordom. 1909 *Westm. Gaz.* 2 Nov. 2/3 There is scarcely a woman in London who has not had to sacrifice the cadences of talk to the Moloch of Motordom. 1910 [see *cracker 4 b]. 1916 W. J. Locke *Wonderful Year* vi. 84 Fashionable motordom halted at the Hôtel des Grottes. 1961 L. Mumford *City in Hist.* xv. 474 The 'progressive' metropolises of motordom, like Los Angeles, exhibit..all the urban evils of the palaeotechnic period.

mo·tored, *a.* [f. Motor *sb.* + -ed².] Provided with a motor.
1928 *Daily Express* 17 Nov. 9/3 The Wright brothers completed their motored glider—the first real airplane—in their bicycle shop at Dayton on Nov. 17th 1903.

motoric (mǒtǭ·rik), *a.* [f. Motor *sb.* and *a.* + -ic.] **a.** Of, pertaining to, or characterized by muscular movement.
1930 *Program 47th Meeting Mod. Lang. Assoc. Amer.* 12 An attempt to illustrate the present status of *Schallanalyse* in its visual, acoustic, and motoric approach. 1958 *New Scientist* 14 Aug. 610/2 A simple motoric rhythm in which the animal repeatedly pulled the brush towards himself. 1971 *Nature* 18 June 470/2 The reasons for favouring this method are not quite clear. Motoric convenience, for example, seems an inadequate explanation, in view of normal writing practice.
b. Of a piece of music or its performance: marked by much movement or energy.
1937 *Sat. Rev. Lit.* (U.S.) 23 Jan. 6 Stravinsky's crisp sentences, often naked in their brevity, recall the 'motoric' rhythms of his *Noces*. 1942 *Ann. Reg. 1941* 338 The Symphony..was an interesting..example of what the continental critics used to call 'motoric music'. 1972 *Daily Tel.* 4 May 14/3 Its determined progress was crisply outlined without excessively motoric playing.

motoring, *vbl. sb.* (In Dict. s.v. Motor *sb.* and *a.*) Also *attrib.*, as *motoring-cap, chocolate, coat, glove, goggles, map, offence, veil.*
1909 *Chambers's Jrnl.* Sept. 588/2 She raised a little gloved hand and patted the hair under a dainty motoring-cap. 1966 D. Holbrook *Flesh Wounds* 130 In handy places here and there the tank men kept their round tins of cigarettes and their bars of motoring chocolate. 1907 G. B. Shaw *John Bull's Other Island* iv. 83 Broadbent enters, soiled and disordered as to his motoring coat. 1931 M. Allingham *Look to Lady* xx. 204 Mr. Campion in a motoring coat and hatless. 1920 W. J. Locke *House of Baltazar* viii. 101 Drawing on his motoring gloves. 1942 *R.A.F. Jrnl.* 3 Oct. 30 There were those first flights..in all the glory of an old flying helmet and motoring gloves. 1903 G. B. Shaw *Man & Superman* iii. 78 Tanner and Straker, in their motoring goggles, leather coats, and caps, are led in from the road by the brigands. 1975 'M. Duke' *Death of Holy Murderer* ii. 37 She removed a pair of antique motoring goggles. 1926–7 *Army & Navy Stores Catal.* 915/1 Contour Motoring Map of British Isles. 1931 R. H. Heaton *Perfect Hostess* 133 A set of motoring maps of England and Scotland. 1974 *Camping & Caravanning* Sept. 12/1 The R.A.C. have introduced two new series of motoring maps. 1933 M. Allingham *Sweet Danger* xvii. 213 The young gents who were wanted for a motoring offence. 1972 *Police Rev.* 17 Nov. 1484/1 Although penalties for motoring offences may be low, it is a truism that purely in financial terms it works out cheaper being a petty criminal than being a forgetful motorist. 1973 P. Moyes *Curious Affair of Third Dog* i. 20 He's not a *criminal*... It was only a motoring offence—but unfortunately he killed somebody. 1955 N. Marsh *Scales of Justice* vii. 160 Illustrations of young women in dustcoats and motoring veils. 1971 *Country Life* 14 Oct. 966/3 Intrepid ladies in large hats, goggles, motoring veils and dust-coats. 1974 I. Murdoch *Sacred & Profane Love Machine* 229 She now donned a floppy green hat and tied it over her head with a sort of motoring veil.

motorism (mǭᵘ·tǫriz'm). [f. Motor *sb.* + -ism.] The use or prevalence of motor vehicles; the world of motoring.
1913 *Chambers's Jrnl.* Feb. 132/2 It is but ten or twelve years since the time when the tide of motorism began to flow. 1930 *Time & Tide* 24 Jan. 101 The death-dealing motorism of the highway. 1952 *Birmingham* (Alabama) *News* 20 Sept. 4 If you want to be in the forefront of motorism these days, it seems you have to have headlamps outside the fenders, mudguards that only partially protect, and if possible wire wheels.

motorist. (Later examples.)
1912 *Collier's* 5 Oct. 30/1 In the new Model C-Six the motorist is constantly in a position of rest and free from care or strain. 1960 *Spectator* 30 Sept. 479 Gibberd gave a motorist's-eye view of the town. 1975 *Times* 10 Sept. 3/2 Motorists..would be..invited to discuss their driving with road safety officials.

motorium (motǭ·riǝm). [mod.L., f. *mōtōrius* moving, f. L. *mōt-*, stem of *movēre* to move: cf. -ory¹.] **1.** *Psychol.* Collectively, the centres in the brain concerned in the function of voluntary muscle; that system of the body capable of initiating and putting into effect muscular movement.
1885 J. Ross *Handbk. Dis. Nervous Syst.* 73 The central ends of the individual sensory mechanisms are unified by a collective centre—the *sensorium commune*—which is the general centre of nervous connections on the afferent side. It may also be supposed that the individual motor centres are unified by a general centre of nervous connections which may be regarded as a *motorium commune*. 1900 Dorland *Med. Dict.* 398/1 *Motorium*, (1) a motor center; especially the common center of motor influences; (2) the motor apparatus of the body. 1924 R. M. Ogden tr. *Koffka's Growth of Mind* 80 The optical sensorium and motorium can not be regarded as two independent pieces of apparatus, since for many types of performance they constitute a unitary organ. 1959 S. Koch *Psychology* II. 76 For certain purposes, in dealing with transactions with the physical world (walking, performing a motor task, driving a car, etc.), the motorium may be considered to be the muscular apparatus of the organism. Even in such cases, however, it may prove to be useful to consider tools as a part of the motorium.
2. *Zool.* A cytoplasmic structure in some ciliate protozoa forming the centre of the neuromotor apparatus.
1914 R. G. Sharp in *Univ. Calif. Publ. Zool.* XIII. 83 We now come to the description of what is believed to be the most interesting structure in the anatomy of this organism [*sc.* the protozoan *Diplodinium ecaudatum*]... The possibility of this structure functioning..as a neuromotor apparatus is suggested and..the designation neuromotor apparatus will be used, and its constituent parts will be described as a motorium or motor mass. 1926 G. N. Calkins *Biol. Protozoa* ii. 104 Certain special types of cytoplasmic kinetic elements such as myonemes, motorium, and conductile fibers, are characteristic of the ciliates. 1940 L. H. Hyman *Invertebrates* I. iii. 65 All these fibrils may center in a small body, the motorium. 1966 *New Scientist* 3 Nov. 242/1 Another interesting feature of *Euplotes* is that, as Taylor first noticed in 1920, some of the cirri are connected by a system of minute fibrils to a structure near the front of the animal. This he called the motorium.

motorization (mǭᵘ:tǒrəizēⁱ·ʃən). [f. next + -ation.] **1.** *Psychol.* The process of making a presentation motorial in character. *rare.*
1901 *Amer. Jrnl. Psychol.* XII. 304 The motorization of an exposed word would suggest another similar in sound. *Ibid.* 309 The single-word motorization does not make so much noise in consciousness.
2. The introduction, use, or possession of motor vehicles; equipment with motor vehicles.

1913 in Webster Add. 1919 *N.Y. Times Mag.* 6 Apr. 11 During the war the motorization of artillery proceeded at an unprecedented rate. 1929 H. Rowan-Robinson *Further Aspects of Mechanization* ii. 9 The motorization of infantry and cavalry divisions furnishes..additional strategic mobility. 1930 *Time & Tide* 13 Sept. 1134 The world moves largely on rubber, which alone makes motorization possible. 1960 H. Hondermarn in E. Davies *Roads & their Traffic* vi. 130 The extent of motorization, expressed by the number of vehicles per 1000 inhabitants, has had to be periodically reviewed. *Ibid.* 132 Fig. 6.1..shows a significant correlation between the degree of motorization and national income in different countries. 1960 G. Mikes *How to be Inimitable* 57 This motorisation has developed into a war between the motorists and the authorities. 1969 *Daily Tel.* 26 Aug. 14/3 Full motorisation of Britain's cities..might go as high as £12,000 million.

motorize (mǭᵘ·tǒrəiz), *v.* [f. Motor *sb.* and *a.* + -ize.] *trans.* **1.** *Psychol.* To convert (visual or auditory sensations or images) into motorial presentations; to apprehend in a motorial manner. Also *intr.* or *absol.* *rare.*
1901 *Amer. Jrnl. Psychol.* XII. 308 The word..seems to be motorized as soon as singly presented. *Ibid.* 309 This.. has reference to readers who motorize.
2. To provide or furnish with a motor or with motor vehicles. Hence **mo·torized** *ppl. a.*; **mo·torizing** *vbl. sb.*
1913 Webster Add., *Motorize*, to substitute motor-driven vehicles or automobiles, for the horses and horse-drawn vehicles of (a fire department, city, etc.). 1922 *Daily Mail* 24 Nov. 6 These machines have gone beyond the stage of motorised pedal cycles and are in all respects real motor-cycles. 1924 *Public Opinion* 8 Aug. 130/3 If one should dream of motorising the entire world on the scale of the United States. 1927 *Glasgow Herald* 5 Apr. 8 Serried ranks of tanks advancing against each other.. with motorised artillery bringing up the rear. 1933 Ade *Let.* 13 Nov. (1973) Now there ain't no income from the darn stuff [*sc.* farm lands]. I think it is entirely because of the fact that the world has become motorized but I am not proposing any remedies. 1938 *Times* 16 May 9/1 The extension of the air arm and the motorizing of units. 1938 *Sun* (Baltimore) 31 Aug. 20/3 The regiment is motorized and will come over the Governor Nice Highway to the city. 1958 *Edmonton* (Alberta) *Jrnl.* 24 June iv. 16/6 Motorized toboggans have made their appearance in the Mackenzie River delta. 1972 *Physics Bull.* July 402/2 The construction of a motorized drill table for physically handicapped operators. 1972 'G. Black' *Bitter Tea* (1973) ii. 26 Angels mightn't be watching over them, but a good third of the motorized police in the State were.

motorless (mǭᵘ·tǫɹlės), *a.* [f. Motor *sb.* + -less.] **1.** Not provided with a motor; performed without the use of a motor; esp. of gliders or flying in gliders.
1897 [see *air-sailer* (*-or*) s.v. *Air *sb.*¹ B. III. 5]. 1908 [see *aerodone* s.v. *Aero- b]. 1930 V. W. Pagé *Henley's ABC of Gliding & Sailflying* (1931) i. 11 They had proved their aëronautical theories with motorless ships which actually stayed aloft a short-time, supported only by air currents. *Ibid.* iii. 49 (*caption*) Outline Drawings..Showing How Bird Form is Approximated in Creations Intended for Motorless Flight.
2. Of a road, etc.: having little or no traffic.
1970 *Daily Tel.* 19 Sept. 12 Experiments with motorless zones in wild and remote country. 1972 *Listener* 21 Dec. 849 We should drive on motorless roads, praying that conditions would be good.

motor-man. (Further examples.)
1904 F. Lynde *Grafters* xxiv. 298 The conductor and motorman of an early morning electric car. 1908 [see *dead man's handle* s.v. *dead man]. 1935 'J. Guthrie' *Little Country* xxv. 367 Ernest..climbed on board a tram... The motorman, instead of moving off immediately, waited a moment. 1956 *Railway Mag.* Nov. 744/1 The back of the motorman's compartment is formed by a glass panel so that passengers can enjoy a good view of the track. 1971 *Railway World* Mar. 127 Someone in High Office..gave orders that motormen should count ten between releasing the air brakes and re-applying power.

motor-minded (stress variable), *a.* [f. Motor *sb.* and *a.* + Minded *ppl. a.*] **1.** *Psychol.* Having a mind in which motor images predominate over visual and auditory ones; thinking largely in terms of movements.
1897 *Psychol. Rev. Monogr. Suppl.* II. 1. 40 This, in itself, does not prove that motor-minded persons are less intelligent readers than auditory- or visual-minded. 1900 *Amer. Jrnl. Psychol.* XI. 297 Consonants were not thought to be generally more important than vowels for word perception. The relative importance of these elements might depend upon the reader's tendency to be motor or auditory minded. 1904 W. James in *Atlantic Monthly* Jan. 98/1 Every one knows that in some of us the material of thought is mainly optical, in others auditory, etc., and the classification of human beings into the eye-minded, the ear-minded, and the motor-minded, is familiar.
2. Interested in motor vehicles.
1927 *Glasgow Herald* 14 Oct. 8 With motor trade exhibitions running..in London (at Olympia) and in Paris.., the motor-minded have got quite a lot to talk about.
So **mo:tor-mi·ndedness,** the state of being motor-minded (sense *1).
1897 *Psychol. Rev. Monogr. Suppl.* II. 1. 41 It is evident that motor-mindedness is far from being an advantage to a reader. 1902 [see *ear-mindedness].

† **motorphobe** (mǭᵘ·tǫɹfǭᵘb). *Obs.* [f. Motor *sb.* + -phobe.] One who has a morbid dread or hatred of motor vehicles.
1905 *Automobile Topics* 27 May 448 The time will come when..the motorphobes will wonder what ever possessed them to act so foolishly. 1911 *Chambers's Jrnl.* Aug. 533/1 A motorphobe was quoted as declaring solemnly in 1906, 'In another ten years there will not be half the autocars on the roads that there are now.'

motorway (mǭᵘ·tǫɹwēⁱ). [f. Motor *sb.* + Way¹.] A specially designated class of highway with two or more lanes in each direction, designed and regulated for use by fast motor traffic. Also *attrib.*, as **motorway box,** a rectangular system of motorways; **motorway madness** *colloq.*, reckless driving on a motorway, esp. in fog.
1903 *Car* 4 Nov. 327 (*title*) Concerning motor-ways. *Ibid.* 327/1 The Motor-way is *bound* to come! 1930 Troubridge & Marshall *John Ld. Montagu of Beaulieu* xx. 254 Where the economic conditions guarantee an adequate volume of traffic, as between large cities, direct roads solely for fast traffic, *i.e.* motorways, must be constructed. 1937 [see *autobahn]. 1955 *Times* 6 July 10/1 Motorways, 345 miles in total length, for motor traffic only, are to be built by the Government. 1959 *Radio Times* 23 Oct. 3/1 A motorway differs from all other types of roads in that it has no crossroads, no traffic lights, no pedestrian crossings. Certain categories of road-users are barred from it altogether; these include pedal cyclists, invalid carriages and 'L' drivers. 1964 *Daily Tel.* 22 Jan. 14 The British are relative newcomers to the motorway age. 1967 *Economist* 4 Nov. 495/1 The more central, shorter, but more expensive 'motorway box' favoured by the GLC. 1971 *Daily Tel.* 14 Sept. 2/6 This supports the 'motorway madness' theory of a false sense of security which makes drivers behave quite differently on motorways than on ordinary roads. 1972 J. Mann *Mrs Knox's Profession* v. 31 She had pulled into the motorway café, hating the look of it. 1973 R. Busby *Pattern of Violence* i. 15 He..gave them a name and the time of arrival of a motorway coach. 1973 *Guardian* 15 Oct. 14/1 Fog is the number one motorway menace with irresponsible.. driving coming second. Mix the two, and the headlines next morning scream 'Motorway madness' yet again.

‖ **motoscafo** (mǭᵘtoskā·fo). Pl. **motoscafi.** [It.] In Italy: a motor-boat; esp. such a boat used as a public conveyance on the Venetian canals.
1942 *Britannica Bk. of Year* (U.S.) 468 Another type developed considerably in scope and design was the motor torpedo boat... Before the war the Italians were proud of their vessels of this category, officially classed as M.A.S. (*motoscafi anti-sommergibili*) but these had accomplished singularly little. 1950 E. Hemingway *Across River* xxxiv. 210, I will get into the *motoscafo*..and..we will not ever see one another again. 1971 P. Loraine *Photographs have been Sent* v. i. 143 The father gave up his gondola and bought himself a nice new *motoscafo*. 1975 'D. Rutherford' *Mystery Tour* iii. 40 There are no motor vehicles in Venice, so..you will be conveyed to the hotel by *motoscafo*.

Motown (mǭᵘ·taun). [Shortening of *Motor Town*, nickname for Detroit, Michigan, U.S.A., an important car-manufacturing city.] Music of a type made popular by Negro musicians and singers from Detroit. Also *attrib.* Cf. *Tamla-Motown.
1970 *Melody Maker* 11 July 19/7 Some of their songs are extremely unusual, for example 'Jesus Buddha Moses Gauranga' which starts with a bit of 2001 string sound and contains one line which is pure Motown. 1971 *New Yorker* 25 Sept. 127 The set included several songs from each of the Band's three albums..and two Motown oldies.

motso, var. *matzah.

mott: see *mot⁴, mott.

mott, var. Moat *sb.*¹
1937 *Times* 16 Nov. 19/6 One pair [of ducks] yearly chose the old moat or 'mott', all among the reeds and the moorhens, [to nest in].

mott, var. Mote *sb.*¹ 4.
1930 H. Williamson *Village Bk.* 328 It had looked like a sheaf of reed motts—the unbruised wheaten stalks used for thatching. 1959 *Times* 19 Aug. 9/7 The 'motts' (or stems) are carefully bumped together, butts down, before being tied in bundles for thatching.

motte. Substitute for etym.: [ad. Amer. Sp. *mata* grove, plant, f. Sp. *mata* bush, clump.] (Earlier and later examples of spelling *mott.*)
1848 C. W. Webber *Old Hicks* v. 52 Our course bearing west of north, over broken prairie, diversified by clumps or motts of scrubby growth. 1962 E. B. Atwood *Regional Vocab. Texas* iii. 42 A small group of trees together, surrounded by open country... *Mott* is rather heavily concentrated in South Central and West Texas.

motte (mǫt), *sb.*² [after F. *motte* mound.] Now the usual spelling of Mote *sb.*² 1, esp. in phr. *motte and bailey*, denoting the principal type or design of castle built in Britain by the Normans, consisting of a fort surmounting a mound (*motte*) at the foot of which was an enclosed Bailey or court.
1884 G. T. Clark *Mediæval Mil. Archit. Eng.* I. ii. 16 This 'mound', 'motte', or 'burh'..was formed from the contents of a broad and deep circumscribing ditch. 1892 J. H. Round *Geoffrey de Mandeville* 336 The *motte*, though its name was occasionally extended to the whole fortress, was essentially the actual keep, the crowned mound. 1900 *Proc. Soc. Antiquaries Scotl.* XXXIV. 269 As these are the proper Norman names, and there are no others, I shall henceforth speak of this type of castle as the motte-and-bailey type. 1912 E. S. Armitage *Early Norman Castles Brit. Isles* vi. 80 The motte-and-bailey type of castle is to be found throughout feudal Europe. 1934 *Proc. Soc. Antiquaries Scotl.* LXVIII. 59 The Peel of Kirkintilloch had been a medieval fortress of the motte-and-bailey type. 1947 T. H. White *Elephant & Kangaroo* (1948) xxi. 169 A Norman castle on a motte. 1962 C. W. Hollister *Anglo-Saxon Mil. Inst.* vii. 142 The motte-and-bailey style was unknown to them, just as it was unknown to their contemporaries on the Continent. 1967 J. B. Nellist *Brit. Archit.* vi. 145/1 Although the bailey might be captured, the motte as a self-contained unit could still offer resistance. 1971 *Country Life* 27 May 1297/1 At Dunsany the present castle was apparently built in succession to an earlier motte and bailey castle, the mound of which exists to the north-east of the present house. *Ibid.* 11 Nov. 1280/4 The building stands on a raised motte; it had a moat on the south-west side.
2. *Comb.*, as **motte-castle,** a (Norman) fort standing on a motte.
1912 E. S. Armitage *Early Norman Castles Brit. Isles* vi. 83 It is rare indeed to find a motte-castle in a wild, mountainous situation in England. 1926 *Archæologia Cambrensis* LXXXI. 223 The earthen mounds of two of the earlier 'motte' castles—moated mounts which once had wooden towers upon their summits.

motted (mǫ·tėd), *ppl. a. rare.* [f. *mott(e *sb.*² + -ed¹.] Situated upon a motte or mound.
1947 Auden *Nones* (1952) 64 Do they sponsor In us the mornes and motted mammelons?

mottledy (mǫ·t'ldi), *a. rare.* [f. Mottled *ppl. a.* + -y¹.] Having a mottled appearance.
1929 E. Bowen *Last September* i. vii. 78 Her grey-and-blue 'mottledy' foulard.

motto. Add: **1. c.** (Earlier examples.)
1848 A. H. Clough *Bothie* ii. 12 You ask me to join you in snapping—What but a pink-paper comfit, with motto romantic inside it? 1850 Dickens *Dav. Copp.* xxxiii. 340 There were crackers..with the tenderest mottos.
4. Freq. *attrib.*, as *motto theme.* (Further examples.)
1934 C. Lambert *Music Ho!* v. 316 The ascription of actual individuality to a recurrent or 'motto theme' and the attaching of symbolic significance to its later transformations, devices wholly at variance with the classic principles of symphonic form, are here perfectly justified. 1935 *Downside Rev.* LIII. 30 Against the noble motto-theme of the introduction is set a restless surging subject in the distant tonality of D minor. 1962 *Times* 13 June 13/1 The symphony's motto-theme. 1964 *Listener* 2 Apr. 570/3 The sixth quartet is more difficult to analyse, though the crucial factor is probably the motto-theme which appears in increasingly contrapuntal guise before each movement. 1966 *Ibid.* 17 Nov. 746/3 The first (*Largo*) begins with imitative entries of the 'motto' and quotes the opening of the first symphony, twice.
5. (Further examples.)
1835 *Southern Lit. Messenger* I. 358, I only ate..a few macaronies and mottoes. 1860 *North-West* (Port Townsend, Washington) 5 July 3/3 Candies, Gum drops, Mottoes.

Motu (mǭᵘ·tū), *sb.* and *a.* [Native word.] **A.** *sb.* A Melanesian people of New Guinea inhabiting the area of Port Moresby. Also, their language. **B.** *adj.* Of or pertaining to this people. Also, **Motuan.**
1880 O. C. Stone *Few Months New Guinea* iv. 46 The inhabitants of Anuapata and the neighbourhood consist principally of the Motu tribe. 1885 W. B. Lawes *Motu Gram. & Vocab.* p. v, In the grammar of the Motu dialect of New Guinea one peculiarity is in the use of letters so much alike as to be scarcely distinguishable..b and p,.. d and t [etc.]... The Motuan language seems to be a strange mixture of Papuan and Eastern Polynesian. 1894 A. C. Haddon *Decorative Art Brit. New Guinea* 236 Port Moresby..is the centre of the Motu tribe... The Motu make great trading voyages to the Gulf of Papua. 1910 C. G. Seligmann *Melanesians Brit. New Guinea* 16 The Koita..have for generations intermarried with the Motu. .. Although the Koita still speak a Papuan language the majority of the males speak Motu, a Melanesian language, and have adopted..certain Motu customs. 1965 *Language* XLI. 169 Motu...is spoken fluently..by a few thousand Koitabuans intermingled among the Motuans themselves. 1967 'E. Lindall' *Time Too Soon* xviii. 197 He spoke in Motuan for the valley folk..and then in English for the white man. 1971 *Guardian* 4 Aug. 5/4 Motu, the language of the Port Moresby district.

‖ **motuca** (motū·kā). Also **motuka.** [Native name.] A Brazilian horse-fly, *Hadrus lepidotus*, of the family Tabanidæ.
1836 H. W. Bates *Naturalist on River Amazons* I. vii. 306 In the daytime the Motúca, a much larger and more formidable fly than the mosquito, insisted upon levying his tax of blood. *Ibid.*, The Motúca is of a bronzed-black colour; its proboscis is formed of a bundle of horny lancets. .. Its puncture does not produce much pain, but it makes such a large gash in the flesh that the blood trickles forth in little streams. 1914 *Chambers's Jrnl.* Oct. 630/1 Hornets dispute with motucas and other biting flies over the luckless intruder. 1927 W. M. McGovern *Jungle Paths & Inca Ruins* 224 Unlike the vampire bat, the motuka inflicted a very painful sting. 1933 P. Fleming *Brazilian*

Adventure I. xvii. 145 There were also *motuca* flies, which looked like a lethal and slightly futuristic form of blue-bottle, and whose bites drew blood and oaths but had no worse effects.

‖ **motu proprio** (mǭᵘ·tu prǫ·prio). [L., 'on his, its, etc., own impulse'.] Of one's own volition, on one's own initiative, spontaneously. Also as *sb.*, an edict issued by the Pope personally to the Roman Catholic Church, or to a part of it. Also *attrib.*
1603 C. Heydon *Defence Judicial Astrol.* xxi. 447 But the Moone and other Planets moove also motu proprio. 1613 J. Chamberlain *Let.* 14 Oct. in T. Birch *Court & Times James I* (1848) I. 278 Signor Gabellione, the Duke of Savoy's ambassador, came *motu proprio* about three weeks since to Ware Park. 1620 N. Brent tr. *Soave's Hist. Councel of Trent* iv. 354 Dispatching the dispensations under the name of Motu proprio, or with other clauses, with which the Chancery doth abound. 1848 Godde de Liancourt & Manning *Pius IX* II. xv. 404 The Pope published, on the 2nd of October at Rome, a new *motu proprio* for the organisation of a Senate. 1911 *Catholic Encycl.* X. 602/2 *Motu proprio*, the name given to certain papal rescripts... The words signify that the provisions of the rescript were decided on by the pope personally, that is, not on the advice of the cardinals or others. *Ibid.*, A favour granted *motu proprio* is valid even when counter to ecclesiastical law. 1938 *Oxf. Compan. Mus.* 596/2 *Motu proprio*... The word is often heard in connexion with church music on account of the Motu Proprio of Pius X, issued in the year of his becoming Pope (1903)... Instruments other than the organ were not to be employed without the bishop's special permission. 1964 P. F. Anson *Bishops at Large* x. 455 The Patriarch of the West (in the person of Pope Pius XI) might have been forced to issue a *motu proprio* denouncing the Patriarch of Glastonbury. 1972 *Daily Tel.* 15 Sept. 8/7 Restatement of celibacy rules for deacons and priests came in a separate *motu proprio* decree.

motza, var. *matzah.

‖ **mou** (mū̆). Also **mow, mu.** [Chinese.] A Chinese unit of area, varying according to locality but usu. equivalent to about 670 square metres (800 square yards).
1876 *Encycl. Brit.* V. 39/1 A ground-rent of 15,000 cash (about £3) per *mow* (a third of an acre). 1924 *Other Lands* Jan. 56/2 Having acquired a tract of land on the north bank of thirty to forty thousand mow it has been very successful. 1937 E. Snow *Red Star over China* ii. 71 Here in Shensi a peasant may own as much as 100 *mou* of land and yet be a poor man. 1955 *Ann. Reg. 1954* 305 'Of the over 80 million mou of the still flooded farmlands, 30 million mou have not yet been drained.'.. A mou was almost exactly one-sixth of an acre. 1957 *Encycl. Brit.* XV. 140/2 Mou... Commonly 806.65 sq. yd. Varies locally. Shanghai = 6,600 sq. ft. (Municipal Council). By Customs Treaty = 920.417 sq. yd., based on ch'ih of 14.1 inches. 1973 *Times* 21 Mar. (China Trade Suppl.) p. xiv/3 Since a *mu* is one sixth of an acre, and a catty is 1.1 lb. these figures come out at 18 cwt and 24 cwt of unhusked rice an acre.

mouche. Restrict † *Obs.* to sense in Dict. and add: **b.** A natural mark on the face similar in appearance to a patch of plaster.
1859 G. A. Sala *Twice round Clock* 143 In a hoop petticoat, a point lace apron, red-heeled *mules*, a *toupet* and a *mouche* on the left cheek. 1959 M. Steen *Woman in Back Seat* i. i. 8 At the angle of the jaw,..there should be a dark brown *mouche*, about the size of a small pea.

mouchette (mū̆ʃe·t). *Archit.* [Fr.] A motif in curvilinear tracery having the shape of a curved dagger.
1927 *Proc. Soc. Antiquaries Scotl.* LX. 388 The space under the arch bears an heraldic shield supported by a composition of *soufflets* and *mouchettes*. 1949 *Archit. Rev.* CV. 4/1 The mouchettes of the Court School are equally distinctive... They change only within certain fixed limits, ..and yet in a very subtle manner the mouchettes create changes of style. 1955 M. Hastings *St Stephen's Chapel* 179 The fan tracery of Henry VII's chapel is no more than an extremely elaborate combination of mouchettes. 1967 J. B. Nellist *Brit. Archit.* ii. 43/1 The elaborate tracery is almost entirely composed of mouchette shapes.

moue. (Later examples.)
1908 A. Bennett *Old Wives' Tale* iv. i. 429 She gave a *moue* and a flounce in reply, and swished out. 1955 *Times* 18 July 3/1 The hero..seeks the male companionship of sea life only to collapse powerless before the frail charm, the irresistible *moue*, and the wide eyed Frenchness of a passenger on the boat. 1974 'J. Le Carré' *Tinker, Tailor* xii. 103 'Why, George..it was you who arranged the deal!' With a quaint moue of professional vanity, Smiley conceded..it.

‖ **moue** (mū̆), *v.* [f. the sb.] **1.** *trans.* **a.** To ogle; stare at. **b.** To utter (words) with a moue.
1909 Hardy in *Daily Chron.* 1 Apr. 4/6 Whither have danced those damsels now; Is Death the partner who doth moue Their wormy chaps and bare? 1938 G. Greene *Brighton Rock* v. v. 213 'Work, work, work,' she moued at them.
2. *intr.* To make a moue.
1938 G. Greene *Brighton Rock* vii. ix. 346 They looked up and moued to each other, as much as to say—'Oh well, she wasn't really worth the trouble.' 1968 *Punch* 7 Aug. 202/3 The girls look the same young flappers to me, moueing and mincing and doing the long-lashed come-hither look.

mouf (mauf), repr. a pronunc. of Mouth *sb.*
The examples are taken from Black English sources in the U.S. but the pronunciation is also characteristic of Cockney speech in London.
1909 *Dialect Notes* III. 350 *Mouf*, mouth: a negroism rapidly gaining ground among the whites, especially in the derivatives, as *mouffle* (mouthful). 1935 Z. N. Hurston *Mules & Men* (1970) i. ii. 49 She got plenty hips, plenty mouf and no brains. 1969 R. Fair in A. Chapman *New Black Voices* (1972) 114 Oh, shut yo mouf up man. 1971 *Black World* Oct. 63/2 Like I got hoof and mouf disease and there ain't no hope at all.

moufflon. Add: Further examples of **mouflon,** which is now the usual spelling.
1843 J. Saunders in C. Knight *London* V. cxvii. 261 In the centre..is a curious little open hut, with projecting eaves... A horned sheep, the mouflon, is confined in it. 1932 J. S. Huxley *Probl. Relative Growth* iii. 88 The adult wild Mouflon ewe is in its proportions but little in advance of the improved Suffolk lamb. 1970 *Ashmolean Mus. Rep. Visitors 1969* 14 Omphalos bronze bowl decorated on the inside with mouflon among the trees. 1974 *Environmental Conservation* I. 9/1 Antelope, mouflon, and ostriches, have been exterminated over large areas.
2. The fur of the moufflon; woollen fabric made from the moufflon's coat. Occas. used erron. of the fur of the Mongolian goat. Also *attrib.*
1905 [in Dict.]. 1910 *Encycl. Brit.* XI. 351/2 Many Mongolian goats with the long hairs pulled out are sold as mouflon. 1949 *Amer. Speech* XXIV. 98 There is a species of goat from the mountains of northern China characterized by pelts of downy underhair... These furs are widely used and are called *moufflons* or *pulled goat*. 1952 C. W. Cunnington *Eng. Women's Clothing* 295 *Moufflon serge* ('18), 'a kind of brushed woollen textile'. 1962 *Guardian* 21 July 4/7 A mist of softest peltry—chinchilla, sables, mink, and millinery moufflon. 1973 *Ibid.* 27 Feb. 11/2 Coat in white mouflon wool.

mought (mǭᵘt). [See May *v.*¹ 4 c β.] A recurring var. of the pa. t. form *might* in representations of regional English (esp. Black English) in the U.S.
1872, 1885 [see May *v.*¹ 4 c β]. 1927 A. P. Randolph in A. Dundes *Mother Wit* (1973) 203 You can't 'speck des 'nigger' bosses to speak up for our rights when it mought cos dey jobs. 1933 J. M. Brewer in *Ibid.* 248/2 Yuh mought as well die wid de chills. 1938 M. K. Rawlings *Yearling* 144 Mought be, we'll find 'em in a pen some'eres.

moujik, muzhik. Add: Also **mouzhik.**
1949 I. Deutscher *Stalin* iii. 67 Eight or nine soldiers out of ten were mouzhiks. 1961 *Observer* 10 Dec. 7/2 The most fascinating Khrushchev of all is the high-powered *mouzhik*, the super *kulak*.

moulage (mū̆·lāʒ). [F. *moulage* action of moulding, moulded reproduction, f. *mouler* to mould.] An impression of a (part of a) person or of an object, the material used, or the process of taking an impression. Also *attrib.*
1902 *Encycl. Brit.* XXX. 788/2 In anatomy and physiology, models are specially employed as aids in teaching and study, and the method of *moulage* or chromoplastic yields excellent impressions of living organisms, and enables anatomical and medical preparations to be copied both in form and colour. 1940 R. Morrish *Police and Crime Detection* xii. 120 In these days an attempt is made to create a complete image of the murdered person by casts. This is called the 'Moulage' system. The results are almost as lifelike as the dummies in Madame Tussaud's. 1947 C. Brooks *Well Wrought Urn* iv. 69 A detective making a moulage of a footprint in wet clay. 1947 J. C. Rich *Materials & Methods Sculpture* v. 96 The late Dr. Alphons Poller was among the first to utilize agar as a negative mold material in fashioning molds from flesh. He developed a moulage system and subsequently wrote a book on the subject. Poller's moulage compounds were patented and the trade names of his negative mold compositions.. were registered. 1957 V. J. Kehoe *Technique Film & Television Make-Up* iii. 35 Moulage is a general name for impression materials some of which can be remelted and re-used (regular hydrocolloids)..and others (known as non-reversible hydrocolloids),..which are alginate non-reusable materials. 1969 R. Mayer *Dict. Art Terms & Techniques* 254/1 A specially prepared moulage plaster.. may be used on delicate or valuable materials. 1973 R. C. Dennis *Sweat of Fear* xi. 81 The detection of murder no longer need depend on fingerprints, blood types and moulages.

mould, *sb.*³ Add: **17.** *mould-cutting, -resisting* adjs. and sbs.; **mould-blowing** *Glass-making*, the blowing of glass inside a mould to give it the required shape; so **mould-blown** *a.*; **mould cavity** (see quots.); **mould-loft,** also *Aeronaut.*; **mould-made** *a.*, of paper, made on a type of machine which produces sheets having characteristics imitating those of hand-made paper, esp. the so-called deckle edge; **mould oil** *Building*, an oil applied to formwork to prevent concrete adhering to it; **mould-runner,** an operative in a pottery responsible for transferring a completed article, still attached to its mould, to the drying-oven; hence **mould-running** *vbl. sb.*
1948 E. B. Haynes *Glass through Ages* 307 Mould-blowing. 1949 P. Davis *Devel. Amer. Glass Industry* iv. 48 Glass for purposes other than glazing..was made by two different processes known technically as 'off-hand blowing' and 'mold-blowing'. 1972 E. Fletcher *Bottle Collecting* iii. 48 Most of the early examples of case bottles to survive have sides which sagged badly after removal from the mould; but the techniques of mould-blowing were soon to improve. 1925 Hodkin & Cousen *Text-bk. Glass Technol.* xxxii. 412 Much of the preliminary work in shaping parisons for mould-blown bottles might be mechanically performed. 1970 *Ashmolean Mus. Rep. Visitors 1969* 15 A clear green glass flask with hexagonal mould blown body decorated with panels of lattice and chevron pattern. 1951 *Gloss. Terms Plastics Industry* (*B.S.I.*) 37 *Mould cavity* (cavity), the female portion of a mould impression. 1971 W. K. V. Gale *Iron & Steel Industry: Dict. Terms* 136 *Mould cavity*, the impression left in a foundry mould after the pattern has been removed. 1947 J. C. Rich *Materials & Methods Sculpture* v. 114 The author has employed dental floss, which is waxed silk thread, for mold-cutting purposes, with good results. 1947 *Jrnl. R. Aeronaut. Soc.* LI. 307/2 The mold loft consisted of a building with a large floor area, the floor being painted a mat black. 1916 H. A. Maddox *Paper* viii. 120 Mould-made imitations of hand-made paper are produced..by several types of apparatus. 1923 —— *Dict. Stationery* 53 *Mould-made paper*, a class of high-grade paper which closely embodies the characteristic features of handmade. The sheets are made on a special machine which forms them singly and imparts four deckled edges... In selling mould-made note paper the stationer is legally compelled to describe it as such. 1938 *Times Lit. Suppl.* 15 Jan. 40/4 The text of the poem [*sc.* the Nonesuch edition of *Comus*] is printed in Fell types.. on Pannekoek mould-made paper, at the Oxford University Press. 1955 S. C. Gilmour *Paper* vii. 64 Nowadays the relatively few mould-made papers that are produced rank as a close second in character and quality to handmades, though not altogether comparable. 1973 S. Jennett *Making of Bks.* (ed. 5) xi. 182 Mould-made Papers are a paradox. They are in effect hand-made papers made by machine. 1939 W. H. Glanville *Mod. Concrete Construction* I. vi. 166 Mould oils of a variety of types are used in the various fields of concrete products manufacture. 1948 L. J. Murdock *Concrete Materials & Pract.* xvi. 240 The requirements of a good mould oil are that it shall prevent sticking, it shall reduce to a minimum adsorption of water by the formwork, and it shall not harm the concrete either by staining or by softening of the surface. 1962 D. F. Orchard *Concrete Technol.* II. xi. 321 Care must be taken to see that the plywood or hardboard does not buckle through expansion due to atmospheric influences or the absorption of water from the concrete; several coats of mould oil or a brush on plastic are a great help in this respect. 1962 Mould-resisting [see *damp-proofing vbl. sb.]. 1863 *1st Rep. Children's Employment Comm.* p. ix, in *Parl. Papers* XVIII. 9 As the potter forms the plate or saucer in the mould, the mould runner runs off with it into the 'store'. 1910 A. Bennett *Clayhanger* i. iv. 29 He was 'mould-runner' to a 'muffin-maker', a muffin being..a small plate, fashioned by its maker on a mould. 1961 M. Jones *Potbank* viii. 34 In the older workshops.. the mould-runner really does plenty of running. 1910 A. Bennett *Clayhanger* i. iv. 31 The labour was much lighter than that of mould-running and clay-wedging.

mould, *sb.*⁴ Add: **c.** *Comb.* (Earlier and later examples.)
1699 Pepys *Let.* 19 Oct. (1926) I. 200, I have found time ..to looke over all my heads; 'tis only mould-spotts some of them are touched with, by being putt together before they were dry. 1944 J. S. Huxley *On Living in Revolution* v. 64 Some [butterfly wings like dead leaves] even go so far as to be marked with imitation mould-spots and holes.

mould (mǭᵘld), *sb.*⁶ = Mould-board¹.
1858 *Trans. Illinois Agric. Soc.* III. 367 In fall-plowing we run the share and mould of the plow under the soil and invert it. 1868 *Rep. Iowa Agric. Soc. 1867* 266 There is no clogging, and the mould and lay are so hardened that they scour readily.

mouldability. (Further examples.)
1938 H. I. Lewenz tr. *Brandenburger's Processes & Machinery Plastics Industry* vi. 97 (*heading*) The effect of fillers on the mouldability of compounds. 1956 J. N. Anderson *Appl. Dental Materials* xix. 228 The water also improves mouldability by acting as a plasticizer. 1970 *Cabinet Maker & Retail Furnisher* 23 Oct. 174/3 Melded fabrics..have a sufficient degree of stretch and 'mouldability' to aid the upholstering of curved areas.

mould-board¹. Add: Also *mould-board plough.*
1808 W. H. Marshall *Rev. Rep. to Board Agric. from N. Eng.* i. 79 The seed is covered in by going once over with a light harrow, or..by a double-mould-board-plough. 1858 *Trans. Illinois Agric. Soc.* III. 366 A bull-tongue or shovel plow put to the same depth, will raise better corn than a mouldboard plow. 1971 *Power Farming* Mar. 31/1 There is a definite place for machines of this type to replace the mouldboard plough under certain circumstances.

moulder, *sb.*¹ Add: **1. b.** (See quot.)
1894 *Gloss. Terms Evidence R. Comm. Labour* 58/1 in *Parl. Papers 1893–4* (C. 7063) XXXVIII. 411 *Moulders*, men in the seed-crushing industry who draw the rolled seed from the fixture wherein it is made hot, and..subject it to a slight pressure.

moulder, *v.* Add: **4.** *intr.* To move *off* in an aimless or lifeless manner. *rare.*
1945 E. Bowen *Demon Lover* 48, I mouldered off by myself..to watch the old clock.

mouldiness. Add: *fig.* A state of boredom or discontent. Cf. *mouldy *a.*² 2 b.
1916 'Taffrail' *Stand By!* 23 Our mouldiness in the morning is merely temporary.

moulding, *vbl. sb.*² Add: **4.** *moulding machine, -plane* (later example), *powder, -sand* (later examples).
1890 *Cent. Dict.*, Molding-machine. 1921 *Daily Colonist* (Victoria, B.C.) 25 Mar. 8/4 To meet the needs of the small foundry with a varied demand, a British firm has, however, introduced an adaptable molding machine which can be quickly and easily adjusted to take molding boxes and pattern plates of any size within a comparatively wide range. 1964 W. L. Goodman *Hist. Woodworking Tools* 52 The remainder include moulding plane irons,.. rebate- and shoulder-plane irons, and plough irons. 1940 *Chambers's Techn. Dict.* 560/1 *Moulding powder*, the finely ground mixture of binder, accelerator, colouring matter, filler, and lubricant which is converted under pressure into the final moulding. 1957 *Which?* Autumn 9/1 Plastic frames should be made of optical sheet. The reason for this is that frames made from moulding powder are not so practical. 1930 *Engineering* 21 Feb. 247/1 Simple tests have been devised for regular daily foundry control of moulding sands. 1969 Bennison & Wright *Geol. Hist. Brit. Isles* xii. 274 The Pebble Beds are followed by the Upper Mottled Sandstone ≡ Moulding Sand... The term moulding sand refers to their widespread use.

mouldwarp. 1. (Later *poet.* examples.)
1916 Blunden *Harbingers* 59 Mouldwarps working late. 1928 A. D. Mackie *Poems in Two Tongues* 51 Alang his pad the mowdie-worps Like sma' Assyrians lie.

mouldy, *a.*² Add: **2. b.** Wretched, boring, depressing, gloomy, sick. *colloq.* or *slang.*
1876 [in Dict., sense 2]. 1896 Farmer & Henley *Slang* IV. 362/1 *Mouldy*,..worthless: *e.g.*, a *mouldy* offer. 1912 F. M. Hueffer *Parnel* i. iii. 93, I slogged like that for Nancy... We could have got along on a major's pay, out there. Just got along! And then the blasted girl goes and gets rotten titles and mouldy houses to her back on the day the bottom drops out of me. 1916 E. V. Lucas *Vermilion Box* 220, I should be mouldy company for you, I fear, because I can't talk. 1916 'Taffrail' *Pincher Martin* x. 174 Since you're all so mouldy, I suppose I must..turn in. 1924 M. Kennedy *Constant Nymph* iv. xxiii. 322 She looked more wan and frail than ever and he exclaimed: 'You look very mouldy.' 1936 —— *Together & Apart* i. 95 Do please come home soon, for it's mouldy without you. 1956 A. Huxley *Let.* 25 Dec. (1969) 814 One feels a bit low and mouldy after those bouts of flu. 1962 *Guardian* 20 Jan. 3/6 Local support for the event had deteriorated, but it did not deserve to be called 'mouldy'. 1972 *Sat. Rev.* (U.S.) 9 Sept. 92/1 The average cabby is a moldy old fascist.
c. *mouldy fig*: a supporter (occas., a performer) of traditional jazz. Also *attrib.* or as quasi-*adj.*
1945 *Esquire* Mar. 10/2 Why do aforementioned connoisseurs insist upon maintaining that the Chicago and New York (white) styles are the real Jazz, when it's perfectly obvious that New Orleans was—and is—the birthplace of the true 'stuff'?.. Sincerely, Moldy Fig, France. 1945 S. Platt in *Ibid.* June 10/3, I wish to protest against the 'Moldy Fig' genre of music lovers. There seems to be some perverse streak in critics such as Avakian or 'Moldy Fig' which prevents them from liking anything but the very oldest available. 1958 G. Lea *Somewhere there's Music* x. 83 Dixie Cats and the rest of the Moldy Figs, okay for them, they don't need to think. 1959 *Hi Fi Rev.* Apr. 79/2 Lines of seething fury were drawn between the [jazz] traditionalists and the boppers who viewed each other as 'moldy figs', on the one hand, and players of 'all them wrong notes', on the other. 1968 *Listener* 4 Apr. 450/3 Readers over 30 will remember the term 'Mouldy Figge' as contemporaneous with Little Jackie Dennis and Suez. 1968 *Blues Unlimited* Nov. 7 Many collectors are mouldy-fig enough to believe that virtually every worthwhile blues singer was recorded at least once in the '20s and '30s. 1973 *National Observer* (U.S.) 6 Oct. 23/1 Charles Keil satirizes the 'moldy-fig' aspirations of earlier blues scholars.
3. *Comb.*, as *mouldy-minded* adj.
1906 Hardy *Dynasts* II. vi. vii. 320 The rawest Dynast ..Will..Down-topple to the dust like soldier Saul, And Europe's mouldy-minded oligarchs Be propped anew.

mouldy (mǭᵘ·ldi), *sb. Navy* and *R.A.F. slang.* [Origin unknown.] **1.** A torpedo.
1916 M. T. Hainsselin *In Northern Mists* xvi. 62 A German submarine..kept one of the bug-traps bailed up.. for a week by waiting..ready to squirt a mouldy at her directly she showed her nose outside... To fire a torpedo at her, of course! 1918 *Yachting Monthly* XXIV. 297/1 When H.M.S. Carraco was torpedoed she received the mouldy right forward. 1928 *Observer* 11 Mar. 17/4 The King of Afghanistan will be given a lesson in torpedo firing and himself discharge a 'mouldy' from one of L22's tubes. 1932 *Flight* 19 Aug. 777/1 At the same time, no doubt, the A.A. gunners on board are gleefully telling all and sundry how they simply riddled the 'Horsleys' with shells before ever a mouldy was dropped. 1943 C. H. Ward-Jackson *Piece of Cake* 43 *Mouldy*, a torpedo..was brought into air force use by the Royal Naval Air Service.
2. A confection sold at naval colleges.
1916 G. Franklin *Naval Digression* xii. 105 The various cadets engaged in stuffing themselves with 'pinkmen', 'mouldies'..and suchlike *vinos y comida*. 1962 Granville *Dict. Sailors' Slang* 78/2 *Mouldy*,..confection popular at Dartmouth.

‖ **moule** (mūl). [Fr.] A mussel, *spec.* in **moules (à la) marinière,** mussels served in their shells and cooked in a wine and onion sauce.
1890 E. Lebour-Fawssett *French Cookery for Ladies* 161 *Moules à la Marinière*... After you have taken your mussels out of the saucepan.., put in three onions,..two sliced carrots, [etc.]. 1928 D. L. Sayers *Unpleasantness at Bellona Club* ix. 106, I was just wondering whether to have *moules marinières* or not. 1950 D. Ames *Corpse Diplomatique* iv. 33 She was almost up to her elbows in a dish of *moules à la marinière*. 1959 *Good Food Guide* 36 The cooking is something above 'plain English' and even includes moules marinières quite frequently. *Ibid.* 162 Take moules Mendip (2/6), pâté (2/6), various vol-au-vent (3/-). 1971 Coombes & Wakelin *Good Househ. Advanced Cooking is Fun* 117 *Mussels* (Moules Marinière): These are best eaten from a bowl or soup plate, with a fork to get the mussel from the shell..and a spoon for the liquor.

Mouli (mū̆·li). Also **mouli.** A proprietary name, shortened from *Moulinette. Also *attrib.*
1937 *Trade Marks Jrnl.* 21 July 862/2 Mouli... All goods included in Class 6. [Machinery of all kinds, and parts of machinery, except agricultural and horticultural machines and their parts.] Mouliware Limited,..London, ..; manufacturers and merchants. 1969 O. John *Dead on Time* i. 11 Homus bathini is made from chick peas.., put..through a mouli. 1972 C. Fremlin *Appointment with Yesterday* vii. 51 The burnt chip-pan. And all that white, sticky stuff in the Mouli-Mixer. 1972 *Times* 12 Aug. 11/1 Pass the soup through a 'mouli' soup mill, this.. keeps back all the stalky bits.

moulin. Add: **2.** A kitchen utensil used for grinding food or reducing it to pulp. (See Mill *sb.*¹ 2 a.)
1959 *Listener* 17 Dec. 1095/1 Put it through a sieve, or a *moulin à legumes*. 1962 *Harper's Bazaar* Aug. 37 Black pepper freshly ground from the *moulin*. 1966 'K. Nicholson' *Hook, Line & Sinker* viii. 92 In the kitchen Mrs. Chilperic..was urging the apple sauce through the moulin.

Moulin-à-Vent (mulæ̃ a vaṅ). Also **Moulin à Vent.** [Fr. place-name.] The name of a Beaujolais wine produced in the commune of Moulin-à-Vent.
1833 C. Redding *Hist. Mod. Wines* v. 112 The first class of Burgundies in the Saone and Loire, are Moulin à Vent, Torins, and Chenas. 1927 A. E. Housman *Let.* 17 Oct. (1971) 254 We had two bottles of white wine (the first probably Pouilly-Fuissé) and then half a bottle of *Moulin-à-Vent*. 1961 'J. Welcome' *Beware of Midnight* vi. 79, I ordered..a bottle of Moulin à Vent. 1967 A. Lichine *Encycl. Wines* 121 Such a quintessential Beaujolais as Fleurie will be more characteristic than the bigger wines—the Morgon and Moulin-à-Vent. 1968 'G. Bagby' *Corpse Candle* vi. 67 There was even a respectable wine, a Moulin-à-Vent.

Moulinette (mū̆line·t). Also **moulinette.** The proprietary name of a type of food mill.
1936 *Trade Marks Jrnl.* 23 Dec. 1585/2 The Moulinette. .. Food strainers and sieves all of ordinary metal. Mantelet & Boucher.., Bagnolet (Seine), France; manufacturers and merchants. 1951 *Ibid.* 7 Mar. 229/1 Moulinette... Hand operated mincing machines for food. Mouliware Limited,.. London,..; manufacturers and merchants. 1951 E. David *French Country Cooking* 19 A purée-maker or food mill, usually called a *moulinette* in France. For soups, fruit and vegetable purées this is absolutely invaluable. 1961 *Listener* 24 Aug. 295/1 If you enjoy sieved greens, fruits, and other purées, a *moulinette*, or rotary sieve..will pay its way.

moulvi(e. (Examples.)
1862, etc. [see Moolvee]. 1864 J. A. Grant *Walk across Afr.* xi. 239 The Seedees, though knowing nothing of the Mohammedan religion, the majority not being circumcised, were much more particular on those occasions, and offered more opinions than a 'moulvie', or Mussulman priest, would. 1972 *Guardian* 1 Dec. 14/3 The great bulk of [Guyanese] Indians are Hindus or Moslems, and pundits, moulvis and parents always viewed such [Christian] schools with suspicion.

mound, *sb.*³ Add: **3. c.** In *Baseball*, 'the slightly elevated ground from which the pitcher pitches' (D.A.).
1914 *Collier's* 7 Feb. 7/2 There's a pitcher who never has to be urged to go to the mound. 1957 [see *bull-pen 1 b]. 1974 *Evening Herald* (Rock Hill, S. Carolina) 18 Apr. 6/3 Mussman went the entire nine inning stint on the mound for Rock Hill and was credited with the win.
5. mound ant *Austral.* = *meat-ant* *Meat *sb.* 6); **Mound City** *U.S.*, a name for St. Louis, Missouri; **Mound of Venus** = *Mons Veneris* (s.v. Mons a, b); **mound region,** a region in which there are many mounds.
1907 Mound Ant [see *meat-ant* s.v. *Meat *sb.* 6]. 1926 *Austral. Encycl.* I. 68/2 Amongst the objectionable species the Mound Ant (*Iridomyrmex detectus*) is prominent; its huge nests are particularly destructive to garden paths. 1935 K. C. McKeown *Insect Wonders Austral.* 5 The Mound Ants form their great gravel nests in the grassy plains, scouring in search of food..to the dead body of some horse or sheep which has perished in time of drought, the marauders issuing from holes in the carcass in long streams, each ant bearing a fragment of flesh in its jaw. 1855 Mayne Reid *Hunters' Feast* i. 5 On the western bank of the Mississippi..stands the large town of St. Louis, poetically known as the 'Mound City'. 1860 Bartlett *Dict. Amer.* (ed. 3) 282 *Mound-City*, the city of St. Louis, so-called from the number of artificial mounds that occupied the site on which the city is built. 1865 R. Beamish *Psychonomy Hand* 35 The mound of Venus, devoid of lines, is the index of chastity, coldness, tranquillity in love. 1963 C. R. Mueller tr. *Büchner's Danton's Death* i. v, in *Compl. Plays & Prose* 20 A woman's thighs will be your guillotine, and her mound of Venus your Tarpeian rock. 1873 J. H. Beadle *Undevel. West* i. 38 This is the centre of the 'Mound Region' of Wisconsin—so called from the many Indian mounds scattered about the valley.

moundy, *a.* Add: (Earlier and later examples.)
1851 W. Kelly *Excursion Calif.* I. vi. 97 Revealing a range of elevated hills stretching north and south, moundy on the surface and where they were broken showing a fine light rabbit sand. 1955 D. D. C. P. Mould *Irish Pilgrimage* iv. 44 Mirage may lift them out of the water, so that they float above the sea, dark moundy masses.

mount, *sb.*² Add: **2. c.** An act of copulation.
1896 Farmer & Henley *Slang* IV. 362/1 *Mount*, an act of coition. 1937 Partridge *Dict. Slang* 226/2 *Do a grind, a mount*, to have sexual intercourse (of men). 1970 *Nature* 12 Dec. 1107/1 In mounts from behind, the mounting cat often had its pelvic region well forward on the back of the mounted cat.
4. b. *collect.* A supply of riding- or draft-horses.
1907 S. E. White *Arizona Nights* i. iii. 53 He kept his own mount of horses, took care of them. 1933 *Amer. Speech* VIII. i. 30/1 *Mount*, a string of horses, usually eight or ten, assigned by the boss to one man.
5*. A stuffed and mounted bird-skin.
1935 *Auk* LII. 281 Since the mounts were similarly posed, it seemed that the male Northern Yellow-throat was discriminating between the sexes primarily on a basis of color pattern. 1938 *Brit. Birds* XXXII. 30 The female mount..had a half-spread tail. 1957 J. W. Moyer *Pract. Taxidermy* v. 34 Tie down the feathers with soft, fine thread or string to hold them in place until the mount is dry.

mount, *v.* Add: **8. f.** To rise on to an obstruction, etc.
1930 *Morning Post* 19 July 12/6 He just managed to avoid a crash by cutting out to his right and in doing so he mounted the footpath.
9. (Further examples.)
c 1884 'Mark Twain' *Speeches* (1923) 109, I renewed my youth, to outward appearance, by mounting a bicycle. 1912 W. Owen *Let.* 1 Feb. (1967) 113, I had arranged to go to the Cyclists... the machine is only £5.19.6!.. It *will* be a *joy-ride* when I am mounted on one of these!
10. (Later examples.) Also of people.
1963 A. Heron *Towards Quaker View of Sex* 54 The young bachelor males of herds where the overlord male jealously protects his harem will mount each other. 1970 *Nature* 12 Dec. 1107/2 A mounting female was frequently immediately mounted by the cat she was mounting, or by another oestrous female. 1970 Masters & Johnson *Human Sexual Inadequacy* 307 The wife once mounted is instructed to hold herself quite still. 1971 'V. X. Scott' *Surrogate Wife* 19, I was a man, mounting a beautiful and passionate woman. 1973 J. Elsom *Erotic Theatre* ix. 174 Men no longer want to mount women simply because, like Everest, they are there.
11. (Further example.)
1891 'Mark Twain' in *Illustr. Lond. News* 26 Dec. 834/1 Everybody else had..'mounted the train', as they say in those regions [e.g. Geneva].
16. g. *to mount an attack, offensive*, etc. Also *fig.*
1952 *N.Y. Times* 3 May 2/4 Striking at Communist targets in excellent flying weather (Thursday) warplanes of the Far East Air Forces mounted 1,283 sorties. 1957 *Times Lit. Suppl.* 20 Dec. 771/1 A British private-army leader would have mounted, or at least planned, an incessant series of operations. 1965 *Listener* 2 Sept. 334/1 Governments mount big campaigns to secure an 'incomes policy'. 1966 *Ibid.* 20 Oct. 579/2, I am mounting a devastating attack on the seriousness of the book. 1972 *Daily Mirror* 12 Oct. 1 An all-out attack is to be mounted against the porn-pushers in Britain's High Streets.
17. d. (Earlier and later examples.) Also, to put on or produce (a radio or television programme).
1870 *N.Y. Times* 11 Oct. 5/5 'The Two Roses' is.. prettily mounted, and nicely, if not greatly acted. 1962 *Listener* 10 May 808/1 It is the first town that approached us and asked us to mount a festival. *Ibid.* 30 Aug. 328/1 His staff..mounted several brisk little propaganda numbers about social evils in Britain. *Ibid.* 20 Sept. 437/1 The Arts Council has mounted..an exhibition which has certain flaws. 1963 *Times* 8 Feb. 14/2 The production is mounted in the later Brechtian manner. 1971 *Daily Tel.* 2 Dec. 12 The BBC is scrapping normal programme schedules..during Christmas to enable it to mount special productions.

mountain. Add: **2. c.** A stockpile, a surplus.
1969 *Times* 10 Sept. 11/4 In Germany..they are beginning to resent it [*sc.* the price for protection], as the sardonic remarks in the supermarkets about the 'Butter Mountain' reveal. 1974 *Daily Tel.* 5 Feb. 2/4 It is intervention buying that leads to the creation of the Common Market's notorious commodity 'mountains'. 1974 *Times* 1 May 4/7 Measures designed to disperse the Community's growing beef surpluses. The beef mountain now stands at more than 70,000 tons. 1975 *Times* 7 Feb. 4/8 The prospect of a 'cheese mountain' in the EEC.
3. b. *to make a mountain* (*out*) *of a molehill*: see Molehill 2.
6. (Further examples.)
In England also applied to a group of Conservatives at the beginning of the 20th c.
1829 H. Hardinge *Let.* 19 June in C. Arbuthnot *Corr.* (1941) 116 It would, if true, keep the high Whigs disunited from the Mountain, & assist our union with our old Tory party. 1965 E. Fellowes in *Political Q.* XXXVI. 257 Among the Supporters of the Government [in the Parliament of 1918] was the National Party led by Sir Henry Page-Croft, and a group of Conservative backbenchers (the 'Mountain'), not so formally organised, but working in concert... This..group in fact often proved a more successful opposition than the Labour and Liberal parties, who shared the Opposition front bench.

7. a. *mountain-echo, slope, -wreath.*
1805–6 Wordsworth *Prelude* (1959) I. 390 Not without the voice Of mountain-echoes did my Boat move on. 1841 Thoreau *Jrnl.* 4 Mar. in *Writings* (1906) VII. 228 Their way is a mountain slope, a river valley's course. 1930 A. Clarke *Coll. Plays* (1963) 63 To look across Kildare in sun and know The far flocks move along the mountain slope. 1928 Blunden *Retreat* 70 Warm furze-perfume, stern mountain-wreath Of pines.

b. *mountain-climber*; *mountain-cresting, -walking* adjs.
1880 'Mark Twain' *Tramp Abroad* xxxvi. 375 We were in the..home of the mountain-climbers. 1951 S. Spender *World within World* iii. 179 Then we came to that extraordinary river-encircled, mountain-cresting city of Toledo. 1821 Shelley *Let.* 22 Oct. (1964) II. 361, I..raised a small turf altar to the mountain-walking Pan.

c. *mountain-clear, -cool* adjs.; *mountain-headed* adj.
1955 P. Larkin *Less Deceived* 36 Their visions mountain-clear. 1919 A. Huxley *Leda* (1920) 1 Brown and bright as an agate, mountain-cool. 1925 Blunden *Eng. Poems* 104 O firmament, O mountain-headed march Of clouds through that blue arch.

e. *mountain-bound, -cradled, -echoed, -guarded, -roofed, -sheltered* adjs.
1860 G. M. Hopkins *Poems* (1967) 3 There is a massy pile above the waste Amongst Castilian barrens mountain-bound. 1954 W. Faulkner *Fable* 158 That thirty-mile-long mountain-cradled saucer. 1860 G. M. Hopkins *Poems* (1967) 6 Then pass'd the wind, and sobb'd with mountain-echo'd woe. 1939 Belloc *Decameron* in *Tablet* 11 Feb. 166/2 Mountain-guarded gardens. 1937 Blunden *Elegy* 90 The rough walls back to Chaucer reach, Near windowless, mountain-roofed, wry-angled. 1924 W. J. Locke *Coming of Amos* xiii. 169 A coast of romantic mountain-sheltered creeks.

8. a. *mountain air* (earlier and later examples), *shelter.*
1807 Southey *Lett. from Eng.* II. xli. 201 The mountain air had made us almost ravenous. 1810 E. Weeton *Let.* 11–15 Nov. (1969) I. 314 The keen mountain air had sharpened our appetites. 1965 *Listener* 30 Dec. 1063/1 The notes of *Colonel Bogey*, played by a military band.., sounding so clearly in the crystalline mountain air. 1922 W. G. Kendrew *Climates of Continents* xxvii. 188 There are almost constant north-west winds, strongest where there is no mountain shelter.

b. *mountain people.*
1881 *Harper's Mag.* Nov. 868/2 They are poor mountain people.

c. *mountain-saddle, wagon.*
1849 F. Parkman *Calif. & Oregon Trail* 145 Though aided by the high-bowed 'mountain-saddle' I could scarcely keep my seat on horseback. 1867 W. H. Dixon *New Amer.* (ed. 6) I. 170, I had the honour of riding in the mountain wagon with an old road-agent.

d. *mountain-mass.* (Cf. quot. 1833 s.v. sense 8 a in Dict.)
1918 A. Huxley *Defeat of Youth* 21 Soon will they lift towards the summer sky Their mountain-mass of clotted greenery.

9. a. mountain-building, the formation of mountains, esp. as a result of folding and thrusting of the earth's crust; **mountain Damara**: see *Damara; **mountain dew,** also used for other, esp. home-made or illicit, whiskies (further examples); **mountain fever** (earlier and later examples); **mountain-folding,** the formation of mountains as a result of folding of the earth's crust; **mountain man,** (*b*) (earlier and later examples); (*c*) *fig.* = Pioneer *sb.* 3; **mountain oyster** = *lamb's fry* (Lamb *sb.* 7); **mountain railway** (earlier example); (*b*) a miniature ascending railway designed for amusement; a scenic railway; a funiculaire; **mountain-schooner,** a wagon used in mountainous country; **mountain-sick** *a.*, suffering from mountain sickness; **mountain slide** (earlier example); **mountain spectre,** a reflection (of persons or things) seen under certain conditions on a mountain (cf. *Brocken); **mountain system,** a group of mountain ranges showing similarity in form, orientation, etc., and assumed to be due to the same general causes; **mountain (standard) time** *N. Amer.*, 'the time of the seventh time zone west of Greenwich based on the 105th meridian and used in west central Canada and the U.S.' (Webster 1961).
1871 J. D. Whitney in *N. Amer. Rev.* CXIII. 238 We cannot separate the phenomena of volcanoes and earthquakes from those of mountain-building and continental growth. 1919 *Jrnl. Geol.* XXVII. 250 Only moderate igneous activity was associated with the mountain building. 1944 J. S. Huxley *On Living in Revolution* p. vii, Periods of mountain-building accompanied by the emergence of more land from the sea. 1971 I. G. Gass et al. *Understanding Earth* i. 26/2 Andesites are typical of the world's greatest volcanoes lying in zones of active mountain building. 1899 R. L. Taylor in B. A. Botkin *Treas. Amer. Folklore* (1944) III. 411 They gathered there on rainy days to talk politics and religion, and to drink 'mountain' dew and fight. 1945 Baker *Austral. Lang.* ix. 168 Illicit whisky, as made in stills in bush areas, is known as *mountain dew*. This is a variation of the Standard English use of the term for genuine Scotch whisky. It is also used in America. 1970 *Times* 15 Oct. 30/3 The distilled spirits industry..wages an expensive propaganda campaign against..mountain dew. 1849 O. W. Lipe *Let.* 15 Aug. in R. P. Bieber *Southern Trails to Calif. in 1849* (1937) 346 There has been much sickness in our company; the disease is mountain fever. *a* 1918 G. Stuart *40 Yrs. on Frontier* (1925) I. 51 A severe attack of mountain fever ..laid me on my back in the wagon. 1970 *Islander* (Victoria, B.C.) 10 May 7/1 An epidemic of mountain fever struck the Kootenays in 1884. 1925 J. Joly *Surface-Hist. Earth* 170 The effects of these conditions on mountain-folding would probably be principally experienced where the geosynclines had forced the continental materials deep into the magma. 1971 *Geogr. Abstr.* A. 445 (*heading*) Fundamental principles of the development of collapses and slips in mountain-folding regions. 1781 *Calendar Virginia State Papers* (1875) I. 494 A late pressing application of General Greene for the aid of the Mountain Men. 1973 R. D. Symons *Where Wagon Led* I. vii. 117 In the United States the 'Mountain Men',—beaver trappers mostly—penetrated the western wilderness at an early date, before and during the middle years of the last century. 1973 *Sci. Amer.* Aug. 113/1 Professor Luria is an authentic pioneer of molecular biology. Even before the first wagon train set out he ventured as a mountain man among bacterial viruses. 1890 *Cent. Dict.* s.v. *Oyster*, *Mountain-oyster*, a lamb's testicle. 1951 E. Paul *Springtime in Paris* xi. 189, I have consumed mountain oysters and prairie dancers that are actually poetic. 1962 *Alberta Hist. Rev.* Autumn 15/2 In the commissariat department [are] 'dope' (butter)... 'Mountain oysters' (calves fries). 1880 'Mark Twain' *Tramp Abroad* xxviii. 256, I..in the distance detected a long worm of black smoke crawling lazily up the steep mountain. Of course that was the locomotive... we had never seen a mountain railway yet. 1910 *Penny Guide Japan-Brit. Exhib.* 23 Mountain Railway. The visitor enters the cars which travel slowly round and upward until the top of the mountain range is reached. 1925 A. Huxley *Those Barren Leaves* II. iii. 106 The switchback, the water-shoot and the mountain railway. 1869 C. L. Brace *New West* xiv. 188 It is more than a hundred miles away from the first link with civilization, and yet coaches, wagons, and the stream of 'mountain-schooners' pour into it unceasingly. 1937 *Discovery* June 171/1 People have been very mountain-sick at this hut. 1830 *Mass. Spy* 25 Aug. (Th.), Mountain slides. 1880 *Encycl. Brit.* XI. 399/2 Mountain spectres are caused by reflexion, and often appear accompanied by chromatic halos. 1882 A. Geikie *Text-bk. Geol.* 918 The Alps offer an instructive example of a great mountain system formed by repeated movements during a long succession of geological periods. 1931 C. M. Nevin *Princ. Struct. Geol.* xi. 289 A mountain system is characterized by folding, faulting, and igneous activities that vary in their complexity and relative importance throughout the zone of deformation. 1953 *Q. Jrnl. Geol. Soc.* CVIII. 2 The great mountain systems, or orogens, are zones of extreme compression of the earth's crust weaving a complex pattern of majestic sweeps around the world. 1968 R. A. Lyttleton *Mysteries Solar Syst.* iii. 91 It can be expected that mountain-systems will have formed on Venus comparable with those on Earth. 1935 *World Almanac* (ed. 50) 115/2 Mountain Standard Time is the local time of the 105th meridian. 1968 *Globe & Mail* (Toronto) 5 Feb. 22/8 Sealed tenders..will be received up to 2 o'clock P.M. Mountain Standard Time. 1891 *Cent. Dict.* s.v. *Time*[1] 19, The time of the 105th meridian (called *mountain time*). 1952 B. Malamud *Natural* 12 It looked around half-past five, but he couldn't be sure because somewhere near they left Mountain Time.

c. mountain bluebird *N. Amer.*, a bluebird of western North America, *Sialia currucoides*, distinguished by a blue breast instead of a red one; **mountain boomer,** substitute for def.: *U.S.*, (*a*) the sewellel or mountain beaver, *Aplodontia rufa*; (*b*) the red squirrel, *Sciurus hudsonicus*; also *transf.* (examples); **mountain buffalo** *U.S.*, (*a*) = *mountain sheep 2; (*b*) a mountain variety of the American buffalo, *Bison bison*; **mountain cock,** (*b*) *U.S.* = Prairie-chicken, *cock of the plains* (*cock *sb.*[1] 10); **mountain devil** (later examples); **mountain goat** (earlier and later examples); see also sense 8 b in Dict.; **mountain herring** (earlier example); **mountain jay** *U.S.*, Steller's jay, *Cyanocitta stelleri*; **mountain lion** (earlier and later examples); **mountain plover** *U.S.*, a small plover, *Eupoda montana*; **mountain ringlet,** an alpine butterfly, *Erebia epiphron*, found in limited parts of the Lake District, Scotland, and western Ireland; cf. Ringlet 4; **mountain tortoise,** a large African tortoise, *Geochelone pardalis*, also known as the leopard-tortoise (Leopard 6 b).
1860 S. F. Baird *Birds N. Amer.* I. 224 *Sialia arctica*, Swainson. Rocky Mountain Blue Bird. 1904 I. G. Wheelock *Birds Calif.* 506 The exquisite coloring of the Mountain Bluebird renders him easily the most beautiful of all Californian birds. 1971 *Islander* (Victoria, B.C.) 13 June 13/1 We were fortunate to see such birds as..a sky-blue mountain bluebird. 1858 D. K. Bennett *Chronol. N. Carolina* 94 The only inhabitants we saw on these high points were pheasants, cross bills, ..and mountain boomers, a sort of squirrel. 1859 H. E. Taliaferro *Fisher's River* 33 A mountain 'Boomer' dressed in a linsey hunting-shirt down to his knees. 1922 H. Kephart *Our Southern Highlanders* (new ed.) 87 Out of a tree overhead hopped a mountain 'boomer'. *Ibid.* 280 They call themselves mountain people, or citizens; sometimes humorously 'mountain boomers'. 1940 *Mt. Hood Guide* 21 The sewellel or mountain beaver, sometimes colloquially called 'mountain boomer'..resembles the porcupine and marmot rather than the beaver. 1958 *Amer. Speech* XXXIII. 265 (*table*) Pejorative designations of rural dwellers in the Upper Midwest..mountain boomer. 1868 *Amer. Naturalist* II. 538 'Mountain Buffalo'... The Bighorn is sometimes called so. *Ibid.*, I saw no difference in the skulls, indicating a different species, or 'Mountain Buffalo' of the hunters. 1884 *Encycl. Brit.* Amer. Suppl. I. 540/2 Buffaloes long inhabiting other localities than the open plains, their natural homes, acquire distinguishable varietal characters. They are known as 'wood-buffalo' and 'mountain-buffalo'. 1892 *Scribner's Mag.* Sept. 277/1 There are, besides the ordinary animal of the plains, the 'mountain buffalo',.. the 'wood buffalo',..and the 'beaver buffalo'. 1805 M. Lewis in Lewis & Clark *Orig. Jrnls. Lewis & Clark Expedition* (1905) II. 124 Saw (near the hills) a flock of the mountain cock, or a large species of heath hen with a long pointed tail which the Indians informed us were common to the Rocky Mountains. 1943 H. Drake-Brockman in *Coast to Coast 1942* 158 There were little barking lizards and mountain-devils and eagle-hawks and white owls in the blow-holes. 1966 *Times* 11 Nov. (W. Austral. Suppl.) p. iv/4 The hideous little mountain devil (*Moloch horridus*) ..trades on its frightening aspect while sustaining a perfectly blameless existence on a diet of ants. 1841 G. Catlin *Lett. on N. Amer. Indians* II. 196 His leggings and shirt were of the mountain goat skin. 1936 D. McCowan *Animals Canad. Rockies* xiv. 119 A herd of mountain goats on an immense buttress of rock. 1966 *Globe & Mail* (Toronto) 7 May 41/2 Mountain goat also abound in the mountainous region..in..south-eastern British Columbia. 1877 C. Hallock *Sportsman's Gazetteer* 350 Williamson's Whitefish; Mountain Herring.—*Coregonus williamsoni.* 1872 *Amer. Naturalist* VI. 398 The great-crested, Woodhouse's and the Canada jays were of frequent occurrence in the mountains, the former being familiarly known as the 'mountain jay'. 1917 T. G. Pearson *Birds Amer.* II. 219 Steller's Jay... [also called] Mountain Jay; Pine Jay; Conifer Jay. 1859 G. A. Jackson *Diary* 1 Jan. in *Colorado Mag.* (1935) XII. 204 Killed a mountain lion today. 1936 D. McCowan *Animals Canad. Rockies* ix. 77 The cougar or mountain lion is a large tawny brown cat with a small head, rather slender body and long round tail. 1972 *Radio Times* 1 June 30/1 Wild life to be found in different parts of Canada including a mountain lion and a wapiti. 1858 S. F. Baird in *Rep. Explor. Route to Pacific* (U.S. War Dept.) IX. 693 Mountain Plover..is only known to inhabit the western countries of North America. 1917 T. G. Pearson *Birds Amer.* I. 267 Mountain Plover... Nest: On the open prairies; a depression in the ground, lined with leaves and grass. 1870 F. O. Morris *Hist. Brit. Butterflies* (ed. 5) 53 (*heading*) Small Ringlet. Mountain Ringlet. 1945 E. B. Ford *Butterflies* xiii. 288 The Mountain Ringlet, *Erebia epiphron*, cannot live in England at an altitude of less than 1800 feet, but in Scotland it is able to descend to about 1500 feet. 1975 *Times* 12 Nov. 14/6 The Mountain Ringlet is an extremely rare butterfly that may still exist in the wild, high mountains of Western Ireland. 1958 L. van der Post *Lost World of Kalahari* i. 26 For the equivalent of cello and bass violin he used the shell of our big dark mountain tortoises. 1966 E. Palmer *Plains of Camdeboo* xiii. 228 Mountain tortoises are said to have a great homing instinct. 1971 D. J. Potgieter et al. *Animal Life S. Afr.* 303/2 Mountain or leopard tortoise..is a widely distributed species, extending from the Sudan and Ethiopia in the north to the Cape in the south.

d. mountain beech, (*b*) (earlier and later examples); (*c*) *N.Z.*, the evergreen tree, *Nothofagus cliffortioides*; **mountain-bell,** an alpine species of *Campanula*; **mountain cherry** *U.S.*, substitute for def.: one of several wild cherries (earlier and later examples); **mountain chestnut oak** *U.S.*, an American oak, *Quercus montana*, with leaves resembling those of the chestnut; **mountain daisy** *N.Z.*, a perennial herb of the genus *Celmisia*, belonging to the family Compositæ; **mountain fern,** substitute for def.: *Thelypteris limbosperma* (earlier and later examples); **mountain hemlock,** a large coniferous tree, *Tsuga mertensiana*, of the family Pinaceæ, native to western North America; **mountain mahogany** *U.S.*, (*a*) = *mahogany birch* (*mahogany 7); (*b*) a shrub or small tree of the genus *Cercocarpus*, esp. *C. ledifolius*, belonging to the family Rosaceæ and native to western North America; **mountain maple** *U.S.*, one of several maples found in upland areas, esp. *Acer spicatum*; **mountain pine,** (*b*) *N.Z.*, a small evergreen tree, *Dacrydium bidwillii*, found in boggy or subalpine regions; **mountain pink,** an alpine species of *Dianthus*; **mountain tea** *N. Amer.*, the wintergreen, *Gaultheria procumbens*, or the beverage made from its leaves.
1884 A. Nilson *Timber Trees New South Wales* 92 *L[omatia] longifolia.*—Mountain Beech.—An erect small tree. 1928 Cockayne & Turner *Trees N.Z.* 154 *Nothofagus cliffortioides* (mountain-beech) is of little value as a timber, for it decays rapidly. 1959 A. McLintock *Descr. Atlas N.Z.* p. xv, The forest is of silver and mountain beech, with traces of red beech and podocarps at low altitude. 1965 *Austral. Encycl.* V. 360/2 *L[omatia] longifolia* (sometimes called mountain beech), has a hard light-coloured ornamental wood, used for turnery. 1923 D. H. Lawrence *Ladybird* 227 Sometimes the hairy mountain-bell, pale-blue and bristling, stood alone. 1813 H. Muhlenberg *Catal. Plantarum Americæ Septentrionalis* 48 *Prunus montana*,..mountain cherry. 1871 *Harper's Mag.* Oct. 707 We must..gather mountain cherries (*Prunus cerasus*). 1801 A. Michaux *Hist. Chênes de l'Amérique* sig. 6[1]v Mountain Chesnut Oak. 1821 T. Nuttall *Jrnl. Trav. Arkansa* i. 42 Much of the *Quercus Prima monticola* (or mountain chestnut oak) presents itself on the mountain. 1857 J. T. Thomson in N. M. Taylor *Early Travellers N.Z.* (1959) 335 Half-way up the mountains some pretty flowers were gathered, amongst which the mountain daisy

deserves notice. 1900 A. Dendy in *Canterbury Old & New* 188 Only rivalled in beauty by the marguerite-like flowers of 'cotton plant' or mountain daisy (species of *Celmisia*). 1959 A. McLintock *Descr. Atlas N.Z.* p. xv, The upper mountain limits were covered with alpine vegetation—mountain daisies (Celmisia), [etc.]. 1840 E. Newman *Hist. Brit. Ferns* 47 (*heading*) Mountain fern. 1879–81 J. Britten *European Ferns* 151 The Mountain Fern, as *L[astrea] Oreopteris* is sometimes called,..is well worthy of cultivation. 1960 P. Taylor *Brit. Ferns & Mosses* 155 The Mountain Fern is widely distributed throughout the northern hemisphere. 1884 C. S. Sargent *Rep. Forests N. Amer.* 572 The timber on these ridges [in Idaho] was often small and scattered..with larch and red fir, balsam, hemlock, and sometimes mountain hemlock. 1969 T. H. Everett *Living Trees of World* 64/2 Another western American, the usually bluish-leaved mountain hemlock.., occurs at high altitudes from Alaska to California. 1974 *Daily Colonist* (Victoria, B.C.) 24 Dec. 32/2 The major reason for establishing the reserve..is to allow biologists to study mountain hemlock. 1810 F. A. Michaux *Hist. Arbres Forestiers de l'Amérique Septentrionale* I. 26 Sweet birch, [ou] Mountain Mahogany dans Virginia. 1832 D. J. Browne *Sylva Amer.* 118 Wherever it grows in the United States, it is known by the name of Black Birch: its secondary denominations are Mountain Mahogany in Virginia, [etc.]. 1875 *Amer. Naturalist* IX. 201 Much more attractive with its glossy foliage and long feathery seeds, is the mountain mahogany *Cercocarpus ledifolius* Nutt. 1951 *Dict. Gardening* (R. Hort. Soc.) I. 440/1 *C[ercocarpus] ledifolius.* Mountain mahogany. Erect slender tree up to 40 ft. 1785 H. Marshall *Arbustrum Amer.* 2 *Acer pennsylvanicum*, Pennsylvanian Dwarf Mountain Maple,..grows naturally upon the mountains in the back parts of Pennsylvania. 1832 D. J. Browne *Sylva Amer.* 102 The mountain maple seldom rises above 20 feet in height. 1969 T. H. Everett *Living Trees of World* 222/2 Similar in size [to the moosewood] and also favoring shaded locations through much of eastern and central North America is the mountain maple (*A. spicatum*). 1933 L. G. D. Acland in *Press* (Christchurch, N.Z.) 4 Nov. 15/7 Mountain... Often used..of a species found in the back country, mostly smaller than their down-country relatives, e.g. mountain pine. 1963 Poole & Adams *Trees & Shrubs N.Z.* 26 *D[acrydium] bidwillii*... Bog pine, mountain pine. Reaching 4 m. Leaves: juvenile linear, sessile, passing abruptly into adult foliage with scale-like, imbricate leaves. 1850 L. Sawyer *Diary* 12 June in *Way Sk.* (1926) 53 We found some grass and some beautiful specimens of what we called mountain pink. It is much smaller than our garden pink, but resembles it somewhat in form, but more in its sweet perfume. 1936 *Discovery* Feb. 46/2 There [sc. on a mountain in Yugoslavia] I found tall mountain pinks of deep crimson, single-flowering. [1765 in D. Campbell *Hist. Prince Edward Island* (1875) 6 The Mountain Shrub and Maiden Hair are also pretty common, of whose leaves and berries the Acadian settlers frequently make a kind of tea.] 1785 H. Marshall *Arbustrum Amer.* 53 *Gaultheria procumbens.* Canadian Gaultheria, or Mountain Tea... The leaves have been used as a substitute for Bohea Tea, whence the name of Mountain Tea. 1804 [see *grouse-berry]. 1830 *Trans. Lit. & Hist. Soc. Quebec* (1837) III. 96 Mountain tea [is] a very small evergreen half-shrubby plant, with strong, shining, leathery leaves. 1832 W. D. Williamson *Hist. State Maine* I. 121 This 'mountain tea' promotes mammillary secretions. 1858 [see Tea *sb.* 6]. 1886 *Harper's Mag.* June 62/1 Another beverage is 'mountain tea' which is made from the sweet scented golden-rod and from winter-green. 1891 M. E. Ryan *Pagan of Alleghanies* 65 They reached the level above the cliff, the level carpeted with mountain-tea and rabbit-berries. 1941 J. Stuart *Men of Mountains* 187, I would love to get out with you and get mountain tea from the knolls. 1964 H. D. Wilson *Tales from Barrett's Landing* 44 To supplement our meals..we ate berries, mountain teas (a leaf that tastes like wintergreen), and seaweed.

mountain ash. 2. b. (Earlier and later examples.)
1884 A. Nilson *Timber Trees New South Wales* 74 *E[ucalyptus] virgata.*—Mountain ash; White-top.—A tree of considerable size. *Ibid.*, *E. micrococca.*—Mountain Ash... Habitat, brush forests, from Illawarra to the Richmond River and New England; also Blue Mountains. 1934 [see *blackbutt]. 1957 *N.Z. Timber Jrnl.* Dec. 59/1 Mountain ash. *Eucalyptus regnans* F. Muell. and *E. delegatensis* R.T.B.... Victoria and Tasmania. Also called Tasmanian oak and Australian oak. Moderately hard, heavy, durable, strong, elastic and resilient.

mountain flax. Add: **1. f.** *N.Z. Phormium colensoi*, a smaller and hardier form of the New Zealand flax, *P. tenax*.
1867 E. Sauter tr. *F. von Hochstetter's New Zealand* vii. 152 We may distinguish about three principal varieties..*Wharariki*, mountain flax, with coarse fibres; little used. 1949 P. Buck *Coming of Maori* (1950) II. v. 167 Relief in decoration was obtained by spacing tags of the mountain flax (*Phormium colensoi*) which turn a distinct yellow in colour.

mountain laurel. = *kalmia.
1759 [in Dict. s.v. Mountain 9 d]. 1810 F. A. Michaux *Hist. Arbres Forestiers de l'Amérique Septentrionale* I. 35 Mountain laurel,..dénomination la plus générale. 1832 D. J. Browne *Sylva Amer.* 191 The Mountain Laurel is a large shrub, which indifferently bears the name Mountain Laurel, Laurel, Ivy, and Calico Tree. 1845 S. Judd *Margaret* I. ii. 8 She got running mosses..and mountain laurel blossoms. 1880 *Harper's Mag.* June 80 The mountain laurel, with its deep green foliage and showy clusters peers above that rocky crag. 1887 [in Dict. s.v. Mountain 9 d]. 1906 J. A. Harrison *George Washington* 91 The bluish-pink masses of the mountain-laurel. 1955 [see *kalmia]. 1973 Robichaud & Buell *Vegetation of New Jersey* 329 Kalmia..latifolia. Mountain laurel.

mountain rescue. Used *attrib.* and *absol.* to designate an organization for rescuing mountaineers (sense 3) in distress.
1956 F. T. K. Bullmore *Dark Haven* xxiv. 182, I think the best thing I can do is to have all leaders of Mountain Rescue Units up to London for a round-table talk. *Ibid.*, I couldn't help thinking of the many..difficulties that these Mountain Rescue Squads would have to face. *Ibid.* xxvi. 189 Roxborough had been making great progress with the Mountain Rescue Organization. To help him with the technical difficulties of mountaineering he had unearthed..an Austrian mountain guide. 1957 R. G. Collomb *Dict. Mountaineering* 108 Mountain districts in Britain and the Alps have their own mountain rescue posts... Mountain rescue teams are local organizations made up of people working in the district. 1959 'G. Carr' *Swing Away, Climber* iv. 74 Telephone Higgs at the Pen-y-gwryd for the mountain rescue stretcher. 1971 O. Norton *Corpse-Bird Cries* v. 85 There was a mountain rescue van there. 1973 C. Bonington *Next Horizon* i. 26 He..had spent some years with the Royal Air Force in Mountain Rescue, and was a good steady goer.

mountain sheep. 1. [Mountain 8 b.] Sheep kept in mountainous regions. Cf. Sheep *sb.* 1 b.
1829 T. L. Peacock *Misfortunes Elphin* xi. 141 The mountain sheep are sweeter, But the valley sheep are fatter. 1910 W. de la Mare *Three Mulla-Mulgars* 17 The little coat of mountain-sheep's wool. 1961 J. Gunston *Profit from Sheep* i. 11 Great experience of local conditions and practice is the only real guide to the proper management of mountain sheep. 1975 *Country Life* 9 Oct. 942/1 The Castlemartin range..is..a custom-built wintering ground for mountain sheep.

2. *U.S.* [Mountain 9 c.] Either of two species of North American wild sheep, the bighorn, *Ovis canadensis*, or the Dall sheep, *Ovis dalli*.
1802 in *Med. Repository 1803* 240 The mountain ram, or sheep, though not very often seen, is to be met with, in considerable numbers, in some parts of the mountains. 1807 P. Gass *Jrnl.* vii. 82 On the top of the highest [bluff] we saw some Mountain sheep, which the natives say are common about the Rocky mountains. 1837 W. Irving *Capt. Bonneville* I. iii. 69 This animal is commonly called the mountain sheep, and is often confounded with another animal, the 'woolly sheep' found more to the northward. 1841 G. Catlin *Lett. on N. Amer. Indians* II. 188 Dressed in a beautiful costume of the mountain-sheep skin. 1846 R. B. Sage *Scenes Rocky Mts.* xiv. 121 In size the mountain sheep is larger than the domestic animal of that name, and its general appearance is in every respect dissimilar—excepting the head and horns. 1918 T. Roosevelt in *Maine my State* (Maine Writers Research Club) (1919) 21 We had a couple of antelope and a yearling mountain sheep. 1947 *Trail & Timberline* Feb. 15/2 We saw moose, bear, caribou, mountain (Dall) sheep, and wolves almost every trip.

mountain side. Add: Also with hyphen or as one word. (Further examples.) Also *fig.*
1910 W. de la Mare *Three Mulla-Mulgars* xiii. 175 The bare, snow-flecked mountain-side. 1913 J. Hulbert in *Granta* 7 Mar. 255/1 No man is helped up the mountain side of toil by undue ragging. 1960 *Farmer & Stockbreeder* 23 Feb. Suppl. 2/1, I was late and began to run, panting up the mountainside with my hand under my heart. *a* 1963 L. MacNeice *Astrol.* (1964) ix. 286 (*caption*) A recent 'end-of-world' forecast was the catastrophe predicted by Indian astrologists in 1962. In Britain, a mountain-side prayer meeting was held to avert this disaster.

mountain snow. 3. (Earlier and later examples.)
1889 'C. E. Craddock' *Despot of Broomsedge Cove* ix. 159 He mechanically noted..how the blooming 'Mountain snow' brushed his mare's fine coat. 1913 Britton & Brown *Illustr. Flora Northern U.S.* (ed. 2) II. 469 *Euphorbia marginata*... In dry soil, Minnesota to Colorado, south to Texas. Introduced into waste places in the Central and Atlantic States. Snow-on-the-mountain... Mountain-snow.

mounted, *a.* Add: **3.** *mounted branch, infantry* (earlier examples), *police.*
1792 *Deb. Congress U.S.* (1849) 18 Sept. 1134 General Harman..detached..the mounted infantry. 1834 R. M. Martin *Hist. Brit. Colonies* I. iii. 147 There is a mounted police officered by [Indian] natives. 1858 E. H. D. Domenech *Missionary Adventures Texas & Mexico* 177 The Indians were once nearly taking prisoners a whole company of mounted infantry. 1859 J. Blackwood *Let.* 24 July in Geo. Eliot *Lett.* (1954) III. 121 The animal is.. under charge of..the veterinary of the Mounted Police. 1878 *Saskatchewan Herald* (Battleford) 25 Aug. 1/2 Shoal Lake has been created the headquarters of the Mounted Police for that district. 1935 N. Mitchison *We have been Warned* iv. 452 Six mounted police went by at the trot, scattering the crowd. 1963 *Calgary Herald Mag.* 4 Oct. 48/4 An old Mounted Police stable may..be seen. 1970 P. Laurie *Scotland Yard* ii. 50 A specialist job: the CID.. Mounted Branch.

7. Of a project, exhibition, or radio or television programme: produced, directed, arranged. Cf. *mount *v.* 17 d.
1895 in *Funk's Stand. Dict.* 1966 *Listener* 30 June 945/2 The carefully mounted [broadcast] series for mentally handicapped children. 1971 *Nature* 9 July 74/1 The medical problems of many developing countries,..should be amenable to all kinds of internationally mounted projects.

Mountie (mau·nti). *colloq.* Also **Mounty.** [f. Mount(ed *a.* + -ie, -y[6].] **1.** A member of the Royal Canadian (formerly North West) Mounted Police.
1914 *Eye Opener* (Calgary) 12 Dec. 3/4 Ketchen, the Mountie,..was easily placated. 1924 A. J. Small *Frozen Gold* vi. 139 A sentence that is at once the badge and boast of the Mounted—'the Mounties never come in without their man'. 1927 *Sunday at Home* 106/2 The Eskimo borrowed the Mounty's gun and shot him. 1971 D. Heffron *Nice Fire & Some Moonpennies* i. 12 We all looked all around us as though there might be a Mountie skulking behind every tree. 1973 *Saturday Night* (Toronto) Feb. 22/1 All I could see was the Mounties' legs.

2. A member of a similar police force outside Canada.
1931 *Skipper* 25 Apr. 112 A detachment of the Camel Corps on the march outside Cairo. These desert 'mounties' keep law and order in Egypt. 1953 R. Campbell *Mamba's Precipice* 125, I wonder why such a smart man with a castle in England has to come out here [*sc.* S. Africa] and work as an ordinary Mountie.

mounting, *vbl. sb.* Add: **1.** (Further examples.)
1828 J. Ebers *Seven Yrs. King's Theatre* xii. 331 The mounting of this, the first performance of the season, afforded me an illustration of the obliging disposition of Madame Biagioli. 1846 *Swell's Night Guide* 43 She undertakes the art of mounting, which she teaches with considerable success. 1959 *Times Lit. Suppl.* 16 Jan. 27/2 Twenty per cent. of the current United States military budget is to be allocated to economic aid and industrial development of the underdeveloped countries, including the mounting of a crash programme for the development of compact and readily transportable atomic power plants. 1962 *Listener* 14 June 1016/2 Ministers of Defence connived at the mounting of an abusive legal action which again jeopardized the journal's financial stability. 1974 H. Waugh *Parrish for Defence* (1975) xiii. 200 It aroused him, and his second mounting was in the nature of an unbridled emotion.

b. *mounting board, bracket, point, ring, test.*
1854 Mounting board [in Dict., sense 2 a]. 1926 *Paper Terminol.* (Spalding & Hodge) 18 *Mounting boards* are made up of a wood pulp centre lined on one or both sides with paper. 1937 E. J. Labarre *Dict. Paper* 177/1 *Mounting test*, a method of testing the absorbing power of blotting paper by allowing the ink or water to 'mount' up a strip of blotting, the 'weight test' determining the weight of liquid absorbed. 1944 *R.A.F. Jrnl.* Aug. 292 Strange is the conversation to the uninitiated ear—Mounting rings..—Double engine changes—these are the phrases heard. 1962 *Which? Car Suppl.* Oct. 131/1 The mounting points for the sun visors..were metal and unguarded. *Ibid.* 138/2 The mounting bracket for the accelerator linkage. *Ibid.* 143/1 Safety belt mounting points..are now fitted as standard equipment.

2. a. (Further examples.)
1914 C. F. Tweney *Dict. Naval & Mil. Terms* 157 *Mountings*, a term applied to the platforms on which heavy naval guns and guns of position for fortresses are mounted. 1962 *Which? Car Suppl.* Oct. 138/2 Flexible gear box rear mounting made gear lever vibrate considerably on rough ground. *Ibid.* 139/2 Fuel pump mounting [was] slightly insecure. *Ibid.*, Central mounting of exhaust system failed.

Mountmellick (mauntme·lik). The name of a town in the Irish Republic used *attrib.* to designate a type of white-work embroidery made there.
1893 E. T. Masters *Gentlewoman's Bk. Art Needlework* iii. 77 The material upon which the original Mountmellick embroidery was executed was a stout make of white satin jean, white knitting-cotton of various sizes being used to form the stitches. The designs are usually somewhat naturalistic in style. 1920 A. K. Arthur *Embroidery Bk.* viii. 85 Bullion knots are used very frequently in Mountmellick work. 1936 M. Thomas *Embroidery Bk.* 187 Contrary to most other forms of white work, there are no open or drawn spaces in Mountmellick Embroidery. The stitches are planned to lie on the surface of the material with as little thread as possible beneath, and to provide the sense of 'stitchery in relief'. *Ibid.* 194 The following is a list of stitches, suitable for Mountmellick Embroidery. 1959 *Chambers's Encycl.* V. 156/1 Mountmellick embroidery is an Irish form of white work, using a variety of stitches but no open spaces. 1960 H. Hayward *Antique Coll.* 191/2 *Mount Mellick embroidery*, a type of white-work, first introduced about 1830 at Mountmellick in Queen's County, Ireland, and revived as a local industry in the 1880s. 1962 B. Morris *Victorian Embroidery* ii. 40 An entirely different type of whitework known as 'Mountmellick embroidery' was popular throughout the Victorian period. 1967 E. Short *Embroidery & Fabric Collage* i. 19 *Mountmellick embroidery.* This embroidery relies on padded stitches, knots, etc., giving a raised surface to the embroidery.

Mounty, var. *Mountie.

mourner[1]. **1. d.** (Earlier and later examples.)
1834 F. Lieber *Lett. to Gent. in Germany* xvi. 312 These tents..are divided lengthwise by a bench about a foot high, and called the mourners' bench. 1845 J. J. Hooper *Some Adventures Simon Suggs* x. 121 Having thus deposited his charge among the mourners, he started out, summarily to convert another soul! *Ibid.* 126 'And then,' continued Simon, 'I had to go yonder—pointing to the mourner's seat. 1857 P. Cartwright *Autobiogr.* xxvi. 403 You must go to the Methodist's despised mourner's bench. 1888 [see *anxious *a.* 2]. 1891 *Harper's Mag.* Jan. 214/2 Everybody else war either convicted o' sin, an' at the mourner's bench [etc.]. 1904 *Charlotte* (N. Carolina) *Observer* 27 July 4 In the city police court a motley crowd of prisoners filled the mourners' benches. 1972 *News & Observer* (Raleigh, N. Carolina) 30 Dec. 4/3 The August meeting, the mourner's bench and the amen corner have gone out with stewards who yelled out 'amen' during the sermon.

mournful, *a.* (and *sb.*). Add: **5. Mournful Maria** = *Mournful Mary* (*b*); **Mournful**

Mary *Forces' slang*, (*a*) a siren; (*b*) *spec.* the siren used at Dunkirk during the 1914–18 war. ? *Obs.*
1918 K. MacLeish *Let.* 6 May in R. D. Paine *First Yale Unit* (1925) II. 162 The French have installed about six new, loud 'Mournful Marys'. 1918 in J. McG. Grider *War Birds* (1926) 185 Everybody knows when to take shelter and Mournful Mary, the siren, goes off automatically ten minutes before. 1920 G. S. Maxwell *Motor Launch Patrol* ix. 151 Above all the voice of the siren—the famous Mournful Mary—kept up a moaning obbligato. 1925 Fraser & Gibbons *Soldier & Sailor Words* 160 *Mournful Maria*, a nickname given to the Dunkirk syren, employed to give warning of enemy air attacks and long range shelling. 1927 E. W. Springs *Nocturne Militaire* 77 Soon Mournful Mary, the siren in Dunkirk, sounded the All Clear.

mourning, *vbl. sb.*[1] Add: **5.** *mourning armlet, badge, bonnet, card, -dress, handkerchief, hat, hat-band* (later example), *millinery, note-paper, picture, tie, veil*; **mourning-band** (later examples); **mourning envelope,** a mourning-bordered envelope; **mourning iris,** *Iris susiana* (see quots.); **mourning jewellery,** jewellery decorated with miniature funereal ornaments or pictures; **mourning-piece** (further examples); **mourning-vein,** a vein of mourning granite; **mourning warehouse,** a warehouse selling mourning clothes, etc. (cf. Warehouse *sb.* 1 e).
1966 *Olney Amsden & Sons Ltd. Price List* 29 Mourning armlets. 1968 Mrs. L. B. Johnson *White House Diary* 8 Oct. (1970) 718 A mourning badge worn after Lincoln's assassination. *c* 1874 D. Boucicault in M. R. Booth *Eng. Plays of 19th Cent.* (1969) II. 211, I was wrong to come here at all. I feel like a mourning band on a white hat. 1966 J. Potts *Footsteps on Stairs* (1967) ii. 19 Enid kept her loneliness hidden, while Martin's was there for all the world to see; he wore it like a mourning band. 1897 *Sears, Roebuck Catal.* 302/2 Ladies' Mourning Bonnet..a very handsome Bonnet and Mourning Veil. 1804 M. Wilmot *Let.* 2 Jan. in *Russ. Jrnls.* (1934) 72 A Mourning Card was presented to the Princess. 1939–40 *Army & Navy Stores Catal.* 301 Mourning note papers, cards,..and envelopes. 1840 *Knickerbocker* XVI. 70 A conclusive proof that the mourning-dress is an empty ordinance of Fashion. 1843 Dickens *Christmas Carol* ii. 65 A fair young girl in a mourning-dress. *a* 1922 L. Luck in J. Burnett *Useful Toil* (1974) I. 72, I had my new mourning dress torn from my back, through trying to part them when fighting. 1862 Mourning envelope [see Note-paper]. 1907 *Yesterday's Shopping* (1969) 332/1 Mourning envelopes. 1939–40 Mourning envelope [see *mourning card* above]. 1897 *Sears, Roebuck Catal.* 226/3 Mourning Handkerchiefs... with neat fast black hem-stitched borders. 1907 *Yesterday's Shopping* (1969) 819/2 Mourning Handkerchiefs. With black border. 1896 *T. Eaton & Co. Catal.* Spring & Summer 46/3 Telegraph orders for Mourning Hats continue to be received. 1899 in A. Adburgham *Shops & Shopping* (1964) xxii. 261 Mourning hat bands. 1883 W. Robinson *Eng. Flower Garden* 158/1 *I[ris] susiana* (Mourning I[ris])... The flowers, which are produced in early summer, are very large and densely spotted and striped with dark purple on a grey ground. 1966 M. Price *Iris Bk.* vii. 78 The celebrated silver and black mourning iris.. is easiest. 1895 *Montgomery Ward Catal.* 179 Real onyx and jet mourning jewelry. 1960 H. Hayward *Antique Coll.* 192/1 *Mourning jewellery* became particularly fashionable in the second half of the 18th cent... Earlier mourning jewellery of the 16th and 17th cent...was of a more gloomy kind. 1896 *T. Eaton & Co. Catal.* Spring & Summer 46/3 Mourning Millinery. 1862 Mourning note-paper [see Note-paper]. 1907 *Yesterday's Shopping* (1969) 332/1 Mourning note paper. 1947 T. H. White *Elephant & Kangaroo* (1948) xxi. 170 Two mourning pictures of her father and mother. 1972 *Village Voice* (N.Y.) 1 June 3/4 Crewel work and other kinds of embroidery, mourning pictures, theorems, and stencil work. 1843 *Knickerbocker* XXII. 189 The parlor..was ornamented..among the rest, [with] the indispensable family mourning-piece. 1967 Mrs. L. B. Johnson *White House Diary* 12 Nov. (1970) 588 On the walls are samplers and 'mourning pieces' and quaint American primitive portraits. 1970 *New Yorker* 28 Nov. 158/2 The picture is one of the earliest examples of the mourning piece—a folk genre that originated at the time of Wash.'s death. 1970 B. Knox *Children of Mist* vii. 155 A black mourning tie knotted neatly at his shirt collar. 1897 Mourning veil [see *mourning bonnet* above]. 1872 *Rep. Vermont Board Agric.* I. 662 The other layers most desirable and most valuable are the dark and light mourning veins. *c* 1860 in A. Adburgham *Shops & Shopping* (1964) vi. 65 The London General Mourning Warehouse. 1885 *List of Subscribers, Classified* (United Telephone Co.) (ed. 6) 157 Mourning Warehouses.

mourning, *ppl. a.* **3. mourning dove,** *N. Amer.*, substitute for def.: a blue-grey pigeon, *Zenaidura macroura carolinensis*, distinguished by its plaintive call; (earlier and later examples); **mourning willow,** the weeping willow.
1841 G. Catlin *Lett. on N. Amer. Indians* I. 158 The mourning or turtle-dove..is not to be destroyed or harmed by anyone. 1880 J. M. Farrar *Five Yrs. Minnesota* 166 Mourning-doves fill every wood with their plaintive notes. 1929 M. de la Roche *Whiteoaks* xvi. 203 High in the pines she heard the plaintive notes of a mourning dove. 1968 *Globe & Mail* (Toronto) 13 Jan. 41/2 Once again Ontario gunners are asking for an open season on mourning doves and once again considerations of sentimentality ..will almost certainly defeat them. 1975 A. Dillard *Pilgrim at Tinker Creek* ii. 33, I saw the backyard cedar where the mourning doves roost. 1813 H. Muhlenberg *Catal. Plantarum Americæ Septentrionalis* 91 Mourning Willow.

mouse, *sb.* Add: **2. b.** (Further examples.)
1785 R. Burns *To a Mouse* in *Poems & Songs* (1968) I. 128 The best laid schemes o' Mice an' Men, Gang aft agley. 1937 J. Steinbeck (*title*) Of mice and men. 1938 *Time* 30 May 48/3 Are we mice or men? 1961 *Times* 24 Nov. 15/1 There is a particularly good comic performance by Mr. Denys Graham as the mouse who for one glorious moment believes he is a man. 1975 L. Deighton *Yesterday's Spy* xxii. 169 She gave a mocking laugh. 'What are you?.. A man or a mouse?'

4. e. (See quot.)
1905 *Jrnl. Franklin Inst.* Mar. 185 A fine wire is sometimes drawn through a duct by a conical piece of wood with a thin leather washer filling the duct, and forced ahead by the air pressure at the rear... This piece of wood is termed the 'mouse'.

10. d. *mouse-gnawn* adj.
1921 W. de la Mare *Veil* 1 From crock of bone-dry crusts and mouse-gnawn cheese.

e. *mouse-poor, -quiet* adjs.
1921 R. Graves *Pier-glass* 50 Baffled, aghast with hate, mouse-poor. 1946 —— *Poems 1938–45* 13 And we remain mouse-quiet when they begin Suddenly in their unpredictable way To weave an allegory of their lives.

f. mouse pox = *ectromelia 2.
1947 F. Fenner in *Austral. Jrnl. Exper. Biol. & Med. Sci.* XXV. 334 Marchal originally proposed the name 'infectious ectromelia' on account of the frequency with which amputation of the extremities occurred. Few subsequent observers have found this at all common...In view of this and of the newly found close relationship of the disease to the mammalian pox diseases, Professor F. M. Burnet has suggested that 'mouse pox' should be used as a synonym for 'infectious ectromelia'. 1948 *Brit. Jrnl. Exper. Path.* XXIX. 77 Mice dying of acute mouse-pox do not shed much virus into the environment. 1955 F. M. Burnet *Princ. Animal Virology* vii. 157 The content of elementary bodies was very clearly shown in ultraviolet micrographs of the..mouse-pox inclusion body. 1970 S. M. Brooks *World of Viruses* v. 47 The so-called variola-like poxviruses cause smallpox,..mousepox,..and turkeypox.

mouse-deer. Substitute for def.: A chevrotain, a small deer-like mammal of the genus *Tragulus*, found in southern Asia, Sumatra, Borneo, and Java. (Later examples.)
1914 *Oxf. Survey Brit. Empire* II. xii. 383 The Indian pig, the mouse-deer..are found throughout the territories. 1965 R. McKie *Company of Animals* x. 153 Joe's mouse-deer are about nine inches high. 1969 Ld. Medway *Wild Mammals Malaya* 106/1 Mouse-deer are true forest mammals.

mouse-eared, *a.* Add: (*c*) *mouse-eared bat*, a brown bat, *Myotis myotis*, with greyish-white underside, found in Europe, parts of western Asia, and, rarely, in Britain.
1910 G. E. H. Barrett-Hamilton *Hist. Brit. Mammals* I. 191 The Mouse-eared bat, should it occur again in Britain, can hardly be mistaken for any other. 1956 *Proc. Zool. Soc.* CXXVII. 201 The animal was identified as a Mouse-eared Bat *Myotis myotis*..and then photographed *in situ* before being touched. 1975 *Observer* 26 Jan. 1/8 Mr Peter Hardy..said mouse-eared bats survived in Britain in only a single colony.

mouse-hole. Add: Hence **mou·sehole** *v. trans.* and *intr.*, to make a narrow passage or tunnel (through something); **mou·seholed** *ppl. a.*
1950 O. Nash *Family Reunion* (1951) 40 Little mouse Blink strategically mouseholed. 1967 L. Deighton *Expensive Place* xx. 137 It was another half-hour before they had broken into the cellars..and then it took twenty minutes more to mousehole through into Datt's house. *Ibid.* 138 Loiseau's men were moving up from the mouse-holed cellars.

mouse-hunt[2]. Delete *rare*[−0] and add later example.
1975 *Country Life* 27 Mar. 761/1 Our adopted cat..no longer seems to want to be off on a nocturnal mousehunt.

mouser. 1. (Later examples.)
1921 T. S. Eliot *Let.* *c* 11 Oct. in *Waste Land Drafts* (1971) p. xxi, Would you be able to house a small cat which we are very fond of?.. It is a very good mouser. 1939 —— *Old Possum's Pract. Cats* 38 No commonplace mousers have such well-cut trousers. 1966 A. Christie *Third Girl* xvii. 178 You're so exactly like a good mouser. A cat sitting over a hole waiting for the mouse to come out.

mousery (mau·səri). [f. Mous(e *sb.* + -ery.] **a.** A place where mice abound; a colony of mice.
1888 F. A. Lucas in *Auk* V. 280 The occasional disturbance of this populous mousery by the visits of Owls. 1925 W. de la Mare *Broomsticks* 109 He never paid the smallest attention to mouse or mouse-hole or mousery.

b. A place where mice are bred or kept.
1935 A. Carrol *Man the Unknown* vi. 207 The mice belonging to one of the strains kept in the mousery of the Rockefeller Institute died of pneumonia in the proportion of fifty-two per cent while subjected to the standard diet. 1946 *All Pets Mag.* Sept. 64/3, I will try and give you a picture of my mousery. I use twelve..breeding units [etc.]. 1949 *Amer. Small Stock Farmer* Sept. 10 Weltytown Mousery. Breeders of white mice for laboratory and research.

mousetrap, *sb.* Add: **1. b.** (Later examples.)
1945 R. Hargreaves *Enemy at Gate* 84 Wounded at the 'mouse-trap' of Dettingen, he was present at Fontenoy. 1966 M. R. D. Foot *SOE in France* viii. 203 All the agents taken in the Villa des Bois mousetrap at Marseilles..were in the noisome Béleyme prison at Périgueux.

d. In various *transf.* and *fig.* uses.
1903 *Dialect Notes* II. 342 *Mouse-trap*, an implement 'for cutting and fishing out rope when matted in the well. It will also take out small pieces of iron or steel, or any small object'. 1929 *Papers Mich. Acad. Sci., Arts & Lett.* X. 309 *Mouse-trap*, a plant at Willoughby, near Cleveland, built for the manufacture of Lewisite. It was so called by the workmen because every one who entered it did so under an agreement not to leave until the end of the war. 1941 *Reader's Digest* Feb. 54 The boys of Britain's R.A.F. have developed a language all their own... 'roller skates' are tanks, and 'mousetraps' submarines. 1953 Rowland & Boyd *U.S. Navy Bureau of Ordnance in World War II* 137 The NDRC project thus became an attempt to combine the respective virtues of the Hedgehog and rockets. By the spring of 1942 the solution was at hand in the Mousetrap, an antisubmarine projector that launched a salvo of sixteen 7".2 rockets ahead of the attacking ship. 1957 *Amer. Speech* XXXII. 194 *Mouse trap*, any device of an unorthodox nature attached to a handgun for the purpose of making it more accurate or, less frequently, of making it function more smoothly.

2. Delete † and add: Now used *joc.* to denote inferior or unpalatable cheese. (Later examples.) Also *mousetrap cheese.*
1936 R. Lehmann *Weather in Streets* III. iii. 299 A portion of mousetrap cheese, extra charge threepence. 1947 F. Smythe *Again Switzerland* iii. 53 The tasty local cheese —what a change from the eternal 'mousetrap'! 1958 *Times Rev. Industry* Sept. 104/3 Statistics are silent as to whether the cheese is 'mousetrap'..a processed type or.. has a character of its own. 1960 *News Chron.* 18 Feb. 3/6, I don't even know what kind of cheese they will be serving. But..mousetrap would be the most suitable. 1972 *Times* 2 Sept. 15/7 Salty butter, red 'mousetrap' cheese. 1975 *Observer* (Colour Suppl.) 27 Apr. 10/1 Although sometimes dismissed as 'mousetrap', Cheddar is much the most popular cheese in Britain.

3. *mousetrap gun, mechanism*; **mousetrap cheese** (see sense *2).
1964 H. L. Peterson *Encycl. Firearms* 325/1 Some collectors called Cochran rifles 'mousetrap guns'. 1969 *Guardian* 25 July 4/4 Watson gave him a smoke grenade which had a 'mousetrap' mechanism.

mousetrap (mau·s,træp), *v.* [f. the sb.] *trans.* To entrap, deceive, fool (a person, etc.).
1890 *Cent. Dict.*, *Mouse-trap*, to catch, as a mouse, in a trap; entrap. 1960 Wentworth & Flexner *Dict. Amer. Slang* 346/1 *Mousetrap*, *v.t.* 1. In sports, to feint an opponent out of position. 2. To fool or mislead by false promises; to entice; to cajole. 1961 *Guardian* 13 Feb. 16/7 The whole system..is close to insane. It is this system which has mouse-trapped Secretary McNamara. 1964 Mrs. L. B. Johnson *White House Diary* 24 July (1970) 183 Bryson Rash used the expression that Lyndon had 'mouse-trapped' Goldwater by leaving him nothing to say about his appointment.

mousquetaire. 2. (Earlier examples.)
1860 M. A. & R. H. M. Wallace-Dunlop *How we spent the Autumn; or, Wanderings in Brittany* 223, I was often struck by the number of fashions we seem to have borrowed from Brittany; for instance, mousquetaire sleeves have their origin there. 1872 *Young Englishwoman* Dec. 651/2 Sleeves with a mousquetaire cuff. 1873 *Ibid.* Oct. 493/2 A mousquetaire hat is of grey felt, turned up with maroon velvet. *Ibid.* 494/1 The redingote tunic of grey poplin, with mousquetaire revers. 1881 C. C. Harrison *Woman's Handiwork* III. 203 Sachets vary in size..to the ..very long ones, meant to contain sixteen-buttoned or mousquetaire gloves.

‖ **moussaka** (mū̆sa·kă, mūsākā·). Also **mousaka.** [ult. ad. Egyptian Arab. *musakk'a* through Turk.: cf. mod. Gr. μουσακᾶς, Rum. *musaca*, Alb., Bulg. *musaka*, etc.] A Balkan and eastern Mediterranean dish of minced beef or lamb, aubergines or potatoes, and onions.
1941 H. D. Harrison *Soul of Yugoslavia* 284 Moussaka is another dish which savours of the east. An earthenware dish is filled with alternative layers of minced mutton and slices of potato or egg plant which have been dipped in batter and lightly fried. Over this is poured a batter of eggs and sour cream and the whole is baked slowly for some hours. 1950 E. David *Bk. Mediterranean Food* 33 (*heading*) Moussaka. Mince 1 lb. of beef or lamb very fine. .. Stir the yolk of an egg into..cream or milk and pour it on top of the Moussaka... The egg and the cream form a kind of custard on the top of the Moussaka. 1957 L. Durrell *Bitter Lemons* 215 We lunched..on a deserted beach to the west of Kyrenia,..eating a *moussaka*. 1960 *Sunday Express* 25 Dec. 8/7 Exotic dishes: stuffed *dolmas*, *mousaka*. 1962 *Listener* 2 Aug. 191/2 Ask six people for a recipe for the Balkan dish *mousaka* and you will get six different answers, but they are all variations on the theme of minced meat (generally lamb) layered with vegetables. 1966 *Punch* 3 Aug. 189/3 We were suspicious when she started us automatically with ouzo, and brought three bottles of retzina with the moussaka. 1972 J. Aiken *Butterfly Picnic* i. 19 Moussaka..is a sort of super-shepherds-pie with a cheese omelette on top.

mousse. Substitute for def.: A frothy dish made with a savoury or sweet purée or other base, stiffened with cream, gelatine, or egg whites, and freq. served chilled. (Further examples.)
1906 *Mrs. Beeton's Bk. Househ. Managem.* xxxiii. 989 Parfaits, mousses, and soufflés differ from ordinary ices, inasmuch as the cream preparation is at once moulded and placed on ice. 1908 *Daily Chron.* 22 July 7/4 A mousse is nothing more than the addition of whipped

cream instead of plain cream when making ices. 1948 *Good Housek. Cookery Bk.* 490 (*heading*) A mousse or soufflé. A very light frozen mixture, usually containing stiffly beaten egg whites. 1960 E. David *French Provincial Cooking* 445 Nearly everyone knows and appreciates the old and reliable formula for a chocolate mousse—4 yolks beaten into 4 oz. of melted bitter chocolate, and the 4 whipped whites folded in. 1965 L. Deighton *Action Cook Bk.* 119 In France a mousse is part of an ice cream. 1970 Simon & Howe *Dict. Gastron.* 269/2 As entrées, main dishes for lunch, or a fine addition to a buffet, savoury mousses of puréed meat or fish, poultry, game or vegetables can be served hot or cold. 1972 K. Stewart *Times Cookery Bk.* ii. 22 (*heading*) Cucumber and cream cheese mousse. *Ibid.*, Allow several hours for mousse to chill. 1975 *Times* 16 Apr. 12 The fruit I can use..for hot crumble or cobbler or cold in a mousse.

‖ **mousseux** (mūsø̄), *a.* [Fr.] = Sparkling *ppl. a.*[1] 3 c (often placed after the sb.). Also *fig.* Also as *sb.*, a sparkling wine.
[1789 A. Young *Jrnl.* 7 July in *Travels* (1792) I. 132, I solaced myself with a bottle of excellent *vin mousseux* for 40s.] 1819 H. Busk *Dessert* 25 The sillery champagne, champagne mousseux, Clear as their taper vase escape the view. 1861 Mrs. Beeton *Bk. Househ. Managem.* 889 There are the sparkling wines (*mousseux*), and the still wines (*non-mousseux*). 1906 W. De Morgan *Joseph Vance* xl. 400, I feel like the contents of a bottle, and am very curious to know what will happen when the bottle is uncorked. Perhaps I shall be *mousseux*—who knows? 1951 R. Postgate *Plain Man's Guide to Wine* v. 92 Not one of the 'mousseux', 'méthode champenoise' or 'gazifiés' possesses the clean, hard flinty taste of champagne. 1968 L. Deighton *Continental Dossier* 29 At St. Péray they make mousseux wine—and say they have made it as long as they have in Champagne. 1973 P. Audemars *Delicate Dust of Death* vii. 95 This is a local *mousseux*... *Mousseux* is drunk after the grapes are gathered—in October and November.

moustache. Add: **6. moustache-lifter,** a device for lifting one's moustache when drinking, sleeping, etc.
1906 *Trans. Asiatic Soc. Japan* XXXIV. II. 122 (*caption*) Ainu moustache lifters. *a* 1930 D. H. Lawrence *Phoenix II* (1968) 261 Before the war, in Germany I used to see advertised in the newspapers a moustache-lifter, which you tied on at night and it would make your moustache stay turned up. 1966 J. S. Cox *Illustr. Dict. Hairdressing* 100/2 *Moustache lifter*, an implement shaped like a paper knife and used by the male hairy ainu of Japan when drinking, to lift his moustache away from the liquid.

Mousterian, Moustierian (mūstiə·riăn), *a.* and *sb.* [ad. F. *moustiérien* (G. de Mortillet *a* 1873, in *Classification des diverses périodes de l'âge de la Pierre* 4), f. *Moustier* (see below) + -ian.] **A.** *adj.* Of or pertaining to the people, culture, and tools, esp. the flint industries, typified by remains found in the Moustier cave in the Dordogne region of France, and properly attributed to the Neanderthal peoples living in Europe and round the Mediterranean; of the Middle Palæolithic period (*c* 70,000–30,000 B.C.) during which these tools were made. Also *absol.*, this culture. **B.** *sb.* A Mousterian man or woman.
1890 T. Wilson in *Rep. U.S. Nat. Museum 1888* 614 The Mousterian implements have been found in the river gravels of Europe. 1896 A. H. Keane *Ethnol.* 86 Moustierian or First Cave Age. *Ibid.* 90 K. Moustierian bed, with typical pointed flint. 1907 *Ann. Rep. Board of Regents Smithsonian Inst. 1906* 374 Glacial period: Mousterian culture. 1912 *Edin. Rev.* Apr. 366 The first discovery of Mousterian man to excite interest was made in 1856 near Düsseldorf. 1912 R. R. Marett *Anthropol.* ii. 45 Those were the days of the Mousterians who dined off woolly rhinoceros in Jersey. 1927 Peake & Fleure *Hunters & Artists* iv. 39 Breuil thinks that the Neanderthal men of Mousterian culture lived in western Europe from the time of the Riss on through the Würm glaciation. 1927 H. G. Wells *Short Stories* 679 These Mousterians are also called Neandertalers. 1928, etc. [see *Aterian *a.*]. 1947 J. & C. Hawkes *Prehist. Brit.* (rev. ed.) i. 14 One of the products of the stimulating interaction between the older traditions was a new flake culture, the Mousterian, which enjoyed its heyday during the final glaciation of the Ice Age. 1949 M. E. Boyle tr. *Breuil's Beyond Bounds of Hist.* 47 A party of courageous hunters is attacking a group of Mammoth with lances and axes of the type known as Mousterian. 1970 J. D. Clark *Prehist. Afr.* iv. 116 Some forty-five feet of stratified occupation waste and cave earth, showing a long sequence of developing Mousterian overlying an industry made on long blades and blade-like flakes. 1974 *Sci. Amer.* June 101/1 The artifacts ..have been assigned to two successive cultural units. The earlier of the two is known as the Mousterian. *Ibid.* 101/2 The Mousterian stone tools from various Ukrainian sites. *Ibid.* 104/3 Like the Mousterians, Upper Paleolithic peoples often buried their dead.

Hence **Mouste·roid, Mou·steroid** *adjs.*, resembling the Mousterian culture or tools.
1946 F. E. Zeuner *Dating Past* ix. 287 Quite probably, many 'mousteroid' industries mentioned in literature are of the same type. 1950 *Proc. Prehist. Soc.* XVI. 176 Among a handful of mousteroid specimens from the cave of Bacho Kiro in Bulgaria a typical specimen of a miniature plano-convex point may be noted. 1969 C. S. Chard *Man in Prehistory* xi. 114 Since it seems desirable to have a broad term..to designate the flake industries of the neanderthaloid phase in the stretch of contiguous territory from western Europe to Mongolia and south to the Near East and North Africa, which all share certain common features..while retaining their individuality, the label 'Mousteroid' would probably be acceptable.

‖ **Moustiers** (mūstie). The name of a small French town, *Moustiers*-Sainte-Marie, in the Basses Alpes, used *attrib.* or *absol.* to designate a type of faience made there.
1863 W. Chaffers *Marks Pott. & Porc.* 96 *Moustiers*... This beautiful ware much resembles porcelain and the early pieces are generally painted in monochrome, in blue, brown, or green; the later specimens are usually in more than one colour. 1870 C. Schreiber *Jrnl.* (1911) I. 99 A very fine Moustier Ware dish. 1872 *Ibid.* 162 The specimens were..not equal to fine Moustiers or Marseilles. 1900 F. Litchfield *Pott. & Porc.* vii. 214 Barber's dish of Moustiers faience. 1962 *House & Garden* Dec. 58/2 We've bought..a French Moustiers blue and white dish. 1970 G. Savage *Dict. Antiques* 281/2 (*caption*) *Moustiers.* Dish painted in high-temperature colours with a pastoral scene within rococo scroll borders. Moustiers faience..*c* 1740.

mousy, *a.* Add: **1.** (Earlier and later examples.)
1853 Mrs. Gaskell *Cranford* ix. 164, I was..most particularly anxious to prevent her from disfiguring her small gentle mousey face with a great Saracen's-head turban. 1936 K. A. Porter *Flowering Judas* 107 He could not bear hearing Miriam called a mousy little nit-wit. 1959 W. Golding *Free Fall* iv. 82 Fair heads and mousy ones. 1959 'O. Mills' *Stairway to Murder* vi. 61 Her skin was sallow, her hair was 'mousey', and her forehead was the lowest Charles had ever seen. 1975 *Times* 15 Feb. 14/2 Hair which is 'light brown' sounds more becoming than hair which is 'mousy'.

4. *mousy-eyed* adj.; *mousy-quiet* (later example).
1909 M. B. Saunders *Litany Lane* I. ii. 13 Only a fold of dark chestnut hair and a hint of red in the lip gave colour—otherwise a little mousy-eyed *gamin* of a thing. 1958 *Observer* 3 Aug. 10/6 Young married business man, cleared by mousey quiet private detective.

moutan (mū·tăn). [Chinese.] The tree peony, *Pæonia suffruticosa*, of the family Ranunculaceæ, a large shrub bearing pale pink flowers, native to China and Tibet, the parent of many garden varieties producing single or double flowers of many colours.
1808 *Curtis's Bot. Mag.* XXIX. 1154 The Moutan, though cultivated in China about fourteen hundred years, is considered in that ancient empire..as rather of modern introduction. 1880 *Encycl. Brit.* XII. 258/1 The Moutans or Tree Pæonies are remarkable for their sub-shrubby habit, forming vigorous plants sometimes attaining a height of 6 to 8 feet. 1963 M. Haworth-Booth *Moutan* i. 11 In Chinese art each month is represented by a flower, and Moutan is specifically the flower for March.

mouth, *sb.* Add: **3. m.** Also, *to put on* (or *up*) *a poor mouth.*
1892 'Mark Twain' *Amer. Claimant* iii. 37 Any selfish tramp..can come and put up a poor mouth. 1949 F. Urquhart *Ferret was Abraham's Daughter* I. vii. 26 'Charity!' Bert shouted. 'Aye puttin' on a poor mouth.'

q. *plum-in-the-mouth* adj. phr.: see *plum *sb.*

4. b. *to shoot off one's mouth*: see Shoot *v.* 23 g in Dict. and Suppl.

20. a. *mouth-aperture, -articulation, -cavity, -gesture, -gymnastics, -heat, -passage, -rim, -sound*; *mouth-honour* (later example); **b.** *mouth-opening* adj.; **d.** *mouth-shrivelled* adj.; **e.** *mouth-blown, -formed* adjs.
1953 K. Jackson *Lang. & Hist. Early Brit.* 573 The degree of mouth-aperture. 1934 J. J. Hogan *Outl. Eng. Philol.* 6 The nasal consonants *n* and *m* are stops in their mouth-articulation, opens in their nasality. 1930 R. Paget *Human Speech* 231 A separate mouth-blown vowel or consonant resonator. 1924 R. M. Ogden tr. *Koffka's Growth of Mind* iii. 132 The most primitive phenomena are figural; as examples..the too cold or too warm milk in contrast with the temperature level of the mouth-cavity. 1964 C. Barber *Ling. Change Present-Day Eng.* iii. 38 A vowel is a voiced sound..in which there is a free flow of air out of the mouth, without any audible friction or any obstruction of the mouth-cavity. 1930 R. Paget *Human Speech* 111 A mouth-formed whistle. 1930 —— *Babel* 60 Making the same mouth-gesture. 1921 H. E. Palmer *Princ. Lang.-Stud.* 89 We must go through a course of mouth-gymnastics. 1942 W. Faulkner *Go Down, Moses* 103 Not even warmed from mouth-heat. 1907 G. B. Shaw *Major Barbara* Pref. 157 The mouth-honor paid to poverty and obedience by rich and insubordinate do-nothings. 1960 *Guardian* 3 May 7/7 A truly mouth-opening recording of Handel's 'Messiah'. 1964 C. Barber *Ling. Change Present-Day Eng.* iii. 47 This [l] is a *lateral* consonant: to produce it, you press the tip of the tongue against the teeth-ridge, thus blocking the centre of the mouth-passage. 1933 *Burlington Mag.* June 265/1 It was customary to bind the mouth-rims of bowls and dishes with metal. 1925 Blunden *Eng. Poems* 89 That old man, face like parchment tanned, Wrinkled, mouth-shrivelled. 1929 W. Faulkner *Sartoris* II. i. 74 All the other mouth-sounds that stood for repose.

21. mouth-board, a wooden instrument to which the mouth is applied, in order to secure a constant position of the head for observation or experiments; **mouth-breather,** a person who breathes through the mouth; **mouth-breeder,** a fish of the families Cichlidæ or Ariidæ which protects its eggs, and sometimes its newly-hatched offspring, by carrying them in its mouth; so **mouth-breeding** *ppl. a.*; **mouth-harp** = Mouth-organ 1; **mouth music,** (*a*) = *mouth-harp*; (*b*) singing without distinct utterance of words; **mouthparts** *Ent.*, the organs surrounding the mouth of an insect or other arthropod, specially adapted to the particular method of feeding of the animal concerned; **mouth root** (earlier and later examples); **mouth rot,** an oral canker sometimes affecting snakes in captivity; **mouth-to-mouth** *a.*, involving the contact of one individual's mouth with another's; *spec.* applied to a method of artificial respiration in which a person places his mouth tightly over the patient's and blows into him every few seconds so as to inflate his lungs; also *absol.*; (for *fig.* sense see 3 j); similarly **mouth-to-nose** *a.*; **mouth wash,** (*a*) (later examples); also *transf.*; (*b*) nonsense, twaddle; **mouth-watering** *a.* (later examples); **mouth-way,** entrance.
1901 E. B. Titchener *Exper. Psychol.* I. II. 245 Materials.—Head-rest, with mouth-board and sighting mark. 1910 *Practitioner* Jan. 69 The child was a mouth-breather and showed signs of adenoids. 1927 *Sunday at Home* June 239/1 The mouth-breeder protects her eggs by carrying them about in her mouth. 1962 K. F. Lagler et al. *Ichthyol.* x. 298 (*caption*) Mouthbreeding catfish. *Ibid.* 301 Some African fishes, *Tilapia*, are called 'mouth-breeders' because the young when they are hatched escape at time of danger into the oral cavity of the female. 1903 Ade *In Babel* 40 I'd walked from Loueyville over to Terry Hut with a nigger that played the mouth-harp. 1968 *Publ. Amer. Dial. Soc.* XLIX. 15 The lack of familiarity with the musical instrument probably accounts for the decline of *juice harp* and *jew's harp* and the student's use of *mouth harp.* 1972 *Village Voice* (N.Y.) 1 June 76/5 She takes out a mouth harp, fills her cheeks with wind, and blows 'the ballad of the shadows'. 1887 *Lantern* (New Orleans) 3 Sept. 3/2 The music was furnished by a kid with a mouth music. 1936 C. Day Lewis *Hope for Poetry* (Postscript) 93 A new language of purely emotive sounds (e.g. the 'mouth-music' of the Hebridean islanders). 1938 L. MacNeice *I crossed Minch* I. iv. 45 An example of the old 'mouth-music'—Port a Beul—to which the islanders used to dance before they had musical instruments. 1973 Boyd & Parkes *Dark Number* ix. 96 He gave way to a brief rendering of the mouth music of his youth. 1869 A. S. Packard *Guide to Study of Insects* 34 We have already treated of the external appendages (mouth-parts) which prepare the food for digestion. 1905 V. L. Kellogg *Amer. Insects* i. 6 Attached to the thorax are three pairs of legs, which are jointed appendages, homologous in origin and fundamental structure with the mouth-parts and antennæ. 1932 *Times Lit. Suppl.* 17 Mar. 206/2 There are, too, such enigmas as *trophia*, for which we guess mouthparts. 1972 L. E. Chadwick tr. *Linsenmaier's Insects of World* 24/1 The variations of the insect mouth and mouthparts are nearly limitless. Here we find an assemblage of instruments such as only a surgeon, an artisan, or a burglar might wish for—from a harmless sucking proboscis, through implements for boring, sawing, pinching, and cutting, to such devilish contraptions as poison syringes or stilettos that snap out to pierce the prey. 1974 *Sci. Amer.* Apr. 103/1 When feeding, adult beetles busily squeeze pellets of moist dung between their mouthparts and suck in the expressed juice. 1784 Cutler in *Mem. Amer. Acad. Arts & Sci.* (1785) I. 457 Goldenthread. Mouth Root... The roots are astringent, and of a bitterish taste. Chewed in the mouth they cure apthas and cankerous sores. 1931 W. N. Clute *Common Names of Plants* 123 The canker-root (*Coptis trifolia*) or mouth-root, as it is called,..continues to hold its place among medicines for the cure of sore mouth. 1961 C. H. Pope *Giant Snakes* (1962) 195 Another threat to captive snakes..is mouth rot, which is also called canker mouth and osteomyelitis. 1965 R. & D. Morris *Men & Snakes* vii. 145 Snakes are very liable to mouth-rot when their jaws are damaged. 1969 A. Bellairs *Life of Reptiles* II. xii. 516 Among the more important bacterial diseases are the oral canker or 'mouth rot' seen in captive snakes and apparently caused by species of *Pseudomonas* and *Pasteurella*. 1909 *Lancet* 13 Mar. 747/2 By 1782 the Royal Humane Society recommended inflation by bellows in preference to the mouth-to-mouth method. 1932 E. Step *Bees, Wasps, Ants* 32 The first eggs hatch, and the legless little [bee] larvae have to be fed, which is a direct mouth-to-mouth exchange between mother and larva. 1941 *Jrnl. Amer. Med. Assoc.* 6 Dec. 1942/2 Mouth to mouth insufflation has many adherents. It permits an exchange of gases under proper conditions as to temperature, content of moisture and carbon dioxide. 1961 *Sunday Times* 17 Sept. 4/5 Mouth-to-mouth resuscitation, by which a life-saver inflates a victim's lungs with his own breath, was approved officially by the British Red Cross Society last week. 1968 K. Weatherley *Roo Shooter* 113 The bush nurse gave him mouth to mouth and got his heart going again. 1970 S. J. Perelman *Baby, it's Cold Inside* 203 Luckily, before Wemyss had to apply mouth-to-mouth breathing, the three of us recovered. 1954 *New Eng. Jrnl. Med.* 6 May 754 Without the mask a mouth-to-nose technic with a handkerchief 'filter' is effective. 1961 *Times* 6 Dec. 9/4 A life-size inflatable doll for demonstrating mouth-to-mouth and mouth-to-nose artificial respiration. 1962 S. Miles *Underwater Med.* xiv. 224 Those who advocate 'mouth to mouth' claim that better inflation is possible, it is more natural and nasal obstruction may be an obstacle if 'mouth to nose' is used. 1920 Ade *Hand-Made Fables* 5 The Cleaners left nothing behind them in Glass Receptacles except Bluing and Mouth-Wash. 1951 M. McLuhan *Mech. Bride* (1967) 60/2 Mouth washes, gargles..are backed by long-standing national advertising campaigns. 1957 C. S. Lewis *Let.* 2 Sept. (1966) 279 Give your imagination a good mouth-wash by a reading..of the *Odyssey*. 1971 *Mod. Law Rev.* XXXIV. 630 Any suggestion that the principle was also

applied can be dismissed as so much mouth-wash. **1960** *Guardian* 7 Oct. 12/5 A mouth-watering bowl of fruit. **1973** *Country Life* 29 Nov. 1796/2 Mouthwatering grills—fish, meat, vegetables. **1920** A. E. W. Mason *Summons* xii. 121 Crossed the road and disappeared into the mouth-way of an alley.

mouth, *v.* Add: **10.** *trans.* To estimate the age of (a sheep) by examining the teeth. *Austral.* and *N.Z.*

1933 *Bulletin* (Sydney) 6 Sept. 24/1 Graziers buy old ewes without troubling to 'mouth' them. *a* **1948** L. G. D. Acland *Early Canterbury Runs* (1951) 404 A competent shepherd should be able to do anything necessary with sheep—draft, shear, mouth, [etc.]. **1972** P. Newton *Sheep Thief* ix. 74, I found the opportunity to mouth several of those double fork sheep—and one was only a four tooth.

mouthful. Add: **b.** Phr. *to say a mouthful*, to make a striking or important statement; to say something noteworthy. *colloq.* (orig. *U.S.*)

[**1790** *Sessions Papers* Sept. 781/1, I never said a *mouth full of ill against her* in my life.] **1922** C. Sandburg *Slabs of Sunburnt West* 7 You said a mouthful. **1924** A. J. Small *Frozen Gold* i. 14 A fight, he says—and he don't know what a mouthful he's said. **1929** A. Conan Doyle *Maracot Deep* vi. 165 He said a mouthful when he asked her to marry him. **1947** L. Hastings *Dragons are Extra* vii. 143 And that says a mouthful. **1973** Wodehouse *Bachelors Anonymous* xii. 153 'Nice nurse?' 'Ah, there you have said a mouthful, Pickering. I have a Grade A nurse.'

mouthing, *vbl. sb.* (Later examples.)

a **1948** L. Acland *Early Canterbury Runs* (1951) 391 s.v. *Race*, Narrow passage in a sheepyard, for drafting, mouthing, branding, etc. **1950** *N.Z. Jrnl. Agric.* Feb. 122 (*caption*) The photograph shows part of the preparation of a breeding flock for the season, which entails mouthing, drafting for wool, age, and other factors.

mouthpiece. Add: **1. b.** The part of a telephone into which one speaks.

1888 *Encycl. Brit.* XXIII. 131/1 The [Reis telephone] receiver consists of an electromagnet made up of a magnetized coil.., with a stout knitting needle for a core. When in use these two instruments are joined in circuit with a battery.., so that under ordinary circumstances a continuous current is flowing through the line. Suppose a sound is then produced in front of the mouthpiece.., the successive variations in the pressure of the air are communicated to the inside of the box, and cause the membrane to vibrate in unison with the sound. **1907** *Sears, Roebuck Catal.* 204/2 Telephone mouthpieces. Male or female thread. **1926** T. E. Lawrence *Seven Pillars* (1935) i. viii. 71 Abdulla went to the telephone..and transferred the mouthpiece to Storrs. **1941** [see *intercom]. **1955** W. Gaddis *Recognitions* ii. v. 509 He turned quickly into the other telephone booth, dialed and stood bent rigid before the mouthpiece.

3. b. (Further example.)

1897 J. L. Allen *Choir Invisible* xiii. 197 Her face concealed by a black velvet riding-mask kept in place by a silver mouth-piece held between her teeth.

4. b. (Earlier and later examples.)

1857 'Ducange Anglicus' *Vulgar Tongue* 13 *Mouth-piece*,..counsel. **1914** [see *fixer 1]. **1926** E. Wallace *More Educated Evans* iv. 91 Mouthpiece!.. Why, all these so-and-so lawyers hang together. **1931** [see *bottle sb.¹ 1 g (*a*)]. **1960** L. Cooper *Accomplices* iii. iii. 164 I'm an Australian citizen and I know my rights and I want to see my mouthpiece before I say any more. **1974** 'P. B. Yuill' *Hazell plays Solomon* xii. 144 The Abreys would get legal aid. The state would fix them up with a good mouthpiece.

mouthy, *a.* Add: **a.** (Later examples.)

1957 M. Spark *Comforters* v. 102 Before they said good night, Eleanor, slurred and mouthy, declared, 'Now, Laurence, take care of Caroline.' **1963** W. H. Boore *Valley & Shadow* ix. 43 Too free you have been..with your talk... Too mouthy all of you. **1968** *Sun Mag.* (Baltimore) 13 Oct. 19/2, I was kind of a big mouth. I wasn't a bully, just mouthy. **1972** *News & Observer* (Raleigh, N. Carolina) 30 Dec. 4/2 Whatever else they are, kids aren't biggity, mouthy [any more].

c. Of a hound: (see quots.).

1946 M. C. Self *Horseman's Encycl.* 457 A hound that is noisy and a babbler is said to be 'mouthy'. **1968** J. Gordon *Beagle Guide* 173 A hound which babbles or is unnecessarily noisy is said to be mouthy.

mouton. Add: **4.** (See quot. 1950.)

1950 Webster Add., *Mouton*, the skin of sheep, usually of Merino Sheep, processed, sheared, and dyed to resemble beaver or a certain type of seal. **1963** *Retail Trading-Standards Assoc. Bull.* Aug./Sept. 3/1 The following names [for furs] may be used notwithstanding the general rule:—*Animal*: Sheep (skin). *Accepted Trade Name*: Lamb or Mouton.

‖ **mouton enragé** (mūtoṅ aṅraʒe). [Fr., lit. angry sheep.] A normally calm person who has become suddenly enraged or violent.

1932 *Times Lit. Suppl.* 27 Oct. 784/4 Her *mouton enragé* of a discarded adorer. **1955** J. Thomas *No Banners* xii. 101 Many of them were peace-loving idealists who, like *moutons enragés*, had been literally goaded into ridding the earth of the Nazi pestilence. **1965** *Economist* 23 Jan. 307/1 The most unenviable reputation as the *moutons enragés* of this Parliament.

Mouton Rothschild (mū·toṅ rǫ·þs,tʃəild). A type of Médoc produced in the Château Mouton-Rothschild, formerly the Château Mouton d'Armailhacq until purchased by the Rothschild family in 1853. Also *ellipt.* as *Mouton*.

[**1860** C. Redding *French Wines & Vineyards* ix. 82 Messrs. Scott,..and d'Armailhac, at Mouton,..are the largest producers... The famous Bran-Mouton is from this district.] **1877** Trollope *Amer. Senator* II. xv. 156, I think you will find that claret what you like... It's a '57 Mouton. **1888** *Encycl. Brit.* XXIV. 605/1 Second Growths. Château Mouton-Rothschild, Pauillac. **1935** M. Morphy *Recipes of all Nations* 102 (*heading*) Growths: Among the best are Château Mouton-Rothschild. **1958** A. L. Simon *Dict. Wines* 115/2 *Mouton-Rothschild, Château*, the first of the second *Crus Classés* of the Médoc (Pauillac). **1974** L. Deighton *Spy Story* xi. 111 What's he going to do..? Put Mouton Rothschild labels on the Algerian? **1974** *Listener* 18 Apr. 487/1 In July 1973 the French Government saw the light and elevated Chateau Mouton Rothschild to one of the top five wines of the Médoc.

‖ **mouvementé** (mūvmaṅte), *a.* [Fr.] Animated, agitated, bustling, full of variety; *spec.* of music: lively.

1918 L. Strachey *Eminent Victorians* 238 The next three years were the most *mouvementés* of his life. **1929** *Theatre Arts Monthly* Mar. 231/2 The chief figure of her time led a life exceedingly *mouvementé*. **1938** *Oxf. Compan. Mus.* 597/1 *Mouvementé*, 'bustling', animated. **1950** A. L. Rowse *England of Elizabeth* vi. 222 We may say that Tudor society was..more flexible and *mouvementé* than is altogether realised. **1961** B. Fergusson *Watery Maze* viii. 201 He had so far had a highly *mouvementé* war, beginning in a Q-ship in Norwegian waters before being appointed navigating officer to the force assembled early in 1941 for the assault on Rhodes. **1963** G. Battiscombe *John Keble* i. 4 Even today..the district round Oxford can hardly be described as *mouvementé*. **1965** *Guardian* 8 Dec. 8/5 Her life has been unusually mouvementé—and erratic.

mouzhik, var. Moujik, muzhik (in Dict. and Suppl.).

movable, moveable, *a.* and *sb.* Add: **A.** *adj.* **7. b.** *Philol.* Designating a consonant or other element affixed to a word, usu. under determined phonetic conditions.

1933 C. D. Buck *Compar. Gram. Greek & Latin* 160 The *ν* movable in forms like λέγουσι(ν), εἶπε(ν), etc., is an added element which, except for a few examples of dat. pl. -σιν in other dialects, is peculiar to Attic-Ionic. **1951** Sturtevant & Hahn *Compar. Gram. Hittite Lang.* (ed. 2) 66, *s* Movable. **1958** Priebsch & Collinson *German Lang.* (rev. ed.) 70 'Movable' *s* is prefixed in Indo-European to many roots. **1973** A. H. Sommerstein *Sound Pattern Anc. Greek* ii. 40 *Moveable Nu.* This is the name given in traditional studies of Greek to a dental nasal which is inserted at the end of certain words that would otherwise end in a vowel, if the following word begins with a vowel or [h] (that is, with a non-consonantal segment), or if a major pause follows.

8. *movable type*: pieces of metal type, individually cast, usu. with reference to early printing.

[**1732** S. Palmer *Gen. Hist. Printing* 5 With the great space so many Pages of Wood must take up, we shall perceive the necessity of inventing moveable Metal Types.] **1770** P. Luckombe *Conc. Hist. Printing* 2 Those who have asserted that Faust was the first inventor of printing, have given for a reason, that they have never seen any book with Guttemberg's name to it; without considering, that their first essays in printing, both by blocks and moveable types,..were anonymous. **1816** W. Y. Ottley *Inquiry Orig. & Early Hist. Engraving* I. 6 The prodigious number of these characters..renders it impracticable for them [*sc.* the Chinese] to print their books with moveable types. **1859** *Abridgements of Specifications relating to Printing* (Patent Office) 15 Where, when, and by whom printing with moveable types was first practised, it seems impossible to determine with any certainty. **1933** T. S. Barber in W. Atkins *Art & Pract. Printing* IV. xx. 239 One of the difficulties encountered in actual printing with movable types was the fact that at times they moved and became displaced when the printer had no desire they should. **1955** S. H. Steinberg *500 Yrs. Printing* i. 22 The available evidence about the invention of printing with movable types cast from matrices is unfortunately less conclusive than might be wished. **1965** J. Moran *Composition of Reading Matter* i. 11 The basis of making pieces of movable type was the punch, matrix and mould. **1974** *Bertram Rota Ltd. Catal.* No. 192. 2 Now the steady advance of film-setting may be heralding the beginning of the end of printing from moveable type.

movant (mū·vănt). *U.S. Law.* [f. Move *v.* + -ant¹: cf. Movent *a.* b.] One who applies to or petitions a court of law or a judge with the intention of obtaining a ruling in his favour. Also *attrib.* or as *adj.*

1927 *Corpus Juris* XLII. 517/1 A 'renewal' of the motion, so-called, is the remedy usually available to the dissatisfied movant whose motion has been denied. **1949** *Corpus Juris Secundum* LX. 38/1 Relief to Movant... In the absence of a prayer for general relief, the moving party usually is confined to the relief asked for in his motion. **1961** *Words & Phrases State & Federal Courts* XXVII. A. 353/1 Intention was that such statute should be remedial in operation, to permit court to correct errors ..into which it has been led..by the party opposing movant. **1972** *N.Y. Law Jrnl.* 22 Aug. 2/3 Motion is granted and the requested subpoena shall issue. Movant shall furnish a copy of the records produced to the other parties. **1973** *Ibid.* 19 July 11/1 The movant fails to establish either defective service of the notice of petition and petition, upon which the default was granted, or a meritorious defense thereto, since nothing is asserted beyond an expression of dissatisfaction with the award. *Ibid.* 1 Aug. 14/2 It is ordered that this motion by defendant..is denied without prejudice to the right of movant defendant to oppose the application of a general preference made by plaintiffs.

move, *sb.* **6.** For *U.S.* read 'orig. *U.S.*' Now usu. *to get a move on.* (Examples.)

1888 *Troy* (Alabama) *Enquirer* 28 July, Get a move on you. **1893** *Columbus* (Ohio) *Dispatch* 7 July, Now is the time for the mover of dead animals 'to get a move on himself'. **1899** A. H. Quinn *Pennsylvania Stories* 138 Come, get a move on. **1906** E. Dyson *Fact'ry 'Ands* xiv. 183 Get er move on, 'r you'll get ther shoot. **1911** C. E. Mulford *Bar-20 Days* x. 107 Come on! Come on!..Get a move on! Will you hurry up! *Ibid.* xx. 198 But why in Jericho don't you fellers get a move on you? **1914** W. G. Lawrence *Let.* 13 Aug. in T. E. Lawrence *Home Lett.* (1954) 569 In ten days time we should get a move on, but we won't go far. **1920** [see *gas sb.¹]. **1937** D. L. Sayers *Busman's Honeymoon* vii. 149, I only hope they're getting a move on out there. **1973** 'D. Mariner' *Beaufort Dossier* vii. 138 What about getting a move on, then! Get out on the flaming roof an' grab them!

8. *Comb.* **move-man** (see quots.).

1923 J. M. Scott-Maxwell *Costing & Price-fixing* 94 Move-men are the men who move the raw material and manufactured parts from the store to the shop, and move all jobs from one machine to another or one department to another. **1955** *Amer. Speech* XXX. 226 *Move man*, the dispatcher in the [aircraft production] shop who moves parts from one operation to the next.

move, *v.* Add: **1. f.** Also *trans.*, to sell; to cause to be sold.

1938 *Sun* (Baltimore) 20 Sept. 10/1 A drastic tax on chain stores has been defeated in a referendum in California, where the function of the chains in 'moving' citrus-fruit surpluses is now more fully appreciated. **1962** *Guardian* 5 Nov. 3/6 Our displays are moving about 25,000 cans a week. **1971** *Ibid.* 24 Dec. 2/5 The boys..are causing the sales lady a good deal of worry... She simply cannot move Indians this season, even with a wigwam. **1975** *Publishers Weekly* 6 Jan. 54/1 Booksellers should easily be able to move this slender 'autobiography' of Lincoln.

g. *Cricket.* To cause (the ball) to swerve.

1956 N. Cardus *Close of Play* 37 We are supposed to be enlightened by news that Lindwall is 'moving' the ball. **1962** *Times* 24 May 4/2 Platt, however, kept plugging away around the good length mark, moving the ball a little either way and generally looking a thoroughly useful practitioner.

16. g. (Further examples.) So, *to move house*. Also, *to move in on*: to take up residence with (someone), esp. so as to inconvenience or annoy; to attach oneself to, put pressure on, or take control of (a person, project, etc.).

1898 G. B. Shaw *You never can Tell* i. 216, I spent my last sovereign on moving in; and I havnt paid a shilling of rent yet. **1924** R. Macaulay *Orphan Island* ii. 22 Your aunt and Martha and myself have recently moved house. **1941** N. Coward *Australia Visited* vii. 47 God help us when the scenery and properties move in on us, when we have to adapt ourselves to new settings and different furniture. **1945** E. Bowen *Demon Lover* 91 Mona moved out..and moved in on Isobel. **1966** *Listener* 6 Jan. 14/2 The society was formed about four years ago... We still have not moved in. **1967** J. Redgate *Killing Season* (1968) ii. xviii. 147 We'll give him a few months to get entrenched in England... Then we'll move in on him. **1968** M. Woodhouse *Rock Baby* xviii. 180 That was the idea, if somebody hadn't moved in on the operation. **1973** 'D. Halliday' *Dolly & Starry Bird* xiv. 205 Sophia deciding to move in on somebody was a sight worth selling tickets for. **1974** R. Rendell *Face of Trespass* 10 If you ever feel like moving house to live among your constituents I'll be happy to oblige.

i. *move on*: also (with hyphen) as *attrib. phr.*

1908 E. J. Banfield *Confessions of Beachcomber* ii. i. 246 He was almost knocked down by a move-on sort of shove. **1971** *Ceylon Daily News* 17 Sept. 1/5 The vagrancy laws and the move-on by-law of the Municipality.

k. To go quickly. *colloq.*

1954 *Amer. Speech* XXIX. 100 It can really move. **1959** M. Gilbert *Blood & Judgement* xvi. 164 'What price the law's delays.' 'They can move when they have to.' **1967** 'G. North' *Sgt. Cluff & Day of Reckoning* ii. 23 You'll have to run to catch him:..when he's in that frame of mind, he can move. **1969** J. Fredman *Fourth Agency* vii. 53 The car was rocketing along... We were moving all right. **1973** W. McCarthy *Detail* ii. 77 Immediately the large man..was out of the apartment and through the fire door. It took him little more than two seconds. God, he can move, Ben thought.

l. To dance or play music energetically or with a strong rhythm; to be exciting or dynamic. *colloq.*

1955 *Down Beat* 6 Apr. 15 The only time it does start to move is in the second chorus, with Charlie Shavers. **1958** G. V. Kennard in R. J. Gleason *Jam Session* 176 'It's got to move,' jazzmen say. If it doesn't 'swing', it's not jazz. **1959** [see *drive v. 25 c]. **1968** [see *groove v. 5]. **1969** N. Cohn *AWopBopaLooBop* (1970) ix. 85 Go to a club one week, go back the next and everyone is moving differently.

19. c. *to move with the times*, to be up to date in one's way of living.

1936 Robinson & Browne *How to live in a Flat* 90 We have no hesitation in offering the following hints to those who—wishing to move with the times..—desire to construct a service-flat for their own use. **1960** L. P. Hartley *Facial Justice* xvii. 144 You must try to move with the times. **1973** *Times* 28 Nov. 22/6 If you change anything, you are accused of eroding history. If you change nothing, you are accused of failing to move with the times.

movement, *sb.* Add: **1. e.** The conveying of cattle from one district to another, often prohibited or restricted during an epidemic of cattle disease.

1869 *Act* 32 & 33 *Vict.* c. 70 § 48 Every local authority and the police of every county..shall..do or cause to be done all things..necessary..for securing, as far as may be, the effectual isolation of infected places in respect of the movement of animals and things. **1878** *Act* 41 & 42 *Vict.* c. 74 § 32 Prohibiting or regulating the movement of animals and persons into, in, or out of an infected place or area. **1954** W. R. Wooldridge *Farm Animals* iv. 74 The Contagious Diseases (Animals) Act..gave adequate powers for the control of all movement of cattle. *Ibid.* vi. 173 Much has been written for and against the policy of suppressing the disease [*sc.* foot-and-mouth] by..rigid control of movement within a wide area.

f. The departure or arrival of an aircraft.

1961 P. W. Brooks *Mod. Airliner* vi. 144, 130 aircraft movements (arrivals and departures) on a peak Summer day. **1969** *Sunday Times* (Colour Suppl.) 17 Aug. 19/1 Many of the major airports are suffering from aerial traffic jams, and the 747 will mean only one movement where there might have been two or more. **1971** *Physics Bull.* Nov. 660/2 Night flying also assumes a significant percentage of the total number of movements.

5. c. *Dancing.* (See quot. 1967.)

1949 Shurr & Yocom *Mod. Dance* v. 135 When the count is finally speeded-up, do *not* omit any of the suspended movements. **1949** *Ballet Ann.* iii. 40 Movement after movement was ruined by the broken line. **1967** Chujoy & Manchester *Dance Encycl.* (rev. ed.) 647/2 *Movement*, the fundamental language of dance; a purposeful change of weight or position, as contrasted with motion, which in dance is considered as purely kinetic, i.e. physical, while movement is metakinetic.

6. b. *in the movement* (earlier and later examples).

1894 *World* 5 Sept. 11/2 We have in *The New Woman* a live play, a play which is distinctly in the movement. **1926** C. Sidgwick *Sack & Sugar* xxvi. 299 She..had quite antiquated Victorian ideas of what English people had nowadays if they were in the movement. She had not got beyond shiny chintzes and overmantels.

d. (With capital initial.) A group of English poets in the nineteen-fifties, their characteristics, or their influence. Also *attrib.*

[**1954** *Spectator* 27 Aug. 261/1 Poets of the Fifties... For better or for worse, we are now in the presence of the only considerable movement in English poetry since the Thirties.] *Ibid.* 1 Oct. 399/1 (*heading*) In the Movement. *Ibid.* 399/2 In these columns some weeks ago Mr. Anthony Hartley remarked upon some of the characteristics of this new Movement of the Fifties—its metaphysical wit, its glittering intellectuality, its rich Empsonian ambiguities. *Ibid.* 400/2 Genuflections towards Dr. Leavis and Professor Empson, admiration for people whom the Thirties by-passed, Orwell above all..are indeed signs by which you may recognise the Movement... The Movement, as well as being anti-phoney, is anti-wet. **1958** A. Alvarez in *Internat. Literary Ann.* I. 99, I have been using the past tense because the Movement has now ended and its poets have gone off on their separate ways... The Movement, for all its limitations and negatives, was immensely valuable. **1958** *Times Lit. Suppl.* 17 Jan. 30/4 The poetry-fanciers say that there are two better Movement poets than either Mr Wain or Mr Amis. *Ibid.* 7 Mar. 127/1 The outstanding performer, or even the 'real poet' of the original Movement is Mr. Philip Larkin. **1959** *Ibid.* 27 Feb. 114/2 Around 1953, Mr. Larkin, Mr. Amis, Mr. Wain and the rest of them burst on a startled world—and 'the Movement' restored to English poetry self-control, precision, and clarity. **1961** *Listener* 24 Aug. 268/1 The decade just ended, a decade dominated by a group of university poets usually referred to as the Movement. **1972** *Times Lit. Suppl.* 14 Apr. 411/3 One need only look at the standard Movement and post-Movement poetry of that period. **1972** D. Timms *Philip Larkin* i. 15, I am reminded here of John Wain's description of the Movement as an *avant garde* that was a rear guard.

10. *movement area, -complex, control, -habit, -illusion, -impulse, -melody, order, -response, -sensation, -study, time*; **movement permit,** permission to move cattle from a particular district; **movement restriction,** a ban on the movement of cattle from a district, esp. when cattle disease is present.

1958 *Chambers's Techn. Dict.* 995/2 *Movement area*, that part of an aerodrome reserved for the take-off, landing and movement of aircraft. **1924** R. M. Ogden tr. *Koffka's Growth of Mind* v. 251 Movement-complexes of grasping and touching..are learnt even earlier than walking. **1956** W. A. Heflin *U.S. Air Force Dict.* 334/2 Movement control. **1969** *Times* 6 Jan. 7/7 Secondary effects of movement control and closed markets left a trail of expense and inconvenience. **1972** L. Lamb *Picture Frame* ii. 21 He arranged with movement control to go a different way. **1920** T. P. Nunn *Education* xiii. 169 Recognition-habits of increasing complexity corresponding to the increasingly complex movement-habits of writing. **1894** Creighton & Titchener tr. *Wundt's Human & Animal Psychol.* ix. 137 It has sometimes been thought that the act of will suffices of itself to explain these subjective movement-illusions. **1924** R. M. Ogden tr. *Koffka's Growth of Mind* iv. 156 The sensory impressions and the movement-impulse of the animal under investigation. *Ibid.* v. 259 In this unification a 'movement-melody' composes itself. **1956** W. A. Heflin *U.S. Air Force Dict.* 334/2 Movement order. **1957** P. Kemp *Mine were of Trouble* iii. 51 Your Movement Order will be sent to your Squadron in a few days. **1973** *Radio Times* 8 Nov. 19/2 It's really like the movement order for an operation. You move up to your start line. **1969** *Times* 27 Jan. 10/8 Brucellosis accredited cattle only will be shown at the Surrey county show... Cattle entered there will require only a movement permit. **1954** L. B. Ames et al. *Rorschach Responses in Old Age* xii. 141 Human movement responses become more passive with increasing deterioration of subjects. **1969** *Times* 6 Jan. 7/7 Hogs kept at home, and sometimes in one field because of movement restrictions, fared less well. **1898** G. F. Stout *Man. Psychol.* I. ii. vi. 192 The distinction between position-sensations and movement-sensations is important. **1951** J. M. Fraser *Psychol.* xv. 187 It is fatally easy to make movement-study sound difficult. **1952** *New Biol.* XIII. 58 The time between the beginning of his movement and the moment at which he starts to return back again is called 'The Movement Time'. **1968** R. N. Singer *Motor Learning & Human Performance* iii. 67 Movement time may include reflex or reaction time, or, as it is usually viewed in research literature, the time a particular act takes to be completed after it has been initiated.

mover¹. 5. Delete ?*Obs.* and add to def.: *spec.* a person migrating westwards; a tenant farmer who moves on after exhausting the fertility of a piece of land. (Earlier and additional examples.)

1810 M. Dwight *Journey to Ohio* (1912) 47 We are.. near a tavern which is fill'd with movers & waggoners. **1878** J. H. Beadle *Western Wilds* xx. 327 Reluctantly the 'movers' consented to his remaining for the night. **1913** J. London *Valley of Moon* 434 The 'movers'..lease, clean out and gut a place in several years, and then move on. **1944** W. Blair *Tall Tale America* 100 Instead of selling the seedlings from these nurseries..he'd..give them to the movers—free. **1945** J. L. Marshall *Santa Fe* 230 'Boomers' and 'movers' tried again and again to take up land, filtering down from Kansas and up from Texas along the Santa Fe track.

6. For ?*U.S.* read *N. Amer.*, and add earlier and later examples.

1838, etc. [see *house-mover* s.v. *house sb.¹ 23]. **1894** *Boston Directory* 1944 J. W. Cook & Son,.. movers of pianofortes, furniture, etc. *Ibid.* 1947 T. G. Buckley, piano and furniture mover. **1956** W. H. Whyte *Organization Man* (1957) 286 It's a hell of a day—the kids are crying..and the movers won't be finished till late. **1968** *Globe & Mail* (Toronto) 3 Feb. 3/9 'The honorable gentleman will be glad to know that I will be getting out as soon as I can get a mover,' the Prime Minister replied.

movie (mū·vi). orig. *U.S.* Also † **movy.** [Abbrev. of *moving picture: see -ie, -y⁶.] **a.** = *moving picture; also, a moving-picture show; a cinema; *pl.* (freq. *the movies*), motion pictures as an industry, an art-form, or a form of entertainment; a cinema or a cinema-show.

1912 *Survey* (N.Y.) 20 Jan. 1628 (*heading*) 'Movies' and the law. **1912** J. Sandilands *Western Canad. Dict.* 30 *Movie*, a moving picture show. **1913** F. A. Talbot *Pract. Cinematogr.* iii. 22 Taking the 'movies' is quite as simple as snap-shot photography with a Kodak. **1913** *N.Y. Even. Post* 10 July 5/7 Guiding the wheel-chair through the entrance gate of the outdoor 'movie'. **1913** *Home Chat* 27 Sept. 578/1 The comparatively small towns [in America] have installed 'movies'—as they call them over there—in their schools. **1914** M. Carrington *Memories* i. 69 A night at the 'movies'. **1916** 'B. M. Bower' *Phantom Herd* xi. 193 Say, do I get it right that you're in the movies. **1918** C. Sandburg *Cornhusker* 51 There is drama in that point... Griffith would make a movie of it to fetch sobs. **1918** G. B. Shaw *Pen Portraits* (1932) 36 Is an evening with Ibsen as popular as an evening with Mary Pickford at the movies? **1919** —— *Heartbreak House* i. 26 You frequent picture palaces... Talk like a man, not like a movy. **1925** F. Scott Fitzgerald *Great Gatsby* (1926) vii. 150 Those big movies around Fiftieth Street are cool. **1929** H. G. Wells *King who was King* i. 15 It was possible for some of us to forget the crude, shallow trade 'movies' we had seen. **1938** E. Bowen *Death of Heart* ii. v. 268 Some of the party wished to go to a movie. *Ibid.* vi. 281 He wanted me to cut off with him somewhere last night, after the movies. **1942** *Short Guide Gt. Brit.* (U.S. War Dept.) 13 The British have theaters and movies (which they call 'cinemas') as we do. **1950** T. S. Eliot *Cocktail Party* iii. 148 You must have been living a quiet life! Don't you go to the movies? **1971** *Mod. Law Rev.* XXXIV. 705 There is so much happening in the field of human rights that there is a great temptation to postpone publication until we see how the movie is going to end.

b. *attrib.* and *Comb.*, as *movie actor, actress, ad, advertisement, buff, business, camera, cowboy, fan, film, hero, industry, king, -land, magazine, -maker, -making, -man, medium, picture, play, projector, queen, scenario, script, show, star, studio, test, theatre*; **movie-goer,** one who goes to the cinema; one who frequently attends film-shows; hence **movie-going** *vbl. sb.* and *ppl. a.*; **movie house,** a cinema; **movie palace,** a (palatial) cinema.

1913 *Writer's Mag.* Dec. 264/2 If you want a chance to pay tribute to some 'movie' actor get a copy of the *Ladies World*, and 'full particulars'. **1935** Wodehouse *Luck of Bodkins* xv. 178 The kid points and says: 'Look, mamma. Movie actors!' And the mother says: 'Hush, dear—you don't know what *you* may come to some day.' **1924** 'J. Sutherland' *Circle of Stars* xi. 114 Mountrose is going to marry an American movie-actress. **1949** M. Lowry *Let.* 1 July (1967) 180 My..brother-in-law, who entertains himself..by listening to the chests of movie actresses. **1951** M. McLuhan *Mech. Bride.* (1967) 82/1 In one movie ad the woman says: 'I killed a man for this kiss.' **1918** N. Angell *Political Conditions Allied Success* p. vii, As the movie advertisement of the war play says: 'You can't put up a good fight until your blood boils.' **1973** *Guardian* 30 Mar. 12/6 Imagine the feelings of a movie-buff if he were told he'd have to get along without Bunuel, Bergman, Chabrol, [etc.]. **1916** 'B. M. Bower' *Phantom Herd* v. 71 There's no art for art's sake in the movie business. **1925** H. L. Foster *Trop. Tramp with Tourists* 118 The company's cinema operator had his movie camera set up in one corner. **1934** *Punch* 14 Nov. 536/2 He and another journalist..bought..a couple of small movie-cameras and all the film there was in Australia. **1961** T. Henrot *Belgium* 185 You may bring in..a small movie-camera with fifteen meters of film. **1926** A. Huxley *Let.* 13 May (1969) 269 One mounts a mule and goes off with a movie cowboy down into the gulf. **1941** Auden *New Year Let.* iii. 68 Some Texas where real cowboys seem Lost in a movie-cowboy's dream. **1913** *Outlook* 5 Apr. 784/1 The 'movie' fan lays himself open to overtired eye nerves. **1952** S. Kauffmann *Philanderer* (1953) iv. 59 There were seven pulps, ten comic magazines, three movie-fan magazines, three confession magazines. **1922** *Atlantic Monthly* June 775 Half the movie films seem almost to have been made for the flapper. **1959** Halas & Manvell *Technique Film Animation* 11 A movie film exposed at normal speed can..be said to analyse every second of live action in terms of a given number of successive phases of movement. **1923** T. Lane *What's Wrong with Movies* 49 The author's name ..means absolutely nothing to the bulk of movie-goers. **1953** *Time* 27 July 76/1 There is..a danger that some impressionable movie-goers, unable to make up their minds which of the stars they prefer, may go quietly hysterical. **1973** *Sat. Rev. Society* (U.S.) Mar. 80/1 *Deep Throat* has grossed nearly a million dollars from moviegoers. **1938** I. Barry tr. *Burdèche & Brasillach's Hist. Motion Pict.* 68 Further *Nick Carter* series..taught the public the habit of regular moviegoing. **1946** A. Warren in W. S. Knickerbocker *20th Cent. English* iii. 312 His conclusions..reassure the parents of movie-going children. **1951** 'J. Wyndham' *Day of Triffids* ii. 43 The great movie-going public. **1958** J. Baldwin in W. King *Black Short Story Anthol.* (1972) 284 A lifetime of moviegoing behind her. **1928** *Manch. Guardian Weekly* 5 Oct. 266/4 The movie hero's knack Of dangling from a trembling wire Across a railway track. **1914** *Automobile Topics* XXXIV. 190/1 Selected bits of the picture..are being shown in the local 'movie houses'. **1932** *Times Lit. Suppl.* 22 Dec. 976/1 Sam, ticket-chopper in a movie-house. **1967** 'La Meri' *Spanish Dancing* (ed. 2) 9 Maria Montero in whose group I danced in vaudeville and the movie houses. **1974** *Times* 19 Jan. 10/1 Local movie houses offering major releases. **1928** H. Crane *Let.* 27 Apr. (1965) 325 'Crashing the gate' ..seems to be exclusively applied to the movie industry. **1917** *Vanity Fair* (N.Y.) Dec. 68 Here we are privileged to behold the interior of a movie king's office. **1914** C. L. Hagen in R. Grau *Theatre of Science* p. xxiv, Is it to be wondered at..that authors, actors, and science await the call to 'movie' land? **1928** *Daily Express* 16 Mar. 4/2 There are few people..better qualified to explain the mysteries of movie-land and of the technical side of films. **1929** E. Wilson *I thought of Daisy* ii. 90 It was a movie magazine called *Photo-Life*. **1955** W. Gaddis *Recognitions* i. v. 180 Movie magazines, simply all sex. **1957** N. Frye *Anat. Crit.* 164 The moviemakers find some difficulty in getting anyone over the age of seventeen into their audiences. **1975** *Listener* 20 Mar. 387 Ingmar Bergman is not everyone's favourite movie maker. **1939** F. Scott Fitzgerald *Lett.* (1964) 415 To be plunged immediately into movie-making. **1973** *Listener* 22 Nov. 716/3 Movie-making depends on movie distributors. **1915** *Pearson's Mag.* Jan. 80 My first action..was to ask a movie-man going home with films, to bring me back a blue serge suit. **1916** 'B. M. Bower' *Phantom Herd* xix. 309 The movie-man that runs this show for the Convention. **1948** *Sun* (Baltimore) 1 Dec. 17/4 Because the most hotly disputed of the series came unexpectedly the movie men did not have their cameras directed at second base. **1951** M. McLuhan *Mech. Bride* (1967) p. vi/1 A film expert, speaking of the value of the movie medium. **1917** *Chicago Defender* 22 Dec., Watch for our new movie palace. **1936** A. Huxley *Olive Tree* 40 When I was last at Margate a gigantic new movie palace had just been opened. **1955** W. Gaddis *Recognitions* i. vi. 220 The woman with me led me down a long street, and we came to a movie palace. **1966** L. Cohen *Beautiful Losers* (1970) i. 22 Let.. no naked flashing breasts lure the dirty laundry of our daily lives into the movie palace. **1916** E. V. Lucas *Vermilion Box* ccxiii. 250 I wish a movie picture could be taken of him. **1917** 'W. Wynning' *Princes St. & Other Otago Rhymes* 11 A girl one sees in 'movie' plays. **1936** Chesterton *As I was Saying* xxxi. 191 The fiction on the film, the partisan version in the movie-play. **1959** *Sears, Roebuck Catal.* Spring & Summer 1396/7 Movie Projectors. **1962** M. McLuhan *Gutenberg Galaxy* 124 The reading of print puts the reader in the role of the movie projector. **1927** U. Sinclair *Oil!* xiii. 314 It's cost me eight million dollars to make a movie queen out of this baby. **1917** H. Crane *Let.* 8 Oct. (1965) 10 Mrs. Walton and I are working out movie scenarios. **1964** M. McLuhan *Understanding Media* (1967) i. v. 65 When the movie scenario or picture story was applied to the *idea* article. **1950** M. Lowry *Lett.* (1967) 2/9 Have written..a detailed movie script. **1973** L. Hellman *Pentimento* (1974) 268 The money I got from the movie script of *The Little Foxes*. **1913** *Technical World* Mar. 17/1 This novel feeling of amused satisfaction over a 'movie' show deepened. **1925** A. P. Herbert *Laughing Ann* 33, I ..wish That life was a little like a movie-show. **1974** M. Z. Lewin *Enemies Within* xxv. 107 I'll..tell Janie about the movie shows at the frat. **1919** H. L. Wilson *Ma Pettengill* ii. 39 [They saw] how much they were paying their president..quoted beside some movie star's salary. **1937** H. G. Wells *Brynhild* vi. 78 Two other of the applicants had 'done publicity' for movie stars. **1973** D. Ramsay *Deadly Discretion* 129 Why do you conceal your beautiful eyes..? You are practising to be a movie star? **1914** *Munsey's Mag.* Jan. 735/2 Then he closed the door and advanced into the 'movie' studio. **1926** A. Huxley *Jesting Pilate* 261 Within the movie studio there shone no sun, only the lamps. **1952** M. McCarthy *Groves of Academe* (1953) x. 197 One of Furness's long-tressed Ritas was promised a movie-test. **1915** *Film Flashes* 13 Nov. 1 It's a long lane that has no movie theatre. **1968** *Michelin Guide N.Y. City* 9 Movie Theaters. —Like the theaters, many first-run movie houses are clustered in the Broadway area.

moviedom (mū·vidəm). orig. *U.S.* Also **Moviedom.** [f. *movie + -dom.] = *filmdom* (s.v. *film sb.).

1916 in *Dialect Notes* (1918) V. 7 Methra Morrell, Moviedom's Greatest Actress. **1927** *Daily Express* 28 Nov. 13 Adolphe Zukor, from whose weakening hand the sceptre of Moviedom has been wrested by Joseph M. Schenk. **1947** *N.Y. Times* 27 July ii. 4/3 Even in its depiction of moviedom, Hollywood has seen fit to glamourize itself. **1973** *Lebende Sprachen* XVIII. 38/2 US moviedom —BE/US film industry, BE cinema industry—Film-industrie.

Movieola, var. *Moviola.

Movietone (mū·vitōun). Also **movietone.** [f. *movie + Tone *sb.*] The proprietary name of a system employed in the making of sound films; a film made by this system; also used allusively of the style of presentation of news-reels formerly produced by the Movietone Company. Also *attrib.*

1927 *Daily Express* 27 Aug. 1/3 The 'movietone' is an invention with the same technical basis as the 'phonofilm'. **1927** *Glasgow Herald* 12 Oct. 11/1 The movietone..is a vast improvement on previous talking films. **1927** G. B. Shaw *Platform & Pulpit* (1962) 178, I am an actual real animal; I am not the latest movietone illusion. **1928** *Trade Marks Jrnl* 4 Apr. 534/1 Movietone... Cinematograph transparencies for exhibition. Fox Case Corporation..City..of New York, United States of America; manufacturers. **1928** *Liberty* 11 Aug. 25/2 George Bernard Shaw as he appears in a strip of Movietone film. Note the sound track on the left margin. **1930** *Nature* 19 July 93/1 With such a sound standard at a known distance from the movietone microphone, a record of pitch and intensity of pure notes covering a wide range of frequencies and intensities may be recorded on the film. *Ibid.*, Fig. 1 is a reproduction of a movietone film of an orchestra and the micro-photometric record of the music recorded. **1939** *Motion Pict. Herald* 28 Oct. 30/1 Movietone News has devoted two complete issues to the war. **1953** L. J. Wheeler *Princ. Cinematogr.* vii. 213 An invention by Theodore W. Case made possible the Fox-Case Movietone news, photographed by cameras fitted with a device capable of recording sound alongside the picture and on the same negative film. The Movietone camera was made possible by the ability of certain gas-filled lamps to vary in brightness in true relationship to the current which is passing through them. **1969** *Listener* 23 Jan. 123/2 A reading from the Old Testament about the death of Moses delivered in a hushed movietone voice.

moving, *vbl. sb.* Add: **5.** *moving business, -van*; **Moving Day,** (*a*) (earlier and later examples); also (with lower-case initials), any other day on which people move to new premises; **moving-man** *N. Amer.* = Mover¹ 6.

1973 *Times* 23 Oct. (Pickfords Suppl.) 1 Three hundred years experience in the moving business has helped keep Pickfords the biggest name in household removals. **1832** J. F. Watson *Hist. Tales N.-Y.* 123 'Moving day' was, as now, the first of May. **1855** *Knickerbocker* XLV. 585 In the southern part of New-Jersey, one who rents or purchases a house or farm usually takes possession of the same on the twenty-fifth day of the present month [*sc.* March], which is therefore denominated 'moving-day'. **1947** *Pasadena* (Calif.) *Star-News* 9 Sept. 16/6 Moving days are ahead for several county departments in the Hall of Records and Hall of Justice, Los Angeles. **1973** Moving day [see *moving-man* below]. **1922** H. L. Foster *Adventures Trop. Tramp* xii. 179 While he shipped the furniture from the old place, I was to go down to the new one to see that the moving-men stole none of it en route. **1965** M. McIntyre *Place of Quiet Waters* ii. 29 'How right you are, Miss,' said the moving-man. **1973** *National Observer* (U.S.) 3 Feb., When our moving men finally showed up ('removals' in their business, whether you are coming or going, and on moving day we hardly knew which we were) we received another language lesson. **1898** *Kansas City* (Missouri) *Star* 21 Dec. 9/2 The moving vans and pie-wagons of New York are changing their pictures. **1925** T. Dreiser *Amer. Trag.* (1926) II. iii. xxvii. 334 A venal moving-van company had revealed her address. **1963** *PMLA* Dec. p. viii/2 U.K. removals lorry: U.S. moving van.

moving, *ppl. a.* Add: **3.** Special collocations: **moving average,** an average derived from a series of values in which the interval contributing to it is of constant size but is moved progressively along the series (usually one value at a time) to give a succession of averages; **moving-coil** *attrib.*, denoting electrical instruments and apparatus in which a coil of wire is situated in a magnetic field, so that either the coil moves when a current flows through it or else a current is generated when it is caused to move; similarly **moving-conductor**; **moving-iron** *attrib.*, denoting electrical instruments and apparatus in which the passage of a current through a fixed coil of wire causes the movement of a piece of iron inside it; **moving map,** a map carried in a ship, aircraft, etc., which is displayed so that as the craft moves its position always corresponds to a fixed point in the middle of the map; **moving pavement,** a section of pavement arranged as a conveyer belt for the carrying of passengers; **moving stair(case),** an escalator; also *fig.*; **moving-target** *attrib.*, applied to radar apparatus or techniques which give an indication only of targets moving relative to the transmitter, signals from stationary targets being eliminated.

1912 W. I. King *Elem. Statistical Method* xv. 168 In determining on the size of groups to be used in calculating a moving average, one should use a period of time approximately equal to the length of the cycle which it is desired to eliminate. **1969** J. Argenti *Managem. Techniques* 174 If demand for a product over the past four weeks has been for 20, 23, 19, 19 units respectively, the average demand has been for 20 units. Now suppose another week passes in which the demand is 22 units, the new moving average is 20·5. This is calculated by dropping the earliest figure (20), adding the latest (22) and averaging again. **1896** W. E. Ayrton *Pract. Electr.* (new ed.) I. ii. 144 A very convenient, portable and accurate moving coil ammeter has been perfected by Mr. Weston, of Newark, America. **1903** G. D. A. Parr *Electr. Engin. Measuring Instruments* iii. 106 The ammeters belonging to the class of moving coil instrument measure current indirectly by indicating the fall of potential down a low-resistance 'shunt'. **1930** *Manch. Guardian* 20 Sept. 15/7 Moving-coil loud speakers are, of course, strongly in evidence. **1933** L. E. C. Hughes *Elem. Engin. Acoustics* v. 77 The three forms of diaphragm action are incorporated respectively in the condenser (Wente) microphone, the moving-coil (Sykes) microphone, and the ribbon (Olson) microphone. **1968** *Radio Communication Handbk.* (ed. 4) ix. 12/2 A miniature moving-coil loudspeaker can sometimes be made to serve as a fairly satisfactory microphone. **1933** K. Henney *Radio Engin. Handbk.* xvi. 421 These have practically been superseded by the moving conductor drive (principally electrodynamic) and the larger portion of all loud-speakers in use today are of this type. **1966** *McGraw-Hill Encycl. Sci. & Technol.* VIII. 360/2 A moving conductor microphone consists of a straight-line conductor located in a magnetic field and coupled to a V-shaped diaphragm acted upon by sound waves. **1908** K. Edgcumbe *Industr. Electr. Measuring Instruments* 27 One of the simplest windings to calculate is that of an ammeter coil such as would be used for a moving-iron instrument. **1940** A. Wood *Acoustics* xviii. 529 The large diaphragm required in lieu of a horn may also be driven by an arrangement of the 'moving iron' type. **1966** *McGraw-Hill Encycl. Sci. & Technol.* XIV. 372/2 The moving-iron instrument, like the electrodynamic voltmeter, when properly calibrated indicates true rms volts. **1969** *Daily Tel.* 6 Mar. 18 The former Ministry of Aviation applied for Patent No. 926448 in 1960 to cover the moving map developed by that Ministry. **1969** *Sunday Tel.* 30 Mar. 1/4 Throughout the world more than 13,500 ships, 2,000 aircraft and some 1,000 helicopters use Decca navigation equipment including many 'moving map' displays. **1970** *New Scientist* 12 Mar. 508/2 The aircraft's position at any instant is shown by a spot of light at the centre of a six-inch moving-map display, showing the area over which the aircraft is flying. **1960** *Times Rev. Industry* Apr. 28/1 Passenger-carrying conveyor belt..known as moving pavements. **1971** *Guardian* 22 June 6/6 Passenger conveyor systems, popularly known as moving pavements, could become a major form of city transport. **1922** Moving stair [see *escalate v. 1]. **1940** L. MacNeice *Last Ditch* 12 And two people with the one pulse (Somebody stopped the moving stairs): Time was away and somewhere else. **1910** Moving staircase [see *escalator 1]. **1927** Haldane & Huxley *Animal Biol.* xiii. 309 The gill cilia are so arranged that all food-particles strained off by the gills are driven up to the dorsal groove. Here they become entangled and stuck in the slime, and are passed on by this sort of moving staircase to be digested in the intestine. **1959** *Chambers's Encycl.* V. 389/1 The escalator or moving staircase was originally developed in America; it was exhibited at the Paris Exposition in 1900, and the first to be installed in Britain (in 1911) was at Earl's Court station, London. **1953** Penrose & Boulding *Princ. & Pract. Radar* (ed. 4) xxiv. 645 The principles on which moving target indication and permanent echo cancellation work. **1966** E. Larsen *Radar works like This* (rev. ed.) 39 An important advance in airport radar is the 'moving target indicator'.., a type of radar equipment which shows only moving aircraft on the PPI. **1972** A. Sundararabu *Fund. Radar* xviii. 485 (*heading*) Moving target indication.

movingness. Delete † *Obs. rare* and add later examples.

1927 Beerbohm *Lett. to R. Turner* (1964) 268, I do so agree about the movingness of the Jews at the Süss funeral. **1930** G. Greene *Two Witnesses* 99 He was touched almost to speechlessness by the movingness of Christ.

moving picture. [f. Moving *ppl. a.* + Picture *sb.*] † **1.** A picture in which objects move, or appear to move, in imitation of their natural motion. *Obs.*

1709 *Daily Courant* 9 May, The most Famous, Artificial and Wonderful Moving Picture that came from Germany. **1709–10** [see Moving *ppl. a.* 1]. **1713** Swift *Jrnl. to Stella* 27 Mar. (1948) II. 647, I went afterwards to see a famous moving Picture, & I never saw any thing so pretty. **1715** *Boston News-Let.* 14 Mar. 2/2 The Italian Matchean, or Moving Picture, wherein are to be seen, Wind-Mills and Water-Mills moving round. **1727** M. W. Montagu *Let.* Mar. (1966) II. 73, I now and then peep upon these things with the same coldness I would do on a moving Picture. **1782** *Town & Country Mag.* Apr. 208/2 With moving pictures Loutherburg displays His graphic skill... He derives new pow'rs, To drive fatigue from exhibition hours. **1822** *Hive* I. 66/1 The illustrations consisted of a series of finely painted scenes, which were made to change their effect and constructions, and accompany, in a rapid and beautiful manner, the progress of the story, embodying, as it were, in a moving picture, almost as vivid as the tablet of a camera obscura, every circumstance as it fell from the speaker's lips. **1899** *Science Siftings* XV. 242/1 A second edition has reached us of the 'Motograph Moving Picture Book'... These moving pictures should prove interesting to children.

2. A cinematographic picture or film. Also *attrib.*

1896 Queen Victoria *Jrnl.* 3 Oct. in *Lett.* (1932) 3rd Ser. III. 87 We were all photographed..by the new cinematograph process, which makes moving pictures by winding off a reel of films. **1897** *Sketch* 13 Oct. 480/2 One guinea and half-a-guinea are being asked for stalls to see the moving pictures of the Corbett-Fitzsimmons fight at the Empress Theatre. **1898** *N.Y. Tribune* 30 Oct. 8/3 The always interesting moving pictures in the biograph..will include..four new scenes. **1906** 'O. Henry' *Four Million* 3 He cursed the moving pictures. **1907** *Pearson's Mag.* Jan. (Advt.), Moving Picture Machines. **1912** F. A. Talbot (*title*) Moving pictures: how they are made and worked. **1913** *Punch* 2 Apr. 253/3 It [*sc.* Mexico] contains a town of 10,000 inhabitants where there is no moving picture palace. **1921** G. B. Shaw *Back to Methuselah* iii. 103 What earthly interest is there in looking at a moving picture of a lot of people merely because they were drowned? *Ibid.* 105 Weve had moving pictures of all four put on the screen today for this American. **1934** W. Saroyan *Daring Young Man* 184 This sense of being out of time has driven thousands of people from their homes into moving-picture theatres where new universes appear before them. **1950** T. S. Eliot *Cocktail Party* i. i. 34 A common interest in the moving pictures Frequently brings young people together. **1959** *Observer* 22 Feb. 19/6 The film is interesting, as a moving-picture record of a movement of our times.

Moviola (mūvi,ōu·lă). *Cinemat.* orig. *U.S.* Also **Movieola,** and with small initial. [f. *movie + -*ola*, after Pianola.] The proprietary name of a device whereby the picture and sound of a cinematographic film are reproduced on a small scale so that the film may be edited or checked.

1929 *Photoplay* (Chicago) Apr. 31/2 *Movieola*, miniature projection machine with earphones used in the cutting room of a talking picture studio for rapid viewing of pieces of film. **1931** B. Brown *Talking Pict.* xi. 276 Lately, a device known as the 'Movieola' has been adopted, ..a miniature peep show projector with a loud speaker. **1933** C. Winchester *World Film Encycl.* 483/1 *Moviola*, a little machine used by the film editors to see and hear the picture. **1944** *Trade Marks Jrnl.* 20 Sept. 455/1 Moviola. .. Projectors, synchronizing apparatus, film measuring machines, sound reading apparatus, sound amplifiers and re-winding apparatus; all for use in editing cinematographic films. Iwan Serrurier, trading as Moviola Company,..Hollywood. **1959** Halas & Manvell *Technique Film Animation* xxv. 302 Music track run on a movieola and each note, phrase and sentence marked with grease pencil. **1970** *Daily Tel.* (Colour Suppl.) 6 Feb. 22/3 For ten weeks, three weeks longer than it took to shoot the movie, he works with the editor at the movieola. **1970** *Times* 5 Dec. 19/6 There was no movieola in those days... You just ran the film through your fingers and cut with a pair of scissors.

mow (mōu), *sb.*⁶ [f. *mow v.¹ 3 d.] In *Cricket*, a sweeping stroke to leg.

1925 D. J. Knight in *Country Life* 15 Aug. 244/1 Leg-side shots. They are the glides,..the mow and the pull. **1926** J. B. Hobbs *Test Match Surprise* xvi. 171 What he intended for a leg glance was nothing more than a 'mow' between square leg and mid-on.

mow, *v.*¹ Add: **3. d.** *Cricket.* To make a sweeping stroke to leg as if mowing the grass with a scythe. Also *trans.*

1844 *Bradford Observer* 8 Aug., Holmes cleverly mowed the ball from the off stump to the leg side. **1868** J. Lillywhite *Cricketers' Compan.* 81 H. M. Mills..might score well if he did not think it necessary to mow at straight long-hops. **1925** D. J. Knight in *Country Life* 15 Aug. 245/1 Supposing there is a deep square-leg, it is better..to kneel down on your right knee and mow or drag the ball round in the direction of long-leg.

4. Combs. containing the verb stem, as **mow-land** *U.S.*, land where grass is grown for mowing; **mow-lot,** a plot of this land.

1845 S. Judd *Margaret* ii. i. 214 She saw..women.. raking and turning hay among alders and willows that yet flourished in their best mow-lands. **1874** *Rep. Vermont Board Agric.* II. 411 The breeding of wrinkled sheep is like a farmer who ridges up his level mow-land and seeds the ridges with an inferior grass. **1845** S. Judd *Margaret* ii. viii. 325, I kept him here in the mow-lot.

mow, var. *mou.

mowhay (mōu·hēi, mōu·i). *dial.* [f. Mow *sb.*¹ + Hay *sb.*² 2.] A stack-yard, a rick-yard.

Given as Devon and Cornwall in *E.D.D.*

1862 R. S. Hawker *Let.* 14 Sept. in C. E. Byles *Life & Lett. R. S. Hawker* (1905) xvi. 362 Now I trust to be able to pay my Farm Wages in Wheat as the custom is, while for the last two years I have had to find money instead. To small Farmers and Vicars the Mowhay is the Farm Bank. **1880** M. A. Courtney in Courtney & Couch *Gloss. Words Cornwall* 38/2 *Mowhay*, an enclosure of ricks of corn or hay. **1959** *Listener* 22 Jan. 174/2 The granite shafts in the old mowhay.

mowing, *vbl. sb.*¹ Add: **3.** *mowing crook, ground* (later examples), *land, machine* (earlier examples).

1943 *Antiquity* XVII. 203 To use a mowing-crook such as one still sees used in connection with the swap-hook in this country. **1654** *Rec. Early Hist. Boston* (1880) VI. 17 Twenty acres more or lesse of mowing ground upon the marsh. **1722** in *Essex Inst. Hist. Coll.* (1906) XLII. 90 [To pay damages for] digging in any Corne field,..mowing ground. **1770** in *Ibid.* (1872) XI. 31 My little mare had provided for herself, by leaping out of a bare pasture into a lot of mowing ground. **1787** G. Washington *Diaries* (1925) III. 222 The same difference was equally obvious on a piece of mowing grd. not far distant from it. **1640** in *Connecticut Hist. Soc. Coll.* (1912) XIV. 357 One parsell called Swamp, now mowing land. **1704** *Proprietors' Rec. Waterbury, Connecticut* (1911) 60 No man shal stak horses in the moing land in said feild. **1816** *North Amer. Rev.* III. 428 At the distance of five or six miles it begins to wind gently through large tracts of fine rich mowing land. **1858** C. L. Flint *Milch Cows* 169 The grasses differ widely; and their value as feed for cows will depend..on the management of pastures and mowing-lands. **1823** *H. R. Doc. 17th U.S. Congress 2 Sess.* No. 36. 6 Improvement in the mowing machine, Feb. 13, [1822], Jeremiah Baily. **1838** H. W. Ellsworth *Valley Upper Wabash* v. 47, I have a plan in view..and that is, to introduce the mowing and grain-cutting machine into this state.

mowra(h), varr. Mahwa (in Dict. & Suppl.).

moxibustion. (Examples.)

1910 E. Playfair tr. *Neuburger's Hist. Med.* I. 71 Moxibustion..also serves as a prophylactic. *Ibid.* 78 Moxibustion and acupuncture were also in Japan the favourite methods of treatment. **1958** *Manch. Guardian* 6 Dec. 4/3 This has a history of three thousand years, and works in various ways: by acupuncture..; by moxibustion..; and by medicaments prepared from herbs and parts of animals. **1965** *New Scientist* 15 July 129/1 Chinese surgery has clearly advanced at a pretty phenomenal rate. Presumably this is why traditional methods like acupuncture, moxibustion (the burning of a herbal mixture on part of the body to transfer the site of irritation from one place to another)..are still respected. **1974** *Sci. Amer.* Apr. 25/1 The years since 1949 have seen the general adoption of certain traditional techniques: the use of herbal preparations, of gymnastics and respiratory exercises, and of two related treatments, moxibustion and acupuncture.

moxie (mǫ·ksi). *U.S. slang.* [f. the name of an American soft drink.] Courage, 'guts', 'nerve'; energy, pep.

1930 D. Runyon in *Collier's* 20 Dec. 32/3 Personally, I always figure Louie a petty-larceny kind of guy, with no more moxie than a canary bird. **1943** M. Shulman *Barefoot Boy with Cheek* xv. 158 We knew you had the old moxie, the old get out and get. **1947** S. J. Perelman *Westward Ha!* (1949) x. 122 Before I could summon up enough moxie to bolt after them,..the attendants herded us into Indian file. **1955** *Publ. Amer. Dial. Soc.* xxiv. 94 If he admits his own limitations, he loses his nerve (blows his moxie). **1959** C. Williams *Man in Motion* iv. 47 She's a real Latin type, dark brown eyes with a lot of moxie in 'em. **1967** *Telegraph* (Brisbane) 6 Mar. 7/2 A girl with all the moxie (outstanding intelligence). **1969** *Maclean's Mag.* Aug. 1/3 The qualities that make business click—qualities such as moxie and timing and stealth. And wealth. **1970** *New Yorker* 29 Aug. 19/1 It takes moxie, skill, and self-reliance. **1975** *Daily Colonist* (Victoria, B.C.) 16 Mar. 17/3, I was very impressed with his all-round moxie. He could snap back at any of them, news reporters, police, and me.

moya (moiā·), *int. Ireland.* Also **maureeyah, mauryah, mor-yah, moyah, moy-yah.** [ad. Ir. *mar dh'eadh*, as if it were so.] Used as an ironic interjection (see quots.).

The spelling in the head-word, which is that of James Joyce, is not otherwise authenticated.

1910 P. W. Joyce *Eng. as we speak it in Ireland* 296 Mor-yah; a derisive expression of dissent to drive home the untruthfulness of some assertion or supposition or pretence, something like the English 'forsooth', but infinitely stronger. **1911** *Jrnl. Co. Kildare Archæol. Soc.* VI. 535 *Maureeyah*, this expression denotes a strong doubt as to the truthfulness of a statement. 'That notorious poacher, Loughlin Murphy, tould the magistrates he didn't know how to set a snare, Maureeyah.' *Ibid.*, *Moyyah*, this is much the same expression as *Maureeyah*... It conveys the same sense as the English saying, 'You may tell that to the Horse Marines'. **1914** Joyce *Dubliners* 151 And the men used to go in on Sunday morning before the houses were open to buy a waistcoat or a trousers—moya! **1922** —— *Ulysses* 327 Beggar my neighbour is his motto. Love, Moya! **1925** *Blackw. Mag.* Dec. 784/2 'Tis a bad pair to beat we are, moyah! **1939** Joyce *Finnegans Wake* 375 The wonder of the women of the world together, moya! **1944** *Béaloideas* XIV. 176 Mauryah. A word implying doubt and irony.

Moyen Age (mwayenaʒ). Also **moyen âge,** etc. [Fr.]. The Middle Ages, the medieval period. Also, a representation of medieval life. Also as quasi-*adj.*: of the Middle Ages, medieval.

1849 Thackeray *Pendennis* I. xxiii. 213 We'll..furnish the oak room with the Moyen-age cabinet and armour. **1852** E. Ruskin *Let.* 10 Jan. in M. Lutyens *Effie in Venice* (1965) ii. 245 The traditions of their families moved with Moyen-age sentiments of Chivalry. *Ibid.* 26 Jan. A coiffure I have made for myself which John says is decidedly Moyen-age. **1864** G. A. Sala *Quite Alone* I. ii. 37 'If a man wants to get on in life, he can't do better than study the History of the Middle Ages'. To which Moyen Age culture Mr. Blunt owed much of his success. **1870** C. Schreiber *Jrnl.* (1911) I. 78, I cannot describe the collection. It was entirely Moyen Age. **1923** R. Fry *Let.* 21 May (1972) ii. 535 Spain..this big rich and on the whole up-to-date civilisation which is yet definitely *moyen age*. **1927** R. H. Wilenski *Mod. Movement in Art* 66 Moyen-âge elements in their art. **1961** M. Levy *Studio Dict. Art Terms* 77 *Moyenage*, paintings depicting reconstructed scenes, etc. from the Medieval period. **1962** R. G. Haggar *Dict. Art Terms* 219/2 *Moyenage*, pictorial reconstruction of the Middle Ages.

Moygashel (moi·gæʃəl). Also **moygashel.** [Name of a village in Co. Tyrone, N. Ireland.] The proprietary name of a type of Irish linen.

1931 *Trade Marks Jrnl.* 11 Nov. 1506/1 Moygashel... Linen piece goods. Stevenson & Son, Limited, Moygashel Mills, Dungannon, County Tyrone, Ireland. **1958** *Woman* 1 Mar. 4/2 (Advt.), Unlined suit, trimmed white pique in a moygashel pebble line tweed. **1966** *Punch* 7 Sept. 363/2, I placed the card in the pocket of my moygashel jacket.

moy-yah, var. *moya *int.*

Mozambican (mōuzæmbī·kăn), *a.* and *sb.* Also † **Mosambican.** [f. Mozambique + -an.] **A.** *adj.* Of or pertaining to Mozambique.

1875 *Encycl. Brit.* III. 758/2 The Libyan Subregion..is then succeeded by the 'Mosambican' Subregion, which continues perhaps to Sofala. **1893** A. Newton *Dict. Birds* ii. 352 The 'Mosambican' Province next follows. **1970** *Guardian* 1 Aug. 9/1 A Mozambican consciousness, an awareness of African history. **1971** *Standard* (Tanzania) 7 Apr. 4/1 Mozambican freedom fighters. **1974** *Daily News* (Tanzania) 13 Sept. 4/1 This bunch of bandits..is desperately trying to oppose the Mozambican People's and the Portuguese People's wish for peace.

B. *sb.* A native or inhabitant of Mozambique.

1971 *Standard* (Tanzania) 7 Apr. 4/1 This news was joyfully received by other Mozambicans. **1973** *Times* 17 Feb. (Mozambique Suppl.) p. ii/2 The identity card carried by any Mozambican is exactly the same document as that carried by anyone in Lisbon.

Mozambique. Add: Hence **Mozambi·quer,** a native or inhabitant of Mozambique.

1803 R. Semple *Walks & Sk. Cape Good Hope* iii. 40 Without the inactivity or dulness of the Mozambiquer or the penetrative genius of the Malay, he [*sc.* the Malabar slave] forms an excellent medium between the two. **1876** F. Boyle *Savage Life* 271 Besides this, we had..two cooks, a Malay and a Mozambiquer.

Mozartean, *a.* Add: (Further examples.) Also as *sb.*, an admirer or adherent of Mozart; an interpreter of his works.

The usual spelling is now *Mozartian.*

1898 G. B. Shaw *Perfect Wagnerite* 114 The Mozartian method. **1947** [see *chromaticism]. **1947** A. Einstein *Mus. Romantic Era* v. 44 Even in the 19th century there were still many..who demanded no more than that of music: they were the 'Mozartians', the opponents particularly of Wagner and Liszt. **1955** *Times* 14 May 4/5 It was a curious whim that brought a Mozartian body all the way from Germany in order to play Beethoven's Eighth Symphony. **1958** *Spectator* 10 Jan. 50/2 Despite one or two very lovely movements, this is only for the insatiable Mozartian. **1959** *Listener* 15 Jan. 131/3 The Mozartian aristocracy has long disappeared [in Prague]. **1960** *20th Cent.* Aug. 141 A mood of carnival that one might call Mozartian. **1975** *Daily Tel.* 1 May 13/8 Once again the young Jamaican Nerine Barrett proved last night a natural-born Mozartian.

Also **Mozartia·na** *sb. pl.* (Ana *suff.*), minor works by Mozart; **Moza·rtianly** *adv.*, in the manner or style of Mozart's music; **Mozartism** (later example); **Mozart-lover**; **Mozart-like, -size** *adjs.*

1925 W. de la Mare *Two Tales* 55 The Mozart-like light and sweetness. **1928** *Observer* 24 June 23 The other three..give the elegant (and in the last movement Mozartianly deep) music an air of special distinction. **1955** H. van Thal *Fanfare for E. Newman* x. 147 But he also thought the Finale of Beethoven's *Eroica*, with its 'gentle Mozartisms', unworthy of the rest. **1958** *Spectator* 10 Jan. 50/2 A rarer contribution to recorded Mozartiana is the fifteen one-movement Church Sonatas for chamber orchestra and organ. **1959** D. Cooke *Lang. Mus.* v. 236 True Mozart-lovers of the nineteenth and twentieth centuries. *Ibid.* 245 Only those who have heard the work played in a small hall, by a Mozart-size orchestra with a handful of strings, can be said to have experienced the truly violent impact of the symphony. **1963** *Punch* 3 Apr. 496/3 *Così fan Tutte* is..brightly and strongly though not always Mozartianly sung.

‖ **mozo** (mōu·þo). *Latin America.* [Sp.] **1.** A male servant or attendant; a groom; a labourer.

1836 C. J. Latrobe *Rambler in Mexico* 49 The remainder were sent in advance under his domestics or mozos. **1847** G. A. F. Ruxton *Adventures Mexico & Rocky Mts.* vii. 48, I at length hired a *mozo* to proceed with me as far as Durango. **1904** Conrad *Nostromo* i. viii. 107 But Captain Mitchell's right-hand man..after looking down critically at the ragged mozo, shook his head. **1923** *Blackw. Mag.* July 46/2 The mozo, the ostler lad,..was a son of the house. **1936** A. Huxley *Eyeless in Gaza* xli. 495 The *mozos* had loaded their baggage on to the pack-mules. **1955** W. Gaddis *Recognitions* iii. iii. 819 It turned out that the Señorito had asked the same question, and fled directly he got this same answer, leaving this mozo behind, to chat with her.

2. A bull-fighter's attendant.

1926 [see *espada]. **1934** H. Baerlein *Belmonte the Matador* iii. 37 Every matador has got a valet, his *mozo de estoques*, whose important days are those on which there is a fight, not only in the dressing of his master but the duties of attendance..at the ring-side, preparing the muletas and handing to his master the swords as he requires them. *Ibid.* 38 Antonio Conde..undertook the part of mozo which he kept throughout the years. **1963** *Parade* (Austral.) Dec. 47/1 Meijas had to start at the bottom of the ladder as a 'mozo', the attendant who waits on a matador. **1967** McCormick & Mascareñas *Compl. Aficionado* iv. 146 *Mozo(s)*, the badly paid and often brave ring-attendants who clean up the mess, accompany the picador on foot, [etc.].

mozza, var. *matzah.

mozzarella (mǫtsăre·lă). Also **mozarella**; *pl.* **moz(z)arelle.** [It.] In full, *mozzarella cheese*. An Italian cheese originally made in the Naples area from buffalo milk.

1911 C. Besana in *Bull. Bureau Agric. Intelligence & Plant-Dis.* (Internat. Inst. Agric., Rome) Nos. 8–10. 2231 With Buffalo milk..'pasta filata' cheeses are made, which are eaten fresh and much esteemed. These are 'Provature', 'Mozarelle', etc. **1912** *Experiment Station Rec.* (USDA) Oct. 475 Provature, or Provole, and Mozzarelle cheeses are made from buffalo milk. **1935** M. Morphy *Recipes of all Nations* 160 Slices of Mozzarella cheese. **1962** *Spectator* 16 Feb. 219/2 Was the mozzarella cheese dripping, positively dripping, fresh? **1962** L. Deighton *Ipcress File* ii. 19 The girl in the delicatessen was small,..and rather delicious. We had been flirting across the mozzarella for years. **1969** *Daily Tel.* (Colour Suppl.) 5 Sept. 33/3 Limpid responsive eyes, and a chin that slopes back from a face as pale and bland as mozzarella. **1974** *Times* 2 Nov. 11/4 Fresh mozarella..for making pizza.

mozzle (mǫ·z'l). *Austral. slang.* Also **moz, mozz.** [ad. Heb. *mazzāl* luck; cf. *shemozzle.] Luck; *spec.* in phr. *to put the moz on*, to inconvenience; to jinx. Also as *v. trans.*, to hinder, interrupt.

1898 *Bulletin* (Sydney) 17 Dec. (Red Page), *Mozzle* is luck... *Good mozzle* = good luck; *Kronk mozzle* = bad luck. **1903** 'T. Collins' *Such is Life* vi. 225 'And how much do you stand to lose, if your mozzle is out?' I asked. **1919** E. Dyson *Hello Soldier!* 32 'Twas rotten mozzle, Neddo. We had blown out every clip. **1941** Baker *Dict. Austral. Slang* 47 *Moz*, to interrupt, to hinder. 'Put the moz on someone', to inconvenience a person. *Mozzle, to*, as for 'moz'. **1956** A. Marshall *How's Andy Going?* 200 'Looking ahead like that never does any bloody good to any man,' observed Pat. 'It puts the moz on him.' **1963** H. Porter *Watcher on Cast-Iron Balcony* 81 Mother is wishing Miss Brewer some female ill, is putting the mozz on her. **1965** F. Hardy *Yarns Billy Borker* xx. 107 'Don't mozz a man,' I tells him. 'You're well named, I'll say that for you, Calamity.' **1974** J. Powers *Last of Knucklemen* 49 Don't let him mozz you, Monk. **1974** K. Stackpole *Not just for Openers* ii. 32 She felt she put the moz on him... She couldn't bear to go in case she was a jinx.

mpalla, var. *impala.

mph ('mh), *int.* Also **mphm.** [Imit.] Used to express an inarticulate sound of disapproval, doubt, or qualified approval. Cf. Umph *int.*

1876 'Mark Twain' *Tom Sawyer* 272 What would people say? 'Mph! Tom Sawyer's gang! pretty low characters in it.' **1949** 'N. Blake' *Head of Traveller* vii. 105 To conceal the way he was killed? Mphm. Don't see how it helps us.

Mpret (bret). Also **Mbret, M'pret.** [ad. Alb. *mbret*, f. L. *imperator* emperor.] The title given to William of Wied, elected ruler of Albania after the declaration of Albanian independence in 1913.

1914 *Conc. Oxf. Dict.* Add., *Mpret*, Albanian ruler [L. *imperator*, emperor]. **1915** *Truth* 27 Jan. 133/2 He and not the Prince of Wied ought to be M'pret. **1921** C. Torr *Small Talk at Wreyland* 2nd Ser. 83 When I first went to Greece, they still spelled Byron's name phonetically Mpairon. They pronounce *b* like our *v*, but *mp* like our *b*—a fact unknown to many of the people who talk about the Mpret of Albania. **1954** M. Hasluck *Unwritten Law in Albania* xv. 154 Esad Pashë Toptani revolted against Prince William of Wied, then *Mbret* (king) of Albania. **1959** *Chambers's Encycl.* I. 221/2 The European ruler selected to be 'mpret' of Albania—the German prince William of Wied.

Mr. Add: **2. d.** One who is entitled to be addressed as 'Mr.'; the word 'Mr.' as a title (in correspondence).

a **1817** Jane Austen *Persuasion* (1818) III. iii. 55 'I have let my house to Admiral Croft,' would sound extremely well; very much better than to any mere *Mr.* ——; a *Mr.* (save, perhaps, some half dozen in the nation,) always needs a note of explanation. **1857** Geo. Eliot *Let.* 2 June (1954) II. 337 Mr. Eliot..may be a relation of Mr. Liggins's or some other 'Mr.' who knows Coton stories. **1882** W. Pater *Let.* 4 Nov. (1970) 43 My dear Sharp, (I think we have known each other long enough to drop the 'Mr.'). **1915** R. Fry *Let.* 26 Aug. (1972) II. 389 Dear Waley (May we drop the Mr).

e. Used with following adj. or sb. to denote an exemplar, a type, or a victor in a contest; esp. **Mr. Big,** the head of an organization of criminals; also, any important man; **Mr. Charley, Charlie** (see *Charley, Charlie 7); **Mr. Clean,** an honourable or incorruptible politician; **Mr. Fix** (see quot. 1950); **Mr. Fixit,** one who does repairs or odd jobs; one skilled at managing difficult problems or situations; a 'trouble-shooter'; **Mr. Right,** a man who would make the ideal husband; a 'Prince Charming'.

See 1958 *Amer. Speech* XXXIII. II. ii. 84–87.
1814 H. Brougham *Let.* June in T. Creevey *Creevey Papers* (1903) I. ix. 194, I was finally decided in favour of publishing to-day by the apprehension of Alexr., &c., coming in a day or two, and taking off the attention of Mr. and Mrs. Bull. **1913** C. Mackenzie *Sinister St.* I. I. ii. 28 Because he had been slow in choosing..he had been called Mr. Particular. **1922** Joyce *Ulysses* 362 Till Mr Right comes along then meet once in a blue moon. **1925** R. Lardner in *Liberty* 9 May 5 (*title*) Mr. and Mrs. Fix-It. **1937** E. H. Sutherland *Professional Thief* v. 129 Since most of the cases of professional thieves in the stores are taken care of by Mr. Fix, it is evident that the store detectives must get an end. **1940** G. Marx *Let.* 10 Oct. (1967) 26, I may motor east..to see your 'Mr. Big'. **1950** H. E. Goldin *Dict. Amer. Underworld Lingo* 139/1 *Mister fix*, a go-between, especially one who shuttles between the underworld and the overworld, handling bribes, ransom payments, etc. **1951** J. Cannan *And All I Learned* vi. 76 Stevens offered her the last cake on a plate... Mildred laughed and replied, 'What about Mr Manners?' but took the cake. **1952** M. Laski *Village* i. 24 In the ordinary way you'd expect someone like Miss Margaret to stay at home and go to tennis-parties and things until Mr. Right came along and she could make a home of her own. **1953** R. Chandler *Long Good-Bye* xxxii. 200 He was Mr. Big, the winner. **1959** A. W. Sherring *Tip Off* iii. 28 Hardly the kind of district one would expect to find Mr. Big of London's underworld. **1962** E. Cleaver in A. Dundes *Mother Wit* (1973) 18/1 To..crown him..Mister..Universe. **1967** P. E. H. Durston *Mortissimo* (1968) xii. 100 He's got very little decent for sale. More of what the Americans call a 'Mr. Fixit'. Mends things. **1968** 'J. Le Carré' *Small Town in Germany* v. 71 Do you a girl as well, would he? Mister Fixit, is that it? **1969** C. Booker *Neophiliacs* vii. 179 Hints of the existence of a powerful 'Mr. Big'. **1970** G. Greer *Female Eunuch* 88 The whole point of a woman's existence is to be exploited by Mr Right. **1970** *Sunday Times* (Colour Suppl.) 20 Dec. 35/3 Still an entrepreneur and a Mr Fix-it, Levis..died from cirrhosis of the liver. **1972** *Village Voice* (N.Y.) 1 June 15/1 Peter M. Flanigan.. became an assistant to the President and acquired a reputation as 'Richard Nixon's "Mr. Fixit" when it comes to powerful business interests'. **1973** J. Mann *Only Security* x. 132 Sylvester..could have modelled as a Mr Average. **1973** *Times* 21 Nov. 8/4 Mr Elliot Richardson, who resigned as Attorney General..is Mr Clean to many Republicans. **1974** E. McGirr *Murderous Journey* 59, I ..asked if I could go through Siskin's papers... He'd been a methodical man... It was more or less the picture of Mr. America. **1974** *Guardian* 28 Jan. 2/5 Mr Shultz himself had never been touched by Watergate... His reputation as a 'Mr Clean'..has led him..to voice a growing sense of unease. **1974** *Observer* 17 Feb. 5/5 Smalls said he had not seen a 'Police 5' TV programme about the Wembley raid, but agreed that Turner's photograph was 'splashed in the papers', accompanied by the title of 'Mr Big'.

mrad: see *M 5 d.

mridangam ('mrida·ŋam). Also **m'ridang(a)**. [a. Sanskrit *mṛidaṃga* a kind of drum.] A double-headed, barrel-shaped drum, once made of clay, now usually of wood, with one head larger than the other, used in southern Indian music. Also *attrib.*

1888 A. J. Hipkins *Mus. Instruments* 87 The Drum with the striped body and leather braces is a kind of M'ridang. The genuine Drum bearing this name is longer in proportion to its diameter, and has one head larger than the other. *Ibid.*, A good Tabla or M'ridang player will earn from 100 to 150 rupees, per month... The M'ridang is considered to be the most ancient of Indian Drums; its origin is popularly ascribed to the god Mahadeo (S'iva). **1891** C. R. Day *Mus. & Mus. Instruments S. India* vi. 138 (*caption*) The Mridang is beaten by the hands, finger-tips, and wrists... The smaller head of the Mridang is struck by the right hand, the larger head by the left. **1921** H. A. Popley *Mus. India* vii. 120 The *Mṛidaṅga* or *Mardala* is the most common and probably the most ancient of Indian drums... The word *mṛidaṅga* or *mardala* means 'made of clay', and probably therefore its body was originally of mud. **1968** *N.Y. Times* 12 Sept. 56 Palghat Raghu on the mridanga and Alla Rakha on the tabla. **1968** *Daily Tel.* 28 Sept. 13/3 Among them was a drum dialogue between Alla Rakha (tabla) and Palghat Raghu (mridangam), a tour de force of fantastic subtlety and eloquence. **1970** *Ibid.* 6 July 12/6 A musical interlude in which Nagarajan on the mridangam drum had a friendly battle..with the dance director. **1971** *Shankar's Weekly* (Delhi) 4 Apr. 24/1 The infinite variety of the mridangam includes some of the elemental sequences that jazz makes capital out of.

Mrs. Add: **1. b.** A wife (with ellipsis of the name of her husband). *colloq.*

1913 R. Brooke *Let.* 22 Nov. (1968) 535 He passed through Fiji lately... Mrs, I gather, is not with him. **1938** M. Allingham *Fashion in Shrouds* vi. 82 Paul Taretan is taking 'three girls from totally different environments', and 'Mrs.' has selected one rather beastly little boy. **1950** J. Cannan *Murder Included* i. 9 Mr and Mrs Scampnell... Mrs has a daughter by her first husband. **1970** A. Morice *Death in Grand Manor* iv. 35 Mr Cornford wasn't so bad.. but Mrs was awful. **1974** J. Montgomerie *Implosion* x. 67 Another picture of Mrs., side-view.

c. *the Mrs.*: one's wife. *colloq.*

The examples happen to be U.S. but the use is widespread: cf. Missis 1.

1920 [see *monkey suit* s.v. *monkey *sb.* 17 a]. **1937** *Amer. Speech* XII. 103 The farmer will often refer to his wife as the Mrs. and he commonly addresses her as Wife. **1944** *Publ. Amer. Dial. Soc.* II. 58 *The Mrs.*, the wife. **1967** E. Bullins *Theme is Blackness* (1973) 96 I'll have the Mrs. call the doctor as soon as I get home. **1973** *Philadelphia Inquirer* (Today Suppl.) 7 Oct. 42/3 You know, when I go home, the Mrs. says to me: 'Well, what happened tonight, night clerk?'

2. a. Also prefixed to the Christian name of the husband (without a following surname).

a **1817** Jane Austen *Persuasion* (1818) III. vi. 103, I shall tell *you*, Miss Anne..that I have no very good opinion of Mrs. Charles's nursery-maid. **1842** Geo. Eliot *Let.* 18 Feb. (1954) I. 126, I imagine, from a message my sister Mrs. Isaac told me of, that you had the idea that I was at Griff. **1857** C. M. Yonge *Dynevor Terr.* II. vi. 81 Is it in the nature of things that she should live in such society as Mrs. Walby's and Mrs. Richardson's? People who call her Mrs. James! **1910** E. M. Forster *Howard's End* xxxiii. 335 Mrs. Charles is expecting her fourth. **1953** A. Christie *Pocket Full of Rye* iv. 28 'Who was at breakfast?' 'Mrs. Fortescue, Miss Fortescue, Mrs. Val Fortescue... Mrs. Val and Miss Fortescue always eat a hearty breakfast.' **1971** E. Lemarchand *Death on Doomsday* vi. 97 We know Mrs Giles has got a key into the public rooms.

c. Prefixed to the surname of (the husband of) an actual or a fictional married woman, or to a generic sb., indicating esp. a woman of a certain occupation or temperament, as **Mrs. Beeton** (abbr. **Mrs. B.**) [the name of Isabella Mary *Beeton* (1836–65), nominal author of a Book on Household Management], an authority on cooking and domestic subjects; also *transf.*; **Mrs. Dale** [the name of the wife of a doctor in a radio serial], a conventional, middle-class woman; so **Mrs. Dale-ish** *a.*, middle-class; **Mrs. Grundy** (see Grundy^{3}); **Mrs. Mop(p** [the name of a charwoman in a radio series], (a nickname for) a charwoman; the typical charwoman; also *attrib.*; **Mrs. next-door** (see *next door 1 b); **Mrs. Thing** [Thing *sb.*^{1} 10 in Dict. and Suppl.], used in place of a married woman's name of which the speaker is uncertain.

1960 *Listener* 7 Jan. 18/2 The modern trick, I am told by Dr. Spock and the other Mrs. Beetons of child care, is not to show anxiety to children. **1970** *Times* 9 Feb. 8/6 Explains Ita Jones, Light's switched-on Mrs B.: 'Though serving up this food may blow some minds, remember that you eat the animal because you love it.' **1954** A. Melville *Simon & Laura* in *Plays of Year* XI. 36 You probably get the idea it's to be a sort of TV 'Mrs Dale'. It's not: we want interesting people who meet other interesting people: people..with a bit of—..glamour, colour, excitement in their lives. **1961** *Times Lit. Suppl.* 24 Nov. 851/1 The setting is suburban, even Mrs. Dale-ish. **1948** J. F. Wolfenden *Public Schools To-Day* v. 99 A great deal of the welfare..of a boarding school depends upon the unsung 'warrant officers and N.C.O.s', from the school messenger to Mrs. Mop. **1950** A. Christie *Murder is Announced* vii. 60 Our Mrs Mopp says he came from one of the big hotels. **1956** 'A. Gilbert' *Riddle of Lady* x. 145 A machine in a Mrs Mopp apron. **1972** *Listener* 27 July 105/1 Today's generation monotonously describe charwomen as Mrs Mops. **1939** N. Marsh *Overture to Death* iii. 41 The other one—Mrs. Thing or whoever she is! **1960** 'R. East' *Kingston Black* xiv. 136 Old Mrs. Thing at the exchange may listen in.

Ms^{2} (miz). Also **Ms.** (with full stop). [A 'compromise' between Mrs. and Miss *sb.*^{2}] A title prefixed to the surname of a woman, regardless of her marital status. Hence as *sb.*, a woman so designated. Cf. *miz 2.

An increasingly common, but not universally accepted, use.

1952 *The Simplified Letter* (Nat. Office Managem. Assoc., Philadelphia) Jan. 4 Use abbreviation Ms. for *all women* addressees. This modern style solves an age-old problem. **1952** *Ibid.* (rev. ed.) June 4 Use abbreviation Ms. if not sure whether to use Mrs. or Miss. **1970** *Daily Tel.* 28 Aug. 14 The American feminists..object to being addressed as Mrs or Miss but admit that Ms, which the New York Commission on Human Rights has adopted for correspondence, [etc.]. **1970** *New Yorker* 5 Sept. 27 'How come no woman heads a super-agency?' demanded Ms. Komisar. **1971** *Publishers' Weekly* 1 Nov. 22 A crowded New York press conference heard this morning that a new magazine, called *Ms.* (pronounced 'Miz'), will begin publication in January. **1972** *Guardian* 29 Mar. 11 Mrs Chisholm (and it is definitely Mrs not Ms). **1973** *Lancet* 24 Mar. 633/2 We thank Ms Payge Hodapp for technical assistance. **1974** *Daily Tel.* 21 May 1/6 The Passport Office yesterday conceded the right to women to call themselves Ms (pronounced Miz) on their passports instead of Mrs or Miss. This followed a month's campaign by Women's Lib. **1975** *Publishers Weekly* 27 Jan. 232/1 *Sanditon* by Jane Austen and Another Lady..Ms. Austen's seventh and unfinished novel. *Ibid.* 19 May 170/1 Samantha Lay, a comely Ms. who devises gimmicks like a 'flying bullpen'. **1975** *Times* 17 Sept. 16/2 A circular..states grandly: 'Female staff in this department may..use the Ms title (usually pronounced Miz) as an alternative to the Miss or Mrs.' *Ibid.* 16/3 The Civil Service Department..has a fair sprinkling of Mss among its own staff.

msasa ('msa·sa). Also **m'sassa, musaasa.** [Mashona.] A Central African tree, *Brachystegia spiciformis*, of the family Leguminosæ, distinguished by compound leaves, racemes of small, fragrant, white flowers, and a spreading crown of branches. Also *attrib.*

1923 *Kew Bull.* 133 'Musaasa'..is by far the most common of the bark trees and is found all over Southern Rhodesia at an altitude of from 3,000 to 6,000 feet. It grows to 30 or 40 feet in height, has a rough outer bark in the large trees but almost smooth in the young trees... The trunks of the 'Musaasa' vary in diameter from 8 to 24 inches. **1949** K. L. Simms *Sun-Drenched Veld* iii. 33 On these islets you can..amuse yourself by trying to pick out the various trees. There is..the crimson-leaved m'sasa.. and the scarlet-flowering kaffirboom. **1951** D. Lessing *This was Old Chief's Country* i. 8 This child could not see a msasa tree, or the thorn, for what they were. **1964** *Listener* 6 Aug. 191/1 The black shadow of the African msasa Squats among the lawn's colonial company. **1973** Palmer & Pitman *Trees S. Afr.* II. 847 The msasa is one of the few trees, growing beyond the borders of the Republic, that is well-known to South Africans. It is the dominant tree in most of the savannah forest of Rhodesia, reaching northwards to Zambia and Malawi, and is famous for the gorgeous pink, wine red, copper and bronze colours of its spring foliage.

mtepe ('mte·pe). [Swahili.] A sailing craft characterized by a square matting sail, used on the East coast of Africa.

1872 R. F. Burton *Zanzibar* I. iv. 73 Various native craft..anchor close in shore... The quaintest and freshest local build is to us the Mtepe, which the Arabs call Muntafiyah. **1938** *Jrnl. R. Anthrop. Inst.* LXVIII. 343 It is only when the Swahili coast is reached that the oculus is again encountered. In this region the only sea-going craft that can be classed as indigenous is the *mtepe*. **1942** *E. African Ann.* 1941–2 27/1 His 'M'tepe' squatted, with many others, on the deep waters beyond the jetty. The tapering mast, raking forwards, was now devoid of its lateen sail.

m' tutor: see *M' = my.

mu (miū). [a. Gr. μῦ, name of the letter *M*, *μ* (see M).] **1.** One micrometre (micron). Usu. denoted by *μ*.

[**1880**: see *micron.] **1888** *Nature* 8 Mar. 438/1 *Micron* is currently used here to express 1/1000 of a millimetre. French botanists call it *μ*, and seldom use its first decimal because they cannot see such a small space. **1900** Dorland *Med. Dict.* 398/2 *Mu*, in micrometry, a micron or micromillimeter. **1957** *Jrnl. Anat.* XCI. 2 The mean nuclear volume in six [vole] oocytes was found to be 5200 μ^3. **1958** *Times Rev. Industry* June 53/2 It [*sc.* the Mercast process] yields castings to extremely close dimensional limits, one example quoted being..a surface finish of 60*μ* an inch or better. **1959** *Listener* 31 Dec. 1161/1 Cell sizes range from a few mu to a few dozens of mu. **1970** Ambrose & Easty *Cell Biol.* i. 7 Bacteria..vary greatly in size from a diameter of 5,000 Å for the cocci up to about 20 *μ* in length for some of the filamentous bacteria.

2. *Electronics.* The amplification factor of a valve.

1918 H. J. van der Bijl in *Physical Rev.* XII. 184 The factor μ_0, which plays a very important part in the theory of operation of the thermionic amplifier, can be more easily and accurately determined with the help of equation (19), which gives the relation between the anode and grid potentials necessary to maintain the current at some convenient constant value. *Ibid.* 192 The voltage amplification *μ* is given by $\mu = \mu_0 R/(R+R_0)$. **1927** B. F. Dashiell *Pop. Guide Radio* vi. 98 The amplification constant of a tube, usually designated as mu, is a measure of the relative effect of changing the grid bias compared to changes in the plate voltage. **1931** *Electronics* Apr. 609/1 Mutual conductance meter... The compensator dial is set for the value most nearly corresponding to the 'Mu' of the tube under test. **1962** *Newnes Conc. Encycl. Electr. Engin.* 860/1 A coarse-pitch grid gives a low magnification and requires greater bias to cutoff. In the variable-mu valve the grid is graded, and this gives a valve whose gain can be altered by changing the bias applied to it.

mu, var. *mou.

‖**Mu'allaqat** (mu,alakā·t), *sb. pl.* Also **Moallakat, Muallakat.** [a. Arab. *mu'allaḳāt*, lit. suspended odes, pl. of *mu'allaqa*.] An anthology of seven pre-Islamic Arabian poems made by the rawi Hammad al Rawija (d. 772). Also in shortened form **Moal.**

1782 (*title*) The Moallakát, or seven Arabian poems.. with a translation..by William Jones. **1834** *Penny Cycl.* II. 219/1 The poems called Moallakat. **1875** *Encycl. Brit.* II. 263/1 Seven of these, known in Arab literature by the title of the *Múallakat* or 'Suspended',..all of them belonging to the 6th century, have become..the..classical standards of Arab poetical composition. **1905** G. Bell *Let.* 12 Feb. (1927) I. 186 At dinner I produced the Muallakat (pre-Muhammadan poems). **1930** R. A. Nicholson *Lit. Hist. Arabs* iii. 101, I will now turn directly to those celebrated odes which are well known by the title of *Mu'allaqát*, or 'Suspended Poems', to all who take the slightest interest in Arabic literature. **1959** *N. & Q.* June 216/2 'The Poem of Amriolkais', the *moál* upon which Tennyson relied for the framework of 'Locksley Hall'. **1964** *Listener* 25 June 1036/1 Behind the *qasidas* of Andalusia lay..the *Mu'allaqat*, the 'hanging odes' of the pre-Islamic oral verse of the Arabian desert.

muccuck, var. *mocock.

much, *a.*, quasi-*sb.*, and *adv.* **A.** *adj.* **2. d.** Delete † *Obs.* and add further examples. Now chiefly *U.S. dial.* and in echoes of quot. 1602 in Dict.

1889 *Kansas City* (Missouri) *Times & Star* 13 Dec., For the latter's fall-down, much thanks. **1890** S. Hale *Lett.* (1919) 253, I have much funny things to tell you. **1928** J. Peterkin *Scarlet Sister Mary* (1929) xxii. 217 How much chillen you had? **1952** E. O'Neill *Moon for Misbegotten* I. 17 You didn't get much thanks from Mike, I'll wager, for your help.

B. *absol.* and quasi-*sb.* **2. e.** (Earlier example.)

1847 A. Brontë *Agnes Grey* xvii. 251, I paid more attention to dress than ever I had done before... This is not saying much, for hitherto I had been a little neglectful in that particular.

g. *a bit much*: see *bit *sb.*^{2} 5. Also in similar phrases with the same sense, as *a trifle much*, *rather much*.

1911 O. Onions *Widdershins* 265, I too smiled... 'It *was* rather much, wasn't it?' I said. **1930** R. Lehmann *Note in Music* v. 200 After all, perhaps it would have been a trifle much, applied to mother. **1964** J. Symons *End of Solomon Grundy* I. i. 27 It's enough to break up any party. I must say I thought it was rather much.

C. *adv.* **1. c.** Also still current in *U.S. dial.*

1916 *Dialect Notes* IV. 347, I don't guess she's much old. **1929** W. Faulkner *Sound & Fury* 268, I..went up front. 'Been much busy?' Earl says. 'Not much,' I says.

d. *not much*: certainly not, far from it, 'not likely'; also occas. (ironically), certainly, 'not half'. *colloq.*

1886 *Harper's Mag.* Dec. 148/1 'Go home?—explain?' he began, more calmly. 'Not much.' **1888** 'R. Boldrewood' *Robbery under Arms* I. xviii. 248 Starlight and I wasn't likely to break down—not much—whatever the jury did. **1904** E. Nesbit *Phoenix & Carpet* xi. 212 'He didn't mean stay and be roasted,' said Robert. 'No boys on burning decks for me, thank you.' 'Not much,' said Cyril. **1911** A. Bennett *Card* x. 255 Do you suppose I was going to let you go by that steamer? Not much. **1928** D. L. Sayers *Unpleasantness at Bellona Club* xvi. 188 'They can get it from Robert or George Fentiman,' warned Wimsey. 'Not much, they won't,' said Salcombe Hardy feelingly. **1970** A. Ross *Manchester Thing* 81 'Got a going over, did you?' 'Not much, I got a going over. Want to see the bruises?' **1973** J. Porter *It's Murder with Dover* i. 1 'I am not asking for any preferential treatment,' said Lord Crouch... Not bloody much! thought the Chief Constable..and tried to work out what His Nibs was up to.

e. *not to be* (or *go*) *much on* (or *for*): (*a*) not to be enthusiastic about (something); not to like or care for; (*b*) not to be useful or effective for (a purpose); to be no good at (something). *colloq.*

1896 *Dialect Notes* I. 417 'I don't go much on that', I don't care much for that. **1908** A. Bennett *Buried Alive* v. 111 'I'm not much for these restaurants,' she said, over grilled kidneys. 'No?' he responded tentatively. 'I'm sorry.' **1928** R. Bradford in B. A. Botkin *Treas. S. Folklore* (1949) III. 485 Angels is all right for singin' and playin' and flyin' round, but they ain't much on workin' de crops and buildin' de levees. **1968** A. Munro in R. Weaver *Canad. Short Stories* 2nd Ser. 262 They may not be much on intellectual conversation but their hearts are in the right place.

5. (*a*) *much-abused, -admired, -branched, -criticized, -debated, -discussed, -dreaded, -needed, -quoted, -travelled, -used, -vaunted*; (*b*) *much talked of* (further example).

1879 W. James *Coll. Ess. & Rev.* (1920) 95 The much-abused subject of mental physiology. **1936** *Discovery* Dec. 382/1 The much abused engine..gave out utterly. **1960** *Farmer & Stockbreeder* 23 Feb. 75/1 The next big price was 3,000gs paid by exporter Mr. James Schofield for Westdrums Winson, Messrs. Boots' much-admired reserve best two-year-old. **1927** Haldane & Huxley *Animal Biol.* i. 35 Most glands..are many-celled tubes or pockets of epithelium, either unbranched or slightly branched..or much-branched like the liver or salivary gland. **1970** T. J. Ross *Film & Liberal Arts* 148 The much-criticized woodenness of her playing is of small matter. **1843** Mill *Logic* I. 9 To this science [*sc.* metaphysics] appertain the great and much debated questions of the existence of matter. **1956** *Nature* 11 Feb. 262/2 The whole of the much-debated line up Borrowdale. **1946** *Ibid.* 23 Nov. 759/2 The much-discussed hypothesis that galactic noise is analogous to the noise associated with solar flares. **1865** G. M. Hopkins *Poems* (1967) 150 You may trust Your footing now to the much-dreaded dust. **1886** E. G. White *Notes of Trav.* in *Hist. Sk. Foreign Missions Seventh Day Adventists* 236/1 We might obtain a little much-needed rest. **1936** *Burlington Mag.* Apr. 202/2 It is a pity not to have given the *Birth of Venus*..a much-needed rest. **1964** *Ann. Reg.* 1963 229 The United States and Canada traditionally supplied much-needed grain. **1927** *Mod. Philology* Nov. 224 The much-quoted example..does not illustrate sound change. **1912** 'Saki' *Chron. Clovis* 196 The County..mustered in full strength to witness the much-talked-of production. **1928** C. Singer *From Magic to Sci.* I. 12 This erudite and much travelled man exhibits great industry. **1890** W. James *Princ. Psychol.* II. xviii. 58 In one it [*sc.* cerebral injury] will throw a much-used brain-tract out of gear; in the other it may affect an unimportant region. **1939** W. S. Churchill *Into Battle* (1941) 150 The magnetic mine..may perhaps be Herr Hitler's much vaunted secret weapon. **1974** *Times* 9 Dec. 12/2 Mr Tanaka..rashly published his much vaunted plan for decentralizing industry.

muchacho (mŭtʃā·tʃo). [Sp.] A boy, young man; a male servant. Also *attrib.* So **muchacha**, a girl.

1591 Garrard & Hitchcock *Arte of Warre* III. 212 The followers of the campe, pages and muchachos, who must be chosen able to fight in a day of seruice, for the defence of themselues and their masters baggage. **1852** *San Diego Herald* 10 Apr. 2/2 Gay, dashing muchachos in boots and spurs. **1877** B. Harte *Story of a Mine* 412 Father Pedro had taken a muchacho foundling for adoption. **1888** 'R. Boldrewood' *Robbery under Arms* II. xi. 190 So the muchacha went back on yer. **1904** Conrad *Nostromo* I. viii. 119 Would not the muchachos of Hernandez like to get hold of this insignificant object..? **1963** *Punch* 23 Oct. 589/3 What we're going on with now, muchachos..is how to catch squid.

muchán, mucharn, varr. Machan.

muchi, var. *moocha.

muchocho, var. *mohoohoo.

muchwhat, *sb.* and *adv.* (Later examples.)

1796 'Tim Bobbin the 2nd' *Plebeian Politics* (1801) 29, I think eawer kese is mitchwhot th' same. **1899** B. W. Green *Word-bk. Virginia Folk-Speech* 246 *Much-what*, nearly; almost. For the most part. 'They are all *much-what*.' **1922** Joyce *Ulysses* 388 It was muchwhat indifferent.

mucification (miūsifikēi·ʃən). *Physiol.* [f. Muc(us + -ification.] Transformation (of epithelial cells) into mucus-secreting cells.

1930 *Proc. R. Soc. Edin.* L. 88 Mucification..is typical of the second phase of the genital cycle; it always accompanies true or pseudo-pregnancy. **1957** *Brit. Jrnl. Radiol.* XXX. 243/1 The intermediate cells did not cornify but instead differentiated into mucus-secreting cells, the mucification being most marked on the third and fourth days after the end of treatment. **1973** *Nature* 9 Feb. 398/1 Progesterone administered for 9 weeks..to.. 3-month-old mice induced a strong intracellular reaction of mucification along the vaginal mucosa.

So **mu·cify** *v. intr.*, to undergo mucification; **mu·cified** *ppl. a.*; **mu·cifying** *ppl. a.*

1930 *Proc. R. Soc. Edin.* L. 95 On sectioning, the wall was found to be thickened and lined with a mucified epithelium. **1941** *Endocrinology* XXVIII. 314, 72 to 96 hours are required for the mucified vagina to become fully cornified after the influence of the corpus luteum hormone has been removed. **1962** *Brit. Jrnl. Cancer* XVI. 648 Castration and progestational hormones have a mucifying effect on the cervico-vaginal epithelium of mice. **1973** *Nature* 9 Feb. 398/1 It is well known that ambivalent cells of the vaginal squamous epithelium..mucify when progesterone acts synergically with oestrogen.

mucilage. 1. c. (Earlier example.)

1859 *La Crosse* (Wisconsin) *Daily Union* 15 Oct. 3/3 Mucilage, sealing wax, playing cards.

mucin. Substitute for def.: **a.** Mucus. **b.** Any of the glycoproteins found in the mucus of various animals. (Add further examples.)

1905 *Chambers's Jrnl.* Jan. 84/2 The mucine which issues from the body of the garden snail. **1938** *Cold Spring Harbor Symp. Quant. Biol.* VI. 91/2 The term 'mucin' is entirely avoided. It should be used only in a physiological sense to denote a viscous fluid of secretory origin. **1966** Schultze & Heremans *Molecular Biol. Human Proteins* I. IV. iv. 769 Salivary mucins are most abundant in sublingual saliva. *Ibid.*, The main sublingual mucin, at least in cattle, differs in composition from submaxillary mucin, as it contains less sialic acid and much more hexose and fucose than the latter product.

muck, *sb.*^{1} **1. d.** Add def.: Soil material consisting of decayed plant remains, similar to peat; in mod. use distinguished from peat by being more thoroughly decomposed (and usu. darker in colour) and having a higher mineral content. (Earlier and later examples.)

1832 H. L. Barnum *Farmer's Own Bk.* 35 On tearing up some handfuls of the ground, this is well blackened of course, and little is thought of looking for the sub-soil, as those invariably do, who have once been deceived by black muck, and these soft beds of leaves. **1839** J. Buel *Farmer's Compan.* ix. 73 Peat earth, or swamp muck, is vegetable food, in an insoluble state, and requires only such a chemical change as shall render it soluable, to convert it into an active manure. **1849** E. Chamberlain *Indiana Gazetteer* (ed. 3) II. 305 The soil is a black muck, based on clay. **1859** S. W. Johnson *Ess. Peat, Muck, & Commercial Manures* 63 Some intelligent farmers call the surface layers of their swamps, which are loose and light in texture, swamp muck, and to the bottom layers, which are more compact and often serviceable as fuel, they apply the term peat. **1889** *Century Mag.* Dec. 217/2 The soil proved to be a wet muck, overlaying sand with boulders. **1897** G. P. Merrill *Treat. Rocks* II. ii. 149 An impure variety [of peat] containing a considerable quantity of siliceous sand, and locally known as 'muck', is used as a fertilizer for 'multching' throughout New England. **1928** *Bull. Amer. Soil Survey Assoc.* IX. 44 Peat has been defined as containing over 65% of organic matter and Muck as containing from 25% to 65%. It does not appear desirable to place such definite limits of composition but rather to base the distinction mainly on the degree of decomposition and secondarily on the content of mineral material. **1930** C. E. Thorne *Maintenance Soil Fertility* ii. 15 Beds of muck and peat..are found on the watershed and in some of the river valleys. *Ibid.*, The distinction between peat and muck is that in these bogs the lower strata, being continually covered with water, may still consist of only partially decomposed fibrous matter, brownish in color, while the surface that has been exposed to the air has more completely decayed, losing its fibrous texture and becoming darker in color. Muck and peat, therefore, may be compared to soil and subsoil. **1943** Millar & Turk *Fund. Soil Sci.* ii. 63 When these organic deposits are yet in the crude, fibrous state, they are frequently designated as peat, but when decay has broken down the plant tissues until the material has something of a loamy consistency, it is called muck. **1971** *Gloss. Soil Sci. Terms* (Soil Sci. Soc. Amer.) 11/2 *Muck*, highly decomposed organic material in which the original plant parts are not recognizable.

3. b. (Earlier and later examples.)

1882 'F. Anstey' *Vice Versâ* xvi. 282 'If you think the tea worth racing like that for, I don't,' said Coggs viciously; 'it's muck.' **1928** W. Ponder *Clara Butt* 138 All I can say is..sing 'em muck! It's all they can understand. **1943** K. Tennant *Ride on Stranger* iv. 34 He had a habit of greeting any new dish with a loud: 'What's this muck?' **1959** I. & P. Opie *Lore & Lang. Schoolch.* ix. 162 School dinners are 'muck', 'pig swill', 'poison', 'slops',.. and Y.M.C.A. (Yesterday's Muck Cooked Again). **1967** *Listener* 20 Apr. 524/3 Is this the kind of muck which the National Film Theatre is going to bring to Norwich?

d. Waste material that is removed during mining or civil engineering operations; *spec.* (*U.S.*), surface material that overlies a placer deposit.

1883 W. S. Gresley *Gloss. Terms Coal Mining* 171 *Muck* (Y[orkshire]), see *Dirt* [= 'clay, bind, or other useless rubbish produced in mining, and which accidentally is sent out of the pit mixed with the coal']. **1897** J. W. Leonard *Gold Fields Klondike* 180 The top 'muck', as it is called by the miners, is, when thawed out, about two-thirds water and one-third sediment. **1908** J. M. MacLaren *Gold* II. 484 The low-level gravels..lie on decomposed schist bed-rock, and are covered by black frozen 'muck' (silt, vegetable matter, and ice, the last forming 75 per cent. of the mass) of a thickness of 2 to 30 feet. **1914** G. Atherton *Perch of Devil* I. 148 His..hands were white with 'muck', a mixture of rock-dust and water. **1959** *Times* 16 Nov. 8/5 About 400,000 tons of spoil (or muck as the mining engineer calls it) will be brought to the surface. **1960** *Vancouver Sun* 4 June 2/4, 150 vertical feet of muck in the stope above had settled gently down over the mule, the string of cars—and our only way out.

e. Phrases. *as muck*: used emphatically following adj.; *like muck*: used negatively with a statement.

1935 J. C. Masterman *Fate cannot harm Me* viii. 154 He would be out any ball and poor old George would be as sick as muck. **1952** [see *drunk *sb.*^{2} 1].

f. *R.A.F. slang.* (*a*) Hostile anti-aircraft fire; (*b*) (see quot. 1943).

1940 Michie & Graebner *Their Finest Hour* iv. 65, I climbed to 12,000 feet, circling along the outside of the searchlights and all the muck [gunfire] that was coming up. **1943** Hunt & Pringle *Service Slang* 46 *Muck*, dirty weather.

g. *Lord Muck*: see *lord *sb.* 14 c. So *Lady Muck.*

1957 I. Cross *God Boy* (1958) xxii. 190 She sat there, sipping away at her tea like Lady Muck. **1966** 'L. Lane' *ABZ of Scouse* 61 Lady Muck of Muck Hall: A woman who puts on airs, has a condescending manner and is regarded as excessively conceited.

4. b. *to make a muck of*, to do (something) badly; to spoil or bungle.

1906 D. F. T. Coke *Bending of Twig* xiv. 222 There'll be nobody much there, so it doesn't matter if you *do* make a muck of it. **1936** R. Lehmann *Weather in Streets* III. i. 265, I would like to paint her, but..would make a muck of it. **1947** 'N. Shute' *Chequer Board* iv. 94 He's made a bloody muck o' things, the way I knew he would. **1970** Y. Carter *Mr. Campion's Falcon* xxi. 159 I've made a muck of it. What the hell can I do now?

5. *muck-headed* adj.; (sense 1 d) *muck-bed, -land, -swamp*; **muck-crome, -crone, -croom** [Crome, cromb *sb.*], a dung-fork; **muck-pot** ? *nonce-wd.* = **muck-spout*; **muck-shifter,** a man who or a machine which removes earth; **muck-shoveller,** (*a*) a farmhand employed in collecting or distributing dung; (*b*) *Austral. slang* (see quot. 1945); so *muck-shovelling* vbl. sb.; **muck soil,** a soil composed of muck (sense 1 d) (see quot. 1928^{1}); **muck-spout** *dial.* and *slang*, one who uses obscene language or who displays a salacious mentality; **muck-spreader,** a machine for spreading dung; **muck-spreading** *vbl. sb.*, the action of distributing dung over a field; **muck-sweat** (later examples).

1872 A. de Morgan *Budget of Paradoxes* 163, I certainly think the words would never have come together except in this way:—I, quart pyx, who fling muck beds. **1823** E. Moor *Suffolk Words & Phr.* 239 A crooked fork for pulling the article out of carts on to heaps we call muck-crome. **1846** Muck-croom [see Crome, cromb *sb.*]. **1969** G. E. Evans *Farm & Village* x. 111, I got an old *muck-crone* (a fork with curved tines), and I used to put this muck-crone under the door and draw a slab of linseed cake through. **1971** *Country Life* 11 Mar. 533/1, I got out o' me punt and stuck a muck-crome, a muck-rake in it [*sc.* a stranded sturgeon]. **1909** H. G. Wells *Ann Veronica* xiii. 272 'Ass!' he went on, still warming. 'Muck-headed moral ass! I ought to have done anything.' **1847** W. Bacon *Let.* 24 Nov. in *Rep. Comm. Patents 1847* (U.S.) (1848) 358 They have been planted the present year, on deep muck lands. **1936** *Sun* (Baltimore) 17 Jan. 3/4 Shattered remnants of the transcontinental airliner which plunged seventeen persons to death in a nearby muckland. **1950** *Daily Progress* (Charlottesville, Virginia) 24 July 1/2 It plunged into the pine-spotted muckland. **1938** Dylan Thomas *Let.* 6 July (1966) 205 It's a crack at young Georgians,..intellectual muckpots leaning on a theory, post-surrealists and orgasmists. **1880** D. W. Barrett *Life & Work among Navvies* II. ii. 43 Navvies themselves speak of one another as muck-shifters, or thick-legs. **1961** *Engineering* 9 June 797 Designed to work under rugged off-highway conditions as a muck-shifter. **1967** G. F. Fiennes *I tried to run a Railway* vi. 63 Muck shifting is easy nowadays. **1970** *Daily Tel.* 5 Nov. 13/6 The whole 'muck-shifting' industry changing the shape of the landscape is experimenting all the time with bigger, and sometimes better, machines. **1945** Baker *Austral. Lang.* v. 98

Muck-shoveller, a tin miner. **1960** *Farmer & Stockbreeder* 23 Feb. 105/3 Of these 32 [farmers], 21 simply wanted a muck-shoveller. **1928** *Bull. Amer. Soil Survey Assoc.* IX. 44 *Muck soil*, soil composed of thoroughly decomposed organic material, with a considerable amount of mineral soil material, finely divided and with few fibrous remains. **1928** F. E. Bear *Theory & Pract. in Use of Fertilizers* xvii. 280 If..more potash can be used to advantage..the ..analysis might well be changed in that direction. This may be the case with muck soils. **1970** *Jrnl. Econ. Entomol.* LXIII. 1283/1 Studies were made to determine the fate of ^{14}C-labeled aldicarb in sand, loam, clay, and muck soils maintained at different moisture levels. *a* **1825** R. Forby *Vocab. E. Anglia* (1830) II. 223 *Muck-spout*, one who is at once very loquacious and very foul-mouthed. **1916** D. H. Lawrence *Let.* c 15 Dec. (1962) 492 And Murry..is a little muck-spout. **1961** *Guardian* 30 May 5/1 In the more developed countries everything from washing machines to muck-spreaders has been mechanised. **1975** *Listener* 29 May 702/2 Gardeners going berserk with mechanical diggers and muck-spreaders. **1903** *Eng. Dial. Dict.* IV. 187/1 [Nottinghamshire] A farmer on being asked in Court when the event occurred, said, 'It wor abaout three weeks afore muck spreading.' **1948** *Brit. Birds* XLI. 358 This is paralleled by the Faeroe farmers' belief that muck-spreading should be completed before the coming of the White Wagtail. **1960** *Farmer & Stockbreeder* 8 Mar. 74/1 An interesting attachment designed for fitting to conventional tipping trailers was this p.t.o.-driven muck-spreading device. **1870** *Rep. Comm. Agric. 1869* (U.S. Dept. Agric.) 270 The soil was..black mud or muck swamp, five feet deep, containing a mixture of sand. **1879** Browning *Dramatic Idyls* 115 Publican Black Ned Bratts and Tabby his big wife too: Both in a muck-sweat. **1922** Joyce *Ulysses* 515 I'm all of a mucksweat. **1953** R. Lehmann *Echoing Grove* 48 Must have a shower. I've been in a muck sweat all day. **1972** J. Porter *Meddler & her Murder* i. 10 There's nothing for you to get into a muck sweat about.

muck, *v.* Add: **1.** Also freq. with *out*, and occas. with *off*, and *absol.*

1851 Muck out [in Dict.]. **1914** Kipling *Divers. Creatures* (1917) 52, I was obligin' Jim that evenin' muckin' out his pig-pen. **1921** [see *cleaner a]. **1950** *Landfall* XIII. 16 They always want one [*sc.* a cup of tea] after they've finished mucking out. **1958** J. Betjeman *Coll. Poems* 252 She can muck out the stables and clean Her snaffle and saddle and bridle. **1952** R. S. Summerhays *Elem. Riding* (ed. 3) iii. 23 The deep bedding has now been 'mucked out'. **1966** M. Torrie *Heavy as Lead* xiv. 169 Sir G. had told me special to muck out the pigs. **1967** C. Watson *Lonelyheart 4122* ix. 91 He would have to be strong, energetic, used to stud work and willing to muck out. **1973** J. Burrows *Like an Evening Gone* ii. 30 I've mucked out the henhouses.

4. (Later example.) Also, to mix *up*.

1909 H. G. Wells *Tono-Bungay* III. i. 279 It's a festering mass of earths and heavy metals... There they are, mucked up together in a sort of rotting sand. **1916** 'Boyd Cable' *Action Front* 109 If it [*sc.* a shell] had fell in the trench, now, and mucked up half a dozen men, there'd have been something to squeal about. **1949** 'J. Tey' *Brat Farrar* xi. 85 You don't want that dazzling outfit of yours to be mucked up.

b. (Earlier and later examples with *up*; later examples with *about*.)

1886 in H. Baumann *Londinismen*. **1922** 'R. Crompton' *Just—William* viii. 161 You seem to have pretty well mucked it up. **1935** 'N. Blake' *Question of Proof* i. 17 Old Simmie will probably muck up the stop-watch like he did last year. **1946** K. Tennant *Lost Haven* (1947) xi. 180 This is a real good stove... She isn't mucked about and cleaned, and that's what makes her a good stove. **1959** 'M. Cronin' *Dead & Done With* vi. 99 'Lena could muck it all up.' 'I don't think she will, so long as she's scared about herself.' **1959** Anon. *Streetwalker* iii. 58 Muck me about and you're out. **1966** *Listener* 17 Nov. 719/2, I was delighted to see Mr Bernard Levin..heading a review.. '*Much Ado About Nothing*, by William Shakespeare, the text mucked about by Robert Graves'. **1971** *Nature* 7 May 65/3 Let us not muck up our language, lest we also muddle our minds. **1973** *Nation Rev.* (Melbourne) 31 Aug. 1446/6 But she went and mucked it all up with her television debut.

5. (Earlier and later examples.) Freq. with *with*, and also with *around*.

1856 H. Phillips *Jrnl.* 26 Sept. (typescript) 41 Cutting firewood and mucking about the house. **1918** H. G. Wells *Joan & Peter* xi. 391 They had long bicycle rides together... They 'mucked about' with Baker's boat. *Ibid.* xiii. 659 He would be climbing trees with Joan, 'mucking about' in the boats with Joan. **1928** D. L. Sayers *Lord Peter views Body* 276 His art..[is] the one thing a genuine artist won't muck about with. **1935** N. Marsh *Enter Murderer* vii. 90 'E was a-mucking arahnd Trixie. **1946** K. Tennant *Lost Haven* (1947) x. 152 We been mucking about and mucking about, and got nowhere. **1957** P. Mansfield *Final Exposure* ix. 121 Why don't you haul him in instead of mucking around asking me bloody silly questions? **1959** 'M. Neville' *Sweet Night for Murder* xiii. 128 They get fed up with their own wives and have to go mucking about with someone new. **1959** *Engineering* 13 Mar. 343/3 Those Americans are mucking about with their journal titles again. **1963** *Truth* (Wellington, N.Z.) 9 July (*heading*) Don't muck about, Mac. **1973** *Times* 12 Dec. 2/8 The other girls..wanted to muck around with boys.

b. *to muck in with*: to share army rations with; to consort or co-operate with; so *to muck in*, (*a*) to eat; (*b*) to help, to 'pull one's weight', to participate. Also, *to muck in to*.

1919 *Athenæum* 1 Aug. 695/2 'To muck in' with anyone is to share rations with him. **1929** F. Manning *Middle Parts of Fortune* I. v. 105 'Martlow and I have mucked in together, since you've been in the orderly-room.' 'Well, the three of us can muck in together now,' said Bourne. *a* **1935** T. E. Lawrence *Mint* (1955) I. viii. 30 'Muck in', we did, yet still looked lean. **1936** F. Richards *Old-Soldier Sahib* xiii. 223 For nine months he had been mucking-in with a youngster who had only arrived in the country the previous winter. **1942** G. Kersh *Nine Lives Bill Nelson* v. 26 The Army was hell because I couldn't muck in there. **1942** Wodehouse *Money in Bank* (1946) iii. 26 When we came to visit here, I understood that that room was reserved for I and my husband. Nobody ever mentioned that we were supposed to muck in with the butler. **1952** M. Laski *Village* vi. 112 We all muck in together and the jobs get done in no time. **1958** J. Cannan *And be a Villain* iv. 89 I'm delighted to muck in, but I'm afraid I'm too conscientious for Mrs Langley. **1959** *Encounter* July 27/1 They want men who will muck in as colleagues. **1966** F. Shaw et al. *Lern Yerself Scouse* 42 *Muck in, yer at yer granny's!* Bon Appetit! **1967** *Guardian* 21 Apr. 7/2 Prince Charles..will be able to muck in to all the student activities. **1970** *Times* 13 Feb. 10/8 The company..all muck in, take small or big parts. **1974** J. Pope-Hennessy *R. L. Stevenson* vi. 128 His readiness to muck in with any of his working-class fellows on boat or train.

c. To search for coal on a coal-tip, a beach, etc.

1935 A. J. Cronin *Stars look Down* I. ii. 22 'It's my duff,' Softley kept whimpering... 'Aw mucked for it, aw did, for my man to hev a fire.'

6. Euphemistically (chiefly in written work) = *fuck *v.*

1929 R. Aldington *Death of Hero* x. 376 Spree be mukked—one of you * * fired his rifle and muckin' near copped me. **1940** E. Hemingway *For whom Bell Tolls* xxii. 273 He may have just mucked off. *Ibid.* xxxv. 369 You're just mucked..you're mucked for good. *Ibid.*, Muck my grandfather and muck this whole treacherous muck-faced mucking country. **1941** *Penguin New Writing* II. 90 'I shall report you to the foreman.' 'Muck the foreman!' I said. **1946** D. Hamson *We fell among Greeks* vii. 85 Another song I used to perpetrate..was 'I'm a man that's in trouble and sorrow.'..'Muck it,..that's a good song, Denys, let's have it again.' **1949** C. Fry *Lady's not for Burning* 92 Youse only has to say muck off, and I goes, wivout argument. **1950** E. Hemingway *Across River* vii. 58 Now muck off..try and have some fun. **1974** R. Adams *Shardik* xxxvi. 300 Come on, now,..you'll get nothing here, so just muck off, there's a good lad.

mu·ck-about. [f. the phrase *to muck about* (Muck *v.* 5).] **a.** A person who 'mucks about'. **b.** The action of 'mucking about'.

1933 H. G. Wells *Bulpington of Blup* ix. 353 Rich old women in Paris—middle-aged muck-abouts—art shops. **1968** L. Berg *Risinghill* 122 The boys..were restless, ready for a 'lark' or a 'muck-about' or a 'giggle'.

muck-a-muck (mɒ·kămɒk). [Chinook jargon.] **1.** Used by or with reference to Amerindians of western North America: food.

1847 J. Palmer *Jrnl. Trav. Rocky Mts.* 150 Muck-a-muck, Provisions, eat. **1852** *Oregonian* (Portland) 25 Dec. 2/3 The aborigine..'put' for the settlement with a sort of legs-do-your-duty-for-the-body-is-in-danger resolution for his *muckamuck*. **1863** *Norfolk Reformer* (Simcoe, Ontario) 8 Jan. 3/1 On arriving as far back as Lytton or Lilooet, there was employment..and 'muca muc', as the Indian name implies. **1880** *Forest & Stream* 11 Nov. 285/2 We should have to come ashore and have some 'muck-a-muck'. **1895** H. S. Somerset *Land of Muskeg* 167 Yes, all kinds of muck-a-muck at McLeod; jam, cake, biscuits.. plenty plenty muck-a-muck, you see. **1915** R. D. Cumming *Skookum Chuck Fables* 18 Perhaps he had bought all his luxuries on jaw-bone from one store while he paid cash for his muck-a-muck in another. **1963** R. D. Symons *Many Trails* 74 Hi-ya tillicum... You plenty muck-a-muck stop.

2. *fig.* Shortening of *high-muck-a-muck.

1912 Kipling *Songs from Bks.* (1913) 159 Shaman, Ju-ju or Angekok, Minister, Mukamuk, Bonze. **1914** *Dialect Notes* IV. 113 *Squeegee*, a person of importance; muckamuck:—used derisively. **1966** H. Kane *Conceal & Disguise* iv. 28 Cape Ulrich was for the muckamucks, the coupon-clippers, the expense account lads, the heavy rich.

mucka-mucka (mɒ·kămɒkă). [Native name.] A name used in Guyana for the large perennial herb, *Caladium arborescens*, native to tropical America and belonging to the family Araceæ. Also *attrib.*

1918 C. W. Beebe *Jungle Peace* (1919) vi. 123 Mucka-mucka was here and there in the foreground. **1955** *Times* 16 July 7/6 They perched, feeding on the leaves of mucka-mucka, a giant variety of arum. *Ibid.* 23 July 7/6, I sat alone and very still in a primitive dug-out after edging it into the heart of a mucka-mucka plant (*Caladium arborescens*) and tying the boat to one of its thick stems. **1958** J. Carew *Black Midas* iii. 35 The weeds, the wild cane, the lilies, the mucka-mucka were all still under the burning sun.

mucked, *ppl. a.* Add: **mucked-about,** spoiled, subjected to unnecessary interference; **mucked-up,** bungled, spoiled.

1930 M. Allingham *Mystery Mile* xxvii. 273, I should be very interested to know how you intend to get away with your reputation all pure and virgie and our Albert's poor little mucked-up corpse lying about. **1930** V. Palmer *Passage* I. iii. 29 It was a mucked-up job. **1966** J. Bingham *Double Agent* iv. 52, I don't remember what I ate. Some mucked-about dish of stew.

mucker, *sb.*^{1} Restrict † *Obs.* to senses in Dict. and add: **3.** Euphemistically (chiefly in written work) = *fucker. Cf. *muck *v.* 6.

1929 R. Aldington *Death of Hero* III. i. 263 Does the old mucker think we're going to run away? **1942** *Penguin New Writing* XV. 19 Suddenly old Bob says: 'I lost my kit, Cockney.' Silly mucker! **1952** M. Tripp *Faith is Windsock* ii. 38, I thought there was twelve hundred muckers in this raid—where's t' other eleven hundred and ninety-nine?

mucker, *sb.*^{2} Restrict *slang* to sense in Dict. and add later examples.

1914 Galsworthy *Mob* I. 8 You're riding for a fall and a godless mucker it'll be. **1916** 'Taffrail' *Pincher Martin* xiv. 270 'Minefield! I thought the one that got you was a floater.' 'Don't you believe it... You can thank your lucky stars you didn't bump one.'..'I'm glad we didn't come a mucker—jolly glad!' **1928** Galsworthy *Swan Song* I. vi. 41 But for you, old girl, I might have gone a holy mucker myself. **1943** M. Aklom *Of Social Significance* in *Best One-Act Plays 1942–43* 166 You know what the Guardianship of Infants Act is. They passed it to prevent legal infants coming matrimonial muckers at the age of indiscretion. **1974** G. Mitchell *Javelin for Jonah* xii. 154, I like old Jimmy boy and I wouldn't want to see him come a mucker.

2. One who, or a machine which, removes muck (*muck *sb.*^{1} 3 d).

1899 *Harper's Weekly* 20 May 498/1 [The] Company.. paid $3. for miners and $2.50 for 'muckers', or underground laborers. **1908** [see *jumbo 1 c]. **1916** C. Sandburg *Chicago Poems* 21 Twenty men stand watching the muckers. Stabbing the sides of the ditch..for the new gas mains. **1923** 'B. M. Bower' *Parowan Bonanza* viii. 94 Now you've staked yourself to the luxury of a mucker, you can leave him in charge. **1927** *Dialect Notes* V. 456 *Mucker*, a shovel man. **1931** 'D. Stiff' *Milk & Honey Route* ii. 24 The rawjawed teameos and muckers..who built that railroad got little or no hard cash. **1956** *Fatal Accidents* (Ontario Dept. Mines) Apr. 2 The mucker was hoisted clear of the bottom timber by 10 feet, to where it might normally have been anchored. **1965** S. G. Lawrence *40 Yrs. on Yukon Telegraph* iv. 23 A mucker's duties are simply to shovel and wheel out rock, and in those days on a wheelbarrow.

mucker, *sb.*^{3} For *U.S. slang* read orig. *U.S. slang*, and add further examples.

1900 Ade *Fables in Slang* 108 They were not Muckers; they were Nice Boys. **1905** D. G. Phillips *Plum Tree* 35 He used to class himself and me together as 'us gentlemen', in contrast to 'them muckers', meaning my colleagues. **1920** F. Scott Fitzgerald *This Side of Paradise* 175 Why is it that the pick of young Englishmen from Oxford and Cambridge go into politics and in the USA we leave it to the muckers? **1936** A. Huxley *Olive Tree* 182 There are the refined and aristocratic Muckers in East Prussia, with their ritual of exhibitionism and long-drawn sexual confessions.

† c. A youthful townsman, as distinct from a member of a college; a young 'townee'. *Obs. U.S. university slang.*

1893 W. K. Post *Harvard Stories* 75 On the first corner of Harvard Street were stationed three or four boys (the occasionally useful Cambridge muckers) employed as vedettes. **1899** A. H. Quinn *Pennsylvania Stories* 168 Del went through his pockets to the great joy of a limited assortment of muckers who were following. **1948** [see *Führer].

2. [Perh. f. *muck *v.* 5 b.] A companion, friend, 'mate'. *slang.*

1947 J. Bertram *Shadow of War* VII. v. 239 What's the griff, mucker? **1954** 'S. Carnegie' *Noble Purpose* 21 McLeod..was a small dark man of lugubrious appearance. His mucker Reed was fat, fair and cherubic. **1963** 'R. Erskine' *Passion Flowers in Italy* v. 56 Well, my father was at Magdalen, but all my muckers seemed to be flocking to the House, so I thought I might as well go there too. **1971** B. W. Aldiss *Soldier Erect* 112 It isn't Taff. It's a mucker of mine down in M/T. Jock McGuffie. He's a real cure. **1972** M. Woodhouse *Mama Doll* ix. 120 'Is that my old mucker?' said Bottle. 'None other,' I said.

3. With advbs.: **mucker-in,** one who 'mucks in' (see *muck *v.* 5 b); **mucker-upper,** a bungler.

1942 A. P. Jephcott *Girls growing Up* iv. 83 When a girl first goes to a factory she may be a 'mucker in' and have to turn her hand to anything. **1972** *Guardian* 18 Feb. 14/1 A raconteur..good value at a gathering where he knows his audience, a mucker-in, a natural wit. **1942** T. Rattigan *Flare Path* I. 9 She's a proper mucker-upper though—she'll go and catch the wrong bus.

mucker, *v.*^{2} Add: **b.** Also with *away*: to squander.

1928 H. G. Wells *Way World is Going* 15 The Western Powers of Europe..muckered away an enormous amount of war gear and money in supporting crazy 'white hopes' against the nascent new thing in Russia.

muckerdom (mɒ·kərdəm). [f. *mucker *sb.*^{3} + -dom.] The world of muckers or 'townees'; muckers collectively. So **mu·ckerish** *a.*, befitting or characteristic of a mucker; **mu·ckerishly** *adv.*; **mu·ckerism,** the characteristic conduct of muckers; unsportsmanlike behaviour.

1893 W. K. Post *Harvard Stories* 254 In five minutes all the best talent in muckerdom will be there with tin-cans and stones. **1904** *Public Ledger* (Philadelphia) 4 June 6 If a player on the opponent's side happened to make a muff or misplay, cheering by the side benefited was.., in the elegant language of the campus, 'muckerish'. **1906** *Outing* Jan. 494/1 This year there was caterwauling and shouting by cadets individually and muckerishly that was so unusual and unpleasant as to make one discredit one's ears. *Ibid.* 494/2 We hope it does not mean an entrance of muckerism into our Army and Navy games. **1952** C. Stead *People with Dogs* 330 It's muckerism, saying we all grunt in monosyllables!

muck-heap. Add: (Later examples.) So **mu·ck-heapy** *adj.*

1881 *Punch* 29 Oct. 193/1 We sincerely hope that..His Grace [of Bedford] drove up Wellington Street, made at once for Mud-Salad Market, and..examined the streets 'all round and about that quarter', whose greasy, muck-heapy state is still a disgrace to the Metropolis in general. **1938** 'N. Shute' *Ruined City* vi. 112 One didn't finance a muck-heap of a country like Laevatia for fun. **1946** *Coast to Coast 1945* 177 Somebody counted fourteen bottles on old Gormie's muck-heap.

mucking, *vbl. sb.* Add: **2. b.** Also *mucking-about.*

1937 Partridge *Dict. Slang* 539/2 *Mucking-about*,..an intimate fondling: low (mostly costers'). **1969** *Listener* 27 Feb. 264/3 'Knowledge is conceived in the hot womb of Violence,' said Auden in his poem on Oxford: perhaps he knew what he was doing when he omitted the phrase in his last mucking-about with the piece.

3. Also with *out.* The action of removing muck, esp. dung from a stable or waste material from a mine.

1641 [see Muck *v.* 1]. **1840** *How to buy a Horse* viii. 160 This Augæan labor is termed 'mucking out'. *Ibid.*, In fact, such an operation as 'mucking out' should, in a well-regulated stable, be an impossibility, for there never should be any 'muck' to take away. **1918** H. L. Carr in R. Peele *Mining Engineers' Handbk.* vii. 260 Mucking, or loading broken rock into hoisting conveyance, occupies 50% of shaft-sinking time. **1932** E. Wilson *Devil take Hindmost* xxi. 218 The men, who had been displaced by new mucking machines (mucking is cleaning out the tunnel after the blast), were to be transferred. **1935** *Mining Mag.* LII. 55/2 Two buckets were used in mucking, one of which was being filled while the other was being hoisted. **1957** *Times* 2 July (Agric. Suppl.) p. vi/3 In a modern fattening house we have also to make provision for minimum labour requirements in feeding, mucking out and weighing. **1960** *New Scientist* 7 Jan. 38/1 'Mucking out', as the removal of the rock fragments is termed, is thus simplified and speeded up. **1961** *Encycl. Brit.* XV. 543/2 The method of mucking is reflected in the choice and design of the haulage system. **1973** K. Giles *File on Death* v. 147 Old Joe keeps the rain channel free on the duck-pond side... Old Joe does a thorough job at mucking-out.

mu·cking, *ppl. a.* [f. Muck *v.* + -ing^{2}.] Euphemistically (chiefly in written work) = *fucking ppl. a. Cf. *muck *v.* 6. Also as quasi-*adv.*

1929 R. Aldington *Death of Hero* III. x. 375 What the muckin' hell are you doing, down there? **1933** H. G. Wells *Bulpington of Blup* vi. 237 Don't be a mucking fool! *Ibid.* 241 Do you think we want to sit round telling ghost stories in this mucking hole? **1935** E. Hemingway *Green Hills Afr.* xiii. 277 And if I ever hit you I'll break your mucking jaw. **1942** *Penguin New Writing* XV. 19 I'll miss the silly mucker... Poor old Bob. Went down with his mucking duffle. **1946** D. Hamson *We fell among Greeks* v. 61 By Christ, it's that mucking dog. **1974** R. Adams *Shardik* xxxvi. 301 You'd better lend him a hand. .. We'll be 'alf the mucking night else. *Ibid.* xlv. 363 The first man peered in his turn. 'He mucking is, too,' he said. 'Aren't you?'

muckluck (mɒ·klɒk). Also **maclock, mucluc, muklek, mukluk.** [Eskimo.] A high boot made of sealskin, canvas, etc. Also *attrib.*

1868 F. Whymper *Trav. Alaska* 136 Their boots vary in length, and in the material used for the sides, but all have soles of 'maclock', or sealskin, with the hair removed. **1901** *Pall Mall Mag.* Jan. 56 We stop on our way..and buy a pair of mucklucks or Esquimaux seal boots. They are water-tight, clumsy, evil-smelling, [etc.]. **1904** E. Robins *Magnetic North* I. 51 Nothing like muck-lucks with a wisp of straw inside for this country. *Ibid.* 176 Stretching out his feet, very comfortable in their straw-lined mucklucks. **1913** R. W. Service *Rhymes of Rolling Stone* 118 Then it's down to chewing muclucs, to the water you can *eat*, To fish you bolt with nose held in your hand. **1924** *Chambers's Jrnl.* Jan. 41/1 He was habited..in anorak and skin breeches and mukluks. **1947** *Chicago Tribune* 11 Dec. 20/3 He was presented with..a pair of mukluk boots made to order for Paul Bunyan. **1962** W. Noyce *To Unknown Mountain* v. 54 Unable to fit his vast feet into any boots but a pair of canvas mucklucks. **1972** *Daily Tel.* (Colour Suppl.) 8 Dec. 10/4 His legs were now swathed almost to the knee in the white sealskin mukluks that are standard issue in the [Canadian National Parks] Service. **1973** *Observer* 29 Apr. 34/8 His mukleks—the Eskimo moccasins with the hard, sealskin soles.

muckna (mɒ·kna). *India* and *Ceylon.* [Hind., f. Skr. *matkuṇa* (among various senses) an elephant without tusks.] A male elephant without tusks, or one having only rudimentary tusks.

c **1780** R. Lindsay in A. W. C. Lindsay *Lives of Lindsays* (1849) III. 194 The *muckna*, or elephant born without teeth, is thought the best. **1848** tr. *W. Hoffmeister's Trav. Ceylon & India* vi. 207 Of the herd of elephants,..the largest, whose height does not much exceed nine feet, is a '*Muckna*' [sic] *i.e.*, an elephant with short, straight tusks, which never grow. **1878** G. P. Sanderson *Thirteen Yrs. among Wild Beasts India* vi. 66 Mucknas breed in the herds, and the peculiarity is not hereditary nor transmitted. **1886** G. Yule in Yule & Burnell *Hobson-Jobson* 454/1, I can distinctly call to mind 6 mucknas that I had ..out of 30 or 40 elephants that passed through my hands. **1890** S. W. Baker *Wild Beasts* I. vi. 226 A tiger sprang from the grass, and seized a large muckna (tuskless male) by the trunk.

mucko (mɒ·ko). *slang.* [? f. Muck *sb.*^{1} 3 + *-o^{2}.] (See quots.)

1937 Partridge *Dict. Slang* 539/2 *Mucko*, orderly man: military. **1943** Baker *Dict. Austral. Slang* (ed. 3) 52 *Mucko*, a sailor. (R.A.N. slang.) **1961** Partridge *Dict. Slang* Suppl. 1191/1 *Mucko*,..esp. a man detailed to serve food to troops aboard ship. **1964** *Punch* 6 May 659/1 No general had..had till then..a mucko–chummo relationship with his men.

muck-rake, *sb.* (Later *attrib.* and *fig.* examples.)

1906 *Sun* (N.Y.) 12 Apr. 8/3 On Saturday the President is to pronounce the formal address at the grave of the Man With the Muck Rake... The Muck Rakers worked merrily for a time in their own bright sunshine, and an unthinking populace applauded their performance. **1906** T. Roosevelt in *Cincinnati Enquirer* 15 Apr. 4/4 The men with the muck-rakes are often indispensable to the well-being of society; but only if they know when to stop raking the muck. *c* **1926** 'Mixer' *Transport Workers' Song Bk.* 18 Then you start your muckrake scandel [*sic*] And good men you dare defame. **1968** *Listener* 28 Nov. 726/2 Frank Harris deplored the dismal fate of Sophia in a Paris boarding-house,..regretting that Bennett..preferred to give her a muck rake instead of a soul.

muck-rake, *v. intr.* Delete *nonce-wd.* and add later *fig.* examples.

1954 *Amer. Scholar* XXIII. 421 McCarthy would enjoy himself if he could muckrake in our barnyard. **1970** *Guardian* 6 Mar. 8/3 They need no prompting to register complaints, muckrake among the police, petition for redress against indecencies.

2. *trans.* [Back-formation f. *muck-raker.] To subject (powerful persons or institutions) to allegations of corruption or other illegal or scandalous behaviour; to discover and publish (such scandals); to examine (political districts) so as to determine the extent of corruption. So **muck-raked** *ppl. a. U.S.*

1910 *N.Y. Even. Post* 10 Dec. 8 Their knowledge of how it feels to be a muck-raked millionaire. **1913** J. London *Let.* 26 June (1966) 388, I..muck-rake the powers that be from one end of the world to the other. **1931** *Time & Tide* 22 Aug. 991/2 Steffens was more successful with his experiments when he was muckraking America's big cities. **1943** M. Flavin *Journey in Dark* 193 The country has been muckraked from one end to the other. **1973** G. Jenkins *Cleft of Stars* vii. 86, I couldn't bring myself to muck-rake details of the guard's murder.

mu·ck-raker. [-er^{1}.] A person who uses a muck-rake. In literary use only *fig.* **† a.** A miser. *Obs.* (Cf. Muckworm 2 a.) **b.** One who seeks out and publishes scandals, allegations of corruption, etc., about prominent people, esp. public officials. orig. *U.S.* **c.** A prurient inquirer into private morals; a writer of pornography.

The source of this figurative use in b is T. Roosevelt's speech (itself reminiscent of Bunyan) cited s.v. *muck-rake *sb.*

1601 A. Dent *Plaine Mans Path-Way to Heaven* 102 We see the world is full of such pinch-pennies, that wil let nothing goe, except it bee wrung from them perforce, as a key out of Hercules hande. These gripple muck-rakers, had as leeue part with their bloud, as their goods. **1906** S. Ford *Shorty McCabe* xi. 233 That's the style you live in when..you've got to be a top-notch grafter that the muck-rakers ain't jungled yet. **1914** R. Brooke *Let.* Apr. (1968) 579 Damn it, we're not muck-rakers or German novelists. **1921** G. B. Shaw *Back to Methuselah* IV. I. 159, I leave them to the chumps and noodles, to the blockheads and the muckrakers. **1950** G. Barker *True Confession* iii. 16 And the muckrakers I have known. **1973** *Guardian* 26 May 1/3 He is a sanctimonious creep..a muck raker. **1974** *Times* 9 May 21/4 In its origins the term 'muckraking' described a tradition of American journalism around the turn of the century which was committed to the exposure of trusts and monopolies and of corruption..; the Muck-rakers..were responsible for progressive reforms.

muck-raking, *ppl. a.* (Later examples.)

1951 M. McLuhan *Mech. Bride* (1967) 7/1 This exciting suspicion about personal plots and dastard motives everywhere led to the popularity of the muckraking press. **1972** J. Philips *Vanishing Senator* (1973) II. i. 48 A muck-raking journalist who's gotten rich on other people's misfortunes.

mu·ck-raking, *vbl. sb.* [f. Muck-rake *sb.* and *v.* + -ing^{1}.] The employing of a muck-rake, in fig. senses. Also *concr.*, the results of psychological or social inquiries.

1911 *N.Y. Even. Post* 25 Jan. 14 The same articles brought President Roosevelt to the defence of the Senate, and led him to apply the word 'muck-raking' to the literature of higher exposure. **1919** 'W. N. P. Barbellion' *Jrnl. Disappointed Man* 211 Any eminently 'right-minded' *Times* or *Spectator* reader will ask: 'Who in Faith's name is interested in your introspective muck-rakings?' **1931** L. Steffens *Autobiogr.* I. xiii. 105 My first essay into muck-raking cost me nothing. **1959** [see *demagoguery]. **1973** 'M. Innes' *Appleby's Answer* xv. 128 Don't imagine I have the slightest wish to be in on your muck-raking.

mu·ck-up. *slang.* [f. phr. *to muck up* (*muck *v.* 4).] A mixing or confusing; a confused situation; a blunder, a fiasco, a mess, a muddle; *Austral.* (see quot. 1945).

1930 *Daily Express* 9 Sept. 8/7 The muck-up of society ..is almost complete. **1934** N. Marsh *Man lay Dead* xii. 217 Only Bathgate's prints on the electric switch and a muck-up of everybody's on the bannister. **1939** N. Monsarrat *This is Schoolroom* I. i. 33 The muck-up the Labour people had left us in, in 1931. **1942** E. Waugh *Put out More Flags* iii. 209 You seem to have made a pretty good muck-up. **1945** Baker *Austral. Lang.* xv. 267 Frigg-up or muck-up, a confusion, a row or argument. **1957** W. Camp *Prospects of Love* III. iii. 159 Mummy..making me do beastly shorthand in case I make a muck-up of the Latin. **1963** B. James *Austral. Short Stories* 46 The 'muck-up' in 4B died down at this announcement. **1967** K. Giles *Death & Mr Prettyman* iii. 72 Old Jabeez made his usual muck-up. **1972** C. Drummond *Death at Bar* ii. 60 The food was probably fry-ups or nasty little continental muck-ups.

mucky, *a.* Add: **1. d.** More widely: grimy, grubby, horrid.

1872 H. T. Dunn *Let.* 10 Sept. in G. Pedrick *Life with Rossetti* (1964) xiii. 121 'The Beatrice' had been got out of the mucky state in which it was, & now looks a very good copy. **1942** *Tee Emm* (Air Ministry) II. 88 Lack of power through mucky plugs..will immediately be seen on the Rev. counter. **1959** I. & P. Opie *Lore & Lang. Schoolch.* ix. 161 Juvenile repugnance continues to be expressed by the old standbys..mouldy, mucky, nasty, no fair, no good, orrid (usual spelling).

2. (Further U.S. examples.)

1840 C. F. Hoffman *Greyslaer* I. I. v. 61 He had laid the logs right down on a piece of deep, mucky soil, made up of old roots, rotten leaves, [etc.]. **1874** *Rep. Vermont Board Agric.* II. 548, I have about five acres of mucky meadow that was mostly covered with alders.

mu·cky-muck. *N. Amer. slang.* [redupl. Mucky *a.*, after *muck-a-muck.] = *muck-a-muck 2. Also *attrib.*

1934 H. Miller *Tropic of Cancer* 194 One of the big mucky-mucks from the other side of the water had decided to make economies. **1941** E. P. O'Donnell *Great Big Doorstep* iv. 61 The Governor and the big mucky-mucks from New York with the oil company, was passing close to the levee. **1968** 'E. McBain' *Fuzz* iv. 60 That big mucky-muck got shot. *Ibid.* 61 All them big mucky-muck doctors. **1968** *Globe & Mail* (Toronto) 13 Feb. 27/1 Orpen was always let out at the members' enclosure, but he never sat with the mucky-mucks.

mucluc, var. *muckluck.

muco-. Add: Also used in *Biochem.* (with broader meaning). **mucoly·tic** *a.*, able to disperse or decompose mucus or mucopolysaccharides; **mucope·ptide** *Biochem.* = *murein; **mucoperio·steum** *Anat.*, a periosteum closely associated with a mucous membrane; **mucopro·tein** *Biochem.*, any of a group of compounds (some of which occur in mucus and other body fluids) whose molecules consist of protein or polypeptides in combination with mucopolysaccharides; **mu:coviscido·sis** *Path.* [Viscid *a.* + -osis], a congenital metabolic disorder, usually causing early death, in which exocrine glands (such as the pancreas) secrete very viscid mucus, which accumulates and blocks the passageways of the body; also called *cystic fibrosis* (*of the pancreas*), *fibrocystic disease* (*of the pancreas*).

1939 *Nature* 9 Dec. 977/2 (*heading*) A mucolytic enzyme in testis extracts. **1940** [see *hyaluronidase]. **1953** *Jrnl. Soc. Leather Trades' Chemists* XXXVII. 80 It should be possible to produce pelt by removing the mucoids with mucolytic enzymes. **1971** *S. Afr. Med. Jrnl.* XLV. 948/1 In patients with acute obstructive airway disease pancreatic dornase and acetyl-L-cysteine (Airbron) showed excellent mucolytic activity. **1959** Mandelstam & Rogers in *Biochem. Jrnl.* LXXII. 655/1 This class of substances from cell walls, which will be referred to as mucopeptides, constitutes a high proportion of the hot trichloroacetic acid-insoluble material from bacteria. **1962** *Jrnl. Exper. Med.* CXV. 60 A rigid mucopeptide structure, composed of *N*-acetylglucosamine, *N*-acetylmuramic acid, and alanine, lysine, glutamic acid, and glycine, has been identified in the cell wall of each of these three groups of streptococci. **1972** Mucopeptide [see *murein]. **1903** Dorland *Med. Dict.* (ed. 3) 419/1 Mucoperiosteum. **1906** *Practitioner* Nov. 712 Should the carious tooth be in the upper jaw, the pus may travel down alongside of the fang of the tooth, around the free border of the bone, make its way through the muco-periosteum on the outer side, and then point as a gumboil. **1974** W. J. Banks *Histol.* xix. 205/3 Because of the intimate relationship of the mucosa to the underlying bone, these structures are referred to collectively as a mucoperiosteum. **1925** *Jrnl. Biol. Chem.* LXV. 700 The carbohydrate radicle isolated from the mucoproteins of the mucus of *Helix aspersa* and *Helix pomatia* belongs to the group of mucoitin sulfuric acid. **1962** D. G. Cogan in A. Pirie *Lens Metabolism Rel. Cataract* 291 The capsule surrounding the lens..stains heavily with periodic acid Schiff (indicating more neutral mucoprotein than most tissues). **1968** Mucoprotein [see *glycoprotein* s.v. *glyco-]. **1971** *S. Afr. Med. Jrnl.* XLV. 948/1 Normal human bronchial mucus is a semiviscid substance composed almost entirely of mucoprotein and mucopolysaccharide fibres. **1945** S. Farber in *Jrnl. Michigan State Med. Soc.* XLIV. 592/2 Until the etiological factors are defined and for the purposes of present convenience only a purely descriptive term suggested by the physical character of the material produced by the mucous glands in this disease may be employed—*muco-viscidosis*. **1949** *New Eng. Jrnl. Med.* 4 Aug. 185 (*heading*) Aureomycin therapy in the pulmonary involvement of pancreatic fibrosis (mucoviscidosis). **1966**

Wright & Symmers *Systemic Path.* I. xxiii. 692/1 Farber has suggested that the disease [*sc.* fibrocystic disease of the pancreas] is a manifestation of a generalized abnormality of mucus secretion, for which he proposed the term 'mucoviscidosis'. **1973** L. C. Carey *Pancreas* ix. 181/2 In mucovisoidosis, steatorrhea is usually present.

mucoid (miū·koid), *sb. Biochem.* [a. G. *mucoid*, f. *mucin* Mucin + *-oid* -oid.] A mucin-like substance; esp. = *mucoprotein* s.v. *muco-.

1900 Gies & Cutter in *Amer. Jrnl. Physiol.* III. p. vi, We believe that continued investigation will show that the diffirences among the mucins, mucoids, and chondroproteids are not as great as their varying physical proproties and behavior have suggested. **1901** *Ibid.* VI. 155 (*heading*) The composition of tendon mucoid. *Ibid.*, Following Cohnheim's suggestion..we use the term 'mucoid', instead of the previously accepted 'mucin', to designate this substance. We agree with Cohnheim that.. the term 'mucin' may be best applied to the glucoproteids elaborated by true secretory cells, and the term 'mucoid' to similar substances in the tissues. **1945** *Adv. Protein Chem.* II. 250 As mucoids we define substances which contain a mucopolysaccharide in firm chemical union with a peptide, where the hexosamine content is greater than 4 per cent. **1953** [see *mucolytic* adj. s.v. *muco-]. **1964** A. White et al. *Princ. Biochem.* (ed. 3) viii. 122 The protein components of mucoids or mucoproteins are combined with large amounts (more than 4 per cent) of carbohydrate, measured as hexosamine. **1967** D. A. L. Davies in D. M. Weir *Handbk. Exper. Immunol.* xi. 405 Ovarian cysts are a rich source of mucoids and a classical source of blood-group mucopolysaccharides. **1974** *Nature* 20 Dec. 711/2 Cervical mucus is a heterogeneous secretion, the most important constituent of which is a hydrogel made of glycoproteic mucoids.

mucoitin (miūkōu·ĭtin). *Biochem.* [f. Muc(in + *-oitin*, after *chondroitin.] A supposed mucopolysaccharide, now generally considered to be a mixture, occurring in combination with sulphuric acid in the mucin of pigs' stomachs, the cornea, and elsewhere. So *mucoitinsulphuric acid.*

1916 Levene & López-Suárez in *Jrnl. Biol. Chem.* XXV. 513 Further work on mucoitin sulfuric acid (as we propose to name this substance) is now in progress. **1925** P. A. Levene *Hexosamines* II. iii. 85 A considerable part of the sulphuric acid radicle had been removed in the process of preparation, so that it was possible to determine the properties of mucoitin and of mucoitin sulphuric acid in the same sample. **1925** [see *chondroitin]. **1964** Brimacombe & Webber *Mucopolysaccharides* vi. 142 It seems likely that the material originally designated as mucoitin sulphate contained hyaluronic acid admixed with sulphated mucopolysaccharides containing 2-amino-2-deoxy-D-glucose residues. **1967** *Chem. Abstr.* LXVI. 9649/1 The chondroitinsulfuric acid and mucoitinsulfuric acid contents were higher in the umbilical cord and myxoid chondroma.

mucopolysaccharide (miū:kopǫlisæ·kăraid). *Biochem.* [f. Muco- + *polysaccharide.] Any of a group of polysaccharides whose molecules contain amino-sugar residues (*spec.* hexosamine residues) and are often found in complexes with protein molecules, and which include as important examples heparin, hyaluronic acid, keratosulphate, and the blood-group substances.

1938 K. Meyer in *Cold Spring Harbor Symp. Quant. Biol.* VI. 91/1 The mucopolysaccharides of our terminology were formerly included as mucins and as mucoids in the group of conjugated proteins. They occur in nature either as free polysaccharides or as protein salts. **1947** *Physiol. Rev.* XXVII. 335 Hyaluronic acid is a mucopolysaccharide acid which in animal tissues seems to bind water in interstitial spaces. It further holds cells together in a jelly-like matrix and serves as a lubricant and shock-absorber in joints. **1963** R. W. Jeanloz in Florkin & Stotz *Comprehensive Biochem.* V. vii. 262 Two groups of substances, which in the past had been described under the name of mucopolysaccharide, namely the glycoproteins and the glycolipids, will be discussed in other chapters. **1967** [see *mucoid *sb.*]. **1971** J. Z. Young *Introd. Study Man* xl. 584 There is polymorphic variation in certain mucopolysaccharide constituents of the red cells of the blood.

mucormycosis (miū:kǫɹmaikōu·sis). *Path.* Pl. -mycoses. [f. Mucor + Mycosis.] = *phycomycosis.

1918 Stedman *Med. Dict.* (ed. 5) 621/2 *Mucormycosis*, a mycosis caused by a fungus or mould of the family *Mucoraceæ*, one of the slime fungi. **1943** *Bull. Johns Hopkins Hosp.* LXXIII. 405 (*heading*) Mucormycosis of the central nervous system. *Ibid.*, These mucormycoses of animals..may be transferred to humans. **1957** *Jrnl. Amer. Med. Assoc.* 9 Mar. 806/2 Rhizopus, rather than Mucor, appears to be the usual cause of human mucormycosis. **1965** Wilson & Plunkett *Fungous Dis. Man* xv. 190 Although the term 'mucormycosis' has long been used synonymously with the newer name 'phycomycosis', species of the genus *Mucor* are not often involved. **1974** Passmore & Robson *Compan. Med. Stud.* III. I. xii. 94/1 Cerebral mucormycosis presents with drowsiness or stupor.

mud, *sb.*[1] Add: **1. e.** *transf.* A 'mud-student' (see Dict., sense 5).

1906 C. G. Grey *Sequel to Story Official Life* 9 Some of the men from across Tweed were very kind to us muds.

f. A liquid (commonly a suspension of clay and other substances in water) that is pumped down the inside of the drill pipe and up the outside during the drilling of an oil or gas well, so as to remove the drill cuttings, cool and lubricate the bit, support the sides of the hole against caving, and prevent the leakage into it of gas or water from the formations encountered; also (with *a* and *pl.*), a kind of mud. Orig. called *mud-laden fluid* and later *mud fluid*.

[**1901** J. G. McIntosh tr. *Neuberger & Noalhat's Technol. Petroleum* xxix. 379 It is..by causing a current of water to circulate continually from the surface of the ground to the bottom of the well, and again from the bottom to the surface, that the mud is continually carried away. **1914** *Times Fuel No.* 104/2 Water and heavy mud forced along the column of the pipe..keep the bit clean and bring the detritus up to the surface.] **1922** D. T. Day *Handbk. Petroleum Industry* I. 249 A column of pure water exerts a pressure of 43 pounds per square inch for each 100 feet in height but with mud-laden fluid this pressure may be increased to 50 or 55 pounds... This lateral pressure forces the mud into sands and porous structures, stabilizes caving formations and effectively shuts off water, gas or oil. **1926** E. R. Lilley *Oil Industry* vii. 174 In the rotary system mud is introduced continuously and the control of gas seldom becomes a serious problem. **1938** J. G. Crowther *About Petroleum* ix. 73 An artificial mud is preferable, as it is more viscous and does not allow the debris to settle. Drilling-mud serves other important functions. **1957** Van der Have & Verver *Petroleum & its Products* ii. 59 Three types of drilling fluid are at present in use: water-base muds, emulsion-type muds and oil-base 'muds'. **1970** W. G. Roberts *Quest for Oil* iv. 45 A great deal of research continues into the making of suitable muds. **1970** *Daily Colonist* (Victoria, B.C.) 31 Dec. 3/5 Workers cut off a blazing oil well with heavy 'drilling mud'. **1974** *Daily Mail* 3 Apr. 23/2 The mud men ..supervise the texture of the mud in the giant tanks that are reservoirs for the special lubricant which is pumped into the ocean bed.

2. † **c.** A fool. *slang. Obs.*

1708 *Memoirs Right Villanous John Hall* 22 *Mud*, a Fool, or Thick skull Fellow. **1886** in H. Baumann *Londinismen.*

d. Opium. *U.S. slang.*

1922 *Dialect Notes* V. 182 *Mud*, obviously so-named from the color and consistency of the drug. **1926** *Flynn's* 16 Jan. 638/2 Some stiffs uses mud but coke don't need any jabbin', cookin' or flops. **1935** *Amer. Speech* X. 17/1 *Hop*, opium. Modern mud, O., pen-yen, tar. **1955** *U.S. Senate Hearings* (1956) VIII. 4161 Opium in the underworld is referred to by various names. For instance, 'mud', 'tar', 'black stuff', 'hop', 'pen yan' and 'yen pocks'. **1974** *Publishers Weekly* 11 Feb. 60/1 Western efforts to open up trade with China in the early to mid-19th century were largely unscrupulous, inspired by the immense profits to be made from 'mud' (opium).

e. Coffee. *slang.*

1925 G. H. Mullin *Adventures Scholar Tramp* iii. 34, I received punk (bread) and a cup of mud (black coffee) or—to use the familiar hobo expression for the combination—duffer. **1931** 'D. Stiff' *Milk & Honey Route* 210 *Mud*, strong coffee mixed with weak milk. **1945** L. Shelly *Jive Talk Dict.* 15/1 *Mud*, coffee or a homely person. **1957** 'N. Culotta' *They're a Weird Mob* (1958) ix. 135 Got another cuppa mud, Joe?

3. *to fling mud* (earlier U.S. example); (*as*) *sick as mud*: depressed, exasperated, furious; (*here's*) *mud in your eye*: a toast, 'good health!'; *one's name is mud*: one is discredited, held in low esteem, ineffective, unlucky [cf. 2 c, above]; hence, passing into senses suggested by Dict. 2 a, b: one is regarded as a pariah, untrustworthy, with the worst reputation; so in allusive phrases; *up to mud* (Austral.): see quot. 1945.

1823 'J. Bee' *Slang* 122 *Mud*—a stupid twaddling fellow. 'And his name is mud!' ejaculated upon the conclusion of a silly oration, or of a *leader* in the Courier. **1880** 'Mark Twain' *Tramp Abroad* 187 These people fling mud at that elegant Englishman..and make fun of him. **1887** *Lantern* (New Orleans) 16 Apr. 2/1 Zeller wants to be Recorder..but his name is mud. **1906** E. Dyson *Fact'ry 'Ands* viii. 92 D'yeh mean to tell me how Hoggy's let you loose agin after you gettin' glorious in his dry-goods, 'n' makin' his name mud all up 'n' down ther town? **1916** C. J. Dennis *Songs Sentimental Bloke* 13 I'm crook; me name is Mud; I've done me dash. **1927** H. V. Morton *In Search of England* iii. 60 'Here's mud in your eye!' said one of the modern pilgrims, tossing down his martini. **1929** J. P. McEvoy *Hollywood Girl* ix. 148 Well, I hope when I'm through I'll have sense enough to know it. Mud in your eye! **1930** K. Brush *Young Man of Manhattan* vii. 87 'Well,' somebody said, 'here's mud in your eye.' **1933** *Bulletin* (Sydney) 1 Nov. 42/2 Still, men today are mostly up to mud. **1935** M. de la Roche *Young Renny* xviii. 152 She hates the thought of his staying on as much as we do. She's as sick as mud about it. **1938** S. V. Benét *Thirteen o' Clock* 335 Mud in your eye! **1942** M. Dickens *One Pair of Feet* vii. 108 Nurse Dickens had no idea of hospital etiquette... Nurse Dickens was too opinionated. Nurse D.'s name, in short, was Mud. **1945** Baker *Austral. Lang.* vi. 128 *Up to mud, up to tripe*,..describe things that are bad, disliked or out of order. **1949** Wodehouse *Mating Season* xxiii. 198 'Skin off your nose, Jeeves.' 'Mud in your eye, sir, if I may use the expression.' **1954** —— *Jeeves & Feudal Spirit* viii. 72 He's as sick as mud about it. He moons broodingly to and fro, looking like Hamlet. **1956** J. Symons *Paper Chase* x. 73 Here's mud in your eye, Eileen. **1957** D. Robins *Noble One* xviii. 169 If tha' doan't put ring on finger shortly, my lad, tha' name will be mud in Mountaindale. **1973** W. M. Duncan *Big Timer* vii. 51 Riordan his name is and so far as I'm concerned from now on it's Mud.

4. a. *mud cabin, floor* (later *fig.* example), *-hut, -land, -puddle, -rush, -scatter, -shoal, -side, -slide, -trap*; **b.** *mud-caked, -chinked, -choked, -greasy, -layered, -moulded, -splashed, -splattered, -stained* adjs.; **c.** *mud-bottomed, -floored, -heaped* adjs.; **d.** *mud-couched, -mattressed, -stuck* adjs.; **e.** objective, as *mud-feeding* adj.; **f.** similative, as *mud-grey* adj.

1908 Hardy *Dynasts* III. vii. iii. 492 Where there is a mud-bottomed stream, the Lasne. **1949** C. Longfield *Dragonflies Brit. Isles* (ed. 2) 132 It inhabits, in England, slow-running, mud-bottomed streams. **1780** A. Young *Tour in Ireland* I. 102 This town appears exceedingly flourishing..yet 40 years ago..there were nothing but mud cabbins in it. **1829** J. MacTaggart *Three Yrs. Canada* II. 243 It is a singular fact..with the Irish, that if they can get a *mud-cabin*, they will never think of building one of wood. **1922** Joyce *Ulysses* 324 Their mudcabins and their shielings by the roadside were laid low by the batteringram. **1912** W. Deeping *Sincerity* xxxix. 281 Grassless fields, mud-caked ponds, and empty wells. **1961** *Times* 11 Jan. 16/4 Their mud-caked forwards battled back into the Eastern Counties' 25. **1946** W. Faulkner *Portable Faulkner* App. 737 Jefferson Mississippi was one long rambling onestorey mudchinked log building housing the Chickasaw Agent and his tradingpost store. **1922** Joyce *Ulysses* 98 Past beds of reeds, over slime, mud-choked bottles, carrion dogs. **1936** W. Faulkner *Absalom, Absalom!* ii. 44 It was that same Akers who had blundered onto the mudcouched negro five years ago. **1926** J. S. Huxley *Ess. Pop. Sci.* xvii. 209 The endostyle degenerates together with the rest of the mud-feeding apparatus. **1960** Koestler *Lotus & Robot* I. i. 28 It had brought..a mud-floor of starved refugees, increased poverty, land hunger, and the threat of civil war. **1951** —— *Age of Longing* I. vi. 108 All men did that who had met in mud-floored rooms before the Great Change. **1921** D. H. Lawrence *Sea & Sardinia* ii. 44 Badly paved, mud-greasy..road. **1923** —— *Birds, Beasts & Flowers* 169 You meet a huge and mud-grey elephant. **1935** W. Empson *Poems* 26 Empty, mudheaped, through which the alluvial scheme Flows temporary as the modern world. **1803** J. Davis *Trav. U.S.* 3, I have entered with equal interest the mud-hut of the negro, and the log-house of the planter. **1858** T. Viele *Following Drum* 125 Half-a-dozen mud-huts neatly thatched with straw..presented a study for an artist. **1941** L. Hellman *Watch on Rhine* III. 166 In every town..and every mud hut in the world, there is always a man who..will fight to make a good world. **1971** *Daily Nation* (Nairobi) 10 Apr. 10 The expression traditional housing is a misnomer... The brick homes in the area will have nothing in common with the traditional mud huts. *a* **1865** Smyth *Sailor's Word-Bk.* (1867) 487 *Mud-lands*, the extensive marshes left dry by the retiring tide in estuaries and river mouths. **1927** *Daily Tel.* 22 Nov. 14/1 This scheme..involves the reclamation of over 400 acres of mudland. **1930** Blunden *Poems* 145 Mud-layered cobble-stones. **1960** S. Plath *Colossus* (1967) 48 Mud-mattressed under the sign of the hag In a clench of blood, the sleep-talking virgin Gibbets..the moon's man. **1906** Hardy *Dynasts* II. v. viii. 288 We are the only phantoms now abroad On this mud-moulded ball! *a* **1841** W. P. Hawes *Sporting Scenes* (1842) I. 183 The thawing mud-puddles. **1912** Kipling *Songs from Books* (1913) 76 As a frog shows in a mud-puddle. **1928** Kipling *Limits & Renewals* (1932) 3 The advance of education and the standard of living would submerge all mind-marks in one mudrush of standardised reading-matter. **1919** J. Masefield *Reynard* 78 Mud-scatters chased him as he scudded. **1842** *Knickerbocker* XX. 309 [He] knew a great deal more about the inconveniences of groping about among mud-shoals in the dark. **1923** G. B. Shaw *Let.* 5 Apr. in *To a Young Actress* (1960) 42, I am here at the mudside (the Bristol channel can hardly be called sea) to recuperate. **1921** *Daily Colonist* (Victoria, B.C.) 29 Oct. 1/7 Mud slides also add to the danger of operating trains until repairs can be effected. **1928** Blunden *Undertones of War* iv. 35 That dugout was a deep one, with a steep mud-slide of an entrance. **1969** *Courier Mail* (Brisbane) 27 Jan. 4/8 Mud-slides buried sleepers alive..as Southern California was deluged by rain. **1908** *Daily Chron.* 28 Sept. 7/4 He sat his mud-splashed saddle, motionless under the moon. **1922** Joyce *Ulysses* 33 The mudsplashed brakes. **1930** J. Dos Passos *42nd Parallel* I. 5 Mudsplattered trouserlegs. *a* **1922** H. Lawson in *Penguin Bk. Austral. Ballads* (1964) 144 And mud-stained, wet, and weary, He goes by rock and tree. **1908** Hardy *Dynasts* III. vii. iv. 493 His horse got mud-stuck in a new-ploughed plot. **1970** *Motoring Which?* July 93/2 Both cars had areas which may well give trouble in a couple of years' time—mud traps around the headlamp cowls of the 3-litre, [etc.].

5. mud balance, a balance designed for measuring the density of drilling mud; **mud-bar** [Bar *sb.*[1] 15], a bank of mud in a river or off an estuary or a shore; **mud-barge,** a barge transporting dredged mud; **mud-bath,** also *transf.* and *fig.*; **mud box** *Naut.*, a box containing a coarse filter used to trap sediment in bilge-water; **mud-boy** (see quot.); **mud-brick,** brick that is made with mud; also *attrib.*; **mud-chute,** a chute down which mud is discharged (in example, *fig.*); **mud-clerk** *U.S.*, an assistant to the purser on a river steamer; **mud-cone** (earlier example); **mud-crack,** a crack formed in drying mud; **mud engineer,** a person responsible for the quality and supply of drilling mud; **mud fever,** a disease of horses, in which patches of the skin on their feet become inflamed and swollen; **mud flap,** a piece of rubber, metal, etc., hung behind each of the wheels of a vehicle to prevent mud, etc., from splashing; mud-flat, (*b*) a mud-bank in a river

which is not tidal; (*c*) *N.Z.* (see quot. 1947); **mud-flinger,** a person who hurls abuse [from *to fling mud* (Mud *sb.*[1] 3)]; so **mud-flinging** *vbl. sb.*; **mud-flow,** a (fluid or hardened) stream or avalanche of mud, e.g. one consisting of soil made fluid by excessive water, one produced by a mud volcano, or a lahar; also, the flow or motion of such a stream; **mud fluid** = *mud *sb.*[1] 1 f; **mud flush,** a flush by means of drilling mud; **mudguarded** *a.*, provided with mudguards; **mud hog** = *mud pump* below; † **mud-laden fluid** = *mud *sb.*[1] 1 f; **mud-lighter** [Lighter *sb.*[1]], a barge for transporting mud; **mud-line,** the line on the sea-bed in front of a coast-line which represents the upper limit at which wave action allows mud to settle permanently on the bottom; **mud logging,** examination of the mud (sense *1 f) coming out of a bore-hole for signs of oil or gas or other indications of the strata being drilled; so **mud logger,** a person responsible for this; **mud-lump** (earlier example); **mud-mask, -pack,** a preparation of fuller's earth applied to the face as a beauty treatment; so **mud-mask** *v. trans.*, to treat with a mud-mask; **mud pie,** also *attrib.* and *fig.*; **mud pilot** (earlier, *fig.*, and later examples); so **mud pilotage**; **mud proof** *a.*, impervious to mud; **mud pump,** a pump for circulating mud (sense *1 f) through the drill pipe and up the bore-hole; **mud room** *N.Amer.*, a cloakroom, *spec.* one in which wet or muddy footwear may be left; **mud runner** *U.S.*, a horse which habitually performs well on a wet racecourse; a *mudder; **mud-shoe** (see quots.); **mud-show** *arch. slang*, an exhibition or performance held in the open air; so **mud showman**; **mud-sill** (*a*) (earlier Amer. example); (*b*) (later examples); **mud-slinging, -throwing** *vbl. sbs.* (the employment of) abuse, calumny or slander; malevolent gossip; so **mud-sling** *sb.*, **-slinger**; **mud-valve** (U.S. example); **mud volcano,** add further examples and substitute for def.: a mound or cone formed of hardened mud discharged from its centre (usu. of much smaller dimensions than a volcano); also *fig.*; (*b*) *U.S.* = *mud-lump*; **mud wing,** a mudguard, a mud flap.

1960 C. Gatlin *Petroleum Engin.* vi. 70/2 The density of drilling muds is normally measured with a mud balance. **1899** C. J. C. Hyne *Further Adventures Capt. Kettle* i. 10 There was a mud-bar with twenty-four feet, but steamers drawing twenty-seven could scrape over, as the bar was soft. **1906** E. Dyson *Fact'ry 'Ands* xviii. 242 Ther toadstools..was growin'..like mussels on er mudbarge. **1926** W. Runciman *Collier Brigs* 77 They..took in chalk ballast from the wharves, and occasionally from mud-barges when they could not get a ready turn at the wharves. **1851** J. Chapman *Diary* 26 June in G. S. Haight *Geo. Eliot & J. Chapman* (1940) 184 M[rs] Hennell..thinks ..that..pure monogamy..will only be reached thro' a previous age of general licence! I don't agree with her that such a mud bath is at all necessary. **1856** D. G. Rossetti *Let.* 10 Dec. (1965) I. 307 Bath has been a mud-bath ever since I came. **1961** *Times* 12 May 19/2 Mr. Beckett's anti-metaphysical mudbath. **1971** *Sunday Nation* (Nairobi) 11 Apr. 5/1 The remaining cars in the Safari braced themselves for a possible mud bath. **1883** A. E. Seaton *Man. Marine Engin.* xii. 231 Mud boxes. Between the directing box and the pump should be fitted a box with a strainer, which shall intercept such solid matter as would derange the pump valves. **1972** *B.S.I. News* Oct. 28/2 Mudboxes intended for the coarse filtration of bilge water which accumulates in ships' machinery spaces. **1958** *Times* 15 May 14/6 The quality of the mud during drilling operations [for oil] is, therefore, very important, and the mudboy, who is responsible for preparing it, is quite a skilled workman. **1810** Z. M. Pike *Acct. Expeditions Sources Mississippi* App. II. 7 Houses would be built entirely of mud-brick (like those in New Spain). **1903** *Speaker* 5 Sept. 527/2 The old town being built of mud-brick had vanished. **1934** F. Stark *Valleys of Assassins* I. i. 29 The castle is a mud-brick square with round towers. **1963** M. Laurence *Tomorrow-Tamer* 227 Then she was gone, shutting quietly behind her the packing-case door of the mudbrick shanty. **1938** H. G. Wells *Apropos of Dolores* iv. 211 It is the most awful gabble—but it is nothing more than the inevitable end of this mud-chute called 'history'. **1872** E. Eggleston *End of World* xxvi. 171 It was natural enough that the 'mud-clerk' on the old steamboat Iatan should have taken a fancy to the 'striker', as the engineer's apprentice was called. **1912** I. S. Cobb *Back Home* 103 Even her two mud clerks, let alone her captain and her pilots, wore uniforms. **1869** 'Mark Twain' *Innoc. Abr.* lviii. 632 Groups of mud cones stuck like wasps' nests. **1895** *Funk's Stand. Dict.*, Mud-cracks. **1917** *Jrnl. Geol.* XXV. 135 (*heading*) Some factors affecting the development of mud-cracks. **1968** R. W. Fairbridge *Encycl. Geomorphol.* 761/1 Mud cracks form largely because of solar radiation. **1970** W. G. Roberts *Quest for Oil* iv. 46 The mud engineer and geologist will be examining the mud all the time to find out what sort of rock is being bored. **1975** *Petroleum Rev.* XXIX. 27/3 The toolpusher was ready,..the mud engineers, the experts in every field were ready. **1872** W. Williams *Princ. & Pract. Vet. Surg.* xxxvii. 623 Mud-fever is occasionally attended with a considerable degree of systemic disturbance. **1901** F. T. Barton *Vet. Manual* 161 (*heading*) Erythema and Mud-rash. (Mud fever.).. Sometimes there is a slight degree of fever, hence the term 'mud fever'. **1928** *Black's Vet. Dict.* 617/2 Mud fever is the popular name for a variety of erythema that attacks the heels and coronets of horses' feet when these parts are subject to long-continued irritation. **1971** G. W. Serth *Horse Owner's Guide Common Ailments* vi. 63 Mud fever affects the legs and under the belly. **1963** *Times* 29 Jan. 3/6 Its front wings and front and rear panels are bolted on for easy repair and the car has rubber cushioned over-riders and mud flaps as standard equipment. **1967** *Autocar* 28 Dec. 29/2 Rear mud flaps should be made compulsory in the interest of safety. **1972** *Guardian* 13 Nov. 8/5 The fitting of mud flaps behind the rear wheels should be made compulsory. **1922** H. Footner *Huntress* 134 The only breaks in the endless panorama of cut-banks, mud-flats, willows, and grass were occasional little inlets. **1943** K. Tennant *Ride on Stranger* ix. 92 The broken grey rocks sloping down to mud flats. **1947** P. Newton *Wayleggo* (1949) 154 *Mud-flats*, a high country man's name for down country flats. Heavy land. **1839** Thackeray *Let.* 1–2 Dec. in A. T. Ritchie *Lett.* (1924) i. 8 Very curious the abuse is of that character. Old Southey is one of the chief mudflingers. **1958** *New Statesman* 22 Feb. 220/2 This latest torrent of mud-flinging was, of course, set off by the letter..from Mr A. H. Milward, Chief Executive of British European Airways. **1964** C. S. Lewis *Discarded Image* iv. 80 This towering vaunt, this philosophic *panache* which goes beyond mere indifference to mud-flinging and actually courts it, is of Cynic origin. **1901** H. W. Monckton in *Q. Jrnl. Geol. Soc.* LVII. 295, I have for several years noted details of landslips in the Drift near Scarborough, and, as in other cases, they may be classed as: (1) Mud-flows. (2) Earth-slips... (3) Falls which, owing to the dryness of the clay, resemble rock-falls. **1902** *Bull. Amer. Mus. Nat. Hist.* XVI. 347 The largest ejected block that we saw was one on the surface of the mud-flow between the rivers Blanche and Sèche and not more than two hundred yards from the sea coast. **1928** *Bull. Geol. Soc. Amer.* XXXIX. 465 (*heading*) Mudflow as a geologic agent in semiarid mountains. **1944** C. A. Cotton *Volcanoes* xiii. 240 The contents of a large crater lake on St Vincent (Antilles) were ejected so as to generate extensive mudflows which rushed down various radial valleys to the sea. *Ibid.* 241 Other destructive mudflows..have been ascribed to the melting of snow and glacier ice by volcanic heat. This has occurred in Iceland. **1963** D. W. & E. E. Humphries tr. *Termier's Erosion & Sedimentation* x. 206 It is probable that earthquake shocks play an important part in initiating submarine mudflows and it is likely that such processes materially assisted in the filling of oceanic basins. **1970** R. J. Small *Study of Landforms* ii. 30 Mud-flows occur in mountain areas after heavy rainfall, in periglacial areas during the thaw season, on the slopes of erupting volcanoes and even in deserts, when a heavily loaded stream-flood is gradually transformed as it loses its water by evaporation and percolation. **1914** Heggem & Pollard *Drilling Wells in Oklahoma by the Mud-Laden Fluid Method* (U.S. Bur. of Mines Techn. Paper No. 68) 23 The mud fluid was then bailed from the inside of the casing and drilling continued. **1921** W. H. Jeffery *Deep Well Drilling* vii. 240 All this can be prevented by moving the casing as occasion requires, and then the mud fluid will rise uniformly on all sides of the casing and the cuttings will have no chance to pack. **1946** *Mod. Petroleum Technol.* 88 When drilling through porous sands it is often necessary to prepare a mud fluid..which leaves on the walls of the well a thin and impervious sheath which seals the pores. **1949** *Our Industry* (Anglo-Iranian Oil Co. Ltd.) (ed. 2) ii. 32 The next essential operation is to remove by a continuous process the debris formed by the action of drilling; this is accomplished by the use of a mud-flush circulation. **1957** Van der Have & Verver *Petroleum & its Products* ii. 59 An accumulation of gas in the mud-flush may lower its density to such an extent that the danger of eruption would increase instead of diminishing. **1923** *Daily Mail* 30 July 6 (Advt.), So well shielded and efficiently mud-guarded that anyone..can ride it in ordinary costume. **1932** *Amer. Speech* VII. 268 Mud-hog. **1939** D. Hager *Fund. Petroleum Industry* ix. 209 Mud pumps, often called 'mud hogs', are of various sizes. **1914** Pollard & Heggem *Mud-laden Fluid Applied to Well Drilling* (U.S. Bur. of Mines Techn. Paper No. 66) 7 In this paper the term 'mud-laden fluid' is applied to a mixture of water with any clay which will remain suspended in water for a considerable time. **1914** F. A. Talbot *Oil Conquest of World* iii. 43 In drilling through sand formations mud-laden fluid is used to seal the sides of the borehole temporarily. **1921** W. H. Jeffery *Deep Well Drilling* vii. 236 Water impregnated with clay, otherwise known as mud laden fluid, is forced by slush pumps down inside the casing,..returning between the casing and the wall of the hole. **1946** *Mod. Petroleum Technol.* 88 The circuit of this mud-laden fluid..begins at the slush-pumps. **1909** *Cent. Dict.* Suppl., Mud-lighter. **1915** W. B. Yeats *Reveries* (1916) 78 He..had nothing to do but work himself into a rage if he saw a mudlighter mismanaged. **1891** *Rep. Sci. Results Voy. H.M.S. Challenger: Deep-Sea Deposits* iii. 185 The greater the extent and depth of the ocean, the greater the depth to which water-movement extends, and consequently the greater is the depth at which the mud-line is formed around the coasts, but the average depth of the mud-line may be taken as approximately about 100 fathoms. *Ibid.* vi. 383 It appears to be most abundant.. in the neighbourhood of what we have termed the mud-line surrounding continental shores. **1963** D. W. & E. E. Humphries tr. *Termier's Erosion & Sedimentation* xi. 226 The maximum amounts of organic matter occur in calm waters, either in estuaries (5–15%) or near the edge of the belt of mud which surrounds the land mass, the 'mud line', where the amount is 6%. **1975** *Daily Tel.* 9 Jan. 25 (Advt.), NEC Gas have a limited number of vacancies for experienced Mudloggers and Instrument Technicians with oilfield experience. **1960** C. Gatlin *Petroleum Engin.* xi. 199/1 Continuous mud logging is..an excellent exploratory tool. **1868** *Putnam's Mag.*, May 591/2 *Mud-lumps*, or more properly Mud-volcanoes, have been known to rise to the height of twenty-five feet. **1928** *Daily Express* 16 June 3/4, I suggested that I should like a mud-mask. The assistant appeared to be alarmed. *Ibid.* 22 Dec. 3/5 Faces have been massaged and mud-masked. **1934** M. Verni *Mod. Beauty Culture* iv. 28/2 Fuller's earth is a powdered clay with healing properties. Some forms are called 'clay packs', or 'mud packs', and are used as masks. **1938** L. MacNeice *Earth Compels* 37 Hot towels for the men, mud packs for the women Will smooth the puckered minutes of your lives. **1971** *New Scientist* 19 Aug. 401/2 The inamorata..wearing a moss-green mud-pack, with hair in rollers. **1927** D. H. Lawrence *Mornings in Mexico* 122 The low, square, mud-pie houses. **1958** Mud-pie [see *dribbly *a.*]. **1856** C. Nordhoff *Merchant Vessel* viii. 94 A Dungeness or *deep-sea* pilot as these are called, in contradistinction to the river men, who are known as mud pilots. **1934** P. H. Godsell *Arctic Trader* 265 The captains of the whaling vessels always took care to keep between the ice pack and the shore, and for this reason were often contemptuously referred to as 'Mud-Pilots'. **1946** G. Millar *Horned Pigeon* xxii. 375 The first guide, a kind of mud-pilot, remained on his bicycle. **1965** R. B. Oram *Cargo Handling* i. 15 In the port of London, a lesser type of pilot..will take over the ship at the lock from the Trinity House pilot... To indicate his inferior status he is known generally as a mud pilot. **1932** S. G. McNeil *In Great Waters* viii. 125, I had to do practically all the mud pilotage myself in the various ports of call. **1897** *Sears, Roebuck Catal.* 255/1 Extra Superfine, strictly rain and mud proof French serge. **1926** E. R. Lilley *Oil Industry* vi. 137 (*caption*) Duplex slush or mud pumps. **1973** J. W. Jenner in Hobson & Pohl *Mod. Petroleum Technol.* (ed. 4) iv. 115 The heart of the fluid circulating system is the mud pumps. **1950** A. E. Burke et al. *Archit. & Building Trades Dict.* 212/1 *Mud room*, in building, a small room or entranceway where members of the family remove their muddy overshoes or rubbers before going into any of the other rooms. **1962** F. Williams *Amer. Invasion* v. 51 The ranch-style houses..with their..sun-rooms, 'bi-level brunch bars', mud rooms [etc.]. **1964** *Calgary Herald* 13 Feb. 16/5 One of the most common places of theft occurrences in the school is in the mud room. **1970** *Globe & Mail* (Toronto) 25 Sept. 36/4 (Advt.), Necessary additions as family room with indoor barbecue,..laundry and mud room..have been architecturally added... **1905** *Evening Sun* (N.Y.) 17 Aug., All the races..were won by the product of stallions that in their day were famous mud runners. **1954** *Mem. Ghost Pine Homesteaders* (Ghost Pine Community Group, Three Hills, Alberta) 119 A chap.. breaking a particularly wet piece of ground, made oval shaped hardwood mud shoes which he fastened to his horses' shoes to keep the horses from miring. **1959** A. Hardy *Fish & Fisheries* xvi. 303 The tractors..will.. have buoyancy tanks so that they are light enough to skim the bottom on their mud-shoes without sinking in. **1969** E. H. Pinto *Treen* 93 Elmwood horse mud shoes, with iron staples,..were used to give a horse better bearing in the cranberry swamps of Wisconsin... Similar devices.. were known as Fen overshoes. **1909** J. R. Ware *Passing Eng.* 178/2 *Mud show*, an agricultural, or other out-door show. **1931** *Amer. Mercury* Nov. 353/2 *Mud show*, the old-time horse-and-wagon circus; now derisive. **1927** K. Nicholson *Barker* 150 *Mud showman*, a carnival man. **1685** *Rec. Early Hist. Boston* (1881) VII. 178 The middle of the wall to lie even with northerlie or outward side of the said Simkins Mudsell in the Old Celler. **1935** S. Lewis *It can't happen Here* xvii. 177 Talking to the dirtiest and tiredest mudsills as warm friend to friend. **1964** *College English* Feb. 333/2 The culturally deprived in our great cities—a more massive social mudsill than the most sanguine dream of Calhoun and other slavocrats ever envisaged, the social converse of Madison Avenue's grey-flanneled dream. **1968** *Times Lit. Suppl.* 4 Apr. 329/3 The mudsill is still Negro, not poor white. **1919** E. Pound *Let.* 1 Feb. (1971) 146 Stings and mud-slings of the ungodly and unco-decorous. **1896** *Advance* (Chicago) 17 Sept. 366 The swarm of caricaturists, libelers and party mud-slingers. **1930** T. S. Eliot tr. *St.-J. Perse's Anabasis* 27 Instigator of strife and discord! fed on insults and slanders, mudslinger! **1884** *Lisbon* (N. Dakota) *Star* 22 Aug., Campaign lies and mud slinging fail to carry the day. **1914** *National Municipal Rev.* (N.Y.) III. 581 This sweeping provision, if constitutional and enforceable, would have the effect of eliminating 'mud-slinging' in political campaigns, perhaps indeed of revolutionizing campaign methods entirely. **1928** *Sunday Express* 27 May 10/1 The social mud-slinging which gives half society its sole virtuous and intellectual amusement. **1952** R. Knox *Hidden Stream* xiii. 119 And that something is not affected, really, by all the mud-slinging which starts, among the more embittered kind of Protestants, the moment the sanctity of the Church is mentioned. **1973** J. Thomson *Death Cap* vii. 98 The accusations..are..not specified, just mud-slinging of a general sort. **1895** W. H. Chambliss *Diary* xxii. 276 They were willing to resort to the most detestable methods of 'mud-throwing'. **1931** *Times Lit. Suppl.* 20 Aug. 633/2 He settles down to protracted mud-throwing with Goodwin Wharton as the target. **1873** 'Mark Twain' & Warner *Gilded Age* 38 The awful thunder of a mud-valve suddenly burst forth. **1868** Mud-volcano [see *mud-lump* above]. **1902** *Ann. Rep. Board of Regents Smithsonian Inst. 1901* 71 Within the region [*sc.* the lower Colorado valley] lie a number of 'mud volcanoes', apparently analogous to the 'mud lumps' of the lower Mississippi. **1914** W. Owen *Let.* 24 May (1967) 252 My face is certainly satisfactory,..free from whelks and knobs and mud-volcanoes. **1953** *Caribbean Q.* III. II. 80 The mud volcanoes..are caused..by the seepage of natural hydrocarbon gases from underground. They have been given their name because..they erupt intermittently, they extrude flows of 'Mud lava', and they build up cones. **1955** *Bull. Geol. Soc. Amer.* LXVI. 1117/1 One mud volcano on an embankment was particularly instructive. Its cone was 7 or 8 feet in diameter. **1967** K. Wilcoxson *Volcanoes* xv. 158 Mud volcanoes are not true volcanoes at all but have features more in common with hot springs and geysers. **1927** *Daily Express* 9 Sept. 11/3 All the six 1928 models are of improved appearance, with..mudwings which not only look better, but also keep off the mud more efficiently. **1959** *Times* 27 Apr. (Rubber Industry Suppl.) p. vi/1 On many goods and passenger vehicles rubber mud-wings are used.

b. mud cat, substitute for def.: *U.S.* one of several species of catfish found in the Mississippi valley (earlier and later examples); also

fig. and *transf.*, an inhabitant of Mississippi, which was sometimes called the *Mudcat State*; **mud catfish** *U.S.*, the bullhead, *Ameiurus nebulosus* (examples); **mud-dauber,** (*a*) for '*Pelopæus*' substitute '*Sceliphron*' (later examples); (*b*) *U.S.* = *mud swallow*; also *transf.* and *fig.*, a travelling workman; **mud duck** *U.S.*, a domestic duck; **mud dweller,** an animal living in a muddy habitat, esp. a water beetle, *Ilybius fuliginosus*; **mud eel** (earlier example); **mud hopper** = *mud-skipper*; **mud-skipper,** substitute for def.: a small Asian, Australasian, or African fish of the family Periophthalmidæ, which is able to scramble over mud and along tree roots, etc. (later examples); **mud snail,** either of two species of pond snail, *Lymnæa glabra* or *L. trunculata*; **mud swallow** *U.S.*, a North American cliff swallow of the genus *Petrochelidon*, which builds jar-shaped nests of mud; also *attrib.* and *fig.*; **mud trout,** a name used in Newfoundland for the brook trout, *Salvelinus fontinalis*; **mud-turtle** (examples of dial. **mud-turkle**); also *transf.* and *fig.* (examples); so **mud-turtle-shaped** *a.*; **mud wasp** *U.S.*, = *mud dauber* (*a*) in Dict. and Suppl.

1819 D. Thomas *Trav. Western Country* 211 The mud cat is covered with clouded spots, and is a very homely fish. **1872** Schele de Vere *Americanisms* 660 Mississippi is occasionally spoken of humorously as the *Mudcat State*, the inhabitants being quite generally known as Mud-cats. **1883** 'Mark Twain' *Life on Mississippi* liv. 532 He didn't really catch anything but only just one small useless mud-cat. **1935** in Z. N. Hurston *Mules & Men* (1970) I. vii. 159 Not no great big trouts nor mud-cats but li'l perches and brims. **1945** B. A. Botkin *Lay my Burden Down* 27 The next is a mudcat; this kind of a fish likes dark trashy places. **1945** *Chicago Daily News* 16 Aug. 10/7 While we are laying down surrender terms for the Japanese, how about a Declaration on Senator 'Dear Dago' Bilbo, the Mississippi mudcat? **1842** J. E. DeKay *Zool. N.Y.* IV. 187 The Mud Catfish..[is] recognized by the scarified and clouded appearance of its skin. **1964** Mud catfish [see *goujon]. **1866** 'Mark Twain' *Sk. New & Old* (1875) 297 The old mud-dobber tackled the piano, and ran his fingers up and down once or twice. **1899** F. Bergen *Animal & Plant Lore* 34 The building of the mud-daubers, or swallows, on the barn or house is a sign of prosperity to the occupants. **1932** E. Step *Bees, Wasps, Ants* 71 Mud-daubers (*Sceliphron*) of warmer countries..build great clusters of mud cells for their eggs and prey. **1945** H. S. Pearson *Country Flavor* 49 There was often a phoebe's home to explore and dozens of mud daubers' nests. **1963** *Amer. Speech* XXXVIII. 271 Mason: *mud dauber.* **1966** C. Sweeney *Scurrying Bush* v. 76 Mud daubers can be a great nuisance, for they build their nests, stuffed with spiders or caterpillars depending on the species, in any hidden place. **1974** A. Dillard *Pilgrim at Tinker Creek* xii. 214 My life inside the cottage is mostly Tinker Creek and mud dauber wasps. **1857** *Spirit of Times* 26 Sept. 54/2 There is duck of every quality, canvas-back, wood, mud, and various other species. **1903** *Forest & Stream* 27 Feb. 150 They are a cross between the mallard and ordinary mud duck. **1920** E. Pound *Let.* 11 Sept. (1971) 158 If you weren't stupider than a mud-duck you would know that every kick to bad writing is by that much a help for the good. **1952** J. Clegg *Freshwater Life* xiv. 221 The Mud Dweller, *Ilybius fuliginosus*, [is] a bronze-coloured beetle with yellow margins to its wing-cases. **1963** *Times* 19 Jan. 10/6 It [*sc.* the Dublin Bay Prawn] feeds, somewhat indiscriminately like all its kind, on its fellow mud-dwellers. **1823** *Charleston* (S. Carolina) *Courier* 7 Mar. 2/4 The British..fairly chased our militiamen across Broad River, to the huge amazement of the mud eels and cats. **1959** *Listener* 29 Oct. 738/3 The skittering of the mud hoppers, those extraordinary little fish that can climb trees. **1965** *Sunday Mail Mag.* (Brisbane) 21 Nov. 14/1 The mud-hopper belongs to a special group of goby (or small fish) and is notable for its strange eyes and for the stiff fins which it seems to be able to use like limbs. **1957** L. Eiseley *Immense Journey* 58 Of all these fishes the mud-skipper *Periophthalmus* is perhaps the strangest. **1972** *Islander* (Victoria, B.C.) 16 Apr. 4/2 [Queensland] After exploring the mangroves for the amphibious mudskipper fish [etc.]. **1926** A. E. Ellis *Brit. Snails* 111 *L[imnæa] glabra*..Mud Snail..inhabits ponds and ditches, and frequently lives in places which are left dry in summer. **1972** *Country Life* 2 Mar. 524/3 The damp and muddy slopes where they [*sc.* liver-flukes] can find their next host ..the species of snail called the mud-snail, or..*Lymnaea truncalata.* **1873** C. G. Leland *Egyptian Sk.-Bk.* 43 Those curious mud-swallow nests of little villages. **1898** M. Deland *Old Chester Tales* 181 Mud-swallows had built their nests in the corners. **1917** T. G. Pearson *Birds Amer.* III. 84 Cliff Swallow... Barn Swallow; Mud Swallow; Republican Swallow. **1917** *Dialect Notes* IV. 332 *Mud-trout*, the brook-trout. **1969** H. Horwood *Newfoundland* 224 Newfoundland's only native trout..is the speckled brook trout (known locally..as a 'mud trout'). **1884** 'Mark Twain' *Huck. Finn* 69 And so you ain't had no meat nor bread to eat all this time? Why didn't you get mud-turkles? **1909** *Dialect Notes* III. 351 *Mud-turkle*, the mud-turtle: chiefly among the negroes. **1946** *Publ. Amer. Dial. Soc.* VI. 21 *Mud turkle*, a small turtle found in muddy bottoms, in either fresh or salt water. Pamlico. Mainly among Negroes. **1796** *Aurora* (Philadelphia) 17 May (Th.), The crocodile throats of the gentle snappers or mud tortles in the Jersey market. **1873** 'Mark Twain' & Warner *Gilded Age* 48 He's in that pilot-house now, showing those mud-turtles how to hunt for easy water. **1896** *Harper's Mag.* Sept. 527 A mud-turtle of a back-settlement lawyer. **1915** Conrad *Victory* vii. 135 Fancy a mud-turtle like you trying to pass an opinion on a gentleman! **1871** 'Mark Twain' *Screamers* 132 A picki-ninny, mud-turtle-shaped craft of a schooner. **1824** *Old Colony Memorial* (Plymouth) 6 Mar. (Th.), A sort of would-be dandy; having the bottom of his waist pinched up to the size of a quart pot, and thus resembling in shape what we call a mud wasp. **1861** *Trans. Illinois Agric. Soc.* IV. 338 The common black and yellow mud wasp (*Pelopæus lunatus*) belongs to this group. **1881** *Amer. Naturalist* XV. 443 Baron Osten Sacken..records the breeding of *A[rgyramæba] cephus*..from the nest of a Texan mud-wasp.

mud, *v.*[1] Add: **2. b.** Also with *off* or *up.* (*a*) *trans.* To seal (porous strata) by causing a layer of mud to be deposited on the sides of a bore-hole. (*b*) *intr.* To become coated in this way. Cf. *mud *sb.*[1] 1 f.

1916 Johnson & Huntley *Princ. Oil & Gas Production* xii. 123 In drilling by the rotary system, usually there is but one size of hole and but one string of casing used, as the sides of the hole are 'mudded up' as drilling proceeds, and caving beds and minor gas and water sands are shut off in this way. **1916** Lewis & McMurray *Use of Mud-Laden Fluid in Oil & Gas Wells* (U.S. Bur. of Mines Bull. 134) 23 When a sand is to be 'mudded off', a comparatively thin mixture is first used. **1921** W. H. Jeffery *Deep Well Drilling* iii. 117 The wire drilling cable now is almost universally used for drilling in deep wet holes and in soft or shale formations that 'mud up'. *Ibid.* xii. 350 The mud fluid under pump pressure has a tendency to 'mud off' an oil or gas producing formation before its paying possibilities may be discovered by the driller. **1924** L. C. Uren *Textbk. Petroleum Production Engin.* ix. 299 It is not always easy to mud an exhausted oil sand, so that it does not continue to absorb fluid. **1926** E. R. Lilley *Oil Industry* vi. 129 Mud fluid is introduced into the hole. This is primarily for the purpose of mudding up the walls of the hole to prevent caving.

‖ **mudalali** (mu·dalali). *India* and *Sri Lanka.* Also mutalali. [Marathi, f. *muddal, mudal* capital, principal, stock; cf. Modeliar, and Malayalam *mutalali.*] A proprietor, a businessman, a rich trader.

1855 in H. H. Wilson *Gloss. Judicial & Revenue Terms* 359/1 s.v. *Mutalali.* **1913** L. Woolf *Village in Jungle* ii. 31 The fat Sinhalese Mudalali, Kodikarage Allis Appu, had supplied grain and curry stuffs on the same terms. **1971** *Ceylon Daily News* 18 Sept. 4/1 Beggar auctions where.. the beggars themselves have their services bought by bidding beggar mudalalis.

mud-bank. [f. Mud *sb.*[1] + Bank *sb.*[1]] A bank of mud in the bed of a river or on the bottom of the sea. Also *transf.*

1774 *Virginia Gaz.* (Williamsburg) 17 Mar. 2/2 A Sloop ..ran aground on a Mud Bank, a little Way up the Creek. **1832** W. D. Williamson *Hist. State Maine* I. 38 From both [islands] a mud bank extends to the main shore. **1908** *Westm. Gaz.* 30 Dec. 8/2 In spite of the employment of many thousands..on..clearing away the muddy remains of the recent snowstorm, the principal roadways.. presented an extraordinary spectacle of mud-heaps, mud-rivers, and mud-banks. **1963** W. Soyinka *Dance of Forests* II. 75 From Limpopo to the Nile coils but one snake On mudbanks, and sandy bed. **1974** P. Dickinson *Poison Oracle* i. 18 The heated air rose..above reed-bed and mud-bank.

mudder (mʊ·dəɹ). *slang* (chiefly *U.S.*). [f. Mud *sb.*[1] + -er[1].] A horse which runs well on a wet or muddy racecourse; *transf.*, a sportsman or team similarly proficient.

1903 *Outing* XLIII. 266/2 'He's a mudder,' he growled, 'and the track today will be like lightning.' **1935** *Amer. Speech* X. 315/2 Grand Slam,..winner of the Arlington futurity.., is a good mudder. **1941** *Sun* (Baltimore) 30 Aug. 13/1 Off-Track Seen. Rain today made the prospect for off-going for the first card, thus giving the 'mudders' an opportunity to strut their stuff. **1942** Berrey & Van den Bark *Amer. Thes. Slang* § 683/1 Football player... *mudder, mudlark*, a player for whom a wet field is no great handicap. **1948** *Time* 1 Nov. 44/3 Halfback Jack Swaner, a superior mudder, had a big day scoring all three touchdowns. **1950** *New Yorker* 11 Nov. 121/2 Cornell's last one [*sc.* fumble] gave Columbia, a remarkably good mudder, the chance to tie the score in the fourth quarter. **1952** *Time* 5 May 71/1 Gehrmann and Druetzler proved no mudders and..Purdue's Denis Johansson.. splashed past the leaders on the last lap. **1960** I. Wallach *Absence of Cello* (1961) 29 Will-o'-the-Wisp in the fifth at Hialeah... He's a mudder... It rained all last night in Hialeah. **1969** *Courier-Mail* (Brisbane) 1 Jan. 1/9 Chance for 'mudders'..after last night's flash storm in Brisbane. **1975** *New Yorker* 24 Mar. 64/1 In my book, Stardust Mel is the best mudder in California. Early last month Mrs. Marjorie Lindheimer Everett's rangy gray gelding splattered through the rain and murk to win.

muddied, *ppl. a.* (Later examples with allusion to quot. 1902 in Dict.)

1912 *New Age* 29 Feb. 416/2 The 'flannelled fools' and 'muddied oafs', who come down from Oxford and Cambridge. **1964** C. Mackenzie *My Life & Times* III. iv. 160 No amount of writing about flannelled fools at the wicket and muddied oafs in the goal by Rudyard Kipling could save the British Empire from ultimate collapse.

muddiness. (Later *fig.* examples.)

1915 D. H. Lawrence *Let.* c 7 July (1962) I. 352, I am rid of all my Christian religiosity. It was only a muddiness. **1964** S. Duke-Elder *Parsons' Dis. Eye* (ed. 14) x. 107 'Muddiness of the iris' is the expression used for indistinctness of the pattern, caused by inflammatory exudates.

mudding, *vbl. sb.* Add: **1. b.** The filling of cracks in the walls of a house or log-cabin with mud. *Canad.*

1898 F. Russell *Explor. Far North* 2 The autumnal 'mudding' was poorly done. **1965** E. L. Myles *Emperor of Peace River* II. iii. 209 On the exterior of the logs treatment was a must if the rooms were to be warm in winter. This treatment consisted of an annual 'mudding', the forcing into the chinks between the logs of a mixture of mud and straw.

muddle, *sb.* Add: **1.** (Earlier example of phr. *in a muddle.*)

1833 J. Constable *Let.* 14 Jan. in *Corr.* (1965) III. 90, I shall be glad when these great pictures are out of doors —but still it's a good thing to be in a muddle.

4. *muddle-thoughted* adj.

1905 E. Phillpotts *Secret Woman* III. v. 250 What a muddle-thoughted man you be—all in a maze!

muddle, *v.* Add: **5.** Also with *up.*

1944 R. Lehmann *Ballad & Source* 104 Sometimes she doesn't remember our names and muddles us up.

6. Also with *at.*

1895 G. B. Shaw *Let.* 1 Mar. (1965) 491, I should muddle at it until I got it right.

b. *to muddle along* = to muddle through; *to muddle through* (further examples).

1899 J. E. Taylor *Let.* 22 Dec. in D. Ayerst *Guardian* (1971) xviii. 245, I suppose we shall muddle along and suffer the natural results. **1910** Belloc *Verses* 86 A gentleman who cannot jest Remarked that we should muddle through. **1931** *Economist* 21 Mar. 599/1 It reveals us as indolent, complacent, mentally lazy, hide-bound by tradition, content to 'muddle along', neglectful of self-help. **1940** L. MacNeice *Poems 1925–40* 287 Muddling through and glad to have no answer. **1948** D. B. Hawkins in R. O'Sullivan *King's Good Servant* viii. 92 You can muddle through only with the aid of sound instincts; without them you make the muddle but you do not get through. **1972** *Village Voice* (N.Y.) 1 June 9/5 In the absence of a national program, America muddles through to produce its energy.

mu·ddledly, *adv.* [f. Muddled *ppl. a.* + -ly[2].] In a muddled or disorganized manner; with confusion of mind.

1918 E. Marsh in *R. Brooke Coll. Poems* p. cxxv, All these people at the front who are fighting muddledly enough for some idea called England. **1935** F. M. Ford *Let.* 27 Sept. (1965) 243 Though I shall read it..it will probably be rather muddledly.

muddle-head. Add: Also, a disorganized, vague mind.

1938 H. Nicolson *Let.* 11 July (1966) 349 About that I think that your dear muddle-head gets confused.

muddle-hea·dedly, *adv.* [f. Muddle-headed *a.* + -ly[2].] In a muddle-headed manner, confusedly.

1909 G. B. Shaw *Pen Portraits* (1931) 232 Furtively, hypocritically, and muddle-headedly. **1968** *Trans-Action* V. viii. 75 The..letters..are simply too muddle-headedly authoritative to let pass. **1973** *Esquire* May 6 Shortsighted obstructionists, muddleheadedly gumming up the wheels of progress.

muddleheadedness. (Later example.)

1972 *N.Y. Rev. Bks.* 30 Jan. 38/3, I charge them both with muddle-headedness for their own views on this subject.

muddlement. (Later examples.)

1910 *Blackw. Mag.* Sept. 422/1 Nor does any amount of sincerity compensate for muddlement. **1948** W. de la Mare *Chardin* 22 A picture, which, without risk of muddlement between the arts, may be said to be quiet, sonorous, shrill, strident or harsh. **1968** J. R. Ackerley *My Father and Myself* xiii. 146, I had suddenly recollected that my birthday was about to fall and foresaw muddlement.

muddler. Add: **1. b.** *Comb.*, as *muddler-through*, one who conducts affairs without system or foresight (see Muddle *v.* 6 b in Dict. and Suppl.).

1930 *New Statesman* 1 Nov. 110/2 That is my complaint against the peace-lovers, the muddlers-through. **1945** G. Millar *Maquis* iv. 66, I am a muddler-through if ever there was one.

2. (Earlier and later examples.)

1880 W. H. Patterson *Gloss. Words Antrim & Down* 70 *Mudler*, a small metal stamper, used in public houses to crush the lumps of sugar in punch. **1955** M. McCarthy *Charmed Life* (1965) 24 She did them [*sc.* Old-Fashioneds] ..in their best glasses..putting in a silver muddler. **1971** *Scope* (S. Afr.) 19 Mar. 77/4 A 'muddler'—for crushing lump sugar and mixing it with bitters or other flavourings in the bottom of a glass.

muddling, *vbl. sb.* Add: Also *muddling-along, -through* (see Muddle *v.* 6 b in Dict. and Suppl.).

1949 Koestler *Promise & Fulfilment* ii. 17 What both Jews and Arabs believed to be a 'diabolic policy' was in fact the traditional muddling-along policy. **1955** *Times* 28 July 3/3 Can it be that..we shall abandon the immemorial practice of muddling through, and discover logic and consistency at last?

mu·ddly, *a.* Also muddley. [f. Muddle *sb.* +

-y[1].] Confused, muddled; passing imperceptibly into: confusing, bewildering.

1909 M. B. Saunders *Litany Lane* II. xvii. 227, I gather it from some of the muddly things he said. **1929** D. H. Lawrence *Let.* 24 Aug. (1962) II. 1184, I can't make out if I have to pay this muddley bill for the gramophone or not. **1938** N. Marsh *Artists in Crime* ii. 14, I won't have that sort of thing—it's too muddly. **1959** —— *Singing in Shrouds* viii. 161 A long muddly argument. **1970** 'O. John' *Diamond Dress* i. 6 I'm sorry to be so muddly.

muddy, *a.* Add: **1. c.** As *sb.* The Missouri or Mississippi. Esp. *The Big Muddy*, the Missouri River.

1825 in S. F. Cooper *Rural Hours* (1850) 481 Ye plains where sweet Big-Muddy rolls along, and Teapot, one day to be found in song. **1859** *Trans. Illinois Agric. Soc.* III. 352 In the winter of '55–6, when one wide sweep of destruction laid dead most of the orchard trees north of the Big Muddy. **1884** 'Mark Twain' *Huck. Finn* lxvi. 130 When it was daylight, here was the clear Ohio water inshore,..and outside was the old regular Muddy. **1948** *Newsweek* 30 Aug. 21/3 We're going clear to the Missouri River and smash this stuff back across the Big Muddy.

4. d. Of a musical sound: blurred, not clearly differentiated.

1962 A. Nisbett *Technique Sound Studio* iii. 55 The balance to seek is one where you get plenty of reverberation, but not so much that the sound becomes muddy or coloured.

5. (Later examples.)

1876 Geo. Eliot *Let.* 25 Feb. (1956) VI. 223, I am rather muddy as to the relation of total sales. **1934** A. Huxley *Let.* 28 Apr. (1969) 380 Pareto...doesn't, like these 'deep' and muddy Germans, invent gratuitous metaphysical entities.

8. *muddy-grey, -minded* (later examples) adjs.; *muddy oaf* [cf. *muddied *ppl. a.*].

1939 'N. Blake' *Smiler with Knife* vi. 96 Her face looked muddy-grey. **1867** Trollope *Last Chron. Barset* II. lxi. 185 Though he knew himself to be muddy-minded and addle-pated, he could see that. **1956** W. H. Whyte *Organization Man* (1957) iv. 33 People who have been the intellectual founders..have not been as muddy-minded. **1934** R. Campbell *Broken Record* ii. 51 Modern international rugby has been going more and more in the muddy-oaf direction.

muddy, *v.* Add: **1.** *fig.* Also with *up.*

1917 E. Pound *Let.* 10 Nov. (1971) 124 You thank your bloomin gawd you've got enough Spanish blood to muddy up your mind, and prevent the current American ideation from going through it like a blighted collander.

Mudéjar (mūde·hāɹ), *a.* and *sb.* Also Mudejar, and with small initial. Pl. Mudéjares. [a. Sp. *mudéjar*, f. Arab. *mudajjan* permitted to remain.] **A.** *adj.* Of, pertaining to, or characteristic of the Mudéjares (see below); *spec.* denoting a partly Islamic, partly Gothic style of architecture and decorative art of the 12th to the 15th century. Also *ellipt.* as *sb.*, this style.

1865 H. O'Shea *Guide to Spain* p. xxix, Moorish architecture may be divided into three periods and styles. 1st. Byzantine–Arabic; 2nd. Mauritane–Almohade; 3rd. Mudejar or Granadine. **1872** M. D. Wyatt *Architect's Note-bk. in Spain* p. ix, I have preferred,..in the binding of this volume, to take its ornament in fac-simile from a beautiful little Mudejar casket. **1909** R. Tyler *Spain* ix. 208 The most complete monuments of the Mudejar style are the two synagogues, El Transito and Santa María la Blanca. **1927** G. G. King *Mudéjar* i. 2 Formerly it was customary to define Mudéjar as a hybrid of oriental and Gothic. **1931** J. B. Trend in Arnold & Guillaume *Legacy of Islam* 15 In later years fine and characteristic work was done by Mudéjar bookbinders. **1938** L. MacNeice *Earth Compels* 31, I was in Spain... Gobbling..the architecture Moorish mudejar churrigueresque. **1946** E. Diehl *Bookbinding* I. vi. 91 Their [*sc.* the Spaniards'] mudéjar bindings, showing the Arab influence, were characterized by interlaced strapwork patterns. **1961** *Times* 30 Sept. 11/3 The elaborately joined wood ceilings produced by the Mudejar (Moorish) craftsmen in Spain. **1972** F. M. López-Morillas in R. Highfield *Spain in 15th Cent.* 197 Aragon, more bound to the Mudéjar tradition of construction in brick, has nevertheless preserved notable examples of Gothic architecture.

B. *sb.* During the reconquest of the Spanish peninsula from the Moors, a subject Muslim who was allowed to retain his laws and religion in return for his loyalty to a Christian king.

1893 H. E. Watts *Spain* vi. 167 It was the *mudejar* who drew the design, a *mudejar* who laid the stones, a *mudejar* who painted the walls. **1901** H. C. Lea *Moriscos of Spain* i. 4 When, in 1212, Alfonso IX..won the great victory of Las Navas de Tolosa and advanced to Ubeda, where 70,000 Moors had taken refuge, they offered to become Mudéjares and to pay him a ransom. **1938** B. Bevan *Hist. Spanish Archit.* xii. 107 In Aragon the Mudéjares were not, as they were elsewhere, a servile minority. **1972** F. M. López-Morillas in R. Highfield *Spain in 15th Cent.* 152 In Valladolid on November 9, 1408, both regents signed a law concerning the *Mudéjares* who lived in Castile.

mu·dhead. [Mud *sb.*[1] 5.] **1.** *U.S. colloq.* A native of Tennessee. *rare.*

1838 [in Dict. s.v. Mud *sb.*[1] 5]. **1949** *Amer. Speech* XX. 27 Buckshine, Mudhead, or Whelp for a Tennessean.

2. *slang.* A fool.

1882 W. D. Hay *Brighter Britain!* I. ix. 234 Shut up, you Milesian mudhead, and listen to me. **1886** [in Dict. s.v. Mud *sb.*[1] 5].

3. The name of a ceremonial clown among the Zuñi people who wears a mud-daubed mask.

1959 E. Tunis *Indians* 129/1 There were other clowns, too, known as *goyemshi*, or Mudheads.

mu·d-hole. Also mudhole, mud hole. [f. Mud *sb.*[1] + Hole *sb.*] **1. a.** A hole containing mud or in which mud collects, esp. as forming a defect or obstacle in a road or highway.

1760 in *Documentary Hist. Amer. Industr. Society* (1910) I. 310 As soon as one Gets out..he Is In a large due or in a mud hole. **1784** A. Ellicott in C. V. Mathews *Andrew Ellicott* (1908) 26 The ground [was] covered with Snow which hid the Mud-Holes. **1857** P. Cartwright *Autobiogr.* xx. 314, I thought of a desperate mudhole about a quarter of a mile ahead;..many wagons had stuck in it. **1911** E. M. Clowes *On Wallaby* xi. 291 The water-holding frog found in the central deserts, which can blow its body out with a sufficiency of fluid to support it for a year or more in a dried-up mud-hole. **1937** *Discovery* May 148/1 There are many different kinds of baths available, supplying various mineral waters; boiling mud holes innumerable; waterfalls over which hot water flows side by side with cold; [etc.]. **1948** *Coast to Coast 1947* 71 Those tourist people came and looked at his boiling spring and his bubbling mudhole. **1973** *Sci. Amer.* Apr. 88/3 They find themselves in an alley where they sink into a bottomless mudhole.

b. *transf.* and *fig.*

1784 in *Pennsylvania Mag. Hist. & Biogr.* (1877) I. 51 The general curse of the country, disunion, rages in this little mudhole [*sc.* Uniontown, Pa.]. **1890** *Cent. Dict.*, *Mud-hole*, a salt-water lagoon in which whales are captured. [Whalers' slang, California.] **1938** S. V. Benét *Thirteen o'Clock* 317 We've all of us been on your back long enough... I know of twice you pulled Jerry Pye out of the mudhole. **1958** J. Carew *Black Midas* x. 231 The thought of returning to the mud-hole where I was born made me shy away.

2. A hole at the base of a boiler, condenser, or other apparatus through which sediment can be removed.

1841 W. Templeton *Locomotive Engine* 16 The mud holes..are for the purpose of removing the sediment and scale that constantly accumulate at the bottom of the water spaces. **1893** Langmaid & Gaisford *Elem. Less. Steam Machinery* xiii. 123 To allow access to the inside of the boiler, for examining and cleaning, manholes and mudholes are provided.

mud-hook. *slang.* [Mud *sb.*[1] 5.] **1. a.** An anchor.

1827 J. F. Cooper *Red Rover* I. ii. 44 He would..fasten her to the spot with good hempen cables and iron mud-hooks. **1874** W. M. Baines *Narr. E. Crewe* vii. 138 Cunningly drop your 'mud-hook' so that you exactly swing with the tide over the right spot. **1884** [in Dict. s.v. Mud *sb.*[1] 5]. **1905** J. C. Lincoln *Partners of Tide* xii. 230 The partners agreed to undertake the job of recovering the lost 'mud-hook'. **1960** M. Sharcott *Place of Many Winds* x. 172 Gusts of wind tore their mud-hooks from the bottom of the anchorage.

b. (See quots.)

1918 'Tommy' *If I goes West!* 25 Forget old Billy Summers with his board at 'crown and anchor'—'The Mud'ook, boys, and now's the time to bet!' **1919** W. H. Downing *Digger Dial.* 34 *Mud-hook*..(2) the anchor in the game of 'Crown and Anchor'. **1943** Hunt & Pringle *Service Slang* 46 *Mudhook*, Army name for the Crown and Anchor board used surreptitiously by members of the forces.

2. a. A foot. **b.** A hand.

1850 L. H. Garrard *Wah-to-Yah* xx. 276 This 'mud-hook', holding out his foot, hasn't a moccasin on for nothin'. *a* **1897** F. B. Lloyd *Sk. Country Life* (1898) xl. 239 When a farmer goes to foolin with figgers he is puttin his mudhooks on powerful slippery ground. **1915** *Dialect Notes* IV. 244 *Mud hooks*, n. pl. Feet. **1941** Baker *Dict. Austral. Slang* 47 *Mudhook*, a foot. (2) A hand. **1952** in Wentworth & Flexner *Dict. Amer. Slang* (1960) 347/2 C'mon, lift them mud hooks!

Mudie (miū·di). The name of Charles Edward *Mudie* (1810–90) used *absol.*, *attrib.*, or in the possessive to designate the lending library opened by him in London in 1842, which continued in business until 1937, or labels, lists, boxes, etc., associated with the library or its contents.

1859 H. C. Robinson *Diary* 26 May (1967) 300, I subscribed to Mudie's Library, as I have long intended. *Ibid.* 30 May (1967) 300, I went to Mudie's for the poems of Macdonald. **1864** Mrs. Gaskell *French Life* i, in *Fraser's Mag.* Apr. 439/2 They have nothing equivalent to 'Mudie' in Paris. **1879** C. M. Yonge *Magnum Bonum* III. xxxii. 675 'It is the advantage of having no Mudie boxes,' said his mother. 'We are taking up our Southey.' **1885** —— *Nuttie's Father* I. xvi. 191 She could only have recourse to Mudie's box to try to drive dull care away. **1888** *Bks. for Presents & Prizes: Catal. Bks. Best Authors* (Mudie's Select Library) Aug. 1 A few Choice Books in Mudie Calf, Extra Morocco, and other Superior Bindings are also kept in Stock. **1890** S. Butler in *Universal Rev.* VIII. 518, I believe I have the smallest library of any literary man in London... I keep my books at the British Museum and at Mudie's. **1908** Mrs. H. Ward *Diana Mallory* I. vi. 113 The Book-Club..embraced some ten families who drew up their Mudie lists in common, and sent the books from house to house. **1922** H. S. Walpole *Cathedral* I. ii. 28 He looked at them all, with their light yellow Mudie labels, their fresh bindings. **1931** 'G. Trevor' *Murder at School* ii. 44 He..said you had written a novel... I must get Mudie's to send it down with their next batch. **1966** Berry & Poole *Ann. Printing* 230/2 'Mudie's Lending Library'..was to become world famous... As the watch-dog of contemporary literary morals, Mudie soon became a ruthless dictator as to what the people should read... 'What will Mudie say?' became the proverbial question when new fiction was under consideration in publishers' offices. **1970** G. L. Griest (*title*) Mudie's Circulating Library and the Victorian novel.

‖ **mudim** (mū·dim). Also 9 mudin. [Malay (now *modin*), prob. ad. Arab. *mu'addin* muezzin.] A junior Muslim official in Malaysia, *spec.* one who performs the operation of circumcision.

1817 T. S. Raffles *Hist. Java* I. vi. 284 A *Kabáyan*.. with the *Kamítuah* and *Múdin* (priest), formed a court for settling petty village disputes. **1858** *Jrnl. Indian Archipelago* 121 The Mudin or professional circumciser performs the operation in a trice. **1900** W. W. Skeat *Malay Magic* vi. 360 Among Malays it [*sc.* circumcision] is performed by a functionary called the 'Mudim'. **1966** D. Forbes *Heart of Malaya* xiv. 194 If the traditional *mudim* carries out the operation [*sc.* circumcision] instead of the hospital doctor, he will rub a salve into the wound.

mudir. (Earlier and later examples.)

1844 A. W. Kinglake *Eothen* xxvi. 383 The appointment of a special Commissioner—they called him 'the Modeer'. **1902** [see *mamur]. **1958** L. Durrell *Balthazar* v. 99 Local mudirs and sheiks.

Mudjur (mu·dʒuəɹ). Also Mujur. The name of a city in central Turkey, used, freq. *attrib.*, to designate rugs made there, usu. with deep borders and prayer arch designs.

1913 W. A. Hawley *Oriental Rugs* xvi. 290 Figures of vandykes, which are seen in some Anatolians and Mudjurs, are also a constant feature of Ladiks. **1931** A. U. Dilley *Oriental Rugs & Carpets* vi. 165 Mujur..prayer rugs [are distinguished] by central panels covered by 'flight-of-stairs roofs'. **1959** *Chambers's Encycl.* III. 137/2 Mujurs, with long pile and glowing reds and greens, can be sumptuous little rugs. **1967** [see *Kir-Shehr]. **1972** P. L. Phillips tr. *F. Formenton's Oriental Rugs & Carpets* 86 Mudjur carpets are different from Yuruk not only in the fact that they are designed as prayer rugs, but also because of the different colour tones.

mudlark, *sb.* Add: **1.** (Examples.)

1801 T. Campbell *Mobiade* in W. Beattie *Life & Lett. T. Campbell* (1849) I. 380 Or fry the mud-lark's odoriferous wing... The poetical name for a pig, principally used in.. Kilmainham jail. **1833** J. Neal *Down-Easters* i. 47, I should like to know..what upon irth he means by..mud-larks that's made into Virginny-ham. **1869** *Overland Monthly* III. 129 A hog clandestinely killed outside of camp and smuggled in..was called a 'slow bear'... 'Mud-lark' signified the same thing. **1923** *Dialect Notes* V. 240 Boiled potatoes an' mud lark.

2. a. (Later examples.)

1959 *Times* 16 Mar. (Port of London Suppl.) p. xvi/1 'Long apron men' and mudlarks who..waited to pick up goods thrown to them by accomplices on board merchantmen. **1975** *Times* 17 May 8/4 Jack Dash..recalling the Mudlarks, river pirates and knife-belted prostitutes.

3. c. Substitute for def.: A black and white Australian bird, the magpie-lark, *Grallina cyanoleuca*, which builds a nest of mud. Add examples.

1898 Morris *Austral Eng.* 278/1 Magpie-Lark..an Australian black-and-white bird..resembling the Magpie in appearance, but smaller; called also Pee-wee, and Mudlark, from its building its nest of mud. **1911** E. M. Clowes *On Wallaby* xi. 290 The mud larks, rather like our water-wagtails, only much larger, come there with the most wanton flutter of broad black and white tails, to disport themselves upon the patch of green at its verge. **1965** *Austral. Encycl.* V. 460/1 The name 'magpie-lark' was presumably bestowed upon it because it runs on the ground like a lark and has pied plumage;..'mud-lark', owing to its preference for the muddy banks of creeks and waterholes.

4. = *mudder. Cf. *mud runner.*

1909 in *Cent. Dict.* Suppl. **1935** A. J. Pollock *Underworld Speaks* 78/2 *Mud lark*, a race horse that excels in mud. **1941** Baker *Dict. Austral. Slang* 47 *Mudlark*, a racehorse that runs well on a muddy course. Also footballers who play on a sodden field. **1975** *Sunday Tel.* (Sydney) 6 Apr. 48 Born Star a Mudlark. Born Star, a two-year-old, yesterday outclassed the field at Sandown in his first start on a rain-affected track.

5. *attrib.*, as *mudlark meet* (see quot.).

1971 *Nat. Geographic* May 719/2 Guernsey invented annual mudlark meets, in which old bangers—near-wrecked automobiles—are raced across the oozing sands at low tide.

mudlarker. (Earlier example.)

1818 'A. Burton' *Adventures J. Newcome* I. 26 Slopmen, Mud-larkers and Crimps.

mudlarking, *vbl. sb.* (Later example.) Cf. *mudlark *sb.* 5.

1960 *Life* (Internat. ed.) 1 Feb. 45/1 The back-bruising sport of mudlarking.., one of England's popular recreations, calls for a special-body car.

‖ **mudra** (mʊdrā·). *Hinduism.* Also moodra. [Skr. *mudrā* seal, sign, token.] One of a large number of symbolic hand gestures used in Hindu religious ceremonies and (hence) in Indian dance. Also, a movement or pose in Yoga.

1811 W. Ward *Acct. Hindoos* II. 26 The ceremonies

called moodra; the names of the different moodras... Certain motions with the hands and fingers. 1832 H. H. Wilson in *Asiatick Researches* XVII. 224 The performance of the *Mudrá*, or gesticulations with the fingers, accompanying the different stages of the ceremony. 1877 M. Williams *Hinduism* ix. 127 The term *Mudrā* is also used in Tāntrism to denote mystical intertwinings of the fingers so as to form symbolical figures. 1899 Max Müller *Six Syst. Indian Philos.* vii. 457, I shall abstain from giving descriptions of the Mudrās (dispositions of upper limbs). 1917 A. Coomaraswamy in Coomaraswamy & Duggirala tr. *Mirror of Gesture* 8 These poses, chiefly of the hands, are spoken of as *mudrās* (seals), and are more or less familiar to students of Hindū iconography. 1948 G. Venkatachalam *Dance in India* xv. 125 There is a regular science of the mudras as there is a science of the mantras. 1952 L. MacNeice *Ten Burnt Offerings* 84 He was..A dancer..a clown, With his gags, his mudras, his entrechats. 1960 Koestler *Lotus & Robot* I. iii. 111 The eight siddhis are promised..in remuneration for the more difficult mudras. 1967 Singha & Massey *Indian Dances* ix. 93 According to..a text on Kathakali mudras, there are twenty-four basic hand gestures... The significance of a mudra changes according to the position of the hand. 1970 D. Brahmachari *Yogasana Vijñana: Sci. of Yoga* 3 There is..no Mudrā (pose) like Khecarī Mudrā (Khechari pose) and nothing that absorbs one's self completely like Nāda (sound). 1972 R. S. Mishra *Fund. Yoga* ii. 36 *Mudras*,..movements of limbs and fingers, etc., according to the circulation of *kuṇḍalinī* force, magnetic force in the body.

mudwalled, *ppl. a.* (Later examples.)
1892 Kipling *Barrack-Room Ballads* 171 Put them in mud-walled prisons. 1961 L. Mumford *City in Hist.* vi. 159 Did the forms of Phidias rise in this barnyard scattering of workshops..mid these mud-walled huts? 1971 *Inside Kenya Today* Mar. 37/1 The mud-walled and grass-thatched office. 1974 *Nat. Geographic* Aug. 255/1 The traditional mud-walled farmhouse with thatched roof has two bedrooms and a kitchen.

mu·d-walling, *vbl. sb.* [Mud *sb.*[1]] (See quots.)
1905 *Tent Catal.* (John Boyle & Co., N.Y.) No. 8. 15 Both grades are provided with sod cloth (mud walling). 1963 *Which?* Apr. 120/2 The outer tent has a strip of heavier, plastic-coated fabric, called mudwalling, round the bottom to protect the tent's walls. 1970 *Ibid.* May 133 The bottom of the walls [of frame tents] is made from plastic (called mudwalling) so that the canvas does not touch the damp ground, and to keep the wind and rain out. 1974 *Camping & Caravanning* Sept. 39/2 (Advt.), 50" PVC Groundsheeting or Mudwalling, 66p yd.

muesli (mü·sli, mü·zli). Also **müesli, musli.** [Swiss-Ger.] A dish, originating in Switzerland, consisting of a cereal (usu. oats) and fruit to which milk is added, often eaten as a breakfast dish.
[1926 Bircher-Benner & Bircher *Fruit Dishes & Raw Vegetables* iv. 29 This dish..is popularly known as 'Birchermüesli'.] 1939 *Ibid.* (ed. 2) iv. 29 The fact that it is cold is never harmful as long as the müesli is well chewed and thus sufficiently warmed in the mouth. 1943 M. Y. Brady *Health for All: Wartime Recipe Bk.* 58 Müesli is a delicious dish, even when modified to suit wartime shortages. 1965 H. Rubinstein *My Life for Beauty* vi. 76, I lived on a diet of Muesli (a mixture of rolled oats, lemon juice, sweetened condensed milk, grated apples and hazel-nuts). 1971 J. Aiken *Nightly Deadshade* vii. 71, I find them finishing their breakfast of musli. 1971 M. Thompson in C. Bonington *Annapurna South Face* App. D. 274 Ideal High Altitude Rations..Readi-Brek *or* Muesli, [etc.]. 1973 [see *instant *a.* 4 c].

muff, *sb.*[2] Add: **3. b.** *slang.* The female pudenda. Also *Comb.*, as **muff-diver,** one who practises cunnilingus.
1699 B. E. *New Dict. Cant. Crew, Muff*, c. a Woman's Secrets. *To the well-wearing of your Muff Mort*, c. to the happy Consummation of your Marriage Madam, a Health. 1935 J. Hargan *Gloss. Prison Lang.* 5 *Muff-diver*, performer of cunnilingus. 1961 Partridge *Dict. Slang* Suppl. 1191/2 *Muff-diver*, a cunilingist [*sic*]. 1972 *Screw* 12 June 5/2 Blowjobs are nice, but blowing into her muff can kill her. 1973 H. Miller *Open City* xiv. 159 The local bookie's got Polaroids of her flashing her muff.
c. A woman or girl, esp. one of low morals; a prostitute. *slang* (orig. *U.S.*).
1914 [see *drag *sb.* 3 e]. 1918 *Dialect Notes* V. 26 *Muff*, a girl. South Idaho and University of Idaho. 1965 H. C. Rae *Skinner* I. i. 10 Flappin' about a muff they found up in the woods.
5. muff-box (further examples); **muff pistol,** the smallest size of nineteenth-century pocket pistol believed by some to have been designed to be carried in a muff (but see quot. 1969).
1834 Muff-box [see *bonnet-box* s.v. *bonnet *sb.* 10]. 1868 *Mich. Agric. Rep.* VII. 363 Henry Fowler [exhibited] ..1 dozen muff and collar boxes, combined in one. 1938 J. N. George *Eng. Pistols & Revolvers* vii. 135 These so-called muff-pistols were not necessarily intended as ladies' weapons. 1956 W. E. Bird *Off-Trail in Nova Scotia* vi. 181 Another item is a muff pistol that was fired with a percussion cap and ball. 1969 F. Wilkinson *Antique Firearms* 135 Some very small examples..are often described as muff pistols, although it is not at all clear on what basis this term was chosen. Certainly there is little or no evidence to suggest that they were intended primarily for ladies or indeed were ever carried in the muff, for both men and women of the period used muffs.

muff, *v.*[3] **1.** (Earlier and later examples.)
[1827 W. Clarke *Every Night Bk.* 84 When one of the fancy dies, the survivors say, that he has..'mizzled'—'morrised'—or 'muffed it'!] 1846 W. Denison *Cricket: Sk. Players* 24 All the best of our players completely muffed their batting. 1950 *Times Lit. Suppl.* 3 Mar. 131/3 It would be impossible for Sir Max Beerbohm to muff a parody. 1968 J. R. Ackerley *My Father & Myself* xiii. 145 A few opportunities occurred. He muffed one of them, I another. 1972 *Jazz & Blues* Oct. 30/1 If Mezz is indeed responsible for this exchange then the author muffs his lines badly. 1973 *Tucson* (Arizona) *Daily Citizen* 22 Aug. 57/3 Third baseman Phil Garner muffed a grounder by Bill Ralston.

muffetee. 2. (Earlier Amer. example.)
1749 in A. M. Earle *Costume Colonial Times* (1894) 165 Men's fine Worsted Gloves and Muffetees.

muffity (mɒ·fiti). [Origin unknown.] A size of Cotswold stone roofing slate.
1914 [see *bachelor 4 d]. 1929 N. Lloyd *Building Craftsmanship* 93 The names and sizes of Stonefield [*sic*] Quarry slates..are Muffity or movedays 8 in.

muffle, *sb.*[4] Add: **3.** *muffle furnace* (later examples); **muffle kiln,** a kiln in which the pottery which is being fired is enclosed within a chamber and thus protected from direct contact with the source of heat.
1949 *Electronic Engin.* XXI. 412 An apparatus was constructed for holding a bar-shaped specimen in the muffle furnace. 1974 *Nature* 27 Sept. 305/2 Several single crystals were placed on a glass slide and heated in a muffle furnace at 180° C for 24 h. 1897 Sparkes & Gandy *Potters* i. 43 The pigments..are all mixed in vehicles of an oily nature, and a most necessary step is to fire this oil out... This is done in..a muffle kiln..fired up to red heat only. 1947 W. B. Honey *German Porcelain* 8 More or less glassy pigments [were] applied over the glaze and fixed by a low-temperature firing in a muffle-kiln. 1971 L. A. Boger *Dict. World Pott. & Porc.* 260/1 Enamel colors are fired at a comparatively low temperature of 700°C to 900°C in a muffle kiln.

muffledly (mɒ·f'ldli), *adv.* [-ly[2].] In a muffled manner.
1903 Conrad & Hueffer *Romance* I. iii. 21 The church clock began muffledly to chime the quarters. 1946 E. O'Neill *Iceman Cometh* (1947) III. 178 He hides his face on his arms, sobbing muffledly.

muffler. Add: **4.** Also (chiefly *U.S.*) = Silencer 2.
1895 in *Funk's Stand. Dict.* 1896 [in Dict.]. 1897 [see *back-firing *vbl. sb.* 2]. 1915 [see *chug *v.*]. 1931 *Economist* 28 Feb. 431/2 There is reason to believe that.. mufflers, chassis springs,..and accessories are all being manufactured or obtained in France. 1964 M. Banton *Policeman in Community* iii. 69 An officer had stopped a driver for a faulty muffler (exhaust silencer) and had given him a ticket. 1973 *Houston* (Texas) *Chron.* 21 Oct. 12/1 Congress is being pressured to ease the auto emission standard that would lead to the installation of catalytic mufflers on new cars sold after Jan. 1, 1975. 1973 J. M. White *Garden Game* 136 There was a horrible grinding and clanking under the car..and I did mental arithmetic about the bill for a new muffler. 1974 *Index-Jrnl.* (Greenwood, S. Carolina) 19 Apr. 5/1 There were also two charges each of muffler violation, disregarding a traffic signal and operating an uninsured vehicle.

mufflered (mɒ·fləɹd), *a.* [f. Muffler + -ed[2].] Wearing a muffler; characterized by the wearing of mufflers.
1927 *Daily Express* 13 Mar. 5/1 From the bridge the mufflered figures glanced down at the business men moving across the quayside. 1963 *Times* 3 May 16/7 Raincoated and mufflered, tropical gear a thing of the past, we steamed into cold drizzle and short, steep seas. 1969 *Punch* 26 Mar. 450/2 The hard collar is wilting at last. In a choice between starched indigence and mufflered affluence, proper pride has had it.

mufti[1]. Add: **1.** (Further examples.) Also *transf.* and *fig.*
1930 W. S. Maugham *Cakes & Ale* iv. 52 He looked a little like a dean in mufti on his summer holiday in Switzerland. 1959 *Listener* 16 Apr. 681/3 A time of extravagance already remote from the mufti sobrieties of the Third Republic. 1966 *Ibid.* 13 Oct. 547/1 Students will observe the tears that, Terry-fashion, the actor (in mufti and communing only with the cameras) cannot hold back in the Deposition scene of *Richard II*. 1971 *Illustr. Weekly India* 4 Apr. 23/2 An old man..was nearly killed by a lathi and was carried away by the police in mufti. 1975 A. Christie *Curtain* xvii. 182 Nurse Craven I saw for the first time in mufti instead of her nurse's uniform.

mug, *sb.*[3] Add: **1. b.** A portrait or photograph of a person, esp. in police records. *slang.*
1887 *Lantern* (New Orleans) 9 July 2/2 He had his mug taken in fireman's clothes. 1889 Clarkson & Richardson *Police!* xxiii. 323 Circulating thieves' photos... Pushing the mugs round. 1940 R. Chandler *Farewell my Lovely* vi. 43 Nulty turned over a photo..and handed it to me. It was a police mug, front and profile.
3. The act of throttling or strangling a person; usu. in phr. *to put the mug on* (someone).
1862 *Sessions Papers Cent. Criminal Court* 26 Nov. 41 Roberts..said, 'You want me for putting the *mug* on, do you? I will put the b—y *mug* on you.' *Ibid.*, *Mug* is slang used by thieves; it means garotting. 1940 *Amer. Speech* XV. 121/1 *To put the mug on* (a mark), to put a strangle-hold on a mark who grows obstreperous after he has been fleeced. 1955 *Publ. Amer. Dial. Soc.* xxiv. 171 A strangle hold is applied... This hold is called..a *mug* on the East Coast.
4. *attrib.* and *Comb.*, as **mug book** *U.S.*, (*a*) (see quot. 1935); (*b*) a book kept by the police containing photographs of criminals; **mug-faker,** a street photographer; **mug shot** *slang* (orig. *U.S.*) = sense 1 b above.
1935 A. J. Pollock *Underworld Speaks* 78/2 *Mug book*, a book published for prominent business and professional men who are induced by high pressure solicitors to vividly write about themselves with youthful photographs. They pay well for this blue book privilege. 1958 J. & W. Hawkins *Death Watch* (1959) i. 19, I couldn't find him in the mug books; his picture isn't there. 1933 'G. Orwell' *Down & Out* xxxii. 236 A mugfaker—a street photographer. 1952 M. Allingham *Tiger in Smoke* i. 16 These old photographers—mugfakers we call 'em—in the street. 1950 in Wentworth & Flexner *Dict. Amer. Slang* (1960) 349/1 Police passed around a mug shot of Willie. 1962 K. Orvis *Damned & Destroyed* xxvii. 200 The police record plate number..indicates that mug-shot was taken in the Receiving Room. 1970 R. Jeffries *Dead Man's Bluff* vii. 66 Check through the mug shots and see if you can find him. 1974 *Daily Tel.* 14 Nov. 2/2 From the mass of information collected, a picture is built up of the personalities in the IRA's command structure, and this is then used in the 'mugshot' booklets carried by soldiers on the streets.

mug, *sb.*[5] Add: **1. b. mug's game,** a thankless task; a useless, foolish, or unprofitable activity. *colloq.*
1910 Belloc *Pongo* xv. 233 One cannot arrest millionaires with impunity... Even in a wild democracy to arrest them is Mug's game. 1918 *Flying* 12 June 427/2 Flying's a mug's game, mater, A fact I know full well. 1930 G. B. Shaw *Apple Cart* II. 77, I am going out of politics. Politics is a mug's game. 1959 T. S. Eliot *Elder Statesman* I. 26 Forgery, I can tell you, is a mug's game. I say that——with conviction. 1973 *Times* 9 Nov. 21/2 Running a reserve currency is a mug's game; and the world is running out of mugs.
c. A person, fellow, chap; *spec.* (*a*) a rough or ugly person; a criminal; (*b*) applied by criminals to someone who is not part of the underworld; (*c*) a policeman.
1890 in Barrère & Leland *Dict. Slang* II. 73 'What are *mugs*?' 'Hard characters... Those are thieves from the First Ward, the fellows that rob immigrants, steal cotton from the bales, [etc.]. 1895 E. W. Townsend *Chimmie Fadden Explains* 15 Dat Mr. Paul is de funniest mug you ever see. *Ibid.* 17 De mug what plays de flute has de music all t'himself when de odder mugs in de orchestra don't do nottin. 1903 H. Hapgood *Autobiogr. Thief* (1904) xii. 267 I'm only stealin' for certain mugs (policemen) and fer those 'igher up, so they can buy real estate. 1921 [see *gee *sb.*[4]]. 1930 [see *jitter *sb.* 1]. 1938 F. D. Sharpe *Sharpe of Flying Squad* i. 13 Underworld men and women speak of all outside their world as 'mugs'. 1960 *Observer* 24 Jan. 5/2 There were recognised prop-men or putters up of jobs, what the mugs called master minds.
2. b. *attrib.*, passing into *adj.*, that is a 'mug' or fool; stupid; easily duped or defeated.
1922 E. Wallace *Flying Fifty-Five* xxxiii. 197 The mug punter was he who dreamed of long-priced winners and refused to bet on the six to four certainty preferring rather the hopeless proposition that started at twenty to one. 1963 T. Tullett *Inside Interpol* xvi. 216 There are still thousands of 'mug' criminals..who invariably make mistakes. 1971 *Sunday Australian* 8 Aug. 5/6 Let's just say I'm a good average mug golfer.

mug, *v.*[3] Add: **2.** Also, to fight, beat up, assault; to strangle; (now the prevailing sense) to attack and rob (a person). Also occas. *intr.*, to fight. Cf. *mugging *vbl. sb.* 1.
In the more usual current sense not *slang*.
1846 *Swell's Night Guide* 37 Most of them can..mug.. alias fight. *Ibid.* 76 She felt inclined to mug her rival, only she thought it would be no bottle. 1859 Hotten *Dict. Slang* 65 *Mug*, to fight, or chastise. 1864 *Ibid.* 183 *Mug*,.. to rob by the garrote. 1866 *London Misc.* 5 May 102 Suppose they had mugged you?..Slogged you, you know. 1904 'No. 1500' *Life in Sing Sing* 250/2 *Mug*,..to strangle. 1948 *N.Y. Times* 15 Aug. 36 The police said the victims were mugged in the hallways of their homes. 1960 I. Wallach *Absence of Cello* 23 She's going into Central Park for her constitutional. I hope she gets mugged. 1966 Wodehouse *Plum Pie* vii. 166 Somebody mugged Sam last night... Yessir, laid him out cold. 1968 A. Diment *Bang Bang Birds* iii. 28 In New York..people conversed about their friends being beaten up and robbed—mugged is the local term. 1971 B. Malamud *Tenants* 198 Lesser.. daily fears that..the writer will be mugged on the subway stairs and lie there unable to crawl home. 1972 *Daily Tel.* 7 Oct. 2/6 Judge Hines, Q.C., jailed three youths for three years for 'mugging' a middle-aged man and stealing £7 from his wallet.
3. To take a photograph of (a person), esp. for police records. *U.S.*
1899 'J. Flynt' *Tramping with Tramps* iv. 395 *Mug*,.. to photograph. 1912 [see *mugging *vbl. sb.* 2]. 1929 M. A. Gill *Underworld Slang*, *Mugged*, photographed. 1934 *Sun* (Baltimore) 5 July 13/6 Attempts of Federal authorities to 'mug' him proved futile. When Robert C. Johnson ..prepared to snap a picture of the prisoner, Kent held his hands before his face. 1960 *Wall St. Jrnl.* 3 Nov. 2 More than 15,000 New Jersey securities salesmen are being fingerprinted and 'mugged' at police stations and private detective agencies over the state, under a new state law. 1972 G. V. Higgins *Friends of Eddie Coyle* xix. 115 We brought him up to the marshal's office and mugged him and printed him.
4. To kiss, to fondle. Chiefly *Austral.* and *N.Z.*
1916 C. J. Dennis *Songs Sentimental Bloke* 126 *Mug*, to,

to kiss. 1932 *Amer. Speech* VII. 334 *Mug*, to kiss. 1957 I. Cross *God Boy* (1958) x. 80 You think there is something funny about them mugging up each other like that?

mug, *v.*[4] Add: (Further examples.) Also *trans.*, to supply with beer or liquor; to buy a drink for (someone).
1854 A. E. Baker *Gloss. Northamptonshire Words* II. 38 Come! mug the girls, and they'll get on with their work. 1939 *Daily Mirror* 14 Mar., Are you going to 'mug' us... Are you going to stand me a drink? 1966 P. Moloney *Plea for Mersey* 23 If ye say to them 'scouse, Mug us dem on de house,' Yerl make Birty and Girty all shirty. 1966 F. Shaw et al. *Lern Yerself Scouse* 42 *Ile mug yer*, allow me to treat you. *Ibid.* 77 Many's the fella dat I use'ter mug.

mug, *v.*[5] Add: Hence **mu·gging** *vbl. sb.*, hard studying, 'swotting'. Usu. with *up*.
1959 I. & P. Opie *Lore & Lang. Schoolch.* x. 179 'Swotting' or 'mugging up' is only considered good form if a person is on the point of taking an exam. 1959 *Daily Tel.* 10 June 10/2 But no one..after yesterday's inauguration of the new electric services from London to the Kent Coast is going to mind mugging up any number of amended arrival and departure times. 1960 *Guardian* 15 July 6/6 Hasty muggings-up and regurgitations of fact.

mug, *v.*[7] For *U.S. slang* read *slang* (chiefly *Canadian* and *Naut.*). (Add further examples.) Also, to have a snack, a meal, or a hot drink.
1917 'Bartimeus' *Long Trick* iv. 78 Coats and mufflers were donned and a bottle of sloe-gin uncorked. 'Mug-up!' cried the Sub. 'Mug-up, and let's get 'appy and chatty.' 1927 G. Bradford *Gloss. Sea Terms* 117/1 *Mug up*, to have a drink of coffee or tea which is always on the galley stove of a fishing schooner. 1929 F. Bowen *Sea Slang* 93 *Mug up, to*, to eat, used principally in the Grand Banks schooners. 1936 A. Stringer *Wife Traders* xv. 214 They fell into the habit of stopping more often to 'mug up' along the trail. 1950 J. Hambleton *Abitibi Adventure* 94 René was just 'mugging up' when the shaggy terrier..sent the blackened tea pail flying. 1972 L. Hancock *There's a Seal in my Sleeping Bag* v. 84 We..mugged up on boiled eggs, toast, jam, and coffee.

Muganda (mugæ·ndă). Also 9 **Mganda.** Pl. *Baganda *sb.* and *a.* [Bantu *ganda* + sing. prefix *mu*.] A native or inhabitant of the former kingdom of Buganda, now a province of Uganda; hence, loosely, any native or inhabitant of Uganda. Also *attrib.* and as *adj.*
1862 J. H. Speke *Jrnl. Discovery Source of Nile* (1863) xi. 299 Turning round with true Mganda impetuosity, he walked away. *Ibid.* 313 Every Mganda will say the first Uganda year dates from the arrival of the first Mzungŭ (white) visitor. 1889 R. P. Ashe *Two Kings Uganda* xxiii. 287 A Muganda mother takes no little pride in her child. *Ibid.* 290 Should any person of position suddenly address a Muganda, he should politely reply, 'Kabaka'. 1905 J. F. Cunningham *Uganda* ix. 185 A Muganda chief looks with loathing on all work in the fields. 1908 *Daily Chron.* 24 Dec. 4/6 Buganda is the country, a Muganda is an individual native, the Baganda are the natives as a whole, and Luganda is the language. 1971 *Sunday Nation* (Nairobi) 11 Apr. 11/7 Possibly he was the greatest Muganda patriot in the four hundred years of the Buganda kingdom's history.

mugearite (mɒ·gi,ărəit). *Petrogr.* [f. *Mugeary*, the name of a village in the Isle of Skye, Scotland + -ite[1].] A dark, fine-grained trachyte which has oligoclase as the main feldspar and also contains olivine, orthoclase, and apatite.
1904 A. Harker *Tertiary Igneous Rocks Skye* xv. 257 The other rock..is, when fresh, a black compact-looking rock without phenocrysts... As this rock..belongs to a peculiar type, we..give it a provisional name..'mugearite', a name adopted from that of Mugeary, the crofter village lying at a short distance north. 1928 *Trans. Geol. Soc. Glasgow* XVIII. 348 A characteristic feature which the mugearites share with the Jedburgh basalts is the presence of close-set parallel somewhat irregular platy joints. 1956 *Trans. R. Soc. Edin.* LXIII. 68 The mugearite lava east of Dunsapie Hill, thought to have been the first of the flows, contains a number of phenocrysts of albitised plagioclase. 1971 *Nature* 23 Apr. 510/2 The lavas [of the Baringo District, Kenya] comprise a succession of alkali-olivine basalts, basanites and mugearites.
Hence **mugeari·tic** *a.*, of or containing mugearite.
1927 *Trans. R. Soc. Edin.* LV. 504 We regard the mugearitic and essexitic intrusions of the Dalmahoy syncline as hypabyssal expressions of the magma which gave rise to the accompanying effusive mugearites and allied basalts. 1973 *Nature* 9 Feb. 375/1 There, the volumes of individual hawaiitic and mugearitic lavas are frequently of the order of 0·1 km³.

mugful (mɒ·gful). Also † **mug-full.** [f. Mug *sb.*[1] + -ful.] The contents of a mug; the amount that a mug will hold (in quot. 1867 = bowlful).
1838 Dickens *Nickleby* (1839) xv. 133 A glass-full of spirits and water for Nicholas, and a cracked mug-full for the joint accommodation of himself and Smike. 1867 J. T. Staton *Rays fro th' Loominary* 110 Aw pusht th' owd woman uv hur bustle in a mugful o dowf [*i.e.* dough]. 1924 J. M. Barrie *Mary Rose* I. 15, I wonder if you would give me a mug of tea. Not a cup, we drink it by the mugful where I hail from. 1973 J. Mann *Only Security* vi. 59 Roger accepted the plastic mugful of tea.

mugger, *sb.*[4] Add: **2.** One who 'mugs' people (see *mug *v.*[3] 2); *spec.* one who commits robbery with violence. orig. *U.S.*
1865 J. H. Browne *Four Yrs. in Secessia* xlv. 340 The Muggers, like most bullies and ruffians, manifested a fine discrimination respecting the party they attacked, selecting those they thought they could rob with little resistance and entire impunity. 1874 Hotten *Slang Dict.* 220 *Maceman*, or *macer*, a welcher, magsman, or general swindler; a 'street-mugger'. 1942 *N.Y. Times* 3 Oct. 17/1 The police were..tracking down three known..muggers who..had received suspended sentences in a mugging case. 1955 *Sun* (Baltimore) 5 Jan. 2/4 Corporal ——..of Cleveland reported to police early today that he had been attacked by two muggers while walking near the Pennsylvania Station. 1965 Wodehouse *Galahad at Blandings* i. 11 Muggers, stick-up men and hoodlums in general he disliked. 1970 *Daily Mail* 21 Feb. 9/5 Clarendon Road in London's Notting Hill area..is the haunt of the Muggers —men who clobber you and steal whatever you have of value. 1973 *Sun* 18 Jan. 7 (*headline*) Muggers attacked detective.

mugger (mɒ·gəɹ), *sb.*[5] A nail, usually of wrought iron, used for protecting the inner soles of mountaineering boots.
1941 C. F. Kirkus *Let's go Climbing!* v. 68 For the inner part of the sole muggers, rough wrought iron hobnails, are excellent. 1954 C. D. Milner *Wedderburn's Alpine Climbing* (ed. 2) ii. 10 The Alpine mugger is a very good nail, but the alternate freezing and thawing of the sole will soon loosen it. 1970 A. Blackshaw *Mountaineering* (rev. ed.) iii. 99 Soft iron nails which grip as a result of the rock biting into them (muggers and clinkers).

mugger (mɒ·gəɹ), *sb.*[6] and *v.* A slang euphemism for some senses of Bugger *sb.* and *v.* (see quots.).
1945 'N. Shute' *Most Secret* vii. 150 What do you think we stopped to pick the mugger up for? 1948 —— *No Highway* ix. 229 The pilot said, 'I don't give a mugger about that, sir. It's plain bloody nonsense.' 1951 —— *Round the Bend* 69 'Well, I'm muggered,' I said in wonder. 1954 'G. Carr' *Death under Snowdon* v. 54 Privileges are for the man who works for 'em, not for the mugger who plants his bottom on 'em because his father owned land. 1962 J. P. Carstairs *Pardon my Gun* i. 10 She's muggered off... Hopped it to Italy.

mugginess. (Earlier example.)
1856 Geo. Eliot *Let.* 29 Dec. (1954) II. 284 Don't you enjoy the frost after that long time of mugginess?

mugging (mɒ·giŋ), *vbl. sb.* [f. Mug *v.*[3] + -ing[1].] **1.** orig. *slang.* The action of *mug *v.*[3] 2; *spec.* robbery with violence. Also *attrib.*
1846 *Swell's Night Guide* 75, I knows that 'ere white-house warment..would chaff—and you knows I'm soon shirty, and then we should have a mugging match. 1866 J. E. Brogden *Provincial Words Lincolnshire* 131, I gave him a sound mugging, he was so chappy. *a* 1876 E. Leigh *Gloss. Words Dial. Cheshire* (1877) 139 'To receive a muggin' is to be beaten. 1939 *Fortune* July 168/3 The vicious art of 'mugging' by which a Negro thug grabs the wayfarer around the neck, from behind, while two others with knives clean out his pockets. 1942 [see *mugger *sb.*[4] 2]. 1949 *Sat. Even. Post* 8 Oct. 171/2 The only things she reads in our newspapers are the murders and the muggings and the obituaries. 1951 *Manch. Guardian Weekly* 28 June 2/2 The increase in petty thieving, in 'muggings' of night-time strollers, and in prostitution. 1971 B. Malamud *Tenants* 64 Bugsy is shot..in Catshit Alley by two white pigs who had cornered him there after a mugging. 1973 *Guardian* 7 June 9/5 Much of the [crime] increase consists of robberies after sudden attacks in the open, known commonly as mugging.
2. The action of *mug *v.*[3] 3; the taking of photographs of persons. Also *attrib.* *U.S. slang.*
1899 'J. Flynt' *Tramping with Tramps* iv. 389 In some cities suspicious characters are arrested on general principles and immediately photographed by the police authorities. Such towns are called 'muggin' joints', and the police authorities 'muggin' fiends'. 1912 A. B. Reeve *Poisoned Pen* (1913) ii. 58 An hour later, at headquarters, after the pedigrees had been taken, the 'mugging' done.. O'Connor led the way into his private office. 1912 A. Train *Courts, Criminals & Camorra* 5 'Mugging' was all right, so long as you 'mugged' the right persons.
3. Kissing, love-making, 'necking'. Cf. *mug *v.*[3] 4.
1924 P. Marks *Plastic Age* xxiii. 271, I hate mugging and petting and that sort of thing... Petting is jazzing love; and I hate it. 1926 K. S. Prichard *Working Bullocks* xxvii. 242 But Deb had never kissed her mother, as she had seen children in the township kiss and cuddle up to their mothers... Her mother did not encourage 'mugging', as she called it. 1970 C. Major *Dict. Afro-Amer. Slang* 83 *Mugging*,..making love.
4. Grimacing; making faces (see also quots.). Also *fig.*
1937 *Amer. Speech* XII. 317/2 *Mugging*, motion picture acting. 1961 *Times* 25 Apr. 20/4 Grimaces and gestures straight out of silent films, properly deserving the name 'mugging'. 1968 *Listener* 25 July 124/2 Paul Daneman made a promising start as Higgins, but dwindled into a kind of verbal mugging.

muggins. Add: **1.** (Earlier and later examples.) Freq. used by a speaker to refer to himself.
1855 *Golden Era* 28 Jan. 2/1 You are a veritable 'Muggins' in [choosing] cigars. 1859 C. E. De Long in *Calif. Hist. Soc. Q.* (1931) X. 167 Spent the evening until late with Jim in having one of our regular old fashioned long talks about women love &c: and both arrived at the conclusion that we were mugginse's. 1973 *Daily Tel.* 29 Aug. 1/3 The letter bomb was not meant for me personally. I was just the muggins who opened it. 1973 E. Lemarchand *Let or Hindrance* 181 'In a nutshell,' Michael said, '..Muggins [*i.e.* himself] has agreed to be in charge.'
2. a. (Earlier examples.)
1855 *Pioneer* (San Francisco) Nov. 358 We returned to the hotel, to engage in the intellectual game of Muggins. Ladies and gentlemen were all decided to be Muggins ere the game had closed. 1865 S. Jex-Blake *Let.* 18 Aug. in M. Todd *Life S. Jex-Blake* (1918) xiii. 165 After the ices we went back to the Hospital, and played a most ridiculous game of cards called 'Muggins', keeping us in roars of laughter half the time.
b. (Earlier U.S. example.)
1868 F. B. Zincke *Last Winter in U.S.* 268 Consoling herself with a kind of dominoes she called 'muggins'.

muggle[3] (mɒ·g'l). *slang* (orig. *U.S.*). [Origin unknown.] *pl.* Marijuana; *sing.* or *pl.*, a marijuana cigarette. Also *Comb.*, as **muggle-head, -smoker,** one who smokes marijuana; so **mu·ggler,** a marijuana addict.
1926 Maines & Grant *Wise-Crack Dict.* 11/1 *Muggle-head*, smoker of Mexican loco weed. 1928 L. Armstrong (*title of gramophone record*) Muggles. 1933 C. de Lenoir *Hundredth Man* i. 10, I found myself on the Mexican border with a bad 'yin', and nothing to relieve me but the native drug marijuana. In New Orleans and other Southern American towns this is known as 'muggles', being sold in the form of cigarettes. 1933 *Fortune* Aug. 90/1 Louis Armstrong, who blew such frenzied tattoos as he has recorded under the titles *Mahogany Hall Stomp*, *Knee Drops*, *Skip the Gutter*, and *Muggles* (named for the Mexican cigarettes drugged with marijuana which have inspired perfectly incredible solos). 1938 *Manch. Guardian Weekly* 2 Sept. 188/3 Many swing players are 'killer-dillers' (first-rate players). Some are 'mugglers' (Marijuana addicts), but very few are 'long-hairs' (people who like classical music). 1946 Mezzrow & Wolfe *Really Blues* (1957) 51 'Ever smoke any muggles?' he asked. 1949 R. Chandler *Little Sister* xxxiv. 248 Desk clerk's a muggle-smoker. 1969 A. Arent *Laying on of Hands* (1971) vi. 50 Offer our guest a muggle. 1972 *Sunday Sun* (Brisbane) 2 July 14 Detectives from the CIB Drug Squad in Brisbane are becoming quite familiar now with words like muggles, griefs, mezz, Mary Jane, jive, tea, rope and loco-weed.

Muggur, var. *Magar[2].

mugo (mu·go). Also **mugho(s).** [a. Fr. *mugho*, It. *mugo* mountain pine.] In full, *mugo pine*. The European mountain pine, *Pinus mugo* (or *P. montana*), a large shrub, or one of its many varieties, ranging from dwarf forms to small trees.
c 1756 P. Collinson in L. W. Dillwyn *Hortus Collinsonianus* (1843) 40 Mr. P. Miller received seeds of the Cembro and Mugos Pines from the mountains of Valencia in Spain, Sept. 12, 1753. 1822 J. M. Good *Study of Med.* I. 315 The Hungarian balsam, or distilled oil of that variety of the *pinus sylvestris* which has been called Mughos. 1861 R. Bentley *Man. Bot.* 659 *P[inus] Pumilio*, the Mugho or Mountain Pine, yields by spontaneous exudation an oleo-resin called Hungarian balsam. 1951 H. S. Conard tr. *Kerner's Background Plant Ecol.* xxiii. 191 The formation of the Rusty Sedge generally occurs in the calcareous Alps as islands in the dark green mugho pine forest. 1966 H. J. Welch *Dwarf Conifers* v. 266 The Mountain Pine (Mugo is a local name) is a very variable species with a distribution in nature throughout Europe from the Pyrenees to the Balkans. 1967 N. T. Mirov *Genus Pinus* ii. 98 It [*sc.* *Pinus montana*] grows there [*sc.* in the Carpathians] in a shrub form and is known as variety *mughus*, or Mugo pine. 1969 T. H. Everett *Living Trees of World* 52/1 The Swiss mountain pine (*P. mugo*) is best known to many Americans in its dwarf variety, the mugho pine (*P. m. mughus*).

‖ **muguet** (müge). [Fr.] Lily of the valley; the smell or scent of lily of the valley. Also *attrib.*
[1873 *Young Englishwoman* May 234/2 Bodice..edged with muguet fringe of violet silk.] 1919 W. L. Dudley *Askinson's Perfumes* (ed. 4) xvi. 208 *Muguet*. Oil of jasmine... Oil of ylang-ylang... Solution of heliotropin. 1928 H. S. Redgrove *Scent* xi. 66 *Hydroxy-citronellal*.., very useful for making artificial lily of the valley (muguet) perfume. 1957 E. Hyams *Into Dream* II. iii. 115 She used a little too much scent: Paul identified it as *muguet*. 1970 *Guardian* 12 May 9/2 Coty have recently reformulated their Muguet des Bois fragrance range.

mug-up (mɒ·gɒp, mɒgɒ·p). *slang* (chiefly *Canadian* and *Naut.*). [f. Mug *v.*[7]] A snack, a meal, or a drink.
1933 E. Merrick *True North* 233 Back at the tent we had a mugup, lashed up and said good-bye to the Indians. 1941 *Beaver* Mar. 12 Where the traveller stops long enough to have a 'mug-up', each dog is given about a pound of frozen fish. 1950 R. Moore *Candlemas Bay* 143 They'd warm up the fishhouse and have a mug-up, before Jeb went to work on his traps for the rest of the daylight. 1961 J. W. Anderson *Fur Trader's Story* xxv. 225 The Eskimos too would be entertained, but because of their numbers they would have a picnic 'mug-up' on deck. 1963 J. T. Rowland *North to Adventure* ix. 123 We had a mug-up of hot coffee and turned in. 1970 R. Price *Howling Arctic* iv. 45 Occasionally they stopped for mug-up.

mugwump, *sb.* For *U.S.* read orig. *U.S.* Add: **1.** (Further examples.)
1925 *N.Y. Times* 10 May, The royal red Indian mugwump, the chief, was copiously red-blooded. 1945 [see *bigwig].
2. (Further examples.) Also, a person who withdraws his support from any group or organization; one who is aloof, independent, or self-important.
1887 J. D. Billings *Hardtack & Coffee* xv. 286 [The mule's] reputation as a kicker is world-wide. He was the Mugwump of the service. 1894 P. L. Ford *Hon. Peter Stirling* 302 I'd have believed anything but that you [*sc.* a Democrat] would be a dashed Mugwump! 1903 G. B. Shaw *Man & Superman* III. 116, I told him I did not care whether he got into parliament or not; so he called me Mugwump and went his way. 1923 H. E. Buchholz *Of what Use are Common People?* vi. 66 Mugwump may be thought of as a fitting term for the man who because of real or imagined superiority separates himself from the group with which he has been associated. *Ibid.* 73 The proposal of the intelligence mugwumps is that the majority of ordinary minds should step out of the picture of the body politic,..while creatures with better than normal minds should be commissioned to do all the thinking. 1946 *Tuscaloosa* (Alabama) *News* 31 Mar. 4/7 A few moments after Secretary Wallace made his pun, he hastened to add that he himself had been a mugwump. 1975 *Sat. Rev.* 22 Mar. 58/3 Among the delicious names taken by, or given to, minor political parties in the United States (apart from Mugwumps and Bull Moose) are..Quids, Locofocos, [etc.].
3. (Earlier and later examples.)
1887 *Courier-Jrnl.* (Louisville, Kentucky) 8 Jan. 4/5 The Mugwump War Department is a horse-power to the Republican machine. 1923 H. E. Buchholz *Of what Use are Common People?* viii. 108 The various mugwump groups that propose changing the form of the present government. 1931 H. F. Pringle *Theodore Roosevelt* I. ix. 115 Roosevelt's gnawing hatred of the Mugwump independents made it impossible for him to withdraw. 1970 *Times* 9 Dec. 16/2 Even a doubtful and controversial conclusion..would have been more useful than this irritating mugwump approach.
Hence **mu·gwumpish** *a.*, characteristic of a mugwump; professing disinterestedness; pretentiously superior; **mugwumpism** (earlier and later examples).
1886 *Congress. Rec.* 3 Mar. 2968/1 That maudlin political sentiment which we recognize, for want of a better, under the name of 'Mugwumpism'. 1898 G. B. Shaw *Our Theatres in Nineties* (1932) III. 297 It [*sc.* *Hamlet*] belongs to a detached residence, a select library, an exclusive circle, to no occupation, to fathomless boredom, to impenitent mugwumpism. 1918 F. Hackett *Ireland* ix. 252 This conviction was accompanied..with many mugwumpish strictures such as 'in the main', 'within certain limits', [etc.]. 1923 *Spectator* 22 Sept. 390/1 Racial, intellectual or moral tests..may turn out to be not aristocratic at all, but merely mugwumpish.

Muhammad, Muhammadan. Now the preferred spelling of Mohammed, Mohammedan *a.* and *sb.*
1961 [see *Black Muslim* (*black *a.* 19)]. 1963 *Cambr. Rev.* 20 Apr. 369 (*title*) Oriental Studies at Cambridge: Muhammadan Law and Urdu Literature.

muhimbi (muhi·mbi, muwi·mbi). Also **muhindi.** [Lunyoro.] An evergreen tree, *Cynometra alexandri*, belonging to the family Leguminosæ and native to East Africa, or the timber obtained from it, which is also called Uganda ironwood.
1906 M. T. Dawe *Rep. Bot. Mission through Forest Districts of Buddu* 43 in *Parl. Papers* (Cd. 2904) LXXX. 41 *Cynometra Alexandri*... Native name, Muhindi: immense and important timber tree. 1939 Eggeling & Harris in L. Chalk et al. *Forest Trees & Timbers Brit. Empire* IV. 27 Two pieces of Muhimbi (Ironwood), buried near a termite hill in 1930, were removed after 16 months, in excellent condition. 1951 *Archit. Rev.* CX. 144/3 Muhimbi, an extremely hard wood from Uganda, is here used for the first time in a floor of any appreciable size in this country. 1956 *Handbk. of Hardwoods* (Forest Prod. Res. Lab.) 158 Muhimbi is a heavy, hard timber of equal weight to East African olive. 1971 F. H. Titmuss *Commercial Timbers of World* (ed. 4) 155 The timber of *Cynometra alexandri*..is now beginning to be much better known under the title of Muhimbi.

muhtar, var. *mukhtar.

muid. 1. a. (Later examples.)
1859 R. J. Mann *Colony of Natal* 124 One farmer in the Umvoti country reaped 120 muids (of [illegible] bushels each) from 30 acres of land which had been sown with 5 muids of seed. 1873 [see *boermeal* (*Boer b)].

muishond (mǒŭ·shǫnt). *S. Afr.* Also 9 **mausehund, mousehund.** [Afrikaans, f. *muis* Mouse *sb.* + *hond* dog; cf. Mouse-hunt[1].] The Cape polecat, *Ictonyx striatus*, which is black with white stripes, or one of several species of mongoose. Also *fig.*
1796 tr. *F. Le Vaillant's New Trav. Afr.* I. 236 It [*sc.* a stranded whale] was attacked..by different species of those small quadrupeds..which, at the Cape, are known under the general name of *muyshond*. *Ibid.* III. 278 My Hottentots of the colony all recognized it as a *muyshond* (mouse-dog), a general name among the inhabitants of the Cape for all the little carnivorous quadrupeds. 1818 C. I. Latrobe *Jrnl. Visit S. Afr.* 36 We observed a tame mongoose or mausehund from Java..which ran about and suffered itself to be handled. 1835 A. Smith *Diary* 5 Jan. (1939) I. 204 A hare and a muishond for sale. 1853 *Edin. New Philos. Jrnl.* LV. 210 Two or three different kinds of Mousehund, or weasel..are commonly seen. 1900 B. Mitford *Aletta* xii. 111 She went round among her fowl-houses, then strolled along the quince hedges to see if any of the hens had been laying out and in irregular places for the benefit of the egg-loving *muishond*, or similar vermin. 1912 F. W. Fitzsimons *Snakes S. Afr.* (new ed.) 30 Within two yards of us was a striped Muishond with his paws firmly planted on an adult Black-necked cobra. 1937 H. Klein *Stage-Coach Dust* xxiv. 208 A muishond (pole-cat) startled the town-bred horse, and away he galloped. 1946 *Cape Argus* 16 Nov. 6/2 Two varieties of muishond now have prices on their heads. 1952 —— 12 Jan. 4/5 His request for tame 'wild' animals has only produced in Cape Town a muishond and a monkey. 1970 *News/Check* (S. Afr.) 12 June 11 An assumption has grown up over the past year or more that South Africa was well and truly isolated, that it remained the *muishond* of the world with whom nobody wanted anything to do. 1974 *Daily Dispatch* (East London, Cape Province) 21 Nov. 10 A rabid muishond has been killed on Arbrook, farm of Mr. George Michau.

muisvoël (mǒŭ·sfōəl). *S. Afr.* Also **muisvoel, muisvogel** (-fōχəl). [Afrikaans, f. as prec. + *voël* bird.] = *mouse-bird* (Mouse *sb.* 10 f).
1822 W. J. Burchell *Trav. S. Afr.* I. xi. 214 Muis-vogel (Mouse-bird). *Colius erythropus* of Linnæus. 1849 A. Smith *Illustr. Zool. S. Afr.: Aves* plate ii, *Chizærhis concolor*..As soon as it was observed, the Hottentots declared it to be a *muis vogel*, or *Colius*, Lin... which was not surprising, since it evinces considerable similarity to birds of that genus. 1867 E. L. Layard *Birds S. Afr.* 221 Of the three species of this genus [*sc.* *Colius*] found in South Africa, and known by the trivial name of Muisvogel or Mouse-bird, this [*sc.* *C. erythropus*] is the only one that is found in the neighbourhood of Cape Town. 1920 R. Y. Stormberg *Mrs. Pieter de Bruyn* 7 It [*sc.* peachblossom] has been vandalised by swarms of 'finks' and wretched little stiff-tailed 'muisvogels' who snip off the blooms most viciously. 1952 *Cape Times* 27 Nov. 2/3 Muisvoëls are destroying whole orchards of early ripening fruit. 1959 *Ibid.* 20 Jan. 7/3 Sparrows and muisvoels are not a pest in areas where our native birds of prey have not been eliminated. 1973 *Eastern Province Herald* (Port Elizabeth) 19 July 17 How many of us city-dwellers..would know the difference between a mossie and a muisvoël?

mujik, var. Moujik, muzhik. (Examples.)
1833 [see Moujik, muzhik 1]. 1934 *Discovery* Dec. 364/2 The mujiks know that Lenin hath given, but Stalin hath taken away. 1963 V. Nabokov *Gift* iv. 216 He began dabbling in propaganda by conversing with mujiks.

‖ **mujtahid** (mudʒtā·hid). Also **mooshtâhed, mujtehid.** Pl. **mujtahids, mujtahidūn.** [Pers. 'one who strives hard to acquire correct and sound views', 'one who has arrived at the highest degree in knowledge of the law'; Arab. 'one who exerts himself'.] In Islamic countries, the title given to a person accepted as an authority on the interpretation of Islamic law. Now only in Iran.
1815 J. Malcolm *Hist. Persia* II. xxiii. 428 That power ..is exercised..by these holy men who are raised, by popular suffrage, to the dignity of Mooshtâhed, or 'high priest'; and who may be deemed at the head of the hierarchy of Persia. 1885 T. P. Hughes *Dict. Islam* 197/2 *Ijtihād* is the deduction made by a single Mujtahid, whilst *Ijma'* is the collective opinion of a council of Mujtahidūn, or enlightened doctors. 1909 *Daily Chron.* 2 Feb. 1/7 The Shah would be well advised to listen to the petition of the Mujtehids and clergy now being presented,..begging for the restoration of the Constitution. 1958 A. Toynbee *East to West* 161 'The veil between us is very thin,' said the old mujtahid. 1961 *Ann. Reg. 1960* 299 The leading *mujtahid* of the day having declared this law to be contrary to the *shari'a* and the Constitution, it seemed likely to remain a dead letter.

Mujur, var. *Mudjur.

mukhtar (mu·ktāɹ). Also **muhtar, muktar.** [ad. Arab. *mukhtār* chosen.] The head of the local government of a town or village in Turkey; a minor provincial official. Hence **mu·khtarship,** the status or office of a mukhtar.
1911 T. E. Lawrence *Let.* 21 May (1938) 104 The Muktar carried off his cousin on his saddle-bow from amid the shrieking women at the spring. 1916 —— *Lett.* (1938) 209 Two Mukhtars and two prominent sheikhs. 1946 [see *dunam]. 1946 Koestler *Thieves in Night* 122 The British rulers had always appointed the most distinguished member of each of the two clans to dual mukhtarship. 1954 F. Stark *Ionia* 65 The mukhtar and the doctor from Pitane..came to give the city's welcome. 1970 *Guardian* 1 Apr. 4/2 The 'muhtar' (local official) of the village of Kakoy. 1974 *Times* 5 Apr. 16/4 The Mukhtar..explained: '..we are Druse and we will never leave our land.'

‖ **mukim** (mū·kim). Pl. **mukim, mukims.** [Malay, ad. Arab. *muḳīm* remaining, resident.] In Malaysia, a Muslim parish, the smallest administrative district.
1839 T. J. Newbold *Pol. & Statistical Acct. Straits of Malacca* I. v. 247 The immediate religious care of the inhabitants of the Mukim (or parish) to which the mosque belongs, devolves upon the Imam, Khatib, and Bilal. 1907 F. A. Swettenham *Brit. Malaya* x. 228 The districts were again sub-divided into *Mukim*. 1911 *Encycl. Brit.* XVII. 482/2 The state is divided into *mukim* or parishes. 1948 *Malayan Pict. Observer* Dec. 1 In Jimah Mukim of Port Dickson District..lies the Chuah Indian Settlement. 1958 J. M. Gullick *Indigenous Political Syst. W. Malaya* ii. 36 A *mukim*, a sub-district which may include several villages ... *Mukim* denotes the area served by a mosque and is sometimes translated 'parish'. 1964 K. G. Tregonning *Hist. Mod. Malaya* vii. 157 By making administration at local *mukim* level the responsibility of people trusted in the kampongs, he created a stable State.

muklek, mukluk, varr. *muckluck.

muktar, var. *mukhtar.

mukti (mɒ·kti, mu·kti). *Hinduism* and *Jainism.* Also 9 **mooktee.** [Skr. *mukti* a setting or becoming free, release, f. *muc* to let loose, free, release.] = *moksha.
1785 C. Wilkins tr. *Bhăgvăt-Gēētā* 140 The Hindoos believe..that an eternal release, which they call *Mŏŏktēē*, is only to be attained by a total neglect of all sublunary things. 1801 *Asiatick Researches* VII. 34 Bhooddha after his death ascended to the Hall of Glory, called Mooktzé, otherwise Nirgoowané. 1807 W. Jones *Works* III. 384 The *Mucti*, or *Elysian* happiness of the *Vèdánta* school is far more sublime. 1811 W. Ward *Acct. Writings, Relig. & Manners Hindoos* IV. x. 315 The shastrus teach that there are four kinds of mooktee, or deliverance. 1832 [see *bhakti]. 1840 [see *moksha]. 1875 M. Williams *Indian Wisdom* iii. 70 The right apprehension of truth..which, if once acquired by the soul, confers upon it final emancipation, whether called Mukti, Moksha,..or Nirvana. 1915 *Encycl. Relig. & Ethics* VIII. 773/2 The evolution of *mukti*, or *mokṣa*, in Jainism cannot be adequately dealt with till its early literature is more fully accessible. 1970 T. W. Organ *Hindu Quest Perfection of Man* ix. 296 *Mukti* is freedom from the individualistic inlook, and the attainment of the divine outlook.

‖ **muktuk** (mɒ·ktɒk). Also **maktuk.** [ad. Eskimo *maktak*.] The skin of any of several species of whales used for food by the Eskimo.
1835 R. Huish *Last Voy. Sir J. Ross Arctic Regions* 701/2 Skin of a whale..Maktuk. 1909 A. D. Cameron *New North* xii. 220 The wedding breakfast consisted of seal-meat, frozen rotten fish, and muktuk (whale meat). 1940 *Beaver* Mar. 25/2 All were chewing *muktuk*—the outer protective skin of the whale—with great enjoyment. 1966 *Star Weekly* (Toronto) 12 Mar. 10/1 The Eskimos love muktuk and eat it raw. It's rather rubbery but doesn't taste bad at all—a bit like hazel nut. 1973 *Nat. Geographic* Mar. 353/1 After the blubber is cut, the Eskimos pare off the tough, rubbery skin and a thin layer of attached blubber—called *muktuk*. A delicacy to the Eskimos, muktuk tastes surprisingly good when boiled and salted. 1974 *N.Y. Times* 30 May 39/2 But the *muktuk*—the skin and outer layer of blubber that is the staple of the Eskimo diet—was fine.

Mukuzani (mukuzā·ni). [Russ.] A red wine from Georgia, U.S.S.R.
1961 *Spectator* 7 Apr. 495 The first Soviet wines to be imported into Britain—Mukuzani, Gurdzhaani and Tsinandali, all from Georgia. 1967 A. Lichine *Encycl. Wines* 470 Some Georgian wines seen abroad are white Myshako Riesling and Ghurdjurni. The Takhetia region is the most successful and the best wines include..the red Mukuzani. 1968 A. H. Gold *Wines & Spirits of World* 467 The wines tend to be full and the red ones dark. *Mukuzani* and *Saperavi* are two dark strong red wines of 14° alcoholic strength from the eastern side of Georgia in the Tiflis region.

Mulana, var. *Maulana.

mulatta. Delete † *Obs.* and add later examples.
1963 *Times* 12 June 16/6 Of the other painters the pioneers were Di Cavalcanti, famous for his sensuous mulatas, and Portinari. 1965 E. Bishop *Questions of Travel* 37 Going out, he met a *mulata* Carrying water on her head.

mulatto, *sb.* and *a.* Add: **A.** *sb.* **4. mulatto-clay** *U.S.*, a dark-coloured clay; **mulatto land** (earlier and later examples); **mulatto loam, mould** = *mulatto land*; **mulatto prairie,** a prairie of mulatto-soil; **mulatto-soil** (later examples).
1741 in *Amer. Speech* (1940) XV. 287/2 A Tract of rich Mulattoe Land, lying in that County. 1788 T. Jefferson *Tour of Amsterdam* in *Writings* (1854) IX. 386 It has a southern aspect, the soil a barren mulatto clay, mixed with a good deal of stone, and some slate. 1883 E. A. Smith *Rep. Geol. Survey Alabama 1881–82* 435 The red or mulatto lands are much the best for cotton. 1837 J. L. Williams *Territory of Florida* 82 The surface is covered with a mulatto or chocolate colored loam. 1838 *Jeffersonian* (Albany) 28 Apr. 88 (Th.), The mulatto mould of the Colorado does not surpass in fatness the alluvial soil of Red River. 1869 *Overland Monthly* III. 130 Then there is the 'chocolate' prairie, and the 'mulatto', and the 'mezquite'. 1819 E. Dana *Geogr. Sk. Western Country* 190 Next to this is very often found a skirt of rich pine land, dark mulatto soil with hickory..characteristic of good land. 1861 *Trans. Illinois Agric. Soc.* IV. 112 He..would not choose the dark prairie mold, but that kind of soil best known in the west as the 'mulatto soil'.
B. *adj.* **1.** (Earlier example.)
1677 *Rec. Court of New Castle on Delaware* (1904) 91 The upholding & detayning of this p[laintiff]'s molatto servant in Maryland.

mulattress. (Earlier and later examples.)
1805 in *Amer. Pioneer* (1843) II. 234 The chief of the

audience is formed of mulatresses and negresses. 1887 *Harper's Mag.* Mar. 609/1 A handsome, strong-limbed, and light-footed mulattress. 1932 *Dict. Amer. Biogr.* VIII. 485/1 He lived openly with a mulattress and was only prevented from marrying her by the law against miscegenation.

mulberry. Add: **3.** (Later examples.)
1951 M. Kennedy *Lucy Carmichael* I. i. 10, I impressed her with my mulberry house-coat. 1971 *Vogue* 15 Oct. 41 Pullover..in Antwerp Brown, French Navy and Mulberry. 1974 R. Harris *Double Snare* xii. 83 Sir Jonathan wore a white doublet and mulberry trunk hose. *Ibid.* xxviii. 226 The colours; dog-rose pink, green, and mulberry.
3*. In full, *Mulberry harbour.* The code name of the prefabricated harbour used in the invasion of the Continent by British and American forces in 1944; also applied to any artificial harbour. Also *attrib.*
1945 *N. & Q.* 15 Dec. 263/2 The word 'Mulberry' was selected as the secret name for the artificial harbour..from its being that which happened to come next in rotation on the Admiralty's List of Ships' Names then available for use. 1946 J. E. Taylor *Last Passage* i. 19 Mulberry is the name given to an artificial harbour erected primarily for the landing of stores off the enemy beaches. 1958 *Listener* 5 June 930/2 Further north—in the island of Schouwen-Duiveland—I saw where it had been necessary to float great mulberry harbour caissons in and sink them in the gaps. *Ibid.* 21 Aug. 258/2 The famous Mulberry Project, the creation of vast artificial seaports on the open coast of Normandy. 1965 R. B. Oram *Cargo Handling* viii. 152 Movable quays of the war time Mulberry pattern. 1972 *Daily Tel.* 11 May 9/3 Vice-Adml Hughes-Hallett told me he..frankly considered the 'embellishments' of the Mulberries quite superfluous.
4. *mulberry wine*; *mulberry-nosed, -red* adjs.; **mulberry-bird,** (*a*) *Austral.* substitute for def.: the southern figbird, *Sphecotheres vieilloti*; (examples); **mulberry molar** *Med.*, a first molar with a small crown that is nodular and pitted, somewhat like a mulberry, as a result of congenital syphilis.
1891 A. J. North in *Rec. Austral. Museum* I. 113 It [*sc.* the Southern Sphecotheres] is fairly common on the Tweed River, where it is locally known as the 'Mulberry-bird', from the decided preference it evinces for that species of fruit. 1966 N. W. Cayley *What Bird is That?* (ed. 4) 21 Southern Figbird... Also called Mulberry-bird and Banana-bird... Flocks may be seen feeding in native fruit- and berry-bearing trees, and sometimes in orchards, where they eat mulberries, figs, and other soft fruits. 1923 *Arch. Dermatol. & Syphilol.* VIII. 794 There is a definite shoulder formation, and the cusps occupy a smaller area ..and..are smaller than normal... The grouping or arrangement of these small cusps have suggested the term 'mulberry' molar. 1941 K. H. Thoma *Oral Path.* v. 242 The mulberry molar is covered on the sides with normal, smooth enamel, but the occlusal surface is pinched together, dwarfed, rough, and hypoplastic, often pigmented. *Ibid.*, Often a supernumerary nodule or pseudocusp appears which clinicians emphasize as an important feature of the mulberry molar. 1961 R. D. Baker *Essent. Path.* ix. 199 Stigmata of congenital syphilis may be found in those who survive. These defects are Hutchinson's teeth, mulberry molars, saddle nose, deafness and keratitis. 1924 R. Graves *Mock Beggar Hall* 6 A dissolute Mulberry-nosed philosopher. 1927 Mulberry red [see *ash *sb.*[1] 1 e]. 1945 W. de la Mare *Scarecrow* 35 His large mulberry-red face and eyes like bits of agate. 1723 J. Nott *Cook's & Confectioner's Dict.* sig. T4 To make mulberry wine. Gather your Mulberries when they are thorough ripe, pick off the Stalks, and press out the Juice. 1971 *Country Life* 23 Dec. 1777/3 There was an abundance of wine, particularly claret, mulberry wine and mead at a peacock feast.

mulctuary, *a.* Delete † *Obs.* and add later example.
1968 *Punch* 25 Sept. 421/2 The taxes used to buy the state industries were collected by savage mulctuary imposts on the rich.

mule[1]. Add: **2. d.** *Naut.* A large triangular sail sometimes used on a ketch. So *mule-rigged* adj.
1932 *Yachting* Sept. 46 'Tidal wave', winner of the race, at the start, with the mule pulling aloft and motor pushing below. 1954 *Ibid.* Apr. 61 The swell rolls the wind out of her spinnaker, but the 'mule' aloft between her masts is pulling like its long-eared namesake. 1954 *Motor Boating* Dec. 29/1 The mule-rigged yawl Flame at the start in Newport. 1964 M. Weeks *Compl. Boating Encycl.* 366/2 *Mule*, a large triangular staysail sometimes used on a ketch. It sets on the main backstay and is sheeted to the mizzenmasthead.
e. *U.S. slang.* (See quots.)
1935 A. J. Pollock *Underworld Speaks* 78/2 *Mule*, person who carries dope for a drug trafficker and passes drug to buyer after a sale has been made. 1951 *Life* 11 June 126/2 He becomes a 'mule' (delivery boy) for a peddler and earns his own heroin by introducing his friends to dope and making customers of them. 1959 'E. McBain' *Pusher* viii. 78, I bought from him a coupla times. He was a mule, Dad. That means he pushed to other kids.
4. c. (Earlier and later examples.)
1801 C. Pye *Provincial Coins* 3 The endless varieties (not unaptly termed *mules*) produced by a combination of dies not originally intended for the same coin. 1961 G. van der Meer in R. H. M. Dolley *Anglo-Saxon Coins* 183 The variety is an 'Intermediate Small Cross/*Crux*' mule. *Ibid.* 184 The type is a mule of the 'Arm-and-Sceptre' type of 'Cnut' (= Harthacnut) and the '*Pacx*' type of Edward the Confessor.
d. A small tractor or locomotive, usually powered by electricity, used for towing canal-boats, moving trailers, etc. (see quots.).
1903 *Electr. World & Engin.* 14 Nov. 795 The 'mule' has two large hooks for the towropes. 1924 *Chambers's Jrnl.* Nov. 731/2 These wire ropes are stretched from the ship to motor-tractors running on rails the length of the docks. Electric 'mules' the tractors are called... These mules both guide and propel the ship. 1928 *Amer. Speech* III. 366 'Mules' are the little gas and electric tractors used in the studios of Hollywood. 1971 M. Tak *Truck Talk* 109 *Mule*, a small tractor used to relocate dollies in a terminal or warehouse.
5. a. (sense 1) *mule-back* (further examples; also *attrib.* and quasi-*adv.*), *-boy, -bray, -cart, -driver, -hoof, -kick, -load, -man, -meat, -path, -power, -race, -road, -skin, -steak, -team, -track, -trail, -train, -wagon, -way*.
1878 *Harper's Mag.* Jan. 283/2 He put it in his broken English, 'On horse-back or mule-back, and many times on foot-back.' 1897 R. M. Stuart *In Simpkinsville* 15 Many's the time..he's rode into town, mule-back, with her settin up in front of him. 1904 'O. Henry' *Cabbages & Kings* i. 16 The mule-back system of transport that prevailed between Coralio and the capital. 1968 D. M. Smith *Medieval Sicily* xviii. 182 All wheat had to be carried to the ports on muleback. 1973 K. Benton *Craig & Jaguar* iii. 29 An estate..surrounded by high mountain ranges.. and until recently only accessible by mule-back. 1958 J. Carew *Black Midas* i. 11, I had started work as a mule boy..trotting beside a mounted overseer. 1960 S. Plath *Colossus* (1967) 20 Mule-bray, pig-grunt. 1847 *Knickerbocker* XXX. 228 Our little mule-cart was but ill-fitted for the passage of so swift a stream. 1929 J. Buchan *Courts of Morning* i. 37 Country mule-carts struggled towards the market-place. 1857 P. St. G. Cooke *Scenes & Adventures* 90 A charge..would have proved disastrous to the mule-drivers. 1909 'O. Henry' *Roads of Destiny* 192 He had been mule-driver..and cattleman. 1880 'Mark Twain' *Tramp Abroad* 486 We found the masonry slightly crumbled, and marked by mule-hoofs. 1930 Blunden *De Bello Germanico* 5 That sudden backward mule-kick which gives troop-trains one of their unique charms. 1968 *Globe & Mail* (Toronto) 13 Jan. 24/3 A flying leap known in dancing as a reverse dolphin dive, and in wrestling jargon as a mule kick. 1810 Z. M. Pike *Acct. Expeditions Sources Mississippi* App. iii. 4 There are taken, to be coined, 100 mule-loads of bullion in silver and gold monthly. 1880 'Mark Twain' *Tramp Abroad* 491 We had plenty of company, in the way of wagon-loads and mule-loads of tourists. 1867 in E. Custer *Tenting on Plains* (1887) 537 Teams of luggage, dogs, horsemen, mulemen, cross and recross at will. 1952 E. F. Davies *Illyrian Venture* iii. 60 Hare gave an order to the mulemen to build the fires. 1970 R. Lowell *Notebk.* 209 The mule-man lost his footing in the clouds. 1846 R. B. Sage *Scenes Rocky Mts.* xxix. 251 We ended our fast.. with a feast of mule meat. 1891 *Century Mag.* Mar. 774 We made our Christmas and New Year's dinner on mule meat. 1834 A. Pike *Prose Sk. & Poems* 25 They would find a mule-path leading from the ford. 1880 'Mark Twain' *Tramp Abroad* 131 Big keel-boats on their way up, using sails, mule power, and profanity. 1914 *Illustr. London News* 12 Dec. 814/3 The mountain gunners can take their gun, by mule-power or man-power, where the Horse and Field cannot follow. 1937 N. N. Puckett in A. Dundes *Mother Wit* (1973) 172/1 On plantations with more abundant mule-power, we find appearing..such.. descriptions as Young Beck,..Leader Kit, and even such regular surnames in muledom as Jane Henkel..and Pol Jones. 1883 'Mark Twain' *Life on Mississippi* xlv. 462 The ladies of New Orleans attend so humble an orgy as a mule-race. 1777 P. Thicknesse *Year's Journey* I. xxiv. 210 The foot-road..is only one thousand three hundred paces;..the mule-road is above four times as far. 1880 'Mark Twain' *Tramp Abroad* 445 We followed the mule-road, a zigzag course. 1970 C. Kopas *Bella Coola* iii. 55 Applied for a contract to build a mule road from Bella Coola to the mouth of the Quesnel. 1897 *Sears, Roebuck Catal.* 228/2 Men's genuine unlined Mule Skin Gloves. Not very pretty, perhaps, but full of real goodness. 1926 *Daily Colonist* (Victoria, B.C.) 7 Jan. 11/1 Brown Muleskin Gauntlets, warmly lined, cuff has fringe and red star. 1854 J. R. Bartlett *Pers. Narr. Explor. Texas* I. v. 113 We might reach El Paso by..taking an occasional mule steak. 1846 S. Magoffin *Diary* in *Down Santa Fé Trail* (1926) 25 His mule team (some eighteen or twenty) were.. passing the little wet creek. 1949 L. G. Green *In the Land of Afternoon* ix. 131 Every twelve miles along the route a fresh mule team awaited the coach. 1908 A. Bennett *Old Wives' Tale* i. i. 2 The ineffaceable mule-tracks that had served centuries before..Watling Street. 1975 P. Somerville-Large *Couch of Earth* x. 181 A traditional mule track that crossed the new road. 1859 *Brit. Colonist* (Victoria, B.C.) 29 Jan. 1/4 A good wagon road to Fort Yale, and a mule trail thence equal to the best in California. 1932 *World Today* Feb. 217/2 Twenty miles of mule-trail were built. 1849 H. Page *Let.* 8 May in E. Page *Wagons West* (1930) 116 The Indians..will not trouble us, so much as they will the mule trains. 1952 E. F. Davies *Illyrian Venture* vii. 129 Chesshire and Kadri went back to find Hare and the mule train at dawn. 1846 P. St. G. Cooke *Jrnl.* 19 Oct. in P. St. G. Cooke et al. *Exploring Southwestern Trails* (1938) 69 There are three mule wagons to each company. 1863 *Harper's Mag.* June 9/2 With utmost speed, in mule-wagons, they started for the Lower Agency. 1850 J. L. Tyson *Diary of Physician in Calif.* 24 [Down] the only pass..was a narrow mule-way. 1930 T. S. Eliot tr. *St.-J. Perse's Anabasis* 65 The parties for upkeep of muleways.
c. mule-beater, a stick used for beating mules; **mule chest** (see quot. 1911); **mule-colt** *U.S.*, a young mule; **mule deer,** substitute for def.: *Odocoileus hemionus*, a black-tailed deer native to western North America, and distinguished by its large ears; (further examples); **mule-ear(ed) rabbit** = Jack-rabbit (in Dict. and Suppl.); **mule-gate,** the space in a spinning-room within which a mule works; **mule-headed** *a.*, stubborn; **mule-killer** *U.S.*, (*a*) (see quot. 1847); (*b*) (see quots. 1890 and 1899); **mule-litter,** a litter borne by mules; **mule rabbit** = *mule-ear(ed) rabbit*; **mule-skinner** (earlier and later examples); also *Canad.*; hence **mule-skinning** *vbl. sb.*; **mule-sweep** = *mule-gate*; **mule-whacker** *U.S.*, a mule-driver.
1909 E. Banks *Mystery F. Farrington* 123 Pedro took up one of the disused mule-beaters, and laid it on him thick and fast. 1911 Blake & Reveirs-Hopkins *Little Bks. Old Furnit.: Tudor to Stuart* vi. 96 In some cases during the transition from the simple chest to the chest of drawers, we find a chest with drawer or drawers below and false drawer fronts to match on the upper portion. The old chest lid is still retained. This type, being a hybrid, is known amongst collectors as the 'Mule Chest'. 1923 J. C. Rogers *Eng. Furnit.* ii. 15 The 'mule chest', or dower chest, with its proverbial bottom drawer. 1972 *Country Life* 21 Dec. 1752/1 A well proportioned 18th century mule chest of well coloured mahogany. 1788 G. Washington *Diaries* (1925) III. 400 Turned..the two yearling Mule Colts..into the Clover Paddock. 1885 *Rep. Indian Affairs* (U.S.) 11 The increase has been 8 horse and mule colts, 50 calves, and 150 pigs. 1805 M. Lewis *Jrnl.* 10 May in Lewis & Clark *Orig. Jrnls. Lewis & Clark Expedition* (1904) II. 21 With the mule deer the horns consist of two beams. 1806 W. Clark *Jrnl.* 11 Mar. in *Ibid.* (1905) IV. 158 The Ears and the tail of this Animale [*sic*] when compared with those of the Common Deer, so well comported with those of the Mule when compared with the Horse, that we have by way of distinction adapted [*sic*] the appellation of the Mule Deer. 1880 *Scribner's Monthly* May 129/1 For meat we have bacon and generally steaks or roasted ribs of elk, mule-deer or mountain sheep. 1936 D. McCowan *Animals Canad. Rockies* xxxi. 265 A full grown Mule deer measures about sixty five inches from nose to tail, stands from forty to forty two inches high at the shoulder and.. weighs from two hundred to two hundred and fifty pounds. 1962 E. Lucia *Klondike Kate* viii. 170 The big handsome mule deer wandered right into the yard. 1972 *Village Voice* (N.Y.) 1 June 75/2 At dusk mule deer venture out to graze under the apple trees. 1855 *Life Illustrated* 10 Nov. 16/3 She will follow the mountain or 'mule-eared' rabbits. 1885 C. A. Siringo *Texas Cowboy* 142, I had just eaten a mule-eared rabbit. 1889 H. H. McConnell *Five Yrs. a Cavalryman* 56 The English hare..is not nearly so large as our jack or 'mule-ear' rabbit. 1892 J. Nasmith *Student's Cotton Spinning* 409 The pillars..are so pitched that they fall into the alleys between the mules and not into the mule-gate. 1884 'Mark Twain' *Huck. Finn* xxix. 278 That mule-headed old fool wouldn't give in then! 1847 *Knickerbocker* XXIX. 161 One might have seen a small French cart, of the sort very appropriately called a 'mule-killer'. 1852 J. Evans *Let.* 30 Apr. in G. N. Jones *Florida Plantation Rec.* (1927) 67, I would Call the New Waggon a nother Mule killer. 1890 *Cent. Dict.*, *Mule-killer*, the whip-tailed scorpion, *Thelyphonus giganteus*. 1899 *Mem. Amer. Folk-Lore Soc.* VII. 63 Mule-killer, devil's war-horse, praying mantis. *Kansas.* 1887 J. D. Billings *Hardtack & Coffee* xvi. 315 Another invention for the transportation of the wounded from the field was the *Cacolet* or *Mule Litter*. 1904 R. J. Farrer *Garden of Asia* 81 Nor does a mule-litter hurry upon the road. 1857 *Spirit of Times* 28 Feb. 414/3 Some of our expedition.. went farther out, for the purpose of fetching in some of the deer, bar, and mule rabbits aforesaid. 1877 S. W. Cozzens *Crossing Quicksands* 80 More commonly known as the 'mule' rabbit, so called on account of its enormous ears. 1870 J. H. Beadle *Life in Utah* 224, I took to the plains..in the capacity of a 'mule-skinner'. 1889 J. McLean *Indians* v. 198 The men..are called bull-whackers and mule-skinners, applied to freighters who drive oxen or mules. 1962 A. Fry *Ranch on Cariboo* 160 I'd the repertoire of a mule-skinner, developed behind a wide variety of knotheaded horses. 1881 E. W. Nye *Bill Nye & Boomerang* 34 A practical knowledge of..mule skinning, vocal music, horsemanship. 1945 B. MacDonald *Egg & I* 50 If only I had studied carpentry or mule skinning instead of ballet. 1971 J. Gray *Red Lights* vii. 158 Nicholas Sheran gave up mule-skinning for coal mining. 1869 *Overland Monthly* III. 9 Here..is a cotton gin..and ponderous wooden wheels, and the mule-sweep underneath. 1873 J. H. Beadle *Undevel. West* iv. 88 The streets were thronged with motley crowds of railroad men and mule-whackers. 1889 H. O'Reilly *50 Yrs. on Trail* 357 The town was full of cow-punchers, mule-whackers, [etc.].

mule[2]. **2.** Delete *Obs.* exc. *Hist.* and add later examples.
1922 S. Lewis *Babbitt* xx. 254 She wore..torn stockings thrust into streaky pink satin mules. 1944 H. Croome *You've gone Astray* v. 43 Kitty was gone in a light clatter of pink mules. 1973 'I. Drummond' *Jaws of Watchdog* xii. 160 The girl..padded softly across the room in her fluffy mules.

muled (miūld), *a.*[2] [f. Mule[1] + -ed[2].] Of a coin: that is a mule (sense 4 c).
1914 *Brit. Mus. Return* 115 in *Parl. Papers* LXXI. 193 A muled groat of Henry VIII combining an obverse of the third coinage with a reverse of the second. 1961 Dolley & Metcalf in R. H. M. Dolley *Anglo-Saxon Coins* 160 *BMC* types III and IV are found muled, though the coins are of the greatest rarity.

Muler, var. *Maler *sb.* and *a.*

Mules (miūlz). The name of J. H. W. *Mules* (d. 1946), Australian sheep-farmer, used *attrib.* or in the possessive to designate an operation developed by him to reduce the incidence of

blowfly strike in sheep by removing folds of skin from the crutch, the area most likely to be affected. So **mulesed** (miū·lzd) *ppl. a.*, of a sheep, treated in this way; **mu·lesing** *vbl. sb.*, the use of this operation.
1933 *Sci. Bull. Council Sci. & Industr. Res.* (Austral.) No. 40. 9 By the surgical removal of the side folds (e.g., Mules' operation), the breech is opened up, urine soiling reduced, and susceptibility to strike decreased. *Ibid.* 103 By careful surgical procedure on the Mules principle, liability to strike should be reduced considerably, because the folds selected constitute the commonest site of strike. 1946 *Queensland Country Life* 18 Apr. 3 Mulesed sheep are much easier to crutch. *Ibid.*, We don't say that mulesing will stop the fly altogether. 1957 *New Biol.* XXII. 101 This experiment showed clearly that the Mules operation gave good protection against crutch strike in wrinkly breeched sheep. 1965 *Austral. Encycl.* II. 37/1 To effect a permanent improvement in the conformation of the crutch, J. H. W. Mules developed a technique for the surgical removal of the folds of skin occurring in the breech of many ewes. The Mules operation, as it is now called, has been modified and developed to a routine procedure. 1970 *Black's Vet. Dict.* (ed. 9) 558/1 Mules's operation.—This involves the removal of a fold of skin from the crutch of Merino sheep and is carried out by Australian sheepmen for the control of blowfly strike. Mulesing is a synonym.

‖ **muleta** (mūlē·tă). *Bullfighting.* [Sp.] The red cloth fixed to a stick which is used by a matador during the faena.
1838 *Q. Rev.* LXII. 417 The matador must not let the bull run on the muleta above two or three times. 1846 R. Ford *Gatherings from Spain* xxii. 308 The matador.. stands before his victim alone... In his left [hand] he waves the *muleta*, the red flag. 1873 J. W. Revere *Keel & Saddle* 111 Taking his stand firmly, he slowly unfolded the scarlet muleta. 1931 R. Campbell *Georgiad* i. 10 But leave the banderillos in their place Nor shake the red muleta in its face. 1932 E. Hemingway *Death in Afternoon* 304 The muleta is used to defend the man; to tire the bull and regulate the position of his head and feet. 1971 *Guardian* 20 July 8/6 Rita Hayworth..flaunting her red hair like a muleta.

muley (miū·li), *a.*³ Also **muly.** [f. Mule¹ + -y¹.] Mulish, stubborn; sulky.
1871 *Yorks. Mag.* I. 28/2 Enah means by and by; *muly*, sulky [etc.]. 1922 J. A. Dunn *Man Trap* v. 68, I got another drink into him, and made a fatal error in doing it, for he turned muley. 1956 *Sat. Even. Post* 21 Apr. 82 With the profits come a steady succession of worries that would discourage any but the most muley.

mulga. 1. Substitute for def.: A small spreading tree, *Acacia aneura*, widespread in dry inland areas of Australia, or one of several related trees found in such regions; also, land covered with vegetation of this kind. Also *attrib.* (Further examples.) Also used *colloq.* in general sense 'uninhabited or inhospitable region'.
1896 H. Lawson *While Billy Boils* 40 Two or three white-washed galvanized-iron roofs start out of the mulga. 1903 'T. Collins' *Such is Life* 79 When he says 'mulga', he means any tree except pine or currajong. 1909 'S. Rudd' *From Selection to City* (1910) xiv. 119 We had listened to many lies about the glory of running wild horses and cattle in the mulga. 1911 C. E. W. Bean *'Dreadnought' of Darling* i. 3 Belts of low mulga scrub. *Ibid.* x. 88 The graceful grey mulga trees..came right down to the river. 1928 V. Palmer *Man Hamilton* 26 Manager of this isolated place in the mulga, a hundred miles away from anywhere. 1936 F. Clune *Roaming round Darling* xxii. 223 A mulgawood causeway was built in front for a distance of thirty yards. 1946 F. D. Davison *Dusty* (1947) 1 They live at the bidding of herdsman and drover.., punching stubborn bullocks through the mulga. 1960 *Times* 5 July 11/6 The country was covered with light mulga scrub. 1964 *Punch* 18 Nov. 750/2 To pack a gun around the mulga. 1967 *Sunday Mail Mag.* (Brisbane) 8 Jan. 6/3, I had fancied myself original in conceiving the idea of taking a hot bath at 9 o'clock at night out in the 'mulga' in the middle of winter, and had borrowed a pair of swim trunks for the purpose. 1968 K. Weatherly *Roo Shooter* 5 On the stony hills, the mulga trees gave no shade. *Ibid.* 114 In this mental state he couldn't stay out in the mulga. 1969 *Northern Territory News* (Darwin) *Focus '69* 97/1 Contractors take the fuel off the Track to places out in the sticks and the mulga.
b. *Comb.*, as *mulga-covered*, *-dotted* adjs.; **mulga madness** *slang* (see quot. 1943); **mulga parrot**, a brightly coloured parrot, *Psephotus varius*, whose plumage includes patches of red, yellow, blue, and green; **mulga scrubber** *slang* (see quot. 1945); **mulga wire** *slang* = *grape-vine 2 a; a rumour, message, or report; a lie or false report.
1936 I. L. Idriess *Cattle King* iv. 27 They rode for thirty miles among the low, mulga-covered ridges that are the foot-hills of the Barrier Range. 1971 *World Archaeol.* III. 170 The mulga-covered flats. 1936 F. Clune *Roaming round Darling* xxii. 226 Steep mulga-dotted headland along Bluff Waterhole. 1943 Baker *Dict. Austral. Slang* (ed. 3) 52 *Mulga madness*, the 'queerness' sometimes developed in lone bushmen or fossickers. 1931 N. W. Cayley *What Bird is That?* 144 Mulga Parrot... Also called Many-coloured Parrot and Varied Parrot. 1941 I. L. Idriess *Great Boomerang* ii. 12 Mulga parrots and blue-bonnets lent a flash of living colour to the light-grey saltbush. 1968 K. Weatherly *Roo Shooter* 115 With the sun would come the parrakeets, mulga parrots, their green wings flashing in the sun. 1945 Baker *Austral. Lang.* iii. 67 The bullock driver and stockman have invented many terms by which to describe their charges. Here are some of the best:..*mulga scrubbers*.., stock that have run wild and deteriorated in condition. 1899 Mulga-wire [see sense 3 below]. 1913 H. Lawson *Triangles of Life* iii, in *Stories* (1964) 3rd Ser. 182 Tom had been out early, or had got what we call a bush telegraphy or mulga wire. 1937 E. Hill *Great Austral. Loneliness* xxiii. 207 Warned by the mulga wires of the approach of a white woman. *Ibid.* 329 Mulga wires. Bush telegraph, gossip of the outback. 1940 A. W. Upfield *Bushranger of Skies* ix. 101 This morning I found Itcheroo squatted before a little fire and sending or receiving a mulga wire. 1964 *Evening Post* (Wellington, N.Z.) 3 Mar. 17 The grapevine or 'mulga wire' in opera circles was instrumental in the return to Wellington of [a]..mezzo-soprano.
3. *slang.* Short for *mulga wire.
1899 *Bulletin* (Sydney) 4 Mar. 15/2 A lie or false report is, in N.S.W., a 'Mulga' or 'Mulga-wire'. 1908 E. S. Sorenson *Quinton's Rouseabout* 186 They'll be that delighted to find it was only a mulga that they'd toast you as 'a jolly good fellow'. 1941 Baker *Dict. Austral. Slang* 47 *Mulga*, a rumour. 1950 K. S. Prichard *Winged Seeds* 297 The troops 've had it all by mulga. They've heard too.

muliebrity. (Further examples.)
1888 B. Harte *Phyllis of Sierras* II. i. 169 This tall.. woman..possessed a refined muliebrity superior to mere liberality of contour. 1911 H. G. Wells *New Machiavelli* II. ii. 206 She was one of those women who are wanting in —what is the word?—muliebrity.

mull, *sb.*¹ Add: **2.** A suspension of a finely ground solid in a liquid, esp. one used in recording the infra-red spectrum of the solid.
1956 *Mineral. Mag.* XXXI. 193 Spectra were obtained on the Perkin–Elmer model 21 double-beam instrument, using both the 'Nujol' mull technique and pressed KBr pellets. 1964 H. A. Szymanski *IR: Theory & Pract. Infrared Spectroscopy* iii. 78 Hydrocarbon oils as well as halogenated hydrocarbons have been used as the liquid in mulls. 1971 Skoog & West *Princ. Instrumental Analysis* vi. 152 The resulting mull is then examined as a thin film between flat salt plates.

mull, *sb.*⁷ Add: Esp. the fine muslin sometimes stuck to the spine of a book before its cover is put on.
1880 J. W. Zaehnsdorf *Art of Bookbinding* xix. 85 In 'throw up' backs, or in 'flexible not to show', a piece of thin linen or stuff called mull (muslin) is glued on the back first, and one piece of paper on the top. 1951 S. Jennett *Making of Books* xi. 173 A length of mull (the open-weave material that can be seen through the endpapers of most books). 1967 V. Strauss *Printing Industry* x. 673/2 The next preparatory step is lining up. It consists in attaching one or more strips of fabric, known as crash—called mull in Canada—and super, as well as a strip of strong paper to the back of the book.
attrib.
1873 *Young Englishwoman* Dec. 559/1 This edging may be worked on cambric or mull muslin. 1910 *Encycl. Brit.* VII. 277/2 (Cotton) The finer kinds, made from Egyptian yarns, are called mull-dhooties. 1960 Cunnington & Beard *Dict. Eng. Costume* 265/1 *Mull muslin*, 19th c., a soft thin muslin, not silky, finer than nainsook.

mull (mɒl), *sb.*¹⁰ *Soil Sci.* [ad. Da. *muld* Mould *sb.*¹ (adopted in this specific sense by P. E. Müller 1879, in *Tidsskrift for Skovbrug* III. 7): etymologically = Mull *sb.*¹] Humus which does not form a distinct layer on top of the soil but is admixed with the underlying mineral soil, which is characteristic of grasslands and hardwood forests and is generally weakly acid to weakly alkaline in reaction. Cf. *mor.
1928 *Ecology* IX. 9 The soil on this..was identified by Dr. Hesselman as a true mull profile. *Ibid.* 10 The mull is deepest under ash and sugar maple. 1931 *Ibid.* XII. 570 Müller distinguished two main types of humus layer called mull..and mor. 1935 *Forestry* IX. 43 Mull or neutral humus..though usually acid in reaction contains sufficient calcium to allow of a crumb or grain structure with a generally 'loose' or 'porous' constitution. 1952 P. W. Richards *Trop. Rain Forest* ix. 218 The humus of rain-forest soils..would appear to resemble the 'mull' rather than the 'mor' of temperate forest soils. 1974 *Encycl. Brit. Macropædia* VII. 537/1 In the temperate regions two major types of organic matter predominate, the mull and the mor.

mull (mɒl), *sb.*¹¹ [f. Mull *v.*³] Mulled wine.
1925 J. Thomas *Bon Vivant's Compan.* 105 (*heading*) Mulls. 1953 D. A. Embury *Fine Art of Mixing Drinks* xiii. 296 A mull, or mulled wine, is simply a spiced and sweetened wine served piping hot. 1959 *Listener* 24 Dec. 1135/3 The mull must be kept hot. 1962 J. Conil *Epicurean Book* xiii. 209 Do not boil the wine, nor allow the mull to reach boiling point. 1972 *House & Garden* Dec. 103/1 The best mulls have as their base an inexpensive, full-bodied red wine.

mull, *v.*¹ Add: **1. b.** *trans.* To convert (solid material) into a mull (*mull *sb.*¹ 2). So **mulled** *ppl. a.*²
1941 *Austral. Jrnl. Dentistry* XLV. 163/2 The writer feels justified in urging all dentists to give up the practice of mulling their amalgams in the bare palm of the hand. *Ibid.* 161/2 It should..be stated that the curves for the mulled specimens are average ones. 1943 *Industr. & Engin. Chem. (Analytical Ed.)* XV. 663/2 Samples of insoluble, infusible materials may be prepared for examination by grinding the substance to as fine a powder as possible, then mulling it thoroughly in a straight-chain hydrocarbon, such as Nujol. 1948 *Rev. Sci. Instruments* XIX. 165/1 The present technique is to mull the sample in a mineral oil such as Nujol. 1956 J. N. Anderson *Appl. Dental Materials* xxv. 359 The dentist..may show symptoms of slight mercurial poisoning if he employs the technique of 'mulling' or 'palming' amalgam in the hand over a period of years. 1964 H. A. Szymanski *IR: Theory & Pract. Infrared Spectroscopy* iii. 78 The name given to this technique [sc. dispersing a solid in a liquid] is *mulling* the sample.

mull, *v.*⁵ Add: **2.** Delete '*colloq. U.S.*' and add earlier example.
1857 J. G. Holland *Bay-Path* xvii. 200 'What do you do with them [troubles]?' 'Let 'em mull.'
b. *mull over* (an idea, etc.), to turn over in one's mind, cogitate upon.
1880, etc. [in Dict.]. 1910 J. London *Let.* 9 Feb. (1966) 299 If I can get from you a suggestion of a motif.. which, after mulling over, I decide I can do, I could..join you. 1949 B. Woolfe in A. Dundes *Mother Wit* (1973) 529 The Rabbit mulls the problem over. 1958 *Times* 20 Aug. 10/7 Bill and I discovered a mutual hobby in fishing. Rods had to be produced for inspection..and experiences mulled over. 1966 *Listener* 3 Nov. 650/2 Of course one mulls over that.
c. *trans.* To consider, ponder upon. *U.S.*
1923 *Dialect Notes* V. 215 *Mull*, v., to ponder over, to cogitate upon. 1949 *Sun* (Baltimore) 27 Dec. 5/1 At last report, the county was mulling a price. 1958 *Wall St. Jrnl.* 29 Oct. 1/5 The idea of a U.S. pledge to facilitate state and local borrowing is mulled by the joint Federal-state committee on swapping services and revenue sources between the U.S. Government and the states. 1972 *Newsweek* 10 Jan. 39/3 Mr. Nixon is mulling two possible tactics. 1972 *Science* 22 Sept. 1081/1 The Germans.. were mulling a public recommendation from their safety advisory committee.

mull (mɒl), *v.*⁸ [Perh. related to other vbl. uses of Mull.] **a.** *trans.* To moisten (leather) during manufacture so as to make it more supple. **b.** *intr.* Of leather: to become soft by moistening. So **mulled** *ppl. a.*³, **mu·lling** *vbl. sb.*³
1931 F. Plucknett *Boot & Shoe Manuf.* xxvi. 166/2 One of the more recent innovations in the lasting-room is the 'mulling chamber', the idea being that if the upper materials absorb a suitable amount of dampness they will stretch easier. 1953 W. Moore in J. H. Thornton *Textbk. Footwear Manuf.* xx. 277 A criticism of mulling is that the fibre of the leather is adversely affected... The mulled upper after lasting is rather damp. 1962 *New Scientist* 12 Apr. 33/1 Almost any known leather, it is believed, can be mulled in fewer than four minutes. 1963 *Times* 7 Mar. 20/1 In previous shoemaking methods the uppers of the shoes had to remain on 'lasts' for five or six days in order to get their shape. The new process does the job in half an hour by means of pumping moisture into the leather in a special mulling machine. 1969 T. C. Thorstensen *Pract. Leather Technol.* xii. 197 After the oils and greases have been distributed uniformly over the surface, the leather is removed from the drum and allowed to mull, once the initial heat has been removed by air cooling.

mullarkey, var. *malarkey.

Muller (mu·ləɹ), *sb.*⁴ The name of J. P. *Muller* (b. 1866), Danish physical educationalist, used *attrib.* and in the possessive esp. to designate a set of bodily exercises published and promoted by him.
c 1912 J. P. Muller *My Syst. for Ladies* 86 In a pamphlet published by the *Ligue des Mères de Famille* (France), the following letter from Mme. A. de Four appeared:—Two years practice of Muller's exercises has given me muscles sufficiently resistive to enable me to leave off wearing the corset. 1921 A. Huxley *Crome Yellow* xxiv. 260 He would have to try and do his Muller exercises more regularly. 1952 M. Gilbert *Death in Captivity* xii. 182 We must therefore maintain bodily suppleness and agility by means of Muller's exercises before breakfast and a regular physical training class at least once a day.

Müller (mü·ləɹ), *sb.*¹ *Min.* Also **Muller.** [f. the surname (supposedly of its discoverer).] *Müller's glass* (also † *Müller glass*, † *glass of Müller*): = Hyalite.
1794 Müller's Glass [see Hyalite]. 1836 T. Thomson *Outl. Min., Geol.* I. 73 Hyalite or Muller glass. 1852 Glass of Muller [see Hyalite]. 1940 G. F. H. Smith *Gemstones* (ed. 9) xxxii. 302 Hyalite, sometimes called Müller's glass after its discoverer, is a remarkable opal which is as clear as glass. 1962 R. Webster *Gems* I. x. 186 Hyalite is a colourless transparent variety of opal that closely resembles melted glass; it is sometimes called Müller's glass.

Müller (mü·ləɹ), *sb.*² *Anat.* The name of Heinrich *Müller* (1820–64), German anatomist, used *attrib.*, in the possessive, and with *of*-adjunct to designate various anatomical structures of the eye which he described, as: **a.** Neuroglial cells, having the appearance of fibres, which form the supportive ground tissue of the retina.
1856 G. V. Ellis *Demonstrations Anat.* (ed. 4) x. 777 Passing vertically through the retina are other fine threads —fibres of Müller. 1886 C. F. Pollock *Normal & Path. Histol. Human Eye & Eyelids* 127 The fibres of Müller are the most prominent parts of a supporting matrix, which pervades the retina. 1932 W. Penfield *Cytol. & Cellular Path. Nervous Syst.* II. ix. 436 (*heading*) Astrocytes of the retina (Müller's cells). *Ibid.* xvi. 767 In other strata the Müller cells are followed with difficulty. 1961 Duke-Elder & Wybar *Syst. Ophthalm.* II. 256 (*heading*) The fibers of Müller. *Ibid.*, On the vitreal aspect of the retina Müller's fibres terminate in large conical or bulbous expansions. 1971 M. J. Hogan et al. *Histol. Human Eye* ix. 462 The mature Müller cell is complex in shape.
b. Each of three smooth muscles, of which one is part of the ciliary muscle, one (the orbitalis) is in the orbit, and one (the superior tarsal muscle) is in the upper eyelid.
1875 *Encycl. Brit.* I. 887/2 The inner part of the [ciliary] muscle forms a ring-like arrangement of fasciculi close to the circumference of the iris, and is often called the annular muscle of Müller. 1890 Billings *Med. Dict.* II. 174/1 *Müller's annular muscle*, ciliary muscle. *Ibid.* 174/2 *Müller's muscle*, smooth muscular fibres found in the membrana orbitalis. *Ibid.*, *Müller's palpebral muscle*, a collection of smooth muscle-fibres in the upper eyelid. 1912 A. Thomson *Anat. Human Eye* i. 13 Corresponding to the bases of the free projecting extremities of the ciliary processes..is a group of fibres..arranged circumferentially; these constitute the circular fibres of Müller. 1921 S. E. Whitnall *Anat. Human Orbit* 82 This muscle, described by H. Müller in 1858, is known as the 'orbital muscle of Müller', or *musculus orbitalis*. *Ibid.* 296 The palpebral involuntary muscles (of Müller). 1950 *Irish Jrnl. Med. Sci.* 39 (*heading*) Some aspects of Muller's [*sic*] orbital muscle. 1964 [see *enophthalmus, -mos]. 1969 Beard & Quickert *Anat. Orbit* 4 Mueller's [*sic*] sympathetic (superior tarsal) muscle can be seen extending from the junction of the muscular levator and the levator aponeurosis to the superior border of the tarsus.

müller (mü·ləɹ), *v.* Also **muller** (mu·ləɹ). [f. the name of Franz Müller, a murderer, who was convicted in 1864 on circumstantial evidence in which a hat was of considerable significance.] *trans.* To alter (a hat) in the manner alleged to have been done by Franz Müller. Also as *sb.*, a type of flat-topped felt hat similar to that associated with Müller.
1864 in Farmer & Henley *Slang* (1896) IV. 384/1 In a small shop not far from Sloane-square, Chelsea, may be seen the following tasteful announcement: Hats muller'd here! 1909 *Daily Chron.* 22 Nov. 4/7 Müller's hat..formed the connecting link in a remarkable chain of circumstantial evidence. Henceforth 'mullers', as they were called, were tabooed. 1934 *Trans. Philol. Soc.* 1933 101 A *Müller hat* is a flat-topped, hard felt.

Müllerian (mülɪə·riăn), *a.*² [f. the name of J. F. T. (Fritz) *Müller* (1821–97), German zoologist, who, in 1878, explained this type of mimicry + -ian.] In *Müllerian mimicry, resemblance*, a form of mimicry (see sense 2 in Dict. & Suppl.) in which insects of different species develop similar patterns of coloration, etc., as a protective device. Also *Müllerian mimic*, an insect exhibiting this type of mimicry.
1899 [see *Batesian *a.*]. 1934 *Discovery* July 195/1 Two or more, often many, inedible forms assume a similar aspect and thus enjoy a common advantage by virtue of uniting to share the toll levied on them by experimental tasting. This is the kind of assimilation which was explained by Fritz Müller and often goes by the name of Müllerian mimicry. 1951 *New Biol.* X. 73 'Müllerian' resemblance is when two or more distasteful organisms resemble one another, thereby deriving collective advantage in that a predator has only to 'learn' one pattern which will suffice to protect all the species which have it. *Ibid.*, Müllerian and Batesian mimics may be involved together in nature in a complex mimetic association, all the members of which wear the same livery. 1968 R. D. Martin tr. *Wickler's Mimicry in Plants & Animals* vii. 78 Most examples of Müllerian mimicry..involve different species with the same complex patterns. *Ibid.* 79 Müllerian mimicry increases in efficiency the greater the number of species in the system and the better the correspondence in pattern. 1975 *Trans. Roy. Entom. Soc.* CXXVI. 632 *Heliconius* species are frequently highly distasteful, and act as models in many Müllerian systems.

Müller-Lyer (mülər lai·ər). Also erron. **Muller-.** The name of Franz Carl *Müller-Lyer* (1857–1916), German sociologist and philosopher, used *attrib.* (and *absol.*) to designate an optical illusion he described (*Arch. f. Anat. u. Physiol.* (*Physiol. Abth.*) (1889) Suppl. 263–70), by which a line with a V-shaped arrowhead at each end appears shorter than an adjacent line of equal length but with the V-shaped portions reversed and pointing inwards.
1899 *Psychol. Rev.* VI. 241 Brentano's unsuccessful attempt to explain the Müller-Lyer illusion by the general fact that acute angles are overestimated and obtuse angles are underestimated. 1938 R. S. Woodworth *Exper. Psychol.* xxv. 646 One of the two Müller-Lyer figures is kept of constant length and used as a Standard, while the other figure is the Variable. 1958 *New Biol.* XXVII. 27 He chose for his experiment the well-known Muller-Lyer illusion. 1969 F. C. Shontz *Perceptual & Cognitive Aspects Body Experience* iii. 41 In certain complex stimuli, like the Müller-Lyer illusion, the organization of a perceptual field may lengthen or shorten the apparent length of a line. 1972 *Jrnl. Social Psychol.* LXXXVII. 144 Samples from Western cultures..are more inclined than Eastern ones to see the Muller-Lyer.

mullet¹. Add: **1, 2.** The name, often with an adjective prefixed to it, is used for any fish of the families Mullidæ, or red mullets, and Mugilidæ, or grey mullets. (Further examples.)
1820, etc. [see *kanae]. 1925 J. T. Jenkins *Fishes Brit. Isles* 123 There are a large number of species of mullets in the genus *Mugil*. 1951 T. C. Roughley *Fish Austral.* 31 The mullets contained in the family Mugilidae range through the temperate regions of the world. 1962 K. F. Lagler et al. *Ichthyol.* ix. 277 Neurosecretory cells in the preoptic nucleus of the hypothalamus lose their secretions rapidly in several marine fishes, including.. mullets (Mugilidae) when the fishes are temporarily placed in hypertonic sea water. 1966 *Encycl. N.Z.* II. 600/2 A familiar sight is the gleaming flash as a mullet leaps out of the water and returns to its element. 1969 A. Wheeler *Fishes Brit. Is. & N.-W. Europe* 465/1 Very young mullet are also common in intertidal pools on the Channel coast.

mullet-head. [? f. Mullet¹.] **1.** *U.S.* A fresh-water fish with a large flat head.
1866 *Harper's Mag.* Sept. 537/1 Dat fish is a mullet-head; it hain't got any brains. 1873 J. H. Beadle *Undevel. West* v. 102 There is a fish called the mullethead, that cannot be intoxicated by any amount of liquor. 1893 W. Forbes-Mitchell *Reminisc. Great Mutiny* vi. 110 That fish, my son, is called a mullet-head: it has got no brains.
2. [Cf. *mull-head* a dull, stupid fellow (*E.D.D.*).] A stupid person. So **mullet-headed** *a.*
1857 *Quindaro* (Kansas) *Chindowan* 6 June 1/3 The men, for the most part sleepy, ignorant, mullet-headed looking wretches. 1884 'Mark Twain' *Huck. Finn* xxxix. 370 They're so confiding and mullet-headed they don't take notice of nothing at all. 1916 *Dialect Notes* IV. 278 Look at that mullet-head of a Sam Smith. He don't know beans. 1935 Z. N. Hurston *Mules & Men* (1970) I. i. 30 Hey, you mullet heads! Get out de way. 1942 Berrey & Van den Bark *Amer. Thes. Slang* § 150/5 Stupid,..mullet-headed. *Ibid.* 433/3 Stupid person,..mullet-head.

mulligan (mɒ·ligăn). *N. Amer.* [Apparently f. a proper name.] **1.** A stew made from odds and ends of food. Also *attrib.*
1904 *Yukon Midnight Sun* (Dawson, Yukon Territory) 10 Jan. 3/4 All the roadhouses served big Christmas dinners and most of them made a mulligan. 1913 *Collier's* 11 Jan. 16/2, I suppose you never have eaten any 'Mulligan'... The recipe calls for canned Willie, spuds, onions, canned tomatoes, all mixed up together. 1914 *Sat. Even. Post* 4 Apr. 10/1 It was a mulligan. Everything was in that stew—meat, potatoes, onions, bread—an appetizing hodgepodge. 1918 C. Sandburg *Cornhuskers* 80 Then they go to the bunk cars and eat mulligan and prune sauce. 1926 J. Black *You can't Win* vii. 83 There was a grand jungle by a small clean river where they..cooked their mulligans. *Ibid.* xv. 198 He's crazy as a bed bug and the best 'mulligan' maker on the road. 1955 R. P. Hobson *Nothing too Good for Cowboy* 138 That's what's good for me, a great beeg vegetable mulligan with feesh in it. 1966 H. Marriott *Cariboo Cowboy* ii. 35 Lots of mulligan stew with beef, carrots and onions. 1971 'R. Macdonald' *Underground Man* xix. 135 A number of plywood tables had been set up on trestles. A hundred or more men were seated at them, eating mulligan stew.
2. (With capital initial.) *Golf.* An extra stroke awarded after a poor shot, not counted on the score card. *colloq.*
1949 P. Cummings *Dict. Sports* 275 *Mulligan*, *Golf*, a handicap of a free shot given after a player makes a bad one. 1952 *Golf Digest* May 7 'It's just a friendly match,' he said. 'Wanna take a mulligan?' 1965 H. Graffis *Esquire's World of Golf* xii. 174 Magnanimously, the hustler will allow his prey a Mulligan off the tee now and then. 1971 *Daily Colonist* (Victoria, B.C.) 7 Feb. 1/2 With millions watching on color television, Shepard took another poke—the extra chance golf duffers call a 'Mulligan'—and connected solidly.

mulling, *vbl. sb.*² [f. Mull *v.*¹ + -ing¹.] The process of rubbing or grinding; the conversion of a solid into a mull (*mull *sb.*¹ 2).
1937 *Austral. Jrnl. Dentistry* XLI. 127 Mixing. This consisted of 1, 1½ or 2 minutes trituration in a standard mortar using about 120 revolutions of the pestle per minute, followed by 1 or 1½ minutes mulling in the hand. 1944 E. R. Riegel *Chem. Machinery* xi. 249 Mulling is the second way in which solids may be mixed with solids. The mulling action is comparable to the rubbing, kneading, or smearing action of the mortar and pestle; no grinding is involved. The machine for mulling consists of a circular stationary pan in which is mounted a special combination of mullers and plows which revolves. 1964 H. A. Szymanski *IR: Theory & Pract. Infrared Spectroscopy* iii. 79 One of the disadvantages of mulling is that it cannot be used very conveniently on rubber, plastics, or resinous materials. 1971 Rosenblatt & Davis *Laboratory Course Org. Chem.* iii. 71 Another method for preparing a solid sample is to grind 5 mg of the solid to a fine dispersion in a drop of suitable mulling agent. 1971 *Materials & Technol.* II. iii. 140 Mulling improves the final shape of the grain by knocking off its brittle corners, and this is carried out by impact crushing, pan milling, or air blasting.

mullion. Add: **2.** *Geol.* Each of a series of ribs or columns of rock (*spec.* those composed of the local rock) on a rock face, usu. formed by folding. So *mullion structure*.
1891 E. Hull et al. *Explanatory Mem. Sheets 3–5, 9–11, 15 & 16 Maps Geol. Survey Ireland* iv. 53 To the north of Carnteena, quartzite of the yellowish-white compact variety, forms the bare rocky hills of Crocknaglogh and Croaghnacreggy and in some parts 'mullion structure' was observed—a peculiar fluting due to the shearing of the rocks. 1937 *Q. Jrnl. Geol. Soc.* XCIII. 597 The individual mullions are not always due to the presence of small folds, as is apparently true of the quartz rods of Beinn Thutaig. 1953 *Proc. Geologists' Assoc.* LXIV. 118 Distinction is drawn between mullions which look like 'clustered columns' and are composed of the local country-rock, and rodding. *Ibid.* 124 A remarkable lone mullion resulting from erosion of an abrupt double fold is found at Poll Cròm. 1963 E. S. Hills *Elem. Struct. Geol.* x. 310 The subdivision of a bed, more particularly a sandstone interbedded with slate, into long parallel slabs with smooth rounded surfaces is known as mullion-structure. 1966 E. H. T. Whitten *Struct. Geol. Folded Rocks* ix. 313 Excellent mullions developed in these quartzites, and in the granite close to the northwestern contact zone.

mullite (mɒ·ləit). *Min.* [f. *Mull*, the name of an island off the west coast of Scotland + -ite¹.] A silicate of aluminium, approximately $Al_6Si_2O_{13}$, that occurs as orthorhombic crystals, is formed naturally or artificially by the action of high temperatures on other aluminium silicates, and is used for making refractory porcelains and bricks.
1924 N. L. Bowen et al. in *Jrnl. Washington Acad. Sci.* XIV. 184 The occurrence in natural rocks necessitates a mineralogical name and for this [sc. the compound $3Al_2O_3.2SiO_2$] we propose the term, mullite, after the locality whence came the rocks in which it is here first identified, the Island of Mull. 1930 *Engineering* 26 Dec. 813/3 The successful application of refractories prepared from..synthetic mullite emphasised the effect of 'mullite' content in promoting chemical resistance. 1941 *Proc. Prehist. Soc.* VII. 64 Porcellanite..probably representing a pisolitic bauxitic clay from the inter-basaltic horizon.. is composed of dark spinel and a light fine-grained aggregate of sillimanite and mullite. 1967 M. Chandler *Ceramics in Mod. World* i. 19 Some of these later combine with each other to form new crystalline phases, including needle-like particles of crystalline mullite. 1974 *Nature* 2 Aug. 403/1 Quartz samples..were heated in a vacuum in outgassed mullite or alumina tubes.
Hence **mullitiza·tion**, conversion into mullite; **mu·llitized** *ppl. a.*
1939 *Ceramic Abstr.* XVIII. 251/1 Mullite is formed rapidly during the firing of clay... The temperature of firing determines the degree of mullitization. 1948 *Jrnl. Amer. Ceramic Soc.* XXXI. 254 (*heading*) Mullitization of kaolinite. *Ibid.* 258 In more completely mullitized instances, e.g., after firing to 1300°C., groups of mullite needles in associated growth simulate hexagonal aggregations.

mullock, *sb.* Add: **1. b.** *fig.* Worthless information, nonsense. *Austral.* and *N.Z.*
1866 R. Burgess *Autobiogr.* (typescript) 127 He said, No b....y fear. I should know it was a lot of mullock they were telling for you are not like this Jew. 1911 *Triad* XVIII. 41 We have a lot of trash and maudlin mullock in these days. 1934 *Bulletin* (Sydney) 20 June 47/1 'Cooney,' I said, 'it is madness to present such mullock to an intellectual audience.'
c. *to poke mullock* (*at*): to deride, ridicule, make fun of. *Austral.*
1916 C. J. Dennis *Songs Sentimental Bloke* 126 *Mullock, to poke*—to deride, to tease. 1916 —— *Moods of Ginger Mick* 74, I own me eyes git brighter When I see 'em pokin' mullock at the everlastin' sea. 1931 V. Palmer *Separate Lives* 210 D'you think I'm going to sit in that galley with Curran and the other blokes all poking mullock at me? 1942 G. Casey *It's Harder for Girls* 153 The chaps poked mullock at me, but it wasn't that that hurt. 1945 Baker *Austral. Lang.* v. 94 *To poke mullock* (also *muck*) *at*, an extension of *poke borak at*, to make fun of a person. 1957 R. Lawler *Summer of 17th Doll* 71 Oh, so that's what you got me in for, is it—to poke mullock. 1962 J. Morrison *Twenty-Three* 86, I heard what you said when you grabbed that rope. Poking mullock at us because we won't go out over an empty hatch.

mullock, *v.* (*Austral.* and *N.Z.* examples.)
1862 *Otago Daily Times* 9 July 5 Whether the flat will pay to resluice at present, from its being so much worked and 'mullocked' and also rather deep, is doubtful. 1893 *Age* (Melbourne) 23 Sept. 14/4 No man could shear 321 sheep in eight hours, although I will admit he might do what we shearers call 'mullock over' that number. 1945 Baker *Austral. Lang.* 94 *To mullock over*, to work shoddily. 1965 J. S. Gunn *Terminol. Shearing Industry* II. 5 *Mullock over*, to rush the work quickly and carelessly, thus turning out badly shorn sheep.

mully-grub (mɒ·ligrɒb). *Austral.* [Var. of Mulligrubs *pl.*] A wichetty grub (cf. Witchetty), the larva of various insects, esp. goat-moths of the family Cossidæ and ghost-moths of the family Hepialidæ, which infest the stems and roots of plants, from which they are extracted and used as food by aborigines and as bait by fishermen.
1959 J. Wright *Generations of Men* 28 A gin had gone over to look for mully-grubs in the bush. 1966 *Courier-Mail* (Brisbane) 26 Nov. 8/8 The woman..had told him frequently to 'go and eat his mully grubs', and to 'go back to the bush where you came from'.

multi-. Add: **1.** (From the adjs. are formed advbs. (e.g. *multiserially* in Dict.) and sbs. (e.g. *multicellularity* below).) **a.** *multicelled*, *-molecular*, *-perforated*, *-stranded*; **multicellular**: hence **mu:lticellula·rity**, the state or condition of being multicellular; **multice·ntric**, pertaining to or having many centres or foci; (of a chromosome or chromatid) having many centromeres; **multicy·clic** *Geol.*, produced by or having undergone many cycles of erosion and deposition; **multidentate**, (*b*) *Chem.* (of a ligand) having more than one point at which it can be attached to a central atom; **multidimensional** (further examples); hence **mu:ltidimensiona·lity**, the property of being multidimensional; **multidime·nsionally** *adv.*, in a manner that involves or requires more than three dimensions; **multifacto·rial**, involving or dependent on a number of factors (*spec.* genes) or causes; hence **multifacto·rially** *adv.*; **multifo·cal**, having or pertaining to several foci, or a range of focal lengths; also as *sb. pl.*, spectacles with multifocal lenses; **multigene·ric**, derived from or involving more than one genus; **multimammate**, having several pairs of mammæ, esp. used to designate the multimammate rat or mouse, *Mastomys natalensis*, a rodent found in tropical Africa; **multipote·ntial** *Med.*, capable of differentiating into any of several kinds of cell or tissue; **multiseriate** (later examples); **multispe·ctral**, operating in or involving several of the regions into which the electromagnetic spectrum is conventionally divided; **multispi·cular** = *multispiculate*; **multista·ble**, (of a system) composed of a number of interconnected subsystems each of which can achieve stability independently of the others; so **multistabi·lity**, the property or state of being multistable.

1968 *Punch* 31 July 157/2 Some urge..has merged it [*sc.* the public] into one worldwide, multicelled acerebral organism. 1972 *Sci. Amer.* Apr. 72/2 The globins were evolving to play several different roles during and after this period, as multicelled organisms arose. 1916 W. Trotter *Instincts of Herd* 18 Looked at in this way, multicellularity presents itself as an escape from the rigour of natural selection. 1972 *Sci. Amer.* Dec. 95/1 His Mesozoa may yet provide us with valuable evidence on the evolution of multicellularity and on the mechanisms of differentiation and development. 1934 Webster, Multicentric. 1941 *Cold Spring Harbor Symp. Quant. Biol.* IX. 153 Di- and multicentric chromosomes. 1971 *Brit. Med. Bull.* XXVII. 68/2 Multi-centric hypotheses have not proved as popular as multi-stage hypotheses, or theories which assume that a change from a normal to a malignant cell takes place in stages. 1972 *Year Bk. Dermatol.* 185 The present study examined the ultrastructural characteristics of multicentric reticulohistiocytosis. 1952, 1954 Multicyclic [see *monocyclic *a.* 5]. 1966 G. H. Dury *Ess. Geomorphol.* 128 The degree to which multi-cyclic bevelling is displayed varies widely. 1972 *Science* 3 Nov. 503/1 A large proportion of the grains appears to be multicyclic, having undergone several cycles of erosion and deposition. 1959 *Nomencl. Inorg. Chem.* (I.U.P.A.C.) 64 A group containing more than one potential co-ordinating atom is termed a multidentate ligand. 1974 *Chem. Rev.* LXXIV. 351/1 This review deals with the synthesis of multidentate macrocyclic compounds. 1956 E. H. Hutten *Lang. Mod. Physics* v. 171 The propagation of the wave must be described as taking place, in most instances, in a multi-dimensional hyper-space, and not in ordinary space. 1963 F. G. Lounsbury in J. A. Fishman *Readings Sociol. of Lang.* (1968) 49 Inflection in some languages is carried into several dimensions of variation, resulting in multidimensional paradigms running into hundreds or even thousands of forms. 1970 *Jrnl. Gen. Psychol.* LXXXII. 172 Anxiety is a vague and multidimensional concept. 1967 J. W. Johnston in T. Hayashi *Olfaction & Taste II* 48 This information could be utilized to investigate the relationship between the multidimensionality of odors and the hypothetical sensor units or sites on the receptor membrane. 1968 W. A. Scott in Lindzey & Aronson *Handbk. Social Psychol.* (ed. 2) II. 251 The dilemma of multidimensionality stems from the acceptance of a measurement model which equates dimension, in the geometrical sense, with psychological notions of attribute magnitude. 1970 *Computers & Humanities* IV. 210 Meylan has to be selective in what he chooses for encoding, and is unable to represent the entire 'multidimensionality' of the melody (e.g., its complete context). 1957 C. E. Osgood et al. in Saporta & Bastian *Psycholinguistics* (1961) 294/2 We may assume at the outset that meanings vary multidimensionally. 1920 W. E. Castle *Genetics & Eugenics* (ed. 2) xxxi. 275 It would be a mistake to cover up our present ignorance concerning the inheritance of these characters by classifying them either as unifactorial or as multifactorial. 1937 *Biol. Rev.* XII. 481 The melanism of *Oporinia dilutata* has a multi-factorial basis. 1960 H. J. Eysenck *Handbk. Abnormal Psychol.* viii. 309/2 Most of the characteristics..are of a quantitatively variable kind, and..regulated by multifactorial and not unifactorial inheritance. 1965 J. Pollitt *Depression & its Treatment* 96 The use of these terms is not in keeping with the view that the ætiology of depression is multifactorial. 1974 *Nature* 10 May 145/2 It can be assumed that the aetiology of cranioschisis is at least complex and multifactorial involving environmental agents as well as genetic factors. 1954 P. M. Sheppard in J. S. Huxley et al. *Evolution as Process* 212 (*heading*) The evolution of multifactorially controlled characters. 1973 *Nature* 22 June 433/1 The word genic is extended from its regular use to include multifactorially determined disorders. 1920 T. P. Nunn *Education* 111 The multi-focal view, to which Thorndike has now transferred his allegiance, holds that our abilities fall into a small number of groups. 1928 S. Duke-Elder *Pract. Refraction* xxiii. 344 In the multifocal lenses, introduced by Gowland of Montreal, in 1922..the reading portion of the lens has a continuous variable curve, there being a gradual accretion of power from the periphery to the reading centre. 1961 *Brit. Jrnl. Surg.* XLIX. 92/1 Low-grade, primary, multifocal osteosarcoma occurring in middle age appears to be a distinct entity. 1962 L. S. Sasieni *Princ. & Pract. Optical Dispensing* v. 125 The accurate fitting of bifocals and multifocals will be considered later. 1971 *Optometry Today* (Amer. Optometric Assoc.) 12 Special optical devices such as multi-focal lenses, contact lenses and other vision aids are used. 1953 *Internat. Code Nomencl. Cultivated Plants* 25 Such [hybrids] as are trigeneric or multigeneric. 1971 J. Z. Young *Introd. Study Man* xxxii. 445 It may be that ultimately some or all the forms now called *Australopithecus* will be referred to as various species of the genus *Homo*... We prefer to keep a somewhat more conservative (multigeneric) classification. 1902 *Ann. & Mag. Nat. Hist.* IX. 219, I fail to find any point whatever by which *M[us] Hildegardeæ* can be distinguished externally from the Machakos member of the multimammate group. 1955 *New Biol.* XIX. 108 House rodents such as the native multimammate mouse..and the black rat, which is replacing it, are the reservoirs [of plague in Central Africa] rather than fully rural species. 1959 *Nature* 12 Sept. 794/2 The multimammate rat is a wild rodent which lives in close contact with man in Africa. 1974 *Ibid.* 13 Sept. 101/3 Lassa virus was isolated from a single murine species, the multimammate rat *Mastomys natalensis*, which is a common commensal rodent in West Africa adapted to life both within houses and in the fields. 1944 S. Brunauer *Adsorption of Gases* I. 6 When the surface can take up only one layer of the adsorbed gas, the adsorption is called unimolecular; when more than one layer it is called multimolecular. 1969 *Science* 12 Dec. 1365/3 Bacteria, viruses, or cells..often called antigens..really are multimolecular mixtures of polydeterminant antigens. 1928 *Funk's Stand. Dict.*, Multiperforated. 1957 *Jrnl. Brit. Interplanetary Soc.* XVI. 207 It is for such reasons as these that the more complicated multiperforated charge..has been developed, which can have small propellent thickness between the perforations without reducing the loading density, and which has little unused propellent at the end of burning. 1913 O. C. Gruner *Biol. Blood-Cells* iv. 191 At certain times the wandering cell becomes stationary... In place of migration..it settles down as an essentially multipotential or indifferent cell. 1939 Dible & Davie *Path.* viii. 152 If the germinal cells for a tumour be split off at this stage, only those tissues can be represented in the growth which are proper to the layer from which these cells have come. Such cells are multipotential but not totipotential. 1966 *Cancer* XIX. 1/1 We have designated these types of tumors as primitive multipotential primary sarcomas of bone:..'multipotential' because each shows differentiation along multiple lines (i.e. bone, cartilage, blood vessels, etc.). 1973 *Nature* 2 Mar. 20/2 The finding of multipotential haemopoietic stem cells (that is, cells capable of becoming any of the mature blood cell types, lymphoid or myeloid) in early mouse embryonic thymus. 1933 *Tropical Woods* XXXVI. 10 Heterogeneous Ray.—A xylem ray composed of cells of different morphological types. (Typically, with the cells of the multiseriate part radially elongated.) 1965 K. Esau *Plant Anat.* (ed. 2) vi. 138 A ray may be one cell wide and one cell high in the beginning; later, the initial divides or more initials are added to the first. The ray thus increases in height and may increase in width if multiseriate rays are characteristic of the plant. *Ibid.* xvii. 485 A well-known example of a multiseriate epidermis is the velamen of air roots of tropical Orchidaceae. 1966 *Proc. 4th Symposium Remote Sensing Environment* (Univ. Mich., U.S. Govt. AD 638-919) 135 Much additional information is needed before multispectral photography and other remote sensors can be..incorporated into operational urban information centers. 1971 *New Scientist* 27 Apr. 256/2 Placing the Earth under the most detailed observation with multispectral cameras. 1973 L. F. Curtis in Cruise & Newman *Photogr. Techniques Sci. Res.* ii. 101 Photographic analysis of the data obtained in multispectral studies may include colour enhancement techniques which aim to emphasize the tonal differences between objects. 1902 *Proc. Zool. Soc.* 210 In one of the two specimens [of sponges] in the collection..there are a few multispicular strands in the otherwise very regular unispicular meshwork. 1952 W. R. Ashby *Design for Brain* xvii. 182 S. 9/4 showed the necessity for ultrastability; is the hypothesis of multistability equally necessary? 1962 A. Battersby *Guide to Stock Control* vii. 65 The effect of this sort of control by interconnected systems is called 'multistability' by cyberneticians. 1952 W. R. Ashby *Design for Brain* xvi. 171 To study the interactions of these two properties we might start by examining the properties of an ultrastable system whose main variables are all part-functions. But it has been found simpler to start by considering a system defined thus: a multistable system consists of many ultrastable systems joined main variable to main variable, all the main variables being part-functions. *Ibid.* xvii. 179 The concept of the 'multistable' system clearly refers primarily to the nervous system. 1966 J. Singh *Gt. Ideas Information Theory, Lang. & Cybernetics* (1967) xiv. 221 Every animal is thus built up of a number of ultrastable subsystems organized with the possibility of dispersion of behavior. Such a multistable system can therefore adapt its behavior to settle in a stable resting state in a much shorter time than a single ultrastable system of the kind considered earlier. 1961 *Nature* 11 Mar. 833/1 Multi-stranded deoxyribonucleic acid as determined by X-irradiation. 1970 *New Scientist* 17 Dec. 496/2 The antitumour effects of multistranded RNA which have been reported recently.
b. *multi-authored*, *-barrelled*, *-centred*, *-componential*, *-consonantal*, *-cored*, *-directional*, *-engined*, *-ethnic*, *-faceted* (later examples), *-geminal*, *-generational*, *-holed*, *-hued* (later examples), *-manned*, *-marbled*, *-sectional*, *-sectioned*, *-secular*, *-segmental*, *-syllabic*, *-tribal*, *-valued*, *-volumed*; **mu·ltiplaned**, having or occupying several planes (cf. *multiplane *sb.* and *a.*); **multi-ta·ped**, recorded from several (magnetic) tapes played simultaneously.
1964 K. Winetrout in I. L. Horowitz *New Sociol.* 159 *The Lonely Crowd* is multi-authored and *Faces in the Crowd* a research team product. 1974 *Nature* 22 Feb. 581/1 It suffers from the usual disadvantage of a multi-authored book in which no editorial attempt has been made to avoid redundancies and overlap. 1944 *Horizon* Oct. 225 'Moaning Minnies' (multi-barrelled mortars). 1957 K. A. Wittfogel *Oriental Despotism* iii. 50 In contrast to the controlled state of multicentered societies, the state of the single-centered hydraulic society was a veritable apparatus state. 1973 *Sci. Amer.* Apr. 90/2 Bruno's belief in a centerless or multicentered universe was derived from..Nicholas of Cusa. 1971 *Archivum Linguisticum* II. 59 Cf. [*make up*] (*sc.* with cosmetics or greasepaint) (verb) and [*make-up*] (noun),..examples whose multicomponential character as idioms is partly revealed by the difference of tonicity or accentuation corresponding to the grammatical distinction indicated. 1974 *Nature* 7 June 571/2 The survival curve of HVS is multicomponential after treatment with heat as well as after treatment with formaldehyde. 1948 D. Diringer *Alphabet* ii. 63 They did not employ it [*sc.* the alphabet] when they could use word-signs or multi-consonantal phonograms. 1965 *Language* XLI. 346 Such multiconsonantal clusters as /mb/ and /ngb/ are merely graphic and represent consonants that are phonemically unitary. 1974 D. Kyle *Raft of Swords* ii. 14 Loop after loop of insulated multi cored copper wire. 1942 *Illuminating Engin. Nomencl. & Photometric Stand.* (Illuminating Engin. Soc. U.S.) (ASA Z7.1–1942) 27 Multidirectional illumination on a surface is that produced by several separated light sources of relatively small area. It is characterized by the fact that a small opaque object placed near the illuminated surface casts several shadows. 1964 *Discovery* Oct. 6/2 Control of a vast and complex multi-directional communications network. 1972 *Geo Abstr.* C. 368 (*heading*) Problems of multi-directional development of automobile transport. 1917 *Jrnl. R. Aeronaut. Soc.* XXI. 320 In the case of most single-engined aeroplanes or seaplanes a fuselage of circular cross-section is admirable... This is not so, however, in the case of multi-engined aircraft, in which the power plant units are placed away from the body of the machine. 1931 C. Day Lewis *From Feathers to Iron* 14 Bodies we have, fabric and frame designed To take the stress of love, Buoyant on gust, multi-engined. 1967 *Jane's Surface Skimmer Systems 1967–68* 121/2 A drive pad at the rear of the primary box enables the engines of a multi-engined aircraft to be coupled together by spanwise shafting. 1967 'Malcolm X' in A. Chapman *New Black Voices* (1972) We Afro-Americans..are not opposed to multi-ethnic associations in any walk of life. 1969 *Alberta Hist. Rev.* Autumn 10/2 Both were born in remote wilderness conditions and embraced multi-ethnic populations. 1965 H. I. Ansoff *Corporate Strategy* (1968) i. 15 The nature of decisions is multifaceted and continually variable. 1971 *Computers & Humanities* VI. 38 Paralleling this trend toward the multifaceted department articulating with almost all the traditional disciplines is the emergence of the divisional course. 1922 Joyce *Ulysses* 404 Recorded instances of multigeminal..births. 1965 *New Society* 1 Apr. 25/2 Subjects from multi-generational households have better scores. 1972 P. Laslett *Household & Family in Past Time* 7 A high proportion of multigenerational extended families among the remaining Dutch peasantry in the 1950s. 1953 *Lancet* 12 Sept. 540/1 A multi-holed catheter had been passed. 1963 A. J. Hall *Textile Sci.* ii. 47 It is pumped at a uniform rate through multi-holed spinnerets into the coagulating bath. 1927 *Daily Express* 13 July 5/2 The hidden wiring of the multi-hued electric lights. 1971 *Sci. Amer.* Oct. 43/1 (Advt.), 'Philippine Birds' contains 569 illustrations..portraying the multi-hued birds. 1973 *Islander* (Victoria, B.C.) 1 July 2/1 Multihued confetti of clouds drifted across the sky. 1961 *Life* 3 Mar. 33/1 The Russians will follow their manned orbiter..with a multimanned moon orbiter..and space station. 1967 *Observer* 5 Nov. 4/5 A multi-manned expedition around the moon and back to earth. 1902 Hardy *Poems of Past & Present* 40 Multimarbled Genova the Proud. 1909 *Westm. Gaz.* 23 Mar. 4/2 The multi-planed helicopter has been abandoned for the biplane. 1959 M. T. Williams *Art of Jazz* (1960) xii. 110 The nation's misery was mirrored in every facet of its multi-planed existence. 1961 Y. Olsson *On Syntax Eng. Verb* ii. 33 A classification can be made of the sentences of the language..: I. multi-sectional (consisting of two or more Sections). 1964 *Gloss. Letterpress Rotary Printing Terms* (*B.S.I.*) 22 Multi-sectioned newspapers or magazines. 1922 Joyce *Ulysses* 655 The multisecular stability of its primeval basin. 1901 C. P. Steinmetz *Theoret. Elem. Electr. Engin.* 122 A closed coil armature, connected with a multisegmental commutator. 1955 *Trans. Philol. Soc.* 1954 68 We may..classify phonemic features by reference to the *types* of unit, of which they are constituents..; in this way we may establish a broad distinction between (*a*) 'uni-segmental' features, i.e. such as combine within single phonemes, and (*b*) 'multi-segmental' features, i.e. such as can only be extracted from more complex units (syllables, morphemes, words, constructions, or sentences). 1909 *Cent. Dict. Suppl.*, Multisyllabic. 1953 N. Tinbergen *Herring Gull's World* ii. 10 The most elaborate call of the Herring Gull's repertoire is the multisyllabic call. 1972 F. Knebel *Dark Horse* (1973) xix. 294 He had long ago passed the point where he sought to impress bewildered laymen by scattering multisyllabic medical terms about. 1955 L. Feather *Encycl. Jazz* vii. 245 The unique tonal effects obtained from these multi-taped, multi-speed novelties sold many millions of Capitol records. 1962 *John o' London's* 16 Aug. 162/4 Not all the tracks are multi-taped. 1949 I. Deutscher *Stalin* iii. 69 The multi-tribal cities of the Caucasus. 1934 *Mind* XLIII. 200 Whatever is a proposition within the two-valued logic is recognized as a proposition within the multi-valued 'logic' and *vice versa*. 1965 N. Chomsky *Aspects of Theory of Syntax* iv. 171 Regarding each of the dimensions of the

system of paradigms as a multivalued feature, [etc.]. 1963 *Times Lit. Suppl.* 10 May 348/4 Multi-volumed books.
2. **mu:ltibillionai·re**, one who is worth many billions of money; **mu:lticollinea·rity** *Statistics*, the existence of a perfect linear correlation between a set of variables when the regression of some dependent variable on them is being investigated; **multi-millionaire** (later examples); **multi-ne·gative** *Printing*, an array of many similar images in negative form used in the printing of small items several at a time by photolithography; similarly **multi-po·sitive**; † **multirota·tion** *Chem.* [a. G. *multirotation* (Parcus & Tollens 1890, in *Ann. d. Chem.* CCLVII. 161)] = *mutarotation.
1906 'Mark Twain' in *North Amer. Rev.* Sept. 322 There would not be any multi-billionaire alive, perhaps, who would be able to buy a full set. 1934 R. Frisch *Statistical Confluence Analysis* xi. 75 There exist two or more independent linear relations between the systematic parts of these variates, but..we are not aware of this multicollinearity. 1972 T. H. & R. J. Wonnacott *Introd. Statistics for Business & Econ.* xiii. 296 Suppose demand for a group of goods is being related to prices and income, with the overall price index being the first independent variable. Suppose aggregate income measured in money terms is the second independent variable. If this is real income multiplied by the same price index, the problem of multicollinearity may become a serious one. The solution is to use real income, rather than money income, as the second independent variable. 1921 Multi-millionaire [see *Corona⁴]. 1972 F. Knebel *Dark Horse* (1973) ix. 120 You got multi-millionaires who don't pay more than two or three per cent of their income in taxes. 1933 N. Montague in W. Atkins *Art & Pract. Printing* III. xii. 95 Multi-negatives have become an economic necessity in the case of printing by photo-lithography such things as stamps, coupons, playing cards, and small labels. 1967 E. Chambers *Photolitho-Offset* vi. 70 The automatic operation can be commenced at any part of the layout and an exposure made to any position on the multi-negative. 1933 N. Montague in W. Atkins *Art & Pract. Printing* III. xii. 96 The 'H.L.' step and repeat projecting machine.. is designed for the production of multi-negatives or positives. 1971 D. Potter *Brit. Eliz. Stamps* ii. 23 In the first stages of printing a multipositive is prepared, from which the printing cylinders are made. 1890 *Jrnl. Chem. Soc.* LVIII. 1084 The authors propose to employ the term 'multi-rotation' to denote the change..of rotatory power which takes place in a sugar solution, from immediately after its preparation in the cold until constant rotation is attained. 1904 J. W. Mellor *Chem. Statics & Dynamics* 224 The influence of acids upon the rate of multirotation of sugars. 1935 Tipson & Stiller in Harrow & Sherwin *Textbk. Biochem.* ii. 43 In the course of time other sugars were examined, and it was found that the rotation is not necessarily halved, so the phenomenon was named multi-rotation.
3. (Compounds of this type are occas. used predicatively, as in quot. 1947 for *multifactor* and quot. 1970 for *multifont*.) *multiaperture*, *-author*, *-band*, *-bed*, *-billion*, *-blade*, *-car*, *-centre*, *-class*, *-component*, *-cycle* (cf. *multicyclic* in *1 a), *-cylinder* (further examples), *-day*, *-element*, *-engine*, *-family*, *-floor*, *-image*, *-jet*, *-lane*, *-lens*, *-man*, *-megaton*, *-microphone*, *-million*, *-part*, *-pin*, *-plate*, *-reel*, *-role*, *-row*, *-screen*, *-seat*, *-step*, *-syllable*, *-track*, *-tube*, *-vane*, *-vitamin*, *-volume*, *-way*, *-word*; **multi-a·ccess**, pertaining to, involving, or being a computer that may be used simultaneously by independent operators at a number of terminals; **multicha·nnel**, employing or possessing many communication or television channels (*channel *sb.*¹ 9 d); also = *multi-track* above; **multici·rcuit**, supplying or containing many electric circuits; **mu·lticore**, having many cores; applied *spec.* to an electric cable (cf. Core *sb.*¹ 11); **multi-ele·ctrode**, possessing or involving several electrodes; applied *spec.* to a valve in which there are two or more sets of electrodes, associated with separate electron beams, within a single envelope; **multifa·ctor** = *multifactorial* in *1 a; **multifa·culty**, (of an institution of higher education) possessing several faculties; **multi-fi·lament**, containing or composed of several filaments; applied *spec.* to a yarn made up of many fine threads; also *absol.*, a multifilament yarn; **mu·ltiflash**, (*a*) *Photogr.* made with two or more flash-bulbs operated simultaneously; (*b*) applied to a process or apparatus for desalinating sea-water in several stages in which water is 'flashed', i.e. converted suddenly into steam, at successively lower pressures; **mu·ltifont**, pertaining to or possessing the ability (of a machine) to read characters of several different 'fonts' (founts) or designs; **mu·ltifuel**, capable of running on more than one kind of fuel; **mu·ltigerm** *Agric.*, (of a sugar-beet variety or seed produced by it) having or consisting of more than one seed in each seed ball (cf. *monogerm* adj. s.v. *mono- 1); **mu·ltigrade** (see quot. 1972); **mu·ltipath** *Telecommunications*, involving (the receipt of) radio signals that have travelled from a single source by several paths; **mu·ltipoint**, having or serving many points; applied *spec.* to a water-heater serving a number of hot-water taps; **mu·ltistart** *Engin.* (see quot. 1940); **mu·ltiwall**, having many walls; applied *spec.* to a sack or large bag made of several layers of strong paper that are usu. attached to one another along the bottom and the side folds.
1966 *Guardian* 4 May 20/3 By the end of the seventies in Britain there will be more than 300,000 people interested in multi-access computers. 1967 *Economist* 4 Mar. 850/3 This will mean hierarchies of powerful and less powerful multi-access computers linked throughout the country with hundreds of terminals and consoles connected to them from commerce, industry and research establishments. 1968 *Times* 1 Nov. 23/5 Multi-access computing needs data-links between users and computers. 1971 R. A. Wisbey *Computer in Lit. & Ling. Res.* 191 The Atlas multi-access system makes it possible to create and manipulate files of information from remote consoles and to use these files for the initiation of jobs on the Atlas computer. 1966 *Electronics* 14 Nov. 25 Honeywell spokesmen will not disclose details of the new design, beyond saying that it used conventional ferrite cores with one hole, not multiaperture cores. 1968 *New Scientist* 15 Feb. 360/2 The elements, called MADs (multiaperture devices), are specially shaped ferrite discs..which act as very fast switches. 1973 *Nature* 9 Feb. 412/1 As is inevitable in any such multi-author work, the individual chapters vary considerably in scope. 1969 *Sci. Jrnl.* June 64 Vertical aerial view of part of the Carrizo Plains, California, at near right was taken using ordinary Ektachrome film and reveals little trace of the phenomena rendered starkly visible in the enhanced multiband photograph at far right. 1971 *Flying* Apr. 90/1 About our portable multi-band radio. 1964 G. L. Cohen *What's Wrong with Hospitals?* i. 24 To the archetypal American patient of the medical magazines, there is no disgrace in entering a 'multi-bed ward'. 1963 *Observer* 31 Mar. 27/3 A background of multi-billion defence spending. 1971 *New Scientist* 28 Jan. 184 What would be the practical and scientific benefits of a multibillion dollar space station? 1910 *Westm. Gaz.* 24 Mar. 5/2 This is the Pitter multi-blade propellor, which..is claimed to give an enormous thrust at low speeds. 1960 *Farmer & Stockbreeder* 1 Mar. 70 The multi-blade fan located in front of the engine has an output of up to 40,000 cu ft per min. 1962 *Economist* 11 Aug. 526/1 The proportion of multi-car families is continuing to rise. 1965 Phillips & Williams *Inorg. Chem.* I. xi. 403 Metals and metal-like hydrides are similarly electron-deficient... The electrons are thought to be in molecular orbitals embracing all the atoms in the structure, i.e. multi-centre rather than two-centre orbitals. 1930 *Engineering* 31 Oct. 541/2 It is still the only multi-channel radio-telephone service in the world. 1955 *Ann. Reg. 1954* 402 There were on exhibition at the radio show ..many television receivers designed for two-band or multi-channel reception. 1962 *Science Survey* III. 315 Indeed one can conceive of such a system in which a multi-channel magnetic tape carries the programme of the plant. 1966 *Punch* 20 July 101/2 The United States, where of course there's multi-channel TV operating more or less round the clock. 1909 *Cent. Dict. Suppl.*, *Multicircuit*, having a multiple circuit... *Multicircuit generator*, an electric generator, as for the supplying of arc-lights, which is connected with and feeds several parallel circuits. 1969 *Jane's Freight Containers 1968–69* 536/2 An a.c. drive motor with single and multi-circuit generators built as one unit. 1970 *Sci. Amer.* Feb. 28/2 One obstacle..was that a multicircuit chip would have to incorporate logic gates of several different types and all would have to function perfectly. 1961 *Guardian* 16 June 6/5 The rise of the anonymous, multi-class mass audience. 1934 Webster, Multicomponent. 1966 D. G. Brandon *Mod. Techniques Metallogr.* iii. 125 Identifying the elements in a multicomponent sample. 1971 C. Placek (*title*) Multicomponent fibers. 1908 *Westm. Gaz.* 6 Feb. 4/2 The high-tension wires are moulded in one, forming a multicore cable. 1957 *Railway Mag.* Nov. 760/2 A prototype length of special 'thin'-type multi-core cable was used for certain indication circuits. 1961 *New Left Rev.* Jan.–Feb. 8/1 A multi-core city is gradually taking shape. 1916 C. A. Cotton in *Geogr. Rev.* I. 39 The diagrams..represent stages in the growth of a composite (i.e., two-cycle or multi-cycle) delta. 1918 —— in *Ibid.* VI. 320 In a previously published article the writer called attention to the occurrence of multicycle fault coasts. 1962 *Gloss. Terms Automatic Data Processing* (*B.S.I.*) 91 *Multi-cycle feeding*, a method of processing punched cards in which several fields of a single card are read sequentially on successive machine cycles... The principal use is as a card-saving stratagem when printing names and addresses. 1966 J. C. Pugh in G. H. Dury *Ess. Geomorphol.* 127 Multi-cycle bornhardts can also be found in the form of dome-on-dome residuals. 1904 *Sci. Abstr.* VII. 1080 (*heading*) Multi-cylinder motors. 1909 *Westm. Gaz.* 10 June 4/2 Of the twenty-five pretty little machines which will career round the twenty-three-mile circuit.., no fewer than seventeen are fitted with four-cylinder engines, which shows..the progress that has been made in the multi-cylinder car. 1969 *Jane's Freight Containers 1968–69* 495/3 Insulated flexi-van multi-cylinder high pressure gas holder. 1973 *Nature* 23 Mar. 217/1 By use [in their motor cycles] of multi-cylinder engines, overhead camshafts and such devices as electric starting motors, they have been able to sell products which are in many ways superior to those of British manufacturers. 1971 *Canad. Antiques Collector* May 10/1 These field studies extend up to a week with the multi-day visitors accommodated economically in camps, lodges, schools and private homes. 1973 C. Bonington *Next Horizon* v. 85 He was not interested in the multi-day epic, or the highly technical rock-climb. 1926 *Wireless World* 26 May 722/1 (*heading*) The multi-electrode valve. 1963 Multi-electrode [see *Dekatron]. 1965 *Math. in Biol. & Med.* (Med. Res. Council) iv. 139 Is it possible to recognize automatically particular waveforms in a single channel of a multi-electrode recording [of an electro-encephalogram]? 1961 Y. Olsson *On Syntax Eng. Verb* iv. 85 The 'genitive' is a multi-element term for PRE. 1970 *New Scientist* 24 Dec. 554/2 The Marconi Company..developed a high-gain multi-element vhf aerial. 1934 Webster, Multiengine. 1961 *Economist* 25 Nov. 830/2 Rolls-Royce immediately started hawking its competing multi-engine system round the international market. 1934 R. Frisch *Statistical Confluence Analysis* vii. 49 We get the special hypothesis regarding the correlation coefficients on which Spearman's two factor and Thurstone's multifactor theory are built. 1947 F. A. E. Crew *Genetics Rel. Clin. Med.* v. 84 There is the possibility that certain abnormal characterisations are themselves multifactor in nature. 1966 A. Battersby *Math. in Managem.* iii. 68 The bonus is to be paid at the rate of one shilling for each ton of production in excess of 160... The manager.. undertakes [in addition] to apply a penalty of 10 shillings for a fall of 1 point in the yield, and a corresponding bonus for an increase. Such a scheme is called a 'multi-factor incentive'. 1961 *Technology* Feb. 36/4 The C.A.T.s ..should become multi-faculty institutions with the proportion of science and social studies increased. 1968 *Economist* 13 Apr. 42/1 The schools are to be reduced by amalgamation from twelve to six, and each is to be closely associated with a multi-faculty institution of the University of London. 1965 G. McInnes *Road to Gundagai* xiv. 249 The enormous multi-family picnics that took place. 1972 *Real Estate Rev.* Winter 73/2 The Federal Housing Administration then began requiring its field offices to report on the number of minorities occupying federally assisted multifamily housing. 1937 W. Watson *Textile Design & Colour* (ed. 4) 437 The diameter of each orifice is about 0·1 mm. for ordinary yarns, and may be as small as 0·06 mm. for fine denier multi-filament yarns. 1940 *Chambers's Techn. Dict.* 562/2 *Multi-filament lamp*, an electric filament lamp having more than one filament in the same bulb, so that failure of one filament will not cause the lamp to be extinguished. 1944 *Textile Manufacturer* July 306/1 We made 150/150 yarn of that type but the multifilaments could not compensate for the inherent harsh character of hosiery made from it. 1968 E. Miller *Textiles* 70 Multifilament yarns give more flexibility and cover, the finer and more numerous the filaments are in the yarn the better it is in this respect. 1945 Webster Add., Multiflash. 1956 *Nature* 21 Jan. 109/2 (*caption*) Multiflash photographs..of a typical water-entry trajectory. 1964 *Listener* 21 May 835/1 The new 'multi-flash' distillation process was first realized in this country, and more than two-thirds of all the installations in the world are British. 1971 *Chem. Abstr.* 17 May 195/2 Multiflash evapn. for concg. solns., and producing fresh water. 1938 *Archit. Rev.* LXXXIII. 118/2 A multifloor scheme was devised which, besides allowing high economy of site area, gave other advantages of even greater moment. 1960 *News Chron.* 7 Oct. 8/5 Multi-floor garages. 1961 *Proc. IRE* XLIX. 185/2 Extensions of the notions described will result in improved multifont recognition and in gains toward the isolation of the character separation problem. 1969 *Computers & Humanities* IV. 83 Of all devices for speeding the preparation of text for machines, none has been awaited..with more confidence than some sort of fast, multifont, full page scanner, more appropriately called an optical character recognition (OCR) machine. 1970 *Ibid.* V. 75 The Scan-Data 300 is genuinely multifont. 1959 *Engineering* 9 Jan. 45/1 (*heading*) Multi-fuel engines. 1965 *Economist* 22 May 941/2 The Wankel could come in here, with possibly only marginal disadvantages over available multi-fuel engines. 1950 *Proc. Amer. Soc. Sugar Beet Technologists* VI. 160 (*heading*) Embryology of mono- and multigerm fruits in the genus Beta L. 1962 *Times* 21 May 18/5 The monogerm seed which is still being bred and selected to take the place of the multi-germ seed now used. 1959 *Motor Manual* (ed. 36) x. 237 The 'multi-grade' engine oils now available ..cover a range of ratings. 1972 *Materials & Technol.* IV. ii. 68 Modern motor oils with good viscosity/temperature characteristics may meet the requirements of several SAE [*sc.* Society of Automotive Engineers] grades and are termed multigrade oils (e.g. 10W/30). 1974 *Drive* Autumn 56 Fast-moving engine parts 'snip up' ordinary multigrade oils like scissors. 1962 *Punch* 17 Jan. 133/3 Technicians..explain..what you can and cannot do with a multi-image lens [in television]. 1969 *Focal Encycl. Film & Television Techniques* 472/2 Multi-image films make use of expanding and diminishing pictures. 1963 *Daily Tel.* 21 Nov. 16 The experimental application of multi-jet techniques for lifting transport aircraft vertically. 1966 *McGraw-Hill Yearbk. Sci. & Technol.* 411/2 The multi-jet configuration permits higher speeds..than single-jet turbines. 1974 *Encycl. Brit. Macropædia* XVIII. 637/1 The large transport aircraft, now almost exclusively of multijet design. 1961 *Times* 12 Dec. 14/3 A conservative British driver might envisage a multi-lane nightmare in a wonderland of flyovers and freeways. 1966 *Times* 28 Feb. (Canada Suppl.) p. viii/5 Vancouver's west end, looking northwards over the multi-lane bridge. 1936 *Discovery* Aug. 237/1 This unique camera..is intended for mapping from a higher altitude than has ever before been practical with multi-lens equipment. 1948 E. Raisz *Gen. Cartogr.* (ed. 2) xix. 194/1 The largest multilens camera is the nine-lens camera of the U.S. Coast and Geodetic Survey. 1951 J. M. Fraser *Psychol.* xv. 189 Wherever a number of people are involved in one task at the same time, a multi-man chart can be used to record the task of each individual. 1967 Multi-man [see *ejection 1 d]. 1958 J. Blish *Case of Conscience* I. vi. 58 The prettiest multimegaton explosion you ever saw. 1965 H. Kahn *On Escalation* x. 198 Multimegaton weapons appeared to be unusable for any rational, and even for many irrational, purposes. 1941 *B.B.C. Gloss. Broadcasting Terms* 19 *Multi-microphone technique*, method of production in which several microphones, with outputs centralized in a mixer, are used either simultaneously or in succession to reproduce sounds from a single studio or hall. 1957 Manvell & Huntley *Technique Film Music* iv. 194 In England some years ago it was fashionable to use a multi-microphone technique, in which..considerable physical

separation between different sections of the orchestra was necessary to enable each microphone to record only that section which it was intended to cover. **1962** A. Nisbett *Technique Sound Studio* iii. 61 If a close balance is adopted in a multi-microphone mix, artificial 'echo' may be added. **1948** *Amer. Jrnl. Roentgenology* LIX. 771/1 Multimillion volt beta rays and multimillion volt roentgen rays in the treatment of cancer. **1966** *Times* 28 Feb. (Canada Suppl.) p. ix/4 It is a multi-million dollar operation. **1974** *Listener* 23 May 652/2 Multi-million pound shopping precincts. **1928** *Daily Express* 4 May 3/3 The powers given under the Electricity Supply Act of 1926 to introduce what are known as multi-part tariffs, whereby a consumer, who is willing to use energy for a number of purposes, may pay a fixed charge yearly and then pay a low running charge per unit. **1936** *Proc. IRE* XXIV. 912 It is..extreme high order harmonic distortion which makes the reception of frequency modulation over a multipath medium far more distorted than the reception of amplitude modulation over the same medium. **1966** *Electronics* 3 Oct. 174 There are three vhf propagation problems that must be considered in designing the Aerocom satellite: fading caused by Faraday rotation, ionospheric fluctuation and sea-water multipath fades. **1970** J. Earl *Tuners & Amplifiers* iii. 75 Multipath interference can adversely disturb stereo reception. **1963** *Thorn Electr. Industr. Group Profile* 16 Multi-pin electrical connectors. **1964** R. F. Ficchi *Electr. Interference* v. 72 The package units are less reliable than the individual units. The dependence of [read on] multi-pin connectors which have a relatively high failure rate is the cause. **1948** *Motor Manual* (ed. 33) vii. 122 The multi-plate clutch, in which two sets of metal discs were arranged alternately in sandwich fashion. One set was attached to the driving shaft, and the other..to the driven shaft. When a spring was permitted to force them all tightly together, they gripped firmly; but when the spring was restrained by the driver, one set of discs could slip round without turning the others. *Ibid.* 135 For top gear, ..there is a separate clutch (sometimes of the multi-plate type) inside the gearbox. **1963** Bird & Hutton-Stott *Veteran Motor Car* 253 Transmission: Multiplate clutch, 4 speeds and reverse, shaft. **1971** *Engineering* Apr. 49 (*caption*) This figure shows the multiplate construction of the differential capacitor probe fitted to the gauging head. **1906** *Trans. Amer. Inst. Electr. Engin.* XXIII. 209 It seems as though the latter class of advantages might be obtained as readily and more simply and cheaply by the use of the multi-point control-switches, used in parallel with the ordinary ones. **1931** *Conc. Househ. Encycl.* 531/2 (*caption*) Automatic water heater for multi-point service. **1966** *Bk. Good Househ.* (Good Househ. Inst.) (rev. ed.) ix. 96/1 Multi-point heaters are now available to fit under a draining-board. **1964** M. Laski in S. Nowell-Smith *Edwardian England* iv. 160 The bathroom (was) served by..a multi-point pressure heater (the first..came on the market in 1899). **1971** B. Scharf *Engin. & its Lang.* x. 82 The cutting tools used on machine tools may be conveniently classified as single-point and multi-point tools. **1929** *Encycl. Brit.* XV. 836/2 Multireel pictures began to appear in 1909... A year later screen history was made by releasing multireel pictures in their full length. **1970** *Times* 4 Sept. (Aviation Suppl.) p. iii/1 They include..the RB 199 turbofan for the Anglo-German MRCA-75 multi-role combat aircraft. **1968** *Chambers's Techn. Dict.* 996/1 *Multi-row radial engine*, a radial aero-engine with two or more rows of cylinders. **1977** *Power Farming* Mar. 11/1 For beans, Herbert had come up with a multi-row harvester which not only reduced the time in the field, but caused less damage. **1967** *Economist* 21 Oct. 267/1 A film in the very latest multiscreen technique extolled the opportunities for everyone in this blossoming province. **1973** *Times* 5 Oct. 25/2 The highlight is the multiscreen show—consisting of what the layman would describe as a cinema screen divided into six parts with different but related films showing on each at the same time. **1958** *Times* 11 Nov. 8/1 The electoral system chosen for Algeria is that of the multi-seat constituency. **1967** *Jane's Surface Skimmer Systems 1967–68* 112/1 Then work began on the Yard's first multi-seat passenger craft. **1940** *Chambers's Techn. Dict.* 564/1 *Multi-start worm*, a worm in which two or more helical threads are used in order to obtain a larger pitch and hence a higher velocity ratio of the drive. **1971** B. Scharf *Engin. & its Lang.* xii. 158 With single-start worms, lead equals pitch, but on multi-start worms, it amounts to pitch × number of starts (i.e. number of thread sections in a plane at right angles to the axis). **1962** G. R. Choppin *Exper. Nucl. Chem.* xi. 182 If the labeled compound is the end product of a multistep synthesis, in order to obtain a higher final yield, it is best to achieve the labeling at as late a point in the synthesis as possible. **1964** *English Studies* XLV. 383 The difference between a normal one-step process and..the multi-step process..is so great that [etc.]. **1964** *Language* XL. 176 Multisyllable words where a syllabic split in the written word occurs in the sequence vowel–consonant–consonant–vowel. **1958** J. Moir *High Quality Sound Reproduction* xviii. 545 The film standards used were those previously adopted for multi-track recording in the studios. **1959** W. S. Sharps *Dict. Cinematogr.* 113/2 *Multitrack recording system*, the system employed for the recording of a multiple sound track. **1969** *Time* 29 Aug. 47 Recording studios also offer new technical means of composing, through such devices as the echo chamber, multi-track recording and tape superimposition. **1935** *Discovery* Feb. 43/2 The multi-tube parachute rocket used for the Hart Mountain experiments. **1959** *Times* 23 Sept. 4 (Advt.), The Oldham Pg battery is unique in possessing a patented sleeve-sealed multi-tube positive plate. **1909** *Cent. Dict. Suppl.*, Multivane. **1963** R. R. A. Higham *Handbk. Papermaking* ix. 25 Situated in the trough [of the hydrapulper] is a multivane rotor, driven by a vertical shaft. **1942** *Endeavour* I. 32/1 On the matter of multivitamin treatment, there are different schools of thought. *Ibid.* 32/2 No multivitamin 'pill'..can at present replace a mixed all-round diet. **1960** *Woman's Own* 19 Mar. 25/3 A course of multi-vitamin tablets are a wise investment. **1968** M. Pyke *Food & Society* ii. 17 But it is different when a manufacturer advertises that..expensive multivitamin tablets protect the well-fed people who can afford them against deficiency diseases they will never experience. **1938** L. M. Harrod *Librarians' Gloss.* 104 *Multi-volume book*, a book in more than one volume. **1964** Y. Malkiel in Householder & Saporta *Probl. Lexicogr.* 15 Multi-volume encyclopedias shot through..with genuinely lexical entries. **1940** *Dict. Paper* (Amer. Paper & Pulp Assoc.) 234 *Multiwall-bag paper*, normally, a 40- or 50-pound..kraft paper. **1946** *Nature* 6 July 32/1 A review of literature on the acoustics of building, preparatory to carrying out experimental work with particular reference to multiwall units. **1955** *Times* 5 July (Paper Suppl.) p. iii/3 In little more than twenty-five years the development of the multi-wall paper sack..has established new standards of efficiency, convenience and cleanliness in the packaging of powdered and granular products. **1967** *Times Rev. Industry* May 82/1 As cement production grew, so did the use of multiwall sacks. **1974** *Columbia* (S. Carolina) *Record* 25 Apr. 9-D/1 The paper in question is used in multiwall bags, common in grocery stores. **1961** *B.S.I. News* Aug. 16/2 It was agreed that multi-way adaptors would not be permitted. **1967** *Gloss. Terms Materials Handling* (*B.S.I.*) iii. 10 *Multi-way valve*, a rotatable chamber having a central inlet, and one outlet which can be positioned in line with any one of several radially spaced conveying lines. **1961** R. B. Long *Sentence & its Parts* i. 15 Many kinds of words and multi-word units function as subjects. **1966** G. N. Leech *Eng. in Advertising* ii. 22 Idioms are treated as multi-word lexical items.

multiar (mʊ·ltiɑːɹ). *Electronics.* [f. Multi- + -ar, of unknown origin.] A circuit which produces an output signal when a varying voltage applied at one input exceeds an adjustable constant voltage applied at a second input, and which consists of a simple regenerative circuit with a diode and a pulse transformer in the feedback loop.

1946 *Jrnl. Inst. Electr. Engin.* XCIII. IIIA. 321/1 The Multiar..may be made to generate a delayed marker by utilizing the time-base waveform itself. *Ibid.* 1197/2 Two or more complete pick offs of the type described can be operated from a single sawtooth without mutual interaction, and for this reason the circuit is called the 'Multiar'. **1953** *Electronic Engin.* XXV. 464 The discriminators consist of ten multiars, each with an individual H.T. supply. **1966** Millman & Taub *Pulse, Digital & Switching Waveforms* xvi. 625 The term multiar is not useful for positive-going signals because multiple oscillations are then obtained. However, a multiar constructed with a p-n-p transistor can be used if the input is a positive-going signal.

mu·lti-colouredness. [f. Multi-coloured *a.* + -ness.] The condition or quality of being multi-coloured.

1953 G. E. M. Anscombe tr. *Wittgenstein's Philos. Investigations* § 47 Multi-colouredness is one kind of complexity.

multicultural (mʊltikʌ·ltiŭrăl), *a.* [f. Multi- 1 + Cultural *a.* 2.] Of or pertaining to a society consisting of varied cultural groups.

1941 *N.Y. Herald-Tribune Books* 27 July 3 A fervent sermon against nationalism, national prejudice and behavior in favor of a 'multicultural' way of life. **1959** *Times* 18 June (Suppl.) p. iv/4 This multi-cultural, multi-lingual society [of Montreal]. **1965** *Economist* 21 Jan. 273/2 For the provision of troops and police [to Rhodesia], consideration should be given to states which are multi-cultural (e.g., Canada) or multiracial (e.g., Trinidad, Jamaica, and New Zealand). **1970** P. K. C. Millins (*title*) Education in a multicultural society. **1973** *Daily Tel.* 20 July 17 Although Britain has a multi-cultural society, where are the black faces among television announcers, newscasters and sports commentators? **1979** *Globe & Mail* (Toronto) 12 May 1/1 Trustee Daniel Leckie, chairman of the work group, puts it this way: 'Morally, legally and pragmatically we are bound to adopt a thoroughly multicultural approach toward public education in Toronto.'

Hence **multicu·lturalism.**

1965 *Prelim. Rep. R. Comm. Bilingualism & Biculturalism* (Canada) iii. 46 The answer they often gave was 'multiculturalism', or, more elaborately, 'the Canadian mosaic'. **1971** *Daily Colonist* (Victoria, B.C.) 9 Oct. 5/2 A policy of multiculturalism within a bilingual framework commends itself to the government as the most suitable means of assuring the cultural freedom of Canadians. **1972** *Times* 12 Dec. 17/6 The country [sc. Canada] faces complex and unresolved problems of..bilingualism and multiculturalism. **1973** *Stornoway Gaz.* 30 June 9/4 A Gaelic Society that has outgrown its original meeting place in just a few short years and enabled the Gaels to meet the new and promising challenge of Multiculturalism.

multidialectal (mʊltidaiălĕ·ktăl), *a.* [f. Multi- + Dialectal *a.*] Proficient in speaking or comprehending more than two dialects. So **multidiale·ctalism.**

1964 *English Studies* XLV. 27 The individual will acquire either a new dialect, or a new accent, or both. [illegible] **1971** *Language* XLVII. 294 They show..that command of heterogeneous structure is not a matter of multidialectalism or performance, but a necessary part of unilingual linguistic competence. **1975** *Amer. Speech 1972* XLVII. 290 Kurath accepts multidialectalism as a linguistic fact.

multidisciplinary (mʊltidisipli·nări), *a.* [f. Multi- + Disciplinary *a.*] Combining many academic approaches, fields, or methods.

1949 [see *family-oriented adj.]. **1967** *Times* 13 June 2/4 Senior member of a multi-disciplinary unit. **1965** *Math. in Biol. & Med.* (Med. Res. Council) p. vii, This is no new problem for a multidisciplinary subject like medical research. **1972** *Physics Bull.* Feb. 81/1 Interplay between different branches..follows the wider multidisciplinary trend between branches of science. **1974** *Nature* 10 May 195/1 This is an outstanding, balanced and humane multidisciplinary book which is much more than an introduction to psychology.

Also **multidiscipli·ned** *a.*

1960 *Brit. Jrnl. Delinquency* I. 151 For the purposes of this Summer School it was agreed that groups should be 'multi-disciplined'. **1963** A. K. C. Ottaway *Educ. & Society* viii. 144 This kind of collaboration has been called the 'multi-disciplined approach'. **1972** *Accountant* 6 Apr. 441/2 The Chartered Institutes should leave the provision of higher qualifications to other multi-disciplined specialist bodies.

multiflora. Substitute for def.: In full, *multiflora rose.* A rose belonging to the species *Rosa multiflora*, which is native to Japan and bears clusters of white or pink flowers, or one of the varieties developed from it; also used *attrib.* to designate a plant bearing several flowers on one stem. (Add earlier and later examples.)

1829 *Western Monthly Rev.* III. 57 It is literally embowered in vines of the multiflora rose. **1839** 'Pencil' *White Sulphur Papers* 82 A small arbor is covered with the multiflora rose and honey-suckle. **1860** *Harper's Mag.* Jan. 189/1 A multiflora rose, tangled with honeysuckle, clambered up the squat chimney. **1913** H. H. Thomas *Rose Bk.* iii. 16 I am making a somewhat apologetic introduction of the multiflora group. **1954** *Times Educ. Suppl.* 12 Nov. p. iv/1 Mermaid, one of the best of modern roses in leaf, flower, and length of flowering season when it is grafted on the *multiflora* stock, deserves to head the analysis of climbing and rambling roses. **1955** C. C. Hurst in G. S. Thomas *Old Shrub Roses* ix. 91 Guillot..planted some of these Japanese Multifloras in his nursery. **1971** N. Young *Complete Rosarian* xii. 217 The main objection to the use of the multiflora stock is the fact that it is less prone to throw up suckers than any of the others. **1979** *Daily Tel.* 14 Oct. 8/1 If you are out for a real display in the ground, you might think of using the new multiflora tulips which have only lately turned up from Holland.

multiform, *a.* and *sb.* Add: **A.** *adj.* (Later examples.)

1903 H. James *Ambassadors* xvi. 146 The question of where, among packed accumulations, so multiform a wedge would be able to enter. **1975** *Times* 21 July 7/5 It includes a model, drawings, and photographs of the multiform playhouse which the Questors Theatre at Ealing hopes to build and which could be adapted to five different types of stage.

B. *sb.* (Later example.)

1913 D. H. Lawrence *Love Poems* 37 You who are twisted in grief like crumpled beech-leaves,..who fall to earth At last like a bean-pod: what are you, oh multiform?

multifu·nctional, *a.* Also **multi-functional.** [Multi- 1 b.] Having or fulfilling many functions.

1941 *Industrial & Engin. Chem.* XXXIII. 351/1 The effectiveness of the multifunctional addition agents is improved by the introduction of metal substituents, particularly in respect to antioxidant value. **1961** *Thorpe's Dict. Appl. Chem.* (ed. 4) VII. 591/2 Multifunctional Additives.—Such compounds combine the properties of power-point depressants, viscosity-index improvers and detergents. **1957** *Essays in Crit.* VII. 21 Shakespeare utilises his multi-functional poetry. **1959** Benn & Peters *Social Princ. & Democratic State* xii. 255 The family was once a multi-functional organization within which men and women found their work, amusements, and religion under patriarchal government. **1972** *Nature* 28 Jan. 187/2 Dr Walter Marshall, director of the Atomic Energy Research Establishment at Harwell, described his establishment as a 'multi-functional' laboratory.

So **multifunction** *attrib.* [Multi- 3].

1967 *Electronics* 6 Mar. 67/2 The company already has a contract for the beam-forming subsystem of the Nike-X multifunction array radar.

Multigraph (mʊ·ltigrɑf). Also **multigraph.** [f. Multi- + -graph.] The proprietary name of a small printing machine which uses specially cast type fitted in to grooves on a rotating cylinder. Also *attrib.* So **mu·ltigraphed** *ppl. a.*, printed by a machine of this type.

1905 *Typewriter Trade Jrnl.* Jan. 17/1 Those who were not able to attend the recent show at Madison Square Garden, missed the opportunity of seeing the 'Multigraph', a machine for producing multiple copies of typewritten work. **1908** *Busy Man's Mag.* May 154/1 The multigraph operated by an office boy or girl really typewrites letters individually. **1921** *Glasgow Herald* 12 Apr. 5 The Gammeter Multigraph, shown by the International Multigraph Co., is an office machine capable of turning out actual printing at over 4000 impressions per hour. **1925** *Daily Tel.* 13 May 20/5 (Advt.), Experienced multigraph operator and stationery buyer required by City firm. **1936** *Brit. Birds* XXX. 32 Explanatory text..is supplied in a multigraphed brochure. **1967** Karch & Buber *Offset Processes* 546 Multigraph—A duplicator which prints from specially cast type.

multigravida (mʊltigræ·vidă). *Obstetrics.*

Pl. **-idas, -idæ.** [f. Multi- after Primigravida.] A pregnant woman who has had at least one previous pregnancy (or at least two) (formerly less specific: see quots. 1890, 1900).

1890 Billings *Med. Dict.* II. 175/1 *Multigravida*, a woman who has been pregnant more than once. **1900** Dorland *Med. Dict.* 399/2 *Multigravida*, a woman who has often been pregnant. **1938** A. L. Mudaliar *Clin. Obstetr.* ii. v. 30 Marked changes take place in the breasts consequent upon pregnancy, and such changes are more obvious in primigravidæ than in multigravidæ. **1943** W. Shaw *Textbk. Midwifery* viii. 146 In normal cases the average duration of the first stage of labour in a primigravida is between twelve and fifteen hours, whereas in a multigravida the duration is usually very much shorter. **1958** *Observer* 3 Aug. 10/5 Nice young upper-middle-class multigravidæ, surrounded by children. **1959** *Amer. Jrnl. Obstetr. & Gynecol.* LXXVII. 104 In multigravidas there is practically no distress caused by difficult labor or head pressure in the second stage (although both are seen where the previous pregnancies have been only abortions). **1972** E. C. Hughes *Obstetr.-Gynecol. Terminol.* 331 A multigravida is a woman who is pregnant and who has been pregnant one or more times previously.

multihull (mʊ·ltihʊl). [f. Multi- 2 + Hull *sb.*²] A boat with more than one hull; freq. *attrib.* Also **mu·ltihulled** *a.*

1956 *Amat. Yacht Res. Soc. Publ.* No. 10. 4 In spite of the fact that multihulled craft have received such little attention..there is enough interest on the part of amateurs to bring about a high degree of development. **1960** R. B. Harris *Mod. Sailing Catamarans* 9 By derivation it [sc. catamaran] could include any multihulled craft. [illegible] this book is concerned only with the twin-hulled sailing craft. [illegible]

multilateral, *a.* Add: **3.** *spec.* Pertaining to or concerning three or more countries, esp. of the trade and financial agreements made by them, or of the control of (part of) their armed forces by a supranational authority.

[illegible] **1963** [see *multinational *a.* and *sb.*].

4. *Educ.* Of a school providing for two or more types of secondary education; also *ellipt.* as *sb.*, a school of this kind. Cf. *comprehensive *a.* 1 d.

1938 *Rep. Consultative Comm. Secondary Educ.* (H.M.S.O.) p. xix, We use the term 'multilateral' to denote a school which [illegible] **1947** [see *comprehensive *a.* 1 d]. [illegible]

Hence **multila·teralism,** the quality of being multilateral (sense *3); **multila·teralist** *a.* and *sb.*, (one) advocating multilateral disarmament; **multila·teralize** *v.*, to embrace in an agreement amongst many parties; so **multilateraliza·tion,** the integration of armed forces under supranational authority; **multilaterally** *adv.*, also, amongst three or more parties.

1928 *Glasgow Herald* 14 June 10/6 M. Briand insisted [illegible] part of the plan to progress towards world economic multilateralism. **1957** *Essays & Stud.* New Ser. X. 7 No one would willingly admit to our vocabulary so unprepossessing and nebulous a word as *multilateralization.* **1960** *News Chron.* 7 Oct. 6/3 How ought we 'multilateralist' Labour MPs to vote on defence? **1960** *Guardian* 11 Oct. 11/1, I am resolutely opposed to urging any member of the Labour party to resign because he is a unilateralist or a multilateralist. *Ibid.* 22 Oct. 1/1 Mr Harold Wilson..declared that the issue was not multilateralism versus unilateralism, but unity versus civil war. **1964** *Listener* 30 July 169/2 The British Government has recently suggested the mixed manning, the multilateralization, of the so-called tactical nuclear weapons now in western Europe. **1965** *Economist* 17 July 207 The Germans still don't want to non-proliferate until they've been multilateralised. **1972** *Sci. Amer.* Apr. 17/2 Government representatives from East and West met.., bilaterally and multilaterally, to discuss increased trade. **1973** *Current Hist.* May 202/1 Yugoslavia supports the convening of a European Security Conference, which would multilateralize the European territorial status quo.

multilayer (mʊ·ltilēˑəɹ), *a.* and *sb.* Also **multi-layer.** [f. Multi- 3, 2 + Layer *sb.*] **A.** *adj.* Composed of or taking place in many or several layers.

1923 *Daily Mail* 28 Apr. 5 Modern amateur practice has run violently in the direction of those compact multilayer coils called 'basket', 'pancake', or 'honeycomb' coils. **1932** J. W. McBain *Sorption of Gases & Vapours by Solids* x. 325 (*heading*) Monomolecular, multilayer and classical thick films. **1960** *Farmer & Stockbreeder* 29 Mar. 77/1 The front-mounted one..can be used for multi-layer stacking in the barn. **1966** R. J. Ross *Television Film Engin.* xii. 452 The multilayer Kodachrome film was first developed in a normal negative developer to produce silver images.

B. *sb.* A structure or film composed of many or several layers, *spec.* of more than one monolayer.

1932 J. W. McBain *Sorption of Gases & Vapours by Solids* x. 326 No such excitement would be aroused by a demonstration of the actual occurrence of multilayers deriving from the second hypothesis. **1963** R. R. A. Higham *Handbk. Papermaking* viii. 214 Laminating machines are used to apply an adhesive to single or multi-layers of paper which are then bonded together by heat to form one ply. **1966** D. G. Brandon *Mod. Techniques Metallogr.* i. 34 Films may be prepared from diele[c]tric multilayers (usually alternating zinc sulphide with magnesium fluoride). **1975** *Nature* 24 July 297/1 Its bottom is formed of multilayers of salt deposits separated by films of sand and gypsum.

So **mu·ltilayered** *a.* [Multi- 1 b].

1935 *Anat. Rec.* LXI. 391 Multilayered follicles were increasing in number and as many as ten or twelve granulosa layers surrounded the ova. **1959** *Biol. Abstr.* XXXIV. 1163/2 (*heading*) The multilayered tube wall as [a] model for the aorta. **1974** *Tetrahedron Lett.* 1573 We have previously described the synthesis of multilayered [2.2] metacyclophanes by way of the Stevens rearrangement.

mu·ltilevel, *a.* Also **multi-level.** [Multi- 3.] Having, involving, or operating on several levels (in any sense). Also *ellipt.* as *sb.*, a multilevel set of apartments.

1952 *Los Angeles Examiner* 19 Mar. iii. 1 (*caption*) Multi-level apartment hotel is planned as the center for the development. **1959** G. Troup *Masers* iv. 50 One multilevel excitation method is that known as 'optical pumping'. **1959** *Sunday Times* 12 Apr. 21/6 America welcomed the one-storey 'ranch' house and the 'split level'. Today, the development builders, who can find new names for old things, are trumpeting the 'multi level'. **1960** *Guardian* 17 Mar. 9/5 Multi-level traffic to separate vehicles from pedestrians. **1961** L. Mumford *City in Hist.* viii. 208 The new marketplaces at Coventry and Harlow, with their upper tiers of shops and offices, are..only a recovery of the admirable Roman multi-level plan. **1964** E. A. Nida *Toward Sci. Transl.* v. 80 When such vertical divisions are multilevel, that is to say, when the cuts extend through several layers, they mark very important distinctions in the language. **1969** *Jane's Freight Containers 1968–69* 121/3 Unloading automobiles from multi-level cars. **1972** *Village Voice* (N.Y.) 1 June 82/3 (Advt.), Unusual multi-levels. Charming bedroom apts. **1973** *Times* 5 Oct. (Safety Suppl.) p. ii, Junctions are multilevel and local roads..are carried over or under their dual carriageways.

So **multi-le·velled** *a.* [Multi- 1 b].

1948 *Hudson Rev.* I. i. 117 But precisely because it is a poem—compact, symbolic, multileveled, all details woven elaborately into the meaning pattern—it cannot be totally accounted for as yet. **1962** M. McLuhan *Gutenberg Galaxy* 103 When only the eye is engaged, the multi-levelled gestures and resonances of Senecan oral action are quite impertinent. **1964** E. Bach *Introd. Transformational Gram.* iv. 58 A multileveled theory such as is necessary for describing a natural language. **1968** A. Lomax *Folk Song Style & Culture* iv. 95 The overall impact of the African style is multileveled, multiparted, highly integrated, multi-textured, gregarious, and playful-voiced.

multilineal (mʊltili·niăl), *a.* [f. Multi- + Lineal *a.*] Having many lines; *spec.* denoting a kinship system which includes relationships derived from parents, grandparents, etc., of both father and mother. Hence **multilinea·lity,** the fact of multilineal kinship.

a **1800, 1882** [in Dict. s.v. Multi- 1 a]. **1943** T. Parsons in *Amer. Anthropologist* XLV. 26 Since the same principle of lack of distinction by sex of intervening relative applies to still higher ascendant generations—the four great- and eight great-great-grandfathers—it is perhaps more accurate to speak of a 'multilineal' than a 'bilateral' system. **1956** R. Firth *Two Stud. Kinship in London* 16 The 'multilineal' American family. *Ibid.* 17 It is in this way that the conversion of bilateralism into multilineality occurs. **1965** G. P. Murdock *Culture & Society* xiii. 179 The 'multilineal' of Parsons..is inappropriate since.. many such systems, including the American, are not in fact 'lineal' in any sense.

multilingual (mʊltili·ŋgwăl), *a.* and *sb.* [f. Multi- + L. *lingu(a* tongue, language + -al.] **A.** *adj.* Speaking, characterized by, or written in many languages.

1838, 1880 [in Dict. s.v. Multi- 1 b]. **1958** *Listener* 2 Jan. 5/2 Indonesia..is multi-racial and multilingual. **1960** E. Delavenay *Introd. Machine Transl.* vi. 96 This Cambridge research leads in the direction of a new kind of bilingual or multilingual dictionaries. **1968** *Amer. Speech* XLIII. 127 How is the language variance in multilingual communities best explained?

B. *sb.* One who speaks many languages. Also **multili·nguist.**

1923 W. J. Locke *Lengthened Shadow* 187 A white-haired white-moustached multilinguist by the name of Soussouki. **1956** *Essays in Crit.* VI. 67 Fr. Jarrett-Kerr is dealing with creative artists who worked in Spanish, Italian, Russian and French—and he confesses himself no multilinguist. **1959** J. C. Catford in Quirk & Smith *Teaching of English* vi. 164 Strictly speaking, the alternative use of more than two languages is 'multilingualism' and the persons involved are 'multilingual' (adj.) or 'multilinguals' (n.). **1964** E. Palmer tr. *Martinet's Elem. Gen. Ling.* v. 160 These are the bilingual speakers, or if they possess more than two codes, the multilinguals, whatever degree of perfection they may achieve in handling each of the said languages. **1968** *Amer. Speech* XLIII. 127 What theoretical model most appropriately describes the individual multilingual's speech? **1974** *Daily Tel.* 10 July 18 An Oxford graduate and multilinguist, Saunders worked at the Treasury and the Coal Board.

Hence **multili·ngually** *adv.*; **multili·ngualism,** the ability to speak many languages; the use of many languages.

1940 *Q. Jrnl. Speech* Oct. 394 Problems confronting the speech therapist in the Union are legion, resulting from racial inequalities, multi-lingualism, [etc.]. **1953** U. Weinreich in *Publ. Ling. Circle N.Y.* I. 1 Unless otherwise specified, all remarks about bilingualism apply as well to multilingualism, the practice of using alternately three or more languages. **1958** *Times Rev. Industry* June 76/1 Working principles are explained multi-lingually through loudspeakers. **1968** J. A. Fishman *Readings Sociol. of Lang.* 12 Multilingualism has long been a topic recognized by sociologists, linguists, anthropologists, and others as shedding light on many aspects of language learning. **1972** *Sci. Amer.* Sept. 78/1 Bilingualism or multilingualism ..cannot be rigidly separated from interdialectal fluctuations. **1972** *Physics Bull.* Dec. 136/2 It [sc. a European *Physics Bulletin*] would have to be printed multilingually, and would be an extremely expensive exercise. **1973** *Archivum Linguisticum* IV. 55 Another way of putting this is to refer to one of the most common questions faced by those wishing to make sense of multilingualism—how many languages are there here, and how many dialects?

Multilith (mʊ·ltiliþ). Also **multilith.** [f. Multi- + Lith(ograph *sb.*] The proprietary name of a small, offset-lithographic, printing machine. Also *attrib.* So **mu·ltilithed** *ppl. a.*, printed by a machine of this kind.

1933 *Profit Making Up-to-Date* (Addressograph-Multigraph Corp.) 81 Multilith greatly reduces the necessity of type composition. **1936** *Business Office Training Course* (Addressograph-Multigraph Corp.) xi. 97 The latest addition to the Multigraph line is the 'Multilith'. This term applies to the simplified process of 'offset lithography' developed by the Multigraph Company. **1937** *Geogr. Jrnl.* LXXXIX. 403 Multilith photographer. **1951** *Archivum Linguisticum* III. 106 From 1942 dates the journal *Studies in Linguistics*..published..at first in mimeographed and later in multilithed form. **1965** *Times Lit. Suppl.* 14 Jan. 25/4 A new multilithed newsletter called *Cultural Events in Africa* is appearing monthly. **1968** *Globe & Mail* (Toronto) 13 Jan. 48/2 (Advt.), Pressman..—knowledge of multilith an advantage. **1974** *Oxf. Univ. Gaz.* 19 Dec. 373/1 (Advt.), Academic, literary, and other typing... Multilith duplicating.

multi-me·dia, *a.* [f. Multi- + *media.] Designating or pertaining to a form of artistic, educational, or commercial communication in which more than one medium is used. Hence as *sb.*

1962 *Times* 26 Feb. (Canada Suppl.) p. xviii/4 The first prong is a multi-media publicity campaign to encourage school children..to obtain adequate educational qualifications. **1962** *Listener* 5 Apr. 603/3 Both Futurists and Dadaists had a keen interest in multi-media art, in breaking down the technical and formal self-sufficiency of the work of art. **1968** *Sun* (Baltimore) 4 July A. 16/3 The notes of one conference we attended a few weeks ago.. show that speakers were using such terms..as..multimedia and multi-mode curriculum. **1970** *Times* 28 Apr. 7 Late night performances of the Military Tattoo and of the multimedia rock musical *Stomp.* **1970** D. Bergen in *Americana Ann.* 420 Multimedia information centers with print and audiovisual materials. **1971** *Black Scholar* Jan. 20/2 As originator of the practice of reading poetry to jazz, he not only stitched backwards and forward in his lineage and idiom, but wrought a new force in the now obscenely exaggerated concept of multi-media. **1974** *News & Press* (Darlington, S. Carolina) 25 Apr. 9/7 Hooser creates a multi-media effect that draws the viewer into the picture.

multimer (mʊ·ltimɑɹ). *Chem.* [f. Multi- + *-mer.] An aggregate of molecules held together by relatively weak bonds, such as hydrogen bonds.

1959 M. Davies in D. Hadži *Hydrogen Bonding* 560 The use of the term 'polymer' for the associated molecules formed from stable monomers by hydrogen bonding, etc., seems objectionable, and it is suggested that the equivalent, self-explanatory term 'multimer' be used in these circumstances. **1967** *Arch. Biochem. & Biophysics* CXX. 158 (*heading*) Studies on protein multimers. **1972** H.-G. Elias in M. B. Huglin *Light Scattering from Polymer Solutions* ix. 399 A unimer of molecular weight M_1 will multimerize to a multimer with a degree of multimerization N̄.

Hence **multime·ric** *a.*; also **mu:ltimeriza·tion,** the formation of multimers; **mu·ltimerize** (or *multi·merize*) *v. intr.*, to associate to form a multimer or multimers; **mu·ltimerized, mu·ltimerizing** *ppl. adjs.*

1971 *Arch. Biochem. & Biophysics* CXLII. 329/1 The experimental basis for supposing that high pressure disaggregates multimeric proteins is already well established. **1972** H.-G. Elias in M. B. Huglin *Light Scattering from Polymer Solutions* ix. 451 It is believed that multimerizations occur preferentially in thermodynamically poor solvents as expressed for polyvinyl chloride solutions. *Ibid.* 398 For the multimerized material, we shall use the following nomenclature. *Ibid.* 400 In light scattering studies of multimerizing solutes, many different concentration dependences of the reciprocal apparent weight average molecular weights may be observed. *Ibid.* 441 Cellulose nitrates multimerize in ethanol at temperatures above 20° C. **1973** *Microbios* VII. 53 (*heading*) The multimeric nature of NADPH-nitrate reductase from *Aspergillus nidulans.* **1973** *Nature* 24 Aug. 528/3 Chapter 9 deals with the problems of aggregation or multimerization.

multimeter (mʊ·ltimītəɹ). *Electr.* [f. Multi- 2 + Meter *sb.*³] An instrument designed to measure voltage, current, and usu. resistance, often over several different ranges of value.

The word *Multimeter* is registered as a proprietary name in the U.S.

1910 *Hawkins' Electr. Dict.* 292/1 *Multi-meter*, a 'universal' electronic measuring instrument designed to serve the purpose of a voltmeter, ammeter, ohmmeter, ground detector and Wheatstone bridge. **1926** *Jrnl. Sci. Instruments* III. 346 The vacuum-tube multimeter has three ranges as a voltmeter, and six ranges as an ammeter for each range of the voltmeter. **1929** *Official Gaz.* (U.S. Patent Office) 5 Mar. 23/2 Rawson Electrical Instrument Company, Cambridge, Mass... *Multimeter* for Electric Meters. Claims use since latter part of 1919. **1967** *Electronics* 6 Mar. 117/1 The Instrumentation division of Fairchild Camera & Instrument Corp. has brought the price of a multimeter down to $249. **1971** *Nature* 12 Feb. 482/1 The 1970 resistance measurements were made with a digital multimeter (Fluke '8100A') to the same accuracy as the 1969 measurements. **1975** *Physics Bull.* Jan. 28/2 Sinclair Radionics new DM2 digital multimeter provides the facilities of the conventional multimeter: ac and dc voltage, ac and dc current ranges and resistance measurements, in a portable instrument.

multimodal (mʊltimōᵘ·dăl), *a.* Also (with hyphen) **multi-modal.** [f. Multi- 1 a + Mod(e *sb.* + -al.] **a.** Of a frequency curve or distribution: having several modes or maxima (*mode *sb.* 7 c). Of a property: occurring among different individuals in accordance with such a distribution.

1902 *Biometrika* I. 305 A frequency distribution with more than one true mode is multi-modal. **1932** J. S. Huxley *Probl. Relative Growth* vii. i. 210 The body-length for males is irregular, multimodal and skew. **1959** Schuell & Jenkins in Saporta & Bastian *Psycholinguistics* (1961) 436/1 The distribution of scores on each test was plotted for the 100-patient sample... Only a few tests yielded irregular multi-modal distributions. **1962** *Lancet* 26 May 1090/1 In both cases distributions in two of the six age-groups are irregular..and of a form which might readily be interpreted as multimodal or bimodal.

b. Characterized by several different modes of occurrence or activity.

1928 D. B. Leary *Mod. Psychol.* 128 The so-called multi-modal theory..assumes that there are several distinct types of intelligent conduct. **1968** W. A. Stewart in J. A. Fishman *Readings Sociol. of Lang.* (1968) 534 When a language has come to be used in more than one country and has..developed multimodal standardization, the form of standardization prevalent in any one country may [etc.]. **1969** *Jane's Freight Containers 1968–69* 32/3 As shippers adjust to the advantages of an integrated, multimodal transportation system.

Hence **multimo·dalism** (*rare*), **mu:ltimoda·lity,** the property or quality of being multimodal.

1902 *Biometrika* I. 305 Much of the multimodalism interpreted in the case of flowers as polymorphism is due either to misinterpretation.., i.e. is not true multimodalism..or..is due to some heterogeneity..introduced by the gatherer. **1932** J. S. Huxley *Probl. Relative Growth* ii. 78 There is a slight tendency to multimodality.. in the male body-length frequency curve. **1972** *Comprehensive Psychiatry* XIII. 391 (*heading*) Impact of a multimodality treatment program for heroin addicts. **1973** D. Nelkin *Methadone Maintenance* i. 24 'Multimodality', an expression coined by Dr. Jaffe, which refers to a concept of rehabilitation that coordinates within a single administrative structure all varieties of approaches to addiction.

multinational (mʊltinæ·ʃənăl), *a.* and *sb.* [f. Multi- + National *a.* and *sb.*] **A.** *adj.* **a.** Comprising or pertaining to many nationalities or ethnic groups. **b.** Possessing branches, factories, offices, etc., in many countries. **B.** *sb.* A multinational company. So **multina·tionally** *adv.*; **multina·tionalism,** the realm of multinational companies.

1926 *Time* 17 May 14 A majority of the multi-national citizens of what is now Hungary. **1940** *Mind* XLIX. 117 Their inequality makes for an intrinsic instability in their relations which causes one to regret the passing of the multi-national Empires, like Austria and Russia, with their stabilising effect. **1957** *Times Lit. Suppl.* 25 Oct. 635/2 The volume contains in fact five separate essays—on the beginnings of the American nation..on the problems of a multi-national nation [etc.]. **1961** *New Scientist* 23 Mar. 733/1 There will be an important multi-national programme of space research, starting in 1962. **1962** *Economist* 15 Dec. 1108/3 Assignments to multinational organisations should be up-graded. **1963** *Guardian* 19 Mar. 11/4 A multinational force would be one composed of Service units already in existence... Multilateral would mean manning.., for example, Polaris surface ships with mixed crews. **1964** *Economist* 17 Oct. 271/2 Very few such companies..seem..to think 'multi-nationally'. **1968** *Listener* 12 Dec. 783/1 The multinational firm transcends national boundaries..through the establishment of subsidiaries in foreign countries. **1971** C. Tugendhat (*title*) The multinationals. **1971** *Rand Daily Mail* 4 Dec. 12/2 There are plans for extra senates and supra-parliaments and multinational assemblies. **1972** *Listener* 18 May 652/2 Yours is a multinational group. What are the prospects for multinationalism? **1973** *Daily Tel.* (Colour Suppl.) 22 June 23/2 The five multi-national pilots—as her name implies the Europa is a truly European airship. **1975** *Bookseller* 18 Jan. 155/1 (Advt.), To manipulate the takeover of the Holmes Motor Corporation by an American multi-national and to destroy a rival spying network.

multi-occupa·tion. [Multi- 2.] Occupation of a house by more than one family, with shared kitchen or sanitary facilities. Hence **multi-o·ccupy** *v. trans.*, to tenant (a house) with more than one family with such shared facilities; to place (tenants) in such a house; **multi-o·ccupied** *ppl. a.* Cf. *multiple occupancy, occupation* (s.v. *multiple *a.* 2 c).

1963 *Guardian* 30 Mar. 2/2 A graduated percentage addition to the rateable value of houses in multi-occupation. **1965** *Ibid.* 3 Sept. 20/8 For every house multi-occupied by Pakistanis..there are many more multi-occupied by Irish. **1967** *Economist* 14 Oct. 177/1 Improvement grants are not available for multi-occupied houses. **1970** *Guardian* 31 July 9/1 The borough council is..planning to multi-occupy its emergency homeless in..a tenement building built in 1878. **1972** *Ibid.* 26 June 7/3 If families are evicted from multi-occupation property, they tend to move to the next street. **1973** C. Mullard *Black Brit.* xiii. 164 There is still no national code in existence. If there were, multi-occupation and other housing problems would be non-issues.

multip, colloq. abbrev. Multipara (in Dict. and Suppl.).

1948 Mencken *Amer. Lang.* Suppl. II. xi. 756 *Primip*, a mother having her first child... At subsequent deliveries she is a *multip*,..or a *para-two, three*, etc. **1967** *Midwives Chron.* Apr. 115/1 The risks to the grande multip cover pregnancy and labour and puerperium.

multipacket (mʊ·ltipækĕt). [f. Multi- + Packet *sb.*] A cargo boat built in two parts, having the propulsion unit and crew quarters at one end and one or more cargo units at the other. Also *attrib.*

1965 *Daily Express* 11 Nov. 17/2 A design for a cargo ship that splits in half has passed Ministry of Transport tests... The design, called the Multi-packet, has been developed by a Liverpool firm. **1967** *New Scientist* 28 Dec. 743/3 The multipacket vessel can enter harbour, deposit a fully laden cargo unit at the quayside, and depart with a different unit—all within a matter of hours. *Ibid.* 744/1 The greatest technical difficulty to be overcome in the new design is that of connecting and disconnecting the multipacket units quickly and easily.

multipara. Add: Pl. **-paras, -paræ.** Freq. used (in contrast to *primipara*) to include pregnant women with a single previous delivery, the forthcoming birth being anticipated in the enumeration. (Earlier and later examples.)

1860 [see Primiparous *a.*]. **1890** tr. *F. Winckel's Textbk. Obstetr.* iii. v. 112 It is by no means easy to distinguish a multipara from a primipara, especially when the previous pregnancy took place many years before. **1938** A. L. Mudaliar *Clin. Obstetr.* ii. vi. 35 In multiparæ the changes in the breasts are not of much diagnostic value, because pregnancy may take place in a lactating woman. **1958** J. R. Willson et al. *Obstetr. & Gynecol.* xxviii. 321 In primigravidas the cervix is usually well effaced before the contractions of true labor begin, but preparation of the cervix in multiparas differs slightly. **1967** J. B. Lawson in Lawson & Stewart *Obstetr. & Gynaecol. in Tropics* i. 5 None of the risks of high parity..are unique to grande multiparae (para 5 and above) but are commoner in this group. **1972** E. C. Hughes *Obstetr.-Gynecol. Terminol.* vii. 333 A multipara is a woman who has given birth two or more times to an infant or infants, alive or dead, weighing 500 gm or more.

multi-pa·rty, *a.* *Polit.* [f. Multi- + Party *sb.* 6.] Comprising several parties or members of several parties; of an electoral or political system which results in the formation of three or more influential parties.

1909 *Englishwoman* Apr. 325 The suffrage societies, nominally non-party and really multi-party in their composition. **1935** R. Bassett *Essent. Parl. Democracy* ii. 43 Under the multi-party conditions prevalent in most continental parliamentary regimes, it is almost certain that no single party will be able to secure a working majority. **1954** B. & R. North tr. *Duverger's Pol. Parties* ii. i. 229 The typology of the multi-party system is difficult to establish. **1956** J. Klein *Study of Groups* 161 A democratic country with a multi-party system. **1958** A. Leiserson *Parties & Politics* ix. 358 The multiparty system operates quite differently in a three-party situation..from the way it does when there are five or six fairly identifiable party groupings. **1964** *Ann. Reg. 1963* 204 Dr Raúl Leoni..stated that his party would try to form a multi-party coalition. **1970** B. M. Barry *Sociologists, Economists & Democracy* v. 125 In a multi-party system, the parties are more precise but it is more difficult to tell what one is doing by voting for one.

Hence **multi-pa·rtism, multi-pa·rtyism,** a political system in which the major interest groups of the electorate are reflected in three or more influential parties. Also *multi-partist* adj.

1946 F. A. Ogg *European Govt. & Politics* (rev. ed.) xlv. 901 The error would lie in failure to remember..the total absence of anything approaching Western multi-partyism. **1954** B. & R. North tr. *Duverger's Pol. Parties* ii. i. 229 In this sense multi-partism is fairly characteristic of Western Europe, Great Britain excepted but Ireland included. *Ibid.* iii. 393 Great Britain and the Dominions, under a two-party system, are profoundly dissimilar from Continental countries under a multi-partist system. **1962** S. E. Finer *Man on Horseback* xii. 236 In Indonesia, where the sole unifying force in government is the personality of Sukarno, his recent activities provide a grotesque commentary on multi-partyism in a new and divided state. **1967** J. J. Linz in Lipset & Rokkan *Party Syst. & Voter Alignments* 259 (*heading*) The transition from extreme multipartism to polarized conflict and civil war. **1971** D. W. Rae *Pol. Consequences Electoral Laws* (rev. ed.) iii. 53 An extreme case of 'multi-partism', with, say, ten parties, each polling about one tenth of the total vote.

mu·ltiphase, *a.* [Multi- 3.] Having or producing two or more phases; in *Electr.* = Polyphase *a.*

1890, etc. [in Dict. s.v. Multi- 3]. **1916** W. H. N. James *Alternating Currents* v. 87 The relative merits of single and multiphase systems depend upon the purpose to which the power is to be applied. **1946** *Nature* 31 Aug. 307/2 The electrolytic polishing of multi-phase metals is usually more difficult than that of single-phase metals owing to differing properties..of the different phases. **1966** D. G. Brandon *Mod. Techniques Metallogr.* ii. 82 Absorption contrast arises from variations in chemical composition and hence provides information about the composition and distribution of phases in a multiphase alloy.

So **mu·ltiphased** *a.*, occurring in several stages.

1970 *Sci. Amer.* Feb. 103/3 The effect of laser irradiation on the cells varied with the stage of mitosis, the multiphased process of cell division.

multiphasic (mʊltifēⁱ·zik), *a.* [f. Multi- 1 a + Phas(e + -ic.] Applied to tests or investigations designed to reveal various phases or aspects of personality, health, etc.

1940 *Psychol. Bull.* XXXVII. 425 The Multiphasic Personality Schedule is a group of 504 items for administration to adults. **1958** M. Argyle *Relig. Behaviour* viii. 87 Brown and Lowe (1915) tested a large number of students on the Minnesota Multiphasic Personality Inventory (MMPI) and found that a group of Bible students—who would be extreme Protestants—scored high on hysteria. **1968** *Daily Tel.* 22 Nov. 25/4 Multiphasic screening could disclose much potential sickness and make possible early treatment. **1971** *Brit. Med. Bull.* XXVII. 6/1 In a multiphasic programme the individual will be subjected to a number of tests within a short space of time.

multiplane (mʊ·ltiplēⁱn), *sb.* and *a.* [f. Multi- 2, 3 + Plane *sb.*³] **A.** *sb.* An aeroplane or glider having several 'planes' or main supporting surfaces placed one above another. Also *attrib.* or as *adj.*

1909 A. Berget *Conquest of Air* ii. ii. 139 Will the aeroplane be a 'monoplane' or 'multiplane'? *Ibid.* 147 The arrangement employed..is 'partitioning'..and applies to multiplane aeroplanes. **1928** *Jrnl. R. Aeronaut. Soc.* XXXII. 145 The method of obtaining the reduction of drag for a multiplane is described. **1939** C. H. L. Needham *Aircraft Design* I. vi. 74 When two or more planes are situated one above the other, so as to form biplane, triplane or multiplane arrangements, the low pressure region over the top of the lower wing is closely adjacent to the high pressure area below the upper wing. **1960** C. H. Gibbs-Smith *Aeroplane* 337 Wenham was the first to build..a multiplane aeroplane, a full-size glider with five wings. **1969** K. Munson *Pioneer Aircraft 1903–14* 154/2 In 1904 appeared the first man-carrying multiplane (Phillips I). This had a framework of twenty rigid blade sustainers. **1973** *Country Life* 11 Oct. 1074 In 1903 a multiplane, somewhat similar to the one built in 1893, was constructed and taken to Essex for trials.

B. *adj.* Involving or occupying several distinct planes or layers (not necessarily horizontal); applied *spec.* in *Cinemat.* to a technique employed to give an enhanced impression of perspective.

1909, etc. [see A above]. **1948** E. Lindgren *Art of Film* vii. 127 Objects in different planes will appear to move at different speeds according to their distance from the camera lens..and the relationship between these various movements can..suggest almost a three-dimensional effect; this is the principle behind the multiplane technique now employed in the making of cartoon films. **1955** J. Caunter *How to do Tricks in Amat. Films* 65 To give the effect of maximum perspective to a background scene you will have to divide it into more than one layer of scenery... This is the multiplane principle. **1959** Halas & Manvell *Technique Film Animation* ii. 28 Now the animation studios have their own specialized cameras.. capable of photographing multi-plane images with back lighting. **1961** G. Millerson *Technique Television Production* xix. 362 Multiplane captions. Here the caption is comprised of several transparent or cut-out layers, and by revealing or obscuring these layers, we can add or remove detail to order. **1969** *Daily Tel.* 7 Feb. 21/6 A clever white multiplane set is used for the projection of slides and movie-film, either bathing the stage in dazzling colour or else enthroning unexpected VIPs.

multiple, *a.* and *sb.* Add: **A.** *adj.* **2. b.** *multiple shop, store*: one of several shops of the same kind belonging to the same firm, opened in different localities. Cf. *chain store.

1903 J. Hazelip *Multiple-Shop Accounts* 1 There is considerable difference in the class of business carried on by multiple-shop firms. **1909** *Westm. Gaz.* 10 May 9/4 Some of the multiple shops..have been retailing at 9s. 4d. sugar which has cost them 14s. 7½d. to buy. **1927** *Daily Tel.* 14 Mar. 4/7 The multiple shops..have reduced their prices in accord with the drop in wholesale prices. **1929** E. Gill *Art-Nonsense* 315 Politics and social guidance are left to ..novelists, multiple-store keepers, manufacturers of motor-cars. **1959** *Times Lit. Suppl.* 4 Sept. 504/3 *An Introduction to Music* is another of what might be called the multiple-stores books which aim to provide everything for the music-lover. **1963** N. Marsh *Dead Water* (1964) ii. 46 There's a rash of boarding establishments and a multiple store. **1965** *Modern Law Rev.* XXVIII. 553 The new multiple shops and shopping centres in both the United States and England. **1972** N. Marsh *Tied up in Tinsel* vi. 140 There's no joy down your way: big multiple stores robbery.

c. *multiple occupancy, occupation* = *multi-occupation.

1971 *Guardian* 11 June 12/3 There is a certain amount of substandard accommodation largely created through multiple occupation. **1973** *Times* 26 July 4/6 Probably 100 to 120 families are breaking the regulations applicable to multiple occupancy houses. **1975** *Times* 30 Aug. 13/6 Perhaps Mr Cutler..would like to relax controls on houses in multiple occupation so that even more people can share a bath.

3. a. *multiple allele* or *allelomorph*: any allele which is located at a genetic locus known to have three or more alleles; *multiple factor*: any gene which acts in concert with other, non-allelic, genes to control the expression of a character; *multiple fission*: the division of a cell into more than two daughter-cells.

1912 E. A. Minchin *Introd. Study Protozoa* vii. 120 In most cases, probably, of multiple fission the nucleus contains a centriole, and..the centriole multiplies by fission a number of times without the nucleus as a whole becoming divided. **1913** *Amer. Naturalist* XLVII. 234 (*heading*) The Himalayan rabbit case, with some considerations on multiple allelomorphs. **1915** R. Pearl *Modes Res. Genetics* i. 21 In recent developments of Mendelian theory it has been a common practice to assume the existence of multiple factors as the causal agents of a single character. **1938** A. F. Shull *Heredity* (ed. 3) xi. 103 (*heading*) Multiple alleles. **1945** E. Altenburg *Genetics* v. 83 Members of several gene pairs which act in a cumulative way on a trait are known as multiple factors. **1948** *Nature* 30 Oct. 684/1 (*heading*) Multiple allelomorphs in colour vision. **1971** J. Z. Young *Introd. Study Man* xl. 585 Fourteen different systems of blood antigen genes are known... The most familiar of them are three multiple alleles known as G^A, G^B, and G, the first two being both dominant to G. These genes are often called simply *A*, *B*, and *O.* **1972** *Proc. Nat. Acad. Sci.* LXIX. 2346/1 Organisms dividing by multiple fission can be used to study the initiation and control of cytokinesis, because those unique events essential to cytokinesis are separated from events of the cell cycle that pertain to DNA synthesis, mitosis, or cytoplasmic growth.

d. *multiple myeloma* [tr. G. *multiples myelom* (J. von Rustizky 1873, in *Deutsche Zeitschr. f. Chirurgie* III. 163)], = *myelomatosis; *multiple sclerosis*, a chronic, progressive, demyelinating disease in which sclerosis occurs in patches in the brain and spinal cord, which chiefly affects young adults, and is often manifested initially as mild attacks with varying symptoms followed later by successive remissions (often long-lasting) and relapses, but typically leading to weakness and paresis of the lower limbs, intention tremors in the upper limbs, disturbed sight and speech, emotional changes, and mental deterioration; also called *disseminated sclerosis.*

[Multiple sclerosis: cf. quot. 1877 in Dict.] **1885** J. Ross *Handbk. Dis. Nervous Syst.* xx. 674 (*heading*) Multiple sclerosis of the brain and spinal cord (disseminated or insular sclerosis). **1897** *Trans. Path. Soc.* XLVIII. 169 (*heading*) General lymphadenomatosis of bones, one form of 'multiple myeloma'. **1904** Multiple myeloma [see *myelomatosis]. **1917** *Jrnl. Nervous & Mental Dis.* XLV. 454 In multiple sclerosis we had been in the habit of considering the disease incurable and although there might be a remission for a number of months or years, it could not be looked upon as a cure. **1957** S. L. Robbins *Textbk. Path.* xxxii. 1311/2 Studies on animals with injections of nervous tissue and adjuvants may yet prove to be important in multiple sclerosis, but no worker has yet reproduced a convincing replica of the disease. **1961** *Lancet* 5 Aug. 290/1 On paper electrophoresis, macroglobulinæmic sera yield patterns which are indistinguishable from those found in multiple myeloma. **1966** Multiple myeloma [see *myelomatosis]. **1971** *New Scientist* 7 Jan. 6/1 Multiple sclerosis is the commonest disease of the nervous system in northern Europe and the USA. **1972** C. E. Seiverd *Hematol. for Med. Technologists* (ed. 4) xxxvii. 667 Myeloma means marrow tumor; thus, multiple myeloma may be interpreted as simply meaning many tumors of the bone marrow. Multiple myeloma is also referred to by the following terms; myelome, plasmocytoma, plasma cell myeloma, myelomatosis, and Kahler's disease.

j. *Statistics.* Involving the joint effect on the variable under investigation of two or more other variables.

1903 *Phil. Trans. R. Soc.* A. CC. 3 (*heading*) On the fundamental theorem in multiple correlation. **1938** A. E. Waugh *Elem. Statistical Method* xi. 318 Problems that involve the determination of the relationship between one variable and several other variables acting together are called problems of multiple correlation. *Ibid.* 321 The multiple-regression equation must obviously be altered so that we can account for changes in all the independent variables. **1958** M. Argyle *Relig. Behaviour* iii. 18 An alternative procedure for estimating the relative weight of different variables influencing percentages is multiple regression. **1971** *Brit. Med. Bull.* XXVII. 22/1 A multiple regression study showed that..both water calcium and rainfall made significant contributions to the variance of cardiovascular death-rates between the towns studied. **1972** T. H. & R. J. Wonnacott *Introd. Statistics for Business & Econ.* xiv. 350 Whereas the partial correlations measure the significance of regressors one by one, the multiple correlation *R* measures the significance of all the regressors at once.

k. *multiple use*: = *multi-use *a.*

1969 *Gloss. for Landscape Work* (*B.S.I.*) v. 16 *Multiple use*, the use of land for more than one purpose, resulting in a modification in the methods of maintenance, for example, water catchment areas also used for cattle grazing or for forestry. **1969** *Jane's Freight Containers 1968–69* 505 (*caption*) SALwall..incorporates a multiple-use double wall.

4. b. *multiple-beam*; **multiple-access** = *multi-access* (*multi- 3); **multiple-aspect,** applied to a colour-light railway signal capable of displaying at least three aspects; **multiple-choice,** applied to an educational or psychological test in which the subject is asked to select his answer from several items; **multiple-disc,** applied to a kind of friction clutch consisting of a row of co-axial discs, fixed alternately to the driving and the driven parts, which may be brought in contact to transmit the drive from the one to the other; **multiple-unit,** of, pertaining to, or designating a train having a number of coaches provided with engines all of which can be controlled by a single driver; also as *sb.*, a coach of this kind.

1966 *New Scientist* 27 Oct. 160/3 This valuable experience..has..enabled us to take in our stride the 'multiple-access' problem of how to serve a large number of control consoles apparently simultaneously. **1970** O. Dopping *Computers & Data Processing* ix. 130 A control program for a multiple access system must be able to identify a subscriber who requires service, find out which program he needs, and put him in the queue, if any. **1932** *Proc. Inst. Railway Signal Engin.* i. 57 (*heading*) Railway colour light signalling in relation to manual block and multiple aspect signals. **1963** Kichenside & Williams *Brit. Railway Signalling* iii. 24 In colour-light areas, a multiple-aspect signal cannot display a 'green——clear' indication unless the next signal is showing at least a 'caution' if not a 'clear' aspect also. *Ibid.* 25 In colour-light areas..every multiple-aspect signal serves as a distant, home and starting signal at the same time. **1970** *Railway World* Dec. 524 Multiple-aspect colour-light signalling with continuous track circuiting will be installed. **1945** *Proc. R. Soc.* A. CLXXXIV. 41 A multiple beam interferometric procedure..can be used as a powerful method for revealing the details of the surface topography of approximately flat crystal planes. **1966** D. G. Brandon *Mod. Techniques Metallogr.* i. 25 This may be done most simply by placing a half-silvered plate in contact with the specimen, when either two-beam or multiple-beam interference patterns may be obtained. **1928** Orleans & Sealy *Objective Tests* xiii. 220 Facility in framing optional answers for a multiple-choice question comes with practice. **1959** J. Barzun *House of Intellect* 268 A basic defect of multiple choice tests: that they call for choices but not for reasons for choices. **1967** *Observer* 26 Nov. 1/4 In the American 'multiple-choice' examinations the candidate marks one of a number of possible responses to a question, instead of writing an essay on it. **1906, 1909** Multiple disc [see *disc-clutch* (*disc *sb.* 8)]. **1967** *Jane's Surface Skimmer Systems 1967–68* 124/1 Flange-mounted converter automatically changing over from multiple-disc clutch..to converter, operation, then again operation through multiple-disc clutch when moving on foils. **1902** *Encycl. Brit.* XXVIII. 93/1 The train operated by two or more motor cars under a common secondary control. This..is called the 'multiple unit system'. **1955** *Oxf. Jun. Encycl.* VIII. 366/2 In Britain 'in-built' motive-power units have proved more popular than separate locomotives. These have 'multiple-unit' control, that is, one driver can control the current supply and braking throughout the whole train. **1967** *Economist* 21 Jan. 202/1 In October 1965 an agreement was reached between the railway management and the unions on single-manning for passenger and long-distance freight locomotives (and multiple-unit trains that have no loco as such). **1970** *Railway World* Dec. 540 At times of peak traffic, nine-car formations composed of two three-car multiple-units and three additional trailers are often seen.

c. Special collocations: **multiple birth** *Med.*, the birth of more than one child at a single confinement; **multiple exposure** *Photogr.*, the repeated exposure of the same frame of a film so as to produce superimposed images; **multiple image,** a composite image comprising two or more superimposed or adjacent images originally distinct (e.g. resulting from the repeated reflection of light, the reception of television signals that have travelled from the transmitter by different paths, or the simultaneous showing of several scenes on a cinema screen); **multiple personality** *Psychol.*, a dissociative condition in which an individual's personality is apparently split into two or more sub-personalities, each of which may become dominant and then is relatively distinct and complete; **multiple pregnancy** *Med.*, a pregnancy which would normally result in a multiple birth; **multiple resistance** *Med.*, resistance of a micro-organism to the action of more than one antibiotic; so *multiple-resistant* adj.; **multiple shift** *Industry*, a double or treble shift (sense 12); freq. *attrib.*; **multiple switchboard** *Teleph.* (see quot. 1932); **multiple twin,** (*a*) *Telephony*, a cable with a number of cores each of which consists of four wires arranged as two twisted pairs twisted together; usu. *attrib.*; (*b*) *Cryst.*, a twinned crystal composed of alternating lamellæ whose relative orientations are all governed by the same twin law; **multiple valve** *Electronics*, a multi-electrode valve.

1826 *Edin. Jrnl. Med. Sci.* II. 366 (*heading*) Memoir upon multiple or twin births. **1841** *Lancet* 9 Jan. 549/1 (*heading*) Statistics of multiple births. **1966** *Amer. Jrnl. Obstetr. & Gynecol.* XCIV. 490 (*heading*) Pregnancies following treatment with human gonadotropins with special reference to the problem of multiple births. **1968** *Guardian* 3 Oct. 9/2 Multiple births, enormously distressing to the mothers.., are becoming rarer. **1971** Hellman & Pritchard *Williams' Obstetr.* (ed. 14) xxv. 657 Some marriages appear to have an inordinately high frequency of multiple births. Greulich reported the case of a 35-year-old mother who in nine births delivered six pairs of two-egg twins and three single children. **1923** F. A. Talbot *Moving Pict.* xxvii. 382 The perfection of the various devices for accomplishing multiple exposures rendered this manifestation of novelty in photography..easy of ready accomplishment. **1968** *Listener* 18 Jan. 81/2 Multiple-exposure photograph of American mechanical claw used for scraping up samples of the moon's surface. **1863** Multiple image [see sense A. 3 e]. **1963** A. F. Abbott *Ordinary Level Physics* xxv. 313 (*heading*) Multiple images formed by a thick glass mirror. **1965** Multiple image [see *freeze *v.* 4 f]. **1971** *Gloss. Electrotechnical, Power Terms* (*B.S.I.*) iii. iii. 23 *Multiple image; double image; ghost*, a defect apparent in reproduction in which an additional outline (ghost), or succession of outlines (multiple image), of prominent features of a picture may be observed displaced from the correct position of the outline by a noticeable amount. **1972** L. D. Giannetti *Understanding Movies* iii. 100 Multiple images, widely seen at Expo 67, will probably be the next modification of editing... Multiple image film-making does not intensify a movie's sense of realism,.. but tends to emphasize the expressionistic aspects of film art. **1901** *Proc. Soc. Psychical Res.* XV. 466 Cases of multiple personality are not very uncommon, but, so far as I know, no attempt has been made to determine the relation which the different personalities bear to each other. **1906** M. Prince *Dissociation of Personality* i. 3 The disintegration resulting in multiple personality is only a functional dissociation of that complex organization which constitutes a normal self. **1942** 'M. Innes' *Daffodil Affair* i. 38 'And did you say dissociation?' 'Yes. What is sometimes called multiple personality.' **1967** M. Argyle *Psychol. Interpersonal Behaviour* iii. 56 Everyone has a number of 'sub-personalities'..—cases of multiple personality are an extreme of what is universal. **1857** Multiple pregnancy [see sense A. 3 d]. **1964** *Obstetr. & Gynecol.* XXIV. 819 (*heading*) Size and number of umbilical vessels, a study of multiple pregnancy in man and the armadillo. **1952** S. K. R. Clarke et al. in *Lancet* 7 June 1132/1 Besides these organisms, called here 'completely resistant', there have appeared staphylococci with what may be called 'multiple resistance'—i.e. resistant to several of these drugs. **1960** *Brit. Med. Jrnl.* 2 Jan. 11/2 It cannot be assumed..that multiple resistance and enhanced virulence are necessarily associated. **1972** *Med. Microbiol. & Immunol.* CLVII. 142 (*heading*) Antimicrobial resistance of the genera *Proteus, Providencia* and *Serratia* with special reference to multiple resistance patterns. **1960** *Brit. Med. Jrnl.* 2 Jan. 17/1, 23 of the fatal infections were due to multiple-resistant strains. **1961** *Lancet* 29 July 248/2 It would prevent the selection and breeding of multiple-resistant strains. **1926** *Rep. R. Comm. Coal Industry* I. 175 in *Parl. Papers* (Cmd. 2600) XIV. 1 (*heading*) Advantages of multiple shifts. **1946** M. Dobb *Wages* (rev. ed.) iii. 62 Where it is practicable to introduce a multiple-shift system—two or three working-shifts a day instead of one—the same economy could be obtained together with the employment of additional workers. **1964** T. W. McRae *Impact of Computers on Accounting* vii. 208 Costs arising out of multiple-shift working. **1891** J. Poole *Pract. Telephone Handbk.* viii. 148 (*heading*) Multiple switch-boards. **1932** T. E. Herbert in E. Molloy *Pract. Electr. Engin.* V. 1865/2 A switchboard in which the subscribers' circuits are repeated at several points so as to make each subscriber accessible to every operator is known as a multiple switchboard. **1969** S. F. Smith *Telephony & Telegr. A* iv. 98 The need for a series type of multiple connexion which these jacks require on a multiple switchboard has certain disadvantages. **1922** *Encycl. Brit.* XXXII. 709/1 In..the 'multiple twin cable' the centre of yarn is dispensed with, and the cable consists of a number of 4-wire cores. **1932** *Amer. Mineralogist* XVII. 360 The diamond multiple twin..is from South Africa. **1959** J. W. Freebody *Telegr.* vi. 173/2 With the multiple-twin type [of cable] the paper insulated conductors are first twisted into pairs and two such pairs are then twisted together to form the two pair core or quad. **1966** *McGraw-Hill Encycl. Sci. & Technol.* V. 208/1 The width of the lamellae in multiple twins of plagioclases is usually about 1 mm and smaller. **1929** *Wireless World* 6 Mar. (*heading*) A Loewe multiple valve for mains operation. **1968** *Radio Communication Handbk.* (ed. 4) ii. 20/1 The trend is to make radio equipment as compact as possible and it is therefore convenient to take advantage of the special multiple valves which have more than one unit contained in a single envelope.

B. *sb.* **4.** *Teleph.* **a.** A section of a multiple switchboard containing one jack for each subscriber.

1905 A. C. Booth in M. Maclean *Mod. Electr. Pract.* VI. iii. ii. 110 The line shown..is already engaged by the insertion of a plug on the first multiple. **1948** J. Atkinson *Herbert & Procter's Telephony* (new ed.) I. x. 195/1 In general, a multiple is designed so that the best compromise between vertical and horizontal reach is obtained when the switchboard is equipped with the ultimate number of lines.

b. *in multiple.* Of calling jacks: (connected) between the same pairs of wires, so enabling the same connection to be made at different points.

1943 A. L. Albert *Fund. Telephony* viii. 174 The calling jacks of each section must be connected in parallel or multiple with the calling jacks of every other section. **1969** S. F. Smith *Telephony & Telegr. A* iv. 98 The bush of the jack will be connected to the exchange battery via the plug inserted in one of the other jacks with which it is connected in multiple.

5. A multiple shop or store (see A. 2 b above).

1951 in Partridge *Dict. Slang* (ed. 4) 1115/1. **1957** *Economist* 5 Oct. 60/1 The multiples have also been ahead in adjusting their range of shoe styles. **1966** *New Statesman* 29 July 156/1 A businessman, stating..that he had sold 'self-owned multiple to avoid capital gains tax'. **1972** *House & Garden* Feb. 100/4 Some of the small merchants ..do a good business with Cape growths. The large multiples..are missing the opportunity.

6. An inexpensive work of art mass-producible by industrial methods.

1968 *Times* 26 Mar. 7 The artist who becomes interested in multiples takes the first step towards involving himself with the demands of technology. **1970** [see *graphic *sb.* 3]. **1971** P. Dickinson *Sleep & his Brother* ii. 36 The one touch of art—a bronze and bulbous paperweight, vaguely post-Brancusi—had the look of one of a large issue of multiples. **1973** J. A. Walker *Gloss. Art, Archit. & Design since 1945* 141 The idea of Multiples was first suggested by Agam and Jean Tinguely. They put their idea to the Parisian gallery dealer Denise René in 1955 but none were produced until 1962.

multiple (mʌ·ltip'l), *v.* *Teleph.* [f. Multiple *a.* and *sb.*] *trans.* To make (a circuit) accessible to operators at more than one point on a switchboard or switchboards; to provide or employ duplicates of (a device) for this purpose.

1906 Bell & Wilson *Pract. Telephony* (ed. 4) ix. 108 At the outgoing end the lines are multipled three times on every two sections, so that every operator has every line almost directly in front of her. **1932** T. Sherratt in J. A. Fleming *Electr. Educator* (ed. 2) III. 1194/1 The trunks are multipled to the contacts of a number of line switches, so that several subscribers can gain access to the same trunks. **1942** Knight & Prickett *Poole's Telephone Handbk.* (ed. 8) xii. 312 Each answering jack and lamp is multipled every four, six, or twelve panels according to traffic requirements. **1969** S. F. Smith *Telephony & Telegr. A* iv. 96 It is sufficient..to multiple the jacks without also multipling the calling indicators or lamps.

multiplet (mʌ·ltiplĕt). [f. Multiple *a.* and *sb.* + -et, after *doublet, triplet.*] **a.** A group of lines in a spectrum that are close together and spaced approximately in accordance with a simple rule; a group of related levels in an atom that differ slightly in energy, esp. a group in which this is due to differing relative orientations of either the electronic spin and orbital angular momenta, giving different values of the quantum number *J* (in the case of fine structure), or the electronic and nuclear angular momenta, giving different values of the quantum number *F* (in the case of hyperfine structure).

1922 M. A. Catalán in *Phil. Trans. R. Soc.* A. CCXXIII. 147 As will be seen later there are many 'groups' of lines in the manganese spectrum with similar structure to that of the foregoing 'group', and for this form of regularity the name 'multiplet' is suggested. **1929** *Trans. Faraday Soc.* XXV. 672 Next, the coupling between spin momenta of the electrons gave a 'resultant' spin vector *S*, which determined the multiplicity. The multiplicity was given by $R = 2S+1$, if *S* was measured in units of $h/2\pi$. Finally, the components of a multiplet were regarded as determined by the coupling between the resultant *L* vector and the resultant *S* vector. **1942** [see *fine structure 1 a]. **1959** *Sci. News* LIII. 87 In order to interpret all the observations, such as the multiplet structure of spectral lines.., it has to be assumed that the electron has, independently of its orbital motion, an intrinsic angular momentum and a magnetic moment. **1967** W. R. Hindmarsh *Atomic Spectra* v. 50 A quantum number *F* can be ascribed to each hyperfine level such that the number increases by one from one level to the next and..the separation between two successive levels is..*AF* where *A* is a constant for a hyperfine multiplet. **1971** A. G. Sharkey in R. I. Reed *Recent Topics Mass Spectrometry* 128 High-resolution mass spectrometry is unique in providing a precise mass from which a molecular formula can be derived. The basis of the high-resolution technique is the ability of the instrument to resolve multiplets, having the same nominal mass but differing in precise mass.

b. A group of sub-atomic particles that are alike as regards the values of the various quantum numbers except for the third component of isospin (and hence charge), which is different for each particle and has one of the $2I+1$ values $0, \pm\frac{1}{2}, \pm 1, \ldots, \pm I$ ($I = 0$ or half-integral); also, a larger group (also called a *supermultiplet*) composed of a number of such charge multiplets, each likewise characterized by a different value of hypercharge (or strangeness) but having the same spin and the same parity.

In quots. 1937 more a *transf.* use of prec. sense.

1937 *Physical Rev.* LI. 106 The structure of the multiplets of nuclear terms is investigated, using as a first approximation a Hamiltonian which does not involve the ordinary spin and corresponds to equal forces between all nuclear constituents, protons and neutrons. The multiplets turn out to have a rather complicated structure, instead of the *S* of atomic spectroscopy, one has three quantum numbers *S*, *T*, *Y*. *Ibid.* 117/1 Every one of these six states can be doubly occupied, with a particle $\tau = 1$ and $\tau = -1$ (neutron or proton). The half sum of the τ is denoted by T_ζ and the different T_ζ from $-T$ to T united into a multiplet. **1954** *Proc. Nat. Acad. Sci.* XL. 490 In discussing baryon states we used the intuitive argument of approximate mass degeneracy in a multiplet. **1956** [see *hypercharge]. **1964** *New Scientist* 20 Feb. 458/3 The particles appear in groups, or 'multiplets', of particles of different charge but very nearly equal mass. **1965** *Ibid.* 18 Mar. 738/3 Just as in the theory of SU(2) the particles are arranged in multiplets, but now they are distinguished within a multiplet by both their charge and hypercharge. The SU(3) symmetry relates not just the proton and neutron one to another, but includes also in one multiplet the six particles known as hyperons. **1970** [see *isospin]. **1970** D. B. Lichtenberg *Unitary Symmetry & Elem. Particles* iii. 34 The $2I+1$ different charge states of a particle with isospin *I* constitute a multiplet. However, since isospin is not an exact symmetry, the different states are not exactly degenerate in energy. It is often said under such circumstances that different members of the multiplet are different particles, rather than different states of the same particle. **1973** L. J. Tassie *Physics Elem. Particles* iv. 39 We now consider the three pions π^-, π^0, π^+ as a charge multiplet or isospin multiplet of multiplicity $2I+1 = 3$ yielding $I = 1$. We identify the $I_3 = +1$ state as the π^+; $I_3 = 0$ as the π^0; $I_3 = -1$ as the π^-. *Ibid.* xi. 134 The arrangement of *N*, Λ, Σ and Ξ into one multiplet, an octet.., is illustrated in Fig. 49.1. The mass splittings between the different isospin multiplets in the octet are about 20 times the mass splitting within isospin multiplets.

multiplex, *a.* and *sb.* Add: **A.** *adj.* **3. b.** More widely in *Telecommunications*, applied to processes and equipment for transmitting two or more independent signals or programmes (to be later separated and recovered) simultaneously over a single wire or channel, and to a composite signal so formed. (Earlier and later examples.)

1873 M. Gally *Brit. Pat.* 1395 17 Apr. 13 My Invention has particular reference to 'multiplex telegraphy', or the employment of a number of operators sending without conflict a number of different messages upon a single wire. **1883** *Jrnl. Franklin Inst.* CXVI. 479 In the multiplex [system], as I myself have seen, an instantaneous change in the resistance of the line equal to five hundred miles, may be made without practically affecting the synchronous movement of the distributors. **1934** [see *coaxial *a.* 2 c]. **1962** A. Nisbett *Technique Sound Studio* 261 Multiplex radio transmissions and the single groove of a stereo disc each contain more than one channel. **1970** J. Earl *Tuners & Amplifiers* ii. 47 Multiplex signal..consists of 19kHz pilot-tone, the A+B mono information and sidebands of the A−B stereo information (the..subcarrier..having been suppressed at the transmitter). **1973** *Mod. Railways* Jan. 12/2 Interlockings within three or four miles of Bristol are operated by individual circuits... Beyond this distance electronic time division multiplex equipment is used.

B. *sb.* **3.** *Electr.* A multiplex system or signal.

1959 J. W. Freebody *Telegr.* ii. 50/2 This printer was used in connexion with the Murray multiplex. **1967** *Technology Week* 20 Feb. 4/2 Initial experiments with time-division multiplex have convinced Comsat Corp. that satellite transmission does not produce insuperable synchronization problems. **1970** *Toronto Daily Star* 24 Sept. 26/1 (Advt.), It..tapes for multiplex and even makes 'sound with sound' recordings. **1973** Hills & Evans *Telecommunications Syst. Design* I. iv. 92 Modern equipment has group and supergroup limiters applied to the signal prior to their assembly into the higher multiplex.

multiplex (mʌ·ltipleks), *v.* [f. the sb.] *trans.* To incorporate into a multiplex signal or system.

1925 C. A. Wright *Telephone Communication* xiii. 317 The terminal equipment required to multiplex a telephone line is costly and its use is, consequently, justified only with long lines. **1955** *Sci. News Let.* 5 Mar. 158/2 You may soon be able to hear programs from your local FM station—minus commercials—in buses..and on the job. This is just one of the possibilities of a system being considered..that would allow FM stations to 'multiplex' their broadcasts. In the process, two separate signals would be sent from the same tower using the same radio frequencies. **1966** *McGraw-Hill Encycl. Sci. & Technol.* V. 352/2 When communication channels are multiplexed by time division, a number of messages is propagated over a common transmitting medium by allocating different time intervals in sequence for the transmission of each message. **1969** *Sci. Jrnl.* Dec. 42/3 A colour TV system multiplexes (by frequency division) video, colour and sound information. **1972** *Sci. Amer.* Sept. 120/3 The T1 carrier of the Bell Telephone System, which multiplexes 24 speech signals, is a typical example of a pulse-code-modulation (PCM) system.

Hence **mu·ltiplexed** *ppl. a.*; **mu·ltiplexing** *vbl. sb.* Also **mu·ltiplexer, -or,** a device which multiplexes.

1955 *Sci. News Let.* 5 Mar. 158/3 Multiplexing is only practical on FM broadcasts because at present there are too many technical difficulties in splitting up AM programs. **1957** *Electronic Engin.* XXIX. 159/1 The information received from, say, two channels of the multiplexor are admitted serially to the clamping circuit. **1964** *Ann. N.Y. Acad. Sci.* CXV. 574 Multiplexers can be sequential sampling devices, or electronic commutators, which sample each channel in turn cyclically. **1969** *Sci. Jrnl.* Dec. 42 This spacing of the different signals over adjacent frequency ranges of the communication channel is called frequency-division multiplexing. *Ibid.* (*caption*) Frequency spectrum of a frequency division multiplexed channel shows how..five carriers..are each surrounded by sidebands. **1970** O. Dopping *Computers & Data Processing* viii. 121 A multiplexor channel has its own facilities for sharing its time between different slow input/output operations. **1973** *Physics Bull.* July 413/3 A multiplexed hologram is produced from the sequence of exposures on the cine film.

multiplication. Add: **6. multiplication constant** or **factor** *Nuclear Physics*, in nuclear fission, the ratio by which the number of neutrons increases during a period equal to the lifetime of a neutron; **multiplication sign,** the sign × placed between two quantities to denote their multiplication; also *fig.*

1956 A. H. Compton *Atomic Quest* 137 These 'delayed' neutrons..make the difference between a multiplication constant of less than one and greater than one. **1962** *Newnes Conc. Encycl. Nucl. Energy* 111/1 The actual number of neutrons per fission varies considerably, but has an average value of 2·5. This is a high enough gain per cycle to suggest that a chain-reacting system of uranium could be made with a multiplication constant of at least unity. **1941** E. Fermi *Coll. Papers* (1965) II. 89 The chain reaction may reach a steady state only when the multiplication factor of the neutrons (including neutron losses due to diffusion outside of the reacting mass) is equal to 1. **1974** *Encycl. Brit. Macropædia* XIII. 305/1 If a fission bomb is to be made, the multiplication factor must be as large as possible. **1907** W. D. Eggar *Elem. Algebra* i. 4 When a number is written in front of a bracket without a sign following it, thus 10(6+5), the multiplication sign is understood, just as when we write 10*a*. **1908** W. Owen *Let.* 20 July (1967) 46 Best love, and a 'multiplication sign' (×) from your ever loving W. **1973** P. M. Aitchison et al. *Form Two Maths.* vi. 67 Remember that a multiplication sign connects factors.

multiplicative, *a.* and *sb.* Add: **b.** (Later example of *sb.*)

1928 H. Poutsma *Gram. Late Mod. Eng.* (ed. 2) iv. 304 The multiplicatives *double, treble, quadruple.*

So **mu·ltiplicatively** *adv.*

1895 in *Funk's Stand. Dict.* **1914** C. D. Broad *Perception* iii. 180 The strengthening of probability takes place multiplicatively. **1973** *Nature* 23 Feb. 513/1 The latter possibility has an attractive feature, namely that Misner's mechanism, and the one described here would operate simultaneously, the two effects combining multiplicatively.

multiplicator. 1. (Later example.)

1923 *Glasgow Herald* 16 Oct. 7 In Cologne order was restored on Friday night... The price 'Multiplicator', which played a large part in creating the agitation, has been lowered from 1,800,000,000..to 400,000,000.

multiplicity. Add: **1. e.** *Physics.* The number of components (whether one or several) in a multiplet; *spec.* (*a*) the quantity $2S+1$, where *S* is the spin quantum number of a term; (*b*) the quantity $2I+1$, where *I* is the isospin of a charge multiplet.

1923 H. L. Brose tr. *Sommerfeld's Atomic Struct. & Spectral Lines* vi. 385 Those terms are defined as similar which have the same multiplicity and the same azimuthal quantum number. **1929** [see *multiplet a]. **1934** H. E. White *Introd. Atomic Spectra* xiv. 248 It is found experimentally that the multiplicities of the levels belonging to a given spectrum will either be all even or all odd. *Ibid.* 249 Spectral terms arising from successive elements in the periodic table alternate between even and odd multiplicities. **1948** G. R. Harrison et al. *Pract. Spectroscopy* x. 241 For two electrons, $S = 0$ or 1, depending on whether

the two spins are antiparallel or parallel. Hence atomic spectra that arise from two electrons (helium and the alkaline-earth metals Be, Mg, Ca, Sr, and Ba) can have multiplicities of 1 (that is, no splitting of levels because of spin) or 3 (splitting of each level into three). **1964** *Sci. Amer.* Feb. 82/2 The number of different charge states in a multiplet, or its 'multiplicity' (*M*), is directly related to the isotopic-spin quantum number *I* by the equation $M = 2I+1$. **1972** *Physics Bull.* Feb. 92/1 A further interesting feature which these multiplets illustrate is the greater bonding of the states of lower multiplicity. **1973** [see *multiplet b].

f. *Biol.* The ratio of the number of infective particles to the number of susceptible cells; usu. in phr. *multiplicity of infection.*

1947 *Proc. Nat. Acad. Sci.* XXXIII. 259 (*heading*) Dependence of the probability of reactivation on the multiplicity of infection. **1957** *Virology* III. 286 Luria (1947) found that cells of *Escherichia coli* B which had been infected with more than one inactivated particle of phage T2, or other members of the T group of coliphages, yielded active phage. He called this 'multiplicity reactivation'. **1964** W. Hayes *Genetics of Bacteria & their Viruses* xvi. 367 If a suspension of virulent phage is added to a growing culture of sensitive bacteria in broth, so that the number of phage particles exceeds the number of bacteria (multiplicity of infection = >1),..the culture becomes clear, as a result of the infection and lysis of virtually all the bacteria. **1974** *Nature* 11 Oct. 542/1 Chang cells in suspension were infected at an input multiplicity of 10. *Ibid.* 543/2 Addition of VSV or EMC virus at a multiplicity of infection of 1.

multiplier. Add: **2. b.** *Econ.* (See quot. 1964.)

1936 J. M. Keynes *Gen. Theory Employment* x. 113 A definite ratio, to be called the *Multiplier*, can be established between income and investment and..between the total employment and the employment directly employed on investment. **1940** G. Crowther *Outl. Money* v. 157 In the upward process, the change in the National Income necessary to restore balance and reverse the trend [downwards] may be very many times larger than the original discrepancy... Thus a small original change results in a large ultimate change in the National Income. The ratio between them is usually known as the 'multiplier'. **1963** *Economist* 9 Feb. 481/2 A direct injection of £500 million of new Government spending can sometimes increase total national spending by £1,000 million (i.e. a multiplier of two). **1964** Gould & Kolb *Dict. Social Sci.* 449/1 *Multiplier*... The term denotes the process (or the index, or coefficient, measuring such a process) whereby initial changes within economic systems (e.g. changes in the levels of investment) have cumulative and, in principle, measurable effects upon the system, its components and its equilibrium.

c. *multiplier effect*, an effect such as could be assessed in terms of the economic multiplier.

1957 R. K. Merton *Student-Physician* 14 Such an experimental institution for medical education would have multiplier-effects. **1963** *Economist* 16 Feb. 619/1 The multiplier effect..of an injection of purchasing power into personal incomes..is unlikely to exceed unity and may be as low as 0·7. **1964** T. W. McRae *Impact of Computers on Accounting* vii. 211 Such improvement will have a 'multiplier effect'. **1965** *Listener* 16 Sept. 405/2 What we can do is to concentrate our help where it will have the highest multiplier effect.

4. (Earlier and later examples.)

1795 T. Cavallo *Compl. Treat. Electr.* (ed. 3) III. 98 With this machine, which..we shall call the Multiplier of Electricity, the accumulation of the communicated power does not advance..so quick as with the doubler. *Ibid.* 106, I shall now describe a few experiments made with this multiplier. **1938** J. B. Hoag *Electron & Nucl. Physics* vi. 127 In another form of multiplier tube, electrons are oscillated back and forth between two surfaces which are good secondary emitters... At each impact of the electrons with the cathode surfaces, the number of electrons is appreciably multiplied. **1940**, etc. [see *electron multiplier* s.v. *electron² 2 b]. **1950** K. Henney *Radio Engin. Handbk.* (ed. 4) xiii. 558 The multiplier phototube has a number of advantages compared with a phototube and separate amplifier. **1973** *McGraw-Hill Yearbk. Sci. & Technol.* 333/1 Because fewer gain stages with higher interstage voltages are needed, the speed of the multipliers has been improved, greatly increasing their usefulness for applications such as laser optical communications.

6. (Examples.)

1969 J. J. Sparkes *Transistor Switching* viii. 200 In the accumulator of this multiplier it is convenient to use flip-flops..which have two separate logic and pulse inputs per flip-flop. *Ibid.* 201 The multiplier circuit operates as follows. **1972** *Daily Tel.* 5 Apr. 10 (Advt.), The world's fastest and smoothest mechanical multiplier. Operates at 360 rpm, lists 10 columns and totals 11.

multiply, *adv.* Restrict *Math.* to terms in Dict. and add: In a multiple manner; many ways or times, or more than once. (Further examples.)

The use of a following hyphen is similar to that described at Well *adv.* VII.

1892 *Mind* I. 353 The rules for the synthesis of multiply-quantified propositions follow immediately from those for the synthesis of singly-quantified propositions. **1922** F. W. Aston *Isotopes* 71 These multiply-charged clusters give most reliable values of mass. **1932** Lewis & Langford *Symbolic Logic* ix. 301 A hierarchy..is involved in multiply-general propositions and functions. **1962** W. & M. Kneale *Devel. of Logic* ii. 112 Two distinct quantifiers are required, as in the statement '*Some* man does not possess *all* knowledge', and multiply general statements are quite common in science and philosophy, e.g. '*Every* event has *a* cause'. **1963** B. Fozard *Instrumentation Nucl. Reactors* ii. 12 In many cases only one electron is separated from the parent atom in an ionising process but cases occur where several electrons are emitted and the atom is said to be multiply ionised. **1966** D. G. Brandon *Mod. Techniques Metallogr.* iv. 204 The ions will be multiply-charged. **1971** *Jrnl. Gen. Psychol.* LXXXV. 165 Almost no studies have been specifically directed to the detection of the relative power of the different kinds of categories into which words seem to be multiply encoded.

multi-ply, *sb.* and *a.* [f. Multi- + Ply *sb.*, *plywood.] **A.** *sb.* Plywood comprising more than three layers. **B.** *adj.* Designating textiles, etc., formed by having several threads or webs twisted together; also denoting paper comprising several layers.

1940 *Chambers's Techn. Dict.* 563/2 *Multi-ply*, plywood formed of more than three layers of wood. **1950** *Gloss. Aeronaut. Terms (B.S.I.)* i. 49 *Multi-ply fabric*, fabric formed of more than one ply. **1951** *Archit. Rev.* CIX. 159 (*caption*) Multi-ply was developed during the war for aircraft propellers. **1962** *Times* 3 Dec. (Agric. Suppl.) p. vii/5 The multi-ply paper sack. **1963** A. J. Hall *Textile Sci.* i. 9 The threads may be single or consist of two, three, or more single threads twisted together to give what are often termed folded or multi-ply threads. **1966** A. W. Lewis *Gloss. Woodworking Terms* 52 Laminated boards.. include plywood, multi-ply, laminboard, [etc.].

multipolar (mʌltipōu·lăɹ), *a.* [f. Multi- + Polar *a.*] Having or pertaining to many poles. **a.** *Anat.* Of a nerve cell: having many processes.

1859 [in Dict. s.v. Multi- 1 a]. **1885** *Encycl. Brit.* XIX. 27/2 The cells of the spinal ganglia are usually rounded; those of the sympathetic more angular; those of the spinal cord multipolar. **1948** A. Brodal *Neurol. Anat.* xi. 354 The autonomic ganglion cells in most of the ganglia of the sympathetic trunk and in the peripheral ganglia are usually multipolar and are characterized by possessing numerous and long dendrites. **1964** J. Z. Young *Model of Brain* xix. 307 Multipolar amacrines are numerous in the cell islands of the optic lobes.

b. *Electr.* Of electrical machinery: having more than one pair of magnetic poles in the system of field magnets.

1884 [in Dict. s.v. Multi- 1 a]. **1903** [see *bar winding* (*bar *sb.*¹ 30)]. **1964** N. N. Hancock *Matrix Analysis of Electr. Machinery* v. 72 To apply the results to multipolar machines it will be necessary to reduce the speed and to increase the torque according to the number of pole-pairs.

c. *Cytology.* Having or involving more than two spindle poles.

1894 *Ann. Bot.* VIII. 313 Nuclear divisions with multipolar spindles. **1896** E. B. Wilson *Cell* ii. 68 The abnormal forms of mitoses are arranged by Hausemann in two general groups, as follows: (1) asymmetrical mitoses,.. and (2) multipolar mitoses. **1912** E. A. Minchin *Introd. Study Protozoa* vii. 120 In cases where the division of the nucleus is of the karyokinetic type, repeated divisions of the centriole result in the formation of a complicated multipolar mitotic figure, leading to a multiple division of the nucleus. **1937** M. J. D. White *Chromosomes* ii. 19 In many cells such as those of cancerous tissues and in Sea-Urchin eggs which have been fertilized several times as a result of polyspermy, multipolar spindles with a number of equatorial planes intersecting one another are found; there may be as many as 12 poles and 6 equatorial planes. **1950** *Hereditas* XXXVI. 393 (*heading*) Multipolar spindles in human cancer cells.

d. Consisting of, or divided into, more than two (political or other) alliances, ententes, parties, etc.

1968 *Guardian* 4 Dec. 4/6 The most profound challenge to American policy will be..to develop some concept of order in a world which is bipolar militarily but multipolar politically. **1969** S. Henig in Henig & Pinder *European Pol. Parties* 502 The country-by-country chapters have indicated a threefold classification between bipolar, unipolar and multipolar party systems. In bipolar systems.. government will be dominated either by the two parties singly..or by both together in a 'big coalition'. In a unipolar system .. the norm is for one party to be easily the largest legislative group... In multipolar systems, neither one nor two parties are dominant even to the extent of always emerging the strongest from succeeding elections. **1975** *Times* 7 Apr. 14/2 Kissinger['s]..concept of a multipolar world, where major powers could compete and cooperate simultaneously.

multipolarity (mʌltipolæ·rĭti). [f. prec. + -ity.] **a.** Multipolar quality or state.

1947 *Hereditas* XXXIII. 468 The anaphases..all showed serious disturbances, lagging chromosomes, multipolarity..and so on. **1950** *Ibid.* XXXVI. 395 Multipolarity seems..to be fairly independent of the chromosome number in that highly polyploid bipolar as well as multipolar configurations which contain low chromosome numbers are found. **1968** *Guardian* 4 Dec. 2/3 NATO..has yet to adjust to the political multipolarity of the late sixties. **1969** *Punch* 5 Mar. 329/2 Henry Kissinger..has already summed up the Soames affair as an example of 'political multipolarity', spoken of the problems of 'a pluralist world'... All good stuff, if we knew what it meant.

b. *Physics.* The highest order of multipole associated with a state or phenomenon.

1955 *Physical Rev.* XCVIII. 1198/1 (*heading*) Multipolarity of gamma transitions in F^{19} and Na^{23} by Coulomb excitation. **1973** *Nature* 31 Aug. 541/1 There have been speculations that other giant resonances with other multipolarities might exist; for example, quadrupole motions in which the nucleon distributions oscillate between prolate and oblate shapes with quadrupole deformations.

multipole (mʌ·ltipōul), *sb.* and *a.* *Physics.* [f. Multi- 2 + Pole *sb.*²] **A.** *sb.* A system of 2^l monopoles ($l = 1, 2, 3, \ldots$) with no net charge or pole strength and no moment of a lower order than *l* (cf. *moment *sb.* 8 c); the dipole ($l = 1$) is often treated as a special case and the quadrupole ($l = 2$) regarded as the multipole of lowest order. Freq. *attrib.* or as *adj.*, esp. designating electromagnetic radiation of the kind produced by a multipole with a moment varying sinusoidally in magnitude.

In quot. 1916 this may not be the precise sense.

1916 *Jrnl. Amer. Chem. Soc.* XXXVIII. 764 One or more electrons are held by sufficiently weak constraints so that they may..in the extreme case pass altogether to another atom, thus producing in the molecule a bipole or multipole of high electrical moment. **1929** *Physical Rev.* XXXIV. 1252 Since *m* must change by ±1,..the transition $j = 0 \to j = 0$ cannot occur for multipole radiation. **1932** J. H. Van Vleck *Theory Electr. & Magn. Susceptibilities* i. 12 The omitted terms are sometimes characterized as representing 'multipoles'. The omitted term in φ of lowest order, for instance, is..Nq/R^3, where *q* is the 'quadrupole moment'. **1950** D. Halliday *Introd. Nucl. Physics* ii. 57 If $|\Delta I| = 1$ the multipole order is 2 and we have..a dipole transition; if $|\Delta I| = 2$, the multipole order is 4 and we have a quadrupole transition. **1952** Blatt & Weisskopf *Theoret. Nucl. Physics* xii. 586 The electric multipole fields have non-vanishing radial components of **E**, and the magnetic multipole fields have non-vanishing radial components of **H**. The angular momentum carried by one quantum of multipole radiation is ħ*l*, with *z* component ħ*m*. Electric and magnetic multipole radiation of order *l*, *m* carry the same angular momentum but differ through their parity. *Ibid.* 587 Any arbitrary field **E**(**r**), **H**(**r**) which obeys the free-space Maxwell equations can be expanded in terms of multipoles. **1962** Corson & Lorrain *Introd. Electromagn. Fields* ii. 52 For the general multipole, characterized by the letter *l*, the potential varies as $1/r^{l+1}$. **1970** [see *moment *sb.* 8 c].

B. *adj.* (Freq. written **multi-pole.**) Designed to close or open several circuits simultaneously.

1927 *Wireless World* 22 June 786/3 It is sometimes recommended that a multi-pole switch should be fitted in order to switch off both batteries. **1934** *Pract. Wireless* IV. 667/3 This multi-pole switch converts the usual panel-controlled push-pull action into a lateral movement. **1962** G. A. T. Burdett *Automatic Control Handbk.* iii. 24 The multi-pole type of control switch is a double break changeover switch especially designed for the switching of squirrel-cage motors.

multi-posi·tion, *a.* [f. Multi- + Position *sb.*] Usable or placeable in more than one position. So **multi-posi·tional** *a.*

1944 *Illustr. Aviation Encycl.* (Aviation Research Associates) 143/2 *Multiposition propeller*: see controllable pitch propeller. **1948** F. J. Murray *Theory Math. Machines* (ed. 2) i. 3 Another method of indicating the digit which appears in a decimal place in a counter is by means of a multiposition switch. **1961** Y. Olsson *On Syntax Eng. Verb* iv. 82 A nucleus counts as multi-positional not only if it is identical in all of the Positions in which it occurs, but also if it is combined with a non-nuclear element. **1969** *Jane's Freight Containers 1968–69* 472 (*caption*) Duramin 30 ft ISO Container with side-loading facility and 'Salwall' lining (for multi-position load anchorage). **1971** *Power Farming* Mar. 69/2, 5 ft convertible hay or silage flail mower, straw chopper and haulm pulveriser. Easily adaptable, multi-position wheels.

multipro·cessing, *vbl. sb.* *Computers.* [f. Multi- 2.] Processing by a number of processors sharing a common memory and common peripherals.

1961 *Communications Assoc. Computing Machinery* Oct. 421/1 Multiprocessing is defined here as the sharing of a common memory and all peripheral equipment by two or more processor units. **1967** C. J. Duncan in Cox & Grose *Organiz. Bibliogr. Rec. by Computer* 11. 47 The decision as to whether to run the composing machine on-line or off-line will..depend upon the size of the machine, what multiprocessing facilities are available, and the precise machine configuration and loading. **1969** P. B. Jordain *Condensed Computer Encycl.* 327 The Honeywell 800 was an early powerful multiprocessing computer. It was equivalent to 8 conventional computers sharing a common memory. A hardware/software executive coordinated the eight processes to ensure efficient operation.

So **multipro·cessor**, a processor capable of performing multiprocessing; usu. *attrib.*

1961 *Communications Assoc. Computing Machinery* Oct. 421 (*heading*) Problems of storage allocation in a multiprocessor multiprogrammed system. **1965** A. G. Favret *Introd. Digital Computer Applic.* xvi. 233 In multiprocessor systems the executive routines must be capable of allocating tasks among different processors or deciding which of several inquiries should be given priority. **1970** O. Dopping *Computers & Data Processing* vi. 97 Two or more computer systems can be interconnected electrically to form a multi-processor system. *Ibid.*, The multi-processor alternative has the advantage that the capacity of the faster machine is always on tap also during input and output operations. **1971** *New Scientist* 25 Feb. 425/2 The 6000 models are well equipped with communications gear, and multi-processor configurations have interconnected memory units and multiplexors as well as processors.

multipro·gramming, *vbl. sb.* *Computers.* [Multi- 2.] The execution of two or more independent programs concurrently.

1959 *Communications Assoc. Computing Machinery* Nov. 15/1 The normal requirement in multiprogramming is to communicate with a particular program and at the same time allow all other programs to proceed... We now desire to stop a program rather than the machine. **1965** A. G. Favret *Introd. Digital Computer Applic.* xvi. 234 Multiprogramming techniques will probably also be used in some industrial process control applications where one computer is utilized to control several different processes on a time-shared basis. **1970** A. Cameron et al. *Computers & Old Eng. Concordances* 25 You can do a run on a 360, of course, without tying up the entire 360 through..something called multi-programming. **1973** Murrill & Smith *Introd. Computer Sci.* vii. 260 The objective of multiprogramming is to maintain two or more programs in memory simultaneously so that the processor can be working on one program while it is waiting for an input or output operation to be completed in the other program.

So **multipro·gram, -pro·grammed** *adjs.* [Multi- 3, 1], designed for or pertaining to multiprogramming.

1959 *Proc. Eastern Joint Computer Conf.* Dec. 75 (*heading*) Arithmetic and control in a multiprogram computer. *Ibid.* 80/2 The amount of work which can be accomplished per second..is, of course, increased enormously in a multi-programmed design. **1961** Multiprogrammed [see *multiprocessor]. **1964** T. W. McRae *Impact of Computers on Accounting* vii. 209 By far the most interesting development..has been the introduction of multiprogramme machines, which can process a number of jobs at the same time. **1967** H. Hellerman *Digital Computer System Princ.* ii. 107 In multiprogrammed systems it [sc. a control program] also resolves all conflicts among problem programs for calls on common facilities and ensures privacy among the several users that may be sharing these facilities. *Ibid.* viii. 345 The major impetus for storage protection arises in multiprogram sharing of processor storage.

multi-pu·rpose, *a.* [f. Multi- + Purpose *sb.*] Serving, or intended to serve, many purposes; performing many duties.

1935 *Sci. Amer.* Sept. 164 It is capable of handling much larger signal voltages than the diodes incorporated in the multi-purpose type of glass tube. **1937** *New Republic* 6 Jan. 292 The multipurpose TVA dams. **1939** *Electronics* June 48/1 (*title*) Multi-purpose Midget Tubes. **1943** J. S. Huxley *TVA* xviii. 134 A co-operative, unified, multipurpose undertaking. **1950** *Mind* LIX. 422 An all-round multi-purpose classification. **1959** *Listener* 12 Nov. 810/2 The use of what is now known..as the 'generalist' or 'multi-purpose' worker at village level. **1960** *Farmer & Stockbreeder* 8 Mar. 68 The only multi-purpose oil to be tested by the National Institute of Agricultural Engineering. **1965** *English Studies* XLVI. 30 While it [sc. þe] is a multivalent form it is not a multi-purpose form. **1972** *Jrnl. Social Psychol.* LXXXVII. 154 The multipurpose role inner city mothers often assume may explain this [observation] in part. **1972** M. Sheppard *Taman Indera* 65 A long multi-purpose head-cloth of red or blue cotton. **1973** *Country Life* 28 June 1904/1 A new multi-purpose mistblower, suitable for crops of all kinds. **1974** *Daily Tel.* (Colour Suppl.) 29 Nov. 72 Many rooms are multi-purpose, and the use to which they are put by family or friends should be analysed before making any decisions.

multi-ra·cial, multiracial, *a.* Of, pertaining to, or comprising several races, peoples, or ethnic groups; characterized by the co-existence or co-operation of individual members of such groups on amicable and equal terms. Also *fig.* So **multi-ra·cially** *adv.*

1923 *Overseas* Sept. 45 The interests of modern civilisation and, I think, Christian ethics, are better expressed in large, bi-racial or multi-racial States,..where racialism is accounted a public curse rather than a civic virtue. **1933** E. B. Reuter in E. S. Bogardus *Social Probl. & Social Processes* 96 The type of accommodation made is of course an individual matter, but the forms that it takes are.. those familiar in other bi-racial or multi-racial political areas. **1947** *Forum* (Johannesburg) X. i. 25/1 We, as a multi-racial society, have had our differences, while sharp antagonisms unfortunately exist today. **1957** L. F. R. Williams *State of Israel* 209 The unifying influence which this hostility is exerting upon Israel's multiracial population. **1957** *Economist* 19 Oct. 204/2 He triumphantly created the first multiracial government in Africa at the height of Mau Mau. **1958** *Times Lit. Suppl.* 10 Jan. 21/2 One feels the pleasant relief of a man living a multi-racial life away from the colour bar. **1959** *New Statesman* 28 Feb. 300/1 But it is his attack on the multi-racial clothing industry—involving the dismissal and replacement of 35,000 non-whites—that has frightened the coloured people especially. **1963** *Economist* 30 Nov. 887/3 Such a multi-racially-run world. **1966** *Listener* 6 Oct. 499/3 Closer contact between..the university worker and the industrial scientist, to make a truly 'multi-racial' commonwealth of scholarship. **1972** T. Lilley *K Section* ix. 40 The Dock Labourers' Union was one of the biggest... Multi-racial, it owed allegiance only to itself.

multi-ra·cialism, multiracialism. [f. prec. + -ism.] The condition or quality of being multi-racial; *spec.*, the conception of a state in which members of different races, peoples, or ethnic groups live on amicable and equal terms. So **multi-ra·cialist, multiracialist,** an advocate of multi-racial governments or societies.

1958 *Economist* 3 May 396/2 Its professed multi-racialism was bound to appear an attempt to entrench separate racial streams. **1959** *Manch. Guardian* 25 July 5/3 The concept of partnership or multi-racialism [in Kenya]. **1960** *Guardian* 25 June 5/5 Mr. Blundell is a convinced multi-racialist. **1962** *Times* 4 Jan. 9/2 Britain is pushing the Federation towards a multiracialism that will end in African control. **1962** *Punch* 28 Nov. 758/1 Uncle Tom, a born multi-racialist. **1964** *Ann. Reg. 1963* 326 In September, another Coloured man..escaped to Swaziland..hoping..to represent the case for multi-racialism in South African sport. **1972** *Guardian* 8 Feb. 13/8 Multi-racialism here [sc. in Senegal] is not just an experiment. It is a proven fact.

mu·ltisided, *a.* [f. Multi- + Sided *ppl. a.*] = Many-sided *a.* So **multisi·dedness.**

1903 A. W. Patterson *Schumann* xvii. 202 The multisidedness of Schumann's genius. **1963** *Times* 30 May 9/6 The first floor elevation has a sawtooth plan with adjoining windows set at an angle to each other so that occupants can look through into each other's rooms, which may disconcert an ambassador seeking privacy; and the resulting multisided rooms may prove awkward in use.

mu·ltistage, *a.* Also **multi-stage.** [f. Multi- 3 + Stage *sb.*] Consisting of, occurring in, or involving several stages (cf. Stage *sb.* IV in Dict. and Suppl.).

1904 *Engineering News* LI. 324/3 The pumps shown and described in the following are multi-stage turbine pumps. **1911** A. M. Greene *Pumping Machinery* xiii. 522 (*heading*) Multi-stage compression. **1920** *Wireless World* Jan. 574/2 (*heading*) The design of multi-stage amplifiers using three-electrode thermionic valves. **1944** [see *intercool* vb. s.v. *intercooling *vbl. sb.*]. **1946** *Nature* 5 Oct. 464/2 It was admitted that these ranges might be attained by the employment of multi-stage rockets. **1962** F. I. Ordway et al. *Basic Astronautics* x. 400 Large boosters..are being designed for use as the initial stage of a multistage space carrier vehicle. **1962** Simpson & Richards *Physical Princ. Junction Transistors* xi. 246 We shall be concerned with typical intermediate stages in a multistage amplifier and not with unusual impedance matching arrangements at the input and output. **1963** R. R. A. Higham *Handbk. Papermaking* vi. 139 Either single- or multi-stage bleaching may be carried out according to the conditions. **1971** *Brit. Med. Bull.* XXVII. 68/2 Multi-centric hypotheses [of the origin of cancer] have not proved as popular as multi-stage hypotheses, or theories which assume that a change from a normal cell to a malignant cell takes place in stages. **1971** B. Scharf *Engin. & its Lang.* xv. 207 Multi-stage turbines may be compound, i.e. they have both impulse and reaction stages.

b. *spec.* in *Statistics.* Applied to sampling carried out in a number of stages, the sample obtained at each stage being divided into smaller units and taken as the population for the next stage of sampling.

1949 F. Yates *Sampling Methods for Censuses & Surveys* iii. 34 Multi-stage sampling introduces a flexibility into sampling which is lacking in the simpler methods. **1966** Kendall & Stuart *Advanced Theory Statistics* III. xxxix. 204 The motive for multi-stage sampling is almost invariably to reduce costs rather than reduce variance directly; the additional resources can, of course, be applied to an increase in sample size. **1973** F. E. Fischer *Fund. Statistical Concepts* ix. 215 The extensive survey work which precedes every national election..is a perfect example of multistage sampling.

So **mu·ltistaged** *ppl. a.* [Multi- 1].

1965 *Times Lit. Suppl.* 25 Nov. 1051/1 It is like a Greek tragedy in which the Furies have been replaced by a multi-staged rocket.

multista·ge, *v.* Also **multi-stage.** [f. prec.] *trans.* To make multistaged. So **multista·ging** *vbl. sb.*, the use of several stages.

1911 A. M. Greene *Pumping Machinery* xiii. 523 Since it is not possible to bring the curve of compression close enough to the isothermal, a method of multi-staging the compression has to be devised. **1944** A. H. Church *Centrifugal Pumps & Blowers* vi. 128 If the total head to be developed by a pump is too large for a single stage to handle, multistaging is resorted to. **1972** *Physics Bull.* Oct. 591/2 In general one would like a higher overall pressure ratio and this can be obtained by multistaging.

multi-sto·ry, -sto·rey, *a.* [f. Multi- + Story *sb.*², storey.] Of many storeys or floors. So **multi-sto·ried, -sto·reyed** *a.*

1918 *N.Y. Times Mag.* 16 June 4 The stenographers.. are here from multi-storied city skyscrapers. **1918** L. Galloway *Office Management* (1919) 50 In a multistory factory building the location of the office within the factory itself vitally affects the efficiency of the clerical force. **1946** *Nature* 3 Aug. 152/2 The framed multi-story construction, to be advocated both on architectural grounds and for blast resistance. *Ibid.* 28 Sept. 439/1 The first remedy was to provide adequate car parks, underground, at ground level, and in multi-storied buildings. **1955** *Times* 31 Aug. 11/6 Rehousing..will create a total overspill of from 35,000 to 54,000, depending on the extent to which multi-storey flats are erected as the old and crumbling tenements come down. **1958** *Manch. Guardian Weekly* 7 Aug. 6/4 The sooner we have more meters the sooner the chaos in our streets will disappear—especially if the profits on metered parking are translated into proper garages and multi-storey parks. **1964** J. Summerson *Classical Lang. Archit.* ii. 15 A building which..is multi-storeyed. **1973** *Times* 28 Feb. (Victoria Centre, Nottingham, Suppl.) p. i/1 The Victoria Centre, Nottingham's new £10m multi-storey shopping centre, is, according to its developers.. the biggest of its type in Europe. **1975** J. Symons *Three Pipe Problem* xviii. 208 The car was in a multi-storey park.

multi-tie·r, *a.* [f. Multi- + Tier *sb.*¹] Having or comprising several tiers or layers. So **multi-tie·red** *a.*; **multi-tie·ring** *vbl. sb.*, arrangement in tiers or layers.

1952 *Flow Q.* Jan. 4 A multi-tier conveyor. **1957** *Textile Terms & Definitions* (Textile Inst.) (ed. 3) 67 *Multi-tier loom*, a loom with a batten having several rows of shuttles. **1958** T. Landau *Encycl. Librarianship* 283/2 The *multi-tier stack*... This system ensures a maximum storage space. Many of these stacks are several tiers high and are connected by book lifts, and occasionally by passenger lifts. **1960** (*title*) The power gain of multi-tiered VHF transmitting aerials (B.B.C. Engin. Monogr. 31). **1965** K. D. Metcalf *Planning Academic & Research Library Buildings* viii. 134/2 Multi-tier book stacks hold themselves up. Each level of stack supports the levels above... A multitier stack provided about as concentrated shelving space as could be found until one or another of the recently developed compact-shelving arrangements began to be used. **1968** *Punch* 27 Nov. 786/2 Dear Public, You've had the Two-Tier System. Now for our fabulous Multi-Tier System, with *Flavorlik*, the new Soft-Centre stamps in *nine* exciting flavours. **1971** *Engineering* Apr. 63/2 If required, the machine will carry out multi-tiering and sleeve-wrapping.

multituberculate (mʌltitiubɜ·ɹkiŭlĕt), *sb.* and *a.* *Palæont.* [ad. mod.L. name of order *Multituberculata* (E. D. Cope 1884, in *Amer. Naturalist* XVIII. 687), f. Multi- + Tuberculate *a.*] A small fossil mammal of the order Multituberculata, characterized by teeth bearing many cusps arranged in two or three rows. Also as *adj.*, pertaining to or resembling an animal of this type.

1884 E. D. Cope in *Amer. Naturalist* XVIII. 687 The extinct marsupials belong to three types, as distinguished by the form of their superior molar teeth. These are tri-tuberculate, quadrituberculate, or multituberculate. **1888** [in Dict. s.v. Multi- 1 a]. **1898** A. S. Woodward *Outl. Vertebr. Palæont.* 248 There is also some reason to suspect that other double-rooted multituberculate teeth from the Rhætic of Europe, commonly claimed as mammalian, may be similarly interpreted [as reptilian]. **1926** *Amer. Jrnl. Sci.* CCXI. 228 (*title*) The Multituberculates as living animals. **1933** A. S. Romer *Vertebr. Paleont.* xiii. 257 The multituberculates had the longest history of any known mammalian order, for they appeared in the Triassic as the oldest known forms of mammals and persisted through into the Lower Eocene. **1969** *Sci. Jrnl.* 38/2 Entire skulls, including the basicranial region, of multituberculates have so far been found only in the Palaeocene genus *Ptilodus*, known from North America.

multitudinosity. (Later example.)

1939 R. A. Knox *Let Dons Delight* vi. 161 For all that, sheer multitudinosity has power to oppress the mind.

multi-u·se, *a.* [f. Multi- + Use *sb.*] Having many uses; intended to serve many purposes. Also **multi-u·ser**, having many users.

1952 *Bull. Amer. Inst. Archit.* VI. iv. 11 The multi-use classroom was used to provide for as many as 4 or 5 different functions within a single space. **1960** *Farmer & Stockbreeder* 29 Mar. 125/1 It has been developed from 10 full years of American 'know-how' and British experience into the ideal multi-use machine for any size of farm. **1964** *Daily Tel.* 30 Jan. 14/2 It rejects the transport theory for the 'multi-user' conception. *Ibid.* 19/1 (*heading*) 'Multi-use' policy 'best for Britain's canals'. **1969** *Jane's Freight Containers 1968–69* 183/1 Berth No. 43—Multi-user container berth. **1971** P. Gresswell *Environment* 181 Examples of multi-use park design, and good maintenance, are the great London parks. **1974** *BP Shield Internat.* Oct. 9/1 (*caption*) The multi-user container berth at Tilbury.

multivalent, *a.* Add: **b.** *Med.* Of an antigen or antibody: having several sites at which it can become attached to an antibody or antigen, respectively.

1934 *Harvey Lect.* XXVIII. 198 In this way it has been possible to show that antigen and antibody in this system, too, are multivalent with respect to each other—that is, that the composition of the precipitate varies according to the relative proportions of the reactants. **1948** Kabat & Mayer *Exper. Immunochem.* iii. 67 The immunochemist.. has found it more useful to..consider the visible agglutination as an indication that the antigen particles are linked together by antibody molecules to form aggregates as a result of the combination of multivalent antibody with multivalent antigen on the particle surface. **1963** Humphrey & White *Immunol. for Students of Med.* vi. 185 The combination in multiple proportions depends upon the fact that both antigen and antibody are multivalent. Antibody valencies are restricted to two per molecule but antigen valencies may exceed 200, although they are more usually around 5 to 10 per molecule.

2. *Cytology.* That is (part of) a multivalent.

1929 *Jrnl. Genetics* XXI. 41 The whole process by which multivalent combinations arise has been followed only in the tulips. **1937** C. D. Darlington *Recent Adv. Cytol.* (ed. 2) iv. 129 The result seems to depend on..the distribution of the chiasmata in the multivalent chromosome. **1967** *Biol. Abstr.* XLVIII. 983/1 (*heading*) Multivalent associations in oocytes of *Triturus helveticus helveticus*.

3. *gen.* Having many applications, meanings, or values.

1933 *Mind* XLII. 484, I propose to call words which can enter sentences in more than one sense multivalent words. **1952** *Essays in Crit.* II. 99 It is unnecessary to heap up detailed parallels for the reader..; our object is to indicate their multivalent mode of functioning. **1963** *Listener* 31 Jan. 213/2 Spenser's allegory is sometimes naïve, often multivalent, and often, indeed, non-existent. **1971** *Archivum Linguisticum* II. 59 Interrogative form considered in the abstract is as multivalent as the flection -*s* and is variously used..for suggestions.., exclamations.., and so on.

B. *sb.* *Cytology.* An association of three or more completely or partly homologous chromosomes during the first division of meiosis.

1929 *Jrnl. Genetics* XXI. 12 At post-diplotene stages associations of three, four and five chromosomes are found, clearly the forerunners of the metaphase multivalents. **1937** M. J. D. White *Chromosomes* v. 82 The frequency of formation of multivalents in polyploids varies a great deal and apparently depends in part on the length of the chromosomes and in part on the rapidity of zygotene pairing. **1959** *Biol. Abstr.* XXXIV. 12/1 Multivalent formation and 'secondary association' in the meiotic stages of autotetraploid races of Lycopersicon Mill. **1971** *Nature* 9 Apr. 390/1 Though most of the chromosomes are associated as bivalents, a few multivalents are present.

multivalence, -valency (further examples); also (after *ambivalence), the property of having many meanings or interpretations.

1933 *Mind* XLII. 49 The solution of these paradoxes lies in the ambiguity... Multivalence would be better. Because it is a multiplicity of use not of meaning. **1937** *Jrnl. Immunol.* XXXII. 119 (*heading*) On the mutual multivalence of toxin and antitoxin. **1940** *Jrnl. Amer. Chem. Soc.* LXII. 2646/1 (*heading*) The bivalence of antibodies and the multivalence of antigens. **1963** Humphrey & White *Immunol. for Students of Med.* vi. 218 The so-called 'sandwich' technique for the detection of antibody ..employs a primary layer of a dilute solution of unlabelled antigen. After reacting for 30 minutes or so, this is rinsed off..and then exposed to specific fluorescein-labelled antibody. Such a method depends upon the multivalency of the antigen. **1965** *English Studies* XLVI. 28 This very multivalence..is the crux of a mythological interpretation, by which such a general mother-goddess may be associated with almost anything on earth and sea.

multivallate (mʌltivæ·lĕ¹t, -ĕt), *a.* [f. Multi- + Vallate *a.*] Having encircling ramparts which form multiple lines of defence. Cf. *bivallate *a.* Also **multivalla·tion.**

1948 *Archaeol. Jrnl.* CV. 50 The best developed barbicans are found in multivallate camps. **1954** *Bull. Board Celtic Stud.* XV. 74 Multivallate 'camps' with various types of lateral or oblique entrance. **1963** [see *bivallate *a.*]. **1963** E. S. Wood *Collins Field Guide Archaeol.* ii. ii. 169 The chief areas of multivallation are the south-west.. and Wessex.

multivalve, *a.* Add: **c.** *Electronics.* Having many thermionic valves.

1920 *Wireless World* Jan. 574/2 More than one valve may be connected in cascade to form a multi-valve amplifier. **1957** *Ibid.* July 310 (*heading*) Multi-valve cathode follower circuits.

multiva·riant, *a.* [f. Multi- + Variant *a.* and *sb.*] Influenced by or taking account of several variables; † *spec.* (of a chemical system) having more than two degrees of freedom (*obs.*).

1902 J. E. Trevor in *Jrnl. Physical Chem.* VI. 136, I would..suggest..that when the variance is successively zero, one, two, three, and more than two, the system be said to be in an Invariant, Univariant, Bivariant, Trivariant, Multivariant, state. **1904** *Ibid.* VIII. 491 The indifferent points of a bivariant or multivariant system form a series analogous to the series of states of equilibrium of a univariant system. **1953** A. K. C. Ottaway *Educ. & Society* iii. 47 Enough should have been said to indicate the multiplicity of factors which must be considered in any analysis of social change. That is why the concept of causation is so dangerous in social science; because we are dealing with multi-variant phenomena. **1959** *Biol. Abstr.* XXXIV. 561/2 (*heading*) The application of multivariant analysis to the physical characteristics of Guarao and Caribe Indians.

multiva·riate, *a.* *Statistics.* [f. Multi- + *variate *sb.*] Involving or having two or more variates or variables.

1928 *Biometrika* XXA. 32 (*heading*) The generalised product moment distribution in samples from a normal multivariate population. **1939** *Proc. Cambr. Philos. Soc.* XXXV. 180 (*heading*) A note on tests of significance in multivariate analysis. **1958** *Economist* 29 Nov. 782 The multi-variate linear regression technique..is satisfactory only if the explanatory variables vary independently of each other or if the degree of interaction between two or more of them can be measured and allowed for. This is certainly not the case in milk production and it was for this reason that the technique was rejected. **1967** C. Berners-Lee in Wills & Yearsley *Handbk. Managem. Technol.* i. 7 Procedures like multivariate regression analysis require a great deal of computation when more than a small number of variables are involved. **1968** R. A. Bradley in *Internat. Encycl. Social Sci.* X. 527/1 Multivariate analysis in statistics is devoted to the summarization, representation, and interpretation of data when more than one characteristic of each sample unit is measured. **1972** *Computer Jrnl.* XV. 215/2 Evaluation of the error matrix is therefore reduced to the evaluation of a multivariate polynomial at a number of points. **1973** *Nature* 16 Mar. 210/3 Altogether, the theory of multivariate distributions poses the most difficult problems in mathematical statistics.

mu·ltiverse. [f. Universe by substituting Multi- for Uni-.] An alternative suggested for the word Universe in order to indicate the absence of order or of a single ruling and guiding power.

1895 W. James *Will to Believe* (1897) 43 Visible nature is all plasticity and indifference,—a moral multiverse, as one might call it, and not a moral universe. **1904** *Daily News* 11 Oct. 3 [Reporting Sir O. Lodge], The only possible alternative was to regard the universe as a result of random chance and capricious disorder, not a cosmos or universe at all, but rather a 'multiverse'. **1920** Chesterton *New Jerusalem* viii. 163 When I told a distinguished psychologist..that I differed from his view of the universe, he answered, 'Why universe? Why should it not be a multiverse?' **1957** *Times Lit. Suppl.* 11 Oct. 602/1 It is precisely Mr Powys's ever-present contact with the vital, or spiritual, principles within the universe which enables him to explore with so uncanny a penetration the deeper problems of that comparatively small section of the universe—or as he would say multiverse—which constitutes man. **1959** N. N. Holland *First Mod. Comedies* 128 Out of this 'pluralistic multiverse', as Robert Oppenheimer has recently called it. **1975** C. Burt *ESP & Psychol.* ii. 34 Modern physics presents us with a heterogeneous multiverse, in place of the homogeneous universe of Newton and Laplace.

multiversity (mʌltivɜ̄·ɹsĭti). Chiefly *U.S.* [f. Multi- + Uni)versity *sb.*] A very large university comprising many different departments and activities.

1963 C. Kerr *Uses of University* i. 42 The multiversity is in the main stream of events. To the teacher and the researcher have been added the consultant and the administrator. Teaching is less central than it once was for most faculty members; research has become more important. **1966** *Economist* 14 May 713/1 For the second year in succession Dr Clark Kerr, the president of the University of California, has been bitterly attacked for permitting too much freedom by a subcommittee of the State Legislature, which finances this great, sprawling 'multi-versity'. **1968** *Guardian* 30 Nov. 2/3 The inevitable transformation of universities everywhere into 'multiversities' is being achieved with appalling birth pangs in the University of California, which a dozen years ago had two campuses and now has 10. **1969** C. Davidson in Cockburn & Blackburn *Student Power* 337 Transforming the academic landscape into what we now call the 'multiversity'. **1969** *Nature* 13 Dec. 1064/1 By the mid-sixties, the result..was clearly an overloading of the university system as a 'multiversity', to use the word of one of its first victims, Clark Kerr. **1971** *Wall Street Jrnl.* 12 Jan. 1/1 They don't like the way it has grown from a sleepy teachers' college of 3,000 students in 1950 to its present status as a giant, cosmopolitan 'multiversity'.

mu:ltivibra·tor. *Electronics.* [ad. F. *multivibrateur* (Abraham & Bloch 1919, in *Compt. Rend.* CLXVIII. 1107), f. *multi-* Multi- + *vibrateur* Vibrator.] A device that consists of two amplifying valves or transistors, each with its output connected to the input of the other, and produces an oscillatory signal rich in harmonics and capable of being triggered and stabilized by an applied sinusoidal signal of slightly higher frequency (the form of the coupling determining the resonant frequency of the device and whether it is astable, bistable, or monostable).

1919 *Sci. Abstr.* B. XXII. 287 The source of oscillations which has been found very successful in operation has been called a 'multivibrator'. It possesses, besides the fundamental wave-length, all the harmonics from 1 up to two or three hundred. **1940** *Amat. Radio Handbk.* (ed. 2) xiv. 219/1 A multi-vibrator operating on 10 kc can be readily locked by a few volts..of 100 kc frequency, and the harmonics of the multivibrator stage will produce carrier frequencies spaced throughout the radio-frequency spectrum at 10 kc intervals with a precision equal to that of the locking source. **1969** J. J. Sparkes *Transistor Switching* iii. 52 The multivibrator can, of course, produce square waves as well as rectangular waves simply by ensuring symmetry between the two halves of the circuit. **1971** J. H. Smith *Digital Logic* iv. 75 Digital systems frequently need a multivibrator, or at least a source of continuous signals.

muly, var. *muley *a.*²

mum, *sb.*³ For *dial.* read *dial.* and *colloq.*, and add further examples.

1876 C. M. Yonge *Three Brides* I. xvi. 274 You'll never do anything with my Governor..you should hear him and the Mum talking. **1955** M. Allingham *Beckoning Lady* vi. 88 Charlie Luke has a mum too, hasn't he? **1956** I. Bromige *Enchanted Garden* iii. iii. 140 This will be our first holiday without the children for a good many years. It'll be good to be with Laurie as a wife and not a mum for a short time. **1960** *News Chron.* 8 Feb. 3/4 The mums from Manchester and Mitcham dressed up in their Sunday best ..had to listen to the principal's report. **1966** Auden *About House* 15 The flesh Mum formulated. **1973** A. S. Neill *Neill! Neill! Orange Peel!* ii. 223 Many married men address their wives as Mum.

b. *Sociol.* Used *spec.* to designate the working-class mother who is a dominating influence in the lives of her children, even in their adulthood and marriage.

1957 Young & Willmott *Family & Kinship in East London* iii. 34 Where Mum plays so large a part in the lives of her descendants, she should be honoured for what she does... Since her status as 'Mum' is so high, it is derogatory to call her by any other name. **1958** M. Kerr *People of Ship Street* ii. 22 The first characteristic which emerges clearly in this group is the central figure of the Mum. Whatever her personal qualities she is the focus of all the family activities. **1958** *Spectator* 30 May 710/2 The Mum restricts her son-in-law's roles and most of the men in the neighbourhood seem to suffer accordingly. **1958** *Listener* 11 Sept. 395/2 Many families who have been moved out of Bethnal Green..are in most cases perfectly wretched. They crave for their old associations, for Mum, the familiar pub, their pets and the old cosy squalor. **1959** *Ibid.* 4 June 981/2 There is, however, some difference of opinion in the field of 'Mum' sociology.

c. Colloq. phr. *to be mum*, etc. = *to be mother* (see *mother *sb.*¹ 3 e).

1962 N. Freeling *Love in Amsterdam* i. 37 'I'll be Mum,' he said, and poured the coffee. **1965** J. Potter *Death in Office* xv. 147, I did the chores as mum..the tea-tray ploy was one of the standard fitness tests for prospective employees.

2. *Comb.*, as **mum figure** *colloq.*, one who represents the attributes of a mother; = *mother figure.

1959 'N. Blake' *Widow's Cruise* 23 I'm a Mum figure—everyone coughs it up in my lap. **1960** *Times Lit. Suppl.* 16 Sept. 589/1 His wife is a shadowy mum-figure who collects for good causes.

mum (mʌm), *sb.*⁴ Abbrev. of Chrysanthemum. Also **'mum.**

1924 A. M. Martineau *Gardening in Sunny Lands* ii. 38 The gardeners call chrysanthemums 'mums'. **1949** [see *cuke]. **1965** J. Philips *Black Glass City* ii. ii. 62 A basket of bright-colored fall flowers..mums and asters. **1967** 'E. Queen' *Face to Face* lxiii. 197 The enormous basket-spray of shaggy 'mums. **1967** *Times* 4 Apr. 18/1 (Advt.), Captivating Meteor Mums... These chrysanthemums..cascade into bloom in August. **1975** M. H. Clark *Where are Children?* i. 16 The window boxes.. were filled with yellow and champagne mums.

mum, vulgar var. Ma'am (examples).

1847 A. Brontë *Agnes Grey* xi. 177 For you know mum he's now't at all to live on, but what he gets fra' th' rector. **1866** Trollope *Belton Estate* (ed. 3) III. vi. 159 The gentleman..was blown up with all the ceremony of which Mrs. Bunce was capable. 'Here he be, mum.' **1973** S. Cohen *Diane Game* (1974) vi. 62 It's me, mum... Are you having dinner in?

mumble-the-peg. Delete restriction of **mumbledepeg** to 7 and **mumblety-peg** to 7, 9 and add **mumbledy-peg** to forms. (Later examples.) Hence **mumblety-pegging** *vbl. sb.*

1931 *Sun* (Baltimore) 18 Feb. 8/7 In my boyhood the spring was ushered in by children rolling hoops, spinning tops, skipping rope, playing mumblety-pegs. **1932** *Ibid.* 17 Sept. 10/7 The Park will attempt next week to match this week's golf tournament with a grand quoit pitchin' and mumblety-peggin' contest. **1938** M. K. Rawlings *Yearling* vii. 64 Jody and Fodder-wing lost interest in the talk and went into a corner to play mumbledepeg. Ma Baxter would never have allowed pocket knives to be flipped into her clean smooth floors. **1963** T. Pynchon *V.* x. 298 They found two more musicians playing mumbledy-peg with clam knives.

mumbo jumbo. Add: **2. b.** Obscure or meaningless talk or writing; nonsense. Also *ellipt.* as **mumbo.**

1896 Farmer & Henley *Slang* IV. 386/2 *Mumbo-Jumbo*, 2. (colloquial).—Unmeaning jargon. **1931** E. Pound *Let.* 27 Dec. (1971) 237 The continuation [of a magazine] can be called Poetry, Second Series, or new series, if that hackneyed term is still big heap mumbo on the lake shore. **1938** *Sun* (Baltimore) 16 Nov. 10/1 The Orson Welles mumbo jumbo of a few weeks ago has revived talk of greater public control over this form of communication. **1955** *Times* 2 May 17/3 A mumbo jumbo of meaningless words and phrases. **1969** M. Pei *Words in Sheep's Clothing* (1970) vii. 54 Mumbo jumbo developed by educators to confound the public in general and inquisitive parents in particular. **1973** *Sunday Express* 29 July 12/3 There is a great deal of unnecessary mumbo talked. **1975** *Times* 14 Aug. 14/5 Labour's elected representatives.. mouth the mumbo-jumbo of capitalism: 'The pound must be kept strong', 'We must all buy British'.

mumbudget. Add: **2.** Phr. *to come mumbudgeting* 'to come clandestinely, secretly' (*E.D.D.*). (See also quot. 1909.) *dial.*

1872 Hardy *Under Greenwood Tree* I. ii. ii. 115 There was this to be said for him, that you were quite sure he'd never come mumbudgeting to see ye, just as you were in the middle of your work, and put you out with his anxious trouble about you. *Ibid.* II. iv. ii. 110 Now, don't come mumbudgeting so close again. **1909** *Daily Chron.* 30 Apr. 6/5 In Mr. Hardy's 'Under the Greenwood Tree'..mumbudgeting..seems to mean rather 'fussily', a sense in which the word budget is still used in the Midlands. **1939** N. Marsh *Overture to Death* xiii. 139 Whatever be the matter with you, then, mum-budgeting so close to my apron strings?

mumchance, *sb.* and *a.* **B.** *adj.* Delete *arch.* and *dial.*, and add later examples.

1897 *Daily Tel.* 11 Mar. 5/5 The man or woman who can sit 'mumchance'..over a well-acted farce do not deserve to be ranked in the noble army of all-embracing playgoers. **1903** G. B. Shaw *Man & Superman* iv. 167, I couldn't sit mumchance and have everything put on me. **1957** L. Durrell *Justine* ii. 133 For my part I remained always stupefied and mumchance at all the avenues opened up by these thoughts. **1971** F. Meynell *My Lives* iv. 42 In a room of awesome vastness with a phalanx of dons at the distant end of it, I was mum-chance.

mu·mchanceness. [f. Mumchance *a.* + -ness.] Silence, reticence. Also **mumchanciness.**

1910 'A. Hope' *Second String* xxi. 435 Perhaps his very

MU-MESON

mum-chanceness was his saving. Glib protestations would have smacked too strongly of the principal to commend the agent. **1920** J. Joyce *Let.* 29 Aug. (1966) III. 17, I am much inconvenienced by their cursed mumchanciness.

mu-meson (miū·mī·zǫn, -me·zǫn). *Nuclear Physics.* Also **mu meson.** [f. *Mu + *Meson³.] The original name for the *muon. (Freq. written *μ-meson.*)

[**1947** Lattes, Occhialini, & Powell in *Nature* 4 Oct. 455/1 There is, therefore, good evidence for the production of a single homogeneous group of secondary mesons. .. It is convenient to refer to this process.. as the μ-decay. We represent the primary mesons by the symbol π, and the secondary by μ. *Ibid.* 455/2 (*heading*) Evidence of a difference in mass of π- and μ-mesons.] **1952** B. Rossi *High-Energy Particles* 566/1 Mu-mesons. **1953** *Jrnl. Brit. Interplanetary Soc.* XII. 203 The final remnants of cosmic-radiation which we observe at sea level are mainly electrons, mu-mesons, gamma rays and a proportionately small component of protons and neutrons. **1969** *Times* 2 Jan. 16/2 The measurements may indicate a means of distinguishing between mu mesons and electrons on grounds other than the difference of mass between the two particles.

Hence **mu-me·sic, -meso·nic** *adjs.* = *muonic *a.*

1954 *Physical Rev.* XCIV. 1619/1 Mu-mesonic energy levels. **1957** *Ann. Rev. Nuclear Sci.* VII. 17 (*heading*) Mu mesonic X-rays. *Ibid.* 495 (Index), Mumesic atoms. **1964** W. E. Jones tr. *A. A. Sokolov's Elementary Particles* vi. 52 Since muons are similar to electrons in their properties.., they may form mu-mesic atoms, where the principal role is played by electrical forces, just as in the normal atom. **1969** *New Scientist* 9 Oct. 63/2 The quadrupole moments derived from mu-mesic X-rays can be used to check the reliability of standard calculations in atomic physics.

Mumetal (miū·metăl). Also **Mu-metal,** and with small initial. [f. *Mu (μ being conventionally used to denote permeability) + Metal *sb.* (and *a.*).] The proprietary name of an alloy of iron that contains approximately 75–78 per cent nickel, 4–6 per cent copper, and 1½–2 per cent chromium by weight and is a useful material for transformer cores and magnetic shields because of its high permeability and low hysteresis loss in weak magnetic fields.

1924 *Trade Marks Jrnl.* 16 Apr. 858 Mumetal... Metallic alloys, unwrought or partly wrought. The Telegraph Construction and Maintenance Company, Limited, .. London, .. manufacturers. **1925** *Jrnl. Iron & Steel Inst.* CXII. 74 The first commercial application of high-frequency melting in Europe was made by a British firm for the preparation of nickel–iron alloys for submarine cables. The research work.. quickly resulted in the perfection of a series of alloys known under the name of 'Mumetal'. **1932** *Discovery* May 142/1 Pure iron, silicon–iron, permalloy, and mumetal are used because they can be magnetized and demagnetized without loss. **1945** *Electronic Engin.* XVII. 384 (*caption*) Recorder unit showing tube enclosed in Mu-metal screen. **1966** *New Scientist* 18 Aug. 351/3 The enclosures were formed of a material (mu-metal) which screens off the Earth's field. **1973** *Physics Bull.* Dec. 719/2 Until now conventional recording heads have had cores of a soft magnetic material, typically ferrite or laminated mumetal.

Mumm (mum). The proprietary name of the champagne produced by the firm of Mumm in Rheims.

[**1851** C. Redding *Hist. Mod. Wines* (ed. 3) vi. 106 The great complaint against Champagne wine has been that it cannot be obtained of a uniform quality... To remedy this evil.. Mumm, Geisler & Co., at Rheims, provided tuns holding twelve thousand litres each.] **1885** *Christie's Auction Catal.* 18 Dec. 17 Three dozens of Champagne, 1868, Mumm. **1907** C. E. Hawker *Chats about Wine* 87 Such names as Cliquot, .. Mumm, Pommery, Roederer, etc., will occur to everyone as carrying with them a guarantee of high-class quality and excellence. **1928** E. Waugh *Decline & Fall* v. 188 The wedding.. was to take place in church with all the barbaric concomitants of bridesmaids, Mendelssohn and Mumm. **1967** 'L. Black' *Two Ladies in Verona* x. 161 A bottle of Mumm Cordon Rouge. I leave the year to you. **1975** J. Symons *Three Pipe Problem* xii. 87 'Glass of Mumm, delicious.'.. 'I shouldn't have thought you could have seen the name.' 'Only the corner of the label. The Mumm label is distinctive.'

mummer (mʌ·məɹ), *v.* [See Mummer 2.] To take part in a mumming. So **mu·mmering** *vbl. sb.* = Mumming *vbl. sb.* 2.

1884 F. Madan in *Bodleian Libr. MS. Eng. Poet. c.* 17 39 Christmas mummering at Ducklington, Oxon. **1964** L. Diack *Labrador Nurse* ii. xiii. 70 From Christmas Day to Twelfth Night.. was the season for 'Mummering' or 'Janny-ing'. There was much dressing up and disguising, and parties went round from house to house to entertain and have fun. **1969** L. J. Chiaramonte in Halpert & Story *Christmas Mumming in Newfoundland* 91 The single male mummer.. does not always accept a drink at the houses he mummers in. **1969** Widdowson & Halpert in *Ibid.* 147 Disguise is a central element in Christmas mummering throughout Newfoundland.

Mummerset (mʌ·məɹset). Also **Mummersetshire.** [Modelled on *Somerset(shire)*, and perh. influenced by Mummer.] An imaginary rustic county in the West of England, and its dialect, invented by actors.

1951 J. B. Priestley *Festival at Farbridge* ii. ii. 219 'Az tew Papular Antertainment,' he drawled in his best Mummerset. **1952** Granville *Dict. Theatrical Terms* 120 An adviser on regional dialect helps the cast to reach a mean which has earned the jocular nickname *Mummersetshire*, or actor's dialect. **1957** [see *honestly *adv.* 2 b]. **1961** *Punch* 15 Mar. 411/2 Stage yokels from Mummerset. **1961** *Times* 19 June 13/4 How long must we wait for a dialect dictionary in one combined volume with pronunciations in Barsetshire and Mummerset? **1965** *Listener* 25 Oct. 640/3 The characters did speak real dialect, and not Mummerset. **1966** D. Blakelock *Eleanor* x. 77 That exaggerated, bogus country dialect known to actors as 'Mummerset'. **1966** C. Mackenzie *Paper Lives* viii. 114 Nowadays you can't be sure if they *are* eggs, even when somebody on television says they are in B.B.C. Mummerset. **1968** N. Marsh *Clutch of Constables* viii. 203, I sat in the Northumberland Arms.. listening to the dullest brand of Mummerset-type gossip. **1970** *Guardian* 26 Mar. 11/3 The straight English is often sloppy and soft; and throughout there is often a damp sprinkling of mummersetshire such as was never heard or spoken. **1971** *Times* 16 Feb. 10/2 Philip Grout's production selflessly follows the spirit of the piece: it is all Mummerset and wagging bottoms, slack jaws and graceless tussles. **1975** *Times Lit. Suppl.* 28 Feb. 220/2 Ordinary people in British war films became more than lovable Cockneys, Mummerset rustics, or Bunny Doyle Northerners full of blunt common sense.

mummied, *a.* Add: **3.** Of a fruit: brown and dry as a result of brown rot disease, caused by a fungus of the genus *Sclerotinia.*

1909 B. M. Duggar *Fungous Diseases of Plants* xi. 190 These mummied fruits are the chief sources of infection the following season, under ordinary conditions. **1935** *Bull. Min. Agric. & Fish.* LXXXVIII. 1 A fruit so infected [with a fungus of the genus *Sclerotinia*], instead of disintegrating becomes dried up and 'mummied'.

mummified, *a.* Add: **2.** Of a fruit: = *mummied *a.* 3.

1928 F. T. Brooks *Plant Diseases* xi. 143 In New Zealand apothecia [of *Sclerotinia laxa*] are only found where mummified fruits have been buried in hard, compact soil. **1973** H. Martin *Scientific Princ. Crop Protection* (ed. 6) xv. 537 For the control of the brown-rot of stone fruits (*Sclerotinia* spp.) the removal and burning of diseased twigs and mummified fruits is of great importance.

mum-mumble (mʌm,mʌ·mb'l). Onomatopœic variant of Mumble *v.*, perhaps under the influence of Mum *v.*

1917 R. Graves *Fairies & Fusiliers* 75 From which the ancient poet was mum-mumbling A song about some Lovers at a Fair. **1956** H. Gold *Man who was not with It* (1965) xxxi. 292 This was our first fret-fingered, mum-mumbling babe.

mummy, *sb.*¹ Add: **3. e.** An apple, plum, or other fruit of the family Rosaceæ, made brown and desiccated by the brown rot disease caused by a fungus of the genus *Sclerotinia.*

1909 B. M. Duggar *Fungous Diseases of Plants* xi. 190 The fruit which has decayed may fall to the ground or hang upon the trees, gradually shrinking with evaporation each to a crumpled dried mass, generally known as a mummy. **1952** E. Ramsden tr. *Gram & Weber's Plant Diseases* ii. 153/2 Similar mummies that have fallen and remained on or near the surface of the soil may very rarely produce clusters of small brown-stalked cup-shaped apothecia.

4. a. *mummy-dust*; **b.** *mummy-dead* adj.; **c.** **mummy brown,** a shade of brown akin to that of the pigment mummy; **mummy-case** (*fig.* examples); **mummy disease,** a disease of mushrooms of uncertain ætiology, indicated by the dying back of young plants, or the distorted shape and hardened gills of older ones.

1886 R. Ridgway *Nomenclature of Colors for Naturalists* ii. 92 *Mummy Brown*, a bright brown color, nearly intermediate in tint between burnt umber and raw umber. **1930** Maerz & Paul *Dict. Color* 168/1 *Mummy Brown*, R[idgway] has a color by this name, 15 c 8, which is simply a somewhat reduced tone of the pigment known as 'mummy'. *a* **1936** Kipling *Something of Myself* (1937) i. 13 A tube of 'Mummy Brown'. **1949** *Dict. Colours for Interior Decoration* (Brit. Colour Council) III. 18/2 Mummy Brown, see Clove Brown. **1922** Joyce *Ulysses* 191 Coffined thoughts around me, in mummy cases, embalmed in spice of words. **1922** L. Mumford *City Devel.* (1946) i. 9 What remained of the provincial town in New England was a mummy-case. **1939** W. B. Yeats *Last Poems* 20 Everything else withered and mummy-dead. **1942** Tucker & Routien in *Res. Bull. Missouri Agric. Exper. Station* No. 358. 3 The common name 'mummy disease' was suggested by Dr. [A. M.] Klingman. It is quite appropriate for the symptoms that develop in white strains or varieties, and with Dr. Klingman's permission, will be used here. **1950** R. L. O. Jackson *Mushroom Growing* vii. 66 The Americans have a similar disease called Mummy Disease, which spreads at an alarming rate, and which they think is caused by a virus, but so far this is of doubtful occurrence in this country. **1969** R. Genders *Mushroom Growing for Everyone* xiv. 198 Mummy disease. For long this has been a source of worry to American growers but only since 1950 has it become known to British growers. **1922** E. Sitwell *Façade* 13 When the moon's hurdy-gurdy wheeze Grinds out her slow mummy-dust.

mummy, *sb.*² Add: Also **mummie.** (Earlier and later examples.)

1784 J. Cullum *Hist. Hawsted* iii. 172 *Mummy*, corrupted from mamma. **1903** *Punch* 30 Sept. 231 Mummy dear, of course Uncle Jack is coming to meet us by a Circle Train, isn't he? **1914** 'Bartimeus' *Naval Occasions* ix. 68 Thank you, mummie darling. **1933** E. A. Robertson *Ordinary Families* ii. 39 'Mummy, did you put in my straw hat?' came Marnie's adenoidal whine from upstairs. 'Oh, mummy, you always say yes. Sure you did? Mummee!' **1974** M. Penoyre *Breach of Security* v. 26 Oh, mummy, you're going out... I'd hoped you were going to ..read me a story.

2. Phr. **mummy's boy** = *mother's boy* (*Mother *sb.*¹ 15 a).

1927 E. Bowen *Hotel* xv. 177 None of us seem to be making much impression on young Ronald... Did you ever see such a Mummy's boy! **1945** E. Taylor *At Mrs Lippincote's* xxi. 180 What a mummy's boy Norman sounds. **1967** 'H. Calvin' *Nice Friendly Town* iv. 49 'Are you a mummy's boy?' 'No,' I said 'but she's a sonny's mum.' **1968** 'P. Hobson' *Titty's Dead* xv. 152 You're not a man at all. Just a mummy's boy. **1975** W. J. Burley *Wycliffe* xi. 173 'What sort of a boy was he?'.. 'Quiet. A bit of a mummy's boy.'

mummy apple, var. *mammee-apple* (Mammee 3).

1905 *Daily Graphic* 16 Jan. 4/4 The mummy-apple, a delicate tree-melon, springs up spontaneously wherever land is cleared. **1911** J. London *Adventure* vii. 85 Mummy apples, which he had regarded as weeds, under her guidance appeared as appetizing breakfast fruit.

mump, *sb.* Add: **4.** A block of peat.

1953 A. Jobson *Household & Country Crafts* vii. 84 Each piece of peat was cut to a uniform size. One man cut the blocks or 'mumps', and another carried them away to dry. **1962** *Amateur Gardening* 5 May 1/4 A 'mump' is a block of peat 10 in × 9 in × 9 in.

mumper. (Later examples.)

1882 *Sydney Slang Dict.* 6/1 *Mumper*, beggar. **1967** *Economist* 30 Sept. 1172/1 This is the total number of 'travellers'—Romanies; mixed-blood 'past-rats' and 'didcois'; 'mumpers', who have no Romany blood; and (mostly Irish) tinkers—recorded by a Ministry of Housing census. **1972** *Countryman* Autumn 86 Beside the gypsies there are many other pickers—tramps, mumpers, all sorts.

mums (mʌmz), shortened form of *mumsy *sb.* and *a.*

1939 L. M. Montgomery *Anne of Ingleside* xxix. 195 It is God who makes everything beautiful, but we can help Him out a bit, can't we, Mums? **1942** J. Cary *To be Pilgrim* cxviii. 254 'How are you really, Johnny?'.. 'Very hungry, Mums darling.' **1942** T. Bailey *Pink Camellia* ii. 18 Say you don't hate me, Mums. **1948** D. Ballantyne *Cunninghams* 8 Dear Mums, I received your ever welcome letter. **1968** 'R. Llewellyn' *End of Rug* (1969) xv. 123 You were always the tough one, Dads... I think I've got more of Mums. She'd cry over a hurt dog.

mumsy (mʌ·mzi), *sb.* and *a.* Also **mumsey.** [f. Mum *sb.*³ + -sy.] **A.** *sb.* A playful variant of Mummy *sb.*² **B.** *adj.* = Maternal *a.* (a playful use).

1876 C. M. Yonge *Three Brides* I. xvi. 274 'Well,' says Mumsey, 'it is not what was thought the thing for ladies in my time.' **1916** *Farmer's Wife* Mar. 248/1 'Dear old motherkins!' 'Good old Mumsy!' **1927** M. Ostenso *Mad Carews* (1928) vii. 93 'Hello, Mumsey!' she greeted her mother. **1953** A. Christie *Pocket Full of Rye* xxvii. 184 Poor Mumsy, she was so devoted to Dad, you know. **1961** 'T. Hinde' *For Good of Company* i. 15 They're all three tucked up in one great big mumsy bed. **1970** 'W. Haggard' *Hardliners* vi. 64 The nurse.. was a West Indian woman, large and mumsy. **1972** J. McClure *Caterpillar Cop* ii. 16 To hell with them and all that crap about mumsy-love.

mumu, var. *muu-muu.

mun (mʌn). Colloq. abbrev. of Money *sb.* Cf. *Mon³.

1896 W. C. Gore *Student Slang* 7 *Mun*, .. money. **1959** M. Steen *Tower* ii. i. 147 You'll adore Kendall—American —tons and tons of lovely *mun*! **1968** J. N. Chance *Rogue Aunt* v. 86 'You should write little tabs, and tie them on,' said Clara... 'Tabs corsts mun,' said Barnskin. **1974** *Sunday Times* 22 Dec. 12/7 They were in this gambling club and he won lots and lots of lovely mun.

Munchausen. Add: **b.** Used *attrib.* and in the possessive with reference to a syndrome in which the patient repeatedly feigns a dramatic or severe illness so as to obtain hospital treatment (see quot. 1951).

1951 R. Asher in *Lancet* 10 Feb. 339/1 Munchausen's syndrome... Here is described a common syndrome which most doctors have seen, but about which little has been written. Like the famous Baron von Munchausen, the persons affected have always travelled widely; and their stories, like those attributed to him, are both dramatic and untruthful. Accordingly the syndrome is respectfully dedicated to the baron, and named after him. **1959** *Perspect. Biol. & Med.* II. 347 The peripatetic medical vagrant, the itinerant fabricator of a nearly perfect facsimile of serious illness—the victim of Munchausen's syndrome. **1967** *Amer. Jrnl. Med.* XLIII. 579/2 This complex of factors.. distinguishes the Munchausen patient from the malingerer, hypochondriac, hysteric, self-mutilator and drug addict, with all of which the diagnosis of Munchausen's syndrome has been confused in the past. **1967** *Cecil-Loeb Textbk. Med.* (ed. 12) 1453/2 The only ones who can be called malingerers with any confidence are some self-mutilating patients and the remarkable pathological liars, picturesquely called examples of the Münchhausen syndrome, who travel from hospital to hospital gaining admission by means of dramatic acts of illness.

Munchi (mu·ntʃi), *sb.* and *a.* Also **Midsi, Mitshi, Munshi.** [Native name.] **A.** *sb.* **a.** A Negro people living in Nigeria near the junction of the Niger and Benue rivers; a member of this people. **b.** The language of this people. **B.** *adj.* Of or pertaining to this people.

1854 S. A. Crowther *Jrnl. Expedition Niger* (1855) ii. 61 There is a tribe on the south bank called Mitshi. They have been represented all along as a wild people and wicked archers. **1883** R. N. Cust *Sk. Mod. Lang. Afr.* I. xi. 231 Michi, Midsi, Mbidsi is the language spoken by a tribe dwelling on the Left or South Bank of the Binué. **1892** A. F. Mockler-Ferryman *Up Niger* iv. 76 The Mitshis are a difficult people to deal with, since they acknowledge no one as head of the tribe, and live in independent families, fearing no one, yet feared by all foreign tribes. *Ibid.*, There was a sort of feeling of relief among our crew when.. we had left the Mitshi country. **1905** C. Partridge *Cross River Natives* v. 89 This angle is peopled by the fierce Munchis or Mitshis, a warlike pagan tribe dreaded by their enemies on account of the deadly poison with which they smear their arrows. **1908** *Daily Chron.* 12 Nov. 5/6 At least one-half of the Munchi country is now open to trade. **1913** F. W. H. Migeod *Lang. West Afr.* II. xxi. 324 The fact of its being found in Munshi is of special interest... This dual connection, having in view the actual numerals which are similar, brings Munshi into touch with languages far to the west. **1933** J. Cary *American Visitor* viii. 96 When you say pagans, do you mean the Munchis.. or the Kukuruku?

munchie (mʌ·nʃi). [f. Munch *v.* + -ie.] **a.** A light meal; the name given to any type of food. *slang.* **b.** In *pl.* Hunger induced after ingesting marijuana; also a snack eaten to satisfy this hunger. *U.S. slang.*

In quot. 1917 a family name for a kind of chocolate.

1917 W. Owen *Let.* 21 Feb. (1967) 437 All I really want is Cigarettes, Munchie, and *plain Cadbury's*. **1959** I. & P. Opie *Lore & Lang. Schoolch.* ix. 163 Food in general is referred to as 'bait'.. 'grub', or 'grubber', 'munchie', [etc.]. **1962** *Austral. Women's Weekly* Suppl. 24 Oct. 3/3 *Munchie*, any type of food. **1971** E. E. Landy *Underground Dict.* 136 *Munchies*, hunger introduced by marijuana—eg. *I have the munchies.* **1971** *Current Slang* (Univ. S. Dakota) VI. 8 *Munchies*, snacks to be eaten after smoking marijuana. **1972** *Dict. Contemp. & Colloquial Usage* (Eng.-Lang. Inst. Amer.) 20 *Munchie*, a snack; food to nibble on. **1973** *Times* 7 Feb. 16/1 There are.. munchies (to be hungry usually after ingesting marijuana).

Munda (mū·nda), *sb.* and *a.* [Native word.] **A.** *sb.* **a.** A member of an ancient Indian people of pre-Aryan stock which was overrun by invading Caucasians and Mongols and which survives in present times as primitive tribes living in north-eastern India. **b.** The name given to the language group which includes the dialects of the Mundas and is believed to belong to the Austroasiatic family of languages. **B.** *adj.* Of or pertaining to the Mundas or their language. Cf. *Kolarian.

1847 B. H. Hodgson *On Aborigines of India* iii. 150 Among the Kóls, I have seen *many* Oraúns and Múndas nearly black. **1854** M. Müller in C. C. J. Bunsen *Christianity & Mankind* III. 437 These people, called themselves 'Munda', which as an old ethnic name, I have adopted for the common appellation of the aboriginal Koles... It is said that the Mundas and Uraons lived peaceably together until the Brahmans reached their country. *Ibid.* 438 The dictionaries of the Munda and Tamulian languages differ more than should be the case with cognate dialects. **1866** *Jrnl. Asiatic Soc. Bengal* XXXV. ii. (Special Number) 26 There are.. 'moondahs' and Santals.. speaking dialects of a language very different from the Dravidian. **1872** E. T. Dalton *Descr. Ethnol. Bengal* vii. 163 The people I am now about to describe comprise the Múndáris or Múndas of Chútia Nágpúr proper. **1877** A. H. Keane tr. *Hovelacque's Sci. of Lang.* iv. 138 *Munda.* The language of the Kols, or Kolhs (south-west of Calcutta), would seem, like Sinhalese, to be independent of the Dravidian group. **1888** *Encycl. Brit.* XXIII. 41/2 There are scattered remnants of a still earlier population of India (Mundas, Kolarians), whose race characteristics.. do not so essentially differ from those of the Dravidians. **1904** G. A. Grierson *Linguistic Survey of India* II. 1 The Mundā order is subject, object, verb, while in Khassi and Mōn it is subject, verb, object. **1912** S. C. Roy *Mundas & their Country* i. 16 The site of the original home of the Mundas will perhaps ever remain hidden from view in the mist of ages. **1933** [see *Austric *a.*]. **1956** J. Whatmough *Language* ii. 30 Munda is now found chiefly in the eastern Himalayas and in isolated regions in the Central Provinces. **1970** Yamada Ryuji *Cultural Formation of Mundas* i. i. 13 The Mundas, together with several other Austroasiatic peoples in India, are called an agricultural people. **1975** *Amer. Speech* 1972 XLVII. 289 The Indo-Aryan languages have acquired various structural features from the Dravidian and Munda languages spoken by the subjected peoples.

mundanity. Delete 'Now *rare*' and add further examples.

1959 *Listener* 30 July 173/2 The outward mundanity of the master's life. **1963** *Movie* July–Aug. 20/2 The presentation of the torture sequences horrifies by its casual mundanity. **1972** *Daily Tel.* 27 Apr. 8/6 Nothing could be further removed from the mundanities of sausage and haddock than the subject of this book of circumferential recollections. **1974** *Nature* 17 May 199/2 It may mean the moral support that leads to a government grant, permission to work abroad for a spell or even such mundanities as the price of an airline ticket.

mung² (muŋ). [See Moong.] In full, *mung bean.* A legume, *Phaseolus aureus*, native to India, where its seeds are an important food.

1866 [see Moong]. **1868** B. H. Powell *Handbk. Econ. Products of Punjab* I. 239/1 Másh, múng and channa (gram), are the pulses most in use. **1884** tr. *A. de Candolle's Orig. Cultivated Plants* v. 346 Green Gram or Múng—.. A species commonly cultivated in India and in the Nile Valley. **1916** C. J. Bamber *Plants of Punjab* 600 *Phaseolus mungo*... Mung... Cultivated for its seeds which are eaten as *dal.* **1955** *New Biol.* XIX. 101 Mung bean preparations given succinate consumed oxygen at one-third of the rate shown by the tissue they were derived from. **1960** J. Organ *Rare Veg.* iv. 79 There are two kinds of bean used for sprouting, the Soy bean (*Soja max*) and the Mung bean (*Phaseolus aureus*). **1969** *Oxf. Bk. Food Plants* 38/2 Green Gram (*Phaseolus aureus*) is often known by the Indian name of 'mung' and is probably native to India... In China and the United States it is also used to produce bean sprouts.. which are popular in certain dishes.

munga² (mʌ·ŋgā). *Austral., N.Z.,* and *Forces' slang.* Also **manga, munger, mungey, mungy.** [Said to be f. F. *manger* to eat.] (See quots.) Also *attrib.*

Quot. 1907 may not belong here.

1907 'Q' *Major Vigoureux* xxii. 218 Annet, Linnet, and Matthew Henry sat.. and watched their friend Jan eat his mid-morning snack—or 'mungey', as it is called in the Islands. **1919** W. H. Downing *Digger Dialects* 34 *Mungy* (Fr., Manger)—Food; a meal. **1925** Fraser & Gibbons *Soldier & Sailor Words* 161 *Mungy wallah*, a man employed in the Cook House. **1929** *Papers Michigan Acad. Sci. & Arts* X. 309/2 *Mungey*, food in general. **1942** C. Barrett *On Wallaby* iv. 70 Those munger vendors did a roaring trade. *Ibid.* vii. 152 There were rush-baskets full of mysterious munger; bricks of sugar, bubbly discs of native bread, and piles of vegetables. **1943** 2 *N.Z.E.F. Times* 25 Oct. 11 He argued quite a lot until munga time. **1947** D. M. Davin *For Rest of Lives* 75 [The Cook] swore he'd give them some good manga to go off with. **1949** 'The Sarge' *Excuse Feet* ix. 102 Herbert felt quite sure she would have put on a much better act for a bit of army *munga*. *Ibid.* 157 *Munga*, food. **1959** S. H. Courtier *Death in Dream Time* vi. 69 Come an' get your munga. **1970** *N.Z. Listener* 12 Oct. 12/4 Certainly do feel hungry. How about a plate of steak and eggs now? Or a rich, smoking mutton chop? A man has to eat. White or brown, everyone scoffs the same munga. **1971** *Sunday Mail* (Brisbane) 24 Oct. 42/4 After the munga Ian showed me his cocktail bit.

mungaree (mʌ̆·ndʒā·ri). *slang.* Also **mungaree, mungare(er, munjari, † numgare.** [ad. It. *mangiare* to eat.] Food.

1861 H. Mayhew *London Labour* III. 139/2 We [*sc.* strolling actors] call breakfast, dinner, tea, supper, all of them 'numgare'. **1889** *Answers* 11 May 374 Broken meat and scraps of bread ('Bull and Munjari' they are called). **1942** C. Barrett *On Wallaby* iv. 64 Chameleons.. are insectivorous and get their own mungaree (food). **1944** L. Glassop *We were Rats* iv. xlvi. 252 Wailed the [Cairene] woman. 'Gibbit *bucksheesh*. Gibbit *mungareer*.'

mungo⁴. Substitute for def.: A fabric made from the short fibres recovered from old hard-woven or felted material. Add further examples.

1961 Blackshaw & Brightman *Dict. Dyeing & Textile Printing* 116 *Mungo*, the poorest grade of *shoddy*, being that obtained from rags, etc., and from materials which have been felted. **1968** [see *Fibro, fibro 1].

mungofa. (Earlier example.)

1836 J. E. Holbrook *N. Amer. Herpetol.* I. 41 *Testudo polyphemus*—Daudin. Synonymes... Gopher and Mungofa, *Vulgo.*

‖ **muni** (mu·ni). *Hinduism & Jainism.* Also **moonee.** [a. Skr. *múni* impulse, eagerness (?), one moved by inward impulse, a seer, saint, etc., f. *man* to think.] An inspired or holy man; a sage; an ascetic or hermit.

1785 C. Wilkins tr. *Bhăgvăt-Gēētā* ii. 41 A man is said to be confirmed in wisdom, when he forsaketh every desire which entereth into his heart, and of himself is happy, and contented in himself... Such a wise man is called a *Mōōnĕĕ.* **1796** [see *dharma]. **1811** W. Ward *Acct. of Hindoos* II. 46 How the moonees instruct their disciples in the different kinds of learning. **1828** H. H. Wilson in *Asiatick Res.* XVI. 18 The *Chárvákas* were so named from one of their teachers, the Muni Chárváka. **1854** M. Müller in C. C. J. Bunsen *Christianity & Mankind* III. 285 The Turanian life is no longer a family life, or the life of a troglodyte Muni. **1866** E. T. Dalton in *Jrnl. Asiatic Soc. of Bengal* XXXV. ii. (Special Number) 160 There is no tradition even of the 'Munis' having sought retreats amongst its rocks or by its waterfalls for their devotional exercises. **1875** Monier-Williams *Indian Wisdom* x. 260 Let him remain without fire, without habitation, feeding on roots and fruits, practising the vow of a Muni. **1949** A. Daniélou *N. Indian Music* I. ii. 39 Matangā Muni considers that the word deshī (worldly) applies to all earthly music. **1969** W. R. Trask tr. *Eliade's Yoga* (ed. 2) viii. 327 Let us recall the *muni* of the *Rg-Veda* who, 'in the intoxication of ecstasy', mounted the 'chariot of the winds'. **1971** *Illustr. Weekly India* 11 Apr. 17/1 The Jain Munis believe that the body is a great source of sin and must be subjugated and won over. **1972** P. Holroyde *Indian Music* iii. 74 The most famous of all their musical theoreticians: Bharata Muni.

Munich (miū·nik). [G. *München.*] The name of the capital of Bavaria used with reference to a meeting of representatives of Germany, Great Britain, France, and Italy on 29 September 1938, when (by the Munich Agreement) the Sudetenland of N. Czechoslovakia was ceded to Germany; also *transf.* as a typical example of dishonourable appeasement. Hence **Munichee·r, Mu·nichite,** advocates of such an appeasement policy. Also **Mu·nichism,** such a policy. Also *attrib.*

1938 H. Nicolson *Diary* 8 Oct. (1966) I. 376 Go up to Leicester. Bertie Jarvis says that I have put the women's vote against me by abusing Munich. **1939** *Ann. Reg. 1938* 76 The Prime Minister followed with his defence of the Munich Agreement. **1939** A. Huxley *After Many a Summer* i. viii. 106 These last months, since the *Anschluss* and Munich, one had found that political discussion was one of the unpleasant things it was wise to avoid. **1939** L. MacNeice *Autumn Jrnl.* 36 Glory to God for Munich. **1941** *Amer. Speech* XVI. 66/1–2 *Munichism*.. 'the spirit of the surrender at Munich; appeasement'. **1942** *R.A.F. Jrnl.* 27 June 13 They do not enter into discussions with dishonourable men..; they have always known the real futility of 'Munichs'. **1942** *Sun* (Baltimore) 5 Aug. 4 (*heading*) Writer finds people hostile to war officialdom, Commons and Municheers. **1944** H. G. Wells *'42 to '44* 76 The misconduct of the war from Munich onward. **1950** A. Wilson *Such Darling Dodos* 88 Tony was not.. unpatriotic, but he had been a great Munichite. **1955** *Ann. Reg. 1954* 24 A further complication.. was the absence of normal diplomatic relations between a number of countries... Someone had to provide a channel of communication, even at the risk of being called a 'Municheer'. **1957** R. W. Zandvoort in *Wiener Beiträge* LXV. 281 *Munich*.. giving rise to the words *Municheers* and *anti-Municheers.* **1958** *Spectator* 6 June 735/2 The pressure put upon President Benes to accept the Munich agreement. **1960** C. Day Lewis *Buried Day* v. 97 The hallucinatory and irresponsible Munich period. **1962** M. Foot *Aneurin Bevan* I. ix. 285 At Bridgwater, Vernon Bartlett, .. won a spectacular victory in the teeth of all the 'peace' propaganda of the Munichites. **1967** *Listener* 5 Oct. 422/3 There are analogies from history—another Munich, another Dunkirk [etc.]—although historians assure us that history never repeats itself. **1973** K. Giles *File on Death* i. 7 Some of the most valuable men in this country were.. Men of Munich, vulgarly so called. Now.. we have little Munichs every week.

municipal, *a.* Add: *municipal law* (see quot. 1959).

1959 Jowitt *Dict. Eng. Law* II. 1201/1 *Municipal law*, that which pertains solely to the citizens and inhabitants of a State, and is thus distinguished from political law and the law of nations. **1965** *Mod. Law Review* XXVIII. 626 There is a useful review of municipal decisions which points to some areas of prospective conflict between community law and municipal law. **1971** *Ibid.* XXXIV. 602 A miscellany of other changes in United Kingdom law will have to be introduced in order to bring municipal law into closer harmony with Community law.

municipalization. (Later examples.)

1958 *Economist* 13 Dec. 967/2 When Labour's election manifesto two weeks ago turned its back on the party's previous extraordinary scheme for the compulsory municipalisation of virtually all rented houses, Conservative propagandists growled angrily like a dog that was being deprived of a bone. **1970** *Internat. & Compar. Law Quarterly* 4th Ser. XIX. ii. 211 The Dutch have.. a system of nationalisation of housing with 85 per cent of all housing in municipal ownership. Nor is this a development occasioned by wartime devastation, as municipalisation can be traced back to a statute of 1905. **1972** *Guardian* 2 Nov. 12/4 Anthony Crosland.. puts the case for the large-scale municipalisation of low-income rented housing. **1974** *Daily Tel.* 12 Jan. 5/2 The 'municipalisation' of privately rented property—except where the owner-occupier shares a house with a tenant—would be encouraged.

municipalize, *v.* (Later examples.)

1955 *Times* 1 June 7/2 The first action of the revolutionaries was to abolish, without compensation, all private ownership of land; a little later most urban dwellings were 'municipalized'. **1973** *Times* 25 Sept. 17/2 The programme of municipalization includes the building societies. .. It also includes the expensive commitment to municipalize rented housing.

municipalizer (miuni·sipălaizəɹ). [f. Municipalize *v.* + -er¹.] One who favours municipal control of public services, institutions, and the like; = Municipalist a.

1908 G. B. Shaw *Commonsense of Municipal Trading* p. ix, The most.. disinterested of them would.. become ardent municipalizers. **1928** *Weekly Dispatch* 24 June 9/3 Within the movement there are (1) prohibitionists, .. (2) municipalisers; (3) advocates of State ownership and control.

municipally, *adv.* Add: **b.** *Comb.*, as *municipally-owned* adj.

1898 E. Howard *To-Morrow* vi. 65 It may be found—especially on municipally-owned land—that the field of municipal activity may grow so as to embrace a very large area, and yet the municipality claim no rigid monopoly. **1972** *Guardian* 4 Dec. 6/8 Hull Corporation.. runs the only municipally-owned telephone service in the country.

‖ **municipio** (miunisi·pio, -tʃi·pio). [Sp. and It.] A municipality; a corporation, a town council. Also *attrib.*

1896 G. Bell *Let.* 14 Apr. (1927) I. 35 The Municipio appeared in splendid gondolas hung with streamers. **1938** E. Hemingway *Fifth Column* (1939) 275 No one will make any trouble forme in Cortina. I know them at the municipio. **1948** K. Davis *Human Society* 317 The *municipios* of Latin America (somewhat similar to our New England townships). **1965** *Language* XLI. 471 All are natives of the large municipio town of Xochistlahuaca, Guerrero. **1972** *Country Life* 2 Mar. 508/1 French is as much in evidence as Italian [in Valle d'Aosta]... it strikes one as odd seeing the word *Mairie* over an Italian Municipio.

munition, *sb.* Add: **2.** Also *colloq.*, the production of munitions; munition-work. *Ministry of Munitions*: a ministry which from 1915 to 1921 controlled the manufacture and supply of munitions. So *Minister of Munitions.*

1915 *Times* 26 May 9/6 The Prime Minister has decided that a new Department shall be created, to be called the Ministry of Munitions. **1915** *Act* 5 & 6 *Geo. V* c. 54 § 4 If the Minister of Munitions considers it expedient.. that any establishment in which munitions work is carried on should be subject to the special provisions, [etc.]. **1917** Hall Caine *Our Girls* i. 11 By permission of Mr. Montague, the Minister of Munitions.. we are at the gates of the great Arsenal. **1917** *Dalton* (Lancs.) *Guardian* 28 Apr. 3/5 He had been sent to munitions, and had not been out to the front. **1924** B. Gilbert *Bly Market* 66, I expect.. you'll be leaving the schooling and go to the munitions. **1925** D. Carnegie *Hist. Munitions Supply in Canada* xxvi. 251 Mr. H. E. Morgan.. was sent out to Canada from the Ministry of Munitions. **1935** A. J. Cronin *Stars look Down* ii. vii. 301 There was a future in munitions... They were going to put up a line of sheds.. filling sheds. **1957** *Encycl. Brit.* XV. 963/1 On the formation of the first wartime coalition government in 1915, a major change in organization was made by setting up a ministry of munitions, with David Lloyd George as the first minister.

5. *munition factory, girl, -maker, -making, work, worker, works*; also *munitions work, worker.*

1909 *Westm. Gaz.* 9 Oct. 2/2 In 1895 he visited the chief firearm and munition factories of France. **1921** G. A. B. Dewar *Great Munition Feat* v. 112 The notion that a munition factory was a place full of shirkers and profiteers is grossly ignorant. **1916** M. Cosens (*title*) Lloyd George's munition girls. **1918** *Daily Mirror* 12 Nov. 6/2 Soldiers, sailors, munition girls and mere civilians clung on anywhere. **1916** *Home Companion* 12 Aug. 16/1 This is my last chat to you, little mother munition-makers. **1916** 'B. Cable' *Doing their Bit* 24 No man or lathe or tool that can be turned to munition-making is possibly doing anything else. *Ibid.* 40 Anything less promising of munition work it would be hard to find. **1918** *Times* 27 Mar. 3/1 These are all chapters in the romance of munitions work in the Midlands. **1915** *Daily Sketch* 18 Aug. 2 (*heading*) Badges for the volunteer munition worker. **1915** W. Owen *Let.* 16 Oct. (1967) 359 Dr. Rayner says I should become a Munitions Worker at Birmingham. **1925** D. Carnegie *Hist. Munitions Supply in Canada* xxvi. 250 One manufacturer said that it cost him approximately $300 to train each munition worker. **1957** *Encycl. Brit.* XV. 964/1 Specialist ladies were set up to deal with the health of munitions workers, especially women. **1917** W. Owen *Let.* 7 Sept. (1967) 491 The other owner of a large Munition Works. **1932** H. Simpson *Boomerang* xvi. 436 The women were haring off to munition-works at five pounds a week. **1940** G. D. H. & M. Cole (*title*) Murder at the munition works.

munition, *v.* Add: **3.** *intr.* To do munition work; to work in a munition factory.

1916 'B. Cable' *Doing their Bit* 23 A man cast for a commission and refused for the ranks a year ago on account of his eyes has 'gone munitioning'.

munitioneer (miuniʃənīᵊ·ɹ). [f. Munition *sb.* + -eer.] A worker in a munition factory.

1916 E. V. Lucas *Vermilion Box* 254 In the need for copper there is quite a good price for engraved plates, and theirs have been weeded out for the munitioneers. **1919** *Athenæum* 23 May 360/1 'Trinitrotoluene', which the munitioneers shortened to T.N.T. **1927** W. Deeping *Kitty* xxvii. 342 The men.. returned to a land that was not full of heroes... The voice of the yellow dog was heard in it, the snarl of the ex-munitioneer. **1940** *New Statesman* 19 Oct. 376 The munitioneer can see no difference in the management or control of his factory.

munitioner. Add: **c.** = *munitioneer; a maker of ammunition.

1917 *Graphic* 30 June 806 (*caption*) The King with his munitioners.

munitionette (miuniʃəne·t). *colloq.* [f. Munition *sb.* + -ette.] A female worker in a munition factory.

1915 *Daily Sketch* 9 Nov. 13/1 (*heading*) Munitionettes who receive threepence an hour. **1917** *Daily News* 17 May 3/1 A shell-shop filled with blue-clad mob-capped cheering munitionettes. **1919** *Punch* 7 May 366/2 Work for the ex-munitionette drawing unemployment pay. **1935** A. J. Cronin *Stars look Down* ii. xiii. 381 He had taken a little flutter with a munitionette from the Wirtley Works.

munitionless (miuni·ʃənlės), *a.* [Munition *sb.* + -less.] Not provided with munitions.

1927 W. S. Churchill *World Crisis 1916–18* i. 224 The mastered agony of the munitionless retreat, the slowly regathered forces;.. has he no share in these?

munitionment (miuni·ʃənmĕnt). [f. Munition *v.* + -ment.] Provision with or supply of munitions; munitions collectively.

1915 H. Belloc in *Land & Water* 29 May 10*/1 If the Austro-German forces under the effect of superior munitionment for the heavy pieces do pierce their opponent's line. **1917** [see *gunning *vbl. sb.* 3]. **1929** J. Buchan *Courts of Morning* i. 133 Science has now created a norm of weapon and munitionment, which is substantially the same for all armies. **1930** H. Jackson *Anatomy of Bibliomania* I. viii. v. 167 Whether they have proved successful as armour or not, they [*sc.* Bibles] are no despicable munitionment of war in other respects.

Munro (mʌnrōᵘ). *Mountaineering.* Name applied to Scottish mountains of at least 3,000 feet after Sir H. T. *Munro*, who published a list of all such peaks in the Journal of the Scottish Mountaineering Club for 1891.

1903 *Jrnl. Scottish Mountaineering Club* VII. 366 The view from the top was magnificent, all the big Munros in the neighbourhood showing up clear and resplendent. **1972** D. Haston *In High Places* i. 3 These [*sc.* 'the bigger hills of Scotland'] are relatively small, mostly around 3,000 feet, and called 'Munros'. **1973** Sillar & Meyler *Skye* x. 188 Between then [*sc.* 1889] and 1891 H. T. Munro —later Sir Hugh Munro—listed Scottish peaks over 3,000 ft... These are known among Scottish mountaineers as 'Munros'.

Munsell (mʌ·nsĕl). The name of the American Albert H. *Munsell* (1858–1918), used *attrib.* in connection with his classification of colours by means of the three quantities hue, 'value' (lightness or brightness), and chroma, each of which can be assigned, for any particular colour, a numerical value on prescribed scales.

1905 H. E. Clifford in A. H. Munsell *Color Notation* 4 The Munsell photometer.. is an instrument of wide range, high precision, and great sensitiveness, and permits the valuations which are necessary in his system to be accurately made. **1913** [see *colour-balance* (*colour *sb.*¹ 17)]. **1937** *Discovery* Oct. 325/2 Other well-known systems besides Ostwald's, such as the American 'Munsell' colour-system. **1950** *Psychol. Abstr.* XXIV. 278/2 The key indicates the Munsell notation for each of the Ridgway color names. **1954** *Archit. Rev.* CXV. 198/2 The following colours from the Munsell range were used: putty-colour on metal panels below windows, [etc.]. **1957, 1959** [see *chroma]. **1964** H. Hodges *Artifacts* xvii. 196 Colour readings using the Munsell Chart are, perhaps, most useful when considering the variation of firing conditions in a large sample of sherds of similar composition. **1971** *Jrnl. Gen. Psychol.* LXXXIV. 24 Ball is beginning to trace saturation changes in monochromatic input, employing Munsell Color Chips as matching targets.

munshi, var. Moonshee.

1849 [see Moonshee]. **1947** R. O. Winstedt *Malays* 152 His father was *munshi* to William Marsden. **1968** D. Duff *Victoria in Highlands* 360 Abdul Karim advanced to become the Queen's India Secretary and was thereafter known as the Munshi Hafiz Abdul Karim... The Munshi's privileged position led to much opposition. **1975** *Times* 26 July 7/6 The language teacher, or munshi.. was larger than I expected.

Munster (mu·nstəɹ). Also **Muenster, Münster.** The name of a town in the Haut-Rhin Department of France used *attrib.* and *absol.* to designate a strongly flavoured, semi-soft cheese made in the Munster valley.

1902 *Encycl. Brit.* XXV. 336/2 Cows are grazed on the S. Vosges in summer and large quantities of cheese (Münster cheese) are made and exported. **1946** G. Millar *Horned Pigeon* xvii. 245 A fat chicken, a bottle of Burgundy, and a Münster cheese. **1947** M. Given *Encycl. Cooking* I. 661 Muenster... Ripened 2 or 3 months to develop pronounced flavour. **1954** H. Smith *Classical Recipes of World* 379 Munster, a French semi-hard, whole-milk fermented cheese. **1960** E. David *French Provincial Cooking* 34 A little bowl of caraway seeds came with the Münster, a strong, rich, creamy textured cheese which, at the right stage of ripeness, is one of the great cheeses of France. *Ibid.* 87 The Münster cheese of Alsace. **1971** *Sunday Times* (Colour Suppl.) 28 Mar. 34/2 The softer a Münster looks the creamier it will be.

munt (munt). *S. Afr.* and *Rhodesian slang.* [ad. Bantu *umuntu*, sing. of *abantu* (see *Bantu *a.* and *sb.*), lit. a person, black person, servant.] A black African: usu. as a term of contempt.

[**1926** G. Callaway *Fellowship of Veld* iii. 25 To the Native the qualities which go to make up *ubuntu*, the qualities which make an *umntu* (person), are largely social. **1937** E. G. Malherbe *Educational Adaptations in Changing Society* xx. 500 Standard of development will be judged by the extent to which the Native has ceased to be an Umuntu and has become a European.] **1948** O. Walker *Kaffirs are Lively* 77 It's the towns that muck the *munt* up. **1953** N. Gordimer *Lying Days* i. iii. 35 Man, there's a whole lota niggers round Ockerts', all over the garden and in the street and everywhere. Just a lot of munts from the Compound. **1962** *New Statesman* 24 Aug. 218/1 The old 'munt', as the African is still widely and insultingly termed. **1964** *Listener* 13 Feb. 257/1 His friends were muttering.. about 'cheeky munts', which is the insulting white Rhodesian term for Africans. **1972** P. Driscoll *Wilby Conspiracy* (1973) xxiv. 309, I used the whistle too, of course, but those two munts in the truck didn't even hear it.

muntjak. Substitute for def.: A small south-Asian deer of the genus *Muntiacus*, which has been introduced into parts of western Europe. (Later examples.)

1939 *Geogr. Jrnl.* XCIV. 429 The Chinese tufted muntjac, or Mickie's tufted deer. **1963** *Times* 13 Mar. 14/7 A muntjac buck ventured into a well-populated corner of Hertfordshire to lie up beneath a hedge of holly in a secluded garden. **1965** D. Morris *Mammals* 384 Muntjacs are usually found singly or in pairs in the thick undergrowth. **1971** *Guardian* 8 Dec. 12/2, I have had the sad task of identifying the corpse of a Muntjac. **1973** *Country Life* 29 Nov. 1763/3 Besides red deer and roe, muntjac have also recently been reported among the Thetford trees.

Muntz (mʌnts). The name of George Frederick *Muntz* (1794–1857), English political reformer and metallurgist, used (now always in *Muntz metal*) to designate a type of brass he patented (*Brit. Pat. 6325* (1832), *11,410* (1846)) that contains about 55–64 per cent of copper (often with 1 per cent or more of lead), can be readily hot-worked, and is used esp. in shipbuilding.

1860 *Chambers's Encycl.* I. 159/1 Muntz sheathing-metal, 16 [parts] copper and 10⅔ zinc. *c* **1865** J. Wylde *Circle of Sciences* I. 376/1 An alloy of copper, .. called Muntz's metal, is.. employed for sheathing ships' bottoms. **1866** H. E. Roscoe *Less. Elem. Chemistry* xxv. 217 The yellow, or muntz metal, used for the sheathing of ships, contains sixty per cent. of copper. **1887** D. A. Low *Introd. Machine Drawing & Design* xiv. 58 The piston.. is attached to a Muntz metal piston rod. **1930** *Engineering* 17 Jan. 68/2 Details of the water cut-offs, consisting of a combination of Muntz metal plates, asphalte and iron sheeting. **1963** *Listener* 17 Jan. 114/2 The Cutty Sark Society is offering an original sheet of Muntz metal [from the hull of the ship] to anyone who pays for a new one. **1969** E. N. Simons *Dict. Alloys* 112 Typical Muntz Metal parts are heavy plates, bolting and valve stems, heat-exchanger tubes, hot forgings, and brazing rods for copper alloys and cast iron.

munyeroo (mʌ·nyərū·). *Austral.* Also **munyeru.** [Aboriginal name.] A small, succulent herb, *Claytonia balonnensis*, of the family Portulacaceæ; also, the paste made from its ground seeds mixed with water, formerly used as food by Aborigines in central Australia.

1896 E. C. Stirling in B. Spencer *Rep. Horn Sci. Exped. Central Austral.* iv. 56 In these districts 'Mūnyeru' takes the place of the spore cases of 'Nardoo'.., which is so much used in the Barcoo and other districts to the south and east. **1934** A. Russell *Tramp-Royal in Wild Austral.* xx. 127 The munyeroo, a form of pigweed, was also plentiful... The natives collect the seed in large quantities and pound it into flour for bread; they also eat the leaves fresh, both as a food and to relieve thirst. **1935** *Bulletin* (Sydney) 1 May 20/2 The succulent leaves of the munyeroo and parakylia, which the birds feed upon, provide sufficient moisture for their needs. **1941** I. L. Idriess *Great Boomerang* xxiii. 174 Some passing shower has fallen here, and the yellow gold is the munyeroo plant in flower. **1956** *Landfall* X. 97 Small black and brown munyeru seeds.

muon (miū·ǫn). *Nuclear Physics.* [f. *Mu(-meson + *-on¹.] A lepton that appears to be almost identical to the electron, except for being unstable and having a mass about 207 times greater, and is the chief constituent of cosmic radiation at the surface of the earth. (Orig. called a mu-meson, but now no longer classed as a meson.)

1953 *Sci. News Let.* 3 Jan. 14/1 About 20 or so particles exist in or can be knocked out of the atoms. Some of these are well-known oldtimers, like the electron... Others are new and stranger, like the pions, the muons, the V-particles. **1958** *Engineering* 4 Apr. 430/2 The fusion of hydrogen and deuterium (*p* – D reaction) catalysed by negative muons. The part played by the muon is to bind the two nuclei in a mesic molecular ion, so bringing them close enough together for fusion to occur. **1961** *Guardian* 21 Jan. 2/4 Muons were first discovered in the naturally occurring cosmic rays and identified precisely in 1947.. at Bristol University. **1968** M. S. Livingston *Particle Physics* iv. 74 Muons are the 'penetrating' component of the ionizing particles in cosmic radiation observed beneath great layers of earth in salt mines. *Ibid.*, Slow μ⁻ muons can also displace electrons in atomic orbits and form temporarily stable atoms... Eventually, the muon decays into an electron and two neutrinos. **1968** *Times* 28 Nov. 14/1 The muons are created by the collision of cosmic rays with matter in the upper atmosphere of the earth. **1972** *Nature* 14 July 86/1 The mass difference of the electron and muon poses one of the inscrutable problems of particle physics. Except for the mass difference the muon does not appear to be any different from the electron. **1974** *Ibid.* 2 Aug. 377/2 Muons are more penetrating than any other charged particles.

b. Special Comb.: **muon number,** a quantum number assigned to sub-atomic particles that is ±1 for muons and their neutrinos and 0 for other particles and is conserved in all known interactions.

1961 *Physical Rev.* CXXIII. 1439/2 The possibility of essentially imaginary relative parities can only arise when some quantum number (here, muon number) is multiplicatively conserved. **1970** D. B. Lichtenberg *Unitary Symmetry* i. 7 Other important symmetries are the invariances under gauge transformations. These symmetries are associated with conservation of charge Q, baryon number B, and two lepton numbers, the electron number L_e and the muon number L_μ.

Hence **muo·nic** *a.*, of, pertaining to, or involving a muon, or an atom having a negative muon orbiting the nucleus.

1955 S. S. Schweber et al. *Mesons & Fields* II. 369 The same considerations lead to the conclusion that the two neutrinos ejected in μ decay are like neutrinos, even though a Dirac theory of neutrinos opens the possibility that one is an anti-neutrino... Such conclusions are important in calculating expectations for the muonic processes. **1965** *New Scientist* 1 July 36/1 Studies of the so-called 'muonic' X-ray spectra do not call for high energies. **1969** *Sci. Jrnl.* July 44/1 In a dense material, a muon is slowed down to a velocity corresponding to 1 keV in about 1 nanosecond, after which it is captured into a 'Bohr orbit', forming a hydrogen-like atom known as a muonic atom. **1972** *Physics Bull.* Mar. 149/2 Muonic molecules are systems in which a negative muon binds together two nuclei. Some examples are $(p\mu^-p)^+$, $(p\mu^-d)^+$, and $(d\mu^-d)^+$, and in practice only such ionic states are of interest. **1972** *Sci. Amer.* Nov. 104/3 The behavior of muonic atoms has provided much information about the structure of the nucleus, particularly about the distribution of protons within nuclei.

muonium (miu,ǫᵘ·niǝm). *Nuclear Physics.* [f. *muon + -ium, after *positronium.] A short-lived neutral system, analogous to an atom, consisting of a (usu. positive) muon bound to an electron.

1957 Friedman & Telegdi in *Physical Rev.* CV. 1681/2, μ^+ can form 'muonium', i.e. (μ^+e^-). **1960** *Physical Rev. Lett.* V. 63/1 The present Letter reports the formation of muonium in pure argon gas. *Ibid.*, Muonium.. can be formed when a positive muon is slowed down in matter and captures an electron from an atom in the stopping material. **1964** S. DeBenedetti *Nuclear Interactions* viii. 593 The absence of muonium annihilation, despite the fact that both muons and electrons behave separately as simple Dirac particles, furnishes further evidence for the basic difference between electrons and muons. **1970** G. K. Woodgate *Elem. Atomic Struct.* ii. 20 The equation therefore applies to all the hydrogenic atoms hydrogen, deuterium, tritium, muonium, positronium, etc.

‖ **muraji** (mură·dʒi). Also, with prefixed ō- 'great'. [Jap.] In early imperial Japan, one of the hereditary titles for a family or clan (cf. *kabane) who claimed their descent from other gods than the divine ancestors of the emperor. Also, **ō-muraji,** the representative of this family when he was entrusted with affairs of state.

1901 F. Brinkley *Japan* I. iii. 53 Such titles as 'great body' (*omi*), 'master of the multitude' (*muraji*), 'honourable intermediary' (*nakatomi*) and so on, were employed as terms of respect, and ultimately passed into use as official titles. *Ibid.*, The head of the clan then came to be distinguished by the prefix *O* (great or senior); as *O-mi* (the senior honourable person), *O-muraji* (the great master of the multitude), and so on. **1931** G. B. Sansom *Japan* i. ii. 37 The *ō-muraji*, territorial administrative officers of high rank who traced their descent from gods other than the divine ancestors of the emperor. *Ibid.* 39 It became the custom to describe the more important members of a clan or corporation by the name of their hereditary office or by some honorific title granted by the court. Thus we have *muraji*, which means 'leader of a group'. **1964** *Japan* (Unesco) (ed. 2) i. 16/2 The clans.. had *kabane* (hereditary family title), to show the status of their families. *Kabane* were divided into *omi, muraji, kimi, atai, obito, miyatsuko* and *fubito.* **1970** J. W. Hall *Japan from Prehist. to Mod. Times* v. 37 Spokesmen chiefs, *Ō-omi* and *Ō-muraji*, were named to serve as chief ministers of state.

mural, *sb.* Restrict † *Obs.* to sense in Dict. and add: **2.** [Short for *mural painting* (Mural *a.*¹ 2 a).] A painting executed on a wall or ceiling as part of a scheme of decoration. So *mural painter.* orig. *U.S.*

1921 *Quill* Mar. 23 I'm doing big things... 'They satisfy' and such outdoor murals. **1929** *Arts & Decoration* Oct. 63 (*heading*) Arresting murals of classic inspiration in the Long Island home of F. Russell. **1931** H. Crane *Let.* 4 Nov. (1965) 385 He's fundamentally a mural painter, and even his smaller paintings have a tremendous *scale.* **1936** R. E. Sherwood *Idiot's Delight* ii. i. 76 He has been out in Australia doing colossal murals for some government building. **1946** H. Feibusch *Mural Painting* 15 H. J. K. Tisdall, the mural painter, helped me with his wide knowledge of technique and materials. **1949** F. Maclean *Eastern Approaches* i. ii. 18 The Customs' building [in the frontier station of Negoreloye].. was a fine big, bright room, decorated with murals depicting scenes from Soviet life. **1954** T. Gunn *Fighting Terms* 43 Turn your closed eyes to see upon these walls A mural scratched there by an earlier man. **1964** Mrs. L. B. Johnson *White House Diary* 17 July (1970) 181 There was a great mural in the Library by Thomas Hare Benton. **1973** F. Taubes *Painter's Dict. of Materials & Methods* 159 Only a certain portion of a large mural can be completed in one day's work. **1974** *Impressionism* (R. Acad.) 32/2 Saint-Sulpice murals, completed 1861, were a revelation for young artists.

mural, *a.*¹ Add: **2.** (Further example.)

1916 H. F. Osborn *Men of Old Stone Age* 316 This Art.. is also mural or parietal.., consisting of drawings, engravings, paintings and bas-reliefs on the walls of caverns and grottos.

muralled, *a.* Restrict *rare* to sense in Dict. and add: **2.** Decorated with murals. Cf. *mural *sb.* 2.

1962 *UCLA Librarian* (Los Angeles) 25 May 106 The oaken walls and muraled ceiling of the drawing room were pulsing with sweet sound. **1967** A. West in *Coast to Coast 1965–6* 218 Stella watched the shadows.. as they bobbed on the muralled wall.

muramic (miūᵊræ·mik), *a.* *Biochem.* [f. L. *mūr-us* wall + Am(ine + -ic.] *muramic acid*: an amino-sugar, $C_9H_{17}NO_7$, which is present in combined form in the cell walls of bacteria and in bacterial spores.

1957 *Nature* 27 Apr. 841/2 An acidic hexosamine, first found in a product from bacterial spores, which Strange has provisionally characterized as 3-O-α-carboxyethyl hexosamine and has named 'muramic acid'. **1969** [see *murein]. **1973** *Physiol. Plant Path.* III. 366 Muramic acid and glucosamine are less stable to acid hydrolysis than are amino acids.

‖ **murchana** (mū·ɹtʃanā). Also **murcchana, murchhana.** [a. Skr. *mūrchanā* modulation.] (See quots.)

1891 C. R. Day *Music in S. India* ii. 23 The s'rutis are arranged in their different svâras, or intervals of the scale, according to the 'murchanas'. Of what these murchanas consisted is very doubtful. *Ibid.* iv. 41 The following short melodies.. show as much as possible the full murchanas of their respective râgas. **1914** [see *jati, jâti]. **1921** H. A. Popley *Music of India* vi. 85 We shall notice.. the graced notes, these being called in the north Mūrchanā. **1954** *Grove's Dict. Mus.* (ed. 5) IV. 457/1 Now it is clear that the *Sa-grâma* and the *Ma-grâma* are modes of one another. All the same, the Indian theory considers them as separate entities and derives seven octave scales (*mûrchanâs*) from each of them. **1968** *Indian Mus. Jrnl.* V. 54 The term '*mūrcchanā*'. In the early sources this refers to an octave-species within a basic tuning-system... One or two 17th-century *śāstra-s*, on the other hand, use the term '*mūrcchanā*'.. to refer to that 'pitch-area' within which the phrases of a *rāga* operate. **1972** P. Holroyde *Indian Music* 274 *Murchhana*... The word has been variously used: (a) It is the ancient name for the early modes in Indian music before the raga system emerged. (b) According to Popley's book,.. murchhana in South India refers to ascending and descending varna or graph-like grace movements. (c) According to Tagore the murchhana is the extending of the note 'to another in the ascending as well as the descending scale (like an *appoggiatura*) without any intermediate break in the disposition of the srutis in the interval'.

murder, *sb.* Add: **1. e.** Phr. *to get away with murder*: see *get *v.* 54 c.

f. An excellent or marvellous person or thing. *U.S. slang.*

1940 *Music Makers* May 37/3 *Murder*, something excellent or terrific... 'That's solid murder, gate!' **1943** M. Shulman *Barefoot Boy* ix. 90 We got on the dance floor just as a Benny Goodman record started to play. 'Oh, B.G.!' cried Noblesse... 'Man, he's murder, Jack.' **1948** H. L. Mencken *Amer. Lang.* Suppl. II. 707 The vocabulary of the jazz addict is largely identical with that of the jazz performer.. anything excellent is *killer-diller, murder* or *Dracula.* **1970** C. Major *Dict. Afro-Amer. Slang* 83 *Murder*, (1930's–40's) excellent; the best.

3. *to cry blue murder* (earlier and later examples); also *blue murder*, a loud or alarming noise, a great commotion, din, or disturbance; used in intensive phrases as *like blue murder*, at a terrific pace, at top speed. *colloq.*

1859 Hotten *Dict. Slang* 8 *Blue-murder*, a desperate or alarming cry. *c* **1874** D. Boucicault in M. R. Booth *Eng. Plays of 19th Cent.* (1969) II. 228 They were standing by and thrying to screech blue murdher. 'Stop their mouths,' said a voice. **1893** G. B. Shaw in *World* 10 May 28/2 What it [*sc.* the slow movement of Stanford's Irish Symphony] does end in is blue murder. **1914** *Evening News* 1 Oct. 2/1 They were off down the road like blue murder. **1921** G. B. Shaw *Back to Methuselah* ii. 84 You couldnt produce it. There would be blue murder. It's out of the question. **1959** 'A. Gilbert' *Death takes Wife* xiii. 164 Corpses don't yell blue murder.

4*. In *fig.* and hyperbolic use: (an act of) destruction or spoliation supposed to be tantamount to murder. Also in weakened senses: a situation or condition that is very unpleasant or undesirable.

1857 Trollope *Barchester Towers* II. ii. 37 This cellar is perfectly abominable. It would be murder to put a bottle of wine into it till it has been roofed, walled, and floored... Goodenough never had a glass of wine that any man could drink. **1878** 'R. Boldrewood' *Ups & Downs* ix. 90 What a murder that one should have all these hundredweights of nails, .. and forests of posts and wallplates to get all over again! **1924** Kipling *Debits & Credits* (1926) 316, I was never keen on bombin' myself... But bombin'-instruction's murder! **1951** J. B. Priestley *Festival at Farbridge* ii. ii. 255 Cook's gone, and it's murder trying to do it all myself. **1956** A. J. Lerner *My Fair Lady* (1958) i. i. 8 By right she should be taken out and hung For the cold-blooded murder of the English tongue! **1960** H. Pinter *Room* in *Birthday Party*, etc. 105 *Rose.* You look cold. *Mrs Sands.* It's murder out. Have you been out? **1965** M. Bradbury *Stepping Westward* i. 64 Private life was simple enough, but the communal centres were murder. **1973** A. Ross *Dunfermline Affair* 69 An old hip injury... Not so bad when I'm just walking.. but murder climbing stairs.

4.** A popular parlour or children's game for a number of participants, which involves a mock murder hunt, led by a 'detective', to find the 'murderer' of one who is playing dead, having been 'murdered' in the dark. Also called **murders, the murder game, murder in the dark.**

1933 Phillips & Westall *Bk. Indoor Games* iv. 276 To give as good an idea of 'Murder' as we can, we will describe in narrative form an actual game. **1934** N. Marsh *Man lay Dead* i. 14 Silly games are played... It's going to be Murders this time. *Ibid.* 20 Are we really going to play the Murder Game? **1937** M. Hillis *Orchids on your Budget* (1938) vi. 103 You can still serve refreshments after bingo played for pennies or the murder game. **1948** C. Day Lewis *Otterbury Incident* v. 59 We'd get him interested if we made a detective job of it—sort of Murder Game. **1964** R. Jeffries *Embarrassing Death* xii. 147 Did you ever play the game 'Murder'?.. Everybody but the murderer must tell the truth: the murderer may lie. **1972** G. Brandreth *Party Games* 24 Murder in the Dark. 6 or more players. A chilling game, definitely not for those with weak hearts.

5. *attrib.* and *Comb.*, as *murder bout, case, charge, film, gun, hunt, story, trial, victim, weapon*; **murder bag** (see quot. 1938); **murder book** (**file, log**), a book or file in which are kept details of a police investigation of a murder; **murder game**: see sense 4**; **murder inquiry,** a police investigation into a case of murder; **murder investigation** = **murder inquiry*; **murder-man,** (*b*) a writer of murder stories; cf. *murdermonger; **murder mystery,** a mysterious murder; *spec.* a murder story in which the murderer's identity is concealed by a complicated plot until the dénouement; **murder one** *U.S. colloq.*, (a charge of) first-degree murder; **murder rap** *slang* (orig. *U.S.*), a charge of murder; **murder room,** after the discovery of a murder, a (nearby) room used as a centre for directing a police inquiry into the crime; **murder squad,** a division of the police appointed to investigate crimes of murder.

1938 F. D. Sharpe *S. of Flying Squad* vi. 65 In the Superintendent's office at Scotland Yard repose two plain cowhide bags... They are the Murder Bags which contain all the tools which a detective is likely to need in solving a major crime. **1962** 'J. Bell' *Crime in our Time* v. iii. 156 He takes with him all the necessary apparatus for a detailed examination on the spot, the so-called 'murder bag'. **1972** 'A. Garve' *Case of Robert Quarry* i. i. 8 Methodically checking the contents of his murder bag. **1972** J. Wainwright *Requiem for Loser* viii. 162 The final write-up would be bound into a single volume, called 'The Murder Book'. **1973** Murder Book [see *murder log* below]. **1906** Hardy *Dynasts* II. vi. vii. 298 Bonaparte and Alexander .. Are closing to a mutual murder-bout. **1930** A. Christie *Murder at Vicarage* xv. 118, I should never have suspected that Hawes would take such a keen interest in the details of a murder case. **1974** M. Birmingham *You can help Me* iii. 52 This is a murder case... To answer a few questions will hardly hurt her. **1937** 'M. Innes' *Hamlet, Revenge!* iii. v. 297 There is nobody.. who would wish to incriminate me in a murder charge. **1974** *Times* 15 Feb. 1/3 (*heading*) Murder charge after London shooting. **1967** W. Keenan *Lonely Beat* iv. 39 He picked up the thick murder file... The reports were in chronological order. **1973** Murder File [see *murder log* below]. **1947** M. Gilbert *Close Quarters* xvi. 244 It was a murder film. **1939** E. S. Gardner *D.A. draws Circle* (1940) v. 57 'What are they, Doug, fingerprints?' 'Yes, on the murder gun.' **1968** *Observer* 28 Apr. 8/2 (*heading*) Murder-hunt police appeal to motorists. **1972** J. Wainwright *Requiem for Loser* viii. 162 The enquiry was still 'The Murder Enquiry'. **1973** Murder enquiry [see *murder log* below]. **1937** 'M. Innes' *Hamlet, Revenge!* iii. i. 220 The scene.. suggested.. a riot rather than a murder-investigation. **1973** R. Lewis *Blood Money* iii. 28 This is a murder investigation. Give me your assistance. **1972** J. Wainwright *Requiem for Loser* viii. 162 The log was still 'The Murder Log'. **1973** —— *High-Class Kill* 123 Bits and pieces which are part of a murder enquiry—the Murder File, the Murder Log.. and the Murder Book. **1889** Murder-man [see *blood-curdler* (*blood *sb.* 18)]. **1900** Ade *Fables in Slang* 198 The Book that begins with a Murder Mystery. **1960** Auden *Homage to Clio* 26 The sin of Gluttony Is ranked among the Deadly Seven, but in murder mysteries One can be sure the gourmet Didn't do it. **1973** A. MacVicar *Painted Doll Affair* vi. 68, I bought two of the paperbacks described as 'murder mysteries'. **1975** *Times* 22 Sept. 11/5 They.. worked out a scheme about two lexicographers involved in a murder mystery. **1971** 'H. Howard' *Murder One* xiv. 177 Murray's going to stand trial charged with murder one. **1972** G. V. Higgins *Friends of Eddie Coyle* xxvi. 162 The three of them're up on murder one, they're gonna be having a hearing this afternoon. **1929** D. Hammett *Dain Curse* (1930) xv. 169 He hasn't a chance in the world of hanging murder-raps on them. **1972** J. Potter *Going West* 57 Sergeant O'Leary said how about pulling her in on a murder rap. **1968** P. N. Walker *Carnaby & Gaolbreakers* xv. 143 I'd like a room set aside as a Murder Room. **1972** G. Sereny *Case of Mary Bell* i. iii. 38 The 'Murder Room' at Newcastle's West End Police.. was a hive of activity all night. **1929** M. A. Gill *Underworld Slang*, *Murder squad*, police who investigate murders. **1958** S. Hyland *Who goes Hang?* xlv. 220 A straightforward fact.. accepted by Macaulay and his murder squad. **1972** C. Drummond *Death at Bar* iii. 84 Sergeant Reed had retired from the pub before the City of London murder squad arrived. **1831** M. Edgeworth *Let.* 16 Mar. (1971) 490 Rogers.. told me.. a capital murder-story. **1929** F. N. Hart *Hide in Dark* i. 27 It's despicable to tell a murder story with the lights on. **1888** Kipling *Let.* 2 May in C. E. Carrington *R. Kipling* (1955) v. 97 He has been concerned in most of the more distinguished murder trials of the past twenty years. **1973** D. WESTHEIMER *Going Public* iii. 42 Lee went to the *Houston Post* and looked through back issues, studying murders and murder trials. **1971** *Guardian* 11 Dec. 10/5 The weekly number of murder victims [in India] is 283. **1959** M. GILBERT *Blood & Judgment* vii. 71 He had turned up.. the murder weapon, ready furnished with a print of the murderer. **1962** K. ORVIS *Damned & Destroyed* xii. 84 Fay's cap of heroin burned like a murder-weapon in my pocket. **1973** R. LEWIS *Blood Money* iv. 36 Frust..confirmed that..the poker..is the murder weapon.

murder, *v.* Add: **1. g.** (Later examples.)
1857 J. HYDE *Mormonism* vii. 181 These men will fight, lie, rob, murder for Mormonism if commanded. **1910** *New Mag.* Nov. 224/2 Yes. I am the man who murders for the king.

h. *colloq.*, as a jocular threat.
1939 JOYCE *Finnegans Wake* III. 460 So don't keep me now for a good boy for the love of my fragrant saint, you villain,..or I'll first murder you. **1942** T. RATTIGAN *Flare Path* II. ii. 66 *Patricia.* You're ill. I'm going to ring up a doctor. *Teddy.* I'll murder you if you do.

5. *slang.* To defeat (an opponent or rival) totally or conclusively, esp. at a game or sport.
1952 G. TALBOT in Wentworth & Flexner *Dict. Amer. Slang* (1960) 349/2 The National Leaguers..eat up south-paws. They murdered them all season. **1973** 'J. PATRICK' *Glasgow Gang Observed* v. 49 Mick had stepped in and challenged Bertie to a 'square-go'. 'Mick murdered him, man,' Tim recalled. **1974** *Observer* 24 Feb. 23/7 If the passing had got any worse, a team of corporals' grandmothers would have murdered them. They ran like fugitives from a church parade.

murderable (mɜ̄·ɹdərăb'l), *a.* [f. MURDER *v.* + -ABLE.] Giving cause for murder, provoking or inviting murder.
1920 D. H. LAWRENCE *Women in Love* ii. 32 A murderee is a man who is murderable. And a man who is murderable is a man who in a profound, if hidden lust, desires to be murdered. **1927** *Sunday Express* 21 Aug. 10/4 This tendency to associate unpopular opinions with murderable offences seems to be an increasing one on both sides of the Atlantic. **1966** *New Statesman* 18 Mar. 377/1 They had made good, and they *were* good; pre-eminently, they had achieved 'security'. And hence the immediate public terror caused by their murder: the least murderable people in the world, if they could be murdered, then anyone could be murdered.

murderee (mɜ̄ɹdərī·). [f. MURDER *v.* + -EE[1].] A person who is murdered. Also, a person whose character and disposition suggest the passive qualities of an easy victim of murder.
1920 D. H. LAWRENCE *Women in Love* ii. 32 It takes two people to make a murder: a murderer and a murderee. And a murderee is a man who is murderable. **1925** *New Yorker* 28 Mar. 13/2 Some day, somebody is going to write a mystery story in which the murderee will be a swell guy. **1928** F. T. JESSE *Trial of S. H. Dougal* 4 The potential murderer has met the born murderee. **1928** A. HUXLEY *Point Counter Point* xii. 209 He's the real type of murderee ..the obvious victim; he fairly invites maltreatment. **1939** —— *After Many a Summer* 20 You're probably the sort of person that invites persecution. A bit of a murderee ..as well as a scholar and a gentleman. **1958** S. HYLAND *Who goes Hang?* xii. 59, I don't *know* who the murdered man, the 'murderee', was. **1970** *Times Lit. Suppl.* 7 Aug. 883/1 This one hinges on the killing of one of those fiendishly attractive girls who, fictionally at least, are so often murderees. **1973** K. GILES *File on Death* vi. 151 Title wasn't much chop however you look at him, a murderee if ever I investigated one. **1974** *Sunday Times* 31 Mar. 52/3 TV Film...about Elizabeth Montgomery set up as possible next murderee in a storm.

murdermonger (mɜ̄·ɹdəɹmɒːŋgəɹ). [f. MURDER *sb.* + MONGER[1].] A dealer in murder, a professional murderer. So *fig.*, a writer of murder stories; fem. (*nonce-wd.*) **murdermongeress.** Also **murder-mongering** *vbl. sb.*, the purveying of news or stories of murder.
a **1889** in Barrère & Leland *Dict. Slang* (1889) I. 139/1 The only one who is annoyed is our own special murdermonger, who has got several blood-curdlers of English extraction up his sleeve. **1900** M. HEWLETT *Richard Yea-and-Nay* II. vi. 299 She knew something of the Marquess, her cousin. Any ally of his must be a murdermonger. **1957** O. NASH *You can't get there from Here* 97, I repeat that one book by this murder-mongeress [*sc.* Agatha Christie] Will last you as long as the Library of Congress. **1967** *Punch* 4 Oct. 524/3 You may think that the enormous sums of money spent on murder-mongering by the newspapers have bred up a lamentable race of scribes.

mureed, var. *MURID.

murein (miū̃ə·rīn). *Biochem.* [f. L. *mūr-us* wall + *-ein*, after PROTEIN.] A polymeric substance whose molecules consist of mucopolysaccharide chains interlinked with short peptide chains and which is an important structural component of the cell walls of many bacteria.
1964 WEIDEL & PELZER in *Adv. Enzymol.* XXVI. 195 The new type of polymer from which bacterial sacculi are tailored. We propose the term 'murein'.., in analogy to the trivial name 'protein'. Various other designations presently in use, like 'mucocomplex' or 'mucopolymer', have little to recommend them because they would fit a variety of chemically rather different structures. The term 'mucopeptide' which has also been proposed, appears to be somewhat misleading for a polymer which is..at least as much a polysaccharide as it is a (poly)peptide. **1969** W. G. MURRELL in Gould & Hurst *Bacterial Spore* vii. 236 All spore mureins so far studied..differ from known cell wall polymers in having..a two- to three-fold excess of glucosamine or muramic acid residues over glutamic acid or DAP residues. **1969** *New Scientist* 10 July 64/1 Murein contains an amino sugar, muramic acid, which is not found in higher organisms, and also several amino acids in the unusual D-configuration. **1972** *Nature* 3 Mar. 10/1 The bacterial cell wall is bounded by a fragile cytoplasmic membrane... This is surrounded by, or interwoven with, a more rigid cell wall..built up from a limited number of constituents and known variously as mucopeptide, peptidoglycan or murein.

Muria (miū̃ə·riă), *sb.* and *a.* Also **Morea, Moria.** [Native word.] **A.** *sb.* A member of a hill people of Bastar in India, a division of the Gonds. Cf. *GOND *sb.* and *a.* **B.** *adj.* Of or pertaining to this people.
1861 *Selections from Records of Govt. of India (Foreign Dept.)* xxx. 8 The Moreas are distributed over the north of the dependency and the vicinity of Jugdulpore, and the Marias to the south and west of it. *a* **1863** S. HISLOP *Aboriginal Tribes of Central Provinces* (1866) i. 22 *Moria Gonds.* These are more civilized than the Márias. *Ibid.* i. 23, I do not possess detailed information regarding the mythology of the Morias. **1877** V. BALL *Jrnl.* 18 Mar. in *Jungle Life in India* (1880) xiii. 620 According to the Dewan, the following are the names of the principal of these races; they are said to possess distinct languages: Bhatra, Muria (=Gond), Purji, Gudwa (or Gudaba), Jhoria, and Mariah or Meriah. **1938** W. V. GRIGSON *Maria Gonds of Bastar* I. iii. 37 In 1931 the tribes enumerated as Gonds in the previous three censuses were separately enumerated under the names Bhattra, Gond, Maria, Muria, Koya and Parja. **1968** P. C. AGARWAL *Human Geogr. Bastar District* xiii. 258 Burha Deo is the principal deity of the Muria tribe. **1971** [see *GOND *sb.* and *a.*].

murid (miū̃ərī·d). Also **mooreed, mureed** [Arab. *murīd.*] A follower of a Muslim pir; a disciple; a member of the second order of the Sufi 'way', aspiring to join the third order.
1815 J. MALCOLM *Hist. Persia* II. xxii. 396 The person who makes the attempt [for the third class of Sufiism] must be a holy mooreed or disciple, who..has already made a progress that has placed him above the necessity of the common usages and forms of established religion. **1885** T. P. HUGHES *Dict. Islam* 421/2 *Murid*,..'one who is desirous or willing'. A disciple of some *murshid*, or leader, of a mystic order. Any student of divinity. **1929** E. D. Ross tr. *H. Lammens's Islām* vi. 129 The *murid* or novice aspiring to be admitted into the congregation. *Ibid.* 135 Admission into a *ṭariqa* is preceded by a period of trial or noviciate, called *irāda*; whence the name *murid* given to the *Ṣūfi* aspirant. **1967** F. RAHMAN *Islam* ix. 154 The absolute authority of the Ṣūfi leader, called Shaykh..over his disciples called *faqir*,..*darwish*, *murid*..or *ikhwān*.

Muridism (miū̃ərī·diz'm). [f. *MURID + -ISM.] A revival movement in Islam encouraging the rising of Muslims against their religious and political opponents.
1866 *Chambers's Encycl.* VIII. 656/2 He was one of the zealous disciples of Kasi-Mollah, the great apostle of Muridism. **1875** C. HENEAGE tr. *Von Thielmann's Caucasus, Persia & Turkey* I. iv. 261 This movement, designated Muridism from the name Murid..borne by the initiated, constitutes one of the greatest events in the modern history of Islamism. *Ibid.* 262 The growth of Muridism was to the Russians a matter of fearful import. **1895** *Macmillan's Mag.* July 217/1 It was called Muridism, and was mainly if not avowedly borrowed from Sufism; it was held to be a strictly orthodox form of the Mahommedan religion. **1931** *Times Lit. Suppl.* 12 Feb. 111/2 The doctrine of Muridism united the Mohamedan sects in a common crusade against the Russians. **1973** HOWAT & TAYLOR *Dict. World Hist.* 1034/2 *Muridism*, Muslim mystical..movement originating in Shirvan in the 18th cent... The movement reached its peak under the third imam..and manifested strong egalitarian tendencies.

murillo, var. *MORILLO.

murine, *a.* and *sb.* Add: **A.** *adj.* **b.** Affecting mice or rats; esp. in *murine typhus* [tr. F. *typhus murin* (C. Nicolle et al. 1932, in *Compt. Rend.* CXCIV. 1706)].
1933 *Biol. Abstr.* VII. 395/1 The name 'historic typhus' is proposed for the African (Old-World) form, and 'murine' typhus for the other 2. **1940** *Jrnl. Exper. Med.* LXXII. 417 In marked contrast to the white mouse and cotton rat, the murine virus induces no symptoms whatever..in other rodents. **1947** J. C. SNYDER in F. H. Top et al. *Communicable Diseases* (ed. 2) lix. 879 Unlike the human body louse, which suffers a fatal infection with typhus, the rat flea is unharmed by the multiplication of murine typhus rickettsiae in its tissues. **1955** *Sci. Amer.* Jan. 74/3 Murine typhus, which occurs in many parts of the world including the U.S., is comparatively mild but closely related to epidemic typhus; an attack by one confers immunity after recovery against both. **1970** PASSMORE & ROBSON *Compan. Med. Stud.* II. xviii. 86/1 *Rickettsia prowazekii* is responsible for epidemic typhus, which is transmitted by lice. *R. mooserii* [sic] causes the milder endemic murine typhus, primarily a disease of rats, amongst which it is spread by the rat flea. *Ibid.* xix. 51/2 Murine typhus is not lethal to rats and hence is not accompanied by deviation of rat fleas to man on the scale found in murine bubonic plague.

murky, *a.* Add: **6.** Phr. *the murky past*: confused or 'dubious' happenings at a previous time; an obscure episode or period. Cf. PAST *sb.* 2.
1917 'CONTACT' *Airman's Outings* 141 At times the R.F.C. pilot, like the man with the murky past, is constrained to have clouds for a covering against attack. **1963** J. CLEUGH *Love locked Out* 11 St. Augustine, tormented by the murky past which had preceded his appointment as bishop of Hippo in North Africa. **1966** *Oxf. Univ. Gaz.* 23 Dec. 432/2 If I may go back into the murky past,..these sort of powers..would..have been used at least to influence planning..described..as precipitate and ill-considered.

murmanite (mɜ̄·ɹmănəit). *Min.* [ad. Russ. *murmanit* (A. E. Fersman 1923, in *Doklady Ross. Akad. Nauk* 63), f. *Murman*, name of a shore on the north of the Kola peninsula in Russia + *-it* -ITE[1].] A hydrated silicate of sodium and titanium with lesser and variable amounts of manganese, zirconium, iron, calcium, and niobium, which is found as violet monoclinic crystals.
1924 *Mineral. Abstr.* II. 263 Various new primary occurrences were found for..a new mineral murmanite from Alluaiva, Angvunsiok, and Sengischorr. **1951** *Ibid.* XI. 240 Recalculation of analyses of murmanite..suggests a formula $(Na_2,Ca)TiSiO_5.xH_2O$ analogous to that of sphene. **1968** I. KOSTOV *Mineralogy* II. v. 298 Murmanite and lomonosovite form a complete isomorphous series and are monoclinic like sphene and fersmanite.

murmur, *sb.* Add: **5.** *Comb.* **murmur diphthong,** a diphthong ending with a weak (murmur) vowel; **murmur vowel,** a glide or weak vowel, a *SCHWA.
1892 H. SWEET *New Eng. Gram.* I. 234 There is another class of *murmur diphthongs* ending in (ə), as in *hear*, *here* (hiə), *fare*, *faire* (fεə). **1933** O. JESPERSEN *Essentials Eng. Gram.* ii. 26 There are three kinds of diphthongs in English: (1) full (long-distance) diphthongs... (2) slow (short-distance) diphthongs... (3) murmur-diphthongs ending in the indistinct central vowel [ə]: [iə] as in *peer*, [εə] as in *pair*, [etc.]. **1965** A. D. CORDTS *Phonics* xii. 228 Today every basic reading system..recognizes the 'short' and 'long' vowel sounds, the diphthongs, the so-called murmur diphthongs and 'digraphs', [etc.]. **1910** Murmur vowel [see *BACK *a.* 1 c]. **1924** W. J. SEDGEFIELD in Mawer & Stenton *Introd. Survey Eng. Place-Names* I. II. i. 8 The vowel that was distinctly pronounced loses under the secondary stress its clear character and becomes either the obscure sound called by phoneticians the 'murmur-vowel' [ə] or an indistinct [i]. **1957** S. POTTER *Mod. Linguistics* i. 27 The inhabitants of Birmingham..call their city [bəːmiŋgəm], but Londoners call it [bəːmiŋəm]. .. Midlanders raise the velum against the wall of the pharynx and make a plosion with the back of the tongue upon it before passing from agma to the murmur-vowel, whereas the people of London keep the nasal pharynx open at this juncture. **1966** A. WIJK *Rules of Pronunciation for Eng. Lang.* iv. 64 The murmur vowel is..very common in both open and closed syllables when the stress falls on the immediately following syllable.

murmur, *v.* Add: **3.** Also with *out*.
1837 DICKENS *Pickw.* xxviii. 304 Gabriel murmured out something about its being very pretty. **1894** A. CONAN DOYLE *Mem. Sh. Holmes* 41 My lips were parted to murmur out some sleepy words of surprise or remonstrance.

murmuration. Delete † *Obs.* and add later example of sense 1.
1908 *Westm. Gaz.* 5 Aug. 2/1 The plaints and murmurations of these Randlords for the grievances which they clamoured to have redressed.

2. Delete 'An alleged term for' from def., and add later examples.
1932 AUDEN in *New Statesman* 16 July 69/1 Patterns a murmuration of starlings Rising in joy over wolds unwittingly weave. **1938** *Times* 6 Jan. 15/5 Great murmurations of starlings are often represented as a peculiar phenomenon of our own times. **1946** M. PEAKE *Titus Groan* 306 The clearing ended where a derelict stone building..held back a grove of leafless elms, where a murmuration of starlings was gathered. **1966** *Sunday Mail Mag.* (Brisbane) 6 Mar. 6/4 Starlings when they're on the wing have to labor under the collective title of murmuration.

murmured, *ppl. a.* Add: **murmured vowel,** = **murmur vowel*; also, a vowel-sound uttered with slow vibration of free vocal chords.
1933 L. BLOOMFIELD *Language* vi. 102 Some languages ..distinguish different voice-qualities, such as *muffled* vowels, *murmured* vowels, with slow vibration of the vocal chords, or *whispered* vowels. **1952** R. JAKOBSON et al. *Preliminaries to Speech Analysis* ii. 26 It is highly questionable whether there are languages in which..there actually is a..distinctive opposition of voiced and murmured vowels. **1966** M. PEI *Gloss. Linguistic Terminol.* 173 Vowels of weakly stressed syllables are sometimes murmured (reduced vowels), and they tend to lose their identity, contrary to fully voiced and whispered vowels.

murning, var. *MURRNONG.

Murphy[2]. Add: **3.** *Murphy's law*: a name humorously given to various aphoristic expressions of the apparent perverseness and unreasonableness of things (see quots.). orig. *U.S.*

1958 *Nation* (N.Y.) 7 June 506/1 There is an old military maxim known as Murphy's Law which asserts that wherever there is a bolt to be turned, someday there will be someone to turn it the wrong way. **1961** LEEDS & WEINBERG *Computer Programming Fund.* viii. 241 What we desire is the presentation of the information in such an accurate and complete form that the reader will be able to use the sub-routine correctly without hesitation or question. Recalling 'Murphy's law'—'If something can go wrong or be misinterpreted, it will'—should be enough stimulus for the goals we desire. **1962** J. GLENN in *Into Orbit* 85 We blamed human errors like this on what aviation engineers call 'Murphy's Law'. 'Murphy' was a fictitious character who appeared in a series of educational cartoons put out by the U.S. Navy... Murphy was a careless, all-thumbs mechanic who was prone to make such mistakes as installing a propeller backwards. **1970** *Sci. Amer.* June 143 Recently..I learned of a governing principle known as Murphy's first law of biology. It states: 'Under any given set of environmental conditions an experimental animal behaves as it damn well pleases.' **1972** *Oxf. Times* 22 Sept. 12/7 That malign influence which presides over human and typographical enterprises, and which is sometimes described as Murphy's Law ('Things are out to get us'). **1973** G. HART *Right from Start* i. 45 According to Murphy's law, if the worst possible thing can happen, it will. **1974** *N.Y. Times Mag.* 8 Sept. 33/1 'If anything can go wrong, it will,' says Murphy's Law.

4. Name given (in full, *Murphy game*) to a confidence trick in which the victim is duped by unfulfilled promises of money or sex, etc. So *Murphy man*, one who practises confidence tricks of this kind; *Murphy* v. trans., to dupe, to swindle by means of such a trick. *U.S. slang.*
1959 *Washington Post* 2 Oct. B 8/1 The 'Murphy game' is..a confidence game... The victim or 'pigeon' is lured by promises of a woman and then given..paper cut to bill size, in an envelope exchanged for the victim's cash. **1965** *N.Y. Times* 6 Apr. 1/3 Mayor Smitherman..and..a Selma lawyer..were 'murphyed' by the Negro confidence man at 2:30 A.M. today. **1965** *Time* 16 Apr. 16 'The Murphy game' is underworld argot for a slick maneuver in which a victim puts his cash in an envelope and gives it to the con man, who makes a fast sleight-of-hand switch and hands back an identical envelope stuffed with newspaper strips. **1966** *N.Y. Times* 4 Sept. IV. 5 Everybody should have a car... How are you going to get it?.. You know, you can get it playing the Murphy. **1968** W. LABOV et al. in A. Dundes *Mother Wit* (1973) 331 The right of a hand of some murphy man. **1970** C. MAJOR *Dict. Afro-Amer. Slang* 83 *Murphy*, a con game played on innocent (especially white) men who are expecting sex with a prostitute (usually black). **1972** T. KOCHMAN *Rappin' & Stylin' Out* 244 An adept hustler who was playing the 'murphy' game on a white trick. **1972** J. MILLS *Report to Commissioner* 52, I thought he was a complainant...some school kid who'd been Murpheyed. *Ibid.* 55 We stop in a bar..and it's filled with pimps, Murphy men, guys like that.

Murphy[3] (mɜ̄·ɹfi). *N. Amer.* The name of William Lawrence *Murphy* (1876–1959), American manufacturer, used *attrib.* in *Murphy bed*, any of various types of folding beds, developed from an original design by Murphy.
1925 *Small Home* Mar. 31 (Advt.), The closet off the living-room has been equipped with a Murphy Bed. **1930** *N.Y. Times* 2 Mar. XIII. 10/2 (Advt.), Apartments Unfurnished... 40s East—Most charming one-room apartment in town; restaurant, telephone service, Murphy beds: $78.50. **1969** L. HELLMAN *Unfinished Woman* vi. 63 They lived in the Murphy-bed, modern apartments that were already the slums off Hollywood Boulevard. **1974** *Apartment Life* May E2/2 The concealed wall bed, best known as a Murphy bed in the 1930s, is coming out of the woodwork again.

murram (mɒ·răm). Also **murrum.** [Native name.] A hard lateritic material associated with soils with impeded drainage in tropical Africa and locally used as road metal.
1925 *Dollar Mag.* Sept. 120 What first catches the attention..of eyes weary of murrum or black cotton soil.. is the soothing greenness. **1932** G. W. ROBINSON *Soils* iii. 59 In certain types of tropical soils, with impeded drainage, the deposition of iron oxides may result in the development of highly indurated concretionary material, known in Africa as 'murram'. **1935** *Soil Res.* IV. 192 Towards the foot of the slopes are murram soils. **1959** E. HUXLEY *Flame Trees of Thika* i. 7 A plain whose soil was largely murram, a coarse red gravel that baked hard and supported only thin, wiry grass, sad-looking thorn trees and tortured-branched erythrinas. **1961** *Engineering* 2 June 765 A variable thickness of murrum (a tough clay bound gravel). **1963** A. SMITH *Throw out Two Hands* vi. 71 The heavens open in their own African fashion to turn the murram roads into sloshy causeways. **1971** D. CREED *Trial of Lobo Icheka* vii. 78 In the town centre the wide murram road was a bedlam of activity.

Murray[1] (mɒ·ri). Any of the series of guide-books, or of time-tables of all railway trains running in Great Britain, issued by John *Murray* (1808–92) or by his successors. Also in the possessive, and *fig.* So **Mu·rray-less** *a.*
1845 THACKERAY *Leg. Rhine* in *G. Cruikshank's Table-Bk.* Sept. 198 Crowds of English,..armed with Murray's guide-books. **1846** R. FORD *Gatherings from Spain* viii. 86 A solitary wanderer..can read the book of Spain.. dwelling on what he likes, and skipping what he does not, as with a red Murray. **1847** [see *BRADSHAW]. **1862** MRS. GASKELL *Let.* 23 July (1966) 927 We stopped all night in a clean hotel... It is not down in Murray, but ought to be for its cleanliness, civility, and moderate charges. **1864** *Daily Tel.* 26 Sept. 5/1, I hope he found his 'Murray' again, for one would not willingly see him wandering..Murray-less and unguarded. **1885** RUSKIN *Praeterita* I. vi. 183 Murray..did not exist in those days; the courier was a private Murray. **1898** H. G. WELLS *Let.* 22 Jan. in *G. Gissing & H. G. Wells* (1961) 77 The more I see of these Murrays the more I settle to Rome. **1925** A. HUXLEY *Along the Road* I. 42 Old guide-books..make excellent travelling-companions. An early Murray is a treasure. **1961** *Economist* 18 Mar. 1047/2 Murray's ABC Time-table (now a pocketbook..) which sells 75,000 monthly in its two editions for Glasgow and Edinburgh. **1961** E. M. FORSTER *Alexandria* (ed. 3) p. xv, I have always respected guidebooks—particularly the earlier Baedekers and Murrays.

Mu·rray[2]. The name of the main river of a large river-system of south-eastern Australia, used *attrib.* to designate plants or animals native to this region, as **Murray cod,** a large carnivorous food-fish, *Maccullochella macquariensis*, of the family Serranidæ; **Murray lily,** a bulbous plant, *Crinum flaccidum*, belonging to the family Amaryllidaceæ and bearing white, lily-like flowers; also called the Darling lily; **Murray perch** = *CALLOP; **Murray (river) pine,** a cypress pine, *Callitris glauca*.
1875 *Encycl. Brit.* III. 112/2 A very fine fresh-water fish is the Murray cod. **1911** C. E. W. BEAN *'Dreadnought' of Darling* xxxiv. 301 The Murray and the Darling and their tributaries contain a certain large, sluggish fish, which rather reminds one of a big carp, very delicious... They call him the Murray cod. **1952** D. STIVENS in *Coast to Coast 1951–2* The bunyip..was about ten feet long with a body like a Murray cod with gleaming golden-yellow scales. **1966** Murray cod [see *CALLOP]. **1877** F. VON MÜLLER *Introd. Botanic Teachings* xx. 119 This showy genus Crinum furnishes also Victoria with a beautiful species, the *Murray Lily* (Crinum flaccidum), not however to be found away from the Murray-River southward. **1962** J. H. WILLIS *Handbk. to Plants in Victoria* I. 326 Murray Lily (Darling Lily, Macquarie Crinum)..on a few sandy inundated flats along the Murray R. flood-plain. **1880** G. WALCH *Victoria in 1880* 124 (Morris), Our noble old 1400-mile river, the Murray..produces..that finny monster, the Murray cod, together with his less bulky, equally flavourless congener, the Murray perch. **1951** T. C. ROUGHLEY *Fish Austral.* 147 In New South Wales it [*sc.* the callop, *Plectroplites ambiguus*] was until recently referred to as golden perch or yellowbelly,..in Queensland and Western Australia as Murray perch. **1877** F. VON MÜLLER *Introd. Bot. Teachings* 38 Frenela or Callitris verrucosa,..known by the name Murray-Pine. **1933** *Bulletin* (Sydney) 4 Jan. 17 Is nothing to be done in methodical re-afforestation with Murray pine... A beautiful timber white-ant and rot resistant, and its dark rosewood heart and soft cream subcortical layers. **1951** *Dict. Gardening* (R. Hort. Soc.) I. 362/1 *C*[*allitris*] *glauca.* Murray River Pine. Up to 100 ft... Widely distributed. The wood is in constant use in Australia. **1966** 'J. HACKSTON' *Father clears Out* 100 The solitary Murray Pine in the college paddock was flinging itself about like a windjammer in a gale. **1967** A. M. BLOMBERY *Guide to Native Austral. Plants* 225 Murray Pine. A tree with spreading light-green branches, and cones tending to be longer than broad when ripe.

Murrayism (mɒ·ri,iz'm). [f. *MURRAY[1] + -ISM.] **a.** A mode of expression reminiscent of a guide-book published by Murray. **b.** A tourist's fondness for Murray's guide-books.
1846 in *Downside Rev.* (1945) LXIII. 216 That Rome should be called the Metropolis of Art..'from the miracles of art which it contains' is a Murrayism. **1865** *Daily Tel.* 27 Sept. 7/3, I saw yesterday in the Tribune a British couple who had brought to the races 'Murray's Handbook'. I think Murrayism can scarcely go further.

Murray Valley. The name of the valley of the Murray River (see *MURRAY[2]), used *attrib.* to designate a severe form of encephalitis and also the mosquito-borne virus that causes it.
1951 *Med. Jrnl. Australia* I. 526/1 A severe human encephalitis of virus origin spread diffusely along the Murray Valley during the early months of 1951. This has been provisionally referred to as Murray Valley encephalitis (MVE). **1955** *Sci. Amer.* Mar. 65/2 The New York laboratory decided to employ the neutralization test to compare each of the 11 new viruses with various other insect-borne viruses, such as the agents of yellow fever, dengue, louping ill, Murray Valley fever, Russian encephalitis, [etc.]. **1966** WRIGHT & SYMMERS *Systemic Path.* II. xxxiv. 1207/1 Epidemics of encephalitis occurred in Australia in 1912, 1918, 1922 and 1926... It is probable that these epidemics were outbreaks of the disease that has become known as Murray Valley fever. **1974** *Sydney Morning Herald* 6 Feb. 7 (*heading*) Mystery of the Murray Valley virus.

murrnong (mɔ̄·ɹnɒŋ). *Austral.* Also **mirr-n'yong, murning.** [Aboriginal name.] A yellow-flowered native herb, *Microseris lanceolata* (also called *M. scapigera*), of the family Compositæ, resembling a dandelion but having clusters of small tuberous roots, formerly used as food by the Aborigines.
1852 J. MORGAN *Life W. Buckley* 85 There was another sort of food very useful to me; this was a particular kind of root the natives call Murning—in shape, and size, and flavour, very much resembling the radish. **1878** R. BROUGH SMYTH *Aborigines Victoria* I. 209 *Murr-nong* or *Mirr-n'yong*, a kind of yam (*Microseris Forsteri*), was usually very plentiful and found in the spring and early summer, and was dug out of the earth by the women and children... The root is small, in taste rather sweet, not unpleasant, and perhaps more like a radish than a potato. **1889** J. H. MAIDEN *Useful Native Plants of Australia* 45 *Microseris Forsteri*... 'Murr-nong' or 'Mirr n'yong', of the aboriginals of New South Wales and Victoria. The tubers were largely used as food by the aboriginals. They are sweet and milky, and in flavour resemble the cocoa-nut. **1965** *Austral. Encycl.* II. 494/1 The most familiar native member of the tribe [*sc.* Compositae] is *Microseris lanceolata* (syn. *M. scapigera*), the 'yam', 'yam daisy', or 'murrnong' of the aborigines, who used its tuberous roots for food.

Murrumbidgee (mɒrɒ̆mbi·dʒī). The name of a river in southern New South Wales used *attrib.* in various Austral. slang collocations, as **Murrumbidgee jam** (see quot. 1943); **Murrumbidgee w(h)aler,** a swagman who camped in the region of the Murrumbidgee or other river, regarded as a classic type of indolent person; so **Murrumbidgee whaling.**
1873 J. C. F. JOHNSON *Christmas on Carringa* 16 Men when on the tramp through the Riverina country often carry a piece of twine and a hook to catch cod or black-fish. This is termed Murrumbidgee Whaling. **1878** 'IRONBARK' *Southerly Busters* 177 Murrumbidgee whalers are a class of loafers who work for about six months in the year —*i.e.*, during shearing and harvest, and camp the rest of the time in bends of rivers, and live by fishing and begging. **1885** *Australasian Printers' Keepsake* 72 [He] eyed Bob very suspiciously, muttering 'spieler' and 'Murrumbidgee whaler'. **1943** BAKER *Dict. Austral. Slang* (ed. 3) 52 *Murrumbidgee jam*, brown sugar moistened with cold tea and spread on damper. **1953** A. RUSSELL *Murray Walkabout* 147 He was a 'Murrumbidgee whaler', the river prototype of the tramping sundowner. **1969** *Sunday Truth* (Brisbane) 2 Nov. 34/1 The so-called Murrumbidgee whalers roamed the inland rivers, particularly for about 50 years from the 1880s, and caught cod which they sold to hotels and farms.

‖ **muru** (mū·rū). *Obs. N.Z.* [Maori.] (See quot. 1863).
1836 J. A. WILSON *Jrnl.* 24 Aug. in *Missionary Life & Work in N.Z.* (1889) III. 48 We were told the events which led to the burning of the mission station, and the *muru* which followed. **1863** F. E. MANING *Old N.Z.* vii. 96 There were in the old times two great institutions..in Maori land—the *Tapu* and the *Muru*. Pakehas..called the *muru* simply 'robbery'... But I speak of the regular legalized and established system of plundering, as penalty for offences. **1905** W. B. *Where White Man Treads* 32 Hiakai.., forgetting the rules of muru, crashed it [*sc.* the tomahawk] into his enemy's skull. **1921** H. GUTHRIE-SMITH *Tutira* xxvii. 267 Neither..was *muru* an institution likely..to foster foresight. **1949** P. BUCK *Coming of Maori* (1950) III. vi. 421 The custom of *muru* (raiding) was sometimes employed by visitors if a death was due to accident.

‖ **murumuru** (mu·rumu·ru). [Tupi.] A Brazilian palm, *Astrocaryum murumuru*, whose stem is covered with black spines. Also *attrib.*
1853 A. R. WALLACE *Palm Trees of Amazon* 101 On the Upper Amazon cattle eat the fruits of the Murumurú, wandering about for days in the forest to procure it. **1860** MAYNE REID *Odd People* 135 These thorns are the spines of the 'murumuru', or 'pupunha' palm. **1927** R. R. GATES *Botanist in Amazon Valley* vi. 134 We stopped again to take on a canoe load of murumuru nuts. They are sometimes burned in the furnaces and they burn rather like charcoal. **1961** P. B. TOMLINSON *Anat. Monocotyledons* II. 142 *Astrocaryum murumuru* (Murumuru Palm), *A. tucuma* (Tucum Palm), and other spp. are included in a list of Brazilian palms yielding vegetable oils.

‖ **murus gallicus** (miū̃ə·rɒs gæ·likɒs). [L. 'Gaulish wall' (Caesar *De Bello Gallico* VII. xxiii).] A type of late Iron Age Celtic fort having stone walls bound by horizontally placed timber frames. Also called *Gaulish* or *Gallic wall* (*fort*).
1947 J. & C. HAWKES *Prehist. Britain* 265 The specially constructed type known as the Gallic Fort in which stone walls are bound with timber (*murus gallicus*). **1947** *Proc. Prehistoric Soc.* XIII. 16 Moreover the distribution of 'murus gallicus' forts is strictly complementary to, and mutually exclusive with, that of the brochs. **1953** R. J. C. ATKINSON *Field Archaeol.* (ed. 2) ii. 62 Walls of *murus gallicus* type have in addition internal timbering. **1963** *Field Archaeol.* (Ordnance Survey) (ed. 4) 71 This work must not be confused with the special type of timber reinforcement (*murus gallicus*) which it resembles and which is reported by Caesar as frequent among the Gaulish forts which he assaulted. **1970** BRAY & TRUMP *Dict. Archaeol.* 233/2 One specialized form of timber-laced rampart, the *murus gallicus* or Gaulish wall, was encountered by Caesar during his campaigns against the Celtic tribes.

Murut (mū·rŭt). Also **Marut.** [f. Bajau *belud* hill.] A member of a primitive Dyak people originally inhabiting the hill country in the interior of North Borneo, although now more widely scattered; the language of this people. Also *attrib.* or as *adj.*
1846 H. KEPPEL *Expedition to Borneo* II. viii. 171 The Murut inhabits the interior of Borneo Proper. They are

not tattooed,..and have a peculiar dialect. **1846**, etc. [see *KAYAN]. **1848** F. S. MARRYAT *Borneo & Indian Archipelago* 111 The whole space was covered with naked savages. These were the Maruts, a tribe of Dyaks who live in the mountains. **1881** J. HATTON *New Ceylon* iii. 81 The interior of the country is inhabited by the descendants of the aboriginal population, called variously Muruts, Dusuns, or Ida'an. **1896** H. L. ROTH *Natives of Sarawak & Brit. N. Borneo* I. p. xx, British North Borneo has been less fortunate with the Muruts. **1923** *Blackw. Mag.* July 49/2 The Muruts are not a warlike or courageous people. **1929** O. RUTTER *Pagans of N. Borneo* ii. 31 Speaking generally, the Muruts may be said to inhabit the hilly country of the interior of North Borneo. **1939** A. KEITH *Land below Wind* xi. 197 Here was a Murut headman in only a loincloth. *Ibid.* 198 Arusap called to them in Murut. **1968** *Encycl. Brit.* XV. 1012/2 The Muruts..are ..abandoning their communal houses in favour of private dwellings.

Mus' (mɒs), dial. abbrev. of MISTER *sb.*[2] or MASTER *sb.*[1]
1875 W. D. PARISH *Dict. Sussex Dial.* s.v. *Master.* 'Master' is quite a distinct title from 'Mr.', which is always pronounced Mus, thus,—*Mus* Smith is the employer. *Master* Smith is the man he employs. **1906** KIPLING *Puck of Pook's Hill* 224 Oh, Mus' Reynolds, Mus' Reynolds... If I knowed all was inside your head, I'd know something wuth knowin'. **1917** —— *Diversity of Creatures* 68 Whoever pays the taxes old Mus' Hobden owns the land.

musa[2] (miū̃·ză). *Radio.* Also **MUSA, Musa.** [See quot. 1937.] A radio aerial consisting of a number of rhombic elements in an end-fire array and giving a beam that is varied in direction by varying the phase relations between the elements.
1937 FRIIS & FELDMAN in *Bell System Techn. Jrnl.* XVI. 340 The word MUSA is coined from the initial letters of 'multiple unit steerable antenna'. **1940** *Ibid.* XIX. 309 The principal parts of the two musa receivers occupy three rows of bays each about 25 feet long and 11½ feet high. **1946** *Nature* 10 Aug. 190/1 Vertical angles were measured on transmissions from Rugby received at Holmdel with 'Musa' equipment. **1966** *McGraw-Hill Encycl. Sci. & Technol.* I. 447/2 The multiple unit steerable antenna, abbreviated MUSA,..has a directional pattern 1° wide at 18 Mc.

musaasa, var. *MSASA.

Musak, erron. var. *MUZAK.
1961 *Times Lit. Suppl.* 30 June 396/2 The comfortable times before Musak (relayed music) and large-scale cultural enterprise took over. **1966** AUDEN *About House* 62 The radio in students' cars, Musak at breakfast.

Muscadet (mu·skadɛ). [f. the name of the Muscadet grape from which it is made.] A white wine made in the Loire valley near Nantes.
1920 A. L. SIMON *Blood of Grape* ix. 221 Muscadet white wines are in no way objectionable, and in no way fine. **1951** R. POSTGATE *Plain Man's Guide to Wine* iv. 88 A white wine called Muscadet, from the grape used in it. **1966** H. JOHNSON *Wine* 103 Nobody claims greatness for Muscadet, and yet it is one of the most useful of the lesser wines of France. **1969** R. HUTCHINGS *Lucky in Jeopardy* 21 In lieu of the Hock the Squire usually selects with a curry meal we had a bottle of *Muscadet*, which went down very refreshingly on our scorched palates. **1971** *Guardian* 13 Aug. 9/4 Three magnums, one of Rosé, one of Burgundy, one of Muscadet, from which you could help yourself.

muscarine. Substitute for entry:
muscarine (mɒ·skărin, -īn). *Chem.* Also † **muscarin.** [ad. G. *muscarin* (Schmiedeberg & Koppe *Das Muscarin* (1869) 2, f. L. *muscārius* (see MUSCARIOUS *a.*): see -INE[5].] A quaternary ammonium base, $C_9H_{21}NO_3$, which is a poisonous alkaloid found in the fungus *Amanita muscaria* and which produces copious secretion by the mucous and sweat glands, nausea, vomiting, contraction of the pupils, and laboured respiration.
1872 *Jrnl. Chem. Soc.* XXV. 830 The physiological action of muscarine is antagonistic to that of atropine. **1878** [in Dict.]. **1910** *Practitioner* June 824 Putrefactive organisms split up proteins into..ptomaines—neurin, cholin, muscarin, cadaverin, [etc.]. **1914** DALE & EWINS in *Jrnl. Physiol.* XLVIII. p. xxv, The choline esters hitherto examined, while showing the peripheral 'muscarine' action in small doses, show in larger doses..a nicotine-like action. **1940** H. A. MCGUIGAN *Appl. Pharmacol.* 576 The action of pilocarpine and muscarine.. seems to be on the parasympathetic nerve endings and not directly on the gland cell. **1951** J. STEINBECK *Log from 'Sea of Cortez'* (1958) 124 Muscarine, the active poisonous principle of *Amanita muscaria* and other fungi. **1951** A. GROLLMAN *Pharmacol. & Therap.* xiii. 246 Muscarine has never been used in medical practice because its powerful parasympathomimetic effects are so generalized as to make it impossible to elicit the desired effects in any given organ. **1961** *Q. Rev.* XV. 154 The story of muscarine revolves round the presentation of three formulae, two erroneous ones $C_5H_{14}O_2N^+$ and $C_8H_{18}O_2N^+$, and the correct one $C_9H_{20}O_2N^+$.

Hence **muscari·nic** *a.*, resembling (that of) muscarine; capable of responding to muscarine; **muscari·nically** *adv.*, in a muscarinic manner.
1941 GOODMAN & GILMAN *Pharmacol. Basis Therapeutics* xix. 339 The nicotinic and muscarinic actions of acetylcholine can be strikingly demonstrated on the intestinal tract of..the tench. **1961** *Nature* 25 Feb. 673/1 Observed differences in the muscarinic activities of the above isomers is [*sic*] therefore explicable in terms of differences in their fit at the muscarinic receptors. **1971** *Ibid.* 16 Apr. 444/1 The muscarinically, weakly active substrates of acetylcholinesterase, acetylthiocholine..and acetylselenocholine..are antiplanar at τ2 and τ3 both in crystals and in solution. **1974** *Canad. Jrnl. Physiol. & Pharmacol.* LII. 332/2 An alternative explanation..is based on the concepts that there exist separate binding subsites on the muscarinic receptor to accommodate the polar and nonpolar side chains of muscarinic agonists [etc.].

Muscatel. Add: **1.** Also applied to wines made in Australia, California, and England.
1867 *Australasian* 9 Feb. 186/1 Australian Wines in the Indian Market... The Muscatel has plenty of body. **1958** A. L. SIMON *Dict. Wines, Spirits, & Liqueurs* 117 *Muscatel*, the name of a sweet fortified wine made in California. Also, in England, that of a cheap, sweet, white, sparkling wine, usually flavoured with elderberry flowers essence. **1967** A. LICHINE *Encycl. Wines & Spirits* 374/1 In California a regrettable quantity of fortified wine known as 'Muscatel' is sold as a substitute for spirits.

muscicole, *a.* Add: Also as *sb.*
1965 B. E. FREEMAN tr. *Vandel's Biospeleology* ix. 110 These two species are commonplace muscicoles.

muscle, *sb.* Add: **3. b.** Force, violence; an act, or the threat, of violence; fear inspired by force; intimidation; a person employed to use or threaten violence (freq. in *collect.* pl.); strength; influence, the exertion of economic or political influence. Phr. *on the muscle* (see quot. 1950). *slang* (orig. *U.S.*).
1930 *Amer. Mercury* Dec. 457/1 *On the muscle*, angry, quarrelsome. 'He busts up to me strictly on the muscle. So I let him have it.' **1931** *Detective Fiction Weekly* 5 Sept. 436/1 When the police drag his name into every gang killing or big shot feud he makes no denial. This circumstance has given Madden that terrorizing thing known in the underworld as 'muscle'. **1935** C. F. COE *G-Man* ii. 26 Winky and Palmy ain't hot for the muscle. **1942** *Detective Fiction Weekly* May 58/1 You were the best muscle in the whole crowd, Julio. **1950** H. E. GOLDIN *Dict. Amer. Underworld Lingo* 143/2 *On the muscle.* 1. By strong-arm methods; engaged in any criminal activity requiring strong-arm methods. 2. Acquired without paying as the result of a reputation for violence... 3. Belligerent. **1959** *Alfred Hitchcock's Mystery Mag.* Feb. 71/1 Skreen visited ..the organization's punitive department. He gave one of the muscles the name of a pusher who'd taken a junkie's credit. **1962** *Cosmopolitan* Aug. 108/3 A female decoy to set Shanley up and adequate muscle to take care of him. *Ibid.* 122/1 'What kind of people were they?'.. 'They were muscle.' **1968** B. TURNER *Sex Trap* v. 30 Free women, a rent-free pad, and all the muscle you need to keep the things running smoothly. **1968** [see *black power* (*BLACK *a.* 19)]. **1973** H. NIELSEN *Severed Key* xvi. 173 The muscle on the trucks..were free-lancers.

4. a. *muscle power*, *strain*.
1937 *Jrnl. R. Aeronaut. Soc.* XLI. 1092 The third part deals with the practical side of flying..autogiros, helicopters, muscle power flight, etc. **1959** *Daily Tel.* 2 Nov. 1/2 There has been increasing interest recently in the possibility of achieving sustained flight by the use of muscle power only. **1968** *Brit. Med. Bull.* XXIV. 189/1 Machines through the ages have been principally concerned with deploying muscle power more effectively. **1935** *Mind* XLIV. 353 We know by inference that muscle-strain in our eyes (for instance) is a part-cause of our headache.

b. *muscle(-)relaxant* adj. and sb., *-relaxing* adj., *-training* sb.
1951 *New Biol.* XI. 101 The muscle-relaxant drugs, which are sometimes given in addition to anaesthetics in surgery. **1968** W. C. BOWMAN et al. *Textbk. Pharmacol.* xxii. 613 Mephenesin is useful as a muscle relaxant in the treatment of tetanus. **1947** *Lancet* 18 Jan. 97/1 (*heading*) Muscle-relaxing action of myanesin. **1960** R. W. MARKS *Dymaxion World of Fuller* 94 All the sanitary and muscle-relaxing effects of other types of bathing could be effected without the use of any bathroom. **1969** D. B. TAYLOR in J. A. Bevan *Essent. Pharmacol.* xvii. 162 The neuromuscular junction of voluntary muscle..is..the site of action of a group of valuable muscle-relaxing drugs. **1869** 'MARK TWAIN' *Innoc. Abr.* 622 Who shall say it is not a muscle-training pastime, climbing the pyramids?

d. muscle-bound *a.*, having the muscles stiff and enlarged, esp. as a result of too much exercise or training; also *fig.*; **muscle car** *N. Amer. slang*, = *HOT ROD; **muscle-feeling** (example); **muscle-flexing** *a.*, demonstrating aggression or strength; **muscle sense,** add: also, the sense of position or movement, kinæsthesis (but see quots.); now more commonly used than *muscular sense*; (examples); **muscle spindle** [tr. G. *muskelspindel* (W. Kühne 1863, in *Arch. f. path. Anat. u. Physiol.* XXVII. 528)], any of numerous small sensory organs within muscle, which consist of a bundle of intrafusal muscle fibres richly supplied with nerves and enclosed in a capsule and which respond to passive stretching and active contraction of the muscle.
1879 W. BLAIKIE *How to get Strong* i. 18 Scarce any man grows earlier muscle-bound [than the coal-heaver], for few backs do so much hard work. **1909** *Westm. Gaz.* 8 June 9/2, I have met many cases where an unintelligent use of such exercises has..so thickened the muscles as to lead to the condition known as being 'muscle-bound'. **1918** *Nation* (N.Y.) 7 Feb. 133/2 There has been a curious sense of leisureliness, of muscle-bound movement. **1946** R. G. COLLINGWOOD *Idea of Hist.* IV. 189 German historians, muscle-bound in their struggle with the facts. **1973** J. WAMBAUGH *Blue Knight* ii. 32, I stopped by the arcade and saw a big muscle-bound fruit hustler standing there. **1969** *Time* (Canad. ed.) 16 May 98/1 It is a hyped-up Mustang—one of Ford's fast-moving contenders in what Detroit calls 'the muscle-car' market, where the best sales pitch is neck-snapping acceleration. **1970** *Wall Street Jrnl.* 16 July 1/6 Two of the insurers' major targets are high-horsepower 'muscle' cars and fragile bumpers that offer little protection in minor accidents. **1974** *Weekend Mag.* (Montreal) 16 Mar. 2/1 But lower drinking ages have crippled the high school dope trade throughout Canada, the big Detroit muscle cars have been displaced by Vivas, MGBs and even Jeeps, and everywhere long hair for boys and short skirts for girls are right out of it. **1975** *Daily Colonist* (Victoria, B.C.) 26 Feb. 17/4 The spokesman said 'muscle cars', today's term for hot rods, had used the stretch up to several times a month. **1890** W. JAMES *Princ. Psychol.* II. xx. 198 Muscle-feeling belongs to that class of general sensations which tell us of our inner states. **1961** *Times* 6 Dec. 17/5 Annoyance at the chest-beating, muscle-flexing, tear-jerking banality of some moments in the second part [of an opera]. **1973** J. ROSSITER *Manipulators* xi. 112 Christou was turned over to you as a sort of muscle-flexing exercise. To prove I could do it. **1895** E. B. TITCHENER in *Amer. Jrnl. Psychol.* VII. 83 *Muskelsinn*, muscle sense. **1933** HEWER & SANDES *Introd. Study Nervous System* (ed. 2) II. i. 71 The sensations produced are those of 'muscle sense', stereognosis, vibratory sense, and a sense of pressure-pain. **1938** R. S. WOODWORTH *Exper. Psychol.* xix. 454 The traditional 'sense of touch' has been broken up by experiment into at least five senses. The first to be split off, early in the nineteenth century, was the muscle sense, kinesthesis. **1968** J. J. GIBSON *Senses considered as Perceptual Systems* vi. 111 The so-called muscle-sense should probably be ascribed no role in detecting the spatial postures and movements of the body, and therefore it had better not be called a sense. **1971** *Jrnl. Gen. Psychol.* Jan. 147 Formerly, kinesthesis was the so-called 'muscle sense'. Now it is known that receptors in joints mediate the appreciation of limb position. **1894** C. S. SHERRINGTON in *Jrnl. Physiol.* XVII. 238 He [*sc.* Kühne] designated the bundles simply in virtue of their shape, muscle-spindles, 'muskelspindeln', the name adopted here as the most suitable of all that have been applied. **1972** P. B. C. MATTHEWS *Mammalian Muscle Receptors* i. 47 With the exception of a few muscles innervated by cranial nerves, muscle spindles have now been found in every striated muscle of every species of mammal in which they have been appropriately sought. **1974** BERGMAN & AFIFI *Atlas Microsc. Anat.* vi. 140 Muscle spindles are found within skeletal muscles. Each spindle is formed of two to ten small muscle fibers, the intrafusal fibers, enclosed within a sheath of connective tissue which is pierced by nerve fibers.

muscle (mɒ·s'l), *v.* [f. the sb.] **1.** *trans.* To move by the exercise of muscular force. *U.S. dial.* and *colloq.*
1913 H. KEPHART *Our Southern Highlanders* xiii. 262 We can muscle this log up. **1969** *New Yorker* 14 June 62/2 Graebner muscles one down the line so fast that Ashe's volley makes a high, awkward parabola. **1974** J. WAINWRIGHT *Hard Hit* 33 The refuse cart is collecting the empties... Three guys..are muscling the bins around.

2. To coerce by violence or by economic or political pressure. *slang* (orig. *U.S. underworld*).
1929 *Chicago Tribune* 18 Jan. 21/4 A certain gentleman in the illicit spirits business was accosted by two sinister characters, who 'muscled' him,..removing from his wallet the sum of $150. **1935** C. F. COE *G-Man* viii. 142 Mebbe it's a new mob. If they're musclin' Rap, it won't be long before they're musclin' us too. **1950** H. E. GOLDIN *Dict. Amer. Underworld Lingo* 143/2 *Muscle out*, to expel by force or threat of force an individual or gang from any racket or area. **1953** in Wentworth & Flexner *Dict. Amer. Slang* (1960) 349/2 If she persists,..she'll be muscled out of the [Free German Youth] movement. **1958** *Time* (Atlantic ed.) 13 Oct. 15/1 The old Union Pacific and Central Pacific railroads had once muscled each other. **1967** *Wall Street Jrnl.* 24 Apr. 32/1 Are aggressive underworld operators beginning to achieve some success in muscling a beachhead among the nation's established securities markets? **1971** *N.Y. Law Jrnl.* 23 Nov. 4/2 Big brewers may be illegally muscling small ones through predatory pricing—that is deliberately using below-cost pricing methods to force competitors out of the market.

3. *intr.* With *in* or *in on* (occas. *into*). To introduce oneself into the business, criminal activities, sphere of influence, etc., of another or others, by force or fraud; to enter forcibly, or uninvited. *slang* (orig. *U.S. underworld*). So **muscling-in** *vbl. sb.*
1929 W. R. BURNETT *Little Caesar* v. iii. 171 If you think you can muscle into this joint you're off your nut. *Ibid.* v. 182 You know, I been watching you ever since you muscled in on Sam Vettori. **1929** E. D. SULLIVAN *Look at Chicago* (1930) iv. 50 When the beer organization, which is trying to muscle in, arrives with its delivery, there is a scene reminiscent of the Fall of Port Arthur. **1931** G. IRWIN *Amer. Tramp & Underworld Slang* 133 *Muscle*, to use force or intimidation so as to secure a share in a 'racket' or graft, or to force one's way into an enterprise or gang by threat of violence. Also 'muscle in'. **1931** F. D. PASLEY (*title*) Muscling in. **1932** WODEHOUSE *Hot Water* vi. 123 You muscle in here, pretending to be the Vicomte de Blissac. **1932** *Observer* 26 June 13/3, I began to feel that he could probably beat Lindrum at billiards and muscle in on Al Capone. **1934** J. O'HARA *Appointment in Samarra* (1935) i. 13 So the feminine members had

muscled in on the smoking-room. **1934** DYLAN THOMAS *18 Poems* 20, I would not fear the muscling-in of love If I were tickled by the urchin hungers. **1940** WODEHOUSE *Eggs, Beans & Crumpets* 57 Beetles tried to muscle in between his collar and his neck. *Ibid.* 99 And no more muscling in and trying to dictate the policy of the 'Uncle Joe To His Chickabiddies' page? **1942** E. WAUGH *Put out More Flags* iii. 162 You're muscling in on my territory. **1955** *Times Lit. Suppl.* 22 July 415/3 The technical secondary school is failing to challenge the grammar school, which is muscling-in on the occupations to which it primarily looks. **1963** N. MARSH *Dead Water* (1964) vii. 182 Far be me from it—I mean it from me—to muscle in where I'm not wanted. **1973** J. WAINWRIGHT *Devil you Don't* 30 'The Ponderosa' was his spread and no cheap, jumped-up, fiddle-foot was gonna muscle in.

4. To make one's way by employing muscular strength.
1961 in WEBSTER. **1972** D. HASTON *In High Places* i. 7 He [*sc.* the leader in a rock-climb] doesn't just find something for his hands and muscle up with scrabbling feet. Climbing should be a balanced set of movements.

mu·scle man. *slang* (orig. *U.S.*). Also **muscle-man, muscleman.** [f. MUSCLE + MAN *sb.*[1]] **a.** A person who employs or threatens violence on behalf of a professional criminal, or who commits crimes of violence without instigation. **b.** Used to denote a paragon of powerful physique. **c.** *fig.*
1929 HOSTETTER & BEESLEY *It's a Racket!* 232 'Muscle Men' are those who 'muscle' their way. **1931** *Times Lit. Suppl.* 19 Feb. 125/1 O'Banion with the comment 'To hell with the Sicilians!' set his muscle-men moving. *Ibid.* 24 Sept. 728/3 Pinkerton's detectives were the most respectable of numerous bands of hired 'muscle men'. **1932** *Sun* (Baltimore) 23 Nov. 10/2 This country wants a frontal attack on the powerful citadel of muscle-men made rich by illegal beer. **1948** *Daily Mail* 21 Jan. 2/5 Princess MacFarlane..makes presents of boots to all the poor children, touring from slum to slum in a fast black saloon packed with muscle-men. **1952** S. KAUFMANN *Philanderer* (1953) xiv. 235 The other kind was the muscle men. You know, right off the farm where they were lifting tractors with one hand. **1953** W. BURROUGHS *Junkie* (1972) vii. 65 Bert was known as a muscleman. He was a heavy-set, round-faced, deceptively soft-looking young man who specialized in strong-arm routines and 'shakes'. **1960** *Times Lit. Suppl.* 27 May 333/3 *St. Petersburg*..was published..under the comparatively liberal tyranny of the Tsar and proscribed by Zhdanov, Stalin's literary muscle-man. **1962** A. HUXLEY *Island* xiii. 206 Little muscle-men and muscle-women—children with tendencies towards aggressiveness and love of power. **1966** 'C. KEITH' *Elusive Epicure* (1968) v. 76 If there is such a thing as a California type it's the big muscleman with a rather simple mind. **1968** P. OLIVER *Screening Blues* iv. 134 With the considerable returns accruing from operating policy wheels the racket came under the control of syndicates with muscle-men and hired gunmen ensuring that their 'rights' were protected. **1975** *Times* 26 Sept. 10/3 Auditions for 'the muscle man with a voice like a bird' [*sc.* Tarzan] will start soon.

Muscovite, *sb.*[1] and *a.* Add: **A.** *sb.* Also, a native or inhabitant of Moscow.
1961 in WEBSTER. **1973** J. SHUB *Moscow by Nightmare* x. 113 Are you one of those Muscovites who feel Leningrad is just a provincial town? **1975** *Times* 8 Mar. 13/3 Muscovites expect journalists to strengthen good will between our countries.

muscovite, *sb.*[2] Add: (Earlier example.) Hence **mu:scovitiza·tion,** conversion into muscovite; **mu·scovitized** *ppl. a.*
1850 J. D. DANA *Syst. Min.* (ed. 3) v. v. 356 Muscovite, *D*[*ana*]. **1909** *Cent. Dict.* Suppl., Muscovitization. **1930** *Trans. Geol. Soc. S. Africa* XXXII. 145 Since, however, albitization on an extensive scale generally accompanies the muscovitization, both processes in part merely involve a mutual exchange of alkali-radicles. **1944** *Q. Jrnl. Geol. Soc.* XCIX. 118 In slices they show..scattered plates of partly chloritized or muscovitized biotite. **1956** E. W. HEINRICH *Microsc. Petrogr.* ii. 28 Deuteric changes are.. muscovitization of biotite [etc.]. **1965** G. J. WILLIAMS *Econ. Geol. N.Z.* xiii. 200/1 The granite in the general area of the Tin Range is more or less muscovitized.

Muscovy. Add: **4. Muscovy duck,** also *ellipt.*, as **Muscovy.** (Further examples.) (*b*) = MUSK-DUCK 2.
1822 J. CAMPBELL *Trav. S. Afr. 2nd Journey* I. xiv. 148 We halted at the side of a lake, when one of our people brought down four wild ducks by one shot, and another found eight Muscovy ducks' eggs, as large as those of a turkey. **1911** C. E. W. BEAN *'Dreadnought' of Darling* ii. 26 They fairly rushed the cargo into her... Japanese onions, condensed milk, tomato sauce, Worcestershire relish, half a dozen hens, and a dozen Muscovy ducks. **1953** *Amer. Speech* XXVIII. 276 In this type of folk naming, the domesticated varieties of ducks also are not wholly neglected, the large Muscovy, with distinctive head adornment, suggesting the scoters, sizable among wild ducks and with unusually shaped and colored bills. All of our three species are known as bay muscoveys [*sic*] in Maryland. **1956** *Nigerian Field* XXI. 108 The presence of the Muscovy duck in Africa has never been satisfactorily explained. The bird is indigenous to the Americas, hence its presence in Africa as a domesticated fowl must be due to the activity of man. **1972** *Country Life* 2 Mar. 489/1 The China goose can't stand up to the muscovy and retreats before him.

muscular, *a.* Add: **1. b.** (Further examples.) *muscular dystrophy*, any of a group of hereditary disorders (or these disorders collectively) marked by the progressive wasting and weakening of some muscles owing (apparently) to some defect of the muscles themselves.
1886 Muscular dystrophy [see *dystrophy* s.v. DYS-]. **1932** W. BOYD *Text-bk. Path.* xxx. 840 The muscular dystrophies have several clinical divisions, but the basic pathology is the same... The large muscles concerned with fixation (shoulder, girdle, hip) are chiefly affected, the small muscles concerned with active movement (hand, etc.) usually escaping. This is the opposite to what occurs in progressive muscular atrophy. **1939** DIBLE & DAVIE *Pathology* lviii. 896 (*heading*) The myopathies or muscular dystrophies. **1961** R. D. BAKER *Essent. Path.* xi. 275 In muscular dystrophy the atrophy is due to changes in the muscle fibers themselves... In muscular atrophy the primary change is in the nerves which supply the muscles or in the cord or brain. **1961** *Lancet* 9 Sept. 601/1 We have recently come across a family where five members suffered from muscular dystrophy, inherited apparently as an autosomal or X-linked dominant. **1973** *Nature* 1 June 287/1 The traditional concept that muscular dystrophy is a 'primary degenerative myopathy' has recently been challenged, and the suggestion put forward that the disease may..have a neural basis.

3. b. *muscular Christianity* (earlier example). *muscular Christian* (later examples); so *muscular-Christian* adj.
1857 *Sat. Rev.* 21 Feb. 176/1 We all know by this time what is the task that Mr. Kingsley has made specially his own—it is that of spreading the knowledge and fostering the love of a muscular Christianity. **1966** *Listener* 27 Oct. 613/2 That muscular Christian, Welldon, who was his headmaster and housemaster at Harrow,..always appreciated his [*sc.* Churchill's] merits. **1970** T. HILTON *Pre-Raphaelites* v. 133 The tone of the place [*sc.* the Working Men's College] was heavily muscular-Christian. **1975** J. BLACKBURN *Mister Brown's Bodies* xviii. 151 Great hulking fellows... Muscular Christians to a man.

muscularis mucosæ (mɒskiulā·ris miukōu·sī). *Anat.* [mod.L., f. mod.L. *musculāris* MUSCULAR + *mūcōsæ*, gen. (?) of MUCOSA: the term may be a shortening of *lamina muscularis mucosæ*.] A thin layer of smooth muscle fibres in certain mucous membranes. Also *ellipt.*, as **muscularis.**
[**1854** BUSK & HUXLEY tr. *Kölliker's Man. Human Histol.* II. 89 We find a dense, continuous, reddish layer 0·022–0·044‴ in thickness, (Brücke) the muscular layer of the mucous membrane.] **1867** *Quain's Anat.* (ed. 7) III. p. cxcix, The deepest layer of the alimentary mucous membrane..is formed throughout by non-striated muscular tissue, and is named muscularis mucosæ. **1892** C. S. MINOT *Human Embryol.* (1897) i. 2 It is commonly asserted that the muscular coat of the uterus is largely made up of the hypertrophied muscularis mucosæ. **1959** N. ANDREW *Textbk. Compar. Histol.* vi. 280 In man the entire length of the intestine shows an inner circular and an outer longitudinal layer of muscularis mucosae. **1974** *Nature* 15 Mar. 238/1 Strips of gastric mucosa (dissected free of the underlying muscularis) were removed from (barbiturate-)anaesthetised adult mongrel dogs.

musculo-. Add: **musculofa·scial** *a.*, of or pertaining to both a muscle and its associated fasciæ; **musculoske·letal** *a.*, of, pertaining to, or involving both muscular and skeletal structures.
1949 *New Gould Med. Dict.* 647/1 *Musculofascial*, consisting of both muscular and fascial elements, as in an amputation flap. **1950** *Jrnl. Amer. Med. Assoc.* 25 Feb. 557/2 (*heading*) Musculofascial pain. *Ibid.* 559/1 A sharp, slender needle may be introduced into musculofascial tissues without harm. **1964** L. MARTIN *Clin. Endocrinol.* (ed. 4) iii. 115 The musculo-fascial cone behind the eye. **1944** DORLAND & MILLER *Med. Dict.* (ed. 20) 947/1 Musculoskeletal. **1962** D. NICHOLS *Echinoderms* iii. 54 Burrowing is apparently effected by musculo-skeletal methods rather than by digging with spines and tube-feet. **1971** *Brit. Med. Bull.* XXVII. 82/2 Rheumatic complaints..are presumed to stem from..the supporting structures of the locomotor or musculoskeletal system. **1972** *Lancet* 2 Sept. 449/1 (*heading*) Musculoskeletal disorders associated with type-IV hyperlipoproteinæmia.

muse, *sb.*[1] Add: **2. d.** In phr. *the tenth Muse*, a muse of inspiration figured as being added to the nine of classical mythology.
1650 A. BRADSTREET (*title*) The tenth Muse lately sprung up in America, or Severall poems, compiled with great variety of wit and learning. *c* **1838** C. J. MATHEWS in M. R. Booth *Eng. Plays of 19th Cent.* (1973) IV. 125 Tell her from me that she's a second Venus, a tenth Muse, a fourth Grace. **1855** A. TROLLOPE *Warden* xiv. 220 It was here that Tom Towers lived, and cultivated with eminent success the tenth Muse who now governs the periodical press. **1957** H. READ (*title*) The tenth Muse: essays in criticism. **1973** *Guardian* 29 June 15/4 Poet confronts poet..they 'talked shop like a tenth muse'.

muse, *v.* Add: **1. d.** To say or murmur meditatively.
1834 A. MARSH *Two Old Men's Tales* II. xviii. 224 'But what can I promise? I who have nothing,' mused she, 'and am now penniless.' **1843** DICKENS *Christmas Carol* 32 'You must have been very slow about it, Jacob.'.. 'Slow!' the Ghost repeated. 'Seven years dead,' mused Scrooge 'And travelling all the time?' **1881** MRS. J. H. RIDDELL *Senior Partner* I. vii. 150 'That's strange,' mused Mr. McCullagh; 'and you getting on for thirty year of age.' **1922** JOYCE *Ulysses* 212 Amused Buck Mulligan mused in pleasant murmur with himself, selfnodding: 'A pleased bottom.'

‖ **museau** (mūzo). *colloq.* [Fr., lit. 'muzzle, snout (of animal)'.] A person's face.
1915 D. H. LAWRENCE *Rainbow* xv. 423 You've got a *museau*, not a face. **1922** —— *England, my England* 228 A young man of twenty-two, with a fresh, jaunty *museau*. **1925** —— *St. Mawr* 7 She, with her odd little *museau*, not exactly pretty, but very attractive. **1955** *Essays in Criticism* V. 79 Lou, with her gipsy wildness, her animal *museau*.

musee. Delete † *Obs. rare*[−1] and add later examples.
Normally in form *musée* as in French.
1861 C. M. YONGE *Young Stepmother* xvii. 233, I shall be most happy to introduce you into my *atelier* and show you my notes on the various *Musées*. **1885** H. JAMES *Little Tour in France* xv. 102 Of course I had time to go to the Musée; the more so that I have a weakness for provincial museums. **1924** R. FRY *Let.* 27 June (1972) II. 553 I've.. seen the much vaunted Musée here [*sc.* in Montpellier]. It's badly lit and worse hung. **1944** *Amer. N. & Q.* Aug. 72/2 In compiling material on the dime museum I could not help being struck with what seems to have been a showman's affectation—the almost universal use of *musee* for *museum*. **1970** B. WHELPTON *Painter's Paris* vii. 107 The Musée Cernuschi..houses a superb collection of the art of China.

2. musée imaginaire (‖ imaʒinɛ̄r') [Fr., lit. 'imaginary museum']: an imaginary collection of all the works of human artifice.
[**1947** A. MALRAUX *Musée Imaginaire* 17 Un Musée imaginaire sans précédent s'est ouvert, qui va pousser à l'extrême l'intellectualisation commencée par l'incomplète confrontation des vrais musées; répondant à l'appel de ceux-ci, les arts plastiques ont inventé leur imprimerie.] **1959** *Twentieth Century* Oct. 262 He [*sc.* the culture-vulture]..haunts the limbo of the *musée imaginaire*, where all cultures and styles coexist. **1960** *Ibid.* Sept. 275 In Audenesque abstraction lies the power to keep us from..the object labelled and dismissed in the *musée imaginaire*. **1962** *Listener* 8 Mar. 431/1 We are no longer bewildered by the 'matter' of civilizations now that the *musée imaginaire* has accustomed us to the quick pounce on the artifact that illumines a lost world. **1967** *Ibid.* 7 Sept. 293/2 Dallagret has gone for..a kind of *musée imaginaire* sealed in polythene.

museless. Delete † *Obs.* and substitute *rare*. Also, uninspired. Add later examples.
1882 G. M. HOPKINS *Let.* 4 Nov. (1955) 159 My mind is dull and museless. **1958** P. L. FERMOR *Mani* i. 6 Nothing in the grace and the enchantment of all this could remind one of museless and unbookish Sparta.

museography. Add: Also = MUSEOLOGY. Also **museogra·phical** *a.*
1935 R. B. PERRY *Tht. & Char. of W. James* I. ix. 176 He had no archaeological or museographical interest. **1953** *Times* 4 Mar. 9/6 An exhibition that is certainly a remarkable example of showmanship or, as it is now called, museography. **1973** *Guardian* 20 Jan. 10/2 French museography has improved.

museology. Add: (Later examples.)
1946 *Nature* 9 Nov. 678/2 Another interesting connexion with the University of Chicago is the establishment of university classes in museology in the Department of Anthropology of the Museum. **1957** G. CLARK *Archaeology & Society* (ed. 3) i. 36 Prehistoric archaeology was virtually a branch of museology. **1973** *Nature* 21–28 Dec. 541/1 Dr. Carmichael also achieved renown as a science administrator in the worlds of academia, museology and government.

Hence **museolo·gical** *a.*
1949 K. CLARK *Landscape into Art* 137 Cézanne's *petite sensation* was transformed into the most cerebral and museological of all styles—cubism. **1960** *Spectator* 4 Nov. 693 Modern methods of reproduction and museological arrangement. **1970** *Times* 18 Mar. 11/7, I..would rather enjoy the image of myself as a kind of museological Hell's Angel or Jesse James.

mu·sery[2]. *rare*[−1]. [MUS(E *sb.*[1] + -ERY.] The work of the Muses.
1869 G. MEREDITH *Let.* 19 Dec. (1970) I. 407 Why, this stuff is not the Muse, it's Musery.

musette. Add: **5.** In full, *musette bag*. A type of canvas haversack. orig. *Mil.*
1923 E. HEMINGWAY *Three Stories* 13 The young gentleman had a musette over his shoulder. **1934** WEBSTER, Musette bag. **1943** [see *flight bag* (*FLIGHT *sb.*[1] 14)]. **1949** A. HAYES *Girl on Via Flaminia* iii. 38 He got up from the bed and went to the table where he had placed the musette bag. He began to unbuckle the straps. **1955** J. THOMAS *No Banners* xxiii. 229 Alfred had the Sten gun in a canvas *musette* tied with string and slung over his shoulder. **1961** J. HELLER *Catch-22* (1962) xiii. 130 Major—de Coverley.. packed his musette bag, commandeered a plane and..had himself flown to that city to rent two apartments. **1965** *Sun* 3 Nov. 11/5 The musette, or featherweight knapsack from the Continent, has largely replaced the saddlebag. **1968** *Amer. Hist. Illustrated* July (rear cover), The diary stayed behind in his musette bag.

museum. Add: **2.** (Further examples.)
1893 M. HOWE *Honor* 320, I was expected to give ten thousand to the Art Museum. **1936** *S.P.E. Tract* XLV. 176 There are..a few English words whose meaning in the United States corresponds more closely to that of their French counterparts than to their meaning in England; e.g...*museum*. **1950** [see *GALLERY *sb.* 6]. **1975** *Times*

27 Sept. 14/6 The first railway museum in Britain was opened..at York in 1928.

b. *attrib.* and *Comb.*, as *museum-goer*, *interest*; **museum piece**, an object suitable for exhibition in a museum; also *transf.* (usu. with derogatory sense).

1930 *Times Educ. Suppl.* 23 Aug. 362/2 April is the general 'peak' month for museum-goers. **1971** *Daily Tel.* 11 June 16 An admission charge would not of course greatly deter the habitual museum-goer. **1933** P. Godfrey *Back-Stage* xiv. 180 The attempt of recent years to stage a music-hall revival has unearthed a few shaky veterans of variety, who have little more than a museum interest for the post-War generation. **1901** *Brit. Chess Mag.* XXI. 351 The more stately carved pieces (named for the sake of distinction 'museum-pieces'). **1908** R. Fry *Let.* 12 Dec. (1972) I. 306 It would be a great Museum piece, but..the price is high for Ribera. **1920** W. J. Locke *House of Baltazar* iii. 31 Quong Ho was admitted to be a museum-piece of discretion. **1923** J. Galsworthy *Captures* 228, I felt as if I had a priceless museum piece which a single stumble might shatter to fragments. **1928** —— *Swan Song* I. xi. 82 The girl and her brother had been museum pieces, two Americans without money to speak of. **1936** L. MacNeice tr. *Aeschylus' Agamemnon* Pref. 8 It is my hope that the play emerges as a play and not as a museum piece. **1949** B. A. Botkin *Treas. S. Folklore* I. i. 4 What saves these relics of the past from being mere museum-pieces is their symbolic and often living relation to the culture of the region. **1955** *Amer. Speech* XXX. 94 *Museum piece*,..an old truck. **1960** H. Nicolson *Let.* 22 Sept. (1968) 385, I had to lunch with such a bore... He is a museum piece; it is like seeing a railway-engine of 1854. **1964** D. Varaday *Gara-Yaka* v. 48 He..handed over his rifle, an ancient muzzle-loading museum piece. **1974** P. Erdman *Silver Bears* xv. 153, I am looked upon as a mildly amusing curiosity—even a museum piece.

muse·umish, *a. colloq.* [f. Museum + -ish[1].] Resembling a museum or its exhibits.

1926 D. H. Lawrence *Let.* (1932) 657, I wouldn't care to live in Assisi..it's too museumish, not enough life in it now. **1937** *Observer* 5 Sept. 13/3, I must confess that I went to 'Ghosts' with some apprehension. Might not these unhappy far-off things seem too far off; might not the echo of battles long ago bring a museumish air to theatre walls?

mush, *sb.*[1] Add: **1.** (Earlier and *attrib.* examples of *mush and milk*.)

1745 E. Kimber *Itinerant Observations Amer.* (1878) 34 The meaner Sort you find little else but Water amongst, when their Cyder is spent, Mush and Milk, or Molasses, Homine,..and Fish, are their principal Diet. **1817** in *Essex Inst. Hist. Coll.* (1866) VIII. 244 They..gave me a supper of mush and milk. **1866** 'Mark Twain' *Lett. fr. Hawaii* (1967) 210 I'm disgusted with these mush-and-milk preacher travels. **1895** *Montgomery Ward Catal.* 534/1 China Oatmeal or Mush and Milk Set..cream pitcher, bowl and plate.

3. d. *slang.* The mouth or face. Freq. pronounced (muʃ).

1859 G. W. Matsell *Vocabulum* 127 *Mush*, the mouth. **1906** E. Dyson *Fact'ry 'Ands* xv. 202 Er stream iv water ..takes Fuzzy fair in the mush, heels him over. **1919** *Dialect Notes* V. 66 *Mush*, the mouth. 'Stop his mush and give us a rest.' **1932** J. T. Farrell *Young Lonigan* I. i. 20 That time he had pasted Weary in the mush with an icy snowball. **1959** I. Jefferies *Thirteen Days* i. 18 He said if anybody opened his mush, he'd kill 'em. **1971** B. W. Aldiss *Soldier Erect* 53 My regret was that I had not given Wally a bunch of fives in the mush while I had the chance. *Ibid.* 194, I hit him smack in the mush. Not very hard but pleasureably hard. My right fist did not hurt a great deal —not as much as his face hurt him. **1974** T. Barling *Shooter Man* i. 12 A big grin all over his ugly mush.

e. *Surfing slang.* The foam produced when a wave breaks.

1969 *Surfer* IX. 57 Hardy rides and cuts out as the shoulder flattens to mush. **1970** *Surf '70* (N.Z.) 13/2 If there is any flat mush the board tends to stop and lose its turning ability. **1971** *Studies in English* (Univ. Cape Town) Feb. 28 When a surfer is tired, he catches a wave all the way in. For the last part of the distance he will be riding the frommel; the soup; white water; the mush.

4. *Radio.* Interference or noise heard as a hissing or the sound (ʃ); *spec.* (see quot. 1924).

1924 *Harmsworth's Wireless Encycl.* II. 1456 *Mush*, term used for the irregular intermediate frequencies set up by an arc transmitter which interfere with the fundamental wave-lengths. **1927** [see *background *sb.* 3]. **1928** *Observer* 1 July 4/4 In listening to very faint signals from a great distance a limit is put by the 'mush' and statics and other noises brought in from the aerial. **1952** *Electronic Engin.* XXIV. 120 Unwanted low frequency fluctuations of high frequency 'mush'. **1971** *Daily Mail* 17 Nov. 9/3 Were subjected to a continuous 'hissing noise', or 'electronic mush'.

5. *attrib.* and *Comb.*, as (sense 1) *mush pan*, *pot*; **mush-head,** a person of a yielding disposition; one lacking in firmness; so *mush-headed* adj.; **mush-ice,** water only partly frozen, ice mixed with water; **mush-sugar,** a mixture of syrup and crystals of sugar.

1890 Barrère & Leland *Dict. Slang* II. 77/2 *Mush-head* (American), a stupid, witless fellow. **1919** H. L. Wilson *Ma Pettengill* ii. 63, I up and told her flat she could never run a boarding-house and make it pay; that no woman could who hadn't learned to say 'No!' and she was too much of a mushhead for that. **1932** *Screenland* Apr. 70/1 She has married the poor little mush-head that had been wished upon her. **1914** R. Cullum *Way of Strong* III. viii. 294 The game isn't worth it, fighting this mush-headed crowd who have to get other folks to think for 'em. **1815** *Niles' Weekly Reg.* IX. 201/2 You may, by digging down three feet, take a pole sixty feet long and with the strength of your hands run it down the whole length, and find no termination of what is called the mush ice. **1907** J. London *White Fang* 202 The fall of the year, when the first snows were falling and mush-ice was running in the river. **1966** R. M. Patterson *Trail* 86 They came..poling and tracking against the slowly drifting mush ice of the fall. **1847** J. S. Robb *Squatter Life* 59 Betsy Jones' Tumble in the Mush Pan. **1940** H. H. Hatcher *Buckeye Country* 173 When his self-made pasteboard hat fell to pieces..he covered his head with his mush pan. **1847** H. Howe *Hist. Coll. Ohio* 432 Johnny, who wore on his head a tin utensil which answered both as a cap and a mush pot, filled it with water and quenched the fire. **1868** *Rep. Iowa Agric. Soc. 1867* 178 When sugar is contemplated, White Imphee is..the best, as all I have made went to thick mush sugar immediately.

mush (mʌʃ), *sb.*[3] [f. *mush *v.*[3]] A journey made through snow with a dog-sledge.

1910 R. Service *Trail of '98* 341, I was still weak from my illness and my long mush had wearied me. **1926** *Glasgow Herald* 8 Mar. 10/7 From Hudson in northern Ontario it is a twelve-day 'mush' for men and dogs over the frozen sub-arctic prairie to the Red Lake district. **1965** *Kingston* (Ont.) *Whig-Standard* 8 Feb. 8/6 It's 'Mush, Mush' time for Owner-driver Vern Zoschke and his Dogs.

mush (muʃ, mʌʃ), *sb.*[4] *Mil. slang.* [? f. Mush *v.*[2]] A guardroom or cell; a military prison.

1917 W. Muir *Observations of Orderly* xiv. 228 If one of those 'hooks' [*sc.* pilferers] were caught, he would be first 'rammed in the mush' (put in the guardroom). **1933** E. Partridge *Words, Words, Words!* III. 204 *Mush*, sometimes spelt and pronounced *moosh*..denotes guard-room or cell(s). **1943** Hunt & Pringle *Service Slang* 46 *Mush*, the *glasshouse* or guardroom.

mush (muʃ), *sb.*[5] *slang.* [Orig. uncertain.] Man, 'chap'; hence also as a term of address.

1936 J. Curtis *Gilt Kid* v. 53 I'm a bit of a coring mush, myself. **1961** J. Maclaren-Ross *Doomsday Book* vii. 76 Long's you don't go laughing in the wrong place, mush. **1968** A. Diment *Bang Bang Birds* v. 57 So this mush is running a string of..houses of ill-repute... So what?

mush (mʌʃ), *sb.*[6] *slang.* Shortened f. Moustache, mustache *sb.*

1967 C. Drummond *Death at Furlong Post* xv. 181 Take my oath it's he..with his hair parted..and the mush shaved off his lip. **1969** K. Giles *Death cracks Bottle* xiii. 172 He read one of these Service ads... You know, a young bloke with a mush telling troops to go plunging into the jungle.

mush (mʌʃ), *v.*[3] [Apparently f. Fr. *marchez* or *marchons*, imp. of *marcher* to advance, the command given to sled dogs; cf. *mush *sb.*[3]] Also const. *on*. **a.** *intr.* To travel on foot through the snow with a dog-sledge (said also of the dogs); *trans.* to drive dogs through the snow.

[**1862** R. Kennicott *Jrnl.* in J. A. James *First Sci. Explor. Russian Amer.* (1942) 130 My dogs are *dogs*! and we will *mouche* very likely, after all.] **1897** *Medicine Hat* (Alberta) *News* 30 Sept. 7/4 It is laughable to hear the driver yell, 'Mush, Mush,' at them. **1900** J. London *Son of Wolf* 5 'Come, mush on there, you poor sore-footed brutes!' he murmured. **1903** *Sun* (N.Y.) 22 Nov., His little boat was cut out, and then he started to 'mush' back over the ice. **1904** S. E. White *Silent Places* xvii. 180 'Mush! Mush on!' shouted Sam. The four dogs leaned into their collars. **1914** R. Cullum *Way of Strong* I. i. 1 Five great sled dogs crouched in their harness. They were waiting the long familiar command to 'mush'; an order they had not heard since the previous winter. **1927** *Brit. Weekly* 13 Jan. 409/2 They were mushing on to a new strike. **1932** *Sun* (Baltimore) 15 Jan. 1/5 Through a raging blizzard McDowell mushed a dog team the eighty miles to Aklavik. **1934** *Beaver* Sept. 26 Constable Lee and his Indian interpreter, Albert, came mushing up with a jingle of bells from Fort Providence to pay their annual visit to trappers in the bush. **1947** *Mazama* Dec. 6/2 Norris left Mt. McKinley Park station on 11 April and mushed his dog team to Base Camp arriving 15 April. **1963** R. D. Symons *Many Trails* 198 And hurry! Hurry! Before it is too late —mush, mush on—the whip cracks hysterically. **1966** *Kingston* (Ont.) *Whig-Standard* 25 Feb. 12/1 There hasn't been so much excitement over sled-dogs in the north since Leonard Sepala mushed through the land of the midnight sun.

b. *transf.* To travel (through snow or ice).

1898 W. N. Robertson *Yukon Memories* 210 You think all the while you are nearing the top, and 'mush on', like viewing a ship at sea. **1906** 'O. Henry' *Four Million* 106 I never got off the train since I mushed out of Seattle, and I'm hungry. **1958** P. Berton *Klondike Fever* 19 He thought nothing of making a present of his trousers to a pantless native and mushing home in his red flannels. **1966** *Globe & Mail* (Toronto) 24 Jan. 17/8, I then struck out to mush to the nearest bus stop.

Hence **mu·shing** *vbl. sb.*

1904 *Prof. Papers U.S. Geol. Survey* No. 20. 15 In 'mushing', the best progress is made in relatively cool weather. **1930** W. N. Robertson *Yukon Memories* viii. 114 There is a lingering feeling that the monotonous *mushing* along has not been devoid of its pleasures. **1966** *Kingston* (Ont.) *Whig-Standard* 25 Feb. 12/1 Wilfrid Charles is regarded as a sure-pop betting cinch to retain the mushing title.

mush (mʌʃ), *v.*[4] *colloq.* [f. Mush *sb.*[1] 3.] Const. *in*. To sink into a soft surface.

1948 N. Shute *No Highway* ii. 41 The landing was a hazardous one because of the alternate thaw and freeze: the skis [of the aircraft] mushed in beneath the icy crust. **1962** J. Glenn in *Into Orbit* 13, I..skimmed right over the top of a rice paddy—almost mushing into it.

musha. (Later examples.)

1898 J. D. Brayshaw *Slum Silhouettes* 15 Oh, musha! the divil a drown ye'll drown, Murty Hooligan, while there's hemp for the sowing. **1933** Davies & Thomson tr. *O'Sullivan's Twenty Years A-Growing* i. 7 Ah, musha, youth is a queer thing.

musher (mʌ·ʃəɹ). [f. *mush *v.*[3] + -er[1].] A person who mushes; one who travels through snow on foot (with a dog-sledge). Also *fig.*

c **1900** *Western Miner* (1963) Mar. 30/1 Way up north there's a rail road...It followed the tracks of the mushers, Where they grasped the rocks for a hold. **1902** L. McKee *Land of Nome* 178, I felt that I had received a very high compliment..when an old-timer in the party..told me that I was a 'musher from hell'. **1925** *Chambers's Jrnl.* July 456/2 Those far northern regions are inaccessible.. except to the most hardy and expert 'mushers'. **1948** *Time* 19 July 34/3 Klondike Mike, the greatest of the mushers, the sourdough who struck it rich and kept his poke, is a living legend. **1973** *Islander* (Victoria, B.C.) 20 May 3/1 Art Fraser, owner and musher of the dog team, was my guide.

mushie (mʌ·ʃi). *Austral. colloq.* shortening of Mushroom *sb.* 1.

1935 O. P. Meyer *Four of Us* 52, I don't see many mushies yet. **1972** *Sunday Mail* (Brisbane) 9 Jan. 5/2 Mushies are seldom sold, and are grown so easily from cow manure..that they're usually given away.

mushily (mʌ·ʃili), *adv.* [f. Mushy *a.* + -ly[2].] In a soft or mushy manner.

1896 'Mark Twain' in *Harper's Mag.* Jan. 296/2 The column of inert mortality sank mushily to the ground. **1910** 'O. Henry' *Strictly Business* xvi. 187 'A cool hundred,' said Fuzzy thoughtfully and mushily. *a* **1935** T. E. Lawrence *Mint* (1955) II. xiv. 138 A black east wind froze the sweat mushily over our skins. **1971** 'D. Shannon' *Ringer* (1972) ix. 147 'That I did,' confirmed Mr. Gebhart mushily. ''Scuse me—get my teeth—.'

mushiness (mʌ·ʃinės). *colloq.* [f. Mushy *a.* + -ness.] The condition of being soft or pulpy; also *fig.*, sentimental character, weakness, sloppiness.

1890 [see Mushy *a.*]. **1893** 'M. Rutherford' *Cath. Furze* II. xiv. 70 Partly this distressing weakness is due to the absence of a clear conviction that we are right;..but frequently it is simple mushiness of character. **1946** A. L. Bacharach *Brit. Music of our Time* xv. 197 One is perhaps disconcerted by an harmonic mushiness—he was especially fond of the now much-abused chords of the seventh and ninth. **1962** *Yale Rev.* June 540 Pound puts Joyce with Hueffer, diametrically opposite to 'the softness and mushiness of the neo-symbolist movement'.

mu·shrat. = Musk-rat 1, 2. *N. Amer.*

1890 *Dialect Notes* I. 74 *Mushrat*, the muskrat or musquash. **1901** R. Connor *Man from Glengarry* 19 The Glengarry men..despised the Murphy gang as sawlog-men; 'log-rollers' or 'mushrats' they called them. **1939** W. Faulkner *Wild Palms* 218 If I took aboard every floating sardine can you sonabitchin mushrats want me to I wouldn't even have room forrard for a leadsman. **1939** T. H. Raddall *Pied Piper* 45 An' I'm goin' to play hell with you, you game-poachin', fish-spearin', wood-stealin' ol' mushrat! **1954** C. Bruce *Channel Shore* 18 Stan was carrying on..talk of mushrats and rabbit-snaring. **1954** *Encounter* Oct. 3/2 All the lesser beasts—coon and possum and beaver and mink and mushrat (not muskrat: mushrat); they were still there.

mushroom, *sb.* Add: **3. c.** *Archit.* A reinforced concrete pillar that broadens out towards the top, with the reinforcing rods passing upwards and outwards into a reinforced concrete slab forming part of the floor above, which is thereby supported by the pillars without the use of beams. Freq. *attrib.*

1907 *Western Architect* May 51/1 The essential feature of this new construction is the formation of a so-called mushroom at the top of each column, by extending its reinforcing rods, laterally, some four feet or more out into the slab in a radial direction and supporting on these, ring rods, which, in turn, carry the lighter reinforcement for the slab construction. **1927** T. P. Bennett *Archit. Design in Concrete* 14 A scheme similar in principle has been employed in a number of factory buildings, and has acquired the name of 'Mushroom' construction. **1936** *Archit. Rev.* LXXX. 113/1 (*caption*) These conditions have been met by an adaptation of the 'mushroom' floor slab type of construction. Each floor consists of a reinforced concrete slab, 10 ins. thick, supported on external wall columns, spaced 21 ft. centre to centre, and two internal columns with mushroom caps. **1941** S. Giedion *Space, Time & Archit.* vi. 374 The American engineer C. A. P. Turner had been experimenting with the mushroom system a year before Maillart, but the Swiss engineer had employed slabs as basic elements in bridges since the beginning of the century. **1963** tr. *Hatje's Encycl. Mod. Archit.* 182/2 His [*sc.* Robert Maillart's] most important invention in the field of high structures was made in 1908 with mushroom slab construction, which he used for the first time on a large scale in 1910. **1969** W. R. Dalzell *Architecture* 41 (*caption*) Reinforced concrete mushrooms support slab constructions.

d. A cloud (of smoke, fire, etc.) that spreads upwards and outwards.

1916 J. Buchan *Greenmantle* xxi. 291 There was the dull shock of an explosion and a mushroom of red earth. **1924** A. J. Small *Frozen Gold* iv. 110 A sudden, abrupt mushroom of smoke spread out above the building on the uprush of air. **1945** *N.Y. Times* 26 Sept. 16/6 At first it was a giant column that soon took the shape of a supramundane mushroom. **1952** G. Wilson *Julien Ware* xvi. 129 He..pulled energetically at the stem until a mushroom of smoke hung above his head. **1954** *Amer. Speech* xxix. 275 *Mushroom*, a fire which, reaching a ceiling and finding no outlet, starts to spread out and burn downward.

e. A mushroom-shaped implement used in darning.

1931 'R. Crompton' *William's Crowded Hours* viii. 172 The cavity through which both her fist and the darning 'mushroom' slipped so unavailingly. **1931** R. H. Heaton *Perfect Hostess* 95 Little comforts for the work-basket... Ribbon threader. Mushroom for darning. **1969** E. H. Pinto *Treen* vii. 134 Cabbage pressers..are larger versions of 'mushroom' darners, usually with the button end about 5 in. in diameter.

6. a. *mushroom lip, minaret, omelette, pickle, sauce* (earlier example), *sculpture*.

1962 D. Harden *Phoenicians* xi. 150 In tombs at Carthage jugs with mushroom or trefoil lips abounded. **1930** R. Campbell *Adamastor* 50 Their mushroom minarets and toadstool towers. **1892** *Encycl. Pract. Cookery* I. xiii. 961/2 Mushroom omelet—Either fresh or canned mushrooms may be used for this. **1954** G. M. Lapolla *Mushroom Cooking* 78 Mushroom omelette. One way of using mushroom stems. **1972** P. D. James *Unsuitable Job for Woman* iv. 137 She made herself a mushroom omelette. **1747** H. Glasse *Art of Cookery* ii. 35 Add a Spoonful of Mushroom Pickle, pickled Mushrooms, or fresh, if you have them. **1911** W. Owen *Let.* 25 Sept. (1967) 85 In the kitchen making mushroom pickle. **1747** H. Glasse *Art of Cookery* ii. 35 To make Mushroom-Sauce for White Fowls of all Sorts. **1970** *Washington Post* 30 Sept. B 1/5 This holy mushroom sculpture of gray volcanic stone is part of the pre-Columbian art exhibit.

c. **mushroom city,** = *mushroom town* (below); **mushroom cloud,** the cloud of smoke that forms above the site of a nuclear explosion, with the characteristic shape of a tall pillar with a broad, flattish top; also *fig.*; (cf. sense *3 d); **mushroom-faker** (earlier example); **mushroom-flap** *obs.*, the cap of a common mushroom when fully opened; **mushroom growth,** rapid growth like that of a mushroom; **mushroom hat** (earlier and later examples); **mushroom-headed** *a.* (later example); **mushroom town,** a town that has sprung up rapidly; **mushroom valve,** a lift valve whose moving element somewhat resembles a mushroom in shape.

1860 H. Greeley *Overland Journey* 140 A rush of three or four hundred, mainly men of broken fortunes from the dead mushroom 'cities' of Nebraska and Kansas. **1948** P. Johnston *Gold Rush* 42/1 By April 1850, Downieville had become a mushroom city of large proportions with a population of about 5000. **1958** *Spectator* 17 Jan. 63/1 If Europe is not to go up in a mushroom cloud we must find an area of agreement with the Soviet Union. **1964** M. Gowing *Britain & Atomic Energy* ix. 267 Dr Penney was one of the scientific observers who saw the mushroom cloud rise from the ruins of Nagasaki. **1965** *Spectator* 5 Feb. 167/1 Hatred grows to mushroom-cloud proportions. **1839** H. Brandon *Poverty, Mendicity & Crime* 164/1 *Mushroom fakers*, itinerant umbrella makers and repairers. **1747** H. Glasse *Art of Cookery* xi. 121 Two Quarts of the large Mushroom Flaps rubbed to Pieces. **1861** Mrs. Beeton *Bk. Househ. Managem.* 226 Choose full-grown mushroom-flaps, and take care they are perfectly fresh-gathered. **1911** E. Wallace *Sanders of River* viii. 108 These secret societies he knew well enough... He knew their mushroom growth; how they rose from nothingness with rituals and practices ready-made. **1931** *Economist* 2 May 944/2 Thus has terminated the career of a bank which enjoyed a mushroom growth immediately after the war. **1879** C. M. Yonge *Magnum Bonum* I. xi. 206 She looked up under her brown mushroom hat. **1888** —— *Beechcroft at Rockstone* iv. 73 A..lady in a mushroom hat. **1912** [see *ablow *adv.* or *pred. a.*]. **1963** *Guardian* 29 Jan. 7/7 Dots..come on giant mushroom hats. **1875** Pitt-Rivers in *Proc. R. Inst. Gt. Brit.* VII. 514 We see [in the plate] the mushroom-headed waddy, with its projecting ridge flattened, then [in a later stage of development] curved. **1819** R. L. Mason *Diary* 9 Dec. in *Narr. in Pioneer West* (1915) 66 Those mushroom towns in a short time will produce their own death. **1938** H. Asbury *Sucker's Prog.* 310 The invasion reached its peak with.. the building of the trans-continental railroads which the gamblers followed step by step, carrying on their thieving business in every mushroom town that sprang up along the route. **1957** *Economist* 12 Oct. 105/2 These peasants turned workers..are the most interesting feature of the whole Polish social landscape. The mushroom town.. displays the problem in a nutshell. **1877** *Design & Work* 7 July 114/3 Exhaust valve, consisting of ordinary mushroom valve. **1924** S. G. Wheeler *Marine Engin.* I. vii. 102 Occasionally steam engines are controlled by 'mushroom' valves like internal combustion engines. **1966** J. A. Dolan *Motor Vehicle Technol.* I. ix. 91 The valves used in motor vehicle engines are called poppet or mushroom valves.

mushroom, *v.* Add: **4.** *intr.* To rise like a mushroom; to expand or increase rapidly; also const. *up*, *out*. Freq. *fig.*

1903 *Sun* (N.Y.) 2 Nov. 3 The flames had gone up the stairs to the very top of the house, and had then 'mushroomed' out, as the firemen say. **1911** *Ithaca Jrnl.* 10 Aug., The flames mushroomed from the shaft on all floors above. **1937** D. L. Sayers *Busman's Honeymoon* v. 97 The loosened soot of centuries came plunging in a mad cascade down the chimney; it met the floor with a soft and deadly violence and mushroomed up in a Stygian cloud. **1947** J. C. Rich *Materials & Methods of Sculpture* ix. 249 If the end [of the chisel] receiving the impact of the carving hammer is too soft, it will quickly 'mushroom'. *Ibid.* ix. 261 The untempered ends of working chisels will frequently begin to mushroom..after repeated hammer blows. **1959** *Daily Tel.* 1 Dec. 11/8 A private Bill, promoted by the L.C.C.; it is intended to secure greater control of clubs which have 'mushroomed' into existence in recent years. **1962** *Listener* 9 Aug. 223/2 His characters are huge, they mushroom out one after the other and yet the plays aren't shapeless. **1972** *Language* XLVIII. 429 The number of publishing companies, both in Europe and North America, which have prepared new editions of primary sources [in linguistics] has mushroomed.

mushrooming, *vbl. sb.* Add: **c.** Growth like that of a mushroom, as regards shape (cf. b) or rapidity. Freq. *fig.* Also as *ppl. a.*

1947 J. C. Rich *Materials & Methods of Sculpture* ix. 251 For fashioning chisels... It is advisable to cut the chisel blanks 1 to 2½ extra inches in length, since the occasional fracturing of a point, a mushrooming of the end receiving the hammer blows, and repeated sharpening will slowly shorten the total length. **1954** *Encounter* June 13/1 Over the past decade there has been a fantastic mushrooming of arts and crafts hobbies. **1957** *Listener* 12 Sept. 379/1 Economic changes..have brought..Africans into new jobs and opportunities, and into mushrooming new towns. **1968** A. Diment *Great Spy Race* x. 186 The mushrooming fireball of exploding petrol. **1968** *Brit. Med. Bull.* XXIV. 189/1 Recent mushrooming of activity and interest in this subject. **1972** K. S. Roundhill *Prescription for Today's Missionary* 11 Positions for foreign teachers in the mushrooming educational programmes..are often available.

mushy, *a.* Add: (Earlier and later examples.)

1839 *Southern Lit. Messenger* V. 209/2, I soon recognized old *Noth Calinur* in their nasal mushy pronunciation. **1880** 'Mark Twain' *Tramp Abr.* xxvi. 267 Mushy, slushy early spring roads. **1952** M. Tripp *Faith is Windsock* ii. 34 Unfortunately reception was indistinct and towns showed vague and mushy, making identification acutely difficult. **1962** *Which?* (Suppl.) July 86/2 Its handling was described as 'mushy' and our drivers said that it tended to 'wallow' in corners.

2. *fig.* Tender, sentimental, insipid. Also as *adv.*

1870 *Nation* (N.Y.) 3 Feb. 67 The death penalty is disappearing, like some better things, before a kind of mushy and unthinking doubt of its morality and expediency. **1876** [in Dict.]. **1913** G. Stratton-Porter *Laddie* vii. 220 They formed a circle around Sally and Peter and as mushy as ever they could they sang, 'As sure as the grass grows around the stump, You are my darling sugar lump,' while they danced. **1928** G. B. Shaw *Intell. Woman's Guide Socialism* 458 You may..be a sharp, cynical sort of person; or you may be a nice, mushy, amiable, goodnatured one. **1951** J. D. Salinger *Catcher in Rye* xvi. 137 She sings it very Dixieland and whore-house, and it doesn't sound at all mushy. **1971** Wodehouse *Much Obliged, Jeeves* i. 7 [Marriage] wouldn't have fitted in with my plans at all, she though physically in the pin-up class, being as mushy a character as ever broke biscuit. **1974** L. Lamb *Man in Mist* i. 12 Some mushy mag. She calls that a book!

3. (See quots. and cf. *mush *sb.*[1] 4.)

1924 *Harmsworth's Wireless Encycl.* II. 1456 A mushy note is one which is not absolutely definite or clear cut, and so hard to read. It is a note received by the heterodyne method when damped wires or modulated continuous waves are being received. **1937** *Printers' Ink Monthly* May 39/3 *Mushy*, poor musical definition of an orchestral pickup.

4. *Surfing.* (See quots.)

1967 J. Severson *Great Surfing* Gloss., *Mushy*, a slow, sloppy wave that has little power. **1969** *Surfer* IX. No. 6, 69 My little board's good for mushy waves, but when it's five foot or over and fast, I use the other. **1972** R. Abbott *Science of Surfing* v. 83 It is sometimes possible to 'beat' breaking sections, although this is easiest if the waves are 'mushy'.

music, *sb.* Add: **11.** *to face the music* (earlier U.S. and later examples); *music while you work*: continuous light music played to workers, esp. in factories; *to make (beautiful) music (together)*: to have sexual intercourse.

1850 *Congress. Globe* App. 4 Mar. 324/3 There should be no skulking or dodging..every man should 'face the music'. **1920** R. Macaulay *Potterism* III. viii. 140 If Gideon didn't shove him, he's nothing to be afraid of in our talk, and if he did he'll have to face the music. **1940** *Radio Times* 21 June 3/2 This coming week there will be, twice every day, half-an-hour's music meant specially for factory-workers..under the title 'Music while you work'. **1958** J. Betjeman *Coll. Poems* 227 Though 'Music while you work' is now our wont, It's not so nice as 'Music while you don't'. **1958** 'J. Byrom' *Or be he Dead* xv. 186 So the old bitch *did* recognize me! Mrs Kernan and I were pretty sure she had. That's why we did a bunk so hastily, leaving Byron to face the bill and the music. **1960** R. Pound *Selfridge* ii. 40 One of the first applications of the music-while-you-work incentive to modern labour. **1967** F. Mullally *Prizewinner* vii. 117 He could say good-bye to any hope of making music later on with the Swede. **1968** S. E. Ellacott *Everyday Things in England 1914–68* xi. 162 The factory system of 'music while you work'..was introduced during the Second World War..in munitions factories. In mechanical, repetitive jobs lively music helped production. **1969** H. Nielsen *Darkest Hour* xxi. 238 You and Buddy can make beautiful music together. **1972** C. Short *Naked Skier* xxi. 115, I think we should make music together.

11*. Flexibility or 'give' in the shaft of a golf club. *rare.*

1890 H. G. Hutchinson *Golf* iii. 57 A heavy head may bring just the right amount of life—of what Tom Morris calls 'music'—out of a very stiff shaft. **1903** W. J. Travis *Pract. Golf* (rev. ed.) ix. 111 The man with a less rapid swing will get equally as long a ball by using a more supple shaft. The more 'music' there is in the shaft, however, the greater is the liability to slice or pull.

12. a. *music cabinet, critic, -man* (further examples), *-master* (later example), *-rack* (earlier example), *-rest* (further example), *-sheet*, *-shop* (earlier example), *-stool* (earlier and later examples), *-track*. **b.** *music-lover, -maker, -making, -publisher, -seller; music-hungry, -loving, -sounding, -stunned* adjs.

1883 *Heal & Son Catal.* Sept. 216 Music Cabinet, Walnut or Ebonised. **1911** *Daily Colonist* (Victoria, B.C.) 22 Apr. 4/4 It does not matter to us what style of piano you may have we can give you a music cabinet style to harmonize. **1901** G. B. Shaw *Three Plays for Puritans* p. v, The difference between the leisure of a Persian cat and the labor of a cockney cab horse is not greater than the difference between the official weekly or fortnightly playgoings of the theatre critic and the restless daily rushing to and fro of the music critic. **1956** Auden & Kallman *Magic Flute* (1957) 58 That is changed Since music-critics learned to feel 'estranged'. **1950** A. Lomax *Mister Jelly Roll* iv. 181 A gang of music-hungry, life-hungry white kids. **1933** *Maclean's Mag.* 15 July 32/4 Music lovers will thrill at the music of His Majesty's Scots Guards, and thirty other bands. **1963** Auden *Dyer's Hand* 507 That great music-lover, Bernard Shaw. **1870** D. J. Kirwan *Palace & Hovel* xxxiv. 493 Theatre going and music-loving people. **1864** Browning *Dramatis Personae* 49 Schumann's our music maker now. **1955** N. Cardus in H. van Thal *Fanfare for E. Newman* iii. 34 The composer's physiology as music-maker. **1946** Music-making [see sense 12 c below]. **1963** *Times* 12 Feb. 14/4 A recital which had more of the air of a friendly music-making at home than the deadly earnest aspiration usually encountered on this platform. **1937** H. G. Wells *Brynhild* v. 61 The brightest and best producers, directors,.. cutters, music-men. **1950** L. Bennett *Anancy Stories & Dial. Verse* 40 Once upon a time Cockroach was a very good music-man an everybody wat have a dance always want Bredda Cockroach fe play fe dem. **1958** *Spectator* 10 Jan. 38/1 Mrs. Legge is what music-men call the flipover. **1845** *Ainsworth's Mag.* VII. 507, I see your ladyship's music-master..in the carriage. **1848** *Knickerbocker* XXXI. 450 Mr. Joseph F. Atwill, one of our most popular music-publishers..is issuing..a musical publication of rare interest. **1885** *List of Subscribers, Classified* (United Telephone Co.) (ed. 6) 157 Music publishers and sellers.. Boosey & Co. **1919** Wodehouse *Damsel in Distress* xvi. 199 She was a stenographer in a music-publisher's office when we first met. **1855** *Knickerbocker* XLV. 136 You hide behind the music-rack while Miss Mince passes. **1922** Joyce *Ulysses* 691 It's musicrest supporting the music.. for voice and piano of *Love's Old Sweet Song*. **1801** T. Busby *Dict. Music*, *Music-seller*, one who buys and sells music..likewise all kinds of musical instruments. **1811** Jane Austen *Sense & Sens.* I. xvii. 216 Booksellers, music-sellers, and print-shops. **1914** W. Owen *Let.* c 3 July (1967) 264 Practically all my Music Sheets..have been stolen! *c* **1760** in Dryden *Alexander's Feast* (title-page), Set to Musick by Mr. Handel. Edinburgh, sold by Robert Bremner, at his Musick-shop, at the Harp and Hautboy. **1925** E., O., & S. Sitwell *Poor Young People* 20 To music-sounding moonlight. **1834** Dickens *Sk. Boz* (1836) 1st Ser. I. 206 Mr. Wisbottle was describing semi-circles on the music-stool. **1902** Conrad *Youth* 314 He swung half round on the music-stool, listening with his fingertips at rest on the keyboard. **1975** *Country Life* 2 Oct. 852/1 The wood's great weight also recommended it for pillar furniture: ranging from tea-poy to music stool. *a* **1930** D. H. Lawrence *Sex, Literature & Censorship* (1955) 114 The dream of our pathetic, music-stunned young girl of today. **1953** K. Reisz *Technique Film Editing* III. xii. 187 It is most important that the dubbing editor should be able to cross fade from the dialogue to the music track at any time.

c. Delete † and add later examples.

1877 G. B. Shaw *How to become Mus. Critic* (1960) 19 Tristan and Isolde, a music drama which, in order of development, is the latest of Wagner's works. **1885** —— *Ibid.* 91 His allegorical music-play Die Zauberflöte. **1903** R. Hughes *Musical Guide* I. 207/2 Music-drama. An opera (particularly of the Wagnerian School) in which the text and the action determine the music, and are not interrupted by set arias, duets, etc. **1935** *Discovery* Sept. 250/2 This..building..is now the scene of the Triennial Music Festivals. **1946** Mezzrow & Wolfe *Really Blues* viii. 117 We had a ritual for these music-appreciation classes. **1946** *Penguin Music Mag.* Dec. 10 Since the early eighteenth century the music festival has been one of the most important and sustaining influences in English music-making. **1947** A. Einstein *Music in Romantic Era* xvi. 227 Wagner..considered himself, justifiably, not as a pure musician, but as a music-dramatist. **1966** *Listener* 5 May 663/2 Puccini's *Madama Butterfly*, his only opera that may lay claim to be called a psychological music drama. **1975** *Country Life* 2 Oct. 846/1 Wagner's great works are music-dramas rather than operas in the old sense.

d. **music case,** (*a*) a container (see Case *sb.*[2] 9) in which the component parts of a fount of music type are arranged before being set; (*b*) a container for sheet music; **music centre,** a stereophonic system combining record-player, radio, and cassette tape recorder in a single unit, usu. with separate loudspeakers; **music-club,** delete † and add earlier and later examples; **music-hall** (further examples in the *spec.* sense); so **music-hallish, music-hally** *adjs.*,

suggestive of a music-hall; **music line,** a line (Line *sb.*[2] 1 e) whose transmission characteristics are good enough for the transmission of music, and along which programme material is usually sent; **music-roll,** (*a*) a receptacle for the carriage of rolled-up sheet music; (*b*) a roll, usually of perforated paper, used in a pianola or player-piano or similar instrument; **music-room** (later examples); **music type,** substitute for def.: a fount of type, including several hundred pieces, used for the typographic printing of music, as distinguished from the use of engraved plates; also *attrib.* (earlier and later examples).

1841 W. Savage *Dict. Art of Printing* 487 All that I shall attempt is to give the plan of a pair of Music Cases, with the characters. **1890** *Cent. Dict.*, Music-case. **1894** *Amer. Dict. Printing & Bookmaking* 383/2 *Music cases*, special cases of a complicated character employed in composing type-music. **1924** *Southward's Mod. Printing* (ed. 5) I. v. 41 Music Cases. This class of case has to be made to suit the founts of different founders, who vary in the number of characters they cast. **1960** K. Amis *Take Girl like You* ii. 27 Jenny..fetched her music case, with books and not music in, from the classroom. **1974** *Hi Fi for Pleasure* Jan. 15 Model G 2601 KL..is a music centre with everything. **1975** *Gramophone* Nov. 958/2 Decca's Compact 4 music centre now incorporates long wave in its tuner section. **1975** *Which?* Dec. 356/3 Most music centres are designed to be sold with one set of loudspeakers only. **1740** R. Verney *Let.* 11 Nov. in M. M. Verney *Verney Lett.* (1930) II. xxix. 185 Mr. Millward on returning..from the inn at East Claydon, where there is a Musicck [*sic*] Club instituted, fell down. **1946** *Penguin Music Mag.* Dec. 7 Gramophone societies and music clubs which flourish all over the country. **1857** J. E. Ritchie *Night Side London* 145 The new Music Hall, Hungerford Market. **1870** D. J. Kirwan *Palace & Hovel* xxxiv. 503 A popular street and music-hall song. **1927** J. Adams *Errors in School* 112 His music-hall cluster. **1959** *New Statesman* 25 Apr. 574/3 The opening of the prison scene is made the occasion for an interminable drunken charade which destroys the balance of an act that is precarious enough in its original form—all for the sake of a primitive music-hall joke repeated at least a dozen times without variation. **1975** J. Wainwright *Square Dance* 248 You dress outrageously and you behave like a music-hall comedian. **1893** M. Beerbohm *Let.* Sept. (1964) 60 A good many common music-hallish people were standing in front of me. **1930** *Observer* 30 Mar. 15 The patriotic song of 1900 should have been sung by someone more robust and music-hallish than the cute and charming Miss Ada-May. **1889** G. B. Shaw *London Music in 1888–89* (1937) 211 The inferior theatre orchestra is music-hally, blatant, thumping, out of tune. **1923** A. Christie *Murder on Links* xxii. 251 Their voices were..rather thin and music-hally, but attractive. **1941** *B.B.C. Gloss. Broadcasting Terms* 20 Music line. **1962** A. Nisbett *Technique Sound Studio* 246 A control line..is a telephone circuit on which programme details may be discussed, and is so called to distinguish it from the broad band (i.e. high quality) 'music line' along which programme is fed. **1890** *Cent. Dict.*, Music-roll. **1890–1** *T. Eaton & Co. Catal.* Fall & Winter 42/1 Music rolls, in leather... Music rolls, in plush. **1906** *Bazaar, Exch. & Mart* Suppl. 3 Oct. 1322/1 Kastner's upright grand auto-piano for sale,..played by hand or by music rolls. **1907** *Yesterday's Shopping* (1969) 399/3 *Music rolls.* American cloth, lined cloth—1/5. **1913** *Strand Mag.* Jan. 13 (Advt.), For all player-pianos 'Songola' music rolls. **1875** 'Mark Twain' *Sketches New & Old* 307 We shall have billard-rooms [*sic*], card-rooms, music-rooms. **1928** 'Sapper' *Female of Species* x. 173 The music-room appeared to be so-called because there was no trace of any musical instrument in it. **1942** E. Blom *Music in England* vi. 92 In 1713 a dancing-master named Hickford opened a music room..where the celebrities of the day appeared. **1975** *Listener* 6 Nov. 608/3 My fourth new stopwatch..fell to the tiled floor of our music-room from my nerveless fingers. **1825** T. C. Hansard *Typographia* Index 11, Music type. **1841** W. Savage *Dict. Art of Printing* 488 (*caption*) C. Hancock's plan of cases for Hughes' music type. **1892** A. Powell *Southward's Pract. Printing* (ed. 4) xxxi. 292 Most of the letter-founders supply music types. **1923** W. Gamble *Music Engraving & Printing* xv. 169 The setting up of music type is a difficult and tedious operation. **1934** C. Lambert *Music Ho!* 11 There are few technical terms and no music-type illustrations. **1968** *Listener* 27 June 845/2 To demonstrate this [*sc.* the shapes of this plain-chant melody] would need music-type.

‖ **musica** (miū·zikă). [L., = music.] Used in special collocations to designate different kinds of music or musical techniques, as **musica ficta,** in contrapuntal music of the 10th to 16th centuries, the introduction by the singer of conventional chromatically altered tones to avoid unacceptable intervals; **musica figurata,** (*a*) contrapuntal music in which the different melodic strands move more or less independently; (*b*) plainsong with decorated melody; **musica plana** = **cantus firmus*; plainsong; **musica reservata** (see quot. 1972).

1740 J. Grassineau *Musical Dict.* 154 *Musica Figuralis, Figurata*, or *Colorata*, figurate music, wherein the notes are of different values, and the motions various, now slow then quick. **1801** T. Busby *Dict. Music*, *Musica ficta*,..the name given by Franchinus and other musical writers, to the first deviations from the old ecclesiastical modes. **1886** W. S. Rockstro *Hist. Mus.* 482 The Rules of ..Musica ficta..enable us to supply, with certainty, the accidental Sharps and Flats, which the older writers omitted. **1940** G. Reese *Music in Middle Ages* (1941) v. 121 Not until the 13th century did *musica plana* etc. come to connote music in free rhythm as distinguished from measured music, as do the English words 'plainsong' and 'plainchant'. **1944** W. Apel *Harvard Dict. Mus.* 467/2 *Musica reservata* denotes the Renaissance-like clarity, balance, and expressiveness, the full-blooded humanism of the Josquin period, in contrast to the abstract and mysterious transcendentalism of the earlier masters. **1954** A. Hughes in *New Oxf. Hist. Music* II. xi. 370 The position which called for such a system as *musica ficta* is as simple as the processes of its working out are complex. It is the problem of the tritone. **1954** G. Reese *Music in Renaissance* ix. 512 The expression '*musica reservata*' has become one of the problems of music history. **1963** R. Donington *Interpretation of Early Music* iv. 67 D minor, which we notate with a signature of one flat, grew out of the Dorian mode when its leading note, C, became habitually sharpened by the workings of *musica ficta*. **1968** *Listener* 23 May 676/2 His [*sc.* Palestrina's] Mass saves music, convinces the Pope that a regression to *musica plana*, plainsong, at the indiscriminate expense of *musica figurata* (the later, florid polyphonic style of the early Flemish composers) would be wrong—wrong metaphysically. **1972** *Composer & Conductor* Aug. 1/1 'Musica reservata'..lends itself to diametrically opposite definitions. Some musicologists describe it as music made suitable for popular use, while others regard 'musica reservata' as literally 'reserved for connoisseurs'.

musical, *a.* (and *sb.*) Add: **A.** *adj.* **1.** (Further examples.)

1790 Sheridan *Let.* in T. Moore *Life R. B. Sheridan* (1825) 468 We had a very pleasant musical party last night at Lord Erskine's. **1791** F. Burney *Jrnl.* Dec. (1972) I. 95 We have spent a charming Musical Evening at Mr. Burney's. **1814** J. Mayne *Jrnl.* 7 Nov. (1909) x. 183 Countess Cardelli, who has agreeable musical parties at her house every Sunday evening. **1885** *List of Subscribers, Classified* (United Telephone Co.) (ed. 6) 157 Musical instrument makers..Chappell & Co. **1926** *Punch* 10 Nov. p. iii (Advt.), Recreation for yourself—Dance Parties—Musical Evenings. **1964** W. L. Goodman *Hist. Woodworking Tools* 92 These small block planes were very useful to violin and other musical-instrument makers.

10. **musical appreciation,** informed response to music; **musical arms,** a modification of the game of musical chairs (see quot.); **musical box** (earlier and later examples); also *transf.* and *fig.*; **musical bumps,** a game similar to musical chairs, in which the competitors sit down on the floor or ground when the music stops and the last person to sit is out of the game; also *fig.*; **musical chairs,** a competitive parlour game in which a number of persons walk to music round a smaller number of chairs and each tries to secure a seat when the music stops, or an outdoor game on the same principle played on horseback; also *attrib.* and *fig.*; **musical clock** (later example); **musical comedy,** a light dramatic piece, on stage or in a cinema, consisting of dialogue, songs, and dancing, connected by a slight plot; also *fig.*; **musical director,** the conductor of the orchestra of a theatre, either for opera or for plays; **musical dramatist,** a composer of music-dramas (see *music *sb.* 12 c); **musical drive,** an exhibition of horsemanship by a military unit in which the horses pull along the military equipment to the accompaniment of music; **musical festival** = Festival *sb.* 1 b; **musical fright** = **musical chairs*; **musical ride** (later example, here equivalent to *musical drive* above); **musical saw,** a hand-saw held between the knees and 'played' with a violin bow; **musical watch,** a watch which incorporates a comb and cylinder mechanism to produce a tune at specified times.

1929 *Jrnl. Abnormal Psychol.* XXIV. 75 It is possible.. in the teaching of 'musical appreciation' to inculcate dogmas about what is symbolized by certain musical combinations. **1947** C. Gray *Contingencies & Other Essays* i. 20 The recognition of the truth..has brought about in recent years the development of an activity known as 'musical appreciation', which aims at fostering a love of music among the populace by means of broadcast lectures, evening classes, elementary handbooks and textbooks, educational gramophone records, and so forth; by teaching music, in a word, in very much the same way as one would teach any other subject in the educational curriculum. **1955** P. Vincent in H. van Thal *Fanfare for Ernest Newman* 174 The courses of musical appreciation that have been taught in schools these last twenty years or so. **1962** M. Sargent *Outline of Music* p. v, I believe this 'Outline' to be more thorough than many popular books on musical appreciation. **1924** D. C. Minter *Children's Parties* x. 137 Musical arms. This game is played in the same way as Musical Chairs, without, however, using chairs. [**1821** M. Wilmot *Let.* 16 Apr. (1935) 105 The musical bonbon box, and other trifles to *amuse* the children.] **1829** A. Royall *Pennsylvania* I. 10 A small wooden musical box..by mechanical invention produced the name of any capital town you called for. **1925** T. E. Lawrence *Let.* 3 Nov. (1938) 486 Our hut now has a little musical box. **1925** E. Sitwell *Poetry & Criticism* 21 No time for darkness there, excepting in the cubes of their musical-box brains. **1932** S. G. Hedges *Indoor & Community Games* vi. 76 *Musical Bumps*... All march round in single file, while the pianist plays... When the music stops everyone must flop down on the floor—and the one who 'bumps' last falls out of the game. **1967** E. Grierson *Crime of One's Own* xi. 92 He..went a few paces and stopped abruptly—the technique one used at the piano to defeat too proficient children at a game of Musical Bumps. **1974** *Listener* 18 July 85/3 England may not be the world champions at soccer, but we are definitely past-masters at musical bumps. **1877** *Cassell's Family Mag.* Dec. 41/1 When there is a piano, 'Musical Chairs' played in the usual way..is quite as interesting. **1916** 'Peter' *Trench Yarns* 25 We had to get the men through the danger zone by a sort of musical-chairs rush. They came slowly up to the entrance, and then dashed in and round the corner into safety behind the bricks. **1933** A. Blewitt *Ponies & Children* viii. 121 Musical Chairs—the posts are stuck in a small circle, one less than the number of competitors, who have to canter in a wide circle, led by a grown-up on horseback. **1939** R. S. Summerhays *Riding for All* xiv. 87 The almost inevitable event at all gymkhanas is musical chairs. **1950** G. Brenan *Face of Spain* x. 202 The Spanish economic system is like a game of musical chairs, in which there are only half as many seats as there are performers. **1973** E. Page *Fortnight by Sea* v. 52 How rapidly the years slipped by, with what speed the game of musical chairs was played, how swiftly one was forced out of one role and into another. **1974** *Times* 16 Jan. 12/3 President Bourguiba..was still in the mood of playing musical chairs with his Prime Ministers. **1939** S. Spender tr. *E. Toller's Pastor Hall* I. 54 Even the musical clock is afraid of being denounced. **1765** R. Cumberland (*title*) The summer's tale: a musical comedy of three acts. **1791, 1904** Musical comedy [in Dict., sense 6]. **1910** 'Saki' in *Bystander* 7 Dec. 484/1 Noted lights of the musical-comedy stage. **1923** A. Huxley *Antic Hay* iii. 35 He looked..positively soldierly in his black jacket and his musical comedy trousers. **1957** *New Yorker* 12 Jan. 34/1 It was her favorite kind of film, a musical comedy in full color. **1967** *Listener* 3 Aug. 138/1, I thought I'd start in musical comedy,..carry on training as a dancer and singer. **1829** H. Foote *Compan. to Theatres* 147 Covent Garden —Season 1828–9... Musical Director and Composer—Mr H. R. Bishop. **1902** W. H. Chantrey *Theatre Accounts* 73 All reasonable requirements from time to time made by the Manager, and the directors of the Musical Director or Stage Manager. **1885** G. B. Shaw *How to become Mus. Critic* (1960) 73 The most subtle and profound of all musical dramatists. **1963** *Listener* 3 Jan. 45/1 Compared with the microcosm created by the greatest musical dramatists his [*sc.* Puccini's] world is limited in subject-matter. **1930** *Times Educ. Suppl.* 31 May 1/3 The musical drive by 'J' Battery, Royal Horse Artillery, was carried out at the gallop. **1852** Dickens *Our Bore* in *Househ. Words* 9 Oct. 75/1 He was at the Norwich musical festival. **1879** 'L. Hoffmann' *Drawing-room Amusements* 24 Musical Fright. **1960** C. Day Lewis *Buried Day* v. 93, I seen this regiment..performing their intricate 'musical ride', with guns and limbers, at the Naval and Military Tournament. **1927** *Melody Maker* Aug. 773/2 In the combination are two performers on musical saw, which novelty, says Mr. Haggleton, always goes down well with his audience. **1946** R. Blesh *Shining Trumpets* (1949) v. 104 Exotic instruments may be utilized as well, such as harmonica, kazoo, jug, washboard, wood blocks and musical saw. **1957** W. C. Handy *Father of Blues* x. 139 The technique..was similar to that used in playing musical saws. **1958** A. Jacobs *New Dict. Mus.* 231 Mayuzumi..has written..music..including..'Tonepleromas 55' for wind and percussion instruments and musical saw. **1973** J. Wainwright *Devil you Don't* 22 Rawlings will have you warbling *La Traviata*, backwards..with musical saw accompaniment. **1899** F. J. Britten *Old Clocks & Watches* 148 Musical watches of large size with moving figures were a favourite conceit among French makers during the latter part of the eighteenth century. **1952** T. P. C. Cuss *Watches* xi. 145 On earlier musical watches—at the end of the eighteenth century—a pinned cylinder took the place of the disc. **1954** *Grove's Dict. Mus.* (ed. 5) V. 1018/2 Musical watches..were made, usually in Switzerland, at the beginning of the 19th century... There might be one tune.. or, more rarely, an air and variation. **1973** *Watches in Usher Coll.* 12 Musical watch with tiny cylinder playing on 13 teeth.

B. *sb.* **3.** A film or a theatrical piece (not opera or operetta) of which music is an essential element.

1938 *Chatelaine* Jan. 50/2 You can guess what it's about without my telling you—it's a musical. **1940** *Illustr. London News* CXCVI. 464/1 Some of these 'musicals' have proved extremely popular. **1944** [see **cover-girl*]. **1954** T. S. Eliot *Confidential Clerk* II. 47 *Lucasta*:..But what about taking me to a concert? *Colby*: Only the other day I invited you... *Lucasta*: To go to see that American Musical! **1959** *Listener* 13 Aug. 259/1 A Novello musical. **1973** *Ibid.* 14 June 815/2 Musicals..exist in a far from splendid isolation, endlessly repeating the tricks which once worked.

musicalization (miūzikăləizēi·ʃən). [f. next + -ation.] The expression or rendering of an art (other than music) in the style or manner of music; the action or process of musicalizing.

1928 A. Huxley *Point Counter Point* xxii. 408 The musicalization of fiction. **1930** H. Read *Julien Benda & New Humanism* 14 The cult of vagueness, the musicalisation of all the arts, the general subjectivism and romanticism. **1971** *Listener* 15 Apr. 473/3 The facts behind this 'musicalisation' of our culture..are fairly obvious. But the underlying motives are..complex.

musicalize (miū·zikăləiz), *v.* [f. Musical *a.* + -ize.] *trans.* To set (a novel, play, or poem) to music; to express or render (an art other than music) in the style or manner of music.

1919 *N.Y. Times* 20 July IV. 2/6 'Kitty Mackaye'.. is another play which is about to be musicalized. **1928** *Sunday Express* 19 Aug. 5/4 He prepared a musicalised version of the..play. **1947** *Penguin Music Mag.* Dec. 72 The art of musicalising films rather than filming music.

1957 Manvell & Huntley *Technique Film Music* 284 Benjamin Britten had proved particularly adroit at musicalizing the rhythm and intonation of speech beyond their actual use in a vocal line. **1962** *Listener* 3 May 761/2 It was Huxley who showed, in *Point Counter Point*, how fiction could be musicalized. **1966** *New Statesman* 20 May 736/2 Usually with Zukofsky you get..a poem intricately musicalised (he is much affected by Bach's contrapuntal patterns).

musicassette (miūzikæse·t). [f. Music *sb.* + Cassette.] A tape cassette of prerecorded music.

1966 *Melody Maker* 8 Oct. 13/1 Musicassettes are with us... What exactly is a musicassette? It's the size of a bar of chocolate..and the tape is permanently housed inside the cassette. **1968** *Sat. Rev.* (U.S.) 25 May 50 As yet, of course, musicassettes are more for the uncritical listener to popular music. **1971** *Hi-Fi Sound* Feb. 12 (Advt.), Announcing the new professional musicassette library starting early in 1971. **1975** *Gramophone* Jan. 1293 (Advt.), A stereo musicassette player need not be expensive.

musicianship. (Later examples.)

1903 A. W. Patterson *Schumann* iv. 45 Surely a sufficient guarantee of his creative musicianship. **1930** *Times Educ. Suppl.* 2 Aug. 338/3 Miniatures full of dainty musicianship. **1959** *Times* 2 Feb. 1/4 The holder of the post will also be responsible for General Musicianship classes. **1970** P. Oliver *Savannah Syncopators* 98 But if blues singers appeal because of these talents as well as their musicianship, they are, like the *griots*, frequently considered as lazy, lacking in industry and job application.

musicogenic (miūzikodʒe·nik), *a. Path.* [f. Musico- + *-genic.] Applied to a form of epilepsy in which attacks are precipitated by the hearing of music, and to the attacks themselves.

1937 M. Critchley in *Brain* LX. 13 The aim of this paper is to describe the occurrence of epileptiform attacks in factual association with the hearing of music... Opportunity has since brought to notice a series of such cases of 'musicogenic epilepsy' (or 'musicolepsia'). **1962** *Jrnl. Amer. Med. Assoc.* 17 Feb. 503/1, 44 patients with musicogenic epilepsy have been reported in the literature. **1972** Boshes & Gibbs *Epilepsy Handbk.* (ed. 2) xiii. 61 Musicogenic seizures are extremely rare; the patient is usually an adult who is musically orientated.

musicological (miūzikǫlǫ·dʒikăl), *a.* [f. *musicology: see -ical.] Of or pertaining to musicology. Hence **musicolo·gically** *adv.*

1915 W. S. Pratt in *Musical Quarterly* I. 1 Several handy derivatives, such as 'musicologist'..'musicological', and the like. **1941** G. Haydon *Introd. Musicology* i. 1 By no means everything written about music qualifies as musicological literature, but only that which exhibits an acceptable quality of scholarship. **1959** D. Cooke *Lang. Music* p. x, We concern ourselves more and more with parochial affairs—technical analyses and musicological *minutiae*—and pride ourselves on our detached, dehumanized approach. **1966** *New Statesman* 23 Sept. 451/2 Musicologically, they are..scrupulous. **1969** *Daily Tel.* 12 Apr. 13 Musicologically sensational and musically highly enjoyable was Alan Hacker's performance on the basset clarinet of Mozart's Concerto, K. 622. **1970** P. Oliver *Savannah Syncopators* 95 Musicological analyses of the blues have scarcely been undertaken to date. **1973** *Listener* 1 Mar. 286/2 Musicologically-inhibited people catalogue Caruso's liberties as heinous sins.

musicologist (miūzikǫ·lŏdʒist). [f. next: see -ologist.] One who studies or practises musicology.

1915 [see *musicological *a.*]. **1927** *Observer* 20 Feb. 18/2 The widest co-operation has..been secured from French, Spanish, German, and other foreign musicologists. **1930** *Music & Lett.* Apr. 138 The distinguished list of English musicologists. **1944** W. Apel *Harvard Dict. Mus.* 474/1 The men who undertook to 'discover Bach'..were musicologists in the true sense of the word. **1955** F. Toye in H. van Thal *Fanfare for E. Newman* xi. 164 Who fifty years ago could have imagined that a musicologist of the calibre of Sir Donald Tovey should sponsor the performance of such an intrinsically popular opera as *Il Trovatore*? **1973** *Oxf. Mag.* 1 June 4/1 The members of the Faculty of Music are for the most part primarily musicologists.

musicology (miūzikǫ·lŏdʒi). [ad. F. *musicologie* or f. Music *sb.* + -ology.] The systematic study of music as opposed to the art of composition and performance; *esp.* academic research in music.

1909 M. S. Logan (*title*) Musicology. **1915** W. S. Pratt in *Musical Quarterly* I. 3 'Musicology', if it is to rank with other comprehensive sciences, must include every conceivable scientific discussion of musical topics. **1919** *Proc. Mus. Assoc. 1918–19* 106 The foundations of Musicology are the documents, manuscripts or printed music of past times. **1928** *Music & Lett.* Apr. 108 The most valuable piece of musicology is the treatment of the six motets. **1937** R. H. Lowie *Hist. Ethnological Theory* xiv. 255 Comparative musicology..rose to a new plane with phonographic recordings. **1941** G. Haydon *Introd. Musicology* i. 7 Musicology, following the lead of philosophy, adopts the epistemological concepts of space and time, and divides its subject matter under the two main headings *systematic* and *historical*. **1958** *Times Lit. Suppl.* 4 July 384/1 Only since the turn of the century has musicology been accepted fully and internationally as an independent branch of scholarship.

musicophile (miū·zikofəil). [f. Musico- + -phil, -phile.] One who professes love of music.

1931 J. Joyce *Let.* 3 Jan. (1957) 298 The wealthy musicophiles in London and New York who control the destinies of opera in those cities.

Musigny (mū·sinʸi). The name of a vineyard in the commune of Chambolle-Musigny in the Côte d'Or department of France used *attrib.* and *absol.* to designate the red Burgundy produced there.

1833 C. Redding *Mod. Wines* v. 112 To recapitulate the wines of the Côte d'Or: the finest Burgundies are Romanée-Conti [etc.]. Of the second class, Chambolle, Musigny [etc.] among red wines. **1907** *Yesterday's Shopping* (1969) 97/1 Burgundy—Musigny—Vintage 1899—Per doz. bots. 52/0. **1932** A. E. Housman *Let.* 31 May (1971) 321 The Sommelier insisted on Chambolle-Musigny 1921, which was most excellent. **1966** H. Johnson *Wine* 191 Musigny..lingers and spreads in your mouth, making what French tasters call 'the peacocks tail' with its bouquet of flavours and flavours within flavours. **1967** A. Lichine *Encycl. Wines & Spirits* 374/2 In a good year the wines, in addition to their..feminine delicacy, have an incomparable bouquet... A fine estate-bottled Musigny can be expected..to profit enormously from imprisonment in the bottle.

‖ **musique concrète** (müzik końkrę̄t). [Fr. (P. Schaeffer *À la recherche d'une musique concrète* (1952)).] = *concrete music* (*concrete *a.* and *sb.* A. 7). Also semi-anglicized as *musique concrete*, and *fig.*

1952 *Time* 10 Nov. 95/1 Last summer the U.S. got a taste of creative recording in France's *musique concrète*, a compilation of natural sounds..recorded on tape, cut and spliced at patterns to make a composition. **1953** *Musical Quarterly* XXXIX. Oct. 608 A special session was devoted to Pierre Schaeffer's *Musique concrète* and Elektronische Musik developed in the studio laboratories of the Cologne Rundfunk. **1954** *Gramophone Record Rev.* Apr. 297 These experiments come to mind after studying the so-called *Musique Concrète* broadcast recently in the Third Programme. **1957** *Electronic Engin.* XXIX. 350 The recent Musique Concrete in France. To quote a reviewer of records 'This is neither music nor concrete'. **1958** *Oxf. Mag.* 8 May 410/1 The *musique concrète* of dishwasher, duplicator and the pneumatic drills of Restoration Oxford, Thalia shuts her pretty mouth. **1962** A. Nisbett *Technique Sound Studio* xi. 191 *Musique concrète* transformations. **1970** *New Scientist* 30 July 221/2 Modern office workers partial to John Cage and *musique concrète*.

musk, *sb.* Add: **4. a.** *musk-perfumed, -scented* (later examples).

1922 Joyce *Ulysses* 11 Phantasmal mirth, folded away: muskperfumed. **1891** O. Wilde *Picture of Dorian Gray* (ed. 2) xv. 272 He bathed his hands and forehead with a cool musk-scented vinegar. **1937** *Discovery* Feb. 53/2 A musk-scented lady. **1957** R. Campbell *Coll. Poems* II. 98 Musk-scented lace and fans of ostrich plume.

f. **musk lorikeet,** an Australian parrot, *Glossopsitta concinna*; = *musk-parrakeet*.

1901 A. J. Campbell *Nests & Eggs Austral. Birds* 596, I have observed many nests of the Musk Lorikeet in Victoria. **1966** Eastman & Hunt *Parrots of Australia* 32 Musk lorikeet... Pleasant strong musky odour associated with this Lorikeet.

muskellunge. Now the usual spelling of Maskinonge, the large North American pike *Esox masquinongy*. (Later examples.)

1946 *Wisconsin State Jrnl.* 18 July 1/4 Gen. Dwight D. 'Ike' Eisenhower and his four brothers all continued their fishing luck Wednesday when they hit the jack pot with a muskellunge each. **1959** E. Tunis *Indians* 40/2 Not to be despised, however, was the big pike called muskellunge by the Algonquins, the name it still bears. **1963** *Globe & Mail* (Toronto) 13 Apr. 5/6 He is also an officer of the Outdoor Writers of Canada and only in one respect might he come into conflict with his companions on that body—by using the U.S. term muskellunge instead of the accepted, original Canadian name, maskinonge. **1970** R. Lowell *Notebk.* 251 Each night, a star, gold-on-black, a muskellunge, Dies in the highest sphere that never dies.

musketeer. Add: *musketeer gauntlet, glove*: see Mousquetaire 2.

1922 Joyce *Ulysses* 458 Fawn musketeer gauntlets. **1968** *Guardian* 26 July 7/3 There were musketeer gloves and gaiters of sequins.

musketoo·ner. [f. Musketoon + -er[1].] = Musketoon 2.

1925 *Chambers's Jrnl.* June 339/2, I felt a tearing blow above my right hip and I knew that the kneeling musketooner had touched me.

muskin[3]. Delete † *Obs. rare*[−1]. and add later allusive example.

1922 Joyce *Ulysses* 395 The embraces of some unaccountable muskin.

musking-place. A place where weasels or kindred animals deposit musk.

1920 *Chambers's Jrnl.* May 291/2 The wolves have their calling-posts, the beavers have their castor-signs, the weasels have their musking-places.

Muskogean (mʌ:skogī·ăn, mʌskōᵘ·giăn). Also **Muskhogean.** [f. *Muskogee.] An Indian language family in south-eastern N. Amer., consisting of the Creek-Seminole, Hitchiti-Mikasuki, Alabawa-Koasati, Apalachee, and Choctaw-Chickasaw languages. Also *attrib.* or as *adj.*

1891 J. W. Powell in *7th Ann. Rep. Bureau Ethnol.* 94 (*heading*) Muskhogean family. **1907** F. W. Hodge *Handbk. Amer. Indians* I. 962/2 The Muskhogean population at the time of first contact with Europeans has been estimated at 50,000. **1932** *Times Lit. Suppl.* 8 Sept. 624/3 Forming part of the Muskhogean culture area east of the Mississippi. **1933** [see *Creek *sb.*[2]]. **1940** *Language* XVI. 142 Curiously enough..certain other Muskogean languages closely related to Muskogee do employ gemination as a grammatical process. **1959** E. Tunis *Indians* 19/2 Remnants of the temple-mound culture survived into historical times among the Muskhogean Indians of the Southeast. **1972** *Language* XLVIII. 845 Hockett exemplifies 3a with 'Arunta..', Muskogean (except Creek-Seminole).., and Wishram'.

Muskogee (mʌskōᵘ·gī). Also **Muscogee, Muskhogee, Muskhokee, Muskogi, Muskohge.** [f. Creek *maskóki*, perh. of Algonquian origin.] The name of a group of N. American Indians, consisting mainly of Creek Indians; also the language spoken by them; also *attrib.*

1775 J. Adair *Hist. Amer. Indians* 161, I had a conversation..with several of the more intelligent Muskohge traders. *Ibid.* 392 A Muskohge warrior .. got to a bramble swamp. *Ibid.* 430 The Cherokees and Muscogees still observe that old custom. **1836** *Trans. Amer. Antiq. Soc.* II. 94 The Muskhogees are the prevailing nation... The Hitchittees..speak a dialect of the Muskhogee. **1868** [see *Creek *sb.*[2]]. **1907** F. W. Hodge *Handbk. Amer. Indians* I. 962/2 The recognized languages of the stock..are as follows:..Muskogee (including almost half the Creek confederacy, and its offshoot, the Seminole). **1948** D. Diringer *Alphabet* I. x. 183 The Muskhokee dialect is the principal dialect of the Muskhogean group; politically, the Muskhokee were the dominant tribe of the Creek Confederacy. **1949** E. A. Nida *Morphology* (ed. 2) vi. 159 Muskogee has some verbs of rather irregular formation which indicate differences of singular, dual, and plural subjects. **1959** E. Tunis *Indians* 70/2 The principal tribe was the Muskhogee, called Creeks by the English.

musk-rat. Add: **1. b.** (Earlier example.)

1845 in C. Cist *Cincinnati Misc.* 240 The inhabitants of ..Delaware [are called] Muskrats.

3. *musk-rat burrow, cap, skin, study*; **musk-rat house** (later examples).

1870 *Amer. Naturalist* IV. 385 This fish, when the bank was carelessly approached, would withdraw to a deserted muskrat burrow. **1916** C. A. Eastman *From Deep Woods to Civilization* 101 We saw that they were colored troopers, wearing buffalo overcoats and muskrat caps. **1962** W. Stegner *Wolf Willow* III. iii. 230 Crouching in the trench in his muskrat cap, he looked like some digging animal. **1939** *Beaver* June 10 Older men remembered when the muskrat houses had been so thick that the marsh looked like a hayfield in coils. **1962** W. Stegner *Wolf Willow* III. ii. 150 They had seen muskrat houses built six feet high in the sloughs—and when the rats built high you could depend on a hard winter. **1642** in *Archives of Maryland* (1887) IV. 99, 2. musk-rat skins. **1796** E. P. Simcoe *Diary* (1911) 315 We..slept well,..but the smell of musk-rat skins..was disagreeable. **1823** in T. L. McKenney *Memoirs* (1846) I. App. 296 Furs and other articles..to wit,..2,500 muskrat skins, [etc.]. **1939** *Amer. Midland Naturalist* XXI. 514 (*title*) Central Wisconsin muskrat study.

musky, *sb.*[2] Add: Also **muskie.** (Later examples.)

1928 *Game Fishes of Canada* 12 The next most definite cult among the anglers are those who devote themselves to muskies. **1936** G. Clark *Which we Did* i. 20 Jimmie Frise..was successfully commissaried by me to our favorite muskie water for the closing day of the season. **1962** *Times* 12 Apr. 7/3 Waterways full of bass, northern pike and muskies. **1968** *Globe & Mail* (Toronto) 3 Feb. 40/1 They recaptured these previously marked fish: 13 walleyes, three smallmouth bass, 11 largemouth bass, but not one musky. **1975** G. V. Higgins *City on Hill* ix. 215, I should spend my days..going after muskies and telling lies when I didn't catch any.

musli, var. *muesli.

Muslim. Now the preferred spelling of Moslem *sb.* and *a.* Also, = *Black Muslim* (*black *a.* 19).

1961 [see **Black Muslim*]. **1971** *Black Scholar* June 52/1 Submission to democratic centralism instead of the egoism that sent him first against his Muslims.

musquash. Add: **3.** **musquash-root** (earlier and later examples); **musquash weed,** substitute for def.: (*a*) = *musquash root*; (*b*) = *musk-rat weed* (examples).

1807 *Massachusetts Spy* 22 July (Th.), Five children were lately poisoned in Scipio (New York) by eating Wild Parsnip or Musquash Root. **1857** [see **beaver-poison*]. **1940** E. Early *New England Sampler* 309 The Indian had told his grandmother..of a decoction of Cowbane (called Musquash Root) that would make a woman forever sterile. **1767** *Massachusetts Gaz.* 21 May 3/1 Persons (especially Children) would do well to beware of this Weed. It is called wild Hemlock by some, and Musquash Weed by others. **1892** *Jrnl. Amer. Folk Lore* V. 91 *Thalictrum polygamum*, muskrat-weed; musquash weed. **1907** A. B. Lyons *Plant Names* (ed. 2) 457 T[halictrum] polygamum. .. Celandine, Muskrat-weed, Musquash-weed.

Musquito, var. *MISKITO *a.* and *sb.*

muss, *sb.*⁴ Add: **1.** (Earlier and later examples.) No longer current.
1830 *Constellation* (N.Y.) 11/3 I've been in many a *scrape* before, but never such a *muss*! 1838 *N.Y. Advertiser & Express* 17 Feb. 4/6 The complainant testified that there was a 'muss' in Chatham Street. 1840 C. F. HOFFMAN *Greyslaer* I. II. iv. 206 That's just what I told Red Wolfert when he showed signs of kicking up a muss. 1864 O. L. JACKSON *Colonel's Diary* (1922) 135 We have just had quite a muss [preceding the battle of Kenesaw Mt.]. 1873 'MARK TWAIN' & WARNER *Gilded Age* 267 That gentleman delivered the young aspirant for a muss one of his elegant little left-handers. 1903 A. H. LEWIS *Boss* 19 It was nothin' but a cheap muss on the pier.
2. (Earlier and later examples.)
1840 C. F. HOFFMAN *Greyslaer* II. III. iv. 142 All this muss is of Wat Bradshawe's cooking. 1842 *Spirit of Times* (Philad.) 22 Jan. (Th.), I upset my table, spilt my ink, and knocked down my books, making a deuced muss. 1970 *Toronto Daily Star* 24 Sept. 40/1 (Advt.), Here's your cartridge... Instant loading. No muss or fuss. 1972 M. MEAD *Blackberry Winter* ix. 103 The first year Miss Abbott, the head of the dormitory apartments, described us as 'a mental and moral muss', and we accepted this with a kind of wicked glee.

muss, *v.*² Add: **1.** (Earlier and later examples.) Hence **mussed** *ppl. a.*; also *mussed-up*.
1837 in *Amer. Speech* (1965) XL. 127 Wish he muss imself [*sic*] up well and arty. 1888 *Detroit Free Press* (Farmer), Neither of us got two winks of sleep during the night on the car... We reached Chicago in a mussed up condition. 1902 C. J. C. HYNE *Mr. Horrocks Purser* 57 If Boy Austen..contrived to get badly mussed up with an undesirable woman [etc.]. 1907 C. E. MULFORD *Bar-20* v. 47, I paid twenty wheels for that [*sc.* a hat] eight years ago, and I don't want it mussed none. 1909 *Philad. Publ. Ledger* 24 June 7/6 (Advt.), Lot 3..Some four hundred soiled or mussed waists, in white lawn & lingerie. 1925 D. H. LAWRENCE *Death of Porcupine* 112 The really quick, Tolstoi loved to kill them off or muss them over. 1930 'R. CROMPTON' *William's Happy Days* iii. 66 Trampling over his doorstep and 'mussing up' everything. 1955 W. GADDIS *Recognitions* II. ii. 351 The wind hit him,.. mussing his hair from behind. 1973 D. WESTHEIMER *Going Public* ii. 28 Only a little mussed, the kid says! I'm not leaving looking like something the cat dragged in.

mussel, *sb.* Add: **4.** *mussel-gatherer* (earlier example), *-opener*, *-pooled*, *-spawn*; **mussel-cracker** (*b*), **-crusher** = *BISKOP; **mussel mud** (later N. Amer. examples).
1905 *East London* (Cape Province) *Dispatch* 6 Nov. 7/4 Judging by the enormous incisors, and the perfect pavement of rounded molars with which the jaws of these white steenbras are armed, these fish live largely upon shell-fish, hence the local name *mussel cracker* and the Durban name *mussel crusher*. 1930 C. L. BIDEN *Sea-Angling Fishes of Cape* xviii. 256 Mussel-Crusher or Mussel-Cracker. 1951 *Cape Times* 7 Feb. 2/5 Twenty-five mussel-crackers and ten blou stompkop were landed. 1953 J. L. B. SMITH *Sea Fishes S. Afr.* (ed. 2) 502 The name 'Musselcracker' is applied in South Africa mainly to 2 fishes, *Sparodon durbanensis* and *Cymatoceps nasutus*, both of which develop massive jaws and powerful teeth. 1973 *Stand. Encycl. S. Afr.* VIII. 14/1 The musselcracker or musselcrusher (*Cymatoceps nasutus*) is one of the best-known angling fishes in South Africa... The jaws are very powerful, with canine teeth in front and molars behind... Their flesh is coarse, but the head is a delicacy. The young are silvery, yellow and black; the adults are dark, and develop a curious fleshy nose. 1905, etc. Mussel-crusher [see *mussel-cracker* above]. 1859 A. J. MUNBY *Diary* 18 July (1972) 38, I met the mussel gatherers..fine young women with brown bare limbs. 1825 *Prince Edward Island Register* (Charlottetown) 23 Aug. 1/1 Plenty of manure [is] at hand, either kelp, seaweed, mussle-mud, [etc.]. 1851 J. F. W. JOHNSTON *Notes on N. Amer.* II. 151 Mussel-mud,..or sea-mud full of mussels, abounds in the Bay of St Andrews... This is an excellent fertilising substance... But the most apparently singular way of using it is to put it, with the mussels still living, into the turnip-drills, where it gives alone an excellent crop of turnips. 1973 *Canadian Antiques Collector* Jan.–Feb. 64/1 Of special value [as fertilizer] was the 'deep, black, stinking mud', better known as 'mussel-mud'—which lay thick in the beds of the Island's many tidal estuaries. 1909 *Daily Chron.* 25 Sept. 7/6 (Advt.), Oyster and mussel opener (young) wanted for evenings. 1946 DYLAN THOMAS *Deaths & Entrances* 9 The mussel pooled and the heron Priested shore. 1902 *Chambers's Jrnl.* May 277/2 Some seasons the mussel-spawn is pretty much in evidence here.

Mussolini (mʊsolī·ni). [Name of Benito *Mussolini* (1883–1945), prime minister of Italy and leader of the Fascist Party in Italy.] One who embodies the characteristics of Mussolini; also *attrib.* So **Mussolini-e·sque**, **Mussoli·nian** *adjs.*, of, pertaining to, characteristic of, or (somewhat) resembling Mussolini; **Mussoli·nism**, the political principles or policy of Mussolini or of the Fascist party in Italy.
1924 *Glasgow Herald* 29 Jan. 9 Signor Mussolini..said that he was unable to accept the antithesis that some people wished to create between Fascism and Mussolinism. *Ibid.* 30 June 11 The working classes needed to be warned against the establishment of 'Mussolinism' in this country. 1928 H. G. WELLS *Way World is Going* ii. 15 The consolidation and reconstruction of China..has not gone on under the direction of some strong-jowled hero of the Diaz or Mussolini type. 1928 *Daily Express* 21 Nov. 19 Mr. [J. W. H. T.] Douglas' stern, unbending look always interests me. So does his Mussolini-esque and frequent call of 'Fight!' 1932 H. NICOLSON *Diary* 15 Mar. (1966) 112 Tom does not wish to end it, being still obsessed by Mussolinian ideas. 1936 [see *HITLER]. 1946 R. CAPELL *Simiomata* II. 86 There are still Mussolinian Italians with the Germans. 1959 *Guardian* 16 Oct. 11/3 He..fails to acknowledge that there were many more sides to 'Mussolinism' than the less creditable later period. 1970 R. A. H. ROBINSON *Origins of Franco's Spain* iii. 134 The policies of Gil Robles meant 'the enslavement, in Mussoliniesque style, of the proletariat'.

Mussulman, *sb.* and *a.* Add: Now falling into disuse. The more usual forms are *Muslim* and *Moslem*.
A. *sb.* **2.** [tr. G. *muselmann*.] pl. Mussulmans, Mussulmen [G. *muselmänner*]. Under the Third Reich, an inhabitant of a concentration camp or extermination camp who had reached a state of physical and emotional exhaustion in which he displayed fatalism and loss of initiative.
[1950 H. NORDEN tr. *E. Kogon's Theory & Practice of Hell* xviii. 214 Not all the prisoners were what came to be known as 'Moslems', men who were physically and mentally broken, who allowed anything to be done to them.] 1953 G. REITLINGER *Final Solution* xvii. 458 Most of those who died were just *Mussulmen*, the camp slang word for a walking skeleton wrapped up in a bit of blanket. The journey from Auschwitz,..for days on end in open trucks, freezing and without food, was alone sufficient to make a *Mussulman*. 1959 M. LEVIN *Eva* 224 These..were the creatures called mussulmans..in the last stage, when they reacted only as brute animals. 1975 N. FREELING *What are Bugles blowing For?* xxvii. 169 The man..has let himself die. It is a phenomenon similar to what was seen in the camps... The Musselmen they were called.

mussurana (musūrā·nā). [ad. Pg. *muçurana* f. Tupi, cord.] A Brazilian colubrid snake, *Clelia clelia*.
1914 T. ROOSEVELT *Through Brazilian Wilderness* i. 14 The most formidable enemy of the many dangerous Brazilian snakes is a non-poisonous, entirely harmless, rather uncommon Brazilian snake, the mussurana. 1934 *Discovery* July 207/2 A cannibal snake, known in Brazil as the Mussurana..though harmless to men and animals, lives entirely on other snakes to whose venom it is immune. 1964 G. VEVERS tr. *Vogel's Reptiles & Amphibians* xiii. 162 In Brazil there is a handsome snake, much-loved and protected, which feeds, although not exclusively, on other snakes, particularly poisonous ones. This is the Mussurana, *Clelia clelia*, which is frequently depicted in the act of eating a Jararaca or other poisonous snake. It reaches a length of 7½ ft., is typically a metallic black, although the young are pale, and it lives on the ground. 1970 *East African Standard* 23 Jan. 6/4 The Mussurana, or snake-eater, feeds on deadly snakes, particularly the lethal Fer-de-lance.

mussy, dial. var. MERCY *sb.* Cf. LAWK, LAWKS, *int.*
1823 E. MOOR *Suffolk Words & Phr.* 243 Mussy on us. 1881 J. C. HARRIS *Nights with Uncle Remus* (1883) xii. 67 For mussy sake gin 'im a walkin'-cane. 1903 P. L. DUNBAR *In Old Plantation Days* 63 Lawd a' mussy, 'pon my soul, an' you one of de faifful of de flock! 1909 *Dialect Notes* III. 344 Lawsy mussy (pon my soul). 1935 *Scribner's Mag.* Feb. 120/2 Hab mussy, Jedus!.. When a bull-bat go lak dat he diggin' a grabe.

must, *sb.*⁴ Add: **b.** Something that must be done, possessed, considered, etc.; a necessity. orig. *U.S.*
1892 *Dialect Notes* I. 205 An article marked with the word *must* is spoken of as a *must*, or emphatically—if there is absolutely no way of keeping it out of the paper—as a *dead must*. 1941 *Amer. Speech* XVI. 67/1 The *dirndl*, a dress which was a 'must' for every woman in 1937. 1948 'J. TEY' *Franchise Affair* xxi. 239 The Feathers—one of the 'musts' of American visitors to Britain—was.. famous. 1951 'J. WYNDHAM' *Day of Triffids* v. 103 It was close on midnight when we had finished adding our own secondary wants to the list of musts. 1957 R. HOGGART *Uses of Literacy* x. 250 These valuable books are *musts* for *you*. 1960 *House & Garden* Mar. 136 Nature in Schwepp-shire is a top must, is actively encouraged. 1973 *Guardian* 30 Mar. 2/4 A film and a song made the Trevi Fountain a 'must' for tourists.
c. *attrib.* and *Comb.* Essential, mandatory, obligatory. *colloq.* (orig. U.S.).
1912 M. NICHOLSON *Hoosier Chron.* 180 His gratification at being able to write 'must' matter for both sides of a prominent journal. 1937 *Amer. Speech* XII. 8 Roosevelt ..was called..The New Deal Caesar who specialized in must-legislation. 1939 *Canadian Forum* June 94/2 This is a *must* book. 1941 *Britannia & Eve* Sept. 4/1 They had come to accept the fact that mails and government-priority 'must' passengers monopolised the flying boats. 1952 *Manch. Guardian Weekly* 27 Nov. 3/2 A list of suggestions which he [*sc.* Senator Taft] had marked as either 'must' or 'optional' legislation. 1959 *New Statesman* 10 Jan. 32/2 A struggle of some interest..is now taking place among those women's periodicals..which are still 'must-reading' for debutantes with any claim to literacy. 1973 *Black World* May 69/2 His forthcoming paper is a 'must-read'.

must, *a.*³ and *sb.*⁵ Add: **A.** *adj.* Quot. 1886 to read:
1887 KIPLING *Departmental Ditties* (1888) 21 The Commissariat *hathee* had—forgive the rhyme—gone *musth*.
B. *sb.* **1.** (Later and *fig.* examples.)
1959 M. PUGH *Chancer* 35 He suffered from the recurring nightmare that his cameras would go on musth and stampede out of the studios at the second he was due on the air. 1972 *New York* 1 May 10/3 In twenty seconds she can go from emotional neutral to the fury of an elephant in musth. 1973 *Nature* 2 Nov. 17/2 The madness of the animal 'on musth' causes other elephants to avoid him.

must, *v.*¹ Add: **3. c.** *if you must know*: used to introduce information provided against the judgement or inclination of the speaker.
[1818 SCOTT *Heart Midl.* xxix, in *Tales of My Landlord* 2nd ser. III. iv. 104 He's in Gaffer Gabblewood's wheat-close, an ye maun ken.] 1861 HUGHES *Tom Brown at Oxf.* II. x. 193 Well, if you must know, I never saw her before yesterday. 1885 *Boy's Own Paper* 28 Mar. 403/2 Well, if you must know, it's our boat. 1902 H. JAMES *Wings of Dove* xxxiii. 511 Well, if you must know—and I want you to be clear about it—I didn't even seriously think of a denial to her face. 1927 C. ASQUITH *Black Cap* 250 'If you must know,' she had said, 'well, you've got dirty nails, haven't you? Look.' 1951 J. D. SALINGER *Catcher in Rye* xix. 173 Girl lives in the Village. Sculptress. If you must know. 1972 'G. HARDING' *Skytrap* ii. 39 If you must know, I wanted to be kissed gently.

musta, must'a' (mʊ·stā), a representation of the colloq. or vulgar pronunciation of *must have* (MUST *v.*¹ 5).
1903 P. L. DUNBAR *In Old Plantation Days* 266 It must 'a' been P'ovidence that directed bofe yo' min's in de same channel. 1927 M. OSTENSO *Mad Carews* (1929) v. 93 You're late... Must'a' stayed at the Phillips' till it was dark. 1938 M. RICHARDSON in B. A. Botkin *Treas. S. Folklore* (1949) III. i. 443 One of the other guards musta told him to bring it in. 1965 J. WAINWRIGHT *Death in Sleeping City* II. 104 They killed—Tommo. They musta—killed Tommo. 1973 *Black World* Jan. 62/2 They laid me out real good... Musta used up most of the insurance money.

mustang. Add: **1.** *attrib.*, as *mustang horse*.
1821 S. AUSTIN *Jrnl.* 5 Sept. in *Texas State Hist. Assoc. Q.* (1904) VII. 300 Mustang horses very plenty saw at least..150. 1941 *Sun* (Baltimore) 30 July 7/6 A Spanish vessel loaded with Mustang horses was shipwrecked years ago.
2. (Earlier and later examples.)
1846 J. GREGG *Diary* 27 Sept. (1941) I. v. 239 There is a large species, called by the Americans, the *Mustang grape* which very much resembles the Muscadine of the western country, except in growing in large bunches or racemes. 1951 *Dict. Gardening* (R. Hort. Soc.) IV. 2250/2 *V[itis] candicans*. Mustang Grape. Vigorous high climber.
3. An officer in the U.S. services who has been promoted from the ranks; in quots. 1847: a volunteer officer as distinct from a regular-army officer.
1847 G. B. MCCLELLAN *Mexican War Diary* 2 Jan. (1917) 23 'Mind, Mr. Smith,' said the old Mustang [*sc.* General Patterson] the night before, 'mind and appear as early as possible, so that you may not delay us'—all this with that air of dignity and importance so peculiarly characteristic of Mustangs. *Ibid.* 4 Jan. 43, I have repeatedly seen a Second Lieutenant of the regular army exercise more authority over the Volunteers..than a Mustang General. [1874 L. P. A. D'ORLÉANS *Hist. Guerre Civile en Amérique* I. 35 Le surnom dérisoire de *mustang*.. qu'il appliquait aux volontaires inexpérimentés avant l'épreuve commune.] 1931 *Leatherneck* Feb. 13 We have three..mustangs, two of whom are..completing their probationary periods. 1939 J. B. CONNOLLY *Navy Men* 172 He was a 'mustang'—up from the enlisted ranks. 1950 *Time* 11 Dec. 22 A mustang who had worked his way up from the ranks in 13 years of service. 1953 M. DIBNER *Deep Six* v. 35 A damned 'mustang'. Never went to the Academy or anything. 1962 *Amer. Speech* XXXVII. 288 *Mustang*, a self-made man, i.e., a Marine officer who started as a private instead of as a second lieutenant. 1971 *N.Y. Times Mag.* 5 Sept. 11 The most decorated enlisted man in the Korean War—the mustang everybody thought was the perfect combat commander.

mustard, *sb.* Add: **1. e.** (Earlier and later examples.)
The use of *keen as mustard* is anticipated by the expression *keenest mustard* in quot. 1658 s.v. KEEN *a.* (*adv.*) 4.
1659 [see STRONG *a.* 15 b]. 1672 W. WALKER *Parœmiologia* 25/1 As keen as mustard. 1679 'T. TICKLEFOOT' *Clod-Pate's Ghost* 4/2 You shall see a man as hot as Mustard against Plot and Plotters. 1916 'B. CABLE' *Doing their Bit* ii. 36 They get as keen as mustard on it. 1932 R. ALDINGTON *Soft Answers* 185 His famous Uncle Harold.. had unfortunately died insolvent, owing to being as keen as mustard on the form of the Empire's horses and the forms of the Empire's barmaids. 1938 N. MARSH *Artists in Crime* ii. 13 He's as keen as mustard,..and he can talk of nothing but his work.
f. The colour of the condiment, usu. a brownish-yellow. Also *attrib.* or as *adj.* Cf. *mustard-coloured*, *-yellow* adjs.
1848 [see *DANDIE DINMONT]. 1922 JOYCE *Ulysses* 226 She stared at the large poster of Marie Kendall... Mustard hair and dauby cheeks. 1923 *Daily Mail* 17 Apr. 7 (Advt.), In shades of Jade, Mustard, Cherry, Nigger or Grey. 1951 E. PAUL *Springtime in Paris* ii. 33 Most of the young coloured men, yellow, brown, snuff, mustard or ginger in hue, had white girls with them. 1965 J. GALE *Clean Young Englishman* III. 101, I wore my demobilisation suede shoes and a mustard ferreting coat. 1969 G. MACBETH *War Quartets* 60, I rose against a full moon, sharp in mustard light. 1971 *Vogue* 15 Sept. 129/2 Coats..colours: orange, mustard, brick, royal.
g. In fig. phrases: that which enhances the flavour of anything, or adds piquancy or zest;

the best of anything. Phr. *to cut the mustard*: to accomplish, to succeed, to make good; to come up to expectations, to meet requirements. *slang* (chiefly *U.S.*).
1903 A. ADAMS *Log Cowboy* xv. 237 For fear they were not the proper mustard, he had that dog man sue him in court for the balance, so as to make him prove the pedigree. 1904 'O. HENRY' *Cabbages & Kings* vi. 101 I'm not headlined in the bills, but I'm the mustard in the salad dressing just the same. 1905 'H. MCHUGH' *You can search Me* 17 Petroskinski is a discovery of mine, and he's all to the mustard. 1907 'O. HENRY' *Trimmed Lamp* 217 Why don't you invite him if he's so much to the mustard? 1907 —— *Heart of West* x. 163, I looked around and found a proposition that exactly cut the mustard. 1909 —— *Roads of Destiny* 99 'She cut the mustard,' he said, 'all right.' 1922 C. SANDBURG *Slabs of Sunburnt West* 7 Kid each other... Tell each other you're all to the mustard. 1923 WODEHOUSE *Inimitable Jeeves* iv. 41 Never before had I encountered a curate so genuinely all to the mustard. 1930 —— *Very Good, Jeeves* iii. 70 Life at this juncture seemed pretty well all to the mustard. 1960 J. MITFORD *Hons & Rebels* ix. 56 Perhaps I could get a job as a maid in somebody's house..but Idden convinced me I would never cut the mustard at this occupation. 1966 'J. HACKSTON' *Father clears Out* 114 Nell was all to the mustard. On her first Sunday afternoon she sat with young Bob Marlow..and, when it was Herby Carter's Sunday, went for a spin with him in his..jinker... For Tom..she sat in the old creek-bed. 1968 *Down Beat* 7 Mar. 18/1 There, in the Apple, his skill was tested in competition with the established ones. If he couldn't cut the mustard, he became part of the anonymous mob. 1974 *Citizen* (Ottawa) 24 Aug. 78 What if it doesn't work out? What if I'm bored with it? What if I'm no good at it? What if I just can't cut the mustard?
h. As quasi-*adj.*: very good, keen, enthusiastic; thorough. Also (*fig.*) hot, pungent.
1925 E. WALLACE *King by Night* ix. 41 That fellow is mustard. 1931 T. LYELL *Slang, Phrase & Idiom* 537 *To be mustard*, to be excellent at anything. It must never be used of the female sex. 1936 WODEHOUSE *Laughing Gas* xvii. 186 Avoid actors. They are mustard. 1959 *Times* 19 June 16/3 He [a batsman] is mustard on anything a trifle short of a length. 1960 L. COOPER *Accomplices* I. vi. 62 He showed me..an extract from a report from..the F.B.I. There wasn't much of it but what there was was mustard. 1966 L. SOUTHWORTH *Felon in Disguise* ix. 131 He could see the couple and what they were doing. 'Blimey, she's mustard,' he thought. 1972 *Daily Express* 4 July 6/8 Britain is particularly hot on calculus. The Russians and the East Germans are mustard on the theory of numbers and on solid geometry. 1973 'B. MATHER' *Snowline* xix. 231 You can wash out South Africa... They're mustard there. You couldn't smuggle in a grain of wheat. 1973 J. WAINWRIGHT *Touch of Malice* 62 'Careful, Charlie,' warned Sanderson quietly. 'He's mustard.' 'Not mustard,' said Ripley. 'Dynamite—'.
2*. a. = *mustard gas* (see sense *3 c).
1918 *Jrnl. Amer. Med. Assoc.* 7 Dec. 1911/2 (*heading*) Mustard (yellow cross) burns. 1937 A. M. PRENTISS *Chem. in War* ii. 50 If sufficient effort is expended in finding more efficient ways of using it, mustard will undoubtedly yield far greater results. 1966 *Science* 5 Apr. 409/1 Mustard, bis(β-chloroethyl)sulfide, shared interest with a series of nitrogenous analogues. 1966 *McGraw-Hill Encycl. Sci. & Technol.* III. 44/2 Wet or perspiring skin absorbs more mustard than does dry skin.
b. *Pharm.* Any of the group of substances that contains mustard gas and the nitrogen mustards (cf. *NITROGEN b).
1946 *Science* 5 Apr. 412/1 The oxidation of pyruvate by brain brei was significantly inhibited by sulfur mustard. 1964 *Brit. Med. Bull.* XX. 91/1 The biological alkylating agents..include the sulphur and nitrogen mustards, epoxides, [etc.]. 1970 *Nature* 6 June 897/1 Quinacrine mustard makes part of the human Y chromosome fluoresce particularly brightly.
3. b. *mustard-keen* (adj.), *-plaster* (earlier and later examples).
1935 G. HEYER *Death in Stocks* iii. 29 I'd had a letter from Arnold, and they were instantly mustard-keen to see it. 1968 'J. J. MARRIC' *Gideon's River* xviii. 162 Singleton's got the frogmen out already, the Warbler says he's mustard-keen. 1810 E. WEETON *Let.* 25 Feb. (1969) I. 235 Mustard plasters and bottles of warm water were applied to the feet. 1869 'MARK TWAIN' *Sketches New & Old* (1875) 46 That mustard-plaster of a newspaper. 1909 *Westm. Gaz.* 23 Dec. 4/1 It had lapels that were far too wide,..being what some tailors called 'the mustard-plaster type'. 1975 D. DELMAN *One Man's Murder* v. 95 Let me go home early to a mustard plaster.
c. **mustard bush, tree** *Austral.* (see quot. 1965); **mustard gas**, dichlorodiethylsulphide, $(ClCH_2CH_2)_2S$, a colourless, oily liquid which is a powerful poison and vesicant, acting directly on the skin, and which was first used in warfare by the Germans in 1917, at Ypres.
1933 *Bulletin* (Sydney) 7 June 25/2 Another hardy northwesterner is the mustard bush. It is of dwarf habit, with a dense growth of leaves relished by sheep. 1965 *Austral. Encycl.* II. 263/1 The mustard-bush or mustard-tree (*Apophyllum anomalum*), an almost leafless tree with small fragrant flowers, is an inland species whose young shoots have a somewhat mustard-like flavour; in spite of its wiry nature it is much relished by stock. 1917 *Nation* (N.Y.) 15 Nov. 524/2 The Germans have just invented a new and particularly powerful weapon in their so-called 'mustard gas'. 1934 *Discovery* Feb. 32/1 As far as can be judged from the sample of casualties officially studied, one in thirty-eight of the British casualties from 'mustard gas' died and about one in four hundred was incapacitated for six months or longer. 1937 J. M. MURRY *Necessity of Pacifism* v. 76 The sword of the Lord is broken in men's hands... The mustard-gas of the Almighty—it cannot be.
1973 G. BEARE *Snake on Grave* xxvi. 160 Less than fifty years ago this place was a sea of blood and mud.. scoured and sterilised by mustard gas.

mustard seed. Add: **1.** *fig.* (Later examples.)
1926 C. VAN VECHTEN *Nigger Heaven* 286 *Mustard-seed*: see high yellow (= mulatto). 1952 DYLAN THOMAS *Coll. Poems* 170 In the mustardseed sun.
3. (Earlier examples.)
1809 T. G. FESSENDEN *Pills Poetical* 8 Her single good gun loaded with mustard seed shot. 1844 *Knickerbocker* XXIII. 440 None of the fine mustard-seed or robin, but the heavy duck-shot.

mustardy, *a.* (Later examples.)
1911 W. J. LOCKE *Glory Clementina Wing* xix. 280 Her ill-fitting mustardy brown stuff skirt. 1936 W. JAMES *Gangways & Corridors* vi. 69 There was a hideous affair of mustardy yellow diagonal serge. 1975 *Times* 5 July 11/3 The plate of crudités, with a mustardy mayonnaise dip.

must-be (mʊ·st,bī). [f. MUST *v.*¹ + BE *v.*] The inevitable, what is fated to happen.
1907 *Cambr. Hist. Eng. Lit.* I. 4 The must-be often helps an undoomed man when he is brave. 1922 D. H. LAWRENCE *Aaron's Rod* 242 A ghastly atmosphere of must-be. 1970 R. LOWELL *Notebk.* 177 Hope is the must-be, the tomb of a small child.

mustelid (mʊ·stilid). [f. mod.L. family name *Mustelidæ*, f. L. *mustēla* weasel, adopted as the name of a genus by Linnæus (*Systema Naturæ* (ed. 10, 1758) I. 582): see -ID³.] A small carnivorous mammal of the family Mustelidæ, which includes weasels, stoats, badgers, mink, and others. Cf. MUSTELINE *sb.*¹
1910 H. F. OSBORN *Age of Mammals* iv. 259 The mustelids were becoming more numerous [in the Miocene]. 1933 A. S. ROMER *Vertebrate Paleontology* xv. 291 Continued emphasis on a flesh-eating diet has..resulted in the retention of well-developed shearing teeth in the greater number of mustelids. 1953 *New Biol.* XIV. 120 Most mustelids and wild dogs and cats do not [have fleas]. 1968 A. S. ROMER *Procession of Life* xv. 247 The mustelids tend to be relatively small in size, short of leg, and primarily forest-dwellers in habitat. 1972 T. A. VAUGHAN *Mammalogy* XII. 201/1 Most mustelids aggressively search for prey in burrows, crevices, or dense cover. 1975 *Nature* 20 Mar. 187/1 Other interesting elements of the fauna include..a new species of gomphothere, a canid, a mellivorine mustelid.

muster, *sb.*¹ **3. g.** Read: *Austral.* and *N.Z.* A collecting of stock (cattle, sheep, etc.) by riding round the scattered herd and driving it together.
1841 S. REVANS *Lett.* I. 90 (MS.), I am not yet confident of the mode in which flock and stock musters will be dealt with by the natives. 1867 LADY BARKER *Station Life in N.Z.* (1870) xx. 173 It is impossible to estimate our loss until the grand muster at shearing. 1884 [see Dict.]. 1892 W. E. SWANTON *Notes on N.Z.* ii. 97 Previous to the shearing, there is the general muster, which means the rounding up and bringing in of all the sheep, good or bad, on the 'run'. 1946 F. D. DAVISON *Dusty* ix. 90 The paddock..was not the easiest in the world to lift sheep from, but Tom had a feeling..that the count would show a clean muster. 1950 *N.Z. Jrnl. Agric.* Oct. 348/3 After the marking muster musterers proceed to the high country and muster wethers. 1956 *Coast to Coast 1955–6* 35 Whole country's gone dead since muster. 1961 B. CRUMP *Hang on Minute Mate* 75 We're starting the shearing muster in a few days. 1963 *Times* 12 Mar. p. xii/7 (*caption*) A muster of merinos for shearing on Benangaroo, a 40,000 acre sheep and cattle property near Coolac, in the southern tablelands of New South Wales.
9. *muster-field*, *-ground* (earlier U.S. example).
1838 B. DRAKE *Tales* 179 Our sons..assembling in the 'muster field', divide themselves into armies, and pelt each other with Buckeye balls. 1798 *Deb. Congress U.S.* 15 May (1851) 1707 At the muster ground on the Commons of Portsmouth.

muster, *v.*¹ **6.** Read: *Austral.* and *N.Z.* To collect stock (cattle, sheep, etc.) together for counting, shearing, drafting, branding, etc. Also with *off* and *up*, and with place as obj.
1846 C. J. PHARAZYN *Jrnl.* 23 June 45 (MS.), Assisted in mustering Ewes and Lambs into Stock-Yard. 1852 J. R. CLOUGH *Diary* 4 May in J. Deans *Pioneers of Canterbury* (1937) 292 Port Philip men just arrived..and two of the survey men mustered the cattle this day. 1858 [in Dict.]. 1860 *St. Leonard's Jrnl.* 31 July, To muster stragglers off Isolated Hill. 1867 LADY BARKER *Station Life in N.Z.* (1870) xvi. 122 It is very difficult to 'muster' these ranges. 1875 [in Dict.]. 1878 E. S. ELWELL *Boy Colonists* 208 They all started from the hut to muster off 'Nob' mountain. 1930 L. G. D. ACLAND *Early Canterbury Runs* vii. 164 The country had never been stocked... [They] had once mustered it..and got about 300 wild sheep off it. 1934 A. RUSSELL *Tramp-Royal in Wild Australia* xxiii. 210 Here we loosed the herd and went east to muster up another section of the run. 1946 F. D. DAVISON *Dusty* vi. 65 There were a lot of sheep to be done, to be mustered from the paddock and brought to yard and shears. 1947 P. NEWTON *Wayleggo* (1949) 12 The musterer's job is to muster the sheep off such country into the respective homesteads. 1950 *N.Z. Jrnl. Agric.* Oct. 349/3 While the hill wethers are being shorn the ewes and lambs are being mustered on the various blocks and are brought in.
absol. (Earlier and later examples.)
1874 A. BATHGATE *Colonial Experiences* vii. 80 A shepherd, while out mustering, descried the errant steed on a small plateau. 1878 E. S. ELWELL *Boy Colonists* 173 Walker's men never mustered beyond the 'Saddle'. 1944 F. CLUNE *Red Heart* 18, I found Bob Buck out mustering.

musterer² (mʊ·stərəɹ). *Austral.* and *N.Z.* [f. MUSTER *v.*¹ (sense *6) + -ER¹.] One who musters sheep or cattle.
1863 E. C. CHUDLEIGH *Diary* 19 Dec. in Richards *Diary of E. R. Chudleigh* (1950) 114 All the musterers dogs have come home. 1872 LADY BARKER in D. M. Davin *N.Z. Short Stories* (1953) 20 Of course Christmas Day would be a complete holiday, and we had invited shearers and musterers, and all the odd hands which flock to a station at shearing-time, to come up to our house. 1892 W. E. SWANTON *Notes on N.Z.* ii. 97 To accomplish this [muster], ..on large 'runs' additional hands called 'musterers' have to be engaged. 1947 [see *MUSTER *v.*¹ 6]. 1961 B. CRUMP *Hang on Minute Mate* 73 I'm short of a couple of musterers, if you're interested. Done any sheep work before? 1970 *Telegraph* (Brisbane) 4 July 2/1 Lawrence Ryan..worked as a musterer on a South Queensland cattle station 20 years ago.

mustering, *vbl. sb.* Add: (Austral. and N.Z. examples.)
1860 G. DUPPA *Station Diary* Oct. in S. S. Crawford *Sheep & Sheepmen of Canterbury* (1949) v. 47 To collect stragglers for a second mustering. 1874 A. BATHGATE *Colonial Experiences* xv. 206 Life on a sheep-station is rather a dull one. There are periodical 'musterings' and other duties to be attended to. 1878 E. S. ELWELL *Boy Colonists* 172 The mode of mustering adopted by Ernest was this. 1911 C. E. W. BEAN *'Dreadnought' of Darling* xviii. 172 There was the operation of mustering going on as clear and plain as if the hillside were a mile distant. 1968 K. WEATHERLY *Roo Shooter* 95 Tales told in a bush pub; tales of droving, mustering.
b. (Further examples.)
1745 J. HEMPSTEAD *Diary* 6 May (1901) 442 The 1st & 2d Company are viewing Arms &c. Mustering Day through the Colony. 1841 *Southern Lit. Messenger* VII. 768/1 Who should I see..but Mr. Jim Guest himself, in..mustering jacket and trowsers, and tarpaulin hat. 1878 E. S. ELWELL *Boy Colonists* 169 In mustering times Ernest used to set some of the huge boulders on the sides of its cliffs rolling. 1930 L. G. D. ACLAND *Early Canterbury Runs* vi. 138 More stories were told of him in shearers' huts and mustering camps than of any other runholder. 1953 O. E. MIDDLETON in C. K. Stead *N.Z. Short Stories* (1966) 186 See him coming in from the back country at mustering-time.

mut, var. *MUTT.

‖ **muta** (miū·tā). [It. *muta* imp. = change.] In *Music*. a direction to the player (see quots.).
1876 STAINER & BARRETT *Dict. Musical Terms* 298/2 *Muta*, a direction to a player on a horn, trumpet, etc., or on drums, to change the *key* of his instrument. 1880 GROVE *Dict. Mus.* II. 439/1 A word often seen attached to Horn parts—'muta in Es', 'muta in B', etc., meaning simply 'change to E♭ or B♭', etc. 1938 *Oxf. Compan. Music* 602/2 *Muta D in C* means 'change tuning from D to C'. 1959 *Collins' Music Encycl.* 446/1 *Muta*,..a direction to the performer found in parts for wind instruments and timpani, indicating either a change of instrument,.. or a change of crook,..or a change of tuning.

muta, var. *MOOTAH, MOOTER.

mutability. Add: **3.** *Biol.* The tendency to undergo, or capacity for undergoing, mutation.
1908 J. A. THOMSON *Heredity* iii. 92 It is possible that the prolific multiplication in a new environment may have had something to do with the awakening of the impulsive mutability [in *Œnothera lamarckiana*]. 1916 *Genetics* I. 606 (*heading*) Mutability in different species of Drosophila. 1929 *Proc. Nat. Acad. Sci.* XV. 834 That gene..does not influence the mutability of the miniature-gamma gene. 1958 *Nature* 11 Oct. 984/1 The sites in each segment have been plotted in order of decreasing mutability for each mutagen. 1974 *Ibid.* 9 Aug. 493/2 There was no significant difference in mutability between the two strains.

mutable, *a.* and *sb.* Add: **A.** *adj.* **3.** *Biol.* Capable of undergoing mutation; liable to undergo mutation frequently.
1905 F. E. CLEMENTS *Res. Methods in Ecol.* 319 *Mutable*, able to produce mutants. 1908 J. A. THOMSON *Heredity* iii. 91 There he [*sc.* de Vries] found his long-looked-for mutable plant, an evening primrose (*Œnothera lamarckiana*). 1928 *Genetics* XIII. 360 To designate a character showing frequent heritable changes, the terms 'mutable', or 'frequently mutating' will be used. 1941 *Nature* 22 Mar. 356/2 We have carried out an experiment with a mutable gene which produces coloured spots on the petals and stems of *Portulaca grandiflora*. 1972 *Molecular & General Genetics* CXIV. 144 (*heading*) The genetics of a new mutable allele at the *white* locus in *Drosophila melanogaster*.

mutagen (miū·tădʒĕn). *Biol.* [f. MUTA(TION + -GEN.] An agent that causes mutation.
1946 *Proc. R. Soc. Edin.* B. LXII. 220 In unpublished experiments with chemical mutagens, a significant change in the rate of induced lethals was recorded. 1952 *New Biol.* XII. 86 Certain chemical substances, or penetrating radiation such as ultra-violet or X-rays, may increase the mutation rate considerably. Those agents, called 'muta-

gens', are at present being intensively studied. 1971 J. Z. YOUNG *Introd. Study Man* xxviii. 393 Another implication of the randomness of the effects of mutagens is that no amount of them is too small to be biologically irrelevant.

mutagenesis (miūtădʒe·nĕsis). *Biol.* [f. MUTA(TION + -GENESIS.] The production or origination of mutations.
1953 *Nature* 21 Nov. 964/1 (*heading*) Chemical mutagenesis in bacteriophage T2. 1961 *Jrnl. Molecular Biol.* III. 121 (*heading*) The theory of mutagenesis. 1972 *Heredity* XXIX. 203 (*heading*) Gamma ray mutagenesis in a strain of *Escherichia coli* deficient in DNA polymerase I.

mutagenic (miūtădʒe·nik). *Biol.* [f. MUTA(TION + *-GENIC.] Causing or capable of causing mutation.
1946 *Proc. R. Soc. Edin.* B. LXII. 211 Muller's classical discovery in 1927 of the mutagenic action of X-rays has provided an extremely useful tool for studying the nature of gene mutation and chromosome breakage. 1952 *New Biol.* XII. 87 All known mutagenic agents also damage the cell. 1959 *New Scientist* 20 Aug. 219/1 Plant geneticists..now combine mutagenic radiations and chemicals with the classic techniques of plant breeding to produce novel plants. 1971 *Nature* 1 Oct. 296/3 Dr Mary Lyon.. argued that the radiation from nuclear fall-out is likely to be less mutagenic than the medical use of X-rays. 1974 *Times* 11 Jan. 16/1 The 'mutagenic' effect of lead compounds.
Hence **mutageni·city**, the property of being mutagenic.
1956 *New Biol.* XX. 30 Different types of radiation were tested for mutagenicity. 1970 *Nature* 30 May 800/1 The concern felt by many scientists about the potential mutagenicity of the ever increasing number of synthetic chemicals in the environment. 1972 *Guardian* 15 Dec. 7/4 Evidence of mutagenicity in a chemical compound is a danger signal.

mutagenize (miū·tădʒĕnəiz), *v.* *Biol.* [f. *MUTAGEN + -IZE.] *trans.* To treat (cells or organisms) with mutagenic agents; also *absol.* So **mu·tagenized** *ppl. a.*
1966 *Virology* XXIX. 339/1 The mutants were produced by mutagenizing with hydroxylamine. *Ibid.* 339/2 A mutagenized phage stock was prepared. 1969 A. M. CAMPBELL *Episomes* ii. 16 From a mutagenized lysogenic culture, some bacteria can be recovered that are no longer lysogenic. 1970 *Nature* 31 Oct. 414/2 Mutagenized stocks of virus. 1972 *Ibid.* 18 Feb. 363/3 Instead of mutagenizing stocks of transforming virus..[they] took mouse 3T3 cells..and..selected for cells which behave at 32° C as typical transformants.

mutalali, var. *MUDALALI.

mutant (miū·tănt), *sb.* and *a.* *Biol.* [ad. L. *mutant-em* pr. pple. of *mūtāre* (see MUTATE *v.*).] **A.** *sb.* An individual (or, formerly, a species or form) which has arisen by or undergone mutation, or which carries a mutant gene (in *Science Fiction*, usu. an individual with freak or grossly abnormal anatomy, abilities, etc.); also, a mutant gene.
1901 *Rep. Brit. Assoc. Adv. Sci.* 848 Two genera of Conifers which are in a period of spontaneous generation, a period in which they do form mutants, to use the terminology of de Vries, which mutants *may* be true to seed. 1919 T. H. MORGAN *Physical Basis Heredity* xx. 249 Of the twelve dominant mutants that have appeared in *Drosophila* each appeared at first in a single individual. 1919 *Jrnl. Exper. Zool.* XXVIII. 363 By aid of these two dominants it is very easy to determine in a single experiment whether a given mutant is in the second or third chromosome. 1929 *Encycl. Brit.* VI. 947/1 There is a mutant or sport of the fruit fly with black coloration of the body. 1930 R. A. FISHER *Genetical Theory Nat. Select.* i. 13 Under the blending theory of inheritance, every individual was regarded as a mutant. 1954 'J. CHRISTOPHER' *22nd Century* 140 You really can produce a breed of telepathic mutants that will support us? 1955 'J. WYNDHAM' *Chrysalids* ii. 21 The nearest approach to decoration was a number of wooden panels with sayings..burnt into them... Blessed is the norm, and... Watch thou for the mutant! 1962 [see *hypomorph* s.v. *HYPO- II]. 1969 *Nature* 18 Oct. 272/1 (*heading*) Vitamin K-deficient mutants of bacteria. 1974 *Encycl. Brit. Macropædia* XII. 756/1 Better strains of *Penicillium* and other industrial fungi..have been raised from artificially produced mutants.
fig. 1971 *Time* 7 June 7/1 An ominous new mutant of the American tragedy in Viet Nam... The troops who became addicted to heroin while serving in SE Asia.
B. *adj.* Having the attributes of a mutant; produced by mutation.
1903 *Amer. Naturalist* XXXVII. 740 The form, habit and behavior of some of the mutant forms discovered by deVries seventeen years ago. 1928 *Genetics* XIII. 389 In this work two mutant characters have been studied. *Ibid.* 392 The sex-ratio in the mutant stock is normal. 1930 R. A. FISHER *Genetical Theory Nat. Select.* i. 11 In domestication..the rigour of Natural Selection [is] relaxed so that mutant types can survive. 1955 *Bull. Atomic Sci.* June 215/3 There is evidence that a mutant gene frequently does produce some disability of the individual who carries it. 1973 *Nature* 9 Feb. 368/2 Mutant cells resistant to particular drugs..are being routinely used today by somatic cell geneticists.

mutarotation (miūtărōtē·ʃən). *Chem.* [f. L. *mūtā-re* to change + ROTATION.] The change of optical activity with time exhibited by the solutions of some compounds (esp. sugars) after being made up. (Orig. called *multirotation* (*MULTI- 2).)
1899 T. M. LOWRY in *Jrnl. Chem. Soc.* LXXV. 213 The essential feature of the phenomenon, however, is the *change* of rotatory power which takes place in the freshly prepared solution, and the term 'mutarotation' may, therefore, be used with advantage to include all those cases in which such a change occurs without reference either to the sign or to the relative magnitude of the initial and final rotations. 1937 *Chem. Rev.* XXI. 65 The brucine salt..was found..to exhibit definite mutarotation. 1971 *Jrnl. Gen. Psychol.* LXXXV. 116 Each quantity of the sugar solution was refrigerated for a day before use at room temperature to allow adequate time for mutarotations.
So **mutarota·tional** *a.*, of, or of the nature of, mutarotation. Also **mutarota·te** *v. intr.*, to exhibit mutarotation; **mutarota·ting** *ppl. a.*
1930 *Jrnl. Chem. Soc.* II. 2618 These two interconvertible, mutarotating forms of mannose are thus assigned different ring structures. 1951 I. L. FINAR *Org. Chem.* I. xviii. 372 These glycosides..do not undergo many of the reactions of the sugars, *e.g.*, they show no reducing properties, they do not mutarotate, etc. 1971 N. L. ALLINGER et al. *Org. Chem.* xxvii. 701 Monosaccharides commonly mutarotate. 1971 M. F. MALLETTE et al. *Introd. Biochem.* v. 145 Because the acetals of carbohydrates are kinetically stable, they do not undergo mutarotational changes. 1973 *Biochemistry* (Easton, Pa.) XII. 2544/1 If the enzyme.. promoted instability by distorting the ring, the mutarotational rate would be enhanced.

mutasarrif, var. *MUTESSARIF.

mutase (miū·tēiz, -s). *Biochem.* [a. G. *mutase* (J. Parnas, at the suggestion of F. Hofmeister, in *Biochem. Zeitschr.* (1910) XXVIII. 284), f. L. *mūt-āre* to change: see *-ASE.] **a.** Any enzyme which brings about a dismutation reaction.
1914 [see *DEHYDRASE a]. 1943 SUMNER & SOMERS *Chem. & Methods Enzymes* xvii. 298 Aldehyde mutase brings about the Cannizzaro reaction. Here, two molecules of an aldehyde undergo an oxido-reduction.
b. Any enzyme which brings about the transfer of a phosphate group from one carbon atom to another in a molecule.
1938 *Jrnl. Biol. Chem.* CXXIV. 552 Mannose-1-phosphate and galactose-1-phosphate, synthetically prepared, were not converted to the respective 6-esters by muscle, liver, or yeast mutase, with or without the addition of accelerating ions. 1964 CORI & BROWN in Florkin & Stotz *Comprehensive Biochem.* XV. vi. 212 In a system consisting of hexokinase, mutase and phosphorylase, glucose could be converted to glycogen, provided that the inorganic phosphate formed in the phosphorylase reaction was removed by barium ions. 1970 R. W. MCGILVERY *Biochemistry* xv. 292 The sequence begins by a freely reversible mutase reaction, in which the phosphate group of glucose-6-phosphate is transferred to form glucose-1-phosphate.

mutassarif, mutasserif, varr. *MUTESSARIF.

mutate, *sb.* Add: **2.** *Chess.* (See quot. 1970.)
1922 HUME & WHITE *Good Companion Two-Mover* 297 We shall see that the term changed mate block (or 'mutate', as Brian Harley has suggested) is often indefinite in that it may include or exclude added mate features according to the individuality of separate problems. 1924 FEAST & WHITE *Simple Two-Move Themes* 62 The changed-mate waiters, or Mutates,..classify more readily. *Ibid.* 66 Every type of mating strategy discussed in Part III may theoretically be treated in Mutate form. 1931 B. HARLEY *Mate in Two Moves* ii. 20 We now come to the last of the big groups into which the two-mover can be divided: the Complete Block both before and after the Key, which changes one or more mates: that is, the Complete Block-Change, or, as I have christened it for short, the Mutate. 1959 B. J. HORTON *Dict. Mod. Chess* 85/1 He [*sc.* Brian Harley] did much to popularize the 'mutate' in his regular chess columns. 1962 K. S. HOWARD *Amer. Two-Move Chess Problem* 11 More than any other type of problem, the attractiveness of a mutate is dependent on the excellence of its construction. *Ibid.* 12 Wurzburg in No. 133 utilizes a mutate setting to show a charming illustration of a Bristol withdrawal key. 1970 J. M. RICE *ABC of Chess Problems* 182 Mutate is the term given to a complete-block problem in which White, with no mere waiting-move at his disposal, is forced to abandon one or more of his set mates and introduce new replies to Black's defences.

mutate, *v.* Add: **2.** *Biol.* **a.** *intr.* To undergo mutation.
1913 *Amer. Naturalist* XLVII. 12 A mutation need not be a loss, and..a recessive may revert in the sense that it may mutate. *Ibid.* 68 There is in *Œnothera lamarchiana* a tendency to mutate in certain definite directions generation after generation. 1926 [see *MUTATING *ppl. a.*]. 1951 E. HAMILTON *Star Kings* xix. 145 They're creatures of this crazy planet... I think they were human once—human colonists who mutated under radioactive influence. 1972 *Sci. Amer.* Apr. 65/2 We have no reason to think the gene for cytochrome *c* mutates more slowly than the gene for hemoglobin. 1974 *Nature* 1 Feb. 261/1 The evolution of enzymes in bacteria is studied by introducing novel substrates and selecting strains that have mutated to be able to use these substrates.
b. *trans.* To cause (genetic material) to undergo mutation.
1961 *Jrnl. Molecular Biol.* III. 142 There are many genetic sites which can be mutated by HA. 1972 *Nature* 14 Apr. 326/1 Clearly, in at least some phage genes certain segments of DNA are more readily mutated than others.
mutated (further examples); **muta·ting** *ppl. a.*, undergoing mutation.
1903 *Amer. Naturalist* XXXVII. 739 This particular factor in distinguishing between fluctuating and mutating variability therefore becomes a safe one, when it is modified to make mutating variability include only newly acquired and transmissible qualities. 1913 *Ibid.* XLVII. 71 Such study should be continued for many years, breeding..from some highly mutating stock. 1919 T. H. MORGAN *Physical Basis Heredity* xx. 248 When these are recessive it is probable..that the actual mutation occurred several generations before the mutated genes came together to produce the mutant character. 1926 *Proc. Nat. Acad. Sci.* XII. 687 The first frequently mutating character observed in *Drosophila virilis* was the body 'reddish', which was found to mutate frequently to wild type. 1945 *Genetics* XXX. 497 From outcrossing flies of the mutating cultures to those of normal cultures (Reedsport) the results..in table 2 were obtained. 1964 *Jrnl. Molecular Biol.* IX. 352 (*heading*) Identification of the altered bases in mutated single-stranded DNA. 1967 *Biol. Abstr.* XLVIII. 3217/1 (*heading*) Notes on mutant forms among Philippine plants. IV. A case of a 'recurrently mutating' banana. 1973 *Guardian* 18 May 18/1 There are reports of mutated fish in the lagoon and radiation sickness from eating the fish.

mutation. **6.** Substitute for def.: **a.** The process whereby detectable and heritable changes in genetic material arise; also, formerly, a process by which de Vries (*Die Mutationstheorie* (1901–3)) supposed a new species to be suddenly produced by a departure from the parent type (in contrast with *variation*). **b.** A change of this kind in the genetic material. **c.** An individual (or, more rarely, an assemblage of like individuals) which has been produced by this process; a mutant. (Further examples.)
1901 *Jrnl. R. Microsc. Soc.* 439 Quite distinct from these are those abrupt..variations..which sometimes occur, and of which de Vries records a remarkable instance in the genus *Œnothera*. For such variations de Vries proposes the term *mutations*. 1919 *Jrnl. Exper. Zool.* XXVIII. 381 In our opinion, the attempted distinctions between 'saltations', 'mutations', and 'variations of slight degree' have led rather to confusion of thought than to clearer thinking. To us these are all a single class, 'mutations', and the term carries no restrictions of degree, covering the most extreme as well as the slightest detectable inherited variation. 1925 *Genetics* X. 117 If one thinks of mutations as being simply inherited changes, it becomes necessary to distinguish changes that involve whole chromosomes.., changes that involve several adjacent genes.., and what have been called 'point-mutations' or 'gene-mutations'. 1928, 1930 [see *gene mutation* s.v. *GENE 2]. 1955 *Sci. News Let.* 25 June 409 Many mutations are lethal. If man-made irradiation increases the mutation rate, the result is sure to be harmful. 1955 *Sci. Amer.* July 74/2 Reproduction is one of the two essential features of life. Mutation is the other. 1957 I. ASIMOV *Naked Sun* (1958) xi. 145 Even the best gene analysis of parents can't assure that all gene permutations and combinations will be favourable, to say nothing of the possibility of mutations. That's our big concern, the unexpected mutation. 1965 A. H. STURTEVANT *Hist. Genetics* x. 62 It is ironic that few of the original mutations observed by de Vries in Oenothera would now be called mutations. 1971 J. Z. YOUNG *Introd. Study Man* xxviii. 392 The cause of mutation is some faulty copying during replication of the DNA. This..will be inherited only if it occurs in the line of the germ cells. Mutations elsewhere in the body are called somatic mutations... The accumulation of somatic mutations may produce some of the diseases of old age. 1972 [see *MENDELIAN *sb.*]. 1974 *Sci. Amer.* Sept. 82 (*caption*) One-gene mutation is responsible for the difference between the beta chain of a normal hemoglobin molecule..and that of hemoglobin *S*, the variant form responsible for sickle-cell anemia.
7. Special Comb.: **mutation mink**, a mink belonging to a mutant strain with a fur colour different from the normal; fur or a garment made from the skin of such a mink; **mutation pressure**, a tendency for recurring mutation (rather than selection) to alter the frequency of a particular allele within a population; **mutation rate**, the rate at which gene mutations occur (see quot. 1971); **mutation theory**, the theory of de Vries concerning the origin of new species (see *6 a).
1942 H. BOCK in *Amer. Fur Breeder* Oct. 14 (*title*) Mutation mink and their use in coats. 1945 R. G. HODGSON *Mink Book* (ed. 2) 41 Mutation mink got their start in the United States..and the Silverblu..was the important mink to give the industry something to think about. 1956 J. G. LINKS *Book of Fur* 166 The mutation mink names Silverblu, Aleutian, Royal Pastel..each representing a different mutation which had been developed since the original Silverblu. 1958 *Vogue* Jan. 12 Natural pale beige mutation mink. 1966 R. SERJEANT *Mink on my Shoulder* xi. 152 To the fur trade and perhaps also to the general public, the mutation mink means a mink of an abnormal colour or fur pattern that can be repeated at will. 1931 *Genetics* XVI. 100 (*heading*) Mutation pressure. 1962 D. J. MERRELL *Evol. & Genetics* xxiv. 237 Theoretically mutation pressure alone could bring about evolution. 1930 R. A. FISHER *Genetical Theory Nat. Select.* vi. 122 If ..the mutation rates..are high enough to maintain any considerable genetic diversity, it will only be the best adapted genotypes which can become the ancestors of

future generations. 1948 *Proc. R. Soc.* B. CXXXV. 168 It follows that human mutation rates are about twice those of *Drosophila* per nuclear division, and about one two-hundredth of those of *Drosophila* per day. 1971 LEVITAN & MONTAGU *Textbk. Human Genetics* xvii. 649 Mutation rates are usually stated in terms of the number of changed genes per locus per generation. 1904 Mutation theory [in Dict., sense 6]. 1912 *Amer. Naturalist* XLVI. 359 We can distinguish and trace the history of these quantitative variations from generation to generation only when the differences between them are of some size. This has led many to think that only variations of some size are inherited (the mutation theory) and others to deny that such variations can be increased in size by selection. 1962 D. J. MERRELL *Evol. & Genetics* xxi. 213 In the very early days of genetics de Vries (1902) proposed the mutation theory of evolution as an alternative to the theory of natural selection.

mutational (miutē·ʃənăl), *a.* *Biol.* [f. MUTATION + -AL.] Of or pertaining to mutation.
1904 *Nature* 25 Feb. 386/2 De Vries's 'mutational' variations. 1931 E. B. FORD *Mendelism & Evolution* iv. 72 This criticism omits all reference to an important condition of the theory: the size of the mutational steps involved. 1955 *Bull. Atomic Sci.* June 210/2 No exposure [to radiation] is so tiny that it does not carry its corresponding mutational risk. 1964 G. H. HAGGIS et al. *Introd. Molecular Biol.* p. xi, Recent refinements in genetic analysis..have made it possible to relate mutational events to modification of nucleic acid structure.
Hence **muta·tionally** *adv.*, as a result of mutation.
1934 in WEBSTER. 1960 *Biochim. & Biophys. Acta* XLIII. 288 (*heading*) The similar properties of tryptophan synthetase and a mutationally altered enzyme in *Neurospora crassa*. 1971 *Nature* 24 Dec. 443/3 Strains of *E. coli* with mutationally altered ribosomal proteins.

mutationist (miutē·ʃənist). [f. MUTATION + -IST.] One who stresses the importance of mutation as a factor in the evolution of new forms or species.
1904 *Science* 10 June 881/2 It might be inferred that Lamarckians and Darwinians are..regarded as believers in adaptiveness as a factor in evolution, and mutationists are necessarily supposed to hold the opposite view. 1909 *Fabian News* XX. 76/1 The mad mutationist who claims that evolution takes place by mutation only. 1911 A. D. DARBISHIRE *Breeding & Mendelian Discov.* 4 The point at issue between the Selectionist..and the Mutationist.

‖ **mutatis mutandis** (miutā·tis miutæ·ndis), *advb. phr.* [L., f. *mutatis*, *mutandis*, ablative pl. respectively of pa. pple. and gerundive of *mūtāre* to change.] 'Things being changed that have to be changed', i.e. with the necessary changes; with due alteration of details (in comparing cases).
1498 *Coventry Leet Bk.* (1909) 595 And like billes, mutatis mutandis, were put In ayenst Gloucestre & Worcestre. 1615 T. BONE *Let.* 9 Nov. in J. P. Collier *Egerton Papers* (1840) 472 The very same (*mutatis mutandis* onely) were put in practize by Foreman. 1666 *Phil. Trans. R. Soc.* I. 289 The like may be fitted to Mars in other positions, *mutatis mutandis*; and so for the other Planets. 1710 SWIFT *Tale Tub* (ed. 5) Author's Apology sig. a1, I know nothing more contemptible in a Writer than the Character of a Plagiary; which he here fixes at a venture, and this, not for a Passage, but a whole Discourse, taken out from another Book only *mutatis mutandis*. 1753 CHESTERFIELD in *World* 14 June 146 The utility of this invention extends, *mutatis mutandis*, to whatever can be the subject of letters. 1817 COLERIDGE *Biogr. Lit.* I. xiii. 288 The actual application of the positions which had so wonderfully enlarged the discoveries of geometry, mutatis mutandis, to philosophical subjects. 1877 *Trans. Connecticut Acad. Arts & Sci.* III. 345 The same may be said, *mutatis mutandis*, of the other symbols of the same type. 1892 [see *HOMOSEXUAL *a.* and *sb.* A.]. 1931 *Times Lit. Suppl.* 23 July 570/2 *Mutatis mutandis*, one might trace a succession from the seventeenth-century divine and Fellow of the Royal Society..to the well-known modern names in philosophy. 1955 *Times* 11 July 13/6 *Mutatis mutandis* these lucid words still apply to the exertions of Russian historians carrying on the same tradition. 1962 S. F. FINER *Man on Horseback* ii. 6 What is said of the army here is to be taken also to apply, *mutatis mutandis*, to the air force and the navy. 1973 *Times* 13 Dec. 18/2 *Mutatis mutandis*, we are going to face a series of enforceable demands by the 'undeveloped' nations, equivalent to the Arabs' demand for higher pay for their oil.

‖ **mutato nomine** (miutā·to nōᵘ·minĕi), *advb. phr.* [L., f. *mutato*, ablative sing. of pa. pple. of *mūtāre* to change + *nomine*, ablative sing. of *nōmen* name.] 'The name being changed': applicable in a transferred context if the name of the person, place, etc., is altered accordingly.
The original phrase (cited in quot. 1621) quotes Horace *Sat.* I. i. 69.
1621 BURTON *Anat. Mel.* (Democritus to Reader) 37 Accounting it an excellent thing..to make our selues merry with other mens obliquities, when as he himselfe is more faultie then the rest, *mutato nomine de te fabula narratur*, he may take himselfe by the nose for a foole. a 1661 FULLER *Worthies* (1662) Lincs. 153 Which Proverb, *Mutato Nomine*, is used in other Counties. 1828 J. S. MILL in *Westm. Rev.* IX. 146 Nobody would pretend that a man unacquainted with the properties of simple substances would be perfectly capable of performing such an analysis, or that the knowledge of the ultimate elements of bodies was of no service to the chemist. The same observations apply, *mutato nomine*, to the logician, and the syllogism. 1840 J. B. FRASER *Koordistan* II. vi. 152 The words of Burns, *mutato nomine*, describe their country exactly. 1860 THACKERAY in *Cornhill Mag.* May 632 A score of such queer names and titles I have smiled at in America. And, *mutato nomine*?

mutator. Restrict † *Obs. rare*⁻¹ to sense in Dict. and add: **2.** *Biol.* Any gene which increases the mutation rate of other genes. Usu. *attrib.*, esp. in *mutator gene*.
1943 *Proc. Nat. Acad. Sci.* XXIX. 138 In determining the abnormal mutation rate we shall consider only those cultures which have given at least one sex-linked recessive, since this is the only indication of the presence of the 'Mutator' gene in the mother. 1945 *Genetics* XXX. 496 Recently, high mutation frequency in a strain of *D. pseudoobscura*, race B, was found to be due to a dominant mutator. 1970 *Jrnl. Molecular Biol.* L. 129 Mutator genes cause a generalized increase in mutation rate. They have been identified in a variety of organisms, including phage and bacteria.

mute, *a.* and *sb.*¹ Add: **B.** *sb.* **3. e.** (Later example.)
1962 WODEHOUSE *Service with Smile* i. 16 That's why she slinks about the place like a funeral mute, is it?
4*. *Cinemat.* A positive or negative film print which has no synchronous sound-track. Also *attrib.* or as *adj.*
1933 A. BRUNEL *Filmcraft* 161 *Mute*, the negative or positive of the pictorial image. 1953 K. REISZ *Technique Film Editing* 281 *Mute negative*, picture negative of a sound film, without the sound-track. *Mute print*, positive print of the picture part of a sound film without the sound track. 1963 E. LINDGREN *Art of Film* (ed. 2) ii. 37 We now have two lengths of negative film, one known as the mute negative (or picture negative or action negative) and the other known as the sound negative. The projection print is made by printing these two negatives on to a single positive film. 1969 W. RUTHERFORD *Gallows Set* ii. 27 We're filming him tomorrow morning... And we're doing a bit of mute, showing him going up to the gate. *Ibid.* iv. 53 There's..a couple of cans on shipbuilding, mostly mute but with a bit of sound.

muted, *a.*² Add to def.: Now in extended use; *spec.* of colour = SUBDUED *ppl. a.* 2.
1939 M. B. PICKEN *Language of Fashion* 101/3 Muted, subdued or toned down. Often applied to colours. 1950 *Britannica Bk. of Year* 682/2 *Muted*, of lighting, subdued. 1958 *Oxf. Mail* 9 July 6/2 Sea colours—aquamarine, sapphire blue, muted green. 1964 *Daily Tel.* 7 Feb. 16/2 'It is unlikely that the development of public expenditure on the scale implied will leave much scope for a reduction in taxation.' This muted sentence..must often come to people's minds as they read of new commitments. 1964 MRS. L. B. JOHNSON *White House Diary* 17 July (1970) 181 My West is muted brown and green and gray. 1968 *Jazz Monthly* Nov. 23/1 Davis uses that distinctive hard-edged smokey-centred muted tone. 1974 *Impressionism* (R. Acad.) 45/2 He uses a very restricted colour-range to convey an overcast snow-effect—muted blues and greens set off against grey and white.

‖ **mutessarif** (mutəsa·rif). Also **mutasarrif, mutassarif, mutasserif, mutesarrif, mutussarif.** [Turk., ad. Arab. *mutaṣarrif* governor of a sanjak.] In the Ottoman Empire and Iraq, a governor of a province.
1875 *Encycl. Brit.* III. 446/2 Batum..is now the seat of a *mutessarif*, or deputy-governor. 1900 G. BELL *Let.* 2 May (1927) I. 87 There has come a telegram from Damascus to bid me say the Mutussarif fears for the safety of your presence. 1914 T. E. LAWRENCE *Let.* June (1938) 174 All the Consuls & Valis & Kaims & Mutasarrifs & Commandants came out also. 1919 R. WEBB *Let.* 28 June in M. Llewellyn Smith *Ionian Vision* (1973) vi. 106 We were gradually getting the bad Valis, Mutessarifs, &c, removed. 1921 *Blackw. Mag.* June 705/2 A few of the political officers remain in the new capacity of Adviser to an Arab Mutassarif. 1933 *Times Lit. Suppl.* 9 Nov. 760/2 It was in Jerusalem that he was able to watch the simple methods of Rauf Pasha, then Mutessarif of El Kuds. 1969 D. WALDER *Chanak* xi. 199 The Mutessarif of Chanak (a combination of English mayor and French prefect) had taken over the Sanjak. 1972 D. DAKIN *Unification of Greece* viii. 116 The *mutessarifs* (provincial governors) were, if Christian, to have Moslem assessors, and, if Moslems, then Christian assessors.

muthafucka, muthafukka, muthafucking, varr. *mother-fucker*, *mother-fucking* ppl. adj. (*MOTHER *sb.*¹ 16 a).
1969 H. R. BROWN *Die Nigger, Die* ii. 29 The dirty muthafucka. 1969 R. D. TAYLOR *Drum Song* in A. Chapman *New Black Voices* (1972) 312 Ain't that blackmuthafukka beautiful. 1971 *Black World* Mar. 54/2 The best muthafucking thing that done ever happened to me.

muti (mū·ti). *S. Afr.* Also 9 **booti, mooti.** [ad. Zulu *umuthi* tree, plant, medicine.] Medicine; a medicinal charm (used esp. by a medicine-man or witch-doctor).
1831 W. B. BOYCE *Jrnl.* 23 June in A. Steedman *Wanderings & Adventures in S. Afr.* (1835) II. 282 He mentioned the report of Mr. Shepstone having sent men on horseback to plant booti (bewitching matter) upon his place. 1891 R. MONTEIRO *Delagoa Bay* viii. 172 They [*sc.* the Kaffirs] don't mind saying some are used as 'mooti', physic. 1911 *State* Dec. 659 (Pettman), He produced a few pinches of powder from the *muti* bag suspended round his neck. 1947 *Cape Times* to May 1/1 It was thought that the plant had been killed out by witchdoctors, who consider it an important ingredient of a powerful muti. 1957 J. PACKER *Nor Moon by Night* xviii. 176 He may have made *muti* of that boy's eyes and tongue. *Ibid.* xxvi. 234 This *muti* is like the bite of a cobra—it makes a man sleep. 1971 *Sunday Times* (Johannesburg) 28 Mar. 20/3 The club which has often used the rolling of the bones to forecast the results of vital matches need some new 'muti' to revive their form.

mutillid (miuti·lid), *sb.* and *a.* [f. mod.L. family name *Mutillidæ*, f. generic name *Mutilla* (Linnæus *Systema Naturæ* (ed. 10, 1758) I. 582): see -ID³.] **A.** *sb.* A solitary, parasitic, fossorial wasp of the family Mutillidæ, including insects also known as velvet ants, whose bodies are covered with fine hair. **B.** *adj.* Pertaining to or resembling an insect of this type.
1910 W. M. WHEELER *Ants* xiv. 244 Existing mutillids, however, present two highly specialized characters..: they are, so far as is known, parasitic and their females are wingless. 1913 [see *CLERID *a.* and *sb.*]. 1971 E. O. WILSON *Insect Societies* xviii. 338/1 Several observers.. have witnessed guard bees protecting their nests against mutillid wasps and ants. 1972 L. E. CHADWICK tr. *Linsenmaier's Insects of World* 299/1 The mutillids chirp with a sound like that of boiling water.

muting, *vbl. sb.* (s.v. MUTE *v.*³). Add: **b.** *Electronics.* The automatic suppression of the output of an amplifier when the input signal falls below some predetermined level. Freq. *attrib.*
1936 W. T. COCKING *Wireless Servicing Man.* xv. 168 In practically all cases muting is obtained by arranging for a valve to apply a large bias to the last I.F. valve, the detector, or the first L.F. valve in the absence of a carrier, so that the stage is rendered inoperative. 1959 H. N. GANT *Mobile Radio Telephones* ii. 51 Receivers for mobile radio telephony have very large amplification and the transmitter to which they are tuned is only on when a message is to be passed, so there are long periods when the receiver produces only noise, in the form of a loud roaring sound. To prevent this annoyance the more complex receivers often incorporate a muting or squelch circuit to cut off the receiver output in the absence of the signal. 1970 J. EARL *Tuners & Amplifiers* ii. 39 The diodes and four transistors..give a.g.c. for the FET in the f.m. front-end, a potential for the tuning meter and inter-station muting.

mutism¹. Add: (Further examples.) Also *fig.*
1873 *Nature* 27 Feb. 323/2, I know of two instances where perfect mutism accompanied the deafness in cats. 1882 SALA *America Revisited* I. ix. 138 Behind the counter was a very paragon of mutism in the shape of an hotel clerk. 1894 A. GRIFFITHS *Secrets Prison House* II. IV. viii. 220 There was no other outlet but confession or obstinate mutism. 1969 P. ANDERSON in Cockburn & Blackburn *Student Power* 215 Given the complete mutism of the past, any such initial attempt will inevitably suffer from errors, lapses, [etc.]. 1975 *N.Y. Times Book Rev.* 8 June VII. 21/3 All we can hope is..that the innumerable poems..will not give us nausea, or a new, science-inspired dose of mutism.
b. In *Psychol.* used (in contrast to *aphonia*) to imply the absence of any ascertainable defect of the vocal organs and hence a cause that is primarily psychological rather than physiological.
1892 D. H. TUKE *Dict. Psychol. Med.* II. 827/2 *Mutism*, dumbness from mental defect or disorder. In addition to the cases of Deaf-Dumbness.., mutism occurs in the course of various mental disorders, as Mental Stupor, Delusional Insanity, &c. 1930 *Internat. Jrnl. Psychoanal.* XI. 185 (*title*) On the physiology of hysterical aphonia and mutism. 1940 *Amer. Jrnl. Psychiatry* XCVI. 1445 In psychotic patients mutism is often associated with other manifestations of negativism. 1961 W. R. BRAIN *Speech Disorders* vii. 107 Mutism is the term applied to a complete loss of speech in a conscious patient in the absence of organic disease of the nervous system. 1972 S. CASHDAN *Abnormal Psychol.* ii. 50 In extreme cases, social withdrawal leads to mutism (refusal to speak) and regression (acting in infantile ways).

muto-. Add: mutoscope (further examples).
1901 *World's Work* Aug. 1057/2 Prints made from the film are mounted consecutively about a cylinder. As the cylinder is revolved the mounted pictures are held back by a stop, and snap past the eye so that the illusion is of a continuous motion picture. Encased in a box and with the automatic penny-in-the-slot attachment the mutoscope is ready for its common commercial use. 1922 JOYCE *Ulysses* 362 Mutoscope pictures in Capel street: for men only. 1969 *Jabez Elliott* (Ringwood, Hants.) *Miscellany Catal.* No. 6. 7 A little booklet or pad..of 82 photographs of two girls dancing. When the leaves are flicked over the dancers spring to life... These little books, called Mutoscopes, were popular in the early days of the Cinema.

muton (miū·tǫn). *Biol.* [f. MUT(ATION + *-ON¹.] The smallest element of the genetic material (supposedly a single pair of nucleotides) which when altered can give rise to a mutant individual.
1957 S. BENZER in McElroy & Glass *Symp. Chem. Basis Heredity* 71 The unit of mutation, the 'muton', will be defined as the smallest element that, when altered, can give rise to a mutant form of the organism. 1959 [see *CISTRON]. 1970 AMBROSE & EASTY *Cell Biol.* x. 356 It seems probable that the smallest unit of recombination (the recon) has in fact the same dimensions as the smallest unit of mutation (the muton), that is a single nucleotide base pair in the DNA molecule. 1971 J. Z. YOUNG *Introd.*

Study Man iii. 53 Mutation may occur by the change of any single base and these are thus the letters of the genetic language (also known as mutons).

mutsha, mutshi, varr. *MOOCHA.

mutt (mɒt). *slang* (orig. *U.S.*). Also **mut.** [abbrev. *mutton-head* (MUTTON 8 b in Dict. and Suppl.] **a.** One who is stupid, ignorant, awkward, blundering, incompetent, or the like; a blockhead, dullard, or fool; also, non-pejoratively, a person, fellow.

1901 'H. McHugh' *Down Line* 79, I knew that Clara Jane would cancel the contract with the mutt that mixed in just as soon as she saw the automobile snap. **1910** O. Johnson *Varmint* 377 Engaged to that Ver Planck fellow that was hanging around. I think he's a mutt. **1915** Wodehouse *Psmith, Journalist* xvi. 116 This ain't him. This is some other mutt. **1920** *Blackw. Mag.* Feb. 176/2 Dougal, the elder brother, was a quiet, inoffensive kind of a mutt. **1926** *Spectator* 1 May 813/1 Not doubting the poor mutt's love for her. **1929** [see *GOOF]. **1942** 'M. Innes' *Daffodil Affair* iv. 146 Why couldn't you keep to the racket, you poor mut, and leave tinkering alone! **1955** W. Gaddis *Recognitions* ii. vii. 597 Got to look up a mutt named Chavenay. **1972** *Police Rev.* 10 Nov. 1444/2 Some male driver—poor mutt!—will be fiddling with the jack. **1973** D. May *Laughter in Djakarta* xii. 203 The poor mutt must have driven it along the bank.

b. A term of contempt applied to a dog, *esp.* a mongrel.

Quot. 1904 refers to a horse.

1904 *Outing* XLV. 170/2 Watch that mut curl up out there. **1906** H. Green *Actors' Boarding House* 335 A fellow can't leave nothin' on his bed without that mutt chawin' it up! **1911** R. W. Chambers *Common Law* x. 310 Now fat old women..Arrive to exercise their various dogs; And 'round and 'round the little mutts all run. **1927** *Ladies' Home Jrnl.* Dec. 4/1 Be careful the mutt doesn't get into a race with a caterpillar some day, and die of heart collapse. **1932** *Sun* (Baltimore) 15 Oct. 20/5 There are people who especially desire a mutt dog. **1949** *Sat. Even. Post* 16 Apr. 44/2 That cat! That mutt! they fight it out And back and forth they shuttle. **1970** *New Yorker* 29 Aug. 50/3 The cast includes a Sheepdog.., a Mutt Bitch. **1972** C. Weston *Poor, Poor Ophelia* (1973) vi. 29 Two barefoot hippies were sharing a bag of potato chips with a happy-looking mutt.

c. Phr. *Mutt and Jeff* [from the names of two characters called *Mutt* and *Jeff*, one tall and the other short, in a popular cartoon series by H. C. Fisher (1884–1954), American cartoonist]. (*a*) A stupid pair of men; stupid dialogue; (*b*) (see quot. 1943); (*c*) as *adj.*, deaf.

1917 E. E. Cummings *Let.* 4 June (1969) 26 By failing to get up..I escaped departing with the bums mutts and jeffs (not to say ginks, slobs, and punks) who came over with us. **1937** M. Huxley in A. Huxley *Lett.* (1969) 426 A sort of Mutt and Jeff on war and peace and religion. **1937** Partridge *Dict. Slang* 545/2 *Mutt and Jeff*, the British War Medal and Victory Medal: military: 1918. **1943** C. H. Ward-Jackson *It's a Piece of Cake* 43 *Mutt and Jeff*, the King George V silver Jubilee and the Edward VIII Coronation medals, or ribbons, worn together; or the 1918 Victory and Overseas medals or ribbons worn together. **1949** *Sun* (Baltimore) 10 Dec. 24/5 Richard ——, identified as the taller man in the recent series of 'Mutt and Jeff' robberies here, yesterday pleaded guilty to armed robbery. **1960** I. Franklyn *Dict. Rhyming Slang* 98/1 *Mutt and Jeff*, deaf. Current in the theatrical world, and formed on the names of the two famous strip-cartoon characters. **1973** *Washington Post* 21 Nov. C2 The women..call the two cops 'Mutt and Jeff'. 'Who is Jeff? The white one'... Mutt is always being kicked around. Jeff makes him do all the work.' **1974** D. Seaman *The Bomb* xx. 200 He silently named them Mutt and Jeff. One [man] stood well over six feet..while the other barely reached to his mate's armpits.

muttering, *vbl. sb.* (Later example of sing. use.)

1971 *Guardian* 14 Jan. 1/3 There may be some feminine muttering that in promoting the 'Woman's Guardian' we are perpetuating the idea of the female ghetto.

muttering, *ppl. a.* (Later examples.)

1917 T. S. Eliot *Prufrock & other Observations* 9 Certain half-deserted streets The muttering retreats Of restless nights in one-night cheap hotels. **1927** Joyce *She weeps over Rahoon* in *Poems Penyeach*, The moongrey nettles, the black mould And muttering rain.

mutt-eye (mɒ·t,ai). *Austral. slang.* [Orig. unknown.] (Cut) corn.

1946 K. Tennant *Lost Haven* (1947) x. 135 Five sacks of potatoes, three of mutt-eyes, another three of pumpkins. **1966** Baker *Austral. Lang.* (ed. 2) iv. 84 Bull's eyes, fried eggs; mutt eyes, corn; frog's eyes, boiled sago.

‖ **mutti** (mu·ti). [G., f. *mutter* mother.] A childish or familiar form of 'mother' (used in German-speaking countries).

1906 M. A. von Arnim *Princess Priscilla's Fortnight* ii. 32 She's a witch—Mutti, she's a witch! **1939** C. Isherwood *Goodbye to Berlin* 169 'Poor little Mummy, little Mutti, little Muttchen,' he crooned. **1967** A. Wilson *No Laughing Matter* iii. 314 And as for Mutti—strange little bent witch-like lady! **1972** *Guardian* 29 July 3/5 An announcement from..the lifeguards. 'Achtung, achtung—morning gymnastics for the ladies..' And off go the Muttis to trim their waistlines, while father relaxes.

mutton. Add: **4.** (Further examples.) So, the genital organs of a woman; copulation. Phr. *to hawk one's mutton*, (of a woman) to seek a lover, to solicit (cf. *hawking* ppl. a. s.v. HAWK *v.*[2]). See also MUTTON-MONGER in Dict. and Suppl.

1864 Hotten *Slang Dict.* (ed. 3) 184 *Mutton*, a contemptuous term for a woman of bad character... In that class of English society which does not lay any claim to refinement, a fond lover is often spoken of as being 'fond of his mutton'. **1937** Partridge *Dict. Slang* 380/2 Hawk one's mutton. **1939** H. Hodge *Cab, Sir?* v. 53 He can't quite believe she hawks her mutton in hexagonal horn-rimmed spectacles. **1964** N. Freeling *Double-Barrel* ii. viii. 65 In the army we used to say, of such and such a girl, nurse, waaf, whatever she was, 'That one hawks her mutton.' **1973** 'J. Patrick' *Glasgow Gang Observed* vii. 73 They're aw cows hawkin' their mutton.

7. *to return to one's muttons* (earlier and later examples); so, *to resume one's muttons*; conversely, *to stick to one's muttons*; *mutton dressed as lamb*: an elderly or middle-aged woman dressed (coiffured, painted, etc.) as though she were young; *to be one's muttons* (N.Z.): see quot. 1941.

1821 M. Edgeworth *Let.* 19 Dec. (1971) 297, I think he is winning..the heart of Lady Caroline—But to return to my muttons. **1895** Kipling *Brushwood Boy* in *Day's Work* (1898) 348 Look at young Davies makin' an ass of himself over mutton-dressed-as-lamb old enough to be his mother! **1903** A. Bennett *Leonora* iii. 72, I shall have to return to my muttons directly. **1922** Joyce *Ulysses* 541 *Passée*. Mutton dressed as lamb. **1930** *Punch* 28 May 606/3 Both houses, having dealt with the Whitsuntide holidays, resumed their muttons. **1933** *Sun* (Baltimore) 3 Mar. 6/7 Let's stick to our muttons, old man radio, and make it music alone. **1937** D. L. Sayers *Busman's Honeymoon* xv. 307 Aggie Twitterton. Runs arter 'im like an old cat... At 'er age! Mutton dressed as lamb. **1940** *National Education* (Wellington, N.Z.) Feb. 17 Milk, however, is small Charlie's muttons. **1941** Baker *N.Z. Slang* vi. 54 The farming community has given us [this century] another useful expression in *our muttons*. When we speak of something *being our muttons* or *a person's muttons* we mean that it is regarded with particular favour, that we like it especially well. **1943** A. Hastings *Bright Conversations* 24 Stick to your muttons, and don't talk tripe. **1967** V. S. Naipaul *Mimic Men* i. iii. 41 Our middle-aged ladies, mutton dressed as lamb, as our barman says. **1974** N. Marsh *Black as he's Painted* i. 31, I digress... Shall we return to our muttons?

7*. *Printing.* = *mutton quad* (MUTTON 8 b).

1938 *Amer. Speech* XIII. 270 An em quad is a space the square of the type body... In the shop..frequently called muttons or monkeys. **1960** G. A. Glaister *Gloss. Book* 122/2 The popular name for an em quad is mutton.

8. b. mutton cloth (see quot. 1957); **mutton-faced** *a.*, having a face suggestive of mutton (as a term of abuse); **mutton fat,** (*b*) in full, *mutton-fat candle*; (*c*) used *attrib.* and *absol.* of jade to designate a creamy white colour valued highly by connoisseurs; **mutton-fisted** *a.*, clumsy, heavy-handed; also *fig.*; **mutton-head** (earlier U.S. and later examples); **mutton-headed** (later examples); **mutton-leg sleeve** = *leg-of-mutton sleeve*.

1923 *Daily Mail* 12 Mar. 2 (Advt.), Stockinette is white mutton cloth in its new form, and everyone knows there is nothing to beat it for any kind of cleaning or polishing. **1957** *Textile Terms & Definitions* (Textile Inst.) (ed. 3) 67 *Mutton cloth*, a plain-knitted fabric of loose texture, usually cotton, made on a multi-feeder circular knitting machine. **1892** Stevenson & Osbourne *Wrecker* xii. 193 'You ——, ——, little, mutton-faced Dutchman,' Nares would bawl. **1864** Le Fanu *Uncle Silas* III. xvii. 259 The imperfect light of our mutton-fat candle. **1912** B. Laufer *Jade* xii. 328 A brown-red tinge passing into light-yellow shades is strewn over a background of a glossy white which the Chinese designate as mutton-fat. **1920** 'K. Mansfield' *Let.* 20 May (1928) II. 34 A flamingo in a cage made of mutton-fat jade. **1935** [see *CH'IEN LUNG]. **1936** J. Goette *Jade Lore* vi. 104 White jade from Central Asia..shows a greater range than does the Burmese. The native connoisseur has always put great emphasis on what he terms mutton fat. **1969** F. Koval tr. *O. Luzzatto-Bilitz's Antique Jade* 46 White jade which is slightly translucent and opalescent is called by the Chinese 'Mutton-Fat Jade'. **1918** Mutton-fisted [see *ham-handed* adj. (*HAM *sb.*[1] 3)]. **1927** *Observer* 27 Nov. 6 A critic of his central sound sense..and mutton-fisted manner of calling a spade a spade. **1934** *Times Lit. Suppl.* 22 Nov. 813/2 But he [*sc.* a hunter] was naturally a little headstrong, and a mutton-fisted stable-boy speedily made of him an incurable puller. **1965** D. Francis *For Kicks* ix. 120, I worked in a slovenly fashion and rode..like a mutton-fisted clod. **1803** T. G. Fessenden *Terrible Tractoration* (ed. 2) iv. 159 And couldst thou, pertinacious B—, But maul these mutton heads, most sadly. **1928** D. H. Lawrence *Woman who rode Away* 209 That *fool*, Joe, standing there like a mutton-head! **1938** H. Nicolson *Let.* 11 July (1966) 349, I suppose that some mutton-head might say [etc.]. **1960** I. Wallach *Absence of Cello* 160 I'm neither an oaf, a boor, or a muttonhead. **1972** 'J. & E. Bonett' *No Time to Kill* xi. 148 Bone-heads, that's what you are. Mutton-heads. Idiots. **1897** G. Bartram *People of Clopton* ii. 49 He were sich a mutton-headed fool theer were no valley in oot-wittin' him. **1934** Wodehouse *Right Ho, Jeeves* xix. 242 She had caused all the trouble by her mutton-headed behaviour in saying 'Yes' instead of 'No'. **1942** E. Paul *Narrow St.* xx. 164 A mutton-headed obstinate father. **1830** *Ladies' Mag.* III. 183 Think of such terms as *mutton leg sleeves*, for example. **1845** *Lowell* (Mass.) *Offering* V. 201 Here is a piece of the first dress I ever saw, cut with what were called 'mutton-leg' sleeves. **1922** Joyce *Ulysses* 431 In..widow Twankey's blouse with muttonleg sleeves buttoned behind.

mutton-bird. Substitute for def.: *Austral.* and *N.Z.* **a.** Either of two species of the genus *Puffinus*, in New Zealand the sooty shearwater *P. griseus*, and in Australia the short-tailed shearwater *P. tenuirostris*. **b.** An Antarctic petrel of the genus *Pterodroma*. Also *attrib.* (Further examples.)

1846 G. H. Haydon *Five Yrs. Australia Felix* 47 (Morris), The mutton-bird, or sooty petrel..is about the size of the wood-pigeon of Europe... it is black in colour. These birds are migratory, and are to be seen ranging over the surface of the great southern ocean far from land... millions of these birds are destroyed annually for the sake of their feathers and the oil of the young. **1905** W. B. *Where White Men Treads* 202 In the olden days we opened the mutton-bird burrows, I heard you remonstrate with our lads. **1911** A. E. Mack *Bush Days* 110 On the surface of the water sat a mutton-bird. **1944** A. Russell *Bush Ways* v. 27 The mutton-birds make underground nesting burrows. **1954** A. Moorehead *Rum Jungle* ix. 133 It is usually in a gale (known as the 'mutton bird gale')..that the main swarm suddenly appears. **1963** B. Pearson *Coal Flat* xxii. 375 Mrs Torrens brought out..a half-kerosene tin of cooked mutton-birds preserved in fat. **1965** *Courier-Mail* (Brisbane) 14 Mar. 3/7 Mutton bird rookeries can be found on the..islands. **1970** *Southerly* XXX. 226 The children saw countless thousands of mutton birds sweeping over the farm, returning from ocean wanderings north of the equator to their nesting colonies on the Bass Strait coasts.

2. mutton-bird scrub, tree *N.Z.*, an evergreen shrub or small tree, *Senecio reinoldii*, of the family Compositæ, bearing dark green leaves and clusters of small yellow flowers.

1898 Morris *Austral Eng.* 311/1 *Mutton-bird Tree*..so called because the mutton-birds, especially in Foveaux Straits, New Zealand, are fond of sitting under it. **1906** Laing & Blackwell *Plants N.Z.* 438 *S[enecio] rotundifolius* is the mutton-bird scrub of Stewart Island. Its leaves are much used by tourists for post cards, the white tomentum of the underside affording a suitable surface for writing. **1963** Poole & Adams *Trees & Shrubs N.Z.* 200 *S[enecio] reinoldii* Endl. Mutton-bird scrub. Shrub or small tree reaching 9 m.

d. *slang.* In full, *mutton-bird eater.* (See quot. 1937). Loosely, any Tasmanian.

1937 Partridge *Dict. Slang* 546/2 *Mutton-Bird* (gen. pl.), a resident in North Tasmania: Southern Tasmanians'. **1941** Baker *Dict. Austral. Slang* 48 *Mutton-bird*, a resident of Northern Tasmania. Also, 'mutton-bird eater'.

So **mutton-birder,** one who catches mutton-birds in season; one who is engaged in the preparation of mutton-birds for the market. Also **mu·tton-birding** *vbl. sb.*

1881 G. Walch *Victoria in 1880* 49 Armed with a piece of stout curved wire, fastened at the end of a long stick, the mutton-birder fishes in the holes for his prize. *Ibid.*, One of the sports of the neighbourhood is 'mutton-birding'. **1900** *N.Z. Illustr. Mag.* II. 519 Mutton-birding on Stewart Island. **1969** *Austral. Encycl.* VI. 234/1 (*caption*) A mutton-birder's hut, with casks of dressed and salted mutton-bird carcasses.

mutton-chop. 2. (Further examples.)

1875 [see *BURNSIDE, BURNSIDE]. **1972** J. Wambaugh *Blue Knight* (1973) vi. 83 Some of the boys had mutton-chops and moustaches. **1973** J. Jordan *Nile Green* xxiv. 97 The mutton-chop whiskered auctioneer. **1975** J. Symons *Three Pipe Problem* iii. 24 A square honest face framed by mutton-chop whiskers.

mutton-fish. 2. Substitute for def.: *Austral.* The flesh of a shell-fish of the genus *Haliotis*, esp. *H. iris*; = *PAUA. (Earlier and later examples.)

1840 J. S. Polack *Manners & Customs of New Zealanders* II. xvi. 176 The *paua*, (mutton-fish) shell. **1843** E. Dieffenbach *Travels in N.Z.* II. x. 239 *Haliotis iris*... Inhabits New Zealand... The 'mutton-fish' of the colonists; eaten boiled, but very tough. **1874** A. Bathgate *Colonial Experiences* xvi. 245 The beautiful iridescent shell of the mutton-fish..is not uncommon in the south. **1944** *Living off Land* ii. 30 In the case of molluscs like mutton fish, cook until most of the natural moisture has boiled out of the shell. **1956** M. West *Gallows on Sand* ix. 98 The ecology of *Haliotus asinina*—mutton-fish to you. **1959** W. G. Mc[illegible] lived on mutton fish.

mutton-monger. Restrict † *Obs.* to sense 2 in Dict. and add later example of sense 1.

1923 J. Manchon *Le Slang* 201 A mutton-monger, un coureur de filles.

mutual, *a.* Add: **1. a.** spec. *mutual aid, deterrence.*

1539, 1894 [in Dict.]. **1912** J. S. Huxley *Individual in Animal Kingdom* v. 153 Mutual aid (though it implies mutual dependence) establishes minimum waste. *Ibid.* vi. 154 The ideals of active harmony and mutual aid as the best means to power and progress. **1927** —— *Religion without Revelation* i. 49 Permanent facts of human existence—..suffering, mutual aid, comradeship, physical and moral growth. **1942** *New Statesman* 10 Oct. 233/1 Lease lend, now officially named 'Mutual Aid', is giving rise to a great deal of discussion. **1954** *Ann. Reg. 1954* 149 Canada had previously made 400 Sabres available to the R.A.F. as part of the mutual aid programme. **1955** *Bull. Atomic Sci.* Jan. 16/1 In the non-Communist world a new formula is gaining more and more popularity—'mutual deterrence' by the consideration of the opponent's retaliatory power. **1966** Schwarz & Hadik *Strategic Terminology* 61 *Mutual deterrence*, situation obtaining between nuclear powers when each is deterred from attacking the other (i.e., launching a first strike) because the damage expected to result from the victim's retaliation (second strike) would be unacceptable.

c. (Earlier and later examples.) Also *mutual admiration gang.*

1851 Thoreau *Jrnl.* 27 Feb. in *Early Spring in Massachusetts* (1881) 16 It is the hip-hip-hurrah and mutual admiration society style. **1920** G. B. Shaw *How to become Musical Critic* (1960) 313 A ridiculous little mutual-admiration gang of snobs. **1953** G. Hough *Romantic Poets* v. 160 Keats would have been better off at this stage..without so many ladies, or with ladies of a different kind; and some of the familiar sonnets suggest a small and rather silly mutual admiration society. **1969** G. Battiscombe *Queen Alexandra* x. 139 Seldom can there have been a more devoted family: the unkind might even have called them a mutual admiration society.

d. (Further examples.) *Spec.* in phr. **mutual fund** *U.S.*, a unit trust; also *attrib.*; hence **mutual funding.**

1950 J. C. Clendenin *Introd. to Investments* 600 (*Index*) Mutual funds (*see* Open-end companies). **1958** *Spectator* 18 July 108/1 In the United States, total investment in Mutual Funds (the American term for unit trusts) stands at over £3,000 million. **1962** S. Strand *Marketing Dict.* 472 *Mutual Funds Market*, the daily buying and selling transactions of the shares of mutual funds. Companies selling mutual funds are listed on the financial pages of newspapers. **1969** *Times* 30 Apr. 28/3 As a Mutual Society all our profits must go to our policyholders. *Ibid.* 5 May (Suppl.) p. vi/3 The average turnover of mutual fund portfolios rose..to 43 per cent. **1972** *Daily Tel.* 4 Nov. 28/4 By lending those funds on mortgage, the [building] societies, which are mutual and non profit-making, pass on the benefit of the short-term cost of money to long-term borrowers. **1973** *Times* 18 May 29/2 Expanding industries such as insurance or mutual funding.

e. *Electr.* Applied to quantities and properties that depend equally and symmetrically on two circuits or circuit elements and represent an effect on either of a certain kind of change in the other; esp. *mutual inductance* (see *INDUCTANCE), *induction*.

1865 J. C. Maxwell in *Phil. Trans. R. Soc.* CLV. 507 To find the coefficient (M) of mutual induction between two circular linear conductors in parallel planes. **1886,** etc. [see *INDUCTANCE]. **1896** F. Bedell *Princ. Transformer* iii. 37 The relation between two circuits is strictly a mutual one, the coefficient of mutual induction having the same value with either circuit as primary or secondary. **1931** Moyer & Wostrel *Radio Handbk.* vii. 395 In most circuits, the coupling between the input and output circuits is adjusted by changing the mutual reactance. **1959** B. J. Ley et al. *Linear Circuit Anal.* iii. 150 The coefficients $Y_{11}, Y_{22}, \ldots, Y_{nn}$ of the principal diagonal terms were called self-admittances and the off-diagonal coefficients Y_{hk} ($h \neq k$) were called the mutual admittances. .. To find the mutual admittance Y_{12} we examine the term $Y_{12}(j\omega)V_2$ in the first equation of Eqs. (3–69). This term represents the phasor current leaving node 1 when nodes 1, 3,.., *n* are grounded and node 2 has a phasor voltage V_2. **1969** A. M. Howatson *Princ. Appl. Electr.* v. 100 The impedance Z_{12} is, physically, the voltage produced in mesh 1 due to unit current flowing in mesh 2. It is called the open-circuit transfer impedance or mutual impedance of mesh 2 to mesh 1, and because $Z_{12} = Z_{21}$ it is also that from mesh 1 to mesh 2.

f. Electronics. *mutual conductance*, the ratio of the change in the anode current of a valve to the change of grid voltage causing it, the anode voltage being held constant; so *mutual characteristic*, a characteristic curve representing the variation of anode current with grid voltage at constant anode voltage.

1918 L. A. Hazeltine in *Proc. IRE* VI. 64 The effectiveness of the audion as a relay depends primarily on the slope of the characteristic curve. This slope, being the quotient of a current by a voltage associated therewith, is of the dimensions of a conductance and may be called the mutual conductance of the grid towards the plate. **1933** K. Henney *Radio Engin. Handbk.* viii. 195 The mutual characteristic, or transfer characteristic of the tube, shows the effect of the grid voltage upon the plate current. **1942** *Electronic Engin.* XIV. 734/1 The alternating potential is applied to control the electron stream at a point where the electron velocity is low, thereby producing a positive feedback which increases the effective mutual conductance of the valve. **1962** D. F. Shaw *Introd. Electronics* xi. 212 The value of g_m is low when I_a is very small but after an initial curvature the mutual characteristics are almost linear and parallel. **1963** B. Fozard *Instrumentation Nuclear Reactors* x. 116 At middle frequencies where the undesirable effects of the various capacitances are negligible the gain is g_mR where g_m is the mutual conductance of the valve and *R* is the effective anode load resistance.

B. *sb.* **a.** = *mutual friend.* **b.** = *mutual fund* (*MUTUAL *a.* 1 d).

1901 Kipling *Kim* xii. 314 The wire came in about what our mutual friend said he had hidden... I meet our mutual at Delhi on the way back. **1971** *Financial Mail* (Johannesburg) 26 Feb. 690/1 Some mutual fund men..are once again pressing the Registrar for permission to go ahead with those property mutuals.

mutualism. Add: **1. a.** Esp. in connection with the theory of non-profit credit and voluntary association for the exchange of services advocated by P. J. Proudhon (1809–65).

1849 C. Dana *Proudhon* (1896) 36 But how can they gain possession of this instrument [*sc.* capital]? By the organization of credit, on the principle of reciprocity or mutualism, if we may use a new word. *Ibid.* 40 Mutualism of credit, or credit at cost. **1929** *Amer. Jrnl. Sociol.* XXXIV. 783 By 'mutualism' Proudhon meant a practice of voluntary association for strictly specified and limited purposes. **1968** *Internat. Encycl. Soc. Sci.* XII. 606/1 Federalism, that is, mutualism transferred into the realm of politics, is the solution.

2. (Further examples.) By some writers applied esp. to such a relationship that (*a*) is not necessary for the survival or reproduction of the organisms involved, or else (*b*) is necessary for one or both of the organisms. Also *transf.*

1949 W. C. Allee et al. *Princ. Animal Ecol.* xxxv. 711/1 Varying degrees of mutualism exist, from a slight benefit to a remarkable interdependence of both species in the partnership. **1953** *Parasitology* XLII. 261 Symbiosis can be broadly divided into the three well-recognized categories: commensalism (where the host is for all practical purposes unaffected by the presence of the symbiote), parasitism (where the host is injured), and mutualism (where the host is benefited). **1956** T. W. H. Cameron *Parasites & Parasitism* 231 Commensalism, in turn, grades into mutualism, which implies a certain benefit by the host from the presence of the invader, but the association is not an essential one. **1962** J. D. Smyth *Introd. Animal Parasitol.* i. 6 An association in which both associates benefit has long been referred to as mutualism by some authors and symbiosis by others... *Symbiosis* could..broadly be used to include all the different kinds of relationship which exist in nature. By usage, however, it has come to be restricted to associations..in which the participating species are dependent on each other for existence... In cases of mutualism, on the other hand, the association is not obligatory to existence. **1962** K. F. Lagler et al. *Ichthyol.* xiv. 436 [Among fishes] instances of mutualism (symbiosis) are not clearly known; this is the relationship in which neither of two species can reproduce or grow in the absence of the other. **1963** C. J. McCall in A. Dundes *Mother Wit* (1973) 420 The relationship is rather one of mutualism, in that it is latently eufunctional for both institutions. **1967** M. E. Hale *Biol. Lichens* v. 70 *Mutualism.* This term describes a mutually beneficial relationship where one or both components may be dependent on the association for survival. **1970** *Nature* 5 Sept. 1001/1 The commonest species, *Remora remora*, is an active cleaner, and the mutualistic relationship between fish and host is strong, but the mutualism is less strong in other species.

mutualist. Add: Also *attrib.*

1909 F. Lawton *3rd French Republic* xiv. 320 From 1852 onwards, the Mutualist movement extended rapidly. **1929** *Amer. Jrnl. Sociol.* XXXIV. 783 Let exchange take place directly within each 'mutualist' association. **1948** M. Nomad in F. Gross *European Ideologies* viii. 329 Proudhon's 'mutualist anarchism', with its panacea of a 'People's Bank'. **1969** A. Ritter *Pol. Thought Proudhon* v. 130 The units of a mutualist society are not only to be equal in power; they are also to differ in their occupations, personalities, ideas, inclinations.

mutualistic (miūtiuăli·stik), *a.* [f. MUTUALIST + -IC.] Exhibiting or characteristic of mutualism (esp. sense 2).

1885 *Anarchist* 15 Sept. 2/2 He shews how two principles pervade human society—one mutualistic, which leads to equality, the other, antagonistic, which destroys it. **1911** J. G. McIntosh *Manuf. Varnishes* III. 291 Giard does not see that the ants are enemies of the cochineal... Their relations are mutualistic and in no way predatory. **1936** *New Phytologist* XXXV. 129 The presence of the fungus in a mycorrhizal association is to be regarded as an example of controlled parasitic attack and has no mutualistic significance. **1967** V. Ahmadjian *Lichen Symbiosis* iv. 78 To say that all lichens are mutualistic is as wrong as saying that all are parasitic. **1970** [see *MUTUALISM 2]. **1974** *Nature* 18 Oct. 574/2 Mortimer Starr (University of California, Davis) produced a new classificatory scheme which has the virtue of both clarifying and thinking about the phenomenon and indicating where further experimental work is required—to decide, for example, whether a symbiotic association is mutualistic or obligately parasitic in this respect.

Hence **mutuali·stically** *adv.*

1949 W. C. Allee et al. *Princ. Anim. Ecol.* xvii. 251 There does not appear to be any development of a strict taxonomic relation like that in so many examples of mutualistically paired species in which pollination is involved. **1967** V. Ahmadjian *Lichen Symbiosis* iv. 78 In some *Collema* lichens the fungal hyphae periodically destroy certain parts of the algal colony while living mutualistically with the rest.

mutuality. Add: **1. c.** A system of organizing conditions of work by agreement between the workmen involved and the employer; also **mutuality system.**

1968 *Sunday Tel.* 20 Oct. 19/2 The employers team pressed the unions to surrender the 'mutuality' system. **1970** *Guardian* 4 Dec. 15 'Mutuality'..means that the rates for each job have to be 'mutually agreed' on the shop floor where they are to be carried out and by the people who will perform the tasks. **1972** *Times* 27 Jan. 14/5 The agreement had that degree of mutuality to bring it within the set-off provisions of section 31. *Ibid.* 7 Nov. 19/3 The management has also conceded a large measure of 'mutuality', the arrangement under which shop floor representatives retain bargaining rights on many shop floor working conditions.

mutualization. (Further examples.)

1930 *Economist* 12 Apr. 825/2 It was created four years ago in connection with the 'mutualisation' of the company. **1951** *Ann. Reg. 1950* 5 It [*sc.* the party programme] was followed, two days later, by proposals for 'mutualization', rather than nationalization, of industrial assurance. **1955** *Times* 28 May 9/1 With the return of the Conservative Government measures of re-nationalization, of fresh nationalization, and of mutualization are no longer risks to be discounted, however vaguely, as contingencies just round the corner. **1960** *Times* 15 Jan. 17/2 Yesterday it [*sc.* the Canada Life Assurance Company] asked to be removed from the list because it had completed its 'mutualization' programme by purchasing all its own shares outstanding.

mutually, *adv.* Add: **1.** (Examples of *mutually exclusive*.)

1874 W. Wallace tr. *Hegel's Logic* ii. 30 Sense..is individual, and as the individual (which, reduced to its simplest terms, is the atom) is also a member of a series, sensible existence presents a number of mutually exclusive units. **1908** A. Bennett *Jrnl.* 5 Jan. (1932) I. 274 Every day's experience shows the folly of mutually-exclusive generalisations concerning two countries. **1969** D. C. Hague *Managerial Economics* vii. 141 To be drawn up correctly, our list of probabilities must be such that if any one event occurs, this automatically rules out the possibility that any other event in the same list could also occur. The events will then be mutually exclusive.

mutuel (mūtūĕl, miū·tiuĕl). Chiefly *N. Amer.* shortening of PARI MUTUEL in Dict. and Suppl. Freq. *attrib.* and in *pl.* Also *transf.* (see quot. 1949).

1908 *Westm. Gaz.* 12 Dec. 16/2 In France last year 320,000,000 francs passed through the 'mutuel'. **1926** *Daily Colonist* (Victoria, B.C.) 23 July 10/5 Returns from the 'mutuels' show that British Columbians are betting as much as ever, or more so, at the Vancouver races. **1938** D. Runyon *Furthermore* vi. 113 Then I see Herbie start for the mutuels windows. **1944** *Sun* (Baltimore) 21 Sept. 17/4 The mutuel machines will do close to a $400,000,000 business around New York tracks this season. **1949** *Amer. Speech* XXIV. 193/1 *Race mutuels*, a number game in which the winning number is derived from the racing results of any well-known race track. **1968** *Globe & Mail* (Toronto) 13 Jan. 40/8 The better class horses would attract more patrons who..would accelerate the mutuels play. **1972** *N.Y. Times* 1 June 56/1 The scramble had handled $659,439..meaning the scramble accounted for 8.45 per cent of the mutuel business. **1974** *New Yorker* 25 Feb. 86/2 Serious horseplayers will be sorry to hear..that the take from the mutuels will again be seventeen per cent.

‖ **mutum** (mū·tŭm). [Native name.] = CURASSOW.

1863 [see CURASSOW]. **1933** P. Fleming *Brazilian Adventure* i. xvi. 137 Sometimes we got a *mutum*, a big black and white turkey with a speckled crest. **1964** A. L. Thomson *New Dict. Birds* 175/2 The Razorbilled Curassow *Mitu mitu*, which is called 'mutum' by the Brazilians, is black with a bluish sheen in both sexes, with brown on the belly, and with a high narrow frontal comb of the same red colour as the bill; it weighs about 7¾ lb. (3500 grams) and inhabits the Amazon region and northern Brazil.

muu-muu (mū·mū). Also **mumu.** [Hawaiian *mu'u mu'u* lit. cut off, from the fact that the yoke was originally omitted.] A woman's loose-fitting dress, usually brightly coloured and patterned, which originated in Hawaii as a local adaptation of the 'Mother Hubbard' dress provided by the missionaries.

1923 C. London *New Hawaii* xv. 250 Not far off..swam a dozen men and women. They wore respectively loincloths and white or red muumuus. **1930** D. Blanding *Hula Moons* 40 Helen being a woman, put on a yellow *mumu*, a cross between a flour sack and an old-fashioned nightie. *Mumus* were designed by the well-intentioned missionary ladies..as a covering for the Hawaiian women, in the early days when a few flowers sufficed for a garment. **1959** M. Jansma *Wandering Malihini* ii. 33 A Hawaiian lady, costumed in an attractive *muumuu*,..sang to us songs of welcome and love. **1960** [see *HOLOKU]. **1962** O. Ruhen *Tangaroa's Godchild* xvi. 230 The missionary..invented the Mother Hubbard, from which the muu-muu is derived, to cover the glorious voluptuous bodies of the earth's most beautiful women. **1964** *Daily Mail* 21 Jan. 3/3 Mu-mus or Mother Hubbards describes the beachwear better. **1964** *Sunday Truth* (Brisbane) 19 July 21/7 The new cafe at Brisbane airport is to give a typical Australian welcome..by means of waiters wearing Hawaiian shirts, waitresses with muu-muus and a background aborigine motif. **1964** *Asia Mag.* 16 Aug. 5 (*caption*) A blonde Caucasian girl wears the long muumuu of the islands. **1966** 'O. Mills' *Enemies of Bride* viii. 76 It was Madeleine, tousle haired, in a short mu-mu with..Ella in a still shorter mu-mu. **1969** C. Armstrong *Seven Seats to Moon* xi. 105 Nanjo with..a blue muumuu concealing her bathing suit..came..and curled herself on the floor.

‖ **muvule** (muviū·li). Also **mvula, mvule, mvuli.** [a. Luganda *muvule*, Kiswahili *mvule*.] The East African name for *IROKO. Also *attrib.*

1911 *Encycl. Brit.* XXVI. 397/1 The largest timber tree [in Tanganyika] is the mvule, which attains vast dimensions, its trunk supplying the natives with the dug-out canoes with which they navigate the lake. **1940** W. J. Eggeling *Indigenous Trees of Uganda Protectorate* 131 Muvule is a tree of great importance to the timber trade of Uganda. **1947** *E. African Ann.* 1946–7 82/1 In the shade of an immense *mvuli* tree, an old man..is sitting. **1962** *Times* 9 Oct. (Uganda Suppl.) p. viii/4 Giant muvule trees (African teak). **1966** B. Kimenye *Kalasanda Revisited* 77 Hunched over the heavy, old-fashioned mvule desk, they set about answering the numerous personal questions. **1966** C. Sweeney *Scurrying Bush* ii. 31 Here and there a *Mvule* tree soared upwards.., the smooth, sheer trunks branchless for the first sixty or seventy feet. **1973** *Daily Colonist* (Victoria, B.C.) 2 May 1/7 Livingstone's heart was buried under a mvula tree in Chipundu.

muvver (mɒ·vəɹ). Representation of a Cockney or childish pronunciation of MOTHER *sb.*[1]

1888 Kipling *Wee Willie Winkie* 5, I don't fink I'll ever want to kiss big girls, nor no one, 'cept my muvver. **1898** J. D. Brayshaw *Slum Silhouettes* 118 A tanner for muvver, an' tuppence for me. **1899** [see *KIP *v.*[4]]. **1924** R. Macaulay *Orphan Island* viii. 79 Going shopping for my muvver. **1931** S. Jameson *Richer Dust* xviii. 516, I kep' the letter you wrote poor muvver after you was married. **1958** J. Townsend *Young Devils* x. 83 Andrews has gone to get 'is muvver's washing.

mux (mɒks), *v. dial.* and *U.S. local.* [Of obscure formation; cf. MUSS *v.*[2], MUCK *v.*, and dial. *mucksy* dirty.] = MUCK *v.* 4.

1806 *Balance* (Hudson, N.Y.) 26 Aug. 272 (Th.), To do observance, make obliging mention, Wink lovingly, mux chastity away. **1859** Bartlett *Dict. Amer.* (ed. 2) 287 To *mux* is much used in New England for *muss*; as, 'Don't *mux* my crinoline.' **1869** Blackmore *Lorna Doone* III. x. 157 By vice of mismanagement on the part of my mother, and Nicholas Snowe, who had thoroughly muxed up everything. **1877** J. M. Bailey *They all do It* 22 Stop muxin' that bread!..you've eaten enough for twenty people. I shan't have you muxing and gauming up the victuals. **1914** *Dialect Notes* IV. 77 *Mux*, to handle, paw over, maul. **1934** *West Virginia Rev.* Dec. 78/1 One may hear *muxed* in certain sections of West Virginia now and then... It is a term synonymous to *messed up*, rather than a form of mixed.

mux (mɒks), *sb. U.S. local.* [f. the vb.] A disordered or muddled state; = MUCK *sb.*[1] 4.

1848 in *Amer. Speech* (1935) X. 41/1 'In a mux.' Confused, disarranged. **1865** E. Stoddard *Two Men* iv. 28, I knew you would come back. Now we are in a mux. **1890** *Cent. Dict.*, *Mux*, work performed in an awkward or improper manner; a botch; a mess; as, he made a mux of it. **1910** *Dialect Notes* III. 454 *Mux*, confusion, 'all in a mux'.

Muzak (miū·zæk). [Cf. MUSIC *sb.*] The proprietary name of a system of piped music for factories, restaurants, supermarkets, etc.; also used loosely, with small initial, to designate recorded light background music generally. Also *attrib.* or as *adj.* Hence (nonce-wds.) **mu·zakal** *a.*, **mu·zakman.** Hence as *v. trans.*, to introduce Muzak to; to equip for the relaying of Muzak; to play in the style of Muzak; so **Mu·zaked** *ppl. a.*

1938 *Trade Marks Jrnl.* 11 May 572/1 Muzak... Instruments and apparatus and parts thereof..for use in the wireless transmission and reception of signals and pictures. Rediffusion Limited,..London,..engineers. **1946** *Ibid.* 27 Mar. 152/1 Muzak... Records of sound, light or electric signals; and instruments and apparatus and parts thereof..for the production of and reproduction from such records. Rediffusion Limited,..London,..manufacturers. **1954** *Official Gaz.* (U.S. Patent Office) 21 Sept. 497/2 Muzak Corporation..*Muzak*. For transmitting from central locations specially programmed background music to stores, restaurants, homes, hotels, banks, [etc.]. **1959** *Observer* 21 June 10/5 Several firms have already arranged to receive Muzak in the factories. *Ibid.*, Executives in the brave new muzakal world will have a new and special privilege: they alone will be able to switch it off. **1960** *Guardian* 9 May 5/7 Canned music, or Muzak..will to-day ooze into yet another corner of the world. **1965** *Listener* 17 June 898/2 We shall have muzak wherever we go. **1968** *Times* 29 Nov. p. xi/1 If muzak be the food of love, no wonder it is commonly to be found..among the frozen mint-flavoured peas and the crinkle-cut chips. *Ibid.* p. xi/2 In my sort of terms this is what makes it all sound the same, though that is heresy to a professional muzakman. **1969** *Listener* 8 May 661/1 Mifune..walks along a beach to the sound of a Bach chorale, muzak-ed. **1970** M. Moorcock *Chinese Agent* xiii. 87 He made the cab stop outside Hennekeys' recently repainted and muzaked pub. **1971** *Daily Tel.* (Colour Suppl.) 11 June 20/3 When subjects were doing a simple manual assembly task, the 'Muzaked' group's output was 17 per cent. up on a 'silent' group. **1971** *New Scientist* 17 June 707/2 The volume has to be kept down to Muzak level so as not to interfere with ..conversation. **1974** *N.Y. Times* 17 Aug. 17/2 'We needed a catchy name and the best known trade name at that time was Kodak,' explained U. V. (Bing) Muscio, current president. 'So we just combined Kodak and music and got Muzak.'

Muzbi, Muz(hu)bee, varr. *MAZHABI.

muzzle, *v.*[1] Add: **3. c.** *transf.* To close (a fishing-net).

1876 F. Francis *Bk. Angling* (ed. 4) xii. 438 The Esk..is a fine river, and would be finer if the Solway stake nets were only muzzled.

9. (Later example.)

1888 'R. Boldrewood' *Robbery under Arms* II. vi. 106 Jim and I jumped off and muzzled him.

muzzy, *a.* Add: **3.** *Comb.*, as **muzzy-headedness,** a fuddled or intoxicated condition.

1930 R. H. Mottram *Europa's Beast* vii. 169 Cocktails were sheer silliness, a short cut to muzzy-headedness.

mvula, mvule, mvuli, varr. *MUVULE.

Mwami (mwā·mi). [Native title, = chief.] The royal title of the former kings of Ruanda and Urundi in Africa.

On independence in 1962 Rwanda (formerly Ruanda) became a republic; Burundi (formerly Urundi) remained a monarchy until 1966 when the Mwami was deposed by a coup.

1890 H. M. Stanley *In Darkest Africa* x. 246 The chief was styled Mwami. **1923** *Glasgow Herald* 30 Jan. 12 The Mwami or Sultan of Ruanda..was assisted in his administration of the country by a state religion, and he was in a sense the personification of the deity Imana. **1959** *Times* 9 Nov. 8/6 The Mwami (Paramount Chief) of Ruanda sent a letter..expressing regret that he is unable to come to Brussels, where he was to have been received by the king, with the Mwami of Urundi, in private audience. **1970** A. T. Grove *Africa South of Sahara* (ed. 2) xi. 201/2 In Burundi the dead Mwami was sewed into the skin of a black bull for burial.

mwchin, var. *MOOCHIN.

Mweru, var. *MERU *sb.* and *a.*

myasthenia gravis (məi,ăsþī·niă græ·vis). *Med.* [mod.L., shortening of *myasthenia gravis pseudoparalytica* (F. Jolly 1895, in *Berliner klin. Wochenschr.* 7 Jan. 7/1), f. MYASTHENIA + L. *gravis* severe, grave.] A rare chronic disease, occurring chiefly in young adults, that is characterized by muscular weakness unaccompanied by atrophy, with temporary paresis following exertion, and is caused by a defect in the mechanism which converts a nervous impulse into a muscular contraction.

1900 Campbell & Bramwell in *Brain* XXIII. 278 The disease has received various names. The earlier cases were published as 'cases of bulbar paralysis without discoverable anatomical changes'... Strümpell introduced the term 'Asthenic bulbar palsy'... Other names which have been employed are 'general profound myasthenia', 'Erb's disease', the 'Erb–Goldflam' and the 'Hoppe-Goldflam-symptom-complex'. Jolly has proposed the term 'myasthenia gravis pseudo-paralytica', and this, or for short 'myasthenia gravis', appears to us the most suitable and convenient hitherto suggested. **1939** W. Haymaker tr. *R. Bing's Textbk. Nerv. Dis.* v. 182 Myasthenia gravis occasionally appears in more than one member of a family. **1961** *Lancet* 5 Aug. 281/2 The clinical diagnosis of myasthenia gravis is usually easy. **1970** *Toronto Daily Star* 24 Sept. 1/4 The twins have myasthenia gravis, a disease that killed their brother Richard last year when he was 6 months old. Excitement of any kind..can set off the choking, gasping attacks. **1974** Passmore & Robson *Compan. Med. Stud.* III. xxxiv. 31/2 Myasthenia gravis is also associated with an immunological disturbance.

Mycenæan, *a.* and *sb.* Add: **a.** *adj.* (Later examples.) **b.** *sb.* Also, the language used by the Greeks of the Mycenæan Age.

1930 Magoffin & Davis *Romance of Archaeol.* ii. 31 Mycenaean civilization could now be seen to be the mainland exotic of which Minoan island civilization had been the stem and the flower. **1956** Ventris & Chadwick *Documents in Mycenaean Gk.* i. iii. 74 There seems as yet to be little certain indication which dissociates Mycenaean from the Aeolic group. **1958** J. Chadwick *Decipherment of Linear B* v. 73 The name Mycenaean, originally a label for the culture of the Greek mainland in the Late Helladic period, is now generally extended to the Linear B script and the dialect it contains. **1965** *Language* XLI. 314 Antonio Tovar..comes out..against the Porzig–Risch thesis that Mycenaean represents a form of 'Southern Greek'. **1968** *Encycl. Brit.* XXII. 975/2 The dialect (conventionally known as Mycenaean) is the oldest Greek known and belongs to the Achaean subdivision. **1968** P. Martin tr. *Pallottino's Meaning of Archaeol.* 160 The people of the Mycenaean civilization..were neither more nor less than the forebears of the Greeks of history. **1971** L. A. Boger *Dict. World Pottery & Porcelain* 235/2 The style of painting on the early Mycenaean vases is almost exactly like Minoan Floral and Marine styles. **1972** M. Magnusson *Introducing Archaeol.* iii. 34 Schliemann was convinced he had found the graves of Homer's Mycenaean royal family, the graves of murdered Agamemnon and his companions.

myceto-. Add: **myceto·phagous** [-PHAGOUS] *a.* = FUNGIVOROUS *a.*

1906 J. B. Smith *Explanation of Terms used in Entomology* 86 Mycetophagous: feeding upon fungi. **1920** *Amer. Naturalist* LIV. 314 Many of these mycetophagous insects undoubtedly show a very close association with certain species of fungi. **1946** C. T. Brues *Insect Dietary* v. 193 Mycetophagous insects..are abundant. **1957** Snell & Dick *Gloss. Mycology* 101/1 Mycetophagous. Fungivorous; eating fungi. **1974** *Sci. Amer.* Apr. 128/3 Mycetophagous Breeding in the Australian Dung Beetle, *Onthophagus Dunningi*, G. F. Bornemissza.

mycetocyte (məisī·tosəit). *Zool.* [ad. G. *mycetocyt* (K. Šulc 1911, in *Sitzungsber. d. k. böhm. Ges. f. Wissensch.* (*Math.-naturw. Classe*) *Jahrg. 1910* iii. 10): see MYCETO- and -CYTE.] Any of the large cells found in some insects, either aggregated into a mycetome or not, which contain symbiotic micro-organisms, esp. yeasts.

1924 *Philippine Jrnl. Sci.* XXIV. 157 In this species there are no specialized 'mycetoms' or 'mycetocytes', as in the aphids. **1946** E. A. Steinhaus *Insect Microbiol.* iv. 198 The number of mycetocytes present in each individual aphid ranges from about 60 to 70. **1952** *Science* 25 Apr. 459/1 All aphids, except very old ones, contain numerous large cells in the body cavity (mycetocytes) containing many more or less spherical intracellular particles. **1973** W. S. Romoser *Sci. Entomol.* iii. 72 In many cases the microbes present are passed from generation to generation. In some insects in which this occurs specialized cells, mycetocytes, and the tissues composed of these cells, mycetomes, can be found associated with the gut, fat body, or, appropriately, the gonads.

mycetome (məi·sītōᵘm). *Zool.* Also **mycetom.** [ad. G. *mycetom* (K. Šulc 1911, in *Sitzungsber. d. k. böhm. Ges. f. Wissensch.* (*Math.-naturw. Classe*) *Jahrg. 1910* iii. 10): see MYCETO- and *-OME.] An organ in some insects consisting of an aggregation of mycetocytes.

1924 *Philippine Jrnl. Sci.* XXIV. 143 (*heading*) Studies on the embryogeny and postnatal development of the Aphididæ with special reference to the history of the 'symbiotic organ', or 'mycetom'. **1946** E. A. Steinhaus *Insect Microbiol.* iv. 190 In 1850 Leydig observed certain organs in aphids which have subsequently been called 'symbiotic organs', 'pseudovitelli', 'green bodies', 'bacteriotomes', or 'mycetomes'. **1946** *Nature* 30 Nov. 795/2 These insects would starve for lack of nitrogenous food were it not for mycetome symbiosis with *Azotobacter*. **1973** [see *MYCETOCYTE].

-mycin (məi·sin), *suffix. Pharmacol.* [f. Gr. μύκ-ης fungus + -IN[1].] Used to form the names of antibiotic compounds derived from fungi, usu. with part of the generic name or specific epithet of the fungus, as *ACTINOMYCIN, *ERYTHROMYCIN, *STREPTOMYCIN, etc.

myco-. Add: **mycomy·sticism,** mystical sensations induced by drugs extracted from mushrooms; so **mycomy·stical** *a.*; **my:cotoxico·sis** [-OSIS], a pathological condition caused by a mycotoxin; **mycoto·xin,** any toxic substance produced by a fungus.

1962 A. Huxley *Island* ix. 141 Neurotheology, metachemistry, mycomysticism. *Ibid.* 156 The local panel of medical and mycomystical experts. **1948** M. V. Gorlenko in *Amer. Rev. Soviet Med.* V. 164 The moulds..are ubiquitous and not particular as to the substrate upon which they grow. This circumstance, as well as the toxicity, suggests that stachybotriotoxicosis is not induced by *Stachybotrys alternans* alone, but may be caused also by other moulds. The disease may be appropriately called mycotoxicosis. **1957** *Rev. Med. & Vet. Mycol.* II. 502 Mycotoxicoses..are stated to be causing great losses among farm animals in the U.S.S.R. **1968** *Prog. Industr. Microbiol.* VII. 156 Alimentary toxic aleukia..appears to be the only mycotoxicosis other than ergotism to have directly affected major segments of human populations. **1962** *Adv. Vet. Sci.* VII. 335 If mycotoxins are indeed present in such grain, sufficient levels of toxin may be added to the mixed feed to be harmful to animals fed thereon. **1974** *Daily Tel.* 26 Mar. 16 The mycotoxins enter the food chain and accumulate in man through his use of milk, eggs, poultry or meat.

mycobacterium (məikobæktīə·riŏm). [mod. L. (Lehmann & Neumann *Atlas und Grundriss der Bakteriologie* (1896) II. 108), f. MYCO- + BACTERIUM.] A saprophytic or parasitic bacterium of the genus so called, which includes those causing tuberculosis, leprosy, and other diseases in man and other animals; also, a bacterium of the family Mycobacteriaceæ.

1909 E. R. Stitt *Pract. Bacteriol.* vii. 66 (*heading*) Study and identification of Bacteria. Mycobacteria and corynebacteria. **1949** H. W. Florey et al. *Antibiotics* I. iv. 211 The sensitivity to antibiotics of the slow-growing mycobacteria, notably various strains of the tubercle bacillus, has been studied. **1962** *Lancet* 1 Dec. 1153/1 Disease resembling tuberculosis..may be produced by mycobacteria other than tubercle bacilli. **1973** R. G. Krueger et al. *Introd. Microbiol.* iii. 62/2 Many types of mycobacteria occur as free-living organisms in soil, water, and dairy products... Some mycobacteria are parasites of animals in which they may cause disease, including tuberculosis.

mycobiont (məikobəi·ɒnt). *Bot.* [f. MYCO- + Gr. βιουντ-, pr. pple. stem of βιοῦν to live, f. βίος life.] The fungal component of a lichen; any fungus which is associated with an alga to form a lichen.

1957 G. D. Scott in *Nature* 2 Mar. 486/2 Three new terms are here proposed... They are:..(2) 'mycobiont', applicable to a fungus in association with an alga in the formation of a lichen. **1961** *Science* 10 Mar. 700/2 In our investigation, 11 pure cultures of mycobionts were tested for active compounds. **1969** *Ecology* L. 744/2 In the laboratory the lichen organization breaks down and either the phycobiont..or the mycobiont..overgrows the other partner. **1973** *Phytochemistry* XII. 2249 (*heading*) Sterols of the mycobiont and phycobiont isolated from the lichen *Xanthoria parietina*.

mycologist. (Later examples.)

1903 *Westm. Rev.* 30 Nov. 2/1 The Director..is a botanist; the Assistant-Director..is a mycologist. **1933** *Jrnl. R. Hort. Soc.* LVIII. ii. 266 Dr. Downson..resigned his appointment as mycologist to the Government of Tasmania. **1967** M. E. Hale *Biol. Lichens* ii. 33 The definition of the term stroma, however, has not yet been settled even among mycologists.

mycophage (məi·kofēidʒ). [f. MYCO- + *-PHAGE.] One who or that which eats fungi.

1958 *Times Lit. Suppl.* 19 Dec. 742/1 Why do the English and the Celts fear all but the cultivated mushroom, while Slavs and Italians are enthusiastic mycophages? **1965** B. E. Freeman tr. *Vandel's Biospeleology* xix. 328 It is generally impossible to classify a cavernicole as a humiphage, xylophage, mycophage, coprophage, or necrophage.

mycophagy. Add: (Later example.) **myco-phagist,** also, an animal that eats fungi; **myco·phagous** *a.* = *mycetophagous* adj. (s.v. *MYCETO-).

1919 *Trans. Brit. Mycol. Soc.* VI. 355 (*title*) The red squirrel of North America as a mycophagist. **1922** *Ibid.* VIII. 84 (*title*) Some observations on the mycophagous propensities of slugs. **1939** *Bull. Torrey Bot. Club* LXVI. 1 (*title*) The Snail *Polygyra thyroidus* as a mycophagist. *Ibid.* 5 *P. thyroidus* is decidedly mycophagous in regard to its food preferences. **1950** A. P. Kelley *Mycotrophy in Plants* xi. 151 As Frank continued to study mycorrhizae, he saw digestion of the mycelium and was led to develop another idea which may be termed 'mycophagy'. According to this concept, the 'fungus-eating plants' are able to draw their victim into the protoplasm,..and finally to digest it.

mycophilic (məikofi·lik), *a.* [f. MYCO- + *-PHILIC.] Fond of mushrooms; feeding upon mushrooms.

1972 *Times Lit. Suppl.* 25 Feb. 211/1 (Advt.), A favourite with mushroom-hunting enthusiasts and mycophilic epicures.

mycophobia (məikofōᵘ·biă). [f. MYCO- + -PHOBIA.] (See quot. 1962.) So **mycopho·bic** *a.*

1958 *Times Lit. Suppl.* 19 Dec. 742/2 The mycophobic hate was back again,..and everything which is not the cultivated mushroom is now banished as 'toadstools'. **1962** *Spectator* 22 June 834/2 Mycophobia, i.e. unreasoning fear of mushrooms.

mycoplasma (məikoplæ·zmă). *Biol.* Pl. **-plasmas, -plasmata.** [mod.L. (J. Nowak 1929 in *Ann. de l'Inst. Pasteur* XLIII. 1349): see MYCO- and PLASMA.] An individual belonging to the genus *Mycoplasma*, comprising a group of pleomorphic, Gram-negative organisms which are much smaller than, and lack the cell wall of, bacteria but, unlike viruses, are capable of growth in artificial media, and which in nature are nearly all animal parasites though only a few are proven pathogens; also called a pleuropneumonia-like organism or PPLO.

[**1955** *Internat. Bull. Bacteriol. Nomenclature & Taxonomy* V. 15 (*heading*) Mycoplasma peripneumoniae. Nowak 1929... The generic name *Mycoplasma* was validly published. The older generic names *Asterococcus*, *Coccobacillus*, and *Micromyces* are unavailable. As next in priority sequence *Mycoplasma* apparently is, therefore, the legitimate name of the genus... It should be noted that..the later *Borrelomyces* is more commonly accepted. However, unless some later name is conserved by international action, *Mycoplasma* probably should be recognized as correct.] **1960** *Ann. Rev. Microbiol.* XIV. 140 (*heading*) Nutrition, metabolism, and pathogenicity of mycoplasmas. **1963** *New Scientist* 25 Apr. 200/2 We know, too, of organisms (the mycoplasmata) which multiply inside cells. **1970** Ambrose & Easty *Cell Biol.* i. 7 The smallest free-living organisms are the mycoplasmas. These minute cells, the smallest of which are only 1,000 Å in diameter, have been isolated from soil and have been identified as the infective agent in various animal infections. **1971** *Nature* 22 Jan. 231/3 Mycoplasmas also share another important characteristic with bacteria and other cells; they are susceptible to infection by viruses. **1973** *Times* 14 July 1/2 In the United States the term 'viral pneumonia' usually means a lung infection with mycoplasma, an organism intermediate between bacteria and viruses.

So **mycopla·smal** *a.*, of, pertaining to, or caused by mycoplasmas.

1960 *Ann. Rev. Microbiol.* XIV. 154 Mycoplasmal infection of the trachea. **1972** *Science* 5 May 504/2 The structure of the mycoplasmal cell. **1974** *Nature* 31 May 422/1 No viral or mycoplasmal contamination was detected.

mycorrhiza (məikorəi·ză). *Bot.* Pl. **-æ** (-ī) or **-s.** Also **mycorhiza.** [a. mod.L. *mycorhiza* (A. B. Frank 1885, in *Ber. Deutsch. Bot. Ges.* III. 129), f. MYCO- + Gr. ῥίζα root.] A symbiotic or slightly pathogenic fungus growing in association with the roots of a plant. Also *attrib.* Hence **mycorrhi·zal** *a.*

1895 W. R. Fisher in *Schlich's Man. Forestry* IV. iii. ii. 376 This altered root with its matted coating of mycelium receives the name mycorhiza, and the fungus has the power of absorbing nutritive matter from the soil and conveying it to the roots of the host. **1898** [in Dict. s.v. Myco-]. **1900** Mycorhizal [in Dict. s.v. Myco-]. **1916** *Nature* 2 Nov. 172/1 The abhorrence of lime by the humus-loving rhododendrons appears to be intimately connected with the mycorrhiza, the symbiotic fungus which lives in association with the roots of the rhododendron and heath family. **1924** J. A. Thomson *Science Old & New* xxvii. 148 In many cases..the partner fungus or mycorhiza confines itself to the outside of the root, forming a dense, absorbing feltwork... The fungus absorbs water and salts and organic materials from the soil and passes these on to the tree; the benefit it gets in return is a supply of carbohydrates from the root. **1926** Tansley & Chipp *Study of Vegetation* ix. 158 Etymologically the word mycorrhiza refers to the fungus–root combination. *Ibid.*, Mycorrhizal fungi may be divided roughly into endotrophic and ectotrophic. **1927** *Forestry* I. 115 Mycorrhiza formation and the mycotrophic habit are of the first importance to trees growing in humus soils. **1934** *Ibid.* VIII. 97 Coralloid clusters of short thick rootlets (mycorrhizas) varying in colour with that of the associated mycelium. **1955** *Jrnl. Ecol.* XLIII. 408 The results of experiments performed with excised mycorrhizal roots may not necessarily be directly applied to mycorrhizas attached to parent trees. **1967** *Punch* 24 May 766/3 The orchid seedling, having no green leaves, depends for its food on the phenomenon of mycorrhiza, an association with a resident fungus that is less of a coming together for mutual benefit than an unending war wherein the opposing forces are so perfectly balanced there is no victor, no vanquished. **1967** M. E. Hale *Biol. Lichens* v. 69 Mycorrhizal fungi associated with the roots of trees. **1973** *Nature* 7 Dec. 366/2 The necessity (or not) of raising nursery seedlings under conditions conducive to mycorrhizal infection, are discussed.

mycotrophy (məikǫ·trŏfi). *Bot.* [a. G. *mykotrophie* (R. Falck 1923, in *Mykologische Untersuchungen und Berichte* II. 49), f. MYCO- + Gr. τροφή nourishment: see -Y[3].] The state of certain plants which have mycorrhizæ growing in association with their roots, possibly as an aid in the assimilation of nutrients. So **mycotro·phic** *a.*

1927 *Forestry* I. 115 Defining mycotrophy as the capacity to assimilate organic compounds through the agency of specific root fungi, Falck reviews the historical evidence bearing on the significance of mycorrhiza. **1927** [see prec.]. **1930** *Nature* 9 July 80/1 The mycotrophic habit has been shown to be of wide occurrence in many families of plants. **1959** J. L. Harley *Biol. Mycorrhiza* i. 6 The mycotrophy of the Ericales is described because it now appears to have something in common with the ectotrophic mycorrhizas of forest trees. *Ibid.* xi. 172 (*heading*) Other mycotrophic plants with septate endophytes.

myctophid (mi·ktŏfid), *sb.* and *a. Zool.* [f. mod.L. family name *Myctophidæ*, f. the generic name *Myctophum* (C. S. Rafinesque *Indice d'Ittiologia Siciliana* (1810) 56), f. Gr. μυκτήρ nostril + ὄφις snake: see -ID[3].] A small, marine, deep-water fish of the family Myctophidæ, distinguished by luminous organs along its sides; also called lantern-fish. Also as *adj.*, of or pertaining to this family of fishes.

1931 J. R. Norman *Hist. Fishes* viii. 151 In the Myctophids or Lantern-fishes (*Myctophidae*) the photophores are fewer in number. **1956** A. Hardy *Open Sea* xii. 233 There are a great many different species of myctophids, but most are very similar in appearance. *Ibid.* 245 (*caption*) A myctophid Lantern-fish *Myctophum punctatum*. **1971** *Nature* 29 Oct. 623/2 The myctophids, or lantern fishes..are among the most common and widely distributed deep-sea fishes known.

myelo-. Add: **myelocœle,** delete entry in Dict. and cf. *MYELOCELE, *MYELOCŒLE; **my:elomonocy·tic** *a.*, designating that form of leukæmia which is characterized by the presence in the circulating blood of myeloid cells of both the monocytic and the myelocytic series; **my:eloproli·ferative** *a.*, characterized by or pertaining to the proliferation of cells of or derived from the bone marrow; esp. in *myeloproliferative syndrome* (see quot. 1974).

1958 Dameshek & Gunz *Leukemia* viii. 161 (*heading*) 'Myelo-monocytic' leukemia. **1960** F. G. J. Hayhoe *Leukaemia* ii. 13 A mixed picture of early monocytic and myelocytic cells is seen, and this variety is called acute myelo-monocytic leukaemia or the Naegeli type of acute leukaemia. **1972** *Nature* 4 Feb. 274/2 High levels of urinary lysozyme activity were observed in acute monocytic and myelomonocytic leukaemia. **1951** W. Dameshek in *Blood* VI. 372 (*heading*) Some speculations on the myeloproliferative syndromes. **1962** *Lancet* 19 May 1044/2 The levels were generally normal in various other conditions—kidney disease,..acute or chronic liver disease, and neoplastic or myeloproliferative disease. **1974** R. P. Custer *Atlas Blood & Bone Marrow* (ed. 2) xix. 383/1 The term 'myeloproliferative syndrome' was suggested by Dameshek to include a heterogeneous group of disorders characterized by proliferation of any or all cell lines indigenous to the bone marrow at some time during the course of the disease.

myeloblast (məi·ĕloblɑst). *Anat.* [a. G. *myeloblast* (O. Naegeli 1900, in *Deutsche med. Wochenschr.* 3 May 289/1): see MYELO- and -BLAST.] Any of the immature cells (approximately 15 microns in diameter, with large nuclei and a small amount of densely staining cytoplasm) which are confined to the bone marrow, appearing in the circulating blood only in pathological states, and which, according to the so-called monophyletic (unitary) theory of hæmopoiesis, are the precursors of all other myeloid cells of the blood and bone marrow, or, according to the diphyletic (dualistic) or polyphyletic theories, are the precursors only of the myelocytes and of cells derived from them.

1904 F. P. Foster *Appleton's Med. Dict.* 1395/1 Myeloblast. **1909** R. J. M. Buchanan *Blood in Health & Dis.* viii. 143 Nägeli termed them 'myeloblasts'. **1911** W. K. Hunter *Rec. Adv. Hæmatol.* 30 Both the neutrophile and the eosinophile myelocyte are, according to Ehrlich, originally derived from a cell to which the name myeloblast has been given. **1961** *Lancet* 19 Aug. 434/2 The leucoerythroblastosis persisted, although granulocyte precursors, occasionally including myeloblasts, became more numerous than the nucleated red cells in the peripheral blood. **1968** Passmore & Robson *Compan. Med. Studies* I. xxvi. 11/2 After birth, the production of the granulocytic white cells takes place only in the red marrow... The first stage of development is the myeloblast.

So **myelobla·stic** *a.*, of, pertaining to, or involving myeloblasts; **my:eloblasto·sis** [-OSIS], the condition of having large numbers of myeloblasts in the bone marrow and circulating blood.

1916 L. F. Barker *Monographic Med.* III. 214 The cells of these myeloblastic proliferations. **1924** *Jrnl. Exper. Med.* XL. 845 (*heading*) Studies on the maturation of myeloblasts into myelocytes and on amitotic cell division in the peripheral blood in subacute myeloblastic leucemia. **1937** Kracke & Garver *Dis. Blood* v. 65 A true myeloblastosis, in which the myeloblast is the predominant cell type in both the bone marrow and peripheral blood, occurs only in the terminal, exacerbation stage of chronic myelosis and in acute myeloblastic leukemia. **1938** H. Downey *Handbk. Hematol.* III. xxv. 2014 They [*sc.* the cells] are clearly of the myeloblastic type. **1965** *New Scientist* 17 June 800/2 The three main leukaemias of poultry—lymphomatosis, erythroblastosis and myeloblastosis—are caused by filterable transmissible agents. **1974** Passmore & Robson *Compan. Med. Studies* III. xxi. 62/1 (*table*) Acute myeloblastic leukaemia.

myelocele (məi·ĕlosīl). *Path.* Also (erron.) **-cœle.** [f. MYELO- + Gr. κήλη tumour, *esp.* rupture, hernia.] **1. a.** = *MYELOMENINGOCELE. **b.** Spina bifida in which tissue of the spinal cord lies exposed over part of its length, without protrusion as a swelling; an area of neural tissue so exposed.

1875 Jones & Sieveking *Path. Anat.* (ed. 2) xii. 290 The protruded sac contains some portion of the spinal cord itself, forming the so-called myelocele. **1922** *Brain* XLV. 44 Myelocele... Here..segments of the medullary folds have remained open and the neural ectoderm lies exposed on the surface of the body as a ribbon of delicate tissue down the mid-dorsal line. **1924** J. A. Foote *Dis. New-Born* x. 140 Myelocele, or meningo-myelocele, is the name given to the tumor when the spinal cord itself is concerned in the embryonic fissure. **1952** J. E. Morison *Foetal & Neonatal Path.* xiv. 247 Especially when the tissues are infected a myelocele may be incorrectly described as a myelomeningocoele. **1963** [see *MYELOMENINGOCELE]. **1974** Passmore & Robson *Compan. Med. Stud.* III. xxxvi. 9/2 The [spinal] cord may be fully exposed, usually for a length of several segments without any covering from the dura mater, forming a myelocoele.

2. Var. *MYELOCŒLE.

myelocœle (məi·ĕlosīl). *Anat.* Also **-cele, -cœl.** [f. MYELO- + *-cœle* (f. Gr. κοιλία cavity of the body: cf. CŒLO-[1]).] **1.** The central canal of the spinal cord.

1885 B. G. Wilder in *N.Y. Med. Jrnl.* 21 Mar. 326/2 (*caption*) Schematic representation of the cavities of the brain and myelon... Encephalocoele. Myelocoele. Neurocoele. **1896** —— in *Jrnl. Compar. Neurol.* VI. 318 (*table*) Myelocele. **1940** *Chambers's Techn. Dict.* 566/1 *Myelocoel*, the central canal of the spinal cord. **1972** T. W. Jenkins *Funct. Mammalian Neuroanat.* ii. 15/1 The lumen of this division (metacoele) joins the myelocoele without any line of demarcation so that both form the fourth ventricle.

2. Var. *MYELOCELE.

myelocyte. Add: **2.** A cell generally confined to the bone marrow (appearing in the circulating blood only in pathological states) which is smaller than the myeloblast from which it derives, which when mature has neutrophil, eosinophil, and basophil cytoplasmic granules, and which is the precursor of the polymorphonuclear leucocyte of the circulating blood.

1891 *Johns Hopkins Hosp. Bull.* II. 87/1 These elements seem never to acquire the power of amoeboid movement which the polynuclear cells possess. They appear to arise in the marrow and have been called by Ehrlich 'myelocytes'. **1911** [see *MYELOBLAST]. **1962** *Gray's Anat.* (ed. 33) 65 Myeloblasts divide and give rise to smaller cells, the myelocytes. **1968** Passmore & Robson *Compan. Med. Studies* I. xxvi. 12/1 The myelocyte is a smaller cell with a round nucleus.

So **myelocy·tic** *a.*, of, pertaining to, or involving myelocytes.

1896 *Boston Med. & Surg. Jrnl.* 2 Jan. 6/1 The shape of the myelocytic nucleus..is generally to be contrasted with that of the 'polynuclear' cell. **1972** *Science* 20 Oct. 304/1 Tumor cells derived from patients with both acute lymphocytic and acute myelocytic leukemia.

myelogram (məi·ĕlogræm). *Med.* [f. MYELO- + -GRAM.] **1.** A radiograph obtained by myelography.

1937 *Arch. Neurol. & Psychiatry* (Chicago) XXXVIII. 1126 The method of injection of air and the roentgen technic are discussed and five air myelograms shown. **1942** *Surg., Gynecol. & Obstetr.* LXXV. 735/2 The myelogram..demonstrated a tumor which had invaded the upper thoracic subarachnoid space. **1974** *Daily Colonist* (Victoria, B.C.) 19 Sept. 2/1, I had a whiplash and am being told a myelogram is the only way to determine if I have a ruptured disc.

2. (A list of) the proportions of the various cells in a sample of bone marrow.

1940 *Biol. Abstr.* XIV. 1635/1 Sternal puncture may also aid in diagnosis through modifications of the myelogram. **1949** *Jrnl. Clin. Path.* II. 8/1 Most workers perform differential counts on marrow films, and by presenting the data in the form of a 'myelogram' express the incidence of the various cell types as percentages. **1956** E. PONDER tr. *M. Bessis's Cytol. Blood & Blood-Forming Organs* vii. 172 Some include megakaryocytes in the myelogram, while others do not.

myelography (məi,ĕlǫ·grăfi). *Med.* [f. MYELO- + -GRAPHY.] Radiography of the spinal chord after injection of a contrast medium (often air) into the subarachnoid space.

1937 *Arch. Neurol. & Psychiatry* (Chicago) XXXVIII. 1126 This report is based on 10 cases of lesions of the spinal canal in which the level of the block was found by air myelography. **1943** *Amer. Jrnl. Med. Sci.* CCVI. 691 If..a protruded intervertebral disk in the lumbar region is suspected air myelography should be used. **1969** N. A. LEWTAS in D. Sutton *Textbk. Radiol.* lxiii. 1134/2 Myelography of spinal tumours is best performed as a pre-operative examination.

Hence **myelogra·phic** *a.*, observed using myelography; **myelogra·phically** *adv.*, by myelography.

1940 H. K. PANCOAST et al. *Head & Neck in Roentgen Diagnosis* xii. 892 If there is a difference in the localization of tumors by neurologic and myelographic findings, the possibility of multiple tumors occurring within the spinal chord should be considered. **1950** *Med. Radiogr. & Photogr.* XXVI. 27 To demonstrate myelographically the lesions.., 6 cubic centimeters of *Pantopaque* were injected into the lumbar subarachnoid space. **1957** *Trans. Amer. Neurol. Assoc.* LXXXI. 171 (*heading*) Myelographically demonstrated cervical intervertebral discs, co-existing with tumors. **1969** N. A. LEWTAS in D. Sutton *Textbk. Radiol.* lxiii. 1142/1 There are four principal myelographic signs of disc prolapse.

myeloid, *a.* Add: Also, of, pertaining to, or involving myeloid cells. (Further examples.)

1927, 1961 [see *ERYTHROID *a.* 2]. **1966** WRIGHT & SYMMERS *Systemic Path.* I. iv. 179/2 Leukaemias may arise from any of the stem cells of the leucopoietic series, and the three most frequent types may be classified as myeloid, lymphatic and monocytic. **1971** *Brit. Med. Bull.* XXVII. 66/1 Leukaemia, in the acute and in the chronic myeloid forms, is the type of malignancy which is most frequently induced by whole-body exposures of several hundred rads.

myeloma (məi,ĕlōu·mă). *Path.* Pl. **-omas, -omata.** [f. MYEL(O- + *-OMA.] A tumour composed of bone-marrow cells (see quots.); *spec.* (*a*) (the tumour found in) myelomatosis; (*b*) a giant-cell sarcoma.

1857 MAYNE *Expos. Lex.* 736/1 *Myeloma*, term for a medullary tumour or enlargement. **1894** *Med. News* (Philadelphia) LXV. 239/1 (*heading*) Myeloma: report of a case. **1900** DORLAND *Med. Dict.* 419/2 *Myeloma*, (1) any medullary tumor; (2) giant-cell sarcoma; (3) a slow-growing tumor of a tendinous sheath containing myeloplaxes. **1902** *Encycl. Medica* XII. 449 Myelomata are rare tumours after the age of twenty-five. **1914** *Lancet* 28 Nov. 1236/2 This large class [*sc.* giant-cell sarcoma] is made up of two groups of cases: (1) the myeloid sarcomata, myelomata of some authors; and (2) the malignant giant-cell sarcomata. **1922** J. EWING *Neoplastic Dis.* (ed. 2) xviii. 256 Giant-cell sarcoma of tendon sheaths and aponeuroses.—The specific structure and benign clinical course of a group of giant-cell sarcomas of tendon sheaths of the hand and feet have long been recognized. They have been fully described, especially by French observers, under the terms 'myeloma' or 'xanthosarcoma' (Gross, Paquet, Reverdin, Heurteaux, Spiess, Lit.). **1948** R. A. WILLIS *Path. Tumours* xliii. 680 The changing views regarding the nature of this tumour [*sc.* osteoclastoma or giant-cell tumour of bone] have resulted in a diverse and confusing terminology—'myeloma', 'myeloid sarcoma', 'tumeur à myéloplaxes', 'benign giant-cell tumour', 'osteoclastoma', and 'chronic (non-suppurative) hemorrhagic osteomyelitis', being only some of the names applied to it. 'Myeloma' and 'myeloid' should be discarded, for the tumours are unrelated to the haemopoietic tumours of bone marrow, and in particular are quite distinct from the lesions of myelomatosis. **1966** WRIGHT & SYMMERS *Systemic Path.* I. iv. 185/2 Sometimes myelomas produce extramedullary masses in the viscera. **1967** *New Scientist* 4 May 276/3 The other source of information..about antibody structure has been the discovery of diseases, called myelomas, in which molecules, chemically similar to the immune globulins but without known antibody activity, are produced in the blood of patients suffering from a cancer-like proliferation of cells in the bone marrow and bloodstream. **1970** PASSMORE & ROBSON *Compan. Med. Stud.* II. xxviii. 4/2 An important tumour affecting the bone marrow is the plasma cell myeloma (myelomatosis or multiple myeloma). In this disease the bone marrow becomes packed with plasma cells.

myelomatosis (məi:ĕlomătōu·sis). *Path.* [f. *myelomat-* (taken as stem of *MYELOMA) + -OSIS.] A malignant proliferation of plasma cells causing numerous accumulations of them to form in the bone marrow and abnormal proteins to be present in the blood and urine.

1904 *Jrnl. Path. & Bacteriol.* IX. 173 At the necropsy the bone marrow of all the bones examined was found to be more or less affected by a diffuse sarcoma-like growth of rounded or polyhedral mononuclear cells—a form of 'multiple myeloma' or 'myelomatosis'. **1906** *Jrnl. Amer. Med. Assoc.* 16 June 1893/2 Hoffmann defines myelomatosis as the multiple development of malignant tumors in the bone marrow, originating in hyperplasia of one of the cellular elements of the mother soil [*sic*: ? *read* cell] (lymphocyte, myelocyte, plasma cell). **1961** *Lancet* 16 Sept. 639/2 Neither in myelomatosis in man nor in any of the plasmacytomas of mice..do we ever regularly observe a direct Coombs test or signs of hæmolytic anæmia. **1966** WRIGHT & SYMMERS *Systemic Path.* II. xxxix. 1419/2 Myelomatosis or multiple myeloma is a neoplastic condition originating from the bone marrow.

myelomeningocele (məi:ĕlomīni·ŋgosīl). *Path.* Also (erron.) **-cœle.** [f. MYELO- + *meningocele* s.v. MENINGO-.] Spina bifida in which tissue of the spinal cord and its investing membranes (the meninges) protrudes through the cleft, forming a rounded swelling of the skin usu. slightly above the base of the spine; the tissue so protruding. Also called *meningomyelocele*; cf. *MYELOCELE.

1889 *Buck's Handbk. Med. Sci.* VIII. 471/2 In myelomeningocele the external wall of the sac consists not of dura mater but of pia mater turned inside out. **1924, 1952** [see *MYELOCELE 1]. **1963** *Cecil-Loeb Textbk. Med.* (ed. 11) 1581/1 Myelocele involves a lesser defect of the meninges [than meningocele] with exposure of the spinal cord, but without cystic protrusion. Myelomeningocele is characterized by an incorporation of the spinal nerve roots and cord in the sac of a meningocele. **1974** PASSMORE & ROBSON *Compan. Med. Stud.* III. xxxvi. 9/2 There may be a developmental failure of the spinal cord and then the flat neural plate with its exposed central canal is liable to be involved with the sac and form a myelomeningocoele.

myelosis (məi,ĕlōu·sis). *Path.* Pl. **myeloses.** [f. MYEL(O- + -OSIS.] **1.** The formation of a tumour of the spinal cord.

1891 in *Syd. Soc. Lex.* **1900** in DORLAND *Med. Dict.*

2. The proliferation of blood-cell precursors in the bone marrow.

1916 L. F. BARKER *Monographic Med.* III. 213 (*heading*) Acute myeloid leukemia. (Acute leukemic myelosis.) **1922** V. ELLERMANN *Leucosis Fowls* i. 9 Schnidde's shorter and correcter words, myelosis and lymphadenosis, are now used in several text-books. **1952** *Blood* VII. 767 The term erythremic myelosis is used to indicate the parallelism with the leukemic myeloses. **1974** PASSMORE & ROBSON *Compan. Med. Stud.* III. xxi. 72/2 Pure erythraemic myelosis, di Guglielmo's syndrome, is rare, and is characterized by intense and bizarre proliferation of red cell precursors, which escape into the peripheral blood. *Ibid.* 73/1 Megakaryocytic myelosis is probably a variant of chronic myeloid leukaemia, as the clinical picture, course and prognosis are similar, and it usually terminates with definite evidence of myeloid leukaemia.

Mylar (məi·lāɹ). Also **mylar.** A proprietary name for a polyester which is the condensation product of ethylene glycol and terephthalic acid and is used in the form of films having high strength and heat resistance.

1954 *Trade Marks Jrnl.* 3 Nov. 1110/1 Mylar... Non-mouldable plastics in the form of films for use as a substitute for glass; and electrical insulating materials. E. I. Du Pont de Nemours and Company.., Wilmington, State of Delaware, United States of America; manufacturers. **1962** *New Scientist* 5 Apr. 796/2 A self-inflating balloon made of aluminized mylar was launched in August, 1960. **1966** *McGraw-Hill Encycl. Sci. & Technol.* VIII. 38/1 The tape used in magnetic recording consists of a plastic base with a coating of magnetic oxide. The base material of Mylar or cellulose acetate varies in thickness from 0·0007 to 0·0015 in. **1973** *Daily Colonist* (Victoria, B.C.) 16 May 1/3 Another possibility..is for astronauts to envelop the spaceship with a giant sheet of mylar, an insulating material.

Myleran (məi·ləræn). *Pharm.* A proprietary name for 1,4-di(methanesulphonyloxy)butane, $C_6H_{14}O_6S_2$, which is a cytotoxic agent with a selective action on bone marrow and is used in the treatment of myeloid leukæmia.

1952 *Trade Marks Jrnl.* 12 Nov. 1049/1 Myleran... The Wellcome Foundation Limited,..London,..manufacturing chemists. **1953** *Lancet* 31 Jan. 207/2 'Myleran'..showed an intense inhibitory effect on the growth of the Walker rat carcinoma 256. **1959** *Progr. Hematol.* II. 227 With Myleran,.. the margin between the therapeutic dose and the toxic dose, producing irreversible aplastic change in the bone marrow, is relatively narrow. **1968** *Internat. Pharmacol. Abstr.* V. 1509 A 26-year-old woman had to be treated with busulphan (Myleran), 2–8 mg. daily during the first to sixth and 19–30th weeks of pregnancy because of chronic myeloid leukemia.

mylonite. Add: Hence **my:lonitiza·tion, myloniza·tion,** the formation of mylonite; **my·lonitize** *v. trans.*, to convert into mylonite; **my·lonitized** *ppl. a.* = MYLONIZED *ppl. a.*

1910 *Trans. Edin. Geol. Soc.* IX. Spec. Part 'The Thrust-Masses in the Western District of the Dolomites' i. 18 The actual shear-plane is occupied by a finely mylonitized, greenish tuffoid material. *Ibid.* 19 The rock material immediately below the inturned ends is Porphyrite finely mylonitized into a tuff. **1913** B. N. PEACH et al. *Geol. Central Ross-shire* (Mem. Geol. Surv. Scot.) iii. 13 It would appear that the mylonisation of the gneiss is probably due to the passage over it of the great mass of rock brought forward by the Moine thrust-plane. *Ibid.* 112/2 (Index), Mylonitisation. **1926** H. H. READ et al. *Geol. Strath Oykell & Lower Loch Shin* (Mem. Geol. Surv. Scot.) iii. 24 The intrusion is truncated by thrust-planes, and its rocks show evidence of shattering and mylonisation. **1956** E. W. HEINRICH *Microsc. Petrogr.* vii. 187 Such rocks as gabbros and peridotites may also be mylonitized. **1970** *New Scientist* 7 May 275/2 The rocks are crushed, smashed and sheared into an extremely fine grained powder, in a process called mylonization. **1971** *Scott. Jrnl. Geol.* VII. 311 At the western edge of the Moine Nappe mylonitization is the earliest recognizable tectonic event..and involves Cambrian rocks.

myo-. Add: **myoche·mistry,** the chemistry of muscle; **myoepithe·lial** *a.*, (of an animal cell) having characters of both a muscular and an epithelial cell; so **myoepithe·lium,** a tissue composed of such cells; *spec.* the contractile cells outside the epithelium of some mammalian glands, e.g. in the breast; **myofi·lament,** any of the elongated threads, revealed by the electron microscope, which are arranged side by side in bundles to form a myofibril and of which there are two kinds in an ordered arrangement, viz. thick filaments composed of myosin molecules and thin filaments composed of actin molecules; also, one of the related filaments of smooth muscle; **myoge·nesis,** the formation of muscular tissue; **myohæmoglo·bin** *Biochem.* = *MYOGLOBIN; **myo-inositol** (usu. with *myo-* in italics) (see *INOSITOL); **myome·trium** *Anat.* [Gr. μήτρα womb], the muscular coat of the uterus, which forms the bulk of the wall of that organ and surrounds the endometrium; tissue from this muscular coat; so **myome·trial** *a.*; **my·oneme** *Biol.* [Gr. νῆμα thread], any of the minute contractile filaments found in the cytoplasm of many protozoa; **myoneu·ral** *a. Physiol.*, neuromuscular; having characteristics of both muscular and nervous tissue; **my·oplasm** *Anat.* [f. mod.L. *myoplasma* (coined in Ger. by P. Schiefferdecker 1905, in *Sitzungsber. d. niederrhein. Ges. f. Natur- und Heilkunde zu Bonn 1904* B. 90)], the cytoplasm or sarcoplasm of a muscle cell; hence **myopla·smic** *a.*; **myota·tic** *a. Physiol.* [Gr. τατικ-ός exerting tension (f. τείνειν to stretch)], applied to a muscular contraction (usu. a reflex) that occurs as a result of the stretching of the muscle; **myotonic** *a.*, delete entry in Dict. and see *MYOTONIC *a.*; **my·otube** *Biol.* [ad. Sp. *myotubo* (J. F. Tello 1917, in *Trabajos d. Lab. de Investig. biol. de la Univ. de Madrid* XV. 122)], a cylindrical cell that develops from myoblasts during the formation of a muscle fibre (see quots. 1960, 1972).

1962 *Lancet* 1 Dec. 1165/1, I still wonder if two hours' instruction in pathological myochemistry will help the candidate. **1968** *Nature* 2 Nov. 433/2 (*heading*) New myochemistry. **1881** F. M. BALFOUR *Compar. Embryol.* II. 550 In all the Coelenterata, except the Ctenophora, the contractile elements of the body wall consist of filiform processes of ectodermal or entodermal epithelial cells. The elements provided with these processes, which were first discovered by Kleinenberg, are known as myo-epithelial cells. **1904** *Nature* 3 Mar. 431/1 At certain stages complete continuity could be observed between motor nerve trunk and the protoplasmic body of the myoepithelial cell. **1973** R. P. GOULD in G. H. Bourne *Structure & Function Muscle* (ed. 2) II. iv. 192 These are known as myoepithelial, basal, or basket cells, and they are able to act like smooth muscle cells and so aid movement of the secretion into the excretory ducts. **1892** F. P. FOSTER *Med. Dict.* IV. 2367/1 Myo-epithelium. **1943** *Amer. Jrnl. Path.* XIX. 474 Myoepithelium can be demonstrated in the male breast of gynecomastia when the ducts are developed. **1961** L. MARTIN *Clinical Endocrinol.* (ed. 3) i. 16 The act of suckling appears to provoke a reflex secretion of oxytocin whereby the myoepithelium surrounding the alveoli of the breast is stimulated to contract and the milk ducts are simultaneously kept open. **1949** *Jrnl. Clin. Invest.* XXVIII. 770/1 (*heading*) Myofilaments and myofibrils of cardiac muscle. **1970** *Nature* 11 Apr. 180/1 Two to eight cells in each complete transverse section contain what appear to be sparse myofilaments and probably represent degenerate muscle cells. **1973** *Hewer's Textbk. Histol.* (ed. 9) x. 119 Although the fibrous proteins actin and myosin can both be isolated from smooth muscle fibres it is not yet clear what their exact relationship is to the longitudinally orientated myofilaments observed in these cells. **1891** *Syd. Soc. Lex.*, Myogenesis. **1921** *Amer. Jrnl. Physiol.* LVIII. 182 (*heading*) Tension of differential growth as a stimulus to myogenesis. **1956** *Biol. Bull.* CXI. 303 This relationship

prompted a re-examination of the situation in amphibian embryos, where it has been claimed that the notochord is essential to somitic myogenesis. **1974** *Nature* 10 May 106/1 In ways not anticipated, myogenesis has yoked in an uneasy alliance muscle biochemistry, molecular genetics and cell differentiation. **1924** Myohemoglobin [see *COPROPORPHYRIN]. **1934** M. BODANSKY *Introd. Physical Chem.* (ed. 3) xviii. 608 Muscle hemoglobin (myohemoglobin, myoglobin) has been isolated in crystalline form. **1953** J. HUNT *Ascent of Everest* 274 Myohaemoglobin is an oxygen-carrying pigment similar to haemoglobin. **1951** H. G. FLETCHER et al. in *Jrnl. Org. Chem.* XVI. 1241 It is suggested that, since the substance [*sc. meso*-inositol] was first discovered in muscle..and was at one time called 'muscle sugar', it is better called *myo*-inositol. **1965** *Adv. Biochem.* XXXIV. 85 As a degradation product of phytoglycolipid, a complex lipid found in a variety of seed phosphatides, 2-α-D-glucopyranosyl-*myo*-inositol was obtained. **1943** L. R. WHARTON *Gynecology* xxxvii. 542 The prognosis in myometrial sarcoma is poor. **1974** PASSMORE & ROBSON *Compan. Med. Stud.* III. xl. 16/2 Drugs used in labour should not depress myometrial contractility. **1900** DORLAND *Med. Dict.* 421/1 *Myometrium*, the muscular substance of the uterus. **1907** *Practitioner* Dec. 792 The myometrium contained many thick-walled blood-vessels. **1956** *Nature* 10 Mar. 478/2 Progesterone..prevents..œstrogen-induced growth of the myometrium. **1965** *Science* 1 Oct. 67/2 Samples of myometrium were obtained from the uteri of five adult females undergoing hysterectomies. **1901** G. N. CALKINS *Protozoa* ii. 38 The outside is covered by living membranes which may become complicated by the addition of muscular fibrils (myonemes). **1973** E. VIVIER in G. H. Bourne *Structure & Function Muscle* (ed. 2) II. iii. 182 The best characterized myonemes have chiefly been identified in ciliates, in which their structure recalls that of smooth muscle fibers of Metazoa cells. **1905** *Jrnl. Physiol.* XXXII. 436 But when plain muscle developes connection with sympathetic nerves it must at the myoneural junction acquire a mechanism that can receive the nervous impulse. **1960** G. H. BOURNE *Struct. & Function Muscle* I. ii. 48 It has been suggested that the iridial muscles are really myoneural elements rather than true smooth muscle cells. **1963** *Lancet* 19 Jan. 153/1 They also showed that the myoneural junction of a neonatal infant behaves in many respects like that of a patient with myasthenia gravis. **1907** *Alienist & Neurologist* XXVIII. 58 The author [*sc.* Schiefferdecker] suggests a number of new terms. The indifferent protoplasm becomes 'myoplasm' as soon as the cell is plainly recognizable as a muscle cell. **1933** *Physiol. Rev.* XIII. 302 Since the myofibrils represent differentiation products of the original myoplasm, it would seem in accord with a more precise terminology to speak of intra- and inter-fibrillar sarcoplasm; or, perhaps even better, the protoplasm of the fibrils (sarcostyles) might be designated sarcoplasm, that of the interfibrillar regions myoplasm. **1952** *Jrnl. Physiol.* CXVIII. 348 The aim..was to measure the myoplasm resistance and the membrane resistance and capacity in Purkinje fibres of the mammalian heart. **1975** *Nature* 10 Jan. 97/2 The action potential on the surface membrane depolarises the T-system within the fibre, which in turn triggers the release of Ca^{2+} into the myoplasm from its internal storage site, the sarcoplasmic reticulum. **1970** T. TOMITA in E. Bülbring et al. *Smooth Muscle* vii. 207 The myoplasmic resistance was measured by means of double micro-electrodes in one arm of a bridge circuit. **1975** *Nature* 10 Jan. 100/2 If the ionic current causing this presumed SR potential change was primarily carried by Ca^{2+} moving from one side of the SR to the other, it is interesting to note how much increase in total myoplasmic Ca^{2+} would correspond to a 135 mV potential change. **1881** W. R. GOWERS *Diagn. Dis. Spinal Cord* (ed. 2) ii. 29 It seems..desirable to discard the term 'tendon-reflex' altogether... They may be termed 'tendon-muscular phenomena', but the intervention of tendons is not necessary for their production; the one condition which all have in common is that passive tension is essential for their occurrence, and they may more accurately be termed myotatic contractions. **1924** LIDDELL & SHERRINGTON in *Proc. R. Soc.* B. XCVI. 240 Gowers, in 1881, proposed for the 'tendon-phenomena', then not commonly accepted as reflex, the term 'myotatic contractions'... His suggested adjective would suitably apply to the reflexes brought forward in this paper... Myotatic reflexes could embrace stretch-reflexes in general, including 'jerk' and 'clonus', which we regard as fractional forms of the complete and fully functional myotatic reflex. **1972** J. A. WILSON *Princ. Animal Physiol.* xi. 438/1 The peak of reflex tension in a myotatic reflex occurs at the time of completion of the stretch movement. *Ibid.* 438/2 When the foot of a spinal dog is raised so that the leg is flexed, the leg muscles respond with an extensor thrust—a myotatic reflex. **1933** M. FERNÁN-NÚÑEZ tr. *S. Ramón-Cajal's Histology* xvi. 280 The periphery of the cells exhibits a striated cortex, which goes on successively enlarging. In the axis resides a string of nuclei, as well as an undifferentiated protoplasmic cord (*D*); the ensemble finally represents tubes of contractile material filled with the protoplasm and nuclei (myotubes of Tello). **1960** L. PICKEN *Organization of Cells* vii. 295 By the fourth day of incubation, the multinucleate coenocytic cells (derived from the spindle-shaped, uninucleate myoblasts) have developed a peripheral layer of myofibrils, embedded in sarcoplasm, and a linear series of nuclei occupying the cell axis. These cylindrical cells are conveniently distinguished as a particular stage in the development of the muscle fibres, namely as 'myotubes'. From the seventh day after their appearance, the myotubes begin to multiply by longitudinal fission... The formation of a secondary is prepared by nuclear fission, leading to the formation of a second linear series of nuclei; round which new myofibrils form. **1972** D. A. FISCHMAN in G. H. Bourne *Structure & Function Muscle* (ed. 2) I. iii. 107 The myotube is here defined as the multinucleated syncitium which results from the cytoplasmic fusion of myoblasts. The term is purely descriptive, for it only implies an immature muscle fiber in which the myofibrils are, in general, circumferentially distributed within the cell, and the nuclei occupy the core or central zone of the syncitium. **1973** *Nature* 3 Aug. 253/3 From the fourth to the sixth day of incubation, motor nerve fibres invade the limb-bud muscles [of the chick embryo], and the first myotubes appear.

myocardiac (məi,okā·ɹdiæk), *a.* [f. MYO- + CARDIAC *a.*] Of or pertaining to the myocardium; myocardial.

1908 *Practitioner* Oct. 610 As the author says, myocardiac deficiency is the chief indication for the Nauheim treatment. **1940** A. HUXLEY *Let.* 7 July (1969) 454, I am still under the weather with..myocardiac weakness.

myoelectric (məi,o,īle·ktrik), *a.* Also (with hyphen) **myo-electric.** [f. MYO- + ELECTRIC *a.*] Applied to (apparatus or techniques using) the currents produced in the body which would normally cause muscular contraction and relaxation.

1955 *Jrnl. Bone & Joint Surg.* XXXVII-B. 506 (*heading*) The use of myo-electric currents in the operation of prostheses. **1963** *Aerospace Med.* XXXIV. 267/1 By attempting to move his arms, the pilot generates muscle action potentials, or myoelectric signals, which may be utilized as a control source. **1964** *New Scientist* 12 Mar. 668/2 A powered prosthetic device with myoelectric control. **1965** D. FRANCIS *Odds Against* xx. 246 The myo-electric arm..worked entirely by harnessing the tiny electric currents generated in one's own remaining muscles. **1972** *Science* 12 May 607/3 The design of myoelectric artificial aids for the physically handicapped.

Hence **myoele·ctrically** *adv.*

1964 *New Scientist* 12 Mar. 671/1 A Russian team..has introduced a myoelectrically controlled hand. **1970** *Ibid.* 5 Mar. 477/1 We constructed in 1966 a myoelectrically controlled hand and arm unit.

myofibril (məi,ofəi·bril). *Anat.* Also in mod. L. form **myofibrilla** (pl. **-fibrillæ**). [f. MYO- + FIBRIL, FIBRILLA.] Any of the elongated cylindrical threads, about one micrometre thick, which arranged side by side in a bundle constitute the contractile components of a striated muscle fibre.

1898 *Jrnl. R. Microsc. Soc.* 64 The muscle-cell produces conducting substance (myofibrils). **1913** DORLAND *Med. Dict.* (ed. 7) 605/1 *Myofibril, myofibrilla*, a muscle-fibril. **1928** *Biol. Abstr.* II. 804/1 The author compares his results on the origin and formation of myofibrillae in striated muscles with those of other investigators. **1933** *Physiol. Rev.* XIII. 318 Thus far it has been assumed that myofibrils are actual morphologic entities. **1960** G. H. BOURNE *Struct. & Function Muscle* I. ii. 34 Hogue (1937) demonstrated cross striated myofibrillae in living cultures of cardiac muscle cells. Microdissection of fresh muscle fibers as a means of demonstrating myofibrils is exceedingly difficult. **1968** *New Scientist* 11 Jan. 70/1 Normal skeletal muscle..is composed of parallel fibres averaging 70 micrometres in diameter. These are in turn made up from hundreds of fine myofibrils, one to two micrometres thick, running the whole length of the fibre. **1970** AMBROSE & EASTY *Cell Biol.* xi. 370 The myofibril is itself composed of fine filaments.

Hence **myofi·brillar** *a.*

1927 *Biol. Abstr.* I. 861/1 Cerebellar ataxia [is due] to loss of 'myofibrillar perceptions'. **1956** *Anatomical Rec.* CXXV. 483 Lacking the regular myofibrillar organization of striated muscles, these filaments appear to span the cell as a loose bundle. **1972** D. A. FISCHMAN in G. H. Bourne *Structure & Function Muscle* (ed. 2) I. iii. 117 Probably no aspect of muscle differentiation has elicited as much interest with biologists as the problem of myofibrillar biosynthesis and assembly.

myogen (məi·ŏdʒĕn). *Biochem.* [a. G. *myogen* (O. von Fürth 1895, in *Arch. f. exper. Path. und Pharm.* XXXVI. 274), f. *myo(sino)gen* MYOSINOGEN.] A mixture of albumins (varying between species) extracted from skeletal muscle plasma; crystalline substances obtained from such mixtures have been designated *myogen A, B, I, II*, etc.

1896 *Jrnl. Chem. Soc.* LXX. II. 48 The proteids in the muscle plasma are three in number, namely paramyosinogen.., myosinogen or myogen 77 to 83 per cent. of the total proteid, and traces of serum albumin. **1939** *Biochem. Jrnl.* XXXIII. 1342 Baranowski..has recently obtained from rabbit muscle extract a crystalline protein, which he has named myogen A. **1970** R. E. SCOPES in E. J. Briskey et al. *Physiol. & Biochem. of Muscle as Food* II. xxii. 472 It is now clear that..myogen A (when prepared from rabbit muscle) consisted of cocrystallized α-glycerophosphate dehydrogenase (α-GPDH) and aldolase... It has since been shown that pig myogen A is a different enzyme, lactate dehydrogenase. *Ibid.*, Myogens I and II from fish sarcoplasms..have been crystallised; ..neither is related to the myogen A of rabbit.

myoglobin (məi,ŏglōu·bin). *Biochem.* [a. G. *myoglobin* (H. Günther 1921, in *Virchows Arch. f. path. Anat. und Physiol.* CCXXX. 150), f. *myo-* MYO + *globin* *GLOBIN.] A red protein responsible for the transport of oxygen in muscle cells, which differs from hæmoglobin in containing only one hæm group and one peptide chain in its molecule (instead of four of each) and in having a much greater affinity for oxygen.

1925 *Proc. R. Soc.* B. XCVIII. 332 This difference does not justify, however, the introduction of the names 'myochrome' or 'myoglobin' for muscle hæmoglobin proposed respectively by Mörner and Günther. **1956** *New Biol.* XXI. 45 Muscles contain myoglobin, a compound related to haemoglobin which takes up oxygen released by haemoglobin..and stores it for use in time of oxygen shortage. **1958** *Manch. Guardian Weekly* 2 Oct. 12 J. C. Kendrew's molecular model of myoglobin, derived from the first successful X-ray analysis of a protein molecule ever to be made, aroused tremendous interest at the Congress [of Biochemistry at Vienna]. **1968** PASSMORE & ROBSON *Compan. Med. Stud.* I. xv. 11/2 Myoglobin can pick up and store in the cells the oxygen brought by the blood even when the circulation is too inadequate to maintain a high oxygen tension in the tissue fluid interposed between capillaries and muscle cells.

myoid (məi·oid), *a.* and *sb.* [f. MY(O- + -OID.] **A.** *adj.* Resembling muscle; composed of muscular tissue.

1857 DUNGLISON *Dict. Med. Sci.* (rev. ed.) 616/1 *Myoid*, an epithet given to tumours composed of fibre cells or muscular fibres of organic life. **1970** *Amer. Jrnl. Anat.* CXXIX. 399 (*heading*) Myoid elements in the mammalian nephron. *Ibid.* 399/1 This procedure revealed myoid features in the nephric tubules. **1973** *Anatomical Rec.* CLXXVII. 525 (*heading*) Myoid cells in the capsule of the adrenal gland and in monolayers derived from cultured adrenal capsules.

B. *sb.* [perh. a different word.] A structural part of the cones and rods of the retina.

1900 *Lancet* 7 July 7/2 The cell-body, traced from the cuticular end, begins as a distinct granular protoplasmic swelling, called the myoid. **1961** S. DUKE-ELDER *Syst. Ophthalmol.* II. iii. 241 In man and other mammals the inner segment of the rod is an elongated slightly barrel-shaped structure, homogeneous in appearance, made up of two elements, an outer ellipsoid and an inner myoid. **1972** *Vision Res.* XII. 1841 (*heading*) Optical function of myoids. *Ibid.*, The myoid is a cylindrical part of the photoreceptor connecting the region of the cell containing its nucleus with the region containing the ellipsoid.

myomectomy (məi,ome·ktŏmi). *Surg.* [f. MYOM(A + *-ECTOMY.] Excision of a myoma, esp. in the uterus.

1886 *Buck's Handbk. Med. Sci.* III. 811/2 The operation of myomectomy..is another product of the fertile mind of Schröder. It has been termed by him myomotomy, but for the sake of greater exactness the other term is to be preferred. **1900** *Lancet* 18 Aug. 501/1 Myomectomy should always be done when it was possible to save the uterus. **1908** *Practitioner* Oct. 608 Vaginal myomectomy. **1974** PASSMORE & ROBSON *Compan. Med. Stud.* III. xxviii. 46/1 Myomectomy is the treatment of choice in cases where it is desirable to preserve the uterus for future pregnancies.

‖ **myon** (myŏn). Also **myen.** [Korean.] A Korean administrative unit approximately equivalent to a rural district or township. Also *attrib.*

1898 I. L. BISHOP *Korea & her Neighbours* II. xxxii. 203 The country is now divided into districts (*Kun*), each *Kun* containing a number of *myen* or cantons, each of which includes a number of *ni* or villages. **1951** C. OSGOOD *Koreans & their Culture* ii. 20 The kun..comprises a series of myŏn, or townships, of which there are said to be currently fourteen in Kanghwa Kun. **1972** P. M. BARTZ *South Korea* 48/2 From provincial governor to rural *myon* chief and city mayor to *dong* head, officials are appointed, not elected. **1972** *Korea Herald* 17 Nov. 1/6 Detailed work on formulation of a plan to construct one factory in each of 770 selected eups and myons throughout the country..began yesterday.

myopic, *a.* (*sb.*) (Later *fig.* examples.)

1955 *Sci. Amer.* Oct. 103/1 Until Maupertuis's death in 1759 Voltaire did not relent in his flood of unmerciful, unscrupulous and myopic ridicule. **1972** G. DURRELL *Catch me a Colobus* v. 82 For some considerable time I had been endeavouring to persuade the BBC to film an animal-collecting trip, but they had been very myopic about the whole thing. **1975** *Times* 3 July 15/2 The Inner London Education Authority..appears to share this myopic vision of the troubles at North London Polytechnic.

myosis. Add: Also **miosis** (now the usual form). (Further examples.)

1890 BILLINGS *Med. Dict.* II. 160/2 *Miosis* (G[erman]), myosis. **1892** A. DUANE tr. *E. Fuchs's Text-bk. Ophthalm.* iv. 257 There is a series of alkaloids which produce either dilatation of the pupil (mydriasis) or its contraction (miosis). [*Note*] From μείωσις, contraction; hence miosis, and not myosis, as it is generally written (Hirschberg). **1910** *Trans. Ophthalm. Soc.* XXX. 163 Miosis on the right side was associated with dilatation of the left pupil. **1947** F. B. WALSH *Clinical Neuro-Ophthalmology* 201 Miosis has already been defined as indicating a pupil 2 mm. or less in diameter. **1964** W. B. SMITH *Allergy & Tissue Metabolism* vii. 78 The pharmacological effects of reserpine administration are: ..miosis (contraction of pupil of the eye), ptosis [etc.].

myotic, *a.* and *sb.* Add: Also **miotic** (now the usual form). (Further examples.)

1892 A. DUANE tr. *E. Fuchs's Text-bk. Ophthalm.* iv. 259 The action of miotics is of shorter duration than that of the mydriatics. **1918** *Jrnl. Amer. Med. Assoc.* 29 June 1992/2 There are contradictory opinions expressed on the effect of miotic drugs in cases of incomplete sphincter. **1946** *Nature* 28 Sept. 432/1 The work ..developed to include other problems, such as the mechanism..of lachrymation, of the action of myotics, etc. **1971** *Lancet* 6 Nov. 1040/2 The use of a miotic..should be regarded as mandatory whatever the mydriatic previously instilled.

myotonia (məi,ŏtōu·niă). *Path.* [mod.L., f. MYO- + Gr. τόν-ος TONE *sb.* + -IA[1].] **1.** The inability to relax voluntary muscle for a period

following its use; any condition characterized by this.

1898 *Jrnl. Nerv. & Mental Dis.* XXV. 509, I think it proper to assign all such cases to the one clinical category of myotonia. *Ibid.* 510 Cases which show symptoms of organic disease..can at once be excluded from the group of myotonias. **1905** *Ibid.* XXIX. 416 Myotonia appears to have been used also in the collective sense. Hochsinger describes as myotonia of infancy the persistent hypertonicity of the flexors, and diagnosticates it from tetany, but if myotonia is used as an inclusive term it is a question if it should not cover tetany itself, while if it is used in the narrow sense it has been preempted by Thomsen's disease. **1948** F. B. CARLSEN tr. *Thomasen's Myotonia* ii. 15 There is a definite difference in the ages at which the myotonia in Thomsen's disease..on the one hand, and dystrophia myotonica on the other, manifests itself. **1963** CAUGHEY & MYRIANTHOPOULOS *Dystrophia Myotonica* p. x, Her father was a London professional man who had myotonia of the grip. **1970** MASTERS & JOHNSON *Human Sexual Inadequacy* viii. 220 Specific evidence has been accumulated of the incidence of both myotonia and vasocongestion in the female's pelvis as she responds physiologically to sex-tension elevation.

2. In the (usu. mod.L) names of specific diseases in which myotonia is a prominent feature, as: **a. myotonia congenita** [A. Strümpell 1881, in *Berliner klin. Wochenschr.* 28 Feb. 121/2], **congenital myotonia** (*rare*), a rare hereditary disease manifested soon after birth and characterized by myotonia but without muscular wasting or other symptoms. Also called THOMSEN'S DISEASE.

1886 *Brain* IX. 113 The eighth article..makes confusion worse confounded by using another term, viz. dysmyotonie congénitale, almost adopting Strumpell's term, congenital myotonia. **1887** *Jrnl. Nerv. & Mental Dis.* XIV. 129 (*heading*) Thomsen's disease (myotonia congenita). **1963** CAUGHEY & MYRIANTHOPOULOS *Dystrophia Myotonica* xviii. 216 Myotonia congenita..was described in detail in 1876 by Dr. Asmus Julius Thomsen who was himself affected and in whose family it had occurred for five generations. **1974** J. T. HUGHES *Path. Muscle* iv. 76 In contrast to myotonia congenita and paramyotonia congenita, dystrophia myotonica is frequently diagnosed.

b. myotonia atrophica [tr. F. *myotonie atrophique* (G. Rossolimo 1902, in *Nouvelle Iconographie de la Salpêtrière* XV. 63)], a hereditary condition characterized esp. by myotonia, muscular wasting, cataracts, frontal baldness, and gonadal atrophy. Also called *dystrophia myotonica*.

1908 *Jrnl. Nerv. & Mental Dis.* XXXV. 269 Myotonia atrophica is an extremely rare affection. **1963** CAUGHEY & MYRIANTHOPOULOS *Dystrophia Myotonica* p. viii, Dystrophia myotonica has been known as myotonia atrophica, Steinert's disease, dystrophia myotonica or myotonic dystrophy. **1974** J. N. WALTON *Disorders Voluntary Muscle* (ed. 3) xv. 595 Dystrophia myotonica (myotonia atrophica) was described by Steinert (1909) and Batten and Gibb (1909).

myotonic, *a.* Substitute for entry s.v. MYO-: **myotonic** (məi,ŏtǫ·nik), *a. Path.* [f. MYO- + TONIC *a.*] Producing, exhibiting, or characteristic of myotonia.

1887 A. DE WATTEVILLE tr. *H. von Ziemssen's Handbk. Therap.* VI. 213 The Myotonic Electrical Reaction. **1887** *Jrnl. Nerv. & Mental Dis.* XIV. 129 Those [cases] which, although showing the principal symptoms, the 'myotonic disorder of the muscles', still..cannot be considered as true cases of Thomsen's disease. **1887** *Lancet* 14 May 973/2 'Myotonic contractions', which has also been suggested, has the advantage of avoiding prejudice of the question whether the condition is always congenital. **1909** *Brain* XXXII. 190 The muscles did not give the myotonic reaction as described by Erb in Thomsen's disease. **1948** F. B. CARLSEN tr. *Thomasen's Myotonia* ii. 18 Myotonic patients may often play the piano without difficulty. **1963** CAUGHEY & MYRIANTHOPOULOS *Dystrophia Myotonica* ii. 20 Brown and Harvey (1939) examined myotonic goats electromyographically.

myrcene (mə̄·ɹsīn). *Chem.* [f. mod.L. *Myrc-ia*, name of a genus of tropical trees and shrubs + -ENE.] 2-Methyl-6-methylene-2,7-octadiene, $C_{10}H_{16}$, a liquid terpene found in bay, hop, and other essential oils.

1895 POWER & KLEBER in *Pharmaceut. Rundschau* XIII. 61/1 As this body is not identical with any of the known terpenes, we propose for it the name *myrcene*. **1926** H. FINNEMORE *Essential Oils* lviii. 528 When 1 part of myrcene was heated..with 3 parts of glacial acetic acid and 1/50 part of 50 per cent sulphuric acid, an oil of lavender-like odour was obtained. **1953** I. A. PREECE *Biochem. Brewing* ix. 254 The myrcene content of a hop sample might well prove to be an outstanding factor in determining the storage characteristics of that sample. **1969** *Daily Tel.* 16 Sept. 21/2 The disaster tank..had contained myrcene, an oil closely allied to turpentine, which in certain circumstances can turn into a sticky residue with a low flashpoint.

myriad, *sb.* and *a.* Add: **B.** *adj.* **3.** *Comb.*: parasynthetic, *myriad-accomplished, -islanded -jewelled, -limbed, -minded* (later examples), *-mirrored, -voiced, -wrinkled*; advb., *myriad-flaking, -times, -tinkling, -wise.*

1909 'MARK TWAIN' *Is Shakes. Dead?* 67 The man who wrote the plays was not merely myriad-minded, but also myriad-accomplished. **1957** R. CAMPBELL *Coll. Poems* II. 106 The Heliades, The myriad-flaking snowstorm of whose boughs Is never still. **1922** JOYCE *Ulysses* 48 Tides, myriadislanded, within her. **1909** E. POUND *Personae* 43 Unless it were to make the halo round each one Appear more myriad-jewelled marvellous. **1923** D. H. LAWRENCE *Birds, Beasts & Flowers* 41 Nude fig-tree... Rather like an octopus, but strange and sweet-myriad-limbed octopus. **1916** BLUNDEN *Harbingers* 12 And wed me with the myriad-minded man. **1968** *Jrnl. Mus. Acad. Madras* XXXIX. 9 Tyagaraja can truly be said to be myriad-minded. **1921** W. DE LA MARE *Veil* 52 Their myriad-mirrored eyes Great day reflect. **1944** BLUNDEN *Shells by Stream* 26 We may glide Over a myriad-times extended sea And land of life abundant. *a* **1918** W. OWEN *Coll. Poems* (1963) 127 The myriad-tinkling flocks. **1883** J. G. WHITTIER *What Traveller Said* in *Bay of Seven Islands* 48, I dread the myriad-voiced strain. **1917** D. H. LAWRENCE *Look! We have come Through!* 120 Its oneness veers Out myriad-wise. **1859** TENNYSON *Idylls of King: Elaine* 156 Then came an old, dumb, myriad-wrinkled man. **1942** W. FAULKNER *Go down, Moses* iii. 75 Then he saw the myriad-wrinkled face.

myriapodal (miriæ·pŏdăl), *a.* [f. MYRIAPODA + -AL.] = MYRIAPOD *a.*

1893 *Rep. U.S. Nat. Mus. 1892* 258 Comparatively little can ever be known concerning the probably great abundance of Insect, Arachnid, and Myriapodal life of former geological time.

myrmeco-. Add: **myrmecologist** (examples); **myrmecophile** (later examples); **myrmecophilism** = *myrmecophily*.

1901 W. M. WHEELER in *Amer. Naturalist* XXXV. 432 It thus becomes necessary to review much that is well known to the myrmecologist. **1972** *Sci. Amer.* Sept. 193/1 Forty-five years ago the American myrmecologist William Morton Wheeler wrote a volume with a very similar title, the fruit of his lifetime study of ants. **1927** H. ST. J. K. DONISTHORPE *Guests of British Ants* p. xv, A large number of creatures do manage to live in or near ants' nests. Such creatures when associated with, and not merely present with ants, are known as myrmecophiles. **1959** E. F. LINSSEN *Beetles Brit. Is.* I. 47 Certain beetles are known as myrmecophiles (ant-lovers) from their association with ants. The name myrmecophilism has been given to this curious habit.

myrmekite (mə̄·ɹmĭkəit). *Petrogr.* [ad. G. *myrmekit* (J. J. Sederholm 1897, in *Bull. de la Commission géol. de la Finlande* I. VI. 113) f. Gr. μυρμηκ-ιά anthill, wart + *-it* -ITE[1].] An intergrowth of plagioclase with drops or worm-like forms of quartz.

1916 *Bull. de la Commission géol. de la Finlande* IX. XLVIII. 1 The intergrowth of plagioclase and 'vermicular' quartz..I have called *myrmekite*. *Ibid.*, I prefer to write *myrmekite*, as analogous to poikilite etc., instead of myrmecite. **1926** H. H. READ et al. *Geol. Strath Oykell & Lower Loch Shin* (Mem. Geol. Surv. Scot.) vi. 154 In slice ..the pinkish medium-grained rock is seen to be composed of quartz, oligoclase, and myrmekite. **1967** *Mineral Mag.* XXXVI. 491 Myrmekite and the other vermicular textures all appear to lack any specific crystallographic or lattice co-relationship between the vermicular and host mineral.

Hence **myrmeki·tic** *a.*, of the nature of or containing myrmekite; **my:rmekitiza·tion,** conversion into myrmekite; a myrmekitic state.

1916 *Bull. de la Commission géol. de la Finlande* IX. XLVIII. 120 Myrmekitic intergrowths may occasionally originate like the minerals crystallising in a molten magma, not as crystalloblastic minerals. *Ibid.* 118 The myrmekitization and the formation of the biotite are here obviously independent phenomena. **1944** *Q. Jrnl. Geol. Soc.* XCIX. 118 The plagioclase is sericitized and its larger plates show myrmekitic margins. **1964** *Amer. Mineral.* XLIX. 50 The myrmekite type I..is formed by the myrmekitization of plagioclase already in the rock. **1972** *Mineral. Mag.* XXXVIII. 573 Microprobe and optical measurements have shown the myrmekitic plagioclase of such intergrowths to be very similar compositionally to non-myrmekitic plagioclase.

myrrhy, *a.* Add: Also *Comb.*, as *myrrhy-threaded*.

1865 G. M. HOPKINS *Poems* (1967) 21 For sackcloth and frieze And the ever-fretting shirt of punishment Give myrrhy-threaded golden folds of ease.

myrtle, *sb.* Add: **3.** (Earlier example.)

1872 [see *HUNTER 5 d].

5. myrtle green (later example); **myrtle-of-the-river,** a large evergreen shrub, *Calyptranthes zuzygium*, of the family Myrtaceæ, native to Jamaica and bearing panicles of white flowers; **myrtle warbler** = *myrtle bird*.

1925 E., O., & S. SITWELL *Poor Young People* 2 The colours most in favour are marine Blue Louise, gris bois, grenate, myrtle green. **1919** *Ann. Rep. Board of Regents Smithsonian Inst. 1917* 384 The myrtle-of-the-river..with opposite glossy leaves and clusters of fruit resembling blueberries. **1924** J. A. THOMSON *Science Old & New* v. 27 Even the names transport us into a land of pure delight—the paradise tree, the myrtle-of-the-river, the marlberry, and the bois-fidèle. **1892** B. TORREY *Footpath-Way* 95 Not so was it with the myrtle warblers. **1947** *Proc. Iowa Acad. Sci.* LIV. 379 Myrtle Warbler..is often a late migrant, for a warbler. **1963** R. D. SYMONS *Many Trails* ii. 17 Small birds that at first glance look like myrtle warblers.

myrtly (mə̄·ɹtli), *a.* [f. MYRTLE *sb.* + -Y[1].] Containing myrtles or redolent of myrtle.

1882 G. F. ARMSTRONG *Garland fr. Greece* 135 Every brake And myrtly jungle seemed to undulate With motions of strange beings.

myself. Add: **5.** Phr. *says I to myself*.

1720, 1825 [see SAY *v.*[1] B. 3 b ¶]. **1920** J. LEE *Penitent in Northern Numbers* 81 As I lay in the trenches at Noove Chapelle,..Sez I to mysel', sez I to mysel':—Billy, me boy, here's the end o' you.

Mysian (mi·siăn), *sb.* and *a.* [f. L. *Mysia*, Gr. Μυσία Mysia + -AN.] **A.** *sb.* **a.** A native or inhabitant of ancient Mysia in north-west Asia Minor. **b.** The language of Mysia. **B.** *adj.* Of or pertaining to ancient Mysia, its inhabitants or its language.

1555 [see *BULGARIAN *a.* and *sb.*]. **1601** HOLLAND tr. *Pliny's Nat. Hist.* v. xxx. 110 Teuthrania, which the Mysians in old time held. **1834** J. S. MILL in *Monthly Repos.* VIII. 837 The most despised of all foreign nations. Witness the phrase Μυσῶν λεία, the spoil of the Mysians, applied to any people so poor in spirit, that even the unwarlike Mysians could plunder them with impunity. **1844** A. W. KINGLAKE *Eothen* iii. 54, I saw, and acknowledged the snowy crown of the Mysian Olympus! **1884** *Encycl. Brit.* XVII. 122/2 Ancient writers all agree in describing the Mysians as a distinct people,..though they never appear in history as an independent nation. *Ibid.* 123/1 The only relic of the Mysian language is a very short inscription,..supposed to be in the Mysian dialect. *a* **1936** A. E. HOUSMAN *Coll. Poems* (1939) 216 Up the Mysian entry wending, Lydians, Lydians, what is yon? **1939** L. H. GRAY *Foundations of Language* 383 Of *Mysian* we may have a scanty inscription of the fourth or third century B.C., consisting of five lines. **1948** D. DIRINGER *Alphabet* 466 Some scholars mention a Mysian, a Cilician and a Cappadocian alphabet. **1954** PEI & GAYNOR *Dict. Linguistics* 142 *Mysian*, an extinct language, once spoken in Asia Minor... A language of undetermined linguistic affinities, classified as Asianic. **1972** W. B. LOCKWOOD *Panorama Indo-Europ. Languages* 174 In Asia Minor, Mysian is said to have been a living language in the sixth century. **1974** *Encycl. Brit. Micropædia* VII. 151/3 The Mysians..were mentioned by Homer as primitive allies of the Trojans.

mysid (məi·sid). [f. mod.L. family name *Mysidæ*, f. the generic name MYSIS (P. A. Latreille in C. S. Sonnini *Buffon's Hist. Nat. Insectes & Crustacés* (1802) III. 36): see -ID[3].] A small shrimp-like crustacean of the family Mysidæ or the suborder Mysidacea; an opossum-shrimp. Also *attrib.*

1941 STEINBECK & RICKETTS *Sea of Cortez* xv. 152 That night we rigged a lamp..and hung it close down to the water... Pelagic isopods and mysids immediately swarmed to the illuminated circle. **1956** A. HARDY *Open Sea* xiii. 254 Many ostracods and copepods, and at least one mysid, are known to flash brightly. **1961** *New Scientist* 15 June 661/1 Small marine crustaceans, called mysids, would orientate themselves preferentially at right angles to the direction of polarization of vertical light. **1965** B. E. FREEMAN tr. *Vandel's Biospeleology* ix. 117 The mysids are mainly marine Crustacea. **1969** A. WHEELER *Fishes Brit. Is. & N.-W. Europe* 133/2 The 'whitebait herring', i.e. those in their first year,..eat crustacea, the copepod *Calanus*, the larvae of acorn barnacles, mysid shrimps, and the eggs and larvae of decapods and amphipods.

Mysoline (məi·sŏlīn). *Pharm.* Also **mysoline.** A proprietary name for primidone.

1949 *Trade Marks Jrnl.* 12 Oct. 901/2 Mysoline... Pharmaceutical substances for human and veterinary use; sanitary substances and disinfectants;..Imperial Chemical (Pharmaceuticals) Limited,..Slough, Buckinghamshire, England; manufacturers and merchants. **1958** J. H. BURN *Lect. Notes Pharmacol.* (ed. 5) 52 Primidone (Mysoline) is the most effective agent for the form of epilepsy known as 'grand mal'. **1966** *Economist* 30 Apr. 512 Other drugs that ICI hope will attract a large market are..the anti-epileptic Mysoline and the long-established anti-malarial Paludrine. **1973** *Clin. Chim. Acta* XLIV. 383 Both mysoline and phenytoin subjects show a more severely raised GGTP level than those treated with phenobarbitone.

Mysore (məi·sōəɹ). The former name of a state in southern India (now Karnataka), used to designate coffee produced there.

1907 *Yesterday's Shopping* (1969) 2 The Mocha and Mysore mixed in equal quantities are recommended for best Coffee. *c* **1938** *Fortnum & Mason Catal.* 7/2 Coffees. ..'Mysore'. **1951** W. H. BRINDLEY in W. G. Copsey *Modern Grocer* (ed. 5) II. viii. 97 The best quality and most widely known coffees from India are Mysores... Mysores are now full-bodied in cup, but often a little rough and dry on the palate. **1965** L. DEIGHTON *Action Cook Bk.* 132/2 Here are some good combinations: Costa Rica with Tanganyika (Chagga); Mocha mixed with Mysore.

Mysorean (məisŏrī·ăn), *sb.* and *a.* **A.** *sb.* A native or inhabitant of Mysore. **B.** *adj.* Of or pertaining to Mysore.

1871 C. M. YONGE *Pioneers & Founders* iii. 61 The Mysoreans complained that the English promises had not been kept. **1932** *Times Lit. Suppl.* 3 Mar. 159/4 A photograph of Mysorean industrial life. **1968** *Jrnl. Mus. Acad. Madras* XXXIX. 74 Kangari Radhakrishnaiya..is believed to have been a Mysorean.

mystagoguery (mi:stăgǫ·gəri). [f. MYSTAGOGU(E + -ERY.] = MYSTAGOGY (in trivial or *transf.* sense).

1927 *Observer* 17 July 6 The mystagoguery of weak human beings. **1963** *Times Lit. Suppl.* 4 Jan. 4/5 But mystagoguery like this will not earn for Indian art the friends it so much needs. **1966** *New Statesman* 2 Dec. 838/1 It works out as practically a capitulation..to the literary mystagoguery of France and the United States.

mysterioso (mistiəriōu·so), *a.* [ad. It. *misterioso* adj., mysterious.] Of music: executed in a mysterious manner. Also *transf.* as *sb.*

1953 in H. E. Vizetelly *New Internat. Year Book 1952* 619 *Mysterioso*, a motion-picture, radio or television program which depends upon the element of mystery for its effect. **1957** S. DANCE in S. Traill *Concerning Jazz* 44 In mysterioso numbers suitable for accompanying the more exotic dances in the night-clubs.., there was evoked what was considered a 'jungle' atmosphere. **1957** MANVELL & HUNTLEY *Technique Film Music* ii. 28 The theme is played twice on strings pizzicato, with suitable mysterioso embellishments. **1967** J. SEVERSON *Great Surfing* Gloss., *Mysterioso*, an 'adornment' or surf trick popular in the late fifties... The stance is with the surfer bending over, hiding his head in his hands.

mysterium (mistiə·riŏm). *Astr.* [f. MYSTER(IOUS *a.* + -IUM.] A hypothetical substance to which a galactic radio emission at 1665 megahertz was attributed until it was identified as an exceptionally strong component of a set of four lines emitted by the hydroxyl radical.

1965 H. WEAVER et al. in *Nature* 2 Oct. 30/1 There is no known identification of the strong emission line at 1,665 Mc/s... We shall speak of this unidentified line as arising from 'mysterium'. *Ibid.* 30/2 Our observations indicate that 'mysterium' is found in strong H II regions, but not in all strong H II regions. **1967** *Science* 25 Aug. 885/1 Mysterium has now been identified as OH. **1974** G. L. VERSCHUUR *Invisible Universe* ix. 77 How do we now know that the mysterium signals are in fact OH lines?

‖ **mysterium tremendum** (mistiə·riŏm treme·ndŏm). [L., = tremendous mystery.] A term used to express the overwhelming awe and sense of unknowable mystery felt by those to whom this aspect of God or of being is revealed.

1923 J. W. HARVEY tr. *R. Otto's Idea of Holy* iv. 12 We shall find we are dealing with something for which there is only one appropriate expression, *mysterium tremendum*. *Ibid.* v. 25 We gave to the object to which the numinous consciousness is directed the name 'mysterium tremendum'. **1957** J. S. HUXLEY *Relig. without Revelation* (rev. ed.) ii. 32 The *mysterium tremendum* of religion. **1963** A. HUXLEY *Let.* 17 Feb. (1969) 949 One..opens oneself up receptively to the *Mysterium tremendum et fascinans* within and without. **1968** *Times* 15 Oct. 11/4 Should we not beware of the mysterium tremendum, the Wrath of God. **1971** *Nature* 13 Aug. 450/2 Their interrelationship [*sc.* that of biology and physical-chemical research] seems to me to represent one of the deepest questions man can ask, a real *mysterium tremendum*.

mystery[1]. Add: **12. d.** Shortened form of *bag of mystery* (*BAG *sb.* 17 b). *slang.* **e.** A girl newly arrived in a town or city; a girl with no fixed address; a young or inexperienced prostitute. *slang.* **f.** A mystery story.

d. *a* **1890** G. HORNCASTLE in Barrère & Leland *Dict. Slang* (1890) II. 79/2 The peelers I scorn and defy, While strings of these *mysteries* I wave round my head, And then to the people I cry, 'Sassidges, oh, sassidges! Oh, beef and pork and German!'

e. 1937 J. WORBY *Other Half* 278 *Mystery*, a girl who is down and out, come to town to look for a job. **1955** C. H. ROLPH *Women of Streets* x. 120 When you're a new girl they call you a 'mystery'. And you're a mystery until you've been here three or four years. Then you become a 'history'. **1960** *Observer* 28 Feb. 23/4 Many teddys, tearaways and mysteries (drifting girls) are put off by the typical orthodox youth club. **1960** C. MACINNES *Mr Love & Justice* 19 'All those men. Maybe two or three a day...' 'Two or three? Are you kidding? What you take me for—a mystery?' **1962** R. COOK *Crust on its Uppers* (1964) iv. 37 'I Saw a Human Monster in My Bedroom, says Teenager'..means the little mystery's woken up when she wasn't supposed to. **1967** M. M. GLATT et al. *Drug Scene* 117 *Mystery*, girl (young) having left home on arrival in London. **1974** G. F. NEWMAN *Price* v. 169 Instead of calling a couple of mysteries, he called a cab.

f. 1973 *Directory of Dealers in Secondhand & Antiquarian Bks. in Brit. Isles 1973–75* 111 Fantasy, scientific romance, Gothic novel, mystery and detective.

13. *mystery-mongering* adj. and sb.; *mystery-piety*; **mystery-bag** = *bag of mystery* (*BAG *sb.* 17 b); **mystery-man,** (*b*) a man about whom little is known; **mystery-religion** (see quot. 1967); **mystery ship,** an armed and camouflaged merchantman used in the war of 1914–18 as a decoy or to destroy submarines; **mystery story,** a detective or crime story; **mystery tour,** a pleasure trip for which there is no advance announcement of the places to be visited; **mystery train,** a train taking passengers on a mystery tour; **mystery trip** = *mystery tour*; **mystery woman,** a woman about whom little is known; **mystery writer,** a writer of mystery stories.

1889 *Sportsman* 2 Feb. 4/1 But the 'mystery-bags' of Sieur X, if we are to believe the common report, were far from being fragrant. This gentleman has been sentenced to six months' imprisonment for 'making sausages of tainted meat'. **1910** *Encycl. Brit.* III. 570/1 'Adventurer', as applied to Disraeli, was a mere term of abuse. 'Mystery-man' had much of the same intention, but in a blameless though not in a happy sense it was true of him to the end of his days. **1933** H. G. WELLS *Shape of Things to Come* I. 112 That Mystery Man of Mystery Men, Sir Basil Zaharoff, the armaments salesman. **1972** G. LYALL *Blame Dead* ii. 10 You sounded a bit of a mystery man in those stories this morning—they'll want to know more. **1901** W. JAMES *Mem. & Stud.* (1911) vii. 150 With all these things, infected by their previous mystery-mongering discoverers, even our best friends had rather avoid complicity. **1912** MRS. R. DAVIDS *Buddhism* i. 20 There is no evidence..that this late recourse to writing was due to any mystery-mongering or esotericism. **1939** DYLAN THOMAS *Let.* 29 Sept. (1966) 240 Censorship and conscription, mystery-mongering and umbrella-worship. **1939** P. S. WATSON tr. *Nygren's Agape & Eros* II. II. iii. 355 Neoplatonism..to a large extent bears the stamp of Mystery-piety. **1913** H. A. A. KENNEDY *St. Paul & Mystery-Religions* III. 69 There are special strains of religious thought and feeling more or less common to all the Mystery-Religions, such as that of regeneration (in some sense) and union or communion with deity. *Ibid.* 86 This connection with Dionysus leads us into the heart of conceptions typical for mystery-religion, the conception of union with the Divine and attainment of undying life. **1925** S. ANGUS *Mystery-Religions & Christianity* ii. 45 A Mystery-Religion was (I) a religion of symbolism which, through myth and allegory, iconic representations, blazing lights and dense darkness, liturgies and sacramental acts, and suggestion..provoked in the initiate a mystical experience conducing to *palingenesia* (regeneration), the object of every initiation. *Ibid.* 50 II. A Mystery-Religion was a religion of Redemption which professed to remove estrangement between man and God. *Ibid.* 52 III. The Mystery-Religions were systems of *Gnosis* akin, and forming a stage to, those movements to which the name of Gnosticism became attached. **1967** D. T. KAUFFMAN *Dict. Relig. Terms* 322/1 *Mystery religions*, secret cults in pre-Christian Greek and Roman culture, as well as in areas of Egypt and Asia... The mystery religions included the Orphic, Eleusinian, Mithraic. **1914** *Daily Mail* (Greater Manchester ed.) 7/4 The grey, gaunt outline of the mystery ship took definite shape. **1925** Mystery ship [see *DECOY *sb.*[2] 6]. **1975** B. MEYRICK *Behind Light* xvi. 207 The crew of Mystery Ship 51 listened as the warning boom of their foghorn echoed..through..the Dover Straits. **1908** CHESTERTON *All Things Considered* 115 Mystery stories are very popular, especially when sold at sixpence. **1932** H. CRANE *Let.* 20 Mar. (1965) 404 Even the suspense of the usual mystery story utilizes that device. **1934** Mystery story [see *crime-story]. **1974** A. PRICE *Other Paths* I. vii. 86 You've been holding on to the book... I can't wait to hear your mystery story. **1947** J. BETJEMAN in *Strand Mag.* Aug. 41 The morning paddle, then the mystery tour By motor-coach inland this afternoon. **1973** C. BONINGTON *Next Horizon* vii. 105 Climbing with Tom Patey was a kind of Magical Mystery Tour, in which no one, except perhaps himself, knew what was coming next. **1933** H. A. PIEHLER *England for Everyman* 35 Recent enterprises include 'mystery' trains for hikers, bound for unknown destinations, and circular tours by special trains through beautiful scenery. **1958** *Listener* 23 Oct. 653/2 An enterprising char-à-banc proprietor advertised 'mystery trips'. **1913** R. C. PRAED (*title*) The mystery woman. **1922** M. ARLEN *Piracy* 321 The *Daily Mail* at once called her a 'mystery woman'. **1974** 'A. GILBERT' *Nice Little Killing* ix. 127 It was something she couldn't afford to be made public... Mrs Brown was the original mystery woman. **1942** *Amer. Speech* XVII. 3 Most of this [*sc.* slang and cant] I have excluded,..because it is already rather thoroughly recorded in the special dictionaries, out of which, no doubt, the mystery-writers took it in the first place. **1973** *N.Y. Times* 1 Aug. 37/1 Good mystery writers have always known that man himself is the greatest mystery of all.

mystery[2]. **2.** (Later examples.)

1889 'MARK TWAIN' *Conn. Yank.* xxxii. 367 A good blacksmith..[offered]..to..teach him the trade—or 'mystery', as Dowley called it. **1957** *Listener* 25 July 141/1 We usually start with some sort of prejudice against the verse-writer who is better known as a writer of prose: there is a (very proper) feeling that the two are different mysteries.

mystic, *a.* and *sb.* Add: Hence **my·sticness.**

1912 F. LAWRENCE *Let.* in A. Huxley *Lett. D. H. Lawrence* (1932) 75 Her weird mysticness throws a veil over her.

mysticismus: see *-ISMUS.

mystico-. Add: *mystico-humanitarian, -oriental, -religious* (earlier example), adjs.

1932 R. CAMPBELL *Taurine Provence* 53 Our modern mystico-humanitarian scientists who are merely barbarians. **1893** A. BEARDSLEY *Let.* *c* June (1971) 50, I am going to do a full page illustration of the wonderful and gorgeous *Song of Songs*, in mystico-Oriental style. **1834** J. S. MILL tr. *Plato's Phædrus* in *Monthly Repos.* VIII. 636 There are two sorts of madness; one coming from human disease, the other from a divine influence. This last we divided into four kinds: viz., prophetic inspiration ..; mystico-religious, (τελεστική,) to Bacchus; poetic, to the Muses; and finally, that of which we are speaking, the inspiration or enthusiasm of Love.

mystificatory, *a.* (Later example.)

1927 C. E. MONTAGUE *Right off Map* viii. 68 This mystificatory drug had been working to some effect on the people of Ria's City.

mystify, *v.*[2] Add: Hence **my·stifyingly** *adv.*

1934 in WEBSTER. **1953** in Botkin & Harlow *Treas. Railroad Folklore* 508 The Delaware and Hudson came tooling up from Pennsylvania with what it mystifyingly called 'Standard gauge; 6 foot and 4 foot, 3 inches'. **1966** *English Studies* XLVII. 19 A conjunction which appears mystifyingly vexing. **1973** *Daily Tel.* 2 Mar. 15/2 *The Wrath of God* (Empire Two, 'X') would seem to parody that novel, mystifyingly, I would say, to most people.

mystique (mistī·k). [F. *mystique*: see MYSTIC *a.* and *sb.*] The atmosphere of mystery and veneration investing some doctrines, arts, professions, or personages; any professional skill or technique which is designed to mystify and impress the layman.

1891 E. DOWSON *Let.* 10 May (1967) 197 Its curious mixture of French technique & the mystique of Rossetti. **1940** *Economist* 14 Sept. 330/1 It is assumed that the only policy which can counter the Hitlerian *mystique* is that of 'European Revolution' which has a *mystique* of its own. **1940** *Time* 21 Oct. 94/2 In the '30s Ernest Hemingway expounded the *mystique* of bullfighting in *Death in the Afternoon*. **1943** H. READ *Politics of Unpolitical* ii. 27 But I do not claim that the principle of equality is a rational doctrine. On the contrary, it is an irrational dogma, a *mystique*. **1949** 'G. ORWELL' *Nineteen Eighty-Four* II. 211 All the beliefs, habits, tastes, emotions, mental attitudes that characterize our time are really designed to sustain the mystique of the Party. **1951** *Observer* 2 Dec. 4/6 The 'mystique' built up around him [*sc.* Stalin] has become a genuine outlet for the Russian religious instinct. **1951** *Sunday Times* 16 Dec. 4 True, the new style [of the Civil Servant] has its own mystique, and the Mr. Alphabet Precis today prides himself on knowing when to write 'Dear' and when 'My dear' [etc.]. **1952** M. MUGGERIDGE in A. Mayor tr. *Ciano's Diaries 1937–8* p. viii, The strange mixture of bombast, lies, cynicism and sincerity which furnished its [*sc.* the Fascist Regime's] mystique. **1952** A. POWELL *Buyer's Market* 141 Money, with its multifarious imagery and restrictive mystique. **1955** H. SPRING *These Lovers fled Away* 316 Talk to me about wine. To me it's something I drink... But to them it's a mystique. **1958** *Times* 25 Apr. 3/1 There is a mystique about violins, especially old violins. **1958** *Spectator* 20 June 813/1 His System has been..damned by lazy actors as a rarefied mystique. **1959** *Times Lit. Suppl.* 15 May 284/5 The particular virtue of this account is that it shows how the task was done and thereby disposes of its *mystique* without destroying the belief that it was a miracle. **1960** *Times* 7 Jan. 13/2 Flying, both the *mystique* and the technicalities of it, plays its part, too, in the Egyptian stories. **1972** *Guardian* 3 Nov. 12/4 The City, of course, talks its own language and this has helped to add to the mystique.

myth, *sb.* Add: **1. b.** (Further examples.) Also, an untrue or popular tale, a rumour. *colloq.*

1854 GEO. ELIOT *Let.* 23 Oct. (1954) II. 179 Of course many silly myths are already afloat about me, in addition to the truth, which of itself would be thought matter for scandal. **1939** J. S. HUXLEY *'Race' in Europe* 28 Napoleon, Shakespeare, Einstein, Galileo—a dozen great names spring to mind which in themselves should be enough to disperse the Nordic myth. The word *myth* is used advisedly, since this belief frequently plays a semi-religious role, as basis for a creed of passionate racialism. **1940** C. S. LEWIS *Problem of Pain* v. 64, I offer the following picture—a 'myth' in the Socratic sense, a not unlikely tale. **1941** H. G. WELLS *You can't be too Careful* v. i. 240 As the New Deal unfolded, American myth and reality began to take on an increasing parallelism with Europe. **1950** *Scot. Jrnl. Theol.* III. 37 To this inner fellowship of disciples the 'mystery' of the Kingdom of God is disclosed, whereas to outsiders this same Kingdom remains veiled in parables, remains, that is, a figure of speech, a colourful vision, an imaginative dream, or, as we might say, a myth. **1959** *Listener* 31 Dec. 1171/2 The theme of *Sacrilege in Malaya*..is that any institution of this kind needs some myth, that is some nonsense, to make it work. **1961** *Ibid.* 2 Nov. 739/2 Disraeli set himself to recreate a national political party out of the wreckage of Peel's following. A new myth had to be evolved. **1963** *Brit. Jrnl. Sociol.* XIV. 27 We use *myth* in a sense a little different from the popular one. To us it does not mean an untrue or impossible tale, but a tale which is told to justify some aspect of social order or of human experience. **1973** *Times* 13 Nov. 6/6 There is a myth going around that there are an awful lot of empty houses in Windsor Great Park. *Ibid.* 4 Dec. 7/4 Egypt's decision to sit at the table with Israel would 'shatter the myth' surrounding Israel's constant call for 'direct negotiations'.

3. *myth-maker* (later examples), *-making* vbl. sb. (examples), ppl. adj. (later example); *myth-addict, -addiction, -criticism, -monger, -pattern, -play, -removal, -stage, -system, -talk, -transcriber*; *myth-bound, -destroying, -haunted, -producing, -provoking* adjs.

1945 KOESTLER *Yogi & Commissar* II. i. 133 Almost every discussion with myth-addicts, whether public or private, is doomed to failure. **1954** —— *Invisible Writing* ii. 31 It does not matter by what name one calls this mental process—double-think, controlled schizophrenia, myth addiction, or semantic perversion. **1964** *Economist* 8 Aug. 551/2 Trying to educate the myth-bound Americans. **1957** N. FRYE *Anat. Criticism* 72 The most conspicuous today being fantastical learning, or myth criticism. **1949** KOESTLER *Insight & Outlook* x. 153 The only [myth] which his myth-destroying genius embodied into his system. **1940** G. BARKER *Lament & Triumph* 33

The Avalon haven I have in the grave Is now myth-haunted by God like Arthur. **1961** *Guardian* 22 Sept. 10/5 The myth-makers are always quick to produce a propaganda image of the Leader. **1972** *Listener* 10 Aug. 183/1 Why did Buonarroti, who had started life as a court page, become a professional revolutionary and myth-maker, to bore historians for the next 130 years? **1881** J. Royce *Let.* 28 Dec. in R. B. Perry *Tht. & Char. of W. James* (1935) I. 791 Ontology, whereby I mean any positive theory of an external reality as such, is of necessity myth-making. **1965** *Times Lit. Suppl.* 25 Nov. 1068/4 The myth-making gestures in her work. **1974** *Listener* 24 Jan. 111/3 There is..no causal connection between art and revolution..to suppose there is one is to take a step towards mythmaking. **1961** *Ibid.* 28 Sept. 479/3 They find their natural allies in the political myth-mongers and the political gangsters. **1951** M. McLuhan *Mech. Bride* 5/2 This urgent appetite to have the cake and eat it, too, is widely prevalent in the myth patterns.. of industrial society. **1957** N. Frye *Anat. Criticism* 282 The scriptural play is a form of a spectacular dramatic genre which we may provisionally call a 'myth-play'. **1954** Koestler *Invisible Writing* xxxvi. 390 An indication of the deep, myth-producing forces that were and still are at work. **1966** *Punch* 26 Jan. 139/1 Author explores the myth-provoking north-west coast of Spain. **1951** Myth-removal [see *demythologize *v.*]. **1950** *Scot. Jrnl. Theol.* III. 39 We have seen that..Christians are to..get beyond the myth-stage of spiritual understanding. **1953** A. K. C. Ottaway *Education & Society* 42 Every society is held together by a myth-system. **1970** *Jrnl. Ecumen. Studies* VII. 822/1 In this essay, Gilkey is the theologian who establishes guidelines for myth-talk. **1924** D. H. Lawrence in *N.Y. Times Mag.* 26 Oct. 3/2 White people always, or nearly always write sentimentally about the Indians—all of them, anthropologists, and myth-transcribers and all.

mythically, *adv.* (Earlier examples.)

1817 Coleridge *Biogr. Lit.* (1907) I. ix. 100 The philosopher who cannot utter the whole truth without conveying falsehood..is constrained to express himself either *mythically* or equivocally. **1846** Geo. Eliot tr. *Strauss's Life of Jesus* II. ii. ix. 425 The two narratives in the Old Testament are to be understood mythically.

mythify (mi·þifəi), *v. rare.* [f. Mythif(ication: see -ify.] To construct a myth or myths about.

1906 *Critic* Feb. 161/1 The truth is that no distinguished actor in modern history has been so recklessly mythified as the great diplomatist [*sc.* Talleyrand]. **1951** Auden *Nones* (1952) 50 We have time To misrepresent, excuse, deny, Mythify, use this event While, under a hotel bed, in prison, Down wrong turnings, its meaning Waits for our lives.

mythistory. (Later example.)

1972 *Times* 8 June 16/4 The reason for Eton's interest in Jane [Shore] is that, in the barnacle-encrustation of legend and mythistory that has grown up around her in the past 500 years, she is said to have used her influence with Edward to save Eton from destruction.

mythless (mi·þlės), *a.* [f. Myth *sb.* + -less.] Without a myth.

1924 C. K. Ogden tr. *Vaihinger's Philos. of 'As If'* 343 The myth..we have lost 'in the abstract character of our mythless existence'. **1936** *Times Lit. Suppl.* 14 Mar. 218/2 In Conrad he finds another writer with an heroic conception of life, but one that is austere, mythless, without veneration for the deep forces of Nature.

mytho-. Add: **my·thomane** = *mythomaniac* in Dict. and Suppl.; also *attrib.*; **mythoma·nia,** the condition or tendencies of a mythomaniac; **mythomaniac,** (*b*) one who has an abnormal or pathological tendency to lie or exaggerate; also as *adj.*; **mytho·nomy** (see quots.); **mythopœ·ia** = Mythopœism; **my:tho-theo·logy,** theology based on myth.

1954 *Encounter* Dec. 77/2 [Socialism is to be] treated as the way to that abolition of 'injustice' whose necessary existence in any human society these mythomanes find intolerable. **1959** *Ibid.* June 79/1 The mythomanes seized the new means of communication. **1962** *Punch* 26 Sept. 464/3 A mythomane tart with a line in imaginary family grandeur. **1975** *Times* 2 May 11/8 *Paper Tiger*..is a modest..entertainment, with David Niven as a pathological mythomane who..finds himself obliged to live up to his fantasies. **1909** *Cent. Dict.* Suppl., Mythomania. **1955** *Antiquity* XXIX. 197 Deception for the mere fun of deceiving—a sort of mythomania. **1958** *New Statesman* 6 Sept. 311/1 As for the formal principle on which the New Critics plumed themselves, it cannot be said that they ever applied it with any consistency or finesse; and lately it has been giving way to mythomania and symbol-hunting. **1973** *Times Lit. Suppl.* 30 Nov. 1476/1 It proves to have no resemblance to the sculpture, and the assertion is seen as a manifestation of mythomania. **1922** W. S. Maugham *Writer's Notebk.* (1949) 186 She is not only a liar, she is a mythomaniac who will invent malicious stories that have no foundation in fact. **1961** *Times* 15 June 17/3 Mary ends up sadder, if less mythomaniac, than she began. **1973** C. McCarry *Miernik Dossier* (1974) 90, I regarded Miernik as a mythomaniac... I did not believe in the existence of the sister. **1882** *Amer. Naturalist* Oct. 829 Mythography... Mythology... Mythonomy. **1890** *Cent. Dict.*, *Mythonomy*, the deductive and predictive stage of mythology. **1939** C. S. Lewis *Lett.* (1966) 163 We now need a new word for 'the science of the nature of myths'... Would 'mythonomy' do? **1959** H. Bloom *Shelley's Mythmaking* i. 8, I do not claim that *all* of Shelley's major and mature poems are mythopoeic, especially in the precise and narrow sense of mythopoeia that I insist upon here. **1970** *Listener* 30 July 154/2 Science is not immune to mythopoeia. **1927** J. S. Huxley *Relig. without Revelation* vi. 191 They possess a vague and elastic mytho-theology. **1932** R. Knox *Broadcast Minds* iv. 72 They had ancestors..to whom that mytho-theology was real.

mythogenic (miþodʒe·nik), *a.* [f. Mytho- + *-genic.] Myth-forming; of or pertaining to the creation of myths.

1964 *Economist* 8 Feb. 511/2 The mythogenic dragons of the golden age of steam. **1970** *Time* 5 Oct. 73 Religion has cut itself off from its 'principal sources of nourishment—the soul, the symbolic and mythogenic process, the psychic energy resources'.

mythographic (miþogræ·fik), *a.* [f. Mythography + -ic.] Of or pertaining to the representation of mythical subjects in art, literature, etc. So **mythogra·phical** *a.*

1939 Tillyard & Lewis *Personal Heresy* v. 120 Between Aristotle and the modern mythographical school of Miss Maud Bodkin, Professor Wilson Knight, and Professor D. G. Jame, we find almost nothing. **1955** R. Graves *Crowning Privilege* iii. 42 The death of Marvan's pig is a mythographic way of recording the murder of inspired poetry by a new-fangled academicism. **1965** *Listener* 2 Dec. 902/1 A mythographic fantasy *The Complaint of Nature*, diffuse and heady with virtuoso language. **1968** J. A. W. Bennett *Chaucer's Book of Fame* i. 17 The attraction of this mythographical tradition for Chaucer results in a Venus very different from the divinity he had found described in the *Roman de la Rose*.

mythologem (miþolǒu·dʒĕm). Also **mutholo-gema.** [ad. Gr. μυθολόγημα.] A mythical story; a fundamental theme or motif of myth.

1884 *Jrnl. Hellenic Stud.* V. 236 The apotheosis of Homer and his marriage with Hebè..do not properly belong to the Homeric era, but to the *mutholegema* of later times. **1939** L. Bloomfield in C. F. Hockett *Bloomfield Anthol.* (1970) 424 Any normal human being could analyze this mythologem of 'metrical convenience'. **1961** *Times Lit. Suppl.* 27 Jan. 56/3 Professor Kerényi.. repeatedly inquired of Mann his views on particular images and mythologems. **1973** G. Pocock *Corneille & Racine* 305 This passion is not merely sexual, but is linked more generally with what, to use a modern mythologem, we would call the unconscious.

mythology. Add: **3.** (Further *transf.* examples.)

1949 'G. Orwell' *Nineteen Eighty-Four* ii. 155 She only questioned the teachings of the Party when they in some way touched upon her own life. Often she was ready to accept the official mythology. **1961** *Listener* 24 Aug. 281/2 This is an antidote to militarism which does not exist in western 'mythology'. **1965** M. Schofield *Sexual Behaviour of Young People* i. i. 9 The frequent articles in the press and the radio and TV programmes tend to create a teenage mythology. **1975** *Times* 22 Sept. 13/2 Not all private [pension] schemes were in fact as generous as popular mythology suggested.

mythopoetic, *a.* Add: (Further examples.) So **mythopoe·tical** *a.*

1900 R. Fry in *Monthly Rev.* Dec. 152 In such periods the magnifying mythopoetical effect, which for us comes only with time, takes place at once, and swells their contemporaries to heroic proportions. **1959** B. Bernardi *Mugwe, a Failing Prophet* iii. 52 It constitutes a typical instance of mythopoetical amalgamation. **1965** *Philos. Rev.* LXXIV. 548 Wheelwright is engaged not in metaphysical but in mythopoetic investigation. **1970** J. O. Love (*title*) Worlds in consciousness: mythopoetic thought in the novels of Virginia Woolf. **1972** *Times Lit. Suppl.* 7 Apr. 381/2 This mythopoetic quality.

mythos. (Later examples.)

1946 'G. Orwell' in *Polemic* Jan. 6 The poisonous effect of the Russian *mythos* on English intellectual life. **1957** N. Frye *Anatomy of Criticism* 52 *Mythoi* or plot-formulas. **1974** *Sat. Rev. World* (U.S.) 2 Nov. 30/2 The same mythos of violence that was indigenous to..the American West.

Mytilenæan, Mytilenean, Mytilenian (mitilī·niăn, mitilenī·ăn), *sb.* and *a.* [f. L. *Mytilenæus*, Gr. Μυτιληναῖος Mytilenæan.] **A.** *sb.* A native or inhabitant of Mytilene, the ancient city and modern capital of the Aegean island of Lesbos. **B.** *adj.* Of or pertaining to Mytilene or its inhabitants.

The form Μυτιληναῖος (L. *Mitylenæus*) also is attested by ancient sources, and is sometimes followed by modern writers in the form *Mitylen(a)ean*. The name of the modern city is frequently transliterated *Mitilini*.

1550 T. Nicolls tr. *Thucydides' Hystory* iii. iii. fol. lxxii, The sayde Mytilenians. **1601** Holland tr. *Pliny's Nat. Hist.* v. xxx. 110 Acheleum..founded first by the Mityleneans. **1790** W. Mitford *Hist. Greece* II. xv. 227 The principal Mitylenæans had sent offers to the Lacedæmonian administration to renounce the Athenian. *Ibid.* 228 The whole Mitylenæan people would go in procession out of the city. **1900** J. B. Bury *Hist. Greece* x. 413 The Mytilenaeans received secret intelligence and postponed the feast. **1911** *Encycl. Brit.* XIX. 887/1 This base coinage ..ceases about 450 B.C., when the Mytilenaean silver begins. *Ibid.* XXI. 73/2 An assembly was held and under the invective of Cleon it was decided to kill all male Mytileneans of military age. **1945** G. B. Grundy *Fifty-Five Years at Oxford* xiv. 221 He could not have supposed that any reader of his history who read the Mytilenian Debate or the Melian Dialogue would regard the Funeral Oration as being a picture of a political..life..at Athens. **1965** T. T. B. Ryder *Koine Eirene* iv. 77 The Athenian people voted a decree of thanks to the Mytileneans for their services in the war against the Spartans. **1968** V. Ehrenberg *From Solon to Socrates* viii. 359 In the Mytilenean debate..Cleon attacks rhetoric. **1972** R. Meiggs *Athenian Empire* iii. 47 We may prefer the different motivation he attributes to the Mytilenaeans. *Ibid.* xvii. 312 A rich Mytilenaean left two daughters. **1974** *Guardian* 11 Nov. 4/6 Life is rough for the Mytilenians at present.

myxamœba (miksămī·bă). Pl. **-æ.** [f. Myxo- + Amœba.] In a slime mould of the division Myxomycota, a cell lacking flagella but capable of amœboid movement; = *myxo-amœba* (s.v. Myxo-). So **myxamœ·boid** *a.*

1887 H. E. F. Garnsey tr. *A. de Bary's Compar. Morphol. & Biol. Fungi* viii. 423 Swarm-cells with purely amoeboid motion have been unnecessarily distinguished by the name of myxamoebae. **1888** [in Dict. s.v. Myxo-]. **1927** Gwynne-Vaughan & Barnes *Structure & Development of Fungi* 45 In early stages of development they [*sc.* Myxomycetes] appear as small, naked uninucleate amoebae, the myxamoebae. *Ibid.*, Eventually multiplication ceases, the zoospores resume the myxamoeboid form and fuse in pairs. **1947** F. A. & F. T. Wolf *Fungi* I. iv. 40 In moist weather the spore walls open to emit swarm cells (myxamoebae), which ingest bacteria and fungus spores, assimilate them, and grow to become a large, multinucleate, naked mass of protoplasm (plasmodium). **1968** H. Harris *Nucleus & Cytoplasm* vi. 118 The experiments..demonstrate the failure of high concentrations of actinomycin D to inhibit decisive events in the differentiation of colonial myxamoebae. **1971** P. H. B. Talbot *Princ. Fungal Taxonomy* viii. 100 A myxamoeba may put out flagella and become a swarmer, or a swarmer may retract its flagella and become a myxamoeba.

myxo (mi·kso). Abbrev. of *myxomatosis.

1953 S. J. Baker *Australia Speaks* 105 Another popular Australian speech habit is to truncate a word and add the suffix *o*. Among words..which fall into this category ..are..*myxo*, the disease myxomatosis, used to kill off rabbits; [etc.]. **1961** Partridge *Dict. Slang* Suppl. 1193/2 Myxo. Myxomatosis: Australian: since ca. 1945... But, since ca. 1950, also current in Great Britain. **1967** J. Morrison in *Coast to Coast 1965–6* 135 The myxo'll look after the rabbits. **1970** M. Tarmey *Skinman* i. 22 They reckon myxo'll spread quite a bit this summer.

myxobacterium (miksŏbæktiə·riŭm). Also **myxobacter.** [ad. mod.L. *Myxobacter* (R. Thaxter 1892, in *Bot. Gaz.* XVII. 403), f. Myxo- + Bacterium.] A slime bacterium of the order Myxobacterales, which includes predominantly saprophytic bacteria having a vegetative state in which the unicellular rods are embedded in slime to produce thin, flat colonies, and forming spores, often in distinct fruiting bodies. So **myxobacte·rial** *a.*, of or pertaining to an organism of this kind.

1932 D. E. Johnson in *Jrnl. Bacteriol.* XXIV. 340 Other cultures had the characteristics of the vegetative stage of Myxobacteria. **1946** *Nature* 23 Nov. 745/1 The lytic effect of certain myxobacteria upon the true bacteria (eubacteria) has been known for some years. **1949** H. W. Florey et al. *Antibiotics* I. i. 40 Three other bacteria.. were inhibitory to *Ustilago*, a non-sporing rod, a motile sporing rod, and a myxobacterium. **1957** R. S. Breed et al. *Bergey's Man. Determinative Bacteriol.* (ed. 7) 854 In the vegetative condition, myxobacters consist of unicellular rods which occur in two characteristic shapes. *Ibid.* 855 The myxobacterial colony, also sometimes designated as a swarm or pseudoplasmodium, consists characteristically of a flat, thin mass of vegetative cells which spreads rapidly. **1961** *New Scientist* 16 Mar. 669/1 The predators [in the soil] include also myxo-bacteria, giant rhizopods, acrasieae and others. **1973** R. G. Krueger et al. *Introd. Microbiol.* iii. 60/1 Myxobacteria grow as individual cells that are typically embedded in a mass of slime they produce during growth. *Ibid.* 60/2 In soil and in dung, myxobacteria are active in the degradation of complex organic materials.

myxococcus (miksŏkǫ·kŏs). [mod.L., f. Myxo- + Coccus.] A myxobacterium of the genus so called. Also *attrib.* So **myxoco·ccal** *a.*, of or pertaining to an organism of this kind.

1892 R. Thaxter in *Bot. Gaz.* XVII. 396 In forms like Myxococcus, in which the rods are somewhat scattered, the first preparation for spore production..consists in the appearance of groups of rods moving with a circular tendency. [*Ibid.* 403 Myxococcus n. gen. – Rods slender, curved, swarming together after a vegetative period.] **1932** *Jrnl. Bacteriol.* XXIV. 337 There is some variation in the ability of Myxococcus cultures to fruit on the same types of media. **1946** *Nature* 23 Nov. 745/1 The growth of the myxococcus concerned..results in the production of a true non-enzymic antibiotic substance. *Ibid.*, An inoculum of myxococcal microcysts. **1973** R. G. Krueger et al. *Introd. Microbiol.* iii. 61/1 Representatives of the *Myxococcus* group behave in a fashion that requires response of vegetative cells to the presence and behavior of other vegetative cells of the same type.

myxomatosis (mi:ksŏmătōu·sis). [f. *myxomat-* (taken as stem of Myxoma) + -osis, as tr. G. *myxomkrankheit* (G. Sanarelli 1898, in *Centralbl. für Bakteriol.* XXIII. 871).] A highly infectious virus disease of rabbits, originally detected in Brazil but now occurring elsewhere, characterized by fever, swelling of the

mucous membranes, and the presence of myxomata; the disease has been artificially introduced into several countries to reduce rabbit populations. Also *attrib.* and *transf.* So **myxo·matized** *ppl. a.*, infected with myxomatosis.

1927 *Proc. Soc. Exper. Biol. & Med.* XXIV. 436 The point of particular interest concerning the myxomatosis of rabbits..is the fact that both epithelial tissue and connective tissue is affected. **1938** *Nature* 16 Apr. 682/2 A field test is now in progress of the virus of myxomatosis, as an agent in reducing rabbit population. **1953** *Times* 20 Oct. 5/4 The Ministry of Agriculture and Fisheries announced yesterday that in the past week myxomatosis, a virus disease of rabbits, was discovered near Edenbridge, Kent. **1953** A. Upfield *Murder must Wait* v. 44 'Was the child raised on cow's milk?'..'No, of course not. It wouldn't do, what with cows feeding off the same grass as myxomatised rabbits and things.' **1955** *Sci. News Let.* 5 Feb. 88/3 Myxomatosis..is threatening to wipe out the European rabbit. **1955** *Sci. Amer.* May 32/1 The myxomatosis virus..was recently introduced in France by a doctor who wished to get rid of the rabbits on his estate, and ..the disease soon spread over most of Western Europe. **1958** *Times* 6 Jan. 5/6 Rabbit numbers are increasing in many parts of the country, although isolated outbreaks of myxomatosis have lately occurred in over 30 counties... It seems that myxomatosis is now in a weakened strain. **1966** [see *columnaris]. **1968** M. Pyke *Food & Society* v. 59 The farmers.., for their own ends, had introduced the disease, myxomatosis, to exterminate the rabbits, which previously kept the brush in check. **1972** R. Adams *Watership Down* ii. 8 He had coolly..stood firm during the terrible onslaught of the myxomatosis.

myxophycean (miksofəi·siăn), *a.* [f. mod.L. name of class *Myxophyceæ*, f. Myxo- + Gr. φυκ-ος seaweed + *-eæ* suffix designating the class (cf. -aceæ).] Belonging or pertaining to the Myxophyceæ or Cyanophyceæ, a class of unicellular or filamentous blue-green algae.

1939 G. W. Prescott in F. R. Moulton *Probl. Lake Biol.* 74/1 The greatest pests [in water blooms] are Myxophycean species. **1957** *New Biol.* XXIII. 96 In many respects there is a great similarity between this Myxophycean poison [of *Microcystis aeruginosa*] and that of the Death Cap mushroom, *Amanita phalloides*. **1964** *Oceanogr. & Marine Biol.* II. 214 This internal environment affects the proportion of the Myxophycean endophytes [of *Codium bursa*].

myxovirus (mi·ksovəiərŏs). *Biol.* [mod.L., f. Myxo- (see quot. 1954) + Virus.] Any of a group of related viruses that includes the influenza virus (see quot. 1966). (At first the term was used to include the paramyxoviruses, later regarded as a separate group.)

[**1954** *Nature* 3 Apr. 621/1 The group name *Myxovirus* for viruses related to influenza was intended to indicate that the viruses in question have a particular kind of affinity for certain mucins.] **1955** *Virology* I. 180 When tests are made with fowl red cells it has not been shown that the same receptors are involved as with other *Myxoviruses*. **1962** C. H. Andrewes in *Adv. Virus Res.* IX. 285 The known myxoviruses are: 1. True influenzas. .. 2. Paramyxoviruses... The viruses of the second group differ from the 'true influenzas' in their larger, more variable size, in the absence of filamentous forms (except for NDV) and in normally producing hemolysins as well as hemagglutinins... 3. Viruses possibly related to myxoviruses—measles, distemper, rinderpest. **1966** J. E. Prier *Basic Med. Virol.* x. 247/1 In effect then, a new definition of the myxovirus group has emerged that is oriented around these biophysical characteristics: (*a*) a size range of 80 to 300 mμ..; (*b*) possession of a coiled ribonucleoprotein inner helix surrounded by an envelope studded with projections; and (*c*) the presence of lipid as an essential constituent of the virus particle. **1968** H. Harris *Nucleus & Cytoplasm* v. 90 The virus used in this work was the 'Sendai' virus, a member of the para-influenza group of myxoviruses. **1974** *Sci. Amer.* Feb. 33/2 Myxoviruses (influenza viruses) and paramyxoviruses..have a set of genes that encodes the manufacture of several viral proteins.

myxy (mi·ksi). Colloq. shortening of *myxomatosis. Also *attrib.* or as *adj.*, suffering from myxomatosis.

1961 R. Jeffries *Evidence of Accused* i. 15 Rabbits.. were slowly returning after the second scourge of myxy. **1962** —— *Exhibit No. Thirteen* x. 97 He'd paid a quid for a myxy rabbit which he'd dropped about the warren. **1973** *Daily Tel.* 22 Oct. 18 It is our custom to shoot any 'myxy' rabbits sitting out in the last stages of this horrible disease.

N

N. Add: **I. 1. b.** (See also *en *sb.*)

c. (Later example.)

1899 J. G. Millais *Breath fr. Veldt* 55 The birds alighting in the background are represented in their usual N-shaped formation.

4. (Further examples.) Also used in place of *bi-*, *di-*, *tri-*, etc., in words (e.g. n-*ary*).

1873 G. Salmon *Treat. Higher Plane Curves* (ed. 2) p. xi, Number of points which determine a *n*-ic. **1903** B. Russell *Princ. Math.* xxxvii. 310 Even an *n*-dimensional series of such terms..is still denumerable. **1924** Galsworthy *White Monkey* ix. 193 For the *n*th time it inspired in him a certain liking and confidence. **1940** W. V. Quine *Math. Logic* 42 The truth table for an *n*-ary mode of composition appears as in Table 3. *Ibid.* 225 A range of *n*-argument functionality..is itself an *n*-adic relation. **1940** W. de la Mare *Pleasures & Speculations* 223 Its children's children to the n-th generation. **1954** I. M. Copi *Symbolic Logic* ix. 306 A dyadic or triadic or *n*-adic relation. **1956** F. Pohl *Alternating Currents* (1966) 85 Spaceships ran from point to point in n-dimensional hyperspace. **1963** *Dict. U.S. Mil. Terms* (U.S. Dept. Defense) 156 *Nth country*,..the next country of a series to acquire nuclear powers. **1964** E. Bach *Introd. Transformational Gram.* vii. 154 Relations..are accordingly named dyadic, triadic, (or, for *n* terms, *n*-adic) or binary, ternary, and so on. *Ibid.* 155 A relation is the set of ordered pairs (triples, *n*-tuples) for which it holds. **1965** Patterson & Rutherford *Elem. Abstract Algebra* v. 156 A vector space, namely the set..of all ordered *n*-tuples $a = [a_1, \ldots, a_n]$ of elements from a field *F*. **1972** H. B. Enderton *Mathematical Introd. Logic* i. 50 An *n*-ary connective symbol combines with *n* wffs to produce a new wff. **1974** *Observer* (Colour Suppl.) 1 Sept. 10/1 (Advt.), This is where the standby duty crews wait, in full flying clothing, ready for the off. N-number of aircrew will always be there—24 hours a day, 365 days a year.

b. In *Physics* and *Chem.* *n* represents the principal quantum number of an electronic orbit in an atom, which determines the energy of the orbit (to the first order) and can take the values 1, 2, 3, ... (corresponding to increasing energy). In molecular spectroscopy *n* was introduced to denote the vibrational quantum number of a diatomic molecule (now usu. replaced by *v*), and later (as *N*) the total angular momentum, apart from spin, of diatomic and polyatomic molecules (see also quot. 1962[1]).

1914 N. Bohr in *Phil. Mag.* XXVII. [507 According to Balmer, Rydberg, and Ritz the frequency of the lines in the line-spectrum of an element can be expressed by the formula $\nu = f_r(n_1) - f_s(n_2)$, where n_1 and n_2 are whole numbers and $f_1, f_2, \ldots$ a series of functions of *n*, which can be expressed by $f_r(n) = (K/n^2)\phi_r(n)$, where *K* is a universal constant and ϕ a function which for large values of *n* approaches the value unity.] *Ibid.* 508 We shall assume that this spectrum is emitted by a system possessing a series of stationary states in which, corresponding to the *n*th state, the energy, omitting the arbitrary constant, is given by $A_n = -hK/n^2$. *Ibid.* 509 For $n = 1$, corresponding to the normal state of the atom, we get [etc.]. **1920** *Sci. Abstr.* A. XXIII. 423 If the nuclei are related in a quasi-elastic manner to the equilibrium positions, then the energy values for a diatomic non-rotating molecule are the same as for a Planck oscillator, namely: $E = nh\omega$, where *n* denotes the quantum number for the atomic oscillation. **1922** A. D. Udden tr. *Bohr's Theory of Spectra* iii. i. 67 This quantum number which will always be denoted by *n* will therefore be called the 'principal quantum number'. **1926** E. C. Kemble et al. in *Bull. Nat. Res. Council* No. 57. 5 Some discrepancies in minor details of notation will be found from chapter to chapter in this report. The following major items, however, have been followed consistently: ..Vibrational quantum number for diatomic molecule: *n*. **1951** D. Bohm *Quantum Theory* xv. 348 *n* = the principal quantum number..defined as one plus the total number of nodal surfaces in the wave function. **1953** *Jrnl. Optical Soc. Amer.* XLIII. 425/1 [Report of Sub-committee of the Joint Commission for Spectroscopy.] Notation for diatomic molecules... *N*. Total angular momentum of the electrons and nuclei, excluding spin... Replaces former *K*. **1962** P. J. & B. Durrant *Introd. Adv. Inorg. Chem.* vii. 207 The angular momentum due to the rotation of the [diatomic] molecule about an axis at right angles to the internuclear axis is denoted by **N**. **1962** J. Potter tr. *Messiah's Quantum Mech.* II. xxi. 964 Each quantum number *n* represents the number of vibrational quanta relative to a particular normal mode of vibration. **1966** G. Herzberg *Molecular Spectra & Molecular Struct.* III. i. 73 For $^2\Sigma$ states each rotational level except the one with $N = 0$ is split into three component levels..corresponding to $J = N + 1$, N, $N - 1$, respectively. **1973** J. Yarwood *Atomic & Nuclear Physics* vi. 188 The quantum mechanical treatment [of the one-electron atom] introduces the integral principal quantum number *n* which..decides the total energy of the system, the quantum number *l* which ..has any integral value between 0 and $(n-1)$, and the azimuthal magnetic quantum number m_l which is an integer having $(2l+1)$ values between $+l$ and $-l$.

6. Add: Soon after their alleged discovery it was concluded that such rays do not exist and Blondlot's work was in error. (Further examples.)

1906 *Nature* 1 Mar. 413/1 It would be interesting to know whether anyone has obtained success in repeating the latest experiment designed to show the objective reality of the *n*-rays. **1925** A. Huxley *Let.* 25 Jan. (1969) 241 You remember the unanimously favourable verdict in favour of N rays? **1941** W. Seabrook *Doctor Wood: Mod. Wizard of Laboratory* xvii. 234 According to Blondlot, the rays were given off spontaneously by many metals. A piece of paper, very feebly illuminated, could be used as a detector, for, wonder of wonders, when the N rays fell upon the eye they increased its ability to see objects in a nearly dark room. **1973** *Nature* 12 Oct. 344/2 The protracted error immortalised in scores of publications can be due only to psychological (self-deception), not to chemical causes. The 'anomalous' water belongs to the category exemplified by the N rays of Blondlot.

†**7.** *Radiology.* A unit of neutron dosage (see quot. 1942). *Obs.*

1942 Aebersold & Lawrence in *Ann. Rev. Physiol.* IV. 36 In this laboratory, a standard Victoreen *r*-meter with a 100r condenser-chamber is used. The unit of fast neutron exposure is arbitrarily taken as that amount which produces the same reading with this meter as a roentgen of x-ray. Thus an *n* of fast neutrons produces inside this 100r Victoreen thimble chamber the same amount of ionization as produced by an *r* of x-rays. **1948** E. Paterson in R. Paterson *Treatment of Malignant Dis. by Radium & X-Rays* xxxiii. 593 The *n* unit has fulfilled a useful function, however, in demonstrating differences in effect between X- or gamma-radiation and neutrons when ratios of the dosage required to produce effects are compared.

II. 1. N, naira (formerly, £N = Nigerian pound); N, nuclear; N.A.A.C.P., National Association for the Advancement of Colored People (*U.S.*); N.A.A.S., National Agricultural Advisory Service; N.A.B., National Assistance Board; N.A.C.A., National Advisory Committee for Aeronautics (*U.S.*); NAD(P), nicotinamide-adenine dinucleotide (phosphate); N.B., New Brunswick; N.B.C., National Broadcasting Corporation (*U.S.*); N.B.G., no bloody good (cf. **N.G.*); N.C.B., National Coal Board; N.C.O., non-commissioned officer; N.C.R., no carbon required, the registered trade-name for paper chemically treated so that the pressure of writing or typing alone produces duplicate copies without the use of carbon paper between sheets; *n.d.*, also N.D., (examples); N.D.C., National Defence Contribution; N.D.P., National Democratic Party (*Canad.*); N.E.D., New English Dictionary; N.E.D.(C.), National Economic Development (Council); N.F., New Franc (nouveau franc); N.F.S., National Fire Service; N.F.T., National Film Theatre; N.G., n.g., no go, no good (the latter is the current use, cf. **N.B.G.*); N.H.I., National Health Insurance; N.H.S., National Health Service; NIR [f. the initials of Russ. *Nauchno-Issledovatel'skaya Rabota* scientific research work], a colour television system developed in Russia, similar to SECAM; N.I.R.A., National Industrial Recovery Act (*U.S.*); N.K.V.D. [Russ. *Naródnyĭ Komissariát Vnútrennikh Del*], Soviet Commissariat of Internal Affairs, replacing the *Ogpu; N.L.F., National Liberation Front; NNI, noise and number index (see *noise *sb.* 3 d); N.O., Naval Officer; N.O.R.A.D., North American Air Defence; N.O.W., National Organization for Women (*U.S.*); n.p. (see quot. 1952); also, not paginated; N.P.D. [G. *Nationaldemokratische Partei Deutschlands*], National Democratic Party of Germany; N.P.L., National Physical Laboratory; N.P.V., Net Present Value; N.R.A., National Recovery Administration (*U.S.*); N.R.D.C., National Research Development Corporation; N.S.P.C.A., National Society for the Prevention of Cruelty to Animals; N.S.P.C.C., National Society for the Prevention of Cruelty to Children; N.S.W., New South Wales (*Austral.*); N.T., Northern Territory (*Austral.*); N.T.S.C., National Television System Committee (*U.S.*); N.U.M., National Union of Mineworkers; N.U.R., National Union of Railwaymen; N.U.S., National Union of Students; N.U.T., National Union of Teachers; N.Y.S.E., New York Stock Exchange. See also *N.A.A.F.I., *N.A.L.G.O., *N.A.S.A., *N.A.T.O., *N.A.T.S.O.P.A., *N.E.P., *Nibmar, *N.I.C., *n-p-n (as main entries).

1960 *Brit. Exports & Exchange Restrictions* (Swiss Bank Corporation) May 108, 1 Nigerian Pound = 20 Shillings. 1£N = 1£stg. **1973** *Times* 1 Oct. (Nigeria Suppl.) p. xi/3 Each member would be paid N120 a month. **1958** *Oxford Mail* 19 July 1/2 Beirut battle group is part of N-weapon division... U.S. paratroopers..equipped with tactical nuclear weapons. **1958** *Observer* 2 Nov. 6/1 Plenty of Aldermaston badges, with their semaphore symbol of ND—nuclear disarmament. **1910** *Crisis* Nov. 12 (*heading*) The N.A.A.C.P. **1974** *Black Panther* 16 Mar. 4/4 The NAACP filed its first suit attacking segregation and discrimination in education on March 15, 1933, against the University of North Carolina. **1948** *Sci. News* VI. 48 The N.A.A.S., which is the only body giving advice on the whole of agriculture. **1971** *Arable Farmer* Feb. 70/1 A nationwide series recently completed by NAAS, of 77 experiments testing nitrogen, phosphate and potash rates for spring barley. **1953** B. Abel-Smith *Reform of Social Security* 27 Malingering is very hard to detect, as the N.A.B. is constantly aware. *Ibid.*, We might keep our whole system on a flat rate basis but take the N.A.B. level as representing the community's present definition of subsistence. **1969** *Listener* 12 June 834/3 In Northborough.. mothers were not allowed to collect the father's weekly contribution..through the NAB. **1922** *Aviation* 13 Nov. 665/1 (*heading*) N.A.C.A. For National Air Policy. **1970** N. Armstrong et al. *First on Moon* i. 17 Vaguely coordinated by something called NACA—the National Advisory Committee for Aeronautics. **1961** *Biochem. Jrnl.* LXXX. 322/1 NAD formation almost certainly involves nucleophilic attack on the α-phosphorus of ATP. **1964** W. G. Smith *Allergy & Tissue Metabolism* viii. 85 The resulting α-β-unsaturated acid is reduced to its saturated counterpart by an enzyme which requires the reduced form of NADP. **1970** R. W. McGilvery *Biochemistry* xvi. 319 In general, NADP is used as an electron carrier for reductive syntheses, such as the formation of fatty acids.., whereas NAD is used more in the processes of energy production. *a* **1912** W. T. Rogers *Dict. Abbrev.* (1913) 133/1 *N.B.*...New Brunswick. **1929** F. W. Wallace *Rec. Canad. Shipping* 2 *Abyssinia*, ship, 833 tons. Built 1868, St. John, N.B. **1973** *Fisheries Fact Sheet* (Environment Canada Fisheries & Marine Service) No. 1. 1 (*caption*) Purse seining in Baie des Chaleurs, N.B. **1950** G. Marx *Let.* 7 Nov. in *Groucho Lett.* (1967) 75, I heard you on the NBC radio show. **1969** *Listener* 13 Nov. 667/1 The Duke of Edinburgh 'met the press' recently on America's NBC network. **1903** R. Bedford *True Eyes* xxxiv. 195 So see here—this place is N.B.G., Billy, me man. Let's do what I say—sell everythin' an' clear. **1929** C. Mackenzie *Three Couriers* ii. v. 180 Crowder messing about with a typewriter? O.K. Crowder outside an office? N.B.G. **1956** A. Wilson *Anglo-Saxon Attitudes* ii. iii. 369 Those Barkers were n.b.g.: I saw quite a lot of their dishonesty at Melpham. **1973** G. Mitchell *Murder of Busy Lizzie* iii. 42 Bang goes our reason for coming here... She said it was N.B.G. and that seems to be just about right. **1948** *Coal* Dec. 12/2 There are many complaints and criticisms of the N.C.B. It would be surprising if there were not. **1955** *Times* 3 Aug. 11/4 An official of the N.C.B. opencast executive said that the prospecting had shown the site to contain about 750,000 tons of good quality coal. **1963** *Times* 7 Feb. 13/4 It is essential not to impair the work at the N.C.B. which has made it possible to raise productivity four times above the national average. **1803–10** *Orderly Bks. of Manx Fencibles* in *Yn Lioar Manninagh* (1890) I. 152 Any party, consisting of 6 men or upwards, must have a N.C.O...appointed to go with them. **1883** *Army Regulations* II. 102 Report on conviction of N.C.O. by civil power. **1915** *Cornhill Mag.* Mar. 388 Had a chat with my N.C.O.s. **1955** *Times* 11 May 6/7 Corporal J. R. Saunders, who receives the George Medal, was the n.c.o. in charge of fire fighting at the storage unit. **1973** K. Giles *File on Death* i. 5 Old top secrets..guarded by superannuated N.C.O.s. **1954** in *Official Gaz.* (U.S. Patent Off.) (1955) 12 July 104/1 NCR no carbon required paper. **1956** *Brit. Printer* Mar. 42/1 Since its appearance on the US market some 18 months ago, ncr paper..is now being manufactured in this country. The initials ncr represent not only the name of the manufacturers—the National Cash Register Co Ltd—but the particular characteristics of the paper (No Carbon Required). **1967** C. Wilson in Wills & Yearsley *Handbk. Management Technol.* 46 Copies of print-outs are obtained with carbon paper or NCR paper. The latter ('no carbon required') has a chemically coated surface activated by percussion. **1834** *William Pickering's Catal. Manuscripts & Bks.* 19 Angler's Guide... Lond. n.d. **1879** *Blackwell's Catal.* No. 1. 10 Chappell's Old English Ditties..n.d. **1917** T. J. Wise *Bibliogr. Brontë Family* vi. 225 Museum of Brontë Relics..A Descriptive Catalogue. (N.D. but circa 1896). **1952** J. Carter *ABC for Book-Collectors* 124 No date (n.d.). This term, unqualified, means that research has failed (or has not attempted) to establish even an approximate date for the book described. **1959** *N. & Q.* CCIV. 292/1 *A Chronicle History of Portsmouth* by Henry Slight (3d. edit., n.p., n.d.). **1937** *Times* 1 June 17/5 The present N.D.C. scheme..imposes a high rate of tax on excess profits only. **1939** *Times* 4 Jan. 17/2 The figures in each case are struck after provisions for income-tax and N.D.C. **1961** *Edmonton* (Alta.) *Jrnl.* 4 Aug. 17/3 The NDP organizers produced mimeographed songsheets complete with Tommy Douglas lyrics. **1972** *Maclean's Mag.* Mar. 12/2 The NDP..refused to have the issue even raised for discussion. **1904** A. S. Palmer in R. C. Trench *On Study of Words* (rev. ed.) 175 Dr. Murray decides in favour of this explanation, of unknown authorship, and so spurious, uncanonical,—*N.E.D.* **1961** *Essays & Studies* XIV. 35 Of the many accounts of the origin and progress of the dictionary, it may suffice to mention here the following: the *Preface* to Vol. I of *N.E.D.*, [etc.]. **1973** *Jrnl. Soc. Bibliogr. Nat. Hist.* VI. 229 The work is known familiarly as NED, OED, or even 'Murray'. **1975** *N. & Q. for Som. & Dorset* Sept. 127 The single reference cited under lug-fall in the *N.E.D.*..takes the definition a little further. **1962** *Daily Tel.* 9 Feb. 12/2 The NEDC will not be a col-

lection of amateurs or mere theorists. *Ibid.* 7 Mar. 12/2 The collective wisdom of 'NED'. **1964** M. Argyle *Psychol. & Social Probl.* xvi. 199 A new government advisory department, like N.E.D.C. or N.I.C. **1960** *British Exports & Exchange Restrictions* (Swiss Bank Corporation) May 55 Only spare parts from O.E.E.C. countries may be obtained through import certificates when the value is less than NF. 5,000. **1971** A. Diment *Think Inc.* ii. 28, I ..added three $100 bills and two NF 1000 notes to the wad. **1942** *Daily Mirror* 25 Feb. 4/5 It will be necessary to enrol men under 41 for whole-time duties in the National Fire Service... Men over 25 when they registered will still be given an opportunity to express preference for the NFS. **1947** *Science News* IV. 47 Four days after she disappeared a fire broke out in the cellar of the Baptist Church and a passing police officer called the N.F.S. **1965** *New Statesman* 19 Mar. 462/3 Ozu..made some 54 films, of which only nine have ever been seen in London and those only by assiduous patrons of the NFT. **1969** *Listener* 24 July 125/1 Members of the NFT poured into their cinema to hear a discussion between Satyajit Ray and Lindsay Anderson. **1838** *Morning Post* (Boston) 25 June 2/3 They then went together to the plaintiff's to try to settle, but it was n.g. **1840** *Daily Pennant* (St. Louis) 20 June (Th.), The bells, boys, and engines tried to get up a fire last night, but it was N.G. **1922** Joyce *Ulysses* 427 Fish and taters. N.g. **1972** 'L. Egan' *Paper Chase* (1973) x. 160 So that little ride on the merry-go-round was n.g. **1934** Foster & Taylor *National Health Insurance* xiv. 191 Enter on debit side of Cash Book and credit side of N.H.I. Fund Current Account. **1958** *Spectator* 20 June 806/1 A doctor..wrote on what he called unjustified claims by his patients upon the N.H.I. **1948** *Lancet* 20 Nov. 823/3 Other doctors do not discriminate between private and N.H.S. patients. *Ibid.* 4 Dec. 904/1 The pharmacist's principal trouble just now is over delay in payment of his accounts under the N.H.S. **1968** *Listener* 11 July 58/3 One of the hospital committee shown arguing in the NHS film whether or not it should refund the cost of a 7s 11d vest. **1975** *Guardian* 5 Nov. 14/5 There will be the equivalent of 2,500 to 3,000 more beds available for NHS use. **1966** *Economist* 12 Feb. 628 This Russian-designed system offers the best prospect of a single acceptable colour television system for the whole of Europe... The Russians call the Soviet system NIR. **1969** Carnt & Townsend *Colour Television* II. vii. 237 In April, 1963, the Non-linear NIR system had already been anticipated by B. W. B. Pethers of the BBC. This was some two years before the NIR system came to light, but at that time it was felt that NIR did not have any significant advantages over the other systems. **1933** *Sun* (Baltimore) 23 Aug. 1/3 'Threats, intimidation, compulsion, boycotts, blacklists and suppression of opinion,' he declared, 'were never contemplated by the N.I.R.A. and, therefore, have no rightful place in the picture.' *Ibid.* 9 Sept. 1/2 The power of propaganda and the press has had few better illustrations than the way in which the so-called 'Nira' is obscuring the other great Administration activities in Washington and in the country. **1942** V. Conolly *Soviet Asia* (Oxf. Pamphs. on World Affairs) 13 The N.K.V.D. (Commissariat of Internal Affairs, former O.G.P.U.). **1945** Koestler *Yogi & Commissar* iii. iii. 208 The deportees were called for..by the so-called 'Executive Troikas' of the N.K.V.D. **1973** T. Allbeury *Choice of Enemies* vi. 21 He was wearing an NKVD uniform. **1965** *New Society* 22 Apr. 259 Nguyen Huu Tho was made president of the NLF. **1971** *Ink* 12 June 8/3 The delegates ..would have been Thieu/Ky (S. Vietnam), Tho (NLF), Ho Chi Minh (N. Vietnam). **1972** H. Evans *Editing & Design: Newsman's English* vii. 147 When the report says the NLF, the deskman should write in 'the National Liberation Front, political arm of the Vietcong..' This does not offend the reader who knows and it helps the rest, which is most of us. There was at one time, anyway, another NLF—in Aden—and at another an FLN, in Algeria and France. **1963** NNI [see *noise *sb.* 3 d]. **1969** *Guardian* 5 Mar. 9/2 Within such an area noise levels would be 40 NNI. **1970** R. D. Ford *Introd. to Acoustics* vii. 144 An external NNI of 55 has been taken as the recommended limit in the vicinity of London Airport and householders exposed to more noise have been offered a special grant for soundproofing their homes. **1914** 'Bartimeus' *Naval Occasions* xvii. 144 That girl to-night—Molly—I suppose she has refused half a dozen N.O.'s. **1916** G. Franklin *Naval Digression* i. xv. 136 Which brings to light another peculiarity of the average N.O.—how he can adopt a friendly 'hail-fellow-well-met' attitude to all and sundry and lose nothing in respect by it, even should they be ex-service men. **1958** M. Dickens *Man Overboard* i. 14 There'll be a lot of sharks about waiting for the innocent N.O. with his touching faith in human nature. **1959** *Roundel* June 3/1 Creation of the North American Air Defence Command by the governments of Canada and the United States was recognition..that air defence of the continent..is a single, common problem. NORAD is a truly integrated, international Command—responsible to both the U.S. Joint Chiefs of Staff and the Canadian Chiefs of Staff Committee. **1970** *Toronto Daily Star* 24 Sept. 31/3 In NORAD we permitted American nuclear bombers to make provocative flights over Canadian territory. **1966** *N.Y. Times* 22 Nov. 44/1 NOW..was formed three weeks ago in Washington to press for 'true equality for all women in America'. **1970** G. Greer *Female Eunuch* 296 When NOW was formed it was read into the Congressional record. **1973** *Times* 2 June 8/3 Betty Friedan..'founding mother' of NOW in 1966. **1882** Halkett & Laing *Dict. Anon. & Pseudon. Eng. Lit.* I. 122/1 Answer to the Declaration published by the Archbishop of Canterbury... n.p., n.d. Octavo. Pp. 12. *a* **1912** W. T. Rogers *Dict. Abbrev.* (1913) 137/1 *n.p.*, no place (no printer's name). **1952** J. Carter *ABC for Book-Collectors* 124 No place, no printer, no publisher (indiscriminately or collectively shortened to n.p.). **1966** *Dawson's of Pall Mall Catal. No.* 162. 12 Cicero..Laelius de amicitia. (n.p., n.d.) (Cologne..about 1467). **1969** N. B. Eales *Cole Library of Early Med. & Zool.* 330 The Complete Dictionary of Arts and Sciences... 2 vols. Fr., [4], vi, n.p. Pls. 1–11, 14–35, 40–42. **1966** *Guardian* 14 Mar. 21/8 The NPD..won up to 10 per cent in some Bavarian towns. **1969** S. Hyland *Top Bloody Secret* ii. 125 The two people who..organised the first conference of the Neo-Nazi party, NPD. *a* **1912** W. T. Rogers *Dict. Abbrev.* (1913) 137/1 *N.P.L.*, National Physical Laboratory (Bushey, by Teddington). **1920** *Flight* XII. 1131/2 With regard to the staffs at the N.P.L. and at the R.A.E., he said he hoped that provision would be made to maintain these on an adequate scale. **1959** *Ann. Reg. 1958* 388 There was a scientific symposium at the National Physical Laboratory (N.P.L.), Teddington. **1964** *Economist* 26 Sept. 1195/1 Then DCF is no more use than NPV. **1969** D. C. Hague *Managerial Econ.* ii. vi. 128 If we..add together the sums of money in Column C, including the minus £1,000 for Year 0, we are left with a net amount of £10. This is known as the net present value (NPV) of the project. **1933** *N.Y. Times* 3 Aug. 16/4 The most vital issue in the whole National Recovery Program is, of course, the question whether we shall end up speaking of it as the NRA or the N.R.A. People are saying NIRA and NRA but there is no denying the fact that there is a self-conscious air about it. **1933** G. Ade *Let.* 25 Aug. (1973) 170 The country editor..need not ignore the N.R.A. or the Farm Relief Board or any of the agencies intended to bring us back to happier times. **1962** *Amer. Speech* XXXVII. 48 Many such letter groups have entered the American language as items of common exchange: *O.K.*... *N.R.A.* **1954** F. A. Buttress *World List Abbrev.* 186 NRDC, National Research Development Corporation, 1 Tilney Street, London, W.1. **1959** *Engineering* 6 Feb. 182/3 If proof were required that NRDC's activities are useful and beneficial, it is provided by their profitability. **1967** *Jane's Surface Skimmer Systems 1967–68* 30/1 The Company has been granted an NRDC licence. *a* **1912** W. T. Rogers *Dict. Abbrev.* (1913) 137/2 *N.S.P.C.A.*, National Society for the Prevention of Cruelty to Animals. **1895** *Civil Rights for Children* (N.S.P.C.C.) (inside back cover), Mary P. Bolton Asst.-Sec., N.S.P.C.C. **1963** *Social Work* Oct. 20/2 Some will emphasise the word 'neglect' and think in terms of an extended N.S.P.C.C. inspectorate. [**1852** T. Cass *Let.* 1 May in J. Deans *Pioneers of Canterbury* (1937) x. 221, I take the opportunity of sending you a few lines by way of N. S. Wales, a cattle vessel leaving in a day or two.] **1889** H. C. Russell (*title*) The source of the underground water in the western districts... (Read before the Royal Society of N.S.W., August 7, 1889.) **1909** B. Stevens *Golden Treasury of Austral. Verse* 336 *Gidya*, a Queensland and N. S. W. aboriginal word for a tree of the acacia species. **1971** *Sunday Australian* 8 Aug. 3/1 Justice ministers in Victoria and NSW, said yesterday they were considering..changes. *a* **1912** W. T. Rogers *Dict. Abbrev.* (1913) 138/1 *N.T.*,..Northern Territory, -ies (Australia). **1930** A. G. Price *Hist. & Problems of Northern Territory* 64 For labour in the N.T., see reports of Gilruth..and Urquhart. **1969** *Northern Territory News* (Darwin) 11 July 3/3 The NT Police Commissioner..wants breath-analyser tests introduced in the Territory. **1957** *B.B.C. Handbook* 136 The N.T.S.C. system now in use for a public service in the United States of America. **1966** *Economist* 12 Feb. 628/2 The Americans manage to transmit colour from coast to coast by NTSC. **1948** *Ann. Reg. 1947* i. v. 90 The negotiations with the N.U.M. on the method of working overtime ..languished. **1914** *Railway Mag.* Nov. 401/1 It was.. agreed that all existing contracts and conditions of service shall remain in operation,..(signed) for General Managers' Committee:—..for N.U.R.:—..for A.S.L.E.&F.:—. **1955** *Times* 2 May 12/7 The union have won the right to put their case without having representatives of the N.U.R. present. **1975** *Guardian* 17 June 26/8 The NUR executive threw the negotiating ball firmly back into Mr Wilson's court. **1924** *University* I. 21 F. G. Connor..who attended as the representative of the English N.U.S., described the constitution and development of the National Union of Students of England and Wales. **1973** J. H. M. Scott *Dons & Students* xii. 145 The vice-chancellors turned it down, effectively siding with the nus against it. **1889** *Schoolmaster* 4 May 634/1 In place of the familiar initials, N.U.E.T. we have the shorter, and let us hope the improved, form of N.U.T... The objects of the N.U.E.T. remain the objects of the N.U.T. **1973** L. Holcombe *Victorian Ladies at Work* iii. 39 The National Union of Teachers..was organized in 1870... In 1911 the N.U.T. elected its first woman president. **1941** *Exchange* Feb. 3/1 'Secondary distribution!' That's a phrase to make the heart of any floor member of the New York Stock Exchange skip a beat. It means—in most cases, where NYSE stocks are being distributed—that just so much business is not passing through the Exchange. **1964** *Financial Times* 25 Feb. 3/7 A special committee set up by the N.Y.S.E. **1973** *N.Y. Law Jrnl.* 4 Sept. 4/2 It might be possible to include all NYSE-listed securities.

b. N.W., north-west, esp. (usu. followed by a numeral) a London postal district.

1615 T. Roe *Jrnl.* 26 Mar. in *Embassy to Court of Gt. Mogul* (1899) I. 3 Wee..saw land N.W. for the Canarye 8 leauges off. **1857** J. A. Symonds *Let.* 1 Feb. (1967) I. 88 Sunday. Harrow *N.W.*... You will wonder what N.W. means on the top of this page. It is in consequence of some Postal arrangements that those letters should be affixed to districts of London. **1899** W. J. Locke *White Dove* (1900) x. 151 Dr. Frodsham..had..[moved] from the house in Weymouth Street into the purer air of the N.W. district. **1965** *Listener* 27 May 792/1 Living with him was his beautiful companion, a Madame de Bargeton of N.W.3, something of a Madame Verdurin also. **1967** *Ibid.* 30 Nov. 733/2 Who could have littered the *Times* with byelines and signed columns and a woman's page so close to NW1's heart? **1971** A. Bennett in D. Nathan *Laughtermakers* iv. 93, I did a series on television with sketches about this area, N.W.1, of which I am a part, and so in a sense it was making fun of myself. But it pinpointed something and since then N.W.1 has been used as a catchphrase to indicate Sunday supplement trendiness which people now find rather suspect.

2. N.B. (later examples).

1861 G. H. Lewes *Jrnl.* 3–6 June in *Geo. Eliot's Lett.* (1954) III. 424 Reposed awhile and chatted (N.B. all the journey we have had fine talk with Trollope). **1965** I. Fleming *Man with Golden Gun* iii. 40 Distinguishing marks: a third nipple about two inches below his left breast. (N.B. in Voodoo and allied local cults this is considered a sign of invulnerability and great sexual prowess.)

3. a. N (also rarely *N*, *n*) (*Chem.*) = *normal *a.* and *sb.* A. 2 b (i).

1863 F. Sutton *Syst. Handbk. Volumetric Anal.* 165 Free iodine is..very readily estimated by solution in iodide of potassium, and titration with N/10 hyposulphite. **1906** *Amer. Chem. Jrnl.* XXXV. 511 A saturated solution of silver chloride..is only about 0·00001N. **1931** J. C. Ware *Analyt. Chem.* iv. 137 Test the solubility of a portion of the precipitate..with 2 cc. 6*n* HNO_3. **1970** M. D. Hawkins *Calculations in Volumetric & Gravimetric Anal.* ii. 35 The solution was titrated against 0·0908N potassium thiocyanate.

b. *n* (also rarely n) (*Chem.*) = *normal *a.* and *sb.* A. 2 b (ii).

1889 G. M'Gowan tr. *A. Bernthsen's Text-bk. Org. Chem.* i. 44 From petroleum have been separated..normal heptane, n-octane, n-nonane, and n-decane. **1938** L. F. Fieser in H. Gilman *Org. Chem.* I. ii. 53 The conversion of *n*-butylcyclopentane into *o*-ethyltoluene on dehydrogenation with palladium charcoal. **1971** N. L. Allinger et al. *Org. Chem.* iii. 28 One isomer, isobutane, has one carbon atom which is bound to three other carbons, and there is no such carbon atom in the other isomer, *n*-butane.

c. In *Physics* *N* is used to designate the series of X-ray emission lines of longer wavelength than the *M*-series obtained by exciting the atoms of any particular element (cf. *M 4* a); these arise from electron transitions to the atomic orbit of fourth-lowest energy, of principal quantum number 4, which is thus termed the *N-shell*, and electrons in this shell *N-electrons*.

1911 [see *M 4* a]. **1923** H. L. Brose tr. *Sommerfeld's Atomic Structure & Spectral Lines* viii. 505 Our relativistic formula of the fine-structure..furnishes us with a principle of sub-division for the multiplicity of M- and N-lines. *Ibid.* 507 Just as the M-shell belongs to the quantum-number 3, so does the N-shell to the quantum-number 4. **1948** Lapp & Andrews *Nuclear Radiation Physics* v. 74 In the heavier elements such as tungsten ($Z = 74$) the *N* shell is saturated with 32 electrons. **1967** G. L. Clark *Handbk. X-Rays* i. 7 At still longer wavelengths *M*- and *N*-series lines have been measured only for heavier elements. **1968** *Physical Rev.* CLXVI. 944/1 Radiationless transitions are favored in low-*Z* elements and always dominate in the *M*, *N*, and higher shells for the ordinary Auger effect. **1970** E. P. Bertin *Princ. & Pract. X-Ray Spectr. Anal.* i. 27 An electron having just enough energy to expel, say, an *L*II electron can also expel *L*III, *M*, and *N* electrons, but not *L*I or *K* electrons.

d. N (*Physics*) = newton.

1951 *Symbols, Signs & Abbrev.* (R. Soc. Symbols Comm.) 14/2 Newton... N. **1973** Kleppner & Kolenkow *Introd. Mechanics* ii. 84 Since $g = 9{\cdot}8\ m/s^2$, the weight of 1 kg mass is 9·8 N.

'n, 'n' (ən), colloq. shortening of And *conj.* Esp. common in **rock 'n' roll*.

1858 O. W. Holmes in *Atlantic Monthly* Sept. 497/1 To beat the taown 'n' the keounty. **1906** E. Dyson *Fact'ry 'Ands* xvii. 233 Ther revolvin' arm was bent out, 'n' it got home a left lead 'n' er right cross. **1923** *Radio Times* 28 Sept. 23/2 Dance Programme..One-step, 'By 'n' Bye'. **1925** C. R. Cooper (*title*) Lions 'n' tigers. **1928** F. Hurst *President is Born* v. 67 God bless Mother 'n' Father. **1959** W. Golding *Free Fall* ii. 58 He was helmeted, assured, delicate at the rudder-bar and joystick in the fish-'n-chip smell of the engine oil. **1968** *Listener* 18 July 93/2 Tony Carruthers has turned the whole building into a prison to assist Charles Marowitz's production of a Canadian play about four boys having sex-'n-violence in a reformatory. **1971** *Black World* June 63/1 The President's tryin' his level best to bring more'n more boys home all the time. **1973** J. Wainwright *Devil you Don't* 30 Dudley was mad.. but good 'n mad.

'n, colloq. reduced form of Than *conj.*

1867 J. T. Trowbridge *Darius Green* in *Oxf. Bk. Children's Verse* (1973) 262 The little chatterin', sassy wren, No bigger 'n my thumb. **1898** J. D. Brayshaw *Slum Silhouettes* 2 That on'y made Bill madder 'n ever. **1903** K. D. Wiggin *Rebecca* i. 8 We've only just started on it,..'it's more 'n two hours'. **1910** C. E. Mulford *Hopalong Cassidy* ix. 63 He hates Greasers worse 'n I do. **1973** 'J. Ashford' *Double Run* viii. 57 The duty would be more 'n Mr. Smith could pay if it's properly declared.

N.A.A.F.I. (næ·fi). Also **NAAFI, Naafi, Naffy.** The Navy, Army, and Air Force Institutes which run canteens, stores, etc., for service personnel; also, a canteen, restaurant, etc., run by this organization. Also *attrib.* Cf. *Naffy.

[**1921** *Times* 7 Jan. 7/3 The joint organization for the administration of canteens for the three Services, hitherto known as the Navy and Army Canteen Board, has been reconstituted with effect from January 1, 1921, under the title 'The Navy, Army, and Air Force Institutes'.] **1927** *Daily Express* 16 Mar. 9/2 Cheering workgirls surrounded the Prince of Wales when he visited the 'Naafi'..headquarters in Kennington yesterday. **1928** J. Fortescue *Canteens in Brit. Army* 73 The general policy of the N.A.A.F.I. is to afford facilities to its members of all three services to purchase commodities of high quality at prices competitive with those of 'multiple' shops. **1940** *War Weekly* 19 Jan. 409/3 N.A.A.F.I., or 'Naffy', as it is affectionately called by the troops. *Ibid.*, The 'Naffy' Institute is a comfortable, red-brick building. **1959** *Times* 17 Jan. 8/4 Common rooms sometimes suggest more the Naafi than the bookman's study. **1959** *Spectator* 23 Jan. 127/3 The NAAFI Is a sort of caafi Where soldiers are rude About the food. **1972** *Oxford Times* 4 Aug. 5 Mr Lewin leaves Oxford for Belfast on Monday to take up an appointment as manager of a NAAFI club in the city. **1973** *Times* 14 May 11/5 Kafka and Brasso make a great double act with a NAAFI piano. **1974** R. Gentil *Trained to Intrude* ii. 20 At that time Dover.., apart from the pubs and the NAAFI, offered very little entertainment to the serviceman.

naan, var. *Nan³.

naartj(i)e, varr. *Nartjie.

nab, *sb.*³ Restrict 'rare' to sense 2 and add further examples of sense 1.
1969 *New Yorker* 17 May 32/3 I talk him into splitting the scene and we start hitchhiking back down Sunset, and just like that the Nabs stop us for bumming rides. **1971** J. Wainwright *Dig Grave* 78 All the nabs in the world were in the downstairs front.

nabam (næ·bæm). *Chem.* [f. *Na*, the chemical symbol for sodium + bisdithio*carbam*ate (see quot. 1950).] A water-soluble powder used as a fungicide, usu. as a spray with zinc sulphate as a stabilizer; $(NaS·CS·NH·CH_2)_2$.
1950 *Phytopathology* XL. 118 The Subcommittee on Fungicide Nomenclature of the American Phytopathological Society, cooperating with the Interdepartmental Committee on Pest Control, has selected common names for two commercially-available fungicidal chemicals which are useful in the control of various destructive plant diseases... Nabam for the fungicidal chemical, disodium ethylene bisdithiocarbamate. **1960** [see *Zineb]. **1967** L. H. Purdy in D. C. Torgeson *Fungicides* I. vi. 206 Nabam reacted with zinc sulfate was the first commercially used spray fungicide for in-the-row treatment control of cotton seedling diseases.

Nabataean (næbătī·ăn), *sb.* and *a.* Also **Nabatean, Nabathæan, Nabathean.** [f. L. *Nabat(h)æ-us*, Gr. *Ναβαταῖ-ος, Ναβαθαῖ-ος* (cf. *Nebāṭu* the native name of the country) + -an.] **A.** *sb.* One of an ancient Arabian people; their language. **B.** *adj.* Of or pertaining to the Nabatæans.
1601 Holland tr. *Pliny's Nat. Hist.* I. xii. xx. 374 The Troglodyte Nabathæans: who onely of the other Arabians, border upon Syria. **1875** *Encycl. Brit.* III. 411/2 Two forms of Shemitic writing (the Palmyrenian.., and the Sinaitic or Nabathæan). **1884** *Ibid.* XVII. 169/1 Nabathæans, a famous people of ancient Arabia. **1897** F. Hommel in H. von Hilprecht *Recent Res. in Bible Lands* 146 Between the decline of the Nabataean Empire and the appearance of Muhammad. **1898** E. Clodd *Tom Tit Tot* vi. 65 Ibn Khaldûn.. describes how the Nabathean sorcerers of the Lower Euphrates made an image of the person whom they plotted to destroy. **1911** *Encycl. Brit.* XXIV. 626/1 The language of this country was Nabataean. **1920** *Public Opinion* 9 July 42/3 Little did the ancient Nabataean imagine that a people called Americans would one day wander among the ruins of their proud city. **1932** *Times Lit. Suppl.* 29 Oct. 832/2 His new work.. is intended to summarize all that is known on the subject of the Nabateans. *Ibid.*, The sources of Nabatean history. **1973** *Times* 9 June 11/7 A fragment of ancient Nabatean pottery. **1974** *Times* 21 Dec. 11/4 The Arava, stretching from the Dead Sea to Eilat, which has been virtually virgin soil since the days of the Nabataeans.

nabbie (næ·bi). Also **nabby.** [? f. Nobby *a.*] A type of Scottish boat used esp. in herring-fishing on Loch Fyne and in the Firth of Clyde, and originally having a raking mast, lug-sail and jib. Cf. Nobby *sb.* 2.
1884 H. Hogarth *Herring Fishery* 4 These boats were round-sterned—from fourteen to sixteen feet keel and about seven feet beam. It was not possible to go any distance to look for herrings in boats of this description. They were known by the name of 'nabbies'. **1907** *Yachting Monthly* iv. 366/1 It may interest 'M.I.N.A.' to know that he is quite correct in his use of the word 'nabbie' as applied to the present-day Loch Fyne type of fishing boat. **1955** *Mariner's Leader* (Stonehaven) 17 June 46 The 'Nabby' boat, *June Rose*, from Fisherrow. She is very well equipped and powered by a 66 h.p. Kelvin Diesel. **1969** P. Norton *End of Voyage* 85 The Loch Fyne skiff or nabbie is a graceful lug-rigged craft.

nabe (nēb). *U.S. slang.* [f. the pronunc. of Neighbo(u)rhood.] A neighbourhood; *spec.*, a local cinema. Also *attrib.*
1935 *Evening Sun* (Baltimore) 8 Apr. 17 On Sunday two powerful [box office pictures] were released to the nabes. **1937** *Amer. Speech* XII. 317/2 *Nabe*, neighborhood motion picture theatre. **1942** Berrey & Van den Bark *Amer. Thes. Slang* § 591 *Nabe*, neighborhood, nabe, neighborhood theater. **1960** *Ibid.* § 587/5 *Nabe*, neighborhood motion-picture theater. **1961** A. Berkman *Singers' Gloss. Show Business* 61 *Nabes*.. neighborhood movie houses. **1970** *New Yorker* 20 June 30/1 They picked an aging star, slapped together a moldy script, and sent the result out to the nabes. **1973** *Time* 8 Nov. 73/1 In Portland, where business is also dragging, the nabes (the trade term for the neighborhood houses) are now closed except on weekends.

Nabeshima (næbĕʃī·mă). The name of a baronial family in feudal Japan used *attrib.* and *absol.* to designate the porcelain produced from the kilns established by this family in 1722 at Okawachi on the island of Kyushu in Japan.
1886 W. Chaffers *Marks Pott. & Porc.* (ed. 7) 412 Nabeshimayaki was made at Okawaji, painted principally in blue with plants, fishes, &c., distinct from the Hiradoyaki. **1902** F. Brinkley *Japan* VIII. ii. 95 The *Nabeshima-yaki*, as the Okawachi manufactures were subsequently called, stands first among Japanese porcelains decorated with vitrifiable enamel. **1937** R. L. Hobson *Handbk. Pott. & Porc. Far East* 163 The early Nabeshima porcelain is a good white ware with lustrous glaze of fine texture. **1965** S. Jenyns *Jap. Porcelain* vi. 230 Within their narrow range the saucer dishes of Nabeshima are technically as perfect as any porcelain that the Japanese were able to produce. **1967** H. H. Sanders *World of Jap. Ceramics* 218 The old Nabeshima style is a foot which rises to a high foot, which has a traditional blue-and-white comb-tooth pattern under the glaze and three underglaze blue patterns on the outside of the bowl. **1971** L. A. Boger *Dict. World Pott. & Porc.* 237/1 Both the underglaze blue and the polychrome enameled Nabeshima are notable for the consummate skill with which the decorations have been applied, particularly the underglaze blue.

‖ **nabi** (nā·bi). Pl. **nebi·im.** [Heb. *nābhī* prophet.] **1.** *Theol.* One inspired to speak the word of God; a prophet, *spec.* a Hebrew prophet of the Old Testament.
[**1765** W. Dodd in *Bible with Commentary* note on Gen. xx. 7 The Hebrew נביא *nabia*, signifies a person that speaks something in an eminent and extraordinary manner.] **1877** A. Milroy tr. *Kuenen's Prophets & Prophecy in Israel* iii. 63 The Nabi is, and cannot but be, an improvisatore. **1885** T. P. Hughes *Dict. Islam* 475/1 *Nabi*, a prophet. One who has received direct inspiration.. by means of an angel, or by the inspiration of the heart.., or has seen the things of God in a dream. *Ibid.* 475/2 According to Muhammadan writers a nabi is anyone directly inspired by God. **1900** W. W. Skeat *Malay Magic* 99 Of the prophets (*Nabi*) there are an indefinite number. **1908** R. J. Wilkinson *Life & Customs* (Papers on Malay Subjects) I. 14 Other religions had prophets of their own who were nevertheless true prophets like *Nabi Isa*, the prophet of the Christians, and *Nabi Musa*, the prophet of the Jews. **1918** *Encycl. Relig. & Ethics* X. 384/1 The true religion of Israel.. traced its origin to those who bore the title *nabî*. *Ibid.* 384/2 In the Hebrew order of the OT books the *nebî'îm* are divided into two sections. **1922** J. Skinner *Prophecy & Religion* i. 4 Amos instances the raising up of *Nebî'îm* as a proof of Yahwe's peculiar love for Israel. **1961** B. Vawter *Conscience of Israel* iv. 75 To him Amos was only a *nabi* like the many who grubbed out a living by devising oracles for clients.

2. *Art.* A member of a group of late 19th-century French post-impressionists, including Bonnard, Vuillard, Denis, and Sérusier, who followed the artistic theories of Gauguin.
1931 C'tess de Lauzanne tr. *Basler & Kunstler's Post-Impressionists* xi. 52 All the 'Nabis', all the 'enflammés' who left the Julian Academy, were attracted by the theories of.. Paul Gauguin. **1945** Goldwater & Treves *Artists on Art* (1947) 379 Sérusier gathered about him a group known as the Nabis which included.. Denis, Bonnard, Vuillard, Ranson, Roussel and later Maillol. **1959** *Listener* 22 Oct. 694/1 Felix Vallotton who worked in Paris and had relations with the Nabis. **1963** L. & E. Hanson *Post-Impressionists* xi. 277 They were.. nicknamed the Nabis.. because most of them were bearded, some were Jews and all were desperately earnest. **1965** *Times* 3 May 6/5 The 'Nabi' phase of French painting. **1969** M. Bullock tr. *Chassé's Nabis & their Period* i. 14 The Nabis soon became known as devotees of 'beautiful greys' and broken colours.. When we try to outline Nabism, we always return in the end to the concept of fantasy and dream. **1971** tr. W. Verkade in J. Russell *Vuillard* 85 One day Sérusier said to me: 'Let's go and see the Nabi, Denis, at Saint-Germain.' **1972** L. Lamb *Picture Frame* ii. 24 I should guess that you have got at least one of the Nabis... If you think it's a Vuillard, it probably is. **1973** *Times* 7 May 7/4 Art nouveau, the Nabis, the Pointillists.
Hence **na·bi'ism, na·bism.**
1922 J. Skinner *Prophecy & Religion* i. 5 Nabi'ism had its unprogressive and degenerate representatives. **1969** [see sense 2 above].

nabla (næ·blă). *Math.* [a. Gr. *νάβλα*: see Nable (and cf. quot. 1879 s.v. Nebel).] = *Del.
The operator was introduced by Sir William Hamilton, who represented it by the symbol ◁. (In 1837 he uses ∇ as a symbol for any arbitrary function.)
[**1837** W. R. Hamilton in *Trans. R. Irish Acad.* XVII. 236 Considering *x* as a function ϕ of a new variable ∇ and performing any operation ∇ with reference to the latter variable, ... **1846** *Proc. R. Irish Acad.* III. 291 The following.. general characteristic of operation $\lhd = i\frac{d}{dx} + j\frac{d}{dy} + k\frac{d}{dz} = \lhd$, in which *x, y, z* are ordinary rectangular coordinates, while *i, j, k* are his [sc. Hamilton's] own coordinate imaginary units, appears to introduce one of great importance in many researches.] **1847** W. R. Hamilton in *Phil. Mag.* XXXI. 291 In the paper designed for Southampton.. the characteristic was written ∇; but this more common sign has been so often used with other meanings, that it seems desirable to abstain from appropriating it to the new signification here proposed. **1853** —— *Lect. Quaternions* vii. 610 Introducing, for abridgment, as a new characteristic of operation, a symbol defined by the formula ◁=(etc.).] **1884** W. Thomson *Notes Lect. Molecular Dynamics & Wave Theory of Light at Johns Hopkins Univ.* x. 112 (MS.), I took the liberty of asking Professor Bell.. whether he had a name for this symbol ∇; and he has mentioned to me *nabla*, a humorous suggestion of Maxwell's. It is the name of an Egyptian harp, which was of that shape. **1892** *Phil. Trans. R. Soc.* CLXXXIII. 431 Physical mathematics is very largely the mathematics of ∇. The name Nabla seems, therefore, ludicrously inefficient. **1939** D. E. Rutherford *Vector Methods* iv. 50 A convenient method of writing grad ϕ is ∇ϕ, where ∇ (pronounced 'nabla') is defined as the vector operator $\nabla \equiv \mathbf{i}\frac{\partial}{\partial x} + \mathbf{j}\frac{\partial}{\partial y} + \mathbf{k}\frac{\partial}{\partial z}$. **1964** [see *Del]. **1969** L. Young *Systems of Units in Electr. & Magn.* v. 63 The symbol nabla, ∇, is a vector differential operator.

nabob. **4.** (Earlier example.)
1803 E. S. Bowne *Girl's Life 80 Yrs. Ago* (1888) 151 Silk nabobs, plaided, colored and white, are much worn, very short waists, hair very plain.

nabocklish (nabǫ·klif), *int. Irish dial.* Also **nabocklesh, nabocklish, nabocklish.** [Ir. *na* not + *bac*, imp. sing. of *bacaim* 'I meddle' + *leis* with it: lit. 'don't meddle with it'.] Never mind! Leave it alone!
1841 C. J. Lever *Charles O'Malley* I. ii. 10 Arrest him! —na bocklish—catch a weasel asleep. **1843** W. Carleton *Traits & Stories Irish Peasantry* (new ed.) I. 341 But, nabocklish! what'll ye have? **1867** P. Kennedy *Banks of Boro* xxi. 229 But, nabocklish, we will find ourselves in the wrong box, maybe. **1917** J. Morley *Recollections* II. v. iii. 222 When I hear of some more malicious or injurious word in politics, I find real comfort in saying to myself 'Nabocklish!' **1939** C. Morley *Kitty Foyle* xxix. 285 Like Pop said sometimes, nabocklish! which is Irish for let's not worry too much.

Nabokovian (næbǫ·kǫ·viăn, nabǫkǭ·viăn), *a.* [f. the name *Nabokov* (see below) + -ian.] Of, pertaining to, resembling, or characteristic of the Russian-born novelist and poet Vladimir Nabokov (b. 1899) or his writings.
1959 *Observer* 1 Nov. 21/6 There is Nabokovian poignancy in leaving such delicate things to be destroyed, as he says with a rarified smile, by 'bored people'. **1965** *Times Lit. Suppl.* 28 Jan. 68/4 Mr. Nabokov's *Eugene Onegin* will be read for its bearing.. It will be read for the brilliant fireworks of his prose and for the beauty of the Nabokovian phrase. **1968** *Punch* 25 Dec. 931/3 Mr. Stegner chooses instead to invest detail with significance, and he overwrites in truly Nabokovian manner. **1972** *Sat. Rev.* (U.S.) 10 June 66/2, I found myself searching for Nabokovian anagrams in the names. **1975** *Times Lit. Suppl.* 31 Oct. 1285/1 The narrative manner similarly alternates between abruptly functional stage or screen direction and a Nabokovian obliquity in which words take on an energy of their own and skitter away from the matter in hand.

nacelle. Restrict † *Obs. rare*⁻¹ to sense in Dict. and add: **2.** [after the same uses in Fr.] **a.** The basket or gondola of a balloon or airship.
1901 *New Penny Mag.* XII. 440 The 'nacelle', or basket, from which.. the aeronaut directs his operations. **1909** *Aero* 13 July 117/2 The dirigible.. has a screw at either end of the nacelle, or car. **1934** *Times* 27 May 13/7 The balloon with which the first ascent was made will be used, but a new nacelle will be constructed.
b. The cockpit of an aeroplane; hence, any streamlined structure on an aeroplane for housing something, esp. an engine.
1914 *Sci. Amer.* 8 Sept. 3/7 (Advt.), The Henry Farman seaplane is a biplane of the pusher type.. The pilot and passenger have comfortable quarters in a 'nacelle' which is built out from the front of the machine. **1915** *War Illustr.* 10 July 494/2 The machine was apparently a gun-carrying 'pusher' biplane of the type in which the gunman sits right out in the nose of the boat-shaped body—or 'nacelle'. **1918** 'Avion' *Aeroplanes & Aero Engines* v. 59 The engine, tanks, crew, controls, and instruments are accommodated in a long streamline body, above which the centre section [of the upper wing] is erected in the usual way. **1920** *Blackw. Mag.* Feb. 165/1 The spirit was not entering the tank, but spilling over the side on to the floor of the nacelle. **1938** C. B. Cantle *Story N. Sea Air Station* i. 56 They could seat two persons in the covered-in nacelle with the pilot in front. **1943** W. L. Cowley *Aerodynamics of Aeroplanes* iii. 53 The engine with a tractor propeller is sometimes placed.. in the nose of a nacelle or engine cover. **1953** R. L. Arnold *Proving Ground* xxix. 309 The planes.. upended to show their twin nacelles.
c. A similarly shaped structure on or in a motor vehicle.
1959 *Motor* 1 June 603/3 The furnishings belong to the polished walnut school, upon which is superimposed a hooded crackle-finish nacelle in front of the driver containing a full but tidily-framed set of instruments. **1967** E. Rudinger *Consumer's Car Gloss.* (ed. 2) 71 *Nacelle*,.. sometimes used to describe the streamlined housing on a car's lights or dashboard instruments. **1969** *Practical Motorist* May 1000/2 The instruments are housed in an oblong speedometer with a small rev-counter perched atop the facia in a separate nacelle. **1970** *Daily Tel.* 7 Apr. 9/1 The Lotus 72's radiator has been moved from the nose and replaced by two smaller radiators mounted in nacelles on the sides of the car.

Nachee, var. *Natchez.

‖ **Nachlass** (na·χlas). [G.] Unpublished writings left by an author after his death.
1842 J. S. Mill *Let.* Apr. in *Wks.* (1963) XIII. 515 She bids me.. to ask what you think of Ottfried Müller's Nachlass as a subject for translation. **1948** *Mind* LVII. 382 The argument is supported throughout by cogent evidence from Kant's works. The quotations from the *Nachlass* and the *Opus Postumum* are especially valuable. *Ibid.* 524 It has the advantage of making his lectures far better material for publication than that found in the average professor's *Nachlass*. **1961** D. Ross in Aristotle *De Anima* 49 The chapter is apparently a series of jottings which an early editor found in Aristotle's *Nachlass*, and put together so that nothing of the Master's should be lost.

Nachschlag (na·χʃlak). *Mus.* [Ger., f. *nach* after + *schlag* blow, note.] A grace note which takes its value from that of the note preceding it.
1879 [see *Agrément 2]. **1880** Grove *Dict. Mus.* II. 441/2 In the works of the great masters the Nachschlag, though of very frequent occurrence, is almost invariably written out in notes of ordinary size. **1915** A. Dolmetsch *Interpretation of Music of 17th & 18th Cents.* iv. 255 C. Ph. E. Bach incidentally mentions the *Nachschlag*, but only to condemn it. **1944** W. Apel *Harvard Dict. Mus.* 476/1 The ornamenting notes constitute a melodic movement away from the preceding note, and are to be performed as a part of this... Thus the Nachschlag is the exact opposite of the appoggiatura. **1960** E. Bodky *Interpretation of Bach's Keyboard Works* v. 180 In bar 20 this version leads to ugly parallel fifths, which Landshoff tries to avoid by changing the short appoggiaturas of bars 19 and 20 into *Nachschläge*.

‖ **Nachtlokal** (na·χtlokal). Pl. **nachtlokale.** [G.] A night-club. Also *attrib.*
1939 E. Ambler *Mask of Dimitrios* v. 95 'She is the proprietress of a *Nachtlokal* called *La Vierge St. Marie*.' '*Nachtlokal*?' He grinned. 'Well, you could call it a night club.' **1954** P. Bottome *Against Whom?* xxiii. 173 Konrad, since it was his free evening, would be at a Nacht-Lokal dancing. **1968** D. Hopkinson *Incense Tree* viii. 94 We went dancing together in the almost total darkness of a 'Nachtlokal'. **1972** J. Freeman *Baby, it's Cold Inside* 227 Some grisly *Nachtlokal* in the Kurfürstendamm.

‖ **nacht und nebel** (naχt unt nē·bəl). [G.] The German for 'night and fog', used, freq. *attrib.*, of a situation characterized by mystery or obscurity.
The name of an infamous decree issued by the Nazis in December 1941.
1947 V. H. Bernstein *Final Judgement* xviii. 240 *Nacht und Nebel* was issued over Hitler's signature, but Keitel issued several covering memoranda and interpretations. Indeed, the name of the Chief of the High Command was so closely identified with the order that it was sometimes referred to as the 'Keitel Decree'. **1965** *Times* 21 Feb. 12/5 This was a *Nacht und Nebel* thirteenth-century Germany, peopled by symbols and based on the orchestra. **1968** A. Marin *Clash of Distant Thunder* (1969) i. 5 The transport camp.. where captured commandos and *Nacht und Nebel* prisoners were isolated. **1971** D. Cory *Sunburst* vi. 100 Country after country going under, disappearing in the 'Nacht und Nebel'. **1972** *Listener* 26 July 123/1 The indescribable verbal miasma, exhalations of *Nacht und Nebel*, that is Scientology's contribution to the encyclopaedia of religious knowledge.

Nachtmaal: see *Nagmaal.

nacker, var. *Knacker *sb.*¹ 3.
1866 [see *Knacker *sb.*¹ 3]. **1958** M. Pugh *Wilderness of Monkeys* 79 Oh, smart boy, eh?.. Festival Hall fiddle! Nackers!

‖ **nada**¹ (nā·da). *Hinduism.* Also **naad** (nād). [Skr. *nādd* sound.] Inchoate or elemental sound considered as the source of all sounds and as a source of creation; the 'inner' sound of the body.
1913 'A. Avalon' *Tantra of Great Liberation* p. xxiii, It is Nāda.. when there is sound in which everything is like a connected or combined disposition of the letters. **1920** *Encycl. Relig. & Ethics* XI. 93/1 The Śāktas base their doctrines on the assumption that through Siva and Sakti there is a drop, Bindu, formed which develops into a female element Nāda (sound), continuing in itself the names of all things to be created. **1926** *Indian Art & Lett.* II. 79 Nāda as indicative of immanent sound in the form of a crescent-moon on His [sc. Siva's] head. **1930** S. N. Dasgupta *Yoga Philos.* iv. 269 This sound in stage of pure varṇas is called nāda.. But each varṇa as it is generated, as the tension of bearing has no power to hold them together. **1940** H. E. Kennedy tr. *Marquis-Rivière's Tantrik Yoga* i. 20 Then there is *Laya Yoga*, based upon the sound of the primordial sound (nāda), and produced by closing the ears. **1943** D. Gascoyne *Poems 1937–1942* 50 The incoherent Nada of the Seer. **1960** Swami Prajñānānanda *Hist. Ind. Music* ii. 36 The *nāda* or sound is considered as a primordial power, and upon this primal ground all the phenomena of Indian music are built. **1960** Koestler *Lotus & Robot* i. ii. 99 The last stages before samadhi: the appearance of an 'inner light', and of various 'inner sounds' or *Nādas*. **1968** *Indian Mus. Jrnl.* VII. 5 Nada, an interior, literally 'resonant sound', but like most Sanskrit words its overtones are more than the literal and precise English. It is much more complex, implying 'vital power'. **1974** P. Holroyde *Indian Music* 274 Naad or Nada.

‖ **nada**² (na·da, nā·ða). [Sp.], = nothing, f. L. (*res*) *nata* thing born; small, insignificant thing.] Nothing; nothingness, non-existence; a state or condition as of non-existence.
1933 E. Hemingway *Winner take Nothing* 23 It was all a nothing and a man was nothing too... He knew it all was nada y pues nada y nada y pues nada. Our nada who art in nada, nada be thy name. **1939** Joyce *Finnegans Wake* 521 Vurry nothing, O pothers!.. It amounts to nada in pounds or pence. **1947** *Horizon* XV. 160 The sleepless man—the man obsessed by death, by the meaninglessness of the world, by nothingness, by nada—is one of the recurring symbols in the works of Hemingway. **1962** *Spectator* 25 May 685/1 This sense of the endless *nada* lying beyond the phenomenological world. **1966** E. Figes *Equinox* 145 A mess, or less than a mess: nothing, nichts, nada, niente, rien. **1974** *Punch* 25 Sept. 493/2 *Hudson*: Will there be anything else, old one? *Mr. Bellamy*: Nada, Hudson. Nada y pues nada y pues nada.

Na-Dene (nā·dene·). Also **Na-Déné, Nadene.** [f. Athapascan *na* cogn. with Haida *na* to dwell, Tlingit *na* people + Northern Athapascan *Dene* people.] The name given to a North American Indian linguistic family consisting of the Athapascan, Eyak, Haida, and Tlingit languages. Cf. *Athapascan, *Haida.
1915 E. Sapir in *American Anthropologist* XVII. 534 (*title*) The Na-Dene languages, a preliminary report. *Ibid.* 558 The most important North American branches: Eskimo, Na-Dene. **1932** D. Jenness *Indians of Canada* ii. 20 Philologists.. have discovered no kinship among any of the eleven Canadian tongues, unless perhaps between Haida, Tlinkit, and Athapaskan, which may belong together under the name of Na-Dene. **1954** Pei & Gaynor *Dict. Linguistics* 143 *Na-Dene*, according to Rivet, a family of North American Indian languages, consisting of the Athapascan, Haida and Tlingit. **1957** *Encycl. Brit.* V. 183/1 The Nadene languages, probably the most specialized of all, are tone languages, and, while presenting a superficially 'polysynthetic' aspect, are built up, fundamentally, of monosyllabic elements of prevailingly nominal significance. **1965** *Canad. Jrnl. Linguistics* Spring 99 This could be exemplified, hypothetically, by a discovery that Haida does not belong with the other Na-Dene languages. **1968** H.-J. Pinnow in *Internat. Jrnl. Amer. Linguistics* XXXIV. 204 (*title*) Genetic relationship vs. borrowing in Na-Dene. **1970** C. Laird *Language in America* (1971) xv. 53 Speakers of Na-Dene.. are today found in the lush Pacific Northwest. *Ibid.*, One of the Na-Dene

Naderism (nā·dəriz'm). [f. name of Ralph *Nader* (b. 1934), American lawyer + -ism.] Public agitation for greater safety and higher quality in consumer goods.
1969 *Britannica Bk. of Year* (U.S.) 800/2 *Naderism*, a protest against defective consumer goods in the manner of the American lawyer Ralph Nader. **1970** *Americana Annual* 725 The phenomenon [sc. consumerism] was also called 'Naderism', because of the successful crusades of a young lawyer, Ralph Nader, against unsafe automobiles, industrial hazards, and environmental pollution. **1971** *Guardian* 2 Oct. 10/8 While Medawar is sold on Naderism for his own operation.. he has not advised the CA to change their style. **1973** *Listener* 22 Mar. 380/2 It need not be doubted that Naderism, like the Civil Rights campaigns, is a product of American affluence.

nævus. Add: Also (chiefly *U.S.*) **nevus.** (Further examples.)
1913 *Jrnl. Amer. Med. Assoc.* 1 Feb. 341/1 (*heading*) Melanotic sarcomas resulting from irritation of pigmented nevi. **1961** R. D. Baker *Essent. Path.* xx. 545 The word nevus has been used to designate skin blemishes of various sorts including 'birthmarks', but in current usage nevus means the ordinary mole of the skin. **1971** *Brit. Med. Bull.* XXVII. 69/1 Trivial congenital defects such as.. pigmented or vascular naevi.

Naffy, var. *N.A.A.F.I.
1937 Partridge *Dict. Slang* 548/2 *Naffy* or *Naffy*; properly *Naafi*; loosely *Naffy*.. The canteen: naval and military: from 1920. **1943** Hunt & Pringle *Service Slang* 47 *Naffy*, the services are supplied with beer and tea by the N.A.A.F.I... There are fully equipped shops and grocery stores in addition to the station canteens and mobile canteens which serve outposts. *Naffytime*, the morning break. **1945** J. Maclaren-Ross *Bitten by Tarantula* 36 He'd dropped his notebook in a Naffy. **1945** Partridge *Dict. R.A.F. Slang* 49 *Naffy Romeo*, a ladies' man—addicted to treating the W.A.A.F. personnel with Naffy refreshments. *Naffy gong*, 1939–45 star (medal). Since late 1943. **1948** A. M. Taylor *Long Way* 57 Two-way steamy fights. **1958** *Listener* 24 Apr. 18 Special with the wartime invasion.. but NAAFI clubs were almost invariably called Naffies. **1959** 'M. Ainsworth' *Murder is Catching* 10 Best if I drop you at the Naffy.

nag, *sb.*³ (Later examples.)
1971 J. Gardner *Every Night's a Bullfight* ix. 261 They had covered all their separate problems, yet that nag in Douglas's mind left him edgy about the box office situation. *Ibid.* x. 293 The added knowledge served to compound his growing nag of worry.

nag, *v.* Add: **3.** Also *transf.* and *fig.*
1921 *Challenge* 28 Oct. 375/1 He nags his brain into a state of consuming doubt, but does not arrive at any conclusion. **1958** L. Durrell *Balthazar* viii. 172 As I examined him a phone started to nag somewhere. **1960** *Times* 20 June 4/1 Jarvis, a fighter still nagged by his shoulder. **1963** *Times* 4 Mar. 3/1 Barrington was a little out of sorts with himself, and after Trueman had let him down by curling the ball into the wind, Reid nagged him out. **1969** *Listener* 6 Mar. 314/1, I am told that R. P. Blackmur used to give a lecture on Jane Austen in which he explored her work in terms of the verb 'to nag' ('she nags out her plots').

naga¹ (nā·ga). *India.* [Skr. *nāgá* serpent, snake.] In Hindu mythology, one of a race of serpent-demons, the offspring of Kadru and the sage Kasyapa, supposed to be the guardians of Patala or the regions under the earth.
1785 C. Wilkins tr. *Bhăgvăt-Gēētā* 155 The *Nāgs* are serpents fabled with many heads. **1810** E. Moor *Hindu Pantheon* 391 On the other side of Nandi, or the bull, is Naga; his hooded head upreared; his length coiled under him. **1821** J. Leyden tr. *Malay Annals* ii. 21 'Make no noise,' said Malin, 'it is some great snake or naga.' **1848** H. H. Wilson in *Asiatick Researches* XVI. 462 Kulika is one of the eight chiefs of the *Nāgas*, or serpents of Pátála. **1852** C. Coleman *Mythol. of Hindus* i. xiii. 254 The fifth lunar day of Sravana is held sacred to the Nagas. **1872** Monier-Williams *Indian Wisdom* xix. 430 All the Nagas are described as having jewels in their heads. **1906** *Encycl. Relig. & Ethics* II. 803/1 Nāgas are figured on numberless sculptures all over India, and in popular tales they and their beautiful daughters play an important part. **1948** P. J. Thomas *Epics, Myths & Legends of India* (ed. 3) x. 86 The Nagas are said to be considered as the superior of their half-brother Garuda. **1967** V. Ions *Indian Mythol.* 117/2 The Nagas called upon Vishnu to save them, and he descended to Patala in the form of Narmada (personification of a river). **1968** B. Walker *Hindu World* II. 388 The Nagas possess secrets little dreamt of by creatures living on the surface, and theirs is a realm of magic and mystery.

Naga² (nā·ga). *India.* [a. Hindi *naṅga*, f. Skr. *nagná* cogn. w. Naked *a.*] A naked mendicant belonging to any Hindu sect, *spec.* such an ascetic belonging to a Dadu Panthi subsect whose members are allowed to carry arms and serve as mercenaries.
1828 H. H. Wilson in *Asiatick Researches* XVI. 80 The Dádú Pant'his are of three classes.. The *Nágas*, who carry arms, which they are willing to exercise for hire, and, amongst the Hindu princes, they have been considered good soldiers. *Ibid.* 133 Nagas. All sects include a division under this denomination... They carry their secession from ordinary manners so far as to leave off every sort of covering. **1879** *Rajputana Gazetteer* II. 147 The Nagas of Jaipur are a sect of militant devotees belonging to the Dádú Panthi sect, who are enrolled in regiments to serve the State; they are vowed to celibacy and to arms. **1921** *Encycl. Relig. & Ethics* IX. 123/1 All these Jaipur Nāgas are vowed to celibacy, and their numbers are replenished by children placed by parents under their charge as disciples. **1939** C. von Fürer-Haimendorf *Naked Nagas* i. 1 Several Englishmen.. whom I told him of my plans to work among the Nagas, thought that I wanted to study people whose nakedness had religious grounds.

Naga³ (nā·ga). *India.* [Of disputed origin: perh. f. Skr. *nága* mountain or f. Skr. *nagná* naked.] **a.** A member of a group of peoples living mainly in the Naga hills which divide Assam from Burma; a native or inhabitant of Nagaland. Also *attrib.* **b.** The Tibeto-Burman language of these tribes.
1837 J. M'Cosh *Topography of Assam* xiv. 156 The next border tribes met with in proceeding westward are the Nagas.. The Nagas go literally naked in their hills. **1841** W. Robinson *Descr. Acct. Assam* 380 The origin of the word Naga is unknown; but it has been supposed by some to have been derived from the Sanskrit.. and applied in derision to the people, from the paucity of their clothing. **1853** *Jrnl. Indian Archipelago* VII. 15 Several of the adjacent languages—Kyen, Heuma, Naga &c., have *v, f*, or both. **1874** *Jrnl. Asiatic Soc. Bengal* XLI. i. 84 The Rajah had first intended to fly to the Naga Hills, but from fear of our army, the Nagas would not afford him an asylum. **1890** Kipling *Barrack-Room Ballads* (1892) 57 We've chivied the Naga an' Looshai, we've give the Afreedeeman fits. **1903** [see *Kachin]. **1928** C. Dawson *Age of Gods* ix. 195 Assam and Manipur, where a degraded form of megalithic culture still survives among the Naga and Kachin tribes. **1939** C. von Fürer-Haimendorf *Naked Nagas* i. 7, I left Vicerregal Lodge enriched.. in my knowledge of Naga cuisine. **1964** R. Perry *World of Tiger* xvi. 270 A man-eater.. killed fifty-two Nagas in eight months. **1971** N. Rustomji *Enchanted Frontiers* xvi. 257 At a mammoth meeting held by the Nagas at Mokokchung in October 1959, known as the Third Naga People's Convention, a demand was formulated for the constitution of the Naga hills as a new State within the Indian Union, to be named Nagaland. **1972** Prakash Singh *Nagaland* i. 9 The appellation 'Naga' was quite foreign to the Nagas themselves until very recent times.. The appellation 'Naga' was actually given to these hill tribes by the plains people. **1973** *Times* 14 Nov. 18/3 The Nagas are a group of 20 tribes of Sino-Tibetan origin, now numbering about half a million.

‖ **nagaika** (nägāi·kă). Pl. **-kas, -ki.** [Russ., of Turkic origin.] A thick plaited whip used by Cossacks.
1842 tr. *J. G. Kohl's Russia* 410 The genuine Cossack of the steppe.. depends only on his well plaited nagaika or whip, with which he rarely fails to cut down a wolf, as a sabre. **1914** N. Jarintzoff *Russia: Country of Extremes* vii. 102 The *nagaiki* flashed in the air, heavily lashing people's faces. **1917** *Daily Chron.* 22 June 1/4 The Cossacks.. drove off the agitators from the station with their nagaikas. **1928** F. Utley tr. *Astrov's Illustr. Hist. Russ. Rev.* I. x. 64 An old, grey-haired police officer.. gave the command: 'Charge with the nagaikas!' (whips).

nagana (nägā·nă). [ad. Zulu *nakane*.] A disease of domestic animals in southern Africa, characterized by fever, lethargy, and œdematous swellings, and caused by the hæmoflagellate parasite *Trypanosoma brucei* which is transmitted by tsetse flies of the genus *Glossina*. Also *attrib.*

1895 D. Bruce (*title*) Preliminary report on the tsetse fly-disease, or nagana, in Zululand. **1896** *Nature* 16 Apr. 567/1 Nagana pursues a much slower course in cattle than in horses. **1904** *Q. Rev.* July 120 The 'fly districts' where nagana disease is rife. **1904** *Brit. Med. Jrnl.* 20 Aug. 368 Nuttall of Cambridge..experimented on the conveyance of the nagana trypanosome from sick to healthy animals. **1925** *Times* 29 Dec. 11/3 It was believed that wild game.. formed a permanent reservoir from which tsetses could convey 'nagana' to domestic stock. **1931** J. S. Huxley *What dare I Think?* i. 33 Human sleeping sickness and nagana disease of cattle, [are] transmitted by tsetse-flies. **1947** [see *horse-sickness* (*Horse *sb.* 27 a)]. **1965** *New Scientist* 26 Aug. 504/3 Wild game don't suffer from nagana, the tsetse-borne trypanosome disease that disastrously affects domestic cattle. **1970** Jubb & Kennedy *Pathol. Domestic Animals* (ed. 2) I. iv. 324/1 *Trypanosoma brucei* is a cause of nagana in most domestic species in Africa, but man is refractory.

nagelfluh (nā·gĕlflū). *Geol.* Also 9 **nagelflue,** and with capital initial. Pl. **-fluhe.** [a. G. *nagelfluh*, f. *nagel* Nail *sb.* + Swiss Ger. *fluh* (S. Ger, dial. *flüe*) rock face.] A massive conglomerate which accompanies the molasse in the Swiss Alps and contains pebbles supposed to look like nail-heads.
1808 R. Jameson *Syst. Mineral.* III. ix. 210 Nagelfluh.—This rock is usually composed of fragments of limestone, more or less rounded, and of various magnitudes, cemented together by a basis of calc-sinter. **1849** *Q. Jrnl. Geol. Soc.* V. 229 If the masses of nagelflue which constitute the Rigi mountain near Lucerne, and the still loftier Speer..near Wesen be included in one group, their thickness must be enormous, certainly exceeding 6000 or 8000 feet. **1879** *Encycl. Brit.* X. 321/2 The well-known nagelflue of Switzerland..can be shown from its fossil contents to be essentially a lacustrine formation. **1912** T. G. Bonney *Building of Alps* i. 25 The gravels, the so-called Nagelfluhe, are occasionally fully a mile in thickness. **1962** L. C. King *Morphology of Earth* iv. 98 The presence of conglomerates and breccias grading into arkoses shows the rapid erosion of adjacent highlands... Thus arose the coarse *Nagelfluh* outwash of the Alps which passes northward into finer grades of sediment.

nagged (nægd), *ppl. a.* [-ed¹.] That is subject to nagging. Hence, downtrodden, wearied, irritated, annoyed.
1893 K. D. Wiggin *Polly Oliver's Problem* ix. 107 Existence was wearing a particularly dismal aspect on that afternoon... He felt 'nagged', injured, blue, out of sorts with fate. **1961** M. Spark *Prime of Miss Jean Brodie* ii. 35 'Mary, don't you *want* to walk tidily?' 'Mary,' said Sandy, 'stop staring at the brown man.' The nagged child looked numbly at Sandy and tried to quicken her pace.

nagging, *ppl. a.* Add: **1.** Also of thirst.
1906 E. Dyson *Fact'ry 'Ands* vi. 67 He sighed frequently; his nagging thirst got at him again.
2. b. (Further examples.)
1946 W. S. Maugham *Then & Now* xxviii. 162 He had not the strength to withstand the nagging arguments of the others. **1953** R. Macaulay *Let.* 11 Aug. (1961) 106, I think nagging doubts..would always from time to time raise their heads and disturb.

naggingly (næ·giŋli), *adv.* [f. Nagging *ppl. a.* + -ly².] In a nagging or insistent way; in a persistently irritating, annoying, or exhausting manner.
1936 P. Fleming *News from Tartary* vi. 98 Things had gone quietly, naggingly against us all the way. **1951** M. Leinster in D. Knight *100 Yrs. Sci. Fiction* (1969) 209 Haynes wished he could talk with him once more—talk sensibly, quietly, without..this naggingly demanding wonderment.

naggish (næ·giʃ), *a.*² [f. Nag *sb.*¹ + -ish¹.] Of horses: suggestive of a nag; small, inferior.
a **1800** *Spirit Farmer's Museum* (1801) 204, I see some here in gay coats, mounted on naggish horses.

naggy (næ·gi), *a.*² [f. Nag *sb.*¹ + -y¹.] Of horses: inferior in size or quality; naggish.
a **1861** T. Winthrop *John Brent* (1883) vii. 54 The little villain's mount was a red roan, a Flat-head horse, rather naggy, but perfectly hardy and wiry.

‖ **Nagmaal** (na·χmāl). *S. Afr.* Also **Nagtmaal**. [Afrikaans, f. *nag* night + *maal* meal.] Also **Nachtmaal** [Du., = nightmeal]. The Lord's Supper, or Holy Communion; also the Communion service, or the occasion on which the service is held.
1835 A. Steedman *Wanderings S. Afr.* I. 184 During '*naacht-maal*' (the administration of the Lord's Supper,) the village becomes a scene of great bustle and activity. **1842** R. Godlonton *Sk. Eastern Districts Cape G.H.* 50 The period for the quarterly administration of the sacrament, (or *nachtmaal*). **1852, 1871** [see *Dutch *a.* 2 b]. **1916** J. Buchan *Greenmantle* iii. 35 It was as slow as a vrouw ooming from *nachtmaal*. **1925** *British Weekly* 25 Nov. 182/1 Shy young Dutchmen in 'nachtmaal' clothes. **1930** R. Campbell *Adamastor* 23 The baboons..all their Simian dignity forgot Would hold a sort of Nagmaal on the spot. **1945** *Cape Argus* 7 Aug., The accident occurred while the ..family were on their way to Nagmaal on Sunday. **1948** *Ibid.* 5 June (Mag. Sect.) 2/4 At nagtmaal the farmers drove in from miles around and opened up their little dorp dwellings. **1953** *Cape Times* 15 Apr. 13/4 They would be utter fools to put her there in broad daylight at *Nagmaal* time. **1974** *Sunday Times* (S. Afr.) 24 Nov. 5 Seldom-worn black suits bought decades before to last a lifetime of weddings, Nagmaals and funerals.

Nago (na·go). [f. Ewe *anagó* a Yoruba Negro.] **a.** A member of an African Negro people, originally Yoruba-speaking, of whom many were taken to the Americas as slaves. **b.** The language of this people, now applied *spec.* to the lingua franca or pidgin form spoken by this people in Brazil. Also *attrib.* or as *adj.*
1775 *Jamaica Gaz.* 25 Mar., [A man] of the Nago country..says his master died when he was a boy and has been run-away ever since. **1793** B. Edwards *Hist. Brit. Colonies in W. Indies* II. 73 Many of the Whidah Negroes are found to be circumcised... It is practised universally by the Nagoes. **1825** R. Bickell *W. Indies* 43 Frank, a Nago, 5 ft 7½ in. no brand-mark, country marks on his face. **1942** D. Pierson *Negroes in Brazil* iii. 73 It is likely that Nagô was not maintained in a pure form but came to be a patois containing numerous elements from other African dialects as well as from Portuguese. **1949** *Caribbean Quarterly* I. i. 11, 33 were Nagoes..from the Slave Coast. **1968** W. J. Samarin in J. A. Fishman *Readings Sociol. of Lang.* 665 As linguistic curios a few others might be cited:..Nago, probably based on Yoruba and ultimately used only in certain Brazilian pseudo-African cults.

nagsman (næ·gzmæn). [f. Nag *sb.*¹ + Man *sb.*¹ 4 p.] A skilled horseman who is employed to train or show horses.
1891 W. A. Kerr *Practical Horsemanship* x. 171 The nagsman who will ride him up and down the yard is pretty certain to be an artist in the saddle, one who, as he gets 'the office', rides either to sell or to buy. **1903** F. M. Ware *First-Hand Bits of Stable Lore* viii. 107 A 'nagsman' handling a green and raw horse may seem..to be rough in his treatment. **1952** R. S. Summerhays *Encycl. for Horsemen* 191/1 *Nagsman*, a horseman who, by his skill, rides to improve a horse, whether as a ride, or on account of some vice or bad manners. **1955** H. Smith *Horseman through Six Reigns* xxi. 203 If the rider's body is still..it is generally a sign that the horse is a good ride, though an expert showman or nagsman can cover up a lot. **1959** *Times* 25 Apr. 9/3 What the P.T. instructor and the drill sergeant do for the recruit..the nagsman does for the colt. **1963** E. H. Edwards *Saddlery* ix. 78 While it might be of assistance to the experienced nagsman, it could be very dangerous in other hands.

Nagtmaal: see *Nagmaal.

nah¹ (nā), a representation of a colloq. or vulgar pronunciation of Now *adv.* Also **na.**
1847 E. Brontë *Wuthering Heights* II. v. 109 He'll hev his lad; und Aw mun tak him—soa nah yah knaw! **1907** Shaw *Major Barbara* ii. 245 Wot prawce Selvytion nah? **1954** E. Pangborn *Mirror for Observers* (1955) ii. iv. 135 'Did you think I was sore?' 'You had a right to be.' 'Nah. Hold everything.' **1959** B. Kops *Hamlet of Stepney Green* i. 108 Na Davey, what can I say to you? **1959** A. Wesker *Chicken Soup with Barley* i. i. 176 Two o'clock they plan to march—nah!

nah² (nā), a representation of a colloq. or vulgar pronunciation of No *adv.*³ Cf. Na *adv.*²
1920 C. Sandburg *Smoke & Steel* 45 Nothin' ever sticks to my fingers, nah, nah, nothin' like that. **1950** *Astounding Sci. Fiction* Nov. 97 'I'll fight you,' offered Borklin. His huge fists closed. 'Nah—why? Wouldn't be a fight, anyway.' **1961** I. Jefferies *It wasn't Me!* iii. 46 'I don't need permission...' 'I think you do.' 'Nah,' I said. **1961** J. Heller *Catch-22* (1962) xxxviii. 393 Havermeyer shook his head thoughtfully. 'Nah, I couldn't do that.' **1966** *New Society* 12 May 9/2 The waiter knows better. 'Nah, you don't want herrings, I'm gonna give you the soup.' **1973** *Time Out* 2–8 Mar. 13/2 Nah, she don't know.

‖ **nahal** (nahā·l). Also with capital initial. [Heb., f. initials of *No'ar Halutzi Loḥem*, Pioneer and Military Youth.] The name of a military youth organization in Israel, used to designate an agricultural settlement manned by members of this organization. Also *attrib.*
1963 D. R. Elston *Israel* ii. 79 *Nahal* members are under no compulsion to continue as agricultural settlers once their period of service is over. **1964** L. Deighton *Funeral in Berlin* xlii. 272 Our people in the *nahals* have got to pack a punch. **1969** *Times* 15 Nov. 7/7 The J.N.F. directly promotes the Israeli war effort by helping to establish *nahal* military-agricultural colonies. **1972** I. T. Naamani *Israel* ix. 131 in 1950, Pioneer and Military Youth (Noar Halutzi Lohem, or Nahal) was founded. This was an organisation of young people who wished to serve in the army as units and later continue as a group to establish for themselves or help others establish agricultural settlements on the borders. Usually the members of Nahal received several months of intensive military training and then were assigned to a frontier village, where..they underwent rigid agricultural tutoring. **1973** *Guardian* 21 Apr. 3/6 It has always been an axiom of Israeli planning that borders must be peopled... The first 12 villages have already been established. All were started as 'nahal' settlements by young soldiers, combining military and agricultural training.

nahcolite (nā·kŏlait, nākōu·lait). *Min.* [f. $NaHCO_3$, its chemical formula + -lite.] Native sodium bicarbonate, found as colourless, transparent, monoclinic crystals.
1928 F. A. Bannister in *Nature* 1 Dec. 866/1 The name nahcolite is proposed for naturally occurring sodium bicarbonate. These incrustations were found lining the walls of a cuniculus near the Stufe di Nerone, Baia, Naples, Italy. **1929** —— in *Min. Mag.* XXII. 60 Dr. L. J. Spencer has very kindly suggested an ingenious name, *Nahcolite*. **1940** *Amer. Mineralogist* XXV. 777 The origin of nahcolite in the Stufe de [*sic*] Nerone and in a lava tunnel at Vesuvius can be easily explained by the action of CO_2 and water vapor upon thermonatrite or trona. **1971** *Prof. Papers U.S. Geol. Survey* No. 750-B. 194 This report describes a new analytical technique for determining the amount of nahcolite in oil shale.

Na'htchi, var. *Natchez.

Nahua, var. *Nahuatl *sb.* and *a.*
1948 D. Diringer *Alphabet* 122 The Nahua civilization of Mexico. **1957** E. Hyams *Speaking Garden* vi. 70 *Tumatl* is a Nahua word. **1962** *Listener* 23 Aug. 270/2 Mexico's own Nahua and Mayan cultures.

Nahuatl (nā·wăt'l), *sb.* and *a.* Also **Nahuat, Nahuatla, Nahuatlac.** [f. Sp., f. Nahuatl.] **A.** *sb.* **a.** A group of peoples of Southern Mexico and Central America which includes the Aztec; a member of these peoples. **b.** The Uto-Aztecan language of this people. **B.** *adj.* Of or pertaining to the Nahuatl people or their language.
1822 J. Black tr. *de Humboldt's Essay on Kingdom of New Spain* I. 138 The Toultecs,..the Acolhuas, and the Nahuatlacs, all spoke the same language as the Mexicans. **1873** A. Lang in *Fortnightly Rev.* May 624 The rite itself was actually practised by the Nahuatls in America. **1877** L. H. Morgan *Ancient Society* ii. vi. 181 The question of the organization of these, and the remaining Nahuatlac tribes of Mexico, in gentes will be considered in the next ensuing chapter. **1891** Z. Nuttall *Atlatl or Spear-thrower of Ancient Mexicans* 6 The Nahuatl text of Sahagun's invaluable Manuscript Historia. **1902** *Encycl. Brit.* XXV. 373/2 The dominant [American] ethnic groups:—Eskimo, on Arctic shores;.. Nahuatla-Maya, Southern Mexico and Central America. **1915** E. Sapir in *American Anthropologist* XVII. Jan.–Mar. 99 The total number of Southern Paiute consonants that have to be directly accounted for in terms of Nahuatl consonants is considerably less. **1921** —— *Language* iv. 70 The compounded object of a verb precedes the verbal element in.. Nahuatl. **1948** C. L. B. Hubbard *Dogs in Britain* xix. 215 The Aztecs (tribal name for the last Nahuatl immigrants into the Mexican valleys). **1971** *Language* XLVII. 737 The language as a whole, once called Mexicano, has more recently been referred to as Nahuatl (in some cases Nahuat). **1973** *Guardian* 23 Mar. 12/6, I was going to learn Nahuatl, the Aztec language, which some of the two-dollar whores could speak.
So **Nahuatlan**, the group of Nahuatl dialects; also as *adj.*
1902 *Encycl. Brit.* XXV. 374/1 The [linguistic] families of..Middle America..:—..Nahuatlan, Mex.; Otomitlan [etc.]. **1933** L. Bloomfield *Language* iv. 72 A *Uto-Aztecan* family has been proposed..to include..the great *Nahuatlan* family in Mexico, including *Aztec*... Beside Nahuatlan, we may mention the *Mayan* family in Yucatan. **1953** *Amer. Speech* XXVIII. 62 The ending *-an*..is used after vowels as well as after consonants..*Tibetan*, *Nahuatlan*, [etc.].

naiad. Add: **2.** *Ent.* = Nymph 3.
1918 J. H. Comstock in *Ann. Ent. Soc. Amer.* XI. 224, I, therefore, propose the restriction of the term nymph to the designating of the early stages of insects with a gradual metamorphosis and the use of the term naiad for designating the immature stages of the Plecoptera, Odonata, and Ephemerida. **1930** W. J. Lucas *Aquatic (Naiad) Stage Brit. Dragonflies* i. 1 The two stages in a dragonfly's life are no less strikingly unlike... Hence arises the necessity for the student of our dragonfly fauna to study and classify the naiads independently of the imagines. **1942** E. O. Essig *College Entomology* i. 11 Naiads of aquatic insects are frequently referred to as nymphs. **1972** Swan & Papp *Common Insects N. Amer.* xv. 171 Metamorphosis in the group [*sc.* the Odonata] is incomplete or gradual. The naiads or nymphs are important predators of mosquito larvae.
3. *Bot.* A submerged aquatic plant with linear leaves, belonging to the family Najadaceæ, which includes only one genus, *Najas* (or *Naias*).
1966 *Rhodora* LXVIII. 216 (*title*) Two new naiads from Illinois and distributional records of the Naiadaceae.

naice (nēs), *a.* Representation of an affected pronunc. of Nice *a.*; freq. joc. or derogatory.
1925 A. Huxley *Those Barren Leaves* ii. iii. 112 'So naice, I always think, these Corner Houses,' says Mrs. Cloudesley. **1941** A. Christie *Evil under Sun* v. 82 Ay am sure it has always been the quayettest [*sc.* quietest] place imaginable! The people who come here are such naice people. **1950** *Spectator* 29 Sept. 354/2 If you want to read about nice people who are really nice and not in the least 'naice', you will enjoy..*Rain on the Wind*. **1962** J. Cannan *All is Discovered* iii. 49 Having agreed..that it was a naice morning. **1973** *Observer* (Colour Suppl.) 17 June 46/4 The girls in well-cut tweeds from the 'naice' public schools.

naïf, *a.* Add: **1.** Also as *sb.*
1893 G. B. Shaw in *World* 10 May 28/2 You will sometimes find some naïf doing this, and verdantly assuming that *his* point of view commands the absolute truth. **1932** T. S. Eliot *Sel. Ess.* 305 He [*sc.* Blake] becomes the apparent naïf, really the mature intelligence. **1968** L. Durrell *Tunc* ii. 36 Somewhere inside she was a naïf—always a bad sign in a woman connected with politics and public life. **1975** *Times Lit. Suppl.* 20 June 701/1 The Bronowski who concluded this was no mathematical naïf. .. He had spent a whole decade as a senior lecturer in mathematics.

b. *Art.* See *Naïve *a.* 1 c.
1947 M. McCarthy in *Partisan Rev.* XIV. 178 As in the case of the *naif* painters, his very faults, the crudity of his conceptions,.. become part of the subject. **1955** —— *Charmed Life* (1956) iii. 58 If Warren had been a carpenter or a plumber, he could have made his marks as a *naif* painter. **1974** *Times* 7 Jan. 8/1 At the age of 39, Haddelsey is one of the world's leading *naif* painters.

nail, *sb.* Add: **13. a.** *nail brush* (earlier example), *enamel, file, lacquer, -pick, polish, scissors* (earlier and later examples, also *transf.*), *varnish*; *nail buffer, clipper(s, polisher.*
1802 M. Edgeworth *Let.* in E. Inglis-Jones *Great Maria* (1959) iv. 65 Enter Miss Linwood who looks not unlike a strolling player... I *wish* she had a nail brush. **1908** *Sears, Roebuck Catal.* 328/1 Solid Silver Nail Buffer, fancy handle, 4½ inches long. **1971** *Petticoat* 24 July 3/4 There are six appliances for the nails—two files, a nail buffer, cuticle stick, nail brush and callous remover. **1945** J. Steinbeck *Cannery Row* xvi. 67 Dora..was there to buy a pair of nail-clippers. **1969** *Sears, Roebuck Catal.* 869/3 Resco Nail Clipper. Helps guard hosiery, furniture, pet's paw-health. **1907** Nail enamel [see *nail polish* below]. **1913** *T. Eaton & Co. Catal.* Spring & Summer 177/3 Lustrite Nail Enamel..will produce a brilliant lustre. **1972** *Vogue* June Special 90 Blueberry Wine Nail Enamel. **1881** *Graphic* 21 May (back cover) (Advt.), Nail file. Cigar cutter. Scissors. **1894** *Country Gentlemen's Catal.* 173/2 Pair nail scissors, button-hook and nail-file. **1922** F. Courtenay *Physical Beauty* 47 When you have shaped the external edge of the nails with a fine pair of scissors, finish with emery or a steel nail file. **1969** *Sears Catal.* Spring/Summer 397 Sapphire-coated Nail File.. Two surfaces..one for shaping, one for finishing. **1966** J. S. Cox *Illustr. Dict. Hairdressing* 101/2 *Nail lacquer*, nitrocellulose dissolved in Butyl Acetate and allied substances forming, when dry, a hard, tough, resistant film on the surface of the nail. **1973** J. Rossiter *Manipulators* viii. 90 A cosmetician's display of nail lacquers and lipstick cases. **1974** *Harpers & Queen* Sept. 50/3 We want to change our nail lacquers as often as our clothes. **1947** J. Steinbeck *Wayward Bus* xvii. 222 He took his gold nail-pick from his pocket and opened it and cleaned his nails. **1907** *Yesterday's Shopping* (1969) 538/3 Diamond Nail Enamel..box 0/10½... Electric Nail Polish..bot. 1/6. **1937** *Discovery* Mar. 86/1 It [*sc.* diatomite] is used in face powders and nail polishes. **1972** 'H. Carmichael' *Naked to Grave* i. 15 A tiny bottle of pink nail polish and another of nail polish remover. **1897** *Sears, Roebuck Catal.* 435/2 Chamois Nail Polisher, fancy Solid Sterling Silver mountings. **1913** E. Wharton *Custom of Country* i. i. 13 Mrs. Heeny, driving her nail-polisher cheeringly. **1854** J. E. Millais *Let.* 5 June in M. Lutyens *Millais & Ruskins* (1967) 222, I have had to operate upon myself with a pair of nail scissors. **1919** W. H. Downing *Digger Dial.* 35 *Nail scissors*, the crossed sword and baton worn as a badge of rank by a General. **1951** *Catal. of Exhibits, South Bank Exhib., Festival of Britain* 60/1 Baby's nail scissors ..Cuticle nail scissors. **1926–7** *Army & Navy Stores Catal.* 495/2 'Elite' Nail Varnish..bot. 1/6. **1936** *Chem. Abstr.* XXX. 7784 Cosmetics in pharmacy... A discussion of face powders, foundation creams,..shampoos and nail varnishes. **1954** *Granta* 6 Nov. 23/2 One uses nail varnish the other not. **1972** *Vogue* Feb. 46/2 Nail Varnish, Revlon's Ultima II. **1974** *She* Jan. 63/4 Nail Varnish Remover pads, 14p for 17.
c. *nail-biting* (further examples).
1941 Koestler *Scum of Earth* 151 But what was the symbolical meaning of all these..nail-biting..figures? **1959** *Ann. Reg. 1958* 190 Sinclair Weeks said..that the outlook was 'far better than the nail-biting pessimists think'. **1973** *Times* 2 Oct. 21/4 The next few weeks will be nail-biting ones for Leonard Grouse.
14. a. *nail boot, factory, machine* (earlier example), *mill*; **nail bomb,** a lethal weapon, used *esp.* by urban guerrillas, made from nails wrapped round a stick of gelignite; **nail-gall,** a nail-shaped gall produced on the leaves of lime and other trees by a mite of the genus *Phytoptus*; **nail-plate** (earlier and later examples); **nail-stubb,** a worn horse-shoe nail; a stub-nail; **nail violin** (see quot. 1959).
1971 *New Scientist* 26 Aug. 483/3 They will assess the effects on the elderly, the sick, children..of the explosives, nail-bomb, indiscriminate rifle and machine gun fire. **1971** *Guardian* 11 Oct. 20/3 A boy of about 10 threw a nail bomb at troops in Belfast yesterday. **1973** R. Clutterbuck *Protest & Urban Guerrilla* x. 111 The same unemployed teenage boys..turned out..almost every evening to throw stones and nail-bombs at army posts and patrols. **1973** 'I. Drummond' *Jaws of Watchdog* x. 135 These revolutionaries were not the ill-shaven backstreet throwers of nail-bombs. They were a disciplined corps. **1923** D. H. Lawrence *Ladybird* 242 The other fellows with sticks and nail-boots had now taken heart and were scrambling like crabs past our hero. **1833** H. Barnard in *Maryland Hist. Mag.* (1918) XIII. 374, I found my old friend..who took me to see..a nail factory [etc.]. **1879** *Encycl. Brit.* X. 46/1 The lime-leaf 'nail-galls' of *Phytoptus tiliae* closely resemble the 'trumpet-galls' formed on American vines by a species of Cecidomyia. **1951** *Dict. Gardening* (R. Hort. Soc.) III. 1345/1 Nail-galls. Galls on leaves of Lime and other trees somewhat resembling tin-tacks driven through the leaf tissues. They result from attacks of species of Phytoptus. **1819** E. Dana *Geogr. Sk. Western Country* 77 Zanesville is ..at the falls, whereon various mills are erected..including..an oil mill, nail machine, and woolen factory. **1850** *Rep. Comm. Patents: Agric. 1849* (U.S. Dept. Agric.) 93 Within its present limits are about fifty cotton factories.. seven rolling, slitting, and nail mills. **1797** in *Essex Inst. Hist. Coll.* (1918) LIV. 107 Agreed with Mr. Allen to work at eight shillings pr. ton..cutting every kind of rods and dubble for iron hoops or nail plates. **1945** *Amer. Speech* XX. 115 Since the cut nail is an American invention, the word *nail-plate* must be of American origin. **1851** H. Melville *Moby Dick* III. xxvii. 176 Look ye, blacksmith, these are the gathered nail-stubbs of the steel shoes of racing horses. **1884** E. Heron-Allen *Violin-Making* v. 108 In the year 1740 a German musician, named Johann Wilde,..invented a curious instrument called a Nail-violin. **1944** W. Apel *Harvard Dict. Mus.* 478/1 There exists a quartet by F. W. Rust for nail violin, two violins and cello. **1959** *Collins Mus. Encycl.* 448/2 *Nail Violin*... It consisted of a semicircular resonator of wood into which were driven U-shaped nails of graduated lengths. The sound was produced by a bow.
b. *nail-driver* (transf. examples).
1823 J. F. Cooper *Pilot* I. viii. 106 The cannon, above which were painted the several quaint names of 'boxer',.. 'exterminator', and 'nail-driver'. **1872** *Life of Bill Hickman* 54 (Th.), I had a nail-driver [*sc.* a horse], very swift, and no end to his bottom.
c. nail-tailed wallaby *Austral.*, a wallaby of the genus *Onychogalea*, distinguished by white stripes on the cheeks and at the top of each limb, and a horny nail near the end of the tail.
1896 F. G. Aflalo *Sk. Nat. Hist. Austral.* i. 43 The Nail-tailed (*Onychogalea*) and Hare (*Lagorchestes*) Wallabies are wanting in Tasmania, though three species of each are found on the mainland. **1896** [in Dict., sense 13 c]. **1970** W. D. L. Ride *Guide Native Mammals Austral.* 53 Three species of rather strikingly marked wallabies are called nail-tailed wallabies because they possess a small dark horny nail, rather like a finger nail, hidden in the dark hair at the end of the slender tail.

nail, *v.* Add: **5. b.** Delete *rare*⁻¹ and add later example.
1902 'Mark Twain' in *Harper's Mag.* Feb. 431/2 Nailing an alibi where it can't be budged.
8. a. Hence also, to arrest. *slang.*
1889 [in Dict.]. **1918** [see *cut *sb.*² 23 d]. **1931** *Amer. Speech* VII. 111 They nailed me right on the border. **1969** C. F. Burke *God is Beautiful, Man* (1970) 29 The cops..nail Ben for havin' the cup.

nailer. Add: **3. b.** [f. *Nail *v.* 8 a.] A policeman or detective. *slang.*
c **1863** T. Taylor in M. R. Booth *Eng. Plays of 19th Cent.* (1969) II. 84 Then there's the Nailer's been after me... What, Hawkshaw, the 'cutest detective in the force? **1935** A. J. Pollock *Underworld Speaks* 79/1 *Nailer*, a uniformed police officer.

nail-head. Add: **1.** Also *fig.*
1855 *Rambler* III. 239 Mr. Forster will have it that the language..scribbled on the rocks of the desert of Sinai, in the Egyptian hieroglyphics, and in the 'nail-head' letters of Assyria and Persia, is all one. **1948** D. Diringer *Alphabet* vi. 358 The vertical strokes ended with wedges or 'nailheads'; this script was therefore termed 'nail-headed'.
2. (Further examples.)
1947 *Sun* (Baltimore) 31 Oct. 3/7 (Advt.), It can look dressed-up or casual with its gold-toned nailheads, oblong gold-toned buckle. **1973** *Philadelphia Inquirer* (Today Suppl.) 14 Oct. 14/1 (Advt.), Zip-front shirt jac with nailhead trim. **1974** *Sumter* (S. Carolina) *Daily Item* 24 Apr. (Belk Stroman Advts. Suppl.) 2 Denim shifts with nailheads, embroidery, polka dot or check trims!

nail-headed, *a.* (Later examples.)
1936 D. Gascoyne *Man's Life is this Meat* 19 In my hand lies the same whispering, nail-headed dude, ever imploring the benefice of a hippograph. **1948** [see *nail-head 1].

nai·l-punch. *Joinery.* [f. Nail *sb.* + Punch *sb.*¹ 2.] A driving punch for nails.
1899 C. G. Wheeler *Woodworking for Beginners* (1900) xvi. 433 *Nail-Set, or Punch*.., for sinking nail-heads below the surface, is quite important, and it is well to have a large one and a fine one. **1951** *Practical Home Woodworking* ii. 36 A nail-punch has a flat 'point'. **1973** *Practical Woodworking* 34 It is sometimes necessary to drive the head of a nail well below the surface and to fill the resulting cavity with putty... Nail sets, or nail punches, as they are sometimes called are designed for this purpose.

nail-rod. Add: **1. b.** (Earlier example.)
1869 *Rep. Comm. Agric. 1868* (U.S. Dept. Agric.) 433 The most convenient method of destroying the bugs is by using a pair of tongs made of nail-rod.
2. (Further examples.)
Quot. 1896 is U.K.
1890 A. J. Vogan *Black Police* 200 He hands our black friend a piece of 'nailrod' with which to charge his evening pipe. **1896** L. Tracy *Final War* xi. 77 Then joy prevailed, and the fumes of nail-rod mingled in the breeze with the strains of 'Annie Laurie'.

Nailsea (nēi·lsī). The name of a town near Bristol, used to designate a style of glassware first made there in the late 18th century. Freq. *attrib.*
[**1897** A. Hartshorne *Old Eng. Glass* xix. 307 Nailsea, near Bristol, was another [source], with darker coloured glass, sometimes blue. **1907** E. Dillon *Glass* 368/2 (*index*) Nailsea glass-works.] **1910** D. Wilmer *Early Eng. Glass* xi. 205 The manufacture of Nailsea glass continued until about the year 1873, some of the latest productions having gilt decoration. **1926** N. H. Moore *Old Glass* i. 170 The Nailsea glass which is of interest to collectors shows specimens very beautiful in color, like the ornamental flasks which were Nailsea's chief product. **1960** H. Hayward *Antique Coll.* 195/1 *Nailsea glass*, the brownish-green glass speckled with white with which the name of Nailsea, near Bristol, is associated, was used on occasions for the making of wine-bottles, and sealed examples exist. **1961** E. M. Elville *Collector's Dict. Glass* 140/1 Also described as 'Nailsea' are the peculiar love-tokens which were..made at all glassworks near coastal towns. **1966** J. Laver *Victoriana* xiv. 173 (*caption*) Bottles decorated with loops of coloured glass, one of many varieties known as 'Nailsea'. **1971** *Guardian* 23 Mar. 32/8 Nailsea style coloured glass for sale..including walking sticks, bells, bottles, bellows.

nai·l-set. *Joinery.* [f. Nail *sb.* + Set *sb.*¹ 33.] = *nail-punch.
1899 [see *nail-punch]. **1927** R. A. Freeman *Certain Dr. Thorndyke* ii. xviii, Carpenters don't fix mouldings on with screws. They use nails and punch them in with a 'nail-set' and stop the holes with putty. **1947** J. C. Rich *Materials & Methods Sculpture* vi. 170 The tools required for chasing may be purchased or fashioned by the sculptor either from tool steel,..or from small steel tools called nail sets. If nail-set tools are used it will be first necessary to remove or draw out the temper, prior to shaping the ends by means of filing into the desired forms. **1973** [see *nail-punch].

naio (nai·o). Also **naeo, naieo.** [Hawaiian.] An evergreen tree, *Myoporum sandwicense*, of the family Myoporaceæ, native to Hawaii and bearing clusters of pink or white flowers; also called bastard sandalwood, as its wood is fragrant. Cf. *ngaio.
1888 W. Hillebrand *Flora Hawaiian Islands* 339 The wood of the 'Naeo' or 'Naieo', most so that of the roots, becomes fragrant on drying. **1915** W. A. Bryan *Nat. Hist. Hawaii* xvi. 222 The bastard sandalwood or naieo is a tree common on the summit of Kaala, and the higher forest belt generally. **1970** S. Carlquist *Hawaii* xv. 275 The naio, *Myoporum sandwicense*, is a common dry forest tree with relatives in New Zealand and on South Pacific islands.

naira (nai·rā). [See quots. 1972.] A unit of currency in Nigeria, equal to 100 kobo.
1972 *N.Y. Times* 9 Aug. 14 On Jan. 1, 1973, Nigeria will scrap her system of Nigerian pounds,..and begin a decimal currency system with units of money called the naira and the kobo... The name is adapted from the word Nigeria. **1972** *Times* 9 Oct. (Nigeria Suppl.) p. viii/4 The new currency being introduced by the Central Bank of Nigeria consists of naira and kobo. Kobo are coins and naira are notes. The origin of the names has not been fully explained, but naira is said to have been derived from the word Nigeria. *Ibid.*, The various denominations of the naira notes, which are now on display in public places, carry pictures showing various aspects of Nigeria's economic activities. The old Nigerian coins which they replace bear the head of the Queen of England. **1974** *Globe & Mail* (Toronto) 2 Oct. 7/3 A number of firms.. were allowed to stay on, provided they raised their capital to at least 200,000 Naira (roughly $350,000).

naïve, *a.* Add: **1. b.** Philos. *naïve realism.* Also *naïf realism.* [The fem. form, despite being adj. to a sb. which in Fr. is masc., has acquired currency in English.] The belief, attributed to non-philosophers, that the world is directly perceived as it really is, as contrasted with philosophical theories of sense-data, the subjectivity of colour, etc. So *naïve realist.*
1882 W. James *Will to Believe* (1897) 290 Even the most *naïf* realism will hardly pretend that the non-table as such exists *in se* after the same fashion as the table does. **1895** F. Thilly tr. *Paulsen's Introd. Philos.* ii. i. 344 We start out from the popular conception whose standpoint is *naïve Realism*. **1897** B. Russell *Ess. Found. Geometry* ii. 93 Those who have done most to further non-Euclidean Geometry —with the exception of Riemann, who was a disciple of Herbart—have usually inherited from Newton a naive realism as regards absolute space. **1909** W. James *Meaning of Truth* ii. 50 The reader will observe that the text is written from the point of view of *naïf* realism or common sense. **1914** C. D. Broad *Perception* i. 1 We are going to begin from the position of naïf realism. It is true that our everyday view of the world is not quite naïvely realistic, but that is what it would like to be. **1932** [see *illusion 4 b]. **1932** H. H. Price *Perception* ii. 26 We ask a Naïve Realist what sort of thing it is whose existence he knows of. *Ibid.* iii. 61 The Naïve Realist conception of 'belonging to' would not have been saved. **1954** A. J. Ayer *Philos. Ess.* vi. 142 The naïve realist is not in error. Naïve realism is not a false theory of perception: it is a refusal to play this sort of game. **1971** A. Flew *Introd. Western Philos.* iii. x. 346 The position supported by the uninstructed 'instincts and propensities of nature' is by instructing philosophers labelled contemptuously, 'Naïve Realism'.
c. *naïve painter*: a painter who has not been trained in a formal manner; so *naïve painting.* Cf. *naïf *a.* 1 b.
1957 *Observer* 3 Nov. 14/4 Naive or primitive painting is a discovery of the twentieth century... The innocent eyes of the often untaught naïve painters give an account of the world quite unhampered by preconceptions of what paintings of it should be like. **1961** *Spectator* 28 July 149 A very intense, fat little book on naïve painters, with Henri Rousseau as the grand point of departure. **1970** *Time* 9 Feb. 54 The United States possesses the oldest, the most original, and just about the most authentic naive painters.

2. *Biol.* and *Psychol.* Not having previously had a particular experience or been the subject of a particular experiment, *esp.* not having taken or received a particular drug; unconditioned. Const. *to*.

1940 *Amer. Jrnl. Psychol.* LIII. 46 (*heading*) Configurational properties considered 'good' by naïve subjects. **1961** *Jrnl. Agric. Sci.* LVII. 401 Assessments of bacon qualities by naïve and experienced judges using photographic reference standards. **1963** *Animal Behaviour* XI. 463 The naïve female cat does not exhibit the complete mating pattern on the first opportunity for sexual behaviour. **1969** *Sci. Jrnl.* Sept. 42/1 When research subjects are volunteers who are naïve to marihuana, an effective placebo is no problem. **1970** *Nature* 11 Apr. 119/1 Interesting social and psychological similarities and differences were revealed among three groups of subjects: marijuana naïve persons (N), persons not naïve to marijuana (NN) and chronic marijuana users (C). **1974** *Ibid.* 19 Apr. 697/1 Twenty experimentally naïve male albino Wistar rats..were anaesthetised with ether.

naïveness (na,ī·vnės). *rare.* [f. NAÏVE *a.* + -NESS.] = NAÏVETY.

1949 KOESTLER *Insight & Outlook* 387 The reconquest of 'naïveness' is one of the periodic revolutions in painting. **1968** *Classical Q.* XVIII. 208 A weakness of the book ..is a certain naïveness of interpretation.

naked, *a.* and *sb.*^{1} **A. 1. b.** Delete †*Obs.* and add further examples.

1654 E. GAYTON *Pleasant Notes Don Quixot* III. iv. 88 As naked and bare as a shorne Sheep, as we say in our English Proverbe. **1860** O. W. HOLMES *Prof. at Breakfast-Table* iii. 39 A friend..had a watch given him ..with a loose silver case... You know them,—the cases that you hang on your thumb, while the..real watch, lies in your hand as naked as a peeled apple. **1879–81** G. F. JACKSON *Shropshire Word-Bk.* 297 W'y yo' bin as naked as a robin. **1890** D. C. MURRAY *John Vale's Guardian* III. xxxviii. 215 Time was I wouldn't ha' married her..without her lands. You can send her now as naked as a robin, if you like. **1939** *N. & Q.* 15 July 42/1 As naked as a needle. **1943** *Amer. Speech* XVIII. 67/2 Naked as a jaybird. **1963** J. T. ROWLAND *North to Adventure* xi. 160 Tell us what you were doing, standing naked as a jaybird. **1974** *State* (Columbia, S. Carolina) 8 Mar. 1-B/1 Just a footnote to ask if there is any truth to the rumor that 'naked as a jaybird' is going to be amended to use 'gamecock'!

e. In printing, *naked forme* (see quots.).

1683–4 J. MOXON *Mech. Exerc. Printing* (1962) 347 *Naked Form*,..is when the Furniture is taken from about all sides of the Form. **1888** C. T. JACOBI *Printers' Vocab.* 86 *Naked forme*, a forme of type waiting for—or stripped of—furniture. **1960** G. A. GLAISTER *Gloss. Bk.* 144/1 A 'naked forme' consists of pages of type secured by page-cord; a 'dressed forme' is one of pages of type with furniture between and around them and the page-cord removed.

6. c. *naked force*, unconcealed, ruthless force.

1963 E. WILSON in *New Statesman* 6 Dec. 847/2 According to the United States State Department, Fidel Castro uses naked force in Cuba.

11. b. (Further examples.) Also *attrib.*

1929 R. GRAVES *Poems* 18 Then it's those naked-light instructions That the muctions plaster up. **1966** *Listener* 18 Aug. 237/1 The naked electric-light bulb had to be left on all day. **1971** *Chambers's Dict. Sci. & Technol.* 787/2 *Naked-light mine*, nonfiery mine, where safety lamps are not required.

15. b. (Further examples.)

1908 *Westm. Gaz.* 11 Aug. 12/1 The other naked-eye planets are too near the sun to be visible. **1930** *Times Educ. Suppl.* 5 Jan. p. iv/3 Within the limit of the naked-eye visibility. **1965** *Listener* 20 May 741/1 Another naked-eye cluster is Præsepe in Cancer, now well placed during the evenings.

17. *naked-footed, -handed* adjs. and advs.; *naked-limbed, -nerved* adjs.; **naked ape,** man, *Homo sapiens*; **naked boys, lady,** see sense A I. 1 c.

1967 D. MORRIS (*title*) The naked ape. **1967** *Spectator* 10 Nov. 577/1 Even before man has destroyed all other animals to make more room for himself, the naked ape may well destroy himself also. **1973** W. BARLOW *Alexander Principle* iii. 33 The 'Naked Ape' has replaced Rousseau's 'Noble Savage'. **1923** D. H. LAWRENCE *Birds, Beasts & Flowers* 16 And yet the soul continuing, naked-footed, ever more vividly embodied. **1848** *Sporting Life* 1 July 210/2 The savage..does not fear to encounter the dangers of his chase naked-handed. **1869** WHITTIER *Hive at Gettysburg* in *Poetical Works* (1898) 380/2 And he who, lone and naked-handed, tore Those jaws of death apart. **a 1930** D. H. LAWRENCE *Phoenix* (1936) III. 164 There the painted women dance..opposite the naked-limbed men. **1933** DYLAN THOMAS *Let.* Nov. (1966) 48 It is typical of the..naked-nerved..to emphasise its brutality.

nakhlite (nā·klait). *Geol.* [f. El *Nakhla* el Baharia, the village near Alexandria, Egypt, where the first known examples fell in 1911 + -ITE^1.] Any of a class of achondrites containing about 75 per cent ferroan diopside and 15 per cent olivine.

1916 [see *HOWARDITE 2]. **1962** B. MASON *Meteorites* vii. 116 The diopside-olivine achondrites or nakhlites are represented by two meteorites, Lafayette (Indiana, U.S.A.) and Nakhla (Egypt). **1971** [see *HOWARDITE 2].

Hence **nakhli·tic** *a.*

1963 *Geochim. & Cosmochim. Acta* XXVII. 1077 Abundances of cadmium have been determined by neutron-activation analysis in 31 meteorites:..0·18 and 1·79 p.p.m. in two nakhlitic achondrites. **1971** P. R. BUSECK in B. Mason *Handbk. Elemental Abundances in Meteorites* 365 (*table*) Type... Nakhlitic.

‖ **nakodo** (nākǭu-do). Also **nakohdo.** [Jap.] In Japan, one who acts as go-between in the arrangement of a marriage.

1890 B. H. CHAMBERLAIN *Things Japanese* 221 The conduct of the affair must be entrusted to a middleman (*nakōdo*)—some discreet married friend, who not only negotiates the marriage, but remains through life a sort of godfather to the young couple. **1895** L. HEARN *Out of East* viii. 258 He was a professional nakōdo, or match-maker, and was..acting in the service of a wealthy rice dealer. **1902** —— *Kottō* x. 123 A majority of Japanese marriages, indeed, are arranged..with the aid of a nakōdo. **1936** K. NOHARA *True Face of Japan* iv. 85 Although he receives monetary compensation for his services, the Nakohdo is not a professional marriage-broker.

nala, var. NULLAH (in Dict. and Suppl.).

N.A.L.G.O., NALGO, Nalgo (næ·lgo). [Acronym f. the initial letters of *N*ational *A*ssociation of *L*ocal *G*overnment *O*fficers.] The name of a trade union of municipal, county, etc., local-government servants.

1909 *Municipal Jrnl.* 29 Oct. 896/2 A change in the Executive Offices of such an organization as the N.A.L.G.O. is always a matter of concern. **1957** *Times* 12 Nov. 8/1 A delegate conference of Nalgo health service branches will be held in London on Sunday. **1962** *Listener* 12 July 74/3 Pompous remarks about Nalgo and the T.U.C. **1975** *Guardian* 5 Nov. 8/8 The social workers, all members of NALGO, say that..the money could be better spent.

nalorphine (næ·lǭɹfīn). *Pharm.* [f. *N*-*al*lylnor*morphine*, another name for the compound (see quot. 1953^2).] A heterocyclic base, $C_{19}H_{21}NO_3$ (or its hydrobromide salt), which is very similar to morphine in chemical structure and is used as an antagonist for that drug and for other narcotics with a similar action.

1953 *Jrnl. Pharmacol. & Exper. Therap.* CIX. 157 N-allylnormorphine (nalorphine) has been shown to have a specific respiratory stimulating effect against the depression caused by morphine,..codeine, methadone. **1953** *Proc. R. Soc. Med.* XLVI. 927 *N*-Allylnormorphine is now officially called nalorphine... A methyl group is removed from the nitrogen [of morphine], hence the term NOR from the German 'Nitrogen ohne radicale'. An allyl group is substituted. **1967** *Martindale's Extra Pharmacopoeia* (ed. 25) 838 Given apart from narcotics, nalorphine may itself cause respiratory depression and disturbing psychotic effects. **1968** J. H. JAFFE in D. H. Efron *Psychopharmacol.* VIII. 854/1 Nalorphine has virtually no abuse potential; its subjective effects are not 'liked' by post-addicts; it does not relieve morphine withdrawal symptoms, but instead precipitates them, and when given over long periods in high doses it produces an atypical physical dependence in which withdrawal symptoms are not associated with craving. **1971** H. W. ELLIOTT in D. H. Clouet *Narcotic Drugs* xxvi. 484 The pharmacological basis for the nalorphine test is an antagonism between nalorphine and the morphine-like drugs, which can be detected by examining the pupil.

naloxone (nælǫ·ksǭun). *Pharm.* [f. *N*-*al*lylnor*oxy*morph*one*, an alternative name for the compound.] A heterocyclic base, $C_{20}H_{23}NO_4$ (or one of its salts), which resembles nalorphine in its chemical structure and antagonism to narcotics (see quot. 1968).

1964 *Amer. Jrnl. Med. Sci.* CCXLVII. 412 The n-allyl derivative of oxymorphone (Numorphan) has recently been synthesized and found to be a potent narcotic antagonist... It has also been suggested that this preparation (Naloxone), differs from other narcotic antagonists in that its administration simultaneously with a narcotic results in antagonism of the respiratory depression but not of the analgesia induced by the narcotic. **1968** *Clin. Pharmacol. & Therapeutics* IX. 215/1 Unlike nalorphine and cyclazocine, naloxone does not produce miosis, subjective effects, tolerance, or physical dependence. **1973** *Nature* 23 Mar. 228/3 Binding was demonstrated by incubating tissue homogenates with tritium-labelled naloxone, a potent antagonist which precipitates withdrawal in humans addicted to opiates.

Nam (næm, nām). Also **'Nam.** Colloq. abbrev. of *Vietnam*.

1969 *Time* 2 May 33/1 'Nam' or 'The Nam' is widely used by U.S. troops to refer to Vietnam. **1973** R. LUDLUM *Matlock Paper* vii. 66 The early days in 'Nam..those weeks of unreported combat. **1973** J. DI MONA *Last Man at Arlington* (1974) 198 [We] were a unit in 'Nam. **1974** *Publishers Weekly* 30 Dec. 93/3 Four Americans caught in Vietnam... The GIs become buddies in Germany... Now in 'Nam' they hope their camaraderie will be closer still.

Nama (nā·mă), *a.* and *sb.* [Hottentot.] **A.** *adj.* Pertaining to or designating one of the four main Hottentot tribes, found in Namaqualand in western South Africa. **B.** *sb.* **a.** A member of this tribe; these people collectively. **b.** Their language, a dialect of Hottentot. Hence **Na·man** *a.* and *sb.*

1864 W. H. I. BLEEK *Reynard the Fox in S. Afr.* p. xxix, To make our available stock of Nama Hottentot literature quite complete. **1881** T. HAHN *Tsuni-‖Goam* i. 3 In the Nama language, one of the Khoikhoi idioms, the Bushmen are called Sā-n (com. plur). *Ibid.* ii. 89, I afterwards made him a present of ammunition, and, as anxious as a Nama is to possess that most precious material, he said: 'No; you want to pay [for] my cow, and I shall not accept it.' **1908** T. G. TUCKER *Introd. Nat. Hist. Lang.* viii. 148 Hottentot dialects: viz. *Nama* (of the Namaqua) to the north-west, [etc.]. **1930** C. G. SELIGMAN *Races Afr.* ii. 34 The customary division of the Hottentots into four main groups—Naman, Korana, Gonaqua, and Old Cape Hottentots. **1944** M. OLDEVIG *Sunny Land* 51 Namas—or Hottentots. *Ibid.* 53 Three different Nama tribes. **1965** [see *HOTTENTOT 2]. **1966** J. H. GREENBERG *Lang. Afr.* (ed. 2) 68 Nama Hottentot indicates the past by an element *go* (in the usual orthography). *Ibid.*, In Nama there is no phonemic distinction between *k* and *g*. **1974** *Times* 14 Jan. 4/6 They have intermarried with Coloured and Nama (Hottentot) people.

Namaqua (nāmā·kwă), *a.* and *sb.* [Hottentot, f. *Nama* (see prec.) + *-qua*, f. *khoi* man.] **1.** = *NAMA *a.* and *sb.*

[**1668** O. DAPPER in I. Schapera *Early Cape Hottentots* (1933) 34 De Hottentots, genaemt *Namaquas*, leggen tegenwoordigh ontrent tachentigh of tuegentigh mylen oost-noord-oostwaerts van de kaep van goeder hope, maer zeer verre te landewaerts in.] **1670** J. OGILBY *Africa* 576 The chiefest people hitherto discover'd in this Southerly part of Africa, are the Gorachouqua's, Goringhaiqua's, Goringhaikona's, Kochoqua's,..Namaqua's, Heusaqua's, Brigoudins, and Hankumqua's. **1790** E. HELME tr. *Le Vaillant's Trav. Afr.* I. vii. 133, I had been assured there were some [*sc.* gazelles] in the country of the great *Namaquois*. **1806** J. BARROW *Voy. Cochinchina* 374 Their dwellings..are constructed on the same principle as those of the Namaqua Hottentots. **1822** W. J. BURCHELL *Trav. S. Afr.* I. 582 To the Hottentot Race are referable the tribes denominated Bushmen, the Namaquas, and Koras or Koraquas, as well as the Hottentots proper. **1822** [see *GRIQUA *sb.* and *a.*]. **1849** E. E. NAPIER *Excursions S. Afr.* I. 53 The Dutch became gradually acquainted not only with numerous hordes of Hottentots, similar to those at the Cape; but likewise with a few varieties of the same race, such as the Namaquas. **1851** J. ANDERSON in G. M. Theal *Yellow & Dark-Skinned People Afr.* (1910) iii. 58 The signs of gender were almost identical in the Namaqua and the Egyptian. **1865** [see HOTTENTOT 2]. **1880** *Encycl. Brit.* XI. 731/2, 3000 Bushmen, 1500 Namaqua. **1930** I. SCHAPERA *Khoisan Peoples S. Afr.* I. ii. 48 The Great Namaqua were always the most northerly group of Hottentots, and probably formed the rearguard of their invasion of South Africa. **1932** *Times Lit. Suppl.* 15 Sept. 642/2 There was a little trouble with some Namaqua guides who suddenly refused to go any further. **1959** *Chambers's Encycl.* VII. 248/1 In south-west Africa many of the 15,000 Namaqua retain their Hamitic speech. **1972** *Stand. Encycl. S. Afr.* V. 607/1 The Nama (also called Naman or Namaqua) consisted of two groups: the Little Nama, living in the North-Western Cape south of the Orange River, and the Great Nama, living in South-West Africa.

2. Namaqua dove, a small, long-tailed dove, *Œna capensis*, found in Africa south of the Sudan; **Namaqua grouse, partridge, sand-grouse,** a terrestrial game bird, *Pterocles namaqua*.

1801 J. BARROW *Acct. Trav. S. Afr.* I. iv. 325 Along the road were numbers of that beautiful little pigeon, called here the Namaaqua dove, not larger than a sparrow. **1864** T. BAINES *Explor. S.-W. Afr.* v. 124 A pretty Namaqua dove fluttered about in abortive efforts to sip the water. **1905** J. DU PLESSIS *Thousand Miles in Afr.* 137 Little black-eyed Namaqua doves. **1947** J. STEVENSON-HAMILTON *Wild Life S. Afr.* xxxiv. 291 Doves... The little namaqua (*Oena capensis*) appears to be a winter migrant to the eastern Transvaal low country. **1972** *Shooting Times & Country Mag.* 27 May 24/2 Mourning, laughing and even tiny, wagtail-sized Namaque [*sic*] doves arrive to raid the crops. **1806** J. BARROW *Trav. S. Afr.* (ed. 2) I. iii. 219 Cape partridges and the Namaaqua grouse were equally plentiful. **1867** E. L. LAYARD *Birds S. Afr.* 277 Namaqua Grouse... Namaqua Patrys of Colonists. **1790** E. HELM tr. *Le Vaillant's Trav. Afr.* II. xxi. 434 The Hottentots of the colonies call them perdrix Namaquais, (Namaquai partridges). **1890** A. MARTIN *Home Life Ostrich Farm* xi. 227 The beautiful little 'Namaqua partridges'..are in reality a kind of grouse. **1936** E. L. GILL *First Guide S. Afr. Birds* 113 Namaqua Sandgrouse, Namaqua Partridge... Much the most widespread of the Sandgrouse in South Africa. **1972** Namaqua partridge [see *KORHAAN]. **1893** H. A. BRYDEN *Gun & Camera S. Afr.* xxii. 477 The Namaqua sandgrouse [drinks] between eight and ten in the morning. **1962** MACKWORTH-PRAED & GRANT *Birds S. Third Afr.* 372 Namaqua Sandgrouse... Birds of the desert fringe. A common Sandgrouse of the drier parts of South Africa especially the western side.

‖ **namaskar** (namaskā·ɹ). *India.* Also **namashkar, namaskara.** [Hindi, ad. Skr. *namaskāra*, the greeting '*namas*', obeisance.] = *NAMASTE *sb.*

1930 C. PARSONS *Mysore City* 88 Indian girls..making the silent and beautiful salutation of the 'namaskāra'—folded hands and bent head and bowing figure. **1948** *U.N. World* (N.Y.) Apr. 35 One by one they filed past, some doing *namaskar* (hands placed together in an attitude of prayer), others bowing low. **1966** J. & R. GODDEN *Two under Indian Sun* iv. 102 Instead of a handshake, he uses the namashkar, the graceful movement that means 'to take the other's dust upon you', hands joined together as if in prayer and raised to the forehead or the breast according to the rank or honour of the person saluted. **1967** SINGHA & MASSEY *Indian Dances* iii. 44 Indians always use joined palms for the namashkar, that is, when they greet anyone, but the position in which they are held indicates the status of the person greeted. **1969** R. SHANKAR *My Music* i. 12/1 The simplest gesture of the *namaskar*, or greeting (putting the hands palm to palm in front of the forehead and bowing). **1972** 'E. PETERS' *Death to Landlords!* viii. 113 Priya offered her *namaskar* shyly.

‖ **namaste** (nā·maste), *sb.* and *int. India.* Also **namasthe.** [Hindi, ad. Skr. *námas*, bowing, obeisance + *te* dat. of *tuam*, you (sing.).] **A.** *sb.* A salutatory gesture made by bringing the palms together before the face or chest and bowing. Also *attrib.*

1948 *Time* 16 Aug. 30 In response Nehru closed palms in front of his chest. This traditional Hindu *namasthe* (greeting) is as much a part of his public manner as was the V sign for Churchill. **1951** J. MASTERS *Nightrunners Bengal* vii. 97 She brought her palms together in front of her face and moved them up to her forehead and down again in the gesture called *namaste*. **1959** M. BRECHER *Nehru* 385 A young man stepped forward from the crowd and greeted Gandhi in the traditional *namaste* salutation, joining his hands palm to palm and bowing slightly. **1960** KOESTLER *Lotus & Robot* I. i. 52 A toothless Brahmin.. walked in and, after a dignified *namaste*, hung a garland round my neck. **1965** E. LINTON *World in Grain of Sand* vii. 114, I shall be waiting for a letter from you. Namastes and love from Savitri. *Ibid.*, I join my hands in 'namaste' to you and Chander Kala. **1973** 'B. MATHER' *Snow Line* vii. 79 The namaste sign..is with the palms together as in prayer, head bowed reverently and nose touching the fingertips.

B. *int.* Used as a respectful greeting.

1967 D. C. COOKE *c/o American Embassy* (1968) ii. 21 '*Namaste*, Sahib,' the Sikh said. **1969** 'E. PETERS' *Mourning Raga* iii. 48 She pressed her hands together in reverence to him, and bowed her head..'Namaste!'

Hence **na·maste** *v. intr.*, to give the *namaste* sign.

1969 J. K. GALBRAITH *Ambassador's Jrnl.* xxvi. 586 After we had shaken hands, waved, *namasted* and got on board, the crowd lined up by the plane and one of the engines failed to start. **1971** R. PARKES *Line of Fire* xi. 101 He namasted gravely. **1971** *Illustr. Weekly India* 18 Apr. 6 The crowds clustered there drew apart, respectfully '*namasteing*', as the big man at the wheel acknowledged their greetings and drove in.

name, *sb.* Add: **1. c.** *Stockbroking.* The ticket bearing the name of the purchaser of stock, handed over to the selling broker on name-day or ticket-day.

[**1891** G. H. STUTFIELD *Rules & Usages Stock Exchange* 59 When the issuer of the ticket is a broker,..he has to insert the name of his client as the person into whose name the Stock is to be transferred.] **1907** POLEY & GOULD *Hist., Law & Pract. Stock Exchange* 178 It is called the ticket or name day because of the passing of tickets or names on that day. **1934** F. E. ARMSTRONG *Bk. Stock Exchange* x. 193 'Names' play an important part in the settlement of Stock Exchange transactions. **1968** J. D. HAMILTON *Stockbroking Today* I. iii. 89 Once in the office the names are sorted by the Names Department so that each name or batch of names matches a certain sale..and where there is more than one ticket they are pinned together.

d. (*or*) *my name is not* ——, appended to a statement as an assertion of its truth.

1803 S. OWENSON *St. Clair* vi. 29 Sir Patrick will make the walls of the old Abbey ring again, or my name is not Michael M'Carty. **1898** J. D. BRAYSHAW *Slum Silhouettes* 220, I tell yer straight, if me an' Kitty don't make Soufend sit up, my name ain't Bill Brown. **1962** C. S. FORESTER *Hornblower & Hotspur* xiii. 174 We'll have a westerly gale, sir, or my name's not William Bush.

e. *to put*, or *write*, *someone's name down for*: to enter someone's name on a list of those interested in sharing in, acquiring, or taking part in a particular commodity or activity (cf. PUT *v.*^{1} 41 i).

1819 M. EDGEWORTH *Let.* 2 Apr. (1971) 193 Lady Jersey..told me she would put down our names and give me some tickets for Almacks. Of the 5 Patronesses she is supposed to rule. **1821** —— *Let.* 30 Oct. (1971) 248 The Colleges are now so full that a young mans name must be written down 3 or 4 years before he can hope to get in. **1824** [see PUT *v.*^{1} 41 i]. **1969** *Guardian* 20 Mar. 20/3 Lord Linley, seven-year-old son of Princess Margaret and Lord Snowdon, had had his name put down for Eton. **1974** *Listener* 25 Apr. 525/3 Frances has her name down for Danesbury [Hospital] just in case something should happen to Ron.

f. *give it a name*: what would you like to drink?

1854 DICKENS *Hard T.* I. vi. 43 What thall it be, Thquire... Thall it be Therry? Give it a name, Thquire! ..have a glath of bitterth. **c 1863** T. TAYLOR in M. R. Booth *Eng. Plays of 19th Cent.* (1969) II. 88, I hope you'll allow me to stand treat—give it a name, gentlemen... Thank you, I never drink with strangers. **1929** J. B. PRIESTLEY *Good Companions* I. vi. 235 The waiter collected orders and told Inigo to give it a name. **1931** T. R. G. LYELL *Slang* 540 'Well, boys, the drinks are on me! Give it a name!' **1951** J. B. PRIESTLEY *Festival at Farbridge* I. ii. 59 What are you drinking? Give it a name, chaps—there's everything here.

g. *to have one's name* (*and number*) *on it*: of a bullet, etc.: to be destined to kill a particular person.

1917 A. G. EMPEY *Over Top* 312 Tommy detests these mortars because..he knows that it is only a matter of minutes before a German shell with his name and number on it will be knocking at his door. **1919** *Athenæum* 18 July 632/2 A soldier refers to the shell that kills him as 'having his name and number on it'. **1925** FRASER & GIBBONS *Soldier & Sailor Words* 163 *Name* (*or number*) *on, to have one's*, said of a bullet that hit a man; *i.e.*, that it was destined for him. **1958** R. STOREY *Touch it Light* in *Plays of Year* XVIII. 376 *Ted.* That farmer don't like us, sir. Ever since that bomb fell on his cowshed. *Og.* He thought it should have fallen on you? *Syd.* It had our name on it all right. **1973** D. FRANCIS *Slay-Ride* xii. 140 The bomb probably had my name on it in the first place.

h. *no names, no pack-drill*: phr. used to indicate that if nobody is named as being responsible, nobody can be blamed. Cf. *pack-drill* s.v. PACK *sb.*^{1} 14 c.

1923 O. ONIONS *Peace in our Time* I. ii. 25 Men had a way of omitting the names of those of whom they spoke; no names no pack-drill. **1926** E. WALLACE *More Educated Evans* vii. 160 There's a certain party—no names no pack-drill—who's fairly doggin' me to get information. **1931** P. MACDONALD *Crime Conductor* I. i. 7 'Meaning?' said Cuthbertson. 'No names,' said Garth Johnson quickly, 'no pack drill!' **1955** M. ALLINGHAM *Beckoning Lady* ii. 32 It just means no name, no pack drill, and always speak well of them as has money to sue. **1962** 'B. GRAEME' *Undetective* iii. 32 'It's a lie, mister. Who told you?' 'No names, no pack drill.'

i. *to have one's name in lights*: to be a well-known actor and so have one's name displayed in lights outside the theatre.

1929 J. B. PRIESTLEY *Good Companions* II. i. 282 His determination to top the bill and have his name in electric lights. **1972** *Guardian* 15 Jan. 8/4, I couldn't wait to get up there with the best of them and see my name up in lights—topping the bill at the Palladium.

j. *the name of the game* (colloq.): the object or essence of an action, etc.

1966 *Legionary* (Ottawa) Oct. 36/1 Where the knight's concerned, quality is the name of the game. **1967** *Maclean's Mag.* Aug. 27/3 And if this means running up against slum landlords, do-nothing local councils or a hostile white community—well, that's the name of the game. **1970** G. JACKSON *Let.* in *Soledad Brother* (1971) 247 We should never make it easy for them—by relaxing —at this stage of the educational process. Examples are crucially important. Well that's the name of the game right now. **1972** *Jazz & Blues* Sept. 7/3 If I can make you feel like you want to holler on your horn then that's the name of the game man. **1972** *Times* 29 Sept. 11/1 The name of the game this week is survival. **1973** *Nature* 6 July 2/1 Call my bluff was the name of the game at last week's meeting of the International Whaling Commission.

5. c. (Later examples of sense 'a celebrity'.) Also *attrib.* or as *adj.* Also in extended use, of a well-known group of people, esp. a jazz band (see *name band* below).

1936 *Variety* 17 June 32/1 The greatest 'names' in the industry, including the cream of its players. **1941** *Sun* (Baltimore) 28 July 11/5 Virtually all of the name horses in the land will be on the scene for the thirty-day meeting. **1943** *Ibid.* 14 Aug. 7/5 At least ten 'name' players, fellows like Gene Sarazen and Craig Wood and Byron Nelson and Walter Hagen, have assured Corcoran that they'd be available. **1945** L. SHELLY *Jive Talk Dict.* 15/1 *Name*,.. most popular band at the moment. **1947** *Sat. Rev.* (U.S.) 11 Oct. 53/1 The growing group of record makers who..turn..to the reservoir of fine performing talent that ..lie outside the galaxy of first-rank star names. **1955** J. BETJEMAN in R. S. Thomas *Song at Year's Turning* 12 His publisher believed that a 'name' was needed to help sell the book. **1960** *20th Cent.* Apr. 342 A big factor in the sale of the more popular 'name' records is the personality cult. **1972** *Times* 12 Dec. 2/6 RIBA circles..had feared that the former Secretary of State, Mr Walker, would appoint a 'name' with glamour to a post where he would have had little real authority or influence. **1973** *Black World* Jan. 28/2 He has concerned himself with the promotional affairs of several 'name' theaters in the Cleveland area. **1973** *Times* 6 Feb. 7/3 There is a narrowing of the opportunities for 'name' designers (couture tailors). **1974** *Times Lit. Suppl.* 18 Jan. 50/3 Professor Eliade is what publishers like to call a 'name', and you cannot ignore him.

d. An underwriter at Lloyd's.

1885 G. VAN DE LINDE *Chartered Accountants' Students' Soc. Lect. Biogr. Lloyd's Policy* 10 The respective partners of Blank & Co.,..head the policy by underwriting it to the extent of £10,000 between them, each name being respectively responsible for the amount against the signature. **1928** WRIGHT & FAYLE *Hist. Lloyd's* xxiii. 422 Let us consider the career of an underwriting 'Name', that is an Underwriting Member of Lloyd's represented by an Agent. **1937** R. STRAUS *Lloyd's* xi. 257 Marine underwriters..offered themselves as 'Names' to those Underwriting Agents who specialised in non-marine risks. **1972** G. LYALL *Blame the Dead* iv. 23 He's a Name. *Ibid.* 24 That means a member; they call them Names. **1973** *Daily Tel.* 16 Oct. 3/7 Discussing evidence given earlier by her father, Dr Dugdale said he was a 'name' several times over at Lloyd's, concerned in shipping and aircraft insurance. He made her a 'name' ensuring her a great deal of capital and a very high income.

9. Also *in name only, only in name*: of a marriage without sexual relations.

1894 W. J. LOCKE *At Gate of Samaria* (1895) xxi. 245 Henceforward Thornton would be her husband only in name. **1972** A. ROUDYBUSH *Sybarite Death* (1974) xxi. 173, I married her..but it never even occurred to me that our marriage would be other than a marriage in name only. **1975** R. PLAYER *Let's talk of Graves* ii. 60 She had hated her husband and been his wife only in name.

b. *in all but name*: of a situation or set of circumstances, existing but not officially acknowledged or recognized.

1934 J. E. NEALE *Queen Elizabeth* xv. 251 In all but name the Papacy was at war with Elizabeth.

15. *attrib.* and *Comb.*, 'bearing a name', as *name-board* (later examples), *card* (later examples), *-label, -ribbon, -tab*; 'well-known', of or pertaining to a name (sense 5 or 7), as *name brand, -worthy* adj.; 'named after, or giving a name to, one', as *name-daughter* (later example), *name-flower*; in *Logic*, as *name-forming, -matrix, -relation, -variable*; in *Linguistics*: consisting of or pertaining to a proper name, as *name-element, -form, -giving, -group, -lore, -stem, -system*; **name-act,** a cabaret act performed by well-known performers; **name band,** a jazz or dance band that has made a name for itself; hence **name band-leader**; **name-calling** *vbl. sb.*, abusive language, mere abuse; hence **name-call** *v.*; **name-dropping** *vbl. sb.*, familiar mention of the names of distinguished people in order to imply one's own importance; also *attrib.*; hence (as a back-formation) **name-drop** *v.*, **name-dropper**; **name-droppingly** *adv.*; **name-part,** also of a book, a ballet, etc.; **name-piece** = **name-poem*, **name-story*; **name-plate,** a metal plate bearing a name; *spec.* one attached to a piece of machinery, or displaying the name of a road or building; also *attrib.* and *fig.*; also as *vb.*; **name-poem,** the poem from which a volume of collected poems is named; **name-story,** the story from which a volume of collected short stories is named; **name-tag,** anything on which a name can be written, to identify the person or object to which it is fixed; **name-tape,** a piece of tape with a person's name woven into it or printed on it, fixed to a person's clothing for identification; hence **name-taped** *ppl. a.*

1942 BERREY & VAN DEN BARK *Amer. Thes. Slang* § 590/15 *Name act*, an act consisting of well-known players. **1949** L. FEATHER *Inside Be-Bop* iii. 21 The Berry Brothers and several other name acts. **1967** *Stage* 2 Mar. 21/4 (Advt.), Top groups required for one night stands. Name acts for winter season cabaret. **1936** *Amer. Mercury* XXXVIII. p.x/1 *Name band*, a band that has gotten the breaks (whether they're good or not). **1938** *Sat. Even. Post* 2 Apr. 9/1 We are to have an orchestra— 'a name-band by all means'. *Ibid.* 9/3 At least $2500 for a name band. **1955** L. FEATHER *Encycl. Jazz* (1956) 122 His son..is also a drummer, heard w[ith] Erskine Hawkins and other name bands including Count Basie, '55. **1963** *Globe & Mail* (Toronto) 8 Jan. 5/1 Although its popularity declined with the passing of the name bands, Toronto's Palace Pier was still a busy, and apparently profitable operation until yesterday. **1958** P. GAMMOND *Decca Bk. Jazz* xix. 236 The year 1951 saw the return to Britain of pre-war 'name' bandleader, Roy Fox. **1939** AUDEN & ISHERWOOD *Journey to War* v. 121 On our left was a little station: we read its name-board, Ling Pao. **1955** J. COPE *Fair House* i. 24 The turn-off to the Boer farm was a gap in the bush at the roadside, no gate or fence or name-board. **1944** *Time* 7 Aug. 38/2 Five times in five minutes the cigaret-counter girl at a Walgreen store in Chicago repeated wearily, 'We have no name brands.' **1973** J. ROSSITER *Manipulators* v. 51 Perhaps.. you've been name-calling somebody. And they didn't like it. **1853** DICKENS *Nobody's Story* in *Househ. Words* Extra Christmas No. 35/2 Such name-calling and dirt-throwing. **1891** Name-calling [in Dict.]. **1947** *Amer. Speech* XXII. 231/1 *Namecalling*, the attempt to put a person or thing in a bad light by attaching to him or it a word with unpleasant connotations. **1965** G. JACKSON *Let.* 16 Mar. in *Soledad Brother* (1971) 69, I have been subjected to the ordeal of hunger, thirst, name-calling, and other uncountable indignities. **1973** R. LUDLUM *Matlock Paper* xxvi. 222, I don't want to be responsible for indiscriminate name-calling, any wide-spread panic. **1975** *Verbatim* Feb. 4/1 The argument is a little uneven, for here it delivers a polemic against name-calling, there against grammar. **1907** *Yesterday's Shopping* (1969) 361/2 (Advt.), Menu and name cards. **1925** W. J. LOCKE *Great Pandolfo* ii. 23 The beautiful lady whose name he had not caught, because, in abstraction, she had turned her name card maddeningly upside down, took little or no interest in him. **1969** A. CADE *Turn up Stone* i. 25 Michael's experience in the Middle East had taught him the importance of the exchange of name cards in many countries. A man without a card was a man without an identity. **1891** R. L. STEVENSON *Let.* Nov. (1899) II. 241, I shal begin to despair of everything but my name-daughter. **1955** J. D. SALINGER *Franny & Zooey* (1962) 25 There's an unwritten law that people in a certain social or financial bracket can name-drop as much as they like just as long as they say something terribly disparaging about the person as soon as they've dropped his name. **1959** I. ROSS *Image Merchants* (1960) v. 94 Newsom does not even have to name-drop. The PR man who can avoid that indulgence has truly arrived. **1969** *Daily Tel.* 22 Aug. 18/3 Mr Walters can name-drop better than most when it comes to generals and film stars. **1947** *San Francisco Examiner* (Pict. Rev.) 7 Sept., Our newest menace. The name dropper. **1959** *Woman's Own* 24 Jan. 31/1 One of my favourite snobs—the name-dropper. **1972** H. KEMELMAN *Monday the Rabbi took Off* xii. 84 He would be likely to point out important people to his son—the wife of the British consul, the American first secretary. He was no name-dropper, but he wanted so much to have his son think well of him. **1950** M. MCCARTHY *On Contrary* (1962) 186 The idea that it's smart to be in step, to be liberal or *avant-garde*, is conveyed through the name-

dropping of a Leo Lerman in *Mademoiselle*. **1951** L. Z. HOBSON *Celebrity* (1953) viii. 119 Rex Stout and Oscar Hammerstein... Conversational spice, he had been thinking; nobody could call it name-dropping. **1966** *Philos.* XLI. 359 Plus a wordy, name-dropping Introduction. **1973** *Times* 7 Feb. 4/5 (*heading*) Solicitors appalled by 'name-dropping' in courts. **1966** *Guardian* 30 Dec. 4/8 He becomes absorbed (name droppingly so) into the ranks of the literati. **1922** E. EKWALL *Place-Names Lancs.* 62 It is probably a Scand. name..as *Brand* is hardly with certainty evidenced as an O.E. name-element. **1932** E. WEEKLEY *Words & Names* ix. 134 From the name-element *mun*, thought, etc., were formed a number of names. **1937** *Harvard Univ. Summaries Ph.D. Theses* 272 The deuterotheme..is by far the more stable name-element in the late Germanic period. **1951** *Traditio* VII. 411 The second theme, *-ferth*..is obviously a metathesized form of *frith* (peace) which occurs in many Germanic names both as a first and second name element. **1907** A. QUILLER-COUCH *Major Vigoureux* ii. 20 Glorious trumpet daffodils!..Major [Narcisse] Vigoureux delighted in them. Were they not his name-flower? **1927** *Observer* 24 Apr. 15 Marigold, its heroine, has the unaffected charm of her name-flower. **1946** B. BLOCH in *Language* XXII. 208 The non-past indicative form of a verb, an adjective, or the copula serves as the *name form*, used to refer collectively to all the members of a paradigm. **1951** TRAGER & SMITH *Outl. Eng. Struct.* 60 The uninflected or name-form is the base. **1970** *English Studies* LI. 445 The name-forms are arranged under OE phonemes. Thus under OE *ǣ* we find head-words like *ǣc, brǣd, rǣp*. **1955** H. LEBLANC *Introd. Deductive Logic* 2 Semiotic quotes..are a name-forming operator. **1956** J. H. WOODGER tr. *Tarski's Logic, Semantics, Metamath.* 161 Quotation marks provide an example of a name-forming functor with *one* sentence argument. **1957** A. N. PRIOR *Time & Modality* 119 Our *x*'s are given..by means of a name-forming operator on intervals. **1864** Name-giving [in Dict.]. **1898** E. CLODD *Tom Tit Tot* vi. 75 Mungo Park thus describes the name-giving ceremony among the Mandingo people. **1940** A. H. GARDINER *Theory of Proper Names* vi. 20 Certain name-givings..do not give rise to proper names. **1970** G. R. STEWART *Amer. Place-Names* p. xii, European scholars rarely concern themselves with the process of name-giving or its motives. **1950** H. L. LORIMER *Homer & Monuments* iv. 125 Apart from the negative evidence of the Pylos tablets, there is the fact..that certain series of signs in both groups form name-groups which also occur at Knossos in the Palace script. **1963** *English Studies* XLIV. 32 Large name-groups with end-variation, e.g. *Cēolwald, -helm, -bald, -ward*, etc. **1910** *Westm. Gaz.* 14 Mar. 11/2 Affixing red name-labels to their seats in the Council Chamber. **1928** D. H. LAWRENCE *Lady Chatterley* iv. 36 I'd be ashamed to see a woman walking round with my name-label on her. **1924** *Daughter of C. Palmore* iii. 36 At one time she is deep in heraldry and name-lore. **1932** E. WEEKLEY *Words & Names* vii. 82 One of the puzzles of name-lore is the process by which the French name *Jacques* ..was early confused with *Jankin* or *Jenkin*. **1940** W. V. QUINE *Math. Logic* iii. 152 Such expressions might be classed as *name matrices*, for they are related to names as statement matrices are related to statements. **1936** *Times Lit. Suppl.* 25 Jan. 73/2 But Reid, after all, is in the 'name-part'. **1961** *Times* 5 Apr. 13/6 With Dame Margot Fonteyn in a memorable account of the name-part [of *Giselle*]. **1924** *Glasgow Herald* 24 Apr. 4 The name-piece of the volume is a genealogical..account of this branch of the..family. **1882, 1896** Name-plate [in Dict.]. **1904** *Electr. Rev.* 3 Sept. 327 The committee recommends that the ratings of generators and motors, except traction motors, be marked plainly on the name-plate. Two types of service are recommended, continuous working and intermittent working, and the name-plate must state to which service it relates. **1908** [see *FINGER-POST *v.*]. **1967** *Gloss. Terms Builders' Hardware* (*B.S.I.*) IV. 17 *Name plate*, a plate..bearing one or more words fixed to a door, gate or cupboard..to convey information concerning contents, premises, business, profession or individuals. **1971** M. TAK *Truck Talk* 110 *Nameplate finders*, cab-mounted spotlights used to find an address on a building at night. **1972** *Village Voice* (N.Y.) 1 June 20/2 Fiction writers.. start getting asked to do book reviews and being invited to name-plate cocktail parties. **1958** BLUNDEN *War Poets* 29 The name-poem is one of the great achievements. **1956** R. CARNAP *Meaning & Necessity* (ed. 2) iii. 96 The method of the name-relation is an alternative method of semantical analysis, more customary than the method of extension and intension. *Ibid.* 97 Following Russell and Church, I used the word 'denotes' for the name-relation in the first version of this book. However, in view of the ambiguity just described, I now prefer to avoid it. **1905** *Daily Chron.* 23 Feb. 6/5 Name-ribbons may have to be changed. **1924** MAWER & STENTON *Introd. Survey Eng. Place-Names* ix. 166 Recent investigation has shown that many Germanic name-stems which are never recorded in England in historic times were still used by the Angles, Saxons and Jutes of the fifth and sixth centuries. **1953** K. JACKSON *Lang. & Hist. Early Brit.* I. 174 The name-stem *Maglocun* which appears both in Ogam and in Latin. **1927** *Observer* 24 Apr. 8/4 There is an air of strain, as if she were attempting—at any rate in the name-story (the others are nearer her usual vein)—to achieve a high-flown style. **1936** *Times Lit. Suppl.* 25 Jan. 76/4 His 'name-story' is of a poor Australian woman who marries an Afghan trader. **1931** C. L'E. EWEN *Hist. Surnames* iii. 50 Fick concluded that the German name-system exceeded in splendour..others of the Aryan group. **1937** *Harvard Univ. Summaries Ph.D. Theses* 271 The aim of this study is..to study the Old English dithematic name against the background of the general Germanic name-system. **1960** V. JENKINS *Lions Down Under* vi. 95 Mother was almost sewing on name-tabs at the airport. **1973** P. O'DONNELL *Silver Mistress* ii. 25 The clothes..had tailor's name-tabs. **1946** W. S. KNICKERBOCKER *20th Cent. English* 333 Each may usurp the business of the other and lose thereby his special name-tag. **1948** H. LAWRENCE *Death of Doll* x. 230 They did not want name tags pinned to their coat sleeves by Nick. **1953** A. UPFIELD *Murder Must Wait* iv. 36 On some of her clothes is a name tag with the initials P.R. overlaid on others which could be J.O. or J.U. **1958** C. WATSON *Coffin scarcely Used* xvii. 163 Purbright watched Gibbins going through pockets. 'Any name tags?' he asked. **1964** G. L. COHEN *What's Wrong with Hospitals?* i. 16 A parallel substitute for the spoken word is the system of name tags prevalent in America and gaining ground in Europe. 'It saves the staff from having to introduce themselves.' **1899** in A. Adburgham *Shops & Shopping* (1964) xxii. 261 Name tapes. **1932** E. BOWEN *To North* v. 45 She stitched name-tapes on to her new summer-term outfit. **1964** —— *Little Girls* II. vii. 152 Her mackintosh, name-taped as St. Agatha's demanded. **1969** *Guardian* 1 Sept. 7/5 Those new-fangled printed heat-adhesive name tapes. **1971** M. MCCARTHY *Birds of America* 49 Peter would have to have a haircut and name-tapes on his clothes. **1955** A. N. PRIOR *Formal Logic* 182 It would not be possible to lay it down..that in any thesis a description may be substituted for a name-variable. **1957** —— *Time & Modality* 46 Q is enriched by name-variables, predicate variables, and quantifiers. **1963** O. WOJTASIEWICZ tr. *Łukasiewicz's Elem. Math. Logic* 103 We shall be concerned with a certain theory of name variables. **1903** *Chambers's Cycl. Eng. Lit.* (new ed.) III. 695/1 *The Growth of Love*,..*Eros & Psyche*.., are amongst his nameworthy poems.

name, *v.*^{1} Add: **1. b.** *to name for*: delete (now only *U.S.*), and add further examples.

1930 ADE *Let.* 20 Aug. (1973) 147 At one time he [*sc.* Peter VanRensselaer] owned thousands of acres in this region and the city of Rensselaer is named for him. **1933** S. HOWARD *Alien Corn* I. 14 We were just saying you must have been named for Wagner's Elsa. **1936** M. DE LA ROCHE *Whiteoak Harvest* vi. 95 You'll have children and perhaps..you'll name a little boy for me. **1957** *Northern Life* June 9/1 Saville Row was named for Col. Sir Geo. Saville, who commanded the garrison of the town [*sc.* Newcastle upon Tyne] in 1776–7 and lived in a house here. **1968** B. FOSTER *Changing Eng. Lang.* v. 226 A very typically American turn of phrase that is showing signs of headway in Britain is the replacing of 'named after' by 'named for'.

5. d. (Later examples.)

1928 [see *LAST *a.* 1 f]. **1972** *Guardian* 11 Feb. 22/4 The Speaker failed to 'name' or suspend Miss Devlin after hitting Mr Maudling. *Ibid.* 15 Mar. 1/4 Mr Charles Loughlin, MP for Gloucestershire West, was named by the Deputy Speaker after telling him that he 'did not give a damn' whether Sir Robert did or did not listen to his point of order. In accordance with the usual custom, the Leader of the House..moved that Mr Loughlin be suspended.

f. To specify officially (someone) by name to whom certain political (usu. Communist) affiliations are imputed, esp. in South Africa under the Suppression of Communism Act, 1950, and in the U.S.A. during the period of McCarthyism. Hence **naming** *vbl. sb.*, **named** *ppl. a.*

1950 *Times* 9 Mar. 5/3 Senator McCarthy has been ordered by Senator Tydings..to name to-morrow the high State Department official who he has alleged in the Senate intervened to protect an employee who was regarded as a bad security risk. **1952** *Economist* 31 May 581/3 He [*sc.* Mr E. S. Sachs] has been 'named' by the Minister of Justice as a Communist under the Suppression of Communism Act. **1956** L. KUPER *Passive Resistance in S. Afr.* II. viii. 188 For most whites, a 'named' person bears a permanent social stigma; he is not acceptable as an employee, or in ordinary intercourse. **1957** S. ADLER *Isolationist Impulse* xv. 460 He [*sc.* McCarthy] said he could name 205, or 57, or 81 Reds (the numbers usually varied with each harangue) in the State Department. *Ibid.* 461 He was unable to substantiate these charges by naming one Communist survivor of the Truman purge. **1958** G. M. CARTER *Politics of Inequality* II. ii. 65 The Minister could then forbid those 'named' to take part in any specified organization. But the 'naming' process is not essential before taking action, for the Minister.. can also prohibit any gathering if it appears to aid the objects of Communism.

g. To cite as co-respondent in a divorce petition.

1971 YELDHAM & CARNE *Rees's Divorce Handbk.* (ed. 4) ii. 27 Unless otherwise directed, where a wife's petition alleges adultery with a woman *named*, the alleged adultress must be made a respondent in this cause. **1971** A. HUNTER *Gently at Gallop* ii. 11 Laing divorced her, naming Berney... Berney was named in another suit, and his first wife petitioned, using that as grounds... He's been named once or twice since then. **1972** *Guardian* 23 Dec. 24/5 The television actress, Linda Thorson..was named yesterday in a divorce suit.

6. e. Phr. *you name it*, *I* (or *we*) *have* (or *have done*) *it* (also with other verbs), everything that you can think of is available, has been done, etc.; also ellipt., *you name it*.

1962 J. BRAINE *Life at Top* xviii. 213 You name the drink, we have it. **1964** M. S. ALLWOOD *American & British* 137 *American.* You name it! *British*.... or whatever you like. **1967** *Field & Stream* Aug. 63/2 Mallards, gadwall, partridge, quail—you name it— they're up here for the season every year. **1968** *Sun* (Baltimore) 18 Sept. A. 14/4 Bear Creek, Back River, you name it; the story is the same. **1969** N. FREELING *Tsing-Boum* vii. 45 What sort of world are they born into anywhere?—hunger, napalm, you name it and we've got it. **1969** *Rolling Stone* 28 June 17/3 I've written every kind of music there is. You name it, I've written it. All except one thing I couldn't do: rhythm and blues. **1972** D. LEES *Zodiac* 53 He's been a smuggler, a gun runner, a dope peddler—you name it. **1973** *Times* 6 Jan. 9/4 Bits of chicken, port, olive—you name it. *Ibid.* 22 Jan. 9/2 At that time the cops knew me. You name it, I'd done it. **1973** *Black Panther* 8 Sept. 17/1 I've seen police call people slur names such as nigger, mother fuckers, bitches, whores..you name it, they had a name for it.

7. c. *to name no names*: to refrain from mentioning the names of the people involved in an incident, etc., in order to protect them; often with the implication that the hearer or reader could supply these names.

1792 F. BURNEY *Jrnl.* June (1972) I. 212 She desired he would name no names, but merely mention that some ladies had been frightened. **1843** DICKENS *Mart. Chuz.* (1844) iv. 46 Naming no names, and therefore hurting nobody but those whose consciences tell them they are alluded to. **1890** KIPLING *Soldiers Three* 12 Av coorse I will name no names, for there's wan that's an orf'cer's lady now, that was in ut. **1908** K. GRAHAME *Wind in Willows* iv. 89 The Wild Wood is pretty well populated by now; with all the usual lot, good, bad and indifferent— I name no names. **1919** BEERBOHM *Seven Men* 203 But now my sense of duty forces me To a departure from my custom of Naming no names. One name I must and shall Name. **1972** L. LAMB *Picture Frame* xiii. 107 You put that tale around, naming no names, at one o' your police smokers, you'll have 'em all rolling in the aisles.

8. *to name the day* (further examples); also *transf.*

1766 GOLDSMITH *Vicar of Wakefield* I. xvii. 171 'Name, then, your day...' ..She again renewed her..promise of marrying Mr. Williams, and, that day month was fixed upon for her nuptials. **1841** DICKENS *Let.* 25 Mar. (1969) II. 243 Chigwell, my dear fellow, is the greatest place in the world. Name your day for going. **1974** *Times* 9 Feb. 16/6 'Heath names the day,' shrilled the billboards yesterday. And an elderly couple on a bus said: 'It's about time he got married.'

nameable, *a.* Add: **2.** (Earlier U.S. example.)

1780 J. ADAMS *26 Lett. Revol. Amer.* (1789) 23 The sixth talk is to shew, 'that no person, in America, is of so much influence, power, or credit, that his death, or corruption, by English money, could be of any nameable consequence'.

name-day. Add: **3.** *Stock Exch.* The day on which the seller of registered securities receives from the buyer a ticket with the name and details of the person to whom the securities are to be transferred.

1885 P. CAMPBELL *Stock Exchange* 25 The second [account day] is called 'Name or Ticket Day', devoted by Brokers and Jobbers to concluding Investment operations by completing and delivering transfers. **1895** A. J. WILSON *Gloss. Colloq. Terms Stock Exchange* 170 Each settlement occupies three days—1st, 'contango day'..; 2nd, 'name day', when the names of buyers of registered securities are passed to the sellers; and 3rd, 'pay day'. **1907** [see *NAME *sb.* 1 c]. **1941** DICE & EITEMAN *Stock Market* (ed. 2) x. 155 The second day of settlement is called 'ticket day' or 'name day'. On this day a broker who wishes delivery on stock that he has bought sends a ticket to his jobber, giving the name of the security. **1970** M. GREENER *Penguin Dict. Commerce* 325 Ticket day..is also sometimes called name day.

namesmanship (nēi·mzmănʃip). *colloq.* [f. NAME *sb.* + *-MANSHIP, after *GAMESMANSHIP.] Skill in the use of influential names of people or objects; skill in name-dropping.

1964 *Amer. Speech* XXXIX. 225 Namesmanship at sea. **1967** *Economist* 28 Oct. p. xxii/2 Namesmanship has been in, in Sweden, since the appearance this summer of Mr Ake Ortmark's book on 'The Power Game in Sweden'.

Namibia (nămi·biă). [f. *Namib* (a desert on the western coast of southern Africa) + -IA^1.] A name given to South-West Africa in 1968 by the United Nations in anticipation of its being released from the mandate granted to South Africa by the League of Nations in 1919 (see quots. 1968). Hence **Nami·bian** *sb.*, a native or inhabitant of Namibia; also *adj.*, of or pertaining to the land or people of Namibia.

1968 *Post* (S. Afr.) 4 Feb. 5 We are Namibians and not South Africans. **1968** *United Nations 22nd Session* Suppl. 16A 12 June 1/2 Resolution 2372... Proclaims that, in accordance with the desires of its people, South West Africa shall henceforth be known as 'Namibia'. *Ibid.* 2/1 The Council shall organize a training programme for Namibians. *Ibid.*, Reaffirms the inalienable right of the Namibian people to freedom and independence. **1970** R. HALL *South-West Afr.* (*Namibia*) 3 The Reverend Michael Scott..has won international respect by the tenacity of his fight..to sustain the rights of the Namibian peoples. **1973** *Times* 8 Mar. 6/3 Namibians prefer to be misgoverned or misruled by themselves rather than by others. **1973** *Guardian* 28 June 13/8 The Namibian strikes last year..secured important concessions. **1974** *Stand. Encycl. S. Afr.* X. 145/1 In United Nations circles the territory [*sc.* S.-W. Africa] is called Namibia, a designation derived from the name of the coastal desert, the Namib.

Namierian (nēimiə·riăn), *sb.* and *a.* [f. the name of Sir Lewis B. *Namier* (1888–1960), British historian + -IAN.] **A.** *sb.* An adherent of the methods and theories of Sir L. Namier concerning the structure of, and research into, political history. **B.** *adj.* Of, pertaining to, illustrating, or concerning the historical theories or techniques of Namier.

1958 *Oxf. Mag.* 27 Feb. 330/2 The Namierian revolution in historical method, now a quarter of a century old, destroyed a long accepted view of the nature of political parties in the 18th century. *Ibid.* 332/1 If the party is of little importance, as the Namierians have proved, then [etc.]. **1961** *Encounter* XVI. 39/2 The general configuration of world history will continue to look much the same, whether one takes the Namierian or pre-Namierian view about the structure of politics in Britain in George III's reign. **1962** *Guardian* 6 July 6/5 The Namierian House of Commons. **1967** *Economist* 6 May 575/3 His expertise with the telescope at one moment and the microscope at another is quite Namierian.

So **Na·mierite** *a.* and *sb.* in same sense. **Namieriza·tion,** the application of Namier's methods and theories to a historical situation; hence **Na·mierize** *v.*; **Na·mierizing** *vbl. sb.*

1948 *Bull. Inst. Hist. Research* XXI. 123, I refer to the process, which I can only call, after its distinguished inventor, the 'Namierizing' of British History: the discovery and exploration of the personal side of parliamentary politics and the parliamentary groups which flourished and manoeuvred, formed, broke and reformed their alliances under cover of debates and constitutional dogmas. **1952** *John o' London's Weekly* 20 June, A really intensely detailed and penetrating piece of analysis is known to historians all over the world as a 'namierisation'. **1957** *Economist* 12 Oct. 120/1 The expression 'to namierise' is a deserved tribute to Sir Lewis's influence. **1958** *Listener* 17 Apr. 661/3 Surely namierite scholarship has.. underlined the great differences..between the political systems of the eighteenth century and our own. **1958** *N.Z. Listener* 5 Sept. 12/4 May one hope that in future volumes he will add a dash of Namierite bitters to the cocktail? **1960** *Observer* 28 Aug. 8/4 Essentially, Namierisation meant a rigorous substitution of accurate detail for the generalisations which had contented older historians. **1961** *Times Lit. Suppl.* 2 June 334/4 This..piece of Namierisation does not offer portraits of any individual members of the group. **1967** *Economist* 22 July 329/3 Are we not, in fact, with the nineteenth century, in danger of much the same kind of difficulties as Namierite zeal landed us in with the eighteenth? **1969** *Listener* 20 Mar. 388/3 A generation ago the Namierites were busy counting heads while the Clapham sect totted up the trade statistics. **1971** *Sunday Times* 30 May 31/4 To 'namierise' is now a well-known historical technique which has been applied— often with far less subtlety—in very different fields. It consists, essentially, of constructing multiple biographies of inarticulate men whose common but unconscious aims may be a force, even a determining force, in history. **1972** *New Society* 13 Apr. 51/2 It [*sc.* the Scarman inquiry into violence in Ulster in 1969] is the Namierisation of a nightmare in which ten people died, [etc.]. **1973** P. A. ALLUM *Politics & Society Post-War Naples* ix. 280 The disappearance of the monarchy left the Vatican as the major institutional support of the *status quo*. In consequence, the Christian Democrats became..a 'Court Party' in the Namierite sense.

naming, *vbl. sb.* Add: **b.** *naming of parts*: the process of becoming acquainted, or of acquainting others, with the essentials of an unfamiliar object or topic.

1946 H. REED *Map of Verona* 22 To-day we have naming of parts. Yesterday We had daily cleaning. And tomorrow morning, We shall have what to do after firing. But to-day, To-day we have naming of parts. **1967** *Listener* 15 June 782/2 The Carnegie Report..begins with a very plain 'naming of parts', which is vital if the different forms of television are to be easily distinguished. **1967** *Punch* 26 July 135/2 We all know the experience, when we take up any new subject..of taking in..an enormous new naming-of-parts, so that we are able to make the distinctions the new subject demands. **1971** *Guardian* 29 Sept. 20/4 Many of the eruptions over topics like sex-education..spring..from a resistance to what may be summed up in the old grammarian's phrase, 'The naming of parts'. **1975** *Nature* 10 Apr. 483/1 (*heading*) Teleost ancestry, or naming of parts. *Ibid.* 483/3 Patterson has been meticulous in naming of parts, that is in tracing homologous structures throughout his series of fishes and naming them accordingly.

namma hole, var. *GNAMMA HOLE. *Austral.*

1893 *Australasian* 5 Aug. 252/4 (Morris), The route all the way from York to Coolgardie is amply watered, either 'namma holes' (native wells) or Government wells being plentiful on the road. **1928** K. S. PRICHARD *Coonardoo* iv. 52 His eyes, namma holes in viscid orbits, glittered at her, as he swung his naked feet.

nam-nam, var. *NUM-NUM.

Namurian (næmiū·riăn), *a.* (*sb.*) *Geol.* [f. *Namur*, name of a town and province of southern Belgium + -IAN.] Of, pertaining to, or designating the lowest stage of the Upper Carboniferous in western Europe, lying above the Dinantian; also as *sb.*, the Namurian stage or epoch.

1915 C. SCHUCHERT *Text-bk. Geol.* II. xl. 729 Kayser adds a third series, the Namurian (Namur, Belgium), which is referred to the base of the Middle Carboniferous. **1931** GREGORY & BARRETT *Gen. Stratigr.* vii. 118 Above the Dinantian [in Belgium] is a coarse sandstone, the great Grès d'Ardenne, corresponding to the Millstone Grit; it is called the Namurian. **1938** A. K. WELLS *Outl. Hist. Geol.* 100 The normal facies of the Namurian Stage is deltaic. **1963** D. W. & E. E. HUMPHRIES tr. *Termier's Erosion & Sedimentation* xiii. 292 By the Namurian, subsidence had practically finished and the last reef phase passed into large biostromes. **1967** D. H. RAYNER *Stratigr. Brit. Isles* vii. 200 It is also the type area of the Millstone Grit, where millstones were once fashioned out of the coarser sandstones... The sandstones..probably represent a lesser proportion of Namurian time than their thickness at first sight suggests.

nan^1. (Later example.)

1922 JOYCE *Ulysses* 46 Lambert Simnel, with a tail of nans and sutlers, a scullion crowned.

nan^2 (næn). *colloq.* [Prob. formed on GRANNY, or a shortening of *NANA^1, NANNA.] A grandmother; occas., a children's nurse.

1940 N. MARSH *Surfeit of Lampreys* (1941) iii. 43 Nanny's hands..made a quick involuntary movement. 'You'll be all right, Nan,' added Henry. **1954** —— *Spinsters in Jeopardy* viii. 150 The new Nanny... I didn't think I had to have a Nan over here. **1955** E. BLISHEN *Roaring Boys* iii. 123 'Been to see my nan,' he growled. **1968** L. BERG *Risinghill* 45 Perhaps it is Nan's importance in Islington.. that gives warmth and security to grandchildren. **1969** *Guardian* 15 Sept. 7/3 Her Mum and her Nan aren't a lot of help. **1975** *New Society* 18 Sept. 631/1 Jackie gets £1 a week off her grandmother, who owns a pub: 'My nan's got tons of money.' **1975** *Country Life* 13 Nov. 1277/2 I had a great, great grandmother who was a witch... Old Nan would remove the spell from a churn.

‖ **nan^3** (nān). Also **naan.** [Hindi.] In Indian and Pakistani cooking, a type of leavened bread.

1967 *Guardian* 8 Dec. 8/4 North Indian or Pakistani [cooking] which depends on the clay oven called a tandoor, into which kebabs..and the special bread called nan are lowered and rapidly cooked. **1971** *Carry Singapore in your Pocket* (Singapore Tourist Promotion Board) (ed. 3) 31 One first-class Kashmiri restaurant..serves..a wide variety of naan and pilau prepared in Kashmiri style. **1972** 'E. PETERS' *Death to Landlords!* iv. 76 The dough-cake type of bread called *nan*. **1973** *Times* 19 June (Bombay Suppl.) p. xv/3 Nan, a leavened bread dotted with poppy seed. **1973** *Sat. Rev. World* (U.S.) 18 Dec. 48/2 *Naan*, bread baked in a special oven by an Indian baker. **1974** R. HOWE in Moraes & Howe *J. K. Galbraith introduces India* 189 It [*sc.* chicken] is eaten with the fingers accompanied by slabs of *nan roti*, a type of bread also made in the *tandur*.

nana^1, nanna (næ·nă). Also **nan-nan.** Forms of address used by a child to a grandmother; occas., a children's nurse.

c 1844 E. PEPYS *Diary* 8 Oct. in *Country Life* (1972) 9 Mar. 577/1, I cut Herbert's hand..he only said 'O, Emy you had better send for Nana.' *Ibid.* 25 Dec. 578/1 Mama rung the bell for Nana. **1869** C. M. YONGE *Release* II. xi. 176 Poor dear Zélie..had been mother's nana. **1899** E. E. CUMMINGS *Let.* 27 Nov. (1969) 3, I am sorry dear Nana but I will be a good boy. **1901** *Punch* 4 Dec. 405/2 Please, Nanna, don't turn on the dark. **1927** W. E. COLLINSON *Contemp. Eng.* 7 Children still use, as in the nineties, and long before, mummy, daddy, nanny or nanna, nursie. [**1928** BARRIE *Peter Pan* I. in *Plays* 18 Nana the nurse.. is a Newfoundland dog.] **1954** *Caribbean Q.* III. I. 5 The role of the *nana* or nurse in preserving folk songs, tales.. herb lore is not to be under-estimated. **1958** M. KERR *People of Ship Street* 48 The granny, or 'nin', or 'nanny', or 'nanna' or 'gran' as she is often called. **1959** D. WALLACE *Richard & Lucy* xv. 210 A mewling toddler, clutching at her grandmother and weeping if Nan-nan were out of sight for more than a minute. **1959** *Woman's Own* 2 May 9/2 Cliffie..lived with his grandparents... Sometimes he would say..'I think I must go home today, thank you. Nana is expecting me.'

nana^2 (nā·nă). *slang* (chiefly *Austral.*). [Perh. f. BANANA.] **a.** A foolish person, a fool. **b.** The head. Also *attrib.*, as in *nana haircut* (see quot. 1941).

[**1894** A. B. PATERSON 'Hughey's Dog' in C. Semmler *The World of 'Banjo' Paterson* (1967) 29 'Off his nanny again,' thought the boss, 'the sooner he goes the better.'] **1941** BAKER *Dict. Austral. Slang* 48 *Nana* (*hair*)*cut*, a utilitarian haircut in which the back of the head is closely shaved. **1965** G. MCINNES *Road to Gundagai* ix. 148 Although he was obviously a gent, he was not a 'tonk' or a 'nana'. **1967** *Coast to Coast 1965–66* 35 If you..horror of horrors, had a *nana* haircut. **1967** C. DRUMMOND *Death at Furlong Post* xiii. 165 It seems if you have a bit of cash it snowballs unless you are a proper 'nana. **1968** J. LOCK *Lady Policeman* iv. 41 Do we stand there like a 'nana' holding up our hands to stop invisible traffic? **1971** *TV Times* (Austral.) 18 Dec. 19/4 It's only the famous ones who get fired... They're the ones..making..statements like 'The management is a nana'. **1972** M. FARHI *Pleasure of your Death* iii. 106 The answer, you 'nana. **1974** *Times* 22 Jan. 14/2 A frank admission that he had made a nana of himself. **1974** *Telegraph* (Brisbane) 22 Apr. 2/4 The baby started crying again. I did my nana and I hit him. **1975** *Australian* 8 Feb. 13 'We've all learned to laugh at ourselves and our predicament,' Trevor England said. 'If we hadn't we'd all be off our nanas.'

nanberry, -bury. (Later examples.)

1876 F. K. ROBINSON *Gloss. Words used in Neighbourhood of Whitby* 129/1 *Nanberries*, s. pl. warty spots on the groin of a horse. **1883** A. EASTHER *Gloss. Dial. Almondbury & Huddersfield* 90 *Nanberry* or *Nanbury*, *sb.* a kind of wart formed on the bag of a cow. **1970** 'J. HERRIOT' *If only they could Talk* xxv. 152 They [*sc.* horses] did get those little dangling growths sometimes—nanberries, the farmers called them. **1973** —— *Let Sleeping Vets Lie* xiii. 113 Oh, and Mr. Ross said would you take some nanberries off a stirk's belly while you're here.

nance (næns). *slang.* = *NANCY^2.

1924 G. C. HENDERSON *Keys to Crookdom* 412 *Nance*, effeminate fellow, sissy man. Nance walk—walk like a woman. **1932** [see *FAG *sb.*^{4}]. **1934** *Amer. Speech* IX. 27/1 *Nance*, an effeminate man. **1935** A. J. POLLOCK *Underworld Speaks* 79/1 *Nance*, a fairy; a male sexual pervert. **1941** B. SCHULBERG *What makes Sammy Run?* viii. 158 They were all so different—a titled Englishman and a famous poet and an aesthetic nance and a tough, drunken ex-reporter. **1957** J. G. COZZENS *By Love Possessed* (1958) II. v. 450 Here; I'll sneak out the back. Nances and I don't get on. **1971** F. FORSYTH *Day of Jackal* xx. 336 We're looking for a fellow who screwed the arse off a Baroness.. not a couple of raving nances.

Nancy. (Later example of *Nancy-story*.)

1967 *Times* 24 June 11/5 Here in Guyana, the term 'a nancy story' is common everyday parlance for a far-fetched yarn or even a downright lie... A 'Nancy Story' is ..one of the very few fragments of African culture which survived transplantation with the slaves.

nancy^2 (næ·nsi). *slang.* Also **nancy-boy.** [orig. *Miss Nancy* (MISS *sb.*^{2} 2 b), f. pet-form of the female name *Ann*.] An effeminate man or boy; a homosexual. Also as *adj.* Hence **na·ncified** *ppl. a.*; **na·ncifully** *adv.*

1904 'No. 1500' *Life in Sing Sing* 251/1 Nancy, effeminate man. **1918** C. MACKENZIE *Early Life Sylvia Scarlett* I. iv. 123 If you can let that nancified milksop mess you about, you can put up with me. **1930** E. WAUGH *Vile Bodies* 36 Where's my Fairy Prince? Powdering hisself again, I suppose... Come here, Nancy, and put away the beauty cream. **1931** R. CAMPBELL *Georgiad* ii. 36 Sharing out my last desires and fancies With tough old suffragettes and ageing nancies. *Ibid.* 42 Grim guards enough to awe the Nancy race. *Ibid.*, Their flag of truce they raise To bribe his laughter with their public praise, And all their little nancy husbands too. **1933** H. S. WALPOLE *Vanessa* IV. 723 But he isn't one of those, you know, or anything like that. Not a bit nancy. **1936** 'G. ORWELL' *Keep Aspidistra Flying* i. 19 A youth..with gilded hair, tripped Nancifully in. **1937** —— *Road to Wigan Pier* ii. 35 You and I..and the Nancy poets..—all of us *really* owe the comparative decency of our lives to poor drudges underground,..with their lungs full of coal dust. **1937** *Times* 25 Jan. 37/2 That nancyfied nonentity in the Foreign Office. **1937** E. LINKLATER *Juan in China* xxii. 288 You're a fop and a nancified civilian. **1947** L. HASTINGS *Dragons are Extra* x. 224 In the early part of the War I had occasional glimpses of an unfamiliar species known comprehensibly, I believe, as the Nancy Left. **1958** L. DURRELL *Balthazar* ii. 31, I can't stand that Toto fellow. He's an open nancy-boy. **1967** K. GILES *Death & Mr. Prettyman* ii. 58 Beautiful smooth dark rum, not like that nancified white stuff you poms put in your cokes. **1974** L. DEIGHTON *Spy Story* xv. 148 Pop music and nancy-boy actors.

Nancy Dawson (næ·nsi dǭ·sən). ?*Obs.* [See quot. 1890.] A sailor's dance or song; a nancy-boy.

1766 C. ANSTEY *New Bath Guide* ix. 64 With what Grace his Gloves he draws on, Claps, and calls up Nancy Dawson: Me thro' ev'ry Dance conducting. **1771** SMOLLETT *Humph. Cl.* I. 176, I can dance a Welsh jig, and Nancy Dawson. **c 1810** W. HICKEY *Mem.* (1960) xxv. 418 The dragoons..marched off..the fifes playing 'Nancy Dawson'. **1840** *Family Mag.* (Cincinnati) 332/2 She sailed through the waltz like an elephant dancing 'Nancy Dawson' in the ring of a menagerie. **1890** BARRÈRE & LELAND *Dict. Slang* II. 81/1 *Nancy Dawson*, a name for a molly, an effeminate youth, apathetic, &c... The original Nancy Dawson was a noted prostitute, on whom there is a song still current among sailors.

NAND (nænd). *Computers.* [f. *n*(*ot*) *and*.] A Boolean function of two or more variables that has the value zero when all of them are unity, and is otherwise unity; 'not..and..'. Usu. *attrib.*

1958 *IRE Trans. Electronic Computers* VII. 181/3 Using a positive logic (+ is identified with true and ONE) the circuit represents the binary NOR function, $\overline{A_1 + A_2} = B$. A negative logic (− is identified with true and ONE) results in the binary NAND function, $\overline{A_1 \cdot A_2} = B$. **1960** N. R. SCOTT *Analog & Digital Computer Technol.* x. 392 Of course, $A \mid B = \overline{A \cdot B}$; so this circuit is sometimes called a NOT AND (or NAND) circuit. **1966** *Electronics* 31 Oct. 42 The circuits used..are transistor-transistor NAND logic. **1971** J. H. SMITH *Digital Logic* iv. 48 Due to practical manufacturing reasons the NAND configuration forms the basis of the majority of integrated circuit systems. *Ibid.*, The action of the NAND is that it gives a *0* output when *all* its inputs are *1* signals.

Nandi (næ·ndi), *sb.*^{1} [Skr., 'the happy one'.] In Hindu mythology, the name of the bull of Siva which is his vahan or vehicle, and symbolizes fertility; also, a figure or statue of Nandi.

1807 H. T. COLEBROOKE in *Asiatick Researches* IX. 425 Near it stands a bull, intended perhaps for the bull called *Nandi*, a constant attendant of Śiva. **1891** M. MONIER-WILLIAMS *Brāhmanism & Hinduism* (ed. 4) xvii. 440 A shrine for the stone image of Siva's bull (nandi). **1910** *Encycl. Relig. & Ethics* III. 311/2 A representation of his sacred animal, the bull Nandi, is usually placed before him [*sc.* Śiva]. **1953** *Antiquity* XXVII. 169 Worth mentioning also are some clay censers, the handle of one being ornamented with a nandi. **1969** *Hindu Weekly Mag.* (Madras) 3 Aug. p. ii/6 The majestic 'Nandi', the conventional vehicle of Lord Siva in front of the 'sanctum' in the forecourt.

Nandi (næ·ndi), *sb.*^{2} and *a.* [Native name.] **A.** *sb.* **a.** An East African people of mixed Nilotic, Hamitic, and Bantu origin which inhabits an area on the Uganda-Kenya border and has given its name to the Nandi plateau;

a member of this people. **b.** The Nilotic language spoken by the Nandi and some neighbouring tribes. **B.** *adj.* Of or pertaining to this people or their language.

1885 J. THOMSON *Through Masai Land* (ed. 3) 469 From this place we could see the high forest region of Nandi... The Wa-nandi are allied in language and customs to the Wa-kamaria and Wa-elgeyo, though much braver and more warlike. **1902** H. H. JOHNSTON *Uganda Protectorate* II. xix. 876 All the Nandi-speaking peoples except the Andorobo make *pottery*. *Ibid.* 882 The Nandi especially believe profoundly in the powers of their medicine men, and follow them implicitly. **1909** C. ELIOT in A. C. Hollis *Nandi* p. xiv, Most of the wild hunting tribes called Dorobo speak a dialect of Nandi. *Ibid.* p. xxi, When a Nandi is ill, it is necessary to discover and propitiate the particular ancestor who has occasioned the disaster. **1909** *Westm. Gaz.* 3 Feb. 4/1 (*heading*) Nandi customs and folklore. **1930** C. G. SELIGMAN *Races Afr.* vii. 163 The Nandi and Suk may be said to live in garden settlements, each man having his own homestead in or near his fields of grain. **1951** in E. E. Evans-Pritchard *Social Anthropol.* i. 12 (*thesis-title*) 'The political organization of the Nandi' (East Africa). **1964** A. N. TUCKER in D. Abercrombie et al. *Daniel Jones* 445 The Nandi Group, comprising Nandi, Kipsigis, [etc.].

b. Nandi bear, a hypothetical animal resembling a bear, said to inhabit parts of East Africa.

1931 C. R. S. PITMAN *Game Warden* xiv. 287 Tales of that elusive monster popularly known in Kenya as the 'Nandi bear' can boast a more substantial basis. **1937** *Discovery* Jan. 2/2 The Wandarobo name 'Keret' or 'Kerit' is applied to the lynx or caracal as well as to the mythical 'Nandi Bear'. **1950** *Sun* (Baltimore) 2 May 13/2 The curious Nandi bear..is often blamed for the killing of humans and domesticated animals in East Africa. Although scientific evaluation is lacking, native and white hunters alike claim to have encountered the beast; and several descriptions of the bearlike creature agree on its shambling gait, shaggy hair, little ears and long snout. **1956** *Nature* 10 Mar. 446/1 This animal may be the dreaded Nandi bear.

Nanga (næ·ŋgă). [Jap., abbrev. *Nanshuga*, f. *nanshu* southern China + *ga* painting, picture.] Used, chiefly *attrib.*, to designate an intellectual style of Japanese painting.

1958 M. L. WOLF *Dict. Painting* 188 *Nanga School*, in Japanese art, a style of painting depicting genre subjects in a manner of Chinese idealism. **1970** J. W. HALL *Japan from Prehist. to Mod. Times* x. 217 The property of the samurai class was the style of 'literati' painting (*bunjinga* or *nanga*). **1970** *Oxf. Compan. Art* 762/1 *Nanga School*, school of painting which arose in Japan at the end of the 17th c.,..and persisted until late in the 19th. *Ibid.* 762/2 The Nanga painters were not all amateurs, but they spurned the professional schools of their day... In general their work represented the art of the intelligentsia as opposed to Ukiyo-e, which was that of the people. **1970** *Ashmolean Mus. Rep. Visitors 1969* 43 A particularly desirable acquisition in connection with the Museum's collection of Nanga painting. **1972** *Times* 18 May 21/6 Ikeno Taiga's *Taigado Gafu*..made £400. This is one of the finest colour-printed Nanga books and extremely rare. It is dated 1803.

Nankeen, Nankin, *sb.* (and *a.*). Add: **6.** (Further examples.)

1784 H. WALPOLE *Descr. Strawberry Hill* 8 A blue and white saucer with a landscape, of fine Nankin china. **1917** D. H. LAWRENCE *Phoenix II* (1968) 113 There were delicate blue Nankin cups. **1921** W. DE LA MARE *Mem. Midget* ii. 12 Dwarf trees..in green Nankin tubs. *Ibid.* xxiv. 157 A little masterpiece: and real old Nankin. **1957** MANKOWITZ & HAGGAR *Conc. Encycl. Eng. Pott. & Porc.* 161/2 Large quantities of porcelain decorated in blue were exported through this town [Nankin], hence the term 'Nankin Blue' or 'Nankin China' or simply 'Nankin'..a synonym for blue-and-white porcelain.

7. A kind of lace.

1865 F. B. PALLISER *Hist. Lace* xv. 199 It was not until 1745 that the blondes..made their appearance. The first silk used for the new production was of its natural colour, 'écrue', hence these laces were called 'nankins' or 'blondes'. **1882** Nankin [see BLOND *sb.* 2 b].

nanna, nan-nan: see *NANA[1].

nannofossil (næ·nofɒsil). *Geol.* [f. as next + FOSSIL *a.* and *sb.*] A fossil of a minute plankton organism.

1963 H. STRADNER in *Proc. 6th World Petroleum Congr.* I. 167/1 Nannofossils (minute calcareous elements of planktonic marine flagellates) were found in sediments younger than Triassic. *Ibid.*, A key for the determination of nannofossil assemblages is given. **1971** *Nature* 3 Sept. 46/1 The ophiolites in those deep-sea troughs were overlain by..radiolarian and nannofossil oozes. **1972** *Sci. Amer.* Dec. 33/2 The white ooze is a typical oceanic sediment, made up almost entirely of the skeletons of microfossils and nannofossils. **1974** *Nature* 13 Sept. 129/2 The chalk contains abundant nannofossils whose preservation is, unfortunately, somewhat marred by calcite overgrowths.

nannoplankton (næ·noplæ:ŋktǫn). *Biol.* Also **nanoplankton.** [a. G. *nannoplankton* (H. Lohmann 1909, in *Verhandl. Deutsch. Zool. Ges.* XIX. 234), f. *nanno-*, used as comb. f. of Gr. νᾶνος, L. *nanus* dwarf + PLANKTON.] Very small forms of plankton.

1912 MURRAY & HJORT *Depths of Ocean* vi. 356 As far as quantity is concerned, the smallest plankton organisms, Lohmann's Nanno-plankton, play a far more important rôle than the whole of the other species caught in our silk nets. **1939** *Nature* 7 Oct. 643/1 There was a nannoplankton consisting entirely of very small flagellates. **1959** *New Biol.* XXIX. 40 Examination of marine nannoplankton, that is the very small plankton organisms which are not retained even by the finest nets, has shown the presence of many colourless forms related to the algae. **1966** R. S. WIMPENNY *Plankton of Sea* iii. 114 The water..is strained through a funnel of fine pipes which filter out the very small organisms of the nannoplankton. **1974** B. H. MCCONNAUGHEY *Introd. Marine Biol.* (ed. 2) i. 7/1 Nanoplankton is, then, simply a size designation that includes parts of both phytoplankton and zooplankton. Strictly speaking, the term nannoplankton is used for organisms ranging from 5 to 60 μm in size.

nanny (næ·ni), *sb.* Also **nannie.** [Appellative use of pet-form of the female name *Ann(e*: see -Y[6].] A child's form of address to a nurse; hence, a children's nurse. Also *transf.*, of a person or institution, etc., considered to be unduly protective or apprehensive. Also *attrib.* and *Comb.*

1795 LADY NEWDIGATE *Let.* in A. E. Newdigate-Newdegate *Cheverels* (1898) 154 Nanny Ashcroft got me yᵉ most delightful & perfect Warm Sea Bath last night..after wᶜʰ I ate my Bason of Milk & went to Bed. **1861** A. HALLIDAY in H. Mayhew *London Labour* (1862) Extra vol. 418/2 An old woman..give me a lodging once for two years. We used to call her Nanny. **1864** *Chambers's Jrnl.* Sept. 506/1 Don't you know I'm a locomotive, and that you should always shunt yourself on to a siding when you hear *me* coming, Nanny? **1912** A. N. LYONS *Clara* i. 3 That little boy was..inured to the coming and going of 'nannies'. **1919** [see *CIG]. **1955** T. H. PEAR *Eng. Social Differences* 273 The effect upon the Nannie-cultured Englishman, of two mothers. **1958** A. WILSON *Middle Age of Mrs Eliot* i. 98 He was so clam and soothing and nannylike that she wanted to hit him. **1959** *Listener* 23 Apr. 735/1 Barristers are handled rather like naughty children by their nanny chief clerks. *Ibid.* 2 July 31/2 They can be nursery memories reproducing folk memory: immemorial nanny-lore. *Ibid.* 9 July 49/1 An extraordinarily powerful old bureaucratic nanny..goes stalking up and down the United States, pouncing on people who are telling commercial fibs. **1961** J. WADE *Back to Life* iv. 35 Nan came in... Her nannie days were past. **1965** B. SWEET-ESCOTT *Baker St. Irreg.* v. 147 There was also a good deal of Nanny work we had to do for Professor Pyke... The professor suffered from the traditional absent-mindedness of his kind, and we were expected to keep track of him. **1972** *Times* 2 Aug. 26/1 (Advt.), Country loving nannie. **1973** 'M. INNES' *Appleby's Answer* xvii. 155 Blabbing for your old nanny—that's what you'll be. **1973** *Times* 31 May 10/2 Kicked upstairs by the knighted nannies of his Institute, Hamo is sent on a world tour. **1973** *Listener* 7 June 742/2 The top authorities..who regulate television should [be]..impervious to the huge army of self-appointed nannies. **1974** *Times* 2 Feb. 18/3 Our natural modesty led us to cheer on the rest with encouraging cries..while deprecating our own dazzling accomplishments in the way that Nanny would have wished. *Ibid.* 29 Mar. 19/6 Surely our English cousins, having themselves retired from being 'nanny to the world', must understand how we feel. **1974** J. MANN *Sticking Place* vi. 106 Lorna took Rachel by the arm in a nanny-like way.

Hence **na·nnified** *ppl. a.*; **na·nnify** *v.*; **na·nniness, na·nnyism; na·nnyish** *a.*

1959 *Punch* 8 Apr. 474/1 Of all the nannyisms that have constrained the English middle classes the most inhibiting has been that favourite injunction about not putting ideas into the child's head. **1960** *Guardian* 1 July 8/1 The ponified and nannified Children's Bookland created by middle-class writers. **1962** *Times* 8 Feb. 9/7 Easy for his opponents to smile tolerantly on Wilberforce and make him seem impossibly nannyish. **1962** *Listener* 13 Dec. 1024/1 Is free speech in this country to be nannified out of existence by official vanity? **1972** J. ROSSITER *Rope for General Dietz* iii. 34 The air of nannyish protectiveness she had possessed in her BEA uniform. **1973** *Times Lit. Suppl.* 8 June 647/2 Mr Davidson's gentle male nanniness, initiating charming boys only to lose them.

nanny (næ·ni), *v.* [f. prec.] To act in the manner of a nanny; to be unduly protective. Hence **na·nnying** *vbl. sb.*

1954 J. TRENCH *Dishonoured Bones* ii. 46 Don't nanny me. **1963** *Economist* 6 July 41/3 Labour..would produce a less compulsively nannying piece of legislation. **1969** *Guardian* 11 June 11/7 Professional women get none of this obligatory nannying: their main grudge is inequalities of opportunity rather than of pay. **1971** J. OSBORNE *West of Suez* i. 35 Devoting all his time to Daddy and nannying him like he does. **1972** E. LONGFORD *Wellington: Pillar of State* iii. 31 In the short run, his five-year plan for nannying France—for such it soon seemed to be—could only damage the popularity of Wellington. **1974** E. DEWAR *Dying Business* ix. 110 There was no need for you to come nannying me.

nannygai (næ·nigəi). *Austral.* Also **nannyghai, nennigai.** [Aboriginal name.] A large marine food fish, *Centroberyx affinis*, found off the south-eastern coast of Australia.

1871 A. OLIVER *Fisheries New South Wales* 5 Fish, common to our coast,..when properly smoked or corned, are far superior to any imported codfish or ling... Nothing can surpass a corned 'moorra nennigai'. **1882** J. E. TENISON-WOODS *Fish New South Wales* 52 The Nannygai. ..Colour, a most beautiful pink, with silver stripes on the body. **1896** *Badminton Mag.* III. 206 A great variety of large game fish..the Nannyghai, the snapper, and a dark species of rock cod. **1896** F. G. AFLALO *Sk. Nat. Hist. Austral.* 199 The large-eyed, crimson Nannygai (*Beryx*), taken on the Schnapper grounds. **1947** *Coast to Coast 1946* 237 Morwong, rock-cod, nannygai and the rest, but not a real red fish among the lot. **1965** *Austral. Encycl.* VI. 240/1 Both [fishes] are wrongly called red snapper; the increasing use of this name in the fish trade for different species of nannygai is deprecated because of confusion with the true snapper. **1966** T. C. ROUGHLEY *Fish & Fisheries of Australia* (rev. ed.) 18 In some quarters..there was a prejudice against the nannygai,..and the New South Wales Fisheries Department therefore condoned the use of the name 'redfish' for it, and the sales were given a helpful impetus.

nanny-goat. Add: **2.** An anecdote. *slang.*

1843 T. C. HALIBURTON *Attaché* 1st Ser. II. vii. 114 The old knight's got an anecdote.., and nanny-goats ain't picked up every day in country. *c* **1895** F. M. FORD *Let.* (1965) 8 That is a subject abt. which I have a nanny goat to commence the next chapter with & it is reserved.

3. *to get* (a person's) *nanny-goat*: to get one's goat (see *GOAT 3 c). Also, *to get* (a person's) *nanny.*

1914 R. LARDNER in *Sat. Even. Post* 15 Aug. 9/3 'Good night, Horseshoes!' he says. That got my nanny up this time. 'Shut up, you lucky stiff!' I says. **1928** E. WALLACE *Double* xi. 122 'She was most kind and gracious—recognised me in an instant. Didn't mention you, by the way,' he dug Dick in the ribs playfully. 'That's got your nanny-goat!' **1972** J. MINIFIE *Homesteader* xvii. 143 Take it easy, old boy...Don't let them get your nanny.

4. A totalisator. *slang.*

1961 PARTRIDGE *Dict. Slang* Suppl. 1194/2 *Nanny-goat*, totalisator... Rhyming on tote. **1970** *Daily Mail* 10 Mar. 8/6 The poor old ailing Tote—the Nanny Goat, as they call it.

nano- (næ·no-, nēi·no-, nā·no-), *prefix.* [f. L. *nan-us*, Gr. νᾶν-ος dwarf + -O.] Prefixed to the names of units to form the names of units 10^9 times smaller, i.e. one thousand-millionth part of them (symbol n), as *nanoamp(ere, -farad, -gramme, -henry* [*HENRY[2]], *-litre -metre, -mole* [*MOLE *sb.*[7]] (hence *-molar* adj.), *-volt, -watt*; **na·noequivalent**, one thousand-millionth of a gramme-equivalent. Also *NANOSECOND.

1947 *Compt. Rend. de la 14me Conf.* (Union Internat. de Chimie) 115 The following prefixes to abbreviations for the names of units should be used to indicate the specified multiples or sub-multiples of these units:..n nano- 10^{-9} ×. **1952** *Wireless World* May 187/2 The prefixes 'pico' and 'nano' became popular in this country fifteen or twenty years ago, mostly through the technical publications of Philips and others, with 'pico' as favourite.

1973 *Nature* 23 Nov. 190/3 The transient currents generated are small (of the order of a few nanoamps) and flow for a very short time (of the order of 10 nanoseconds). **1962** *Flight Internat.* LXXXII. 634/2 Designers are talking in terms of nanoAmperes. **1967** *McGraw-Hill Yearbk. Sci. & Technol.* 146 Accuracy and precision [of coulometric analysis]..range from a few hundredths of a per cent at the hundred microequivalent level..to approximately 10% at the five nanoequivalent level. **1951** *Wireless World* Nov. 458/1 This is undesirable..if another metric value, the nF (nano-farad) could be accepted for one-milliardth of a farad... Originally introduced—so far as I know—in Germany and also used in other Continental countries before the war, this *nano* abbreviation is now, in the Indonesian PTT, as normal as *km* for length of wire. **1951** *Nature* 8 Dec. 1008/2 Most microanalytical needs are satisfied by the subunits milligram and microgram. This sequence has now been extended by the nanogram (ng = 10^{-9} g) and picogram (pg = 10^{-12} g). **1964** W. G. SMITH *Allergy & Tissue Metabolism* vii. 76 In man, 10^9 platelets contain 60 nanograms (0·06 mg.) of serotonin. **1975** *Nature* 11 Sept. 141/2 Nanogram quantities of DNP-D-GL induced anti-DNP antibody formation, while larger amounts prevented responses to DNP-conjugates, as previously reported. **1975** WILLIAMS & WILSON *Biologist's Guide to Princ. & Techniques Pract. Biochem.* i. 16 Response of isolated organs to nanogramme quantities of active substances has been obtained by this technique. **1968** *New Scientist* 29 Feb. 484/1 The capacitors..are 28 microfarads, 5 nanohenries, 2 milliohms. **1974** *Nature* 26 Apr. 774/2 The use of injection experiments..demands careful quantitative control over the injection of nanolitre volumes of material. **1963** *Calibration & Test Services* (Nat. Bureau of Standards Misc. Publ. 250) 41 Transmittances of these disks at wavelengths from 365 to 390 nm (nanometer, 10^{-9} meter)..will also be determined on request. **1970** *Sci. Jrnl.* Feb. 49/3 Usually points closer than one nanometre (10^{-9}m) can be separated with the electron microscope. **1971** *Sci. Amer.* Sept. 89/2 Blue light at a wavelength of 450 nanometers (a nanometer is a billionth of a meter). **1969** *Nature* 18 Oct. 221/1 They are of such high biological potency that nanomolar concentrations can produce well-marked effects. **1968** *McGraw-Hill Yearbk. Sci. & Technol.* 387/2 Conduction is blocked in isolated, desheathed frog sciatic nerves by a solution [of tetrodotoxin] containing about 3 nanomoles per liter. **1971** *Sci. Amer.* Sept. 98/2 Plants will increase photosynthesis with increasing concentration of carbon dioxide to at least three times the normal concentration of 12·5 nanomoles per cubic centimeter (·03 percent by volume). **1968** *New Scientist* 22 Aug. 391/1 The other problem is how to take just a few microvolts from the national standard of one volt, and to infer its correctness to a few parts of a nanovolt (10^{-9}V). **1968** *Sci. Amer.* Mar. 17 (Advt.), The new COS/MOS units..operate on nanowatts of power in the quiescent state.

nanophanerophyte (næ:nofæ·nəɹofəit). [a. F. *nanophanérophyte* (C. Raunkiær 1905, in *Oversigt K. Danske Videnskabernes Selskabs Forhandl.* 352), f. prec. + *PHANEROPHYTE.] A shrub or sub-shrub between 25 cm and 2 m

in height, bearing its resting buds above the surface of the soil.

1913 W. G. SMITH in *Jrnl. Ecol.* I. 17 Nanophanerophytes less than 2 m. high. **1932** FULLER & CONARD tr. *Braun-Blanquet's Plant Sociol.* xii. 293 The round, woody cushions..are not to be classed with the cushion forms, because of the lignified shoots, but rather among the semishrubs and nanophanerophytes. **1960** N. POLUNIN *Introd. Plant Geogr.* iii. 94 Refinements may be used such as the subdivision of phanerophytes into nanophanerophytes (shrubs) in which the renewal buds lie less than 2 metres above ground, microphanerophytes [etc.].

nanoplankton, var. *NANNOPLANKTON.

na·nosecond (see *NANO-). [f. *NANO- + SECOND *sb.*[1]] A unit of time equal to one thousand-millionth of a second.

1959 W. C. G. ORTEL in *IRE Trans. Electronic Computers* III. 265 (*heading*) Nanasecond logic by amplitude modulation at X band. [*Note*] Although the term *millimicrosecond* has attained some currency, the metric system provides the standard prefix *nano-*, (abbreviated n-), by which the decimal multiplier 10^{-9} may be denoted conveniently without multiple prefixes. **1960** *Times* 17 Feb. 2/6 Pulses of nanosecond duration are being studied by means of fast recording techniques and spectrographic methods. **1964** *Evening Standard* 30 Oct. 19/3 The latest computers are now doing their sums in 'nano-seconds'... According to General Sarnoff, former head of the Radio Corporation of America, 'A nano-second is to a second what a second is to 30 years'. **1966** [see *GIGA-]. **1967** *New Statesman* 14 July 47/3 It takes a nanosecond for an electric signal to travel a foot. **1968** *Dataweek* 24 Jan. 1/3 The 9400 would appear to be slower than the IBM System/360 Model 25 (memory cycle of 600 nanoseconds per two bytes against IBM's 180 nanoseconds per 64 bytes) but faster than the ICT 1902A (3 microseconds per 24-bit word). **1972** *Sci. Amer.* Jan. 121/1 Present atomic clocks keep time to better than 100 nanoseconds per day.

Nansen passport. [f. name of Fridtjof *Nansen* (1861–1930), Norwegian diplomat and explorer, who was responsible for the issue of the papers described below.] A document of identification issued after the war of 1914–18 to a stateless person ineligible for a passport. Also *absol.*

1925 *Measures to help Refugees: Rep. 5th Comm. to 6th Assembly League of Nations* 1 After lengthy negotiations, the Refugee Service has secured recognition for the Nansen Passports from forty Governments in the case of Russians and from twenty-eight in the case of Armenians. **1932** J. B. C. WATKINS tr. *Sörensen's Saga F. Nansen* 285 Nansen called together representatives from the various governments to a meeting at Geneva in July 1922. Thirty-one were represented, and they accepted Nansen's proposal for an identification certificate for each refugee which could be used as a passport. Fifty-two governments have recognized these certificates, which are stamped with Nansen's picture and known as 'Nansen passports'. They may be obtained by Armenian, Chaldean, Turkish, and Syrian refugees. **1944** H. G. WELLS *'42 to '44* 50 The practical organization of the Nansen passports that were for a time a resort for the multitude of people who had lost their national standing through changes of boundaries and similar dislocations. **1958** *Spectator* 14 Feb. 209/2 The 'Nansen passport' was one result. **1975** L. DICKSON *Radclyffe Hall at Well of Loneliness* xvii. 215 Souline had only a 'Nansen', a letter of identity not a passport, allowing her to reside in France. *Ibid.* 220 The long process of obtaining for Souline British naturalization papers, a first step in which was to exchange her French 'Nansen' for an English 'Nansen'.

‖ **nant** (nænt). [W.] A brook; a valley.

1862 G. BORROW *Wild Wales* II. xxxii. 368, I saw a small house close by a nant or dingle. **1883** G. ALLEN *Flowers & their Pedigrees* vi. 184 There was once a time when great glacial sheets spread over the combs and glens of Snowdonia, as they spread to-day over the nants of Chamounix. **1923** *Chambers's Jrnl.* 26 May 409/1 A hill-road scrambles through orchards and vineyards and across dashing *nants* (mountain torrents) to a lofty ledge of pasture-land.

Nantgarw (nænt,gæ·ru). Also **Nantgarrow** (nænt,gæ·ro). The name of a village in Glamorgan, used to designate a translucent soft-paste porcelain produced between 1813 and 1920 at the Nantgarw pottery founded by William Billingsley and Samuel Walker.

1820 W. W. YOUNG *Diary* 28 Oct. in E. M. Nance *Pott. & Porc. Swansea & Nantgarw* (1942) 524 Put Advertisement for the Nantgarw China in *The Cambrian*. **1820** *Cambrian* 28 Oct. in *Ibid.* xiii. 395 (Advt.), For Sale by Auction... At the China Manufactory at Nantgarw on Wednesday the 8th day of November 1820. A Quantity of Nantgarw Porcelain; a considerable portion of this is enamelled and gilded in a superior style. **1849** L. W. DILLWYN *Let.* 5 June in J. Marryat *Coll. Hist. Pott. & Porc.* (1850) ix. 186/2, I believe that all the China with granulated fracture was marked 'Nantgarrow'. **1880** J. RANDALL *Hist. Madeley* 206 Employing agents in Paris to buy up Sevres china in white for the purpose of having it painted in London, as Nantgarw had been. **1948** W. D. JOHN *Nantgarw Porc.* iii. 53 Two glazes were used on the Nantgarw porcelain. **1964** M. KELLY *March to Gallows* i. 10 It wasn't an environment for highflown escape. Nantgarw china and French song, for example. **1973** *Times* 16 Oct. 4/8 The Welsh porcelain brought particularly competitive prices, with one Nantgarw plate painted with fruit and flowers at £714.

Nantucketer (næntʊ·kėtəɹ). [f. *Nantucket*, the name of an American island off the coast of Massachusetts + -ER[1].] A native or inhabitant of Nantucket.

1851 H. MELVILLE *Moby Dick* II. xli. 279 Now some Nantucketers rather distrust this historical story of Jonah and the whale. **1882** E. K. GODFREY *Island of Nantucket* 24 The Nantucketers seem to have a mania for bell-ringing. **1914** R. A. DOUGLAS-LITHGOW *Nantucket* xv. 296 A Nantucketer does not pull, he always 'hauls', he does not tie or fasten anything, he 'splices' it. **1935** *Amer. Speech* X. 38/1 The calm superiority of the Nantucketer is shown in the case of the school boy who began his composition thus: 'Napoleon was a great man..but he was an off-islander.' **1948** *Sat. Even. Post* 26 June 46/3 Russells had also played a part, along with Howlands and Rotches, who had been Nantucketers, in committing New Bedford to the gigantic gamble on whaling. **1974** *Sci. Amer.* Mar. 119/1 The yellowfin-tuna fishery on the Pacific Equator, worked only recently by American and Japanese longliners, is plainly disclosed in the Nantucketers' logbooks.

nap, *sb.*[3] Add: **1. d.** Blankets or other covering used by a person sleeping in the open air. *Austral. slang.*

c **1905** in Stewart & Keesing *Old Bush Songs* (1957) 249 My nap is rather thin, But my rig is pretty good. **1918** C. FETHERSTONHAUGH *After Many Days* 279 That night he could not catch the donkey, and he had to camp without any 'nap' (blankets). **1926** L. C. E. GEE *Bushtracks & Goldfields* 41 The quarts were boiled and an after-dinner supply of coffee made, the 'nap' spread out on soft sandy spots, the two travellers reclined with pipes. **1933** R. B. PLOWMAN *Man from Oodnadatta* 2 The blackboy's nap (blankets, underproof sheet, etc.) filled in the rear compartment. **1934** A. RUSSELL *Tramp-Royal in Wild Austral.* vii. 54 Rolled in our 'nap' with the 'break' at our heads and the camp fire at our feet, we slept snug and comfortable. **1936** —— *Gone Nomad* iii. 15 Here I hobbled out my horse and built a fire, placing beside it my nap and sweat-soaked saddle for pillow. **1945** *Coast to Coast 1944* 157 If you carry enough nap, you goes hungry; if you carry enough tucker you sleeps cold.

nap, *sb.*[5] Add: **2.** *nap hand*, a hand which will probably take all five tricks in the game of nap; a strong hand. Also *fig.*

1862 C. C. ROBINSON *Dial. Leeds* 371 He's a nap hand at his traade! **1899** *Captain* I. 369/2 He showed me the way to deal myself a 'nap' hand, no matter who shuffled the cards. **1955** *Radio Times* 22 Apr. 9/2 Both these races (run over the Rowley Mile) test three-year-olds in the spring and complete the 'nap hand' of classics.

b. (Later *fig.* examples.) In *Austral.* also used neg.

1938 F. S. ANTHONY *Me & Gus* 8 That's why the girls go nap on you the way they do. **1955** P. WHITE *Tree of Man* (1956) vii. 80, I never went nap on the priests meself. **1957** D. NILAND *Call me when Cross turns Over* viii. 184 But he never went nap on the city, he said, and now he had done with it for good. **1959** *Times* 22 Dec. 3/2 That is the sort of thing that poses another problem within the selectors' main one—whether to go nap on the ability that they know a man has in him, or whether to go entirely on the evidence of trials. **1961** A. UPFIELD *Bony & White Savage* vii. 59 The woman who runs the bookshop knows we don't go nap on the sexy stuff. **1967** N. MARSH *Death at Dolphin* x. 256 When you get one of your hunches..I reckon it's safe to go nap on it.

c. A tip that a horse or greyhound is the most likely to win; *spec.* (see quot. 1937). In full, *nap selection*. Also, the horse or greyhound so tipped; a bet on such a horse or greyhound. *colloq.*

1895 *Starting Price* 30 Mar. 1/2 Our 'Outsider's' nap of Docker for the Hainton Stakes. **1926** *Westm. Gaz.* 20 July 1/4 The Whip, who yesterday gave Lightstep, Nap (won 3–1),..continues to hold a strong lead in Naps over the selections of the other racing critics. **1927** W. E. COLLINSON *Contemp. Eng.* 30 He stars this one, and the horse so starred is the nap selection. **1937** E. RICKMAN *On & off Racecourse* ix. 195 Every racing writer gives a single 'nap' or starred selection each day. It is his idea of the most promising bet the programme affords. **1960** *Which?* Mar. 60/1 The figures in the table are based on the correspondent's 'nap' selection—the word comes from a card game—for each day's racing, the horse that he thinks is the best bet. **1971** *Post* (S. Afr., Cape ed.) 9 May 16/5 (Advt.), Information from 'Horseman' includes jackpots, naps, accumulators, duplas, quinellas, doubles.

nap (næp), *v.*[4] [f. *NAP *sb.*[5] 2 c.] *trans.* To recommend (a horse or greyhound) as a likely winner.

1927 *Daily Express* 22 June 17/7 Great Chum napped for White City Cup Final. **1973** *Listener* 28 June 864/3 You start napping odds-on chances, and they get beaten just as easily as the others.

Napa (næ·pă). Also **nappa.** The name of a county, town, and valley in California, U.S.A., used *attrib.* and *ellipt.* to designate (*a*) leather prepared from sheep- or goat-skin by a special tawing process; (*b*) wine grown in the valley.

1883 R. L. STEVENSON *Silverado Squatters* 34 (*heading*) Napa wine. **1895** *Montgomery Ward Catal.* 289/3 Men's Smoke Napa Gauntlet. **1897** C. T. DAVIS *Manuf. Leather* (ed. 2) 275 The staking machine..can be adjusted to any kind of leather, including napa. **1903** L. A. FLEMMING *Pract. Tanning* 51 Coloring black Napa. Black Napa leather is generally colored blue on the flesh side. **1921** B. E. ELLIS *Gloves & Glove Trade* 58 'Nappa' gloves are made from tawed leathers. **1961** *Housewife* Apr. 59/1 Fawn nappa leather handbag. **1971** M. MCCARTHY *Birds of America* 157 Several glasses of Napa Valley wine. **1974** *Country Life* 21 Mar. 688/1 Slip-on mules in coloured nappa leather for £7.

napalm (næ-, nēi·pām), *sb.* orig. *U.S.* [f. *NA(PHTHENATE + PALM(ITATE (see quot. 1946).] **1. a.** A thickening agent consisting essentially of aluminium salts of naphthenic acids and of the fatty acids of coconut oil. **b.** A thixotropic gel consisting of petrol and napalm (or some similar agent), used in flame-throwers and incendiary bombs; jellied petrol.

1942 in L. F. Fieser *Sci. Method* (1964) iv. 47 Novello and Harris will arrive..on Monday morning and deliver 7 small bombs and 7 large ones, each loaded with 100 grams of white phosphorus and with Napalm Polymer Gum prepared in gasoline. **1942** H. C. HOTTEL *Stud. Fuel Projection from Nozzles* (U.S. Govt. Res. Rep. PB 23769, 8 June) Summary, Very striking and extremely promising, however, is the enormous increase in range obtainable by the use of special materials, specifically a starch-water suspension and an incendiary gasoline gel known as Napalm. The latter can be projected three times as far in a one-eighth inch jet as can a conventional hydrocarbon fuel. **1946** L. F. FIESER et al. in *Industr. & Engin. Chem.* Aug. 769/1 It was next found (January 29, 1942) that a combination of aluminium naphthenate with the same 'aluminum palmitate' could be easily incorporated into gasoline to form a promising gel, and we termed this naphthenate-'palmitate' combination a Napalm gel. Subsequently it developed that the supposed 'aluminum palmitate' was actually the aluminum soap of the total fatty acids of coconut oil, and that the specific gelling quality is due to a high content of lauric, not palmitic, acid. **1951** KIRK & OTHMER *Encycl. Chem. Technol.* VI. 580 The U.S. uses a mixture of aluminum soaps, generally referred to as Napalm, as a thickening agent. A fuel of any desired consistency may be prepared by mixing the proper amount of Napalm with gasoline. **1952** R. CUTFORTH *Korean Reporter* xix. 174 He was no longer covered with a skin, but with a crust like crackling which broke easily. 'That's napalm,' said the doctor. **1953** *Armed Forces Chem. Jrnl.* (U.S.) July 8/2 The thickening agent, known popularly as napalm, gelled the gasoline to a honey-like consistency. **1964** L. F. FIESER *Sci. Method* ii. 27 On Feb. 14 [1942], we reported to NDRC development of two lines of gels that could be prepared by stirring with gasoline at room temperature. To one, made from aluminum *na*phthenate and 'aluminum *palm*itate', I gave the name Napalm. *Ibid.* 32 The most interesting sample, Napalm X-104, arrived on April 13, 1942... The material is a brownish, dry, nonsticky powder. When an amount of Napalm powder sufficient to produce a 12% solution is poured into gasoline and given one stir, solvation occurs rapidly and the swollen solvated particles soon fill the container with material of applesauce consistency... After aging for a few hours without attention, the gel reaches its final form, in which it is tough, strong, and sticky. **1966** *Chem. & Engin. News* 14 Mar. 24/3 An Air Force contract to supply 100 million pounds of the new napalm, which contains 50% polystyrene. *Ibid.*, Napalm-B, besides the 50% polystyrene, contains 25% benzene and 25% gasoline. It is replacing the soap-jelled gasoline napalm formulations of World War II and the Korean action. **1966** *Daily Tel.* 16 Aug. 17/3 Mrs. Anne Kerr, Labour M.P. for Rochester and Chatham, said in Tokyo today that the United States should withdraw from South Vietnam 'rather than drop napalm on innocent peasants'. **1967** *Freedomways* VII. 121 We saw a man who was a victim of napalm, and two women who had been tortured in Diem's prisons. **1968** [see *INCENDIJEL]. **1968** V. W. SIDEL in S. Rose *Chem. & Biol. Warfare* iii. 46 The adhesiveness, prolonged burning time and high burning temperature of napalm favour third-degree burns, and such burns are likely to be deep and extensive. **1970** K. LYLE in A. Chapman *New Black Voices* (1972) 293 I-am-wishing for an hour of napalm on ALL Junior Chambers-of-Commerce.

2. *attrib.* and *Comb.*, as *napalm bomb, -bombing, burn, gel, jelly.*

1945 *N.Y. Times* 17 Aug. 2/7 Only a few [new weapons] can be mentioned. Among them were:..the Napalm bomb—jellied gasoline with a detonator in a plane's detachable fuel tank. **1957** S. JAMESON *Cup of Tea for Mr. Thorgill* v. 41 All these years he has done harm—from a position of complete safety and comfort, like a pilot dropping napalm bombs. **1963** *Daily Tel.* 22 Aug. 10/2 The United Arab Republic..admits to using napalm bombs against the Royalist tribes. **1973** *Black Panther* 13 Oct. 14/2 He also charged that the Portuguese had been attacking them with tanks, napalm bombs and highly sophisticated and dangerous weapons supplied mostly by NATO countries. **1955** G. GREENE *Quiet American* 197 What I detest is napalm bombing... The poor devils are burnt alive, the flames go over them like water. **1959** *Times Lit. Suppl.* 23 Oct. 614/3 Fontane sees the war at close quarters—the heat, the dysentery, the napalm-bombing. **1952** *Times* 14 July 7/6 The napalm burns are not necessarily the most painful of wounds. **1968** *Listener* 8 Aug. 170/2 A television shot of a patient suffering from napalm burns in Vietnam. **1946** Napalm gel [see *1]. **1968** V. W. SIDEL in S. Rose *Chem. & Biol. Warfare* iii. 46 The napalm gel proved far superior to the original rubber-based gel. **1966** *Daily Tel.* 2 Nov. 14/8 Pictures showing the full horror of burns in Vietnam caused by the diabolical napalm jelly of phosphorus and petrol.

napalm (næ-, nēi·pām), *v.* [f. prec.] *trans.* To attack or destroy with napalm. Also *fig.* Hence **na·palmed** *ppl. a.*, **na·palming** *vbl. sb.*

1950 *N.Y. Times* 26 Sept. 2/7 Troops were napalmed when they were found hiding in caves at the dead end of a canyon. *Ibid.* 6 Nov. 2/2 Capt. Warren Nichols..dived down in a napalming attack on a concentration of Red troops. **1952** DYLAN THOMAS *Let.* 8 Oct. (1966) 378, I think it has something to do with what Our Side gives to people after it has napalmed them. **1966** *Punch*

12 Jan. 64/2 The nobility of our democracy obliges us to succour the people..not to napalm them down. **1966** *New Statesman* 26 Aug. 292/2 There is no patience left in Vietnam. It has been napalmed and shot away. **1967** *Observer* 10 Dec. 5/3 The Americans are very busy bombing, rocketing and napalming air fields. **1968** *Economist* 27 Apr. 21/1 The nationalists have no intention of occupying these stockades, since the villages that contain them would then be napalmed or razed. **1968** *Guardian* 1 Oct. 8/4 Napalmed peasants in Vietnam. **1969** R. PETRIE *Despatch of Dove* III. 155 How to get rid of poverty? Blow it up! Napalm it! **1972** *Village Voice* (N.Y.) 1 June 62/2 (Advt.), They said they wanted to napalm draft records instead of people.

nape, *sb.*[1] Add: **1.** (Later *attrib.* example.)

1922 JOYCE *Ulysses* 256 Stooping her fair pinnacles of hair, stooping, her tortoise napecomb showed.

2. b. (Later example.)

1884 G. B. GOODE *Fisheries U.S.: Nat. Hist. Aquatic Anim.* 201 George's fish are very fat fish with white 'napes'.

naphtha. Add: **b.** *naphtha lamp* (earlier example); *naphtha-bearing* adj.; **naphtha engine** (see quot.); **naphtha launch**, a launch powered by a naphtha engine.

1909 *Westm. Gaz.* 2 Dec. 9/2 The leasing of certain areas of naphtha-bearing land contrary to existing regulations. **1892** P. BENJAMIN *Mod. Mechanism* 270 Naphtha-engines, which utilize naphtha both as the fuel under the boiler and as the fluid to be vaporized in the boiler and used in the engine, have recently come into somewhat extensive employment as motors for light launches. **1877** *Scribner's Monthly* Nov. 3/1 It consists of a large reflector behind a common naphtha lamp and mounted upon the bow of the boat. **1887** *Forest & Stream* 8 Dec. 395/3 The success which the naphtha launch has attained in a very short time must be taken as very strong proof of the inherent excellence of the machine itself. **1903** *Mobile* (Alabama) *Advertiser* 2 Jan. 6/3 The big naphtha launch Stella arrived from up the Alabama river yesterday afternoon. **1923** H. E. WILLIAMS *Spinning Wheels & Homespun* xv. 182 There have been times when..skimming away in little naptha [*sic*] launches..has seemed an enviable lot.

naphthalene. (Now the regular spelling.)

naphthaleneacetic (næ:fþălīnăsī·tik), *a.* [f. NAPHTHALENE + ACETIC *a.*] *naphthaleneacetic acid*: either of the two crystalline compounds, $C_{10}H_7CH_2COOH$, obtained by replacing one of the hydrogen atoms of naphthalene by an acetic acid group; *spec.* α-naphthaleneacetic acid, which has the action of an auxin and is used to stimulate the rooting of plant cuttings, to initiate flowering of the pineapple, to prevent premature drop of fruit, and to improve the colour of apples.

1917 *Chem. Abstr.* XI. 3876/2 (Index), Naphthaleneacetic acids. **1936** *Contrib. Boyce Thompson Inst.* VIII. 78 Treating cuttings or shoots of *Ilex, Taxus, Pachysandra, Hibiscus, Acer*, and *Chrysanthemum* with preparations of indoleacetic, indolepropionic, indolebutyric, or naphthaleneacetic acids induced earlier rooting, increased the number of roots, and roots emerged from a greater area of stem tissue as compared with control cuttings. **1942** *Science* 22 May 536/1 Low concentrations of α-naphthaleneacetic acid (the compound used most extensively) applied as foliage sprays induced formation of inflorescences in advance of the normal period, but high concentrations..delayed flowering far beyond that of the controls. **1968** F. C. STEWARD *Growth & Organization in Plants* viii. 394 The application of auxin during the dark period eliminates the photoperiodic induction due to the dark period in a short-day plant like cocklebur. Auxins, whether IAA or a synthetic substitute like naphthalene acetic acid..are all active in this respect.

naphthaquinone, var. *NAPHTHOQUINONE.

naphthene. Add: [ad. F. *naphtène* (Pelletier & Walter 1840, in *Jrnl. de Pharm.* XXVI. 561).] The substance orig. called by this name is now regarded not as a single compound but as a mixture of hydrocarbons, and the term is now *Obs.* in this sense. (Further example.)

1854 THOMSON *Cycl. Chem.* 378 Naphthene. $C_{16}H_{16}$. That portion of mineral naphtha which comes over second in distillation.

b. [ad. F. *naphtène* (coined afresh by Markovnikov & Oglobine 1884, in *Ann. de Chim. et de Physique* II. 447).] Any of a class of saturated cyclic hydrocarbons (including cyclopentane and cyclohexane) that are present in or obtained from petroleum.

1884 *Jrnl. Chem. Soc.* XLVI. 1276 Hydrocarbons of the C_nH_{2n} series, called by the authors naphthenes, constitute the principal part of Caucasian petroleum. **1921** J. S. CHAMBERLAIN *Textbk. Org. Chem.* 38 The saturated cyclic hydrocarbons known as naphthenes are characteristic of Russian (Baku) petroleum and are also found in that from Galicia. **1951** C. R. NOLLER *Chem. Org. Compounds* xxxix. 759 The saturated alicyclic hydrocarbons frequently are called cycloparaffins or cyclanes although petroleum technologists usually call them naphthenes because cyclopentane..and cyclohexane..and their homologs have been isolated from the naphtha fraction of petroleum. **1964** N. G. CLARK *Mod. Org. Chem.* v. 78 The naphthenes (a term used in the petroleum industry for the cyclic paraffins) are mainly cyclopentane and cyclohexane, and their simple derivatives.

naphthenic (næfþī·nik), *a. Chem.* [ad. F. *naphténique* (Markovnikov & Oglobine 1884, in *Ann. de Chim. et de Physique* II. 476), f. *naphtène* NAPHTHENE: see -IC.] **1.** *naphthenic acid*: any of the carboxylic acids obtained in the refining of petroleum, esp. one derived from an alicyclic compound (as cyclohexane or cyclopentane); also, an unspecified mixture of such acids.

1894 *Jrnl. Chem. Soc.* LXVI. i. 532 The methyl salts of the lower natural naphthenic acids were prepared and fractionated with a view to isolate the methyl salt of heptanaphthenic acid (hexahydrobenzoic acid). **1932** *Jrnl. Amer. Chem. Soc.* LIV. 240 We observed that naphthenic acids are formed when linoleic acid is cracked under pressure. **1942** *Chem. Rev.* XXX. 100 The mixture of acids now known as naphthenic acid. **1946** S. B. ELLIOTT *Alkaline-Earth & Heavy-Metal Soaps* ii. 33 Naphthenic acids..are the raw materials from which some of the most important soaps are manufactured. **1957** VAN DER HAVE & VERVER *Petroleum* xiii. 383 The free naphthenic acids also find certain applications, for instance as wetting agents for the pigments used in printing inks and paints. **1966** *McGraw-Hill Encycl. Chem. Technol.* X. 47/2 Naphthenic acid production is about 16,000,000 lb annually.

2. Of, pertaining to, or containing naphthenes.

1931 *Engineering* 2 Jan. 1/2 Cyclohexane, a hydrocarbon of the naphthenic series. **1969** *Sci. Jrnl.* Nov. 32/2 The basic structure of bitumen is that of clusters of naphthenic and aromatic rings... It is called a naphthenic crude oil. **1974** *Sci. Amer.* Dec. 7 (Advt.), Up to this point, the best traction drive lubricants were naphthenic oils.

Hence **na·phthenate**, any of the salts of a naphthenic acid, some of which are used as paint driers, as fungicides, and in lubricants.

1899 *Jrnl. Chem. Soc.* LXXVI. i. 423 (*heading*) Properties of naphthenates and their qualitative distinction from salts of fatty acids. **1924** *Chem. Abstr.* XVIII. 2688 Al naphthenate forms a transparent, colorless, and porous mass which can be powdered; it has a water-proofing action. **1942** *Chem. Rev.* XXX. 103 In the case of castor machine oils, where increased..viscosity..is desired, various naphthenates have been used. **1960** E. L. DELMAR-MORGAN *Cruising Yacht Equipment & Navigation* xiii. 158 It is usually possible to see when power kerosene is deteriorating, by the fact that it becomes bright green in colour, as a result of the formation of copper naphthenate formed by the interaction of the fuel and the metal of a copper or brass tank. **1973** *Biol. Abstr.* LV. 6965/2 Potassium naphthenate significantly increased the ascorbic acid content of green pods of bush bean plants.

naphthoquinone (næfþokwi·nōᵘn). *Chem.* Also **naphtha-**. [f. NAPHTH(ALENE + -O + QUINONE.] Each of the six compounds, $C_{10}H_6O_2$, obtained (theoretically) by replacing two of the ⩾CH groups of naphthalene by carbonyl groups, *spec.* 1,4- (or α-)-naphthoquinone, a volatile yellow solid whose molecule forms part of the structure of vitamin K.

1870 *Jrnl. Chem. Soc.* XXIII. 446/2 (Index), Naphthoquinones. **1914** H. T. CLARKE *Introd. Study Org. Chem.* xxxiv. 417 The *ortho* quinone, β-naphthaquinone,..is a red crystalline solid, which decomposes at 115–120°. **1942** *Ann. Reg. 1941* 344 The four forms of vitamin K proved to be naphthoquinone derivatives. **1951** I. L. FINAR *Org. Chem.* I. xxix. 593 Theoretically, six naphthaquinones are possible... Only three are known, the 1:2-, 1:4- and 2:6-, but it appears that derivatives of 2:3-naphthaquinone have been prepared. **1966** *McGraw-Hill Encycl. Sci. & Technol.* XI. 194/2 The naphthoquinones are prepared by oxidation of the corresponding aminonaphthols.

napkin, *sb.* Add: **1. c.** A rectangular piece of towelling or absorbent material used as a baby's undergarment by folding, drawing up between the legs, and fastening at the waist.

1845 MRS. GASKELL *Let.* (1966) 824 Meta is so neat & so knowing, only, handles wet napkins very gingerly. **1861** MRS. BEETON *Bk. Househ. Managem.* 1021 Soiled baby's napkins should be rolled up and put into a pan, when they should be washed out every morning. **1961** *Brit. Med. Dict.* 1212/1 The skin..is the more easily affected by the free ammonia liberated through the interaction of acid urine and badly washed napkins. **1974** *Janet Frazer Catal.* Spring & Summer 292/2 Fully-bleached terry napkins. Soft and absorbent... Size 24" × 24"..£3·35 (dozen).

4. napkin-ring (earlier and later examples).

1686 in *Narragansett Hist. Reg.* III. (1884–5) 105, 18 Napkins & 9 Napkin Rings. **1972** *Country Life* 21 Dec. 1755 This woven design, using sterling silver ribbons interwoven..is very attractive. There are napkin rings (ideal for a wedding present)... The napkin ring costs £20.

Napoleon. 2. For 'top-boot' read 'long boot'.

3. (Later examples.)

1880 *Harper's Mag.* May 917/1 The artillery is almost entirely the old brass Napoleon. **1897** *Outing* XXX. 80/1 These gun companies were each supplied with one 12-pounder Napoleon gun and one Gatling gun.

4. b. = *NAP *sb.*[5] 2 c.

1895 *Starting Price* 23 Mar. 1/1 With ruinous 'all day wires' and extortionately priced 'Napoleons' we will have nothing to do.

5. One resembling Napoleon I, esp. in having gained supremacy in his own sphere through ruthless ambition. Also *transf.*

1821 SHELLEY *Let.* 25 Sept. (1964) II. 353 He was a little Napoleon..with a dukedom instead of an empire for his theatre. **1866** J. BLACKWOOD *Let.* 2 Aug. in Geo. Eliot *Lett.* (1956) IV. 293 He is a Napoleon in his Trade is True-fitt. **1879** R. L. STEVENSON *Trav. with Donkey* 144, I had travelled hitherto through a dull district, and in the track of nothing more notable than the child-eating Beast of *Gévaudan*, the *Napoleon Buonaparte* of wolves. **1894** A. CONAN DOYLE *Mem. Sherlock Holmes* 260 He is the Napoleon of crime, Watson. He is the organizer of half that is evil and of nearly all that is undetected in this great city. He is a genius, a philosopher, an abstract thinker. **1907** I. ZANGWILL *Ghetto Comedies* 353 This, then, was the notorious multi-millionaire, 'the Napoleon in dollars'. **1932** Q. D. LEAVIS *Fiction & Reading Public* 312 Northcliffe, being the Napoleon of the Press, naturally disliked having to play second fiddle to the advertiser. **1939** T. S. ELIOT *Old Possum's Pract. Cats* 35 The Cat who all the time Just controls their operations: the Napoleon of Crime! **1969** *N.Y. Rev. Bks.* 18 Dec. 33/2 We had Napoleons of finance, Napoleons of industry, Napoleons of the betting ring. **1973** 'I. DRUMMOND' *Jaws of Watchdog* vi. 79 'Royston,' said Colly. 'What kind of a name is that for a Napoleon of crime?' **1975** *Times* 2 June 10 A Time Profile—General Vo Nguyen Giap—Communist Napoleon who conquered South Vietnam.

6. A type of rich cake made from layers of puff pastry filled with cream, custard, or jam. Also *attrib.*

1892 *Encycl. Pract. Cookery* I. 246/1 *Napoleon Cake.* Lay in a Napoleon-cake pan..a layer of puff paste. spread over that a layer of pastry cream, cover with puff paste, glaze the top with sugar, and bake. **1896** E. TURNER *Little Larrikin* xxviii. 342 In the centre were five white-iced Napoleons, ornamented with devices cut from silver tea-paper, this being the nearest the funds would stretch to wedding-cake. **1956** E. STARKIE *Diary* Sept. in J. Richardson *Enid Starkie* (1973) xxviii. 204 There was..every kind of cake, biscuit and sweet—the most succulent cake was called Napoleon! **1961** J. HELLER *Catch-22* (1962) xxiv. 249 Prune and cheese Danish from Copenhagen, éclairs, cream puffs, Napoleons and *petits fours* from Paris, Reims and Grenoble, [etc.]. **1968** A. BINKLEY *What shall I Cry?* 139 'You got more of those cakes?'..Lenni handed him a box in which one napoleon remained. **1969** W. S. KUNICZAK *Sempinski Affair* (1970) vii. 69, I..felt my resolve melting like the Napoleon pastries on my plate. **1975** C. NESBITT *Little Love & Good Company* xvi. 201 A plate of luscious Napoleons oozing thick whipped-cream.

7. Used *attrib.* and *absol.* of brandy of supposed great age or special merit.

1930 J. DOS PASSOS *42nd Parallel* II. 207 They were sitting over Napoleon brandy in big bowlshaped glasses and cigars. **1967** C. CHURCHILL *World of Wines* x. 208 'Napoleon Brandy' ranks high among some of the more transparent promotional myths of the industry. **1967** A. LICHINE *Encycl. Wines* 214/2 The worst deception in Cognac brandies is the fraud and fakery connected with alleged 'Napoleon' brandy—but a distinction must be made between brandies pretending to derive from Napoleon's time, and the Napoleon style. **1968** C. FORSYTE *Murder with Minarets* vii. 42 They sat over their Napoleon after dinner. **1968** T. PARKER *People of Streets* 48 Could I have a glass of Napoleon brandy please? **1975** D. BEATY *Electric Train* 147 Liqueurs—two Napoleon brandies. Warm on her stomach.

Napoleonism. Add: (*a*) *spec.* the assumption of absolute control over subject peoples or countries; (*c*) conduct or behaviour resembling that of Napoleon.

1966 *New Statesman* 28 Oct. 828/1 The military life had the advantage of absorbing what could easily have been insidious at this stage: the premature dream of glory, the stupefying effects of youthful Napoleonism. **1969** *Sunday Times* 6 Apr. 27 Personal incompatibility..and the traditional Latin American leader's custom of renouncing ultimate power as soon as he had won it (for fear of accusations of 'Napoleonism'), were less significant than the internal political and social divisions. **1970** *Guardian Weekly* 6 June 2 Without our nation standing up against Russia's modern day mad dog Napoleonism Europe itself would not remain independent through the 1970s.

napoo (nāpū·), *int.*, *a.*, and *v. colloq.* Also **na poo, napooh.** [Corruption of F. (*il n'y e*)*n a plus* there is no more.] **a.** *int.* Finished; gone; done for. **b.** *adj.* Finished; good for nothing; dead. **c.** *v. trans.* To finish, kill, or destroy.

1915 'I. HAY' *First Hundred Thousand* 302 You say 'Na pooh!' when you push your plate away after dinner. ..'Poor Bill got na-poohed by a rifle-grenade yesterday.' **1917** W. J. LOCKE *Red Planet* xvi. 194 Instinctively I stretched out my hand. He laughed. 'Napoo. You must take it as gripped.' **1919** J. B. MORTON *Barber of Putney* xv. 253 'Can't do nothing,' said Curly, ''e's napoo.' *Ibid.* xviii. 301 Even if they themselves were na-pooed, they'd hate to think of the lousy Boche living in their home. **1925** N. VENNER *Imperfect Impostor* i. 6 If you haven't got a job to do, you're a washout. You might as well be napood right off. **1927** W. DEEPING *Kitty* xvi. 205 A man's phrase—a war-phrase—seemed to trickle into his head. Everything was na poo, a wash-out. His marriage—. **1936** F. CLUNE *Roaming round Darling* vi. 52 All the boys about here..have looked for money in the gullies, and the only thing they ever come across was a rum-keg—empty—'*Napoo*', like the Diggers used to say! **1943** J. B. PRIESTLEY *Daylight on Saturday* xxix. 228 You're as good as dead—just waitin' to stiffen. Fini—napoo! **1973** L. MEYNELL *Thirteen Trumpeters* v. 81 Prudence..fell down dead in the croupier's bag. Fini. Napoo.

nappa, var. *NAPA.

nappe (næp). [a. F. *nappe* table-cloth, NAPE *sb.*[2]] **1.** *Hydraulics.* A sheet of water falling over a weir or similar surface.

1892 MARICHAL & TRANTWINE tr. Bazin in *Proc. Engineers' Club Philadelphia* IX. 231 We now proceed to study the form of the sheet of water passing over the weir. [*translator's footnote*] For want of a convenient English equivalent, we shall designate this sheet by its very appropriate French name, the *nappe*, a name applied primarily to a table-cloth, the form of which..is well imitated by the sheet of water passing over the weir. *Ibid.* [*text*], The upper surface of the nappe has already been studied by certain experimenters. **1945** W. P. CREAGER et al. *Engin. for Dams* II. xi. 357 If the area below the lower nappe is filled with masonry, the shape of the sheet and the discharge will not be changed appreciably. **1966** F. M. HENDERSON *Open Channel Flow* vi. 177 If the nappe is contained within parallel walls downstream of the weir, it may well enclose the air between itself, walls, and floor.

2. *Geol.* A sheet of rock which has moved sideways over neighbouring strata, usu. as a result of overthrusting.

Earlier used in F. as *nappe de recouvrement* sheet of overlaying (e.g. *Bull. Soc. vaudoise des Sci. nat.* (1893) XXIX. 252).

1922 *Q. Jrnl. Geol. Soc.* LXXVIII. 87 In deciding upon the basal limit to be assigned to any particular nappe, one generally chooses some prominent thrustplane; failing this, one is entitled to select the axial plane of some recumbent anticline or syncline, according to local convenience. **1932** [see *ALLOCHTHONOUS *a.*]. **1944** A. HOLMES *Princ. Physical Geol.* vi. 81 Tear faults are commonly developed in nappes, where they naturally arise if one part of a nappe has been driven forward further than the adjoining parts. **1970** LINTON & MOSELEY in *Cambr. Anc. Hist.* (ed. 3) I. i. 15 The Gulf of Oman, in which a series of 'nappes' or horizontally displaced rock sheets have been successively driven from north-east to south-west.

napper[4]. (Further examples.)

1916 'TAFFRAIL' *Pincher Martin* xii 232 Yer kin be drownded in yer barth, or git a chimney-pot dropped on yer napper in a gale o' wind. **1936** F. CLUNE *Roaming round Darling* ix. 78 Half a score of Dubboites..waited with suspense and open mouths for a brick to fall back and crack the thrower on the napper. **1936** W. LAWSON *When Cobb & Co. was King* v. 81 Fell on his napper on a log. **1947** [see *CONK *sb.*[1] c]. **1959** G. M. WILSON *Shadows on Landing* vii. 78 If anyone ever asked for an orangeade bottle on his napper, Fruity did.

nappy, *sb.*[2] (Later examples.)

1916 *Daily Colonist* (Victoria, B.C.) 1 July 2/1 (Advt.), If it's Cut Glass you want, then we have what you want. Berry Bowls, Bon-Bon Dishes, Nappies, [etc.]. **1969** *Northwest* (*Sunday Oregonian Mag.*) 14 Dec. 24/1 Faceted knife rests, nappies (those flat bottomed, slope-sided little dishes for relish and a myriad of other uses), vases, celery dishes, whatever your budget would support—these offered marriage fitting recognition.

nappy (næ·pi), *sb.*[3] *colloq.* = *NAPKIN *sb.* 1 c. Also *fig.*

1927 W. E. COLLINSON *Contemp. Eng.* 7 Mothers and nurses use pseudo-infantile forms like pinny (pinafore), nappy (napkin). **1938** S. SPENDER *Trial of Judge* ii. 49 The babe's scream till the nurse brings its nappy. **1955** M. F. GILBERT *Sky High* 118 He's got about as much sense as a baby that kicks its nappy off and then cries because it's sleeping in a puddle. **1959** *Economist* 6 June 921/2 An able minister..will need anything up to six months to get out of his administrative nappies. **1961** *Which?* Apr. 84/1 There are various types of nappy used in this country, including a T-shaped one made of gauze.. and disposable nappies. **1972** LD. ROBENS *Ten Year Stint* xvi. 322, I had been a Minister when John Eden was in his political nappies, but had never experienced such peculiar behaviour. **1973** *Guardian* 3 Feb. 24/2 The greatest pollution in this land..was..committed against women 'from the nappy to the coffin'.

b. *attrib.* and *Comb.*, as *nappy-changing, towel*; **nappy pin**, a curved safety pin used for fastening nappies; **nappy rash**, infantile erythema caused by persistent contact with urinary ammonia from a damp nappy; **nappy service**, a commercial laundry service for babies' nappies.

1970 D. BAGLEY *Running Blind* ii. 17 No nappy-changing for you. **1973** *Guardian* 23 May 9/3, I draw the line at nappy-changing, not because I am a man but because..I don't like surprises of that nature. **1966** *Olney Amsden & Sons Ltd. Price List* 1 Nappy Pins. Snaplock curved with enamelled safety cap. **1967** H. W. SUTHERLAND *Magnie* i. 2 He saw Bridie stop and look across at him, a nappy pin in her mouth. **1972** J. MANN *Mrs. Knox's Profession* xv. 110 He wielded the tin of powder and the nappy pins with voluntary attentiveness. **1959** *Which?* Sept. 104/2 Do the synthetic detergents cause dermatitis, or don't they? And does a baby get nappy-rash if its nappies are washed in them? **1966** 'G. NORTH' *Confounding Sgt. Cluff* xvi. 128 An infant red in the face, tormented by nappy-rash, howled in a pram. **1967** *New Scientist* 25 May 449/3 What started out as an effective treatment for nappy rash in infants may now be an essential instrument of modern warfare. **1972** J. WILSON *Hide & Seek* ii. 34 A baby who had teething troubles and nappy rash. **1959** *Manch. Guardian* 8 July 5/5 The nappy service cost 2s 6d a dozen, three times a week. **1970** K. GILES *Death in Church* ii. 47 Auntie..switched me on to Nappy Service, six dozen delivered every morning. **1940** C. DAY LEWIS tr. *Virgil's Georgics* iv. 89 Her sisters washed his hands Duly with pure water and fetched the nappy towels.

nappy, *a.*[1] Add: **2.** Fuzzy, kinky; used colloquially and freq. derogatorily of Negroes' hair. So **nappy-head**, a Negro; **nappy-haired, -headed** *adjs. U.S.*

1950 A. LOMAX *Mister Jelly Roll* (1952) 80 Light-skinned Downtown shared the bandstand with 'real black and nappy-headed' Uptown. **1956** S. LONGSTREET *Real Jazz* xviii. 150 To call a man '*nappy*' is to say his hair is kinky—a real insult. **1962** E. CLEAVER in A. Dundes *Mother Wit* (1973) 16/1 Good hair, bad hair, nappy hair. **1966** K. L. MORGAN in *Ibid.* 606/2 They would come up with black nappy-haired babies. **1967** J. TAYLOR in A. Chapman *New Black Voices* (1972) 175 One of the new breed of nappy-haired, bangle-wearing nationalists. **1968** J. PULLUM in P. Oliver *Screening Blues* iii. 92 Joe Pullum's high-pitched 'black gal, black gal, what makes your nappy head so hard?' **1971** *Black World* June 71/2 Her hair..was in the bushy style that the freedom riders had brought. They called it 'natural'; Bojack called it nappy. **1973** *Ibid.* Apr. 63 All them ol' nappy-heads runnin' up there tryin' to pull his clothes off.

nappy, *a.*[2] Add: **3.** Of a horse: awkward, disobedient.

1924 G. BROOKE *Horse-Sense* v. 36 There are horses that refuse from temper, generally called 'nappy horses'. **1952** R. S. SUMMERHAYS *Elem. Riding* (ed. 3) xii. 72 A 'nappy' horse is a horse which can be described as one who will not go forward, will often go sideways and sometimes backwards. **1960** *Times* 23 July 9/2 Do not give the pony more than 20 minutes of this or it will sicken him, and..he may turn nappy and start refusing. **1963** BLOODGOOD & SANTINI *Horseman's Dict.* 134 *Nappy (to be)*, for a horse to be inclined to shy or refuse suddenly and without warning. **1973** J. WHITE *Norfolk Child* 131 My father..was utterly patient with every horse he rode, however nappy.

Nara (nā·ra). [The name of a town in Central Honshu, Japan, capital of Japan A.D. 710–84.] Used *attrib.* to designate the Buddhist sculpture of the period of Japanese history during which Nara was the capital.

1902 F. BRINKLEY *Oriental Series: Japan* VII. i. 19 As to sculpture, the point of excellence to which it had been carried is attested by several statues which form part of the Nara temple relics. **1903** K. OKAKURA *Ideals of East* 112 The stone Buddhas of the Tin Tal in Ellora.. are beautiful, with a self-contained grandeur and harmony of proportion. In them we find the sources of inspiration of the Tāng and Nara sculptures. **1955** PAINE & SOPER *Art & Archit. Japan* iv. 26 The earliest clay images of the Nara period are the groups of figures dating from 711 in the Pagoda of Hōryūji. **1970** *Oxf. Compan. Art* 609/2 During the Heian period..the trend was away from the realism of the Nara period.

narc (nāɹk). *U.S. slang.* Also **nark.** [abbrev. NARCOTIC *sb.* or *a.*] A federal, state, or local narcotics agent. Cf. *NARCO.

1967 *N.Y. Times* 17 Oct. 44 Most speed freaks get to a point where they're seeing narks in the trees with cameras. **1968–70** *Current Slang* (Univ. S. Dakota) III–IV. 85 *Narc*, an official of the Narcotics Control Board. *Ibid.*, *Narco, or narc*, narcotics officer. (Drug users' jargon). **1970** E. TIDYMAN *Shaft* (1971) iv. 50 The police didn't frighten him. The Narcs didn't frighten him. **1972** J. WAMBAUGH *Blue Knight* (1973) vii. 114 A narco cop nailed him..the narcs busted in the pad. **1973** *Black Panther* 20 Oct. 15/1 Then I made contact for the ring with the narc agent. **1975** *New Yorker* 19 Feb. 96/3 Bo, a rookie detective..is so confused by the Department's manipulations that he doesn't guess that she is an undercover narc.

narcissine, *a.* Add: **2.** Resembling, or of the nature of, Narcissus (see *NARCISSUS 2); loving or admiring oneself.

1813 E. S. BARRETT *Heroine* II. xxii. 77 Flinging off my bonnet, I shook my narcissine locks over my shoulders. **1911** BEERBOHM *Zuleika D.* ii. 15 Yet there was nothing Narcissine in her spirit. Her love for her own image was not cold æstheticism.

narcissism (nāɹsi·siz'm). esp. *Psychol.* [f. the name of *Narcissus*, a beautiful youth who fell in love with his own reflection in a fountain (Ovid *Metam.* III. 370; subsequently referred to by Havelock Ellis in *Alienist & Neurologist* (1898) XIX. II. 280; the term *Narcissismus* was used by Näcke in *Die sexuellen Perversitäten* (1899)): see -ISM.] Self-love and admiration that find emotional satisfaction in self-contemplation; occas. (erron.) narcism.

1822 COLERIDGE *Let.* 15 Jan. (1971) V. 196 Of course, I am glad to be able to correct my fears as far as public Balls, Concerts, and Time-murder in Narcissism. **1905** H. ELLIS *Stud. Psychol. Sex* IV. iii. 187, I have referred to the developed forms of this kind of self-contemplation..and in this connection have alluded to the fable of Narcissus, whence Näcke has since devised the term Narcissism for this group of phenomena. **1922** J. RIVIERE tr. *Freud's Introd. Lect. Psycho-Anal.* xxiv. 347 It is probable that this *narcissism* is the universal original condition, out of which *object-love* develops later without thereby necessarily effecting a disappearance of the narcissism. **1938** H. A. MURRAY *Explorations in Personality* iii. 180 Narcism (or Egophilia) is technical for self-love. **1946** [see *INTROJECTION 3 a]. **1955** H. MARCUSE *Eros & Civilization* (1969) viii. 169 The striking paradox that narcissism, normally understood as egotistic withdrawal from reality, here is connected with oneness with the universe, reveals the new depth of the conception. **1969** P. LOEWENBERG in B. B. Wolman *Psychoanal. Interpretation of Hist.* vi. 182 He [*sc.* Herzl] regressed to the stage of narcissism in which his only sexual object was his own ego and its fantasies. **1970** HINSIE & CAMPBELL *Psychiatric Dict.* (ed. 4) 487/2 *Narcism*, a shortened (and incorrect) form of narcissism.

narcissist (nāɹsi·sist). [f. as prec. + -IST.] A person affected with narcissism. Also *attrib.*

1930 B. RUSSELL *Conquest of Happiness* I. i. 22 A narcissist,..inspired by the homage paid to great painters, may become an art student. **1934** C. DAY LEWIS *Hope for Poetry* vi. 32 We find, instead of criticism, long narcissist ramblings, the 'reviewer's' views on everything under the sun. **1953** *Essays in Crit.* III. 89 A narcissist indulgence in a 'phantasmal world' of self-reflection. **1958** *Times Lit. Suppl.* 24 Jan. 46/3 The doctrine here is that Shelley is a self-absorbed narcissist, yearning emptily and 'regressively' for death.

narcissistic (nāɹsisi·stik), *a.* [f. as prec. + -ISTIC.] Of or pertaining to narcissism; marked by excessive love of self.

1916 E. JONES tr. *Ferenczi's Contrib. Psycho-Anal.* 174, I was just striving to make clear to a patient her excessive ambition, arising from narcissistic fixation. **1922** J. RIVIERE tr. *Freud's Introd. Lect. Psycho-Anal.* xxiv. 347 Thus it appeared that auto-erotism was the sexual activity of the narcissistic phase of the direction of the Libido. **1935** S. SPENDER *Destructive Element* xv. 267 It was not a narcissistic desire to recover lost charms and innocence. **1964** *Economist* 23 May 801/2 Students..would be vastly improved if they were a lot less frowsty and narcissistic. **1973** L. BELLAK et al. *Ego Functions* II. x. 159 Kernberg has discussed 'Barriers to Being in Love' from the standpoint of narcissistic personality features.

narcissistically (nāɹsisi·stikăli), *adv.* [f. prec.: see -ICALLY.] In a narcissistic manner.

1925 T. DREISER *Amer. Trag.* (1926) I. I. xv. 112 She narcissistically painted her flight. **1962** *Times* 3 May 17/2 Some literary men seem almost narcissistically concerned with themselves and their art. **1968** *Daily Tel.* 13 Nov. 15/5 If you don't go for seeing yourself narcissistically reflected in the shimmering walls, there is the view of lacy trees outside. **1974** D. MEIRING *President Plan* x. 84 The father [was] bound narcissistically within the mirrors of his supposed professional importance.

Narcissus. Add: **2.** [See *NARCISSISM.] The name of a youth in classical mythology who died of self-love after seeing his reflection in water and was turned into the flower, used chiefly *attrib.* and *Comb.* allusively for: one who admires himself exclusively, one who resembles Narcissus in handsomeness.

1606 SHAKES. *Ant. & Cl.* II. v. 96 Go get thee hence, Had'st thou Narcissus in thy face to me, Thou would'st appeere most vgly. **1767** W. KENRICK tr. *Misc. Wks. J. J. Rousseau* II. 121 (*title*) Narcissus, or the self-admirer. A comedy. **1860** A. J. MUNBY *Diary* 10 Feb. in D. Hudson *Munby* (1972) 49 His face..seemed to me weak and self-conscious; a Narcissus face. **1891** O. WILDE *Pict. Dorian Gray* i. 4 This young Adonis, who looks as if he was made of ivory and rose-leaves. Why, my dear Basil, he is a Narcissus. **1928** R. CAMPBELL *Wayzgoose* ii. 44 Here was a man—I thought it was a mess, A bottle-nosed Narcissus of the Press. **1929** D. H. LAWRENCE *Pornography & Obscenity* 26 The most emancipated bohemians..are still utterly..enclosed within the narcissus-masturbation circle. **1930** —— *A Propos Lady Chatterley* 23 Poor, self-conscious, uneasy, narcissus-monk as he was. **1935** A. HUXLEY *Let.* 20 Sept (1969) 397 Why is it that when one enters that [*sc.* theatrical] world, one always finds oneself with crooks, imbeciles, narcissus complexes? *a* **1963** C. S. LEWIS *Poems* (1964) 89 So should I quickly die Narcissus-like of want. **1964** M. MCLUHAN *Understanding Media* (1967) v. 66 The frontiers between forms that snap us out of the Narcissus-narcosis.

3. narcissus fly, a hover-fly, *Lampetia* (= *Merodon*) *equestris*, whose larva infests the bulbs of narcissus and other plants, causing them to rot.

1903 F. V. THEOBALD *First Rep. Econ. Entomol.* 107 A correspondent..sent the larvæ of the Narcissus Fly, from Chertsey, with the following note: 'They play havoc with the narcissus bulbs and are evidently the maggot of some fly.' **1926** *Jrnl. Econ. Entomol.* XIX. 249 The observations were made on the larvae of the narcissus fly. **1951** COLYER & HAMMOND *Flies Brit. Isles* xii. 166 Our last example from the group is *Merodon equestris* F., the 'Large Bulb Fly' or 'Narcissus Fly', which is a serious pest to the horticulturalist. **1966** *Punch* 28 Sept. 485/3 There are in Britain some two hundred and twenty known varieties of hover fly, of which only the Narcissus flies are destructive (to narcissi, naturally).

narco (nā·ɹko). U.S. slang abbrev. of *narcotic* or *narcotics*, used esp. (freq. *attrib.*) = *NARC (see also quots.).

1955 *Amer. Speech* XXX. 87 *Narco*, the narcotic hospital in Lexington, Kentucky. **1958** J. & W. HAWKINS *Death Watch* 43 This informant is a thief, a narco or a four-bottle bum. **1960** *Sat. Rev.* (U.S.) 6 Feb. 11/3 The Beat Generation has marihuana and the ritual of dodging the 'narcos'—the narcotics squad. **1961** RIGNEY & SMITH *Real Bohemia* p. xvi, *Narco, the*, federal narcotics agents. **1964** *Manhunt* Mar. 67/2, I feel very strongly about helping narco violators. *Ibid.* 68/1 You've gone to bat on forty-three narcos *without a single acquittal*! **1968** *Wall St. Jrnl.* 19 Feb. 1/1 Students also have agitated against university acquiescence in the presence on the campus of 'narcos'—police agents seeking to make arrests for violations of narcotics laws, whose basic premises many of the students question. **1970** K. PLATT *Pushbutton Butterfly* (1971) xi. 126 Tina was in the drug racket... The narco squad might have had reason to suspect her. **1971** 'D. Shannon' *Ringer* (1972) iii. 58 The pedigrees varied from burglary to narco dealing to rape. **1972** [see *narc].

narco- (nā·ɹko). *Psychol.* [f. Gr. νάρκη numbness.] Prefixed to a sb. to indicate that use is made, in the treatment specified by the sb., of a drug (usu. a barbiturate such as amylobarbitone or thiopentone sodium) which, while inducing relaxation, facilitates the remembering and verbalizing of repressed emotional experiences. So *narco-analysis, -hypnosis* (also *-hypnotic* adj.), *-therapy; narco-synthesis*, the acceptance into the conscious self of repressed emotional experiences revealed by the use of drugs.

1936 J. S. Horsley in *Jrnl. Mental Sci.* LXXXII. 416 Narco-analysis is an eclectic technique based on the observation that a combination of narcosis with psychotherapy is quicker and sometimes more effective than the formal methods of analytical psychology. **1943** —— (*title*) Narco-analysis. *Ibid.* i. 3 Clearly any title which suggests that a particular narcotic is essential to the technique is undesirable. Therefore I devised the term 'Narco-Analysis' which is inclusive of all such methods. **1973** E. Rudinger *Treatm. & Care in Mental Illness* (rev. ed.) 161 Narco-analysis or abreaction is a treatment that is also available privately, but the drugs used and the nursing home fee have to be paid in addition to the psychiatrist's fee. **1949** A. Huxley *Let.* 21 Oct. (1969) 605 The world's rulers will discover that infant conditioning and narco-hypnosis are..efficient. **1974** Hawkey & Bingham *Wild Card* v. 51 McElroy was sufficiently promising a subject to be questioned under narco-hypnosis. **1949** A. Huxley *Let.* 9 Dec. (1969) 611 The latest in..narco-hypnotic techniques. **1945** Grinker & Spiegel *Men under Stress* iii. v. 102 Here he underwent pentothal narcosynthesis and ventilated his intense anxiety. **1955** A. Huxley *Let.* 10 Jan. (1969) 720 Mescalin..acts in the opposite way to narcosynthesis. **1958** [see *abreaction]. **1963** *Times* 9 Mar. 8/1 It is believed to be the first instance in which narcosynthesis—a technique to help patients release suppressed or forgotten information—has ever been used for this purpose. **1966** P. Polatin *Guide to Treatm. in Psychiatry* v. 50 Narcotherapy is most effective for acute anxiety states, conversion reactions and dissociative phenomena such as amnesias and fugues. **1968** S. Loebl *Exploring the Mind* xv. 184 Hypnotherapy and narcotherapy, as these procedures are called, have by now become part of psychiatry's ever-growing therapeutic arsenal.

narcolepsy. Add: (Earlier and later examples.) [First formed as F. *narcolepsie* (Gélineau 1880, in *Gaz. des Hôpitaux* LIII. 626/2).]

1880 *Jrnl. Nervous & Mental Dis.* VII. 737 Dr. Gelineau [*sic*], *Gaz. des Hopitaux* [sic], Nos. 79 and 80, describes at length, and discusses a case in his practice which..seems to be one of a new disease, for which he proposes the name 'Narcolepsy'. **1928** *Brain* LI. 78 All narcolepsy is..symptomatic, but at present we do not always know of what pathological or physiological condition it is thus symptomatic. **1969** *Sunday Times* 2 Nov. 10/3 It [*sc.* amphetamine] is the best drug for narcolepsy (uncontrollable attacks of sleeping). **1972** H. L. Williams in Brady & Nauta *Princ., Pract. & Positions Neuropsychiatric Res.* 454 It has been observed that patients with idiopathic narcolepsy frequently move directly from waking to REM sleep.

narcolept (nā·ɹkŏlept). [Back-formation from next.] = *narcoleptic *sb.*

1957 R. A. Heinlein *Door into Summer* (1960) iv. 54 I'm forever deprived, like a narcolept on a honeymoon. **1973** *Times Lit. Suppl.* 19 Oct. 1268/5 Like that later narcolept Edgar Allan Poe.

narcoleptic (nāɹkole·ptik), *a.* and *sb.* [ad. F. *narcoleptique*; cf. Epileptic *a.* and *sb.*] **A.** *adj.* Characteristic of or affected with narcolepsy.

1904 G. S. Hall *Adolescence* I. 264 Sometimes a sense of fatigue, lassitude, and sleepiness, rarely narcoleptic, may supervene. **1912** *Jrnl. Nervous & Mental Dis.* XXXIX. 131 (*heading*) Adipose pituitary syndrome of Lannois with narcoleptic fits. **1949** *Progress Neurol. & Psychiatry* IV. 414 Narcoleptic sleep,..major convulsions, or other 'epileptic' symptoms. **1960** *Arch. Gen. Psychiatry* III. 422/1 This narcoleptic patient is an intelligent 42-year-old married Negro male shipping clerk. **1972** *Ibid.* XXVI. 462/1 The narcoleptic sleep attack is more imperative but of shorter duration than diurnal hypersomniac sleep.

B. *sb.* A narcoleptic person.

1928 *Brain* LI. 84 The sleep of the narcoleptic is to be likened in all respects to normal, healthy sleep. **1957** *Proc. Mayo Clinic* XXXII. 324 A serious problem among the narcoleptics is the tendency to drowse when driving. **1972** *Arch. Gen. Psychiatry* XXVI. 457/2 Hypersomniacs do not fall asleep while eating, speaking, or walking, as narcoleptics do.

narcosis. Add to def.: Also a psychologically therapeutic sleep artificially prolonged by the use of drugs. So *narcosis therapy*.

1936 [see *narco-analysis* s.v. *narco-]. **1943** Curran & Guttmann *Psychological Med.* viii. 97 Probably the most humane method is to give continuous narcosis for the first few days after the drug [*sc.* morphine] has been withdrawn; but relapse is very frequent. **1945** Grinker & Spiegel *Men under Stress* iv. xvii. 394 At the present time, aside from the method of narcosis therapy (continuous sleep), the use of intravenous barbiturates is dignified by three different terms.

narcotic, *sb.* Add: **b.** In extended use: any drug which affects the mind in some way and is prohibited or under strict legal control in many countries owing to the social problems associated with its misuse, but which tends nevertheless to be extensively sold and used illegally. orig. *U.S.*

1926 *Rep. Drug Addiction in Calif.* (Calif. State Narcotic Comm.) 13 When used in this report, the term 'narcotics' or 'drugs' includes all of the following: Cocaine, opium, morphine, codeine, heroin, alpha eucaine, beta eucaine, flowering tops and leaves, extracts, tinctures, and other narcotic preparations of hemp or loco weed, Indian hemp, peyote, or chloral hydrate or any of the salts, derivatives or compounds of the foregoing substances. **1955** *U.S. Senate. Hearings* (1956) VIII. 4160 There are three groups into which the narcotics set forth in division 10 of the Health and Safety Code are divided: 1. Opiate group. 2. Cocaine, marihuana, and lophophora. 3. Synthetic group. **1972** *Daily Tel.* (Colour Suppl.) 11 Aug. 12/1 The Americans prefer the word 'narcotics', which although not entirely accurate—not all banned drugs cause narcosis—is better understood, so it will be used from here on. **1974** A. Goth *Med. Pharmacol.* (ed. 7) xxv. 300 In some state statutes the legal category 'narcotics' embraces the opiates, opiate-like drugs, marihuana, and cocaine. Medically defined, however, the term narcotic refers only to drugs having both a sedative and an analgesic action and is essentially restricted to the opiates and opiate-like drugs. *Ibid.* xxv. 304 Although marihuana is not medically a narcotic, it is classified legally as a narcotic in some states for purposes of control. **1974** *Encycl. Brit. Macropædia* V. 1049/2 Prejudice and ignorance have led to the labelling of all use of nonsanctioned drugs as addiction and of all drugs, when misused, as narcotics.

2. *attrib.* (freq. in *pl.*).

1926 *Proc. 1st World Conf. Narcotic Educ.* 111 How the police catch the crook or the narcotic dealer;..it is only when someone tells the police who the narcotic addict or peddler is that the police get them. *Ibid.* 227 The narcotic squad [in Philadelphia] was under the command of Captain Van Horn. **1929** *Narcotic Educ.* Jan. 46/1 To get sufficient evidence against John Smith it is necessary that a Government agent purchase narcotics from John Smith, but John Smith will only sell..for..$2,500 in advance... Where can a narcotic agent be found who has $2,500 to use? *Ibid.* Apr. 73/1 Out of the gray, neutral shadows cast by the murder of Arnold Rothstein, federal authorities forged a sturdy weapon expected to crush the most gigantic narcotic ring in the history of the illegal drug trade. **1951** *Manch. Guardian Weekly* 28 June 2/2 The State Legislature at Albany ordered an inquiry, which met last week. And this week the American Legion has followed it up with a 'narcotic clinic' of its own. **1953** *Conf. Drug Addiction among Adolescents* (N.Y. Acad. Med., Comm. Public Health Relations) II. 159 He..devised a plan..for the preventive and follow-up care of young narcotics addicts. **1953** W. Burroughs *Junkie* (1972) iii. 37 They found out the narcotics squad had a warrant for him sworn out by the State Inspector. **1963** *Listener* 4 Apr. 585/1 They were the bookies, and money lenders..and the narcotic pedlars. **1972** *Daily Tel.* 4 Sept. 3/1 A national Narcotics Bureau is to be set up in London next month to coordinate the activities of police drug squads in this country. **1973** *N.Y. Law Jrnl.* 19 June 4/5 The defendant argued that the offence itself, sale of dangerous drugs..made it 'appear' that he was a narcotics addict.

‖ **narikin** (na·rikin). [Jap.] In Japan, a wealthy parvenu.

1920 *Glasgow Herald* 2 Sept. 7 'Narikin', or mushroom millionaires, have spent their rapidly amassed fortunes on extravagant living. **1923** in J. Manchon *Le Slang* 203. **1933** *Times Lit. Suppl.* 16 Nov. 784/3 *Narikin*..as commonly used is..a term of reproach, applied to those *nouveaux riches* who made rapid fortunes as profiteers of the War. **1946** R. Benedict *Chrysanthemum & Sword* (1947) iv. 95 Narikin is often translated 'nouveau riche' but that does not do justice to the Japanese feeling... A narikin is a term taken from Japanese chess and means a pawn promoted to queen.

naringin (nāri·ndʒin). *Chem.* [ad. G. *naringen* (E. Hoffmann, at the suggestion of Flückiger, 1879 in *Arch. der Pharm.* CCXIV. 140), f. *naringi*, given as Skr. for Orange *sb.*[1], *a.*: see -in[1].] A bitter glucoside, $C_{27}H_{32}O_{14}$, of a tricyclic alcohol which is found in shaddock, grapefruit, and certain types of orange.

1879 *Jrnl. Chem. Soc.* XXXVI. 468 Naringin, the hesperidin of de Vrij, which is found in the fully developed buds of *Citrus decumana* to the amount of 2 per cent., is obtained from the residue left on distillation of the ethereal oil. **1926** *Biochem. Jrnl.* XX. 1305 Naringin, the specific glucoside of the grape-fruit, was found present in the rind and in alcoholic extracts of it. **1957** E. V. Miller *Chem. Plants* viii. 113 Naringin is found in large quantities in the juice of immature grapefruits..and to a certain extent in the segment walls of mature fruits. **1970** *New Scientist* 1 Jan. 21/1 The most promising [sweetening] compounds are made from naringin isolated from grapefruit, hesperidin from sweet oranges and neohesperidin from Seville oranges.

nark, *sb.* For **1865** *Slang Dict.* (ed. 2) read **1860** in Hotten *Dict. Slang* (ed. 2), 179, and add further examples.

1916 G. B. Shaw *Pygmalion* i. 110 It's a—well, it's a copper's nark, as you might say. What else would you call it? A sort of informer. **1933** Auden *Dance of Death* 37 Quick under the table, it's the 'tecs and their narks, O no, salute—it's Mr. Karl Marx. **1936** J. G. Brandon *Pawnshop Murder* x. 90 Police 'narks', 'noses', and all such kindred brethren. **1954** [see *grass *sb.*[1] 11*]. **1968** *Listener* 19 Dec. 810/3 Don't do it.., he's a copper's nark. **1975** *Times* 9 Jan. 4/4 If it was thought we were coppers' narks it could endanger the lives of our film crews.

b. A policeman.

1891 F. W. Carew *No. 747* vi. 65 If you don't turn up my fair share, I'll put the narks upon you. S'elp me never, I will. **1937** M. Allingham *Dancers in Mourning* xxvii. 327 There's a bunch o' narks at either end of the lane. **1959** A. Sillitoe *Loneliness of Long-Distance Runner* 27 Don't let that gate creak too much or you'll have the narks tuning-in. **1959** I. & P. Opie *Lore & Lang. Schoolch.* xvii. 369 There are, in the London area, at least thirty nicknames current among boys... The Law, Nark or Narker, Nobby, [etc.]. *a* **1966** M. Allingham *Cargo of Eagles* (1968) v. 74 I've 'appened on a little something wot the official narks 'aven't cottoned to yet.

2. *slang* (freq. in *Austral.* and *N.Z.*). **a.** A person who is annoying, unpleasant, obstructive, or quarrelsome. Cf. Knark.

1846 *Swell's Night Guide* 68 They are the rankest narks vot ever God put guts into, or ever farted in a kickses case. **1898** *Bulletin* (Sydney) 17 Dec. (Red Page), An informer or mar-plot is a *nark* or *Jonah*. **1906** E. Dyson *Fact'ry 'Ands* i. 12 'Yeh know, Feathers, she's no bally nark; er bloke kin trust 'er,' he said. **1908** *Austral. Mag.* 1 Nov. 1251 *Nark*, a spoil-sport. **1918** *Aussie* Sept. 2/2 He appealed in vain to Madame... 'Orright, then, Madame, I can be a nark, too! I know where there's..fat, juicy frogs, and you wont get one of 'em!' **1925** Fraser & Gibbons *Soldier & Sailor Words* 164 *Nark*, a bad-tempered man; a spoil-sport... Also, a man eager to curry favour by running about and doing odd jobs for a superior. **1928** V. Palmer *Man Hamilton* 94 'Oh, don't be a nark, Miss Byrne,' he coaxes her. **1933** F. Clune *Try Anything Once* 81 Lieutenant Hennessy fulfilled the most exacting requirements of what a 'nark' should be. **1937** Partridge *Dict. Slang* 551/2 *Nark*... A person on inquiry from head office: London clerks', managers', etc.: from before 1935. **1943** *Penguin New Writing* XVIII. 63 If I said anything..he'd go crook and tell me not to be a nark. **1959** I. & P. Opie *Lore & Lang. Schoolch.* xviii. 391 'It is a way to get your own back on an old nark who has spoilt a game,' declares an Edinburgh 12-year-old. **1965** J. S. Gunn *Terminol. Shearing Industry* ii. 6 *Nark*, a troublemaker who interrupts the rhythm of the shed. A nark shearer only hurts himself, but a shedhand or roller who does not keep the board clear can slow up the work of all men handling the fleece.

b. An annoying or unpleasant thing or situation; a source of astonishment or vexation. *slang* (freq. in *Austral.* and *N.Z.*).

1923 J. Manchon *Le Slang* 203 *Nark*.. (rare), rancune, une dent qu'on a contre quelqu'un. **1926** R. R. Terry *Shanty Bk.* II. 25 When they went to church to say 'I will', the drummer got a nark. [*footnote*] *Nark*, a disagreeable surprise. **1937** N. Marsh *Vintage Murder* 114 It's a blooming nark, dinkum it is. **1947** *Book* (Christchurch, N.Z.) IX. 23 'It's a nark, isn't it,' she said. 'I thought you'd get by without the op.' **1948** R. Alley *Gung Ho* 16 Typhoid, malaria, and all the narks. **1966** P. Moloney *Plea for Mersey* 51 Not hardened junkies, when deprived of dope, Ere felt such anger, ere got such a nark As Scouseville driver seeking space to park.

nark, *v.* For **1865** *Slang Dict.* (ed. 2) read **1859** in Hotten *Dict. Slang* 67 and add:

2. *trans.* and *intr.* To annoy, exasperate, infuriate; to complain, grumble. Hence **narked** *ppl. a.*; **na·rking** *vbl. sb.* *slang* (freq. in *Austral.* and *N.Z.*) and *dial.*

1888 J. Dalby *Mayroyd of Mytholm* II. 45 That's just what he's ta'en to him for, just to nark Mayroyd. **1888** S. O. Addy *Gloss. Words Sheffield* 155 *Narked*, vexed, angry... 'He wor narked about it.' **1896** H. Lawson *Shearing of Cook's Dog* in *While Billy Boils* 167 The cook usually forgot all about it in an hour... But this time he didn't; he was 'narked' for three days. **1899** *Bulletin* (Sydney) 25 Feb. (Red Page), Tom was gettin' narked, so I tries to get him to turn in. **1908** E. J. Banfield *Confessions of Beachcomber* I. v. 174 He'll be a bit narked at having wasted a whole bloomin' day. **1916** *Anzac Book* 142 But wot narks us more than any is to 'ear the sergeant say: 'The sea's too rough to land our stores; There ain't no jam today!' **1916** J. B. Cooper *Coo-oo-ee* xix. 299, I thought you would. That's very like the Tippinses to nark like that. **1930** *Bulletin* (Sydney) 19 Mar. 23/1 It was this that narked me most—'e couldn't see a joke. **1932** L. A. G. Strong *Don Juan & Wheelbarrow* ix. 162 Is it true..that So-and-So is narked as the Cottage 'Ospital be over to 'Arraton, 'stead of here? **1940** F. Sargeson *Man & his Wife* (1944) 49 That got Tom narked. He told George he ought to be ashamed of himself for telling things like that. **1945** *Daily Sketch* 20 Apr. 2/2 Like you, I am all for personal liberty—and so no doubt are the inmates of Wormwood Scrubs—but the fact remains that so long as there is not economic freedom for everybody, what is the point of constantly narking about State-planning? **1947** D. M. Davin *Gorse blooms Pale* 78 It narked her that everyone was eating out of her hand except me. **1958** *Times Lit. Suppl.* 15 Aug. p. xxxii/3 This naturally brings out the worst in their opponents and in the resultant narking and name-calling the 'legitimate contention' is lost sight of. **1962** [see *half *adv.* 3]. **1966** *Crescendo* Apr. 29/2 The narked comments of some respected musicians. **1968** *Daily Progress* (Charlottesville, Virginia) 11 July C.14/1 In Britain, to nark is to nag, scold, annoy or irritate. Mrs. Ewing, a vivacious blonde, narks mostly about home rule for Scotland. She has managed to irritate some members of Parliament and her election victory annoyed Prime Minister Harold Wilson's Labor party. **1970** *New Scientist* 23 July 190/2 The Chinese have naturally been somewhat narked at being left out of the fun. **1973** *Daily Tel.* 2 Nov. 15 If you feel especially narked about something, you can turn it into a theory of human behaviour.

3. *trans.*, usu. with *it*. To cease, desist, stop, terminate. Freq. in imp. *slang*.

1889 *Sporting Times* 29 June 1/3 And as terseness of expression was an art she'd studied well, She determined that her lady friend should nark it. **1925** in Fraser & Gibbons *Soldier & Sailor Words* 164. **1933** *Bulletin* (Sydney) 1 Feb. 20 It was 'ard luck when a bloke narked the show. **1936** J. Curtis *Gilt Kid* xvii. 175 Nark it for God's sake... You'll get us done, yelling around the gaff like that. **1943** M. Harrison *Reported Safe Arrival* 12 There was a sharp interchange of: 'Nark it!' and 'Oo says so?' I think there had been an interchange of more than words had not the Orderly Officer then appeared. **1943** *R.A.F. Jrnl.* Aug. 30 Nark it, Flight,..you sound like a penny uplift. **1959** I. & P. Opie *Lore & Lang. Schoolch.* x. 199 Saying by the one being tortured: "Ere, nark it.' **1973** N. Graham *Murder in Dark Room* xxiii. 162 'Nark it,' I said. 'I want a little bit of information from you.'

narker (nā·ɹkəɹ). *slang.* [f. Nark *v.* + -er[1].] An informer; a policeman; one who complains or disparages.

1932 Nordhoff & Hall *Mutiny on Bounty* iii. 39 The captain's 'narker' or spy among the men. **1937** *Sunday Dispatch* 7 Feb. 22/6 The narkers are at it again with their fancy explanation of our Test mistakes. **1959** [see *nark *sb.* 1 b]. **1959** A. Sillitoe *Loneliness of Long-Distance Runner* 28 I'd..screw my eyes up like I was on my way to the hospital,.. 'Cancer,' I'd manage to say to Narker, which would make his slow punch-drunk brain suspect a thing or two. **1971** *Daily Tel.* 3 May 8/4 His motto will be to celebrate not denigrate, and I commend this to the legion of glib narkers who tend to monopolise the screen.

narks (nāɹks), *sb. pl.* Colloq. abbrev. of nitrogen *narcosis*.

1962 S. Miles *Underwater Med.* vii. 100 Unromantic British divers simply call it 'Narks'. **1967** 'J. Palmer' *Above & Below* i. 9 It's lucky the ship lies in such shallow water. We shan't get the 'narks'.

narky (nā·ɹki), *a. slang.* [f. *nark *sb.* 2 + -y[1].] Irascible, vexed, bad-tempered, sarcastic.

1895 *Leeds Mercury Weekly Suppl.* 13 July 3/8 Doan't let's get narky ower it. **1898** B. Kirkby *Lakeland Words* 105 He were a bit narky ower t' trottin' do. **1943** M. Harrison *Reported Safe Arrival* 53 Harry was interested. 'Garn!' he said. 'Yer kiddin', Perfess [Professor]!... Shakespeare, eh? Full of swearin', eh? I allus thought 'e wz fer school-kids and blokes like yew, Perfess. I don' mean nothin' narky: on'y 'e ain't the sort of bloke me and de Pen 'ud settle dahn to read a basin of, if yer get me meanin'?' **1958** L. Durrell *Balthazar* ii. 33 Below the belt... Dirty. Cruel. Narky. **1958** [see *ducky *sb.* 2]. **1970** *New Society* 5 Mar. 399/1 Shelter has incurred a lot of narky criticism since it announced its intention of setting up a chain of housing advisory centres. **1973** *Irish Times* 2 Mar. 14/5 My husband is narky in the house. If I was to bring heaven down it would not satisfy him.

‖ **Narodnik** (narǫ·dnik). Also **narodnik.** Pl. **Narodniki, Narodniks.** [Russ., f. *narod* people + *-nik.] A supporter of the type of populist agrarian socialism originating amongst the Russian intelligentsia in the late 1860s which regarded the peasants and intelligentsia as the only revolutionary forces and denied the revolutionary role of the working class; one who tries to educate politically communities of rural or urban poor while sharing the conditions of their lives. Also *attrib.* and *transf.* Hence **Naro·dnikism,** the theory of making political power a reality for the masses.

1885 E. Noble *Russ. Revolt* 204 In the spring of 1877 the members of a revolutionary society called the 'Narodniki' (Party of the People) 'went to the people', establishing a large number of propaganda centres along the line of the Volga. **1904** G. Drage *Russ. Affairs* i. 51 A party arose who called themselves *Narodniki* (Nationalists). **1921** M. P. Price *My Reminisc. Russ. Revolution* ii. 27 The Narodniks therefore preached 'back to the land'. **1929** L. Krassin *Leonid Krassin* vii. 58 The Narodniki, the first organised agrarian party, welcomed..common possession of the soil. **1950** E. H. Carr *Bolshevik Revolution* I. i. i. 4 For the past thirty years [*sc.* 1868–98] the leading Russian revolutionaries had been the *narodniks*—a composite name for a succession of revolutionary groups believing in the theory of peasant revolution and in the practice of terrorism against members of the autocracy. *Ibid.* iii. 52 The peasantry remained for the Mensheviks an essentially anti-revolutionary force; any revolutionary policy which counted on its support was a reversion to the *narodnik* heresy of a peasant revolution. **1965** *New Statesman* 20 Aug. 240/1 Last year's Mississippi Summer Project..was..composed of young Southern Negroes and Northern students on the staff of the Student Non-Violent Co-ordinating Committee... They were the original *narodniks* of the movement. **1966** *Economist* 3 Sept. 887/2 A new *narodnik* movement, a fresh attempt by the intellectuals to 'go to the people'. **1969** J. Saul in Ionescu & Gallner *Populism* 135 It may seem useful to lump together Russian Narodnikism and North American Populism..because both represent largely rural responses to..'capitalism' or 'modernization' or 'industrialization'. **1970** G. Jackson *Let.* 25 Mar. in *Soledad Brother* (1971) 197 The dialectic between Narodnik and Nihilist should never break down. **1971** *Graphic* (Durban) 7 May 12/2 Her aunts were active in the populist narodniki movement [in Russia].

narp (nāɹp). *slang.* [Origin unknown.] A shirt.

1839 H. Brandon *Poverty, Mendicity & Crime* 164/1 *Narp*, a shirt. **1949** in Partridge *Dict. Underworld* 464/1.

narra[1] (na·rā). *S. Afr.* Also **nara(s).** [ad. Hottentot *'narab*.] A spiny shrub, *Acanthosicyos horrida*, of the family Cucurbitaceæ, found in arid regions of south-western Africa, distinguished by thorns, which replace the leaves of young plants, and yellow flowers; also, the large globular fruit of this plant. Also *attrib.*

1838 J. E. Alexander *Expedition Interior Afr.* II. iii. 68 The 'naras was growing on little knolls of sand; the bushes were about four or five feet high, without leaves, and with apposite thorns on the light and dark green striped branches. **1853** F. Galton *Narr. Explorer Trop. S. Afr.* i. 21, I have mentioned above the 'Nara, a prickly gourd, which..is the staple food of these Hottentots. **1881** T. Hahn *Tsuni-‖ Goam* 101 !Naras.—This fruit is a *Cucurbitacea*, almost as large as a new-born child's head. The flesh of it is eaten raw, and the seeds are kept for the dry season, when there is no fruit. The seeds taste almost like almonds. **1946** L. G. Green *So Few are Free* (1948) xii. 165 The narra, a member of the melon family..sends its roots down through the dunes for fifty feet, if necessary, to find moisture... The ripe narra is full of edible seeds, which are treated in many ways by the Hottentots. Boiled, they make a porridge. Tough pancakes are formed by the narra fluid and stored for months. Narra beer may be brewed from the syrupy juice. **1959** G. Jenkins *Twist of Sand* xiii. 267 The trails of naras creeper would provide some sort of fuel. **1973** *Stand. Encycl. S. Afr.* VIII. 40/2 Naras (properly: narra)... Spiny, much-branched, prostrate shrub of the family *Cucurbitaceae*, found growing on sand-dunes in the dry coastal strip of South-West Africa.

narra[2] (nā·rā). [Tagalog.] The Filipino name for the south-east Asian tree, *Pterocarpus indicus*, of the family Leguminosæ, or its timber; = Amboyna. Also *attrib.*

1859 J. Bowring *Visit to Philippine Islands* xv. 270 (*table*) Narra, or Naga, or Asang... Buildings, furniture, doors and windows. **1890** J. Foreman *Philippine Islands* xix. 371 Narra..gives logs up to 35 feet long by 26 inches square. It is the Mahogany of the Philippines. **1911** *Philippine Bureau of Forestry Bull.* No. 10. ii. 35 The members of the Narra family mentioned here have simply or doubly compound leaves. *Ibid.*, Narra is found throughout the Philippines. *Ibid.* 36 Narra is the most common commercial name for the wood in the Philippines. **1929** *Amer. Speech* IV. 284 Of the native [Filipino] words, a few have already been accepted... *narra* and *camagon*, hard woods used in fine cabinet work. **1947** J. C. Rich *Materials & Methods Sculpture* x. 292 *Narra* is a Philippine wood related to the Andaman Islands' Padouk. It has a yellow to reddish-brown color and is not always available commercially. **1971** F. H. Titmuss *Commercial Timbers of World* (ed. 4) 48 This [*sc.* amboyna, *Pterocarpus indicus*] is one of the most important export species of the Philippines (the wood being known as narra on the North American market).

Narragansett (nærăgæ·nsĕt). Also **Nar(r)haganset, Nar(r)ohiganset(t),** etc. [Algonquian *Naiaganset* people of the small point (of land).] **1.** An Algonquian people orig. living in New England; a member of this people. Also *attrib.*

1622 *Relation Eng. Plantation Plimoth, New Eng.* 44 They told us that if they were Narrohigganset men they would not trust them. **1637** T. Morton *New Eng. Canaan* i. xiv. 45 The cause why these other Salvages of the Narohigansets, came into these parts, was to see what strength we were of. **1682** M. Rowlandson *Narr. Captivity among Indians* 5, I was sold to him by another Narrhaganset Indian. **1764** T. Hutchinson *Hist. Colony Mass.-Bay* I. 138 At first, the Naragansets gave kind words to the messengers. **1809** 'D. Knickerbocker' *Hist. N.Y.* II. v. iv. 29 He has been secretly endeavoring to instigate the Narrohigansett (or Narraganset).. Indians. **1848** *Trans. Amer. Ethnol. Soc.* II. i. 77 (*table*) *Families*... Algonkins. *Languages* or *Dialects*... Narragansets. **1850** R. G. Latham *Nat. Hist. Varieties Man* 329 Algonkins... Narraganset.—Extinct. In 1674, in Rhode Island. **1871** C. M. Yonge *Pioneers & Founders* i. 8 The Pequots were..at war..with the Narragansets, or river Indians. **1933** L. Bloomfield *Lang.* 72 The *Algonquian* family..includes the languages..of New England (..*Narraganset, Mohican*, and so on). **1945** C. M. Webster *Town Meeting Country* 11 The Narragansetts on the west shores of Narragansett Bay. **1951** Hepburn & Logan *New England* 84/1 The Old Narragansett Church is worth seeing.

2. In full, **Narragansett pacer** (or **pacing mare**): a breed of pacing horse (now extinct) which originated in Rhode Island.

1777 J. Adams in J. & A. Adams *Familiar Lett.* (1876) 272 We are looking about for American curiosities to send across the Atlantic as presents... Narraganset pacing mares, mooses, wood ducks,..have all been thought of. **1809** 'D. Knickerbocker' *Hist. N.Y.* II. vii. iii. 194 A crafty man..bargained him out of his..charger, leaving in place thereof a villainous, spavined, foundered, Narraganset pacer. **1826** J. F. Cooper *Last of Mohicans* I. ii. 18 Giving her Narraganset a smart cut of the whip. **1943** E. Forbes *Johnny Tremain* 92, I never saw a horse his color before... Narragansett breed. **1946** M. C. Self *Horseman's Encycl.* 286 A noble breed of horses was developed, namely, the Narragansett Pacer. They are said to have had a smooth gliding motion..and were very popular..with the Virginia plantation owners. **1963** Bloodgood & Santini *Horseman's Dict.* 144 *Pacer*... Specifically: a Standardbred pacing horse descended from the Thoroughbred English stallion 'Messenger' and from the native, so-called Naragansett pacing mares.

narratage (nærā·tĕdʒ). [f. Narrate *v.* + -age.] A technique used in films, plays, and on television in which one of the characters has the role of storyteller.

1948 E. Lindgren *Art of Film* vi. 112 Other films.. employ a method known as *narratage*, in which one of the characters, usually a minor character in the film, is depicted as telling the story. **1954** *Encounter* Aug. 55/1 Here, still surviving in an age of redundant dialogue and, worse still, explanatory narratage, are stories really told in pictures.

narrate, *v.* Add: **1. b.** To speak the commentary of (a broadcast, film, exhibition, etc.).

1974 *Anderson* (S. Carolina) *Independent* 24 Apr. 1B/2 Rear Adm. Rice narrated a slide show depicting historical portions of Anderson. **1975** *Daily Tel.* 30 Oct. 6/6 The Prince of Wales introduces and narrates 'One Day in November', a £20,000 colour film about the work of the Royal British Legion.

narrative, *a.* Add: **1.** Also in *Painting*. Also *narrative line*, a consecutively developed story.

1902 R. Fry *Let.* 10 Oct. (1972) I. 196 Already he has Giovanni Bellini's *farmito* colours..and he has too the pure narrative style..of the great Venetians. **1962** R. G. Haggar *Dict. Art Terms* 221/1 Tissot painted admirable narrative pictures. **1962** *Listener* 22 Feb. 335/2 That rich outpouring of 'narrative' painting which began in England with the Bayeux tapestry, continued through the missals and Books of Hours, to be picked up again by Hogarth, Rowlandson, and Gillray. **1962** *Times* 1 Nov. 8/4 Supporting this narrative line is a cross section of Alf's social background. **1972** *Guardian* 16 Feb. 12/3, I pictured a story with every western cliche in it... Why have a straight narrative line when everybody knows the story?

narrator. Add: **b.** One who speaks a commentary in a broadcast or a film; hence also, a character who relates part of the plot of a play to the audience.

1941 *B.B.C. Gloss. Broadcasting Terms* 20 *Narrator*, person whose role is to deliver, in either his own or an assumed character, narrative passages in a radio-dramatic programme. **1948** E. Lindgren *Art of Film* vi. 112 There may be..portions [of film]..in which we see the action of the story, but hear the words of the narrator, which thus become..a form of commentary. **1955** *Radio Times* 22 Apr. 25/1 Bob Hope comperes the Annual Awards... Narrator, Leslie Mitchell. **1960** *Times* 3 Oct. 16/2 For Ahlsen clearly employs the 'narrator' technique in order to allow his protagonist not only to act out the dramatic scenes..but also to speak his thoughts aloud in monologue form if not actually fire them into the auditorium. **1961** *Times* 23 May 15/1 It is the incidentals of Brecht's method—the narrator-characters half in, half outside the action..which have..proved capable of easy assimilation. **1973** D. Gifford *Brit. Film Catal. 1895–1970* 15 *Credits*..Associate Producer.. Director..Narrator [etc.].

‖ **narrischkeit** (nā·riʃkəit). [Yiddish *naarishkeit, narrishkeit*, ad. G. *närrischkeit*, f. *närrisch* foolish.] Foolishness, nonsense; that which is of no consequence.

1892 I. Zangwill *Childr. Ghetto* I. 179 What *Narrischkeit!* Why should he die? **1963** *Punch* 4 Dec. 803/1 This piece of *narrischkeit* currently being serialised.

narrow, *a.* and *sb.* Add: **A.** *adj.* **1. c.** *narrow axe* U.S., an axe used primarily for chopping; opp. Broad-ax; *narrowback* U.S., a citizen of the United States of Irish ancestry (see also quot. 1941); *narrow band* Electr., a band (*band *sb.*[2] 14) of frequencies lying within a narrow range; freq. *attrib.* (with hyphen); *narrow boat*, a canal boat, *spec.* one not exceeding 7 feet in width or 72 feet in length; *narrow-cut* adj., applied to filters which transmit only a narrow band of wavelengths; *narrow-range* attrib., restricted in incidence or scope.

1641 *Public Rec. Colony of Connecticut* (1850) I. 444 A broad axe, 2 narrow axes, wimbell & chessells. **1854** Thoreau *Walden* 46, I went on for some days cutting and hewing timber..with my narrow axe. **1941** J. Smiley *Hash House Lingo* 39 *Narrow back*, dishwasher, cashier. **1957** *N.Y. Times* 29 Sept. X3 William Joseph Patrick (Pat) O'Brien, a Milwaukee-born Irishman or narrowback. **1966** *Publ. Amer. Dial. Soc.* 1964 xlii. 32 Irish informants use..*narrowback* for a second generation Irishman who has neither the need, the desire nor the physical equipment to do the work his father had to do. **1975** G. V. Higgins *City on Hill* ii. 56 You went out and married the same kind of commoner he always was himself, and a narrowback to boot. **1956** *Nature* 28 Jan. 178/2 A narrow-band tuned amplifier on 16 kc./s. **1962** A. Nisbett *Technique Sound Studio* 261 *Music line*, broad-band circuit for carrying programme (including speech), as distinct from a telephone line which may occupy only a narrow band. **1971** M. S. Ghausi *Electronic*

Circuits vii. 437 When extremely high Q and excellent stability are required in narrow-band bandpass filters, the use of crystal filters is practical. **1951** *Oxf. Jun. Encycl.* IV. 37/2 The narrow boat, or 'monkey boat'.., is possibly the commonest craft on inland waterways in England... because the locks on the group of canals in the Midland counties..can only pass a boat about 7 feet wide and 70 feet long. **1972** *Country Life* 13 Jan. 93/1 Only the deep draught of narrow boats can keep the canals open in the absence of proper dredging. **1975** *Times* 8 Mar. 10/6 Some of the older narrow boats with a three-foot draught are constantly ploughing through mud, for the canals are silted up. **1964** L. A. Mannheim tr. *Clauss & Meusel's Filter Pract.* 63 (*caption*) Absorption curve of a deep red filter (narrow-cut type). **1969** *Focal Encycl. Photogr.* (rev. ed.) 615/1 Special narrow-cut tricolour filter sets are used to make indirect colour separation negatives from colour transparencies. **1932** Faucett & Maki *Study of Eng. Word-Values* 7 It is fair to require of all students of English a mastery of wide-range words but.. it is unfair standardizing of procedure to pass or fail students on their knowledge or lack of knowledge of narrow-range words. **1964** E. Uldall in D. Abercrombie et al. *Daniel Jones* 279 The narrow-range 'smooth' contours..vary most often from one sentence to another.

e. Denoting a type of phonetic transcription in which separate symbols are used to denote all identifiable features (phonemic and non-phonemic) of an utterance; opp. *broad*.

1877 [see *broad *a.* 5 c]. **1908** H. Sweet *Sounds of English* 10 In comparing the sounds of a variety of languages..we require a 'narrow', that is, a minutely accurate notation covering the whole field of possible sounds. **1933** *Amer. Speech* VIII. ii. 49/2, I have never been able to understand how there might be any advantage whatever of *broad* over *narrow* transcription for English. **1961** Y. Olsson *On Syntax Eng. Verb* ii. 20 The editing [when linguists transcribe] would not imply a greater crime than a phonetician using a broad, instead of a narrow, notation. **1964** D. Abercrombie *Eng. Phonetic Texts* 35 It is..convenient to use 'broad' as an equivalent of *simple phonemic*, and 'narrow' for any departure from this, either in the direction of comparative or in the direction of *allophonic*, or both together.

3. e. *Econ.* (See quots.)

1935 *Economist* 12 Oct. 712/2 Technically, markets remain 'narrow', and day-to-day price movements are correspondingly exaggerated. **1940** G. Crowther *Outl. Money* vii. 267 The market is at all times 'narrow', that is, quotations are available only for half a dozen of the most important currencies. **1962** S. Strand *Marketing Dict.* 474 *Narrow market*, in the stock market, a dull trading session, generally limited to a few fields. **1965** J. L. Hanson *Dict. Econ.* 288/2 *Narrow market*, a term used more particularly of stock exchange securities of which there is only a small supply available in the market.

7. *narrow-beamed, -billed, -bodied, -gutted, -slitted, -slotted*.

1927 *Observer* 14 Aug. 7/4 Agile, narrow-beamed cars, with plenty of acceleration and 'safety-first' brake-power, are his ideal. **1895** *Funk's Stand. Dict.*, Narrow-billed. **1909** A. E. Mack *Bush Calendar* 9 Birds breeding in August... Chalcococcyx basalis. Narrow-billed bronze cuckoo. **1953** D. A. Bannerman *Birds Brit. Isles* I. 299 There is much to be said in favour of retaining these broad-billed buntings, as distinct from the narrow-billed species, under the specific name *tschusii*. **1949** M. Mead *Male & Female* vi. 133 The slender, narrow-bodied of the Arapesh, Tchambuli, Swede, Eskimo, and Hottentot. **1963** *Times* 1 Feb. 14/5 This narrow-bodied look was extended to coats. **1903** *Dialect Notes* II. 299 (Cape Cod dial.) *Narrow gutted*, mean, ungenerous. **1903** *Eng. Dial. Dict.* IV. 228/2 *Narrow-gutted*, of horses..weak in the loins. *Ibid.*, Not[tinghamshire]. A narrow-gutted brute. **1924** D. H. Lawrence *Let.* 29 Oct. (1962) II. 816 A narrow-gutted 'artist' with a stutter. **1965** G. J. Williams *Econ. Geol. N.Z.* xviii. 283/2 Unless the seam is narrow-gutted as at Kawakawa,..it seems certain that coal does not exist above the greywacke. **1905** J. London *Jacket* (1915) xi. 90 His eyes..were cunning and.. narrow-slitted. **1923** Kipling *Irish Guards in Gt. War* I. 234 The narrow-slotted pill-boxes.

narrow, *v.* Add: **1.** Also with *down*.

1906 L. J. Vance *Terence O'Rourke* ii. ii. 214 Then it narrowed down to a mere contest of endurance.

2. Also with *down* and *in*.

1885 J. Morris *Kotaka* xi. 106 The entire force and volume of the Fuzikawa being here narrowed down to the width of the gorge. **1885** J. W. Dawson *Egypt & Syria* ii. 39 It is just where the broad expanse of alluvium..is narrowed in by that great promontory. **1889** R. L. Stevenson *Master of Ballantrae* iv. 122 The family was now so narrowed down (indeed, there were..just the father and the two sons) that it was possible to break the entail.

narrow gauge. Add: **2.** *Cinemat.* (See quot. 1959.) Cf. Gauge, gage *sb.* 1.

1951 G. H. Sewell *Amateur Film-Making* (ed. 2) xii. 115 It is not impossible that one day research will place in the hands of the narrow-gauge worker the means of obtaining fully stereoscopic pictures without undue difficulty. **1953** L. J. Wheeler *Princ. Cinematogr.* x. 283 It is interesting to note that the B.K.S. has officially adopted the term 'narrow gauge' [film] in preference to 'sub-standard'. **1959** W. S. Sharps *Dict. Cinematogr.* 100/1 Ciné film is designated according to its width, 35 mm. film..being said to be of 'standard gauge'...the only other professional size of film..is 16 mm. wide and this is termed 'sub-standard', or 'narrow-gauge'.

narrow-mouthed, *a.* **1.** (Later examples.)

1910 W. de la Mare *Three Mulla-Mulgars* xvi. 247 Sipped slowly from their gurgling, narrow-mouthed bags or bottles. **1967** Karch & Buber *Offset Processes* vi. 248 When mixing solution 'A', 'B' or 'C' it is advisable to use a one gallon, narrow-mouthed bottle.

narsarsukite (nāɹsāɹsu·kəit). *Min.* [f. *Narsarssauk, -suk*, name of a plain in SW. Greenland + -ite[1].] A silicate and fluoride of sodium, iron, and titanium, $Na_2(Ti,Fe)Si_4(O,F)_{11}$, which is found as yellow tetragonal crystals.

1900 G. Flink in *Meddelelser om Grønland* XXIV. 154 The new mineral..was the first that attracted my attention on Narsarsuk. None of the other new minerals has been found at so many different localities on the plateau or is so largely distributed there. It may therefore, more than any other mineral, be said to be characteristic of the place. On this ground I propose for it the name narsarsukite. **1935** *Amer. Mineralogist* XX. 598 The narsarsukite was seen only in the quartz veins in the roof rock and upper part on the intrusive. **1968** I. Kostov *Mineral.* ii. 300 Baotite is a typical example of silicates with tetragonal rings of silicon tetrahedra $[Si_4O_{12}]$. Narsarsukite has similar rings but they are not isolated as in baotite but linked together along the tetragonal axis in a tube-like manner.

nartjie (nā·ɹtʃi, na·rkʸi). *S. Afr.* Also **naartj(i)e, naatje(e), naretje.** [Afrikaans, f. Tamil *nārattai* citrus.] The tangerine or mandarin orange. Also *attrib.* and *transf.*

1790 E. Helme tr. *Le Vaillant's Trav. Afr.* I. ii. 23 The citrus and oranges (specially the sort called *naretyes* [sic]) are excellent. **1796** tr. *Le Vaillant's New Travels* I. 190 There is a kind of orange..called at the Cape *naretjes*, which, notwithstanding the smallness of its size, is sold at a higher rate. **1870** *Cape Monthly Mag.* Oct. 219 The 'nartjie', a dwarf orange. **1890** A. Martin *Home Life Ostrich Farm* xi. 211 Dutchmen..often bringing with them very good onions, oranges, *naatjes* or mandarines, nuts, dried peaches and figs. **1908** *Westm. Gaz.* 14 Aug. 4/3 The only orange now on the market here is the Cape 'naatjee'. **1909** *Daily Chron.* 15 June 4/3 The little Tangerine..and its bigger relative the Mandarin are called Naartjes in Natal. **1923** *Chambers's Jrnl.* Apr. 269/2 Oranges, nartjes and grape-fruit ripen during June, July and August. **1937** S. Cloete *Turning Wheels* iii. 46 Others handled the seeds they had brought with them—..the pips of nartjes. *Ibid.* xiii. 210 Martha, your heart is like a naartjie. **1947** [see *dennebol]. **1949** A. Wilson *Wrong Set* 41 Oranges of all sizes..from the tiny nartjies, through tangerines and green mandarins to the great navel oranges. **1953** N. Gordimer *Lying Days* iii. xxvii. 276 'You've been eating a naartjie,' I said. **1966** E. Palmer *Plains of Camdeboo* ii. 33 Fanny's pot-roasted venison and pou, her van der Hum made of brandy, syrup and naartjie peel, her pickled peaches,.. are still remembered. **1967** *Punch* 5 July 17/1 Accept and enjoy a naartjie off the tree. **1970** *Rand Daily Mail* 28 Feb. 7/4 Many South African words of Dutch or Afrikaans origin are now part of English usage even in other countries—..naartjie and mealie. **1971** *Sunday Express* (Johannesburg) 28 Mar. 11/4 Couldn't the orange [of air-hostesses' uniforms] be toned down a little and the blue be slightly deeper? No wonder the poor girls are referred to as naartjies. **1972** D. Francis *Smokescreen* vii. 85 The *naartjies* turned out to be like large lumpy tangerines with green patches on the skin. **1974** *Sunday Times* (Johannesburg) 28 July 4/2 The record for the number of naartjies and oranges brought into the ground by spectators is held jointly by Ellis Park, Loftus Versfeld and Kings Park, a total of 200,000 citric fruit units being introduced at each ground. **1975** *Darling* (S. Afr.) 9 Apr. 95 You only feel a naartjie riding in the back of a truck with three drums of pig swill and a stack of moulting lucerne bales for company.

nary, *a.* Add: (Further examples.) Now more widely used, and almost always followed by the indef. article. Also (*U.S.*) *nary (a) red* or *cent*, not a cent.

1746 N. Wright *Jrnl.* 6 Aug. in *New England Hist. & Geneal. Reg.* (1848) II. 209 The Indians..escaped them, and there was no 'spile dunne on nary side'. **1821** *Mass. Spy* 14 Feb. (Th.), He asked her whether she was most fond of writing prose or poetry. 'Nary one,' says she, 'I writes small hand.' **1849** *Picayune* (New Orleans) 6 May 2/6 I'm goin' tew get my breakfuss yere, and not pay 'nary red' till I dew! **1856** *Knickerbocker* XLVII. 99 'Aint you gwine for to give us three dollars?' 'Nary a red!' sung out *Hart*. **1856** 'J. Phoenix' *Phoenixiana* 125 Playin at billiards and monte Till they've nary red cent to ante. **1872** G. P. Burnham *Mem. U.S. Secret Service* p. vii, *Nary red*, out of pocket; 'broke' of ready funds. **1895** S. Crane *Red Badge of Courage* x. 102 There's too much depending on me fer me t' die yit. No, sir! Nary die; I *can't*! **1906** *Dialect Notes* III. 147 (N.W. Arkansas) Nary a red. **1917** 'H. H. Richardson' *Fortunes R. Mahony* iv. x. 379 She would discover a thousand drawbacks to his scheme, but nary a one of the incorporeal benefits he dreamed of reaping from it. **1920** E. O'Neill *Beyond Horizon* i. ii. 36 They've been thick as thieves all their lives, with nary a quarrel I kin remember. **1932** *New Yorker* 9 Apr. 54/1 We flit hither and thither among the gadgets,..with nary a thought for the morrow. **1941** *Time* 6 Jan. 16/1 From the National Defense Advisory Commission,..there was nary a peep. **1971** M. Torrie *Bismarck Herrings* xi. 156 'Were many turned out during her time?' asked Timothy. 'Nary a one,' replied Mrs Baines. **1973** *New Society* 30 Aug. 522/3 You can wander around the cavernous vaults of the Law Courts in the Strand these days and come across nary a person. **1973** *People's Jrnl.* (Inverness & Northern Counties ed.) 1 Dec. 4/4 And there was nary a stranger on hand to be impressed! **1974** *Verbatim* Dec. 6/2 The computer that did all the dogwork gets nary a mention.

N.A.S.A., NASA, Nasa (næ·sā). [Acronym f. the initial letters of *N*ational *A*eronautics and *S*pace *A*dministration, set up in 1958.] A body responsible for organizing research in extraterrestrial space conducted by the United States.

1958 *Aviation Week* 14 Apr. 25/1 The President's recommendations and proposed legislation to create NASA, using NACA as a nucleus, were sent to Congress before the Easter recess. **1958** D. Eisenhower in *N.Y. Times* 30 July 10/7 The present National Advisory Committee for Aeronautics..will provide the nucleus of the N.A.S.A. **1965** *Economist* 4 Sept. 873/2 The Manned Orbiting Laboratory, a project which President Johnson has decided to entrust not to Nasa but to the Air Force. **1969** *Times* 13 June 7/6 N.A.S.A.'s Apollo applications programme has reached an advanced stage of discussion.

nasalism. (Further examples.)

1887 *Proc. R. Soc. Edin.* XXXII. 349 The Yankee nasalism is another familiar instance of the same kind. **1937** *Evening News* 23 Mar. 4/2 'We just dote on a good English accent,' she said, 'and we've got to get away from nasalism, our teacher says.'

nasalizable (nēɪ·zăləizăb'l), *a.* [f. Nasalize *v.* + -able.] Capable of being pronounced nasally.

1872 S. S. Haldeman *Pennsylvania Dutch* 11 This vowel being nasalisable.

Nasara: see *Nasrani.

Nascapee, Nascapi, varr. *Naskapi *a.* and *sb.*

nascence. (Later example.)

1901 *Science* 21 June 983/1 Formations often disappear through the agency of fires, floods, mankind, etc., in which cases new formations may arise by nascence.

nascent, *a.* **2. c.** Add: (Further examples.) *spec.* applied to hydrogen that has just been released from a compound by electrolysis or chemical action (marked by its great reactivity and reducing power).

1807 H. Davy in *Phil. Trans. R. Soc.* XCVII. 11 It was natural to account for both these appearances, from the combination of nascent oxygene and hydrogene respectively. **1826** —— in *Ibid.* CXVI. 388 Nascent hydrogen was not, as had been generally believed, the cause of the appearance of metals from metallic solutions. **1959** *Proc. Nat. Acad. Sci.* XLV. 1441 Thus there is a protein component which is transiently associated with the ribosomes and has all the characteristics which would be expected in a compulsory precursor of the soluble proteins. It appears that this nascent protein is a polypeptide strand which is formed on the ribosome and is subsequently released as soluble protein. **1965** Phillips & Williams *Inorg. Chem.* I. xi. 412 Nascent hydrogen is an unstable form of hydrogen which has sufficient life or kinetic stability for it to be able to react with Po before it is transformed into normal H_2 molecules. Similarly, the hydrides of phosphorus, arsenic, and antimony can be made by the action of nascent hydrogen on soluble compounds of the element. **1974** *Nature* 1 Nov. 74/2 Experiments..have shown that nascent RNA chains have 5'-triphosphate termini.

nascently (næ·sĕntli), *adv.* [f. Nascent *a.* + -ly[2].] In a nascent manner; incipiently.

1890 W. James *Princ. Psychol.* I. ix. 258 Other processes discharge that ought as yet to be but nascently aroused. **1920** *Times Lit. Suppl.* 29 Apr. 264/2 Notions of the type of mana or orenda are of 'a nascently philosophic order'. **1936** White & Renner *Geogr.* xxiv. 517 The water present in volcanic and geyser eruptions is at least partially of magmatic origin. It is, in that event, nascently formed by the union of hydrogen and oxygen atoms, as these gases are liberated from hot magmatic materials deep in the earth.

nashi, nashki, varr. *naskhi *sb. pl.*

Nashiji (nāʃī·dʒi). Also **Nashidji** and with lower-case initial. [Jap., lit. 'pear ground'.] A Japanese lacquer containing gold or silver flakes. Also, the technique of decorating with this lacquer. Also *attrib.*

1881 J. J. Quin in *Trans. Asiatic Soc. of Japan* IX. 6 The kind of powdered gold lacquer called *nashiji*, from its resemblance to the spotted skin of a pear, was introduced about the beginning of the 10th century. *Ibid.* 7 The groundwork of this box is thick *nashiji*, inlaid with various coloured shells. **1889** J. J. Rein *Industries of Japan* iii. iii. 352 Nashi-ji gets its name 'pear ground' from its use in a kind of surface decoration with coarse gold powder or its bronze substitute, which is said to be an imitation in colour of the Japanese pears. *Ibid.* 368 Nashi-ji is one of the most frequent and popular modes of surface decoration. The fine particles of gold dust and foil have at first a brownish yellow colour, but always with age become brighter and more brilliant, because of the greater transparency of the lacquer varnish. **1904** S. Hartmann *Japanese Art* vii. 247 At the beginning of the tenth century, the Nashidji lacquer, of a yellowish orange colour, sprinkled with gold. **1911** J. F. Blacker *ABC of Japanese Art* ix. 176 Now all *Nashiji* on articles intended for exportation is applied by workers in plain lacquer. The best *Nashiji*, the best *Togi-dashi*, the best *Honji*, have the first twenty-two processes alike. **1957** *Encycl. Brit.* XIII. 575/2 In Japanese lacquer, the following are the chief processes used:—*Nashiji* (pear-skin), small flakes of gold or silver sunk to various depths in the lacquer [etc.] **1975** *Country Life* 10 Apr. 907/1 The chests..were..in *Hiramakie* in gold and grey lacquer on a *Nashiji* ground... *Nashiji* (pearskin) is a technique which can be traced back to AD 905, employing small flakes of gold and silver foil at different levels in the lacquer.

Nasho (næ·ʃo). *Austral. slang.* Abbrev. of 'national serviceman'; also, compulsory military training (discontinued in 1972).

1962 C. Rohan *Delinquents* 52 They..had supper with a couple of national servicemen... 'I'm not keen on Nashos,' she said. **1966** B. Beaver *You can't come Back* (1968) 5 Sam, the new one, was just eighteen and due for his Nasho training. **1966** *Bulletin* (Sydney) 23 Apr. 12/3 The bulk of the Nashos—how the Army loathes that term—have little time for the 'protests'. **1970** *Telegraph* (Brisbane) 8 July 16/3 (*heading*) RSL [*sc.* Returned Services League] wants full Nasho details.

Nashville (næ·ʃvil). The name of the capital city of Tennessee, U.S.A., used *attrib.* and *absol.* to designate a type of country-and-western music originating there.

1963 *Broadcasting* (U.S.) 28 Jan. 74/3 Bill Fuller.. said, 'We play your Nashville numbers in Ireland and they like it because it's our kind of music.' *Ibid.* 78/2 Composer [Harlan] Howard..operates without the benefit of musical education. His 'Pick Me Up on Your Way Down' and scores of others are pure Nashville. **1968** *Sing Out!* Oct./Nov. 57/2 The songs are pleasant, though rarely distinctive (but, of course, that's Nashville). **1969** *Rolling Stone* 28 June 19/1 Along parallel lines a conservative approach to maintaining the Nashville sound could eventually backfire. **1973** *Nation Rev.* (Melbourne) 31 Aug.–6 Sept. 1459/1 The music is pure Nashville. **1975** *Hi-Fi Answers* Feb. 60/2 Musical styling ranges from the almost pure Nashville sound of 'Can't Go Back' to the rocking gospel tune of 'Let It Flow' and the convincing train beat of 'Sunshine Special'.

‖ **nasi** (nā·si). [Malay.] Cooked rice. Used usu. in *Comb.* to designate *spec.* dishes, as: **nasi br(i)yani** (briyā·ni) [Malay *beriani, biani*], a type of pilau; **nasi goreng** (gǫre·ŋ), savoury fried rice with a variety of ingredients including meat and/or fish; **nasi Padang**, rice served as in Padang in Sumatra.

1894 N. B. Dennys *Descr. Dict. Brit. Malaya* 331 *Rice* (*padi* and *nasi*). **1900** W. W. Skeat *Malay Magic* v. 226 The rice is..then boiled so that it becomes *nasi* (cooked rice). **1958** *Catal. County Stores, Taunton* June 4 Nasi Goreng (Fried Rice with Pork)—a tin 4/-. **1963** *Times* 16 Feb. 5/5 *Nasi goreng* contains boiled rice, pork, curry powder, soy sauce, sliced omelette and tomatoes. **1971** *Carry Singapore in your Pocket* (Singapore Tourist Promotion Board) (ed. 3) 30 Malay and Indonesian type food is mainly of Sumatran origin and known as Nasi Padang. *Ibid.* 31 One first-class Kashmiri restaurant.. serves nasi bryani. **1972** T. Lilley *K Section* xli. 225 Lunch?..It's only nasi-goreng, but it'll fill a hole. **1972** *New Nation* (Singapore) 25 Nov. 7/5 The nasi briyani was of excellent consistency. **1972** R. McKie *Singapore* 109 Popular Indonesian and Malay food includes nasi Padang, after the town of Padang in Sumatra.

nasion (nēɪ·ziǫn). *Anat.* [a. F. *nasion* (P. Broca *Instruct. gén. pour les Recherches anthropol.* (ed. 2, 1879) iii. 143), f. L. *nās-us* nose: see *-ion[2].] The centre of the frontonasal suture.

1889 *Buck's Handbk. Med. Sci.* VIII. 202/1 (*caption*) *N*, Nasion (junction of the nasal and frontal). **1934** F. Stark *Valleys of Assassins* I. ii. 193 The face is broken off just at the nasion, and only the ends of the molars are present. **1937** Glaister & Brash *Medico-Legal Aspects Ruxton Case* x. 154 The greatest discrepancy..can easily be accounted for by the difficulty of determining the exact position of nasion and prosthion in a portrait. **1971** *Nature* 30 July 311/1 The nasal length [of Olduvai hominid 24] from nasion to the left lateral lip of the nasal aperture is 42·5 mm and nasion to nasospinale 36·0 mm.

Naskapi (næ·skāpi), *a.* and *sb.* Also **Nascapee, -i, Nascopi(e), -upi, Nasquapee, Nescaupick.** [Native name.] **A.** *adj.* Pertaining to or designating an Indian people inhabiting northern Quebec and the interior of Labrador. **B.** *sb.* **a.** This people; a member of this people. **b.** Their language.

1774 G. Cartwright *Jrnl.* 18 June (1792) II. 9 Soon after our arrival there, two canoes of Nescaupick Indians came. **1779** —— *Jrnl.* 7 Mar. (1792) II. 418 My Nescaupick sled..was drawn by the two men, a bloodhound, and a Newfoundland dog. **1849** J. McLean *Notes 25 Years' Service Hudson's Bay* I. 305 The Nascopies, or mountaineers of Labrador, speak a mixture of Cree and Sauteaux, the former predominating. **1863** H. Y. Hind *Explor. Labrador Peninsula* I. iii. 34 On Father Laure's map..the Nasquapees are stated to occupy the country north of this lake under the name of Les Cuneskapi. **1908** *Catholic Encycl.* III. 229/2 To the east are the Micmac, Malecite, Abnaki, Nascapi, and the Montagnais of Labrador. **1934** D. Jenness *Indians of Canada* (ed. 2) xviii. 270 Two tribes (*Montagnais* 'Mountaineers'; *Naskapi* 'rude, uncivilized people') were the first to come into close contact with Europeans. *Ibid.* 271 Much of the Naskapi territory was open plateau covered with grasses and lichens. **1957** *Encycl. Brit.* XV. 741/2 North of the St. Lawrence, the Montagnais and Nascapee resemble the southern Algonkin. **1966** C. F. & F. M. Voegelin *Map N. Amer. Indian Lang.* (*caption*) Algonquian Family. 1. Cree-Montagnais-Naskapi. **1974** *Country Life* 4 Apr. 780/3 A painted deerskin coat of the Naskapi tribe.

‖ **naskhi** (næ·ski), *sb. pl.* Also **nashi, nesk(h)i, niskhi,** etc. and with capital initial. [Arab. *naskī*, f. *nasaka* to copy.] The normal cursive Arabic script. Also *attrib.* or as *adj.*

1771 W. Jones *Gram. Pers. Lang.* 15 Our books are printed in the Niskhi hand, and all Arabick manuscripts, as well as most Persian and Turkish histories, are written in it; but the Persians write their poetical works in the Tâlík, which answers to the most elegant of our Italick hands. **1777** J. Richardson *Dict. Pers., Arab., & English* p. iii/1 In this character the Alcoran was originally written: it was afterwards improved under the denomination of Cufik; and continued in use till the appearance of the Niskhi, in the tenth century of our era... The Niskhi, ..with some variation and corruption, is the same which now prevails in Arabia, Persia, India, and other Eastern countries. **1807** in W. Jones *Wks.* XIII. 416 *Zafar Námeh*. A most elegant history of Taimur, written in the Niskh character. **1820** J. G. Jackson *Acct. Timbuctoo & Housa* 350 The Arabs have various modes of writing, the principal of which is that used by the Koreish, the most learned of all the Western tribes, and is denominated the *Niskhi*, or upright character: if this is understood, the others may be easily comprehended. **1849** F. Madden tr. *Silvestre's Universal Palæogr.* I. 52 The Arabic Neskhi alphabet, already in use among the Arab tribes, was forced upon Persia with the Coran, and took the place of the ancient cuneiform and other writings. **1854** A. Sprenger *Catal. Arab., Pers. & Hindústany MSS. in Libraries of King of Oudh* I. 8 Copies are not frequent, yet there are five in the Moty Mahall. The best is written in Naskhy. **1879** C. Rieu *Catal. Pers. MSS. in Brit. Mus.* I. 7/2 The Coran in Arabic, with a Persian version written in small Nashki. **1880** *Encycl. Brit.* XIII. 117/2 Arabic epigraphy begins with the rise of Islam. Two systems of writing were used concomitantly, the Cufic or uncial, and the Neski or running hand. **1893** [in Dict. s.v. Cufic *a.*]. **1948** D. Diringer *Alphabet* 271 The two main branches of the Arabic script, Naskhi and Kufic. **1959** *Chambers's Encycl.* XII. 291/1 The earliest form of the North-Arabic script was the Kufic...in this script however many of the signs were inconveniently alike, and crushing the so-called *Nashi*-script, in which diacritical points are used to distinguish similar letters, began to take its place as early as the 7th century A.D. **1970** *Oxf. Compan. Art* 634/1 Naskhi inscriptions are common in architecture and the decorative arts.

naskh-ta'lyq: see *nastalik.

Nasmyth (nēɪ·smiþ). The name of James *Nasmyth* (1808–1890), Scottish engineer, used *attrib.* and in the possessive to denote a form of hammer or pile-driver he invented, in which the falling weight is raised by steam pressure on a piston attached to it.

1845 *Jrnl. Franklin Inst.* XL. 283 Nasmyth's steam hammer for pile driving. *a* **1877** Knight *Dict. Mech.* III. 2348/2 Plate LXII shows the new Nasmyth steam-hammer at Woolwich, England. It is at present the largest and most powerful in the world. **1902** L. F. Vernon-Harcourt *Civil Engin.* iv. 56 Nasmyth's pile-driver is the oldest and best-known form of these machines. **1928** R. Bridges in H. Bradley *Coll. Papers* 43 A friend whose mind I used to compare to a Nasmyth's hammer, which can weld a ton of iron or crack a nut without crushing its kernel. **1945** R. Hargreaves *Enemy at Gate* 278 Bringing, in effect, a Nasmyth hammer to the task of cracking a walnut. **1963** A. F. Burstall *Hist. Mech. Engin.* vi. 207 Considerable improvements took place in forging, particularly after the invention of Nasmyth's steam hammer..in 1839.

Nasmyth's membrane (nēɪ·smiþ). *Anat.* [f. the name of Alexander *Nasmyth* (d. 1848), British dentist, who described the membrane in 1839.] A transient membrane covering the crown of young teeth; the primary enamel cuticle.

1853 *Q. Jrnl. Microsc. Sci.* I. 162 The layer is about 1-40th of an inch thick, and consists of an external delicate structureless Nasmyth's membrane. **1906** *Brit. Dental Jrnl.* XXVII. 1017 The interest at the time was centred on the condition of its Nasmyth's membrane. **1973** *Gray's Anat.* (ed. 35) 1235/2 The newly erupted enamel is covered by a two-layered structure (reduced enamel epithelium and primary cuticle). This is called Nasmyth's membrane.

naso-. Add: **nasoci·liary,** applied to a branch of the ophthalmic nerve that supplies the skin and mucous membrane of the nose, the eyelids, and parts of the eyeball, and communicates with the ciliary ganglion; **naso-phary·ngeal** *a.*, of or pertaining to the nasopharynx, or the nose and the pharynx; **na·sopharynx,** the uppermost part of the pharynx, lying above the soft palate and communicating with the posterior nares.

1895 *Quain's Elem. Anat.* (ed. 10) III. ii. 236 The nasal nerve (oculo-nasal or naso-ciliary) enters the orbit between the heads of the external rectus muscle. **1968** Passmore & Robson *Compan. Med. Stud.* I. xxi. 10/2 The ophthalmic nerve..breaks up into its lacrimal, frontal and nasociliary branches while still within the wall of the cavernous sinus. **1872, 1884** Naso-pharyngeal [in Dict.]. **1892** P. McBride *Dis. Throat, Nose & Ear* 281 Naso-pharyngeal Catarrh associated with Crust Formation.—As we have seen, crusts may exist in the naso-pharynx in cases of atrophic rhinitis. **1973** *Sci. Amer.* Oct. 32/3 As the term indicates, nasopharyngeal carcinoma originates in the nasal cavity, the pharynx and the rear of the oral cavity. **1877** Nasopharynx [in Dict.]. **1902** D. J. Cunningham *Text-bk. Anat.* 981 This sheet [*sc.* the soft palate]..cuts into the cavity of the pharynx.., and, falling short of the posterior wall, incompletely divides it into two, namely, an upper part or naso-pharynx (pars nasalis), and a lower part or pharynx proper, which is further subdivided..into the oral pharynx..and the laryngeal pharynx. **1935** Imperatori & Burman *Dis. Nose & Throat* xxiv. 283 Most often neoplasms in the nasopharynx really have their origin in the nose but in their growth extend back into the nasopharynx. **1973** *Daily Colonist* (Victoria, B.C.) 3 Nov. 1/3 Snoring results from vibration of the soft tissues of the nasopharynx.

nasogastric (nēɪzogæ·strik), *a.* *Med.* [f. Naso- + Gastric *a.*] Reaching or supplying the stomach via the nose; effected via such a route.

1958 *Med. Ann. District of Columbia* XXVII. 331 (*heading*) Hypertonic dehydration complicating high protein nasogastric tube feeding. **1961** *Lancet* 2 Sept. 519/2 The uncritical use of naso-gastric suction in all upper abdominal cases should be reconsidered. **1964** *Ibid.* 26 Dec. 1351/1 The indwelling polyvinyl nasogastric tube scarcely disturbs the infant. **1974** Passmore & Robson *Compan. Med. Stud.* III. 10/1 Most nasogastric feeds are based on milk.

nasonite (nēɪ·sǫnəit). *Min.* [f. the name of Frank Lewis *Nason* (1856–1928), U.S. geologist + -ite[1].] A silicate and chloride of calcium and lead, $Ca_4Pb_6(Si_2O_7)_3Cl_2$, which forms hexagonal crystals and is found as white or grey granular masses.

1899 Penfield & Warren in *Amer. Jrnl. Sci.* clviii. 346 Hand specimens usually present a mottled or spotted appearance owing to numerous inclusions of yellow axinite and brown garnet, which are scattered rather uniformly through the massive nasonite. **1935** *Prof. Papers U.S. Geol. Survey* No. 180. 93/2 Nasonite commonly forms the matrix of glaucochroite crystals, and in the few specimens seen is also associated with garnet, axinite, and barite. **1971** *Amer. Mineralogist* LVI. 1175 If one imagines two unit cells of apatite superposed in the *c* direction..and rotates one of them of 60° around the 6_3 axis, one obtains the unit cell of nasonite.

nasospinale (nēɪ:zospinā·li). [app. mod.L., f. Naso- + late L. *spīnāle*, neut. of *spīnālis* Spinal *a.*

So etymologized, e.g., in R. Martin *Lehrb. d. Anthrop.* (1914) 513. But cf. F. *naso-spinal* adj., used and prob. coined (as the fem. *naso-spinale*) by P. Broca (*Mém. de la Soc. d'Anthrop. de Paris* (1875) II. 72) as the epithet of a line 'de la racine du nez à la base de l'épine nasale'.]

The point of intersection of a line joining the lowest points on the anterior nasal apertures with the midsagittal plane.

1920 H. H. Wilder *Lab. Man. Anthropometry* i. 47 (*heading*) Nasospinale... This point is defined as a point (usually within the bone substance), where a line tangent to the two lateral curves of the lower margin of the piriform aperture crosses the median line. **1933** *Jrnl. R. Anthrop. Inst.* LXIII. 44 The definition of the nasospinale we have given above is an elaboration of that which was first given by Broca. **1949** H. Sicher *Oral Anat.* i. 97 The nasal profile angle is the angle formed by the line from nasion to nasospinale with the Frankfort horizontal. **1971** Nasospinale [see *nasion].

Nasquapee, var. *Naskapi *a.* and *sb.*

Nasrani (næzrā·ni). Also **Nasrany,** etc. Pl. **Nasara,** etc. [Arab. *Naṣrānī*, pl. *Naṣārā* cogn. with Nazarene *a.* and *sb.*] A Christian, so called by Muslims; = Nazarene *sb.* 1 b.

1583 J. Newbery *Let.* 15 July in Purchas *Pilgrimes* (1625) II. ix. xviii. 1643 In Aleppo I hired two Nastraynes, and one of them hath the Indian tongue. **1615** G. Sandys *Relation* III. 153 On the twentieth of March with the rising Sunne we departed. A small remainder of that great Caruan; the Nostraines (so name they the Christians of the East) that rid vpon Mules and Asses being gone before. **1792** R. Heron tr. *Niebuhr's Trav. Arabia* II. xxvi. 242 In Arabia, the Christians are called *Nassara* or *Nusrani*. **1820** J. G. Jackson *Acct. Timbuctoo & Housa* 510 It is not correct to assert that *Nasari is a general term* applied to infidels in Muhamed; it is applied to Christians only. **1826** Denham & Clapperton *Narrative* p. xxxi, You were the first man whose hand I ever touched—but they all said it did not signify with you, an *Insara* (a Christian). **1836** E. W. Lane *Account* II. iv. 52 The kelbs before removal from their original places are called *Nasa'ra* (or Christians, in the singular, *Nusra'nee*). **1844** J. H. Drummond Hay *Western Barbary* xi. 76/2 The inhabitants rushed out from their houses to have a sight of the Ensara. **1888** C. M. Doughty *Trav. Arabia Deserta* II. xvii. 514 When I responded,..Sâlem exclaimed, 'Ullah! how truly the Nasrâny speaks!' **1905** *Jewish Encycl.* IX. 195/1 In the Koran also the Christians are called 'Al-Naṣara'. This name may be traced back to Nazareth, Jesus' birthplace. **1912** *Catholic Encycl.* XIV. 681/1 The St. Thomas Christians now prefer to be called *Nasrani* (Nazarenes), the designation given by the Mohammedans to all Christians. **1926** R. Bell *Origin Islam* v. 149 If we could be sure that *Ṣâbi'in* denoted the Christians of South Arabia, Naṣârâ would then denote specially those of the north. But we shall not..go far wrong in taking it as meant to denote Christians in general. The word Naṣârâ is apparently derived from *Nazaraioi*, which is mentioned as the name of a Jewish-Christian sect. **1936** F. Stark *Southern Gates Arabia* iv. 40 About fifty children pursued, calling 'Nasrani' in a monotonous but not insulting way. **1937** R. Bell tr. *Qur'ân* I. xxii. 318/2 Those who have believed, those who have judaised, the Ṣâbi'în, the Naṣârâ, the Magians, and the Polytheists—

verily Allah will distinguish between them on the day of resurrection. **1958** L. Durrell *Balthazar* iv. 87 He, a city-bred Alexandrian—almost a despised *Nasrany*—could out-shoot, out-talk and out-gallop..them. **1963** *Times* 14 Feb. 14/7 They drew aside and conferred among themselves since a *Nasrani* or Christian is seldom met in their deserts.

Nass (næs). The name of a river in British Columbia, Canada, used *attrib.* to denote a tribe of the Tsimshian people who inhabit the basin of this river, and whose native name is *Niska. Also as *sb.*, the language of this tribe.

1829 J. S. Green in *Missionary Herald* (Boston, Mass.) (1830) XXVI. 344 The Nass Indians live on the continent. *Ibid.*, The Nass to my ear is harsh and disagreeable. **1911** F. Boas *Handbk. Amer. Indian Lang.* I. 290 The system of vowels of Tsimshian is nearly the same as that of the Nass dialect. *Ibid.* 292 Many *u*-sounds of Tsimishian are *i* or *ē* in Nass. [**1934** D. Jenness *Indians of Canada* (ed. 2) ii. 18 The Tsimshian of the Nass and Skeena rivers. *Ibid.* xxi. 336 The Tsimshian ('people inside of the Skeena river'), were divided into three groups, the Tsimshian proper around the mouth of Skeena river, the Gitksan ('Skeena River people') farther up the same stream, and the Niska who inhabited the basin of Nass (Niska) river.]

nassa (næ·să). [mod.L. (J. B. P. A. de Monet, Chevalier de Lamarck 1799, in *Mém. Soc. d'Hist. Nat. Paris* I. 71), f. L. *nassa* basket fish-trap.] The shell of a marine gastropod of the genus once so called, now included in the genus *Nassarius*; a basket shell; = Dog-whelk. Also *attrib.*

1853 L. Reeve *Conchologia Iconica* XVI. s.v. *Nassa*, plate 1 The rough Nassa... A very elegantly coronated species, with the spine rising into a sharp turret. **1971** *World Archaeol.* III. 136 String bands to which cowrie or nassa shells are attached. **1974** S. P. Dance *Encycl. Shells* 156/1 *Nassarius fossatus* Gould. Giant Western Nassa. Shell ovate. Conical spire, apex pointed.

Nassauvian (năsǭ·viăn). Also **Nassavian.** [f. *Nassau*, capital city of the Bahama islands + -ian.] A native or inhabitant of Nassau. Also *attrib.* or as *adj.*

1924 F. Holmes *Bahamas during Gt. War* xii. 121 On May 29th [1917], greatly to the satisfaction of all Nassauvians, the Lighting Restrictions were..suspended. **1928** R. A. Curry *Bahamian Lore* vi. 64 Along the sidewalks of Bay Street you will meet..the Nassauvian in white clothes. **1961** I. Fleming *Thunderball* xii. 126 A discussion about the merits of Nassavian labour... The Chief of Immigration and Customs was a sleek Nassavian with quick brown eyes. **1973** *Listener* 16 Aug. 202/2 Nassauvians generally have a care for their visitors' feelings.

nassella (næse·lă). [mod.L. (E. Desvaux in C. Gay *Historia Física y Política de Chile, Botánica* (1853) VI. 263), f. L. *nassa* net + *-ella*, fem. of *-ellus*, diminutive suffix.] A coarse grass of the genus so called, native to Chile, but also found in other countries to which it has been introduced, esp. New Zealand, where it has become a troublesome weed. Also *attrib.*

[**1936** H. H. Allan *Introd. Grasses N.Z.* 118 *Nassella trichotoma*, a rather graceful tussocky species, somewhat resembling *Festuca novae-zelandiae* in habit, but with very different spikelets. Recently collected by Professor A. Wall on the Waipara River bed, and in the Omihi Valley, where it appears to be thoroughly established.] **1946** *Nature* 21 Dec. 920/1 Weed problems..such as nassella, constitute a serious threat to good pasture land [in New Zealand]. **1950** *N.Z. Jrnl. Agric.* June 515/1 Nassella tussock... This insidious pest crept almost unawares on to some of our good pastoral country. **1966** G. W. Turner *Eng. Lang. Austral. & N.Z.* viii. 172 In scientific language Latin botanical names often become general. The recently troublesome Nassella tussock has no more popular name. It is controlled in accordance with the Nassella Tussock Act of 1946 by Nassella Tussock Boards.

Nasserite (næ·sərəit). [f. the name of Gamel Abdel *Nasser* (1918–70), President of Egypt from 1956 to 1970 + -ite[1].] A supporter of Nasser. Also *attrib.* or as *adj.* So **Na·sserism,** the political principles or policy of Nasser; **Na·sserist,** a Nasserite; also *attrib.* or as *adj.*

1958 *Times* 18 Mar. 11/3 The argument..between the west-looking policy pursued by President Chamoun and M. Solh and the Nasserite policy espoused by some of the Opposition. **1958** *Observer* 20 July 10/2 We were prepared (as Sir Anthony Eden seemed to be at the time of Suez) to overthrow 'Nasserism' by force. **1958** *Economist* 26 July 300 It is obvious and well understood here that the coup in Iraq means a further link in the encirclement of Israel by Nasserist forces and thus increases Israel's military danger. **1958** *Times* 16 Dec. 9/6 The others are the Nasserists, the Communists, and Brigadier Kassem. **1959** *Daily Tel.* 24 Apr. 12/5 Five visiting Iraqi officers are reported to have recently investigated public feeling here, much to the concern of the Nasserites. **1965** M. Spark *Mandelbaum Gate* iv. 105 Are you a nationalist?.. A Nasserite? **1966** *Economist* 5 Mar. 889/1 If this is not potential 'nasserite material', what is? **1968** *Ibid.* 3 Feb. 23/3 One hears a lot about 'Young Turks', 'Nasserists', and so on these days. **1969** J. Mander *Static Society* viii. 225 We might say that Peron invented 'Nasserism' a decade before Nasser. **1971** *Guardian Weekly* 10 Apr. 7/1 Officially Sadat upholds the Nasserist orthodoxy. He told the Palestinians..that there was no such thing as an Egyptian strategy or a Syrian strategy, but only an Arab strategy. **1973** Howat & Taylor *Dict. World Hist.* 1047/2 *Nasserism*, Western term used to denote first the policies of Egypt..under President Abd Al-Nasir, and second, a generalized movement among Arabs.

‖**nastalik** (næstā̆l*ī*·k). Also **nastaliq, nestalik, nestaliq,** etc. [Pers., f. Arab. *naskī* *naskhi + *ta'līq* hanging.] A Persian cursive script, characterized by rounded forms and elongated horizontal strokes. Also called *talik.

1795 W. Ouseley *Pers. Miscellanies* I. i. 7, I must here remark, that in India the Talik hand is generally called Nustaleek... Although used occasionally by the Arabian, and commonly by the Turkish penmen, yet it seems to be more particularly a favourite of the Persians. **1809** C. Stewart *Descr. Catal. Oriental Library of Late Tippoo Sultan of Mysore* 57/1 Octavo, Nastâlik Character, beautifully written. **1854** A. Sprenger *Catal. Arab., Pers., & Hindústány MSS. in Libraries of King of Oudh* I. 89 Myr 'imád of Qazwyn a most exquisite calligraph particularly in Naskh-ta'lyq. **1879** C. Rieu *Catal. Pers. MSS. in Brit. Mus.* I. 5/2 The texts..are given in Arabic, and mostly accompanied with an interlinear Persian version in Nestalik. **1908** Maulavi Abdul Muqtadir *Catal. Arab. & Pers. MSS. in Oriental Public Library at Bankipore: Pers. Poetry* 5 Written in a perfect Nasta'liq, in four columns, with one gold and two ornamental rules. **1913** *Encycl. Islām* I. 391/2 It was probably not until this later period that the *nesta'lik* arose (said to be a contraction of *naskhi* and *ta'lik*), a variation of *ta'lik*, from which it does not differ in any essential features. **1919** D. C. Phillott *Higher Pers. Gram.* i. 36 The *nasta'liq*.. is a combination of the *naskh*..or ordinary hand and the *ta'liq*..: it is a beautiful hand, chiefly used by the Persians for well-written manuscripts; but the modern Arabs call the Persian writing generally *ta'liq*. **1948** D. Diringer *Alphabet* 595/2 (*index*) Nesta'liq script; *see* Ta'liq. **1954** A. F. L. Beeston *Catal. Addit. Pers. MSS. in Bodl. Libr.* 92/1 Main text and index in neat Naksh, the other pieces in Nasta'līk. **1957** *Encycl. Brit.* XVII. 593/1 In this academy [*sc.* the Academy of Artistic Book-Production at Herat] the Persian texts were no longer written in *naskhi*, but in *nastaliq* an offshoot of the old cursive character, invented by the celebrated Mir Ali of Tabriz. **1966** Hosking & Meredith-Owens *Handbk. Asian Scripts* ii. 20 After 1600 there is a steady deterioration in *Nasta'līk* which becomes larger and coarser, whereas *Naskhi* gradually improves after this date. **1973** 'D. Jordan' *Nile Green* xxiv. 99, I vaguely heard him expounding the difference between *naskhi* and *nasta'liq* script.

nastic (næ·stik), *a. Bot.* [a. G. *nastisch* (E. Strasburger *Lehrbuch der Botanik* (ed. 6, 1904) I. ii. 221), f. Gk. *ναστ-ός* pressed together + -ic.] Of movements of parts of plants: uninfluenced by the direction of an external stimulus.

1908 W. H. Lang tr. *Strasburger's Text-bk. Bot.* (ed. 3) 270 When..the movement results from the internal disposition of the structure it is spoken of as nastic. **1912** *Ibid.* (ed. 4) 300 Nastic movements..are curvatures which bring about a particular position in relation to the plant and not to the direction of the stimulus. **1929** J. C. Bose *Growth & Tropic Movements Plants* xx. 216 If the movement be nastic, then the closure or the opening movement will remain the same, whether the organ be held in normal position or upside down. **1951** M. Abercrombie et al. *Dict. Biol.* 150 Nastic movements are classified according to nature of stimulus, e.g. photonasty, response to alteration in light intensity, thermonasty, response to alteration in heat intensity, seismonasty, response to shock. **1965** Bell & Coombe tr. *Strasburger's Textbk. Bot.* 380 Nastic movements, depending on growth differences on the two sides of the organ concerned, are seen particularly clearly in the perianth segments of the many flowers which open or close according to the temperature.

Nastrayne, obs. var. *Nasrani.

nasturtium. Add: ¶ Used jocularly in place of *aspersion*.

1922 Joyce *Ulysses* 315 Don't cast your nasturtiums on my character. **1933** D. L. Sayers *Murder must Advertise* i. 22 He's been a long time in the firm and doesn't like any nasturtiums cast at it. **1970** *Guardian* 2 Feb. 9/4 No nasturtiums are cast upon anyone's actual hormones.

nasty, *a.* Add: **1. b., 4.** (Further examples.)

1914 G. B. Shaw *Misalliance* 16 Thats what theyre like: theyve nasty minds. With really nice good women a thing is either decent or indecent. **1930** W. S. Maugham *Cakes & Ale* xvii. 162 'Don't bother about it,' she said. 'He's got a nasty mind.' **1933** [see *hot *a.* 6 c]. **1942** Berrey & Van den Bark *Amer. Thes. Slang* § 295/1 *Nasty look*, a reproving look. **1961** [see *equalizer d]. **1969** M. Allingham *Case-Book* 29 One doesn't have to have a nasty mind to wonder.

7. Phrases: *a nasty piece* (or *bit*) *of work* (or *goods*): an unpleasant or contemptible person; *something nasty in the woodshed*, a traumatic experience or a concealed unpleasantness in a person's background.

1923 'Bartimeus' *Seaways* vii. 110 Nasty Bit of Work. I'd go and bash his head for two pins. **1928** 'M. Hoffe' *Many Waters* ii. iv. 62 Edith shows in Mr. Rosel. He is really a rather nasty piece of work. **1932** S. Gibbons *Cold Comfort Farm* x. 141 When you were very small.. you had seen something nasty in the woodshed. **1945** G. B. Grundy *55 Yrs. at Oxf.* v. 87 Among the many pupils I had..there was only one I disliked. He was what is called a 'nasty piece of work'. **1949** 'M. Innes' *Journeying Boy* ix. 109 Nasty bit of work, isn't he?.. Specialises in blackmailing adolescents. **1952** A. Christie *Mrs McGinty's Dead* xxvii. 184 She was a nasty bit of goods all right—children know. **1953** 'H. Cecil' *Nat. Causes* xix. 223 He was a nasty piece of work all right. A real blackmailer. **1959** *Sunday Times* 5 July 6/6 He enjoyed a temperate childhood: nothing nasty in his woodshed. **1959** *Listener* 3 Sept. 350/1 Although the idea is no longer entirely respectable (rather akin to 'something nasty in the woodshed'), we [etc.]. **1960** D. Fearon *Murder-on-Thames* xiv. 116 He was a nasty bit of work. I don't know that he actually had a criminal record, but the blighter stank of black market. **1968** B. Bainbridge *Another Part of Wood* ii. 70 They had all, Joseph, brother Trevor, the younger sister,..come across something nasty in the woodshed, mother or father or both, having it off with someone else. **1975** A. Christie *Curtain* v. 46 You do not like him?..What you call the nasty bit of goods.

8. *Comb.*, as *nasty-minded* adj.; *nasty-mindedness*.

1921 D. Canfield *Brimming Cup* xiv. 236 A nasty-minded remark from somebody who didn't know what he was talking about. **1935** *Times Lit. Suppl.* 3 Oct. 613/3 A nasty-minded boy. **1972** 'J. & E. Bonett' *No Time to Kill* vii. 96, I don't want everybody to know about Greg and me. It's not that I'm ashamed... But nasty-minded people make it sound dirty. **1940** M. Marples *Public School Slang* p. x, For some reason food inspires a particular kind of satirical nasty-mindedness. **1960** *Encounter* Mar. 75/1 The nasty-mindedness of the Freudians.

nasty (nɑ·sti), *sb.*[1] [f. the adj.] **1.** (Freq. with capital initial.) Used as a jocular alteration of 'Nazi'.

1935 'R. Crompton' *William—the Detective* vi. 121 I'm jolly well not going to be called Her Hitler... I'll be called Him Hitler... Now I'm Him Hitler an' we four are the nasties. **1939** *Airman's Gaz.* Dec., All aircraft off duty being allowed to..view the stupendous, side-splitting entertainment provided by the Nasty Air Force and our Navy. *Ibid.*, All races won by the Nasties with a three mile start each time. **1940** H. G. Wells *Babes in Darkling Wood* iv. iii. 349 If it helps beat them Nastys,..I'll sit in the 'ole pond all day. **1942** *R.A.F. Jrnl.* 30 May 19 Ole Nasty can hit a hay-stack all right, but it's all he can hit. **1943** J. B. Priestley *Daylight on Saturday* xxi. 163 Cripes, ol' Nasty give us Phoney War all right, didn't 'e?

2. A nasty person; something nasty.

1935 'L. Luard' *Conquering Seas* 39 Fair is foul and foul is fair when Jack nasties step aboard. **1959** *Listener* 2 Apr. 609/2 For a Silver Wedding party there was such a grouping of nasties that one seemed to be involved in a misanthropist's nightmare. **1968** *Saturday Night* (Toronto) Feb. 27 Nasties—they're the newest social force, waiting in the wings to displace the last 1960s social force, the Flower People. They've always been around, the Nasties—disguised as merely unpleasant people. **1970** *Guardian* 8 Aug. 21 Nice nasties are de rigueur these days. **1971** *Ibid.* 22 Feb. 9/1 You come up against all manner of nasties in the woodshed: inadequacy, fear, alcoholism, ignorance, poverty, and hopelessness. **1973** H. Miller *Open City* xiv. 153 Here was a big nasty in a Crombie coat, standing right in the middle of her own living room. She was becoming very frightened. **1974** H. MacInnes *Climb to Lost World* ix. 145, I was convinced that there must be nasties in this territory, but..I found only one spider. **1975** *Country Life* 30 Jan. 257/3 It is the business of museums to present us with nasties as well as with fine things.

nasty (næ·sti), *sb.*[2] *Bot.* [a. G. *nastie* (E. Strasburger *Lehrbuch der Botanik* (ed. 6, 1904) I. ii. 221), f. Gk. *ναστ-ός* pressed together.] A nastic movement.

1936 J. B. Hill et al. *Bot.* ix. 228 Nasties are responses of bilaterally symmetrical organs like leaves and flower petals. **1955** *Sci. Amer.* Feb. 101/2 Nasties (or, more euphoniously, nastic movements) are among a plant's more beautiful motions: a typical example is the opening of a flower. They are the result of differing responses of different parts of the plant structure to the same external stimulus. **1965** Bell & Coombe tr. *Strasburger's Textbk. Bot.* 361 If, however, the direction of the movement is quite independent of that of the stimulus, but is determined solely by the structure of the organ, the movements are designated nasties.

nasute, *a.* Restrict *Obs.* to senses in Dict. and add: **3.** *Zool.* Nose-shaped or having a pronounced proboscis, esp. in reference to a caste of soldier termites of the genus *Nasutitermes*. Also as *sb.*, a termite of this caste.

1884 J. Hall in *Geol. Mag.* Dec. 560 In other forms [of lamellibranchiate molluscs], the anterior extremity becomes nasute or rostrate. **1926** R. J. Tillyard *Insects Austral. & N.Z.* xi. 103 In the family Termitidae the soldiers are sometimes of the nasute type, i.e. they are provided with pear-shaped heads more or less produced anteriorly, above the mandibles, into a narrow, snout-like process. **1934** *Discovery* Nov. 307 (*caption*) Earthen mound built by nasute termites in northern Australia. **1963** A. W. Leftwich *Student's Dict. Zool.* 160 Nasutes. Soldier-termites: specialized forms within the genus *Nasutitermes*, able to defend the colony by discharging an acrid secretion from glands situated at the end of a long snout or rostrum. **1970** F. J. Gay in *Insects of Australia* (Commonwealth Sci. & Industr. Res. Organization) xv. 292/1 *Nasutitermes* is widely distributed throughout Australia, particularly in the northern half of the continent. The soldiers are the typical nasute type.

nasutus (nĕɪsiu·tǝs). Pl. **nasuti.** [L. *nāsūt-us*, f. *nāsus* nose.] = *nasute *sb.*

1869 A. S. Packard *Guide to Study of Insects* 587 There also occur among workers [of certain termite species] certain individuals (Nasuti) which have the front of the head prolonged into a horn. **1895** D. Sharp in *Cambr. Nat. Hist.* V. xxi. 371 The prolongation and form of the head of these Nasuti may be fairly described as adaptation to useful ends. **1920** *Bull. U.S. Nat. Mus.* CVIII. 71 Two specimens [of *Nasutitermes costaricensis*] from Texas..; a dealated female and two nasuti.

Nat[1] (næt). *India.* Also **Nut.** [Skr. *naṭa* dancer, actor, tumbler.] A member of an itinerant class of entertainers, fortune-tellers, and the like, found esp. in northern India, but coming from no tribe in particular. Also **Na·ti,** the argot spoken by these people; also *attrib.* or as *adj.*

1801 D. Richardson in *Asiatick Researches* VII. 457 Strictly speaking, these people might be denominated *players* or *actors*, from their Persian name of *Bazee-gur*,.. *juggler* or *tricker*; but the appellation of *Nut* extends to several tribes, and properly belongs to many more. **1855** H. H. Wilson *Gloss. Judicial & Revenue Terms* 369/2 Nata, Naṭ, or Nuṭ,..a dancer, an actor, a tumbler, a public performer; applied also to a tribe of vagrants who live by feats of dexterity, sleight of hand, fortune-telling and the like. **1896** W. Crooke *Tribes & Castes N.-W. Provinces & Oudh* IV. 66 These Nats say that they came originally from Ratanpur and Bilâspur in the Central Provinces. *Ibid.* 70 Their domestic ceremonies are of the usual Nat type. **1908** *Encycl. Relig. & Ethics* I. 451/1 The ritual of the Nats, a tribe of wandering acrobats, is more remarkable. **1917** Kipling *Eyes of Asia* (1918) 8 The nature of the enemy in this war is like the Nat (juggler) who is compelled to climb a pole for his belly's sake. **1922** G. A. Grierson *Ling. Survey India* XI. 121 Any tribe may be represented among the people acting as Naṭs. *Ibid.* 122 The great majority of Natī slang words..have been taken from the common Aryan vocabulary of Northern India.

nat[2] (nāt). [Burmese *nāt*, f. Skr. *nāthá* lord, protector.] **a.** In the animistic native religion of the Burmese people, a spirit or demon, a supernatural being.

1819 F. Hamilton *Acct. Kingdom Nepal* i. i. 57 The Bhotiyas..worship all the spirits, that by the Burmas are called Nat. **1826** J. Conder *Mod. Traveller: Birmah, Siam & Anam* 82 Carved images..are to be seen.., the supposed representatives of different *nats* or demons. **1828** *Asiatick Researches* XVI. 468 In Ava and Siam..in the existence of *Nats*, it is admitted, that other animated creatures than man and animals exist. **1858** C. T. Winter *Six Months Brit. Burmah* i. 13 The Nats who guard the royal city, palace, and umbrella. **1878** C. J. F. S. Forbes *Brit. Burma* viii. 223 A man going on a journey through a forest, comes to a large and conspicuous tree..and places..an offering to the Nát of the tree. **1923** *Blackw. Mag.* Aug. 149/1 We had been talking..of folk-lore, superstitions, witches, djinns, nats, spookes, ghouls and other inventions of primitive man. **1934** 'G. Orwell' *Burmese Days* xiv. 209 Sacrificing to the local god. Nats, they call them—a kind of dryad. **1959** C. Ogburn *Marauders* (1960) iv. 115 These were intended to propitiate, entice, or exorcise the nats, or forest spirits. **1968** O. Wynd *Sumatra Seven Zero* i. 10 'Rubies..only come out of the ground when their protecting "nats" permit it.'..'What is a "nat"?'..'A Burmese spirit of the earth and air.' **1974** *Times* 30 Apr. 16/7 A Burmese journalist..asked if I would like to come with him to an evening of 'Nat' Dancing. To the Burmese the Nats are the multitudinous spirits that inhabit all natural phenomena and can exercise at will power over people and objects... The monks are tolerant of Nats: a good many of them are probably Nat-conscious themselves.

b. *Comb.*, as *nat-worship, -worshipper*.

1833 A. Judson *Let.* 29 Nov. in F. Wayland *Mem. A. Judson* (1853) II. 56 The best outward test is to have refrained from rum, nat-worship, &c. **1910** *Encycl. Relig. & Ethics* III. 21/1 The practical everyday religion of the Burmese peoples is Animism, called generally in Burmese 'Nat-worship'. **1965** M. Nash *Golden Road to Modernity* ix. 320 There is a system of animistic belief (nat worship) which is integrated with Buddhism and gives a villager a belief system reaching from hut to heaven and beyond. **1906** J. G. R. Forlong *Faiths of Man* I. 257 Its [*sc.* Burma's] population in 1894 was.. Nāt worshipers..420,000. **1923** *Blackw. Mag.* Feb. 183/2 They are all Nat or spirit worshippers.

Nat[3] (næt). [Abbrev. of National *a.* and *sb.*, Nationalist.] A member of (*a*) the National Party in South Africa; (*b*) the Scottish or Welsh Nationalists.

In S. Afr. also = 'National'.

1934 W. Saint-Mandé *Halcyon Days Afr.* ii. 24 Labour had done right to join forces with the Nats., for the Smuts government would gradually have reduced the white workers to slaves. **1958** *New Statesman* 22 Mar. 367/2 The movement..cannot lose momentum. The Nats must know this, unless the strain of continually having to defend an indefensible position has lost them their sanity. **1967** *Economist* 18 Mar. 1008/2 Both Scotland and Wales could in time do well. But the start of their national lives would be bleak... One cannot take seriously the present Nats who advocate freedom and promise prosperity in one breath. **1968** *Guardian* 20 Sept. 13/2 Jeremy Thorpe's overtures to the Celtic Nationalists... The Liberals would back the Nats on devolution. **1970** B. Knox *Children of Mist* i. 23 Scot Nats? He'd met all kinds. **1971** *Sunday Times* (Johannesburg) 28 Mar. 2/2 The Transvaler, official organ of the Nat Party in the Transvaal. **1972** M. Sinclair *Norslag* ix. 73 'Weren't you with the...' 'Yes, I was an ardent Nat then.'..MacCaig was already neatly pigeonholed. **1974** *Times* 22 Apr. 7/1 The white people..will be voting overwhelmingly for Mr. Vorster's National Party... The average voter will..continue to vote 'Nat'. **1974** *Sunday Post* (Glasgow) 28 Apr. 5/5 And there are other Labour strongholds where it wouldn't need much of a swing to put in the Nats.

Natal (natā·l), *sb.*[2] The name of a province of the Republic of South Africa used *attrib.* in: **1. Natal sore** *Path.* = *Oriental sore* s.v. Oriental *a.* and *sb.* A. 3 b.

1852 C. Barter *Dorp & Veld* ii. 13 The Natal sore, a very painful boil. **1855** J. W. Colenso *Ten Weeks in Natal* 245 The Natal sores or boils..are such as I have known among my parishioners in Norfolk. **1915** O. S. Ormsby *Pract. Treat. Dis. Skin* 360 As distinguished from the Natal sore, which was chiefly found in the lower part of that country, the veldt sore was most abundant in the high, barren table-lands. **1951** G. Panja in R. B. H. Gradwohl *Clin. Trop. Med.* xxxiii. 641 Tropical Phagedenic Ulcer... Synonyms... Natal sore.

2. The names of plants and animals found there, as **Natal francolin,** *Francolinus natalensis*; **Natal lily,** a bulbous plant of the family Amaryllidaceæ, *Clivia miniata*, which bears clusters of scarlet flowers, or one of several species of *Crinum*; **Natal mahogany,** either of two evergreen timber trees, *Kiggelaria africana*, of the family Flacourtiaceæ, or *Trichilia emetica*, of the family Meliaceæ; **Natal plum,** one of several shrubs or trees bearing fruit resembling a plum, esp. a spiny evergreen shrub or small tree, *Carissa macrocarpa*, of the family Apocynaceæ, which bears tubular, white, fragrant flowers and an edible purple fruit.

1906 Natal francolin [see *coast-partridge* (*coast *sb.* 14)]. **1947** J. Stevenson-Hamilton *Wild Life S. Afr.* xxxii. 274 The Natal francolin (*Francolinus natalensis*).. may be recognized by its red bill and legs, and white breast with V-shaped black markings. **1855** *Cape of Good Hope Almanac & Ann. Reg. for 1856* 283 The Natal lily is the perfection of beauty and fragrance. **1859** R. J. Mann *Colony of Natal* viii. 152 Most places are commonly covered with another very beautiful amaryllid..which is termed *par excellence* the 'Natal lily'. The flowers of this striking plant are large white pink-ribbed bells, hanging in enormous bunches round the summit of the flower-stalk. **1962** Watt & Breyer-Brandwijk *Medicinal & Poisonous Plants S. & E. Afr.* (ed. 2) 1378/1 Crinum sp... Natal lily. **1904** H. Stone *Timbers of Commerce* 3 (*heading*) Natal Mahogany. Kiggelaria Dregeana. **1907** T. R. Sim *Forests & Forest Flora Cape Good Hope* 128 *Kiggelaria africana*, Wild Peach, Natal Mahogany.. does best in open forest. *Ibid.* 161 This tree [sc. *Trichilia emetica*]..is known in the Transkei as Cape Mahogany, Manuti Mahogany, or Natal Mahogany. **1973** Palmer & Pitman *Trees S. Afr.* II. 1071 *Trichilia emetica* Vahl. Woodland mahogany, Natal mahogany... This is one of the widespread trees of Africa. **1859** R. J. Mann *Colony of Natal* viii. 158 The *Amatungulu* (Natal plum) is the berry of an evergreen periwinkle (*Vinca*) growing as a small shrub on the sea-coast lands. The fruit is about the size of a damson. **1876** H. Brooks *Natal* v. 168 A plant.. bearing a really valuable fruit which is familiarly known as the Natal plum. **1970** *Country Life* 17 Dec. 1230/3 Scarlet ixoras and the spiny carissa (Natal plum) are used as evergreen hedges. **1973** Palmer & Pitman *Trees S. Afr.* III. 1901 *Carissa macrocarpa*... Amatungulu, Natal plum..is a common and often conspicuous species in coastal bush, on sand dunes and on the edges of coastal forest... Although it is often a low bush, it can grow into a small tree up to 4 m high, many-branched, spiny, with dense evergreen foliage.

Natalian (natā·liăn), *a.* and *sb.* [f. *Natal *sb.*[2] + -ian.] **A.** *adj.* Of or pertaining to the province of Natal in South Africa. **B.** *sb.* A native or inhabitant of Natal.

1867 R. J. Mann in *Intellectual Observer* X. 186 In the year 1842,..the Dutchmen within the Natalian territory became subjects of the British Crown. **1870** A. Lindley *After Ophir* ii. 27 The Natalians carry their *penchant* for genteel and sonorous titles nearly as far as the Sicilians. **1878** [see *Jimmy 1]. **1897** J. Bryce *Impressions S. Afr.* xviii. 359 The Natalians have.. become the less energetic. **1928** R. Campbell *Wayzgoose* i. 10 Seldom do suns such striking talent show As when they set Natalian woods aglow. **1935** S. Desmond *Afr. Log* xxxix. 191 Throughout the day I see the lovely Natalian landscape roll past. **1960** *Guardian* 14 Oct. 12/5 Natalians are subjects of the Queen. **1973** *Sunday Times* (S. Afr.) 18 Feb. 2 An independent Natal has long been the dream of many White Natalians.

Nataraja (na:tără·dʒă). *India.* [Hindi, lit. 'prince of dancers' f. Skr. *naṭa* dancer, actor + *rājan* prince, king (see Raja).] A name of Siva, the Hindu god of creation and dissolution, in his role as lord of the dance, when he symbolizes cosmic energy. Also, a figure depicting Siva as lord of the dance.

1911 V. A. Smith *Hist. Fine Art India & Ceylon* vii. 249 The place of honour may be given to the spirited images of Siva as 'Natarāja', 'Lord of the Dance', the first of their kind to be found in Ceylon. *Ibid.* vii. 251 (*caption*) Bronze Siva Nataraja, 3 feet high. **1916** H. Krishna Sastri *S. Indian Images Gods & Goddesses* iv. 77 Natarāja ..is the well-known dancing form of god Siva. **1917** Coomaraswamy & Duggirala tr. *Mirror of Gesture* 15 The image of Śiva as dancer (Naṭarāja, Naṭeśa) and actor is everywhere conspicuous in Śaiva literature. **1918** A. Coomaraswamy *Dance of Śiva* 66 Explorers of the infinitely great and infinitely small, we are worshippers of Naṭarāja still. **1968** *Indian Mining & Engin. Jrnl.* V. 40 God in his aspect of Naṭarāja, or Lord of the Dance. **1972** P. Holroyde *Indian Mus.* i. 42 That first artist who fashioned the Nataraja into a symbolic equation of energy.

natatorium (nĕɪtătōᵊ·riǝm). *N. Amer.* [L. *natātōrium* a place for swimming.] A swimming-pool.

1890 in Webster. **1897** E. Bellamy *Equality* x. 60, I propose..that we go over to the natatorium and take a plunge. **1900** *Amer. Jrnl. Sociol.* Jan. 473 A considerable number of Wellesley college girls..patronize the Brookline natatorium. **1907** *Amherst College Catal.* 1907–08 104 In the natatorium, swimming is taught by a competent instructor. Every student who on entering college cannot swim, is required to learn the first year. **1911** *Daily Colonist* (Victoria, B.C.) 2 Apr. 2/4 Pearl Moore came to her death by drowning in a pool at the Washington natatorium. **1921** *Ibid.* 6 Oct. 1/5 Plans have been completed for the immediate construction of an up-to-date natatorium in Vancouver. **1966** *Punch* 21 Dec. 921/2 Only in Texas, perhaps, will a man in search of somewhere to swim pick up the Yellow Pages and look under Natatoriums; the directory refers him to Swimming Pools. **1971** *Parks & Recreation* (U.S.) VI. 31 This modern natatorium demonstrated that using a vinyl-coated nylon air-supported structure as the roof in the winter and completely removing it in the summer is a desirable and effective arrangement.

natch (nætʃ), *adv.* Colloq. abbrev. of Naturally *adv.* orig. *U.S.*

1945 in L. Shelly *Jive Talk Dict.* 15/1. **1946** *Sun* (Baltimore) 24 Sept. 3/3 (Advt.), Natch! Mom's gettin' both of us Hen self-starters... Baltimore's most popular 'first shoes for babies'. **1953** *New Yorker* 10 Jan. 21/3 Their disapproving papa..inserted a dime, which, he said, would otherwise have gone to buy them a Coca-Cola... Well, natch, this brought a twenty-dime return. **1957** P. Wildeblood *Main Chance* 132 'You don't mean to say,' she whispered tragically, 'that we're going to eat?' 'Why, natch. We're going to have another drink first, though.' **1962** J. Wain *Strike Father Dead* 60 We went in: Saloon Bar, natch. Lucille may have been Public Bar in her heart, but she was Saloon Bar on the outside. **1968** *N.Y. Times Bk. Rev.* 23 June vii. 1/3 The banning—in Boston, natch—of 'Fanny Hill' back in 1821. **1970** *Daily Tel.* (Colour Suppl.) 3 July 5/2 A crowd of opulent-looking masochists..were actually savouring this culinary surrealism and, natch, paying through the nose for it. **1971** A. Diment *Think Inc.* ii. 20 They blamed you, natch. **1973** *Times* 29 May 8/5, I went down to Bath to celebrate (along with the coronation of King Edgar, natch) the re-opening of the Assembly Rooms.

Natchez (næ·tʃez, nā·tʃī). Also **Nachee, Na'htchi.** [Native name.] A N. Amer. Indian people of Mississippi; a member of this people; also, their language. Also *attrib.* or as *adj.*

1775 J. Adair *Hist. Amer. Indians* 86 In the year 1747, a Nachee warrior told me, that while one of their prophets was using his divine invocation for rain..he was killed with thunder on the spot. **1845** *Cherokee Advocate* (Tahlequah, Okla.) 24 Apr. 4/3 The tradition has been widely recorded, that the dominion of the Natchez once extended even to the Wabash. **1848** *Trans. Amer. Ethnol. Soc.* II. i. 77 (*table*) Families... Natchez. **1850** R. G. Latham *Nat. Hist. Varieties Man* 340 Disinterred skulls from the Natchez area... Such is a brief notice of the customs of the Natchez. **1890** J. G. Frazer *Golden Bough* II. 382 The chief solemnity of the Natchez, an Indian tribe on the Lower Mississippi, was the Harvest Festival or the Festival of New Fire. **1891** *7th Ann. Rep. Bureau Amer. Ethnol. 1885–86* 96 The Shetimasha, the language of which is known from a vocabulary to be totally distinct..from the Na'htchi. *Ibid.* 97 The missionary..affirmed the affinity of the Taensa language to that of the Na'htchi... The Taensa language is considered to be a branch of the Na'htchi. **1949** M. Mead *Male & Female* 396 This was done..by certain women in Natchez Indian society. **1959** E. Tunis *Indians* 82/2 The Natchez were Muskhogeans who lived on the Mississippi River. **1965** *Canad. Jrnl. Ling.* Spring 99 An 'Eastern Group' comprising Siouan-Yuchi and Natchez-Muskogian. *Ibid.* 145 My Natchez informant..told about the great removal.

nation, *sb.*[1] Add: **4. b.** *two nations*: phr. used of two groups within a given nation divided from each other by marked social inequality; hence *one nation*, a nation which is not divided by social inequalities.

1845 Disraeli (*title*) Sybil, or The two nations. *Ibid.* I. ii. v. 149 Two nations; between whom there is no intercourse and no sympathy; who are..ignorant of each other's habits, thoughts, and feelings..; who are formed by different breeding, are fed by different food,.. and are not governed by the same laws... The Rich and the Poor. **1892** *Youth's Compan.* 8 Sept. 446/1 (*heading*) Salute to the Flag... I pledge allegiance to my Flag and the Republic for which it stands: one Nation indivisible, with Liberty and Justice for all. **1971** *Guardian* 19 Nov. 12/2 There are two nations now within schools... Social inequality is growing. **1973** *Times* 15 Oct. 17/6 The Disraelian doctrine of 'One Nation' has..been in the past little more than an ideal or a pretence. **1974** *Times* 16 Oct. 14/6 God bless the squire and his relations; full speed astern to the Two Nations.

5. d. (Earlier examples.)

1650 *Archives of Maryland* (1883) I. 260 The Ports adjoyning are very much pestered with great Concourse of

Indians of several nations. **1709** J. Lawson *New Voy. Carolina* 199 Two Nations of Indians here in Carolina were at war together. **1722** D. Coxe *Descr. Carolana* iv. 49 Near the Bottom of the Bay..is the fair River of the Miamihas (so call'd because upon it lives Part of a Nation bearing the same Name). **1740** in *South Carolina Hist. Soc. Coll.* (1887) IV. 83, I desire also that you will send me..the Indian presents, with power to distribute them, for much Depends on the Nations.

9. *attrib.* and *Comb.* (see also sense 1 a *ad fin.*), as **nation-building,** the creation of a new nation, *spec.* a newly independent nation; hence **nation-builder; nation-state,** a sovereign state the members of which are also united by those ties such as language, common descent, etc., which constitute a nation; **nation-wide** *a.*, as wide as a nation; extending over, reaching, or affecting the whole nation; also as *adv.*

1907 *Collier's* 12 Jan. 7/2 Next week's issue will be our annual Automobile Number, and, in addition to general news and illustrations concerning the modern vehicle, it will give some lucid arguments for the automobile as a nation-builder. **1920** N. M. Butler *Is Amer. Worth Saving?* xvi. 285 (*heading*) Alexander Hamilton, nation-builder. **1933** P. S. Cleary (*title*) Australia's debt to Irish nation-builders. **1967** *Freedomways* VII. 167 These are the new lessons in old African history that are giving many present-day African nation builders a new consciousness of past achievements. **1913** N. M. Butler in *Educational Rev.* (N.Y.) Apr. 405 These six men are.. the moving forces of the constructive nation-building of the American people. *Ibid.* 406 The most prominent in the galaxy of our nation-building heroes. **1928** *Daily Tel.* 4 Sept. 10/4 For such an enterprise of nation-building peace..is the essential condition. **1931** *Economist* 7 Mar. 486/1 One of the urgent needs of India is that the Provinces should have funds available for so-called 'nation building' services. **1971** *Sunday Nation* (Nairobi) 11 Apr. 3/2 The Ambassador advised the students to study hard and return home after graduating to play their part in nation-building. **1973** *Express* (Trinidad & Tobago) 7 Apr. 12/4 Unless we are prepared to..rid ourselves of our petty differences and general smallmindedness, we are not ready for nation building. **1918** J. A. R. Marriott *European Commonwealth* ii. 18 The ultimate genesis of the world conflict of to-day is sought..in..the existing European polity..based upon the recognition of the rights of a large number of Nation-States, entirely independent and nominally coequal. **1935** Huxley & Haddon *We Europeans* i. 11 The nation-state is a modern conception and product, the result of certain peculiar social and economic circumstances. *Ibid.* vi. 187 With the sixteenth century, nation-states of the modern type began to appear. **1945** H. Kohn *Idea of Nationalism* i. 19 Nationalism demands the nation-state; the creation of the nation-state strengthens nationalism. **1950** Theimer & Campbell *Encycl. World Politics* 301/2 Absolutism paved the way for the modern nation-state marked by sovereignty and the repudiation of any superior authority. **1959** *Encounter* July 75/1 Egypt is turning into a modern nation-state. **1971** *Black Scholar* June 29/1 As the cultural revolution and students become more politically sophisticated, the question of an independent black nation-state will become a popular demand. **1973** *Listener* 10 May 616/1 The nation state requires the idea of an entity along with other entities that are equal. **1915** *Munsey's Mag.* May 708/2 The nation-wide primary is coming. **1925** E. S. Jones *Christ of Indian Road* iii. 72 A year ago began a struggle in South India that has had nation-wide consequences. **1928** *Daily Express* 27 Apr. 1 In deciding to publish this most moving..narrative the 'Daily Express' embarks on a venture which will command nation-wide attention. **1958** *Economist* 1 Nov. 413/1 The one real issue of nation-wide importance which this campaign has produced concerns the trade unions. **1960** *Farmer & Stockbreeder* 15 Mar. 72/1 Perhaps F.M.C. could have done more if it hadn't started off on so wide a scale—nation-wide in fact. **1972** *Daily Tel.* 21 Jan. 1/7 Mr. Ian Smith, the Rhodesian Prime Minister, is to make a nationwide broadcast today. **1975** *Lamp* (Exxon Corporation) Winter 3/2 Now being telecast nationwide..each of these television messages pays tribute to an outstanding man or woman.

nation, *adv.*, *a.*, and *sb.*[2] Add: **A.** *adv.* (Earlier and later examples.)

1771 *Trial of Atticus* 26 He is a nation bawdy creature to talk. **1884** 'Mark Twain' *Huck. Finn* xix. 186 Looky here, Bilgewater,..I'm nation sorry for you, but you ain't the only person that's had troubles like that.

B. *adj.* (Earlier and later examples.)

1765 in Bartlett *Dict. Amer.* (1877) (ed. 4) 419, I believe, my friend, you're very right, They'll get a nation profit by it. **1962** A. Jobson *Window in Suffolk* vi. 100 What a nation fule he wur tew be shure.

C. *sb.* (U.S. examples.)

1775 *Yankee Doodle* in O. E. Winslow *Amer. Broadside Verse* (1930) 141/1 A swamping gun..makes a noise like father's gun, Only a nation louder. **1880** *News & Press* (Cimarron, New Mexico) 23 Dec. 1/7 'Well, I've got the g.b.' 'The geebee, Thomas! What in the nation is that?' 'I've got the grand bounce.' **1884** 'Mark Twain' *Huck. Finn* xiii. 113 Why, how in the nation did they ever git into such a scrape? **1918** J. C. Lincoln *Shavings* 213 Now how in the nation did I get it Wood?

national, *a.* and *sb.* Add: **A.** *adj.* **1.** (Further examples.)

1743 Bolingbroke *Remarks Hist. Eng.* (1780) ii. 28 A Spirit of Liberty will be always and wholly concern'd about national Interests. **1914** G. B. Shaw *Dark Lady of Sonnets* 132 The very stupid people who cannot see that a National Theatre is worth having for the sake of the National Soul. **1927** 'A. Hope' *Memories & Notes* xiv. 239, I went to Boston in company with Peter Dunne, whose 'Mr. Dooley' had already made him what the papers in the States call a 'national figure'. **1931** *Hansard Commons* 28 July 2217, I should take no exception to a small sum towards the preservation of such things as Hadrian's Wall and the other national monuments. *c* **1935** E. E. Cummings *Let.* 7 Jan. (1969) 134 Here lies a national hero (Who governed by fits and by starts). **1939** *Hansard Commons* 2 Sept. 232 In the national interest, having in mind the great inroads that war must..make on the youth of our nation, we should see.. that some of our youth are left to carry on in the future. **1966** Tacheron & Udall *Job of Congressman* vii. 198 And then, you see, some of your friends back home say, 'Why, he is becoming a national figure.' **1966** *Oxf. Compan. Amer. Hist.* 560/1 National monuments include buildings, statues, homesteads, battlefields, cemeteries, and sites of historic, scenic, or political significance. **1969** Plano & Olton *Internat. Relations Dict.* vi. 128 When a state bases its foreign policy solely on the bedrock of national interest with little or no concern for universal moral principles, it can be described as pursuing a realistic in contradistinction to an idealistic policy. **1973** *Times* 17 Dec. 15/3 The NUM..cannot be totally indifferent to the national interest. **1973** *N.Y. Times* 22 June 35 One of the most successful codebreaking organizations of those post-war years was the National Security Agency. **1975** *Atlantic Monthly* Jan. 32/2 The Department of Defense could give no adequate definition of 'national security'.

d. Designating a B.B.C. radio service which operated during the 1930s. Also *ellipt.*

1931 *Daily Express* 22 Sept. 13/3 Tonight..on the National begin..the new poetry readings. **1938** W. Goatman *By-Ways of B.B.C.* viii. 81 Control-Room engineers are responsible..for controlling three separate programmes—National, Regional, and Empire. **1956** B. Paulu *Brit. Broadcasting* vii. 145 From 1930 to the outbreak of World War II, there was the National Service, uniform throughout the country, consisting mainly of London programs of national appeal. **1965** A. Briggs *Golden Age of Wireless* 36 By 1934..there was more light entertainment on the National than on the Regional service. **1971** D. G. Bridson *Prospero & Ariel* viii. 177 Back in the thirties, there had been two nation-wide broadcast programmes, the National and the Regional.

e. In the war of 1939–45 used to designate foodstuffs made to official specifications for nation-wide distribution, as *national flour, loaf*, etc.

1940 *Hansard Commons* 25 Jan. 810/2 All butter sold by the Ministry of Food is..described as 'National Butter'. **1941** *New Statesman* 16 Apr. 157 The bakers are equally reluctant to change their habits. They shake their heads with a mournful smile at mention of the national loaf. **1941** *Lancet* 15 Nov. 605/1 When the Medical Research Council drew up its second memorandum on national flour in the spring of 1941 it offered the nation something new. **1943** *Daily Tel.* 20 Dec. 3 National Milk-Cocoa—the new food drink evolved by the Ministry's scientific advisers—is already available to all factories and industrial undertakings in limited quantities. **1945** *ABC of Cookery* (Ministry of Food) vi. 20 National Household Milk is dried skim milk. *Ibid.* xix. 69 The best foods for iron are liver, eggs,..national or wholemeal bread and oatmeal. **1956** W. Thompson *Time off my Life* xix. 146, I couldn't stand the stuff they called 'national loaf' in the war.

2. spec. *national football* (Austral.) = *Australian rules; national game* (see quots.).

In quots. 1930 and 1963 the game is two-up.

1828 *Oscotian* (ed. 2) I. 35 The day or two previous.. gave employment for our youth, in adjusting the arrangements, necessary for the noble and national game, of Cricket. **1869** W. G. Beers (*title*) Lacrosse: the national game of Canada. **19..** *Primer Austral. Football* (Austral. National Football Council) 1 Youngsters who are keen to learn and understand the fundamentals of their National Football. **1912** Hilton & Smith *Royal & Anc. Game of Golf* 34 The Scots themselves have yielded to the softening influences of the South..and the change is reflected in their national game. **1930** L. W. Lower *Here's Luck* 70 He had a small piece of wood in his hand, on which were balanced two pennies. The national game was in progress. **1958** National football [see *Australian rules*]. **1963** F. Hardy *Legends from Benson's Valley* 108 The demise of the bookmakers increased attendance at the Sunday sessions of Australia's national game.

5. National Assistance, a form of welfare payment combining Unemployment Assistance and Public Assistance, begun in 1948, administered by the **National Assistance Board,** and replaced in 1965 by Supplementary Benefits; **national bank** (earlier example); **national cake** (see *cake *sb.* 7 b); **national character,** personality or cultural characteristics which seem to be wide-spread enough in a particular nation or racial group for generalizations to be made concerning either the whole group or individuals belonging to it; so **national characteristic; National Front,** a political group in Britain with extreme reactionary views; **National Government,** a coalition government, esp. one in which party differences are subordinated to the national interest in times of crisis; also in the sense of a government free from external domination (see quot. 1943); **national guard** (further examples); also (with capitals), in the United States, a militia force which may be used by its own state, e.g. for law enforcement, or by the Federal government as part of the U.S. army; **national health,** health as it concerns the nation as a whole; **National Health Service,** the comprehensive health service provided in Great Britain, initiated in 1946 and financed by taxation; freq. *ellipt.* as *National Health*; **National Hunt Committee,** the body which controls steeplechasing and hurdle-racing in Great Britain; freq. *ellipt.* and *attrib.* as *National Hunt*; **national income**: see *income *sb.*[1] 6 a; **National Insurance** = *insurance 4 e; also *attrib.*; **National Mark,** a mark designating grade for use on British agricultural produce; also *attrib.*; **national minority,** a minority group, belonging historically to another nationality, which feels itself or is felt to be culturally or racially separate from the majority in a country; **national park** = Park *sb.* 2 b; **national product,** the monetary value of all goods and services produced in a country in one year; cf. *gross national product* (s.v. *gross *a.* and *sb.*[4] A. 6 c); **National Savings,** a method of saving through investment in British Government securities, started in the war of 1914–18 as National War Savings; also applied to similar schemes in other countries; so *National Savings Certificate*, etc.; **national school,** a school provided for under a system of state-aided education, esp. one of the type set up in Ireland after 1831 under the National System of Education; **national service,** a statutory obligation to serve in the armed forces for a specified period; hence **national serviceman,** one who is performing national service; **National Socialism,** the name adopted by Adolf Hitler for his doctrines of nationalism, racial purity, anti-Communism, and the all-powerful role of the State; so **National Socialist,** a member of the National Socialist Workers' Party led by Adolf Hitler after 1920; = *Nazi *sb.*; also *attrib.*, of or pertaining to this party or to the doctrines of National Socialism; **national theatre** (freq. with capital initials), a theatre endowed by the State; also ellipt., *the National*; **National Trust,** a trust for the preservation of places of historic interest or natural beauty in England, Wales, and N. Ireland, founded in 1893, incorporated in 1907, and supported by endowment and private subscription. Also *national grid.

1948 *Act 11 & 12 Geo. VI* c. 29 §2 (*title*) National Assistance Act, 1948. *Ibid.*, The National Assistance Board (hereafter in this Act referred to as 'the Board') shall exercise their functions in such manner as shall best promote the welfare of persons affected by the exercise thereof. **1958** *New Statesman* 27 Sept. 398/3 Few people are at a worse disadvantage than the man who is fresh from prison and is also homeless. National Assistance methods..must sometimes seem designed to assist such a man back into prison in the shortest possible time. **1959** J. Braine *Vodi* viii. 124 There isn't any dole for him. It's bankruptcy first and then National Assistance. **1968** J. Lock *Lady Policeman* xviii. 152 The National Assistance Board telephoned. **1969** *Listener* 12 June 834/3 The women he has studied are the very poorest, depending almost entirely on National Assistance. **1838** *Democratic Rev.* I. 52 That portion of the plan.. which involved a present non-committalism on the question of a National Bank. **1778** National character [in Dict., sense A. 2]. **1863** *Home & Foreign Rev.* III. 549 Next to those who form the national taste and fix the national character, the greatest geniuses are those who corrupt them. **1908** Kipling *Lett. of Travel* (1920) 144 'Hustle' does not sit well on the national character. **1967** M. Argyle *Psychol. Interpersonal Behaviour* iv. 82 If the people from some cultural group are consistently aggressive, [etc.].., we *can* say that their norms are different; but it may be more useful to say that their level of motivation is different, and to regard this as a feature of their 'national character'. **1960** Duijker & Frijda *National Character* i. 4 The existence of a stereotype might give rise to a national characteristic. **1967** *Spearhead* Nov.–Dec. 6 (*caption*) On October 7th there took place..an event which may well prove to have historic significance in British politics. This was the first Annual General Meeting of Britain's new party, The National Front. *c* **1970** *Facts: National Front* 2/3 When the National Front comes to power it will know well how to deal with the murderer and the thug. **1973** C. Mullard *Black Brit.* iii. viii. 91 White fascist groups like the National Front are lumped together with progressive, power-demanding groups like the Black People's Alliance. **1931** J. Reith *Diary* 24 Aug. (1975) i. 106 A wire..informed me that the Labour government had resigned, that MacDonald was prime minister of a National government. **1931** *Economist* 17 Oct. 692/1 Sir John Simon..stated that..'if the broad base National Government finds it [*sc.* a tariff] necessary, I am not going on that ground to refuse to support the National Government.' **1936** J. A. Spender *Gt. Brit.: Empire & Commonwealth* IV. xliii. 512 National Government [in 1915] was the logical corollary of the party-truce. **1943** *Ann. Reg. 1942* 143 With the security of India threatened, we.. should bring her people into full moral support..by conceding forthwith the demands for a 'National Government', with virtual independence of British authority.

1969 T. A. Neal *Democracy & Responsibility* xviii. 223 Strong arguments may be marshalled to explain MacDonald's formation of the National Government. *Ibid.* xix. 230 In November of the same year [*sc.* 1935], the National Government was re-elected, with a total of 428 supporters in the House of Commons. **1793** *Observer* 27 Jan. in M. Miliband *Observer of 19th Cent.* (1966) i. 4 The train moved on with a slow pace from the Temple to the Boulevards, which were planted with cannon, and beset with National Guards, drums beating, trumpets sounding, and colours flying. **1847** *Santa Fé* (New Mexico) *Republican* 18 Dec. 3/1 Some National Guards that were at San Antonio had a small fight. **1868** *N.Y. Herald* 4 July 6/1 The First division of the National Guard will parade. **1909** *World Today* Oct. 1099 While many of the American states designate their amateur soldiery as 'National Guards', the venerable Bay State still sticks to the thoroughly Yankee caption of 'militia'. **1966** *Oxf. Compan. Amer. Hist.* 530/1 *Militia*, now officially termed the National Guard in the U.S., is the body of armed forces within the states, formed by enlistments. **1973** *Freedomways* XIII. 14 The forces of state power (the police, the army, the National Guard) will be used. **1908** *National Health* I. 1. 18/2 (*heading*) The Women's National Health Association of Great Britain. **1911** Lloyd George *Insurance of the People* 2 A few weeks ago I had the honour of introducing in the House of Commons a measure dealing with proposals for securing the national health. **1935** *Economist* 7 Sept. 456/2 The most important event..in the field of health during the reign was the creation of the scheme of National Health Insurance in 1912. **1946** *Act 9 & 10 Geo. VI* c. 81 (*title*) National Health Service Act. **1952** A. Christie *Mrs McGinty's Dead* xii. 96 Nowadays even if you've got a chilblain you run to the doctor with it so as to get your money's worth out of the National Health. **1958** J. Cannan *And be a Villain* i. 23 A doctor w'at was too taken up by 'is posh paying patients to trouble with a National 'Ealth kid. **1958** R. Graves in *Times Lit. Suppl.* 15 Aug. p. x/5 What is even worse, any slight mania or differentiating oddness..they now take along to the National Health psychiatrist, pleading to be de-thinged. **1958** *Listener* 30 Oct. 699/2 The workings of the National Health Service. **1966** J. S. Cox *Illustr. Dict. Hairdressing* 102/1 *National Health wig*, a wig or other similar postiche of necessity supplied by a Wigmaker for a client whose health is said to be affected by lack of hair, to the order of the Ministry of Health. **1967** M. Argyle *Psychol. Interpersonal Behaviour* ix. 174 This [*sc.* counselling] is the most widely practised kind of psychotherapy—it is given by National Health Service psychiatrists in England. **1967** *Melody Maker* 27 May 5 With his bushy sideboards and National Health specs he resembled an animated Victorian watchmaker. **1973** J. Porter *It's Murder with Dover* ii. 13 Gritting his National Health teeth, [he] limped off. **1866** C., J., & E. Weatherby *Racing Calendar for 1866. Races Past* p. vi, Advertisements and reports of Steeple Chases under the Grand National Hunt Rules are included in the Sheet Calendar. **1873** C., J., E., & J. P. Weatherby *Racing Calendar, Steeple Chases Past, for Season 1872–73* p. xiii, Any person running a horse in contravention of this Rule shall (at the discretion of the Grand National Hunt Committee) be disqualified from ever running a horse where these rules are in force. *Ibid.* p. xxv, The following form of certificate is that approved by the National Hunt Committee. **1898** A. E. T. Watson *Turf* ix. 173 Horses are usually put to jumping because for some reason or another their career on the flat has ceased to look promising, it having been so continually proved that failures under Jockey Club Rules were brilliant successes under the Rules of the National Hunt. **1902** *Encycl. Brit.* XXIX. 333/1 None other than thoroughbred horses are nowadays ever found in races run under the rules of the National Hunt Committee, the body which governs the sport of steeplechasing. **1935** *Encycl. Sports* 590/1 National Hunt Rules fix the opening of the steeplechasing season as July 1. **1963** Bloodgood & Santini *Horseman's Dict.* 152 *Point-to-point*, cross-country race for qualified hunters..ridden by amateurs.. over a flagged natural hunting country; under the jurisdiction of the National Hunt (England). **1967** *Everyman's Encycl.* VI. 482/2 In 1866, as a result of the efforts of Lord Suffolk, Lord Coventry and the duke of Beaufort, the National Hunt Committee was formed as the authoritative governing body over steeplechasing. **1975** *Oxf. Compan. Sports & Games* 496/1 Over the years, the National Hunt season has gradually become longer, and the 1973–4 season began early in August and will end on 1 June. **1878** Blackley in *19th Cent.* Nov. 834 (*title*) National Insurance: A cheap, practical and popular means of abolishing poor rates. **1911** National insurance [see *insurance 4 e]. **1913** *Q. Rev.* Apr. 510 The labourer has gained something in a pecuniary sense from.. national insurance. **1922** *Encycl. Brit.* XXX. 998/2 A few doctors..had started a new organization in opposition to the British Medical Association, called the National Insurance Practitioners Association. **1931** *Times Lit. Suppl.* 20 Aug. 626/3 The indebtedness of the national insurance fund has already risen. **1967** E. Rudinger *Wills & Probate* 60 Matthew's father-in-law had also been receiving the state old age pension, or national insurance retirement pension, as it is properly called. **1968** *Brit. Med. Bull.* XXIV. 207/1 National Insurance records giving the dates on which patients left and returned to work. **1973** *Times* 13 Jan. 19/7 The National Insurance Scheme covers benefits payable by the State (through the Department of Health and Social Security) and covers pensions, widows' benefits, maternity benefits and sickness and unemployment benefits, together with certain minor benefits. **1928** *Times* 8 Aug. 12/3 The Minister of Agriculture and Fisheries has appointed a Committee, to be known as the National Mark Committee,..empowered to authorize the use of grade designation marks. *Ibid.*, An artistic and striking 'National Mark' has been designed,..a silhouette map of South Britain with a circle inset bearing the words around the margin, 'Produce of England and Wales'. **1929** *Daily Express* 7 Nov. 3/4 There has been a greatly increased demand for British beef since the National Mark scheme was introduced. **1931** *Times* 17 Nov. (Suppl.) p. xiv/1 One mark, the National Mark, which is a trade mark registered in the name of the Minister of Agriculture, is being used for all graded produce of England and Wales. **1934** *Daily Tel.* 31 Dec. 14/4 The National Mark Scheme began eight years ago to encourage home producers. **1937** *Food Manufacture* Oct. 340 The National Mark Scheme for Fruit Products..has now been extended to include.. fruit juices. **1921** R. W. Seton-Watson in H. W. V. Temperley *Hist. Peace Conf. Paris* IV. 266 National minorities shall enjoy equal rights. **1934** C. A. Macartney (*title*) National states and national minorities. **1945** E. E. Brooke tr. *Azcárate y Florez's League of Nations* p. vii, When those responsible for reconstructing Europe find themselves once again confronted by the difficulties which national minorities have created in the past. **1970** V. Van Dyke *Human Rights, U.S. & World Community* ii. v. 98 A recognition of 'the right of members of national minorities to carry on their own educational activities'. **1868** J. D. Whitney *Yosemite Bk.* i 22 The Yosemite Valley..has been made a National public park and placed under the charge of the State of California. **1871** [see Park *sb.* 2 b]. **1872** *Publ. Colonial Soc. Mass.* VIII. 377 Congress, by an Act approved March 1, 1872, has set apart a tract of land near the head-waters of the Yellowstone River... The reservation so set apart is to be known as the 'Yellowstone National Park'. **1903** [see Park *sb.* 2 b]. **1933** *Discovery* Feb. 68/2 The creation of the great American national parks..has saved the bison from extinction. **1949** *Act 12 & 13 Geo. VI* c. 97 An Act to make provision for National Parks and the establishment of a National Parks Commission. **1950** R. N. Hutchins *National Parks & Access to Countryside Act* i. i. 3 New Zealand..has 10 National Parks and over 200 reserves and public domains. *Ibid.*, In England and Wales, the designation of a National Park will not affect the ownership of the land,..nor will any additional facilities for public access be made available merely by such definition. **1962** *Listener* 1 Mar. 373/1 The Planning Board or committee in charge of a national park has a statutory duty.. to 'preserve and enhance' the natural beauty of the landscape. **1974** *Times* 1 Apr. (Yorkshire & Humberside Suppl.) p. ix/1 Within reasonable driving distance of the Humber the choice lies between three national parks, a noble sprinkling of country estates and several attractive cathedral cities. **1945** S. Kuznets *National Product in Wartime* i. ii. 7 Estimates of peacetime national product assume that economic activity is to produce goods to satisfy ultimate consumers; that production is for man, not man for production. **1962** *Listener* 20 Dec. 1041/1 The French national product rose at a rate of no less than 8 per cent. a year. **1964** Gould & Kolb *Dict. Social Sci.* 454/1 If a series of national product estimates for several years is divided by a price index, each year's national product being divided by the price index for that year, the resulting series is known as *deflated* or *real* national product. [**1917** (*title*) National War Savings Committee... First annual report, in *Parl. Papers 1917–18* (Cd. 8516) XVIII. 703.] **1919** *Saving* 3 Dec. 138/2 The Model Schemes suggested by the National Savings Committee are of various forms. **1922** *Encycl. Brit.* XXXI. 369/2 The system of linking up National Savings certificates with local finance becomes, in effect, a national credit bank spread over the whole country. **1932** *Discovery* Nov. 374/2 The much abused Post Office ..has actually allowed it [*sc.* the Institute of Industrial Psychology] to redesign the National Savings Certificate. **1941** *Picture Post* 3 May 35/3 Buy National Savings Stamps—they cost 6d. and 2/6 each...When you have the right amount you convert them into National Savings Certificates—15/- each. **1964** A. Wykes *Gambling* x. 238 Offering a lottery ticket worth one mark..for every eight marks deposited in national-savings accounts. **1972** *Times* 21 Nov. 19/6 Historically national savings have been treated as a poor relation to other investment media. **1838** *Digest Evidence before Comm. Houses Lords & Commons, 1837, on National Syst. Educ. Ireland* ii. 25 She was highly gratified; indeed she was astonished; she could not have believed that in the national school the Bible was used at all. **1966** *New Statesman* 27 May 774/3 But certain schools, to be called perhaps National Schools, should be placed directly under the Department of Education. **1968** T. Kinsella *Nightwalker* 65 But the authorities Used the National Schools to try to conquer The Irish national spirit. **1972** M. E. Collins *Ireland Three* vi. 49 The National Schools were free, but not compulsory... The National Schools, though in theory of mixed religion, were in practice, separate, and children of each religion went to their own schools. **1916** R. Fry *Let.* 25 Jan. (1972) II. 393, I don't think Ha talked to me of national service. **1916** *Hansard Commons* 19 Dec. 1352 It is proposed to appoint at once a director of National Service, to be in charge of both the military and civil side of universal national service. **1939** *Ibid.* 2 Sept. 240 National Service (armed forces) Bill, 'to make provision for securing and controlling the enlistment of men for services in the armed forces of the Crown; and for purposes connected with the matter aforesaid', presented, accordingly, read the First time; and ordered to be printed. **1940** *Ann. Reg. 1939* 397 The National Service (Armed Forces) Act, which rendered liable for service in the forces all male persons between eighteen and forty-one. **1940** H. G. Wells *Babes in Darkling Wood* i. i. 21 Romeo and Juliet weren't called on for national service. **1942** G. G. Slack *Liability for National Service* i. i. 1 The Government powers of control over persons are, in relation to their liability for National Service, exercised by the Minister of Labour. **1944** J. S. Huxley *On Living in Revolution* 21 It is both probable and desirable that some form of National Service will continue after the war is over. **1958** M. Kerr *People of Ship Street* iii. 29 The boys in between 18 and 20 are of course doing National Service. **1970** M. Moorcock *Chinese Agent* xiii. 87 He had avoided being called up for his National Service for two years and, when he had been called up, he'd managed to desert. **1949** *Times* 20 Oct. 5/5 The Secretary of State for War gave an assurance that no national service man would be posted to the Far East with less than 18 weeks' service. **1957** *B.B.C. Handbk.* 170 British National Servicemen. **1957** C. Hunt *Guide to Communist Jargon* vii. 19 Those who have taken up soldiering as a profession as opposed to National Servicemen, who have not. **1973** D. Lees *Rape of Quiet Town* vii. 113 A number of former national service men who had never heard a shot fired in anger. **1931** *Times Lit. Suppl.* 16 Apr. 296/3 Such is the doctrine, not inaptly named National Socialism, which by uniting Moscow and Paris in a common anathema has appealed at once to the quiet steady man and to the boisterous patriot. **1933** J. J. Bronowski in *Granta* 19 Apr. 358/2 National Socialism pictures itself as persecuted, not only by the political world but by art, science and philosophy. **1923** *Times* 23 Aug. 9/6 At the conclusion of a National Socialist meeting last night Herr Hitler's storm troops.. attempted to march through Munich. **1931** *Daily Tel.* 21 May 13/1 The National-Socialists (Hitlerites) are now definitely committed to a Hohenzollern restoration. **1933** J. J. Bronowski in *Granta* 19 Apr. 358/1 The National Socialist revolution in Germany is most deeply founded, has its most bigoted and best organised support in the German universities. **1942** E. Barker *Reflections on Govt.* xiii. 369 We have here the ultimate roots of the National Socialist doctrines of the Leader and the Folk which he leads and expresses. **1942** L. B. Namier *Conflicts* 40 The depression..everywhere brought new political forces to the surface, violent and brutal—National Socialists and National Radicals. **1967** D. Eisenberg *Re-emergence Fascism* i. 59 The judge..hit out at Colin Jordan's National Socialist movement. **1973** *Black Panther* 27 Oct. 8/3 On another occasion the National Socialist Party, also known as the 'White People's Party', passed out literature in front of her house pledging to 'get rid of all the niggers'. **1889** G. B. Shaw *London Music 1888–89* (1937) 150 Miss Reimers.. upset even my gravity..by asking me..why the splendid, the intellectual, the free English people had no national theatre. **1901** Beerbohm *Around Theatres* (1924) I. 245 The offer made (or *not* made) by Mr. Carnegie to endow with us a national theatre has duly revived that periodic cry to which I have alluded. **1928** G. Hughes *Story of Theatre* xiv. 215 The Burgtheater [in Vienna]..had been founded in 1741 as a Court Theatre..and changed in 1776..into a National Theatre (in imitation of the Comédie Française). **1961** E. Waugh *Unconditional Surrender* 306 The foundation stone was solemnly laid for a National Theatre. **1972** *Listener* 18 May 644/3 The new National Theatre rising on the South Bank. Until the coming of the National, the British theatre had survived..without the alien office of dramaturgs. **1974** *Country Life* 24 Jan. 125 The National has been running successfully for ten years and plays generally to good houses. **1893** *Times* 17 Nov. 9/6 'The National Trust for Places of Historic Interest or Natural Beauty' is an association which held its first meeting yesterday. **1894** *Spectator* 7 July 12/2 Men who love their glens and their hill-tops..would far rather see them go to the National Trust than be put up to auction. **1896** O. Hill *Let.* 26 Oct. in C. E. Maurice *Life O. Hill* (1913) x. 538, I suppose you do not happen to know any gentleman likely to do for, and accept, our National Trust secretaryship? **1907** *Hansard Commons* 21 Aug. 758 The following Bills received the Royal Assent:—..14. National Trust for Places of Historic Interest or Natural Beauty. **1930** J. S. Huxley *Bird-Watching* vi. 115 And he [*sc.* the bird-lover] can help by supporting such bodies as the National Trust and the Royal Society for the Protection of Birds, which are saving wild bits of country from being built over or otherwise developed, or reserving them as actual sanctuaries, inviolate to the birds. **1934** *N. & Q.* 16 June 415/2 East Riddlesden Hall..has just been publicly handed over to the National Trust. **1938** D. Garnett in T. E. Lawrence *Lett.* 873 His cottage at Clouds Hill now belongs to the National Trust. **1958** *Listener* 9 Oct. 556/2 On the preservation of historic buildings many voices must obviously be heard:..art-historians and archivists, the National Trust. **1971** P. Gresswell *Environment* 166 The National Trust..was founded..by Octavia Hill, Sir Robert Hunter and Canon Rawnsley.

B. *sb.* **4. b.** (See quot. 1904.)

1904 J. Westlake *Internat. Law* I. 3 All the members of a state, whether sovereign, subjects or citizens, are denoted by the convenient name of its nationals. **1929** *Times* 31 Jan., The official thanks of the German Government..for the rescue by aeroplane of their nationals in Kabul. **1931** *Times* 22 Sept. 7/4 There is no evidence of any substantial export of capital by British nationals. **1953** *Stroud's Judicial Dict.* (ed. 3) III. 1855 In English courts phrases which refer to the national law of a propositus are prima facie to be construed not as referring to the law which the courts of that country would apply in the case of its own nationals domiciled in its own country with regard..to property in its own country. **1955** *Times* 3 June 6/2 The two Governments have agreed to take measures for the conclusion of a treaty for the purpose of settling questions of citizenship, and with regard to the repatriation of nationals of one contracting party residing in the territory of the other party. **1969** Plano & Olton *Internat. Relations Dict.* xi. 266 *Statelessness*, the condition of an individual who is not recognized by any state as one of its nationals.

5. = *Grand National*: see *grand *a.* 12.

1909 *Westm. Gaz.* 26 Mar. 12/1 A horse that had never run a National. **1931** *Times* 12 Mar. 5/1 Several National horses have been entered at Hurst Park. **1972** L. P. Davies *What did I do Tomorrow?* ii. 26 The only time he backed horses was for the Derby or the National.

6. A national as opposed to a local newspaper; usu. *pl.*

1938 C. Hunt *You want to be a Journalist?* iii. 29 It is when staff increase—on county papers, dailies published from the great cities, London nationals—that specialisation develops. **1960** H. L. Lawrence *Children of Light* v. 68 Some of the nationals had rushed their own men to Keniston, but a flattering number had wired Johnny for coverage. *a* **1966** M. Allingham *Cargo of Eagles* (1968) i. 19 She must have sold the idea to the local papers because the nationals picked it up a year or two back. **1966** *Listener* 28 July 129/3 In thirty years' time there could probably be only four nationals left. **1973** J. Porter *It's Murder with Dover* iv. 36 We'll make the front page of the nationals with this!

national grid. [f. National *a.* + Grid.]

1. The grid (sense *8) that interconnects the major power stations and distribution centres in Great Britain; any similar grid in another country. Also *transf.*

1930 *Times* 22 Mar. 19/2 There will be no great rush of new business consequent on the completion of the national 'Grid' system, as was contemplated in some quarters. 1943 [see *GRID 8 b]. 1955 [see *GRID 8 a]. 1967 *Times* 13 Dec. 4/2 Mr. Wilson had said that comparative calculations were not possible for the gas industry, because there was not yet an equivalent of the national grid.

2. The metric co-ordinate system and reference grid used by the Ordnance Survey and printed on its maps, having a false origin west of the Isles of Scilly and a true origin at 2°W., 49°N.

1938 *Final Rep. Dept. Comm. Ordnance Survey* (Ministry of Agric. & Fisheries) 4 We recommend that a National grid should be super-imposed on all large-scale plans, and on smaller scale maps, to provide one reference system for the maps of the whole of the country. 1952 *Proc. Prehist. Soc.* XVIII. 3 The National Grid position on the 6-inch map Sheet XLIV, N.W. is 832642. 1963 *Atlas of Britain* (Clarendon Press) map 1 The National Grid is used in this Atlas with the sanction of the Director General, Ordnance Survey and of H.M. Stationery Office; the Irish Grid is used with the sanction of the Ordnance Survey of Northern Ireland. 1969 C. B. M. LOCK *Mod. Maps & Atlases* i. 32 The Ordnance Survey maps are now prepared on the Transverse Mercator, which enables the National Grid reference system to be easily operated. 1971 *Nature* 5 Feb. 375/1 A combination of these two surveys showed that before 1900 forty-four species..occurred in only one or two 10 km squares of the national grid; by 1930 the number was fifty-nine species.

nationali·stically, *adv.* [f. NATIONALISTIC *a.*: see -ICALLY.] In a nationalistic manner; on nationalistic lines.

1913 H. W. ROBINSON *Relig. Ideas Old Testament* 32 The redemption is differently conceived and nationalistically applied. 1924 *These Eventful Years* (Encycl. Brit.) II. 126 The nationalist problem is renewing itself in the succession states, which are nationalistically as varied as was Old Austria. 1951 *Current Hist.* Sept. 141/2 The nationalistically minded Japanese might be easily made to believe that all the West has done in Japan in the past is nothing but a plot to use the Japanese for its own purposes.

nationality. 3. Add to def.: *spec.* a legal relationship between a state and an individual involving reciprocal rights and duties. Also with reference to the legal device by which ships, aircraft, and companies acquire the protection of the state in which they are registered.

1880 W. E. HALL *Internat. Law* II. v. 188 The more important states recognise..that the child of a foreigner ought to be allowed to be himself a foreigner, unless he manifests a wish to assume or retain the nationality of the state in which he has been born. 1907 L. A. ATHERLEY-JONES *Commerce in War* vi. 345 Every merchant vessel is expected to carry on board some official documents vouching for her nationality. 1928 E. M. BORCHARD *Diplomatic Protection of Citizens Abroad* iii. 555 With the rise of the modern state in Europe..nationality became the test of civil and political status. 1961 N. BAR-YAACOV *Dual Nationality* 3 Having wide discretion to formulate their nationality laws according to their own interests, States adopt different methods for acquisition of nationality, the result being that two States may simultaneously confer their nationality on the same individual. 1964 GOULD & KOLB *Dict. Social Sci.* 456/2 The normal way in which nationality is acquired is through birth.... Nationality may also be granted to a person who is originally foreign or stateless. This process is known as naturalization.

5. Also occas., a racial or ethnic group.

1952 S. SELVON *Brighter Sun* v. 88 Whenever he saw a couple of different nationalities he used to hail out to them, and tell Stella that that was the way to live, especially in Trinidad. *Ibid.*, He used to say that all this business about colour and nationality was balls. 1964 GOULD & KOLB *Dict. Social Sci.* 244/1 In the Soviet Union, *nationalities* is more frequently applied to the diverse national-ethnic units who make up the membership of the Union. 1971 *Times* 17 Dec., 'Nationality', in the sense of citizenship of a certain state, must not be confused with 'nationality' as meaning membership of a certain nation in the sense of race.

native, *sb.* Add: **3. b.** (Later examples.)

1818 *London Guide & Stranger's Safeguard* 6 The practices of 'shouldering' passengers, on their own account—doing *the natives* out of articles of life, which they bring to town to dispose of—..bring them [*sc.* coachmen] to 'take care of things', for which there is no immediate owner. 1823 'J. BEE' *Slang* 124 *Natives*, silly people generally; the untravelled population of any town, wrapt up in incipient simplicity are natives. 1975 D. DELMAN *One Man's Murder* ii. 46 The house, Odum Harborage, corrupted to Odum Garborage by irreverent natives, was a notable mishmash even for Long Island.

c. (Further examples.)

Sense 4 also is current in Australia (of the Aborigines). 1863 R. HENNING *Let.* 21 Sept. (1966) 141 They were natives, and a little colonial, as might be expected. They had just left school in Melbourne. 1886 [see *AUSTRALIANA]. 1895 A. B. PATERSON *Man from Snowy River* (1896) 43 They were long and wiry natives from the rugged mountain side. 1966 G. W. TURNER *Eng. Lang. Austral. & N.Z.* iii. 62 Early writers called them *natives* or *Indians*, but *Indians* fell entirely from use, and the word *natives* was required by Europeans born in Australia, who formed an Australian Natives' Association in 1871.

4. For second part of def. read: now *esp.* one belonging to a non-European race in a country in which Europeans hold political power. (Further examples.)

1896 F. C. SELOUS *Sunshine & Storm Rhodesia* i. 5 No one could have recognised..in the quiet, submissive native..the arrogant savage of old times. 1921 G. PAGE *Jill on Ranch* viii. 184 The native is a strange child, and he needs sympathetic dealing... Make a boy laugh and you can do anything with him. 1924 E. M. FORSTER *Passage to India* II. ii. 33 Whether the native swaggers or cringes, there's always something behind every remark he makes. 1931 E. O'NEILL *Mourning becomes Electra* III. ii. 238 The natives dancing naked and innocent without knowledge of sin. 1934 G. B. SHAW *On Rocks* II. 72 *Sir Bemrose*: If a Conservative Prime Minister of England may not take down a heathen native when he forgets himself there is an end of British supremacy. *Sir Arthur*: For Heaven's sake don't call him a native. You are a native. *Sir Bemrose* (*very solemnly*): Of Kent, Arthur: of Kent. Not of Ceylon. 1944 *Living off Land* ii. 43 The corkwood..blossoms are nourishing. The natives chew these... but..it should be left well alone by the white man. 1948 *Times Lit. Suppl.* 9 Oct. 569/3 'Native' is a good word that may not now be employed without giving deep offence. 1950 M. CHAPPELL *Rhodesian Adventure* xiii. 143 There was nothing here when the pioneers came. Save bushveld and natives and wild animals. 1950 J. C. FURNAS *Anat. Paradise* ii. 24 The meaning of 'Native' can be approximated. It means: Darker. Productive of quaint handicrafts... Greedy for beads..and alcoholic drinks. Suspect of cannibalism. Addicted to drumbeating and lewd dancing. More or less naked. Sporadically treacherous. Probably polygynous and simultaneously promiscuous. Picturesque. Comic when trying to speak English or otherwise ape white ways. 1950 D. LESSING *Grass is Singing* viii. 178 When a white man in Africa by accident looks into the eyes of a native and sees the human being (which it is his chief preoccupation to avoid), his sense of guilt, which he denies, fumes up in resentment and he brings down the whip. 1951 J. MASTERS *Nightrunners Bengal* i. 9 We of the Company's service *live* here all our working lives. We do our work and enjoy ourselves and lord it over the country entirely by the good will of the average native... If you even think of them insultingly, of course they know it and resent it. 1975 C. ALLEN *Plain Tales from Raj* xix. 195 The regimental cook slaughtered a cow... The natives got to know about this and nearly stoned the camp.

c. In U.S. and Canada, a North American Indian.

1636 *Public Rec. Colony of Connecticut* (1850) I. 1 None..shall trade w[i]th the natives or Indians any peece or pistoll or gunn. *a* 1772 J. WOOLMAN *Jrnl.* in *Works* (1774) I. viii. 153 My meditations were on the alterations in the circumstances of the natives..since the coming in of the English. 1846 R. B. SAGE *Scenes Rocky Mts.* xxxiii. 287 Skins furnish to the natives a favorite material for arrow-cases. 1856 R. M. BALLANTYNE *Snowflakes & Sunbeams* vii. 72 This is the trading-store. It is always recognisable, if natives are in the neighbourhood, by the bevy of red men that cluster round it, awaiting the coming of the store-keeper. 1951 R. BULIARD *Inuk* 316 The company nowadays certainly does give help to the natives, in the forms of loans, gifts, and medicine.

d. In phr. *to astonish the natives* (orig. *U.S.*), to shock, or otherwise profoundly impress, public opinion.

1807 *Salmagundi* 27 June 233 He was determined to '*astonish the natives a few*'. *Ibid.* 238 Unfortunate Straddle! may thy fate be a warning to all young gentlemen who come out from Birmingham to *astonish the natives!* 1842 *Ainsworth's Mag.* I. 302 At last, having astonished the natives,..he rolled off to bed. 1852 [in Dict., sense 4]. 1901 M. E. RYAN *That Girl Montana* 96 Much of her afternoon was spent..fashioning a party gown with which to astonish the natives.

e. *to go native*: see *GO *v.* 44 a.

native, *a.* Add: **12. c.** *native son* U.S., a native of a particular state.

1833 *Southern Patriot* (Charleston, S. Carolina) 27 July 2/6 Col. William Drayton..a native son of Carolina..left our shores. 1850 *Ex. Doc. 31st U.S. Congress 1 Sess. Senate* No. 76. 3 The native sons of the United States living in New Mexico knew their right to equality of privilege. 1864 *Weekly New Mexican* (Santa Fé) 25 Nov. 3 Lieut.-Colonel Chaves..is one of New Mexico's favorite native sons. 1913 *Dialect Notes* IV. 27 You can't get a job in California unless you are a native son. 1916 'B. M. BOWER' *Phantom Herd* 27 [He] backed out of the way of the Native Son, who sprawled himself over the table corner. 1949 *Sun* (Baltimore) 7 Sept. 12/4 A bare victory in Ohio could not be taken as a decisive popular verdict for Mr. Taft's philosophy because the Taft philosophy had been supplemented by loyalty to a native son and that element could not be counted as a safe factor in general calculations. 1974 *News & Reporter* (Chester, S. Carolina) 22 Apr. 2-A/1 The man is a native son, is old enough and has resided in the state long enough.

13. c. *native bear* = *KOALA (later examples); *native cat* = DASYURE (earlier and later examples); *native companion*, the *BROLGA, *Megalornis rubicundus* (later examples).

1863 R. HENNING *Let.* 10 Aug. (1966) 137 It was not rats but a large 'native cat' that had so alarmed John... They are pretty little creatures with soft spotted fur, about the size of a kitten. 1896, 1911 Native companion [see *BROLGA]. 1911 E. M. CLOWES *On Wallaby* xi. 279 You may yet meet with..a lumbering native bear, like nothing on earth so much as a child's woolly toy, really the most ingratiating creature. Standing about two feet high, and covered with soft, thick fur, it has an odd, blunt, wistful sort of nose. 1916 J. B. COOPER *Coo-oo-ee* xii. 174, I saw for the first time a native bear on the bough of a black butt. 1928 'BRENT OF BIN BIN' *Up Country* ii. 25 Bert's heelers..were tethered in a number of kennels placed around the fowl-houses as protection against native cats, which could devastate a fowl roost in one attack. 1934 T. WOOD *Cobbers* xvi. 189 Native companions—strange white stalky birds on stilts whose courtship dance is a marvel. 1946 K. TENNANT *Lost Haven* (1968) xiv. 237 The native companions..beating the water with their stumpy wings to frighten the little fish. They seemed to have their legs fastened on backwards at the joints. 1966 G. W. TURNER *Eng. Lang. Austral. & N.Z.* iii. 41 Koalas were more often called native bears in the early years. 1968 *Times* 23 Jan. (Austral. Suppl.) p. xiii/3 He..caught instead a pair of dibblers..believed to be extinct and of importance as a link between the smaller phascogales and the larger native cats.

d. Also, in the names of New Zealand plants; (further examples). Also **native bush** *N.Z.*, woods or forests made up of indigenous trees and shrubs.

1826 Native cherry [see *five corner(s)]. 1884 A. NILSON *Timber Trees New South Wales* 125 *X[ylomelum] pyriforme*.—Wooden Pear; Native Pear. 1889 J. H. MAIDEN *Useful Native Plants Austral.* 286 *Ricinocarpus pinifolius*, Desf., Native Jasmine. This plant yields abundance of seeds, like small castor-oil seeds. They yield an oil. 1891 R. WALLACE *Rural Econ. Austral. & N.Z.* xxii. 294 *Panicum decompositum*, R. Br.—Barley grass, native millet, umbrella grass. Throughout Colonies, except Tasmania. 1898 MORRIS *Austral Eng.* 5/1 Emu A[pple]—*Owenia acidula*, F. v. M.; called also Native Nectarine. 1905 Native nectarine [see *emu-apple]. 1908 E. J. BANFIELD *Confessions of Beachcomber* I. i. 20 Strong and spicy are the odours..in the jungle, the so-called native ginger, nutmeg..and many others. 1926 *Trans. N.Z. Inst.* LVI. 662 The word 'native' has been prefixed to almost as many names as the words 'New Zealand'. 1928 'BRENT OF BIN BIN' *Up Country* iii. 43 Its floor was spread with glowing embers from the bark of the native apple tree, specially suitable for the purpose. 1930 L. G. D. ACLAND *Early Canterbury Runs* vi. 125 The glorious view, native bush and trim gardens make it [*sc.* Peel Forest] one of the most beautiful homesteads in Canterbury. 1933 *Bulletin* (Sydney) 29 Mar. 25/1 Some of the potential plants of these pastures are rib grass, coolah grass..and native wheat grass. *Ibid.* 7 June 25/2 Another ornamental plant is the native jasmine, which grows to about 6 ft. high. This bush is often completely denuded of foliage by stock. 1934 *Ibid.* 15 Aug. 21/4 The thirsty traveller in the western parts of N.S.W. and Queensland welcomes the sight of a native nectarine (*Owenia acidula*) almost as much as a fresh-water spring. 1947 A. VOGT in D. M. Davin *N.Z. Short Stories* (1953) 377 Further back, even here, the hills were thick with native bush. 1959 A. MCLINTOCK *Descr. Atlas N.Z.* 80 Stewart Island..is a popular tourist resort with unspoiled native bush,..and a wealth of birdlife. 1966 W. S. RAMSON *Austral. Eng.* v. 82 In some cases the resemblance to an English species was sufficient for the choice of name, but a qualification was introduced through the use of a particularizer, as, for instance, in..native pear. 1967 A. M. BLOMBERY *Guide Native Austral. Plants* 313 *S[antalum] acuminatum*. Quandong, Native Peach... A shrub or small tree with light greyish-green, narrow, lanceolate leaves and bright-red fruits.

14. (Further examples.)

1891 R. WALLACE *Rural Econ. Austral. & N.Z.* xiv. 215 Mrs. Donnelly is perhaps one of the most able lawyers in relation to native affairs that New Zealand possesses, and..has earned..the reputation of a Maori Portia. 1914 R. FRY *Let.* (1972) II. 378 Are you going to Tunis? If so..he'll give you Arab dinners in a lovely Arab house in the middle of the native quarter. 1950 M. CHAPPELL *Rhodesian Adventure* ix. 101, I believe the original native name for the place where Salisbury stands was 'Harari'. 1961 G. GREENE *In Search of a Character* I. 14 The 'method' here seems to be to drive around the native town until a likely girl is seen. 1975 C. ALLEN *Plain Tales from Raj* i. 25 The native bazaar was..out of bounds..to all European children.

15. Special Combs., as *Native American*, a North American Indian; also *attrib.* or as *adj.*; *native location* S. Afr., = LOCATION 5 (in Dict. and Suppl.); *native oven* (*N.Z.*) = *COPPER MAORI; a similar oven in Australia; *native question*, the question of relations between colonizers and the indigenous population of a country; *native reserve*, an area of land set aside by a colonial power for the exclusive use of the indigenous population; *native state, Native State*, during the period of British dominion of India, the term used to designate a state outside British territory which was governed by a native ruler; also called 'Indian State', 'princely state'.

1956 A. HUXLEY *Let.* 20 Oct. (1969) 809 Thank you for your most interesting letter about the Native American churchmen. 1973 *Black Panther* 7 Apr. 4/1 Appearing at the awards in Brando's behalf was the beautiful, gracious, and now famous Native American woman, Sacheen Littlefeather, who, dressed in the traditional garments of her people, read a prepared statement. 1973 *New Society* 19 July 137/1 Services at a Native American Church, a denomination that combines Indian and Christian beliefs. 1974 *Black Panther* 19 Jan. 3/2 In a vain attempt to cover the highly-political nature of the trial, the government has accused the two Native Americans of crimes such as burglary, larceny, and auto theft rather than accuse them of the real charges: of standing up for the dignity and culture of Indian peoples. [1855 W. C. HOLDEN *Hist. Natal* viii. 176 The plan of government devised was, to preserve the Natives distinct from the whites; and, for this purpose, large tracts of country were set aside, under

the designation of 'Locations for the Natives'. On these Locations the Natives were to be collected, and governed by their own laws, through the medium of their own chiefs.] 1866 in *Towards Dict. S. Afr. Eng.* (1971) 51 Crime has considerably increased during the year; which is in a great measure to be attributed to the scarcity of food in the native locations. 1881 *Convention of Pretoria* in J. Nixon *Compl. Story Transvaal* (1884) 348 Article 21, Forthwith, after the taking effect of this Convention, a Native Location Commission will be constituted. 1928 R. L. BUELL *Native Probl. Afr.* I. I. iii. 50 Each South African city has its native location in which the native population must supposedly live, and in which houses are usually rented from the municipality. 1955 *Problems & Tensions in S. Afr.* 374 In a few Native locations there is adequate provision for good schools, health centres, stores and churches. 1832 A. EARLE *Narr. Residence N.Z.* xxviii. 96 On a spot of rising ground, just outside the village, we saw a man preparing a native oven, which is done in the following simple manner:—A hole is made in the ground, and hot stones are put within it, and then all is covered up close. 1834, 1889 [see *COPPER MAORI]. 1905 W. B. *Where White Man Treads* 101 (*heading*) The haangi—(native oven). 1911 C. E. W. BEAN *'Dreadnought' of Darling* xxvi. 227 Menindie..swarms with native ovens and other relics of the blacks. 1917 H. W. WILLIAMS *Dict. Maori Lang.* 41/2 *Hāngi*..., native oven, consisting of a circular hole in the ground, in which the food was cooked by heated stones. 1900 (*title*) Native question in South Africa. 1924 E. M. FORSTER *Passage to India* II. ii. 32 Adela, who meditated spending her life in the country, was a more serious matter; it would be tiresome if she started crooked over the native question. 1927 W. PLOMER *I speak of Afr.* 255, I am the only one here who doesn't depend for a living on the native question. 1971 *Oxf. Hist. S. Afr.* II. viii. 403 In 1914 'the Native question' was mentioned only in passing. 1928 R. L. BUELL *Native Probl. Afr.* I. v. 71 All land was simply declared public land,..which alienated it to European settlers after establishing in several cases, notable in Natal, native reserves. 1950 M. CHAPPELL *Rhodesian Adventure* ix. 101 The whole area is a native reserve and looks no different today than it has for centuries. 1953 P. ABRAHAMS *Return to Goli* iv. 106 The result is that nowhere else in Africa is land-hunger as acute as it is in the Union's 'Native Reserves'. 1966 M. M. COLE *S. Afr.* (ed. 2) xlv. 687 The Native Reserves in the Republic are incapable of supporting all the Bantu population in agriculture. 1784 *Act 24 Geo. III* c. 25 § 15 Treating or negociating with any of the Native Princes or States in India. [*Ibid.* §35 To negociate or conclude any Treaty of Peace..with any Indian Prince or State.] 1823 J. MALCOLM *Mem. Cent. India* II. xvi. 280 The present condition of our empire in India requires..in the exercise of political control and superintendence over Native States, a school..distinct from other branches of the service. 1883 J. S. COTTON in Cotton & Payne *Colonies & Dependencies* I. iii. 23 The native states are sometimes called feudatory—a convenient term to express their vague relation to the British crown. 1886 KIPLING *Departmental Ditties* (ed. 2) 7 Rustum Beg of Kolazai—slightly backward Native State—Lusted for a C.S.I. 1894 W. LEE-WARNER *Protected Princes India* i. 2 The most cursory examination of the Native states brings to light a confusing variety in their size, their origin, and their development. 1931 P. KENDALL *India & British* viii. 223 Hyderabad, the largest..of all the Native States, absorbed Golconda centuries ago. 1963 M. A. RAHIM *Ld. Dalhousie's Administration* i. 4 As regards the condition of Indian states, there were many so-called independent native principalities or states, but actually..they were firmly controlled by the British Government.

nativism. Add: **1.** (Earlier and later examples.)

1845 *Congress. Globe* 18 Dec. 66 In the City of New York nativism had its origin in the disputes of the Tammany party. 1965 *Guardian* 6 Aug. 5/3 The revived nativism of white, rural, Protestant America. 1968 G. W. STOCKING *Race, Culture & Evolution* xi. 297 This reaction was related to the national outburst of nativism which..was to lead finally in 1923 to the passage of legislation which closed the era of mass immigration by non-'Nordics'.

b. *Anthrop.* (See quot. 1972.)

1964 GOULD & KOLB *Dict. Social Sci.* 458/1 Examples of nativism would include peyotism and cargo cults. Another example is the Ghost Dance of the American Indians. 1972 D. DAVIES *Dict. Anthropol.* 132/1 *Nativism*, the movement of societies back toward a reaffirmation of their native tribal cultures as a reaction when acculturation seems to be threatening their tribal identity and culture.

2. b. *Psychol.* and *Linguistics.* The doctrine that certain capacities or abilities, esp. those of sense perception, are innate rather than acquired; the theory that in the development of language an inherent connection exists in the mind between sound and sense.

1924 R. M. OGDEN tr. *Koffka's Growth of Mind* i. 1 And even to-day no agreement has been reached between the rival theories of *empiricism* and *nativism*, the first of which emphasizes the influence of environment, and the second the influence of heredity. 1968 D. PRICE-WILLIAMS in J. Clifton *Introd. Cultural Anthropol.* xii. 312/1 Visual illusions have given cross-cultural psychologists a considerable body of data with which to assess the factors of nativism and empiricism in perception. 1970 *Language* XLVI. 139 In the revival of Cartesian philosophical views of language, a number of other moribund doctrines have been brought to spectral life, too: mentalism, nativism (the innate-endowment component in child language studies). 1972 *Encycl. Psychol.* II. 308/1 *Nativism*, the belief that human behavior and particularly human perceptual mechanisms are inborn and determined by genetics, as opposed to the belief that they are the result of learning and experience.

nativist. Add: **2. b.** *Psychol.* and *Linguistics.* One who holds the doctrine of innate capacities, or of an inherent connection between sound and meaning in language.

1924 R. M. OGDEN tr. *Koffka's Growth of Mind* iii. § 5. 76 To the empiricist the observed development [of fixation] is regarded as a process of learning; while the nativist regards it as a process of maturation. 1930 W. LEOPOLD in J. T. Hatfield et al. *Curme Vol. Ling. Stud.* 106 It might be possible to find..a tie of union even between views as contrasting as those of Wundt and Marty, of nativists and teleologists, in the philosophy of language. 1971 *Jrnl. Gen. Psychol.* LXXXV. 18 The argument of nativists that the phenomenal experience is not found to be as fluid or flexible as would be expected under an empirical approach.

nativistic, *a.* (Further examples.)

1922 O. JESPERSEN *Lang.* xxi. 415 A closely related theory is the nativistic, nicknamed the *ding-dong*, theory, according to which there is a mystic harmony between sound and sense. 1943 *Amer. Anthropologist* XLV. 230 The term 'nativistic' has been loosely applied to a rather wide range of phenomena... We may define a nativistic movement as, 'Any conscious, organized attempt on the part of a society's members to revive or perpetuate selected aspects of its culture'. 1946 D. MCCARTHY in L. Carmichael *Manual of Child Psychol.* x. 501/2 Sapir..proposes a nativistic theory of phonetic symbolism which has given rise to considerable controversy. 1958 F. M. KEESING *Cultural Anthropol.* xvi. 406 'Nativistic' movements, including the new religious cults spoken of earlier. 1966 M. PEI *Gloss. Ling. Terminol.* 177 *Nativistic theory*, the view that a mystic harmony or connection exists between sound and meaning, and that human speech is the result of an instinct of primitive man. 1968 D. PRICE-WILLIAMS in J. Clifton *Introd. Cultural Anthropol.* xii. 312/1 The results support the Gestalt position and thus the nativistic viewpoint. 1974 *Listener* 7 Mar. 294/1 A nationalist—or as the officials prefer to call it—'nativistic' movement..in the South Pacific.

nativize (nē̆i·tivəiz), *v. Linguistics.* [f. NATIVE *a.* + -IZE.] *trans.* To render native; *spec.* **a.** To adapt (a loan word) to the phonetic structure of the native language. **b.** To develop (a pidgin language) into a creole used as a first language. Hence **na·tivized** *ppl. a.*, **na:tiviza·tion.**

1933 L. BLOOMFIELD *Lang.* xxv. 454 From the completely nativized ['šowfr] *chauffeur*, we have the back-formation *to chauffe* [šowf]. 1940 *Proc. Amer. Philos. Soc.* LXXXII. 15 'Scandalous examples of Great Russian chauvinism' have often interfered with what is called the nativization of the Soviet apparatus. 1970 *Language* XLVI. 66 A Nupe speaker will consistently 'nativize' [Cɔ] as [Cʷa] and [Cɛ] as [Cʸa]. *Ibid.*, The position supported by this evidence is that the nativization of foreign sounds is a valid indicator of what rules have been internalized. 1971 I. F. HANCOCK in J. Spencer *Eng. Lang. W. Afr.* 113 When a pidgin supplants a 'full' language, changes must occur... Therefore in becoming *nativised* and thereby *creolised*, it expands its vocabulary, produces more explicit grammatical constructions and becomes more fixed in pronunciation.

N.A.T.O., NATO, Nato (nē̆i·to). [Acronym f. the initials of *N*orth *A*tlantic *T*reaty *O*rganization, set up in 1949.] A military alliance of the United States and Canada with certain European nations. So **Na·to,-ish** *a.*, supporting N.A.T.O.; **Na·toism,** adherence to N.A.T.O.; **Na·toist,** a supporter of N.A.T.O.

1950 *Newsweek* 14 Aug. 40/3 Nato is the newest synthetic word in the international gobbledygook and stands for the North Atlantic treaty organization. 1962 H. O. BEECHENO *Introd. Business Stud.* i. 5 Britain had also put her military forces under a unified command in NATO (the North Atlantic Treaty Organization). 1965 *Economist* 28 Aug. 770/3 The Socialist People's party has accused Labour of being too capitalistic and too Nato-ish. 1965 *New Statesman* 24 Dec. 992/2 De Gaulle will not budge on his slightly neutralist type of Natoism. 1966 *Ibid.* 18 Mar. 366/2 The opposition..includes a lot of outright Natoists (Mollet and even Mitterrand also support Nato, with reservations). 1975 *Guardian* 5 Dec. 3/1 Britain is interested in buying several of the aircraft provided that other European NATO countries also join in.

natriuresis (nē̆i:tri-, næ:tri,jurī·sis). *Med.* [f. NATRI(UM + Gr. οὔρησις urination.] The excretion of abnormally large amounts of sodium in the urine.

1957 H. W. SMITH in *Amer. Jrnl. Med.* XXIII. 625/2 An increase or decrease in sodium excretion will be designated as natriuresis and antinatriuresis. 1974 *Nature* 11 Jan. 109/1 The natriuresis produced by the intra-renal release of kallikrein may be effected by vasodilator and permeability effects on the renal vasculature and tubules.

Hence **natriuretic** (-e·tik) *a.*, causing or pertaining to natriuresis.

1957 *Amer. Jrnl. Med.* XXIII. 625/1 The natriuretic effects of the infusion of large volumes of saline solution are in part controlled by the absence of natriuresis after the infusion of iso-oncotic albumin. 1974 *Nature* 11 Jan. 109/1 The data on sodium excretion place substance P amongst the most potent natriuretic substances so far described.

natro- (nē̆i·tro, næ·tro-), comb. form of NATRIUM, used in *Min.* to form the names of sodium-containing minerals (as NATROLITE), and also prefixed to existing mineral names to indicate the (often partial) substitution of sodium for some other metal in that mineral, as **natroa·lunite,** any of the range of basic sulphates of sodium, potassium, and aluminium, $(Na,K)Al_3(SO_4)_2(OH)_6$, which have the proportion of sodium greater than that of potassium and are found as white hexagonal crystals; *spec.* the sodic end-member of this series; **natrocha·lcite** [L. *chalcītes* CHALCITES], a basic hydrated sulphate of sodium and copper, $NaCu_2(SO_4)_2(OH).H_2O$, which is found as emerald-green monoclinic crystals at Chuquicamata in Chile; **natroda·vyne** [ad. It. *natrodavyna* (F. Zambonini 1910, in *Atti d. R. Accad. d. Sci. fis. e mat.* XIV. VI. 188)], a silicate and carbonate of aluminium and sodium, found as colourless hexagonal crystals; **natroja·rosite,** a basic sulphate of sodium and iron, $NaFe_3(SO_4)_2(OH)_6$, found as yellow or brown hexagonal crystals; **na:tromontebra·site** [a. F. *natromontebrasite* (F. Gonnard 1913, in *Bull. de la Soc. franç. de Min.* XXXVI. 120)], a basic fluoride and phosphate of sodium, lithium, and aluminium, $(Na,Li)Al(PO_4)(OH,F)$, with more sodium than lithium, which is found as whitish triclinic crystals.

1902 HILLEBRAND & PENFIELD in *Amer. Jrnl. Sci.* CLXIV. 220 The name natroalunite might be employed to designate the two varieties of alunite from Colorado..where the proportion of the soda to the potash molecule is 7:4. 1937 *Amer. Mineralogist* XXII. 944 Small amounts of finely granular natroalunite have attacked the andalusite ores. 1968 I. KOSTOV *Mineral.* II. 493 Natroalunite is a sodium analogue [of alunite] $NaAl_3(SO_4)_2(OH)_6$. 1969 *Mineral. Abstr.* XX. 146/1 The first occurrence of natroalunite and natrojarosite from Argentina is noted from the area around La Flecha gorge. 1908 PALACHE & WARREN in *Amer. Jrnl. Sci.* CLXXVI. 343 The collection..was obtained from exhausted copper veins and includes..natrochalcite (a new mineral). 1944 *Econ. Geol.* XXXIX. 274 The SO_3 concentration is so reduced that chalcanthite, natrochalcite, and krohnkite become unstable and antlerite takes their place. 1968 I. KOSTOV *Mineral.* II. 514 Natrochalcite is monoclinic (*C2/m*) occurring as acute pyramidal green crystals with perfect {001} cleavage. 1913 *Mineral. Mag.* XVI. 367 Natrodavyne... Hexagonal crystals, rich in faces, hitherto referred to nepheline or to davyne. 1963 *Canad. Mineralogist* VII. 632 Dry material of the natrodavyne composition was prepared by mixing α-cristobalite, γ-alumina and Na_2SiO_3 in the correct stoichiometric proportions for the formation of nepheline, and then adding the required weight of anhydrous Na_2CO_3. 1902 HILLEBRAND & PENFIELD in *Amer. Jrnl. Sci.* CLXIV. 219 It has seemed to the writers best to designate the new compounds described in this article as natrojarosite and plumbojarosite, the names signifying their relation to a well known species. 1938 *Amer. Mineralogist* XXIII. 757 At Chuquicamata natrojarosite is associated with chalcanthite, kroehnkite and sulphur. 1969 Natrojarosite [see *natroalunite* above]. 1971 *Norsk Geol. Tidsskrift* LI. 195 Natrojarosite is reported as a secondary mineral from Forvik antimony deposit, Helgeland, North Norway. 1915 *Mineral. Mag.* XVII. 355 Natromontebrasite... To replace the name Natramblygonite or Natronamblygonit of W. T. Schaller.., since the mineral is a hydrofluophosphate rather than a fluophosphate. 1955 *Amer. Mineralogist* XL. 1141 Fremontite.., in which Na:Li = 1·7:1 and OH > F, becomes natromontebrasite and possesses the dubious distinction of having been referred to under three different names within 40 years.

N.A.T.S.O.P.A., NATSOPA, Natsopa (næt-sō̆u·pă). [Acronym f. the initial letters of *Na*tional *S*ociety of *O*perative *P*rinters and *A*ssistants.] A trade union composed of printing workers; pl., the members of this union.

From 1975 retitled the National Society of Operative Printers, Graphical and Media Personnel.

1917 *Natsopa Jrnl.* May 1/1 Our intentions..will no doubt go to swell the number with which the road to a certain place is paved, but that does not matter,..because all Natsopas will go the journey to another place and will require wings to convey them. 1922 *Times* 11 Feb. 4/1 In the January number of *Natsopa*, the official journal of the society, the executive council gave notice that..the office of general secretary would be declared vacant. 1926 *Glasgow Herald* 3 May 11 The Natsopas..at Carmelite House took exception to the leading article. 1935 F. D. KLINGENDER *Condition Clerical Labour Brit.* ii. 48 The following remark made by the secretary of the clerical section of Natsopa. 1958 *Oxford Mail* 6 Aug. 1/1 NATSOPA indicate in a statement that..the question of craft and non-craft unions [etc.]. 1963 *Times* 11 Feb. 11/2 Severe disruption of the production of the *Daily Mirror* took place last night as a result of unofficial action by some members of the Natsopa union.

natter, *v.* Add: **1. b.** To chatter, to chat (in an aimless manner). Also with *about, away, of, on*. (Now the usual sense.)

1943 HUNT & PRINGLE *Service Slang* 47 *Natter*, to chide or chatter in an irritatingly aimless fashion. 1949 M. ALLINGHAM *More Work for Undertaker* xxvii. 310 The shares Campion keeps nattering about. 1952 J. CANNAN *Body in Beck* i. 9 No human voices..nattering of sodomy and of being bumped by Cat's. 1954 C. P. SNOW *New Men* 213 You're saying we ought to find a bogus reason for putting him in the street—just because some old women might natter. 1958 *Sunday Times* 26 Jan. 17/3 They..nattered away for an hour about nothing. 1959 C. MACINNES *Absolute Beginners* 74 She nattered on. I gave up. 'Well—you win,' I told her. 1972 R. MAUGHAM *Escape from Shadows* iv. 169 Seeing me look like a village idiot when he nattered away at me in Arabic. 1973 *Times* 12 Nov. 10/6 Women who..natter about discrimination.

2. (Later example.)

1946 J. B. PRIESTLEY *Bright Day* ix. 274 If you've got summat in you that wants to be let out an' goes on natterin' at you day an' night, then you let go of everything else an' get it out.

So **na·tterer**[1]; **nattering** *vbl. sb.* and *ppl. a.* (later examples in various senses); **na·ttery** *a.* = *nattered* ppl. adj.

1825 J. T. BROCKETT *Gloss. North Country Words* 146 *Nattry*, ill natured, petulant. '*Nattry faced*.' 1873 J. STANDING *Echoes Lancs. Vale* 17 One o' thoose nattery owd maids 'at con olez tell so mitch better heaw to bring a family o' childer op nor thoose 'at have 'em. 1900 *Eng. Dial. Dict.* II. 657/2 s.v. *Gnatter*, *Lin[colnshire]*. Eh! Miss, she is such a natterer; she is always nattering about. 1923 *Sunday At Home* Mar. 335/2, I 'ate them skinny owd women—always bad tempered and nattery that kind is. 1937 M. ALLINGHAM *Dancers in Mourning* xvii. 217 Her energy, her constant nattering at one. 1942 *Tee Emm* (Air Ministry) II. 64 Your C.O. tears you off a strip for nattering too much over the R/T. 1949 H. PAKINGTON *Young W. Washbourne* 36 It was no longer a dear old pouch, but a nattering, irritating little pouch that twanged upon the strings of his conscience. 1953 E. SIMON *Past Masters* II. vi. 116 I'm sick and tired of all this lily-livered nattering..behind closed doors. 1956 'N. SHUTE' *Beyond Black Stump* vii. 207 To kill the nattering of hope that lingered on. 1959 H. HOBSON *Mission House Murder* xviii. 116 'Did this girl talk?' 'Not much...she wasn't one of the natterers.' 1959 G. MITCHELL *Man who grew Tomatoes* i. 23 Do you hold your tongue, now. Like a nattering old mawther, you are! 1966 J. WAINWRIGHT *Crystallised Carbon Pig* x. 47, I was nattery and on edge.

na·tter, *sb.* [f. the vb.] Grumbling, nagging talk; (now esp.) aimless chatter; a chat, a talk.

1866 W. GREGOR *Dial. Banffshire* 119 *Nyatter*, peevish chattering, grumbling. 1943 HUNT & PRINGLE *Service Slang* 47 *Natter party*, a Conference which leads nowhere. 1945 PARTRIDGE *Dict. R.A.F. Slang* 40 *Natter can*, a person—especially a 'Waaf'—prone to talk too much. 1947 *Forum* (Johannesburg) X. 1. 23/2 So it is that words like 'interdenominisationalism' and 'polyphiloprogenitive', with which we are wont to sprinkle our normal natter, sound like the mouthings of the village idiot. 1951 *News Chron.* 8 Nov. 6/1 From the swarm he singled out one bird... 'That's Joey,..he usually comes for a natter when there's nothing else doing.' 1955 'N. SHUTE' *Requiem for Wren* iii. 57 I've got a natter on with the Americans tomorrow evening. 1959 G. FREEMAN *Jack would be Gent.* v. 94 I'd give anythin' to 'ave a natter with some of them blokes. 1966 'L. LANE' *ABZ of Scouse* 74 *Owd natterbag*, a scolding woman. 1967 N. FREELING *Strike Out* 28 The natter of silly women. 1974 E. LEMARCHAND *Buried in Past* vi. 102 We wanted a natter with you... You can fill us in as nobody else can.

Natterer[2] (næ·tərəɹ). The name of Johann *Natterer* (1787–1843), Austrian naturalist, used in the possessive in **Natterer's bat,** *Myotis nattereri*, a greyish-brown bat with a white underside, found in Europe and Asia. Also *absol.*

1889 J. E. HARTING in *Zoologist* XIII. 245 The present distribution of Natterer's Bat in the British Islands cannot be stated in a few words. 1910 G. E. H. BARRETT-HAMILTON *Hist. Brit. Mammals* I. 178 Natterer's Bat ranges through boreal and temperate Europe and Asia. 1960 *Times* 14 June 14/6 My first encounter with Natterer's bat was a night to remember. *Ibid.*, The most up-to-date bat book today speaks of the Natterer's flight as 'slow and steady'.

Nattier blue (na·tye). [f. the name of the Fr. painter Jean-Marc *Nattier* (1685–1766).] A soft shade of blue much used by Nattier and associated *spec.* with his work.

[1909 *Westm. Gaz.* 4 May 5/3 We have quoted the painter Nattier for the soft shade of blue he used.] 1912 *Queen* 4 May p. xvii, The bonnet is fashioned of Nattier blue satin. 1918 W. J. LOCKE *Rough Road* xxi. 261 His own bedroom with the satinwood furniture and nattier blue hangings. 1923 [see *BISQUE[2] 3]. 1928 *Times* 9 May 10/3 A train of Nattier-blue satin and silver lace. 1963 *New Yorker* 1 June 34 She wore Nattier-blue with a collar so high it almost reached her ears. 1969 *Queen* 17–30 Sept. 91/1 Nattier blue silk curtains.

Natufian (natū·fiăn). *Archæol.* [f. the name of Wādi an-*Natuf*, seventeen miles north-west of Jerusalem.] Name coined by Prof. D. Garrod for a late Mesolithic culture the type-site of which was discovered by her at Wādi an-Natuf. Also *attrib.*

1932 D. A. E. GARROD in *Jrnl. R. Anthrop. Inst.* LXII. 257 (*heading*) A new mesolithic industry: the Natufian of Palestine. *Ibid.* 261 It was abundantly clear that we were dealing with a microlithic culture that would not fit exactly into any of the pigeon-holes already existing, and I therefore decided to give it a label of its own, adopting the name Natufian from the Wady en-Natuf at Shukba. 1949 [see *food-gatherer* (*FOOD *sb.* 8)]. 1960 K. M. KENYON *Archaeol. in Holy Land* ii. 36 The Natufians of Mount Carmel, and of rock-shelters on the eastern and western slopes of the Judaean hills, lived mainly by hunting. 1960 tr. *S. Moscati's Face of Anc. Orient* i. 13 The Natufian civilization brings two principal innovations: the harvesting of wheat and barley, and the beginnings of the domestication of animals. 1970 BRAY & TRUMP *Dict. Archaeol.* 159/1 The shrine at the base of the Tell at Jericho was built during the early Natufian phase and the descendants of the Natufians built the earliest Neolithic town at the site.

natural, *sb.* Add: **1.** (Later Amer. example.)

1748 in *Maryland Hist. Mag.* (1911) VI. 229, I..am become a Natural of the country or country born as some call themselves.

6. b. (Later example in *sing.*)

1922 JOYCE *Ulysses* 396 Any female..with the desire of fulfilling the functions of her natural.

9. b. Also in other gambling games, any combination or score that immediately wins the stake. Also *fig.*

1762 GOLDSMITH *Citizen of World* I. 165 He had something in his face gave me as much pleasure as a pair-royal of naturals in my own hand. 1929 E. WALLACE *Red Aces* xi. 110 Somebody would draw a six to these, and the banker would have a 'natural'—which means, I understand, that he would win. 1930 J. LAIT *Big House* 15 Dean Ward Kent arrived at the big house with a 'natural' staring him in the face, for that is what the crap-shooting inmates call a seven-year 'stretch'. 1962 K. ORVIS *Damned & Destroyed* xv. 109 The dice bounced to a natural.

13. *colloq.* Short for *natural life* (NATURAL *a.* 9 b).

1893 G. L. GOWER *Gloss. Surrey Words* 27 *In my natural*, phrase for 'in my life', 'at any time'. 1898 J. D. BRAYSHAW *Slum Silhouettes* 220 Yer never see sich a 'owlin' swell as Cocky was in yer born natural. 1911 L. STONE *Jonah* II. ii. 161, I niver 'eard anythin' like that, in my natural. 1913 C. MACKENZIE *Sinister St.* I. I. iv. 46, I never worked so hard in all my natural. 1925 WODEHOUSE *Carry on, Jeeves!* iii. 59, I didn't want to have England barred to me for the rest of my natural. 1967 J. PORTER *Chinks in Curtain* xviii. 185, I couldn't stay like that for the rest of my natural.

14. A person naturally endowed *for* (a role, etc.); one having natural gifts or talents; also, a thing with qualities that make it particularly suitable (*for* some purpose.)

1925 *Hearst's International* June 80/2 The fight was what promoters call a 'natural'. 1929 D. HAMMETT *Red Harvest* xiii. 132 'So you and Noonan are trying to paste his brother's death on me?' 'It doesn't need pasting. It's a natural.' 1930 *Publisher's Weekly* 21 June 2971/2 Mystery fans will devour it; and you can sell it also to anyone who likes a finely written and witty novel. A possible natural. 1933 F. BALDWIN *Innocent Bystander* (1935) xiii. 260 But she's a natural... I watched her walk across the stage..and the audience rose to her. 1939 *Sun* (Baltimore) 2 Jan. 1/8 The Hopkins and Murphy appointments are regarded as 'naturals' for early-session debates. 1946 *Coast to Coast 1945* 125 This is a natural, son. I can pick this one. 1948 F. BROWN *Murder can be Fun* (1951) ix. 132 Hell, it was a *natural* for publicity for a writer. 1955 *Observer* 24 July 13/7 The sort of play which should have been a natural for television. 1958 *Times Lit. Suppl.* 15 Aug. 455/1 But the theme of how the Labour Party was born of the Labour Representation Committee is, as the film-makers say, a 'natural'. Poor Party makes good. 1964 *McCall's Sewing* i. 10/2 You're a natural for high fashion. 1966 *Listener* 24 Nov. 780/1 (Advt.), These five talks..diversified with 64 photos and 8 maps..make up what will be a 'natural' at Christmas, for young and old. 1971 B. MALAMUD *Tenants* 154 I'm not a natural. This present play is my last, I've decided. 1975 *Sunday Times* (Colour Suppl.) 23 Feb. 26/4 He was a natural, and gradually began to pick up something of a reputation. 1975 *Times* 25 Feb. 3/6 Railways are a natural for process control and on-line computer systems.

16. *Bullfighting.* A type of pass made with the cape. Also with Spanish pronunc. (naturāl).

1932 E. HEMINGWAY *Death in Afternoon* xviii. 198 The greatest pass with the muleta, the most dangerous to make.., is the natural. In this the man faces the bull with the muleta held in his left hand, the sword in his right. 1959 V. J. KEHOE *Aficionado!* xiv. 174/1 A natural is always in the direction of the arm that makes it. 1967 MCCORMICK & MASCAREÑAS *Compl. Aficionado* i. 24 A *natural* is a pass; it too may be called a suerte. 1973 *Sat. Rev.* (U.S.) 25 Sept. 29/2 Taking the smaller *muleta*, he ran off several fine *naturales*, bringing the bull tightly round him.

17. *Archæol.* Undisturbed terrain, below the level of cultivation or other working; virgin rock or soil.

1946 R. J. C. ATKINSON *Field Archaeol.* 210 *Natural rock* or '*natural*', the undisturbed material upon which the soil lies. 1950 *Notes on Archaeol. Technique* (ed. 3) 13 Many avoidable mistakes have been made through failure to identify the real 'natural' (undisturbed) layer of a site. Before the excavation proper is started, dig a cutting in undisturbed ground... In one half, stop on the 'natural'; in the other dig well into it. The 'natural' can then be studied in plan and section. 1954 M. B. COOKSON *Photogr. for Archaeologists* i. 14 A sharp right-angle where the last archaeological layer meets 'natural', balks swept at the end of a day's work, will repay the trouble taken a hundredfold.

18. A hair-style among Negroes in which the hair is not straightened or bleached; *spec.*, an Afro haircut. *U.S.*

1969 *Ebony* Feb. 27 There's a lean young cat wearing a natural who knows where it's at and tells it like it is. 1971 B. MALAMUD *Tenants* 42 She wore a natural of small silken ringlets, and a plain white mini with purple tights. 1971 *Black Scholar* Apr.–May 17/1 He has a Black is Beautiful bumper sticker on his car; he has a natural and even wears a dashiki to work. 1973 E. BULLINS *Theme is Blackness* 150, I love you, baby... I sure dig that sexy natural.

natural, *a.* Add: **I. 1.** Add to def.: esp. in phr. *natural law*: in political and legal philosophy and theology, doctrines based on the theory that there are certain unchanging laws which pertain to man's nature, which can be discovered by reason, and to which man-made laws should conform; freq. contrasted with positive laws; also (with hyphen) *attrib.* (Add further examples.)

1899 W. R. INGE *Christian Mysticism* viii. 306 Wordsworth..shows his affinity with the modern spirit in his firm grasp of natural *law*. 1915 tr. *Aquinas's Summa Theologica* II. I. Question 91. Article 2 This participation of the eternal law in the rational creature is called the natural law. 1934 E. BARKER tr. *Gierke's Natural Law Theory of Society* I. III. ii. 111 (*heading*) The natural-law view of the purposes of society and its various groups. 1950 A. VERDROSS-DROSSBERG in *Contemp. Pol. Sci.* (Unesco) 598 They are a residue of the ideas of natural law, since the unanimous agreement among civilized nations on a legal principle shows that the latter satisfies the elementary requirements of the legal conscience of mankind. 1951 A. PASSERIN D'ENTRÈVES (*title*) Natural law: an introduction to legal philosophy. *Ibid.* 15 The belief in natural law, both as a recognition of a law common to humanity and as an assertion of the fundamental rights of man, was..the distinguishing mark of political thought in Western Europe. 1967 *Encycl. Philos.* V. 451/1 The ideal or ethical law, which is contrasted with positive law..is regarded by natural-law theorists..as grounded in something..more enduring than the mere practical needs of men. 1970 W. E. VOLKOMER *Passionate Liberal* iii. 64 The immediate reason for Frank's conversion to natural law would appear to lie in the rise of totalitarianism.

2. c. Also of wind instruments, as *natural trumpet* (see quot. 1959).

1910 K. SCHLESINGER *Instruments Mod. Orchestra* I. xviii. 83 The *natural trumpet* in which the length and pitch are varied by means of crooks. 1959 *Collins' Mus. Encycl.* 450/1 *Natural horn, natural trumpet*, a horn or trumpet which is not provided with any method, such as valves, of altering the length of the tube, and can therefore sound no other notes than those of the harmonic series above its fundamental, except as stopped notes. 1966 P. BATE *Trumpet & Trombone* vi. 99 (*heading*) Natural trumpets: medieval to modern.

e. Also *Zool.*, applied to systems of classification based on the characteristics of the animals concerned, and the groups resulting from a classification of this type.

1841 T. R. JONES *Gen. Outl. Animal Kingdom* i. 2 The apparatus of digestion appears to be among the least efficient for the purpose of a natural division [of the animal kingdom]. *Ibid.* 3 The researches of this profound physiologist [*sc.* John Hunter]..did much to approximate a more natural method of classification. 1945 *Bull. Amer. Mus. Nat. Hist.* LXXXV. 4/1 It is understood that a 'true' or 'natural' classification has, by intention, quite a different basis and expression. 1970 *Nature* 19 Sept. 1272/2 It is always helpful for students to understand thoroughly the natural orders into which insects are divided.

f. *natural order*, the order apparent in the constitution of matter and operation of forces in nature.

1697 M. EARBERY *Answer to Tractatus Theologico Politicus* 18 The one [*sc.* Human Reason] is founded on the Natural Order of things, and therefore subject to those Imperfections, which are common to all the Works of Nature. 1895 F. THILLY tr. *Paulsen's Introd. Philos.* I. ii. 322 The intellectual law of causality is the basis of our belief in the natural order. 1934 *Encycl. Social Sci.* XI. 284/1 The stoics and certain Roman jurists interpreted natural law as that law which conformed to the natural order of the universe. 1941 W. TEMPLE *Citizen & Churchman* v. 83 The 'natural order' by which is meant the consideration of the various departments of life in the light of the essential function of each. 1948 BERGIN & FISCH tr. *Vico's New Sci.* § 2. 3 The philosophers, contemplating divine providence only through the natural order. 1951 C. C. GILLESPIE *Genesis & Geol.* vi. 169 Revealed truth, though indispensable to belief, could be apprehended inductively, by inferring a moral order parallel to natural order. 1953 C. E. RAVEN *Nat. Relig.* i. 2 If grace is radically contrasted with the beauty and truth and goodness of the natural order, then any belief in a real Incarnation is impossible.

g. *natural selection*: see SELECTION 3 b in Dict. and Suppl. Hence *natural selectionist*, a supporter of the theory of natural selection.

1913 G. B. SHAW *Quintessence of Ibsenism Compl.* p. xv, Capitalism, built up by generations of Scotch Rationalists and English Utilitarians, Atheists, Agnostics and Natural-Selectionists..is proclaimed the bulwark of the Christian churches. 1916 —— *Androcles & Lion* p. lxxi, The efforts of Natural Selectionists..to reduce evolution to mere automatism.

h. *natural deduction*, in Logic, the name given to a method devised separately in 1934 by G. Gentzen (1935 *Math. Zeitschrift* XXXIX) and S. Jaśkowski (1934 *Studia Logica* I) whereby formal proofs are obtained solely by the application of rules of inference without appeal to axioms.

1950 W. V. QUINE *Methods of Logic* (1952) § 28. 166 The method set forth in the present pages is of a type known as *natural deduction*, and stems, in its broadest

outlines, from Gentzen and Jaśkowski (1934). 1954 I. M. COPI *Symbolic Logic* iv. 119 The methods of proof so far assembled (techniques for 'Natural Deduction', as they are sometimes called) permit the demonstration of all logically true propositions constructed out of truth-functional connectives and the quantification of individual variables. 1966 *Amer. Philos. Q.* III. 27 (*title*) Natural deduction rules for obligation. 1969 *Aristotelian Soc. Suppl. Vol.* XLIII. 53 The expression 'natural deduction' was introduced, I surmise, under the influence partly of the name bestowed on Herbrand's 'theorem of deduction' and partly of the French expression 'la déduction naturelle'. 1973 B. A. BRODY *Logic* iii. 103 Two..methods for showing that inferences are valid. The first, the natural deduction technique, starts from the results of this section.

3. d. (Later examples.) Esp. in phr. *natural causes.*

1889 A. B. HICKS *Hints to Medical Men concerning Certificates of Death* 6 Deaths which may be due to either natural causes, or to neglect or gross carelessness. *c* 1900 H. A. JONES in M. R. Booth *Eng. Plays of 19th Cent.* (1969) II. 382 When I heard this story was being circulated I thought it would be better to take no notice and let it die a natural death. 1921 A. CHRISTIE *Mysterious Affair at Styles* xiii. 292 He strenuously, and quite uselessly, upheld the theory of 'Death from natural causes'. 1970 P. MOYES *Who saw her Die?* viii. 105 The death certificate says 'Natural Causes'. That's the doctor's verdict. 1974 'J. LE CARRÉ' *Tinker, Tailor* xii. 104 Disappeared... May have died of course. One does *tend* to forget the natural causes. 1975 S. BRETT *Cast* xi. 104 Assisting her investigations into a perfectly natural death as if it were murder.

c. *natural childbirth*, methods of relaxation and of physical co-operation with the natural process of childbirth, first advocated by G. D. Read in 1933, intended to counteract the muscular tension, resulting from fear, and consequent pain suffered in childbirth esp. by the women of civilized peoples; also *attrib.*

1933 G. D. READ (*title*) Natural childbirth. 1948 H. HEARDMAN (*title*) A way to natural childbirth. 1960 *Guardian* 6 July 5/1 There are still many doctors and hospitals which do little or nothing to teach expectant mothers about the various methods of 'natural childbirth'. 1964 W. MARKFIELD *To Early Grave* (1965) x. 181 Inez, six and a half months gone, was at a natural childbirth class. 1965 W. LAMB *Posture & Gesture* ix. 124 One is the practice of 'natural childbirth', still contentious, but with an organized following. 1971 D. D. MOIR *Pain Relief in Labour* ii. 11 The Grantly Dick Read or Natural Childbirth method dates from 1935 and is based on the idea that fear, tension and pain are linked together. 1974 'E. LATHEN' *Sweet & Low* vii. 76 It was not easy..to become an instant swinger after ten years..of natural childbirth.

6. b. *spec.* in phrs. *natural foundation* (see quots. 1906, 1963); *natural gas* (orig. *N. Amer.*), inflammable gas occurring underground, consisting chiefly of methane and other simple paraffins and often found associated with petroleum; *natural glass*, any of various naturally occurring substances which resemble glass in appearance, having solidified too quickly to crystallize.

1825 *Canad. Courant* (Montreal) 17 Dec. 1/5 This is undoubtedly the first attempt which has ever been made to apply natural Gas to so extensive and useful a purpose. 1846 S. F. SMITH *Theatr. Apprenticeship* 102 Many of the stores and shops in the village are lighted with natural gas! 1887 *Encycl. Brit.* XXIII. 813/2 The use of natural gas for illumination, and even for metallurgical purposes, has lately become a matter of importance... The gas obtained from wells or bore-holes was used for illumination in Fredonia, N.Y., as early as 1821. 1906 H. Y. MARGARY in G. A. T. Middleton *Mod. Buildings* I. III. i. 71/1 Natural foundation is the name applied to such as are formed on the soil itself, and it is applicable when the soil is practically incompressible. 1917 *Jrnl. Geol.* XXV. 540 Natural glass is, of course, varied in composition in comparison with the various types of igneous rocks, yet the average obsidian is probably not much more varied than some of the amorphous minerals. 1930 *Economist* 22 Mar. 654/2 Interest is being taken in utilities and oils with natural gas possibilities. 1938 E. G. WARLAND *Building Construction* I. i. 1 Natural foundation beds should be incompressible or equally yielding over the whole area and not subjected to atmospheric or other influences which may alter its nature or powers to resist the loads to be placed upon them. 1942 *Chem. Abstr.* XXXVI. 6452 (*heading*) Transformation of natural glasses into crystalline rocks by subjection to high gas and water-vapor pressures. 1957 *Ann. Reg.* 1956 76 The Government's proposal for financing the natural gas pipeline from Alberta to Ontario. 1963 *Gloss. Gen. Building Terms* (*B.S.I.*) 18 *Natural foundation*, soil requiring no preparation or other foundation to support a building or structure. 1971 W. VOGEL *Struct. & Crystallization Glasses* i. 13 In the earliest times of world history, quartz porphyry in particular, and other extrusive rocks, solidified on rapid chilling as natural glasses, e.g. pitchstone, perlite, obsidian or pumice. 1972 *Guardian* 17 Feb. 9 People ran from their homes yesterday because of a gas explosion in Elgin Street, Sheffield... The area had been converted to natural gas about six months ago. 1974 *Sat. Rev. World* (U.S.) 2 Nov. 29/1 Calgary, Canada's oil and natural-gas capital.

d. Esp. in *Forestry*: *natural regeneration*, the growth of young trees from seed of those already established.

1902 B. E. FERNOW *Econ. Forestry* vii. 167 There is also a choice of producing the new crop by seeds falling from the trees of the old crop, by 'natural regeneration'. 1928 R. S. TROUP *Silvicultural Syst.* ii. 11 Natural regeneration..may be obtained either (1) from seed already on the area, or (2) from seed disseminated from trees outside and usually adjoining it. 1946 *Q. Jrnl. Forestry* XL. 18 A good deal of natural regeneration goes on at Bedgebury and had the seedlings been left..a Scots Pine forest could have been formed on the whole of the Pinetum area without any trouble. 1964 W. E. HILEY *Forestry Venture* iii. 64 The artificial weeding of natural regeneration, in which the young trees are irregularly scattered, is almost an impossible task.

e. Special collocations: *natural area* (see quot. 1964); *natural break*: see *BREAK *sb.*[1] 8 k; *natural cement*: a cement obtained by calcining naturally occurring argillaceous limestone; *natural food* (see quot. 1972); *natural language*: any naturally evolved language, as opposed to artificial languages constructed (*a*) for universal or international communications, or (*b*) for formal logical or mathematical purposes; *natural region*: each of a number of regions of the earth's surface characterized by a certain uniformity and individuality of character (see quots. 1905, 1937); *natural resources* (see quot. 1956); *natural seasoning* = *air-seasoning* vbl. sb. (*AIR *sb.*[1] B. II).

1932 *Amer. Jrnl. Sociol.* XXXVIII. 339 It is more likely that census tracts near the central business districts of Philadelphia would conform more closely to natural areas than in West Philadelphia. 1964 GOULD & KOLB *Dict. Social Sci.* 458/2 A *natural area* is a territorial unit whose distinctive characteristics—physical, economic, and cultural—are the result of the unplanned operation of ecological and social processes. 1970 G. A. & A. G. THEODORSON *Mod. Dict. Sociol.* 271 *Natural area*, a territorial area with some common, unifying characteristic. The term has been used primarily in human ecology, and usually refers to an area that emerges without planning from the operation of ecological processes. 1882 *Chem. News* 27 Oct. 187/2 (*heading*) Japanese soils—a natural cement. *Ibid.*, The chemist to the Geological Survey Department of the Japanese Government, was led to look for a natural cement. Such cements are formed by mixing burnt lime with substances of volcanic origin, generally tufas. 1921 W. H. WARREN *Engin. Construction* II. iii. 38 In America, Rosendale cement is largely used; it is a natural cement, being first found at Rosendale, Ulster County, N.Y. It is slower setting, weaker, and cheaper than Portland cement. 1966 *McGraw-Hill Encycl. Sci. & Technol.* II. 627/1 Natural cement, a naturally occurring argillaceous limestone, calcined and pulverized, is slower-setting and less uniform in quality than portland cement. 1934 H. C. SHERMAN *Food & Health* xvii. 160 Natural foods,..nature's wholes of the kinds to which our own bodies have been adjusting themselves throughout our evolutionary history. 1956 I. ORGA *Cooking with Yogurt* 8 Yoğurt is a natural food and..only needs the simplest equipment. 1963 B. T. HUNTER *Natural Foods Cookbk.* p. xv, Many thoughtful people are..seeking out the good old flavors, textures and nutrients of the natural foods their grandparents enjoyed. 1970 U. M. CAVANAGH *Cooking & Catering Wholefood Way* 9 The emphasis is on natural whole foods such as stone-ground wholewheat flour, brown sugar, natural unpolished rice and honey. 1972 *New York* 8 May 49 *Natural food*..refers to food after the growing stage; food that is unprocessed, not treated with preservatives, artificial colorings or flavorings... Ideally the rule is 'nothing added, nothing taken away'. [1668 J. WILKINS *Ess. Real Character & Philosophical Lang.* I. i. 2 There is scarce any subject that hath been more thoroughly scanned and debated amongst Learned men, than the Original of Languages and Letters. 'Tis evident enough that no one Language is natural to mankind.] 1774 LD. MONBODDO *Orig. & Progress of Lang.* II. III. xiii. 445 If we understand the sign, we have in effect the definition of the thing, then is the language truly a philosophical language, and such as must be universal among philosophers... It may also be said to be a *natural* language..since it follows the order of the human mind in forming the ideas of which language is the expression. 1864 MAX MÜLLER *Lect. Sci. Lang.* 2nd Ser. ii. 58 A grammatical framework, too, is wanted before the problem of an artificial language can be considered as solved. In natural languages the grammatical articulation consists either in separate particles or in modifications in the body of a word. 1871 S. P. ANDREWS *Primary Synopsis Universology & Alwato* vi. 95 The ideas themselves are the most subtle and embarrassing, and *natural* language then exactly *echoes* this embarrassment. As we descend to more *feasible* domains the words will become correspondingly more *feasible*. 1888 H. A. STRONG tr. *Paul's Princ. Hist. Lang.* xxiii. 501 The artificial language of a large area has a tendency to become dialectically differentiated..in much the same degree as the natural language within a particular territory. 1889 *Literary World* 22 June 209/1 The progress of education..will enable each to divest himself of the crudities of his natural language. 1933 L. BLOOMFIELD *Language* xxviii. 506 The political difficulty of getting any considerable number of people all over the world to study, say, Esperanto, will probably prove so great that some natural language will outstrip it. 1956 J. H. WOODGER tr. *Tarski's Logic, Semantics, Metamath.* viii. 267 Philosophers who are not accustomed to use deductive methods..are inclined to regard all formalized languages with a certain disparagement, because they contrast these 'artificial' constructions with the one natural language—the colloquial language. 1962 U. WEINREICH in Householder & Saporta *Probl. Lexicography* 30 Much less can we claim for natural-language lexicography that the definiens should be literally substitutable for the definiendum in normal discourse. 1963 L. LOEVINGER in H. W. Baade *Jurimetrics* 14 The abstract is recorded in natural language stating the significant index terms. 1970 A. CAMERON et al. *Computers & O.E. Concordances* 6 University courses in natural-language programming are now widely available for undergraduates. 1973 M. DUMMETT *Frege* xiii. 463 A theory of truth which attempts to display the role of the notion of truth..is not a completely separate enterprise from an account of the word 'true' as used within natural language. 1905 J. HERBERTSON in *Geogr. Jrnl.* XXV. 302 What are the characteristic and distinguishing elements of the areas which we may term natural regions? *Ibid.* 309 We may divide the world up into the following types of natural regions:—1. Polar... 2. The cool temperate regions... 3. The warm temperate regions... 4. (*a*) The west tropical deserts... 5. Lofty tropical or sub-tropical mountains... 6. Equatorial lowlands. *Ibid.*, A natural region should have a certain unity of configuration, climate, and vegetation. 1937 *Geography* XXII. 253 'Natural regions' has been used to cover two distinct types of unit-areas of the earth's surface: (i) those which are marked out as possessing certain common physical characteristics—e.g., a certain kind of structure and surface relief, or a particular kind of climate,—and (ii) those regions which possess a unity based upon any significant geographical characteristics, whether physical, biological or human..as contrasted with areas marked out by boundaries imposed..without reference to any geographical unity of the areas. 1971 *Biol. Conservation* IV. 247 (*heading*) Towards a system for classifying natural regions of the world and their representation by National Parks and Reserves. 1870 Natural resources [see *RESOURCE 1 b]. 1921 *Daily Colonist* (Victoria, B.C.) 24 Mar. 13/3 We have toasted our natural resources and talked of the wonderful possibilities of the Province. 1956 J. C. SWAYNE *Conc. Gloss. Geogr. Terms* 100 *Natural resources*, any materials or conditions existing in nature which may be capable of economic exploitation. 1936 R. R. RIVERS *How to buy Timber* iii. 15 Small, specially cut pieces of wood (usually Pine) are inserted crossways between each board in order to let in the air, and so season. That method is called 'air-drying', or 'natural seasoning'. It is a slow process. 1966 A. W. LEWIS *Gloss. Woodworking Terms* 87 The two main methods of seasoning are natural or air seasoning, and kiln drying or artificial seasoning.

II. 8. c. *natural right(s)*, in Western political philosophy, esp. since the 18th century, doctrines derived from concepts of the nature of man and the relationship of the individual to the state whereby certain rights are formulated (see quots.) which the state ought to safeguard.

1689 tr. *B. de Spinoza's Treat. Theol. Pol.* xvi. 343 In Democratical Government, no man so transfers his own Natural Right to another, as for ever after to be excluded from consultation, but only transfers it upon the major part of the Society, of which he still makes one. 1791 T. PAINE *Rights of Man* 111 The end of all political associations, is, the preservation of the natural and imprescriptible rights of man; and these rights are liberty, property, security, and resistance of oppression. *Ibid.*, The exercise of the natural rights of every man, has no other limits than those which are necessary to secure to every *other* man the free exercise of the same rights. 1796 *Encycl. Brit.* XVI. 244/1 *Natural* rights are those which a man has to his life, limbs, and liberty; to the produce of his personal labour; to the use, in common with others, of air, light, and water, &c. 1925 A. D. LINDSAY *Karl Marx's Capital* iii. 60 The labour theory of value is..like all theories of natural right, a revolutionary doctrine. *Ibid.* 61 Theories of natural right are always to this extent misleading—that they are statements of ideals which pretend to be statements of fact. 1939 E. BENEŠ *Democracy* i. 7 The French Revolution became by the declaration of human rights the expression of the whole school of philosophy which for centuries fought for the recognition of the so-called '*natural rights*'—that is, for the innate rights of man, the equality of human beings. 1955 *Philos. Rev.* LXIV. 175, I shall advance the thesis that if there are any moral rights at all, it follows that there is at least one natural right, the equal right of all men to be free. 1971 A. R. BALL *Mod. Pol. & Govt.* vii. 124 The justification for these individual rights was to be found in the theories of natural rights, rights that were beyond the competence of any government interference.

d. *natural frequency*, the frequency at which a mechanical or electrical system oscillates when not subjected to any external forces.

1908 J. A. FLEMING *Elem. Man. Radiotelegr.* i. 33 If..oscillations are maintained which have a frequency different from the natural frequency of [the] circuit, they are called forced oscillations. 1922 GLAZEBROOK *Dict. Appl. Physics* II. 961/1 If the natural frequency is nearly equal to that of the applied force..we have the phenomenon known as resonance. 1962 F. I. ORDWAY et al. *Basic Astronautics* ix. 370 Very serious problems arise if the lowest natural frequency of the vehicle approaches the control frequency. 1971 L. T. AGGER *Introd. Electr.* xxiv. 435 The resonant frequency..in the present simple case is the same as the natural frequency.

12. a. For † *Obs. rare* read 'Now *Hist.*', and substitute for def.: in *Old Med.*, that one of the three spirits (SPIRIT *sb.* 16) which was held to be produced in the liver and pass thence to the heart (see quots.). (Earlier and later examples.)

1477 Spirit Naturall [see SPIRIT *sb.* 16]. 1888 *Encycl. Brit.* XXIV. 95/1 The blood-making organ, the liver, separates from the blood subtle vapours, the natural spirits, which, carried to the heart, mix with the air introduced by respiration, and thus form the vital spirits. 1928 C. SINGER *Short Hist. Med.* ii. 56 Galen believed that food-substance from the intestines was carried as 'Chyle' by the portal vein to the liver. There it was converted into blood and endowed with a particular pneuma, the Natural Spirit, which bestowed the power of growth and nutrition. *Ibid.* iv. 126 We have already traced the wrecking of the Galenic physiology. With its destruction, the old ideas concerning the three types of spirit, natural, vital, and animal, went by the board. **1945** D. Guthrie *Hist. Med.* v. 77 In the liver the blood, endowed with Natural Spirit, passed to the right ventricle, whence it was distributed to nourish all the tissues and organs, and also to the lungs, in order that impurities might be exhaled.

IV. 18. a. (Examples of *attrib.* and pl. use of *natural science.*) Hence *natural-scientific* adj.

1890 W. James *Princ. Psychol.* I. vii. 184 It is highly important that this natural-science point of view should be understood at the outset. **1924** R. M. Ogden tr. *Koffka's Growth of Mind* i. § 4. 15 The behaviour of an animal as it takes place is something to be determined as a natural-scientific event. **1944** H. A. Hodges *Wilhelm Dilthey* iii. 49 Natural-science psychology..takes the mind as a thing among things and studies its processes from a causal point of view. **1949** M. Fortes *Social Struct.* p. ix, Their theme was the comparative study of human society by the methods of the natural sciences. *a* **1963** L. MacNeice *Astrol.* (1964) viii. 262 He began his huge and gallant undertaking while he was still a natural-sciences student at the University of Geneva. **1970** G. E. Evans *Where Beards wag All* xx. 231 When he [sc. the archaeologist] comes..to the question *Why?*, he is forced to look beyond the natural sciences, and his aids are more likely to come from anthropology, history, geography, [etc.].

c. Also *natural scientist.*

1895 F. Thilly tr. *Paulsen's Introd. Philos.* I. i. 59 Natural scientists..are inclined to pursue the former path. **1951** E. E. Evans-Pritchard *Social Anthropol.* iii. 48 This is not the procedure of natural scientists, which most writers of this persuasion—and that means most English social anthropologists—consider themselves to be. **1975** *Notes & Rec. R. Soc.* XXIX. 193 My first intimation of Johnson's views on natural science and on natural scientists came from the *Life of Milton.*

19. Of wool, cotton, silk, etc.: having a colour characteristic of the natural state when unbleached and undyed. Also *natural-coloured* adj. Hence as *sb.* to denote a shade of off-white or creamy beige.

1854 *Morning Post* 7 July 1/5 (Advt.), Aberdeen Linsey Woolseys in granite, heather, and natural wools. *c* **1860** in A. Adburgham *Shops & Shopping* (1964) vii. 74 Vicuna,..woven in its natural colour,..is admirably adapted for Ladies' Cloaks and Gentlemen's Costumes. **1895** *Montgomery Ward Catal.* 9/3 Dress Linen..in the 'natural' flax color only. *Ibid.* 12/3 Plain Habutai Silk in natural (cream) color only. **1927** T. Woodhouse *Artificial Silk* 85 Natural-coloured artificial silk yarns. **1930** *Daily Express* 6 Oct. 9/6 (Advt.), Real Italian hand embroidery on pale cream linen, beautifully worked in blue, gold, rose, or natural. **1941** R. Stout *Red Threads* i. 6 He must have the natural kasha, the one with nubs, by tomorrow afternoon. **1954** [see *alizarin]. **1971** *Vogue* 15 Oct. 144/3 Dress;..colours: cocoa, blue, natural.

20. *natural shoulder* U.S. (see quot. 1973). Freq. *attrib.*

1957 *Men's Wear* (N.Y.) 8 Feb. 69/2 Natural shoulders mark the topcoats as well as the suits. **1958** *Ibid.* 21 Feb. 70/2 What is one man's 'Ivy' is another man's 'natural shoulder garment'. **1962** 'I. T. Ross' *Old Students never Die* i. 13 In a natural-shoulder suit now, instead of a sloppy sweatshirt. **1969** *New Yorker* 11 Oct. 128/1 (Advt.), We don't make outerwear with padding. All our coats are natural shoulder. **1973** *Esquire's Encycl. 20th Cent. Men's Fashions* 670/1 *Natural shoulder.* Term applied to a straight-hanging jacket with medium-width, lightly padded shoulders and a center vent. With this style, favored by university men and others, were worn pleatless, trim-cut trousers.

naturalism. Add: **2.** Also, the view that moral concepts can be analysed in terms of concepts applicable to natural phenomena.

1894 J. Seth *Study of Ethical Princ.* III. ii. 398 We are offered..a new version of the 'Ethics of Naturalism', far superior to the old Utilitarian version, superior because so much more scientific. **1903** G. E. Moore *Principia Ethica* ii. 40, I have thus appropriated the name Naturalism to a particular method of approaching Ethics. **1945** K. R. Popper *Open Society* I. v. 60 Ethical naturalism..has recently been used for confusing the whole issue by advertising certain reactionary, and allegedly 'natural' rights as 'natural laws'. **1952** R. M. Hare *Lang. Morals* ii. 30 Professor G. E. Moore's celebrated 'refutation of naturalism'. *Ibid.* v. 92 Naturalism in ethics, like attempts to square the circle..will constantly recur so long as there are people who have not understood the fallacy involved. **1967** *Encycl. Philos.* III. 69/1 According to ethical naturalism, moral judgments just state a special sub-class of facts about the natural world.

naturalist. 1. (Further examples.)

1952 R. M. Hare *Lang. Morals* v. 92 It is therefore no answer..to claim that a 'naturalist' might if he pleased define 'good' in terms of some characteristics of his choice. **1964** [see *contextualism 1].

naturalistic, *a.* Add: **1. b.** (Further examples.) Spec. *naturalistic fallacy* (see quot. 1903).

1894 J. Seth *Study of Ethical Princ.* III. ii. 398 In spite of his professed impartiality between matter and mind, Spencer does not hesitate to offer such a materialistic or naturalistic interpretation of the moral life. *Ibid.*, A naturalistic scheme of morality, the correlation of the ethical with the physical process. **1903** G. E. Moore *Principia Ethica* i. 10 Ethics aims at discovering what are those other properties belonging to all things which are good. But far too many philosophers have thought that when they named those other properties they were actually defining good; that these properties, in fact, were simply not 'other', but absolutely and entirely the same with goodness. This view I propose to call the 'naturalistic fallacy'. *Ibid.* vi. 201 The naturalistic fallacy has been quite as commonly committed with regard to beauty as with regard to good: its use has introduced as many errors into Aesthetics as into Ethics. **1934** C. D. Broad *Five Types Ethical Theory* vii. 257 Those theories which hold that ethical characteristics *can* be analysed without remainder into non-ethical ones may be called..*Naturalistic Theories.* **1936** A. J. Ayer *Lang., Truth & Logic* vi. 157 We have already rejected the 'naturalistic' theories which are commonly supposed to provide the only alternative to 'absolutism' in ethics. **1965** *Philos.* XL. 308 The attack on the naturalistic fallacy..has been welcomed.

‖ **natura naturans** (natiū·rä næ·tiŭrænz). *Philos.* [L. *nātūra* birth, constitution, etc., f. *nāt-*, pple. stem of *nasci* to be born + med. L. *nāturans* pres. pple.] Nature creating; the essential creative power or act. Also **natura naturata** (-ā·tā), nature created; the natural phenomena and forces in which creation is manifested. Also *transf.* Cf. Nature *v.*[1] 2.

1605 [see Nature *v.*[1] 2]. *c* **1818** Coleridge *On Poesy & Art* in *Biog. Lit.* (1907) II. 257 If the artist copies the mere nature, the *natura naturata*, what idle rivalry!..Believe me, you must master the essence, the *natura naturans*, which presupposes a bond between nature in the higher sense and the soul of man. **1855** Bagehot in *National Rev.* I. 57 A school of 'common-sense poets',..who proceed to describe what they see around them, to describe its *natura naturans*, to delineate its *natura naturata.* **1903** R. A. Duff *Spinoza's Pol. & Ethical Philos.* i. 3 It might be shown that Spinoza did, and could, make no claim to terms such as substance,..idea, *natura naturans* and *natura naturata*..., nor even to the definitions which he gives of them. **1933** R. Tuve *Seasons and Months* iv. 152 Matfre treats first of the nature and power of God ('natura naturans'), Who has created all 'Natura naturata'. **1950** F. Copleston *Hist. Philos.* II. iv. xix. 198 Averroes's..metaphysical scale reaches from pure matter..to pure Act, God, as the highest limit, between these limits being the objects composed of potency and act, which form *Natura naturata.* (The phrases of the Latin translation, *Natura naturans* and *Natura naturata*, reappear eventually in the system of Spinoza.) **1958** E. F. J. Payne tr. *Schopenhauer's World as Will* II. xvii. 175 Such an *absolute system of physics* as described above, which would leave no room for any *metaphysics*, would make *natura naturata* (created nature) into *natura naturans* (creative nature).

nature, *sb.* Add: **2. a.** Also, (one's) *better nature.*

1848 Lytton *Harold* III. xii. ix. 375 His own better nature which, ere polluted by plotting-craft, and hardened by despotic ire, was magnanimous and heroic, moved and won him. **1919** M. K. Bradby *Psycho-Anal.* III. x. 129 Any repressed desire which belongs to my undeveloped 'better nature'. **1949** D. Smith *I capture Castle* xiii. 246 By the time Stephen got home, my better nature had asserted itself and I was terribly worried about his feelings. **1965** M. Frayn *Tin Men* i. 9 Appeal to their better natures.

4. e. *the nature of the beast.*

1678 J. Ray *Coll. Eng. Proverbs* (ed. 2) 77 It's the nature o' th' beast. *c* **1683** J. Verney *Let.* in M. M. Verney *Mem.* (1899) IV. vii. 254 I'me very Sorry John my Coachman Should be soe great a Clowne to you..but t'is the nature of the Beast. **1748** Richardson *Clarissa* III. 218, I might as well have preserv'd the first; for I see it is the *nature of the beast.* **1893** Kipling *Many Inventions* 254 'Twas the nature av the baste to put the comether on the best av thim. **1969** V. Gielgud *Necessary End* v. 48 Barry Compayne never made bones about..the number of girls that he had 'laid'... Anthea had chosen deliberately to put down such exploits to 'the nature of the beast'.

f. *nature and nurture, nature–nurture*, heredity and environment as influences on, or the determinants of, personality (see quot. 1874). Also *attrib.*

[**1610** Shaks. *Tempest* IV. i. 188 A borne-Deuill, on whose nature Nurture can neuer sticke.] **1874** R. Galton *Eng. Men of Sci.: their Nature & Nurture* i. 12 The phrase 'nature and nurture' is a convenient jingle of words, for it separates under two distinct heads the innumerable elements of which personality is composed. Nature is all that a man brings with himself into the world; nurture is every influence from without that affects him after his birth. **1914** F. W. Mott (*title*) Nature and nurture in mental development. *Ibid.* 1 The problem of nature and nurture in mental development is one that has recently acquired importance. **1933** L. Hogben (*title*) Nature and nurture. **1946** *Brit. Jrnl. Psychol.* May 159 The particular nature-nurture ratio value, or the physiological or anatomical associations which it possesses. **1952** C. P. Blacker *Eugenics* 267 Unconscious prejudices can throw the nature-nurture controversy into different perspectives. **1965** R. B. Cattell *Sci. Analysis of Personality* ii. 50 The nature-nurture ratios are not fixed and immutable laws, but statements which may change with culture patterns and the ranges of racial, genetic difference within the given population. **1972** *Times* 2 Sept. 14/3 The argument between the 'nature' and 'nurture' schools of thought. **1974** *Science* 5 July 20/2 The disagreement about the causation of autism... First, there is the usual nature–nurture controversy.

7. Restrict † *Obs.* to sense a and add later example of sense b.

1922 Joyce *Ulysses* 373 Frightened she was when her nature came on her first.

11. a. *balance of nature*: see *balance *sb.* 13 c.

d. *from nature*: see From *prep.* (*adv., conj.*) 13.

12. f. (*one of*) *nature's gentlemen*: a natural gentleman, a person who is a gentleman by nature. Hence in similar phrases, and in extended use: by temperament.

1841 Thackeray *Second Funeral of Napoleon* iii. 67 In the matter of gentlemen democrats, cry pshaw! Give us one of nature's gentlemen, and hang your aristocrats! *a* **1882** Trollope *Autobiogr.* (1883) I. iii. 53 If I say that a judge should be a gentleman..I am met with a scornful allusion to 'Nature's Gentlemen'. **1898** A. J. Munby *Diary* 23 Mar. in D. Hudson *Munby* (1972) 423 A splendid woman, full of rustic health & vigour, & one of Nature's ladies. **1902** G. B. Shaw *Admirable Bashville* ii. 309 You need not be an idle gentleman. I call you one of Nature's gentlemen. **1929** A. Huxley *Let.* 9 May (1969) 311 Now..I can write a letter. It will be a poor one for all yours, because I am not one of nature's letter-writers. **1969** L. Durrell *Spirit of Place* (1969)... One of nature's innocents. He couldn't dissemble if he tried. **1973** *Guardian* 18 June 9/6 We all know the kind of women who are just nature's doormats and..put up with anything.

13. d. *all nature*, everything, everyone, all creation; *like all nature*, like anything, like blazes. *U.S. colloq.*

1819 *Mass. Spy* 3 Nov. 3/1 Father and I have just returned from the balloon—all nature was there, and more too. **1847** *Woodstock* (Vermont) *Observer* 17 Feb., They said too Twould shock like all nature..what stories they tell. **1825** J. Neal *Bro. Jonathan* II. 93 Hurra for you—that beats all nater! **1840** C. F. Hoffman *Greyslaer* II. iii. xiv. 254 The poor critter would have been sucked under, smashed on the rocky bottom, and dragged off like all nater. **1878** Mrs. Brown in *Atlantic Monthly* Oct. 473/1 Cuff would prance round..and seem to think he..had the charge of all natur'. **1892** J. C. Duval *Early Times in Texas* vii. 82 'Well, I declar, boys,' said he, 'if this don't beat all natur.'

14. a. (*d*) (Examples.)

1802 C. Wilmot *Let.* 3 Jan. in *Irish Peer* (1920) 23 My first impression was amazement, at beholding the women in this Town of the first rank. **1970** *Brewer's Dict. Phr. & Fable* (ed. 12) 747/1 *In a state of nature*, nude or naked.

15. a. *nature-cure* (*attrib.* examples), *-folk*, *-lover*, *-loving* (adj.), *-mystic*, *-mysticism*, *-myth* (*attrib.* example), *poem*, *poet*, *poetry*, *ramble*, *-symbol*, *-symbolism*, *walk*, *-worship* (earlier and later examples), *-worshipper*, *-writer*, *-writing* (vbl. sb.); **Nature Conservancy**, an organization responsible for the conservation and study of flora and fauna in Britain, which runs nature reserves, research stations, etc.; **nature conservation**, the preservation of wild fauna and flora and the habitat necessary for their continued existence; **nature-faker** orig. *U.S.*, a person who falsifies reports of natural phenomena, esp. animal behaviour; so **nature-faking** *vbl. sb.* and *ppl. a.*; **nature-name**, a toponym embodying an allusion to a natural feature or phenomenon; **nature-notes**, comments on natural history; **nature reserve**, a tract of land managed in order to preserve its fauna, flora, physical features, etc.; **nature sanctuary**, an area in which the fauna and flora are protected from any disturbance; **nature strip** *Austral. local* (see quot. 1966); **nature study**, the study of natural objects and phenomena, esp. as a subject taught in schools; an example of this; so (*rare*) **nature-student**; **nature trail**, a path linking features of interest, esp. in relation to local natural history, which are described and interpreted by explanatory notices, printed leaflets, or a guide.

1949 *Times* 12 Feb. 3/4 Mr. Herbert Morrison announced in the House of Commons yesterday that arrangements have been completed for forming a Nature Conservancy, and that a separate committee will supervise activities in Scotland... The conservancy—'a more convenient title than conservation board'—would be responsible for the whole of Great Britain. *Ibid.* 5/3 The Nature Conservancy is to have a general charge over all matters relating to the native fauna and flora of Britain. **1959** *News Chron.* 4 Dec. 7/6 The Nature Conservancy has had to abandon plans to establish a warden on Dungeness. **1971** O. Norton *Corpse-Bird Cries* ii. 25 Nature conservancy, or Snowdonia national park, or something. They've been out a nature trail at Llyn Coedig. **1974** M. Blackmore in Warren & Goldsmith *Conservation in Pract.* xxvii. 427 On March 23rd, 1949, it [*sc.* the Government] created the Nature Conservancy by royal charter as a new research council. **1943** *Nature Conservation & Nature Reserves* (Brit. Ecol. Soc.) 7 The whole problem of nature conservation requires to be viewed against the human or social background. **1948** *Times* 30 Apr. 6/3 Mr. H. Morrison..said that the Government accepted in principle the recommendations..calling for the establishment of a Nature Conservation Board. **1953** *Rep. Nature Conservancy to 1952* 3 Though the Act as a whole does not extend to Scotland, Part III and such other Sections as relate to nature conservation are so extended. **1968** C. Burke *Elephant across Border* v. 203 You can do more for nature conservation by shocking people..than a

whole heap of discussion groups and bird magazines can do. **1974** M. Blackmore in Warren & Goldsmith *Conservation in Pract.* xxvii. 423 Nature conservation illustrates the gradual evolution and development of modern ideas. **1906** *Chambers's Jrnl.* 24 Nov. 832/1 At Dr. Lahmann's nature-cure sanatorium,..care is taken to cook vegetables so as to retain the nutritive and soluble salts. **1965** C. Watson *Flaxborough Crab* xvii. 180 What have I to do with this—this nature cure chicanery? **1906** *Everybody's Mag.* June 770 (*heading*) Roosevelt on the nature fakirs. **1909** *Daily Chron.* 8 Dec. 6/4 A President..who never..'goes for' Congress or nature-fakers or millionaires. **1949** *Natural Hist.* Mar. 131/2 Many nature fakers have obtained free meals..through the gullibility of newspaper reporters. **1921** *Daily Colonist* (Victoria, B.C.) 29 Apr. 17/1, I should be sorry to have Mr. John Burroughs catch me nature-faking. **1923** Kipling *Land & Sea Tales* 85 To say that William did not sleep a wink that night would be what has been called 'nature-faking'. **1947** *Sports Afield* Dec. 6/3 It was apparent to me that the writer colored his material, particularly in regard to the nature faking episode. **1927** Peake & Fleure *Peasants & Potters* i. 8 They had settled down into a routine, as had many nature-folk the world over before European industrialism touched them in the last century. **1909** *Chambers's Jrnl.* July 426/2 Many an angler and nature-lover is a veritable 'prisoner of Hope'. **1937** *Discovery* Jan. 32/1 The book is a most suitable gift for nature lovers of all ages. **1960** *Islander* (Victoria, B.C.) 17 Apr. 4/1 It is the perfect spot in summertime for lazing on the beach..more than this, it is a nature lover's paradise. **1913** *Eng. Illustr. Mag.* June 254 It is scarcely possible to find a mountain track or woody dell..which has not fascinated and inspired this nature-loving poet. **1927** J. S. Huxley *Relig. without Revelation* iv. 123 The nature-mystic. **1958** *Economist* 8 Nov. Suppl. 11/1 Tolkien was a visionary possessed of a powerful and discerning mind. To regard him as a lone eccentric or a pre-Romantic Nature-mystic is to under-estimate his stature as a Christian humanist. **1899** W. R. Inge *Christian Mysticism* viii. 302 The true Nature-Mysticism is prominent in St. Francis of Assisi. **1932** C. Williams *Eng. Poetic Mind* ii. 13 Wordsworth was..not ever a Nature-mystic..; he left that to his commentators. **1965** E. E. Evans-Pritchard *Theories Primitive Religion* i. 4 Other anthropologists—if we may include Max Müller and the rest of the nature myth school under this heading—were busy explaining religion in terms of personification of such natural phenomena as sun, sky, and rain. **1960** P. H. Reaney *Orig. Eng. Place-Names* ii. 30 Farnborough 'fern-clad hill', Hertford, 'stag-ford'..were originally nature-names from which later settlements near-by took their names. **1906** M. Cawein (*title*) Nature-notes and impressions in prose and verse. **1935** A. Thirkell..journey of aerial nature-study. **1905** F. H. Shoosmith (*title*) The 'Kingsley' nature poetry books I (*heading*) The 'Kingsley' Nature Poetry Books. **1946** G. Orwell *Coll. Ess.* (1968) I. 55 I wrote bad and usually unfinished 'nature poems' in the Georgian style. **1906** A. Mackie *Nature Knowl. Mod. Poetry* v. 55 (*heading*) Wordsworth as a nature poet. **1923** A. Huxley *Along Road* i. 66 A 'nature poet' (the expression is somehow rather horrible, but there it is). **1932** L. MacNeice *Mod. Poetry* i. 8 Rooted, as nature-poets should be, in their subject. **1905** F. H. Shoosmith (*title*) The 'Kingsley' nature poetry books for schools. **1936** F. R. Leavis *Revaluation* v. 186 'Nature poetry', Victorian or Georgian, pays at the best only an equivocal tribute to his [*sc.* Wordsworth's] greatness. **1934** A. Thirkell *Wild Strawberries* iii. 231 There would be a Nature Ramble at a good brisk pace in Lord Pomfret's grounds. **1969** 'O. Mills' *Dusty Death* xi. 121 We used to..go to one of the cookery classes, or go on a nature ramble. **1915** E. Lankester *Diversions of Naturalist* ii. 17 A society has been founded for the formation of 'nature-reserves' in the British Islands. **1973** *Handbk. Soc. Promotion of Nature Reserves* 8 Woodwalton Fen, Huntingdonshire. This nature reserve..consists of about 360 acres of primitive fenland, a relic of the once extensive Huntingdonshire fens, and is rich in plant and insect life. **1949** *Act* 12 & 13 *Geo. VI* c. 97 § 15 The expression 'nature reserve' means land managed for the purpose—(*a*) of providing..special opportunities for the study of, and research into, matters relating to the fauna and flora..or (*b*) of preserving flora, fauna or geological or physiographic features of special interest in the area. **1959** *News Chron.* 4 Dec. 7/6 A warden to see that building operations cause as little harm as possible to wild life and plants in what is a bloody nature-reserve. **1967** N. Freeling *Strike Out* 39 The sand dunes..have been made into a sort of nature-reserve. **1969** *Times* 3 Mar. 5/8 In a general discussion on establishing nature reserves on farms, the question of unrestricted access and possible vandalism clearly worried some farmers. **1972** *Country Life* 6 Jan. 22/3 Nature-reserve management. **1932** V. E. Shelford in *Ecology* XIII. 202 Reports from the Advisory Board and other members of the committee showed them unanimously in favor of nature sanctuaries to which only persons conducting scientific, artistic or literary work of a serious nature are to be admitted. *Ibid.*, Nature sanctuaries..an account of the desire that they should be kept inviolate. **1953** *Country Life* 6 Apr. 838/3 Where..a marsh has been reclaimed, one cannot expect..in the shape of a nature sanctuary. **1966** Baker *Austral. Lang.* (ed. 2) 401 *Nature strip*, a strip of lawn beside the footpath outside Melbourne homes in 'garden suburbs'. **1973** *Listener* 25 Jan. 118/1 The ground in front of the house—what the Australians call a 'nature strip'. **1902** *Pall Mall Mag.* Aug. 485 Few of these nature-students are men of leisure. **1896** L. C. Miall (*title*) Round the year. A series of short nature-studies. **1897** J. H. Comstock (*title*) Insect life: an introduction to nature-study and a guide for teachers, students and others. **1902** *Chambers's Jrnl.* Oct. 683/1 The Nature-study Exhibition which was held at the Botanical Gardens, London, this autumn. **1928** D. Patton (*title*) Nature study for beginners. **1953** G. Bell *Black Marigolds* i. 19 Bugs got his nature study prize as expected. **1972** J. Wilson *Hide & Seek* ii. 34 Alice taught Mary how to read and how to add up..and they did nature study and learnt about cavemen. **1927** H. Crane *Let.* 12 Sept. (1965) 305 Pocahontas is the mythological nature-symbol chosen to represent the physical body of the continent. *Ibid.* 307 The mother who died... Her succession to the nature-symbolism of Pocahontas. **1926** F. E. Lutz (*title*) Nature trails. **1927** *58th Ann. Rep. Amer. Mus. Nat. Hist. 1926* 106 Dr. Frank E. Lutz,..has conducted for some time a Station for the Study of Insects, in Harriman State Park, near Tuxedo, N.Y. In connection with this outdoor station, in 1925 he established and developed a Nature Trail in the vicinity. *Ibid.* 107 This Nature Trail has been a wonderful stimulus to the present-day movement toward the emphasis of the outdoor museum and hundreds of nature trails have been made in various parts of the country, and also in foreign countries. **1950** W. Hillcourt *Field Bk. Nature Activities* 47 The best location for a nature trail is a park, a camp, a grove adjacent to the school grounds. **1963** *Rep. Nature Conservancy* vii. 103 During the [National Nature] Week the Conservancy set up Nature Trails at Castor Hanglands and Studland Heath National Nature Reserves. **1969** M. Pugh *Last Place Left* iv. 23 All the work of..laying out nature trails had been undone. **1974** *Country Life* 14 Mar. 583/3 An excellent nature trail has been laid over this land. **1932** R. Lehmann *Invitation to Waltz* I. vii. 80 She saw two figures..James and Miss Mivart, returning from their nature walk. **1964** O. Blakeston *Fingers* i. 9 Drilling the catechism class..and giving the children a yearly 'nature walk' as a treat. **1850** R. W. Mackay *Progress of Intellect* I. iii. 151 The elements of personification, as well as Pantheism, are in all Nature-worship. **1960** C. Day Lewis *Buried Day* vii. 147, I took up nature worship now because it was a poetic thing. **1929** A. Huxley *Do what you Will* 158 St. Francis is often hailed as the first nature-worshipper..in Europe since..the Greeks. **1946** Blunden *Shelley* 137 A seer and a nature-worshipper. **1931** —— *Votive Tablets* 259 Some of the..nature-writers mentioned above. **1969** *Times Lit. Suppl.* 16 Jan. 61/1 Thomas's nature writing was good of its kind and time.

b. Restrict *Sc.* to uses in Dict. and add *nature food.*

1847 E. Walker *Diary* in C. M. Drury *Elkanah & Mary Walker* (1940) viii. 205 The year has been fruitful in nature food. **1971** *Sunday Express* (Johannesburg) 28 Mar. 5/6 He tended the children himself, using nature foods prescribed by Mr. Peter Dowling, a practising naturopath.

nature, *v.*[1] Restrict † *Obs.* to sense 1 and add later example of sense 2 (used as *vbl. sb.*).

1880 G. M. Hopkins *Note-bks. & Papers* (1937) 312 The whole function, the naturing, the selving of that nature.

naturelle (natürɛ·l), *a.* and *sb.* [a. Fr. *naturelle* fem. of *naturel* natural. Cf. *natural *a.* 19 above.] (Of) a pale pink or beige colour; skin-colour(ed).

1873 *Young Englishwoman* Feb. 78/1 A dark blue marine velvet hat..with black pompom, and long plume naturelle. **1907** *Yesterday's Shopping* (1969) 537/1 Poudre de Riz..in 4 shades (Blanche, Rachel, Naturelle, Rose). **1927–8** *T. Eaton & Co. Catal.* Fall & Winter 367 Pompeian Face Powder..fine and clinging, in White, Rachel, Naturelle or Flesh.

‖ **nature morte** (natür mort). [Fr.] = Still life. Also *transf.*

1921 R. Fry *Let.* 14 Dec. (1972) II. 518 He's bought..one superb *nature morte*, and he's two lovely Renoirs. **1923** A. Huxley *Antic Hay* xvi. 225 Her face, painted in two tones of red, white, green, blue and black, is the most tasteful of *nature-mortes.* **1938** L. Durrell *Let.* in *Spirit of Place* (1969) 53 He hasn't painted a stroke..not a bloody *nature-morte* even; and when he admits it he looks rather *nature-morte* himself. **1947** *Horizon* Jan. 17 The fixed elements of cubist nature-morte came to a new life. **1960** *Times* 19 Jan. 13/2 By the religious-minded these *nature morte* passages can be interpreted as a humble hymn to the Creator. **1963** M. McCarthy *Group* vi. 123 Pale and lifeless..a veritable *nature morte.* **1970** *Oxf. Compan. Art* 1096/2 Dutch *still-leven*, which..denotes simply a motionless (*still*) aspect of nature (*leven*)... means exactly the same as the French term *nature morte*, which dates from the 18th c.

nature-printing. Add: (Later examples.) **nature-printed** *ppl. a.* (later example); **nature-print** *sb.* (example).

1950 W. Blunt *Art Botanical Illustr.* xi. 141 Some of these works illustrated by nature printing have considerable charm, especially where the process has been used to record grasses, ferns and delicately formed plants. **1967** Cave & Wakeman *Typographia Naturalis* i. 2 It is appropriate that the earliest description of the original technique of nature printing and the oldest extant nature print should both be by Leonardo da Vinci. *Ibid.* 11 Other nature printed illustrations to books were produced in Germany in the late eighteenth and early nineteenth centuries.

naturism. Add: **4.** A movement for, or the practice of, communal nudity in private grounds.

1933 *Gymnos* Nov. 18/1 This book..is the first serious attempt to link Nudism with..Naturism, and Feminism. **1961** *Daily Tel.* 30 Oct. 11/2 Delegates..at the annual conference of the British Sun Bathing Association...agreed..to substitute 'naturism' for 'nudism'. **1973** *Guardian* 28 June 6/1 The Central Council for British Naturism has launched a publicity campaign..[for] official 'naturist beaches'.

naturist. Add: Also *attrib.*

1950 G. Brenan *Face of Spain* v. 113 He is a vegetarian and a firm adherent of the Naturist clinic in Malaga, with its theories of opposites and harmonies in foods. **1971** M. McCarthy *Birds of America* 184 A naturist diet of fruit and raw vegetables.

2. A practitioner of naturism (sense *4). Also *attrib.*

1929 M. Parmelee *Nudity in Modern Life* i. 15 We have all heard of so-called 'naturists', who insist that one should discard everything artificial such as clothing, books, cooked food, etc. **1930** *Observer* 27 Apr. 12/5 Advocates of the health cure of complete nudity..spent a holiday in a naturist colony on an island in the Seine. **1958** *New Statesman* 15 Mar. 330/3 Nudist camps (or 'as we prefer the word "naturist"') were in violent competition. **1963** *Daily Tel.* 20 Mar. 22/5 The description 'a nudist camp', according to the naturist terminology, is defunct... Instead club members are asked to use the expression 'sun club' or 'naturist club'. **1973** [see *naturism 4].

naturopathy (nēitiŭrǫ·pǎþi). [f. Natur(e *sb.* + -o + -pathy (cf. Hydropathy).] A theory of the nature of disease and a system of therapeutic practice founded on the supposition that diseases can be cured by natural agencies.

1901 L. Staden in *Kneipp Water Cure Monthly* Jan. 30/2 There is no doubt that you can get cured without operation by Naturopathy. **1925** [see *Kellogg]. **1948** [see *cultish *a.*]. **1971** *Sunday Express* (Johannesburg) 28 Mar. 5/5 Mr. Fell asked the Supreme Court to reverse the magistrate's findings and to accept that naturopathy is a legitimate form of medicine.

Hence **na·turopath**, one who advocates or practises naturopathy; **naturopa·thic** *a.*

1901 *Kneipp Water Cure Monthly* Jan. 30/2 (*heading*) Naturopathic Adviser. *Ibid.* Nov. 311/1 L. Staden, Naturopath. **1928** S. Lief *Nat. Cure versus Med. Sci.* v. 14 The crux of the Naturopathic contention is that the suppression of every acute malady—after the Allopathic procedure—lays the foundation for some chronic malady. **1937** *Evening News* 20 Jan. 9/1, I think that everybody will admit that we owe the present-day benefits of fasting to the teaching of naturopaths. **1956** *Spectator* 28 Oct. 647 A number of naturopathic practitioners. **1973** *Times* 8 May 18/8 The conference of the British Naturopathic and Osteopathic Association. **1973** *Nation Rev.* (Melbourne) 31 Aug.–6 Sept. 1434/5, I do have sympathy to those people who advocate cures for homosexuality as chiropractor and naturopath.

‖ **Naturphilosophie** (natū·rfilǫsǒfī:). Also **natur-philosophie.** [G., f. *natur* nature + *philosophie* philosophy.] The name given to the theory put forward, esp. by Schelling (1775–1854) and other German philosophers, that there is an eternal and unchanging law of nature, proceeding from the Absolute, from which all laws governing natural phenomena and forces derive. Hence **natur-philosoph, -er**, one who adheres to the theory of Naturphilosophie.

1817 Coleridge *Biogr. Lit.* I. ix. 148 In Schelling's '*Natur-Philosophie*',..I first found a genial coincidence with much that I had toiled out for myself. *a* **1834** —— in K. Coburn *Inquiring Spirit* (1951) 118 The Natur-Phil[osophen] are apt to mistake the new name of a thing..for additional insight. *Ibid.* 251 There are the passages that annoy me in the *Natur-philosophen.* **1846** J. D. Morell *Hist. View Philos.* II. v. 109 For the method by which Schelling accounts for the three dimensions in space, we refer the reader to a little work called the Elements of Schelling's Natur-Philosophie. **1892** W. Wallace tr. *Hegel's Logic* (ed. 2) 430 The formalism of *Naturphilosophie* may teach *e.g.* that understanding is electricity. **1920** A. N. Whitehead *Concept of Nature* ii. 47 A *Natur-philosoph* raises nature to independence, makes it construct itself, and he never feels the necessity of opposing nature as constructed (i.e., experience) to real nature, or of correcting the one by means of the other. **1946** M. R. Cohen *Pref. to Logic* vi. 102 Since the failure of the romantic *Naturphilosophie* to derive infallible knowledge of nature *a priori*, it has become generally evident that all our factual knowledge is only probable. **1957** G. S. Carter *100 Yrs. Evolution* iii. 16 The German abstract or transcendental zoology, usually called Natur-philosophie..owed its origin to Goethe more than to any other biologist. *Ibid.* 19 The theories of the natur-philosophers might have been. **1965** *Listener* 3 June 833/1 One of the important forces which shaped the intellectual milieu of early Victorian England was German *Naturphilosophie.* **1974** *Sunday Times* (Colour Suppl.) 9 June 52/4 In Germany *Naturphilosophie* often coincided with extreme nationalism.

Naugahyde (nǭ·gāhəid). Also (erron.) **naugahide.** [f. *Nauga(tuk*, the name of a town in Connecticut, U.S.A., where rubber is manufactured + *-hyde*, modified form of Hide *sb.*[1]] The proprietary name of a material consisting of a fabric base coated with a layer of rubber or vinyl resin and finished with a grain like that of leather, which is used in upholstery.

1937 *Official Gaz.* (U.S. Patent Office) 7 Dec. 41/2 United States Rubber Products, Inc.... *Naugahyde* for upholstery material, more specifically fabric material which has been treated with rubber and other substances for producing artificial leather. **1971** *Flying* Apr. 49/1 He is..a pilot's pilot..for whom 'cockpit' means not an air-conditioned cell with air-conditioning, but a bucket seat and a parachute. **1973** H. Nielsen *Severed Key* v. 58 The apartment was..finished in contemporary motel naugahyde. **1974**

Anderson (S. Carolina) *Independent* 22 Apr. 5B/3 (Advt.), Several Console Stereos. Modern sewing machines. One Duncan Phyfe sofa, covered in Red velvet. One 2 piece den set, covered in naugahyde.

naught, *v.* Add: **b.** To bring to naught; to annihilate. Also **nau·ghting** *vbl. sb.*

1913 E. Underhill *Mystic Way* 137 It is the final disestablishment and 'naughting' of the separate will. **1930** C. Williams *Poetry at Present* 85 The word death generally suggests a 'naughting' of all that we know. **1958** C. Pepler *Eng. Relig. Heritage* IV. i. 225 Concentrating only on the outgoing features of simplicity in the naughting of self. **1958** *Times Lit. Suppl.* 10 Oct. 581/3 The Self, the One, in whom a Western mystic..seeks to lose his particular warped and transient self is not Himself naughted, is not naught.

naughty, *a.* Add: **2. b.** Also, *naughty naughty*: a reprimand used to a child; also used jocularly designating disapproval of something, *spec.* concerning sex.

1882 *National Police Gaz.* (U.S.) 4 Nov. 10/1 Those naughty naughty parsons up and at it again. **1938** I. Goldberg *Wonder of Words* viii. 150 To a child..we say, 'Nightie-nightie'. Or, if it has been mischievous, 'Naughty-naughty'. **1940** 'G. Orwell' *Inside Whale* 133 From a mere account of the subject-matter of *Tropic of Cancer* [by Henry Miller] most people would probably assume it to be no more than a bit of naughty-naughty left over from the 'twenties. **1946** —— *Crit. Ess.* 127 The naughty-naughty touches in Dali's autobiography.

3. b. *the naughty nineties* (see quot. 1970).

1925 R. Le Gallienne *Romantic '90s* iv. 162 'The '90s' are usually spoken of as if they had only one colour: the 'yellow' '90s, or the 'naughty' '90s, or the 'decadent' '90s. **1930** Sellar & Yeatman *1066 & All That* lix. 111 Oscar Wilde..was the leader of a set of disgusting old gentlemen called 'the naughty nineties'. **1937** *Jrnl. R. Aeronaut. Soc.* XLI. 128 He is carrying us back into the nineties of last century..(today I note referred to as the 'naughty nineties'!). **1939** *Burlington Mag.* Apr. 200/2 He was essentially a 'naughty 'nineties' figure. **1970** *Brewer's Dict. Phr. & Fable* (rev. ed.) 748/1 *Naughty Nineties, The*, the 1890s in England, when the puritanical Victorian code of behaviour and conduct gave way in certain wealthy and fashionable circles to growing laxity in sexual morals, a growing cult of hedonism, and a more light-hearted approach to life.

naughty (nǭ·ti), *sb.* [f. the adj.] **1.** *to do the naughty* (slang): to behave in a sexually promiscuous way. Similarly *to go naughty.*

1869 F. Hall in D. Lyndesay *Works* IV. 498 The wealth of the prelates keeps our daughters unwedded. And some of them go naughty. **1902** Farmer & Henley *Slang* V. 20/2 Shop and working girls in large towns sometimes say they work for their living, but do the naughty for their clothes. **1937** Partridge *Dict. Slang* 553/1 *Naughty, do the*, play the whore; to coït (of women only).

2. *Austral.* and *N.Z. colloq.* or *slang.* (An act of) sexual intercourse. Hence as *v. trans.*, to have sexual intercourse with.

1959 D. Niland *Big Smoke* vii. 169 The woman giggled. ..'Come on, what about it?' Ocker shook his head, grinning. 'I'd like to oblige, but I can't... It's in me contract, no leaving the job for a naughty.' **1961** Partridge *Dict. Slang* Suppl. 1195/2 *Naughty, v.* To coït with... 'He naughties her.' **1962** *Times Lit. Suppl.* 12 Oct. 793/4 Would you please whisper in the ear of the young lady who reviewed *The Stuart Case* in your issue of August 10 that 'to have naughty'..is throughout the South Seas the polite and strict analogue of 'to have sexual intercourse'. **1963** F. Hardy *Legends from Benson's Valley* 11, ..smiled, remembering his oft-repeated remark: 'I get a lot of knock backs but I get a lot of naughties.' **1967** F. Sargeson *Hangover* vii. 55 He read: 'We'll naughty anyone. No, not on your life, not after Coral.' **1969** *Private Eye* 25 Apr. 12 What bloody fun have I had? Eight kiddies and two more on the way and I haven't even negotiated a naughty!

naughty (nǭ·ti), *adv.* [f. Naughty *a.*] In a naughty or improper manner.

1898 J. D. Brayshaw *Slum Silhouettes* 142 He looked reg'lar dook. He'd a pair o' lavinder-coloured bell-bottom trowsis, cut werry naughty. **1919** Mencken *Amer. Lang.* 228 The child behaved naughty. **1922** Joyce *Ulysses* 441 Naughty cruel I was.

naujakasite (nauyākā·zəit). *Min.* [f. *Naujakas-ik*, a point on the southern coast of Tunugdliarfik Fiord, Greenland + -ite[1].] A hydrated silicate of sodium, iron, and aluminium (with substitution by other metals), $(Na,K)_6(Fe^{II},Mn,Ca)(Al,Fe^{III})_4Si_8O_{26}.H_2O$, which is found as white, platy, monoclinic crystals.

1933 O. B. Bøggild in *Meddelelser om Grønland* XCII. ix. 7 Among the minerals collected by G. Flink on his mineralogical journey in 1897 there is a specimen labelled 'Chorite?' by Flink. The locality is stated to be Naujakasik... The specimen containing the naujakasite was probably found loose on the ground. **1960** *Mineral. Abstr.* XIV. 370/2 Naujakasite, formerly only known from a loose boulder, has now been found in situ at Tupersuatsiaq and the northern part of the Ilímaussaq batholith. **1967** *Meddelelser om Grønland* CLXXXI. vi. 14 Naujakasite alters easily through several stages to analcite.

naupliar (nǭ·pliǝɹ), *a.* [f. Naupli(us + -ar[1].] = Nauplial *a.*

1961 in Webster. **1963** [see *copepodid *a.*]. **1975** *Sci. Amer.* Mar. 80/2 In some instances there is enough wax in the egg to carry *E*[*uchaeta*] *japonica* through the entire six naupliar stages. **1975** *Nature* 17 Apr. 591/2 The freshwater layer of the lake has been colonised by a calanoid copepod, *Pseudoboeckella* sp. (average concentration recorded 0.5 individuals l⁻¹, including naupliar stages).

Nauruan (nɑu·ruǎn), *sb.* and *a.* [f. *Nauru*, an island in the western Pacific + -an.] **A.** *sb.* A native or inhabitant of Nauru. **B.** *adj.* Of or pertaining to Nauru.

1921 R. D. Rhone in *Nat. Geogr. Mag.* Dec. 563/2 The Nauruans have never been cannibals, but they had the reputation of being savage warriors. *Ibid.* 576/1 In the Nauruan legends the coconut..either owes its eyes and mouth to human ancestry or man owes his eyes and mouth to the coconut. **1925** *Windsor Mag.* Mar. 398/2 A middle-aged Nauruan, clad only in a *ridi*, presents an impressive and ponderous appearance... But..nothing could be more charming than a Nauruan maiden. **1935** A. F. Ellis *Ocean Islands & Nauru* v. 41 From time immemorial a special kind of fish has been cultivated by the Nauruans in this lagoon. **1951** L. Mason in O. W. Freeman *Geogr. Pacific* x. 296 About a thousand Nauruans were transferred in 1943 to Truk... A postwar sequel in Nauruan history has to do with its new status as a Trust Territory. **1963** *Austral. Encycl.* VI. 257 In the past most of the Nauruans were skilled fishermen. **1969** *Age* (Melbourne) 24 May 2/4 The Acting Chief Secretary (Mr. Manson) warned yesterday that Victorians taking part in the Nauruan pools could face penalties of up to $100.

nausea. Add: Also with pronunc. (nǭ·ziǎ).

4. Special Comb.: **nausea gas**, a gas used to induce nausea in people.

1966 *Guardian* 10 May 12/4 US Army planes dropped nearly three and a quarter tons of nausea gas on the suspected jungle headquarters of the Vietcong yesterday. **1970** *Globe & Mail* (Toronto) 28 Sept. 17/1 Soldiers fired nausea gas to drive back a stone-throwing mob..in Belfast.

nauseate, *v.* Also with pronunc. (nǭ·zɪ,eɪt).

nauseous, *a.* Also with pronunc. (nǭ·ziəs).

Naussie (nǫ·si, nǫ·zi). *colloq.* [f. N(ew *a.* + *Aussie *sb.*] = *New Australian* (*Australian *sb.* 2 b).

1959 [see *Australian *sb.* 2 b]. **1959** Baker *Drum* 129 *Naussie*, a N(ew) Aussie.

nautic, *a.* and *sb.* Add: **B.** *sb.* **b.** A sailor, esp. of the Royal Navy.

1909 *Westm. Gaz.* 3 June 4/2 'Nautics' love the spray of the waves more than they do the dust of the roads. **1943** B. J. Hurren *Eastern Med* vi. 76 A complete reversal of policy was now thrust upon an eager company of flying nautics. **1951** P. Brickhill *Dam Busters* xix. 244 A certain..personality at Bomber Command..when he heard the *Tirpitz* was sunk, [said] 'That's one in the eye for the Nautics!' **1973** *Sunday Tel.* 4 Mar. 38/1 The Army did as they had been done by in the first-half. Now it was the Nautics who were under hideous pressure.

navaid (næ·vēɪd). [f. Nav(igational *a.* + Aid *sb.*] Any navigational device in an aircraft, ship, etc.

1956 S. Rosenberg *Systems Analysis Approach to Choice of Long Distance Navaid* (Rome Air Devel. Technical Note 56–279, U.S. Govt. Res. Rep. AD 97714, Aug.) 1 Increased interest has been displayed recently on the need for a single, standard long-distance navigational aid (navaid) capable of meeting the latest..operational objectives. **1960** *Electronics Weekly* 21 Sept. 2 (*heading*) Ocean-based navaid to be tested by US. **1971** *Electronics & Communications Abstr.* XI. 47 An analytical study is presented of a new navaid system designed to assist the navigation of aircraft in remote and low traffic areas by facilitating homing and orbiting on ground beacons. **1971** *Flying* Apr. 83/2 When ranked against the ..expenditures proposed for terminal and en-route traffic control (not counting navaids and landing aids), the FSS money looks like a couple of drops they might spill while pouring.

Navajo (næ·vǎho), *sb.* and *a.* Also **Nabajo, Nabeho, Navaho, Navajoe.** [f. Sp. *Apaches de Navajó*, f. Tewa *Navahu* large field.] **A.** *sb.* **a.** An Athapascan Indian people of Arizona, Utah, and New Mexico; a member of this people. **b.** The Athapascan language of this people. **B.** *adj.* Of, pertaining to, or characteristic of this people.

[**1629** Zárate-Salmeron in *Papers Archaeol. Inst. Amer.* (1892) Amer. Ser. IV. 294 La nacion de los Indios Apaches de Nabajù.] **1780** in *New Mexico Hist. Soc. Publ.* (1918) No. 21, 36 The war cruelly made upon them [*sc.* pueblos] by the Utes and Navajos. **1822** J. Fowler *Jrnl.* 8 Mar. (1898) 123 The Spanierds Have Sent 700 men against the nabeho Indeans. **1834** A. Pike *Prose Sk. & Poems* 99 An Indian girl with her Nabajo blanket, black, with a red border. **1850** [see *Jicarilla]. **1873** J. H. Beadle *Undevel. West* xxv. 524 John H. Van Order acted as interpreter from English into Spanish, and Jesus Alviso from Spanish into Navajo. Nearly all the employes understood a little Navajo, but not enough to interpret. **1907** [see *Apache 1]. **1910** G. W. James *Grand Canyon* iii. 18 Gray Navaho rugs cover the brown floor. **1921** [see *Apache 1]. **1929** D. H. Lawrence *Pansies* 39 A Navajo woman, weaving her rug in the pattern of her dream. **1951** see *beat-up *sb.* and *a.*]. **1957** P. Worsley *Trumpet shall Sound* 243 The Navaho..were one of the few Indian tribes who remained unaffected by the Ghost Dance of 1890. **1972** A. Fowles *Double Feature* ix. 173, I stared, instead, at a Navajo rug hanging on the wall. It was worked in small black triangles of black and red, worked in a stark zigzag pattern on a cream background. **1973** T. Allbeury *Choice of Enemies* xxviii. 155, I can get by quite happily in Navajo. **1975** *Guardian* 21 Jan. 2/6 The more culturally intact tribes such as the Navajos, the nation's largest.

naval, *a.* Add: **2. d.** *naval brass*, a type of brass containing about 60 per cent copper, 39 per cent zinc, and one per cent tin, used for bolts and other small fittings of ships.

1881 *Calvert's Mechanics' Almanack 1882* 27 In 1874, an alloy, composed of 62 parts of copper, 37 of spelter, and one of tin, was proposed by Mr. Farquharson... The new alloy is specified for all ships built for the Admiralty, and the details now given may be of service to anyone using naval brass. **1928** S. G. Wheeler *Marine Engin.* II. xvii. 510 Naval brass can be rolled or forged, but unless it can be worked in this way has poor strength when cast. **1964** S. H. Avner *Introd. Physical Metall.* xii. 353 Leaded naval brass with the addition of 1·75 Pb for improved machinability is used for marine hardware. **1969** D. K. Allen *Metall.* xiii. 438/2 Naval brass or Tobin bronze..has increased resistance to salt water spray and is used for condenser plates, welding rod, propeller shafts, and marine hardware.

e. *naval base*, a securely held seaport from which naval operations can be carried out.

1906 F. T. Jane *Heresies of Sea Power* II. ii. 126 These are they who assign the first and second places to the fleet; the shore and the shore forces come but a bad third. The advocates of naval command of naval bases may be found among these. **1941** A. J. Marder *Brit. Naval Policy 1880–1905* x. 183 England's key position in the Mediterranean, though strongly fortified, could hardly be called a naval base at this time. **1957** P. Mackesy *War in Mediterranean 1803–10* 13 Malta..was the only naval base possessed by England from which the Toulon fleet could be watched and maritime command exercised in the central Mediterranean. **1969** *Islander* (Victoria, B.C.) 9 Nov. 16/2 In the early days Esquimalt was a British naval base, but merchantmen also used the harbor extensively.

f. *naval bank holiday*: (see quot. 1961). *colloq.*

1916 G. Franklin *Naval Digression* vii. 220 We had a typical 'naval bank holiday' on Boxing day—coaling ship. **1948** Partridge *Dict. Forces' Slang* 124 *Naval bank holiday*, coaling the ship. **1961** —— *Dict. Slang* Suppl. 1195/1 *Naval bank holiday*, a day spent in coaling the ship.

4. d. *naval brigade*, a landing force; a reinforcement force for land troops.

1883 Melton & Oliphant *Cruise of U.S.S. Galena* vii. 80 On several occasions our Naval Brigade was landed upon the breakwater and exercised in marching and counter-marching. **1884** *Naval Encycl.* 510/1 It has been customary in the service to give the name of 'naval brigade' to even a single ship's company, although by rights it refers to a larger organization. **1901** J. Blake *How Sailors Fight* xii. 241 Naval brigades took part in the operations with Buller, with Methuen, and with Lord Roberts; but of all it is probable that at the siege of Ladysmith the services rendered were most valuable. **1904** J. S. Corbett *Eng. in Mediterranean* II. xxiv. 119 The real attack was made from the centre with five battalions of infantry, the naval brigade, and the three troops of British horse. **1937** H. Fitch *My Mis-spent Youth* iii. 14 As a Captain, Beresford had been in charge of the naval brigade during the River War which led up to the battle of Omdurman.

navalist (nēɪ·vǎlist). [f. Naval *a.* + -ist.] One who stresses the importance of having a strong navy. Also *attrib.*

1920 *Glasgow Herald* 30 Dec. 6/3 Mr. Daniels's rather flamboyant allusions to the American naval programme would be utilised by our domestic navalists. **1927** *Observer* 20 Mar. 16/4 'Neon'..is a good old-fashioned navalist and an obsolete politician. **1961** A. J. Marder *From Dreadnought to Scapa Flow* I. 158 The insistence of navalist opinion..that in an era of international strife it was imperative that the private armament firms be kept in efficient condition. **1963** R. A. Hough *Hunting of Force Z* iii. 64 Navalists and airmen and their supporting politicians and suppliers bandied about the old dogmas and statistics. *Ibid.* 72 Bombardment from the air, navalists the world over calculated, was no more than another threat to be added to that of the torpedo-boat and submarine.

‖ **navarin** (navaræ̃). [Fr.] A mutton stew made with small onions and potatoes. Also **navarin printanier** (præ̃tanie), a navarin made with spring vegetables.

1877 E. S. Dallas *Kettner's Bk. of Table* 309 Navarin is a stupid word which has arisen from a desire to get rid of the unintelligible and misleading name, Haricot de mouton, without falling back on the vulgar phrase, Ragoût de mouton. It was at first selected with a thought of punning upon the *navet* or turnip, which is so prominent in the Haricot de mouton. **1907** A. Escoffier *Guide Mod. Cookery* xv. 448 Navarin Printanier... Transfer the pieces of mutton..to another saucepan with..small, new onions..new trimmed carrots..new turnips..new potatoes..fresh peas, and..French beans. **1951** E. David *French Country Cooking* 111 *Navarin Printanier.* This is a ragoût of lamb or mutton to which spring vegetables give special character... As soon as they [*sc.* peas] are cooked the *Navarin* is ready. **1963** *Economist* 30 Nov. 931/2 A basic recipe like *navarin.* **1966** *Vogue* Nov. 154/3 Dishes of the day may be Navarin Printanier, daube of lamb..or Chicken à la King. **1970** Simon & Howe *Dict. Gastron.* 275 *Navarin*, the French culinary

name for a ragoût of mutton..made either with small onions and potatoes or with different vegetables such as carrots, turnips, new potatoes and green peas in which case *à la printanière* is added to its name.

Navarrese (nævārī·z), *sb.* and *a.* Also **Nava·rran.** [f. *Navarr(e* a province of northern Spain, formerly a kingdom which included part of south-west France + -ese.] **A.** *sb.* The people of Navarre; a native or inhabitant of Navarre. **B.** *adj.* Of or pertaining to Navarre.

[**1699** J. Stevens tr. *Mariana's Gen. Hist. Spain* VIII. iii. 122 At this time the Count of Toulouse, came in with fresh supplies to assist the Navarrois.] **1846** R. Ford *Gatherings from Spain* xiii. 147 The Navarrese drink their Peralta, the Basques their Chacolet. **1855** C. M. Yonge *Lances of Lynwood* xiv. 219 The swarthy Navarrese mountaineer. **1915** C. C. Martindale *In God's Army* I. 122 His servant, Miguel, was a Navarrese of bad character. **1932** E. Hemingway *Death in Afternoon* xii. 125 Navarrese bulls are almost a different race, smaller and usually of a reddish color. **1943** E. A. Peers *Spain in Eclipse* I. i. 25 The difficult country north of Reinosa had been negotiated by the Navarrans. *Ibid.* ii. 46 Navarran troops entered Barbastro and a rather slower advance began against Lérida. **1956** P. Kemp *Mine were of Trouble* ii. 26 A young Navarrese officer. *Ibid.* iv. 76 Father Vicente, the Company Chaplain, a stern-faced, lean Navarrese.

nave, *sb.*[1] Add: **1. b. nave plate** = *hub-cap.

1962 *Which? Car Suppl.* Apr. 67/1 The Austin A60 had a grease gun, nave plate remover and box spanner. **1968** *Radio Times* 13 June 29/3 A rattle from the tail end [of a car],..sounding..like a loose wheel nut bouncing about inside the nave plate.

navel, *sb.* Add: **1. a.** *to contemplate* (or *regard*) *one's navel*: to engage in meditation or contemplation; to be complacently parochial or escapist; so *navel-contemplation.* Cf. *navel-contemplator*, *navel-gazer.

1921 D. H. Lawrence *Let.* 2 May (1962) II. 650 *Your* Nirvana is too much a one-man show: leads inevitably to navel-contemplation. **1933** E. O'Neill *Days without End* (1934) I. 21 His letters..extolled passionless contemplation so passionately that I had a mental view of him regarding his navel frenziedly by the hour and making nothing of it! **1966** *Listener* 24 Nov. 770/1 One sits in a New York traffic jam, contemplating, as it were, the city's navel, and the conclusion is inevitable that death from a combination of congestion and suffocation is not far off. **1974** *Times* 27 June 18/3 To fight off the navel-contemplation mood induced by our move of office. **1975** *Times* 2 June 12/8 Lift our eyes for a moment from the contemplation of our own unlovely navels and look out to where..our fellow human beings live.

4. navel-cord = Navel-string; **navel-gazer** = Omphalopsychite; also *transf.* (cf. *navel-contemplator*); so **navel-gazing** *vbl. sb.*; **navel-stone**, a stone that marks a navel (sense 2).

1922 Joyce *Ulysses* 385 Our grandam, which we are linked up with by successive anastomosis of navelcords. **1952** L. MacNeice *Ten Burnt Offerings* 37 Crystal-gazers, navel-gazers. **1963** *Kenyon Rev.* XXV. 549/1 This piece of navel-gazing also reveals a dangerous and sometimes excessive self-consciousness. **1972** *Publishers' Weekly* 10 July 27/2 David Obst has no monopoly on national navel-gazing. **1917** *Encycl. Relig. & Ethics* IX. 492/2 Zeus, wishing to ascertain the exact centre of the earth, sent forth two eagles to fly simultaneously at equal speed from its eastern and western ends. They met at Delphi, and there in Apollo's temple was set up in commemoration the holy Navel-stone..to mark earth's central point. **1922** A. E. Housman *Last Poems* 50 Mute's the midland navel-stone beside the singing fountain.

navette (nævε·t). [Fr., lit. = little boat, f. med. L. *naveta*, dim. of *navis* ship. Cf. Navet[2].] A cut of jewel in the shape of a pointed oval; a jewel cut in such a shape. Also *attrib.* Cf. Marquise 4.

1908 H. C. Smith *Jewellery* 246 Many of the best-known collections contain examples of these 'nef' or 'navette' pendants. **1944** *Times* 21 Mar. 2/2 A flower pendant, the six petals formed of diamond navettes with a single ruby centre. **1945** A. Selwyn *Retail Jeweller's Handbk.* xv. 217 Fancy shapes, such as the three-cornered, the marquise or navette.., the pear-shaped..or pendeloque, make unusual jewels, and are generally suggested by the natural form of the diamond itself. **1949** G. F. H. Smith *Gemstones* (rev. ed.) xii. 153 The commonest distorted shapes are the marquise or navette..and the pendeloque or drop-form. **1963** *Guardian* 15 Nov. 10/5 The leading trend today in diamond cutting is the navette—shuttle-shape an enormous pendant *navette* diamond. **1967** *Times* 21 Feb. 21/7 (Advt.), A navette-shaped diamond single-stone ring. **1973** *Country Life* 18 Oct. (Suppl.) 71 Sunburst brooch..brilliant baguette and navette diamond set.

Nav. House, Nav House, *colloq.* name for the Navigation School in H.M. Dockyard, Portsmouth.

1924 'Nauticus' *Sea Ways & Wangles* viii. 48 You will then be sent to the Navigation School or 'Nav. House' as it is more familiarly called. This, as you might expect, is situated ashore, near the main gate of Portsmouth Dockyard. *Ibid.* 52 However you are not likely to become Navigator of a flagship until you have been through another course at the 'Nav. House' to qualify you for first class ships. **1948** Partridge *Dict. Forces' Slang* 124 *Nav House, the*, the R.N. Navigation School, Portsmouth.

navicert (næ·visɔ̄ɹt). [f. L. *navis* ship + Cert(ificate *sb.*] A consular certificate granted to a neutral ship testifying that her cargo is correctly described in the manifest. Hence as *v. trans.*, to authorize with a navicert.

1923 C. E. Fayle *Seaborne Trade* II. xx. 304 The 'Navicert System', as it was called, from the code word employed, became, in fact, one of the chief instruments in the prevention of enemy trade. **1939** *Times* 3 Nov. 5/1 Arrangements are approaching completion for introducing the navicert system, which was much used in the last war. **1940** *Times* 21 Mar. 5/5 On Tuesday a vessel came through with no fewer than 200 navicerted items of cargo. **1940** R. W. B. Clarke *Britain's Blockade* 10 A plan is in operation for issuing compulsory navicerts to all ships coming to Europe and North Africa. **1941** *Ann. Reg. 1940* 7 To make matters easier for neutrals, the 'navicert' system introduced in the last war was being greatly extended. **1941** *Times* 2 Apr. 3/5 After the British Government had agreed to navicert two gift cargoes of flour. **1967** D. L. Busk *Craft of Diplomacy* ii. 46 During the last war..the British issued what were called 'navicerts'. These guaranteed shippers that 'clean' items of cargo would not be confiscated by the Royal Navy as contraband destined for the enemy.

naviculoid (nǎvi·kiŭloid), *a.* [f. L. *navicula*, dim. of *nāvis* a ship; (in sense 1) adopted as a generic name by J. B. G. M. Bory de St. Vincent in *Dictionnaire classique d'Histoire naturelle* (1827) XI. 472/2: see -oid.] **1.** Of, pertaining to, or resembling a diatom of the genus *Navicula.*

1894 P. T. Cleve in *Kongelige Svenska Vetenskaps-akademien Handlingar* XXVI. II. 10 It is by no means an easy task to construct an artificial key of such numerous and variable forms as the naviculoid diatoms. **1916** G. S. West *Algæ* I. 90 In some of the larger naviculoid diatoms the cell-wall is destitute of pores. **1946** *Nature* 26 Oct. 588/1 It was found that the form of the protoplasmic bodies was returning to the naviculoid.

2. = Navicular *a.* 3 b.

1961 F. W. Butcher *New Illustr. Brit. Flora* II. 172 Seed..naviculoid, rugose, black.

navigable, *a.* Add: **2. b.** (Earlier and later examples.) Also of airships. Now *Obs.* exc. *Hist.*

1783 in S. Stubelius *Balloon* (1960) 151 We have received a prodigious number of letters relative to the aerostatic machine of Mess. de Montgolfier. Some of these propose methods for rendering this machine truly navigable. **1835** *Mechanics' Mag.* 15 Aug. 374/2 This has been the case with steam-carriages, steam-boats, and other machines, and why should a navigable balloon be excepted? **1887** *Nature* 13 Jan. 260/1 Captain Renard has recently sent in to the French Academy an account of his experiments with his so-called navigable balloon, *La France*, at Meudon. **1907** *Cornh. Mag.* May 611 We shall not go very far wrong if we say that the limits of speed attainable with navigable balloons are not widely different from those attainable in marine navigation. **1908** H. G. Wells *War in Air* ii. 41 There were several navigable gas air-ships, not to mention balloons, in the air.

Hence as *sb.*, a navigable balloon. Cf. *dirigible *sb. Obs.* exc. *Hist.*

1882 W. N. Hutchinson in *United Service Mag.* II. 262 This principle of diminishing buoyancy by diminishing bulk is as applicable to the ordinary balloon..as to the navigable, but..the strain on the material..of the navigable..would be trifling. **1908** H. G. Wells *War in Air* i. 18 They started ironclads, they started submarines, they started navigables. **1933** —— *Shape of Things to Come* I. § 7. 70 That primitive 'navigable' the Zeppelin.

navigate, *v.* Add: **1. c.** *U.S.* To walk steadily; to keep on one's course.

1843 'J. Slick' *High Life* (1844) I. vii. 109 It warn't no easy matter tu navigate so as not tu git a second ducking, for every nigger in York seemed to be out a washing winders. **1846** *Spirit of Times* 11 July 234/3 Well, by this time I began to think of navigating. **1881** R. T. Cooke *Somebody's Neighbors* 88 What are you navigating round me for? **1904** *Sun* (N.Y.) 9 Aug. 10 She was so drunk that she could barely navigate. **1908** G. H. Lorimer *Jack Spurlock* 117 While he could navigate successfully..he could only just stuttah. **1930** *Randolph Enterprise* (Elkins, W. Virginia) 13 Feb. 1/1 The fellow was..hardly able to navigate as he was carrying a heavy load of Prohibition poison.

3. b. *transf.* and *fig.*

1901 G. B. Shaw *Capt. Brassbound's Conversion* I. 215 Spiritually a little weatherbeaten, as having to navigate his creed in strange waters crowded with other craft. **1934** Dylan Thomas *18 Poems* 12 Sleep navigates the tides of time.

c. *absol.* and *transf.* of a motor vehicle.

1965 I. Fleming *Man with Golden Gun* v. 67 Mary Goodnight had insisted on coming along, 'to navigate and help with the punctures'. **1967** J. Caird *Murder Scholastic* viii. 93 Once they were in the car, David said: 'I'll navigate. Turn left at the school gate'. **1967** L. Meynell *Mauve Front Door* ix. 117 We all piled into it [*sc.* the car]. Three of us in the front seat, Ray and Zena behind. Tessa navigated. **1971** 'H. Howard' *Million Dollar Snapshot* vii. 108 Zombie tossed me the keys. He said, 'You drive, I'll navigate.'

6. *trans.* To manage, direct, sail, or fly (a balloon, airship, aeroplane, or the like) in the air. Hence *spec.* to plot and supervise the course of (an aircraft or spacecraft).

1784 *Universal Mag.* LXXIV. p. ii, By imitating the action of..wings, sails, oars, and a rudder, we may be able to navigate a Globe [*sc.* a balloon] in any direction we please. **1877** *Design & Work* III. 603/1 To navigate it [*sc.* an airship] in England, and navigate it to Zanzibar. **1920** *Blackw. Mag.* July 5/1 The pilot of an aeroplane is almost wholly occupied with navigating his craft. *Ibid.* 13/2 If we can succeed..in building and in navigating a few score of serviceable dirigibles. **1922** *Encycl. Brit.* XXX. 43/1 Not only had the flying-boats on the water to be navigated but the pilot and observer had to 'navigate' a bomb to its desired target. **1951** *Oxf. Jun. Encycl.* IV. 289/2 Aeroplanes are navigated first by careful planning before the flight, and then by an effort to keep the course planned throughout the journey. **1958** C. C. Adams et al. *Space Flight* xiii. 326 Seven stars must be known to properly navigate a ship [*sc.* a spaceship].

b. To travel or fly through (the air).

1901 *Chambers's Jrnl.* Mar. 207/2 Count Zeppelin's airship,..with a row of seventeen balloons..for navigating the air, has also pointed cigar-like ends. *Ibid.* July 479/2 In the meantime his efforts to navigate the air have, as a matter of course, resulted in the inauguration of many rival schemes. **1907** *Cornh. Mag.* May 609 Grotesque and fantastic schemes for navigating the air were put forward. **1927** C. L. M. Brown *Conquest of Air* 8 Stories of wizards and witches who navigated the upper air with the assistance of tubs and broomsticks.

navigating, *vbl. sb.* Add: Also in aeronautical and motoring uses.

1950 M. Laski in *Contact* May–June 50/2, I must warn you, too, of the danger of letting navigating become an end in itself, of sitting with one's eyes glued to the map until it becomes a substitute for the country outside. **1966** 'E. Peters' *Piper on Mountain* iii. 51 He enjoyed driving, but to him navigating was a chore. **1971** *Gloss. Electrotechnical, Power Terms* (*B.S.I.*) III. vi. 22 A radio navigating system. **1973** A. Ross *Dunfermline Affair* 103 Charlie drove, and I did the navigating.

navigation. Add: **1. d.** The action or practice of travelling through the air by means of aircraft; flying. More fully, *aerial navigation.* ? *Obs.*

1804 [see *aerial navigation* s.v. *aerial *a.* 5]. **1835** *Mechanics' Mag.* Aug. 290/1 The first experiment of this new system of aërial navigation will be made from London to Paris, and back again. **1870** tr. *F. Marion's Wonderful Balloon Ascents* II. ix. 163 The idea of aerial navigation by means of an apparatus heavier than the atmosphere. **1910** *Blackw. Mag.* July 12/2 The navigation of the air. **1920** *Act* 10 & 11 *Geo. V* c.80 § 1 Limited to aircraft of any special description, or engaged in any special kind of navigation.

2. b. The art or science of directing the movements of aircraft or spacecraft, esp. in regard to a craft's position and course. More fully, *aerial* (or *air*) or *celestial navigation.*

1922 *Encycl. Brit.* XXX. 14/1 Aerial navigation, as distinct from piloting with the ground in view, developed tardily everywhere, though first in Britain. **1931** *Times* 19 Feb. 17/1 A haze..limited visibility to five miles. Navigation was undoubtedly difficult. **1951** *Oxf. Jun. Encycl.* IV. 291/1 Some methods of navigation employ both radar and radio at the same time... Some of the radar and radio devices..can be used both for the main part of a flight and for the actual landing. **1964** F. I. Ordway et al. *Basic Astronautics* ix. 385 (*heading*) Celestial navigation. *Ibid.*, The platform of a celestial navigation system includes an automatic sextant. **1974** *Encycl. Brit. Macropædia* I. 376/2 The flight schools taught the essentials of air navigation and blind flying to thousands of military pilots during World War II.

c. The action or practice of navigating a car (*navigate *v.* 3 c) or other vehicle.

1944 *Return to Attack* (Army Board, N. Z.) 9/1 Battalions and regiments entered the wilderness in battle order and exercised in desert navigation. **1950** [see *navigator 1 c]. **1973** H. Miller *Open City* xiii. 149 I'm relying on you for navigation..just keep me on the road.

7. a. (Later example.)

1916 A. Bennett *Lion's Share* i. 7 Probably the largest yacht that had ever threaded that ticklish navigation.

8. In senses relating mainly to air navigation: *navigation beacon, instrument, log*; **navigation light**, one of a set of lights shown by a ship or aircraft at night; **navigation satellite**, a satellite whose orbit is accurately known and made available, so that observations of its position may be used for navigational purposes.

1941 J. Masefield *Nine Days Wonder* 3 In war-time, when the navigation-beacons are extinguished. **1959** F. D. Adams *Aeronaut. Dict.* 117/1 *Navigation instrument*, an aircraft instrument that indicates, or that is used to ascertain, information relating to the position of the aircraft in flight, or to the direction in which it is flying. **1922** Navigation light [see *landing light*]. **1939** T. L. Stocken *Oncoming Ships* 99/1 Lights..mistaken for navigation lights. **1943** C. D. Lane *Boatman's Manual* x. 355 In addition to the navigation light, the lightship carries a riding light forward. **1951** *Oxf. Jun. Encycl.* IV. 4/1 Small lights attached to the wing tips are navigation lights, similar to those used by ships. **1966** Navigation light [see *collision course*]. **1970** V. Canning *Great Affair* vii. 117 A helicopter..came low over the villa... It wasn't showing any navigation lights. **1937** Navigation log [see air speed (*air *sb.*[1] B. III. 1)]. **1961** *Ann. Reg. 1960* 386 Transit IB (U.S.A.)..Navigation satellite. Not transmitting. **1968** *Listener* 27 June 825/1 On 21 April..the U.S. military authorities launched a navigation satellite.

navigational, *a.* (Further examples.)

1920 *Act* 10 & 11 *Geo. V* c.67 §69 Lighthouses, light vessels, buoys, beacons, or other navigational marks.

1903 *Proc. Internat. Illumination Congr. 1931* II. 995 The navigational light on a light-vessel is fitted in a lantern which, in the older types, is made in halves bolted together around the mast. **1951** *Oxf. Jun. Encycl.* IV. 291/1 Another radio navigational aid is the radio compass. **1954** *Communications & Electronics* I. i. 60 The progress of modern aviation is becoming increasingly bound up with developments in communications and navigational aids. **1962** D. Slayton in *Into Orbit* 21 Carpenter..told us what he was doing about navigational aids, although the word 'navigation' was really a misnomer in our case... When you've been..tossed into either a ballistic trajectory or into an orbit, your course is already set. There is nothing you can do to change it. **1966** *Electronics* 3 Oct. 177 Teldix..supply instrumentation, navigational platforms and sensors. **1971** *Gloss. Electrotechnical, Power Terms (B.S.I.)* iii. vi. 22 *Navigational radar*, radar equipment installed on a craft as an aid to its navigation. **1973** *Fisheries Fact Sheet* (Environment Canada Fisheries & Marine Service) No. 1. 4/2 Small boats..equipped with..navigational aids and radio.

navigator. Add: **1. b.** One who navigates an aircraft or spacecraft.

1784 *Universal Mag.* LXXIV. 20/1 But they soon lost sight of our aerial navigators. **1825** [in Dict., sense 1 transf.]. **1834, 1915** [see *air navigator* s.v. *air *sb.*¹ B. III. 4]. **1929** T. E. Lawrence *Let.* 12 July (1938) 663 The Navigator of the airship will be getting his W/T bearings & time signals all the way, and will plot his course exactly. **1930** *Daily Express* 6 Oct. 3/5 The commander [of the airship] must have known, and the navigator. **1943** [see *air-bomber* (*air *sb.*¹ B. III. 3)]. **1951** *Oxf. Jun. Encycl.* IV. 290/2 Once in flight, the navigator's task is continually to 'fix' his position by observations. *Ibid.* 291/2 A pilot or navigator..cannot choose any course he pleases, as his is not the only aircraft in the sky. **1960** F. Gaynor *Dict. Aerospace* 161 *Navigator*, a crew member who plots and directs the movement of a space ship from within the ship.

c. One who navigates a motor vehicle.

1950 M. Laski in *Contact* May–June 26/1 It is..possible to take an extended motor-trip abroad without ever coming into contact with the Art of Navigation... Far better to become an Accomplished Navigator, and be free of all roads everywhere. **1964** W. Markfield *To Early Grave* (1965) vii. 120 A groan went round the car. '—we are a little, little bit *fahrblunged*.' 'Our navigator,' fumed Levine. **1968** M. Carroll *Dead Trouble* ii. 25, I sat in the car, content and lazy. Lisa needed no navigator. She seemed to know just where she was going. **1971** 'D. Rutherford' *Clear the Fast Lane* 37 Grant would be responsible for everything to do with the car..whilst Ritchie, who would act as his navigator, would be responsible for maps, routes and the latest road information. **1971** M. Tak *Truck Talk* 110 *Navigator*, the co-driver of a two-man operation who reads the road maps while going through unfamiliar towns.

navy¹. Add: **5. c.** A navy revolver.

1867 *Harper's Mag.* June 131/1 Judge put hand under pillow, drew out 'navy', and fired—*through a looking-glass!* **1875** 'Mark Twain' *Sk. New & Old* 122 She turned on that smirking Spanish fool like a wild cat, and out with a 'navy' and shot him dead in open court. **1931** G. F. Willison *Here they dug Gold* 92 Early boom towns and mining camps generally prefer the Colt 'Navy' (·36). **1968** R. F. Adams *Western Words* (ed. 2) 204 *Navy*, a westerner's term for the Navy Colt revolver.

d. A type of tobacco. Also, cigarette ends, etc., as picked up by tramps.

1872 *Kansas Mag.* 177/1 Another pull at the bottle,..a chaw of navy, and the repast is finished. **1876** G. H. Tripp *Student-Life Harvard* 399 Hawes had smoked 'navy' in it all the year of Sam's probation. **1889** J. W. Riley *Pipes o' Pan* 40, I draw my plug o' navy, and I climb the fence. **1926** *Amer. Speech* I. 652/1 *Navy*, cigar end or 'butts' found on side-walk. **1931** 'D. Stiff' *Milk & Honey Route* 214 *Navy*, butts of cigarettes and cigars. **1934** *Amer. Ballads & Folk Songs* 383 The higher you pitch, the sweeter my navy tastes. **1960** Wentworth & Flexner *Dict. Amer. Slang* 351/2 *Navy*,..a cigar end or butt found on a side walk. Perhaps from 'navy' = a type of chewing tobacco.

e. = navy rum.

1946 J. Irving *Royal Navalese* 121 *Navy*, Service issue; Service ways. Most usually it is a sobriquet for the rum-ration—e.g., 'I'll trade my *Navy* for a turn out of watch!'. **1962** Granville *Dict. Sailors' Slang* 80/1 *Navy, tot of*, measure of Navy rum offered to a guest by one who has saved his ration for some special occasion.

6. Articles as supplied to the navy; *navy biscuit, bread, jacket, -plug, revolver* (earlier examples); **navy bean** = Haricot *sb.* 2; **navy bullet,** a bullet used with a navy revolver; **navy catapult** (see quot.); **Navy Cut,** proprietary name of a kind of tobacco; **navy register** (example); **navy stroke,** the style of rowing practised in the navy; **navy-yard** (later U.S. examples).

1897 *Sears, Roebuck Catal.* 15/3 Beans, small Navy, hand picked. **1903** A. Adams *Log of Cowboy* xii. 77 Our supply of flour and navy beans was running rather low. **1951** *Good Housek. Home Encycl.* 349/2 Haricot Beans (called 'navy beans' in U.S.A.). **1955** W. Gaddis *Recognitions* ii. iii. 426 She went out..and left Gwyon staring into a plate of white navy beans. **1972** *Arable Farmer* Feb. 55/2 The navy bean crop must remain a matter for research and speculation for at least a year or two. **1973** *Times* 30 Apr. 9/1 The baked beans that pass over the shelves of Britain's supermarkets..originate in the United States. Navy beans grown in the Michigan area are shipped to Britain..for processing and canning. **1975** *New Society* 3 July 6/1 The Scottish Horticultural Institute are developing plants less susceptible to colder northern conditions; and research is going on into the production of a navy bean (baked bean) which can be grown in this country. **1867** 'T. Lackland' *Homespun* ii. 216 The people not only want the Word, but they want it as hard and dry as a navy biscuit. **1831** *Constellation* 54/1 Ephraim Treadwell..has for sale..Pilot and Navy Bread. **1848** *Rep. Comm. Patents 1847* (U.S.) 374 The 'navy bread' is usually made out of the coarser particles of the meal. **1873** J. Miller *Life amongst Modocs* 312 Was it possible that this man..could still live with a navy bullet through his body fired at two feet distance. **1914** C. F. Tweney *Dict. Naval & Mil. Terms* 161 *Navy Catapult*, a device for launching hydro-aeroplanes from a ship by means of compressed air. **1907** *Yesterday's Shopping* (1969) 67/3 Cigarettes... Virginia... Navy Cut, mild. **1911** *Trade Marks Jrnl.* 6 Sept. 1309 Player's Navy Cut... Manufactured tobacco. The Imperial Tobacco Company (of Great Britain and Ireland), Limited,..Bristol. **1959** E. Burgess *Divided we Fall* vii. 96 Harry was filling his briar with his favourite navy-cut. **1972** 'G. North' *Sgt. Cluff rings True* i. 15 Harrison inhaled contentedly... 'One third navy-cut, two thirds herb mixture.' **1840** C. Mathews *Politicians* ii. i. 30 He..had boasted out of doors he could and would save his life with a word as easily as hem-stitch a navy-jacket! **1864** in *Maryland Hist. Mag.* (1926) XXI. 300 He..had on his navy jacket with bright buttons and pants of the same dark blue. **1870** T. B. Aldrich *Story Bad Boy* 245 Between the beer and the soothing fragrance of the navy-plug, I fell into a pleasanter mood. **1909** 'O. Henry' *Roads of Destiny* xxi. 357 It seems that the only maritime aid I am to receive from the United States is some navy-plug to chew. **1916** 'Taffrail' *Pincher Martin* xiii. 242 'Care for a bit of navy plug?' He..never dreamt of boarding a trawler without a couple of inches of strong navy plug tobacco in his pocket. **1841** *Southern Lit. Messenger* VII. 4/1 Statistics..furnished by the Navy Register will show [etc.]. *a* **1861** T. Winthrop *Canoe & Saddle* (1883) iii. 21 This machine..is called a six-shooter, an eight-inch navy revolver. **1903** Kipling *Traffics & Discov.* (1904) 143 Aren't they rowing Navy-stroke, yonder? **1842** *Knickerbocker* XIX. 107 The General landed at the navy-yard. **1886** *Harper's Mag.* Sept. 619/1 The fact of establishing a navy-yard. **1936** Mencken *Amer. Lang.* (ed. 4) 239 What we call..a navy-yard is a *dock-yard* or *naval-yard*. **1946** E. O'Neill *Iceman Cometh* i. 75 De booze dey dish out around de Brooklyn Navy Yard.

naw, var. of No. Add: Also in *U.S.* use.

1906 *Dialect Notes* III. 147 Naw sir, we didn't do it. **1930** J. P. Burke in *Amer. Mercury* Dec. 455/1 Naw, we don't cut hooch any more; we make the bunk with malt. **1949** C. Himes *Black on Black* (1973) 277 'Naw suh,' Lemuel said. **1959** N. Mailer *Advts. for Myself* (1961) 52 'Does one of you want to go?'.. 'Naw.' **1971** *Black World* June 54/2 Naw, that scene was out. **1973** *New Yorker* 21 Apr. 62/3 'You didn't tear her clothes?' 'Naw, nothing like that.'

Naxalite (næ·ksălait). [f. *Naxal(bari* (the name of a place in W. Bengal) + -ite¹.] A name given in India to supporters of Chinese-type communism (see quot. 1969¹); also *attrib.* Hence **Na·xalism,** Chinese-type communism as practised in parts of India.

1969 *Times* 9 Jan. 9/1 'Naxalites' are in the headlines and seem to be at work in West Bengal, Andhra, Kerala, and half a dozen other states. The name comes from Naxalbari in West Bengal where Indian communists started arming peasants in the spring of 1967. **1969** *Amrita Bazar Patrika* 5 Aug. 5/5 Classes in Calcutta University..could not be held on Monday following a strike call given by the Naxalite-dominated Students' Action Committee. **1970** *Guardian* 2 July 4/4 Three Naxalites were lynched in the Midnapore district of West Bengal after they had killed a school teacher... The first major sign of popular resistance to Naxalite violence. *Ibid.* 4/6 Nepal..is the place where Naxalites and their Chinese mentors meet. **1970** *Guardian Weekly* 26 Dec. 16 The menace of Naxalism, the Maoist movement in West Bengal, may be less virulent than before. **1971** *Illustr. Weekly India* 11 Apr. 33/2 The authorities saw in it an urgent need to stamp out Naxalism before it became too late to do so. **1972** 'E. Peters' *Death to Landlords!* iii. 58 There are Naxalite bosses who are themselves greedy and tenacious landlords. *Ibid.* v. 88 The Naxalites probably quote the Baghavadgita [*sic*], too. **1974** *Daily Tel.* 18 Mar. 1/6 Four people were killed, one of them a policeman, when Naxalite Communists ambushed a police patrol in a Calcutta suburb yesterday.

Naxian (næ·ksiăn), *sb.* and *a.* [f. Gr. Νάξιος, L. *Naxius* + -ian.] **A.** *sb.* An inhabitant of Naxos, a Greek island in the Cyclades group, or Naxos, a part of Sicily colonized from the island of Naxos. **B.** *adj.* Of, pertaining to, or characteristic of the island or colony of Naxos. So **Naxiote** *sb.* and *a.*

1601 P. Holland tr. *Pliny's Nat. Hist.* II. xxi. xviii. 101 The Naxian Cypirus hath a quicker sent: the Phœnician Cypirus smelleth but a little. **1797** *Encycl. Brit.* XII. 779/2 The Ionians..in time..possessed the whole island; whence the Naxians are, by Herodotus, called Ionians, and ranked among the Athenian colonies. **1835** Mitford & Davenport *Hist. Greece* (new ed.) II. i. 4 The expelled Naxians..consented to guide a Persian army against a Grecian island. **1862** G. Grote *Hist. Greece* (new ed.) II. ii. xxii. 522 The oligarchy of Chalkis..sent out..settlers, Chalkidian and Naxian, who founded the Sicilian Naxos. *Ibid.* V. ii. lviii. 169 The Naxians cordially received the armament, which then steered southward along the coast of Sicily to Katana. **1885** J. T. Bent *Cyclades* xiv. 334 The Naxiotes were aghast when they heard that they were to be ruled by a Jew. *Ibid.* 348 This fortress..is..the acropolis of the Naxiote valleys. **1911** *Encycl. Brit.* XIX. 318/1 Four Naxian ships took part in the expedition of Xerxes, but deserted and fought on the Greek side at Salamis in 480. **1913** J. B. Bury *Hist. Greece* (ed. 2) xv. 674 The Naxians were the first Sicilians to welcome the deliverer of Sicily to her shores. **1946** R. Capell *Simiomata* i. 33 The Naxian pilot's presence on board was overlooked. *Ibid.* 36 A biggish party of German prisoners has been brought in to join our Naxians. **1968** V. Ehrenberg *From Solon to Socrates* v. 130 The Naxians took to the hills, but a few were taken prisoner. **1972** R. Meiggs *Athenian Empire* vii. 123 The earlier date will..suit the Naxian evidence better.

nay, *adv.*¹ and *sb.* Add: **B.** *sb.* **1.** (Earlier U.S. examples in sense: a negative reply or vote.)

1774 in *Vermont Hist. Soc. Coll.* (1870) I. 8 Passed in the affirmative—all yeas, no nays. **1807** J. Turner *Let.* 23 Jan. in J. Steele *Papers* (1924) II. 492 A Bill has passed the H.R repealing the duty on salt with only 5 nays. **1871** *Trans. Illinois Agric. Soc.* VIII. 5 Mr. Dalton demanded the yeas and nays.

‖ **naya paisa** (nəi·yă pəi·sā). Pl. **naye paise.** [Hindi, = new pice.] A unit of currency in India, equal to 1/100 rupee.

1956 *India: Reference Ann.* (Ministry of Information, Govt. India) xi. 127 The Indian rupee..will be divided into 100 units, each unit being called *naya paisa.* **1957** *Times* 2 Apr. 10/6 India's new decimal coinage was introduced today [*sc.* 1 Apr.]...All business transactions will now be calculated in rupees and *naye paise*, 'new coins', of the value of one hundredth of a rupee. **1957** *Whitaker's Almanack 1958* 974/1 An issue of 600m. coins of one, two, five, and 10 *naye paise* denominations was begun. **1963** *Times* 30 Jan. 12/7 King Edward rupees and everything since even to *naya paisa.* **1969** H. R. F. Keating *Inspector Ghote plays Joker* viii. 114 He plonked down fifteen naye paise, asked if he could use the phone. **1972** 'E. Peters' *Death to Landlords!* xi. 158 At the cost of a few *naye paise* they acquired three satisfied business contacts.

Nayar, var. Nair.

1911 *Encycl. Brit.* XIX. 318/2 *Nayar*.., a caste or tribe on the W. coast of S. India, who form the dominant race in Malabar. Traditionally they are soldiers, but many have taken to professions... Their total number in all India in 1901 was just over one million. **1922** *Edin. Rev.* Jan. 187 Even the Nayar, when he addresses a Brahman, must use the language of deprecation. **1955** M. Gluckman *Custom & Conflict in Afr.* iii. 68 The father's rôle seems to have been reduced to a minimum among the Nayar castes of Malabar in India in the past.

nay-sayer. Delete *rare*⁻¹ and add further examples. Also, one who votes against something.

1939 Joyce *Finnegans Wake* 108 Naysayers we know. **1961** C. MacInnes *England, Half English* 47 In this decade we witness the second Children's Crusade, armed with strength and booty, against all 'squares', all adult nay-sayers. **1962** *Times* 5 Apr. 7/3 Mephistopheles the eternal nay-sayer. **1970** *New Statesman* 12 June 835/2 Food scientists should try to be 'nay-sayers against profit curves'. **1972** *Guardian* 28 Nov. 12/6 The Republican Party began the Nixon years as an unrepresentative minority of nay-sayers. **1974** *Publishers Weekly* 5 Aug. 57/3 Further to refute the nay-sayers, he cites poll after poll in which Americans say they are happy despite their feelings that all is not well with the country.

nay-saying. (Later examples.)

1952 B. Wolfe *Limbo* (1953) iv. 222 His mind brimming with..the necessity for some nay-saying gesture against EMSIAC. **1972** *Jrnl. Social Psychol.* LXXXVI. 220 Subjects who obtained scores of 0, 1, and 11, 12 were dropped from the analysis as representing extremes of yeasaying or naysaying.

‖ **nazar** (næ·zāɹ). *India.* [Hindustani.] = Nuzzer.

1855 H. H. Wilson *Gloss. Judicial & Revenue Terms* 373/2 *Nazr, nazar, najar*,..a present, an offering, especially one from an inferior to a superior, to a holy man, or to a prince [etc.]. **1885** G. C. Whitworth *Anglo-Indian Dict.* 225/2 *Nazar*,..a present, an offering; especially one made by an inferior on his presentation to a superior. **1922** *Glasgow Herald* 11 Feb. 10 Twelve Nobles, each offering 'Nazar', or the symbolic tribute of gold. **1934** E. L. Tottenham *Highnesses of Hindostan* iv. 110 The deputations from the various States were presenting their gifts—the *poshaks* and *nazars*—of valuable cloth and plate. **1956** K. Fitze *Twilight of Maharajas* ii. 41 All Durbaris made their act of homage and obeisance, accompanied by the presentation of coins, known as *nazar*, which..found their way in due course to the Palace treasury. **1973** M. Bence-Jones *Palaces of Raj* x. 169 The civilians offering *nazars* of a few rupees laid on a white handkerchief.

Nazarene, *a.* and *sb.* Add: **A.** *adj.* **2. b.** Of or pertaining to, or characteristic of, the Church of the Nazarene (sense *B. 1 c).

1910 *New Schaff-Herzog Encycl. Relig. Knowl.* VIII. 453/2 As official organs of the church the *Nazarene Messenger*, Los Angeles, Cal.,..and the *Holiness Evangel*, Pilot Point, Tex., are recognized. **1958** M. Argyle *Relig. Behaviour* iv. 33 Many of these sects—the Pentecostal, Holiness, Nazarene churches and others—have increased enormously in proportion to their size during this period [*sc.* 1926–1953]. **1968** [see C.O. (*C III. 3)]. **1968** *War Resistance* II. xxiv. 26 The Nazarene Church is a plain building with simple furniture. *Ibid.* 28 Many do leave for USA, Canada or Australia where there are Nazarene Communities.

4. Of, pertaining to, or characteristic of the school of artists called Nazarenes. Cf. B. 4 below.

1950 *Chambers's Encycl.* XIV. 380/1 In many respects Wackenroder was a forerunner of the Nazarene school of painters. **1952** W. Gaunt *Victorian Olympus* i. 29 The master was found in the Nazarene painter, Jacob Eduard von Steinle. **1959** *Listener* 29 Jan. 217/3 The Bayreuth *Parsifal* has escaped from the old-fashioned 'Nazarene' presentation only to fall into a kind of fashionable 'grimness'. **1965** *Ibid.* 9 Sept. 382/2 For a short time after settling in London in 1846, Brown painted in the Nazarene style. **1970** T. Hilton *Pre-Raphaelites* i. 20 *Wycliffe*, with its light, flat tones and dispersed composition, looks very like a Nazarene painting.

B. *sb.* **1. c.** A member of the Church of the Nazarene, a Protestant sect formed at the beginning of the 20th cent.

1898 P. F. Bresee in T. L. Smith *Called unto Holiness* (1962) v. 121 It is now somewhat more than two years since..the Nazarenes, putting the old things behind them, went out to follow in the footsteps of Him whose name they bear. **1962** K. S. Latourette *Christianity in Revolutionary Age* V. i. 14 On the Pacific coast denominations of recent American origin loomed larger... the Four-Square Gospel, the Pentecostals, the Churches of God, and the Nazarenes. **1962** T. L. Smith *Called unto Holiness* iv. 75 Men of the Green Mountain State have been significant leaders among New England Nazarenes. **1968** *War Resistance* II. xxiv. 26 According to a report in the mid-20's there were 12–15,000 Nazarenes in Yugoslavia.

4. A name given to members of a group of German artists called the Brotherhood of St. Luke, founded in 1809, who aimed to restore to art the religious quality found in mediæval painting; cf. Pre-Raphaelite *sb.* Also called **Nazarite.** Sometimes in pl. **Nazarener.** Also *transf.*, of a style of musical composition.

1880 J. B. Atkinson *Schools of Mod. Art in Germany* ii. 11 Enemies in the opposite camp nicknamed the new saints 'Nazarites' and 'Pre-Raphaelites'. **1882** J. B. Atkinson *Overbeck* iii. 74 The type is ascetic and æsthetic after the pre-Raphaelite pattern affected by the Nazarites. **1889** Armstrong & Graves *Bryan's Dict. Painters & Engravers* (rev. ed.) II. 460/1 Schadow..went..to Rome, and joined the 'Nazarenes'. **1911** *Encycl. Brit.* X. 375/1 The group of artists who styled themselves *Nazarener*. **1928** E. Waugh *Rossetti* ii. 29 Nazarene, Florentine, and Crusader fused into one shadowy figure, glowing and distorted. **1942** *Archit. Rev.* XCI. 29 (*caption*) There is nothing elegant in this interior except for the reredos by William Dyce, the 'Nazarene' amongst English Early-Victorian painters. **1947** A. Einstein *Mus. Romantic Era* xii. 161 The 'Nazarenes' of church music were replaced by the 'Caecilians'. *Ibid.*, Nazarene-like church music. **1965** *Listener* 9 Sept. 382/2 He..had come under various influences, including that of the Nazarenes,..a group of German artists who had founded a 'pre' Pre-Raphaelite movement..in the eighteen-tens. **1973** *Country Life* 20 Sept. 779/2 One of his [*sc.* William Dyce's] Nazarene-type paintings.

Nazarenism (næ·zărīniz'm). [f. Nazarene *a.* and *sb.*] **1.** The principles or doctrines of early Christianity.

1892 T. H. Huxley *Let.* in J. S. Huxley *Ess. Pop. Sci.* (1926) 161, I have a great respect for the Nazarenism of Jesus—very little for later 'Christianity'. **1923** *Expository Times* Nov. 73/2 Here..the story of Jesus ends, and the story of Nazarenism begins.

2. The characteristics or artistic principles of the Nazarene school of artists.

1895 E. Dowson et al. tr. *Muther's Hist. Modern Painting* I. vi. 229 It was of no avail to him [*sc.* Philip Veit] that he mingled with his Nazarenism a certain air of the world. **1947** A. Einstein *Mus. Romantic Era* iv. 36 The Protestant church turned back to Bach..; the Catholic went back still further, to Palestrina and his time, in whose style it fostered a new musical Nazarenism, comparable to the Pre-Raphaelitism of the English school of painters. **1960** R. H. Boothroyd tr. *Novotny's Painting & Sculpture Europe 1780–1880* vi. 69 In the strange mixture of styles pervading his paintings of saints there appear a kind of return to the beginnings of Nazarene and Romantic principles and a breaking down of the narrow confines of academic Nazarenism reminiscent even of Blake.

Nazi (nā·tsi, nā·zi), *sb.* and *a.* [repr. pronunc. of *Nati-* in G. *Nationalsozialist.*] **A.** *sb.* A member of the National Socialist (Workers') Party in Germany, led by Adolf Hitler from 1920 and in power from 1933–45; a member of a similar organization; a person who believes in the aims of Nazism or similar doctrines and in the methods necessary to achieve them. Also *Comb.* So **Na·ziphil(e,** a person sympathetic to the ideology of Nazism.

1930 *Times* 19 Sept. 10/1 Herr Hitler, the leader of the victorious National-Socialists (Nazis), has very carefully refrained from saying anything. **1931** W. Lewis *Hitler* 57 The Democrats..have not been able to deal with the Nazi because of his Mastery of the Street. **1934** D. Teilhet *Talking Sparrow Murders* xiv. 202 The police are Nazi-controlled even in Heidelberg. **1938** *Sun* (Baltimore) 6 Sept. 1/3 The center of Santiago was kept in turmoil after Chilean Nazis..seized the National University. **1939** *Ann. Reg. 1938* 195 Added a number of minor Naziphile persons to the Cabinet. *Ibid.* 234 The chief change made by the new Premier in the Cabinet was to replace Dr. Homan, a strong Naziphil, with Count Paul Teleki. **1942** W. S. Churchill *End of Beginning* (1943) 222 The horde of divisions provided by Finland, Rumania, Hungary, and others of the Nazi-ridden or Fascist-ridden states. **1946** *R.A.F. Jrnl.* May 169 Lancasters..carried the war from one end of Nazi-controlled Europe to the other. **1956** A. H. Compton *Atomic Quest* i. 7 The Nazis saw in the atomic bomb the possibility of a new weapon of decisive importance. **1967** D. Eisenberg *Re-emergence of Fascism* iii. 127 The American Nazi and his uniformed 'storm-troopers' have frequently demonstrated outside the White House against racial integration in the South. **1974** *Times* 21 May 1/3 Mr Begin described Arab terrorists as 'the new Nazis' who made children their targets..adding: 'We must arm the population to fight these Nazis.'

B. *adj.* Of, pertaining to, or connected with the National Socialist Party in Germany or a political organization with similar aims, beliefs, or methods elsewhere. So **Na·zi-ish** *a.*

1930 *Times* 16 Sept. 16 (*caption*) Herr Hitler, the leader of the National Socialists, speaking at the last big Nazi election meeting. **1935** C. Isherwood *Mr. Norris changes Trains* vii. 105 The local Nazi storm-troop. **1939** W. S. Churchill *Into Battle* (1941) 108 There is the Nazi-Fascist ideology, and the Communist ideology. **1939** 'N. Blake' *Smiler with Knife* x. 154 There's nothing Nazi-ish about it... No talk of labour camps and so on. **1949** Koestler *Promise & Fulfilment* i. v. 54 To put the callous policy of the Mandatory Power on a par with the barbarity of Hitlerism, as the Jewish terrorists did in their slogan of the 'Nazi-British', is..unjust and stupid. **1967** W. Soyinka *Kongi's Harvest* 64 The carpenters end with a march down-stage with stiff mallet-wielding arms pistoning up in the Nazi salute. **1973** *Guardian* 29 Mar. 16/4 'Nazi' has become an indiscriminate political cliché applied to insensitive bureaucrats, Americans in Vietnam, IRA Provos, British paras in Ulster, Black September, Zionists, *et al.* **1974** R. Thomas *Porkchoppers* xviii. 162 'You mean Peter Majury?' '*Jawohl*,' Gayan said and made a Nazi salute.

Nazidom (nā·tsidəm, nā·zi-). [f. prec. + -dom.] The concepts and institutions of the Nazis.

1933 *Time* 20 Nov. 19/3 To make sure of a real 'intellectual eruption' last week—an utter blasting of Nazidom's foes—the Chancellor and his chief henchmen have been shouting themselves hoarse. **1935** *Times* 25 June 15/3 Professor Karl Barth..has been finally dismissed by the German Minister of Education. Into the details of his dispute with Nazidom there is no need to enter. **1941** *Ann. Reg. 1940* 387 Do not think that Hitler's Nazidom is going to be easily overthrown. **1947** A. L. Rowse *End of Epoch* xv. 181 The threat to ourselves and the rest of Europe involved by Nazidom. **1971** D. E. Westlake *I gave at the Office* (1972) 179 The story of his escape from Nazidom took slightly longer than a dental appointment.

Nazify (nā·tsifəi, nā·zi-), *v.* Also **nazify.** [f. as prec. + -fy.] *trans.* To cause or force to adopt Nazism or similar doctrines. Hence **Nazifica·tion; Na·zifying** *vbl. sb.*; **Na·zified** *ppl. a.*

1933 *Times* 7 June 13/3 Dr. Dollfuss..is resisting..the combined efforts of the German and Austrian Nazis to nazify Austria. **1933** *Christian Cent.* 20 Sept. 1164 (*heading*) Nazification of German Protestantism continues. **1934** *Times Lit. Suppl.* 26 July 524/2 It may, of course, be said that the process of 'Nazifying' has not had time to achieve its full results. **1934** *Sun* (Baltimore) 24 Sept. 8/2 Such events do not augur well for the future of the 'unified', *i.e.*, Nazified, Evangelical Church of Germany. **1939** *Ann. Reg. 1938* 199 Preparations for the plebiscite..fully occupied the party and the nazified administration during the preceding weeks. **1941** *Ibid. 1940* 229 Their policy of..nazifying the population gradually. **1961** R. Seth *How Resistance Worked* vi. 67 Throughout 1941 he issued a number of laws by which he intended to nazify Norway. **1972** C. Short *Naked Skier* xvi. 79 This Nazified policeman was on the side of the Countess and against us. **1973** E. Osers tr. *Waldheim's Austrian Example* iii. 45 At first the new regime made a point of cultivating public opinion, but as the process of Nazification spread to all spheres of public life less and less attention was paid to it.

Nazism (nā·tsiz'm). Also **Naziism, Nazi-ism.** [f. as prec. + -ism.] The political doctrines evolved and implemented by Adolf Hitler and his followers, esp. those relating to racial superiority, the all-powerful state, and the cult of the leader. So **Na·zist, Nazi·stic** *adjs.*

1934 *Times* 28 Feb. 14/3 Finally it is urged that the disinterested support given by Italy to Nazism should counsel higher respect on her part for Italian foreign policy. *Ibid.* 16 Mar. 20/4 A clever essay on the cultural aims of Nazi-ism. **1938** *Ann. Reg. 1937* 171 Chancellor Schuschnigg rejected the Italian suggestion that he should introduce a representative of Naziism into the Austrian Government. **1938** *Times* 17 Mar. 15/5 In Belgium there is already a semi-Nazist party of Germanic nationality looking for help abroad. **1938** *New Statesman* 25 June 1053/2 Quite recently..a new Nazistic party called the Christian Nationalist Socialist Front has been formed. **1940** A. Huxley *Let.* 24 Apr. (1969) 453 The doctrines of Nazism, Communism, nationalism..are manifestly idiotic. **1957** J. S. Huxley *Relig. without Revelation* (rev. ed.) iii. 63 Nazism was inherently self-destructive because of its claim to world domination by a small group. **1971** *Nature* 16 July 205/2 Feelings about the scientific community's relationship to Naziism. **1973** *Guardian* 29 Mar. 16/4 Nazism was the ultimate degradation of mankind, singular even in the history of vileness.

N'Dama (n,dā·mă). Also **N'dama, Ndama.** [Native name.] A West African breed of cattle, usually fawn or light red in colour and bearing lyre-shaped horns; an animal belonging to this breed. Also *attrib*

1938 J. L. Stewart in *Vet. Rec.* 9 Oct. 1291/2 The present day N'Dama is almost identical with the Longhorn of the ancient Egyptian drawings. *Ibid.* 1292/1 The N'Dama cattle until a few years ago were found in the Fouta Djalon mountains of French Guinea but have now spread over a large part of West Africa as this breed is especially good for stock improvement. **1952** *Ann. Trop. Med. & Parasitol.* XLVI. 128 The N'damas were born and bred at the Nigerian Agricultural Department's farm at Ilorin. **1956** *Nature* 21 Jan. 132/2 The West African N'Dama breed of cattle shows a remarkable tolerance to trypanosomiasis. **1973** *Guardian* 20 June 12/3 The Zebu cattle of Senegambia..[are] less resistant than the tougher Ndama breed.

Ndebele (n,dĕbī·li). [Native name, f. *n-* sing. prefix + *Tebele*.] The name of a Zulu people branches of which are found in Rhodesia and in the Transvaal; also a member of this people; (also **Sin-debele**), the language of this people.

The Rhodesian Ndebele are better known as the Matabele (*ma* = pl. prefix).

1913 J. O'Neil (*title*) A grammar of the Sindebele dialect of Zulu. **1919** [see *Matabele]. **1930** C. G. Seligman *Races Afr.* viii. 187 The Zulu-Xosa, chiefly in the coastal region south and east of Drakenberg Mountains,..include ..the Amandebele (Matabele) of Matabeleland. **1937** A. W. Hoernlé in I. Schapera *Bantu-Speaking Tribes S. Afr.* iv. 86 Among the Southern Transvaal Ndebele, each ..clan has a special species of animal..which may not be named or eaten or used by the members of that clan... The Rhodesian Ndebele..have introduced a new hierarchy of rank resulting from their conquest of indigenous tribes in their new homes. **1951** P. Abrahams *Wild Conquest* 175 'This one brought the news.' 'An Ndabele?' 'No, a Barolong.' **1971** *Rand Daily Mail* 27 Mar. 6/2 Our Zululand and Ndebele beadwork..has always been regarded as the best in the world. **1973** *Drum* (Johannesburg) 8 Jan. 18/3 My mother language is Shona and I can speak English and Ndebele.

‖ **Ndugu** (n,dū·gu). [Swahili, = relative, brother.] A general form of address in Tanzania.

1973 *Black World* May 47 San Francisco Afrikans (brothers) Afrikan Afrikans (*ndugu*) West Indian Afrikans (Hey man). **1974** *Sunday Tel.* 16 June 36/3 A [Tanzanian] Government directive says that mister, honourable, excellency and all other honorifics will be replaced by the Swahili word *ndugu*, which means brother. **1974** *Daily News* (Tanzania) 13 Sept. 3/1 One of the passengers Ndugu Mohamed Ismaili had this to say: 'We have a transport problem about buses going to Lindi.' *Ibid.* 3/4 'Ndugu' (to trained nurse) 'you're going to work in the ward for four hours today.'

ne', ne (ne), colloq. shortening of Never *adv.*

1934 J. D. Carr *Eight of Swords* ii. 28 His daughter and my son—hurrumph, ne' mind. **1949** C. Himes *Black on Black* (1973) 277 Ne you mind.

nealie (nī·li). Also **nelia.** *Austral.* [Etym. unknown.] = *needle-bush* (*b*) (*Needle *sb.* 14).

1889 [see *needle-bush* s.v. Needle *sb.* 14]. **1933** *Bulletin* (Sydney) 15 Nov. 21/4 Two miles farther on..was a mile-wide stretch of nelia scrub. **1936** I. L. Idriess *Cattle King* xxxviii. 326 Where had all the mulga gone, the white-wood, and acacias, and turpentine bush, the black oak and nelia, and bullocky bush, the thick shrub life that had carpeted this country when he rode through here a boy? **1965** [see *needle-bush* s.v. *Needle *sb.* 14].

Neanderthal (nĭæ·ndəɹtāl). Also **Neandertal.** The name of a valley near Düsseldorf in western Germany, used *attrib.* in **Neanderthal man, skull,** etc., to designate the Middle Palæolithic fossil hominid *Homo neanderthalensis*, first identified from a skull found there in 1856, and also known from other remains in Europe, Africa, and Asia. Also *absol.*, *fig.* (with reference to appearance), and as *adj.*

1861 G. Busk tr. D. Schaaffhausen in *Nat. Hist. Rev.* Apr. 167 The considerations which have led us to compare the Neanderthal cranium with those of the most ancient races are still further confirmed by the discovery ..of skulls exhibiting a yet closer correspondence with it. **1863** T. H. Huxley *Evidence Man's Place in Nature* iii. 142 The posterior lobe of the brain of the Neanderthal man must have been as much flattened as I suspected it to be. In truth, the Neanderthal cranium has most extraordinary characters. **1864** *Q. Jrnl. Sci.* I. 88 It is my intention to confine myself to the consideration of the Neanderthal fossil with reference to its place in Nature. *Ibid.* 90 The Neanderthal skull is of an elongated oval form. **1899** A. H. Keane *Man Past & Present* 9 Certain skulls from South Australia seem cast in almost the same mould as the Neanderthal. **1922** *Encycl. Brit.* XXX. 145/2 Neanderthal man is now revealed as an uncouth race with an enormous flattened head, very prominent eye-brow ridges and a coarse face. **1923** A. L. Kroeber *Anthropol.* ii. 24 Neandertal man was short. *Ibid.* xv. 472 Whenever the origin of a people remains obscure, be they Neandertals, Alpines, Sumerians,..or what not, some one propounds the convenient hypothesis of deriving them from this vast interior land [*sc.* central Asia]. **1928** *Weekly Dispatch* 13 May 12/3 The pictures which show modern influence are even more unpleasant than the streaky bacon sunsets..or the most realistic neanderthal clergymen which otherwise decorate the walls. **1939** 'N. Blake' *Smiler with Knife* iii. 57 You know what the other side says—..'Woman's place is in the kitchen'—all the rest of that Neanderthal tommyrot. **1966** E. Palmer *Plains of Camdeboo* vii. 117 Neanderthal Man lived here in Southern Africa. **1966** M. Woodhouse *Tree Frog* xii. 86 A fourteen-stone Neanderthal with a sub-machine gun appeared. **1970** *Guardian* 9 July 10/4 The Neanderthal men of the military and industrial establishments. **1971** *Observer* 23 May 24/6 Neanderthal grunts over the cornflakes..were the best most husbands could manage. **1973** B. J. Williams *Evolution & Human*

Origins xi. 176/2 In the material that follows I shall use the term Neandertal *sensu lato*, that is, in the broad sense. The term will designate the early stages of the within-species evolution of *Homo sapiens*. *Ibid.* 185/2 In 1958 part of a Neandertal skull was recovered from a cave at Ma-pa, southern Kwangtung. **1973** A. Price *October Men* xi. 164 He is a creature from the Dark Ages, a man of violence. A Neanderthal. **1973** D. Ramsay *Deadly Discretion* 97 How about your pal Ivan? Does he have sensitive feelings under that Neanderthal exterior?

Hence **Nea·ndertha:ler,** a Neanderthal man; **Neandertha·lian** *a.* and *sb.*; **Neandertha·lic** *adj.* (in quot. *fig.*).

1913 *Science Progress* VIII. 278 The Neandertaler differed from *Homo sapiens* in having an extremely receding forehead, with enormously developed brow-ridges. **1920** H. G. Wells *Outl. Hist.* I. ix. 51/2 The Tasmanians were not racially Neanderthalers. *Ibid.* x. 58/2 The tremendous advance they [*sc.* later Palæolithic men] display upon their Neanderthalian predecessors. **1920** H. Reinheimer *Symbiosis* iii. i. 205 Some of them, e.g., the Neanderthalians, reached the stage of the savage beast, marked by chronic acromegaly. **1955** *Sci. News Let.* 19 Feb. 122/2 With succeeding generations Neanderthalers became adapted physically to withstand the cold. **1967** *Boston Sunday Herald* 26 Mar. i. 29/3 Neanderthalic sadists forced near-dead men to run when they scarcely had the strength to put one foot ahead of another. **1971** *Times* 11 Feb. 14/7 He pointed out that Neanderthalers, living as they did in the ice age, might easily have suffered from a deficiency in vitamin D.

Neanderthaloid, *a.* Add: Also as *sb.*

1971 *Nature* 23 Apr. 489/2 Although Swanscombe had long been considered closely affiliated with modern *Homo sapiens* the multivariate analysis indicated a closer resemblance to the neanderthaloids.

neanic (nī,æ·nik), *a.* *Zool.* [f. Gk. νεανικ-ός youthful.] Designating the early stages of the growth of an animal, esp. the pupal stage of an insect.

1892 Buchman & Bather in *Zool. Anzeiger* XV. 421 Hyatt... Nealogic. Here proposed..Neanic. Literary equivalent..Adolescent. *Ibid.* 430 Neanic. During this stage specific characters and all other morphological features present in the adult, appear and undergo development. **1903** *Amer. Naturalist* XXXVII. 519 At this stage [in the growth of *Sycotypus canaliculatus*], the early neanic, the lines of growth are well marked and of nearly equal strength with the revolving lines, the two together giving the shell surface a reticulated appearance. **1906** J. B. Smith *Explanation Terms Entomol.* 87 Neanic: referring to the pupal stage. **1938** *Nature* 20 Aug. 341/1 It is helpful to distinguish between the very early, or embryonic, stages of development and the later, or neanic stages, during which the young gradually assumes the characteristics of the adult. *Ibid.* 10 Sept. 461/1 In the neanic phase the organism exhibits a combination of less stable characters. **1971** F. E. Eames *Davies's Tertiary Faunas* (ed. 2) i. i. 90 Neanic. Youthful stage in ontogeny.

neanthropic (nī,ænþrǫ·pik), *a.* Also **neoanthropic,** and with capital initial. [f. Neo- + Anthropic *a.*] Of, pertaining to, or designating the single extant species of man as distinguished from extinct forms known from their fossil remains.

Quot. 1916 may represent an independent coinage.

1894 J. W. Dawson *Meeting-Place Geol. & Hist.* i. 17 The modern, or anthropic [period], is..divisible into two sections—the early modern, or palanthropic, sometimes called quaternary, or post-glacial, and which may coincide with the antediluvian period of human history; and the neanthropic, extending onward to the present time. [*Note*] I have preferred..to call the earlier races of men palaeocosmic and the later neocosmic,..while the periods to which they belong are respectively the Palanthropic and Neanthropic. **1916** G. E. Smith *Primitive Man* 18 Thus the new spirit of man and modern man himself are revealed in the Upper Palaeolithic period. This Neoanthropic phase, as I have called it, thus begins in the Aurignacian period. **1939** McCown & Keith *Stone Age Mt. Carmel* II. xxii. 362 In Neanderthal skulls the great wing and the orbital plate of the malar tend to be wide; in Neanthropic skulls they tend to be narrow. **1959** J. D. Clark *Prehist. S. Afr.* iv. 77 The 'neanthropic' or modern (*Homo sapiens*) form of man. **1973** B. J. Williams *Evolution & Human Origins* xi. 175/1 The neanthropic (new man) line included fossils that were more or less modern throughout time.

neap, *a.* Add: **1.** Also *neap rise* (see quot.).

1888 *Encycl. Brit.* XXIII. 369/2 The height between high-water mark at neap tide and mean low-water mark at spring tide is called the neap rise.

Neapolitan, *a.* and *sb.* Add: **A.** *adj.* **b.** *Neapolitan ice*, a block of ice cream made in layers of different colours and flavours; also *transf.* and *fig.*; *Neapolitan system* (see quots.).

1895 'M. Ronald' *Century Cook Bk.* xxii. 498 *Neapolitan ice-cream.* This cream is molded in brick form in three layers of different flavors and colors. **1911** *Daily Colonist* (Victoria, B.C.) 26 Apr. 7/1 (Advt.), Ice Cream Bricks, Neapolitan, Vanilla, Strawberry, Pineapple, [etc.]. **1933** A. Huxley *Let.* 29 Apr. (1969) 369 Mexico was..very curious. Such a strange Neapolitan ice, with its layers of Indian, mestizo, white. **1954** —— *Let.* 25 Oct. (1969) 714 We have to think of the mind in terms of a stratified Neapolitan ice, with a peculiar flavour of consciousness at each level. **1969** R. & D. De Sola *Dict. Cooking* 158/1 *Neapolitan ice cream*, ice-cream brick containing several flavors arranged in contrasting color layers, usually chocolate, vanilla, and strawberry. **1959** Reese & Dormer *Bridge Player's Dict.* 151 *Neapolitan System*,..one of the principal Italian systems, played by Forquet and Siniscalco and others. It is a one-club system, with a series of artificial responses. **1962** *Listener* 22 Nov. 886/2 West, playing the Neapolitan system, opened Two Diamonds. **1964** *Official Encycl. Bridge* 374/2 *Neapolitan*, a system..played in many World Championship events by a group of Neapolitan players.

B. *sb.* **d.** A dialect or language of Naples.

1598 Florio *Worlde of Wordes* Epistle Dedicatorie, How shall we, naie how may we ayme at the Venetian, at the Romane, at the Lombard, at the Neapolitane, at so manie, and so much differing Dialects, and Idiomes, as be vsed and spoken in Italie, besides the Florentine? **1901** M. Carmichael *In Tuscany* 99 Had Dante been born not at Florence but in Venice and written in Venetian, had Petrarch been born not at Arezzo but in Naples and written in Neapolitan, there would have been two classical languages in Italy today. **1936** G. F.-H. & J. Berkeley *Italy in Making* II. xviii. 276 As a general rule he spoke French or Neapolitan, not Italian. **1973** A. Price *October Men* xi. 157 The man's Italian was..faultless... There was even..the hint of Neapolitan in it.

near, *adv.*² Add: **1. e.** *so near and yet so far*: describing a person who or thing which is unattainable despite its apparent proximity.

1755 W. Hay tr. *Martial's Sel. Epigrams* i. 15 In the whole town no soul can be So near, and yet so far from me. **1850** Tennyson *In Mem.* xcv. 145 He seems so near and yet so far. **1863** 'Ouida' *Held in Bondage* I. i. 25 The long sunny future that stretched before us in dim golden haze,—so near and yet so far from our young longing eyes. **1863** *Harper's Mag.* Dec. 93/2 Thou art so near, and yet so far! **1920** R. Macaulay *Potterism* i. 8 In June and July 1914 the conversation turned largely and tediously on militant suffragists, Irish rebels, and strikers... It was a curious age, so near and yet so far, when the ordered frame of things was still unbroken, and violence a child's dream. **1939** Joyce *Finnegans Wake* 213 I've lost it!..So near and yet so far! **1962** E. Cleaver in A. Dundes *Mother Wit* (1973) 12/1 The actual conditions to which they aspired were..all around them, as it were 'so near and yet so far'.

8. (Further examples.) Esp. in colloq. phrases: (*as*) *near as dammit* (or *damn-it*), *as near as makes no difference* (or *matter*), extremely near; virtually.

1894 G. F. Northall *Folk-Phrases* 9 As near as damn it. As near as fourpence to a groat. As near as two ha' pennies for a penny. **1897** Conrad *Nigger of Narcissus* iv. 91 You were as near hanging as damn-it tonight. **1911** A. Bennett *Hilda Lessways* v. i. 338 That first night..I was as near as dammit to letting out the whole thing and chancing it. **1931** [see *dammit]. **1937** D. L. Sayers *Busman's Honeymoon* viii. 171 You left here some time after 6.10 by that clock, which was right as near as makes no difference. **1961** [see *dammit]. **1961** I. Jefferies *It Wasn't Me!* iv. 54 They near as dammit showed interest. **1970** P. Moyes *Who saw her Die?* xiv. 183 Funny, isn't it, Miss Threep ending up in the hospital with Billing? Well, not *with* him, of course, but as near as makes no matter.

21. Delete † *Obs.* and add further examples. Now very commonly prefixed to adjs. in the sense 'almost, nearly, approximately', and to sbs. in the sense 'something that is nearly the same as, or is a substitute for, the thing specified'; 'artificial'; *spec.* **near-beer** orig. *U.S.*, a beverage resembling beer; beer with a very low alcoholic content; also *attrib.*; **near-print** (see quot. 1943); **near-seal** *N. Amer.*, any fur treated and dyed to resemble sealskin; **near-silk,** artificial silk.

1928 *Daily Express* 27 Apr. 12/4 The aspirant painters ..drink near-absinthe instead of beer. **1909** *N.Y. Even. Post* (Semi-Weekly ed.) 23 Aug. 2 The refusal of the Cities Commission to prohibit the sale of imitation beer, commonly known as 'near beer'. **1921** *Daily Colonist* (Victoria, B.C.) 27 Mar. 7/5 The thirty-six near-beer bars in operation here at present pay into the city treasury a total of $5,400 a year in the form of licence fees. **1963** *Times* 14 Feb. 4/7 A hostess in a Soho near-beer club told a jury..that..customers were charged 15s. for two soft drinks made from blackcurrant juice. **1952** *Manch. Guardian Weekly* 4 Sept. 3 Add the near-certainty of Massachusetts. **1822** W. Sewall *Diary* 27 Sept. (1930) 89, I am much troubled with a bowel complaint. This must have been caused by drinking freely of near cider. **1973** L. Snelling *Heresy* i. i. 11 Andover, Yale, fine athlete.., captained the Yale crew, near-effortless Phi Beta key, et cetera. **1926** *Ladies' Home Jrnl.* Apr. 24 The decision was based on..two broken engagements, one near-engagement..and several flirtations. **1949** Koestler *Promise & Fulfilment* ii. ii. 218 Each of them means a cherished near-escape story to some member of the commune. **1973** C. Bonington *Next Horizon* xix. 261 Twenty-four hours later the shock of my near-escape really hit me, and manifested itself mainly in a sense of horror at letting down my family. **1942** *Daily Tel.* 22 May 5/3 If the harvest collection should break down through a failure of fuel supplies, then near-famine conditions..may not be far away. **1938** *New Statesman* 3 Dec. 904/2 He must now embrace..not merely those propertied groups which are reliably republican, but near-Fascists of the Flandin type as well. **1926** *Scribner's Mag.* Sept. 34/2 (Advt.), Some more piquant revelations of the great and near-great in English and Continental Society. **1855** W. Whitman *Leaves of Grass* 37 The laughing-gull..laughs her near-human laugh. **1943** *Times* (Weekly ed.) 10 Feb. 5 A British army pitch-forked on to Mars would first smile at and then discover near-human characteristics in any local inhabitants that might appear. **1964** M. Critchley *Developmental Dyslexia* xiv. 81 Many of them..become items within the community of illiterates or near-illiterates. **1955** Koestler *Trail of Dinosaur* 253 The trouble with all near-miracles..is the unpredictability of their timing. **1957** C. S. Lewis *Let.* 6 Mar. (1966) 275, I married..a very sick, save by near-miracle, a dying woman. **1949** Koestler *Promise & Fulfilment* iii. i. 391 Transport cooperatives which have..a near-monopoly of cross-country bus services. **1964** *Oceanogr. & Marine Biol.* II. 326 A temperature of 25° C represents near-optimum conditions. **1939** *Ann. Reg. 1938* 261 The condition of the inhabitants was one of near-panic after this merciless destruction. **1956** *Nature* 4 Feb. 239/2 Carbons in which neighbouring ordered regions lie in near-parallel orientation. **1963** C. R. Cowell et al. *Inlays, Crowns & Bridges* ii. 5 The principle of 'near parallelism'—that is, opposing walls must be nearly parallel with only enough divergence to make withdrawal of the pattern or impression possible. **1962** Near-perfect [see *attitude 2 c]. **1934** Dylan Thomas *Let.* 9 May (1966) 124 You've brought 'conventional' poetry..to a point of near-perfection. **1943** E. H. Thompson *A.L.A. Gloss. Library Terms* 106/2 *Processed*, reproduced by duplicating processes other than ordinary printing, as by mimeograph, multigraph, rotoprint, multilith, etc. Also called *Near-print.* **1956** Wilson & Tauber *University Library* 7 Journals, government documents, near-print, newspapers, manuscripts..have to be secured on an unprecedented scale. **1969** *N.Z. National Bibliogr.* II. p. viii, The growing volume of mimeographed and near-print editions has presented particular difficulty. **1949** *Word Study* Apr. 2/2 Cat..gut. Near rhyme..is the most generally accepted name. **1912** J. H. Moore *Ethics & Educ.* 109 That state of near-savagery when any low-browed irresponsible..is allowed to go out and shoot to death everything that has the breath of life in it. **1902** G. H. Lorimer *Lett. Merchant* 184 He leads the nag out..and examines every hair of his hide, as if he expected to find it near-seal. **1906–7** *T. Eaton & Co. Catal.* Fall & Winter 4/1 Near Seal Jacket, made of finest quality skins, closely resembling real seal in appearance. **1919** Mencken *Amer. Lang.* 159 Many characteristic Americanisms of the sort to stagger lexicographers—for example, *near-silk*—have come from the Jews. **1937** D. Canfield *Fables for Parents* (1938) 251 Nude-coloured, near-silk stockings with a run down one leg. **1911** E. Ferber *Dawn O'Hara* iii. 34 Assuming a near-smile, she entered the room. **1950** Koestler et al. *God that Failed* 63 The next five years were for me years of near-starvation. **1962** W. Nowottny *Lang. Poets Use* vii. 169 The treachery of near-synonyms which can slant meaning in a new direction. *Ibid.* 168 Ambiguities (drawn variously from near-synonymity and from rhetorical repetitions involving syntactical similarity). **1910** *Century Mag.* Apr. 891 Clothes and the Man. A near-true story. By Edith Rickert. **1951** S. Spender *World within World* iii. 138 These writers wrote with a near-unanimity, surprising when one considers that most of them were strangers to one another. **1964** *Language* XL. 206 Many of these 'near-universals' involve structural relationships. **1930** *Cambridge Daily News* 25 Sept. 3/2 Never having worn even near-wool within rubbing distance of my skin.

near, *a.* Add: **2. a.** Freq. in phr. *nearest and dearest*; also *absol.* as *sb.*, one's closest and most beloved relatives or friends.

1596 Shakes. *1 Hen. IV* iii. ii. 123 Why, Harry, doe I tell thee of my Foes, Which art my neer'st and dearest Enemie? **1610** J. Chamberlain *Let.* 23 Jan. (1939) I. 296 The neerest and deerest frends he hath know not what to guesse of this humor. **1654** [in Dict.]. **1822** T. Creevey *Let.* 23 Dec. in *Creevey Papers* (1963) xi. 186 Brougham arrived here on Saturday, on his way—or rather *out* of his way—to his nearest and dearest. **1839** Dickens *Let.* 21 Jan. (1965) I. 493 For those who are nearest and dearest to me I can realise little more than a genteel subsistence. **1926** F. M. Ford *Man could stand Up* i. ii. 38 Look how you let in your nearest and dearest—those who have to sympathise with you. **1959** J. Burke *Echo of Barbara* ix. 92 One never does know much about one's nearest and dearest. **1975** 'C. Aird' *Slight Mourning* iv. 42 It's always family or friends who do you in... Nearest and dearest, that's who it'll be.

3. c. Of a motor vehicle; usu. in *near side* (freq. *attrib.*).

1926 *Times* 6 May 3/7 The defendant..went to the rear of the car and attempted to rip open the near side wheel with a knife. **1927** [see *dip-stick* s.v. *dip *sb.* 11]. **1959** *Times* 15 Dec. 13/5 The windscreen wipers were so arranged that the nearside blade flicked rain from the offside blade across the driver's line of vision. **1973** 'H. Howard' *Highway to Murder* i. 12 It felt like they'd got a flat on the nearside front wheel.

4. a. *near space*: space in the immediate vicinity of the earth; inner space.

1962 *New Statesman* 24 Aug. 219/1 The Russians 'have been busily preoccupied with the near-space environment and have ably demonstrated their..de-orbiting accuracies'. **1967** *Economist* 20 May 791/3 The practical uses of 'near-space' in weather observation and in geological surveys of the earth.

c. *near-term*, used *attrib.* = *short-term*; occurring in or pertaining to the near future; opp. *long-term *a.*

1958 *Listener* 18 Sept. 407/1 To turn to Australia. There the near-term position is much more comfortable, unemployment is less than two per cent. **1965** H. I. Ansoff *Corporate Strategy* (1968) iv. 50 If the profitability picture urgently needs a near-term boost, a more modest goal ..will be acceptable. **1971** *Flying* Apr. 51/2 This..certainly does not suggest near-term improvement.

d. *nearest neighbour*, the member of a series or array nearest to that being considered; freq. *attrib.* (usu. with hyphen).

1937 *Jrnl. Appl. Physics* VIII. 654/1 In a liquid the number of nearest neighbors and the interatomic distances are roughly the same as in the crystalline material. **1945** A. F. Wells *Structural Inorg. Chem.* iii. 110 In each of these close-packed arrangements, cubic and hexagonal, each atom has twelve equidistant nearest neighbours, six in its own plane and three in each adjacent layer. **1961**

Jrnl. Biol. Chem. CCXXXVI. 864 (*heading*) Frequencies of nearest neighbor base sequences in deoxyribonucleic acid. **1972** *Computers & Humanities* VII. 40 The statistical analysis of pattern in such horizontal distributions is another area to which some attention is now being given, with nearest-neighbour analysis the most frequently adopted method.

e. *near money*, a deposit, bond, etc., that can easily be converted into ready money.

1948 G. Crowther *Outl. Money* (rev. ed.) ii. 65 Bank deposits, in those circumstances, might be something very close to money, but still not quite within the definition. They would be 'near-money'. *Ibid.* 66 We have already come across another example of 'near-money' in the form of the Bill of Exchange. **1968** *Economist* 23 Nov. 83/1 Ottawa itself is relying on a $2 billion 'near-money' savings bond issue at 6¾ per cent to meet its current excessive needs.

6. b. *near miss*, a shot that only just misses a target; also *transf.* and *fig.*; *near thing*, something barely effected; a narrow escape.

1751 Fielding *Amelia* IV. xi. ii. 123 You certainly know..how hard Colonel Trompington is run at your Town, in the Election of a Mayor; they tell me it will be a very near Thing, unless you join us. **1894** Somerville & 'Ross' *Real Charlotte* II. xxiii. 120 'That was a near thing,' remarked Mr. Hawkins complacently, as a slight grating sound told that they had grazed one of these smooth-backed monsters. **1930** W. Gibson *Hazards* 8 A near thing! But he caught the plane: 'twas well He did not miss it. **1940** *Life* 9 Sept. 120/2 The other was a near miss amidships. **1940** *Illustr. London News* 28 Dec. 829/2 They came back with direct hits on a mine-sweeper and a supply ship, and a near miss on a destroyer to their credit. **1957** F. Hoyle *Black Cloud* i. 17 You mean, Dave, that there's no chance of the cloud missing the solar system, of it being a near-miss, let us say? **1957** *Listener* 21 Nov. 853/2 For those children who are near-misses as well as for those who make the grade. **1964** D. Varaday *Gara-Yaka* xxi. 190 When one moving coil lashed over the crocodile's head, Mulembe snapped at it with spiky-toothed jaws which almost bit through... It was a near thing for the snake. **1967** G. F. Fiennes *I tried to run a Railway* iv. 38 It had a lot of bomb damage to repair from a near miss whose crater was still unfilled. **1972** 'E. Lathen' *The Longer the Thread* xiv. 129 The taxi-drive..was..marked by enough near-misses to put from his mind any thoughts but survival. **1973** *Guardian* 7 Mar. 2/1 French and Scandinavian pilots..had reported 11 'near misses' over France to their airlines since Friday.

nearabout (nīə·ɹăbaut), *adv.* Also **near about, nearabouts, near 'bout.** [f. Near *adv.*² + About *adv.* and *prep.*, Abouts *adv.* and *prep.*] † **a.** In this vicinity; nearby. *Obs.* **b.** *dial.* Nearly, almost; approximately.

c **1400, 1634** [see Near *adv.*² 1 c]. [**1702** *Rec. Early Hist. Boston* (1882) VIII. 20 The front of the Old house..is neer abt Eleven foot from the street.] **1708** *Ibid.* 81 The mouth of the Said creek bears Neer about North East from ye Knowl of Trees. **1834** W. A. Caruthers *Kentuckian in N.Y.* II. 206 Yes, I believe everybody's married, nearabouts, as far as I can learn. *a* **1878** [see Near *adv.*² 5 b]. **1907** A. Quiller-Couch *Major Vigoureux* xxii. 251 The tide bein' nearabouts on the top of the flood. **1928** 'M. Chapman' *Happy Mountain* iii. 25 All the fields were nearabout flat. **1938** M. K. Rawlings *Yearling* i. 9 'I near about give you out, son,' he said. **1941** J. Faulkner *Men Working* i. 19 We can make twelve bales of cotton, near 'bout, every year.

near by, *adv.*, *prep.*, and *a.* **A.** *adv.* **1.** Delete 'Now chiefly *dial.* and *U.S.*' and add further examples.

1931 *News Chron.* 25 Aug., There had been feverish activity at the Trade Union Congress office nearby. **1965** E. Gowers *Fowler's Mod. Eng. Usage* (ed. 2) 382/1 *Near by* has been long established as an adverb, and there is no good reason for those who draft police notices to prefer *in the vicinity.*

C. *adj.* Delete 'Chiefly *U.S.*' and add further examples.

1939 Joyce *Finnegans Wake* 448 Take a good longing gaze into any nearby shopwindow. **1954** *Manch. Guardian Weekly* 26 Aug. 4 He went to the near-by police station. **1973** *Times* 24 May 27/3 There was a slackening interest in nearby metal.

Near East. [Near *a.* 4, East *sb.*] A region comprising the countries of the eastern Mediterranean, sometimes also including those of the Balkan peninsula, south-west Asia, or north Africa. (Also *Nearer East.*) Also *attrib.* Cf. *Far East, *Middle East.

[**1869** *Wesleyan-Methodist Mag.* (Sixpenny ed.) July 312 (*heading*) Peeps at the Near East (*i.e.* Spitalfields in London).] **1891** J. L. Kipling *Beast & Man in India* iv. 84 There was once a time when in the nearer East..he [*sc.* the ass] was held in high honour. **1898** W. Miller (*title*) Travels and politics in the Near East. **1902** D. G. Hogarth (*title*) The Nearer East. **1909** [see *Middle Eastern *a.*]. **1910** *Chambers's Jrnl.* Dec. 800/1 In the Near East the keynote of cookery is disguise. The Turk brings his oriental love of mystery with him to the dinner-table. **1912** *Review of Reviews* July 70/1 (*heading*) The Near East problem. **1920** *Sat. Rev.* 16 Oct. 320 He took very little notice of Balkan intrigues, because the Near East was not his business. **1923** E. Wharton *Son at Front* 10 Poor little circumscribed Paul Dastrey, whose utmost adventure had been..an occasional six weeks in the Near East. **1936** *Discovery* Sept. 264/1 The Wellcome Archaeological Expedition to the Near East. **1959** *Listener* 1 Jan. 31/2 Interest in the lands and peoples of the nearer east. **1973** 'D. Jordan' *Nile Green* xliv. 283 Sue..told the Near East Desk, who..sent a cable to the Cairo Embassy.

Hence **Near-Ea·sterly** *adv.*, **Near-Ea·stern** *a.*, **Near-Ea·sterner.**

1906 *Q. Rev.* Jan. 284 Lord Salisbury and his successor have..skilfully withdrawn England from the Near-Eastern entanglements. **1909** *Ibid.* Apr. 654 (*heading*) The Near Eastern question. **1909** *Daily Chron.* 25 Aug. 3/6 Near-Easterly. Bosnia and Herzegovina have figured recently in European politics. **1925** A. J. Toynbee *Survey Internat. Affairs 1920–23* i. 3 They appear..to have refrained from interfering in any way regarding Near Eastern affairs. **1951** E. Paul *Springtime in Paris* ii. 24 There were a few North Americans, hordes of English, many South Americans, and a few Near-Easterners. *Ibid.* xii. 215 Helen Hatounian put an Armenian record on the gramophone... The afternoon stillness..was rent with the shrill voice of a near-eastern soprano. **1959** *Listener* 15 Jan. 145/2 His predilection for the interval of the augmented second (of the near-Eastern scale).

nearmost, *a.* Delete *dial.* and add further examples.

1913 E. H. Barker *Wayfaring in France* 468 It is almost a shriek when the wind strikes the nearmost crests [of trees]. **1952** *Times Lit. Suppl.* 7 Mar. 168/4 You are my nearmost, you who have travelled the farthest.

nearshore (nīə·ɹʃōəɹ), *a.* [Near *a.* + Shore *sb.*¹] Situated or occurring (relatively) close to a shore; pertaining to or involving the study of such a zone.

1896 B. Willis in *U.S. Geol. Survey Geol. Folio* xxxiii. 2/2 This formation..represents the near-shore, muddier sediment of those times. **1936** *Rep. Comm. Sedimentation* (Nat. Research Council) 3 The main subjects that are contemplated are (1) relation of oceanography to sedimentation, including both physical and biological oceanography, (2) pelagic deposits, (3) inland sea deposits, (4) near-shore deposits, [etc.]. **1957** *Gloss. Geol.* (Amer. Geol. Inst.) 196/1 *Nearshore zone*, in beach terminology an indefinite zone extending seaward from the shore line somewhat beyond the breaker zone. It defines the area of nearshore currents. **1966** R. W. Fairbridge *Encycl. Oceanogr.* 614/1 Nearshore oceanography may be taken to include the study of all aspects of the area lying between the backshore zone and the outer edge of the continental shelf. **1973** *Nature* 6 Apr. 393/1 The Galana Boi beds..range from coarse-grained fluvial deposits to fine-grained near-shore lacustrine silts. **1974** *Ibid.* 15 Feb. 452/1 Deposition in a shallow, nearshore marine environment.

near-si·ghtedly, *adv.* [f. Near-sighted *a.* + -ly².] In a near-sighted manner.

1909 *Daily Chron.* 11 Oct. 7/1 Dr. Shuttleworth blinked near-sightedly throughout the time he was in the witness box. **1971** B. Malamud *Tenants* 209 With furrowed brows he reads the marriage contract over and over, then nearsightedly reads it again.

neat, *sb.* Add: **2. b.** *neat beast, -beef, cattle* (earlier examples), *leather* (= Neat's Leather), *stock.*

1624 in *Essex Inst. Hist. Coll.* (1914) L. 235 All my Cattell nowe upon the farme..as neat bests, horse bests, and swine. **1727** *Rec. Smithtown, N.Y.* (1898) 82 It is agreed on that the pounder shall have for pounding a horse four pence, for a net best four pence. **1755** in S. M. Hamilton *Lett. to Washington* (1898) I. 135 Not under twelve shillings and sixpence per Hundred Neet Beef. **1619** *Jrnl. House of Burgesses, Gen. Assembly Virginia* (1915) 13 No man without leave from the Governour shall kill any Neat cattle whatsoever. **1648** *Archives of Maryland* (1887) IV. 390 Certaine neate-cattle to the number of 27. **1776** in *New Hampsh. Hist. Soc. Coll.* (1889) IX. 263 Mens Neat Leather Shoes of the best common sort. **1850** *Rep. Comm. Patents: Agric. 1849* (U.S. Dept. Agric.) 94 It is estimated that there are in this country..fifteen thousand two hundred and eighty five neat stock. **1869** *Rep. Comm. Agric. 1868* (U.S. Dept. Agric.) 427 The present winter (1868) he feeds forty-three head of neat stock, equivalent to thirty-four mature animals. **1882** *Rep. Maine Board Agric.* XVI. 265 The way is to fence off such a piece, and allow no neat stock or horses to run in it at any time.

neat, *a.* and *adv.* Add: **A.** *adj.* **3. d.** *neat cement*: a mortar made from cement and water only, without the addition of sand.

1932 T. Corkhill *Conc. Building Encycl.* 142 *Neat*, a term applied to cement mortar without sand. **1947** J. C. Rich *Materials & Methods Sculpture* xi. 328 Neat cement is a mixture of cement and water. It is not recommended for sculptural use save as a retouching medium. **1964** H. F. W. Taylor *Chem. of Cements* I. 2 Mechanical or physical determinations, such as strength tests, are usually made with an aggregate present, as determinations of this type on neat cement pastes can give misleading results.

7. d. Semi-proverbial phr. *neat* (*but*) *not gaudy* and variants. Also *absol.* and *fig.*

[**1602**: see Gaudy *a.*² 3.] **1700** S. Wesley *Epistle to Friend concerning Poetry* 5 Style is the Dress of Thought; a modest Dress, Neat, but not gaudy, will true Critics please. **1806** C. Lamb *Let.* 26 June (1935) II. 14 A little thin, flowery border round, neat, not gaudy. **1838** Ruskin in *Archit. Mag.* Nov. 484 That admiration of the 'neat but not gaudy', which is commonly reported to have influenced the devil when he painted his tail pea-green. **1849** Thackeray *Pendennis* I. xiii. 116 'You seem to like my dressing-gown, sir,' he said to Mr. Tatham. 'A pretty thing, isn't it? Neat, but not in the least gaudy.' **1887** *Lippincott's Mag.* July 116, I have sent, I say, just such manuscript as editors call for, fair, clean, written on one side, not with a pencil, but with a good gold pen, stamps enclosed for return if declined; the whole thing 'neat, but not gaudy, as the monkey said' on the memorable occasion 'when he painted his tail sky-blue'. **1892** *Society* 6 Aug. 757/1 Tennyson when in a rage is neat and not gaudy. **1974** L. Deighton *Spy Story* xxi. 222 If Toliver complains to the Home Secretary you say it was the C.I.A. doing it. Neat, but not gaudy.

11. b. *colloq.* Excellent, desirable, attractive.

1934 J. T. Farrell *Calico Shoes* 54 A girl in a two-piece bathing suit without brassière walked by them. 'Oh, baby, you can make me so happy!' Don sing-songed. 'Neat!' Jack appraised. **1942** Berrey & Van den Bark *Amer. Thes. Slang* § 29/4 Excellent; first-rate... neat. **1947** *Sat. Rev.* (U.S.) 10 May 26/1 Each of these adjoining rooms has a radio in it, which they find 'neat' and I don't. **1972** D. Westheimer *Over Edge* (1974) i. 10 'I could drive you on into Idyllwild if you want...' 'That would be neat.' **1974** *Washington Post* 24 Feb. H. 13/5 I've passed up some neat dinner invitations.

neaten, *v.* Add: (Later examples.) Often const. *up*.

1942 Berrey & Van den Bark *Amer. Thes. Slang* § 4/4 Arrange; Put in order; Clean up..*neaten up*. **1954** *Daily Progress* (Charlottesville, Virginia) 11 Mar. 10/1 (*heading*) Shrubs and fences neaten up yards. **1970** 'H. Pentecost' *Plague of Violence* (1972) ii. ii. 92 If the killers had tried to neaten up the room after a struggle, they couldn't have put everything back in place just as it had been. **1972** *Daily Tel.* 27 May 9/7 Finally, pulling and weaving neatens the whole nest.

neatnik (nī·tnik). *slang* (chiefly *U.S.*). [Modelled on *beatnik.] One who is neat in his personal habits, as opposed to a *beatnik.

1959 *N.Y. Times* 30 Aug. 67/1 The beatniks and the neatniks had at each other this week. **1960** *Ibid.* 3 Jan. 48 (Advt.), Seeing how you're a Neatnik, you'll be buying things like soap and ties and stuff from now on. **1961** *Britannica Bk. of Year* 537/1 The reaction of one section of beatniks to the appearance of others produced first *Washed Beatnik* and then *Neatnik*. **1962** *Amer. Speech* XXXVII. 146 *Neatnik*..one who is neat in his personal habits. **1969** *Sears Catal.* Spring/Summer 16 A new look in Rally-back Jeans that can be worn by Neatniks of any age.

neb, *sb.* Add: **3. b.** (Further example.)

1893 R. L. Stevenson *Catriona* xi. 116, I couldna see the nebs of my ten fingers.

d. 'The pole of an ox-cart' (E.D.D.); *neb ox*, a draught ox.

1710 *New Hampsh. Probate Rec.* (1907) I. 650 All my household goods and four Cows, and a yoak of neb Oxen..to be for her own proper use. **1865** 'G. Hamilton' *Skirmishes* ii. 7 Men left their oxen standing by the nebs.

‖ **nebbich** (ne·biχ), *sb.* and *a.* Also **nebbish, nebbishe, nebbisher, nebish.** [Yiddish.] A nobody, a nonentity. As *adj.*, innocuous, ineffectual, luckless, hapless, etc. Also as *int.*, an expression of commiseration, dismay, etc.

1892 I. Zangwill *Childr. Ghetto* I. 46 '*Achi nebbich*, poor little thing!' cried Mrs. Kosminski, who was in a tender mood. **1907** —— *Ghetto Comedies* 205 '*Nebbich*, the poor little children!' cried Natalya, horrified. **1959** A. Wesker *Chicken Soup with Barley* iii. i, in *New Eng. Dramatists* I. 222 It's ach a nebish Harry now. It's not easy for him... Other men get ill but they fight. **1960** *Commentary* June 530/1 The *nebbish*, the cynic, the sophisticate. *Ibid.* 539/1 The sad *nebbishe* Podolsky is the owner of the building. **1960** F. Raphael *Limits of Love* i. i. 11 Your daughter ends up by..marrying a good for nothing, nebbisher nobody. **1962** B. Abrahams tr. *Life Glückel of Hameln* i. 4 All the pleasures and riches he, *nebbich*, was denied here. **1968** *Times* 6 Apr. 21 The central character is so nebbish he has not even a name. **1969** *Atlantic Monthly* Sept. 57 Paranoid psychopaths who, after nebbish lives, suddenly feel themselves invulnerable in the certain wooing of sweet death. **1973** *Jewish Chron.* 9 Feb. 5/1 The kings [in this Jewish chess-set] are dead, long live the nebbishes (the deprived, signifying the decline of royal power). **1975** *New Yorker* 3 Feb. 77/2 Mr. Antonacci is both antic and affecting as the jumpy, craven *nebbish* Honey Boy, and John Bottoms is superb in several roles.

‖ **Nebbiolo** (nebiǭ·lo), *sb.* and *a.* Also **Nebiolo.** [It.] A black grape of northern Italy; the red table wine made from this grape. Also *attrib.* or as *adj.*

1833 C. Redding *Hist. Mod. Wines* ix. 246 At Asti, the plants called *Passaretta* and *Malvasia Nebiolo* produce *vins de liqueur*, with the smell of the raspberry. **1875** [see *Barolo]. **1957** *Encycl. Brit.* XXIII. 664/2 Good honest red drinking wine comes from northern Italy, usually marketed under vine names (Nebiolo, Grignolino, Freisa, Barbera). **1958** A. L. Simon *Dict. Wines* 117/2 *Nebbiolo*, one of the best black grapes grown in the North of Italy for the making of red table wines. Many of the wines made from Nebbiolo grapes are sold as *Nebbiolo* wines. **1959** W. James *Word-Bk. Wine* 132 *Nebbiolo*. The vine of this name produces some of the best red wines of northern Italy. **1970** Simon & Howe *Dict. Gastron.* 275/2 *Nebbiolo*, a black grape grown in Piedmont and Lombardy and used to make red table wines. Among these are Barolo and Barbaresco, two of Italy's best-known wines. **1970** *Sat. Rev.* (U.S.) 12 Sept. 69/1 In Piedmont and Lombardy they have harvested the Nebbiolo grapes (a variety named because it ripens late in the autumnal *nebbia*, or mist). **1970** *House & Garden* May 138/3 Nebbiolo is a red wine named after the most aristocratic of the Piedmont grape varieties.

‖ **nebelwerfer** (nē·bəlvę·ɹfəɹ, -wə̄ɹfəɹ). [G., f. *nebel* mist, fog, haze + *werfer* thrower, mortar, f. *werfen* to throw.] A German six-barrelled rocket mortar. Also *attrib.*

1943 *Hutchinson's Pict. Hist. War* 4 Aug.–26 Oct. 191 These German gunners have certainly chosen a well-concealed position in which to hide their nebelwerfer (fog-thrower) battery. **1946** C. Wilmot in R. W. Zandvoort et al. *Wartime Eng.* (1957) 163 The nebelwerfers—those many-barrelled mortars that put down the rocket-propelled bombs with the high wailing sobbing note. **1948** A. M. Taylor *Lang. World War II* (rev. ed.) 136 *Nebelwerfer*, a German rocket thrower similar to the Russian Katyusha. **1961** W. Vaughan-Thomas *Anzio* vi. 112 The Scots Guards captured a German officer who had driven into their lines while looking for sites for the unpleasant *Nebelwerfers*, or six-barrel mortars.

nebenkern (nē·bĕnkəɹn). *Cytology.* Pl. **-kerne.** [a. G. *nebenkern* (O. Bütschli 1871, in *Zeitschr. f. wissensch. Zool.* XXI. 527), f. *neben* near + *kern* kernel, nucleus.] Any of various cytoplasmic structures associated with or resembling the nucleus (see quots.).

1898 *Jrnl. R. Microsc. Soc.* 624 (*heading*) 'Nebenkern' in spermatogenesis of Pulmonata. **1901** G. N. Calkins *Protozoa* i. 14 He [*sc.* Bütschli] held..that in addition to the macronucleus there is a second and a smaller nucleus in Infusoria, and to this he gave the name *Nebenkern* or micronucleus ('76). **1917** *Q. Jrnl. Microsc. Sci.* LXII. 416, I have been unable to find any body in the spermatid formed from 'spindle fibres' or 'yolk granules', and I do not intend to use the term 'nebenkern', which has been, and still is, used without discrimination for almost any granule or body in a cell. For example, Hegner..lately draws attention to the 'granules of Blochmann' in the wasp and two ants, which have also been called 'nebenkerne', quite regardless of whether or no they are of the same nature as the original 'nebenkern' of Bütschli. **1928** E. L. Opie in E. V. Cowdry *Special Cytol.* I. ix. 246 The name 'nebenkern' or 'paranucleus' has been given to structures situated within the cytoplasm [of the pancreatic acinous cell] and staining like nuclear material... Other chromatin-like bodies designated 'nebenkerne' are spherical and have an even contour. **1958** *Internat. Rev. Cytol.* VII. 223 In other invertebrates..the mitochondrial *nebenkern* is transformed into a spiral wrapping that extends for the greater part of the length of the sperm tail. **1962** D. W. Bishop *Spermatozoan Motility* 157 In insects, gastropods, and other invertebrates the conspicuous *nebenkern* formed during spermiogenesis arises by fusion of mitochondria. **1965** E. N. Willmer *Cells & Tissues in Culture* II. xi. 531 The round or oval nucleus of the hair-cell [in the organ of Corti] occupies the third quarter of the cell. The infranuclear part contains a small number of mitochondria and a group of granular cisternae. Occasionally, the granular cisternae might be arranged in a circular fashion forming a 'Nebenkern' which recalls the so-called Retzius Body.

Nebraskan (nĕbræ·skăn), *a.* and *sb.* [f. *Nebraska*, name of a state in the central U.S. + -AN.] **A.** *adj.* **1.** Of, pertaining to, or from Nebraska.

1875 E. A. Curley *Nebraska* xxii. 289 The Mormon fields, irrigated as they are, are much smaller than those of Nebraskan farmers. **1884** *N.Y. Weekly Tribune* 2 Apr. 10/4 The advantages which the Nebraskan Mennonites secure by co-operative purchasing of their implements, machinery, and supplies of all kinds, are manifestly..great. **1933** Blunden & Norman *We'll shift our Ground* 155 Some thesis writer, Nebraskan or Wisconsinese. **1973** S. Dobyns *Man of Little Evils* (1974) x. 96 The story concerned a Nebraskan couple who had been..robbed.

2. *Geol.* Of, pertaining to, or designating the first Pleistocene glaciation in North America, now generally identified with the Günz glaciation in the Alps. Also *absol.*, the Nebraskan glaciation or the deposits it produced.

1909 B. Shimek in *Bull. Geol. Soc. Amer.* XX. 408 The tough, impervious, bluish-black fill which has been known as the sub-Aftonian or pre-Kansan drift in Iowa, is here so well developed, reaching an exposed thickness of more than 15 feet.., that it can no longer be regarded as merely a remnant, but should rank with other well developed drift sheets... This leaves it without a name, and in view of this fact, and of the wide distribution of this formation, the name *Nebraskan* is proposed for it... The name Nebraskan was suggested to the writer by Professor Calvin. **1930** *Science* 22 Aug. 194/1 The Nebraskan is probably a million years old. **1934**, **1957** [see *Illinoian *sb.* and *a.* B]. **1966** *Science* 11 Nov. 771/3 The Early Pleistocene record in Nebraska is at present interpreted as indicating two stadial advances of the continental glacier during Nebraskan time and three stadial advances during Kansan time. **1970** [see *Günz].

B. *sb.* A native or inhabitant of Nebraska.

1888 [see *Minnesotan]. **1900** *Congress. Rec.* 27 Jan. 1241/2 He aided in the nomination of Mr. Bryan for President, and was an enthusiastic supporter of the brilliant Nebraskan. **1948** *Time* 26 Apr. 13/2 Ohioans seemed as friendly to him as Nebraskans. **1969** I. Kemp *Brit. G.I. in Vietnam* iii. 49 The medic, 'Jake' Jacobs..was a pale, thin Nebraskan, recently married.

Nebuchadnezzar (nebiŭkădne·zăɹ). [So called in allusion to *Nebuchadnezzar* King of Babylon, d. 562 B.C.] A very large wine-bottle (see quots.).

1913 A. Huxley *Let.* 11 Nov. (1969) 55 Fireworks ensue, then (children dismissed) supper and afterwards, the most magnificent Nebuchadnezzars, and finally a good form of blind man's buff. **1962** [see *Balthazar]. **1967** *Punch* 27 Sept. 484/3 In the thirtieth of fifty-one chapters, Patrick Forbes lists the eleven different sizes of champagne bottle, from the quarter-bottle to the Nebuchadnezzar, eighty times as big. **1967** A. Lichine *Encycl. Wines* 317/2 Since wine ages more slowly in large bottles, there are many magnums and double magnums in the Lafite cellars, as well as Imperials, which hold eight bottles, and Nebuchadnezzars (ten bottles). **1970** Simon & Howe *Dict. Gastron.* 80/2 *Nebuchadnezzar*, 20 bottles.

nebula. 3. Add to def.: In mod. use the term is applied to (*a*) a cloud of gas or dust situated within the interstellar space of a galaxy (usu. our own) and appearing as either a bright or a dark cloud (according to whether or not there are stars present to make it luminous); (*b*) a galaxy (usu. one other than our own). (Further examples.)

1924 H. Dingle *Mod. Astrophysics* xvi. 302 A dark nebula is really dark, and not merely too faint for its light to be seen on account of its great distance from us. **1930** R. H. Baker *Astron.* xi. 465 Frequently a 'nebula' turned out to be a star-cluster, thereby encouraging the opinion, in former times, that all nebulae are really clusters of stars. *Ibid.*, Modern investigations have shown that nebulae, as distinguished from ordinary star-clusters, fall into two classes having entirely different characteristics, namely, the galactic nebulae and the extra-galactic nebulae. **1963** B. & J. Lovell *Discovering Universe* ix. 114 The Milky Way system is typical of many extra-galactic spiral nebulae, and very much akin to M31 in Andromeda. **1968** P. Moore *Sky at Night* II. xxxii. 232 With other nebulæ, there are no convenient stars—and so the nebulæ cannot be seen directly, but betray themselves because they blot out the light of the stars beyond... Mention should also be made of the nebulæ which shine by pure reflection. Such is the nebula in the Pleiades. *Ibid.* 235 The old term of 'spiral nebula' has become obsolete, to be replaced by 'spiral galaxy'; a proper nebula is gaseous, and belongs to the Galaxy in which we live. **1971** D. W. Sciama *Mod. Cosmol.* iii. 39 In this way he [*sc.* E. P. Hubble] obtained a distance of 800 000 light-years for the Andromeda nebula, and similar values for other spiral nebulae. Now that these nebulae are well established as stellar systems outside our own, we shall henceforth call them galaxies. **1974** F. W. Cole *Fund. Astron.* xiii. 358/1 The Orion nebula is 1600 LY distant and about 30 LY in diameter. In our galaxy, other emission nebulae of the same type are about the same size. Vastly larger emission nebulae are known outside our galaxy; an excellent example is in M33, a spiral galaxy in Triangulum.

nebulium. Add: (Further examples.)

It is now considered that no such element exists, the lines formerly attributed to it having been identified with those produced by known elements.

1937 J. W. T. Spinks tr. *Herzberg's Atomic Spectra & Atomic Struct.* iv. 157 Bowen..first showed that the nebulium lines, which had been observed in the spectra of many cosmic nebulae but were long a complete mystery, were to be explained as forbidden transitions between the deep terms of O^+ (^4S, ^2D, ^2P), O^{++} (^3P, ^1D, ^1S), and N^+ (^3P, ^1D, ^1S). **1940** *Astrophysical Jrnl.* XCII. 408 The free electrons, liberated from hydrogen by photo-ionization, may excite the 'nebulium' lines by inelastic impact. **1965** Phillips & Williams *Inorg. Chem.* I. ii. 43 Transitions between these states in atomic spectra are also 'forbidden', but they have been observed for the isoelectronic N^+ and O^{2+} ions in the spectra of cosmic nebulae (so called nebulium lines). **1973** L. Oster *Mod. Astron.* xiii. 205 For a long time, in fact until 1928, the origin of some of the strongest lines was unclear, and, in desperation, they were ascribed by some astronomers to a hypothetical element called nebulium.

nebulization (nebiŭlaizē¹·ʃən). *Med.* [f. Nebulize *v.* + -ation.] **a.** The conversion of a liquid into a mist or spray. **b.** Medical treatment using a nebulizer.

1906 *Index-Catal. Library Surg.-General's Office, U.S. Army* 2nd Ser. XI. 366/1 (*heading*) Nebulizers and nebulization. **1949** *Dis. Chest* XVI. 410 The technique designed ..by us is based on nebulization of the surface anesthetic solution. **1968** *Jrnl. Asthma Res.* V. 253 Each child was subjected to 15 minutes of nebulization, three times daily. **1969** *Daily Tel.* 25 June 15/5 You can have different products incorporated into your nebulisation shower: products to help ease pregnancy scarring, correct dry or greasy skin, [etc.].

nebulize, *v.* Restrict *rare* to sense 2 and add to sense 1: Hence **ne·bulizing** *vbl. sb.* (Further examples.)

1915 Wright & Smith *Text-bk. Dis. Nose & Throat* i. 58 There are a number of mechanical devices for nebulizing medicated solutions, which have a certain amount of value in tracheal and bronchial affectations. **1944** *Proc. Soc. Exper. Biol. & Med.* LVII. 257/1 The nebulizing unit..consists of a container and a metal atomizer. **1949** *Dis. Chest* XVI. 412 Oxygen or air may be used to nebulize the solution. **1958** *Ann. Allergy* XVI. 628 The liquid is disrupted by the air jet in the nebulizing area and forms an aerosol. *Ibid.* 631 The DeVilbiss No. 40 model nebulizes about twice as much water as the Vaponefrin model. **1973** *Amer. Rev. Respiratory Dis.* CVIII. 511/2 The small deposition of ^{99m}Tc in the lungs in these normal subjects even during slow, deep breathing raises serious doubts that effects claimed for aerosol therapy are related to nebulized fluid.

‖ **nécessaire** (nesesę̄r), *sb.* Also **necessaire.** [Fr. (see Necessaire *a.*).] A small case, sometimes ornamental, for small articles, as pencils, scissors, tweezers, articles of cosmetics, etc.

1800 E. Hervey *Mourtray Family* III. ix. 177 A chance of his travelling *necessaire*, and all the apparatus of his toilet, being burned. **1854** Thackeray *Newcomes* I. xxviii. 266 Gousset empty, tiroirs empty, nécessaire parted for Strasbourg! **1876** Geo. Eliot *Dan. Der.* I. i. ii. 29 Gwendolen..thrust necklace, cambric..and all into her *nécessaire*. **1960** *Times* 16 Feb. 20/7 An old English gold and agate necessaire. **1967** V. Nabokov *Speak, Memory* (rev. ed.) xiii. 253 The handful of jewels which Natasha, a farsighted old chambermaid,.. had swept off a dresser into a *nécessaire*. **1973** *Times* 11 Dec. 18/5 A George III gold and enamel *nécessaire* attributed to James Cox.

necessary, *a.* and *sb.* Add: **A.** *adj.* **I. 1. a.** *necessary evil* (earlier and later examples).

1547 W. Baldwin *Treat. Morall Phylosophie* III. xv. sig. O5ᵛ A woman is a necessary euyll. **1765** Johnson in Shakes. *Plays* I. Pref. p. lxix, Notes are often necessary, but they are necessary evils. **1776** T. Paine *Common Sense* 1 Society in every state is a blessing, but Government even in its best state is but a necessary evil; in its worst state an intolerable one. **1815** C'tess Granville *Let.* 18 July (1894) I. 54 The humiliation of now having him is great, but he is reasonable about it and thinks it a necessary evil. **1863** *Country Gentleman* 16 Apr. 254/3 The manuring of the vines is regarded as 'a necessary evil'. **1927** E. O'Neill *Marco Millions* I. iv. 51 Don't waste pity. Her kind are necessary evils.

d. *necessary condition* = Condition *sb.* 4; cf. *sufficient condition.

1817 Coleridge *Biog. Lit.* I. ix. 136, I began to ask myself; is a system of philosophy..possible? If possible, what are its necessary conditions? **1859** Mill *On Liberty* i. 9 The consent of the community..was made a necessary condition to some of the more important acts of the governing power. **1949** A. Pap *Elem. Analytic Philos.* x. 212 If a sufficient condition is complex—as it almost invariably is—then it may consist in a conjunction of necessary conditions. **1965** E. J. Lemmon *Beginning Logic* i. 28 Whenever it is the case that only if *P* then *Q*, *P* is a necessary condition for *Q*.

necessitarian, *sb.* and *a.* (Earlier and later examples.)

1796 F. A. Nitsch *Gen. View Kant's Princ. concerning Man* 17 The Necessitarians..make a considerable party in the philosophic world. **1912** Kipling *Songs from Books* (1913) 154 (*title*) The necessitarian. **1952** C. P. Blacker *Eugenics* 267 Unconscious prejudices can throw the Nature–Nurture controversy into different perspectives. Those who believe in predestination as an article of religion have a necessitarian bias which might incline them to over-estimate the role of heredity. **1956** E. H. Hutten *Lang. Mod. Physics* vi. 212 Ever since Hume the necessitarian interpretation [of determinism] has been rejected. **1966** M. R. D. Foot *SOE in France* viii. 183 By one of those accidents that baffle the necessitarians, the only sentry awake at the moment of the drop was a newcomer who did not know where the alarm telephone was.

necessitated, *ppl. a.* **2.** (Later example.)

1857 A. & M. Ward *Husband in Utah* xvii. 194 Mrs. Farrow informed me of several sisters, who having inherited money from Eastern quarters, were immediately assailed by the necessitated priest.

nece·ssitator. [f. Necessitate *v.* + -or.] = Necessitater.

1904 Hardy *Dynasts* I. vi. iii. 118 O Great Necessitator, heed us now!.. Quicken the issue as Thou knowest how.

necessity. Add: **14.** *attrib.* **necessity-operator** *Logic*, a word or symbol signifying that the proposition to which it attaches is a necessary truth.

1957 A. N. Prior *Time & Modality* 57 Church's necessity-operator is not an operator of this sort. **1968** Hughes & Cresswell *Introd. Modal Logic* ii. 24 We shall call *L* the necessity operator. **1973** M. J. Cresswell *Logics & Lang.* ii. 32 The simplest is the necessity operator, or as Dana Scott calls it, the universal necessity operator.

neck, *sb.*¹ Add: **1. e.** Phrases. *to get* (*catch, take*) *it in the neck*: to be hard hit (by something); to be severely reprimanded or punished. Conversely, *to give it in the neck*: to assault or reprimand (someone) severely.

1882 *National Police Gaz.* 25 Nov. 3/3 An 'Artless' Young Girl Gives it to Her 'in the Neck', as the Sports Say. **1887** [see *dub *sb.*⁴]. **1903** A. Adams *Log of Cowboy* xi. 175 Old Nat will get it in the neck this time, if that old girl dallies with him as she did with us. **1908** H. G. Wells *War in Air* ii. 58 They'll get it in the neck in real earnest one of these days, if they ain't precious careful. **1914** D. O. Barnett *Let.* 31 Dec. (1915) 30 You probably don't know what a village looks like when it has caught it in the neck. **1923** Wodehouse *Inimitable Jeeves* iii. 30 Something always comes along to give it you in the neck at the very moment when you're feeling most braced about things in general. **1927** F. Niven *Wild Honey* iii. 21 If you sit.. facing ahead, that's called 'punching the breeze'. If you sit..looking back, it's called 'taking it in the neck'. **1928** 'Sapper' *Female of Species* x. 169 I'd never forgive myself if one of you took it in the neck. *a* **1930** D. H. Lawrence *Phoenix II* (1968) 259 Give it the blue-bottles [*sc.* policemen] in the neck! **1955** *Times* 11 July 12/7 Do they belong to an unlucky generation that has got it in the neck before the law can catch up with the swift development of civilian aviation, and insist upon silenced airliners? **1973** *Guardian* 18 June 10/2 It's the poor old vicar who gets it most in the neck... He runs the risk of losing the best-kept-village competition because..the churchyard is looking its shaggiest. **1974** *Times* 12 Dec. 14/4 Giscard..

apparently caught it in the neck from his continental colleagues.

2. d. In Racing, *to win by a neck*, i.e. by the length of the horse's neck. Also *fig.* So *a neck*, such a distance separating two horses at the end of a race. Also of greyhounds.

1823 'J. Bee' *Slang* 94 Won by..indeed 'a neck'. **1865** 'Mark Twain' *Sk. New & Old* (1875) 32 She'd always fetch up at the stand just about a neck ahead. **1873** J. Blackwood *Let.* 7 June in Geo. Eliot *Lett.* (1956) V. 421 There was a grand [golf] match...my man the young champion Tom Morris came in winner by a neck. **1886** [see Head *sb.* 1 c]. **1930** *Daily Express* 6 Oct. 17/6 Three-quarters of a length; neck. Beggarman fourth. **1931** T. R. G. Lyell *Slang* 544 The worst of it is that I only lost by a neck; the other fellow beat me by three marks! **1975** *Times* 21 July 7/4 If Juliette Marny had not cocked her head..a few strides from the post Piggott thought the margin of success would have been half a length rather than a neck.

e. *N.Z.* The wool shorn from the neck of a sheep.

1950 *N.Z. Jrnl. Agric.* Oct. 311 (*caption*) Frames hinged to a wall [in the shearing shed] can be very useful to support a wool pack for bellies, necks, etc.

3. b. Also implying insolent speech or presumptuous behaviour, esp. in phr. *to have a neck.*

1893–4 R. O. Heslop *Northumb. Words* II. 494 *Neck*, forwardness, impudence. 'What a neck ye hev efter aa'!' **1933** *Punch* 25 Jan. 108/3 I'm afraid I was so overcome by his barefaced 'neck' that it never occurred to me to call him back. **1935** G. Heyer *Death in Stocks* iii. 34 He'd had the infernal neck to say I wasn't going to marry the man. **1942** L. A. G. Strong *Unpractised Heart* xii. 77 And then you have the sheer neck, the bloody effrontery to say you think there's more in life than I do. **1960** J. Symons *Progress of a Crime* v. 34 If that doesn't beat anything for hard neck.

e. *to speak* (*talk*) *through* (*the back of*) *one's neck*: to use extravagant or inaccurate words or language; *to stick* (or *put*) *one's neck out*: to expose oneself to danger, reprisal, criticism, etc.; *up to the neck*: fully concerned or deeply implicated in (a transaction, freq. illegal); also *ellipt.*, occupied without intermission; *to breathe down* (*someone's*) *neck*: to be close behind (someone); to keep a close or oppressive watch upon; *dead from the neck up*: see *dead *a.* 32 h. Also in further miscellaneous uses.

1388 Wyclif *Gen.* xlv. 14 And whanne he hadde biclippid, and hadde feld [*other MSS.* falle] in to the necke of Beniamyn, his brother, he wepte, the while also Benjamin wepte in lijk maner on the necke of Joseph. **1611** Bible *Gen.* xlv. 14 And he fel vpon his brother Beniamins necke, and wept: and Beniamin wept vpon his necke. **1738** Swift *Polite Conv.* I. 47 If ever I hang, it shall be about a fair Lady's Neck. **1896** [see Oafish *a.*]. **1899** E. W. Hornung *Amat. Cracksman* 199 'Don't talk through yer neck,' snarled the convict. 'Talk out straight, curse you!' **1903** A. H. Lewis *Boss* 174 Still I must say you went in up to your neck on sparks and voylets. **1907** *Strand Mag.* June 672/1 We are not slow to tell them they are 'talking through the back of their neck'. **1909** R. Brooke *Let.* 3 Nov. (1968) 192 Your offer is splendid and noble. I fall upon your neck. **1912** Kipling *Songs from Books* (1913) 153 So back I go to my job again, Not.. quite so ready to sob again On any neck that's around. **1923** Conrad *Rover* xii. 203 It's the very spot for hatching treacheries. One feels steeped in them up to the neck. **1923** *Pall Mall Gaz.* 13 Apr. 3/3 Anybody who gets up in this House and talks about universal peace knows he is talking through the back of his neck. **1926** *Univ. Mag.* (Univ. Va.) Oct. 16/2 Absolutely original slang at the University of Virginia includes..*to stick one's neck out.* **1933** *New Republic* 22 Nov. 45/2 Instead, there is a general disposition now to regard him as a fat-headed fellow.. who 'put his neck out' and got what he deserved. **1935** *Planning* III. xlix. 2 Opinion in administration and in industry, among people who are up to the neck in current problems, is far more advanced than opinion among some of those who occupy comfortably detached positions. **1936** R. Chandler in *Black Mask* June 31/2 You sure stick your neck out all the time. **1941** Wodehouse *Berlin Broadcasts* in *Performing Flea* (1961) i. 266 Algy didn't know a thing about it and was almost certainly talking through the back of his neck. **1946** K. Tennant *Lost Haven* (1947) vii. 97 There were big black moths in the wardrobe; not to mention a beastly big mountain breathing down the back of your neck. **1950** H. Hastings *Seagulls Over Sorrento* in *Plays of Year* IV. 64 We've stuck our necks out—we're looking for trouble, see? **1955** R. Frost *Small Hotel* in *Ibid.* XIII. 189 I'll be out of here on my neck as soon as they've got what they want out of me in this court business. **1955** A. L. Rowse *Expansion Eliz. Eng.* ii. 64 Three mayors..were up to their neck in the trade. *Ibid.* viii. 302 The conclusion she [*sc.* Elizabeth] drew from that was not to put out her neck again. **1957** A. Huxley *Let.* 18 Jan. (1969) 816 Selznick is up to his neck in his forthcoming..film and probably won't be able to read the piece for some days. **1959** *Listener* 5 Mar. 414/1 It was likely that he would be thrown out on his neck very quickly. **1959** *Times* 19 May 5/5 Because Kent were always breathing down their necks, Hampshire could never really establish themselves. **1961** *Debates Senate Canada* 5 July 1021/1 So I shall try not to stick my neck out on the legal aspect too much; although, as I say, even from a legal standpoint it does seem rather simple to me. **1961** A. Wilson *Old Men at Zoo* iii. 162 But you shouldn't worry. You can never do the best work that way. Of course with Falcon and Sanderson round your neck, I'm not surprised. **1965** J. Porter *Dover Two* xi. 141 MacGregor rushed..away, delighted to be able to pursue his own line of investigation and..not to have Dover breathing down his neck all the time. **1971** 'F. Clifford' *Blind Side* iv. ii. 157 'I haven't seen him for a couple of days... He's been up to his neck.' 'Who with?' 'Same man.' **1971** A. Price *Alamut Ambush* xiii. 157 You don't have to stick your neck out, David—I'll stick mine out. And it'll be a pleasure! **1973** J. Fleming *You won't let me Finish* ix. 79 You're in it up to the neck whether you like it or not. **1973** *Times* 27 Feb. 14/4 He.. unfortunately began a sentence 'If I disagree with my local party..' whereupon a heckler added loudly 'You'll be out on your neck.' *Ibid.* 24 Apr. 11/7 Shakespeare.. gives the troupe a chance to try something new without the Academie breathing down its neck.

6. c. *neck and crop*: (later examples). Also *attrib.*

1932 Kipling *Limits & Renewals* 398 That does not excuse the neck-and-crop abruptness..of..our expulsion. **1963** A. Ross *Australia 63* v. 110 Titmus,..trying to force an in-swinger away, was bowled neck and crop. **1967** [see *decolonization].

9. a. *neck and neck*: (further examples).

1799 *Sporting Mag.* XIII. 309/1 In this way, neck and neck, whipping and spurring, all the speed of the horses, and all the skill of the jockies exerted, they rode up to the ending post. **1835** Dickens *Let.* 2 May (1965) I. 58 We came in literally neck and neck. **1901** *Chambers's Jrnl.* June 361/2 There a horse fell or staggered, and was instantly recovered. Now we were a few yards ahead, again neck-and-neck with the 'Quicksilver'; and so we raced on until we approached the old bridge at Bow. **1926** P. Guedalla *Palmerston* vii. 320 The republicans, the Orleans princes.., and the President might soon be running neck and neck. **1955** *Times* 23 June 9/4 Production ran neck-and-neck in the studios, but the second version..reached the public screen last.

b. *attrib.* (Earlier and later examples.)

1828 M. R. Mitford *Our Village* III. 204 The strength and luck of the parties were so well balanced, that it produced quite a neck-and-neck race, won only by two notches. **1952** E. F. Davies *Illyrian Venture* v. 84 Nicholls and I had a race across the plain, with a neck-and-neck finish.

9*. *neck-to-knees*: (see quot. 1941). Also *neck-to-knee* attrib. *Austral.*

1941 Baker *Dict. Austral. Slang* 49 *Neck-to-knees*, bathing costumes covering the body from the neck to half-way down the thigh. **1965** G. McInnes *Road to Gundagai* xiv. 261 Refusing to wear the regulation 'neck-to-knee' bathing togs.

11. b. For second part of def. read: *neck of the woods* (orig. *U.S.*), a settlement in wooded country; a small or remotely situated community; a district, neighbourhood, or region. Also *neck of timber*, and *ellipt.* (Earlier and later examples.)

1839 *Spirit of Times* 15 June 175/2 In this neck of the woods. **1874** [see *beatenest *a.*]. **1931** 'Grey Owl' *Men of Last Frontier* 15 A man may be soaking wet, half-frozen, hungry, and tired, landed on some inhospitable neck of the woods, vowing that a man is a fool to so abuse himself. **1955** M. Gilbert *Sky High* vi. 76 They don't come to live in this particular neck of the woods. **1967** *Listener* 19 Oct. 501 Some jerk has applied for a job as the new Cyril Connolly. Perhaps you would look him over, he lives in your neck of the woods. **1973** R. D. Symons *Where Wagon Led* I. vi. 95 Lee said there were springs about two miles up, and any cattle in what he called 'this neck of the woods' would probably be there. **1973** J. Wainwright *Devil you Don't* 21 In this neck, I say what. I also say when.

15. a. *neck-buckle, -gear, -ribbon, -rope, -wear.* **c.** *neck-hold.*

1767 in *Essex Inst. Hist. Coll.* (1917) LIII. 298, 7 pair silver Sleeve Buttons, together with Neck-Buckles, etc. **1888** *Judge* (U.S.) 29 Sept. 401/1 Neck-gear will, as always, cause the torture of dudes and dukes. **1890** Hardy *Melancholy Hussar* ii, in *Three Notable Stories* 167 His head would probably have been bent..but for his stiff neck-gear. **1912** W. Owen *Let.* 3 Apr. (1967) 127 In Church with a neck-gear such as Wordsworth wore. **1905** *Daily Chron.* 23 Feb. 3/5 By means of a peculiar 'neck-hold' he can render his man unconscious. **1852** A. Cary *Clovernook* 97 She selected a white muslin which she thought would do if she only had a new neck-ribbon. **1883** *Century Mag.* Aug. 572/2 Partly to rescue the rest of her raiment from the shower which had ruined her neck-ribbon. **1777** *Horae Subsecivae* (MS.) 302 *Neck-rope*, a wooden bow to come round the neck of a bullock, and fastned above to a small beam, by which bullocks are fastned with a cord or rope in the linney. **1822** J. Fowler *Jrnl.* 18 June (1898) 159 In the evening the Indeans [s]tole all the neck Roaps of our Horses. **1879** Webster Suppl. 1569/1 *Neck-wear*, a collective term for cravats and collars. (*Colloq.*) **1887** *Harper's Mag.* May 947/2 He waited at the corner of the block,..affecting an interest in the neckwear of a furnisher's window. **1910** *Westm. Gaz.* 15 Apr. 4/1 Similar good results have followed upon the use of looser neck-wear. **1924** *Barnsley Brit. Drapery Stores Sale Catal.*, Lace, embroideries and neckwear. **1959** *Sears, Roebuck Catal.* Spring & Summer 145/2 Neckwear... Collar and Cuff Set... Nylon Dickey.

16. neck-canal *Bot.*, in ferns and bryophytes, a central channel in the neck of the archegonium, made up of **neck canal cells; neck-cell** (later examples); **neck-lock,** (*a*) in a wig, a sausage-curl (see also quot. 1966); (*b*) in *Judo*, a form of strangle-hold; **neck-oil** *slang*, alcoholic drink, chiefly beer; **neck-rein** *v.* (see quot. 1946); also as *sb.*; so **neck-reining** *vbl. sb.*; **neck-roll,** (*a*) in *Gymnastics*, a swing of the body backwards to rest on the back of the neck: (*b*) (see quot. 1966).

1887 W. Hillhouse tr. *Strasburger's Handbk. Pract. Bot.* xxv. 275 The neck [of the archegonium] is traversed by the neck-canal, which is composed of a series of neck canal-cells, the walls between which are dissolved, and the disorganized contents of the four neck canal-cells are thus fused into a connected string. **1938** G. M. Smith *Cryptogamic Bot.* II. ii. 16 Marchantiales typically have six rows of neck cells. *Ibid.* 17 During the course of development of the neck, the primary canal cell divides to form four neck canal cells. **1957** *New Biol.* XXII. 116 In effecting fertilization the spermatozoid has to traverse a lengthy 'neck-canal', a distance perhaps 200 times its own length. **1965** Bell & Coombe tr. *Strasburger's Textbk. Bot.* 524 The ventral and neck canal cells may be regarded as gametes which have become functionless. **1761** W. Hogarth *Works* (1833) II. facing p. 177 Triglyph membretta or necklock. **1906** Miyake & Tani *Game of Ju-Jitsu* ix. 59 L may make direct opposition to the neck-lock in several ways. **1925** Kelly & Schwabe *Historic Costume* vi. 201 The twisted central neck-lock is a survival of the late seventeenth century dildo. **1962** E. G. Bartlett *Judo & Self-Defence* 56 The Single Wing Necklock... Stranglehold.., pass your left hand under his left armpit and up behind his head, so as to press his head forward. **1966** J. S. Cox *Illustr. Dict. Hairdressing* 169/1 *Neck lock*, the vertical curl at the neck of a barrister's wig. **1860** Hotten *Dict. Slang* (ed. 2) 179 *Neck*, to swallow. *Neck-oil*, drink of any kind. **1880** C. H. Poole *Attempt towards Gloss. Words Stafford* 16/2 *Neck-oil*, ale. A word I once heard at Walsall. **1919** H. Jenkins *John Dene of Toronto* i. 27 They'd be attacked all along the three thousand miles route, and would go down like neck-oil on a permit night. **1936** F. Richards *Old Soldier Sahib* iv. 75 He inquired if we were fond of a drop of 'neck-oil', which like 'purge' was a nickname for beer. **1970** *Private Eye* 2 Jan. 12 A chance encounter..leads Barry to consume a lot of nice neck-oil. **1935** H. D. Chamberlin *Riding & Schooling Horses* iv. 153 'Bearing' or 'Neck Rein'. The right hand is carried just over the crest of the neck, and acts towards the left front. **1940** W. Fawcett *Young Horseman* xii. 151 'Neck reining'... Here the horse is taught with voice, hand and heel to turn away from the side on which he feels the rein against his neck. *Ibid.*, It is very handy to have a horse which is trained to 'neck rein'. **1946** M. C. Self *Horseman's Encycl.* 288 Neck reining a horse is turning him by use of the indirect rein. That is, in turning to the left the reins are carried to the left with no backward pull nor any direct pull on the left rein; the right rein, coming against the neck, gives the signal for the turn. **1953** G. Brooke *Introd. Riding* vii. 76 What in polo language is called 'neck-reining', e.g. when turning or diverging to the right, while applying the left leg, the hand is carried over to the right. **1959** E. Collier *Three against Wilderness* ii. 23, I..neck reined the gelding northward. **1920** *Royal Navy Handbk. Physical & Recreational Training* v. 180 'Tricks of Ground Work'... Neck roll (backwards) to Long-arm Balance. Back Handspring,[etc.]. *a* **1935** T. E. Lawrence *Mint* (1955) II. xii. 133 Hand-springs and neck-rolls. **1946** G. Millar *Horned Pigeon* xiii. 177 Instead of hitting something very solid,..I found myself doing neck rolls down a granite chip embankment. **1966** J. S. Cox *Illustr. Dict. Hairdressing* 169/1 *Neck roll*, (1) The ends of the natural hair worn in a roll at the nape. (2) A postiche worn at the nape. Also called neck piece.

neck, *v.*¹ **2.** Read: *slang.* To drink, to swallow. (Later examples.)

1860 [see *neck-oil* (*Neck *sb.*¹ 16)]. **1889** E. Peacock *Gloss. Words Manley & Corringham, Lincolnshire* (ed. 2) 366 He neck'd a good share o' beer that neet o' th' jewbilee. **1899** C. Rook *Hooligan Nights* i. 13 He wasn't selling 'is meat over-quick, 'cos 'alf the time he was necking four-ale in the pub 'cross the way. **1929** J. Masefield *Hawbucks* 135, I do wish..you'd chuck necking Scotch the way you do.

4. a. *trans.* To clasp (a member of the opposite sex) round the neck; to fondle. **b.** *intr.* To engage in holding and fondling, or to embrace and caress, a member of the opposite sex. Hence **necking** *vbl. sb.*; also *attrib.*, as *necking party.*

1825 A. Crawfurd *Tales my Grandmother* I. 138 Let's see nae mair o' Peter Wallett's neckin' an' touslin' here. **1842** Allnutt *Diary* (MS.) 10 Newcastle... I came rather suddenly upon a man who unceremoniously put his arm round a young lady, and..said..'I was only a-necking on her a little bit, Sir.' **1877** G. Fraser *Wigtown & Whithorn* 272 When sufficiently near him, she necked her supposed partner, greeting him with the following affectionate salute. **1890** J. Service *Thir Notandums* xi. 82 I'm muckle mista'en if I haena seen him neckin' wi' the said Betty. **1922** *Dialect Notes* V. 148 *Necking*, dancing with cheeks together, also known as 'parking'. **1923** *Cosmopolitan* Nov. 72/3 The necking parties in dark nooks about the deck at night. **1924** P. Marks *Plastic Age* xiv. 149 Some of those janes certainly could neck, and they were ready for it any time. **1928** *Daily Tel.* 4 Sept. 7/5 High school children..whose favourite pastime is 'necking' in motor-cars in dark roads with the lights turned off. **1930** *Punch* 26 Mar. 341/3 *Necking-Control.* The Ministry of Transport is now building a chain of illuminated posts... Red means 'stop', amber 'get ready to love' and green 'go'.—*Indian Paper.* **1932** E. Waugh *Black Mischief* v. 179 It's pretty dim for me, floundering about half the day, ..while you neck with the chap who's cut me out. **1935** Wodehouse *Blandings Castle* xii. 296 Do you know who that is that this necker is necking?.. My girl. No less. **1938** E. Bowen *Death of Heart* II. vi. 278 A spot of necking with Daphne. **1940** J. O'Hara *Pal Joey* 59, I was even surprised I could neck her at all. **1950** G. Barker *True Confession* iv. 24 That this rapscallion Was necking with his legal bride. **1958** *Times Lit. Suppl.* 15 Aug. p. x/3 For the active young non-reader, sport, cars, dancing and necking are the prime immediacies. **1970** G. Greer *Female Eunuch* 181 The best behaved teenager necks. **1971** *Petticoat* 24 July 39/5 In necking and/or petting, a boy may ask his partner to 'make love to him'. **1974** 'J. Le Carré' *Tinker, Tailor* xiii. 115 A loving couple necking in the back of a Rover.

5. To fasten *to* or *together* by means of ropes put around the neck. *U.S.*

1857 D. E. E. Braman *Information Texas* iv. 73 The usual practice of farmers whenever they want work oxen, is to..neck together, with ropes, as many pair of..steers as they desire. **1923** J. H. Cook *50 Yrs. on Old Frontier* 21 Each of them had to be 'necked' to a gentle one, to be led for a time. **1930** J. F. Dobie *Coronado's Children* iii. 102 Every animal in the pen had been roped and [illegible] necked to an old brindle ox. **1933** J. V. Allen *Cowboy Lore* i. 9 *Necking*, in range terminology... On the range an unruly cow or one with roving proclivities will often be necked or tied to a more tractable animal.

6. *intr.* To undergo a local reduction in width when subjected to tension. Usu. with *down.*

1938 J. Newton *Introd. Metall.* iv. 105 The contraction begins to concentrate at some one point on the bar, [and] the bar begins to 'neck down'. **1942** *Industr. & Engin. Chem.* Jan. 56/2 During drawing each filament 'necks down' and takes a smaller diameter. **1964** *Mod. Textiles Mag.* Jan. 67/2 When an undrawn nylon filament is stretched beyond its elastic limit, it suddenly (and irreversibly) 'necks down' to a fraction of its original diameter. **1969** P. J. Vincent in P. D. Ritchie *Physics of Plastics* ii. 84 It may also happen that a specimen does not neck [illegible] softening.

necked, *a.* Add: **3.** (Further examples.)

1956 R. J. C. Atkinson *Stonehenge* v. 154 Two main groups of population, distinguished by the forms of their pottery as the Bell-Beaker and Necked-Beaker cultures. *Ibid.*, The Necked-Beaker culture..is an indigenous development in England. **1967** *Antiquaries Jrnl.* XLVII. 174 The latter comprised sherds of two pots—part of the body of a probable necked..beaker and part of the base of a beaker of indeterminate type—probably a necked variety. **1972** *Sci. Amer.* Dec. 34/3 The Taslan process feeds a filament into a rapidly moving airstream at the necked region of a nozzle.

4. [perh. f. *Neck *v.*¹ 6.] Reduced in width as a result of having been subjected to tension.

1959 C. E. Birchenall *Physical Metall.* vi. 136 *A* is the cross-sectional area in the necked region. **1964** R. E. Reed-Hill *Physical Metall. Princ.* xvii. 555 Fracture begins at the center of the necked region on a plane that is macroscopically normal to the applied tensile-stress axis.

Necker¹ (ne·kəɹ). [name of Louis Albert *Necker* (1786–1861), Swiss naturalist and mineralogist, who described the phenomenon in 1832 (*Phil. Mag.* I. 329).] *Necker('s) cube*: a line drawing of a transparent cube in which parallel sides are drawn with the same length, so that one seems to look successively down at the top and up at the bottom as the perspective reverses.

1901 E. B. Titchener *Exper. Psychol.* I. ii. ix. 309 The Instructor should have a few prepared as large wall-diagrams: Schröder's stair-figure, Necker's cube,.. the Müller-Lyer figure. **1938** R. S. Woodworth *Exper. Psychol.* xxv. 628 Many line drawings readily suggest three dimensions and are called figures of ambiguous or reversible perspective. The Necker cube and the Schröder staircase are the best known. **1966** R. L. Gregory *Eye & Brain* xii. 231 Certain figures..are ambiguously seen in depth, for example the Necker cube. **1975** *Sci. Amer.* Jan. 81/ [illegible] Rubin's well-known vase, Schröder's stairs and Necker's cube [illegible] M. C. Escher's work on the regular subdivision of the plane.

necker² (ne·kəɹ). [f. *Neck *v.*¹ 4 + -er¹.] One who indulges in caresses and fondling.

1923 Mencken *Amer. Lang.* (ed. 3) 373 *Necker*, one given to cheek-to-cheek dancing. **1925** H. L. Foster *Trop. Tramp with Tourists* xvi. 300 Listen, you would-be necker! **1935** [see *Neck *v.*¹ 4]. **1947** L. Brown *Say the Word* 81 But they [*sc.* young gentlemen] are not known, I think, as neckers. **1966** D. Shannon *With a Vengeance* (1968) i. 15 Well, it's a body... Found about half an hour ago... By, I gather, a couple of neckers.

neckful (ne·kful). *colloq.* [f. Neck *sb.*¹ + -ful 2.] As much (punishment, vituperation, etc.) as one can endure; cf. phr. *to get* (or *catch, take*) *it in the neck* (*Neck *sb.*¹ 1 e). So **ne·ck-full** *a.*, full up to the neck.

1920 R. Graves *Country Sentiment* 71 On pay-day nights, neck-full with beer, Old soldiers stumbling homeward here. **1950** W. Hammond *Cricketers' School* ii. 30 There was plenty of time..to give them a neckful of their own medicine.

neck-handkerchief. (Earlier Amer. example.)

1642 *Archives of Maryland* (1887) IV. 95, 9 plaine neckcloths and 5 plaine neckhandkercheifs.

necking, *sb.* Add: **3. necking-cord,** in a drawloom (see quot. 1910).

1910 L. Hooper *Hand-Loom Weaving* 328 *Necking cords*, cords joining pulley cords and leashes in a mounting. **1958** A. Hindson *Designer's Drawloom* vi. 26 The doubled part of the cord stretching from top to bottom of the Simple frame is known as the Simple cord and the divided sections from the top of the Simple frame to where they meet the cords from the shafts as the Necking cords.

4. *Archæol.* A circlet around a projection (as the boss of a shield).

1946 *Antiquity* XX. 24 A decorative necking of tinned bronze connects the boss with its wide ornamental border.

necking (ne·kiŋ), *vbl. sb.* [f. Neck *v.*¹ + -ing¹.] (In Dict. s.v. Neck *v.*¹) (In Suppl. s.v. Neck *v.*¹ 4.)

3. Also **necking down.** A local reduction in width occurring when a sample is subjected to tension.

1957 *Textile Terms & Definitions* (Textile Inst.) (ed. 3) 67 *Necking*, the sudden reduction in diameter occurring on stretching an undrawn filament. **1959** C. E. Birchenall *Physical Metall.* vi. 145 The fcc [*sc.* face-centred cubic] metals, like copper and aluminium, undergo duplex slip and necking without a distinct fracture stage. **1966** *McGraw-Hill Encycl. Sci. & Technol.* VIII. 309/1 Even the tension test, which is in general the most satisfactory, gives some difficulty because of the instability that leads to necking down. **1974** [illegible] *Analysis Metall. Failures* ii. 15 When failure occurs, the necking leads to a variety of fracture surfaces, depending on the material's ductility.

necklaced, *a.* (Later example.)

1968 G. Jones *Hist. Vikings* iv. i. 322 Frey was a god of fruitfulness and sexuality. His necklaced sister was his genial counterpart.

ne·ck-line. [Neck *sb.*¹] † **1.** = Neck *sb.*¹ 10 d. *Obs.*

1671 J. Lacy tr. *Tacquet's Mil. Archit.* 34 Look in the Table for the Neck line, you'l find it to be 160 feet. **1810** C. James *New Mil. Dict.* (ed. 3) II, *Neck-line*, an old term in fortification, signifying the gorge.

2. The shape of the neck of a garment; the line of the top of a woman's garment at the front of the neck.

1906 A. M. Smith *Cutting Out Sewing* xxviii. 217 Construction lines... The 'neck line' is the 'waist' line and the upper line the neck line. **1935** [see *Chest *sb.* 5]. **1961** *Times* 26 Feb. 14/4 The bride..wore a gown of ivory-tinted satin with a square neckline. **1964** *McCall's Sewing* viii. 116/2 Neck-line slips smoothly without pulling or gaping. **1974** *Country Life* 17 Jan. 106 Sweater with a sweetheart neckline... Sporting pullover..with a fashionable square neckline.

3. (See quot. 1966.)

1931 G. A. Foan *Art & Craft of Hairdressing* iii. 126/1 Note the flat top, ragged crown, high neck line, and straight side-pieces. **1966** J. S. Cox *Illustr. Dict. Hairdressing* 169/1 *Neckline*, (a) That part of the neck where the hair growth begins... (b) The outline at the neck of a cut and shaped hair style. **1972** I. Parkinson *Princ. & Pract. Hairdressing* viii. 113 Necklines may be curved, straight, pointed, or graduated high or low.

4. *Archæol.* An ornamental line around the neck of a vessel.

1939 *Antiquaries Jrnl.* XIX. 394 They show minor differences, additional necklines and bosses on the shoulder [illegible].

neck-tie. b. Read: *attrib.*, *spec.* in **necktie-party,** a lynching or hanging. (Also *slang*.)

1871 *Harper's Mag.* Nov. 949/2 Mr. Jim Clemenston, equine abductor, was made the victim of a neck-tie sociable. **1882** in *Nat. Geogr. Mag.* (1929) Aug. 247 Ef found within the Limits of this City after Ten O'Clock p.m. this Night, you will be Invited to attend a Grand Neck-Tie Party. **1893** [in Dict.]. [illegible] the victim of a neck-tie party.

necro-. Add: **necrobacillo·sis** (pl. **-o·ses**) *Path.* [Bacill(us + -osis], any of several conditions in animals, esp. domestic animals, and occas. in humans, characterized by diffuse or localized necrotic lesions caused by the bacterium *Sphærophorus necrophorus* (also called *Bacteroides funduliformis*, etc.); **necrophi·lia, necro·phily** = *necrophilism*; hence **necrophi·lic, necrophili·stic** *adjs.*; **ne·crophile, necrophi·liac, necro·philist** *sbs.*, one who is morbidly attracted to corpses; also *attrib.* and *fig.*; **necrophilous** *a.*, (*b*) of, pertaining to, or resembling necrophilism; also *fig.*; **ne·crophobe,** one who has a horror of death or of dead bodies; **necrophobia** (examples).

1907 *Ann. Rep. Bureau Animal Industry, U.S. Dept. Agric. 1905* 18 The presence of only one of the morbid conditions noted may be the starting point of an enzootic outbreak of necrobacillosis. **1933** R. A. Kelser *Man. Vet. Bacteriol.* (ed. 2) xvii. 284 *Actinomyces necrophorus* is the etiological factor of a variety of 'necrobacilloses' among domestic animals. [illegible] necrophorus (*Bacteroides funduliformis*) causes calf diphtheria and other animal diseases. In man it is the cause of various infections grouped together as necrobacillosis. **1892** C. G. Chaddock tr. *Krafft-Ebing's Psychopathia Sexualis* iii. 67 Following the preceding horrible group of perversions..come naturally the necrophiles. **1932** V. Woolf *Let. to Young Poet* 20 The large and highly respectable [illegible]

necromancer. Add: **2.** A silver or pewter dish with closely fitting lid and wide rim (see quots.).

1747 H. Glasse *Art of Cookery* iii. 31 Take a large Pewter or Silver Dish, made like a deep Soop Dish, with an Edge about an Inch deep on the inside, on which the Lid fixes (with a Handle at top) so fast that you may lift it up full, by that Handle without falling; this Dish is called a Necromancer. [illegible]

necromant. (Later example.)

1887 A. Lang *Myth, Ritual & Relig.* I. 105 The power of..Sorcerers and Necromants.

necromantically, *adv.* Delete † *Obs. rare*⁻¹ in Dict. and add later examples.

1963 *Times* 1 May 13/6 Voodoo..is seen as a religion of fear in which people become like animals, being possessed by demons and necromantically meddling with the spirits of the dead. **1965** *New Statesman* 12 Nov. 749/1 The most marvellous, necromantically speaking, is probably the title piece, where a boy learns the hazards of magic.

necromantist. For † *Obs. rare*⁻¹ read *rare* and add later example.

1910 *Daily Chron.* 8 Apr. 4/4 A sheaf of conjectures.. which have been drawn from the various necromantists.

necropolitan, *a.* (Later examples.)

1914 C. Mackenzie *Sinister St.* II. iv. v. 965 Always in contrast with these necropolitan streets, these masks of human dwellings, were Michael's own thoughts thronged with fancies of himself and Lily. **1916** A. Huxley *Burning Wheel* 48 The necropolitan ground. **1931** A. Gibbs *New Crusade* 78 The long arm of coincidence was in his case the long arm of the law, white sleeved, and it was raised against the further progress of her vehicle to allow a large necropolitan car to come swinging serenely from the Embankment to the left over Westminster Bridge. **1960** *Times Lit. Suppl.* 20 May p. xi/2 The curiously necropolitan conventions of the worst comic strips.

necropsy (ne·krǫpsi, nekrǫ·psi), *v.* [f. the sb.] *trans.* To perform a necropsy on. So **necro·psied** *ppl. a.*

1927 *Arch. Path.* III. 985 It is of interest to compare in some detail the death rates at ages in the necropsied population as a whole, with the life table death rates of a general population, and with that portion of the necropsied population which had some malignant tumor at death. **1958** *Amer. Jrnl. Path.* XXXIV. 863 Kidneys from an unselected series of 200 necropsied patients were sectioned and stained by a modification of the Bowie technique. **1966** *Internat. Encycl. Vet. Med.* II. 964 In 7 per cent. of 52 severely affected pigs necropsied there was inflammation or ulceration of the stomach wall. **1971** *Nature* 16 Apr. 460/2 All surviving animals were killed and necropsied after 60 weeks.

necrosis. 2. *Bot.* (Later examples.)

1901 H. M. Ward *Disease in Plants* xxvi. 240 Necrosis. —This is a general term for cases where the tissues gradually turn brown or black in patches which die and dry up... Necrosis is often due to frost. **1951** L. L. Pyenson *Elem. Plant Protection* xvi. 302 Necrosis. The browning and death of tissues is a characteristic effect of some viruses. *Ibid.* 303 Internal symptoms may show a necrosis of the phloem. **1958** *U.S. Dept. Agric. Yearbk. 1957* 763/1 Necrosis. Death associated with discoloration and dehydration of all or parts of plant ograns, such as leaves.

necrotizing, *ppl. a.* (s.v. Necrotize *v.*). Add defs.: **a.** Undergoing or becoming affected with necrosis.

1899 [see Necrotize *v.*]. **1966** [see *granulomatosis].

b. Causing necrosis.

1901 *Ann. Rep. Bureau Animal Industry, U.S. Dept. Agric. 1900* 276 Dorset and de Schweinitz described the isolation of a necrotizing acid which they obtained from tuberculous cultures. **1957** Smith & Jones *Vet. Path.* xxiv. 831 Each of these latter infections produces a minimum of exudate, but their toxins are no less deadly and the latter two are also necrotizing. Many streptococci also produce necrotizing (lytic) toxins.

nectarious, *a.* (Earlier examples.)

1771 Smollett *Humph. Cl.* III. 4 My dairy flows with nectarious tides of milk and cream. **1791** W. Enfield *Hist. Philos.* II. ii. 53 He [*sc.* Apuleius]..drank freely of.. the nectarious but unfathomable deep of philosophy.

nectarivorous (nektări·vōrəs), *a.* [f. L. *nectar* (Gr. νέκταρ) nectar + -i- + *vor-us* devouring + -ous, after Carnivorous *a.*, etc.] Of birds or insects: feeding on the nectar of flowers.

1898 *Ann. Rep. Board of Regents Smithsonian Inst. 1896* 421 The nectarivorous insects localize their action upon these nectaries.

necton, var. *nekton.

ned¹ (ned). *Sc. slang.* [? f. *Ned*, a familiar abbrev. of the name *Edward*; cf. *Teddy boy.] Hooligan, thug, petty criminal. Also used as a general term of disapprobation.

1959 *Times* 18 Dec. 5/3 He can..give gloriously funny imitations of Glasgow charwomen, tram drivers, and neds. **1964** B. Gaston *Drifting Death* iv. 57 Lomax..was a tuppeny-ha'penny little ned, not even attached to one of the big mobs. **1968** H. C. Rae *Few Small Bones* II. i. 71 Even the bloody neds from the newspapers were getting critical with their wisecracks. **1969** B. Knox *Tallyman* ii. 21 Millside had the worst pockets of unemployed and unemployable,..and some of the toughest hooligan 'neds' in the city [*sc.* Glasgow]. **1971** —— *To kill a Witch* i. 10 Most were neds, the city's [*sc.* Glasgow's] verbal shorthand for petty thugs, second-rate criminals and professional layabouts. **1973** P. Malloch *Kickback* xvi. 100 He was a ned. You could always spot them. There was something about them that no trained policeman would ever miss.

Ned² (ned). Short for *Neddy 3.

1961 *Guardian* 19 Dec. 14/1 His National Economic Development Council (or 'Ned', as it is ominously being called). **1963** *Ann. Reg. 1962* 3 The National Economic Development Council (already familiarly known as Ned or even Neddy). **1964** *New Society* 23 Jan. 23/2 Each little Ned..is going to have a highly individual character, according to the wishes of the industry involved.

neddy. Add: **1. b.** A fool, a simpleton.

1823 'J. Bee' *Slang* 124 *Neddy*—sometimes 'ass-neger', other names for jackass—the living emblem of patience and long suffering. **1853** Thackeray *Newcomes* (1854) I. i. 4 All types of all characters march through all fables:.. victims and bullies; dupes and knaves; long-eared Neddies, giving themselves leonine airs. **1854** A. E. Baker *Gloss. Northamptonshire Words* II. 49 What a neddy you must be, to do that! **1963** L. Deighton *Horse under Water* xlix. 212 'I'm sorry,' he said, 'you must think I'm a terrible neddie.'

c. *Austral. slang.* A horse, esp. a racehorse.

1900 [see *bag *sb.* 17 c(*a*)]. **1918** B. Cronin *Coastlanders* 74 A hot cinder lit on my neddie's rump. **1965** W. Dick *Bunch of Ratbags* 40 My old man was backing the neddies as usual.

3. [Properly a different word.] *colloq.* name for the National Economic Development Council. Also, one of its sub-committees. Also *attrib.*

1962 *Engineering* 1 June 729/2 Since poor Neddy was formed everyone seems to be jumping on to his band wagon. **1963** *Times* 11 June 6/6 The great significance of Neddy is that it is the first time in this country that a concerted effort has been made by the Government, management and the unions to set the country moving on a course which can be steadily sustained. **1964** *Financial Times* 8 Sept., We shall deal with them [*sc.* problems]..most of all through N.E.D.C., which we shall retain, and the regional and industrial Neddies. **1966** *Times* 14 May 17/6 Two little neddies are due to meet, wool on Tuesday and mechanical engineering on Thursday. **1968** W. Davis *Three Years Hard Labour* I. iv. 50 The so-called 'Little Neddies'. These were Councils for individual industries, designed to bridge the gap between Whitehall and industry. **1968** *Times* 18 Apr. 21/7 A Neddy-sponsored questionnaire sent out to 2,000 firms in the wool trade, is the first phase of a £80,000 survey designed to establish the competitiveness of the British wool textile industry. **1969** *Times* 30 Apr. 25/4 New chairman for building Neddy... A new chairman for the building and civil engineering economic development committees has been found.

‖ **Nederlands** (nē·dəɹlants). *S. Afr.* [Afrikaans, ad. Du. *Nederlandsch.*] The Dutch language.

1926 *Spectator* 21 Aug. 278/2 Africaans resembles, in vocabulary, the Dutch of the seventeenth and early eighteenth centuries almost more than modern Nederlands does. **1959** *Cape Times* 8 June 8/5 Nederlands is an old and highly-developed language with a wide literature.

Ned Kelly (ned ke·li). The name of the most famous Australian bush-ranger (1857–1880), used allusively to designate one of reckless courage or unscrupulous business dealings. *Austral. colloq.* See also *game *a.*¹ 2.

1941 Baker *Dict. Austral. Slang* 41 *Kelly, Ned*, any person of buccaneering business habits. **1945** 'R. Rene' *Mo's Memoirs* 24 He was game as Ned Kelly, and he'd ride anything. **1953** D. Cusack *Southern Steel* 41 Is that kid game? Game as Ned Kelly. **1953** R. Braddon in I. Bevan *Sunburnt Country* 129 Such a feat of bluff is known to Australians as a 'Ned Kelly'... It is phrases such as 'do a Ned Kelly' that lend so much verve and colour to the Australian serviceman's vocabulary. **1958** H. D. Williamson *Sunlit Plain* 90 In fact, to pay him his due compliment, he was as game as Ned Kelly. **1965** J. O'Grady *Aussie English* 62 Included in this Ned Kelly category are..characters who overcharge for mediocre work and services, and the bloke who sells you a second-hand, guaranteed, 'every bit as good as new..' vehicle, which falls to pieces in the first hundred miles. *Ibid.*, To say that a man is 'as game as Ned Kelly', on the other hand, is to praise him highly. It means that he..is brave to the point of recklessness in the face of any odds. **1966** D. Crick *Period of Adjustment* 66 'Are you game?' 'As Ned Kelly.' **1973** *Guardian* 19 Mar. 7/7 Sporty boys now..peer out of the windowed skull of the full-face [crash] helmet dubbed Ned Kelly. **1974** *Courier-Mail* (Brisbane) 20 Aug. 12/5 Mr. Bizzell said the council offered him $740 for 74 perches of land, including a 33 perch block he had levelled... 'They are just Ned Kellys,' he said. 'They certainly won't put it back on the market without making a handsome profit.'

2. *Austral.* rhyming slang for 'belly'.

1945 Baker *Austral. Lang.* xv. 271 Here are a few examples of undisguised rhyme that seem to be Australian:..*Port Melbourne Pier*, an ear; *Ned Kelly*, the belly. **1960** J. Franklyn *Dict. Rhyming Slang* 100/1 *Ned Kelly*, belly. **1970** *Private Eye* 27 Mar. 16 If I don't get a drop of hard stuff up me old Ned Kelly there's a good chance I might chunder in the channel.

‖ **née** (ne), *a.* [Fr., fem. of pa. pple. of *naître* to be born.] Born: placed before a married woman's maiden name.

1758 M. W. Montagu *Let.* 27 Nov. (1967) III. 192 The advantage of being casually admitted in the train of Madame de B., *née* O. **1831** M. Edgeworth *Let.* 30 Apr. (1971) 529 This Abroad & at Home is by Mrs. Eaton née Waldy. **1848** Thackeray *Van. Fair* xlviii. 429 The interview between Rebecca Crawley, née Sharp, and her Imperial Master. **1919** T. S. Eliot *Sweeney Among Nightingales* in *Poems*, Rachel née Rabinovitch Tears at the grapes with murderous paws. **1955** *Times* 2 July 8/7 He married, after divorce proceedings, Mary Barrie (*née* Ansell), by whom he was, himself, subsequently divorced. **1973** *Times* 3 Nov. 16/6 Mrs Fanny Harwood, nee Fanny Pain..was born in 1889.

neechee, var. *nitchie.

need, *sb.* Add: **10. c.** *Psychol.* A state of physiological or psychological want that consciously or subconsciously motivates behaviour towards its satisfaction.

1929 J. B. Miner tr. *Piéron's Princ. Exper. Psychol.* iii. 54 These instincts are generally designated by a special name..which expresses in a measure the imperious character of the tendencies; we say that they are needs. **1935** K. Koffka *Princ. Gestalt Psychol.* viii. 329 But needs are..states of tension which persist until they are relieved. **1936** *Jrnl. Psychol.* III. 27 Two commonly used terms for a motivational process are *drive* and *need*, and, since I cannot see that one is to be preferred to the other, I shall..use them interchangeably. Need is a concept to account for certain objective and subjective facts. **1961** F. H. Sanford *Psychol.* viii. 200/2 The need for achievement, referred to in the literature and in the following paragraphs as *n ach.* **1964** L. J. Bischof *Interpreting Personality Theories* II. iii. 146 In studying the need structure of man, Murray found that he required criteria in order to establish that a need existed.

15. *need-achievement, condition, -disposition, pattern, -push*; *needs analysis, test.*

1971 F. H. Farley in H. J. Eysenck *Readings in Extraversion–Introversion* III. xlv. 406 The personality variables of anxiety and need-achievement were considered. **1960** N. Maier in Kaplan & Wapner *Perspectives Psychol. Theory* 153 Like all need conditions, social needs select goal-oriented behaviors. **1951** Parsons & Shils *Toward Gen. Theory Action* I. i. 18 The child's development of a 'personality'..is to be viewed as the establishment of a relatively specific, definite, and consistent system of need-dispositions. **1958** D. Emmett *Function, Purpose & Powers* 30 From the point of view of any given actor in the system it is both a mode of the fulfilment of his own need-dispositions and a condition of 'optimizing' the reactions of other significant actors. **1947** G. Murphy *Personality* III. xvi. 395 It would seem that mood or need patterns can intensify and enrich the world of images. *Ibid.* 992/1 *Need pattern*, total organization of the needs of the organism. **1951** Parsons & Shils *Toward Gen. Theory Action* III. iii. 308 Identification does involve..locomotion away from some other region of valenced activity because of the stronger need-push to get to the region of love and approval. **1969** J. Argenti *Managem. Techniques* 175 Needs analysis, then, consists of systematically examining the requirements of each job and comparing these with the skills of the incumbent of, or an applicant for, the job. **1932** *Ann. Reg. 1931* I. iv. 102 By the regulations issued by the Ministry of Labour in October, the task of applying the 'needs test' to applicants for transitional benefit had been left to the Public Assistance Committees. **1940** *Economist* 29 June 1106/2 The chief objection is.. that it introduces another needs test. It is now possible that in the same household there will be a means test for an unemployed member..and a means test for pensions of war.

need, *v.*² Add: **7. a.** *spec.* In colloq. phrases implying that something is completely unnecessary or unwanted, as *who needs it?* [tr. Yiddish *ver darf es?*], *to need* (*something*) *like a hole in the head*: see *hole *sb.* 11. orig. *U.S.*

1951, etc. [see *hole *sb.* 11]. **1960** *Mademoiselle* Jan. 34/2 Popular idiom deals best with racial prejudice: 'Who needs it?' **1960** *N.Y. Post* 24 Feb. 56/5 Who needs them? **1962** *Sat. Even. Post* 31 Mar. 70 (*heading*) Good news—who needs it? **1963** *TV Times* 11 Jan. 8 It was so easy to say: 'Education? Who needs it?' **1968** *Melody Maker* 23 Nov. 11/3 They envision themselves wearing berets..and crawling about the rubble, throwing Molotov cocktails. 'But who needs Che Guevara? It's not like that.' **1968** M. Woodhouse *Rock Baby* xvii. 164 A twenty-two-year-old bomb disposal expert? I needed a twenty-two-year-old bomb disposal expert like I needed four more thumbs and a teen-age brain surgeon. **1973** R. Hayes *Hungarian Game* xxxi. 185, I needed a cat like I needed a nervous breakdown. **1974** *New Yorker* 17 June 92 True, he's one damn hell of a fine human being. But who needs him?

needful, *a.* Add: **4. a.** (Further examples.) Esp. in phr. *to do the needful.*

1710 J. Lovett *Let.* 1 Apr. in M. M. Verney *Verney Lett.* (1930) I. xii. 210 Waiting on proper persons and doing the needful in all places. **1822** M. Edgeworth *Let.* 27 Jan. (1971) 338, I resolved to write..only 3 or 4 lines just to say the needful. **1831** Scott *Jrnl.* 24 Apr. (1946) 164 Young Clarkson had already done the needful—that is had bled & blisterd severely, and placed me on a very restrictd diet.

needle, *sb.* Add: **1. d.** (Earlier and later examples.)

a **1530** T. Lupset *Compendious Treat. Dyenge Well* (1534) 35 For as harde a thynge it is to plucke through the smale nedels eie a greatte caboull rope, as to brynge a ryche man in at heuens wycket. **1872** Besant & Rice *Ready-Money Mortiboy* III. xiii. 234 A single-hearted.. rich man, for whom the needle's eye is as easy to pass, as for the poorest pauper. **1925** A. Huxley *Those Barren Leaves* I. i. 11 Those roaring lions at Lady Trunion's..had no hope of passing through the needle's eye. **1929** H. W. Nevinson *The English* vi. 43 They are well fitted to carry on the traditions of Victorian vulgarity, and to prove yet again that no camel will ever get through the needle's eye. **1940** V. W. Brooks *New England* xx. 414 People solemnly chewed their food very fine and slowly to be slender enough to pass through the eye of the needle. **1957** F. R. Scott (*title*) Eye of the needle.

2. b. A slender, usu. pointed, indicator on a dial or other measuring instrument, *spec.* on a speedometer.

1928 Kipling *Limits & Renewals* (1932) 60 She preferred cars to her own feet... Her place was at his left elbow, nose touching his sleeve, until the needle reached fifty. **1937** D. L. Sayers *Busman's Honeymoon* xiv. 244 He let the needle drop back to twenty-five and they dawdled on through the lanes. **1958** 'Castle' & 'Hailey' *Flight into Danger* ii. 31 The altimeter needle on the winking instrument panel steadily registered a climb of five hundred feet a minute. **1962** J. Glenn in *Into Orbit* 42 The periscope..gives you an horizon-to-horizon view of the earth below so you can check your actual attitude against the needles. **1973** 'S. Harvester' *Corner of Playground* II. v. 118 He drove faster, watching the needle flick up to a hundred.

3. b. *spec.* A hypodermic needle used to inject drugs; the use of, or addiction to, injected drugs, esp. in phr. *on the needle*,

NEEDLE

engaged in, or addicted to, injecting drugs; also *rarely*, a morphine-addict; a dose of a drug for injection. *slang* (orig. *U.S.*).

1929 M. A. Gill *Underworld Slang* (s.v. *don't*), *Don't break the needle*, don't use all the dope. **1936** L. Duncan *Over Wall* i. 21, I saw and became familiar with the hopheads or cokes—the cocaine addicts on the snow; the needles or hypes—morphine users. **1943** *N.Y. Times* 9 May II. 5/6 He's got a band that don't need a five o'clock needle like some other bands. **1953** W. Burroughs *Junkie* (1972) x. 94 'You've been hooking that spot so much it's about to get infected,' he said, pointing to a needle welt. **1955** W. Gaddis *Recognitions* I. v. 196 I've heard about her... On the needle. A schiz. **1957** C. MacInnes *City of Spades* II. iv. 102 He's using all his dope allowance now... You know who put him on the needle and supplied him? **1968** R. Jeffries *Traitor's Crime* i. 8 'When d'you get your fixes?' asked Elwick. 'In—in the evenings.' 'How much?' 'Fifteen bob a needle.' **1973** *Listener* 6 Sept. 306/1 Middle Britain thinks..one puff on the joint leads to the needle.

d. A thin pointed or tapering rod used to secure fine adjustment in closing apertures, as in valves.

1884 Knight *Dict. Mech.* Suppl. 632/2 In order to regulate the supply of oil [from a needle lubricator], a metallic feed-rod (needle) passes through the tube, and rests upon the shaft to be lubricated. **1909** *Chambers's Jrnl.* Nov. 698/1 The gas-regulator can be adjusted to the fiftieth part of an inch, with dead centralisation of the needle. **1927** G. W. C. Kaye *High Vacua* iv. 52 The needle readily beds itself into its seating, and very little pressure is needed to close the valve completely. **1965** C. M. Van Atta *Vacuum Sci. & Engin.* viii. 328 The principal feature of the design [of the needle valve] is the slowly tapering needle fitting snugly into a carefully reamed conical seat.

e. The small pointed jewel or piece of metal, wood, etc., which rests in the groove of a gramophone record when it is being played and communicates the undulations to the pick-up or diaphragm; also, a similar device used to cut the groove; = Stylus 2.

1902 *Encycl. Brit.* XXXI. 679/2 The marker..instead of being a stiff needle coming from the centre of the membrane or glass plate, is now a lever. **1930** A. B. Wood *Textbk. Sound* 438 The vibrations of the diaphragm cause a needle to cut grooves on the surface of a prepared cylinder or disc. **1949** Frayne & Wolfe *Elem. Sound Recording* xiii. 240 Motion of the needle can be utilized to apply a force to a piezoelectric crystal and thus generate a voltage. **1957** *Records & Recording* Nov. 20/1 These grooves..must be tracked with absolute accuracy by the pickup needle—nowadays it is more usually called a stylus. **1973** D. Ramsay *Deadly Discretion* 190 The concerto came to an end. The needle began to click against the ungrooved portion of the record.

12. b. (Earlier and additional examples.) Also (sometimes without *the*), anger, bad temper, enmity; esp. *to get the needle*, to become angry or upset, to lose one's temper.

1874 Hotten *Slang Dict.* 235 To 'cop the needle' is to become vexed or annoyed. **1884** [see *bird *sb.* 5 b]. **1890** Barrère & Leland *Dict. Slang* II. 84/2 It gives a man the needle when he hasn't got a bob, To see his pals come round and wish him joy. *Ibid.*, To get the needle is to feel very nervous and funky. **1923** *Daily Mail* 1 Aug. 8/2 It may be, of course, that there was too much 'needle' (to employ a boxing term which means bad spirit) about this contest. **1929** H. A. Vachell *Virgin* viii. 141 The silly ass got the needle, 'cos she asked for the ring. **1959** *Times* 8 June 3/1 Perhaps it was this very lack of needle, this air of unreality in the late evening of Saturday..that failed to see Davies home to a victory. **1967** *Time* 22 Dec. 48 A needle from [Bob] Hope becomes an emblem instead of a scar. **1970** G. F. Newman *Sir, You Bastard* v. 130 He's got the needle with you. You've got to go very careful.

13. b. *needle-painted* (later example).

1910 *Westm. Gaz.* 2 Feb. 5/4 An exhibition of needle-painted wild flowers of South Africa was opened yesterday.

c. *needle-nosed, -sharp* adjs.

1955 *Sci. News Let.* 26 Mar. 196/2 An eight-foot-tall, needle-nosed rocket. **1973** *Times* 4 June 1/3 The crash came at the end of the last of three passes the needle-nosed plane made to show off its qualities. **1923** J. Galsworthy *Captures* 161 That fellow was needle-sharp, though not always correct in his conclusions. **1973** *Times* 5 Oct. (Safety Suppl.) p. i/3 Everything depends on needle-sharp reflexes.

14. needle-and-pin, rhyming slang for Gin *sb.*[2]; **needle-and-thread,** rhyming slang for 'bread'; **needle bearing,** a bearing using needle rollers; **needle beer** *U.S. slang* (see quot. 1928); **needle board** (see quots.); **needle-bush,** substitute for def.: *Austral.* (*a*) either of two shrubs or small trees of the genus *Hakea*, *H. leucoptera* or *H. vittata*, of the family Proteaceæ; (*b*) the nealie or nelia, *Acacia rigens*, of the family Leguminosæ; (later examples); **needle-cast,** a fungus disease of conifers, causing the leaves to go brown and drop off; **needle contest** = **needle match*; **needlecord,** a finely ribbed cut-pile fabric; **needle-felt, needlefelt** = **needle-loom* (b); **needle fight** = **needle match*; **needle game** = **needle match*; **needle gap** *Electr.*, a pair of needle-shaped electrodes placed in line, between which an electric discharge can take place when the potential difference between them exceeds a value dependent on the size of the gap; **needle gate,** a dam or sluice consisting of several thin spars which are placed vertically one after the other into a frame; **needle ice,** ice formed into thin needle-like crystals just beneath the soil surface and often pushing up through it; **needle-loom, needleloom,** (*a*) (see quot. *a* 1877); also *attrib.*; (*b*) carpeting made of felt attached to a base of rubber, hessian, or other material; **needle lubricator, oiler,** a form of lubricator in which the supply of lubricant is controlled by a needle fitted in the supply tube and resting on the shaft to be lubricated; **needle mark,** a mark made by a hypodermic injection; **needle match,** a match or contest that arouses much interest and excitement; a crucial or keenly fought match; a contest in which the contestants have a grudge against each other; a dispute; **needle paper,** a stout black paper orig. used for wrapping up needles; **needle roller,** a roller in the form of a long, thin, sometimes tapered cylinder, used in roller bearings; freq. *attrib.*; **needle scar,** a scar made by a hypodermic injection; **needle shower, spray,** a shower-bath of strong fine jets of water; also *fig.*; **needle-threader,** a device for threading needles; **needle time,** an agreed time during which gramophone records are allowed to be broadcast; **needle track** = **needle mark*; **needle valve,** a valve which works by means of a narrow pointed rod fitting into a conical seating and is operated either automatically or by a screw; **needleweaving** (see quots.); **needle-wood** = **needle-bush* (*a*); also *attrib.*

1937 Partridge *Dict. Slang* 555/2 *Needle and pin*, gin. **1859** Hotten *Dict. Slang* 144 *Needle and thread*, bread. **1935** A. J. Pollock *Underworld Speaks* 80/1 *Needle and thread*, bread. **1930** *Automotive Industries* LXIII. 869/1 The needle bearing offers particular advantages for certain applications in high speed engines, as on crankpins. **1946** *R. Air Force July 1943 to Sept. 1944* 134 (*caption*) Bomber Command launched an attack on a needle-bearing factory..in France. **1972** R. C. Gunther *Lubrication* xiii. 408 Needle bearings are suited for slow speeds, or for oscillating and intermittent motion which permits the rollers to return to their required position upon load relief. **1928** *Flynn's* 1 Apr. 19/2 On the way back you can get your needle beer—near beer shot with alki or ether. **1936** J. Dos Passos *Big Money* 81 He..had a session with the helpwanted columns over some glasses of needle beer. **1897** T. R. Ashenhurst *Pract. Treat. Weaving & Designing Textile Fabrics* 63 The pressure thus bestowed upon the crosswires keeps them in position through the needle-board. **1889** *Cent. Dict.*, *Needle-board*, in the Jacquard loom, a perforated board or plate through which the points of the needles presented to the cards pass, and the perforations of which act as guides for the needles when the latter are actuated by the cards. The needle-board holds all the needles in proper relation with the prism or cylinder to which the cards are attached, and with the perforations in the cards. **1961** Webster, *Needle board*, a board covered with very short fine wires that is used for pressing pile fabrics. **1909** A. E. Mack *Bush Calendar* 4 All through the bush the needlebush showed white blossoms amongst its spiky leaves. **1944** *Living off Land* iii. 47 One of the main water supplies of the aborigines came from the roots, principally those of the mallee, the needle-bush. *Ibid.* 48 The needlebush, a dark green shrub with sharp-pointed needles in place of leaves. **1965** *Austral. Encycl.* VI. 266/2 Needlebush or Needlebush, popular names for *Hakea leucoptera* and *H. vittata*, dry-country shrubs or small trees possessing rigid acicular leaves. *Ibid.*, The so-called 'nealie', *Acacia rigens*, is sometimes referred to as needle-bush, because of its long terete phyllodes. **1895** W. R. Fisher *Schlich's Man. Forestry* IV. 408 This sudden shedding of pine needles is the characteristic of the disease so widely spread in Germany, and termed *Schütte*, or needle-cast. **1964** W. E. Hiley *Forestry Venture* iv. 81 We are bothered to some extent by needle cast caused by a fungus (*Phaeocryptopus gaeumannii*). **1922** *Daily Mail* 21 Nov. 11 There is also a 'needle' contest, recently arranged, between two stable-lads. **1967** *Times* 21 Jan. 41 And then what about that 'needle' contest, the University match? **1959** *Manch. Guardian* 26 June 5/3 A needlecord in a dove-like grey was excellent for slacks or jackets. **1973** 'D. Halliday' *Dolly & Starry Bird* xvi. 236 Charles had on a needlecord shirt with a flowery pattern. **1957** *Textile Terms & Definitions* (Textile Inst.) (ed. 3) 67 *Needle felt*, felt produced by the needleloom process. **1927** *Daily News* 25 May 8/1 England's native champion ..went down in a needle fight with Samuel Robinson, an experienced golfer. **1970** *Globe & Mail* (Toronto) 28 Sept. 21/2 In soccer there is a word for a tense match, it is called a needle game and this one fitted into that category. **1916** C. C. Garrard *Electr. Switch & Controlling Gear* viii. 563 One source of uncertainty can be removed if needles are used for the measuring gap, in which case the result obtained is termed the 'equivalent needle gap'. **1927** *Ibid.* (ed. 3) viii. 641 Let us assume that this [line] is protected by a needle gap with a breakdown voltage.. of 66,000 volts. **1962** *Newnes Conc. Encycl. Electr. Engin.* 701/2 Needle gaps have been used for measuring voltages of a few kilovolts as they have larger, and therefore more convenient, spacings than sphere gaps at these low voltages. **1969** H. M. Wilson *Irrigation Engin.* (ed. 6) 230 Simple flash-board or needle gates should be used only where the pressure upon them is low. **1928** *Engineering News-Record* 7 Feb. 262/1 The ice columns, or 'needle ice', formed on bare clayey soils are familiar to most people living in regions where the nights are cold enough for heavy frosts. **1939** H. H. Bennett *Soil Conservation* xii. 284 Where there is sufficient soil moisture, a freeze will produce layers of needle ice, or spew frost, which will lift the overlying soil and vegetation as much as several inches. **1968** R. W. Fairbridge *Encycl. Geomorphol.* 381 Such 'needle ice' is sometimes called pipkrake. *a* **1877** Knight *Dict. Mech.* II. 1512/1 *Needle-loom*, one in which the weft is carried by a needle instead of a shuttle. The usual form of loom for narrow wares, such as ribbons, tapes, bindings, etc. **1966** *Good Housek. Home Encycl.* (ed. 4) 92/2 Needleloom carpetings have a rubber or plastic back. **1957** *Textile Terms & Definitions* (Textile Inst.) (ed. 3) 66 The needleloom process is essentially a method of attaching a lap or batt.. of loose fibrous material to a base, e.g. fabric, paper, rubber and/or plastic material. **1973** *Observer* 28 Oct. 10/2 Almost all the new tufted carpets (which have, to a large extent, superseded the rubber-backed felt called needleloom) are made almost entirely of rayon. **1969** A. J. Hall *Standard Handbk. Textiles* (ed. 7) iii. 161 For the manufacture of needle-loom carpets a machine is used which comprises rows of vertical needles. **1884** Knight *Dict. Mech.* Suppl. 632/2 Needle lubricator. **1887** D. A. Low *Introd. Machine Drawing & Design* vii. 32 In the block illustrated the journal is lubricated by a needle lubricator. **1949** N. Marsh *Swing, Brother, Swing* iii. 206 He hasn't been long on the injection method... Curtis had a look for needle-marks and didn't find any. **1971** 'D. Shannon' *Murder with Love* (1972) iv. 69, I doubt very much whether he's really hooked... No needle marks on him. **1923** *Daily Mail* 16 Jan. 3 There will be a 'needle' match in Sheffield if Barnsley beat Swindon and visit the Wednesday. **1951** L. A. G. Strong *Darling Tom* xvii. 136 (*heading*) Needle match. Family quarrel will be fought out at Olympia. **1962** *Listener* 1 Nov. 731/3 More enjoyable was an off-beat needle match between Hans Keller, who felt that Gershwin was a neglected genius, and Deryck Cooke, who didn't. **1965** D. Francis *For Kicks* iii. 41, I.. watched Paddy and one of Granger's lads engage in a needle match of dominoes. **1909** *Westm. Gaz.* 3 Apr. 14/2 If needle-paper of the required kind is not available a very excellent substitute can be prepared by placing good stout paper in a solution of gelatine and glycerine to which has been added some good strong black colouring. **1975** *Sci. Amer.* May 118/2 Black needle paper and white typewriter bond differ by about a factor of 15 in reflectance all across the spectrum. **1935** *Jrnl. R. Aeronaut. Soc.* XXXIX. 470 The crankshaft was mounted on roller bearings, and the connecting rod big ends were mounted on needle roller type split bearings. **1961** *Engineering* 20 Oct. 533/1 An open propeller shaft provided with Hardy Spicer needle-roller joints transmits the drive to the rear axle. **1959** R. R. Slaymaker *Mech. Design & Analysis* ii. xv. 299 Loose needle rollers are now universally used throughout the automotive industry to serve as bearings in planetary gear systems. **1974** 'A. Haig' *Peruvian Printout* 37 A rubber-wheeled trolley which glided on needle roller bearings. **1962** K. Orvis *Damned & Destroyed* vii. 52, I made a mental note of the needle-scars. **1931** A. J. Cronin *Stars Look Down* ii. xix. 183 After that a needle shower and a hard rub down. **1973** 'H. Howard' *Highway to Murder* iii. 34 Ten minutes under a needle-shower washed the clammy heat out of me. **1974** R. McGirr *Wanderers' Journey* 6, I.. shivered under a cold needle shower. **1967** *Freedomways* VII. 153 A scattered needle-spray of unrelated, often ephemeral, facts and events which confuse the readers more than they inform them. **1970** H. McLeave *Question of Negligence* (1973) vi. 54 In the shower room he... received the needle spray on his head and body. **1889** *Cent. Dict.*, *Needle-threader*. **1964** *McCall's Sewing* v. 66/1 If you have an eye-sight problem, or just find threading needles a chore, use a needle threader. **1962** *Sunday Express* 30 Dec. 1/6 'Needle time'—the number of hours given to records. **1970** *B.B.C. Handbk.* 103 An agreement with Phonographic Performance Ltd. provides for the right to broadcast commercial gramophone records 'live', the B.B.C.'s various radio and television services being allocated fixed periods of 'needle time' in return for an annual lump sum payment. **1959** A. K. Lang in H. Q. Masur *Murder Most Foul* (1971) 69 The kid had been a user; they'd know that from the gear in her purse and the needle-tracks in her arm. **1973** J. Martin *95 File* 80 He checked her arms but found no needle tracks. **1903** *Sci. Amer. Suppl.* 24 Jan. 22697/2 The inventor's idea, in designing the carbureter, was to do away with the needle-valve usually employed for controlling the flow of gasoline. **1925** H. E. Owen in E. F. Norton *Fight for Everest 1924* 362 The pressure gauge was connected close to the mouths of the cylinders on the back (of the breathing apparatus), and the rate of flow regulated by a needle-valve close beside it. **1971** *Sci. Amer.* Sept. 221/1 All gases are admitted through needle valves to a manifold that connects to the laser. **1932** C. C. Minter *Mod. Needlecraft* 16/1 Needleweaving is a form of embroidery worked on the threads of the material when the threads in the opposite direction (i.e. either the warp or the weft) have been withdrawn. **1967** E. Short *Embroidery & Fabric Collage* ii. 48 Needleweaving. This is a variation of drawn thread work which could be used effectively on accessories such as handbags and belts. **1911** C. E. W. Bean *'Dreadnought' of Darling* xv. 141 The pretty grey needle-wood. **1936** F. Clune *Roaming round Darling* xxv. 286 Plenty of mulga, needlewood, belah, budda, and broad leaf-box. **1941** I. L. Idriess *Great Boomerang* i. 1 Upon a needlewood-tree a crow waited. *Ibid.* iii. 22 It [*sc.* the gold] lay by a needlewood bush. **1959** A. Upfield *Bony & Black Virgin* xiii. 111 He sat in the shade of the needlewood tree, or rather its trunk, for the narrow leaves give but scant shelter.

needle, *v.* Add: **1. b.** (Further examples.) Also, to goad; to provoke into anger.

1881 G. R. Sims *Dagonet Ballads* 77 There, he's off! the

young warmint, he's needled; whenever I talks about work He puts on his cap and he hooks it. **1898** G. B. Shaw *Our Theatres in Nineties* (1932) III. 358 Old Indian women get 'fairly needled' at the spectacle of their houses and crops being burnt. **1941** *Time* 7 Apr. 22/3 Some 20 Manhattan reporters gave the Ambassador a going-over for 50 minutes... He did not let it appear that he knew he was being needled. **1958** J. Wain *Contenders* 154 It was that bit about forgetting his business worries that needled Ned. **1959** M. Pugh *Chancer* 177 He was needling me, needling me this night, and I wouldn't provoke. **1972** D. Haston *In High Places* ii. 36 Once again we'd needled each other into a state of open warfare.

3. *U.S. slang.* (See quots.) Also **nee·dled** *ppl. a.* (cf. *needle beer*, *needle *sb.* 14).

1929 *Amer. Speech* IV. 387 Many Kansans..buy the ordinary non-alcoholic near-beer, and add a little alcohol to each bottle. The resulting mixture is called..*needled beer*. **1929** Hostetter & Beesley *It's a Racket!* 233 *Needle*, to inject alcohol or ether into any liquid, such as beer, to make it stimulating. **1930** *Amer. Mercury* Dec. 456/2 *Needle*, to make near-beer intoxicating by injecting ether or alcohol. 'This beer knocks you for a loop. It's needled with ether.' **1931** D. Runyon *Guys & Dolls* (1932) iv. 79 It is sleeping so sound that I am commencing to figure that Butch must give it some of the needled beer he is feeding us.

needleman. Add: **2.** (Spelt as two words, or hyphenated.) A drug-addict, esp. one who is addicted to injecting drugs. *U.S. slang.*

1925 *Flynn's* 11 July 128/1 So surely was Howard a needle man—that is, a hopeless drug addict. **1955** [see *hype *sb.*[1]].

needle-point. Add: **1.** (Examples corresponding to *needle *sb.* 3 e).

1929 *Radio Times* 8 Nov. 444/2 The needle-armature is so light that the needle point actually *feels* its way along the record groove. **1949** B. Semeonoff *Record Collecting* v. 22 The term 'tracking' refers to the path traced by the needle-point as it travels inward towards the centre of a record.

2. (Earlier and additional examples.) Also *attrib.*

1865 F. B. Palliser *Hist. Lace* iii. 28 Lace is divided into point and pillow. The first is made by the needle on a parchment pattern, and termed needle point. **1882** *Encycl. Brit.* XIV. 189/2 A technical peculiarity in making needlepoint lace is that a single thread and needle are alone used to form the pattern. **1902** Jourdain & Dryden *Palliser's Hist. Lace* (rev. ed.) xiii. 195 'Needle point' is the name by which point d'Alençon was alone known in England during the last century. **1967** E. Short *Embroidery & Fabric Collage* ii. 39 Cut work.. later developed into needlepoint lace. *Ibid.*, This delicate embroidery with its contrasts of cut work, eyelets, fine needlepoint fillings, [etc.].

needless, *a.* Add: **2.** Freq. in phr. (*it is*) *needless to say* (or *add*), often used parenthetically.

1770 A. Young *Six Months' Tour North of Eng.* I. 185 It is almost needless to add upon the course of crops in question, that the turnips ought..to be fed off the land by sheep. **1770** Baretti *Journey London to Genoa* I. xx. 148 It is needless to say that thousands and thousands have migrated to other places. **1826** *Kaleidoscope* 31 Jan. 247/3 The Squire was hard hit by this nonchalance, and (as the newspapers say), 'it is needless to add', acted upon Sheridan's suggestion. **1885** Rider Haggard *K. Solomon's Mines* vi. 88 That night we covered nearly five-and-twenty miles, but, needless to say, found no more water. **1902** R. J. Mecredy in A. C. Harmsworth *Motors* vii. 122 Needless to say, the shoulder F is thereby raised. **1930** *Sunday Times* 12 Oct. 5/5 Needless to say, such a visitor is immensely impressed and at once enrols for the Pelman Course in the particular language in which he is interested.

needling, *vbl. sb.* Add: **2. c.** The action of annoying, irritating, or goading (see Needle *v.* 1 b in Dict. and Suppl.). Also as *ppl. adj.*

1941 *Sun* (Baltimore) 10 Jan. 12/7 The word 'needling' ..is being used more and more frequently in the sense of using sharp bits of persuasion to bring a person to adopt a desired course. **1945** *Ibid.* 17 Feb. 7/3 P. K. W...was plainly irked by some sharp needling of his group. **1956** W. H. Whyte *Organization Man* (1957) 246 It was Keefer, with his clever mind, his needling of authority, who led the ordinary people..astray. **1958** *Spectator* 10 Jan. 33/1 Their needling and often impertinent questions. **1959** *Times Lit. Suppl.* 27 Nov. 698/5 She undergoes, still buoyant, the familiar needlings of interrogation. **1962** *Listener* 7 June 999/1 When the next careful, needling letter arrived from Samuel, the black temper broke loose again. **1971** C. Bonington *Annapurna South Face* xi. 128 '..Anyway, if you want to go out in front on this trip, you'd better prove you can keep going.' This kind of needling was the ideal treatment for Mick. **1973** *Islander* (Victoria, B.C.) 28 Oct. 2/1 He was the great complainer of his time, and..Victoria became better place because of his needling.

needs, *adv.* **d.** (Further examples of proverbial use.) Cf. proverbial uses under sense c in Dict.

1835 Southey *Doctor* III. lxxxiii. 77 'Needs must go when the Devil drives', says the proverb: but the Devil shall never drive me. **1853** T. C. Haliburton *Sam Slick's Wise Saws* I. xiii. 267 'Needs must when the devil drives, so here goes,' and off he went for the water. **1886** [see Drive *v.* 1 b]. **1916** E. Wallace *Clue of Twisted Candle* (1918) xi. 133 But needs must when the devil drives, as the saying goes. **1956** G. Durrell *My Family* xiv. 189, I think we had better have a cab. An extravagance, of course, but needs must where the devil drives, eh?

need-to-know (nī:d,tŭnōu·). [Need *v.*[2] 8.] Used, freq. *attrib.*, to denote a principle or practice, esp. in counter-espionage, whereby people are kept ignorant of things which they do not need to know.

1954 *Amer. Documentation* V. 120 In most security controlled report systems, the dissemination of information is regulated by 'need-to-know' or 'compartmentalization' principles. **1956** W. A. Heflin *U.S. Air Force Dict.* 342/2 *Need-to-know*, n. A criterion used in security procedures that requires a person requesting classified information to establish his need to know such information in terms of his mission. **1966** J. Bingham *Double Agent* ix. 137 Mr. Ryan had then hastily told him to keep that to himself. Nothing further had been said. Mr. List had been naturally dismayed at this gross breach of the need-to-know principle. **1969** A. Marin *Rise with Wind* i. 7 You will notice that there are some gaps in the Clay material. Part of the information is..strictly on a 'need to know' basis. **1970** K. Benton *Sole Agent* xviii. 193 There is the rule about need-to-know. They won't *need* to know the details of your contacts with MI5. **1971** D. Bagley *Freedom Trap* vi. 127 They worked on the 'need to know' principle, and an escapee didn't need to know how he escaped—just that he had done so. **1975** *Observer* 12 Jan. 1/1 The CIA has instructed the companies to limit all knowledge of the exercise to the spymaster's traditional 'need-to-know' criterion. **1975** N. Luard *Robespierre Serial* iv. 27 There was no need for Carswell to be filled in on all the ramifications... On the strict need-to-know basis they simply didn't concern him.

neejee, var. *nitchie.

Néel (nē[i]·ĕl). The name of Louis Eugène Félix *Néel* (b. 1904), French physicist, used *attrib.* to designate certain phenomena connected with his work on magnetism, as **Néel point** or **temperature,** the transition temperature for an antiferromagnetic or ferrimagnetic substance, above which it is paramagnetic (analogous to the Curie point for ferromagnetics); **Néel spike,** a sharply pointed triangular domain extending from a small hole or inclusion in a magnetic substance diagonally in relation to the field direction in the surrounding area; **Néel wall,** a type of domain boundary that Néel predicted should occur in thin layers of magnetic material, in which the rotation of the field direction in passing from one side of the wall to the other occurs in the plane of the film.

1949 *Rev. Mod. Physics* XXI. 572/2 Measurement of the ratio of length to width of the Néel domains offers a possible experimental method for determining..the surface energy density of a Bloch wall. **1951** *Jrnl. de Physique et le Radium* XII. 311 (*heading*) Néel 'spikes' around holes in a crystal surface. **1952** *Physica* XVIII. 714 In warming the salt through the region of the Néel temperature a continuous but rapid change occurs from a state in which the spins of the Cu^{++} ions are highly ordered to a state of comparative disorder. **1960** *Jrnl. Appl. Physics* XXXI. 303S/1 The smooth parts of the wall at the thin and thick ends of the film have to be interpreted as pure Néel and Bloch walls, respectively. **1963** J. S. Smart in Rado & Suhl *Magnetism* III. ii. 66 The Néel point..of a ferrimagnet or an antiferromagnet can be determined in a number of different ways, of which magnetic susceptibility is probably the most common and specific heat measurements one of the most accurate. **1966** Carey & Isaac *Magn. Domains* ii. 24 Néel walls become energetically more favourable than Bloch walls for thin films. *Ibid.* 28 In films of thickness less than about 200 Å no cross-tie walls occur but normal Néel walls are observed. *Ibid.* viii. 142 Consideration of the magnetostatic energy associated with large inclusions in iron crystals led Néel to predict the existence of closure domains..now termed Néel spikes. **1973** *Physical Rev.* B. VII. 287 (*heading*) Heat-capacity measurements on manganese dibromide tetrahydrate near its Néel temperature.

neem. (Further examples.)

1876 *Cornh. Mag.* Sept. 320 There was Beena..standing apart under a *nim* tree. **1911** J. Frazer *Golden Bough: Magic Art* (ed. 3) I. v. 293 In order to procure rain people of low caste in the Central Provinces of India will tie a frog to a rod covered with green leaves and branches of the *nīm* tree. **1937** *N. & Q.* 8 May 338/1 The 'nimb' was used by Terence Mulvaney to make 'a thundering big poultice av neem leaves..'. The Indian sais (or groom) still uses the neem leaf as a poultice for galls. **1949** H. W. Florey et al. *Antibiotics* I. xiv. 586 The nim tree (*Melia azadirachta*, *M. indica*) has been cultivated throughout India on account of its medicinal properties. **1969** T. H. Everett *Living Trees of World* 210/2 The neem or nim tree..common in India and Ceylon, is greatly valued for its bitter, antiseptic resin, which is used in medicines, soaps, lotions and toothpaste. *Ibid.* 211/1 The neem tree is a graceful evergreen, up to 50 feet tall; it thrives in dry climates. Its pinnate leaves have an odd number of curved, pointed, toothed, shiny leaflets. The numerous small, fragrant white flowers occur in loose panicles..followed by small yellow berries. **1971** R. Russell tr. *Ahmad's Shore & Wave* i. 9 There were only tracks on Gipsies' Hill, winding their way through the trees of..bitter-leaved neem among rocks.

neencephalon (ni,ense·fălǫn). *Anat.* [a. G. *neencephalon* (L. Edinger *Vorlesungen über den Bau der nervösen Zentralorgane des Menschen und der Tiere* (ed. 7, 1908) II. xvi. 242): see *neo- 1 e and Encephalon.] The phylogenetically younger part of the brain, comprising the cerebral cortex and related structures. So **neencepha·lic** *a.*

1917 *Jrnl. Compar. Neurol.* XXVIII. 216 The emergence of the true cortex (neencephalon, or suprasegmental apparatus of the brain). *Ibid.* 217 Here the picture is uncomplicated by the great neencephalic systems. **1948** *Brit. Jrnl. Psychol.* XXXIX. 71 The infant is born without mental life... Only gradually does the neencephalon exert its influence. **1972** *Encycl. Psychol.* II. 310/2 *Neencephalon*, the 'new brain' or 'harmonious prolongation' lying above the 'old brain' or paleencephalon. A fairly pure neencephalic area of the brain is the six-layer cerebral cortex (isocortex) in mammals... Under narcosis or the influence of alcohol the neencephalic systems are the first to cease to operate.

ne'er, var. New-year. *Sc.*

nefast, *a.* For *rare*$^{-1}$ read *rare*, and add further example.

1887 R. L. Stevenson *Let.* Oct. (1899) II. 71 In good case and spirits, as I am now, after a most nefast experience of despondency before I left.

nefedyevite (nefedye·vəit). *Min.* Also †**nefediewite, nefedievite.** [ad. Russ. *nefed'evit* (P. Puzyrevsky 1872, in *Zapiski imperat. S.-Peterburgsk. mineral. Obshch.* VII. 15), f. the name of V. V. *Nefed'ev*, 19th-cent. Russian mineralogist: see -ite[1].] A white or pinkish aluminosilicate of magnesium and calcium belonging to the clay family and similar to montmorillonite.

1873 *Jrnl. Chem. Soc.* XXVI. 1210 Nefediewite; a new mineral... This amorphous mineral, very much like lithomarge, occurs, together with fluorspar, in the limestone of Nertschinsk. **1938** *Mineral. Abstr.* VII. 104 Greenish-grey unctuous clay interbedded among Cretaceous limestones, the bulk of which consists of a fibrous crystalline aggregate..and in composition..corresponds to nefedievite. **1961** *Doklady Earth Sci.* CXXXV. 1296/1 In view of the great structural similarity between true montmorillonite and nefedyevite, doubling of the unit cell along the *c* axis can be expected.

neg (neg), colloq. abbrev. of *negative* or *negatively*, esp. = Negative *sb.* 8.

1874 W. H. Jackson *Diaries* (1959) 275 While Bob was saddling and packing up, I made a couple of negs. **1909** *Cent. Dict.* Suppl., *Neg.* An abbreviation (*a*) of *negative*; (*b*) of *negatively*. **1948** M. Allingham *More Work for Undertaker* xv. 185, I told you it was like a neg. Black shadows and everything else a sort of chilly grey. **1959** H. Hamilton *Answer in Negative* i. 8 You don't want to keep negs in the same place as pix..because there's always a risk of fire. **1971** *Hi-Fi Sound* Feb. 71/1 It is not always so difficult to identify 'pos. and neg.' but long practice has made it a habit with me to tie a knot in each negative connection. **1972** *Amat. Photographer* 12 Jan. 29 The resultant negs had a remarkable degree of density.

negate, *v.* Add: **b.** *Gram.* To render negative in sense.

1930 W. Empson *Seven Types of Ambiguity* vii. 269 *Not* may negate *going* or *weeping*. That the ear expects *did go* may mean that all nature wept for Polonius. **1961** R. B. Long *Sentence & its Parts* iv. 105 Even the words that commonly negate clauses do not always do so. **1972** R. Quirk et al. *Gram. Contemp. Eng.* 382 *Not* here functions as a predeterminer in the italicized noun phrase; but it has the effect of negating the whole clause.

negater (nĭgē[i]·tər). *Computers.* Also **negator.** [f. Negat(e *v.* + -er[1]. Cf. Negator.] = *inverter 2 c.

1962 *Gloss. Terms Automatic Data Processing* (*B.S.I.*) 60 Negater. **1963** *New Scientist* 14 Nov. 387/3 The system employs a 'negator' element—a logical element which reverses, for example, the polarity of the voltage, or the direction of the magnetic flux. **1971** J. H. Smith *Digital Logic* iv. 65 The British standard gives little guidance to the use of a digital buffer amplifier and the reader is recommended to use a single input NOR symbol as an inverting buffer (this is defined as a negater) and the amplifier symbol for the non-inverting buffer.

negation. Add: **1. c.** Also as a logical operation in Computing; = *inversion 2 k. (Further examples.) Also **negation-sign,** the sign or symbol used to indicate negation.

1948 McKinsey & Tarski in *Jrnl. Symbolic Logic* XIII. 1 As regards constants, they are three in number: the negation sign, the conjunction sign, and the possibility sign. **1949** E. C. Berkeley *Giant Brains* iii. 34 The simplest computing operation is negation. **1955** A. N. Prior *Formal Logic* III. ii. 253 We regard..his negation-sign as meaning impossibility. **1959** E. M. McCormick *Digital Computer Primer* v. 64 The NOT logical operation (negation) results in an output which is opposite to its single input. **1962** T. C. Bartee et al. *Theory & Design Digital Machines* iii. 23 Some authors refer to *x'* as *not x* or as the negation of *x* corresponding to our complement of *x*. **1965** Hughes & Londey *Elem. Formal Logic* x. 67 We used the Law of Double Negation to insert a pair of negation signs. **1969** F. M. Hall *Introd. Abstr. Algebra* II. xi. 313 The negation, or

complement, of *A*, written *A'*, is the statement 'it is false that *A*', or briefly 'not *A*'.

negatival (negătəi·văl), *a.* [f. Negative *sb.* + -al.] Negative; negativistic; characterized by negation.

1936 J. J. Cohen *Psychotherapy* ii. 7 Psychophonism, or mental logography, is a word which I have coined to distinguish a psychotherapeutic system of negatival auto-suggestion from all rival systems which, all, start from an entirely wrong foundation, this in their being all alike positival in their psychology. **1966** J. E. Buse in C. E. Bazell *In Memory of J. R. Firth* 56 Phrases are classed as..negatival if they commence with a negative.

negative, *sb.* Add: **1. d.** Used quasi-*advb.*, orig. in radio communication, = No *adv.*[3] *colloq.*

Quot. 1946 perhaps illustrates sense 2 a of the adj.

[**1946** J. Irving *Royal Navalese* 121 Orders for a Church Parade 'Dress for Officers No. 3, negative swords'.] **1955** *Amer. Speech* XXX. 118 *Negative*.., I refuse; I disagree; no (in answer to a question). (For reasons of clarity, any negative expression is expressed over the radio as *negative*...) **1961** E. Waugh *Unconditional Surrender* I. i. 29 'Any result of my application for the return of my typist?' 'Negative,' said Mr Oates. **1972** C. Kearey *Last Plane from Uli* vii. 81 'Any snags, Captain?' 'Negative, she's running like a clock.' **1972** P. Cleife *Slick & Dead* ix. 69, I shook my head. 'Negative,' I said.

8. b. A mould for, or reverse impression of, a piece of sculpture or the like. Cf. next, sense 11 a.

1911 A. Toft *Modelling & Sculpture* x. 195 The mould or negative is next coated with a preparation of plumbago or black-lead, and placed in a bath where the metal is deposited into it. **1947** J. C. Rich *Materials & Methods Sculpture* i. 18 The 'negative' is the term applied to the hollow containing form or mold into which the positive, temporarily plastic casting material is poured. *Ibid.* v. 95 If a plaster negative is fashioned over an earth-clay model, the original should not be too dry. **1961** J. Challinor *Dict. Geol.* 134/1 *Negative*, a fossil in the form of an impression. **1973** D. Cowley *Working with Clay & Plaster* 81 (*caption*) Plaster negative taken from a positive plaster cast.

c. A disc similar to a gramophone record but having ridges in place of grooves.

1918 H. Seymour *Reproduction of Sound* 17 In 1900 he applied the vacuous deposit system in electrolysis to the production of record negatives. **1931** A. Nadell *Projecting Sound Pict.* xiv. 240 This metal plate.. constitutes a 'negative' with which any number of 'positive' records may be stamped. **1974** *Encycl. Brit. Macropædia* XVII. 52/1 Berliner did not, however, contemplate using this etched master as the record to be played; rather, a negative was made from the master by electroforming.

negative, *a.* Add: **2. a.** *negative flag* (see quots.).

1909 *Daily Chron.* 18 Aug. 7/5 If it is hoisted superior to the flag called the Negative flag, it signifies that the man is drowned. **1916** 'Taffrail' *Carry On!* 24 If the 'Negative flag', white with five black crosses, had been displayed, he would have known that the worst had happened, and that a life had been lost. **1948** R. de Kerchove *Internat. Maritime Dict.* 487/1 *Negative flag*, a single-letter signal consisting of letter 'N' of the International Code of signals. Means 'No'.

7. b. (Further examples.) *negative booster*, a booster used to lower the potential of the station end of a negative feeder to below earth potential; *negative feeder*, a wire which connects the rails forming the negative connection for an electric traction vehicle to the bus-bars at a substation.

1890 *Proc. R. Soc.* XLVII. 543 How far the positive charges in the polarising layer and the negative charges projected away from the kathode are *alone* sufficient to account for the whole current, cannot be decided at present. **1902** Negative electron [see *electron[2] 1 a]. **1909** P. Dawson *Electr. Traction on Railways* xv. 475 The negative booster consists simply of a rotating machine driven at a constant speed by an independent motor. *Ibid.* 476 The booster is usually designed so that the E.M.F. produced in the armature is sufficient to cancel out the loss of voltage in the negative feeder. **1932** R. Rawlinson in E. Molloy *Pract. Electr. Engin.* V. 1598/2 The negative booster is so connected as to reduce the station end of the negative feeder to below earth potential. **1933** *Discovery* Mar. 69/1 The negative electron, the massless unit charge of electricity, was isolated first in the Cavendish Laboratory by Sir J. J. Thomson in 1897. **1956** *Ann. Reg. 1955* 402 The bevatron, at Berkeley, made possible the discovery of a new atomic particle—the negative proton. **1974** *Encycl. Brit. Macropædia* VI. 666/1 The magnitude of the negative charge *e* was obviously of basic importance and a scale parameter for the whole of atomic physics.

d. *negative glow*, the luminous region in a discharge tube between the Crookes dark space and the Faraday dark space.

1890 *Proc. R. Soc.* XLVII. 557 This is the dark interval separating the positive part of the discharge from the negative glow. **1939** H. J. Reich *Theory & Applic. Electron Tubes* xi. 369 The brightness of the negative glow decreases toward the anode, and the glow gradually merges into another relatively dark region, the Faraday dark space. **1971** J. F. Waymouth *Electr. Discharge Lamps* iv. 71 In the negative glow, the rate of ion production required to supply a cathode ion current of 10% of the total may be many times what it is in the positive column.

8. c. Misc. special collocations: *negative capability* (see quot. 1817); now also used (in the light of Keats's other observations on the nature of the creative artist) for *empathy; *negative catalysis* (Chem.) = *inhibition 3 b; *negative catalyst* = *inhibitor 2 c; *negative eugenics*, an attempt to prevent the birth of children to persons considered unfit to be parents; cf. *dysgenic *a.*; *negative feedback*, feedback that tends to diminish or counteract the process giving rise to it; *negative g* or *G*, (a force resulting from) the deceleration of a vehicle, esp. an aircraft or spacecraft; *negative growth*, the cessation or reversal of growth, esp. in lower animals, in response to starvation or other unfavourable conditions; *negative income tax*, a scheme whereby low-paid workers receive a government subsidy to raise their pay to subsistence level; also *negative tax*; *negative* (*phase*) *sequence* (Electr. Engin.), a three-phase system in which the voltages or currents in each phase reach their maxima in the opposite order (i.e. red–blue–yellow) from the positive sequence; *negative resistance*, (the property of) a device in which an increase in the potential difference between its terminals causes a drop in the current flowing through it; *negative transfer*, habits or methods learned for one task which interfere with or transfer negatively to the learning of a subsequent task; cf. *transfer *sb.* 3, *positive *a.* 8; *negative transference*, the transfer or imputing to a doctor of feelings of hostility, or of resistance and constraint, that may be aroused in the patient through fear of giving expression to his repressed emotions; also, the transferring to a relationship of negative emotions which persist from a previous relationship or experience; cf. *transference 1, *positive *a.* 8.

1817 Keats *Let.* c 21 Dec. (1958) I. 193, I had not a dispute but a disquisition, with Dilke, on various subjects; several things dovetailed in my mind, & at once it struck me, what quality went to form a Man of Achievement especially in Literature & which Shakespeare possessed so enormously—I mean *Negative Capability*, that is when man is capable of being in uncertainties, Mysteries, doubts without any irritable reaching after fact & reason—Coleridge, for instance, would let go by a fine isolated verisimilitude caught from the Penetralium of mystery, from being incapable of remaining content with half knowledge. **1964** S. Barnet et al. *Dict. Literary Terms* 97 Negative capability is sometimes identified with empathy, sometimes with objectivity. **1975** *Studies in Eng. Lit.: Eng. Number* (Tokyo) 167 The poem is representative of the poet [*sc.* Wallace Stevens] in that it is a poem of 'Negative Capability', his basic mental attitude. **1904** *Jrnl. Chem. Soc.* LXXXVI. ii. 113 The simultaneous effect of a positive catalyst (copper sulphate) and a negative catalyst (mannitol or stannic chloride) has been studied, and the experiments support the view that negative catalysis consists in a counteracting of the effect of positive catalysis. **1940** Glasstone *Text-bk. Physical Chem.* xiii. 1121 Negative catalysis in gas reactions is probably also to be ascribed to the breaking of reaction chains. **1966** *McGraw-Hill Encycl. Sci. & Technol.* II. 546/2 Such substances, formerly called negative catalysts, are now known to be consumed in the process... Accordingly such materials are not true catalysts. **1968** M. M. Jones *Ligand Reactivity & Catalysis* i. 10 Negative catalysis resulting from the removal of catalytically active metals by a chelating agent, for example, the inhibition of hemoglobin by carbon monoxide. **1908** F. Galton in *Nature* 22 Oct. 645/2 Little or nothing will be said relating to what has been well termed by Dr. Saleeby 'negative' eugenics, namely,.. hindering the marriages and the production of offspring by the exceptionally unfit. **1914** C. W. Saleeby *Progress Eugenics* i. 20 It is no less necessary to discourage parenthood among defective individuals, and to this, with Galton's approval, I gave the name of negative eugenics, calling his own scheme positive eugenics. **1931** J. S. Huxley *What dare I Think?* iii. 93 Negative eugenics is concerned with preventing degeneration. **1974** J. R. Baker *Race* iv. 57 He [*sc.* Madison Grant] was harsh in his schemes for negative eugenics... He favoured the forcible sterilization of criminals, diseased and insane persons, and 'worthless race-types', and the enactment of laws against race-mixture. **1934** H. S. Black in *Bell Syst. Techn. Jrnl.* XIII. 5, $1/(1-\mu\beta)$ will be used as a quantitative measure of the effect of feedback and the feedback referred to as positive feedback or negative feedback according as the absolute value of $1/(1-\mu\beta)$ is greater or less than unity. Positive feedback increases the gain of the amplifier; negative feedback reduces it. **1956** *Science* 11 May 848/1 (*heading*) Evidence for a negative-feedback mechanism in the biosynthesis of isoleucine. **1966** 'A. Hall' *9th Directive* x. 97, I put in fifty or sixty shots,..gradually allowing the negative feedback data to correct the aim. **1967** M. Argyle *Psychol. Interpersonal Behaviour* iii. 62 Such unstable sequences are known as cases of 'positive feedback' (unstable vicious circles) and can be contrasted with self-correcting 'negative feedback', which is also a common feature of social performance. **1952** R. L. Christy in White & Benson *Physics & Med. Upper Atmosphere* 510 Certain unusual attitudes of the aircraft in which irregular accelerations, including negative g, are encountered. **1955** *Aeroplane* 25 Nov. 794/1 The need for a negative-g stressing case for civil aircraft was questioned. **1962** J. Glenn in *Into Orbit* 71 We also made runs to simulate the forces of deceleration, or negative Gs. **1932** J. S. Huxley *Probl. Relative Growth* iii. 87 Here [*sc.* in the shore crab Ocypoda] the low point of growth, or 'negative growth-centre', is also in the merus. **1957** G. E. Hutchinson *Treat. Limnol.* I. iii. 217 The life of such lower organisms as are capable of negative growth may be greatly prolonged. **1964** A. E. Needham *Growth Process in Animals* iii. 29 In the lower animals negative growth or degrowth is commonly reversible, sometimes to a remarkable degree. It is a normal response to starvation and to some other conditions. **1967** *Yale Law Jrnl.* Nov. 1 (*title*) Is a negative income tax practical? **1969** *Daily Tel.* 17 Jan. 27/1 The use of 'negative income tax', by which low-wage earners receive a PAYE handout. **1973** *Guardian* 30 Mar. 6/4 The Government's proposed tax credit system—the 'negative income tax' proposed in a Green Paper last year. **1930** M. G. Malti *Electr. Circuit Analysis* xv. 244 We shall call a negative phase sequence such a sequence of the phases that phase 1 leads phase 3 by 120° and leads phase 2 by 240°. **1896** Frith & Rodgers in *Phil. Mag.* XLII. 410 [Prof. Ayrton concluded that if an attempt were made to measure the resistance of the arc by altering the P.D. between the carbons and finding the corresponding alteration of current produced, the resistance found by taking this ratio must be *negative*.] *Ibid.*, All these experiments..lead to the conclusion that the arc has a *negative* resistance. **1932** W. L. Everitt *Communication Engin.* xviii. 479 A negative resistance can be used in a circuit to counteract the effect of positive resistance and so cause a combination of inductance and capacitance to oscillate. **1942** C. L. Amick *Fluorescent Lighting Manual* ii. 21 The 'negative-resistance' characteristic of fluorescent lamps means that the voltage drop across the lamp decreases as the arc current goes up. **1974** G. J. Angerbauer *Electronics for Mod. Communications* xvi. 319 This negative-resistance effect depends upon secondary emission from the plate. **1973** J. R. Neuenswander *Mod. Power Syst.* ix. 175 The revolving field..rotates *with* the rotor while the negative-sequence armature currents..set up a revolving field rotating at the same speed but in opposite direction to the rotor. **1963** M. S. Gordon *Econ. Welfare Policies* vi. 117 A few economists, appalled at the piecemeal character of our approach to welfare policies, have espoused the so-called 'negative tax' proposal. **1921** F. N. Freeman *Exper. Educ.* ii. 47 There is strong negative transfer from Set 2 to Set 3. **1933** M. Viteles *Industr. Psychol.* III. xx. 427 The study of transfer effect has indicated that under certain conditions the practice of similar tasks may induce a negative transfer or interference with the acquisition of skill. **1938** R. S. Woodworth *Exper. Psychol.* viii. 176 When an act is carried over but impedes the learning of a second act we obviously have positive transfer and a negative transfer effect. **1950** O. Mowrer *Learning Theory & Personality Dynamics* vii. 193 If 'reinforcement' learning were alone operative, the initial 'mis-training' given to the experimental-group subjects should have produced negative transfer. **1966** J. T. & K. W. Spence in C. D. Spielberger *Anxiety & Behavior* 303 An investigation..that involved a type of negative transfer design. **1916** C. E. Long tr. *Jung's Coll. Papers Analytical Psychol.* ix. 270 If it is a 'negative' transference, you can see nothing but violent resistances which sometimes veil themselves in seemingly critical or sceptical dress. **1924** J. Riviere et al. tr. *Freud's Coll. Papers* II. xxviii. 319 One is forced to distinguish 'positive' transference from 'negative' transference, the transference of affectionate feeling from that of hostile feeling and to deal separately with the two varieties of the transference to the physician. **1954** R. W. Pickford *Analysis of Obsessional* i. 20 A technique for manipulating the positive and negative transferences to the patient's advantage. **1960** L. Pincus *Marriage* ii. 91 The distress, pain and aggression of a negative transference. **1964** Zaleznik & Moment *Dynamics Interpersonal Behavior* viii. 268 The negative transference reactions where the individual experiences hatred toward a person in the present because of a past relationship. **1968** A. J. Mandell in J. Marmor *Mod. Psychoanal.* xi. 283 Side reactions may begin to take the place of verbalized negative transference.

10. b. *negative after-image*, an after-image of complementary colour or brightness to that of the original impression.

1899 L. Hill *Man. Human Physiol.* xxxv. 439 Look steadfastly at a piece of white paper placed on a sheet of black paper, and then look up at a dark wall. You will now see a black spot on a whitish ground. This is a negative after-image... Look steadfastly now at a piece of red paper; a negative after-image appears on looking at the ceiling. This image will not be red, but of the complementary colour greenish blue. **1967** D. A. Schreuder in J. B. de Boer *Public Lighting* iv. 167 Complete adaptation [to the luminance of the field of vision] requires a certain length of time. The slowest, and hence in practice the most important process is the disappearance of the negative after-images.

11. a. Of, pertaining to, or designating a mould or reverse impression. Cf. *negative *sb.* 8 b.

1911 G. H. Wilson *Man. Dental Prosthetics* ii. 55 An impression is a negative likeness of an object or part taken in a plastic material, from which a cast or positive likeness may be produced. **1939** M. Hoffman *Sculpture Inside & Out* xii. 214 While working, it is useful to squeeze the wax often against the negative mold, thereby verifying just what the effect will be. **1940** J. Osborne *Dental Mech.* i. 1 The technique necessary for the accurate construction of a model, or positive likeness of the patient's mouth, from an impression or negative likeness. **1947** J. C. Rich *Materials & Methods Sculpture* v. 90 Negative molds are of two varieties: those that are flexible and those that are rigid. *Ibid.* 96 Agar-base negative mixtures are a recent casting development. *Ibid.* 122 Wax is rarely used as a negative material in sculpture. **1966** D. Z. Meilach *Creating with Plaster* iv. 58/2 Grease the negative mold very well with a separating medium. **1975** *N.Y. Times* 29 Nov. 19/2 The work, unusual for a cast bronze in that it has a negative impression on the back corresponding to the subject on the front, was apparently designed that way so castings could easily be made from it.

b. Having the character of a negative (*negative *sb.* 8 c).

1949 Frayne & Wolfe *Elem. Sound Recording* xiv. 264 A negative matrix or 'stamper' must be made from the original record. **1974** *Encycl. Brit. Macropædia* XVII. 54/1 To make copies of the recording, dies must be made the surface of which is a negative replica of the master-record surface.

12. *Theol.* = *apophatic *a.*

1956, **1961** [see *apophatic *a.*]. **1964** C. S. Lewis *Discarded Image* iv. 70 It is the 'negative Theology' of those who take in a more rigid sense, and emphasise more persistently than others, the incomprehensibility of God.

III. 13. Comb.: **negative-going,** increasing in magnitude in the direction of negative polarity; becoming less positive or more negative; **negative–positive** *a.*, pertaining to or designating a photographic process, device, etc., employing or producing both negative and positive film, or employing negative film to produce a positive image, or vice versa.

1959 J. M. Pettit *Electronic Switching* v. 136 The plate waveforms in Fig. 5–15 are square waves, except for the negative-going spikes and the exponential rise. **1969** J. J. Sparkes *Transistor Switching* v. 121 The logic inputs.. predetermine the direction the circuit will tend to switch next time a negative-going voltage step (from *ECC* to zero) is applied to the pulse inputs. **1936** C. E. K. Mees *Photogr.* 212/1 Negative–positive process, cine film. **1938** G. H. Sewell *Amateur Film-Making* ii. 12 The 'negative–positive' method consists of recording the negative image on one piece of film, and 'printing off' the positive image on to another piece of material in the manner described above. **1958** *Newnes Compl. Amat. Photogr.* 3 The next step [after daguerreotype], technically, was the development by Fox Talbot of a negative–positive process... Light-sensitive paper was the negative, reversed on to a second sheet of sensitised paper for the positive. *Ibid.* v. 85 Negative–positive films are designed to produce colour negatives, from which colour prints can be made. **1967** E. Chambers *Photolitho-Offset* x. 145 The line positive is made on 'lith' film from the negative–positive combination.

negativism. Add: **2.** *Psychol.* Resistance to attempts to impose a change of activity or posture, characteristic of various neuro-psychiatric disorders.

1892 D. H. Tuke *Dict. Psychol. Med.* II. 724/2 As soon as we attempt to produce passive movements of any part of the patient's body, we meet almost always with a powerful resistance; the groups of muscles antagonistic to the attempted movement commence to contract energetically—this has been termed the symptom of negativism. **1902** A. R. Defendorf *Clin. Psychiatry* 63 Negativism consists in the reaction to stimuli which are [*sic*] the reverse of the normal reaction. **1916** A. A. Brill tr. *Freud's Wit & its Relation to Unconscious* III. vi. 278 This..behavior of antagonistic relationships is probably not without value for the understanding of the symptom of negativism in neurotics and in the insane. **1947** G. Murphy *Personality* II. ix. 218 Negativism appears..at *any* period in which the external interference is insufficient to redirect behavior but is sufficient to necessitate an extra effort in continuing one's own activities. **1969** *Sci. Jrnl.* June 72/2 The patient continued to maintain his negativism and said 'I don't know'. **1971** *Publishers' Weekly* 30 Aug. 209/3 Solutions to the problems of fear, tension, guilt, loneliness, and negativism.

negativist. Add: Also *attrib.* or as *adj.*

1927 J. S. Huxley *Relig. without Revelation* i. 52 They have..rejected the whole ground of divinity.., and so been forced into a negativist attitude, compelled to satisfy their natural and normal religious needs in other ways and other spheres. **1958** W. Stark *Sociol. of Knowl.* 317 The main representatives of this doctrine which..we wish to label the *negativist* theory, are.. Nietzsche and..Pareto.

negativistic (ne:gătivi·stik), *a.* [f. Negativ(ism + -istic.] Of, pertaining to, or characterized by negativism.

1902 A. R. Defendorf *Clin. Psychiatry* 64 The negativistic patient shows great equanimity, he seldom defends himself..but merely resists. **1916** C. E. Long tr. *Jung's Coll. Papers Analytical Psychol.* vi. 200 Disposing causes of negativistic phenomena are: the *ambitendency* by which every impulse is accompanied by its opposite. **1930** E. & F. Jensen tr. *Adler's Educ. Children* xii. 213 If an adolescent shows himself very negativistic in regard to the other sex, we will find, if we trace back his life, that he was probably a fighting child. **1957** M. Millar *Soft Talkers* 186, I do hope your negativistic attitude towards me personally won't interfere with your better judgement. **1970** H. McLeave *Question of Negligence* (1973) x. 84 The negativistic attitude of Fairchild, the ministry sneak with the almost psychotic habit of saying no to everything.

negativity. (Earlier and later examples.)

1854 Geo. Eliot tr. *Feuerbach's Essence Christianity* ii. 42 It is true that thus, negativity, as the speculative philosophers express themselves—*nothing* is the cause of the world. **1953** R. Lehmann *Echoing Grove* 41 Had decided to my own satisfaction what it meant, in psychological terms: not tension, not active boredom—simply negativity. **1956** D. Gascoyne *Night Thoughts* 17 The Tyrant Negativity has usurped power. **1975** *Times* 24 Oct. 3/3 Rabies virus was not found in a man who died of the disease after intensive vaccine treatment... 'The negativity of the finding..was perhaps the light at the end of the tunnel.'

negaton (ne·gătǫn). *Physics.* [f. Negat(ive *a.* + *-on[1].] = *negatron 2.

1928 H. D. Hubbard *Explanatory Key to Periodic Chart of Atoms* 20 The positive atom of electricity, the 'proton' is 1845 times heavier than the negative. (Since 'electron' is ambiguous, the term 'positon' is suggested as more explicit, and 'negaton' would avoid ambiguity as the corresponding name for the negative electron.) **1938** *Encycl. Brit. Bk. of Year* 404/1 The prediction of the possibility under certain conditions of the materialization of radiation quanta into a positon and a negaton. **1955** O. Klein in W. Pauli *Niels Bohr* 115 Assuming..that there are no negaton–positon interaction terms. **1955** R. D. Evans *Atomic Nucleus* xviii. 569 The Sixth General Assembly of the International Union of Pure and Applied Physics (Amsterdam, July 8 to 10, 1948) unanimously recommended the use of the terms *positon* and *negaton* as a means of distinguishing between positive and negative electrons. However, common usage, as seen in the periodical literature, still tends to retain the 'r'.

negator. Add: **2.** A word expressing negation; = Negative *sb.* 2.

1961 R. B. Long *Sentence & its Parts* iv. 100 *Not* is by no means the only adjunct with negator force... Negative subjects and negative complements can serve as negators. **1966** G. N. Leech *Eng. in Advertising* xviii. 159 The universal negators *no*, *never*, etc. **1973** *Archivum Linguisticum* IV. 14 Of these auxiliaries, only one negator /no/ can occur independently. **1974** L. Todd *Pidgins & Creoles* iii. 34 In the 1830 examples the negator occurs in the same position as in the pidgins and creoles we have examined.

3. See *negater.

negatron (ne·gătrǫn). [f. Nega(tive *a.* + *-tron.] **1.** *Electronics.* A kind of valve that exhibits negative resistance, having two anodes (one either side of the cathode) and a control electrode between the cathode and one of the anodes.

1919 J. Scott-Taggart *Brit. Pat.* 166,260 17 Sept., The present invention relates to an electron discharge device, hereinafter to be termed a negatron, which is capable of being used as a negative resistance. **1964** J. Groszkowski *Frequency of Self-Oscillations* iii. 70 The negatron belongs to the voltage-controlled negative resistors.

2. *Physics.* The ordinary, i.e. negatively charged, electron (as distinct from the positron). Cf. *negaton.

1934 C. D. Anderson et al. in *Physical Rev.* XLV. 353/1 To remove the ambiguity in the definition of the term 'electron'..the terms 'negatron' and 'positron' are here used. **1955** R. D. Evans *Atomic Nucleus* xviii. 576 Consider the collision of an incident negatron..with an atomic electron. **1961** G. R. Choppin *Exper. Nucl. Chem.* vii. 100 When the gamma energy is greater than 1·02 MeV, a certain probability exists for the creation of a negatron..and a positron. **1971** Stern & Lewis *X-Rays* iii. 47 Triplet production can occur when the photon interacts with an electron and not with a nucleus. Three particles result, the original electron as well as the positron and negatron pair.

negentropy (nege·ntrǫpi). [f. Neg(ative *a.* + Entropy.] Negative entropy, as a measure of order or information.

1950 L. Brillouin in *Amer. Scientist* XXXVIII. 594/1 Every observation in the laboratory..is made at the expense of a certain amount of negative entropy (abbreviation: negentropy), taken away from the surroundings. **1956** —— *Sci. & Information Theory* p. xii, We prove that information must be considered as a negative term in the entropy of a system; in short, information is negentropy. *Ibid.* ix. 117 An isolated system contains negentropy if it reveals a possibility for doing mechanical or electrical work. **1958** *Archivum Linguisticum* X. 137 The calculation of redundancy, by the method taken over from information theory, requires the three terms 'negentropy', 'relative negentropy' and redundancy. **1966** S. Beer *Decision & Control* ix. 188 Hence the energy transferred through the system is exactly balanced by the information flowing the opposite way; to speak of a change in entropy is *ipso facto* to speak of an equivalent change in negentropy. **1970** H. C. Shands *Semiotic Approaches to Psychiatry* xxiii. 386 Knowing systems lose information as physical systems gain entropy. The continuous input of 'negentropy'.. is required to maintain balance in either context. **1971** *Sci. Amer.* Sept. 183/2 Taking commonly accepted average values for the temperatures of the sun and the earth, the 1·6 × 10^{14} megawatt-hours of energy radiated to outer space carries with it the capability for an entropy decrease, or 'negentropy flux', of 3·2 × 10^{23} joules per degree K. per year, or 10^{46} bits per second.

négligé. Add: Also **negligee, negligée.** (Further examples.) Cf. Negligee in Dict. and Suppl.

1908 R. W. Chambers *Firing Line* xviii. 313 Cecile, in distractingly pretty negligee, waved him audacious adieu from her window. **1913** E. M. Forster *Let.* 1 Jan. in *Hill of Devi* (1953) 24 Her dress was on the negligée side, but she had not been intending to receive. **1917** 'Taffrail' *Sub* iv. 99 The members of the mess had purchased pictures, most of them of beautiful ladies in rather négligé raiment. **1945** S. Lewis *Cass Timberlane* (1946) xliii. 308 Jinny weaved in, much too pretty in her négligé. **1954** A. Gardiner *Theory of Proper Names* (ed. 2) p. v, The bulk of this book is an exact reprint of a..booklet published..in 1940. Doubtless owing to the circumstances of the times, but possibly also to the original publication's somewhat négligé apparel, this passed almost unnoticed.

negligee. Add: **1.** Also, an informal garment worn by men in the 18th century.

1795 tr. *C. P. Moritz's Trav.* 87 In the morning, it is usual to walk out in a sort of negligée [*sic*], or morning-dress, your hair not dressed, but merely rolled up in rollers, and in a frock and boots.

b. *negligee shirt* (N. Amer.), a soft-fronted shirt worn by men.

1895 *Montgomery Ward Catal.* 278/3 Men's negligee shirts. **1921** *Daily Colonist* (Victoria, B.C.) 19 Oct. 7/2 (Advt.), Men, if you want the best shirt..buy..these well made Cambric Negligee Shirts, in neat stripes.

c. *transf.* A shroud. *U.S.*

1927 *Amer. Mercury* May 33/1 The corpse is not a corpse nor does it wear a shroud. It is the body, or the remains;..and the garment in which it is wrapped, when there is one aside from ordinary clothing, is a negligee. **1963** J. Mitford *Amer. Way of Death* i. 26 Florence Gowns Inc...exhibited their line of 'streetwear type garments and negligees'..at a recent convention of the National Funeral Directors Association.

3. A woman's dressing-gown, usu. made of flimsy, semi-transparent fabric trimmed with ruffles, lace, etc.

1930 M. Story *Individuality & Clothes* III. 311 *The negligée*, this garment for intimate occasions may be.. silky, lacy, colorful, and dainty. **1942** H. Pepin *Mod. Pattern Design* x. 219/1 The boudoir negligée is not to be confused with the 'hostess gown'. **1952** S. Kauffmann *Philanderer* (1953) i. 12 She shrugged and walked lazily over to the kitchenette, her filmy negligée heightening the illusion that she floated. **1960** A. Clarke *Later Poems* (1961) 80 Rosalind, in a négligée, Began to sketch me as I lay Naked. **1967** B. Patten *Little Johnny's Confession* 51 A thousand negligées, pyjamas, nightgowns.

negligent, *a.* and *sb.* Add: **B.** *sb.* **2.** A type of wig worn in the 18th century.

1753 in F. W. Fairholt *Costume in England* (1885) II. 320 The pigeon's wing, the comet, the cauliflower,..the rose, the crutch, the negligent, the chancellor, [etc.]. **1762** Goldsmith *Life R. Nash* 74 Nash..had seen flaxen bobs succeeded by majors, which in their turn gave way to negligents. **1971** J. Woodforde *Strange Story False Hair* vii. 46 *Men's—eighteenth century*..Negligent.

‖**négociant** (negosiaṅ). [Fr., = merchant, used *ellipt.* for *négociant des vins.*] A wine merchant.

1910 'Saki' *Reginald in Russia* 104 The little Lemberg *négociant* plucked up heart. **1961** W. E. Massee *Wines & Spirits* 51 A shipper who owns vineyards in one town but ships from another must put the word *négociant* on the label. **1970** *House & Garden* May 38/3 The principal *négociants* of Burgundy are very hospitable to properly introduced visitors who have a serious interest in wine. **1974** *Harpers & Queen* Sept. 73/1 It's a good thing for a young man to have the M[aster of] W[ine]... Continental *négociants* and growers now recognise and respect it.

negotiate, *v.* **4.** (Later examples.)

1909 *Q. Rev.* Oct. 492 Some rival..had 'negotiated'—this we believe to be the sporting phrase—the same 150 miles in forty-seven hours. **1922** H. Titus *Timber* xvi. 151 Pelly negotiated the *cuspidor* safely. **1973** E. Lemarchand *Let or Hindrance* xiii. 157 Toye negotiated the narrow entry, and they arrived in a small enclosed space in which several cars were already parked.

Negress. Add: Now customarily written with a capital initial.

In recent years felt by some to have 'racist' connotations.

1970 J. Updike *Bech: a Book* 113 'In *Travel Light*, for example, you keep calling Roxanne a Negress.' 'But she was one.' He added, 'I loved Roxanne.' 'The fact is, the word has distinctly racist overtones.' **1974** *Times Lit. Suppl.* 21 June 671/1 Blacks all over the world find the term 'Negress' offensive. What is wrong with saying 'a Black woman'?

transf. (Later example.)

1920 E. Sitwell *Wooden Pegasus* 92 The negress Night, within her house of glass Watched the processions pass.

Negri (nē[i]·gri). *Path.* The name of Adelchi *Negri* (1876–1912), Italian physician, used *attrib.* (and formerly in the possessive) to designate eosinophil cytoplasmic inclusion bodies (first described by him in 1903) found in the neurons of the brains of human beings and animals infected with rabies, the demonstration of which provides the most certain diagnosis of that disease.

1904 *Brit. Med. Jrnl.*, *Epitome of Current Med. Lit.* 30 Apr. 72/2 (*heading*) The minute structure of Negri's bodies. **1905** *Jrnl. Amer. Med. Assoc.* 2 Sept. 744/1 It is just as possible that the Negri bodies are the result of the infection as that they are the cause of rabies. **1906** *Jrnl. R. Microsc. Soc.* 626 (*heading*) Demonstrating Negri's corpuscles. **1952** [see *chromatoid *a.*]. **1970** *Sci. Jrnl.* Apr. 38/1 Diagnosis of rabies by the recognition of Negri bodies in the brain of an affected animal.. requires the animal to be dead. **1974** Passmore & Robson *Compan. Med. Stud.* III. xii. 96/2 Smears and tissue sections should be searched for Negri bodies which stain pink with polychrome stains.

negrification (nīːgrifikēⁱ·ʃǝn). Also **Negrification.** [f. Negro on model of words in *-fication*, as *pacification*. Cf. Nigrification.] The action or fact of making Negro in character; placing under the control of Blacks. So **ne·grify** *v. trans.*

1929 *Nation* (N.Y.) 9 Jan. 56 At the beginning of 1928 it seemed as if an end would be made to the slave traffic and the 'negrification' of Cuba, when the Government decided to restrict the output of sugar to 4,000,000 tons per year. **1961** *Spectator* 9 June 835 Through the ceremony the negroes 'negrify' themselves. **1962** *Economist* 7 Apr. 28/2 Some young Belgian technicians have complained of the effects of 'negrification'. **1966** P. Green tr. *Escarpit's Novel Computer* vi. 76 Suppose..you were asked to join a revolt..to save the Western world from materialism and Arabo-Marxist negrification. **1972** *Daily Tel.* 7 Mar. 11/4 'Black' in Trinidad means black... Hence the need for 'Negrification' of public and private employment.

negritic (nīgri·tik), *a.* [f. Negro + -itic. Cf. Nigritic *a.*] Of or pertaining to Negroes or Black peoples; nigritic.

1878 C. Keary *Dawn of Hist.* 220 The reader may consult an interesting paper by Professor Huxley..for some further views concerning the extension of the Negritic family. **1926** *Contemp. Rev.* Apr. 529 The one class that had kept itself pure from negritic intermixture. **1947** E. Hooton in H. Gladwin *Men out of Asia* p. xi, I am fairly sure that the earliest arrivals here were non-Mongoloids carrying archaic White strains ('Australoids', if you like) probably mixed with Negritic elements. **1950** *Cold Spring Harbor Symposia on Quantitative Biol.* XV. 260/1 Using the living Andamanese as a basis for reference, the negritic migrants may be characterized as of very short stature, dark skin color, woolly hair form, moderate round-headedness, low nasal relief, and a very short and narrow face.

negritize (nī·gritǝiz), *v.* Also **Negritize.** [Irreg. f. Negro or *negrit(ic *a.* + -ize.] *trans.* To make Negro or nigritic in character or appearance.

1901 *Ann. Rep. Board of Regents Smithsonian Inst. 1899* 513 Not one fact is in evidence from which we may conclude that a single neighbouring people known to us has been Negritized. **1930** C. G. Seligman *Races Afr.* v. 112 The less negritized type are of a slight, rather graceful build.

Negrito. (Earlier and later examples.)

1814 J. Maver tr. *Martinez de Zuñiga's Hist. View Philippine Islands* I. p. xii, It is generally allowed that the language spoken by the Papuans..and Negritos of the Philippines, and adjacent islands, is totally different from the Malayan. **1928** *Times Lit. Suppl.* 9 Feb. 90/4 The Negritoes..live at a primitive level, using wind-breaks and not houses. **1958** J. Slimming *Temiar Jungle* ii. 22 Originally much of this part of the Nenggiri was Negrito country. **1965** C. Shuttleworth *Malayan Safari* i. 16 Negritoes—a small negroid people with crinkly hair. **1969** *Age* (Melbourne) 24 May 12/5 Mr. Robinson..takes it for granted that two 'waves' of Negritos, the Kartans and the Tartangans, preceded the Aborigines to Australia. **1972** *Guardian* 22 Sept. 9/3 At the top of the mud bank was a tiny village of palm shelters, just high enough to sit in and here I met the Negritos, the oldest inhabitants of Malaysia, a short, negroid nomadic group.

Negritude (nī·gritiud). Also with small initial and in Fr. form *Négritude*. [a. F. *négritude* Nigritude.] The quality or characteristic of being a Negro; affirmation of the value of Black or African culture, identity, etc.

1950 *French Rev.* XXIII. 383 Their [sc. pre-1939–45-war young French Colonial Negro authors'] writing would be different, so different that only a new term could describe it; hence they invented the word: *négritude*. **1960** *Guardian* 29 July 4/2 The deeper cultural manifestations of colonialism. The best of this..kind of analysis seems..to be coming from those who have been colonised themselves—in Africa..from 'Présence Africaine's' explorations of *négritude*. **1960** *Observer* 20 Nov. 7/1 A movement for what French Pan-Africanists call 'Negritude'—the recognition of the Negro personality in world civilisation. **1961** *Ibid.* 29 Oct. 13/7 Senghor has been one of the leading prophets of *Négritude*, a literary and philosophical movement which expresses in an almost mystical way the African identity over against Western materialism. **1962** *Times Lit. Suppl.* 10 Aug. 596/4 The recent African Writers Conference..was significant for its hostility to négritude. **1963** *Internat. Year Bk.* 274/1 Aimé Césaire, a Martinique poet, invented the term, 'négritude', to describe the poetry that he and Haitians Jacques Roumain and Léon Laleau, and Léon Damas from Guiana were attempting to write. The word referred to the elevating of Africa as a place toward which all people of Negro blood aspired spiritually, but this Africa was not so much a geographical location as a condition of the mind. **1965** *Time* 27 Aug. 19 The wholehearted attempt by other Negroes to emphasize their Negroid features and hair texture shows their pride in their 'negritude'—a word currently in fashion in Negro communities. **1966** *New Statesman* 18 Nov. 730/1 Negritude is the least characteristic thing about Senator-elect Edward Brooke... On election night..Walter Cronkite told..television listeners that Brooke was 'five-fourths White'. **1969** N. Hare in A. Chapman *New Black Voices* (1972) 435 The debate was kicked off by leading Negritude theoretician, Leopold Senghor. *Ibid.* 436 Negritude.. permitted an escape into excessive glorification of the past and the traditional..so that one found difficulty in incorporating the techniques of the present and the future or in turning them effectively against the oppressor. **1971** *Black Scholar* Apr.–May 8 The laws and order of this nation are contrary to the black man's nature, contrary to our Negritude. **1972** M. Riofrancas in J. Pinkham tr. *Césaire's Discourse on Colonialism* 72 (*tr. interview*) How did you come to develop the concept of Negritude? A[imé] C[ésaire]. I have a feeling that it was something of a collective creation. I used the term first, that's true. .. It was really a resistance to the policy of assimilation. **1975** *Times Lit. Suppl.* 7 Mar. 247/2 (Advt.), Negritude has been defined by Senghor as 'the sum of the cultural values of the black world as they are expressed in the life, the institutions, and the works of black men'.

Negro. Add: In the nineteenth and twentieth centuries also applied (now somewhat less frequently because of the increasing use of the word *Black*) to individuals of African ancestry born in or resident in the United States or in other English-speaking countries. (Now customarily written with a capital initial.) (Further examples.) Cf. *New Negro, Nigger *sb.* (in Dict. and Suppl.).

1876 tr. *O. Peschel's Races of Man* 464 Narrow and more or less high skulls are prevalent among the negroes. **1906** *Harper's Weekly* 2 June 763/2 Professor Booker T. Washington, being politely interrogated..as to whether negroes ought to be called 'negroes' or 'members of the colored race' has replied that it has long been his own practice to write and speak of members of his race as negroes, and when using the term 'negro' as a race designation, to employ the capital 'N'. **1911** E. C. Semple *Influences Geogr. Environment* ii. 38 It is generally conceded by scientists that pigment is a protective device of nature. The Negro's skin is comparatively insensitive to a sun heat that blisters a white man. **1930** *N.Y. Times* 7 Mar. 22/5 (*heading*) 'Negro' with a capital 'N'... Major Robert R. Moton..has written..that his people universally wish to see the word 'Negro' capitalized... In our 'style book' 'Negro' is now added to the list of words to be capitalized. It is not merely a typographical change; it is an act in recognition of racial self-respect. **1938** F. Boas *Gen. Anthropol.* iii. 104 In a strict sense a race must be defined as a group of common origin and of stable type. In this sense extreme forms like the Australians, Negroes, Mongolians, and Europeans may be described as races because each has certain characteristics which set them off from other groups, and which are strictly hereditary. **1965** S. S. Smith *Ess. Causes of Variety of Complexion & Figure* p. lvii, It remains hazardous..to offer summary findings as to skeletal differences between whites and negroes. **1970** R. D. Abrahams *Positively Black* ii. 33 By espousing the term 'black' for themselves, they are also arguing implicitly that 'Negro' is a status term imposed by whites to underline the white's sense of the place of blacks in the American system. **1970** C. Major *Dict. Afro-Amer. Slang* 84 *Negro*, another way of calling [a] person an *Uncle Tom*. **1971** *Black Scholar* Jan. 53/2 His protagonist, a white-skinned Negro..decides to leave the black race. *Ibid.* Apr.–May 9 The United States of America has..deprived me and my brothers and sisters, the 30 to 60 million so-called Negroes, better known as Asiatic Black people, of life, liberty, and the pursuit of happiness. **1973** *Black World* May 37/2 Upon spotting the Afro-American, the Ghanaians shouted out, 'Hey, Negro!' The other.. retorted angrily, 'I'm a Black Man, not a Negro. Don't call me Negro.'

d. (Earlier example.) Also in *Combs.* with a language name, as *Negro-English*, *-French*, etc.

1704 S. Knight *Jrnl.* (1825) 38 You speak negro to him. I'le ask him. **1808** T. Ashe *Trav. Amer.* 79 The husband..had lived long enough in Virginia to pick up some Negro-English. **1819** R. L. Mason *Narr. in Pioneer West* (1915) 56 Negro-French is the common language of this town. **1862** 'E. Kirke' *Among Pines* 132 Not to weary the reader with a long repetition of negro-English, I will tell in brief what I gleaned from an hour's conversation with the two blacks. **1932** W. L. Graff *Lang.* 436 As a result of European trade a number of creolized trade languages have developed along the Atlantic Coast. They are chiefly Negro-Portuguese, Negro-English, and Negro-French. **1964** *Language* XL. 291 Sranan (also known as..Surinam Negro-English). **1971** J. Spencer *Eng. Lang. W. Afr.* 9 A pidginised form of Portuguese (often referred to as Negro-Portuguese).

2. *Negro-baiting* vbl. sb. and ppl. a., *-breaker*, *-breaking*, *equality*, *-hate*, *-holder* (earlier example), *-monger*, *question*, *-rank* adj., *slavery*, *-stale* adj., *-stealer*, *-stealing*, *-trader*, *-white* adj.

1949 *Sat. Rev. Lit.* (U.S.) 24 Sept. 6 Something about Negro-baiting in the South. **1951** Koestler *Age of Longing* x. 183 You are a Negro-baiting, half-civilised nation. **1845** F. Douglass *Narr. Life F. Douglass* x. 73 Mr. Covey enjoyed the most unbounded reputation for being a first-rate overseer and negro-breaker. **1855** —— *My Bondage & my Freedom* xv. 216 His proficiency in the art of negro breaking. **1856** *Illinois State Register* (Springfield) 26 June 3/3 The cry for negro equality is on their lips. **1905** N. Davis *Northerner* 52 You think I might be nice to Mr. Falls, negro equality and all? **1862** *N.Y. Tribune* 21 Apr., Southern negro-hate, being based on Slavery, is kept within bounds; that of the North, being mainly a hypocrisy or imitation, is affected & exaggerated to caricature. **1780** J. Jones *Lett.* (1889) 47 The negro holders in general already clamour against the project. **1741** T. Jones *Let.* 1 July in *Colonial Rec. Georgia* (1906) IV. 678 This exposes them to the Envy and Hatred of our Negro-Mongers. **1832** *Reg. Deb. Congress U.S.* 2 Apr. 2348 The South must be threatened with the negro question. **1949** *Time* 31 Oct. 84/3 The South gradually transformed 'the Negro question' into a fanatical folk bias, coloring its segregated religion, its sex attitudes, its every moment in life. **1942** Negro-rank [see *Negro-stale* adj. below]. **1831** M. Holley *Texas* (1833) 87 The question of negro slavery..is one of great importance. **1942** W. Faulkner *Go Down, Moses* 199 It all seemed to stand there about them, intact and complete and visible in the drafty, damp, heatless, negro-stale negro-rank sorry room. **1827** *Western Monthly Rev.* I. 69 It will be the refuge of Negro-stealers and the Elysium of rogues. **1819** *Niles' Reg.* XVI. 160/1 Sentence of death has been pronounced on a fellow in North Carolina, for negro stealing. **1732** in *Rhode Island Hist. Soc. Coll.* (1923) XVI. 108, 4 Negro Traders then on board. **1873** 'Mark Twain' & Warner *Gilded Age* vii. 78 The Hawkins hearts had been torn to see Uncle Dan'l and his wife pass from the auction-block into the hands of a negro trader. **1956** J. C. Furnas *Goodbye to Uncle Tom* ii. 70 Our town then had the largest Negro–white ratio in the North. **1961** *Times* 2 Dec. 11/5 Eight Negroes, mostly of mixed Negro–white descent.

3. *Negro minstrel* (earlier and later examples).

1855 C. E. De Long in *Calif. Hist. Soc. Q.* (1929) VIII. 346 A negro minstrel performance at home went to it and took some girls. **1884** *Century Mag.* Mar. 688/1 At that time the negro-minstrel was not a black-faced singer of sentimental songs but a man who sang and jumped Jim Crow..and other genuine plantation songs. **1915** *Scribner's Mag.* June 754 Time was when the Negro-minstrels held possession of three or four theatres in the single city of New York. **1970** *Oxf. Compan. Mus.* (ed. 10) 675/1 Towards the end of the nineteenth century Negro Minstrels were a feature of every considerable British coast resort.

5. *Negro quarters* (later examples).

1813 E. Gerry Jr. *Diary* 26 June (1927) 144 Mr. Carrol has 1000 slaves, whose huts, called negro quarters, constitute a small town around the mansion. **1913** W. P. Eaton *Barn Doors & Byways* 167 The old foundation stones show that the house was once one hundred and ten feet long, with a gigantic kitchen and outstanding negro quarters.

6. (Further examples.) Spec. of any art form; associated with or characteristic of Negroes. Of clothes, fabrics, etc.: designed to be worn or used by Negroes.

1732 *South Carolina Gaz.* 1 Apr., He had on..blue Negro Boots. *Ibid.* 30 Sept. 4/2 Just imported, white and blue Negro Cloth. **1769,** etc. Negro cloth [in Dict., sense 7]. **1786** *Maryland Jrnl.* 26 Sept., Fine and coarse broadcloths; coatings; Negro cottons. **1818** *Amer. Beacon* (Norfolk, Va.) 19 Dec. 1/4 Negro cotton. 10 Bales just received. **1841** *Picayune* (New Orleans) 3 Mar. 3 Negro Blankets, in store and for sale. **1844** J. Cowell *30 Yrs. among Players* 66 [Blakeley] was the first to introduce negro singing on the American stage. **1847** F. A. Kemble *Let.* Dec. in *Rec. Later Life* (1882) III. 279 Do you remember that delightful Negro song, the 'Invitation to Hayti'? **1855** A. M. Murray *Lett.* (1856) 395 This morning we have had some negro music. **1912** *Chambers's Jrnl.* Jan. 23/1 Negro songs have always been popular among us, and deservedly so. **1925** W. S. Braithwaite in A. Locke *New Negro* 39 It was the stirring year of 1917 that heard the first real masterful accent in Negro poetry. **1936** A. Locke *Negro & his Mus.* i. 1 Negro music is the closest approach America has to a folk music, and so Negro music is almost as important for the musical culture of America as it is for the spiritual life of the Negro. *Ibid.* ii. 15 The Negro dance has the feature, characteristic of Russian, Polish and other Slavic folk-dances, of sudden changes of the pace and daring climaxes of tempo. **1960** A. E. Keep tr. *Leuzinger's Afr.* (1962) i. i. 13 Negro art attracts and enthrals us by its emotional vigour and clarity of form. *Ibid.* vi. 52 Is negro art primitive? No—if the word 'primitive' is understood to mean something crude, barbaric and contemptible. Yes—if by 'primitive' we mean something honourable, as the term is applied, for instance, to the Fauvists in European painting. **1963** *Times Lit. Suppl.* 18 Jan. 43/2 The rock 'n roll and Negro-jazz rhythms of his American cycle of poems. **1970** R. D. Abrahams *Positively Black* ii. 51 Distrust of even one's closest friends is a constant theme of Negro life and Negro fictions.

7. Negro Renaissance (see quot. 1973); **Negro spiritual**, an American Negro religious song; **Negro State**, any of the Southern States of America in which slavery was legal; **negro yam**: for '*Dioscorea alata*' substitute '*Dioscorea sativa*'; (earlier and later examples).

1925 A. Locke *New Negro* p. xi, We speak of the offerings of this book..as culled from the first fruits of the Negro Renaissance. **1952** B. Ulanov *Hist. Jazz in Amer.* (1958) x. 103 The Negro poets who won such a large audience for their work, good, bad, and indifferent, in the intense days of the so-called Negro Renaissance. **1964** J. H. Clarke *Harlem* 16 The stock market collapse of 1929 marked..the end of the period known as the Negro Renaissance. **1973** Baskin & Runes *Dict. Black Culture* 324 *Negro Renaissance*, a creative outpouring in art, music, and literature in the 1920s, giving expression to the discontent of the Negro... Writers of the twenties displayed considerable talent..in developing Negro themes in a highly personal way. **1867** *Atlantic Monthly* June 685/1, I had for many years heard of this class of songs under the name of 'Negro Spirituals'. **1928** *Observer* 22 July 21/1 As important..is their singing of negro spirituals and 'work songs'. **1949** *Oregonian* (Portland) 10 Aug. 8/4 He found time to write books on his hobbies, on alligators and on Negro spirituals. **1970** *Oxf. Compan. Mus.* (ed. 10) 1064/2 The words of Negro spirituals are for the most part adaptations of passages from the Bible. **1780** in *Essex Inst. Hist. Coll.* (1877) XIII. 220 You did not carry home contemptible Ideas enough of the negro States or of this great Braggadocio. **1809** *Deb. Congress U.S.* 20 Jan. (1853) 1152 The Potomac the boundary—the Negro states by themselves! **1696** H. Sloane *Catal. Plantarum Jamaica* 219 Negro Country Yam. **1707** —— *Voy. Jamaica* I. 140 Negro Country Yams. This has a great Root a Foot broad... They being cut into pieces and boiled or rosted are eaten by Negros, Slaves, or Europeans, instead of Bread. **1756** P. Browne *Civil & Nat. Hist. Jamaica* 359 The Negro Yam... The Yam. Both these plants are cultivated for food, the roots, which grow very large, being mealy and easy of digestion. **1814** J. Lunan *Hortus Jamaicensis* II. 308 This [sc. *Dioscorea sativa*] is commonly called negro yam. **1864** A. H. R. Grisebach *Flora Brit. W. Indian Islands* 789/2 Yams, Negro country: *Dioscorea alata*. **1953** *Caribbean Q.* II. iv. 32 *Dioscorea sativa* the so-called negro-yam, may have been indigenous, for it..sometimes grows wild; but more probably the wild specimens were originally escapes from cultivation. **1971** *Jamaican Weekly Gleaner* 3 Nov. 34/3 (Advt.), Negro yams, yellow yams, sweet potatoes.

Negrodom. (Earlier and later examples.)

1847 *Congress. Globe* 13 Feb. App. 376/1 Our measures have given all that wide region to the empire of negrodom. **1942** Z. N. Hurston in A. Dundes *Mother Wit* (1973) 31/2 Neither the top nor the bottom of Negrodom. *Ibid.* 32/1 A flight away from Negrodom.

negrofy, *v.* (Earlier example.)

a **1790** B. Franklin *Autobiogr.* in *Writings* (1905) I. 391 Finding he was likely to be negrofied himself, he..grew tir'd of the contest.

negro-head. Add: **2.** (Earlier example.)

1809 'D. Knickerbocker' *Hist. N.Y.* II. vi. ii. 88 He.. thrust a prodigious quid of negro head tobacco into his left cheek.

5. = Niggerhead 2.

1910 F. Wood-Jones *Coral & Atolls* xxiv. 284 There are many enormous 'negro heads' upon the windward barrier flats of the Southern islands of the Cocos-Keeling atoll. **1943** Baker *Dict. Austral. Slang* (ed. 3) 54 *Nigger-head*, an anthill-like peak of coral showing above water. Also called 'negro-head'. **1963** D. W. & E. E. Humphries tr. *Termier's Erosion & Sedimentation* xiii. 280 Reefs are broken into blocks, the large mushroom-shaped fragments are thrown up onto the beaches, where they form 'negro-heads'.

Negroish, *a.* (Earlier example.)

1789 J. Morse *Amer. Geogr.* 65 The children, by being brought up, and constantly associating with the negroes.. contract a negroish kind of accent and dialect.

Negroism. Add: **1.** (Earlier and later examples.)

1847 *Congress. Globe* 29th Congress 2 Sess. App. 323/2 He..thanked God that he voted against that Wilmot proviso. It smelt rank of negroism. **1935** Z. N. Hurston *Mules & Men* (1970) 17 When I pitched headforemost into the world I landed in the crib of negroism. **1970** D. Caute *Fanon* vi. 80 The vague, cosmic cults of Negroism and Arabism.

2. (Earlier and later examples.)

1859 Bartlett *Dict. Amer.* (ed. 2) p. viii, The term 'Americanisms'..may then be said to include the following classes of words:..8. Negroisms. **1930** G. B. Johnson in B. A. Botkin *Treas. S. Folklore* (1949) iv. iii. 697 *Ax*, ask. Not a Negroism, but a usage which was once good English.

Negroization (nīːgroˌǝizēⁱ·ʃǝn). [f. Negro + -ization.] A making or becoming Negro in character.

1898 A. J. Butler tr. *Ratzel's Hist. Mankind* III. 258 From them Rohlfs expects an ever-increasing 'negroisation' of the Libyan oasis.

Negroland. Add earlier and later examples and example referring to the Southern States of the U.S.A.

1756 J. Wesley *Works* (1872) IX. 209 Either in Negroland or round the Cape of Good Hope. **1836** F. A. Kemble *Let.* 5 Oct. in *Rec. Later Life* (1882) I. 66 The nearest town to this estate, Brunswick, is..a wretched hole, where I am assured it will be impossible to obtain a decent lodging for me... The owner will go..without us, on his expedition to Negroland. **1931** *Times Lit. Suppl.* 26 Feb. 145/1 Africa south of the Sahara is by no means coterminous with negroland.

Negroness (nī·gronės). [f. Negro + -ness.] Negro qualities and characteristics in the aggregate.

1946 *Sat. Rev. Lit.* (U.S.) 29 June 42 The native suffixes *-hood*, *-ness*, and *-wright* have given us *Hottentothood* and *sahib-hood*; *at-easeness*, *Negroness*, and *tasklessness*; and *filmwright*. **1958** J. Kerouac *Subterraneans* 31 The old eccentric lady not any more conscious of her *Negroness*. **1965** *Times Lit. Suppl.* 25 Nov. 1047/5 He had to confront being 'Baldwin' without atrophying into an idol, and his method is to transcend enforced initial Negroness, even to avoid being 'a Negro writer', or any 'Negro' trap. **1971** G. S. Holt in T. Kochman *Rappin' & Stylin' Out* (1972) 155 So identifying someone as 'Jew' or 'Negro' devastates the category of Negroness or Jewishness more so than saying Negro *writer*, Jewish *doctor*. **1973** *Black World* Mar. 19 Bakara asks: 'Where is the Negro-ness of a literature written in imitation of the meanest of social intelligences to be found in American culture, *i.e.*, the white middle class?'

‖ **negroni** (nigrō·ni). Also **negrone.** [It.] A drink made from gin, vermouth, and Campari; a glass of this.

1950 E. Hemingway *Across River* vi. 34 They were drinking *negronis*, a combination of two sweet vermouths and seltzer water. **1952** P. Bonner *SPQR* (1953) xx. 174, I called the barman for a *negrone*. **1960** I. Fleming *For your Eyes Only* v. 146 Bond nodded. 'A Negroni. With Gordon's, please.' **1960** B. Marshall *Divided Lady* ii. vi. 129 Before I had finished my third *negrone*. **1972** J. D. Buchanan *Professional* xvi. 158 Summoning the waiter he ordered a Negroni.

Negrophil. Add further examples of spelling *-phile*; also example of *attrib.* use.

1934 R. Campbell *Broken Record* iii. 58 Pringle obtained and asserted the freedom of the Press... He had a negrophile bias and misrepresented much. **1936** H. Preece in A. Dundes *Mother Wit* (1973) 36/2 Our indictments of the professional Negrophiles need not even be so personal. **1958** *Times Lit. Suppl.* 10 Jan. 21/1 Many of the first explorers had an objectivity and sympathy with the indigenous people which would startle a modern negrophile.

Negrophilism. (Earlier and later examples.)

1846 *Congress. Globe* 18 May 838/1 The gentleman from Ohio.., the advocate of *negro-philism*. **1859** *Harper's Mag.* Oct. 694/2 If negrophilism seeks to substitute the Chinaman or the Indiaman for the African, it will neglect all the lessons of experience. **1860** A. B. Longstreet in *U.S. Ann. Treas. Rep.* 475 A man..of more negrophilism than brains. **1889** Farmer *Americanisms* 386/2 *Negrophilism*, the anti-slavery movement.

Negrophilistic (nīːgrofili·stik), *a. rare.* [f. Negrophil + -istic.] Having Negrophil characteristics.

1899 *Leisure Hour* Dec. 168/2 The Colonists have always despised the volume [Thomas Pringle's South African poems] because of the Negrophilistic character of some of the pieces therein.

Negrophobe (nī·grofōᵘb). [f. Negro + -phobe.] One who has a violent aversion to or hatred of Negroes.

1900 *Spectator* 15 Sept. 329/2 Negrophiles may be.. wiser as well as better men than negrophobes. **1962** C. L. Barnhart in Householder & Saporta *Probl. Lexicogr.* 179 A vulgar, offensive term of contempt, as used by Negrophobes. **1974** *Guardian* 23 Jan. 12/2 Edward Long, historian, statistician and so-called absentee sinecurist and negrophobe.

Negrophobia. (Earlier and further examples.)

1819 *Niles' Reg.* XVI. Suppl. 173/2 The gentleman from Kentucky..has charged us..with being under the influence of *negrophobia*. **1859** N. P. Willis *Convalescent* 173 His hatred of colored people amounted to a negrophobia. **1945** Mencken *Amer. Lang.* Suppl. I. 408 Whenever a new suffix appears in the United States, it is put to use. An example is..*-phobia*, borrowed from the psychiatrists, and made to do heavy duty in a multitude of nouns designating violent aversions *e.g.*,..*negrophobia*..with attendant adjectives in *-phobic*.

Nehru (nēᵊ·ru). [f. the name of Jawaharlal *Nehru* (1889–1964), first Prime Minister of independent India from 1947 to 1964.] A garment of the style often worn by Nehru, consisting of a long narrow jacket with a high, stand-up collar. Usually *attrib.*, as *Nehru coat*, *collar*, *jacket*, *suit*, etc.

1967 *N.Y. Times* 24 May 46 Variations on the narrow 'Nehru' jacket with the stand-up collar continue. **1967** *Time* 25 Aug. 38 There came the groom,..wearing baggy trousers, a white, Nehru-collar tunic with red trim and cowboy boots. **1968** *N.Y. Times* 6 June 50, I wouldn't wear Nehrus or those damn beads either. **1968** *Punch* 24 July 107/2 In, says an informed source, are 'the Nehru coat and the Nehru suit'.·Gandhi's turn next? **1969** C. F. Burke *God is Beautiful, Man* (1970) 47 A new chain to wear with his turtle neck sweater and Nehru jacket. **1971** 'G. Black' *Time for Pirates* ii. 33 The man is tall for a Chinese and very thin. He wears a white Nehru jacket over black cotton jeans. **1973** *Caribbean Contact* Feb. 9/4 The Ambassador for the People's Republic of China stands in a corner with the interpreter, wearing a black, Nehru-style shirt.

neighborite (nēⁱ·bǒrǝit). *Min.* [f. the name of Frank *Neighbor* (see quot. 1961) + -ite¹.] A fluoride of sodium and magnesium, $NaMgF_3$, which occurs as colourless orthorhombic crystals, isostructural with perovskite.

1961 E. C. T. Chao et al. in *Amer. Mineralogist* XLVI. 379 The mineral is named 'neighborite' in grateful acknowledgement of the continued friendly interest and helpfulness of Mr. Frank Neighbor, District Geologist of the Sun Oil Co. at Salt Lake City. *Ibid.* 380 Neighborite was first found as clusters of pink and brown rounded grains in a dark brown to grayish black dolomitic oil shale of the Green River formation of South Ouray, Utah. **1967** *Doklady Earth Sci.* CLXXIV. 140/1 In the USSR neighborite was detected in alkalic metasomatite developed in the Western Urals from porphyry tuff and argillaceous carbonate sediments of Proterozoic age. **1971** *Amer. Mineralogist* LVI. 1520 Neighborite is synthesised from NaF and MgF_2 via a eutectic solidification process in which the product is a two-phase mixture of NaF and $NaMgF_3$.

neighbourhood. Add: **3. c.** *Math.* (i) The set of points whose distance from a given point is less than, or is less than or equal to, some non-zero, usu. small, value (see also quot. 1921).

1891 G. L. Cathcart tr. *Harnack's Introd. Study Elem. Differential & Integral Calculus* I. v. 26 We have attained the conception of the Region or Neighbourhood of a point. By it we mean an arbitrarily small but still always finite interval at both sides of the value *x*. **1921** H. S. Carslaw *Introd. Theory Fourier's Series & Integrals* (ed. 2) iii. 52 Sometimes the neighbourhood is meant to include the point $x = a$ itself. In this case it is defined by $|x-a| \leq h$. **1939** M. H. A. Newman *Elem. Topology of Plane Sets of Points* ii. 20 If *a* is a point of a space with the metric ρ, and ϵ any positive number, the set of all points, *x*, satisfying $\rho(x, a) < \epsilon$ is called a spherical neighbourhood, and more particularly an ϵ-neighbourhood of *a*. **1956** E. M. Patterson *Topology* ii. 22 A convenient way of describing continuity is to introduce the idea of neighbourhood... The set of points whose coordinates *x* satisfy $|x-x_0| < \epsilon$ is called the ϵ-neighbourhood of x_0; it consists of all points whose distance from x_0 is less than ϵ. **1971** M. Gemignani *Introd. Real Analysis* iii. 32 We denote the *p*-neighborhood of *a* by $N(a, p)$. (Thus, $N(a, p)$ consists of all real numbers within distance *p* of *a*.) *Ibid.*, Not only is every *p*-neighborhood an open interval, but every open interval is a *p*-neighborhood.

(ii) Any open set containing a given point or non-empty set; also, any set containing such an open set.

1934 C. C. Krieger tr. *Sierpiński's Introd. Gen. Topology* ii. 33 We shall understand by a neighbourhood of an element *a* any open set containing *a*. **1946** E. Lehmer tr. *Pontrjagin's Topological Groups* ii. 28 We shall give a method of defining a topological space by means of neighbourhoods rather than by means of the operation of closure. This method is rather important and is often used as the foundation of the axiomatic treatment of the concept of a topological space. **1964** W. J. Pervin *Found. Gen. Topology* iii. 45 A set is open iff it is a neighbourhood of each of its points. **1967** I. Adler *New Look at Geom.* xii. 372 The sets that belong to this special class of subsets in a topological space are called the open sets of the space or the neighborhoods of the space. **1968** S. Moran tr. *H. Schubert's Topology* 86 An open set is a neighbourhood of every one of its subsets.

6. c. In urban planning and development, a small sector of a larger inhabited area with an integrated community provided with its own shops and other facilities.

1951 *Social Aspects Town Devel. Plan* (Univ. Liverpool, Social Sci. Dept.) ii. 25 It must be emphasised that the essence of a Neighbourhood from the point of view of the planner and sociologist alike, is the opportunity it provides for people to meet together, to share the burdens of daily life, and to co-operate in an endeavour to overcome their common problems. **1961** *Listener* 2 Nov. 702/2 People are beginning..to insist that their local authorities give proper consideration to all those aspects of life which make a neighbourhood different from an estate. **1961** *Observer* 3 Dec. 23/3 The principle, so depressingly practised in the first batch of British new towns, of grouping low-density housing in so-called neighbourhoods, punctuated by random acres of open space and served by small shopping centres. **1973** *Country Life* 6 Dec. 1952/1 One of London's greatest attractions is..its village-like localities. In planning jargon these are called neighbourhoods or..environmental areas.

7. *neighbourhood bookie*, *bookstore*, *centre*, *council*, *grocery*, *market*, *park*, *road*, *school*, *shop*, *shopping centre*, *store*, *unit*; **neighbourhood friendly** *U.S. colloq.*, a well-known local shop, a neighbourhood shop.

1971 *Black Scholar* June 6/1 A man of some kind is usually around. He may be a boyfriend, an uncle or just the neighborhood bookie. **1973** *N.Y. Law Jrnl.* 17 Apr., Each appellant managed a large neighborhood bookstore. **1961** *Listener* 28 Sept. 470/3 The idea that a neighbourhood centre was a sort of rag-bag where you put all the social functions if you could not find anywhere else to put them. **1973** *Guardian* 24 Dec. 13/6 The idea of neighbourhood councils or 'community councils' or 'urban parish councils' is based on the simple proposition that there is no urban equivalent of the rural parish council and that there should be. **1970** J. Hansen *Fadeout* (1972) xx. 167 But those envelopes were there. Nine of them. The kind you buy at your neighbourhood friendly. In packs of a dozen. **1966** B. H. Deal *Fancy's Knell* v. 67 The supermarkets had killed the neighborhood grocery. **1938** *Richmond* (Va.) *News-Leader* 28 Sept. 1/3 The rush of tobacco to market has operated greatly to the advantage of the smaller towns of the belt. Leaf that once was sold in the big centers of North Carolina and Virginia is overflowing into the neighborhood markets. **1961** L. Mumford *City in Hist.* xvi. 502 The neighborhood park, conceived either as a Greenbelt around the neighborhood,..or as a ribbon of internal green. **1972** *Village Voice* (N.Y.) 1 June 16/3 Proceeds of the fair will help the association purchase two vacant lots to be developed into a neighborhood park. **1835** W. G. Simms *Partisan* II. xxviii. 266 A small track—a common wagon or neighbourhood road—wound into the forest. **1843** 'R. Carlton' *New Purchase* i. 89 Notice here, a neighborhood road does not imply necessarily much proximity of neighbours. **1842** *Southern Lit. Messenger* VIII. 65/1 As this was what is called a 'neighborhood school', the pupils necessarily came from a great distance. **1967** *New Yorker* 31 May 88/2 School integration and the preservation of neighborhood schools have been white-collar movements. **1973** *Times* 19 Apr. 18/1 The tendency for neighbourhood schools to develop on class lines. **1966** B. H. Deal *Fancy's Knell* v. 67 The shopping centers had killed the neighborhood shops. **1968** *Guardian* 26 Mar. 9/3 Her customers..used to shop in the West End, but are thankful to find a first-rate fashion shop nearer home. The success of such 'neighbourhood shops'..is an interesting trend. **1961** *Listener* 28 Sept. 471/1 Strong neighbourhood shopping centres. **1949** E. S. Gardner in *Argosy* Apr. 108/3 In that district of small neighbourhood stores, he's in a position to keep irregular hours. **1974** *Amer. Speech* 1971 XLVI. 76 *Grocery store*,.. neighborhood store. **1943** *Archit. Rev.* XCIII. 91/1 Five such residential units make up into one neighbourhood unit of approximately a thousand families. **1953** P. C. Berg *Dict. New Words* 114/1 *Neighbourhood unit*, one of the residential areas in a planned town, containing about 10,000 inhabitants, complete with schools, shops, and a community centre of its own. **1961** E. A. Powdrill *Vocab. Land Planning* iii. 40 It is by now a fundamental concept of town and country planning that the rehabilitation of existing towns and the building of new towns should be based on the 'neighbourhood unit' principle. **1966** *Listener* 19 May 729/3 It was interesting to discover how much Patrick Abercrombie's neighbourhood-unit concept..had influenced Moscow planners.

b. Attrib. phr. *the*, *your*, etc., *friendly neighbourhood*, applied to a well-known and popular local person or thing; also *ironical. colloq.*

1955 W. Gaddis *Recognitions* ii. ii. 366 Just tell Mummy to ask about *Cuff* next time she visits her friendly neighborhood druggist. **1968** *Peace News* 25 Oct. 7 (*heading*) Your friendly neighbourhood senior detective officer. **1971** *Guardian* 4 June 5/5 Their friendly neighbourhood stockbroker gave his talents free. **1973** 'R. MacLeod' *Nest of Vultures* vii. 154, I feel like I've just made a date with the friendly neighbourhood vampire. **1974** *Times* 22 Mar. 21/4 (*heading*) Your friendly neighbourhood fuel cell.

neighbourize (nēⁱ·bǝrǝiz), *v. rare.* [f. Neighbour *sb.* + -ize.] *intr.* To associate with others as neighbours; to act in a neighbourly fashion.

1899 G. B. Burgin *Bread of Tears* i. ii. 43 We thought we'd just neighbourise, and happen in to hear what it says.

Neil Robertson stretcher (nīᵊːl rǫ·bǝːtsǝn). [App. f. a proper name.] (See quots. 1941, 1967.)

1941 N. Hammer *Warwick & Tunstall's First Aid* (ed. 18) xviii. 242 The Neil Robertson stretcher (the hammock stretcher) is made of split bamboo sewn onto stout canvas, on the principle of Gooch splinting. It surrounds and encloses the patient completely and rigidly. **1959** *Times* 16 Mar. (Port of London Suppl.) p. xvi/3 All boats carry a Neil- Robertson stretcher designed to remove casualties from such awkward places as the hold or engine room of a ship. **1967** J. Wainwright *Worms must Wait* lxii. 154 The 'Neil-Robertson' stretcher was a contraption of stout canvas, bamboo, leather straps, buckles and rope. Its purpose was (to quote the book of words) 'to lift casualties in any position through small hatches, man-holes, sewer ventilators and for lowering from heights'. **1974** P. McCutchan *Call for S. Shard* xiv. 134 Tuball and his plastered leg safely strapped into a Neil Robertson stretcher with his crutches attached.

‖ **neinei** (nē·nē). *N.Z.* Also **nene.** [Maori.] A New Zealand shrub or small tree of the genus *Dracophyllum*, esp. *D. latifolium*, of the family Epacridaceæ, distinguished by clusters of long, narrow leaves; also called *grass-tree*.

1858 S. P. Smith in N. M. Taylor *Early Travellers N.Z.* (1959) 355 Some of us..found what was to us quite a new kind of tree, called *Nei nei*. **1879** J. von Haast *Geol. Provinces Canterbury & Westland, N.Z.* 78 An undescribed, superb, tree-like *Dracophyllum*..began to appear here... The natives call it Nene. **1882** W. D. Hay *Brighter Britain!* ii. 197 The Neinei..is but a small tree. **1949** P. Buck *Coming of Maori* (1950) ii. x. 262 The wood [of this flute] is *neinei*. **1966** *Encycl. N.Z.* I. 498/2 Some of the larger-leaved species [of *Dracophyllum*] like nei-nei or *D. latifolium* are true forest plants.

neither. A. *adv.* **3. b.** Delete † *Obs.* and add further examples.

1909 L. M. Montgomery *Anne of Avonlea* x. 98 'Davy declares he never saw her since I left.' 'Neither I did,' avowed Davy. **1926** J. Black *You can't Win* xiii. 165 'I wouldn't plead guilty to anything if I were you,' I advised him. 'Me, neither,' said his partner. **1973** R. Busby *Pattern of Violence* iv. 61 'Can you place either of them?' The young detective shook his head. 'Me neither.'

nekoite (ne·koˌǝit). *Min.* [anagram of Okenite.] A triclinic form of the hydrated calcium silicate $Ca_3Si_6O_{15}.6H_2O$ found at Crestmore, California, which differs from the okenite of other localities in its optical and X-ray properties.

1956 Gard & Taylor in *Mineral. Mag.* XXXI. 20 The Crestmore material must be regarded as a new species. Because of the likeness to okenite, the anagram *nekoite* is suggested. **1962** *Ibid.* XXXIII. 70 Nekoite was discovered at Crestmore, California, by Eakle (1917), who identified it as okenite. **1971** *Acta Crystallogr.* B.XXVII. 473/2 When fibre bundles of nekoite are heated to temperatures between 200 and 250°C water is lost reversibly, the apparent limiting water content of the crystal from the 'immediate weight-loss' curve corresponding to the composition $3CaO.6SiO_2.3H_2O$.

nekro-: see *necro-.

nekton (ne·ktǫn). *Zool.* Also **necton.** [ad. G. *necton*, *nekton* (E. Haeckel 1890, in *Jenaische Zeitschr. Naturw.* XXV. 251, 252), f. Gr. νηκτόν, neut. of νηκτός swimming, f. νεῖν to swim.] A collective name for aquatic animals that are able to swim and move about independently. Cf. *benthos, Plankton. So **nekto·nic** *a.*

1893 G. W. Field tr. *Haeckel's Planktonic Stud.* in *Rep. U.S. Comm. Fisheries 1889–91* 580 We must distinguish the actively swimming nekton from the passively drifting plankton. **1895** *Natural Sci.* July 31 The Plankton, Nekton, and Benthos form three well-marked communities of organisms. **1903** *Amer. Geologist* XXXI. 211 It has been asserted..that slowly creeping organisms preceded the planktonic and nektonic forms. **1909** *Chambers's Jrnl.* Dec. 784/1 The Nekton are those animals which are capable of vigorous swimming movements, and are able to migrate freely from one part of the sea to another. **1923** *Nature* 17 Mar. 374/2 The author concludes that the kerogen has arisen from 'nectons and kelps' which have 'been repeatedly buried by ash and detritus from submarine volcanoes'. **1923** *Glasgow Herald* 17 Nov. 4/2 The animals of the sea are sometimes divided into the Nekton and the Plankton, the swimmers and the drifters. **1956** *Nature* 25 Feb. 375/2 Only one nektonic species apart from *Thyrsites* itself has been found to occur in Bass Strait at the time of year in question. **1969** *Austral. Law Jrnl.* XLIII. 430 The traditional biological classification of fish and fish like creatures into drifters or plankton, swimmers or necton, and the bottom dwellers or benthos. *Ibid.*, The nectonic species of fish are divided into demersal and pelagic. **1974** Lucas & Critch *Life in Oceans* ii. 25 The nekton comprises all those animals, such as squid, cuttlefish, fishes, seals and whales, which are able to swim more or less powerfully and which are therefore independent of water movements for transport. **1975** *Nature* 17 Apr. 591/2 The stomachs of the fish contained the remains of nektonic, planktonic and benthic organisms.

‖ **nekulturny** (nekultu·rni), *sb.* and *a.* Also **nekulturniiy.** [Russ. *nekul'turnyĭ* uncivilized.] **A.** *sb.* One who is by Russian standards considered unenlightened, a boor. **B.** *adj.* Not having cultured manners, boorish. Cf. *kulturny.

1959 I. R. Levine *Real Russia* xxiv. 347 It is *nekulturny* for a man..to put feet up on a desk, to cross legs, or to keep hands in trouser pockets. **1960** W. Miller *Russians as People* vi. 139 To accuse someone of being 'uncultured' (*nekulturny*) is not a light matter... It may signify that you do not clean your teeth, that you never read a book, or that you are pushing rudely or giving way to a coarse expression of opinion. **1962** J. Wade *Running Sand* xiii. 166 To dance so close to a woman was uncultured, *nekulturny*, a sign of western decadence. **1965** *New Statesman* 16 July 71/3 What bothers Washington is that the President is so determinedly *nekulturniiy*. **1967** J. Fores *Desirable Dictator* iv. 97 We are not gangsters, Mr. C.I.A. We leave that kind of *nekulturny* behaviour to the West.

nelia, var. *nealie.

nelly² (ne·li). Also **Nelly, Nellie.** [A fem. Christian name.] **1.** Slang phr. *not on your Nelly* [f. rhyming slang *Nelly Duff* = *puff* = breath of life], 'not on your life' (see *life *sb.* 3 d), not likely.

1941 *New Statesman* 30 Aug. 218/3 *Not on your Nelly Duff*,..not likely. **1961** *John o' London's* 14 Dec. 663/1 You might have thought Mr. Samuel Bronston would have rested... Not, as they say, on your nelly. **1961** Partridge *Adventuring among Words* xi. 55 The trouble begins when part, usually the latter part, of the rhyming phrase is omitted, as so often it is, as in..'not on your Nellie' for '...Nellie Duff!' = 'not on your *puff*' = 'not on your life' or 'most certainly not' or, less politely, 'like hell, I will!'. **1966** *Times* 15 May 9 That would mean me investing in another man's career. Not on your Nelly! **1968** *Manch. Guardian Weekly* 11 Apr. 15/1 So the Liberals dropped Acton and Dudley and concentrated on Warwick and Leamington? Not on your Nellie! **1972** *Times* 24 June 11/4 Ooh, no, not on your nelly, ah, fearless Francis! **1974** *Globe & Mail* (Toronto) 21 Sept. 35/2, I appear to be giving away most of the plot? Not on your nelly. That's only the beginning.

2. A cheap wine. *Austral. slang.*

1945 Baker *Austral. Lang.* ix. 166 Here is a group of indigenous terms used to describe cheap wines: *Africa speaks*,..*nelly*, *nelly's death* [etc.]. **1952** A. G. Mitchell in *Chambers's Shorter Eng. Dict.* Suppl., *Nelly*,..(*slang*) cheap wine.—phr. *on the nelly*, given to drinking cheap wine.

3. A weak-spirited or silly person; a homosexual. Also as quasi-adj., of feminine appearance, effeminate. Cf. *nice-nellyism.

1961 *Sunday Times* 17 Sept. 41/4 [Henry] Livings's latest work, 'Sacred Nit'... This same play, now called 'Big Soft Nellie', opens at the Oxford Playhouse. **1962** *Sunday Times* 29 Apr. 12/4 See, what you've got to do is get on the same wavelength as the nellies who write in. **1967** 'T. Wells' *Dead by Light of Moon* (1968) iv. 45 You don't suppose it could have anything to do with that Strangler business, do you? Not that I'm a nervous Nelly type. **1970** K. Platt *Pushbutton Butterfly* (1971) xvi. 182 He..puffed daintily on a long cigarette as he watched the nellies cruising to the 'tearoom'. **1971** *Psychiatry* XXXIV. 187/2 Some guy came up to me and tried to take me for some money, and I knew it, and he said, 'You know, you're very nellie.' **1972** B. Rodgers *Queens' Vernacular* 141 *Nelly*..outrageously effeminate; coy, silly. **1973** C. Wittman in P. Brown *Radical Psychol.* xix. 459 There is a tendency among 'homophile' groups to deplore gays who play visible roles—the queens and the nellies.

4. Slang phr. *sitting next to* (or *by*, *with*) *Nelly*, learning an occupation on the job by observing how others do it.

1963 J. U. Fraser *Psychol.* (ed. 2) xiv. 173 Observation of a skilled operator (often characterized as 'sitting next to Nellie'). **1966** *Guardian* 30 July 6/5 The generous 'General Trainees Scheme' which allows them to spend up to two years in different departments [of the BBC]—'sitting with Nellie', is the union phrase. **1972** *Listener* 10 Aug. 180/2 Journalists are the casual labourers of the intellectual world... Most training still consists of sitting next to Nellie. **1975** *New Society* 17 July 130/1 It was then made compulsory for doctors from overseas to 'sit by Nellie' for a month. Immigrant doctors had to complete satisfactorily a month's supervision under a consultant supervisor.

Nelson (ne·lsǝn). [App. f. a proper name.] **1.** *Wrestling.* The name of a class of holds in which one or both arms are passed under the opponent's from behind and the hand(s) applied to his neck; often with words prefixed to indicate the precise form of the hold, as *double nelson*, **half nelson*, *(three-) quarter nelson*. Also *fig.*

1889 W. Armstrong *Wrestling* 233 Probably the most dangerous move in Lancashire and Cornwall and Devon wrestling..is what is called the 'Double Nelson'. **1893** *Lippincott's Mag.* Feb. 211 Among the many holds the Nelson is the most popular one with wrestlers, while the half-Nelson and half-walch-lock are next in order. **1897** [see *hammer-lock* (*hammer *sb.* 7)]. **1900** A. E. T. Watson *Young Sportsman* 644 The principal chips associated with catch as catch can wrestling are the double Nelson, the half Nelson, the heave, the Lancashire lock, the flying mare and the three-quarter Nelson. **1930** P. MacDonald *Link* ix. 168 They lose Dinwater—or lose half the Nelson they've got on him, so immediately they switch on to you.

2. The name of Admiral Lord *Nelson* (1758–1805) used *attrib.* in **Nelson cake** (see quot. 1966); **Nelson eye**, a blind eye: usu. *transf.* and *fig.* (cf. *eye *sb.*¹ 5 e); **Nelson knife**, a combined knife and fork for the use of a one-armed person; **Nelson's blood**, Navy rum; **Nelson touch**, a stroke, action, or manner characteristic of Nelson.

1909 *Daily Chron.* 16 Dec. 4/7 The Nelson cake consists of two thin pieces of reputed pastry, with a dark agglomeration between them of currants and sweet mush. **1966** F. Shaw et al. *Lern Yerself Scouse* 42 *Nelson cake*, a cake made from compressed, broken biscuits, pastry remnants etc. with dried fruit added: the whole soaked in syrup or burnt sugar and stacked in great piles. **1965** *New Society* 7 Jan. 12/3 The police made no attempt to arrest him. The State simply turns a Nelson eye to his presence. **1970** B. Turner *Another Little Death* xxiv. 149 It's often possible for the police to turn a Nelson eye to misdemeanours confessed to them if they can catch a blackmailer by doing so. **1973** J. Leasor *Mandarin Gold* i. 13 He makes most out of opium. Turning a Nelson eye to a trade he should be stamping out. **1902** *Chambers's Jrnl.* 4 Oct. 692/2 The mention of knife *and* fork attached to the one wrist is at first slightly puzzling; but probably the combination was what is called the 'Nelson knife', after its most distinguished user. The handle is like that of an ordinary table-knife; but the end of the blade, instead of being rounded off in the ordinary way, turns up at a right angle in its own plane, and is divided into four fork-prongs. **1974** *Times* 27 Nov. 17/6 A 'Nelson' knife..has a curved blade..and also teeth at the end of the blade... Nelson knives are a Government issue to those..who have lost an arm but they can also be purchased. **1925** Fraser & Gibbons *Soldier & Sailor Words* 166 *Nelson's Blood*, rum. Old Navy, and probably derived from the old story of the sailors on board the *Victory* tapping the cask in which Nelson's body was brought home and drinking the spirits. **1968** *Telegraph* (Brisbane) 26 June 45 To preserve Nelson's body it was placed in the ship's rum ration—that's why rum is now often referred to as 'Nelson's blood'. **1970** A. Draper *Swansong for Rare Bird* ix. 76 After all the Nelson's blood he'd stashed away the night before, I was surprised to see Gorgeous already there. **1805** Ld. Nelson *Let.* 25 Sept. in C. Oman *Nelson* (1947) xix. 607, I am anxious to join the fleet, for it would add to my grief if any other man was to give them the Nelson touch, which *we* say is warranted never to fail. *Ibid.* 1 Oct. 609 When I came to explain to them the '*Nelson touch*', it was like an electric shock. **1898** H. Newbolt *Island Race* 26 But cared greatly to serve God and the king, And keep the Nelson touch. **1963** *Economist* 12 Jan. 90/1 There was a 'Nelson touch' about the bloodless occupation of Jadotville. **1968** B. Turner *Sex Trap* xiii. 116 A little of the old Nelson touch is a big help in getting on with other departments. **1971** D. Cory *Sunburst* i. 20 Only the British, in Intelligence matters, retain the Nelson touch. *What's Fedora doing in Spain?*.. Up goes the old telescope. *Fedora? I see no Fedora.*

Nelsonian (nelsōᵘ·niǎn), *a.* [f. *Nelson 2 + -ian.] Belonging to the time of Nelson; pertaining to, characteristic of, or relating to Nelson. Hence **Nelsonia·na**, a collection of papers, relics, etc., relating to Nelson.

1909 *Daily Chron.* 24 Feb. 3/5 A volume of Nelsoniana, which realised £145. **1913** *Q. Rev.* Apr. 461 We have only to go back to the eighties to find ships..with their..guns still mounted on the Nelsonian trucks worked by handspikes. **1931** *Times Lit. Suppl.* 17 Sept. 699/4 The Nelsonian mannerisms of words, dress and movements. **1958** *Listener* 2 Oct. 519/1 This valuable addition to Nelsoniana is the hundredth volume published by the Navy Records Society. **1969** *Sunday Times* 2 Nov. 8/7 Every editor knows how to turn a Nelsonian eye on a lack of O and A levels if he has a promising journalist before him. **1971** B. Callison *Plague of Sailors* iv. 134 It was conceivable that Their Lords of the Admiralty might have been prepared to turn a Nelsonian blind eye to H.M. Ships shooting up the odd foreign cargo boat.

Nelsonic (nelsǫ·nik), *a.* [f. *Nelson 2 + -ic.] Pertaining to, relating to, or characteristic of Nelson.

1846 R. Ford *Gatherings from Spain* xvi. 188 They are ill-bred enough, in spite of the Montpensier marriage, and the Nelsonic achievements of Monsieur de Joinville, to consider the words as synonymes. **1909** *Daily Chron.* 13 Sept. 3/3 These Nelsonic qualities in Wolfe do not..come out very clearly in the letters. **1922** *Q. Rev.* Apr. 361 Orders of this nature..are always dangerous in the absence of the Nelsonic spirit. **1972** V. Gielgud *Black Sambo Affair* xxviii. 220 Various persons at the airports were given the hint to turn the Nelsonic blind eye.

nelumbium. In etym. for '(of De Jussieu, 1789)' substitute '(A. L. de Jussieu *Genera Plantarum* (1789) 68)'. Substitute for def.: An aquatic plant of the genus *Nelumbo*, belonging to the family Nymphæaceæ, formerly called *Nelumbium*, and including the East Indian or sacred lotus, *N. nucifera*, which has white or pink flowers, and the American lotus, *N. pentapetala*, which has fragrant yellow flowers. (Earlier and later examples.)

1806 *Curtis's Bot. Mag.* XXIII. 903 The Nelumbium is no longer found in Egypt. **1853** *Harper's Mag.* Nov. 751/2 The broad prairies..are also diversified by lakes, their surfaces shaded from the hot sun by the broad-leafed nelumbium. **1916** L. H. Bailey *Stand. Cycl. Hort.* IV. 2117/1 Nelumbiums are bold plants, suitable for large ponds. **1951** *Dict. Gardening* (R. Hort. Soc.) III. 1359/1 The Nelumbiums need rich soil.

nelumbo (nīlɒ·mbo). [mod.L. (M. Adanson *Familles des Plantes* (1763) I. 73); see Nelumbium.] = Nelumbium (in Dict. and Suppl.).

1794 E. Darwin *Botanic Garden* (ed. 4) ii. 169 With sweet loquacity Nelumbo sails, Shouts to his shores, and parleys with his gales. *Ibid.*, Linneus[*sic*], who has enlisted all our senses into the service of botany, has observed this rattling of the Nelumbo. **1804** J. E. Smith *Exotic Botany* I. 61 A carved horn of a rhinoceros, sent to Linnæus from China,..is now before me... The whole inverted base of the horn is carved into an elegant leaf of *Nelumbo*, rising from the water. **1818** C. Abel *Narr. Journey in China* v. 103 Fields of *Nelumbo* rearing high its glossy leaves and gorgeous flowers..spread at our feet. **1850** S. F. Cooper *Rural Hours* 275 It is chiefly in our western waters that the Nelumbo is found. **1884** tr. *A. de Candolle's Orig. Cultivated Plants* ii. i. 75 The nelumbo of Indian origin has ceased to grow in Egypt. **1916** L. H. Bailey *Stand. Cycl. Hort.* IV. 2117/1 American Lotus, or Nelumbo... A bold and useful plant for colonizing. **1961** F. Perry *Water Gardening* (ed. 2) vii. 75 Nelumbos are reproduced by means of long creeping rootstocks. *Ibid.* 76 [In India] the stranger is presented with fruit and flowers laid in a simple basket fashioned from Nelumbo leaves.

nemasperm (ne·măspɔ̄ːm). *rare*⁻¹. [f. Gr. νῆμα thread + Sperm *sb.*] = Spermatozoon.

1922 Joyce *Ulysses* 411 Must we accept the view of Empedocles of Trinacria that the right ovary..is responsible for the birth of males or are the too long neglected spermatozoa or nemasperms the differentiating factors.

nematic (nemæ·tik), *a.* *Physical Chem.* [ad. F. *nématique* (G. Friedel 1922, in *Ann. de Physique* XVIII. 277), f. Gr. νῆμα, νήματ- thread: see -ic.] Applied to (the state of) a mesophase in which the molecules all have the same orientation but are not arranged in well-defined planes (as they are in a smectic phase). Also as *sb.*, a nematic substance.

1923 *Chem. Abstr.* XVII. 3267 A complete treatise upon the types and properties of matter existent between the true amorphous and cryst. states. The condition formerly designated as liquid crystals or anisotropic liquids is divided into 2 classifications: smectic.., and nematic. **1940** S. Glasstone *Text-bk. Physical Chem.* vii. 505 Nematic liquids are closer to true anisotropic liquids than are the smectic phases. **1971** *Physics Bull.* June 357/2 One of Marconi's exhibits at this year's Physics Exhibition demonstrated the effective use of organic nematic liquid crystals for visual display purposes. **1972** *Sci. Amer.* Mar. 39/1 (Advt.), We hereby announce the commercial availability of a multicomponent nematic that might suffice to call off some chemists. **1974** *Nature* 25 Jan. 178/3 Nuclear magnetic resonance spectra of molecules oriented in thermotropic and lyotropic nematics.

nematicide (ne·măt-, nemæ·tisǝid). [irreg. f. Gr. νῆμα, νήματ- thread + -cide.] = *nematocide* s.v. Nemato-.

1933 [see *nematology]. **1952** *Nature* 8 Mar. 420/2 Work has already been published..on the value of chlorphenol as a nematicide. **1960** *Jrnl. R. Horticultural Soc.* LXXXV. ii. 54 Trial of various nematicides has shown that treatment with methyl bromide gives a yield almost double that of the control plots. **1972** P. Sivapalan in J. M. Webster *Economic Nematology* xiii. 308 In order to improve the trend of the decline in yield in the tea fields infested with *Pratylenchus loosi*, the field is ..treated with suitable nematicides.

Hence **nematici·dal** *a.* = *nematocidal* adj. s.v. *Nemato-.

1950 *Helminthol. Abstr.* XVI. 122 Ethylene dibromide is highly nematicidal since 0·1 c.c. will eradicate nematodes from a gallon of soil. **1973** C. A. Anderson in G. Zweig *Analyt. Methods for Pesticides* VII. x. 253 Dasanit® exhibits both insecticidal and nematicidal activity.

nemato-. Add: **nematoci·dal** *a.*, characteristic of a nematocide; **nematogen**, add: [ad. Fr. *nématogène* (E. van Beneden 1876, in *Bull. Acad. R. Belgique* 2 Sér. XLI. 1195)]; (later examples).

1943 *Phytopathology* XXXIII. 1174 The nematocidal effect of chloropicrin was tested in the field against the root-knot nematode, *Heterodera marioni*. **1973** *Biol. Abstr.* LVI. 6884/2 Nematocidal activity, similar to that of trichothecin, was detected in mycelial extracts and volatile substances of *Arthrobotrys*, a predacious fungus. **1899** *Zoologischer Anzeiger* XXII. 171 The mother of vermiform embryos is called a nematogen, the mother of infusoriform embryos a rhombogen. **1972** *Sci. Amer.* Dec. 98/2 Two different wormlike stages of the Mesozoa have been described: 'nematogens' were reported to produce nematogens and 'rhombogens', whereas rhombogens were reported to produce infusoriform dispersal larvae and occasionally to both produce larvae and revert to the nematogen stage. Observations on laboratory populations of Mesozoa make this distinction untenable.

nematodiriasis (neːmătodirǝi·asis). [f. *Nematodir(us + *-iasis.] A disease of young lambs caused by the larvæ of nematodes of the genus *Nematodirus* and characterized by diarrhœa, loss of weight, and dehydration.

1957 Fraser & Stamp *Sheep Husbandry & Diseases* (ed. 3) x. 384 This disease for which the rather ugly word 'nematodiriasis' might be used is characterised by its sudden and dramatic appearance in low-ground flocks during late May and early June. **1967** *New Scientist* 13 July 63/3 Nematodiriasis, a worm-caused disease of young lambs. **1971** *Farmers Weekly* 19 Mar. 66 (Advt.), Frantin Paste halts the scours associated with nematodiriasis and other worm infestations in unweaned lambs.

nematodirus (neːmătodǝi·rɒs). [mod.L. (B. H. Ransom 1907, in *U.S. Dept. Agric. Bur. Anim. Ind.* No. 116. 4), f. Nemato(de *a.* and *sb.* + Gr. δειρή neck.] A parasitic nematode worm of the genus so called, belonging to the family Trichostrongylidæ, which is found in the intestine of many mammals and causes disease in young lambs. Also *attrib.*

1915 *Parasitology* VIII. 146 The *Nematodirus* larvae are able to withstand complete desiccation. **1934** *Jrnl. Agric. Sci.* XXIV. 207 *Nematodirus* larvae can regain the surface after being turned in by the plough. **1956** *Vet. Rec.* 21 July 471 (*title*) Some observations on Nematodirus Disease in Northumberland and Durham. **1957** Fraser & Stamp *Sheep Husbandry & Diseases* (ed. 3) x. 384 Until recently the worm nematodirus was thought to be relatively harmless to sheep. *Ibid.* 385 Nematodirus worms may be present in large numbers. **1960** *Farmer & Stock-breeder* 9 Feb. 64/3 Border farmers..who have never seen nematodirus and who have long ceased to drench for worms. **1963** *Times* 20 May 16/5 Forecasts could now be made of conditions likely to induce..fluke and nematodirus infections.

nematology (nemătǫ·lŏdʒi). [f. Nemat(ode *a.* and *sb.* + -ology.] The study of nematodes. So **nematolo·gical** *a.*, of or pertaining to this study; **nemato·logist**, a person making such a study.

1926 C. W. Stiles in Yorke & Maplestone *Nematode Parasites of Vertebrates* p. v, A few genera..were of some slight interest in human and veterinary medicine, but this fact played a distinctly secondary *rôle* in nematological studies... Zenker..and others, pointed out the far-reaching medical and economic bearings of nematology. **1933** T. Goodey *Plant Parasitic Nematodes* i. 2 In the United States of America these small forms are often called 'nemas'; a word introduced by the late Dr. N. A. Cobb together with several cognate terms such as nematology, nematologist, nematized, and nematicide, &c. **1960** *Times* 7 July 3/6 Research appointments in..plant nematology. **1961** G. Thorne *Princ. Nematology* p. ix, The writer is indebted to the..U.S. Department of Agriculture, for the opportunity to work on nematological problems in many parts of the United States. **1963** *New Scientist* 5 Sept. 490/1 The comparative effectiveness of four nematicides ..has recently been under investigation by the Nematology Department at Rothamsted Experimental Station. **1971** *Nature* 15 Jan. p. xvii (Advt.), A nematologist is required to work in some aspects of the biology of *Trichodorus* species. **1974** *Ibid.* 19 Apr. 715/3, I commend this book to all thoughtful nematologists, plant pathologists and ecologists.

nembie (ne·mbi). Also **nebbie, nemish, nemmie, nimby.** *U.S. colloq.* contractions of *Nembutal.

1950 [see *goof ball]. **1953** W. Burroughs *Junkie* (1972) ii. 23 This night Herman was knocked out on 'nembies' and his head kept falling down onto the bar. *Ibid.* iv. 40 Next day I was worse and could not get out of bed. So I stayed in bed taking nembies at intervals. *Ibid.* 158 *Nembies, Goof Balls, Yellow Jackets*...Nembutal capsules. **1969** R. R. Lingeman *Drugs from A to Z* 182 Nembutal... Slang names: nebbies, nemmies, nemish, yellow jackets. **1971** E. E. Landy *Underground Dict.* 139 Nembutal... Nemish, nemmie, nimby.

Nembutal (ne·mbiutæl). Also **nembutal.** [f. *Na*, chemical symbol for sodium (f. Na(trium) + initial letters of the elements of 5-*e*thyl-5-(1-*m*ethyl*but*yl)barbiturate, chemical name of the compound + *-al* (as in *Veronal*, *barbital*).] A proprietary name of pentobarbitone sodium, a short-acting barbiturate used as a hypnotic and an anticonvulsant. Also, a capsule of Nembutal.

[**1930** *Anaesthesia & Analgesia* IX. 215/2 Sodium ethyl 1-methyl butyl barbiturate (sodium embutal).] **1930** *Druggists Circular* Dec. 73/2 Nembutal, '844', a new pre-anesthetic sedative and hypnotic,..described as being effective in allaying the apprehension and fear of patients about to undergo an operation. **1931** *Lancet* 10 Jan. 74/1 (*heading*) Nembutal as a basal hypnotic in general anæsthesia. **1931** *Official Gaz.* (U.S. Patent Off.) 21 Apr. 560/1 Nembutal. For hypnotic sedative anesthetic compound. **1936** *Discovery* July 206/2 Another modern method of lessening nervous apprehension is to give the patient a sleeping draught the night before, and then in the morning, sometime before the operation, he is given a capsule of Nembutal or Pronocton. **1937** *Trade Marks Jrnl.* 15 Dec. 1474/1 Nembutal... Medicated preparations for use as hypnotics and sedatives. Abbott Laboratories Limited.., Montreal. **1953** W. Burroughs *Junkie* (1972) xiii. 128, I drank it all with two nembutals and slept several hours. *Ibid.* 158 Nembutal is a barbiturate used by junkies 'to take the edge off' when they can't get junk. **1959** N. Marsh *Singing in Shrouds* ix. 183 I've given her a nembutal. She's asleep in bed. **1962** A. Pirie *Lens Metabolism Rel. Cataract* 429 Rabbit lenses were obtained from the laboratory stock of Dutch rabbits killed by overdose of Nembutal. **1971** J. Wright *Coll. Poems* 274 This was the dream that woke me From nembutal sleep into the pains of grief.

nembutalized (nembiū·tǝlǝizd), *ppl. a.* *Vet.* [f. *Nembutal + -iz(e + -ed¹.] Anæsthetized with Nembutal.

1940 *Amer. Jrnl. Physiol.* CXXIX. 55 Two of the nembutalized dogs included in group II..were adrenalectomized. **1972** *Brain Res.* XXXVI. 350 Presynaptic inhibition in nembutalized cats.

nem. diss. (Further examples.)

1870 *Brewer's Dict. Phr. & Fable* 610/1 *Nem. diss.*, without a dissentient voice. (Latin, *nem'ine dissent'ientĕ.*) **1945** R. Hargreaves *Enemy at Gate* 123 The thanks of both Houses were duly agreed *nem. diss.*

Nemedian (nėmī·diǎn). [f. *Nemed*, name of a legendary invader of Ireland + -ian.] One of the followers or descendants of Nemed of Scythia, who settled in Ireland, and who were later driven out by the Fomorians.

1876 *Encycl. Brit.* V. 299/2 We find four successive colonies mentioned in the following order:—*Nemedians*, *Firbolgs*, *Tuatha Dé Danann*, and *Milesians*. The *Nemedians* are said to have occupied the country during only two hundred years..owing to..their final overthrow, by a people..called Fomorians. **1898** [see *Fomorian]. **1911** *Encycl. Brit.* XIV. 757/2 After undergoing great hardship the Nemedians succeeded in destroying the fortress.. but the Fomorians received reinforcements... Of the Nemedians only thirty warriors escaped. **1950** *Funk's Stand. Dict. Folklore* II. 788/1 *Nemedians*, an early people who invaded Ireland; descendants of Nemed, son of Agnoman of Scythia; the third group to land in Ireland, following next after Partholan and his people. **1971** *It* 2–16 June 24/1 The Partholyans, Nemedians..are all remembered as being of semi-human stock.

nemesia (nĕmī·ʒiă). [mod.L. (E. P. Ventenat *Jardin de la Malmaison* (1803) 41), f. Gr. νεμέσιον a name used by Dioscurides for a similar plant.] An annual or perennial herb of the genus so called, native to southern Africa, belonging to the family Scrophulariaceæ, and usually bearing racemes of flowers of various colours.

1815 F. Pursh *Donn's Hortus Cantabrigiensis* (ed. 8) 196 Nemesia, [English name] Nemesia. Germander-leav. [*sic*]. **1829** J. C. Loudon *Encycl. Plants* 526 Nemesia, Vent[enat. English name] Nemesia...fœtid...horned. **1838** *Edwards's Bot. Reg.* XXIV. 39 (*heading*) Many-flowered Nemesia. **1892** *Gardeners' Chron.* 3 Sept. 269/3 There are many plants that vary to the extent of two or three colours in a wild state, but it is exceedingly rare to find them indulging in such wholesale variation as this Nemesia does. I have seen sixteen varieties of colour. **1927** *Observer* 20 Mar. 24/3 Modern florists and men of science..have made a rainbow out of a single colour (as in the nemesia). **1949** L. G. Green *In Land of Afternoon* v. 72 Hybrid gladioli, nemesias, even the Hottentot fig, were all startling novelties. **1970** C. Lloyd *Well-Tempered Garden* ii. 97 There is a freshness about a display of nemesias.

nemesism (ne·mĭsiz'm). *Psychol.* [f. Nemes(is + -ism.] Feelings of frustration turned inward and expressed by aggression directed against the self. So **nemesi·stic** *a.*, of or connected with nemesism.

1938 S. Rosenzweig in H. A. Murray *Explorations in Personality* vi. 588 The psycho-analyst might appropriately call the turning of aggression upon the individual's own self 'nemesism' from the name of the Greek goddess of vengeance. Nemesism could then be thought of as the counterpart of narcism. **1945** J. C. Flügel *Man, Morals & Society* vii. 78 We propose here to adopt Rosenzweig's suggestion and to use the term 'nemesism' as an alternative and technical term for 'aggression turned against the self'. *Ibid.* 87 The channel for the discharge of the children's aggressiveness towards the natural outer objects being blocked, there may be no alternative for them but to become nemesistic and intropunitive. *Ibid.* xi. 143 This nemesistic urge. **1946** [see *introjection 3 a]. **1968** P. McKellar *Experience & Behav.* ix. 248 Nemesism is a term coined by Flügel.

|| **nemine contradicente** (ne·mini kǫ:ntrăditʃe·nti). See Nem. Con. Also **nemine con.**
1662 J. Davies tr. *A. Olearius's Voyages & Travels of Ambassadors* iii. 77 Where, *nemine contradicente*, it was declar'd. **1718** Vanbrugh *Let.* 7 Aug. in *Athenæum* 30 Aug. (1890) 290/2 Amongst many material things in our conversation it was Nemine Contradicente agreed, That your Grace had writ a most Tyrannical letter. **1762** *Monthly Review* XXV. 363 The very vote that should be passed in our favour, would, in all likelihood, be a new grievance; as we should possibly see the honest English resolution, dated, *Die Mercurii, Feb.* 10mo, and concluded with a *Nemine Contradicente*. **1766** H. Brooke *Fool of Qual.* II. x. 145 They concluded, *nemine con.* to get as speedily as they might from the ministers of darkness. **1962** *Listener* 10 May 798/2 The decision, although not strictly unanimous, can be described as *nemine contradicente*. **1966** *Rep. Comm. Inquiry Univ. Oxf.* I. 316 A special resolution..shall be published..and shall be deemed to have been approved *nemine contradicente*.

nemish, nemmie: see *Nembie.

nemmind, nemmine (ne·məin(d), nem-əi·n(d)), representation of a colloq. pronunciation of *never mind*. Chiefly *U.S.*
1914 *Dialect Notes* IV. 77 *Nemmind!*..Contraction of *never mind!* **1935** R. Bass in A. Dundes *Mother Wit* (1973) 389 Nemmine dat. **1960** N. Hilliard *Maori Girl* iii. ix. 241 'You better not do that. They might put you in jail.' 'Nemmine. I get all the baby things, that's all I'm worried about.'

nemo (nī·mo). *U.S.* [Etym. unknown.] (See quots.)
1937 *Printers Ink Monthly* May 39/3 *Nemo*, any program originating outside of the broadcasting studios. A collection of 'remote'. **1942** Berrey & Van den Bark *Amer. Thes. Slang* §619/2 A program originating outside the studio and transmitted to the studio for broadcasting. .. Nemo. **1950** Chester & Garrison *Radio & Television* 527 *Nemo*, a remote, a program originating away from the studio. **1960** O. Skilbeck *ABC of Film & T.V.* 89 *Nemo* (American), a telecast from location: an outside broadcast. **1961** H. B. Jacobson *Mass Communications Dict.* 222 *Nemo*, a program originating outside of the station's studios.

nenadkevichite (nenæ·dkĕvitʃəit). *Min.* [ad. Russ. *nenadkevichit* (Kuz'menko & Kazakova 1955, in *Dokladȳ Akad. Nauk SSSR* C. 1159), f. the name of K. A. *Nenadkevich* (b. 1880), Russian mineralogist: see -ite[1].] A hydrated silicate of sodium, calcium, niobium, and titanium, $(Na,Ca)(Nb,Ti)Si_2O_7.2H_2O$, which is found as pale yellow orthorhombic crystals.
1955 *Mineral. Abstr.* XII. 569 (*heading*) Nenadkevichite—a new mineral. **1965** *Doklady Acad. Sci. U.S.S.R. Earth Sci. Sect.* CLX. 118/1 Among the natrolite and feldspar are..large (to 10 cm and more) dull flakes of nenadkevichite of a pale to pinkish yellow color. **1972** *Mineral. Abstr.* XXIII. 225/2 Localities are described representing the alkalic complex of Augusta County [Virginia]... Associated with the alkalic rocks are nenadkevichite, astrophyllite, and bastnäsite.

nenadkevite (nenæ·dkĕvəit). *Min.* [ad. Russ. *nenadkevit*, f. as prec.] Any member of a range of isomorphous, basic, hydrated silicates of uranium(IV), uranium(VI), magnesium, calcium, thorium, and lead.
1955 *1st Internat. Conf. Peaceful Uses Atomic Energy: U.S.S.R. Sci. & Techn. Exhibition* (Eng. ed.) 7 Of interest in the second group of silicates containing tetra- and hexavalent uranium is the new mineral species nenadkevite... This mineral is deposited from hydrothermal solutions in the form of veins. **1960** *Mineral. Abstr.* XIV. 401/1 Rarer uranium deposits include those in which the ore mineral is the silicate nenadkevite.

|| **nene** (nē[1]·nē[1]). Also **néné.** [Hawaiian.] = *Hawaiian goose* (*Hawaiian *a.* and *sb.* II).
1902 H. W. Henshaw *Birds of Hawaiian Islands* 103 Upon the island of Hawaii the haunts of the nene..are the uplands. **1915** [see *Hawaiian goose*]. **1945** *Condor* XLVII. 34 Direct influences were brought to bear upon the Nene by white men directly and indirectly through altered environment. **1958** [see *Hawaiian goose*]. **1962** *Punch* 1 Aug. 177/3 Various animals whose survival is in question, with the accent on big game rather than..the Hawaiian *ne-ne*. **1972** *Shooting Times & Country Mag.* 1 July 19/2 The Wildfowl Trust established the principle with the Ne-ne or Hawaiian goose. **1975** *Nat. Geographic* Mar. 411/1 A beautiful, unique species of goose called the nene (nay-nay) was being brought back from near extinction here.

nene, var. *Neinei.

Nenets (ne·nets). Pl. **Nentsi, Nentsy.** [a. Russ. *Nénets*, pl. *Néntsy*.] A Samodian (formerly Samoyedic) people inhabiting the far north-east of Europe and the north of Siberia; a member of this people; their language.
[**1886** *Encycl. Brit.* XXI. 251/1 The names assumed by the Samoyedes themselves are Hazovo and Nyănyăz.] **1944** Gregory & Shave *U.S.S.R.* iv. 162 Occupying territory lying roughly between that of the Palæo-Asiatic groups and the groups of Finnish origin are the peoples commonly known in the past as the Samoyedes, although now more correctly named Nentsi and Ostyak Samoyedes. *Ibid.* viii. 286 The..Nentsi are nomadic tent dwellers. **1954** Pei & Gaynor *Dict. Ling.* 189 Samoyedic..consists of Samoyed, called *Nenets* by the speakers themselves (with various local dialects), Yurak, Kamassin and Tagvy. **1957** *Encycl. Brit.* XX. 598/1 The earliest known inhabitants of Siberia were the Finno-Ugrian Nentsy. **1962** *Amer. Speech* XXXVII. 68 The phonemic systems of five languages of the Samoyed subgroup (Nenets, Nganasan, Enets, Sel'Kup, Kamas). **1967** D. S. Parlett *Short Dict. Lang.* 107 The Northern language [of Samoyedic] has three dialects: Yurak (Nenets) with Taiga and Tundra variants, Yenisei (Enets) and Tavgi (Nganasan). **1968** S. P. Dunn tr. *Diószegi's Pop. Beliefs & Folklore Trad. Siberia* 123 An 'Enetsized' Nenets called Yar. **1968** *Encycl. Brit.* XVI. 210/2 The Nenets people, a Finno-Ugrian group formerly known as the Samoyed, make their living chiefly by reindeer herding. **1972** *Language* XLVIII. 206 Soviet scholars usually distinguish three languages: Nenets ('Yurak Samoyed'), Enets..and Nganasan.

neo-. Add: **1. a.** *neo-American* adj., *-Aramaic* adj., *-Aristotelian* adj. and sb., *-Aristotelianism*, *-Babylonian* adj., *-baroque* adj., *-behaviourism*, *-behaviourist*, *-istic* adjs., *-Bloomfieldian* adj., *-Buddhist* adj. and sb., *-Buddhistic* adj., *-capitalism*, *-capitalist* adj., *-Catholic* (adj.), *-Christian* (adj.), *-Confucian* adj., *-Confucianism*, *-conservatism*, *-conservative* adj. and sb., *-critical* adj., *-criticist*, *-Dada*, *-Dadaism*, *-Darwinian* (sb.), *-Darwinism*, *-Darwinist*, *-Dixieland* adj., *-Edwardian* adj., *-Elizabethan* adj. and sb., *-expressionism*, *-expressionist* adj. and sb., *-Fascism*, *-Fascismo*, *-Fascist* adj. and sb., *-feminist*, *-Firthian* adj. and sb., *-Freudian* adj. and sb., *-Freudianism*, *-Gaullism*, *-Gaullist* adj. and sb., *-Georgian* adj. and sb., *-Georgianism*, *-German*, *-Gothicism*, *-Hegelian* adj and sb., *-Hegelianism*, *-Hittite* adj. and sb., *-Humboldtian* adj. and sb., *-imperial* adj., *-imperialism*, *-imperialist*, *-isolationism*, *-isolationist* adj. and sb., *-Kantian* (adj.), *-Kantism*, *-Keynesian* adj. and sb., *-Keynesianism*, *-Lamarckian* sb., *-Lamarckism*, *-liberal* adj., *-Maoist* adj., *-modal* adj., *-modalism*, *-modality*, *-nationalism*, *-Nazi* adj. and sb., *-Nazism*, *-Nietzschean* adj., *-Norman* adj., *-orthodox* adj., *-orthodoxy*, *-paganism*, *-Palladian* adj., *-populism*, *-positivism*, *-positivist* adj. and sb., *-primitivism*, *-Pythagorean* (sb.), *-realism*, *-realist* adj. and sb., *-realistic* adj., *-revisionist* adj. and sb., *-rococo* adj., *-Romantic* (sb.), *-scholastic* adj. and sb., *-scholasticism*, *-slave*, *-slaver*, *-slavery*, *-Stalinism*, *-Stalinist* adj. and sb., *-symbolist* adj. and sb., *-theosophical* adj., *-Thomism*, *-Thomist* adj. and sb., *-Tory*, *-Toryism*, *-traditionalism*, *-Tudor* adj., *-Victorian* adj. and sb., *-vitalism*, *-vitalist* adj. and sb., *-vitalistic* adj.
1904 *Amer. Naturalist* XXXVIII. 682 The Neo-American nomenclature is adopted—with synonymic citation where the generic name is unfamiliar to the ordinary reader. **1970** I. Reed in A. Chapman *New Black Voices* (1972) 334 So sez d neoamerican hoodoo Church of free spirits. **1948** D. Diringer *Alphabet* 289 The east-Syrian or neo-Aramaic dialect. **1931** *Times Lit. Suppl.* 16 July 561/1 The attempt to apply the neo-Aristotlean [*sic*] rules was..doomed to failure. **1947** *College English* VIII. 403 We should not leave the new critics without notice of the neo-Aristotelians who do not accept fellowship with the new critics, nor with the neo-Thomists. **1953** R. S. Crane *Languages of Criticism* 149 It is not a question of..looking upon ourselves, in any exclusive sense, as forming an 'Aristotelian' or 'Neo-Aristotelian' school. **1962** W. S. Scott *Five Approaches of Lit. Criticism* 183 He also arraigns the 'Neo-Aristotelians' for going outside the poem—not to history, psychology, or morality, but to a theory of the genre. **1952** R. S. Crane *Critics & Criticism* 2 A particular philosophic dogma the nature of which has been defined..as Aristotelianism or neo-Aristotelianism. **1915** L. W. King *Hist. Babylonia* II. 278 Nebuchadnezzar ..established the Neo-Babylonian empire on a firm basis. **1931** *Times Lit. Suppl.* 31 Dec. 1054/1 Reliefs in glazed brick of the neo-Babylonian revival. **1966** *Daily Tel.* 9 Nov. 17/3 'The Rake's Progress'..is either one of the most enigmatic or the most straight forward, according to the view you take of its neo-rococo or neo-baroque idiom. **1970** *Canadian Antiques Collector* June 10/1 His execution of neo-baroque designs to some degree anticipates the Victorian penchant for period revivals. **1967** *Philos. Rev.* LXXVI. 97 The stimulus-response (S-R) theories of neo-behaviorism. **1964** *Listener* 14 May 785/2 Watsonian Behaviourism is..the foundation on which the..neo-Behaviourist systems of Professor Guthrie, Clark Hull, and Skinner were built. **1952** *Mind* LXI. 349 In *The Concept of Mind* Professor Gilbert Ryle has presented us with a neo-behaviouristic theory. **1964** *Language* XL. 265 Weinreich rejects..the hard-boiled neo-Bloomfieldian behaviorism of such philosophers as Carnap and Quine. **1891** *Literary World* 17 Apr. 357/1 The extracts we have given serve to show the dogmatic assertiveness of the Neo-Buddhist philosophy. **1972** *Times of India* 28 Nov. 11/3 A majority of them come from the Scheduled Castes. Many say they are neo-Buddhists. **1896** A. H. Keane *Ethnology* xiii. 346 There is something similar in the Neo-Buddhistic teachings. **1968** *Economist* 20 July 25/3 By neocapitalism they mean the new type of industrial economy dominated by the very large industries. *Ibid.*, An educational system which they look upon as the instrument of 'neocapitalist' oppression. **1842** Mill *Let.* 10 Jan. in *Wks.* (1963) XIII. 497 Our neo-Catholic school at [Oxf]ord. **1956** K. Clark *Nude* viii. 333 The neo-Catholic doctrines of his [*sc.* Rouault's] friend Leon Bloy, by which in the lukewarm, materialistic society of 1900, absolute degradation came closer to redemption than worldly compromise. **1910** Wyndham Lewis *Lett.* (1963) 44 He [*sc.* Masefield] makes Pompey a sort of Tolstoyan or neo-Christian hero. **1948** *Mind* LVII. 535 Dr. Fung..does not, like many Chinese and most Occidental expositors, make Confucius teach medieval Neo-Confucian doctrine. *Ibid.*, Fung continues Chinese traditions by attempting to bring up to date medieval Neo-Confucianism. **1960** *Encounter* Nov. 78/1 If anything can render neo-conservatism intellectually respectable, it is writing of this kind. **1964** K. Winetrout in I. L. Horowitz *New Sociology* 151 They can turn to Russell Kirk and his fellow neo-conservatives. **1971** *Time* 23 Aug. 31 Judaism and Christianity have always placed primacy in man. Now this primacy is being attacked by what I call the neoconservative ecological approach to life. **1960** *Times Lit. Suppl.* 3 June 357/2 Though it may be short on neo-critical knowhow, it will provide valuable ammunition for the literary gamesman. **1925** Neo-criticist [see *Empirio-critical *a.*]. **1962** *Times* 8 Mar. 16/7 The neo-dada work of which Rauschenberg's 'combine-paintings' (i.e. collages) are the only example here. **1966** *Times* 16 July 7 Perhaps a change in the air encouraged..the recent revival of surrealism and neo-Dadaism. **1895** G. J. Romanes *Darwin* II. 10 The Neo-Darwinians strain the teachings of Darwin. **1902** J. M. Baldwin *Development & Evol.* 135 The possible truth of either of the current doctrines of heredity, called Neo-Darwinism and Neo-Lamarckism respectively. **1970** *Observer* (Colour Suppl.) 15 Feb. 18/1 All that neo-Darwinism does at present, the critics say, is to pronounce that 'what survives, survives'. **1895** G. J. Romanes *Darwin* II. 28, I am not a Neo-Darwinist, and so have no desire to make 'natural selection' synonymous with 'natural causation'. **1972** *Listener* 3 Aug. 138/2 You have said one or two things which would be regarded by neo-Darwinists as treasonable. **1955** L. Feather *Encycl. Jazz* vii. 235 The conventional neo-Dixieland jazz of the Eddie Condon variety. **1947** 'N. Blake' *Minute for Murder* iv. 84 The hair brushed up from the back in the neo-Edwardian manner. **1909** *Westm. Gaz.* 7 Aug. 7/1 Mr. Figgis is enlightened enough to be a neo-Elizabethan. **1924** R. Graves *Mock Beggar Hall* 29 Don't cramp my neo-Elizabethan manner. **1958** *Listener* 3 July 23/3 Conditions would exist that would justify a renewal of the nexus between Church and State; there might be a sort of neo-Elizabethan settlement. **1962** *Guardian* 20 Feb. 5/2 An antiquated system more in keeping with Victoria than with neo-Elizabethan England. **1961** Neo-Expressionism [see *Neo-Liberty]. **1955** P. Heron *Changing Forms of Art* 171 The occasional gaunt landscapes by young French neo-expressionists such as Bernard Buffet or Raymond Guerrier. **1968** *N.Y. Times* 20 Jan. 25 Landscape and other subjects are handled by Miss Rosenberger as a series of neoexpressionist clichés. **1946** *Commonweal* (N.Y.) 1 Feb. 398/2 (*title*) Neo-fascism: Italian sample. **1973** *Guardian* 20 June 14 As neofascism under the leadership of Giorgio Almirante makes new gains, suspicions and fears are once again rife in Italy. **1928** *Observer* 29 Jan. 17/1 Herein lies an incident of neo-Fascismo which has no controversial element. **1944** *Birmingham* (Alabama) *News* 28 June 8/5 The Germans had dissolved the Neofascist movement in Northern Italy. **1944** H. McCloy *Panic* 143 She—or rather he—is a neo-Fascist and Nazi sympathizer. **1972** *Guardian* 17 Feb. 2/5 Seven people are now in Milan prison in connection with recent neo-Fascist activities. **1969** *Harper's Mag.* Nov. 28 'Are you still against abortion?' Ellen Willis asks, alluding to a position I had once taken which gave me, for a time, some small notoriety among the rest of the neo-feminists. **1964** R. H. Robins *Gen. Linguistics* p. xix, Several of the principles taught by [J. R.] Firth..are continued in the work of those who are coming to be known as 'neo-Firthians'. **1964** *Language* XL. 305 The term 'neo-Firthian linguistics' may be suggested to characterize the recent work of M. A. K. Halliday, R. M. Dixon, and others. **1948** L. Schneider *Freudian Psychol. & Veblen's Social Theory* v. 166 Some of the work of the neo-Freudians..suggests that the strain in Freud above referred to has not quite been eliminated. *Ibid.* 173 This lacuna in neo-Freudian literature is..quite evident in Fromm's *Escape From Freedom*. **1959** Neo-Freudian [see *Biologistic *a.*]. **1958** W. J. H. Sprott *Human Groups* iv. 64 The 'neo-Freudianism' of Karen Horney. **1958** *Spectator* 6 June 720/2 A neo-Gaullism that has little to do with anything he [*sc.* General de Gaulle] has sought to represent. **1958** *Ibid.* 30 May 677/1 All the neo-Gaullists who have been utilised by M. Soustelle to create the situation in which the General [*sc.* de Gaulle] is invited to act as supreme arbitrator. **1958** *Times* 19 Dec. 8/1 The innovation was proposed by the leader of the neo-Gaullist Union for the New Republic. **1915** *Truth* 24 Mar. 459/2 Essentially early nineteenth century in character, he was hopelessly out of his element in this neo-Georgian era. **1923** Hardy *Coll. Poems* (ed. 2) 525 The launching of a volume of this kind in neo-Georgian days by one who began writing in mid-Victorian, and has published nothing to speak of for some years, may seem to call for a few words of excuse or explanation. **1923** W. Deeping *Secret Sanctuary* xix. 195 The Jinkses with their drink-philosophy, their blatant motor-bike, their stout insensitiveness, were Neo-Georgians. **1933** *Archit. Rev.* LXXIV. 130/2 The Kirkgate and the Briggate..cast off their Georgian glory and assumed the Jacobean,..the Perpendicular and the neo-Georgian, in Leeds phorpres brick and stone and terra-cotta. **1955** J. Betjeman in R. S. Thomas *Song at Year's Turning* 12 The 'wain and stook' pastoral poetry of the neo-Georgians. **1970** *English Studies* LI. 270 The insipidly 'pleasant' strain which.. became a conditioned reflex in the 'neo-Georgian' ripplings of the Squire–Turner–Freeman school. **1940** *Scrutiny* IX. 289 The 'simplicity' of Auden's later manner merges into neo-Georgianism. **1940** *Tablet* 4 May 421/1 This is the feat attempted only by the neo-Germans of the two Great Wars. **1964** *Daily Tel.* 3 Feb. 12 It savours of all the neo-Gothicism of Sir Walter Scott. **1885** W. James *Let.* 13 Aug. (1920) I. 254 Why don't you have a special 'Neo-Hegelian Department' in 'Mind',..

which educated readers skip? **1928** *London Aphrodite* Aug. 6 This analysis is carried on so as to involve anyone at all who admits Time as a philosophic symbol, whether he be an academic neo-Hegelian like Bosanquet, or an intelligent one like Croce. **1917** *Encycl. Relig. & Ethics* IX. 300/2 Neo-Hegelianism is a title which has been given to that current of thought inspired by Hegel and the idealists of Germany which began to make itself felt in British and American philosophy in the third quarter of the 19th century. **1971** *Jrnl. Gen. Psychol.* Apr. 253 Philosophical tendencies have also influenced the practice of psychology in the course of the nineteenth century: in particular neo-Kantianism, neo-Hegelianism, [etc.]. **1925** J. B. Bury et al. *Cambr. Anc. Hist.* III. p. vii, Dr Hogarth ..discusses the successors of the Hittites, and the history and art of the Neo-Hittite states of North Syria. **1952** O. R. Gurney *Hittites* i. 40 The language and the religion of these 'Neo-Hittite' inscriptions are not those of the Hittites of Hattusas. **1964** G. Roux *Anc. Iraq* xvii. 223 We reach, in the extreme north of Syria, the realm of the people called 'Hieroglyphic Hittites' or, more simply, 'Neo-Hittites'. *Ibid.* 224 It is important to emphasize that there was no break in the transmission of Hittite culture in those regions, and that the term 'Neo-Hittite' is no more than a convenient appellation. *Ibid.* xviii. 245 The Neo-Hittite and Aramaean princes whom Ashurnasirpal had caught by surprise had had time to strengthen themselves. **1952** H. Basilius in J. A. Fishman *Readings Sociol. of Lang.* (1968) 447 (*title*) Neo-Humboldtian ethnolinguistics. *Ibid.* 458 The basic point of departure for the Neo-Humboldtians is always Wilhelm von Humboldt's idea that language is simply 'the human being approaching the objective idea'. **1962** *Language* XXXVIII. 319 Karl-Otto Apel provides an excellent illustration of the interconvertibility of Neo-Humboldtian and Existentialist terminology. **1968** *N.Y. Times* 17 Mar. iv. 12 This analysis could embrace not just neoimperial China and neoimperial Russia. **1962** *Listener* 5 Apr. 583/2 It does..make sense to regard this neo-imperialism of the left as an extension of its traditional Little Englanderism. **1971** *Yearbk. World Affairs* XXV. 136 Neo-imperialism is more broadly defined to include the creation and maintenance of an economic sphere of influence by one or more developed countries embracing one or more countries at a very much lower stage of economic development. **1967** *Atlantic Monthly* Jan. 55, I believe that, in fact, we are in danger of seeing the isolationists of the 1920s and 1930s replaced by the neoimperialists, who somehow imagine that the United States has a mandate to impose an American solution the world around. **1952** *Britannica Bk. of Year* 666/2 *Neo-isolationism*, an isolationism in which military preparedness plays a part. **1971** *Atlantic Monthly* Jan. 4 Disillusionment about the Indochina War and preoccupation with domestic ills will bring on neoisolationism in the United States. **1950** *N.Y. Times* 30 Dec. 12/1 The neo-isolationists..propose to retreat to the Western hemisphere, with some outlying bases. **1967** *Listener* 21 Dec. 813/1 People who think we ought not to be in Asia are neo-isolationists. **1971** *Atlantic Monthly* Jan. 4 This month's magazine offers a considerable dose of anti-neoisolationist nourishment: two articles on Egypt and the Middle East situation, reports on Berlin, Canada, and China policy; as well as a lighthearted exercise in the ways of England. **1886** S. H. Hodgson *Let.* 23 Nov. in R. B. Perry *Tht. & Char. W. James* (1935) I. 640 Bradley's *Logic*..belongs to that Neo-Kantian line of thought from which, except as mental gymnastic, I hope nothing. **1958** W. Stark *Sociology of Knowledge* p. x, Heinrich Rickert was a member of the neo-Kantian school. **1904** W. James *Ess. Rad. Emp.* (1912) i. 5 If neo-Kantism has expelled earlier forms of dualism, we shall have expelled all forms if we are able to expel neo-Kantism in its turn. **1959** *Encounter* Jan. 52/1 Neo-Keynesian models. **1968** *Time* 25 Oct. 31 [Senator Hubert H.] Humphrey pledges to continue neo-Keynesian policies that have helped stimulate the nation to 7½ years of unprecedented growth in jobs, wages and production. **1970** *Times* 7 Sept. 19 It was also recognized that monetary policy could be effectively exploited for purposes of stabilization policy, and this position is accepted by Neo-Keynesians and monetarists alike. **1965** *Times Lit. Suppl.* 25 Nov. 1039/3 It was only towards the end of his life that the neo-Keynesianism of his professional advisers prevailed. **1899** J. A. Thomson *Sci. of Life* xvi. 228 The Neo-Lamarckians have added breadth and subtlety to Lamarckism. **1910** *Contemp. Rev.* Jan. 107 This important factor of direct action, which has been brought so much into prominence by the Neo-Lamarckians. **1902** Neo-Lamarckism [see *neo-Darwinism* above]. **1966** *Economist* 14 May p. xii/3 The restrictive view of the common market as a strictly neo-liberal free trade affair would win the day. **1970** *Times* 21 Feb. 7 If a neo-Maoist China is to take shape after the cultural revolution and the ninth congress one can say that it is still very murky and that its institutional life is not in sight yet. **1959** D. Cooke *Lang. Mus.* ii. 76 The austere nature of modal and neo-modal music. **1930** *Music & Lett.* XI. i. 62 Polytonality or atonality on the one hand, and neo-modalism on the other, must set the bounds of our next enquiry. **1958** *Times* 6 June 4/4 The neo-modalism of Vaughan Williams nor the neo-primitivism of Carl Orff. **1947** *Penguin Music Mag.* Sept. 11 They needed a stronger purgative, needed a new instrument—neo-modality, atonality, polytonality, quarter-tonality. **1968** *N.Y. Times* 11 May 13 Signs that 'middle-class radicalism and neo-nationalism' were growing in West Germany. **1950** Neo-Nazi [see *neo-Nazism* below]. **1952** *Time* 9 June 30 Threateningly resurgent neo-Nazis. **1952** *Harper's Mag.* Dec. 33/1 Fissures are appearing in Communist and neo-Nazi parties. **1973** D. Bagley *Tightrope Men* xxv. 172 Who the hell cares what happens to a lot of neo-Nazis? **1974** *Washington Post* 18 Jan. A29/5 The facade of the Euram Building looks almost neo-Nazi. **1950** *Britannica Bk. of Year* 739/2 *Neo-Nazism*, a rightist movement designed to revive former nazi principles and beliefs (1947; neo-Nazi, 1938). **1966** *Daily Tel.* 15 Aug. 17/8 Deep concern about an increase in neo-Nazism and other forms of bigotry was expressed at the end of the International Conference on Jewish–Christian Co-operation. **1974** *Times* 22 Oct. 5/1 National Front members.. have been embarrassed..by Mr Tyndall's vulnerability to the charge of neo-Nazism. **1920** E. Pound *H. S. Mauberley* 25 Mildness, amid the neo-Nietzschean clatter. **1958** *Listener* 19 June 1025/3 The neo-Nietzschean oracles of Colin Wilson. **1960** J. Betjeman *Summoned by Bells* v. 46 St. Andrew's first, with neo-Norman apse. **1958** *New Statesman* 6 Sept. 296/2 Serious neo-orthodox theologians regard the Billy Grahams and Norman Vincent Peales with revulsion. **1952** *Britannica Bk. of Year* 666/2 *Neo-orthodoxy*, certain new interpretations, especially on original sin, the righteousness of God, and the Bible as a source of revealed religious truth to which man must respond in every situation by a decision involving obedience or disobedience. **1880** J. McCarthy *Hist. Own Times* IV. 542 Pre-Raphaelitism..has got mixed up with æstheticism, neo-paganism, and other such fantasies. **1940** *Burlington Mag.* Sept. 79/2 The Neo-palladian architecture in England from the twenties of the eighteenth century onwards. **1972** *Newsweek* 27 Mar. 3 Once dismissed as a racist demagogue, Alabama Gov. George Wallace brings a quirky but potent neopopulism to 1972 Presidential politics—as witnessed by his triumph in the Florida Democratic primary. **1975** *Times Lit. Suppl.* 31 Jan. 116/5 In small, localized communities, the tyranny of social control and of majority pressures can be all the more dangerous, as the neopopulism of Mid-Western small towns and the American South amply illustrates. **1935** *Mind* XLIV. 540 (*title*) Some metaphysical assumptions and problems of neo-positivism. **1963** R. Carnap in P. A. Schilpp *Philos. of R. Carnap* 866 Since Lenin's book against Mach's philosophy, no author in the Soviet Union has dared to discuss neo-positivism with sympathy. **1935** *Mind* XLIV. 540 Neo-positivists who define the class of possible operations by reference to *future* acts of verification can do so only by assigning some metaphysical status to time. **1964** *English Studies* XLV. Suppl. 109 Present-day British philosophers of the analytical or neo-positivist schools have therefore one criterion, namely the contribution to positive knowledge. **1927** *Observer* 24 Apr. 14/4 Post-impressionism, Expressionism, Neo-primitivism. **1971** *Times Lit. Suppl.* 31 Dec. 1622/1 He would have denounced their neo-primitivism, the Rousseauian belief that poverty and roughness are closer to nature than austerity and civilized habits, and therefore more authentic and morally pure. **1891** *Chambers's Encycl.* VII. 436/2 Neopythagoreans may be divided into two groups. **1916** *Mind* XXV. 314 The theory of neo-realism that colours, shapes and sounds must be accepted at their face value whatever the difficulties. **1955** *Times* 4 July 12/3 The earnest talker on films, the *cinéaste*, finds his brother in the balletomane and his cousin in the first-nighter..and his conversation is full of words like montage, neo-realism, and audio-visual correspondence. **1917** A. S. Pringle-Pattison *Idea of God* x. 191 The opening years of the twentieth century have been marked..by a strong attack on the fundamental tenet of Mentalism on the part of thinkers who call themselves Realists or Neo-Realists. **1931** G. F. Stout *Mind & Matter* 171 The Neo-Realist holds that when we see a star the star itself becomes in one aspect of its being an immediate content of our actual sense-experience. **1940** *Mind* XLIX. 122 The 'neo-realist' positions of Russell, Alexander, and Whitehead. **1974** *Times* 11 Dec. 14/4 *Downpour*, a gentle neo-realist study of the difficulties of a young teacher. **1955** *Times* 26 May 13/3 The neo-realistic Italian cinema has made English audiences familiar with the poverty and crime that flourish in the back streets of Italian cities. **1970** *Times* 26 Feb. 6 One of the Maoists' main targets is the Marxist Communist Party, branded by Peking as neo-revisionist to distinguish it from the pro-Moscow Communist Party. **1970** *Guardian Weekly* 18 July 5 The Indian Marxists have often been denounced by Radio Peking as 'neo-revisionists'. **1926** A. Huxley *Two or Three Graces* 76 That sham *dix-huitième* language, those neo-rococo sentiments. **1899** Beerbohm *More* 95 The Neo-Romantics, the dalliers with pretty sentiment, would paint admirable sign-boards. **1947** A. Einstein *Music in Romantic Era* xv. 221 That crude, scribbling materialism with which the French neo-Romantics amuse themselves. **1915** *London Q. Rev.* Jan. 49 The 'Neo-scholastic' movement would have attracted no attention outside Catholic circles, had it not been for the literary activity of the Louvain professors. **1930** *Times Lit. Suppl.* 26 June 524/4 It is true that there are certain neo-Scholastics and certain naturalists who divide into neo-Realists and critical Realists influenced both by William James and Mr. Bertrand Russell. **1915** *London Q. Rev.* Jan. 46 It is strange that a movement with so much verve as 'Neo-scholasticism' has attracted so little attention in Britain. **1934** C. S. Lewis *Let.* 7 June (1966) 157 Beware of people who are at present running what they call 'neo-scholasticism' as a fad. **1970** G. Jackson *Let.* 4 Apr. in *Soledad Brother* (1971) 211 Part of the credo of the neoslave, the latter-day slave,..is to shuffle away from any situation that becomes too difficult. *Ibid.* 214 The neoslaver destroyed the uneconomic plantation, and built upon its ruins a factory. **1965** —— *Let.* 30 Mar. in *Soledad Brother* (1971) 70 Our people react in different ways to this neoslavery. **1969** *Guardian* 30 Sept. 12/2 Those who are allowed to speak to the [Czech] nation speak with the accents of neo-Stalinism. **1974** *Times* 27 May 4/2 The [Yugoslav] climate changed and the attacks began to centre on neo-Stalinism. **1960** *Guardian* 28 July 9/6 The neo-Stalinists..are now prepared to make their stand known in public. *Ibid.* 9/7 This is also the neo-Stalinist view in China. **1973** A. Mann *Tiara* i. 3 Things are beginning to simmer again in Moscow, and the neo-Stalinists or hard-liners..have recovered a good deal of influence. **1930** *Times Educ. Suppl.* 3 May 197/2 The schools or coteries of the last few decades—the symbolists, neo-classicists, neo-symbolists, synthetists..and so forth. **1960** *Guardian* 19 Aug. 4/5 How to combine the neo-symbolist image..with the language of rational discourse. **1907** W. de Morgan *Alice-For-Short* viii. 75 A neo-theosophical reincarnationism without so much as a single Himalayan Brother to back you up! **1964** P. F. Anson *Bishops at Large* ix. 367 He [*sc.* Steiner] founded his own neo-theosophical body. **1915** *London Q. Rev.* Jan. 51 Neo-scholasticism is essentially 'Neo-Thomism'. **1968** P. B. Austin *On being Swedish* xxiii. 168 Neo-Thomism is also making its voice heard. **1928** H. Crane *Let.* 17 Apr. (1965) 323 This method is reversed—as with the neo-Thomists. c**1928** R. Fry *Lett.* (1972) II. 632 He [*sc.* Herbert Read]'s one of this neo-Thomist lot with a whole bag of metaphysical nostrums on his back. **1947** Neo-Thomist [see *neo-Aristotelian* adj. and sb. above]. **1945** 'G. Orwell' *England Your England* (1953) 56 All Neo-Tories are anti-Russian, but sometimes the main emphasis is anti-American. *Ibid.*, The real motive force of Neo-Toryism..is the desire not to recognise that British power and influence have declined. **1962** *Economist* 19 May 674/1 Neo-traditionalism was now so strong that Buganda became its own political 'party'. **1932** Auden *Orators* iii. 103 In the neo-Tudor club-house the captains frown. **1960** J. Betjeman *Summoned by Bells* vi. 57 And not far off the neo-Tudor shops. **1929** Galsworthy *Modern Comedy* p. x, Such subsidiary neo-Victorians as the self-righteous Mr. Danby. **1959** Neo-Victorian [see *Betjeman]. **1902** *Encycl. Brit.* XXXI. 712/2 These efforts..by their unfortunate designations of Vitalism and Neo-vitalism give rise to entirely false conceptions. **1899** J. A. Thomson *Sci. of Life* 9 The rise of a school of 'neovitalists', who have helped to save the science from self-conceit by their emphasis on the partial nature of all physiological analysis. **1920** A. S. Pringle-Pattison *Idea of God* (ed. 2) p. vi, I have added some detailed criticism of recent neo-vitalist statements from which I wish to dissociate myself. **1966** C. G. Hempel *Philos. of Nat. Sci.* vi. 72 All that the neovitalist doctrine enables us to do is to make the *post factum* pronouncement. **1902** *Encycl. Brit.* XXXI. 712/2 All the so-called neo-vitalistic efforts..have nothing to do with the older vitalism.

b. neoarsphe·namine, a bicyclic arsenic compound, $H_2NC_6H_3(OH)As{:}As(OH)C_6H_3\text{-}NHCH_2SO_2Na$, which is a derivative of arsphenamine and is a toxic yellow powder formerly much used in the treatment of syphilis; **neohespe·ridin** [a. G. *neohesperidin* (Kolle & Gloppe 1936, in *Pharmazeut. Zentralhalle* LXXVII. 425)], a bitter compound, $C_{28}H_{34}O_{15}$, which is a glycoside of a flavone and is found in Seville oranges; **neosa·lvarsan** [a. G. *neosalvarsan* (E. Schreiber 1912, in *München. med. Wochenschr.* 23 Apr. 905/1)] = *neoarsphenamine* above; **neosti·gmine** [Physo)stigmine], the aromatic quaternary ammonium ion $(CH_3)_2N{\cdot}CO{\cdot}O{\cdot}C_6H_4N^+(CH_3)_3$ or its bromide or methylsulphate salts, which are white crystalline compounds used in treating myasthenia gravis and other muscular complaints.
1918 *Public Health Rep.* (U.S.) XXXIII. 1003 Previous to the year 1914, all of the arsphenamine (salvarsan) and neo-arsphenamine (neosalvarsan) on the market was manufactured by a single German firm. **1951** A. Grollman *Pharmacol. & Therap.* xxx. 698 Neoarsphenamine is easier to prepare and administer than arsphenamine; causes fewer reactions; and is less irritant to the tissues. **1969** *Radiation Res.* XXXIX. 579 Neoarsphenamine, a sulfhydryl-binding agent, sensitizes hypoxic suspensions of *E. coli*..to irradiation. **1936** *Chem. Abstr.* XXX. 6509 The new glucoside '244', for which the name *neohesperidin* is suggested, is split on hydrolysis in 2 phases in accordance with the following scheme. **1962** *Anal. Biochem.* IV. 110 The isolation and identification of neohesperidin, a 7-rhamnoglucoside of hesperetin..from a commercially produced grapefruit flavonoid preparation containing mostly naringin. **1970** Neohesperidin [see *Naringin]. **1912** *Chem. Abstr.* VI. 2111 Neo-salvarsan is more easily sol. than salvarsan, thus simplifying this part of the salvarsan therapy and preventing the effects produced by NaOH. It is more easily tolerated than salvarsan; intestinal disturbances and collapse are nearly entirely absent. **1932** Schamberg & Wright *Treatment of Syphilis* x. 159 The product known as '914' or 'neo-salvarsan' (neoarsphenamine) possessed the very desirable property of being more freely soluble in water and being neutral in reaction. **1970** Passmore & Robson *Compan. Med. Stud.* II. xx. 2/1 Salvarsan tended to be toxic and was soon replaced by a less toxic compound called neosalvarsan, which was universally employed until it was superseded by penicillin in 1945. **1943** *Proc. Soc. Exper. Biol. & Med.* LIV. 254/1 Oral administration of 0·2 mg/kg of neostigmine bromide to mice reduces mortality in severe anoxia but larger doses are not prophylactic. **1951** A. Grollman *Pharmacol. & Therap.* xiii. 253 Neostigmine, like physostigmine, exerts its action by inactivating cholinesterase. **1973** *Exper. Eye Res.* XV. 35 The increase in blood flow in the ciliary processes may be due to a direct effect on the muscarinic receptor of neostigmine, which is a quaternary compound, but it may also be due to an ability of neostigmine to release acetylcholine from the nerve terminals besides protecting it.

Spec. (after *neoparaffin*: see Dict.), denoting compounds and radicals in which one carbon atom is linked to four others, as *neohexane*, *neohexyl*; **neope·ntane**, an isomer, $C(CH_3)_4$, of pentane which is an easily condensable gas and which is found in small amounts in petroleum; **neope·ntyl**, the radical $(CH_3)_3\text{-}C{\cdot}CH_2$— derived from neopentane.
1876 *Phil. Mag.* I. 207 (*caption*) Neohexane. **1942** V. J. Clancey *Chem. & Aeroplane* iv. 64 Neo-hexane, with an octane value of 94, has been produced for use in blending with normal aviation fuels. **1957** G. I. Brown *Introd. Org. Chem.* vi. 83 Ethylene and iso-butane react.. at 500°C and 250 atmospheres to form neo-hexane. **1968** R. O. C. Norman *Princ. Org. Synth.* viii. 263 The simplest example of an efficient alkylation is the preparation of neohexyl chloride..from t-butyl chloride and ethylene. **1938** *Encycl. Brit. Bk. of Year* 145/1 The highly symmetrical neopentane molecule. **1966** *McGraw-Hill Encycl. Sci. & Tech* IX. 249/1 Two alkanes, neopentane (dimethylpropane) and neohexane (2,2-dimethylbutane) are named unambiguously by using the prefix 'neo'. **1876** *Phil. Mag.* I. 209 The paraffin- or alcohol-monad radicals propyl, butyl, isobutyl, pentyl, isopentyl or amyl, and neopentyl, &c. **1933** *Jrnl. Amer. Chem. Soc.* LV. 3803 Because of the importance of the neopentyl system in rearrangements the preparation and properties of the parent hydrocarbons have been studied. **1969** A. Nica *Theory & Pract. Lubrication Syst.* vi. 168 The neopentyl polyol esters were used in gas turbine engines when the temperatures..prohibited the utilisation of diesters.

c. In anatomical terms designating parts of the brain which are considered to be of relatively recent development phylogenetically, as *Neencephalon, *Neocerebellum, *Neocortex, *Neopallium, *Neostriatum, *Neothalamus. Cf. *Palæo-, paleo- b.

2. ne·oblast *Zool.* [-blast], any of the specialized cells in annelid worms by the division of which a lost portion of the body can be regenerated; **neoge·nesis**, chiefly in scientific use: the formation of something new; the renewed formation of something formed previously; **Neo-Gre·c** [F. *grec* Greek], a modern style of architecture based on classical Greek architecture; **neomorpho·sis** *Biol.*, a type of regeneration (see quot.); **neona·tally** *adv.*, soon after birth; **neopho·bic** *a.*, fearing or disliking what is new; **neota·ntalite** *Min.* [ad. F. *néotantalite* (P. Termier 1902, in *Bull. de la Soc. franç. de Minéral.* XXV. 37)], a mineral found as yellow octahedra in Allier, France, that was formerly thought to be a tantalate and niobate of iron, manganese, and sodium but is now thought to be an impure microlite; **neote·nin** *Ent.* [*neoten*(*ia* s.v. Neo- 2] = *juvenile hormone* s.v. *Juvenile *a.* 4.
1891 H. Randolph in *Zoologischer Anzeiger* XIV. 154 The new mesoderm is formed in great part from specialized cells in the region of the peritonaeal epithelium of the ventral longitudinal muscles, on each side of the ventral nerve cord between it and the ventral row of setae. These cells, which I propose to call neoblasts, are distinguishable from the cells of the peritonaeum by their great size and by the presence of a cell body. **1930** *Jrnl. Linn. Soc.* XXXVII. 186 The cells concerned in the phagocytosis and replacement of tissues are derived from neoblasts in the ventral body-wall. **1972** *Jrnl. Morphol.* CXXXVII. 217/2 The confusion between peritoneal cells, so-called neoblasts, and oocytes in polychaetes has a considerable historical basis. **1903** *Ibis* III. 15 But the opposite hypothesis, that we have in this singular small Owl a case of neogenesis—i.e. the *ex-abrupto* formation of a new type with sufficient differential characters to constitute, if maintained, a new species,—can, I believe, be upheld. The term neogenesis was first used to explain this sudden origin of new forms from old-established species, if I am not mistaken, by my friend and colleague Prof. Paolo Mantegazza, many years ago; it has been since used, more or less in the same sense, by the late Prof. Cope and by others. **1946** *Nature* 10 Aug. 202/1 The situation is essentially different in coma, in the sense that although neogenesis of sugar occurs (high blood sugar during fast) storage of glycogen in the liver is no longer possible. **1959** *Ann. N.Y. Acad. Sci.* LXXXIII. 507 (*heading*) Neogenesis of human hair follicles. **1972** *Gloss. Geol.* (Amer. Geol. Inst.) 477/1 *Neogenesis*, the formation of new minerals, as by diagenesis or metamorphism. **1931** C. H. Reilly in W. Rose *Outl. Mod. Knowl.* 989 The next stage, generally called the Neo Grec, was probably due to the indirect influence of the Gothic revivalists. **1939** *Archit. Rev.* LXXXV. 52/3 No better background could be imagined for the parade ground manoeuvrings of red-coated guards than these stern, unadorned neo-grec façades. **1901** T. H. Morgan *Regeneration* i. 24 In one case [of heteromorphosis] the new part is not only different from the part removed, but is also an organ that belongs to a different part of the body (or it may be unlike any organ of the body). This we may call 'neomorphosis'. **1945** *Ann. Amer. Acad. Political & Social Sci.* CCXXXVII. 140 Puerperal fatality increases sharply when the infant is either stillborn or dies neonatally. **1974** *Nature* 7 June 564/2 No significant increase in tumour incidence has been observed in either neonatally thymectomised..mice or congenitally athymic..mice. **1925** *Glasgow Herald* 20 June 4 The greeting extended to the steamship was quite as neophobic. **1971** W. Hanley *Blue Dreams* iii. 22 Her unfailingly neophobic response to those attempts. **1903** *Mineral. Mag.* XIII. 374 Neotantalite... Minute, regular octahedra resembling pyrochlore, in kaolin from dép. Allier, France. **1932** *Mineral. Abstr.* V. 185 The several minerals (..microlite, neotantalite, atopite, [etc.]..) of this group [*sc.* pyrochlore-romeite]..crystallize as small octahedra and have the same type of crystal-structure. **1973** *Ibid.* XXIV. 282/1 'Neotantalite' is a microlite $A_2B_2O_6(OH,F)$ with large deficiences in the *A* ions which may be Ba, Pb, U, or Ca. Microprobe analyses show that the Fe and Mn included in the first description..are impurities. **1954** V. B. Wigglesworth *Physiol. Insect Metamorphosis* iv. 56 The 'inhibitory hormone'..has.. been called for preference the 'juvenile hormone'. If a Greek term is preferred it might be called 'neotenin'. **1966** tr. *V. J. A. Novák's Insect Hormones* 79 (*heading*) The juvenile hormone (neotenin), JH (= inhibitory hormone, Wigglesworth, 1935; = status quo hormone, Williams, 1952; = das Larvalhormon, Weber, 1954). **1968** *New Scientist* 16 May 354/1 In larval insects, moulting and metamorphosis are controlled by two hormones produced by epithelial endocrine glands:..neotenin from the corpora allata controls the change towards the adult form.

neoanthropic, var. *Neanthropic *a.*

neocerebellum (nīˌoserī·be·ləm). *Anat.* [mod.L., f. *Neo- 1 e + Cerebellum.] The phylogenetically youngest portion of the cerebellum, comprising mainly its lateral lobes ('hemispheres').
1925 W. H. F. Addison tr. *Villiger's Brain & Spinal Cord* (ed. 3) 80 The hemispheres constitute the neocerebellum, and the vermis and flocculus constitute the palaeocerebellum. **1954** Jansen & Brodal *Aspects Cerebellar Anat.* vi. 384 We may appropriately close this discussion by commenting briefly on the much debated concepts paleocerebellum, comprising the vermis and the flocculus, and neocerebellum, comprising the cerebellar hemispheres... From the point of view of gross morphology these concepts may appear well founded. On the basis of fiber connections, however, a paleocerebellum and a neocerebellum *sensu strictiori* are not distinguishable. **1974** *Encycl. Brit. Macropædia* XII. 990/2 This new part of the cerebellum, or neocerebellum, coordinates skilled movements initiated at cortical levels.

So **neocerebe·llar** *a.*, of or pertaining to the neocerebellum.
1914 *Jrnl. Nerv. & Mental Dis.* XLI. 540 (*heading*) Neocerebellar hemiatrophy. **1948** A. Brodal *Neurol. Anat.* v. 126 The neo-cerebellar syndrome. Here homolateral hypotonia and atactic movements, asynergic and clumsy, appear, and when the dentate nucleus is involved tremor also develops. **1958** *Jrnl. Neurophysiol.* XXI. 3 Stimulation of some neocerebellar areas, however, resulted in equilibratory changes.

neo-cla·ssic, -cla·ssical, *a.* [Neo-.] Of, pertaining to, or characteristic of a style of art, architecture, music, literature, etc., that is based on or influenced by classical style or by a style that has become established as 'classical'; *spec.* of such a style in 18th-century literature, late-18th-century art and architecture, or 20th-century music.
1877 *Contemp. Rev.* Feb. 360 The imagination of the men of Spenser's time was affected by his use of the neo-classical mythology of the Renaissance. **1881, 1882** [see Neo- 1 a]. **1923** C. Gray *Contingencies* (1947) ii. 65 We find throughout almost all the work of Brahms..the perpetual striving after the ideal of a grandiose, neo-classic art. **1927** *P.M.L.A.* Mar. 237 This linguistic attitude, expressing itself in grammars of arbitrary rules, seems to be but another manifestation and *survival* of that tendency called the neo-classic creed of literary criticism. **1933** *Archit. Rev.* LXXIV. 79/2 Many is the ill-built block of London flats whose internal planning has been sacrificed for some ponderous neo-classical façade. **1934** C. Lambert *Music Ho!* ii. 101 Stravinsky's neo-classical period,..apart from the adoption of eighteenth-century forms and titles, is chiefly noticeable for its attempt to create melody by synthetic means. **1944** *Burlington Mag.* Apr. 97/1 A fully developed neo-classic sculptor. **1964** J. Summerson *Classical Lang. Archit.* v. 37/1 The Panthéon is the first major building which can be called neo-classical—'neo-classical' being the expression which has come to be used for architecture which..tends towards the rational simplification advocated by Cordemoy and Laugier and..seeks to present the orders with the utmost antiquarian fidelity. **1966** *English Studies* XLVII. 150 In the period up to the Civil War the neo-classic Happy Husbandman becomes the neo-Stoic Contemplator of the world. **1974** *Times Lit. Suppl.* 15 Mar. 259/4 Throughout the book—as for example in the emphasis on jobs rather than on the ownership of capital as a key determinant in occupational mobility—the authors assume the validity of neo-classical economics as taught in the United States.

Hence **neo-cla·ssicism**, neo-classic style or principles; **neo-cla·ssicist**, a follower or exponent of neo-classic style or principles.
1893 Neo-classicism [see Neo- 1 a]. **1930** *Times Educ. Suppl.* 3 May 197/2 The schools or coteries of the last few decades—the symbolists, neo-classicists, neo-symbolists, [etc.]. **1934** C. Lambert *Music Ho!* ii. 73 *Pulcinella*.. marks the beginning of the movement sometimes dignified with the name of neo-classicism. *Ibid.* iv. 245 It may seem contradictory to condemn composers like Honegger for basing their work on the contemporary scene after complaining that the neo-classicists are so out of touch with contemporary life. **1943** *Philological Q.* XXII. 143 The fairy tales about 'neo-classicism' and 'romanticism' in the eighteenth century which have so long been allowed to come between us and the direct appreciation of eighteenth-century texts. **1944** *Burlington Mag.* Apr. 97/2 He..gives no hint of any sympathy with the principles of the Neo-Classicists. **1947** *Penguin Music Mag.* Sept. 24 Both men [*sc.* Schumann and Mendelssohn] turned to a sort of neo-classicism before they had done. **1965** *Times Lit. Suppl.* 25 Nov. 1063/2 The 'lower order' was one which allowed Gay his particularly delicate critical assimilation of the vulgar writers and the 'trivial' moderns, while himself remaining, in theory at least, in the camp of the neo-classicists. **1972** *Listener* 21 Sept. 361/3 What was then called the 'True Style' which we now tend to call Neo-Classicism.

ne:o-colo·nialism. [f. Neo- + Colonialism.] The acquisition or retention of influence over other countries, esp. one's former colonies, often by economic or political measures. So **ne:o-colo·nial** *a.* and *sb.*; **ne:o-colo·nialist** *a.* and *sb.*; **ne:o-colo:nializa·tion**; **ne:o-colo·nialized, ne:o-co·lonized** *adjs.*
1961 *New Statesman* 20 Jan. 82/1 The most dangerous type of colonialism is neo-colonialism. **1961** *New Left Rev.* July–Aug. 12/1 One of the effects of EEC is therefore to deepen the split between independent and neo-colonial Africa. **1961** *Economist* 16 Dec. 1113/2 Casuistry worthy of the most devious neo-colonialist. **1963** *Ann. Reg.* 1962 98 Accepting the Indonesian Communist Party's definition of 'Malaysia' as a 'neo-colonialist' device. **1964** K. Nkrumah *Consciencism* v. 111 Just as a liberated territory can be produced by the application of D (na > pa), so a neo-colonized territory can be produced by the application of D (pa > na). **1965** —— *Neo-colonialism: the Last Stage of Imperialism* p. ix, In place of colonialism as the main instrument of imperialism we have today neo-colonialism. The essence of neo-colonialism is that the State which is subject to it is, in theory, independent and has all the outward trappings of international sovereignty. In reality its economic system and thus its political policy is directed from outside. **1966** *Economist* 26 Mar. 1226/1 Libya is not nearly so neo-colonialised as a country like Algeria with its government-to-government oil and development agreements with France. **1967** *N.Y. Times* 5 July 12 The 'neocolonials' (the Belgians) and the 'imperialists' (the Americans). **1969** *Listener* 3 July 12/2 You'll always have people moving from a stage of neo-colonialisation back to capitalism. **1971** *Black Scholar* Apr.–May 9 The black people in the United States..are de facto neocolonial subjects. **1972** *Sci. Amer.* Apr. 19/3 Strong elements of neocolonialism persist in the economic relations of the rich and poor countries. **1973** J. Rex *Discovering Sociology* ix. 109 The inhabitants of the former British and French empires.. addressed themselves to the twin tasks of dealing with their own poverty and fighting neo-colonialism. **1973** *Times* 1 Jan. 14/4 One of the black speakers..mentioned Europe in passing as a 'massive neo-colonialist conspiracy to castrate black men everywhere'.

neocortex (nīˌokǫ·ɹteks). *Anat.* Also (with hyphen) **neo-cortex.** [mod.L., f. *Neo- 1 e + Cortex.] The phylogenetically youngest portion of the cerebral cortex, which is co-extensive with the neopallium.
1909 C. U. A. Kappers in *Arch. Neurol. & Psychiatry* IV. 162 Just as in the pallium there can be distinguished three territories according to the connections which they exhibit, so the cortical structures occurring in them should be distinguished, according to the same principle, into a palæo-cortex, archi-cortex and neo-cortex. **1947** H. C. Elliott *Textbk. Nerv. Syst.* vii. 87/1 This first, or paleo-cortex..is soon overshadowed by the development of a general, or neocortex. **1948** [see *Neopallium]. **1964** J. Z. Young *Model of Brain* xiv. 234 Although it is notoriously dangerous to try to speak of the locality of 'engrams' in a mammal, there can be little doubt that they reside largely in the neocortex. **1972** T. W. Jenkins *Functional Mammalian Neuroanat.* xvi. 250/2 The area of the canine neocortex is 84·2 per cent of the entire hemispheric area. **1972** *Science* 19 May 804/1 We have used the development of the neocortex..as an index of generation length in extinct taxa.

Hence **neoco·rtical** *a.*, of or pertaining to the neocortex.
1909 *Arch. Neurol. & Psychiatry* IV. 163 The upper part of the lateral cortex layer has certainly already a neo-cortical character. **1971** *Nature* 11 June 397/1 All rats were killed 4 h later and the RNA was extracted from hippocarpal, thalamic and medial neocortical tissue. **1971** *Daily Colonist* (Victoria, B.C.) 7 Oct. 5/1 There are other species like ourselves who through evolutionary quirk if not neocortical ascendancy lack biological commands to insure reasonable populations.

neodymium (nīˌodi·miəm). *Chem.* [mod.L., f. G. *neodym* (C. A. von Welsbach 1885, in *Monatshefte f. Chem.* VI. 490), f. *neo-* Neo- + *di*)*dym* Didymium: see -ium.] A metallic element that is a typical member of the lanthanide series, forms red compounds in which it has a valency of three (some of the salts being used for colouring glass and for glazes), and can also have a valency of two or four. Atomic number 60; symbol Nd.
1885 *Jrnl. Chem. Soc.* XLVIII. ii. 1113 By repeated crystallisation of a mixture of the double nitrates of lanthanum and didymium with ammonium, the lanthanum salt was obtained pure, whilst the didymium salt separated into the salts of two new elements, neodymium and praseodymium. **1923** *Glasgow Herald* 16 Feb. 11 Number 61 is a rare-earth metal in the midst of that troop of strange elements with stranger names found chiefly in Scandinavian minerals; its neighbours are neodymium and samarium. **1971** *Sci. Amer.* June 20/3 At present solid-state ruby and neodymium-glass laser materials are used to obtain the highest peak-power output in pulsed operation. **1974** *Encycl. Brit. Micropædia* VII. 253/2 Of the rare earths, only cerium and yttrium are more plentiful than neodymium.

neoglacial (nīˌoglē[i]·siăl, -glē[i]·ʃăl), *a.* Also **Neo-.** [f. Neo- + Glacial *a.*] Of or pertaining to a neoglaciation; also *absol.*, a neoglacial period.
1960 *Amer. Jrnl. Sci.* CCLVIII. 325 Some Neoglacial ice bodies are judged to have attained a length of nearly two miles. **1970** *Sci. Amer.* June 103/3 In 1794, George Vancouver observed the position of ice in Glacier Bay. At that time glaciers in southern Alaska had begun to retreat from a maximum neoglacial position attained earlier in the 18th century. **1974** *Nature* 20/27 Dec. 680/1 During the Neoglacial, the three major advance phases of alpine and polar glaciers around 4,590–5,260, 2,100–2,940 and 40–460 yr b.p. were exactly contemporaneous with the three major post-Wisconsin volcanic phases.. in New Zealand, Japan and southern South America.

neoglaciation (nī:oglē[i]sië[i]·ʃən). Also **Neo-.** [f. Neo- + Glaciation.] A minor, short-lived increase in glaciation following the major glacial retreat at the end of the Ice Age.
Though freq. attributed to F. E. Matthes, the term appears not to have been used by him in print.
1951 J. H. Moss *Early Man in Eden Valley* v. 62 In the Sierra Nevada, Matthes (1939) has described a similar set of small, very fresh moraines fronting the existing glaciers... They contrast sharply with the older moraines slightly farther down the valley. Matthes suggested the name Neo-glaciation for the minor glacial pulsation represented by these moraines, and pointed out that it probably postdates the so-called Climatic Optimum. **1954** *Jrnl. Geol.* LIV. 340/1 Most of the moraines described in the literature which have been correlated with the Neoglaciation are close to, or in contact with, existing ice masses and are devoid of all vegetation. **1970** *Sci. Amer.* June 102 Historical records of the latest glacier fluctuations during neoglaciation are available from many alpine regions.

ne:o-gramma·rian, ne:ogramma·rian. [f. Neo- + Grammarian.] A member of the *Junggrammatiker. Also as *adj.*
1885 *Encycl. Brit.* XVIII. 782/2 This younger school (often branded with the name of Neo-Grammarians, 'Junggrammatiker', by its opponents real and imaginary) is marked by certain distinct tendencies. **1933** L. Bloomfield *Lang.* xx. 354 The neo-grammarian insists..that his hypothesis..sorts out the resemblances that are due to factors other than phonetic change. *Ibid.* 355 The opponents of the neo-grammarian hypothesis claim that resemblances which do not fit into recognized types of phonetic correspondence may be due merely to sporadic occurrence or deviation or non-occurrence of sound-change. **1947** E. Sturtevant *Introd. Linguistic Sci.* vii. 70 In the 1870's a number of scholars announced..that phonetic laws have no exceptions. The earliest declaration..seems to have been made by August Leskien in 1876, but the discovery really belonged to a group, who, from that time to this, have been called the neo-grammarians (*Junggrammatiker*). **1965** *Language* XLI. 188 Mention of the neogrammarians..can elicit so much emotional noise that no one can hear what you are saying. **1972** *Ibid.* XLVIII. 437 The subsequent section illustrates ..the Neogrammarian position. *Ibid.*, Arens refers to Saussure's association with the Neogrammarians.

ne:o-impre·ssionism. Freq. with capital initials. [ad. F. *néo-impressionnisme*, f. Neo- + Impressionism.] A movement or style in art, originated by the French painter Georges Seurat (1859–91), and characterized by a systematic use of divisionism. Cf. *Pointillism. So **ne:o-impre·ssionist** *a.* and *sb.*
[**1886** F. Fénéon in *L'Art Moderne* 19 Sept. 302/1 La vérité est que la méthode néo-impressionniste exige une exceptionnelle délicatesse d'œil.] **1892** *Mag. of Art* p. xxxv/1 Though neo-impressionism has, indeed, asserted itself in the exhibitions of the Twenty, symbolism and realism also hold their own. *Ibid.*, M. Camille Pissaro, a neo-impressionist in his pictures, betrays Japanese influence in his woodcuts. **1901** *Sat. Rev.* 23 Feb. 240/1 Impressionism in France had..passed through the phases of luminism, vibration, pointillisme, independence and neo-impressionism, all comparatively short-lived extreme phases. **1903** *Studio* XXIX. 112/1 Coteries of artists.. have..tried to 'go one better', the most formidable and temporarily successful being that of the 'Néo-Impressionists'. **1908** R. Fry *Let.* Mar. (1972) I. 299 These neo-Impressionists follow straight upon the heels of the true Impressionists. **1914** A. J. Eddy *Cubists & Post-Impressionism* (1915) 27 Neo-Impressionism was the logical outcome of Impressionism. It was simply the attempt to paint light in still more scientific fashion, by the use of the primary colors laid on in fine points in such a manner that at the proper distance the points fuse and produce the tone desired. **1938** *Burlington Mag.* June 289/2 A confirmed disciple of the neo-impressionists. **1944** *Ibid.* Apr. 104/1 A history of the whole Neo-Impressionist movement. **1968** *Michelin Guide N.Y. City* 37 Adepts of Neo-Impressionism, like Seurat. **1972** *Country Life* 23 Nov. 1369/2 These artists' [*sc.* Futurists'] neo-Impressionist brushwork and Expressionist colour is invariably awful. **1975** *Physics Bull.* Feb. 59/2 Our start is the beginning of the 20th century, taking for granted the legacy of Leonardo Da Vinci, photography and the effects of the theories of colour on the neoimpressionists.

ne:o-La·tin. Also **Neo-Latin.** [f. Neo- + Latin *sb.*] **a.** = Romance *sb.* 1. **b.** Latin in use since the end of the Renaissance. Also *attrib.* or as *adj.* Hence **ne:o-La·tinist**, a writer of neo-Latin.
1850, 1880–1 [see Neo- 1 a]. **1946** H. Jacob *On Choice of Common Lang.* 16 Idiom Neutral is considered a neo-Latin rather than an autonomic system. **1951** [see *Copulative *a.* 1]. **1964** *Archivum Linguisticum* XVI. 4 The author's uniquely panoramic view of the neo-Latin languages. **1965** J. Lawlor in J. Gibb *Light on C. S. Lewis* 79 Lewis consistently turned his neo-Latin authors into sixteenth-century English. **1966** *English Studies* XLVII. 150 Miss Røstvig stresses the importance of the hitherto neglected neo-Latinist Casimir Sarbiewski. **1970** B. M. H. Strang *Hist. English* 130 These learned written sources, usually referred to as *Neo-Latin*, are not confined to the donation of whole words. *Ibid.*, The following suffixes come from Neo-Latin.

Neo-Li·berty. Also **Neoliberty.** [f. Neo- + Liberty *sb.*] A movement or style in architecture originating in Italy, a revival of *art nouveau*.
1959 *Archit. Rev.* CXXV. 232/2 Paolo Portoghesi seems to have been the first to call the style of the Retreat by the apt term 'Neoliberty' as late as the end of 1958. **1961** *Listener* 16 Feb. 300/1 There is Neo-Liberty in Italy, there is quite a lot of Neo-Gaudí, and there is Neo-Expressionism.

neo-li·nguist. Also **neolinguist.** [f. Neo- + Linguist.] A member of a school of linguistics which arose in opposition to the neo-grammarians, rejecting the claim that phonetic laws have no exception, and maintaining that linguistic change results from individual innovation. So **neo-lingui·stic** *a.*; **neo-lingui·stics** *sb. pl.*
[**1925** M. Bartoli (*title*) Introduzione alla neolinguistica.] **1937** J. Orr tr. *Iordan's Introd. Romance Linguistics* i. 29 Here..is..the source of the misunderstanding between the neo-linguists..and the Italian neo-grammarians of to-day, with regard to their attitude towards Ascoli. *Ibid.* iii. 273 The so-called Neo-linguistic School, whose tenets..are little more than a somewhat hard-and-fast formulation of certain of Gilliéron's ideas. **1944** *Jrnl. Amer. Oriental Soc.* LXIV. 177 The *areal* theory of linguistics..has been developed and brought into a system by the Italian neolinguistic school. **1946** *Language* XXII. 273 Bartoli's 'Neo-Linguistics' has both a negative and a positive side. **1953** J. H. Greenberg in A. L. Kroeber *Anthropology Today* 265/1 The reconstructions of the neo-linguistic school are not generally accepted by other scholars. **1972** *Language* XLVIII. 439 The Neo-Humboldtians in Germany and the Neolinguists in Italy.

neolocal (nīˌolōᵘ·kăl), *a.* *Anthrop.* [f. Neo- + Local *a.*] Denoting a place of residence chosen by a newly-married couple which is independent of parental or family ties. Hence **neolo·cally** *adv.*
1949 G. P. Murdock *Soc. Structure* i. 16 When a newly wedded couple, as in our own society, establishes a domicile independent of the location of the parental home of either partner..residence may be called neolocal. **1958** F. M. Keesing *Cultural Anthrop.* x. 264 The married couple may.., as in the modern urban society, set up a home apart from the parents of both: neolocal (new-place). **1967** J. Deetz *Invitation to Archaeology* 95 We Americans reside *neolocally*, apart from both parents. **1973** *Times Lit. Suppl.* 6 July 774/5 Many young married couples must live 'neolocally': that is, away from the extended family households characteristic of other kinds of area.

neologism. Add: **1. c.** *Psychol.* An invented or concocted word or word-sound without recognizable meaning, freq. interpolated in otherwise correct sentences, and used by persons in a variety of neuropsychiatric disorders.
1905 A. J. Rosanoff tr. *Rogues de Fursac's Man. Psychiatry* ii. 46 Neologisms the meaning of which may remain absolutely enigmatical to the patient himself. *Ibid.* viii. 200 Neologisms are frequent in the period of dementia. **1906** J. H. Macdonald tr. *Bianchi's Textbk. Psychiatry* iii. xiii. 680 In the typical form [of mania].. neologisms and symbols are found in great number. **1932** Cannon & Hayes *Princ. & Pract. of Psychiatry* 378 The verbal repetition of these 'new' words—neologisms or senseless words invented by himself. **1960** R. F. C. Hull tr. *Jung's Coll. Wks.* III. i. 25 Word-formations, which are so bizarre that they immediately bring to mind the neologisms of dementia praecox. **1969** W. Mayer-Gross et al. *Clin. Psychiatry* (ed. 3) v. 286 Other patients refer the origin of neologisms to hallucinatory experiences.

neologistic, *a.* (Further examples.)
1935 *Mind* XLIV. 524 Philosophers who have absorbed a glut of new systems of logistic during the last decade.. must face the invention of yet another neologistic language with something like dismay. **1936** *Theology* XXXII. 73 The Catholic is quite sure that there is a God, and that in no neologistic sense.

ne:o-Malthu·sian, *a.* and *sb.* Also **Neo-Malthusian.** [f. Neo- + Malthusian *a.* and *sb.*] **A.** *adj.* Of or pertaining to the belief that the size of population should be controlled, *spec.* by the use of contraceptives. **B.** *sb.* An advocate of birth control or the limitation of population. So **ne:o-Malthu·sianism.**
1885 Neo-Malthusianism [see Neo- 1 a]. **1901** J. A. Godfrey *Sci. Sex* ii. vi. 253 The main points to be proved by the Neo-malthusian are, therefore, that it is advisable for society to have some means of checking the increase of population in advance, and that the means now at hand are both harmless and effective. **1910** G. B. Shaw *Brieux: a Preface* 31 Just about forty years ago the propaganda of Neo-Malthusianism changed the bearing of children from an involuntary condition of marriage to a voluntary one. *Ibid.*, The expectation of the Neo-Malthusians that the regulation of births in our families would give the fewer children born a better chance of survival..has no doubt been fulfilled in some cases. **1911** Havelock Ellis *Stud. Psychol. Sex* VI. xii. 594 James Mill was the pioneer in advocating Neo-Malthusian methods. **1934** A. Huxley *Beyond Mexique Bay* 255 You cannot teach primitive Indians to practise the Neo-Malthusian techniques and expect them to remain primitive Indians. **1934** H. G. Wells *Experiment in Autobiogr.* II. vii. 436 The spreading knowledge of birth-control,—Neo-Malthusianism was our name for it in those days—seemed to justify my contention that love was now to be taken more lightly than it had been in the past. **1962** *Punch* 3 Jan. 52/1 Some neo-Malthusians have been heard to suggest that the bomb is Nature's Way..of checking..the..over-spawning of our species. **1967** *Listener* 20 July 94/2 Mr Eversley..dispelled any doubt in the minds of those listeners who felt guilty about having '*two* children and a motor-car'—apparently anti-social luxuries in the neo-Malthusian wave.

Ne:o-Melane·sian. [f. Neo- + Melanesian *sb.*] (See quots.)
1961 Webster, *Neo-melanesian*, an English-based pidgin language used in New Guinea and the Solomon islands. **1962** *Listener* 20 Sept. 418/2 This word 'neo-Melanesian' is an indication of the changed status that pidgin English has in south-east Asia. The name is given to the pidgin that is spoken particularly in places like New Guinea, and a translation of the Bible has appeared in neo-Melanesian; grammars of neo-Melanesian and a dictionary of neo-Melanesian have appeared. **1972** W. B. Lockwood *Panorama Indo-European Lang.* 120 An important language used in the South Seas, technically known as Neo-Melanesian, but more popularly as Beach-la-Mar... Beach-la-Mar came to denote the Pidgin English spoken between East Australian and Melanesian tripang fishermen... Neo-Melanesian is..of obvious political importance. **1973** *Sunday Times* (Colour Suppl.) 10 June 42/2 Pidgin is taught as a subject at Brisbane University and is referred to by some American scholars as 'Neo-Melanesian'.

neomorph (nī·omǫɹf). [f. Neo- + *-morph.]
1. *Biol.* An anatomical structure or feature that is of recent origin phylogenetically.
1886 W. N. Parker tr. *Wiedersheim's Elem. Compar. Anat. Vertebrates* 33 It is uncertain whether the dermal skeleton present in Armadillos..is to be derived directly from that of Reptiles, or whether it is to be considered as formed independently, that is, as a new acquisition or 'neomorph' (Gadow).
2. *Genetics.* A mutant allele which effects a different character from that effected by the wild-type allele.
1932 H. J. Muller in *Proc. Sixth Internat. Congress Genetics* I. 245 Somewhat different from the negatively acting, competing mutant genes, or antimorphs, is the class which I am provisionally terming 'neomorphs'. A good example is the dominant mutant, Hairy wing, near the left end of the X chromosome. **1946** *Nature* 12 Oct. 520/1 Among spontaneous mutations which are known to occur at this locus some would be expected to be neomorphs. **1967** *Evolution* XXI. 850 (*heading*) Latent neomorphs and the evolution of dominance.
Hence **neomo·rphic** *a.*, of, pertaining to, or being a neomorph (in either sense).
1903 *Rep. Brit. Assoc. Advancement of Sci.* 1902 631 Madagascar has yielded a Physa (*P. lamellata*) with a neomorphic gill, a character shared by species of Planorbis (*P. corneus* and *P. marginatus*). **1922** W. Garstang in *Jrnl. Linnean Soc.: Zool.* XXXV. 99, I propose in future to use *palæogenetic* and *neogenetic* when referring to ontogenetic processes, and *palæomorphic* and *neomorphic* when contrasting primitive and modified types of structure. **1932** H. J. Muller in *Proc. Sixth Internat. Congress Genetics* I. 245 (*heading*) Neomorphic mutations. **1940** G. R. de Beer *Embryos & Ancestors* xiv. 89 A palaeomorphic character may of course..have made its first appearance in early stages of ontogeny. Conversely, characters of more recent origin, to which Garstang applies the term neomorphic, may have originated in terminal ontogenetic or adult stages. **1966** E. A. Carlson *Gene* xiii. 113 In this neomorphic class, Muller placed Bar eyes.

neomycin (nīˌomai·sin). *Pharm.* [f. Neo- + *-mycin.] An antibiotic that is a mixture of two stereoisomers produced by a selected strain of *Streptomyces fradiæ*, is active against many strains of Gram-positive and Gram-negative bacteria, and is used (as the sulphate) in lotions and injections for treating a wide variety of infections and orally as an intestinal antiseptic; also, either of the two constituent isomers (*neomycin B* and *C*) or an inactive degradation product of them (*neomycin A*).
1949 Waksman & Lechevalier in *Science* 25 Mar. 305/2 In search for new compounds, particular attention was paid to those that would be effective against streptomycin-resistant bacteria, notably against the streptomycin-resistant strains of *M*[*ycobacterium*] *tuberculosis*. The discovery of such an agent, designated as neomycin, is reported here. **1949** R. L. Peck et al. in *Jrnl. Amer. Chem. Soc.* LXXI. 2590/2 Evidence has been obtained that the neomycin activity is due to more than one chemical entity; hence one may define it as a 'neomycin-complex'. The substance isolated as described herein, has been designated neomycin A. **1960** M. E. Florey *Clinical Applications of Antibiotics* iv. v. 140 Neomycin.. was expected to replace streptomycin since it was much less liable to induce resistance in pathogenic organisms. **1963** *Lancet* 19 Jan. 161/2 The infant [*sc.* a *Salmonella heidelberg* excretor] was treated with neomycin, and thereafter all specimens were negative. **1970** Passmore & Robson *Compan. Med. Stud.* II. xx. 30/1 Neomycin and kanamycin reduce ventilation during general anaesthesia by a neuromuscular blocking action. **1974** M. C. Gerald *Pharmacol.* iii. 45 The antibiotic neomycin is used to sterilize the gastrointestinal tract prior to abdominal surgery.

neon. Substitute for def.: **1.** One of the inert or noble gases, which is present in low concentration in the earth's atmosphere and is used at low pressure in discharge tubes, where it emits an orange-red glow. Atomic number 10; symbol Ne. (Add further examples.)
1905 *Chem. News* 5 May 204/1 A fair quantity of the mixture of neon and helium was prepared by liquefying air. **1935** *Industr. & Engin. Chem.* Jan. 116/1 Neon is characterized by its high electrical conductivity and light-emissive powers when an electrical current is discharged through it. **1966** *McGraw-Hill Encycl. Sci. & Technol.* IX. 36/1 Neon is used as the current-carrying agent in lightning arrestors; virtually no current is carried at voltages below the breakdown potential of the neon, but

when lightning strikes, the neon is ionized and allows the current to flow to ground. **1966** Cotton & Wilkinson *Adv. Inorg. Chem.* (ed. 2) xxiii. 598 Helium, neon and argon have so far not been brought into chemical combination..and it seems unlikely that they are capable of reaction.

2. A neon lamp or tube; neon lighting. Also *fig.*

1934 S. Gold *Neon* 18 Whilst the customer may want his name expressed in neon and surrounded by a border, it is up to the sign-man to supply it with 'finish'. **1957** J. Braine *Room at Top* vi. 54 Too clean and well-lighted. They'll be installing neons soon. **1958** *Spectator* 30 May 687/3 The dialogue [of a play] is stuffed as full of metaphors as a copywriter's prose and Tony Richardson's production sets up each phrase in neon. **1969** A. Glyn *Dragon Variation* v. 162 To the left it was the glare of the rest of the Strip, the great river of neon, the jazziest in the world. **1974** R. Butler *Buffalo Hook* iii. 26 The neons were flashing over the plushy restaurants.

3. a. *attrib.* and *Comb.*, as *neon advertisement, glow, lighting, strip, wilderness, world; neon-blazing, -bright, -coupled, -filled, -lighted, -lit* adjs.

1972 D. Haston *In High Places* iv. 57 There was a neon advertisement by the entrance. **1962** K. Orvis *Damned & Destroyed* xxvii. 206 The city's wide, neon-blazing Sunset Strip. **1958** *Spectator* 25 July 133/3 A neon-bright café. **1946** *Nature* 21 Sept. 414/1 The rectified signal from the second detector was amplified by a neon-coupled two-stage D.C. amplifier. **1935** Miller & Fink *Neon Signs* iii. 37 (*caption*) Relative light energy from a neon-filled tube. **1966** *McGraw-Hill Encycl. Sci. & Technol.* IX. 36/1 A very small wattage produces visible light in neon-filled glow lamps. **1935** Miller & Fink *Neon Signs* iii. 37 Why is it, then, that the neon glow has such a powerful effect? **1945** A. Huxley *Time must have Stop* xxx. 276 The neon glow from those technological New Jerusalems beyond the horizons of the next revolution. **1936** C. Rouse *Old Towns* i. 21 Inside, a neon-lighted glass sign directs you to 'Ye olde Beamed Tudor Cocktail Bar'. **1954** *Encounter* Feb. 37/2 It begins, in the West, with the neon-lighted brilliance of the Kurfürstendamm. **1913** *Trans. Illum. Engin. Soc.* (U.S.) 371 (*heading*) Neon lighting. **1933** *Times* 5 Dec. (Electricity Supply Number) p. xxiii/7 Neon lighting can be seen in almost every town of any size in the country. **1958** *Times Lit. Suppl.* 31 Jan. 63/2 His wrath is only excited by the Subtopia, the neon lighting, the petrol fumes, the 'soulless concrete' of our time. **1954** Neon-lit [see *aspirin]. **1958** *Spectator* 14 Feb. 197/1 The actress, the best-selling author, or the famous film star who..remain tragically single, trapped in their own accursed, neon-lit achievement. **1972** R. Busby *Reasonable Man* iii. 24 They left the car outside the store and went..into the brash, neon-lit interior. **1939** 'J. Struther' *Mrs. Miniver* 243 This part of the town was almost unrecognizable—a street of angular lettering and neon strips. **1973** 'H. Howard' *Highway to Murder* xiii. 149 Over the dressing chest a neon strip shone down on jars and bottles. **1953** *Encounter* Nov. 8/1 The neon wilderness of noise and music, fun and sin, boredom and high, desperate spirits. **1959** *Listener* 10 Dec. 1048/3 The American neon-wilderness. **1959** *New Statesman* 28 Feb. 302/3 Living in a neon-world of semi-legality, its leadership organises conferences and 'stay-at-homes' as if keeping the government informed about its intentions was a rule in the revolutionary game.

b. Special Comb.: **neon fish** = **neon tetra*; **neon lamp, light,** a lamp in which an electric discharge is passed through neon (giving an orange-red coloured light) or a mixture of neon with other gases (giving other colours); **neon sign,** a sign incorporating a neon light (usu. a neon tube), and often serving as an advertisement on a building; **neon tetra,** a small characin, *Hyphessobrycon innesi*, native to the Amazon, and remarkable for its colouring, which is dark green and white with a shining blue-green stripe; **neon tube,** a neon light in the form of a tube.

1938 L. MacNeice *Zoo* 182 The tiniest fish here are the neon fish. **1911** *Chem. Abstr.* V. 1024 In contradistinction to the Hg vapor lamp the neon lamp's light is rich in red rays. **1931** *B.B.C. Year-bk.* 447/1 A special type of neon lamp is now used in television receivers to convert the electrical impulses back into light impulses, dependent for its action on a linear relation between light response and applied electrical potential and on an absence of time-lag. **1940** L. MacNeice *Last Ditch* 15 And the neon-lamps of London Stain the canals of night. **1966** Ainsworth & Robinson in Hewitt & Vause *Lamps & Lighting* xix. 309 Neon lamps using the positive column as the light source do exist however, the conventional 'neon lamps' used in advertising displays being of this type. **1913** *Trans. Illum. Engin. Soc.* (U.S.) 376 The neon light is physiologically excellent on account of its dull luminescence. **1931** H. G. Wells *Work, Wealth & Happiness of Mankind* (1932) v. 211 This Neon light has great penetrating power in a fog. **1958** J. Betjeman *Coll. Poems* 231 So up I rose and went along To that old village alehouse where In neon lights is written 'Bear'. **1972** J. Potter *Going West* 38 'Aloha', the airport tower announced in neon lights... Aloha meant welcome. **1927** *Advertising & Selling* 28 Dec. 34/3 Neon signs overcome the handicap of high first cost by lowered current consumption. **1934** *Times* 19 Feb. 13/5 In two of our quiet residential streets flaring neon signs have been put up to announce that the houses on which they are placed are hotels. **1958** J. Betjeman *Coll. Poems* 232 The neon sign's a work of art and visible for miles. **1973** *Black World* Mar. 62 Pulsating rock music seemed to control the flashing neon signs. **1936** W. T. Innes in *Aquarium* V. 82/2 The recent feat of bringing some wonderful new Characins, Neon Tetras (*Hyphessobrycon innesi* Myers), from Germany to the Shedd Aquarium in Chicago..in less than 60 hours. *Ibid.* 135/2 Regarding the Neon fishes... Four of these fishes were dead when the airship arrived... The red coloring from the fishes had made bright red blotches on the paper. *Ibid.* 136/2 The very good descriptive popular name for the fish, 'Neon Tetra' or 'Neontet' was originated by M. Lepaut, of Paris... The name is so fitting that Mr. Lew Willumsen, on seeing the new importation, independently hit upon the same name for them. **1952** D. Gohm *Tropical Fish* 100 The Neon Tetra is generally regarded as the most beautiful of all aquarium fish. **1962** *Listener* 22 Nov. 852/2 One [*sc.* tropical fish] which seemed to be lit up by a greenish blue light running from his head to his tail—this was a neon tetra. **1971** *Ceylon Observer Mag. Ed.* (Colombo) 19 Sept. 2/6 (Advt.), Goldfish, Angelfish, Neontetras. **1904** *Electr. World & Engin.* XLII. 583/2 The wavelength of the oscillator's vibration can..be measured by isolating a complete wave on the helix by means of a sliding earthed saddle using a neon tube as indicator. **1936** *Discovery* Nov. 364/2 The manufacture, erection, and maintenance of all neon tubes for advertising and display purposes. **1945** R. C. Walker *Electronic Equipment & Accessories* xi. 211 Fig. 187 shows an application of the neon tube as a voltage stabiliser in a valve-rectifier circuit. **1962** A. Nisbett *Technique Sound Studio* 272 Stroboscopes work best with neon tubes.

neonate (nī·onē̆ᶦt). *Med.* [f. Neo- + L. *nāt-us* born.] A recently born individual; *spec.* an infant less than four weeks old. Also *attrib.* or as *adj.*

1932 M. B. McGraw in *Child Development* III. 292/1 Most 'partunates'[1] display a decidedly helpless response to the force of gravity. [*Note*] [1] A term indicating infants who are just born... It covers about the first fifteen or thirty minutes of life since it includes the time during and immediately following parturition. When the umbilical cord is dressed and the baby is taken to the maternity nursery, then he becomes a 'neonate'. **1936** *Q. Rev. Biol.* XI. 70 (*heading*) Problems in the classification of neonate activities. **1951** L. Carmichael in S. S. Stevens *Handbk. Exper. Psychol.* viii. 289/1 The fetus shows most of the specific patterns of response that can be elicited in the neonate. **1962** *Lancet* 12 May 1026/2 Neonates are capable of withstanding quite profound hypothermia for a short period of time. **1967** *Nature* 10 June 1099/2 The dentate gyrus is considerably more mature in the neonate guinea-pig than in the rat. **1973** *Jrnl. Genetic Psychol.* CXXII. 320 The overall body proportions of the typical neonate, 2-year-old, 6-year-old, [etc.].

neonatology (nī͵onēᶦtǫ·lŏdʒi). *Med.* [f. prec. + -ology.] The branch of medicine concerned with the disorders and problems of recently born infants. Hence **neonato·logist,** an expert or specialist in neonatology.

1960 A. J. Schaffer *Diseases of Newborn* 1/1 We trust we shall be forgiven for coining the words 'neonatology' and 'neonatologist'. We do not recall ever having seen them in print. The one designates the art and science of diagnosis and treatment of disorders of the newborn infant, and the other the physician whose primary concern lies in this specialty. *Ibid.* p. vi/1 The situation of the pediatrician practising neonatology differs but little qualitatively from his everyday posture with respect to older infants and children. **1967** *Obstetrics & Gynecol.* XXX. 890 The obstetric trainee would assume the role of a pediatric resident in the field of neonatology. **1972** *Daily Colonist* (Victoria, B.C.) 3 Aug. 26/5 The intensive care unit..combines the disciplines of both fetology and neonatology. **1975** *Sci. Amer.* Jan. 51 (Advt.), Most neonatologists contend that many of these tragedies can be prevented by specially-trained perinatal medical teams using intensive care techniques.

neoned (nī·ǫnd), *a.* [f. Neon + -ed².] Illuminated by neon lighting.

1945 *Tomorrow* Mar. 46 The shadow went, he hardly noticed that he was already moving again toward the neoned glare. **1968** *Punch* 29 May 778/3 We..came upon an imperfectly neoned structure which might have passed for a church hall. **1973** J. Wainwright *Devil you Don't* 34 The neoned entrance to the *Roll-a-Ball Arcade*.

neopallium (nī͵opæ·liŏm). *Anat.* Also (with hyphen) **neo-pallium.** Pl. **-pallia.** [mod.L., f. *Neo- 1 e + Pallium.] The phylogenetically youngest portion of the pallium of the brain, which appears first among the more advanced reptiles and which among the mammals has become the largest part of the brain. Cf. *neocortex.

1901 G. E. Smith in *Jrnl. Anat. & Physiol.* XXXV. 431 It is only one of the three histological formations which constitute the true pallium; and, as it is the latest of these to reach the height of its development, we may call it the 'new pallium' or..'neopallium', in contradistinction to the 'old pallium' of the Sauropsida and earlier Vertebrata, which is *chiefly* formed of the other two pallial areas. **1907** *Arch. Neurol.* III. 51 Maps have been obtained of these neopallia which are fairly full. *Ibid.* 52 The term neopallium is employed throughout in the sense suggested by Elliot Smith to indicate the variable area intercalated between the 'basal pallium' or pyriform lobe and the marginal pallium, or hippocampus. **1922** *Glasgow Herald* 23 Dec. 4/2 When our ancestors, with their free hands, their enlarged cerebral cortex or neopallium, and their capacity for co-operative action, had resources sufficient to enable them to stand up to Carnivores.., they left the trees and became once more terrestrial. **1948** A. Brodal *Neurol. Anat.* x. 323 With the development of the neopallium or neo-cortex, the two more primitive areas are finally pushed medially, and in man are found entirely on the medial aspect of the hemisphere. **1973** *Gray's Anat.* (ed. 35) vii. 921/1 In the course of evolution visual, auditory and other conduction paths have been transferred, through the thalamus, to the pallium of the cerebral hemispheres... Consequently, each cerebral hemisphere enlarges as a result of the formation of an additional region, the neopallium.

Hence **neopa·llial** *a.*, of or pertaining to the neopallium.

1907 *Arch. Neurol.* III. 112 This appears to be the lowest grade of conscious neopallial association—*i.e.*, of psychic function. **1921** Tilney & Riley *Form & Functions Central Nervous Syst.* xlii. 772 (*heading*) Neopallial projection system. **1953** G. A. G. Mitchell *Anat. Autonomic Nerv. Syst.* v. 66 This neopallial responsibility for autonomic activities is shared by subcortical centres in the hypothalamus. **1962** E. C. Crosby et al. *Correl. Anat. Nerv. Syst.* vii. 411/1 Neopallial cortex makes up the major portion of the cerebral cortex in man.

neophilia (nī͵ŏfi·liă). [f. Neo- + *-philia.] Love for, or great interest in, what is new; a love of novelty. So **neophi·liac,** a person characterized by neophilia; also **neophi·li(a)c** *a.*; **neo·phily.**

1932 B. Malinowski in R. F. Fortune *Sorcerers of Dobu* p. xxvii, Terminological neophily..is a habit to which I have always been hostile. **1966** R. & D. Morris *Men & Apes* vii. 217 There is a perpetual struggle going on inside the brain, between the fear of the new (neophobia) and the love of the new (neophilia). The neophobic urges keep the animal out of danger, while the neophilic urges prevent him from becoming too set in his ways. **1969** C. Booker (*title*) The neophiliacs. **1971** R. Petrie *Thorne in Flesh* xi. 142 From where Thorne had sat, without neophiliac needs, he had seen..not spangles but sweating flesh. **1972** *Daily Tel.* 2 Aug. 2/2 The exaltation of novelty (neophilia) had been turned into a cult. *Ibid.*, Neophiliacs suffer from a collective fantasy which leads them to describe every change as inevitable and an improvement on what preceded it.

neophyte. Add: **3.** *Bot.* A plant found in an area in which it has not been recorded before.

1916 B. D. Jackson *Gloss. Bot. Terms* (ed. 3) 248/1 Neophyte..a newly introduced plant (Rikli). **1970** *Watsonia* VIII. 157 Besides these three widespread neophytes, some other exotic taxa occur as ornamentals in gardens and parks.

neopilina (nī͵opiləi·nă). [mod.L. (H. Lemche 1957, in *Nature* 23 Feb. 414/1), f. Neo- + *Pilina*, name of a similar fossil genus of molluscs.] A primitive, deep-sea mollusc of the monotypic genus so called, belonging to the order Monoplacophora.

1957 H. Lemche in *Nature* 23 Feb. 416/1 It would seem probable..that one of the main food-sources of *Neopilina* is radiolarians. **1959** *Listener* 12 Feb. 300/2 The recent discovery of *Neopilina*..is scientifically quite as important as that of the coelacanth. **1968** A. S. Romer *Procession of Life* v. 101 Belief as to the primitive nature of these little ancient molluscs was abundantly confirmed by the discovery by a Danish oceanographic vessel a few years ago of specimens of a modern descendant of *Pilina*, living in deep oceanic waters off the west coast of South America; American exploration has resulted in further finds of this small mollusc, appropriately named *Neopilina*... Little *Neopilina* has aroused little public interest, but scientifically it is far more important than *Latimeria*... *Neopilina* sheds much light on the origin of a whole animal phylum. **1973** P. Tasch *Paleobiol. Invertebrates* viii. 325/2 *Neopilina* is segmented (that is, has internal metamerism) in a way similar to that of annelid worms and arthropods.

neoplasia (nī͵oplēᶦ·ziă). *Biol.* and *Med.* [f. Neo- + Gr. πλάσις formation: see -ia¹.] The formation of neoplasms; the state or condition of having neoplastic growth.

1890 C. P. Mitchell *Philosophy Tumour-Dis.* i. 7 (*heading*) Explanation of the self-dependence of tumours: neoplasia identified with genesis in the Protozoa. **1926** *Glasgow Herald* 22 May 7/2 No one who had made a broad study of neoplasia of all varieties could support either of those theses. **1947** *Radiology* XLIX. 358/2 The late effects resulting from exposure to penetrating radiations..consisted of generalized atrophy and neoplasia of hemopoietic organs. **1962** *New Scientist* 12 Apr. 48/2 These processes do resemble in certain respects neoplasia. **1970** *Nature* 18 Apr. 290/1 The general belief in clinical circles is that bladder tumours are..produced as the result of chemical constituents in the urine which act on the urinary epithelium and produce neoplasia.

ne:o-pla·sticism. Freq. with capital initials. [ad. F. *néoplasticisme*, f. Neo- + Plasticism.] A movement or style in art originated by the Dutch painter Piet Mondrian (1872–1944), characterized by the use of primary colours and abstract forms. So **neopla·stic** *a.*²; **neoplasti·cian; neopla·sticist** *a.*

[**1920** P. Mondrian (*title*) Le Néo-plasticisme.] **1933** *Gallery of Living Art Catal.: A. E. Gallatin Coll.* (N.Y.), Mondrian, the neo-plastician. **1934** J. J. Sweeney *Plastic Redirections in 20th Cent. Painting* 40 Piet Mondrian, member of the Dutch 'de Stijl' group and founder of the Neoplastic school, can..be taken as a producer of..'abstractions'... Yet, Mondrian has said, 'It is a great error to envisage a Neoplastic work as a total abstraction from life.' *Ibid.* 69 Such groups as the Purists, the Neoplasticians, and the Constructivists. **1935** *Art Digest* 15 Jan. 9/1 The Gallery of Living Art at New York University..has reopened after redecoration... In the rearrangement, emphasis has been placed on the work of the Cubists and on Constructivism and Neo-plasticism. **1936** A. H. Barr in *Cubism & Abstract Art* (N.Y. Mus. Mod.

Art) 140 (*heading*) Abstract art in Holland: *de Stijl* and Neo-Plasticism. *Ibid.* 144 This project is clearly a three-dimensional projection of a Neo-Plasticist painting. **1970** *Oxf. Compan. Art* 770/2 Neo-plasticism..was distinguished from Cubism in that it did not make use of figurative elements even as a starting-point for abstraction.

Neoplatonic, *a.* Add: Hence as *sb.* = Neoplatonist.

1840 E. Cox tr. *Döllinger's Hist. Church* i. 71 The Neoplatonics endeavoured, therefore, to unite the different systems of philosophy, especially the Pythagorean, Platonic, and Aristotelean, in one body with the principles of oriental learning. **1879** *Dublin Rev.* Apr. 524 On the Catholic side no theologian has followed Scotus Erigena: his system may be called an offspring of the Neoplatonics.

neoprene (nī·oprīn). [f. Neo- + *-prene*, after *isoprene, *chloroprene.] Any of various synthetic rubbers made by polymerizing chloroprene (2-chloro-1,3-butadiene, $CH_2{:}CCl{\cdot}CH{:}CH_2$) and useful for their resistance to oil, heat, and weathering and their higher strength than natural rubber.

1937 *Du Pont Mag.* Feb. 16/3 As of December 15th the name *Neoprene* has been adopted to describe our chloroprene rubber formerly sold under the trade-mark 'DuPrene'. **1939** *Jrnl. R. Aeronaut. Soc.* XLIII. 150 In the matter of fuel tank construction, there are just about as many welded tanks as riveted tanks, the latter using duprene or neoprene as a leak preventative. **1950** *Archit. Rev.* CVIII. 411 Exterior butt joints are sealed with a flexible neoprene tube which is compressed when the panels are bolted together. **1957** H. L. Fisher *Chem. Natural & Synthetic Rubbers* ix. 101 The neoprenes are available in several types of latexes. **1959** *Observer* 27 Sept. 4/3 Diving-suits, made out of 'expanded neoprene' (a kind of foamed rubber), are now coming on the market. **1959** J. P. Munn in M. Morton *Introd. Rubber Technol.* xiii. 339 Neoprene Type KNR is a soft, chemically plasticizable type, suitable for the production of cement, paints, and in spreading compounds. **1969** *New Scientist* 28 Aug. 430/3 The pyrotechnical compound is ignited by applying a soldering iron or propane gas flame to a small ignition hole. The heat makes the neoprene gasket expand to form a continuous seal. **1974** 'J. Graham' *Bloody Passage* i. 24, I took down a neoprene wetsuit in black and pulled it on.

neostriatum (nī͵ostrəi͵ēᶦ·tŏm). *Anat.* [mod. L., f. *Neo- 1 e + *striatum.] The phylogenetically younger portion of the corpus striatum, consisting essentially of the caudate nucleus and the putamen.

1909 *Arch. Neurol. & Psychiatry* IV. 163 As the nucl. med. thal. (and also the nucl. rotund.) receive trigeminal fillet-fibres, it results that the first neocortical tactile region is one of oral sensibility (just as the first neostriatum). **1929** *Physiol. Abstr.* XIV. 398 (*heading*) Morphogenesis of neo- and palæostriatum in man. **1972** T. W. Jenkins *Functional Mammalian Neuroanat.* xvi. 237/1 Functionally the neostriatum (caudate nucleus and putamen) differs from the paleostriatum (globus pallidus). **1974** *Nature* 1 Feb. 284/1 Brains from other animals prepared similarly were dissected into neostriatum, nucleus accumbens, [etc.].

Hence **neostria·tal** *a.*, of or pertaining to the neostriatum.

1936 C. U. A. Kappers et al. *Compar. Anat. Nervous Syst. Vertebr.* II. ix. 1475 It..forms a hypopallial (or neostriatal) area. **1958** L. Hausman *Clin. Neuroanat.* xxxiii. 287 (*heading*) The efferent neostriatal connections.

Neo-Synephrine (nī͵osine·frin, -īn). *Pharm.* Also **neo-synephrin(e, neosynephrine.** [f. Neo- + *synephrine* (f. Syn- + *epin)ephrine.] A proprietary name for phenylephrine.

1934 *Jrnl. Amer. Med. Assoc.* 16 June 2024/1 Neo-synephrin hydrochloride is a vasoconstrictor which is active when administered orally. **1937** *Amer. Jrnl. Ophthalmol.* XX. 176/2 In ophthalmoscopy neo-synephrine has the properties of other epinephrines. It produces mydriasis as quickly, as fully, and as safely as any mydriatic. **1950** *Trade Marks Jrnl.* 2 Aug. 711/1 Neo-Synephrine... Vasoconstrictors and antispasmodics being pharmaceutical preparations. Winthrop-Stearns, Inc... City and State of New York, United States of America; manufacturing chemists. **1972** *Obstetr. & Gynecol.* XL. 23/1 The nonsuture technic using a neosynephrine injection procedure for hemostasis..is recommended. **1974** M. C. Gerald *Pharmacol.* vi. 109 Drugs such as..phenylephrine (Neo-Synephrine) activate the adrenergic receptor directly.

neotechnic (nī͵ote·knik), *a.* [f. Neo- + Gr. τεχνικ-ός (see Technic *a.* and *sb.*).] Denoting, or belonging to, the most recent stage of industrial development. Hence **neote·chnics,** neotechnic technology.

1915 P. Geddes *Cities in Evolution* iv. 64 Simply substituting *-technic* for *-lithic*, we may distinguish the earlier and ruder elements of the Industrial Age as Paleotechnic, the newer and still often incipient elements disengaging themselves from these as Neotechnic. **1927** Neotechnics [see *geotechnics* (*geo-)]. **1934** [see *eotechnic *a.*]. **1963** *Punch* 21 Aug. 284/1 The forms, symbolism and idiom of neotechnics. **1967** *Punch* 27 Sept. 451/1 The country would be rejigged to meet the challenge of the neotechnic revolution.

neotene (nī·otīn). *rare*⁻¹. [Cf. next.] A species (or member of a species) in which the period of immaturity is indefinitely prolonged.

1959 Auden *Homage to Clio* (1960) 25 The neotene who marches Upright and can subtract.

neoteny (ni͵ǫ·təni). *Zool.* [ad. G. *neotenie* (J. C. E. Kollman 1884, in *Verh. Naturf. Ges. Basel* VII. 391), f. Gr. νέος young + τείνειν to extend.] **a.** The retention of juvenile characteristics in adult life. **b.** The possession of sexual maturity by an animal still in its larval stage. Cf. *neoteinia, -tenia* (Neo- 2).

1901 H. Gadow in *Cambr. Nat. Hist.* VIII. iii. 65 These cases of neoteny are therefore instances of more or less complete retardation, or of the retention of partially larval conditions. **1920** *Conquest* Apr. 278/2 Neoteny..here means the abnormal time-extension of youthful characters. **1932** J. S. Huxley *Probl. Relative Growth* vii. 237 It is clear that changes in rate-genes could as easily lead to the opposite of recapitulation as to recapitulation. Many examples of neoteny would fall under this head. **1962** D. Nichols *Echinoderms* xiv. 178 An interesting embryological phenomenon known as neoteny, in which, by the acceleration of development of the gonads, an animal becomes sexually mature while still retaining the larval body form. **1965** *New Scientist* 14 Jan. 86/1 The retention of the puppy characteristic (neoteny) of floppy ears. **1971** J. Z. Young *Introd. Study Man* xxxiv. 479 Such a change, technically called neoteny (or paedomorphosis), has in fact occurred often in the course of the evolution of diverse animals.

Hence **neote·nic, neo·tenous** *adjs.*, **neo·tenously** *adv.*

1901 H. Gadow in *Cambr. Nat. Hist.* VIII. iii. 64 Not unfrequently typical neotenic and overgrown specimens occur side by side with others which have completed their metamorphosis. **1930** G. R. de Beer *Embryol. & Evol.* 27 Some animals have become permanently committed to this neotenous state. **1932** J. S. Huxley *Probl. Relative Growth* ii. 67 Workers [*sc.* termites] have been derived from soldiers by a suppression of their final development into the normal big-jawed type—..in fact, they are neotenic. **1957** *New Biol.* XXIII. 103 Animals which breed as juveniles or larvae are described as 'neotenous'. **1963** R. P. Dales *Annelids* iii. 72 The fusiform typhloscolecids...were possibly derived neotenously from phyllodocid stock. **1965** B. E. Freeman tr. *Vandel's Biospeleology* xiv. 233 All the troglobious urodeles are neotenous. They do not undergo metamorphosis and thus retain their larval characters throughout life.

neoterical, *a.* Delete ? *Obs.* and add later *arch.* example.

1941 E. R. Eddison *Fish Dinner* (1972) xvi. 266 For the more mockery, let it arise from the sea: a very neoterical Anadyomene.

neothalamus (nī͵opæ·lămŏs). *Anat.* [mod. L. (L. Edinger *Vorlesungen über den Bau der nervösen Zentralorgane des Menschen und der Tiere* (ed. 7, 1908) II. xv. 234), f. *Neo- 1 e + Thalamus.] The phylogenetically younger portion of the thalamus, which includes its lateral part.

1917 *Jrnl. Compar. Neurol.* XXVIII. 217 These parts of the thalamus..are termed by Edinger the neothalamus by reason of their functional relationship with the neopallium. **1921** Tilney & Riley *Form & Functions Central Nerv. Syst.* xxxi. 567 The neothalamus is an addition to the pars thalamica appearing in mammals only. **1973** *Gray's Anat.* (ed. 35) 893/1 The anterior and medial part of the thalamus, together with some of the smaller groups of nuclei, are often regarded as phylogenetically the older regions, and designated paleothalamus in contrast with the lateral part or neothalamus, which reaches its greatest development in anthropoid apes and man.

So **neotha·lamic** *a.*, of or pertaining to the neothalamus.

1909 *Arch. Neurol. & Psychiatry* IV. 163 The upper part of the lateral cortex layer has certainly already a neo-cortical character, as is proved by the fact that a certain amount of fibres, originating in a neo-thalamic nucleus of the fillet.., end in it. **1936** C. U. A. Kappers et al. *Compar. Anat. Nervous Syst. Vertebr.* II. ix. 1481 Here for the first time there are projection fibers from neothalamic centers.

neotocite. Read: (nī͵otō͞ᵘ·kəit). *Min.* Also † **neotokite.** [ad. G. *neotokit* (N. Nordenskiöld *Verzeichn. d. in Finland gefund. Min.* (1852) (Dana)), f. Gr. νεότοκ-ος new-born, recent (f. νέος new + τόκος offspring, childbirth) + G. *-it* -ite¹.] A hydrated silicate of manganese and iron, approximately $Mn_2Fe_2Si_4O_{12}.6H_2O$, which is found as black amorphous masses.

1854, etc. [in Dict.]. **1921** *Jrnl. Washington Acad. Sci.* XI. 27 Some specimens of the bementite rock are cut by thin veinlets of dark brown to black amorphous neotocite. **1932** *Amer. Mineralogist* XVII. 18 Areas of neotocite, several centimeters across are characterized by glistening black color and conchoidal fracture. **1968** I. Kostov *Mineralogy* 351 The amorphous silicate of manganese known as neotocite, occurring as black, opal- or coal-like masses is found as a weathering product of rhodonite and as a volcanogenic-sedimentary mineral.

neotype (nī·otəip). [f. Neo- + Type.] **1.** (In Dict. s.v. Neo- 2.)

2. *Taxonomy.* [a. F. *néotype* (A. E. M. Cossman *Essais de Paléoécologie Comparée* (1896) II. 2 and (1904) VI. 9), f. Neo- + *type *sb.*¹ 8.] A specimen designated as a type, in the absence of any other type material.

1905 C. Schuchert in *Bull. U.S. Nat. Mus.* No. 53. 13 The term is here used as redefined by Cossman in 1904..as follows: Neotype for the specimen afterwards taken as the type of a species when the original type has been destroyed. **1961** G. G. Simpson *Princ. Animal Taxonomy* i. 32 Neotype: a substitute for a type that has been lost or (in some usages) is otherwise inadequate for the type role. (Neotypes have not hitherto been recognized in the Rules, but there are proposals to incorporate them therein.) **1966** *Internat. Code Bot. Nomenclature* ii. 19 A neotype is a specimen or other element selected to serve as nomenclatural type as long as all of the material on which the name of the taxon was based is missing.

N.E.P., NEP, Nep⁴ (nep). [f. the initial letters of *New Economic Policy.*] A programme initiated in the Soviet Union in 1921 for the revival of the wage system and private ownership in industry. Hence **ne·pman, NEP-man,** one engaged in this programme.

[**1922** *Communist Rev.* June 107/2 The basic task of 'Nepa' (New Economic Policy),..is to establish a close connection between the socialist economy..and the economy of the peasant masses.] **1924** *Glasgow Herald* 2 Jan. 7 The wholesale arrest of thousands of persons, including many 'Nepmen' supporters of the new economic policy. **1927** *19th Cent.* Nov. 655 In 1921 the economic catastrophe threatened to bury under its ruins even the Soviet Government. It was then that Lenin proclaimed the New Economic Policy or 'N.E.P.' This 'Nep' was nothing else than a whole series of concessions to life—it was the victory of life over the deadly system inaugurated by the Soviet Government. **1929** *Daily Tel.* 22 Jan. 10/7 The Nep..is continually re-asserting itself, in spite of the persecution and restrictions under which the Nepmen suffer. **1949** E. Pound *Pisan Cantos* lxxiv. 25 But in Russia they bungled and did not apparently Grasp the idea of work-certificate And started the N.E.P. with disaster And the immolation of men to machinery. **1951** Koestler *Age of Longing* vi. 98 There were rich and poor again; though the rich were now called NEP-men. **1961** C. Cockburn *View from West* ix. 127 The British upper middle classes..richer and happier than ever—the great NEP-men of the 1950s. **1971** W. H. McNeill in A. Bullock *20th Cent.* 48/1 Extreme economic disruption even compelled Lenin to abandon Communist principles to the extent of allowing limited private trading (NEP or New Economic Policy, 1921). **1973** *Times Lit. Suppl.* 3 Aug. 910/3 The wickedness, real or alleged, of Nepmen became a favourite theme of official propaganda.

Nepalese (nepǭlī·z), *a.* and *sb.* Also **Nepaulese.** [f. *Nepal*, name of a country on the north-eastern frontier of India + -ese.] **A.** *adj.* Of, pertaining to, or connected with Nepal. **B.** *sb.* A native or inhabitant of Nepal; these people collectively. Also, the language of this people.

The form *Nepali* (*a.* and *sb.*) also occurs, esp. as the name of the language of the Nepalese.

[**1811** W. Kirkpatrick *Acct. Nepaul* vii. 207 The Nepaulians most commonly barter it for the rock salt and borax of Tibet.] **1819** F. Hamilton *Acct. Nepal* 1. i. 39 The following account of the Nepalese, or rather Newar, architecture, I have taken from papers communicated by Colonel Crawford. The Nepalese possess a great advantage in having an excellent clay for making bricks and tiles. **1848** J. D. Hooker in L. Huxley *Life J. D. Hooker* (1918) I. xiii. 251 The Sikkim Rajah, whose territories were once the prey of the Nepalese. *Ibid.* 263 Accepting the invitation of Major Thoresby, the Nepaulese Resident. **1884** *Encycl. Brit.* XVII. 343/1 In all matters of domestic policy the Nepalese brook no interference. **1885** G. C. Whitworth *Anglo-Indian Dict.* 226/2 *Nepāli*, an inhabitant of Nepal or an emigrant from that country. The principal Nepāli tribes are the Gurkhās, the Gurungs, the Newārs and Limbus, the Kirātis, the Bhutiās, the Lepchās. Also the name of the language of the country, which is a dialect of Hindi. **1908** T. G. Tucker *Introd. Nat. Hist. Lang.* 187 *Gujarāti, Sindhi, Punjābi, Kashmiri*, and *Nepāli* in the regions which their names imply. **1911** *Encycl. Brit.* XX. 453/1 Khas-kurā,..passes under various names. The English generally call it Nēpālī or Naipālī (i.e. the language of Nepal), which is a misnomer, for it is not the principal form of speech of that country. **1920** *Mission News* June 45/2 He has translated..the whole of the Bible into his own Nepali language. **1927** *Chambers's Jrnl.* June 370/1 The bungalow..remained in charge of a Nepali. **1931** *Times* 16 Mar. 22/5 Nepalese bronzes and brass work. **1961** *Listener* 10 Aug. 211/1, I had once known Nepali almost as well as I know English. **1964** R. Perry *World of Tiger* iv. 52 The Nepalese stockmen were only too well aware of this habit. **1969** *Sunday Tel.* 12 Jan. 7/3 Many of the hippies..left Nepal..in 1968. The Nepalese didn't like them. **1971** C. Bonington *Annapurna South Face* i. 13 He spoke fluent Nepali. *Ibid.* iii. 40 The Internal Departures office, an open shed crammed with Nepalis and hippies. **1973** *Times* 14 Apr. (Nepal Suppl.) p. ii/2 It was only after the unification of the country under the royal house of Gorkha that Nepali became the *lingua franca* of the whole of Nepal. *Ibid.* p. ii/3 Traditionally Nepali Hindus are extremely conservative.

neper (nī-, nēᶦ-pəɹ). Also †**napier.** [f. *Neperus*, Latinized form of the name of John *Napier* (or *Neper*) (see Napier's bones).] A unit used in comparing the power levels in

two communication circuits or the intensities of two sounds: their difference in nepers is equal to half the natural logarithm of their ratio.

1924 K. S. Johnson *Transmission Circuits for Telephonic Communication* vii. 46 The transformer loss (expressed in napiers). **1929** [see *bel]. **1931** [see *decineper* (*deci- 1)]. **1954** *Electronic Engin.* XXVI. 91 The short line with which we are concerned in this article is defined as one for which the overall attenuation coefficient is small, say not greater than 0·2 neper, although the line may be several wavelengths long. **1972** J. M. Taylor tr. *Meyer & Neumann's Physical & Appl. Acoustics* i. 11 One neper corresponds to 8·69 dB.

nepeta (nĕpī·tă). [L. *nepeta* catmint, perh. f. *Nepeta* ancient name of Nepi, a town in central Italy, adopted as the name of a genus by Linnæus (*Genera Plantarum* (1737) 170) and earlier botanists.] An annual or perennial herbaceous plant of the genus so called, belonging to the family Labiatæ; = Catmint.

1915 H. H. Thomas *Bk. Hardy Flowers* 304 Some of the Nepetas are little more than weeds, though a few kinds are attractive. **1939** R. E. Clarkson *Magic Gardens* ii. 21 If you like, the gray santolina (but only for wide edgings), the nepetas, French thyme. **1961** *Times* 28 Jan. 9/3 Interplanted with the nepeta were daffodils. **1975** *Country Life* 6 Mar. 561/2 Blue anchusas..rising from misty blue nepeta.

nephanalysis (nefănæ·lĭsis). *Meteorol.* Pl. **-ses.** [f. Gr. νέφ-ος cloud + Analysis.] An analysis of the amounts and kinds of cloud present over an area; *esp.* a chart showing this in symbolic form.

1945 F. A. Berry et al. *Handbk. Meteorol.* xi. 882 Synoptic studies of clouds (nephanalysis) indicate the possibility of a visual weather analysis based on cloud observations. *Ibid.* 905 Observations plotted on maps for nephanalysis include clouds, weather, [etc.]. **1961** *N.Y. Times* 21 Sept. 31/5 Meanwhile the meteorologists were translating the cloud pictures into diagrammatic meteorological charts, known as 'nephanalyses'. **1965** *Q. Jrnl. R. Meteorol. Soc.* XCI. 526 The latest Tiros 9 weather satellite is on a polar orbit and facsimile nephanalyses are received twice daily at Bracknell for an area close to or covering the British Isles. **1973** Barry & Perry *Synoptic Climatology* ii. 52 With the nimbus and essa satellites, global coverage is now available on a daily basis and nephanalyses are prepared from the photography.

nephelauxetic (nefĕlǭkse·tik), *a.* *Physical Chem.* [f. Gr. νεφέλ-η cloud + αὐξητικ-ός Auxetic *a.*] Causing an expansion of the *d*-electron cloud of a central atom, i.e. a decrease in the inter-electron repulsion parameter; *nephelauxetic series*, a series of ligands arranged in order of their nephelauxetic effects.

1958 Schäffer & Jørgensen in *Jrnl. Inorg. & Nuclear Chem.* VIII. 147 We shall propose to call this series the nephelauxetic (cloud expanding) series, as it corresponds to a development of the *d* shell in the region of the ligands. The neo-greek word was constructed by Professor Kaj Barr from the University of Copenhagen. **1965** *Molecular Physics* X. 7 The red shift of the ligand field bands in metal complexes with respect to the free ions has been explained..as being due to the nephelauxetic effect. **1966** Phillips & Williams *Inorg. Chem.* II. xxviii. 402 The nephelauxetic (cloud-expanding) series.., it is suggested, represents the spread of the electron away from its parent nucleus on increase of covalent character in the binding of a given cation. **1973** *Chem. Soc. Rev.* II. 177 The *cis*-influence of ligands increases in accordance with their position in the nephelauxetic series.

nepheline. Add: **nepheline-syenite** [ad. G. *nephelin-syenit* (H. Rosenbusch *Mikrosk. Petrogr.* (1877) II. 203)], a rare plutonic rock which resembles syenite in containing alkali feldspars such as orthoclase as essential minerals and usu. dark ferromagnesian minerals also (commonly amphibole or pyroxene), but differs in containing nepheline as an additional essential mineral, in being rich in soda but always lacking quartz, and in the frequent occurrence of rare minerals as accessories.

1892 F. H. Hatch *Text-bk. Petrol.* (ed. 2) vi. 143 The syenites and diorites are plutonic granitoid rocks, less acid than granite, and consequently containing little or no free silica. They may be divided into—Syenites. Nepheline-syenites. Diorites. **1938** A. Johannsen *Descr. Petrogr. Ign. Rocks* IV. 78 The nepheline-syenites are generally gray rocks with a greasy luster, usually moderately dark... Yellows, reds, and blues are not unknown. **1967** M. J. Coe *Ecol. Alpine Zone Mt. Kenya* 11 The lava cooled slowly and produced a material that is highly crystalline. These rocks are called nepheline-syenite and contain large Felspar crystals up to 3 cm long as well as smaller nepheline crystals.

nephelinite. Add: Hence **nephelini·tic** *a.*, containing or characteristic of nephelinite.

1909 J. P. Iddings *Igneous Rocks* I. ii. ii. 386 Rocks rich in feldspar have a trachytic texture; those rich in nephelite have a nephelinitic texture. **1944** C. Palache et al. *Dana's Syst. Min.* (ed. 7) I. 733 Pegmatites of the contact zone of the nephelinitic intrusive of the Chibina tundra, Kola. **1974** *Nature* 8 Feb. 354/1 An older series of basaltic-trachytic shield volcanoes, together with smaller nephelinitic centres.

nephelinization (nefĕlinəizēᶦ·ʃən). *Petrol.* [f. Nephelin(e + -ization.] The alteration of a rock to one in which nepheline is an essential constituent.

1943 *Science* 26 Mar. 286/2 It is believed that the nephelinization is post-folding, since the flow marble contains fragments of all rocks except those containing nepheline. **1969** *Bull. Brit. Mus.* (*Nat. Hist.*) *Mineral.* II. 214 The nepheline syenites and ijolites are capable of desilicating the enclosing rock envelope—nephelinization. *Ibid.*, The nephelinization at Tundulu has been shown..to be related to the main ring-dyke of foyaite.., while at Dorowa the nephelinized fenite..surrounds a body of foyaite and ijolite.

So **ne·phelinized, ne·phelinizing** *ppl. adjs.*

1943 *Science* 26 Mar. 286/2 (*heading*) The nephelinized paragneisses of the Bancroft region, Ontario. **1946** *Jrnl. Geol.* LIV. 167/2 The nephelinizing solutions..carried..soda and alumina, with quantities of volatiles, including H_2O, Cl, P, and others. **1959** W. W. Moorhouse *Study Rocks in Thin Section* xv. 311 The nephelinized gneisses include a wide variety of melanocratic to leucocratic rocks. **1969** Nephelinized [see above].

nephelo-. Add: **nephelometer,** (*b*) any of various instruments by which the turbidity of a suspension, culture, etc., can be measured or compared with a standard by means of the light scattered (at right angles) by it; **nephelometric,** *a.*, add: or to nephelometry (examples); **nephelome·trically** *adv.*, by nephelometry; **nephelo·metry** *Chem.*, the technique of quantitative analysis using a nephelometer.

1895 T. W. Richards in *Proc. Amer. Acad. Arts & Sci.* XXX. 385 Since the opalescence was so faint that one could only with difficulty see it at all under ordinary conditions, a piece of apparatus, which may be named a 'nephelometer' (νεφέλη, a cloud), was devised for detecting it. **1906** McFarland & L'Engle in *Medicine* (Detroit) XII. 249/1 It occurred to us that uniformity in the number of bacteria..could be secured with reasonable accuracy by some means of measuring and standardizing the turbidity of the fluid containing them. We therefore devised a simple instrument, for which we suggest the name *nephelometer*.., by which it became easy to secure any desired degree of turbidity. **1936** F. D. & C. T. Snell *Colorimetric Methods of Anal.* viii. 91 In the usual nephelometer the opaque tubes with clear glass bottoms used in colorimetry have been replaced with clear glass tubes with opaque bottoms. **1969** *Atmospheric Environment* III. 561 The multi-wavelength adaptation of the integrating nephelometer makes possible the local measurement of the wavelength dependence of the extinction coefficient due to scatter of atmospheric air. **1974** *Times-Herald-Record* (Middletown, N.Y.) 12 May, (*caption*) Researchers set up a light scattering instrument called an integrating nephelometer atop Mount Beacon. The tests were part of an air-pollution study. **1905** *Jrnl. Amer. Chem. Soc.* XXVII. 485 The application of these observations to the nephelometric analysis of a very dilute silver chloride solution is obvious. **1929** P. A. Kober in J. H. Yoe *Photometric Chem. Anal.* II. vi. 71 Precipitants for the production of nephelometric suspensions are as varied as the substances precipitated. **1971** P. R. Hesse *Textbk. Soil Chem. Anal.* xii. 314 The only difference between turbidimetric and nephelometric measurement of sulphate is in the measurement of the final turbidity. **1905** *Jrnl. Amer. Chem. Soc.* XXVII. 507 When in a day or two the precipitate had settled and the mother-liquor had become clear the latter was examined nephelometrically. **1971** P. R. Hesse *Textbk. Soil Chem. Anal.* xvi. 423 Silver is extracted from soil by acid digestion and is classically determined nephelometrically as the chloride. **1876** T. P. Blunt in *Chem. News* 7 Jan. 7/2 It would appear that the usefulness of colorimetry, and also of judgment by turbidity, which may provisionally be termed 'nephelometry', might be widely extended. **1906** *Amer. Chem. Jrnl.* XXXV. 113 In order to save the time of those attempting nephelometry, the precautions for several typical cases are here collected. **1929** J. H. Yoe *Photometric Chem. Anal.* II. iii. 18 When colorimetry or nephelometry is used at all, in routine analytical work or in research problems, it is apt to be used almost continuously. **1966** *McGraw-Hill Encycl. Sci. & Technol.* IX. 40/1 The advantage of nephelometry [over turbidimetry] is its greater sensitivity, accuracy, and precision in the determination of small amounts of turbidity.

nepheloid, *a.* Delete *rare*—⁰ and add examples referring to the ocean; chiefly in *nepheloid layer*, a layer about a kilometre thick in the deep water of the western North Atlantic and elsewhere that is turbid owing to suspended mineral matter.

1965 Ewing & Thorndike in *Science* 12 Mar. 1291/1 Strong background haze in some of our ocean bottom photographs has been difficult to explain by any cause other than clouded (nepheloid) water. *Ibid.* 1291/2 The degree to which the water is nepheloid. *Ibid.* 1291/3 A sample of 200 liters obtained from the nepheloid layer was found..to contain some organisms and about 0·50 g of suspended lutite. **1972** Ettreim & Ewing in A. L. Gordon *Stud. Physical Oceanogr.* II. 123 The turbulence associated with the vigorous bottom currents of the western North Atlantic maintains a nepheloid layer with an average thickness of one kilometer. **1974** *Nature* 6 Sept. 43/2 The thickness of this nepheloid layer is several orders of magnitude greater than the characteristic thickness of the turbulent Ekman layer at the ocean floor.

nephometer (nefǫ·mītəɹ). [ad. F. *néphomètre* (L. Besson 1906, in *Annuaire de la Soc. météorol. de France* LIV. 241), Gr. νέφ-ος cloud + -ometer.] = *nephelometer* s.v. Nephelo-.

1910 J. Moore *Meteorol.* (ed. 2) xvii. 218 Spherical Mirror Nephometer.—This instrument..permits the cloud-percentage (nebulosity) to be measured without any fear of an error in the number of the tenths. The description of, and the method of using, this new nephometer have been given by the inventor, M. L. Besson. **1959** R. E. Huschke *Gloss. Meteorol.* 388 *Nephometer* (also called *nephelometer*), a general term for instruments designed to measure the amount of cloudiness.

nephrectomize, *v.* Add: (Further example.) So **nephre·ctomized** *ppl. a.*, deprived of a kidney by surgery.

1949 Florey & Jennings in H. W. Florey et al. *Antibiotics* II. xl. 1286 The slopes of the curves showing the disappearance of the two penicillins from the plasma were roughly parallel for both normal and nephrectomized dogs. **1953** Nephrectomized [see *hypokalæmic* adj. s.v. *hypo- II]. **1972** *Science* 9 June 1146/3 Rats..either had their ureters ligated or were bilaterally nephrectomized.

nephridiopore (nefri·diopō͞ᵊɹ). [f. Nephridium + Pore *sb.*¹] The excretory opening of a nephridium.

1888 F. E. Beddard in *Q. Jrnl. Microsc. Sci.* XXVIII. 397 There are more than one pair of nephridiopores in each segment of the body. **1946** *Nature* 9 Nov. 665/2 Overton observed an initial loss of weight on handling the worm, and attributed this to the expulsion of fluid through the nephridiopores. **1963** R. P. Dales *Annelids* v. 98 The fluid which issues from the nephridiopores of an earthworm does not have the same composition as the coelomic fluid.

nephridiostome (nefri·diostō͞ᵘm). [f. Nephridium + Stoma.] = *nephrostome* s.v. Nephro-.

1902 *Encycl. Brit.* XXXIII. 882/2 The Hesionidæ have compound organs, serving both as excretory and as genital ducts, formed by the grafting of the coelomostome on to the nephridiostome. **1963** R. P. Dales *Annelids* i. 29 In many other polychaetes..and in oligochaetes and leeches, an open funnel or nephridiostome is formed.

nephro-. Add: **ne:phrocalcino·sis,** deposition of concretions of calcium compounds in the kidneys; **nephrocyte** *Zool.* [ad. G. *nephrocyt* (A. Korotneff 1894, in *Mitt. Zool. Stat. Neapel* XI. 344)], a cell in insects which stores or excretes waste products; **nephromi·xium** (pl. **-mixia**) *Zool.* [Gr. μίξις mingling], in certain polychaetes, an organ formed by the fusion of the coelomoduct and the nephridium; **nephropathy,** for 'diseases' read 'disease' and add examples; hence **nephropa·thic** *a.*; **ne·phropexy** *Surg.* [Gr. πῆξις fixing], the operation of securing an abnormally movable kidney; **nephrosclero·sis,** thickening and hardening of the walls of the blood vessels of the kidney, which is often associated with hypertension and can lead to renal failure; **nephro·stomy** *Surg.* [ad. F. *néphrostomie* (Guyon & Albarran 1898, in *Rev. de Chir.* XVIII. 1052), f. Gr. στόμα mouth], the operation of making an opening from the surface of the body directly into the pelvis of the kidney; **ne·phrotome** *Zool.* [ad. G. *nephrotom* (J. W. van Wijhe 1889, in *Arch. f. mikrosk. Anat.* XXXIII. 465): see *-tome], a block of tissue at the edge of a somite, giving rise to the excretory organs; **nephroto·xic** *a.*, having a toxic effect on the kidneys; so **ne:phrotoxi·city,** the property or effect of being nephrotoxic; **ne·phrotoxin,** (*a*) a nephrotoxic antibody produced by injecting kidney tissue into an animal; (*b*) any nephrotoxic substance.

1934 F. Albright et al. in *Amer. Jrnl. Med. Sci.* CLXXXVII. 60 The initial disturbance, however, is not an inflammation, but presumably a deposition of calcium. Therefore, the term chronic nephro-calcinosis would seem preferable. **1951** A. C. Allen *Kidney* xi. 366/2 Parenchymal nephrocalcinosis rarely interferes with renal function to any significant degree. **1974** Passmore & Robson *Compan. Med. Stud.* III. xxiii. 51/1 In nephrocalcinosis calcium is deposited in the basement of renal tubules in cortex and medulla, in tubular lumina and in small foci in the interstitial tissue. **1895** *Jrnl. R. Microsc. Soc.* 165 He [*sc.* Korotneff] finds colossal cells, whose possible function is suggested in the title 'nephrocytes'. **1932** Borradaile & Potts *Invertebrata* xiv. 390 Nitrogenous end products are found in the nephrocytes (cells found commonly associated with the fat body and the pericardium [of insects]). **1969** R. F. Chapman *Insects* xxv. 494 Nephrocytes, or pericardial cells, are cells occurring singly or in groups in various parts of the body. They may be very large, as in dipterous larvae, or small and numerous and usually they contain more than one nucleus. **1900** E. S. Goodrich in *Q. Jrnl. Microsc. Sci.* XLIII. 742 The ordinary wide-mouthed segmental

organs of the Polychæta, formed by the fusion of the nephridium with the genital funnel, may be called Nephromixia... Kindly suggested to me by Professor E. Ray Lankester. **1900** E. R. Lankester *Treat. Zool.* II. ii. 37 The composite organ thus formed may be termed a 'nephromixium' or 'nephromix', in reference to its hybrid composition. **1932** Borradaile & Potts *Invertebrata* ix. 229 Nephromixia may take on the functions of coelomoducts where these do not exist independently. **1963** R. P. Dales *Annelids* i. 30 In other polychaetes the coelomoducts may be grafted on to the stem of the nephridium to form a nephromixium, which may be used both as a genital and as an excretory duct. **1917** Dorland *Med. Dict.* (ed. 9) 655/1 Nephropathic. **1973** *Nature* 3 Aug. 289/2 In the nephropathic form of cystinosis, patients appear normal at birth. **1916** Nephropathy [see *nephrosis b]. **1956** Root & White *Diabetes Mellitus* xv. 189 Diabetic nephropathy is commonly first indicated by recurring edema, persistent albuminuria and in many cases an increase in the plasma lipids. **1968** A. Walsh tr. *J. Hamburger's Nephrology* I. xvi. 529/1 Even in the group of nephropathies of toxic origin, mercury poisoning takes second place..after carbon tetrachloride. **1897** *Jrnl. Amer. Med. Assoc.* 11 Dec. 1190/1 Nephropexy is a legitimate and established procedure in all cases in which it can be established that the kidney is not only displaced, but is at the same time the direct cause of the manifold symptoms which such a condition may and often will produce. **1932** Ball & Evans *Dis. of Kidney* xiii. 394 Nephropexy has proved a sufficient method of treatment of mild degrees of hydronephrosis associated with excessive mobility of the kidney. **1968** Nephropexy [see *fixation 3 c]. **1890** Billings *Med. Dict.* II. 199/2 Nephrosclerosis. **1926** H. Elwyn *Nephritis* xvi. 280 Arteriosclerosis of the kidney with its end stage of primary contracted kidney was classified under the term nephrosclerosis. **1951** A. C. Allen *Kidney* xiii. 397/1 The color of the kidneys with malignant nephrosclerosis is brownish or greyish red. **1966** *McGraw-Hill Encycl. Sci. & Technol.* VII. 343/2 Nephrosclerosis is only part of a generalized disorder of arteries. **1900** Keen & DaCosta *Amer. Yr.-bk. Med. & Surg.* II. 184 MM. Guyon and T. Albarran discussed the subject of nephrotomy at the French Surgical Congress of 1898. They restrict the term nephrotomy to the making of an incision into the kidney; but when the pelvis of the kidney is opened through the kidney-substance, and the wound is kept open and a fistula is formed, they call the operation nephrostomy. *Ibid.*, Nephrostomy is employed, in the first place, to relieve renal retention, septic or aseptic. **1932** *Jrnl. Amer. Med. Assoc.* 8 Oct. 1229/2 Nephrostomy on only one side makes the care of the urinary drainage apparatus much easier. **1967** S. Taylor et al. *Short Textbk. Surg.* xxv. 357 If the patient is too ill for a major operation nephrostomy or drainage of the kidney only may be possible, with secondary nephrectomy later, a very difficult operation. **1895** Gadow & Abbott in *Phil. Trans. R. Soc.* B. CLXXXVI. 166 Concerning the segmentally arranged mesodermal products (emitting nephrotomes and gonotomes) the following subdivision is adhered to. **1949** A. S. Romer *Vertebrate Body* xii. 378 In our embryological story we noted that in every trunk segment the mesoderm includes, on either side, a nephric region, often segmentally distinct as a nephrotome, a small discrete block of tissue interposed between somite and lateral plate. **1974** M. Hildebrand *Analysis of Vertebrate Structure* xiii. 308 Relatively few nephrotomes are incorporated [in the pronephros]. **1902** Vaughan & Novy *Cellular Toxins* (ed. 4) vii. 144 The blood of animals in which one ureter has remained tied for some time becomes laden with a nephrotoxic substance. **1937** Nephrotoxic [see *nephrotoxin* below]. **1973** *Jrnl. Pharmacol. & Exper. Ther.* CLXXXVI. 593/1 Administration of the nephrotoxic agent uranyl nitrate to rats specifically stimulated organic base accumulation by renal cortical slices. **1961** *Lancet* 22 July 179/1 The new preparation..appeared..to be devoid of..nephrotoxicity. **1970** Passmore & Robson *Compan. Med. Stud.* II. xxxii. 4/2 Mercury poisoning is the oldest and best understood type of nephrotoxicity. **1902** *Jrnl. Chem. Soc.* LXXXIV. ii. 443 Repeated injections into the rabbit..of kidney cells from the dog..provoke the appearance in the rabbit's blood of a substance, nephrotoxin, which is most harmful to the kidney cells of the dog. **1937** *Jrnl. Exper. Med.* LXV. 564 The nephrotoxic effect induced by anti-kidney serum is dependent upon a relatively organ specific antibody, nephrotoxin. **1961** A. G. White *Clin. Disturbances of Renal Function* v. 153 Among the nephrotoxins encountered most frequently in clinical practice is carbon tetrachloride. **1970** R. C. Muehrcke in F. W. Sunderman *Lab. Diagnosis Kidney Dis.* xxxvii. 444/1 Other nephrotoxins such as sulfonamides produce their adverse pathopharmacological effects through a hypersensitivity reaction.

nephron (ne·frǫn). *Anat.* [a. G. *nephron* (H. Braus *Anat. d. Menschen* (1924) II. 351), f. Gr. νεφρός kidney.] Each of the numerous filtration units in the kidney, which consist of a tube divided (in higher forms) into a glomerulus, a proximal convoluted tubule, a loop of Henle, a distal convoluted tubule and a collecting tubule, and through which the glomerular filtrate passes, undergoing selective reabsorption and emerging as urine.

1932 *Anat. Record* CIV. 185 The nephron of the sculpin was nevertheless compared with that of the toadfish. **1937** *Amer. Jrnl. Anat.* LXI. 21 (*heading*) Observations upon the structure of the nephron in the common eel. **1965** *New Scientist* 24 June 868/2 A human kidney contains approximately one million units called nephrons, each consisting of a thin tube about 20 to 50 micrometres wide and 50 millimetres long. **1974** Passmore & Robson *Compan. Med. Stud.* III. xxii. 11/1 Such surviving nephrons would be in a state of continuous osmotic diuresis. **1975** A. Dillard *Pilgrim at Tinker Creek* viii. 133 The nephron..is a filtering structure which produces urine and reabsorbs nutrients.

nephrops (ne·frǫps). [mod.L. (W. E. Leach 1816, in *Trans. Linn. Soc.* XI. 344), f. Nephr(o- + Gr. ὤψ eye.] = *Dublin (Bay) prawn* (*Dublin); = *Norway lobster* (s.v. Norway¹ in Dict. and Suppl.).

[**1830** *Edin. Encycl.* VII. 398/2 This last [*sc.* *Astacus Norvegicus*] is considered as a distinct genus by Mr. Leach, under the name of Nephrops, from the kidney shaped eye.] **1961** *New Statesman* 23 June 1000/3 She is fishing off Barra, both seine-netting and after nephrops (scampi), a new and at present a paying type of fishery. **1971** *Nature* 29 Jan. 299/1 *Nephrops* tends to stay hidden away during the day, coming out during the night.

nephrosis (nefrō͞ᵘ·sis). *Path.* Pl. **-oses.** [f. Nephr(o- + -osis.] †**a.** (See quot.) *Obs.* exc. as in *b.

1900 Dorland *Med. Dict.* 428/1 *Nephrosis*, any disease of the kidney.

b. [a. G. *nephrose* (F. Müller 1905, in *Verhandl. d. Deut. Path. Ges.* IX. 65).] A syndrome characterized by œdema, albuminuria, a fall in the plasma albumin, and usu. an increase in the blood cholesterol, formerly attributed to degeneration of the renal tubules but now known to arise from increased permeability to protein of the glomerular capillary basement membranes.

1916 L. F. Barker *Monogr. Med.* III. 954 (*heading*) Toxic degenerative tubular nephropathies without marked glomerular involvement (the so-called nephroses). **1926** H. Elwyn *Nephritis* v. 74 Tubular degenerations occurring in the course of other diseases, when presenting definite renal symptoms, may be similarly designated by the term nephrosis of the particular disease in which they occur. **1929** *Amer. Jrnl. Path.* V. 619 Lipoid nephrosis is to be regarded as a form of glomerulonephritis in which the glomeruli are damaged but their capillaries are only partially obstructed. **1946** E. T. Bell *Renal Dis.* vi. 217 The tubular atrophy in lipoid nephrosis is due to disuse and anemia. **1973** J. Brod *Kidney* xix. 405/1 Unlike glomerulonephritis pure nephrosis is encountered mostly in children under 5 years of age.

nephrotic (nefrǫ·tik), *a.* *Path.* [f. prec.: see -otic.] Of, associated with, or suffering from nephrosis.

1928 *Amer. Jrnl. Path.* IV. 633 Every case of glomerulonephritis has a certain nephrotic element. **1940** *Jrnl. Clin. Invest.* XIX. 317/1 The albumin was much reduced in plasma, but unlike a nephrotic serum with a similar diminution of albumin, the globulin increase was most striking in the γ-fraction. **1954** *Amer. Jrnl. Physiol.* CLXXVIII. 329/2 The results demonstrated either an unchanged or a diminished hepatic synthesis of cholesterol by the nephrotic rat. **1966** *McGraw-Hill Encycl. Sci. & Technol.* III. 141/2 An example of edema formation secondary to a decrease in the colloid osmotic pressure of the plasma proteins is nephrotic edema.

b. *nephrotic syndrome* = *nephrosis b.

1939 *Arch. Internal Med.* LXIII. 646 Throughout the last fifteen months of the patient's illness all the cardinal signs of the nephrotic syndrome were repeatedly found. **1961** *Lancet* 5 Aug. 290/2 Specimens were obtained from..2 [patients] with the nephrotic syndrome. **1974** Passmore & Robson *Compan. Med. Stud.* III. xxii. 30/2 In the case of toxic or allergic nephrotic syndrome the offending agent should be removed.

nephsystem (ne·fsistĕm). *Meteorol.* [f. Gr. νέφ-ος cloud + System.] (See quot. 1959.)

1945 F. A. Berry et al. *Handbk. Meteorol.* xi. 903 Cloud forms and cloud species over large areas form synoptic entities (called nephsystems) usually surrounded by fair-weather areas of clear sky, cumulus humilis, or cirriform clouds. **1959** R. E. Huschke *Gloss. Meteorol.* 113 *Cloud system* (or *nephsystem*), an array of clouds and precipitation associated with a cyclonic-scale feature of atmospheric circulation. **1974** E. C. Barrett *Climatol. from Satellites* iv. 116 Especially heavy and prolonged rain may be expected from major nephsystems within the meteorological tropics.

nepionic (nī͵pi͵ǫ·nik), *a.* *Zool.* [f. Gr. νήπιος child + *-onic* as in *embryonic.*] Larval; in an early stage of development.

1889 A. Hyatt in *Smithsonian Contrib. Knowledge* XXVI. ii. 9 The first larval or næpionic stage of a Nautiloid was, therefore, represented by the apex of the conch in that order. **1893** —— in *Proc. Boston Soc. Nat. Hist.* XXVI. 94 Larval or young: nepionic. Here first used as a substitute for 'næpionic'. **1898** A. S. Packard *Text-bk. Ent.* 706 As regards the organization of larval (nepionic) as compared with imaginal forms, the nymphs and larvæ of insects are, with the exception of many Diptera, nearly as perfectly developed as the adult. **1903** *Amer. Naturalist* XXXVII. 517 The nepionic stage of shell growth begins with the second whorl.

‖ **nepitella** (nepite·lă). [It.] = Calamint.

1926 D. H. Lawrence *Sun* iv. 13 The paths were all grown high with grass and flowers and nepitella.

nepman, NEP-man: see *N.E.P. (as main entry).

nepotistic (nepǒti·stik), *a.* [f. Nepotist + -ic.] Characterized by, or given to, nepotism.

1936 *Fortune* Jan. 133 Mr. Aldrich's sudden elevation, whereas it may have been dramatic, was certainly not nepotistic. **1949** Koestler *Promise & Fulfilment* ii. v. 296 The one-time pioneers have, as so often happens, developed into a somewhat nepotistic coterie of the ancients. **1955** W. Gaddis *Recognitions* ii. i. 305 The Viareggio, a small Italian bar of nepotistic honesty before it was discovered by exotics.

Neptune. 3. Add to def.: and lying beyond Uranus. (A more distant planet, *Pluto¹, was discovered in 1930.)

Neptunism (ne·ptiŭniz'm). *Geol.* [f. Neptun(e + -ism.] = Neptunianism.

1905 A. Geikie *Founders of Geol.* (ed. 2) viii. 257 Powerful as an advocate for the Vulcanist doctrines in opposition to the prevailing Neptunism of his time, he wrote some excellent monographs on the geology of different parts of Italy. **1951** C. C. Gillispie *Genesis & Geol.* ii. 46 The appeal of Neptunism is easier to understand than its acceptability. **1953** S. F. Mason *Hist. Sci.* xxxiii. 321 In the period 1790–1830..the Vulcanist view became associated with the theory that rock strata had gradually evolved, and Neptunism with the theory that the strata were formed suddenly and catastrophically. **1966** *Mercian Geol.* I. 291 The geological concepts Plutonism and Neptunism have, until recently, been thought to belong more properly to the igneous and sedimentary environments respectively.

Neptunist. 2. For '*attrib.*' read '*attrib.* or as *adj.*', and add further examples.

1905 A. Geikie *Founders of Geol.* (ed. 2) viii. 262 He would have run some risk of being regarded as having gone over to the Neptunist camp. **1951** C. C. Gillispie *Genesis & Geol.* ii. 44 The Neptunist synthesis explained stratification by postulating that all rock formations had been precipitated from an aqueous solution and suspension. **1965** M. Smith *Essent. Mod. Geol.* vi. 101 Considerable controversy existed in the eighteenth century until the Vulcanists, led by Nicholas Desmarest, finally overcame the opposition of the Neptunists.

neptunite (ne·ptiŭnəit). *Min.* [ad. Sw. *neptunit* (G. Flink (at the suggestion of N. O. Holst) 1893, in *Geol. Förening i Stockholm Förhandl.* XV. 196): see Neptune and -ite¹. (So called because of its occurrence with Ægirite, Ægir being the Scandinavian god of the sea.)] A silicate of sodium, potassium, iron, manganese, and titanium, $(Na,K)_2(Fe^{II},Mn)TiSi_4O_{12}$, which is found as black monoclinic crystals.

1895 *Mineral. Mag.* XI. 100 Owing to the close similarity of the crystal angles to those of sphene, neptunite is placed in the titanite group. **1926** *Mineral. Abstr.* III. 102 Neptunite has been found at eight spots in the Kola peninsula. **1950** *Mineral. Mag.* XXIX. 27 The main point of interest in the new material is the presence here and there of the rare mineral neptunite in the syenite, a mineral previously unknown in the British Isles. **1966** *Doklady Acad. Sci. U.S.S.R.* (*Earth Sci. Sect.*) CLXVI. 121/2 The pyroxenic structure of neptunite is typically apparent from the endless zigzag baroque columns, in which Ti and Mn octahedra connected by common edges alternate in pairs.

neptunium (neptiū·niŏm). *Chem.* [mod.L., f. Neptun(e + -ium.] † **1.** [coined in G. (R. Hermann 1877, in *Jrnl. f. prakt. Chem.* XV. 105).] A supposed element similar to tantalum found in a sample of tantalite from Haddam, Connecticut, U.S.A. *Obs.*

1877 *Jrnl. Chem. Soc.* XXXII. 167 Neptunium is distinguished from tantalum by the fact that its fluoride forms an easily soluble double salt with potassium fluoride. **1877** *Potter's Amer. Monthly* Sept. 238/2 Hermann calls it neptunium.

2. An artificially produced transuranic element (traces of which have subsequently been found in nature) which is a silvery metal and whose longest-lived isotope has a half-life of about 2¼ million years. Atomic number 93; symbol Np.

The word does not occur in the article by McMillan and Abelson (*Physical Rev.* (1940) LVII. 1185) in which they announced the discovery of the element, though McMillan is often said to be the coiner (e.g. quot. 1945; cf. also *Jrnl. Amer. Chem. Soc.* (1948) LXX. 1128 (an abridgement of a secret report of 1942)).

1941 *Sci. News Let.* 30 Aug. 135/1 The uranium outpost was passed some years ago by Prof. Enrico Fermi..with his discovery of the radioactive element No. 93, now called neptunium. **1945** *Chem. & Engin. News* 10 Dec. 2190/3 Element 93 was given the name neptunium by McMillan after Neptune, the planet immediately beyond Uranus, which gives its name to uranium. **1946** *Electronic Engin.* XVIII. 88 On bombarding the 238 isotope of uranium with a neutron of resonant energy value, a new isotope of uranium..is formed. This isotope is unstable..and emits one electron to become a new element..called neptunium. **1950** M. Lowry *Let.* 6 Mar. (1967) 200 Oddly enough I put neptunium in but abandoned it for niobium... I just took the elements out of the dictionary. **1968** *New Scientist* 23 May 410/2 Neptunium-237 is an important isotope because it is the precursor of plutonium-238. **1974** *Encycl. Brit. Micropædia* VII. 261/3 Neptunium is chemically reactive and similar to uranium with oxidation states from +3 to +6.

neral (nīᵊ·ræl). *Chem.* [f. *Ner(ol + Al(dehyde.] A colourless oily aldehyde, $C_{10}H_{16}O$, which is the *cis* form of citral (*citral² b) and gives nerol on reduction.

1939 [see *citral¹]. **1953** [see *geranial]. **1973**

NERAM

Zoon Suppl. I. 56 The main components of the secretion [of four species of *Prosopis*] have been identified as the two isomers of citral, geranial..and neral.

neram (nera·m). [Malay.] A large evergreen tree, *Dipterocarpus oblongifolius*, of the family Dipterocarpaceæ. Also *attrib.*

1927 F. W. FOXWORTHY *Commercial Timber Trees of Malay Peninsula* 41 *D[ipterocarpus] oblongifolius*..is known as Neram and is not used commercially. **1940** E. J. H. CORNER *Wayside Trees of Malaya* I. 211 The *Neram* is the big tree that arches over the rocky rivers in the eastern and northern states of Malaya. **1965** R. MCKIE *Company of Animals* i. 24 Neram trees, with trunks that weigh many tons and sprawling root systems ..reached thirty degrees from the jungle banks to meet above the centre of the stream.

nerf (nɔ̄ɹf), *v. slang* (orig. *U.S.*). [Origin unknown.] *intr.* In drag-racing, to bump another car. Hence **ne·rf-bar, ne·rfing-bar,** a bumper fitted to a car used in drag-racing.

1953 BERREY & VAN DEN BARK *Amer. Thes. Slang* (1954) § 728/1 *Nerfing*, bumping another car out of the way. **1955** *Hot Rod Mag.* May 28 The nerf bar itself is mounted in a 'slip tube' that is welded permanently to the reworked bumper irons. **1960** WENTWORTH & FLEXNER *Dict. Amer. Slang* 352/2 The nerfing bar that supports the bumper on most cars. **1962** *Punch* 17 Oct. 561/1 A custom-built nerfing bar (bumper). **1969** R. E. JENNINGS *Automotive Dict.* 158/1 *Nerf bar:* see 'Nerfing Bar'. *Nerfing bar*, small, lightweight, vertical bumpers normally used on race cars, hot rods, and custom cars. *Ibid.*, *Nerf*, to bump, shove, or push a car during a racing event with another racer. Nerfing is very popular on short tracks.

nerine (nīrəi·nī). *Bot.* Also **nerina.** [mod.L. (W. Herbert 1820, in *Curtis's Bot. Mag.* XLVII. 2124), f. L. *Nērīnē* (Virg. *Ecl.* vii. 37), Gr. νηρηίς a water nymph; see NEREID.] A South African bulbous plant of the genus so called, belonging to the family Amaryllidaceæ and including *Nerine bowdenii*, widely cultivated for its autumn-blooming pink flowers, and the Guernsey lily, *Nerine sarniensis*. Also *attrib.*

1820 *Curtis's Bot. Mag.* XLVII. 2124 Rose-coloured Nerine... Nerine is probably confined to South Africa. **1837** W. HERBERT *Amaryllidaceæ* 285, I have no hesitation in saying that it is a Nerine. **1886** G. NICHOLSON *Dict. Gardening* II. 446/2 When in flower, Nerines are amongst the most beautiful of greenhouse bulbous plants. **1923** *Chambers's Jrnl.* Dec. 786/2 The scarlet or rose-red nerine (the Japanese spider lily) appeared next. **1929** *Amateur Gardening* 3 Aug. 292/3 Nerines, or Guernsey lilies, are attractive flowering bulbs. **1949** L. G. GREEN *In Land of Afternoon* v. 69 They [*sc.* lilies growing on the Guernsey coast] were identified as the nerinas of the Table Mountain ledges. **1955** K. A. THOMPSON *Great House* iv. 116 A Nerine lily glistening like rose-crystal. **1961** *Amateur Gardening* 14 Oct. 19/3 Nerines are often looked upon as bulbs to grow in favoured gardens. **1974** R. L. FOX *Variations on Garden* 165, I prefer its [*sc.* the amaryllis's] cousin, the Nerine or Guernsey lily.

neritic (nīri·tik), *a.* [a. G. *neritisch* (E. Haeckel 1890, in *Jenaische Zeitschr. Naturw.* XXV. 253), perh. f. NERIT(A + -IC.] Of, pertaining to, or inhabiting the region of water bordering coasts, down to a depth of a hundred fathoms.

1891 MURRAY & RENARD *Rep. Deep-Sea Deposits* iv. 251 The organisms living in mid-ocean in the great oceanic currents are quite different from those in the surface waters near land, and Haeckel proposes to designate the former oceanic Plankton, and the latter neritic Plankton. **1909** [see *holoplankton* (*HOLO-)]. **1913** J. MURRAY *Ocean* vii. 136 The neritic area surrounds all continents and islands. **1926** [see *bathyal adj.]. **1957** *Sci. News* XLIII. 71 Certain environments, such as the neritic zone of the ocean, are much more commonly represented in the fossil record than others. **1967** *New Scientist* 16 Mar. 546/2 Marine fish from the 'neritic' zone between the low-water mark and the edge of the continental shelf. **1974** LUCAS & CRITCH *Life in Oceans* i. 24 The pelagic division is divided into the region inshore of the continental edge, known as the 'neritic province', and the remainder, called the 'oceanic province'.

Nernst (nɛ̄ɹnst, nɔ̄ɹnst). [Name of Walther Hermann *Nernst* (1864–1941), German physical chemist.] **a.** Used *attrib.* to designate an electric incandescent lamp invented by Nernst in which an unenclosed rod or wire consisting of a mixture of rare earths and other metallic oxides (as magnesia or zirconia) is made hot and luminous by the passage of an electric current (after being first brought to a conducting state by heating), and which is used esp. as a source of infra-red radiation.

1899 *Chambers's Jrnl.* 25 Mar. 269/2 For some time there have been rumours of an electric lamp on an entirely new principle, and..the contrivance was recently exhibited at the Society of Arts, London. It is known as the Nernst incandescent electric lamp, and its chief peculiarity is that it employs a rod of refractory earth in place of the usual carbon filament, and that this material is not enclosed in a glass exhausted of air. **1912** W. S. FRANKLIN *Electric Lighting* v. 134 There are five important kinds of glow lamps as follows:... (*e*) The Nernst lamp in which the glower is a small rod of porcelain-like material. **1950** L. J. BRADY in M. G. Mellon *Analytical Absorption Spectrosc.* viii. 444 One serious objection to the Nernst glower is the frequent mechanical failure of the source due to the poor bonding of the platinum leads to the element itself. **1962** R. E. DODD *Chem. Spectrosc.* i. 16 For use in the infra-red region, mainly beyond 2μ, the hot body takes the form of a rod of semi-conducting lanthanon oxides, a Nernst glower. *Ibid.* 17 The Nernst filament is only suitable for the near infra-red and visible.

b. Used *attrib.* with reference to a thermo-magnetic effect investigated by Nernst, in which a temperature gradient in a metal subject to a magnetic field at right angles to the gradient gives rise to an e.m.f. in a direction at right angles to both.

1901 M. G. LLOYD in *Amer. Jrnl. Sci.* CLXII. 57 It has already been proposed† to call the galvano-magnetic temperature-difference, the thermo-magnetic temperature-difference, and the thermo-magnetic potential-difference by the respective names, Ettingshausen effect, Leduc effect and Nernst effect. [*Note*] † Thesis: The Transversal Thermo-magnetic Effect in Bismuth, M. G. Lloyd, Philadelphia, 1900; Beiblätter, 24, p. 1014. **1911** *Physical Rev.* XXXIII. 300 Both the Hall electromotive force and the Nernst electromotive force seem to be proportional to the intensity of magnetization in the plate rather than to the magnetic field. **1960** E. H. PUTLEY *Hall Effect & Related Phenomena* ii. 27 In the Nernst effect, electrons attempting to diffuse down a temperature gradient are deflected by a magnetic field but a transverse electric field is set up to balance out the Lorentz force. *Ibid.* 28 The units for the Nernst coefficient are $cm^2\ sec^{-1}\ (°K)^{-1}$ or $m^2\ sec^{-1}\ (°K)^{-1}$.

c. Used *attrib.* and in the possessive to designate a theorem in thermodynamics enunciated by Nernst: the change in entropy accompanying a chemical reaction between pure crystalline solids tends to zero as the temperature at which it occurs tends to absolute zero. (Also called the third law of thermodynamics, esp. in more generalized formulations.)

1913 J. R. PARTINGTON *Text-bk. Thermodynamics* xvii. 484 The required information is furnished by a hypothesis put forward in 1906 by W. Nernst, and usually called by German writers 'das Nernstsche Wärmetheorem'. We can refer to it without ambiguity as Nernst's Theorem. **1928** J. K. ROBERTS *Heat & Thermodynamics* xviii. 354 Such a wide range of reactions is considered that we are certainly justified in regarding the Nernst Heat Theorem as being established as a first approximation. *Ibid.* 355 If the Nernst Theorem as originally stated should prove to be only a first approximation, the theorem does not on that account become less important. **1971** G. SOCRATES *Thermodynamics & Statistical Mech.* vii. 139 An alternative statement of Nernst's heat theorem is: It is impossible to reduce the temperature of any system to absolute zero in a finite number of operations.

nerol (nīə·rǫl). *Chem.* [a. G. *nerol* (Hesse & Zeitschel 1902, in *Jrnl. f. prakt. Chem.* LXVI. 498): see NEROLI and -OL.] An oily unsaturated primary alcohol, $C_{10}H_{18}O$, which is present in neroli and some other essential oils and is used in perfumery, having a fragrance similar to but finer than that of geraniol, with which it is stereoisomeric.

1903 *Jrnl. Chem. Soc.* LXXXIV. I. 189 Nerol, $C_{10}H_{18}O$, is an oil which boils at 225–227° under 765 mm. pressure.. and is distinguished from geraniol by its odour of roses. **1949** E. GUENTHER *Essent. Oils* II. 174 Nerol and its acetic ester have been found in several volatile oils, for instance, in oil of helichrysum (30–50 per cent), neroli, bigarade, petitgrain, rose, linaloe, lavender, bergamot, Ceylon citronella, etc. **1949** T. F. WEST et al. *Synthetic Perfumes* iii. 19 Nerol is a superior odorant [to geraniol] in exquisite rose compositions as it provides them with that velvety blossom top-note of freshly cut dark-red roses. **1974** *Jrnl. Chem. Soc.: Perkin Trans. I* 1637 Treatment of geraniol or nerol with fluorosulphonic acid at low temperatures gave good..yields of a novel iridoid ether.

nerts (nɔ̄ɹts), repr. a colloq. or euphemistic pronunc. of *nuts* (*NUT *sb.*[1] 7 e).

1932 *Amer. Speech* VII. 334 (Johns Hopkins jargon) Nerts—exclamation of incredulity or disgust. (Nuts?) **1932** *Sun* (Baltimore) 28 Sept. 10/6 The first couple of months are devoted to getting the right title, suggestions running all the way from 'Alors, Mes Enfants' to 'Nerts'. **1933** M. ALLINGHAM *Sweet Danger* xviii. 222 'Life's very beautiful, isn't it?' 'Speaking as a soul not yet mated, nerts,' said Amanda. **1937** HEMINGWAY *To have & have Not* III. vii. 128 Oh, nerts to you. **1937** B. HOWARD in *New Statesman* 9 Jan. 52/1 Heaven knows that no little word of mine can possibly be heard above the deepening hosannas, but all the same, I shall say it, and it is Nerts. Nerts to everybody, all round, except the authoress. **1948** D. BALLANTYNE *Cunninghams* II. xiii. 222 Nerts, you did your part bringing them into the world.

nerve, *sb.* Add: **8. b.** Phr. *to live on one's nerves:* to lead an emotionally exhausting life.

1927 *Daily Tel.* 1 Nov. 7 The correspondence about the dramatic version of 'The Secret Agent'..is almost interminable. One sees that Conrad lived on his nerves, as the French say, and that he took a secret delight in parading his petty cares. **1932** AUDEN *Orators* I. 16 Dare-devils of the soul, living dangerously upon their nerves.

d. (Earlier and later examples.)

Quots. ?1792 and 1839 perhaps belong in sense 8 b.

? **1792** I. PIGOT *Let.* in A. Leslie *Mrs Fitzherbert* (1960) ix. 87 She says her spirits are so damped and her nerves so bad, she must go out to..soothe her mind by change of scene and country. **1839** DICKENS *Let.* 5 Mar. (1965) I. 519 Recovering from an attack 'on the Nerves'. **1914** G. W. YOUNG *From Trenches* vii. 143 The control of the population is admirable in its restraint. We have no 'nerves' here yet. **1920** H. CRANE *Let.* 23 Nov. (1965) 46 I'm sorry to know you are having such an ordeal of 'nerves'. **1948** A. CHRISTIE *Taken at Flood* I. i. 20 The poor girl was blitzed and had shock from blast... She's a mass of nerves.

e. Phr. *to get on one's nerves:* to (begin to) affect one with irritation, impatience, fear, or the like.

1903 'C. E. MERRIMAN' *Lett. from Son to his Self-Made Father* vi. 75 I've worried a lot since you went away. The business seems to have got on my nerves. **1908** H. G. WELLS *War in Air* iv. 125 This flying gets on one's nerves. *Ibid.* v. 174 All this looking down and floating over things and smashing up people, it's getting on my nerves. **1910** *Chambers's Jrnl.* Mar. 155/2 Sometimes I hate this accursed country...It gets on one's nerves at times. **1920** B. C. CRONIN *Timber Wolves* ii. 34 Hotel hogs, I call 'em. Come in and jolly a chap as if they owned the whole joke. Gets on your nerves when you've been out of your bed all night. **1941** A. L. ROWSE *Tudor Cornwall* xi. 276 The siege..got on people's nerves. **1959** T. S. ELIOT *Elder Statesman* III. 99, I remember, when I came home for the holidays How it used to get on my nerves, when I saw you Always sitting there with your nose in a book. **1972** J. WILSON *Hide & Seek* vii. 118 Alice and I are really close, even if we do get on each other's nerves sometimes.

f. *war of nerves:* the use of hostile or subversive propaganda to undermine morale and cause confusion and uncertainty; psychological warfare.

1940 *Ann. Reg. 1939* 81 The British public..did not allow the 'war of nerves' organised by the Nazi Government to interfere in the least with its August holiday. **1953** E. SIMON *Past Masters* IV. ii. 223 War of nerves... Best thing to do is take no notice. **1974** P. WRIGHT *Lang. Brit. Industry* ii. 26 Recent threats of conflicts have produced, e.g., *cold war* and *war of nerves*.

10. b. *colloq.* Audacity, impudence, cheek. Esp. in phr. *to have a nerve* or *to have the nerve to.*

1887 *Lantern* (New Orleans) 6 Aug. 3/3 Oh, this is a nerve, sure. **1890** BARRÈRE & LELAND *Dict. Slang* II. 84/2 *Nerve* (Eton), impudence. *a* **1911** D. G. PHILLIPS *Susan Lenox* (1917) I. xx. 352 More money!.. You *have* got a nerve!—when factories are shutting down everywhere. **1921** *Daily Colonist* (Victoria, B.C.) 5 Apr. 4/2 No one had the nerve to claim this should be done, because it would have been laughed out of court immediately. **1925** WODEHOUSE *Sam the Sudden* xxv. 219 'You mean to tell me that you had the—the nerve—the insolence—.' He gulped. **1929** [see *HONESTLY *adv.* 2 b]. **1930** V. SACKVILLE-WEST *Edwardians* vi. 220 The cabby exclaimed that the young toff had a nerve and no mistake. **1942** *Q. Jrnl. Speech* Feb. 6/2 In low brow programs..we note..'you've got a nerve'. **1975** S. BRETT *Cast* xiii. 136 Joanne Menzies looked at him coolly. 'You've got a nerve.'

11. a. *nerve-channel, -end* (*fig.* examples), *-network, -pill, -test, -tester, -tip, -world.*

1902 *Chambers's Jrnl.* 15 Feb. 166/1 Muscular repose means that the muscles are relaxed and the nerve-channels free. **1951** J. M. FRASER *Psychol.* ii. 13 A message.. will be sent along a nerve-channel to the brain. **1953** R. LEHMANN *Echoing Grove* 159 In the silence Dinah's nerve ends crept, contracted, listening for the guns, the sirens. **1969** *Listener* 22 May 733/3 In Debussy's case this led to Impressionism, which lived at the nerve ends. **1947** *Mind* LVI. 57 In the nervous system the 'variables' are mostly impulse-frequencies at various points in the nerve-network. **1972** *Sci. Amer.* Feb. 88/2 The major part of the sympathetic nerve network stimulates the secretion of norepinephrine. **1907** *Yesterday's Shopping* (1969) 505/1 Carter's little liver pills... Nerve pills. **1975** M. CRICHTON *Great Train Robbery* xiii. 78 He was forced to down two Carter's Little Nerve Pills and some tincture of opium for his pain. **1909** *Westm. Gaz.* 29 May 8/3 (*heading*) A zoological nerve-test. **1894** *Strand* Feb. 119/1 A visit to this place is the finest and most complete nerve-tester in the world! **1890** W. JAMES *Princ. Psychol.* I. vi. 150 This must be because the number of sensations from the elementary nerve-tips affected was too small. **1936** F. R. LEAVIS *Revaluation* v. 184 The sentiments and attitudes of the patriotic and Anglican Wordsworth..are external, general and conventional; their quality is that of the medium they are proffered in, which is..not felt into from within as something at the nerve-tips, but handled from outside. **1890** Nerve-world [see *mind-world*].

b. Objective, as *nerve-lacerating, -racking, -shaking* (earlier and later examples), *-shattering, -testing, -wracking; nerve-wrackingly* adv.; instrumental, as *nerve-drawn, -racked, -ridden* (further examples), *-shattered, -worn, -wracked.*

1937 V. WOOLF *Years* 388 A queer face; knit up; nerve-drawn; fixed. **1911** W. OWEN *Let.* 17 Sept. (1967) 83 The nerve-lacerating speech of the pompous vigilator. **1933** A. N. WHITEHEAD *Adventures of Ideas* xvi. 245 A red-irritation is prevalent among nerve-racked people and among bulls. **1812** SHELLEY *Let.* 27 Dec. (1964) I. 347 My removal from your nerve racking & spirit quelling metropolis. **1908** *Westm. Gaz.* 22 Feb. 2/3 The nerve-racking work of the telephone-girls. **1909** *Chambers's Jrnl.* Mar. 205/2 In all the large towns in Britain there is also far too much nerve-racking, unnecessary noise. **1973** *Times* 30 June 13/5 My own King Charles's head is the use of 'nerve-wracking' for 'nerve-racking'. **1930** R. MACAULAY *Staying with Relations* xxi. 313 Adrian..had already shown himself a husband too impatient and too nerve-ridden to endure for long so much irrational excitement at his side. **1939** D. CECIL *Young Melbourne* vi. 150 He [*sc.* Byron] was a raw, nerve-ridden boy of genius. **1820** KEATS *Let.* ?Feb. (1958) II. 262 The medicine I am at present taking..is of a nerve-shaking nature. **1933** J. CARY *Amer. Visitor* vii. 88 Uli shrank away from the white men and uttered another nerve-shaking yell. **1929** *Observer* 17 Nov. 11/3 Jean Jacques Bernard's interesting

play..about a nerve-shattered and unreasonably jealous husband. **1975** C. DENNIS *Somebody just grabbed Annie!* [illegible] **1909** *Westm. Gaz.* 28 Sept. 4/2 The work of the engine-driver and fireman is..by no means of the nerve-shattering description that some would have us believe. [illegible]

12. nerve block, blocking, inactivation of the nerves supplying a particular area of the body, esp. by use of a local anæsthetic; **nerve cell** (earlier example); **nerve-centre** (later *fig.* examples); **nerve current,** a signal propagated along a nerve; a series of nerve impulses; **nerve-doctor,** a specialist in the treatment of nervous diseases; **nerve gas,** any poisonous gas or vapour that has a weakening or paralysing effect on the nervous system, esp. for use in warfare; also *fig.*; **nerve impulse,** a wave of excitation in a nerve fibre accompanied by a brief, temporary change in electrical potential, [illegible] which constitutes transmission of a stimulus along the fibre; **nerve net,** a diffuse network of neurons found in cœlenterates, echinoderms, and other organisms, which conducts excitations in all directions from the area stimulated; **nerve-path,** a route (assumed to be inborn or developed through use) by which a specific sensory stimulus or motor response is propagated through the nervous system; **nerve-patient,** a patient suffering from a nervous disorder; **nerve physiology** = *neurophysiology* s.v. *NEURO-; so **nerve-physiologist; nerve-route** = *nerve-path* above; **nerve-sick** *a.*, suffering from nervous illness; **nerve-specialist** = *nerve-doctor* above; **nerve war** = *war of nerves* (*NERVE *sb.* 8 f); also *fig.*

1923 *Jrnl. Amer. Med. Assoc.* 29 Sept. 1079/1 (*heading*) The value of sacral nerve block anesthesia in obstetrics. [illegible] *Animal Biol.* v. 123 (*caption*) Diagram to illustrate the course of nerve-impulses concerned in a spinal reflex. **1966** *McGraw-Hill Encycl. Sci. & Technol.* IX. 52/2 The rate of conduction of nerve impulses varies in proportion to the diameter of the nerve fiber. **1971** J. Z. YOUNG *Introd. Study Man* ii. 28 This is best known for nerve cells, whose signals are the nerve impulses (action potentials) [illegible] **1904** C. S. SHERRINGTON in *Nature* 8 Sept. 460/1 We can distinguish two main types of [nervous] system according to the mode of union of the conductors.—(i.) the nerve-net system, such as met in Medusa and in the walls of viscera, and (ii.) the synaptic system, such as the cerebro-spinal system of Arthropods and Vertebrates. [illegible] *Nervous System* i. 3 The older view of the nerve net as a conduction apparatus for diffusing nerve impulses by protoplasmic continuity of the nerve cell processes has largely yielded to the conception of a synaptic system, even in coelenterates. **1968** [illegible] *Physiol.* x. 214 The nerve [illegible] the Coelenterata are [illegible] organized central nervous [illegible] *Psychol.* I. xi. 458 A [illegible] kept constantly open during [illegible] J. S. HUXLEY *Essays in* [illegible] secretions of ovaries pick out [illegible] nerve-paths appropriate to [illegible] paths appropriate to males. [illegible] (*heading*) Age changes of [illegible] *Chambers's Jrnl.* Dec. 818/1 [illegible] obtains an increasing number [illegible] per year. **1890** W. JAMES [illegible] another way the psychologist [illegible] of a nerve-physiologist. [illegible] *Sci. 1859* II. 166 (*heading*) [illegible] nerve-physiology. **1890** W. [illegible] The conception of *all* action [illegible] reflex] is the fundamental [illegible] physiology. **1927** J. S. [illegible] ion 135 How, precisely, [illegible] psychology and nerve- [illegible] us. **1890** A. HILL tr. *H.* [illegible] *Organs* 162 One must be [illegible] object to nerve-routes, [illegible] an internode..in length. [illegible] *Adventures of Ideas* xiv. 276 Also [illegible] be forming themselves in [illegible] FAULKNER *Sartoris* III. 173 [illegible] led over with a fine cold [illegible] *Staying with Relations* xv. [illegible] of quarrelling, nerve-sick [illegible] **1889** Nerve-specialist [see [illegible] in 'W. N. P. Bar- [illegible] persuaded him..to see [illegible] **1941** *Argus* (Melbourne) [illegible] *nerve war,* grousing or com- [illegible] **1942** *Ann. Reg. 1941* 212 [illegible] nese blackmail, cajolery, [illegible] *Daily Tel.* 26 Aug. 6 [illegible] The nerve war has reached [illegible] at possible. **1946** F. WIL- [illegible] *People* iii. 61 They [*sc.* the [illegible] ves that the message had [illegible] of a clever piece of nerve [illegible] make their defence forces [illegible]

nerver (nɔ̄·ɹvəɹ). [f. NERVE *sb.* or *v.* + -ER[1].] Something, esp. a drink, that gives one 'nerve' or courage.

[illegible] 117/2 *Nerver,* eins für den [illegible] *Times's Gaz.* 10 Aug. 3/2 His [illegible] mums, hydrocyanic acid, [illegible] capital 'nerver' in its way. [illegible] *That Fall* iii. 33 You'll pull [illegible] nerver of pale sherry and [illegible]

nervily (nɔ̄·ɹvili), *adv.* [f. NERVY *a.*: see -LY[2].] In a nervy manner; boldly; agitatedly.

[illegible] *Daily Express* 15 June 9/4 [illegible] as he would like to have [illegible] ones roved nervily from [illegible] *Comforters* iii. 63 Caroline [illegible] opened.

nerviness. Delete *rare*⁻⁰ and add: **2.** The state or character of being nervous (sense 9) or nervy (sense *5).

[illegible] The home men had only [illegible] to reduce the Australian [illegible] **1924** J. SUTHERLAND [illegible] footstep he started... [illegible] that even Barbara com- [illegible] J. RATHBONE *Kill Cure* [illegible] all right, but there was a [illegible] not quite like.

nervo-. Add: † **nervo-electric** *a.* = *neuro-electric* adj. s.v. *NEURO-; so **nervo-electricity.**

[illegible] *us Electricitatis in Motu* [illegible] fluidi excursum per [illegible] **1860** A. C. GARRATT [illegible] *Therapeutics* iv. 216 (*heading*) [illegible] nervo-electric batteries. [illegible] can be thus incessantly [illegible] indeed it be..'an electric [illegible] charging itself by the nerves [illegible] at brief intervals, when the tension of the nervo-electricity, developed, reaches a certain point'?

nervous, *a.* Add: **2. c.** (Examples of *nervous energy.*) Also, *nervous tension.*

1849 THOREAU *Week Concord Riv.* 338 They..handled their paddles unskilfully, but with nervous energy and determination. **1933** *Burlington Mag.* Feb. 54/1 Every- where the nervous tension is relaxed, the rhythms are less vital and less consistent. **1936** *Discovery* Nov. 357/2 His nervous tension is surely not lessened if precautions are taken against his actually watching the operation [on himself]. **1951** M. MCLUHAN *Mech. Bride* (1967) 33/1 Books on how to relax would seem just about to cancel out books on how to build up nervous tension for success drive. **1953** J. S. HUXLEY *Evolution in Action* iv. 91 Behaviour is always the result of a flow of something—what many psychologists and most laymen call nervous energy. Unfortunately, the physiologists are driven to say 'excitation', because *energy* is another word the physicists have taken over from ordinary speech and given a restricted scientific meaning. Perhaps the term 'neurergy' would serve.

7. b. *nervous breakdown:* see *BREAK-DOWN 1 c; *nervous fever* (earlier and later U.S. examples). Also *nervous exhaustion, nervous headache.*

1787 H. MORE *Let.* June (1925) 120 A very tedious nervous headache has made me less than ever qualified to traffic with you. **1807** A. PUTNAM in *Danvers Hist. Soc. Coll.* (1917) V. 57 Mr. David Tapley is very sick with the nervous fever. **1857** J. A. SYMONDS *Let.* 16 Nov. (1967) I. 125, I must..not work much these trials since yesterday I had the same sort of nervous headache. **1924** H. CRANE *Let.* 23 Sept. (1965) 190 The sneezing and nervous fever.. begin to subside. **1927** 'R. CROMPTON' *William—in Trouble* viii. 207 He's suffering from *nervous exhaustion.* **1973** E. BERCKMAN *Victorian Album* 142 The thought of any considerable threat to her happiness..is always enough to give me a nervous headache.

9. Also, *nervous wreck. colloq.*

1899 [see WRECK *sb.*[1] 7 b]. **1906** W. JAMES *Let.* 9 May (1920) II. 251, I did n't hear one pathetic word uttered at the scene of disaster, though of course the crop of 'nervous wrecks' is very likely to come in a month or so. **1932** H. CRANE *Let.* 22 Apr. (1965) 412 It certainly has about made a nervous wreck of me. **1936** J. BUCHAN *Island of Sheep* I. vi. 110 He started at every noise. He was the very model of a nervous wreck.

nervuration (nɔ̄ɹviurēⁱ·ʃən). [f. NERVURE + -ATION.] = NEURATION.

1899 D. SHARP in *Cambr. Nat. Hist.* VI. v. 319 In the aberrant moths of the genus Costina the nervuration is unusually complex.

nervy, *a.* Add: **2. b.** (Earlier and later examples.) *U.S. colloq.*

1896 ADE *Artie* viii. 75, I just received your nervy letter. **1904** E. ROBINS *Magnetic North* II. 118 Feeling that it is a little 'nervy'..to walk into another man's house uninvited. **1948** *Pauls Valley* (Okla.) *Daily Democrat* 2 May 4 Wouldn't it be rather nervy to ask her help?

5. Having one's nerves disordered; easily excitable, 'jumpy'; = NERVOUS *a.* 9.

1891 R. T. COOKE *Huckleberries gathered f. New England Hills* 319, I expect I be sort o' nervy, what with takin' a journey and the thought o' seein' Melindy. **1906** *Sat. Rev.* 3 Mar. 254 They are very 'nervy' in Russia. **1916** *Daily Mail* 21 Sept. 7 (Advt.), When you are weak, anæmic, 'nervy', run-down, [etc.]. **1923** W. DEEPING *Secret Sanctuary* viii. 77 You can explain that your brother was shell-shocked, and that he is still a bit nervy. **1932** A. CHRISTIE *Peril at End House* iii. 40, I have been worried to death. Everybody's been telling me I'm nervy. **1973** J. BURROWS *Like Evening Gone* v. 63 Greta was grey as paper and peevish and nervy.

Nescafé (ne·skæfe). The proprietary name of a brand of instant coffee; a drink made of Nescafé. Also *colloq.* abbrev. **Nes;** [cf. *CAFF] **Nescaff.**

1946 *Trade Marks Jrnl.* 4 Sept. 475/2 Nescafé... Preparations of coffee in powder form. Nestlé's Milk Products Limited,..London,... Manufacturers. **1959** [see *ABLUTE *v.*]. **1962** A. LEJEUNE *Duel in Shadows* xi. 156 'Only Nescaff, I'm afraid,' said Fiona Stewart-Long. **1966** *Observer* 19 June 29/5 Silk shantung dress in Nescafé brown. **1966** *New Statesman* 24 June 939/3 Maurice Ronet as a devil-may-care sadist, and George Segal in Nescafé as an old Algerian buddy. **1968** H. READE *Comeback for Stark* xii. 175 Sure you won't have a cup of nes? I prefer it to anything except the real bottled yellow. Take it cold in the summer. **1968** *Listener* 18 Apr. 515/1 How many spoonfuls of Nescafé did the Chancellor of the Exchequer have in his elevenses that vital morning? **1970** J. FLEMING *Young Man, I think you're Dying* ii. 33, I could make you a cuppa hot Nescaff, eh? How about it? **1971** 'P. HOBSON' *Three Graces* viii. 54 Downstairs for a cup of Nes. **1975** N. FREELING *What are Bugles blowing For?* iv. 17 We'll have some coffee—or to be accurate, Nes. *Ibid.* v. 21 La Touche seemed to be waking up... Shock wearing off, or the stimulus of Nescafé.

Nescaupick, var. *NASKAPI *a.* and *sb.*

nesh, *a.* **1. a.** (Further example.)

1915 R. C. THOMPSON *Pilgr. Scrip* 71 The road from the bridge is like an English lane with blackberry hedgerows.. and a nesh track for a morning gallop.

Nesite (ne·səit). Also **Nesian, Nesitic.** [f. Hittite *našili, nešumnili,* f. *Nešaš* name of an ancient Hittite city in Asia Minor (identified by B. Hrozný 1929, in *Archiv orientální* (Prague) I. 273–99) + -ITE[1].] **a.** A name given to the official language used in Hittite documents, suggested as an alternative to *KANE-

SIAN. **b.** A member of the Hittite people who used this language.

The meaning of *našili* and *nešumnili* in Hittite texts has been disputed, but the view of Hrozný (above, and 1931, in *Journal Asiatique* CCXVIII. 318–19) is now generally accepted, that the words relate to the former Hittite capital Nešaš (or Nesa) and designate the Indo-European Hittite aristocracy and its language (see also quot. 1954).

[**1920** A. H. SAYCE in *Jrnl. Royal Asiatic Soc.* 605 Literary Hittite is once called 'Luvian', *Lûi-li*, 'in Luvian', taking the place of the more usual *nâsi-li.* **1921** M. BLOOMFIELD in *Jrnl. Amer. Oriental Soc.* XLI. 197 The supposedly I. E. Hittite [in the Boghazköi inscriptions] seems, according to both authors [*sc.* Forrer and Hrozný], to be well entitled to the name Kanesian, named after the city of Kaneš. But this latter designation is never indicated by an ethnical adjective...Instead there occurs, more frequently than the mention of Kaneš, the ethnical designation Nâšili, which Forrer takes to be the same as Kanesian. **1929** B. HROZNÝ in *Archiv orientální* (Prague) I. 296 Les Nêsites sont donc en effet les fondateurs du grand empire hittite... La langue hittite doit donc être nommée..la langue nêsite.] **1933** E. H. STURTEVANT *Compar. Gram. Hittite Lang.* i. 28 Hrozný.. may be correct in thinking that the word *Nesumnili*..refers to official Hittite and therefore in calling the latter Nesite. **1949** F. O. STEIN tr. *L. Matouš's Bedřich Hrozný* 26 Later, in a paper published in Volume I of '*Archiv orientální*' (Oriental Archives)..., Hrozný shewed that the Indo-European Hittites originally called themselves *Nesites*, from their ancient city *Neshash*, and that therefore their language should more correctly be called *Nesitic*, and not Hittite, as had always been taken for granted. **1954** O. R. GURNEY *Hittites* (ed. 2) 122 While Hrozný adopted the name 'Nêsite' (i.e., the language of the city Nesa), Forrer preferred 'Kanisisch' (i.e., the language of Kanesh). It is now generally agreed that 'Nesite' or 'Nesian' (derived from the Hittite adverb *našili* or *nešumnili*) is indeed the true name of the language. **1959** *Chambers's Encycl.* VII. 155/1 The language..was seen to be mainly Indo-European in structure and vocabulary. In the Hittite texts it is sometimes called the speech of Nesas, and has therefore been termed Nesite. **1964** G. ROUX *Ancient Iraq* xiv. 191 The Luwians..settled to the west of Cilicia, along the coast,..and the so-called Nesites in Cappadocia... Centuries later, those Nesite-speaking invaders conquered the centre of the Anatolian plateau. **1972** W. B. LOCKWOOD *Panorama Indo-Europ. Lang.* xiv. 262 The terms used are *nasili* (or *nisili*) and *nesumnili* meaning respectively 'in the language of Nesa' and 'in the language of those of Nesa', and Hittite has since occasionally been called Nesite.

nesk(h)i, varr. *NASKHI *sb. pl.*

‖ **nespola** (ne·spǫlă). [It.] = MEDLAR.

1883 LADY MONKSWELL *Jrnl.* 29 Apr. in *Victorian Diarist* (1944) 104 On the entrance terrace were lemon trees in pots, trees of magnolia (not in flower) & nespola. **1920** D. H. LAWRENCE *Let.* 7 May (1932) 504 Nespoli look like apricots, and taste a bit like them—but they're pear-shaped. They're a sort of medlar. **1922** *Blackw. Mag.* Feb. 157/2 Florindo threw a number of nespole with deadly accuracy.

nesquehonite (neskwihōᵘ·nəit). *Min.* [f. *Nesquehon-ing* (see quot. 1890) + -ITE[1].] Magnesium carbonate trihydrate, $MgCO_3.3H_2O$, found as colourless monoclinic crystals.

1890 GENTH & PENFIELD in *Amer. Jrnl. Sci.* CXXXIX. 122 The clear crystals..proved on examination to be a new mineral, having the composition $MgCO_3.3H_2O$, to which we have given the name Nesquehonite, after the locality where the mineral was found, the Nesquehoning Mine being one of the best known in Pennsylvania. **1934** *Mineral. Abstr.* V. 431 Galleries in the Cogne magnetite mine in serpentine rocks, at an elevation of 2500 metres in the Piedmontese Alps, have a constant temperature of 4°C., and this has favoured the formation of lansfordite.. and nesquehonite. **1957** G. E. HUTCHINSON *Treat. Limnol.* I. x. 665 At ordinary temperatures the carbonate that precipitates when saturation is reached at from 10^{-3} to 10^{-1} atm. of CO_2 is $MgCO_3.3H_2O$, or nesquehonite. **1971** *Contrib. Mineral. & Petrol.* XXXIII. 88 The 'cauliflower' crusts mainly consist of hydromagnesite and/or nesquehonite with traces of monohydrocalcite and calcite.

nessberry (ne·sberi). [f. the name of Helge *Ness* (d. 1928), American horticulturist + BERRY *sb.*[1]] A variety of *Rubus*, produced by crossing a dewberry and a raspberry, introduced by Helge Ness in 1921.

1925 H. NESS in *Bull. Texas Agric. Exper. Stat.* No. 326. 19 This new fruit has been named the Nessberry by the authorities of the College. *Ibid.*, The Nessberry is of an excessively strong growth, some of its branches attaining a length of 10 to 15 feet in a single season. **1948** J. S. SHOEMAKER *Small Fruit Culture* (ed. 2) iii. 232 Nessberry..has a delightful flavor and thrives well in East Texas, but it proved unpopular because of poor picking quality and difficulty in propagation. **1952** *New Biol.* XIII. 46 Among the newer crops which have arisen by allopolyploidy..the most notable are the various berries related to blackberry and raspberry... Two typical examples are the nessberry and loganberry. **1972** BROOKS & OLMO *Register of New Fruit & Nut Varieties* 164 Nessberry... Fruit: larger than Logan; size variable; hemispherical; skin deep to blood red.

Nesselrode (ne·səlrōᵘd). The name of Karl R. *Nesselrode* (1780–1862), Russian statesman, used *attrib.* in *Nesselrode pie, pudding,* etc., an iced dessert made with chestnuts, cream, preserved fruits, etc., and freq. flavoured with rum. Also ellipt. *Nesselrode.*

1845 E. ACTON *Mod. Cookery* xx. 441 *Nesselrode Cream*, ..Chestnuts..sugar..isinglass..cream..vanilla [etc]. *Ibid.* 461 Nesselrode pudding. We give Monsieur Carême's own receipt..as it originated with him. **1877** E. S. DALLAS *Kettner's Bk. of Table* 311 (*heading*) Nesselrode pudding was invented..by Mony, cook to the famous Count Nesselrode. **1894** E. SKUSE *Compl. Confectioner* 155 Nesselrode or Ice Pudding. Prepare a custard of one pint of cream, [etc.]. **1952** S. J. PERELMAN *Ill-Tempered Clavichord* (1953) 12, I wouldn't bolt that Nesselrode if I were you. **1965** *Harper's Bazaar* Nov. 136/2 Make a Nesselrode pudding such as Françoise made for Monsieur de Norpois the first time he went to dine with the Prousts. **1972** *New Yorker* 15 Apr. 35/2 Toward the back of the shop are two refrigerated cases—one for birthday cakes, the other for ice-cream cakes and for desserts composed of perishable whipped cream, such as Nesselrode pie.

Nessie (ne·si). *colloq.* Also **Nessy.** [f. *Ness* (see below) + -IE or -Y[6].] A name for a monster or monsters supposedly inhabiting Loch Ness in northern Scotland. Cf. *LOCH NESS.

1945 *Sun* (Baltimore) 25 June 2 'Nessie', the unidentified Loch Ness monster which used to disport itself in those Scottish waters before the war..has reappeared, according to several women residents. **1960** in T. Dinsdale *Loch Ness Monster* (1961) v. 76 According to substantiated reports, Nessie was seen a number of times last spring..and a few times in the autumn. **1968** *New Scientist* 26 Dec. 729/1 The trouble with Nessie was that she became an immediate newspaper sensation. **1971** *Ibid.* 10 June 641/2 Nessy has lost her mythical charm and has become the target of serious-minded technocrats. **1974** *People's Jrnl.* (Inverness & Northern Counties ed.) 29 June 2/6 They sighted Nessie as they returned from a delivery at Inverfarigaig. **1975** *Nature* 11 Dec. 467/3 Should the Nessies wish to breathe quite frequently, they would not be detected easily if the nostrils were at the topmost point to break surface.

Nessler (ne·slər). *Chem.* The name of Julius *Nessler* (1827–1905), German agricultural chemist, used *attrib.* and in the possessive with reference to a delicate test for ammonia he devised; as **Nessler('s) reagent** or **solution,** an alkaline solution of potassium mercuric iodide, which gives a yellow or brown colour or precipitate when added to an aqueous solution containing ammonia; **Nessler('s) tube,** a glass cylinder marked with a line indicating a certain volume and depth, used in nesslerization and other colorimetric procedures.

1865 *Jrnl. Chem. Soc.* XVIII. 125 The distillate is divided into two equal portions; one of these is submitted to Nessler's test for ammonia. **1868** *Ibid.* XXI. 87 Mr. Chapman..recommends that the ammonia determination should be made by the application of Nessler's solution directly to the water. *Ibid.*, Waters containing chalk in solution become turbid on the addition of the Nessler test. **1873** *Chem. News* 11 July 13/2 The time required for the development of the Nessler colour. *Ibid.*, One sample of Nessler reagent gives its maximum of colour almost immediately, and another takes a quarter of an hour or an hour. **1913** CUMMING & KAY *Text-bk. Quantit. Chem. Anal.* IV. 156 Measure into a 100 c.c. Nessler tube a portion of the solution containing from 0·1 to 3 mgrm. of iron. **1946** *Thorpe's Dict. Appl. Chem.* (ed. 4) VII. 579/1 The action of ammonia on an alkaline solution of potassium mercuri-iodide..is the basis of the Nessler test for ammonia. **1963** SKOOG & WEST *Fund. Analytical Chem.* xxviii. 656 In its simplest form, colorimetry consists of visual matching of the color of the solutions of the substance with a set of standards. For such a procedure flat-bottomed tubes called Nessler tubes are frequently employed. **1970** Nessler's reagent [see *NESSLERIZATION].

nesslerize, *v.* Add: Also **Nesslerize.** Substitute for def.: *trans.* To treat with Nessler's reagent in order to test for the presence of ammonia; also, to test for (ammonia) by this means. Also *absol.* (Earlier and later examples.) So **ne·sslerized** *ppl. a.*, **ne·sslerizing** *vbl. sb.*

1873 *Chem. News* 11 July 13/2 Whether the Nesslerising takes a couple of minutes, or whether it takes an hour, is a matter of vital importance to those persons who are working the ammonia process of water analysis. **1884** *Jrnl. Amer. Chem. Soc.* VI. 122 Such a tube was found.. to be unsuitable for comparison with the color of Nesslerized liquids. **1890** *Amer. Chem. Jrnl.* XII. 426 Standards and distillates are made ready for nesslerisation during the day, and all stand on the same table overnight. In the morning they are nesslerised and compared. **1916** *Jrnl. Biol. Chem.* XXVI. 473 It was hoped that it would be possible to Nesslerize directly the ammonia produced by the destructive digestion of the urine. *Ibid.* 481 It is better when working in urine to dilute the final Nesslerized solution to 200 cc. instead of to 100 cc. **1956** *Jrnl. Lab. & Clin. Med.* XLVII. 645 In one the clot is dried and weighed to the nearest microgram; in the other it is nesslerized and assayed colorimetrically. *Ibid.* 646 If a microbalance is not available, the fibrin may be estimated by digesting it in sulfuric acid and hydrogen peroxide, nesslerizing, and determining the color density.

nesslerization (examples).

1890 [see above]. **1916** *Jrnl. Biol. Chem.* XXVI. 475 This sediment does not interfere with the Nesslerization of the ammonia, but it must be removed before the color comparison is made. **1964** *Oceanogr. & Marine Biol.* II. 148 They determine ammonia in the digest by Nesslerization. **1970** *Nature* 12 Sept. 1136/2 L-Glutaminase and L-asparaginase assays were performed..as previously described except that the colour formed by nesslerization of ammonia was measured 1–3 min after addition of Nessler's reagent.

Nessus (ne·səs). Name of the centaur slain by Hercules and in whose blood was soaked the tunic which consumed Hercules with fire, used allusively in *Nessus robe, Nessus shirt, shirt of Nessus* of any destructive or expurgatory force or influence.

1606 SHAKS. *Ant. & Cl.* IV. xii. 43 The shirt of Nessus is vpon me. **1835** CARLYLE *Lett. to his Wife* (1953) 108 It is now almost my sole rule of life: to clear myself of Cants and formulas, as of poisonous Nessus' shirts. **1905** S. J. WEYMAN *Starvecrow Farm* xxxii. 297 Remorse is the very shirt of Nessus. It is of all mental pains the worst. **1924** R. GRAVES *Mock Beggar Hall* 10 The Nessus-robe that beauties wear, Burning away their beauty. **1931** *Times Lit. Suppl.* 5 Mar. 162/2 The Nessus-robe of vice clings festering about him. **1957** E. SITWELL *Coll. Poems* 414 Then the heart that was the Burning-Bush May change to a Nessus-robe of flame. **1961** M. WEST *Daughter of Silence* iv. 116 Accept the guilt, know yourself for what you are, wear the knowledge like a Nessus shirt on your own back and bear the pricks and the poison with as much dignity as you can muster.

Nessy, var. *NESSIE.

nest, *sb.* Add: **2. d.** *Mil.* An emplaced group of machine-guns.

1914 E. A. POWELL *Fighting in Flanders* v. 120 Other wagons..contained 'nests' of nine machine-guns. **1949** 'G. ORWELL' *Nineteen Eighty-Four* I. 8 A maze of barbed-wire entanglements, steel doors and hidden machine-gun nests. **1959** R. POSTGATE *Every Man is God* xviii. 157 The Germans come out of the machine-gun nest holding up their hands.

6. (Further examples.)

1807 R. SOUTHEY *Lett. from England* I. 155 Here is also a nest of tables for the ladies, consisting of four, one less than another, and each fitting into the one above it. **1924** R. KEABLE *Recompence* iv. 76 A delightful nest of occasional tables. **1973** J. LEASOR *Host of Extras* i. 17 A nest of spanners and some lengths of wire that might come in useful in starting an engine. **1975** *Country Life* 20 Feb. Suppl. 32d/2 A rare nest of five rosewood 'quartetto' tables..English, circa 1820.

8. *nest-burrow, -factory, -mate, -material, -relief, -scrape, -site; nest-raiding.*

1948 *British Birds* XLI. 341 During our short stay we saw no trace of nest burrows of the Great Shearwater. **1961** G. DURRELL *Whispering Land* ii. 55 Once the parent bird [*sc.* a penguin] reached the edge of the colony it had run the gauntlet of several thousand youngsters before it reached its own nest-burrow and babies. **1908** *Westm. Gaz.* 19 Aug. 5/3 The Hungarian Government go so far as to pay large sums of money in subsidies to artificial nest factories, these nests being fixed in the forests by the thousand and regularly looked after. **1913** G. STRATTON-PORTER *Laddie* v. 146 If a young bird failed to get the bite it wanted, it sometimes grabbed one of its nestmates by the bill..and tried to swallow it whole. **1941** J. S. HUXLEY *Uniqueness of Man* x. 214 The nestling cuckoo.. does not know why he is murdering his fellow nest-mates. **1923** —— *Essays of Biologist* iii. 121 We invariably find the seizing of nest-material in the beak as a part of courtship. **1953** N. TINBERGEN *Herring Gull's World* vii. 68 We often wondered how another gull knew whether a grass-pulling gull was in an aggressive mood or merely collecting nest-material. **1937** *British Birds* XXXI. 205 The evidence for nest raiding is so scanty as to lead to the conclusion that it is not a usual habit. **1953** D. A. BANNERMAN *Birds Brit. Isles* I. 15 It is in the early morning ..that the crow carries out most of its nest-raiding. **1923** J. S. HUXLEY *Essays of Biologist* iii. 112 After the eggs [of the Louisiana Heron] are laid both sexes brood, and there is a nest-relief four times in every twenty-four hours. **1953** N. TINBERGEN *Herring Gull's World* xvi. 136 The variation in the behaviour of the birds at nest-relief is primarily due to variations in the intensity of the incubation urge. *Ibid.* ii. 11 When the birds of a pair make a nest-scrape together..they make a queer rhythmical sound. **1964** *Oxf. Bk. Birds* 48/1 The nest scrape [of the kestrel] may be in a hollow tree. **1930** J. S. HUXLEY *Bird-Watching* iii. 53 Each breeding pair occupies and defends against intrusion a considerable area around its nest-site. **1964** A. L. THOMSON *New Dict. Birds* 531/1 Summer migrants may select a nest site and begin building on the same day.

nest, *v.* Add: **2. c.** *U.S. colloq.* To squat. (Cf. *NESTER 2.)

1918 C. E. MULFORD *Man from Bar-20* xi. 114 Not satisfied with nestin' on a man's range, you had to start a little herd.

4. b. (Later example, not in pa. pple. Also, examples referring to abstract entities; cf. *NESTED *ppl. a.* 2.)

1925 N. E. ODELL in E. F. Norton *Fight for Everest 1924* 362 Two saucepans that nest into one another. **1961** D. V. HUNTSBERGER *Elem. Statist. Inference* ix. 230 Situations of this sort, where every classification is nested within the next larger one, are called nested or hierarch[ic]al classifications. **1967** KLERER & KORN *Digital Computer User's Handbk.* I. i. 16 Groups of loops need not be strictly nested and may have parts which are done in parallel or parts which are done in series. **1968** J. LYONS *Introd. Theoret. Linguistics* vi. 233 In subordinative constructions one modifier may be recursively 'nested' within another. **1970** P. M. SHERMAN *Techn. Computer Programming* iii. 43 In many problems one loop is nested within another one. This occurs when a loop contains a box with a repetitious process that itself can be drawn as a loop.

nestalik, nestaliq, varr. *NASTALIK.

nest-box. 2. (Later examples.)

1893 *Burpee's Farm Ann.* 64 These gourds are useful for many household purposes, such as..nest-boxes, soap

and salt dishes. **1946** *Q. Jrnl. Forestry* XL. 83 The habit of maintaining a territory also leads to birds being well scattered over the area, and as a consequence nest boxes should not be too close together. **1963** *Times* 8 June 12/7 Your Correspondent recalls a horrid stench from a nest-box after seven or eight young great tits had flown. **1971** *Country Life* 2 Sept. 541/1 Many bird-lovers put up two or three nest-boxes in their gardens.

nested, *ppl. a.* Add: **2.** Such that each item or constituent contains or is contained within another similar item in a hierarchical arrangement. **a.** Of concrete objects (cf. NEST *v.* 4 b).

1921 *Glasgow Herald* 1 June 7 Two or three 'nested' dishes, the biggest one large enough to take two plates side by side. **1969** 'S. TROY' *Swift to its Close* vi. 81 A little man carrying three nested chairs brushed past. **1972** *Sci. Amer.* Apr. 32/1 The magnetic field in the compressor is supplied by a nested set of copper coils.

b. Of abstract entities.

1959 E. L. LEHMANN *Testing Statistical Hypotheses* vii. 290 (*heading*) Nested classifications. **1963** ERVIN & MILLER in J. A. Fishman *Readings Sociol. of Lang.* (1968) 80 Preliminary evidence suggests that nesting is learned early... It is difficult to assess the evidence, however, because nested constructions are usually delineated by markers which are largely lacking in the early grammatical systems. **1965** N. CHOMSKY *Aspects of Theory of Syntax* i. 12 The phrases *A* and *B* form a nested construction if *A* falls totally within *B*, with some nonnull element to its left within *B* and some nonnull element to its right within *B*. **1967** [see *NESTING *vbl. sb.* 2]. **1968** J. F. HART et al. *Computer Approximations* iv. 68 Consider a polynomial given in the conventional power form $P_n(x) = \sum_{i=0}^{n} a_i x^i$. It can be written in nested form $P_n(x) = (\ldots(a_n x + a_{n-1})x + \ldots + a_1)x + a_0$. **1970** P. M. SHERMAN *Techn. Computer Programming* iii. 43 Some flowcharts contain three or more nested loops. **1973** A. M. COHEN et al. *Numerical Anal.* ii. 21 The computation of a polynomial function is usually done by means of 'nested multiplication'. **1973** *Computers & Humanities* VII. 226 Procedures for handling nested loops and subscripted variables.

nest-egg, *sb.* Add: **1.** (Later examples.)

1939–40 *Army & Navy Stores Catal.* 989/2 Poultry nest eggs. **1973** *Country Life* 18 Oct. 1129/3 What happened to the old-fashioned nest-egg, the white pot egg one could once..buy to ensure that hens would lay where they were supposed to lay?

4. (Later examples.)

1968 J. R. ACKERLEY *My Father & Myself* 41 This nest-egg towards the cost of the farm. **1975** *Radio Times* 24 Apr. 12 (Advt.), Williams & Glyn's Nest Egg is a regular savings plan.

nester. Delete *rare* and add: **1.** (Earlier example.)

1887 *Ibis* 95 It [*sc. Cisticola cursitans*] is both an early and late nester.

2. *N. Amer.* An opprobrious term for a person who settles permanently in a cattle-grazing region as a farmer, homesteader, etc. (Cf. *NEST *v.* 2 c.)

1880 *Ft. Griffin* (Tex.) *Echo* 3 Jan., [A sheep man is] a tramp, an ingrate, a 'Nester', and a liar. **1907** C. E. MULFORD *Bar-20* xix. 192 Ain't th' Panhandle full of nesters (farmers)? **1918** —— *Man from Bar-20* iii. 27 He had found the ruins of a burned homestead..and he guessed that it had been used by 'nesters'. **1970** *Alberta Hist. Rev.* Summer 9/2 The one small cloud on this otherwise sunny horizon was the would-be farmer, or as he was more inhospitably known to the cattlemen—the 'sod-buster' or 'nester'. **1973** R. D. SYMONS *Where Wagon Led* p. xx, You plant a garden and someone'll see it an' we'll have nesters.

nesting, *vbl. sb.* Add: **b.** *nesting-burrow, -call, -cover, -ground, -habit, -hole, -territory, -tree;* **nesting-box** = NEST-BOX 2 (in Dict. and Suppl.); also *fig.*

1873 *Young Englishwoman* May 225/1 The space beneath the nesting-boxes should be partitioned from the rest of the cage. **1895** W. SCHLICH *Man. Forestry* IV. ii. 148 The wooden nesting-boxes..are made out of half-inch boards, and tarred. **1933** H. NICOLSON *Diary* 5 Jan. (1966) 131 Up Broadway and Madison to the hotel. Nesting-boxes. **1938** *Brit. Birds* XXXI. 331 Noticing..a Blue Tit.. entering about dusk a nesting box..I began watching. **1970** *New Yorker* 8 Aug. 56/3 We're building nesting boxes and doing everything else we can think of to encourage the kestrels. **1937** *Brit. Birds* XXX. 237 On Rona the thrift-grown banks..are riddled with nesting burrows of Leach's Petrels. **1959** VAN TYNE & BERGER *Fundamentals of Ornithology* x. 277 The Turquoise-browed Motmot.. may dig nesting burrows over 5 feet long. **1924** J. A. THOMSON *Sci. Old & New* xxi. 116 The selection is marked by the bird's remaining near the chosen spot and giving the nesting-call to the mate. **1936** *Brit. Birds* XXIX. 28 Elstree reservoir..is about 80 acres in extent with relatively little suitable nesting-cover for the Great Crested Grebe. **1921** *Daily Colonist* (Victoria, B.C.) 27 Oct. 10/3 The purchasing of suitable marsh and swamp lands to furnish nesting grounds for the birds in the North. **1961** O. L. AUSTIN *Birds of World* (1962) 38/2 They [*sc.* diving petrels]..are nocturnal on the nesting grounds. **1936** *Brit. Birds* XXX. 98 The average size of the broods of the Swallow..and other questions connected with their nesting-habits..bring the enquiry to a close. **1964** *Oxf. Bk. Birds* 102/2 They [*sc.* rock doves] build on ledges in caves and crevices and have..nesting habits like other doves. **1936** *Brit. Birds* XXIX. 379 The height above ground of the nesting-hole varies according to the choice there is of decayed wood. **1961** O. L. AUSTIN *Birds of World* (1962) 173/2 Guatemalans have always claimed that the Quetzal selects a nesting hole with two entrances ...Recent studies have shown this to be pure legend. **1923** J. S. HUXLEY *Ess. Biologist* vii. 291 The sense of being a trespasser so often shown by a bird that has ventured upon the nesting-territory of another. **1959** VAN TYNE & BERGER *Fundamentals of Ornithology* x. 268 Small nesting territories of colonial and some non-colonial birds—a small area round the actual nest. **1974** *Country Life* 14 Feb. 285/3 Fulmars..lay claim to sites, even when there is frost... Ravens, too, patrol a nesting territory. **1935** *Brit. Birds* XXVIII. 347 Return to nesting-trees.—First seen on February 18th.

2. (The making of) a nested arrangement (see *NESTED *ppl. a.* 2).

1957 D. D. MCCRACKEN *Digital Computer Programming* xvii. 207 The truncated series is factored: $e^x = 1 + x/1! + x^2/2! + x^3/3! + x^4/4! + x^5/5! + x^6/6! = 1 + x[1 + \frac{1}{2}x[1 + \frac{1}{3}x[1 + \frac{1}{4}x[1 + \frac{1}{5}x[1 + \frac{1}{6}x]]]]]$. The factored series is then evaluated 'from the inside out', which is also called nesting. **1958** C. F. HOCKETT *Course in Mod. Linguistics* xxi. 189 The presence of certain attributive constructions in the nesting precludes the occurrence of certain others at a more inclusive level: we can say *this fresh milk*, but not *fresh this milk*. **1963** [see *NESTED *ppl. a.* 2 b]. **1965** *Language* XLI. 71 Internal layering and multiple nesting characterize many syntagmemes. **1966** Y. BAR-HILLEL in *Automatic Transl. of Lang.* (NATO Summer School, Venice 1962) 11 The same degree of nesting is also assigned to the terminal expression as analysed by this tree. **1967** KLERER & KORN *Digital Computer User's Handbk.* I. iii. 75 Another technique to reduce operations is by nesting when increasing powers of a variable are to be calculated. For example, the expression $AX^3 + BX^2 + CX + D$ may be written in nested form as $((B + Ax)x + C)x + D$ which requires less arithmetic. **1972** HARTMANN & STORK *Dict. of Lang. & Linguistics* 150/2 *Nesting*, the embedding of a phrase or clause within an endocentric phrase to modify its head word.

nesting, *ppl. a.* Add: **2.** Of a table, chair, etc.: that forms part of a set of similar articles which can be fitted into one another (cf. NEST *sb.* 6).

Passing into the *vbl. sb.*

1934 H. READ *Art & Industry* II. 89 Another chair designed by Alvar Aalto. The plywood process combined with tubular steel to make nesting-chairs. **1934** A. WOOLLCOTT *While Rome Burns* 29 He [*sc.* Harpo, a French poodle] was gracious, for instance, toward Erich Maria Remarque, a dachshund who fitted under him like a nesting table. **1958** J. CANNAN *And be a Villain* iii. 57 In the meagre lounge, half-heartedly modernized with nesting chairs and formica-topped tables. **1959** N. MAILER *Advts. for Myself* (1961) 436 Hand-worked nesting tables. **1961** G. MILLERSON *Technique Telev. Production* viii. 150 (*caption*) Nesting sets may be placed one within the other, the inner one being struck to reveal the outer. **1969** *Guardian* 26 Sept. 11/7 Nesting tables..5 gns. each or £15. 10s. per set of three. **1970** *Washington Post* 30 Sept. B8/5 (Advt.), Front-zipped cotton luggage... Six nesting sizes. **1975** *Harpers & Queen* May 145/2 Gladstone bags in woven rush; 3 nesting sizes.

nestle-chick = NESTLE-COCK.

1932 R. MACAULAY *They were Defeated* I. 32, I couldn't spare her—my last nestle-chick, and the best. **1937** —— *I would be Private* 36 It was to have been her..nestle-chick.

net, *sb.*[1] Add: **2. a.** *spec.* In Games and Sports: a piece of netting used as part of the equipment for the game; esp. *Cricket,* the netting used to divide off practice wickets; hence in *pl.*, a name for such a wicket. Also, the safety net used by acrobats.

1845 'N. FELIX' *Felix on Bat* I. I. 7 The way to secure much practice..is to procure a large net, about twenty yards long and six feet in height, [etc.]. **1856** [in Dict.]. **1884** E. M. BUTLER in F. Gale *Life R. Grimston* (1885) xii. 191 In the evenings he would stand behind the nets when the eleven were practising, and coach them very thoroughly. **1889** [in Dict.]. **1899** *Captain* II. 127/1 The most difficult shot for a goal-keeper to stop is a low one that crosses him into the corner of the net... Beware lest..you shoot outside the net. **1903** G. B. SHAW *Man & Superman* IV. 144 There is no tennis net nor set of croquet hoops. **1905** A. BENNETT *Tales of Five Towns* II. 250 We should..flash past each other in mid-flight.. and soar to opposite platforms again, amid frenzied applause. There were no nets. **1915** W. S. MAUGHAM *Of Human Bondage* xvii. 67 They practised at nets in the summer. **1937** [see *HOOP *sb.*[1] 8 c]. **1947** *Sun* (Baltimore) 8 Nov. 12/2 Although Friends took the ball on several long runs down the field, the Bryn Mawr defense was too effective to be penetrated and Goalie Jo Nelson found little to do in the nets. **1955** *Times* 6 July 3/6 With his left side heavily plastered he batted in the nets before the Yorkshire game resumed. **1958** [see *GRAFT *v.*[2]]. **1958** J. BETJEMAN *Coll. Poems* 50, I adore you, Pam, you great big mountainous sports girl, Whizzing them over the net. **1960** N. STREATFEILD *Look at Circus* iv. 80 In any properly run circus under the trapezes or high tight-rope act a net is stretched. **1961** J. S. SALAK *Dict. Amer. Sports* 295 *Net* (ice hockey), the goal. **1965** *Men's Hockey* ('Know the Game' Series) 6 Nets are attached to the goal-posts, cross bar and ground behind the goal. Goal-boards..are placed at the foot of the goal-nets. **1967** C. B. MILLS *B. Mills Circus* viii. 109 Fritzi Bartoni..did one very difficult trick in which she fell forward from a trapeze and caught herself by her heels on the cross bar and she had been doing this without a net. **1975** *Times* 4 Dec. 10/1 Their goalkeeper..rolled out a ball that rebounded..to Summerill who sent Hart away to run the ball into an empty net.

c. *Lawn Tennis.* = LET *sb.*[1] 2. *colloq.*

1904 J. P. PARET et al. *Lawn Tennis* 344 *Net,*..also (same as 'let'), a ball that touches the net and goes into the proper court. **1929** D. MACKAIL *How Amusing!* 450 Clampson served a let. They actually—yes, in the twentieth century and the Centre Court—they distinctly called it a 'net'.

4. c. A network; *spec.* (*a*) a network of spies; (*b*) a broadcasting network.

1919 J. BUCHAN *Mr. Standfast* iii. 63 By the middle of 1915 most [enemy spies]..had been gathered in. But there remained loose ends, and..somebody was very busy combining these ends into a net. **1952** *Brewer's Dict. Phr. & Fable* (rev. ed.) 644/2 *On the Old Boy net*, to arrange something through a friend (originally, someone known at school) instead of through the usual channels. **1959** *Listener* 22 Oct. 668/2 Television programmes are to be exchanged among the East European countries and the Soviet Union... A coaxial cable is already being laid on the Soviet-Polish section of the net. **1961** H. B. JACOBSON *Mass Communications Dict.* 222 *Net*, abbreviation for radio network, which is merely a group of stations joined by wires to release a given program simultaneously. **1966** 'H. TALBOT' *Catch me Traitor* I. 14, D. I. 6 was probably short of yet another carefully built up net in the East German People's Republic. **1966** *Electronics* 17 Oct. 129 Capt. D. A. Jones..cautioned that 'our national security could be affected by the dissemination of data by these geodetic nets'. **1967** M. CHILDS *Taint of Innocence* (1968) ii. 97 She's got her own net going in that nest of vipers and I'd stake a lot on it. **1969** J. ELLIOT *Duel* I. i. 25, I had a regular weekly spot on a national net—thirty-six in a year. **1969** A. MARIN *Rise with Wind* xiv. 168 Whenever a net was surfaced..the word went around the *intelligence community.* **1971** C. BONINGTON *Annapurna South Face* iii. 33 He eventually got an agreement that the set in Katmandu should be kept in the Nepalese Army headquarters so that they would have control of the wireless net. **1972** 'W. HAGGARD' *Protectors* iv. 42 Shay was after the gold, on information received through the usual net.

5. a. *net-fishing* (later examples), *game, -man* (earlier and later examples), *-owner, -pocket, shot;* **d. net-bag** (earlier and later examples); **net-cord,** the cord passing along and supporting the top of a net, esp. a tennis net; so **net-cord (stroke),** in lawn tennis, a stroke which hits the net cord but still crosses the net; **net curtain,** a curtain made of net, usu. now fixed permanently across windows to ensure privacy; hence **net-curtained** *a.*; **net-drifter,** earlier term for *net-layer*; also *attrib.*; **net-fish** *v. intr.*, to fish with a net; **net-layer,** a vessel which lays anti-submarine nets; hence **net-laying** *vbl. sb.*; **netminder** = *goal-tender* (*GOAL *sb.* 6); **net-player** *Lawn Tennis* and *Badminton*, a player who advances close to the net; hence **net play; net-practice,** cricket practice at the nets.

1598 J. FLORIO *Worlde of Wordes* 61/1 *Carnero, Carnièro*, a net-bag to carie meat in, a hawking bag. **1934** *Discovery* Sept. 261/1 In those days the wild fowler had not learnt the trick of curving the pipes, and so had to trust to frightening the duck into the net bag, or some such contraption. **1951** COLYER & HAMMOND *Flies Brit. Isles* 324 This instrument can often be used under conditions where a net-bag is liable to be torn, e.g. on thorn or bramble. **1970** *New Yorker* 28 Nov. 51/1 On the third bed lay a bulging net bag. **1844** Net-cord [in Dict.]. **1887** *Boy's Own Paper* 3 Sept. 778/1 If you are tall enough, take it before it descends to the level of the net-cord. **1904** J. P. PARET et al. *Lawn Tennis* 345 Net-cord stroke. **1959** *Times* 1 Sept. 3/3 It was the third game when twice MacKay got away with net cords. **1961** F. C. AVIS *Sportsman's Gloss.* 258 *Net cord stroke*, one that causes the ball to contact the net, the ball afterwards going into the proper court. **1931** M. ALLINGHAM *Look to Lady* xxii. 224 There was a thin net curtain over the windows, but the light inside rendered it transparent. **1935** A. FREMANTLE tr. *Wynne Diaries* I. i. 2 The Curé of Berkheim's house is very pretty. There is in it a sofa and net curtains which he himself has made. **1967** E. SHORT *Embroidery & Fabric Collage* iii. 83 On net curtains, permanently in place across a window, designs of a more pictorial type would be quite suitable. **1972** J. MANN *Mrs Knox's Profession* x. 78 The local style was for picture windows, sometimes net curtained. **1973** C. EGLETON *Seven Days to Killing* xviii. 188 The net-curtained window. **1919** R. BACON *Dover Patrol 1915–17* I. vi. 157 The procedure of the mine-net drifters was simple. As each division arrived at its mark-buoy, the rear boats proceeded to shoot their nets, and, after steaming a definite time interval, each succeeding pair stopped and shot their nets to complete the line. *Ibid.*, In addition to the above, the 10th, 11th, and 12th Divisions of net-drifters anchored parallel to the West Hinder and shot their nets. **1923** E. K. CHATTERTON *Auxiliary Patrol* iii. 41 Net-drifter bases were quickly established also at Peterhead..and Larne. **1925** *Home Waters* (Naval Staff Monogr., Historical, XIII.) IV. vii. 126 By the beginning of March [1915] the term 'Net Drifter Flotillas' was in use to designate units of drifters working indicator nets, or other anti-submarine devices. **1891** *Chambers's Encycl.* VIII. 256/1 He may be arrested if he is net-fishing, but not if he is fishing in another way. **1968** J. ARNOLD *Shell Book of Country Crafts* 228 As no new licences are being granted, net fishing on the upper stretch of the river will die out. **1961** F. C. AVIS *Sportsman's Gloss.* 259 *Net game*, that kind of play taking place in the vicinity of the net, and thus the opposite of base line game. **1930** *Economist* 8 Mar. 513/1 Five submarines, one submarine depôt ship, two sloops and one netlayer. **1957** *Observer* 1 Dec. 1/1 H.M.S. Protector, a converted net-layer. **1923** E. K. CHATTERTON *Auxiliary Patrol* xv. 229 One hundred and two ships were secretly assembled, and the actual operation of net-laying was carried out in an hour and a half. **1847** C. LANMAN *Summer in Wilderness* xxvi. 160 A false movement of the net-man will cause the canoe to be swamped. **1915** [see *cross-court* adj. s.v. *CROSS- B]. **1934** R. GRAVES *Claudius the God* xxx. 518 He was disarmed and a net-man was standing over him with his trident raised. **1942** BERREY &

Van den Bark *Amer. Thes. Slang* §662/2 *Net minder*,..a goalkeeper. **1961** J. S. Salak *Dict. Amer. Sports* 296 *Net minder* (ice hockey), the goaltender. **1968** *Globe & Mail* (Toronto) 15 Jan. 19/1 Eddie Shack spoiled Montreal net-minder Rogatien Vachon's chance for a shutout when he scored. **1973** *Cleveland* (Ohio) *Plain Dealer* 6 Apr. 10-C/1 Cheevers has come on strong to maintain his reputation as one of the best netminders in hockey. **1901** *Chambers's Jrnl.* Sept. 585/1 The estuary limits..have been fixed more in the interests of net-owners than in those of the salmon. **1961** J. S. Salak *Dict. Amer. Sports* 296 *Net play* (badminton), the player returning the bird from a position close to the net. **1919** S. Lenglen *Lawn Tennis for Girls* 53 The net player..has a free hand at the net. **1930** *Amer. Speech* VI. 118 California's net players. **1902** *Chambers's Jrnl.* Jan. 48/2 The younger man..swung the captive ashore in the net-pocket. **1899** *Captain* I. 378/2 Net practice is good in moderation, but nothing is so good as practice games. **1975** *Country Life* 19 June 1625/3, I waited to greet the Australians for their first net practice. **1961** J. S. Salak *Dict. Amer. Sports* 296 In mixed doubles the lady is usually responsible for the net shots.

net, *a.* Add: **3. a.** Also as *sb.*

1910 *Gt. Central Railway Rep.* 11 Feb. 7 Deducting from our receipts..our expenses of £1,488,474, we have a net of £796,956. **1961** H. B. Jacobson *Mass Communications Dict.* 222 *Net*, the amount paid to the advertising medium by the advertising agency after deducting the agency commission. **1969** *N.Y. Rev. Books* 2 Jan. 6/4 Its net after taxes..after allowing for sales of assets,..was only $9 million.

b. *spec.* in phr. *net book*, a book sold at the net price; *Net Book Agreement*, the formal arrangement between publishers and booksellers, binding the latter not to sell below the net price; also, an agreement between publishers and public libraries to adhere to a set discount; *net* (*and*) *net*, used to indicate that the price of a book so marked cannot be subject to any discount whatever; *net system*, the system laid down in the Net Book Agreement for the sale of net books.

1890 *Bookseller* 6 Mar. 244/2 With the hearty co-operation of the retail trade, the *net* system could easily be introduced. **1893, 1894** [in Dict.]. **1911** *Encycl. Brit.* XXII. 630/1 The Booksellers' Association signed an agreement to charge the full published price for every net book. *Ibid.* 630/2 The net system was gradually introduced, net books and discount books being issued side by side. *Ibid.* 631/2 In 1901 the net system, as adopted in Great Britain, was partially introduced into America. **1938** L. M. Harrod *Librarians' Gloss.* 105 *Net book agreement*, an agreement drawn up in 1929..between the Publishers' Association, the Booksellers' Association, and the Library Association, enabling rate-supported libraries and other libraries..to receive a discount of 10 per cent on all new books purchased. **1939** M. Plant *Eng. Bk. Trade* xx. 441 Net books were to come, but not in that way. It was Frederick Macmillan who was to take the lead. He wrote to the *Bookseller* in 1890 suggesting the publication of net books. **1952** *Bookseller* 5 Jan. 27 This list is..published..at quarterly intervals..with Classified List, May 6/– net net. **1959** L. M. Harrod *Librarians' Gloss.* (ed. 2) 187 *Net book*, one which..may be sold to a signatory to the agreement at a recognized discount. **1960** G. A. Glaister *Glossary of Book* 274/2 The 1901 Net Book Agreement, now replaced, was based on the right of the Publishers Association to act on behalf of its publisher members in ensuring that the conditions of supply of net books were observed. *Ibid.* 275/2 *Net net*, an indication that a book so marked or described must be sold at the full published price, and that the publisher will not allow any discount to the book trade... Confined almost exclusively to reference works published at the lowest possible sum for the benefit of the trade. **1961** M. C. Turner *Bookman's Gloss.* (ed. 4) 113 When an English book carries no trade discount, it is sometimes called a 'net net' book. **1963** Kenneison & Spilman *Dict. Printing* 133 *Net net*, term indicating that a book will not be sold to the book trade at any discount. **1972** C. Bingley *Business Bk. Publishing* iv. 135 The validity of the British Net Book Agreement does not extend beyond the U.K.

c. *net reproduction rate*: a reproduction rate representing the average number of girls born to each woman of a population who can be expected to reach their mothers' age at the time of birth, calculated from the average fertility rates and death rates of each age-group during the period considered.

1928 R. R. Kuczynski *Balance Births & Deaths* I. iii. 46 The net reproduction rate, as we may call it, was 1·435. **1952** C. P. Blacker *Eugenics* viii. 170 Gross and net reproduction rates have been published for France since 1806. **1972** *Nature* 7 Apr. 270/1 In a population which is just barely reproducing itself, the Net Reproduction Rate should, of course, be 1·0. **1973** *Sci. Amer.* Apr. 19/1 Under conditions of high mortality the difference between the net reproduction rate and the gross reproduction rate is large.

d. *net worth* (see quot. 1930).

1930 *Economist* 17 May 1108/2 'Net worth' (that is, the stockholders' equity, made up of preferred and common stock outstanding, surplus account and undivided profits, taken at the beginning of the fiscal year). **1955** *Times* 13 July 13/1 Some estimates..have now been made..of the net worth of consumers. **1964** Gould & Kolb *Dict. Social Sci.* 465/2 The expression *net worth* is becoming obsolete. The British now frequently use the expression *total equity*.

e. *net national product* (see quot. 1964 and *national product*, *National *a.* 5).

1945 S. Kuznets *National Product in Wartime* I. ii. 13 Net national product consists of (a) flow of goods to consumers, (b) net nonwar Capital formation, (c) net war output, i.e., net additions to the inventory of war goods. **1945** S. E. Harris *Inflation & American Economy* i. 31 (*table*) Output rises. The percentage rise of 1939–43..Net national product. **1962** *Listener* 25 Oct. 679/1 The proportion of the national product going to property and enterprise..has fallen relative to that going to wages and salaries. **1964** Gould & Kolb *Dict. Social Sci.* 453/2 In defining national product..one has a choice about whether or not to subtract from the total an allowance for the depreciation of capital goods that occurs during the period... If one does make such a subtraction, the results are called net national product... Net national product measures the value of goods and services produced, after *setting aside* whatever is required to maintain the stock of capital goods as it was at the beginning of the period.

4. *Comb.*, as *net-priced* adj.

1909 *Daily Chron.* 16 July 3/2 In his new net-priced series. **1973** *Writers' & Artists' Year Bk.* [illegible] to sell their net-priced books [illegible]

net, *v.*¹ Add: **1. a.** To fasten *down* with a net.

1909 H. G. Wells *Tono-Bungay* [illegible] I contracted my sausage gas-bag by netting it down.

2. a. (Later example.) Hence in *colloq.* or *slang* use, to acquire (cf. Land *v.* 3 b).

1900 Ade *Fables in Slang* [illegible] the Bazaar was pleased to learn that the Sixty-cent Vase had Netted over Seven Hundred Dollars. **1922** Galsworthy *In Chancery* [illegible] Some stars [illegible] **1936** H. L. Mencken *Amer. Lang.* (ed. 4) [illegible] favorite verbs [illegible] three letters [illegible] 10 Feb. 52/2 [illegible] mishap, a fall that netted her a broken arm and an injured ankle.

4. *trans.* and *absol.* In ball games in which a net is employed: to send (a ball) into the net.

1906 *Peel City Guardian* [illegible] meant to force the custodian into the net, and not the ball. **1907** *Ibid.* [illegible] netting. **19**[illegible] for Birmingham [illegible] 2/2 Raymond [illegible] a succession of returns. **1942** Berrey & Van den Bark *Amer. Thes. Slang* [illegible] *Times* 1 Oct. [illegible] Bradford City [illegible] Delman *Sudden Death* [illegible] on a shoddy return [illegible] away. **1975** [illegible] ford knocked up [illegible] season when [illegible]

Ne·therla·ndic, *a.* [f. *Netherland* + -ic.] = Netherlandish *a.*

1902 *Encycl. Brit.* [illegible] for the sake of his early Netherlandic manner. **1972** E. Krispyn in [illegible] Paul Boon, [illegible] and controversial [illegible] Netherlandic [illegible] story of Reynard the Fox. **1973** [illegible] *letter* Mar. 1 [illegible] but with the exception of [illegible] ture', which [illegible] groups in [illegible] **1975** *Times Lit. Suppl.* [illegible] [series of books] [illegible] including a [illegible] landic) section.

Netherlandish, *a.* Add: **B.** *sb.* The language of the Netherlands; Dutch.

1890 *Chambers's Encycl.* [illegible] Netherlandish [illegible] **1944** *Britannica* [illegible] Spanish, E[illegible] X. Apr. 91 [illegible] hâtive,' bar[illegible] from Nether[illegible] that lg. [illegible] Netherland[illegible] *N.Y. Times* [illegible] educated and [illegible] 'Vlaams', or Flemish; [illegible] Dutch, or Netherlandish.

netherward(s, *adv.* Add: Also (in form *netherward*) as *adj.*

1865–6 W. Whitman *When Lilacs last in Door-Yard Bloom'd* 6, [illegible] netherward [illegible] I. vii. 149 [illegible] fervour had [illegible] netherward Egdon.

netherworld (ne·ðəɹwɜːld). [f. Nether *a.* 6 + World *sb.*] = Underworld 4.

1930 *Sat. Evening Post* [illegible] sibly deny [illegible] P. Heron [illegible] symbol of the artist [illegible] years: for it [illegible] vast netherworld [illegible] scious. **197**[illegible] word is always [illegible] it with 'netherworld' (the netherworld of deceit, subversion and espionage'). **1975** *Times Lit. Suppl.* 13 June 679/1 Occasionally he even descended from Baltimore into this netherworld of benighted yokels and hillbillies.

netsman. (Further examples.)

1937 H. Pope *St. Augustine of Hippo* vi. 245 One of the netsmen has just been bitten by a crab. **1962** *Punch* 21 Feb. 328/2 Who but a lunatic or a salmon netsman would be astir? **1973** *Nature* 27 July [illegible] Many were caught by netsmen off other neighbouring salmon rivers. **1975** *Times Lit. Suppl.* [illegible] The salmon netsmen of the east coast of Scotland were the first to complain of the damage done by seals.

netter. Add: **3.** *U.S. colloq.* A lawn-tennis player.

1932 *Sun* (Baltimore) 6 Sept. 14/2 (*heading*) Coast netter stops invader. [illegible] **1974** *Anderson* (S. Carolina) *Independent* 23 Apr. 6A/3 The Anderson College netters [illegible]

netting, *vbl. sb.* Add: **1.** (Earlier example.)

1785 E. Sheridan *Jrnl.* 20 June (1960) [illegible]

b. *netting-case*, *-cotton*, *-needle* (earlier and later examples).

1790 F. Burney *Diary* Jan. (1905) IV. [illegible] **1817** Jane Austen *Northanger Abbey* [illegible] **1865** Mrs. Gaskell *Wives & Daughters* [illegible] **1832** in A. Adburgham *Shops & Shopping* (1964) [illegible] **1847** C. Brontë *Jane Eyre* [illegible] **1918** N. Duncan [illegible] **1973** [illegible] the netting-needle, on which a certain amount of twine is wound, and the spool.

nettle, *sb.*¹ Add: **2. d.** *to cast* (*throw*) *one's frock* (*or cassock*) *to the nettles* [= Fr. *jeter le froc aux orties*], to renounce the clerical life; also *transf. rare.*

1916 W. J. Locke *Wonderful Year* [illegible] **1928** —— *Rough Road* [illegible]

4. a. *nettle beer* (further examples), *-field.*

1910 A. Bennett *Clayhanger* [illegible] **1953** [illegible] **1969** G. B. Schaller *Year of Gorilla* [illegible]

b. *nettle-weed U.S.*, a plant of the nettle family.

1843 'R. Carlton' *New Purchase* I. xix. 159 [illegible] **1867** T. Lackland *Homespun* [illegible]

network. Add: **2. c.** (Further examples.) Also without *of*.

1869 *Bradshaw's Railway Manual* XXI. 346 [illegible] **1964** A. Battersby *Network Analysis* ii. 13 It is not the purpose of this book to discuss such methods of charting in detail..but rather to point out why their inadequacies have made necessary a new method of charting: arrow diagrams, otherwise known as networks. **1966** S. Beer *Decision & Control* ix. 191 Naturally, the critical path through a converging network can be discovered under any criterion or balance of criteria. **1966** *Economist* 3 Dec. 1042/2 If one wanted to

work out what was involved in doing something, an engineer would use a bar chart. (The 'hard way' in our charts showing the networking of a cup of tea.) **1967** S. Woodgate in Wills & Yearsley *Handbk. Management Technol.* v. 73 Network planning techniques first came into general use in 1957. *Ibid.* 74 Since 1957 hundreds of variations of network planning systems have proliferated and many of these variations have perished while undergoing the supreme test of operational effectiveness. *Ibid.* 98 *Network*, a schematic representation of events and activities which shows their inter-relationships. **1968** *Listener* 4 July 13/2 They propose..a 'network' system which would allow variety and versatility to be developed. **1970** O. Dopping *Computers & Data Processing* xxii. 345 Network planning helps to reveal all the possibilities for parallel operations, identify critical activities, and follow up the job... A network is built up of the different *activities* which constitute the project. **1970** *Canadian Jrnl. Linguistics* XV. 103 Once we express a grammar in terms of a relational network, intermediate symbols become completely superfluous. *Ibid.* 108 One can organize types of grammar in a two-by-two table. There are re-write grammars and network grammars.

e. A system of cables for the distribution of electricity to consumers; *spec.* one in which interconnections are such that each consumer is supplied by more than one route; hence, any system of interconnected electrical conductors or components, sometimes including a source of e.m.f., that provides more than one path for the current between any two points.

1883 *Jrnl. Soc. Telegr.-Engineers & Electricians* XII. 551 (*heading*) On a method of calculating the total horse-power expended in a network of conductors (such, for instance, as a system of street mains). **1914** J. W. Meares *Electr. Engin. in India* iii. 47 The terms 'network' and 'distributor' are also applied to any system of distributing mains. **1930** Dannatt & Dalgleish *Electr. Power Transmission & Interconnection* v. 110 Any network can be divided into conducting polygons or meshes. **1940** *Proc. IRE* XXVIII. 415/1 Figs. 3 and 4 represent mathematically identical networks, even though they refer to physically different systems. **1962** *Newnes Conc. Encycl. Electr. Engin.* 201/1 Distribution systems can be classified into the following basic categories: (*a*) Radial system. (*b*) Ring system: (i) h.v. feeders ring. (ii) l.v. distribution ring. (*c*) Network system. **1966** *McGraw-Hill Encycl. Sci. & Technol.* IX. 72/1 A network may be solved when it is possible to set up a number of independent equations equal to the number of unknown quantities.

f. A broadcasting system, consisting of a series of transmitters capable of being linked together to carry the same programme; also, in a more general sense: a nation-wide broadcasting company; the broadcasting companies as a whole. Also *attrib.*

1914 W. A. Du Puy *Uncle Sam's Modern Miracles* 170 This great network of stations may not merely hurl forth its messages. **1933** *B.B.C. Year Bk.* 37 In step with the gradual nationalisation of the network, local autonomy has declined. **1941** *B.B.C. Gloss. of Broadcasting Terms* 20 *Network cue*, cue consisting of a standardized phrase which constitutes a signal for the performance of technical operations at switching centres and transmitting stations forming part of a network. **1956** *Newsweek* 7 May 59 On its three wholly owned TV stations, exclusive of network operations, CBS reported a total net investment as of Dec. 31, 1953, of $3,322,023. **1957** *Listener* 21 Nov. 853/3 His hour-by-hour account of what the networks were saying. **1961** *Washington Post* 17 Feb. A14 The Columbia Broadcasting System's handling of its excellent TV show 'The Spy Next Door' on Wednesday night at 10 is a good illustration of what is amiss with network television. **1966** *Publ. Amer. Dial. Soc.* xlii. 46 A Southern rustic's imitation of 'network' pronunciation. **1969** *New Statesman* 11 Apr. 506/1 If there were 100 commercial stations..on the air all day they would either have to negotiate..or have back-up programmes from some central concern—virtually a network outfit. **1970** *New Yorker* 28 Feb. 29/2 The more general avoidance of controversial issues which has been noticeable among politicians and on the networks and in the press. **1972** *Daily Tel.* 14 Aug. 7/6 The plans for Radios 1 and 2 are designed to put the networks into an impregnable position when the commercial companies start. **1972** *Sci. Amer.* Sept. 117/1 The radio and television broadcasting networks provide the people of the world with on-the-spot news reports, entertainment and educational programs through a billion radio and television receivers. **1972** J. L. Dillard *Black English* vi. 261 Many a preacher of this type [*sc.* 'storefront'] presents a better performance on Sunday than is ever to be seen on network television.

g. A structure proposed for glass in which the non-oxygen atoms (usu. silicon) are joined together in a three-dimensional array by oxygen atoms.

1932 W. H. Zachariasen in *Jrnl. Amer. Chem. Soc.* LIV. 3842 As in crystals, the atoms in glasses must form extended three dimensional networks. **1947** *Jrnl. Soc. Glass Technol.* XXXI. 117 A quantitative factor may be deduced which is..an indication of the extent to which these atoms may be expected to take part in the glass network. **1971** *Oxidation of Metals & Alloys* (Amer. Soc. for Metals) 42 Oxides of the form..M_2O_3 can form a network if the oxygens form triangles around each metal atom.

h. An interconnected group of people; an organization. Also *attrib.*

1947 *Science News* IV. 37 He [*sc.* the habitual criminal]..is matching his brain and slender resources against all the might of a vast police network with infinite resource. **1954** [see *girl *sb.* 2 a]. **1957** E. Bott *Family & Social Network* iii. 62 Intermediate degrees of conjugal role-segregation and network-connectedness. *Ibid.* 92 Although the networks of husband and wife are distinct, it is very likely, even at the time of marriage, that there will be overlapping between them. **1960** *Analog Science Fact/Fiction* Dec. 68/1 As long as they're allowed to think they haven't been spotted, they may lead the way to other spies or spy net-works. **1960** L. Pincus *Marriage* i. 15 Supports and satisfactions..normally provided by participation in..social and kinship networks are sought, and often not found, within the resources of marriage. **1972** *Daily Tel.* 19 June 2/1, I was paid..about £500 for infiltrating the IRA network in London. **1973** *Times* 15 Aug. 21/7 'Multinational organizations are becoming increasingly integrated..', especially in the case of 'network' companies having subsidiaries in many countries. **1974** 'J. Le Carré' *Tinker, Tailor* xxii. 196 Moscow..was busy denouncing him for blowing the San Francisco network.

5. *attrib.* and *Comb.*, as **network analyser,** an assembly of inductors, capacitors, and resistors used to model an electrical network and facilitate its analysis; **network analysis,** (*a*) *Electr.*, calculation of the currents flowing in the various meshes of a network; (*b*) *Work Study* (see quot. 1968; cf. sense 2 d above); **network former,** (a substance containing) an atom which can become part of the network of a glass; so **network-forming** *a.*; **network modifier,** (a substance containing) a metal ion which can occupy an interstice in the network of a glass; so **network-modifying** *a.*; **network structure** *Metallurgy*, the structure of an alloy in which one component forms a continuous network around the grains of the other component; **network theorem,** any of various theorems about the currents and voltages in an electrical network that can be used to determine their values in any particular case.

1930 H. L. Hazen et al. in *Trans. Amer. Inst. Electr. Engin.* XLIX. 1102/1 The following paper deals with an improved form of network-computing device... The Network Analyzer, as this device is called, is installed in the Electrical Engineering Research Laboratory. **1945** *Physical Rev.* LXVII. 48/1 The tests reported here..indicate that it is practical to use for this purpose [*sc.* determining eigenvalues and eigenfunctions of the Schrödinger equation] existing a.c. network analyzers. **1974** *Encycl. Brit. Macropædia* VI. 625/2 Since World War II, digital computers have largely replaced network analyzers. **1930** *Trans. Amer. Inst. Electr. Engin.* XLIX. 1102/1 Several types of experimental computing devices or miniature systems for network analysis have been developed. **1945** H. W. Bode *Network Analysis & Feedback Amplifier Design* xvi. 360 Expressions having the mathematical form of a reflection coefficient appear frequently in network analysis. **1962** *Operations Res.* X. 728 Network analysis—also known as PERT (program evaluation and review technique), PEP (program evaluation procedure), CPM (critical-path method), CPS (critical path scheduling), and arrow diagramming—for the planning and control of research and industrial projects and programs has been the subject of many published articles. **1966** *Economist* 3 Dec. 1043/1 The Post Office Savings Bank has made massive use of network analysis to control its move to Glasgow. **1968** *Gloss. Terms Project Network Analysis* (*B.S.I.*) 5 *Project network analysis*, a group of techniques for presenting information to assist the planning and controlling of projects. The information, usually represented by a network, includes the sequence and logical interrelationships of all project activities. **1969** *Power System Protection* (Electr. Council) I. iii. 201 Network analysis by manual computation may involve a very considerable amount of labour. **1973** A. Parrish *Mech. Engineers' Ref. Bk.* xx. 73 Network analysis is a general term which is used to embrace a whole series of similar planning methods dealing with project control. **1947** *Jrnl. Soc. Glass Technol.* XXXI. 114 For some years it has been customary to classify the various constituents of glass into network formers, such as SiO_2 and B_2O_3; intermediates..; and modifiers. **1966** C. R. Tottle *Sci. Engin. Materials* iii. 83 Aluminium occurs in the natural glass obsidian, as a network former, in the proportion of one aluminium to five silicon ions, with sodium and potassium to modify the network. **1950** Kirk & Othmer *Encycl. Chem. Technol.* V. 723 Oxygen ratio is defined as the total number of oxygen ions to the total number of network-forming ions. **1971** Fehlner & Mott in *Oxidation of Metals & Alloys* (Amer. Soc. for Metals) 43 Ionic transport by anions is expected in network-forming oxides. **1961** *Progress Ceramic Sci.* I. 7 Systematic measurement of elastic properties of glasses, containing various proportions of the single network former SiO_2, and a single univalent network modifier, would be of great value in assessing the role of the network former in determining theoretical strength. **1971** *Materials & Technol.* II. vi. 335 Aluminium, when in tetrahedral coordination..is undoubtedly a network former, and when in octahedral configuration it is a network modifier. **1950** Kirk & Othmer *Encycl. Chem. Technol.* V. 732 The 'holes' between these tetrahedral groups are considered to be occupied by the network-modifying ions. **1966** C. R. Tottle *Sci. Engin. Materials* iii. 83 Healing glass containing metallic ions that are network-modifying tends to move them into positions that are network-forming and so varies the colour. **1939** E. C. Rollason *Metall. for Engin.* vii. 101 If sufficient time is allowed for the diffusion phenomena all the ferrite is precipitated, whilst the pearlite occupies the centre forming a network structure. **1940** E. J. Teichert *Ferrous Metallurgy* III. xiv. 317 Slow cooling will induce the Widmanstatten and rapid cooling will promote the network structure. **1965** A. D. Merriman *Concise Encycl. Metall.* 594 When the structure of an alloy is such that one constituent occurs partly or completely surrounding the crystals of another constituent, then the appearance of an etched section taken across the grains shows as a network structure. **1930** Dannatt & Dalgleish *Electr. Power Transmission & Interconnection* vii. 220 (*heading*) Network theorems. **1966** *McGraw-Hill Encycl. Sci. & Technol.* IX. 75/2 In the solution of specific problems, much time can often be saved by making use of special methods or relations, known as network theorems.

network (ne·twɒɹk), *v.* [f. the sb.] **a.** To cover with a network.

1887 *Courier-Jrnl.* (Louisville, Kentucky) 24 Jan. 8/1 It is only a question of time when railroads will net-work the Pan-handle. **1914** *Cycl. Amer. Govt.* III. 139/1 Whole regions are networked, and one can go by trolley car from the Atlantic to the Middle West. **1928** E. Wright *Great Horn Spoon* xviii. 217 Gourds and several varieties of squashes networked the compound with their vines.

b. To broadcast simultaneously over a network of radio or television stations.

1940 [implied in *networking* vbl. sb. below]. **1952** *Sun* (Baltimore) 12 Feb. 14/3 Matthews beat Murphy in a bout networked out of Madison Square Garden. **1957** *Times* 28 Aug. (Radio & Television Suppl.) p. iii/3 This practice of networking the major items will continue. **1958** *Observer* 21 Sept. 16/6 The first sample of *This Wonderful World* to be networked in England came up to high expectation. **1968** *Times* 13 Nov. 1 (Advt.), Early in 1969 Thames Television are networking twelve one hour documentaries about one man. **1973** *Times* 8 Jan. 3/2 His series on the castles of Wales is being networked.

Hence **ne·tworked** *ppl. a.*; **ne·tworking** *vbl. sb.* and *ppl. a.*

1940 Porterfield & Reynolds *We present Television* iv. 149 Television, now that a practicable means of networking has been developed, has been supplied with the final implement necessary for the creation of what will eventually be a nation-wide service. **1956** *Newsweek* 7 May 59 The television networking business is a complicated and delicate business. **1962** *Rep. Comm. Broadcasting 1960* 159 in *Parl. Papers 1961–2* (Cmnd. 1753) IX. 259 They must..use an old recorded networked programme. **1968** *Listener* 8 Aug. 187/1 The new companies..operate within a networking system still dominated by Granada, ATV and Thames. **1970** *New Statesman* 4 Sept. 281/3 A small [television] company can rely on getting full networking for just about four programmes a year... This figure of about four hours' worth of fully networked television time ought to be placed beside..the fact that STV is obliged..to do nine hours' output of its own every week. **1971** *Writing for B.B.C.* 65 Scotland contributes to all the networked series described under the 'Drama' heading... Plays intended for networking..should not be so Scottish that they cannot be readily understood by listeners in the other parts of the British Isles.

Neuchâtel (nɜ̃ʃatel). The name of a town and a canton in western Switzerland, used *attrib.* or *absol.* to designate the white wine, or the less common red one, made there.

1903 S. Weyman *Long Night* i. 8 The landlord..vanished, to return..with four tall glasses and a flask of Neuchatel. **1967** A. Lichine *Encycl. Wines & Spirits* 516/1 The Neuchâtel white wines, grown in a chalky soil, are light and sprightly. *Ibid.* 516/2 Neuchâtel is the most widely exported wine of Switzerland. **1969** N. Denny tr. *Veraldi's Spies of Good Intent* viii. 114 My brother's a wine-grower and he keeps the best of his Neutchatel [*sic*] for me.

‖ **Neue Sachlichkeit** (noi·ə za·χliχᵛkait). [G., 'new objectivity'.] A movement in the fine arts, music, and literature, which developed in Germany during the 1920s and was characterized by realism and a deliberate rejection of romantic attitudes.

1929 G. F. Hartlaub *Let.* 8 July in *Arts* (1931) XVII. 237 The expression *Neue Sachlichkeit* was in fact coined by me in the year 1924. A year later came the Mannheim exhibition which bore the same name. The expression ought really to apply as a label to the new realism bearing a socialistic flavor. **1938** C. Fullman tr. *Thoene's Modern German Art* 87 The new group..wanted super-concrete precision; after the elimination of objects, journalistic reporting of time and place. This was the starting-point for German *vérisme*, the 'Neue Sachlichkeit', and the art of such men as Max Beckmann, Otto Dix, George Grosz and their like. **1950** B. S. Myers *Modern Art in Making* xix. 351 There appeared a movement since classified as *Neue Sachlichkeit*, or New Objectivity, begun in 1920 by Otto Dix and George Grosz. **1954** *Grove's Dict. Mus.* VI. 53/1 *Neue Sachlichkeit*.., a term that came into fashion in Germany between the two world wars. It describes a tendency in some composers of that time to write music entirely detached from sentiment and free from any pictorial suggestion. **1959** J. Willett *Theatre of Bertolt Brecht* vii. 194 A deliberately impersonal quality marks much of the art of that time [*sc.* the late 1920s], as the words 'Neue Sachlichkeit' themselves imply. **1964** *Listener* 17 Sept. 445/1 After its first performance in 1923..[Hindemith's] *Marienleben* was hailed as a shining example of *Neue Sachlichkeit*, a new matter-of-fact contrapuntal style then emerging, which represented a conscious reaction against the hyper-romantic attitude prevalent earlier this century. **1970** *Oxf. Compan. Art* 772/2 *Neue Sachlichkeit*, a movement in German art and literature that arose in the mid 1920s. It represented a sharp reaction against experimental and idealistic art of any sort... Artists strove for an 'honest objectivity' (Sachlichkeit), depicting with matter-of-fact literalness their everyday experience.

Neufchâtel (nɜ̃ʃatel). The name of a small town in Normandy, NE. of Rouen, used *attrib.* or *absol.* to designate the soft, white cheese originally made there.

c **1865** E. C. Gaskell *Let.* 6 Oct. (1966) 778 Lunch..chocolate, cold meat, bread & butter Neufchatel Cheese & grapes. **1902** [see *bondon]. **1951** *Good Housek. Home Encycl.* 568/2 *Neufchâtel Cheese*, a soft whole-milk cheese prepared in Normandy and other parts of France. **1966** P. V. Price *France: Food & Wine Guide* 283 *Neufchâtel*, the name of the best cheese produced in the Pays de Bray.

It may be eaten when it has only been kept for a few days, in which case it is creamy, soft and mild..; if allowed to ferment and ripen it..is fairly strong.

Neumann (noi·man). The name of Johann G. *Neumann*, 19th-cent. Austrian geologist(?), used *attrib.* to designate narrow bands, lines, or lamellæ parallel to crystallographic planes which are seen in α-iron (ferrite) subjected to a sudden shock, and are usu. attributed to twinning (investigated by Neumann *c* 1848).

1886 *Proc. Amer. Acad. Arts & Sci.* XXI. 498 On different sections of meteorites Widmanstättian figures and Neumann lines can be exhibited in every gradation, from the broadest bands to the finest markings. **1914** *Jrnl. Soc. Chem. Industry* 15 Aug. 791/2 The Neumann lamellæ occurring in any given grain are parallel to not more than 6 planes corresponding to the crystalline orientation of the grain itself. **1922** *Chem. Abstr.* XVI. 3845 (*heading*) Neumann bands as evidence of action of explosives upon metal. **1946** *Jrnl. Iron & Steel Inst.* CLIV. 129P/1 The Neumann band is primarily a shearing or faulting movement operating along the pre-existent planar disjunctions of the mosaic structure, and..as a secondary operation twinning may be completed, as is known to be possible in ferrite. **1966** C. R. Tottle *Sci. Engin. Materials* iii. 64 In alpha iron (b.c.c.) impact loading at room temperature can produce shear twins, called Neumann bands (which are found, incidentally, in metallic meteorites).

neurally (niū̆ə·răli), *adv.* [f. Neural *a.* + -ly².] **a.** By a nerve or nerves.

1902 W. James *Var. Relig. Exper.* p. vii, All states of mind are neurally conditioned. **1974** *Nature* 13 Dec. 591/2 Neurally-evoked responses.

b. As regards one's 'nerves'.

1936 L. C. Douglas *White Banners* vi. 141 He was too far spent, neurally, to offer any resistance.

neuraminic (niū̆ərămi·nik), *a. Chem.* [tr. G. *neuraminsäure* neuraminic acid (E. Klenk 1941, in *Zeitschr. f. physiol. Chem.* CCLXVIII. 51): see Neur-, Amine and -ic.] *neuraminic acid*: a crystalline carboxylic acid, HOOC·CO·CH_2·CHOH·$CHNH_2$·$(CHOH)_3$·CH_2OH (or a cyclic form of this), acyl derivatives of which are the sialic acids.

1942 *Chem. Abstr.* XXXVI. 3204 (*heading*) Neuraminic acid, the cleavage product of a new brain lipoid. **1956** *Nature* 17 Mar. 524/2 Klenk *et al.* have also investigated the amount of ganglioside in human brain by determining the content of neuraminic acid. **1966** *McGraw-Hill Encycl. Sci. & Technol.* II. 12/1 Neuraminic acid, an important constituent of carbohydrate-protein complexes of animal cells and tissues, has been reported in polymeric form in cultures of *Escherichia coli*.

Hence **neura·minate,** a salt or ester of neuraminic acid.

1970 R. W. McGilvery *Biochemistry* xxiv. 585 Part of the neuraminates contain an additional 4-O-acetyl group.

neuraminidase (niū̆ərămi·nidēiz, -s). *Biochem.* [f. *Neuramin(ic *a.* + -id⁴ + *-ase.] (See quot. 1957.)

1957 A. Gottschalk in *Biochim. & Biophys. Acta* XXIII. 646 We propose the name 'neuraminidase' for the enzyme and define its action as the hydrolytic cleavage of the glycosidic bond joining the keto group of neuraminic acid to D-galactose or D-galactosamine and possibly to other sugars. **1960** —— in P. D. Boyer et al. *Enzymes* (ed. 2) IV. xxvii. 472 The wide distribution of neuraminidase among microorganisms inhabiting the respiratory and intestinal tracts would suggest that the enzyme has been evolved as a vital mechanism to guard the microbe against solitary confinement in a coating of host mucin. *Ibid.* 473 Neuraminidase has not been met in animal tissues. It is not a metabolic enzyme but a tool to facilitate accommodation of a parasite or symbiont within its host. **1970** *New Scientist* 8 Jan. 51/1 Influenza viruses possess another surface protein, the enzyme neuraminidase. *Ibid.* 31 Dec. 608/3 The enzyme neuraminidase, injected into mice as a post-coital contraceptive, seemed to be the most effective base for a 'morning-after pill' of any yet tried.

neurasthenia. (Further examples.)

1922 J. Riviere tr. *Freud's Introd. Lectures on Psycho-Anal.* xxiv. 325 We distinguish three pure forms of actual neurosis: neurasthenia, anxiety-neurosis and hypochondria. Even this classification has been disputed. The terms are certainly all in use, but their connotation is vague and unsettled. **1926** J. I. Suttie tr. *Ferenczi's Further Contrib. Theory & Technique Psycho-Anal.* ii. 35 The obsessional neuroses are nowadays mostly ranked among the 'neurasthenias'. **1965** Rosen & Gregory *Abnormal Psychol.* xii. 250/1 Neurasthenia is observed with such frequency in housewives who are bored and feel neglected by their husbands that it has often been called 'housewives' neurosis'. **1968** C. Rycroft *Crit. Dict. Psychoanal.* 96 *Neurasthenia*, obsolescent medical and psychiatric term for a state of excessive fatiguability and lack of vigour. **1974** M. Sainsbury *Key to Psychiatry* xiv. 234 Much disagreement is in evidence through the years regarding the status of neurasthenia as a nosological entity.

neurastheniac (niū̆ərǽsþī·niæk). *rare.* [f. Neurasthenia: see -ac.] = Neurasthenic *sb.*

1904 *Lancet* 18 June 1737/2 The 'neurastheniacs' that present themselves at the clinic for diseases of the nervous system.

neurilema, neurilemma. Add: Also neuro-. **c.** The thin outer sheath that is seen with the light microscope surrounding the axon (and the myelin sheath, if present) of an individual peripheral nerve fibre; also called *sheath of Schwann*. (The usual sense: senses a, b are now *rare* or *Obs.*) (Further examples.)

1852 Neurolemma [in Dict. s.v. Neuro-]. **1874** A. E. J. Barker tr. *Frey's Histol. & Histochem. of Man* 307 The existence of an envelope on the nervous tube [*sc.* fibre] is easily inferred... This neurilemma may be seen not infrequently as [etc.]. *Ibid.* 317 The nerves of the brain and spinal cord..become clothed with a delicate envelope at their exit from the nervous centres. This covering receives another addition from the *dura mater* of connective-tissue bundles in its passage through the latter, and..constitutes what was formerly known as 'neurilemma', but which we will designate from henceforth 'perineurium'. **1890** *Gray's Anat.* (ed. 12) 42 The tubular sheath of the funiculi, called the neurilemma or perineurium. **1892** E. A. Schäfer *Essent. Histol.* (ed. 3) xviii. 83 Outside the medullary sheath is a delicate but tough homogeneous membrane, the primitive sheath or nucleated sheath of Schwann... The primitive sheath is also known as the neurolemma. **1930** Maximow & Bloom *Text-bk. Histol.* xii. 251 Some European writers, like Cajal, apply the term 'neurilemma' to the epineurium and not to the sheath of Schwann of the individual fibers. **1966** Wright & Symmers *Systemic Path.* II. xxxiv. 1252/1 Neurolemmomas arise from the cells of the neurolemma, or nerve sheath. It is believed that the cell of origin is the Schwann cell. **1968** Bloom & Fawcett *Textbk. Histol.* (ed. 9) xii. 321/1 The myelin is actually part of the Schwann cell, consisting of spirally wrapped layers of its surface membrane... The outer membrane..and the protein-polysaccharide boundary layer on its outer aspect were resolved with the light microscope as a single layer, which has traditionally been called the neurilemma.

Hence also **neurile·mmal** *a.* (now the usual adj.).

1903 *Rep. Brit. Assoc. Adv. Sci. 1902* 782 From the microscopic study of the distal portions of divided nerve-trunks we arrived at the conclusion that the activity of the neurilemmal cells has some relation to the development of the new nerve-fibres. **1937** E. E. Hewer *Text-bk. Histol.* 96 Within the central nervous system there is no neurilemmal sheath. **1954** T. L. Peele *Neuroanat. Basis Clin. Neurol.* i. 13/2 The neurolemmal sheath of the spinal nerves, and of those cranial nerves possessing it, terminates (or begins) a short distance from spinal cord or brain stem. **1973** H. M. Ráliš et al. *Techniques Neurohistol.* i. 19 The neurilemmal sheath plays an important role in the nutrition and protection of nerve fibres.

neurilemoma, neurilemmoma (niū̆ərilemōᵘ·mă). *Path.* Also neuro-. [f. Neurilem(a, neurilemm(a + *-oma.] A tumour formed by proliferation of the neurilemma.

1935 A. P. Stout in *Amer. Jrnl. Cancer* XXIV. 752 Therefore it seemed advisable to take this phrase 'nerve sheath tumor'..and to construct from it a new name for the neoplasm. After consultation with Dr. G. F. Laidlaw and Dr. F. H. Vizetelly, editor of the *New Standard Dictionary*, the word 'neurilemoma' has been constructed and it is proposed to use this term for the tumor under discussion. *Ibid.* 753 Neurilemomas can be found sometimes in cases of von Recklinghausen's disease. **1948** J. M. Beattie et al. *Text-bk. Path.* (ed. 5) II. xxix. 1310 On the supposition that this word is derived from the Greek words νεῦρον, *neuron*, a nerve-fibre.., λέμμα, lemma, a skin or sheath..we ventured to alter the spelling of Dr. Purdy Stout's term 'neurilemoma' to 'neurilemmoma'... Dr. Purdy Stout..has persuaded us of the error of our ways...It is a pleasure..to take this opportunity of altering our spelling of the word back to..'neurilemoma'—though we are still in some doubt as to whether..it might not with propriety take the form 'Neurolemoma'. **1948** R. A. Willis *Path. Tumours* liv. 828 Masson's and Stout's names 'Schwannoma' and 'neurilemmoma', and Mallory and Penfield's name 'perineural fibroblastoma', express the two opposing views regarding the histogenesis of this distinctive nerve-sheath tumour. **1954** Ackerman & Del Regato *Cancer* (ed. 2) vii. 202 Neurilemomas [*ed.* 1 (1947): neurilemmomas] have also been reported to arise in the nasal fossa. **1962** *Lancet* 29 Dec. 1354/2 The preoperative diagnosis rested between a neurilemmoma of the posterior tibial nerve and compression of the nerve by the retinaculum. **1966** Neurolemmoma [see *neurilema].

neurinoma (niū̆ərinōᵘ·mă). *Path.* Pl. **-omas, -omata.** [mod.L. (J. Verocay 1910, in *Beitr. z. path. Anat. u. z. allgem. Path.* XLVIII. 65), f. Neur- + Ino- + *-oma.] = *Neurilemoma, neurilemmoma.

1913 in Stedman *Med. Dict.* (ed. 2) 600/1. **1917** H. Cushing *Tumors of Nervus Acusticus* vii. 195 Though the term 'fibrosarcoma' predominates, other terms, which have been met with in cases which appear to be unquestionably acoustic tumors, are as follows: steatoma of the older writers,..neurinoma (Verocay), fibroblastoma (Borst). **1928** W. G. MacCallum *Text-Bk. Path.* (ed. 4) liv. 966 Such neurinomata occur only in peripheral nerves in which there are the sheath cells of Schwann. Of this character is the so-called acusticus tumor which is a neurinoma (formerly called neurofibroma). **1932** W. Penfield *Cytol. & Cellular Path. Nervous Syst.* III. xix. 968 These tumors contain long slender fibers of even caliber which tend to be arranged in parallel sheaves. Verocay (1910) considered these fibers to be nerve fibers and the type cell to be the sheath of Schwann cell... He therefore suggested the term neurinoma, meaning nerve fiber tumor. **1935** *Amer. Jrnl. Cancer* XXIV. 751 It is proposed..to limit the discussion to that specific encapsulated tumor composed of highly differentiated tissues which is so characteristic of the nerve sheath within which it develops...Neurinoma is the term most widely used,..but there is the insuperable objection to it that it means 'nerve fiber tumor'. Whatever else..all authorities are convinced that this is not a nerve fiber tumor but a nerve sheath tumor. **1948** A. Brodal *Neurological Anat.* vii. 240 Likewise it may happen that patients affected with a neurinoma of the VIIIth nerve tell a story of their sufferings starting with a pain situated deep in the ear. **1961** R. D. Baker *Essent. Path.* xx. 554 Neurinomas (neurilemmomas) of the skin are rare.

neuristor (niuri·stəɹ). *Electronics.* [f. Neur- + *-istor*, as in *resistor*, *transistor*.] Any device that is effectively a transmission line along which a signal will travel without attenuation (generally with a supply of energy along its length).

1960 H. C. Crane in *IRE Trans. Electronic Computers* IX. 370/1 A novel device and its properties have been hypothesized and possible digital systems employing it are briefly outlined in this note. The device, termed Neuristor, may be used to synthesize all digital logic functions... A neuristor may be visualized as a one-dimensional channel along which signals may flow. A signal propagates along the channel in the form of a discharge... A discharge signal has the following properties: 1) attenuationless propagation, 2) uniform velocity of propagation, 3) a refractory period. These characteristics are somewhat similar to the gross properties of transmission of discharge pulses by neurons in the nervous system—hence the name neuristor. **1963** *New Scientist* 7 Feb. 285/1 The concept of devices called 'neuristors', which behave like the axon or connection process of the neuron in the brain, is exciting to medical workers. **1965** *Electronics Lett.* I. 134/1 In most cases, neuristors which can propagate pulses have been considered, but one which can only propagate a discharge has also been studied. **1972** *Nuclear Instruments & Methods* CIV. 593 (*heading*) On the possibility of using superconducting neuristor line as particle detector.

neurite (niū̆ə·rəit). *Anat.* [f. Neur- + *-ite*, after *dendrite 3.] An axon or dendrite (formerly, an axon only).

1894 P. A. Fish in *Jrnl. Compar. Neurol.* IV. 173 For the axis-cylinder process the term neurite is proposed and for the other processes of the cells, retain the word dendrites. **1898** *Jrnl. R. Microsc. Soc.* 64 Nerve-cells with a long neurite (Purkinje's cells). **1924** R. M. Ogden tr. *Koffka's Growth of Mind* ii. 53 At its end this neurite divides into a fine net-work. **1956** *Nature* 28 Jan. 185/2 Individual nerve cell bodies and a part of their neurites 100–200µ long were dissected out by micromanipulation. **1973** *Gray's Anat.* (ed. 35) vii. 750/1 The neurites..may be divided into those which conduct broadly towards the cell body..and another one which may branch more or less profusely, and conducts away from the cell body.

neuro-. Add: **neuroana·tomy,** the anatomy of the nervous system; hence **neu:roanatomi·cal** *a.*, **neuroana·tomist**; **neurobio·logy,** the branch of biology which deals with nervous tissue; hence **neu:robiolo·gical** *a.*, **neurobio·logist**; **neu:robiota·xis** *Biol.* [Bio- + Taxis], a tendency of nerve cells, in the course both of development and of evolution, to remain close to their source of stimulation by migrating; so **neu:robiota·ctic** *a.*; **neuroblasto·ma** (pl. **-omas, -omata**) *Med.* [*-oma], a tumour composed of neuroblasts; *spec.* a malignant tumour composed of small cells with darkly staining nuclei and little cytoplasm, which is common in infants and usually appears in the adrenal gland; **neuroche·mistry,** (the study of) the chemical composition and processes of nervous tissue; hence **neuroche·mical** *a.*, **-che·mist**; **neu:rocircula·tory** *a. Med.*, of or pertaining to the nervous system and the circulatory system jointly; chiefly in *neurocirculatory asthenia* = *irritable heart* s.v. *Irritable *a.* 2 b; **neurocra·nium** *Anat.* = Cranium 1 a; **neu·rocrine** (-krəin) *a. Physiol.* [a. F. *neurocrine* (Masson & Berger 1923, in *Compt. Rend.* CLXXVI. 1750), after *olocrine* holocrine, *endocrine* endocrine, etc.], secreting or secreted directly into nervous tissue; **neu·rocyte** [-cyte], a neurone, a nerve cell; hence **neurocyto·ma** (pl. **-omas, -omata**) *Med.* [*-oma], a tumour composed of nerve cells; **neu:rodermati·tis** (pl. **-dermatitides**) *Path.* [ad. F. *névrodermite* (Brocq & Jacquet 1891, in *Ann. de Dermatol. et de Syphiligraphie* II. 98)], lichen simplex chronicus or atopic dermatitis, esp. when aggravated by a nervous disorder; also more widely, = next; **neu:rodermato·sis** (pl. **-oses**) *Path.*, any skin disorder that is nervous or psychosomatic in nature; **neuro-effe·ctor** *a. Physiol.*, pertaining to or composed of both a nerve and an effector; also as *sb.*, a neuro-effector system; **neuroele·ctric** *a.*, of or pertaining to the electrical phenomena and properties of the nervous system; cf. *nervo-electric* adj. s.v. *Nervo-; also **neuroele·ctrical**

a. (*rare*), **-electri·city; neuroembryo·logy,** the science which deals with the development of the nervous system in embryos; hence **neu:roembryolo·gic** (chiefly *U.S.*), **-lo·gical** *adjs.*; **neuroembryo·logist,** a specialist or expert in neuroembryology; **neuroe·ndocrine** *a.*, involving both nervous and endocrine participation; **neu:roendocrino·logy,** the study of the interactions between the nervous system and the endocrine system; hence **neu:roendocrinolo·gical** *a.*, **-endocrino·logist**; **neurofi·bril, -fi·brilla,** any of the fibrils visible within the body of a nerve cell using light microscopy; hence **neurofi·brillar, -fi·brillary** *adjs.*; **neurofi·lament,** a long filamentary structure, typically about 100 ångströms in diameter, visible in the cytoplasm of neurones under electron microscopy; **neurogla·ndular** *a.*, involving or possessing both glandular and nervous tissue or functions; **neurohæ·mal** (*U.S.* **-hemal**) *a.*, designating any of the organs, esp. among insects, which are composed of a group of nerve endings closely associated with the vascular system and are believed to have a neurosecretory function; **neurohisto·logy,** the histology of the nervous system or of nervous tissue; so **neu:rohistolo·gic** (chiefly *U.S.*), **-lo·gical** *adjs.*; **neurohisto·logist,** an expert or specialist in neurohistology; **neu:ro-interme·diate** *a. Anat.*, used to designate the posterior lobe (pars nervosa) of the neurohypophysis together with the adjacent pars intermedia (which is usu. regarded as part of the adenohypophysis); **neuroki·nin** *Med.*, a kinin reported to have been obtained from subcutaneous fluid in the scalp during attacks of migraine; **neu·rokyme** *Psychol.* [ad. G. *neurokym* (O. Vogt 1895, in *Zeitschr. f. Hypnotismus* III. 300), f. Gr. κῦμ-α wave] (see quot. 1908); **neurolingui·stic** *a.*, of or connected with the application of neurology to linguistic research; so **neuroli·nguist,** an expert or specialist in neurolinguistics; **neurolingui·stics** *sb. pl.* (const. as *sing.*), neurological linguistics; **neuromuscular** *a.*, add to def.: having characteristics of both nervous and muscular tissue; being or pertaining to a junction between a nerve fibre and a muscle fibre (earlier and later examples); **neuromy·al** *a.* [Myo-] = prec.; **neu:ropharmaco·logy,** the study of the action of drugs on the nervous system; hence **neu:ropharmacolo·gic** (chiefly *U.S.*), **-lo·gical** *adjs.*, **neu:ropharmacolo·gically** *adv.*, **neu:ropharmaco·logist**; **neurophysin** (-fəi·sin) *Biochem.* [ad F. *neurophysine* (J. Chauvet et al. 1960, in *Biochim. & Biophys. Acta* XXXVIII. 266), f. *neuro*(*hypo*)*physe* *Neurohypophysis], any of a group of proteins which are found in the neurohypophysis in complexes with oxytocin and vasopressin and are believed to act as carriers for these hormones during their passage from the hypothalamus; **neurophysiolo·gic** *a.* (chiefly *U.S.*), neurophysiological; **neurophysiolo·gically** *adv.*, with respect to neurophysiology; **neurophysio·logist,** an expert or specialist in neurophysiology; **neu·roplasm,** the cytoplasm of a neurone; occas. applied to that of the cell body, as distinguished from *axoplasm; hence **neuropla·smic** *a.*; **neuropsycho·sis** *Psychol.*, a neurotic condition in which certain features characteristic of a psychosis can be recognized; **neuroradio·logy** *Med.*, radiology of the central nervous system; hence **neu:roradiolo·gic** (chiefly *U.S.*), **-lo·gical** *adjs.*; **neuroradio·logist,** a specialist or expert in neuroradiology; **neu·roscience,** any of the sciences (as neurochemistry or psychology) which deal with the nervous system or mental phenomena; such sciences collectively; hence **neurosci·entist,** a specialist in a neuroscience; **neurosecre·tion,** (*a*) the process of secretion by a (specially adapted) nerve cell; (*b*) the substance secreted in this process; so **neurosecre·tory** *a.*, performing, produced by, or pertaining to neurosecretion; **neurosema·ntic** *a.*, of or connected with a neurological approach to semantics; **neurose·nsory,** pertaining to or involving both nervous and sensory properties; *spec.* applied to a sensory nerve cell, esp. one in which the cell boby (usu. situated in epithelium) is the receptor and has a single process by which impulses are transmitted to a ganglionic neurone or an effector organ; **neu·rosurgeon,** one who practises neurosurgery; **neurosu·rgery,** surgery of the nervous system; hence **neurosu·rgical** *a.*; **neurosy·philis** *Path.*, syphilis involving the central nervous system; hence **neu:rosyphili·tic** *a.* (also as *sb.*, a neurosyphilitic person); **neurote·ndinous** *a.*, of or pertaining to a nerve fibre and a tendon, esp. the termination of a nerve in a tendon; **neurotoxi·city,** (*a*) toxicity towards the nerves; (*b*) poisoning by a neurotoxin; **neurotu·bule,** a microtubule of a neurone; **neurovi·rulence,** virulence towards the nervous system.

1904 *Biol. Bull.* VI. 90 These facts point toward either a specialization or a modification in function which is of interest because of its bearing upon certain neuro-anatomical facts. **1971** H. A. Whitaker in W. O. Dingwall *Survey of Linguistic Science* 148 It is plausible..that there is a universal neuroanatomical substrate for the language system. **1931** H. L. Hollingworth *Abnormal Psychol.* iv. 68 This condition would be difficult to understand if it were not that the neuro-anatomist had mapped out the course of the various sensory pathways in the spinal cord. **1963** *Zeitschr. für Zellforschung* LX. 815 The basis of the neuron doctrine was established by the work of a few eminent neuroanatomists. **1974** *Nature* 6 Sept. 83/1 For the neuroanatomists of the nineteenth century, the cerebellum was a source of fascination. **1900** Dorland *Med. Dict.* 441/1 Neuro-anatomy. **1913** *Official Publ. Cornell Univ.* IV. xvi, Strauss, Israel. A.B., M.D., Instructor in Neuro-Anatomy, 1911. **1931** H. L. Hollingworth *Abnormal Psychol.* iv. 77 We may illustrate the 'fictional' or hypothetical notion of explanation also in the field of neuro-anatomy. **1971** *Sci. Amer.* May 89/3 In view of the basic neuroanatomy of the visual system, this means that the visual effects must have been the result of neuronal activity in the brain rather than in the eye itself. **1959** *Internat. Rev. Neurobiol.* I. p. vii, Progress in neurobiological research must maintain a delicate balance between the fascination of basic explanation of clinical and physiological phenomena by means of chemical and physical concepts on the one hand and the pressing needs for the development of new and effective treatments of disease on the other. **1971** *Nature* 5 Mar. 25/1 The neurobiological approach to mental health. **1957** H. Read *Tenth Muse* xiii. 110 If one reads a neurobiologist such as J. R. Smithies one has the feeling, perhaps deceptive, that the problems discussed by Professor Ayer are being discussed more realistically. **1971** J. Z. Young *Introd. Study Man* p. vii, What the neurobiologist finds out about the brain must surely be relevant to fundamental views of the nature of all this knowledge. **1906** *Index-Catal. Library Surg.-General's Office, U.S. Army* 2nd Ser. XI. 622/1 (*heading*) Neurobiology. **1960** *Times* 22 Sept. 4/2 Researches cover neurobiology—such as 'eye movements and optogyral illusion'. **1972** *Sci. Amer.* Sept. 51/1 Thus neurobiology has now shown why it is human—and all too human—to hold Euclidean geometry and its non-intersecting co-planar parallel lines to be a self-evident truth. Non-Euclidean geometries of convex or concave surfaces, although our brain is evidently capable of conceiving them, are more alien to our built-in spatial-perception processes. **1908** C. U. A. Kappers in *Jrnl. Physiol.* XXXVII. 143 Another feature of the autonomic system might find its explanation in the neurobiotactic influence of the axon stimulation. **1948** A. Brodal *Neurological Anat.* vii. 192 The neurobiotactic influences (Kappers) of which the above-mentioned fact is an example, are also revealed in other morphological features. **1908** C. U. A. Kappers in *Jrnl. Physiol.* XXXVII. 140 Referring for further details concerning these phenomena of neurobiotaxis to my former papers on this subject, I only wish to state here that..it is obvious that the motor cells migrate in the direction whence they get the greatest quantity of stimuli. **1961** P. Glees *Exper. Neurol.* vi. 168 In accordance with his hypothesis of neurobiotaxis..Kappers (1920) believes that the dendrites grow from the cell towards the source of their afferent stimuli. **1910** J. H. Wright in *Jrnl. Exper. Med.* XII. 556 The essential cells of the tumor are considered to be more or less undifferentiated nerve cells or neurocytes or neuroblasts, and hence the names neurocytoma and neuroblastoma. **1925** *Arch. Neurol. & Psychiatry* (Chicago) XIV. 193 In our Brigham Hospital records..many of them have in the past been designated neurocytomas or neuroblastomas. **1948** R. Greene *Pract. Endocrinol.* iv. 128 The neuroblastomata are highly malignant tumours. **1974** *Nature* 17 May 224/1 The C1300 mouse neuroblastoma has been useful for studies of the differentiation and trophic interactions of nerve cells. **1949** Koestler *Insight & Outlook* v. 68 Their neurochemical substratum [*sc.* that of our emotional impulses] cannot be worked off in overt activities. **1963** *Lancet* 12 Jan. 79/1 If their effect on the brain could be evaluated, understanding of the neurochemical changes in depression might follow. **1972** *Sci. Amer.* Nov. 28/3 The hypothesis proposed that hypothalamic control of the secretory activity of the anterior pituitary could be neurochemical. **1958** *Neurology* VIII. (Suppl. 1) 27/2 Many investigators using such methods would not consider themselves neurochemists, but they certainly contribute to the growing body of neurochemical data. **1971** *Nature* 22 Mar. 130/3 Undoubtedly, it [*sc.* myelin] now has the added attraction to neurochemists that it can be obtained in bulk from the mammalian nervous system. **1955** *Neuropharmacology* I. 11 There are enough problems and enough possibilities for an entire science in the field of 'neuro-chemistry' or/and 'neuro-pharmacology'. **1969** *Nature* 11 Oct. 118/2 A significant development in neurochemistry has been the finding that thiol groups and disulphides are involved in the functioning and activity of neurones. **1918** 'T. Lewis' [*i.e.* B. S. Oppenheimer et al.: see *Jrnl. Amer. Med. Assoc.* (1918) 21 Sept. 994] in *Military Surgeon* XLII. 410 An appreciable number of soldiers present a well-defined symptom-complex, in which certain nervous and circulatory symptoms are associated with an increased susceptibility to fatigue. The descriptive name of Neuro-Circulatory Asthenia ('N.C.A.') is suggested for this syndrome in preference to Disordered Action of the Heart ('D.A.H.'). **1953** R. A. McFarland *Human Factors in Air Transportation* vi. 305 The effect of smoking on neurocirculatory efficiency..may be of particular importance to the airman since this test is often used to appraise fitness. **1971** Neurocirculatory [see *irritable heart* (*Irritable *a.* 2 b)]. **1907** W. N. Parker tr. *Wiedersheim's Compar. Anat. Vertebrates* (ed. 3) 75 The portion of the skull which is situated along the main axis in continuation of the vertebral column and which encloses the brain, is known as the brain-case or cranium (neurocranium). **1942** Grove & Newell *Animal Biol.* xv. 257 Other cartilage is laid down at the sides and above the brain until it becomes enclosed in a cartilaginous box, the brain box or neurocranium. **1972** *Nature* 24 Mar. 143/2 A sulcus on the outer face of the neurocranium in front of the supposed foramen for the vagus nerve is interpreted as the lateral occipital fissure. **1925** *Physiol. Abstr.* IX. 544 The secretion of these 'neurocrine' glands acts directly on the nerves. **1947** *Phil. Trans. R. Soc.* B. CCXXXII. 394 The relationship of the pars distalis of the pituitary and the possible neurocrine secretion by the hypothalamic nuclei to the water balance of the body are still debatable points. **1962** *Science Survey* III. 264 There exist equally interesting relationships between odours and animal behaviour on a different and perhaps more profound level..which are very likely mediated by the vegetative nervous system and the neurocrine and other endocrine glands. **1890** Billings *Med. Dict.* II. 209/1 *Neurocyte*, nerve-cell. **1894** P. A. Fish in *Jrnl. Compar. Neurol.* IV. 174 For the equivalent of a nerve unit including the nerve cell with all its processes to the uttermost filament, the term neurocyte is suggested. It has not been possible to trace the word to its originator. It is in use in the French language and is included in the dictionary of the New Sydenham Society and Gould's New Medical Dictionary with the simple definition: a nerve cell. **1932** W. Penfield *Cytol. & Cellular Path. Nervous Syst.* III. xviii. 941 Between the neurocytes are numerous smaller cells. **1910, 1925** Neurocytoma [see *neuroblastoma* above]. **1948** R. Greene *Pract. Endocrinol.* iv. 127 The most malignant tumours are the neuroblastomata or neurocytomata. **1966** Wright & Symmers *Systemic Path.* II. xxxiv. 1246/1 A ganglioneuroma (neurocytoma) of the brain is a slowly growing tumour composed of neurons. **1896** *Amer. Year-bk. Med. & Surg.* 715 (*heading*) Vitiligo, lichen ruber planus, and chronic circumscribed neurodermatitis. **1935** Neurodermatitis [see *neurodermatosis* below]. **1947** *N.Y. State Jrnl. Med.* XLVII. 1889/2 MacKenna and others have suggested that the neurodermatitides may have a corresponding symbolic meaning. **1954** *Bull. Muscogee County Med. Soc.* Aug. 9 A large list of diseases has been included under the term 'psychosomatic',..it includes..certain skin diseases—notably the so-called neurodermatitides. **1974** Passmore & Robson *Compan. Med. Stud.* III. xxxi. 13/1 The condition [*sc.* lichen simplex chronicus] is sometimes referred to as neurodermatitis because it is most frequently encountered in obsessional and anxious individuals. **1909** *Cent. Dict.* Suppl., Neurodermatosis. **1911** M. Morris *Dis. Skin* (ed. 5) iv. 54 (*heading*) Neuroses of the skin. Classification of neuro-dermatoses. **1935** *Jrnl. Amer. Med. Assoc.* 5 Oct. 1099/1 A group of eighty patients presenting typical clinical examples of the neurodermatoses..was selected.., with diagnoses as follows: neurodermatitis (dry type), nineteen; pruritus ani or vulvae, five; neurodermatitis (exudative type), twenty-eight; dyshidrosis, four; [etc.]. **1941** S. H. Kraines *Therapy of Neuroses & Psychoses* xiv. 313 Clearing up of these neurodermatoses is more difficult than removal of many other emotionally conditioned physical symptoms. **1961** *Lancet* 12 Aug. 369/1 MacKenna very artistically described neurodermatosis when he stated that 'in some cases the skin is an organ of stress which bears the brunt of nervous agitation, acting as the canvas on which the perturbation of the mind is painted.' **1935** *Q. Rev. Biol.* X. 335/2 According to the data available, the following is the sequence of functionally related events which probably occur in the electrical excitation of a neuro-effector system. **1937** Cannon & Rosenblueth *Autonomic Neuro-Effector Systems* p. viii, Previous researches on autonomic neuro-effectors and the occurrences at their synapses. **1973** *European Jrnl. Clin. Pharmacol.* VI. 92/1 Further investigations are possible..of the possible role of these [neuronal] pools in the control of NA exchanges at the level of the sympathetic neuro-effector junction. **1973** *Science* 16 Feb. 693/2 (*heading*) Neuroeffectors controlling mucus release by the leech. **1849** G. Bird *Lect. Electr. & Galvanism* i. 24 The neuro-electric theory of Galvani. *Ibid.* 25 Valli..believed the neuro-electric fluid to be secreted by the capillary arteries supplying the nerves. **1956** L. S. Frishkopf in *Technical Rep. Res. Lab. Electronics Mass. Inst. Technol.* No. 307 (*title*) A probability approach to certain neuroelectric phenomena. **1965** *Math. in Biol. & Med.* (Med. Res. Council) iv. 131 Modern electrophysiological techniques permit the recording of several types of neuroelectric potentials, and the patterns of voltage-versus-time traces provide the electrophysiologist's basic data. **1974** *Nature* 15 Feb. 481/1 An exactly similar method is widely used in studying EEG visual evoked responses and other neuroelectric phenomena. **1914** *Practitioner* June 838 The immediate cause of an epileptic attack is a neuro-electrical brain storm. *Ibid.* 831 The chemical generation of nerve force (neuro-electricity) in the human body. *Ibid.* 832 The grey matter of the brain must be the site of generation of this neuro-electricity. **1933** *Science* 18 Aug. 132/1 A notable beginning was made in neuro-embryologic study of behavior. **1950** Hamburger & Levi-Montalcini in P. Weiss *Genetic Neurol.* 129 The *in vitro* culture of nerve cells made important contributions to the solution of neuro-embryological problems. **1970** D. Bodian in F. O. Schmitt *Neurosciences: 2nd Study Program* xiii. 139/2 The considerable insight gained in recent times through analysis of neuroembryological processes. **1950** Hamburger & Levi-Montalcini in P. Weiss *Genetic Neurol.* 141 The neuroembryologist is largely concerned with the further elaboration of these elementary patterns of the early neural tube. **1970** M. V. Edds in F. O. Schmitt *Neurosciences: 2nd Study Program* v. 51/2 The roster of experimental neuroembryologists active since Harrison opened the field early in the century includes some of the most able developmental biologists. **1933** *Science* 18 Aug. 137/1 The whole subject of neuro-embryology of higher vertebrates should be reexamined. **1950** Hamburger & Levi-Montalcini in P. Weiss *Genetic Neurol.* 131 The material used in experimental neuroembryology has been confined largely to teleosts, amphibians, and the chick embryo. **1974** *Nature* 22 Mar. p. xi/1 (Advt.), By combining the facts of neurophysiology, neuroembryology, and behaviour, a new theory is built up. **1922** P. Fridenberg in L. F. Barker et al. *Endocrinol. & Metabolism* II. 769 (*heading*) Neuro-endocrin control of intra-ocular tension. **1944** Neuro-endocrine [see *hypothalamico-hypophysial *a.*]. **1959** T. Lidz in S. Arieti *Amer. Handbk. Psychiatry* I. xxxii. 650/1 The hypothalamus..is now understood to form a critical juncture in a circular feedback system that mediates and regulates neural impulses concerned with emotions and neuro-endocrine activity. **1973** *Folia Biol.* (Cracow) XXI. 329 It has been suggested that alterations in the..secretory pattern of the neuroendocrine components are due to the action of stress. **1963** *Annales d'Endocrinol.* XXIV. 198 (*heading*) Introduction to the neuro-endocrinological study of the pineal gland. **1974** *Nature* 17 May 213/1 Recent neuroendocrinological findings of most of the leading laboratories in Europe were discussed informally. **1969** *Britannica Yearbk. of Sci. & Future 1968* 389 These organic compounds, which were thought to act as chemical mediators (neurohumors) at synaptic junctions..were of interest to neuroendocrinologists. **1922** P. Fridenberg in L. F. Barker et al. *Endocrinol. & Metabolism* II. 757 (*heading*) Ophthalmic neuro-endocrinology. **1961** *Lancet* 19 Aug. 442/2 In 1952 he was appointed senior lecturer in experimental neuroendocrinology at the Institute of Psychiatry. **1967** E. Bajusz (*title*) An introduction to clinical neuroendocrinology. **1898** *Jrnl. R. Microsc. Soc.* 64 The general conception [of Prof. Apáthy] may be briefly stated. The nerve-cell is analogous to the muscle-cell, producing conducting substance (primitive fibrils, neurofibrils), as the muscle-cell produces contractile substance (myofibrils). **1970** A. Peters et al. *Fine Struct. Nervous Syst.* iv. 62/2 The precise correlation between the classical neurofibrils of silver preparations and the structures seen in electron micrographs remains uncertain. **1973** Neurofibril [see *neurofibrillary* adj. below]. **1902** *Jrnl. Nerv. & Mental Dis.* XXIX. 435 (*heading*) The neuro-fibrillae in nerve cells and nerve fibers of the retina. **1963** R. P. Dales *Annelids* vii. 138 Hess suggested that the irregularly shaped refractory body directed the light on to a dense reticulum of neurofibrillae which joined to form the basal nerve fibre. **1902** *Jrnl. R. Microsc. Soc.* 542 (*heading*) Neurofibrillar theory. **1949** Neurofibrillar [see *neurofibrillary* adj. below]. **1971** *Jrnl. Compar. Neurol.* CXLIII. 395/2 Neurofibrillar boutons appear following section of the axons. **1902** *Jrnl. Nerv. & Mental Dis.* XXIX. 435 The author concludes..that a neuro-fibrillary structure of nerve cells and their processes..is abundantly proven. **1949** B. W. Lichtenstein *Textbk. Neuropath.* iv. 46 Neurofibrillary abnormalities occur in a variety of disease states. The normal configuration of the neurofibrillar apparatus is well seen in preparations impregnated with silver according to Bielschowsky's method. **1973** H. M. Rális et al. *Techniques Neurohistol.* iv. 89 Methods for neurofibrils may also be used to demonstrate..the neurofibrillary tangles which are found in certain pathological conditions. **1955** Palay & Palade in *Jrnl. Biophysical & Biochem. Cytol.* I. 78 Neurofilaments.—Fine, long threads, 60 to 100A in diameter and of indefinite length, traverse the cytoplasmic matrix [of the neuron] between masses of Nissl substance and other organelles. **1965** *Progress Brain Res.* XIV. 57 Electron microscopical studies have shown that the neurofibrils of light microscopists are made up of fine, long, apparently non-branching structures approximately 100 Å in diameter. These are the neurofilaments. **1968** G. A. Horridge *Interneurons* i. 11 Throughout the animal kingdom..many axons and dendrites of nerve cells have tubules in the axoplasm; others, such as the squid giant axon, have neurofilaments that are thinner and less obviously tubular. **1969, 1970** Neurofilament [see *neurotubule* below]. **1909** *Cent. Dict.* Suppl., *Neuroglandular*, having the characteristics of sensory and glandular organs: as, the neuroglandular pit of some *Nemertini*. **1941** *Jrnl. Compar. Neurol.* LXXIV. 106 Neuroglandular cells are described in three species of cockroaches.... The main center of neurosecretory activity is found to be the suboesophageal ganglion. **1943** H. Read *Education through Art* 26 Temperament..is closely connected with the neuro-glandular system and the relations of the cortex to the sub-cortex. **1964** J. Z. Young *Model of Brain* xviii. 296 There must be some common principle involved to produce these similar neuroglandular arrangements in completely independent phyla. **1953** Carlisle & Knowles in *Nature* 29 Aug. 405/1 It seems preferable to call these organs by some purely topographical name which does not denote any function, actual or supposed. The adjective 'neurohæmal' seems to us to be the most appropriate topographical name denoting the common feature of these organs. The organs may thus be referred to collectively as neurohæmal organs, while the adjective may be combined with any of the pre-existing names for these various organs, as, for example, 'the post-commissural neurohæmal organ' and 'dorsal neurohæmal lamella'. **1967** C. A. G. Wiersma *Invertebr. Nervous Systems* x. 125 (*heading*) Correlation of propagated action potentials and release of neurosecretory material in a neurohemal organ. **1973** *Nature* 12 Oct. 288/2 Thus another useful criterion for the definition of a neurosecretory neurone—that it ends in a neurohaemal organ—loses its generality. **1957** *A.M.A. Arch. Path.* LXIII. 3/2 Histologic analysis of these alterations does not require special neurohistologic methods. **1940** *Jrnl. Anat.* LXXIV. 413 (*heading*) Observations on the neurohistological basis of cutaneous pain. **1973** H. M. Rális et al. *Techniques Neurohistol.* iv. 82 (*heading*) Neurohistological staining methods. **1901** *Buck's Handbk. Med. Sci.* (ed. 2) II. 336/1 Until further knowledge has been gained it is safer for the neurohistologist to work with the various methods [etc.]. **1968** Clarke & O'Malley *Human Brain & Spinal Cord* ii. 87 Two opposing groups of neurohistologists arose. On the one hand were those who believed that the nerve cells and their processes..constituted independent units in contiguity with other units but not in continuity... Their opponents..considered the cells and fibres to be in direct continuity with one another by way of a network to which the fibres contributed. **1897** *N.Y. Med. Jrnl.* 15 May 652/2 The most important contributions of Golgi in the domain of neuro-histology consisted in (1) the invention of the silver method of staining; (2) the recognition within the central regions of cells of different types..; and (3) the discovery of lateral branches from the axis-cylinder process. **1940** *Jrnl. Anat.* LXXIV. 426 The neuro-histology of this area establishes that in the human skin pain is subserved by fine nerve fibres bearing free nerve endings. **1973** H. M. Rális et al. *Techniques Neurohistol.* v. 146 This final chapter..mentions histochemical methods used in neurohistology as well as some applications of autoradiography in the study of nerve tissue. **1926** G. R. de Beer *Compar. Anat. Pituitary Body* ii. 28 The anterior lobe consists only of the pars anterior, but the posterior lobe, which always contains the partes intermedia and nervosa, may or may not also be associated with the pars tuberalis, since many authors fail to distinguish between the latter and the pars intermedia. In order to avoid ambiguity the term *neuro-intermediate lobe* may be used to include the pars nervosa and the pars intermedia, since they are always in the closest morphological association. **1965** Lee & Knowles *Animal Hormones* ii. 28 In fishes the pituitary gland is conveniently divided into three portions... Closely associated with the posterior portion of the adenohypophysis (pars intermedia), and extending into it, is the pars nervosa; the term neuro-intermediate lobe is often applied to this region. **1973** *Nature* 28 Sept. 207/2 Pituitary control of sebaceous gland activity has generally been assumed to be a function of the anterior lobe. The possibility that the neurointermediate (NI) lobe is involved was first suggested when we found that its removal led to a decrease in sebum secretion. **1960** L. F. Chapman et al. in *Trans. Assoc. Amer. Physicians* LXXIII. 263 Specimens collected from the head during the headache attacks contained a substance that could be distinguished from serotonin,..acetylcholine and histamine... The heat stabilized substance had many of the properties of bradykinin, Kallidin, or 'plasma kinin'... This polypeptide has been labeled 'neurokinin' and has been found..to be released during neuronal excitation. **1969** J. Pearce *Migraine* vii. 39 More recently the polypeptides and kinins have been examined more critically, because of the claims of isolation of neurokinin from the scalp tissue fluid in migraine attacks. **1908** W. McDougall in *Brain* XXXI. 247 This distinction between chemically stored or potential nervous energy and the liberated active nervous energy is, I feel sure, one of the first importance for neurological speculation... Oscar Vogt..has proposed to mark it by calling the freed nervous energy 'neurokyme'... I adopt Vogt's term. **1926** —— *Outl. Abnormal Psychol.* v. 104 All mental activity involves the discharge of neurokyme from the sensory to the motor side of the brain. **1944** W. Brown *Psychol. & Psychotherapy* (ed. 5) v. 57 McDougall regards the passage of nervous energy (neurokyme) across the synapses of the cerebral cortex as the physical correlate of the psychical process. **1961** *Studies in Linguistics* XV. 70 Ideally, the neurolinguist would have thorough training in scientific linguistics and in neuroanatomy and neurophysiology. **1936** Neuro-linguistic [see *neuro-semantic* adj. below]. **1961** *Studies in Linguistics* XV. 70 Neurolinguistic work has certainly been carried out under other names, by people who work with aphasia, by neurosurgeons and neurologists, [etc.]. **1970** J. Laver in J. Lyons *New Horizons in Linguistics* iii. 61 The healthy adult brain is not itself accessible to neurolinguistic experiment. There is thus no possibility of directly observing the neural mechanisms involved in constructing a neurolinguistic program. **1961** E. C. Trager in *Studies in Linguistics* XV. 70 Neurolinguistics is the term proposed here for a field of interdisciplinary scientific study which does not as yet have a formal existence. Its subject matter is the relationship between the human nervous system and language. **1970** J. Laver in J. Lyons *New Horizons in Linguistics* iii. 61 In neurolinguistics the subdisciplinary boundary between phonetics and linguistics, which has always been of doubtful validity, is largely disappearing. **1973** *Tuscaloosa* (Ala.) *News* 10 Apr. 5/4 She is interested in neurolinguistics and studies of aphasia. **1975** *Canad. Jrnl. Linguistics* XX. 94 *A Study in Neurolinguistics* is perhaps the first publication in the form of a monograph in the field of 'language and the brain' with the term 'neurolinguistics' in its title. **1864** *Jrnl. Mental Sci.* X. 37 There appears to be a neuro-muscular, as well as a purely mental retentiveness. **1892** J. A. Thomson *Outl. Zool.* iii. 36 In some Cœlenterates it is possible that some of the external cells combine contractile, nervous, and even other functions. Under this impression many call them 'neuro-muscular'. **1904** *Jrnl. Physiol.* XXX. 494 If our conception of this neuro-muscular junctional tissue is correct the name nerve-ending is obviously a misnomer. **1937** *Physiol. Rev.* XVII. 538 It may be concluded that synapses and neuro-muscular junctions are essentially similar, there being close contact but not protoplasmic continuity. **1948** *Federation Proc.* VII. 452/1 The fundamental change which accounts for the neuromuscular block produced by curare itself is a decrease in the end-plate potential. **1950** J. H. Burn *Lect. Notes Pharmacol.* (ed. 2) 14 Neostigmine restores neuromuscular transmission. **1963** R. P. Dales *Annelids* vi. 119 The failure to respond is not due to failure of the giant fibre itself, but to the relay or to the neuromuscular junction. **1926** *Physiol. Rev.* VI. 564 Comparable data on muscle and the neuromyal junction. **1965** *Jrnl. Pharmacol. & Exper. Therap.* CXLVII. 350 (*heading*) Pharmacological actions of oxamides and hydroxyanalinium compounds at frog neuromyal junction. **1913** *Ibid.* V. 107 We now approach the subject of the susceptibility of the central heat-regulating mechanism to more specific neuro-pharmacologic influences. **1973** *Folia Biol.* (Cracow) XXI. 331 The CAH-positive cells of the brain in the cockroach undergo most of the alterations..after the administration of various types of neuropharmacologic agents. **1959** *Jrnl. Pharmacol. & Exper. Therap.* CXXVI. 312/1 Thioridazine hydrochloride..and chlorpromazine hydrochloride..have been examined for neuropharmacological properties in mice and rats. **1971** *New Scientist* 9 Dec. 119 These various poisons have proved novel tools in neuropharmacological research, especially in elucidating the mechanism of nervous conduction. **1971** *Nature* 24 Sept. 285/2 Other neuropharmacologically-active or serotonin-related drugs were also tested. **1966** *Sci. News Let.* 1 Jan. 6 Rats..given the compound..retained what they had learned longer, Dr. N. P. Plotnikoff, an Abbott neuropharmacologist, reported. **1973** *Nature* 14 Dec. p. xvi/1 (Advt.), We seek a Technician to join a team of neuropharmacologists. **1955** Neuro-pharmacology [see *neurochemistry* above]. **1973** *Nature* 28 Sept. p. xx/3 (Advt.), General experience of either neurochemistry, neurophysiology, or neuropharmacology is essential. **1960** *Biochim. & Biophys. Acta* XXXVIII. 266 From sheep posterior hypophysis a complex was obtained containing 90% of the oxytocic and vasopressic activity of the gland. The complex is an association of oxytocine and vasopressine with a protein, neurophysine [*sic: Eng. summary of article in Fr.*]. **1970** *Biochem. Jrnl.* CXVI. 908/2 The neurophysins of the pig form a group of proteins of different electrophoretic mobilities but all possessing the capacity to bind oxytocin and [8-lysine]-vasopressin. **1973** *Nature* 2 Mar. 63/1 This system is characterized immunochemically by the neurophysins, the specific carrier-proteins for vasopressin and oxytocin. **1937** *Surgery* I. 132 The vast literature on recent neurophysiologic research. **1972** *Science* 12 May 607/1 Any changes in the action potentials of trained motor units..must reflect neurophysiologic changes of the single neuron supplying the motor unit. **1962** C. L. Buxton *Study of Psychophysical Methods for Relief Childbirth Pain* vii. 60 Attempts were made also to explain neurophysiologically how it might be possible for fear and tension to increase the pain and length of labor. **1971** *Jrnl. Gen. Psychol.* LXXXIV. 141 In the early experimentation, there was no way to isolate a channel neurophysiologically. **1949** Koestler *Insight & Outlook* iv. 44 The demonstration or refutation by the neuro-physiologist of the existence of corresponding mechanisms in the central nervous system. **1966** I. Asimov *Fantastic Voyage* i. 17 The extension of the technique could be of great importance to the neurophysiologist. **1973** *Sci. Amer.* July 96/3 Between 1900 and 1920 Charles S. Sherrington, the foremost neurophysiologist of the time, applied the technique of electrical stimulation to study how the cerebrum controlled movement. [**1892** *Syd. Soc. Lex.*, *Neuroplasma*, Kupffer's term for a fluid which he supposed to lie between the fibrils of the cylinder-axis of a nerve.] **1894** Gould *Dict. Med.* 869/1 Neuroplasm. **1896** E. L. Billstein tr. *Stöhr's Text-bk. Histol.* II. i. 81 Each fibrilla [of the axon] represents a special conducting path and is cemented to neighboring fibrillae by a small amount of finely-granular interstitial substance—neuroplasm. **1960** L. Picken *Organization of Cells* vii. 291 In the light of electron microscope studies..it is likely that the axoplasm differs from the neuroplasm rather in the relative abundance and orientation of the various components..than in absolute composition. **1970** *Nature* 5 Sept. 1006/2 It is generally believed that neuroplasm is constantly synthesized in the cell body and moves as a gel down the axon (and probably also along the dendrites). **1909** *Cent. Dict.* Suppl., Neuroplasmic. **1965** *Acta Neuropath.* IV. 33 Neuroplasmic swellings were found within dendrites. **1970** P. A. Weiss in F. O. Schmitt *Neurosciences: 2nd Study Program* lxxiii. 840 (*heading*) Neuronal dynamics and neuroplasmic flow. **1900** Dorland *Med. Dict.* 442/2 *Neuropsychosis*, nervous disease complicated with mental disorder. **1918** A. A. Brill tr. *Freud's Totem & Taboo* iii. 158 The system formation is most ingenious in delusional states (paranoia) and dominates the clinical picture, but it also must not be overlooked in other forms of neuropsychoses. **1924** J. Riviere et al. tr. *Freud's Coll. Papers* I. 59 (*heading*) The defence of neuropsychoses. **1936** A. Myerson in *Amer. Jrnl. Psychiatry* XCIII. 281, I formally introduce the concept of the neuropsychosis. The neuropsychosis comes into being by an intensification of the symptomatology of the neuroses. **1964** Taveras & Wood *Diagnostic Neuroradiol.* 2 The use of gamma rays from radioactive isotopes for scanning, and the use of heat waves in thermography, ..are being incorporated in neuroradiologic clinical practice. **1962** *Brit. Jrnl. Radiol.* XXXV. 501/1 Problems of neuroradiological nomenclature and the radiographic projections were discussed. **1955** *Brit. Jrnl. Surg.* XLIII 8/1 A very high proportion of successful angiograms justifies a wider trial. Even if it may not be equally successful in other hands I believe that it will become an important part of the armamentarium of neuroradiologist and neurosurgeon. **1961** *Lancet* 30 Sept. 746/1 Dr. J. L. G. Thomson, the neuroradiologist in Bristol, and all of the neurosurgical team are now performing about 700 angiographies a year. **1938** Wakeley & Orley (*title*) A textbook of neuro-radiology. **1964** Taveras & Wood *Diagnostic Neuroradiol.* 1 During the last fifteen years, angiography has arrived at its appropriate place of importance in diagnostic neuroradiology. **1963** (*title of periodical*) Neurosciences research program bulletin. **1964** *New Scientist* 10 Sept. 643/1 Man's search for the physical basis of mental processes has evolved a number of disparate neurosciences. Some of these, such as neurology, neuroanatomy, neurophysiology, neurochemistry, neuropathology, and psychology, have advanced to the status of mature sciences. **1970** *Nature* 5 Sept. 1006/1 During the past few years neuroscience, comprising the sciences of brain and behaviour, has been differentiating, integrating, regrouping. **1974** *Times Lit. Suppl.* 18 Oct. 1151/3 It will transform the established neuro-sciences until they become increasingly able to comprehend the problems of behaviour, possibly even of mind. **1967** R. B. Livingston in G. C. Quarton et al. *Neurosciences: Study Program* 500/1 Neuroscientists are drawn to this field..by a desire to learn more about ourselves as human beings. **1974** *Nature* 1 Mar. p. v/1 (Advt.), This new book by the well-known neuroscientist Elliot S. Valenstein. **1941** *Jrnl. Compar. Neurol.* LXXIV. 93 Neurosecretion is present in both vertebrates and invertebrates. **1961** *Biol. Abstr.* XXXVI. 1201/1 (*heading*) Neurosecretions in the insect. **1963** R. P. Dales *Annelids* viii. 166 Neurosecretion may be of major importance in the co-ordination of the annelid

body. **1968** *New Scientist* 16 May 355/1 It is quite possible that neurosecretions are the 'oldest' hormones in the animal kingdom. **1973** *Proc. Indian Acad. Sci.* B. LXXVII. 148 Involvement of neurosecretion in some of the physiological activities of this scorpion will be reported here. **1940** *Nature* 17 Feb. 264/1 (*heading*) Neurosecretory cells in the ganglia of Lepidoptera. **1956** *Ibid.* 17 Mar. 532/1 In the cockroach, activity rhythms may have a neurosecretory basis. **1963** R. P. Dales *Annelids* viii. 166 In *Nephthys* the cells in the blood greatly increase and take up neurosecretory material from the back of the brain during posterior regeneration following amputation. **1968** H. O. Hofer in G. H. Bourne *Struct. & Function of Nervous Tissue* I. xi. 471 The neurosecretory substances act as hormones, long-range and long-acting; and they are not directly transmitted, but are released in a circulating body fluid as acting agents. **1936** A. Korzybski in *Amer. Jrnl. Psychiatry* XCIII. 29 By using the term *evaluation* as a fundamental term, we bridge methodologically and linguistically the exact sciences with other sciences, psychiatry included. We gain thereby powerful neuro-linguistic and neuro-semantic direct methods for education and psychotherapy. **1946** S. A. Hayakawa in W. S. Knickerbocker *Twentieth Century English* 47 In accounting for human behavior it postulates the 'neuro-semantic environment'—the environment, that is, of dogmas, beliefs, creeds, knowledge, and superstitions to which we react as the result of our training—as a fundamental and inescapable part of our total environment. **1929** C. U. A. Kappers *Evolution Nervous Syst. Invertebr., Vertebr. & Man* 3 The different forms of nervous conductors are three: the neuro-sensory cell, which generally retains its place in an epithelial layer, the primitive or asynaptic ganglion cell and the polarized or synaptic neurone, both of which are nearly always located under the epithelium. **1940** O. Lowenstein *Parker & Haswell's Text-bk. Zool.* (ed. 6) I. i. 36 The photosensitive cells in the retina of the vertebrate eye and the olfactory receptor cells situated in the epithelium coating of the vertebrate nose have the structure of neuro-sensory cells. **1946** L. A. White in W. S. Knickerbocker *Twentieth Century Eng.* 93 The animal hearing them understands them..by virtue of his own inborn neuro-sensory equipment. **1962** D. Nichols *Echinoderms* iii. 43 Besides the general scattering of neurosensory cells over the asteroid body, there are five light-sensitive optic cushions, one at the base of each terminal tentacle. **1974** *Sci. Amer.* Nov. 14/3 Reader in neurosensory physiology. **1925** *Arch. Neurol. & Psychiatry* (Chicago) XIV. 192 It..is important, more especially for the neurosurgeon..that his clinical experiences should be correlated with a more detailed classification of the gliomas than is customary. **1972** *Oxford Times* 26 May 6/7 Mr R. Gye, consultant neurosurgeon, explained the many uses to which the blanket could be put. **1904** *Alienist & Neurologist* XXV. 404 (*heading*) Neurosurgery. Trigeminal neuralgia treated by intraneural injections of osmic acid. **1937** *Surgery* I. 132 With the later days of Victor Horsley in England and the early days of Harvey Cushing in America, neurosurgery may truly be said to have been born. **1966** *Lancet* 24 Dec. 1400/1 A speciality like neuro-surgery requires an extra year's training. **1932** *Glasgow Med. Jrnl.* CXVIII. 137 (*heading*) The work of a neuro-surgical clinic. **1955** A. Huxley *Let.* 25 Sept. (1969) 767 Penfield says, absence of evidence, in the present state of neurosurgical knowledge, proves nothing. **1974** *Nature* 13 Dec. 582/2 Recordings of unit activity during neurosurgical operations have demonstrated neuronal activity in cortex and subcortical structures. [**1877** *Med. Times & Gaz.* 10 Nov. 511/1 Nerve-syphilis appears to affect with preference those persons in whom there is the neuropathic constitution.] **1878** *Boston Med. & Surgical Jrnl.* XCVIII. 278 Neuro-Syphilis.—As nervous diseases of syphilitic origin are more amenable to treatment than the corresponding idiopathic ones, a correct diagnosis may at times be sufficient to save a life otherwise lost. **1915** *Ibid.* CLXXIII. 996/1 (*heading*) The significance of changes in cellular content of cerebrospinal fluid in neurosyphilis. **1946** *Nature* 17 Aug. 243/2 Penicillin sodium in saline solution is effective to a greater or lesser degree in all aspects of neurosyphilis studied. **1974** Passmore & Robson *Compan. Med. Stud.* III. xiii. 6/1 In all forms of neurosyphilis the results of treatment depend on the number of neurones already destroyed. **1877** *Med. Times & Gaz.* 10 Nov. 511/1 Neuro-syphilitic affections belong generally to the later portions of the secondary stage, or to the tertiary period of the complaint. **1918** *Jrnl. Amer. Med. Assoc.* 28 Sept. 1023/2 We can..control the majority of the early infections of the [cerebrospinal] fluid and greatly limit the number of neurosyphilitics in the future. **1921** *Ibid.* 2 July 3/2 In eight of the twenty-one neurosyphilitic partners the type of neurosyphilis was the same as in the original patient. **1954** D. Nabarro *Congenital Syphilis* viii. 282 Unless neurosyphilis is actively sought by routine C.S.F. investigations upon patients, many neurosyphilitics will be overlooked. **1972** *Afr. Jrnl. Med. Sci.* III. 195 Cases..diagnosed as neurosyphilitic on clinical grounds should be given the benefit of adequate penicillin therapy. **1901** *Gray's Anat.* (ed. 15) [52] (*heading*) Organ of Golgi (neuro-tendinous spindle) from the human tendo Achillis. **1920** S. W. Ranson *Anat. Nervous Syst.* v. 72 Somewhat analogous structures [to the neuromuscular end organs] are the neurotendinous end organs or tendon spindles where myelinated nerve-fibers end in relation to specialized tendon fasciculi. **1962** E. C. Crosby et al. *Correlative Anat. Nervous Syst.* ii. 87/2 The dendritic endings may be of neuromuscular or neurotendinous type (that is, muscle spindles or tendon spindles). **1949** *Jrnl. Compar. Neurol.* XCI. 339 Only such procedures as bear specifically on the comparative neurotoxicity of this drug for various animals will be set forth here. **1959** *Arch. Internat. de Pharmacodynamie et de Thérapie* CXXII. 98 Rats on a diet deficient in pyridoxine developed signs of neurotoxicity when administered IDPN. **1968** W. C. Bowman et al. *Textbk. Pharmacol.* xxviii. 722 All of the potent compounds which cause neurotoxicity are inhibitors of butyryl cholinesterase. **1971** *Nature* 20 Aug. 525/2 Between them the two proteins manifest ferocious neurotoxicity. **1948** De Robertis & Schmitt in *Jrnl. Cellular & Compar. Physiol.* XXXI. 3 Although unequivocal proof of the structure of the fibers cannot yet be given, the available evidence is consistent with the view that they are tubular, possessing a thin wall of relatively high electron density and a core of low density. To facilitate description they will be called 'neurotubules'. **1969** *Nature* 15 Nov. 710/1 The chief axoplasmic components, extending beyond the neurone cell body, are neurofilaments and neurotubules. **1970** P. A. Weiss in F. O. Schmitt *Neurosciences: 2nd Study Program* lxxiii. 845/2 In contrast to the straight neurotubules, the neurofilaments, 70–100 Ångström units in diameter, show a more wavy course. **1961** *Lancet* 23 Sept. 717/2 The Cox strains..exhibit a much higher level of monkey neurovirulence. **1973** *Nature* 26 Jan. 248/2 The vaccine was used on a limited scale but was withdrawn after Dick had claimed that he had detected a reversion to neurovirulence.

neuroepithelial (niūᵊːroˌepiþī·liăl), *a.* *Histology.* Also **neuro-epithelial.** [f. *neuroepitheli(um + -al.] **a.** Of or pertaining to neuroepithelium (sense *a); applied *spec.* to an epithelial nerve cell that is a sensory receptor.

1889 McKendrick & Stoehr *Text-bk. Physiol.* II. xi. i. 448 In certain medusæ, the transition between a neuro-epithelial cell and a ganglionic or nerve cell may be traced. **1907** G. A. Piersol *Human Anat.* 1463 The inner lamella [of the retina] may be subdivided..into the neuro-epithelial and the cerebral layers. **1952** Habel & Biberstein tr. *Trautmann & Fiebiger's Fund. Histology Domestic Animals* iii. 32 The neuroepithelial cells occur in taste buds, the olfactory epithelium, the cristae and maculae staticae, the organ of Corti, and the retina. **1966** T. S. & C. R. Leeson *Histology* xiv. 267/2 The neuroepithelial taste cells are distributed between the supporting cells and number only 4 to 16 in each taste bud. **1974** P. Constantinides *Functional Electronic Histology* vi. 81/2 (*heading*) Neuro-epithelial synapses.

b. Composed of or derived from neuro-epithelium (sense *b).

1948 R. A. Willis *Path. Tumours* lii. 818 While it is possible that the tumours containing undoubted embryonic neuro-epithelial tissue..were indeed pure embryonic neuro-epithelial growths, this remains uncertain. **1949** B. W. Lichtenstein *Textbk. Neuropath.* iv. 47 One generally recognizes two varieties of interstitial tissue in the fully developed nervous system—the neuroglia of neuro-epithelial origin and the microglia of mesenchymal origin. **1971** J. Minckler *Path. Nervous Syst.* II. clvii. 2111/1 Very primitive tumors occasionally occur in peripheral nerves which are regarded as neuroepithelial.

neuroepithelium (niūᵊːroˌepiþī·liə̆m). *Histology.* Also **neuro-epithelium.** [mod.L., f. Neuro- + Epithelium, as tr. G. *neuro-epithel* (G. Schwalbe in Graefe & Saemisch *Handb. d. ges. Augenheilkunde* (1874) I. 358).] **a.** The sensory epithelium in organs of special sense, such as the eye or the nose.

1885 W. Stirling tr. *Landois's Text-bk. Human Physiol.* II. 960 The layer of rods and cones or neuro-epithelium of Schwalbe. **1899** F. H. Gerrish *Text-bk. Anat.* 43 Sensory epithelium, or neuro-epithelium cells, are found in close relation with the filamentous terminals of the nerves devoted to taste, smell, hearing, and sight. **1929** S. Duke-Elder *Recent Adv. Ophthalm.* (ed. 2) iii. 104 As the primary optic vesicle invaginates, two layers are formed... The inner layer shows two well-differentiated regions: a marginal layer.., and deep to this the cells proper of the primitive neuro-epithelium. **1943** Fischer & Wolfson *Inner Ear* x. 319 There was a circular fold of neuro-epithelium segmenting the utricle. **1952** Habel & Biberstein tr. *Trautmann & Fiebiger's Fund. Histol. Domestic Animals* ii. 32 (*heading*) Sensory epithelium (neuro-epithelium).

b. Embryonic ectoderm that develops into nervous tissue.

1889 *Nature* 10 Jan. 260/2 The epithelium of this ciliated groove..is the only part of the primary central cylinder which is ciliated, and which does not form ganglion elements, and hence it is the only part which is not neuro-epithelium. **1940** S. A. K. Wilson *Neurol.* II. lxxii. 1176 In the epiblast of the embryo there appears a medullary tube lined by primitive neuroepithelium. **1958** R. A. Willis *Borderland Embryol. & Path.* iii. 115 This undifferentiated prospective nervous tissue, from which neurones, neuroglia and ependyma will all arise, is neuro-epithelium. **1971** —— in J. Minckler *Path. Nervous Syst.* II. cxlvi. 1938/2 Malignant teratomas in adults..may contain nervous tissue at all stages of development, from early neuroepithelium to its fully differentiated derivatives.

neurofibromatosis (niūᵊːrofəibrōᵘmătōᵘ·sis). *Path.* [f. *neurofibromat-* (taken as stem of *neurofibroma*, s.v. Neuro-) + -osis.] Any condition characterized by multiple neurofibromas; *spec.* a condition in which multiple (often very numerous) palpable neurofibromas occur on the peripheral nerves (also called *von Recklinghausen's disease*).

1899 *Lancet* 29 July 271/2 The nature of Recklinghausen's disease or, as it has sometimes been called, generalised neuro-fibromatosis. **1900** A. Thomson (*title*) On neuroma and neuro-fibromatosis. **1966** Wright & Symmers *Systemic Path.* II. xxxiv. 1236/1 The rare condition that is sometimes referred to as central neurofibromatosis, or Wishart's disease, may occur in association with neurofibromatosis of the peripheral nerves (von Recklinghausen's disease), or it may exist without the latter. **1973** *Daily Colonist* (Victoria, B.C.) 4 Oct. 2/3 A rare condition called neurofibromatosis is featured by multiple bumps.

neurogenesis (niūᵊrodʒe·nĭsis). *Biol.* [f. Neuro- + -genesis.] The development of nervous tissue.

1900 in Dorland *Med. Dict.* **1908** *Jrnl. R. Microsc. Soc.* 27 Arguments based on embryonic neurogenesis. **1928** R. M. May tr. *Ramón y Cajal's Degeneration & Regeneration Nervous Syst.* I. xvi. 381 The ideas of Harrison..are perfectly applicable to normal neurogenesis as well as to nervous regeneration. **1967** M. V. Edds in G. C. Quarton et al. *Neurosciences: a Study Program* 232/1 Faced with the limitations imposed by working with the embryo, investigators of neurogenesis have often turned to older organisms with the fortunate capacity of regenerating amputated nerve fibers.

neurogenetic (niūᵊrodʒĕne·tik), *a.* [f. Neuro- + -genetic.] = *neurogenic *a.* 2.

1889 *Jrnl. R. Microsc. Soc.* 494 The primordial cells of the embryonic brain are therefore neurogenetic, giving rise to the special nerve-cells. **1928** R. M. May tr. *Ramón y Cajal's Degeneration & Regeneration Nervous Syst.* I. xvi. 383 In the neurogenetic process the following factors come into play.

neurogenic (niūᵊrŏdʒe·nik), *a.* [f. Neuro- + *-genic.] **1. a.** Of a theory: implying or assuming control (esp. of the heart-beat) by the nervous system.

1901 *Buck's Handbk. Med. Sci.* (ed. 2) III. 111/1 The neurogenic theory is based on the following considerations: Muscular tissue..depends upon its motor-nerve impulses to set it into action. **1922** C. P. Howard in L. F. Barker *Endocrinol. & Metabolism* I. 313 Biedl has come to the conclusion that the neurogenic theory was based upon inconclusive experiments.

b. Caused or controlled by (a disorder of) the nervous system.

1904 [see *myogenic* adj. s.v. Myo-]. **1949** Koestler *Insight & Outlook* v. 69 They [*sc.* digestive disorders] could be called neurogenic rather than psychogenic. **1961** *Lancet* 19 Aug. 409/1 The book..ends with a detailed account of the surgical treatment of neurogenic disorders of the entire urogenital tract. **1974** *Nature* 8 Mar. 106/3 In the singing katydids..relatively high rates of sound pulses..are controlled by synchronous, or neurogenic, muscles in which each contraction is initiated by a nerve impulse.

c. *neurogenic bladder*, abnormal functioning of the bladder owing to disturbances of nervous control.

1930 *Urologic & Cutaneous Rev.* XXXIV. 541/2 (*heading*) Treatment of the neurogenic bladder with the slow sinusoidal current. **1952** P. A. Herbut *Urological Path.* I. iii. 323 In neurogenic bladder the sphincteric mechanism ceases to exist as an efficient functioning structure and the detrusor mechanism dictates the behavior of the bladder. **1966** Wright & Symmers *Systemic Path.* I. xxv. 782/1 The neurogenic bladder is in effect a functional obstruction of the urinary tract.

2. Of or pertaining to neurogenesis; giving rise to nervous tissue.

1915 J. A. Nelson *Embryol. of Honey Bee* viii. 127 The neurogenic area..comprises two longitudinal thickenings, the primitive swellings..which are separated by a median furrow. **1928** R. M. May tr. *Ramón y Cajal's Degeneration & Regeneration Nervous Syst.* I. xvi. 381 Some of the neurogenic factors mentioned by Harrison seem to us highly doubtful. **1941** Johannsen & Butt *Embryol. of Insects & Myriapods* xv. 242 A thin layer of elongated cells, the neurilemma, probably arising from the outlying ganglion cells, covers the neurogenic tissue dorsally.

Hence **neuroge·nically** *adv.*, by the nervous system or by nerves.

1960 *Arch. Neurol.* III. 229/1 The result is a sterile inflammatory reaction, neurogenically induced. **1967** *Jrnl. Pharmacol. & Exper. Therap.* CLV. 37/2 Phentolamine blocked with equal facility..the norepinephrine- and neurogenically evoked constriction. **1971** *Nature* 28 May 264/2 Increased sensitivity at the receptor level could play a role in enhancement of neurogenically mediated responses.

neurogram (niūᵊ·rogræm). [f. Neuro- + -gram.] An enduring structural change postulated as being produced in the nervous system by experience and as being the physiological basis of memory.

1914 M. Prince *Unconscious* v. 131, I have been in the habit of using the term *neurograms* to characterize these brain records... Richard Semon..has adopted the term Engramm with much the same signification that I have given to Neurogram. **1921** M. Garnett *Educ. & World Citizenship* v. 42 Examples of neural dispositions, or neurograms, are furnished by the functional systems of nervous arcs. **1939** —— *Knowledge & Character* iv. 54 The same neurones and systems of neurones may form part of several different neurograms.

neurography. Add: **3.** A name proposed for: all the neurograms of an individual, considered collectively.

1921 M. Garnett *Educ. & World Citizenship* v. 63 In the course of life's experience, then, an individual's neurography—if we may so describe all his neurograms, however distributed and arranged—tends to become organised into interest-systems. **1939** —— *Knowl. & Character* xi. 215 That our neurographies do tend to correspond to the realm of facts..is so important that we must insist further upon it.

neurohormone (niūᵊrohǭ·ɹmōᵘn). *Physiol.* [f. Neuro- + *hormone.] Any hormone or neurotransmitter released by the nervous system.

1941 A. Grollman *Essent. Endocrinol.* i. 2 This definition [of true hormones] excludes the..neurohormones,..for these are..local in their effects or are transported by diffusion. **1955** J. H. Welsh in Pincus & Thimann *Hormones* III. iii. 99 In this chapter an attempt will be made to see what all neurons have in common with respect to production, transport, storage, and release of regulator substances. It is proposed that the term 'neurohormone' be used to designate these substances. We may well continue to speak of 'neurohumors' and of 'neurosecretory materials', but it appears highly desirable to have an exclusive term, and 'neurohormone', previously used in this sense without precise definition,..seems highly appropriate. *Ibid.*, The term neurohormone..may be defined as an organic compound produced by neurons and released at their endings to act as a chemical messenger or hormone, either locally or at a distance. **1965** J. Pollitt *Depression & its Treatment* iv. 53 Monoamines are neurohormones regarded as essential for normal activity of the brain, and the two substances of this group playing particularly important roles in mood regulation, are serotonin..and noradrenalin. **1972** *Sci. Amer.* Nov. 28/3 The hypothesis that pituitary function is controlled by neurohormones originating in the hypothalamus was soon well established. **1974** *Nature* 15 Mar. 238/1 Many hormones, including noradrenaline, the neurohormone of the sympathetic nervous system.

So **neurohormo·nal** *a.*, involving both the nervous system and the endocrine system; of or pertaining to a neurohormone.

1949 Koestler *Insight & Outlook* v. 69 Digestive disorders..may be explained in direct neurohormonal terms. *Ibid.* viii. 127 The neurohormonal excitation may persist for a while like the duffed-down pain. **1962** *Science Survey* III. 266 The neuro-hormonal chain of events leading to the olfactory pregnancy block has to some extent been explored. **1965** Lee & Knowles *Animal Hormones* ii. 32 The ejection of milk at suckling..is a neurohormonal reflex. **1974** *Sci. Amer.* Sept. 8/3 What magic gives Grinspoon and Singer the power to know that a behavioral condition..is not due to, say, a neurohormonal deficiency?

neurohumoral (niūᵊrohiū·mŏrăl), *a.* *Physiol.* [a. F. *neuro-humoral* (H. Fredericq 1927, in *Compt. Rend. des Séances de la Soc. de Biol.* XCVII. (Réunion plénière) 3), f. *neuro-* Neuro- + *humoral* Humoral *a.*] Of, involving, or being a neurohumour. Cf. *humoral *a.* 1 a.

1929 P. Bard in C. Murchison *Foundations Exper. Psychol.* xii. 449 (*heading*) The neuro-humoral basis of emotional reactions. **1947** *Jrnl. Endocrinol.* V. 136 The possibility of a neurohumoral transmission of stimuli has been tentatively suggested on many occasions. **1966** G. B. Koelle in Rodahl & Issekutz *Nerve as Tissue* 291 The neurohumoral theory is most consistent with the known facts concerning transmission at the vast majority of synaptic and neuro-effector junctions. **1971** *Nature* 31 Dec. 570/1 Neurohumoral agents from the hypothalamus were discharged into the portal vessels for transmission to the anterior lobe of the pituitary to affect the secretion of hormones.

neurohumour (niūᵊ·rohiūməɹ). *Physiol.* Also (*U.S.*) **-humor.** [f. Neuro- + Humour, humor *sb.* Cf. next.] A neurohormone, esp. a transmitter.

1933 *Proc. Soc. Exper. Biol. & Med.* XXX. 556 That this stripe should be dark is due to the fact that the severance of the nerve fibers excites the discharge of an expanding neurohumor whereby the melanophores are made to enlarge and thus to darken the skin. **1948** *Special Publ. N.Y. Acad. Sci.* IV. 292 Fries..has obtained considerable evidence pointing to the conclusion that xanthophores are also doubly innervated and believes that neurohumors are involved. **1959** J. H. Welsh in A. Gorbman *Compar. Endocrinol.* 123 Although it is convenient to refer to the products of ordinary neurons as neurohumors and to those of neurosecretory cells as neurosecretory substances, their common role as regulatory agents must be recognized. **1971** *Compar. Biochem. & Physiol.* XXXVIII. A.239 The neurohumor, serotonin, has been shown to be functionally involved in hibernation in the ground squirrel.

neurohypophysial (niūᵊːrohəipofi·ziăl), *a.* *Med.* Also **-physeal** (-fi·ziăl, -fisī·ăl). [f. next after *hypophysial *a.*

In sense a, f. Neuro- + Hypophys(is + -al.]

a. Pertaining to the neural tube and the hypophysis (in the embryo). ?*Obs.*

1893 *Q. Jrnl. Microsc. Sci.* XXXV. 301 The process of constriction by which the tube, or, as it may at once be called, the neuro-hypophysial canal, comes to be entirely separated from the cerebral vesicle has therefore now commenced. *Ibid.* 306 The hypophysis and the ganglion ..have been gradually differentiating themselves from the common neuro-hypophysial tube. **1929** *Ibid.* LXXII. 79 The whole of the adult nervous system [in Ascidians] is a new development proliferated from a minute remnant of the embryonic neural canal. The region in which this proliferation takes place is usually included in the neurohypophysial canal.

b. Of or pertaining to the neurohypophysis.

1934 *Physiol. Abstr.* XIX. 481 (*heading*) The central effect on blood pressure of the neurohypophyseal circulation hormone. **1939** *Res. Publ. Assoc. Res. Nerv. & Mental Dis.* XX. 444 There is some question..that the antidiuretic substance in the urine is of neurohypophysial origin. **1963** Montgomery & Welbourn *Clin. Endocrinol.* viii. 336 Any condition which damages the integrity of the neurohypophyseal system causes diabetes insipidus. **1973** C. Ezrin et al. *Systematic Endocrinol.* iii. 21/2 Recovery follows because damage to the supraoptic neurohypophyseal tract is mild enough to be reversible. **1973** *Nature* 12 Oct. 287/2 This also means that the rat now has two neurohypophysial hormones and two neurophysins.

neurohypophysis (niūᵊrohəipǫ·fisis). *Med.* [f. Neuro- + Hypophysis.] The portion of the hypophysis that develops from the embryonic brain, comprising the posterior lobe and usually the infundibulum and median eminence, and differing from the adenohypophysis in having little glandular activity and in being richly supplied with nerve fibres, which originate in the hypothalamus (being the means by which the latter exercises its control of the hormonal activity of the anterior lobe) and release the hormones oxytocin and vasopressin following their production in the hypothalamus.

1912 H. Cushing *Pituitary Body & its Disorders* i. 2 The tip becomes thickened into the infundibular body (neurohypophysis, or pars nervosa). *Ibid.* 3 The neurohypophysis itself is connected with the tuber cinereum by a stalk which varies in length in different species. **1939** *Res. Publ. Assoc. Res. Nerv. & Mental Dis.* XX. 24 It is convenient to regard the neurohypophysis as being divisible into two major divisions, namely, the neural lobe and the neural stalk. **1954** L. C. Martin *Clin. Endocrinol.* (ed. 2) i. 2 The neural division or neuro-hypophysis consists of the pars nervosa or infundibular process, the pituitary stalk or infundibular stem, and the median eminence of the tuber cinereum. **1968** A. van Tienhoven *Reproductive Physiol. Vertebr.* viii. 249/1 Some of the hypothalamic hormones, e.g. oxytocin and vasopressin, are stored in the neurohypophysis. **1968** Passmore & Robson *Compan. Med. Stud.* I. xxv. 15/2 The neurohypophysis is a point of contact between neurological and hormonal control mechanisms. **1972** *Sci. Amer.* Nov. 24/3 This double organ..is the pituitary gland, or hypophysis. The part that migrated from the brain is the posterior lobe, or neurohypophysis; the part that migrated from the pharynx is the anterior lobe, or adenohypophysis.

neuroleptic (niūᵊrole·ptik), *a.* (*sb.*) *Pharm.* [ad. F. *neuroleptique* (Delay & Deniker 1955, in *Bull. de l'Acad. Nat. de Méd.* CXXXIX. 145), after *psycholeptique* *psycholeptic *a.*] Able to reduce nervous tension; tranquillizing; also as *sb.*, a neuroleptic drug; a tranquillizer.

1958 *Science* 10 Jan. 59/1 Others [*sc.* terms for psychopharmacologic drugs] are *ataraxic*.., and *neuroleptic* and *neuroplegic*, indicating diminutions in the intensity of nerve functions. **1959** *Jrnl. Clin. & Exper. Psychopath.* XIX. 286 We believe..the higher dosage of neuroleptics, combination of neuroleptics with other medications, or combinations of neuroleptics is frequently responsible for the unfavorable reactions. **1965** J. Pollitt *Depression & its Treatment* v. 67 This drug [*sc.* chlorpromazine] is neuroleptic, producing Parkinsonism. **1968** *New Scientist* 21 Nov. 417/1 Schizophrenics treated over long periods with neuroleptics have sometimes shown symptoms typical of endogenous depression. **1971** *Nature* 26 Nov. 224/2 All patients remained on their previous neuroleptic medication throughout the study. **1972** *Encycl. Psychol.* II. 310/2 After acute application in healthy subjects, even in low doses, neuroleptics only seldom induce motor relaxation and emotional stabilization.

neurologically (niūᵊrolǫ·dʒikăli), *adv.* [f. Neurological *a.* + -ly².] From a neurological point of view; as regards neurology.

1936 *Acta Med. Scand.* LXXXVIII. 479 Nothing neurologically abnormal was found. **1971** *New Scientist* 5 Aug. 335/1 Psychiatric disorders in neurologically handicapped children. **1974** *Sci. Amer.* July 106/1 Neurologically speaking, brains whose organization was essentially human were already in existence some three million years ago.

neuromast (niūᵊ·romæst). *Zool.* [f. Neuro- + Gr. μαστ-ός breast.] An organ of sensory perception forming part of the lateral line system of fishes and larval or aquatic amphibians.

1912 J. S. Kingsley *Compar. Anat. Vertebrates* 167 Distally the fibres terminate in peculiar collections of sense cells known as sense hillocks or neuromasts occurring in the inner ear and in the lateral line organs of the ichthyopsida. **1937** *Proc. R. Soc.* B. CXXIII. 474 The distinction between pit organs and canal neuromasts is purely conjectural. **1957** O. Lowenstein in M. E. Brown *Physiol. Fishes* II. ii. 156 Fundamentally the end-organs of the lateral line system consist of groups of secondary sensory cells called neuromasts. **1962** K. F. Lagler et al. *Ichthyology* iii. 105 Each ampulla contains sensory nerve endings (neuromast cells) in a gelatinous terminal cup. **1974** M. Hildebrand *Analysis Vertebr. Struct.* xvi. 395 The lateral line system..is present in fishes and in both larval and aquatic adult amphibians. It consists of thousands of microscopic organs called neuromasts.

neuromotor (niūᵊ·romōᵘtəɹ), *a.* [f. Neuro- + Motor *sb.* and *a.*] Pertaining to or involving both the nervous system and motor activities; applied *spec.* to a system of minute ectoplasmic fibrils connecting some of the cirri to the motorium in some ciliate protozoa.

1914 [see *motorium 2]. **1940** L. H. Hyman *Invertebrates* I. iii. 169 It is probable that at least some of the fibrils of the neuromotor system are conductile and serve to coordinate ciliary activities..; but others are presumably of a supporting nature. **1959** W. Andrew *Textbk. Compar. Histology* xiv. 539 The neuromotor system of Stylonichia appears to be very similar to that of Oxytricha... Some authors have ascribed a neuromotor function to certain fibrils connected to the basal granules of flagellates. **1966** *New Scientist* 3 Nov. 242/1 These neuromotor fibres have since been generally accepted as indicating the presence of an elementary intracellular 'nervous system'. **1966** *Amer. Speech* XLI. 226 Speech production by control of vocal tract configuration changes treats..phonemic distinction in terms of neuromotor commands. **1971** *Language* XLVII. 51 We have now reached the stage where automatic phonetic and phonological rules take over, converting the sequences of segments into actual neuro-motor commands to the muscles in the articulation of the utterance.

neuron. Add: **2.** (Earlier and later examples.) *Obs.*

1893 [see *dendron]. **1897** *Med. Chron.* VII. 234 We notice that Mr. Schäfer uses the word 'neuron' in a somewhat different sense from that applied to it by Waldeyer, who used it—and we believe that in his sense it is extensively, if not universally, used—to denote a nerve cell with all its processes. Mr. Schäfer uses it to designate the axis-cylinder process of a nerve cell. **1930** Hartridge & Haynes *Histol.* 114 Nerve processes are of two kinds, long and short. The long are called axons, axis cylinders,..and neurons.

3. [a. G. *neuron* (W. Waldeyer 1891, in *Berliner klin. Wochenschr.* 13 July 691/1).] (Earlier and later examples.)

The spellings *neuron* (niūᵊ·rǫn) and *neurone* (niūᵊ·rōᵘn) are both still widely current.

1891 *Brain* XIV. 569 [*Abstr. of Waldeyer*, 1891] Thus a nerve element, a nerve entity, or 'neuron', as I propose to call it, consists..of the following pieces:—(*a*) a nerve cell, (*b*) the nerve process, (*c*) its collaterals, and (*d*) the end-branching. **1896** L. F. Barker in *Johns Hopkins Hosp. Bull.* VII. 201/1 Van Gehuchten has adopted Waldeyer's word, spelling it in French 'le neurone', and French writers generally employ it. The leading investigators in Spain and Italy have also adopted the same term... The question arises, how is Waldeyer's form to be anglicized? Would it be justifiable to bring it into English through the French and to spell it *neurone*, pronounced neurōne, or could it be brought into English directly from the Greek and be so spelled and pronounced? **1904** E. B. Titchener tr. *Wundt's Physiol. Psychol.* I. vi. 326 This spatial connexion may..consist either in the immediate proximity of neurones lying upon the same side of the brain, or in the union of distant areas by association fibres. **1939** W. E. Le Gros Clark *Tissues of Body* xii. 294 Within the central nervous system impulses are conducted from one part to another along a chain of neurones. **1950** A. Huxley *Themes & Variations* i. 94 The gulf between thought..and..neurones and electric charges is just as wide as that which in Biran's day divided thought from fluids and fibres. **1964** *New Scientist* 10 Sept. 643/1 The brain is estimated to contain ten thousand million nerve cells (neurons) and ten times as many glial cells. **1971** J. Z. Young *Introd. Study Man* xxii. 296 Neurons are probably lost steadily in the later decades of life. **1973** H. M. Rális et al. *Techniques Neurohistol.* i. 16 The nerve cell can make contact with a large number of other neurones through the multiple branches of dendrites.

4. Special Comb: **neuron(e theory**, the theory (now generally accepted) that the nervous system is composed of individual cells which, though effectively in contact with one another, are structurally distinct units all derived from a single neuroblast in the embryo.

1897 *Cincinnati Lancet-Clinic* LXXVIII. 565/1 (*heading*) The histological basis of the neuron theory. **1939** W. E. Le Gros Clark *Tissues of Body* xii. 320 At one time the neurone theory excited considerable controversy, and even to-day some anatomists question its validity. **1972** M. L. Barr *Human Nervous Syst.* ii. 9/1 The Neuron Theory, as opposed to the view that nerve cells form a continuous reticulum, was advanced by His on the basis of embryological studies, by Forel on the basis of the response of nerve cells to injury, and by Ramón y Cajal from his observations with silver staining methods... Wholly convincing evidence in support of the Neuron Theory had to await the introduction of electron microscopy.

neuronal (niurōᵘ·năl), *a.* [f. Neuron + -al.] Of or pertaining to a neurone or neurones.

1901 *Brit. Med. Jrnl.* 29 June 1610/2 (*heading*) Changes in the neuronal centres in beri-beric neuritis. **1946** *Nature* 9 Nov. 647/2 Nachmansohn deals with the theory (his own) that acetylcholine is released at the neuronal surface during the passage of an impulse. **1953** *Brit. Jrnl. Psychol.* XLIV. 304 The final verification of an explanatory hypothesis in psychology will not come from its direct identification in terms of neuronal systems. **1971** Chin-Wu Kim in W. O. Dingwall *Survey of Linguistic Sci.* 32 Lenneberg..is of the opinion that the orders of neuronal firings are adjusted so as to achieve a temporal coincidence at the neuro-muscular juncture. **1974** [see *neuronophagia].

Hence **neuro·nally** *adv.*, by a neurone or neurones.

1960 *Exper. Neurol.* II. 364 (*heading*) Effects of various drugs on activity of the neuronally isolated cerebral cortex. **1975** *Nature* 6 Feb. 448/1 The contractile responses are neuronally mediated.

neurone: see Neuron 3 in Dict. and Suppl.

neuronic, *a.* (s.v. Neuron). Add: (Further examples.) Now a less common word than *neuronal *a.*

1906 *Athenæum* 1 Sept. 246/2 The controversy on the neuronic theory still continues. **1931** *Brit. Jrnl. Psychol.* XXII. 141 If there be some such automatic regulating device in the neuronic mechanism, we must assume that it is in operation in the intact animal. **1964** S. Duke-Elder *Parsons' Dis. Eye* (ed. 14) xxii. 334 The disease is a primary lipid neuronic degeneration of the whole of the central nervous system.

neuronophagia (niūᵊːrǫnŏfēⁱ·dʒiă). *Med.* Also anglicized as **-phagy.** [f. Neuron + -o + Gr. -φαγία eating (*sb.*).] The destruction of neurones by phagocytes.

1911 Stedman *Med. Dict.* 584/1 Neuronophagia, neuronophagy. **1932** M. Bielschowsky in W. Penfield *Cytol. & Cellular Path. Nervous Syst.* I. iv. 153 The neuroglia cells not only multiply, but some of them invade the marginal hollows and vacuoles. This process is included under the rather ill-chosen name of neuronophagia. **1958** J. G. Greenfield et al. *Neuropath.* ii. 62 Neuronophagy is specially common in all forms of encephalomyelitis which affect the grey matter. **1974** Passmore & Robson *Compan. Med. Stud.* III. xxxiv. 128/2 The occurrence of perivascular lymphocytic cuffing may distract attention from the more subtle isolated neuronal necrosis, each dead cell being surrounded by a ring of satellite phagocytes (neuronophagia).

neuropathologic (niūᵊːropæþolǫ·dʒik), *a.* Chiefly *U.S.* [f. Neuropatholog(y + -ic.] = Neuropathological *a.*

1950 *Veterinary Bull.* XX. 278 The neuropathologic diagnosis of hog cholera. **1973** *Neurology* XXIII. 561/1 Neuropathologic studies of 10 cases of Huntington's disease are known to us.

neuropathologist (niūᵊropăþǫ·lŏdʒist). Also **neuro-pathologist.** [f. Neuropatholog(y + -ist] = Neuropathist.

1876 Van Duyn & Seguin tr. *Wagner's Man. Gen. Path.* i. 5 Physicians have distinguished themselves as humoralists or solidists. Among solidists the neuropathologists and cellular pathologists must be separately designated. **1903** *Proc. Soc. Psychical Res.* XVIII. 23 Conservatives in anthropologic science will immediately say that Myers used the concept of the 'subliminal' far too broadly, and that the only safe demarcation of the term is that of the neuro-pathologists. **1932** M. Bielschowsky in W. Penfield *Cytol. & Cellular Path. Nervous Syst.* I. iv. 147 In his method of study the neuropathologist must amalgamate investigation of localization together with investigation of the pathological process. **1973** *Neurology* XXIII. 561/1 The striking clinical disparity between the manifestations of Huntington's chorea in children and those in adults is a challenge to the physiologist and neuropathologist.

neuropil (niūᵊ·ropil). *Neurology.* Also **neuropile** (-pəil). [prob. a shortening of next: cf. Pile *sb.*⁵] **a.** A network of interwoven unmyelinated nerve fibres and their branches and synapses; hence, esp. in organisms with simple nervous systems, a structure composed of, or a region in which is concentrated, such a network. **b.** *rare.* An ultimate branch of a nerve fibre.

1899 L. F. Barker *Nervous System* xxiii. 271 This investigator..isolated the neuropil of the second antenna of Carcinus—in other words, he removed the ganglion cells of the neurones supplying the antenna, but left their processes and side branches. **1900** tr. A. Bethe in J. Loeb *Compar. Physiol. Brain* iii. 45 It was easy to decide this question by separating the ganglion-cells with their axis-cylinder process from the motor neurons without injuring the neuropiles. **1934** *Jrnl. Compar. Neurol.* LIX. 95 In some parts of the [amphibian] brain much branched and contorted unmyelinated axons form a dense entanglement which is termed neuropil. **1941** Johannsen & Butt *Embryol. Insects & Myriapods* ii. 21 The neuropile (Punctsubstanz, fibrillar substance, nerve fibers) develops in the ganglia in the older stages. **1953** *Nature* 29 Aug. 405/1 The neurohæmal organs of the heart ligaments [in higher Crustacea] are..neuropiles. **1964** R. M. & J. W. Fox *Introd. Compar. Entomol.* vi. 179 A ganglion..has a central mass of medullary tissue (neuropile) consisting of the intermixed axons and dendrons of motor and internuncial neurons and the terminal arborizations of sensory dendrons. **1974** *Sci. Amer.* Jan. 41/1 In addition to the major branches running through roots and connectives there are many such processes within the neuropil, the region in the center of the ganglion where synaptic connections are made. **1974** *Nature* 4 Oct. 428/1 The ganglia..have a central fibrous neuropile containing most of the synapses.

†**neuropilema** (niūᵊropəilī·mă). *Obs.* Also **neuropilem.** [ad. G. *neuropilem* (W. His 1890, in *Arch. f. Entwickelungsges.* (*Anat. Abth.*) Suppl. 113), f. Gr. νεῦρο-ν nerve + πίλημα felt.] = *neuropil a.

1891 *Brain* XIV. 568 Nerve-felt, neuro-pilema, His. **1900** A. Hill tr. *Obersteiner's Anat. Cent. Nervous Organs* (ed. 2) iii. 156 The last and finest ramifications of the cell-processes, collaterals and arborescent systems alike, are closely interwoven in the grey substance, forming not so much a network as a felt-work, the neuropilema or neurospongium. **1900** *Jrnl. Nerv. & Mental Dis.* XXVII. 476 He [*sc.* Waldeyer] acknowledges that his ideas of the mode of conduction are based on the view that no anastomosing nerve networks occur, but only a nerve feltwork (neuropilema, His). **1902** Baldwin *Dict. Philos. & Psychol.* II. 175/2 *Neuropilem*, a meshwork of nervous arborizations forming a system of intercommunication between various neurocytes. **1929** C. U. A. Kappers *Evolution Nervous Syst. Invertebr., Vertebr. & Man* 43 In the higher Molluscs..they are arranged peripherally to the neuropilema as in Annelids.

neuropsychiatry (niūᵊrosəikəi·ătri). *Med.* [f. Neuro- + Psychiatry.] Psychiatry which relates mental or emotional disturbance to disordered brain function; neurology and psychiatry as a single discipline.

1918 M. W. Brown (*title*) Neuropsychiatry and the war. **1945** *Times* 11 Jan. 2/4 Smith was examined by Major Thomas March, chief of the section of neuropsychiatry at a United States service hospital. **1955** *Psychiatric Q.* XXIX. 392 The brain has been enthroned once again as the organ of thought, and a new era of neuropsychiatry is about to unfold. **1971** *New Scientist* 5 Aug. 335/1 This careful and scholarly investigation is of great help to the understanding of this difficult aspect of neuropsychiatry; it also..stresses the necessity of close cooperation between paediatricians, neurologists, and psychologists.

So **neu:ropsychia·tric** *a.*; **neuropsychi·atrist**, an expert or specialist in neuropsychiatry.

1918 M. W. Brown *Neuropsychiatry & the War* 113 A plea..for the establishment also of medico-legal centers to collaborate with the medical and neuro-psychiatric centers. **1922** *N.Y. State Jrnl. Med.* XXII. 512/2 The neuropsychiatrist..has to deal with the structure and functions in health and in disease of the most highly organized part of man, namely the multi-neuronic integrate known as the nervous system. **1952** *Sun* (Baltimore) 28 Feb. 11/3 The neuropsychiatrist conducts his interviews in an office apart from the hospital buildings. **1953** *A.M.A. Arch. Neurol. & Psychiatry* LXX. 428 (*heading*) Neuropsychiatric aspects of infantile eczema. **1971** *New Scientist* 5 Aug. 335/1 For anyone who has struggled with the diagnosis of mild conditions in psychologically affected children *A Neuropsychiatric Study in Childhood* comes as very welcome help.

neuropsychology (niūᵊrosəikǫ·lŏdʒi). *Med.* [f. Neuro- + Psychology.] The field of study concerned with the relationship between behaviour and the mind on the one hand, and the nervous system, esp. the brain, on the other; neurological psychology.

1893 Dunglison *Dict. Med. Sci.* (ed. 21) 751/1 *Neuropsychology*, neurology including psychology. **1955** *Sci. Amer.* 72/2 Neuropsychology is Lashley's specialty, and..its studies of cerebral function and neurological structure have been conducted under his personal direction. **1963** *Neuropsychologia* I. 1 Under the term 'neuropsychology', we have in mind a particular area of neurology of common interest to neurologists, psychiatrists, psychologists, and neurophysiologists. This interest is focused mainly..on the cerebral cortex. Among topics of particular concern to us are disorders of language, perception and action. **1964** R. L. Isaacson *Basic Readings Neuropsychol.* p. vii, Since the work of Lashley, neuropsychology has become a major branch of physiological psychology. **1973** *Daily Tel.* 23 Aug. 7/1 A high degree of specialisation by each of man's 'two brains'—the cerebral hemispheres—have [*sic*] been shown by recent work in neuro-psychology.

So **neu:ropsycholo·gical** *a.*; **neuropsycho·logist**, an expert or specialist in neuropsychology.

The mod. use appears to be independent of quots. 1851, where the meaning is rather less specific.

1851 T. Laycock tr. *Unzer's Princ. Physiol.* p. ii, The neuro-psychological essays he inserted in it are frequently referred to in the present work. *Ibid.* p. vi, Sylvius..followed Descartes, while Willis was influenced..by the doctrines of Paracelsus. There was, however, yet another neuro-psychologist. **1949** D. O. Hebb (*title*) Organization of behavior: a neuropsychological theory. **1970** *Sci. Amer.* Mar. 67/3 Of the various lesions in the second block of the brain those in the tertiary zones are particularly interesting to us as neuropsychologists. *Ibid.* 72/3 The neuropsychological approach provides a valuable means of dissecting mental processes as well as diagnosing illness. **1973** B. Haigh tr. *Luria's Working Brain* 12 The author..attempts to fit the facts obtained by neuropsychological studies of individual brain systems into their appropriate place in the grand design of psychological science. **1974** *Nature* 31 May 495/1 Thus was born the 'frontal lobe syndrome', the study of which in modern times owes much to the Russian neuro-psychologist Alexander Luria.

neurotic, *a.* Add: **3. b.** Characteristic of a neurosis or a neurotic. Also *fig.*

1918 A. A. Brill tr. *Freud's Totem & Taboo* iii. 143 Neurotics live in a special world in which..only the 'neurotic standard of currency' counts. **1948** L. MacNeice *Holes in Sky* 31 The taut-necked donkey's neurotic-asthmatic-erotic lamenting. **1972** *Newsweek* 10 Jan. 3/1 The Nixon Administration has blundered again—having become enslaved by its neurotic emphasis on seeking re-election in '72 at all costs. **1975** M. Babson *There must be Some Mistake* xix. 161 My husband simply disappears..and I was foolish enough to get upset about it. Neurotic of me, wasn't it?

neurotically (niurǫ·tikăli), *adv.* [f. Neurotic *a.* + -al + -ly².] In a neurotic manner; as the result of a nervous disorder. Also *fig.*

1890 W. James *Princ. Psychol.* II. xxviii. 685 In many cases..the parental alcoholics are themselves degenerates neurotically. **1919** M. K. Bradby *Psycho-Anal.* 78 Many..thrust their qualms into the unconscious and become neurotically deaf, blind or what not. **1924** R. Graves *Mock Beggar Hall* 21 A row of post-war villas, neurotically built, Standing each at different curious angles to the road. **1961** C. Willock *Death in Covert* iii. 69 He had a dark, keen, neurotically intelligent face. **1972** *Daily Tel.* 20 Apr. 10/6 Christopher Dowson, who slides into a house-party uninvited.., yearns neurotically after an unchanging situation.

neuroticism (niurǫ·tisiz'm). [f. Neurotic *a.* + -ism.] The condition or state of being neurotic; a tendency towards neurosis; esp. as a factor showing liability to neurosis included in certain types of personality assessment.

1900 *Daily Chron.* 5 June 4/5 The holiday season has been darkened by the reports of suicides which suggest..the neuroticism of Paris. **1902** W. James *Var. Relig. Exper.* i. 25, I think that I may let the matter of religion and neuroticism drop. **1922** *Glasgow Herald* 5 Oct. 5 The brilliant neuroticism of recent novels. **1952** H. J. Eysenck *Sci. Study of Personality* ii. 58 We have extracted a general factor of 'neuroticism' from the intercorrelations of fifteen tests for normal and neurotic groups separately. **1957** P. Lafitte *Person in Psychol.* iii. 34 Neuroticism and psychoticism are defined with scrupulous statistical care in terms of factorial scores..: but neither is directly related to ordinary behaviour. **1959** *Times Lit. Suppl.* 31 July 445/1 Mr. Johnston is subtle and satisfying when he shows them needling one another into neuroticism. **1973** *Jrnl. Genetic Psychol.* CXXII. 197 Iranian female subjects had scored higher than their male counterparts on neuroticism but lower on both extraversion and psychoticism.

neu·rotransmi:tter. *Physiol.* [f. Neuro- + Transmitter.] A substance which is released at the end of a nerve fibre by the arrival of a nerve impulse and, by diffusing across the synapse or junction, effects the transfer of the impulse to another nerve fibre, a muscle fibre, or some other receptor.

1961 *Lancet* 2 Sept. 530/1 In these sections appears much of the most recent work on the metabolic as well as the neurotransmitter actions of these amines. **1965** Lee & Knowles *Animal Hormones* ix. 123 Acetylcholine is also released in the ganglia of the autonomic nerves and at the endings of the parasympathetic nerves, acting as a neurotransmitter substance. **1971** *New Scientist* 17 June 669/1 The neurotransmitter serotonin stimulates salivation in the salivary gland of the housefly. **1974** *Sci. Amer.* June 59/1 Neurotransmitters make the heart beat faster or slower and make muscles contract or relax. They cause glands to synthesize hormone-producing enzymes or to secrete hormones.

So **neurotransmi·ssion**, the transmission of nerve impulses.

1961 *Harvey Lect.* LV. 43 (*heading*) Neurotransmission in the adrenergic nervous system. **1973** *Nature* 15 June 426/1 Reviews on biochemical aspects of neurotransmission.

neurotrophic (niūᵉrotrōᵘ·fik, -trǫ·fik), *a.* [f. Neuro- + *-trophic.] Of or pertaining to the control of cells exerted by nervous tissue, esp. in relation to cellular nutrition.

1887 T. W. Mills in *Med. Record* 22 Oct. 529/2, I propose..to substitute for mechanical explanations as applied to cardiac pathology, neuro-trophic ones. **1899** [in Dict. s.v. Neuro-]. **1935** *Amer. Jrnl. Cancer* XXIV. 416 Every epithelial cell is normally under constant neurotrophic control. **1970** *Nature* 28 Feb. 824 (*heading*) Neurotrophic control of protein synthesis in the regenerating limb of the newt, *Triturus*. *Ibid.*, We have examined the chemical nature of the neurotrophic action in a system which is clearly nerve-dependent, the regenerating salamander limb.

neurotropic (niūᵊrotrōᵘ·pik, -trǫ·pik), *a.* [f. Neuro- + *-tropic.] **1.** *Path.* Tending to attack or affect the nervous system preferentially.

1903 *Buck's Handbk. Med. Sci.* (ed. 2) VI. 270/2 Thus the poisons which have at the same time neurotropic and lipotropic effects..will have a much more marked influence upon the nervous system in an emaciated animal than in one which is very fat. **1922** *Med. Jrnl. Australia* II. 261/2 Vaccinia is the least neurotropic, while poliomyelitis is the most exclusively neurotropic. **1933** *Amer. Jrnl. Cancer* XIX. 648 Thus the tumor employed is not neurotropic. **1939** *Lancet* 4 Mar. 497/1 (*heading*) A neurotropic strain of human influenza virus. **1961** R. D. Baker *Essent. Path.* viii. 136 Venoms of snakes are hemolytic or neurotropic. **1973** *Nature* 26 Jan. 247/1 The MV strain of [poliomyelitis] virus..was then shown to be so adapted to monkey brain passage and so neurotropic that it would no longer grow in cells from other organs.

2. *Anat.* Of or pertaining to the control or regulation of nerves, esp. in relation to their growth and regeneration.

1912 *Jrnl. Nerv. & Mental Dis.* XXXIX. 774 Is the comparative incapacity of the nervous system to regenerate due to some incapacity of the mechanical obstacles, absence of preestablished routes, or to some peculiar defect in the Schwann cells, the chief if not the only elaborators of neurotrophic [*sic*] material? **1928** R. M. May tr. *Ramón y Cajal's Degeneration & Regeneration Nervous Syst.* I. 374 Several neurotropic currents may act simultaneously on the same cone of growth. *Ibid.*, This fact bears out the view that the neurotropic material is an enzyme segregated by the living cells, instead of a product of the decomposition or disintegration of the nervous tissue. **1968** A. F. W. Hughes *Aspects Neural Ontogeny* i. 33 For Ramon y Cajal the association of a regenerating fibre with the bands of Büngner was an instance of neurotropic attraction. **1968** S. Sunderland *Nerves & Nerve Injuries* xxxii. 405 Whether or not the tissues of the distal stump exert a neurotropic influence on the regenerating process..is still unsettled.

neurotropism (niūərotrōᵘ·piz'm, niūərǫ·trǫp-iz'm). [ad. G. *neurotropismus* (J. Forssman 1900, in *Beiträge z. path. Anat. u. z. allgem. Path.* XXVII. 408): see *NEUROTROPIC *a.* and -ISM.] **1.** *Anat.* The supposed attraction (or repulsion) exerted by one mass of nervous (or other) tissue upon another mass of nervous tissue which is in the process of growing or regenerating.

1905 Gould *Dict. New Med. Terms* 380/1 *Neurotropism*, the attraction or repulsion exercised upon regenerating nerve-fibers. A substance is said to have positive neurotropism when these regenerating nerve fibers have a tendency to grow toward and into it. **1912** *Jrnl. Nerv. & Mental Dis.* XXXIX. 774 (*heading*) Influence of neurotropism on regeneration. **1955** B. H. Willier et al. *Analysis of Devel.* VII. i. 356/2 Such 'distance action', commonly referred to as 'neurotropism' and assumed to be a form of either galvanotropism or chemotropism, has been invoked to explain oriented nerve growth in the embryo.., as well as during later nerve regeneration. **1968** A. F. W. Hughes *Aspects Neural Ontogeny* i. 34 The two principles which have been proposed as forces which direct the growing nerve fibre, the 'contact-guidance' of Weiss.., and the neurotropism of Ramon y Cajal, both rest on inference from the behaviour of growing fibres under various circumstances, and at present on nothing more.

2. *Path.* The tendency of a virus or other pathological agent to attack the nervous system preferentially.

1911 Stedman *Med. Dict.* 585/2 *Neurotropism*, the attraction of certain pathogenic microorganisms, poisons, and nutritive substances, toward the nerve-centers. **1925** *Jrnl. Exper. Med.* XLII. 523 His theories of dermotropism and neurotropism of viruses. **1933** *Amer. Jrnl. Cancer* XIX. 647 (*heading*) Neurotropism of neoplasms in the mouse. **1940** *Nature* 2 Nov. 596/1 A curious characteristic of the yellow fever virus..is its manifestation of two types of virulence, namely, 'viscerotropism', meaning that it attacks such viscera as the liver, kidneys and heart, and 'neurotropism', meaning that it damages the nervous system. **1959** *New Scientist* 19 Mar. 620/1 Sabin himself admits that there is some return of virulence—neurotropism he calls it, the power to cause paralysis.

neurula (niūə·rŭlă). *Embryol.* Pl. **neurulæ.** [mod.L., f. NEUR- + *-ula*, as in *BLASTULA, GASTRULA.] An embryo at the time when it is developing a neural tube from the neural plate. So **neurula·tion**, the formation of a neurula.

1888 *Amer. Naturalist* XXII. 470 The neurulation is normal. **1909** *Cent. Dict.* Suppl., Neurula. **1926** J. S. Huxley *Essays Popular Sci.* xviii. 259 The late gastrula or neurula is bisected as a whole. **1973** *Nature* 2 Mar. 55/2 In amphibian neurulae following initial removal of much of the totipotent blastula, abnormally few cells create a whole pattern. *Ibid.* 56/2 Gastrulation and neurulation movements are unaffected by mitomycin C.

neuston (niū·stǫn). [a. G. *neuston* (E. Naumann 1917, in *Biol. Zentralbl.* XXXVII. 99), f. Gr. νευστόν neuter of νευστός swimming, f. νεῖν to swim, after *plankton*.] A collective term for minute organisms inhabiting the surface layer of fresh water. Also *attrib.*

1928 K. E. Carpenter *Life in Inland Waters* ii. 35 Neuston.—This term, applied..especially to minute forms, such as bacteria and Protista, which float against the surface-film, may be extended to include all types especially associated with the film. **1939** *Nature* 28 Jan. 139/1 It [*sc.* the book under review] divides the plankton into plankton proper and the so-called 'neuston'—a term new to us—which means the organisms the peculiar province of which is the surface-film. **1957** G. E. Hutchinson *Treat. Limnol.* I. vi. 418 The organisms may be associated with the surface film and the resultant coloration should then be termed *neuston color* rather than section color. **1968** *New Scientist* 26 Sept. 669/2 Neuston are the organisms which live in the surface layer of [*sic*] film of water. **1974** A. Mayr-Harting tr. *Rheinheimer's Aquatic Microbiology* iii. 31 In the neuston—that is, the living community which develops during calm weather on the surface of lakes at the water-air interface—a characteristic microflora develops. **1974** *Nature* 4 Jan. 30/2 Thirty-seven surface tows were made with a neuston net to collect particulate pollutants quantitatively.

Neustrian (niū·striăn), *sb.* and *a.* [f. med.L. *Neustria* + -AN.] **A.** *sb.* A native or inhabitant of Neustria, the western part of the Frankish empire in the Merovingian period. **B.** *adj.* Of or pertaining to Neustria or its inhabitants.

1794 W. Beckford *Hist. France* I. vi. 104 The Austrasians..assisted by the Neustrians..drove him from the throne, and raised Childeric..to the sovereignty. **1874** [see *AUSTRASIAN *a.* and *sb.*]. **1918** A. Hassall *France Mediaeval & Modern* i. 6 In spite of the efforts of Ebroin, the Neustrian Mayor,..the tendency of Austrasia and Burgundy towards independence was too strong to be resisted. **1924** *Public Opinion* 27 June 618/2 The descendants of the Neustrian Franks now garrison the conquered Rhineland. **1927** O. M. Dalton *Gregory of Tours's Hist. of Franks* I. II. ii. 165 We are told that Charles addressed his Neustrians in 'Roman'. *Ibid.* 168 The Neustrian army of Charles was harangued in the Latin vulgar tongue. **1951** B. & R. North tr. *Martin's Making of France* iv. 37 Aquitanians fought with Neustrians, and Austrasians with Burgundians in battles in which the unity of Roman Gaul was undone. **1968** *Encycl. Brit.* XVI. 302/1 In the later Merovingian period the names Neustria and Francia were used interchangeably by Neustrian writers to designate the kingdom of which they were subjects, thus betraying their conviction that Neustria formed the heart and core of the Frankish lands. **1975** F. Heer *Charlemagne & his World* i. 16 After Ebroin was murdered..Pepin defeated the Neustrians in battle. He crushed Neustria as an independent state by confiscating the lands of the great Neustrian magnates and rewarding his own followers with them.

neutral, *a.* and *sb.* Add: **A.** *adj.* **2. c.** *neutral corner*: (see quot. 1954).

1952 *Amateur Boxing* ('Know the Game' Series) 29 When a boxer is 'down', his opponent must immediately retire to the farther neutral corner where he shall remain until ordered to resume boxing by the Referee. **1954** F. C. Avis *Boxing Reference Dict.* 89 *Neutral Corners*, the two corners of the ring not occupied, between the rounds, by the boxers and their seconds.

3. c. *neutral tint* (see also quot. 1911). Also *neutral orange* (see quots. 1934 and 1969).

1835 G. Field *Chromatography* xx. 170 Several mixed pigments of the class of gray colours are sold under the name of Payne's gray, neutral tint, &c. **1869** T. W. Salter *Field's Chromatogr.* (new ed.) xi. 253 Neutral orange or Penley's Neutral Orange, is a permanent compound pigment composed of yellow ochre and the russet-marrone known as brown madder. **1911** M. Toch *Materials for Permanent Painting* xiv. 145 Neutral tint..is a complex mixture of ultramarine, sienna, lamp black or ochre and lamp black, and..is an excellent color which is perfectly permanent. *Ibid.*, Neutral orange..has many of the characteristics of mars orange, but sometimes is made by mixing a brilliant yellow, free from lead, with a bright oxide of iron. **1924** F. W. Weber *Artists' Pigments* 98 Payne's Gray, like Neutral Tint, is prepared by various color manufacturers. **1934** H. Hiler *Notes on Technique Painting* ii. 113 Neutral orange, Penley's orange, a mixture of cadmium yellow with Venetian red. It is only used in water colour. **1969** R. Mayer *Dict. Art Terms & Techniques* 262/1 *Neutral orange*, a prepared artists' color made of mixed pigments. The best grades would contain cadmium orange or deep cadmium yellow and light red.

e. *Phonetics.* Of the central, usu. unstressed, vowel sounds [ə] and [i], produced with the tongue in a rest position and having indefinite quality.

1891 L. Soames *Introd. Phonetics* I. iii. 50 The obscure vowel 'a' in attend.., sometimes called the *natural* or the *neutral vowel*. **1948** J. R. Firth *Papers in Ling. 1934–51* (1957) ix. 131 The weak, neutral, or 'minimal' vowel... The term neutral suits it in English, since it is in fact neutral to the phonematic system of vowels in southern English. **1956** D. Abercrombie *Problems & Principles* iii. 33 The ubiquitous English 'neutral' vowel ə. **1965** W. S. Allen *Vox Latina* 4 The so-called 'neutral' vowel of standard southern British English, as at the end of *sofa*. **1972** R. A. Palmatier *Gloss. Eng. Transformational Gram.* 103 *Neutral vowel*, ..the mid central vowel 'schwa' [ə]—or the high central vowel 'barred i' [i]—to which insufficiently stressed vowels are reduced.

f. *Philos.* Belonging neither to the mental nor to the physical; esp. as *neutral monism*, the theory that there is but one substance of existence of which mind and matter are varying arrangements. So *neutral monist*.

1904 W. James in *Jrnl. Philos., Psychol. & Sci. Methods* 1 Sept. 487 Matter we know, and thought we know, and conscious content we know, but neutral and simple 'pure experience' is something we know not at all. **1912** R. B. Perry *Present Philos. Tendencies* II. iv. 79 It is evident that Mach's view can only mean a reduction of both the physical and the mental order to a manifold of neutral elements. **1914** B. Russell in *Monist* XXIV. Apr. 161 'Neutral monism'—as opposed to idealistic monism and materialistic monism. *Ibid.* 171 Neutral monists..infer that the mental and the physical are composed of the same 'stuff'. **1920** S. Alexander *Space, Time & Deity* II. III. viii. 216 Nor are we free to suppose that there is a neutral non-mental world containing illusions amongst other neutral objects, neither mental nor physical. **1925** C. D. Broad *Mind & its Place in Nature* iii. 133 It might be so on a Double-Aspect theory, or on a theory of Neutral Monism. **1944** J. E. Boodin in P. A. Schilpp *Philos. B. Russell* xv. 495 It seems to me that Russell's neutral monism is an illusion. Our sensory awareness—'sensation' as Russell calls it—is real. But it is not identical with a mathematical equation. **1973** A. Quinton *Nature of Things* III. 237 There is the kind of phenomenalistic or neutral monist theory intimated by Hume and Mill. *Ibid.* xi. 318 Minds, for the neutral monist, are literally composed of impressions.

4. f. Electr. Engin. *neutral point*, the point in an electrical system which has the same potential that the junction of equal resistances would have if they were connected at their other ends to the lines making up the system; (see also sense 4 c in Dict.); *neutral wire*, a wire connected to a neutral point (and usu. also to earth).

1896 R. Robb *Electr. Wiring* iii. 44 The taps are taken off from the neutral wire and either of the other two in such a way that the loads on the two sides of the system will balance as nearly as possible. **1907** J. F. C. Snell *Distribution Electr. Energy* 19 A usual practice is to require the middle of the star, or neutral point, to be earthed. **1930** *Engineering* 7 Mar. 321/3 The necessary apparatus for earthing the neutral point of a three-phase system. **1962** *Newnes Conc. Encycl. Electr. Engin.* 543/1 If no neutral point on a system is earthed and one of the line conductors becomes accidentally earthed, the other two line conductors will immediately assume line voltage to earth and the neutral will assume the line-to-neutral voltage. **1972** Smith & Hosie *Basic Electr. Engin. Sci.* ix. 238 The current in the neutral wire of a balanced 3-ph[ase], 4-wire star-connected system is zero and the conductor may be omitted.

g. *neutral-density* (Photogr.): applied to a filter that absorbs light of all wavelengths to the same extent and so causes no change in its colour.

1938 K. Henney *Color Photogr.* iii. 58 By the use of a neutral-density filter of the proper transmission, in addition to the colored separation filters, it is possible to lengthen the exposure of the green and the red filters. **1962** M. L. Haselgrove *Photographers' Dict.* 151 Mainly used in sensitometry and color work, a neutral density filter is sometimes resorted to in black and white photography when the camera is loaded with a film of very high speed. **1965** M. J. Langford *Basic Photogr.* xi. 195 Other, colourless, filters include neutral density (grey) filters for reducing image brightness without affecting colour reproduction.

6. Denoting various classes of dyes. **a.** Applied to various basic azine dyes: *neutral blue*, a brown crystalline compound, $C_{24}H_{20}N_3Cl$, which dyes cotton a dull blue but has poor fastness; *neutral red*, a dark green crystalline compound, $C_{15}H_{16}N_4.HCl$, which forms a red solution in water or alcohol and is used as an acid–base indicator and for staining granules and vacuoles in living cells; *neutral violet*, a dark green crystalline compound, $C_{22}H_{24}N_6.HCl$, which is occasionally used as a biological stain or an indicator.

1889 *Cent. Dict.* s.v. *blue*, Neutral blue. **1890** Thorpe *Dict. Appl. Chem.* I. 229/2 The commercial product [*sc.* toluene red], which contains a certain amount of impurities, is sold under the name of 'neutral red'... A similar product is..sold under the name of 'neutral violet'. **1905** Cain & Thorpe *Synthetic Dyestuffs* xviii. 133 The only other Eurhodine of importance is Neutral violet.., prepared by the oxidation of a mixture of *p*-amidodimethylaniline and *m*-phenylenediamine. *Ibid.* 140 Neutral blue..and its derivatives, Basle blue B..and Azine green GB..are important tannin cotton dyes. **1914** *Chem. Abstr.* VIII. 723 (*heading*) Neutral red as indicator in determination of the alkalinity of the serum. **1930** *Stain Technol.* V. 133 Dyes employed in the vital staining are:..neutral red, neutral violet, fuchsin, [etc.]. **1950** *Proc. Linn. Soc.* CLXII. 69 He [*sc.* the French cytologist Paiat] saw there no net: on the contrary, he saw separate spheres, and these he called neutral red vacuoles, because he was accustomed to colour them in the living cell with neutral red. **1952** K. Venkataraman *Chem. Synthetic Dyes* II. xxv. 768 A further distinction has been made between Rosindulines..and *iso*Rosindulines, such as Neutral Blue,..in which the auxochrome is in the benzene part of the naphthophenazine nucleus. **1967** *Oceanogr. & Marine Biol.* V. 364 She ascertained the region of epithelial elongation using Neutral Red marks. **1971** E. Gurr *Synthetic Dyes* 110 Neutral violet is only occasionally used as a biological stain.

b. Applied to textile dyes which can be used directly in an approximately neutral dye-bath. So *neutral-dyeing* ppl. adj.

1892 Collin & Richardson tr. *Nietzki's Chem. Org. Dyestuffs* 14 The salts of certain azosulphonic acids may be termed neutral dyestuffs. They may be directly fixed on vegetable fibres. **1920** J. M. Matthews *Application of Dyestuffs* v. 156 A separate group of dyestuffs is frequently made of the eosin or phthalein dyes. This group includes certain of both the acid and basic dyes, which..might be termed 'neutral' dyes... They are applied in neutral or slightly acid bath, and are largely used for the dyeing of silk. **1955** A. J. Hall *Handbk. Textile Dyeing & Printing* iii. 49 Acid wool dyes are sometimes divided into two groups, *viz.* acid dyeing and neutral dyeing. **1955** H. E. Woodward in H. A. Lubs *Chem. of Synthetic Dyes & Pigments* iii. 151 Many neutral-dyeing dyes are aggregated more than level-dyeing acid dyes. **1962** [see sense A. 6 c below].

c. *Histology.* [Introduced in this sense by P. Ehrlich 1880, in *Zeitschr. f. klin. Med.* I. 557.] Applied to biological stains or dyes precipitated on mixing an acid dye and a basic dye.

1893 *N.Y. Med. Jrnl.* LVII. 2/2 Other granules react only to basic, or still others only to neutral colors. **1925** H. J. Conn *Biol. Stains* viii. 86 Compound dyes of this sort are sometimes referred to as neutral dyes or neutral stains. This terminology, of course, does not indicate that they are neutral in reaction any more than do the corresponding terms acid and basic dyes. A dye chemist, in fact, uses the term neutral dye in an entirely different sense. **1958** J. R. Baker *Princ. Biol. Microtechnique* xiv. 263 The granules of polymorphonuclear leucocytes are coloured by both the components of the neutral dye, and that is why Ehrlich called them 'neutrophil'. He regarded the specific dyeing of these granules as an important property of the neutral dyes, not to be obtained without their use. **1962** E. Gurr *Staining Animal Tissues* 31 Neutral 'dyes' should not be confused with 'neutral' or compound stains. The latter are formed by chemical union between a basic and an acid dye.

B. *sb.* **4.** *Electr.* A neutral point or conductor (cf. *A. 4 f above).

1900 *Jrnl. Inst. Electr. Engin.* XXIX. 538 Each of these boards receives from the main generator board five cables, a pair of 'outers' for lighting, a similar pair for power, and a common neutral. **1930** H. P. Seelye *Electr. Distribution Engin.* vi. 98 On many systems, the neutral of the primary circuit, whether or not it is brought out as a four-wire circuit, is grounded at the substation. **1973** [see *LIVE *a.* 4].

5. = *Idiom Neutral* (s.v. *IDIOM 5). Also *attrib.*

1907 W. J. Clark *Internat. Lang.* II. v. 99 Members of the academy..carry on their business by means of circulars, drawn up, of course, in Neutral. **1922** A. L. Guérard *Short Hist. Internat. Lang. Movement* II. vi. 137 International words were selected..and were altered only in order to conform to Neutral spelling. **1928** O. Jespersen *Internat. Lang.* I. 49 Occidental... It forms in that respect a continuation of Neutral and especially of Rosenberger's Reform-Neutral. **1947** H. Jacob *Planned Auxiliary Language* ii. 45 The most favoured systems [based on ethnic languages] were Esperanto, Neutral, Novlatin, and Universal.

6. A position of the driving and driven parts in a gear mechanism in which no power is transmitted. Also *fig.*

1912 G. Harris et al. *Audel's Answers on Automobiles* 442 With clutch still disengaged, the transmission lever is moved from neutral to first speed position. **1925** *Morris Owner* Jan. 1154/1 Don't lose your head and start the car in gear. Take things quietly, put the lever in neutral, [etc.]. **1926** 'J. J. Connington' *Death at Swaythling Court* xiii. 250 The Colonel slipped his gear into neutral. **1958** *Spectator* 22 Aug. 251/2 A time-waster filling the hours when the brain is in neutral. **1962** J. Braine *Life at Top* xix. 222 She turned the ignition key; the car jerked forward convulsively, then stopped. 'Put it in neutral, first.' **1971** R. Dentry *Encounter at Kharmel* iii. 42 Pepper threw the gear stick into neutral, applied the handbrake firmly, switched off. **1973** 'D. Jordan' *Nile Green* xxxv. 174 Sue's face went into neutral. **1975** T. Allbeury *Special Collection* xix. 132 The Special Collection [Operation] has been put into neutral by the Presidium.

neutralism. Add to def: *spec.* a policy of maintaining neutrality and attempting conciliation in conflicts between major world powers. (Further examples.)

1951 *Here & Now* (N.Z.) May 5/1 The second feature of French opinion which seems important today is..the anxiety to keep out of future wars and is summed up in the term 'neutralism'. **1955** *Times* 6 July 11/3 Russia might take a favourable view of some at least of Japan's requests in order to stimulate Japanese 'neutralism' and the adoption of a foreign policy less friendly to the western Powers. **1958** *Listener* 27 Nov. 864/2 The interests of the Arabs require the political neutralization of their countries. This should mean genuine neutrality, and not neutralism; that is, exacerbating and exploiting the rivalries of the great powers, for immediate political advantage. **1959** *Times* 26 Oct. 11/2 Neutralism differs from neutrality in that it is an attitude of mind in time of peace rather than a legal status in time of war. **1963** M. Brecher *New States of Asia* iv. 112 *Neutralism* has in common with non-alignment an expressed desire to remain aloof from bloc conflict. But neutralism goes much further, for it involves a positive attitude towards bloc conflicts. A neutralist state assumes an obligation to help reduce tensions between blocs with a view to maintaining peace or bringing about peace, and more particularly to prevent the outbreak of war.

neutralist. (Further examples.)

1915 *Morning Post* 1 Feb. 8/6 A meeting of neutralists, held here to-day, was broken up by Republicans, who shouted 'Long live the war!' **1920** *Glasgow Herald* 26 May 9/2 The Neutralist elements..hate the very idea of celebrating Italy's entrance into the war. **1957** *Economist* 28 Dec. 1114/2 They have been supported by perhaps an equal number of pacifists, near-pacifists and neutralists. **1963** *Ann. Reg. 1962* 351 The congress called for the withdrawal of American influence from South Vietnam and the formation of a neutralist Government there. **1975** *Nature* 6 Feb. 482/2 Neutralists, however, are delighted to find that this molecule of a living fossil underwent changes at least as rapidly as the homologous molecules in highly-evolved species.

neutralistic (niūtrăli·stik), *a.* [f. NEUTRAL *a.* + -ISTIC.] Characterized by a neutral attitude.

1949 *Mind* LVIII. 57 The element in, for example, a sensation of blue..will be susceptible to Mr. Gallie's neutralistic analysis.

neutrality. Add: **4. a.** (Later example.)

1883 J. S. Stallybrass tr. *Grimm's Teutonic Mythol.* II. 883 Out of the Goth. fairguni's neutrality unfolded themselves both a male *Fiörgynn* and a female *Fiörgyn*.

neutralization. Add: **1. b.** (Earlier example.)

1817 Coleridge *Biog. Lit.* I. ii. 45 This is one instance among many of deception, by the telling the half of a fact, and omitting the other half, when it is from their mutual counteraction and neutralization, that the *whole* truth arises.

c. *Ophthalm.* Combination of a lens with one or more of known power, as a means of finding its power.

1897 J. Thorington *Retinoscopy* iv. 34 This movement, at such a point in neutralization, may give a hint as to the presence of astigmatism. **1946** Dickinson & Hall *Contact Lenses* vii. 82 The lenses should be checked for accuracy of focus and curve. In the absence of a refractionometer, neutralization may be employed. **1974** Jalie & Wray *Pract. Ophthalmic Lenses* 5 The measurement of the focal power of lenses can be done in many ways. Two main methods are used for ophthalmic lenses: (1) Neutralization. (2) Vertex power measuring instruments.

d. *Electronics.* The cancellation of internal feedback in an amplifier, etc. (see *NEUTRALIZE *v.* 3 c).

1923 *Wireless World* 21 Apr. 69/2 Neutralisation is secured provided coils L and L_1 are correctly coupled. **1943** F. E. Terman *Radio Engineers' Handbk.* v. 469 Cross neutralization is used with push-pull amplifiers... It can be thought of as a form of neutrodyne that takes advantage of the fact that the voltages on the two sides of a push-pull amplifier are of opposite polarity, thus giving the phase relations required for neutralizing. **1961** Gray & Graham *Radio Transmitters* iii. 64 Many systems of neutralization are based on forming a balanced bridge with the tube grid-to-plate capacity forming one of the arms of the bridge. **1962** Simpson & Richards *Physical Princ. Junction Transistors* vii. 132 It is possible to determine the elements of an external feedback circuit which will accomplish the required neutralization. Since the transistor is thereby made into a one-way device, in which an input signal can influence the output but the reverse is not true, the compensation process is often called 'unilateralization'. **1968** [see *NEUTRALIZE *v.* 3 c].

3. *Linguistics.* The levelling out of certain phonemic or morphemic distinctions in particular contexts.

1942 C. F. Hockett in *Language* XVIII. 10 Any talk of neutralization or cancellation or archiphonemes confuses the facts without adding anything. **1947** K. L. Pike *Phonemics* 243/2 *Neutralization of oppositions*, the occurrence in some environment of a segment phonetically similar to and mutually exclusive with two other contrasting segments. **1949** A. Martinet *Phonology as Functional Phonetics* 7 Care should be taken not to mistake the non-appearance of a phoneme in a given position with a neutralization. **1962** *Amer. Speech* XXXVII. 69 Review of extant phonemic analyses of Modern Icelandic..and proposal of a third solution hinging on the Prague concept of neutralization. **1964** E. Palmer tr. *Martinet's Elem. General Linguistics* iii. 69 Where an archiphoneme is manifested there is said to be neutralization. **1968** *Language* XLIV. 475, I understand neutralization to mean the lack of consonance between form contrasts and function contrasts in all parts of a single linguistic sub-system. **1972** Hartmann & Stork *Dict. Lang. & Ling.* 151 *Neutralisation*, the cancelling of a phonemic opposition in certain positions. **1973** *Word 1970* XXVI. 102 Morphological neutralization occurs in such examples as *alemães* and *alemãs* (both pronounced [the same]).

neutralize, *v.* Add: **3. b.** *Ophthalm.* To annul the refractive power of (a lens) by combination with one or more other lenses (of known power).

1902 Taylor & Baxter *Key to Sight Testing* xxxvi. 243 A deep convex, if it could be made infinitely thin, would practically neutralize a concave of the same power. **1962** L. S. Sasieni *Optical Dispensing* (ed. 2) xv. 354 When plastic lenses have to be neutralized, the trial lenses should on no account be allowed to touch the outer surface. **1974** Jalie & Wray *Pract. Ophthalmic Lenses* 8 It may be necessary for unknown lenses of high focal power to be neutralized by a combination of two neutralizing lenses.

c. *Electronics.* To cancel internal feedback in (an amplifier stage, valve, or transistor), esp. that due to interelectrode capacitance, by providing an additional external feedback voltage of equal magnitude but opposite phase.

1924 Moyer & Wostrel *Pract. Radio* viii. 123 The adjustment of each neutralizing capacity is made by tuning to the radio current of some transmitting station, turning out the filament of the vacuum tube to be neutralized..and adjusting the capacity until all the sounds in the telephone receiver disappear. **1948** A. L. Albert *Radio Fund.* ix. 367 Triodes used in radio-frequency amplifiers must be neutralized, otherwise feedback from the plate to the grid through the interelectrode capacitance may cause oscillations. **1962** Simpson & Richards *Physical Princ. Junction Transistors* vii. 132 We now consider a general determination of the circuit constants required to neutralize a *CB* or *CE* amplifying stage. **1968** Romanowitz & Puckett *Introd. Electronics* x. 410 Push-pull circuits operating at high frequencies also require neutralization. They may be neutralized by a criss-cross connection of two capacitors, one from each tube plate, to the end of the input parallel circuit that is connected to the grid of the other tube.

5. In motor rallying, to exempt (a section of the course) from having to be covered at a set average speed, so that that section has no effect on the result of a race.

1902 *Encycl. Brit.* XXXI. 13/2 Deducting the Swiss portion of the route (which was neutralized), the distance was 615 miles. **1903** *Sci. Amer. Suppl.* 20 June 22958 A number of different villages were neutralized, and the chauffeurs were given from 5 to 25 minutes to make the passage. **1971** P. Browning *Rally Manual* vi. 54 My immediate reaction was to neutralize the frontier crossing—that is to say..disregard the time taken to pass between the two officials. **1972** H. S. Villard *Great Road Races 1894–1914* vi. 115 The Taunus Circuit took in eight neutralized control areas..and was eighty-seven miles in length.

neutralizing, *vbl. sb.* (Later examples, in sense *3 b of the vb.)

1902 Taylor & Baxter *Key to Sight Testing* xxiii. 148 The lenses used for testing should not be used for neutralizing. **1962** L. S. Sasieni *Optical Dispensing* (ed. 2) xv. 355 The usual practice in neutralising is to hold the neutralising lens against the front surface of the spectacle lens and with the front surface towards the observer.

neutretto (niūtre·to). *Nuclear Physics.* [f. *NEUTR(INO + It. *-etto*, dim. suffix (see -ET).] †**a.** A neutral pion. *Obs.* **b.** A neutral particle having low rest-mass. *rare.*

1938 Arley & Heitler in *Nature* 23 July 159/1 Neutrons would have a much smaller penetrating power and there is no process known by which neutrinos could produce heavy electrons in sufficiently large numbers. We therefore think that we have to deal in these experiments with the neutral counterpart of the heavy electron, for which we propose the name neutretto. *Ibid.*, A neutretto (denoted by Y°) can be transformed into a heavy electron, $Y^\pm$, during a collision with a proton (P) or neutron (N) and vice versa: $Y^\circ + N \rightleftarrows P + Y^-$ or $Y^\circ + P \rightleftarrows N + Y^+$. **1939** *Physical Rev.* LV. 24 These non-ionizing particles must be much more penetrating than photons. This high penetrating power suggests their identification with the neutrettos (neutral particles having mass and other properties similar to the barytron) postulated by Heitler. **1947** *Sci. News* IV. 123 A neutral meson is also expected to appear; the romantic name Neutretto is ready waiting for it. **1948** *Rev. Mod. Physics* XX. 550/1 If the neutral particle is a neutrino, the electron energy should be almost exactly one-half the rest energy of the meson, i.e., 50 Mev. If the neutral particle is considerably heavier than an electron (neutretto) the energy of the decay electron should be correspondingly smaller. **1952** R. E. Marshak *Meson Physics* vi. 208 If one is willing to consider the possibility of a new neutral decay particle of small rest mass (we shall call it a neutretto and denote it by μ^0), then an alternative decay scheme would be $\mu^\pm \to e^\pm + \nu + \mu^0$. **1963** *New Scientist* 1 Aug. 255 The terms 'electron neutrino' and 'muon neutrino' respectively for 'neutrino' and 'neutretto' are probably more common at present in distinguishing these two particles.

neutrino (niutrī·no). *Nuclear Physics.* [a. It. *neutrino* (E. Fermi 1933, in *La Ricerca sci.* II. 491), f. *neutro* NEUTER *a.* and *sb.*, neutral + *-ino*, dim. suffix.] Either of two stable, uncharged leptons (associated respectively with the electron and the muon) which have zero or negligible mass and an extremely low probability of interaction with matter; also, the antiparticle of either of these, one of which is produced (along with an electron and a proton) in the beta decay of a neutron.

1934 *Sci. Abstr.* A. XXXVII. 383 A quantitative theory of the emission of β-rays is explained. This admits the existence of the 'neutrino', a new particle proposed by Pauli having no electric charge and mass of the order of magnitude of that of the electron or less. **1938** *Ann. Reg. 1937* 354 The neutrino was generally accepted as a useful working hypothesis, but at least one attempt was made to show that the beta-ray spectrum could be explained without assuming its existence. **1948** *Sci. News* VI. 79 Beta decay is very slow compared to the times taken by other nuclear reactions, and one must, therefore, expect the reverse process, i.e., the capture of a neutrino, also to be very rare. On the Fermi theory, a neutrino could indeed go very many times across the interior of the earth and still have practically no chance of hitting anything. **1956** *Time* 2 July 46/3 For 20 years nuclear physicists have used neutrinos (small, uncharged particles) in their calculations... But no known apparatus has ever detected neutrinos... Last week from the Atomic Energy Commission came big news. Neutrinos do exist. **1958** *New Scientist* 25 Dec. 1567/1 Pauli produced the sorely needed explanation for the variability of the energy of beta-particles thrown out by radioactive materials. He postulated the simultaneous emission of a ghost particle which carried away part of the energy but which had no charge and virtually no mass and so was not observable directly. The neutrino..has been fully vindicated and now occupies a position of great importance in contemporary theory. **1962** *Physical Rev. Lett.* IX. 36/1 The neutrinos we have used produce μ mesons but do not produce electrons, and hence are very likely different from the neutrinos involved in β decay. *Ibid.* 42/2 The most plausible explanation for the absence of electron showers..is then that $\nu_\mu \neq \nu_e$; i.e., that there are at least two types of neutrinos. This also resolves the problem raised by the forbiddenness of the $\mu^+ \to e^+ + \gamma$ decay. **1968** M. S. Livingstone *Particle Physics* iv. 74 The muon decays into an electron and two neutrinos: $\mu^- \to e^- + \bar{\nu}_e + \nu_\mu$. **1969** *Times* 20 Feb. 17/5 Neutrinos are emitted as by-products in a great many nuclear reactions. **1974** *McGraw-Hill Yearbk. Sci. & Technol.* 306/1 The neutrino is the only known particle that has only weak interactions. Thus, the neutrino is a unique tool in the study of the weak forces, since the interactions are free of the effects of the strong and the electromagnetic interactions, which are many orders of magnitude stronger. **1974** *Sci. Amer.* Dec. 115/1 At proton accelerators muon neutrinos are produced about 100 times more copiously than electron neutrinos.

neutro-. Add: **neutrope·nia** *Med.* [*-PENIA], the presence of an abnormally low concentration of neutrophils in the blood; hence **neutrope·nic** *a.*, suffering from neutropenia; **neutrophil(e** *a.*, add: [a. G. *neutrophil* (P. Ehrlich 1880, in *Zeitschr. f. klin. Med.* I. 558)], and substitute for def.: that can be stained with neutral dyes (*NEUTRAL *a.* and *sb.* A. 6 c); not strongly stained by acid or basic dyes; also as *sb.*, a neutrophil cell;

(further examples); so **neutrophi·lic** *a.*, neutrophil; of or pertaining to neutrophil cells.

1931 *Arch. Otolaryngol.* XIII. 864 The term agranulocytosis has come into the literature because of its brevity. The name is not strictly correct... At present, the name applied by Tuerk, 'malignant neutropenia', seems to be more appropriate. **1973** *Acta Haematol.* L. 223 Neutropenia early in haemodialysis is probably due to the return into the circulation of leucocytes damaged by the first contact with the dialyzing surface. **1961** Webster, Neutropenic. **1963** *Federation Proc.* XXII. 671/1 (*heading*) Potentiation of arthus reactivity in neutropenic rabbits. **1973** *Jrnl. Infectious Dis.* CXXVIII. 248 Both normal and neutropenic mice were treated with gentamicin. **1893** *N.Y. Med. Jrnl.* LVII. 3/2 The neutrophile granules of the polynuclear cells are not stained by this solution. **1954** Neutrophil [see *DRUMSTICK 2 e]. **1973** *Daily Tel.* 2 Oct. 19/2 Neutrophil cells in a healthy person are usually six per cent. of the total number of white blood cells. *Ibid.*, A medical expert said last night: 'An increase of neutrophils to 12 per cent. in a two-year-old is a thoroughly satisfactory reaction to an ear infection.' **1893** *N.Y. Med. Jrnl.* LVII. 4/2 The uninuclear leucocyte does not show neutrophilic granulations. **1962** Luntz & Wright in A. Pirie *Lens Metabolism Rel. Cataract* 321 Group A... These show mainly neutrophilic infiltration.

neutrodyne (niū·trodəin). *Radio.* [f. NEUTRO- (taken as repr. *neutralize, -ization*) + *-DYNE.] A type of high-frequency valve amplifier in which neutralization was first employed to prevent oscillation throughout a range of frequencies. Freq. *attrib.*

1923 *Wireless World* 21 Apr. 68/2 Professor L. A. Hazeltine has designed receivers in which the valve capacity coupling is neutralised. The capacity coupling is neutralised with the aid of fixed condensers, and the receivers are called by him 'Neutrodyne' receivers. **1924** *Mod. Wireless* II. 590/2 In the receivers to be described here, a fixed coupling is used between the anode coil and the neutrodyne coil. **1924** Moyer & Wostrel *Pract. Radio* viii. 124 The neutrodyne works best on an antenna but may be used on a loop. **1936** R. S. Glasgow *Princ. Radio Engin.* ix. 256 The neutrodyne circuit was used quite extensively in broadcast receiving sets prior to the general introduction of screen-grid tubes in 1929. **1943** [see *NEUTRALIZATION 1 d]. **1964** Ghirardi & Dines *Radio & Television Receiver Circuitry* (rev. ed.) iii. 79 A modification of the famous Hazeltine 'neutrodyne' circuit is perhaps the most widely used in commercial receivers that employ triodes.

Hence **neu·trodyning** *vbl. sb.* and *ppl. a.*, neutralizing (see *NEUTRALIZE *v.* 3 c).

1924 *Mod. Wireless* II. 590/3 The reaction is controlled entirely by this neutrodyning condenser. **1926** R. W. Hutchinson *Wireless* 217 Several methods have been devised for the control of this oscillation; neutrodyning, as it is termed, is one of the best. **1943** C. L. Boltz *Basic Radio* xiv. 226 This [instability] was decreased by using condensers to provide negative feed-back to nullify the positive feed-back. These were called neutralizing condensers, and the process was one of neutralization or neutrodyning.

neutron (niū·trǫn). *Physics.* [f. NEUTRAL *a.* and *sb.* + *-ON[1].] An electrically uncharged sub-atomic particle whose mass (939·6 MeV) is very slightly greater than that of the proton, which can decay into a proton, an electron, and an antineutrino (as in beta decay), and which is a constituent (with the proton) of all atomic nuclei except that of the common isotope of hydrogen; it is now usu. regarded as a particular state of a nucleon.

Before its discovery in 1932 it was conceived as a close association of a proton and an electron. Rutherford (who communicated Glasson's 1921 paper to the Royal Society) discusses this concept in a paper of 1920 cited by Glasson, but without using the word *neutron*. Harkins (of Chicago) seems to have coined the term independently. (The use in quot. 1899 is unrelated to the current use.)

[**1899** W. Sutherland in *Phil. Mag.* 5th Ser. XLVII. 273 If the electrons are distributed through the æther, we must suppose that in æther showing no electric charge each negative electron is united with a positive electron to form the analogue of a material molecule, which might conveniently be called a neutron.] **1921** W. D. Harkins in *Ibid.* 6th Ser. XLII. 309 Any complex atom has a mass and weight 0·76 per cent. less than the hydrogen atoms (neutrons) from which it may be assumed to be built. *Ibid.* 315 The term neutron represents one proton plus one electron. **1921** J. L. Glasson in *Ibid.* 597 In the ordinary atom of hydrogen we have a single electron separated from the nucleus by a distance of the order of 10^{-8} cm. It is here contemplated that a more intimate union of the two is possible... Such a particle, to which the name neutron has been given by Prof. Rutherford, would have novel and important properties. It would, for instance, greatly simplify our ideas as to how the nuclei of the heavy elements are built up. **1930** E. Rutherford et al. *Radiations from Radio-Active Substances* xvii. 523 The existence of a neutron, i.e. a close combination of a proton and electron, has been suggested. **1932** J. Chadwick in *Nature* 27 Feb. 312/1 (*heading*) Possible existence of a neutron. *Ibid.*, These results..are very difficult to explain on the assumption that the radiation from beryllium is a quantum radiation, if energy and momentum are to be conserved in the collisions. The difficulties disappear, however, if it be assumed that the radiation consists of particles of mass 1 and charge 0, or neutrons. The capture of the α-particle by the Be^9 nucleus may be supposed to result in the formation of a C^{12} nucleus and the emission of the neutron. **1938** R. W. Lawson tr. *Hevesy & Paneth's Man. Radioactivity* (ed. 2) ix. 94 We must regard these protons and neutrons as the ultimate constituents of the nuclei, and consider the α-particle as a kind of 'molecule' of the 'nuclear atoms' possessing especial stability, and arising from the union of 2 protons and 2 neutrons. **1950** Glasstone *Sourcebk. Atomic Energy* xi. 287 One of the most striking arguments for the wave-particle duality of matter..has been provided by the diffraction of neutrons. **1951** S. Dushman *Fund. Atomic Physics* xiii. 227 The U-235 nucleus, in splitting, emits high-speed neutrons which may produce fission in other U-235 nuclei, thus initiating a chain reaction. **1968** M. S. Livingstone *Particle Physics* ii. 25 The nucleus could now be conceived as a closely packed assemblage of protons and neutrons, with the atomic charge number *Z* given by the number of protons and the atomic weight number *A* by the total of protons and neutrons. **1972** *Physics Bull.* June 349/1 Although it is comforting and often convenient to consider the proton and neutron as elementary particles with no internal substructure, they are in fact particles in a state of continual change... The neutron divides its time between being a neutron and a composite proton-negative pion system, while remaining electrically neutral. **1973** A. J. MacLeod *Instrumental Methods Food Analysis* vii. 684 Isotopes of an element possess different numbers of neutrons in their nuclei, but are otherwise identical.

2. *attrib.* and *Comb.*, as *neutron absorption, bombardment, flux; neutron-absorbing* adj.; **neutron activation**, the process of making a substance radioactive by irradiating it with neutrons; freq. *attrib.*, esp. in *neutron activation analysis*, activation analysis in which this is employed; **neutron bomb**, a bomb that would produce large numbers of neutrons but little blast, and would consequently be harmful to life but not destructive of property; **neutron capture**, the absorption of a neutron by an atomic nucleus; **neutron chopper**, a device for converting a continuous beam of neutrons into a pulsed beam by passing it through a rotating slotted disc or cylinder; **neutron diffraction**, diffraction of a beam of neutrons; **neutron excess**, the excess of the number of neutrons in an atomic nucleus over the number of protons; **neutron number**, the mass number of a nucleus minus its atomic number, taken as being the number of neutrons it contains; **neutron radiography**, radiography in which the radiation employed is a beam of neutrons; **neutron star** *Astr.*, a hypothetical extremely dense kind of star composed predominantly of closely packed neutrons, which is believed to have a mass similar to that of the sun but a diameter of only a few kilometres; cf. *PULSAR; **neutron therapy**, radio-therapy in which the radiation employed is a beam of neutrons.

1947 *Physical Rev.* LXXII. 16 Introduction of a neutron-absorbing substance into a pile decreases the reactivity. **1964** M. Gowing *Britain & Atomic Energy* 29 The Paris group also worked out the idea of controlling the reaction by introducing neutron-absorbing material to limit the multiplication of neutrons. **1947** *Physical Rev.* LXXII. 16 (*heading*) Method for measuring neutron-absorption cross sections by the effect on the reactivity of a chain-reacting pile. **1951** *Analyst* LXXVI. 644 (*heading*) The estimation of tantalum in mixtures by neutron activation analysis. **1960** *Nature* 16 Jan. 196/2 (*heading*) Neutron activation analysis of ancient Roman potsherds. **1965** D. Gibbons in Lenihan & Thomson *Activation Analysis* x. 68 Phosphorus and barium can be determined, rapidly and non-destructively, in lubricating oil in the range 0·01–1% and 0·05–2% using 14 MeV neutron activation and γ-ray spectrometry. **1966** *Encycl. Industr. Chem. Analysis* I. 52 With the high neutron flux of a nuclear reactor, neutron activation analysis (NAA) provides the most sensitive means of detection of low concentrations known today for most of the elements of the periodic system. A typical limit of detection is about one part per billion (ppb) in a 1-g sample. **1973** A. J. MacLeod *Instrumental Methods Food Analysis* vii. 695 Of the three methods described for separation and analysis of the products of neutron activation, gamma ray spectrometry is by far the most common. **1960** *Congress. Rec.* 12 May 10138/3 Although there have been a few fragmentary references to the neutron bomb in the press, I was told..that the matter was classified. **1962** L. Deighton *Ipcress File* xxiv. 155, I guessed that it was a neutron bomb that they were about to explode. **1967** *New Scientist* 14 Sept. 534/1 Whenever the possibility of a neutron bomb was discussed in the late 'fifties and early 'sixties it was with particular reference to the military interest in the application of tactical nuclear firepower in a controlled and highly selective fashion. **1937** G. Gamow *Struct. Atomic Nuclei* x. 193 These phenomena (of induced activity by neutron bombardment) were observed by Fermi for many very heavy elements also. **1964** M. Gowing *Britain & Atomic Energy* ii. 55 It was known that in some nuclei fission occurred spontaneously, without neutron bombardment. **1945** H. D. Smyth *Gen. Acct. Devel. Atomic Energy Mil. Purposes* i. 17 [By 1940] neutron-capture cross-sections had been measured. **1959** *Listener* 19 Nov. 872/1 At slow or 'thermal' speeds neutron capture by nuclei of Uranium 238 is less important. **1966** Phillips & Williams *Inorg. Chem.* II. xxxv. 635 Neutron-capture efficiencies are usually measured in terms of nuclear cross-sections, for which the whimsical unit the 'barn' has been devised. **1950** Glasstone *Sourcebk. Atomic Energy* xi. 306/2 Three main techniques have been developed for making measurements with neutrons of specific energies; these are the time-of-flight velocity selector..; the mechanical selector, sometimes referred to colloquially as a 'neutron chopper'..; and the crystal spectrometer. **1971** *New Scientist* 8 July 72/1 Another rotating component developed at Harwell has been the so-called 'neutron chopper'... When it rotates, the beam is chopped into pulses of neutrons of the same energy level. **1949** *Physical Rev.* LXXVI. 1256/2 (*heading*) Detection of antiferromagnetism by neutron diffraction. **1973** J. Yarwood *Atomic & Nuclear Physics* xiii. 381 Whereas X-ray diffraction cannot readily lead to a knowledge of the positions of hydrogen atoms in a crystal because the scattering is dependent on the number of orbital electrons.., neutrons are strongly scattered by hydrogen nuclei. The technique of neutron diffraction therefore supplements that of X-ray diffraction. **1955** R. D. Evans *Atomic Nucleus* iii. 99 (*N–Z*) is called..the 'neutron excess'. **1947** *Science* 9 May 491/1 Facilities will include..another pile with 100 times the neutron flux of the first. **1971** *Engineering* Apr. 34/2 The neutron flux reaching the sample is in excess of 10^7 neutrons cm^{-2} sec^{-1}. **1955** R. D. Evans *Atomic Nucleus* iii. 99 Nuclei having the same neutron number are isotones. **1973** *McGraw-Hill Yearbk. Sci. & Technol.* 181/1 Theoretical calculations..have shown that the same r-process which synthesizes the actinides..may also synthesize the predicted superheavy elements with atomic number $Z \sim 114$ and neutron number $N \sim 184$. **1948** H. Kallmann in *Research* I. 254/1 The first successful experiments in neutron radiography are due to the author and his collaborator E. Kuhn, who in the years 1935–39 developed and partly tried out the fundamental methods of neutron radiography. **1962** *Sci. Amer.* Nov. 107/2 Neutron radiography has become practical with the advent of nuclear reactors and particle accelerators, which can provide a source of neutrons of the required intensity. **1971** *Engineering* Apr. 37/1 An advantage of using neutron radiography is the possibility of detecting small amounts of low density materials, within sections of high density materials such as lead or steel. **1934** Baade & Zwicky in *Physical Rev.* XLV. 138/2 We advance the view that supernovae represent the transitions from ordinary stars into neutron stars, which in their final stages consist of extremely closely packed neutrons. **1968** *New Scientist* 16 May 331/1 Neutron stars, if they exist, could have densities as much as ten million times higher even than a white dwarf. **1970** B. Lovell in *Times* 19 Aug. 9/7 It is now generally accepted that the pulsars are the remnants of stars which have collapsed to form neutron stars. **1974** *Nature* 13 Sept 99/3 The neutron star represents the most extreme density of matter in the observable Universe (greater densities lead to unobservable black hole matter). **1947** *Radiology* XLVIII. 431/1 (*heading*) Possible progress in the radiotherapy of cancer (neutron therapy, Joliot therapy, alpha therapy, beta therapy..). **1974** *Nature* 25 Jan. 173/1 Because of Hammersmith Hospital's success in the use of neutron therapy for the treatment of cancer a second British hospital is to be supplied with a compact cyclotron. *Ibid.*, The effectiveness of neutron therapy depends on the ability of the neutrons to destroy cancer cells even in the absence of oxygen.

neutronic (niutrǫ·nik), *a. Physics.* [f. prec. + -IC.] Of, pertaining to, or employing neutrons.

1934 [see *ISOMER 2]. **1952** *Proc. Nat. Acad. Sci.* XXXVIII. 450 Particles which are heavy in the sense used here may be said to have unit neutronic charge. The neutronic charge should play the same role for the conservation law of heavy particles as the electric charge plays in the conservation law of electric charges. **1955** *Sci. News Let.* 28 May 349/1 Drs. Fermi and Szilard originally filed application for a patent on the 'neutronic reactor', on Dec. 19, 1944. **1958** *Times* 2 Jan. 2/4 Neutronic calculations on the steady state performance of thermal reactors. **1971** *Nature* 26 Nov. 217/2 If the reflecting box is expanding, all enclosed quantal systems will be decaying and emitting quanta of [*printed* on] all sorts of electromagnetic, electronic, neutronic and so on, waves.

névé. 1. (Earlier example.)

1843 J. D. Forbes *Travels through Alps of Savoy* 31 The part of a glacier covered with perpetual snow is what I understand to be meant by the term *névé* in the writings of the modern glacialists, although that term is vaguely defined.

never, *adv.* Add: **1. d.** (Further examples.)

1896 A. Morrison *Child of Jago* xii. 120 'I never,' protested Dicky stoutly. **1926** A. Bennett *Lord Raingo* II. lxxi. 322 She faintly annoyed him by her ingenuous exclamations: Oh my! Well, I never! Well I never did! *a* **1930** D. H. Lawrence *Phoenix II* (1968) 21 'I never, I never!!' he declared, more emphatically. **1939** L. M. Montgomery *Anne of Ingleside* xxv. 171 'I've an idea Bruno has gone back there.' 'Six miles? He'd never!' said Jem. **1950** [see *INJUN b]. **1972** N. Marsh *Tied up in Tinsel* ii. 49 'A booby-trap.' 'I never!' Mervyn burst out. 'My God..I swear I never.' **1974** N. Bentley *Inside Information* xv. 151 'There's a fellow..got a gun—a pistol.' 'Never!'

e. With suppression of the personal pronoun as subject.

1874 Hardy *Far from Madding Crowd* I. vi. 76 Never heard the man's name in my life. **1968** *Listener* 7 Nov. 610/2 He said: 'Never heard of it.'

2. *never fear, never mind* (further examples); *to make* (or *pay*) *no nevermind* (U.S.), to make no difference; to pay no attention.

1795 tr. *C. P. Moritz's Trav.* 261, I do not recollect to have heard any expression repeated oftener than this *never mind it!* A porter..fell down, and cut his head..'O, never mind it!' said an Englishman who happened to be passing by. **1849** G. E. Jewsbury *Sel. Lett. to Jane Welsh Carlyle* (1892) 344 Dear child, the solution will come to you, never fear. **1935** H. Davis *Honey in Horn* xvi. 264 That ain't no nevermind to me, though. **1946** Mezzrow & Wolfe *Really the Blues* (1957) xii. 225 He pays it no nevermind. **1968** *Guardian* 27 Dec. 8/1 We still have to adapt to Prime Ministers and Presidents, never mind astronauts, who have the essential quality of ordinariness. **1971** B. Malamud *Tenants* 177 Those are old books of his he wrote long ago, says Willie. Both been published. Then it makes no nevermind if we burn them.

3. c. *never say die* (see quot. *a* 1865); also *attrib.*

a **1865** Smyth *Sailor's Word-Bk.* (1867) 497 *Never say die*, an expressive phrase, meaning do not despair, there is hope yet. **1971** *Scope* (S. Afr.) 19 Mar. 10/2 Israel is a land that lives by the maxim: 'Never say die.' **1974** *Country Life* 5 Dec. 1717/2 The mental stamina, and the never-say-die spirit.

6. c. (Further examples.)

1917 E. Wallace *Just Men of Cordova* ii. 32 One never-to-be-forgotten occasion. **1925** R. Graves *Welchman's Hose* 59 In the compilation Of their Grand Dictionnaire de la Langue Française, The full, the final, never-to-be-gainsaid. **1935** L. MacNeice *Poems* 56 That never-to-be-touched Vision is your mistress.

7. a. *never-broken, -come, -conquered* (later example), *-contented, -dreamt, -ended* (earlier example), *-erased, -glutted, -lost, -quelled, -rebuked, -satisfied* (later example), *-tarnished, -tracked.*

1817 Coleridge *Biog. Lit.* II. xv. 16 A series and never broken chain of imagery. **1873** J. R. Lowell *Cathedral* in *Poetical Wks.* 452/1 Never-broken secrecies of sky. **1892** W. B. Yeats *Let.* Nov. (1954) 218 The ever-coming never-come light of that ideal peace and freedom. **1951** L. MacNeice tr. *Goethe's Faust* 235 Now even my army, I fear, must needs Obey the conquering, never-conquered woman. **1845** Poe *Fairy-land* in *Raven & Other Poems* 86 Those butterflies, Of Earth, who seek the skies, And so come down again (Never-contented things!). **1951** L. MacNeice tr. *Goethe's Faust* 302 Make to *this* good soul concession—Only once misled by pleasure To a never-dreamt transgression. **1855** D. G. Rossetti *Let.* 25 Nov. (1965) I. 282 One of his neverended stories was about an anonymous letter. **1855** W. Whitman *Leaves of Grass* 83, I see your rounded never-erased flow, I see neath the rims of your haggard and mean disguises. **1843** J. R. Lowell *Prometheus* in *U.S. Mag. & Democratic Rev.* Aug. 149 This bitter peak, This never-glutted vulture, and these chains. **1957** A. Miller *Coll. Plays* (1958) iv. 27 His terror springs from his never-lost awareness of time and place. **1860** W. Whitman *Leaves of Grass* 368 Those with a never-quell'd audacity. **1950** D. Gascoyne *Vagrant* 12 The serene, robust air as of never-rebuked gaiety. **1940** C. Day Lewis tr. *Virgil's Georgics* I. 16 That crop..Will answer at last the prayers of the never-satisfied Farmer. **1944** Auden *For Time Being* iii. 29 Present to the speculative eye an ever-shining, never-tarnished proof of her amazing unheard-of power to combine and happily contrast. **1848** J. R. Lowell *Growth of Legend* in *Poems* 2nd Ser. 71 The lake's frore Sahara of never-tracked white.

b. *never-diminishing, -ebbing, -eldering, -hastening, -intermitting, -lifting, -moving, -pardoning, -rejecting, -sinking, -stopping.*

1898 'Mark Twain' *Man that Corrupted Hadleyburg* (1900) 326 The war of epithets crashes along with never-diminishing energy for a couple of hours. **1866** J. G. Whittier *Our Master* in *Tent on Beach* (1867) 143 Immortal Love, forever full, Forever flowing free, Forever shared, forever whole, A never-ebbing sea! **1876** G. M. Hopkins *Wreck of Deutschland* xviii, in *Poems* (1967) 57 Tears; such a melting, a madrigal start! Never-eldering revel and river of youth, What can it be, this glee? **1950** W. de la Mare *Inward Companion* 13 With never-hastening feet Time pursues the Infinite. **1849** J. S. Mill in *Westm. Rev.* LI. 34 The immense majority are condemned..to a life of never-ending, never-intermitting toil. **1885** W. B. Yeats *Island of Statues* I. iii, in *Dublin Univ. Rev.* May 84/1 Where their sinewy might is strung In the never-lifting dark. **1913** J. Masefield *Daffodil Fields* 24 The stars did house Their lights like lamps upon those never-moving boughs. **1923** R. Graves *Whipperginny* 54 This never-pardoning life we live May earn God's blackest punishment. **1849** J. R. Lowell *Day in June* in *National Anti-Slavery Standard* 8 Mar. 162/1 O never-rejecting roof of blue. **1849** Thoreau *Week Concord Riv.* 244 The unwearied, never sinking shore. **1931** R. Graves *To Whom Else?* 9 What drew the legs along Was the never-stopping, And the senseless frightening Fate of being legs.

c. *never-anxious, -quiet.*

1889 W. B. Yeats *Wanderings of Oisin* I. 5 And always never-anxious sleep. **1913** J. Masefield *Daffodil Fields* 2 The never-quiet joy of dancing daffodils.

8. never-fail, (*a*) a person who never fails (one); (*b*) an Australian grass, *Eragrostis setifolia*, used as pasture in areas of low rainfall; **never-was,** a person who has never been great, distinguished, useful, or the like; also **never-waser, -wozzer.**

1850 H. C. Watson *Camp-Fires of Revolution* 188 Morgan's one of the never-fails. **1936** F. Clune *Roaming round Darling* xiii. 114 He has a marvellous collection of native grasses, nardoo, Mitchell, neverfail, and a dozen others. **1964** *Austral. Encycl.* IV. 367/1 *E[ragrostis] setifolia* ('never-fail') is a drought resistant species of the inland. **1967** *Coast to Coast 1965–66* 191, I had been riding through..the high grained heads of the grasses, the spinifex and the neverfail. **1911** J. C. Lincoln *Cap'n Warren's Wards* xv. 238 One of 'em's a used-to-be, and the other's a never-was. **1923** 'B. M. Bower' *Parowan Bonanza* i. 14 'Nope, I'm a never-was,' Bill retorted shamelessly. **1938** L. MacNeice *I crossed Minch* I. iii. 36 You ninny, you, you automaton, You Never-Was, you As-Good-As-Gone! **1891** *Sportsman* 1 Apr. 2/6 He is one of the 'has beens' or else one of the 'never wasers', as Dan Rice, the circus man, always called ambitious counterfeits. **1915** A. S. Neill *Dominie's Log* xiv. 155 The average married woman is a 'has been' in thought, while not a few are 'never wasers'. **1931** Wodehouse *Big Money* viii. 176 It's always been half-way between a may-be and a never-waser. **1974** *Economist* 9 Nov. 6/1 With respect, it is silly, on the strength of a typical remark of some anonymous Brussels diplomat, no doubt a strong advocate of the absurd procedure rules of the council of ministers, to describe them as 'has beens' or 'never wasers'. **1929** Neverwozzer [see *HAS-BEEN *sb.*].

9. (Further examples.)

1882 A. J. Boyd *Old Colonials* 202 My soliloquy ends with the inquiry, 'What on earth is to be done in this wretched Never-never country?' **1890** 'R. Boldrewood' *Colonial Reformer* 174 But here it seems to be the Never-Never country, and no mistake. **1916** J. B. Cooper *Coo-oo-ee* iii. 39 He had not forgotten the palship that is often made between men tramping along the bush distances that cover the sunburnt tracks to the Never-Never. **1942** C. Barrett *On Wallaby* iii. 43 Tim..owned a copper show in the Never Never country near the West Australian border. **1963** V. B. Cranley *27,000 Miles through Australia* v. 34 It was far beyond Yuendumu along the great desert traverse..back in the Never-Never, as they call those wastes. **1969** 'A. Garve' *Boomerang* ii. 71 His intention was to enjoy this trip..not to 'do a perish' in the Never-Never.

b. *Never(-Never) Land*, an imaginary, illusory, or Utopian place; freq. with allusion to the ideal country in J. M. Barrie's *Peter Pan* (see quots. 1904 and 1908).

1900 *N.Y. Dramatic Mirror* 3 Nov. 16/1 At Wallack's on Tuesday evening Sarah Cowell Le Moyne supplemented The Greatest Thing in the World with the initial performance of The Moment of Death; or, The Never, Never Land, a drama in one act and three scenes, by Israel Zangwill. **1904** J. M. Barrie *Peter Pan* (1928) I. 34 *Wendy.* Where do you live now? *Peter.* With the lost boys... They are the children who fall out of their prams when the nurse is looking the other way. If they are not claimed in seven days they are sent far away to the Never Land. **1907** *Canadian Mag.* XXIX. 135/1 But instead of the unreal *never-never-land*..the scene is dear old England. **1908** J. M. Barrie *When Wendy grew Up* (1957) 17 Do they ever wish they were back in the Never Never Land? *Ibid.* 28 The dear Never Never Land. **1938** Auden & Isherwood *On Frontier* I. i. 24 Dream of your never-never land, where the parks are covered with naked cow-like women, quite free. **1938** L. MacNeice *Mod. Poetry* v. 80 William Morris..looking wanly..back to a medieval Never-Never Land. **1958** *Spectator* 8 Aug. 203/1 It was no longer the real India they wanted to escape to; it was the Never-Never Land of the East. **1961** *Times* 1 Nov. 13/1 This commercial never-never land. **1968** Mrs. L. B. Johnson *White House Diary* 4 Apr. (1970) 647, I was back at the White House by 3 o'clock from my brief visit to that beautiful never-never land—Mrs. Merriweather Post's home. **1971** *Nature* 30 July 287/1 The result is that the report of the committee under Sir Frederick Dainton on the reorganization of civil research,..has disappeared into never-never land. **1975** *Times* 16 Oct. 13/8 Sending the hero and heroine at the end into an azure never-land that is clearly some distance from both Dorset and London.

c. *never-never* adj., *colloq.* (or *joc.*), denoting a system of paying for articles by periodic instalments over an extended period; = *hire-purchase*; also *ellipt.* as *sb.*, and as *never*.

1926 E. Wallace *More Educated Evans* ii. 39 Her uncle..drove a taxi which he..had purchased on the 'never never' system. You pay £80 down and more than you can afford for the rest of your life. **1939** 'N. Shute' *What Happened to Corbetts* viii. 261 We could have the radiogram... Even if we had to put it on the Never-Never. **1957** F. King *Man on Rock* i. 7, I can't even afford to pay the never-never on a wireless. **1960** *News Chron.* 29 Apr. 6 Twenty per cent is a small deposit for hire purchase, and the most reputable 'never-never' firms have been asking that. **1967** M. Procter *Exercise Hoodwink* iii. 21 'I'm getting it on the never. Anybody can do that.' 'Not a new Rover.' **1973** J. Wilson *Truth or Dare* ii. 24 They've still not paid off their mortgage, you know, and I wouldn't mind betting that Rover of theirs is on the never-never.

d. *never-never* adj., unrealistic, unrealizable, imaginary. Also applied to a person who says 'never, never'.

1928 D. H. Lawrence *Lady Chatterley* xiv. 243 'So when you did get a woman who wanted you..you got a bit too much of a good thing.' 'Ay! Seems so! Yet even then I'd rather have her than the never-never ones.' **1950** [see *CLOUD-CUCKOO-LAND]. **1952** Dylan Thomas *Let.* 21 Nov. (1966) 384 A day's life in a small town in a never-never Wales. **1955** *Bull. Atomic Sci.* Jan. 36/2 Norman Thomas, who had the courage to deplore our never-never attitude to the recognition of Red China, did so with an acute sense of his own isolation. **1956** H. Gold *Man who was not with It* (1965) xiii. 113 She predicted the never-never happiness of others. **1958** *Sunday Times* 16 Nov. 21/7 The atmosphere of some never-never hotel is certainly caught.

10. Colloq. phrases: *never a dull moment!*: see *MOMENT *sb.* 1 c; *never again!*, a phrase expressing emphatic refusal to repeat an experience, etc.

1873 Hardy *Pair of Blue Eyes* II. iv. 51 Thank you. But never again. **1901** Ade *Forty Modern Fables* 161 And everybody said, 'Never Again.' **1915** T. F. A. Smith *Soul of Germany* 298 The oft-quoted phrase is applicable to the case: Never again!

never (ne·vəɪ), *sb. Naut. slang.* [f. the adv.] In phr. *to do a never*: to shirk; to loaf.

1946 J. Irving *Royal Navalese* 121 Never, to do a, to dodge work. **1948** Partridge *Dict. Forces' Slang* 124 *Doing a never* means—in the Navy—shirking work. **1961** F. H. Burgess *Dict. Sailing* 74 Doing a Never, loafing on a job.

never-fading, *a.* (Later example.)

1916 Blunden *Pastorals* 32 Nor where the never-fading rainbow plays.

nevermore, *adv.* Add: **B.** as *sb.*

1951 Koestler *Age of Longing* II. v. 257 The evermore of desire and the nevermore of satiety. **1952** R. Campbell tr. *Baudelaire's Poems* 50 It's by such charms the Nevermore Intoxicates us in the Now.

‖**Nevers** (nevẹ̄r). The name of a city in central France, used freq. *attrib.* to describe a type of deep blue-ground faience in the style of Italian majolica, made there from the latter part of the 16th century to the 18th.

1863 W. Chaffers *Marks Pott. & Porc.* 93 *Nevers*, fine Faience, in the Italian style, decorated in colours, sometimes like Faenza ware; also in blue *en grisaille*, as well as in a deep blue ground of enamel with splashes or spots of white. **1900** F. Litchfield *Pott. & Porc.* vii. 219 Specimens of Nevers are difficult to identify, owing to the similarity of their characteristics to those of the Rouen and other similar faiences. **1960** H. Hayward *Antique Coll.* 198/1 *Nevers faience*, the chief Nevers pottery was founded early in the 17th cent. by Italian potters... Nevers glass. **1960** R. G. Haggar *Conc. Encycl. Cont. Pott. & Porc.* 68/1 Nevers faience decorated with pseudo-oriental or oriental motifs in white, or white, yellow and orange. **1971** L. A. Boger *Dict. World Pott. & Porc.* 241/2 The so-called Bleu Persan or Décor Persan was introduced during the second half of the 17th century and is probably the most important Nevers creation. **1972** *Times* 21 June 16/4 A charming seventeenth-century Nevers glass picture depicting Diana surprised by Actaeon.

neves (ne·vẹs). *Back-slang.* Also **nevis.** [Reversed form of 'seven'.] Seven years' hard labour.

1901 Farmer & Henley *Slang* V. 31/2 *Nevis*,..Seven. ..Nevis-stretch = seven year's hard. **1958** F. Norman *Bang to Rights* 22 Your f——ing lucky, I'm doing a bleeding neves.

nevus, var. NÆVUS (in Dict. and Suppl.).

nevvy, nevy, colloq. abbrevs. of NEPHEW.

1847 Dickens *Dombey* (1848) xv. 149 'He might die a little sooner for the loss of—' 'Of his Nevy,' interposed the Captain. **1903** Wodehouse *Tales of St. Austin's* 138 Yes, prarper good runner, his nevvy. **1940** M. Marples *Public School Slang* 133 Nunky and nevvy, uncle and nephew, are quoted by Wrench's WB as current before 1901 at Winchester; they are certainly now obsolete. **1959** E. Pound *Thrones* civ. 91 Sammy's nevvy got the gold out of the palace bed-room.

nevyanskite (nevyæ·nskəit). *Min.* Also † **newjanskite.** [ad. G. *newjanskit* (W. Haidinger *Handbuch der bestimm. Min.* (1845) 558), f. *Newjansk* Nevyansk, a city in Russia: see -ITE[1].] A variety of iridosmine containing about 35 to 50 per cent of osmium.

1854 J. D. Dana *Syst. Min.* (ed. 4) II. vi. 20 At a high temperature..Newjanskite is not decomposed and does not give an osmium odor. **1938** *Mineral. Abstr.* VII. 162 Ruthenium-bearing nevyanskite was found in alluvial deposits associated with serpentine of the Great Laba river, northern Caucasus. **1963** *Mineral. Mag.* XXXIII. 714 Iridosmine for the hexagonal phase (with Nevyanskite and Sysertskite as varieties) and Osmiridium for the cubic phase.

new, *a.* and *sb.*[2] Add: **A.** *adj.* **2. a.** *a new one* (*spec.* an anecdote or a joke; also, a circumstance not previously encountered). Usu. const. *on* (a person). orig. *U.S.*

1887 *Lantern* (New Orleans) 17 Dec. 2/3 Isn't this a new one on you, Messrs. Police? **1900** Ade *Fables in Slang* 74 The Pew-Holders didn't ever admit among themselves that the Preacher had rung in some New Ones. **1930** D. L. Sayers *Strong Poison* xx. 256 'I warn you..that you still have to establish evidence as to means and opportunity.' 'I know that. Tell us a new one.' **1931** T. H. Dey *Leaves from Bookmaker's Bk.* iii. 72 Charles Austin, Wilkie Bard and Ernie Lotinga too, are excellent private raconteurs and George Robey has always got a 'new one'. **1939** C. Isherwood *Goodbye to Berlin* 72 That's a new one on me... I never suspected the boy of having a mind at all. **1940** J. Cary *Charley is My Darling* xv. 77 Ginger reflects a moment and says then: 'It's a new one on me', meaning that he has not yet broken the law. **1952** 'M. Innes' *Private View* iii. 45 He said there was blood pouring through his ceiling. That was a new one on the station sergeant. **1971** 'A. Gilbert' *Tenant for Tomb* vi. 93 Her brother?.. That's a new one on me.

4. (Further examples.) *new breed*: see BREED *sb.* 2 c in Dict. and Suppl.; *new order*, a new regime or government; *spec.* (cf. G. *die neue Ordnung*), Hitler's plan for the reconstitution of the States of Europe on the basis of a National-Socialist regime. Of a coinage: replacing a former monetary unit, e.g. *new franc, new penny.*

1842 New order [see ORDER *sb.* 16]. **1917** Kipling *Divers. Creatures* (1917) 333 He and his friends had helped the World a step nearer the Truth, the Dawn, and the New Order. **1936** *Discovery* Sept. 295/1 The higher motives of a new order. **1940** *Times* (Weekly ed.) 27 Nov. 16 Every effort of the Quisling Government to induce the Norwegian people to believe that the 'new order' in Norway means the real liberation of the people and the erection of a new and happier Norway, freed from party strife and class distinctions, seems to have fallen on very deaf ears. **1941, 1944** New order [see *CO-PROSPERITY SPHERE]. **1944** G. B. Shaw *Everybody's Political What's What?* viii. 67 There is much talk at present of a New Order to follow the war. **1960** *Whitaker's Almanack 1961* 864/2 The *New Franc*, worth 100 old francs, came into use on Jan. 1, 1960, in metropolitan France and Algeria. **1966** H. Yoxall *Fashion of Life* xxiv. 221 A *menu gastronomique* at twenty-five new francs. **1966** *New Statesman* 16 Dec. 896/3 The government has opted for a pound divided into

100 'new pennies'. 1969 *Times* 21 July p. i/3 In 1971 our coins will be ½p, 1p, 2p, 5p, 10p, and 50p, each new penny equalling 2·4 of our present pennies. 1972 'J. FRASER' *Death in Pheasant's Eye* xvi. 163 Game bird soup at twenty new pence a helping. 1973 D. LEES *Zodiac* 47 I've won twenty-five thousand new francs—it's a fortune. 1973 *Listener* 22 Feb. 258/3 If the Nazis considered it worth their while to try to subvert in their propaganda, surely we should be at least as discerning in our exposure of the New Order.

b. *new bug* (slang), a new boy. Cf. *BUG sb.² 4 c. Also (rare) *new tick*. Cf. *NEW BOY, *NEW GIRL.
1900 FARMER *Public School Word-Bk.* 139 New-bug, a new boy. 1934 'G. ORWELL' *Burmese Days* v. 79 New-tick Flory does look rum. 1936, 1960 New bug [see *BUG sb.² 4 c]. 1971 'M. INNES' *Awkward Lie* v. 90 It mayn't be too bad an idea for the new bugs.

c. *new thing* (freq. with capital initials): something avant-garde or innovative; *spec.* a type of experimental jazz music of the 1960s dispensing with the normal harmonic and rhythmic framework. Also *attrib.* Hence *new thinger*. *slang* (orig. *U.S.*).
1928 [see *MUCKER v.² b]. 1962 *Down Beat* 12 Apr. 20 Coltrane's cohort Eric Dolphy, a member of that group of musicians who play what has been termed 'the new thing'. 1966 *New Statesman* 25 Mar. 438/3 'Pure feeling, pure expression, pure movement' (I quote from *Changes*)!, an extended manifesto of the New Thingers. 1966 *New Yorker* 3 Oct. 214 The newest rock 'n' roll groups are tuning in on the new thing. *Ibid.* 217 Charles Lloyd..wears new-thing clothes (an Army officer's jacket, tinted glasses, and bell-bottom trousers). 1967 *Melody Maker* 28 Jan. 15 Near the end, Dolphy had adopted a determinedly 'new thing' attitude, becoming more and more abstract in his playing, especially on alto. 1970 C. MAJOR *Dict. Afro-Amer. Slang* 84 *New thing*, in jazz, an aggressive and original attitude and feeling;..in black writing, the 'new thing' trend is best indicated through.. *The New Black Poetry.*

5. a. *the New Humanism*, in the U.S., a school of cultural thought based on the pragmatic philosophy of Dewey and others and emphasizing man's superiority to the natural order through the use of his reason; so *New Humanist*, a proponent of the New Humanism; *the New Kingdom*, a name given collectively to the Eighteenth, Nineteenth, and Twentieth Dynasties, which ruled Egypt from the sixteenth to the eleventh centuries B.C.; *the new mathematics*, a system of teaching mathematics to younger children in which an emphasis is laid on investigation and discovery on their part and topics are included that are not in the traditional school curriculum (as set theory, symbolic logic, and number systems); usu. abbreviated to *the new maths* (U.S. *math*); also without *the*; *the new psychology*, a term denoting new and major fields of psychological investigation, such as experimental psychology in the 19th century and esp. those theories in the 20th century that recognize the irrational and unconscious motivations of human behaviour.
1899 W. JAMES *Talks to Teachers* ii. 20 In the light of some of the expectations that are abroad concerning the 'new psychology', it is instructive to read the unusually candid confession of its founder Wundt. 1920 A. G. TANSLEY (*title*) The new psychology. *Ibid.* i. 9 Before we consider the developments of the New Psychology we must first glance at the causes of this failure of the older psychology... Not very many years ago the subject-matter of psychology was almost entirely limited to what is called the 'content of consciousness'. 1928 N. FOERSTER *Amer. Criticism* v. 236 A better way to consider the reconstruction proposed by the new humanism would be to examine its fundamental assumption. 1928 C. DAWSON *Age of Gods* viii. 173 The Late Minoan [corresponds] to the New Kingdom [in Egypt]. 1930 *Proc. Brit. Acad.* XVI. 414 The 'new psychology' with its perhaps exaggerated stress on the hidden roots of our conscious convictions and purposes in the depths of our unconscious mind, had not yet brought all reasoning into suspicion of being merely the 'rationalization' of irrational impulses. 1930 C. H. GRATTAN *Critique of Humanism* 6 In attacking the New Humanism, then, I am not casting aspersions upon the attitude which more than any other will lead to the good life. 1930 K. BURKE in *Ibid.* 169 The men whom the New Humanists in America recognize as their colleagues in France are advocates of Catholicism. 1942 S. R. K. GLANVILLE *Legacy of Egypt* 105 His tomb..bore witness to a period when the art of the New Kingdom had..reached its highest point. 1957 *Antiquity & Survival* II. 122/1 One of the raids carried out by one of the New Kingdom Pharaohs along the Palestinian coast. 1958 *Time* 3 Feb. 48/2 (*heading*) The new mathematics. 1958 I. ADLER (*title*) The new mathematics. [*In the text referred to as* modern mathematics.] 1960 *Math. Teaching* July 21 The self-styled Public Relations Officer for the New Mathematics..declared himself the herald of New Mathematics. 1961 A. H. GARDINER *Egypt of the Pharaohs* iii. 40 It is a matter of some surprise that the much less pleasing sandstone should have supplanted it [*sc.* limestone] from the New Kingdom (c. 1500 B.C.) onwards. 1963 J. S. BRUNER in Z. P. Dienes *Exper. Study of Math.-Learning* p. xii, In recent years there has been much misguided debate about the introduction of 'the new mathematics' into the schools. 1965 R. WELLEK *Hist. Mod. Criticism* IV. 59 His theory of commonplaces..is practically the same as that of the American new humanists. 1966 MEYER & HANLON (*title*) Fun with the new math. 1967 *Punch* 8 Mar. 332/1 If we realised..how much the New Maths derived from a foreign professor attending to the babbling of infants.. our suspicions would be even deeper and darker. 1971 R. A. PARKER in J. R. Harris *Legacy of Egypt* (ed. 2) i. 22 A papyrus known as the Turin Canon..listed the kings of Egypt from the earliest dynasties down to the end of the Hyksos period, that is just prior to the New Kingdom. 1973 *Sci. Amer.* Apr. 101/1 The triumph of the 'new math' in the elementary schools of America during the past decade. 1973 H. WHITNEY in E. Choat *Preschool & Primary Math.* 9 After centuries with little change in the mathematics curriculum in schools, we find ourselves in an era of 'new maths', typified by the acquisition of concepts.

b. *new entry*: a recruit; *collect.*, persons who have recently qualified or become eligible to do something. Also *transf.* and *attrib.*
1919 W. LANG *Sea Lawyer's Log* 5 Approaching him with diffidence our spokesman modestly announced us as 'new entries'—the Navy does not deal in 'recruits'. 1958 *Listener* 20 Nov. 812/1 If the electoral behaviour of the new entry [of voters] differs from that of those who have joined the great majority, there will be a swing one way or the other in consequence. 1962 *Economist* 18 Aug. 623/3 Competition seems to have prevented prices from ever going near their new-entry ceiling.

d. In names of inhabitants of countries, provinces, etc. whose names include the word *New*. Cf. NEW ENGLANDER, NEWFOUNDLANDER, *NEW MEXICAN, NEW ZEALANDER.
1874 C. M. YONGE *Life J. C. Patteson* I. viii. 214 The little New Caledonian remained at Taurarua. 1890 W. A. FOSTER in R. Reid *Canadian Style* (1973) ii. 40 All we need is a sentiment which shall make us feel not as Ontarians or Quebeckers, Nova Scotians or New Brunswickers, but as Canadians proud of our country as a whole. 1911 *Encycl. Brit.* XIX. 469/1 Many New Caledonians having black skins and woolly hair with Polynesian superiority of limb. 1957 *Ibid.* XVI. 696/2 The newcomers spoke languages of the Austronesian..family, which were in time adopted by most of the coastal New Guineans. 1973 *Country Life* 13 Sept. 745/1 Shell jewellery still valued for its magical powers by the New Guineans.

6. Used of colour names.
1897 *Sears, Roebuck Catal.* 360/3 Colors for Artists... Neutral Tint, New Blue, Olive Lake. *Ibid.* 361/1 Water Colors... Dark Green, New Rose, Flesh, New Violet. 1927 [see *DAWN sb. 1 b]. 1948 F. A. STAPLES *Watercolor Painting* iv. 49 New Blue, bright blue transparent.

f. Recently formed; *spec.* (see quot. 1958): said of deposits of ice or snow, esp. in polar regions.
1860 *Jrnl. R. Dublin Soc.* Jan. 374, I have before stated that we were frozen in on the 7th of September, 1857; the new ice then forming around us was of specific gravity 1·0235 (30°). 1918 FINCH & HAWKS *Water in Nature* vi. 126 Nansen has given us particulars of thickness attained by the ice of the Polar seas... 'This formation of new ice on the underside was owing to the layer of fresh water which, by reason of the surface thaw on the ice, now floated above the cold, salt water.' 1935 *Handbk. Weather, Currents & Ice, for Seamen* (Meteorol. Office) vii. 102 Between the Arctic Pack and the fast ice is a moving belt consisting partly of new ice and partly of broken ice from the pack and from the fast ice. 1958 ARMSTRONG & ROBERTS *Illustr. Ice Gloss.* 94 *New snow*, a recent snow deposit in which the original form of the ice crystals can be recognized; usually the daily new snowfall, measured in the morning. 1966 T. ARMSTRONG et al. *Illustr. Gloss. Snow & Ice* 28 *New ice*, a general term for floating ice recently formed. It includes frazil ice, grease ice, slush, shuga, ice rind, milas and pancake ice.

8. (Later examples.) Cf. *New Australian* (*AUSTRALIAN sb. 2 b).
1807 SOUTHEY *Lett. from England* II. xxix. 27 The heretical sects in this country... form a curious list! Arminians, Socinians, Baxterians, Presbyterians, New Americans, Sabellians, [etc.]. 1907 H. A. KENNEDY (*title*) New Canada and the New Canadians. 1919 *Ladies' Home Jrnl.* Sept. 35 The New Americans. 1939 J. M. GIBBON *Canadian Mosaic* ix. 231 Anthology of verse written by New Canadians. 1940 *Chatelaine* Jan. 40/1 Native-born and new Canadians work together for the nation's good. 1965 *Listener* 10 June 860/2 The idealist who saw his vision of the New Man shattered by harsh reality. 1970 *Guardian* 14 Apr. 11/5 The machines..are obviously the New Men, the ultimate revolutionaries. 1970 *Toronto Daily Star* 24 Sept. 7/6 If New Canadians prefer to learn English from Eaton's or Simpson's catalogue, there must be something wrong?

d. *spec. new rich*: in recent use, common as a translation of the French *nouveaux riches*, persons who have recently acquired wealth. Also as *adj.* Hence *new poor*, recently impoverished persons; *new money*, a fortune recently acquired, funds recently raised; so by metonymy, the new rich.
1886 *Harper's Mag.* Oct. 795/2 There are..the sons of the 'new rich' who are like men drunk with new wine. 1920 *Punch* 6 Oct. 279 *Exhausted War Profiteer.* 'Deer forests for the 'idle rich' be blowed! The 'new poor' can 'ave 'em for me.' 1923 'B. M. BOWER' *Parowan Bonanza* xiii. 157 You've never seen me look New-rich, have you, Bill? 1926 A. BENNETT *Lord Raingo* I. xxxv. 165 He had demonstrated publicly..that he belonged to the type of the new rich. 1942 H. C. BAILEY *Dead Man's Shoes* x. 46 Not a drop of fizz for you. I am the new poor and proud of the title. 1958 *New Statesman* 23 Aug. 222/2 How typical is his new-rich business man, seen leaving an elaborate Tudoresque house..who seemed genuinely to feel no animosity either towards the class to which his father, a jobbing gardener, had belonged or towards the class in which, economically at least, he now finds himself. 1961 'W. HAGGARD' *Arena* iii. 25 They had most of the new money: all the new men used them, the takeover boys, the property men. 1967 M. PROCTER *Exercise Hoodwink* xvii. 119 He moves with a pricey crowd, though I wouldn't call 'em classy. Small business people. New money. 1969 *Triumph* (U.S.) Mar. 25/1 The new poor, the misplaced workers, who are a by-product of our technology. 1970 *Daily Tel.* 2 June 20/3 'New money' raised in the United Kingdom during May by the issue of marketable securities was £32 million. 1973 R. LUDLUM *Matlock Paper* xvi. 139 The blooded first families..migrated just a little west to avoid the new rich. 1975 *Daily Tel.* 30 Aug. 13/4 The amount of new money raised in August at £120·1 million was the lowest since last January.

9. (Later examples.)
1903 MRS. H. WARD *Lady Rose's Daughter* xiv. 230 He was a good deal of a politician, himself a 'new man', and on the side of 'new men'. 1936 M. MITCHELL *Gone with Wind* iii. 52 Gerald was a 'new man', despite his nearly ten years' residence. 1943 D. W. BROGAN *English People* iii. 81 Lord Reading was the first Marquess of Reading, a 'new man'. 1962 *Spectator* 30 Mar. 392 It strikes fear, not into the manual worker so much as into the hearts of the new men.

10. a. *new-old* (later examples).
1926 D. H. LAWRENCE *Plumed Serpent* xvii. 279 They seized upon the new-old thrill, with a certain fear, and joy, and relief. 1932 H. CRANE *Let.* 31 Mar. (1965) 405 An environment not half so strange and distractingly new-old curious as this. 1961 *Guardian* 14 June 10/6 These new-old problems that face our daughters.

b. *new-face* = *modern-face(d)* adjs. (*MODERN a. 2 g); *new time*: in the Stock Exchange: of dealings, having the settlement postponed to the next settling-day; of prices, quoted for the next settling-day before the previous settlement is completed; also in more general contexts; *new wheat disease*: a pyelonephritis affecting young poultry, possibly caused by a virus; also called *blue comb (disease)* (*BLUE a. 13) or *pullet disease* (*PULLET 3).
Some of these expressions also occur in non-*attrib.* contexts.
1900 H. HART *Cent. Typogr.* 120 These are the first examples of what are called nowadays 'new-face' types. a 1910 'MARK TWAIN' *Autobiogr.* (1924) I. 198 Any other old-time or new-time palace on the continent of Europe. 1912 *Century Mag.* Jan. 476 Open Letters... On the New-time Negro. 1922 New-time [see *CARRY-OVER]. 1927 *Daily Express* 27 Sept. 10/1 The price for 'new time' was about 15s. 6d., compared with a making up price of 14s. *Ibid.* 10/3 The 'new time' price at one time touched the new record of 4½. 1950 New wheat disease [see *blue comb (disease)* s.v. *BLUE a. 13]. 1957 L. ROBINSON *Mod. Poultry Husbandry* (ed. 4) xx. 677 Since outbreaks usually occur in late summer and autumn, the disease has been called 'New Wheat disease'. There is no evidence, however, that new wheat is responsible. 1964 *Financial Times* 25 Feb. 17/1 First Dealings... 'New time' dealings may take place two business days earlier.

d. *new style*: in Chronology, see STYLE sb. 27; gen., forming attributive compounds.
1914 'I. HAY' *Lighter Side School Life* vii. 193 The new-style parent breaks right away from tradition—kicks over the traces, in fact. 1937 B. H. L. HART *Europe in Arms* iii. 26 These new-style formations were not shown to the foreign officers and military publicists. 1961 *Times* 18 May 19/4 The new-style Arab. 1965 *New Statesman* 7 May 707/2 Such a department, like the rest of a new-style Transport House, would have to have a long-term career structure.

B. *absol.* or as *sb.* **5.** A naval cadet during his first term in a training-ship.
1909 J. R. WARE *Passing Eng.* 181/2 *New* (*Britannia* training ship), fresh arrival, last addition. Used in the plural. 1914 'BARTIMEUS' *Naval Occasions* ix. 63 The path of the 'New' in those days was by no means strewn with roses. 1953 J. MASEFIELD *Conway* (rev. ed.) IV. 224 The 'News' very rarely ventured into the upper room. 1962 GRANVILLE *Dict. Sailors' Slang* 81/1 *New!*, HMS *Britannia*'s equivalent of the public school's 'Fag!'

new, *adv.* Add: **3. a.** *new-baked, -bought, -crowned, -cut, -dated, -fulfilled, -gnarled, -nurtured, -scored, -skeined, -washed.*
1793 W. BLAKE *Visions Daughters Albion* in *Compl. Writings* (1972) 191 The new wash'd lamb ting'd with the village smoke, & the bright swan. 1821 BYRON *Don Juan* v. clvi. 289 The new-bought virgin, made her blush and shake. 1865 G. M. HOPKINS *Poems* (1967) 21 New-dated from the terms that reappear, More sweet-familiar grows my love to thee. 1877 —— *Ibid.* 68, I hear the lark ascend, His rash-fresh re-winded new-skeinèd score In crisps of curl off wild winch whirl. 1889 W. B. YEATS *Wanderings of Oisin* III. 35 Whiter than new-washed fleece. 1909 KIPLING *Actions & Reactions* 51 Over their heads in the branches Of their new-bought ancient trees. 1913 —— *Songs from Books* 43 My new-cut ashlar takes the light. 1918 E. SITWELL *Clown's Houses* 14 In new-washed air. 1922 JOYCE *Ulysses* 155 Steam of newbaked jampuffs. 1934 E. BLUNDEN *Choice or Chance* 37 Yet one I knew who most of all seemed sent Among earth's flowers for new-fulfilled content. 1935 W. EMPSON *Poems* 27 With plump or splash on the new-nurtured field To Reason's arm they proper homage yield. 1942 *R.A.F. Jrnl.* 3 Oct. 7 The wheels kicking up a frenzied cloud of new-cut grass. 1943 L. B. LYON *Evening in Stepney* 20 A script of blackening girders, New-scored by old despair. 1948 E. POUND *Pisan Cantos* (1949) lxxx. 107 Nor is the white bud Time's inquisitor Probing to know if its new-gnarled root Twists from York's head or belly of Lancaster. a 1963 C. S. LEWIS *Poems* (1964) 69 A seminal breeze from the far side Calls to their new-crowned race.

b. *new-awakened, -kerned, -landed.*
1851 H. MELVILLE *Moby Dick* I. xxiii. 169 One Bulkington was spoken of, a new-landed mariner, encountered in New Bedford. 1908 HARDY *Dynasts* III. VII. ii. 486 The peaceful produce of the grange,..new-kerned apples, hairy gooseberries green. 1918 E. SITWELL *Clown's Houses* 7 The new-awakened flower-strange hair.

4. *new-awakened, -crushed, -desired.*
1878 O. WILDE *Ravenna* 13 Thou hast not drunk this wine From grapes new-crushed. 1917 D. H. LAWRENCE *Look! We have come Through!* 131 Now here was I, new-awakened. 1959 BLUNDEN *Hong Kong House*, Except when they, of one tree tired Into another new-desired Over the lawn and scattered playthings chose to glide.

6. b. With active forms of intransitive verbs, as *new-dapple, new-nestle.*
a 1885 G. M. HOPKINS *Poems* (1967) 193 As sure as what is most sure, sure as that spring primroses Shall new-dapple next year. a 1889 —— *Ibid.* 185 Say it is áshboughs: whether on a December day and furled Fast ór they in clammyish lashtender combs creep Apart wide and new-nestle at heaven most high.

7. *new-dawning.*
1820 J. TRUMBULL *Poetical Wks.* II. 108 Prove to the world in these new-dawning skies, What genius kindles and what arts arise.

b. With pres. pples. of intransitive verbs that are const. with predicative adverbs, as *new-looking, -seeming.*
1905 J. JOYCE *Let.* c 12 Oct. (1966) II. 121 A good new-looking..suit. 1928 D. H. LAWRENCE *Lady Chatterley* xv. 276 The man's face, that was smooth and new-looking with love. 1951 S. SPENDER *World within World* 87 Raindrops..through which the sun..would gleam with a new-seeming whiteness. 1969 *Sears Catal.* Spring/Summer 15 Enamel colors oven-baked to stay new-looking longer.

New Aca·demy. Also **new Academy.** [ACADEMY 2.] A name given to schools of philosophy founded in Athens by the successors of Plato as Heads of the Academy and developing some of its principles, *spec.* (*a*) that founded by Arcesilaus (316/15–241/0 B.C.) in the third century B.C. (now more usually called **Middle Academy*); (*b*) (more usually) that founded in the second century B.C. by Carneades of Cyrene (214/13–129/8 B.C.) and developing the mainly sceptical philosophy of Arcesilaus. Also *attrib.* So **New Acade·mic** *sb.* and *a.*
1659 T. STANLEY tr. *Sextus Empiricus's Pyrr. Hyp.* in *Hist. Philos.* III. IV. 33 Those of the new Academy, though they say all things are incomprehensible, differ from the Scepticks... Now the new Academicks, before Phantasie which is simply credible, prefer that which is credible and circumcurrent. 1702 S. PARKER tr. *Cicero's De Finibus* v. 280 For all that, I'll venture to dissuade you from following the new Academicks, and bespeak you in behalf of the Old ones. 1744 W. GUTHRIE tr. *Cicero's Morals* p. v, It was by Sincerity alone that the New Academy could hope to recommend that Moderation which was their peculiar Distinction. *Ibid.*, Thus, the New Academy, by admitting a Degree of Comprehension, preserv'd a Principle of Agency. *Ibid.* 321 This very Cicero is of the New Academy, tho' it appears to me to be the old one, at least we may look upon Plato to have been of the old. 1853 C. D. YONGE tr. *Cicero's Academic Questions* p. xxix, Arcesilaus, or Arcesilas, flourished about B.C. 280... On the death of Crantor he succeeded to the chair of the Academy, in the doctrines of which he made so many innovations that he is called the founder of the New Academy. 1856 W. A. BUTLER *Lect. on Hist. of Ancient Philosophy* II. vi. i. 318 The sceptical side of Platonism represented by the New Academy, the doctrinal by the Neo-platonists. 1874 J. S. REID *Cicero's Academica* p. lvi, Cicero..merely attaches himself to the New Academy. 1902 BALDWIN *Dict. Philos. & Psychol.* II. 496/2 The New Academy, under Philo of Larissa (about 100 B.C.) and Antiochus, turned to dogmatism and eclecticism. 1946 F. C. COPLESTON *Hist. Philos.* I. xxxvii. 414 The founder of the Third or New Academy was Carneades of Cyrene. 1967 P. MERLAN in *Camb. Hist. Later Greek & Early Medieval Philos.* 61 Plutarch even has kind words for the scepticism of the New Academy.

Newar (niwā·ɪ). A member of one of the castes of Nepal, of Mongol or partly Mongol origin. Also *attrib.* So **Newa·ri**, a language of partly Tibetan origin spoken by the Newars.
1811 [see *MAGAR]. 1819 F. HAMILTON *Acct. Kingdom Nepal* I. 1, 49 They afterwards settled in the valley of Nepal, and are the people now called Newars. 1874 [see *BAEL, BEL]. 1877 D. WRIGHT *Hist. Nepal* App. vi. 306 (*heading*) Newari songs. 1893 A. L. WADDELL in *Indian Antiquary* XXII. 293 The Newārs are the aborigines of Nepāl Proper, that is, of the valley in which the present capital Khatmāndū stands. ... Latin and Newari. 1971 J. PEMBLE *Invasion of Nepal* i. 6 Jaya Yaksha Malla..patronized native letters and raised Newari to the status of an official language. 1972 W. B. LOCKWOOD *Panorama Indo-European Lang.* 208 Spoken in 1952–4 by close on 400,000 persons in Nepal, Newari is the only Himalayan language of the Tibeto-Burman group to have developed a considerable literature. The Newars have evolved a high degree of material culture. 1973 *Times* 14 Apr. (Nepal Suppl.) p. ii/5 The three towns of Katmandu, Patan and Bhadgaon are the creation of Newars. 1974 M. PEISSEL *Great Himalayan Passage* xiv. 206 The Newars do not like anyone to die in their houses, so they turn the dying outside on mats. *Ibid.*, ... the windows of traditional Newar houses have to be entirely closed by wooden lattices, as these are believed to stop evil spirits entering.

Newark (niū·ǝɹk, nū·ǝɹk). The name of a city in New Jersey, U.S.A., used *attrib.* in or *Comb.* in various special collocations (see quots.).
1787 in T. F. De Voe *Market Book* (1862) I. 181 (New) Jersey, a Burlington ham and Newark cyder. 1804 A. F. M. WILLICH *Domestic Encycl.* III. 111 *Newark Pippin*... It is said to have been imported from France. 1858 S. F. AVERY *Harp of Thousand Strings* 189 A well-known sporting character of the city had made a bet that he would drive his 'Newark waggon' (a light carriage with two horses) the whole length of Broadway. 1890 R. TOMES *Bazaar Bk. Decorum* 136 He will be sure to detect the Newark cider in your Champagne bottle. 1910 A. E. BOSTWICK *Amer. Public Library* 43 In the type of two-card system known as the 'Newark' system,..an additional record of the date is made on a flap attached to the inside of the book. 1961 T. LANDAU *Encycl. Librarianship* (ed. 2) 261/1 *Newark charging system*, the method of recording book issues which is most widely used in America.

New Art, new art. = *art nouveau* (*ART sb. VI. c).
1903 W. B. YEATS *Let.* 14 May in *Florence Farr, Shaw, Yeats* (1946) 58, I have had a letter from the editor of the 'Daily News' asking permission to interview me on the Theatre and the New Art. 1906 M. H. BAILLIE-SCOTT *Houses & Gardens* i. 13 A bizarre striving after originality and eccentricity of design,..posing as the 'new art'. 1909 *Cottage Furniture* (Studio) 78 ... The 'new art' overmantel smothered in rococo photograph frames. 1909 H. G. WELLS *Tono-Bungay* III. i. 166 Beautiful jam-pots! Get one of these new art chaps to design all the things they make ugly now. 1921 L. WEAVER *House & its Equipment* 53 The superficial cleverness of the New Art movement. 1914 C. MACKENZIE *Sinister Street* II. ii. vi. 617 New Art flower vases. 1936 G. GREENE *Brighton Rock* i. iii. 46 He touched a little buzzer, the New Art doors opened. 1969 J. GLOAG *Short Dict. Furniture* (ed. 2) 464 New Art was characterized by the free use of naturalistic motifs: chairs, tables, cabinets and bedsteads were contorted by writhing plants, exuberant blossoms, and intricate arabesques, while surfaces were punctuated by inserted patches of hammered copper and coloured enamel, and pierced by heart-shaped apertures.

new ball. *Cricket.* A previously unused ball, such as is brought into use at the beginning of an innings or after a prescribed number of overs; freq. used *attrib.* (with hyphen) of an opening bowler or of the type of bowling (usu. fast) in which the new ball is employed. Also *fig.*
1956 K. ALSTON *Test Commentary* 12 The most devastating pair of new-ball bowlers in the world. 1960 J. FINGLETON *Four Chukkas to Australia* vi. 65 We saw a magnificent piece of new-ball bowling. 1963 *Times Lit. Suppl.* 18 Jan. 37/3 One has the impression (in reading a novel) of being for the new ball to become available. 1971 *Times* 15 Feb. 8/1 Edrich and Luckhurst..batted with the greatest of ease against some modest new-ball bowling. 1975 *Cricketer* May 23/2 It was the finest display of new-ball bowling I ever saw.

newberyite (niū·bĕri,əit). *Min.* [ad. G. *newberyit* (G. vom Rath (at the suggestion of G. Ulrich) 1879, in *Sitzungsber. der niederrhein. Ges. in Bonn* XXXVI. 8), f. the name of J. Cosmo *Newbery* (1838–95), Australian mineralogist) + *-it* -ITE¹.] A hydrated magnesium acid phosphate, $MgH(PO_4).3H_2O$, which is found as colourless orthorhombic crystals.
1879 *Mineral. Mag.* III. 108 Newberyite... Found at Skipton Caves, Victoria, with Hannayite. 1928 *Amer. Mineralogist* XIII. 397 (*heading*) Newberyite and other phosphates from Ascension Island. 1957 *Mineral. Mag.* XXXI. 456 Newberyite, together with pyrrhotine and vivianite, was found in cracks of mammoth tusk buried in clay. 1966 *Amer. Mineralogist* LI. 1764 Subsequent submergence and removal of the source of ammonia permitted the replacement of struvite by newberyite in a subaqueous environment.

new-born, *ppl. a.* Add: Also **newborn, new born. 3.** *absol.* as *sb.* A new-born individual; chiefly in *the new-born* (usu. with pl. or collective sense).
1879 A. JACOBI in A. H. Buck *Treat. Hygiene & Public Health* I. i. 75 Landau's essay on 'The Melæna of the Newborn, with Notices on the Obliteration of the Foetal Blood-Vessels', proves that..but little attention has been paid to the subject. 1907 *Jrnl. Amer. Med. Assoc.* 31 Aug. 775/1 Occurrences of vaginal hemorrhages in the new-born are..rare. 1929 R. HURWITZ tr. *Bernfeld's Psychol. of Infant* 2 The new-born is not capable of maintaining life for itself. 1937 *Amer. Jrnl. Dis. Child.* LIV. 1215 Tetany of the new-born..may occur at any time during the neonatal period. 1968 PASSMORE & ROBSON *Compan. Med. Stud.* I. xxxix. 6/2 The distribution and amount of brown fat in the newborn varies among different species. 1974 *Nature* 6 Dec. 514/3 A doubtful study..indicating that newborns can orient visually to a voice source in space.

new boy. Also **new-boy, newboy.** [NEW *a.* 8.] A schoolboy during his first term(s) at a school, esp. one at a preparatory school or English public school. Also *transf.*, a (young) man newly come into a given set of circumstances. Cf. *NEW *a.* 4 b, *NEW GIRL.
1847 DICKENS *Dombey* (1848) xli. 410 Here is the table upon which he sat forlorn and strange, the 'new boy' of the school. 1847 *Punch* 25 Dec. 255 (*caption*) Here's a New Boy, Johnny Russell. Now you see that nobody bullies him. 1905 R. BROOKE *Let.* 12 Mar. (1968) 18 Such brave souls can be content with the admiration of one another, and of the new boys. 1935 H. NICOLSON *Diary* 21 Nov. (1966) 229 He says the rather dramatic circumstances of my election may arouse some jealousy in that old hen the H. of C. I must do the new-boy for six months at least. 1948 PARTRIDGE *Dict. Forces' Slang* 124 *New boy*, a new member of a ship's ward-room. 1953 A. HUXLEY *Let.* 5 Apr. (1969) 667 He had been part of my Order of Things for almost fifty years, ever since we first met as newboys at our preparatory school in the autumn of 1903. 1970 E. PACE *Saberlegs* (1971) x. 87 He was a new boy here, a novice 'philanthropoid', as people in the foundation business called themselves. 1973 *Washington Post* 13 Jan. A. 22/2 Neither the anxious war critic Mr. Hughes nor the cautious new boy Mr. Clements was familiar with the unequivocal policy statement..made..just two years ago. 1974 'J. LE CARRÉ' *Tinker, Tailor* i. 10 Roach was a new boy... Thursgood's was his second prep school.

new broom. [In allusion to the prov. *new brooms sweep clean* (SWEEP *v.* 13 b).] One newly appointed to a position who vigorously makes changes in personnel or procedures; one who effects fundamental or numerous alterations. Also (with hyphen) *attrib.* and as *v. trans.* and *intr.*
[1776 G. COLMAN (*title*) New brooms! An occasional prelude. *Ibid.* 15, I am glad he is gone. *Catcall.* Glad! *Phelim.* To be sure I am glad. *New Brooms*, you know.] 1855 [see BROOM *sb.* 3 b]. 1925 [see *CLEAN *v.* 6 c]. a 1930 D. H. LAWRENCE *Phoenix II* (1968) 114 The Reverend Mr. Flewitt is newly arrived on the circuit, and wants to sweep the chapel very clean of sin, being a new broom. 1938 *Ann. Reg. 1937* 93 The War Ministry, where Mr. Hore-Belisha was proving himself a veritable 'new broom'. 1951 N. MARSH *Opening Night* v. 114 Our little stranger.. seems to be new-brooming away. 1963 *Economist* 10 Aug. 496/3 Lord Hill of Luton, new-brooming his way through the Independent Television Authority. *Ibid.* 7 Sept. 809/1 The new government's new-broom mentality. 1969 *Listener* 14 Aug. 205/1 This seems to me exactly what is happening at the BBC, new brooms bustling about tidying up, rationalising and sorting out on very ill-thought-out principles.

ne·wbuilding. [tr. Da. *nybygning* new building, new ship: NEW *adv.* 8.] A newly constructed ship; the construction of ships.
1948 *N.Y. Maritime Register* 12 May 3/2 The newbuildings include a number of large fast cargo liners. 1957 *Times* 19 Dec. 16/6 It is the intention of that company to arrange a time charter on its newbuilding. 1968 *Marine West* Mar.–Apr. 2 The extra thick plates and oversize scantlings once used in tanker new buildings. 1968 *Marine Digest* 13 July 23 There is a steady trickle these days of newbuilding contracts for sale.

Newcastle¹ (niū·kɑs'l; also with main stress on second syllable). [Name of a city (in full *Newcastle upon Tyne*) in the north of England.] **1.** Phr. *to carry coals to Newcastle*: see COAL *sb.* 13.
2. Used *attrib.* in various special collocations: **Newcastle brown,** a strong brown ale; **Newcastle disease,** an infectious, often fatal, virus disease of poultry, first recorded in Britain near Newcastle in 1927 and characterized by lethargy followed by paralysis and difficulty in breathing; also called **fowl pest*; **Newcastle glass,** a type of colour-free glass manufactured in Newcastle; also *ellipt.*; **Newcastle pottery,** a type of coarse pottery manufactured around Newcastle.
1972 J. WAINWRIGHT *Requiem for Loser* ii. 43 A beer?.. There's a can o' Newcastle Brown in the fridge. 1973 'J. PATRICK' *Glasgow Gang Observed* iii. 29 Pints of lager, heavy beer and bottles of Newcastle Brown were ordered. 1927 T. M. DOYLE in *Jrnl. Compar. Path. & Therapeutics* XL. 144 In order to facilitate description we propose to refer to it [*sc.* a virus disease of fowls] as 'Newcastle disease'. 1938 *Poultry Keepers' Year Book* v. 178 Newcastle Disease is notifiable to the Ministry of Agriculture... Recognisable by dribbling from beak and sudden death of many birds at the same time. 1955 *Sci. News Let.* 21 May 326/3 Newcastle disease virus, cause of a frequently fatal epidemic in poultry..can also cause eye inflammation in humans. 1968 *New Scientist* 5 Dec. 561/1 The infection—the dreaded Newcastle disease—is normally so lethal to chickens that a strict quarantine is imposed on imported poultry. 1972 *Daily Colonist* (Victoria, B.C.) 9 Jan. 35/5 Vaccination of hens against Newcastle disease is being urged by the B.C. Poultry Commissioner. 1767 in *Sc. Nat. Dict.* (1968) VII. 17/1 All the windows to be glazed with Newcastle crown glass, bedded and back-pottied, and all hung with paces. 1779 COWPER *Let.* 26 May in *Corr.* (1904) I. 153, I shall be obliged..if you will inquire at a glass-manufacturer's how he sells his Newcastle glass, such as is used for frames and hothouses. 1883 J. W. MOLLETT *Illustr. Dict. Art & Archæol.* 225/1 *Newcastle glass*, a crown glass, held the best for windows from 1728 to 1830... It was of an ash colour,..and frequently

warped. 1923 H. J. POWELL *Glass-Making in Pottery* vii. 93 In later years John Tyzack's warehouse for Newcastle glass near the Old Swan Stairs was well known. In 1691 Newcastle cut window-glass was sold at 13s. per 100 feet. 1965 P. M. HUBBARD *Hive of Glass* i. 9 It was..[a] quite faultless Newcastle light baluster..ten inches high, the rounded perfect bowl perched on a breathless series of knobs. 1972 *Country Life* 28 Dec. 1783/2 The great rarities in this sale were two Dutch engraved Newcastle glasses of the mid-18th century. 1874 L. W. *Eng. Pott. & Porc.* 40 The principal marks on Sunderland and Newcastle pottery are (stamped in the clay or printed in transfer). 1971 R. C. BELL *Tyneside Pottery* II. 95/2 The Willett Collection in the Brighton Museum contains two rare earthenware mugs marked 'Newcastle Pottery'.

Newcastle². The title of Henry Pelham-Clinton, fifth Duke of *Newcastle-under-Lyne* (1811–64), used as the name of a classical scholarship at Eton College, established by him in 1829. Also *attrib.*
1832 *Eton College Mag.* 25 June 7 It was on the second of April..that we went up for the last Newcastle scholarship. 1845 J. PATTESON *Let.* in C. M. Yonge *Life J. C. Patteson* (1874) I. ii. 46 Do not distress yourself about this unfortunate failure as to the Newcastle. 1875 H. C. MAXWELL-LYTE *Hist. Eton College* xix. 369 One Newcastle Scholar is elected annually after a competitive examination open to Oppidans and Collegers alike. 1879 C. M. YONGE *Magnum Bonum* II. xxiv. 484 But you did like Eton so, and you were going to get the Newcastle and the Prince Consort's Prize. 1884 E. W. HAMILTON *Diary* 5 Nov. (1972) II. 725 Dined at Dilke's—Chamberlain, Sir L. Playfair, Lord Advocate, L. Lawson, Romer, and Kennedy (the Newcastle Scholar of my day). 1922 S. LESLIE *Oppidan* vii. 83, I would rather get the Newcastle than make a century at Lord's. 1953 H. NICOLSON *Diary* 3 Jan. (1968) 236, I feel as if I had got a fourth prize in scripture when I should have liked the Newcastle. 1959 *Chambers's Encycl.* V. 422/1 The classical tradition in Eton education, strengthened by the institution of the Newcastle scholarship in 1829, remains strong to-day. 1975 *Times* 18 Oct. 7/5 My [*sc.* Harold Macmillan's] brother Daniel had won the Newcastle at Eton.

new-comer. (Further examples: used of things.)
1886 W. B. YEATS *Mosada* III. 8 Yonder a leaf Of apple blossom circles in the gloom, Floating from yon barred window. New-comer, Thou'rt welcome. 1930 *New Statesman* 27 Dec. 357/2 The Gum-trees or Eucalypts are newcomers to California. 1966 *B.B.C. Handbk.* 23 A newcomer to many schools is 'radio-vision'.

New Commonwealth. [*COMMONWEALTH 4 c.] *collect.* Those countries which have achieved self-government within the British Commonwealth since 1945, opp. the old Dominions; also used *attrib.* of persons from (or whose parents came from) such countries; a genteelism for persons considered 'non-white'.
1960 *New Commonwealth* Oct. 626/3 The Rt. Hon. Ian Macleod, M.P., Secretary of State for the Colonies, will be the guest-of-honour at the next New Commonwealth luncheon, to be held at Claridges. 1964 *Listener* 6 Feb. 219/1 A respected voice from the new Commonwealth, that of Professor Rajan of the Indian School of International Studies, declared that [etc.]. 1964 S. A. DE SMITH *New Commonwealth & its Constitutions* iv. 137 In a new Commonwealth..country there is a presumption against leaving the power of appointment exclusively in the hands of the executive. 1970 *Guardian* 10 Mar. 7/3 The number of 'new' Commonwealth immigrants admitted during the whole year fell by very nearly one-third. 1973 *Times* 4 Oct. 4/7 About 547,000 people whose parents were born in the New Commonwealth live in Greater London. 1975 *Times* 28 Feb. 3/2 The total New Commonwealth population in Britain. 1976 *Times* 27 Jan. 4/3 The total number of births [in Britain] to New Commonwealth and Pakistan mothers fell.

new-create, *v.* (Later examples.)
a 1930 D. H. LAWRENCE *Last Poems* (1932) 77 Then I have been dipped again in God, and new-created. 1952 R. CAMPBELL tr. *St. John of the Cross's Poems* 25 And have been new-created There where your mother first was violated.

New Criticism, new criticism. [NEW *a.* 5.] An approach to the analysis of literary texts, associated *spec.* with American critics who subscribed to the procedures outlined by John Crowe Ransom (see quots. 1941), which concentrates on the linguistic organization of a text with particular emphasis on irony, ambiguity, paradox, etc. So **New Critic; New Critical** *a.*
1941 J. C. RANSOM (*title*) The new criticism. *Ibid.* p. vii, He [*sc.* R. P. Blackmur] is nevertheless a 'new' critic in the sense of this book. *Ibid.* i. 3 Discussion of the new criticism must start with Mr. Richards. 1948 *Poetry* LXXIII. 153 The Father of the New Criticism is probably I. A. Richards. 1948 [see *CONCRETE *a.* and *sb.* A. 4 b]. 1951 R. P. BLACKMUR in *Hudson Rev.* III. iv. 501 The one thing the 'new critics' shared was skill in analysis. 1952 M. MCCARTHY *Groves of Academe* (1953) x. 212 The true attitude of Eliot, he suspected, was manifest in his disciples, who in all their voluminous New Criticism had given Joyce scarcely a word of exegesis. 1955 J. WAIN *Interpretations* 215 Poetic difficulty..is not identical with grammatical difficulty..nor is the 'New Critic'..simply the old notemaker writ large. 1956 A. WILSON *Anglo-Saxon Att.* I. iii. 59 Hardy was back from New York with some frightfully funny stories about the New Criticism boys. 1960–1 M. SPILKA in *Modern Fiction Studies* VI. iv. 285 The concept of the author is conspicuous, in New Critical thought, by its absence. 1963 P. WEST *Mod. Novel* III. ii. 291 Robert Penn Warren..one of the old New Critics. 1969 *Encycl. Brit.* VI. 781/2 In 1941 John Crowe Ransom coined the term 'new criticism' for what is in certain essential respects a return to the Renaissance rhetoricians' principle that a poem is an arrangement of words, to be apprehended in their interaction. 1973 *Observer* 8 Apr. 37/5 The so-called New Critics in America made a great deal of their capacity to submit poems or prose-passages to detailed, word-by-word analysis. 1973 *College English* XXXIV. 573 The pattern is the same as he moves from one caricature of a 'New Critical' assumption to the next.

new deal, new dealer: see *DEAL *sb.*² 4 d. So **New Dealing, New Dealish** *a.*, **New Dealism.**
1934 *Sun* (Baltimore) 23 Oct. 1/2 In many sections Republican candidates have gone New Dealish. 1936 *Rocky Mt. News* (Denver, Colorado) 7 May 10/3 Two towers of progressive strength, more 'New Dealish' than most Democrats, will be lost to the administration in the next Congress. 1939 in *Amer. Speech* (1941) XVI. 309/1 New Dealism..is nothing but the American form of Communism. 1962 R. B. FULLER *Epic Poem on Industrialization* 21 Lugubrious New Dealing Of a mixed pack Of bright new cards And dirty old ones. 1965 *Punch* 27 Oct. 593/1 Somehow President Johnson has managed to inject his countrymen with a new, and better, shot of New Dealism. 1965 *New Statesman* 29 Oct. 634/3 By that time Mr Johnson may have done all the New Dealing he thinks necessary. 1973 *New Yorker* 13 Jan. 56/2 Melvin Laird.. saved everyone's dignity by declaring that the name—Family Security System—sounded 'too New Dealish'.

Newdigate (niū·digĕt). The name of Sir Roger *Newdigate* (1719–1806), M.P. for Oxford University, used *attrib.* and *absol.* to designate an English verse prize founded by him at Oxford University in 1805, or the poetry associated with this prize.
1852 *Prospective Rev.* VIII. 523 These [*sc.* the mechanical parts of rhythm and metre] any industrious person will find in any collection of the Newdegate forms [ed. 1965 Newdigate poems]. 1856 C. M. YONGE *Daisy Chain* II. v. 383 You should try for the Newdigate Prize. *Ibid.* viii. 414, I am not going to have the Newdigate prizeman shown as brother to a scare-crow. 1860 J. A. SYMONDS *Let.* 15 June (1967) I. 245, I am so glad you are pleased about the Newdigate. 1929 LD. BIRKENHEAD *Hundred Best Eng. Ess.* 386 He subsequently entered Christ Church, Oxford, and won the Newdigate Prize. 1931 'G. TREVOR' *Murder at School* i. 12 So far he seemed to have done nothing in life except win the Newdigate. 1969 G. SMITH in A. Huxley *Lett.* 11, 1915... His Byronic poem on Glastonbury fails to gain the Newdigate Prize. 1975 *Times* 25 Aug. 5/7 John Buchan..won the Newdigate and was President of the Union.

New England. [See NEW ENGLANDER.] **a.** Used to denote a form of U.S. speech characteristic of New England, and *attrib.* of persons, produce, etc., native to New England; of mentality, idiom, etc., marked by the characteristics of New England.
1638 J. UNDERHILL in *Mass. Hist. Soc. Coll.* (1837) 3rd Ser. VI. 5 Let the clamor be quenched.., that New England men usurp over their wives. 1655 *Deeds Suffolk County, Mass.* (1883) II. 166 Thirty quintalls he recieued.. was New England fish. 1709 W. BYRD *Secret Diary* (1941) 13 Parson Ware sent to me for a pint of canary, he being sick of the gripes with the New England rum. 1715 S. SEWALL *Diary* (1882) III. 56 Gave Mr. Short's daughter a New-England shilling. 1787 M. CUTLER in Parker & Cutler *Life M. Cutler* (1888) I. 195 [You are] acquainted with the institution of a Company in the New England States by the name of the Ohio Company. 1839 *Southern Lit. Messenger* V. 112/2 Noah Webster..will ere long succeed in giving us a New England tongue which shall not be intelligible in Britain. 1842 DICKENS *Let.* 29 Apr. (1974) III. 217 A New England Poet buzzed about me on the Ohio, like a gigantic Bee. 1845 J. F. COOPER *Chainbearer* II. xiv. 466 The supercilious feeling of the New Englandman can very easily be traced to his origin in the mother country. 1850 W. C. FOWLER *Eng. Gram.* II. iii. 92 To pass over the local peculiarities of smaller districts, there are certain generic dialectal differences which characterize, 1. New England. 2. The Southern States. 3. The Western States. 1905 R. FRY *Let.* Jan. (1972) I. 228 The Philadelphia Quakeress..said 'You know I can't bear it because I've got a New England conscience.' 1917 T. S. ELIOT *Prufrock* 34 The barren New England hills. 1934 WEBSTER p. xlviii/1, Another variety of short *o* is the 'New England short *o*'. This is acoustically intermediate between ō (nōte) and ŭ (nŭt), being practically an ŭ sounded with rounded lips. 1935 A. C. BAUGH *Hist. Eng. Lang.* 446 In the English language spoken in America three major dialects can be distinguished: New England, Southern, and General American. 1951 *Language* XXVII. 425 The division into Northern, Midland, and Southern (instead of the older New England, General American, and Southern) does not come as a shock. 1952 S. KAUFFMANN *Philanderer* (1953) iii. 33 They had just been graduated from a small New England college. 1975 *Daily Colonist* (Victoria, B.C.) 26 Oct. 30/8 New England peachblow [glass] is made in one layer shading from red to white. 1975 P. ORGAN *House on Cheyne Walk* v. 31 We could stand a little New England uprightness here. One tires of the bohemian life.

b. In special collocations, as **New England aster,** a large Michaelmas daisy, *Aster novæ-angliæ*; **New England boiled dinner,** a **boiled dinner*, esp. one including corned beef; **New England mayflower,** a prostrate, evergreen shrub, *Epigæa repens*, belonging to the family Ericaceæ, native to the eastern half of North America, and better known as trailing arbutus; **New England theology,** a movement in American Congregationalism, also affecting other American Protestant bodies, which repudiated much Calvinist doctrine.
1814 J. BIGELOW *Florula Bostoniensis* 199 *Aster Novæ Angliæ.* New England aster... A tall, and very beautiful plant. Stem three feet high, brown, very hairy. 1931 W. N. CLUTE *Common Names of Plants* 140 The New England aster (*Aster Nova* [sic] *Angliæ*), which lingers long in the fields and fence corners, is further distinguished as last-rose-of-summer. 1968 PETERSON & MCKENNY *Field Guide to Wildflowers Northeastern & Northcentral N. Amer.* 356 New England Aster. *Aster novæ-angliæ.* Our most showy wild aster, deeper violet than the others; rarely rose-colored. 1936 F. M. FARMER *Boston Cooking-School Cook Bk.* (new ed.) 277 New England boiled dinner. Served warm, unpressed corned beef with cabbage, beets, turnips, carrots, and potatoes. 1975 *Times Lit. Suppl.* 20 June 703/2 We have a New England boiled dinner as well as pot-au-feu. 1855 *Harvard Mag.* I. 232 Most admired of our spring flowers is the Ground Laurel, *Epigæa repens*, commonly called Trailing Arbutus, or New England Mayflower. 1952 A. G. L. HELLYER *Sanders' Encycl. Gardening* (ed. 22) 178 *Epigæa... repens*, 'American Ground Laurel', 'New England Mayflower', 'Trailing Arbutus', white, fragrant. 1899 G. N. BOARDMAN (*title*) A history of New England theology. 1967 D. T. KAUFFMAN *Dict. Religious Terms* 329/1 *New England theology*, term for the movement in New England in the latter part of the eighteenth century and the first half of the nineteenth century to tie Calvinism to human reason and experience.

New-Englandism. (Earlier and later examples in sense: An idiom or mode of expression characteristic of New Englanders.)
1831 *Boston Transcript* 7 Sept. 2/3 Mr Pickering, however, thinks *hub* a New-Englandism only. 1948 MENCKEN *Amer. Lang.* Suppl. II. 198 Dr. R-M. S. Heffner, who was born in 1892 at Bellefontaine, in the west central part of the State [*sc.* Ohio], testifies that New Englandisms were common there in his boyhood.

new-fallen, *a.* **1.** (Later example.)
1926 J. FERGUSSON in *Oxford Poetry* 23 Like a shield New-fallen on a stricken field.

Newfie (niū·fi, nū·fi). *colloq.* [Hypocoristic, f. NEWFOUNDLAND, NEWFOUNDLANDER: see -IE.] Newfoundland; a Newfoundlander. Also *attrib.* or as *adj.*
1942 BERREY & VAN DEN BARK *Amer. Thes. Slang* § 49/4 Newfie, New Foundland [sic]. *Ibid.* § 385/5 Newfie, Newfier, a Newfoundlander. 1945 W. H. PUGSLEY *Saints, Sinners & Ordinary Seamen* 81 This certainly is a change after those winters off Sydney and Newfie. 1958 *In Flight* (Montreal) Summer 8/1 Cod fishing has..long been synonymous with Newfoundland, but today the canny 'Newfies' are no longer putting all their economic eggs in one basket. 1963 R. I. MCDAVID *Mencken's Amer. Lang.* 166 Colonial days, when rum was actually the chief tipple of American dipsomaniacs.... It still is in Newfoundland, where *screech* or *Newfie screech* designates the cheapest grade. 1965 *Globe & Mail* (Toronto) 29 Jan. 7/6 Nobody in Newfie..underestimates Joey Smallwood's abilities as a propagandist. *Ibid.* 7 Dec. 6/4, I described a rail trip (on the famous Newfie Bullet) through the great dead heart of the island-province. 1971 S. E. MORISON *European Discovery Amer.: Northern Voy.* vi. 180 The rocky surface of eastern Newfoundland is full of small depressions..; swarms of mosquitoes breed therein and make life miserable for all but most hardened 'Newfies'. 1972 *Evening Telegram* (St. John's, Newfoundland) 24 June 10/4 It was Newfie entertainment at its best.

Newfoundland. Add: (Earlier examples of *Newfoundland dog.*)
1773 M. RISHTON *Let.* 25 Apr. in F. Burney *Early Diary* (1889) I. 204 We intend getting a very large Newfoundland dog before we leave this place. 1779 W. COXE *Sketches of Swisserland* v. 49 How we contrived to arrange ourselves, our servants, a large Newfoundland dog, and the baggage, in so narrow a compass.
b. (Earlier example.)
1827 DISRAELI *Vivian Grey* V. VII. v. 29 They were instantly saluted by an immense Newfoundland.

New Frontier. [cf. FRONTIER *sb.* 4 b.] A new approach to reform and social betterment; *spec.* the name given to a programme of social improvement advocated by J. F. Kennedy, President of the U.S. from 1961 to 1963. Hence **New Frontiersman, new frontiersman,** a proponent of this programme.
1934 H. A. WALLACE (*title*) The new frontier. 1960 J. F. KENNEDY in *N.Y. Times* 16 July 1/8 The New Frontier of which I speak is not a set of promises—it is a set of challenges. 1962 *Listener* 21 June 1057/2 Even the most ardent new frontiersman..would agree that Mr McCormack was probably not exaggerating. 1963 *Guardian* 22 Feb. 11/4 The energy and sophistication of the New Frontiersmen. 1964 *New Statesman* 10 Apr. 574/1 Baulked of any hope of keeping up with the new frontiersmen of science, it is determined to keep tabs on the new backwoodsmen of architecture. 1969 C. BOOKER *Neophiliacs* vii. 162 Kennedy's New Frontier America. 1973 R. THOMAS *If you can't be Good* (1974) ii. 14 The doctoral thesis I had been researching when summoned to the New Frontier.

Newgate. Add: **Newgate hornpipe,** a hanging; **Newgate novel,** a picaresque novel of the

second quarter of the nineteenth century; so *Newgate novelist, Newgate school.*
1829 W. MAGINN in Partridge *Dict. Slang* (1937) 558/1 Toeing a Newgate hornpipe. 1854 C. KNIGHT *Old Printer & Modern Press* vi. 281 The host of penny Newgate novels ..may continue to be sold; but, as far as we can trace, there are no novelties in this once popular literature of the gallows. 1959 *Brno Studies in English* I. 104 The 'Newgate novelists' of the 'thirties (Edward Bulwer, W. H. Ainsworth, and Charles Whitehead) represent a literary school, which is generally called 'the Newgate School' or 'Bulwer's school'. 1963 K. HOLLINGSWORTH *Newgate Novel 1830–47* i. 14 A series of novels having criminals as prominent characters aroused widespread attention. Contemptuous critics at the time called them Newgate fiction, and later writers have grouped them under the label of the Newgate novel. 1965 S. MARCUS *Dickens: from Pickwick to Dombey* ii. 67 In *Catherine*..Thackeray..identified Dickens with the Newgate novelists. 1970 G. M. FRASER *Royal Flash* 188 And then it would be the Newgate hornpipe for Flashy, with the whole damned crew of Sons of the Volsungs hauling on the rope. 1971 R. L. WOLFF *Strange Stories* I. 24 The vogue of the 'Newgate' novel—as the new genre was called after the famous prison, where some of the most affecting scenes usually took place, and after the Newgate Calendar of crime—ran fast and furious.

new girl. [NEW *a.* 8.] A schoolgirl during her first term(s) at a school, esp. one at an English boarding-school; also *transf.*, a (young) woman newly come into a given set of circumstances. Also (with hyphen) *attrib.* or as *adj.* Cf. *NEW BOY.
1853 MRS. GASKELL *Ruth* I. i. 16 Most new girls get impatient at first; but it goes off. 1924 D. MOORE *Fen's First Term* v. 59 A dozen people charged hospitably at the new girl with biscuits and milk. 1936 M. KENNEDY *Together & Apart* I. 130 New girls are always sort of freaks. 1951 N. MITFORD *Blessing* I. viii. 93 She was a new girl, she must watch her step. 1968 'R. LLEWELLYN' *End of Rug* (1969) xix. 151, I went from school to school... I was rather new-girl at times, of course. A little lonely.

new ground. [GROUND *sb.* 16.] **a.** Ground which has been cleared and cultivated only recently. *U.S. local.* **b.** A part of a goldfield unexploited until recently. *Austral.* (Cf. *to break (new) ground* s.v. *GROUND *sb.* 11 b.)
1624 J. SMITH *Gen. Hist. Virginia* IV. 126 We haue ordinarily foure or fiue [barrels of produce an acre], but of new ground six, seuen, and eight. 1771 in *Maryland Hist. Mag.* (1919) XIV. 134 Our new ground tob[acc]o here has been housed 3 or 4 days past. 1862 *Burrangong* (New South Wales) *Courier* 13 Aug. 2/3 The rush to the Three Mile Diggings..is..going ahead in a most satisfactory manner, and a large extent of new ground has lately been taken up. 1868 *Mining Surveyors & Registrars' Rep.* (Victoria, Dept. Mines) Sept. 36 At Barkly some new ground has been opened south of the main lead. 1915 *Dialect Notes* IV. 186 *New-ground*, virgin land prepared for cultivation. 1937 *Shenandoah* (Va.) *Nature Jrnl.* I. III. 11/1 Each year the acre and half of rough rocky mountain and perhaps a little 'new ground' patch were tilled by hand. 1949 H. HORNSBY *Lonesome Valley* 44 The next time he looked back Chester's place was like a newground that had been burned over. 1953 *Amer. Speech* XXVIII. 251 [Bedford Co., Pa.] *New ground*... Pronounced with a heavily accented *new* and as if it were a single word, *newground*. 'There are lots of teaberries this year out on John Bussard's new ground.' In general use.

New Hall. The name of a site at Shelton, Staffordshire, used to designate china and porcelain produced there.
1829 S. SHAW *Hist. Staffordshire Pott.* viii. 201 The partners..settled the manufactory at the New Hall, Shelton..; on which account the Porcelain had the appellation of *New Hall China*. 1896 E. A. DOWNMAN *Eng. Pott. & Porc.* (new ed.) 132 As no distinguishing mark was used before 1820, New Hall porcelain is difficult to identify. 1969 [see *BRISTOL 2 b].

newie (niū·i). *colloq.* Also **newy.** [f. NEW *a.* + -IE.] **a.** Something new, as a new joke, story, or suggestion. **b.** A person without previous experience in professional entertainment. **c.** A song recently issued on a gramophone record. **d.** = *NEW BALL.
1947 *Amer. Speech* XXII. 157 On the 'Can You Top This?' radio show of November 16, 1946, one of the principal characters said, 'Here's a *new-y*.' His reference was to a joke. 1951 CUSACK & JAMES *Come in Spinner* 495 'Yeah! I seem to have heard your propositions before.' Kim leaned over and took the bottle from her. 'Put that down and listen to me. You haven't heard this one, this is a newy.' 1961 A. BERKMAN *Singers' Gloss. Show Business* 61 *Newies* (Var.), novices; neophytes. 1966 *Melody Maker* 26 Nov. 2 Dave Dee's newie 'Save Me' has an 'African sound with cowbells'. 1971 *West Indian Weekly* 12 Nov. 14/4 The Staples Singers are turning out some incredible soul discs of which their newie, 'Respect Yourself', is a classic in the message song genre. 1972 *Shout* Mar., When this newie turned up..I snatched up a copy. 1975 *Saturday Night* (Toronto) July/Aug. 29/1 Mostly, synectics involves sticking big pieces of paper up around the walls, writing down every bright remark that anyone comes up with, and then winnowing these down to a few golden newies. 1975 *Daily Tel.* 6 Sept. 14/5 But did I say 'ball?' How old-fashioned can one get? 'Ah, the new cherry,' mutters Trevor Bailey with nostalgia in his voice. 'It's the newie,' exclaims Brian Johnston.

New Jersey. The name of one of the eastern states of the United States, used *attrib.* in the names of plants native to the region, as **New Jersey pine** = *Jersey pine* (*JERSEY²); **New Jersey tea** = *Jersey tea* (*JERSEY²).
1759 [see *Jersey tea* (*JERSEY²)]. 1785 H. MARSHALL *Arbustrum Americanum* 27 American Ceanothus, or New-Jersey Tea-tree,..is a low shrub, growing common in most parts of North America. 1818 W. P. C. BARTON *Compendium Floræ Philadelphicæ* II. 183 *P[inus] inops*... New Jersey Pine. Scrub Pine. Pitch Pine. A low straggly, and very common species particularly in Jersey. 1832 D. J. BROWNE *Sylva Amer.* 234 New Jersey Pine. 1877 *Rep. Vermont Board Agric.* IV. 159 Riley recommends persons..to plant a small patch of New Jersey Tea (*Ceanothus*)..as a decoy near the strawberry bed. 1941 R. S. WALKER *Lookout* 59 The New Jersey tea or redroot ..grows profusely on the summit as well as on both sides of the mountain. 1958 G. A. PETRIDES *Field Guide to Trees & Shrubs* 139 New Jersey Tea, *Ceanothus americanus* L... A low shrub. Leaves egg-shaped to triangular... Flowers white in dense heads.

New Left, new left. [Phr. coined by C. Wright Mills (1916–62), American political sociologist.] The name of a movement originated by young radicals opposed to the philosophy of the 'old' liberal society; now applied to many movements of protest. Also *attrib.* Hence **New Lefter, New Lefty,** a supporter of radical policies; **New Leftish** *a.*, **New Leftist** *a.* and *sb.*
1960 *New Left Rev.* Jan.–Feb. 1 Our hope is that people in the New Left will feel, with a special urgency, the poverty of ideas in the Labour Movement. *Ibid.* Mar.–Apr. 70/2 The 'anarchism' of young New Lefters. *Ibid.* Sept.–Oct. 15/2 Stop the new-leftist dogs from yapping. 1960 New Lefty [see *HOBOHEMIA]. 1961 *New Left Rev.* Jan.–Feb. 51/1 What can..be called 'new left' student opinion. 1966 *Economist* 3 Sept. 887/1 The phrase 'new left' is applied to all sorts of protesters in the United States, to Provos in Amsterdam, to dissatisfied intellectuals in eastern Europe. 1967 *Philosophy* XLII. 287 The emphasis is rather New-Leftish and Gramscian. 1967 *Time* 28 Apr. 15 The New Leftists have a mystical faith in the purity and wisdom of the poor, 'uncorrupted' by the Establishment. 1968 *Listener* 3 Oct. 442/1 The would-be Monday Clubber and the New Lefter will both shut up when they hear of the accident in the Headrow. 1971 *Daily Tel.* 18 Nov. 15/3 Berkeley City Council—first and only city government in America to represent the 'New Left'—has passed an astonishing resolution..to offer asylum to military deserters. 1973 *Times Lit. Suppl.* 23 Nov. 1422/5 Whose sympathies are clearly more towards the New Left than towards bureaucratic communism. 1975 *Listener* 13 Feb. 209/1 All over Europe in the Sixties you could see a proliferation of political sects—Marxists, Trotskyists, Maoists, Castroists... This New Left..was, above all, a manifestation of youth.

new light. [LIGHT *sb.* 6 d.] **1. a.** Novel religious views or doctrines (see LIGHT *sb.* 6 d).
1650–1785 [see LIGHT *sb.* 6 d]. 1806 T. G. FESSENDEN *Democracy Unveiled* II. 181 Altho' not bless'd with second sight, Divine inflation, or new light.
b. Any of the religious sects or doctrines of the 'New Lights'.
1750 J. BIRKET *Voy. N. Amer.* (1916) 4 There is two Presbyterian meeting houses here, one of the Newlight, and one of the old. 1819 J. G. LOCKHART *Peter's Lett. to his Kinsfolk* III. lxii. 100 The Old Light Antiburghers enjoy the ministrations of..Dr. McCrie... The New Light.. are ruled *in spiritualibus* by Dr. Jamieson. 1850 W. H. FOOTE *Sk. Virginia* 373 In his discourse he..read a hue and cry, for the arrest of 'the new light'. 1874 [see LIGHT *sb.* 6 d]. 1943 J. MACLEOD *Scottish Theology* 229 As a term, New Light came especially to be used in connection with the change that took place in the thinking of the Seceders towards the end of the 18th century.
2. A person holding 'new lights' or novel (religious) doctrines; a revivalist; a member of any of various schisms from several Protestant churches in Scotland and N. Amer. during the eighteenth and nineteenth centuries.
a 1734 R. WODROW *Analecta* (1842) II. 169 You have brought in a stranger, one of the neu-lights, among us. 1743 J. HEMPSTEAD *Diary* 30 Mar. (1901) 407 All come to settle the disorders that are subsisting among those called New Lights which follow Mr Davenport. 1750 J. BIRKET *Voy. N. Amer.* (1916) 22 There is Nineteen different places of Worship in the Town (to wit) thirteen of the Independents Presbyterians & newlights &Ca. 1796 GROSE *Dict. Vulgar T.* (ed. 3) *New Light*, one of the new light; a methodist. 1807 R. MCNEMAR *Kentucky Revival* 29 These ..taught as an important truth, that the will of God, was made manifest to each individual..by an inward light... Hence they received the name of 'New-Lights'. 1828 J. MCGREGOR *Sk. Maritime Colonies Brit. Amer.* II. 465 Let us leave abstract points of Christian doctrine to theological disputants, and the raving of *new lights*. 1847 R. DAVIDSON *Hist. Presbyterian Church in Kentucky* 219 In the Great Revival, Mr. Stone was conspicuous..[in] the subsequent formation of societies, known under the various names of New Lights, Christians, Arians, Marshallites, and Stoneites. 1872 [see *HICKORY 4 a]. 1888 J. M. BARRIE *Auld Licht Idylls* iii. 60 The congregation..had split, and as the New Lights (now the U.P.'s) were in the majority, the Old Lights..had to retire to the community. 1949 *William & Mary Q.* Jan. 43 The New Brunswick group, the 'new lights', wished to give eloquence in preaching precedence over formal knowledge. 1959 *Chambers's Encycl.* XII. 309/1 The original seceders..formed themselves into a body independent of the state church in 1733. The new group of separatists were divided in 1747 by the anti-Jacobite burgess oath into burghers and antiburghers. Subsequently each of these bodies split into Old and New Lights (1799 and 1806 respectively).
b. *transf.* = *CAMPBELLITE 2.
1877 *Bull. U.S. Nat. Museum* No. 9. 21 *Pomoxys annularis*... Throughout Kentucky it is known as the 'New Light', and sometimes as 'Campbellite'. 1884 [see *CAMPBELLITE 2].
3. *attrib.* or as *adj.* **a.** Belonging to or holding the views of the 'New Lights'.
1732 SWIFT *Advantages repealing Sacramental Test* 10 The Quarrel between Old and New Light-Men, is managed with more Rage and Rancour, than any other Dispute. 1742 J. HEMPSTEAD *Diary* 20 Dec. (1901) 402, 2 of them Newlight Exhorters begun their meeting. 1744 [see LIGHT *sb.* 6 d]. 1793 'T. THRUM' *Look before ye Loup* 3, I took the advice of a *newlight* neighbour upo' this knotty point. 1807 R. MCNEMAR *Kentucky Revival* 46 Taking what is called the *New-light* doctrine, as the rudiments of divine truth, they proceeded to consider the nature of justification, reconciliation to God, etc. 1837 W. JENKINS *Ohio Gazetteer* 373 There are in this county..eight christian (or newlight)..and five dunkard churches. 1874 [see LIGHT *sb.* 6 d]. 1883 P. SCHAFF *Relig. Encycl.* II. 1634/2 [New England divines] announced a few principles, which were called 'New-Light Divinity', or 'New Divinity'. 1949 *Canad. Hist. Rev.* Mar. 75 Henry Alline, the New Light evangelist, was shattering the less sedate Dissenting congregations of the Nova Scotian out-ports.
b. *transf.* Novel, newfangled.
1831 J. N. CATRON *Let.* 1 May in N. N. Scott *Mem. H. L. White* (1856) xii. 249 A union of N. Light Federalists with Kentucky republicans..cannot last. 1833 J. B. WYETH *Oregon* 4 What the *new-light* doctrine of Phrenology calls the disposition bump of *Inhabitiveness*. 1839 *Spirit of Times* 26 Oct. 399/3 *Abbreviations*... We heard of this 'new light system' being carried so far as to be adopted by a lady.
Hence **New Lightism,** new-light doctrines.
1755 in *Essex Inst. Hist. Coll.* (1916) LII. 78 He seems a grave, close heavy Man, not given to talk & deeply immerged in New Lightism. 1857 P. CARTWRIGHT *Autobiogr.* 32 B. W. Stone stuck to his New Lightism.

new look. [*LOOK *sb.* 2 f.] **1.** (Freq. with capital initials.) A style of women's clothes introduced in Paris by Christian Dior in 1947, characterized esp. by skirts longer and fuller than those previously worn. Also applied to more recent new styles.
1947 *Time* 15 Sept. 87 What was going on [in fashion]? The search for the 'New Look'. What was the New Look? No one knew precisely. 1948 *Economist* 14 Feb. 285/2 Clothing in styles which the New Look has superseded. 1949 *Ann. Reg. 1948* 59 The 'new look' had swept the great cities. 1957 *Daily Mail* 25 Oct. 10/2 When Christian Dior launched the New Look in 1947 it met with an enthusiasm which amounted to hysteria. 1958 *Spectator* 8 Aug. 187/1 The [fashion] trade now bitterly resents St. Laurent's new look.
2. In wider use: a change in policy or procedure; a renovated or up-to-date presentation or appearance. Also (with hyphen) *attrib.*
1948 *Melody Maker* 3 Apr. 3/3 (*heading*) The 'New Look' in jazz has come to stay. 1952 *Economist* 27 Sept. 765/2 (*heading*) Czech trade unions reorganized... The 'new look' given to, and by, the trade union officials at once brought them into conflict with the rank and file of the movement. 1952 *Science News Let.* 25 Oct. 258/1 A 'New Look' for veterans with smashed noses due to war-time injuries is being developed. 1955 *Ann. Reg. 1954* 169 On January 12 the Secretary of State, Mr. Dulles, announced what came to be called the 'New Look' policy. 1956 E. J. RUPPELT *Report on Unidentified Flying Objects* 85 In early 1949 the term 'new look' was well known. The new look in women's fashions was the lower hemlines, in automobiles it was longer lines. In UFO circles the new look was cuss 'em. 1958 J. CANNAN *And be a Villain* iii. 61 How different life looked now, and how strange that the begetter of this New Look was..the depraved and disgraced Jonathan. 1959 *News Chron.* 2 Nov. 5/1 The new-look colours and markings which are to be on all B.E.A. airliners. 1960 *Guardian* 28 Apr. 6/6 The 'new look' which..the Glasgow business man who now runs the Tourist Board, is giving to tourist centres. 1966 *Melody Maker* 15 Oct. 19 A slightly new-look Johnny Parker band restored to full vigour. 1974 S. GULLIVER *Vulcan Bulletins* 107 It's the new-look Libyan justice. Or rather, it's medieval but recently revived.
Hence **new-loo·kish** *a.*
1950 *Christian Sci. Monitor* 20 Feb. 10/1 Visitors recently returned from Moscow report that it is the most New Lookish of any Russian city.

newly, *adv.* Add: **1. c.** (Further examples.)
1690 LOCKE *Ess. Hum. Understanding* III. x. § 32 He that, in a newly-discovered country, shall see several sorts of animals and vegetables, unknown to him before, may have as true ideas of them, as of a horse or stag. 1831 MILL *Let.* 22 Oct. (1910) I. 19 They have formed a plan for a new colony,... on the coast of Southern Australia, near the place where the newly discovered navigable river discharges itself into the sea. 1834 —— in *Monthly Repos.* VIII. 163 The many are..likely to make a most dangerous use of their newly-acquired power. 1848 S. M. GRAY *Let.* 11 Apr. in M. Lutyens *Ruskins & Grays* (1972) xi. 107 Many compliments [were] paid to the newly married couple. 1937 *Burlington Mag.* Apr. 173/2 Prince of a newly-constituted State. 1968 R. A. LYTTLETON *Myst. Solar Syst.* v. 166 It endows the newly-formed comets with orbits that are almost parabolic. 1974 J. WAINWRIGHT *Evidence I shall Give* viii. 35 Lennox was bringing the newly-arrived Sugden up to date on the murder.
d. newly-wed, newlywed, a person newly married.

1918 *Cosmopolitan* Feb. 90/2 It seemed that a Newlywed can live on Marmalade for about three months. 1932 Auden *Orators* iii. 103 To-day may mean division for the newly-weds. 1935 *Discovery* Mar. 91/1 Generation upon generation of newly-weds are tempted to 'come and live in leafy ——'. 1938 J. I. Rodale *King's English on Horseback* 146/1 Newlyweds. First-nighters. 1959 G. Freeman *Jack would be Gentleman* viii. 163 'Quite the blissful newly-weds, aren't we?'..'Well, we are newly-weds.' 1973 *Guardian* 12 Mar. 9/6 The advertisement..was one of a series which featured a newly-wed, an insurance man and a saleslady.

Newmania (niūmēi·niă). [f. the name of J. H. *Newman* (see NEWMANISM) + -IA¹, punning on MANIA.] Enthusiastic support for Newmanism. Also *attrib.* Also **Newmanic** (niūmæ·nik) *a.*, pertaining to or characteristic of J. H. Newman or his views.
1838 J. Romilly *Diary* 9 Dec. (1967) 158 A very objectionable sermon..as faulty as the 'Newmania' sect at Oxford. 1849 F. D. Maurice *Let.* 9 Mar. in J. F. Maurice *F. D. Maurice* (1884) I. xxv. 518 Froude's hero..adopted the Newmanic theory. He gave God credit for being a tyrant. 1900 W. Tuckwell *Reminisc. of Oxford* ii. 17 Men who formed in Oxford what was known as the Noetic school.., provoking by their political and ecclesiastical liberalism the great revolt of the Newmania. 1958 *Spectator* 7 Feb. 167/3 The Newmania [in Oxford] was succeeded by the Railway Mania.

Newman–Keuls (niū:măn,köls). *Statistics.* The names of D. *Newman* (of the Dept. of Statistics, University College, London) and M. *Keuls* (of the Institute of Horticultural Plant Breeding, Wageningen, Holland), used *attrib.* to designate a test they devised for assessing the significance of differences between the means of different sets of observations, using the ranges of the sets contributing to the means.
1955 D. B. Duncan in *Biometrics* XI. 26 The Newman–Keuls Test. A test proposed by Newman..in 1939 and again by Keuls..in 1952 succeeds very simply in raising all of the low protection levels of the multiple *t* test. This test is equivalent to a multiple *t* test preceded by several preliminary range tests. 1962 B. J. Winer *Statistical Princ. Exper. Design* iii. 85 The level of significance for the Newman–Keuls procedure is considered individually with respect to each test. 1970 *Jrnl. General Psychol.* LXXXII. 159 The Newman–Keuls test..was run on the means for the different diagnostic classes.., with results indicating significantly more cartoon enjoyment for normals than for the neurotic and sociopathic Ss. No other differences reached significance.

Newmarket. Add: **1.** *Newmarket boot*; *Newmarket coat* (earlier example).
1837 B. Webster *My Young Wife & my Old Umbrella* in *Acting National Drama* I. 7 Green Newmarket cut coat. 1933 J. Buchan *Prince of Captivity* ii. iv. 248 She had been gardening..and wore Newmarket boots. 1958 *Times* 6 Oct. 13/1 The foot which would be warm and practical with sheepskin or Newmarket boots.

New Mexican, *a.* and *sb.* [f. *New Mexico*, one of the United States.] **a.** *adj.* Of or belonging to New Mexico. **b.** *sb.* A native or inhabitant of New Mexico.
1834 A. Pike *Sketches* 137 To an American, the first sight of these New Mexican villages is novel and singular. *Ibid.* 170 Even the New Mexicans call him a great rascal. 1893 C. F. Lummis *Land of Poco Tiempo* 294 Twenty miles south of the New Mexican hamlet of Manzano..is..the pueblo of Abó. 1940 E. Fergusson *Our Southwest* 228 Few families are so completely urbanized as not to have a little ranch somewhere, and New Mexicans care for their own as long as they can.

new-mown, *ppl. a.* Add: *new-mown hay*, as the name of a scent.
1890–1 *T. Eaton & Co. Catal.* Fall & Winter 42/2 Jockey club, new-mown hay, lily of the valley, 25 c. per bottle. 1926–7 *Army & Navy Stores Catal.* 490/2 Perfumes..Lilas Blanc, New Mown Hay, Oeillet Blanc. 1971 *Guardian* 17 Aug. 7/2 Their new Meadowsong fragrance range..based on..newmown hay, honeysuckle, clover.

New Negro, new negro. *U.S.* **a.** During the period of slavery, a Negro brought from Africa to the New World. **b.** An artist belonging to the *New Negro Movement*, the efflorescence of Negro writing, etc., during the 1920s. **c.** A Negro claiming equal status with a white American.
1701 C. Wolley *Two Years' Jrnl. N.-Y.* 32 In Barbados new Negro's (i.e. such as cannot speak English) are bought for twelve or fourteen pound a head. 1732 *Calendar State Papers Amer. & W. Indies* (1939) 55 Except proof were made that they were all new negroes and not been above 6 months in America. 1860 S. Mordecai *Virginia* 350, I do not speak of the New Negros, as the imported Africans were called, but of their descendants. 1922 A. P. Randolph in A. Dundes *Mother Wit* (1973) 400/1 It does not occur to the Old South that there is a 'New Negro'; that the 'Uncle Toms' are passing. 1925 A. Locke in *Survey* (N.Y.) Graphic No., 1 Mar. 632/1 The Negro is being carefully studied, not just talked about and discussed. In arts and letters, instead of being wholly caricatured, he is being seriously portrayed and painted. To all of this the New Negro is keenly responsive as an augury of a new democracy in American culture. 1953 S. A. Brown in A. Dundes *Mother Wit* (1973) 40/1 We go then to what is called the New Negro Movement. *Ibid.* 43/2 She wouldn't let anybody kick her around (something like the New Negro). 1963 in J. H. Clarke *Harlem* (1964) 40 For a racial dilemma in Negro art, a racial solution was necessary. This came in the mid-twenties from the inspiration of the New Negro Movement with its crusade of folk expression in all of the arts. *Ibid.* 49 Many of the New Negroes were unwilling victims of an inverted racialistic nationalism,..priding themselves that they could sing, paint and write as well as their white-skinned patrons.

Newnhamite (niū·nămait). [f. *Newnham* (see below) + -ITE¹.] One who is, or has been, a student at Newnham College, one of the Cambridge colleges for women.
1896 [see GIRTONIAN]. 1907 R. Brooke *Let.* 28 Nov. (1968) 115 Most of the committee are Newnhamites, strange wild people. 1913 J. Vaizey *College Girl* xx. 282, I am proud to be a Newnhamite. 1936 M. V. Hughes *London Girl of Eighties* vi. 119 Miss Rogers, a large and genial Newnhamite.

New Orleans (niū ǭ·ıliănz, ǭlī·nz). The name of a city in SE. Louisiana, giving its name to various commercial and natural products; used *spec.* to designate a style of jazz which originated there.
1807 C. Schultz *Travels* (1810) I. 132 Kentucky and New-Orleans boats from one dollar to one and a half a foot. 1849 G. G. Foster *N.Y. in Slices* 82 The grocery-keeper..buys a barrel of common New Orleans molasses at twenty-five cents per gallon. [1905 (*title of tune composed by Jelly Roll Morton*) New Orleans Blues. 1922 (*name of jazz band*) New Orleans Rhythm Kings.] 1935 *Swing Music* July 120/1 The expressions 'Chicago Style', 'New Orleans Style' were certainly invented by the American musicians, and not at all by European hot fans. 1938 D. Baker *Young Man with Horn* i. v. 42 Jeff's band..had two styles of playing, known to the present trade as Memphis style and New Orleans style. *Ibid.* 43 Memphis style is sometimes called 'take your turn', and New Orleans has everybody in at the same time. 1952 B. Wolfe *Limbo* (1953) ii. ix. 102 Program of jazz recordings... Maybe New Orleans had come back..to a world it..belonged to. 1955 R. Blesh *Shining Trumpets* (ed. 3) iii. 58 They prophesy New Orleans jazz as clearly as they recall Africa. 1958 Osborne & Creighton *Epitaph G. Dillon* i. 14 Do you mind if we do without New Orleans just for the moment? 1965 G. Melly *Owning-Up* xi. 128 What the revivalists thought of as 'New Orleans Jazz' was the music of Armstrong, Morton and Oliver—New Orleans musicians but based on, and recorded in, Chicago, during the Prohibition era. What the traditionalists meant by New Orleans Jazz..was the music played by musicians who had never left the city. 1973 J. Drummond *Bang! Bang! You're Dead!* xiii. 34 One of the early New Orleans rags.

New Realism, new realism. *Philos.* Chiefly *U.S.* Doctrines of Realism as revived at the beginning of the twentieth century to refute certain tenets of Idealism, emphasizing the existence of objects in the external world independently of the way they are subjectively experienced. (For the contemporary movement in Britain, cf. *REALISM 1 c.) So **New Realist** *sb.* and *a.*, an adherent or advocate of, pertaining to, New Realism.
1906 J. S. Mackenzie in *Mind* XV. 308 (*heading*) The New Realism and the Old Idealism. *Ibid.*, Some of the leading supporters of the new Realism (especially Mr. Moore and Mr. Russell) connect it with an extremely nominalistic type of Logic. 1920 W. R. Sorley *Hist. Eng. Philos.* xii. 297 Forms of what is called the new realism seem to have been started independently in the United States and in this country. 1929 C. I. Lewis *Mind & World-Order* ii. 39 Immediacy is thus emphasized..by the American new-realists. 1947 *Mind* LVI. 290 As everyone knows, the revolt against Idealism, initiated chiefly by Moore and Russell in this country and by the New Realist group in the United States, proved quite prodigiously successful. 1966 F. Copleston *Hist. Philos.* VIII. v. xvii. 387 The new realists were at any rate agreed on the truth of a basic tenet.., 'things known are not products of the knowing relation nor essentially dependent for their existence or behaviour upon that relation'.

news, *sb.* (*pl.*) Add: **2. c.** *no news is good news* (further examples).
1616 James I *Let.* 13 May in A. J. Kempe *Loseley Manuscripts* (1835) 403 No newis is better then evill newis. 1850 F. E. Smedley *Frank Fairleigh* x. 101 Arguing..(on the 'no-news being good-news' system) that I should have heard again if anything had gone wrong, I dismissed the subject from my mind. 1916 'Taffrail' *Pincher Martin* xviii. 336 They could not bring themselves to believe that 'no news was good news'. 1941 A. Huxley *Let.* 27 Nov. (1969) 473 Matthew never writes; but we interpret no news as good news.
e. *that* (or *it*) *is news to me*: I did not know that. *colloq.*
1898 S. Weyman *Castle Inn* xvi. 159 For the rest, which this gentleman says, about who she is and her claim..it is news to me. 1919 D. Ashford *Young Visiters* viii. 69 Ethel he said blushing a deep red I always wished to marry you some fine day. This is news to me cried Ethel still peevish. 1943 J. B. Priestley *Daylight on Saturday* xii. 76 If you've had any trouble with your husband, I'm sorry, but it's news to me. 1968 S. B. Hough *Sweet Sister Seduced* xv. 79, I had thought we were in tune with one another..that my reactions were her reactions. It was news to me, as she told me in round phrases, that in fact they weren't. 1974 M. Z. Lewin *Enemies Within* xxxiv. 154 'I'm going to Chicago shortly.' News to me. But not a bad idea.
f. *bad news*, used to designate something or someone unpleasant, unlucky, or undesirable (see quots.). *colloq.*
For the literal use cf. quot. 1821 in Dict.
1926 Maines & Grant *Wise-Crack Dict.* 5/2 *Bad news*, piece of pasteboard handed by the waiter after a meal. 1929 M. A. Gill *Underworld Slang*, *Bad news*, shot gun. 1930 *Amer. Mercury* Dec. 454/1 *Bad news*, trouble. 'Sucker, stay out of me district! It's bad news if you don't.' 1935 A. J. Pollock *Underworld Speaks* 120/1 *The bad news*, the bill (check) in a restaurant, speakeasy or night club. 1942 Berrey & Van den Bark *Amer. Thes. Slang* § 256/1 *Difficulty, trouble*.., bad news. 1963 I. Fleming *On H.M. Secret Service* i. 13 Their waiter..had simply put them in the category of 'bad news' and hoped they would soon be on their way. 1964 *Amer. Speech* XXXIX. 189 *Bad news*, 'a poor social evening, a wasted night'. 1968 'J. Welcome' *Hell is where you find It* iv. 67 'Where is she?' I said... 'Listen, I don't know. She's bad news, that one.' 1973 'E. McBain' *Let's hear It* v. 69 Bikies had begun drifting into the area, sporting their leather jackets and their swastikas... The bikies were bad news. 1973 H. Miller *Open City* xviii. 197 Any kind of witness would be bad news on a job with such a tight specification. 1974 D. Gray *Dead Give Away* iii. 31 Milly these days was plain bad news. Her fascination had evaporated.
g. A person, place, etc., regarded as a topic of discussion or note.
1912 Kipling *Diversity of Creatures* (1917) 192 The great Baron Reuter himself..flashed that letter in full to the front, back, and both wings of this scene of our labours. For Huckley was News. 1946 E. Waugh *When Going was Good* v. 260 Abyssinia was News. Everyone with any claims to African experience was cashing in. 1965 *Listener* 23 Sept. 452/2 The reading boom..has made poets news, and it has made them think about being news. 1974 V. Gielgud *In Such a Night* vii. 58, I am not what is commonly called 'news'. But..my wife is 'news' in the biggest possible way.
5. c. A television or radio programme in which the news is announced and sometimes discussed; a newsreel. Also *attrib.*
1923 *Radio Times* 28 Sept. 9/1 10.0. — Time signal, general news bulletin. Broadcast to all stations, followed by London News and Weather Report. 1925 *Daily Herald* 23 June 4/3 To hear the news from your favourite announcer is like buying your favourite newspaper. 1932 G. Greene *Stamboul Train* ii. i. 66 Janet Pardoe said that she wanted to see the news and they both stayed [in the cinema]. 1939 T. S. Eliot *Family Reunion* ii. i. 97 And now it is nearly time for the news We must listen to the weather report And the international catastrophes. 1940 'G. Orwell' in *World Rev.* (1950) June 21, I went to the pub to hear the 9 o'clock news. 1947 Auden *Age of Anxiety* (1948) i. 17 Now the news. Night raids on Five cities. 1953 M. Laski *Victorian Chaise Longue* 64 It was that programme just before the news. 1968 'D. Rutherford' *Skin for Skin* iv. 95 Crisp..glanced at his watch. 'Five minutes till news time... Can you look after her while I listen to the news?' 1972 D. Delman *Sudden Death* (1973) iv. 121 It was news time so..I turned on the TV. 1973 J. Drummond *Bang! Bang! You're Dead!* xxxvi. 126 The ginger-headed Crabbe was watching the nine o'clock news.
6. a. *news-dealer* (earlier and later examples), *-gatherer* (later examples).
1861 *Chicago Tribune* 15 Apr., We..are now prepared to furnish News Dealers and Booksellers with Every Paper, Periodical and Book. 1966 R. Ellison in A. Chapman *New Black Voices* (1972) 407 One newsdealer in Harlem. 1963 *Punch* 20 Feb. 273/1 He is a seer rather than an exclusive news-gatherer. 1971 *Guardian* 20 Dec. 11/1 The Guardian's specialist correspondents are not only newsgatherers but also distinguished commentators.
b. *news-gathering*; *news black-out, -board, bulletin, conference, editor, feature, film, item, magazine, -matter, media, -print* (earlier example), *story, summary, value.*
1944 *Sun* (Baltimore) 17 Aug. 1/6 The whole sector east of the Falaise bottleneck was under an Allied news blackout. 1945 [see *BLACK-OUT 2]. 1974 Hawkey & Bingham *Wild Card* xxiii. 192 To reduce the risk of panic..a news blackout was requested. 1922 Joyce *Ulysses* 116 A stately figure entered between the newsboards. *Ibid.* 218 He passed Grogan's the tobacconist against which newsboards leaned. 1915 (*title*) News Bulletin (Aero Club Amer.). 1923 [see sense 5 c above]. 1925 News bulletin [see *BULLETIN 2 b]. 1973 C. Egleton *Seven Days to Killing* xx. 211 Julyan sat..listening to the transistor radio... The music faded to give way to the news bulletin. 1966 'G. Black' *You want to die, Johnny?* i. 11 At a news conference in Los Angeles..Lil had said to television cameras..: 'Boots and I think God is a drag.' 1972 *N.Y. Times* 3 Nov. 1/1 The Prime Minister's announcement at a televised news conference was a rejection of demands..that the Liberal Prime Minister resign. a 1883 G. W. Bagby *Old Virginia Gentleman* (1910) 190 Pollard he declared was 'the best news editor in the whole South'. 1931 *Daily Express* 16 Oct. 11/3 Before I die, I wish to see the countenance of my own news editor when I stand before him admitting a similar circumstance. 1951 *Oxf. Jun. Encycl.* IV. 300/2 The news editor also receives a great deal of information from Government departments. 1974 *Times* 16 Nov. 15/5 The issue at the *Kentish Times* is..whether six journalists styled 'district editors'..are in fact the editors of newspapers... Each..works under the direction of the news editor of the *Kentish Times*. 1912 *International* (N.Y.) Apr. 79/1 It is wonderful what a variety of cultured subjects are concentrated in the Gould article—economics

..and heart interest 'news feature' as the daily papers would say. 1973 'S. Harvester' *Corner of Playground* ii. i. 81 She loved news-features about herself. 1940 J. Reith *Diary* 16 Jan. (1975) v. 238 Very bothered about a news film..in which Hore-Belisha was cheered and Gort received in silence. 1941 E. Niggeman *Let.* 2 Jan. in H. Nicolson *Diaries & Lett.* (1967) 136, I also went to a News film and saw the film of London's fire. 1965 *B.B.C. Handbk.* 65 The BBC's own newsfilm cameramen. 1974 *Times* 9 Dec. 13/2 Producers of television..want..access to Parliament for the making of news film on the big occasions. 1918 W. G. Bleyer *Profession of Journalism* 27 The Allied governments abroad and our courts at home have struck a hard blow at the Hearst news-gathering concern. *Ibid.* 114 The Associated Press is the child of the first effort at cooperative news-gathering ever made. 1966 *B.B.C. Handbk.* 52 General news-gathering facilities have been enlarged. 1972 H. Evans *Newsman's English* i. 1 This news-gathering is a prodigious if familiar achievement. 1844 *Knickerbocker* XXIV. 179 News-items, matters of information, actual discoveries. 1930–1 G. Ade *Let.* Dec.–Jan. (1973) 149 Here are some news items which have not been printed but which come from pretty reliable sources. 1938 News-item [see *cover design* s.v. *COVER sb.¹ 8]. 1958 *New Statesman* 20 Dec. 871/1 The second intrusive news-item concerns the budget. 1923 (*title*) Time, the weekly news-magazine. 1953 *Encounter* Nov. 5/1 He shifted to the weekly news-magazine, *Der Spiegel*. 1968 *Listener* 4 Apr. 442/3 His job was to compile the early morning news magazine which I shall present at ten past seven. 1923 O. G. Villard *Some Newspapers & Newspapermen* 88 His rivals and critics accuse him..of going to the very edge of the salacious in some of his news-matter and fiction. 1959 *Times* 14 Jan. 12/5 The setting and make-up of newsmatter. 1962 *Amer. Speech* XXXVII. 44 The news media in Moscow relayed to the American press the sensational story of Gagarin's space flight. 1972 J. Mann *Mrs. Knox's Profession* xiv. 105 The news media had got on to the story very quickly. 1973 *Sat. Rev. Soc.* (U.S.) May 69/3 Survey results can be disseminated through the news media. 1843 *Knickerbocker* XXII. 283 The news-prints kept their works and worth before the public eye. 1932 News story [see *MONEY sb. 6 b]. 1974 M. G. Eberhart *Danger Money* (1975) xiii. 136 She can't stop the news stories but perhaps she can soften them. 1941 *B.B.C. Gloss. Broadcasting Terms* 20 *News summary*: (1) Brief statement of salient news items, broadcast at a fixed time. (2) Brief statement of principal news items, broadcast as a preface in a news bulletin. 1973 A. MacVicar *Painted Doll Affair* ii. 33, I turned on the dashboard radio... The pop music was interrupted by a news summary. 1906 J. London *Let.* 8 Apr. (1966) 198 But what I did propose to you was 'events of large news-value'. 1926 T. Beer *Mauve Decade* 172 He had no 'news value'—Julian Ralph invented the phrase in 1892 although it would be long before it became sacred. 1941 [see *EPHEMERAL a.]. 1960 G. Butler *Death lives Next Door* vii. 149 Ezra could see at once that he had news value. 1972 P. Black *Biggest Aspidistra* iii. viii. 229 Lord Hill..justified the occasional trivialisation as better than some academic selection of news values.
c. news agency, (*a*) a business that sells newspapers and periodicals; (*b*) a business organization that collects and supplies news to subscribing newspapers, broadcasters, etc.; **news-boat** (earlier example); **news break,** a newsworthy item (see also quot. 1969); **news butch(er)** *U.S. colloq.* [*BUTCHER sb. 3 b], a seller of newspapers, sweets, etc., on a train; **news cinema,** a cinema which shows a succession of short films, cartoons, and newsreels; **news desk,** the department of a newspaper office responsible for collecting and reporting the news; **news-dick** *literary nonce-wd.*, a news-hawk, a news-hound; **news flash** (see *FLASH sb.² 1 d); **news-girl,** a girl who sells newspapers; **news-hawk, -hound** *colloq.* (orig. *U.S.*), a newspaper reporter; **news peg,** a news story that forms the basis of an editorial, interview, cartoon, etc.; **newsprint,** cheap paper made from mechanical and chemical wood-pulp, and used chiefly for newspapers (see also 6 b); also **news-printing** *vbl. sb.*; **news-reader,** a person who reads the news on radio or television (see also 6 a); hence **news-reading** *vbl. sb.*; **newsreel, news reel,** a short cinema film dealing with news and current affairs; also *attrib.*; **news room,** (*b*) the office in a newspaper, radio, or television station, etc., where the news is processed; **news-stand** (earlier and later examples); **news theatre** = **news cinema*; **news ticker,** a telegraphic recording instrument which automatically prints the news on to a tape; **news-work** (earlier and later examples).
1873 F. Hudson *Journalism* 521 News agencies..branched out and extended into colossal news companies as a..necessity of the age. 1883 *Encycl. Brit.* XX. 405/2 The demand for such reporting had led, on the passing of the telegraphs into the hands of the state, to the formation of news agencies which undertook to supply the provincial newspapers. 1887 *Postal Laws* (U.S.) 147 In admitting second-class publications sent from a news agency, postmasters will observe the following [rules]. 1890 [see *BURGLAR v.]. 1933 J. Buchan *Prince of Captivity* iii. ii. 291 A statement to an international news agency. 1959 *New Statesman* 25 Apr. 564/3 Chancellor is a brilliant journalistic administrator who deserves a very large part of the credit for the wartime and post-war recovery of that famous world news-agency, but he has had no previous experience in actual magazine or newspaper publication. 1974 J. Banning *How I fooled the World* iv. 23 One difference..between working for a newspaper and working for a news agency was that a news agency correspondent is much more vulnerable. 1830 *Boston Transcript* 1 Sept. 2/2 The news-boat, T. H. Smith, belonging to the Associated Morning Papers, boarded the packet ship Caledonia,..25 miles outside Sandy Hook, and before she was boarded by any other news-boat. 1954 D. Dodge *Lights of Skaro* vi. 218 Filing coverage on one of the news breaks of our time. 1959 J. Thurber *Years with Ross* v. 80 Newsbreaks, those garbled..items from American journals..which conveniently fill out..*New Yorker* columns. 1969 *New Yorker* 11 Oct. 43/1 We've just received a letter..enclosing three newsbreaks (those little items we print at the bottom of the page) for our consideration. 1930 A. Henderson *Contemporary Immortals* 54 In connection with his profession as 'news butch' or seller of newspapers, candy and the like, he established a printing press and a small laboratory upon the train. 1894 *Daily Ardmoreite* (Ardmore, Okla.) 1 Jan. 3/1 Ben R. Wheeler, an old time and popular news butcher on the Santa Fe..is in the city. 1930 J. Dos Passos *42nd Parallel* 294 He got a concession as news-butcher on the daily train. 1947 L. M. Beebe *Mixed Train Daily* 85 The news butcher..still carries as stock in trade the immortal volume of senescent anecdotes, Thomas W. Jackson's *On a Slow Train Through Arkansas*. 1935 *Punch* 14 Aug. 192/1 Trousers go wrong the moment you move in them. The news-cinemas and photographs in the papers tell you that. 1965 M. Stewart *Airs above Ground* i. 16 There was an hour to Angy's train and we wanted somewhere to sit, so we went to the news cinema. 1950 *Kemsley Man. Journalism* i. 67 Touch must constantly be kept with the picture news desk which controls the photographers. 1962 A. Lejeune *Duel in Shadows* i. 11 The staff of the Night News Desk could be seen in their usual state of harassment. 1973 *Times* 21 Sept. 5/3 He asked to be put through to the news desk. 1974 *New Society* 7 Feb. 308/1 Wrote for British newspapers through more than three decades of lack of interest on London foreign newsdesks. a 1953 Dylan Thomas *Quite Early One Morning* (1954) 80 Two typewriter Thomas the ace news-dick. 1904, 1938 News flash [see *FLASH sb.² 1 d]. 1972 J. Wilson *Hide & Seek* vi. 111 He was in control now. News-flashes couldn't scare him. 1974 E. Ambler *Dr. Frigo* i. 56 There was a television news flash... The announcer didn't get your father's name quite right. 1868 *Putnam's Mag.* Apr. 518/1 A few years ago, a news-girl was as rare a sight as a Dodo. 1870 *Scribner's Monthly* I. 115 Old and young are enlisted in the street-vending service from the grey-haired grandsire..to the tiny news-girl. 1937 B. Board (*title*) Newsgirl in Palestine. 1931 *Amer. Speech* VI. 283 *Newshawk*.., used for 'reporter'. 1935 [see *big time* (*BIG a. B 2)]. 1940 *Illustr. London News* CXCVI. 544/1 News-hawks reading the tape-machines in New York and California. 1970 E. K. Walker in W. King *Black Short Story Anthol.* (1972) 54 Bull, flanked by his sergeant, the newshawk, and his cameramen, walked to the ambulance. 1974 N. Freeling *Dressing of Diamond* 17 Just ringing the police..is equivalent to inviting the newshawks. They'll be here in half an hour. 1918 *Hatchet* 7 Apr. 48/2 'Got what all figured out,' queried the news hound eagerly. 1926 *Time* 12 July 22/3 In a jazzed age no news hound delved through the reference 'morgue' of his paper to turn up the great story. 1936 E. Ambler *Dark Frontier* x. 163 'What have you been doing with yourself?' 'Trying to be a good newshound.' 1973 'S. Harvester' *Corner of Playground* i. iv. 41 You newshounds never spare yourselves once you've gotten latched onto a story. 1974 H. MacInnes *Climb to Lost World* ii. 35 Our trip was getting more like Conan Doyle's *Lost World* expedition every day, and I was obviously being cast as Ed Malone of the *Daily Gazette*. 'That stalwart news-hound of the cleft sticks', as Neil put it. 1960 *20th Cent.* Apr. 357 These jousts don't seem to need a news peg. 1960 *New Statesman* 15 Oct. 556/2 The BBC's interview programme *Face to Face*..subjects selected public figures to sustained personal questioning before the cameras—without necessarily any particular topical news-peg. 1909 *Westm. Gaz.* 3 June 2/2 The duty of 5 dollars a ton on news-print. 1935 *Geogr. Jrnl.* LXXXVI. 354 A large mill on the Wirral which manufactures newsprint. 1967 Karch & Buber *Offset Processes* xi. 485 Newsprint.., (Use) Newspaper, hand-bills, posters. 1974 *Publishers Weekly* 18 Feb. 44/2 Newsprint remains a headache for most paperback publishers [in Britain] and one was shocked recently to find himself quoted a price almost twice what he usually pays. 1937 *Tablet* 2 Oct. 436/2 The growth of large newspaper combines makes competition increasingly difficult, because the large proprietors are in fact deeply interested in the allied industries like the manufacture of news printing. 1925 *Daily Herald* 23 June 4/3 Instead of receiving a shock at a national calamity, the news reader breaks it to you in a calm and quiet voice. 1959 P. McCutchan *Storm South* i. 6, I hadn't realized it was News-time... I heard the news-reader's voice coming over. 1973 *Times* 15 Jan. 14/8 Mr William Alexander Moyes, the former BBC news reader and announcer, has been found dead at his flat. 1975 J. Wood *North Kill* xii. 181 A BBC news-reader was announcing the details. 1951 in M. McLuhan *Mech. Bride* (1967) 8/3 The Editors of *Time* hope to give..a clearer picture of the world of news-gathering, news-writing, and news-reading. 1971 T. F. Mitchell in *Archivum Linguisticum* II. 38 Although the implication of spoken utterance is less assured for some written functions than others, nevertheless rehearsals.., newsreading, lecturing and public address, though not 'colloquial' language, illustrate speech with the implication of writing and vice versa. 1916 *Wells Fargo Messenger* V. 39/3 Some companies issue their news reels twice a week. 1928 *Manch. Guardian Weekly* 26 Oct. 335/1 There are four motion picture newsreel cameramen, and four 'still' photographers. 1934 *B.B.C. Year-Bk.* 60 There has been another development of this new service by 'special correspondent' in the five-minute topical talk which in turn has now been extended to the 'News Reel'. 1944 L. MacNeice *Christopher Columbus* 15 The radio play..is competing with the Soviet art-cinema rather than with Hollywood or the standardised news-reel. 1949 *Radio Times* 15 July 15/2 Radio Newsreel, a summary of events of the past week. 1973 'D. Jordan' *Nile Green* xliii. 218 We were safe..bowling across the desert like old newsreels of the Eighth Army chasing Rommel into the sea. 1929 C. N. Warren *News Reporting* i. 2 Ed Markham, reporter, entered the *Times* local news room, ready to start his work for the day. 1959 *Times* 5 May 13/5 The news-room scoop is almost a thing of the past—an exciting aspect of newspaper life remembered only by the journalist whose career began in the heady Fleet Street days before the war. 1973 *Scotsman* 13 Feb. 15/6 (Advt.), The Scotsman Publications Limited require a newsroom typist... You will be typing our news stories as phoned in live by 'The Scotsman' reporters. 1872 E. Eggleston *Hoosier Schoolm.* viii. 77 You can buy trap-doors..dirt-cheap at the next news-stand. 1926 G. Ade *Let.* 29 Nov. (1973) 116 You..finally met out west the news-stand girl who had been your friend. 1932 E. Wilson *Devil take Hindmost* iii. 19 Communist publications, sold openly on news-stands. 1973 'J. Ashford' *Double Run* vi. 42 Ryan crossed the main hall of Charing Cross station to the Smith news-stand in the centre. 1973 *Philadelphia Inquirer* (Today Suppl.) 14 Oct. 7/2 The Digest..has lowered its newsstand price a dime. 1933 J. B. Priestley *Wonder Hero* iii. 95 He paid his shilling and entered one of the little News Theatres. 1961 S. Chaplin *Day of Sardine* ii. 54 We nipped into the News Theatre one night and saw one of these Barbecue affairs, in colour. 1974 E. Lemarchand *Buried in Past* ix. 152 Enquiries at steak houses and news theatres in the Tottenham Court Road area. 1902 News ticker [see Ticker² b]. 1933 Balmer & Wylie *When Worlds Collide* v. 46 The news-ticker carried, as additional information, only the effect of the announcement on the markets in Europe. 1967 Mrs. L. B. Johnson *White House Diary* 1 Aug. (1970) 550 It was 11 when we got back..and Lyndon wanted me to go into his office with him to take that last look at the news ticker. 1820 *Rep. Comm. Working on Newspapers* in E. Howe *London Compositor* (1947) xv. 378 The Committee commenced their labours by tracing the Regulations for News Work back to a certain period. 1971 *Library* XXVI. 302 If we can trust the 'oral testimony' reported in 1820, long galleys were in use in newswork as early as 1770.

news-boy. (Earlier and later examples.)
1764 in O. E. Winslow *Amer. Broadside Verse* (1930) 205 The news-boy's christmas and new-year's verses. 1939 T. S. Eliot *Family Reunion* i. i. 43 The hidden shall be exposed and the newsboy shall shout in the street.

new·scast, *sb.* [Modelled on *BROADCAST sb. 2.] A broadcast of the news on radio or television. Also as *vb.* So **new·scaster,** (*a*) a person who reads out the news on radio or television; (*b*) (see quot. 1966); **new·scasting** *vbl. sb.*
1930 *Observer* 28 Sept. 21 Graham MacNamee, the news-caster of our American newspaper newsreel, takes the part of an unseen dramatist. 1934 M. Weseen *Dict. Amer. Slang* xii. 169 *Newscast*, to broadcast news by radio; a radio report of news. 1943 *Amer. Speech* XVIII. 147 The -*cast* of *broadcast* long since cast loose to attach itself in *newscast*. 1956 *Ann. Reg.* 1955 358 I.T.A. news was planned, intentionally, as something different from the traditional, wholly dignified, and impartial B.B.C. news, and was given instead from a personal angle, from less orthodox sources, by a skilful team of 'newscasters'. 1958 *New Statesman* 26 July 106/2 ITN one evening played martial music..during the newscasting between shots. 1966 *Times* 8 Mar. 4/5 An electric newscaster is to be erected in Piccadilly Circus. A newscaster was one of three illuminated signs for the Circus approved by the Minister..last month... Headline news bulletins, containing up-to-the-minute news from all over the world, will be flashed direct from the Press Association. Interspersed with the news bulletins will be advertising 'commercials' in four colours. 1970 *Toronto Daily Star* 24 Sept. 30/4 When the CBC hits it lucky, nobody can beat its newscasts. 1972 J. Wainwright *Requiem for Loser* viii. 171 He, too, momentarily disbelieved the B.B.C. newscaster... The wireless set continued to broadcast the news item. A lot of people listened to the newscast. 1972 'G. Black' *Bitter Tea* (1973) xi. 180 The American style of newscasting, a breathless outpouring of words. 1973 *Daily Tel.* (Colour Suppl.) 9 Mar. 21/4 We've always had great success using newscasters for our appeals, and we prefer to use them, to make it a current affairs thing. The BBC don't see it our way, quite. 1973 *Daily Tel.* 3 Dec. 14/2 The first business to become a victim, presumably unintentionally, of the Government's ban on outdoor illuminated display advertising is Newscaster Publicity, which runs the newscaster on the outside of the Swiss Centre in Leicester Square.

new school. a. An advanced or liberal faction of a party or organization, *spec.* applied to the section of the Presbyterian Church of the United States which was separated from the rest of that Church between 1838 and 1869. Chiefly *attrib.*
1806 T. Fessenden *Democracy Unveiled* i. 113 That were not justice in arrears, These New School folks would lack their ears. 1816 *Emigrant's Guide* 17 Local politicians assume various appellations, such as New School and Old School Democrats, Snyderites, Clintonians, and many others, mostly derived from the name or principles of some popular demagogue. 1837 W. Jenkins *Ohio Gaz.* 317 The public buildings consist of..two presbyterian churches, one of the old, and one of the new school. 1837 J. M. Peck *Gaz. Illinois* (ed. 2) i. 72 McDonough College..is identified with the interests of the 'old school' Presbyterians, as the Illinois College at Jacksonville is with the 'New School' Presbyterians. 1872 R. G. McClellan *Golden State* 406 The Old and New School Presbyterians have..2,600 members and 3,500 Sunday-scholars. 1884 P. Schaff *Relig. Encycl.* III. 2306/1 The 'New-Haven Theology'..was one of the most influential of the types of so-called 'New School Divinity'. 1959 *Chambers's Encycl.* XI. 184/1 The controversies of the old country were reflected in the colony, for instance, in the divisions of the Philadelphian presbyterians into the Old and New sides or schools. 1961 K. S. Latourette *Christianity in Revolutionary Age* III. 87 The congregations which had sprung up from the Plan of Union, the 'New School' Presbyterians, were made up mainly of New Englanders and their descendants. 1967 D. T. Kauffman *Dict. Relig. Terms* 329/2 *New School*, term for that part of American Presbyterianism in mid-nineteenth century which favored liberal positions in theology, church government, and social issues such as slavery.

b. *transf.*
1837 *Southern Lit. Messenger* III. 107 As I once read medicine..under a disciple of the 'new school' (vulgarly called steam doctors). 1838 J. F. Cooper *Eve Effingham* II. iii. 84 But they evidently inclined to the opinion that the new school of pews was far better than the old. 1974 *Times* 18 Apr. 19/6 We conclude that the New School are mistaken in relying on some 'black box' stability in the relationships between the various sectoral balances.

newsie (niū·zi). Chiefly *U.S.* and *Austral. colloq.* Also **newsy.** [f. News *sb.* (*pl.*) + -IE, -Y⁶.] = NEWS-BOY.
1875 *North Alabamian* (Tuscaloosa, Ala.) 1 July 3/2 A newsboy on the M. and C. road was cruelly beating a dog which had jumped on the train, when its owner suddenly appeared at the car door, knocked 'newsy' off and commenced to pay him in his own coin. 1889 *Kansas City Times & Star* 12 Jan., Nearly 900 'newsies' applied for licenses and badges. 1904 *N.Y. Times* 16 July 7 He approached the 'newsy' and offered to buy a paper. 1916 C. Sandburg *Chicago Poems* 42 The newsies are pitching pennies. 1953 Baker *Australia Speaks* iv. 105 *Newsie*, a paper seller. 1962 *John o' London's* 4 Jan. 7/2 To be polite the newsie took a couple of swigs of it. 1962 J. Onslow *Bowler-Hatted Cowboy* i. 10 One early morning as the 'newsy' passed my berth, I leant out to ask him quietly for some fruit. 1969 'E. Lathen' *Murder to Go* (1970) xvi. 161 The newsie in the lobby.

newsily (niū·zili), *adv.* [f. Newsy *a.* + -LY².] In a newsy manner or style.
1949 'N. R. Nash' *Young & Fair* 17 Have a nice summer? Patty (*Newsily*): Oh, fine. We got here in August. 1967 *Listener* 3 Aug. 136/3 Twenty years ago it [sc. a newspaper] was laid out rather newsily too.

news-letter. Delete 'Now only *Hist.*' and add further examples. Also, a periodical sent or handed out to subscribers, members of an organization, etc.
1880 'Geo. Eliot' *Let.* 19 Sept. (1956) VII. 325 Your 'news-letter' was very welcome, and I was especially glad to know that Gertrude tries to keep up her health by little devices of change. 1914 *Writer's Mag.* Jan. 313/2 Some of the journals take daily news letters. 1936 *Time* 29 June 26/2 He runs a publishing house with offices in Chicago and Philadelphia, keeps his friends informed in periodic news-letters which he calls 'Rainbow-Graphs'. 1961 'E. Lathen' *Banking on Death* (1962) xi. 91 Martin seems to be a persistent reader of stock market newsletters. 1972 *Physics Bull.* Sept. 562/2 The purpose of the new Group will be..to circulate newsletters to Group members and to promote the general advancement and dissemination of knowledge in the field of neutron scattering.

news-man. Also **newsman.** Add: **1.** Also, a newspaper reporter, a journalist.
1953 *Manch. Guardian Weekly* 7 May 2. 1958 *Church Times* 24 Jan. 4/3 None the less, the theatre critic or the theatre newsman, each with his seat in the stalls on most nights in the year, must sit through many weary hours. 1968 *Globe & Mail* (Toronto) 17 Feb. 7/5 Some newsmen were concerned about Mr. Trudeau's talk of a government information service. 1970 N. Armstrong et al. *First on Moon* iii. 64 Deke Slayton went outside to tell newsmen, standing in the glare of television camera lights, that the crew would soon be out. 1973 *Nature* 2 Feb. 300/1 A high-ranking newsman from the BBC. 1974 *Anderson* (S. Carolina) *Independent* 23 Apr. 3B/3 The Envoy Towers was evacuated and among its dazed occupants a newsman counted three dozen persons, children included, bleeding from head and body cuts sustained when windows blew in.

newspaper, *sb.* Add: **b.** *newspaper account, advertisement, advertising, agent, article, boy* (further examples), *carrier, clipping, column, correspondence, corresponding, critic, cutting, directory, editor, hack, kiosk, letter, man* (further examples), *office, owner, paragraph* (earlier example), *postage, press* (earlier and later examples), *printing, proprietor, reader, reporter, round, seller, selling, stand, syndicate, woman* (further examples), *wrapper.*
1851 D. B. Woods *Sixteen Months at Gold Diggings* 199 Divesting the newspaper accounts from California of certain expressions bordering rather too much upon the hyperbolic order, they amount to the fact that the outcrops of certain veins [of goldbearing quartz] have been removed. 1936 *Discovery* Dec. 384/2 It was not until February, 1626, that the first newspaper advertisement appeared. *Ibid.*, The next step was newspaper advertising. 1874 'G. Hamilton' *Twelve Miles* ii. 30 The religious newspaper-agents bore into your house like worms of the dust. 1832 J. S. Mill in *Tait's Edin. Mag.* II. 343 Books are run through with no less rapidity, and scarcely leave a more durable impression than a newspaper article. 1858 *Missouri Democrat* 23 Oct. 2/2, I have not seen the letter, and but very few of the newspaper articles on the subject. 1972 D. Wainwright *Journalism made Simple* iv. 109 A magazine article needs an attractive and if possible startling opening sentence, like a good newspaper article. 1920 M. Beer *Hist. Brit. Socialism* II. iv. xiii. 249 He played the part of a newspaper-boy. 1974 P. Lovesey *Invitation to Dynamite Party* iii. 32 Newspaper-boys..bawled their wares. 1851 C. Cist *Sk. Cincinnati in 1851* 50 Occupations... Newspaper publishers, 9; Newspaper carriers, 23. 1926 *Daily Colonist* (Victoria, B.C.) 9 Jan. 14/3 With the neck broken and a deep gash in the head, the body of William Merchant, fourteen-year-old newspaper carrier, was found last night. 1906 W. Churchill *Coniston* ii. xi. 374 She had brought a note from her father... Two newspaper clippings fell out of it. 1958 C. Watson *Coffin scarcely Used* iv. 33, I am here now instead of concocting a mysterious message from newspaper clippings. 1974 J. Banning *How I fooled World* iv. 23 Rosemary was putting teleprinter and newspaper clippings into the filing cabinet. 1843 M. Fuller *Summer on Lakes* (1844) vi. 185 Has ever Art found..a richer theme..sketched carelessly in the newspaper column of to-day? 1868 'Mark Twain' *Lett. to Publishers* (1967) 14, I have cut my newspaper correspondence down a good deal. *Ibid.* 15 If you can stand an advance, I wish you would, and relieve me of this newspaper corresponding until July. 1859 B. Bodichon *Let.* 28 June in *Geo. Eliot Lett.* (1954) III. 103 The book could not have succeeded if it had been known as hers; *every newspaper critic would have written against it* (!!!). 1907 *Yesterday's Shopping* (1969) 436D Albums for Newspaper Cuttings. 1886 'Mark Twain' *Lett. to Publishers* (1967) 206 Please take a glance..at your Newspaper Directory and tell me the aggregate number of dailies in the U.S., big cities and all. 1785 *Daily Universal Register* 1 Jan. 4/1 A Newspaper Editor..should rest himself on truth and facts. 1837 H. Martineau *Society in America* I. i. iii. 151 The majority of newspaper editors made themselves parties to the act, by refusing, from fear, to reprobate it. 1972 C. Wintour *Pressures on Press* i. 6 It is essential for newspaper editors to be concerned with accuracy. 1885 Newspaper-hack [see *CIPHERER 2]. 1894 E. L. Shuman *Steps into Journalism* 65 One of the most prolific newspaper hacks in Chicago once remarked that he did not consider a man..a reporter unless he could make good reading out of anything. 1975 N. Luard *Robespierre Serial* xvii. 151 Carswell..stopped at a newspaper kiosk. 1868 'Mark Twain' *Lett. to Publishers* (1967) 15 In order to give to the book the amount of attention it really requires I shall have to cut loose from everything but one, and sometimes two, newspaper letters a week. 1954 G. Kersh in D. Knight *100 Yrs. Sci. Fiction* (1969) iv. 217 I'm a war correspondent, and newspaper-man, and so I have the right to ask impertinent questions. 1966 *Listener* 21 July 79/1 For newspapermen throughout the world Washington has always been the Mecca of journalism. 1972 J. Mosedale *Football* v. 71 Newspapermen..confessed they could not look on him objectively. 1834 J. S. Mill in *Monthly Repos.* VIII. 173 The priest of the nineteenth century..sets up his pulpit in a newspaper office. 1915 R. Fry *Let.* 27 Aug. (1972) II. 390 Those who encounter the enemy in the newspaper offices are the most bloodthirsty. 1966 Harris & Spark *Pract. Newspaper Reporting* iii. 34 Different newspaper offices mean different things when they refer to 'district reporting'. 1975 A. Fraser *Whistler's Lane* x. 159 I'll take the bus into Clitheroe...that's where the nearest newspaper office is. I want to look up the files. 1959 *Chambers's Encycl.* IX. 845/2 The fears of journalists in Britain that similar systems of chain newspapers, all propagating the views of one newspaper owner, might have ill effects both on the journalistic profession and on the newspapers. 1961 A. Clarke *Later Poems* 92 The leader of the Dublin capitalists was William Martin Murphy, a newspaper owner and ruthless clericalist. 1798 *Deb. Congress U.S.* 5 July (1851) 2107 The gentleman from Connecticut..had communicated to the House..a number of newspaper paragraphs. 1812 *Niles' Reg.* I. 361/1, I..admit your publication to be a newspaper and to be rated at Newspaper Postage. 1829 J. S. Mill *Let.* 7 Nov. in *Wks.* (1963) XII. 38 You know in how low a state the newspaper press of this country is. 1837 H. Martineau *Society in America* I. i. iii. 75 Of all newspaper presses, I never heard any one deny that the American is the worst. 1959 *Chambers's Encycl.* IX. 843/1 In America only the periodical press is national and therefore has priority for advertising purposes over the newspaper press. 1824 J. Johnson *Typographia* II. 651 (*heading*) Newspaper printing offices. 1847 H. Howe *Hist. Coll. Ohio* 241 Kenton..now contains..1 newspaper printing office. 1974 *Encycl. Brit. Macropædia* XV. 243/1 The practice resulted in considerable overmanning in newspaper printing departments. 1885 *List of Subscribers, Brighton* (S. of Eng. Telephone Co.) 14 Printers and Newspaper Proprietors. 1933 J. Buchan *Prince of Captivity* i. iv. 114 Falconet was..a newspaper proprietor on a large scale. 1974 *Encycl. Brit. Macropædia* XV. 243/2 Many of Scripps' methods were adopted by his rivals and by newspaper proprietors in other countries. 1935 *Discovery* Aug. 244/1 To the average newspaper-reader, Mongol is but a generic term. 1959 *Chambers's Encycl.* IX. 842/2 The steadily growing body of potential newspaper readers which increased educational facilities had brought into being. 1813 *Theatrical Inquisitor* II. 213 Newspaper critics and reporters..have had a prodigious addition to their necessary employments. 1910 'O. Henry' *Strictly Business* 87 The newspaper reporters dug out of their trunks the old broad-brimmed hats and leather belts. 1963 L. E. & B. Ryan *So you want to go into Journalism* i. 27 Newspaper reporters make modest salaries. 1948 C. Day Lewis *Otterbury Incident* iv. 48, I was a bit late, the newspaper-round taking longer than I expected. 1973 J. Wayne *Brown Bread & Butter* x. 183 Was she eating breakfast, wasn't the newspaper round too much? 1927 C. Parsons in *Oxford Poetry* 24 Beneath me in the windy stir Newspaper-sellers advertise The death of a philosopher By unintelligible cries. 1974 *Country Life* 28 Nov. 1656/3, I had a word..with a newspaper seller outside the building. 1957 J. Kerouac *On Road* (1958) 58 The little midget newspaper-selling woman. 1893 W. K. Post *Harvard Stories* 31 At a news-paper stand he bought all the picture papers. 1889 W. D. Howells *Hazard of New Fortunes* II. 120, I told 'em I hadn't much practice with Go-devils in the newspaper syndicate business. 1891 'Mark Twain' *Lett. to Publishers* (1967) 275 McCluny..the manager of the newspaper syndicate. 1925 F. Scott Fitzgerald *Great Gatsby* vi. 120 Ella Kaye, the newspaper woman. 1954 D. Dodge *Lights of Skaro* ii. 55 She was a very good newspaperwoman, and not too scrupulous. 1971 B. Graham *Spy Trap* i. 13 Christ! A bloody newspaper woman! He told her everything! 1873 *Brit. Postal Guide* 1 Jan. 21 Every Head Postmaster is required to keep, for sale to the public..newspaper wrappers bearing an impressed halfpenny stamp, and Post Cards. 1926–7 *Army & Navy Stores Catal.* 363/2 *Newspaper wrappers*. Size 12⅛ × 4⅞ in., plain, gummed.
c. newspaper English, the style of English used in newspapers; journalese; **newspaper stamp,** a stamp tax imposed on newspapers between 1711 and 1855.
1888 *Harper's Mag.* May 962/2 The phrase 'newspaper English' has come to have a significance which is not flattering to newspapers. 1942 P. G. Perrin *Writer's Guide & Index* 606 Good newspaper English is simply informal English applied to the daily recording of affairs. It is a style written to be read rapidly and by the eye—tricks of sound outside the headlines are out of place. 1947 Partridge *Usage & Abusage* 38/2 *Betrothal* [for 'engagement'] and *betrothed* are current in American newspaper-English. 1835 J. S. Mill in *London Rev.* I. 513 It was understood..that the ministry intended to take off the newspaper stamps. 1956 J. E. Gerald *British Press under Govt. Econ. Controls* i. 5 The newspaper stamp duty..was allowed to lapse in 1855.
2. *Underworld slang.* (See quots.)
1926 Maines & Grant *Wise-Crack Dict.* 11/2 *Newspaper*, crook's term for thirty days in jail. 1931 G. Irwin *Amer. Tramp & Underworld Slang* 134 *Newspaper*, a thirty days' gaol sentence. 1949 Partridge *Dict. Underworld* 467/2 *Newspaper*, a thirty-days jail sentence. ..The time it takes an illiterate to read one.

newspaper (niū·zpeipəı), *v. U.S. rare.* [f. the sb.] To work on a newspaper, to do newspaper work. Cf. Newspapering *vbl. sb.* in Dict. and Suppl.
1943 *Time* 8 Mar. 64 He had newspapered in Hawaii. 1959 *Time* (Atlantic ed.) 6 July 11 Who newspapered in Chicago.

newspaperdom. (Further examples.)
1933 *Times Lit. Suppl.* 30 Mar. 223/4 *Shooting the Bull*..is a tongue-in-the-cheek march through newspaperdom. 1946 *Sun* (Baltimore) 9 May 12/3 [A] new era of independent journalism in country newspaperdom.

newspapered, *ppl. a.* (Further examples.)
1855 Dickens in *Household Words* 29 Sept. 193/1 Every house was shut up and newspapered. 1926 W. W. Bishop *Backs of Books* 229 We are the most newspapered and magazined nation on earth, I suppose. 1973 E. Williams *Emlyn* i. 7, I saw my two brothers sitting at the table neatly newspapered to save the cloth.

newspapering, *vbl. sb.* (Later examples.)
1911 E. Ferber *Dawn O'Hara* iii. 29, I would fall to thinking of those years of newspapering—of the thrills of them, and the ills of them. 1968 L. J. Braun *Cat who turned on & Off* (1969) i. 11 Interviewing artists, interior decorators and Japanese flower arrangers was not Qwilleran's idea of newspapering. 1971 *Daily Colonist* (Victoria, B.C.) 26 Sept. 2/1, I care because newspapering is my business—make that my life, not business.

newspaperish, *a.* (Earlier and later examples.)
1825 M. Wilmot *Let.* 26 Sept. (1935) 225 And so ends my story, which is a stupid newspaperish sort of thing, tho' it was exceedingly interesting and amusing and pretty to look at at the time. 1929 A. Noyes *Return of Scare-Crow* iv. 54, I know that it's all very noble and distinguished and broad-minded and generally newspaperish.

new·spaperland. [f. Newspaper + Land *sb.*] = Newspaperdom.
1910 *Daily Chron.* 22 Mar. 8/2 As he is one of those who sit in the high places of newspaperland his voice is an authority that must not be overlooked. 1931 G. B. Shaw in *Ellen Terry & Bernard Shaw* 338 In newspaperland a slander is interesting news, always welcomed by news-editors. 1931 T. H. Pear *Voice & Personality* ii. 20 In newspaper-land, too, persons in danger when sleeping, escape in 'night attire'.

new·spaperless, *a.* [f. Newspaper + -less.] Without a newspaper.
c 1889 'Mark Twain' *Speeches* (1923) 152 This was a newspaperless globe. 1909 *Westm. Gaz.* 10 Aug. 7/2 Newspaperless Sweden... No newspapers..are published here this morning. 1920 *Glasgow Herald* 1 Sept. 6 The demand for a distantly printed issue imported into Liverpool was a revelation of the eagerness with which a newspaperless public will snatch at anything conveying a semblance of the news of the day.

Newspeak (niū·spīk). Also **newspeak.** [f. New *a.* + Speak *v.*] The name of the artificial language used for official communications in G. Orwell's novel *Nineteen Eighty-Four*, freq. applied to any corrupt form of English, *spec.* the propagandist and ambiguous language of some politicians, broadcasters, etc. Also *attrib.*
1949 'G. Orwell' *Nineteen Eighty-Four* i. 51 Syme was a philologist, a specialist in Newspeak. Indeed, he was one of the enormous team of experts now engaged in compiling the Eleventh Edition of the Newspeak Dictionary. *Ibid.* ii. 133 Do you know the Newspeak word *goodthinkful*? *Ibid.* App. 299 Newspeak was the official language of Oceania and had been devised to meet the ideological needs of Ingsoc, or English Socialism. In the year 1984 there was not as yet anyone who used Newspeak as his sole means of communication, either in speech or writing. 1950 A. A. Roback *Personality in Theory & Practice* i. 27, I do not think it necessary to resort to 'Newspeak' in order to write scientifically. 1959 *New Statesman* 2 May 602/2 This cynical 'newspeak' naming of Nationalist legislation has, in recent months, been matched by a remarkable change in the language used by their press and politicians. 1961 Y. Olsson *On Syntax Eng. Verb* vi. 148 Even George Orwell's progressive Newspeak still preserves a few 'clumsy remnants of a bygone past'. 1963 *Guardian* 8 Mar. 10/6 Mr John Snagge, Head of Presentation at the BBC, is asking applicants for jobs as announcers to read aloud the 'Guardian's' leading article about 'newspeak', a method of speech, said by the writer to have become common among broadcasters, in which fullstops are put in the middle of sentences. 1966 *Punch* 27 July 140/1 Accusing the Prime Minister of 'the same old excuses', it [sc. the *Daily Telegraph*] labelled 'redeployment' as 'new-speak', which would be 'victimisation of the workers' in any but a Labour Government. 1972 *Guardian* 17 Feb. 14/5 The Orwellian Newspeak style. 1972 *Times Lit. Suppl.* 11 Aug. 935/2 The new party line, directed this time against 'rootless cosmopolitans'—newspeak for Jews. 1975 *Ibid.* 31 Jan. 115/4 A Khrushchevian panache which still makes a refreshing contrast with the computerized newspeak that passes for political discourse among many of his successors.

newst, var. *NOUST.

newsworthy (niū·zwɜ̄ıði), *a.* [f. News *sb.* (*pl.*) + Worthy *a.*] Of sufficient interest to the general public to warrant mention in the news. Also **new·sworthily** *adv.*; **new·sworthiness.**
1932 *Time* 15 Feb. 4/2 *Time* is grateful for support and criticism of its policy, and repeats its promise to cause the minimum of offense in respect to newsworthy oaths. 1936 L. C. Douglas *White Banners* v. 88 The manner of the taciturn chemist's departure had doubled the insurance his widow would have received in the event of his having died less newsworthily. 1940 *Harper's Mag.* Nov. 588a The Nazi offensive..is, judging from our newspapers, apparently the only newsworthy topic on Latin America. 1951 L. Z. Hobson *Celebrity* (1953) x. 145 They were finding his speech a shade less newsworthy than they were supposed to. 1957 *Observer* 29 Sept. 10/7 His lot is harder than Mr. Colvin's, if not so newsworthy. 1960 'M. Cronin' *Begin with Gun* vii. 82 A celebrated French *modiste* who had become eminently newsworthy..had consented to be interviewed. 1961 *Guardian* 22 Mar. 10/4 Publicity-conscious people..make 'newsworthiness' a determinant of their actions. 1967 G. Playfair *Prodigy* vii. 167 His newsworthiness had dwindled. 1969 *Listener* 24 July 113/2 Not that by living life more newsworthily one could hope to provide more than the raw material. 1970 G. F. Newman *Sir, You Bastard* iii. 89 They might get a lot of newsworthy items, but no exclusives. 1972 M. Williams *Inside Number 10* xiv. 345 While the general public will know the names of the political heads of the more newsworthy ministries, there is rarely any knowledge of the names of the permanent features of the departments, the Civil Service heads. 1973 *Guardian* 13 Apr. 11/2 The newsworthiness of the [Black Panther] movement.

newsy, *a.* Add: **2.** Likely to create news.
1959 *Vogue* Nov. 83 A newsy new car, the red sleek Daimler. 1971 *Daily Tel.* 28 July 11 Tweed sprouts all over the collection, best in separates such as..the ponchos worn over slim skirts or (far more newsy) skin-tight tweed trousers.

newsy, var. *NEWSIE.

New Thought, new thought. [Thought¹ 1.] A theory of the nature of disease, a system of therapeutic practice, and a religious sect believing in these, founded on principles formulated by Phineas P. Quimby of Portland, Maine, U.S.A. Hence **New Thoughter,** one who holds and practises 'new thought'; a member of the sect following the principles of Quimby. Also *transf.*
1887 W. H. Holcombe *Condensed Thoughts Christian Sci.* (ed. 6) 45 New thought always excites combat in the mind with the old thought, which refuses to retire. 1899 H. W. Dresser *Voices of Freedom* ii. 22 The term 'New Thought', now the accepted appellation of a doctrine which has differentiated itself from..mental science..and become the representative teaching of those who..are not worshippers of personality, are not bound to certain books, but remain independent. *Ibid.* 23 The New Thought..is both a philosophy of life and conduct and a mode of healing. 1902 W. James *Var. Relig. Exper.* iv. 94 A current..has recently poured over America..to which..I will give the title of the 'Mind-cure movement'. There are various sects of this 'New Thought', to use another of the names by which it calls itself. 1907 —— in *American Mag.* Nov. 58/2 To relax, to say to ourselves (with the 'new thoughters') 'Peace! be still!' is sometimes a great achievement of inner work. a 1910 —— *Some Probl. Philos.* (1911) i. 18 Compare Prentice Mulford and others of the 'new thought' type. 1937 D. Canfield *Fables for Parents* 128 Being a Congregationalist and not a Christian Scientist or a New Thoughter. 1942 A. Huxley *Let.* 30 Dec. (1969) 485 The strengthening and extension of such religions as Christian Science, Theosophy, New Thought, 'I Am', in which the stress is wholly on powers, personal advantages and future time. 1949 *Sun* (Baltimore) 7 Sept. 12/4 Loss of that strength would weaken the old-line men as against the groping 'new-thoughters' in the party. 1967 D. T. Kauffman *Dict. Relig. Terms* 330/1 *New Thought*, American religious movement,..emphasizing 'the divinity of man and his infinite possibilities..'. Groups are now found in many parts of the world.

Newton (niū·tən). [Name of Sir Isaac *Newton*: see Newtonian *a.* and *sb.*] **1.** *Newton's rings*: a set of concentric circular fringes seen surrounding the point of contact when a convex lens is placed on a plane surface (or on another lens), which join points where the intervening thin layer of air has the same thickness and are caused by interference between light reflected from its upper and lower surfaces. When used *attrib.* also *sing.*
[1807 *Phil. Trans. R. Soc.* XCVII. 180 (*heading*) Experiments for investigating the cause of the coloured concentric rings, discovered by Sir Isaac Newton, between two object-glasses laid upon one another. 1809 *Ibid.* XCIX. 299 In order completely to account for the Newtonian rings.] 1835 *Phil. Mag.* VII. 363 In examining Newton's rings I was induced to place a convex lens..between two surfaces of plate-glass, in order to effect the superposition of the rings. 1904 A. Schuster *Introd. Theory of Optics* iv. 72 The colours observed in Newton's rings are the colours of thin films, the film being the layer of air included between the lens and the plate. 1969 S. G. & H. Lipson *Optical Physics* xii. 382 Lens surfaces can be tested to a high degree of accuracy by forming Newton's rings with a known surface. *Ibid.* 383 Coefficients of thermal expansion of quite small crystals can be measured by counting the fringes that pass through the centre of a Newton's ring system as the crystal is heated.
2. (Written **newton.**) The unit of force in the metre-kilogramme-second system (now incorporated in the S.I.): the force that would give a mass of one kilogramme an acceleration of one metre per second per second; 100,000 dynes (approximately the weight of 102 grammes or 3·6 oz.). Abbrev. N.
1904 D. Robertson in *Electrician* 22 Apr. 25/1 The writer suggests that the name 'Newton' be given to the unit of force (10^8 dynes). 1919 *Nature* 4 Sept. 13/2 A fourth matter..which was discussed was the proposal to adopt in future legislation for metric countries the M.K.S. system of units... On this system the unit of force is the 'Newton'. 1935 Hartshorn & Vigoureux in *Ibid.* 7 Sept. 397/2 In the M.K.S. system..no name has yet been assigned to the unit of force... G. Giorgi has provisionally used the word 'vis'... We venture to suggest that 'newton' would be more appropriate. 1942 H. Howe *Introd. Physics* iv. 57 The awkwardly large numbers in Ex. 2 would have been avoided had mks units been used; the answer would have been F = 300 newtons. 1962 Corson & Lorrain *Introd. Electromagnetic Fields* ii. 29 In these units Coulomb's law is written as $\mathbf{F}_{ab} = (1/4\pi\epsilon_0)(Q_aQ_b/r^2)\mathbf{/r}_1$, where the force **F** is measured in newtons; the charges Q_a and Q_b, in coulombs..; and the distance r, in meters. 1970 *Daily Tel.* 14 May 18 In the S I the force is measured not in kilogrammes, but in newtons—spelt with a small 'n' but abbreviated with a large one, 'N'. One newton (reinforcing the legend) is roughly the pull of gravity on one apple.

Newtonically (niūtǫ·nikăli), *adv. rare.* [f. the name of Sir Isaac *Newton* (see Newtonian *a.* and *sb.*) + -ically.] In the manner of Sir Isaac Newton.
1869 J. S. Mill *Let.* 30 Jan. (1910) II. xii. 181 In regard to the Darwinian hypothesis,..Darwin has found (to speak Newtonically) a *vera causa*. 1953 K. Britton *John Stuart Mill* v. 182 His [sc. Mill's] preference for a mechanical view of forces..arises in part from his insistence upon 'speaking Newtonically' in opposition to the Baconian language of Macaulay and the empiricists.

New Town, new town. A planned urban area designed to ease the congestion of a nearby large city, usu. one with special provision for housing, employment, and amenities for a delimited population. Also *transf.* Also *attrib.*, as *New Town blues*, despondency or anxiety suffered by a person resident in a New Town. Hence **New Towner; New-Townish,** *a.*
1918 (*title*) New towns after the war. 1946 *Hansard Commons* 17 Apr. 2710 *New Towns Bill* 'to provide for the creation of new towns by means of development corporations, and for purposes connected therewith' presented by Mr. Silkin. 1948 'J. Tey' *Franchise Affair* xxiii. 266 The little house on the outer rim of the 'new' town. 1958 *Engineering* 21 Feb. 227/1 Suburban railway peak problems could be eased by more New Towns. 1958 *Spectator* 8 Aug. 191/1 An immense new-town bicycle shed. 1958 *Listener* 23 Oct. 655/2 His analysis of the Hellenistic 'new town', Priene. 1959 *Manch. Guardian* 6 Aug. 4/2 The new-town voters. 1960 *News Chron.* 2 July 3/5 He was born in St. Pancras, London, and is now a Hemel Hempstead new towner. 1962 *Britannica Bk. of Year* 546/1 In Great Britain a topical term was New Town Blues, used to describe the dissatisfied, discontented condition of people from large cities who found the recreational amenities of the new towns inadequate. 1962 *Harper's Bazaar* Aug. 58/1 Castries, the capital..is entirely modern and rather New-Townish. 1964 *Daily Tel.* 4 Feb. 14/2 The new environment, though brighter, cleaner and more commodious, is said to lack the social attraction of the old. Hence the so-called complaint of the 'new town blues'. 1965 *Economist* 6 Nov. p. x/2 New towners also pay rates at the level set by the local authorities in whose territory they lie. 1970 *Americana Ann.* 176 The high hopes that 'new towns'..could provide a solution to congestion, high population densities, and the erosion of public services have collapsed. 1971 P. Gresswell *Environment* 174 A visit to one or more of the new towns is essential to anyone interested in architecture and environment. 1973 *Guardian* 30 May 7/1 The fears of the old-towners are certainly understandable. The population has declined... Runcorn new town has been growing. 1975 *Ibid.* 20 Jan. 15 Milton Keynes..is the latest (possibly the final?) refinement of new town theory.

Newtown (niū·taun). Also **Newton.** The name of a town on Long Island in New York state, used *attrib.* or *absol.* to designate two varieties of dessert apple first introduced there, the **Newtown Pippin** and the **Newtown Spitzenburg.** So **New-towner,** a Newtown apple.
1760 G. Washington *Diary* 30 Mar. (1925) I. 147 Grafted 10 of the New Town Pippin from Collo. 1770 in *Maryland Historical Mag.* (1918) XIII. 69 Things sent by the wagon 4 Barrills of Apples Russetins, Golden Pippins, Newtown Pippins & Pairmains. 1785 G. Washington *Diary* 12 Nov. (1925) II. 435 Received two New Town and 2 Golden Pippin trees. 1803 A. F. M. Willich *Domestic Encycl.* (Amer. ed.) III. 110 Newton Pippin. *Ibid.* IV. 182/2 The New-town Pippin, or New-York Rennet, a noble American apple. 1817 W. Coxe *View of Cultivation of Fruit Trees* 126 Newton Spitzemberg [*sic*]. This apple is in some parts of this State [sc. Pennsylvania] called the English, or Burlington Spitzemberg: it was brought from Newton on Long-Island. 1840 *N.Y. Mirror* 4 Apr. 327/2 Dealers in Newton pippins and maple candy tell you..they can't afford their accommodations so low as they can be afforded in Chatham and Church streets. 1846 J. F. Cooper *Ravensnest* I. i. 27 Their *poire beurrée*, here at Paris..will not compare with the Newtowners we grow at Satanstoe. 1860 R. Hogg *Fruit Manual* 18 Newtown Pippin, D. — Medium sized, roundish, rather irregular, and obscurely ribbed... Requires a wall in this country. December to April. 1863 *Rep. Comm. Agric. 1862* (U.S. Dept. Agric.) 168 Newtown Spitzenberg... A very hardy tree; good bearer; fruit of superior quality; keeps and bears transportation well. 1877 E. S. Dallas *Kettner's Bk. of Table* 34 Newtown Pippin... Named after Newtown, Long Island. This apple is imported from America. *Ibid.* 36 Kitchen Apples... Newtown Spitzemberg [*sic*]... Named by William Cobbett 'the matchless'. 1913 J. London *Valley of Moon* 364 Every year he goes to England, and he takes a hundred carloads of yellow Newton pippins with him. 1925 B. D. Drain *Essent. Systematic Pomology* v. 78 Newtown Spitzenburg Group. Medium size, round-oblate, red-striped, high quality. 1953 Brooks & Hesse *Western Fruit Gardening* 97 Yellow Newtown is most widely grown in the coastal districts... The fruit is medium to large, roundish to flattened, green to yellow.

new wave. Also **New Wave.** = *nouvelle vague. Also (with hyphen) *attrib.* Hence **new wavish** *a.*
1960 *News Chron.* 19 July 6/8 A 'new wave' is emerging here [in Spain], too, with an up-to-date philosophy. *Ibid.* 6 Aug. 6/7 Cy Grant will sing 'Carnival' from the French new-wave movie. 1960 *Guardian* 15 Oct. 5/2 The Italian neo-realists..and..the Frenchmen of the 'New Wave' have all been pursuing the same course. 1961 *Sunday Times* 12 Feb. 11/8 Blanchflower is the very crest of the New Wave among professional footballers. 1962 *Spectator* 23 Feb. 242/1 Her central situation concerns a new-wave actress. 1963 *Punch* 20 Feb. 285/1 Perhaps she's New Wavish enough to scorn love. 1967 *Economist* 18 Mar. 1008/2 The new-wave nationalists have not bothered to think out what sort of nationalism they want. 1972 *Newsweek* 10 Jan. 22/1 As New York's new-wave mayor in 1966, he had portrayed himself as a reform insurgent battling the city's 'power brokers'. 1974 *Publishers Weekly* 25 Nov. 47/1 Malzberg is rather on the New Wavish side, writing 'speculative' rather than straight science fiction. 1975 *Times Lit. Suppl.* 21 Nov. 1374/2 The defining characteristic of the New Wave, and its ambiguous legacy to all films made since 1958–61, was selfconsciousness... The New Wave brought the film director to the public's immediate attention as a potential cultural hero.

new-‿wed, *a.* and *sb. arch.* [New *adv.* 3.] **A.** *adj.* Recently married. **B.** *sb.* = *newly-wed* (*newly *adv.* 1 d).
1886 Kipling *Departmental Ditties* (ed. 2) 21 Jones had left his new-wed bride to keep his house in order. 1893 W. B. Yeats *Celtic Twilight* 94 The new-born or the new-wed moves henceforth in the bloodless land of Faery.

new woman, etc.: see Woman *sb.* 1 i.

New World, new world. [World *sb.* 11.] Used *attrib.* to designate phenomena characteristic of, or territories pertaining to, the western hemisphere. Hence **New-Worlder, New-Worldling,** a native or inhabitant of the western hemisphere; **newworldward** *adv.*, towards the western hemisphere.
[1823 Byron *Don Juan* ix. xxxix. 24 How the new worldlings of the then new east.] a 1855 C. Brontë *Professor* (1857) II. xviii. 7 Some natural and graphic touches disclosed to the reader the scene of virgin forest and great, New-World river—barren of sail and flag. 1863 H. W. Bates *Naturalist on Amazons* II. v. 324 All the New World genera of apes..are represented in the Amazon regions. 1867 O. W. Holmes *Guardian Angel* 216 There is a double proportion of oxygen in the New-World air. 1876 H. Melville *Clarel* I. ii. xxvi. 256 Oh, that a New-Worlder should talk so! 1901 *Munsey's Mag.* XXIV. 530/1 These and other causes have acted and reacted to

bring about, in this new world metropolis,..a crowding together of the poor. **1930** *Tablet* 24 May 686/1 Old Worldlings in Europe..the New-Worldlings of the U.S.A. **1935** E. E. Cummings *Let.* 29 Jan. (1969) 134 Delighted to learn that you're casting your anchor and blowing newnewworldward!!!!! **1944** *Amer. Fern Jrnl.* XXXIV. 69 (*heading*) The New World species of Azolla. **1951** M. McLuhan *Mech. Bride* (1967) 77/1 A new-world man towering tree and confident in an untroubled generation. **1957** P. J. Darlington *Zoogeography* vi. 346 Whether one of them [*sc.* a primitive primate] reached South America and evolved the New World monkeys..is unknown. *Ibid.* 390 Erethizontidae, New World porcupines. **1966** *New Statesman* 23 Sept. 435/1 Great Victorian relics, long since peeled of their paint, electric-light bulbs and courteous service—especially toward dollar-bearing New Worlders. These are some recent experiences of myself and a pair of Montreal acquaintances. **1968** *Times* 15 Nov. 8/7 The New World monkeys, or Ceboidea, include such species as the marmosets and the howler monkeys. **1970** *Nature* 24 Oct. 382/1 In Jamaica the fifth instar larvae of the New World hawkmoth, *Erinnyis ello*, exhibit four basic colours.

newy, var. *NEWIE.

new-year. Add: **1. a.** *to see the new-year in*: see *YEAR 7.

2. *New-year('s) honours list*: see *honour(s) list* (*HONOUR *sb.* 10).

3. Also with ellipsis of *day*. *N. Amer.*

1845 *Knickerbocker* XXV. 128 Stay away on New-Year's and you stay away all the year. **1909** *Springfield* (Mass.) *Weekly Republ.* 4 Nov. 1 The general elections are not now expected until after New Year's. **1951** J. Steakley in R. Weaver *Canad. Short Stories* (1960) 383 The Christmas holidays were haunted for me by fear of what would happen when I went back there after New Year's. **1973** F. Decker in S. Terkel *Working* iv. 190 She's just lucky he's home Christmas and New Year's.

New York (niū yǭ·ɹk, niū yǭ·ɹk). The name of a city and state in the United States used *attrib.* in various special collocations (see quots.).

1774 S. Sewall *Diary* 14 Apr. in *Mass. Hist. Soc. Coll.* (1879) 5th Ser. VI. 440, I had my New York Biscuit to eat. **1771** T. Pennant *Synopsis Quadrupeds* 367 New York Bat with a head shaped like that of a mouse. **1842** J. E. DeKay *Zool. N.Y.* I. 140 Index, New York Weasel. **1843** J. Torrey *Flora State of New-York* II. 497 *Aspidium Noveboracense*...New York Shield-fern. Moist woods and thickets. **1862** *Amer. Ann. Cycl.* 1861 307/2 The highest, lowest, and average quotations..at the New York Stock Exchange for the stocks most largely dealt in. **1892** *Scribner's Mag.* Sept. 361/1 A fellow-pupil dictated to him Latin and Greek, and he printed the text in New York point. **1901** C. Mohr *Plant Life Alabama* 316 *Dryopteris noveboracensis*...New York Shield Fern. Alleghenian and Carolinian areas. **1943** H. W. Shimer *Origin & Significance Plant Names* 47 New York fern. *Dryopteris noveboracensis*. **1955** R. Blesh *Shining Trumpets* (ed. 2) xi. 275 New York jazz, even at its best, was inept. **1956** B. Cobb *Field Guide to Ferns* 50 The Marsh Ferns, the Massachusetts Ferns, and the New York Fern have upright, narrow, oblong leaves... These five species of the genus *Thelypteris* have been moved from one genus to another... Elsewhere they are classified in the genus *Dryopteris*. **1974** *Guardian* 23 Jan. 4/7 The poultry industry..called in a slim blonde to wheel to the platform a 90lb New York dressed turkey—New York dressed being a dying species.

New Yorker. Add: (Further examples.) Also *attrib.*, pertaining to or characteristic of the magazine *The New Yorker* (founded 1925).

1859 G. H. Lewes *Let.* 6 Sept. in *Geo. Eliot's Let.* (1954) III. 148 To-day a letter has come from the editor of a 'Parish Magazine'—and really G. E. was almost more likely to be tempted by that audience than by the New Yorkers. **1884** Matthews & Bunner *In Partnership* 127 'Are you a New Yorker, sir?' 'From the north of the State.' **1902** *Chambers's Jrnl.* July 450/1 The New Yorker defends this wretched state of affairs by a peculiar argument. **1926** H. L. Wilson *Somewhere in Red Gap* 398 The New Yorker was now sunk deep in a trance. **1934** *Fortune* Aug. 75/1 No advertising man is believed, by the editors, ever to have understood a New Yorker joke. **1948** *Hearst's International* May 175/1 Literary critics and editors of other magazines are always referring to 'The New Yorker style of writing'. **1948** *N.Y. Star* 30 June 14/1 The Board of Transportation is appealing to New Yorkers to put up patiently with the condition. **1951** R. Hoggart *Auden* iii. 68 We should notice in this to the part of the mind which is tickled but put on its guard by the 'New Yorker' profiles. **1959** *Times Lit. Suppl.* 2 Jan. 4/2 He surveys the established Old Guard..the new 'realists'..., the New Yorker School. **1974** *Times* 4 Mar. 14/8 Brenda Bedanky, a New Yorker, dressed to resemble a man trying to look like Marlene Dietrich.

Hence **New Yo·rkerish** *a.*, characteristic or reminiscent of the magazine *The New Yorker*; **New Yo·rkerism**, an idiom, expression, or word peculiar to *The New Yorker*.

1948 H. F. Pringle in *'48: Magazine of Yr.* Apr. 87/2 His [*sc.* H. W. Ross's] sense of comedy is properly New Yorkerish and fantastic. **1951** *Time* 22 Oct. 102/2 Many a New Yorkerism (e.g., Cartoonist Carl Rose's 'I say it's spinach, and I say the hell with it') has become a part of the language. **1961** *Punch* 4 Jan. 81/3 The drawings.. [are] well suited to the captions, many of which have a *New Yorker*ish elliptical quality. **1967** *Guardian* 20 Oct. 7/4 Just another New Yorkerish monologue. **1970** D. L. Emblen *Peter Mark Roget* xv. 276 Punch in its *New Yorkerish* way, picking up a slip in some Scottish paper.

New Yo·rkese. [-ESE.] The regional form of English used in New York City. Cf. *MANHATTANESE *sb.*

1894 *Harper's Mag.* Oct. 695/1 'Cafe'..is New Yorkese for dram-shop. **1938** F. M. Ford *Let.* 11 Sept. (1965) 242 ...Latest New Yorkese which, I know, is always welcome in Hampstead. **1951** Trager & Smith *Outl. Eng. Structure* I. 23 The sequence /ɔy/ has always been known to him as 'New Yorkese' or 'Brooklynese'. **1973** E.-J. Bahr *Nice Neighborhood* vi. 71 'So continue,' said Don... He sometimes lapses into New Yorkese.

New-Yo·rkish, *a.* [-ISH¹.] = next.

1913 F. H. Burnett *T. Tembarom* xxiv. 305 How thoroughly New Yorkish it was that he should march into a fashionable shop and see that he got..the worth of his money! **1962** *John o' London's* 19 Apr. 385/3 That most New Yorkish of all musical comedies, *Guys and Dolls*.

New-Yo·rky, *a.* [-Y¹.] Suggestive or characteristic of New York.

1908 E. Wharton *Hermit* 150 To be compared to her! to be accused of being 'New-Yorky'! **1969** *Guardian* 25 Jan. 6/6 The 'Partisan', New York-y too, but out of a cosmopolitan liberal intelligentsia. **1973** R. L. Simon *Big Fix* (1974) ix. 66 The accent was nasal and New Yorky, the voice scratchy.

New Zealand. 1. The name of an Australasian country, used *attrib.* to designate plants native there, as **New Zealand flax**, an evergreen plant, *Phormium tenax*, of the family Agavaceæ, cultivated for the fibre it produces or the ornamental value of its tufts of long, stiff, pointed leaves; **New Zealand passion flower**, a climbing plant, *Tetrapathæa tetrandra*, of the family Passifloraceæ; **New Zealand spinach**, an annual herb, *Tetragonia tetragonioides* (*T. expansa*), of the family Aizoaceæ, cultivated for its thick leaves which are used as a substitute for spinach.

1811 W. Aiton *Hortus Kewensis* (ed. 2) II. 284 New Zealand Flax. Nat[ive] of New Zealand. Introd[uced] about 1789, by the Right Hon. Sir Joseph Banks. **1832** *Penny Mag.* 14 Dec. 365/1 The plant known by the name of New Zealand Flax..was thought to be a native of New Zealand... Since 1789 it has been cultivated by Sir Joseph Banks in 1797 did not succeed, but the New Zealand Flax was introduced through the medium of the same enlightened individual in 1789, and thence has been liberally distributed. **1881** W. Robinson *Eng. Flower Garden* 195/1 New Zealand Flax (*Phormium tenax*). **1910** [see *TREASURE *a.*]. **1939** *Times* 11 Jan. 10/3 He saw 2/4 The harakeke or New Zealand flax..doesn't resemble its European cousin. **1853** J. D. Hooker *Bot. Antarctic Voy.* II. *Flora Novæ-Zelandiæ* I. 72. The New Zealand Passion-flower is a perfectly smooth climbing plant, with alternate, simple, petiolate leaves, axillary tendrils, and small axillary panicles of green flowers. **1951** Laing & Blackwell *Plants N.Z.* (ed. 5) 281 *Tetrapathæa tetrandra* (The New Zealand Passion-flower). A slender climber, with glossy leaves. **1822** J. Anderson in *Trans. Hort. Soc.* IV. 458 (*heading*) Account of a new Esculent Vegetable called *Tetragonia*, or New Zealand Spinach. **1867** E. Sauter tr. *F. von Hochstetter's New Zealand* vii. 157 New Zealand spinach,..was first brought into notice by Captain Cook, who used it largely as an anti-scorbutic. **1896** New Zealand [see *ICE-PLANT]. **1951** *Dict. Gardening* (R. Hort. Soc.) III. 1369/2 In New Zealand Spinach is an excellent substitute for the common Spinach for use during the hot dry months. **1973** *Parade* (Melbourne) Sept. 35/1 I have seen this New Zealand spinach..growing on the sandhills along the beach..north of Perth.

2. New Zealand rabbit, also **New Zealand black**, **New Zealand red**, **New Zealand white**, various American breeds of domestic rabbit. Also *absol.*

1914 Roth & Coleman *Rabbit Culture* 95 Standard of the New Zealand Rabbit... Larger than the Belgian Hare and of a beautiful reddish buff color... Everybody's friend wherever known. **1917** C. P. Gilmore *N.Z. Red Rabbit* 83 The New Zealand Reds are..business rabbits for general or utility purposes. **1961** *Times* 24 Jan. 8/3 The New Zealand is practically a new rabbit in the American fancy. **1920** F. L. Washburn *Rabbit Bk.* ii. 43 The New Zealand Red. This rabbit is a close second to the Belgian and Flemish in the race for popularity. **1921** C. A. Kieran *Rabbit & Cavy Bk.* (ed. 4) 21 A new breed..called the White New Zealand..having the type of the New Zealand, but a pure white coat with pink eyes. **1953** C. Goodchild *Keeping Rabbits* ii. 80 The New Zealand Red was first imported into this country around 1920. It was originated in the U.S.A., and was a breed much favoured by the Californian breeders for the frying trade. **Ibid.** 82 Although the New Zealand White has only recently been adopted in this country, they have been one of the most popular breeds in the U.S.A. for over a quarter of a century. **1965** *Amer. Rabbit Breeders' Assoc. Official Guide Bk.* 32 The New Zealand Rabbit had its origin in the United States in the early 1900's. *Ibid.* 33 The standard of perfection for all New Zealands is identical with the exception of the color of the fur, eyes and nails... The Black New Zealand is the newest variety of the New Zealand breed. Blacks were started in California in 1949. **1971** *Guardian* 28 Dec. 1/6 Half the consignment [of rabbits] will be New Zealand Whites which have red eyes. **1973** M. I. Faivre *How to raise Rabbits* i. 11 The New Zealand breed may be divided into three distinct categories: New Zealand Red, New Zealand Black, and New Zealand White... Although all three types are excellent meat rabbits, the New Zealand White fur is in greatest demand by garment makers because it takes a variety of dyes successfully.

New Zealander. Add: **a.** (Earlier example.) Also (occas.) *attrib.* **b.** (Earlier and later examples.)

1773 Capt. J. Cook *Jrnl.* (1961) II. 268 The New Zealanders cut and scar themselves on the same account. **1837** *Sydney Gaz.* 11 Dec., in R. McNab *Old Whaling Days* (1913) ix. 167 They saw the chief officer, James George Bailey, a New Zealander. **1841** M. Edgeworth *Let.* 14 Mar. (1971) 586 A New Zealander boy with large head and frizzled black hair and face as yellow as dirty gold. **1886** J. A. Froude *Oceana* xx. 323 The Australian, the New Zealander, the Californian will have as much in them..of the ancient 'Merry England' as the severely earnest Northern American. **1966** Mrs. L. B. Johnson *White House Diary* 20 Oct. (1970) 430 We departed in a New Zealand airplane with the Prime Minister and Mrs. Holyoake to the tune of a thousand New Zealanders singing.

New Zea·landism. [-ISM.] An idiom or word peculiar to the English spoken in New Zealand.

1957 *N.Z. Listener* 22 Nov. 4/3 'Creek' and 'paddock' are New Zealandisms, because they mean something quite different in the English of England. **1964** A. S. C. Ross in Ross & Moverley *Pitcairnese Lang.* 13 The two letters which I published in New Zealand about possible new zealandisms in Pitcairnese.

nexal, *a.* Add: **2.** *Linguistics.* Of or pertaining to a nexus (sense *1 c).

1933 *English Studies* XV. 148, I don't wish to speak to you ever again..is a more polite formula for: I wish never to speak to you again. We have here a very curious case of 'cleaving of *never*' and 'nexal attraction'. **1939** V. Mathesius in *Mélanges de Linguistique et de Philologie offerts à J. van Ginneken* 82 Professor Jespersen has repeatedly called attention to the difference between what he calls *special negation* and what he calls *nexal negation*.

next, *a.*, *sb.*, and *adv.* Add: **A.** *adj.* **2. c.** *the next man (one, person, etc.)*: the average man; a typical man; the next comer. *orig. U.S.*

1857 *Lawrence* (Kan.) *Republican* 18 June 2 The Judge ..will probably talk as long to a crowd without tiring them as the next man. **1897** Kipling *Captains Courageous* i. 5 Guess I've as good right here as the next man. **1900** Ade *More Fables in Slang* 173 Lute was just about as Nifty as the Next One. **1902** S. G. Fisher *True Hist. Amer. Revol.* 146 We do not surrender our property to save our lives as well as th' nex' man. **1925** E. Caldwell *God's Little Acre* iii. 37 Will can dig as good as the next one, if he wants to. **1938** G. Greene *Brighton Rock* iii. iv. 135 He'd been a loyal old geezer, he hadn't a bigger heart than the next man. **1941** L. Bromfield in *Hearst's International* May 131/2 She thought: 'I'll behave as well as the next woman.' **1957** *Sunday Times* 5 Mar. 15/2 Cecil Beaton..can appreciate the 'excruciatingly bad taste' of a Lancashire living-room as well as the next designer. **1962** L. Peters *Snatch of Music* ix. 155, I can read a paperback translation with the next man. **1966** A. E. Lindop *I start Counting* xviii. 222, I can take a hint as well as the next person—and I know when I'm not wanted. **1973** *Sunday Bulletin* (Philadelphia) (Discover Suppl.) 14 Oct. 8/1, I feel you owe me a smidgin more than the next person.

10. (Further example.)

13. c. *to get* (*or be*) *next* (*to* or *on*): to become acquainted or intimate with, to come to know; to find out about, to understand, to become worldly-wise, to acquire self-knowledge. Also *to put next* (*to*): to acquaint (one) (with). *U.S. slang.*

1896 [see *BRASH *a.*² b]. **1896** Ade *Artie* xvi. 146 I've been next, I'll tell you those. **1900** — *More Fables* 86 She knew that the Treasurer of the Shoe Factory was Next to all these Boarding School Tactics. **1902** 'O. Dy' *Fables of Elkie* 85 'Do you get next to what I'm sayin'?' **1903** *Lit. Digest* 9 May ... I apprehended that a wink to the Wise is sufficient'. **1908** B. Tarkington *Gentleman from Indiana* ... Then along comes Paul Potter, and puts me next on how to write a play. **1908** K. McGaffey *Sorrows of Show-Girl* 31 ... [the] wise guy is a party who is 'next'. You had better drop in your penny and get next to yourselves. **1920** W. M. Raine *O'Connor* 215 Mrs. Mackenzie will put you next to the etiquette wrinkles where you are shy. **1930** Wodehouse *Gentleman of Leisure* vii. 66 Sure, he will.. He'll be good. He's next to the game, Sure. *Ibid.* x. 98 Boss, what's doin' here? **1923** E. O'Neill *Hairy Ape* ii. 24 Yuh don't know... **1923** E. C. Bentley *Trent's Last Case* vi. 93 'Has he any friends?' interjected Trent. Mr. Bunner [an American] glanced at him sharply. 'Somebody has been putting you next, I see that.' **1936** J. Tully *Bruiser* iv. 41 She took me for a hundred before I got next. **1950** A. Lomax *Mr. Jelly Roll* 19 If you could shoot a good agate.., I'm telling you were liable to get next to that broad. **1957** R. Stout *If Death ever Slept* (1958) vi. 73 Maybe you can get a lead to it through Brigham. Get next to him. **1969** 'H. Pentecost' *Girl Watcher's Funeral* (1970) II. iii. 114 He found out I was an actor... He told me he could put me next to some guy who was making underground films. **1973** *Black World* Sept. 4/2 If he can't get next to what we're about, we'll just have to school him.

next door. Add: **1. b.** By extension, the occupant of the adjoining house; so *Mrs. next-door*. Also, *next-door-but-one*, the occupant of the house two doors away.

1855 Dickens *Prince Bull* in *Household Words* 17 Feb.

51/1 One answered, 'I will if next door will;' and another, 'I won't, if over the way does.' **1933** D. C. Peel *Life's Enchanted Cup* xi. 127 It..occurred to him that our extra pair of guests might belong to one of the 'next doors'. **1935** *Punch* 27 Nov. 590/1 Next-door-but-one was rather a strain. Have you ever imagined canvassing through a speaking-tube? **1951** W. Morum *Gabriel* i. ii. 20, I showed it to Mrs. Next-door. *Ibid.* vii 95 Aunt Amy came in carrying a roast chicken... 'One of next-door's,' she said. **1969** Willmott & Young *Family & Class in London Suburb* ix. 109 Next door but one has gone when we go on holiday. Last year she broke her arm and I used to go in and help her. **1962** *Guardian* 30 July 4/5 Most of us have had to pick up Mrs Next-Door's torn-up letters.

2. b. *boy next door*: see *BOY *sb.*¹ 8; (*the*) *girl next door*: see *GIRL *sb.* 2 h.

nexus. Add: **1. c.** In Jespersen's terminology, a group of words containing a verb, or a predicative (with ellipsis of verb); a predicative relation or a construction treated as such. Freq. *attrib.*

1924 O. Jespersen *Philos. Gram.* vii. 97 If now we compare the combination *a furiously barking dog*..with *the dog barks furiously*, it is evident that the same subordination obtains in the latter as in the former combination... We shall call the former kind *junction*, and the latter *nexus*. *Ibid.* ix. 122 A nexus-object is often found: 'I found *the cage empty*'... which is easily distinguished from 'I found *the empty cage*' where *empty* is an adjunct. *Ibid.* 126 The subject-part (primary) of a Latin nexus-subject may be an accusative-with-infinitive or a clause. *Ibid.* x. 138 Nexus-substantives are also often convenient in cases where idiomatic usage does not allow a dependent clause, as after *upon* in 'Close upon his resignation followed his last illness and death'. *Ibid.* xxiv. 329 So we may therefore call questions of this kind (ex. yes-or-no) *nexus-questions*. **1928** — *Internat. Lang.* ii. 130 To form so-called 'abstracts' (i.e. in my terminology predicative nexus-words) from adjectives we use the suffix *-ess*. **1933** [see *JUNCT]. **1952** R. Aiken *Commonsense Gram.* xvii. 222 The clause is basically a nexus performing a single function with reference to a sentence. **1933** O. Jespersen *Analytic Syntax* 16 Nexus-substantive, e.g. work, kind-ness, etc., Nexus substantives denote something predicable... *Mod. Eng. Gram.* VI. v. 47 Something looking like a nexus-tertiary is combined with its S[ubject] as the real subject of the sentence. **1951** A. Gardiner *Speech & Lang.* 261 Jespersen has given to this subject-predicate relation..the name of 'nexus'. **1957** S. Potter *Mod. Linguistics* iii. 71 General or nexus-questions which may be answered by 'yes' or 'no'. **1966** *English Studies* XLVII. 55 Those which are nexus-objects (e.g. I believe Williams the murderer = Williams to be the murderer).

Nez Percé, Nez Perce (nez pɜ̄ɹs). [Fr., lit. 'pierced nose'.] A member of a group of North American Indians; also, the language of this people. Also *attrib.*

1812 in *S. Dakota Hist. Coll.* (1908) IV. 157 The..Nez Perce nation have a tradition that the human race sprung from this dog [sc. the prairie dog] and the beaver. **1836** *Ibid.* 86 The Nez Percé, who are estimated by others at about 1,000 souls, are divided into 10 bands. **1831** *Ibid.* 86 in *Tyler Hist. Calaveras Co.* ... We found the Nez Perces, a fine tribe of Indians, the Flatheads. **1841** G. Catlin *Lett. on N. Amer. Indians* II. 108 The Nez Percé Indians, who..inhabit the country of the Columbia,..are seldom known to flatten the head. **1910** F. W. Hodge *Handbk. Amer. Indians* II. 67/1 Practically the only rupture in these relations was the Nez Percé war of 1877. **1926** L. A. Clare tr. *Lévy-Bruhl's How Natives Think* iv. 157 With the Nez-percés, we assume different forms according to whether the subject or object is advancing or retreating. **1937** R. H. Lowie *Hist. Ethnol. Theory* ix. 133 The Nez Percé myths dictated by an old native woman to her college-bred son form another notable instance. **1949** *Pacific Discovery* May–June 16/1 According to some it is derived from the Nez Percé word meaning 'muddy water'. **1969** E. Tunis *Indians* 112/1 The northern Shoshone, the Nez Percé, and some other tribes clearly differentiated and transferred to reasonable facsimiles of Plains Indians. **1965** *Canad. Jrnl. Linguistics* X. 78 Languages of sure affiliation... Nez Perce (Penutian). **1968** *Times* 8 Oct. ... [see *Sahaptin]. **1969** *Lawrence* XLV. 45 Sahaptian..has two members: Nez Perce and Sahaptian. Nez Perce was spoken in parts of present-day Oregon, Washington, and Idaho. **1973** *Times Lit. Suppl.* 23 Nov. 1425/1 Joseph, chief of the Nez Percé, non-treaty chiefs, was restricted to a Washington reservation.

ngaio (nai·o). Also 9 **ngaiho.** [Maori.] An evergreen shrub or small tree, *Myoporum laetum*, of the family Myoporaceæ, native to New Zealand and bearing clusters of white flowers. Also *attrib.*

1853 J. D. Hooker *Bot. Antarctic Voy.* II. *Flora Novæ-Zelandiæ* I. 205 *Myoporum lætum*... Nat[ive] name 'Ngaio'. Colenso). 'Cultivated in England.] **1861** A. S. Atkinson *Jrnl.* 6 Mar. in *Richmond-Atkinson Papers* (1960) I. xii. 693 Sat down in the shade of the ngaiho's which bordered the beach. **1873** *Descr. Catal. Exhibits from N.Z. Vienna Exhib.* 24/1 *Myoporum lætum*. (Ngaio). A small ornamental tree. Wood light, white, and tough. Used for gun stocks. **1876** *Trans. N.Z. Inst.* IX. 206 A common New Zealand shrub, or tree, which may be made useful for shelter, viz. the Ngaio. **1921** H. Guthrie-Smith *Tutira* xii. 102 In this light bush, tawa..mahoe or hinahina (*Melicytus ramiflorus*), ngaio (*Myoporum laetum*)..were the most common trees and shrubs. **1946** *Jrnl. Polynesian Soc.* LV. 161 Fancy ngaio not being in the Dictionary; ngaio, so reminiscent of the story of Ngaio and her translation to the moon with her calabash and the ngaio tree she clung to for support and stay. **1959** *Listener* 30 Apr. 769/2 The New Zealand 'properties'—ngaio, bluegum, etc.—enter naturally into what he [*sc.* J. K. Baxter] has to say. **1966** *Encycl. N.Z.*

II. 681/2 Ngaio grows to a height of about 30 ft and is a much-branched, rounded tree. The leaves are bright green and somewhat fleshy... They are thickly studded with oil glands in which bacteria live. Flowers are small and appear as little clusters in the axils of leaves.

Ngala, var. *LINGALA.

nganga, var. *MGANGA.

‖ **ngarara** (narā·rā). [Maori.] A name used for various extinct, unidentified, New Zealand lizards; also, in Maori mythology, a lizard-like monster. Also *attrib.*

1874 J. W. Stack in *Trans. N.Z. Inst.* VII. 296 Ngarara burrows were frequently met with on the plains. *Ibid.*, A description of the Ngarara... **1882** W. D. Hay *Brighter Britain* II. 225 There is a little emerald-green lizard in the bush, called by the Maori ngarara. It is dreadfully tapu. **1901** A. A. Grace *Tales of Dying Race* 190 'The Ngarara—you never heard of him?' said the old woman. 'He was—a terrible beast, that one. Big body, eight feet long; big webbed feet; big wings like a bat's, with which he flies and catches fish; long tail like a *taniwha* lizard's, but bigger; skin like the bark of the red pine.' **1905** W. B. *Where White Man Treads* 98 This [*sc.* the Maori's] existence was burdened with the knowledge of huge reptilian monsters, ngarara on land, and taniwha in the water. **1949** P. Buck *Coming of Maori* (1950) I. iv. 61 The crew and passengers of the [canoe] Mangarara consisted of reptiles and insects. The reptiles [ngarara] were lizards. **1966** *Encycl. N.Z.* I. 48/2 Less improbable were Maori tales of the *ngarara*, lizards which were larger than the tuatara... Certain prominent Maoris ..claimed not only to have seen but also to have handled and eaten them. It seems that the *ngarara*, which frequented manuka scrub, varied in size from 2 to 3 ft in length and from 6 to 12 in. in girth. There was also a smaller *ngarara*, about 18 in. long, found in streams. The Maoris attributed the disappearance of the large *ngarara* to scrub-fires and the attacks of cats and..perhaps the Norwegian rat.

Ngbaka (ŋ,bā·kă). [Native name.] A Bantu language of the northern parts of Zaire.

1949 E. A. Nida *Morphol.* (ed. 2) 63 Ngbaka, a language of the Belgian Congo. **1965** *Language* XLI. 347 A few short Ngbaka texts.

Ngbandi (ŋ,ba·ndi). [Native name.] A Bantu language of the Central African Republic and northern Zaire.

1955 J. H. Greenberg *Studies in African Linguistic Classification* 12 An enumeration of the membership of the Niger-Congo family by tentative genetic subfamilies follows:..14. Eastern Branch: Gbaya-Manjia,..Sango-Yakoma-Ngbandi. **1964** E. A. Nida *Toward Sci. Transl.* ix. 199 In Ngbandi, a language of northern Congo, tense distinctions are generally indicated by tonal differences on the subject pronouns, while aspectual differences are signaled by differences of tone on the verbs. **1967** W. J. Samarin *Grammar of Sango* 17 Vernacular Sango, one of the dialects of the Ngbandi complex (which itself is a language of the Adamawa-Eastern group of Greenberg's Niger-Kordofanian), at one time came to be used as a lingua-franca. **1969** *Language* XLV. 659 Sango is a pidginized variant of a Ngbandi language of the Central African Republic.

‖ **ngege** (ŋ,ge·ge). [Native name.] A cichlid food fish, *Tilapia esculenta*, found in Lake Victoria in E. Africa. Cf. *TILAPIA.

1928 *Times* (Weekly ed.) 12 Jan. 54/2 His nets were full of the prime fish of the lake, the carplike ngege. **1932** *Discovery* Jan. 16/1 The fish of economic importance is called the ngege, a carp-like species which is found in Lake Victoria.

Ngoko (ŋ,ōu·ko). [Native name.] The popular written or spoken form of modern Javanese.

1893, 1925 [see *KROMO]. **1948** D. Diringer *Alphabet* vii. 424 Ngoko, the language of the commoner. **1963** R. T. McVey *Indonesia* x. 443 Semar..speaks to the god in Ngoko (low Javanese) and is answered in high Javanese.

‖ **ngoma** (ŋ,gōu·mă). [Swahili *ngoma, goma*, drum, dance, music.] In East Africa, a dance, a social gathering where dancing is general, a night of dancing.

1926 *Glasgow Herald* 27 Jan. 10 They..console them for the temporary absence of their dancing partners at ngomas. **1935** E. Hemingway *Green Hills of Africa* I. i. 16 That is what you should see. The big *ngomas*. The big native dance festivals. **1947** *E. African Ann.* 1946–7 57/1 Most include in their headdresses a 'halo' of giraffe or zebra hair, and one I saw at a Victory ngoma (dance) had his horned headdress crowned by a..teapot. **1960** *Spectator* 29 July 179 They would keep it up far into the night, drinking and drumming; a jolly social party not like the ngomas I used to see which always had a hint of magic and, it seemed, of menace. **1966** C. Sweeney *Scurrying Bush* v. 65 It was not until I went to a *ngoma*, or beer-dance,..that I appreciated the tremendous volume of noise that human beings could produce. **1971** *Standard* (Dar es Salaam) 7 Apr. 1/8 There will be dances in all dancing halls,..competitions in ngomas and singing, [etc.].

Ngoni (ŋ,gōu·ni), *sb.* and *a.* [Native name.] **A.** *sb.* An African people belonging to the Nguni branch of the Bantu. (Groups of Ngoni migrated from their original home in Zululand in about 1830 and are now found in Malawi, Tanzania, Zambia, etc.) **B.** *adj.* Of or pertaining to this people.

1883 R. N. Cust *Mod. Lang. Africa* II. xii. 300 Bands of marauders, or Nomads, are met with as far North as the neighbourhood of the Victoria Nyanza under the names of..Ba-Ngoni, Wa-Ngoni, and Ma-Ngoni. The uniform testimony of travellers is that they speak Zulu. **1891** W. A. Elmslie *(title)* Introductory Grammar of the Ngoni (Zulu) language, as spoken in Mombera's kingdom. **1911** *Encycl. Brit.* XXVIII. 1050/2 In the 19th century various Zulu hordes successively invaded and overran a great part of east-central Africa... Throughout these regions they are variously known as Ma-Zitu, Wa-Ravi, Wa-Ngoni (Angoni), etc. **1948** J. A. Barnes *Marital Culture of Ft. Jameson Ngoni* i. 4 The principal difference between the Ngoni type and the surrounding types lies in the Ngoni form of centralised chieftainship, inherited from father to son. **1950** — *Marriage in a Changing Society* ix. 71 The Ngoni regard uxorilocal residence as part of the obligation a man has towards his wife's parents. **1966** C. G. Seligman *Races of Africa* (ed. 4) ix. 141 Reflex movements northwards had produced disorder long before Arab raids and Ngoni incursions spread ruin and desolation.

Nguni (ŋ,gu·ni), *sb.* and *a.* [Zulu.] **A.** *sb.* A subdivision of the Bantu people which includes the Zulu-Xhosa tribes; also the languages spoken by this group, i.e. the Zulu-Xhosa-Swazi languages. **B.** *adj.* Of or pertaining to this group of peoples or languages.

1929 A. T. Bryant *Olden Times in Zululand & Natal* i. 3 The natives of South-Eastern Africa we distinguish as of three separate families, which we call respectively Ngūni, in Zululand, the Transkei, Natal, and the Cape; Sutū..; and Tonga. *Ibid.* 5 Captured Bushwomen became common in their homes... And the children..adopted..in a Bantuized form, much of the slave-girl's speech... Hence the clicks in Ngūni speech. **1940** M. Gluckman in Fortes & Evans-Pritchard *African Political Systems* i. 29 In the earlier period of Nguni history, political allegiance tended to coincide with kinship affiliation. **1950** Radcliffe-Brown & Forde *African Systems of Kinship & Marriage* 58 In the Nguni tribes the personal name that a woman has in her own family, as a daughter, may not be used by her husband's family. **1957** C. G. Seligman *Races of Africa* (ed. 3) viii. 168 This group [*sc.* Eastern Southern Bantu] consists of two main subdivisions Nguni and Tsonga. The former include the Cape Nguni of the Ciskei and Transkei.. together with the 'Fingo', fugitive remnants of tribes broken up in Natal..; the Natal Nguni, or 'Zulu' of Natal and Zululand, with their offshoot the Ndebele (Tebele) of Southern Rhodesia; the Swazi of Swaziland and the Eastern Transvaal; and the 'Transvaal Ndebele' of Central and Northern Transvaal. **1970** W. Smith *Gold Mine* xxxiv. 88 Basuto is also one of the fighting tribes of the N'guni group.

ngwee (ŋ,gwē¹·). Pl. **ngwee.** [Chibemba, lit. 'bright'.] A small coin of Zambia.

1966 *Guardian* 10 Mar. 12/4 Zambia..is to go over, by January 1968, to a decimal currency based on the kwacha (worth 10s of the present pound), divided into 100 ngwee. **1969** *Reporter* (Nairobi) 16 May 12/4 Zambia's white farmers..appear fairly satisfied with the 38 ngwee per pound average. **1975** *Stand. Encycl. S. Afr.* XI. 573/1 The kwacha was equivalent to half of the pound and itself was made up of 100 ngwee.

niacin (nəi·ăsin). *Biochem.* [f. *NI(COTINIC *a.* + AC(ID *a.* and *sb.* + -IN¹.] **a.** = **nicotinic acid*. **b.** The pellagra-preventing vitamin, which can be either nicotinic acid or nicotinamide.

1942 *Cooperative Consumer* 28 Feb. 5/3 'Niacin' is the new name for 'nicotinic acid', the ingredient of enriched bread which was first discovered as a potent preventive and cure for pellagra. The new name was found to be necessary because some anti-tobacco groups warned against using enriched bread because it would foster the cigaret habit... Federal Security Administrator Paul McNutt..approved the name. **1956** A. Huxley *Let.* 25 Dec. (1969) 814 Are there any published papers on the use of niacin in the treatment of high cholesterol conditions? **1968** Pfeiffer & Murphree in D. H. Efron *Psychopharmacology* 696/2 The majority of schizophrenic patients given niacin or niacinamide had fewer hospital readmissions..when compared to non-vitamin treated control groups. **1970** *Nature* 16 May 665/2 Niacin is now acceptably defined as a blanket word for the acid and amide. **1972** *Materials & Technology* V. xix. 676 The best dietary sources of niacin are yeast, liver, lean meat, poultry, and legumes.

niacinamide (nəi,ăsi·nămaid). *Biochem.* [f. prec. + AMIDE.] = *NICOTINAMIDE.

1951 *Addendum to Brit. Pharmacop.* 23 Injection of Nicotinamide. *Synonym.* Niacinamide Injection. **1955** W. W. Denlinger *Compl. Boston* II. 34 The need for niacinamide, calcium pantothenate and pyr[i]doxine..has not yet been established as pertains to the nutrition of dogs. **1968** [see prec.]. **1972** *Arch. Dermatol.* CV. 574/2 In our patient, oral and parenteral administration of niacinamide..led to rapid clearing of the pellagrous dermatitis.

Niagara. 1. (Earlier and later examples.)

1841 F. A. Kemble *Let.* 28 Dec. in *Rec. Later Life* (1882) II. 153 Such a Niagara of information did surely never pour from the lips of mortal man! **1909** *Chambers's Jrnl.* June 383/1 In the savage blizzards of a frozen Sahara this [ice-]drift becomes a roaring, hissing, blinding Niagara of snow, rising hundreds of feet into the air. **1912** I. S. Cobb *Back Home* 321 Rivers of red pop had

already flowed, Niagaras of lager beer and stick gin had been swallowed up. **1931** A. Huxley *Let.* 24 Aug. (1969) 352 We are reading *Monte Cristo* aloud. What a book! I had never read it before: it is a kind of Niagara! **1970** P. Laurie *Scotland Yard* iii. 86 A Niagara of tinted hair. **1974** *Times* 9 Jan. 6/5 Mr Nixon was swept towards what the White House once called a 'Niagara' of accusations last spring.

nialamide (nəi,æ·lămaid). *Pharm.* [f. the proprietary name *Niamid* by insertion of *-al.*] A crystalline hydrazide, $C_{16}H_{18}N_4O_2$, which is a monoamine oxidase inhibitor used as an antidepressant.

1959 *Proc. Soc. Exper. Biol. & Med.* CI. 832 (*heading*) Pharmacological studies with nialamide, a new antidepressant agent. **1960** A. Carlsson et al. in J. R. Vane et al. *Adrenergic Mechanisms* 434 Fig. 1 shows the action of nialamide, which is a very potent and long-acting monoamine oxidase inhibitor though its action sets in fairly slowly. **1971** *Brit. Med. Bull.* XXVII. 28/2 First nialamide..and then tranylcypromine..began to be used for the treatment of depression.

Niam-Niam (ni·ăm,ni·ăm). [Dinka, lit. 'great eaters'.] = *ZANDE.

1861 J. Petherick *Egypt, Soudan, & Central Africa* xxvi. 469 Attached to the girdle, a strong leather sheath containing a knife, hilt downwards, is worn by every Neam Nam. *Ibid.* 473 The Neam Nam recognise no superior chief; but, like the Dôr, the tribe is divided into numerous chieftainships. **1873** E. E. Frewer tr. *Schweinfurth's Heart of Africa* II. xiii. 3 The name Niam-niam is.. so universally incorporated into the Arabic of the Soudan, that it seems unadvisable to substitute for it the word 'Zandey', the name by which the people are known among themselves. **1891** A. W. Buckland *Anthrop. Studies* v. 59 In Africa, the typical home of the stalwart Negro.. we find the Bushman and Hottentot in the South, and the Akkas and Niam-Niams in the centre, very small in stature and yellow in colour. **1902** *Encycl. Brit.* XXXI. 230/1 The Niam-Niam, or Zandeh people, as they call themselves..are now found to stretch, with interruptions, from the White Nile above the Sobat confluence to the Shari affluent of Lake Chad. **1931** J. G. Leyburn *Handbk. Ethnography* 176/2 Niam-Niam. Between the Welle, a tributary of the Ubangi, and the Nile. **1966** R. & D. Morris *Men & Apes* viii. 238 The Niams-Niams, apparently, still adorn themselves with colobus monkey skins wound round the waist.

‖ **niaouli** (ni,ā,uli). [Native name.] An evergreen tree, *Melaleuca quinquenervia*, of the family Myrtaceæ, native to New Caledonia. Also *attrib.* and *transf.*, a personification of New Caledonia.

1921 *Public Opinion* 29 July 109/2 A mat of pandanus leaves served for his sail and a paddle of niaouli wood for its helm. **1943** *Amer. Speech* XVIII. 14 It would be.. impossible to root them [*sc.* English words] out of the Niaouli's everyday language. *Ibid.* 15 Our expression 'to go bush' has its counterpart in Niaouli pidgin... Niaouli is the name of a tree which..has come to be regarded as a national symbol [of New Caledonia]. **1965** *Univ. Iowa Stud. Nat. Hist.* XX. vii. 52 (*caption*) This frequent burning maintains the enormous area of fire-resistant niaouli.

nib, *sb.*¹ Add: **3.** (Further examples.)

1901 J. Black *Carp. & Builder Series: Slating & Tiling* 13 The ordinary pantile..is provided on the underside with a small projection known as a nib. **1940** *Chambers's Techn. Dict.* 578/1 *Nib*, a small projection, sometimes continuous, formed on the under-side at the top of each tile, enabling the tile to be hung on battens. **1955** *Railway Mag.* May 307/2 The main feature of the detachable nibs in the relay baseboards is that disconnection points are available without the necessity of providing independent terminal boards for each relay. **1962** *Gloss. Terms Glass Industry* (*B.S.I.*) 43 *Nib*, a small protrusion at the corner of a piece of flat glass due to faulty cutting. **1968** *Gloss. Formwork Terms* (*B.S.I.*) 17 *Kicker* (nib), a small concrete upstand cast above floor level to position wall or column forms for the next lift and to assist the prevention of grout loss.

7. A speck of solid matter in a coat of paint or varnish.

1940 in *Chambers's Techn. Dict.* **1958** *Listener* 11 Sept. 399/1 You can now tidy it [*sc.* the undercoat of paint] up with fine sandpaper—just enough to remove any dust nibs or brush marks. **1965** W. N. Lapper in *Applic. Surface Coatings* (Oil & Colour Chemists' Assoc.) iii. 37 A coagulation of pigment can cause 'nibs' or bittiness in the film. **1968** *Pract. Motorist* Feb. 611/3 Once the first coat is fully dry rub it down very gently with wet-or-dry (grade 320) to remove any 'nibs' and runs.

nib, *sb.*³ (Later example.)

1936 Wodehouse *Laughing Gas* viii. 81 You don't run to an English butler in Hollywood unless you are a pretty prominent nib.

nibble, *v.* Add: **1. e.** To produce by nibbling.

1867 A. J. Evans *St. Elmo* xxi. 296 Just see what a hole the pretty little wretch has nibbled in my new Swiss muslin dress!

2. b. (Later *fig.* examples.) Also *absol.*

1921 W. J. Locke *Mountebank* xvi. 208 Moignon was in touch, on his behalf, with powerful American agencies... Moignon had said: 'They are nibbling for the winter.' **1973** *Times* 20 Mar. 21/3 Since the Broadspeed Turbo Bullit..was announced in January, motor manufacturers have been nibbling at the idea.

d. *Cricket.* To play (indecisively) at a ball bowled outside the off stump.

1926 P. F. Warner *Fight for Ashes* 16 Bardsley.. showed a distinct weakness in nibbling at good-length balls outside the off-stump. **1932** E. Blunden *Face of England* 71 'Tom's out.' 'He shouldn't have nibbled at that.'

nibbled (ni·b'ld), *ppl. a.* [f. Nibble *v.* + -ed¹.] That has been nibbled or cropped.

1865 R. D. Blackmore *Cradock Nowell* (1866) I. viii. 68 Over the nibbled sward..came wandering the lightest foot that ever passed. **1905** J. B. Firth *Highways Derbyshire* viii. 119 The Dove flows between closely nibbled hill slopes. **1949** *U.S. Employment Service, Dict. Occup. Titles* 891 May file and grind excess metal and rough edges of nibbled parts to specified dimensions.

nibble-nip, *v.* [f. Nibble *sb.* or *v.* + Nip *v.*¹] *intr.* To give a nibble, a trifling nip (only *fig.*). So **nibble-nipped** *ppl. a.*

1883 G. Meredith *Poems & Lyrics of Joy of Earth* 26 Haggard Wisdom, stately once, Leers fantastical and trips: Allegory drums the sconce, Impiousness nibblenips. **1937** *Sunday Times* 10 Jan. 8/3 The conversation distinguished, sedate, and rather wintry (even Newman had at first felt nibble-nipped) of the Oriel Common Room.

nibbler. Add. **1. b.** *Engin.* A type of metal-cutting tool in which a rapidly reciprocating punch knocks out a line of overlapping small holes from sheet or plate.

1939 *Jrnl. R. Aeronaut. Soc.* XLIII. 144 Practically all of the standard means of cutting are used in all factories. These are the oxy-acetylene torch,..the nibbler; the saw; and the uni-shears. **1958** *Engineering* 28 Mar. 409/2 A nibbler for sheet up to 16 s.w.g. **1961** *Aeroplane & Astronautics* CI. 272/3 On display is a complete range of tools, including drills, screwdrivers, nutrunners, grinders, rivet and bolt millers, nibblers and shears.

nibcocked (ni·b,kǫk't), *a. rare*⁻¹. [f. Nib *sb.*¹ + Cock *sb.*¹ 20 + -ed².] ? Having a penis like the point of a pen.

1939 Dylan Thomas *Let.* Mar. (1966) 226 The English poets now are such a pinlegged, nibcocked, paperhearted crowd you could blow them down with one bellow out of a done lung.

nibful (ni·bful). *rare.* [f. Nib *sb.*¹ + -ful.] As much as a nib can hold.

1930 V. Woolf *Writer's Diary* 29 Apr. (1953) 158, I have just finished, with this very nib-ful of ink, the last sentence of *The Waves*.

niblick. (Earlier, *attrib.*, and later examples.)

1857 H. B. Farnie *Golfer's Manual* (1947) iii. 18 It is called a *Niblick*. **1886** H. G. Hutchinson *Hints on Golf* xiii. 33 In the typical niblick shot the ball lies in a heel-mark or other cup in the sand, with the face of the bunker in front. **1909** *Bystander* 13 Oct. 90/2 Herd..saved an apparently lost hole by a most masterly niblick pitch. **1955** R. Browning *Hist. Golf* 145 Even the niblicks were originally wooden clubs: the first iron-headed niblicks were excessively short in the blade. **1961** M. Spark *Prime of Miss Jean Brodie* v. 141 Sandy gave a hack with her niblick.

ni·blick, *v. Golf.* [f. the sb.] *trans.* To hit (a ball) with a niblick.

1909 *Westm. Gaz.* 15 Jan. 4/2 If bunkered..he would have to niblick the ball out sideways.

Nibmar, NIBMAR (ni·bmāɹ). [f. the initials of '*n*o *i*ndependence *b*efore *m*ajority *A*frican *r*ule'.] The policy of opposing recognition of the minority government which proclaimed the independence of Rhodesia in 1965.

1966 *Time* 23 Sept. 31 They demanded that he agree to something called NIBMAR. **1966** *Economist* 17 Dec. 1222/1 They suspect Britain is on the point of making a settlement with Mr Smith which would snatch from them the apple of Nibmar. **1970** *Times* 12 Nov. 6 Britain, he said, had never accepted a commitment to Nibmar from the United Nations.

nibong, var. Nibung (in Dict. and Suppl.).

nibs. Add: Esp. *his nibs, His Nibs*, an employer, a superior; a self-important person. (Further examples, including occas. uses with other possessive pronouns.)

1846 *Swell's Night Guide* 57 She flokessed his nibs, and hooked it off to his crib. **1877** *Brooklyn Monthly* Oct. 21/2 Salute the hostess by saying: 'Cully, how's his nibs?' **1882** G. W. Peck *Sunshine* 131 A respectable merchant was going to the opera with a friend from the country, when a couple of sirens met them and one said to the other, 'Look at his nibs.' **1906** E. Dyson *Fact'ry 'Ands* ii. 18 They're settin' her nibs t'-day. **1906** [see *dilly *a.*¹]. **1919** G. S. Gordon *Let.* 30 June (1943) 115 We get on to the *Caesar*, and find their nibs strolling the quarterdeck after dinner. **1928** 'Brent of Bin Bin' *Up Country* ii. 33 Gifts..including a splendid meerschaum pipe for his Nibs. **1933** M. Allingham *Sweet Danger* xxi. 218 What if 'Is Nib's boy friends spot us before? **1944** [see *Hampshire b]. **1957** H. Croome *Forgotten Place* 175, I wish I could just lie on a bed and smoke, like His Nibs. **1967** A. Wilson *No Laughing Matter* i. 21 His father smiled. 'Trust His Nibs to have noticed that deficiency.' **1973** A. Hunter *Gently French* i. 9 Since when were you on first-name terms with His Nibs? **1974** O. Manning *Rain Forest* II. i. 141 Her nibs don't like me calling him 'old bugger'. There's a snobby bitch, if you like!

nibung. Substitute for def.: A Malaysian palm, *Oncosperma filamentosa*. (Earlier and later examples.)

The usual sp. is now *nibong*.

1779 T. Forrest *Voyage to New Guinea* ix. 121 We made very good curry; stewing it with the heart of the aneebong, or cabbage tree. **1783** W. Marsden *Hist. Sumatra* 77 The *neebong* or cabbage tree, a species of palm, grows wild in too great abundance to require being cultivated. **1820** J. Crawfurd *Hist. Indian Archipelago* I. iv. 417 The *nibung* is the true mountain cabbage. **1839** T. J. Newbold *Straits of Malacca* I. iv. 139 Thatch of Atap, and floors of split nibong, called lantei. **1898** Conrad *Tales of Unrest* 275 Two tall nibong palms.. leaned slightly over the ragged roof. **1907** F. Swettenham *Brit. Malaya* vii. 151 The floor..is of planks, *nibong*, or split bamboos. **1954** R. H. Holttum *Plant Life in Malaya* ii. 22 Other kinds of palm have many trunks; for example, the Sago palm and the Nibong. **1966** *Listener* 6 Oct. 502/3 Nearly 14,000 people live in wooden houses built on stilts of tough nibong palms.

N.I.C., NIC, Nic (nik). [Acronym f. the initial letters of '*N*ational *I*ncomes *C*ommission'.] The name of a body giving advice to the government on economic policy. Cf. *Nicky *sb.*²

1962 *Daily Tel.* 27 July 24/5 He saw the N.I.C. primarily as a means of 'educating public opinion'. **1962** *Times* 27 July 13/2 (*heading*) To Neddy add Nic.

Nicaraguan (nikăræ·giuăn), *a.* and *sb.* [f. Nicaragua: see -an.] **A.** *adj.* Of, pertaining to, or characteristic of Nicaragua. **B.** *sb.* A native or inhabitant of Nicaragua.

1847 R. G. Dunlop *Trav. Central America* iii. 105 The Caroline, a schooner chartered by the Nicaraguan government. **1852** E. G. Squier *Nicaragua* I. ii. ii. 77 The average rate of duty exacted under the Nicaraguan tariff, is about 21 per cent. ad valorem. *Ibid.* 78 All its inhabitants were, and with the exception of a few.., still are Nicaraguans. **1868** F. Boyle *Ride across Continent* I. p. xxi, The universal legend of the surrounding peoples—Indians, Caribs, Nicaraguans, and Costa Ricans,—declares the Guatuso race to be distinguished by fair hair and blue eyes. *Ibid.* iii. 120 He stayed with us through all our Nicaraguan experiences. **1884** *Encycl. Brit.* XVII. 478/2 The Nicaraguan fauna differs in few respects from that of the other Central-American states. **1923** Ld. Charnwood *Theodore Roosevelt* vii. 151 The Nicaraguan route would be cheaper for a canal. **1934** [see *Honduran, Hondurean *a.* and *sb.*]. **1946** E. A. Peers *Fool of Love* vii. 121 The Nicaraguan, Rubén Darío, took up the picturesque phrase. **1973** *Time* 25 June 17/1 Nicaraguan Ambassador Guillermo Sevilla-Sacasa said sympathetically [to Julie Nixon Eisenhower]: 'Your father still has one friend.'

nice, *a.* Add: **14.** *spec.* Of a cup of tea.

1899 R. Whiteing *No. 5 John St.* iv. 38 Her sex's universal restorative... 'You shall have a nice cup of tea.' **1928** R. Knox *Footsteps at Lock* v. 41 You'd have got a nice cup of tea down at the Gudgeon. **1937** A. P. Herbert *Nice Cup of Tea* (song), I like a nice cup of tea in the morning, For to start the day you see. **1937** 'G. Orwell' *Road to Wigan Pier* v. 88 There is generally a cup of tea going—a 'nice cup of tea'. **1961** I. Fleming *Thunderball* iv. 38 The dimity world of the Nice-Cup-of-Tea. **1974** L. Deighton *Spy Story* xxi. 221 'I'll pour him some tea,' said Dawlish. 'There's nothing so reviving as a nice cup of tea.'

15. a. *nice girl* (of an adult). Freq. somewhat derisive.

1876 C. M. Yonge *Womankind* xvi. 126 Though a well managed, innocent and select rink is quite possible, 'nice' girls would do well to abstain from those where a chance public shares the sport. **1901** W. D. Howells *Heroines of Fiction* I. 12 They imagined the heroine who was after all a Nice Girl; who still remains the ideal of our fiction. **1905** E. Wharton *House of Mirth* I. xiv. 239 He had never wanted to marry a 'nice' girl: the adjective connoting..certain utilitarian qualities..apt to preclude the luxury of charm. **1910** *National Police Gaz.* 16 July 3/1 That's what tells and it pulls the nice girls down with a sudden rush that takes their breath away. **1933** E. O'Neill *Ah, Wilderness!* (1934) iii. i. 89 You're a darned nice girl. **1938** N. Marsh *Artists in Crime* viii. 108 She tries to talk 'Slade'..but the original nice-girl gush oozes out. **1968–70** *Current Slang* (Univ. S. Dakota) III–IV. 86 *Nice Girl*, a sexually permissive girl. **1975** I. S. Black *Man on Bridge* ii. 26 'She's pretty.'..'More than that, she looked what used to be called a nice girl.'

Phr. (Further examples.)

1839 Dickens *Let.* 5 Mar. (1965) I. 521 A capital bed, and all as nice as nice could be. **1937** G. & I. Gershwin *Nice Work if you can get It* (*song*), Holding hands at mid-night 'Neath a starry sky, Nice work if you can get it, And you can get it if you try. **1938** *Sun* (Baltimore) 20 June 8/2 Ruth said, 'Nice going, kid,' and that simple compliment pleased the young Cincinnati pitcher more than all of the other praises he received. **1954** R. Bissell *High Water* xxiii. 279 'Nice going, George,' I said. **1958** *Listener* 2 Oct. 492/1 The Frenchman..may well reply with impatience: 'Nice work if you can get it.'

d. (Earlier and later examples.)

1836 Dickens *Let.* 29 Dec. (1965) I. 217, I have been clearing off all the rejected articles to-day, and nice work I have had. **1892** I. Zangwill *Childr. Ghetto* I. i. xi. 248 Well, you're a nice friend of his, I must say. **1896** E. Turner *Little Larrikin* xviii. 209 Aren't you going to stop and see Clem off?..you *are* a nice one. **1939**

L. M. Montgomery *Anne of Ingleside* v, S'posin' he et a lot of the little green apples..and got nice and sick?

nicely, *adv.* Add: **3. c.** (Later examples.)

In quots. 1935 and 1938 = 'merry from the effects of drinking'.

1935 G. Heyer *Death in Stocks* xvi. 178 It was quite obvious he'd been at a pub all the time, because he was quite nicely. **1938** N. Marsh *Death in White Tie* v. 55 I'm not inebriated..but I am..a little exalted. What I believe is nowadays called nicely thank you. **1949** —— *Swing, Brother, Swing* ix. 207 'How are you, Mr. Fox?' 'Nicely, thank you, sir. And you?' **1954** A. S. C. Ross in *Neuphilologische Mitteilungen* LV. 43 Possible negative non-U answers are *I'm doing nicely, thank you* and (*Quite*) *sufficient, thank you*. **1943** *Times* 31 May 10/8 Aurelia Weatherbournes generally do quite nicely, thank you, on council estates.

Nicene, *a.* (and *sb.*) Add: Cf. *Isnik.

nice-nellyism (nəis,ne·li,iz'm). *N. Amer. slang.* Also **nice-nellieism.** [f. *nice Nelly*, a conventional name for a respectable woman, a prude + -ism.] Prudery; genteelism; excessive prudishness of speech or behaviour: usually applied adversely. Cf. *Nelly² 3.

1936 *New Republic* 28 Oct. 337/1 Perhaps, it is true, as charged that the British press is displaying a brand of Nice Nellyism in refusing to mention the subject [*sc.* the divorce of Mrs Simpson, later Duchess of Windsor]. **1942** *Sun* (Baltimore) 17 June 10/7 Mr. Adams accused the editor [of a new book of soldier songs] of Nice Nellieism in dealing with the songs. **1947** *Sat. Rev.* (U.S.) 15 Feb. 9/2 It takes more than..nice-Nellyism in the name of patriotism to obliterate that spirit. **1952** *New Yorker* 18 Oct. 159/1 Mr. Pyles attributes much of the nice-nellyism that blighted polite speech and writing during the nineteenth century to Webster's Puritan prudishness. **1956** *N.Y. Times Book Rev.* 30 Sept. 2/2 None of the words which Ned Sheldon..found so obnoxious seems to me acutely distasteful, with the exception of 'funeral parlor', which carries nice-nellieism to the nth degree. **1960** I. Wallach *Absence of Cello* (1961) 174 'Experience', as absurd a nice-Nellyism for copulation as she could conceive. **1973** *Saturday Night* (Toronto) Oct. 15/1 The public had been made comatose by the greyness of Mackenzie King and the nice nellyism of Middle Powermanship.

nicey, nicy (nəi·si), *a. colloq.* [f. Nice *a.* + -y¹.] Nice. Also as *sb.*, a nice person or thing.

1859 Trollope *Bertrams* I. vi. 120 The musty fusty people, and the nicy spicy people, and the witty pretty people do severally assemble and get together as they ought to do. Bertram's next-door neighbour was certainly of the nicy spicy order. **1879** C. M. Yonge *Burnt Out* viii. 132 Oh, you're a tell-tale-tit! Catch me giving you nicies again! **1914** *Conc. Oxf. Dict.* Add., *Nicy*..(nursery), sweet, lollipop. **1922** Joyce *Ulysses* 363 Go home to nicey bread and milky and say night prayers with the kiddies. **1959** *Listener* 3 Dec. 1008/1 Young nasties after lots of lolly, and young nicies to foil them at last.

niche, *sb.* Add: Also with pronunc. (nīʃ). **2. c.** = *mihrab 2. In full, *prayer niche*.

1911, etc. [see *mihrab 2]. **1962** C. W. Jacobsen *Oriental Rugs* 251 The prayer niche is typical and each is in this shape unless, by chance, it is one that has three prayer niches.

3. c. *Ecol.* The position of a plant or animal within its community.

1927 C. Elton *Anim. Ecology* v. 63 It is therefore convenient to have some term to describe the status of an animal in its community, to indicate what it is doing and not merely what it looks like, and the term used is 'niche'. *Ibid.* 64 The 'niche' of an animal means its place in the biotic environment, its relations to food and enemies. **1937** [see *biotope]. **1960** N. Polunin *Introd. Plant Geogr.* xiv. 431 The large lianes comprise..by far the more numerous synusiae (groups of plants of similar life-form, each filling much the same ecological niche and playing a similar role). **1962** *Listener* 13 Sept. 388/2 Divergence is related to the existence of ecological niches. *Ibid.*, On the existence of niches, he [*sc.* Darwin] was already clear in 1837... Even more apposite is his question on how a niche is entered. **1965** B. E. Freeman tr. *Vandel's Biospeleology* i. 6 A number of terrestrial planarians are endogeans, occupying a similar niche to the earthworms. **1974** Bennett & Humphries *Introd. Field Biol.* (ed. 2) ii. 8 Each animal species has its own typical food relationships with other species, so in a given community at a given time each is said to have its characteristic food niche. *Ibid.* 12 Its habitat niche can be thought of as the sum total of its many effective environments throughout life. *Ibid.* 13 The term 'niche' on its own is much misused in ecological writing to mean food-niche, habitat-niche, habitat or microhabitat.

Nichrome (nəi·krōum). Also **nichrome.** A proprietary name of various alloys of nickel with chromium (10–20 per cent) and sometimes iron (up to 25 per cent).

1911 *Jrnl. Amer. Chem. Soc.* XXXIII. 190 The nichrome triangles..when used with city gas..lost, at most, one milligram an hour. **1933** *Official Gaz.* (U.S. Patent Off.) 4 July 18/1 Driver-Harris Company, Harrison, N.J. ..Nichrome. For alloys of nickel, chromium, and iron in the form of wire, ribbon, strip, [etc.]. **1937** *Jrnl. Brit. Interplanetary Soc.* IV. 7 The nozzles were machined from nichrome steel rod, threaded at the lower end so that they could be screwed into the exhaust end of the combustion chamber. **1947** J. C. Rich *Materials & Methods of Sculpture* ii. 46 For low-temperature firing, electric kilns equipped with nichrome wire heating elements are very efficient. **1950** *Jrnl. R. Aeronaut. Soc.* LIV. 20/1 For general purposes, the wire used in strain gauges is a straightforward cupro-nickel, or nickel-chrome alloy, such as Eureka, Brightray, Nichrome. **1951** *Trade Marks Jrnl.* 7 Mar. 228/1 'Nichrome'... Unwrought and partly wrought alloys composed mainly of nickel-chromium; and rods, tubes, bars, wire, sheets and pellets, all being goods..made from alloys composed mainly of nickel-chromium. British Driver-Harris Company Limited,.. Manchester,..manufacturers. **1964** S. H. Avner *Introd. Physical Metallurgy* xii. 387 Some nominal compositions are 80Ni-20Cr (Chromel A, Nichrome V, and others) used as electric heating elements for household appliances and industrial furnaces; 60Ni-16Cr-24Fe (Chromel C, Nichrome, and others) used as electrical heating elements for toasters..and hot-water heaters. **1968** E. R. Petty *Physical Metall. Engin. Materials* xiii. 276 There are two main groups of heater wire, one based on 80% nickel-20% chromium (Nichrome) with or without amounts of iron, and the other based on ferritic chromium steels... The former alloy is well known as the basis of the Nimonic high temperature alloys.

‖ **nicht wahr** (niχᵛt vār). [G., lit. 'not true'.] Is it not true?

1924 M. Kennedy *Constant Nymph* xii. 167 We shall do very well without one [*sc.* a maid] for a little, *nicht war?* **1941** M. Treadgold *We couldn't leave Dinah* vi. 106 They are nice liddle horses, *nicht wahr*, Karl? **1948** W. Stevens *Let.* 25 Oct. (1967) 621 One has a sense that the world was never less new than now... Nicht wahr? **1967** 'G. Carr' *Lewker in Tirol* v. 69 Today it is so hot that tempers are not good—*nicht wahr?* **1971** J. Henderson *Copperhead* (1972) xiii. 162 He was going over to the other side. Such is truly a form of suicide, *nicht wahr?*

nick, *sb.*¹ Add: **1. d.** *Squash* and *Real Tennis*. (See quot. 1961.)

1890 J. M. Heathcote et al. *Tennis, Lawn Tennis, Rackets, Fives* i. iv. 69 When the odds of touch-no-walls.. are given, a ball returned by the giver of odds, which makes a nick, is counted for the striker. **1926** C. Arnold *Handbk. on Squash Rackets* iii. 34 Winners can also be made by what is known as a dead service nick, the ball being made to meet the back or side wall and the floor at the same time. **1961** J. S. Salak *Dict. Amer. Sports* 296 *Nick* (court tennis)—the junction of the wall and the floor, or a return when the ball, as it drops or falls, touches the wall and the floor simultaneously. **1963** *Times* 8 Jan. 3/5 This Binns did with his usual touch strokes, boasting with precision and, on his forehand drops, turning his racket over like a butterfly net to impart top spin and make the ball die in the nick. **1973** R. Hawkey *Beginner's Guide Squash* iv. 41 One must avoid..allowing a service to drop into the 'nick' between the back wall and the floor.

13. (Earlier and later examples.)

1824 D. Blaine *Canine Pathol.* (ed. 2) 109 Some rearers of game fowls..are favourable to breeding from the third remove, which they call a nick. **1927** J. E. Platt *Thoroughbred Race-Horse* iv. 31 All the well-known nicks and blending of bloods must be observed, and the leading lines of sires and dams carefully considered. **1973** *Country Life* 15 Nov. 1545/1 The nick owed little to studies of genetics and line breeding [of harriers].

14. Phrases. *in good nick* (slang, orig. dial.), in good condition; (*in the*) *nick* (slang), (in) prison, (at a) police station.

1882 *Sydney Slang Dict.* 6/2 *Nick* (*The*), gaol. **1905** Wright *Eng. Dial. Dict.* Suppl. 151/1 *In good nick*, in good condition. **1949** F. Sargeson *I saw in my Dream* 61 They [*sc.* tennis courts] seemed to be in good nick. **1952** 'N. Shute' *Far Country* 5 She's in good nick. **1957** *Railway Mag.* June 431/1 It does not mean..that an 'A4' in good 'nick'..is not capable of performances well up to the pre-war standard. **1957** P. Wildeblood *Main Chance* 122 'Arrest me,' said Ron. 'Go on, take me to the nick.' **1959** *Listener* 9 Apr. 645/3 Sundry knaves have been routed and put in 'the nick'. **1962** R. Cook *Crust on its Uppers* i. 20 The boys down at Chelsea nick. **1968** *Listener* 8 Aug. 178/2 A second-'and British slave in good nick. **1968** J. Lock *Lady Policeman* i. 7 An address much nearer than the 'nick'. *Ibid.* 8 Back at the nick the station officer was very cross. **1969** [see *bust *sb.*² f]. **1970** P. Laurie *Scotland Yard* iii. 66 The Inspector calls the nick. **1971** 'F. Clifford' *Blind Side* ii. ii. 81 Reports are that he's in fair enough nick.

nick, *v.*² Add: **4.** (Later examples of *intr.* use.)

1876 *Rep. Vermont Board Agric.* III. 132 There is another strain or peculiarity among these Canadians, that seems to nick well with the Morgans. **1974** *Country Life* 3–10 Jan. 43/3 The colour-marking Hereford 'nicks' well with virtually all beef and dairy breeds.

7. d. Of a ball in squash, real tennis, etc.: to strike the floor and wall simultaneously.

1898 W. Morgan *'House' on Sport* 251 By volleying the service you prevent the ball 'nicking', i.e., so pitching in the angle formed by floor and wall as to be unplayable. **1926** C. Arnold *Handbk. on Squash Rackets* iii. 34 It [*sc.* the ball] should be made just to nick on to the side wall and floor and there lie dead. **1960** *Times* 29 Nov. 17/3 Oddy was nicking the side wall more often. **1973** *Times* 31 Jan. 8/7 The pace was ferocious, the 22-year-old Jehan hitting a stream of nicked winners.

8. a. Add to def.: to arrest, to put in gaol. Hence **ni·cking** *vbl. sb.*

1836, etc. [in Dict.]. **1958** *Encounter* May 11/2 He'd got nicked for ponceing off his old woman. **1959** 'M. Cronin' *Dead & Done With* x. 152 They nicked your chum for killing his wife. **1962** *John o' London's* 25 Jan. 82/2 'Arrest' has many Cant synonyms including..*nick*. **1968** J. Lock *Lady Policeman* iii. 30, I don't sleep rough any more. I've learnt a thing or two since you nicked me. **1970** G. F. Newman *Sir, You Bastard* i. 28 Even on leave he looked for possible nickings. **1973** J. Wainwright *Devil you Don't* 32, I am talking to you, copper..either nick me..or close that bloody door.

c. (Later examples.) Also, to rob.

1901 'J. Flynt' *World of Graft* 220/1 *Nick*, to make a 'touch'. **1903** J. London *People of Abyss* xxiii. 280 'At ten we 'ops the wag; at thirteen we nicks things; an' at sixteen we bashes the copper.' Which is to say, at ten they play truant, at thirteen steal, and at sixteen are sufficiently developed hooligans to smash the policemen. **1914** Jackson & Hellyer *Vocab. Criminal Slang* 62 *Nick*... To surreptitiously extract something from the person. **1916** H. L. Wilson *Somewhere in Red Gap* vi. 236 'I did hear that you'd had your pocket picked.'..'That's right. ..Some lad nicked me for my roll and return ticket.' **1954** Wodehouse *Jeeves & Feudal Spirit* ix. 81 Despite this, you succeeded in nicking him for what must have been a small fortune. **1966** J. Porter *Sour Cream* xiii. 169, I had Azatov's own pass which I had nicked from him at the airport. **1973** *Courier Mail* (Brisbane) 21 June 7/4 Nicking toys from chain stores. **1974** S. Gulliver *Vulcan Bulletins* 29 The Libyans will try to nick Javits' shipment.

13. b. *Austral.* To slip away, depart hurriedly.

1896 E. Turner *Little Larrikin* xxiii. 274 Trying to induce the driver of the motor, for whom he had a friendship, to promise at the end of the journey to 'nick away and come too'. **1928** 'Brent of Bin Bin' *Up Country* viii. 120 Bert and I could just nick down to Mungee. **1959** Baker *Drum* ii. 129 *Nick, do a*, to decamp, slip away unnoticed. Also, *nick off*.

14. Substitute for def.: Of animals, to mate with excellent results. (Later examples.)

1942 R. B. Kelley *Sheep Dogs* 51 When the progeny of a bitch by a particular dog are outstanding the parents are said to have 'nicked'. **1959** *New Scientist* 22 Oct. 737/2 Where the offspring's performance is conspicuously superior to that of its parents the mating [of poultry] is said to 'nick'. **1971** *Daily Tel.* 4 Jan. 7/6 Breeders..know from bitter experience that matings do not always 'nick' and that..they are sure to suffer many a disappointment.

nickel, *sb.* Add: **2. c.** Five dollars' worth of marijuana. *U.S. slang.*

1967 *Boston Sunday Herald* 26 Mar. 1/2 Nickel bags of marijuana (in hippie lingo a 'nickel' is $5 worth). **1968–70** *Current Slang* (Univ. S. Dakota) III–IV. 86 *Nickel*.., one-eighth to one-fourth of an ounce of marijuana costing about five dollars. Five dollars. (Drug users' jargon).

3. a. *nickel-candy, -cigar, -faced* (also *-face* vb.), *-facing, -plated* adj. (also *fig.*); **nickel-and-dime,** (*a*) rhyming slang for 'time'; (*b*) *adj.*, designating a store in which articles are cheaply priced; also *transf.* and *fig.*; **nickel bag** *U.S. slang*, a bag containing, or a measure of, five dollars' worth of a drug, esp. heroin or marijuana; **nickel-in-the-slot** *a.*, of a machine, etc.: operated by the insertion of a nickel; **nickel note** *U.S. slang* (see quots.); **nickel nurser** *U.S. slang*, a miser.

1935 A. J. Pollock *Underworld Speaks* 80/1 *Nickel and dime*, time. **1960** J. Franklyn *Dict. Rhyming Slang* 100/2 *Nickel and dime*, time. **1970** *New York* 16 Nov. 42/2 Pinned to its banks are proud, homely nickel-and-dime towns..with sides made of asphalt shingles. **1972** *Times* 16 May 1/3 The first Mrs Wallace had worked in a nickel-and-dime store. **1974** 'E. Lathen' *Sweet & Low* viii. 84 You've got a lot of members who do nickel-and-dime business. **1967** Wentworth & Flexner *Dict. Amer. Slang* 672/1 The usual quantities or sizes of 'bags' are: 'trey' = $3 worth (esp. cocaine and heroin); 'nickel bag' = $5 worth (esp. marijuana, cocaine, or heroin). **1971** *Black World* Apr. 38/2 Black men and women and their children exchange expensive gifts of death—in small nickel bags. **1973** *Ibid.* June 79/2 If..he gets high and blurts it out to a stranger in some bar that he got his nickel bag from Joe, the pusher, then Joe's livelihood is endangered. **1972** *Village Voice* (N.Y.) 1 June 13/4 Nickel candy is 12 cents. **1894** R. Kipling *Let.* 28 July in C. Carrington *Rudyard Kipling* (1955) ix. 217 There's a smell of horse-piss, Italian fruit-vendor, nickel cigars. **1894** *Amer. Dict. Printing & Bookmaking* 402/1 Electrotypes are often nickel-faced when they are to be used with colored inks, as copper injures the color. **1964** E. A. D. Hutchings *Printing by Letterpress* I. xii. 207 The usual copper shell can be nickel-faced after the plate has been finished. This type of plate is termed a nickel-faced electrotype. **1892** A. Powell *Southward's Pract. Printing* (ed. 4) lxxii. 696 (*heading*) Electrotyping.. finishing the plate..nickel facing. **1946** W. H. Church in H. Whetton *Pract. Printing & Binding* xiii. 169/1 No further finishing of the surface of the plate can be undertaken after the nickel facing is applied, without risk of the nickel peeling off when the plate is in use. **1889** *Tacoma* (Washington) *News* 13 Dec. 3/5 The latest nickel-in-the-slot scheme is really a stroke of genius and is destined to revolutionize cheap literature in this country. **1893** *Harper's Mag.* Mar. 494 [In Jacksonville] there were the same..nickel-in-the-slot machines [as in Asbury Park]. **1901** *Daily Colonist* (Victoria, B.C.) 20 Oct. 10/3 So long as..there is no means of obtaining..official reports,..so long will mining stock investment remain on the level of 'nickel-in-the-slot' gambling. **1926** *Amer. Speech* I. 652/1 *Nickel note*, five dollar paper bill. **1970** C. Major *Dict. Afro-Amer. Slang* 85 *Nickel note*, five-dollar bill. **1926** Maines & Grant *Wise-Crack Dict.* 11/2 *Nickle nurser*, one who has a passion for seeing that his nickles don't stray. **1945** L. Shelly *Jive Talk Dict.* 29 *Nickel nurser*, tightwad. **1884** *Encycl. Brit.* XVII. 488/2 The manufacture of cooking utensils and other useful articles out of..nickel-plated iron. **1885** 'Mark Twain' in *Century Mag.* Dec. 194 He had some pathetic little nickel-plated aristocratic instincts. **1910** *Daily Chron.* 10 Dec. 9/5 This five-shilling watch..is made in cases of nickel or of gunmetal nickel-plated, with a crown bow. **1970** T. Hughes *Crow* 39 The tears are nickel-plated. **1974** R. C. Dennis *Conversations with Corpse* iii. 19 A revolver..

or a pistol? Nickel-plated or blue steel? Regulation or snub-nosed?

b. nickel-antigorite [ad. G. *nickel-antigorit* (H. Strunz *Mineral. Tabellen* (ed. 3, 1957) 323)], a nickelian variety, (Mg, Ni)$_3$Si$_2$O$_5$(OH)$_4$, of antigorite; **nickel-chlorite** [ad. G. *nickelchlorit* (H. Strunz *Mineral. Tabellen* (ed. 3, 1957) 317)], a basic silicate and aluminate of magnesium, iron, nickel, and aluminium, (Mg,Fe,Ni,Al)$_6$(Si,Al)$_4$O$_{10}$(OH)$_8$, which has been synthesized but whose natural occurrence is uncertain; **nickel-iron**, any alloy of nickel and iron; freq. *attrib.*; **nickel-skutterudite**, an arsenide of nickel and cobalt, (Ni,Co)As$_3$, with nickel predominating, found as white or grey isometric crystals; also, the cobalt-free compound NiAs$_3$; **nickel spinel** [ad. G. *nickelspinell* (H. Strunz *Mineral. Tabellen* (ed. 3, 1957) 137): see SPINEL], an artificially produced oxide of nickel and aluminium, NiAl$_2$O$_4$.

1961 *Mineral. Mag.* XXXII. 972 Nickel-antigorite... An unnecessary name for nickelian antigorite. **1968** *Proc. Indian Acad. Sci.* B. LXVII. 178 The 'a' spacings of this mineral can also stand a fair comparison with nickel-antigorite. Therefore, this sample could be either nepouite or nickel antigorite. **1961** *Mineral. Mag.* XXXII. 979 Nickelchlorite... Some of the natural nickel silicates may be members of the chlorite group. **1969** *Clays & Clay Minerals* XVII. 133 Nickel-chlorite has been obtained by the co-precipitation of nickelous hydrous oxide and montmorillonite at an OH/Ni ratio of 2·0. **1875** *Geol. Mag.* Decade II. II. 27 Nickel-iron, containing 17·5 per cent. of nickel, constitutes 3·5 per cent. of the stone a less quantity than is found in the Pultusk meteorites. **1946** *Thorpe's Dict. Appl. Chem.* (ed. 4) VII. 458/1 Silicon-iron and nickel-iron.. are noteworthy. *Ibid.*, The nickel-iron alloys ('Permalloys') are used in composites for telephone and radio frequencies. **1971** I. G. Gass et al. *Understanding Earth* viii. 116/1 Nickel-iron is practically absent from terrestrial rocks. **1974** *Gloss. Electrotechnical, Power Terms* (B.S.I.) i. iii. 14 *Nickel-iron sleeve*, longitudinally split sleeve of nickel-iron alloy, the use of which enables a winding to have a higher impedance at speech frequencies. **1892** Waller & Moses in *School of Mines Q.* XIV. 51 This would be a mineral of the type of skutterudite CoAs$_3$... If further analysis confirms these results, the name *Nickel-Skutterudite* is suggested. **1935** *Amer. Mineralogist* XX. 723 Microscopic study of material from the Bullard's Peak district, New Mexico, shows native silver associated with nickel-skutterudite. **1968** I. Kostov *Mineralogy* ii. 134 The analogous NiAs$_3$ (nickel-skutterudite) contains Ni 20·71%, As 79·29%. **1961** *Mineral. Mag.* XXXII. 973 Nickelspinel. **1963** *Jrnl. Amer. Ceramic Soc.* XLVI. 583/2 The other ternary phase assemblage, liquid, NiO, and nickel spinel, exists over wide ranges of temperature (1450° to ~1775° C) and composition (~20 to 100 mole % NiAl$_2$O$_4$).

nickel, *v.* Add: **b.** To foul (the bore of a gun) with nickel off the bullet-casing; *intr.*, to become fouled. **nickeling**, (*b*) the fouling of the bore of a gun with nickel; metallic fouling.

1928 E. S. Farrow *Dict. Mil. Terms* 407 *Nickeling*, in gunnery, metallic fouling caused by a portion of the cupro-nickel of the envelope of the bullet being left on the surface of the bore. **1930** G. Burrard *Notes on Sporting Rifles* 31 When a barrel has once been nickelled it is always liable to nickel again very quickly. *Ibid.* 30 Nickelling at first is impossible to detect with the eye. **1958** J. A. Barlow *Elements of Rifle Shooting* (ed. 5) i. 6 It may be as well here to touch on the problem presented by metallic fouling, generally known as nickelling... Nickelling is mainly due to small particles of the bullet envelope being cut off by roughness or excrescence in the bore.

nickelian (nikī·liăn), *a. Min.* [f. Nickel *sb.* + *-ian 2.] Of a mineral: having a (small) proportion of a constituent element replaced by nickel.

1930 W. T. Schaller in *Amer. Mineralogist* XV. 571 Nickel—nickelian. **1951** C. Palache et al. *Dana's Syst. Min.* (ed. 7) II. 503 A nickelian variety (2·5 per cent NiO) occurs in the Mt. Diablo district. **1974** *Proc. Brit. Macrophacia* XII. 41/1 There is simply a gradually thickening strand of iron oxides—a mixture of goethite, hematite, and nickelian magnetite.

nickeline, *sb.* (in Dict. after Nickel *sb.*). Add: (See quot. 1971.) (Further example.)

1971 *Mineral. Mag.* XXXVIII. 104 Recommendations of the Commission [on New Minerals and Mineral Names, of the International Mineralogical Association] on minerals for which more than one name is in common use: ..Nickeline, not niccolite or nickelite.

nickelodeon (nikĕlō·diən). *U.S.* [f. Nickel *sb.* 2; app. after Melodeon.] **1.** A theatre or motion-picture show for which the admission fee is a nickel; a place containing automatic machines to provide amusement, which can be used for a nickel. Also *attrib.*

1921 *Ladies' Home Jrnl.* June 79/1 It is this class which first patronized the old nickelodeon, and undoubtedly it imposed its tastes and its traditions on the picture makers. **1927** F. Hurst *Song of Life* 292 The nickelodeons and the gewgaw shops of the most terrific city in the world. **1930** *Time & Tide* 27 Sept. 1206 The film was.. handed over by the scientists to the 'nickelodeons' of America. **1938** *Encycl. Brit. Bk. of Year* 432/2 The old nickelodeon programmes. **1939** C. Morley *Kitty Foyle* 68 A dance floor and a nickelodeon piano. **1955** G. Greene *Quiet American* 188 It must have belonged to the same era as the nickelodeon. **1973** *Publishers Weekly* 10 Sept. 43/2 The development of American movies from nickelodeon days to the 1970s.

2. A 'jukebox'; a machine that automatically plays selected gramophone records on the insertion of a coin. Also *attrib.*

1938 *Florida Review* Spring 25/1 The requisites of a place entitling it to the name *juke* are..presence of the nickelodeon, and..of the dance-floor. **1949** *Sat. Even. Post* 15 Jan. 87/1 A nickelodeon at the end of the street emits a tinny piano tinkle. **1957** J. Frame *Owls do Cry* 76 Putting money in the nickelodeon. **1973** *Daily Colonist* (Victoria, B.C.) 18 Aug. 8/2 The former shepherd boy who used a $4,000 nickelodeon as a springboard to the top.

nicker (ni·kəɹ), *sb.*⁶ *slang.* [Origin unknown.] One pound sterling.

1910 *Sessions Papers* 1 June 128, I suppose this has cost you a couple of 'nickers'. **1939** [see *case]. **1960** D. Lessing *In Pursuit of English* ii. 66 It's a little matter. A hundred nicker. And it'd double itself in a year. **1966** F. Shaw et al. *Lern Yerself Scouse* 34 Five nicker, five pounds; five pound note. **1975** J. Symons *Three Pipe Problem* xv. 138 Who said there'd be trouble? Anyway, it's a hundred nicker.

nickie tam, nickie tom, varr. *Nicky tam.

nicking, *vbl. sb.*¹ Add: **1. d.** A method of pruning in which an incision is made below the base of a bud in order to curb its growth.

1949 C. R. Thompson *Pruning of Apples & Pears* ii. 53 The top bud should be prevented from growing vigorously by making a knife incision at its base, a treatment referred to as 'nicking'. **1972** G. E. Brown *Pruning of Trees* ii. 24 Nicking is carried out below selected buds in order to reduce their vigour.

nickpoint, var. *Knickpoint.

nick-tailed (ni·k,tēild), *a.* Having the tail nicked.

1841 *Southern Lit. Messenger* VII. 219/1 Brenda, mounted on Paul Clifford, nick-tailed sorrel pacer. **1853** J. G. Baldwin *Flush Times Alabama* 97 The horse, a nick-tailed trotter. **1867** G. W. Harris *Sut Lovingood* 19 A nick tailed, bow necked, long, poor, pale sorrel horse.

Nicky (ni·ki), *sb.*² [f. *N.I.C. (main entry), infl. by *Nicky*, dim. of *Nicholas*, a male name.] Colloq. name for the National Incomes Commission. Also *attrib.*

1962 *Daily Tel.* 27 July 1/1 The Commission, likely to be known as 'Nicky', will consider pay questions in both private and public sectors. **1964** *Ibid.* 10 Aug. 20/5 (heading) 'Nicky' report for TUC. **1963** *Ann. Reg. 1962* 18 Nicky was expected to run the rule over wages claims.

nicky tam (ni·ki tæm). *Sc.* Also **knicky tom, nickie tam, nickie tom,** and with hyphen. [f. Knicker² + -y⁶ + *tam* (see Taum).] (See quot. 1965.)

1901 *Aberdeen Jrnl. N. & Q.* IV. 17/2 *Knicky-toms*, garters worn over trousers. **1917** E. S. Rae *Private John Macpherson* 34 An' Geordie, ma foreman, a decenter lad Ne'er wore nickietoms. **1931** in E. MacColl *Scotland Sings* (1953) 56, I buckit roond my nappin' knees a pair o' Nicky Tams. **1965** *Sc. Nat. Dict.* VI. 422/3 *Nicky-tams*... One of a pair of straps, or a piece of string or twine, tied by farmworkers over the trousers-legs immediately below the knee to keep the legs clear of the soil and dust, etc. **1967** *Listener* 12 Oct. 479/3 The Scots word for 'a tying worn below the knee to keep the bottom of the trouser-leg tilted clear in dirty work or to exclude dust'. It is a nicky-tam. **1972** *Daily Mail* 19 July 6/5 What we do about to lose the English word 'yorkers' or what the Scots called 'Nicky Tams'.

Nicobarese (nikǒbāri·z), *sb.* and *a.* [f. *Nicobar* (see below) + -ese.] **A.** *sb.* **a.** The people of the Nicobar Islands in the Bay of Bengal. **b.** The Mon-Khmer language of this people. **B.** *adj.* Of or pertaining to this people. Also **Nicoba·rian** *sb.* and *a.*

1790 G. Hamilton in *Asiatick Res.* II. 344 The people of Carnicobar have a tradition among them, that several canoes came from Andaman..and that the crews..killed several of the Nicobarians. **1846** *Jrnl. Asiatic Soc. Bengal* XV. 368 The Nicobarian Pigs appear to have been derived from the Chinese domestic species. **1859** *Sel. Rec. Govt. India Home Dept.* No. 25. 61 Their language bears no affinity to that of the Nicobarians. **1875** F. A. de Roepstorff *Vocab. Dial. Nicobar & Andaman Is.* (ed. 2) 14 The Nicobarese they all want, yet they like very much to barter with foreigners. **1884** *Encycl. Brit.* XVII. 486/2 Nicobarese is a wholly uninflective tongue. **1889** E. H. Man (title) Dictionary of the central Nicobarese language. **1924** G. Whitehead *In Nicobar Is.* ii. 47 The Nicobarese keep a great number of domestic pigs, which they feed on coco-nut. **1972** W. B. Lockwood *Panorama of Indo-European Lang.* 229 Nicobarese..is spoken by 15,000 islanders. *Ibid.*, The population of the Nicobars has been swollen in the last two decades by settlers from all parts of India, forming a polyglot element..about as numerous as the native Nicobarese.

Nicodemite. (Later example.)

1921 *Outward Bound* Apr. 29/2 This is no time to play the Nicodemite.

Nicol². Add: Also **nicol**. (*Nicol* (or *nicol*) *prism* is now more usual than *Nicol's prism.*) (Earlier and additional examples.)

1863 E. Atkinson tr. *Ganot's Elem. Treat. Physics* vii. viii. 485 The Nicol's prism is one of the most valuable means of polarising light. **1906** *Jrnl. Chem. Soc.* LXXXIX. ii. 1150 The nicols were rotated. **1937** *Discovery* Aug. 242/2 A Nicol prism (which is two portions of a rhomb of Iceland spar cemented together with Canada balsam). **1955** W. Gaddis *Recognitions* iii. v. 874 If we can fix a microscope up with polarized light and put a particle of the pigment under it, we can see whether it's isotropic or anisotropic, for that we use nicol prisms.

Niconian, var. *Nikonian *sb.* and *a.*

nicotinamide (nikoti·nāməid). *Biochem.* [f. *nicotin(ic *a.* + Amide.] **a.** The amide, (C$_5$H$_4$N)CONH$_2$, of nicotinic acid, which can be converted into the acid *in vivo* and so can replace it in the diet.

1895 *Jrnl. Chem. Soc.* LXVIII. i. 391 On heating ethylic nicotinate with alcoholic ammonia..it is converted into nicotinamide, which melts at 121°. **1951** in M. McLuhan *Mech. Bride* (1967) 91/1 (Advt.), It is nicotinamide, an important component in the familiar Vitamin B Complex pill. **1957** A. Huxley *Let.* 12 Dec. (1969) 837 The Italians have been using massive doses of Nicotinamide (a variant on nicotinic acid) in psychological cases for some time. **1968** Passmore & Robson *Compan. Med. Stud.* I. v. 14/2 In Great Britain, the U.S.A. and many other countries, nicotinamide must by law be added to white flour used for bread making.

b. *Comb.* **nicotinamide-adenine dinucleotide**, a compound of adenosine monophosphate and nicotinamide mononucleotide which is a co-enzyme for the oxidation of a wide variety of substrates *in vivo*; NAD; co-enzyme I; diphosphopyridine nucleotide.

1961 *Biochem. Jrnl.* LXXX. 323/1 The rate of formation of nicotinamide-adenine dinucleotide from adenosine triphosphate..and nicotinamide mononucleotide..was measured. **1969** J. R. Holum *Introd. Org. & Biol. Chem.* xi. 381 Nicotinamide (niacinamide) is a vitamin used by the body in making the coenzyme nicotinamide adenine dinucleotide.

nicotinate (ni·kotineit). [f. *nicotin(ic *a.* + -ate⁴.] The anion, or a salt or ester, of nicotinic acid.

1879 *Jrnl. Chem. Soc.* XXXVI. 809 If..potassium nicotinate is treated with phosphorus pentachloride, energetic action ensues. **1934** *Jrnl. Amer. Chem. Soc.* LVI. 2426/1 Folkers prepared ethyl nipecotate..by the hydrogenation of ethyl nicotinate. **1970** R. W. McGilvery *Biochemistry* x. 185 Nicotinate and its enzyme occur in all organisms.

nicotine. Add to def.: in small doses it has a stimulating action, but in larger amounts it blocks the actions of autonomic ganglion cells and skeletal muscle fibres. (Further examples.)

1915 W. S. Maugham *Of Human Bondage* xlvii. 231 She had long, beautiful hands, with fingers deeply stained by nicotine. **1940** H. A. McGuigan *Appl. Pharmacol.* 589 When nicotine is injected intravenously, an enormous rise in blood pressure occurs. **1951** A. Grollman *Pharmacol. & Therapeutics* xv. 285 Nicotine produces extreme nausea and vomiting when taken even in comparatively small quantities. **1966** *McGraw-Hill Encycl. Sci. & Technol.* IX. 100/2 Poisoning has occurred from accidental ingestion of insecticide sprays containing nicotine.

b. *attrib.* and *Comb.*, as *nicotine-brown, -free, -like, -stained* adjs.; *nicotine poisoning* (further example).

1945 Dylan Thomas *Let.* 30 July (1966) 278, I can hear..my uncle Bob drinking tea and methylated spirits through eighty years of nicotine-brown fern. **1967** *Daily Tel.* 20 Jan. 17/7 The pleasure of nicotine-free smoking. **1914** Nicotine-like [see *muscarine]. **1948** J. H. Burn *Lect. Notes Pharmacol.* 15 Nicotine is a substance which possesses the nicotine-like actions of acetylcholine. **1951** A. Grollman *Pharmacol. & Therapeutics* xv. 282 In the treatment of nicotine poisoning, artificial respiration should be instituted. **1936** C. Day Lewis *Friendly Tree* 12 A plump man with a nicotine-stained moustache. **1967** A. Marshall in *Coast to Coast 1965–66* 121 He rubbed his chin with a nicotine-stained finger.

Hence **ni:cotiniza·tion**, subjection to the action of nicotine; **ni·cotinized** *ppl. a.*, containing or drugged with nicotine. (Delete entry for **nicotinic** *a.* and see next.)

1873 Nicotinized [in Dict.]. **1911** *Jrnl. Physiol.* XLIII. 181 The partial contraction suggests an anodic inhibition in the nerve cell like that which occurs in certain nicotinized striated muscles. **1940** *Ibid.* XCIX. 73 Acetylcholine injections have been observed to produce vasoconstriction in the nicotinized as well as in the normal perfused guinea-pig lungs. **1945** *Amer. Jrnl. Physiol.* CXLIV. 192 The nicotinization was maintained by adding nicotine (2 mgm./liter) to the perfusion fluid of the heart. **1966** *Punch* 9 Feb. 215/1 Today nicotinisation is complete with our own men taking the first breath of the day through a tube, lighting up automatically at the appearance of food, and thickening the bedroom air well into the early hours of the morning.

nicotinic (nikoti·nik), *a. Chem.* and *Biochem.* [f. Nicotin(e + -ic.] **1.** *nicotinic acid* [tr. G. *nicotinsäure* (H. Weidel 1873, in *Ann. d. Chem. und Pharm.* CLXV. 330)]: a white crystalline heterocyclic acid, (C$_5$H$_4$N)COOH, which is widely distributed (usu. in the form of a complex of its amide) in foods such as yeasts, wheat germ, and meat, is formed when nicotine is oxidized, and can be synthesized in the body from tryptophan; it is a B vitamin, deficiency of which causes pellagra in man; 3-pyridinecarboxylic acid. (Cf. *niacin.)

1873 *Jrnl. Chem. Soc.* XXVI. 509 Acetyl chloride has no action on nicotinic acid. **1913** J. Walker *Org. Chem.* 279 The β acid, or nicotinic acid, may also be produced by the oxidation of the alkaloid nicotine. **1942** *Industr. & Engin. Chem. (Analytical Ed.)* XIV. 663/1 The dietary position of cereals and cereal products in relation to their pellegra [*sic*]-preventive attributes has recently been given considerable prominence by the inclusion of nicotinic acid or niacin in the list of required components of enriched flour. **1968** Passmore & Robson *Compan. Med. Stud.* I. v. 14/2 Like thiamine, nicotinic acid is present in the whole-wheat grain in good quantities, but most is removed by the millers in preparing fine flours or polished rice.

2. Resembling (that of) nicotine; capable of responding to nicotine.

1941 [see *muscarinic *a.*]. **1961** A. Goth *Med. Pharmacol.* iii. 44 The nicotinic receptors may be divided into the ganglionic receptors, which are hexamethonium sensitive, and the skeletal muscle receptors, which are sensitive to *d*-tubocurarine. **1970** *Nature* 5 Dec. 917/1 The pharmacological actions of acetylcholine..outside the central nervous system could be divided into two categories, nicotinic and muscarinic. **1973** *Jrnl. Pharmacol. & Exper. Therap.* CLXXXV. 649 The data..suggest that both stimulant drugs and competitive antagonists interact on a one-to-one basis with the nicotinic receptor of the guinea-pig lumbrical muscle.

nictate, *v.* Delete 'Only in *nictating membrane*' and add further example.

1960 V. Nabokov *Invitation to Beheading* iv. 40 Emmie was still squatting,..her long, pale, almost white lashes nictating as she looked across the table-top at the door.

nictitation. Add: Also *transf.*

1962 V. Nabokov *Pale Fire* 160 A couple..whose blundering Cadillac half entered my driveway before retreating in a flurry of luminous nictitation.

nicy: see *nicey *a.*

nidation (nəidēi·ʃən). *Physiol.* [f. Nid(us + -ation.] † **a.** (See quots.) *Obs.*

1874 J. H. Aveling in *Obstetr. Jrnl.* II. 210 The act of nidation consists of the periodical development of the mucous membrane lining the interior of the body of the uterus. **1892** *Syd. Soc. Lex.*, *Nidation*,..Aveling's term for the monthly renewal of the epithelium of the mucous lining of the womb during the intermenstrual period.

b. = *implantation 6.

1892 *Syd. Soc. Lex.*, *Nidation*,..also, a term for the reception of the fertilised ovum in the uterine mucous membrane. **1921** B. M. Anspach *Gynecol.* iv. 72 These [changes] have for their purpose suitable nidation and nourishment of the ovum. **1966** *New Statesman* 17 June 880/1 Surely an abortifacient is something which interferes with an established pregnancy and not something which prevents the nidation of a fertilised ovum. **1970** *Sci. Jrnl.* June 63/2 It is apparently by this thickened trophoblastic plate that the penetration of the maternal tissues—called nidation or implantation—is accomplished.

niddy-noddy (ni·di,nǫ·di), *sb. Hist.* [Prob. f. Niddy-noddy *adv.* and *v.*] A frame on which to skein and measure wool yarn.

1890 G. S. Hall in *Proc. Amer. Antiquarian Soc.* VII. 111 It was taken from the spindle sometimes on a niddy-noddy held in the hand, at two rounds per yard, but more commonly on a reel, in rounds of two yards each. **1927** M. N. Rawson *Candle Days* ii. 28 The graceful 'swifts', the 'niddy noddy', or hand reels, was also of home construction in wood. **1968** *Beaver* Winter 43/1 A cross between the hand looms..has been found practical, as has been..the English niddy-noddy and butterfly yarn winders. **1969** E. H. Pinto *Treen* 318 The earlier device, which was in general use..was the cross reel, in English speaking countries almost universally known as a niddy-noddy. *Ibid.*, The niddy-noddy, which was held in one hand by the central stem, was wound with a waving motion, to the rhythm of a song, the opening line of which ran, 'Niddy-noddy, niddy-noddy, two heads and one body.'

Niderviller (nidęrvilē̜r). The name of a town in Lorraine, east France, used, freq. *attrib.*, to designate the porcelain and faience made there from 1754.

1857 H. G. Bohn *Guide Knowl. Pottery, Porcelain* 480 *Niderviller.* Hard paste. Manufacture of François Lanfray,..about 1790. Stencilled in blue. **1863** W. Chaffers *Marks Pott. & Porc.* 208 *Niderviller*, the letter N., for Niderviller, occurs on a set of plates, on one of which is the double C, and on another the letter N. **1903** M. L. Solon *Hist. Old French Faïence* 112 The Niderviller faïence is amply represented in the Nancy Museum. **1948** A. Lane *French Faïence* ix. 39 The Niderviller figures belong to a class peculiar to the Lorraine region. **1960** R. G. Haggar *Concise Encycl. Cont. Pott. & Porc.* 329 The flower painting on Niderviller faience was of high quality and exploited the range of tints which 'purple of Cassius' crimson could be made to yield. **1963** *Times* 16 Feb. 4/4 A Niderviller part dinner-service of 37 pieces, painted with landscapes, the borders with butterflies, ladybirds and other insects, made c1770. **1971** H. Wynter *Introd. European Porc.* iii. 105 Niderviller characteristics. Hard-paste.. Shapes and decoration..similar to rococo Strasbourg.

nidicolous (nidi·kǒləs), *a. Ornith.* [f. mod.L. *Nidicolæ* (H. F. Gadow in A. Newton *Dict. Birds* (1894) 629) (f. L. *nīdus* nest + *col-ere* to inhabit) + -ous.] Of a bird: bearing young which are helpless at birth and remain in the nest until they are sufficiently developed to live without parental care. So **ni·dicole** *sb.*, a bird of this type.

1900 H. F. Gadow in *Encycl. Brit.* XXVI. 257/1 Order Sphenisciformes.—Nidicolous, marine. Flightless, wings transformed into rowing paddles. **1927** A. L. Thomson *Birds* ii. 155 Birds may..be divided into two main types, those having nidifugous or 'nest-quitting' young, and those having nidicolous or 'nest-dwelling' young. **1945** S. Smith *How to study Birds* iv. 74 Birds whose young are born naked, helpless and blind (the so-called nidicolous birds). **1962** C. J. Welty *Life of Birds* xvii. 319/2 The altricial bird, is born naked, or nearly so, is usually blind, and it too weak to support itself on its legs... Such birds remain confined to the nest for some days or weeks. They are therefore called nidicoles or nest-dwellers. **1974** I. C. J. Galbraith tr. *Dorst's Life of Birds* I. xiii. 247 The parents are entirely responsible for feeding nidicolous young.

nidifugous (nidi·fiŭgəs), *a. Ornith.* [f. mod. L. group name *Nidifugæ* (H. F. Gadow in A. Newton *Dict. Birds* (1894) 629) (f. L. *nīdus* nest + *fug-ere* to flee) + -ous.] Of a bird: bearing young which are well developed at birth and leave the nest almost immediately. So **ni·difuge** *sb.*, a bird of this type.

1900 H. F. Gadow in *Encycl. Brit.* XXVI. 257/2 Order Colymbiformes.—Plantigrade, nidifugous, aquatic. **1927** [see *nidicolous *a.*]. **1945** S. Smith *How to study Birds* iv. 74 Birds whose young are born strong, with plenty of down, and able to run within a few hours of their birth are called nidifugous birds. **1962** C. J. Welty *Life of Birds* xvii. 318/2 Nidifuges are often ground-nesting species that, as adults, are good runners or good swimmers and feed either on the ground or in the water. These precocial birds include such forms as the ostrich and its relatives, loons, grebes, ducks, [etc.]. *Ibid.* 319/2 (caption) The one-day-old, precocial, nidifugous chick of the Ruffed Grouse. **1974** I. C. J. Galbraith tr. *Dorst's Life of Birds* I. xiii. 245 Nidifugous birds also have visual means of communication.

nid-nod, *a. poet.* [f. Nid-nod *v.*] That nid-nods.

1912 W. de la Mare *Veil* 88 Of whispering boughs, and feathery, nid-nod grass. **1937** —— *This Year, Next Year*, Poppy, cornflower, nid-nod wheat, The sheaves are ripe for rick. **1941** —— *Bells & Grass* 50 The nid-nod daffodil.

nidor. (Later example.)

1923 *Blackw. Mag.* Feb. 159/2 A nidor was to him an agony impossible to endure.

nielsbohrium (nīlzbō·riəm). *Chem.* [f. *Niels Bohr* (see *Bohr) + -ium, as ad. Russ. *nil'sborii*, a name used by G. N. Flerov and co-workers (e.g. in Flerov & Zvara *Report D7-6013* [Joint Inst. Nuclear Res., Dubna, U.S.S.R., 1971] 56), though no explicit coinage of the word has been traced in the literature available.] (A name proposed for) an artificially produced transuranic element, of atomic number 105. (The name *hahnium has also been proposed for it.)

1973 *Nuclear Sci. Abstr.* XXVIII. 2109/2 Proposed names for No. 104 and 105 are Kurchatovium (Ku) and Nielsbohrium (Ns). **1975** *Nature* 27 Mar. 288/2 As a sign of the rival claims element 104 and 105 have been christened Rutherfordium and Hahnium in the USA and Kurchatovium and Nielsbohrium in the USSR.

Niemann–Pick disease (nīmăn,pi·k). *Path.* Also **Niemann–Pick's disease.** [f. the names of Albert *Niemann* (1880–1921) and Ludwig *Pick* (1868–?1944), German physicians, who described the disease in 1914 and 1926 respectively.] A rare, inherited metabolic disorder, usu. fatal in childhood, which is characterized by the accumulation of lipid in the body cells of a lipid (sphingomyelin).

1928 *Jrnl. Amer. Med. Assoc.* 7 Apr. 1166/1 (*heading*) Niemann–Pick disease. *Ibid.* 30 June 2027/2 The stored material in Niemann–Pick's disease is phosphatid. **1942** M. M. Wintrobe *Clin. Hematol.* xii. 521 The cells of Niemann–Pick's disease are round, oval, or polyhedral and are filled with small round hyaline droplets in clusters which give the appearance of foam or a honeycomb. **1966** Wright & Symmers *Systemic Path.* I. v. 146/1 Niemann–Pick disease is considerably rarer than Gaucher's disease. **1973** *Sci. Amer.* Aug. 90/3 Another disorder involving enlargement of the liver and the spleen and mental retardation, Niemann–Pick disease, was shown..to involve the accumulation of the phospholipid sphingomyelin.

Niersteiner (nīə·ɹʃtəinəɹ). [f. *Nierstein* a town in west central Germany near Mainz + -er, G. adj. suffix.] A much esteemed white Rhine wine produced at Nierstein.

[**1825** Scott *Talisman* in *Tales of Crusaders* IV. xi. 200 He invited them to a goblet of nierenstein.] **1833** C. Redding *Hist. Mod. Wines* vii. 207 The wines of Bischeim are very pleasant wines; those of the next strength are ..Rüdesheimer, and Niersteiner. **1855** T. McMullen *Handbk. Wines* xi. 112 Some are of opinion that the Marcobrunner, Rüdesheimer, and Niersteiner, possess more fulness and body. **1907** *Yesterday's Shopping* (1969) 59/3 Still Hock. Niersteiner Pettenthal—Per doz. bots. 30/0. **1930** [see *Steinberger]. **1961** *Guardian* 3 Mar. 12/3 She was eating alone, with a bottle of Niersteiner. **1967** A. Lichine *Encycl. Wines & Spirits* 447/2 Of some 550 parcels, about two dozen produce the soft, elegant, full-bodied, peak wines known around the world as Niersteiners. Look for Riesling on the label: the best Niersteiners are Riesling wines.

Nietzschean (nī·tʃiăn), *sb.* and *a.* Also **Nietzschian.** [f. the name of the German philosopher, Friedrich *Nietzsche* (1844–1900) + -an.] **A.** *sb.* A follower, admirer, or imitator of Nietzsche; one who holds or supports Nietzsche's principles or views, esp. his theories of the superman, and the division of humanity into masters and slaves. **B.** *adj.* Of, pertaining to, or characteristic of Nietzsche or his views.

1904 *To-Day* XLIV. 49/1 This moustache..was Nietzschean at the root, to end in the twirl of the sub-officer. **1908** G. B. Shaw *Sanity of Art* 6 The Nietzscheans were only too glad to see Tolstoy catching it. **1908** M. A. Mügge *Friedrich Nietzsche* iii. 331 There is no doubt that Goethe was a Nietzschean. [illegible]

Hence **Nie·tzscheanism; Nie·tzscheite; Nie·tzscheism.**

1908 H. L. Mencken *Philos. Nietzsche* iii. ii. 278 We have a hero who calls himself a disciple and offers Nietzscheism as a substitute for Christianity. **1908** M. A. Mügge *Friedrich Nietzsche* [illegible]

Nife (nəif). Also **nife.** [f. *Ni* + *Fe*, chem. symbols for nickel and iron (L. *ferrum*) respectively.] Nickel-iron; *spec.* [a. G. *Nife* (E. Suess *Das Antlitz der Erde* (1909) III. ii. xxiv. 626)], the earth's core or the material composing it.

1909 H. B. C. & W. J. Sollas tr. *Suess's Face of Earth* IV. xv. 544 We assume the existence of three zones. [illegible] ..the Ni-Fe cell.

niff (nif), *sb.*¹ *colloq.* and *dial.* Also **nif.** [Origin unknown.] Resentment, offence. Freq. in phr. *to take a niff*, to take offence.

1777 *Horae Subsecivae* 303 Let her alone, her've o'ny a-got a bit of a niff her'll zoon come o' that again. **1865** R. Hunt *Pop. Romances W. of Eng.* 2nd Ser. 78 The woman took a nif, and for a long time never spoke to our John. **1903** in *Eng. Dial. Dict.* IV. 267/2 You're taking the niff. **1914** *Dialect Notes* IV. 77 *Niff*, a quarrel, grudge, or spite. **1946** *Amer. Speech* XXI. 308 To 'take a niff' at a person was to conceive a violent dislike for him.

niff (nif), *sb.*² *slang.* [Perh. f. S)niff *sb.*] A disagreeable smell; a whiff.

1903 in *Eng. Dial. Dict.* **1921** *Chambers's Jrnl.* Mar. 202/1 They found themselves within an outer circle of bee hive huts, fires that had died to red glimmers, and—a 'niff', if I may thus gently put it. **1960** D. Fearon *Murder-on-Thames* ii. 27 It wouldn't be nice for Rachel if some niff of ancient scandal caught up with her poor papa. **1975** *Draconian* Christmas 16922/1 The customary Oxford autumn niff, usually readily recognisable, redolent as it is of bonfires and long grass.

niff (nif), *v.*¹ *colloq.* and *dial.* [Origin unknown.] To quarrel, to be offended. So **niffed** (nift) *ppl. a.*

1875 W. D. Parish *Dict. Sussex Dial.* 79 *Niff*, to quarrel; to be offended. **1880** Courtney & Couch *Gloss. Words in Use in Cornwall* 40/1 She's gone away niffed. **1893** D. Jordan *Forest Tithes* 99 Ye wunt feel niffed like when we meets ye, if we gives ye plenty o' elber-room, mister. **1906** E. Phillpotts *Portreeve* iii. xii. 310 Then—just because you'm niffed about something—you lift your hand to her to let out your bile. **1924** Lawrence & Skinner *Boy in Bush* viii. 119 At last Monica..was niffed. She thought him a muff. **1927** W. E. Collinson *Contemp. Eng.* 116 Anger is expressed by such phrases as these in ascending order of intensity: he was niffed or peeved, he got shirty or hairy, he got his rag out, [etc.].

niff (nif), *v.*² *slang.* Also **nif.** [See *niff *sb.*²] *intr.* To have a disagreeable smell.

1927 [see *hum *v.*²]. **1934** Wodehouse *Thank you, Jeeves* iii. 34 I've started breeding mice and puppies. And, of course, they nif a bit. **1938** —— *Code of Woosters* viii. 177 Scotties are smellie... You will recall how my Aunt Agatha's McIntosh niffed to heaven while enjoying my hospitality. **1950** A. Baron *There's no Home* v. 57 This ol' street may niff a bit, but it don't smell as bad as the water out of polluted wells. **1967** K. Giles *Death in Diamonds* iv. 66 It smelled... 'Niffs, don't it?' said one of the youths. **1974** Wodehouse *Aunts aren't Gentlemen* xvii. 145 'Nasty slinking-looking bleeder.'.. 'He don't half niff.'

niffy (ni·fi), *a. slang.* [f. *niff *sb.*² + -y¹.] Having a disagreeable or strong smell. Also as *sb.* Hence **ni·ffiness.**

1903 *Eng. Dial. Dict.* IV. 267/2 *Niffy*, adj. odorous. **1925** Fraser & Gibbons *Soldier & Sailor Words* 209 *Niffy*, a, a strong nasty smell. **1934** 'Baron Corvo' *Desire & Pursuit of Whole* ix. 81 The niffy silted-up little Rio della Croxe. **1934** Wodehouse *Right Ho, Jeeves* x. 113 The garden was full of the aroma of those niffy white flowers. **1946** —— *Money in Bank* iv. 31 Well anyway, Stinker, putting aside for the moment the question of your niffiness, wasn't it notorious that you couldn't tell the truth without straining a ligament?

nifontovite (nifǫ·ntǒvəit). *Min.* [ad. Russ. *nifontovit* (Malinko & Lisitsyn 1961, in *Doklady̆ Akad. Nauk SSSR* CXXXIX. 188), f. the name of P. V. *Nifontov*, 20th-cent. Russian geologist: see -ite¹.] A hydrated borate of calcium, CaB$_2$O$_4$·2·3H$_2$O, found as colourless monoclinic crystals.

1961 *Mineral. Mag.* XXXII. 990 Nifontovite... Small anhedral grains in skarn deposits in the Urals. **1971** *Soviet Physics—Crystallogr.* XVI. 186/1 The description and investigation of a group of endogenic calcium metaborates: calciborite, korzhinskite, uralborite, nifontovite, frolovite, and pentahydroborite.

niftily (ni·ftili), *adv.* [f. Nifty *a.* + -ly².] In a nifty manner.

1919 *Amer. Mag.* LXXXVII. 93/1 That was a clever girl you had against you tonight. I don't believe in pacifism much, myself, but she used it very niftily for her argument. **1960** P. Mortimer *Saturday Lunch with Brownings* 202 She lay down on her stomach, lifting up the chest with one hand while niftily tucking the carpet down under its foot with the other. **1971** *New Yorker* 13 Nov. 126/3 (Advt.), Our famous Whale Tie..in a niftily contemporary 4½ inch width. **1974** D. Smith *Look Back with Love* xi. 106, I..particularly liked one story..in which the hero fought his way through an army by niftily manipulating a chair.

niftiness (ni·ftinės). *colloq.* [f. Nifty *a.* + -ness.] The quality of being nifty; smartness, cleverness.

1923 M. Watts *Luther Nichols* 27 His fixed purpose was to keep it so or to increase its niftiness. **1974** *Punch* 13 Feb. 268 Some have even been moved to congratulate him on the niftiness of his own footwork—particularly when it comes to the waltz.

nifty, *a.* Add: (Later examples.) Also, clever, nimble, adroit.

1907 C. E. Mulford *Bar-20* ix. 107 I've heard of Smith of Topeka, an' he's mighty nifty with his hands. **1916** H. L. Wilson *Somewhere in Red Gap* v. 213 Hetty.. looking so fresh and nifty and feminine. **1921** Wodehouse *Indiscretions of Archie* xiii. 141 Don't you think it's a nifty scheme? **1923** —— *Inimitable Jeeves* iii. 32 Roville ..is a fairly nifty spot where a chappie without encumbrances in the shape of aunts might spend a somewhat genial week or so. **1930** Chesterton *Four Faultless Felons* iii. 196 He's awfully nifty with his fingers. **1933** G. Ade *Let.* 6 Apr. (1973) 166 You..tell a nifty little story at the finish. **1938** D. Smith *Dear Octopus* ii. i. 52 I'm very nifty on a step-ladder. **1949** N. Marsh *Swing, Brother, Swing* iv. 57 Now that's quite a nifty little idea. **1958** J. Cannan *And be a Villain* i. 19, I..got the niftiest white overalls. **1968** *Globe & Mail* (Toronto) 3 Feb. 35/4 Bruce Kelly scored a tying goal..on a nifty pass from John DeDiana. **1973** R. Hayes *Hungarian Game* xxxi. 184 'Try that coffee now. Tell me if you like it.'.. It was good. 'It's nifty,' I said. **1975** *Observer* 19 Jan. 22/4 Duncan was nifty on occasions, indeed scored an immaculate goal, but was at other times rather daintily ineffective.

nifty (ni·fti), *sb. slang.* [f. the adj.] A joke, a witty remark or story.

1925 Wodehouse *Carry on, Jeeves* vi. 145 Every time I started to pull a nifty, Sir Roderick swung round on me with such a piercing stare that it stopped me in my tracks. **1929** D. Mackail *How Amusing!* 237 He..released no less than six of the wisecracks or nifties which he had been carefully hoarding for his next story. **1957** O. Nash *You can't get there from Here* 65 Had he sought his answers just below where Broadway bisects the lower Fifties, He would have come up with some nifties.

nig, Nig, *sb.*³ Add: (Earlier and later examples.) Now only in derogatory use.

c **1832** T. D. Rice *Jim Crow* x, De Nigs in ole Virginny Be so black dey shine. **1840** *Picayune* (New Orleans) 20 Sept. 2/2 Two little nigs..had a most scientific set-to at the corner. **1905** E. W. Pringle *Woman Rice Planter* 160 Her manner is what the 'nigs' call 'stiff'. **1916** J. B. Cooper *Coo-oo-ee* xvi. 245 He never wipes the glass slobbered over by dozens of dirty nigs! Gosh, it's a good place to get out of! **1939** J. Cary *Mister Johnson* 162 You don't know wot it costs us, you nigs, to tidy up things for you. **1961** C. Willock *Death in Covert* i. 11 Like many of these nigs, he could shoot. Dammit, back home in steaming wogland he probably did nothing else. **1974** R. Gadney *Something worth fighting For* xii. 85 Judd read National Front puts Britain First. Someone had scribbled Nigs Out.

Niger² (nəi·dʒəɹ). Also written **niger. 1.** The name of a West African river, used *absol.* or *attrib.* to designate a type of morocco produced in regions near the river and used for bookbinding.

1898 C. Eyre in *Bookbinding by Women* 4 From Chiswick there are a number of examples of the very beautiful Niger-morocco bindings. **1901** D. Cockerell *Bookbinding* xix. 278 The leather that I have found most useful is the Niger goatskin, brought from Africa by the Royal Niger Company. *Ibid.* 279 It is to be hoped that before long some of the manufacturers interested will produce skins as good in quality and colour as the best Niger morocco. **1930** *Times Lit. Suppl.* 18 Dec. 1081/3 (Advt.), Bound in whole natural niger goat-skin. **1952** J. Carter *ABC for Book Collectors* 124 True niger, which comes from West Africa, is a soft skin with an unemphatic, variable grain. It is locally tanned and dyed... The slight variations of grain and colour which give niger its character are seldom achieved in the imitations of it. **1963** B. C. Middleton *Hist. Eng. Craft Bookbinding Technique* xi. 122 Niger goatskin..was popularized by Douglas Cockerell more than sixty years ago.

2. Niger seed, the seeds of *Guizotia abyssinica*, of the family Compositæ, native to West Africa and cultivated elsewhere for the oil obtained from its seeds; also, the plant itself (cf. Ramtil); **niger (seed) oil,** the oil produced from niger seeds.

1884 *Encycl. Brit.* XVII. 746/1 Niger oil is the produce of the seeds (properly achenes) of *Guizotia oleifera*, a plant native of the east coast of Africa, but cultivated throughout India and to some extent in Germany. The fruits, which are small, tooth-like in form, and shining black in colour, contain from 40 to 45 per cent. of oil... In Western countries niger oil is principally employed in soap-making and as a lubricant. **1889** G. S. Boulger *Uses of Plants* 138 *Guizotia abyssinica*, Cass., Niger or Ramtil seeds, came into the English market about 1851. It is a native of Tropical Africa, but is cultivated in India and Germany. It is used in Europe for soap and lubricating oil. **1917** *Chambers's Jrnl.* May 293/2 Niger-seed oil is used as a substitute for linseed-oil when the latter is scarce. **1944** *Living off Land* ii. 40 Niger-seed..looks like a sunflower. **1974** F. N. Howes *Dict. Useful & Everyday Plants* 175 Niger seed. *Guizotia abyssinica*, an oil seed and much favoured for cage birds. **1974** G. Usher *Dict. Plants used by Man* 288/2 The plant [*sc. Guizotia abyssinica*] is cultivated for the seeds which yield an oil (Niger Seed Oil, Ramtilla Oil, Werinnua Oil). The plant is cultivated mainly in India, but also in E. Africa, W. Indies and Germany... It is used for soap-making and cooking fats.

Niger–Congo (nəi·dʒəɹ,kǫ·ŋgo). [f. the names of the rivers *Niger* and *Congo*.] A group of languages which includes those spoken by most of the indigenous peoples of western, central, and southern Africa.

1955 J. H. Greenberg *Studies in African Linguistic Classification* 8 To the entire family consisting of the West Sudanic nucleus inclusive of Bantu, plus this eastward extension, I have preferred to adopt a new name of a non-committal geographic nature, Niger–Congo, from the two great rivers in whose basins these languages predominate. **1961** F. G. Cassidy *Jamaica Talk* iii. 31 The Niger–Congo languages are characterised by differences of meaning which depend upon pitch—or 'tone'—and stress. **1970** R. Finnegan *Oral Lit. in Africa* iii. 55 In the opinion of some recent scholars, even this large Bantu group is only one subdivision within a much larger family, the 'Niger–Congo' group, which also includes most of the languages of West Africa. **1972** *Language* XLVIII. 273 Accounting for the various kinds of syntactic evidence in the Bantu languages, such as the consistent use of prefixes which often correspond to suffixes of the Niger–Congo languages, will require considerable historical study.

Nigerian (nəidʒɪə·riăn), *sb.* and *a.* [f. *Nigeria* (see below): see -an.] **A.** *sb.* A native or inhabitant of Nigeria, a republic in West Africa occupying the basin of the lower Niger. **B.** *adj.* Of or pertaining to Nigeria or its inhabitants.

1860 W. Cole *Jrnl.* 6 May in *Life in Niger* (1862) 170 This is the general mode in which the Nigerians make known their losses. **1905** F. L. Shaw *Tropical Dependency* xlv. 425 The names, alas, of more than one of the first small Nigerian group are engraved now upon tombstones on that border of the Empire which they helped to make. *Ibid.* li. 481 There are now in all..about 400 white men in the Northern Nigerian service. **1908** *Daily Chron.* 31 Jan. 3/3 When she came back from her first journey still hale and happy, she was accepted as a full-fledged Nigerian. **1923** *Blackw. Mag.* Apr. 514/2 His early days have gone down in Nigerian history. **1954** R. St. B. Baker *Sahara Challenge* vii. 73 This is the secret of the successful growing of groundnuts, which has been known to the Nigerians for many generations. *Ibid.*, We slept just outside a Nigerian resthouse. **1973** *Listener* 14 June 782/1 There is a confidence and exuberance about Nigerians which lifts the spirits. **1973** *Guardian* 19 June 18/4 The Niger herdsmen are paid in Nigerian pounds. They buy Nigerian goods, recross the border and sell the goods.

Nigerianization (nəidʒɪə:riănəizēi·ʃən). [f. *Nigerian *sb.* and *a.* + -ization.] The process of making Nigerian, *spec.* the transfer of posts in government and industry from foreigners to native Nigerians.

1954 *Rep. Lagos Conf.* 80 in *Parl. Papers 1953–4* (Cmd. 9059) XI. 122 We are determined to press forward with the Nigerianization of the Civil Service. **1955** *Times* 23 Aug. 7/4 The 'Nigerianization' of the public service had not kept pace with the speed of the advance of self-government. **1959** *Times* 4 Dec. 9/3 He..pursued a vigorous policy of Nigerianization, particularly in the Post and Telegraphs Department. **1960** *Economist* 8 Oct. 132/1 Nigerianisation will go on, as it ought to do. **1961** *Aeroplane* CI. 450/2 A programme of development has recently been agreed by the board of Nigeria Airways... Steps will be taken leading eventually to the complete 'Nigerianization' of all activities. **1973** *Times* 1 Oct. (Nigeria Suppl.) p. viii/3 Within Nigeria British business continues to play a highly significant role, despite the effects of Nigerianization.

Nigerianize (nəidʒɪə·riănəiz), *v.* [f. *Nigerian *sb.* and *a.* + -ize.] To make Nigerian in character. Hence **Nige·rianized** *ppl. a.*

1960 *Guardian* 4 Nov. 10/3 The politicians..are themselves fully 'Nigerianised'. **1966** J. P. Mackintosh *Nigerian Govt. & Politics* iv. 175 The Eastern and Western Government were composed of men determined to Nigerianize their Public Services with the utmost speed. **1969** I. F. Nicolson *Administration of Nigeria* ix. 251 There are several justifications for the somewhat arbitrary choice of the year 1948 as the starting point of the process of transforming alien 'administocracy' into the apparatus of modern, federal, parliamentary government, complete with Nigerian Ministers and 'Nigerianized' career public services. **1971** J. S. Coleman *Nigeria: Background to Nationalism* iv. 103 The first major effort to Nigerianize the clergy..was made by the Church Missionary Society.

nigerite (nəi·dʒĕrəit). *Min.* [f. *Niger-ia*, the name of the country in which it was discovered + -ite¹.] A basic oxide of zinc, iron, magnesium, tin, and aluminium, (Zn,Fe^{II},Mg)-(Sn,Zn)$_2$(Al,Fe^{III})$_{12}$O$_{22}$(OH)$_2$, found as brown or red hexagonal platelets.

1946 Jacobson & Webb in *Bull. Geol. Surv. Nigeria* No. 17. 11 The dykes to the south-east of Oke Oloke are of special interest on account of the discovery of a new tin-zinc mineral, nigerite, which occurs in quartz–andalusite–sillimanite veins associated with the albitized pegmatites. **1947** *Times* 2 Dec. 6/3 Specimens of nigerite, a new tin mineral, recently analysed and described by the Museum's Department of Mineralogy, have been given. **1974** *Mineral. Mag.* XXXIX. 837 (*heading*) Nigerite in the tin-tantalum pegmatites of Amapá, Brazil.

nigga (ni·gă). Also **niggah.** Repr. a Southern U.S. pronunciation of Nigger *sb.* Cf. *nigguh, niggur.

Now virtually restricted to publications in which Black English vernacular is set down.

1925 L. R. Harris in A. Dundes *Mother Wit* (1973) 563/2 Howdy niggahs,..how's you all dis mawnin'. **1937** C. Himes in *Black on Black* (1973) 139 Niggah, ef'n yo is talkin' tuh me, Ah ain' liss'nin'. **1969** in A. Dundes *Mother Wit* (1973) 655/1 They end up losing all of their money to that big nigga who is supposed to be the epitome of 'nigga-ness'. **1973** *Black World* Apr. 58 My wandering niggah done slipped up on me. **1974** *Ibid.* Dec. 23 In the grammaticality of the mid nighttime sky when them sweet blue nigga dialects of the flesh rise.

nigger, *sb.* Add: **1.** Except in Black English vernacular, where it remains common, now virtually restricted to contexts of deliberate and contemptuous ethnic abuse. (Further examples.)

1931 D. L. Sayers *Five Red Herrings* i. 11 Waters.., like all Englishmen, was ready enough to admire and praise all foreigners except dagoes and niggers. **1934** G. B. Shaw *On Rocks* ii. 70 Pandranath: you are only a silly nigger pretending to be an English gentleman. *Ibid.* 71, I am called nigger by this dirty faced barbarian whose forefathers were naked savages worshipping acorns and mistletoe..whilst my people were spreading the highest enlightenment yet reached by the human race from the temples of Brahma... You call me nigger, sneering at my colour because you have none. The jackdaw has lost his tail and would persuade the world that his defect is a quality. **1936** M. Mitchell *Gone with Wind* 401 'You're a fool nigger, and the worst day's work Pa ever did was to buy you,' said Scarlett slowly... There, she thought, I've said 'nigger' and Mother wouldn't like that at all. **1937** C. Himes in *Black on Black* 132 Uncle Tomism, acceptance, toadying—all there in its most rugged form. One way to be a nigger. Other Negroes did it other ways—he did it the hard way. The same result—*a nigger*. **1948** [see *coon *sb.* 2 c]. **1948** G. Greene *Heart of Matter* i. i. 3, I hate the place. I hate the people. I hate the bloody niggers. Mustn't call 'em that you know. *Ibid.* ii. i. 179 A clerk knocked and said, 'There's a nigger for you, Wilson, with a note.' **1949** B. A. Botkin *Treas. S. Folklore* p. xxiii, In turning his laughter on himself as well as the whites, the Negro has taken over the objectionable word 'nigger' (though not 'darky') and made it a term of praise or blame. **1964** L. Hughes in J. H. Clarke *Harlem* iv. 251 A klansman said, 'Nigger, Look me in the face—And tell me you believe in The great white race.' **1966** *Stage & Television Today* 6 Oct., When 'Ten Little Niggers' opened for a week's run at Birmingham Theatre on Monday members of the co-ordinating committee against racial discrimination (CARD) staged a protest demonstration outside. **1968** C. Brown in A. Dundes *Mother Wit* (1973) 232/1 Perhaps the most soulful word in the world is 'nigger'. **1969** D. L. Lee in A. Chapman *New Black Voices* (1972) ii. 286 Change, stop being an instant yes machine. Change. Niggers don't change they just grow. That's a change; Bigger and better niggers. **1971** *Black World* Apr. 56 Who the hell you think, nigger? **1972** D. Onyeama *Nigger at Eton* iii. 83, I remember that in conversation, some boys occasionally used 'nigger' in reference to black people. I never dreamt that it was a racial name and generally used with contempt; I just reckoned it was a harmless slang word for a black man. **1973** *Black World* Aug. 61/1 Even credit-card niggers didn' really trust banks. **1973** *Times Lit. Suppl.* 14 Dec. 1536/1 You do not reduce to only a ticket-collector, only an asthmatic, only a voter, only a politician, only a pools-winner, only a nigger.

b. (Further Austral. examples.)

1941 I. L. Idriess *Great Boomerang* xxii. 169 The cranky nigger who was on the job broke the only shovel. **1946** K. Tennant *Lost Haven* (1968) 6 No grandson of mine,..is going to be brought up by them thieving, godless, nigger Detwinters. *Ibid.* xviii. 312 When that old nigger went past she turned up her nose.

c. *to work like a nigger*, to work exceptionally hard. orig. *U.S.*

1836 C. Gilman *Recoll. Southern Matron* in *Southern Rose* 23 July 186/1, I have toiled night and day, I've worked like a nigger, and more than any nigger. **1861** Geo. Eliot *Let.* 13 Apr. (1954) III. 404 Charles..will.. work like a nigger at his music. **1880** 'Mark Twain' *Tramp Abroad* 40 He laid into his work like a nigger. **1889** E. Dowson *Let.* 19 May (1967) 80, I have simply worked like a nigger this week. **1920** R. Fry *Let.* 20 June (1972) II. 481, I have worked like a nigger to arrange it [*sc.* an exhibition] well. **1931** D. L. Sayers *Five Red Herrings* ix. 99 Dalziel has been working like a nigger all day, getting him identified by his family and by the station-master at Pinwherry and by the people at Larne.

d. *a nigger in the woodpile* (or *fence*): a concealed motive or unknown factor affecting a situation in an adverse way. orig. *U.S.*

1852 in *Kans. Hist. Quarterly* (1942) XI. 235 No 'nigger in the wood pile' here..; white men are at the bottom of this speculation. **1862** *Congress. Globe* 3 June 2527/1 [These gentlemen] spoke two whole hours..in showing—to borrow an elegant phrase, the paternity of which belongs, I think, to their side of the House,—that there was 'a nigger in the wood-pile'. **1897** *Congress. Rec.* 18 Feb. App. 61/1 Like a great many others ignorant of facts, he finds 'a nigger in the wood pile' when there is neither wood pile nor nigger. **1911** Woodrow Wilson in *Outlook* 11 Aug. 944 If you go through the schedules you will find some nigger in every wood pile. **1930** *Cambridge Daily News* 24 Sept. 7/6 Unless..there is a nigger in the wood pile,..the shares ought to be worth a mild flutter at round 8s. 6d. **1952** A. Christie *They do it with Mirrors* xii. 109 Well now, let's have your point of view. Who's the nigger in the woodpile? The G.I. husband? **1958** *Listener* 13 Feb. 285/1 'The starry heaven that we know.. is inside us.'.. The nigger in the woodpile is to be found in the word 'know'. **1958** 'A. Gilbert' *Death against Clock* 72 The nigger in the woodpile on this occasion being an elderly spinster of decided views. **1960** *Daily Tel.* 16 Jan. 8 This seems to be the nigger in the woodpile—the woodpile being an industrial recovery and activity remarkable by any standard. **1974** M. Gilbert *Flash Point* ii. 19 It wouldn't have been easy to spot... It's taken Jonas himself all this time to spot the nigger in the woodpile.

1850 *California Courier* (San Francisco) 4 Sept. 2/6 The majority of the papers, however, think that there 'is a nigger in the fence' somewhere. **1888** B. Harte *Phyllis of Sierras* i. iii. 90 Ef he aint scooped up by Jenny Bradley he'll guess there's a nigger in the fence somewhere. **1911** H. Quick *Yellowstone Nights* xi. 286 He's always looking for a nigger in the fence.

2. d. Used in Comb. to denote a dark shade of colour, as *nigger-brown* (also *ellipt.*), *-grey, -pink.*

1914 *Lady's Pictorial* 4 July p. v (Advt.), Soft Taffeta Hat..In Black,..Nigger, Mole, and White. **1915** *Home Chat* 2 Jan. 11/1 Nigger-brown cloth. **1917** *Ibid.* 3 Nov. 139/2 Nigger or, as it is now called, 'Zulu', is also to be seen. **1922** D. H. Lawrence *England, my England* 116 She was wearing a wide hat of grey straw, and a loose, swinging dress of nigger-grey velvet. **1923** [see *desert-brown adj.]. **1930** J. Dos Passos *42nd Parallel* i. 124 On each table there were niggerpink and vermilion paper

flowers. **1960** V. Williams *Walk Egypt* 89 A dry-goods store showed a dress of 'nigger-pink'. **1973** *Times* 12 Nov. 4/4 Decorations in autumnal colours, that is, coral pink and what used to be nigger brown.

3. a, b. (Examples.)

1867 J. A. Hosmer *Trip to States by Yellowstone & Missouri* 58 The boat..struck the bar; they then began to work with the spars and nigger, and at two o'clock we got off. **1878** J. H. Beadle *Western Wilds* 378 Then oaths, spars, 'nigger-engine' and all the other available machinery were set in operation. **1882** *Harper's Mag.* Jan. 175/2 One of the 'nigger' engines is suddenly called into service to tighten a two-inch rope, or wind up a discarded cable. **1900** *Atlantic Monthly* LXXXV. 103/2 'Carriages', bright with red and green lanterns..rush to and fro, seizing the logs as they come from the 'kickers' and 'niggers'. **1910** S. E. White *Rules of Game* i. v. 32 When the car had flown back to its starting-point, the 'nigger' rose from obscurity to turn the log half-way round. **1929** *Encycl. Brit.* XIV. 482/1 A steam or air 'nigger' (mechanically operated steel arms) helps to place the log in the proper position.

4. c. *slang.* (See quot. 1960.)

1934 *Tit-Bits* 31 Mar. 12/1 The film world has a colourful compilation of expressions unlike those in other walks of life. 'Niggers' are not men of colour, but blackboards used to 'kill' unwanted reflections from the powerful lights. **1937** A. Buchanan *Film Making* iii. 52 'Niggers' are wooden oblong screens used to 'nigger-off' or shield light from faces, or shadows on walls, and so 'Her face needs a nigger' is not such an alarming statement as it sounds to the uninitiated. **1957** Manvell & Huntley *Technique Film Music* ii. 32 In silent film days we had all the apparatus we needed in the studios, with miles of cables, banks of arcs, and great screens (known in the industry as 'niggers') to reflect light and help shadow effects. **1960** O. Skilbeck *ABC of Film & TV* 89 *Nigger*, an adjustable Mask on a stand, used on the Floor to shield the camera from, or to achieve effects with, lights.

5. *nigger-breaker, -chaser, -dealer, -driver,* (earlier example), *-lover, -stealer, -trader*; also *nigger-loving* adj.

1845 F. Douglass *Narr. Life F. Douglass* 57 All of this added weight to his reputation as a 'nigger-breaker'. **1921** C. E. Mulford *Bar-20 Three* xvi. 217 Most likely they'll be nigger-chasers, th' way some folks'll be steppin' lively to get out of th' way. **1853** F. W. Thomas *John Randolph* 285 You know Robinson the nigger-dealer, who has the pen down town. **1833** J. Neal *Down Easters* I. 70 When the nigger-drivers falls out among themselves. **1909** R. E. Knowles *Attic Guest* xiii. 178 'Then you can take what you deserve, curse you for a nigger-lover,' I heard the Colonel retort madly. **1924** *American Mercury* Feb. 135/2 Shorty—Nigger-lover! *He throws the money in her face.* **1930** J. Dos Passos *42nd Parallel* ii. 140 'Niggerlover,' yelled Joe in her ear... Janey began to cry. **1958** *Church Times* 21 Nov. 6/4 You can call a man a *Kaffir-Boetie* in Johannesburg and a nigger-lover in the Southern States; but both mean precisely the same thing and have the same accent. **1959** *Encounter* Dec. 45/2 They nick our boys and let the Spades go!..Nigger-lovers! **1972** *Guardian* 3 July 8/2 Black Mountain College..was a seat of free love, communism, and nigger-lovers. **1914** S. Lewis *Our Mr. Wrenn* 3 Mrs Zapp was too conscientiously dolorous to be much cheered by the sympathy of a nigger-lovin' Yankee. **1962** *Guardian* 3 Oct. 1/6 It is still wise not to..admit you represent such a Nigger-loving Red publication as the 'Guardian'. **1975** J. Rathbone *Kill Cure* iii. iv. 106 You nigger-loving bitch. **1839** R. M. Bird *Adventures R. Day* I. xxv. 181, I was 'a kidnapper, a Georgeeye nigger-stealer'. **1884** 'Mark Twain' *Huck. Finn* xxxiii. 314 Only I couldn't believe it. Tom Sawyer a nigger-stealer. **1853** F. W. Thomas *John Randolph* 285 He's not in favor of these regular nigger-traders is he? **1884** 'Mark Twain' *Huck. Finn* viii. 60 But I noticed dey wuz a nigger trader roun' de place considerable lately en I begin to get oneasy.

6. a. *nigger-blood, boy, -lips, mouth*; also *nigger-blooded, -skinned* adjs.; *nigger-dead* adj.; *nigger minstrel*, one of a group of entertainers performing songs and dances typical of or based on plantation life in the Southern U.S., freq. by white men with blackened faces; also ellipt., *nigger*.

1833 J. Neal *Down-Easters* I. 66 If there's a drop of nigger-blood in 'em, they'll always show it in their temper. **1932** W. Faulkner *Light in August* v. 96 'Take your black hand off of me, you damn niggerblooded—' The hand shut down again. **1825** J. Neal *Bro. Jonathan* III. 207 Nobody there, I guess, but a nigger boy. **1970** R. D. Abrahams *Positively Black* ii. 26 'Nigger boy,' he said to me, 'how'd you like to meet your maker right now?' **1970** W. Ford in O. Coombs *We speak as Liberators* 43, I would tell of being Black and Proud and Black and Loud and Black and Bowed and Black and niggerdead. **1922** Joyce *Ulysses* 219 From the hoardings Mr. Eugene Stratton grinned with thick niggerlips at Father Conmee. **1860** A. J. Munby *Diary* 17 Mar. in D. Hudson *Munby* (1972) 56 My 'comic' friend suggested that I should apply to certain 'niggers', who had just bowed themselves off the stage... In this den were two or three men with blackened faces, taking off their shabby nigger costume. **1860** E. Cowell *Diary* 13 Sept. in *Cowells in America* (1934) 155 A Company of 'nigger minstrels' are to be here tomorrow night. **1917** A. Waugh *Loom of Youth* ii. i. 93 And do you think he really imagines he is doing any good to his form by giving that nigger minstrel entertainment up there? **1959** I. & P. Opie *Lore & Lang. Schoolch.* i. 13 Nellie Bligh..was the heroine of a mid-nineteenth-century nigger minstrel song by Stephen Foster. **1922** Joyce *Ulysses* 365 The dark one [*sc.* girl] with the mop head and the nigger mouth. **1938** Nigger-skinned [see *fawn-eyed adj.].

b. *nigger culture, dialect, melody* (earlier and later examples), *music, show, song.*

1970 J. B. Cole in A. Chapman *New Black Voices* (1972) iii. 493 When blacks refer to 'Nigger culture', they often very explicitly speak of soul and style. **1834** *Knickerbocker* III. 445 And I would say too, that although *mighty smart*, and a *mighty smart chance*, *mighty big*, and *mighty little* was excellent 'nigger' dialect, yet it was not so refined, as an orator might use. **1846** *Ibid.* XXVIII. 244 Captain Marin would give a touch from a sea-song, or a specimen of a 'nigger-melody'. **1857** J. D. Borthwick *Three Yrs. in Calif.* xii. 212 My entertainers, producing two violins, favoured me with a selection of Nigger melodies. **1894** G. du Maurier *Trilby* i. iii. 219 He..can even scream with laughter at..a nigger melody. **1948** A. Lomax in A. Dundes *Mother Wit* (1973) 472/1 If you like this nigger music. **1856** C. E. De Long in *Calif. Hist. Soc. Q.* (1930) IX. 60 Went to a nigger show. **1884** 'Mark Twain' *Huck. Finn* xxvi. 261 They never go to the circus, nor theatre, nor nigger shows, nor nowheres. **1909** R. E. Knowles *Attic Guest* xiii. 156 They'll be flaunting that Uncle Tom's Cabin nigger show under your noses. **1844** *United Service Jrnl.* XLIV. 21 He was never at rest,—now singing a nigger song on the deck. **1851** H. Mayhew *London Labour* I. 273/1, I sell ballads and manuscript music.., which is 'transposed'..from the nigger songs.

c. nigger cloth = *negro cloth* (Negro 7); **nigger corner** *U.S.*, a part of a public building to which Negroes were confined; **nigger fish**, a small grouper, *Cephalopholis fulvus*, found in the West Indies and off the coast of Florida; = Cony, coney *sb.* 7 b; (earlier example); **nigger goose** *N. Amer. local* (see quots.); **nigger heaven** *U.S. slang*, the top gallery in a theatre; **nigger heel** *Naut.* (see quot.); also **nigger-heeled** *a.* (see quot. 1961); **nigger hunt**, the organized pursuit of Negroes for the purpose of attacking them; so **nigger-hunter, -hunting**; **nigger lice** *U.S.*, informal name of the prickly awns of various species of plants, esp. of the genus *Desmodium*; **nigger luck**, exceptionally good luck; **nigger shooter** *U.S. slang*, a catapult; **nigger-stick** *U.S. slang* (see quot. 1974); **nigger toe** *U.S.*, a Brazil nut.

1857 *Chambers's Jrnl.* 3 Jan. 3/2 The garments of.. copper-coloured nigger cloth. **1860** J. G. Holland *Miss Gilbert's Career* iv. 61 You see he sells some of his nigger cloth for goods. **1955** W. Foster-Harris *Look of Old West* i. 38 Typically the cloth was linsey-woolsey... 'Nigger cloth' it was called. It had been much used for slaves' garments, [etc.]. **1894** 'Mark Twain' in *Century Mag.* June 233/1 In the 'nigger corner' sat Chambers. **1876** G. B. Goode *Catal. Fishes Bermudas* 60 The red form corresponds to *Terranus ouatalibe*, and is known as the Nigger-fish. **1917** T. G. Pearson *Birds of Amer.* I. 97/2 The Cormorants have many local names, such as 'Shag', 'Lawyer', and 'Nigger Goose'. **1941** R. Faherty *Big Old Sun* 313 You can eat curlew, or kill duck or coot or niggergoose if they come flying out yonder. **1947** *National Geographic Mag.* Sept. 339/1 A large flight of cormorants, called there [*sc.* in North Carolina] 'nigger geese', passed close aboard. **1957** W. L. McAtee *Folk-Names Canad. Birds* 5 Double-crested Cormorant.. nigger goose (in allusion to its colour and its goose-like appearance, especially when in flight in the V-formation so closely associated with the common Canada goose). **1878** A. Daly in J. F. Daly *Life A. Daly* (1917) 249 There is a 'Nigger Heaven' (as the third tier is called in Troy) here, & as 'tis very capacious I have been liberal with my pencilled passes. **1931** 'D. Stiff' *Milk & Honey Route* xiv. 151 These..entertainments..have raised their prices beyond the reach of the hobo, unless he wants to go to 'nigger heaven'. **1973** A. Dundes *Mother Wit* 222 This extension or transformation of the 'Nigger Heaven' stereotype has no doubt contributed to the continued currency of Harlem folk speech. **1901** *Rudder* XII. 302/2 The after leech would take an incurve or 'nigger heel', as sailmakers call it. **1922** C. G. Davis *How Sails are Made* (ed. 2) 63 A double-bighted sail would, if not carefully handled and hauled out too hard on peak and clew, become 'nigger-heeled', as a hollow leech was called. **1934** *Yachting Monthly* LVII. 11/1 We were watching a hawse-fallen sloop beating in under a badly nigger-heeled mainsail. **1961** F. H. Burgess *Dict. Sailing* 150 *Niggerheeled*, said of the leech of a sail that curves inward of a line from peak to clew, and is therefore not roached. **1834** *Chambers's Edin. Jrnl.* III. 135/3 When a slave runs away..a party is made up for a nigger hunt. **1940** E. Caldwell *Trouble in July* x. 165 This ain't no nigger-hunt—this here's a jawing match! **1959** *Encounter* Mar. 87 The young 'nigger-hunters' of Notting Hill. **1834** *Chambers's Edin. Jrnl.* III. 135/3 In Kentucky..nigger-hunting is a favourite sport. **1958** *Encounter* Dec. 4/2 They had then armed themselves,..and had gone on what they described..as a 'nigger-hunting expedition'. **1971** in C. Mullard *Black Britain* (1973) iv. xi. 142 The racists, who were no doubt nigger hunting, as they call it, felt that this was easy meat. **1933** *Sun* (Baltimore) 11 Sept. 6/7 The iron weed with its deep color, and..great clumps of 'nigger lice'. **1940** H. L. Mencken *Happy Days* 43 Sometimes a black-hearted boy would sneak into the adjacent brickyard, which was covered in large part with Jimpson weeds, plantains and other such vegetable outlaws, and return with a large ball of nigger-lice. **1946** *Sun* (Baltimore) 2 Oct. 12/3 When I spoke the other day of 'nigger lice', I was referring to the tick trefoil, its scientific name. **1851** R. Glisan *Jrnl. Army Life* (1874) 90, I occasionally made him a little envious by my nigger-luck, as he is pleased to term it. **1900** R. H. Savage *Midnight Passenger* (1901) 135 It has been a great stroke of nigger luck. **1909** *Dialect Notes* III. 352 You can't beat me playing dominoes. It's jest your nigger-luck that gets away with me. **1876** E. W. Heap *Diary* 26 Feb. in *Publ. Amer. Dial. Soc.* (1969) lii. 53, I had a job on hand making Nigger shooters for Dr's children. **1883** Sweet & Knox *On Mexican Mustang through Texas* 339 Just about the time people have got used to tops buzzing about their ears, the 'nigger-shooter' mania breaks out. **1901** Ade *Forty Mod. Fables* 172 All you wanted to do was to tear out with those Toughs and kill Birds with Nigger-Shooters. **1963** R. I. McDavid *Mencken's Amer. Lang.* vi. 284 In the South it [*sc.* a slingshot] is still sometimes called a *nigger-shooter*. **1971**

Guardian 18 Sept. 11/7 Conditions inside American prisons... Prisoners live their lives at the end of gun barrels and what are often known as 'nigger-sticks'. 1973 *Black Panther* 15 Sept. 17/3 They were attacked and brutally beaten by 50 to 60 guards armed with tear gas, plexiglass shields and four-foot long 'nigger sticks'. 1974 *Guidelines to Volunteer Services* (N.Y. State, Dept. Correctional Services) 42 *Niggerstick*, officer's baton. 1896 *Dialect Notes* I. 421 *Nigger toes*: for *Brazil nuts*. 1958 J. M. Lacy in A. Dundes *Mother Wit* (1973) 597/2 He buys..nuts called 'nigger toes'. 1973 *Times* 27 Aug. 5/8 In Virginia brazil nuts are called *nigger-toes* and chewed with great relish.

nigger, *v.* Add: **1. b.** (Earlier and later examples.)
1833 S. Smith *Life & Writings J. Downing* 22 He laid sticks across the large logs.., and *niggered* them off with fire, and then roolled them up in piles. 1843 'R. Carlton' *New Purchase* I. xx. 188 In addition to the 'niggering off', it became necessary, as the cold increased, to chop off logs. 1905 M. G. Sherk *Pen Pictures of Early Pioneer Life in Upper Canada* 49 To save the time and labor of cutting the fallen trees into lengths for being drawn together by the oxen, they were often 'niggered'.
2. To work 'like a nigger'; to work very hard; quasi-*trans.* with *it*.
1857 J. Hyde *Mormonism* v. 120 Many of the people express satisfaction in seeing these 'better-dressed fellers' obliged to 'nigger it' as well as themselves. 1878 J. H. Beadle *Western Wilds* 349 Was it not more of an honor to be the 'bishop's fourth'..than the 'slavey' of a poor mechanic, to 'nigger it on love and starvation'?
3. *refl.* To make (oneself) resemble a Negro, by blackening the face.
1881 M. Crommelin *Miss Daisy Dimity* I. ii. 21 Jemmy the third, was 'niggering' himself, by adorning his rosy cheeks with black.

niggerdom. (Earlier examples.)
1862 *Congress. Globe* 28 Mar. 1414/2 New England, where they hate niggerdom worse than the devil. 1868 *Good Words* 1 Oct. 603/2 The conquering nigger..caught many of the Aborigines, blacked them over, and sent them off to proclaim the glories of Niggerdom.

niggerhead. Add: **1.** For *U.S.* read *N. Amer.* (Later examples.) Also, *N.Z.*, the tussocks formed in swampy ground by species of *Carex*, esp. *C. secta*.
1882 T. H. Potts *Out in Open* 76 Penetrating the dead massy root of an old plant of nigger-head (*carex virgata*). 1904 J. Lynch *Three Yrs. Klondike* 41 We plunged into a mire of muddy water and 'nigger-heads'. 'Nigger-heads' are detachments of dark moss about a foot in diameter, lifting their heads just above the water or marshy subsoil. 1910 R. W. Service *Ballads of Cheechako* 19 And there was the little lone moose trail... By muskeg hollow and niggerhead it wandered endlessly. 1921 H. Guthrie-Smith *Tutira* xii. 103 The outer edges of these marshes were rough with nigger's-head (*Carex secta*). 1947 J. H. Brown *Outdoors Unlimited* 314 The ptarmigan cackled in the manner of a Bronx cheer as it flew to a nearby nigger-head. 1950 *N.Z. Jrnl. Agric.* Apr. 356/3 Excellent crops and pastures..where before only nigger-heads, rushes, and swamp plants were flourishing. 1958 P. Berton *Klondike* 44 The great clumps of grass 'nigger-heads' that marked the mouth of Rabbit. 1961 C. Vyvyan *Arctic Adventure* xxi. 126 We had to negotiate about a mile of open country across nigger-heads.
b. *U.S.* A spherical prickly cactus belonging to the genera *Ferocactus* or *Echinocactus*.
1877 H. C. Hodge *Arizona* 244 The kind [of cactus] commonly called the nigger head is round, of the size of a cabbage, and covered with large, crooked, catlike thorns. 1881 [see *barrel cactus]. 1940 E. C. Jaeger *Calif. Deserts* (rev. ed.) 181 Closely allied to this is the Mohavean niggerhead. 1966 E. Y. Dawson *Cacti of Calif.* 51 (*heading*) Nigger Heads (*Echinocactus polycephalus*).
c. *U.S.* The black-eyed Susan, *Rudbeckia hirta*, a yellow composite flower with a dark centre.
1893 S. F. Price *Flora of Warren County, Kentucky* 15 *Rudbeckia..fulgida*... Cone Flower. 'Nigger-head'. 1931 W. N. Clute *Common Names of Plants* 45 A number of composites with yellow rays and dark centers are commonly known as niggerheads, though the more polite term is black-eyed Susan. 1966 *Publ. Amer. Dial. Soc.* 1964 xlii. 21 Niggerhead... The black-eyed Susan (*Rudbeckia hirta*).
2. A rock, stone, lump of coral, etc. (Earlier and further examples.)
1847 H. Howe *Hist. Coll. of Ohio* 569 It was a saw mill, with a small pair of stones attached, made of boulders, or 'nigger heads', as they are commonly called. 1885 in *Amer. Speech* (1961) XXXVI. 295 The term 'Nigger head' is used by the Kanawha miners to designate a hard, heavy, impure coal often resembling cannel. 1901 *Chambers's Jrnl.* Sept. 634/1 He tightened his grip on the reins as he caught the dim outline of a treacherous nigger-head stone. 1908 E. J. Banfield *Confessions of Beachcomber* I. ii. 57 Nothing was left of the big ship save some distorted fragments of iron jammed in among the nigger-heads of coral. 1916 C. Sandburg *Chicago Poems* 41 A boy passes and throws a niggerhead that chips off the end of the nose [of a statue]. 1948 E. N. Dick *Dixie Frontier* 4 Bears rolled 'nigger head' stones over and ate the grubs and field mice. 1956 M. L. West *Gallows on Sand* vii. 89 We moored the skiff to a niggerhead, one of those jutting stumps of dead coral which are found all over the reefs, and which have the look of a frizzled skull on top of a stumpy neck.
3. (Earlier and later examples.)
1843 J. Lumsden *Amer. Memoranda* (1844) 14 My next communication will probably contain full details of the methods adopted by the Virginian planters in the manufacturing of the nigger-head, ladies'-twist, [etc.]. 1860 *Nor'Wester* (Red River Settlement) 28 June 4/5 After that I would smoke half a plug of 'nigger-head tobacco'. 1884 'Mark Twain' *Huck. Finn* xxi. 194 You borry'd store tobacco and paid back nigger-head. 1936 *Beaver* Mar. 7/2 It is probably the lineal descendant of the nigger-head tobacco used in the Indian trade years ago, and as it came in ropes it was sold by the inch. 1956 Crate & Williams *We speak for Silent* 3 Groceries—particularly tea and 'nigger-head' (a trade-tobacco for smoking and chewing)—are his more necessary 'luxuries'.
5. (Later example.)
1946 W. S. Knickerbocker *20th Cent. English* 149 Niggerhead... After the Civil War it was used for a person in favor of full political equality for Negroes.
6. A type of fabric (see quot. 1950).
1892–3 *T. Eaton & Co. Catal.* Fall & Winter 10/2 In the plain cloth jackets the materials are beavers, niggerheads, serges and worsted. 1950 *'Mercury' Dict. Textile Terms* 366/1 *Niggerhead Curl*, a fancy dress cloth made from spiral yarn warp and mixture weft (cotton and wool).
7. (See quot.)
1927 G. Bradford *Gloss. Sea Terms* 119/1 *Niggerheads*, a name for bollards, and sometimes applied to winch heads.

ni·ggering, *vbl. sb.* [f. Nigger *v.* + -ing[1].] The action of the verb in various senses.
1843 'R. Carlton' *New Purchase* I. xx. 188 Niggering belongs mainly to very large timber, and pertains rather to the science of log-rolling than of preparing fuel. 1894 'R. Andom' *We Three & Troddles* xix. 174 'Busking' be it known is the technical term for amateur niggering. 1948 E. N. Dick *Dixie Frontier* 126 Morning and evening dry limbs were laid in the widening gap until the log was burnt into length. After about a week the fires had done their work. This was called 'niggering off'.

niggerish, *a.* (Earlier example.)
1825 J. Neal *Bro. Jonathan* II. 67 Ye great niggerish lookin', wap-sided haw.

niggerism. (Further examples.)
1844 *St. Louis* (Missouri) *Reveille* 24 Nov. 2/4 Scrub and whitewash your spiritual niggerism, or you will forever rest in the valley of *Sheol*! 1856 *Illinois State Register* (Springfield) 19 June 2/1 For every democrat who deserts to niggerism, one hundred old line whigs join the democracy against it. 1873 *Porcupine* 19 Apr. 38/3 For the occasional dash or suspicion of niggerism in the mangling of the words, common custom and coincidence will quite account. 1970 G. Jackson *Let.* 10 June in *Soledad Brother* (1971) 33 He and my mother went to great pains to impress on me that it was the worst form of niggerism to hook and jab..at other blacks.

niggerize (ni·gəraiz), *v.* [f. Nigger *sb.* + -ize.] *trans.* To make Negro in character; to treat (a Negro) with contempt. Also *intr.*, to act or dress up as a Negro.
1969 V. Ferdinand in A. Chapman *New Black Voices* (1972) 379 Blk people have done it to the english language, they have niggerized it. 1973 *Times* 13 Feb. 16/4 As people, we are tired of being niggerized, ostracized..and ossified. 1974 'J. Marks' *Mick Jagger* 92 Mick niggerising with a fake, greasy dialect and a very heavy debt to Bo on 'I'm a King Bee'.

nigger's head. *Naut.* An ornamental knot; = Turk's head 2.
1925 *Glasgow Herald* 3 Oct. 4 The glittering ship's bell with its pendant of brightly painted niggershead knotting.

niggery, *a.* (Earlier and later examples.)
1855 'Q. K. P. Doesticks' *Doesticks, What he Says* xxiii 204 Coffee, which I sweeten with niggery brown sugar. 1935 E. Hemingway *Green Hills Afr.* (1936) iv. ii. 241 The long, clean niggery legs. 1936 M. Mitchell *Gone with Wind* 427 The faint niggery smell which crept from the cabin increased her nausea. 1940 C. McCullers *Heart is Lonely Hunter* (1943) i. iii. 43 Portia had a certain kind of niggery craziness, but she was O.K.

nigget (ni·dʒėt). *dial.* Also **nidget.** [Origin unknown.] A small insect; *spec.* one used by a witch or sorcerer as a familiar.
1875 W. D. Parish *Dict. Sussex Dial.* 79 *Nidget*, a little bug. 1915 *Times* 3 Sept. 5/2 'Oh, what are niggets?' 'Why those creepy-crawly things that witches keep all over them. She was sitting down with her niggets all round her, feeding them with little bits of grass all chopped up.' 1925 R. O. Winstedt *Shaman, Saiva, & Sufi* ii. 21 But the best known of these familiars..is of the nigget type and takes the shape of a house-cricket.

niggle, *sb.* Add: **2.** A complaint or criticism, esp. one that is petty or trifling; a worry, annoyance; nagging or irritation.
1886 F. T. Elworthy *West Somerset Word-Bk.* 512 Her's always 'pon the niggle way un. 1956 I. Bromige *Enchanted Garden* iii. iii. 144, I even feel a few little niggles of uneasiness myself. 1960 *Guardian* 11 June 1/5 The poor quality of contemporary furniture..can be the only niggle. 1966 *New Statesman* 8 July 51/3 How much of us will be recognisable in the pages of the history books of 2066? This egoist's niggle spiralled up into my mind. 1972 M. Gilbert *Body of Girl* xii. 106 If the boys in blue can get in a niggle at you, they will. 1974 *Times Lit. Suppl.* 15 Feb. 158/3 In view..of the fact that his book should..go into a second edition.., one or two minor niggles may conveniently be ventilated.

niggle, *v.*[2] Add: **1. d.** (Earlier and later examples.)
1844 W. Barnes *Poems Rural Life* 1st Coll. 330 *Niggle*, to complain of trifles from ill temper or bad humor. 1929 W. Deeping *Roper's Row* xxviii. 312 He would niggle at his food. 1966 'L. Lane' *ABZ of Scouse* 75 *Niggle*, to question; to raise objections. 1974 *Sunday Post* (Glasgow) 21 Apr. 16 (Advt.), Your accounts director won't niggle at the bill at the Ormonde Restaurant (just a high-speed lift away from your room).
3. d. To annoy, irritate; to criticize, nag.
1886 F. T. Elworthy *West Somerset Word-Bk.* 512 *Niggle*, same as to nag. To aggravate. Her'd niggle anybody's live out o' them, nif they'd let her to. 1959 I. & P. Opie *Lore & Lang. Schoolch.* x. 178 A short-tempered person is spoken of as being..niggled or niggly. 1968 A. Diment *Great Spy Race* iv. 47 He is liable to start demanding mass executions when niggled. 1971 A. Price *Alamut Ambush* 6 He was mildly niggled that Maitland had found himself some other pressing engagement.

niggliite (ni·gli,əit). *Min.* [f. the name of Paul *Niggli* (1888–1953), Swiss mineralogist + -ite[1].] A mineral containing platinum and tin (or perhaps tellurium), found at Insizwa in Cape Province, Republic of South Africa.
1936 D. L. Scholtz in *Publ. Univ. Pretoria* 2nd Ser. I. 184 (*heading*) Mineral F (niggliite). 1955 *Amer. Mineralogist* XL. 693 Niggliite..was found in a concentrate which was obtained by panning large amounts of oxidized sulphide ore from Waterfall Gorge, Insizwa, South Africa. 1968 I. Kostov *Mineralogy* 115 Various other formulae have been given to niggliite, PtTe, PtSn, and Pt_3Sn_4; its structure appears to be that of niccolite. 1972 *Mineral. Mag.* XXXVIII. 796 The close similarity of niggliite to synthetic PtTe..led Groeneveld Meijer (1955) to suggest that they were identical. Ramdohr (1960) suggests that the formula is Pt_2Sn_3 but in Ramdohr (1969) the composition is given as 'PtTe, perhaps a mixed crystal with isostructural PtSn'. Scholtz..reports that re-examination showed niggliite is essentially PtSn. *Ibid.* 798 Niggliite from the type locality has been shown to be an antimonian-bismuthian variety of PtSn.

niggly (ni·gli), *a.* [f. Niggle *v.*[2] + -y[1].] = Niggling *ppl. a.*; also, irritable, short-tempered.
1840 W. Harcourt *Let.* in A. G. Gardiner *Life W. Harcourt* (1923) I. ii. 24, I think his trees *niggly* as you would say, and he teaches an odd doctrine about trees. 1862 C. C. Robinson *Dial. Leeds* 40 Ah doant want to live soa as fowks could cawal us niggly. 1898 B. Kirkby *Lakeland Words* 107 He was as niggly ower a penny as many a yan is ower a pund. 1952 M. Tripp *Faith is Windsock* iii. 43 Well, aren't you a niggly old bastard? 1959 [see *niggle *v.*[2] 3 d]. 1967 E. Short *Embroidery & Fabric Collage* iv. 106 Care must be taken to avoid niggly detail and a temptation to imitate a pencil line or brush stroke exactly. 1973 A. Ross *Dunfermline Affair* 65 'What about..going to get the bloody stuff?' Thomson said. He was niggly, and showed it.

niggra (ni·grǎ). *U.S.* (See quot. 1960.) Cf. *nigra.
1960 Wentworth & Flexner *Dict. Amer. Slang* 354/2 *Niggra* = nigger. A pronunciation used by Southerners of Southern breeding and ancestry. Conjuring up the period of Negro slavery, the pronunciation is even more derog. than 'nigger'. 1968 *Guardian* 25 Oct. 2/3 We know our niggras and we love them. 1969 P. Cross in A. Dundes *Mother Wit* (1973) 654/2 They had the niggras on their plantation.

nigguh, niggur (ni·gə), varr. Nigger *sb.* *U.S. slang.* Cf. *nigga.
1968 *Amer. Speech* XLIII. 217 The mountain trapper..was *this Injun,..this niggur*, or *this hoss*. 1969 G. Brooks in A. Chapman *New Black Voices* (1972) 202 Lord! Forgive these nigguhs that know not what they do. 1970 J. Cortez in O. Coombs *We speak as Liberators* 17 Love Lives And I wanta taste myself inside Mmmmmmmm that pure nigguh pain. 1973 *Black World* Sept. 70 His sound now turns to nigguh notes.

night, *sb.* Add: **4. f.** An evening or night devoted to the performance of a play, or of music by a specified composer or artist, or celebrations in honour of a particular person, etc.; freq. with defining word prefixed, as in *first night* (see First *a.* (*sb.*) and *adv.* C 2). Cf. quot. 1711 under sense 4 b in Dict.
1707 *Muses Mercury* Jan. 4 This Prologue was forbidden to be spoken the second Night of the Representation of the *Prophetess*. 1784 in C. B. Hogan *London Stage 1660–1800* (1968) v. 760 *Il Curioso Indiscreto*... This Night, the last of performing before the Holidays, will not be counted a Subscription Night, but the Tickets admitted as usual. 1793 in *Ibid.* v. 1584 *Othello*... Paid Music 4 Nights £35 19s. 4d. 1842 Dickens *Let.* 12 Nov. (1974) III. 368 Mrs. Dickens begs me..to say that if you can oblige her with your box at Covent Garden on any of Miss Kemble's nights, she will be very thankful. 1847 *Punch* XIII. 60 (*caption*) Melancholy scene at the opera on a Jenny Lind night. 1859 G. A. Sala *Twice round Clock* 260, I have brought you to her Majesty's Theatre, and this is unfortunately a Verdi night. 1861 Geo. Eliot *Let.* 6 Oct. (1954) III. 456 We are enjoying a great, great pleasure—a new grand piano; and last evening we had a Beethoven night. 1959 *Observer* 18 Jan. 14/4 The Burns Night circles the globe like a sputnik. 1969 M. R. Booth *Eng. Plays of 19th Cent.* II. 346 *Mrs. Dane's Defence* ran for 209 nights on its first appearance. 1970 *Listener* 10 Sept. 326/3 The years of the romantic biographers and the Beethoven pianists, of Beethoven nights at the old Queen's Hall.
g. *the night*, the first occasion on which a play, entertainment, etc., is publicly per-

formed; freq. in phr. (*it will be*) *all right on the night*, an expression of optimism that a performance will go well even if it is given publicly, even if rehearsals are unpromising; also *transf.*
1911 O. Onions *Widdershins* 26 I've not got on very well with it. But it will all right on the night, as you used to say. 1938 R. G. Collingwood *Principles of Art* xiv. 322 In the rehearsal of any given passage..the actors may move and speak exactly as they will 'on the night'. 1939 S. Box in J. W. Marriott *Best One-Act Plays of 1939* (1940) 296 ...God-a-mercy! PROMPT!... *Nurse.* Ah, well! 'Twill be all right on the night. 1949 *Economist* 23 July 172 The hope that the Atlantic Pact would 'turn out all right on the night'. 1967 J. Gardner *Madrigal* ii. 22 Boysie..began to think that everything would be all right on the night. 1973 E. Lemarchand *Let or Hindrance* xiii. 165 Penny may fly off the handle, but she's always all right on the night.
h. (Also *'night.*) Ellipt. for Good night 1. *colloq.*
1912 Mulford & Clay *Buck Peters, Ranchman* viii. 92 'Good-night. I'm goin' to roost.' 'Night, Dave.' 1922 Joyce *Ulysses* 589 Corny Kelleher. Good night, men. *The Watch.* (*Saluting together.*) Night, gentlemen. 1933 A. Thirkell *High Rising* vii. 151 'Good night. We're going to bed now.' 'Night,' said Stoker. 1967 K. Giles *Death & Mr. Prettyman* i. 21 'Seven ack emma tomorrow if you can manage.' 'Night, sir,' Honeybody lumbered off. 1972 'H. Howard' *Nice Day for Funeral* v. 80 'I could sleep standing up.' 'Then let's call it a day.' 'Night.'
5. a. *night out* (earlier examples); also, an evening or night spent in enjoyment or revelling away from one's home; a spree (cf. Out *adv.* 15 b); so *night off*, a night free from work or one's usual duties.
1885 A. Daly in A. Nicoll *Hist. Eng. Drama 1660–1900* (1959) V. 333 [play-title] A night off; or, a page from Balzac. 1890 W. Booth *In Darkest Eng.* ii. v. 190 On the weekly Church service or 'night out' with nowhere to go. 1908 G. Sanger *Seventy Years a Showman* xi. 90 For these people Lansdown Fair was, as they put it, their 'night out'. 1910 *Blackw. Mag.* Jan. 149/1 Mr. Lloyd-George declined to deliver a speech on the ground that it was 'the Prime Minister's night out'. 1916 E. V. Lucas *Vermilion Box* 129 We have done this pictures here, of course, and I go there regularly on my night out. 1943 J. B. Priestley *Daylight on Saturday* xii. 152 'I'm staying to-night.' 'Then we can't have our night out,' she cried. 1947 N. Cardus *Autobiogr.* 27 My friend's 'night off' occurred he bought a ticket. 1961 H. Pinter (*title*) A night out.
13. a. *night-air* (earlier example), *-ascent*, *-attack* (earlier and later examples), *-car*, *-city*, *-class*, *-damp*, *-duty* (further examples; freq. *attrib.*), *-fancy* (further examples), *-fear*, *-fight*, *-flight* (further examples), *-haunt*, *-hospital*, *-lunch*, *-music*, *-noise*, *-nursery*, *-perfume*, *-raid*, *-rate*, *-rehearsal*, *-road*, *-school* (further examples), *-self*, *-smell*, *-sound*, *-speech*, *-talk*, *-town*, *-train* (earlier and later examples), *-wonder*, *-world*.
1788 Ld. Auckland *Jrnl.* 25 June (1861) II. 55 He is afraid of the night air. 1866 *Chambers's Jrnl.* Oct. 644/2 One night-ascent has been made in this way. 1844 *Knickerbocker* XXIII. 117, I knew that Indians in a night attack make sure by imitating the cry of some animal. 1952 R. Campbell tr. *Baudelaire's Poems* 127 Like enemies preparing night-attacks. 1817–19 Wordsworth *MS.* in E. de Selincourt *Prelude* (1959) 136 The night-calm over sea and land. 1970 T. Hughes *Crow* 55 Seeing the night-city's lights. 1838 A. Beardsley *Let.* July (1971) 24 Two hours' daily study is quite sufficient for me, so, as you suggest, I mean to attend night classes. 1936 N. Coward *Fumed Oak* i. in *Tonight at 8.30* II. i. 47 Your father was a gentleman, which is more than your husband ever will be. 1851 H. Melville *Moby Dick* III. xliv. 250 The unheeded night-damp gathered in beads of dew upon the stone-carved coat and hat. 1921 A. Christie *Mysterious Affair at Styles* viii. 170 She had kindly offered to remain on night duty. 1959 [see *beigel]. 1966 J. Bingham *Double Agent* vii. 99 The phone rang. It was the Night Duty Officer with a deciphered message from Vienna. 1823 C. Lamb *Elia* 158 My night-fancies have long ceased to be afflictive. 1904 W. H. Hudson *Green Mansions* xvi. 292 Half-delirious night-fancies. 1823 C. Lamb *Elia* 148 (*heading*) Witches, and other night-fears. 1923 Kipling *Irish Guards in Great War* I. 78 The 3rd Coldstream Guards..beat off that attack in a night-fight. 1918 *War Illustr.* 13 July 372/2 Making a night-flight over London. 1973 L. Cooper *Tea on Sunday* xxxi. 228, I must get a plane to Milan tonight..a night flight. 1859 A. J. Munby *Diary* 20 Mar. in D. Hudson *Munby* (1972) 28 A large & still flourishing crop of secret dens & night haunts. 1950 G. Greene *Third Man* xi. 96 He might have been..in some third-rate night haunt in any other shabby capital. 1963 'Night' hospital [see *day-hospital* s.v. *day *sb.* 23]. 1964 G. L. Cohen *What's Wrong with Hospitals?* vi. 96 He started a night hospital... Executives and professional men who would come by night to be treated. 1933 M. Peel *S.S. Utah* 12 Anybody know where the night lunch hangs out? 1945 *Seafarers' Log* 17 Aug. 6/5 Men coming back from shore leave are not able to get into night lunch. 1910 W. de la Mare *Three Mulla-Mulgars* ix. 128 It pleased Battle mightily, this night-music—music of all the kinds they knew, white man's, Jaqquamusic, Nugga-music, and Mulla-mulgars'. 1952 M. J. Ward (*title*) A little night Music. 1954 J. R. R. Tolkien *Fellowship of Ring* I. xii. 214 He lay tossing and turning and listening fearfully to the stealthy night-noises. 1844 T. Webster *Encycl. Domestic Econ.* xxvi. i. 1189 Night nurseries require little furniture beyond bedding, and utensils for washing and bathing. 1937 Night nursery [see *buddy *sb.*]. 1974 J. Pope-Hennessy *R. L. Stevenson* i. 34 Louis's night-nursery..was a small room to the east of the day-nursery. 1918 E. Sitwell *Clown's Houses* 7 Tulip-trees Spilling night-perfumes on the terrace. 1932 Auden *Orators* 120 The shamming dead, the night-raid, the feinted retreat. 1939 R. A. Knox *Let Dons Delight* ix. 243 My host returned, voluble..[illegible] the cheap night-rate for telephoning. 1975 J. R. L. Anderson *Death in North Sea* iv. 73 'Can you get a helicopter out tonight?' 'I can, but..night-rates for the crews make it rather expensive.' 1812 *Dramatic Censor for 1811* 212 The Public are respectfully informed, that it being found absolutely necessary to have a general night rehearsal of the new Burlesque Tragic Drama, there will be no performance in the Theatre this evening. 1866 M. Mackintosh *Stage Reminiscences* 99 We rehearsed the piece, without music, after which a night rehearsal, including the orchestra, was called. 1934 W. de la Mare *Fleeting* 42 The empty night-road to the sea. 1780 *New Jersey Jrnl.* 22 Nov. 4/2 Wanted, to be bound, a Boy..he shall be..sent to night school if required. 1858 Night-school [see *college lecture* s.v. *college 6]. 1923 E. J. Bamm *Nice Neighbourhood* i. 14, I met Don at night school. He was learning to be an accountant. 1922 D. H. Lawrence *Fantasia of Unconscious* xv. 271 The night-self is the very basis of the dynamic self. 1965 *Punch* 3 Nov. 665/1 The struggle in him between the day-self and the night-self. 1926 C. Morgan *Sparkenbroke* iv. iv. 344 From the darkness of the garden came the soft patter of invisible rain and the earthy night-smell. 1904 W. H. Hudson *Green Mansions* xx. 272 Nor had I any choice but to..listen to the night sounds of the forest. 1953 S. Beckett *Watt* 33 Listening to the little nightsounds in the hedge behind him. 1954 J. R. R. Tolkien *Fellowship of Ring* II. vi. 352, I fear nothing but the night-speech of plants and stones. 1910 W. de la Mare *Three Mulla-Mulgars* iv. 71 Soon the long-billed river-birds began their night-talk across the water. 1922 Joyce *Ulysses* 422 The Mabbot street entrance of nighttown. 1846 C. Brontë *Let.* 4 Sept. The night train. 1954 T. S. Eliot *Confid. Clerk* i. 30 We took the night train, and got the Channel crossing. 1906 Hardy *Dynasts* III. vi. vii. 315 It is where sky-fires flame and flit, Or solar craters spew and spit, Or ultra-stellar night-webs knit! 1900 E. Pound *Persona* 22 The strange night-wonder of your eyes Dies not. 1939 Dylan Thomas *Map of Love* 57 The clean winter sounds of the nightworld.
b. *night-express*, *-ferry*, *-latch*, *-refuge*, *-shelter*, *-sock*, *-stand*, *-suit*, *-wrapper*.
1877 J. Blackwood *Let.* 25 June in *Geo. Eliot's Lett.* (1956) VI. 390 We propose to come down by the night Express. 1854 O. S. Fowler *Home for All* 113 The front door..secured..with a night-latch and two keys. 1967 Karch & Buber *Offset Processes* vi. 321 Turn night latch lever to 'night latch' position. 1872 B. Jerrold *London* xxiii. 185 We are in the receiving house of a night refuge. 1960 *Guardian* 30 Nov. Night shelter. 1888 Night-wrapper. 1966 Night sock. 1974 Night-stand. 1862 Night-suit.
c. *night-editor*, *guard* (later example), *nurse*, *-patrol*, *people*, *police*, *porter* (later examples), *sister*, *tourist*.
1868 M. H. Smith *Sunshine & Shadow in New York* xxxviii. 639 Henry Winson is city editor, and Governor Carr is night editor. 1873 Night editor [see *day editor* s.v. *day *sb.* 24]. 1949 *Time* 30 May 69/1 Stewart..had been night editor, sports editor and state editor of the Scripps-Howard *Press*. 1973 R. L. Simon *Big Fix* (1974) xx. 174 Ask for the night editor. 1913 W. Owen *Let.* 1 Sept. (1967) xvi. 128 'Night Guard,' said the Lieutenant curtly. 1942 Berrey & Van den Bark *Amer. Thes. Slang* §675/6 Night nurse. 1951 Night-patrol. 1971 B. Patten *Irrelevant Song* 64 Leave it out among the night-patrols and the lovers. 1957 *N.Y. Post* 20 Sept. M4 Night people, the professor and his wife used to retire at about 2.30 or 3 A.M. 1963 *Times* 8 Jan. 10/4 The 'night people', cleaners, maintenance men, and so on, who occupy the London Underground after the last train has gone. 1877 E. S. Phelps *Story of Avis* 153 To recommend to the Faculty a stricter régime of night police for those boys. 1887 in C. E. Pascoe *Joyous Neighbourhood Covent Garden* 126 (Advt.), Rougemont hotel, Exeter... Night porter on patrol. 1963 N. Marsh *Dead Water* (1964) ix. 231 The night porter was reading behind his desk. 1969 H. MacInnes *Salzburg Connection* viii. 118 A little late, perhaps, to telephone but he knew the night porter there. 1886 E. C. E. Lückes *Hospital Sisters & their Duties* vi. 138 The Night Sister's object is to help the Day sister by giving the supervision to her patients and Nurses which the latter cannot exercise both night and day. 1972 F. Norton *Duties of Sisters in Small Hospitals* v. 54 The routine duties of the Night Sister consist in taking the Day report, making periodical visits to the wards, supervising the admission of accidents and of emergency operation cases. 1934 P. Bottome *Private Worlds* 3 The door of the night-sister's room was open. 1958 Night-sister [see *garbo]. 1973 'E. Peters' *City of Gold & Shadows* xii. 191 The night sister on duty was an old friend. 1963 L. Deighton *Horse under Water* xxv. 125 Night Tourist Aeroplane. 1971 F. Purser *Holy Father's Navy* xvii. 88 The last plane will have gone, anyway, unless there are night tourist flights.
d. *night snake*, a name used for several nocturnal African snakes.
1931 K. L. Ditmars *Snakes of World* ix. 252 The so-called Bush Snakes or Night Snakes are rear-fanged species. 1954 J. A. Pringle *Common Snakes* 12 Olive Night-Snake..non-venomous. 1957 V. F. M. FitzSimons *Snakes S. Afr.* 112 Aurora or Night-snake.
e. *night-primrose* = *evening-primrose* (Evening *sb.* 5 b, in Dict. and Suppl.) (earlier and later examples); *night stock* = *night-scented stock* (sense *14).
1759 P. Miller *Gardeners Dict.* (ed. 7) s.v. *Œnothera*, Tree Primrose... The Tree Primrose opening in the Evening, many Persons call it the Night Primrose. 1924 Sitwell *Tryst* 21 You shall have all the beauty Like a pale night-primrose. 1928 D. H. Lawrence *New Poems* 33 The night-stock oozes scent.
14. night-adapted *a.* = *dark-adapted* ppl. adj. s.v. *dark *sb.* 6; **night adder,** a nocturnal, venomous, African viper of the genus *Causus*, esp. *C. rhombeatus*, a grey snake with darker patches, common in southern Africa; **night-blooming cereus,** one of several tropical plants belonging to the genera *Hylocereus* and *Selenicereus* of the family Cactaceæ, esp. *H. undulatus*, having very large, fragrant, white flowers that open only at night; **night-blue** (further examples); also *attrib.* or as *adj.*; **night bomber,** an aircraft that drops bombs at night; also, the pilot of such an aircraft; hence **night-bombing**; **night-bound** *a.*, bound, confined, or impeded by night or darkness; **night-box** = *boîte de nuit; **night chain,** a chain for securing a door at night; **night-climber,** one who climbs on buildings at night, esp. at the Universities of Oxford and Cambridge; so **night-climbing**; **night clock,** a clock which is illuminated so that it can be seen in the dark; **night-club,** † (*a*) *U.S.* = *night-stick; (*b*) a club or similar establishment that opens at night, providing food, drink, and entertainment; also *attrib.*; hence **night-club** *v. trans.* and *intr.*, to take or go to a night-club; **night-clubber,** one who frequents night-clubs; **night-clubbing,** the frequenting of night-clubs; **night-clubby** *a.*, characteristic or fond of night-clubs; **night-coach,** (*a*) a coach that travels at night; (*b*) *U.S.*, a commercial aircraft providing a night service; **night crawler** *N. Amer.*, a large earthworm, esp. one caught at night to be used as bait in fishing; **night cream,** cosmetic cream that is applied to the face at night; **night-driving,** the driving of a motor vehicle at night; also *attrib.*; so **night-drive** *v. intr.*, to drive a motor vehicle at night; **night effect,** irregularity of the strength and apparent direction of received radio waves of certain frequencies that is especially marked at night, owing to the reception of polarized waves reflected by the ionosphere; so **night error**; **night eye,** (*a*) *U.S.* = Chestnut *sb.* 6; (*b*) an eye adapted for seeing in the dark (usu. *pl.*); **night-fighter,** a fighter (*fighter 3) used, or designed for use, at night; also, the pilot of such an aircraft; hence **night-fighting**; **night-flowering cereus** = *night-blooming cereus; **night-fossicker** *Austral. Hist.*, a nocturnal thief of gold dust or quartz; so **night-fossicking**; **night-herd** *N. Amer.*, the herding or guarding of cattle at night; hence as *v. intr.* and *trans.*, to herd or guard (cattle) at night; **night-herder,** one who night-herds; **night-herding,** the work of a night-herder; **night-horse,** (*a*) a horse used for work at night; (*b*) a punning alteration of Nightmare *sb.*; **night-lark,** a person who goes about at night; **night-life,** manifestations of life at night; *spec.* the activities of, or urban entertainments open to, pleasure-seekers at night; **night-lifer,** one who enjoys night-life; **night-office** *R.C.Ch.*, (until 1971) the part of the canonical office performed during the night hours; **night op** or **operation,** a military operation at night; **night paddock** *Austral.* and *N.Z.*, a field where stock, esp. dairy cows, are kept overnight; **night parrot** *Austral.*, a nocturnal green and yellow ground parrot, *Geopsittacus occidentalis*; **night rider,** one who rides by night, esp. on horse-back; *spec.* in *U.S.*, one of various gangs of mounted men who commit acts of violence in order to intimidate or punish (see also quots.); so *night-riding* vbl. sb. and ppl. adj.; **night-safe** (see quot. 1930); **night-scented stock,** a small annual herb, *Matthiola tristis* or *M. bicornis*, of the family Cruciferæ, whose fragrant lilac flowers open at night; **night-side,** (*a*) the dark or bad aspect of a person or thing; (*b*) *Shetland dial.*, in phr. *in the night-side*, in the evening; (*c*) (see quot. 1927); (*d*) the side of a planet that is facing away from the sun and is therefore in darkness; **night-sight,** (*b*) a rifle-sight designed for shooting at night; (*c*) = Night-vision 2 a; **night spot,** a night-club or similar place open to pleasure-seekers at night; **night starvation,** hunger at night; also *transf.*, lack of sexual gratification; **night-stick** orig. *U.S.*, a stick or truncheon carried by a policeman or the like, esp. at night; also *attrib.*; **night-stop,** a place where one stops for the night; the action of stopping at such a place; hence as *v. intr.*, to stop for the night; **night storage heater** or **radiator,** an electric heater in which heat can be accumulated at night and released during the day; so *night-stored* ppl. adj.; **night-watchman,** (*a*) a person employed to keep watch at night; (*b*) in *Cricket*, a batsman who goes in to bat just before the end of a day's play.

1961 I. Jefferies *It wasn't Me!* x. 132 When the moon was up I let an hour pass to make sure my eyes were night-adapted. 1972 J. Poyer *Chinese Agenda* (1973) xi. 144 Just enough light for Gillon's night-adapted vision. 1834 Night adder [in Dict., sense 13 d]. 1915 *Chambers's Jrnl.* July 437/2 The night-adder, as its name implies, is most in evidence after sundown. 1947 J. Stevenson-Hamilton *Wild Life S. Afr.* xxxvi. 329 The Night Adder (*Causus rhombeatus*). *Ibid.*, I have often seen my cats eating night adders which they have caught and killed. 1966 C. Sweeney *Scurrying Bush* vii. 98 The night adder was writhing in some discomfort. 1832 J. Lindley *Introd. Bot.* IV. 402 Of a night (*nocturnus*); which appears during the night, and perishes before morning; as the flowers of the night-blooming cereus. 1890 *Harper's Mag.* Mar. 613/1 My wife has a sweet face of her own, but one bearing the same relation to Miss Jasmine's as that existing between a sprig of mignonette and a night-blooming cereus. 1936 *Times Lit. Suppl.* 20 June 521/2 The flowering of the night-blooming cereus..is very lovingly described. 1971 E. L. Wardman *Bermuda Jubilee Garden* 199/2 Night-blooming cereus. A vine-like plant with triangular stems... Useful for clambering over a wall or rock face and even growing up a tree... Its very beautiful flowers open only after sunset and shrivel and die before morning. 1908 *Paris Fashions* 18 Jan. 23/2 Costume of night-blue cloth. 1938 L. MacNeice *I crossed Minch* ii. viii. 106 The sky part of the landscape was night-blue. 1956 D. Gascoyne *Night Thoughts* 45 That profound night-blue abyss of starry vacancy. 1970 R. P. Warren *Incarnations* 17 On night-blue the tetter of cloud-scud. 1918 *Flying* 4 Sept. 221 (*caption*) A British night-bomber photographed by searchlight. 1919 R. H. Reece *Night Bombing with Bedouins* 57 These calculations are all important to the long-distance night bomber. 1936 *Economist* 11 Jan. 85/2 The Fairey Company is an air-frame concern which is chiefly interested in the 'Hendon' night bomber. 1975 *Listener* 13 Mar. 335/1 Scheduled air services began on 25 August 1919..using ten- and 12-seated converted night bombers. 1942 W. S. Churchill *End of Beginning* (1943) 118 These two great night-bombing raids mark the introduction of a new phase in the British air offensive against Germany. 1925 A. S. M. Hutchinson *One Increasing Purpose* i. xxvi. 161 As if the phrase were a path on which, nightbound and groping, he suddenly had stumbled. *Ibid.* ii. ii. 203 A finger-post whose word the nightbound traveller hates to obey yet may not disbelieve. 1954 L. MacNeice *Autumn Sequel* 123 From my seat I see my night-bound double, slumped apart On a conveyor belt. 1938 *New Statesman* 23 July 154/1, I have very rarely been overcharged in France (except in the grotesque night-boxes of Montmartre). 1973 E. McGirr *Bardel's Murder* iv. 108 Night boxes.. came and they went, and the more crowded the more successful. 1904 E. Glasgow *Deliverance* 45 He had fastened the night-chain and shot the heavy bolt. 1973 'E. McBain' *Let's hear It* xiv. 205 Kling heard the night chain being slipped off, the lock turning. 1968 J. M. White *Nightclimber* iv. 28, I had repeated the whole series of safaris among the tiles and chimneypots pioneered by the Night-climbers. *Ibid.* 27, I didn't realise you were interested in night-climbing. 1911 F. J. Britten *Old Clocks & Watches* (ed. 3) v. 266 A night clock..is of ebony on oak, and the top lifts off to allow the insertion of a lamp. Showing through a curved slit in the upper part of the dial is a disc with perforated hour numerals so that the time can be seen at night. The light would also shine through a keyhole-shaped aperture above which serves as a pointer. 1972 *Times* 7 Nov. 25/4 (Advt.), A rare 17th century night clock, by Edward East, London. 1882 J. D. McCabe *New York* 383 The entire force on duty at the station dashed into the street, armed with their long night clubs. 1894 W. J. Locke *At Gate of Samaria* (1903) xxvii. 319 They went together to East End music-halls,.. night clubs in the West End, where ladies are admitted free on a member's introduction. 1906 R. Machray *Night Side of London* i. 21 Finishing up, perhaps, at some night-club, or in some other den. 1915 Night club [see *cabaret[1] 2 b]. 1928 F. B. Young *My Brother Jonathan* ii. viii. 355 A life of night-clubs and jazz-bands. 1938 *Amer. Speech* XIII. 194 Pleasure seekers at first went to night clubs; now, at least in the columns of the Broadway gossips, they simply *night-club*. 1965 *New Statesman* 9 Apr. 557/2 The sharp limitation in the circumstances in which businessmen can wine and dine and nightclub other businessmen. 1972 P. Driscoll *Wilby Conspiracy* (1973) xi. 138 That bitch of a nightclub singer. 1974 *Listener* 31 Jan. 131/1 The foreboding, the mounting menace, that we can trace through, say, the night-club songs of the Weimar republic. 1952 B. Ulanov *Hist. Jazz in Amer.* (1958) xv. 177 From the general hubbub of night clubs and the particular cries and grunts of night-clubbers..Bubber made his music. 1953 'S. Ransome' *Hear No Evil* (1954) xv. 138 Bendley, a confirmed night-clubber. 1936 R. Lehmann *Weather in Streets* i. v. 90 No, she never married... Does some little odd jobs and goes lunching and dining and night-clubbing. 1941 Hermer & May *Havana Mañana* p. xii, There have been no books about Havana that show tourists how to get more than their money's worth out of stopping, eating, sightseeing and nightclubbing. 1947 *Sun* (Baltimore) 4 Aug. 1/2 Senator Pepper (D. Fla.) today tabbed as possibly 'base judgment' the 1943 night clubbing activities which Elliott Roosevelt will be asked to explain tomorrow to senators investigating Howard Hughes's wartime plane contracts. 1971 D. Bagley *Freedom Trap* viii. 181 Too much damned night-clubbing. 1933 G. B. Shaw *Political Madhouse in Amer.* 48 You have become a wonderful night clubby sort of nation. 1958 *New Statesman* 6 Sept. 314/2 A nightclubby girl 'with a spurious American accent' is a minor character in the recent *Sober as a Judge* by Henry Cecil. 1844 Night-coach [in Dict., sense 13]. 1959 *Wall St. Jrnl.* (Eastern ed.) 7 Oct. 8/4 Standard first class Miami–New York fare one-way is $80.80, regular daycoach is $54.55 and regular nightcoach is $46.80. 1960 *Daily Progress* (Charlottesville, Va.) 6 Jan., A National Airlines night coach flying non-stop from New York to Miami crashed with 34 persons aboard. 1924 *Collier's* 2 Feb. 3/1 He could stay up till 10 and hunt night crawlers in the garden with a lantern. 1948 *Esquire* Mar. 88 Members of the Huck Finn school of fishing..have been looking for some way to enliven the almost impossibly sluggish night crawler. 1951 T. Capote *Grass Harp* (1952) ii. 64 Night-crawlers slithered away from its lurching light. 1971 B. Malamud *Tenants* 12 Lesser tried to scare off the nightcrawlers on his floor.. by playing loud his hi-fi at night. 1973 *Islander* (Victoria, B.C.) 5 Aug. 13/2 Sympathy and practical help poured in from every direction. Small donations of native worms—night crawlers—were the most practical help. 1926–7 *Army & Navy Stores Catal.* 496/2 Day Cream... Massage Cream... Night Cream. 1963 D. Gray *Murder in Mind* iii. 23 Except for a thick night cream, she took so little [care] of her face. 1956 *This Week* 29 July 11/2 If you must night-drive, keep the dash-lights as dim as possible—this particular glare is hypnotic. 1929 *Times* 31 Oct. 21/3 Those who have used this latest device under actual night-driving conditions are unanimous in their praise. 1936 *Discovery* Oct. 302/1 The application of a sheet of.. Polaroid..should remove one of the greatest inconveniences of night driving. 1962 L. S. Sasieni *Optical Dispensing* xiii. 327 There have been introduced from time to time certain glasses said to increase vision in low illumination—and so-called night-driving glasses. There are many arguments as to the merits and demerits of tinted glasses for night driving. 1914 R. Stanley *Text-bk. Wireless Telegr.* x. 114 The difference between day and night effects on the transmitted ether energy might possibly be caused by a change in the position of the upper conducting layer of atmosphere. 1932 F. E. Terman *Radio Engineering* xvi. 591 Since the sky wave is always strongest at night the errors that result from downcoming horizontally polarized waves are frequently referred to as 'night effects' although they are always present to some extent in daytime. 1962 J. H. & P. J. Reyner *Radio Communication* viii. 333 A phenomenon which considerably affects D.F. work is what is known as 'night effect'. 1921 *Flight* XIII. 664/1 Aircraft when within an area northward of the parallel of latitude 51° 10′ 00″ N., and westward of the meridian of longitude 8° 30′ 00″ W., should not ask for bearings from Carnsore, as such bearings..will probably be unreliable on account of the effect of the coastline, the night error in particular being of considerable magnitude. 1936 *Jrnl. R. Aeronaut. Soc.* XL. 161 These wave-lengths are subject to night error which affects the accuracy of bearings taken on medium wave direction finders. 1948 *Sat. Even. Post* 29 May 116/1 Six photographs are taken—a front view, side view and close-ups of the horse's four 'chestnuts', or 'night eyes', which are the rough protrusions of scaly, hardened skin that are on the inner side of each leg. 1954 J. R. R. Tolkien *Fellowship of Ring* ii. ix. 402 It was dark, but not too dark for the night-eyes of Orcs. 1956 I. Fleming *Diamonds are Forever* viii. 80 'The Jockey Club are going to change to photos of the night eyes [to help identify race horses].' 'What are night eyes?' 'They're those callouses on the inside of a horse's knees. The English call them 'chestnuts'. Seems they're different on every horse. Like a man's fingerprints.' 1957 'J. Wyndham' *Midwich Cuckoos* ii. 21 The last customers to be persuaded out of The Scythe and Stone had lingered for a few minutes to get their night-eyes and gone their ways. 1963 M. A. Stoneridge *Horse of your Own* ii. 75 Inside each of the horse's legs you will notice a sort of horny excrescence called Chestnuts or Night Eyes, which are vestigial toes (remember, prehistoric horses had four or five of them). Night Eyes vary in form with no two alike; they are as individual as human fingerprints and recorded as part of the identification system for thoroughbred race horses and registered trotters. 1941 *Aeronautics* Jan. 41/3 The effective range of fire of night fighter aircraft will have to be increased. 1941 *Times* (Weekly ed.) 5 Feb. 15 The defence by night-fighter aeroplanes was the most difficult task of anti-aircraft defences. 1942 *Tee Emm* (Air Ministry) II. 62 That..is the gist, or guts, of the night-fighter pilot's training. *Ibid.* 85 Particularly important this for night-fighters. 1947 J. Mulgan *Report on Experience* vi. 80 Meeting sometimes night-fighters or flak coming back over the coast defences. 1933 *Meccano Mag.* Feb. 109/1 Night fighting consists chiefly of individual attacks at close range. 1947 Crowther & Whiddington *Science at War* 60 The radar air interception equipments, used in the early night-fighting battles. 1789 W. Aiton *Hortus Kewensis* II. 152 Great Night-flowering Cereus. Nat[ive] of Jamaica and Vera Cruz. 1853 C. R. Read *What I heard, saw, & did at Austral. Gold Fields* 150 (Morris), The man was what they called a night fossicker, who slept, or did nothing during the day, and then went round at night to where he knew the claims to be rich, and stole the stuff by candle-light. 1889 *Cent. Dict.*, Night-fossicking. 1884 R. Aldridge *Life on Ranch* 62 When on night-herd the men usually keep singing all the time as they ride round. 1903 A. Adams *Log of Cowboy* ii. 11 Forrest night herded them using five guards. *Ibid.* vii. 97 We night herded as usual. 1955 R. P. Hobson *Nothing too Good for Cowboy* viii. 71 Simrose and Rob came in from their night-herd shift. 1962 W. Stegner *Wolf Willow* iii. ii. 163 Everything he said or played or sang during his hours on the night herd was meant seriously. 1963 R. D. Symons *Many Trails* 37 We certainly did not propose to night herd. 1873 J. H. Beadle *Undevel. West* v. 98 The 'night-herder' Billy Keyes, and two other drivers..were Gentiles. 1891 Night herder [in Dict., sense 13 c]. 1888 *Chambers's Jrnl.* Apr. 221/1 This is called night-herding. 1890 Night-herding [in Dict., sense 12 b]. 1908 *Sat. Even. Post* 24 Oct. 10/1 A long-eared, reddish, sleepy-eyed..mule frequently used in night-herding. 1933 J. V. Allen *Cowboy Lore* iv. 71 David went from night-herding to using a sling. 1908 *Sat. Even. Post* 4 July 22/3 They made Blackie a night horse, for his sure-footedness was remarkable. 1925 J. Farnol *Loring Mystery* xlii. 283 'Talking o' bed,' quoth Mr. Shrig .., 'do you ever dream —— d'ye ever have the night-'orse?' 1929 *Amer. Speech* V. 67 Usually the 'kept horses' or 'herding horses' are also 'night horses', those used for 'night herding'. 1937 *Dialect Notes* VI. 618 The night horse is one staked near the cowpuncher's bed for immediate use in some such emergency as a stampede. 1959 C. MacInnes *Absolute Beginners* 81 The capital was a night-horse dream. 1967 *Coast to Coast 1965–66* 195 The smell of the night-horse nearby and the cattle a little beyond will tie you to reality. 1895 G. Meredith *Amaz. Marriage* I. xxiii. 258 Night-larks of different classes, both sexes. 1852 H. Melville *Pierre* xvi. i. 322 All the garish night-life of a vast thoroughfare, crowded and wedged by day, and even now, at this late hour, brilliant with occasional illuminations. 1913 H. L. Mencken *Let.* 17 Aug. (1961) 32 The title 'Night Life in Vienna'..has the air of a lure held out to the Puritanical and dirty-minded. 1927 G. Ade et al. *Let.* 4 Mar. (1973) 119 Our fellow-passengers ..were ashore last night, dancing and hunting up a second-rate African imitation of night life in Paris. 1929 D. L. Moore *Pandora's Letter Box* iii. 53 'Night life'—to use the popular expression for habitual nocturnal dancing and drinking. 1972 *Sat. Rev.* (U.S.) 25 Mar. 68/3 The Kabarett, with its sharp political satire, was part of Berlin night life. 1967 W. Soyinka *Kongi's Harvest* 13 A few night-lifers pick up their drinks and go in. 1767 A. B. *Short Acct. Life Mary of Holy Cross* 84 In the Time of St. Peter Damian, only the Clergy rose to the Night Office. 1909 M. B. Saunders *Litany Lane* i. iii. 34 A small chapel in which the brothers held their short night-office. 1957 *Oxf. Dict. Chr. Church* 960/1 *Night office*, another name for Mattins, the liturgical office prescribed for the night. 1916 W. Owen *Let.* 6 Apr. (1967) 388 We had 'Night Ops.' yesterday till 9.30! *Ibid.* 10 Apr. 390 We had Night Operations again. 1930 *Bulletin* (Sydney) 29 Jan. 62/4 He strode up to the big, heavily-capped yards in the corner of the night-paddock. 1950 *N.Z. Jrnl. Agric.* Jan. 71/2 Night paddocks on dairy farms showed least response [to potash topdressing]. 1966 *Te Reo* IX. 54 The apparently Australian innovation lies..in the adoption of the refined terminology of the *home paddock* and the *night paddock*. 1917 *Chambers's Jrnl.* Feb. 89/1 Even on moonlight nights I could never catch a glimpse of the flying night-parrots. 1934 *Bulletin* (Sydney) 9 May 21/1 The night parrot..continues to hide itself from human ken, and is always referred to either as the elusive parrot or mystery bird. 1965 *Austral. Encycl.* VII. 27/2 A second remarkable parrot of the ground is one known as the night parrot, sole member of the genus *Geopsittacus*, which is, or used to be, distributed widely throughout the drier parts of the inland region. It is closely associated with the spiny spinifex or porcupine-grass, hiding in the thick clusters by day, feeding on their seeds at night. 1877 J. M Wells *Chisolm Massacre* x. 118 They said that night-riders had shot into the houses of the colored people. 1879 *Congress. Rec.* 20 May 1480/1 There was much said..of kuklux, white leagues, and night-riders... There are..no night-riders in the State of Louisiana. 1882 F. W. P. Jago *Anc. Lang. & Dial. Cornwall* 226 *Night-riders*, Piskey (Fairy) people who have been riding Tom (the name of a horse) again. 1906 Kipling *Puck of Pook's Hill* 10 Leprechauns, night-riders, pixies, nixies. 1907 *Lit. Digest* 28 Dec. 976 The first appearance of the night riders was in November, 1906, when they destroyed some tobacco-barns and small factories in Todd County. 1911 J. Masefield *Jim Davis* ii. 12 They were the night-riders or smugglers. 1936 J. G. Miller *Black Patch War* 18 The Night Rider burnt his warehouse and his purchase with it. 1948 E. N. Dick *Dixie Frontier* 94 Patrols, called patterols by the slaves, were organized by the whites, and these night riders endeavored to enforce the regulations. 1970 *New Yorker* 12 Dec. 166/3 Night riders who fire shotgun blasts at the home of an anti-war leader..are routine. 1973 J. Cleary *Ransom* viii. 184 As a plainclothesman he had travelled on late trains out of Sydney to the suburbs, riding shotgun as it were against hooligans... Night riders, he guessed, were the same everywhere. 1875 *Chicago Tribune* 6 Nov. 3/6 To-night..there is to be a 'night riding' and shooting..to arouse a degree of uneasiness in the darky's mind and cause him..not to go to the election. 1909 *Chambers's Jrnl.* Feb. 104/1 Night riding began as soon as the farmers' associations were organised. 1952 C. Day Lewis tr. *Virgil's Aeneid* x. 215 The gentle Moon's Night-riding horses were pacing halfway across the heavens. 1930 W. Thomson *Dict. Banking* (ed. 7) 501/2 *Night safes*. In order that customers may deposit cash or cheques after a bank has closed for the day or for the week-end, night safes were introduced in 1928. The entrance to these safes is in the outside wall of the bank, the opening being fitted with a locked cover to which customers who wish to avail themselves of the safe are supplied with a key. 1959 *Times* 18 Feb. 11/5 Night safes don't cash cheques. 1975 'M. Yorke' *Small Hours* iii. 31 Ray had tracked down one of the night-safe depositors. [1824 H. Phillips *Flora Historica* I. 336 We have no plant that exhales so delightful a fragrance in the night as the..Night-smelling Rocket, or Night Odorous Stock.] 1849 Night-scented stock [see *night-scented*, sense 12 c]. 1870 W. Robinson *Wild Garden* ii. 54 Night-scented Stock... May be established on the sunny sides of old ruins and walls. 1914 E. A. Bowles *My Garden in Summer* xiv. 259 The old Night-scented Stock of one's great-grandmother, is another half-hardy indispensable. 1972 *Country Life* 23 Mar. 690/1 Seeds of small annuals such as night-scented stock..I broadcast by scattering them over the ground. 1848, 1855 Night side [in Dict., sense 13]. 1898 *Shetland News* 10 Dec., If Willie id been some boys, diel wird he'd sung i' da nicht side. 1900 *Ibid.* 26 May, Dis kirn is no laek ta brak i' da night side. 1927 *Amer. Speech* II. 242/2 On papers having both morning and afternoon editions one also hears references to the two divisions of the staff as the 'day side' and the 'night side'. 1951 A. C. Clarke *Sands of Mars* viii. 94 It was a live programme, beamed to Mars from somewhere on the night-side of Earth. 1970 *Nature* 7 Mar. 925/1 On the nightside of Venus at altitudes between 750 km and 1,450 km the presence of a light ion was inferred. 1973 G. Greene *Honorary Consul* v. iii. 285 Suppose the night-side of God swallows up the day-side altogether? Suppose it is the good side which withers away? 1974 *Nature* 4 Jan. 24/1 The orbit is near-polar with the north-going passes on the nightside at about 2230 LT, and the south-going passes on the dayside at about 1030 LT. 1915 *Chambers's Jrnl.* Apr. 269/2 The night-sight does not interfere with the ordinary front-sight for daylight shooting. 1971 *Guardian Weekly* 6 Nov. 12 Our marksman..saw him clearly through his night-sight. 1972 B. Everitt *Cold Front* xv. 143 My night-sight is good and I drove on side-lights only. 1973 'D. Halliday' *Dolly & Starry Bird* ii. 29 At night the full lights in the cupola never go on. They would spoil the plate and ruin your night sight. 1936 *Swing Music* Mar. 9/1 The management at the Friar's Inn, well-known Chicago night spot, was very anxious to feature this new type of music. 1947 *Sun* (Baltimore) 4 Aug. 1/2 He was the guest of John Meyer, publicity man, in a costly round of New York night spots. 1959 F. Usher *Death in Error* i. 18 They went to a night spot.. where they drank champagne. 1973 *Express* (Trinidad & Tobago) 26 June 4/6 A flash fire..swept through a New Orleans night spot. 1936 'G. Orwell' *Keep Aspidistra Flying* xii. 311 What they asked for was a really telling slogan; something in the class of 'Night-starvation'..that would rankle in the public consciousness. 1949 Partridge *Dict. Slang* (ed. 3) Add. 1119/1 *Night starvation*, sexual deprivation, lack of sexual intimacy. 1970 *Southerly* XXX. 286 David's head is in the right place—night starvation—Horlicks—Venus de Milo. 1971 D. Lees *Rainbow Conspiracy* viii. 117 It wasn't as if I was suffering from night starvation. Val was easily one of the best screws in the business. 1974 *Harpers & Queen* Sept. 135/1 My bread rolls she secretes..against night starvation. 1887 W. E. S. Fales *Brooklyn's Guardians* ii. 30 In the wealthier thoroughfares the brawls were of so frequent occurrence that the sight of two watchmen..pounding each other with their night-sticks, occasioned no comment. 1893 S. Crane *Maggie* xi. 102 The officer made a terrific advance, club in hand. One comprehensive sweep of the long night stick threw the ally to the floor. 1904 'No. 1500' *Life in Sing Sing* i. 10 Big clubs, heavier and more formidable than a policeman's night stick. 1905 *N.Y. Times* 15 July, San Juan Hill and the Gut were under nightstick law until early this morning. 1932 *New Yorker* 4 June 23 There's a lot of law at the end of a nightstick. 1963 T. & P. Morris *Pentonville* x. 209 Hospital officers do not carry night sticks. 1973 J. Wambaugh *Blue Knight* i. 18 A beat cop has to be big..or..somebody'd take the night-stick off him and shove it up his ass. 1975 *Daily Tel.* 30 June 1/5 Police chased the demonstrators down the streets [of Delhi], stopping occasionally to swing at anyone in their way with heavy wooden nightsticks. 1791 J. Byng *Diary* 3 July in *Torrington Diaries* (1935) II. 360 Dunnington, a small market-town, where I dined..in a quiet house, but not a night-stop. 1951 'N. Shute' *Round the Bend* 232 Next day took us to Karachi, where we night-stopped before going on to Bahrein. 1959 *New Statesman* 8 Aug. 151/1 When coming back from India took five days, with agreeable night-stops along the way, it was quite pleasant. 1972 C. Kearey *Last Plane from Uli* ii. 34 'Are you night-stopping, sir?' the Immigration Chief asked curiously. 'No, I'm taking off for Lagos as soon as the refuelling's finished.' 1975 *Daily Tel.* (Colour Suppl.) 4 Apr. 16/1 Had the supervisors declared the plane unserviceable, another 747..due to night-stop in London would have been allocated to take her place. 1963 *Good Housek. Setting up Home* iii. 30 Electric night storage heaters. These are electric heaters which..absorb and store heat during the night when off-peak rates for electricity are available, releasing it during the day. 1970 *House & Garden* Mar. 94/3 The eight-hour-charge night-storage radiators..are able to store enough heat to give an even heat output for the rest of the day. 1973 *Guardian* 23 May 9/6 Night storage heaters are metal boxes filled with bricks wired for electric heating. 1962 *Daily Tel.* 20 Aug. 16/2 (*heading*) Night-stored heat from electricity. 1863 A. D. Whitney *Faith Gartney's Girlhood* xxviii. 261 Michael Garvin, the night-watchman,..had left. 1874 Night-watchman [in Dict., sense 13 c]. 1926 J. Masefield *Odtaa* xiii. 214 There must be at least a caretaker or night-watchman. 1949 Partridge *Dict. Slang* (ed. 3) Add. 1110/1 *Night-watchman*, a (usually a second-rate) batsman sent in to 'hold up an end' until the close of play. 1971 *Sunday Times* 2 May 24 The fifth [ball] was caught by Turner off the nightwatchman Wasim Bari's glove and shoulder.

night-bird. **2.** (Later examples.)
1870 D. J. Kirwan *Palace & Hovel* xxxii. 481 When the dancing places..close, this door remains open to catch all stray night birds who can find no other resting place. 1871 G. Easton *Trav. Amer.* 41 In Boston the cars run till midnight, and after that time one car runs every hour through the principal thoroughfare for the accommodation of 'Night Birds'. 1900 [see *brunch]. 1939 Joyce *Finnegans Wake* 438, I have every reason to know that rogues' gallery of nightbirds and bitchfanciers. 1974 'J. Le Carré' *Tinker, Tailor* xxxv. 313 It was nearly midnight... He was a night bird.

night-cap. Add: **3.** Also, a non-alcoholic drink taken at bedtime.
1930 *Daily Tel.* 9 Apr. 11/7 (Advt.), 'Ovaltine'... The world's best 'night-cap' to ensure sound, natural sleep. 1959 *New Statesman* 1 Aug. 141/3 A greater number of housewives than ever before realised the goodness of Bovril..for use as a beverage during the day, a night-cap, and for adding to savoury dishes. 1975 'C. Aird' *Slight Mourning* ii. 18 Sloan's own nightcap was usually milky coffee.
4. The final event in one day's series of sporting contests; *spec.* the second of two baseball games played by the same two teams on a single day. *N. Amer. colloq.*
1939 Webster Add., *Nightcap*, the final race or contest of a day's sports. 1941 *Sun* (Baltimore) 22 Sept. 1/1 The Cardinals pulled the opening game out of the fire with a daring bit of base running..to win 6–5, and took the nightcap, 7–0. 1942 Berrey & Van den Bark *Amer. Thes. Slang* § 675/6 Second game of a 'double-header',.. *nightcap*. 1969 *Internat. Herald Tribune* (Paris) 6 Nov. 13/1 In the nightcap, Baltimore's Wes Unseld..sat out much of the game because of foul trouble. 1970 *Toronto Daily Star* 24 Sept. 17/1 In the nightcap, Jerry Reuss permitted seven hits. 1974 *Union* (S. Carolina) *Daily Times* 22 Apr. 6/2 The Orioles edged New York 6–5 in 13 innings in the opener of a doubleheader and then protested the nightcap, which they lost 3–0.

night-dress. Add: **2.** *attrib.*, as **night-dress bag, case,** a bag or other container in which a night-dress can be placed when it is not being worn.
1975 *Times* 25 Nov. 2/6 He tried to punch her and hit a toy dog nightdress bag instead. 1907 *Yesterday's Shopping* (1969) p. xlvi/4 Night Dress Cases. 1932 D. C. Minter *Mod. Needlecraft* 218/2 A triangular embroidered night-dress case. 1969 'G. North' *Procrastination of Sergeant Cluff* xv. 151 The night-dress case in the shape of a toy Koala-bear that growled when she lifted it.

ni·ght-flying, *vbl. sb.* [f. Night *sb.* + Flying *vbl. sb.*] Flying in an aircraft by night. Also *attrib.* Hence **ni·ght-fly** *v. intr.*
1907 A. C. Johnson *How to find Time at Sea* (ed. 6) 9 Steering by the stars, for night flying, night marching and night boat-work. 1918 *War Illustr.* 13 July 372/2 The petrol or electric flares, which are placed on all night-flying aerodromes for landing purposes. 1927 *Daily Tel.* 31 May 13/2 We night-fly regularly in America. It's the ideal time for flying in the States. 1935 [see *directional *a.* 5]. 1945 *Tee Emm* (Air Ministry) V. 42 He thought the obstructions were left over from the previous night's night-flying taxi path. 1973 'A. Hall' *Tango Briefing* xviii. 222 'Is there any chance of a flarepath on that strip?' 'No. They don't night-fly.'

nightglow (nai·tglōu). *Meteorol.* [f. Night *sb.* + Glow *sb.*] The faint light emitted by the upper atmosphere at night.
1951 Roach & Pettit in *Jrnl. Geophysical Res.* LVI. 325 Many investigators..have reported that the intensity of the green nightglow line at 5577 Å has a maximum near the observer's local midnight. *Ibid.*, The general term 'airglow' is used for radiations originating in the upper atmosphere. We use the term 'nightglow' to connote airglow at night to distinguish it from similar radiations which may be present (though not observable) in the daytime. 1955 *Sci. Amer.* Sept. 150/2 These studies show that the nightglow is faintest at the zenith overhead and grows in intensity down the sky until it reaches a maximum about 10 degrees above the horizon. 1967 R. W. Fairbridge *Encycl. Atmospheric Sci. & Astrogeol.* 9/1 It has become usual to distinguish these differing kinds of airglow by the names 'twilight glow' and 'dayglow', in distinction to the 'nightglow'. 1972 *Sci. Amer.* Jan. 78/2 The nightglow can be observed with a photometer or a spectrometer.

night-hawk. **1. b.** Substitute for def.: A predominantly nocturnal bird of the genus *Chordeiles*, esp. *C. minor*, belonging to the nightjar family, Caprimulgidæ. (Earlier and later examples.)
1793 W. Bentley *Diary* 22 Aug. (1909) II. 48 We observed a great number of the birds, called here [*sc.* at Charlestown, N.H.] night hawks, playing in the air. 1913 W. P. Pritchard *Barn Doors & Byways* 119 The naturalists tell us that the night-hawks nest on top of the Manhattan skyscrapers. 1962 O. L. Austin *Birds of World* 163/2 The plaintive buzzing calls of Nighthawks are now commonly heard above the traffic noises of many American cities as the birds course high over the rooftops in their buoyant, dancing search for flying insects. 1968 R. M. Patterson *Finlay's River* 123 The dew was starting to fall..and the thrum of the night-hawk was sounding from the upper air. 1974 D. Sears *Lark in Clear Air* vi. 73 A night-hawk beeped high and out of sight.
2. (Further examples.) Also, one who stays up late at night; a person who goes out or works at night; *spec.* = **night-herder* (see also quots.).
1868 'Mark Twain' *Let.* 8 Jan. (1920) 80 Jack Van Nostrand, Dan and I, (all Quaker City night-hawks,) had a blow-out at Dan's house. 1903 W. D. Coburn *Rhymes from Round-up Camp* (rev. ed.) 18 Cotton-Eye, the night-hawk, Was then a top cow-hand. 1915 *Dialect Notes* IV. 209 *Night-hawk*, a thief or harlot. 'Those night-hawks ought to be taken up and sent home if they don't know enough to go.' 1929 F. Bowen *Sea Slang* 95 *Night hawks*, night watchmen stewards. 1934 A. Woollcott *While Rome Burns* 180 Your correspondent, a nighthawk of parts in those days, was within ear-shot. 1948 *Sierra Club Bull.* (San Francisco) Mar. 22 Ed Thistlethwaite..was our night hawk. It was to be his job..to get up and watch the dawn in the high and relatively high pasture lands to which the stock had been pushed, then to round them up and bring them down to work. 1972 J. Metcalf *Going Down Slow* vi. 108 As the hour wore on to eleven p.m. she excused herself and retired, always calling them 'night-hawks'.

nightie, nightie-night(ie): see *nighty *sb.*, *nighty-night *int.*

nightingale[1]**.** Add: **1. c.** Also *Cambridgeshire nightingale*. The edible frog, *Rana esculenta*, which was introduced into East Anglia early in the nineteenth century.
1881 *Brewer's Dict. Phr. & Fable* (ed. 12) 615/2 Cambridgeshire nightingales, edible frogs. 1975 *Country Life* 20 Feb. 455/2 The lakes, canals and meres of East Anglia became well stocked with [edible] frogs... Locals called these invaders Cambridgeshire nightingales.
4. nightingale floor, in Japan, a floor that emits a high-pitched sound when it is trodden on.
1959 S. Sitwell *Bridge of Brocade Sash* v. 113 As for the squeaking,..the Japanese..will not allow it to be a 'nightingale floor', and leave it at that, but have to remark it is so constructed that 'at every step the boards emit a sound resembling that of *uguisu*, Japanese bush warbler'. 1964 I. Fleming *You only live Twice* v. 62 This..is what the Japanese call a 'nightingale floor'... Imagine trying to get across here without being heard.

nightingale[2]**.** Add: **2.** *attrib.*, as **Nightingale ward,** a type of hospital ward designed to accommodate several patients in one room.
1964 G. L. Cohen *What's Wrong with Hospitals?* ii. 37 Long after the austere open dormitory had been abandoned abroad, Britain doggedly went on building 'Nightingale wards'. 1970 *Guardian* 10 Sept. 13/1 Cubicles instead of Nightingale wards mean that nurses escape being constantly overseen.

night letter. [f. Night *sb.* + Letter *sb.*[1]] **a.** In full **night telegraph letter,** a cheap-rate inland telegram delivered overnight. (Said in the 1945 P.O. Guide 'to be suspended'. The similar *overnight telegram* service was introduced in 1955.) Cf. *lettergram.
1912 *Post Office Guide* Apr. 94 A service of night telegraph letters is in operation between London and Aberdeen and between London and Belfast. The rate is 6*d.* for 36 words or less and ½*d.* for every 3 words beyond 36. Night Telegraph letters are delivered (except on Sunday) by the first post next morning. 1912 *Times* 29 May 3/4 Night telegraph letters may be posted prepaid at the rates given above. 1938 E. T. Crutchley *G.P.O.* vii. 146 There are various by-products of the telegraph system about which the ordinary man-in-the-street remains strangely ignorant. For instance there is the Night Telegraph Letter which provides a means of communication after the usual posting hour, and at a cheaper rate than by the ordinary telegraph service. 1940 *Post Office Guide* Aug. 413 Night telegraph letters, which are accepted at any time and delivered by first postal delivery the next weekday or by special messenger shortly afterwards. Minimum charge 1*s.* 3*d.* for 36 words. 1953 H. Robinson *Britain's Post Office* xix. 258 A successful effort was made [*sc.* in the 1930s] to give the public a knowledge of the advantages of night-telegraph letters.
b. In full **night letter telegram,** a cheap-rate overseas telegram (see quots.). (Discontinued after 1949.)
[1913 *Post Office Guide* Jan. 883 The Night and Week-end Cable Letter Services are not available viâ the Commercial or French Telegraph Cable Companies' routes... Night Cable Letters are accepted on the condition that they will not be delivered before the day after they are received at the Cable Companies' Stations.] 1914 *Ibid.* Apr. 896 (*heading*) Night and week-end letter telegrams for Canada, Newfoundland and the United States. *Ibid.*, Night Letter Telegrams are accepted on the condition that they will not be delivered before the day following that on which they are received at the Telegraph Companies' Stations. 1934 'E. M. Delafield' *Provincial Lady in America* 234 She gets such quantities of night letters and cables from abroad. 1949 *Post Office Guide* July 229 Night Letter Telegrams (NLT) may be sent to North and South America, British territories in Africa, Egypt, West Indies, certain British territories in the Far East, India, Pakistan and many other places outside Europe at one-third the rates for ordinary telegrams. They are normally delivered on the morning following the day of handing in. The minimum charge is as for 25 words. 1966 N. Freeling *King of Rainy Country* 33 The War Office in London ..promised to get him off a night letter. 1973 T. Tobin in Ade *Lett.* 1 George Ade sent the above night letter to Ashton Stevens in reply to the drama critic's request for an autobiographical sketch to insert in his *Chicago Herald*

and Examiner column, April 30, 1930. The night letter to Stevens illustrates the image Ade manifested throughout his career.

nightman. Add: **2.** (Earlier and later examples.) Also, one who works illegally at night; a burglar.

1851 H. Melville *Moby Dick* III. xlix. 293 The solitary night-man at the fore-mast-head. 1928 *Amer. Mercury* May 78/1 *Rowdy-dowdy*..was borrowed from the more aristocratic *night-men*, who use it in this manner: 'Charge on a town, make as many clouts on the *kiester* (safe) as necessary, and then battle the irate citizens in a *rowdy-dowdy* get-a-way.' 1957 M. Banton *W. Afr. City* v. 87 The people hated any type of investigator because so many of them were 'night men' (i.e. made an illegal living after dark). 1960 *Times* 29 Sept. 7/1 A company taxi is usually driven by both a day-man and a night-man.

nightmare, *sb.* Add: **2. b.** (Further examples.)

1909 *Chambers's Jrnl.* Feb. 75/2 From tip to tip of its outstretched arms this nightmare of the deep measured 56 feet. 1956 A. L. Rowse *Early Churchills* 32 A great deal of genuine learning is displayed, a nightmare of authorities cited in the Elizabethan fashion. 1975 M. Babson *There must be some Mistake* xiii. 100 'It's a nightmare,' Karen agreed. '..I wake up and it's still there.'

3. *nightmare-dreamer, -land*; *nightmare-like* adj. and adv.; *nightmare-ridden* adj.

1954 Koestler *Invisible Writing* vii. 76, I am a chronic nightmare-dreamer. 1957 E. Hyams *Into Dream* 246 For twenty-four hours he had been living in wonderland, nightmareland. 1847 J. R. Lowell *Let. from Boston* in *Pennsylvania Freeman* 1 Jan. 3/3 His words burn as with iron-searers, And nightmare-like he mounts his hearers. 1919 Wodehouse *Damsel in Distress* xv. 176 This blister had become the one great Fact in an unreal nightmare-like universe. 1926 C. Plumb in *Oxford Poetry* 36 Plagued, nightmare-ridden by a million lusts. 1961 *Times* 10 Nov. 18/7 Schoenberg's nightmare-ridden territory.

nightmarey, var. Nightmary *a.*

1851 [see Nightmary *a.*]. 1934 H. Nicolson *Diary* 24 Mar. (1966) 172, I am..very apprehensive and nightmarey.

nightmarishly, *adv.* (Further examples.)

1915 A. Bennett *These Twain* (1916) II. xv. 297 The longer Hilda regarded, the more nightmarishly numerous seemed the doors. 1934 A. Huxley *Beyond Mexique Bay* 249 Then, nightmarishly proliferating, appear the Pittsburgs and Birminghams, the Osakas and Calcuttas of this unhappy world. 1973 *Nation Rev.* (Melbourne) 31 Aug. 1453/5 Backdrops as nightmarishly accurate as the best of Goya.

night-night (nəi·t,nəi:t), *int.* Also **night, night.** [Cf. *night *sb.* 4 h.] = Good night I.

1896 G. F. Northall *Warwickshire Word-Bk.* 158 *Night-night*, good-night: spoken to children. 1905 R. Fry *Let.* 9 Jan. (1972) I. 229 Night, night, dear heart. Thanks for your cheering letter. 1945 W. de la Mare *Burning-Glass* 24 Night-night, my Precious!

night-owl. Add: **2.** A person who is up or out-of-doors late at night. orig. *U.S.*

1847 W. T. Porter *Quarter Race Kentucky* 163 You no-souled, shad-bellied, squash-headed, old night-owl you! 1880 'Mark Twain' *Tramp Abroad* 270 He calculated to be off before night-owls like me turned out of bed. 1963 M. McCarthy *Group* xii. 286 Her father, who was a night owl, was still awake. 1975 *Listener* 11 Sept. 343/3 Jazz lovers have to be night owls..for only two programmes feature early evening jazz.

nights, *adv.* Add: (Further examples.) Also *colloq.*

1786 *Exchange Advertiser* (Boston) 19 Oct. (Th.), Not a flute that has a hole in it, but that is employed very successfully nights. 1861 O. W. Norton *Army Lett.* (1903) 29 To-morrow we do guard duty. It is tiresome work. No sleep nights. 1938 T. Wilder *Our Town* 34 From my window up there I can just see your head nights when you're doing your homework over in your room. 1964 *Panorama* (Brisbane) Sept. 7/1, I don't know who writes the lyrics for their songs, but professionals can still sleep nights. 1970 *Globe & Mail* (Toronto) 25 Sept. 12/1 (Advt.), She wanted a place that would stay up nights to serve her.

night-scape, nightscape (nəi·tskē^i^p). [f. Night *sb.* + Scape *sb.*[3]] = Night-piece.

1915 T. Burke *Nights in Town* 14 There is a short street in Walworth Road..which is as perfect as any nightscape ever conceived by any artist. *Ibid.* 272 Even as a child I was conscious of the call of these wicked nightscapes. 1949 *Archit. Rev.* CV. 248 (*caption*) Nightscape, suspended animation. 1958 L. Durrell *Mountolive* xiii. 250 The youthful figures of himself and Leila moved..to the punctuation of a single soft finger-drum across a violet night-scape.

night-shift. 1. (Later examples.)

1863 Queen Victoria *Let.* 8 Apr. in R. Fulford *Dearest Mama* (1968) 192 Poor, dear Alice had the same night shift on which I had when you were all born! 1923 Conrad *Rover* v. 76 If she had a petticoat over her night-shift, that was all. 1958 *New Yorker* 13 Dec. 69/2 (Advt.), The beautiful night shift..by Barbizon in..nylon satin.

night-soil. Add: **2.** *attrib.* and *Comb.*, as *night-soil cart, collector, man.*

1967 O. Wynd *Walk Softly, Men Praying* ix. 145, I had to steer past a chain of night-soil carts. 1960 *Spectator* 19 Feb. 244 The nightsoil collectors of Lagos's slums were so thoughtless as to go on strike. 1957 M. Banton *W. Afr. City* viii. 158 Loko men are ridiculed by other tribes for working as night-soil men. 1960 C. Achebe *No Longer at Ease* ii. 16 As soon as the night-soil-man passed swinging his broom..and trailing clouds of putrefaction the boy quickly sprang to his feet and began calling him names. 1965 J. R. Hetherington *Selina's Aunt* 42/2, I then learned that he *worked* for the council—in the capacity of night-soilman.

night telegraph letter: see *night letter a.

night-time. Add: Also *attrib.*

1935 T. S. Eliot *Murder in Cathedral* ii. 76 The night-time heaping of the ashes. 1955 *Sci. News Let.* 26 Feb. 138/1 Sirius, the dog-star, brightest of all the stars in the nighttime sky.

night-times (nəi·t,təi:mz), *adv.* Chiefly *dial.* [f. Night-time + -s.] At night; during the night.

1884 'Mark Twain' *Huck. Finn* 41 Just tramp right across the country, mostly night-times. 1886 F. T. Elworthy *West Somerset Word-Bk.* 513, I goes to work, but I goes to school night-times. 1902 'L. Hope' *Garden of Kama* 18 We rested, night-times, on the sand By the rare waters of this weary land. 1940 *Sat. Even. Post* 20 Jan. 36/3 Cut the rest of his fence, nighttimes.

night-vision. Add: Also as two words. **2. a.** (Later examples.)

1946 V. Tempest *Near the Sun* vii. 53 The medical people were..confined to assisting the overcoming of..the effects of height and lack of oxygen,..the difficulties of night vision and the psychological approach to flying. 1961 *Housewife* (Ceylon) Dec. 19 Vitamin A..assists our eyesight, especially 'night vision'. 1969 'P. Kavanagh' *Such Men are Dangerous* (1971) iii. 49 He might even wait until dark. Fine. My night vision was always good. 1973 J. Rossiter *Manipulators* xxvi. 249 Ferris would have been watching his progress through his night-vision glasses.

ni·ght-walk, *v.* [f. Night *sb.* + Walk *v.*[1]] *trans.* To walk or travel across (a place) by night. Also *transf.*

1845 Thoreau *Jrnl.* 6 Aug. in *Writings* (1906) VII. 377 Fallen spirits who once in human shape night-walked the earth. 1899 Kipling *Five Nations* (1903) 8 Consider what toils we [*sc.* cruisers] endure, Night-walking wet sea-lanes, a guard and a lure.

night-walking, *vbl. sb.* Add to def.: Also, sleep-walking.

1621 [in Dict.]. 1943 A. M. Maynard in D. Ibberson *Our Towns* 135 Look for signs of bitten finger-nails, night walking or excitability.

nighty, *sb.* Add: Also **nightie.** (Earlier and later examples.) Also *attrib.* and *Comb.*

1871 'S. May' *Prudy keeping House* 98 After a nice bath..the little one was dressed in her nightie. 1894 S. Hale *Lett.* (1919) 286 A blind I opened (thereby drenching me and my nighty) banged and smashed a big pane. 1908 *Daily Chron.* 14 Sept. 5/5 So I folded up my nightie and went into the street. 1913 R. Brooke *Let.* in *Coll. Poems* (1918) p. lxxx, We may only find each other in a whiter world, nighty-clad, harped, winged, celibate. 1934 R. Nichols *Fisbo* 36 Frivolous and frolicsome and flighty As the naughtiest flapper in her newest nightie. 1968 'O. Mills' *Sundry Fell Designs* iii. 29 Vicky's got a mangy old nightie-case she won't go to bed without. 1972 'G. Black' *Bitter Tea* (1973) x. 157 The hospital nighty..felt slightly scratchy.

ni·ghty-night, *int.* Also **nightie-night(ie), nighty-nighty.** [See -y[6].] = Good-night I.

1876 A. F. Parker *Gloss. Words Oxfordshire* 118 *Nighty-nighty*, good-night, a phrase used by *very old* people. 1888 *Texas Siftings* 7 Jan. 4/1 His nibs wants yer ter fire in yer stuff ter-morrow by eleven sharp. Nighty-night, dovey. 1929 E. Bowen *Joining Charles* 193 Mrs. Moysey would say, 'Well nightie-night I suppose,' and get finally up. 1957 L. Stern *Midas Touch* ii. xviii. 136 Nighty-night, Barbara. Sleep tight. 1959 I. & P. Opie *Lore & Lang. Schoolch.* iii. 52 To one colloquially saying good-night 'Nightie, nightie', their good-humoured riposte is 'Pyjama, pyjama'. 1971 *Southerly* XXXI. 103 They kissed each other 'nightie night'.

nig-nog[1] (ni·g,nǫg). *slang.* [Perh. f. Nigmenog fool.] A foolish person; hence, a raw and unskilled recruit. Cf. *ning-nong.

1953 *Punch* 9 Dec. 692/3 All must be represented on a strict basis of proportion of the number of citizens for whom they cater: Football-pool promoters (six representatives), barrow-boys (two representatives), share-pushers, erks, nig-nogs, [etc.]. 1962 A. Wesker *Chips with Everything* I. iii. 17 A straight line, you heaving nig-nogs, a straight line. 1967 *Times* 30 Nov. 10/8 'Nig-nog' was used on the railways and elsewhere long before coloured immigrants appeared... It is usually taken as a mildly contemptuous but good-humoured name for an unskilled man or novice.

nig-nog[2] (ni·g,nǫg). [Redupl. shortened form of Nigger *sb.*] A coarsely abusive term for a Negro.

1959 M. Pugh *Chancer* 85 First lot, and look lively. Lot of nig-nogs off the trees. 1971 J. Gardner *Every Night's a Bullfight* xiii. 405 I'm talking about you and your precious Juliet, your beloved Carol bloody Evans that nig-nog tart. 1972 D. Onyeama *Nigger at Eton* x. 199 The word 'wog'..was one racial name which I always seemed to fear at Eton. Together with 'nig-nog', it was the term of abuse which..I did not, to start with, understand the meaning of. 1974 *Times* 14 Feb. 16/8 I'm not going to vote until they get me a house and get rid of the nignogs. 1975 J. Symons *Three Pipe Problem* v. 36 He wanted to send the nig nogs and the Pakis back where they belong, in the jungle.

nigra (ni·grä). *U.S.* Also **nigrah** (ni·grä). [f. Negro.] Variant form of 'Negro', used principally in the Southern States. Cf. *niggra.

1944 *Amer. Speech* XIX. 166 In the South it is commonly heard as *nigrah*, and not only from white lips. Indeed *nigrah* is also used by Northern Negroes, including some of the most eminent. 1959 *New Statesman* 6 June 800/1 In the autumn of 1956 I asked a young plantation owner in Mississippi if he had noticed any change in his relations with Negro employees since the Supreme Court rulings against segregation. 'Well,' he said slowly, 'I guess you can say the nigra ain't loyal any more.' 1965 L. Whitten *Progeny of Adder* (1966) 69 The guy has some kind of funny accent, it ain't Jewish or Italian and it ain't Nigra or Southern. 1969 F. Richards *Risky Way to Kill* (1970) xi. 140 'Pretty little thing, as nigras go, Mrs. Prender said.' '"Nigras"? Like that, Henderson?' 'Way it sounded to me.' 'It's a Southern variant,' Heimrich said. 'Between "nigger", which they're beginning—some of them are beginning—not to use so much and "Negro", which a lot of them can't get used to.' 1973 L. Heren *Growing up Poor in London* v. 109, I was with Martin Luther King Jr. on his 1961 freedom ride, and the church in which he preached in Montgomery, Alabama, was attacked. This led to a stopover of three or four days, during which several well-meaning whites, in spite of all the evidence, sought to persuade me that their niggers—or rather, nigras—were happy, well loved, and free to do whatever they wanted.

Nigritian, *a.* (Earlier examples.)

1757 J. Dyer *Fleece* iv. 128 Sailing the western coast of Afric's realms, Of Mauritania and Nigritian tracts. 1856 J. L. Wilson *Western Africa* i. ii. 30 The inhabitants of Northern Guinea are known as the Nigritian family, from their supposed descent from the great Negro families living in the valley of the Niger.

Nigritic, *a.* (Examples.)

1883 R. N. Cust *Sk. Mod. Lang. Afr.* I. iv. 54 Krapf..became aware of the existence of two distinct Groups of Languages on the East Coast, about the line of the Equator, and called them Nilo-Hamitic, or Nilotic, and Nigro-Hamitic or Nigritic respectively. 1938 M. Hailey *African Survey* iii. 76 Dr. Westermann..proposes a four-fold division: (*a*) Nigritic, or Old Sudanic. 1963 C. J. McCall in A. Dundes *Mother Wit* (1973) 420 'Hoodoo represents the syncretistic blend of Christian and Nigritic religious traditions.

nigromancer, -mancy. Delete *obs.* and add examples (quot. 1970 involves a pun on 'Negro').

1922 E. R. Eddison *Worm Ouroboros* xxxi. 388 Yet was it apparent to one so deeply learned in nigromancy and secrets astronomical that this thing was fatal. 1970 I. Reed in A. Chapman *New Black Voices* (1972) 329 He who meddles w/ nigro-mancers courts his demise.

nihilism. Add: **2.** (Further examples.)

1909 A. M. Ludovici tr. *Nietzsche's Will to Power* I. 5 Nihilism, i.e., the absolute repudiation of worth, purpose desirability. *Ibid.* 16 The extremest form of Nihilism would mean that *all* belief—all assumption of truth—is false: because no real world is at hand. 1964 P. Roubiczek *Existentialism* vii. 125 Thus it is no cause for surprise that Sartre lands himself in complete nihilism. 1972 M. Natanson in L. Embree *Life-World & Consciousness* 302 Value-free science as it has been interpreted by many contemporary social scientists does endorse a kind of *cultural nihilism*.

c. *Psychol.* In some forms of severe mental illness, the belief that the outside world, the patient's self, or parts of either, have ceased to function or to exist.

1889 *Jrnl. Mental Sci.* XXXIV. 137 Though not a mental affection in the strict sense of the word, nihilism is a psychical factor very liable to devolve into insanity. 1927 Henderson & Gillespie *Text-bk. Psychiatry* viii. 160 The most characteristic involutional qualities lie in the content of the psychosis, especially in the apprehension, hypochondriasis and nihilism. 1957 E. Mayer in P. A. Schilpp *Philos. K. Jaspers* xi. 451 Jaspers vividly describes nihilism as a symptom of mental illness or..as a manifestation of ultimate situations in which a human being can find himself in depressions and schizophrenias. 1965 J. Pollitt *Depression & its Treatment* iii. 33 The most extreme form of hypochondriacal delusion, referred to as 'nihilism'.

nihilistic, *a.* Add: **c.** *Psychol.* Of or characterized by delusions of nihilism (first described by J. Cotard *Du délire des négations* in *Archives Neurologiques* (1882) IV. xi. 152). See *nihilism 2 c.

1925 Strecker & Ebaugh *Pract. Clin. Psychiatry* iii. 45 Nihilistic ideas deny the existence of things. As an aspect of a depressive trend there may be self-depreciation. 1927 Henderson & Gillespie *Text-bk. Psychiatry* xi. 294 The feature of the depression is the frequency with which absurd nihilistic ideas are expressed. Patients claim that they are dead, that their blood has ceased to circulate,..that their bodies are utterly destroyed. 1944 Lichtenstein & Small *Handbk. Psychiatry* ii. 34 Nihilistic delusions, in which there is a more or less complete denial of

reality and existence. 1967 A. T. Beck *Depression* I. ii. 38 A typical nihilistic delusion is reflected in the following statement: 'It's no use... The world is empty. Everybody died last night.' 1974 M. Rohde tr. *Cotard's Nihilistic Delusions* in Hirsch & Shepherd *Themes & Variations European Psychiatry* xiv. 354, I would tentatively suggest the name 'nihilistic delusions' (*délire de négations*) to describe the condition of the patients to whom Griesinger was referring.

‖ **nihil obstat** (ni·hil ǫ·bstæt). [Lat., 'nothing stands in the way'.] Words appearing on the title-page or elsewhere in the preliminary pages of a Roman Catholic work indicating that it has been approved as free of doctrinal or moral error. Also *fig.*

1886 in P. Soulier *Life St. Philip Benizi* p. iv, Nihil obstat: Guglielmus T. Gordon, *Congr. Orat. Presbyter.* 1932 J. L. Stocks in *Hibbert Jrnl.* XXX. 622 He loves beauty, he admires character, he feels the thrill of poetry and art. He believes in God. For all this the utmost that he can get out of science is a *nihil obstat*. 1933 *Times Lit. Suppl.* 21 Dec. 904/2 The selection [of anthems] carries with it the *nihil obstat* of two such eminent Church musicians as Sir Walford Davies and Dr. Henry Ley. 1938 *Mind* XLVII. 93 The parallel to the *Nihil Obstat* of Roman Catholic censorship is obvious. 1955 *Times* 6 July 11/4 Authoritarians wish that one dictionary enjoyed dictatorial rights. Reference to it then would finally close any argument; its *nihil obstat* would give the green light to *imprimatur*. 1958 *Spectator* 22 Aug. 260/1 This is a lucid, judicious book, with a '*Nihil obstat*' discreetly tucked away in the title pages. 1973 *Times* 11 Aug. 2/8 Mgr. Guazzelli said: 'The *nihil obstat* and the imprimatur were duly signed, and the censor and myself, as the responsible bishop, acknowledge this fact.'

-nik (nik), *suff.* from Russian (cf. *kolkhoznik, *Narodnik, *sputnik) and Yiddish, appended to sbs. and adjs. to denote a person or thing involved in or associated with the thing or quality specified, as **beatnik, *folknik, *no-goodnik*, *nudnik, **peacenik*. Often with humorous or pejorative connotations.

1945 A. Kober *Parm Me* 17 That stuck-upnick fomm the lodge, Sister Leshinsky..she's a regella Yenkee. 1958 *Amer. Speech* XXXIII. 154 On learning that a dog was in the Soviet Moon, the Detroit *News* (and almost every other paper)..referred to the satellite as *Muttnik*... From then on there was no end of -niks. 1959 *Observer* 14 June 22/7 It happened that Mr. Werth arrived in Columbus, Ohio, just as the Russian Sputnik soared into the cosmos; before he left the American flopnik had burnt out on its launching pad. 1965 *Newsweek* 1 Nov. 31/3 The crowded headquarters of the young draftniks and Vietniks pulse with an almost religious fervor. 1965 *Time* 12 Nov. 4 Those guitar-plunking protestniks whose St. Joan is Baez. 1966 *Economist* 5 Mar. 883/1 These protestants represent only a small faction, no more important politically than the nuclear disarmers were in Britain or the Vietniks are in America. 1966 *Sat. Rev.* (U.S.) 22 Oct. 59 Despite the alarums of the computerniks..the book would appear to be here to stay. 1968 B. Foster *Changing Eng. Lang.* ii. 110 This borrowing [*sc.* sputnik]..has given a new lease of life to the suffix *-nik* which had already made its appearance, at any rate in the U.S.A., as a loan from Yiddish... New creations..have usually been..humorous..; thus a device which failed to go into orbit was derided in..1957 as a Kaputnik (*Daily Express*), a Flopnik (*Daily Herald*), a Puffnik (*Daily Mail*), and a Stayputnik (*News Chronicle*). 1968 L. Rosten *Joys of Yiddish* 265 *-Nik* lends itself to delightful *ad hoc* inventions. A *sicknik* would be one who fancies 'sick' or 'black' humor. A *Freudnik* would be an uncritical acolyte of the father of psychoanalysis. And recently homosexuals began to refer to heterosexuals, with some amusement, as 'straightniks'. 1973 *Indexer* VIII. 227/2 Publishers and computerniks can create decadent search systems.

nikau (nī·kau). [Maori.] A New Zealand palm tree, *Rhopalostylis sapida*, or its leaves, once used as a sort of thatch. Also *attrib.*

1831 H. Williams *Jrnl.* 15 Nov. in H. Carleton *Life of Henry Williams* (1874) I. 100 We..erected a tent, with a shed of nikau for the boys. 1844 C. Chapman *Let.* 30 Nov. in A. Drummond *Married & gone to N.Z.* (1960) 70 Another tree called the Nikau is said to be very good; I have tasted a little bit and found it very nice. 1847 *N.Z. Jrnl.* No. 191. 105/1 From Teapu southward the country is..wooded with rata and nikau, or cabbage palm, down to the high water mark. 1853 J. M. Richmond *Let.* 11 Nov. in *Richmond-Atkinson Papers* (1960) I. iii. 133 It [*sc.* their dwelling] is in fact a roof on the ground, thatched with nikau, a palm, the only one in N.Z. 1891 R. Wallace *Rural Econ. Austral. & N.Z.* xvi. 242 The tree fern and the common Nikau Palm of the country dotted the hill-sides and gave..a semi-tropical appearance to the landscape. 1905 W. B. *Where White Man Treads* 16 He could supplement his dry fare with..the tender leaves of the nikau. 1920 J. Mander *Story N.Z. River* 315 He built a nikau whare. 1935 J. Guthrie *Little Country* iv. 98 Nikaus, rata vine, and lycopodium..brought to the hall the charms of the New Zealand bush. 1949 P. Buck *Coming of Maori* (1950) II. i. 87 The middle part of the top of the trunk of the *nikau* palm, just under the leaf spread, was eaten raw. 1959 M. Shadbolt *New Zealanders* 66 He explained how easy it was to live off the bush if you wanted, eating..the juicy white insides of nikau palms. 1966 *Encycl. N.Z.* II. 690/1 The Maoris used the nikau leaves in their whares.

Nike (nəi·kī). [Gr. νίκη victory.] **1.** In Greek art: a winged statue representing Nike, the goddess of victory.

1867 H. M. Westropp *Handbk. Archæol.* 195 *Nike.—Victory.* Victory is represented in a short tunic, with wings, and usually carries a palm. She is also represented writing on a shield, and frequently sacrificing a bull. 1924 A. D. Sedgwick *Little French Girl* iv. iii. 327 She's a Nike..on the prow of a Greek ship. 1960 R. Carpenter *Greek Sculpture* v. 147 The running Nike is an interesting study in formal drapery. 1968 V. Ehrenberg *From Solon to Socrates* vii. 315 There are the figures of a number of winged *Nikai*.

2. Any of a range of surface-to-air guided missiles developed by the U.S., initially as defensive weapons.

1952 *Britannica Bk. of Year* 431/2 Limited production for operational trials of the U.S. army's Nike missile for the anti-aircraft artillery was begun in 1951. 1955 *Times* 4 July 9/7 Short-range guided weapons such as the American Nike are useless for the air defence of Britain. 1961 E. Burgess *Long-Range Ballistic Missiles* vii. 194 On a recent tour of a Nike-Zeus installation the writer saw..the tremendous complexity of the anti-missile missile system. 1970 J. W. R. Taylor *Rockets & Missiles* ii. 56 Typical of first-generation anti-aircraft missiles was America's two-stage Nike-Ajax, made up of a liquid-propellent second stage and a solid-propellent first-stage booster. 1974 *Encycl. Brit. Micropædia* VII. 344/1 Nike Zeus, first antimissile missile, was about 50 feet long... A further development was Nike X, which had a fixed radar antenna that could be electronically scanned. Nike Cajun, a sounding rocket, was capable of lifting a 50-pound payload of scientific instruments to a height of 90 miles.

nikethamide (nike·þămәid). *Pharm.* [f. *ni*cotinic acid di*ethylamide*, its chemical name, by alteration.] A colourless or yellowish oily liquid or crystalline solid, $(C_5H_4N)CON(CH_2CH_3)_2$, which is used as a respiratory stimulant and analeptic.

1940 *Jrnl. Amer. Med. Assoc.* 20 Jan. 249/2 The Council consulted Ciba Pharmaceutical Products, Inc., and agreement was reached on the name 'Nikethamide'..as the nonproprietary name for the substance introduced in medicine under the proprietary name of Coramine. 1951 A. Grollman *Pharmacol. & Therapeutics* viii. 168 Because of its analeptic action, nikethamide is used in acute respiratory depression of anesthesia, alcoholic intoxication, and overdosage with hypnotics. 1969 [see *Coramine].

Nikonian (nikō^u^·niăn), *sb.* and *a.* Also **Niconian.** [f. *Nikon*, the name adopted by Nikita Minin (1605–81), 6th Patriarch of Moscow + -ian.] **A.** *sb.* An orthodox member of the Russian Church who accepted the reforms introduced by Nikon. **B.** *adj.* Of or pertaining to Nikon or his reforms. Hence **Niko·nianism.**

1874 J. H. Blunt *Dict. Sects, Heresies* 373/1 *Niconians*, a name applied by Russian dissenters to the orthodox members of the Established Church who accepted the reforms introduced by the Patriarch Nicon in the year 1654. 1877 D. M. Wallace *Russia* xx. 308 Believing these 'Nikonian novelties' to be heretical. 1888 S. M. Kravchinskii *Russ. Peasantry* II. 454 They [*sc.* the Stranniky] look upon their co-religionists..with the same disgust and abhorrence as they lavish on the Niconians. 1957 *Encycl. Brit.* XIX. 695/2 A long struggle began between the faith of the Old Believers and the Nikonianism, as the official church was now generally called.

nil[2]. Add: **1. a.** (Later examples.)

1936 A. W. Clapham *Romanesque Archit.* i. 5 Of all the various structures referred to or described by Isidore of Seville..the surviving remains are almost nil. 1952 C. Day Lewis tr. *Virgil's Aeneid* viii. 172 Visibility was reduced to nil. 1955 *Times* 6 July 4/4 The British Isles Rugby Union tourists beat South West Africa by nine points to nil in their match here to-day.

2. *Comb.*, as **nil-grade** *Philol.*, the most reduced form of weak ablaut grade, in which the vowel disappears; **nil norm** *Econ.*, during a wage-freeze a standard under which no wage increases are normally allowed.

1922–3 *Modern Philology* XX. 197 The nil-grade of the root vowel is represented by OE *-tyllan* (in *fortyllan* 'to seduce'). 1928 *Language* IV. 163 It follows that *šd* is the nil-grade of ***sed-* 'sit'. 1935 *Mod. Lang. Notes* L. 523 Pre-English uncontracted *Lāamkō* (with nil-grade of the *k*-suffix). 1972 *Language* XLVIII. 9 The difficulty disappears if we start from a root *k^w^erp*, whose nil-grade developed into Aryan *kṛp-* and Goidelic *kri*. 1966 *Sunday Times* 14 Aug. 8 For all the traditional wage demands some principle does need to be hammered out to establish who will be permitted to breach the nil norm. 1967 *Economist* 23 Dec. 1204/1 Britain is still supposed to be operating in conditions of a 'nil norm' for wage increases. 1968 *Manch. Guardian Weekly* 1 Feb. 8 The £1 a week rise would not qualify as an exception to the nil norm laid down in the summer.

b. Passing into *adj.*, containing, reporting, or consisting of nothing.

1959 *Punch* 27 May 704/3 The thoughtless might suppose that hail-watching was dull work, with long blank intervals offering nothing better than a nil return. 1972 *Guardian* 14 Mar. 2/8 They..returned because of nil visibility. 1974 L. Lamb *Man in Mist* ix. 55 In view of my constable's nil report, *could* it be that nothing ever happened?

‖ **nil admirari** (nil ædmirā·ri). [L.] An attitude of indifference to the distractions of the outside world, advocated by Horace (*Epistles* I. vi. 1): *nil admirari prope res est una..quae possit facere et servare beatum*, 'to wonder at nothing is just about the only way a man can become contented and remain so.' Hence, a person adopting this attitude. Also *attrib.* or as *adj.*

Also used erron. with the meaning 'to admire nothing'.

1748 Chesterfield *Let.* 27 Sept. (1774) I. 345 This book..will both divert and astonish you; and at the same time, teach you *nil admirari*. 1749 Fielding *Tom Jones* III. vii. i. 8 The famous *Nil admirari* of Horace, or in the English Phrase, *To stare at nothing*. 1785 Boswell *Jrnl. Tour Hebr.* 118 But Dr. Johnson has much of the *nil admirari* in smaller concerns. 1821 Byron *Don Juan* v. c. 185, I ne'er could see the very Great happiness of the 'Nil Admirari'. 1822 Shelley *Let.* 18 June (1964) II. 715 The *nil admirari*..seems to me a bad sign in a young person. 1857 Trollope *Barchester T.* III. xvi. 277 Very many men now-a-days..adopt or affect to adopt the *nil admirari* doctrine. 1866 Mrs. Gaskell *Wives & Daughters* I. xxii. 287 Every inflexion of the voice breathed out..admiration! And this from the *nil admirari* brother. 1935 J. Buchan *House of Four Winds* vi. 139 Mr Glynde's *nil admirari* countenance..registered surprise. 1951 N. Annan *Leslie Stephen* x. 278 *Nil admirari* was his precept, the salute to genius his practice. 1961 M. Kelly *Spoilt Kill* iii. 193 Why was I always preaching *nil admirari*?

‖ **nil desperandum** (nil despəræ·ndŏm), *int.* Also **ni:hil despera·ndum.** [L.] An exhortation to have hope in difficult circumstances and not to despair, deriving from Horace *Odes* I. vii. 27 *nil desperandum Teucro duce et auspice Teucro*, 'no need to despair with Teucer as your leader and Teucer to protect you'.

[1617 T. Adams *Souldiers Honour* sig. A3v, Nil desperandū Christo Duce, & Auspice Christo.] 1628 R. Burton *Anat. Melancholy* (ed. 3) III. ii. 527 *Nihil desperandum*, there s hope enough yet. 1749 Fielding *Tom Jones* III. viii. vi. 189 All I have is at your Service, and at your Disposal... Nil desperandum est Teucro duce et auspice Teucro. 1774 J. Adams *Wks.* (1851) IV. 12 *Nil desperandum* is a good motto. 1872 E. Braddon *Life in India* iii. 75 But *nil desperandum* was the cry of the Vauxhall partisans. 1921 W. de la Mare *Memoirs of Midget* xx. 142 *Nil desperandum*, Mr. Crimble. And you know what they say about fish in the sea. 1955 *Times* 5 May 15/4 There is..a key-note running through the essays and magazine articles here reprinted—'a note of *nil desperandum*'. 1974 J. Mann *Sticking Place* iii. 46 Edward had thought the job his for the asking... Well, *nil desperandum*..he might yet be moving in.

Nile (nəil). The name of a river in North Africa, used *attrib.* to designate animals native to the region, as **Nile crocodile**, the African crocodile, *Crocodylus niloticus*; cf. Crocodile 1; **Nile monitor** = *iguana 2; **Nile perch**, a large food fish, *Lates niloticus*, found in the rivers and lakes of north and central Africa.

[1860 T. H. Huxley in *Jrnl. Linn. Soc.* (*Zool.*) IV. 8, I could find neither in the British Museum, nor in the Museum of the Royal College of Surgeons, any authentic skeleton of this, the so-called Nilotic Crocodile.] 1898 J. Anderson *Zool. Egypt* I. 1 The Nile crocodile was described, so long ago as 1699, by Oligeus Jacobæus. 1915 A. M. Reese *Alligator & its Allies* i. 39 The African or Nile Crocodile..is found throughout the continent of Africa. 1974 *BP Shield Internat.* Oct. 28/4 Nile crocodiles are a prominent feature of the lake fauna. 1900 H. A. Bryden *Animals of Africa* xv. 173 The Nile Monitor, a big lizard, attaining a length of six feet,..is often miscalled an 'iguana'. 1964 C. Willock *Enormous Zoo* vii. 125 The Nile monitor, an immense and really powerful lizard which sometimes reaches four feet in length, has a banquet when crocodile hatching time comes around. [1905 A. Sedgwick *Student's Text-bk. Zool.* II. 235 *Lates* C. and V., *L. niloticus*, the perch of the Nile.] 1931 J. R. Norman *Hist. Fishes* ix. 162 (*caption*) Skeleton of the Nile Perch. 1949 L. J. McCormick in Vesey-Fitzgerald & Lamonte *Gamefish of World* 367 In colour and form the Nile perch looks like an overgrown cod. 1965 A. J. McClane *Standard Fishing Encycl.* 616/2 Various species of the Nile perch are distributed in the big rivers and lakes of tropical Africa. 1974 *Encycl. Brit. Macropædia* XIV. 46/2 The nile perches (Latidae) have been found as mummies in ancient tombs in Egypt.

Nile-blue, *sb.* and *a.* Add: (Earlier and later examples.) Also *ellipt.* **Nile.**

The ellipt. examples may represent *Nile-green*.

1873 *Young Englishwoman* Feb. 77/2 Dark maroon velvet and Nile blue silk rep. 1895 *Montgomery Ward Catal.* 7/3 Plain Crepe Picardie... Colors: Lilac,..nile, black [etc.]. 1926 *Daily Colonist* (Victoria, B.C.) 21 July 16/6 (Advt.), A 4-ply worsted wool in shades of pink..nile, [etc.]. 1950 Maerz & Paul *Dict. Color* (ed. 2) 199/1 Nile Blue.

2. Also **Nile blue.** A tetracyclic quaternary ammonium ion which is an azine dye and is used as a biological stain to colour fatty acids blue; also, a salt (usu. the sulphate) of this ion.

1888 *Jrnl. Soc. Dyers & Colourists* IV. 96/1 On wool and silk, Nile blue is applied in a neutral bath, and yields red shades of blue. 1920 *Physiol. Abstr.* VI. 617 The affinity of Nile blue sulphate for fat is small. Oleic acid and its esters are the most stainable. 1942 V. Hamburger *Man. Exper. Biol.* 46 A small particle of agar stained with Nile blue sulphate or neutral red is pressed against the surface of the blastula for a short period. 1971 Ayyangar & Tilak in K. Venkataraman *Chem. Synthetic Dyes* IV. iii. 107 It has been shown that Rhodanile Blue is nothing but a mixture of Rhodamine B and Nile Blue.

Nile-green (earlier and later examples).

1871 *Scribner's Monthly* June 209/1 'Nile green' will turn some people into oranges, though twenty empresses ordain its adoption. 1944 R. Chandler *Lady in Lake* xvi. 93 The nile green tiles of the bathroom floor. 1969 'I. Drummond' *Man with Tiny Head* xv. 176 He used his palms against the nile-green hull. 1973 'D. Jordan' (*title*) Nile green.

Nili (nī·li). [Acronym f. Heb. *Netzach Israel lo Ishakare*, the strength of Israel will not lie (1 Sam. xv. 29).] A Jewish espionage group in Turkish-ruled Palestine during the 1914–18 war.

1930 H. B. Samuel *Unholy Memories of Holy Land* viii. 119 Aronson..leapt to prominence in the war by organizing the celebrated Nili Society..whose 'intelligence' was largely instrumental in the success of Lord Allenby's campaign. 1938 W. B. Ziff *Rape of Palestine* vi. 66 Of..great importance was the voluntary intelligence service rendered by the celebrated Nili Society all over the Holy Land. 1959 A. Engle *Nili Spies* ix. 99 Nili became the name of the espionage group, adding a new word to the Hebrew language. *Ibid.* xi. 113 The actual details of Nili's work were never known... But the existence of a group of people carrying on espionage and plotting rebellion could not be concealed. 1973 *Jewish Chron.* 2 Feb. 12/4 The Nili Group..sprang up during the First World War, when Turkish treatment of the Palestine Jewish community raised questions of its survival.

Nilo-. Used as combining form of *Nile* in names of language groups common to inhabitants of the Nile area and of some other specified area. Cf. next.

1932 W. L. Graff *Lang.* 434 *Kanuri*, of the Nilo-Chadian group. 1938 [see *Nilo-Hamitic *a.*]. 1939 L. H. Gray *Foundations of Language* 401 The Nobades (the supposed ancestors of the modern Nilo-Chadian Nuba). *Ibid.* 402 The divisions of *Sudano-Guinean*, according to Delafosse, may now be enumerated:..(2) *Nilo-Abyssinian* (fifteen languages) with evident traces of classes and class-pronouns and tones, and including Shiluk, Dinka, etc. *Ibid.*, (5) *Nilo-Congolese* (nineteen languages) with traces of classes, and including Mangbetu, Mbuda, etc. 1966 J. H. Greenberg *Languages of Africa* (ed. 2) 130 To the new grouping which consists of Songhai, Saharan, Maban, Fur and Coman in addition to Chari-Nile, the name Nilo-Saharan is given. 1969 *Language* XLV. 665 Of the 33 languages and language groups surveyed in *Hdbk* 66, 21 fall into Greenberg's Nilo-Saharan family.

Nilo-Hamite (nəi:lo,hæ·məit). [f. *Nilo-Hamitic *a.*: see -ite[1].] A member of the Nilo-Hamitic group of peoples.

1932 C. G. & B. Z. Seligman *Pagan Tribes Nilotic Sudan* i. 5 There can be no doubt as to the persistent influence, physical and cultural, of the Hamites on the Nilotes and Nilo-Hamites. 1953 G. W. B. Huntingford *Southern Nilo-Hamites* 9 The majority of the Southern Nilo-Hamites are pastoralists practising..a little agriculture. 1955 P. H. Gulliver *Family Herds* iii. 55 Another tribe of the northern Nilo-Hamites, the Lotuko, makes the same distinction [of kinship terms]. 1966 C. G. Seligman *Races of Africa* (ed. 4) vii. 102 The Nilo-Hamites..speak Nilotic languages with Hamitic elements, and are predominantly pastoral Negroids. 1974 J. R. Baker *Race* xiii. 226 The Aethiopids have hybridized secondarily with Nilotid Negrids to give rise to tribes referred to under the general title of Niloto or Nilo-Hamites (or sometimes Half-Hamites).

Nilo-Hamitic (nəi:lo,hæ̆mi·tik), *a.* Also **Niloto-Hamitic, Nilohamitic.** [f. *Nilo- + Hamitic *a.*] **a.** Used (originally by German philologists) to designate the groups of languages spoken by East African peoples of mixed Negro and Hamitic descent. **b.** Of, pertaining to, or designating any one or all of the peoples who speak a language belonging to this group.

1883 R. N. Cust *Sk. Mod. Lang. Afr.* I. iv. 54 Krapf..became aware of the existence of two distinct Groups of Languages on the East Coast, about the line of the Equator, and called them Nilo-Hamitic, or Nilotic and Nigro-Hamitic or Nigritic respectively. 1920 G. W. Murray in *Jrnl. R. Anthrop. Inst.* L. 328 The Hamito-Semitic influence on these first three languages is so strong that we may certainly call them Niloto-Hamitic with Westermann. 1938 W. M. Hailey *Afr. Survey* iii. 77 The Nilotic..comprises tongues spoken in the Upper Nile region as far south as Lake Victoria. Two main branches are distinguished, Nilo-Sudanic..and Nilo-Hamitic. 1955 J. H. Greenberg *Studies in African Linguistic Classification* 66 The term Nilo-Hamitic has been used by different writers with widely varying meanings. 1955 P. H. Gulliver *Family Herds* i. 1 In East Africa..it so happens that many of the pastoral peoples are of the Nilo-Hamitic strain, which tends to give them a more striking physique than their Bantu neighbours. 1957 C. G. Seligman *Races of Africa* (ed. 3) vii. 161 Their kinship is rather with the Nilo-Hamitic tribes of East Africa. 1967 M. J. Coe *Ecol. Alpine Zone Mt. Kenya* 1 For centuries the peak of Mount Kenya has had a magical and religious significance for the Bantu and Nilohamitic peoples.

Nilome·tric, *a.* [f. Nilometer + -ic.] Of or pertaining to a nilometer, or the measurement of the height of the Nile.

1921 G. A. F. Knight *Nile & Jordan* ix. 96 The Second Cataract where Nilometric markings with his cartouche are recorded.

Nilot. Add: Also **Nilote.**

1932 C. G. & B. Z. Seligman *Pagan Tribes Nilotic Sudan* i. 13 The Nilotes..are essentially proud, aloof, tenacious of their old beliefs and ideas, intensely religious, and by far the most introvert of the peoples of the Sudan. 1934 D. Westermann *African Today* ii. 24 The Masai and tribes related to them..show strong Hamitic influence but are on the other hand relatives of the Nilotes. 1966 C. G. Seligman *Races of Africa* (ed. 4) vii. 110 The second great group of hamiticized Negroes are the Nilotes..geographically limited to the Nile Valley or its immediate neighbourhood.

Nilotic, *a.* Add: **2.** Of, pertaining to, or designating the group of East African Negro tribes including the Dinka, Luo, Nuer, and Shilluk, or the sub-group of Sudanic languages spoken by them.

1915 A. Werner *Lang. Families of Africa* iii. 41 These..may perhaps be grouped together as a 'Nilotic' sub-division of the Sudan family. 1930 C. G. Seligman *Races of Africa* vii. 173 The Acholi and Belanda no doubt are of Nilotic origin. 1932 C. G. & B. Z. Seligman *Pagan Tribes Nilotic Sudan* p. xxiii, In the Nilotic languages the *p* and *f* approach each other and seem to be interchangeable. 1949 tr. *L. Homburger's Negro-Afr. Lang.* ii. 42 Central Nilotic languages have no *s*, *z*; in them Arab *suk* becomes *šuk*. 1966 J. H. Greenberg *Lang. of Africa* (ed. 2) 85 The closest relatives of these languages [*sc.* the Nilo-Hamitic languages] are the group of languages traditionally called Nilotic... The term Nilotic may be extended to include both the Nilotic and Nilo-Hamitic languages. 1973 *Sci. Amer.* July 74/2 The Sebei, a Southern Nilotic people who live on the northern slopes of Mount Elgon in Uganda. 1973 A. Mann *Tiara* xvii. 158 The indefinite tyranny of the Northern Arabs over the Nilotic and Nilo-Hamitic Southerners.

Nilotic (nəilǫ·tik), *sb.* **a.** = Nilot. **b.** A subdivision of the Sudanic group of languages.

1924 *Chambers's Jrnl.* Jan. 11/2 The Nilotics are in nearly every case the aggressors. 1938 W. M. Hailey *Afr. Survey* iii. 77 The third family included in the Negro division—namely Nilotic—comprises tongues spoken in the Upper Nile region as far south as Lake Victoria. 1963 *Times* 4 June 14/6 Hieroglyphics, and Demotics, and Nilotics, and Cryptics, and Cufics.

nilpotent (nilpō^u^·tĕnt), *a.* *Math.* [f. Nil[2] + L. *potent-*, *potens* powerful, Potent *a.*[1]] Becoming zero when raised to some positive integral power (see also quot. 1949).

1870 B. Peirce in *Amer. Jrnl. Math.* (1881) IV. 104 When an expression raised to the square or any higher power vanishes, it may be called nilpotent. 1937 A. A. Albert *Mod. Higher Algebra* (1938) iv. 87 Two nilpotent matrices of index two are similar in **F** if and only if they have the same rank. 1949 S. Kravetz tr. *Zassenhaus's Theory of Groups* iv. 111 A group **G** is said to be nilpotent if the ascending central series contains the whole group as a member. 1971 I. T. Adamson *Rings, Modules & Algebras* xxiii. 194 We say that *I* is a nilpotent ideal if there exists a positive integer *n* such that I^n is the zero ideal. 1974 T. W. Hungerford *Algebra* vii. 100 We obtain a sequence of normal subgroups of *G*, called the ascending central series of *G*: $\langle e\rangle < C_1(G) < C_2(G) < \ldots$ A group *G* is nilpotent if $C_n(G) = G$ for some *n*. *Ibid.* 102 A group *G* is said to be solvable if $G^{(n)} = \langle e\rangle$ for some *n*... Every nilpotent group is solvable.

nim (nim), *sb.*[2] Also **Nim.** [Orig. uncertain: perh. suggested by Nim *v.* or G. *nimm* (imp. of *nehmen* to take).] A game in which two players alternately take one or more objects from any one of several heaps, the aim being to compel one's opponent to take the last remaining object (or, sometimes, to take it oneself).

1901 C. L. Bouton in *Ann. Math.* III. 35 Nim. A game with a complete mathematical theory... The writer has not been able to discover much concerning its history, although certain forms of it seem to be played at a number of American colleges, and at some of the American fairs. It has been called Fan-Tan, but as it is not the Chinese game of that name, the name in the title is proposed for it. 1939 Uspensky & Heaslet *Elem. Number Theory* i. 16 It should be interesting to present an application of the binary system..to the theory of the game of Nim. 1955 *Sci. News Let.* 26 Feb. 134/2 A 17-year-old senior at Newton High School..can boast of having an electronic player that almost never loses in the ancient game of wits, known as 'nim'. 1968 Corlett & Tinsley *Pract. Programming* iv. 66 In a game of Nim..two players move alternately and take any number of matches from one pile, the winner taking the last match. If a player can set up a winning position, he cannot lose unless he makes a mistake in a subsequent move.

nim, var. Neem (in Dict. and Suppl.).

nimble, *a.* (and *adv.*). Add: **7. nimble-pimble** *v. intr.* (*nonce?*), to behave in a sentimental or trifling manner (toward); **nimble Will,** substitute for def.: a slender grass, *Muhlenbergia schreberi*, found in the central United States and sometimes used for pasture.

1927 D. H. Lawrence *Let.* 6 Feb. in E. & A. Brewster *D. H. Lawrence: Reminisc. & Corr.* (1934) 115, I feel an infinite disgust at the idea of having to be there while the fools nimble-pimble at the dialogue. 1816 D. Thomas *Jrnl.* 10 July in *Travels through Western Country* (1819) 168 He pointed out to me a grass, of which I had heard much, known through all the western country by the name of nimble Will. 1817 S. R. Brown *Western Gazetteer* 109 This is the short, nutritious grass called 'nimble will', which has completely overspred with astonishing celerity, almost every spot of waste or uncultivated ground. 1865 *Trans. Illinois Agric. Soc.* V. 863 The *Muhlenbergia diffusa*, or Nimble Will, is a common grass, which is rather known as a troublesome weed. 1894 J. M. Coulter *Bot. W. Texas* III. 523 Nimble Will... Dry hills and woods, northern Texas and northward.

8. *nimble-throated* adj.; *nimble-tongued, -winged* (later examples).

1930 Blunden *Poems* 310 The girls are quicker, more nimble-throated. 1951 *Essays in Criticism* I. ii. 165 She [*sc.* Mrs. Behn]..is nimble-tongued indeed at the expense of a military fop who was ready to damn her play. 1960 S. Plath *Colossus* (1967) 37 Each thumb-size bird Flits nimble-winged in thickets.

nimble, *v.* **3.** (Later example.)

1938 E. Bowen *Death of Heart* ii. ii. 191 She nimbled in with the tray.

nimble-witted, *a.* (Later example.)

1922 V. Woolf *Jacob's Room* iv. 86 The talkative, nimble-witted people have taken themselves to towns.

nimbostratus (nimbostrā·tŏs, -strē^i^·tŏs). *Meteorol.* [mod.L., f. Nimb(us + -o + Stratus.] A form of cloud, which usually occurs as a thick, low, extensive layer, which is grey and often dark, and from which rain, sleet, or snow falls (not necessarily reaching the ground) unaccompanied by lightning, hail, or thunder.

Quot. 1887 represents a different sense. The present use appears to have originated independently in 1932, and quot. 1909 is difficult to account for.

1887 R. Abercromby *Weather* iii. 112 Weilbach..gives three varieties [of nimbus]—..and nimbo-stratus, the rain-cloud, in rear of cyclones, which we have designated cumulo-nimbus [i.e. a 'rocky cumulus cloud from which rain falls in squalls or showers']. 1909 *Cent. Dict.* Suppl., *Nimbo-stratus*, same as *nimbo-pallium* [*sc.* 'a broad sheet of cloud from which rain is falling']. 1932 *Internat. Atlas Clouds* (Internat. Meteorol. Comm.) (abridged ed.) I. 14 In the present Atlas it was intended to give the cloud (*a*) the new name of Nimbostratus, which is a better name than nimbus for a continuous layer which is formed by evolution from altostratus. 1940 W. J. Humphreys *Physics of Air* (ed. 3) 295 The nimbostratus, formerly called nimbus, is any thick, extensive layer of formless cloud from which rain or snow is falling or seemingly on the point of falling. 1957 J. I. M. Stewart *Use of Riches* I. ii. 23 Charles pointed to the horizon. 'Nimbostratus. There's going to be rain.' 1967 R. W. Fairbridge *Encycl. Atmospheric Sci. & Astrogeol.* 687/1 Nimbostratus generally grows out of altostratus, thickening downward.

‖ **nimbu-pani** (ni·mbu,pā·ni). Also **nimboo-pani, nimbo-pani.** [Hindi; cf. Punjabi *nimbū* lemon, lime, Pali *pāna-* drink.] An Indian non-alcoholic drink comprising lemon-juice or lime-juice with sugar and ice or water.

1961 Mrs. B. Singh *Indian Cookery* 184 Nimboo pani or fresh lime juice and a little sugar dissolved in water is a very popular drink in India during the hot weather. 1968 P. Lal *Indian Recipes* 259 Rose-flavoured nimbu pani... Squeeze the juice out of the lemons and add the sugar... Add..iced water and the rosewater. Pour into serving glasses and serve with crushed ice. 1971 *Illustr. Weekly India* 4 Apr. 6/1 Portugal does not believe in *nimbu-pani*. *Ibid.* 25 Apr. 8/3 Nuzzling into a glass of nimboo-pani placed by his side. 1973 'B. Mather' *Snowline* xvi. 197 Nimbo-pani..is the juice of fresh limes squeezed on to cracked ice and sugar.

nimby: see *nembie.

niminy-piminy, *a.* (Earlier examples.)

1786 J. Burgoyne *Heiress* iii. ii. 55 *Lady Emily*... You have only, when before your glass, to keep pronouncing to yourself nimini-primini. *Miss Alscrip.* Nimini-pimini-imini, mimini—oh, it's delightfully enfantine. 1786 G. White *Let.* 25 Mar. in R. Holt-White *Life & Lett. G. White* (1901) II. 154, I hope you practice every day at your Glass; and that you are by this time perfect mistress of '*Nimini pimini*'.

nimite (ni·məit). *Min.* [f. *N*ational *I*nstitute for *M*etallurgy + -ite[1].] A basic silicate, aluminate, and oxide of nickel, magnesium, iron, and aluminium (with nickel as the dominant cation), $(Ni,Mg,Fe,Al)_3(Si,Al)_2O_5(OH)_4$, which belongs to the chlorite group and is found as yellowish-green monoclinic crystals near the Scotia talc mine in Transvaal.

1968 Hiemstra & De Waal in *S. Afr. Nat. Inst. Metall. Res. Rep.* No. 344 (*title*) Nickel minerals from Barberton. II. Nimite, a nickelian chlorite. 1970 *Amer. Mineralogist* LV. 21 The exothermic reaction..at 921°C most probably is due to recrystallization of the nimite to a substance that by X-ray diffraction analysis proved to be mainly spinel.

nim-nosed (ni·m,nō^u^zd), *a.* *rare*[−1]. [f. *nim* (*dial.* shortening of Nimble *a.*) + Nosed *a.*] Quick-nosed, swift to pick up the scent.

1936 J. Masefield *Let. from Pontus* 68 The keeper with his nim-nosed dog.

Nimonic (nimǫ·nik). Also **nimonic.** A proprietary name of various nickel-based alloys similar to those known by the name *Inconel.

1941 *Trade Marks Jrnl.* 9 Apr. 136/2 *Nimonic*... Cast and wrought alloys of nickel sold in the form of bars, sheets, rods,..and other shaped pieces. Henry Wiggin & Company Limited,..London,..manufacturers. 1947 *Official Gazette* (U.S. Patent Office) 22 July 529/2 Henry Wiggin & Company Limited, London, England..*Nimonic*. 1947 A. W. Judge *Mod. Gas Turbines* iv. 79, 1270°C (abs.)..is about the limiting value for the turbine blades when made in Nimonic 80 alloy. 1957 *Technology* Aug. 213/1 Ordinary high-temperature alloys, like the nimonics, have reached the limits of development. 1966 [see *Inconel]. 1968 [see *Nichrome]. 1968 D. R. Cliffe *Technical Metall.* xvii. 357 Nimonic 75 contains approximately 80:20 Ni:Cr, with 0·2 to 0·6% Ti and up to 0·15% C. Nimonic 80 and 80A have, in addition, 0·5–1·8% Al, and Nimonic 90 and 95 have 15–21% Co replacing nickel. 1971 *New Scientist* 25 Mar. 667/1 Scientists within the UKAEA had been studying a selection of nickel-based alloys, and in particular a nimonic alloy called PE16.

‖ **n'importe** (nænport'), *phr.* [Fr.] It does not matter, it is immaterial.

1775 H. Walpole *Let.* 16 Sept. (1904) IX. 257 *N'importe*; we know many sages that take great pains to pass their time with less satisfaction. 1779 [see Something *sb.* 1 b]. 1837 Dickens *Let.* c 14 June (1965) I. 271 This is a bad look-out, but n'importe—we will mend it. 1856 S. O. Beeton *Let.* in N. Spain *Mrs Beeton* (1948) I. vi. 97 N'importe, a little more than a month, and I hope and think you will be a happy little wife. 1909 *Manch. Guardian* 24 Nov. 10/1 Dresses by the 'Maison N'Importe' and all the rest of it. 1929 R. Hughes *High Wind in Jamaica* vii. 149 He imagined..rescuing Rachel—or Laura, *n'importe*—from new and complicated dangers. 1952 'M. Cost' *Hour Awaits* 113 'I'm so sorry, Fanchon.' '*N'importe!* You and Albert Augustus were cut out for tight-rope walking from birth.'

Nimzo-Indian (ni:mzo,i·ndiăn), *a.* *Chess.* [f. *Nimzowitsch* (see next) + Indian *a.*] Designating a form of Indian defence popularized by A. Nimzowitsch, in which Black plays his king's bishop to Kt5 instead of fianchettoing it.

1935 *Chess* I. 103/1 Nimzo-Indian Defence... White simply sought to *combat* and not to *refute* the Nimzo-Indian Variation. 1957 *Griffith & White's Mod. Chess Openings* (ed. 9) 261 The Nimzowitsch-Indian Defence, 'Nimzo-Indian' for short, was first worked out in detail by Aron Nimzowitsch, the Latvian-Danish Grandmaster. 1958 *Listener* 13 Nov. 803/3 If you prefer positional play..you will probably do best to specialize in some form of the Ruy Lopez or English with White, and with the French and Nimzo-Indian Defences with Black.

Nimzowitsch (ni·mzovitʃ). *Chess.* Also **Nimzovitch**, etc. The name of A. *Nimzowitsch* (1886–1935), Latvian-born chess-player, used *attrib.* and in the possessive to designate various methods of opening play introduced or popularized by him.

1925 *Griffith & White's Mod. Chess Openings* (ed. 4) 57 Nimzovitch's Defence. 1932 *Ibid.* (ed. 5) 105 Nimzovitch's Attack. 1957 Nimzowitsch-Indian [see *Nimzo-Indian *a.*]. 1961 *Times* 25 Mar. 7/2 The game started with exactly the same variation of the Nimzovitch defence that was played in the first and third games. 1972 *Times* 2 Sept. 4/3 This had started with 1.P–Q4 and had been defended by Fischer, satisfactorily enough with a Nimzowitsch defence.

nin[2] (nin). [ad. W. *nain* grandmother.] In Liverpool working-class use: grandmother.

1958 M. Kerr *People of Ship Street* 48 The granny, or 'nin', or 'nanny', or 'nanna' or 'gran' as she is often called. 1966 P. Moloney *Plea for Mersey* 22 Every true wacker has three relations, viz. 'Me Mar, Me Nin, an me Anti-Mury'.

nincompoopiana (ni:nkǫmpūpi,ā·na). [f. Nincompoop + *-iana.] (See quots.)

1895 Beerbohm in *Yellow Bk.* Jan. 279 Long before this time there had been in the heart of Chelsea a kind of cult of Beauty... 'Nincompoopiana' the craze was called at first, and later 'Æstheticism'. 1970 *Sunday Times* 18 Oct. 49/3 'Nincompoopiana' began in the 1880s and was triggered off by the aesthetic movement which rebelled against the pretty and the respectable, and by the 'new woman'.

nine, *a.* and *sb.* Add: **A.** *adj.* **2. b.** *attrib.* *nine o'clock news*: see *news *sb.* (*pl.*) 5 c. Also *ellipt.*

1942 T. Rattigan *Flare Path* ii. i. 117 Anyone hear the nine o'clock? I clean forgot the time. 1952 M. Laski *Village* i. 11 Since the King had spoken on the nine o'clock.

c. (Further examples.)

1635 F. Quarles *Emblemes* v. iii. 254 'Tis not the sacred wealth of all the Nine Can buy my heart from Him. 1757 J. Dyer *Fleece* iv. 145 My Muse..Be thou the first of the harmonious Nine From high Parnassus. 1803 [see Muse *sb.*[1] 1]. 1852 M. Arnold *Empedocles on Etna* 69 'Tis Apollo comes leading His choir, The Nine. 1933 Kipling in *Times* 23 Feb. 16/1 He called the obedient Nine to aid The varied chase. And Clio kissed.

d. *nine-to-five, nine-till-five*, used *attrib.* (*a*) of a person working between 9 a.m. and 5 p.m.; so *nine-to-fiver, nine-till-fiver*; (*b*) of an occupation pursued between these hours, or of the mentality concerned with such an occupation, or with work only between such hours. So *nine-to-five (it)* v., to work between such hours.

1959 *Manch. Guardian* 16 June 5/1 These people are just workers like everybody else..they're all nine-to-fivers. 1960 *News Chron.* 13 Sept. 4/5 An electronic computer, working the nine-to-five shift. 1960 F. Raphael *Limits of Love* i. vi. 80 What do you want him to be? A nine till fiver? 1961 *Oxf. Mag.* 4 May 318/2 The so called '9 to 5' atmosphere and routine. 1962 R. Cook *Crust on its Uppers* i. 25 The game we play, it's got its risks, but it's..better than nine-to-fiving it. 1965 *Listener* 10 June 856/2 Some 60 per cent. of the students are home-based—a statistic to which the president of the students' union attributes nine-till-five attitudes. 1966 O. Norton *School of Liars* i. 19 'I hate having my corn measured by some-one else's bushel. Especially when it's a tinpot suburban bushel. The nine-to-five world.' That amused him. 'But you're a nine-to-fiver's wife.' 1969 *Times* 13 Mar. 20/3 (Advt.), 9 to 5 men come cheaper by the dozen. 1972 *Nature* 25 Feb. 412/3 The contract people tend to be exclusively '9 to 5' scientists whereas the scientists on grants have the traditional university attitude of ignoring the clock. 1972 F. Warner *Lying Figures* ii. 9, I couldn't stand a nine-to-fiver. 1975 P. G. Winslow *Death of Angel* i. 41 Wants a nine to five... That's her type, not a policeman coming in at all hours.

4. c. *nine-hours' day*: a working day of nine hours. So *nine-hours' movement*. Cf. *nine-hour* in Dict., sense 4 a.

1859 *Times* 5 Aug. 3/3 It appears that the 'Conference of the United Building Trades'..is established for the special purpose of carrying the nine hours' movement. 1862 *Leisure Hour* 28 June 413/2 They agreed upon an effort to shorten working time, and fixed upon a nine-hours' day. The phrase includes nine hours' *actual* work.

5. *nine-banded, -spotted* adj.

1909 *Biol. Bull.* XVII. 181 (*title*) A case of normal identical quadruplets in the nine-banded armadillo [*sc.* *Dasypus novemcinctus*]. 1964 G. Durrell *Menagerie Manor* i. 34 The nine-banded armadillo..trots about his cage. 1861 *Trans. Illinois Agric. Soc.* IV. 347, I found numerous specimens of a nine-spotted lady-bird (*Coccinella novemnotata*, Herbst,) under dry cow-dung. 1972 Swan & Papp *Common Insects N. Amer.* xx. 410 Nine-spotted lady beetle... In California..they are heavy feeders on aphids in alfalfa.

B. *sb.* **4. b.** (See also *long nine.)

6. b. *up to the nines* (earlier and later examples).

1859 Hotten *Dict. Slang* 68 *Nines*, 'dressed up to the *nines*', in a showy or recherché manner. 1928 Galsworthy *Swan Song* i. viii. 63 Women then were defended up to the nines. 1963 N. C. E. Kenrick *Story Wiltshire Regiment* ix. 86 The 99th's sartorial perfection at this time [*c* 1850] is said to have given rise to the expression 'Dressed up to the nines' as the other Regiments in Aldershot were constantly trying to achieve the same standard. 1965 *Listener* 20 May 742/2 So there they are, whenever a concert is given by their own orchestra, dressed up to the nines and bursting with pride.

c. *nine-nine-nine*, a telephone number dialled in an emergency in order to obtain a connection to the ambulance service, fire brigade, police, etc.; also written *999*.

1937 *Rep. Proc. 14th Conf. ASLIB* 76 The first burglar caught by dialling 999—the new telephone alarm signal. 1939 G. Greene *19 Stories* (1947) 159 He chose..a telephone box and dialled..999. 1954 I. Murdoch *Under Net* x. 136 'Better call the police if you ask me.'..'You go and dial nine nine nine.' 1954 M. Procter *Hell is City* i. i. 16, I stopped..and dialled nine-nine-nine. *a* 1956 W. de le Mare *Compl. Poems* (1969) 713 Dial 999, and gain at once, Safety from fire, police and ambulance. 1966 'W. Haggard' *Power House* viii. 89 She staggered to the telephone..dialled Nine Nine Nine. 1966 'A. York' *Eliminator* v. 85 Should something happen to him I don't see Lucinda dialling Nine-Nine-Nine. 1973 *Express* (Trinidad & Tobago) 17 Mar. 10/4 We have introduced the 999 system with four cars to cover Morvant, San Juan and Santa Cruz areas. 1974 'M. Innes' *Appleby's Other Story* vii. 55 He put through a 999 call to the police.

nine-holes. Add: **1. c.** *in the nine-hole(s*: in a difficulty. *U.S.*

1863 'E. Kirke' *My Southern Friends* 76 He owned har [*sc.* a slave] till he got in the nineholes one day, and sold har to the Gin'ral. 1877 *Congress. Rec.* 3 Nov. 230/1 We have put the gentleman in the 'nine-holes'; and there we intend to keep him. 1890 *Ibid.* 12 June 6002/1 The bill..has passed the Senate, and, to use a Western expression, it will put me 'in the nine-hole' if I do not get it through. 1906 B. L. Ridley *Battles & Sk. Army Tennessee* 295 The only time he ever got Johnston apparently in 'a nine hole' was at Resaca, on May 15, 1864.

ninepence. Add: **1.** (Earlier and further examples.)

1546 [see Shilling 5]. 1607 R. C. tr. *Estienne's World of Wonders* xv. 81 Hauing brought their twelue-pence to nine-pence, and their nine-pence to nothing. 1659 J. Howell *Eng. Proverbs* 11/1 As fine as fippence, as neat as nine pence. 1670 J. Ray *Coll. Eng. Proverbs* 206 As like as nine pence to nothing. 1850 F. E. Smedley *Frank Fairlegh* l. 444 Well, let her say 'no' as if she meant it,..and then it will all be as right as ninepence. 1884 *Temple Bar* Aug. 525 The trick of alliteration is often useful to give point to old proverbs. In such familiar sayings as 'fine as fivepence', 'nice as ninepence', 'to lie by the legend', its importance is most curious. 1968 'C. Aird' *Henrietta Who?* iv. 37 A rare old state it was in..but your mother..had it right as ninepence in next to no time.

b. (*nobbut, no more than*) *ninepence in the shilling*: of imperfect intelligence, mentally retarded. *dial.* and *colloq.*

1889 E. Peacock *Gloss. Words Manley & Corringham, Lincolnshire* (ed. 2) 370 *Nine-pence-to-the-shilling*, below the average in common sense. 'How's Mr..? Thaay do saay as he's nobut nine-pence-to-th'-shilling.'—M. F., *Scotton*, 1876. 1951 J. B. Priestley *Festival at Farbridge* iii. ii. 521 She's got a husband who's..ninepence in the shilling, a bit barmy. 1957 M. Kennedy *Heroes of Clone* ii. iii. 97 She's ninepence in the shilling. 1964 'A. Gilbert' *Fingerprint* v. 77 While they might be sympathetic, it was more likely they thought her child no more than ninepence in the shilling.

ninepenny, *sb.* and *a.* Add: **A.** *sb.* **2.** Ale that costs ninepence a gallon.

1886 Hardy *Mayor of Casterbridge* I. xiii. 160 I'm in such a low key with drinking nothing but small table ninepenny this last week or two.

niner. Add: **2.** Formerly, a senior naval cadet in the training-ship *Britannia*.

1914 'Bartimeus' *Naval Occasions* ix. 63 Jerry had to submit to strange indignities and stranger torments at the hands of Olympian 'Niners' (Fourth-term Cadets).

nineteen, *a.* (and *sb.*). **2. b.** (Earlier and later examples.)

1785 E. Sheridan *Jrnl.* 7 Aug. (1960) ii. 63 The Mother good humour'd and Civil but talks nineteen to the dozen. 1916 'Boyd Cable' *Action Front* 187 They must be charging, I think, or our front line's fallen back, because the rifles is going nineteen to the dozen. 1936 A. Christie *Murder in Mesopotamia* v. 39 Presently Mr Coleman bustled in and took the place beyond Miss Johnson... He talked away nineteen to the dozen. 1956 V. H. Collins *Bk. Eng. Idioms* 228 *Talk nineteen to the dozen*... Why 'nineteen'? The obvious numeral would be the round number 'twenty'. Possibly 'nineteen' was chosen just because, not being what might have been expected, it seemed to give a more striking effect.

nineteen eighty-four. The year-date (freq. written in form *1984*): the title of an apocalyptic novel (1949) by 'George Orwell' portraying a society in which government propaganda and terrorizing destroy consciousness of reality; freq. used allusively. Also *1984-ish*. Cf. *doublethink, double-think, *Newspeak.

1959 Benn & Peters *Social Principles* x. 225 Our antipathy to nineteen eighty-four methods may be due in part to the evil ends with which we associate them, or with the physical and mental cruelty involved in particular techniques. 1959 *Times* 6 May 15/7 We had a constant 1984-ish feeling as we worked in these vast abandoned castles. 1961 L. Mumford *City in History* xvii. 527 Whether they extrapolate 1960 or anticipate 2060 their goal is actually '1984'. 1968 *Listener* 29 Feb. 261/1 David Brinkley..pointed to the spokesman who claimed, 'We had to destroy that village to save it,' and commented that 1984 Newspeak had come to 1968. 1970 *Time* 16 Nov. 14 The political uses of television advertising and packaging of candidates were heralded by..doom-sayers as the ominous forerunner of 1984.

nineteenth, *a.* and *sb.* Add: **A.** *adj.* (Earlier and later *Comb.* examples.)

In quot. 1951 used in a quasi-*adj.* phr.

1872 W. F. Butler *Gt. Lone Land* (ed. 2) xvi. 241 Terrible deeds..never perhaps more sickening than now in the full blaze of nineteenth-century civilization. 1923 J. W. Harvey tr. *Otto's Idea of Holy* p. vii, A fair expression of the limitations and bias of the nineteenth-century mind. 1951 C. P. Snow *Masters* xxxiv. 272 His views were eccentric for an old man, but his manners had stayed gentle and nineteenth century. 1956 R. C. Zaehner in A. Pryce-Jones *New Outl. Mod. Knowledge* 65 Nineteenth-century optimism, then, is comparable to Buddhism in the emphasis it lays on individual effort in the battle for salvation.

2. *the nineteenth hole*: the bar-room in a golf club-house. Also *ellipt.* and in extended use. *slang* (orig. *U.S.*).

1901 W. G. van T. Sutphen (*title*) The nineteenth hole, being tales of the fair green. 1926 *Daily Colonist* (Victoria, B.C.) 11 July 4/3 The immeasurable distance of the nineteenth hole on standard courses is altogether beyond our physical capacity on these hot and golden Summer days. 1928 *Daily Express* 3 Jan. 9/2 Most courses have been completely unplayable, except at the nineteenth hole. 1948 'J. Tey' *Franchise Affair* i. 7 A good chap who played a very steady game and occasionally, when it came to the nineteenth, expanded into mild indiscretions. 1956 [see *ear-bash *v.*]. 1971 *Good Food Guide* 317 The Golf Tavern Nineteenth Hole.

nineteenth-ce·nturyism. The distinctive spirit, outlook, or character of the nineteenth century; a feature or trait suggestive of the nineteenth century.

1846 Tennyson in H. Tennyson *Alfred Ld. Tennyson* (1897) I. 238 They seem to be very clever and full of a noble 19th century-ism (if you will admit such a word). 1891 'L. Malet' *Wages of Sin* II. v. iii. 214 There is another of your perverted nineteenth-centuryisms!

ninety. Add: **1. b. ninety-day wonder** *U.S. Services' slang*, a graduate of a ninety-day officers' training course; an inexperienced, newly-commissioned officer; also *attrib.* or as *adj.*

1917 R. Lord in *Captain Boyd's Battery, A.E.F.* (1919) ii. 23 Two tents of Shavetails (i.e... Ninety-Day Wonders ..) have been attached to us for instruction purposes. 1919 K. M. Cortelyou et al. *From Arizona to Huns* 106 'A ninty [*sic*] day wonder' fresh from O.T.S. 1921 F. L. Field *Battery Book* 15 The inexperience of officers was a circumstance of tremendous significance. The fact that we used to call them 'ninety-day wonders' indicates our attitude..at the time. 1926 L. H. Nason *Chevrons* 136 You ninety-day wonder!.. Haven't you got brains enough to know this brook runs east and west? 1928 A. C. Havlin *Hist. Company A, 102nd Machine Gun Battalion* 1 As usual, they were dubbed 'ninety-day wonders'. 1956 E. N. Rogers *Queenie's Brood* 307 The draft law and the ninety-day-wonder officers came in for much discussion also. 1970 W. C. Woods *Killing Zone* 5 A pale punk kid to run my company, another ninety day wonder.

2. b. *spec.*, the years between 1890 and 1899. Also *attrib.*

1911 W. G. Blaikie Murdoch *Renaissance of Nineties* 34 Subtlety and delicacy are not more prominent in the painting of the nineties than in the period's literature. *Ibid.* 45 If Life was the watchword of the men of the nineties, Freedom was the motto on their banner. 1913 H. Jackson *Eighteen Nineties* i. 27 Max Beerbohm, in a delightful essay which could only have been written in the Nineties. *Ibid.* ii. 40 *The Yellow Book* and *The Savoy*.. were the favourite lamps around which the most bizarre moths of the Nineties clustered. There were few essential writers of the Nineties who did not contribute to one or the other. 1920 B. Muddiman (*title*) The men of the nineties. 1938 L. MacNeice *Modern Poetry* i. 7 The nineties poets did to some extent criticize life. 1954 A. S. C. Ross in *Neuphilologische Mitteilungen* LV. 32 It was apparently U and was certainly thriving in the nineties. 1971 P. Muir *Victorian Illustrated Books* viii. 194 (*heading*) Other Nineties people. 1972 *Country Life* 23 Nov. 1414/3 The Nineties in Paris were not all decadence, can-can and absinthe. 1975 *Times* 5 Dec. 13/4 A mock-innocent return to the 'nineties.

ninetyish (nəi·nti,iʃ), *a.* [f. Ninety *a.* and *sb.* + -ish[1].] Of, belonging to, or characteristic of 'the nineties' of the nineteenth century; resembling, or suggestive of, what was then current. So **ni·netyism,** the spirit of 'the nineties'; **ni·netyishness,** ninetyish characteristics.

1909 *Westm. Gaz.* 2 Mar. 2/2 What the *Standard* had hoped was that 'there might be a return to the rule of the 'nineties, when the seat was won or lost by a margin between two and five hundred votes'. Certainly there is nothing ninety-ish about Saturday's figures. 1918 E. Marsh *Rupert Brooke* 13 He entertained a *culte*..for the literature that is now called 'ninetyish'—Pater, Wilde and Dowson. 1931 *Times Lit. Suppl.* 23 Apr. 327/4 Thus 'Proteus' in the *New Statesman* describes me as 'engagingly ninety-ish'. 1941 L. MacNeice *Poetry of Yeats* iv. 66 He ceased to be 'ninetyish' with the Nineties. How saturated with 'Ninetyism' he had been can be seen [etc.]. 1959 *Times* 13 Feb. 13/4 Consider the subjects, or titles, of *Verklärte Nacht, Pierrot Lunaire, Wozzeck, Lulu,* all in varying measure products of a decadent romanticism, ranging from 'ninetyishness' to post-Freudian psychopathology. 1973 *Guardian* 28 Mar. 10/1 Firbank seemed negligible..his fictions artificial, *naughty* and Ninetyish.

ning-nong (ni·ŋ,nɒŋ). *dial.* and *Austral. slang.* Also **ning-nang.** [Origin unknown.] A fool, a stupid person. Cf. *nig-nog[1].

1832 *Whitehaven Poll Bk.* 34 He looks parlish like a ning-nang. 1864 Hotten *Slang Dict.* 188 *Ning-nang*, horse-coupers' term for a worthless thorough-bred. 1881 J. Sargisson *Joe Scoap's Jurneh through three Wardles* ii. 189 Wad teh believe't noo, t'Ning-nang can nowder read ner write. 1957 'N. Culotta' *They're a Weird Mob* (1958) i. 15, I 'ave ter get landed with a bloody ning nong who doesn't know where he's bloody goin'. 1973 *Telegraph* (Brisbane) 14 Mar. 71/1 Even ning-nongs can win prizes on Channel o's daily quiz show.

Ningre Tongo (niŋgre tɒŋgo). [prob. Taki-Taki, ad. Eng. Nigger *sb.* + Tongue *sb.*] An English-based creole language of Surinam, less divergent from English than **Jew Tongo*.

1933 L. Bloomfield *Lang.* xxvi. 474 Ningre Tongo or taki-taki is spoken by descendants of slaves along the coast. 1939 L. H. Gray *Foundations of Language* 37 The languages are termed *creolised*, examples being the *Taki-Taki* (or *Ningre Tongo*). 1968 [see *Jew Tongo* (*Jew *sb.* 3 c)].

ningyoite (ni·ŋgyo,əit). *Min.* [f. *Ningyo*, the name of a pass in Tottori Prefecture, Japan + -ite[1].] A hydrated phosphate of calcium and uranium, $CaU^{IV}(PO_4)_2.1\frac{1}{2}H_2O$, in which there is some replacement of uranium by lanthanons and which is found as brown or brownish-green orthorhombic crystals.

1959 T. Muto et al. in *Amer. Mineralogist* XLIV. 634 The new mineral, ningyoite, is named for the locality, Ningyo Pass. *Ibid.* 649 The synthetic ningyoite ignited at 600°C. in air for 5–10 minutes still retained the original orthorhombic structure. 1962 *Mineral. Jrnl.* III. 306 Ningyoite contains almost all the rare earths and their abundance ratio resembles that of apatite. 1970 *Econ. Geol.* LXV. 470 For the formation of ningyoite, fixation of much sulfur in comparison with carbonaceous matter should occur before the penetration of uranium-carrying solutions.

ninhydrin (ninhəi·drin). *Chem.* [prob. partially f. the chemical name, *triketohydrindene*, and app. first formed (as a trade name) in Ger.] A brown crystalline compound, $C_9H_6O_4$, which forms coloured products with amines and is particularly used for detecting and estimating amino-acids; triketohydrindene hydrate; 1,2,3-indantrione hydrate. Freq. *attrib.*

[1912 *Warenzeichenblatt* Nov. 2500/2 Ninhydrin... Farbwerke vorm. Meister Lucius & Brüning, Aktiengesellschaft, Höchst a/M... Geschäftsbetrieb: Chemische Fabrik. Waren: Pharmazeutische und therapeutische Präparate.] 1913 *Chem. Abstr.* VII. 3765 Ninhydrin is a valuable reagent for the detection of non-biuret dialyzable amino acids. 1915 *Jrnl. Biol. Chem.* XX. 218 This discovery [of Ruhemann's] was confirmed and extended by Abderhalden, who applied it to the detection of pregnancy and cancer, triketohydrindene hydrate now being a commercial product under the name of 'ninhydrin'. *Ibid.* XXIII. 382 Free from soluble nitrogen capable of giving the ninhydrin test for amino-acids. 1949 Abraham & Heatley in H. W. Florey et al. *Antibiotics* I. ii. 107 The positions of the separated amino acids on the paper were made visible by the coloration they gave with ninhydrin. 1962 Dardenne & Kirsten in A. Pirie *Lens Metabolism Rel. Cataract* 416 In these hydrolysed lens extracts we detected some 30 non-protein-bound substances which could be determined by the ninhydrin reaction. 1974 W. J. Burley *Death in Stanley St.* x. 161 [We] decided to try a new test for latents. It's supposed to be better than the old ninhydrin test for some surfaces.

ninon (ninoṅ). [Fr.] A light-weight fabric, used esp. in dresses, made in a plain weave from silk, rayon, or nylon.

1911 *Daily Colonist* (Victoria, B.C.) 15 Apr. 24/1 (Advt.), Three large shipments just to hand making our stock replete with every desired line of silks..Bengaline,..Tamaline..Ninon. 1913 W. J. Locke *Stella Maris* xiii. 160 Dressed in a soft grey ninon gown. 1923 *Daily Mail* 24 Feb. 11 Novel ninons are patterned with designs resembling finely worked embroideries. 1929 D. L. Moore *Pandora's Letter Box* xiv. 254 Everything must be of silk, georgette, triple ninon and so forth. 1948 G. L. Fraser *Textiles by Britain* 165 *Ninon*, fine and closely woven plain-weave semi-transparent dress cloth made either in silk or rayon. 1968 Hollen & Saddler *Textiles* (ed. 3) xvi. 140/1 *Ninon* is a filament sheer widely used for curtains.

ninth, *a.* and *sb.* Add: **B.** *sb.* **2.** Also, a note eight diatonic degrees above or below another note. **b.** In full, *ninth chord.* The chord of such notes; a ninth added to a triad. (Later examples.)

1845 *Encycl. Metrop.* IV. 799 As to pentachords, such as what have been called the major and minor ninth, and compound sharp sixth,..they are..only chords of the seventh..with a fifth note violently forced in. 1876 Stainer & Barrett *Dict. Mus. Terms* 303/1 *Chord of the major ninth*, a chord formed by a combination of thirds starting with the dominant or fifth of the scale, called by some writers the 'added ninth', because it consists of a chord of the dominant seventh, with the addition of the ninth; by others the 'dominant ninth', because it occurs on a dominant bass. *Ibid.* 303/2 *Chord of the minor ninth*.. consists of a dominant, its major third, major (perfect) fifth, minor seventh, and minor ninth. 1926 A. Niles in W. C. Handy *Blues* iii. 16 The device..of ending up the tune on the diminished seventh chord..lately sometimes on the chord of the ninth. 1934 [see *eleventh *sb.* 2]. 1949 L. Feather *Inside Be-Bop* ii. 50 By the 1920's it had become fashionable to end on a seventh or ninth chord. 1952 R. Eldridge in B. Ulanov *Hist. Jazz in Amer.* (1958) xix. 238, I was full of ideas. Augmented chords. Ninths. 1969 *Rolling Stone* 28 June 17/3 In those days country music was very loose in both meter and lyrics... No one had ever heard of a ninth chord.

niobian (nəi,ōᵘ·biăn), *a.* *Min.* [f. Niob(ium + *-ian 2.] Of a mineral: having a (small) proportion of a constituent element replaced by niobium.

1962 [see *ilmenorutile]. 1972 *Nature* 31 Mar. 215/1 Terrestrial dysanalyte is a niobian perovskite which contains about 26% CaO and no detectable ZrO_2.

niocalite (nəi,okæ·ləit). *Min.* [f. Nio(bium + Cal(cium + -ite[1].] A basic silicate and fluoride of niobium and calcium, $Ca_4NbSi_2O_{10}(OH,F)$, found as yellow orthorhombic crystals.

1956 E. H. Nickel in *Amer. Mineralogist* XLI. 785 A new mineral species, for which the name 'niocalite' is here proposed, has been identified in rock from the Oka district, 20 miles west of Montreal, Quebec. 1958 *Canad. Mineralogist* VI. 264 The niocalite occurs as randomly oriented, coarse, prismatic crystals up to one centimetre in length, and occasionally even longer, embedded in the calcite. 1966 *Soviet Physics—Doklady* XI. 197/2 The Patterson projections show that niocalite does not have a screw axis.

niopo (ni,ōᵘ·po). [Native name.] A narcotic snuff used by certain South American Indian tribes, prepared from the seeds of the tropical American trees, *Piptadenia peregrina* and related species. Also *attrib.*

1860 Mayne Reid *Odd People* 134 Snuffing the niopo is not exclusively confined to the Mundrucu. *Ibid.*, The niopo-taker who has one [*sc.* a device for taking snuff, made from the forked bone of a bird], esteems it as the most valuable item of his apparatus. 1900 Dorland *Med. Dict.* 444/1 Niopo-snuff... An intoxicating snuff made from the seeds of *Piptadenia peregrina*, a tree of tropical America. 1966 *New Scientist* 21 Apr. 156/1 Some natives of tropical America use a snuff made from seeds of the tree *Piptadenia peregrina*... This is called cohoba, niopo or parica. 1969 R. R. Lingeman *Drugs from A to Z* 204 A hallucinogenic snuff is made by South American Indians by pulverizing the seeds of the legumes *Piptadenia peregrina, P. colubrina,* and *P. macrocarpa*... Among different tribes it is known variously as yopo, niopo, cohoba, and huilca. These tribes are centered in the Orinoco basin of Colombia and Venezuela and in the Peruvian Andes.

nip, *sb.*[1] Add: **I. 1.** *fig.* phr. *to put in the nips*, to cadge, to ask for a loan. Cf. *nip *v.*[1] 2 c. *Austral.* and *N.Z. colloq.*

1919 [see *nip *v.*[1] 2 c]. 1937 Partridge *Dict. Slang* 564/1 *Put the nips in(to*, to ask a loan (from a person): Australian and New Zealand: from ca. 1908. 1949 L. Glassop *Lucky Palmer* 230 You can't put the nips into old Alf. He's got death adders in his pockets. 1955 D. Niland *Shiralee* 41 He was here yesterday, too. Put the nips into me for tea and sugar and tobacco in his usual style. 1963 B. Pearson *Coal Flat* x. 190 'The woman's getting too serious,' he thought; 'she's putting the nips in.' 1973 F. Huelin *Keep Moving* 48 Parsons, priests, doctors, lawyers and professional people generally were legitimate prey, and we had no scruples about 'putting the nips' into them.

4. *Cricket.* Restrict † *Obs.* to sense in Dict. and add: **b.** The quality in bowling which causes the ball to rise sharply from the pitch.

1963 A. Ross *Australia 63* iii. 75 Coldwell had been varying pace and direction skilfully and only that lack of final nip which Bedser possessed prevented him from being even more troublesome.

6. For (U.S.) read (chiefly U.S.). (Earlier and later examples.) *nip and tuck folder*: see quot. 1964.

1832 J. K. Paulding *Westward Ho!* I. 172 There we were at rip *and* tuck [*sic*], up one tree and down another. 1847 W. T. Porter *Quarter Race in Kentucky* 16 (Th.), It will be like the old bitch and the rabbit, nick and tack every jump. *Ibid.* 123 (Th.), Then we'd have it again, nip and chuck. 1857 *Knickerbocker* L. 498 (Th.), [I got the trout off the fire] by the head, and the dog got him by the tail, and it was nip and tuck, pull Dick, pull devil. 1890 Barrère & Leland *Dict. Slang* II. 87/1 *Nip and tuck* (Cornwall), a close contest. An old term in wrestling. 1906 J. London *Let.* 1 Dec. (1966) 226 This means..loss of money in the first months of the voyage, during which time things will be just nip and tuck with me. 1948 *Economist* 8 May 764/2 It is nip and tuck whether such a last great achievement of the bipartisan foreign policy can be ratified before..the Presidential race. 1964 *Gloss. Letterpress Rotary Print. Terms* (*B.S.I.*) 25 *Nip and tuck folder*, a type of folder in which the fold at right angles to the run of the web is formed by a blade thrusting the web between folding jaws. 1968 *Economist* 9 Nov. 28/3 The Vice President said that there were still several important states nip-and-tuck. 1974 *State* (Columbia, S. Carolina) 5 Mar. 6–A/1 It was nip and tuck the rest of the way with two straight Cunningham baskets to open the fourth quarter knotting the count, 40–40, with 7:28 left to play.

8. c. The narrow gap or area of contact between two rollers; the rollers themselves.

1884 W. S. B. McLaren *Spinning* 250 Nip of rollers, the point where a pair of rollers touch each other, and where, consequently, they hold or nip the wool. *Ibid.* iii. 35 The lowest rollers..have their nip below the level of the suds. 1946 A. J. Hall *Stand. Handbk. Textiles* iv. 171 The fabric receives a light squeeze as it passes between the nip of the mangle rollers. 1969 W. R. R. Park *Plastics Film Technol.* ii. 12 Aluminium foil is made by passing hot sheet through a series of hot, highly polished, precision finished metal roll nips. 1972 *Materials & Technol.* V. 515 The opening between the two rolls—the nip—is adjusted in such a way that when pieces of rubber are fed into the nip they are gripped and squeezed through the opening.

III. 10. *Geol.* A low cliff cut along a gently sloping coastline by wave action; also, a notch cut along the base of a pre-existing coastal cliff by wave action.

1897 *Geogr. Jrnl.* IX. 542 Where the aggradation begins at the shoreline at the foot of the earlier formed 'nip'. 1919 D. W. Johnson *Shore Processes & Shoreline Development* v. 259 If the lagoon waves are too feeble, the nip may be entirely lacking. 1939 A. K. Lobeck *Geomorphology* x. 347 The smaller waves advance landward and cut low cliffs in the weak material of which the land is usually composed. Thus a nip is produced. 1942 C. A. Cotton *Geomorphology* (ed. 3) xxix. 409 Erosion may be so rapid that in cliffs of tough, unjointed rock a nip is cut—that is, a notch along the base, above which the cliff overhangs. 1958 Sparks & Kneese tr. *Guilcher's Coastal & Submarine Morphology* iii. 64 That notches or nips at the base of cliffs are due to mechanical erosion is often much more improbable.

nip, *sb.*[2] Add: **2.** *Comb.* **nip bottle,** a miniature bottle of spirits (literally, one containing enough for one drink); **nip joint** *U.S.*, an establishment illegally selling (small quantities of) spirits.

1915 A. D. Gillespie *Lett. from Flanders* (1916) 150 The etcetera now includes goggles, respirators, and 'nip bottles' of chemicals. 1939 *Sun* (Baltimore) 18 Aug. 24/7 A hearing by the Liquor Board on the advisability of banning the sale of miniature, or 'nip' bottles of alcoholic drink. 1936 *Ibid.* 28 Jan. 9/1 The bills were opposed by.. Raye O. Lawson,..who said the legislation would increase bootlegging and 'nip joints'. 1938 *Ibid.* 10 Sept. 4/1 The court ordered restored to the owners a house which had been padlocked on conviction of a tenant for operating a 'nip joint'.

Nip, *sb.*[6] and *a.* [Abbrev. of *Nipponese.] (A) Japanese. *slang.*

Usually abusive.

1942 *Time* 9 Feb. 23/3, I visited a command post in one sector where they had just rounded up a bunch of Nips. 1942 *R.A.F. Jrnl.* 31 Oct. 13 The Nip pilots. 1947 J. Bertram *Shadow of War* i. ii. 12 The Nips keep bombing the airfields. 1965 R. T. Bickers *Scent of Mayhem* iii. 30 La belle Pauline was a secretary at the Nip Embassy in Paris. 1971 J. Osborne *West of Suez* i. 27 Few little Nips popping away with cameras. 1973 *Islander* (Victoria, B.C.) 19 Aug. 12/2 Who hadn't quite made up their minds about what should be done with Hitler and Mussolini and the Nips.

nip, *v.*[1] Add: **1. d.** (Earlier and later examples.)

1850 S. Osborn *Stray Leaves Arctic Jrnl.* (1852) 72 Penny had passed a long way inside of the spot the steamers had been beset and nipped in. 1937 *Beaver* June 13/2 The 'Fort James', a Company schooner, was 'nipped' in the ice at Tuktuk. 1966 T. Armstrong et al. *Gloss. Snow & Ice* 29 *Nip*, ice is said to nip when it forcibly presses against a ship which is *beset*. A vessel so caught, though undamaged, is said to have been nipped.

2. c. To borrow, to obtain by wheedling. Cf. *nip *sb.*[1] 1 *slang* (chiefly *Austral.*).

1919 W. H. Downing *Digger Dial.* 35 *Nip*, to cadge (or 'Put in the Nips'). 1925 Fraser & Gibbons *Soldier & Sailor Words* 209 *To nip*, to cadge.

4. Restrict † to senses in Dict. and add: **c.** To defeat narrowly (in a sporting contest). *U.S.*

1942 Berrey & Van den Bark *Amer. Thes. Slang* §649/3 Defeat..nick, nip, outbeat, outwin. 1942 *Sun* (Baltimore) 29 Apr. 19/1 Miss Goshen, under Eddie DeCamillis, came along in the stretch to take the second by two lengths. Sobriquet got up to nip Spare the Rod for second honors. 1951 *Amer. Speech* XXVI. 230/2 Oregon nips St. Mary's. 1966 *N.Y. Times* (Internat. ed.) 22 Apr. 12/1 The Pirates nipped the Reds, 3–2. 1969 *Eugene* (Oregon) *Register-Guard* 3 Dec. 1D/5 The Ducks.. will continue their three-game..against highly rated Florida State, which nipped touring Oregon State 69–68 Tuesday night. 1974 *Anderson* (S. Carolina) *Independent* 19 Apr. 4B/2 Danny Ford banged out four hits and knocked in two runs as Augusta College nipped Erskine, 6–5, here Thursday.

5. Also with *off*.

1902 *Chambers's Jrnl.* Nov. 742/2 Small establishments for the treatment of manufacturing refuse..are nipped off by a rise in the price of fuel.

12. (Later examples. In quot. 1919 *trans.*) Freq. *fig.*, as to cut *in*, and in extended use, to move informally or unobtrusively, often quickly, *away*, *out*, etc. Occas. without adv.

1908 H. G. Wells *War in Air* v. 155 She [*sc.* a ship] had..nipped in between the *Susquehanna* and the *Kansas City*. 1909 A. Quiller-Couch *True Tilda* xiii. 169 'If they catch up with us we must nip into a gateway,' panted Tilda. 1919 *Times Lit. Suppl.* 27 Feb. 107/3 'The light-hearted snottie' who nipped in his piquet boat across the knife-edged ram of a fast-travelling cruiser. 1920 *Blackw. Mag.* Jan. 111/2 Your friend..nips in and takes up the running, and you are out of the hunt. 1920 D. H. Lawrence *Let.* 9 Apr. in C. Mackenzie *My Life & Times* (1966) V. 177 Nip over here for a short while! 1926 T. E. Lawrence *Seven Pillars* lviii. 355 So, watching the time, one or two of the quicker youths nipped across to drag back the saddle-bags. 1930 M. Allingham *Mystery Mile* xxi. 199 'Shall we nip off?' said Knapp nervously. 1930 W. S. Maugham *Bread-Winner* 115 If Uncle Alfred wants us to get out we'd better nip before Daddy comes back. 1930 J. B. Priestley *Angel Pavement* vi. 299 Now then,.. just nip back for the plates. 1947 *People* 22 June 7/5 Meantime, Club No. 2..nipped in, handed the 'very famous footballer' the thousand smackers.., and clinched the transfer. 1955 M. Gilbert *Sky High* viii. 115 If you nip along now..you could catch her before the practice starts. 1962 C. Oman *Mary of Modena* v. 187 All the visiting English who could manage it quietly nipped off to see the Prince of Wales. 1969 *Listener* 14 Aug. 205/2 Nipping out for a smoke during the odd bit of Schoenberg. 1973 A. Mann *Tiara* xiii. 118 Piccoli's will still be open. Shall I nip down and get pictures of all these types?

b. Of a cricket ball: to come sharply *off* the pitch; also *absol.*

1899 *Captain* I. 516/1 Another..makes the ball nip off the pitch like a marble off a granite wall. 1903 P. F. Warner in H. G. Hutchinson *Cricket* xiv. 399 Matting on the bare grassless ground favours the batsman, though I am inclined to think that a really good bowler ought always to be able to make the ball 'nip' a bit.

nipa. 2. In def. substitute 'native to coastal regions in tropical Asia and Australia' for 'native to the coasts and islands of the Indian seas'. (Earlier and later examples.)

1779 T. Forrest *Voy. New Guinea* i. 16 She was covered almost entirely with the leaves of a certain Palm tree, called Nipa, such as the natives cover houses with on the south-west coast of Sumatra, and in almost all Malay countries. 1783 W. Marsden *Hist. Sumatra* 46 Those [people] of the lowest class have their fillet of the leaf of the *neepah* tree. 1817 [see *attap]. 1926 E. Blatter *Palms Brit. India* 553 Nipa is the vernacular name of the palm in the Philippines. 1954 R. H. Holttum *Plant Life in Malaya* ii. 23 One of the most peculiar of all palms is Nipah, which grows on muddy river banks near the sea. 1962 B. Harrisson *Orang-Utan* iv. 139, I had boarded ship at the mouth of the Rejang River where it had anchored alongside mud, mangrove and nipah-palms. 1966 D. Forbes *Heart of Malaya* xiii. 161 The wooden beach bungalows. thatched with the nipa palm,..were all empty. 1969 J. M. Gullick *Malaysia* iv. 160 *Nipah* is a type of palm whose fronds are plaited for use as thatch. 1973 *Telegraph* (Brisbane) 2 Apr. 12/1 They tied the two women up in a nipa (thatched grass) hut.

Nipkow disc (ni·pkɒf). *Television.* [f. the name of Paul *Nipkow* (1860–1940), Polish electrical engineer, who invented it in 1884.] A scanning disc used in some early television transmitters and receivers having a line of small apertures near the circumference arranged in a spiral of one complete turn, so that on each revolution of the disc an area is scanned equal in height to the radial distance between the first and last apertures.

1934 J. H. Reyner *Television* xiv. 161 Until recent times most of the Continental systems..have been of the Nipkow disc type, and in that main respect similar to the Baird system. 1962 G. A. T. Burdett *Automatic Control Handbk.* xxi. 6 There are three types of scanning device in existence,..the Nipkow disc,..the flying-spot scanner and the pick-up tube. *Ibid.*, A development of the Nipkow disc has recently been produced in the U.S.A. for use as a slow-scan device for the transmission of documents. 1974 *Encycl. Brit. Macropædia* XVIII. 105/2 Until the advent of electronic scanning, all workable television systems depended on some form or variation (*e.g.*, mirror drums, lensed disks, etc.) of the mechanical sequential scanning method exemplified by the Nipkow disk.

nipper, *sb.*[1] Add: **3.** (Later examples.)

1968 *Courier-Mail* (Brisbane) 21 Nov. 3/5 The men, employed by the Public Works Department, claimed the job foreman would not let their 'nipper' patronise the sandwich shop of his choice. A nipper is a boy or man who boils the billies, runs messages, buys lunches, and does similar jobs. 1971 R. Roberts *Classic Slum* viii. 125 The nippers, carters' helps. The nipper looked after the horse and sat guard over goods at the tail end of the vehicle.

b. (Earlier, *attrib.*, and later examples.) Also, a girl; a child of either sex; the smallest or youngest of a family.

[1847 Dickens *Dombey* (1848) xxiii. 240 Florence endeavoured to believe that the Captain was right; but the Nipper..shook her head in resolute denial.] 1859 Hotten *Dict. Slang* 68 *Nipper*, a small boy. 1875 W. D. Parish *Dict. Sussex Dial.*, *Nipper*, a common nickname for the youngest member of the family, or for one who is unusually small for his age. 1901 M. Franklin *My Brilliant Career* x. 81 By George, you're a wonderful-looking girl!.. You are such a little nipper. 1928 J. Mason *Before Mast in Sailing Ships* 128 Next to Clarke was a Scottish lad by the name of Nisbet, from Inverness-shire. He was the smallest, and was called the 'Nipper'. 1941 *Lilliput* Mar. 371/1 A family party..arrived on the scene: Mother, aunty, two nippers—a girl and a boy. 1959 I. & P. Opie *Lore & Lang. Schoolch.* ix. 170 Little 'uns..midge, nipper, penguin, pint-size. 1967 *Courier-Mail* (Brisbane) 3 Feb. 6/8 The establishment of 'nipper clubs' for boys aged from eight to 13 would guarantee the future of the surf club movement. 1969 Visct. Buckmaster *Roundabout* ii. 35 He [*sc.* a butler] was..always in domestic trouble from the arrival of what he would call 'another little nipper'. 1972 *Times* 3 June 19/1 When I was a nipper at school in Glasgow [etc.].

4. c. (Earlier and later examples.)

1821 D. Haggart *Life* 94 That's one of the bulkies from Dumfries, wanting to clap the nippers on me. 1876 J. S. Ingram *Centenn. Exposition* viii. 235 The curious part of this [hardware] exhibit consisted in the police nippers. 1910 *Encycl. Brit.* X. 296/1 Several recently invented appliances are used as handcuffs, *e.g.* snaps, nippers, twisters. They differ from handcuffs in being intended for one wrist only... The nippers can be instantly fastened on the wrist. 1918 *Outlook* (N.Y.) 25 Sept. 126/1 A newly appointed policeman..has to buy ..a pair of nippers. 1939 *Fortune* July 104/1 At 2145 one of the detectives put nippers on the prisoner's wrist. 1950 H. E. Goldin *Dict. Amer. Underworld Lingo* 145/2 We hit a scorf joint (went into an eating place), and the dick (detective) took the nippers off.

nippet, nippit, Sc. varr. Nipped *ppl. a.* (chiefly in senses: miserly; scanty; starved; of restricted mental attitude; bitter, sarcastic).

1535 [see Nipped *ppl. a.*]. 1808 Jamieson, *Nippit*, adj. 1. Niggardly, parsimonious... 2. Too small, scanty. 1814 C. I. Johnstone *Saxon & Gaël* I. ix. 121 Na, na, I ne'er liket to be nippit or pinging, gie me routhrie o' a'thing. 1857 H. S. Riddell *St. Matthew* xxv. 24, I kennet thee that thou art ane nippet man, sheerin' whare thou hestna sawn, an' getherin' whare thou hestna strinklet. *c* 1860 in *Scotsman* (1912) 13 Sept., 'She is a puir nippet creater' —A poorly-fed child. 1875 N. Elliott *Nellie Macpherson* 165 As regards the langidge, ye maunna be ower nippet on that point. 1924 A. Dodd *Poppies in Corn* 33 Ye're naewise mean or nippit. 1925 *Glasgow Herald* 23 May 4/4 Nannie..rejected nothing—but oatcakes. The misguided and 'nippet' woman at the Sluices once offered these, conspicuously ill-baked. 1935 D. Kirkwood *My Life of Revolt* 222 Philip Snowden's views were precise, narrow, and moulded by the immediate circumstances. The Scotsmen used to call him 'nippet'. 1962 *Buchan Observer* 24 July 2 Nippet words an' soor ill-naiter Gart her man an' bairns behave!

nippily, *adv.* Delete *rare*[−1] and add later examples.

1972 *Guardian* 10 Nov. 11/1 The new 3-Door Renault 5 ..[will] take you nippily through town traffic snarl-ups. 1974 *Times* 3 Jan. 23/3 The 128 [Fiat] handles nippily, with good roadholding.

nippiness (ni·pinės). [f. Nippy *a.* + -ness.] Nimbleness, agility.

1923 U. L. Silberrad *Lett. J. Armiter* ii. 50 Jethro is smitten with admiration of her agility; 'nippiness' he calls it. 1932 *Daily Express* 2 July 11/2 Perry's volleys, Hughes' return of service and overhead smashing, and the perpetual nippiness of our two men, gave us the victory.

nipping, *ppl. a.* (and *adv.*). Add: **5.** *nipping-roller.*

1920 *Discovery* Mar. 88/1 The padded goods are well squeezed through nipping-rollers, and then dried and 'backed'. 1964 *Gloss. Letterpress Rotary Printing Terms* (*B.S.I.*) 25 *Nipping rollers*, a pair of rollers adjustable to exert pressure to set a fold. 1967 V. Strauss *Printing Industry* vi. 382/2 The cutting cylinder cuts the web after it passes the nipping rollers.

nipple, *sb.* **4. c.** nipple cactus, for '*Mamillaria*' substitute the correct spelling '*Mammillaria*'. (Examples.)

1876 *Encycl. Brit.* IV. 625/2 Mammillaria.—This group..is called Nipple Cactus. 1971 D. Weniger *Cacti of Southwest* 122/1 *Mammillaria similis*... 'Nipple Cactus'.

Nippon (ni·pɒn). [Jap., f. *ni(chi* the sun + *pon, hon* source.] **1.** The Japanese name for Japan.

1727 J. G. Scheuchzer tr. *Kæmpfer's Hist. Japan* I. i. iv. 58 This Empire is by the Europeans call'd *Japan*. The Natives give it several names and characters. The most common, and most frequently us'd in their writings and conversation, is *Nipon*, which is sometimes in a more elegant manner, and particular to this Nation, pronounc'd Nifon... It signifies, *the foundation of the Sun*. 1859 K. Cornwallis *Two Journeys to Japan* I. vii. 187 It was against the laws of Nipon—in speaking of their own country this was the invariable term and pronunciation. 1890 B. H. Chamberlain *Things Japanese* 174 Our word 'Japan', and the Japanese *Nihon* or *Nippon* are alike corruptions of *Jih-pên*,..literally 'sun-origin' a name given to Japan by the Chinese. 1914 M. Klein *By Nippon's Lotus Ponds* v. 62 Nippon's old custom wills that the..wife or..servant girl closely seal up the house with..big rain-doors. 1940 E. Pound *Cantos* lviii. 74 Sinbu put order in Sun land, Nippon, in the beginning of all things. 1942 *R.A.F. Jrnl.* 13 June 5 The sons of Nippon had become aware that their destiny was to establish a New Order in the East. 1975 O. Sela *Bengali Inheritance* xviii. 160 The Indians in Nippon dominated territory.

2. Nippon vellum = *Japanese vellum* (*Japanese *a.* b).

1926 *Brit. Weekly* 3 June 185/3 They will simultaneously issue special editions on nippon vellum. 1958 J. R. Biggs *Woodcuts* 91 *Nippon Vellum* and *Jappon Vellum*.. are imitation vellum papers whose smooth, creamy kindly surface is just right for some prints.

Nipponese (nipǒnī·z). [f. *Nippon + -ese.] **a.** The Japanese people; an individual Japanese. **b.** The Japanese language. Also *attrib.* or as *adj.*

1859 K. Cornwallis *Two Journeys to Japan* I. 205 Beyond..was to be seen the houses of the town of Napa.. wherein were moored several large junks, native and Niponese. 1860 R. H. Dana *Jrnl.* 24 Apr. (1968) III. 1027 This island, Yeso, is a conquest of the Japanese (Niponese). 1927 E. Pound *Let.* 9 Nov. (1971) 214 At present it is the scattered fragments left by a dead man, edited by a man ignorant of Japanese. Naturally any sonvbitch who knows a little Nipponese can jump on it or say his flatfooted renderings are a safer guide to the style of that country. 1931 *Lit. Digest* 16 May 45/1 Your Nipponese..looks as tho nothing short of an earthquake would startle him. 1935 J. Joyce *Let.* *c* 18 Feb. (1966) III. 343 He wore a kimono and scarlet vest. I suppose the Nipponese evening dress. 1942 [see *crash-dive vb.]. 1944 *Sun* (Baltimore) 25 Nov. 2/4 Nipponese planes were shot down. 1948 A. Keith *Three came Home* i. 33 The war was coming to the East, and..the Nipponese were coming to Borneo. 1973 'S. Harvester' *Corner of Playground* i. viii. 66 The apparent end of traditional Nipponese spiritual values under the steam-roller of post-war materialism.

Nipponian (nipōᵘ·niăn), *a.* [f. *Nippon + -ian.] Of or pertaining to Japan, Japanese. So **Ni·pponism,** the furtherance of Japanese nationalistic interests.

1909 *Daily Chron.* 19 Aug. 4/6 The best English account of the conflict from the Nipponian point of view. 1914 *Encycl. Relig. & Ethics* VII. 489/1 The cry of 'Nipponism' ..was raised in a somewhat extravagant fashion.

Nippy (ni·pi), *sb.* [f. the adj.] Formerly, a waitress in one of the restaurants of J. Lyons & Co. Ltd., London; hence, any waitress.

1925 *Punch* 11 Feb. 167/2, I can't mike up me mind weyver to be a lidy's 'elp or a 'nippy'. 1941 J. Smiley *Hash House Lingo* 39 *Nippy*, waitress. 1948 G. V. Galwey *Lift & Drop* vi. 161 His hands stuck out in front of him like a Nippy carrying a tray. 1973 *Country Life* 22 Nov. 1736/1 Biba's shiny black tea boxes have three Nippies in gold decorating the label. 1974 W. Foley *Child in Forest* ii. 251 The big brightly lit clean teashop in Oxford Street where I worked as a Nippy.

Nirvana. Add: **b. Nirvana principle** *Psychol.*, in psychoanalytic theory, the attraction felt by the psyche for a state of non-existence,

which Freud connected with the death-instinct, countering the tensions set up by the pleasure principle.

1920 B. Low *Psycho-Analysis* (ed. 2) iii. 73 It is possible that deeper than the Pleasure-principle lies the Nirvana-principle, as one may call it—the desire of the newborn creature to return to that stage of omnipotence, where there are no non-fulfilled desires, in which it existed within the mother's womb. 1924 J. Riviere tr. *Freud's Econ. Problem in Masochism* in *Coll. Papers* II. xxii. 256 For this tendency that has been presumed by us Barbara Low has suggested the name Nirvana-principle, which we accept. But we have unquestioningly identified the pleasure-pain-principle with this Nirvana-principle. *Ibid.* 257 The *Nirvana*-principle expresses the tendency of the death-instincts. 1936 *Brit. Jrnl. Psychol.* Jan. 286 In the biological-physical conception of a tendency to death (for which they would reserve the name 'Nirvana principle') there is no room for Eros. 1973 L. Bellak et al. *Ego Functions* ii. xiv. 212 Do these data invalidate the Nirvana principle, or can that tenet be modified to include the seemingly contradictory clinical observations?

Hence **Nirva·na-ing** *vbl. sb.*, experiencing the state of nirvana; **Nirva·nist,** one who experiences this state.

1898 G. B. Shaw *Perfect Wagnerite* 106 He rested himself as a Pessimist and Nirvanist. 1921 D. H. Lawrence in *Reminisc. & Corr.* (1934) 23 Nirvana-ing is surely a state of continuing as you are. 1922 —— *Aaron's Rod* xxi. 309 And you've never got to think you'll dodge the responsibility of your own soul's self, by loving or sacrificing or Nirvanaing. *a* 1930 —— *Sex, Lit. & Censorship* (1955) 101 Even the Nirvanists consider man as a fixed entity.

‖**Nisan** (ni·săn, nī·săn). Also † **Nysan.** [Heb. *Nīsān*.] The first month of the Jewish ecclesiastical year and the seventh of the civil year, formerly called Abib.

1382 Wyclif *Neh.* (2nd text) ii. 1 Forsothe it was doon in the monethe Nysan, in the twentithe ȝeer of Artaxerses, kyng. 1535 Coverdale *Neh.* ii. 1 In the moneth Nisan of the twentieth yeare of kynge Artaxerses. 1611 Bible *Neh.* ii. 1 And it came to passe, in the moneth Nisan,..that wine was before him. 1737 W. Whiston tr. *Josephus' Works* 9 Moses appointed that Nisan, which is the same with Xanthicus, should be the first month, for their festivals; because he brought them out of Egypt in that month. 1846 Geo. Eliot *Let.* *c* Apr. (1954) I. 213 The evening of the 13th Nisan..would be what Strauss calls the *Vorabend* des Pascha. 1934 T. S. Eliot *Rock* i. 35 In Shushan the palace, in the month Nisan, He served the wine to the King Artaxerxes. 1972 C. Raphael *Feast of History* i. 32/1 A hand-written Hebrew-Russian calendar..open at Nisan, the month of Passover.

nisei (nī·sēⁱ). Also **nissei,** and with capital initial. Pl. **nisei.** [Jap., f. *ni-* second + *sei* generation.] An American born of Japanese parents. Also *attrib.*

1943 S. Menefee *Assignment: U.S.A.* 191 The War Relocation Authority, after a delay of many months, finally began to release those Nisei, or American-born Japanese. 1945 Mencken *Amer. Lang.* Suppl. I. 608 The designation *nisei*..for Japanese of American birth was seldom heard, before Pearl Harbor, save on the Pacific Coast... *Nisei* is sometimes spelled *nissei*. 1948 *Newsweek* 30 Aug. 20/1 The 29-year-old Nisei claimed that she had taken a job with the Tokyo radio merely 'for the experience'. 1957 *New Yorker* 16 Nov. 125/1 It would be difficult for a Japanese born student to..date a nisei girl. 1972 J. Ball *Five Pieces of Jade* v. 51 The bespectacled, crew-cut, Babbitt-looking Nisei detective. 1973 *Publishers Weekly* 3 Sept. 48/3 This diary, kept for eight months in 1942 by a 26-year-old Nisei.

nish, var. Nesh *a.*

1925 *Dialect Notes* V. 237 *Nish*.., delicate. 1963 *Amer. Speech* XXXVIII. 299 [Newfoundland] *Nish, adj.* (1) Tender. (2) Easily injured. 1964 L. E. F. English *Historic Newfoundland* 31 *Nish*, tender, easily injured.

nisin (nəi·sin). *Pharm.* [See quot. 1947 and -in[1].] A mixture of closely related polypeptides produced by the bacterium *Streptococcus lactis* which is active against Gram-positive bacteria and is used in some countries as a food preservative.

1947 Mattick & Hirsch in *Lancet* 5 July 5/1 We have already given some account..of the action in vivo of an inhibitory substance*, isolated from lactic streptococci, against streptococcal infections of the mouse. [*Note*] *Group N Inhibitory Substance: (G) Nisin. 1959 *Times* 9 Mar. (Britain's Food Suppl.) p. viii/2 Nisin is an easily digested protein. 1963 *B.S.I. News* Feb. 13/2 The bacteriological examination of butter, milk, cheese and ice-cream, and for the detection and assay of nisin. 1972 *Sci. Amer.* Mar. 18/3 A number of countries have permitted such antibiotics as tetracyclines, nystatin, nisin..as direct or indirect additives to chilled or raw fish.

Niska: see *Nass.

1895 F. Boas in *Rep. Brit. Assoc. Adv. Sci.* 569 The customs of the Nīsk·a and those of the Tsimshian..are practically identical. *Ibid.* 583 The Nīsk·a does not differ very much from the Tsimshian. 1973 *Times* 27 Dec. 6/4 There is also a portrait mask of an old woman (*c* 1825–1850) from the Niska of British Columbia, of wood inlaid with abalone shell.

niskh(i), varr. *naskhi *sb. pl.*

Nissen (ni·sən). [Name of the inventor, Lt.-Col. Peter Norman *Nissen* (1871–1930).] Used *attrib.* and *absol.* of a tunnel-shaped hut made of corrugated iron with a cement floor.

1917 E. F. Wood *Note-Book of Intelligence Officer* 224 Recently I met the inventor of the now famous Nissen hut. 1932 Auden *Orators* iii. 114 These nissen huts if hiding could Your eye inseeing from Firm fenders were. 1942 *R.A.F. Jrnl.* 16 May 3 In Nissen huts, with orders coming through from Flight Commander all day and night. 1948 G. Greene *Heart of Matter* iii. i. 191, I would never go back there, to the Nissen hut, if it meant that she were happy. *Ibid.* 192 The rain hammered on the Nissen roofs. *Ibid.* 222 He's living in one of the Nissens now. 1954 W. Faulkner *Fable* (1955) 86 It was not the Nissen walls which trembled. 1973 J. Wainwright *Pride of Pigs* 72 There's a glorified Nissen hut—a bit of a youth club—on the site.

Nissl (ni·s'l). *Med.* The name of Franz *Nissl* (1860–1919), German neurologist, used *attrib.* and in the possessive to designate esp. a methylene blue stain (*Nissl('s) stain*) used for the cell bodies of neurones; the application of this stain (*Nissl('s) method*); and hence the small cytoplasmic structures (*Nissl('s) bodies, granules*) revealed in the cell bodies of neurones by this method. Also **Nissl degeneration,** degeneration of the cell bodies of neurones, accompanied by disappearance of their Nissl bodies; **Nissl('s) substance,** the Nissl bodies collectively.

1898 *Jrnl. Mental Sci.* XLIV. 730 The Nissl bodies.., which stain deeply with basic dyes. 1899 *Jrnl. R. Microsc. Soc.* 448 (*heading*) New Nissl method. 1899 L. F. Barker *Nervous System* 110 He thinks it very wrong that these should be thrown all together and designated either 'Nissl's substance' or 'tigroid substance'. 1901 *Jrnl. Exper. Med.* V. 551 The principles of Nissl's method are extremely simple. *Ibid.* 553 A definite reaction to the Nissl stain. *Ibid.* 554 He finds the Nissl granules embedded in a homogeneous coagulum-like mass. 1905 Gould *Dict. New Med. Terms* 129/1 Staining with Nissl's stain (methylene-blue). 1911 Stedman *Med. Dict.* 586/2 Nissl degeneration. 1933 *Amer. Jrnl. Anat.* LIII. 153 The cells are indeed very light; the Nissl substance is scanty. 1943 Strong & Elwyn *Human Neuroanat.* iii. 28/1 The significance of the Nissl stain.. lies in the fact that..each type of nerve cell always presents the same appearance or 'equivalent picture' in normal conditions. 1966 Wright & Symmers *Systemic Path.* II. xxxiv. 1147/1 In the cytoplasm of the neuron, certain polygonal, basiphile structures, known collectively as Nissl substance, are present. 1968 Passmore & Robson *Compan. Med. Studies* I. xiv. 2/2 Nissl granules are present in the perikaryons of all neurones. 1971 *Sci. Amer.* July 48/2 The Nissl method made it possible to outline and identify well-defined regions in the brain. 1971 J. Z. Young *Introd. Study Man* xi. 147 Nerve cells are characterized by a large amount of ribonucleic acid in the cytoplasm—the basophilic Nissl bodies. 1974 Passmore & Robson *Compan. Med. Studies* III. ii. xxxiv. 33/2 The parent cell bodies undergo central chromatolysis (Nissl degeneration).

nit, *sb.*[1] **2.** Delete †*Obs.* and add further examples. Now esp., a stupid or incompetent person. *colloq.*

Influenced by *nitwit.

1941 Baker *Dict. Austral. Slang* 49 *Nit*, a simpleton or fool. 1961 *Sunday Times* 17 Sept. 41/4 Livings's latest work, 'Sacred Nit'. 1962 *Melody Maker* 11 Aug 6/1 I could see he wasn't very impressed with this nit sitting across the table. 1963 [see *coot *sb.*[1] 4]. 1967 S. Knight *Window on Shanghai* xxiii. 97 You ask a crazy question —Xmas ???!!! The Chinese certainly enjoy celebrating, but for goodness sake! You nit, Mum—can you imagine them trooping to church on December 25? 1972 P. Cleife *Slick & Dead* ii. ix. 70 If you think..I would be willing to allow you..to board my aircraft..then you must be a nit.

4. *Comb.*, as *nit-comb.*

1943 D. Ibberson *Our Towns* iii. 72 A square-toothed steel nit-comb..is too expensive for the poorest. 1959 F. Donaldson *Child of Twenties* iii. 39 My mother washed her head again and again, and combed her hair with a nit-comb.

nit (nit), *sb.*[2] *Austral.* [Origin unknown; but cf. Nix[1] 3, a parallel use.] **1.** A word used as a signal that someone is approaching.

1899 H. Lawson *If I could Paint* in *Stories* (1964) 3rd. Ser. 416 I'd call it 'Nit! There's Mother.' 1911 L. Stone *Jonah* 8 Suddenly there was a cry of 'Nit! 'Ere's a cop!' and the push bolted like rabbits.

2. Usu. in phr. *to keep nit*, to keep watch, to act as guard. Hence **ni·tkeeper,** one who acts as sentinel, usu. while some illegal activity is being carried on.

1935 *Bulletin* (Sydney) 22 May 21/1 That outlaw the sulphur-crested cockatoo is not the only bird to post a 'nit-keeper' when transgressing against society. 1940 I. L. Idriess *Lightning Ridge* 20 Bill kept nit for his elder brother who was courting a girl, and earned a shilling. 1943 D. Stewart *Ned Kelly* ii. i. 97 We keep nit and keep quiet. Ned's good watch-dogs! 1947 N. Lindsay in B. James *Austral. Short Stories* (1963) 6 On condition of you kids scooting up the flat to keep nit, case the old man comes down this way. 1952 T. A. G. Hungerford *Ridge & River* 10 Send two men a couple of hundred yards up and down the track to keep nit. 1963 F. Hardy *Legends from Benson's Valley* 108 An elaborate network of nitkeepers on all sides frustrated the new policeman for three weeks. 1971 D. Ireland *Unknown Industrial Prisoner* 77 They had transgressed the unwritten law that you didn't let yourself go to sleep while you were keeping nit for your mates.

nit (nit), *sb.*[3] *U.S. colloq.* [Of obscure origin; perh. a corruption of Naught *sb.*, Nought *sb.*: cf. *nit *adv.*] None; nothing.

1910 'O. Henry' *Strictly Business* v. 66 'You fool... Why did you do it? 'The Stuff,' explained Thomas briefly. 'You know. But subsequently nit. Not a drop.'

nit (nit), *sb.*[4] *Physics.* [a. F. *nit* (formally adopted in 1948 at the 11th meeting of the Commission internationale de l'Éclairage, and published in its *Recueil des travaux* (1950) 145), f. L. *nit-ēre* to shine.] A unit of luminance equal to one candela per square metre.

1953 J. W. T. Walsh *Photometry* (ed. 2) v. 136 On the c.g.s. system the unit [of luminance] is the stilb, equal to one candela per sq. cm. or the nit, equal to one candela per sq. metre. There is no name for the corresponding unit on the British system; the candela per sq. inch or per sq. ft. is generally employed. 1965 G. A. Fry in R. Kingslake *Appl. Optics* II. i. 18 One nit of luminance is equal to one lux of illuminance per steradian of solid angle. 1965 R. Kingslake in *Ibid.* v. 198, 1 ft-L is equal to 3·43 nits. 1969 *Amateur Photographer* 19 Mar. 80/3 A brightness (or luminous intensity) of 1 Candela per square metre is termed a Nit. Therefore, for example, 60 Candelas per square centimetre equals 600,000 Nits.

nit (nit), *adv.* [? f. Yiddish.] = Not *adv.* Cf. *nit *sb.*[3]

1895 W. C. Gore in *Inlander* Nov. 63 *Nit*.., not; sometimes an emphatic not. 1896 *Dialect Notes* I. 421 *Nit*, a decided negative, much stronger than *no*.

nit (nit), *v.*[2] *Austral.* [Cf. *nit *sb.*[2]] To escape, decamp; to hurry away.

1882 *Sydney Slang Dict.* 10/2 *Nit*, get away (usually from a foe), make tracks. 1897 W. T. Goodge *Hits! Skits! & Jingles!* (1899) 150 And to 'nark it' means to stop it, And to 'nit it' means to fly! 1941 Baker *Dict. Austral. Slang* 49 *To nit*, to decamp, get away (from a foe).

nital (nəi·tæl). *Metallurgy.* Also **Nital.** [f. Nit(ric *a.* + Al(cohol.] An etchant consisting of a few per cent of concentrated nitric acid in ethyl or methyl alcohol.

1925 M. A. Grossmann tr. *Heyn's Physical Metallogr.* iv. 144 Alcoholic Nitric Acid (Martens)... Abbreviation HNO_3/Alc., or nital. 1936 *Metals Handbk.* (Amer. Soc. for Metals) 557 Nital brings out ferrite junction lines clearly while both Nital and Picral etch pearlite clearly. 1950 *Engineering* 24 Mar. 339/1 The barrel is etched with 7 per cent. nital solution. 1963 B. Harocopos *Princ. Structural Metall.* v. 55 Occasionally nital will produce pits in the vicinity of non-metallic inclusions, and in such cases amyl alcohol may be used as a solvent instead. 1975 *Nature* 21 Aug. 635/1 The Nital etch did not reveal recrystallisation in the taenite.

nitchie (ni·tʃi). *Canad.* Also **neche, neechee, neejee, nichi, nichiwa, nidge, nitchee, nitchy.** [Algonquian.] Originally (among North American Indians), a friend; hence as a (usu. derogatory) term for a North American Indian.

1791 J. Long *Voy. & Trav. Indian Trader* 268 Neejee, or neecarnis, friend, or companion. 1838 A. Jameson *Winter Stud. & Summer Rambles Canada* III. 83 Thus, one man addressing another says 'nichi' or 'neejee', my friend. 1850 J. J. Bigsby *Shoe & Canoe* II. 161, I sallied forth, and found the Nidges loading the canoes, drest in their best. 1852 C. P. Traill *Canad. Crusoes* vi. 180 While she called Louis, 'Nee-chee', or friend. 1857 J. Palliser *Jrnls.* (1863) 52 Our Indian friend..to whom we had given the name Nichiwa, or friend. 1878 C. Hallock *Amer. Club List & Sportsman's Gloss.* p. viii, *Nitchee*, a common word among Indian tribes signifying brother. 1903 R. Cullum *Devil's Keg* xxii. 242 A neche was leisurely cleaning up round Labiche's store. 1910 R. W. Service *Ballads of Cheechako* 118 Then came I to a land I knew no man had ever seen, a haggard land, forlornly spanned by mountains lank and lean; The nitchies said 'twas full of dread, of smoke and fiery breath. 1930 J. Beames *Army without Banners* 131 Have to see if I can dig up a Nitchie an' trade whitefish for tobacco or something. 1947 *Beaver* Mar. 4 'These confounded nitchies,' he was wont to exclaim, 'are lazy, good-for-nothings.' 1956 *Saturday Night* 8 Dec. 16/2 And since when have Americans or British freely discussed Clear Grits, Digby chickens, Socreds, the Land of Little Sticks, separate schools, nitchies, longlinermen? 1960 in E. Fowke et al. *Canada's Story in Song* 127 They leave their homes on starving pay to take the nitchies' lives. 1973 R. D. Symons *Where Wagon Led* i. viii. 132 'Quick, you fellows,' he said, 'them Nitchies are crawling up all around.'

nite (nəit), *sb.*[2] An arbitrary respelling of Night *sb.* Also *attrib.* and *Comb.* Hence **ni·tely** *a.* and *adv.*

A widespread vulgarism.

1931 *Amer. Speech* VI. 379 Write *rite* (for right) and *nite* (for night). 1934 B. J. Thompson in *Catholic World* Aug. 523 *Nite* connotes speakeasies, gin, cheapness and vulgarity. 1960 *Punch* 27 Apr. 584/1 Didn't you know? It's Rock Nite at the Darby and Joan. 1961 A. Berkman *Singers' Gloss. Show Business* 62 Nite Club, Nite Spot. 1968 *Blues Unlimited* Nov. 6 Both he and Myers were discovered by Johnny in Jackson 'nite-spots'. 1970 *Toronto Daily Star* 24 Sept. 32/2 (Advt.), Nitely dancing to an excellent European trio. 1970 *Globe & Mail* (Toronto) 25 Sept. 9/1 (Advt.), Train for Court Reporting..Special nite classes. 1971 *Times* 25 Aug. 11/7 (Advt.), Where it's at in Yorkshire... Mood with good food, nitely til 2... And for a romping, Bavarian-style nite out, visit the Intercon Bier Keller in Wakefield. 1971 *Leader* (Durban) 7 May 4/3 (*caption*) Elaine Meyers needs no introduction to nitelifers. *Ibid.*, Around the nitespots. 1973 *Black World* June 61 Sister Habiba knew how to give parties alright: three flights up—Saturday nite. 1974 *Marlboro Herald-Advocate* (Bennettsville, S. Carolina) 18 Apr. 7/8 (Advt.), Free parking in paved lot in rear of store. Open all day Wednesday. Open Fri. nite 'til 6:30.

nitery (nəi·təri). *U.S. slang.* Also **niterie.** [f. *nite *sb.*[2] + -ery.] A night club.

1934 M. H. Weseen *Dict. Amer. Slang* x. 147 *Nitery*, a night club. 1935 *Vanity Fair* (N.Y.) Nov. 38/1 We'll never catch a wire in a decent nitery. 1938 *Amer. Speech* XIII. 239/1 The most recent market nose dive played hob with what Abel Green calls 'the niteries'. 1946 *Sat. Rev.* (U.S.) 19 Oct. 20/3 His reportage is accurate in capturing the contemporary nitery scene. 1955 Priestley & Hawkes *Journey down Rainbow* 129 All darkened niteries and dimmed hot spots. 1958 [see *dream *sb.*[2] 4 h]. 1962 *New Statesman* 19 Jan. 95/3 This negro comedian, who refuses to be winkled out of the grimy little niterie in which he made his reputation. 1967 *Boston Sunday Herald* 30 Apr. (Show Guide) 14/3 Our story begins in a narrow strip of niteries on 52nd Street. 1972 *Daily Tel.* (Colour Suppl.) 1 Dec 62/3 The Body Shop..is a nitery of the topless/bottomless variety, and reminds one that this is..LA's own porno-strip. 1972 *Guardian* 13 Dec. 9/1 El Cubano (the New Niterie with the Olde Worlde Flavor).

Nitinol (ni·tinɒl). [f. *Ni*, chem. symbol for nickel + *Ti*, chem. symbol for titanium + the initial letters of *N*aval *O*rdnance *L*aboratory, Silver Spring, Maryland, U.S.A., the place of work of the metallurgists who discovered the alloy and invented its name.] An alloy of nickel and titanium; *esp.* one composed of equimolecular proportions of these elements, which has the property that after deformation it will return to its former shape when heated to a certain transition temperature.

1968 Buehler & Wang in *Ocean Engin.* I. 105 (*heading*) A summary of recent research on the Nitinol* alloys and their potential application in ocean engineering. [*Note*] *Name derived from Ni-Ti-NOL. Prefix numeral value (e.g., 55-Nitinol) indicates the nominal nickel content in weight per cent, balance titanium. *Ibid.*, The 55-Nitinol was found to be a single phase alloy containing only TiNi phase.., while nickel-rich TiNi alloys showed a $TiNi_3$ phase that co-existed in varying quantity with the TiNi phase. Nominal 55-Nitinol..exhibits a very unusual 'mechanical memory'. *Ibid.* 106, 55-Nitinol and 60-Nitinol must be considered as separate and distinct alloy types. 1969 *Nature* 29 Nov. 844/2 An object made of 'Nitinol' can be deformed to an arbitrary shape, briefly heated to 'fix' that shape, and then possesses a 'memory'. 1971 *Sci. Amer.* Mar. 48/1 At present the only commercial use of Nitinol is in a heat-shrinkable tube coupling.

nitkeeper: see *nit *sb.*[2] 2.

†**niton** (nəi·tɒn). *Chem. Obs.* [f. L. *nit-ēre* to shine + *-on[2].] = *radon.

1911 Gray & Ramsay in *Proc. R. Soc.* A. LXXXIV. 550 To show its relation to gases of the argon series, it should receive a similar name...The name 'niton', Nt, which has been used in this paper is suggested as sufficiently distinctive. 1912 *Bath & Wilts. Chron.* 12 Mar. 4/2 Solid niton causes the glass or silica tube in which it is necessary to confine it to glow with a brilliant light. 1926 G. Birtwistle *Quantum Theory of Atom* xv. 154 The conclusions..agree with Bohr's general scheme, which gives niton orbits up to quantum number 6. 1947 *Science News* V. 157 In the six years 1894–1900 there were added to the periodic table five new elements, constituting a group soon to be completed by niton (the radioactive emanation).

nitpicker (ni·tpikəɹ). Also **nit-picker.** [f. Nit *sb.* + Picker[1].] A pedantic critic; one who searches for and over-emphasizes trivial errors. Hence **ni·tpicking, nit-picking** *vbl. sb.* and *ppl. a.* Hence (as back formations) **ni·t-pick** *v.* and *sb.*

1951 *Collier's* 24 Nov. 67/1 Two long-time Pentagon stand-bys are *fly-speckers* and *nit-pickers*. The first of these nouns refers to people whose sole occupation seems to be studying papers in the hope of finding flaws in the writing, rather than making any effort to improve the thought or meaning; nit-pickers are those who quarrel with trivialities of expression *and* meaning, but who usually end up without making concrete or justified suggestions for improvement. 1956 *Time* 16 Jan. 17 The members of the Cabinet commented on the draft of the message, then commented upon one another's comments. 'No nit-picking,' Vice President Nixon adjured his colleagues, but the Cabinet eventually sent out to the President a file of verbatim reaction that piled 1½ inches high. 1959 *Washington Post* 3 July A 12/2 When the nit-pickers and the parliamentary horse-traders had finished with it, the program had shrunk to much smaller proportions. 1961 *Flight* LXXX. 525 Contributions have not sought to attract the jackdaws nor the 'nitpickers'. 1962 W. Schirra in *Into Orbit* 34 We all tried to avoid nit-picking with each other on these things. 1964 *New Statesman* 14 Feb. 261/3 Some of the..modern buildings..which provide a real feast for art-historical nit-pickers. 1968 *Listener* 4 July 22/1 Knox's essay was a stylistic send-up of German *apparatus criticus* nit-picking in the Bible and the classics. 1970 P. St. Pierre *Chilcotin Holiday* 93 Let us bring it down to the point where no nitpicking critic can disagree. 1970 *New Scientist* 10 Sept. 542/1 If I am giving an impression of nit-picking, I can only apologise. 1971 'D. Shannon' *Ringer* (1972) vi. 108 Don't nitpick. 1972 *Times Lit. Suppl.* 18 Feb. 179/1 A savage, malicious, and nitpicking attack on Malone's great Variorum edition of 1790. 1972 *Guardian* 10 June 10/2 A nit-picking approach would be dangerous and impractical. 1972 *N.Y. Times* 19 Dec. 65/7 Every niggling detail is carefully nitpicked. 1973 *Times* 28 Nov. 4 Nit-picking is an occupational activity of MPs. 1974 *Sat. Even. Post* Jan.-Feb. 32/3 Protect yourself and your leaders from preoccupation with the trivial and the picayune: let people control their own time; don't nitpick procedures. 1975 *Time Out* 1 Aug. 3/3, I don't argue about the inaccuracies (though they're so nit-picking as to infer that everything about the film must be defended at all costs). 1975 *Times Lit. Suppl.* 15 Aug. 922/4 Professor Laqueur's nitpicks force me to comment.

Nitralloy (nəi·trăloi). Also **nitralloy.** [f. *nitr(iding* vbl. sb. (s.v. *nitride *v.*) + Alloy *sb.*] A proprietary name for any of a range of alloy steels specially manufactured for nitriding and usu. containing (among other elements) about one per cent of aluminium and 0·2 to 0·5 per cent of carbon.

1928 *Machinery* XXXI. 478/2 Nitralloys are suitable for crankshafts, camshafts, timing gears, wrist pins, worms,..etc. 1930 *Flight* XXII. 170g/1 The Hispano-Suiza patents do not cover the actual hardening process, but its application to cylinder liners of an alloy steel known as Nitralloy, which is especially suitable for this form of heat treatment. 1952 J. Wulff et al. *Metall. for Engineers* x. 182 Steels (especially Nitralloy, which contains aluminium) may be case-hardened by heating in ammonia at 450–540°C. 1966 *McGraw-Hill Encycl. Sci. & Technol.* XIII. 314/1 The special alloy steels used for nitriding are known as nitralloy. 1972 *Trade Marks Jrnl.* 1 June 1059/1 Nitralloy...Unwrought and partly wrought nitriding steel. Firth Brown Limited,..Sheffield,..manufacturers and merchants.

nitrate, *sb.* Add: **2. b.** Cellulose nitrate (i.e. nitrocellulose), used as a base for cinematographic films. Usu. *attrib.* (see *3).

1949 W. H. Offenhauser *16-mm Sound Motion Pictures* iii. 20 The weight of nitrate, however, is but a fraction of one per cent of the total weight of the film. 1971 L. B. Happé *Basic Motion Picture Technol.* 348 *Nitrate*, cellulose nitrate, a highly inflammable material once used as film base.

3. *nitrate base, film, reduction*; *nitrate-reducing* adj.; **nitrate reductase,** an enzyme or group of enzymes which brings about (the second step in) the reduction of nitrate to nitrite.

1936 A. B. Klein *Colour Cinematography* I. iii. 172 Standard nitrate base for motion picture film is coated with three layers of emulsion containing dyes. 1925 *Chem. Abstr.* XIX. 728 Addn. of 16·6% camphor to one to [*sic*] the strongest nitrate films increased the tensile strength from 3·43 to 8·56 kg. per sq. mm. 1975 *Oxford Jrnl.* 6 June 8 The old one [*sc.* projector room] was fire-proof..because of the highly inflammable nitrate film that was used. 1927 S. A. Waksman *Princ. Soil Microbiol.* vii. 180 The bacteria, which reduce nitrates only to nitrites or to ammonia, but not to nitrogen gas (elementary form and oxides), may be best spoken of as nitrate reducing bacteria, reserving the term denitrifying bacteria for the other organisms. 1967 *Oceanogr. & Marine Biol.* V. 194 Inactivation of the nitrate-reducing system occurred on prolonged incubation at elevated pressures. 1939 *Chem. Abstr.* XXXIII. 2176 Reduction of $—NO_3$ to $—NO_2$ by this bacterium [sc. *E. coli*] involves 2 enzymes, an ordinary reductase and a special nitrate reductase... The special reductase transfers the H [from an intermediate acceptor] to the $—NO_3$. 1964 *Oceanogr. & Marine Biol.* II. 216 Nitrate reductase is found in various green seaweeds. 1974 *Arch. Biochem. & Biophysics* CLX. 269 The enzymatic complex nitrate reductase from *Chlorella fusca* is inactivated by simple thiols. 1919 *Jrnl. Bacteriol.* IV. 267 (*heading*) The use of the nitrate-reduction test in characterizing bacteria. 1973 *Jrnl. Gen. Microbiol.* LXXV. 419 The function of cytochromes in the metabolism of bifidobacteria seems to be associated with nitrate reduction.

nitrazepam (nəitrēⁱ·zĭpæm). *Pharm.* [f. Nitr(o- + Az(o- + *-ep(ine* (suffix designating an unsaturated seven-membered ring containing nitrogen) + Am(ide.] A tricyclic, yellow, crystalline compound, $C_{15}H_{11}N_3O_3$, which is a rapidly acting non-barbiturate hypnotic. Cf. *Mogadon.

1965 [see *Mogadon]. 1970 Passmore & Robson *Compan. Med. Stud.* II. v. 35/1 Nitrazepam is also used as a night sedative and induces sleep without the hang-over effects attributed to the barbiturates.

nitre, *sb.* Add: **1. d.** A sediment produced during the refining of maple syrup.

1872 *Rep. Vermont Board Agric.* I. 219 The gritty sediment from maple syrup, commonly termed 'nitre'. 1882 *Ibid.* VII. 65 The higher the tree is tapped the more of nitre or malate of lime is found. 1949 *Highway Traveler* Feb. 39/1 Strainers..through which the hot syrup is passed to remove the 'nitre', or 'sugar sand', a fine gritty substance, before it is canned.

5. *nitre works.*

1775 *South Carolina Hist. Soc. Coll.* (1858) II. 66 If he was assisted with a sufficient sum..he says he could bring the nitre works to a great degree of perfection.

Nitrian (ni·triăn), *a.* [f. the name *Nitria*: see -an.] Of, pertaining to, or designating the desert region of Nitria in Egypt, *spec.* the Christian hermit monks, renowned for asceticism, who lived there in the fourth century.

1867 C. M. Yonge *Pupils of St. John* ix. 149 Christians ..are said to have preferred the Nitrian valley because of the words of Jeremiah—'though thou wash thee with nitre'. 1892 I. G. Smith *Christian Monasticism* vii. 186 In the famous monastery of St. Gall, in Switzerland, as in the Nitrian monasteries of the fifth century, the whip.. was suspended from a pillar in the chapter-house. 1904 J. O. Hannay *Wisdom of Desert* 6 Journeying still southwards over about forty miles of utterly desolate land, he would come to a long valley extending east and west between two ranges of mountains..covered with.. dangerous rocks. This is the famous Nitrian desert... At the end of the fourth century the Nitrian mountains were dotted over with hermits' cells. 1923 T. E. Lawrence *Let.* 14 May (1938) 416 This sort of thing must be madness... It's terrible to hold myself voluntarily here, and yet I want to stay here till it no longer hurts me... It's a lurid flash into the Nitrian desert: seems almost to strip the sainthood from Anthony. 1958 L. Durrell *Balthazar* iv. 80 His mind winged away like a swallow across the dunes into the Nitrian desert itself.

nitridation (nəitrəidēⁱ·ʃən). *Chem.* [f. Nitrid(e + -ation, after *oxidation*.] A reaction analogous to oxidation but involving nitrogen or its compounds rather than oxygen, water, etc.; *spec.* = *nitriding *vbl. sb.*

1911 Browne & Walsh in *Jrnl. Amer. Chem. Soc.* XXXIII. 1728 One..purpose of this research has..been to study the phenomena of ammono-oxidation, or nitridation. 1929 *Chem. Abstr.* XXIII. 3888 (*heading*) Study of surface-hardening of steel by nitridation. 1931 *Jrnl. Amer. Chem. Soc.* LIII. 1478 Browne.. applied the term 'nitridation' to all those reactions which for the ammonia system of compounds are essentially similar to oxidation reactions for the familiar water compounds... Nitridation reactions are today classed under the broader conception of oxidation reactions. 1974 *Jrnl. Materials Sci.* IX. 1362/2 Nitridations under identical conditions in tubes of two types of commercial sintered alumina resulted in the formation of greatly increased proportions of the beta phase.

nitride (nəi·trəid), *v.* [f. the sb.] *trans.* To convert into a nitride or nitrides; *spec.* to heat in the presence of ammonia or other nitrogen-containing gas so as to form nitrides near the surface and improve the hardness and corrosion resistance.

1928 *Machinery* XXXI. 479/1 The parts to be nitrided are placed in a gas-tight box. 1958 Aitchison & Pumphrey *Using Steel Wisely* i. ix. 155 A number of high-alloy steels—such as those employed for aero-engine exhaust valves—are nitrided with very beneficial results. 1966 *McGraw-Hill Encycl. Sci. & Technol.* XIII. 314/1 It is possible to nitride some of the common alloy steels which do not contain appreciable amounts of aluminium, although the hardness in this case is much lower than with nitralloy. 1972 *Nature* 28 Jan. 219/2 The alloys were nitrided in pure, dry nitrogen at 7 atmosphere pressure at 1,050°C.

Hence **ni·trided** *ppl. a.*, **ni·triding** *vbl. sb.*; *nitriding steel*, steel made with a composition to fit it for nitriding.

1928 *Machinery* XXXI. 478/2 Standard electric furnaces can easily be adapted to the nitriding process. 1928 *Jrnl. Iron & Steel Inst.* CXVII. 855 Photo-micrographs show the nitrided case after nitriding for 90 hrs. at 900°F. 1931 *Ibid.* CXXIV. 615 With the exception of aluminium, all metals present in nitriding steels are transition elements, the nitrides of which have a metallic character. 1952 Kirk & Othmer *Encycl. Chem. Technol.* IX. 26 Nitriding improves the fatigue endurance limit of the heat-treated nitriding steels. 1967 *Trans. Inst. Engin. & Shipbuilders Scotl.* CX. 32 A photograph of a frigate's fine-tooth coupling with barrelled, crowned and nitrided teeth. 1967 M. Chandler *Ceramics in Mod. World* vi. 171 Silicon nitride..is formed in the nitriding process, which consists of taking the element silicon, pressing it into the required form, and then heating to a temperature of about 1400°c in an atmosphere of nitrogen.

nitridizing (nəi·tridəiziŋ), *ppl. a.* *Chem.* [f. Nitrid(e + -iz(e + -ing[2].] Bringing about nitridation.

1911 Browne & Walsh in *Jrnl. Amer. Chem. Soc.* XXXIII. 1728 In case the hydrogen pernitrides in liquid ammonia possess properties similar to those of hydrogen peroxide in aqueous solution it is to be expected that under proper conditions they should act in a sense as oxidizing agents, or more strictly as nitridizing agents. 1930 *Ibid.* LII. 2430 Mercuric nitride and bismuth nitride have been shown to have the properties of nitridizing agents.

nitrifiable, *a.* Add: (Earlier example.) Hence **ni:trifiabi·lity,** the property of being nitrifiable.

1827 *Phil. Mag.* I. 177 Tufa, chalk and nitrifiable materials act in nitrification both as absorbents of water and air, and as presenting a base which solicits the formation of nitric acid. **1884** *Jrnl. Chem. Soc.* XLV. 651 Evidence of the nitrifiability of rape-cake.

nitrification. 1. (Earlier and later examples.)
1827 [see *NITRIFIABLE *a.*]. **1897** *Bull. Minnesota Agric. Exper. Station* No. 53. 7 Although a corn crop takes more nitrogen from the soil than a wheat crop, the cultivation of the corn crop favors nitrification (production of nitrates from humus) and results in leaving more available nitrogen in the soil. **1926** TANSLEY & CHIPP *Study of Vegetation* vii. 122 This process of nitrification..carried out by special bacteria, is of the first importance in humous soils. **1966** *McGraw-Hill Encycl. Sci. & Technol.* IX. 111/1 A well-aerated, fertile, neutral to slightly alkaline soil will provide optimum conditions for nitrification.

nitrifier (nəi·trifəi,əɹ). *Biol.* [f. NITRIFY *v.* + -ER¹.] An organism or soil which nitrifies.
1903 *Lancet* 6 June 1590/1 The bacterial organisms themselves are..the real nitrogen bringers or nitrifiers. **1931** E. ASHBY tr. *Lundegårdh's Environment & Plant Devel.* viii. 241 Mycorhiza occur more commonly in soils which are poor nitrifiers, than in soils in which the nitrification is good. **1973** *Communications Soil Sci. & Plant Analysis* IV. 280 Greater populations of nitrifiers were found by the NPN method than by silica gel plating for most soils.

nitrify, *v.* Add: **1.** (Earlier and later examples.) *spec.* To convert (ammonia) into nitrite or nitrate. Also *absol.*
1827 *Phil. Mag.* I. 174 The putrescent blood was at the distance of two feet from the carbonate of lime, which it is pretended that it nitrified. **1932** FULLER & CONARD tr. *Braun-Blanquet's Plant Sociol.* viii. 239 Probably all communities of the Molinion nitrify abundantly. **1957** G. E. HUTCHINSON *Treat. Limnol.* I. xvi. 837 A considerable part of the ammonia so formed is nitrified. **1962** H. S. MCKEE *Nitrogen Metabolism in Plants* iv. 112 These heterotrophs nitrify more completely than the autotrophic species.
2. (Earlier examples.)
1827 *Phil. Mag.* I. 176 Chevraud met with compact chalks which did not nitrify. **1878** *Jrnl. Chem. Soc.* XXXIII. 46, 10 grams of a vegetable soil, known to nitrify with ease, were washed with water.
nitrifying *ppl. a.* (earlier and later examples). Quot. 1827 is *vbl. sb.*
1827 *Phil. Mag.* I. 175 Why..does it exhibit no nitrifying power without the cooperation of carbonate of lime? **1900** *Jrnl. Chem. Soc.* LXXVIII. II. 97 Experiments with amides, proteids, wine, &c., showed that the nitrifying organisms are not able to attack organic nitrogen; the nitrogen must first be converted into ammonia. **1932** FULLER & CONARD tr. *Braun-Blanquet's Plant Sociol.* viii. 236 The nitrogen compounds in the soil must generally be converted into nitric acid... But this transformation can only take place with the cooperation of nitrifying bacteria. **1969** F. E. ROUND *Introd. Lower Plants* i. 4 The nitrifying bacteria operate in the nitrogen cycle by utilizing ammonia ions which they convert into nitrite (nitrile bacteria), which is then excreted to be absorbed by nitrate bacteria which convert the nitrite into nitrate.

nitrile. Add: **2.** Special Comb.: **nitrile rubber,** any of the copolymers of acrylonitrile with butadiene in various proportions, which have properties resembling those of natural rubber and are used esp. when oil resistance is necessary, as for fuel hoses, containers, and adhesives.
1947 *Mod. Plastics* Oct. 91/2 The particular difference between nitrile rubber and the more commonly used Buna-S or GR-S synthetic rubber is the use of acrylonitrile instead of styrene. **1954** W. L. SEMON in G. S. Whitby *Synthetic Rubber* xxiii. 818 One of the largest uses for nitrile rubber is in the form of latex, in which form it finds application as an adhesive, as a modifier for other water-dispersed resins, and as an impregnant for paper, textiles, and leather. **1972** *Materials & Technol.* V. xiv. 481 The first nitrile rubbers were hot polymerized; later on cold rubbers also became available, which are easier to process.

nitritoid (nəi·trəitoid), *a. Med.* [f. NITRIT(E + -OID.] Resembling the effect of nitrite; said of the crisis which may follow the administration of arsphenamine, which mimics the effect of nitrite poisoning.
1921 in STEDMAN *Med. Dict.* (ed. 6) 677/2. **1943** *Ann. Allergy* I. 144 Urbach was able to prevent nitritoid crisis from arsphenamine, for example, by a similar technique.

nitro-. Add: In sense d also used without hyphen as quasi-adj.
b. nitrofu·ran, any of the furans having a nitro group attached to one of the carbon atoms of the furan ring, some of which (with the nitro group attached to a carbon atom next to the oxygen atom) are used as bacteriostatics; **nitrome·thane,** an oily liquid, CH_3NO_2, which is used as a solvent, as a rocket fuel, and in the production of nitro-compounds; **nitrophe·nol,** any compound containing a nitro and a phenolic hydroxyl group; esp. any of the three possible compounds, $C_6H_4(NO_2)OH$, obtained by substituting a nitro group for one of the nuclear hydrogen atoms of phenol, *spec.* 2- (or *ortho-*)*nitrophenol*, a yellow crystalline compound used as a dyestuff intermediate, and 4- (or *para-*)*nitrophenol*, a colourless or yellow crystalline compound used in the manufacture of phosphorus-containing pesticides and azo-dyes; **nitroto·luene,** any of the four possible compounds, $C_7H_7NO_2$, obtained by substituting a nitro group for one of the hydrogen atoms of toluene, two of which (the *ortho* and *para* isomers) are used as intermediates for dyestuffs.
1930 *Jrnl. Amer. Chem. Soc.* LII. 2550 In connection with the preparation of aminofurans and their diazo-compounds, it was necessary to have a series of readily accessible nitrofurans and their derivatives. **1950** *Jrnl. Pharmacol. & Exper. Therap.* XCVIII. 163 Of the nitrofurans fed in this study, the ones which result in appreciable antibacterial activity in the urine are characterized by a side-chain of the semicarbazone, semi-oxamazone, or closely related type in the 2-position of the furan ring. **1959** *Times* 7 Dec. (Agriculture Suppl.) p. vii/4 To prevent coccidiosis in chickens, nitrophenid, a sulfonamide, or a nitrofuran is added to the feed. **1970** W. H. PARKER *Health & Dis. in Farm Animals* xiii. 180 When the vaccine is used antibiotics and nitrofurans must only be used..for calves showing actual symptoms. **1872** *Jrnl. Chem. Soc.* XXV. 804 Nitromethane is a heavy oil, of a peculiar odour; it boils at 99°. **1950** *Sci. News* XV. 78 A third group..is the monergols, which contain the oxygen needed for their own combustion. Members of this group, such as nitromethane (CH_3NO_2), tend to be unstable, and research is now being intensively carried out with the object of making them safe to handle. **1972** *Materials & Technol.* IV. xv. 548 Nitromethane is used as a solvent for cellulose esters and vinyl resins. **1852** Nitrophenol [in Dict.]. **1905** CAIN & THORPE *Synthetic Dyestuffs* xviii. 149 The formation of these compounds [*sc.* the Nigrosines] is brought about by heating crude nitrophenol..with aniline and aniline hydrochloride. **1949** P. W. VITTUM tr. *Fierz-David & Blangey's Fund. Proc. Dye Chem.* I. 148, *o-* and *p-*nitrophenols are the starting materials for *o-* and *p-*phenetidine and anisidine. **1972** *Materials & Technol.* IV. xv. 554 The nitrophenols are extremely hazardous materials; not only as a fire risk but also because the polynitrophenols are explosive. **1871** *Jrnl. Chem. Soc.* XXIV. 871 (*table*) Nitrotoluene. **1915** *Dyestuffs & Coal-Tar Products* i. 24 2-Nitrotoluene may be..reduced to *o-*azoxytoluene, which is then acidified..and reduced to tolidine sulphate. **1964** N. G. CLARK *Mod. Org. Chem.* xix. 377 A substituent in the side-chain of toluene may be designated 'α-'; for example, α-nitrotoluene, $C_6H_5.CH_2.NO_2$. **1972** *Materials & Technol.* IV. xv. 551 *para-*Nitrotoluene, a brownish-yellow solid.., is employed in the manufacture of *para-*toluidine.
d. nitrobacte·rium [ad. F. *nitrobactérie* (S. Winogradsky 1891, in *Ann. de l'Inst. Pasteur* V. 92)], any nitrifying bacterium; *esp.* one of the genus *Nitrobacter*, which oxidizes nitrites to nitrates; **nitro-cellulose** (further examples); now usu. without hyphen; **Ni·trochalk, ni·tro-chalk,** the proprietary name of a fertilizer consisting of a mixture of ammonium nitrate and calcium carbonate; **nitro-compound,** change stress to *ni·tro-co:mpound* and add to def.: any compound containing a nitro group; (further examples); **nitro group,** the radical $-NO_2$, present in nitric acid; **ni·trolim(e** [LIME *sb.*¹], calcium cyanamide, or a mixture of it with carbon, obtained by treating calcium carbide with nitrogen and used as a fertilizer.
1891 *Jrnl. R. Microsc. Soc.* 680 M. Winogradsky, who at one time ascribed the nitrifying faculty to a single species of bacteria.., has..satisfied himself that morphological differences exist in these organisms, and they are now classed together in a group of 'Nitrobacteria', the common characteristic of which is the oxidation of the ammoniacal nitrogen. **1906** E. W. HILGARD *Soils* ix. 146 The oxidation of the nitrites into nitrates by..rod-shaped bacilli, named nitrobacteria. **1965** B. E. FREEMAN tr. *Vandel's Biospeleology* xix. 335 The nitrobacteria have the effect of mineralising proteins. **1911** E. C. WORDEN *Nitrocellulose Industry* I. xiii. 459 The nitrocellulose silks dissolve in concentrated sulphuric acid. **1931** *Economist* 28 Feb. 431/2 There is reason to believe that nitro-cellulose, lacquers, oils..are all being manufactured or obtained in France. **1955** F. D. MILES *Cellulose Nitrate* vi. 221 On account of its capacity to swell in nitroglycerine and to absorb it, nitrocellulose is an almost indispensable component of both the two principal classes of explosive—blasting explosives..and propellant explosives. **1962** F. T. DAY *Introd. to Paper* v. 53 Nitrocellulose finishing is now an established process for printing work, the smooth and polished surface being obtained by coating on the machine. **1972** *Materials & Technol.* V. xi. 336 Nitrocellulose paints and varnishes dry very fast to give hard, flexible, and reasonably durable films. **1927** *Daily Express* 7 Dec. 12/4 To replace Chilean nitrate we shall make..nitrochalk, a rich mixture of nitrogen and calcium. **1936** *Trade Marks Jrnl.* 4 Nov. 1355/1 Nitro-Chalk... Artificial fertilisers for soils... I.C.I. (Fertilizer & Synthetic Products) Limited,.. London,..manufacturers. **1954** *Jrnl. Brit. Grassland Soc.* IX. 323 (*heading*) The influence of 'Nitro-Chalk' on established lucerne leys. **1966** WEBSTER & WILSON *Agric. in Tropics* viii. 197 It would seem wise to consider other types of nitrogenous fertilizer, such as nitro-chalk (15·5 per cent N) or ammonium nitrate.., as alternatives to the long-continued use of sulphate of ammonia. **1928** ADKINS & MCELVAIN *Elem. Org. Chem.* x. 101 Many of these nitro compounds are of considerable importance. **1962** P. J. & B. DURRANT *Introd. Adv. Inorg. Chem.* xix. 679 The nitro compounds are made by the action of silver nitrite on the alkyl iodide. **1886** E. F. SMITH tr. *V. von Richter's Chem. Carbon Compounds* 79 The nitro-group always exerts such an acidic influence upon hydrogen linked to carbon. **1938** C. D. HURD in H. Gilman *Org. Chem.* I. vii. 628 The peculiar activity of the fourth nitro group in $C(NO_2)_4$ should be mentioned. **1964** J. W. LINNETT *Electronic Struct. Molecules* iv. 65 There is much evidence which suggests that the nitro-group (NO_2) is a very stable group in many different molecules. **1908** *Trans. Faraday Soc.* IV. 104 A great outcry was, and still is, made warning farmers against the use of calcium cyanamide, popularly known as nitrolim, or at least advising that it should be employed with the utmost caution. **1909** *Jrnl. Chem. Soc.* XCVI. 893 (*heading*) Formation of 'Nitrolime'. **1923** J. HENDRICK *Farmer's Raw Materials* x. 156 When nitrolim is applied to the soil its nitrogen quickly turns to ammonia, and the ammonia in turn changes to nitrate. **1962** J. H. WHITE *Inorg. Chem.* xxi. 331 Nitrolime is a soluble fertilizer, rich in nitrogen.

nitro. Add pronunc. (nəi·tro). **2.** Abbrev. of NITROGLYCERINE. *slang.*
1935 N. ERSINE *Underworld & Prison Slang* 54 *Nitro*, nitroglycerine. **1950** R. CHANDLER *Let.* 18 May in *R. Chandler Speaking* (1966) 80 Opening a good safe (without a time lock) requires expensive and heavy tools, the finest drills either to drill out the lock or to get in the nitro if he is a peterman. **1972** J. GODEY *Three Worlds* (1973) iii. 30 They had an old-time safe... I hit it with a fat charge of nitro.

nitrofurantoin (nəitrofiŭræ·nto,in). *Pharm.* [f. NITRO- + *fur-furyl* (f. FURFUR-OL + -YL) + HYD)ANTOÏN in the chemical name of the compound, 1-(5-*nitrofur*furylideneamino)-hyd*antoïn*.] A yellow crystalline bicyclic compound, $C_8H_6N_4O_5$, which is an antibacterial agent used in treating infections of the urinary tract.
1953 *Antibiotics & Chemotherapy* III. 151 A new antibacterial nitrofuran, Furadantin, Eaton, brand of nitrofurantoin. **1963** *Amer. Rev. Respiratory Dis.* LXXXVIII. 712/2 Most tubercle bacilli with increased resistance to isoniazid also manifest increased susceptibility to nitrofurantoin. **1972** M. ROWLAND in Melmon & Morrelli *Clin. Pharmacol.* ii. 30/2 The most common side effect associated with nitrofurantoin is gastrointestinal irritation and occasional emesis due to a central action of the drug, probably dependent on high concentrations of nitrofurantoin in blood.

nitrofurazone (nəitrofŭ·rāzōᵘn). *Pharm.* [f. NITRO- + *FUR(FURALDEHYDE + *SEMICARB)AZONE in the chemical name of the compound, 5-*nitrofur*furaldehyde semicarb*azone*.] A yellow crystalline cyclic compound, $C_6H_6N_4O_4$, which is an antibacterial agent used locally on burns, wounds, and skin infections, and in veterinary medicine.
1947 F. K. OLDHAM et al. *Essent. Pharmacol.* xxiv. 323 5-nitro 2-furaldehyde semicarbazone (nitrofurazone) has been recently introduced under the name of furacin as a dressing for wounds and chronic ulcers. **1957** *Veterinary Rec.* LXIX. 1415/2 In Great Britain the drug nitrofurazone has been widely used and flocks of domestic poultry very commonly are given this drug continuously as a preventative measure against Eimeria infections. **1972** *Cancer Research* XXXII. 2623/1 Nitrofurazone chemotherapy was usually initiated after both surgery and radiotherapy had failed to arrest the metastatic invasion of testicular cancer to the lungs and its subsequent growth.

nitrogen. Add: **b.** *nitrogen-containing* adj.; **nitrogen cycle** *Biol.*, the cycle of changes whereby nitrogen is interconverted between its free state in the air and combined states in organisms and the soil; **nitrogen fixation,** the conversion of free, gaseous nitrogen into a combined form; cf. *FIXATION 2 d; so **nitrogen fixer,** a nitrogen-fixing organism; **nitrogen-fixing** *ppl. a.*, bringing about nitrogen fixation; **nitrogen mustard,** any of the group of substances containing the group $-N(CH_2CH_2Cl)_2$, which are cytotoxic as a result of their alkylating property and some of which are used in treating neoplastic diseases, such as Hodgkin's disease and lymphosarcoma; **nitrogen narcosis,** a narcotic state, common among divers, induced by breathing air under pressure.
1956 *Nature* 4 Feb. 234/1 A nitrogen-containing acid. **1967** *Oceanogr. & Marine Biol.* V. 168 If some of the nitrogen-containing gases (evolved during the digestion procedure) escape..the values obtained may be too low. **1908** HALL & DEFREN tr. *Abderhalden's Text-bk. Physiol. Chem.* x. 198 The discovery that ordinary nitrogen can be directly assimilated closes the chain of the nitrogen cycle, which had apparently been broken open by the discovery of the denitrifying organisms. **1931** E. ASHBY tr. *Lundegårdh's Environment & Plant Devel.* viii. 242 The nitrogen cycle..lies to a great degree, though not exclusively, in the soil. **1966** *McGraw-Hill Encycl. Sci. & Technol.* IX. 111/2 The nitrogen cycle comprises the processes of ammonification, nitrification, denitrification, and nitrogen fixation. **1895** LAWES & GILBERT *Rothamsted Exper.* 164 Recently acquired knowledge in regard to nitrogen fixation. **1938** *Nature* 12 Nov. 878/1 Diverse workers..have previously affirmed nitrogen fixation by Blue-green Algæ. **1970** *Sunday Times* (Colour Suppl.) 16 Aug. 13/1 It is likely that Germany would have collapsed before 1918 but for the nitrogen fixation plants built in 1914–16. **1912** E. J. RUSSELL *Soil Conditions & Plant Growth* iv. 83 In all these cases leguminous plants are present in greatest extent where the gains in nitrogen are greatest, but they are not necessarily the only nitrogen fixers. **1932** FULLER & CONARD tr. *Braun-Blanquet's Plant Sociol.* viii. 235 Anaerobic and aerobic nitrogen fixers often occur associated together in the same soil. **1968** *Jrnl. Gen. Microbiol.* L. 487 As these small colonies grew abundantly on nitrogen-poor agar, it was thought possible that some of them might be nitrogen-fixers. **1899** F. H. KING *Irrigation & Drainage* 233 The nitrogen-fixing tubercles were already developed. **1929** WEAVER & CLEMENTS *Plant Ecol.* x. 249 Both nitrifying and nitrogen-fixing bacteria thrive in the humus. **1966** *New Statesman* 2 Dec. 826/1 The first new component isolated from the Du Pont nitrogen-fixing system. **1970** *Nature* 25 July 378/1 The nitrogen-fixing enzyme complex, nitrogenase, catalyses various reductions as well as the formation of ammonia from dinitrogen. **1946** *Jrnl. Amer. Med. Assoc.* 21 Sept. 132/1 Although indications and contraindications for the use of the nitrogen mustards remain to be established definitively, it is felt that these agents are deserving of further clinical trial. **1951** *New Biol.* XI. 97 In the treatment of leukaemia..several different classes of chemicals show some value. There are some compounds, such as the nitrogen mustards, which imitate the action of radiation on the cell, and dislocate nuclear division. **1970** PASSMORE & ROBSON *Compan. Med. Stud.* II. xxix. 5/1 Chlorambucil..can be given by mouth and is the slowest acting and least toxic of all the nitrogen mustards in clinical use. **1937** *U.S. Naval Med. Bull.* XXXV. 379 It is well known that if pressure is applied too quickly the diver becomes dizzy and often is so dazed as to require several minutes to orient himself. Consequently, if the cause were nitrogen narcosis, the difficulty would increase with exposure rather than decrease. **1962** *Listener* 29 Mar. 562/1 Free deep-sea divers—that is to say, divers wearing cylinders filled with compressed air—may suffer at varying depths from a mild intoxicating effect, or 'rapture of the deep', known as nitrogen narcosis. **1972** *Aerospace Med.* XLIII. 1079 (*heading*) Diver performance: nitrogen narcosis and anxiety.
nitrogenize *v.*, add to def.: *spec.* = *NITRIDE *v.*; also **nitrogeniza·tion,** the action or result of nitrogenizing.
1896 *Jrnl. Iron & Steel Inst.* L. 161 Iron specially nitrogenised by the action of ammonia is materially altered in character. **1903** *Lancet* 6 June 1590/1 The increased nitrogenisation of the soil by the widened use of phosphatic manures. **1916** *Engineering* 3 Mar. 218/2 A sample of iron was taken, nitrogenised at 600 deg., ..and then thrown out from the Heraeus furnace into very cold water. **1922** *Ibid.* 29 Sept. 413/1 To obtain positive evidence of nitrogenisation both unannealed and annealed specimens of armco iron were analysed for nitrogen. **1926** *Chem. Abstr.* XX. 2138 In Ni and Mn steels nitrogenization produced hardly any increase in hardness.

nitrogenase (nəitrǫ·dʒĕnēiz, -s). *Biochem.* [f. NITROGEN + *-ASE.] The enzyme which combines with molecular nitrogen as the first step in biological nitrogen fixation.
1934 D. BURK in *Ergebnisse der Enzymforschung* III. 24 Nitrogenase is the specific enzyme within the azotase system that combines directly with N_2 with characteristic affinity. **1950** *Federation Proc.* IX. 548/1 The remarkable effect of molecular nitrogen appears to require activation of nitrogenase which in some way results in simultaneous activation of the dehydrogenases which are thereby enabled to compete successfully with the hydrogenase systems. **1975** *Nature* 3 Jan. 7/3 It is now a tacit assumption of all model studies that a metal ion, most probably molybdenum, is responsible for binding dinitrogen in nitrogenase.

nitrogenate (nəitrǫ·dʒĕnēit), *v.* [f. NITROGEN + -ATE³.] *trans.* To combine with nitrogen; *spec.* = *NITRIDE *v.* So **nitro·genated** *ppl. a.* Also **nitrogena·tion,** the action or process of nitrogenating.
1926 *Jrnl. Iron & Steel Inst.* CXIII. 600 Parts to be case-hardened by nitrogenation must be completely machined and finished off before case-hardening. **1927** *Ibid.* CXV. 893 Brinell tests were made on a chrome-aluminium steel nitrogenated by heating in ammonia. **1938** *Chem. Abstr. Third Decennial Index* 6205/3 Nitrogenation, see *nitridation*. **1967** *Biol. Abstr.* XLVIII. 1394/2 Ways of controlling nitrogenation [of soil] are..treated. **1969** KUNII & LEVENSPIEL *Fluidization Engineering* xv. 517 (*heading*) Conversion of a particle in the nitrogenation of calcium carbide. **1971** *Biol. Abstr.* LII. 8948/2 (*heading*) Tests of the 'Pro-Milk Automatic Apparatus' for use in the serial determination of nitrogenated matter in milk.

nitron (nəi·trǫn). *Chem.* [ad. G. *nitron* (Busch & Mehrtens 1905, in *Ber. d. Deut. Chem. Ges.* XXXVIII. 4049), f. *nitro-* NITRO- + *-on*.] A heterocyclic compound, $(C_6H_5)_3C_2HN_4$, which is used in gravimetric analysis as a precipitant for nitrate, perrhenate, and some other ions.
1906 *Jrnl. Chem. Soc.* XC. I. 118 As a test for nitrates nitron is even more delicate than was stated previously. **1939** A. I. VOGEL *Text-bk. Quantitative Inorg. Analysis* i. 165 The following acids form slightly soluble salts with nitron..perchloric, thiocyanic,..and nitrous acids. **1971** *Mikrochim. Acta* 478 Small amounts of gold can be extracted into dichloroethane as nitron tetrachloroaurate.

nitronium (nəitrōᵘ·niə̆m). *Chem.* [mod.L. (coined in Ger. by A. Hantzsch 1925, in *Ber. d. Deut. Chem. Ges.* LVIII. 943): see NITRO- and *-ONIUM.] †**a.** The cation which was supposed to be formed by ionization in pure nitric acid and formulated as $[NO(OH)_2]^+$ or $[N(OH)_3]^{2+}$. *Obs.*
1925 *Chem. Abstr.* XIX. 2312 The fairly high cond[uctivity] of pure HNO_3 is probably due not to ionization of H^+ and NO_3^-, but to formation of 'nitronium nitrate' $[NO(OH)_2][NO_3]$ or $[N(OH)_3][NO_3]_2$.
b. The cation NO_2^+, which is the nitrating agent in mixtures of concentrated nitric acid with another strong acid and is formed by the dissociation of nitric acid itself.
1946 R. J. GILLESPIE et al. in *Nature* 5 Oct. 480/1 The freezing point evidence..provides an unambiguous proof that the cation into which nitric acid is actually converted in sulphuric acid is..the nitronium ion, NO_2^+. **1961** *Aeroplane* CI. 549/2 New oxidizers such as nitronium perchlorate are being applied experimentally in propellents containing well established binder systems. **1972** S. J. WEININGER *Contemp. Org. Chem.* xv. 371 Salts of the nitronium cation containing nonnucleophilic anions may be prepared and..used as nitrating agents.

nitrophilous (nəitrǫ·filəs), *a. Bot.* [f. NITRO(GEN + -PHILOUS.] Of a plant: growing best in a habitat rich in nitrogen. So **ni·trophile** *sb.*, a plant of this type.
1909 GROOM & BALFOUR tr. *Warming's Oecol. Plants* xvii. 68 Nitrophilous plants..thrive best in soil where compounds of ammonium and nitric acid are abundant. **1932** FULLER & CONARD tr. *Braun-Blanquet's Plant Sociol.* iv. 63 Here [*sc.* among ecologically specialized forms] belong many mycotrophic species, saprophytes, nitrophiles. *Ibid.* viii. 240 Nitrophilous communities are very widely distributed in dry, subtropical regions. **1964** V. J. CHAPMAN *Coastal Vegetation* iv. 97 These [*sc.* marsh plants] are probably nitrophiles. **1967** M. E. HALE *Biol. Lichens* iv. 58 This phenomenon [*sc.* lichens growing on guano] has given rise to a large literature on nitrophilous or ornithocoprophilous lichens. **1972** A. MITCHELL in Leigh & Noble *Plants for Sheep in Australia* v. 42 In eastern Victoria..the native perennial grasses could be maintained in the sward with the introduced clovers without the same tendency to invasion by nitrophilous weeds.

nitrosamine (nəitrōᵘ·sămīn). [ad. G. *nitrosamin* (O. N. Witt 1875, in *Ber. d. deut. Chem. Ges.* VIII. 857): see NITROSO- and AMINE.] Any of the class of compounds containing the group $>N-NO$, which can be prepared by the action of sodium nitrite and a strong acid on secondary amines.
1878 O. N. WITT in *Jrnl. Chem. Soc.* XXXIII. 202 The term 'nitrosamine' I apply to any substituted ammonia which contains, instead of at least one atom of hydrogen, the univalent nitrosyl group,—NO, in immediate connection with the ammoniacal nitrogen. **1912** *Chem. Abstr.* VI. 2319 (*heading*) The nitrosamine of *p-*nitro-*o-*anisidine as a red resist under aniline black. **1942** FUSON & SNYDER *Org. Chem.* x. 118 The nitrosamines are really amides of nitrous acid and as such would be expected to be neutral compounds. **1954** *Chem. Abstr.* XLVIII. 7305 (*heading*) Printing of cotton with nitrosamine dyes using a neutral developer. **1975** *Daily Colonist* (Victoria, B.C.) 15 May 3/2 Animal research studies have shown that nitrate and nitrite preservatives may produce cancer-causing chemicals, called nitrosamines, in the stomach.

nitrosate (nəi·trosēit), *v. Chem.* [f. NITROS(O- + -ATE³.] *trans.* To introduce a nitroso group into (a compound).
1920 F. A. MASON tr. *G. von Georgievics's Text-bk. Dye Chem.* 66 The monoazo dyes derived from resorcinol can be nitrosated, but the mordant dyes obtained possess too slight tinctorial powers to be of much technical value. **1952** K. VENKATARAMAN *Chem. Synthetic Dyes* II. xxv. 783 Rhoduline Pure Blue 3G is prepared by nitrosating diethyl-*m*-phenetidine, and condensing with *m*-diethylaminophenol. **1973** *Nature* 12 Oct. 326/1 The mutagenicity of MG nitrosated in human gastric juice was studied by the spot test.
Hence **ni·trosated, ni·trosating** *ppl. adjs.* Also **nitrosa·table** *a.*, capable of being converted into a nitroso compound; **nitrosa·tion,** the process of converting into a nitroso compound.
1920 F. A. MASON tr. *G. von Georgievics's Text-bk. Dye Chem.* 66 On nitrosation, *p*-nitroso derivatives are obtained where the *para* position is free. **1968** R. O. C. NORMAN *Princ. Org. Synthesis* xi. 385 Nitrosation is limited to the very reactive nuclei of phenols and tertiary aromatic amines. **1970** R. PRICE in K. Venkataraman *Chem. Synthetic Dyes* III. vii. 371 Certain of the complexes which are disclosed are derived from nitrosated *o*-hydroxyarylazoresorcinol dyestuffs. **1972** *Science* 7 July 66/3 These are also the nitrosating species for secondary amines and alkylureas. *Ibid.* 67/3 Addition of ascorbate to certain foods containing nitrate or nitrosatable compounds might be worth considering. **1973** *Nature* 12 Oct. 326/1 Nitrosated MG was found to be mutagenic at up to sixteen-fold dilution of the original reaction mixture. **1974** *Ibid.* 8 Nov. 179/1 Ascorbate might be used to block *in vivo* formation of N-nitroso compounds from nitrosatable chemicals.

nitroso-. Add: Also used without hyphen as quasi-*adj.* **nitroso group,** the nitrosyl radical, $-N:O$.
1885 *Jrnl. Soc. Dyers & Colourists* I. 177/2 Tetramethyltolylenediamine seems to give a similar nitroso compound. **1906** J. M. MATTHEWS tr. *Alexeyeff's Gen. Princ. Org. Syntheses* ii. 58 (*heading*) Substitution of the nitroso group. **1911** I. W. FAY *Chem. Coal-Tar Dyes* v. 71 Naphthol green B. is a nitroso dye which, unlike the other members of this class, does not require a mordant upon the fabric to be dyed. **1938** C. D. HURD in H. Gilman *Org. Chem.* I. vii. 635 This tendency for nitroso compounds to change to oximes..is apparent even in the nitroso alcohols or acids. **1971** N. L. ALLINGER et al. *Org. Chem.* xxii. 604 Aliphatic nitroso compounds which bear at least one hydrogen on the carbon α to the nitroso group are isomerized rapidly and irreversibly by acid or base to oximes.

nitrous, *a.* Add: **2. g.** *nitrous vitriol*, a solution of oxides of nitrogen in sulphuric acid produced in the Gay-Lussac tower in the lead-chamber process.
1879 *Chem. News* 30 May 237/2 In a paper about to be published, I [*sc.* G. Lunge] have proved that denitration by hot water or steam is insufficient when the nitrous vitriol, by faulty work, contains nitric acid. **1933** W. T. READ *Industr. Chem.* xi. 165 A portion of the acid from the coolers is sent to the top of the cold tower. As it passes down the cold tower it picks up the oxides of nitrogen in the form of nitrosyl sulfuric acid and becomes..'nitrous vitriol'. **1954** KIRK & OTHMER *Encycl. Chem. Technol.* XIII. 472 The Glover tower receives the hot burner gas, and is fed at the top with the nitrous vitriol from the Gay-Lussac tower, and with 52° Bé. (65%) acid from the chambers.

nitto (ni·to), *v. Criminals' slang.* [Cf. *NIT *sb.*²] (See quot. 1959.)
1959 J. GOSLING *Ghost Squad* ii. 25 'Nitto' means 'stop' or 'Be quiet'. **1962** D. WARNER *Death of Bogey* iv. vi. 169 You guys better nitto. The Sparrow's got a line to your run-in.

nitty-gritty (ni:ti,gri·ti). *slang* (orig. *U.S.*). [Etym. unknown.] The realities or basic facts of a problem, situation, subject, etc.; the heart of the matter. Also *attrib.* or as *adj.*
1963 *Time* 2 Aug. 14/2 The Negroes present would know perfectly well that the nitty-gritty of a situation is the essentials of it. **1963** *Wall St. Jrnl.* 12 Sept. 14/1 Says W. C. Patton, field secretary for..the National Association for the Advancement of Colored People. 'Now we're down to the nitty-gritty, the hard core who've never been interested in politics.' **1967** *Freedomways* VII. 186 All those 'nitty gritty' actions and styles which set Negroes off from the rest of American society. **1967** *N.Y. Times* 27 June 20 He's not afraid to get down to the nitty-gritty of unpleasant problems. **1968** *Times* 15 Nov. 17/2 To get down to what the American will call the 'nitty-gritty' of the matter—the heart, sir, the heart. **1969** *Listener* 25 Sept. 420/2 The Animals were already into the nitty-gritty of blues history. **1973** *Computers & Humanities* VII. 163 Most of the Harris work covers the nitty-gritty problems of subject analysis. **1974** *Financial Times* 6 Mar. 36/2 Mr Wilson is expected to appoint a trade union MP or two as junior Ministers at the Department of Employment to make up for Mr Foot's lack of experience in the 'nitty-gritty' of trade union negotiations. **1975** *Publishers Weekly* 13 Jan. 56/3 He still can startle the reader with his abrupt shifts from nitty-gritty reality to far-out fantasy.

nit-weed (ni·twīd). [f. NIT *sb.* + WEED *sb.*¹] A North American herb, *Hypericum gentianoides*, of the family Guttiferæ, having wiry stems, scale-like leaves, and yellow flowers; also called orange grass and pine-weed.
1818 A. EATON *Man. Bot.* (ed. 2) 422 Nit-weed, false john's wort... On the sandy plain west of Ball's spring, New Haven. **1843** J. TORREY *Flora N.Y.* I. 89 Ground Pine. Nitweed. Pine-weed. **1907** A. B. LYONS *Plant Names* (ed. 2) 414 Sarothra... Orange-grass, Pine-weed, Bastard Gentian, Ground Pine, Nit-weed, False Johnswort.

nitwit (ni·twit). *colloq.* [Perh. f. NIT *sb.* 2 (in Dict. and Suppl.) + WIT *sb.*] A stupid person, a person of little intelligence. Also *attrib.* and *transf.*
1922 *Dialect Notes* V. 142 Nit-wit. **1926** L. NASON *Chevrons* v. 170 Listen, nit-wit. The rocket you want is a yellow smoke. **1928** *Daily Express* 1 June 9 The Vice-President announced loudly that he wanted a large cup of coffee with his dinner, and none of these 'nit-wit, pee-wee, demi-tasses'. **1930** *Musical Times* Nov. 987 Music..of the type that the nitwits who write..to the *Radio Times* call dry and highbrow. **1933** *Punch* 8 Feb. 157/2 *Barbara.* It's awful explaining to a nitwit. When you're out you put down. *Me.* Put what down? *Barbara.* Your cards, idiot! **1958** *Times Lit. Suppl.* 11 Apr. 193/3 Like some nitwit from a slapstick comedy, Turvey may be outwitted, but he is never really worsted. **1975** J. DRUMMOND *Slowly the Poison* II. 168 For God's sake, Beryl, don't be such a nitwit.

ni·twitted, *a.* [f. prec. + -ED².] Lacking in intelligence. Hence **nitwi·ttedness,** stupidity.
1931 *Observer* 6 Dec. 11 Many of the American films are just as shoddy, just as nitwitted. **1942** A. CHRISTIE *Body in Library* ix. 85 That half-baked nitwitted little slypuss. **1952** *Canad. Forum* XXXII. 13 These gems of phony surprise, of noxious nitwittedness.

nitwittery (nitwi·təɹi). [f. *NITWIT + -ERY.] Imbecility, stupidity, lack of intelligence; foolish behaviour.
1936 *Punch* 29 Apr. 504/1 Eight of the stories are adventures which befell members of that singular focus of nitwittery, the Drones Club. **1949** C. FRY *Lady's not for Burning* 96 Last, vulgarity, cruelty, trickery, sham And all possible nitwittery. **1965** *New Statesman* 6 Aug. 194/1 It does not advance dramatic art to..tolerate nitwittery for the sake of a Cause.

Niuean (ni,ū·ăn, niŭ·iăn), *sb.* and *a.* [f. *Niue*, native name (= 'world') of an island in the S. Pacific + -AN.] **A.** *sb.* **a.** A native or inhabitant of Niue. **b.** The language of the Niuean people. **B.** *adj.* Of or pertaining to Niue, the Niueans, or their language.
[**1893** *Jrnl. Polynesian Soc.* II. 11 The words were obtained from the New Testament translated into Niue by the Rev. Frank Lawes. *Ibid.* 13 The missionaries..'induced' two of the Niue youths to accompany them.] **1901** *Jrnl. R. Anthrop. Inst.* XXXI. 138 The Tongan mode of warfare was frontal attack by desperate charges; the Niuéan, a series of feints intended to frighten the enemy, and entice him into ambush. **1902** B. THOMSON *Savage Island* p. v, I went to the Niuéans in the name of the Queen and Empress whom the world is still lamenting. **1924** R. W. WILLIAMSON *Social & Pol. Syst. Cent. Polynesia* II. xiv. 53, I have..referred to Tregear's dictionary and to a Niuean vocabulary provided by Williams. **1954** K. B. CUMBERLAND *Southwest Pacific* vi. 270 It was native missionaries, Samoans mainly, who converted the Niueans. **1962** H. LUKE *Islands S. Pacific* xii. 222 In the latter half of the [19th] century..Niuean men would voluntarily leave the island..to work phosphate for good wages on Malden Island. **1968** *Encycl. Brit.* XVIII. 205/2 The best known of the dialects [of the Polynesian language] include Hawaiian,..Tongan, Tahitian, Niuean,..and Tuamotuan. **1973** *Guardian* 21 May 13/4 The Cook Islanders and Niueans are New Zealand citizens, have free right of entry... The Fijians and Tongans need individual permits.

nivation (nəivēi·ʃən). *Geol.* [f. L. *niv-*, *nix* snow + -ATION.] Erosion of the ground beneath and at the sides of a snowbank, mainly as a result of alternate freezing and thawing.
1900 F. E. MATTHES in *21st Ann. Rep. U.S. Geol. Survey* II. 183 These névé effects..I shall, for the sake of brevity, speak of as effects of nivation. **1918** *Proc. Nat. Acad. Sci.* IV. 288 (*heading*) The importance of nivation as an erosive factor, and of soil flow as a transporting agency, in northern Greenland. **1957** G. E. HUTCHINSON *Treat. Limnol.* I. i. 59 A few cases may be found where nivation, or the freezing and thawing of water round patches of snow, has produced small closed depressions in jointed or fractured rocks. **1968** R. W. FAIRBRIDGE *Encycl. Geomorphol.* 123/2 When a deep snowdrift fails to melt away during summer, periodic freezing and thawing of the constantly moistened ground around and beneath it leads to the breakup of the rock particles which are then removed by meltwaters. This process is known as nivation.

ni:velliza·tion. *rare.* [f. F. *nivel-er* to level + -IZATION.] A making level or equal.
1879 VIGFUSSON & POWELL *Icelandic Reader* 469 There is a nivellization of all vowels as to their quantities. **1947** R. WELLEK in H. Smith *Columbia Dict. Mod. Europ. Lit.* 186/1 There was after the war a most impressive expansion—with the concomitant dangers in a break with tradition, in commercialization and nivellization.

Nivernois (nivẽ́rnwa). [f. the title of Louis Jules Mancini Mazarini, Duc de Nivernais or *Nivernois* (1716–98), f. Nivernais, *Nivernois*, a former province of central France.] A tricorn hat with a wide brim, fashionable in the late eighteenth century. Also *attrib.*
1765 in Cunnington & Beard *Dict. Eng. Costume* (1960) 147/1 He wears this large umbrella-like hat. This is the Nivernois. **1766** C. ANSTEY *New Bath Guide* x. 68 What with my *Nivernois*' Hat can compare? **1770** E. P. *Art of Dressing Hair* 8 For they to shining Balls the Camp prefer'd, Nor e'er of Powder and Pomatum heard, Of silken Suits, or Nivernois genteel. **1960** CUNNINGTON & BEARD *Dict. Eng. Costume* 147/1 *Nivernois hat*, a tricorne hat with broad spreading brim rolled over a flat crown; known as the 'Nivernois cock'. **1969** R. T. WILCOX *Dict. Costume* (1970) 248/1 *Nivernois*, a diminutive tricorne worn by the English Macaronies with the cadogan wig in in 1770's.

‖**Nivose** (nivōᵘ·z). [F. *Nivôse*, ad. L. *nivōsus* snowy: see NIVOSITY.] The fourth month of the French revolutionary calendar, extending (in 1793–94) from 21 Dec. to 19 Jan.
1802 C. WILMOT *Let.* 3 Jan. in *Irish Peer* (1920) 21 Sunday—3rd Jany. 1802. 13 Nivose, An 10. **1838** H. NICOLAS *Chronol. Hist.* (ed. 2) 184 The French Republicans ..used the number of the day of each month of their Calendar. For example: 1,2,..30 Nivose. **1884** *Encycl. Brit.* XVII. 266/2 He arrested and transported one hundred and thirty persons, whom he knew to be innocent of the plot... This is Nivose, an act as enormous as Fructidor, and with a perfidy of its own. **1957** *Ibid.* IX. 804/1 The winter months were *Nivôse*, the snowy, *Pluviôse*, the rainy, and *Ventôse*, the windy month. **1972** R. COBB *Reactions to French Revol.* iv. 134 On 7 Nivôse year VI (27 Dec. 1798), the *Ministre*..reported to his colleague.

nix¹. Add: **1.** (Further examples.) Also, = No *adv.*³; not possibly.
1909 *Dialect Notes* III. 352 *Nit, nix(y*, adv. Variants of *no*. Slang. **1926** J. BLACK *You can't Win* vi. 67 'I'll go to the farmhouse..and buy something.' 'Nix, nix,' said one; 'buy nothin'.' **1929** A. CONAN DOYLE *Maracot Deep* 14 If I pull down fifty bucks a week it's not for nix. **1932** D. L. SAYERS *Have his Carcase* xxiii. 309 As for getting an experienced actor and giving him a show in the part—nix! **1951** GREEN & LAURIE *Show Biz* 570/1 *Nix*, no, veto, thumbs-down. **1959** 'J. R. MACDONALD' *Galton Case* (1960) xvi. 132 'He..wanted his old job back. Nix.' A gesture of his spread hand swept Lemberg into the dust-bin.
c. Phr. *nix on*, enough of, have done with, no more of. *colloq.*
1902 'H. MCHUGH' *It's up to You* iii. 55 We decided before we stepped on the Pullman that it would be nix on the sweetheart talk. **1923** R. D. PAINE *Comrades of Rolling Ocean* iv. 62 Camp Stuart at ten o'clock. Nix on that kid stuff. **1941** BAKER *Dict. Austral. Slang* 49 *Nix on it!*: Stop it! Lay off!
d. = No *a.*; none, negligible. (Also directly from G. *nix* (or *nichts*), in bilingual conversation.)
1906 E. DYSON *Fact'ry 'Ands* vii. 84 No man can reasonably expect t' live ther life iv er hindependent gent on er nix income. **1928** H. CRANE *Let.* 31 Jan. (1965) 315 One can generally 'place' people to some extent; but out here it's mostly nix. **1945** G. MORGAN *Only Ghosts can Live* xii. 144 'Nix Fish-Tins.' *Gefangener*—German for '*Nicht verstehen*': I don't get you. **1971** *Daily Tel.* (Colour Suppl.) 12 Nov. 21/2 Oh, I just said battery kaput, nix lights, nix motor... And we fix. But all European trucks stop for each other here.
3. (Earlier example.) Also, as a children's 'truce-word'.
1860 in HOTTEN *Dict. Slang* (ed. 2). **1959** I. & P. OPIE *Lore & Lang. Schoolch.* viii. 152 Nicks or nix. Prevailing term [as a truce-word] in Warwick.

nix, *v.* [f. NIX¹.] **1.** *trans.* To cancel, forbid, refuse. Freq. as imp., beware, cease (doing something).
1903 H. HUTCHINS *Autobiogr. Thief* viii. 180, I started in to talk about old times in the stir and..he answered me by saying 'Nix', which meant 'Drop it'. **1914** [see *EYEFUL *sb.* c]. **1934** M. H. WESEEN *Dict. Amer. Slang* xxi. 371 *Nix*, to refuse an offer; a refusal; to deny a request; a denial. *Nixy* is a variant. **1945** in Wentworth & Flexner *Dict. Amer. Slang* (1960) 356/1 The blue-penciler nixed the story. **1961** *New Left Rev.* July/Aug. 53/1 Every time somebody nixes..paid work to fulfil an unpaid commitment..my faith..is..restored. **1969** R. V. BESTE *Next Time I'll Pay* ii. 22 He could have been more explicit... If he had been his holiday would have been nixed, that was for sure. **1973** *Tucson* (Arizona) *Daily Citizen* 22 Aug. 11/1 (*heading*) Nude bathing nixed. **1974** *Maclean's Mag.* Dec. 19/1 It was the inner voice that nixed the deal—the savings contract that the salesman was pushing ensured that I wouldn't break even until after the first eight years.
2. *nix out* (*on*): *U.S. slang* in various senses (see quots.).
1940 *Music Makers* May 37/3 *Nix out*, to eliminate, get rid of. Ex. 'I nixed that chick out last week'. 'I nixed my garments' (undressed). **1945** L. SHELLY *Jive Talk Dict.* 29/2 *Nix out*, to erase. **1946** MEZZROW & WOLFE *Really Blues* (1957) 84 The owner nixed big crowds out. **1969** Nix out on [see *FADE *sb.*¹]. **1970** C. MAJOR *Dict. Afro-Amer. Slang* 85 *Nix out*, to throw away.

nixie¹. (Later poet. examples.)
1918 D. H. LAWRENCE *New Poems* 62 It seems to me The woman you are should be nixie, there is a pool Where we ought to be. **1952** R. CAMPBELL tr. *Baudelaire's Poems* 214 If your gaze the gaze transfixes Of satyresses or of nixies.

nixie² (ni·ksi). Also **nixy.** [f. NIX¹ + -Y⁶, -IE.] **1.** = NIX¹ 1 b. Also *attrib. U.S.*
1890 *Cent. Dict.*, Nixy². **1901** *Congress. Rec.* 17 Jan. 1145/6 These poor 'nixie' clerks in the postoffices of this country. **1904** *Springfield* (Mass.) *Republ.* 29 Oct. 4 He was made what is known in the office as a 'nixy' clerk.. one who looks up misdirected letters. **1905** *N.Y. Even. Post* 8 Feb. 5 What the railway postal clerks most dread is the class of mail they know as 'nixies'. **1929** *Lit. Digest* 5 Oct. 67/1 The similarity in appearance of the letters N.Y. and N.J...is responsible for many letters reaching the 'Nixie' division. **1949** *Amer. Speech* XXVI. 136/1 *Nixie*, a piece of mail so damaged or illegible that it can go no farther in the mails. The *Nixie* section rewraps and tries to discern the scribbled addresses on mail. **1956** *Daily Progress* (Charlottesville, Va.) 19 Sept. 28/1 'Nixie' is mail that can't be delivered because the address is incorrect, illegible, or insufficient.
b. = NIX¹ 1 a (in Dict. and Suppl.).
1906 E. DYSON *Fact'ry 'Ands* xviii. 249 Er storm centre ..redooced..land values t' next' t' nixie.
2. *adv.* No, certainly not. Freq. as expletive.
1886 H. BAUMANN *Londinismen* 120/1 Nixey..nein, nicht. **1903** *Pedagogical Seminary* X. 373 Nixy... Don't you believe it. Not much. **1914** G. ATHERTON *Perch of Devil* I. 108 They're all right to marry,..but to sacrifice your life for, nixie.

Nixonian (niksōᵘ·niăn), *a.* [f. the name of Richard M. *Nixon* (b. 1913), Vice-President of the United States 1953–61 and President 1969–74 + -IAN.] Of, pertaining to, or resembling Richard M. Nixon.
1959 *Listener* 10 Sept. 376/1 Thomas Edmund Dewey was a Nixonian sort of man. **1970** *Harper's Mag.* July 30/3 It was not, after all, critics of the Administration who defined Nixonian language for us—Don't watch what we say, watch what we do—it was the Administration spokesmen themselves. **1971** *Between Lines* (Newtown, Pa.) 1 Nov. 1/1 Voluntarism, the theme of the new Nixonian-economics, puts capitalism to its most severe test. **1972** *N.Y. Times Book Rev.* 29 Oct. 46/3 Herblock analyzes Nixonian techniques such as the Happening or Non-Event in which the Vice President is photographed standing at a podium addressing the President and full Cabinet and telling them what a great job they are doing. **1973** *Nation Rev.* (Melbourne) 31 Aug. 1445/1 Change indeed in nixonian Washington!
Hence **Nixonite** (ni·ksǫnəit), a supporter of Nixon or his policies; **Nixonization** (ni:ksǫnəizēi·ʃən), development of a Nixonian character; subjection to the influence of or domination by supporters of Nixon; **Nixonomics** (niksǫnǫ·miks) *sb. pl.* [f. *econ*)*omics* (ECONOMIC *sb.* 2 c)], the economic policies of Nixon.
1958 *Time* 17 Nov. 22 A staunch Nixonite. **1960** *Guardian* 10 Nov. 10/7 There were some Nixonites present. **1969** *Demo Memo* (Washington, D.C.) 29 Oct. 1 Walter Heller..challenges the soundness of 'Nixonomics'. **1971** *New Statesman* 16 Apr. 516/3 Alternatives to the Hobson's choice of possibly counterproductive street protest or mute acquiescence in the Nixonisation of the war. **1971** *N.Y. Times* 19 Sept. 2E/2 The President dredged up a term he had always treated with derision as a symbol of economic blundering by his Democratic predecessors and converted it to the dynamics of the new Nixonomics. **1972** *Newsweek* 16 Oct. 89/1 Phase two of the new Nixonomics has changed nearly all the rules—and as one result, quite a few companies these days..are going out of their way..to cut profits down. **1973** *Black World* Mar. 68 There is certain to be greater Nixonization of government at every level. **1975** *Maclean's Mag.* Jan. 6/1 Richard Nixon was able to take dramatic economic action over a single weekend in August, 1971 (remember the new Nixonomics?) without a word of sass or backtalk from Congress.

nkonze, var. *KONZE.

Nkrumahism (n,krū·mă,iz'm). Also **Nkrumaism.** [f. the name of 'Kwame *Nkrumah*' (Francis Nwia Kofi) (1909–72), Ghanaian political leader: see -ISM.] The principles and policies associated with Kwame Nkrumah; adherence to or support of these principles. So **Nkru·ma(h)ist** *a.* and *sb.*, (supporting, one who supports) the policies and principles of Nkrumah; of or pertaining to these policies and principles. Also **Nkru:mahiza·tion,** the making of changes in accordance with his policies and principles.
1960 *Sunday Times* 28 Aug. 5/1 Orgy of phrases... These phrases..have cropped up constantly..'Nkrumahism' (off-stage among Ghanaians). **1962** *Listener* 23 Aug. 272/2 Bishop Roseveare had 'tried to subvert the Nkrumahist revolution'. **1964** *Economist* 1 Feb. 396/2 The forming of regional groups is not inimical to pan-African unity (as Nkrumahism says it is). **1968** *Listener* 11 July 53/1 In 1964, 'local conditions'—that is to say, Kwame Nkrumah—demanded that the University appoint certain known 'Nkrumahists' to key positions. The sole specific intellectual content of 'Nkrumahism' was the implied infallibility of Kwame Nkrumah. *Ibid.*, At the University we had the choice of accepting 'Nkrumahisation' or 'sheltering behind political freedom.' **1973** *Ann. Reg. 1972* 247 There were those who saw the new military regime in Accra as returning to a form of 'Nkrumaism without Nkrumah'. *Ibid.*, General Nathan Aferi..was reported as saying..'There are Busiaites and Nkrumaists and uncommitted.'

No., N°. Add: (Further examples.)
No longer restricted to the designation of individual things and persons.
1840 DICKENS *Let.* 4 Feb. (1969) II. 18, I am curious to see how the idea of the first No. of my projected work strikes you. **1905** *Strand Mag.* Apr. 376/1 George II.. made a present of No. 10 [Downing Street] to his Prime Minister, Sir Robert Walpole, and his successors. **1924** J. BUCHAN *Three Hostages* xvii. 242 It was eventually arranged that the district-visitor should call at No. 4 the following afternoon. **1937** N. MARSH *Vintage Murder* v. 47 He was advancing a No. 4 company in St. Helens. I was selling tickets for the worst show in England. **1972** *Oxford Times* 15 Sept. 18/1 Conversion of house into 2 No. self contained flats. **1974** M. GILBERT *Flash Point* ii. 14 It's No. 276 Coalporter Street.
b. *No. 9* (mil.): see quot. 1917.
1911 *Pharmaceutical Formulas* (ed. 8) 922/2 (*heading*) Army Pill No. 9. **1917** A. G. EMPEY *Over Top* 301 'No. 9.' A pill the doctor gives you if you are suffering with corns or barber's itch or any disease at all.

no, *a.* Add: **I. 1. a.** (Later examples.) Still *dial.* or illiterate.
1896 [see *CON *v.*]. **1897** [see *CLASS *sb.* 5 b]. **1968** *Listener* 20 June 796/3 He's not going to be put in no poorhouse.
3. b. (Earlier U.S. examples.)
1832 J. P. KENNEDY *Swallow Barn* II. xvii. 220 Which ..would produce a cure 'in almost no time'. **1868** G. G. CHANNING *Early Recoll. Newport R.I.* 143 The money was ..arranged as to facilitate the payments in 'no time', understood in my day, to mean the shortest period.

II. 5. b. (Further examples.)
In quot. 1948 the sense is 'without the use of the hands'.
1940 *Ann. Reg. 1939* 253 A vote of no-confidence was carried in the Lower Chamber. **1948** D. BALLANTYNE *Cunninghams* I. xviii. 92 Ralph showed off, riding no-hands and skidding in the loose metal. **1954** F. C. AVIS *Boxing Reference Dict.* 75 *No contest*, a declaration of the referee that the fight is null and void, usually because both contestants are making no serious efforts at boxing. *Ibid.*, *No-count*, a slipping to the floor of the ring but getting up again before the count begins. **1956** J. G. PORTER in A. Pryce-Jones *New Outl. Mod. Knowledge* 142 Adaptable as he [*sc.* man] is, can he exist for any length of time under conditions of no-gravity? **1957** L. F. R. WILLIAMS *State of Israel* 158 The four years of the legislature's statutory life (which is, of course, always liable to be shortened by a vote of no-confidence). **1960** *Times* 5 July 18/2 Edwards has had 10 contests and won nine of them, featuring rather unluckily in a no-contest (or double disqualification). **1973** *Houston* (Texas) *Chron. Mag. People, Places, Pleasures* 14 Oct. 8/4 Agnew..pleaded no contest—in effect, guilty—to cheating on his income tax.
d. Used in various colloq. phrases, as *no strings*, no conditions or obligations; also *attrib.* (cf. *STRING *sb.*); *no stuff*, no joking; *no sweat*, no bother, no trouble.
1909 'O. HENRY' *Options* (1916) 50 I've told you..my oral sentiments, and there's no strings to 'em. **1946** MEZZROW & WOLFE *Really Blues* (1957) 376 *No stuff*, no kidding. **1952** A. HUXLEY *Let.* 12 Oct. (1969) 658 The thing should start in a small way, but with adequate equipment, no strings and no red tape. **1955** *Amer. Speech* XXX. 118 *No sweat, no strain* n. phr. used adjectivally, easy, no trouble, no difficulty. **1960** *Ibid.* XXXV. 122 *No sweat*, the GI's reaction to an unpleasant, but necessary task. **1963** *Daily Mail* 11 Nov. 8/8 Mumblemouth especially knows how to blow flicks that cop bread, no-sweat style (knows how to succeed in movies without really trying). **1965** *Economist* 6 Mar. 980/3 In effect, the family doctors will get a no-strings pay rise averaging 9 per cent. **1970** *Times* 18 Aug. 15 Following the February £13m no-strings pay deal..union officials have been conducting a wages and conditions survey of motor plants in Britain. **1970** C. MAJOR *Dict. Afro-Amer. Slang* 85 *No stuff*, expression that implies sincerity. **1972** 'H. HOWARD' *Nice Day for Funeral* iii. 45 He respects me as a person. No strings. **1972** *Publishers Weekly* 16 Oct. 17/1 Mrs Wallach complains that she cannot use plastic book jackets on books with maps on the inside covers. No sweat! We paste the book pocket..on the next inside page, [etc.]. **1973** K. GILES *File on Death* vi. 153 No sweat, mate... We're not looking for trouble.
e. Denoting the complete emptying of the mind described in Buddhist, and esp. Zen, philosophy as *no-mind, no-thought*, etc.
1934 D. T. SUZUKI *Essays in Zen Buddhism* III. ii. 84 'Mind is still subject to measurement. Who is the Buddha?' 'No-mind is he.' **1949** —— *Zen Doctrine of No-Mind* 29 When..the seeing of self-nature has no reference to a specific state of consciousness, which can be logically or relatively defined as a something, the Zen Masters designate it in negative terms and call it 'no-thought' or 'no mind', *wu-nien* or *wu-hsin*. **1956** A. HUXLEY *Adonis & Alphabet* 34 No-thought not-thinks about the world in terms of no-things. *Ibid.*, In Zen the virgin consciousness was called *Wu-nien* or *Wu-hsin*—no-mind or no-thought. **1959** C. C. CHANG *Practice of Zen* ii. 59 The so-called No-mind (Chinese: *Wu hsin*) is not like day, wood, or stone, that is, utterly devoid of consciousness; nor does the term imply that the mind stands still without any reaction when it contacts objects or circumstances in the world. It..is natural and spontaneous at all times... There is nothing impure within it; neither does it remain in a state of impurity. When one observes his body and mind, he sees them as magic shadows or as a dream... When he reaches this point, then he can be considered as having arrived at the true state of No-mind. **1959** D. T. SUZUKI *Zen & Jap. Culture* iv. 74 All things are accomplished when one attains a mind of 'no-mind-ness' according to the great Zen master. **1960** A. KOESTLER *Lotus & Robot* II. x. 240 It [*sc.* Zen] proclaims to be the philosophy of no-mind (Wu-hsin), of no-thought (Wu-mien)..and of 'going ahead without hesitation'. **1966** P. KAPLEAU *Three Pillars of Zen* II. v. 201 Mindlessness, on the other hand, or 'no-mindness' as it has been called, is a condition of such complete absorption that there is no vestige of self-awareness.
6. b. (Further examples.) Also, denoting absence of necessity for. See also *NO-FINES *a.* (*sb.*).
1902 R. MACHRAY *Night Side of London* ii. 23 The clubs, both high-class and no-class, are not all closed. **1930** E. POUND *XXX Cantos* vii. 27 Brown-yellow wood, and the no colour plaster. **1936** 'J. BEYNON' *Planet Plane* 58, I didn't think we were going to hit the no-gravity zone so soon. **1939** No class [see *billiard-hall* (*BILLIARDS 2)]. **1955** *N.Y. Times* 13 Feb. III. 8/1 The favorite fabric is the no-iron type. **1958** *Economist* 1 Nov. 435/2 Nobody really doubts that 'no deposit' business will also be done. **1961** P. WHITE *Riders in Chariot* viii. 233 For Chrisake! Who am I to know what is up to every no-hope Jew that comes to the country? **1963** B. FOZARD *Instrumentation Nuclear Reactors* xi. 132 The reading of the voltmeter may be corrected to zero under no-signal conditions. **1969** *New Statesman* 18 July 80/1 'It's a gas, man, it's a rave,' says a no-bra girl. **1970** *Globe & Mail* (Toronto) 26 Sept. 1/5 (*caption*) Empty cans and no-deposit bottles lie around a tree. **1971** *Flying* (N.Y.) Apr. 18/2 The evidence seemed clear that the no-accident day had been moved from Thursday to Tuesday. **1972** *Guardian* 2 Nov. 10/3 A no-hope telephonist with an invalid mother, an illegitimate child and a bad communication problem. **1973** *Times* 19 Mar. 21/1 The United States Justice Department filed both civil and criminal suits. Ford entered a 'no contest' plea and last month was fined a total of $7m. **1973** GAGNON & SIMON *Sexual Conduct* (1974) x. 291 The no-bra look is serving both males' fantasies and a return to naturalness.
7. b. (Later example.)
1916 JOYCE *Portrait of Artist* (1969) i. 50 His nocoloured eyes looking through the glasses.

no, *adv.*¹ **1.** (Later *Sc.* examples.)
a **1894** R. L. STEVENSON *Weir of Hermiston* (1896) viii. 245 Oh, my dear, that'll no dae! **1931** A. J. CRONIN *Hatter's Castle* II. xiii. 435 What was't he said, 'a loyal wife and a devoted mother,' wasn't no? **1973** *People's Jrnl.* (Inverness & Northern Counties ed.) 28 July 4/5 Who says the Scots are a dour lot? No' us anyway! **1975** M. RUSSELL *Murder by Mile* iii. 22 What's holding ye up?..Was the tyre no' checked?

no, *adv.*² *no better than* (one) *should* (or *ought to*) *be*: see *BETTER *a.* 5.

no, *adv.*³ and *sb.* Add: **A.** *adv.* **1.** (Further examples.)
1853 *Harper's Mag.* Feb. 402/1 This is the gentleman who can't say no. **1853** T. C. HALIBURTON *Sam Slick's Wise Saws* I. v. 119 You first of all force yourself into my cabin, won't take no for an answer, and then complain of oncivility. **1861** G. H. LEWES *Let.* 20 Aug. in *Geo. Eliot Lett.* (1954) III. 446 She allows herself to be preyed upon dreadfully by the boys—she can't say No. **1930** W. S. CHURCHILL *My Early Life* iv. 74 Come on now, all you young men... Don't take no for an answer, never submit to failure. **1961** *Family Jrnl.* Dec. 15/3 'But he would not take "No" for an answer,' she went on. **1961** *Listener* 21 Dec. 1065/2 He was made Minister of Labour in a season when the Government's economic policy meant saying 'no' to wage demands. **1974** M. BUTTERWORTH *Man in Sopwith Camel* viii. 89 I'm warning you that I'm not taking no for an answer. **1975** C. STORR *Chinese Egg* vii. 41 'I can manage. You keep sitting down.' 'I shan't say, No. It's a long drag up to St. Monica's.'
3. c. *no, you don't*: see *DO *v.* 29 b.

No,² Nō: see *NOH.

noa (nōᵘ·ă). [a. Hawaiian (Maori, Tahitian) *noa* (something) free from taboo, ordinary.] An expression substituted for a taboo word or phrase. Freq. in *Comb.*
1925 O. JESPERSEN *Mankind, Nation & Individual* ix. 169 This harmless word that has been substituted, the Polynesians call *noa*. *Ibid.* 179 Both the above Danish names for devil are originally noa-names. *Ibid.* 184 Certain words and expressions are taboo for certain persons and must therefore be replaced by other words, by noa-words. **1951** S. ULLMANN *Words & their Use* III. iii. 75 If a word is struck by a taboo ban, it must be replaced by a harmless alternative, a so-called *noa* term. **1956** *Trans. Philol. Soc. 1955* 1 The special terms used by the [Faroese] fishermen to replace ordinary words when these became tabooed are designated as noa words or noa names, etc. Noa denotes the opposite of taboo, in the present context it stands for what is allowed in contrast to that which is forbidden. **1975** W. B. LOCKWOOD *Lang. Brit. Isles* 55 [Cornish] *pajerpaw* 'newt', lit. 'four foot'..clearly a noa name.

no-account, *a.* orig. *U.S.* Of no account, importance, value, or use; insignificant, worthless. (See also *NO-'COUNT *a.*)
1845 *Spirit of Times* 1 Feb. 583/2 I'll just tell you that the land I'm after is a d–d, little, no-account quarter section, that nobody would have but me. **1886** *Pall Mall Gaz.* 29 Sept. 4/1 We submit to be..treated as no-account people in all affairs of State. **1898** H. S. CANFIELD *Maid of Frontier* 109 It seems to me..that Charlie is gettin mighty no account. **1900** CONRAD *Lord Jim* v. 47 The other two no-account chaps spotted their captain, and began to move towards us. **1902** G. H. LORIMER *Lett. Merchant* vi. 69 A mailing-clerk so no-account as to be writing personal letters in office hours. **1936** 'J. TEY' *Shilling for Candles* v. 51 'You mean she thinks he's a wrong 'un?' 'No. Just no account.' *Ibid.* vii. 71 We both felt no-account and were afraid people'd find it out. **1952** C. DAY LEWIS tr. *Virgil's Aeneid* XI. 246 Are we, we no-account souls, to litter the plains? **1973** M. MACKINTOSH *King & Two Queens* xii. 180 I'm only a no-account Irisher, but I like to pay my debts.
B. *sb.* A 'no-account' person.
1896 'MARK TWAIN' in *Harper's Mag.* Sept. 523 Who ever had anything agin that poor trifling no-account? **1936** J. DOS PASSOS *Big Money* 169 The child of a no-account like Fred. **1970** W. GARNER *Puppet-Masters* xxx. 222 Why is a no-account like Lindsay collected from a police court by a Special Branch cop? **1973** *Philadelphia Inquirer* (Today Suppl.) 14 Oct. 43/2 Others—those Gypsies most respected and strongest in the community—will not challenge him because they believe he is a 'no-account'.

Noah's Ark. Add: **5.** *U.S.* An orchid of the genus *Cypripedium*, esp. the pink *C. acaule* and the yellow *C. calceolus* var. *pubescens*.
1826 W. DARLINGTON *Florula Cestrica* 95 C[*ypripedium*] *pubescens*... Noah's Ark. Yellow Mocasin [*sic*] flower. **1898** C. A. CREEVEY *Flowers of Field* 296 Stemless Lady's Slipper. Noah's Ark. Moccasin-flower. *Cypripedium acaule*. **1949** E. L. PALMER *Fieldbk. Nat. Hist.* 148/2 It bears a number of common names such as Noah's ark, squirrel's shoes, camel's foot, nerveroot, old goose, Indian moccasin, and two lips.
6. *Rhyming slang.* **a.** A police informer. **b.** *Austral.* A shark. (See also quots. 1941, 1960.)
1898 *Bulletin* (Sydney) 17 Dec. (Red Page), An informer or mar-plot is a nark or a Jonah or a Noah's Ark. **1941** BAKER *Dict. Austral. Slang* 49 *A Noah's ark*, a dull, witless fellow. A rhyme on 'nark'. **1941** 'V. DAVIS' *Phenomena in Crime* xix. 254 A stoolie, Noah's Ark, a grasshopper, a nark or informer. **1952** *Chambers's Shorter Eng. Dict.* 800/1 *Noah's ark* (*slang*), a shark. **1960** J. FRANKLYN *Dict. Rhyming Slang* 101/1 *Noah's ark*, (1) nark (an informer), (2) park. (1) has been in use in England since the first decade of the 20 C. It is also used in the form of a Spoonerism '*oah's Nark*, the first word having the inference, '*whore's*' and when thus inverted it is the supreme expression of contempt.
7. (See quot. 1968.)
1945 *Archit. Rev.* XCVII. 59/3 In the 1920's an import from Germany, simpler in erection, faster and thus more thrilling, threatened the supremacy of the switch-back... This new machine—the Noah's Ark..had decorative features which continued the tradition of coarse splendour. **1965** *Observer* (Colour Suppl.) 29 Aug. 10/2 The showmen are setting up their stalls and heavy riding machines: Dodgems, Noah's Arks, Waltzers, Shows, and Hurricane Jets. **1968** D. BRAITHWAITE *Fairground Archit.* 153/2 The term 'Noah's Ark' referred to a circular ride having an undulating track..and a variety of fixed animals carried on segmental platforms, all contained within a static enclosure.

nob, *sb.*² Add: **1.** *bob a nob*: see *BOB *sb.*⁸ 2.

no ball, no-ball, *sb.* Add: **2.** (Earlier and later examples.)
1876 *Haygarth's Cricket Scores* V. 176 Mr. Hankey was bowled by a no ball..; he afterwards carried out his bat. **1955** *Times* 6 Aug. 4/3 The loping run, all arms and legs, and the no-ball were not forgotten. **1974** *Sunday Tel.* 3 Mar. 33/7 Hayes, his middle stump removed by a Holder no-ball.., remained to fight another day. **1975** *Cricketer* May 19/2 There were at least three occasions when he bowled batsmen with no-balls.
3. *fig.* (also *attrib.*).
1922 [see *BEAVER⁴]. **1939** *John o' London's* 9 June 361/2 Mr. Chance's solution is much too complicated, and he does not play the game quite fairly, because he has two men who are related and can pass for each other, which.. has been considered a 'no-ball' in detective fiction for a long time. **1966** *Sunday Times* (Colour Suppl.) 4 Dec. 73/2 GI Jargon. *No-ball Target*, Air Corps nickname for a German rocket launching site.

no-ball, *v.* (Earlier examples of both senses.)
1862 *Baily's Mag.* Oct. 201 John Lillywhite..also 'No balled' the third, fourth, fifth, sixth, and seventh delivered by Willsher, who thereupon walked off the ground. **1867** G. H. SELKIRK *Guide to Cricket-Ground* iv. 60 Dean, as umpire, no balled Mr. C. D. Marsham three times in one over.

nobbily, *adv.* (Examples.)
1859 HOTTEN *Dict. Slang* 69 *Nobby*, fine or showy; *nobbily*, showily. **1877** 'MARK TWAIN' in *Atlantic Monthly* Nov. 591 Two hundred Bermudians..all of them nobbily dressed, as the poet says. **1880** *Punch* 25 Dec. 299/1 There wasn't a chap in the room so good-looking or nobbily drest.

no·bbiness. [f. NOBBY *a.* + -NESS.] Smartness; affected elegance.
1909 *N.Y. Even. Post* 7 Aug. 2 He unwittingly errs.. on the side either of a certain scrubbiness or of an even more unfortunate 'nobbiness'. **1935** *Amer. Speech* X. 10/1 Around 1900 when the present-day elderly good-people were criminal Dapper Dans, their lingo was the last word in linguistic nobbiness.

nobble, *v.* Add: **1.** (Later examples.)
1933 WODEHOUSE *Heavy Weather* iii. 34 At any moment ..the bounder was liable to come sneaking in, mask on face and poison-needle in hand, intent on nobbling the favourite. **1951** *News Chron.* 18 Dec. 1 Lord Rosebery confirms today that his horse which was nobbled was Snap.
b. (Earlier and later examples.) Also in somewhat weakened sense: to reduce the efficiency of (a person, etc.) by some means.
1856 LD. CLARENDON *Let.* in H. R. C. Wellesley *Paris Embassy* (1928) 103 Morny..seems to have talked with enthusiasm about the Empress-mother, and to have been quite nobbled by her. **1912** F. M. HUEFFER *Panel* I. ii. 31 'I want to point out to Miss Delamere that you can't reform the theatre without reforming the conventional idea about marriage.'.. 'Oh, I see,' the major said amiably, 'you want to nobble her before she makes any business arrangements with my uncle.' **1939** H. NICOLSON *Diary* 20 Sept. (1967) 36 Margesson insisted that..he [*sc.* Churchill] must be 'nobbled' by having a department which would occupy all his time. **1960** J. FINGLETON *Four Chukkas to Australia* xvii. 142 He was left bereft when this particular style of 'nobbling' the opposition was outwitted in the first Test. **1963** *Times* 13 May 6/6 Unions felt that if they were to take part in the work of the National Incomes Commission they would be 'nobbled at the start.' **1973** 'M. UNDERWOOD' *Reward for Defector* v. 39 What about the rest of the delegation?.. No chance of nobbling one of them?
3. Also, to strike; to kidnap; to 'steal'.
1841 C. H. HARTSHORNE *Salopia Antiqua* 517 *Nobler*, a man whose duty it is to remind inattentive youths in church, of their misbehaviour, by '*nobling*' them, or hitting them on the head with a wand. **1865** MILTON & CHEADLE *N.-W. Passage by Land* xv. 306 His son had succeeded in 'nobbling' a brace of partridges, knocking the young birds out of the trees with short sticks, missiles they used with great dexterity. **1922** JOYCE *Ulysses* 497 You once nobble that, congregation, and a buck joy ride to heaven becomes a back number. **1928** E. WAUGH *Decline & Fall* vii. 66 'What sort of job?' I says. 'Nobbling,' he says, meaning kidnapping. **1932** H. J. MASSINGHAM *World without End* 296 Off he goes to 'bibble' a

mug of scrumpy and 'nobble' a hunk of cheese. **1968** M. Woodhouse *Rock Baby* viii. 64 We've got this Shackleton we've nobbled off Coastal Command.

nobbler². Add: (Further Austral. and N.Z. examples.) Also, a small glass or container for liquor.
1852 'G.F.P.' *Gold Pen & Pencil Sk.* xiv. (Morris), The summit gained, he pulls up at the Valley, To drain a farewell nobbler to his Sally. **1853** J. Sherer *Gold-Finder of Australia* 177, I have only had two noblers (as they are called) since I came to the place, and paid 1s. 6d. per nobler. **1859** F. Fowler *Southern Lights & Shadows* 52 (Morris), To pay for liquor for another is to 'stand' or to 'shout'... The measure is called a 'nobbler' or a 'breakdown'. **1873** Trollope *Austral. & N.Z.* II. xi. 201 A nobbler is the proper colonial phrase for a drink at a public-house. **1885** R. C. Praed *Austral. Life* 103 Having accepted at my hands the customary 'nobbler', he would sit down for half-an-hour, talking. **1908** D. Ferguson *Bush Life in Austral. & N.Z.* (ed. 4) xxxv. 274 Nor did their thirst for ardent spirits appear to be in the least moderated by the price of the beverages..the good old colonial charge of one shilling per 'nobbler'. **1936** M. Franklin *All that Swagger* xix. 181, I took a nobbler of your poisoned grog. **1949** D. M. Davin *Roads from Home* 216 He..was pouring it into two nobblers he had fished out of the pocket. **1957** D. Niland *Call me when Cross turns Over* ii. 30 He poured himself nobbler after nobbler and drank them straight. **1971** *Walkabout* (Austral.) Nov. 73/1 Whisky costs around 300 rupiahs, or some 75 cents, for a generous nobbler.

nobbut, *adv.* **1.** (Later *dial.* examples.)
1929 J. B. Priestley *Good Companions* I. v. 196 It's nobbut Thursday, isn't it? Well, it seems like months. **1957** 'B. Buckingham' *Boiled Alive* xi. 61 There was nobbut a bunch of dirty foreigners here. **1963** *Times* 25 May 9/7 Mr. Vernon Horsfall still makes clogs. 'But it's a mak a finished is t'trade, you know. It's nobbut farmers and folk in weaving sheds.'

nobby, *sb.* Add: **2.** Also used more widely around the Irish Sea, and by the Royal Navy.
1936 E. Vale *Seas & Shores England* ii. 19 The Morecambe Bay fishermen with their specially evolved cutter-rigged smack called a *nobby* have been for generations famous throughout the three western seas of Britain. **1948** R. de Kerchove *Internat. Maritime Dict.* 491/1 *Nobbie, nobby.* 1. A round-sterned, two-masted, lug-rigged fishing boat found on the south coast of Ireland... 2. A pointed-stern fishing boat of the Mersey estuary rigged with a jib, a dipping lug foresail, and a standing lug mizzen. **1953** J. Masefield *Conway* (rev. ed.) iv. 209 We had three sailing dinghies—and the nobby—a heavier boat of about eighteen feet. **1970** E. J. March *Inshore Craft Gt. Britain* II. viii. 280 The early smacks, 'nobbies' to use the local name, were about 36 ft long, and drew 4 ft of water. **1973** W. Elmer *Terminol. Fishing* i. 26 In the west, an impressive pattern is formed by the distribution of the *prawners* and *shrimpers* of the Cumberland and Lancashire coast, and the Lancashire *nobby.*
3. *Austral.* Black opal found as a silica drop (the characteristic form at Lightning Ridge, N.S.W.).
1924 T. C. Wollaston *Opal* i. ii. 10 Characteristic forms of the Black Opal are locally known as 'Nobbies'...pseudomorphs after sponges and corals. **1948** E. F. Murphy *They struck Opal* 140 Nobbies are..scattered here and there like shells on the beach. **1963** A. Lubbock *Austral. Roundabout* 79 These petrified..bubbles are called 'nobbies'; and they are prised out..by the opal digger. **1967** S. Lloyd *Lightning Ridge Bk.* Introd., Dug out a cleanskin nobby. It was a bonza stone and a whopper too.

Nobel (nōᵘbe·l). [The name of Alfred *Nobel* (1833–96), Swedish chemist and engineer, inventor of dynamite and other high explosives.] **Nobel prize,** one of five prizes, established by the will of Alfred Nobel, which are awarded annually to the person or persons adjudged by Swedish learned societies to have done the most significant recent work in physics, chemistry, medicine, and literature, and to the person or persons adjudged by the Norwegian parliament to have rendered the greatest service to the cause of peace. So *Nobel award, bequest, laureate* (see *laureate sb. 1 d), *prize-man, prize-winner, prize-winning* adj. Also *ellipt.*
A sixth prize, for economics, was first awarded in 1969.
1900 *Sci. Gossip* Nov. 164/2 Just before going to press we have been furnished..with copies of the official statutes and regulations of the Nobel Bequest. *Ibid.* Dec. 195/1 Each candidate for a Nobel prize must be proposed in writing by some one qualified to make such proposal. **1904** *To-Day* 28 Dec. 252/2 (*heading*) The Nobel Prizemen. **1932** *Discovery* Oct. 327/2 Ross..was..awarded the Nobel Prize for Medicine. **1956** Nobel award [see *Guggenheim]. **1958** *Listener* 6 Nov. 749/1 His great discovery, a Nobel-prizewinning matter. **1962** *Ibid.* 8 Nov. 775/1 Deservedly a Nobel Prize winner, O'Neill..was perhaps the greatest twentieth-century dramatist writing in English. **1968** J. D. Watson *Double Helix* xxii. 163 Though the odds still appeared against us, Linus had not yet won his Nobel. **1969** *Times* 29 Sept. 10/8 An award for economics is, I hear, to be added to the list of Nobel prizes for peace, literature, physics, chemistry, and medicine. It has been endowed after a bequest from the Royal Bank of Sweden, which has celebrated its 300th anniversary this year. **1973** D. Robinson *Rotten with Honour* 66 Our own people rate this man's work at Nobel standard. **1973** *Times* 19 Oct. 21/2 Professor Wassily Leontief, of Harvard University, yesterday won the Nobel Prize for Economics. He will receive an award of £41,000. **1975** J. Aiken *Voices in Empty House* ii. 73 August's lab assistants gave him this set when he won his Nobel.

Nobelist (nōᵘbe·list). [f. *Nobel + -ist.] A winner of a Nobel prize.
1941 *Sci. News Let.* 30 Aug. 135/2 Prof. Enrico Fermi, Nobelist now working at Columbia University. **1965** *Amer. N. & Q.* Sept. 9/2 Joseph Breuer as Nobelist—Is there any evidence that Joseph Breuer (1842–1925), physiologist, was ever seriously considered as a candidate for a Nobel Prize? **1972** *Impact of Science on Society* (Unesco) XXII. iv. 282 Nobelists everywhere have emerged from the laboratory corner, the literary alcove or the private study to become involved with the plight of the race.

nobelium (nōᵘbī·-, nōᵘbe·liəm). *Chem.* [f. *Nobel + -ium.] An artificially produced transuranic element, the longest-lived isotope of which has a half-life of about three minutes. Atomic number 102; symbol No.
1957 P. R. Fields et al. in *Physical Rev.* CVII. 1461/2 We suggest the name nobelium, symbol No, for the new element in recognition of Alfred Nobel's support of scientific research and after the institute where the work was done. **1957** *Times* 10 Sept. 11/1 He compared the building up of new elements beyond uranium to the ascent of Everest... In the making of nobelium, the advanced camp had been curium 244, made in the materials testing reactor in Idaho, United States. **1963** *Sci. Amer.* Apr. 70/2 The investigators proposed the name 'nobelium' for element 102, and the name was accepted by the Commission on Atomic Weights of the International Union of Pure and Applied Chemistry. The acceptance turned out to be premature. All attempts..to duplicate the Stockholm experiment have failed. **1974** *Encycl. Brit. Micropædia* VII. 368/2 Radiochemists have shown nobelium to exist in aqueous solution in both the +2 and +3 oxidation states.

nobility. Add: **1. c.** *pl.* Instances of nobleness of nature.
1921 R. Hichens *Spirit of Time* iv. 71 He pointed to the nobilities, the self-sacrifice,..the marvellous examples of courage.
d. The property (of an element) of being noble or relatively unreactive. Cf. Noble *a.* 7 b in Dict. and Suppl.
1907 [see *noble a. 7 b]. **1974** *Sci. Amer.* Aug. 48/2 The supposed 'nobility' of the elements that make up Group Zero in the periodic table was first compromised in 1962, when Neil Bartlett..synthesized xenon hexafluoroplatinate.
4. b. (Later example.)
1927 [see *divinely adv. 2].

noble, *a.* and *sb.*¹ Add: **A.** *adj.* **4.** *noble savage,* primitive man, conceived of in the manner of Rousseau as morally superior to civilized man.
1672 [see Savage sb. 2]. **1914** C. Mackenzie *Sinister Street* II. iii. vi. 628 Every new writer who commands any attention drags out the old idol of the Noble Savage and invites us to worship him. Only now the Noble Savage has been put into corduroy trousers. **1933** J. Cary *Amer. Visitor* vii. 72 Her publisher..belonged to the most modern school of anthropologists and believed in the Golden Age, the noble savage, and all the other resuscitated fancies of Rousseau. **1947** *English Studies* XXVIII. 1 He is prevented from depicting the enemies..in..sinister colours by his interest in the romantic dream of the noble savage. **1954** W. S. Maugham *Ten Novels* i. viii. 201 Let us not forget that *Typee* is a glorification of the noble savage, uncorrupted by the vices of civilization, and that Melville looked upon the natural man as good. **1971** G. Steiner *In Bluebeard's Castle* iii. 52 The myth of the noble savage had interiorized a powerful hierarchic dogma. **1972** *Daily Tel.* 11 Dec. 11/1 They believe in the moral superiority of primitive over civilised man. As a potent idea, the Noble Savage died 100 years ago. But it lives for the Ardens in..the Indian peasants, the exploited Irish.
7. b. *spec.* of a metal: resisting oxidation; relatively unreactive. Hence of any element: low in the electrochemical series. (Further examples.)
1907 E. S. Merriam tr. *Danneel's Electrochem.* v. 134 Metals whose solution pressure is less than that of hydrogen..have a negative potential. The same thing is meant when we speak of the 'nobility' of the metals; silver is more noble than zinc, and zinc is less noble than hydrogen, etc. **1938** R. W. Lawson tr. *Hevesy & Paneth's Man. Radioactivity* (ed. 2) xxii. 218 They can only be contemplated for those radio-elements which are to some extent electrochemically noble, and hence especially for the isotopes of lead, bismuth, and polonium. **1956** E. C. Potter *Electrochem.* x. 234 We may summarize this mode of corrosion..by saying 'it is unwise to permit a metal to contact an aqueous solution of a salt of a metal more noble than itself'. **1973** *Nature* 20 July 137/1 After a beta transformation the daughter element is electrochemically more noble than the mother element. **1974** *Sci. Amer.* Jan. 33/3 If..one puts the corroding metal in contact with a 'nobler' (less active) metal on which the cathodic reaction can proceed more easily, the corrosion current and hence the rate of dissolution of metal can be increased significantly.
(ii) *noble gas*: = *inert gas* (*b*) s.v. *inert *a.* 1 c. So *noble liquid,* one of the noble gases in the liquid state.
noble gas is now the term officially preferred by the International Union of Pure & Applied Chemistry.
1902 J. I. D. Hinds *Inorg. Chem.* xviii. 151 The name *Noble Gases* has been given by Erdmann to the several rare and inactive elements which have recently been discovered. **1927** [see *inert a. 1 c (b)]. **1950** *Electronic Engin.* XXII. 108 Electron tubes filled with a noble gas such as argon, neon or helium are now widely employed. **1971** *Nomencl. Inorg. Chem.* (I.U.P.A.C.) (ed. 2) 11 The use of the collective names..alkaline-earth metals (Ca to Ra), and noble gases [ed. 1 (1959): inert gases] may be continued. **1971** *Nature* 29 Oct. 617/1 We are working towards the development of a thin multiconductor chamber filled with a noble liquid. **1974** *Sci. Amer.* Aug. 48/3 Soon after Bartlett's announcement several other noble-gas compounds were made, chief among them the xenon fluorides (XeF_2, XeF_4 and XeF_6) and xenon trioxide (XeO_3).
f. *noble rot* = **pourriture noble.*
[**1924** H. W. Allen *Wines of France* ii. 114 The Botrytis produces a grey mould, which gives to other wines a most unpleasant taste, but in Sauternes that mould is the *pourriture noble,* the 'noble rottenness', which bestows on the grapes and the wines made from them their extraordinary richness of sweetness and perfume.] **1935** Schoonmaker & Marvel *Compl. Wine Bk.* i. 12 It is necessary to leave the grapes on the vine until they are..over-ripe, sugary and shrunken, until that so-called 'noble rot' (*la pourriture noble*) has set in. **1959** W. James *Word-Bk. Wine* 133 The noble rot also produces glycerine, which gives the wine a fine liqueur-like oiliness. **1965** A. Sichel *Penguin Bk. Wines* 194 To every hectolitre (22 gallons) of the fermenting must of the Furmint and Harslevelü grape they [*sc.* Hungarians producing Tokay] add 3, 4, 5 or 6 *puttony* of the grapes concentrated by the 'noble rot'. **1973** *Country Life* 15 Nov. 1535/3 The German vintage..lacking the 'noble rot' which produces the more luscious *auslese* and *beerenauslese* wines. **1975** P. Vandyke Price *Taste of Wine* 53/1 The primarily sweet wines may have some grapes affected by noble rot, but there is a profundity and a distinctive after-taste to those that are chiefly made with nobly rotten grapes.
B. *sb.* **1. c.** A leader or protector of men hired to replace striking workers. *U.S. slang.*
1930 *Amer. Mercury* Dec. 456/2 *Noble,* a guard for strike breakers. 'Me work? Don't be foolish. I'm a noble, I am.' **1937** *N.Y. Times* 22 Dec. 22 *Noble,* a lieutenant of strike operations usually in charge of a detachment of guards, sluggers and finks. **1950** H. E. Goldin *Dict. Amer. Underworld Lingo* 145/2 *Noble* (rare), a guard hired to protect strike-breakers. **1960** Wentworth & Flexner *Dict. Amer. Slang* 356/2 *Noble.* 1. A strike-breaker's guard. 2. The boss of a gang of strike-breakers; a chief fink.

nobleite (nōᵘ·b'ləit). *Min.* [f. the name of Levi F. *Noble,* 20th-cent. U.S. geologist + -ite¹.] A hydrated calcium borate, $CaB_6O_{10}.4H_2O$, found as colourless, monoclinic crystals at Furnace Creek, Death Valley, California.
1961 R. C. Erd et al. in *Amer. Mineralogist* XLVI. 560 The naturally occurring hydrous calcium borate, $CaO.3B_2O_3.4H_2O$..is named nobleite in honor of Dr. Levi F. Noble, geologist in the U.S. Geological Survey since 1909, in further recognition of his fundamental contributions to geologic knowledge of the Death Valley region. **1964** *Ibid.* XLIX. 1549 Tunellite..is isostructural with the Ca analogue, nobleite. **1968** I. Kostov *Mineralogy* 435 Colemanite Group. The following minerals belong to this group:..nifontovite..tunellite..nobleite.

‖ **noblesse oblige** (noble·s oblī·ʒ), *phr.* [Fr.] Phrase suggesting that noble ancestry constrains (to honourable behaviour); privilege entails responsibility. Also in extended use. Also as *sb.* and *attrib.*
1837 F. A. Kemble *Let.* 1 Aug. in *Rec. Later Life* (1882) I. 86 To be sure, if 'noblesse oblige', royalty must do so still more. **1864** J. H. Newman *Apol.* i. 10 Do you think I can let you go scot free instead of myself? No; *noblesse oblige.* Go to the shades, old man. **1873** C. M. Yonge *Pillars of House* IV. xxxix. 150, I always regarded you as a sacred personage, condemned to *noblesse oblige,* and all that! **1896** Beerbohm *Happy Hypocrite* in *Yellow Bk.* Oct. 12 *Noblesse oblige*..an aristocrat should be very careful of his good name. **1903** Chesterton *Robert Browning* v. 114 When someone excused coarseness..on the ground of genius, he said, 'That is an error: Noblesse oblige.' **1923** D. H. Lawrence *Birds, Beasts & Flowers* 65 Blessed are the pure in heart and the fathomless in bright pride; The loveliness that knows *noblesse oblige.* **1932** J. M. S. Tompkins *Popular Novel in England 1770–1800* iv. 153 The Magdalens of gentle blood..descend..to the family vault. It is a clear case of *noblesse oblige.* **1968** *Listener* 6 June 713/2 *Noblesse oblige* Toryism..is giving way to..managerial Conservatism. **1973** J. Shub *Moscow by Nightmare* xiii. 150 He was pleased to inform the servants of his whereabouts; it simplified their reporting for the KGB. *Noblesse oblige.*

Nobodaddy (nōᵘ·bodædi). [Blend of Nobody + Daddy.] Used by William Blake, and others after him: a disrespectful name for God, esp. when regarded anthropomorphically. Also *transf.,* someone no longer held in esteem. Also *attrib.*
c **1793** W. Blake in *Compl. Writings* (1972) 171 (*title*) To Nobodaddy. *Ibid.* 187 Then again old Nobodaddy swore He ne'er had seen such a thing before, Since Noah was shut in the ark. **1921** G. B. Shaw *Back to Methuselah* p. lxiii, It did not occur to us that Old Nobodaddy, instead of being a ridiculous fiction, might be only an impostor. *Ibid.* p. lxvi, The moment Nobodaddy was slain by Darwin, Public Opinion, as divine deputy, lost its sanctity. **1922** Joyce *Ulysses* 203 Whether these be sins or virtues old Nobodaddy will tell us at doomsday leet. *Ibid.* 388 But the braggart boaster cried that an old Nobodaddy was in his cups it was muchwhat indifferent and he would not lag behind his lead. **1926** A. MacLeish (*title*) Nobodaddy. **1962** *Listener* 29 Nov. 932/1 Goethe is in danger of turning into a Nobodaddy—a booming, boring member of that depressed class, the Illustrious Dead. **1976** *Guardian* 8 Jan. 12/7, I believe we see here [*sc.* in the poetry of Tristan Tzara] the great Nobodaddy Dadaist losing out to the truthful inclusiveness of our women's contemporaries.

nobody. 1. Delete † and add earlier and further examples of proverbial use.
1611 Cotgrave *Dict.* s.v. *Ouvrage,* Euerie bodies worke is no bodies worke. **1725** Defoe (*title*) Every-body's business is no-body's business. **1828** Macaulay in *Edin. Rev.* Sept. 103 The business of every body is the business of nobody. **1829** Cobbett *Advice to Young Men* vi. 345 Public property is never so well taken care of as private property; and this, too, on the maxim, that 'that which is every body's business is no [*sic*] nobody's business'.
2. Phr. *nobody's business* (see *business 16 f); *nobody's fool,* a person who cannot be taken advantage of.
Further examples s.v. *business 16 f.
1839 *Spirit of Times* 8 June 163/1 As to eating, jist go to Snowden's, and the way you can git good things is nobody's business. **1923** H. C. Witwer *Fighting Blood* xi. 323 He's a little too big..for us... And, another thing, Ryan is nobody's fool. **1940** N. Marsh *Surfeit of Lampreys* (1941) xv. 232 They've displayed a surprising virtuosity. They're nobody's fools. **1942** [see *death sb. 17 c]. **1956** *Times* 12 Sept. 1 What she could do with a pencil, notebook, and typewriter was simply nobody's business. **1959** 'A. Fraser' *High Tension* x. 103 He smiled slightly, and I made a note that he was nobody's fool. **1962** 'E. Peters' *Funeral of Figaro* i. 32 'He can sing like nobody's business,' said Stoker positively. **1975** *Times* 20 Sept. 9/7 Poirot..adds..'Never do I pull the leg.' That, alas, is not true. He teased poor Hastings like nobody's business.
b. (Examples with the indef. article.)
1886 G. Meredith *Let.* 15 Nov. (1970) II. 838 In origin I am what is called here a nobody. **1922** Joyce *Ulysses* 316 And who was he, tell us? A nobody. **1950** G. B. Shaw *Farfetched Fables* 67, I replied that if he did not realize that without them he would be a nobody he was no gentleman. **1961** *New Eng. Bible* 2 Cor. xii. 11 In no respect did I fall short of these superlative apostles, even if I am a nobody. **1975** *Listener* 4 Dec. 752/2 Out there, he could become a 'somebody'; in London, he felt he was a 'nobody'.
3. nobody-crab, a marine arachnid of the order Pantopoda (or Pycnogonida), having a small body and four pairs of very long, thin legs; = *pycnogonid* (s.v. Pycno-), Sea spider 1 b.
1881 [see *pycnogonid* (s.v. Pycno-)]. **1935** *Discovery* Sept. 282/1 Those queer creatures the Pycnogonida, the so-called No-body Crabs, real Tom Noddies, with only enough body to hold together the legs, in which are situated the vital organs. **1945** T. H. Savory *Spiders Brit. Isles* (ed. 2) 20 Sea-spiders or nobody-crabs are found only in the sea, where they range from the littoral regions to the depths of the ocean.

no bon (nōᵘ bon), *adj. phr. slang.* [No *adv.*¹ + Bon *a.*] No good.
1918 W. Owen *Let.* 29 May (1967) 554 Music—no bon! Painting—nah pooh! **1929** *Papers Mich. Acad. Sci., Arts, & Lett.* X. 310/2 *No bon,* no good.

no-ca·lorie, *a.* [f. No *a.* + Calorie.] Of a food substance or diet: free from, or very low in, calories. Abbrev. **no-cal.**
1961 S. Price *Just for Record* x. 101 Morosely munching their no-calory diet. **1969** *Guardian* 6 Oct. 11/4 Young mothers..whose entire diet consists of cottage cheese, yogurt, and gallons of 'no-cal' pop. **1971** *Petticoat* 17 July 5 They are no-cyclamate, no-calorie, no-aftertaste sweeteners.

no can do, *colloq. phr.* [f. No *adv.*¹ + Can *v.*¹ + Do *v.*] It is not possible; it is not within the power of (the speaker).
1914 'Saki' *Beasts & Super-Beasts* 289 'Sorry, my dear, no can do,' said Suzanne. **1915** A. D. Gillespie *Lett. from Flanders* (1916) 69 'Nap poo'..was once 'il n'y a plus', but now it's used like the Chinese 'no can do' for everything. **1923, 1958** [see *can v.¹ 8 c]. **1962** 'E. Ferrars' *Busy Body* v. 53 Sorry, no can do—not tonight.

nocardiosis (nōᵘkāɹdiōᵘ·sis). *Path.* [f. mod.L. *Nocardi-a,* the name of a genus of actinomycetes (f. the name of E. I. E. *Nocard* (1850–1903), French veterinary pathologist + -ia¹) + -osis.] Infection with, or a disease caused by, an actinomycete of the genus *Nocardia,* esp. *N. asteroides,* which is occas. pathogenic in humans, producing local lesions or more often generalized disease beginning in the lungs but tending to spread to other organs, esp. the brain, and often proving fatal.
1907 J. H. Wright in Osler & McCrae *Mod. Med.* I. xvi. 341 It has been pointed out that if the name Actinomyces be not used for these microörganisms then the only permissible generic term to apply to them..is Nocardia, and disease processes produced by them should be called nocardiosis. **1920** *Arch. Dermatol. & Syphilol.* II. 137 (*heading*) Nocardiosis cutis resembling sporotrichosis. **1949** *Amer. Jrnl. Path.* XXV. 1 (*heading*) Pure granulomatous nocardiosis: a new fungus disease distinguished by intracellular parasitism. **1963** C. W. Emmons et al. *Med. Mycol.* ix. 75 Nocardiosis is recognized in the dog oftener than in other domesticated animals. **1974** Passmore & Robson *Compan. Med. Stud.* III. i. xii. 90/2 Nocardiosis and mucormycosis are infections usually acquired by inhaling dusts.

‖ **noceur** (nosör). [Fr.] A reveller, rake, libertine; one who stays up late at night.
1908 Nevill & Jerningham *Piccadilly to Pall Mall* iv. 157 The French *noceur* is only too pleased to show himself in the company of some well-known 'horizontale'. **1918** J. Agate *Buzz, Buzz!* i. 11 In Ibsen the characters who bother themselves about the arts are invariably humbugs or hypocrites or *noceurs.* **1931** W. Rothenstein *Men & Memories* I. ix. 94 Though I could, on occasion, sit up most of the night, I was not a *noceur.*

noci- (nōᵘsi), comb. form of L. *noc-ēre* to do harm, used in a few terms, esp. in neurology, as **nocice·ptive** *a. Physiol.* [(Re)ceptive *a.*], (of a stimulus) painful; responding to or caused by such a stimulus; hence **nocice·ptor** [(Re)ceptor], any sensory receptor for painful stimuli; **nocife·nsor** *a.* [L. *dē-fens-,* ppl. stem (in Defensive *a.,* etc.): cf. L. *dēfensor* defender] (of a nerve) concerned in the transmission of the sensation of pain.
1904 C. S. Sherrington in *Nature* 8 Sept. 463/1 In this reaction the reflex arc is (i) the receptive neurone..(nociceptive) from the foot to the spinal segment, (ii) perhaps a short intraspinal neurone, and (iii) the motor neurone..to the flexor muscle. *Ibid.,* Stimulation (nociceptive) of the foot causes flexion of its own leg and extension of the opposite. **1956** *Nature* 18 Feb. 340/1 In the lightly anæsthetised preparations, nociceptive stimuli, twisting the pinna and electrical stimulation of the reticular activating system produced transient changes in the cortical blood-flow. **1961** *Lancet* 2 Sept. 546/2 Dr. W. Koll..described his work on the action of various pharmacological substances on nociceptive spinal reflexes. **1974** *Sci. Amer.* Jan. 43/1 The nociceptive (*N*) cells require still stronger mechanical stimuli. **1906** C. S. Sherrington *Integrative Action Nervous Syst.* ix. 330 The reaction initiated by a noci-ceptor..is to be regarded as consummatory. **1967** *Jrnl. Physiol.* CXC. 541 Seventy-four fibres conducting between 6 and 37 m/sec were classified as nociceptors because they responded only to damaging mechanical stimulation of the skin. **1936** T. Lewis in *Clin. Sci.* II. 402 It will be evident that any system of nerve fibres, which in the exercise of its function gives rise to no obvious and distinctive external manifestations, will tend to escape recognition... The need to postulate a new system of nerves has arisen to explain hitherto unrecognised phenomena. The nerves of this system are at present unnamed. Because they are associated with local defence against injury I propose to call them the 'nocifensor nerves'. **1964** J. Z. Young *Model of Brain* xiii. 209 They are presumed to be nocifensor (pain) fibres, coming either direct from the periphery or after synapse in the sub-vertical lobe.

no-city, *a.* Also **no-cities.** [No *a.*] Of a strategy in a nuclear war: (see quot. 1966).
1962 *Economist* 8 Dec. 1014/1 Many Americans are sceptical of this 'no cities' doctrine. **1965** H. Kahn *On Escalation* vi. 124 If the war were conducted as a no-city war. **1966** Schwarz & Hadik *Strategic Terminol.* 102 *No-cities strategy,* strategy based on the principle that only the enemy's forces are to be targets of nuclear strikes, and that enemy population centers are to be avoided completely or as much as possible.

‖ **nockerl** (nǫ·kěrl). Pl. **nockerln.** [Austrian G., little dumpling.] A small, light dumpling, made with a batter including eggs, and usually fried; **Salzburger nockerl,** a sweet version of this dumpling, using extra eggs which make the mixture puff up when it is cooked.
1855 E. Acton *Mod. Cookery* (rev. ed.) xxxii. 620 (*heading*) A Viennese soufflé-pudding, called Salzburger Nockerl. **1954** G. Beer *Austrian Cooking* 89 A *Nockerl* is tiny to medium-sized, oval in shape and not really a dumpling at all. The only reason why I continuously refer to them as such is for want of a better name. *Ibid.* 90 The *Nockerln* must not brown. *Ibid.* 106 Salzburger Nockerl. ..Remove *Nockerln* carefully with a palette knife... Dust liberally with icing sugar... The centre should be light and creamy, the outside golden brown and puffed. **1958** J. Grossinger *Art Jewish Cooking* (1960) 31 Soup Nockerl... Beat the eggs, water and salt together. Mix in the flour and baking powder. Drop by the teaspoon into boiling salted water. Cook until they rise to the surface.

no claim(s, *attrib. phr.* [No *a.*] Designating a discount allowed on a motor-vehicle insurance premium if no claim has been submitted during the preceding year(s).
1933 G. W. Gilbert *Motor Insurance* iii. 66 (*heading*) No claim bonus. This is allowed..only at a flat rate of 10 per cent of the renewal premium. If more than one cycle is insured the bonus is payable 'per cycle'. *Ibid.* iv. 75 No Claim Bonus is paid..if the policy as a whole runs free of claim for one or two or three succeeding years. **1955** *Times* 24 Aug. 7/3 The 'tariff' companies are credited with the intention of still further increasing the 'no-claims' bonus. **1968** D. Devine *Sleeping Tiger* i. 15 'Take it up with the insurance company, would you?..'It's not worth losing your no-claim bonus. I'll pay for it to be re-sprayed myself.' **1970** *Motoring Which?* July 91/1 Before no-claims bonus and other discounts, fully comprehensive insurance will probably cost you around £45 to £90 for the 3-litre. **1975** *Times* 24 Sept. 3/2 The no-claim bonus is no longer, because of inflation, a sufficient incentive.

no comment: see *comment *sb.* 2 c. Also (with hyphen) as attrib. phr.
1966 S. B. Jackman *Davidson Affair* iii. 22, I thought he was strictly a no-comment man. **1971** 'L. Black' *Death has Green Fingers* iii. 32 The police simply didn't know any more yet. When they did, the 'no comment' answers would begin. **1971** E. Fenwick *Impeccable People* xix. 103 Ben wore his no-comment expression.

no-'count, *a.* Aphetic form of *no-account *a.* Cf. *'count. Also as *sb.*
1853 'P. Paxton' *Stray Yankee in Texas* 282 (Th.), Yes, Massa, dem no 'count calves done fool me again. **1885** H. Jackson *Zeph* iii. 82 Ye miserable, mean-spirited, no-'count critter! **1902** A. D. McFaul *Ike Glidden* xviii. 145 It wasn't enough for your sickly, no-'count mother to waste my grub and money in idleness. **1920** C. E. Mulford *Johnny Nelson* xiii. 135 Judgin' from th' way those no-'count hosses was pullin' when they come over th' hill..I reckoned you got th' hides. **1936** M. Mitchell *Gone with Wind* iv. 65 Dey is de shiflesses, mos' ungrateful passel of no-counts livin'.

noct-. Add: Also used in place of Nocti- in words in which the following element begins with a vowel; **nocta·mbulate** *v. intr.,* to walk at night; **nocta·mbulator,** one who walks at night; **noctambula·tory** *a.* = *noctambulous* adj.; **noctu·ria** [-uria], the condition of being aroused from one's sleep abnormally often by the need to urinate.
1955 H. Spring *These Lovers fled Away* 206 Now and then I would noctambulate through the city. **1962** V. Nabokov *Pale Fire* 221 If you..pull up the window, and..roll out..there is always the chance of knocking clean through into your own hell a pacific noctambulator walking his dog. **1913** C. Mackenzie *Sinister Street* I. ii. vii. 259 Conversations with Brother Aloysius were sufficiently thrilling journeys, and Michael was always ready to follow his footsteps as one might follow a noctambulatory cat. **1911** Dorland *Med. Dict.* (ed. 6) 558/2 *Nocturia,* excessive urination at night. **1928** Eisendrath & Rolnick *Text-bk. Urology* xlvi. 707 The patient has noticed that..there has been an increase in the number of times, he or she, experienced the desire to void urine. This may have occurred only at night i.e. nocturia, or during the day. **1961** *Lancet* 12 Aug. 335/2, 1 patient was troubled by nocturia. **1971** Golden & Maher *Kidney* ii. 36 Nocturia may thus be an early symptom of renal failure.

nocti-. Add: **noctilucent** *a.* (examples); *spec.* in *noctilucent cloud,* a cloud of a kind occas. seen at night in summer in high latitudes, which occurs at a height of about 80 kilometres (at the mesopause) and which some authorities believe is composed purely of cosmic dust and others of ice condensed round cosmic dust particles; cf. **mother-of-pearl cloud.*
1890 *Cent. Dict.,* *Noctilucent,* shining by night or in the dark; noctilucid; as, the noctilucent eyes of a cat. **1910** W. L. Moore *Descriptive Meteorol.* xi. 198 Certain clouds that are seen about midnight in summer have for twenty years received considerable attention from Abbe and others..; sometimes they are called nacreous..at other times noctilucent, because they shine at night. **1936** N. Shaw *Man. Meteorol.* (ed. 2) II. iii. 44 The lower type, known as Perlmutter or iridescent clouds, show brilliant prismatic colours and occur at heights between 20 and 30 kilometres. They differ from noctilucent clouds by appearing almost exclusively in winter. **1940** *Chambers's Techn. Dict.* 581/1 *Noctilucent* (*Zool.*), phosphorescent, light-producing. **1956** *Nature* 18 Feb. 308/1 The technique of observing noctilucent clouds has been improved recently. **1963** *New Scientist* 25 July 169/2 There has been considerable dispute in the past about whether the noctilucent clouds are composed of ice particles or dust particles.

noctuary. (Later example.)
1910 *Chambers's Jrnl.* Sept. 594/2 My sceptical friends..say I kept myself awake on purpose to write this noctuary.

nocturnal, *a.* Add: **1.** *nocturnal emission:* involuntary ejaculation of sperm during sleep.
1928 H. B. English *Student's Dict. Psychol. Terms,* *Nocturnal emissions,* loss of semen during sleep. **1948** A. C. Kinsey et al. *Sexual Behavior Human Male* xv. 516 In the male, nocturnal emissions or wet dreams are generally accepted as a usual part of the sexual picture. **1951** M. McLuhan *Mech. Bride* (1967) 47/1 The [sexual] 'outlets' are..'nocturnal emission', [etc.]. **1958** M. Argyle *Relig. Behaviour* x. 123 *Nocturnal emissions* for Kinsey's devout males were insignificantly more frequent, for his devout females sex dreams to the point of orgasm were less frequent.

nod, *sb.*¹ Add: **1.** In phrases which imply approval, as *to get* or *give the nod.* Chiefly *U.S.*
1948 W. O'Sullivan in *Thrilling Sports* July 55/1 Rebel felt sure of his surmise on the hidden-crew game when his bunch got the nod to start against the highly regarded Tiger crew. **1953** *Wall St. Jrnl.* 23 Apr. 1/3 Paul L. Troast got the G.O.P. nod, beating his nearest rival, State Sen.

Malcolm Forbes, by more than 53,000 votes. **1962** *New Yorker* 17 Nov. 43/2 Industry has at last given literature the nod. **1967** *Boston Globe* 20 May 2/2 (*heading*) Desalting funds get U.S. nod. **1970** *New Yorker* 28 Nov. 151/1 We will not be surprised if the museum gives this piece the nod. **1973** *Bulletin* (Sydney) 25 Aug. 30/2 Perhaps he sees himself—if he gets the nod—as a natural successor to Sir Kevin Ellis in the Speaker's chair of the Assembly. **1975** *Cleveland* (Ohio) *Plain Dealer* 31 Mar. 2-D/2 The five outfielders certainly will include regulars Charlie Spikes, George Hendrick, and Oscar Gamble, meaning the two other jobs will be fought for among Ken Berry, Leron Lee and Rick Manning, with Berry and Lee probably getting the nod.
Prov. (Earlier and later examples.) (See also Wink *sb.*¹ 2.)
1794 Godwin *Caleb Williams* I. viii. 171 A nod is as good as a wink to a blind horse. **1935** T. S. Eliot *Murder in Cathedral* i. 24 My Lord, a nod is as good as a wink. A man will often love what he spurns. **1974** N. Freeling *Dressing of Diamond* 200 All right; a nod's as good as a wink... You've got these people in mind.
d. *on the nod* (later examples).
1907 'Ian Hay' *Pip* ix. 286 He looked all round the room, and I *knew* he knew everything in it had been got on the nod. **1934** *Bulletin* (Sydney) 25 July 46/4 Drunks with determined minds to get bacon, bread, cheese, on the nod. **1945** B. Naughton in C. Madge *Pilot Papers* I. 106 Edith ..got them a house,..and Edith filled it with furniture on the 'nod'.
e. *on the nod,* with a merely formal assent; by abstention from voting.
1959 *Times* 14 Mar. 4/2 The Bill..was given a second reading 'on the nod' by the House. **1969** *Sunday Times* 12 Jan. 4 The agenda, usually the cause of great friction, was accepted 'on the nod'. **1973** C. Mullard *Black Britain* iii. vii. 85 The late Lord (Learie) Constantine, then a member of the Board, opposed the appointment of John Lyttle on the nod, and urged that the job should be advertised in the press.
2. c. A state of drowsiness brought on by narcotic drugs. *Esp.* in phr. *on the nod.*
1942 Berrey & Van den Bark *Amer. Thes. Slang* § 509/22 *Play the nod,* to be drowsy as a result of over-indulgence in narcotics. **1946** B. Treadwell *Big Bk. of Swing* xvi. 125/1 *Nod,* tired feeling, sleepy, fatigued. **1951** *Life* 11 June 126/2 Instead of a warming, bright 'charge', he merely becomes comatose and lethargic (goes on the nod in junkie parlance). **1953** W. Burroughs *Junkie* (1972) iv. 43 Don't ever invite him to your home... He'll go on the nod in front of your family. He's got no class to him. *Ibid.* xiv. 147 When he arrived in Mexico, I gave him half a grain of M and he went on the nod. **1962** K. Orvis *Damned & Destroyed* v. 37 While I was on the nod. **1965** *New Statesman* 20 Aug. 248/3 In addict's language—'going on the nod' (becoming senseless). **1969** H. Waugh *Young Prey* (1970) xxiii. 180 Once you went into the nod, the surroundings no longer mattered.

nod, *v.* Add: **2. d.** (See quot.)
1968–70 *Current Slang* (Univ. S. Dakota) III–IV. 86 *Nod,* to drift in and out of consciousness while under the influence of a drug.

nodal, *a.* Add: **1.** Similarly *nodal surface.*
1937 J. W. T. Spinks tr. *Herzberg's Atomic Spectra & Atomic Struct.* i. 41 For $n > 1$, ψ goes once, or more than once, through the value zero before the exponential decrease sets in; that is, on certain spherical surfaces about the nucleus, the ψ function is always zero. These are the nodal surfaces of the ψ function corresponding to the nodes of a vibrating string.

nodality (nodæ·lIti). [f. Nodal *a.* + -ity.] **a.** The degree to which a place is a point of convergence for routes, roads, or the like.
1897 *Geogr. Jrnl.* IX. 78 A higher degree of 'nodality', to use Mr. Mackinder's term, is found where several such furrows [*sc.* shallow valleys in the chalk] meet to form a well-marked though by no means deep hollow. **1902** H. J. Mackinder *Britain & British Seas* xix. 330 A spot upon which more numerous land and water roads converge, as in a defile past some natural obstacle, may be said to have a higher degree of nodality. **1953** J. M. Houston *Social Geogr. Europe* ix. 205 In cities of long history, such as Paris, it is easy to point out the geographical advantages of the site and the elements of nodality, whether at a bridge-head, a crossing of routes or at a transfer point.
b. The number of nodes of an oscillation.
1905 *Trans. R. Soc. Edin.* XLI. 602 In any given lake, seiches of all degrees of nodality, i.e. uninodal, binodal, trinodal, etc., are possible. **1957** G. E. Hutchinson *Treat. Limnol.* I. v. 299 In the case of Loch Earn, periods have been identified which may belong to a nodality as high as the sixteenth.

nodalize (nōᵘ·dăləiz), *v. rare.* [f. Nodal *a.* + -ize.] *trans.* To make nodal in form or arrangement; to concentrate in a node.
1915 D. H. Lawrence *Rainbow* xv. 412 For what purpose were the incalculable physical and chemical activities nodalised in this..speck under her microscope?

nodding, *vbl. sb.* Add: **a.** Also = *nod *sb.*¹ 2 c.
1970 H. E. Roberts *Third Ear* 10/2 Nodding out, a drug stupor. **1972** C. Weston *Poor, Poor Ophelia* (1973) xxiv. 149 He's high on something. Nodding, I think they call it.
b. (Further examples.)
1825 H. Wilson *Mem.* (ed. 2) II. 108 Having only a sort of bowing, nodding acquaintance with him. **1959** I. & P. Opie *Lore & Lang. Schoolch.* viii. 122 Childhood is on nodding terms with the supernatural. **1969** *Times* 22 July p. ii/6 Nobody volunteers that they are not at least on nodding terms with their hero. **1972** H. C. Rae *Shooting Gallery* ii. 74 Three thousand students in this establishment. I know, on nodding terms, about forty of them. **1972** *Times* 7 Aug. (Jamaica Suppl.) p. ii/5 A health care plan..should give many Jamaicans at least a nodding acquaintance with health services.

node, *sb.* Add: **3. c.** *Anat.* Any small mass of differentiated tissue; also applied to the interruptions (called *nodes of Ranvier*: see *Ranvier) of the myelin sheath of myelinated nerve fibres.
1885 *Encycl. Brit.* XIX. 23/2 The medullary sheath shows at certain intervals interruptions called the 'nodes of Ranvier'. **1892**, etc. [see *lymph node* (*lymph 5)]. **1902** *Trans. Chicago Path. Soc.* V. 151 (*heading*) Are the hemolymph nodes organs *sui generis*? **1917** *Amer. Jrnl. Anat.* XXI. 375 The frequent close association of lymph and hemal nodes makes differentiation of the early stages in development naturally very difficult. **1962** *Gray's Anat.* (ed. 33) 89 At its [*sc.* the primitive streak's] headward end a further area of exceptionally active growth forms a knob-like thickening which is termed the primitive node. **1968** Passmore & Robson *Compan. Med. Stud.* I. xxviii. 9/1 The part of the heart with the highest spontaneous rate..provides the source of excitation of the whole heart, and this pacemaker is normally the sinuatrial (SA) node. *Ibid.* 10/1 Situated low in the right atrium, near the opening of coronary sinus and tricuspid valve, is another group of cells with a specialized conducting function, the atrioventricular (AV) node.
6. a. (Further examples.) Hence also, a point at which a spherical harmonic or similar function has the value zero.
1905 *Trans. R. Soc. Edin.* XLI. 601 At these points, which are called nodes, the level of the surface [of the lake] is unaltered by the seiche. **1927** T. M. MacRobert *Spherical Harmonics* iii. 59 The middle point..is always at rest, and this point is called a node, the end-points also being nodes. For the *n*th mode there are $n-1$ nodes, as well as the two end nodes. **1957** G. E. Hutchinson *Treat. Limnol.* I. v. 299 While the periods and positions of the nodes depend on the form of the lake basin, the amplitude of the seiche depends only on the source of energy that generates it. **1964** J. W. Linnett *Electronic Struct. Molecules* i. 6 For $n=3$, the $l=0$ function would be multiplied by a radial function varying with *r*, the distance from the nucleus, which had two nodes (i.e. it may be considered, rather approximately, as made up of 3 half-waves).
b. Delete the def. in Dict. and transfer the quots. to 6 a.
c. *Electr.* A point on an aerial or in a circuit where the current or voltage is zero.
1915 A. E. Seelig tr. *Zenneck's Wireless Telegr.* ii. 25 In a lineal oscillator..the current amplitude is greatest at the middle and zero at the ends of the oscillator. In other words, there are 'current nodes' at each end and a 'current anti-node' at the middle... The 'potential' or 'voltage anti-nodes' occur at each end of the oscillator, the 'potential node' being at the middle. **1947** E. K. Sandeman *Radio Engin.* I. xvi. 645 At the voltage anti-nodes (current nodes) the incident and reflected voltages are in phase and therefore add, while the incident and reflected currents are 180° out of phase and therefore subtract. **1968** *Radio Communication Handbk.* (ed. 4) xiii. 3/1 The positions of maxima are usually known as current (or voltage) anti-nodes or loops and the intermediate positions as nodes or zeros.
7. b. A point or vertex of a network or graph (sense 1). Also *node point.*
1864 H. J. S. Smith in *Rep. Brit. Assoc. Adv. Sci. 1863* I. 768 Let an infinite plane area be divided by two systems of parallel lines into similar and equal parallelograms. The vertices of these parallelograms we shall call nodes. **1892** G. B. Mathews *Theory of Numbers* I. iv. 124 Let the plane of reference be divided up into a system of equal and similar parallelograms..; such a system will be called a net,..and each joint, where two lines cross, a node. **1941** *Proc. Cambr. Philos. Soc.* XXXVII. 194 Let *N* be a network (or linear graph) such that at each node not more than *n* lines meet (where $n > 2$), and no line has both ends at the same node. **1957** N. Chomsky *Syntactic Structures* vii. 68 The phrase structure of a terminal string is determined from its derivation, by tracing segments back to node points. **1964** J. J. Katz in Fodor & Katz *Struct. of Lang.* 526 The amalgam is assigned to the set of paths associated with the node (i.e., the point at which an *n*-ary branching occurs) that immediately dominates the sets of paths from which the paths amalgamated were drawn. The amalgam provides one of the meanings for the sequence of lexical items that the node dominates. **1964** A. Battersby *Network Analysis* ii. 14 The beginning and end of a job are events; they are represented as numbered circles called nodes. *Ibid.* 19 The nodes which represent the events are numbered successively from the beginning to the end of the network. **1966** S. Beer *Decision & Control* ix. 194 In the first place, the nodes in the network which now represent individuals are not very tidily placed in ranks of equivalent importance. **1967** *Electronics* 6 Mar. 132/1 The feedback loop is formed by connecting the amplifier's inverting input..to potentiometer R_3. When power is first applied..the voltage at the common node rises quickly to 6·6 volts. **1971** [see *graph *sb.*¹ 1]. **1972** *Sci. Amer.* June 52/3 By last month the network..included 24 computer centers ('nodes' in the terminology of the system), ranging from Massachusetts to California. **1975** *Language* LI. 388 The use of the empty node in the derivation of truncated passives is not motivated empirically, but rather is a way of retaining a transformational analysis of passives.

nodular, *a.* Add: **4.** *Metallurgy.* Of cast iron: containing all the graphite in the form of small spheroids rather than flakes, which results in increased strength and ductility and can be brought about by adding a suitable inoculant. Cf. *spheroidal *a.*
1947 *Machinery* LXX. 420/1 (*heading*) Nodular cast iron. **1950** [see *inoculation 2]. **1967** A. H. Cottrell *Introd. Metallurgy* xxv. 520 In nodular cast irons the structure is developed directly, in an as-cast grey iron, by adding a small amount of magnesium or cerium alloy to the metal in the ladle.

nodularity (nǫdiŭlæ·rIti). [f. Nodular *a.* + -ity.] The state or condition of being nodular.
1948 L. Martin *Clin. Endocrinol.* x. 200 Nodularity so extreme as to be obviously pathological. **1953** *Brit. Jrnl. Urology* XXV. 27 Radial striation on the fractured surface of a calculus is often associated with nodularity of the external surface. **1971** *Amer. Jrnl. Public Health* LXI. 241 (*heading*) Thyroid nodularity in southwestern Utah school children exposed to fallout radiation.

nodulate (nǫ·diŭlēɪt), *v. Bot.* [f. Nodul(e + -ate³, or back-formation from Nodulated *a.*] **a.** *trans.* To produce root nodules on.
1939 *Soil Sci.* XLVII. 67 A velvet bean culture..produced nodules on six plant species included within the cowpea group yet did not nodulate three other plant members. **1973** *Nature* 17 Aug. 460/1 The *Trema* root nodule bacterium is the first of the rhizobia known to effectively nodulate a non-legume.
b. *intr.* Of a plant: to undergo nodulation.
1956 *Biol. Rev.* XXXI. 111 The degree to which an individual plant can be induced to nodulate earlier by root exudates depends wholly upon its normal nodulating habit in their absence. **1968** Gibbs & Shapton *Identification Methods for Microbiologists* 61 Only about 1200 species have been examined for the presence of nodules and not all of these nodulate.
Hence **no·dulating** *vbl. sb.* and *ppl. a.*
1944 *Proc. Soil Sci. Soc. Amer.* IX. 95 (*heading*) The nodulating performance of three species of legumes. **1960** *Dissertation Abstr.* XX. 3003/1 Experiments using..nodulating and non-nodulating soybeans were established. **1971** M. Alexander *Microbial Ecol.* xi. 259 The nodulating habit is particularly prominent among genera of Podocarpaceae, a family that includes trees important as timber.

nodulated, *ppl. a.* (Further examples: cf. prec.)
1932 E. B. Fred et al. *Root Nodule Bacteria & Leguminous Plants* ii. 17 In 1905 Norton & Walls... suggested that the wild Leguminosae, nearly all of which are nodulated in Maryland, are important agents for adding nitrogen and humus to the soils. **1957** *Canadian Jrnl. Microbiol.* III. 120 In the nodulated plants the nitrogen content was fully twice the seed nitrogen. **1971** M. Alexander *Microbial Ecol.* xi. 259 In addition to the plants cited, several gymnosperms are nodulated, and they may be capable of assimilating N_2.

nodulation. (Further examples: cf. *nodulate *v.*)
1924 *Soil Sci.* XVII. 439 Fellers found that under many conditions nodulation was increased by an application of phosphorus, calcium, or potassium. **1973** *Nature* 17 Aug. 459/1 Roots of *Trema aspera* Decaisne plants growing between rows of tea in..New Guinea showed abundant nodulation, similar to the nodules found on many tropical legumes.

nodule. Add: **3.** Esp., one formed on the root of a leguminous plant by symbiotic bacteria. Also *attrib.* (Later examples.)
1890 *Proc. R. Soc.* XLVII. 104 The limited growth in pot 1..is coincident with the entire absence of nodule-formation. **1922** *Encycl. Brit.* XXX. 72/1 The so-called 'nodule' organisms (*Pseudomonas radicicola*)..live in symbiosis with the leguminous plants. **1965** Bell & Coombe tr. *Strasburger's Textbk. Bot.* 303 In the development of the nitrogen-fixing nodules on the roots of the Leguminosae the first process is a normal infection of the rootlets by various races of the aerobic *Rhizobium leguminosarum*..living saprophytically in the soil.

nodulize (nǫ·diŭləiz), *v. Metallurgy.* [f. Nodul(e + -ize.] *trans.* To convert (esp. finely-divided iron ore) into nodules. So **no·dulized** *ppl. a.,* **no·dulizing** *vbl. sb.*
1905 *U.S. Patent* 794,673 1/1 A new and useful Improvement in Methods or Processes of Purifying and Nodulizing Metalliferous Materials. **1905** *Iron Age* LXXVI. 590/1 The National Metallurgic Company has erected what it terms an ore purifying and nodulizing plant. **1906** *Jrnl. Iron & Steel Inst.* LXXI. 359 One rotary kiln..has produced an average of 75 to 90 tons of nodulised desulphurised pyrites cinder per day of twenty-four hours. **1921** W. Gowland *Metallurgy of Non-Ferrous Metals* (ed. 3) 139, 62 per cent. of the concentrate is nodulized, 14 per cent. sintered, and 24 per cent. reduced direct. **1968** E. N. Simons *Outl. Metallurgy* i. 3 Nodulizing..passes the ore too fine for use through a rotary kiln, in which it is heated with a relative movement between an internal rotary drum and an outer shell. Nodulizing has some advantages over sintering. *Ibid.* iii. 34 The graphite is present as tiny round balls, i.e. it is nodulized.
Hence **noduliza·tion,** the process of nodulizing.
1915 *Chem. Abstr.* IX. 590 Nodulizing iron ores..by subjection first to a reducing flame at a temp. only slightly below that required for nodulization and then to an oxidizing flame which suddenly raises the temp. of the material above the nodulizing point. **1921** W. Gowland *Metallurgy of Non-Ferrous Metals* (ed. 3) 138 In the pro-

cess of nodulization the proportion of sulphur has been reduced from 28 to 18 per cent. **1968** E. N. Simons *Outl. Metallurgy* iii. 34 Nodulization or spheroidization renders the iron stronger and more ductile.

nodum (nōᵘ·dǝm). *Ecology.* Pl. **noda.** [mod. L., perh. f. L. *nōdus* knot] (See quot. 1955.)
1955 M. E. D. Poore in *Jrnl. Ecol.* XLIII. 243 The additional term Nodum has been introduced to apply to an abstract unit [of vegetation]..of uncertain status. *Ibid.* 259 The term nodum is being adopted to apply to abstract vegetation units of any category. It corresponds to the term taxon in systematics. *Ibid.,* Both the *Rhacomitrium-Carex bigelowii* and the *Nardus* noda..are clearly defined communities. **1961** *Watsonia* V. 76 The nodum is only developed on south- to west-facing slopes of 10 to 15°, fully exposed to wind and salt spray. **1971** *New Phytologist* LXX. 1155 The noda are related to environmental factors. *Ibid.,* A strongly marked relationship between the percentage of maritime species in a nodum and the sodium-organic ratio is noted. **1973** H. J. B. Birks *Past & Present Vegetation of Skye* iv. 24/2 A community is any aggregation of plants constituting a certain spatial whole... It is regarded as synonymous with Poore's general term 'nodum'. *Ibid.* 25/2 Where the vegetational unit is rather heterogeneous, it is termed a *nodum.* The word is italicised when it is used to refer to a specific vegetational unit that does not merit the rank of association.

noegenesis (no,idʒe·nėsis). [f. Gr. νόη-σις Noe(sis + -genesis.] The generating of new knowledge from experience through the apprehension of experience, the eduction of relations and the eduction of correlates; the obtaining of knowledge thus. So **noegene·tic** *a.,* of, pertaining to, or concerned with, noegenesis.
1923 C. Spearman *Nature of 'Intelligence'* iv. 61 'Noegenesis'. Another basal property of the manifestations of all the principles is that they, and they alone, are generative of new items in the field of cognition. If then, it be desired to depict these three principles summarily, taking into account both their noetic and their generative properties, we must compound some such name as 'noegenetic'. **1931** F. Aveling in W. Rose *Outl. Mod. Knowl.* viii. 336 The noegenetic principles, though accounting for the origin of knowledge, do not cover the entire cognitive field. **1936** F. Banks *Conduct & Ability* xii. 232 Clearly some word like *eduction* was needed to distinguish this noegenetic process from the 'merely associative'. *Ibid.* xv. 308 The third principle of noegenesis will also be found adequate to explain the inventions of science. **1938** *Mind* XLVII. 378 Very proper complaint is registered to the effect that Psychologists have always neglected the problem of the perception of relations. Here again Prof. Spearman may reflect with pride on his own contribution in the shape of the 'noegenetic laws'. **1972** L. S. Hearnshaw in Cox & Dyson *20th-Cent. Mind* I. vii. 235 Spearman contributed a powerful and erudite analysis, and though once again his theory of noegenesis, or creative intelligence, has not altogether stood up to subsequent scrutiny, it proved practically useful. **1973** H. J. Eysenck *Inequality of Man* ii. 45 Spearman (1927) tried to lay down in his 'noegenetic laws' the essence of this cognitive ability.

noem (nōᵘ·em), **noeme** (nōᵘ·īm). *Philos.* and *Linguistics.* [f. Gr. νόημα thought.] A term used to denote that aspect of a unit of speech which is concerned with meaning as distinguished from sound.
a **1866** J. Grote in *Jrnl. Philol.* (1872) IV. 55 It will be necessary for me to make use of one or two new-coined words, which I will begin by defining as accurately as I can... When I mean words as thought I shall use the term *noem.* *Ibid.* 56 Logically, a noem may be called a concept, a notion, or what we will; but I would have the term bear simply a relation to language, and mean the thought-word, that, whatever it is, which the sound stands for. **1926** L. Bloomfield in *Language* II. 161 The meaning of a glosseme is a noeme. **1940** *Mod. Lang. Q.* June 178, I have recently encountered the *enthymeme,*..the *noeme,* the *philosopheme,* [etc.]. **1966** M. Pei *Gloss. Ling. Terminol.* 180 *Noeme,* the meaning of a glosseme.

noema (nōᵘ,ē·mă, nōᵘ,ī·mă). *Philos.* Pl. **noemata** (no,ī·mătă). [ad. Gr. νόημα thought.] A term first used by Husserl for that which is perceived or thought as the self experiences it. (See quot. 1966.)
1931 W. R. B. Gibson tr. *Husserl's Ideas* III. iii. § 88. 258 Corresponding at all points to the manifold data of the real (*reellen*) noetic content, there is a variety of data displayable in really pure (*wirklich reiner*) intuition, and in a correlative '*noematic content*', or briefly '*noema*'—terms which we shall henceforth be continually using. **1957** B. Lonergan *Insight* p. xxv, The *noēma* or *intentio intenta* or *pensée pensée,* illustrated by the lower contexts, *P, Q, R,*... and by the upper context that is Gödel's theorem. **1966** A. Gurwitsch *Stud. Phenomenology & Psychol.* vii. 132 What has just been described by these allusions is the *noema* of perception—namely the object just (exactly so and only so) as the perceiving subject is aware of it. *Ibid.* 133 Noemata are not to be found in perceptual life alone. There is a noema corresponding to every act of memory, expectation, representation, [etc.]. **1972** H. Dreyfus in L. Embree *Life-World & Consciousness* 138 If the noema were an abstract entity, we could indeed reflect on the higher order noemata involved in reflecting on the first-order noema.

noematic (no,imæ·tik), *a.* [f. Gr. type νοηματικός, f. νόημα thought.] **a.** Of or pertaining to a noem. **b.** Of, pertaining to, or connected with, a noema.
a **1866** J. Grote in *Jrnl. Philol.* (1872) IV. 55, I shall use the adjectives *phonal* and *noematic*; and I shall give the name of phonal or noematic *schematisms* to modifications of the primary noems and phones. **1931** [see *noema]. **1934** *Mind* XLIII. 310 Kant obviously contemplated the possibility of a noematic 'given'. **1943** M. Farber *Found. Phenomenology* xvi. 526 The *noematic* side (referring to what is correlatively involved, as 'that which is experienced, thought, etc.'). **1957** B. Lonergan *Insight* II. xiv. 415 The pure forms of noetic experience terminating in noematic contents. **1973** J. N. Mohanty in Carr & Casey *Explor. Phenomenology* 212 An awareness of the truth of its own noematic content.

noem-noem, var. *num-num.

no end. (See End *sb.* 21.) (Further examples.)
1871 F. C. Burnand *More Happy Thoughts* (ed. 2) xix. 143 He 'makes no end of stuff', or 'loses no end of stuff'. on the Derby. **1909** R. E. Knowles *Attic Guest* 54 You'll have no end of fun with him. **1912** *Chambers's Jrnl.* Dec. 769/1 'I really must show this to Champneys,' thought Michael; 'it will please him no end.' **1955** *Essays & Studies* VIII. 5 A few clean strokes of Occam's razor would have helped Mr. Jackson no end. **1958** H. Babcock *I don't want to shoot an Elephant* 8, I often walk fifteen miles a day while hunting... This puzzles my wife no end. **1970** *New Yorker* 3 Oct. 90/2 Thomas had been impressed no end by the sight of Klüver ..fixing an art-and-technology malfunction with a pair of pliers.

no entry. [Entry 1.] Used *attrib.* to designate an area to which entrance is forbidden; also, to designate a sign marking this instruction.
1967 *Guardian* 15 Dec. 3/5 Bollards and no-entry signs to deflect vehicles. **1968** *Ibid.* 13 Aug. 5/3 Potential suicides swerve into 'No entry' streets. **1969** S. Coulter *Embassy* xviii. 195 They had reversed the traffic along the Avenue Gabriel. No Entry signs—white-barred red discs—were now planted in concrete bases on either sidewalk. **1971** *Guardian* 30 Oct. 20/1 Mrs. Joy Johnson..demolished a 'No Entry' sign while driving a double-deck bus.

noesis. Add: Pl. **noeses.** **c.** *Philos.* In phenomenology, that which is involved in the act of thinking or perceiving. Cf. *noema.
1931 W. R. B. Gibson tr. *Husserl's Ideas* III. iii. § 92. 268 Our glance, instead of passing through the noeses of perception..goes through a noesis of remembrance into a world of memory. **1957** B. Lonergan *Insight* p. xxv, There is also the *noēsis* or *intentio intendens,* or *pensée pensante* that is constituted by the very activity of inquiring and reflecting, understanding and affirming, asking further questions and reaching further answers. **1967** B. Jager tr. *Kockelmans's Edmund Husserl's Phenomenological Psychol.* v. 244 A consideration and analysis of the related *noeses,* of the different acts in and through which the ego is conscious of something. **1970** E. Pivčević *Husserl & Phenomenology* vi. 67 'Noesis'..must of course be strictly distinguished from the sensory content of an experience.

noetic, *a.*² and *sb.* Add: **A.** *adj.* **1.** (Further examples.)
1907 W. James *Pragmatism* v. 166 This is the hypothesis of *noetic pluralism,* which monists consider so absurd. **1925** *Brit. Jrnl. Psychol.* Oct. 106 In the following pages the word 'noetic' is used for brevity's sake instead of the more elaborate though more correct 'noegenetic'. *Ibid.* 115 The tests..measure the 'Noetic' and the 'Generative' powers of the subject. **1957** J. S. Huxley *Relig. without Revelation* (rev. ed.) ix. 204 Some new term..in which communicable mental activities play a predominant rôle. If so, I suggest the term *noetic.* *Ibid.,* Religions are thus noetic organs of evolving man. **1972** *Encycl. Psychol.* II. 327/2 *Noetic superstructure,* a term applied by strata theorists..to individual thought processes (power of abstraction, judgment, logical reasoning, etc.) that take place, together with voluntary activities, above the endothymic basis.
b. In phenomenology, that which is concerned with or pertains to the act of thinking or perceiving (see *noesis c). Cf. *noematic *a.* b.
1931 [see *noema]. **1943** M. Farber *Found. Phenomenology* xvi. 526 This applies to the *noetic* side ('I think', 'I experience', etc.) as well as to the *noematic* side. **1969** R. McKeon in R. Klibansky *Contemp. Philos.* III. 105 The difference of knower and known disappear in the description of the experiencing (or the 'noetic') and experienced (or the 'noematic').
B. *sb.* **1.** (Later example.)
1931 W. R. B. Gibson tr. *Husserl's Ideas* II. iv. § 59. 176 So too..we cannot suspend general Noetics, which expresses our essential insight into the rationality or irrationality of the judging activity generally.

noetical, *a.* Delete †*Obs.* and add later example.
1931 W. R. B. Gibson tr. *Husserl's Ideas* III. iii. § 87. 256 It is a long..way that leads from meaning, from ontological and noetical insights.

no-fault. *U.S.* Used *attrib.* to designate a form of motor vehicle insurance (see quot. 1973¹).
1967 *Wall St. Jrnl.* 7 Nov. 1 Support for no-fault auto insurance in the U.S. **1968** *Economist* 13 July p. xviii/2 American motor insurance is now to be put under the microscope. And a no-fault compensation system is being discussed. **1973** C. L. Barnhart et al. *Dict. New Eng.* 318/7 *No-fault,* of or relating to a form of automobile insurance by which an accident victim is compensated for damages or expenses by his own insurance company, whether the accident was his fault or not. **1973** *Washington Post* 13 Jan. A.22/2 Governor Holton's call for a strong no-fault insurance bill. **1974** *Aiken* (S. Carolina) *Standard* 22 Apr. 1-A/2 The bill, to come up today, would assure that victims are paid for their losses regardless of who is at fault. It would require every car owner to buy insurance to protect himself and his family against losses arising from traffic accidents. The no-fault bill heads the Senate agenda for the week as Congress returns from a 10-day recess.

no-fines (nōᵘfəi·nz), *a.* (*sb.*) [f. No *a.* + pl. of *fine* sb. (Fine *a.* B).] Applied to concrete made from an aggregate composed of lumps larger than about 9 mm. (and smaller than about 18 mm.), which results in increased porosity and better thermal insulation. Also as *sb.*
1946 A. F. Boyd *Rep. Tests No-Fines Concrete* (Commonwealth Exper. Building Station, Sydney) 1 Little technical information was available on the strength of no-fines concrete as applied to building construction. **1954** *Archit. Rev.* CXVI. 410/3 Hills themselves also made no-fines walling blocks about 7 in. thick..for facing their steel-framed houses. **1955** *Times* 5 July 17/3 (Advt.), We handed over during the year some 19,500 permanent dwellings, most of them built by the Wimpey 'No-Fines' technique, which continues to make a wide appeal to those wanting speed combined with a high standard of construction. **1960** M. Bowley *Innovations in Building Materials* ix. 209 Dense concrete properly made is impervious to moisture..but provides poor heat insulation... 'No fines', on the other hand, is more pervious..but provides better heat insulation. **1969** E. C. Adams *Sci. in Building* III. viii. 139 No-fines is made by mixing a coarse aggregate with cement and water. *Ibid.,* No-fines concrete is not subject to capillarity, and can..provide a weatherproof wall in one thickness..if rendered externally.

no-frill(s. [*frill *sb.*¹ 1 e.] Used *attrib.* to designate a lack of ornamentation or embellishment.
The more usual form is *no-frills.*
1960 *Farmer & Stockbreeder* 16 Feb. 3/2 A no-frill piggery. **1967** 'E. Lathen' *Murder against Grain* iv. 48 The no-frills exterior of Halloran's garage. **1969** *Listener* 27 Mar. 436/2 *World in Action* has an aggressive investigatory policy combined with tight editing and a no-frills presentation. **1969** *Daily Tel.* (Colour Suppl.) 3 Oct. 20/4 The 61st is a no-frills working regiment that spends most of its time in khaki field uniforms.

nog, *sb.*² **2.** (Earlier examples.)
1851 A. O. Hall *Manhattaner* 10, I tremble to think of the juleps, and punches, and nogs, and soups. **1881** A. W. Tourgée *Zouri's Christmas* in *Royal Gentleman* viii. 527 Then he tried to drain the glass, but a part of the foamy nogg remained in it despite his efforts.

Nogai, var. *Nogay.

‖ **nogaku** (nō·gaku). [Jap. *nōgaku,* f. *nō* talent, accomplishment, *Noh, No² + *gaku* music.] The Japanese dramatic form called Noh; the genre of the Noh drama.
1916 Joly & Tomita *Jap. Art & Handicraft* 7 The *Nō Gaku* originated in the Oyei era close upon the beginning of the XVth century. **1932** B. L. Suzuki (*title*) Nōgaku: Japanese Nō Plays. **1938** D. T. Suzuki *Zen Buddhism & its Influence on Jap. Culture* I. v. 101 The beginning of *haiku,* nōgaku theatre,..is to be sought in them. **1948** *Introd. Classic Jap. Lit.* (Kokusai Bunka Shinkokai) p. xvi, The earliest theatrical performances were those of the *nōgaku* (Noh drama), and the *nō-kyōgen* (Noh farces). **1959** W. P. Malm *Jap. Music* iv. 105 (*heading*) Nohgaku, the music of the noh drama.

Nogay (nǫ̆gai·). Also † **Na-, Nogai.** **a.** A Turkic-speaking people inhabiting the north-western Caucasus region of the U.S.S.R. **b.** The language of this people, a member of the Kipchak branch of Turkic languages. Also *attrib.* or as *adj.* So **Nogai·an** *a.*
1652 Nagaian [see *Kirghiz *sb.* and *a.*]. **1840** *Penny Cycl.* XVI. 248/1 The Nogays are a Tartar or Turkish nation, dispersed over the steppes which extend between the lower course of the river Dnieper and Mount Caucasus. **1886** [see *Kazakh]. **1888** [see *Kirghiz *sb.* and *a.*]. **1948** D. Diringer *Alphabet* 568 The Arabic character has also been adopted for..Nogai Turkish, spoken by nearly 200,000 people to the north-west of the Caucasus and in the Crimea. **1958** *Everyman's Encycl.* IX. 268/1 *Nogay,* Turkic-speaking people of N. Caucasus, who live in the Groznyy oblast and Circassian Autonomous Oblast. **1974** *Encycl. Brit. Micropædia* VII. 376/1 Nogay belongs to the northwestern, or Kipchak, division of the Turkic languages, along with such languages as Kazakh and Tatar.

noggin. Add: **1. b.** The head. orig. and chiefly *U.S.*
1866 J. Finley *Hoosier's Nest* 90 But Matty's top-knot..Wasn't there, for his noggin was bald. **1885** 'Phudge Phumble' *Adventures Greenhorn in Gotham* 25 The full force of it against his noggin. **1931** D. Runyon *Guys & Dolls* (1932) 286 She smacks Rusty Charley on the side of the noggin with the bat. **1951** *Landfall* V. 202 'Thanks, chum,' said Vic, 'You used your noggin.' **1957** *New Yorker* 2 Nov. 34/3 The Right Honourable Our Man Stanley, his Lock hat set firmly on his noggin, dropped by

the office the other day and deposited the following dispatch. **1972** J. E. Franklin in W. King *Black Short Story Anthol.* 354 Her hair was short,..and a dozen of those little twig-plaits tucked under and pinned looked like knots rising on her noggin. **1975** P. G. Winslow *Death of Angel* iv. 86 A rap on the back of the noggin that knocked her out.

c. A pail or bucket. *local U.S.*

1885 'C. E. Craddock' *Prophet Gt. Smoky Mts.* x. 175 Mirandy Jane, seated on an inverted noggin, listened tamely to the conversation. **1889** —— *Despot of Broomsedge Cove* xviii. 324 Isabel sat idle on an inverted noggin.

no-go. Add: (Further examples.) Freq. *attrib.*

1946 *Ann. Computation Lab. Harvard Univ.* I. 26 If..the calculator finds that all multiples of the divisor are greater than the dividend or the remainder under consideration, a zero or 'no-go' is entered in the quotient counter and a new comparison made one column to the right. **1961** *Flight* LXXX. 542/1 A self-checking capability was therefore built into the machine so that, whenever a 'No Go' condition occurred, this part..carried out a self-checking routine to ensure that the fault was in the aircraft and not in the tester. **1962** A. Shepard in *Into Orbit* 170 When a particular station is on a 'No Go' status, the appropriate symbol lights up to indicate which of the several systems there is non-operative. **1966** *Electronics* 17 Oct. 93 Usually one test was sufficient to provide a go or no-go indication. **1969** *Guardian* 14 June 7/3 The astronauts would have had to end the mission there for, in space jargon, this was potentially a 'no-go' situation.

b. Used *attrib.* of an area: impossible to enter (because of barricades, etc.); to which entry is forbidden for specified persons, groups, etc.

1971 *Guardian* 13 Nov. 1/6 For journalists and others, the Bogside and Creggan estates are 'no-go' areas, with the IRA in total effective control. **1972** *Times* 24 May 16/1 The UDA organized the Protestant 'no go' areas in Belfast last weekend. **1972** *Guardian* 29 May 8/1 The Duke of Norfolk has decreed the Royal Enclosure at Ascot a 'no-go' area for the miniskirted or hotpanted lass. **1972** *Times* 3 June 2/7 Gypsy Council Names Four 'No-Go' Districts. The first four English boroughs likely to be designated as 'no go' areas for gypsies were named yesterday. **1972** *Daily Tel.* 11 Nov. 2/5 As a result, soldiers have been injured in rioting in the former 'no go' areas. **1974** *Times* 12 Feb. 2/8 Since the dispersal of the IRA leadership from inside the city to Donegal, after the British Army moved into the 'no-go areas' in 1972.

no-God. Used *attrib.* of a set of beliefs, etc., based on atheism. Also (*nonce-wd.*), a personified non-deity. So **no-Goddism, no-Goddite.**

1931 Belloc *Essays of Catholic* viii. 174, I have seen one very monstrous Myth reach maturity..and..explode. That was the Myth of Natural Selection. The enthusiasm which supported it and gave it the atmosphere in which to grow was no-Goddism. **1933** Chesterton *St. Thomas Aquinas* iv. 126 The Bolshevist No-God movement in the twentieth century. **1940** L. MacNeice *Last Ditch* 15 The horses ride..To a place where God and No-God Play at pitch and toss. **1952** R. A. Knox *Hidden Stream* vii. 60 If your atheist friend was a really reasoned no-Goddite, he would hoot with laughter.

no-good. orig. *U.S.* [A sb. use of the phr. *no good*: see Good C. 5 g.] **1.** A useless or valueless person or thing. Also *attrib.* Hence **no-good-er,** (*U.S. slang*) **no-goodnik** [*-nik], a no-good person.

1908 E. J. Banfield *Confessions of Beachcomber* ii. i. 250 A no-good-boy wantonly brought about a big wind. **1924** A. J. Small *Frozen Gold* i. 14 I'll learn you half-suckled no-goods what it means. **1931** M. Allingham *Look to Lady* v. 59 A pack of crazy no-goods—strutting about in funny clothes. **1936** S. P. Spivey in P. Oliver *Screening Blues* (1968) vi. 246 Oh you dirty no-gooder, you don't mean me no good. **1944** *This Week Mag.* 22 Oct. 21/2 Any newspaper reader of the late '20's would remember this no-gooder. **1944** S. J. Perelman *Crazy like Fox* (1945) 9 A parasite, a leech, a bloodsucker—altogether a five-star nogoodnick! **1947** D. M. Davin *Gorse blooms Pale* 28 Taking that no-good Calaghan girl to every dance. **1948** J. Steinbeck *Russ. Jrnl.* (1949) 100 The slovenly, no-good girl. **1958** 'A. Gilbert' *Death against Clock* 67 He recognized her almost at once for what she was, a gold-digger, a no-good. **1958** R. Graves in *Times Lit. Suppl.* 15 Aug. p. x/2 And the committee would, I am sure, be a snuggery of all the no-goods and do-goods whom I have spent half my life successfully avoiding. **1958** *Listener* 9 Oct. 569/2 A splendid comic performance by Alberto Sordi as a no-gooder from the city. **1959** *News Chron.* 19 June 8/2 Holly..gets herself involved with a no-good hoofer in a low night club. **1960** *Sunday Express* 10 July 15/6 He is a lazy no-good. **1960** *Guardian* 23 Sept. 11/7, I know their type... Bums and nogoodniks, the lot of them. **1968** *N.Y. Times* 3 Mar. 37 Lew Archer's job is to find a 17-year-old girl who has run off with a 19-year-old no-goodnik. **1971** *Black Scholar* Sept. 38/2 It was snowin' when your no good daddy left me for that hussy. **1973** W. M. Duncan *Big Timer* x. 67 This no-good whom I do not even see for weeks on end.

2. Phr. *no good to Gundy*: (see quots.) *Austral.*

The explanation in quot. 1945 seems improbable.

1919 W. H. Downing *Digger Dial.* 35 *No good to gundy*, of no advantage. **1945** Baker *Austral. Lang.* iv. 90 *No good to Gundy*, an elaboration of the simple 'no good', has been current since 1907 or before, and probably had its origin in America... The origin is more likely to be found..in the old U.S. phrase, *according to Gunter.* Gunter was a noted mathematician who gave his name to works of precision and accuracy. **1966** G. W. Turner *Eng. Lang. Austral. & N.Z.* vi. 118 The unexplained phrase *no good to Gundy*, meaning simply 'no good' perhaps appeals because of its alliteration.

Noh, No² (nōᵘ). Also **Nō.** [Jap.] The traditional Japanese masked drama evolved from the rites of Shinto worship and substantially unchanged since the 15th c.; the oldest form of drama in Japan. Freq. *attrib.*

1871 A. B. Mitford *Tales of Old Japan* I. 156 A kind of classical opera, called Nô, which is performed on stages specially built for the purpose. **1899** [see *kyogen]. **1911** *Encycl. Brit.* XV. 169 Briefly speaking, the Nō was a dance of the most stately character, adapted to the incidents of dramas. **1917** W. B. Yeats *Lett.* (1954) IV. 631 The music to my Noh play *The Dreaming of the Bones.* **1927** E. Pound *Let.* 9 Nov. (1971) 214, I wonder if Iwasaki is trained in No or if you and he want to undertake revision of my redaction of Fenollosa's paper on the Noh (or No; better I think spelled with the 'h'). **1938** D. T. Suzuki *Zen Buddhism & its Influence on Jap. Culture* i. iv. 70 The Noh-play 'Kokaji' gives us some idea about the moral and religious significance of the sword among the Japanese. **1958** [see *kyogen]. **1960** B. Leach *Potter in Japan* vii. 155 A grey, wooden, thatched, open-air Nō stage where the monks still perform under an August moon. **1964** I. Fleming *You only live Twice* i. 15 'Grey Pearl'..was so thickly made up that she looked like a character out of a No play. **1973** *Times* 20 Mar. 9/5 The bare events of the tragedy are retained and acted out in trance-like, Noh-play fashion. **1974** S. Marcus *Minding the Store* (1975) xiv. 281 Billie gave me..a Noh mask from Japan.

no hit. 1. *Cricket.* (See quot. 1835.)

This rule no longer applies.

1827 *Sussex Weekly Advertiser* 9 July, The umpire called no hit. **1835** *Hoyle's Games* 426 When the striker shall hit the ball, one of his feet must be on the ground, and behind the popping crease, otherwise the umpire shall call no hit.

2. *Baseball.* (See quots. 1961.) So **no-hitter.**

1948 *Richmond (Va.) Times-Dispatch* 15 Mar. 17/4 Even without this new delivery Blackwell..came within two outs of pitching two successive no-hitters. **1949** [see *left field* (*left a. 3 c)]. **1955** A. Budrys in D. Knight *100 Yrs. Sci. Fiction* (1969) 255 Walker's a good pitcher, all right—but he didn't pitch any no-hitters. And he only won eighteen games. **1961** J. S. Salak *Dict. Amer. Sports* 298 *No-hit no-run game*, a game in which the pitcher permits the opposing team no base hits and no runs for the entire game or for at least the first nine innings of the game. **1961** Webster, *No-hitter*, a no-hit game in baseball. **1973** *Tucson (Arizona) Daily Citizen* 22 Aug. 60/2 Bahnsen said he was aware he had a no-hitter going and started bearing down in the seventh inning. **1974** *News & Reporter* (Chester, S. Carolina) 22 Apr. 10-A/8 Blackwell struck out 13 batters in the contest and pitched five innings of no-hit ball before Burton ended his hopes for a no-hitter.

no-ho·per. *Austral. slang.* [f. No *a.* + Hoper.] **a.** A horse with no prospect of winning, an outsider. **b.** A useless or incompetent person, one from whom no good can be expected.

1943 Baker *Dict. Austral. Slang* (ed. 3) 54 *No-hoper*, an outsider. (Racing slang.) **1945** T. Ronan *Strangers on Ophir* (1966) 79 There were actually eight runners in it: the three favourites, three no-hopers hardly up to hack-race standard, [etc.]. **1953** [see *drongo 3]. **1957** D. Niland *Call me when Cross turns Over* i. 27 This son was not a no-hoper. **1959** Baker *Drum* (1960) iii. 33 The suggestion that a heart of gold beats in the breast of every no-hoper. **1959** *Times Lit. Suppl.* 25 Dec. 753/2 Shanty towns where the 'no-hopers' drink their big cheques away. **1967** J. Morrison in *Coast to Coast 1965–66* 137 He's a bit of a no-hoper. **1971** *Sunday Australian* 8 Aug. 5/1 He prefers that they go north to the mining towns to..staying in the cities where they are dragged down to the level of no-hopers they pick up with.

nohow, *adv.* Add: **1. b.** (Earlier and later U.S. examples.) Esp. in phr. *no how you (they) can fix it.* Cf. Fix *v.* 14 c.

1833 J. Hall *Harpe's Head* 91 (Th.), They don't raise such humans in the Old Dominion, no how. **1835** W. G. Simms *Partisan* 506 It won't be an easy journey, ma'am, no how, I tell you. **1836** [see *fix v. 14 c]. **1840** *Knickerbocker* XVI. 19, I mean my name ain't G. Washington Mortimer, no how. **1843** 'R. Carlton' *New Purchase* I. xviii. 141, I..couldn't read a chapter in the Bible no how you could fix it. **1929** H. W. Odum in A. Dundes *Mother Wit* (1973) 187 Boys jes' natchelly tired an' don't want to work no-how. **1970** R. P. Warren *Incarnations* 48 Did he merely blow, and never Was rightly your husband, no how. **1973** C. Williams *Man on Leash* (1974) x. 146 You won't be thinkin' about yore hairdo nohow.

nohowish, *a.* (Earlier and later examples.)

1826 Disraeli *Vivian Grey* II. xii. 171, I was altogether no-howish by the time I got home. *a* **1897** [see *all A. 10 a]. **1935** J. M. Murry *Between Two Worlds* viii. 110 Within a fortnight I was feeling nohowish.

‖ **noia** (noi·ā). [It.] Boredom, weariness, ennui.

1944 'Palinurus' *Unquiet Grave* i. 22 The secret of happiness lies in the avoidance of Angst (anxiety, spleen, noia, guilt, fear, remorse, cafard). **1962** *Encounter* XIX. 69 Let us hear what Moravia has to say about *noia*, the fundamental..situation of all his books. **1967** *Listener* 2 Nov. 583/2 The revenge play is the pretext for all that Renaissance introspection and *noia* and *Angst.*

Noilly Prat (nwaʸi pra). [Name of the manufacturers.] The proprietary name of a French vermouth; a drink of this.

1906 *Hatch, Mansfield Price List* May 32 Liqueurs... Vermouth (Noilly, Prat)..2/7-litre. **1907** *Yesterday's Shopping* (1969) 100/3 Vermouth... French, Noilly Prat —Per bot. 2/7. **1950** D. Ames *Corpse Diplomatique* i. 10, I found Dagobert..sipping a Noilly Prat. **1964** E. Bowen *Little Girls* i. iv. 53 We are also right at the end of the Noilly Prat. **1965** *Harper's Bazaar* Jan. 80/3 Haut Savoie, historically the cradle of French vermouth, though Noilly-Prat has always been made well to the west of it. **1966** J. Pearl *Crucifixion of Pete McCabe* (1967) v. 65 He uncorked the Noilly Prat vermouth. **1974** D. Winsor *Death Convention* v. 40, I want a very dry martini. With Noilly Prat.

noir. Add: **2. a.** The black numbers in the game of roulette.

1850 *Bohn's Hand-bk. Games* 348 The other chances are also designated on the green cloth.., 'le pair, le passe, et le noir'. **1928** M. Carol *How to play Roulette* ii. 21 Among other divisions, or spaces for the stakes, you will find *Passe, Pair, Manque, Impair, Noir, Rouge.* **1939** T. King *Twenty-one Games to play for Money* 31 Even chances are given when a stake is placed on:..Noir, meaning any black number that turns up. **1964** A. Wykes *Gambling* ix. 215 The European betting table is divided into six areas labelled *pair, impair, passe, manque, rouge,* and *noir* (even, odd, high, low, red, black). **1971** P. O'Neil-Dunne *Roulette for Millions* iv. 36 You may be regarded as socially inferior in French casinos, if you do not understand the following French expressions: Rouge: Red Noir: Black Impair: Odds [etc.].

b. The black colour in the game of *rouge et noir* (cf. *Rouge et Noir* s.v. Rouge *sb.*¹ 4).

1850 *Bohn's Hand-bk. Games* 343 The first parcel of cards played, is usually for *noir*, the second for *rouge.* **1850** [see *inverse sb. 3]. **1928** M. Carol *How to play Roulette* iv. 56 The even money chances are Rouge or Red, Noir or Black, Coleur [*sic*] and Inverse. **1964** A. Wykes *Gambling* vii. 171 (*caption*) The dealer lays out two rows of cards (*le noir* and *le rouge*) until each totals 31 or more. Players bet that one or the other row will be nearer to 31 by placing chips on *rouge* or on *noir.*

noise, *sb.* Add: **2. d.** An utterance, usu. in phr. *to make noises*: to express (something) vocally; freq. with defining adj. prefixed.

1951 N. Marsh *Opening Night* vii. 152 Dr. Curtis said: 'I'd better go and make professional noises at him.' **1955** *Times* 21 July 8/6 If this is so, 'why then the noise about the 12 German divisions in W.E.U. and N.A.T.O.?' **1956** N. Marsh *Off with his Head* (1957) v. 91, I suppose I ought to make a polite noise. **1965** *N.Y. Times* 15 Sept. 42 Left-wing Liberals have made neutralist noises in the past. **1967** *New Scientist* 22 June 718 General Electric and Alcoa, for example, are making noises about getting into city building. **1969** S. Hyland *Top Bloody Secret* iii. 232, I made the right kind of encouraging noises. **1971** *Guardian* 14 May 24/2 There is a temptation to see the hand of Tate and Lyle and Mr Cube in any political noises from the sugar trade. **1971** P. Worsthorne *Socialist Myth* iii. 32 The Labour Party cannot make the classical patriotic noises as convincingly as the Tories. **1973** *Times* 8 Jan. 3/3 Although the city council is, as they see it, making more friendly noises, its policies on development and road building..set it on a collision course.

3. c. *noises off*: sound effects, usu. loud or confused, produced off the stage but to be heard by the audience at the performance of a play. Also *transf.*

1924 H. A. Vachell *Quinney's Adventures* 46 As he did so, he heard what is called in stageland a 'noise off'. 'Put them in your pocket,' commanded madame, in a hurried whisper. **1932** Wodehouse *Hot Water* ii. 66 He's got a job with the British Broadcasting Company... He does the noises off. **1934** *B.B.C. Year-bk.* 44 Plays, for example, need their effects, which in their turn call for the studio and staff allocated specially for producing 'noises off'. **1937** N. Marsh *Vintage Murder* xi. 122 'And what..is Scotland Yard's part in the proceedings?' 'Noises off, Mr Ackroyd,' replied Alleyn good-humouredly. **1940** 'G. Orwell' *Inside Whale* 38 The aristocracy and the big bourgeoisie exist in his books chiefly as a kind of 'noises off', a haw-hawing chorus somewhere in the wings. **1949** F. Maclean *Eastern Approaches* iii. v. 359 After some time had elapsed, there were 'noises off' from which those of us who remained concluded that the attention of the enemy was fully engaged elsewhere. **1958** *Times* 7 July 9/4 There are 'noises-off', however, that add to the harmony. **1972** *Daily Tel.* 31 May 10/4 In fact the rustlings of the wind-blown trees were perfect noises-off for Shakespeare's uninhabited island.

d. *noise and number index*: a quantity used in evaluating aircraft noise in terms of its intensity and duration.

1963 *Final Rep. Comm. Probl. Noise* 218 in *Parl. Papers 1962–3* (Cmnd. 2056) XXII. 657 During the Social Survey made in 1961 in the vicinity of London (Heathrow) Airport, measurement of noise levels and studies of the numbers of aircraft likely to be heard were made... The results have been combined..to form a Noise and Number Index (NNI). **1971** *Physics Bull.* Nov. 660/3 An exposure index for aircraft noise has been developed from this survey, called the noise and number index. This index is a combination of the average noise level measured at a point on the ground and the number of times a person is exposed to aircraft noise during a given period of time.

6. b. Also (without preceding *make* or *keep*) used of a person, esp. in phr. *the* (or *a*) *big noise*: a person of great importance. orig. *U.S.*

1908 G. H. Lorimer *Jack Spurlock* vii. 153 A lot of people are beginning to think that Teddy's a mere noise. **1911** R. W. Chambers *Common Law* vi. 169 Well, sister,

take it from muh, she thinks she's the big noise in the Great White Alley. **1927** T. E. Lawrence *Let.* 8 Feb. (1938) 506 Drill Parades bi-weekly when a big noise draws near—Sir Sam. **1931** Galsworthy *Maid in Waiting* vi. 42 Saxenden is a big noise behind the scenes in military matters. **1939** C. Day Lewis *Child of Misfortune* iii. iii. 296 Elderton was the big noise in the Home Office. **1942** J. B. Priestley *Black-out in Gretley* iii. 32 He's rather a big noise here. Landed man really, but has a seat on our Board, and a local J.P. **1957** M. Kennedy *Heroes of Clone* i. v. 50 Say you don't want him. You're the big noise here.

d. *Joc.* phr. *to make a noise like*, to pretend to be.

1920 'Sapper' *Bull-Dog Drummond* v. 126 Make a noise like a sturgeon, and he'll think it's caviare. **1928** D. L. Sayers *Lord Peter views Body* 87, I s'pose I'd better make a noise like a hoop and roll away. Night, night, everybody. **1961** Partridge *Dict. Slang. Suppl.* 1199/1 *Noise like a—, make a*..dates from ca. 1908. Baden Powell, in his *Scouting for Boys*, instructed scouts in danger of detection to take cover and make a noise like a (say) thrush.

6*. In scientific use, a collective term (used without the indef. article) for: fluctuations or disturbances (usu. irregular) which are not part of a wanted signal or which interfere with its intelligibility or usefulness.

1923 *Telegraph & Telephone Jrnl.* IX. 119/2 The variations in noise were plotted, and their effect at times was to reduce the intelligibility to 20 or 30 per cent. **1930** *Proc. IRE* XVIII. 253 Circuit for measuring noise on the plate side of a vacuum tube. *Ibid.* 259 The ratio of signal to noise in the input circuit. **1932** F. E. Terman *Radio Engin.* vi. 207 The output currents obtained..in the absence of a signal voltage produce what is commonly referred to as 'noise' when flowing through a..loud-speaker, and it is also common practice to apply the term 'noise' to the corresponding radio-frequency currents obtained in the output of a radio-frequency amplifier, although these lie above the range of audible frequencies. **1940** Zworykin & Morton *Television* vi. 194 If the noise is appreciable compared with the picture signal, it appears in the reproduction as a myriad of constantly changing bright specks. **1953** J. B. Carroll *Study of Language* vii. 201 It is..necessary to study the effect of the signal-to-noise ratio on the efficiency of communication, noise being defined as that part of a received transmission which is extraneous to the original message. **1962** A. Nisbett *Technique Sound Studio* 262 In all electronic components and recording or transmission media the signal must compete with some degree of background noise. **1966** *New Scientist* 16 June 714/3 In this way a radar echo which may otherwise be hidden by 'noise' is rendered visible. **1968** J. Lyons *Introd. Theoretical Linguistics* ii. 88 The distortions produced in one's handwriting in a moving train can be attributed to 'noise'. **1970** O. Dopping *Computers & Data Processing* i. 21 Information theory deals largely with what happens when a random interference ('noise') is superimposed on the desired signal. **1973** *Computers & Humanities* VII. 160 Knisbacher uses a generalized (context-free) grammar in his algorithm [for machine translation] but avoids the 'noise' of too many resultant analyses for each sentence by simulating context sensitivity within that context-free framework. **1974** *Nature* 10 May 192/1 As normally viewed, displays of video noise ('snow') have the appearance of fields of small speckles which seem to dart about at random.

7. *noise abatement, control, level, measurement, meter, pollution, reduction, suppression*; *noise-free, -measuring* adjs.; **noise check** *Motor Rallying*, the use of a decibel metre to ensure that cars do not make too much noise; **noise contour**, an imaginary line or surface joining points where the noise level is the same; **noise factor** or **figure** *Electronics*, a quantity representing the additional noise introduced by a signal-processing device such as an amplifier (see quots.); **noise filter** *Electronics*, a filter for selectively reducing noise; **noise limiter** *Electronics*, a circuit or device for selectively reducing certain types of noise, esp. by momentarily reducing the output or the gain during peaks of greater amplitude than the desired signal; **noise storm** *Astr.*, a radio emission from the sun consisting of a succession of short bursts or pips in the megahertz range that lasts for a period of hours or days and is associated with sunspots.

1923 *Health* II. 438 A real want, a very great want, and a very immediate want is a Noise-Abatement Society. **1973** *Scotsman* 13 Feb. 8/3 It would be very hard to sustain a reasonable argument against them on noise abatement grounds. **1960** S. Turner *Rallying* vi. 68 One other sort of check which you must treat with respect is a noise check. **1963** [see *dust-raising* adj. (*dust sb.¹ 8 b)]. **1963** P. Drackett *Motor Rallying* iii. 37, I was remarkably unimpressed by the secrecy displayed by noise-check marshals on one big rally. **1971** *Physics Bull.* Nov. 656/1 During design studies of rotorcraft the noise of various designs is assessed by predicting an appropriate noise contour (usually 90 PN dB) and comparing it with the design noise target. **1973** *Times* 25 Apr. 19/6 If there were a prospect of drastically curtailing operations from Heathrow and Gatwick..before the time when the noise contours will start contracting through the increasing use of quieter aircraft, then there would be some point..in building Maplin. **1960** *McGraw-Hill Encycl. Sci. & Technol.* IX. 120/1 The first step in noise control is an analysis of the nature and extent of the problem. **1937** A. G. Tynan in *Radio Engin.* July 21/2 Such a factor is very easily arrived at by multiplying the noise-signal ratio..by the sensitivity of the receiver in microvolts. This may be conveniently christened the noise factor. **1952** *Wireless World* June 224/1 There are various slightly different definitions of noise factor (in America, 'noise figure'). *Ibid.*, The noise factor is 3. It means that the result of amplifier noise is to make the noise 3 times as bad as in the ideal case where the signal source is the sole noise generator. *Ibid.*, In an ideal amplifier or receiver the noise factor..would be 1. **1962** *Rep. Comm. Broadcasting 1960* 333 in *Parl. Papers 1961–2* (Cmnd. 1753) IX. 259 The Band V tests have also shown that the noise factor of receivers in this Band is at present relatively high. **1944** *Proc. IRE* XXXII. 420/2 The noise figure F of the network is defined as the ratio of the available signal-to-noise ratio at the signal-to-generator terminals to the available signal-to-noise ratio at its output terminals. **1952** Noise figure [see *noise factor* above]. **1968** *Wireless World* Dec. 455/2 Noise figure = Total noise output power/Noise output power due to sources only. *Ibid.* 457/2 Negative feedback..has no effect whatever on the noise figure of an amplifier at any given frequency. *Ibid.* 458/2 A noise figure at 1,000 Hz of 0·02 dB. **1960** *McGraw-Hill Encycl. Sci. & Technol.* IX. 130/1 The tone control of a radio or record player can act as a noise filter, as when high-frequencies are cut down to reduce record noise. **1967** E. L. Gruenberg *Handbk. Telemetry & Remote Control* ix. 15 The characteristics of the noise filter are determined by the sampling rate, the overall system accuracy, and the allowable crosstalk between successive samples. **1934** *Discovery* Dec. 348/1 Standard practice in noise-free construction is now available for architects. **1966** D. G. Brandon *Mod. Techniques Metallogr.* 162 Careful design of the collection system for backscattered electrons yields practically noise-free images. **1925** *Sci. Amer.* June 422/3 The limit of radio..is the static—what the radio engineers call the 'noise level' of the disturbances in the ether. **1932** V. O. Knudsen *Archit. Acoustics* ix. 257 The average street noise level in New York varies from about 47 to 80 db. **1959** *Daily Tel.* 24 Apr. 20/7 It must be shown that helicopter operations can be carried out at a noise level tolerable to the public. **1972** R. Adams *Watership Down* xix. 111 Nowadays, among fields and woods, the noise level by day is..too high for some kinds of animal to tolerate. **1939** *Wireless World* 5 Jan. 15 (*heading*) Noise limiters suppressing interference in the receiver. **1954** E. Molloy *Radio & Television Engineers' Ref. Bk.* xxxiii. 8 Noise limiters tend to be less efficient when used in receivers possessing extremely high selectivity. **1934** *Discovery* Dec. 345 Efforts..are being made to provide standards of noise-measurement. *Ibid.* 347/1 The expert with his noise-measuring gear. **1964** R. F. Ficchi *Electr. Interference* vii. 120 About twenty years ago, a new approach to the problem of measurement was taken when it was stated that the most practical type of noise-measuring instrument is, essentially, a radio receiver with an indicating means. **1931** *Proc. IRE* XIX. 1953 This instrument, which has been called a 'circuit noise meter', consists of an amplifier, a frequency weighting network, a rectifier, and an indicating meter. **1950** *Jrnl. R. Aeronaut. Soc.* LIV. 699/1 The microphone..has been calibrated by the makers in conjunction with the noise meter. **1970** *Britannica Bk. of Year 1969* 798/3 *Noise pollution*, pollution consisting of annoying noise (noise pollution caused by automobile traffic, a jet airplane, or a vacuum cleaner); called also *sound pollution.* **1970** *Sci. Jrnl.* Mar. 5 The greatly improved noise pollution characteristics of VTOL compared not only with conventional aircraft (CTOL) but with short take off and landing craft (STOL). **1972** J. Maddox *Doomsday Syndrome* iv. 98 'Noise pollution' is a phrase in everyday use. **1931** *Proc. IRE* XIX. 1763 The noise reduction advantage of the arrays..is some 15 decibels over that of a nondirectional antenna. **1959** W. S. Sharps *Dict. Cinematogr.* 114/2 In photographic recording and reproducing, noise reduction is a process whereby the average transmission of the sound track of the print..is decreased for signals of low level and increased for signals of high level. **1947** C. W. Allen in *Monthly Notices R. Astron. Soc.* CVII. 391 All solar noise storms..in Figs. 2 and 3 coincide with near meridian passage of spots. **1974** G. L. Verschuur *Invisible Universe* iii. 36 Sunspots also produce characteristic radio bursts, called storm bursts or bursts of type I. These are very short-lived intense pips of radio emission, each lasting only a fraction of a second, although thousands of these might be emitted every hour and the noise storm (a string of bursts) may last for days. **1933** *Practical Wireless* II. 37/2 (*heading*) Noise suppression on the short waves. **1968** *IEEE Jrnl. Quantum Electronics* IV. 644 (*heading*) Noise suppression in the Argon FM Laser.

noiseless, *a.* Add: **b.** Characterized by a virtual absence of noise (sense 6*).

1931 A. Nadell *Projecting Sound Pictures* xiv. 244 The recently developed 'noiseless recording' is achieved by passing a 'bias' current through the value ribbons, in addition to the usual speech current. The effect of the bias is to eliminate the former spacing of 2, or 1½ mils, and to keep the ribbons always closed when no speech current is passing. **1944** *Phil. Mag.* XXXV. 394, P must be regarded as a noiseless shunting resistance. **1957** *Sci. Amer.* Feb. 82/2 The maser would..be useful in astronomy and cosmology. As a noiseless amplifier it would eliminate the noise generated in the circuits of radio telescopes. **1972** J. M. Taylor tr. *Meyer & Neumann's Physical & Applied Acoustics* ix. 351 In variable-area recording..the bright side of the sound track is darkened for small amplitudes in the region that does not contribute to sound generation (noiseless recording).

noisy, *a.* Add: **1. b.** *noisy scrub bird Austral.*, a ground-dwelling bird, *Altrichornis clamosus*, found in parts of Western Australia.

[**1848** J. Gould *Birds Austral.* III. tab. 34 (*heading*) *Altrichia clamosa*, Gould. Noisy Brush-bird.] **1901** A. J. Campbell *Nests & Eggs Austral. Birds* I. 504 The Noisy Scrub Bird..lives in the thickets of undergrowth. **1933** *Bulletin* (Sydney) 1 Feb. 21 One of the rarest and most curious of Australian birds is the 'noisy scrub-bird'..whose home, if perchance the secretive creature survives, is in the dense forest of the Leeuwin Peninsula. **1966** *Evening Standard* 14 July 18/2 One of Australia's rarest birds, the noisy scrub bird, has forced the Western Australian State Government to abandon plans to establish a new town. The town..was to have been in the heart of the noisy scrub bird country near Albany, 300 miles south of Perth. *Ibid.* 18/3 Between 1889 and 1961 the bird was neither seen nor heard. But then it reappeared... The noisy scrub bird is extremely elusive. The male has a call like the crack of a whip. **1972** G. Durrell *Catch me a Colobus* x. 218 In Australia they rediscovered the Noisy Scrub Bird which had been thought to be extinct.

2. b. (In the sense of *noise sb. 6*.)

1944 *Phil. Mag.* XXXV. 394 If [the resistance] P were thermally 'noisy'. **1949** *Proc. R. Soc.* A. CXCVII. 480 (*facing caption*) Noisy background chiefly due to imperfections of illuminating objective. **1951** *Sci. Amer.* Aug. 14/3 Transistors..were 'noisy'. **1973** *Nature* 5 Oct. 258/1 In the galactic plane, the 1,420 MHz line is noisy, which decreases its attractiveness.

‖ **noix** (nwa). [F., lit. 'nut'.] (See quot. 1961.)

1845 E. Acton *Mod. Cookery* ix. 210 That part of the fillet [of veal] to which the fat or udder is attached;.. Called by them [*sc.* French cooks] the *noix.* **1893** *Encycl. Pract. Cookery* II. 699/2 Beat the noix well, trim it to a nice shape, and lard with thin fillets of fat bacon. **1906** A. Filippini *Internat. Cook Bk.* 162 (*heading*) Noix of veal, braisée, fermière. **1923** C. H. Senn *Cent. Cookery Bk.* (new ed.) 565 Noix de veau à la Viart..1 small cushion or kernel part of veal [etc.]. **1960** E. David *French Provincial Cooking* 371 A noix of veal, the cut approximating to the topside of the leg in beef. **1961** N. Froud et al. tr. *Montagné's Larousse Gastronomique* 993/1 The *noix* of veal is the topside (rump), the fleshy upper part of the leg, cut lengthwise. **1967** A. Wilson *No Laughing Matter* iii. 193 They hungrily devoured the noix de veau and carrots at lunch. **1974** *Times* 2 Nov. 11/5 Everything from stewing steak to dainty noix de veau.

no-knock. Used *attrib.* (occas. *absol.*) of a search or raid by the police made without permission or warning. *U.S.*

1970 *Atlantic Monthly* Oct. 57 John Mitchell puts on a happy face and suggests that the name of the 'no-knock' law be changed to something more felicitous, like 'quick-entry'. **1971** *New Yorker* 10 Apr. 30 The 'no-knock' and 'preventive-detention' provisions of the District of Columbia Crime Control Act have violated, respectively, the public's right to be secure against unreasonable searches and seizures and the traditional presumption of innocence. **1973** *Black Panther* 8 Sept. 8/3 The California State Supreme Court last week refused to decide in favor of a no-knock rule. In a four to three decision the court held that a search warrant specifically providing that a police officer can enter a home without announcing himself cannot be issued. **1973** *Houston (Texas) Chron. Mag. People, Places, Pleasures* 14 Oct. 4/6 Can't the people see the implications of a thing like 'no knock'? It's a Gestapo tactic.

no-licence. A stipulation that no licence is granted, e.g. for the sale of liquor. Freq. *attrib.*

1908 *Daily Chron.* 24 Apr. 4/3, I have been a police officer for thirty-seven years, and am consequently quite familiar with the conditions prevailing before the no-license law..came into effect. **1909** *Ibid.* 27 Feb. 5/7 A no-license resolution, which means that no certificate for the sale of excisable liquors shall be granted. **1921** E. O'Ferrall in Murdoch & Drake-Brockman *Austral. Short Stories* (1951) 163 I've decided to reform and join the No-licence crowd. *Ibid.*, I'm goin' t' vote No-licence! **1969** *Austral. Encycl.* V. 312/1 There was to be no compensation for licensees in the event of a vote for no-licence.

no-limit. Used *attrib.* of a game, betting, etc., in which no limits are laid down.

1915 *Munsey's Mag.* LIV. Apr. 485/1 Poker game tonight... We'll have a real game this time—a no-limit game. **1954** F. C. Avis *Boxing Ref. Dict.* 75 *No-Limit Boxer*, a heavyweight, from the fact that there is no restriction of weight among the 'heavies'. **1962** K. Orvis *Damned & Destroyed* xv. 110 Play on the no-limit table was completed. **1973** *Times* 12 Apr. 12/6 Ladbroke..aside from the £50m spending spree planned on hotels, is quite happy to welcome no-limit betting.

nolition. (Later example.)

1857 in *Appleton's Illustr. Handbk. Amer. Trav.* 237 Do not allow..the nolition of your womankind..to prevent you.

nolle, *v.* (Later example.)

1910 *Springfield (Mass.) Weekly Republ.* 24 Nov. 10 (*headline*) Case against Haskell Nolled.

‖ **nolle** (nǫ·li), *sb. U.S.* Abbrev. of Nolle prosequi.

1871 E. Eggleston in *Hearth & Home* 23 Dec. 1010/3, I now enter a *nolle* in his case..and I ask that this court adjourn. **1878** J. H. Beadle *Western Wilds* xxxi. 507 He had been indicted along with the others, and a *nolle* entered.

‖ **nolle pros.** = *nolle sb.

1895 *Denver Times* 5 Mar. 1/3 John Doyle was dismissed on a nolle pros in both cases against him.

nolle prosequi. Add: (Later examples.) Also (quot. 1971) *joc.*

1964 *Mod. Law Rev.* XXVII. 268 The jury failed to agree whether there was a trade dispute. The Attorney-General entered a *nolle prosequi*, but wished it to be 'clearly understood' that strikes not in pursuit of trade disputes had no statutory protection. **1971** Wodehouse *Much Obliged, Jeeves* x. 104 When an aunt has set her mind on a thing, it's no use trying to put in a nolle prosequi.

nolle-pros(s, *v.* Add: Also **nolle prosse, nol-pros.** (Earlier and later examples.)

1880 G. A. Pierce *Zachariah* 436 Judge Spalding informed Zach..that the case could be 'nolle prossed' when it came up. **1905** *Springfield* (Mass.) *Weekly Republ.* 15 Dec. 2 The court heard petitions for a new trial, and upon these being granted the cases were nol prossed and the brothers set free. **1926** D. L. Colvin *Prohibition in U.S.* 505 In the two years 14,567 cases were nolle-prossed or dismissed. **1974** D. Tidyman *Dummy* ix. 127 The State would move to *nolle prosse* the case against Donald Lang.

no-load (nōᵘ·lōᵘd), *attrib. phr.* [No *a.* 6 b.] **1.** *Electr.* Corresponding to or involving an absence of any load.

1907 G. W. O. Howe tr. *Thomälen's Text-bk. Electr. Engin.* viii. 180 The no-load loss for a given excitation can generally be looked upon as a constant loss. **1951** W. Sluckin *Princ. Alternating Currents* vii. 175 Calculation of the..iron losses is based on the results of the so-called open-circuit or no-load test. **1966** N. Jones *Basic Electrotechnology* xi. 150 The motor exhibits a practically constant speed characteristic, the full-load speed being about 5% less than the no-load speed.

2. Of shares: sold at net asset value. Also as *sb.*

1964 P. Wyckoff *Dict. Stock Market Terms* 174 *No load fund*, a mutual fund which charges little or no commission (*load charge*) to the buyer of its shares. No sales organization is involved. **1968** *Maclean's Mag.* Oct. 22 A handful or so Canadian funds are 'no-loads'—are offered without any sales charge and are generally available through investment dealers. **1969** *Times* 5 May (Suppl.) p. vi/3 No-loads are more likely to be performance funds because they have no salesmen to sell them. **1972** *National Observer* (U.S.) 27 May 8/6 (Advt.), The Scudder Special Fund is a no-load mutual fund seeking above-average growth of capital and may invest in securities with above-average risk. You pay no sales commission when you invest in this Fund.

‖ **nolo contendere** (nōᵘ·lo kǫnte·ndĕre). [L., 'I do not wish to contend'.] 'A plea by the defendant in a criminal prosecution that without admitting guilt subjects him to a judgment of conviction as in case of a plea of guilty but does not preclude him from denying the truth of the charges in a collateral proceeding' (Webster, 1961).

1872 G. P. Burnham *Memo. U.S. Secret Service* p. vii, *Nolo contendere*, don't wish further to contend. **1960** *Times* 9 Dec. 11/2 Counsel for these companies told the court that their pleas of Guilty and *nolo contendere* were not an admission of all allegations on the indictment, but were made with a desire to terminate what would otherwise be protracted and expensive litigation. **1972** *N.Y. Law Jrnl.* 24 Nov. 3/1 [He] entered a plea of nolo-contendere to ten counts of securities fraud of an indictment charging him with sixteen counts of securities law violations.

‖ **nom.** Add: **c.** *nom de Dieu* (noṅ də dyö), a mild oath.

1867 'Ouida' *Under Two Flags* III. ix. 228 'Nom de Dieu, Miladi!' she swore in her teeth. **1922** E. E. Cummings *Enormous Room* viii. 176 *Nom de dieu*, I thought vaguely. Am I or am I not completely asleep? **1961** *Amer. Speech* XXXVI. 35 Another basic oath is *nom de Dieu.*

d. *nom de théâtre* (noṅ də teˌatr'), a stage name.

1874 W. P. Lennox *My Recollections* I. viii. 189 After leaving the army, he appeared on the regular boards under the *nom de théâtre* of Calcraft. **1935** Wodehouse *Blandings Castle* xi. 278 A short list of names, one of which she proposed as a *nom de théâtre* as soon as her screen career should begin. **1952** Granville *Dict. Theatrical Terms* 122 *Nom de théâtre*, a stage name. These were adopted at the time when connection with the stage was not considered respectable. **1974** *Times* 6 Apr. 14/2 Erik Weisz..in 1890..set himself up as a stage musician under the *nom de théâtre* of Harry Houdini.

e. *nom de vente* (noṅ də vāṅt), a name assumed by a buyer at an auction who wishes to remain anonymous.

1955 *Times* 21 May 5/4 This, it is understood, is the *nom-de-vente* for a member of the nobleman's family. **1959** *Times* 26 Feb. 12/5 All three bought by Felham, a *nom de vente.*

nomadism. (Later examples.)

1963 *Punch* 13 Nov. 721/1 Centuries of fierce nomadism. **1964** Gould & Kolb *Dict. Social Sci.* 724/2 Usages of transhumance..agree in contrasting the term with *pastoral nomadism.* **1974** *Environmental Conservation* I. 12/1 If traditional nomadism finally disappears..it will be impossible to recreate it.

nomadization. Add: **b.** A making or becoming nomadic in character or nature.

1920 H. G. Wells *Outl. Hist.* 606/2 What we now call democracy, the boldness of modern scientific inquiry and a universal restlessness, are due to this 'nomadization' of civilization.

nomadize, *v.* Add: **b.** *trans.* To make nomadic in character.

1902 D. G. Hogarth *Nearer East* 156 The incomers 'nomadised' the south-east. *Ibid.* 272 The southern oases are the most 'nomadised'.

nomady (nōᵘ·mădi). [f. Nomad *sb.* and *a.* + -y³.] The state, condition, or life of a nomad.

1909 Beerbohm *Yet Again* 126 Had nomady been my business, had I been a commercial traveller or a King's Messenger. **1920** —— *And Even Now* 263 The Bohemian, as tending always to nomady, feels that [etc.].

no man. Add: **1. b.** [see No *adv.*³ and *sb.*] A man who says 'no'; one who is accustomed to disagree or to refuse requests in a resolute manner. *colloq.*

1953 Berrey & Van den Bark *Amer. Thes. Slang* (1954) § 400/3 Obstinate person..'no' man. *Ibid.* § 421/1 Critic; opposer..'no' man (*vs.* 'yes man'). **1959** *Listener* 30 July 168/1 He [*sc.* Metternich] was the great 'No-man'. One can make a long list of the changes he did not like. **1961** *Sunday Express* 19 Feb. 8/4 Prince Philip..attacked what he called the 'no-men' who stood in the way of those with energy and imagination. **1967** *Listener* 26 Jan. 143/1 The only role of the abominable no-men of Whitehall was to frustrate him at every turn.

2. a. *no man's land.* (Earlier and additional examples.) See also quots. 1966 and 1972.

1598 Stow *Survey of London* 356 Ralph Stratforde Bishop of London, in the yeare 1348, bought a peece of ground called no mans land, which he inclosed with a wall of Bricke. **1876** H. Brooks *Natal* 234 In 1866..the Government of the colony took possession of 'No-man's-land'. **1902** *Chambers's Jrnl.* Jan. 34/1 The independent warlike tribes formerly sandwiched in a No Man's Land between Afghanistan and India. **1909** *Westm. Gaz.* 21 Aug. 14/1 This place has a higher attraction..for it is no-man's-land, eligible for building on, threatened, but as yet unoccupied. **1910** *Chambers's Jrnl.* Aug. 495/1 These cottages had been built..on ground between two roads, which was a kind of 'no man's land' and rent free. **1921** C. A. W. Monckton *Some Experiences New Guinea Resident Magistrate* xxvi. 306 The country we were camped in was a sort of 'no man's land' or border land lying between the Baruga tribe and their mountain enemies. **1966** G. E. Evans *Pattern under Plough* xiii. 136 The belief that a piece of land in the parish should be left untilled. In English villages this is sometimes called *Jack's Land* or *No-Man's Land.* **1972** P. Newton *Sheep Thief* 187 *No Man's Land*, those areas of wild rough country, bounding the Southern Alps, which are not stocked. **1975** *Country Life* 19 June 1637/1 Until the Dutchman Vermuyden came on the scene..to control ..the river Great Ouse..much of the region was a marshy no-man's-land through which..the only means of transport was by boat.

c. *fig.*

1870 H. A. Nicholson *Man. Zool.* I. 6 Some observers have established an intermediate kingdom, a sort of no-man's-land, for the reception of those debatable organisms. **1888** H. A. Strong tr. *Paul's Princ. Hist. Lang.* ii. 28 This 'no man's land' forms a boundary wall through which the influences cannot pass from one side to the other. **1892** [in Dict., sense 2 a]. **1954** M. Beresford *Lost Villages* v. 166 The crucial years lie half in a no-man's-land of English historical research. **1959** *Wall St. Jrnl.* 9 Nov. 16/4 One aim of the new law was to abolish the so-called 'no-man's-land' between Federal and state authority in labor cases. **1973** *Archivum Linguisticum* IV. 51 This is indeed a No Man's Land, virgin territory one might say. **1974** 'H. Carmichael' *Motive* iv. 53 One question chased another..questions that got lost in a no-man's-land of conjecture. **1974** *Kingston* (Ontario) *Whig-Standard* 11 Jan. 7/1 Dossers are..men and women squeezed into a no-man's-land by an increasingly affluent society.

d. *Mil.* The terrain between the front lines of armies entrenched opposite one another.

1908 *Blackw. Mag.* Dec. 761 Here and there in that wilderness of dead bodies—the dreadful 'No-Man's-Land' between the opposing lines—deserted guns showed up singly or in groups. **1915** G. Adam *Behind Scenes at Front* 101 Perilous work it is repairing wire in the No Man's Land between trenches. **1936** M. Plowman *Faith called Pacifism* 97 The order I received..to go out across No-man's-land, and cut the throat of the nearest German lad on sentry. **1973** R. Hill *Ruling Passion* iii. ii. 173 They were keeping steadily on neutral ground and she was finding it a pleasant experience. Like football in no-man's-land during a Great War Christmas.

e. *Lawn Tennis.* (See quot. 1931.)

1931 Doeg & Danzig *Elem. Lawn Tennis* 65 If you hesitate you are caught in what is known as 'No Man's Land', the territory between the service line and the baseline. **1969** *New Yorker* 14 June 68/3 Graebner, in no man's land, drives the ball far into Ashe's backhand corner.

no-mar (nōᵘ,māɹ), *attrib. phr.* [f. No *a.* + Mar *v.*] That is designed not to mar, spoil, etc.

1970 *Globe & Mail* (Toronto) 26 Sept. 52/4 (Advt.), Trays..of stain-resistant woodgrain, with raised frame and no-mar, plastic tipped feet. **1973** *Washington Post* 13 Jan. A. 19/1 (Advt.), Round extension table with..walnut bronze no-mar top table.

nomen (nōᵘ·mĕn). [L.] **1.** *Taxonomy.* **nomen nudum** (niū·dəm) [L., naked name], a Latin name which has no standing because it was introduced without publication of the full description demanded by the rules governing botanical and zoological nomenclature. Also occasionally used for a popular name that cannot be attached to a definite species (see quot. 1957).

1906 *Verhandl. Int. Bot. Kongr. Wien 1905* viii. 205 A genus or any other group of higher rank than a species, named or announced without being characterised conformably to article 37 cannot be regarded as effectively published (*nomen nudum*). **1925** A. S. Hitchcock *Methods of Descriptive Systematic Bot.* xii. 114 A new name appearing without description is called a *nomen nudum.* **1957** *N. & Q.* Feb. 83/2 Since, however, no such variety has been described or bred, Chub-eel (× *Aminguilla*) remains a *nomen nudum.* **1963** Davies & Heywood *Princ. Angiosperm Taxonomy* viii. 290 It is permissible..to use a name for a new species that already exists as a *nomen nudum*—i.e. has never been validated by a description and therefore has no standing under the Rules [of botanical nomenclature]. **1969** E. Mayr *Princ. Systematic Zool.* xiii. 347 A name published without satisfying the conditions of availability is generally called a *nomen nudum*... A *nomen nudum* has no standing in zoological nomenclature and is best never recorded, not even in synonymy.

2. *Philol.* In phrases: **nomen actionis** (æktiōᵘ·nis), a noun of action; **nomen agentis** (age·ntis), an agent noun.

1928 C. Bergener *Contrib. to Study of Conversion of Adjs. into Nouns* ii. 18 When a nomen agentis is contrasted with an absolute past participle of the corresponding verb (as in Galsw., Flower I, ch. vi: *the tortured does not salute her torturers*). **1930** T. Sasaki *On Lang. Bridges' Poetry* iii. ii. 80 The use of the gerund and the noun of action (nomen actionis)..have in common certain grammatical functions. **1932** A. Gardiner *Theory of Speech & Lang.* iii. 107 In all these languages..the equivalent of 'speech' is a *nomen actionis* for the activity of which the most evident symptoms are articulation and audibility. **1963** F. T. Visser *Hist. Syntax Eng. Lang.* I. iv. 357 With a plural subject the ending *-end* of the nomina agentis appeared as *-ende.* **1966** *English Studies* XLVII. 53 This is an old trait in Germanic where the pres. participles had the character and regimen of nomina agentis. **1970** *Archivum Linguisticum* I. 9 One other class of words seems to have had fluctuating secondary stress: these are the nomina agentium in *-ari, eri, -iri.*

nomenclatorial, *a.* Delete *rare*⁻¹ and add further examples.

1897 *Nature* 19 Aug. 364/1 To distinguish those [references] that relate to habits and biology from those that are systematic and nomenclatorial. **1935** *Jrnl. Heredity* XXVI. 461/1 It has been..a continuing nomenclatorial tragedy that parents..label helpless infants..with the cognomen of the political (or other) hero popular at the moment of birth. **1946** F. E. Zeuner *Dating the Past* viii. 271 This [*sc.* a difference in dating Chinese geological formations] appears to be due to nomenclatorial rather than factual differences. **1953** *Parasitology* XLII. 260/2, I wish also to make reference to another nomenclatorial problem. **1970** *Nature* 5 Sept. 1065/2 Fortunately, the informational content of volume eight of *Primates* is in no way affected by these nomenclatorial niceties.

nomenclatural, *a.* (Later examples.)

1937 N. N. Puckett in A. Dundes *Mother Wit* (1973) 167/1 The enticement may have been the ceremony [of baptism] rather than a delusion of nomenclatural grandeur. **1939** [see *homotype 1]. **1951** G. H. M. Lawrence *Taxon. Vascular Plants* ix. 196 In many countries the nomenclatural patterns and practices were set by botanists of considerable prestige. **1961** G. G. Simpson *Princ. Animal Taxon.* i. 29 Personal preferences and emphases also frequently affect rearrangements of supraspecific taxa..although this is not an important cause of nomenclatural change. **1972** *Nature* 24 Nov. 239/1 The difference between them is more apparent than real and stems from nomenclatural rather than biological considerations. **1974** *Ibid.* 1 Nov. 85/3 He has kindly and lucidly drawn attention to the intricacies of the correct nomenclatural procedures to be used.

Hence **nomenclatu·rally** *adv.*

1944 S. A. Cain *Found. Plant Geogr.* xxx. 473 The first described form in a group is the species and the later forms are varieties, nomenclaturally. **1967** R. E. Blackwelder *Taxonomy* xx. 423 Nomenclaturally it [*sc.* the type specimen] serves as an anchor for the name.

nomic, *a.*² Add: **2.** *Philos.* and *Math.* That pertains to or is concerned with a discoverable scientific or logical law. Hence **no·mically** *adv.*

1892 K. Pearson *Gram. of Sci.* iii. 114, I shall, for convenience, however, speak of natural law in the old sense, or, as a mere routine of perceptions, as law in the *nomic* sense. Law in the nomic sense is thus no product of the reason, but a pure *order* of perceptions. **1905** *Nature* 30 Mar. 517/2 The correlation..is..of nomic heteroscedasticity. **1921** W. E. Johnson *Logic* I. iv. 61, I should propose that *nomic* (from νόμος, a law) should be substituted for *necessary* as contrasted with *contingent.* Thus a nomic proposition is one that expresses a pure law of nature. *Ibid.*, The nomically possible. **1959** K. R. Popper *Logic of Sci. Discovery* 434 A 'necessary conditional' or a 'nomic conditional'. **1961** E. Nagel *Structure of Sci.* iv. 51 The distinction between accidental and nomic universality can be brought out in another way. **1973** N. Rescher *Conceptual Idealism* iv. 59 This nomological necessity of laws is generally called 'nomic necessity'.

nominal, *a.* and *sb.* Add: **A.** *adj.* **1. a.** (Further examples.) (See also quots.)

1924 O. Jespersen *Philos. of Gram.* ix. 120 Here we first encounter the so-called nominal sentences, containing a subject and a predicative, which may be either a substantive or an adjective. **1928** H. Poutsma *Gram. Late Mod. English* (ed. 2) I. i. 2 Predicates are of two kinds, viz.: (*a*) verbal predicates..; (*b*) nominal predicates, i.e. such as are made up of a copula or link-verb and a nominal. **1929** *Ibid.* x. 557 *And* is used to..bring about a junction of a sentence with an undeveloped nominal- or infinitive-clause. **1939** L. H. Gray *Foundations of Lang.* 230 There is only one case in which the sentence does not require a verb, i.e., the nominal sentence of the type of the Latin *omnia praeclara rara* 'all splendid things (are) rare'... Contrary to popular opinion, the nominal sentence does not omit the copula, and none is to be supplied. **1949** *Archivum Linguisticum* I. 183 The hybrid nature—nominal-verbal—of the participle. **1954** Pei & Gaynor *Dict. Ling.* 147 *Nominal sentence*, a sentence in which the principal part..is a noun or nominal form. **1973** *Archivum Linguisticum* IV. 3 Some verbal forms occur as nominal modifiers, others not.

5. d. (Earlier and later examples.)

1869 *Bradshaw's Railway Manual* XXI. 369 The nominal capital now consists of 92,000 original shares of 5*l* each; 38,000 preference shares of 4*l* each: and 32,000 obligations of 4*l* each. **1964** *Lebende Sprachen* IX. 98/2 The amount of capital stated in the memorandum of association which a joint-stock company may issue is called *nominal* or *authorized capital.*

6. (See quot. 1970².)

1966 *Aviation Week & Space Technology* 5 Dec. 30/1 The mission is to launch the 800-lb. Prime vehicle to effect a nominal re-entry at 400,000 ft. following injection at 26,000 fps. **1970** N. Armstrong et al. *First on Moon* vi. 124 An example of misuse is our use of the word 'nominal', which most of the English-speaking world interprets as meaning small, minimal—and we usually use it in the sense of being average or normal. **1970** R. Turnill *Lang. Space* 94 *Nominal*, a favourite word, meaning within prescribed limits; anything from 'perfect' to acceptable. **1972** *Daily Colonist* (Victoria, B.C.) 26 July 3/1 As one engineer said, 'She is phenomenally nominal'—nominal being space jargon for operating-as-planned.

B. *sb.* **5.** *Gram.* (See quot. 1972.)

1928 H. Poutsma *Gram. Late Mod. Eng.* (ed. 2) I. i. 2 A nominal (a noun or adjective) or a nominal equivalent (i.e. a word or word-group doing duty as a nominal). **1935** H. Straumann *Newspaper Headlines* 49 Words of these formal characteristics will be called nominals. **1961** *Amer. Speech* XXXVI. 159 It is customary to describe the English nominal as consisting of a sequence of constituents: predeterminers, determiners, adjectives, the noun head, and finally certain postnominal modifiers, such as relative clauses. **1968** J. Lyons *Introd. Theoretical Linguistics* 347 Let us..draw a distinction between what we will call *first-order* and *second-order* nominals in English, and say that only second-order nominals may occur in sentences whose underlying structure is Nominal + Time. **1972** Hartmann & Stork *Dict. Lang. & Ling.* 151/2 *Nominal*, a name given..to a word which functions as a..noun, but does not have all the formal characteristics of a noun (i.e. in English the distinction between singular and plural and between common and possessive cases). **1974** *Nature* 25 Oct. 705/1 One could defend the thesis that sentences are remembered by forming an associative structure linking representations of their nouns and verbs by arguing that the associative machinery deals with nominals derived from verbs in a very similar way to verbs themselves.

nominalist. Add: Also as *adj.*

1965 *Listener* 9 Dec. 942/1 How much more nominalist, in a way, the whole society is.

nominalistic, *a.* Add: (Later example.) Hence **nominali·stically** *adv.*

1932 C. C. J. Webb *John of Salisbury* 154 In philosophy he favoured the views of the Nominalistic school. **1939** *Mind* XLVIII. 246 To save himself he would have to read 'satisfaction' nominalistically.

nominalize, *v.* Delete † *Obs. rare*⁻¹ and add later examples. Hence **nominali·zable** *a.*; **nominaliza·tion**; **no·minalized** *ppl. a.*; **no·minalizer**; **no·minalizing** *ppl. a.*

1951 Z. S. Harris *Methods in Structural Linguistics* xvi. 277 There is no noun class in Hidatsa, only a stem class (neither noun nor verb), a class of nominalizing suffixes, and a class of verbalizing suffixes. **1955** T. Burrow *Sanskrit Lang.* iv. 125 Such transference is common in nominalised adjectives throughout the system. **1960** R. B. Lees in *Internat. Jrnl. Amer. Linguistics* XXVI. (*title*) The grammar of English nominalizations. **1965** N. Chomsky *Aspects of Theory of Syntax* ii. 87 Nominalizable adjectives. **1965** *Canad. Jrnl. Linguistics* X. 117 The nominalizer *-ma* is attached directly to the future durative suffixes. *Ibid.* 160 These articles are obligatory with every noun or nominalization appearing as subject or object of a verb. **1967** Z. Vendler *Linguistics in Philos.* v. 124 In technical terms, we will end up with a list of nominalized sentences. **1968** J. Lyons *Introd. Theoretical Linguistics* 265 The constituent-string is transformed ('nominalized') into an N[oun] P[hrase] of the form A[djective] + N[oun]. **1972** A. Makkai *Idiom Structure* 140 *The king stepped down*, nominalizable as **the downstepping of the king.* **1972** *Science* 23 June 1306/2 When a surgeon uses a phrase such as the [*sic*] 'the bleeding of the vessels', the only admissible origin for the nominalization 'bleeding' would be 'the vessels bled' and not 'someone (that is, the surgeon) bled the vessels'. **1972** J. W. Bresnan in *Language* XLVIII. 334 It must be noted that *-ation* is not an isolated case... Many other nominalizing affixes affect internal stress: contínue, continúity; oppóse, opposítion; [etc.]. **1974** H. Esau (*title*) Nominalization and complementation in modern German.

nominate, *ppl. a.* Add: **1. d.** *Taxonomy.* (See quot. 1967.)

1948 A. L. Rand *Mammals Eastern Rockies* 100 The Alberta form is the nominate subspecies [of badger]: *Taxidea taxus taxus* Schreber. **1967** R. E. Blackwelder *Taxonomy* xix. 398 A nominate taxon is the subordinate taxon which contains the type of the subdivided 'higher' taxon. **1968** Perring & Sell *Critical Suppl. Atlas Brit. Flora* 146 The question of the use of the nominate subspecies in this case [*sc. Eleocharis palustris*] is complex and disputed. **1971** *Nature* 10 Dec. 360/1 The European stock [of the Atlantic salmon], binominally named by Linnaeus, must become the nominate subspecies and, therefore, should bear the name *Salmo salar salar* Linnaeus. **1974** *New Phytologist* LXXIII. 802 The presence of the nominate subspecies in some cases..and not in others..is not explained.

nominate, *v.* Add: **5.** In horse breeding, to choose (a mare) as suitable for mating *to* a particular stallion.

1950 H. Wynmalen *Horse Breeding* vi. 109 The considerations set out in the preceding chapter will be found helpful in selecting a suitable stallion to which to nominate our mare. **1972** *Harper's Bazaar* Apr. 63/1 [The mare] gave the stallion she was nominated to such a hell of a time he wouldn't touch her with a barge-pole.

nomination. Add: **6.** In horse breeding, the planned mating of a particular mare and a particular stallion.

1912 *Bloodstock Breeders' Rev.* I. 169/2 (*heading*) Free nominations for mares. *Ibid.*, The Board, in 1911, allotted for award by County Committees 777 nominations for the free service of mares by the King's Premium stallions. **1927** J. E. Platt *Thoroughbred Race-Horse* iv. 29 It is necessary to book nominations for two, three, or even four years ahead. **1950** H. Wynmalen *Horse Breeding* vi. 110 If the horse is of a type and character that pleases you.. you are not likely to go wrong in fixing upon your nomination. **1969** G. E. Evans *Farm & Village* x. 118 They gave them assisted nominations to encourage horse-breeding. **1974** *Country Life* 14 Mar. 584/1 The highest price was given for a nomination to the American horse, Never Bend.

nominative. Add: **B.** *sb.* **3.** *nominative absolute.*

1858 C. P. Mason *Eng. Gram.* 97 This adverbial relation may be sustained... By a substantive (accompanied by some attributive adjunct) in the nominative absolute; as '*The sun having risen*, we commenced our journey'. **1916** E. A. Sonnenschein *New Eng. Gram.* iii. 44 The nominative absolute construction is an equivalent of an adverb-clause: *We sitting*, as I said, the cock crew loud. **1949** Bailey & Horn *Eng. Handbk.* vii. 266 Begin some sentences with a nominative absolute... Dr. Carver having proved his point, sweet potatoes and peanuts were planted in abundance. **1963** Pence & Emery *Gram. Present-Day Eng.* ii. 62 The nominative absolute is a perfectly proper construction as far as grammar is concerned... Although a nominative absolute has no *grammatical* function in the statement in which it appears, it should have a logical function. **1972** Hartmann & Stork *Dict. Lang. & Linguistics* 152/2 *Nominative absolute*, absolute construction.

‖ **nominativus pendens** (nǫ:minătəi·vŏs pe·ndenz). [L.] (See quot. 1926.)

1867 Wharton *Law-Lexicon* (ed. 4) 647/2 *Nominativus pendens*, a nominative case grammatically unconnected with the rest of the sentence in which it stands. **1926** Fowler *Mod. Eng. Usage* 611/2 *Nominativus pendens*.., 'hanging nominative'. A form of anacoluthon in which a sentence is begun with what appears to be the subject, but before the verb is reached something else is substituted in word or in thought, & the supposed subject is left in the air. **1929** C. Mackenzie *Gallipoli Memories* viii. 124 And I would have turned a gerund into a participle here and there..and probably there would be a vile *nominativus pendens.* **1929** Blunden *Near & Far* 12 While the pallid herd Of Grecians limit their pedantic gaze To some prodigious *nominativus pendens.* **1963** F. T. Visser *Hist. Syntax Eng. Lang.* I. i. 61 This subject is sometimes called 'nominativus pendens' or 'dangling subject'.

‖ **nominatum** (nǫminēⁱ·tŏm). [L., neut. of *nōminātus*, pa. pple. of *nōmināre* Nominate *v.*] 'The thing that is named by a sign, word, or linguistic expression' (Webster, 1961).

1947 R. Carnap *Meaning & Necessity* iii. 96 The customary method of meaning analysis regards an expression as a *name* for a (concrete or abstract) entity, which we call its *nominatum.* **1949** H. Feigl tr. G. Frege in Feigl & Sellars *Readings in Philos. Analysis* 89 A proper name (word, sign, sign-compound, expression) expresses its sense, and designates or signifies its nominatum. We let a *sign express* its sense and *designate* its nominatum. **1962** W. & M. Kneale *Devel. of Logic* viii. 496 'Nominatum' is too obviously artificial. **1966** S. Ceccato in *Automatic Transl. of Lang.* (NATO Summer School, Venice 1962) 77 However, since thought is always constituted by correlations made up of correlata in a given order, it is clear that the second example requires a repetition of the nominatum of 'water' which is not required in the first example.

no-mind: see *no a. 5 e.

nominee. Add: **4.** One in whose name, though he is not the owner, a stock or registered bond certificate is registered. Also *attrib.*

1869 *Bradshaw's Railway Manual* XXI. 157 The directors..be authorised to offer such new shares to the holders of ordinary stock..or to the nominees of such holders. **1930** *Economist* 22 Mar. 634/2 Again, many securities registered in the names of British banks or even individuals, as nominees, are really the property of foreigners. *Ibid.* 15 Nov. 896/1 Our own figures for 1929 are on the low side owing to the growth of nominee holdings in recent years. **1960** *Times* 24 Oct. (Financial Review) p. xvii/4 The nominee companies endeavour to facilitate the handling and supervision of portfolio investment. **1964** *Financial Times* 12 Mar. 19/5 Nominee..is a holder who, as respects the exercise of any rights in respect of a security, is not entitled to exercise those rights except in accordance with instructions given by the owner. **1975** *Guardian* 24 Feb. 20/7 Mr Marcus Lipton, the Labour MP for Lambeth Central will, however, ask the Government this week to outlaw nominee shareholding.

‖ **nominis umbra** (nǫ·minis ʊ·mbrā), *phr.* [L. (Lucan i. 135)] The shadow or appearance of a name; a name without substance.

1856 Bagehot in *National Rev.* Apr. 363 Taylor's theorem will go down to posterity,..but what does posterity know of the deceased Taylor? *Nominis umbra* is rather a compliment; for it is not substantial enough to have a shadow. *c* **1874** W. James in R. B. Perry *Tht. & Char. W. James* (1935) I. xxxi. 525 The British school say that the laws are nil—*nominis umbra.* **1896** [see *bookful a. 2]. **1959** Jowitt *Dict. English Law* II. 1230/2 *Nominis umbra*, the shadow of a name, *e.g.*, a one-man company.

nomism (nōᵘ·miz'm). [f. Gr. νόμ-ος law + -ism.] (See quots.)

1905 *Jewish Encycl.* IX. 326/1 *Nomism*, that religious tendency which aims at the control of both social and individual life by legalism, making the law the supreme norm. **1917** *Encycl. Relig. & Ethics* IX. 380/1 'Nomism' or 'legalism' is the name given to the view that moral conduct consists in the observance of a law or body of laws.

nomisma (nōᵘ·mi·zmă). Pl. **nomismata.** [ad. Gr. νόμισμα money, f. νομίζ-ειν to use customarily, f. νόμος usage, custom.] = Bezant i.

1908 W. Wroth *Catal. Imperial Byzantine Coins Brit. Mus.* p. lxxiv, Under Alexius I, the nomisma..was issued in several different metals simultaneously. **1957** *Encycl. Brit.* XV. 699/1 The principal coin, the *nomisma*, popularly called bezant, was derived from the Roman gold solidus. **1959** E. Pound *Thrones* xcvii. 24 Vasa Klipped for the people, Lycurgus, nomisma, And 'limitation is the essence of good nomisma'. **1960** H. Hayward *Antique Coll.* 200/1 *Nomisma*, the gold coin of the later Byzantine Empire. Usually scyphate in form with types of the Emperor, Christ, the Virgin and saints.

nomogram (nǫ·mŏgræm). [ad. F. *nomogramme*, f. Gr. νομο- Nomo- + F. *-gramme* -gram.] A diagram representing a relationship between three or more variables in the form of a number of (straight or curved) scales, so arranged that the value of one variable corresponding to given values of the others can be read off by means of one or more straight lines drawn to intersect the scales at the appropriate values.

1909 in *Cent. Dict.* Suppl. [with Fr. reference]. **1910** R. K. Hezlet in *Jrnl. R. Artillery* XXXVI. 460 Fig. 5 shows..the 'nomogram' constructed for the formula on this plan. [*Note*] Coining an equivalent to the French 'nomogramme'. **1918** *Inst. Automobile Engin. Proc.* XII. 191 When..this 'intersection' diagram is changed to a 'nomogram' or 'alinement' diagram, all these difficulties disappear and the reading is rendered much more accurate. **1920** *Chem. Abstr.* XIV. 2440 (*heading*) Nomograms for the functions $\log p = -(Q_s/4{\cdot}57T) + 1{\cdot}75 \log T + C$ and $\log P = -(Q_s/4{\cdot}57T) + 1{\cdot}75 \log T - aT + 3{\cdot}3$. **1938** *Jrnl. R. Aeronaut. Soc.* XLII. 8 Nomograms for picking the best airscrew diameter. **1947** *Electronic Engin.* XIX. 100 Line drawings, tables, charts and nomograms are liberally used. **1957** G. E. Hutchinson *Treat. Limnol.* I. ix. 582 (*caption*) Nomogram for determining O_2 saturation at different temperatures and altitudes. **1965** A. S. Levens *Graphical Methods Res.* ii. 39 There are two general types of nomograms: one recognized as a Cartesian coordinate chart (or concurrency chart) and the other as an alignment chart.

nomograph (nǫ·mŏgrɑf). [f. Nomo- + -graph.] = prec.

1909 in *Cent. Dict.* Suppl. **1921** *Chem. Abstr.* XV. 3233 The methods of calcg. and plotting nomographs are explained. **1939** *Nature* 4 Mar. 371/2 In the third chart a third series of horizontal and vertical lines leading to a final nomograph either confirms the horizontal shear strength as adequate or shows the correction required. **1965** A. S. Levens *Graphical Methods Res.* ii. 81 The nomographs provide a means for the rapid evaluation of the relative merit of various materials for the design of ablating shields on re-entering vehicles. **1975** G. J. King *Audio Handbk.* ii. 30 The intrinsic noise of the power amplifier is always fed to the loudspeaker, and the attenuated preamplifier noise adds in vector sum (see the nomograph in Fig. 1.5).

nomography (nŏmǫ·grăfi). [ad. F. *nomographie* (M. d'Ocagne *Nomographie. Les calculs usuels effectués au moyen des abaques*, etc. (1891)), f. Gr. νομο- Nomo- + F. *-graphie* -graphy: cf. *nomography* s.v. Nomo-.] The technique of using or devising nomograms.

1909 in *Cent. Dict.* Suppl. [with Fr. reference]. **1913** R. K. Hezlet (*title*) Nomography. **1918** *Inst. Automobile Engin. Proc.* XII. 187 Nomography can be an aid to efficiency in the drawing office and factory. **1951** *Engineering* 3 Aug. 160/1 Although nomography is by no means a new branch of mathematics, the use of nomograms seems never to have become widespread. **1965** [see nomographic *a.* below].

Hence **nomo·grapher,** one who employs nomography; **nomogra·phic** *a.*, involving or being a nomogram; **nomogra·phically** *adv.*, by means of a nomogram.

1915, 1918 Nomographic [see *alignment 6]. **1947** Douglass & Adams *Elem. Nomography* p. v, This book uses the term 'nomographic chart' in the popular sense to refer to alignment diagrams. *Ibid.* p. vii, However, the successful rule-of-thumb or short-cut nomographer is not to be criticized for his approach. 1965 W. H. Burrows *Graphical Techniques for Engin. Computations* p. ix, A modification of existing theory..pointed the way towards the development of a new theory of nomographic representation, the hyperbolic coordinate method from which have grown generalizations and extensions covering the entire field of nomography. 1965 A. S. Levens *Graphical Methods Res.* ii. 81 The amount of material given up during ablation..can be determined nomographically for heat input rates from 100 to 5,000 Btu/ft.²-sec.

‖ **nomoli** (nǫ·moli). Also **nomori, numori.** [Mende.] (See quots.)

1910 T. J. Alldridge *Transformed Colony* xxix. facing p. 286 (*caption*) Numori... Steatite figures found in caves and supposed to be of very great antiquity. 1925 T. N. Goddard *Handbk. Sierra Leone* iii. iii. 56 Another superstition of much interest is that associated with the steatite stone figures known locally as 'Nomori'... A Nomori is venerated for its supernatural powers and the good luck it is supposed to bring. 1926 F. W. H. Migeod *View Sierra Leone* xiv. 180 Owing to the demand of Europeans for steatite images which, of unknown origin, have frequently been dug up in various parts of Mende country, the manufacture of them has begun again... Their name in Mende is Nomoli. 1951 K. L. Little *Mende of Sierra Leone* xi. 223 The steatite *nomoli* which the farmer turns up when he is hoeing in old bush are looked upon as the genii's handiwork and can be used to make the rice prosper. 1954 R. Lewis *Sierra Leone* xvii. 161 Small soapstone figures of men and animals called Nomoli, which are found only in Sierra Leone, are not the artistic productions of any existing tribe, though they are greatly valued as bringers of good luck and fertility; at the time of planting rice they are whipped to make them 'work'. They are, apparently, the handiwork of an earlier bush culture and are found in the gravel of rivers or in the bush. They have, however, been convincingly counterfeited by Mende craftsmen for European collectors. 1961 *Times* 12 Aug. 10/5 Called 'Nomoli' by the Mende, these [figures] are credited with a supernatural origin.

nomological, *a.* (in Dict. s.v. Nomo-). Add: (Later examples.) Also as *sb.* Hence **nomolo·gically** *adv.*

1947 H. Reichenbach *Elements of Symbolic Logic* viii. § 61. 360 We shall call these formulas *nomological formulas.* The term 'nomological'..is chosen to express the idea that the formulas are either laws of nature or logical laws... The term 'nomological' is therefore a generalization of the term 'tautological'. 1961 E. Nagel *Structure of Sci.* iv. 51 Such use is characteristic of all nomological universals. *Ibid.* 53 They contend only that genuine nomologicals are logically necessary. 1967 *Philosophy* XLII. 134 My theory permits me nomologically to derive the statement. 1970 J. N. Findlay tr. *Husserl's Logical Investigations* I. xi. 230 One could say..that these sciences are *nomological*, in so far as their unifying principle, as well as their essential aim of research, is a law. 1973 *Jrnl. Genetic Psychol.* CXXIII. 11 The beginnings of a nomological network may have been laid in this experiment. 1973 [see *nomic *a.*²].

nomori, var. *nomoli.

nomos (nǫ·mǫs). *Theol.* [ad. Gr. νόμος usage, custom, law.] The law; a law of life.

1938 P. S. Watson tr. *Nygren's Agape & Eros: Part 2* I. 35 The *Commandment* of love was easier to grasp, and it led back to the Old Testament level, so that Agape was again brought under the scheme of Nomos. 1948 J. L. Adams tr. *Tillich's Protestant Era* iv. 56 The words 'autonomy', 'heteronomy', and 'theonomy' answer the question of the *nomos* or the law of life in three different ways. 1954 *Scottish Jrnl. Theol.* VII. 260 The Church follows Christ as justified, reconciled, and risen with Him, and is therefore differently related to the *nomos*-form of this age.

nomothetic, *a.* Delete *rare* and add: That pertains to or is concerned with the study or discovery of general (scientific) laws, esp. as contrasted with idiographic study.

[1894: see *idiographic* adj. s.v. *idio-.] 1904 *Jrnl. Philos., Psychol. & Sci. Methods* 21 Jan. 42 This is the same distinction as that made by Windelband under the names nomothetic and idiographic sciences. *a* 1943 R. G. Collingwood *Idea of Hist.* (1946) 166 Windelband..laid it down that history and science were two different things. ..There were two kinds of science..: nomothetic science, which is science in the common sense of the word, and idiographic science, which is history. 1948 *Brit. Jrnl. Psychol.* XXXVIII. 160 (*heading*) Nomothetic and idiographic [perspectives in psychology]... Two aims: the discovery of general laws of mind, or the description..of the unique undivided personality... Well-balanced psychologists ought to keep in mind both the nomothetic and idiographic aims. 1955 F. H. Allport *Theories of Perception* xi. 263 Conditions that are truer to life than those in which the exact or 'nomothetic' laws of nature are sought. 1964 *Listener* 16 Apr. 617/2, I [*sc.* G. Freeland] have coined..the term 'nomothetic corrections'. This group of corrections is intimately tied up with scientific laws. 1970 *Daily Tel.* 19 Sept. 9/7 Is musicology—or rather *musikwissenschaft*, which means science of music—a nomothetic or an idiographic discipline? Ought it to emulate physics and concern itself with establishing generally applicable laws or must it, like history, describe what is unrepeatable?

non-, *prefix.* Add: The prefix has continued to be one of the great formative elements in English. The sections that follow contain a selection of the more frequently occurring formations during the last hundred years or so.

1. Prefixed to nouns of action, condition, or quality, as *non-achievement, -activity, -ambiguity, -aspiration, -candidacy, -class, -commitment, -communication, -commutativeness, -commutativity, -comparability, -comprehension, -conservation, -contemporaneity, -creativity, -deposition, -derivability, -disclosure, -discrimination, -employment, -equilibrium* (later examples; see also *4), *-equivalence, -finality, -happening, -implication, -inclusion, -independence, -influence, -invitation, -involvement, -occurrence, -pathogenicity, -planarity, -possession, -reality, -reception, -recognition* (earlier and later examples), *-recurrence, -simultaneity, -stationarity, -stoich(e)iometry, -support, -transitivity, -uniformity, -uniqueness, -universality, -validity* (later examples).

1934 Webster, Nonachievement. 1946 *Nature* 17 Aug. 236/2 The residual carbon dioxide in the inhibited oxidation gases..is shown in the graph by the non-achievement of 100 per cent inhibition. 1847 W. Smith tr. *Fichte's Characteristics Present Age* 6 Every one would accept the proof of *non-existence* at a particular time, as equivalent to the proof of *non-activity* at the same time. 1949 A. Pap *Elem. Analytical Philos.* iii. 74 How could we..insure non-ambiguity of a proper name? 1958 E. Fischer-Jørgensen in Saporta & Bastian *Psycholinguistics* (1961) 140/1 The non-aspiration of final /t/ and the non-closure of final /d/ should be considered as part of the syllable boundary. 1968 *Gainesville* (Florida) *Sun* 16 Jan., All this randomly built superstructure of his non-candidacy could..come tumbling down. 1971 *New Yorker* 4 Dec. 47 In addition to repeatedly asserting his non-candidacy, [Senator Edward] Kennedy had made some forceful speeches in recent months. 1940 W. V. Quine *Math. Logic* 123 The distinguishing feature of a non-class *y* is this: '*x* ∈ *y*' amounts, for every object *x*, to '*x* = *y*'. 1973 'Trevanian' *Loo Sanction* (1974) 11 A dour young man..flaunting his non-class by drinking beer. 1959 *Brno Studies in English* I. 133 This is not so much non-commitment..as withdrawal. 1962 W. Nowottny *Lang. Poets Use* iv. 94 A tone of non-commitment is established. 1960 *News Chron.* 28 July 6/6 Critics talk a lot about the problem of non-communication amongst my characters. 1973 *N.Y. Times* 11 Aug., One is struck by the talent of the lawyers for non-communication. 1940 *Mind* XLIX. 459 It is not always the case that *ab* = *ba*. This non-commutativeness is quite essential to its use in physics. 1964 E. A. Power *Introd. Quantum Electrodynamics* vi. 83 The order of the factors **p** and **A(q)** in the first term of (56) is immaterial despite the non-commutativity of $\mathbf{p}_\alpha$ and $\mathbf{q}_\alpha$ in the quantum theory. 1909 W. James *Pluralistic Universe* 335 The only real relations in comparison are absolute identity and absolute non-comparability. 1973 *Computers & Humanities* VII. 158 A. A. Lyne..spoke on 'The Problem of Non-Comparability of Word-Frequency Counts'. 1910 *Daily Chron.* 3 Mar. 4/4 To non-comprehension of the true meaning of that golden rule..can be traced practically every defect of our system. 1962 *Times* 15 Feb. 15/2 Across a vast abyss of non-comprehension. 1941 *Physical Rev.* LIX. 441 A natural consequence of this nonconservation of spin angular momentum is a dependence of the scattering cross section upon the spin magnetic quantum number of the incident beam. 1968 M. S. Livingston *Particle Physics* vii. 146 As another example of nonconservation of parity in a weak interaction, it shows a correlation between spin and direction for the product photon. 1950 F. E. Zeuner *Dating Past* (ed. 2) iv. 78 This evidence bears out what has been said on p. 54, that the land rose faster than the sea-level in the northern Baltic region, and that the non-contemporaneity of the maximum in different parts of the Baltic may be due to this fact. 1969 Bennison & Wright *Geol. Hist. Brit. Isles* i. 5 The noncontemporaneity of geological events..makes some scale of reference imperative. 1957 P. Suppes *Introd. Logic* 154 Criterion of Non-creativity. 1965 B. Mates *Elem. Logic* xi. 190, ϕ will satisfy the criterion of non-creativity. 1843 Mill *Logic* I. iii. ix. 499 The presence or absence of an uninterrupted communication with the sky causes the deposition or non-deposition of dew. 1969 Bennison & Wright *Geol. Hist. Brit. Isles* i. 8 The older strata may be tilted..during the period of non-deposition. 1957 A. N. Prior *Time & Modality* ii. 14 If we wish to express the non-derivability of *CMpp*. 1964 G. Kreisel in Benacerraf & Putnam *Philos. of Math.* 157 Hilbert emphasized the consistency problem which is so to speak the weakest non-derivability result. 1908 *Westm. Gaz.* 24 July 8/4 Defendants resisted the plaintiff's claim under the policy upon the grounds of the misstatement and non-disclosure. 1950 *Mind* LIX. 248 There is always a possibility..of non-disclosure or lies. 1923 *19th Cent.* Jan. 135 Shinto favours racial nondiscrimination. 1958 *Listener* 16 Jan. 88/1 To say that the Communists, who claim world brotherhood,..practise non-discrimination in the fullest sense on the racial side, [etc.]. 1843 Mill *Logic* II. v. iv. 398 It is not the nature of the faculties..but the non-employment of them. 1966 A. Battersby *Math. in Management* iii. 74 The quick answer 'use machinery wherever possible because it is cheaper' is not necessarily the best when the non-employment of labour is unjustified on social or moral grounds. 1938 J. Newton *Introd. Metallurgy* vi. 162 The iron–carbon equilibrium diagram..is the best diagram to illustrate the meaning of equilibrium and non-equilibrium. 1971 *Contrib. Mineral. & Petrol.* XXXII. 165 Some magnetites..that show a wide variation of Cr content also give very erratic oxygen isotopic results, suggesting non-equilibrium. 1894 J. N. Keynes *Stud. & Exerc. Formal Logic* (ed. 3) ii. iii. 111 To establish their non-equivalence we may proceed as follows. 1955 A. N. Prior *Formal Logic* 212 The connected non-equivalence of $\Pi x M\phi x$ and $M\Pi x\phi x$. 1896 W. Caldwell *Schopenhauer's System* viii. 370 Religion alone pretends to answer the question, why it is that non-finality and non-attainment and illusoriness seem to characterise all human experience and all human life. 1971 D. Crystal *Linguistics* 133 The ..sentence ends with a rising movement.., giving the impression of non-finality. 1892 W. James *Coll. Ess. & Rev.* (1920) 319 Our conscious states are temporal *events*,..the conditions of whose happening or non-happening from one moment to another, lie certainly in large part in the physical world. 1968 *Gainesville* (Florida) *Sun* 13 Feb., The [TV] medium's extraordinary ability to suggest that a true happening resides in a non-happening. 1878 H. MacColl in *Proc. London Math. Soc.* IX. 180 Thus in a non-implication, as in an implication, the rule is *Transpose and change signs.* 1887 J. N. Keynes *Stud. & Exerc. Formal Logic* (ed. 2) ii. ii. 77 The implication or non-implication of existence in propositions. 1932 Lewis & Langford *Symbolic Logic* ix. 281 Facts like these about non-implication and independence. 1972 M. A. Boden *Purposive Explanation in Psychol.* viii. 318 The logical characteristics of indeterminacy,..and non-implication of embedded clauses. 1885 J. Veitch *Institutes of Logic* § 383 *All* are thus excluded, through the non-inclusion of *any*. 1952 *Mind* LXI. 526 We interpret the non-inclusion of a conjunction as the exclusion of one *or* the other of the conjuncts. 1874 W. Wallace tr. *Hegel's Logic* 293 A mean which combines in itself the centrality with the non-independence of the objects. 1965 Hughes & Londey *Elem. Formal Logic* xviii. 131 (*heading*) PM (IV)—independence and non-independence. 1838 W. Howitt *Rural Life of Eng.* I. i. v. 79 Those who ..believe in the nonsense of the economists on the non-influence of absenteeism. 1907 W. James *Pragmatism* iv. 138 All these systems of influence or non-influence may be listed under the general problem of the world's *causal unity.* 1908 *Daily Chron.* 24 July 4/3 Invitations would be robbed of their grace, and non-invitations be invested quite unnecessarily with the air of a slight. 1940 *Economist* 11 May 853/1 Japan's policy, for all her declarations of 'non-involvement' (a new type of non-belligerency?) is bound to be that of fishing in troubled waters. 1970 *Guardian* 31 Dec. 7/8 A devastating..argument for non-involvement as a starting-point for social research. 1807 R. Kirwan *Logick* I. 178 Their occurrence or non-occurrence being consonant to common observation. 1963 J. Lyons *Structural Semantics* i. 8 The non-occurrence of the large set of grammatical, but nonsensical, sentences. 1971 *Nature* 12 Mar. 97/8 It should..be possible to improve greatly the efficacy of live cholera vaccines by preparing them from bacteria in which non-pathogenicity has been induced by a very specific genetic mutation. 1967 A. Battersby *Network Analysis* (ed. 2) xi. 177 There are, however, some minor ones [*sc.* obstacles], including integer restrictions and non-planarity. 1843 Mill *Logic* I. i. ii. 52 The non-possession of any given attribute is also an attribute, and may receive a name as such. 1959 D. Cooke *Lang. Mus.* ii. 69 The feeling is..of non-possession, non-acceptance, need. *a* 1866 J. Grote *Treat. Moral Ideals* (1876) 360 We may assume four degrees of reality or non-reality in the moral ideals. 1909 W. James *Pluralistic Universe* iv. 234 Zeno..has no alternative but to say that our intellect repudiates motion as a non-reality. 1924 W. E. Johnson *Logic* III. i. 7 A verbal formula that..gives the true test of non-reality. 1878 Trollope *Is he Popenjoy?* II. ii. 20 Mr. Holdenough had made no attempt after the reception—or rather non-reception—awarded to his wife. 1963 *Times Lit. Suppl.* 1 Mar. 152/4 His [*sc.* Chopin's] reception—or non-reception—by the English. 1838 J. S. Mill in *London & Westm. Rev.* Aug. 484 His non-recognition of them does not put them out of existence. 1862 Bagehot *Coll. Works* (1965) II. 257 Its non-recognition by what is called the public. 1946 *Nature* 9 Nov. 679/1 A further factor which stems progressive action is the non-recognition of the educational potentialities of museums. 1965 A. J. P. Taylor *Eng. Hist. 1914–45* x. 373 In February 1932 Stimson, the American secretary of state, announced that the United States would not recognize territorial changes brought about by force... 'Non-recognition' became a League [of Nations] principle. 1847 A. De Morgan *Formal Logic* xiii. 278 The demand for non-recurrence of words arises from the public (I beg its pardon) not knowing how to read. 1911 J. Ward *Realm of Ends* xv. 335 In place of the existing certainty of evil, there will be an even chance of its non-recurrence. 1855 H. Spencer *Princ. Psychol.* ii. v. 121 The simultaneity or non-simultaneity of certain events. 1904 *Mind* XIII. 214 The non-simultaneity of perception and object, where internal perception is concerned, need cause no difficulty. 1953 C. E. Bazell *Linguistic Form* 44 Non-simultaneity has different degrees of validity. 1965 *Math. in Biol. & Med.* (Med. Res. Council) IV. 158 Non-stationarity of the EEG signals during each of the 10-second epochs does not pose the same limitations. 1974 *Jrnl. of Business* (Chicago) XLVII. 518 (*heading*) Investigations of nonstationarity in prices. 1962 Simpson & Richards *Physical Princ. Junction Transistors* iii. 35 In the latter materials any unbalance (non-stoichiometry) between constituents degrades physical properties such as mobility and lifetime to a marked extent. 1909 *Daily Chron.* 20 Jan. 6/2 The Englishwoman makes a charge of non-support against her husband. 1939 *Mind* XLVIII. 532 Obviously true of non-transitivity since that is 'ambiguous' with respect to transitivity or intransitivity. 1974 *Sci. Amer.* Oct. 120/2 If the nontransitivity is so counter-intuitive as to boggle the mind, we have what is called a nontransitive paradox. 1902 *Encycl. Brit.* XXVIII. 456/2 Non-uniformity of convergence of the series does not necessarily imply discontinuity in the sum. 1962 Simpson & Richards *Physical Princ. Junction Transistors* vi. 116 For high-power transistors, other factors such as the non-uniformity of current distribution across the base region may become important. 1957 J. D. O'Connor in Saporta & Bastian *Psycholinguistics* (1961) 102/1 The acceptance of the nonuniqueness of phonemic solutions. 1974 *Physics Bull.* Dec. 587/3 Questions of nonuniqueness and stability in nonlinear continuum mechanics. 1846 J. D. Morell *Hist. View Philos.* I. ii. 102 He attempts to prove their non-universality. 1893 W. Minto *Logic* 70 The expression of Quantity, that is, of Universality or non-universality, is all-important. 1952 W. V. Quine *Methods of Logic* § 9.50 Interchange of equivalents..preserves consistency, nonvalidity, nonimplication, and

nonequivalence. 1964 P. F. Anson *Bishops at Large* viii. 271 The validity or non-validity of Anglican orders.

2. Prefixed to agent-nouns and designations of persons and objects, as *non-actor, -addict, -body, -breeder, -candidate, -Christian, -clitic, -cognitivist, -dancer, -drinker, -driver, -electrolyte, -fact, -friend, -Jew, -joiner, -linguist, -motorist, -musician, -owner, -participant, -people, -philosopher, -reader, -scientist, -self, -sentence, -slaveholder, -socialist, -speaker, -structuralist, -swimmer, -theist, -theorem, -visualizer, -writer*; **non-secre·tor,** a person whose saliva and other secretions do not contain his blood-group antigens.

1937 *Times* 17 Aug. 8/3 The air of reality, the judicious use of non-actors, the social consciousness of the Russian film. 1955 *Publ. Amer. Dial. Soc.* xxiv. 34 The natural revulsion which nonaddicts feel toward addicts. 1933 Dylan Thomas *Lett.* (1966) 72 The life of the non-body..is capable..of creating an artistic progeny. 1949 G. Ryle *Concept of Mind* vi. 189 Perhaps it is because of the absurdity of such collocations that so many people have felt driven to describe a person as an association between a body and a non-body. 1944 J. S. Huxley *On Living in Revolution* ix. 99 About 17,000 breeding pairs: with the non-breeders, about 40,000. 1950 *Chem. Engin. Progress* XLVI. 112/1 The nonbreeders are pointed essentially toward power production. 1967 *Time* 22 Sept. 22 (*heading*) The non-candidates. 1972 *Daily Colonist* (Victoria, B.C.) 12 Jan. 16/5 Miami—Sen. Edward Kennedy, a steadfast noncandidate, was one of a dozen Democrats whose names have been entered in the..primary. 1671 A. Woodhead *Consid. Council of Trent* xv. 240 Nor may such separation be understood from Infidels, Heathens, or non-Christians only. 1941 W. Temple *Citizen & Churchman* iv. 68 Children grow up in it who are schoolfellows of non-Christians. 1949 Nonclitic [see *additive *sb.*]. 1952 *Mind* LXI. 546 The non-cognitivists are usually regarded as a relatively homogeneous group. 1963 R. Carnap in P. A. Schilpp *Philos. R. Carnap* 1008 Both cognitivists and non-cognitivists agree that beliefs play a very important role in the origin of attitudes and decisions. 1965 Noncognitivist [see *imperativism]. 1863 Trollope *Rachel Ray* I. vii. 136 Now the room was partially cleared, the non-dancers being pressed back. 1929 L. Bettany in W. J. Temple *Diaries* p. lxxiv, Bored as a Whig, as a non-drinker, as a scholar, and as an opponent of blood sports, by the only society he could generally enter. 1975 E. Mosbacher tr. *Herlin's Commemorations* 4 A tall, thin, fair man.., a non-smoker and non-drinker. 1953 *Essays in Criticism* III. 422 The customary impression of the non-driver, that handling a car by night involves merely the switching on of headlights, and the turning of a wheel. 1891 *Jrnl. Chem. Soc.* LX. ii. 971 (*heading*) Freezing points of dilute aqueous solutions of non-electrolytes. 1964 N. G. Clark *Mod. Org. Chem.* i. 4 With salts very much in mind, he probably associates inorganic chemistry with electrolytes... However,..this branch of chemistry embraces..many compounds which neither dissolve in water nor produce ions, i.e. typical non-electrolytes. 1926 Fowler *Mod. Eng. Usage* 576/1 Utopia, the realm of non-fact or the imaginary. 1964 *Listener* 9 Jan. 48/2 Our opinions..should not be based on non-facts, or dubious facts. 1971 *Harper's Mag.* Sept. 57 'There are people here who are not our friends.' The nonfriends were expelled. 1973 *Jrnl. Genetic Psychol.* CXXIII. 70 Cliques of friends and nonfriends. 1967 W. Q. de Funiak *American-British Dict.* 115 *Non-Jew*, Gentile. 1973 R. Perry *Nowhere Man* 6 For a non-Jew to have risen so high [in Israel] was testimony enough to his ability. 1964 R. Miliband in I. L. Horowitz *New Sociology* 78 Mills was a determined non-joiner, with an intense dislike of togetherness. 1974 *Listener* 14 Feb. 210/2 It was an efficient way of getting people into a group, turning non-joiners into joiners. 1936 *Language* XII. 93 Non-linguists..constantly forget that a speaker is making noise, and credit him, instead, with the possession of impalpable 'ideas'. 1965 *English Studies* Feb. 58 A terminology as simple and as familiar to the non-linguist as possible. 1908 *Westm. Gaz.* 31 Dec. 4/1 Non-motorists can join on equal terms. 1936 *Discovery* Apr. 108/2 Even the non-motorist may enjoy the adventure, as a comfortable bus service has been established. 1949 *Penguin Music Mag.* Feb. 40 This is utterly meaningless to a non-musician, and obvious and therefore unnecessary to a musician. 1912 Belloc *Servile State* 3 This principle of compulsion applied against the non-owners. 1948 *Mind* LVII. 228 The political opinions of automobile-owners are not very symptomatic of the political opinions of non-owners. 1883 'Mark Twain' *Life on Mississippi* 379 Could you come nearer to reproducing it [*sc.* the siege of Vicksburg] to the imagination of a non-participant? 1971 *Jrnl. Gen. Psychol.* LXXXV. 266 All three categories (doublers, singlers, nonparticipants) are compared. 1967 B. Kaufman in S. Henderson *Understanding New Black Poetry* (1973) iii. 206, I am writing these lines to the responsible non-people. 1969 *Guardian* 26 Sept. 11/2 His shiny, plastic world of nonpeople attending nonevents. 1897 W. James *Will to Believe* 210 In all this the philosopher is just like the rest of us non-philosophers. 1956 F. C. Copleston in H. D. Lewis *Contemp. Brit. Philos.* 122 By the term 'ordinary man'..I mean simply the non-philosopher. 1938 S. Beckett *Murphy* 162 Murphy, a strict non-reader. 1964 M. Critchley *Developmental Dyslexia* iv. 16 Monroe, too, preferred the old-fashioned phonic approach, saying darkly that it was better to be a slow reader than a non-reader. 1897 W. James *Will to Believe* 320 We all, scientists and non-scientists, live on some inclined plane of credulity. 1959 C. P. Snow *Two Cultures* i. 5 The non-scientists have a rooted impression that the scientists are shallowly optimistic. 1950 *Sci. News* XV. 111 Most people secrete the appropriate blood group substances (antigens) in bodily secretions such as saliva and tears. About one-seventh of the population, however, do not do this, and are called 'non-secretors'. 1971 J. Z. Young *Introd. Study Man* xl. 586 Those who get duodenal ulcers include more group O than would be expected. Non-secretors are more liable to the disease than secretors. 1874 J. Cunningham *New Theory of Knowing* 98 The bodily organism is the 'debateable land' between self and non-self. 1949 Koestler *Insight & Outlook* viii. 120 Our understanding of other personalities, of the nonself by the self. 1970 *New Scientist* 19 Feb. 345/1 When a foreign protein enters the body the immune system recognizes the invader as non-self and produces antibodies which destroy it. 1940 Bryant & Aiken *Psychology of Eng.* 21 The phrase is any other group of words with a unitary idea, within a sentence or a nonsentence. 1966 M. Gross in *Automatic Transl. of Lang.* (NATO Summer School, Venice 1962) 124 Certain strings on *V* are clearly understood by speakers of *L* as sentences of *L*, others are clearly recognized as nonsentences. 1832 T. J. Randolph *Speech on Abolition of Slavery* 16 The burden of this defence..is to fall upon the less wealthy class of our citizens; chiefly upon the non-slaveholder. 1895 G. B. Shaw *Let.* 15 Dec. (1965) 578 The third division falling to Horobin, a non socialist. 1964 M. A. K. Halliday et al. *Linguistic Sciences* 63 A non-speaker of the language. 1971 J. Z. Young *Introd. Study Man* xxxvi. 518 The selection of those individuals who were competent in such skills became unusually severe at the expense of the non-speakers. 1952 *Archivum Linguisticum* IV. 89 To the non-structuralist at least, the book appears to throw less light on the essential nature of the French language than on the present state of structural linguistics. 1932 Blunden *Face of England* 159 Our young men are non-swimmers. 1962 J. F. Powers *Morte d'Urban* xiii. 285 Paul, the non-swimmer, had probably been thinking of the return voyage when he pinched the life preserver. 1908 *Lit. Guide* 1 Sept. 141/2 This false charge—viz., that non-theists are morally inferior persons. 1944 W. Temple *Let.* 6 Mar. (1963) 148 But there is a Christian way of living and it is possible for non-theists to appreciate it and try to follow it. 1954 I. M. Copi *Symbolic Logic* vi. 185 A purely formal criterion for distinguishing between theorems and non-theorems of the system. 1972 A. Church in Rudner & Scheffler *Logic & Art* x. 198 The expected consequences of any assertion, even of a non-theorem. 1924 G. B. Shaw *St. Joan* p. xxi, Some other people see imaginary diagrams and landscapes ..and are thereby able to perform feats of memory and arithmetic impossible to non-visualisers. 1953 A. Huxley *Let.* 21 June (1969) 676, I am a non-visualizer, and got very little in the way of imagery. 1940 V. Woolf *Writer's Diary* 29 Mar. (1953) 330 All the detail that seems to the non-writer so easy..to me is torture. 1958 *New Statesman* 15 Nov. 664/2 It might even lure some of the great non-writers in the historical profession into putting pen to paper.

b. *spec.* Prefixed to a sb. with an implication of pretence, to denote a person who, or thing which, is not wholly, adequately, or genuinely what is specified by the sb.; used esp. of forms of art or literature, freq. with the implication that they are new or unconventional forms of the things specified (cf. *anti-¹ 2 c); as *non-answer, -architecture, -book, -budget, -conversation, -country, -debate, -drama, -film, -information, -issue, -lecture, -music, -news, -newspaper, -novel, -play, -policy, -problem, -sherry, -story.* See also *non-art, *non-event, *non-hero.

1966 *Listener* 21 July 84/1 Public men have learned to apply the technique of the non-answer, a technique in many cases mastered and perfected at the Commons dispatch-box. 1968 *Gainesville* (Florida) *Sun* 11 Mar. 8 They were non-answers—words calculated solely as a physical response to solve the immediate need for some kind of a response generated in a news conference. 1960 *20th-Cent.* Oct. 357 Manchester..is a city with no architecture, only an inert mass of *building*. I believe that the adherents of the Betjemanesque cult of Victorian bad taste find things to admire in examples of Manchester non-architecture. *Ibid.* May 441 Technical language and presentation are necessary; they do not excuse so many non-books. 1972 *Scholarly Publishing* III. 101 Non-books are of several kinds; but their differences notwithstanding, they all share two common features: they tend to be written for market rather than for intellectual reasons, and they are all means rather than ends. 1967 *Observer* 16 Apr. 10/2 The crescendo of the roll call of defeat..came just two days after the Chancellor of the Exchequer's non-Budget. 1960 *News Chron.* 28 July 6/8 Pinter is..a writer over-occupied with the externals of behaviour. The non-conversations..can stale into nothingness. 1970 *Gainesville* (Florida) *Sun* 24 Sept. A6 Jordan was and remains a non-country, created out of sandscape by Britain to pay off a dynastic debt. 1971 *Britannica Bk. of Year* (U.S.) 779/3 *Noncountry*, a country whose size, location, and economic resources are inadequate for economic self-sufficiency and development. 1970 *Times* 21 May 8 The debate on the amendments to end the Indo-China war was the usual non-debate. 1968 *Sat. Rev.* (U.S.) 20 Jan. 33 Most of the noise is made by Mailer, who, as principal player in a turgid nondrama.. barks sometimes like a dog and sometimes like a seal. 1963 *Listener* 31 Jan. 201/1 Their [*sc.* the Situationists'] violent and derisory manifestations—such as the non-film *Hurlements en faveur de Sade*—are all intended to create situations..such as will produce certain kinds of human behaviour, usually disgust or disorientation. 1970 *Times* 16 Apr. 5 The ritual exchange of non-information. 1965 *New Society* 9 Sept. 5/3 The brain drain is a non-issue. 1972 *Time* 20 Nov. 28 Unfortunately for the country, there were a lot of non-issues in this campaign. 1953 E. E. Cummings (*title*) I: six nonlectures. 1958 *Publ. Amer. Dial. Soc.* xxx. 41 There has always been a tendency in some circles to regard jazz as non-music. 1969 *Listener* 3 July 26/1 Easily the most interesting and successful work in the programme was the one which kept consistently and coherently to the realms of non-music, and could thus be listened to without difficulty on its own terms. 1963 Non-news [see *non-event]. 1972 *Wall St. Jrnl. Index* 681/1 President Nixon clears desk with sweep of proclamations and orders, other (yawn) nonnews items. 1973 *Nation Rev.* (Melbourne) 31 Aug. 1443/4 There's a new type of newspaper occurring in the world: the *non newspaper*. 1961 *Guardian* 27 Oct. 6/5 Henry Miller's two remarkable non-novels. 1963 *New Yorker* 8 June 120 A non-novel, 'Why Not Try God?' 1968 *Ibid.* 28 Dec. 2 Dustin Hoffman does his brave best to make us believe that this non-play by Murray Schisgal is a touching comedy about the ignominy of young manhood. 1962 E. Cleaver in A. Dundes *Mother Wit* (1973) 20/1 The fallacious stupidity of the non-policy of segregation. 1969 *Guardian* 7 Nov. 12/2 The apostles of relevance..bring massive expertise to bear on non-problems. 1967 *Economist* 5 Aug. 477 The non-sherries have carved themselves out healthy chunks of the fortified wine market on the basis of what they really are, more than by pretending to be the real Spanish thing. 1967 *Gainesville* (Florida) *Sun* 19 Mar., The latest non-story out of New Orleans on the Kennedy assassination 'plot'. 1968 *Listener* 11 July 49/2 The House of Lords vote on Rhodesia..was the great 'non-story' of recent weeks. 1971 M. Russell *Deadline* xiii. 157 This promises to be the non-story of the month. 1973 L. Heren *Growing up Poor in London* vii. 179 The coming and going of ships were regularly reported by local stringers, but tarantula spiders among the bananas and other non-stories were always available.

3. Prefixed to adjectives, as *non-academic, -actinic, -adaptive, -addictive, -æsthetic, -alcoholic, -alphabetic, -alphabetical, -ambiguous, -analytic, -antigenic, -aquatic, -aqueous, -articular, -assertive, -assimilable, -atomic, -auditory, -automatic, -axiomatic, -basic, -behavioural, -binary, -biodegradable, -British, -calcareous, -causal, -causative, -cellulosic* (also as *sb.* (*sc.* fibre)), *-chemical, -Christian, -chromosomal, -circular, -classical, -cognitive, -coherent, -coital, -collinear, -combustible, -commercial, -communicative, -Communist, -communistic, -commutative, -compact, -competitive, -complementary, -complex, -compound, -conceptual, -concrete, -conscious, -conservative, -consonantal, -constructive, -contiguous, -continuous, -contrastive, -controversial, -corrosive, -critical, -crystalline, -cumulative, -cyclic, -deductive, -degenerate, -delinquent, -demonstrative, -denominational, -denumerable, -derivative, -descriptive, -designative, -deterministic, -diffusible* (also *-able*), *-directive, -discriminatory, -dispersive, -dispositional, -disruptive, -dissipative, -dominant, -dramatic, -economic, -electrolytic, -electronic, -elementary, -emotional, -emphatic, -English, -enumerative, -enzymatic, -enzymic, -equivalent, -ethical, -exclusive, -existential, -explanatory, -explosive, -extensional, -factual, -fatty* (cf. *non-fat* in *4), *-final, -finite, -finitist, -fissile, -fissionable, -formal, -functional, -generic, -genetic, -geometrical, -German, -Germanic, -grammatical, -gravid, -habitual, -hæmolytic, -Hermitian, -historical, -homogeneous, -identical, -ideological, -illusory, -imitative, -independent, -indigenous, -individual, -Indo-European, -inductive, -industrial, -infectious, -infective, -inferential, -inflammable* (later examples), *-inflected, -inflexional, -initial, -insightful, -integral, -intellectual, -intuitive, -involved, -irritant, -Jewish, -judg(e)mental, -kosher, -lethal, -lexical, -linguistic, -literary, -local, -magnetic, -manual, -marine, -material, -mathematical, -measurable, -mechanical, -medical, -Mendelian, -mental* (also as *sb.*), *-mentalistic, -metabolizable, -metaphorical, -metaphysical, -metrical, -migratory, -military, -mimetic, -minimal, -modal, -molecular, -monetary, -motile, -musical, -mystical, -national, -native, -Nazi, -negative, -negligible, -negotiable, -null, -numerical, -nutritious, -observational, -official, -operational, -organic, -pathogenic, -perceptual, -perfective, -perishable* (also as *sb.*), *-permissive, -phallic, -phenomenal, -philosophical, -phonemic, -phonetic, -physical, -planar, -poetic, -poisonous, -polarizable, -porous, -predicative, -predictable, -pregnant, -prepositional, -prescriptive, -productive, -professorial, -progressive, -propositional, -psychological, -public, -purposeful, -purposive, -quantitative, -racial, -radioactive, -reactive, -realistic, -reciprocal, -recurrent, -redundant, -relational, -relative, -relativistic, -religious, -renewable, -representational, -representative, -reproducible, -revolutionary, -Roman, -roman, -routine, -saponifiable, -sceptical, -sectarian, -selective, -sensational, -sensory, -sensuous, -sentential, -sharp, -shrinkable, -simultaneous, -social, -socialist, -spatial, -spherical, -stationary, -statistical, -statutory, -steady, -structural, -subjective, -suppurative, -syllogistic, -symbolic, -symmetric(al, -synchronous, -synonymous, -syntactic, -systematic, -tautologous, -technical* (later examples),

-teleological, -temporal, -terminal, -theatric(al, -theistic, -theological, -theoretical, -tonal, -totalitarian, -toxic, -traditional, -transitive, -turbulent, -typical, -ultimate, -uniform, -urban, -utilitarian, -vacuous, -valid, -veridical, -visual, -vital, -vocal, -vocalic, -vocational; **non-cli·nical,** not clinical; *esp.* not accompanied by directly observable symptoms; **non-dime·nsional,** (of a quantity) = *dimensionless *a.* 1 d; (of an equation) composed of dimensionless terms; **non-homo·logous,** not homologous; *spec.* in *Genetics* (cf. *homologous *a.* 4 b); **non-Oh·mic** *Electr.*, not obeying or in accordance with Ohm's law; **no:n-stoich(e)iome·tric** *a. Chem.*, containing or representing atoms of the different elements in numbers that do not bear a simple ratio to one another.

1918 *Nation* (N.Y.) 28 Mar. 335 The real trouble with American universities has been their inability to 'get it across' to the 'man in the street', the 'roughnecks', and all those other fearsome beasts that inhabit the non-academic jungles. 1937 *Mind* XLVI. 124 That ease, forcefulness, and felicity of expression which made him so welcome a speaker, alike to academic and to non-academic audiences. 1957 C. S. Lewis *Let.* 16 Mar. (1966) 275 In *writing* I do regard all non-academic works..as being leisure occupations. 1975 *Times* 6 Dec. 3/2 To chop £500,000 off the budget of a college means fewer jobs for staff, academic and non-academic. 1909 *Westm. Gaz.* 16 Oct. 14 The real essential in comfort..is the use of one of the scientifically prepared light screens which gives us all the possible non-actinic light. 1935 *Discovery* Apr. 97/1 The non-actinic paper in which photographic plates were wrapped. 1926 J. S. Huxley *Ess. Pop. Sci.* xviii. 276 An artificial and non-adaptive tendon. 1955 *Sci. Amer.* Apr. 93/1 The modification of T2 is 'nonadaptive'; that is, the modified virus cannot grow in the host that changed it. 1967 *Economist* 7 Oct. 20/1 The 'non-medical', non-addictive drugs such as marijuana and LSD. 1934 C. Lambert *Music Ho!* iv. 251 Music for organized and non-aesthetic action such as military marches and fox-trots. 1957 T. Veblen *Theory of Leisure Class* 164 The limits..are fixed by requirements of a non-aesthetic kind. 1907 *Yesterday's Shopping* (1969) 59/2 Hop Bitter Beer, non-alcoholic. 1909 [see *beerless *a.*]. 1966 B. Kimenye *Kalasanda Revisited* 58 There he was..nursing a glass of non-alcoholic, fizzy liquid. 1960 E. Delavenay *Introd. Machine Translation* vi. 90 A meaningful group of signs, alphabetic and non-alphabetic (spaces, hyphens, etc.). 1973 *Computers & Humanities* VII. 151 A provision must be made for the concurrent appearance of any number of non-alphabetic characters in any position. 1948 D. Diringer *Alphabet* 21 The crudest forms of writing.. are non-alphabetical. 1964 E. Palmer tr. *Martinet's Elem. General Linguistics* iv. 114 A non-ambiguous form. 1971 J. Anderson in A. J. Aitken et al. *Edin. Stud. Eng. & Scots* 76, I want now to turn to the (non-ambiguous) distinction hinted at above. 1890 W. James *Princ. Psychol.* II. xxii. 351 In certain stages of the hypnotic trance the subject seems to lapse into the non-analytic state. 1939 *Mind* XLVIII. 202 The class *k* of non-analytic declarative sentences formulable in a given language. 1949 A. Pap *Elem. Analytical Philos.* viii. 152 A synthetic statement would simply be defined as a non-analytic statement. 1934 Webster, Nonantigenic. 1948 J. H. Burn *Lect. Notes Pharmacol.* 75 The protein must, of course, be non-antigenic. 1951 Whitby & Hynes *Med. Bacteriol.* (ed. 5) vi. 69 Enzymic digestion of antitoxic sera is also used, and by this means even the antibodies themselves may be split into non-antigenic fractions without loss of activity. 1963 D. W. & E. E. Humphries tr. *Termier's Erosion & Sedimentation* xvii. 338 Sediments of nonaquatic origin, such as those of deserts, can receive rain water. 1904 R. A. Lehfeldt *Electro-Chem.* I. i. 84 These laws..are only very partially applicable to non-aqueous solutions. 1957 G. E. Hutchinson *Treat. Limnol.* I. iii. 202 The only other non-aqueous natural liquids are the complex mixtures of hydrocarbons and other organic compounds known as petroleums. 1967 G. M. Wyburn et al. *Conc. Anat.* iv. 111/1 The [parotid] gland occupies the nonarticular part of the mandibular fossa of the temporal bone. 1901 A. Sidgwick *Use of Words* viii. 224 Some statement of theirs is ambiguous, and therefore..non-assertive. 1940 *Kenyon Review* II. 282 The poetic statement..falls between the two poles of full assertion..and pure non-assertive tension between plurisigns. 1920 H. G. Wells *Outl. Hist.* 563/1 There has hitherto existed in the States no organization for and no tradition of what one may call non-assimilable possessions. 1972 P. Green *Shadow of Parthenon* 70 Everything that conventional historiography finds non-assimilable, from science to tragedy. 1943 *Mind* LII. 26 It is supposed that..all complex or non-atomic assertions are just aggregates of atomic propositions linked by 'and', 'on', or 'if—then'. 1955 Koestler *Trail of Dinosaur* 239 The value of an atomic stockpile against non-atomic, local aggression. 1890 W. James *Princ. Psychol.* I. ii. 50 The momentary loss..of our *non*-auditory images.. makes us mentally deaf. 1938 *Times Lit. Suppl.* 8 Oct. 649/3 Calvin was not hostile to art as such, but merely to the use of any non-auditory art in connexion with worship. 1953 C. E. Bazell *Linguistic Form* i. 1 Non-automatic alternations between phonemes differing in respect of one minimal feature alone are seldom more frequent than similar alternations between phonemes differing in respect of several features. 1964 E. Bach *Introd. Transformational Gram.* iii. 44 The number suffixes show different forms for these two classes (nonautomatic). 1896 L. T. Hobhouse *Theory of Knowl.* 480 We shall have to discard any other principle..as non-axiomatic in character. 1966 *Jrnl. Philos.* LXIII. 435 We are not inclined to admit intuition or nonaxiomatic knowledge about sets. 1939 *Mind* XLVIII. 489 The discrimination of basic from non-basic predicates. 1964 R. H. Robins *Gen. Linguistics* 235 Non-basic endocentric constructions. 1946 C. Morris *Signs, Language & Behavior* 201 The explanation has the merit of not invoking non-behavioral factors. 1953 *Mind* LXII. 231 We have not yet enquired whether a machine is essentially distinct from a man in non-behavioural respects. 1947 *Math. Tables & Other Aids to Computation* Oct. 361 It is convenient to use as many selections as there are basic binary places in the computer, or in nonbinary machines as many as on-off signal channels. 1962 F. I. Ordway et al. *Basic Astronautics* vi. 287 By taking an inventory of nonbinary, solar-type stars with low angular momenta, we may at the same time be taking an inventory of stars with planetary systems. 1967 *New Scientist* 31 Aug. 440/1 Organic substances which are largely or totally nonbiodegradable are currently being used in both domestic and industrial applications. 1971 *Ibid.* 27 Apr. 260/1 Pollution created by non-biodegradable products has created immense environmental problems. 1955 *Times* 2 May 18/4 The pressing need of giving non-British immigrants to the Dominions some idea of British standards and culture. 1934 Webster, Noncalcareous. 1946 *Nature* 21 Sept. 421/2 The non-calcareous type of surface waters, chiefly found in mountainous areas of the north and west, requires little or no treatment in the best cases. 1964 V. J. Chapman *Coastal Vegetation* vi. 139 Two main types of dune, i.e. calcareous and non-calcareous. 1911 W. James *Some Probl. Philos.* xii. 201 Non-causal sequences can be reversed; causal ones follow in conformity to rule. 1939 *Mind* XLVIII. 422 But very often in determining the dates of events we use non-causal Law-propositions. 1959 *Archivum Linguisticum* XI. 15 *Feed* can also have the non-causative sense: 'we feed at eight'. 1931 *Jrnl. Physical Chem.* XXXV. 469 The walls [of Corallines] are non-cellulosic. 1959 *Times Rev. Industry* Sept. 4/1 Man made fibres..have seen the development of non-cellulosic fibres. 1964 *Financial Times* 25 Feb. 2/2 Output of all man-made fibres, including non-cellulosics, in the U.K. was 15 per cent. higher in 1963. 1964 N. G. Clark *Mod. Org. Chem.* vii. 119 This and other non-chemical uses of acetylene today account for only about 20 per cent of the amount consumed. 1974 *Verbatim* Dec. 1/1 The elements for which there is likely to be a nonchemical ..name. 1936 T. S. Eliot *Ess. Anc. & Mod.* 132 There may always be schemes, initiated by non-Christian and non-Catholic minds with purely temporal motives and aims, to which we can give unqualified support. 1965 *English Studies* Feb. 3 Such scenes in earlier non-dramatic narrative, where the roots are probably non-Christian. 1975 *Times* 18 Dec. 10/8 The beauty of ancient non-Christian mystical texts. 1960 *New Biol.* XXXI. 109 The special value of *Paramecium* for genetics depends not on its use in the analysis of heredity by Mendelian methods but on its revelation of other hereditary or quasi-hereditary systems, based on non-chromosomal or cytoplasmic elements. 1922 A. D. Udden tr. *Bohr's Theory of Spectra* iii. iii. 100 The non-circular orbits will correspond to a firmer binding than the circular orbits having the same value for the principal quantum number. 1935 *Mind* XLIV. 373 This interpretation..shows the appeal to the divine veracity to be both non-circular and epistemologically necessary. 1926 S. R. James *Seventy Years* ix. 72 In those days the non-classical masters [at Eton College] who held houses were known as 'Dames'. 1955 A. N. Prior *Formal Logic* 258 A non-classical calculus developed by the Russian logician A. Kolmogorov in 1925. 1934 Webster, Nonclinical. 1960 *Farmer & Stockbreeder* 9 Feb. 77/1 The prohibition has been made since the discovery that a form of non-clinical rhinitis exists in pigs in Ireland. 1971 *Brit. Med. Bull.* XXVII. 33/1 The range of conditions is gradually progressive, from normal veins to veins that are elongated, tortuous and dilated but which give rise to no symptoms or to one symptom referrable to the legs; these have been termed 'non-clinical' varicose veins. 1890 W. James *Princ. Psychol.* II. xvii. 9 Plato's earlier pupils used to admit Sensation's existence, grudgingly, but they trampled it in the dust as something corporeal, non-cognitive, and vile. 1963 R. Carnap in P. A. Schilpp *Philos. R. Carnap* 45 The view that these sentences and questions are non-cognitive was based on Wittgenstein's principle of verifiability. 1962 F. I. Ordway et al. *Basic Astronautics* v. 207 (*caption*) Coherent and non-coherent converters. 1967 M. Chandler *Ceramics in Mod. World* iii. 98 The glaze at this stage is a feeble layer of noncoherent particles. 1971 G. H. Bourne *Ape People* x. 255 Among humans the amount of sexual play..and..noncoital sex in general seem to be related to the social level..and intelligence in the individual concerned. 1973 I. Singer *Goals Human Sexuality* i. 35 Claiming that all coital orgasms are physiologically indistinguishable from noncoital ones. 1914 S. E. Rasor tr. *Burkhardt's Theory of Functions* i. 24 Three points *A, B, C* in the plane such that *oA, oB, oC* are non-collinear. 1918 Veblen & Young *Projective Geom.* II. iii. 96 The triads of points..are all triads of noncollinear points. *Ibid.* 97 A, B, C are any three noncollinear points. 1966 Ogilvy & Anderson *Excursions in Number Theory* vi. 69 How many non-collinear points in a plane can be spaced at integral distances each from each? 1969 *Gloss. Terms Fire* (*B.S.I.*) 1. 7 *Non-combustible*, not capable of undergoing combustion. 1975 *B.S.I. News* May 10 The term 'incombustible' as a synonym for 'non-combustible' is deprecated in BS 4422: Part 1: 1969 and should not be used in standards. 1901 *Edin. Rev.* Apr. 492 The 'dumping' abroad at non-commercial prices of goods manufactured in excess of the requirements of the protected home market. 1933 Dylan Thomas *Let.* Jan. (1966) 14 It is prose..but an utterly non-commercial prose. 1973 *Listener* 30 Aug. 295/3 This concern to tailor the news to the listener has even spread..to National Public Radio, the non-commercial radio network. 1953 C. F. Hockett in Saporta & Bastian *Psycholinguistics* (1961) 63/2 Information theory does not deal with the way in which a source maps a noncommunicative stimulus into communicative output. 1920 B. Russell *Pract. & Theory Bolshevism* v. 76 If they elected a non-Communist representative he could not obtain a pass on the railway. 1972 'G. Black' *Bitter Tea* (1973) vi. 82 Singapore is now the real focal point for the whole of non-Communist Chinese life. 1852 Mill *Pol. Econ.* (ed. 3) I. ii. i. 258 The two elaborate forms of non-communistic Socialism known as St. Simonism and Fourierism. 1863 *Rep. Brit. Assoc. Adv. Sci. 1862* ii. 7 An algebra of non-commutative symbols. 1964 E. Bach *Introd. Transformational Gram.* vii. 159 Ordinary arithmetical multiplication and addition form groups of a special character called commutative (or Abelian) groups, since they obey the commutative law: $x + y = y + x$; semi-groups are in general noncommutative. 1954 *Proc. Nat. Acad. Sci.* XL. 1150 The quotient *G/K* of a simple noncompact group with center reduced to {*e*} by a maximal compact subgroup carries a Hermitian symmetric structure invariant under *G* if and only if *K* has a nondiscrete center. 1959 E. M. Patterson *Topology* iii. 63 The transformation operates on a non-compact space. 1964 E. Bach *Introd. Transformational Gram.* v. 114 It is only at the lowest levels that the noncompact nature of the morphemes appears. 1906 *Daily Chron.* 10 Aug. 8/7 This club is extending the idea of holding a long-distance non-competitive run to Bath and back. 1935 *Discovery* Mar. 64/1 We have never ceased advocating this non-competitive use of available idle labour. 1953 C. E. Bazell *Linguistic Form* 40 Non redundancy (non-complementary distribution). 1961 R. B. Long *Sentence & its Parts* i. 22 But the distinction between complementary and noncomplementary contained modifiers can generally be disregarded. 1936 *Mind* XLV. 316 The third class.. actually includes non-complex ideas revived by imagination. 1972 *Jrnl. Social Psychol.* LXXXVI. 36 Complex groups perform better than noncomplex groups. 1887 J. N. Keynes *Formal Logic* (ed. 2) vi. 381 The proposition..is supposed to be non-compound. 1954 I. M. Copi *Symbolic Logic* iv. 66 We must develop methods for analyzing non-compound statements. 1897 W. M. Urban *Hist. Princ. Sufficient Reason* iv. 41 The postulate of ground..is not..unlogical, and non-conceptual, as Schopenhauer later maintained. 1929 A. N. Whitehead *Process & Reality* 344 Physical feelings form the non-conceptual element in our awareness of nature. 1964 F. Bowers *Bibliogr. & Textual Criticism* i. v. 27 Not as words that have meaningful values—but, instead, as impersonal and non-conceptual inked prints. 1928 E. & C. Paul tr. *Marx's Capital* ii. iii. 117 The money serves as an ideal, as a non-concrete, means of purchase. 1912 J. S. Huxley *Individual in Animal Kingdom* iii. 84 To discriminate between conscious and non-conscious brains. 1972 *Computers & Humanities* VII. 111 We propose to do a feasibility study to see if the role of language in the formation of nonconscious belief systems can be ascertained. 1867 Thomson & Tait *Treat. Nat. Philos.* ii. 280 It is easy to arrange a system artificially, in connexion with a source of energy, so that its forces of position shall be non-conservative. 1962 Corson & Lorrain *Introd. Electromagn. Fields* vi. 219 We shall now extend the discussion to include time-dependent magnetic fields and the nonconservative electric fields which accompany them. 1965 H. Kahn *On Escalation* xii. 241 Almost everybody looks for ways to improve a situation, and 'non-conservative' behavior..becomes very possible. 1964 R. H. Robins *Gen. Linguistics* 157 Opposed acoustic features:..consonantal, non-consonantal. 1908 *Westm. Gaz.* 8 Aug. 14/1 Non-resistance is for the non-constructive man. 1964 G. Kreisel in Benacerraf & Putnam *Philos. of Math.* 158 Brouwer ignores non-constructive mathematics altogether. 1964 *Language* XL. 24 Noncontiguous dialect areas. 1865 G. M. Hopkins *Jrnls. & Papers* (1959) 103 All figures must be composed of continuous or of non-continuous lines or of both. 1973 *Archivum Linguisticum* IV. 4 A past time auxiliary which refers to recent non-continuous actions. 1964 R. H. Robins *Gen. Linguistics* 130 The equally non-contrastive [t] and [pʰ]. 1928 J. Reith *Diary* 13 Apr. (1975) i. 100 Winston Churchill..said he would like to.. speak for 15 minutes the next night, factual and non-controversial. 1943 *Mind* LII. 274 This covers only the elementary, non-controversial portions of the system. 1960 E. L. Delmar-Morgan *Cruising Yacht Equipment* v. 66 A modern sextant is..made of non-corrosive light metal alloy. 1969 *Jane's Freight Containers 1968–69* 472/2 Floor screws, flat head, self-tapping with non-corrosive finish. 1893 G. B. Shaw *Let.* 3 Mar. (1965) 386 Assuring you, in our personal, non-artistic and non-critical relations, of my unmitigated defiance. 1967 A. Battersby *Network Analysis* (ed. 2) iii. 26 The systematic analysis of a network sorts out the individual jobs into two main classes, critical and non-critical. 1851 *Phil. Mag.* I. 178 It is..desirable that a complete theory of magnetic induction in crystalline or non-crystalline matter should be established independently of any hypothesis of magnetic fluids. 1966 D. G. Brandon *Mod. Techniques Metallogr.* ii. 64 In electron microscopy the contrast is most commonly caused by elastic scattering only, that from non-crystalline replica materials being due to non-coherent scattering, that from thin crystalline films being diffraction contrast. 1908 *Westm. Gaz.* 15 Aug. 15/1 Home Railway Preference stocks are non-cumulative, 1954 G. I. M. Swyer *Reproduction & Sex* xiv. 173 An efficient spermicide..should be non-toxic, non-irritating, and non-cumulative;..it should preferably be non-absorbable. 1946 F. E. Zeuner *Dating the Past* xi. 341 There are superimposed cycles, or non-cyclic changes of a considerably longer duration. 1962 Corson & Lorrain *Introd. Electromagn. Fields* v. 203 If the vectors are permuted in noncyclic order, the sign is changed. 1914 B. Russell *Our Knowl. External World* ii. 34 The ultimate result of the introduction of the inductive method seems not the creation of a new kind of non-deductive reasoning. 1968 R. A. Lyttleton *Mysteries Solar Syst.* vii. 250 An intricate subjective psychological process must be concerned, of a non-deductive character. 1900 *Trans. Amer. Math. Soc.* I. 462 This represents the most general non-degenerate quadric. 1929 Non-degenerate [see *degenerate *a.* B. 3 a]. 1968 C. G. Kuper *Introd. Theory Superconductivity* xi. 183 The *n*-fold degenerate pair state has been split into an (*n* − 1)-fold level and a non-degenerate lower state. 1941 *Manch. Guardian Weekly* 14 Mar. 214/4 Now is my chance..to synthesize the ideology and emotional attitudes of juveniles and adolescents (cross-sectional and non-delinquent) during their vacational activities. 1970 *Jrnl. General Psychol.* LXXXII. 265 Delinquent and nondelinquent subjects. 1944 *Mind* LIII. 344 We must admit non-demonstrative principles of inference which are not derivable from experience. 1965 P. Caws *Philos. of Sci.* xxvi. 193 The most common name for such non-demonstrative inference is induction. 1908 *Daily Chron.* 19 Nov. 4/4 The Churches..together with all the ethical and non-denominational societies

should be summoned to co-operate. 1953 E. Smith *Guide to Eng. Traditions & Public Life* 17 A non-denominational chapel was commonly provided for religious services. 1905 *Ann. Math.* VI. 175 An example of a non-denumerable class is the class of all non-terminating decimal fractions. 1937 *Mind* XLVI. 59 The properties of integers are non-denumerable. 1968 J. J. C. Smart *Between Sci. & Philos.* vii. 245 A non-denumerable infinity of points. 1902 W. James *Let.* 29 July in R. B. Perry *Tht. & Char. W. James* (1935) II. 650, I may maximize unduly the non-derivative character of these forces, which you minimize. 1956 J. Whatmough *Language* 39 The derivative *manly* as compared with the simple 'non-derivative' word *man*. 1892 A. Sidgwick *Distinction* 182 Normally all proper (or non-descriptive) names have once been general (or descriptive). 1925 L. P. Smith *Words & Idioms* iii. 83 The non-descriptive, non-explanatory, and purely identifying term of the French Précieuses, *je ne sais quoi*. 1944 *Mind* LIII. 241 A formal distinction can be made..between *descriptive* symbols on the one hand and *nondescriptive* or logical symbols on the other. 1946 C. Morris *Signs, Language & Behavior* 116 Knowledge about the adequacy of non-designative signs is a powerful factor in their control. 1952 *Mind* LXI. 459 These theories proposed that ethical terms function like 'hurrah' and 'blast', as non-designative expressions. 1960 *Encounter* XV. v. 42 A markedly non-deterministic interpretation of psycho-analysis. 1965 A. Shimony in M. Black *Philos. in America* 260 As a result of the measurement there is a non-deterministic transition from the initial state to a final state. 1885 Non-diffusible [see Particulate *a.*]. 1951 Whitby & Hynes *Med. Bacteriol.* (ed. 5) iv. 57 The toxins of bacteria which do not produce exotoxins are termed endotoxins and are thought to be firmly bound to the bacterial proteins in a non-diffusible form. 1962 H. Heath in A. Pirie *Lens Metabolism Rel. Cataract* 362 Nondiffusable high polymers. 1967 *Oceanogr. & Marine Biol.* V. 380 Part of the calcium content of the blood is in the form of non-diffusible compounds of calcium with protein. 1956 A. A. Townsend *Struct. Turbulent Shear Flow* v. 91 By substituting in the equation of mean motion.., the non-dimensional equation is obtained. 1956 *Nature* 24 Mar. 548/2 The rather bewildering array of non-dimensional numbers that confronts the student in this subject [*sc.* combustion]. 1971 *Ibid.* 13 Aug. 447/3 The system is then arranged so that the equations of motion have the same form whether they are dimensional or non-dimensional. 1951 *Mind* LX. 398 The Rogerian non-directive technique is of no avail. 1954 J. A. C. Brown *Social Psychol. of Industry* 78 'Non-directive' interviewing in which the interviewer listens rather than talks. 1961 *Ann. Reg. 1960* 467 The expansion of world trade on a multilateral non-discriminatory basis. 1962 W. B. Thompson *Introd. Plasma Physics* ii. 11 These oscillations have the unusual character of being non-dispersive. 1966 D. G. Brandon *Mod. Techniques Metallogr.* 155 Such needle counters hold great promise in the field of 'non-dispersive' analysis. 1941 *Mind* L. 339 So it does not seem as if ϕ can be a non-dispositional, non-relational characteristic. 1949 A. Pap *Elem. Analytical Philos.* xii. 289 The dispositional predicate is said to be reducible, by means of the reduction-sentence, to non-dispositional predicates. 1970 *Manch. Guardian Weekly* 24 Oct. 9 The colleges on their side would agree to permit non-disruptive occupations and boycotts of classes. 1940 *Chambers's Techn. Dict.* 582/1 *Non-dissipative network*, a network designed as if the inductances and condensers are free from dissipation, and with components of minimum loss. 1964 R. F. Ficchi *Electr. Interference* iv. 38 A cavity resonator is described as a cylindrical box with perfect conductivity and containing within itself a nondissipative medium. An oscillation, once started within the box, should continue indefinitely. 1961 *Amer. Speech* XXXVI. 204 The reasons for the persistence of non-dominant languages. 1972 *Jrnl. Social Psychol.* LXXXVII. 92 Ring electrodes were slipped onto two fingers of the subject's nondominant hand. 1947 C. Gray *Contingencies* v. 120 Avoiding..the pitfall of non-dramatic music-making into which so many of his contemporaries and predecessors had fallen. 1965 *English Studies* Feb. 3 It is possible to find such scenes in earlier non-dramatic narrative. 1920 B. Russell *Pract. & Theory Bolshevism* ii. 123, I do not think that non-economic factors can be neglected. 1965 J. D. Chambers in Glass & Eversley *Population in Hist.* xiii. 344 There is more reason to relate the change to non-economic causes. 1853 *Phil. Mag.* V. 398 Nothing is known with certainty regarding the non-electrolytic resistance of liquid conductors. 1964 R. F. Ficchi *Electr. Interference* v. 55 Capacitors, both electrolytic and nonelectrolytic. 1964 *Times Rev. Industry & Techn.* Jan. 22/2 The operation was put straight on to a computer without an intervening non-electronic stage. 1964 *Language* XL. 191 The simple words which by our definitions are nonelementary. 1965 Hughes & Londey *Elem. of Formal Logic* xxiii. 177 Even the full Lower Predicate Calculus..raises problems of a non-elementary nature. 1941 *Mind* L. 168 Expressed in non-metaphoric and non-emotional words. 1961 R. B. Long *Sentence & its Parts* vi. 147 This occurs in subordinate-declarative clauses following ideas of appraisal, emotional or non-emotional. 1901 W. James *Let.* (1920) II. 153 The optimism and healthy-mindedness are yours... But the moderate and non-emphatic way of putting things is not. 1966 *English Studies* XLVII. 257 The normal non-emphatic character of the subject. 1955 *Publ. Amer. Dial. Soc.* xxiv. 20, I have on tape some argots which might at first appear to be non-English. 1941 *Mind* L. 129 We know the truth of a universal non-enumerative proposition, *All A is B*, only by apprehending, in a particular instance, that the universal A-ness necessitates the universal B-ness. 1956 J. O. Urmson *Philosophical Analysis* v. 63 The case becomes still more favourable to general facts if we take the non-enumerative universal proposition of traditional logic—'All As are B'. 1960 *New Biol.* XXXI. 26 A mutation (change) of a single chromosomal gene can lead to a corresponding chemical change in a non-enzymatic protein. 1964 *Oceanogr. & Marine Biol.* II. 347 Luciferin..is a non-enzymatic protein which luminesces in the presence of calcium ions. 1934 Webster, Nonenzymic. 1946 *Nature* 28 Sept. 433/2 Mackworth finds the other —SH enzymes..much less sensitive, requiring concentrations of the order of *M*/1,000, and this was also found by Morgan and Dixon for —SH groups in non-enzymic proteins. 1965 Peacocke & Drysdale *Molecular Basis Heredity* x. 115 Work on non-enzymic proteins. 1885 J. Veitch *Institutes of Logic* § 560 What are equivalent, or non-equivalent, to a common third term, are equivalent or non-equivalent to each other. 1973 J. J. Zeman *Modal Logic* v. 80 The systems which have an infinite number of non-equivalent irreducible modalities will generally be called 'absolutely strict' systems, while systems containing only a finite number of non-equivalent modalities will, in general, be called 'systems of complete modalization'. 1925 C. D. Broad *Mind & its Place* xi. 487 Purely ethical characteristics..cannot be identified with or defined in terms of non-ethical or 'natural' characteristics. 1964 E. A. Nida *Toward Sci. Transl.* v. 93 It might seem strange at first glance that in the above treatment of *jwok* in Anuak we have not indicated such traditional componential contrasts as ethical vs. non-ethical and secular vs. sacred. 1847 A. De Morgan *Formal Logic* iv. 75 We may say that each is *participant*, or *non-exclusive*, of the other. 1963 Auden *Dyer's Hand* 402 Friendship is a nonexclusive, nonpossessive relationship. 1903 B. Russell *Princ. Math.* I. li. 450 The consideration of existence itself leads to non-existential propositions, and so contradicts the theory. 1949 A. Pap *Elem. Analytical Philos.* iv. 85 Such discourse about connections of properties without any regard to particulars exemplifying them, is often called 'non-existential'. 1901 A. Sidgwick *Use of Words* iv. 119 Some kind of non-explanatory verbiage. 1966 *Jrnl. Philos.* LXIII. 446 The author's typical non-explanatory explanations. 1946 F. E. Zeuner *Dating the Past* xii. 381 The *quality* of the species-steps during the period of explosive evolution differs from those of the period of ordinary, non-explosive evolution. 1971 I. G. Gass et al. *Understanding Earth* xxi. 301/2 Non-explosive fissure eruptions. 1949 *Mind* LVIII. 63 We need a modal or non-extensional language which contains such words as 'necessarily' or 'accidentally'. 1955 A. N. Prior *Formal Logic* 269 Modal operators (and non-extensional operators generally). 1934 Webster, Nonfactual. 1936 *Mind* XLV. 359 According to Mr. Ayer every significant, non-factual proposition is analytic; every factual proposition is synthetic. 1934 Webster, Nonfatty. 1935 *Discovery* June 170/1 For analytical purposes, milk is divided into three parts: fat or cream; non-fatty solids..; and water. 1953 C. E. Bazell *Linguistic Form* iv. 47 The opposition in French between 'close e' and 'open e' in non-final syllables is similarly marginal. 1973 *Word 1970* XXVI. 103 In such compounds, nonfinal primary stress is supposed to become secondary. 1923 J. S. Huxley *Ess. Biologist* vii. 264 Man's ideals are in themselves unlimited, non-finite. 1971 *Computers & Humanities* V. 307 Stylistic uses of the non-finite verb clause in Shakespeare's plays. 1940 *Mind* XLIX. 248 The theory that the non-finitist symbols used in pure mathematics are to be interpreted as 'ideal elements'..would provide a fruitful starting-point for a consideration of certain aspects of scientific method. 1956 J. H. Woodger tr. *Tarski's Logic, Semantics, Metamath.* 260 On account of its non-finitist nature the rule of infinite induction differs fundamentally. 1909 Webster, Non-fissile. 1953 Non-fissile [see *fissile *a.* 2]. 1961 J. E. Radford *Nuclear Energy Simplified* ii. 29 The non-fissile isotope U²³⁸..has now become exceedingly valuable because plutonium can be transmuted from it. 1950 *Chem. Engin. Progress* XLVI. 110/1 An immediate concern should be to make use of the relatively large amount of nonfissionable U-238. 1974 *Sci. Amer.* June 36/1 The fusion neutrons can also be used to convert nonfissionable isotopes of uranium or thorium to fissionable isotopes. 1903 B. Russell *Princ. Math.* I. iii. 41 This is simply an instance of the non-formal principle of dropping a true premiss. 1964 R. H. Robins *Gen. Linguistics* v. 194 Those [definitions of the word unit] relying on non-formal, extra-grammatical criteria.. are of little value. 1926 J. S. Huxley *Ess. Pop. Sci.* xviii. 301 The embryonic or non-functional period. 1965 *English Studies* XLVI. 420 The attempt to save the *Gesetz* by saying that in all these cases the alliteration of the verb is non-functional can hardly be taken seriously. 1959 *Brno Studies in English* I. 42 The non-generic indefinite article. 1932 J. S. Huxley *Probl. Relative Growth* vii. 207 The bulk of the observable differences in size and proportions between the existing Scottish strain of Red deer and (a) the existing Carpathian strain and (b) the sub-fossil Scottish type are non-genetic and of no taxonomic significance. 1964 *Oceanogr. & Marine Biol.* II. 314 Others imply a nongenetic character of the changes involved or a combination of both genetic and non-genetic components. 1926 A. Huxley *Essays New & Old* 131 The grandest sight I saw in non-geometrical Holland was Zaandam. 1936 *Mind* XLV. 466 Some non-geometrical kind of analysis. 1861 Mill *Repr. Govt.* xvi. 292 East Prussia..must..be either under a non-German government, or the intervening Polish territory must be under a German one. 1964 *English Studies* XLV. 395 The use of *Textbuch* in the non-German sense may be due to carelessness of translation. 1937 *Jrnl. Eng. & Gmc. Philol.* 4 Oct. 463 Rules are given..and a comparison made between the vocabulary of German and English and the non-Germanic languages. 1965 *English Studies* XLVI. 419 This [*sc.* Gregorian chant] is of non-Germanic origin. 1955 J. L. Austin *How to do Things with Words* (1962) i. 3 The mistake of taking as straightforward statements of fact utterances which are *either* (in interesting non-grammatical ways) nonsensical *or else* intended as something quite different. 1965 *Canad. Jrnl. Linguistics* XI. 47 The moment a 'non-grammatical' sentence is uttered or written, it ceases to be 'non-grammatical'. 1961 L. C. Martin *Clinical Endocrinol.* (ed. 3) i. 15 Oxytocin has little or no effect either upon the non-gravid human uterus or upon the gravid uterus until the time of parturition. 1967 *Oceanogr. & Marine Biol.* V. 161 The perivisceral fluid..constitutes only 25% of the mass of a small (non-gravid) sea urchin. 1964 C. Barber *Present-Day Eng.* vi. 142 This habitual/non-habitual criterion is not typical. 1946 *Nature* 31 Aug. 294/1 It is non-hæmolytic and does not produce any pyrogenic effect on intramuscular injection in man. 1964 W. G. Smith *Allergy & Tissue Metabolism* v. 61 They also caused the formation of hæmolytic lysolecithin-like substances and non-hæmolytic SRS. 1955 W. Pauli *Niels Bohr* 37 This group distinguishes two kinds of complex (non-Hermitian) fields..which are multiplied by gauge-transformations with opposite phase factors. 1970 G. K. Woodgate *Elem. Atomic Struct.* 207 We find it convenient to use two non-Hermitian operators. 1896 W. Caldwell *Schopenhauer's System* ix. 508 A metaphysical analysis of the world must naturally always be taken in a timeless or ideal (non-historical) sense. 1935 *Mind* XLIV. 235 The historical essays..are mostly concerned with Greek philosophy..; and two of the essays are ostensibly non-historical. 1957 L. Fox *Two-Point Boundary Probl.* vii. 190 A change..in the non-homogeneous part of the equation. 1927 *Jrnl. Heredity* XVIII. 269/1 One might suppose that a portion of one chromosome has become attached to a member of a non-homologous pair. 1956 *Nature* 25 Feb. 386/1 'Centric fusion' of non-homologous autosomes. 1971 *Perceptual & Motor Skills* XXXII. 639 Synchrony of bimanual wrist movements by 12 normal Ss was compared when homologous muscles (e.g., left and right wrist flexors) and when non-homologous muscles (e.g., left flexors and right extensors) were simultaneously active. 1971 *Chromosoma* XXXV. 247 Rieger (1957) reported a high level of pairing between non-homologous chromosomes in haploid forms of *Antirrhinum majus*. 1890 R. Adamson in W. S. Jevons *Pure Logic* p. xi, Non-identical contents presented to it. 1965 *Language* XLI. 280 The nonidentical subject constituents. 1964 Rousseas & Farganis in I. L. Horowitz *New Sociology* 270 The non-ideological liberal is uncommitted. 1975 *Times* 25 Nov. 8/8 The [Chinese] authors..have been writing cultural books, largely non-ideological, since the 1930s. 1932 H. H. Price *Perception* v. 110 Some few and as it were privileged expanses (normal or non-illusory ones). 1928 L. P. Smith *Words & Idioms* 114 The non-imitative arts, architecture and music. 1973 *Jrnl. Genetic Psychol.* CXXIII. 145 The reinforced group emitted significant more physical, verbal, and nonimitative aggression than the nonreinforced group. 1874 W. Wallace tr. *Hegel's Logic* 208 The non-independent and changeable. 1965 Hughes & Londey *Elem. of Formal Logic* xviii. 131 A4..can be proved from the other axioms... We express this fact by saying that A4 is a non-independent axiom. 1894 C. Dixon (*title*) The nests and eggs of non-indigenous British birds. 1964 *Language* XL. 93 Regional and nonindigenous influence. 1926 D. H. Lawrence *Plumed Serpent* vii. 128 Men, dark, collective men, non-individual. 1949 A. Pap *Elem. Analytical Philos.* iv. 88 The use of non-individual variables commits the logician to the use of names of abstract entities, like properties, classes or propositions. 1934 Priebsch & Collinson *German Lang.* ii. v. 251 A residue of untraced words which have led some scholars..to postulate a non-Indo-European strain in the early vocabulary. 1965 W. S. Allen *Vox Latina* i. 13 Many of the words..are probably non-Indo-European. 1899 *Jrnl. Inst. Electr. Engin.* XXVIII. 17 A practically non-inductive wire resistance. 1957 G. Ryle in C. A. Mace *Brit. Philos. in Mid-Cent.* 259 Pure mathematics is a non-inductive..science. 1938 *Mind* XLVII. 247 It is this aspect of Dr. Myers' essay that is of interest to the non-industrial reader, though there is much of value in it about the prevention of accidents in industrial occupations. 1909 Webster, Noninfectious. 1956 *Nature* 21 Jan 131/1 The purified protein alone was found to be non-infectious. 1965 McKeown & Brown in Glass & Eversley *Population in Hist.* xii. 291 Infectious and non-infectious cases. 1956 *Nature* 18 Feb. 304/1 Extracts from infected plants contain a range of specific proteins that resemble the virus..but are not infective... Beans appear to contain an unusually large proportion of non-infective protein. 1971 *Brit. Med. Bull.* XXVII. 9/1 It may have a predominantly non-infective allergic origin. 1897 B. Russell *Essay on Foundations of Geometry* ii. 61 This element must be non-inferential. 1935 *Mind* XLIV. 48 The last phrase suggests a less positive stand in regard to the non-inferential apprehension of the objective. 1936 *Discovery* Apr. 106/2 Helium gas is used as a non-inflammable substitute for hydrogen in balloons and airships. 1969 *Jane's Freight Containers 1968–69* 476/2 Both achieved by in-situ foamed non-inflammable urethane. 1837 J. S. Mill *Let.* Sept. in *Wks.* (1963) XII. 350 Our non-inflected language. 1928 *Mod. Lang. Rev.* Apr. 136 Then we shall have two genitives in English, one inflected and one non-inflected. 1924 H. E. Palmer *Gram. Spoken Eng.* 81 All other classes of qualificatives..form the comparative..and superlative by the non-inflexional method. 1940 *Mind* XLIX. 436 The syntax of the Aryan languages differs fundamentally from that of non-inflectional languages, in particular Chinese. 1953 K. Jackson *Lang. & Hist. in Early Britain* ii. 286 He gives substitution by AS. *i, y,* or *e* in all cases, apparently not considering Pr.W. stressed *ī* in non-initial syllables. 1965 *Language* XLI. 453 Medial (i.e. noninitial and nonfinal) dissyllabic sequences. 1962 *Listener* 20 Sept. 436/2 The differences between insightful and non-insightful problem-solving behaviour. 1968 M. S. Livingston *Particle Physics* ii. 18 The non-integral atomic weights were explained as isotopes. 1897 W. James *Will to Believe* 11 Evidently..our non-intellectual nature does influence our convictions. 1946 *Mind* LV. 51 This, so far as it depends on denying the possibility of other sorts of non-intellectual intuition than the sensible, is certainly dogmatic. 1957 N. Chomsky in Saporta & Bastian *Psycholinguistics* (1961) 41/2 Our ultimate aim is to provide an objective, non-intuitive way to evaluate a grammar. 1960 *Suppl. Oxf. Med. Treatment of Disease* VIII. 804 Non-involved. 1973 S. Fisher *Female Orgasm* xv. 434 A signal for potential future sexual difficulties is..a distant, non-involved relationship with the father. 1946 *Nature* 14 Dec. 876/2 Physiological effect on man: non-irritant. 1963 J. Osborne *Dental Mechanics* (ed. 5) i. 9 It [*sc.* the impression material] should be non-irritant to the mouth tissues. 1935 Huxley & Haddon *We Europeans* iii. 96 In each country the Jewish population overlaps with the non-Jewish. 1973 *Guardian* 17 Apr. 15/8 The Passover meal, to which non-Jewish passengers like me were invited. 1965 M. Morse *Unattached* i. 44 His impartial and non-judgemental attitude. 1972 *Jrnl. Social Psychol.* LXXXVI. 224 Low authoritarian subjects are..flexible and nonjudgmental. 1949 Koestler *Promise & Fulfilment* iii. iii. 317 The Cabinet has capitulated to the

[illegible]

need be no 'intrinsic difference'..between veridical and non-veridical experiences, then how can you tell, at any given time, whether your perceptual experience is veridical or non-veridical? **1890** W. James *Princ. Psychol.* II. xx. 212 Berkeley..concluded that distance could not possibly be a visual sensation, but must be an intellectual 'suggestion' from 'custom' of some non-visual experience. **1971** J. Z. Young *Introd. Study Man* xxx. 421 Some lemurs have scent-glands on the fore-limbs; these are common among non-visual mammals. **1933** *Discovery* Sept. 278/1 It is, in fact, not an *aposeme* to frighten an enemy away..but an *episeme*, to distract attack from a vulnerable to a fortified or non-vital region. **1936** J. R. Kantor *Objective Psychol. Gram.* xiii. 184 There is still left the question of interjectional acts or even non-vocal behavior. **1946** C. Morris *Signs, Language & Behavior* 192 That there are non-vocal languages is not generally disputed. **1964** R. H. Robins *Gen. Linguistics* iv. 157 Opposed acoustic features: vocalic, non-vocalic. **1930** *Times Educ. Suppl.* 15 Feb. 70/2 What was known as non-vocational education. **1963** *Higher Educ.* (Cmnd. 2154) 317 Courses of a mainly non-vocational character.

4. Prefixed to a sb. (or vbl. sb.) forming a phrase used attributively, as *non-citizen, -class, -combat, -copyright, -corridor, -county, -craft, -dollar, -equilibrium* (see also sense 1 in Dict. and Suppl.), *-fat* (cf. *non-fatty* in *3), *-food, -image, -jazz, -kernel, -language, -league, -narrative, -pedigree, -print, -profit, -protein, -speech, -structuralist, -teaching, -title, -vintage*; **non-association** (see quot. 1940).

Occas. such phrases are used predicatively (as in quot. 1956 for *non-protein*).

1909 *Installation News* II. 180 The standard qualities listed are:— (1) A class equivalent to what is generally designated throughout the trade as Non-Association Cable. (2) A grade equivalent to that manufactured by [illegible]

[illegible] carbohydrates, non-protein nitrogenous compounds (NPN)). **1954** F. G. Cassidy *Robertson's Devel. Mod. Eng.* iv. 54 The phonetician..must include the entire range of speech sounds... Pike would have him investigate non-speech sounds too. **1953** C. E. Bazell *Linguistic Form* 5 The earlier distinction of morpheme and semanteme is obsolete except in non-structuralist circles. **1959** *New Statesman* 21 Feb. 259/3 Hours are wasted each week in administrative, non-teaching duties such as special staff meetings, frequent house meetings, tutorial group meetings. **1973** *N.Y. Times* 28 Jan. 21/2 This is a nontitle fight and there is no weight restriction on heavyweights. **1953** D. Parry *Going Up——Going Down* vi. 318 At dinner Clive ordered up four bottles of non-vintage champagne. **1959** *Listener* 5 Mar. 422/1 Bratby is here, of course, a non-vintage work half way between the 'larder' period and what is happening at his current exhibition. **1975** *Country Life* 2 Oct. 838/3 No non-vintage champagne may be sold until it has been a year in bottle.

5. b. Prefixed to an infinitive forming a phrase used attributively in the sense 'that does not ——', 'that is designed not to ——', as *non-crease, -crush, -dazzle, -shrink.* Cf. *non-skid *a.*, *non-slip *a.*, *non-stick *a.*

1944 A. Thirkell *Headmistress* ix. 200 Utility non-crease (though they were neither) ready-made dresses. **1969** *Punch* 5 Feb. 195/1 You *should* wear non-crease clothes. **1924** *Times Trade & Eng. Suppl.* 29 Nov. 247/2 It must be remembered that the heavy weight non-crush costume linens absorb an enormous quantity of flax in spinning. **1961** *Times* 4 Dec. (Agric. Suppl.) p. iii/5 An optional extra to be combined with a non-crush cab. **1962** *Times* 5 May 9/4 A non-dazzle protective covering for our remaining pictures. **1965** *Economist* 27 Feb. 929/2 An alternative pair of non-dazzle lamps.. [for] meeting other traffic. **1962** J. T. Marsh *Self-Smoothing Fabrics* xv. 256 The anti-shrink effect so often ignored by later publications on non-shrink processes. *a* **1963** L. MacNeice *Astrology* (1964) vii. 237 The Zodiac..signs are featured..in advertisements for.. non-shrink cottons.

6. a. Prefixed to ppl. adjs., as *non-aspirated, -associated, -centralized, -ciliated, -classified, -clogging, -coloured, -committed, -corroding, -defining, -dividing, -fabricated, -fattening,*

[illegible]

[illegible] Jakobson & Halle in Saporta & Bastian *Psycholinguistics* (1961) 348/2 Consonants: palatalized vs. non-palatalized. **1898** H. E. Byers in W. A. Morgan '*House*' *on Sport* 206 The non-playing men alternately shout their approbation and make heavy bets. **1958** *Atlantic Monthly* Nov. 184/2 One solution is to..subsidize a trip to the movies for the nonplaying members of the family. **1959** *Times* 11 Sept. 9/3 The British team's non-playing captain. **1876** Preece & Sivewright *Telegraphy* iii. 60 A non-polarized relay..is not much used in England. The forms of relay more largely used are called polarized, because their armatures are either permanent magnets or are maintained in a magnetized condition. **1946** *Nature* 30 Nov. 794/2 The erythrocytes become dipoles.. and form chain-like aggregates... Sedimentation-rates of such red-cell aggregates will be greater than those of single non-polarized blood cells. **1967** *Economist* 8 July 107/3 Tenants in some rural areas had to prostrate themselves at a safe, non-polluting distance on the one day they saw their Brahmin landlord, rent day. **1974** *Listener* 14 Feb. 212/2 What is needed is..cycleways and encouragement for people to use this non-polluting form of transport. **1946** *Nature* 21 Dec. 900/1 The processes listed are mainly brought about by non-proliferating suspensions of bacteria or yeasts. **1910** *Daily Chron.* 7 Mar. 1/7 They did not mix up recurring and non-recurring expenditure. **1973** *Ann. Rep. Curators Bodl. Libr. 1971–2* 49 A total balance of £1,614 from special non-recurring grants for stock-piling. **1957** A. N. Prior *Time & Modality* 61 Genuinely non-referring names..are excluded. **1963** F. T. Visser *Hist. Syntax* I. iii. 236 Another [construction] of very recent introduction, in which *it* is non-referring. **1946** *Nature* 21 Sept. 422/1 The anti-reflexion or non-reflecting thin transparent film deposited on the surface of the lens. **1962** L. S. Sasieni *Optical Dispensing* x. 2~1 Coating the lens surfaces with a non-reflecting coating. **1974** *Country Life* 21 Mar. 659/2 The black non-reflecting surface. **1972** P. Black *Biggest* [illegible]

[illegible]

1949 M. Mead *Male & Female* xvi. 330 The division of labour among non-English-speaking immigrants. **1975** *Nature* 20 Feb. p. xiv/1 (Advt.), The work..involves.. language correction of contributions written by non-English-speaking authors. **1933** *Radio Times* 14 Apr. 112/2 This pleasant, gentle laxative..is non-habit-forming. **1949** M. Lowry *Let.* 16 Feb. (1967) 169 One or other of the non-habit-forming barbiturates should be often used in preference to strychnine or chloral. **1974** M. C. Gerald *Pharmacol.* xi. 198 Virtually all of these nonbarbiturate compounds have been initially promoted as being nonhabit-forming. **1967** Cox & Grose *Organiz. Bibliogr. Rec. by Computer* iv. 85 These are a set of words which the user defines as non-information carrying words. **1937** E. Snow *Red Star over China* vi. 233 Some $70,000 in non-interest-bearing loans had been invested by the Government in the co-operatives. **1959** E. Pound *Thrones* ciii. 85 Trying (T.C.P.) 50 years later to keep some of the non-interest-bearing etc. in circulation as currency. **1933** *Planning* I. viii. 11 A non-profit-making organisation for this purpose could probably obtain a large interest in British farming. **1971** *Guardian* 16 Dec. 1/8 A railway porter..has been told to stop his non-profit making early-morning hot drink service to passengers.

8. Prefixed to adverbs, as *non-adaptively, -denumerably, -enzymatically, -enzymically, -inferentially, -relativistically, -spatially, -uniformly.*

1931 *Brit. Jrnl. Psychol.* XXII. 126 The reaction is not adaptive. It is one in which the past..does not function; or if it does function at all, it does so inefficiently, non-adaptively. **1945** E. T. Bell *Devel. of Math.* (ed. 2) viii. 171 Non-denumerably infinite classes of rational numbers. **1965** B. Mates *Elem. Logic* viii. 141 The question thus arises whether one could find a consistent set of sentences that is satisfiable only by interpretations having non-denumerably infinite domains. In view of the Löwenheim-Skolem theorem, the answer is negative. **1953** *Adv. in Protein Chem.* VIII. 169 Labelled amino acids may [illegible] to be non-inferentially discoverable. **1957** Bethe & Salpeter *Quantum Mechanics of One- & Two-Electron Atoms* I. xix. 97 We treat the electron nonrelativistically throughout. **1934** *Mind* XLIII. 299 What confronts the mind, in this sense..is (perhaps non-spatially) *there* to be found, or haply to be found out, by searching. **1893** Harkness & Morley *Treat. Theory of Functions* iii. 72 A non-uniformly convergent series may be continuous. **1969** *Listener* 2 Jan. 11/1 Stars generally spin non-uniformly, rotating faster near the equator than at the pole.

non-, form of *nona- used before a vowel.

nona- (before a vowel **non-**), comb. form of L. *nōnus* ninth. **a.** In scientific words, as Nonagon, *nonapeptide, *nonic *sb.* and *a.* **b.** *spec.* in *Chem.*, used in naming the alkane with nine carbon atoms in its molecule, Nonane, and compounds and radicals which contain a similar carbon chain, as Nonoic *acid*, Nonyl.

non-aggre·ssion. [Non- 1.] Absence of the will, desire, or intention to show aggression on the part of nations, governments, or politicians; freq. *attrib.*, esp. in phr. *non-aggression pact.*

1903 J. Morley *Life of Gladstone* III. x. viii. 50 Then he would go on—'..—My name stands in Europe as a symbol of the policy of peace, moderation, and non-aggression.' **1925** A. Toynbee *Survey Internat. Affairs* 26 The proposed pledge of non-aggression would not have power to deprive the Allies of the right conferred on them by the Treaty of Versailles. *Ibid.* 31 Mr. Lloyd George broached to the French delegation his project for a Pact of Non-aggression. **1935** *Ann. Reg. 1934* [21] The Polish Ambassador in Moscow signed a Protocol extending the Non-Aggression Pact between the two countries till December 31, 1945. **1941** Koestler *Scum of Earth* 19 The classical form for non-aggression treaties had been so far to promise neutrality. **1955** *Ann. Reg. 1954* 106 The two countries [*sc.* India and China] based their relationship on..mutual non-aggression.

non-ali·gned, *a.* [Non- 6.] Of a country, government, etc.: pursuing a policy of non-alignment. Also *absol.* as *sb.* Hence **non-ali·gnedness, non-ali·gner.**

1960 *New Left Rev.* Nov.–Dec. 10/2 The non-aligned powers have grown in strength. **1962** *Guardian* 7 Nov. 18/7 The concept of political non-alignedness. **1963** *Economist* 28 Dec. 1325/1 India's professional non-aligners. **1965** *Times Lit. Suppl.* 25 Nov. 1056/4 The attitudes of allies and of the non-aligned. **1966** *Guardian* 14 Oct. 12/2 The Canadians..believe that there is a real strategic danger of a nonaligned block composed of Austria, Switzerland, and France cutting Europe into two. **1970** *Guardian Weekly* 30 May 6 The two groups in Indonesia opposing each other are the 'nonaligneds' and the 'interventionists'. **1973** *Caribbean Contact* Feb. 9/3 The need..for the Non-Aligned world to probe beyond the detente between the big powers.

non-ali·gnment. [Non- 1.] Lack or absence of alignment; *spec.* in *Politics*, absence of political or ideological affiliations with other nations, esp. with the most powerful nations. Also *transf.*

1934 in Webster. **1955** *Times* 6 May 11/4 He extolled 'non-alignment' and co-existence. **1963** M. Brecher *New States of Asia* iv. 111 *Non-alignment*..is a political status. It refers to a state that declares itself aloof from bloc conflicts; nothing more. It proclaims itself free from *a priori* alliances, notably military entanglements with any bloc or Great Power anywhere in the world. **1970** E. Bullins *Theme is Blackness* (1973) 153 My non-alignment with reality is quite evident and explainable.. my being Black and non-committed to any political truths. **1973** 'D. Jordan' *Nile Green* xx. 81 The Egyptians ..prefer to hang on to what's left of their non-alignment status and not rely on the Russians.

non-alle·rgic, *a.* [Non- 3.] Not allergic; *spec.* not causing allergy.

1936 H. Goodman *Cosmetic Dermatol.* xxv. 368 In the field of cosmetics and beauty-application preparations, the words *allergy* and *nonallergic* are mentioned more and more often. **1938** *Words* Jan. 11/2 'Cosmetics are non-allergic.' Donna Ray broadcast Oct. 24 [19]37. **1954** L. Martin *Clin. Endocrinol.* (ed. 2) vii. 150 The eosinophil count will..be lowered by..50 per cent or more in non-allergic subjects. **1964** W. G. Smith *Allergy & Tissue Metabolism* iii. 42 Inflammation of non-allergic origin. **1972** *Islander* (Victoria, B.C.) 18 June 7/3 Dacron fibrefill II is non-allergic and moth, vermin and mildew resistant.

non-American, *a.* (*sb.*). [Non- 3.] Not American. Occas. as *sb.*, one who is not an American.

1904 G. Saintsbury *Hist. Criticism* III. 640 The poet who seems to some possibly rash non-American persons to divide with Poe the prize due to the worthiest in American poetry, was also a critic—less of the professional kind, much more *borné*, but more concentrated. **1924** S. P. Sherman *Genius of America* 213 He laughed at all the non-American world. **1936** *Mind* XLV. 15 Insuffi-

[illegible]

nonane. Add: Hence **nonano·ic** *a.* = Nonoic *a.*; **nonano·ate**, the anion, or an ester or salt, of nonanoic acid.

1947 K. S. Markley *Fatty Acids* xv. 407 Armstrong and Hilditch described the method for the disruptive oxidation of methyl oleate in acetone solution which they claim yields 80% to 90% of the theoretical amount of nonanoic acid. [illegible]

nonapeptide (nǫnăpe·ptaid). *Biochem.* [f. *nona- + *peptide.] Any peptide composed of nine amino-acid residues.

1960 *Helv. Chim. Acta* XLIII. 1358 The synthesis of a nonapeptide exhibiting bradykinin-like properties is described. [illegible]

non-art. [Non- 2 and *2 b.] Something that is not art; *spec.* a form of art which avoids artifice or which rejects conventional modes and methods.

1936 L. Durrell *Spirit of Place* (1969) 49 [illegible]

non-A·ryan, *a.* and *sb.* [Non- 3, 2.] **A.** *adj.* Of a language or a person: not Aryan or of Aryan descent; used *spec.* of Jews in Nazi Germany.

[illegible]

B. *sb.* One who is not an Aryan; used *spec.* of Jews in Nazi Germany.

[illegible]

no-na·tion, *a. dial.* [f. No *a.* 6 b + Nation *sb.*[1]] (See quots.)

[illegible]

non-atta·ched, *ppl. a.* [Non- 6.] Not attached; *spec.* unconcerned or uninvolved with material things. So **non-atta·chment.**

[illegible]

non-bank, *a.* [Non- 4.] Not connected with, or transacted by, a banking house; of an institution: not being a bank.

1946 C. C. Abbott *Managem. Federal Debt* v. 87 Should a boom or a price inflation develop and conditions call for a restriction of deposits, this policy of relying upon increases of bank credit to take up obligations not wanted by nonbank holders would aggravate the situation. **1961** *Ann. Reg. 1960* 485 The Bank of England commented..that..in the light of the need to attract purchases of stock by 'non-bank' investors, 'a somewhat higher level of gilt-edged yields would now be appropriate'. **1961** *Economist* 25 Nov. 822/3 The gilt-edged market (which provides the Exchequer with non-bank finance). **1964** Gould & Kolb *Dict. Soc. Sci.* 438/2 Like every other nonbank institution, savings and loan associations..must keep accounts with commercial banks. **1973** *N.Y. Law Jrnl.* 4 Sept. 3/5 (Advt.), We are..the largest non-bank firm—providing complete corporation securities services.

non-belli·gerent, *a.* and *sb.* [Non- 3, 2.] **A.** *adj.* Not actively engaged in hostilities; not aggressive. **B.** *sb.* A country which abstains from active involvement in a war but which more or less openly favours one side. Hence **non-belli·gerence, -belli·gerency,** the status or attitude of a non-belligerent country.

1909 *Westm. Gaz.* 14 June 5/3 When the defence has been on anything like a great scale, the non-belligerent defenders have very largely exceeded in number the armed defenders. **1940** *War Illustr.* 5 Jan. 556 'The position assumed by Italy on September 1,' he [*sc.* Count Ciano, the Italian Foreign Minister] said, 'was a position of non-belligerency strictly in conformity with the German intention of localizing the conflict.' **1940** *Manch. Guardian Weekly* 1 Mar. 162 Today all the Balkan states are neutral, Italy is a 'non-belligerent', Turkey is a firm friend, Russia is uncertain. **1946** *Ann. Reg. 1945* 76 The Spanish Government had publicly followed a policy not of neutrality but of non-belligerence. **1953** P. C. Berg *Dict. New Words* (ed. 2) 115/1 A non-belligerent attitude between contending parties or powers. **1975** *Times* 19 Aug. 13/2 A formal pledge of non-belligerency is something which President Sadat cannot or will not give.

non-biolo·gical, *a.* [Non- 3.] Not belonging to biology or forming part of its subject matter; not occurring in, involving, or pertaining to living organisms.

1931 J. S. Huxley *What dare I Think?* i. 32 There have been Boards of Pest Control which were not too anxious to find their occupation gone with the going of their particular pest. Leaving such non-biological or hyper-biological considerations on one side, there have been many pests which have so far baffled research. **1956** *Nature* 21 Jan. 110/2 Photochemistry in higher plants and in non-biological systems. **1957** G. E. Hutchinson *Treat. Limnol.* I. xvi. 866 The existence of non-biological mechanisms of oxidation of ammonia. **1962** F. I. Ordway et al. *Basic Astronautics* vi. 247 An understanding of much corollary evidence and supposition from nonbiological fields. **1971** I. G. Gass et al. *Understanding Earth* ix. 125/1 All known extra-terrestrial environments and many geological environments are still non-biological.

Hence **non-biolo·gically** *adv.*

1974 *Nature* 6 Sept. 43/1 A rapid, routine, new tool for the differentiation of biologically and nonbiologically produced dicarboxylic acids.

non-black, *a.* and *sb.* [Non- 3, 2.] **A.** *adj.* Not black; *spec.* (also *non-Black*), of, pertaining to, or being a person or persons who are not black. **B.** *sb.* A non-black person.

1926 D. M. Liddell *Handbk. Non-Ferrous Metallurgy* I. ix. 283 No radiation pyrometer has been found to be very useful for measuring temperatures of non-black bodies. **1930** *Social Sci. Abstr.* June 1011 In the non-black land the rush toward individual appropriation was naturally less than in the black soil belt. **1970** *Guardian Weekly* 22 Aug. 7 At the University of Natal, non-white students..refer to whites as 'non-blacks'. **1971** *Black Scholar* Jan. 27/2 The argument goes that black artists..must move beyond race in their work to some mythical human (non-black) understanding. **1973** *Black World* Apr. 20/2 Expertly directed by a non-Black, the musical played to standing-room audiences. **1973** S. Henderson *Understanding New Black Poetry* 9 Is it possible that, given a Black Poetic Structure, a non-Black can create in this form—as whites play jazz, for example?

non-ca·pital, *a.* [Non- 3]. Of a murder or other offence: for which a convicted person is not punishable by the death penalty.

1898 C. Ilbert *Government of India* i. 15 The Company were in the habit of procuring for each voyage a commission to the 'general' in command, empowering him to inflict punishments for non-capital offences, such as murder or mutiny. **1956** *Hansard Commons* 15 Nov. 1237/1 The distinction between non-capital murders and capital murders which is made in this part of the Bill must lead to many inconsistencies and anomalies. **1964** Morris & Blom-Cooper *Calendar of Murder* viii. 328 The legislature..sought..to mitigate considerably the incidence of the death penalty..by declaring certain categories of murder non-capital. **1968** *Globe & Mail* (Toronto) 13 Feb. 3/8 Two men were jointly charged yesterday with the non-capital murder of..a taxi driver.

non-Ca·tholic, *sb.* and *a.* [Non- 2, 3.] **A.** *sb.* One who is not a Roman Catholic.

1793 [in Dict. s.v. Non- 2]. **1859** Mill *Liberty* iv. 154 What do Protestants think of these perfectly sincere feelings, and of the attempt to enforce them against non-Catholics? **1971** J. C. Heenan *Not Whole Truth* i. 12 His church was half-filled every Monday evening with enquirers who came to hear his talks to non-Catholics.

B. *adj.* That is not Roman Catholic.

In a somewhat wider sense in quot. 1936.

1823 C. Butler *Continuation of A. Butler's Lives of Saints* p. lxii, Their anti-christian and non-catholic adversaries. **1867** F. Oakeley in H. E. Manning *Ess. Relig. & Lit.* 2nd Ser. 130 The world persists in thinking that the Catholic and non-Catholic ideas are in diametrical opposition. **1936** T. S. Eliot *Essays Ancient & Modern* 132 There may always be schemes, initiated by non-Christian and non-Catholic minds..to which we can give unqualified support. **1966** *Publ. Amer. Dial. Soc.* xlii. 35 To recall many abusive terms for non-Catholic Christians.

nonce. Add: **4. nonce-word** (examples); similarly *nonce-borrowing, -combination, -form, -formation, -meaning.*

1954 U. Weinreich in Saporta & Bastian *Psycholinguistics* (1961) 385/1 At the time of his utterance, it is a 'nonce-borrowing'. **1943** *Amer. Speech* XVIII. 301 A number of them..also meet the condition of not being independent words used in some nonce-combination. **1962** H. A. Gleason in Householder & Saporta *Problems in Lexicography* 88 A dictionary-maker need not include a non-idiomatic nonce-form. **1957** *Archivum Linguisticum* IX. 122 It clearly functions morphemically as everyday nonce-formations testify. **1943** C. L. Wrenn *Word & Symbol* (1967) 97 The most surprisingly beautiful result of Spenser's experimenting in poetic language is in the use..of the word *Cheuisaunce*, which may be described as having acquired for special purpose what I would call a nonce-meaning. **1884** Nonce-wd. [see Anotherness]. **1927** *Englische Studien* Nov. 99 If an alternative explanation presents itself, topographical nonce-words ought to be avoided. **1957** R. W. Zandvoort *Handbk. Eng. Gram.* I. ii. 43 Some of them are nonce-words, i.e. spontaneous creations by a speaker or writer, coined for the occasion.

non-ce·ntral, *a.* [Non- 3.] **1.** *Statistics.* Having or corresponding to a non-zero mean.

1928 *Proc. R. Soc.* A. CXXI. 670 This interpretation of the distribution..is seen to replace the χ^2 distribution of the analysis of variance for cases in which the sum of squares corresponding to n_1 degrees of freedom is derived theoretically for non-central deviations with fixed central displacements. **1949** *Biometrika* XXXVI. 202 In the case of the well-known tests using χ^2, t and F, the evaluation of their power functions involves the use of what have been called non-central distributions. **1968** P. A. P. Moran *Introd. Probability Theory* vii. 323 The quantity χ^2 in (7.121) is then said to have a non-central χ^2-distribution.

2. *Physics.* Of a force: not central, i.e. not in general directed along the line joining the bodies it acts between.

1948 *Physical Rev.* LXXIII. 1403/1 (*heading*) The effect of non-central forces on the collisions of high energy neutrons with protons. **1962** *Encycl. Dict. Physics* V. 610/1 The magnetic forces due to electric currents also have this non-central character. **1966** D. L. Livesey *Atomic & Nuclear Physics* ix. 414 There are noncentral forces between neutron and proton. **1970** G. K. Woodgate *Elem. Atomic Struct.* vii. 107 In the central-field approximation we neglected the residual electrostatic term in the Hamiltonian which represents a non-central force.

Hence **non-centra·lity,** the property of being non-central.

1959 H. Scheffé *Anal. of Variance* 412 The ordinary or central chi-square distribution is the special case of the noncentral distribution when the noncentrality parameter $\delta = 0$. **1969** H. O. Lancaster *Chi-Squared Distribution* xi. 231 It is required to determine the parameters of non-centrality of the test of independence in two dimensions.

non-committal, *sb.* (*a.*). **b.** (Earlier and later examples.)

1829 H. Orne *Lett. of Columbus* 18 The non-committal system prevailed. **1973** *Archivum Linguisticum* IV. 18 The same applies in..(5), which may be interpreted as a 'non-committal' statement.

non-communicating, *vbl. sb.* and *ppl. a.* Add examples in the sense: 'not communicating or conversing'.

1917 A. S. Pringle-Pattison *Idea of God* iii. 57 It is impossible to treat the world of religious belief and the world of fact..as if they were two non-communicating spheres. **1960** K. Amis *New Maps of Hell* (1961) i. 20 Some excellent stories have been written about non-communicating aliens, from *The War of the Worlds* onwards. **1972** *Where* May–June 158/2 The parent who protested that her LEA was sending her non-communicating child 200 miles away from home and so she would not be able to visit, when the child could not write or use a phone, had a real case.

non-concur, *v.* Add: (Later example.)

1823 W. Tudor *Life of James Otis* 239 The Council non-concurred this resolve.

b. *intr.* To fail or refuse to concur; to disagree. Usu. const. *in* or *with.*

1855 *Chicago Times* 3 Mar. 3/5 The House non-concurred with the Senate amendment. **1862** *Congress. Globe* 9 July 3214/1, I hope the house will non-concur in that amendment of the Senate. **1907** *Springfield* (Mass.) *Weekly Republican* 20 June 1 The Senate has non-concurred with the House amendments. **1911** Parton & Manning in C. E. Persons et al. *Labor Laws* 53 The House went on record as favorable to the resolution. The State nonconcurred. **1974** *Spartanburg* (S. Carolina) *Herald* 18 Apr. A. 2/4 Arthur was one of several House members to take the floor Tuesday to urge the legislators to nonconcur in the Senate amendments.

nonconforming, *ppl. a.* Add to def.: Also in non-religious contexts; *spec.* (see quot. 1961).

1899 'Mark Twain' *Man that corrupted Hadleyburg* (1900) 273 Worn officially, our nonconforming swallow-tail is a declaration of ungracious independence in the matter of manners. **1961** E. A. Powdrill *Vocab. Land Planning* iv. 67 When industry of this nature [*sc.* general industry] is located in an area the primary use of which is residential, shopping, or other non-industrial uses, it is said to be 'non-conforming', in that it does not conform, in usage, to the primary use. **1972** *Jrnl. Social Psychol.* LXXXVI. 16 When a non-conforming response was made by an *S* in the low or high reward condition..he automatically received a coin. **1973** *N.Y. Law Jrnl.* 26 July 2/2 The effect of such a zoning law change is to make the property a non-conforming use.

nonconformist. 2. Delete † *Obs.* and add later example.

1952 [see *feather-bedding *vbl. sb.*].

3. Esp. in phr. *nonconformist conscience.*

1890 *Times* 28 Nov. 8/6 The *minimum* demand of the great Nonconformist party is the unconditional abdication of Mr. Parnell, and his immediate retirement from Parliamentary life... Nothing less will satisfy the Nonconformist conscience. **1893** O. Wilde *Lady Windermere's Fan* iii. 90 There is nothing in the whole world so unbecoming to a woman as a Nonconformist conscience. **1894** Beerbohm in *Yellow Bk.* III. Oct. 250 The Nonconformist Conscience makes cowards of us all. **1931** *Times Lit. Suppl.* 10 Sept. 670/1 The difficulties of governing Ireland and keeping the Nonconformist conscience quiet.

4. A moth, *Lithophane lamda*, found in northern Europe, Russia, and North America.

1869 E. Newman *Illustr. Nat. Hist. Brit. Moths* 428/2 The Nonconformist... Their colour is bluish-gray in some specimens, prettily varied with darker and lighter gray in others. **1961** Edelsten & Fletcher *South's Moths Brit. Is.* (ed. 4) 228 The Nonconformist (*Lithophane lamda*, Fab.).

non-conju·nction. [Non- 1.] **1.** *Cytology.* The failure of homologous chromosomes to pair at meiosis. *rare.*

1925 J. Belling in *Jrnl. Genetics* XV. 256 Non-conjunction. This term is used instead of non-conjugation, because in most cases it cannot be told whether conjugation, that is, parasynapsis, has taken place or not, with regard to the chromosome pair in question... Non-conjunction cannot be distinguished from true non-disjunction, in the later stages.

2. *Logic.* The relation of the terms in a proposition asserting the negative of a conjunctive proposition ('not both..and..').

1926 H. M. Sheffer in *Isis* VIII. 229 The authors consider one of the cardinal improvements in the new edition to consist in their substitution, for the two operations of propositional *negation* and *disjunction*, of the single operation of *non-conjunction*. **1956** A. Church *Introd. Math. Logic* (rev. ed.) I. ii. 134 Sheffer uses the sign of disjunction, V, inverted as a sign of non-disjunction; he introduces non-conjunction only in a footnote and uses no special sign for it. *Ibid.* 135 Taking non-conjunction as the only primitive connective, give definitions of the singulary and remaining binary connectives. **1957** *Encycl. Brit.* XIV. 306/1 We shall use..the sign | to denote non-conjunction ('$p|q$' to mean 'not both p and q'). *Ibid.*, If the sign of non-conjunction (Sheffer's stroke) is taken as primitive, all the other connectives can be defined from this one. **1965** Hughes & Londey *Elem. Formal Logic* viii. 54 The following are valid: (i) $(p \uparrow q) \equiv \sim (p.q)$, (ii) $(p \downarrow q) \equiv \sim (p \vee q)$, which suggests the names 'non-conjunction' and 'non-disjunction' for '↑' and '↓' respectively.

non-consumption. (Later example.)

1961 L. E. Davis et al. *Amer. Econ. Hist.* iii. 54 The net..addition to a nation's capital stock..must be the result of nonconsumption of national product.

non-conti·ngent, *a.* [f. Non- 3 + Contingent *a.* 4.] That does not happen by chance or depend on a variable.

1890 J. Rankine *Erskine's Princ. Law Scotl.* (ed. 18) iii. vii. 405 The running of prescription is suspended during the minority of the person who is *verus dominus*, being lawfully vested with a non-contingent title to compete for possession, or of him who is the true creditor. **1939** Kurtz & Edgerton *Statistical Dict. Terms & Symbols* 99 *Mathematically independent*, unaffected by the happening of another event or by the size of another variable. Also called independent, non-contingent, and uncontingent. **1971** *Jrnl. General Psychol.* Apr. 326 *Ss* were given either 30%, 50%, or 70% noncontingent positive *E*-administered reinforcement. **1973** *Jrnl. Genetic Psychol.* Sept. 16 The final design contained two levels of noncontingent preference value within subjects.

non-contri·butory, *a.* [Non- 3.] Of a pension, a pension-scheme, or the like: not involving contributions from the pensioner or beneficiary.

1911 *Q. Rev.* July 198 The provision for old age in our system is non-contributory and wholly paid by the State. **1935** *Planning* II. xlii. 5 The first of these is the non-contributory State pension of 10s a week available at 70 subject to a declaration of means. **1975** *Times* 1 Dec. 12/8 The question of moving from a non-contributory to a contributory [pensions] scheme has been reopened in Whitehall.

non-co-opera·tion. [Non- 1.] Failure or refusal to co-operate; *spec.* used as a means of resistance or protest, esp. against the British in India before 1947. Also *attrib.*

1795 [in Dict. s.v. Non- 1]. **1860** Ruskin *Unto this Last* (1862) iv. 141 The most accurately nugatory labour is..that of which not enough is given to answer a purpose effectually... Also, labour which fails of effect through non-co-operation. **1920** A. Besant (*title*) Gandhian non-co-operation; or, shall India commit suicide? **1922** *Telegraphic Corr. India* 3 in *Parl. Papers* (Cmd. 1586) XVI. 578 The origins of the non-co-operation movement. **1937** J. M. Murry *Necessity of Pacifism* vi. 91 The only possible creative issue for the German Socialist movement was by way of non-violent non-co-operation. **1946** J. S. Huxley *Unesco* i. 7 The non-cooperation or even withdrawal of a number of nations. **1973** C. Mullard *Black Britain* iv. xiii. 156 This resistance has been passive in the tradition of civil disobedience and of Martin Luther King's philosophy of nonco-operation.

Hence **non-co-o·perate** *v. intr.*, to refuse to co-operate; **non-co-o·perating** *ppl. a.*, **non-co-o·perative** *a.*, that refuses or fails to co-operate; **non-co-o·perator,** one who practises or advocates non-co-operation.

1921 *Daily Tel.* 28 Sept. 9/1 The leaflet..asks the Moplahs actively to non-co-operate with the Government. **1922** J. T. Gwynn *Let.* 17 May in *Indian Politics* (1924) iii. 17 A Non-Co-operating politician. *Ibid.*, Non-Co-operative propaganda. *Ibid.* 18 The first-raters may be looked for either among the Co-operators or among the practising Non-Co-operators. **1928** *Observer* 19 Feb. 17/3 The non-co-operators expected to win in the division. **1958** *Punch* 29 Jan. 187/2 The water-fowl are non-co-operative, bobbing about indistinguishably on the far side of the water.

non-cro·ssover. *Biol.* [Non- 2, 4.] A gamete or individual which does not exhibit the results of crossing-over between any two genetic loci. Also *attrib.*

1916, 1919 [see *cross-over 4]. **1962** *Lancet* 12 May 1026/1 In the three families there were six children who could be considered non-cross-overs. **1971** Levitan & Montagu *Textbk. Human Genetics* x. 379 To whom the *AB/ab* parent contributed non-crossover gametes. *Ibid.*, The former could identify a noncrossover gamete of the *AB/ab* parent.

noncurantist, *a.* Add: (Further example.) Also **noncurance,** indifference; **noncurant** *a.*, indifferent.

1904 F. Rolfe *Hadrian the Seventh* iii. 84 He began to undress..with the noncurance of one accustomed to swim in Sandford Lasher. *a***1913** —— *Desire & Pursuit of Whole* (1934) viii. 78 The most elegant young stranger.., exquisitely noncurant in carriage. *a***1913** —— *Nicholas Crabbe* (1958) iii. 28 In the matter of money, he was extravagant, chimerical, quixotic, noncurant, to a degree. *Ibid.* xxiii. 158 Some unknown boy, noncurant, saucy, shrilly whistling. **1928** *Times Lit. Suppl.* 12 July 510/4 He contrasts the Italian measures against manifestations in favour of Enosis (union with Greece) with the completely noncurantist tolerance of them by the British authorities in Cyprus. **1931** *Ibid.* 5 Nov. 870/3 Some of the [Masonic] symbols and ceremonies discarded in the days of uninquiring ignorance and self-satisfied noncurance.

non-destru·ctive, *a.* [Non- 3.] That does not involve destruction, esp. of an object or material that is tested.

1929 *Trans. Amer. Soc. Steel Testing* XVI. 771 (*heading*) Nondestructive testing. *Ibid.*, A number of independent investigators have been working on various methods of nondestructive inspection. **1930** *Mining & Metallurgy* XI. 308 (*heading*) Destructive and non-destructive tests of welds. **1946** *Nature* 19 Oct. 539/2 It was pointed out that the X-ray method is the only non-destructive way of determining surface residual stresses. **1959** *Jrnl. Iron & Steel Inst.* CXCI. 299/2 The economic importance of non-destructive testing is emphasized. **1962** *Gloss. Terms Automatic Data Processing* (*B.S.I.*) 63 *Non-destructive reading*, a reading process which does not change the record of the data which has been read. **1962** *B.S.I. News* June 12/1 Points covered in the section on welding are the quality of welds, tests on welds..and non-destructive examination. **1971** *Engineering* Apr. 37/1 Facilities at Harwell have become available to industry through the Nondestructive Testing Centre. **1975** G. Bram *Manufacturing Technol.* ii. 66 Research and development of methods of non-destructive testing has increased considerably over the past thirty years.

Hence **non-destru·ctively** *adv.*

1930 Stoughton & Butts *Engin. Metallurgy* (ed. 2) iii. 54 (*heading*) Tests sometimes made non-destructively. **1945** A. N. Sachanen *Chem. Constituents of Petroleum* iii. 130 Petroleum fractions are completely and non-destructively hydrogenated so that the ring structure of the naphthenes formed corresponds to the original ring structure. **1969** P. B. Jordain *Condensed Computer Encycl.* 336 Magnetic tapes, disks, drums, and cards are usually read nondestructively. **1971** *Engineering* Apr. 33/1 To be able to examine critically the insides of welds or castings non-destructively has improved fabrication technology.

non-dire·ctional, *a.* (*sb.*) [Non- 3.] Lacking directional properties; *esp.* equally sensitive, intense, or the like, in all directions. † Also as *sb.*

1903 *Nature* 22 Oct. 610/1 In Prof. Henrici's algebra the products of two vectors α, β are:— ($\alpha\beta$) a non-directional or 'scalar',..and [$\alpha\beta$] a vector perpendicular to the plane drawn through α and β. **1931** *Proc. IRE* XIX. 1754 When the noise comes from more than one direction the effects produced by sunrise or sunset upon the noise intensity as received on a nondirectional antenna become somewhat involved. **1946** H. Jacob *On Choice of Common Language* iii. 107 In addition to the primarily spatial directives, *for, of,* and *till* are used on the same analogy, as non-directional prepositions. **1964** *Oceanogr. & Marine Biol.* II. 448 The usual response of untrained fish in aquaria to non-directional sounds is one of non-orientated quickened swimming, often momentary. **1967** E. Short *Embroidery & Fabric Collage* iii. 59 This non-directional type of design is important in an article which may be carelessly thrown down.

non-disju·nction. [Non- 1.] **1.** *Cytology.* The failure of one or more pairs of homologous chromosomes (at meiosis) or sister chromatids (at mitosis) to separate and move away normally from the equatorial plate during nuclear division, usu. with the result that one of the daughter nuclei has too few chromosomes and the other too many.

1913 *Jrnl. Exper. Zoöl.* XV. 587 (*heading*) Non-disjunction of the sex chromosomes of Drosophila. **1925** *Jrnl. Genetics* XV. 251 Non-disjunction was long ago observed in the diploid *Oenothera*. **1961** *Lancet* 5 Aug. 319/2 The mosaicism reported here could have originated in several ways. Mitotic nondisjunction in a normal diploid embryo could result in 1 cell with forty-eight chromosomes (trisomy for chromosome 21 and chromosome 19 or 20) and 1 non-viable cell with forty-four chromosomes. **1962** *Ibid.* 15 Dec. 1270/1 Three types of meiotic non-disjunction have been recognised. **1971** Levitan & Montagu *Textbk. Human Genetics* iii. 63 Changes in the total number of chromosomes most frequently result from the occasional failure of chromosomes to move to opposite poles or disjoin during anaphase of cell division. Such a failure is called non-disjunction.

Hence **non-disju·nctional** *a.*

1913 *Jrnl. Exper. Zoöl.* XV. 589, 10 per cent of the eggs of such a female matured in a non-disjunctional manner. *Ibid.* 590, I counted the offspring of the non-disjunctional male only with respect to white and pink. **1930** *Genetics* XV. 11 These seven non-disjunctional males were red. **1971** Levitan & Montagu *Textbk. Human Genetics* iii. 66 Nondisjunctional error in meiosis could involve *all* 23 pairs of chromosomes.

2. *Logic.* The relation of the terms in a proposition asserting the negative of a disjunctive proposition ('neither..nor..').

1956, 1965 [see *non-conjunction 2].

non-disti·nctive, *a.* *Linguistics.* [Non- 3.] Not distinctive (see *distinctive *a.* 1 b).

1916 Jones & Plaatje *Sechuana Reader* p. xiv, The consonant sounds c, ɫ, ɥ, and the vowel sound ʉ are probably 'non-distinctive'... By this we mean that the substitution of the sounds t, l, w, u respectively..would probably never change the meaning of any word. **1933, 1942** [see *distinctive *a.* 1 b]. **1964** L. S. Hultzén in D. Abercrombie et al. *Daniel Jones* 92 There is of course a great deal of grammatically non-distinctive variation in utterance.

none, *pron.*, *a.*, and *adv.* Add: **C.** *adv.* **3. a.** Freq. with *too* and an adjective forming combs. used *attrib.*

1928 H. Crane *Let.* 31 Jan. (1965) 315 Crowds of ambitious but none-too-successful strumpets of movie-dom. **1941** *Amer. Speech* XVI. 57/2 This none-too-accurate article on the DAE. **1963** D. Ballantyne in C. K. Stead *N.Z. Short Stories* (1966) 151 His none-too-searching questions had already disclosed that the psychologist..was sophisticated enough to take an irreverent view of Freud. **1970** 'M. Hebden' *Mask of Violence* (1971) i. 3 He was dressed in shabby jeans and a none-too-clean shirt. **1974** *Country Life* 26 Dec. 2016/2 Its class of none-too-hardy evergreens.

b. (Later examples.)

1852 J. B. Jones *Col. Vanderbomb* xv. 198 Our adventurers slept none that night. **1890** J. Service *Thir Notandums* i. 3, I would weary nane. **1906** *Advocate of Peace* Mar. 52 Has civilization advanced none from the barbaric days of the 5th century? **1956** B. Holiday *Lady sings Blues* (1973) xviii. 151, I had never cared what the hell people thought, and jail hadn't changed that none. **1973** E. Bullins *Theme is Blackness* 65 Now, now, don't you worry none about that, Mother. We'll find a way.

non-ele·ctive, *a.* [Non- 3.] Not appointed by election.

1909 W. S. Churchill in *Westm. Gaz.* 9 Oct. 9/4 The claim of the House of Lords—that is, the claim of a non-elective and unrepresentative Chamber—to make and to unmake Governments. **1910** *Blackw. Mag.* Aug. 283/1 The independent opinion of men formed in the comparatively 'dry light' of the non-elective chamber has a value of its own.

non-e·mpty, *a.* *Math.* and *Logic.* [Non- 3.] Not empty; having at least one member or element.

1937 *Mind* XLVI. 375 There is another non-empty sub-set..of sentences of C_1 which are nonsensical but not α-nonsensical. **1956** E. M. Patterson *Topology* vi. 113 A complex *K* is said to be connected if it is not the union of two non-empty subcomplexes which have no simplexes in common. **1957** P. Suppes *Introd. Logic* 198 At least one of the three small regions is non-empty. **1965** P. Caws *Philos. of Sci.* xvii. 130 The restriction must be added that the domain be non-empty. **1971** G. Glauberman in Powell & Higman *Finite Simple Groups* i. 54 Let $\mathscr{A}$ be a non-empty set of subgroups of *S*.

Hence **non-e·mptiness,** the property of being non-empty.

1937 A. Smeaton tr. *Carnap's Logical Syntax of Lang.* 261 We introduce conditions which require for symmetrical, reflexive, and transitive relations the property of non-emptiness. **1950** W. V. Quine *Methods of Logic* (1952) 80 The bar..can be made to lie across a boundary and thus indicate non-emptiness of a compound region. **1957** P. Suppes *Introd. Logic* ix. 198 Another kind of symbol is needed for non-emptiness. **1965** Hughes & Londey *Elem. Formal Logic* xlvi. 339 Each schema..is valid under the hypothesis of the non-emptiness of some one term.

none·ntitize, *v.* [f. Nonentity + -ize.] *trans.* To make into a nonentity.

1903 G. B. Shaw *Man & Superman* i. 21 We're beaten——smashed—nonentitized, like her mother. **1913** R. W. Service *Rhymes of Rolling Stone* 176 Sober am I nonentitized; drunk am I more than half a god.

non-entitous, *a.* (Further example.) Also **non-entitious.**

1951 W. Sansom *Face of Innocence* (1954) 77 The girls nubile and the men nonentitous. **1954** C. E. M. Joad *Folly Farm* vi. 143 Henry, the husband, a decent civil servant, utterly non-entitious—another character without a face..without a body of any kind. **1968** *Listener* 15 Aug. 212/3 Much of the rebuilding is elephantine and nonentitious.

non-entry. 2. (Later example.)

1962 H. R. Loyn *Anglo-Saxon Eng.* ix. 359 A factor that determined entry or non-entry in Domesday Book.

none-so-pretty. Add: **1.** (Later examples.) Also as *adj.*

1771 in A. M. Earle *Costume Col. Times* (1894) 173 None-so-Pretty Tapes. **1804** W. Clark in *Orig. Jrnls. of Lewis & Clark Exped.* (1905) VI. 271 For Indians Presents..10 pieces Nonsoprettys. **1969** R. T. Wilcox *Dict. Costume* 248/1 *None-so-pretties*, fancy, decorative tapes used for trimming garments in the eighteenth century in the American Colonies.

non est. (Further examples.)

1945 Mrs. Belloc Lowndes *Let.* 21 July (1971) 261 Eggs are practically *non est*, unless one has hens. **1959** R. Fuller *Ruined Boys* i. v. 34 The weights you put on it at the end of the arm thing were *non est*.

nonesuch, *sb.* and *a.* Add: **1. c.** *Nonesuch* (or *Nonsuch*) *chest*: a type of wooden chest made in the 16th and 17th centuries, inlaid with stylized designs supposedly representing views of Nonsuch Palace (see 1 b in Dict.); also *Nonesuch ornament.*

1905 F. S. Robinson *Eng. Furnit.* v. 71 A somewhat rare type of chest..called a 'Nonesuch' chest. **1920** *Connoisseur* Dec. 205 The 'Nonsuch chest'..possesses the flat surface unrelieved by mouldings, which is characteristic of many of the earlier inlaid pieces. **1960** H. Hayward *Antique Coll.* 200/1 *Nonesuch chest*, chest, dating from the second half of the 16th cent., decorated with an inlaid design representing an ideal building resembling the Palace of Nonesuch. **1970** D. Ash *Dict. Eng. Antique Furnit.* 107/2 Ornament of this kind probably originated in Italy, but quite possibly reached England from Germany without any connection with the palace from which its name is derived. Some pieces of furniture bearing Nonsuch ornament and long regarded as English may well have been imported. **1973** *Country Life* 7 June 1690 English oak chest, with..bands of stylised zig-zag ornament reminiscent of so-called Nonesuch chests... Circa 1630.

2. b. (Later U.S. examples.)

1821 A. Royall *Lett. from Alabama* (1830) li. 123, I went to hear this none such. **1927** 'S. Rohmer' *Moon of Madness* 18 He was a poisonously handsome none-such. *Ibid.* 26 Da Cunha danced perfectly, with all the sensuous grace of a none-such.

nonet. Add: **2.** *Nuclear Physics.* A multiplet (sense *b) of nine sub-atomic particles.

1963 S. Okubo in *Physics Lett.* V. 166/2 We have to take into account of [*sic*] the ninth boson ϕ together with the octet. This means that we are dealing with a nonet rather than an octet, and a singlet. **1970** *New Scientist* 29 Oct. 211/3 To reconcile the A2 meson with current theory, it is necessary to show that the other members of its SU(3) nonet display similar double structure.

nonetheless (nʌnðəle·s, nɒ·nðəles), *adv.* Also none the less, none-the-less. [See None *adv.* 1 b, Less *adv.* 1 a.] Nevertheless.

1847 Dickens *Dombey* (1848) xlii. 419, I thank you none the less. **1875** [see None *adv.* 1 b]. **1924** G. B. Shaw *St. Joan* v. 66 But you will be none the less alone: they cannot save you. **1930** *Principal's Rep., Armstrong College (Newcastle)* 1929–30 4 The elevation of Sir William Noble..to the peerage..has been a matter of pride and pleasure..which nonetheless has been qualified subsequently. **1955** *Times* 13 July 7/5 Later than *Nabucco*.., it is, nonetheless, not such a good opera. **1972** *Nature* 8 Dec. 327/1 Although some research fields seem to grow from a single, germinal experiment, others take root in accidents set off by workers stumbling from one preserve to another—and are none-the-less interesting for that.

non-Europe·an, *a.* and *sb.* [Non- 3, 2.] **A.** *adj.* Not European.

1907 W. James *Pragmatism* v. 182 At this stage of philosophy all non-European men without exception have remained. **1923** D. H. Lawrence *Birds, Beasts & Flowers* 31 A set, stoic endurance, non-European. **1967** *Guardian* 15 May 6/6 Another such Russian idea..is that all 'non-European' nuclear weapons should be banned from European soil and seas.

B. *sb.* A person who is not European; esp. in South Africa: a non-white.

1926 M. Nathan *South Africa from Within* viii. 166 According to this [*sc.* a 1921 census in British S. Africa], there were 1,519,488 Europeans.., and 5,409,092 non-Europeans. **1935** A. M. Carr-Saunders in Huxley & Haddon *We Europeans* viii. 242 In the Union of South Africa there are under 2 million Europeans and about 6 million non-Europeans. **1939** J. S. Marais *Cape Coloured People* p. vii, This book is..a study..of the relations between the Europeans and that group of non-Europeans with whom the former have been longest..in contact. **1958** *Ann. Reg. 1957* 95 The so-called 'open' universities, Cape Town and the Witwatersrand, had a long tradition of admitting non-Europeans. **1970** *Cape Times* 28 Oct. 18/2 (Advt.), Personal loans, for Europeans and non-Europeans.

non-eva·luative, *a.* [Non- 3.] Of or pertaining to that which does not evaluate but is concerned only with fact.

1937 *Jrnl. Philos.* XXXIV. 237 It is a basic rule of valuational logic that an evaluative conclusion can not be deduced from non-evaluative, that is to say factual, premises. **1958** W. Stark *Sociol. of Knowl.* i. iv. 152 'Knowledge' in the context of gnosio-sociology is a non-evaluative term. **1973** 'J. Patrick' *Glasgow Gang Observed* xxi. 219 Adolescents..did not know how to react to the non-evaluative, non-judgmental approach of the street worker.

non-eve·nt. [*Non- 2 b.] An unimportant or unexciting event, *spec.* one which was expected or intended to be important; occas., something that did not happen.

1962 *Spectator* 30 Nov. 853/1 The chief non-event of Gilmour's campaign was also, he says, the product of the 'serious' press. **1963** I. Gilmour in *Hansard Commons* 7 May 272 Drugged by their normal diet of non-news stories and non-events, the newspapers tend to lose their heads when..faced with..a news story. **1965** *Listener* 30 Sept. 480/2 For all its admirable achievements in science and technology and the arts, the Soviet Union is still a great place for the non-event and the non-person, a category in which Mr. Khrushchev is at present languishing. **1966** K. Amis *Anti-Death League* 186 To find this view supported by events, or as now by non-events, was depressing. **1967** *Listener* 23 Nov. 679/2 Here was a non-event to compare with those other annual and totally insignificant English non-events—Ascot, the Boat Race and the Queen's Speech. **1972** *Guardian* 22 Jan. 15/3 The exchange of signatures is a non-event..it is neither truly historic nor of much practical significance.

non-existence. 1. (Later examples.)

1854 K. H. Digby *Compitum* VII. iii. 255 Books of dogmatic scepticism, and expositions of the non-existence of virtue and honour. **1968** P. A. P. Moran *Introd. Probability Theory* vi. 269 The fact that this is not analytic at t = 0 corresponds to the non-existence of the moments in this case.

non-fe·rrous, *a.* [Non- 3.] Containing no iron; of or pertaining to metals other than iron.

1887 [see *ferrous *a.* 2]. **1909** *Westm. Gaz.* 4 Sept. 4/3 The non-ferrous metal trades. **1936** *Times Educ. Suppl.* 13 June p. iv/1 A non-ferrous survey ship..is being fitted out and will probably start work in the Southern Indian Ocean. **1940** Laing & Rolfe (*title*) Non-ferrous foundry practice. **1963** H. R. Clauser *Encycl. Engin. Materials & Processes* 120/1 Nonferrous centrifugal castings are produced from copper alloys, nickel alloys, and tin and ead-base bearing metals. **1971** *Nature* 23 July 215/1 Britain's imports of non-ferrous metals cost £600 million annually.

non-fi·ction. [Non- 2.] Prose writings other than fiction (see Fiction 4). Also *attrib.* or as *adj.*, esp. in **non-fiction novel,** a novel written about real situations or characters.

1909 *Westm. Gaz.* 2 June 5/2 In Capetown the percentage of non-fiction to the total number of volumes is 58. **1922** Holliday & Van Rensselaer *Business of Writing* 174 Successful works of non-fiction not infrequently are later issued in cheaper forms by the original publisher. **1930** *Times Educ. Suppl.* 31 May 248/3 One of the most pleasing features was an increase of 344 issues of non-fiction literature. **1951** L. Z. Hobson *Celebrity* (1953) xvi. 255 In this bad slump, nonfiction's the only thing selling—apart from one or two novels a year. **1965** *Vogue* 15 Oct. 94 [Truman] Capote is an experimenter, an adventurer. His newest experiment is *In Cold Blood*, a unique book, for it is the first non-fiction novel, a precise documentary, in many ways brilliantly composed. **1967** *New Mexico Q.* Autumn 243 (*heading*) The 'non-fiction' novel. **1969** *Times* 30 Oct. 10/7 Non-fiction paperbacks.

Hence **non-fi·ctional** *a.*, of, pertaining to, or characteristic of non-fiction.

1903 *Library World* Mar. 227 Mr. Doubleday's allusions to the expedients that have been adopted to advertise the non-fictional wares remind me of another point. **1911** A. Bennett *Let.* 28 July (1966) I. 159 This clause would prevent me from publishing anything whatever non-fictional until they issued the book. **1938** D. Baker *Young Man with Horn* ii. i. 92 The author's right, as author, to furnish himself, as hero, with everything he lacked in his non-fictional life. **1966** G. N. Leech *Eng. in Advertising* v. 54 Credibility is a dominant consideration in both fictional and non-fictional indirect address. **1967** *Britannica Bk. of Year 1966* 803/3 *Nonfictional novel*, a completely factual narrative characterized by the use of fictional techniques; *nonfictional novelist.*

non-fi·gurative, *a.* [Non- 3.] Not figurative; *spec.* in art = *abstract A. 4 d. So **non-figura·tion.**

1927 O. Jespersen *Mod. Eng. Gram.* III. xiv. 294 *Play* in the non-figurative sense has often *to*: he played Beethoven to us. **1934** *Burlington Mag.* Aug. 94/1 Cubism or non-figurative art is dismissed as an ephemeral movement. **1955** P. Heron *Changing Forms of Art* iii. 40 Until painting became non-figurative, the spatial configuration which the painter registered upon his canvas consisted of forms that could be read as illusionistic references to real objects; objects, that is, that were external to the picture. *Ibid.* 41 In 1945 pure non-figuration began to predominate in Paris. **1961** *Guardian* 16 Nov. 7/4 Abstract pattern deriving from Islamic ornament..to transcend..Western non-figuration. **1962** *Listener* 19 July 94/1 The non-figurative painter avoids stereotypes by means of marks and shapes, the interpretation of which is arbitrary, since the truly non-figurative is infinitely ambiguous.

non-flam, *a.* [Non- 3.] That is not inflammable. Hence as *sb.*, something that is not inflammable. So **non-fla·mmable** *a.*

1906 *Chambers's Jrnl.* 24 Nov. 831/2 A British firm, the patentees of 'Non-Flam' goods, of Aytoun Street, Manchester, points out that however efficacious Professor Doremus's prescription may be, it will need to be applied to the curtains every time they are washed. **1908** *Daily Chron.* 20 Oct. 3/6 It would be a good thing if the Government would..prohibit the use of any but 'non-flam' or other permanently incombustible flannelette for infants' garments. **1909** *Ibid.* 27 Feb. 4/7 'Non-Flam'..will not flare when brought into contact with fire. **1927** *Daily Tel.* 25 Oct. 14 In France from Jan. 1 next no celluloid film may be used unless made from this so-called 'non-flam' film. **1961** *Daily Mail* 20 July 9/4 A resolution urging the association to do more to promote the use of non-flammable clothing was accepted. **1968** *Punch* 17 Jan. 77/2 (*caption*) You've sure got a point there, Senator—non-flam draft cards. **1970** N. Armstrong et al. *First on Moon* iii. 65 It was made of foam and covered with nonflammable tape. **1975** L. Deighton *Yesterday's Spy* xxx. 221 Non-flam helium would not have burned.

non-fraterniza·tion. [Non- 1.] An absence of fraternization (see *fraternization b). Also abbrev. (*slang*) **non-frat.** Also *attrib.*

1945 F. Gillard in Hawkins & Boyd *War Report* (1946) xx. 383 The Germans don't like our non-fraternization rule, and they don't like the way in which our troops are obeying these rules. *Ibid.* 384 The attitude of these people makes our non-fraternization policy more necessary than ever. **1945** Non-fraternisation [see *fraternization b]. **1945** Non-frat [see *fratting *vbl. sb.*]. **1957** *Wiener Beiträge* LXV. 280 Soldiers abbreviated the long words *fraternisation* and *non-fraternisation* to *frat* and *non-frat.*

nong (nɒŋ). *Austral. slang.* Also **nong-nong.** [Origin unknown.] A fool; a stupid person. Cf. *ning-nong.

1953 Baker *Australia Speaks* vii. 171 *Nong*, a simpleton or fool. **1959** —— *Drum* viii. 61 'Nong' (or the duplicated form 'nong-nong') was equivalent not only to dill, but also to drube, dope, and drongo. **1961** *Sunday Mail Mag.* (Brisbane) 5 Feb. 5/1 Nicholson and Fenton were not complete nongnongs. **1970** M. Kelly *Spinifex* vii. 113 Todd whirled, rifle at hip, 'You nong!' 'That's what they call me,' Reid commented affably. **1972** *Telegraph* (Brisbane) 5 Dec. 3/4 Mrs. Margaret Whitlam today emphatically denied that she had referred on Saturday night to journalists as 'nongs'.

nongenary (nɒndʒī·nări). [f. L. *nongēnārius* containing nine hundred, after Centenary *sb.* 3.] A nine hundredth anniversary, or the celebrations connected with it.

1926 Fowler *Mod. Eng. Usage* 72/2 A list of not intolerable forms is here offered... Centē'nary,..octingē'nary, nongē'nary, millē'nary. **1965** E. Gowers *Fowler's Mod. Eng. Usage* (ed. 2) 83/1 Nor is anyone, however passionate a supporter of etymological correctness, likely to try to introduce *sescenary, septingenary, octingenary,* or *nongenary*. **1966** *New Statesman* 8 Apr. 511/2 The painters are in no way concerned with the nongenary of the Battle of Hastings.

non-grea·sy, *a.* [Non- 3.] Not greasy.

1907 *Yesterday's Shopping* (1969) 537/1 Violet Oatmeal Skin Cream... Non-greasy and without stickiness. **1937** *Discovery* Oct. 297 An offset—a print made in a non-greasy powdered chalk. **1967** E. Chambers *Photo-litho-Offset* xvi. 242 Gravure inks are difficult in make-up, being very fluid, finely ground, non-greasy and drying mainly by evaporation. **1974** *Harpers & Queen* Sept. 112 Some foundations are non-greasy, ideal for oily skins.

no·n-hero. [*Non- 2 b.] One who is the opposite or the reverse of a hero; one who is not genuinely a hero. Cf. *anti-hero.

1940 *Sun* (Baltimore) 7 Sept. 10/7 In my list of non-heroes, Carol of Rumania stands close to the top. **1948** E. E. Cummings *Let.* 27 Jan. (1969) 181 Since our (as he's doubtless already boasted to you) non-hero was once a candidate for honours. **1959** *Times Lit. Suppl.* 13 Nov. 657/1 These two novels..take lovable, weak, defeated men as non-heroes. **1972** *Sat. Rev.* 17 June 52/3 Our existential nonhero wanders through the story.

nonic (nōᵘ·nik), *sb.* and *a.* *Math.* [f. *non(a- + -ic.] **A.** *sb.* A nonic curve or equation. **B.** *adj.* Of the ninth order or degree.

1879 *Amer. Jrnl. Math.* II. 233 (*heading*) Nonic. **1894** *Phil. Trans. R. Soc.* A. CLXXXV. 103 The nonic was proved..to have only three real roots. **1958** *New Biol.* XXV. 16 By an elementary transformation I have thrown his rather formidable nonic equation into a form which allows numerical tabulation.

nonillion. (Earlier example.)

1690 [see Billion 1].

non-intervention. Add: (Further examples.) Hence **non-interve·ne** *v. intr.*, not to intervene; **non-interve·ner,** one who does not intervene or who advocates non-intervention; **non-interve·ning** *ppl. a.*

1849 S. P. Chase *Let.* 30 July in J. Schuckers *Life of Chase* (1874) xii. 102 For what will be the cost to the Democracy of the alliance with the slaveholders in a presidential campaign?.. It is non-intervention upon the subject of slavery. **1901** K. D. Best *Victories of Rome* (ed. 4) p. xii, The State..is exhibiting to a non-intervening circle of Catholic kingdoms the newest and latest development of Liberté, Egalité, Fraternité. **1937** H. Nicolson *Diary* 27 July (1966) 309 Anthony Eden..says that the French really did take the initiative in non-intervention [in Spain] and were not put up to it by us. **1937** A. P. Herbert in *Punch* 8 Sept. 275/1 The Non-Intervention Committee are meeting, The Non-Interveners are meeting again. *Ibid.* 275/2 So let every soldier go non-intervening—Oh, fly with me Southward and non-intervene! **1939** S. Spender *Poems for Spain* 10 Satire, like Edgell Rickword's on Non-Intervention, in which the fundamental ideas are turned against the politicians. **1942** 'A. Bridge' *Frontier Passage* iv. 53 Mr. Crumpaun..was playing bridge with the elderly clergyman and two Non-Interveners. **1959** *Ann. Reg. 1958* 51 A discreet non-intervention in the debate was observed by the Leader of the Party. **1969** *Guardian* 29 Aug. 8/4 Mr Chamberlain..threatened to..'non-intervene' on Hitler's side against Czechoslovakia, as they were doing in favour of General Franco in Spain. **1975** *Times* 16 Oct. 15/2 The Soviet Union feels free to propagate its ideas in the west, so the west should not be shy about propagating its ideas in the east... The passages on non-intervention are aimed..at..physical intervention.

non-io·nic, *a.* and *sb.* [Non- 3.] **A.** *adj.* Not ionic; *spec.* (esp. of a detergent) not dissociating into ions in aqueous solution.

1943 *Chemical Industries* LII. 327/1 Non-ionic surface active agents, as their name implies, are not ionizable and owe their effectiveness to a proper balance between certain hydrophilic (polar) and lyophilic (non-polar) groups in their molecules. **1959** *Spectator* 4 Sept. 294/1 Together with the nonionic 'Nonidet' products, they have scores of applications throughout industry both for cleaning and maintenance. **1963** A. J. Hall *Textile Sci.* vi. 296 Surface active agents or surfactants can be usefully classified into three main groups—anion-active, non-ionic, and cation-active. **1964** M. Hynes *Med. Bacteriol.* (ed. 8) vii. 90 It depends on such forces as attraction between polar non-ionic groups..and van der Waal[s] attraction. **1965** Phillips & Williams *Inorg. Chem.* I. v. 171 The difference between the equilibrium lattice energies U and U_i is seen to be approximately equal to the additional non-ionic binding energies.

B. *sb.* A non-ionic substance or detergent.

1952 *Mod. Sanit. & Build. Maint.* IV. 63/1 Considerable interest has been shown as to just what happens when iodine is dissolved in nonionics... It has been observed that nonionics render iodine soluble in aqueous solutions providing the amount of iodine does not exceed 25% based on the nonionic. **1972** *Materials & Technol.* V. x. 308 In several of the different classes of these non-ionics, it is possible to obtain products ranging from completely oil-soluble to completely water-soluble.

non-iron, *a.* [Cf. *non- 5 b.] Of clothes, fabrics, etc.: that does not require ironing after being washed.

1957 *Woman* 16 Nov. 25/4 Cottons and rayons with a special finish—that is ones that have been made crease-resistant, drip-dry or non-iron cannot be dyed at home. **1971** *Vogue* 15 Sept. 43/1 Thick soft American towels, non-iron sheets. **1975** *Times* 27 Sept. 16/1 Easily washable and non-iron fabrics.

nonlea·ded, *ppl. a.* [f. Non- 6 + *leaded *ppl. a.* e.] Of petrol: containing no added tetraethyl lead to counteract 'knocking'.

1955 *Sci. News Let.* 22 Jan. 54/1 The best way to remove carbon monoxide from auto fumes appears to be the use of nonleaded gasoline and a catalytic converter, Dr. W. L. Faith, chief engineer of the Southern California Air Pollution Foundation, said. **1971** *McGraw-Hill Yearbk. Sci. & Technol.* 212 The two most desired types of gasoline components for nonleaded gasolines are highly branched paraffins and the common aromatic components such as benzene, toluene, and xylene. **1971** J. D. Roberts et al. *Org. Chem.* iii. 58 The manufacture of engines capable of running on nonleaded gasoline is also being undertaken.

non-li·near, *a.* [Non- 3.] Not linear, in sense 3 of the adj.; involving terms of an equation that are not of the first degree; involving or possessing the property that the magnitude of an effect or output is not linearly related to that of the cause or input.

1844 *Phil. Trans. R. Soc.* CXXXIV. 282 The integration of non-linear differential equations. **1905** *Drapers' Company Res. Mem. Biometric Ser.* II. 21 In the cases of non-linear regression..I have..had to deal with, I find that parabolæ of the 2nd or 3rd order will suffice as a rule to describe the deviation from linearity. **1925** F. C. Mills *Statistical Methods* xii. 432 (*heading*) Non-linear correlation. **1930** T. E. Shea *Transmission Networks & Wave Filters* ii. 43 Waves would tend to lose character in propagation, because of what is called non-linear distortion. **1949** O. G. Sutton *Sci. of Flight* ii. 43 The inertia terms in these equations contain both squares and products of the velocities and the equations are thus non-linear. **1968** Fox & Mayers *Computing Methods for Scientists & Engineers* i. 7 The Runge–Kutta method, or one of its variants, is a useful general-purpose routine for non-linear first-order equations. **1969** *Sci. Jrnl.* Apr. 53/3 Non-linear optics..is the study of the behaviour of materials subjected to light of such intensity as to change some of the parameters of the material. **1973** [see *linearize *v.*].

b. *Linguistics.* = *suprasegmental adj.

1947 C. F. Hockett in *Internat. Jrnl. Amer. Ling.* XIII. 258/2 Features of stress or tone, for example, which normally stretch over more than a single vowel or consonant, have been called non-linear or suprasegmental, in contrast to the linear or segmental vowels and consonants. **1947** —— in *Jrnl. Amer. Oriental Soc.* LXVII. 258/1 Since it is inconvenient to transcribe other than linearly, we derive the symbols for inclosure between solidi from the symbols defined above in such a way as to eliminate the need for non-linear notation.

Hence **non-li·nearly** *adv.*

1943 *Jrnl. Amer. Chem. Soc.* LXV. 381/2 The ratio π/c of osmotic pressure to concentration has been found..to vary non-linearly with concentration. **1964** R. F. Ficchi *Electrical Interference* iii. 24 Components should not be permitted to operate nonlinearly since this causes harmonics to generate. **1973** *Jrnl. Genetic Psychol.* CXXII. 325 Body proportions..vary nonlinearly with figure age.

non-linea·rity. [Non- 1.] The property of not being linear; *esp.* lack of proportionality between two related quantities (as input and output).

1929 *Physical Rev.* XXXIII. 633 (*heading*) Non-linearity of photoelectric response near long wave-length limit. **1937** J. Orr tr. *Iordan's Introd. Romance Linguistics* iv. 379 Non-linearity of structure is shown to be present when a sign 'cumulates' more than one function. **1948** *Proc. IRE* XXXVI. 37/1 Any transmission system containing vacuum tubes is more or less nonlinear, with the degree of nonlinearity increasing with increasing signal level. **1965** *Math. in Biol. & Med.* (Med. Res. Council) vi. 266 In some cases, such as an excessively branching. syncytial network, the gross non-linearities may be completely smoothed out and virtually linear polarizing current–voltage relations could then be obtained. **1967** *Oceanogr. & Marine Biol.* V. 30 Part of the non-linearity in the latter equations arises from the assumption that the bottom friction is proportional to the square of the bottom current. **1973** *Sci. Amer.* July 29/3 The presence of such nonlinearity in the control characteristics gives rise to additional frequencies at double, triple and quadruple the frequency of the basic oscillation.

non liquet. (Later example.)

1952 *Archivum Linguisticum* IV. ii. 99 Hofmann..has also returned a verdict of *non liquet* on the question of etymology.

non-li·terate, *a.* *Anthropol.* [Non- 3.] Denoting a person or culture that has no written language. Hence as *sb.*, a non-literate person.

1948 M. J. Herskovits *Man & his Works* ii. v. 75 The..form, *nonliterate*, simply describes the fact that these people do not have written languages... Nonliterate, because it is colorless, conveys its meaning unambiguously, and is readily applicable to the data it seeks to delimit, is thus to be preferred to all the other terms we have considered. **1956** R. Redfield *Peasant Society & Culture* i. 6 Among these non-literates there was no history to learn. **1958** A. R. Radcliffe-Brown *Method in Soc. Anthropol.* ii. iii. 155 The customs of ancient times might be better understood in the light of the resemblances they show to customs of non-literate peoples of later times. **1970** G. A. & A. G. Theodorson *Mod. Dict. Sociol.* 276 Sometimes the term *preliterate* is used as synonymous with nonliterate; however, preliterate is associated with certain theoretical assumptions..that are no longer accepted.

non-natural, *a.* and *sb.* **A.** *adj.* **1.** (Earlier example.)

1621 Burton *Anat. Mel.* i. ii. 86 Necessary..are those six non-naturall things, so much spoken of amongst Physitians.

non-naturalism. Add: **2.** *Philos.* A theory of ethics which opposes naturalism; intuitionalism in ethics.

1939 *Mind* XLVIII. 464 The practitioners of a certain kind of ethical theory, which is dominant in England and capably represented in America, and which is variously called objectivism, non-naturalism, or intuitionism, have frequently charged their opponents with committing the naturalistic fallacy.

non-naturalist. Add: **b.** *Philos.* Of or pertaining to non-naturalism.

1966 *Amer. Philos. Q.* III. 299/2 Such a derivation, if valid, might be..a counter-example to the non-naturalist thesis.

B. *sb.* [Non- 2.] *Philos.* An adherent of non-naturalism.

1939 *Mind* XLVIII. 5 It is not indeed the only sense of good which the non-naturalist will hold incapable of reduction to naturalistic terms. **1966** *Amer. Philos. Q.* III. 305/2 Both non-naturalists and their critics seem to be seriously mistaken.

Hence **non-naturali·stic** *a.*; **non-naturali·stically** *adv.*

1902 W. James *Var. Relig. Exper.* vi. 143 Brahmans, Buddhists, Christians, Mohammedans, twice-born people whose religion is non-naturalistic. **1939** W. D. Ross *Foundations of Ethics* i. 6 If you define 'good' as meaning 'such that it *ought* to be desired', you are putting forward a non-naturalistic definition. Consequence theories also may be either naturalistic or non-naturalistic. **1940** *Mind* XLIX. 228 In the Introduction Ross divides attempted definitions of ethical terms into 'attitude-theories' and 'consequence-theories', and then sub-divides each of these into a naturalistic and a non-naturalistic sub-class. **1952** *Mind* LXI. 544 Any moral system is taken to be grounded non-naturalistically which is deduced from principles held to be certain and inviolable.

non-net, *a.* [Non- 3.] Of a book: not subject to the normal conditions of sale of net books.

1939 in F. D. Sanders *Brit. Bk. Trade Organisation* 79 The sale of books published at non-net prices is not covered by the terms of the Net Book Agreement. **1951** B. N. Langdon-Davies *Practice of Bookselling* ii. 21 In respect of non-net books the only condition is that the bookseller may not sell them at less than the price he gave for them. **1963** 'R. Findlater' *What are Writers Worth?* 11 'Non-net' books are mainly educational ones, sold in bulk for use in schools by contractors rather than through the bookshops. **1972** *Bookseller* 2 Dec. 2545/2 Many of them [*sc.* textbooks] will be non-net, and not have prices printed on them.

non-Newto·nian, *a.* [Non- 3.] Not Newtonian, in sense 1 of the adj.; used esp. in connection with the flow of fluids (see quot. 1937).

1913 *Phil. Mag.* XXV. 157 The Einstein transformation equations, and the other principles of non-Newtonian mechanics. **1937** *Rep. Progress Physics* III. 22 It is rather with impure liquids, semi-liquids and semi-solids, that we have to deal in every-day life. Of these a large proportion may be described as 'non-Newtonian', i.e. the rate of shear is not proportional to the shearing stress, and they do not obey Poiseuille's or Stokes's law. **1943** *Amer. Speech* XVIII. 220 We are..on the threshold, he says, of a third stage: the 'non-aristotelian', non-Euclidean, non-Newtonian stage. The complete revolution Einstein has produced in physics is only part of a general revolution that must eventually affect all human knowledge. **1946** *Nature* 17 Aug. 245/1 The measurement of viscosity of non-Newtonian liquids. **1971** *Biorheology* VIII. 79 The 'overall viscosity'..exhibited non-Newtonian behavior.

non-no·rmal, *a.* [Non- 3.] Not normal; *spec.* in *Statistics* (cf. *normal *a.* and *sb.* A. 2 e).

1929 *Biometrika* XXI. 124 (*heading*) On the distribution of the ratio of mean to standard deviation in small samples from non-normal universes. **1966** *English Studies* XLVII. 194 In this last phrase [*sc.* 'Mallice domestique'] the contrasting adjective 'domestique', given emphasis by its non-normal position, brings into full prominence the 'hatred' which the noun contains. **1968** P. A. P. Moran *Introd. Probability Theory* vii. 300 The distributions..converge to a proper distribution, *F*(*x*), which can be chosen to be non-normal if the theorem is false. **1973** J. J. Zeman *Modal Logic* ix. 146 The PC connectives behave in non-normal worlds precisely as they do in normal ones.

So **no:n-norma·lity,** the property or state of being non-normal.

1935 *Proc. Cambr. Philos. Soc.* XXXI. 230 The test of significance of a correlation coefficient has been shown..to be little vitiated by non-normality. **1962** E. S. Keeping *Introd. Statistical Inference* viii. 208 If there is reason to suspect non-normality, from the nature of the data, it is advisable to try a transformation. The logarithm of the variate, or the square root,..may be more nearly normal. **1968** *Brit. Med. Bull.* XXIV. 212/1 The emphasis placed so far on non-Normality is somewhat misleading, in that other kinds of departure from the usual statistical model are often more important.

non-nu·clear, *a.* [Non- 3.] That is not nuclear (in various senses); *spec.* not possessing nuclear weapons; esp. in *non-nuclear club*, a group of nations that agrees not to possess nuclear weapons. Hence as *sb.*, something non-nuclear; *spec.* a nation that has no nuclear weapons.

1920 W. W. Strong *New Philos. of Mod. Sci.* 103 Present day theories of the atom assume that the positive charge on the nucleus is equal in magnitude to the charge of the non-nuclear or planetary electrons. **1932** H. H. Price *Perception* viii. 258 If the sense-datum is a non-nuclear one, we cannot possibly move *towards* it. **1950** *New Biol.* VIII. 20 The unequal distribution of some non-nuclear hereditary substance. **1956** *Bull. Atomic Sci.* June 204/1 The nonnuclear nations may have the feeling that they alone constitute the stakes in the great struggle for world hegemony that the East is waging with the West. **1957** *Ibid.* Sept. 260/2 The U.S. on June 26 also proposed a reduction of nonnuclear weapons. **1959** *Daily Tel.* 6 July 6/2 His [*sc.* Mr. Gaitskell's] promise to try to form a non-nuclear club under British leadership. **1961** Y. Olsson *Syntax Eng. Verb* vi. 95 The most restricted collocability gives the nucleus and the most general collocability gives the non-nuclear elements. *Ibid.* 127 The N^v-expansion contains no non-nuclears. **1966** *Economist* 13 Aug. 634/1 Keeping the devices out of non-nuclear hands (and sight), but providing the service at a much lower cost than the non-nuclears would face. **1967** *Observer* 26 Feb. 10/6 The current leader of the non-nuclear club is Germany. **1968** Mrs. L. B. Johnson *White House Diary* 1 July (1970) 692 The United States, Great Britain, and the Soviet Union, as well as fifty-eight non-nuclear nations, signed the Treaty.

nonny bag, var. *nunny bag.

no-no (nōᵘ·nōᵘ). *colloq.* (orig. *U.S.*). [Redupl. No *sb.*] Something that must not be done, used, etc.; something that is forbidden, impossible, or not acceptable; a failure.

1942 Berrey & Van den Bark *Amer. Thes. Slang* §310/1 *No-no*,..something that should not be done. **1968** *Washington Post* 30 Sept. A.20/2 What she was doing was a no-no, the protectors of our health announced... She must stop baking things for sale. **1970** *Islander* (Victoria, B.C.) 25 Oct. 16/3 Colonial Secretary Young at Victoria had to remind them that the Admiralty timber reserve was a 'no no'. **1972** *New Yorker* 17 June 29/1 Our people in Accounting tell me that cash payments are a no-no. **1972** *Observer* 17 Sept. 34/8 *Nationwide*..has my strict instructions never to touch the subject of rock music again: its piece on Bob Dylan was a total no-no. **1975** *Sunday Advocate-News* (Barbados) 15 June 3/1 Plants that require a great deal of moisture are no-noes unless you have your own well.

non-object. 1. [Non- 2.] **a.** Something which is not a material body.

1914 C. D. Broad *Perception* i. 8 We must accept the possibility of real non-objects, though there cannot be apparent non-objects.

b. The condition of not being a grammatical object.

1964 E. A. Nida *Toward Sci. Transl.* v. 109 Semantic markers are primarily of two types: (1) those which mark positive–negative dichotomies, e.g. nonobject vs. object and verbal vs. nonverbal.

2. *attrib.* [Non- 4.] **a.** Not corporeal. **b.** Not functioning as grammatical object.

1963 F. T. Visser *Hist. Syntax of Eng. Lang.* I. iv. 429 For its [*sc.* 'the self's'] use in non-object relations see 1959 Carstensen p. 226. **1971** *Gloss. Electrotechnical, Power Terms* (*B.S.I.*) iv. ii. 9 *Non-object (perceived) colour*, colour perceived as non-located in depth, such as that perceived as filling a hole in a screen.

non-obje·ctive, *a.* [Non- 3.] Not objective; *spec.* in art = *abstract A. 4 d. Hence **non-obje·ctivism, non-obje·ctivist, non-objecti·vity.**

1905 W. James in *Jrnl. Philos., Psychol. & Sci. Methods* II. 286 That..is enough to save them from being classed as absolutely non-objective. **1936** *Amer. Mag. of Art* Mar. 154 (*heading*) Non-objectivity at Charleston. **1936** *Design* June 13 (*heading*) Definition of non-objective painting. **1937** H. Read *Art & Society* vii. 260 A heightened sensibility to the purity of form..can best be induced and refined by the creation and appreciation of non-objective works of art. **1946** D. D. Runes *Encycl. Arts* 680/1 The non-objectivist uses his medium, line and color..to create freely following his intuition. **1952** *Time* 16 June 61 He enjoys swimming against the current of non-objectivism. **1958** *Times* 24 Nov. (Canada Suppl.) p. xvi/6 In our contracting world few countries can escape the impact of the international artistic trend towards non-objectivity. **1959** *Listener* 3 Dec. 980/2 The non-objectivist Alexander Rodchenko painted three canvases, one red, one blue, one yellow, and announced 'the death of art'. **1960** E. H. Gombrich *Art & Illusion* viii. 286 Even nonobjective art derives some of its meaning and effects from the habits and mental sets we acquired in learning to read representations. **1962** J. Söderlind in F. Behre *Contrib. Eng. Syntax* 110 Non-objective *of*-groups offer some difficulties.

no-no·nsense, *a.* [The phr. *no nonsense* (Nonsense *sb.* 1 c) used as an adj.] That does not tolerate foolish or extravagant conduct; sensible, business-like, realistic, practical.

1928 *Sat. Even. Post* 12 May 25/1 From a no-nonsense business man he has become a romantic. **1943** J. B. Priestley *Daylight on Saturday* xiii. 92 Her own breezy, no-nonsense line. **1957** *Observer* 27 Oct. 16/3 The tunes are vigorous no-nonsense affairs. **1959** W. McGivern *Savage Streets* (1960) i. 7 A blunt, no-nonsense sort of person. **1963** [see *go-anywhere* adj. s.v. *go *v.* VIII]. **1973** J. Wainwright *Pride of Pigs* 113 Its bare-brick walls were painted white. Its floor was no-nonsense concrete.

non-o·rientable, *a.* *Math.* [Non- 3.] Of a surface: such that a figure in the surface can be continuously transformed into its mirror image by taking it round a closed path in the surface; not orientable.

1949 S. Lefschetz *Introd. Topology* ii. 76 The combination of the first type is called nonorientable. **1952** P. Nemenyi tr. *Hilbert & Cohn-Vossen's Geom. & Imagination* vi. 306 The classification of surfaces into two-sided and one-sided surfaces is identical with the classification into orientable and non-orientable. *Ibid.*, A surface is non-orientable if and only if there exists on the surface some closed curve *s* which is such that a small oriented circle whose center traverses the curve continuously will arrive at its starting point with its orientation reversed. **1965** S. Barr *Exper. Topology* ii. 22 The Moebius strip is what is called non-orientable—which is less open to misinterpretation than saying it is 1-sided.

Hence **no:n-orientabi·lity,** the property of being non-orientable.

1949 S. Lefschetz *Introd. Topology* ii. 83 A closed surface..is completely characterized by the value of its Betti number..and by its orientability or nonorientability. **1964** H. Levy *Projective & Related Geometries* v. 388 We shall derive two properties of L_2 that are strongly suggestive of its nonorientability.

nonose (nǭ·nǭs, -z). *Chem.* [a. G. *nonose* (E. Fischer 1890, in *Ber. d. Deut. Chem. Ges.* XXIII. 934): see *Nona- and -ose².] Any monosaccharide having nine carbon atoms in the molecule, esp. when these are all in an unbranched chain.
1890 *Jrnl. Chem. Soc.* LVIII. 1234 (*table*) Nonose. 1915 *Jrnl. Biol. Chem.* XXIII. 327 When it was found later that the gluconose was not fermentable, the configurations of the two nonoses became of great interest. 1945 *Adv. Carbohydrate Chem.* I. 7 Distinctly different nonoses were obtained in the two researches by the reduction of the respective lactones. 1972 Brimacombe & Webber in Pigman & Horton *Carbohydrates* (ed. 2) IA. xiv. 497 Extension of the cyanohydrin synthesis to this nonose gave a decitol.

non-parame·tric, *a. Statistics.* [Non- 3.] Not involving any assumptions as to the form or parameters of a frequency distribution.
1942 J. Wolfowitz in *Ann. Math. Statistics* XIII. 264 Most of these developments have this feature in common, that the distribution functions of the various stochastic variables which enter into their problems are assumed to be of known functional form, and the theories of estimation and of testing hypotheses are theories of estimation of and of testing hypotheses about, one or more parameters..the knowledge of which would completely determine the various distribution functions involved. We shall refer to this situation..as the parametric case, and denote the opposite case, where the functional forms of the distributions are unknown, as the non-parametric case. 1956 S. Siegel (*title*) Nonparametric statistics for the behavioral sciences. 1957 Kendall & Buckland *Dict. Statistical Terms* 87 Distribution-free inference or distribution-free tests are sometimes known as non-parametric but this usage is confusing and should be avoided. It is better to confine the word 'non-parametric' to the description of hypotheses which do not explicitly make an assertion about a parameter. 1962 E. S. Keeping *Introd. Statistical Inference* x. 251 Non-parametric tests are generally simple to apply... In those cases where a parametric test would also be applicable a non-parametric test will naturally be less powerful than the parametric one. 1966 *Lancet* 31 Dec. 1434/1 To avoid making assumptions concerning normality and homogeneity, the non-parametric one-way analysis [of] variance by ranks..was carried out. 1972 *Jrnl. Social Psychol.* LXXXVIII. 204, I have typically used a nonparametric correlation to illustrate the association between awareness and experimental performance.

non-partisa·n, *a.* [Non- 3.] Not partisan. Also as *sb.*, one who is not a partisan. Hence **non-partisa·nship.**
1885 *Century Mag.* Apr. 823 A citizens' ticket, largely non-partisan in character, was run for certain local offices. 1888 *Voice* 9 Feb., The non-partisans have for some time been making Mr. Johnson's position very uncomfortable for him. 1888 G. H. Carrow (*title*) Non-partisanship; or, do not take temperance into politics. 1932 *Times Lit. Suppl.* 11 Feb. 95/4 Critical, non-partisan, and bright in style, this book is a veritable compendium of the theories, claims, and actual achievements of psychology. 1952 G. Sarton *Hist. Sci.* I. xii. 319 His nonpartisanship, objectivity, and honesty. 1957 *Economist* 7 Sept. 850/2 For most of the time, he has put on a determined show of nonpartisanship. 1964 *Ann. Reg.* 1963 215 The latter..dealt in a relatively non-partisan spirit with contemporary Western culture. 1974 *Nature* 25 Jan. 169/1 The inferior quality of the recent Commons debate on the fuel shortage does not bode well for informed non-partisan discussions.

non-passive, *a.* and *sb. Gram.* [Non- 3, 2.] **A.** *adj.* Not in the passive voice.
1962 C. L. Barber in F. Behre *Contrib. Eng. Syntax* 26 Of these [finite verbs], 28% are passive, and 72% non-passive. 1963 F. T. Visser *Hist. Syntax of Eng. Lang.* I. iv. 562 In the non-passive construction the predicative adjunct is usually preceded by *for*: 'Hig hæfdon hyne for ænne witegan'. 1965 *Language* XLI. 107 The non-passive aorist in Greek.
B. *sb.* A construction not in the passive voice.
1962 C. L. Barber in F. Behre *Contrib. Eng. Syntax* 27 A much smaller number of passives—only 3%, against 97% non-passives.

non-past. *Gram.* [Non- 2.] A tense that is not the past tense; usually, the present tense, or the present and future tenses. Also *attrib.* or as *adj.*
1946 *Language* XXII. 213 The non-past indicative of the copula, *dá*, is replaced by a zero alternant (i.e. drops out). 1951 Trager & Smith *Outl. Eng. Structure* II. 60 Verbs are inflected for 3D Person SG. Non-past, Past, Past Participle, Present Participle. 1962 S. E. Martin in Householder & Saporta *Problems in Lexicography* 157 In many parts of the world, verbs are usually entered under the plain present (or non-past) form. 1966 *Amer. Speech* XLI. 201 The present, which many structuralists have preferred to call the 'nonpast'. 1971 *Archivum Linguisticum* II. 25 It is important to notice that [in Classical Latin] both potential and unreal conditions could refer to past and to non-past time. *Ibid.* 27 It is possible to visualize, in Old Spanish, the maintenance of the category of 'non-past, unreal', a category which..existed in Latin. 1972 *Language* XLVIII. 473 One feature of emphatic transformations from verbal bases is the replacement of the verb form of the base clause by an emphatic form; this is a special form in the non-past or past corresponding to the simple forms in the same tenses.

non-pa·trial, *sb.* and *a.* [Non- 2, 4 + *Patrial *a.* (*sb.*) 3.] **A.** *sb.* One who is not a patrial. **B.** *adj.* Applied to a person who is not a patrial.
The 1971 Immigration Act does not use the term 'non-patrial'.
1971 *Telegraph* (Brisbane) 25 Jan. 5/3 Under the Government's new Immigration Act, everyone who regards himself as a UK citizen is to be defined as a 'patrial' or 'non-patrial'. 1971 *Daily Tel.* 11 Oct. 1/5 The Police Federation asked the Government four months ago to drop the proposal in the Bill for 'non-patrial' Commonwealth citizens to register with and report to the police, as aliens do. 1972 *Guardian* 25 Nov. 10/2 A non-patrial Commonwealth citizen who wants to work here. 1973 C. Mullard *Black Britain* II. v. 63 Others—'non-patrials'—do not have an automatic 'right of abode'.

non-perio·dic, *a.* [Non- 3.] Characterized by or exhibiting a lack of periodicity; without regular recurrence; = *aperiodic *a.*
1843 *Scientific Mem.* III. 221 (*heading*) On the non-periodic variations in the distribution of temperature on the surface of the earth. 1902 *Encycl. Brit.* XXVII. 161/2 The non-periodic comets appearing since 1700 are nearly all in the hands of computers. 1932 W. L. Graff *Language & Languages* 17 Many so-called musical tones are nonperiodic. 1965 *Math. in Biol. & Med.* (Med. Res. Council) iv. 141 If the time occupied by one complex cycle is assumed to increase to infinity, then a Fourier Integral ..may be developed which describes, again in terms of sines and cosines, the waveform of a transient or non-periodic function.

no·n-person. [Non- 2.] A person who is regarded as nonexistent or unimportant; someone who is ignored, humiliated, or forgotten.
1959 *Times Lit. Suppl.* 2 Oct. 555/3 The belief that all Africans, students and dockers alike, are the same; the notion that they are 'non-persons', with a kind of animal anonymity which relieves whites of their inhibitions. 1965 [see *non-event]. 1968 *Daily Tel.* 7 Nov. 22/4 Pets are in highest demand in the wealthy Protestant West, with all its rejects and 'non-persons'. 1973 R. Hayes *Hungarian Game* xxiii. 145 Neither AVH nor KGB had a watch on Mityas' house. He'd become such a non-person that not even they worried about whom he saw.

non-pe·rsonal, *a.* and *sb.* [Non- 3, 2.] **A.** *adj.* Not personal. **B.** *sb.* Something that is not personal, or not a person; *spec.* a pronoun representing a non-personal noun.
1902 W. James *Var. Relig. Exper.* xx. 500 The contention of the survival-theory that we ought to stick to non-personal elements exclusively. 1925 Grattan & Gurrey *Our Living Lang.* 189 With Non-Personals the idea of possession is commonly not present—for example —I have recovered this book and mended *the* (not *its*) back. 1928 O. Jespersen *Internat. Lang.* II. 128 If there is an apparatus called in this way, the word belongs to the non-personal class: *telegrafe* (the apparatus), *telegrafa* vb, *telegrafo* telegraphing. 1933 L. Bloomfield *Lang.* ix. 146 *It* has the class-meanings of substantives, singulars, non-personals. *Ibid.* xv. 253 The English definite or third-person pronouns..differ.., in the singular, for *personal* and *non-personal* antecedents: personal *he, she*, versus non-personal *it*. 1957 R. W. Zandvoort *Handbk. Eng. Gram.* IX. ii. 315 The addition of *-ish* to other personal nouns: *boyish, girlish* (= proper to the nature of), and to a few non-personal nouns (*feverish*). 1965 *English Studies* XLVI. 228 It perfectly focusses the gap between a personal standard and the non-personal sonorities of Roman 'honor'.

non-plu·ral, *sb.* and *a.* [Non- 2, 3.] **A.** *sb.* The fact or condition of being only one in number.
1941 N. K. Smith *Philos. of D. Hume* xxiii. 500 Tending to equate 'unity' with 'the simple', in the sense of the non-plural.
B. *adj. Linguistics.* Not in the plural form.
1958 L. Krasner in Saporta & Bastian *Psycholinguistics* (1961) 81/1 When used to reinforce nonplural responses, both stimuli tended to increase the frequency of such responses. 1961 R. B. Long *Sentence & its Parts* viii. 184 The 'singular' forms are really nonplural rather than singular in the strict sense. 1970 *Jrnl. Gen. Psychol.* July 7 Greenspoon noted that both of these reinforcing stimuli effected an increase in the emission of nonplural responses.

non-po·lar, *a.* [Non- 3.] Not polar; *spec.* in *Chem.* and *Physics*, (composed of molecules) having no electric dipole moment.
1892 *Electrician* 29 Jan. 329/1 He [*sc.* G. Forbes] had carried out some encouraging experiments with a dynamo..built on his 'non-polar' principle. 1896 Fisher & Schwatt tr. *H. Durège's Elem. Theory Functions* vii. 126 A non-polar discontinuity. 1913 *Jrnl. Amer. Chem. Soc.* XXXV. 1443 Polar and non-polar bonds may..appear in the same formula. *Ibid.*, There are two distinct types of union between atoms: polar, in which an electron has passed from one atom to the other, and non-polar, in which there is no motion of an electron. *Ibid.*, Characteristics of the non-polar type [of substance] are the ability to form chain compounds, and the existence of separable isomers. 1951 *New Biol.* X. 124 Water molecules, being polar, that is, with electric charges not distributed uniformly, are attracted by other polar molecules or parts of molecules but have no affinity for non-polar groups. 1965 Phillips & Williams *Inorg. Chem.* I. iv. 135 Polar molecules will have higher.. boiling points than otherwise similar non-polar molecules.

1967 E. Chambers *Photolitho-Offset* i. 8 Lithographic etches produce non-polar deposits not capable of absorbing an acid.

non-poli·tical, *a.* and *sb.* [Non- 3, 2.] **A.** *adj.* Not political. **B.** *sb.* One who is not involved in politics.
1860 Dickens *Uncommercial Traveller* (1958) xvii. 172 If the prisoner..had committed every non-political crime in the Newgate Calendar, nothing would have been easier than..to obtain his release. 1888 E. W. Hamilton *Diary* 27 July (1972) I. 311 It appears that H.M., regarding the matter as 'non-political', has also taken the advice of Lord Cranbrook and Northcote about it. 1920 B. Russell *Practice & Theory Bolshevism* I. ii. 25 I was able..to find out how the whole system appears to the ordinary non-political man and woman. 1926 R. Macaulay *Crewe Train* (1967) II. iii. 67 She enjoyed her mother's party, though she found at it too many non-politicals. 1958 *New Statesman* 1 Feb. 131/1 Even in South Africa (I would agree) it was all right for us to live our non-political lives. 1962 *Guardian* 22 Dec. 3/1 The other non-politicals contented themselves with brief speeches.

non-prolifera·tion. [Non- 1.] The prevention of an increase in the number of countries possessing nuclear weapons. Freq. *attrib.*
1965 *Times* ... It would, however, only be prepared to subscribe to an east-west agreement on non-proliferation when the problem of nuclear security within the Atlantic Alliance had been satisfactorily settled. 1965 *Newsweek* 26 July 36/1 The price for a Russian agreement on a non-proliferation pact. 1972 *Friend* 1 Feb. 114/2 The premier..is committed under the Non-Proliferation Treaty to negotiate an end to the nuclear arms race. 1974 *Times* 10 May 15/8 Some Afro-Asian countries maintained that the non-proliferation treaty was..a plot to maintain the hegemony of those who already had the bomb.

non-provi·ded, *a.* [Non- 6.] Of schools or education: that is not provided (see Provided *ppl. a.* 4 b).
1918 A. Quiller-Couch *Foe-Farrell* 44 He..zigzagged off into 'Education: Provided' and 'Non-Provided', lunging and blundering with the Church Catechism and the Rate-book. 1921 *Act* 11 & 12 *Geo. V* c. 51 §29 *marg.*, Conditions as to use of premises of non-provided schools. 1960 R. Williams *Border Country* 162 The entrance to the school, Glynmawr Non-Provided.

non-ra·ndom, *a.* [Non- 3.] Not random. Hence **non-ra·ndomly** *adv.*; **non-ra·ndomness.**
1942 *Genetics* XXVII. 531 (*heading*) The non-random distribution of mutants in the progeny of certain flies. 1943 J. S. Huxley *Evolutionary Ethics* 73 A non-random sample of the population. 1958 *Oxf. Univ. Gaz.* 27 Jan. 524/2 A model of non-randomness in stochastic experiments. 1961 L. F. Brosnahan *Sounds of Language* v. 90 Sounds..occur in apparently non-random distributions which unite languages of different language groups. 1966 D. G. Brandon *Mod. Techniques Metallogr.* 251 Spatial non-randomness may be either orientation-dependent or orientation-independent. 1972 *Science* 5 May 545/1 Alleles of polymorphic loci, in the absence of direct selection, will tend to become nonrandomly associated with closely linked overdominant loci. 1972 *Jrnl. Soc. Psychol.* Aug. 223 A non-random sample of collective behaviour. 1973 *Nature* 16 Mar. 166/1 Human chromosomes are segregated non-randomly into hybrid clones.

non-ra·tional, *a.* [Non- 3.] Not rational. Also *absol.* as *sb.* Hence **non-rationa·lity.**
a 1871 G. Grote *Eth. Fragm.* (1876) v. 187 Aristotle classifies the phenomena of the..non-rational soul into three. 1896 W. Caldwell *Schopenhauer's System* i. 16 His experience of life was such as to bring the non-rational side of things prominently before his mind. 1902 W. James *Var. Relig. Exper.* iii. 74, I do not yet say that it is *better* that the subconscious and non-rational should thus hold primacy in the religious realm. 1910 — *Some Probl. Philos.* (1911) viii. 136 The only alternative allowed by such writers..is the non-rationality—and no philosopher can permit himself to do that. 1951 R. Firth *Elements of Social Organization* vii. 245 The possibilities of intellectual and emotional satisfaction to be derived from religious belief are..the resort of that belief to non-rationality. 1957 J. S. Huxley *Religion without Revelation* ... which is basically non-rational. 1972 *Listener* 21 Dec. 854/2 Reliance on the non-rational means that courses of action are followed without regard for their long-term consequences.

non-refere·ntial, *a.* [Non- 3.] Of hypothesized mental events, such as the awareness of thought or of an image: having no reference to anything beyond themselves.
1925 C. D. Broad *Mind & its Place* vi. 305 Purely inspective situations would belong to the latter class. So would pure sensation, the mere awareness of an image, etc... I am inclined to think that pure sensations, etc., are ideal limits rather than actual facts... Let us call situations of the..second kind 'non-referential'. 1943 *Mind* LII. 176, I shall say something now about the non-referential interpretation of 'copy'. 1971 J. Z. Young *Introd. Study Man* x. 125 All the words of the mental language are said to be 'non-referential'.

non-refle·xive, *a.* [Non- 3.] **1.** *Gram.* Not reflexive (Reflexive *a.* 5).
1949 O. Jespersen *Mod. Eng. Gram.* VII. 170 As a reflexive use of the *self*-forms we must also reckon the predicative in 'he is quite himself again' = 'in a normal

condition', though NED takes it to be the emphatic, i.e. non-reflexive use. 1965 D. Ward *Russ. Lang. Today* v. 139 As a general rule, such nouns are formed from imperfective non-reflexive verbs. 1968 J. Lyons *Introd. Theoretical Linguistics* 361 Many languages..have a set of reflexive pronouns distinguished for person and number..; others..draw a distinction between reflexive and non-reflexive objects only in the third person.
2. *Philos.* Of a relation which may, but need not, hold between a thing and itself. Cf. *Irreflexive *a.*, *Reflexive *a.* 3.
1947 H. Reichenbach *Elem. Symbolic Logic* iii. 120 The nonreflexive functions comprise the irreflexive and the mesoreflexive functions. 1954 I. M. Copi *Symbolic Logic* v. 145 Relations which are neither reflexive nor irreflexive are said to be non-reflexive. ... 1965 Hughes & Londey *Elem. of Formal Logic* xxxix. 274 Every dyadic relation must be either reflexive or irreflexive or non-reflexive.

non-resident, *sb.* Add: **3.** A person who uses some of the facilities of a hotel without residing there.
1910 *Bradshaw's Railway Guide* Apr. 1081 Sea and fresh water baths in the hotel..are open to non-residents. 1966 *Harper's Bazaar* Oct. 120/1 Even at the Dorchester..the bar is closed at 11 p.m. to non-residents. 1971 'A. Gilbert' *Tenant for Death* vii. 132 A printed invitation to non-residents to use the bar and the dining-room.

non-resident, *a.* **3.** *U.S.* Of land: owned by a person who does not reside on it.
1838 E. Chamberlain *Indiana Gaz.* (ed. 3) 315 The large amount of non-resident lands has hitherto retarded the improvement of the country. 1881 *Mich. Gen. Statutes* (1882) 385 Such notice to owners of such non-resident lands, shall be served by posting up the same in three public places.

non-resistance. Delete 'Now only *Hist.*' and add further examples.
1838 W. L. Garrison in *Liberator* (Boston) 28 Sept. 154/5 We shall adhere to the doctrine of non-resistance and passive submission to enemies. 1861 W. Rowntree *War & Christianity* 6 The gospel principle of non-resistance. ... 1934 G. B. Shaw *On Rocks* 155 It is easy to suggest that they [*sc.* evildoers] should be reformed by gentleness and shamed by non-resistance. 1971 R. V. Sampson *Tolstoy: the Discovery of Peace* vi. 109 It is very rare for a writer to allow Tolstoy's arguments in support of the doctrine of non-resistance to evil—the essence of Tolstoy—to be heard.

non-resistant, *a.* and *sb.* **B.** *sb.* (Earlier example.)
1755 in S. M. Hamilton *Lett. to Washington* (1898) I. 91 The fighting Faction..threaten to put to Death all the Non-resistents—Dunkers, Moravians, Dutch and Quakers.

non-resi·ster. [Non- 2.] = Non-resistant *sb.*
1851 J. A. Quitman in J. F. H. Claiborne *Life & Corr. J. A. Quitman* (1860) II. 142 ...

non-restri·ctive, *a.* [Non- 3.] Not restrictive; *spec.* in *Gram.*, denoting a word, phrase, or clause that does not restrict or limit the meaning of the word or words to which it is added.
1924 O. Jespersen *Philos. of Gram.* viii. 112 Next we come to non-restrictive adjuncts as in *my dear little Ann!* 1927 — *Mod. Eng. Gram.* III. iv. 85 Restrictive or defining clauses..give a necessary determination to the antecedent, and thereby make it more precise.. Non-restrictive or loose clauses..might be discarded without serious injury to the precise understanding of the sentence as a whole. 1953 J. S. Huxley *Evolution in Action* vi. 142 What is difficult is to discover just how any one step is effected, still more to distinguish desirable from undesirable change, and restrictive from non-restrictive improvement. 1966 G. N. Leech *Eng. in Advertising* xiv. 132 Proper nouns do occasionally combine with modifiers of non-restrictive force: 'fair Helen'..'beautiful Britain'. 1971 *Archivum Linguisticum* II. 14 To separate non-restrictive ... relative clauses from the preceding clause.

non-retu·rnable, *a.* [Non- 3.] That may not be returned; *spec.* of containers, bottles, etc.: that may not be returned empty to the suppliers. Also *absol.* as *sb.*
1903 T. E. Young *Insurance* x. 265 The premiums in the event of death happening prior to the adult age agreed upon were non-returnable. 1907 *Yesterday's Shopping* (1969) p. x/1 All empties, except 'non-returnable cases'..must be returned within a reasonable period. 1926–27 *Army & Navy Stores Catal.* p. ix/1 Goods sent by Rail or by Carter Paterson are packed in non-returnable cases for which no charge is made. 1961 *New Scientist* 16 Mar. 700/2 The milk bottle..is about five times cheaper than the non-returnable carton. 1968 *Economist* 27 Jan. 59/3 Now that the non-returnable bottle—and breathalysers—are here to stay, grocery stores are the growth point in drink sales. 1971 *Wall St. Jrnl.* 5 Mar. 26/4 About 70 such laws were debated last year by various local, state and federal bodies. Bowie, Md., actually banned nonreturnables. 1971 *Nature* 6 Aug. 359/2 Potential customers should..make appropriate non-returnable deposits as marks of their good faith. 1973 *Times* 11 Dec. 6/4 It is absurd for continuance of non-returnable bottle production still to be allowed.

nonrewa·rd. *Psychol.* [Non- 1.] In learning experiments: deliberate withholding of an expected reward. Hence **nonrewa·rded** *ppl. a.*
1951 N. E. Miller in S. S. Stevens *Handbk. Exper. Psychol.* 458/1 The cues produced by nonreward. *Ibid.*, The interspersing of a number of nonrewarded trials during training with primary drive and reward (i.e. partial reinforcement) may increase the resistance of a learned reward to experimental extinction. 1970 *Jrnl. General Psychol.* Oct. 185 The role of frustrative nonreward in instrumental escape conditioning. 1973 *Jrnl. Genetic Psychol.* Mar. 89 Association with frustrative nonreward alters the affective value of the stimulus.

non-ri·gid, *a.* [Non- 3.] Not rigid; *spec.* denoting an airship which has no framework to support the envelope, and whose shape is maintained solely by the pressure of the gas inside. Also occas. as *sb.*, an airship of this kind.
1909 A. Berget *Conquest of Air* I. v. 108 These airships are non-rigid. 1909 *Daily Chron.* 4 Aug. 1/7 ... 1930 *Daily Express* 6 Oct. 2/4 He..was captain of non-rigid airships in home waters and the East Mediterranean. 1961 *B.S.I. News* Dec. 20/2 Rigid plastics; semi-rigid plastics; non-rigid plastics.

non-sche·duled, *a.* orig. *U.S.* [Non- 6.] Not scheduled; *spec.* of an airline: operating without fixed or published flying schedules; or of or pertaining to such an airline. So **non-sched, -sked** (nɒnske·d), a non-scheduled airline; also *attrib.* or as *adj.*
1946 *Harper's Mag.* Oct. 324/1 The non-scheds naturally look with envy at the mail subsidies now granted to the major airlines. 1949 *Aviation Week* 30 May 7/1 The nonsked may barely break even as a result. 1949 *Industrial Arts Index* June 719/1 Air carriers; non-scheduled operations. 1960 *Wall St. Jrnl.* 12 Jan. 30/2 The ultimate outcome of the case apparently will affect several other suits brought by 'non-skeds' which seek an additional $154.6 million in damages. 1965 *Flight* LXXX. 645/1 The Flying Tiger Line has leased an L.1049H Super Constellation to Trans-International Airlines Inc., a Florida-based 'non-sked'. 1971 *Flying* Apr. 113/2 Cargo and non-scheduled operations. 1975 *Publishers Weekly* 3 Mar. 73/2 This is like being stuck on a nonsched junket to the most obvious European spots.

nonsense, *sb.* Add: **1. a.** *to knock the nonsense out of*: see *Knock *v.* 6 e.
b. Delete † and add later examples; also *concr.*; also, a muddle, fiasco, esp. in phr. *to make a nonsense (of)*.
1803 Scott *Let.* 24 Sept. (1932) I. 204, I daresay I shall go on scribbling one nonsense or another to the end of the chapter. 1942 E. Waugh *Put Out More Flags* iii. 222 Everyone said, 'Lynn made a nonsense of the unpleasantness.' *Ibid.* 224 'It was all rather a nonsense' said the subaltern, in the classic phraseology of his trade which comprehends all human tragedy. 1957 *Listener* 30 May 873 Size is essential for efficiency. A small steel works..is an economic nonsense. 1961 L. Cooper *Accomplices* vi. i. 71 He's made a complete nonsense of the whole thing and got the sack. ... 1966 *Movie* June 62 Ambitious nonsenses like *The Entertainer* and *Flame in the Streets*. 1966 A. Prior *Operators* xv. 235 An occasion that could make a nonsense of any Op. Bad intelligence, weather, flak. 1970 *New Statesman* 16 Jan. 88/3 'Structure' is one of those useful nonsenses in the vocabulary of the fiction critic, the mystery ingredient with the secret formula.
4. Used, usu. in the spelling *non-sense*, without connotations of absurdity: something that is not sense or that differs from sense.
Not clearly separable from sense 1 (or 2) at first.
1932 T. S. Eliot *Music of Poetry* 14 His [*sc.* Lear's] non-sense is not vacuity of sense: it is a parody of sense, and that is the sense of it. 1949 *Amer. Speech* ... Men struggle over non-sense questions. 1965 *English Studies* XLVI. 322 A 'steep ford' is quite obviously non-sense.
6. a. *nonsense song* (later examples); **nonsense syllable**, a syllable formed by putting a vowel between any two consonants, used in memory experiments and tests; **nonsense verse** (earlier and later examples); also *nonsense verse*, comically nonsensical or whimsical verse; *spec.* a limerick; **nonsense word**, a word which has no accepted meaning; also = *nonsense syllable.
1968 *Radio Times* 28 Nov. 65/2 A nonsense song from Virginia, 'Poor Howard'. 1975 *Listener* 6 Mar. 305/2 When Charlie finally opened his mouth in *Modern Times*, out came a nonsense song no one needed to understand. 1890 W. James *Princ. Psychol.* I. xvi. 667 Mr. Burnham..suggested..one ought to test one's self *à la* Ebbinghaus on series of nonsense-syllables. 1924 R. M. Ogden tr. *Koffka's Growth of Mind* i. 29 Impressing certain material (preferably nonsense-syllables in an ordered series) upon the observer. 1971 *Jrnl. General Psychol.* LXXXV. 139 Ss were allowed to work 10 nonsense-syllable examples. 1799 *Public Characters* II. 423 Although few men in England could equal him in writing some sense, yet many might surpass him in writing *nonsense verses*. 1862 Keats *Let.* 17 Sept. (1931) II. 438 ... 1961 [see *Limerick]. 1938 A. Davidson *Edward Lear* ii. 18 It is possible that these extemporized nonsense-verses were in the limerick form. 1940 *Essays & Studies* XXV. 7 The nonsense verse which was well worth having—who had believed, else, that Housman had it in him either to be happy or to write nonsense. 1929 J. B. Watson *Psychology* iv. 136 Nonsense words or syllables are made by separating two consonants with a vowel as *ver, gas* and *mex*. 1933 B. Malinowski in M. Black *Importance of Lang.* (1962) 72 Have we to imagine that magical speech starts from sheer nonsense words..? 1954 J. R. R. Tolkien *Fellowship of Ring* I. vi. 130 Out of a long string of nonsense-words (or so they seemed) the voice rose up loud and clear and burst into this song.
c. *Biol.* Denoting a codon that does not specify any amino-acid, or a mutation which gives rise to such a codon (and may thus prematurely bring about a blockage of polypeptide synthesis).
[1957 F. H. C. Crick et al. in *Proc. Nat. Acad. Sci.* XLIII. 418 We shall assume that there are certain sequences of three nucleotides with which an amino acid can be associated and certain others for which this is not possible. Using the metaphors of coding, we say that some of the 64 triplets make sense and some make nonsense.] 1961 [see *missense *a.*]. 1964 G. H. Haggis et al. *Introd. Molecular Biol.* xii. 325 They then calculated that the sixty-four triplets could be divided..into two classes, with twenty meaningful triplets in one class and forty-four nonsense triplets in the other. 1970 J. McCarthy *Episomes* ix. 222 Nonsense mutations of *lacZ*..produce reduced levels of the permease coded by *lacY*. 1974 *Encycl. Brit. Micropædia* IV. 469/2 It is now known that there are triplets coded for one amino acid and that there are nonsense codons that do not code for the ...

nonsense (nɒ·nsəns), *v.* [f. the *sb.*] *intr.* To talk nonsense.
1822 M. Edgeworth *Let.* 30 May (1971) 404 We have given most elegant repasts as I shall have the honor of describing to you soon when we are all together nonsensing round Lucys sofa. 1909 R. A. Wason *Happy Hawkins* 67, I nonsensed a while, tryin' to get her to laugh at me.

non-se·quence. *Geol.* [Non- 1.] An interruption in the deposition of adjacent conformable strata that was too short (in geological terms) for significant erosion to take place and consequently has to be inferred from a gap in the fossil record.
[1895 S. S. Buckman in *Q. Jrnl. Geol. Soc.* LI. 391 Non-sequential strata, when the sequence is incomplete, but the planes of the deposits are practically parallel.] 1904 *Ibid.* LX. 349 (*heading*) Evidence of non-sequence between the Keuper and Rhætic series in North-West Gloucestershire and Worcestershire. 1933 *Ibid.* LXXXIX. 163 ... The position immediately overlying strata with *Dolymorphites bifidus* would indicate that they mark a considerable non-sequence. 1969 Bennison & Wright *Geol. Hist. Brit. Isles* i. 7 Naturally non-sequences are unprovable in barren unfossiliferous rocks..or in rocks so altered that fossils have failed to record.

non-signi·ficant, *a.* [Non- 3.] Not significant. Also *absol.*
1902 S. H. Butcher *Aristotle's Theory of Poetry & Fine Art* (ed. 3) x. 377 In Plautus the number of names etymologically significant and appropriate largely preponderates over the non-significant. 1905 C. D. Roberts *Barbara Ladd* 219 His wife was a non-significant, abundant, gently acquiescent pudding of a woman. 1923 J. S. Huxley *Ess. Biologist* iv. 139 These differences may be biologically speaking non-significant, mere accidents of the primary difference. 1973 *Jrnl. Genetic Psychol.* CXXIII. 10 A nonsignificant correlation of −.09 was obtained. 1974 *Nature* 22 Nov. 279/1 Unobserved objects (black holes, dead white dwarfs and so on) must only make up a non-significant percentage of the galactic mass.

non-sked: see *non-scheduled *a.*

non-skid, *a.* and *sb.* [Non- 4 or *5 b.] **A.** *adj.* That does not skid, or is designed to prevent skidding. **B.** *sb.* A tyre, substance, etc., designed to prevent skidding. So **non-ski·dding** *ppl. a.*
1908 *Sears, Roebuck Catal.* 169/1 We have added to the tire a link chain non-skid tread, the most practical and scientific tread ever made. 1908 *Daily Report* 12 Sept. 8/4 It is absolutely incumbent upon all owners of such tyres to make use of this sort of non-skid in the winter. 1909 *Chambers's Jrnl.* June 404/1 A perfect non-skidding wheel. 1911 *Daily Colonist* (Victoria, B.C.) 2 Apr. 19/1 (Advt.), Michelin 'Semelle' Non-Skids, unlike so-called non-skids made wholly of rubber, do not wear down into ordinary smooth treads. 1920 *Motor Cycle* 29 July 129/2 Non-skid chains. 1925 *Public Opinion* 11 Dec. 588/3 We want light-coloured, waterproof, nonskid surfaces. 1937 [see *autostrada]. 1944–45 *Jane's Fighting Ships* (1946) 459 Armoured flight deck is..covered with non-skid surface material. 1972 *Country Life* 15 June 1577/2 A useful top oddments shelf with non-skid surface. 1972 *Islander* (Victoria, B.C.) 25 June 2/2 Many busy weeks and months were spent painting and putting non-skid on the decks.

non-slip, *a.* [Non- 4 or *5 b.] That does not slip, or is designed to prevent slipping. So **non-sli·pping** *ppl. a.* and *vbl. sb.*
1904 G. B. Shaw *Common Sense of Municipal Trading* vii. 56 To ride bicycles through greasy mud..on tires advertized as 'non-slipping'. 1909 Webster, Nonslip. 1912 *Chambers's Jrnl.* Jan. 78/2 The gyroscope non-slipping device has been subjected to many exacting tests, but has established its capabilities to overcome skidding. 1933 *Archit. Rev.* LXXIII. 219 The staircase is in concrete, and the steps are rendered non-slip on the nosings. 1956 [see *granolithic *a.*]. 1959 *Times* 27 Apr. (Suppl. Rubber Industry) p. vi/7 Special non-slipping mineral grain-filled flooring is used for safety walks and service corridors. 1973 *Times* 19 Oct. 14/3 (*caption*) Non-slip. They are unlikely to get pushed across the kitchen floor.

non-smo·ker. [Non- 2.] A person who does not smoke.
1846 in H. C. Webster *Railways for All* (1950) xv. 114 Smokers and non-smokers should be apart, but why should smokers exclusively be indulged with the use of the best carriages on the line? 1860 *Pharmaceutical Jrnl.* II. III. 190 Dividing the pupils..into..the smokers and the non-smokers, it is shown that the smokers have proved themselves in the various competitive examinations far inferior to the others. 1891 W. Morris *News from Nowhere* vi. 37 Bob is always telling me that we non-smokers are a selfish lot. 1898 K. Baedeker *Spain & Portugal* p. xv, Every train is bound to have a first-class compartment..and another for non-smokers. 1921 G. B. Shaw *Back to Methuselah* II. 84 They believe the tee-totallers and non-smokers live longer. 1973 K. Giles *File on Death* i. 10 No car, two suits a year, non-smoker, allows more than he should afford on drink.
b. A compartment in a railway carriage, or other conveyance, where smoking is not permitted. *colloq.*
1962 J. Braine *Life at Top* v. 74 All the first-class compartments seemed to be full; finally I got a seat in a non-smoker. 1971 'L. Black' *Death has Green Fingers* iii. 29 When the woman angrily lit a cigarette, he pointed out to her..that she was in a non-smoker.

non-smo·king, *a.* [Non- 4, 6.] Denoting a railway compartment, carriage, etc., in which smoking is not permitted.
1891 H. Spencer *Let.* 27 Mar. in G. S. Layard *Mrs. Lynn Linton* (1901) xx. 278 One may persist in putting down smoking in a non-smoking compartment of a railway carriage. 1909 *Daily Chron.* 17 Sept. 4/7 A well-known clergyman was travelling in a non-smoking compartment of a train. 1922 C. E. Montague *Disenchantment* (1924) vi. 86 Trains are divided into 'non-smoking' and 'smoking'. 1936 *Punch* 5 Feb. 141/3 Discovering as we finish lighting our pipes that we are in a Ladies Only Non-Smoking carriage. 1963 *Mod. Railways* May 359/1 The sliding door to the non-smoking compartment is often left open by passengers.
b. Denoting a person who does not smoke.
1946 *R.A.F. Jrnl.* May 178 Les, non-smoking but not teetotal,..will never fail to stand by for a duty. 1972 J. Mosedale *Football* v. 65 The non-smoking, non-drinking Graham, of whom it was said, 'His idea of a big night is eating chocolate ice cream.' 1975 V. Canning *Kingsford Mark* v. 74 Are you still non-smoking and non-drinking?

non-spe·cialist, *a.* and *sb.* [Non- 3, 2.] **A.** *adj.* That is not specialist. **B.** *sb.* One who is not a specialist.
1919 W. Lewis *Caliph's Design* III. 53, I am..only giving so much general matter..as is necessary to remind a non-specialist public of the rough points of the position. 1933 *Discovery* Sept. 286/2 He has just written a remarkable book, where the non-specialist will find much of interest. 1958 *Times Lit. Suppl.* 24 Jan. 46/2 An editor whose speciality is to be non-specialist. 1971 J. Spencer *Eng. Lang. W. Afr.* p. viii, If our explorations prove to be of interest to the non-specialist, while at the same time stimulating the specialist to embark upon further research, we shall be very well satisfied. 1975 *Listener* 28 Aug. 280/3 The performance criteria were clear to non-specialists.

non-speci·fic, *a.* [Non- 3.] Not specific; esp. in *Med.*, not specific as regards cause or effects. Hence **no:n-specifi·city.**
1938 R. W. Lawson tr. *Hevesy & Paneth's Man. Radioactivity* (ed. 2) xv. 147 Non-specific adsorption forces do not play a part in this process. 1939 H. Kogan et al. in Saporta & Bastian *Psycholinguistics* (1961) 534/2 Gestalt psychology..left unaltered the postulate of its non-specificity. 1942 Darlington & Mather *Genes, Plants & People* (1950) iv. 37 The difference between activity and inertness would then be the difference between specificity and non-specificity. 1964 W. G. Smith *Allergy & Tissue Metabolism* ii. 29 Promethazine produces considerable antagonism, but the effect is probably a non-specific one. 1971 *Brit. Med. Bull.* XXVII. 61/1 This group of diseases might be referred to as 'chronic non-specific lung disease'. 1971 J. Z. Young *Introd. Study Man* xxvi. 372 The first organism included only one enzyme, a non-specific phosphokinase. 1975 I. K. Martin *Regan & Manhattan File* 88 He made some non-specific enquiries to the FBI.

non-sta·ndard, *a.* [Non- 3.] Not standard; esp. of language that is not standard (see *Standard *a.* 3).
1924 *Simplified Practice Recommendation* (U.S. Nat. Bureau Standards) XVI. 61 These tables include certain illustrative combinations of standard, and nonstandard, widths and thicknesses. 1927 *Reliable Poultry Jrnl.* Feb. 33 Whatever may be said of the policy of showing the non-Standard colors, there can be no doubt..that a class of all Standard colored birds makes a much better impression on the lay visitor than a class of mixed shades of color. 1933 L. Bloomfield *Language* iii. 48 'Bad' or 'vulgar' or, as the linguist prefers to call it, *non-standard* English. 1962 A. Battersby *Guide to Stock Control* viii. 77 Another multiplier of stock items is the non-standard product. 1971 *Engineering* Apr. 117 Diameters can be altered slightly to produce non-standard sizes. 1973 *Archivum Linguisticum* IV. 107 It might well only give evidence of a socially determined Nonstandard English.
b. *Math.* Involving infinitesimals and infinities, as quantities which are not defined in the real number system but can be rigorously accommodated in a model which includes that system.
1961 A. Robinson in *Proc. K. Nederlandse Akad. v. Wetensch.* A. LXIV. 432 (*heading*) Non-standard analysis. *Ibid.* 437 Let $f(x, t)$ be a standard function defined in a standard set S_1, and let $g(x) = f(x, \omega)$, where ω is non-standard, e.g. infinite or infinitesimal. 1966 — *Non-Standard Analysis* p. vii, In the fall of 1960 it occurred to me that the concepts and methods of contemporary Mathematical Logic are capable of providing a suitable framework for the development of the Differential and Integral Calculus by means of infinitely small and infinitely large numbers. I first reported my ideas in a seminar talk at Princeton University (November 1960) and, later, in an address at the annual meeting of the Association for Symbolic Logic (January 1961) and in a paper published in the Proceedings of the Royal Academy of Sciences of Amsterdam... The resulting subject was called by me Non-standard Analysis since it involves and was, in part, inspired by the so-called Non-standard models of Arithmetic whose existence was first pointed out by T. Skolem. 1972 *Sci. Amer.* June 86/2 We define the instantaneous velocity not as the ratio of infinitesimal increments, as L'Hôpital did, but rather as the standard part of that ratio; then *ds, dt* and their ratio *ds/dt* are nonstandard real numbers. We have as before $ds/dt = 32 + 16dt$, but now we immediately conclude, rigorously and without any limiting argument, that v, the standard part of ds/dt, equals 32.

non-sta·rter. [Non- 2.] One who does not start something; esp. one who fails to start in a race or other contest; hence, a useless or ineffective person or thing; an impracticable idea.
1909 in Webster. 1932 *Brit. Jrnl. Psychol.* July 31 Besides the 33 non-starters, only two made no attempt at this test. 1934 Wodehouse *Right Ho, Jeeves* ii. 27, I reminded myself that this non-starter and I had been at school together. 1942 'A. Bridge' *Frontier Passage* iii. 49 That's one reason why non-intervention is such a non-starter. 1952 A. J. Ayer *Philosophical Essays* (1954) iii. 42 A statement..might be a non-starter, in the sense that the question of its truth or falsehood did not arise. 1973 C. Mullard *Black Britain* III. ix. 109 People began to dub the council a nonstarter.

non-stick, *a.* [*Non- 5 b.] That does not stick or allow sticking; *spec.* of a cooking utensil: to which food does not adhere.
1958 *Listener* 4 Dec. 967/1 Silicones are used..for..coating cooking pans to make them 'non-stick'. 1960 *Farmer & Stockbreeder* 1 Mar. 152/3 (Advt.), Men's non-stick P.V.C. plastic oilskin coats. 1968 *Guardian* 9 July 7/3 The cost of the non-stick oven lining adds £6 10s to the price of a cooker. 1973 'D. Jordan' *Nile Green* ii. 14 Sticky non-stick frying pans in the sink.

non-stop, *a.*, *sb.*, and *adv.* [Non- 4 or *5 b.] **A.** *adj.* That does not stop; *spec.* of a railway train or other conveyance: that travels between two (usually distant) places without stopping at intermediate ones; of a journey, etc.: made or done without a stop; of a variety show or the like: in which there is no interval between the various acts.
1903 *Work* 11 July 364/1 The L. & N.W. Railway long non-stop run..presents no difficulty. 1904 *Windsor Mag.* Dec. p. ix (Advt.), Awards gained for Glasgow to London non-stop trials. 1910 *Punch* 15 June 441/2 *Platelayer* (to passenger who has jumped from the London-Plymouth Non-stop Express). Jumped aht? did yer.—Wot for? 1914 *Whitaker's Almanack* 1915 822/1 Britannia trophy awarded to Capt. Longcroft, R.F.C., for non-stop flight Montrose to Farnborough. 1923 Wodehouse *Inimitable Jeeves* xv. 192, I was fairly tired, having swung a practically non-stop shoe from shortly after dinner till two a.m. 1932 *Sunday Express* 3 July 17/6 Of all the people in non-stop variety, the one I should most hate to annoy is Veronica, the dancer. She can do fifty high kicks in twenty-five seconds. 1938 *Times Lit. Suppl.* 15 Jan. 44/1 A job as chorus girl in a very undressed non-stop revue. 1957 *Economist* 28 Dec. 1120/2 The sum total of his economics is that a non-stop increase in money wages is..a Good Thing. 1967 G. F. Fiennes *I tried to run a Railway* vii. 86 Averages of 75 m.p.h. for the non-stop business expresses. 1973 'I. Drummond' *Jaws of Watchdog* i. 9 The party will be out to sea. A seventy-two hour trip, way the hell out, a non-stop party. 1973 *Country Life* 23 Aug. 476/2 The Windmill Theatre..became famous as the home of live, non-stop revue.
B. *sb.* A non-stop train; a non-stop journey or run; (see also quot. 1925; *Obs. slang*).
1909 *Westm. Gaz.* 8 Sept. 2/1 The innovation and growth of 'non-stops' upon the Metropolitan and District electric lines. 1911 *Motor Cycle* 27 Apr. 428/1 He..has now made six successive non-stops. 1922 V. Woolf *Jacob's Room* ix. 183 The hordes crossing Waterloo Bridge to catch the non-stop to Surbiton. 1925 Fraser & Gibbons *Soldier & Sailor Words* 210 *Nonstop, a*, a trench expression used of a long range shell passing high overhead. 1957 *Railway Mag.* Jan. 22/2 Thus the non-stops to and from Sheffield were suspended. 1958 J. Betjeman *Coll. Poems* 202 They ..caught the first non-stop to Willesden Green.
C. *adv.* Without stopping.
1920 'Ixion' *Motor Cycle Remin.* 15 The beastie [*sc.* a motor cycle] did the outward journey non-stop. 1927 *Daily Tel.* 14 June 11/3 The second attempt to fly non-stop to India. 1942 *R.A.F. Jrnl.* 16 May 9 The test pilot..flew the aircraft non-stop from Moscow to Baku and back. 1953 A. Upfield *Murder must Wait* xvii. 147 We were talking non-stop. 1963 *Ann. Reg.* 1962 447 One station in New York (WBAI) gave the whole *Der Ring des Nibelungen* in 17 hours non-stop. 1965 W. Sovinka *Road* 21 Have you known any other driver take an oil-tanker from Port Harcourt to Kaduna non-stop since Muftan died? 1968 M. Bragg *Without City Wall* x. 112 Both he and Richard went on non-stop about London, their friends, politics, anything that came up.

non-subscriber. 2. (Further examples.)
1816 T. J. Howell *Stranger in Shrewsbury* 65 The public libraries in Shrewsbury are wretchedly furnished except one, and to *that* entrance is almost denied to strangers and non-subscribers. 1884 *List of Subscribers* (London & Globe Telephone Co.) [3] Subscribers..must not allow non-subscribers to use their instruments for the transmission of messages.

non-sylla·bic, *a.* and *sb. Linguistics.* [Non- 3, 2.] **A.** *adj.* **a.** = *asyllabic *a.*; *spec.* denoting a speech-sound that does not constitute the predominant sonority of the syllable in which it occurs. **b.** Not syllable-timed. **B.** *sb.* A non-syllabic speech-sound.
1909 Webster, Nonsyllabic, *a.* 1933 L. Bloomfield *Language* vii. 120 Some of the phonemes are more sonorous than the phonemes (or the silence) which immediately precede or follow... Any such phoneme is a *crest of sonority* or a *syllabic*; the other phonemes are *non-syllabic*. Thus the [e] in *red* and the [r] in *bird* are syllabics, but the [r] in *red* and the [d] in *red* and *bird* are non-syllabics. 1947 K. L. Pike *Phonemics* 30/1 The acoustic quality is quite similar to that which is found when they are acting as nonsyllabics. 1957 H. Whitehall in N. Frye *Sound & Poetry* 143 The first distinction would be drawn between syllabic and non-syllabic rhythms. 1961 R. B. Long *Sentence & its Parts* xvii. 382 Such monosyllabic complexes employing nonsyllabic suffixes as *truth* and *fourth*. 1965 *Language* XLI. 32 Since the first element of these clusters always represents the syllable peak.., I have not separately indicated the nonsyllabic nature of the second element. *Ibid.* 476 According to their function in the syllable, the consonants are non-syllabics or syllabics.

non-thema·tic, *a.* [Non- 3.] **1.** *Linguistics.* = *athematic *a.* 1.
1933 E. H. Sturtevant *Compar. Gram. Hittite Lang.* vi. 217 There is also, as in I[ndo-] E[uropean], a distinction between thematic and non-thematic conjugation. 1955 T. Burrow *Sanskrit Lang.* vii. 318 The verbs are divided into two major types, (a) non-thematic..and (b) thematic.
b. Not conveying new information; constituting part of the rheme; = *rhematic *a.*
1959 *Brno Studies in English* I. 43 The indefinite article and its zero plural variant..play an important part in that co-operation of means which may render non-thematic even such a subject as occurs at the very beginning, or very near the beginning, of the sentence.
2. *Mus.* = *athematic *a.* 2.
1946 G. Abraham in A. L. Bacharach *Brit. Mus.* 55 *Neptune, the Mystic*, which shares with *Saturn* the distinction of being one of the most remarkable purely impressionistic, practically non-thematic pieces of music ever written.

non-trea·ty. *N. Amer.* [Non- 4.] Used *attrib.* or *absol.* to designate an American Indian who is not subject to a treaty made with the Government.
1878 *Saskatchewan Herald* (Battleford) 18 Nov. 2/2 The non-treaty Indians invade them. 1885 J. L. Onderdonk *Idaho* 15 Councils were held at Lapwai to apportion lands to the various chiefs of the Non-treaties. 1945 *Beaver* Mar. 12/1 Some four hundred Chipewyans and one hundred Crees, the latter being mixed treaties and non-treaties. 1956 C. Relander *Drummers & Dreamers* 121 The intruders occupied the scurfy land along the Columbia where the non-treaty people lived. 1964 W. Duff *Indian Hist. Brit. Columbia* I. 70 The generalization, often heard, that the Indians of British Columbia are 'non-treaty' Indians is not wholly true.

non-tri·vial, *a.* [Non- 3.] Not trivial; significant; *spec.* in *Math.*, such that not all variables or terms are zero.
1915 R. D. Carmichael *Diophantine Analysis* ii. 52 Prove that a non-trivial solution does not exist when *b* is a square number. 1940 *Mind* XLIX. 366 This view has non-trivial consequences. 1948 F. D. Murnaghan *Introd. Applied Math.* i. 12, *n* plane vectors are linearly dependent when, and only when, there exists a nontrivial linear combination of the *n* vectors which is the zero vector. 1953 C. E. Bazell *Linguistic Form* 19 It is only when there are relations of conversion between constructions that complementary distribution is non-trivial. 1971 *Amer. Jrnl. Physics* XXXIX. 486/1 Our aim..is to derive a list of all values of $\Omega > 0$ for which there exist nontrivial q satisfying (11) and (12). 1973 *Sci. Amer.* Feb. 81/2 The ratio of those who received nontrivial injuries was 70 percent higher among the unbelted than among the belted.
So **non-tri·vially** *adv.*

1940 W. V. Quine *Math. Logic* 222 We can speak non-trivially of *the x* of *y*. 1971 Powell & Higman *Finite Simple Groups* i. 2 A conjugate class of *G* that interests *S* non-trivially.

nontronite. Add: *Nontronite* rather than *chloropal* is now the usual name for the mineral. (Further examples.)
1928 *Amer. Jrnl. Sci.* CCXV. 10 A large number of names have been proposed for the hydrous silicates of ferric iron; and of these nontronite, chloropal, pinguite, fettbol, and graminite are recognized by most mineralogists as belonging to the species nontronite. 1935, etc. [see *montmorillonite]. 1945 *Prof. Papers U.S. Geol. Survey Paper* No. 205-B. 28/1 The name chloropal has a slight priority over nontronite. However, the material has been shown to be definitely crystalline by both optical and X-ray methods, and the name chloropal, suggesting a deeply colored opaline mineral, is unsuitable. Dana (1892, p. xliii) has laid down the rule that a name having priority may properly be set aside when its signification is glaringly false. Mineralogists for some time have used the name nontronite as the most suitable for this mineral. 1963 M. H. Hey *Index Min. Species* (ed. 2) App. 117 *Nontronite*... The older name Chloropal has passed out of common use.

non-U (nǫn,yū·, nǫ·n,yū), *a.* and *sb.* [f. Non- 3 + *U 4.] **A.** *adj.* Not upper-class; not characteristic of upper-class people; esp. with reference to linguistic usage. **B.** *sb.* Non-U persons or characteristics collectively; non-U language. Hence **non-U-ness.**
1954 A. S. C. Ross in *Neuphilologische Mitteilungen* LV. 21 (*title*) U and non-U. *Ibid.*, In this article I use the terms *upper class* (abbreviated: U), *correct, proper*,..to designate usages of the upper class; their antonyms (*non-U, incorrect, not proper*,..) to designate usages which are not upper class. *Ibid.* 22 As a boy I heard *not quite a gent* ..used by non-U speakers. *Ibid.* 23 The non-U slang phrase *He's left his visiting card*. *Ibid.* 27 The U-rules for ending letters are very strict; failure to observe them usually implies non-U-ness. *Ibid.* 45 *Pardon!* is used by the non-U in three main ways. 1956 *Times* 13 Jan. 9/4 The cosh is, of course, definitely non-U, whereas the old-fashioned blunt instrument was perfectly *comme il faut*. 1957 O. Nash *You can't get there from Here* 16 The Wicked Queen said 'Mirror, mirror on the wall' instead of 'Looking glass, looking glass on the wall'... So the Wicked Queen exposed herself as not only wicked but definitely non-U. 1958 *Times Lit. Suppl.* 18 July 411/2 They phrase their advice just seriously enough to encourage the uninitiated, or non-U, yet just facetiously enough to reassure the *cognoscenti*. 1960 R. Lister *Decorative Cast Ironwork* iv. 141 Letter racks, tobacco jars, spittoons ('salivaria' in non-U). 1964 H. Kökeritz in D. Abercrombie et al. *Daniel Jones* 140 These pronunciations..might..be classified as U rather than non-U. 1967 M. Argyle *Psychol. Interpersonal Behaviour* iii. 47 Others may be divided simply into the 'saved' and the 'not-saved', U and non-U, or into finer sub-groups. 1973 *Listener* 29 Nov. 755/2 Empty-headed chat, mixed with non-U euphemisms.

non-ultra. (Earlier examples.)
1611 Coryat *Crudities* sig. b 3ᵛ, To give the *Non vltra* of him in a word, he is so Substantive an Author as will stand by himselfe. *a* 1666 Evelyn *Diary* 8 Feb. an. 1645 (1955) II. 354, I left this Ode..as the Non ultra of my Travells.

non-union. Add: **1.** (Further examples.)
1909 *Daily Chron.* 17 Feb. 1/6 Dr. Jameson..warned his hearers of the dangers of non-union, which might possibly lead to internecine war. 1963 *Lancet* 12 Jan. 86/2 There are only two causes of delayed union or non-union of fractures.
2. (Further examples.)
c 1926 'Mixer' *Transport Workers' Song Bk.* 28 Your ranks are broken through By the carping critic 'ring-necks'. And non-union alienage. 1964 S. M. Miller in I. L. Horowitz *New Sociol.* 309 The importance of organizing the non-union and rapidly growing white-collar workers. 1973 L. Snelling *Heresy* II. i. 61 Even if they stayed away I'm sure you could shoot the thing [*sc.* a film] with non-union people. 1974 *Times* 10 May 11/6 The *Daily News*..has continued to be printed..by non-union staff.
Hence **non-u·nionize** *v. trans.*, to make non-union in character; to supply with non-union employees in place of union members; also **non-u·nionized** *ppl. a.*
1926 *Glasgow Herald* 26 July 7 A motion was carried condemning the action of those Scottish newspaper proprietors who had non-unionised their offices since the strike. 1961 W. Sheed *Middle Class Education* xviii. 334, I resent every last butler of yours. And when I think of those non-unionized chambermaids—cheezt! 1968 *Economist* 20 July 19/2 If no union can attain 50 per cent of the votes, the plant remains non-unionised. 1974 *Black Panther* 16 Mar. 15/4 Many non-unionized enterprises were forced to close down as well.

non-uti·lity, *a.* [Non- 3.] Not utility or not for utility; used chiefly of clothes that are of better quality than *utility garments.
1948 *News Chron.* 7 Apr. 1 Children's non-utility garments. *Ibid.*, Non-utility tables, desks, chairs. 1950 *Engineering* 6 Jan. 11/3 'Nonutility generation' is defined as 'generation by producers..for the purpose of supplying electric power required in the conduct of their industrial or commercial operations'. 1957 R. Campbell *Portugal* 82 In Portugal there are still the regional dresses of the people (strictly non-utility). 1960 I. Jefferies *Dignity & Purity* iv. 55 My suits happened to be new and very non-utility.

non-va·lent, *a. Chem.* [f. Non- + *-valent.] **a.** Not capable of entering into chemical combination; inert.
1908 *Athenæum* 28 Mar. 390/1 All the 'non-valent' elements—I should prefer to describe them as the elements which are inert at atmospheric pressure and temperature —should sublime.
b. Having a formal oxidation state of zero.
1950 N. V. Sidgwick *Chem. Elements & their Compounds* II. 1429 Nickel is of course non-valent in the carbonyl $Ni(CO)_4$. 1962 P. J. & B. Durrant *Introd. Adv. Inorg. Chem.* xxiv. 981 None of the elements [of sub-Group VI T] forms any compounds in which it is non-valent.

non-va·nishing, *ppl. a. Math.* and *Physics.* [Non- 6.] Not becoming zero, or not zero.
1907 M. Bôcher *Introd. Higher Algebra* v. 54 The non-vanishing *r*-rowed determinant stands in the upper left-hand corner of the matrix. 1922 G. N. Watson *Theory Bessel Functions* xix. 636 Schlömilch series with non-vanishing coefficients. 1956 *Physical Rev.* CIV. 257/1 The particle has a nonvanishing spin. 1962 Corson & Lorrain *Introd. Electromagn. Fields* ii. 76 In Eq. 2–197 it must include all regions of space in which the field intensity **E** is nonvanishing.

non-ve·rbal, *a.* [Non- 3.] **a.** Not employing or including words; unskilful in the use of words.
1927 B. Russell *Outl. Philos.* xxiv. 271 This keeps us in the verbal realm, and does not get us outside it to some realm of non-verbal fact. 1934 H. C. Warren *Dict. Psychol.* 181/2 *Non-verbal test*, a type of mental (generally intelligence) test, in which no words are used in the test content, but the directions for giving the tests may be either verbal or by pantomime. 1953 W. Burroughs *Junkie* (1972) xv. 152 What I look for in any relationship is contact on the non-verbal level of intuition and feeling, that is, telepathic contact. 1972 J. L. Dillard *Black English* i. 33 There may be many reasons, most of them cultural, why a child may appear 'non-verbal' in certain contexts.
b. Not employing or including a verb; not a form of a verb.
1932 A. H. Gardiner *Theory of Speech & Lang.* iv. 219 Old Egyptian dispensed with the copula in more than one common sub-class of nominal or, as I prefer to call them, non-verbal sentences. 1965 D. Ward *Russ. Lang. Today* iv. 102 Non-verbal predicators include не and нет, as in Он не там *He is not there*. *Ibid.* v. 123 Пред- is productive, however, in non-verbal derivatives. 1973 *Archivum Linguisticum* IV. 49 In respect of their association with a verbal or non-verbal form in the sentence, the suffixes often alternate more or less freely.
So **non-ve·rbalized** *ppl. a.*, not verbalized; **non-ve·rbally** *adv.*, not verbally.
1949 Koestler *Insight & Outlook* xxiv. 342 It is in the very nature of this type of nonverbalized experience that ordinary language..is insufficient to convey it. 1955 J. L. Austin *How to do Things with Words* (1962) ix. 118 The means for achieving its ends non-verbally must be conventional. 1958 B. Bernstein in J. A. Fishman *Readings Sociol. Lang.* (1968) 233 The child is born into a world in which personal qualifications are established non-verbally in the sense that the personal qualifications are left out of the structure of the sentences. 1961 *Countryman* LVIII. III. 462 Most of his waking life the berger is communicating with animals in grunts, cries and other non-verbalized sounds.

non-vi·able, *a.* [Non- 3.] Not viable; *spec.* (*a*) of living beings or living matter: = *inviable *a.*; (*b*) not economically viable.
1879 [in Dict. s.v. Non- 3]. 1946 *Nature* 9 Nov. 678/1 The cells become non-viable. 1956 A. Huxley *Adonis & Alphabet* 204 The wall, which now divides the ancient city from the new, the non-viable state of Jordan from the non-viable state of Israel. 1959 *Times* 6 Mar. 11/5 Her Majesty's Government..is struggling..to build up a primitive people in a non-viable country to a point where they may one day hope to achieve a genuine form of democracy. 1959 *Economist* 18 Apr. 218/1 Agricultural policy is belatedly growing more rational, which means tougher towards the 'non-viable' farm. 1971 J. Z. Young *Introd. Study Man* xxvii. 385 Producing a new version that is not a non-viable freak. 1974 *Nature* 1 Nov. 64/2 The consideration that cells defective for such processes should be non-viable.

non-vi·olence. [Non- 1.] The principle or practice of abstaining from the use of violence.
1920 M. K. Gandhi in J. Nehru *Autobiogr.* (1936) xii. 83, I believe that non-violence is infinitely superior to violence, forgiveness is more manly than punishment. 1936 [see *conscientious *a.* 2]. 1942 Koestler in *Horizon* June 384 His [*sc.* Gandhi's] slope..gradually made him slide down to his present position of non-violence towards the Japanese aggression. 1968 M. Luther King *Trumpet of Conscience* iv. 77 Nonviolence is no longer an option for intellectual analysis, it is an imperative for action. 1974 *Peace News* 1 Feb. 3/1, I knew myself pretty well and knew a little about nonviolence.
So **non-vi·olent** *a.*, characterized by, believing in, or practising non-violence; **non-vi·olently** *adv.*, in a non-violent manner.
1920 M. K. Gandhi *Non-Violence* 9 Mar. (1922) in *Young India 1919–22* 288 We must not intend harm to the English or to our co-operating countrymen, if and whilst we claim to be non-violent. 1923 C. M. Case (*title*) Non-violent coercion. 1937 J. M. Murry *Necessity of Pacifism* i. 22 A Socialism of voluntary community and absolute but non-violent resistance to war. 1948 A. Huxley *Ape & Essence* (1949) 2 Gandhi..just couldn't do anything but resist oppression non-violently. 1971 M. McCarthy *Birds of America* 82 Peter replied non-violently. 'It's for my mother.' 1973 J. Ferguson (*title*) The politics of love: the New Testament and non-violent revolution.

non-vo·coid. *Linguistics.* [Non- 2.] A speech sound that is not a vocoid; = *contoid *sb.*
1943 K. L. Pike *Phonetics* vii. 143 All nonvocoids are contoids. 1952 A. Cohen *Phonemes of English* 37 Pike suggests using the terms vocoids and non-vocoids in phonetic, and vowels and consonants in phonemic analysis.

non-vo·latile, *a.* [Non- 3.] Not volatile. **a.** Of substances (cf. Volatile *sb.* and *a.* B. 3).
1866 Watts *Dict. Chem.* IV. 230 They possess a bitter taste, are non-volatile, and do not suffer any change when exposed to air. 1903 *Phil. Mag.* V. 577 In other cases.. the new matter is itself non-volatile. 1957 G. E. Hutchinson *Treat. Limnol.* I. viii. 553 Evaporating the water to dryness, so obtaining the total solids per unit mass or volume, and then igniting the residue to obtain the non-volatile solids per unit mass or volume. 1971 I. G. Gass et al. *Understanding Earth* 114 It is widely believed that the average composition of meteoric matter provides the best information on the relative abundances of the non-volatile elements.
b. *Computers.* Of a store or storage (cf. *volatile *a.*).
1950 W. W. Stifler et al. *High-Speed Computing Devices* xiv. 313 The storage unit is said to be nonvolatile because it can sustain loss of power without losing all the information stored. 1962 *Gloss. Terms Automatic Data Processing* (*B.S.I.*) 65 A non-volatile store..retains its content when the power supplies are switched off normally, but the content may be lost if power failure occurs. 1969 P. B. Jordain *Condensed Computer Encycl.* 563 Non-volatile storage (almost all magnetic; some optical and some punched-hole media) will hold onto its data through any mishaps.
Hence **non-volati·lity,** the property of not being volatile.
1964 N. G. Clark *Mod. Organic Chem.* v. 67 The highest members [of the paraffin series], e.g. the waxes, because of their non-volatility, possess no smell.

non-white, *a.* and *sb.* Also **non-White.** [Non- 3, 2.] **A.** *adj.* Not having a white skin; of or pertaining to people who are not white. **B.** *sb.* A non-white person. Also *collect.*
1921 *Sci. Amer.* 20 Aug. 125 The States that have the greatest proportion of non-white residents. 1952 B. Davidson *Rep. S. Afr.* I. i. 27 No serious South African will argue any longer (at least in private) that *apartheid*, the complete and geographical segregation of the white from the non-white at all levels, can work. 1953 P. Abrahams *Return to Goli* II. iii. 57 They are the most prejudiced and colour conscious of all the non-White groups. 1956 *Time* 9 Jan. 61/2 The rarity of skin cancer among non-whites..is thought to be caused by a 'true racial difference in susceptibility'. 1959 V. Packard *Status Seekers* (1960) vi. 86 If white residents of an area are in no hurry to leave, but non-whites are eager to come in, the pressure of non-white demand may bid up the price of houses. 1973 E. Bullins *Theme is Blackness* 3 America's educational institutions are predicated upon keeping Black men ignorant of themselves and the other nonwhite four-fifths of the world. 1973 *Black Panther* 5 May 5/2 Eleven non-whites and five civil rights organizations last week charged the San Francisco Police Department with discriminatory employment practices against non-whites and women.

no·n-word. [Non- 2.] A word that is not recorded or not established.
1961 *Life* 27 Oct. 4 (*heading*) A non-word deluge. 1963 *Punch* 16 Jan. 95/1 The aesthetically displeasing non-word 'annoyment'. 1967 D. G. Hays *Introd. Computational Linguistics* v. 87 But for that to be practical, the memory would have to be large enough so that every string of letters shorter than the longest word in the language could serve as a cell index, and almost all the cells in the memory would be empty—since almost every letter string is a nonword. 1970 *Sun* (Gainesville, Florida) 8 Jan., The School Board's petition to the Supreme Court contained an error, the non-word 'predominately'.

non-woven, *a.* and *sb.* [Non- 3, 2.] **A.** *adj.* Not woven (see quot. 1968). **B.** *sb.* A non-woven fabric or material.
1945 *Forbes* (N.Y.) 15 Oct. 16/2 The new non-spun and non-woven cloth will be cheaper, not only because it eliminates so many operations but also because it requires much less of the basic fiber to cover a given area. 1959 A. J. Hall *Standard Handbk. Textiles* iii. 156 It is obvious that non-woven fabrics can never entirely take the place of woven or knitted fabrics. 1963 — *Textile Sci.* i. 11 The characteristics of non-woven fabrics are capable of wide variation. 1968 J. Ironside *Fashion Alphabet* 242 *Non-woven fabrics*. A 'non-woven' ought..to refer to any fabric not made on a loom, but the generally accepted meaning..is that it is any fabric which is not woven, knitted or spun but built up by interlocking fibres by means of chemical bonding agents, and (for fusible fibres) by mechanical means, chemical action, moisture and heat. ..Paper is also included in this definition. 1970 *Cabinet Maker & Retail Furnisher* 23 Oct. 173/2 The chemically or adhesively bonded non-wovens use bonding elements that behave as glueing binders. 1973 *Times* 28 Aug. 17/7 That other branch of the disposable paper industry, nonwovens, is being addressed in the same optimistic tones.

non-ze·ro, *a.* [Non- 4.] Not equal to zero.
1905 *Proc. London Math. Soc.* III. 273 This assumption is valid in the case of functions of finite (non-zero) order.

NOODLE

1925 C. E. Cullis *Matrices & Determinoids* III. I. xxvii. 440 Two sets of non-zero positive integers whose sums are respectively *m* and *n*. 1951 *Physical Rev.* LXXXIII. 1235/1 Small but nonzero errors in the results. 1970 G. K. Woodgate *Elem. Atomic Struct.* vii. 134 The admixture of the $^3P_1$ level accounts for the non-zero intensity of the line 2,537 Å.

noodle, *sb.*[1] Add: **b.** *slang.* The head. cf. Noddle *sb.*[1] 2 b, 3.

1914 Jackson & Hellyer *Vocab. Criminal Slang* 62 *Noodle*,..the human head; brains; savoir faire; mentality. 1923 T. Fane *What's wrong with Movies* vi. 101 To the masses the cinema is only an entertainment, and using their noodles has long been classed..as one of life's hard labors. 1945 M. Trist in *Coast to Coast 1944* 207 Take no notice... She's off her noodle.

noodle (nū·d'l), *sb.*[3] *Jazz.* [Origin unknown.] A trill or improvisation on an instrument. Cf. *noodle *v.*[2] below.

1926 Whiteman & McBride *Jazz* x. 220 'Noodles', that is, fancy figures in saxophone such as triple trills, often crowd out the melody. 1958 *Jazz Review* Nov. 25 My one complaint is that Monk here allows too many of his favourite piano 'noodles' (all pianists seem to have them).

noodle (nū·d'l), *v.*[1] *Austral.* [Origin unknown.] **a.** *intr.* and *trans.* To search for opals (in opal dumps or 'mullock'). **b.** *trans.* To obtain (an opal) in this way. So **noo·dling** *vbl. sb.*[1]

1902 *Geol. Survey* No. 177 (Queensland Dept. Mines) 20 Some splendid opal is found..by turning over and searching the old heaps and mullock—'noodling'. 1931 M. S. Buchanan *Prospecting for Opal in Austral.* 10 Produced ten thousand pounds, with what was 'noodled' or picked up from the dumps. 1940 I. L. Idriess *Lightning Ridge* (1948) xiii. 79 Send the dirt up in the bucket where his mate..would carefully 'noodle' it, seeking tell-tale potch and colour. 1962 R. Webster *Gems* I. x. 197 The practice of picking over old mine dumps for overlooked opal is another form of 'mining'. Such procedure is termed 'noodling'. 1963 A. Lubbock *Austral. Roundabout* 79 Anyone can poke about..and 'paddle' or 'noodle' in the gravelly tailings of the mines. 1967 *Sunday Mail Mag.* (Brisbane) 8 Jan. 6/1 If you do the work by hand, sitting on the mullock heap like a shag on a rock, and patiently sift through the dirt, you are 'noodling'.

noodle (nū·d'l), *v.*[2] orig. *U.S.* [f. the sb.; cf. *noodle *sb.*[3]] *intr.* To improvise or play casually on a musical instrument, esp. in jazz; to play an elaborate or decorative series of notes; also occas. *trans.* Freq. as **noo·dling** *vbl. sb.*[2] Also *fig.*

1937 *Printers' Ink Monthly* May 39/3 *Noodling*, the tuning up of musical instruments with practice runs, trills, scales, etc. 1941 *Sun* (Baltimore) 19 July 8/3 There is something rather exciting about the notion of 'Old Man River', already a folk song, being poured out by the whole choir of woodwinds, with the strings noodling away in the background like rolling waters. 1946 R. Blesh *Shining Trumpets* (1949) ix. 212 Shield's clarinet part..revolves in little runs around lead notes (this is called 'noodling'). 1957 *Nugget* Dec. 5 Every time a jazz musician noodles a passage these days he is followed by a show of bravura on an open Underwood fingered by a jazz writer. 1960 H. O. Brunn *Story of Orig. Dixieland Jazz Band* 164 Larry Shields will go down in history as the father of the 'noodling' style and possessor of one of the most powerful clarinet tones on record. 1966 Auden *About House* 44 In the half-dark, members of an avian orchestra Are already softly noodling, limbering up for An overture at sunrise. 1970 *Globe & Mail* (Toronto) 26 Sept. 25/2 Each piece [of writing] has the feel of jazz, the sense of building, of noodling around here, improvising there and finally resolving.

noodler (nū·dlə*ɹ*). *Austral.* [f. *noodle *v.*[1]] One who 'noodles' (see *noodle *v.*[1]).

1940 I. L. Idriess *Lightning Ridge* xvi. 96 All stones thus found were regarded as the legitimate prey of the noodlers. 1945 *Walkabout* (Austral.) 1 Mar. 15 Other noodlers use a wire-screen and a shovel. 1967 S. Lloyd *Lightning Ridge Bk.* xv. 136 A champion noodler. First to use shear blade to cut opal dirt on dumps.

noodly (nū·dli), *a.* *rare*⁻¹. [Presumably f. Noodle *sb.*[1] + -y[1].] Stupid, silly.

1922 Joyce *Ulysses* 646 A big foolish nervous noodly kind of a horse.

noogenesis (nōᵘ,odʒe·nėsis). [ad. F. *noogénèse*, f. Gr. νόο-ς mind + Genesis.] The coming into being of the stage or sphere of the mind, according to Pierre Teilhard de Chardin's theory of evolution (see quot. 1959).

1959 B. Wall tr. *Teilhard de Chardin's Phenomenon of Man* III. i. 181 Psychogenesis has led to man. Now it effaces itself, relieved or absorbed by another and a higher function—the engendering and subsequent development of all the stages of the mind, in one word noogenesis. 1966 C. F. Mooney *Teilhard de Chardin* ii. 43 'Noogenesis' will therefore be spiritual and social, that is to say, it will concern itself with the development of individuals as persons and with society on the level of interpersonal relationships.

noogenic (nōᵘ,odʒe·nik), *a.* [ad. F. *noogénique*, f. Gr. νόο-ς mind + *-genic.] Pertaining to or connected with noogenesis, or to that which exists in the sphere of the mind.

1959 B. Wall tr. *Teilhard de Chardin's Phenomenon of Man* 307 Without doubt, the 'noogenic' forces of compression, organisation and interiorisation, under which the biological synthesis of reflection operates, do not at any moment relax their pressure on the stuff of mankind. 1964 I. Lasch tr. *V. E. Frankl's Man's Search for Meaning* (rev. ed.) ii. 102 Existential frustration can also result in neuroses. For this type of neuroses, logotherapy has coined the term 'noögenic neuroses' in contrast to..psychogenic neuroses. Noögenic neuroses have their origin..in the noölogical [from the Greek 'noos' meaning mind] dimension of the human existence. 1968 *Time* 2 Feb. 38 Dr Frankl makes great play with words beginning with noo-, from the Greek *noös* (mind), as in noodynamics and noögenic neuroses.

Noogoora (nūgū·rǎ). The name of a Queensland sheep station, used *attrib.* in **Noogoora burr** to designate a composite plant, *Xanthium chinense*, treated as a noxious weed in Australia because of its hooked burrs which become entangled in wool.

1883 F. M. Bailey *Synopsis Queensland Flora* 259 *X[anthium] strumarium*...known as 'Noogoora Burr', and supposed injurious to stock. 1900 —— *Queensland Flora* III. 837 Noogoora Burr. A tall wide-spreading annual... A widespread weed of warm countries. 1934 *Bulletin* (Sydney) 12 Sept. 22/3 Through not obeying a Lands Department order to clear his holding of Noogoora burr, a North [Queensland] grazier has been served with a notice of forfeiture. 1965 *Times* 7 June 3/1 A cerambycid beetle..[is] to be liberated in Queensland..against the important weed Noogoora burr. 1961 *Coast to Coast 1959–60* 41 When we saw how he looked at our cattle-book, we knew he was a dairy farmer, because his eyes classified us—with drought, fire, the pear, flood and Noogoora burr. 1963 *Austral. Encycl.* IX. 227/1 One of the worst offenders [sc. plants whose burrs cling to wool] is noogoora burr (*Xanthium chinense*) which causes annual wool losses of at least £500,000. Since the 1860s it has occupied sheep country from the Gulf of Carpentaria to Sydney... Noogoora burr, indigenous to South America, is supposed to have been introduced with cotton seed.

nook (nŭk), *sb.*[2] *U.S.* representation of an abbreviated form of 'nuclear (weapon)'. Cf. *nuke *sb.*[2]

1964 *Daily Mirror* 24 Aug. 4/5 The generals should be allowed to decide whether to use tactical nuclear weapons, or as the current ugly phrase has it: 'Where and when to put in the nooks.' 1967 *Word Study* Oct. 8/1, I heard a grim, affectionate diminutive: 'nook' (nŭk). 'If the bird is carrying nooks.'

nooky, *a.* (Earlier example.)

1813 M. Edgeworth *Let.* 19 Apr. (1971) 19 Railed-in nice gardens, little nooky green spots.

nooky (nu·ki), *sb.* *slang.* Also **nookie**. [Orig. uncertain; perh. f. Nook *sb.* 3.] Sexual intercourse; hence also, a woman or girl considered as a sex object.

1928 Hecht & MacArthur *Front Page* ii. 46 *Mollie Malloy enters. She is a North Clark Street tart.* [illegible] Hendriks said he'd picked up with a skirt that was a warm baby and he was getting his nookie every night. 1932 J. T. Farrell *Young Lonigan* vi. 142 Schreiber was a good guy, but you know he liked his nookey, and he was always mixed up with some woman or other. 1948 N. Mailer *Naked & Dead* (1949) ii. v. 181 A woman that likes her nookie ain't gonna be satisfied with one man after she gits used to him. 1953 *Landfall* VII. 29 But I've had my bit of nookie on that coast... A sheila of thirty, hot and spry. 1960 A. West *Trend is Up* (1961) iii. 95 [illegible] who cared what the woman was like if the lay was good. 1964 W. Markfield *To Early Grave* (1965) vi. 168 He starts telling me how tough it is for him to break in another nookie, how he's tired of having to make up reading lists for them. 1972 M. Woodhouse *Mama Doll* vii. 83 All the money he'll make as well as the nooky.

noon, *sb.* Add: **6.** *noon-dew*, *-height*, *-reek*, *-rest* (later example), *-scape*, *-top*; *noon-clear*, *-fierce*, *-lod*, *-manned*, *-slight*, *-wide*, *-wild* adjs.; *noon-aglow* adv.

1910 De La Mare *Flora* 13 Her billowing summits heaving noon-aglow. 1874 Hardy *Far from Madding Crowd* II. i. 3 [illegible] 1933 E. Sitwell *Gardeners & Astronomers* 9 [illegible] 1954 W. Faulkner *Fable* (1955) [illegible] 1869 D. G. Rossetti *Let.* 26 Aug. (1965) II. 720 [illegible] 1940 Auden *Another Time* 37 [illegible] 1946 Dylan Thomas *Deaths & Entrances* [illegible] 1922 Joyce *Ulysses* [illegible] 1936 [illegible] 1925 Dylan Thomas [illegible] 1935 *Times to Dance* 25 Buoyed, embayed in heaven's noon-wide reaches. 1936 L. B. Lyon *Bright Feather Fading* 41 Battlement that once glowed Noon-wild is warier lit.

b. noon-basket *U.S.*, a lunch-basket; **noon-halt**, a halt made in the middle of the day; **noon-house** *U.S.*, a house used for rest and meals at midday; now *Hist.*; **noon-mark**, a mark which indicates when it is noon; midday; now *Hist.*; **noonshine**, joc. form of Nuncheon; **noon-spell** *U.S.*, a rest taken in the middle of the day.

1865 A. D. Whitney *Gayworthys* vi. 71 Don't you remember what we used to say at school, when we opened our noon-baskets? 1843 J. C. Frémont *Rep. Exploring Expedition* 15 At our noon halt, the men were exercised at a target. 1854 J. R. Bartlett *Pers. Narr. Explor. Texas* II. xxxvii. 395 On our return we made a noon halt on the banks of the river. *a* 1918 G. Stewart *On Frontier* I. 115 John Dickery rode ahead from our noon halt to try to kill a sage hen. 1845 S. Judd *Margaret* I. 110 Several elderly men and women retired to what was called a 'Noon House', a small building..where they ate dinner and had a prayer. 1891 A. Earle *Sabbath* 102 There might have been seen a hundred years ago, by the side of many an old meeting-house in New England, a long, low, mean, stable-like building... This was the 'noon-house', or 'Sabba-day house'... It was a place of refuge in the winter time, at the noon interval between two services. 1854 B. F. Taylor *Jan. & June* i. 131 The sun..has reached the noon-mark on the threshold. 1889 R. T. Cooke *Steadfast* xxv. 275 Goodness! its most noon-mark and I haven't took a step towards dinner. 1948 *Amer. N. & Q.* Nov. 121/2, I should like to know whether..the term 'noon mark' was once common. 1808 Jane Austen *Let.* 20 June (1952) 195 The Moores came..between one & two o'clock, &..after the noonshine which succeeded their arrival, a party set off for Buckwell. *Ibid.* 24 Oct. 228 The tide is just right for our going immediately after noon-shine. 1839 C. M. Kirkland *New Home* xlv. 300 Even the 'noon-spell' shines no holiday for the luckless subjects of her domination. 1887 J. Kirkland *Zury* 18 Wait till noon-spell, then we'll see! 1889 R. T. Cooke *Steadfast* ii. 30 Its high about noonspell now.

noon-time. Add: (Later (*fig.*) example.) Also as *adv.*, **noontimes** *U.S.*, at mid-day.

1938 T. Wilder *Our Town* 21 Quarter of nine mornings, noontimes, and three o'clock afternoon's, the hull town can hear the yelling and screaming from those schoolyards. 1951 E. Paul *Springtime in Paris* i. 3, I concluded that he was posting the Dove bills noontimes, on a volunteer basis. 1965 *Times Lit. Suppl.* 25 Nov. 1036/3 The noon-time of the genre [sc. the musical] was over.

noosphere (nōᵘ·osfiə*ɹ*). [a. F. *noösphère*, f. Gr. νόο-ς mind + Sphere *sb.* 7.] The name given by Pierre Teilhard de Chardin in his theory of evolution to the stage or sphere characterized by the emergence of consciousness and mind which follows the stage of the establishment of human life (see quot. 1959). Also *fig.* Hence **noosphe·ric** *a.*

1953 J. S. Huxley *Evolution in Action* iv. 110 It provides a new kind of environment for life to inhabit. It needs a name of its own: following Père Teilhard de Chardin, the French paleontologist and philosopher, I shall call it the *noösphere*, the world of mind. 1959 B. Wall tr. *Teilhard de Chardin's Phenomenon of Man* III. i. 182 Much more coherent and just as extensive as any preceding layer, it is really a new layer, the 'thinking layer', which..has spread over and above the world of plants and animals. In other words, outside and above the biosphere there is the noosphere. 1962 M. McLuhan *Gutenberg Galaxy* 32 This externalization of our senses creates what de Chardin calls the 'noosphere' or a technological brain for the world. 1965 *Times Lit. Suppl.* 6 May 350/5 Linguistic anthropologists are at least as agnostic about language which pull that phenomenon down from its cold noosphere back into the warm current of social living. 1966 *New Statesman* 6 May 659/2 Stock fictional characters in some kind of noösphere organisation run by Conchis for his own enlightenment. 1967 *New Scientist* 26 Jan. 227/2 In practice we all act as if the mental aspect of Chardin's noösphere really is a guiding and determining factor in human existence. 1970 *Sci. Amer.* Sept. 53/3 Just before his death on January 6, 1945, he wrote..'I think that we undergo not only a historical, but a planetary change as well. We live in a transition to the noosphere.' By noosphere Vernadsky meant the envelope of mind that was to supersede the biosphere.

noost, var. *noust.

Nootka (nū·tkā, nu·tkā), *sb.* and *a.* [f. *Nootka* Sound, an inlet on the coast of Vancouver Island, British Columbia.] **A.** *sb.* **a.** A North American Indian people of north-western Washington state and Vancouver Island; a member of this people. **b.** The Wakashan language of the Nootka people.

[1841 *Jrnl. R. Geogr. Soc.* XI. 221 The second or Southern Family of the insular tribes may be also denominated *Nootka-Columbian*, from the two places in which they have had most intercourse with Europeans.] 1846 H. Hale *Rep. U.S. Exploring Exped.* VI. 198 (*heading*) The North-Oregon division. All the tribes north of the Columbia..belong to this division... The *Nootkas*..also belong to it. 1868 G. M. Sproat *Scenes Stud. Savage Life* iv. 22 The men (the Nootkahs) are below the middle height, with thick-set limbs, broad faces..and rough, coppery, and tanned skins. 1875 H. H. Bancroft *Native Races of Pacific States* I. iii. 176 The Nootkas are of less than medium height..but rather strongly built. 1910 F. W. Hodge *Handbk. Amer. Indians* II. 82/1 The Nootka form one branch of the great Wakashan family. 1915 E.

Sapir *Abnormal Types of Speech in Nootka* 1 In Nootka there are special words used in speaking of obscene matters or to the presence of women. 1934 *Language* X. 122 In Nootka a monosyllabic word may end in a consonant or a long vowel, but never in a short vowel. 1959 *Chambers's Encycl.* I. 339/2 The tribes of the coast and islands of north-west Canada comprise..southward, coast Salish, Nootka, Chinook, [etc.]. 1972 *Language* XLVIII. 274 Languages like Nootka, in which it seems difficult to distinguish between nominal and verbal roots.

B. *adj.* **1.** Pertaining to or designating this people or their language.

1846 H. Hale *Rep. U.S. Exploring Exped.* VI. 220 We might..add to the synopses and map the *Nootka Family*, comprising the tribes of the west coast. 1854 [illegible] 1875 H. H. Bancroft *Native Races of Pacific States* I. iii. 177 [illegible] 1890 J. G. Frazer *Golden Bough* II. iii. 113 Amongst the Nootka Indians of British Columbia, when a bear had been killed, it was brought in and seated before the head chief. 1915 E. Sapir *Abnormal Types of Speech in Nootka* 3 More specialized Nootka examples to be given presently. 1955 P. Drucker *Indians of Northwest Coast* 12 They [sc. the Chinook Indians] traded slaves from the Californian hinterland up the coast for Nootka canoes.

2. Designating trees native to the Pacific coast of North America, as **Nootka (false-) cypress**, **Nootka Sound cypress**, a coniferous tree, *Chamaecyparis nootkatensis*, of the family Cupressaceæ, also known as the yellow cedar; **Nootka fir** = *Douglas fir* (*Douglas[1]).

1892 A. C. Apgar *Trees Northern U.S.* 195 Nootka Sound Cypress... [illegible] 1897 G. B. Sudworth *Nomencl. Arborescent Flora U.S.* 79 Yellow Cypress. Nootka Sound Cypress. 1957 H. Hadfield *Brit. Trees* 116 The Nootka cypress [illegible] 1966 T. H. Everett *Living Trees of World* [illegible] 1974 *Country Life* 12 Dec. [illegible] 1866 *Treas. Bot.* [illegible] Nootka fir. [illegible] 1977 *N.Z. Timber Jrnl.* [illegible]

So **Noo·tkan** *a.* and *sb.*

1841 *Jrnl. R. Geogr. Soc.* XI. 224 Whoever will compare the list of Nootkan words..with the Tlaoquatch vocabulary..will find that there is very little difference between them. *Ibid.* 225 The Kawitchen tribe..appears..to be a mixed race, compounded of Shahaptan and Nootkans. 1848 *Jrnl. Ethnol. Soc. London* I. 234 Nootkans. 1973 *Amer. Speech* 1969 XLIV. 232 Chinook jargon [has]..Nootkan and Algonkian lexical elements.

no parking. [*parking *vbl. sb.*] Used *attrib.* and *absol.* to designate an area, place, etc., where vehicles may not be parked; also, to designate a sign bearing this instruction.

1946 *Rep. Departmental Comm. Traffic Signs 1944* (Ministry of Transport) 15 We do not consider it desirable that the words 'Halt' or 'No Parking' should be marked on the carriageway. 1956 F. Castle *Violent Hours* (1966) x. 97 The car was..parked in a no-parking zone. 1964 I. Fleming *You only live Twice* vii. 51 A smart Toyopet saloon waiting in a no-parking area. 1964 J. Masters *Trial at Monomoy* ii. 49 That's a No Parking zone. 1968 S. B. Hough *Sweet Sister Seduced* iv. 35 He reached the grass, the No Parking signs. 1971 A. Price *Alamut Ambush* i. 12 A parked car is just a parked car if it's not in a 'no parking' zone. 1973 J. Sinclair *Norslag* x. 85 A 'No Parking' sign by the kerb.

nope (nōᵘp), *adv. slang* (orig. *U.S.*) Extended form of No *adv.*[3] Cf. Yep.

1888 *N.Y. Life* 12 May, Cover 1/2 'Suppose you will be a literary man, like your father, when you grow up.' 'Nope,' said the little boy..'Literary nuthin'! I'm goin' to be a ten-thousand-dollar cook.' 1891 *Harper's Mag.* Nov. 970/1 The professor, wishing to express negation, made use of the objectionable form 'nope'. 1926 C. Mulford *Orphan* ii. 24 Nope, I reckon not—seven husky Apaches are too much for one man to go up against. 1927 Sinclair Lewis *Elmer Gantry* 22 Nope. 1928 E. Wallace *Down Under Donovan* x. 129 'Have you been in Europe before?' 'Nope,' she replied shortly. 1948 R. Stout *Red Threads* iii. 30 'You don't paint in oils, do you?' 'Nope.' 1956 E. Pound tr. *Sophocles' Women of Trachis* 27 Nope, no proof without data. 1964 Mrs. L. B. Johnson *White House Diary* 8 Apr. (1970) 103 And General George Marshall said, 'I'll be damned if I will have people calling me "Marshal Marshall". Nope, we'll just have five-star Generals.' 1971 H. C. Rae *Marksman* i. viii. 72 'Anybody asking for me?' 'Nope.'

no-place, *sb.* and *adv.* [f. No *a.* + Place *sb.*] **A.** *sb.* A place which does not exist.

1929 E. Bowen *Last September* xi. 131 To be enclosed in nonentity, in some ideal no-place. 1934 L. B. Lyon *White Hare* 12 Now is the new page turned and the new light falling From the no-place of the spirit. 1958 *Listener* 28 Aug. 304/2 Neither to accept nor to rebel Is that no-place which I call hell. 1963 *Colorado Portfolio* No. 1 31/3 And in some noplace in some notime he would take out a tiny book..and write.

B. *adv.* (usu. *colloq.*) Nowhere; (in double negative) anywhere. Chiefly *U.S.*

1934 in Webster. 1942 Partridge *Usage & Abusage* 211/1 *No place* is illiterate for *nowhere*, as in 'The jewel was no place to be found'. 1956 H. Gold *Man who was not with It* (1965) ix. 74 'Where are you going?' I..cried out: 'Noplace!' 1969 M. Pugh *Last Place Left* xxv. 187 You're going no place until Herb gets here. 1972 D. E. Westlake *I gave at the Office* (1972) 191 We was in two trucks, packed in tight so you couldn't hardly sit down noplace.

nor, *conj.*[1] Add: **II. 6.** As *sb.* or *adj.* *Computers.* (Written in capitals.) A Boolean function of two or more variables that has the value unity if all the variables are zero and is otherwise zero; 'neither..nor..'. Usu. *attrib.*

1957 *Trans. Amer. Inst. Electr. Engin.* LXXVI. I. 263/1 The transistor NOR circuit was introduced because of the desire to have a single logic element which would combine well in logic configurations with a minimum concern for matching and element loading. A NOR element has a signal output only if there are no input signals. *Ibid.*, P-n-p and n-p-n transistor NOR circuits..cannot be readily intermixed. 1958 [see *NAND]. 1969 J. J. Sparkes *Transistor Switching* iv. 69 For NOR gates the Boolean expression is first reduced to its simplest form containing the product of sums. 1971 J. M. Smith *Digital Logic* iv. 41 The NOR function is used extensively, especially in resistor-transistor logic systems. 1972 *IEEE Trans. Computers* XXI. 153/2 Optimal networks consisting of NOR–OR gates (each gate produces the NOR and/or the OR of its inputs) are tabulated for all Boolean functions of three variables.

nor- (nǭɹ), *prefix.* *Organic Chem.* [f. Nor(mal *a.* and *sb.*) 1 a.] Prefixed to the names of organic compounds to denote the replacement of one or (esp. in terpenes) all the (methyl) side-chains by hydrogen atoms.

The use arose from the designation of a compound $C_8H_{12}O_4$ as 'normal opianic acid', of which opianic acid was supposed to be a dimethyl derivative: 1868 Matthiessen & Foster in *Jrnl. Chem. Soc.* XXI. 358 We therefore propose to call the compound $C_8H_6O_5$ monomethyl-normopianic acid, or..if the contraction is admissible, methyl-noropianic acid.

norbo·rnane [*bornane* f. Born(eol + -ane], the crystalline bicyclic hydrocarbon, C_7H_{12}, which is obtained by replacing the three methyl groups of bornane (camphane) by hydrogen atoms and is the parent compound of a number of terpenes; bicyclo[2.2.1]heptane; **nordihy·droguaiare·tic acid** [f. *guaiaretic acid*, f. Guaia(c + -ret(in + -ic)], a compound, $C_{18}H_{22}O_4$, found in the resin of the creosote bush..; a crystalline phenol, $C_{14}H_{18}(OH)_4$, which is used as an antioxidant for oils and fats; **norephe·drine**, the crystalline, optically active compound, $C_9H_{13}NO$, which is obtained by replacing the methylamino group of ephedrine by an amino group and has actions and uses resembling those of ephedrine; **norethi·sterone**, a crystalline compound, $C_{20}H_{26}O_2$, which has actions similar to those of progesterone and is used as a component of oral contraceptives and in the treatment of amenorrhoea and functional uterine bleeding; (β-)**norni·cotine** [a. F. *nornicotine* (M. & M. Polonovski 1927, in *Compt. Rend.* CLXXXV. 1333)], an optically active liquid alkaloid, $C_9H_{12}N_2$, found in tobacco and with resembling nicotine; **norsyne·phrin(e** = *octopamine; **nortesto·sterone**, the 19-norsteroid derived from testosterone.

1925 *Chem. & Engin. News* 1 Mar. 930/3 Corresponding to these six types [of bicyclic compounds] are the four parent ring systems norbornane..norcarane..norpinane..and northujane. 1962 E. L. Eliel *Stereochem. Carbon Compounds* x. 302 Norbornane, despite its easy accessibility, is an appreciably strained system. 1974 *Jrnl. Chem. Soc. Perkin Trans. I* 322 (*heading*) Inhibition of ring expansion reactions in the norbornane system by neighbouring methyl groups. [1919 *Chem. Abstr.* XIII. 991 Schiff's norguaiaretic acid is shown to be nordihydroguaiaretic acid.] 1944 *Oil & Soap* XXI. 33/1 An investigation was made of the antioxidant properties of nordihydroguaiaretic acid. 1947 *Jrnl. Amer. Pharm. Assoc.* (Sci. Ed.) XXXVI. 81/1 One of..the most effective bacteriocidal agents present in [the shrub] *Larrea divaricata* Cav. has been characterized as nordihydroguaiaretic acid. 1968 M. K. Schwitzer *Margarine & Other Food Fats* viii. 323 Nordihydroguaiaretic acid..was permitted to be used in the United States in December, 1943 at the rate of 0·01 per cent. The antioxidant effect is very marked in animal fats... It is less effective in vegetable fats. 1971 P. Tooley *Fats, Oils & Waxes* iv. 125 Oxidation can cause serious deterioration in a lipophilic producing unpleasant rancid odours and taste... This can be largely prevented by the use of antioxidants such as..nordihydroguaiaretic acid. 1929 *Chem. Abstr.* XXIII. 2431 Reduction of PhCH(OH)CH(NO_2)Me with Zn dust and dil. H_2SO_4 gave norephedrine..and nor-ψ-ephedrine. 1943 *Jrnl. Lab. & Clin. Med.* XXVIII. 704 Norephedrine and pseudonorephedrine together constitute propadrine, of which the commercial preparation contains also a small amount of ammonium chloride. 1973 *Biochem. Jrnl.* CXXXVI. 769/1 Norephedrine is extensively metabolized in the rabbit but not in man. 1956 *Proc. Soc. Exper. Biol. & Med.* XCI. 419/2, 10 mg nor-ethisterone given twice daily represents a reproducibly effective dose in women for the production of marked progestational changes. 1964 Norethisterone [see *ethinylœstradiol]. 1973 *Nature*

20 Aug. 351/1 If inferences can be drawn from this preliminary study using mouse foetal hearts, it would seem that the concentration of norethisterone acetate in the pill should be kept to a minimum. 1927 *Chem. Abstr.* XXI. 3905 (*heading*) β-Pyridyl-α-pyrrolidine (nornicotine). 1936 *Jrnl. Econ. Entomol.* XXIX. 854 Macht & Davis..observed that β-nornicotine was more toxic for certain vertebrates..than natural nicotine. 1951 A. Grollman *Pharmacol. & Therapeutics* xv. 261 Nornicotine, which has similar actions to nicotine, lacks only the CH_3 group in the pyrrolidine nucleus. 1968 Larson & Silvette *Tobacco Suppl.* I. 682/1 Nornicotine inhibited development of the chick-embryo, and was even more potent than nicotine in causing hydrops. 1959 Norsynephrine [see *octopamine]. 1952 *Experientia* XI. 1135 Sympathomimetic amines (tyramine, norsynephrine, adrenaline, noradrenaline). 1960 *Biochim. & Biophys. Acta* XLIII. 566 The enzyme responsible for norsynephrin formation is dopamine β-oxidase. 1950 *Jrnl. Chem. Soc.* 367 This β-unsaturated ketone has now been isomerised..to the αβ-unsaturated ketone..., which is a 19-nortestosterone. 1955 H. Hirschmann *Hormones* III. ii. 544 19-Nortestosterone on the other hand, is inferior to testosterone as an androgen but equal to it in its myotrophic..effect. 1973 *Jrnl. Appl. Physiol.* XXXV. 378/1 Nortestosterone esters are more potent than testosterone in their capacity to stimulate erythropoietic activity in mice.

b. Denoting the contraction of a chain or ring of carbon atoms by one methylene group, as **norchola·nic acid** [*cholanic* f. Chol- + -an + -ic], a crystalline, optically active compound, $C_{23}H_{38}O_2$, obtained by shortening the side-chain of cholanic acid by one methylene group; **norpi·nic acid**, a crystalline dicarboxylic acid, $C_8H_{12}O_4$, existing in *cis* and *trans* isomers and differing from pinic acid in having a carboxyl group attached to the cyclobutane ring in place of a carboxymethyl group.

1927 *Chem. Abstr.* XXI. 590 Oxidation of [*sc.* cholanic acid] by CrO_3 in AcOH gives the next lower homolog, norcholanic acid. 1958 C. W. Shoppee *Chem. Steroids* iv. 191 Similar reactions have been carried out in the norcholanic acid..series. 1909 *Jrnl. Chem. Soc.* XCV. 1170 Our experiments have, however, shown that this ring, in norpinic acid, exhibits quite unusual stability. 1953 J. L. Simonsen *Terpenes* II. iii. 129 By the action of dilute hydrochloric acid at 180° [on the *cis*-isomer], an equilibrium mixture of *cis*- and *trans*-norpinic acids is obtained. 1969 A. R. Pinder *Chem. of Terpenes* v. 91 The first preparation and the physical properties of norpinic acid suggested it was derived from cyclobutane.

2. Denoting the normal (unbranched chain) isomer of the compound to whose name it is prefixed, as **norleu·cine** [ad. G. *norleucin* (E. Abderhalden et al. 1913, in *Zeitschr. f. physiol. Chem.* LXXXVI. 455)], an optically active, crystalline, non-essential amino-acid, $CH_3(CH_2)_3CH(NH_2)COOH$; **norva·line** [ad. G. *norvalin* (Abderhalden & Kürten 1921, in *Fermentforschung* IV. 328)], an optically active crystalline amino-acid, $CH_3(CH_2)_2$·$CH(NH_2)COOH$.

1913 *Jrnl. Chem. Soc.* CIV. I. 1049 The acid, for which the name norleucine is proposed, has been obtained from glycine and hexylbromide to form polypeptides. 1929 *Jrnl. Biol. Chem.* XCVII. 342 Norleucine has been isolated from spinal cord protein. 1963 D. M. Greenberg *Metabolic Pathways* II. xiv. 111 Norleucine and norvaline..[are not] known to occur naturally in proteins. 1971 *Nomencl. Org. Chem.* (I.U.P.A.C.) (ed. 2) C-42 193 (*table*) Norleucine... Norvaline. [*Note*] Use of 'nor' in these names to denote a 'normal' (unbranched) chain conflicts with more recent usage. 1921 *Jrnl. Chem. Soc.* CXX. I. 547 dl-α-Amino-n-valeric acid..to which the name norvaline is given, is prepared by the usual methods from α-valeric acid. 1933 *Ann. Rev. Biochem.* II. 79 Further evidence of the occurrence of norvaline in proteins has been furnished by Abderhalden & Heyns..who report its occurrence in steer horn. 1969 *Biochem. Jrnl.* CXV. 621 Kinetic studies of glutamate dehydrogenase were made with wide concentration ranges of the coenzymes NAD⁺ and NADP⁺ and the substrates glutamate and norvaline.

nor (nǭɹ), *a.* *Organic Chem.* [f. prec.] Applied to compounds or groups of compounds conventionally named by adding the prefix *nor- to the name of a parent compound; esp. with a prefixed numeral indicating the methyl group lacking or a prefixed capital letter indicating the ring which is contracted as compared with the parent compound.

1940 *Chem. Abstr.* XXXIV. 5455 Diminution from a 6- to a 5-ring by loss of C atoms..in rings A, B and C gives rise to the corresponding A-nor, B-nor and C-nor compds. 1946 *Jrnl. Biol. Chem.* CLXIII. 589 The nor, bisnor, and etio homologues of 3,11-dihydroxy-12-ketocholanic acid have been prepared. 1954 *Jrnl. Amer. Chem. Soc.* LXXVI. 4092/2 The efficacy of this 19-nor analog..in human females has been established. 1972 *Adv. Steroid Biochem. & Pharmacol.* III. 93 Two of the 19-nor steroids, norgestrel and allylestrenol, appear to show differences from norethisterone and lynestrenol.

noradrenaline (nǭræ̆dre·nălin, -ǣdrĭnălin). *Med.* Also **-in.** [f. *nor- 1 a + *adrenaline.] An amine, $(HO)_2C_6H_3$·CHOH·CH_2·NH_2, related to adrenaline, having a hydrogen atom in place of the methyl group; 1-(3,

4-dihydroxyphenyl)-2-aminoethanol; *spec.* the lævorotatory isomer, which is the transmitter substance of sympathetic nerves and at some synapses in the central nervous system, is also synthesized in the adrenal medulla, and whose acid tartrate is used, usu. by intravenous infusion, as a vasoconstrictor for raising blood pressure.

1932 *Arch. Internat. de Pharmacodynamie et de Thérapie* XLI. 366 This compound..is referred to occasionally as nor-adrenaline. 1948 *Science* 6 Aug. 135/2 The facts..point to the existence of two highly active sympathomimetic substances: nor-adrenaline and adrenaline. 1962 A. Huxley *Island* xiii. 206 The power generated by fear or envy or too much noradrenalin. 1966 [see *nor *a.* 2 b]. 1971 J. Z. Young *Introd. Study Man* xiii. 629 Carnivorous animals are said to secrete large amounts of noradrenaline, which makes them aggressive. 1973 *Daily Colonist* (Victoria, B.C.) 30 June 12/1 Both humans and rats secrete a large amount of noradrenaline when they shiver.

Hence **noradre·nergic** *a.* [after *adrenergic, *cholinergic *adjs.*], (of a nerve-cell) stimulated by or liberating noradrenaline; **noradrene·rgically** *adv.*

1963 *Arch. Internat. de Pharmacodynamie et de Thérapie* CXLII. 257 The highly vasoconstriction, spleen contraction and cardioacceleration, which occur in response to a threshold stimulation of the corresponding postganglionic noradrenergic nerve supply, are not sensitized by intravenous infusion of noradrenaline. 1973 *Naunyn-Schmiedebergs Arch. Pharmakol.* CCLXXIX. 53 An α-receptor-mediated feed-back control of noradrenaline release, previously demonstrated in postganglionic sympathetic nerves, also operates in central noradrenergic neurones. 1975 *Nature* 22 Aug. 659/2 Peripheral, noradrenergically innervated tissue.

norate (nǭ·rē̍t), *v.* *U.S. dial.* Also **norrate.** [? Corruption of Narrate *v.*] **a.** To announce; to spread (information) by word of mouth. **b.** To denigrate. So **nora·tion, norration.**

1853 J. W. Page *Uncle Robin* 231 Der's some folks who tells de people, dat, dar masters in dis country [*sc.* in the South], when der niggers runs away, puts out a noration, dat dey will give four hundred dollar' to anybody who will bring one o' der runaway niggers to um, dead or live. 1895 *Dialect Notes* I. 373 'We will norate the preaching' (*i.e.* announce the services to be held). 1905 *Ibid.* III. 89 [Arkansas] A French specialist has given out the 'noration' that kissing is not a hurtful process. 1914 *Ibid.* IV. 110 Norate... Also, norration, noration. *n.* 1921 C. Campbell *Southern Highlander & his Homeland* 143 [In the Southern Appalachians] a man wishing to hold a public meeting has it *norated*, that is, the announcement of it spread by report. 1938 *Amer. Speech* XIII. 6/2 [S.E. Arkansas] It is norated about that he is a crook. 1941 H. Skidmore *Hawk's Nest* 22 Norratin' it round bout them wanten men to work. 1949 H. Hornsby *Lonesome Valley* 10 'You're planning to norate the word,' he said... 'More than apt they'll expect me to do the preaching.' 1954 *Publ. Amer. Dial. Soc.* xxi. 33 [South Carolina] *Norate*, *v.t.*, to depreciate (scandalize).

norbergite (nǭ·ɹbəɹgəit). *Min.* [f. *Norberg*, the name of the village in Sweden near which it was discovered + -ite[1].] A basic silicate and fluoride of magnesium, $Mg_3SiO_4(F,OH)_2$, which is found as pink or whitish orthorhombic crystals.

1926 P. Geijer in *Geol. Fören. Stockholm Förhand.* XLVIII. 84 Norbergite. Only in massive aggregates, crystal system unknown... The proposed name is derived from the name of the district, since that of the mine is unsuitable for an international nomenclature. 1928 *Amer. Mineralogist* XIII. 349 The norbergite occurs in a coarsely crystalline limestone from the Nicoll Quarry, and grains or crystals are commonly several millimeters across. 1968 *Mineral. Mag.* XXXVI. 966 The polyhedral distortions in norbergite are smaller than in forsterite in accord with the decreased number of shared edges.

nordenskiöldine (nǭɹdenʃö·ldīn, -ʃkiˌö·ldīn). *Min.* [ad. Sw. *nordenskiöldin* (W. C. Brögger 1887, in *Geol. Fören. Stockholm Förhand.* IX. 255), f. the name of Baron Nils Adolf Erik *Nordenskiöld* (1832–1901), Swedish geologist and explorer: see -ine[5].] A borate of calcium and tin, $CaSnB_2O_6$, found as colourless or yellow rhombohedral crystals.

1890 *Jrnl. Chem. Soc.* LVIII. 1078 Nordenskiöldine, named after the celebrated traveller, is of great rarity in the Norwegian veins. 1935 *Mineral. Abstr.* VI. 46 An ore pipe in marble near a granite contact at Arandis, South-West Africa..consists of the minerals tin as margin cassiterite, tourmaline, and small colourless plates of the rare mineral nordenskiöldine. 1966 *Doklady Earth Sci.* CLXIV. 131/1 Macroscopically, the nordenskiöldine synthesized is a dense white mass in which hexagonal, heavily tabular crystals up to 0·005 mm in size are discernible under the microscope.

Nordic (nǭ·ɹdik), *a.* and *sb.* [ad. F. *nordique* (J. Deniker 1898, in *L'Anthropologie* IX. 127), f. *nord* North: see -ic.] **A.** *adj.* Of or pertaining to the Scandinavian people or their languages; *spec.* of or pertaining to a physical type of northern Germanic peoples characterized by tall stature, bony frame, light colouring, and dolichocephalic head.

In Nazi doctrine the 'Nordic race' was regarded as essentially 'superior' to other races.

1898 W. Z. Ripley in *Pop. Sci. Monthly* Oct. 744 A direct physical relationship between the three [peoples], referring them all to a so-called nordic race, is confirmed by the very latest and most competent authority [sc. J. Deniker]. 1921 *Contemp. Rev.* Jan. 56 All the talk about Nordic supremacy is vanity when we look at the facts in Europe. 1929 Chesterton *Thing* xiv. 113 Englishmen who now call themselves Nordic used to call themselves Teutonic. 1937 [see sense B. 1 below]. 1938 G. Heyer *Blunt Instrument* xi. 197 You ought to have seen me giving my impression of a Nordic public-school man with a pipe at the Wild Bottoms... ... 1939 H. G. Wells *Holy Terror* iv. ii. 419 The new generation of Germans were ashamed of the Hitler period and the Nordic legend. 1940 —— *All Aboard for Ararat* i. 24 The third, Japhet, was what the Germans would consider a Nordic type, all milk and roses. 1957 M. Bellow *Europe & Europeans* iv. 86 The Nordic languages, especially Old Norse, borrowed important words from Anglo-Saxon. 1959 *Chambers's Encycl.* XIV. 319/1 The term viking is Nordic. 1966 W. P. Lehmann in Birnbaum & Puhvel *Anc. Indo-Europ. Dialects* 16 The most striking innovation common to Gothic and the Nordic languages is the development of a stop in geminate j and w clusters. 1968 G. Jones *Hist. Vikings* 69 As Wessén says, at the beginning of his history of the Swedish language..'we have then come to the *Primitive Nordic language*, the parent tongue of the present Nordic vernaculars, common to the Scandinavian countries down to the beginning of the Viking Age'. 1970 Foote & Wilson *Viking Achievement* 5 Some of the Nordic provinces had sent out tribes in the Migration Age to join the more southerly Germanic peoples as they cut their way through the old domains of the Roman Empire. 1973 G. Brade *Snake on Green* vii. 36 A crowd of Nordic drunks at a table near him was singing. 1974 *Encycl. Brit. Micropædia* VII. 366/1 *Nordic Council*, organization of the Nordic states of Denmark, Finland, Iceland, Norway, and Sweden for the purpose of consultation and cooperation on matters of common interest.

b. Of a skiing competition involving cross-country or jumping events.

1954 *Brit. Ski Yr. Bk.* XV. 30 The greatest of all of the Nordic competitions—Holmenkollen. 1960 *Ski-ing* ('Know the Game' series) 24/2 The Nordic events include cross-country races and jumping competitions. 1966 S. Eriksen *Come Ski with Me* i. 4 A Norwegian had won the Nordic Combination and the 18 km. cross-country in Holmenkollen. 1969 H. Heller *Ski* xiii. 170 Nordic skiing is the original and basic form of ski-ing... The fundamental techniques are self-evident—the diagonal stride, double poling with and without a single pole and uphill strides—on skier words, walking on skis. 1972 *Evening Telegram* (St. John's, Newfoundland) 24 June 23/1 Rolf Kjaernsli of Norway has been named nordic skiing program director to develop nordic and cross-country skiing in Canada.

B. *sb.* **a.** A person of the Nordic type.

1925 *Contemp. Rev.* May. June 1105 The tall blonde race of northern Europe, sometimes called 'Teutons', but more scientifically 'Nordics'. 1928 Chesterton *Money for Nothing* ii. 32 Well, all I can say is..it's no life for a refined Nordic. 1936 H. G. Wells *Anat. Frustration* xv. 176 It is for the treatment of the Jews that we are most frequently urged to condemn Hitlerism... Their [sc. the Jews'] racial purity is as much a falsehood as the racial purity of the 'Nordics'. 1937 A. Huxley *Ends & Means* xiii. 242 Hitlerian theology affirms that there is a Nordic race, inherently superior to all other. Hence it is right that Nordics should organize themselves for conquest.

2. The northern branch of the Germanic languages.

1955 T. Burrow *Sanskrit Lang.* 8 Germanic..may be divided into East Germanic or Gothic (extinct), Nordic or Scandinavian, and West Germanic. 1962 *Scandinavian Studies* XXXIV. 26 (*title*) Proto-Scandinavian and Common Nordic. 1972 in Van Coetsem & Kufner *Toward Gram. of Proto-Germanic* 78 The close relationship of 'Nordic' and 'Gothic'.

Nordicism (nǭ·ɹdisiz'm). [f. *Nordic *a.* and *sb.* + -ism.] **a.** The state or condition of being Nordic; the characteristics of the Nordics. **b.** The belief in or doctrine of the cultural and racial supremacy of the Nordic people. Hence **No·rdicist**, one who believes in the supremacy of the Nordic people.

1926 H. H. Dickson *Jefferson Davis* (1931) ii. 24 The modernism of the North and the Nordicism of the South came more and more into conflict. 1924 *Glasgow Herald* 7 Apr. 8 Doubtless we would have seen a transfer of 'Nordicism'. 1925 *Nation* (N.Y.) CXX. 516/1 The transition from Aryanism to Nordicism in Germany. 1930 R. Hughes *High Wind in Jamaica* vii. 151 The Nordicism of captain and mate kept the rest looking clean enough. 1934 A. Toynbee *Study of Hist.* I. ii. 221 The Nordicists claim it [sc. the monopoly of the unique magical ability in Mankind] for all White Men with fair hair and blue eyes. 1957 G. Clark *Archaeol. & Society* (ed. 3) viii. 259 Much..was corrupted by doctrines which stemmed directly from Gustaf Kossinna (1858–1931), an ardent exponent of Pan-Germanism and Nordicism.

nordite (nǭ·ɹdəit). *Min.* [ad. Russ. *nordit* (V.I. Gerasimovsky 1941, in *Doklady Akad. Nauk SSSR* XXXIII. 496): see quot. 1941 and -ite[1].] A silicate of sodium, strontium, manganese, calcium, and lanthanides, found as light brown orthorhombic crystals.

1941 V. I. Gerasimovsky in *Compt. Rend. (Doklady) de l'Acad. des Sci. de l'URSS* XXXIII. 496 A more detailed study of the mineral has shown that it cannot be identified with any of the minerals already known, and therefore it was given a new name—nordite (because of its northern origin). 1958 *Chem. Abstr.* LII. 12701 The predominance of La over Ce in nordite is an exception to the Oddo-Harkins rule. 1970 *Amer. Mineralogist* LV. 1167 The structure of nordite is closely related to the structures of mellilite and datolite-gadolinite, and can be considered as an unusual combination of both.

nordmarkite (nǭ·ɹdmāɹkəit). [f. *Nordmark*, name of an area in Sweden + -ite[1].] **1.** *Min.* A brown manganesian variety of staurolite.

1868 J. D. Dana *Syst. Min.* (ed. 5) v. 389 *Manganese-Staurolite*, *Nordmarkite* (anal. 28); from Nordmark in Wermland, Sweden, of chocolate-brown color, with H. = 6·5, G. = 3·54, and presenting the usual crystalline form. Its easy fusibility is reason for believing this variety the distinctive name *Nordmarkite*. 1968 I. Kostov *Mineralogy* 289 Staurolites rich in manganese are termed nordmarkite.

2. *Petrogr.* [a. G. *nordmarkit* (W. C. Brøgger 1890, in *Zeitschr. f. Kryst. und Mineral.* XVI. I. 55).] A syenite composed mainly of microperthite, with lesser amounts of quartz and usu. oligoclase and biotite, which has a trachytoid or granitic texture.

1895 *Mineral. Mag.* XI. 115 The corresponding abyssal or plutonic rock is nordmarkite. 1928 *Mineral. Abstr.* III. 498 The differentiation of the Lung Wang Miao akerite has given rise to a whole series of granitic derivatives: nordmarkite, alkali-granite-aplite, and quartz porphyry. 1942 *Amer. Jrnl. Sci.* CCXL. 362 Nordmarkite porphyry from a dyke at Blindern, by Oslo, Norway. 1970 *Meddelelser om Grønland* CXC. II. 15 The nordmarkites are grey or fawn in colour and are fairly coarse-grained rocks. *Ibid.* 16 In most of the nordmarkites the ferromagnesian minerals occur in small clusters.

Hence **nordmarki·tic** *a.*, composed of, or having the nature of, (the rock) nordmarkite.

1947 *Mineral. Mag.* X. 90 At Dorowa [S. Rhodesia] nordmarkitic granite occurs with syenite in the outer ring. 1953 *Q. Jrnl. Geol. Soc.* CIX. 161 Keratophyres, kersantites and nordmarkitic rocks occur on both sides of the outcrop of the Moine thrust-plane.

nordstrandite (nǭ·ɹdstrăndəit). *Min.* [a. F. *nordstrandite* (D. Papée et al. 1958, in *Bull. Soc. chim. de France* 1306/2), f. the name of Robert A. van *Nordstrand*, 20th-cent. U.S. chemist: see -ite[1].] One of the phases of aluminium hydroxide, $Al(OH)_3$.

1962 *Nature* 20 Oct. 265/1 The nordstrandite varies from almost colourless to coral pink and reddish-brown, the coloration being largely due to inclusions of finely divided goethite. 1968 *Mineral. Abstr.* XIX. 176/2 Nordstrandite occurs in all (6) Hungarian bricklays examined and is the only phase in the Debrecen clay.

norepinephrine (nǭ·repine·frin). *Biochem.* [f. *nor- + *epinephrine.] = *noradrenaline.

1948 *Jrnl. Pharmacol. & Exper. Therap.* XCII. 369 Norepinephrine ('Arterenol'), a racemic compound,..is 1·5 times more pressor than racemic epinephrine. 1951 A. Grollman *Pharmacol. & Therapeutics* xi. 205 Norepinephrine is a vasoconstrictor and unlike epinephrine does not decrease the peripheral resistance or increase the cardiac output. 1962 J. Glenn et al. *Into Orbit* 115 The flow of norepinephrine [*sic*] from his adrenal glands during the flight had been more than 2½ times what it had normally been. 1971 *Nature* 2 Apr. 330/1 Some, if not all, states of mental depression may be associated with a deficiency in norepinephrine..at functionally important receptor sites in the brain.

norethynodrel (nǭrepi·nōdrĕl). *Pharm.* [f. *nor- + Eth(ane + *-yn(e + *-odrel*, of unknown origin.] A synthetic hormone, $C_{20}H_{26}O_2$, with actions and uses similar to those of norethisterone, with which it is isomeric.

1957 *Endocrinology* LX. 804 Norethynodrel has been examined for its ability to produce endometrial gland development in the estrogen-primed spayed or immature rabbit. 1962 *New Scientist* 7 June 506/3 Most contraceptive pills consist of a mixture of two steroids obtained from yams. One is norethynodrel..which is believed to inhibit ovulation. The other is an oestrogen. 1970 Passmore & Robson *Compan. Med. Stud.* II. xii. 6/1 Shift of the double bond in the A ring of norethisterone from the 4–5 to the 5–10 position produces the isomer, norethynodrel.

Norfolk. Add: **b. Norfolk dumpling,** (*b*) a plain dumpling made from bread dough; **Norfolk jacket** (earlier and later examples); **Norfolk reed,** the common reed, *Phragmites australis*, grown in East Anglia for use as thatching material; **Norfolk spaniel,** a name formerly used for the English springer spaniel, a breed once associated with the estates of the Duke of Norfolk; **Norfolk suit,** a suit with a Norfolk jacket and knee breeches; **Norfolk terrier,** the drop-eared variety of the Norwich terrier (see *Norwich).

c 1600, etc. [see Dumpling 1 a]. 1747 H. Glasse *Art of Cookery* ix. 112 *Norfolk Dumplings*. Make a good thick

batter... Eat them hot. 1877 *Cassell's Dict. Cookery* 458/1 Norfolk Dumplings... When bread is made at home, take a little of the dough.., make it up into small balls.., drop them into fast-boiling water... Send melted butter, sweetened and flavoured with lemon juice, to table with them. 1933 C. H. Senn *Century Cookery Bk.* (ed. 10) 1013 *Norfolk Dumplings*,—Make an ordinary bread dough... Serve with a boat of rich gravy or other suitable sauce. 1972 *Mrs. Beeton's Family Cookery* 449 Norfolk Dumplings..Boiling water, Salt, Bread, Dough... Make the dough... Cream the yeast with the sugar and add the warm water and the melted fat... Serve with..jam, treacle, golden syrup, *or* butter and sugar. 1866 J. Macgregor *Thousand Miles in Rob Roy Canoe* i. 9 The 'Norfolk jacket' is a loose frock-coat, like a blouse, with shoulder-straps, and belted at the waist, and garnished by six pockets. 1969 *Queen* 17–30 Sept. 76 Norfolk jacket in cream leather; long side vents; belt..45 gns. [1925 E. G. Blake *Roof Coverings* ii. 18 The reeds are grown principally in the eastern counties, especially on the Norfolk Broads.] 1952 *Oxf. Jun. Encycl.* VI. 440 A well-bedded thatch of Norfolk reed..has been known to last as long as 50 years. *Ibid.*, Although Norfolk reed is by far the most durable, it is also the most expensive. 1965 P. Wayre *Wind in Reeds* ix. 111 Norfolk reed is in much demand for thatch, being most durable and of good length—often exceeding eight feet. 1971 *Country Life* 18 Nov. 1403/1 The great tithe barn at Tisbury, in Wiltshire, has just been re-thatched with Norfolk reed. 1859 *Field* 5 Mar. 180/3 Will any of the readers of *The Field* kindly inform me what spaniels are considered the best for woodcock-shooting—the Sussex, Clumber, Norfolk, &c.? 1867 'Stonehenge' *Dogs Brit. Is.* I. iii. 40 The Norfolk is one of the four descriptions of spaniels known as 'springers'. *Ibid.*, The Norfolk spaniel is now seldom to be obtained mute. 1945 C. L. B. Hubbard *Observer's Bk. Dogs* 65 The old name of Norfolk Spaniel..is obsolete. 1896 *Junior Army & Navy Stores Catal.* p. xxxvi/2 Norfolk Jackets.. Suits. 1913 C. Mackenzie *Sinister St.* I. I. vii. 112 An exciting Monday spent in buying a Norfolk suit and Eton collars. 1938 J. Cary *Castle Corner* iii. 116 He was seen every day walking about Knockeen with his solemn, fierce air and his smartest norfolk suits. 1964 *Kennel Gaz.* 419/2 On September 22, 1964, the General Committee of the Kennel Club agreed to register drop-eared Norwich Terriers as Norfolk Terriers, a separate breed and not a variety. 1968 C. G. E. Wimhurst *Bk. Terriers* xxi. 154 The Norfolk Terrier is one of the smallest of the terriers. 1971 F. Hamilton *World Encycl. Dogs* 463 The Norfolk Terrier was, until September 1964, the drop-eared Norwich Terrier. *Ibid.* 467 There is little or no difference in character between the Norwich and the Norfolk.

d. *ellipt.* A Norfolk jacket. In pl., a Norfolk suit.

1902 E. Nesbit *Five Children & It* ii. 47 Nine pockets in my Norfolks. *Ibid.* viii. 215 A giant little boy—in Norfolks like my brother's. 1904 *T. Eaton & Co. Catal.* Spring & Summer 87/1 Brownie Norfolks, made of dark green Tweed. 1969 R. T. Wilcox *Dict. Costume* 248/1 The coat of the Duke of Norfolk's hunting suit, the first 'Norfolk', appeared in the 1880's with knickerbockers, a revival of knee breeches for day wear. 1970 S. J. Perelman *Baby, it's Cold Inside* 28 Where'd you get that Norfolk?.. Brooks hasn't carried that model in years.

Norfolk Island (nǭ·ɹfŏk əi·lænd). The name of a South Pacific island about five hundred miles north-west of New Zealand, used *attrib.* in **Norfolk (Island) pine** to designate a large conifer, *Araucaria heterophylla* (formerly *excelsa*), of the family Araucariaceæ, native to this island.

[1778 J. R. Forster *Observations made during Voyage round World* v. 174 In the opposite, or Westernmost part of the South-Sea, lies a small isle, which has obtained the name of Norfolk-Island... Peculiar to this isle, and to the Eastern end of Caledonia, we found a species of coniferous tree, from the cones probably seeming to be a cypress: it grows here to a great size, and is very heavy but useful timber.] 1803 A. B. Lambert *Descr. Genus Pinus* I. 87 Norfolk Island Pine... This tree..is the tallest at present known. 1836 *Agriculturist's Manual* (P. Lawson & Son) 354 The Norfolk Island Pine..was first discovered by the celebrated circumnavigator Captain Cook, in his second voyage, on Norfolk Island and New Caledonia. 1854 J. C. Patteson *Let.* 24 Aug. in C. M. Yonge *Life J. C. Patteson* (1874) I. v. 175, I shall write to you very often, and send you ferns and seeds, and tell you about the Norfolk Island pines. 1920 C. Coltman-Rogers *Conifers* vi. 227 The Norfolk Island Pine..is fast becoming quite a common corner-window side-show in many English homes. 1933 *Bulletin* (Sydney) 27 Dec. 20/4 Almost as well known overseas as the eucalypts is the Norfolk Island pine. 1940 F. Sargeson *Man & his Wife* (1944) 91 A very old settled place with a row of Norfolk pines planted along the beach. 1966 *Times* 28 Mar. (Austral. Suppl.) p. xv/2 The home unit blocks are growing taller than the two Norfolk pines. 1968 *Southerly* XXVIII. 172 Halfway up the west face of the valley was a double row of Norfolk Island pines. 1974 A. Mitchell *Field Guide Trees Britain* 57 One tender species, the Norfolk Island Pine, *A[raucaria] heterophylla* (*excelsa*), frequent as a pot plant and growing to 30 m at Tresco, Isles of Scilly.

norgestrel (nǭɹdʒe·strĕl). *Pharm.* [f. *nor- + *pro)gest(ogen + *-rel*, prob. after *norethynodrel.] An artificial steroid hormone, $C_{21}H_{28}O_2$, which has actions similar to those of progesterone and is used in some contraceptive pills.

1966 R. A. Edgren et al. in *Internat. Jrnl. Fertility* XI. 389 Norgestrel..is a highly potent, totally synthetic progestogen chemically related to norethisterone and norethynodrel. 1974 *Nature* 8 Mar. 98/3 Ovranette contains a progestogen called norgestrel which is not converted to oestrogen in the body. In fact it is actively anti-oestrogenic, reversing some of the changes brought about by oestrogen.

‖**nori** (nǭə·ri). [Jap.] A Japanese food prepared from fronds of a seaweed of the genus *Porphyra*, eaten either fresh or dried, when they stick together to form small sheets. Cf. *laver *sb.*[1] 2.

1892 E. Arnold *Japonica* ii. 86 Large slices of broiled *tai*, and *tsubo* or *nori*, sea-weed,..of which the Japanese are fonder than the foreigner is likely to prove. 1966 P. S. Buck *People of Japan* (1968) xiv. 167 Rice covered with shredded egg and *nori*. 1973 A. Broinowski *Take One Ambassador* xi. 174 We all collected *nori*, the seaweed along the beach.

no-right (nōᵘ,rəit). [f. No *a.* + Right *sb.*[1] 7.] In Jurisprudence, an obligation not to prevent the exercise of a privilege.

1913 W. N. Hohfeld in *Yale Law Jrnl.* XXIII. 32 As indicated in the above scheme of jural relations, a privilege is the opposite of a duty, and the correlative of a 'no-right'. 1923 —— *Fundamental Legal Conceptions* i. 39 The correlative of X's right that Y shall not enter on the land is Y's duty not to enter; but the correlative of X's privilege of entering himself is manifestly Y's 'no-right' that X shall not enter. 1938 M. Radin in *Harvard Law Rev.* LI. 1150 The phrase 'no right' was subjected to a great deal of critical and destructive comment. A '"no-right"', one critic once declared, 'might be an elephant.' 1972 W. A. Wilson in *Juridical Rev.* Aug. 162 The correlative of a liberty is a no-right. The jural opposite of a right is a no-right.

nork (nǭɹk). *Austral. slang.* [See quot. 1966.] A woman's breast. Usu. *pl.*

1962 C. Rohan *Delinquents* 157 Hello, honey, that sweater—one deep breath and your norks will be in my soup. 1966 Baker *Austral. Lang.* (ed. 2) x. 215 *Nork*, a female breast, usually in plural. (Ex Norco Co-operative Ltd., a butter manufacturer in N.S.W.). The form *norg* is reported from Melbourne. 1969 *Private Eye* 21 Nov. 14 She's a top model with norks out to here. 1972 I. Hamilton *Thrill Machine* iii. 17 She's only a body, all she's got is big norks. 1973 P. White *Eye of Storm* 593 Hits herself in the eye with an independent nork it isn't any laughing matter.

Norland[2] (nǭ·ɹlănd). The name of the *Norland* Institute (see below), now the Norland Nursery Training College, used *attrib.* to designate the methods of child care taught there, a nurse trained in these methods, or a nursery following them. Also *ellipt.*, a Norland nurse.

[1892 *St. James's Gaz.* 28 Sept. 12/2 It will be interesting to watch the experiment being made by Mrs. Walter Ward, who has just opened the Norland Institute for the training of kindergarten nurses. 1893 *Farm, Field & Fireside* 10 Feb. 460/1 The Norland Institute is situated at 9, Norland Place, Holland Park Gardens, London, W. It was started by Mrs. Walter Ward (Miss Emily Ford) for the training of ladies as children's nurses on Froebelian principles.] 1894 *Nursing Record & Hospital World* 7 Apr. 234/2 As a Norland Nurse, I have now had opportunity of seeing the value of the work to both employer and employée. 1899 *West-End* 23 Aug. 14/1 [The nurse] must receive a certificate, and do a further three months as probationer in a family.., or at some special Institution, before she dons her pretty neat brown uniform, and sallies forth as a fully-fledged Norland nurse. 1945 N. Streatfeild *Saplings* ii. 19 Lena would have liked..a young Norland or the equivalent, looking smart in her uniform. 1959 *Times* 29 Sept. 2/7 Norland/Trained Nurse required. 1972 J. Gathorne-Hardy *Rise & Fall of Brit. Nanny* vi. 178 From the first a Norland Nurse was forbidden to hit a child... This enlightened view..has remained a key feature of the Norland training. 1975 *Harpers & Queen* May 139/2 When they were eighteen months old they went to a Norland nursery. 1975 *Times* 4 Aug. 5/6 Each weekend they go to Barnwell with [the baby] Alexander and his Norland nanny.

norm. Add: **a.** (Further examples.)

1911 R. Brooke *Coll. Poems* (1918) 154 All of the accents upon all the norms!—And ah! the stress on the penultimate! We never knew blank verse could have such feet. 1941 J. P. Marquand *H.M. Pulham, Esquire* iv. 44 Beatrice considered that I was utterly characteristic, completely true to type; once she called me a norm. 1961 S. R. Herman in J. A. Fishman *Readings Sociol. of Lang.* (1968) 505 Frequently the newcomer arrives with the belief that the prevailing norm about using Hebrew to the exclusion of other languages is more rigidly observed than is actually the case... Soon, however, the newcomer becomes aware of the wide range of deviations from the norm. 1964 M. Argyle *Psychol. & Social Probl.* iii. 38 One particular respect in which a group equilibrium develops is in the formation of norms—shared patterns of behaving, feeling and thinking. All social groups develop norms, particularly about matters connected with the group's main purposes and activities. ..When group members deviate from the norms, various kinds of persuasion, pressure and sanctions are exerted in order to make them conform. 1965 *Economist* 13 Feb. 645/2 The other vital point will be the 'norms' or 'guiding lights' [for incomes] recommended for each year.

b. (Earlier and later examples.) Also defined analogously for other quantities. [Introduced as L. *norma* by Gauss 1832, in *Commentationes Recentiores Soc. R. Scient. Gottingensis* VII. Class. math. 98.]

1856 W. R. Hamilton *Notebook* in Halberstam & Ingram *Math. Papers Sir W. R. Hamilton* (1967) III. 657, *a* + *ib* is said to be a complex number, when *a* and *b* are integers, and $i = \sqrt{-1}$; its norm is $a^2 + b^2$; and therefore the norm of a product is equal to the product of the norms of its factors. 1932 Turnbull & Aitken *Introd. Theory Canonical Matrices* iv. 38 This fundamental Hermitian inner product of $\bar{x}$ and x is often called the norm of the complex vector x... The square root of the norm, taken with positive sign, $(\bar{x}x)^{\frac{1}{2}}$, is sometimes denoted by $|x|$. 1949 A. Albert *Solid Analytic Geom.* i. 3 The norm of a vector *P* is defined to be the inner product $P \cdot P = x_1^2 + \ldots + x_n^2$. 1952 C. Møller *Theory of Relativity* iv. 99 It then follows..that the 'square of the magnitude of the four-vector', or the norm of the vector, $\sum_i a_i^2 = \sum_i a'^2_i$, is an invariant. 1967 MacLane & Birkhoff *Algebra* v. 187 Each quadratic field $Q(\sqrt{d})$, with the elements $\sigma = r + s\sqrt{d}$, has as automorphisms the identity and $\sigma \mapsto \bar{\sigma} = r - s\sqrt{d}$... The product $\sigma\bar{\sigma} = r^2 + s^2d$ is called the norm $N(\sigma)$ of σ.

(ii) The positive square root of the quantity defined in sense b; more generally, a quantity defined on a vector space over the real or complex field which represents a generalization of the concept of length or magnitude and has the properties that $\|u\| > 0$ if $u \neq 0$ ($\|u\|$ being the norm of the vector *u*), $\|u\| = 0$ if $u = 0$, $\|au\| = |a|\,\|u\|$ (*a* being a real number), and $\|u+v\| \leq \|u\|+\|v\|$ (*v* being another vector).

1921 *Proc. Nat. Acad. Sci.* VII. 84 The notion of norm or numerical value of a complex quantity, $c = a + b\sqrt{-1}$, namely, $|c| = \sqrt{(a^2 + b^2)}$, as it arises in algebra, has a more or less immediate generalization to more extensive matric systems. 1951 P. R. Halmos *Introd. Hilbert Space* i. 13 The norm of a vector α in the inner product space..coincides with the absolute value of the complex number *a*. 1955 L. F. Boron tr. *I. P. Natanson's Theory of Functions of Real Variable* I. vii. 199 Let $f(x) \in L_p$... The number $\|f\| = \sqrt[p]{(\int_a^b |f(x)|^p dx)}$ is called the norm of the function $f(x)$ (considered as an element of L_p). 1962 Kacinskas & Counts tr. *L. S. Pontryagin's Ordinary Differential Equations* iv. 152 The maximum modulus of this function..will be called its norm. 1965 Patterson & Rutherford *Elem. Abstract Algebra* v. 184 A norm can be defined in a vector space having an inner product by writing $\|x\| = \sqrt{(x \cdot x)}$... There exist norms which cannot be expressed in this way in terms of an inner product. In the case where an inner product exists, it is clear that the length of a vector satisfies the requirements for a norm. *Ibid.* 185 Let $M_{2,2}(\mathbf{C})$ be the vector space of 2 × 2 matrices over the complex field. Then the mapping $\left[\begin{smallmatrix} a & b \\ c & d \end{smallmatrix}\right] \to \max(|a|, |b|, |c|, |d|)$ defines a norm in $M_{2,2}(\mathbf{C})$. The mapping $\left[\begin{smallmatrix} a & b \\ c & d \end{smallmatrix}\right] \to \sqrt{(|a|^2+|b|^2+|c|^2+|d|^2)}$ defines another norm. 1970 F. A. Matsen *Vector Spaces & Algebras* i. 11 The norm N_c of *c* is defined by $N_c = (cc^*)^{1/2} = \sqrt{(a^2 + b^2)}$.

c. In Communist countries, a standard unit of work prescribed.

1935 S. & B. Webb *Soviet Communism* II. ix. 706 They [*sc.* piece-work rates] are, in some cases, even progressive, the rate rising by stages for output beyond the norm. 1952 *Manch. Guardian* 6 June, Stakhanovite women miners in the Donetz basin are performing four, nine, and eleven norms each. 1959 *Times* 12 Mar. 13/6 This moulding process may vary from the crudest regimentation and subordination..to factory-like specification and 'norms' to mere pilotage..towards the Marxist haven.

2. *Petrol.* A hypothetical mineral composition of a rock calculated by assigning the compounds present to certain relatively simple minerals in accordance with prescribed rules.

1902 [see *mode *sb.* 5 b]. 1932 A. Johannsen *Descr. Petrogr. Ign. Rocks* II. 272 (*heading*) Table 135. Norms of rhyolites. 1973 *Jrnl. Petrol.* XIV. 35 The norms used are not based on stoichiometric formulae..but approach the compositions of the constituents as actually found in natural rocks. *Ibid.* 250 The C.I.P.W. norms which accompany the analyses were calculated using an Fe_2O_3/FeO ratio estimated to be appropriate for the quartz-fayalite-magnetite-buffered charges.

3. *attrib.* and *Comb.*

1934 *Archit. Rev.* LXXVI. 42/1 It represents a return to the machine, regarded realistically and as mechanism to produce mechanical norm-types. 1958 M. Argyle *Relig. Behaviour* v. 57 All these findings can be regarded as instances of social learning, mediated by the usual processes of persuasion, imitation and norm-formation. 1961 J. N. Findlay *Values & Intentions* ix. 399 The religious object..must tend more and more towards the pattern of a detached, suprapersonal, norm-setting *mind*. 1964 I. L. Horowitz *New Sociology* 32 Why do people choose rapid industrialization with its attendant psychological turmoil over social stability and norm adherence? 1966 *Mathematical Rev.* XXXI. 8/2 Since norm-sentences are neither true nor false, logical connectives undergo reinterpretation. 1969 J. F. Szwed in Halpert & Story *Christmas Mumming in Newfoundland* 116 During socialization, each person acquires an awareness of social sanctions and at the same time becomes 'norm-oriented'.

norm (nǭɹm), *v.* *Math.* [f. the sb.] **1.** *trans.* = *normalize *v.* 3 a. ?*Obs.*

1931 P. Dienes *Taylor Series* viii. 274 We 'norm' the mapping with respect to $u = a$ by requiring that $a_1k = 1$. 1941 R. V. Churchill *Fourier Series* iii. 38 The functions of the set are normed by dividing each function $g_n(x)$ by $[N(g_n)]^{1/2}$.

2. *trans.* [back-formation from *normed *a.*] To define a norm on (a space).

1959 L. F. Boron tr. *Naimark's Normed Rings* i. 73 The space $X/\mathfrak{M}$, normed by formula (1), will be called a normed factor-space. **1964** D. E. Brown tr. *Kantorovich & Akilov's Functional Analysis in Normed Spaces* ii. 51 The metric spaces..are also linear sets and can be normed. **1972** A. G. Howson *Handbk. Terms used in Algebra & Analysis* xxii. 111 The vector space $\mathscr{L}(E,F)$ can be normed in the following way.

Hence **no·rming** *vbl. sb.*

1967 L. Rédei *Algebra* I. iv. 307 In different rings norming is carried out in different ways.

normal, *a.* and *sb.* **A.** *adj.* **2. b.** Substitute for def.: (i) Of a salt: containing no acidic hydrogen.

1860 W. A. Miller *Elements Chem.* (ed. 2) II. x. 338 The most usual form of salt, in which 1 atom of a protoxide is united with 1 atom of an acid to form the normal salt. **1869** [in Dict.]. **1915** P. W. Oscroft *Adv. Inorg. Chem.* iv. 38 It must not be understood that normal salts are always neutral bodies with regard to their action with litmus; this is far from being the case. **1965** B. J. Moody *Comp. Inorg. Chem.* xiii. 188/2 Both sodium sulphate and trisodium orthophosphate are normal salts.

(ii) Of (the concentration of) a solution: having one gramme-equivalent of solute per litre.

1863 F. Sutton *Syst. Handbk. Volumetric Anal.* 19 The normal solutions prepared on the gramme system are equally applicable for that of the grain, and *vice versâ.* **1892** [in Dict.]. **1915** P. W. Oscroft *Adv. Inorg. Chem.* v. 48, 25 c.c. of a caustic potash solution required 15 c.c. of a normal sulphuric acid solution for neutralization. **1955** C. R. N. Strouts et al. *Analyt. Chem.* I. xii. 258 A solution of one-tenth normal strength is designated as N/10. **1967** R. Fulton *Course in Titrimetric Anal.* ii. 7 Normal solutions are more widely used [than molar solutions].

(iii) (Composed of molecules) containing an unbranched chain of carbon atoms in an alkane molecule or alkyl radical.

1869 C. Schorlemmer in *Proc. R. Soc.* XVII. 373, I had obtained the normal propyl alcohol by this method. **1871** —— *Ibid.* XIX. 487 The first group, which I called normal paraffins, contain the carbon atoms linked together in a single chain. **1876** [in Dict.]. **1932** I. D. Garard *Introd. Org. Chem.* ii. 20 At average room temperature, those normal paraffin hydrocarbons containing four carbon atoms or less are gaseous, and those from five to seventeen, liquids. **1968** J. A. Monick *Alcohols* iii. 86 If the [parent] hydrocarbon consists of an unbranched carbon chain, the equivalent primary alcohol is called normal.

c. *Physics.* Of, pertaining to, or being a mode of vibration in which every particle executes simple harmonic motion at the same frequency and in phase (or 180° out of phase).

1867 Thomson & Tait *Treat. Nat. Philos.* ii. 274 There are in general..*i* distinct determinate displacements, which we shall call the normal displacements, fulfilling the condition, that if any one of them be produced alone, and the system then left to itself for an instant at rest, this displacement will diminish and increase periodically according to a simple harmonic function of the time, and consequently every particle of the system will execute a simple harmonic movement in the same period. **1877** [see *mode *sb.* 4 c]. **1927** Toft & Kersey *Theory of Machines* xiv. 362 Any type of oscillation other than a normal mode may be considered as being the sum of a number of motions each of which is a normal mode. **1942** Synge & Griffith *Princ. Mech.* vii. 209 The periods and frequencies of normal modes are called normal periods and normal frequencies. **1962** P. J. & B. Durrant *Introd. Adv. Inorg. Chem.* viii. 229 There may be several normal vibrations of different frequencies characteristic of a given molecule. **1971** *Amer. Jrnl. Physics* XXXIX. 484/2 The major purpose of this paper is to determine..the normal-mode longitudinal-vibration frequencies of the chain.

d. *Geol.* Applied to a fault and to faulting in which the relative downward movement occurred in the strata situated on the upper side of the fault plane. So *normal-faulted* adj.

1876 A. H. Green *Geol.* xi. 382 The direction of the hade in a normal fault. **1878** J. LeConte in *Amer. Jrnl. Sci. & Arts* XVI. 99 Thus arise two distinct slips: In the one, the more common or normal, the strata drop on the hanging-wall side of the fissure, in the other or reverse fault, the strata on the hanging-wall side is slidden up and over the other side by the sheer force of the horizontal pressure. **1902** *Jrnl. Geol.* X. 873 Orographic blocks may..display an arrangement in zigzags or *en échelon*, which it is difficult to explain upon any other basis than that of normal faulting. **1944** A. Holmes *Princ. Physical Geol.* vi. 79 Normal faults involve an extension of the faulted beds. **1974** Flint & Skinner *Physical Geol.* iv. 293/2 Normal faults are caused by tensional forces that tend to pull the crust apart, and also by forces tending to expand the crust by pushing it upward from below. *Ibid.* 294/1 In the Earth's crust are many zones that have been deformed repeatedly by normal faulting. **1975** *Nature* 1 May 22/2 This part represents a tensional arm..bifurcating into normal-faulted shear zones south of Sinai.

e. *Statistics.* = *Gaussian *a.* b.

1893 K. Pearson in *Nature* 26 Oct. 615/2 As verification note that for the normal probability curve $3\mu_2^2 = \mu_4$ and $\mu_3 = 0$. **1894** —— in *Phil. Trans. R. Soc.* A. CLXXXV. 72 A frequency-curve, which, for practical purposes, can be represented by the error curve, will for the remainder of this paper be termed a normal curve. **1897** *Proc. R. Soc.* LXII. 176 A random selection from a normal distribution. **1920** [see *Gaussian *a.* b]. **1928** T. C. Fry *Probability & its Engin. Uses* viii. 244 Both the Binomial and the Poisson Laws, under suitable conditions, approach the Normal Law as a limit. **1938** A. E. Waugh *Elem. Statistical Method* vi. 94 Many phenomena of biology, economics, psychology, education, etc., even though not exactly normal in distribution, can be described roughly by the normal curve. **1951** Dixon & Massey *Introd. Statistical Anal.* v. 63 Many practical problems have statistical answers based on the 'assumption' that the distribution of the population is normal... The truth of this assumption may be checked by plotting the sample cumulative-percentage points on normal-probability paper. **1968** *Brit. Med. Bull.* XXIV. 211/1 One of these [alternatives for small samples] consists simply in plotting the individual sample values..on 'Normal-probability' graph paper—this is a special graph paper whose vertical scale has been 'stretched' in such a way that the S-shaped cumulative Normal curve is transformed graphically into a straight line.

f. *Med.* Of a saline solution: containing the same concentration of sodium chloride as the blood.

1895 *Jrnl. Physiol.* XVIII. 50 Neither can the injection of normal saline be of much benefit. **1924** L. Clendening *Mod. Methods Treatment* ii. 153 Normal salt solution given in subcutaneous areolar tissue (beneath the breasts or in the thighs) or intravenously is very frequently used in surgical shock. *Ibid.*, In normal salt solution with glucose any concentration could be used, without hemolysis. **1970** F. N. Douglas *Essentials Pharmacology in Clinical Nursing* iii. 22 Physiologic normal saline (0·9 per cent) is used for treating dehydration in the absence of acidosis.

g. Physics. *normal state* = *ground state* (*ground *sb.* 18).

See also sense *2 i.

1914 [see *N I. 4 b]. **1922** A. D. Udden tr. *Bohr's Theory of Spectra* ii. ii. 32 All the atoms exist in that stationary state in which the value of the energy is a minimum. This state I shall call the normal state. **1952** R. W. Ditchburn *Light* xvii. 550 The atom very quickly makes a transition back to the normal state re-emitting the radiation. **1963** G. F. Lothian *Electrons in Atoms* iii, Mercury atoms have quantized energy levels ..above the normal or 'ground' state.

h. *spec.* = heterosexual.

1914 E. M. Forster *Maurice* (1971) xxii. 106 Against my will I have become normal. I cannot help it.

i. *Physics.* Pertaining to or characteristic of a substance that is not in the superconducting state.

1927 *Nature* 3 Dec. 818/2 The resistance became normal at a certain critical value of the magnetic field. **1938** D. Shoenberg *Superconductivity* i. 4 Apart from the loss of resistance the metal appeared to have identical properties both in the superconducting and normal states. **1955** H. B. G. Casimir in W. Pauli *Niels Bohr* 119 The theory describing the normal state when the absolute temperature tends towards zero is extremely simple. **1968** C. G. Kuper *Introd. Theory Superconductivity* i. 6 An upper bound for the resistance in the superconducting state can be established. If R_n is the resistance of the specimen when normal, then $R/R_n < 10^{-15}$.

j. *normal forest*, a collection of trees at various stages of development, organized to provide a regular yield of timber.

1928 R. S. Troup *Silvicultural Systems* i. 1 With the object of ensuring future sustained yields the ideal of the normal forest has been created. Such a forest contains a regular and complete succession of age-classes..in correct proportion, density, and distribution... The normal forest can hardly be said to exist in reality; rather it should be regarded as an ideal to be aimed at. **1962** C. E. Hart *Practical Forestry* vi. 109 The term normal forest is used for a forest or woodland or group of woodlands containing a regular and complete succession of age classes, from the youngest to the oldest.

k. *Physics.* Applied to a component of a superfluid that is regarded as not having the properties of a superfluid and as co-existing at the atomic level with a component that does have them, in a proportion that decreases with decreasing temperature.

1947 L. Tizas in *Physical Rev.* LXXII. 842/2 The density [of the Bose-Einstein liquid]..will be subdivided into two parts: $\rho = \rho_n + \rho_s$ where ρ_n is the density connected with the 'molecules' of the gas and ρ_s refers to the 'background' in which the molecules are moving. The subscripts refer to 'normal' and 'superfluid', a terminology which will be explained below. *Ibid.* 853/2 In this 'anomalous' region [*sc.* between about 1°K and 2·19°K] the liquid is a mixture of a normal component (like helium I) and a superfluid component. **1967** J. Wilks *Properties of Liquid & Solid Helium* iii. 39 Two baths of liquid helium II..at slightly different temperatures..are connected by a fine capillary. The capillary almost completely inhibits the flow of the normal fluid, on the other hand the superfluid having zero viscosity, may pass freely. **1975** *Nature* 10 Apr. 480/3 At a finite temperature the superfluid phase [of ^{4}He] behaves like a mixture of two fluids: a 'normal' component, behaving like an ordinary viscous liquid..; and a 'superfluid' component,..closely associated with the atoms in the Bose condensate.

3. (Later examples.) Also *ellipt.* = *normal school.*

1888 *Nat. Educ. Assoc. U.S. Addresses & Proc. 1887* 478 We say that normal-school training is as essential to good teaching as the work of a medical school to the physician. *Ibid.* 502 A course of normal instruction. **1925** in P. W. Slosson *Great Crusade* (1931) xv. 431 Unlawful for any teacher in any of the universities, normals, and all other public schools..to teach..that man has descended from a lower order of animals. **1939** H. G. Wells *Holy Terror* iv. i. 359 An increasing number of women are taking up professions now; at architecture, catering, various industries, normal teaching..they are practically as good as men or better. **1960** P. E. Burrup *Teacher & Public School Syst.* iv. vii. 275 'Normal Schools' for the preparation of teachers..have been replaced by teachers' colleges, liberal-arts colleges, and universities, each with a college of education. **1960** Curtis & Boultwood *Introd. Hist. Eng. Educ.* xii. 277 In that year [*sc.* 1836] a normal school of design was established. *Ibid.* 278 The Normal School of Design became the Royal College of Art in 1896.

4. Philos. *normal form* (see quot. 1950).

1948 *Mind* LVII. 173 He [*sc.* Boole] introduced two notions which are of the greatest importance, namely, that of a *truth-function* and that of a *normal form.* **1950** L. M. Hammond et al. tr. *Hilbert & Ackermann's Math. Logic* i. §3. 11 (*heading*) Normal form for logical expressions. *Ibid.* 12 Any combination of sentences can be brought into a certain normal form by means of equivalence transformations;..this normal form consists of a conjunction of disjunctions in which each component of the disjunction is either an elementary sentence or the negation of one. **1952** *Mind* LXI. 564 The procedure is to be applied to normal forms, reduction to which is a matter of propositional logic. **1965** E. J. Lemmon *Beginning Logic* 189 Normal forms have a certain interest in connection with the truth-table method, since they provide an independent test as to whether a wff is tautologous, contingent, or inconsistent; and they are also used in certain proofs of the completeness of the propositional calculus. **1973** J. J. Zeman *Modal Logic* viii. 117 The normal-form theorem holds for systems LS1° and LS1.

B. *sb.* **4.** (Later example.)

1957 M. Spark *Comforters* v. 114 She snapped back at him. And so, in his need for their relations to return to a nice normal, he said peaceably, [etc.].

5. A normal variety of anything; that which, or a person who, is healthy and is not impaired in any way.

1894 W. Bateson *Study of Variation* 17 For the belief that such races are descended from the putative normal scarcely ever rests on proof. **1901** *Amer. Jrnl. Psychol.* XII. 235 The blind rats learned the original task as well as the normals. **1908** *Daily Chron.* 14 Oct. 4/4 We might divide them [*sc.* criminals] into three groups:— Normals, Juveniles and children; and The degenerate. **1916** J. S. Haldane *Organism & Environment* (1917) iv. 102 The normals of anatomy are not mere physical structures, nor are the normals of physiology mere averages: they are manifestations of the life of an organism regarded as a whole. **1940** *Psychol. Bull.* XXXVII. 425 Scales may be successively discovered and standardized on a reservoir sample of normals. **1964** M. Critchley *Developmental Dyslexia* vii. 40 Measuring the reaction time..in normals and in dyslexics. **1973** *Nature* 12 Jan. 99/1 The response of lymphocytes from normals or leukaemics to low (7μg ml.$^{-1}$) doses of PHA.

b. A heterosexual person.

1966 *New Statesman* 29 Apr. 623/3 He [*sc.* Coward]'s working for the same kind of audience—Knightsbridge normals—and still going as near the knuckle as he thinks they can abide. **1971** M. McCarthy *Birds of America* 304 A female *clocharde* had reason to shrink from 'normals'.

normalcy. Add: Now freq. (esp. in U.S.) used in the senses of Normality.

1920 W. G. Harding in F. L. Allen *Only Yesterday* (1931) ii. 41 America's present need is not heroics but healing; not nostrums but normalcy; not revolution but restoration. **1929** G. N. Clark in *S.P.E. Tract* xxxiii. 417 If..'normalcy' is ever to become an accepted word it will presumably be because the late President Harding did not know any better. **1932** G. K. Chesterton *Sidelights* ii. xiv. 182 Life in a modern town, whatever else it is, is not Normalcy. **1939** *John o' London's* 9 June 369/1 That insistent normalcy of men who cannot afford to permit themselves to be thrown off balance. **1951** M. McLuhan *Mech. Bride* 47/1 Professor Kinsey's surveys, with their economist-like normalcy charts. **1957** V. J. Kehoe *Technique Film & T.V. Make-up* i. 17 On stage, where strong lights and distance of the actors from the audience wash out and flatten the features, make-up restores to the face the look of normalcy in both color and contour. **1965** *New Statesman* 7 May 733/1 A kind of spectral normalcy.

normalism (nǭ·ɹmăliz'm). *rare.* [f. Normal *a.* and *sb.* + -ism.] The quality or state of being normal.

1891 F. W. Bain *Antichrist* ii. 113 The planing away of all gnarled and knotty characteristics, the reducing each individual to precisely the same external appearance. This is the essence and the consequence of the impulse to normalism.

normality. Add: **1. b.** *spec.* in *Statistics* (cf. *normal *a.* and *sb.* A. 2 e).

1928 *Amer. Jrnl. Psychol.* XL. 348 We can change 0·93 into a P.E. [*sc.* probable error] by multiplying by 0·845 (assuming normality of distribution). **1938** A. E. Waugh *Elem. Statistical Method* vi. 95 In most statistical problems there is no a priori reason for expecting normality of distribution—no reason for believing in advance that the data will be distributed as are the coefficients of the expansion $(\frac{1}{2}+\frac{1}{2})^n$. **1968** *Brit. Med. Bull.* XXIV. 211/1 The assumption of Normality is central to the most powerful statistical techniques. **1974** *Nature* 22 Mar. 288/1 A weighted least squares analysis of the two sets of correlations..provides (given normality) a test of goodness of fit of the model.

2. *Chem.* The concentration of a solution as a proportion of the normal concentration.

1903 *Sci. Abstr.* VI. 315 Boric acid was agitated..with an excess of aqueous hydrochloric acids of different normalities. **1928** A. W. Wellings *Volumetric Analysis* ii. 41 The normality of the acid solution will be 0·1N. × 18·6/50. **1966** *McGraw-Hill Encycl. Sci. & Technol.* III. 361/2 In double-decomposition reactions normality may be an ambiguous concept unless referred to a specific reaction. **1972** *Nature* 8 Sept. 69/3 At higher concentra-

tions of the salt..the viscosity was considerably smaller (1·60 at a normality of 0·50).

normalizable (nǭ·ɹmăləizăb'l), *a.* [f. Normaliz(e *v.* + -able.] Capable of being normalized.

1939 V. Rojansky *Introd. Quantum Mech.* i. 37 If *u* and all of its derivatives approach zero when $|x| \to \infty$, and if $\int_{-\infty}^{\infty} \bar{u}u\,dx$ is finite, we call *u* a normalizable function (this use of the term is not standard in the literature). **1955** O. Klein in W. Pauli *Niels Bohr* 116 There may be some reason to expect that the most obvious formulation of the interaction problem on these lines should already give rise to a normalizable theory, the effect of the 'wrong' states being limited to further infinite contributions to the mass and charge of the electron. **1968** *Physics Bull.* Nov. 373/2 A normalizable state ϕ_0.

normalization. Add to def.: More widely, the process of normalizing (in any sense). (Further examples.)

1916 *Jrnl. Iron & Steel Inst.* XCIV. 10 It is not suggested to abandon the term 'normalise', but to define normalisation as a treatment which will give equalisation and not metal of abnormal variations. **1929** [see *normalize *v.* 3 a]. **1944** Gregory & Simons *Heat-Treatment Steel* xx. 273 The steels not requiring normalization after forging should be cooled off in lime, then annealed. **1959** E. M. McCormick *Digital Computer Primer* xi. 155 To facilitate normalization, many computers have a special instruction..that counts the positions the number must be shifted left to be normalized. This count is then used to modify the exponent. **1967** C. L. Wrenn *Word & Symbol* p. xii, The problem of normalisation for Old English still requires scholarly attention. **1972** Bergman & Bruckner *Introd. Computers & Computer Programming* vi. 169 The only exception to the normalization rule is zero; there are many ways in which we can represent zero. **1972** P. W. Williams *Numerical Computation* ix. 162 One method of normalization is to divide all the elements of a vector by the largest element so that vectors have unity as the largest element. Alternatively, each element could be divided by the sum of the squares of the elements of the vector in which case vectors have unit length. **1973** *Amer. Speech 1969* XLIV. 220 Two sociolinguistic problems: immigrant bilingualism and language normalization.

b. *Psychol.* The subconscious process whereby the mental image of a shape, pattern, etc., is changed to resemble something more familiar; also *attrib.* So **normali·zing** *vbl. sb.*

1929 J. J. Gibson in *Jrnl. Experimental Psychol.* XII. 3 The first kind [of change] he [*sc.* Wulf] calls *normalizing* (Normalizierung), *i.e.* a change (presumably in the reproductions) in the direction of a familiar object. **1935** K. Koffka *Princ. Gestalt Psychol.* xi. 499 Normalizing occurs when the reproductions approach successively a familiar form. *Ibid.*, Autonomous changes occur against the forces of normalization and pointing. **1970** L. Zusne *Visual Perception of Form* vii. 312 Making the reproduced figure resemble some well-known shape (normalizing). **1971** J. Hochberg in Woodworth & Schlosberg *Experimental Psychol.* (1972) xii. 469 The normalization effects might thus simply be instances of the figural after effects produced by satiation.

c. *Politics.* The achieving of 'normal' or stable political relationships between two countries, freq. between a major power and a weaker or dependent country.

1938 *Times Review of 1937* p. vii/1 The 'normalization' of Polish-German relations. **1955** *Times* 22 Aug. 5/3 The Yugoslav Press has been complaining for some time that the process of 'normalization' with Albania was lagging behind other east European countries. **1956** *Ann. Reg. 1955* 260 'Normalization' of relations with the satellites. **1962** *Daily Tel.* 18 Sept. 12/2 President Tito can be well pleased. 'Normalisation' has been attained without compromise of Jugoslavia's freedom of action, in defence of which he broke with Stalin. **1968** *Economist* 7 Sept. 33/1 It will permit Mr. Dubcek, his party leadership and government to resume the road towards what is laughingly called 'normalisation' just so long as they stick to Moscow's interpretation of normality.

normalize, *v.* Add: **1. b.** *intr.* To become normal.

1923 *Contemp. Rev.* Mar. 366 If a rise in the price of tin should follow on the already normalising price of materials.

2. *Metallurgy.* To heat (steel) to above the transformation range (about 700°C or more) and allow to cool in still air at room temperature, so as to remove any effects of strain-hardening, produce a finer grain structure, and improve the mechanical properties and machinability.

1902 *Jrnl. Iron & Steel Inst.* LXI. plate XIV (*caption*) Mr. Stead's Austenite normalized. **1916** *Ibid.* XCIV. 26 (*heading*) Effect of heating to 850°C., quenching in oil and tempering at 550°C., compared with the same steels simply normalized by heating to 1000°C. and cooling in air. **1937** R. T. Rolfe *Steels for User* vi. 114 Since quenching in a liquid medium was for a long time prohibited, these castings were either annealed or normalised, followed by a tempering process. **1970** E. N. Simons *Dict. Ferrous Metals* 155 The rate of cooling differs according to the mass of the piece, a thin cross-section cooling more rapidly than a thick, so that in the large masses the core is annealed rather than normalized. **1971** B. Scharf *Engin. & its Lang.* ii. 13 Steel is often normalised before hardening or machining.

3. a. Chiefly *Math.* and *Physics.* To multiply (a series, function, or variable) by a factor that makes the norm or some associated quantity (as an integral) equal to a particular value, usu. unity.

1921 *Proc. Lond. Math. Soc.* XX. 125 The sequence $\{\psi_n(x)\}$ being normalised and orthogonal. **1929** Condon & Morse *Quantum Mech.* i. 30 Of course, when $\Psi\bar{\Psi}$ is to be used as probability it has to be so normalized, by choice of a constant multiplier for Ψ, that $\int \Psi\bar{\Psi}\,dv = 1$... The same normalization is required from the distributed charge standpoint to express the fact that the total charge is *e*. **1934** W. V. Houston *Princ. Math. Physics* ix. 134 A function can be normalized by multiplying it by a constant. When it is normalized, $\int_0^a R_1^2\,dx = 1$. **1956** *Math. Tables & Other Aids to Computation* X. 2 The fundamental sequence was normalized so that $-1 \leqslant u_j \leqslant 1$. **1961** Powell & Crasemann *Quantum Mech.* ix. 286 A vector of unit norm is said to be normalized... Every nonzero vector can be normalized. **1975** *Nature* 13 Feb. 563/1 To normalise variations among tissues, responses are reported as percentage of the maximum increase, calculated for each preparation.

b. *Computers.* To express (a number in floating-point representation) in the standard form as regards the position of the radix point, which is usually immediately preceding the first non-zero digit.

1946 *Ann. Computation Lab. Harvard Univ.* I. 495 The quantity to be normalized lies in storage counter A. **1957** D. D. McCracken *Digital Computer Programming* xvii. 205 The first two steps convert the code number into an unnormalized floating point form. The third does nothing but normalize it, i.e., it brings the first nonzero digit into position three of the accumulator. **1962** Huskey & Korn *Computer Handbk.* xv. 23 It is desirable to obtain as many significant figures in the answers as possible. In floating-point systems this can be accomplished by normalizing the nonzero numbers at each step. **1973** H. Dinter *Introd. Computing* v. 167 The decimal exponent will be a value that is adjusted for normalizing the number. For example, if the number 26·5 is to be expressed in normalized form, it will be presented as $\cdot 265 \times 10^2$.

normalizer (nǭ·ɹmăləizəɹ). [f. Normalize *v.* + -er¹.] Someone or something that normalizes.

1926 *Heating & Ventilating Mag.* (U.S.) Oct. 110/1 (*caption*) The Sjostrom Atmospheric Normalizer. *Ibid.* 110/2 When the Normalizer is operating it moves about 800 cu. ft. of air per minute. **1946** *House Beautiful* Nov. 290 If you live in a hard-water area, it helps to use a good water normalizer..in the water. **1960** P. Dorf tr. M. M. Guxman in J. A. Fishman *Readings Sociol. of Lang.* (1968) 773 The puristic strivings of the German normalizers of the 17th and 18th centuries were ridiculed, the striving of many normalizers both in Western as well as Eastern countries to counteract the appearance of the new vital tendencies of the colloquial variety of the language was underscored. *Ibid.* 777 It is a known fact that the French normalizers of the 15th and 16th centuries oriented themselves towards the language of Paris, but towards the form in which it was spoken at court.

normalizing (nǭ·ɹmăləiziŋ), *vbl. sb.* [f. Normaliz(e *v.* + -ing¹.] The action of the vb. Normalize (in Dict. and Suppl.).

1909 *Jrnl. Iron & Steel Inst.* LXXIX. 350 (*heading*) Normalising. **1935** A. B. Kinzel in *Symp. Welding Iron & Steel* (Iron & Steel Inst.) II. iii. 424 Complete normalising or stress-relieving of the entire structure should be considered essential if high internal stress exists. **1970** O. Dopping *Computers & Data Processing* vii. 112 In computers with built-in operations for floating-point arithmetic..normalizing usually occurs as a special operation.

normally, *adv.* Add: **5.** *Statistics.* In accordance with the normal distribution.

1928 T. C. Fry *Probability & its Engin. Uses* viii. 243 It is absurd to speak of a man of negative height:..it simply cannot occur. Yet if height were distributed normally, the second property of the Law would assign a finite probability to this absurdity. **1951** Dixon & Massey *Introd. Statistical Anal.* v. 63 Often a research worker has sufficient data and enough experience..to be able to specify the type of transformation of measurement which will give a normally distributed variable. **1971** *Nature* 3 Sept. 19/1, I take as my datum the fact that Englishmen are normally distributed in height with a mean of 5 feet 8 inches.

normalness (nǭ·ɹmălnės). *rare.* [f. Normal *a.* and *sb.* + -ness.] = Normality.

1854 Geo. Eliot tr. *Feuerbach's Essence Christianity* xvi. 159 The agreement of others is therefore my criterion of the normalness, the universality, the truth of my thoughts. **1972** *Language* XLVIII. 314 We suspect that for some speakers, some of the sentences that we have claimed to be 'odd' will sound quite normal. This is to be expected if the oddness or normalness of a given stress contour depends partly on the speaker's ability to provide a satisfactory context.

Normandy (nǭ·ɹmăndi). [Name of a region of northern France.] *Normandy butter*: butter made in Normandy. Also *ellipt.*

1902 J. T. Law *Grocer's Man.* (ed. 2) 632/2 Normandy butter. **1962** L. Deighton *Ipcress File* ii. 19, I..bought..some Normandy butter and garlic sausage. **1973** *Guardian* 26 Jan. 9/1 'Butter is half the price here, even French butter,' she said, packing pounds of best cream Normandy into her shopping trolley. **1975** R. Butler *Where all Girls are Sweeter* i. 5 There was still a slab of Sainsbury's Normandy butter.

b. *Normandy vellum*, a strong, hand-made paper designed to imitate the qualities of parchment.

1935 S. Beckett *Echo's Bones*, This edition is limited to 327 copies of which 25 on Normandy vellum signed by the author are numbered 1 to XXV. **1968** *Amer. N. & Q.* VI. 158/1 There were only 212 copies [of the magazine *Pages*] of which 12..are on Normandy vellum.

Normanist. Add: **b.** (See quot. 1970.) Also as *adj.*

1943 G. Vernadsky *Anc. Russia* vii. iv. 276 Let us then turn to an appraisal of the results of the tournement of the *Norsephiles* ('Normanists') and *Mysosnorses* ('Anti-Normanists'), so famous in the annals of the Russian historiography. *Ibid.* 277 The whole argument of the first Normanists was based on the premise that the very name of Rus spread from north to south and not otherwise. **1968** G. Jones *Hist. Vikings* iii. iv. 247 The most readable 'Normanist' statement is still V. Thomsen's *Relations*. G. Vernadsky presents a summary of evidence for the 'anti-Normanist' view. *Ibid.* 264 To the disengaged it may well seem that the Normanist case has been as over-presented by a majority of Scandinavian historians as it has been played down by a majority of the historians of Russia. **1970** Foote & Wilson *Viking Achievement* vi. 220 Scholars have tended to side with one of two factions; the first firmly believing in the importance of the Scandinavian element in the founding of Russian towns and of the Russian state itself, the second discounting this influence almost completely (these two schools of thought are labelled in technical literature by the unfortunate terms 'Normanist' and 'anti-Normanist'). *Ibid.*, Discussion of the validity of this evidence is at the basis of the Normanist controversy. *Ibid.* 222 Both Normanists and anti-Normanists are more or less agreed that Ladoga was at one time dominated by Scandinavian traders.

normative, *a.* Add: **a.** (Later examples.) Deriving from, expressing, or implying a general standard, norm, or ideal. Also *absol.*

1897 J. H. Gulliver et al. tr. *Wundt's Ethics* I. 1 The normative point of view considers objects with reference to *definite rules*, which find expression in them, and to which they are at the same time required to conform. *Ibid.*, From the normative point of view, it is the purpose of the inquirer to estimate the relative values of facts. **1912** L. Bloomfield in *Jrnl. Eng. & Germ. Philol.* XI. 622 The scientific study of language has nothing to do with the normative (i.e. purely pedagogic) purpose of teaching people of ('fixing') the use or the better use of a literary language. **1931** M. R. Cohen *Reason & Nature* iii. iv. 403 Instead of refuting the normative standpoint of the old natural law, these writers substitute an unconscious natural law of their own. **1932** E. C. Tolman *Purposive Behav.* v. xxiii. 389 Relations to pre-individual or 'normative psychology'. *Ibid.* 397 Normative psychologies..tend to take an average (normal) individual or..sample of individuals..and then to apply all sorts of different stimuli. **1941** [see *genetic *a.* 1 f]. **1954** *Mind* LXIII. 264 As examples of 'normatives' he gives 'I ought to go' [etc.]. **1963** H.-N. Castañeda in Castañeda & Nakhnikian *Morality & Lang. of Conduct* vii. 279 A complete elucidation of the nature and structure of normatives requires an understanding of the logic of imperatives. **1968** A. Etzioni in Lindzey & Aronson *Handbk. Social Psychol.* (ed. 2) V. xliii. 547 The world situation warrants an approach that would emphasize common interests—above all, peace—as the key normative principle. **1970** R. C. Zaehner *Concordant Discord* iii. 46 It is the custom of nature mystics to assume that their own experience, even though it last only for a few moments, must be normative of all such experiences. **1973** *Archivum Linguisticum* IV. 107 It is hazardous to state correlations between social status and speech behaviour, especially if it is done in this normative fashion.

b. In special collocations: *normative grammar*: grammatical rules set up as a fixed standard to which language in use must conform; also, a treatise setting out such rules; so *normative grammarian*; *normative science*: a discipline such as ethics or æsthetics which aims at evaluation as well as description; *normative system*: a system based on what is established as the norm.

1901 H. Oertel *Lect. on Study of Lang.* ii. 87 Normative or didactic grammar sets up a certain standard as correct. **1933** L. Bloomfield *Language* i. 7 This gave the authoritarians their chance: they wrote *normative grammars*, in which they often ignored actual usage in favor of speculative notions. **1966** L. J. Cohen *Diversity of Meaning* (ed. 2) i. 9 The late eighteenth century was the first period to treat words' meanings as raw material for historians as well as normative grammarians. *Ibid.*, Some bold spirit had to cry down the pretensions of normative grammar and lexicography. **1968** *Word* XXIV. 387 Normative grammarians come into being when a society begins to concern itself with its language. **1895** F. Thilly tr. *Paulsen's Introd. Philos.* 26 Three normative sciences are added: Logic, practical philosophy based on psychology, and technology based on physics. **1921** W. E. Johnson *Logic* i. xiv. 224 Any study of which imperatives constitute the subject-matter has been called a normative science. **1934** Cohen & Nagel *Introd. Logic* vi. 110 Logic has..been defined as the normative science which studies the norms distinguishing sound..from unsound thinking. **1953** G. E. M. Anscombe tr. *Wittgenstein's Philos. Investigations* i. 38ᵉ F. P. Ramsey once emphasized in conversation with me that logic was a 'normative science'. **1966** *Mathematical Rev.* XXXI. 8/2 (*title*) The principles of a logic of normative systems. **1967** C. Margerison in Wills & Yearsley *Handbk. Management Technol.* 31 The normative system is a central factor in securing group participation. *Ibid.* 32 Various studies have indicated how the group's activity can be structured and led to bring about a normative system consonant with the organizational goals. **1973** *Jrnl. Genetic Psychol.* Mar. 56 Internalization was defined as correspondence between the child's normative system (i.e., his generalized expectations) and the parent's behavior or own normative system.

2. *Petrol.* Of or pertaining to the norm of a rock.

1902 W. Cross et al. in *Jrnl. Geol.* X. 604 The standard minerals which make up the norm are to be called the normative minerals, not the normal ones, since the latter adjective has the meaning of usual or common. **1909** J. P. Iddings *Ign. Rocks* I. ii. iii. 420 Normative minerals are frequently not the normal ones in certain rocks, though they may be the normal minerals in many others. **1962** *Jrnl. Petrology* III. 353 The normative components in basalts closely approximate the mode in most cases. **1973** *Ibid.* XIV. 35 (*heading*) Normative composition.

Hence **no·rmatively** *adv.*; **no·rmativeness**; **no·rmativist**; **normati·vity.**

1945 Normativity [see *facticity]. **1948** J. Towster *Pol. Power in U.S.S.R.* 16 The normativists (Kelsen etc.), too, say that every law is state (public) law. **1953** J. G. Peristiany in *Durkheim's Sociol. & Philos.* p. xv, The generality of a fact..is significant in this context only in relation to its generality, in the sense of its normativeness. **1957** *Archivum Linguisticum* IX. i. 75 Among the normativists both Gaertner/Passendorfer and Slonski admit of both genders. **1958** W. J. H. Sprott *Human Groups* ix. 144 Normativeness as such is a prerequisite for continuous interaction. **1964** Gould & Kolb *Dict. Soc. Sci.* 495/2 This rules out..any items of conduct which are not founded on past experience (in their words 'normatively regulated'). **1970** J. N. Findlay tr. *Husserl's Logical Investigations* I. ii. 78 Has not the concept of normativity got an inherent relation to a guiding aim. **1973** *Nature* 23 Nov. 228/2 The intent of these very simple models is neither to forecast, even implicitly, the levels reached in either population or food, or to forecast 'normatively', to set a target and then to deduce what is needed to achieve it.

normed (nǭɹmd), *a. Math.* [f. Norm + -ed².] Having a norm.

1935 *Trans. Amer. Math. Soc.* XXXVIII. 360 We now add the permanent assumption that $\mathfrak{B}$ is 'normed', that is, that there is associated with $\mathfrak{B}$ a rule assigning to every $\xi \in \mathfrak{B}$ a number $\|\xi\|$ called the 'norm' of ξ, and satisfying [etc.]. **1947** *Ibid.* LXII. 193 A normed linear vector space is a Banach space..if it is furthermore complete. **1959** L. F. Boron tr. Naimark (*title*) Normed rings. **1962** C. Wexler *Analytic Geom.* ii. 29 The most important normed linear spaces are those that have one additional property, that of completeness. **1972** A. G. Howson *Handbk. Terms used in Algebra & Analysis* xxii. 111 A space with a norm is known as a normed space; it is..a metric space with metric $d(x, y) = \|x - y\|$.

no·rmlessness. [f. Norm + -less + -ness.] Without any relevant standard or norm.

1944 H. P. Fairchild *Dict. Sociol.* 206/1 *Normlessness*, the absence of any appreciable norm. **1951** R. K. Merton *Soc. Theory & Soc. Structure* ii. iv. 128 There develops what Durkheim called 'anomie' (or normlessness). **1961** B. R. Wilson *Sects & Society* iv. v. 350 The individual is sheltered from the wider social anomie..the normlessness of the modern world. **1964** I. L. Horowitz *New Sociology* 20 All agree that anomie is defined by a 'condition of normlessness'. **1969** R. Blackburn in Cockburn & Blackburn *Student Power* 198 Seeman suggests reducing the concept [*sc.* alienation] to five psychological dimensions (feelings of normlessness, meaninglessness, powerlessness, isolation and self-estrangement).

normo- (nǭ·ɹmo), comb. form of L. *norma* (see Norma) used in several biological and medical words, esp. in physiology, to express the condition of being close to the average in respect of any particular character which varies (often contrasted with Hyper- and Hypo-). **normochro·mic** [Gr. χρῶμ-α colour] *a.*, having the normal amount of hæmoglobin; (of anæmia) characterized by the presence of red blood cells with the normal content of hæmoglobin; **no·rmocyte** [-cyte], an erythrocyte which is normal (esp. in size); so **normocy·tic** *a.*, of or pertaining to a normocyte; (of anæmia) characterized by the presence of erythrocytes which are normal in size, etc., but reduced in numbers; **normoglycæ·mia** (*U.S.* **-e·mia**), a normal concentration of sugar in the blood; so **normoglycæ·mic** *a.*, characterized by normoglycæmia; **normote·nsive** [*hyper-, *hypo)tensive *adjs.*] *a.*, having, or being, a normal blood pressure; hence as *sb.*, one who has such a blood pressure; **normothe·rmic** *a.* [Gr. θέρμ-η heat], characterized by or occurring at a normal body temperature; **normovolæ·mia** (*U.S.* **-e·mia**) [Vol(ume *sb.* + Gr. αἷμα blood], the condition of having a normal volume of circulating blood in the body; so **normovolæ·mic** *a.*, characterized by or pertaining to normovolæmia.

1935 Whitby & Britton *Disorders of Blood* iii. 48 The technical terms used are, firstly, 'hypochromic', 'normochromic' and 'hyperchromic', which indicate whether the cells contain an amount of hemoglobin which is less than, equal to, or more than normal. **1958** Normochromic [see *hypochromic *a.* 1]. **1974** *Nature* 7 June 551/1 Clinically the disease is characterised by proteinuria, corneal opacities and normochromic anaemia with decreased erythrocyte life span. **1900** *Buck's Handbk. Med. Sci.* (rev. ed.) I. 273/1 The red cells..are spoken of as microcytes, normocytes, and megalocytes, according to their size. **1935** Whitby & Britton *Disorders Blood* iii. 49 Hypochromia may be found with microcytes and normocytes. **1969** W. R. Platt *Color Atlas & Textbk. Hematol.* ii. 28/1 The erythrocyte (normocyte or red blood cell) measures 6–8 μ in diameter. **1911** Dorland *Med. Dict.* (ed. 6) 559/2 Normocytic. **1935** Whitby & Britton *Disorders of Blood* iii. 48 The technical terms used are..'microcytic', 'normocytic' and 'macrocytic', which indicate whether the [red] cells are smaller than, equal to, or larger than normal. **1971** *Indian Jrnl. Med. Res.* LIX. 427 In 60% of the cases the anemia was of the normocytic normochromic type. **1932** Dorland & Miller *Med. Dict.* (ed. 16) 871/2 Normoglycemia. **1961** *Lancet* 16 Sept. 637/2 To maintain normoglycæmia a [diabetic] woman of 81 needed 45 units P.Z.I. before the course but only 15 units after it. **1973** *Nature* 17 Aug. 447/2 Direct injection of pancreatic islets into the portal vein resulted in normoglycaemia and normal urine volumes in the five rats studied. **1933** *Stedman's Med. Dict.* (ed. 12) 733/2 Normoglycemic. **1961** *Diabetes* X. 322/1 The ketonuria may..represent the relatively benign normoglycemic ketosis of hunger or starvation which usually responds to increase in the amount of carbohydrate in the diet. **1969** *Hormone & Metabol. Res.* I. 266 (*heading*) Insulin levels during pregnancy or obesity in normoglycemic women with a positive history of diabetes mellitus. **1941** Dorland & Miller *Med. Dict.* (ed. 19) 982/1 *Normotensive*, marked by normal blood pressure. **1948** *Federation Proc.* VII. 41/2 (*heading*) The immediate pressor effect of desoxycorticosterone acetate in hypertensive and normotensive subjects. **1953** *Lancet* 12 Sept. 541/2 None of the other normotensives exceeded 0·9 ml. per minute. *Ibid.* 543/2 Ten normotensive women. **1962** *Ibid.* 26 May 1092/1 Possible sources of irregularity in the distribution of blood-pressure include..the observer's subconscious distinction between normotensive and hypertensive levels of pressure. **1972** *Aerospace Med.* XLIII. 1225 (*heading*) Effect of hydrochlorothiazide on $+G_z$ tolerance in normotensives. **1959** *Surg., Gynecol. & Obstetr.* CIX. 721 (*heading*) Normothermic perfusion and replantation of the excised dog kidney. **1960** *Amer. Jrnl. Physiol.* CXCIX. 163/1 The heart was cooled while maintaining the rest of the body normothermic. **1974** *Nature* 22 Feb. 568/2 Obviously the destruction of cells by growth of virus may also be different in hypothermic and normothermic conditions. **1925** Normovolemia [see *hypervolæmia* s.v. *hyper- IV]. **1966** *Ann. Surg.* CLXIV. 51 (*heading*) Effects of adrenergic blocking agents on renal blood flow in normovolemia and experimental hypovolemia. **1947** *Acta Cardiologica* II. 134 This method..permitted comparisons of sequential cardiovascular changes which occurred in the same animal during oligemic shock and normovolemic shock (i.e., shock with essentially normal blood volume). **1966** *Amer. Jrnl. Physiol.* CCXI. 878/1 Normovolemic anemia was produced by removing from a femoral artery 40 ml/kg blood while infusing an equal volume of 6% dextran in saline into a femoral vein.

Norn, *sb.*¹ Add: Also pl. Nornir.

1859 G. W. Dasent *Pop. Tales from Norse* p. xli, The worshippers of Odin and the Nornir were gradually converted into votaries of the Virgin Mary. **1875** *Encycl. Brit.* I. 211/1 The three principal Norns or Nornir are Urd, past time; Verdandi, present time; and Skulld, future time. **1959** R. W. V. Elliott *Runes* vii. 105 Other possibilities occur to one: such as that the figures represent the Nornir, the 'three fatal sisters' of Northern mythology.

Norn, *a.* and *sb.*² Delete *dial.* and add to def.: **a.** *adj.* Also, of or pertaining to Norn. **b.** *sb.* Also formerly spoken on parts of the northern mainland of Scotland. Also *the Orkney* (or *Shetland*) *Norn.* (Further examples.)

1932 *Times Lit. Suppl.* 6 Oct. 712/3 Jakobsen's great dictionary of the Norn language in Shetland. **1956** 'H. MacDiarmid' *Stony Limits & Scots Unbound* 48 The old Norn words. **1966** E. W. Marwick in J. Shearer et al. *New Orkney Book* iv. 25 At the beginning of the eighteenth century some of them were still speaking the old Norn language. **1966–69** *Saga-Book* XVII. 11 In Appendix II he printed specimens of the Orkney Norn of the fourteenth and fifteenth centuries. **1972** W. B. Lockwood *Panorama Indo-Europ. Lang.* vii. 122 Norse obliterated the Pictish of Orkney and Shetland and hastened its demise on the mainland. Known locally as Norn, the language lived on there and across the Pentland Firth in Caithness until modern times. *Ibid.* 127 On the Scottish mainland generally, Norn was ousted by Gaelic during the Middle Ages, except in the North-East of Caithness where it appears to have lingered on until the sixteenth century. **1970–73** *Saga-Book* XVIII. 382 The section on the Orkney and Shetland Norn..is most noteworthy in this respect.

Norse, *sb.* and *a.* **A.** *sb.* **2.** Also, any native or inhabitant of ancient Scandinavia. (Further example.)

1972 W. B. Lockwood *Panorama Indo-Europe. Lang.* vii. 126 As many as 5,000 Norse are believed to have lived in Greenland.

3. For '(sometimes loosely used to include early Swedish and Danish)' read 'Also, the North Germanic language which was the immediate ancestor of the Scandinavian languages'. (Further example.)

1927 E. V. Gordon *Introd. Old Norse* p. xvi, The structure of Gothic..reveals its affinity with Norse, but the differences between the oldest surviving Gothic..and Norse of the same period are too great for Gothic to be included in the Norse group of tongues.

B. *adj.* Also, of or pertaining to ancient Scandinavia or the Norse peoples as a whole. (Further examples.)

1885 *Encycl. Dict.* V. i. 216/1 *Norse*,..of or pertaining to ancient Scandinavia or its inhabitants; Norwegian. **1927** E. V. Gordon *Introd. Old Norse* p. xv, The home of the oldest Norse culture and the oldest Norse traditions was Sweden. *Ibid.* p. xxxv, The oldest Norse poetry preserved traditions which belonged not merely to the Norse peoples but to the Germanic race as a whole. **1970–73** *Saga-Book* XVIII. 208, I may insert a word of warning on Dr Marwick's old-fashioned and narrowly correct use of the term 'Norse'—to him it translates *norsk*, Norwegian and then especially *landsmål*.

Hence **No·rseness**, the state or quality of being Norwegian or Scandinavian.

1961 *Listener* 7 Sept. 363/3 Ibsen's characters are always about to take flight into fantasy but their manner, their stubborn Norseness, holds them fast in a two-dimensional theatre. **1968** G. Jones *Hist. Vikings* iii. iv. 264 The adoption of Slavonic customs had quietly eroded the Norseness of the Rus.

norseller (nǭ·ɹsɛ'ləɹ). Also **orseller.** [f. Norsel *sb.* and *v.* + -er¹.] A person who fits nets with norsels.

1921 *Dict. Occup. Terms* (1927) §398 *Net orseller, norseller*, attaches orsells or norsells (short lines about ten inches long) to top and bottom of fishing net at regular intervals.

Norseman. Add: Also, any native or inhabitant of viking-age Scandinavia.

1864 *Chambers's Encycl.* VI. 792/2 The first Danish Norsemen made their appearance on the eastern and southern coasts of England in 787. *Ibid.* 793/2 The Swedish Norsemen directed their expeditions chiefly against the eastern coasts of the Baltic. **1935** *Ibid.* VII. 527/2 *Northmen*, or Norsemen, were the sea-rovers who came from the north—Denmark, Norway, Sweden. **1943** G. Vernadsky *Anc. Russia* vii. iii. 273 It seems highly probable that..a band of Norsemen—more exactly, of Swedes—established their control over the lower Don and Azov area. **1948** *Oxf. Jun. Encycl.* I. 137/1 *Danes.* These are a Scandinavian people, as are their neighbours the Swedes..and Norwegians... The ancestors of all three were the Norsemen or Vikings. **1968** G. Jones *Hist. Vikings* iv. iii. 395 Ireland was different. The Norsemen (part Danes, part Norwegians) had established a number of important trading towns in the southern half of the island. **1970** Foote & Wilson *Viking Achievement* i. 14 Saxon political influence went hand in hand with Saxon efforts to convert the Danes to Christianity. The archbishopric of Hamburg-Bremen..was intended as a missionary outpost with prime responsibility for bringing the heathen Norsemen to the true faith.

norsethite (nǭ·ɹsĕþəit). *Min.* [See quot. 1959² and -ite¹.] A carbonate of barium and magnesium, $BaMg(CO_3)_2$, that is found as whitish crystals with a vitreous or pearly lustre, similar to calcite and dolomite.

1959 C. Milton et al. in *Bull. Geol. Soc. Amer.* LXX. 1646 Norsethite, $BaMg(CO_3)_2$, was found in dolomitic, black oil shale below the main trona bed in the Westvaco trona mine. *Ibid.*, Norsethite is named in honor of Mr. Keith Norseth, engineering geologist of the trona mine at Westvaco, Sweetwater County, Wyoming. **1967** *Amer. Mineralogist* LII. 1770 The occurrence of norsethite, $BaMg(CO_3)_2$, in large quantities in a newly discovered zinc-lead-copper deposit in South West Africa, is reported. It is one of the major gangue minerals of the deposit, is closely associated with calcite and is apparently of hydrothermal origin. **1968** I. Kostov *Mineralogy* ii. xi. 540 Norsethite is probably trigonal-trapezohedral with distinct rhombohedral $\{10\bar{1}1\}$ cleavage.

norsteroid (nǭɹsti^ə·roid, -ste·roid). *Biochem.* [f. *nor- + *steroid.] A steroid lacking a methyl side chain (esp. the one containing the carbon atom numbered 19), or having one of its rings contracted by one methylene group.

1950 *Jrnl. Chem. Soc.* 367 The synthetic approach to nor-steroid hormones has now been greatly simplified by Birch and Mukherji's method of reduction. **1954** *Ibid.* 1984 Values for 2-substituted A-norsteroids..may be considered. **1970** *Indian Jrnl. Biochem.* VII. 116/1 Since the introduction of 19-norsteroids as the synthetic progestational agents, it has become necessary to know their role as regulators of metabolic processes.

nortes, *sb. pl.* Add: Also in *sing.*

1926 *British Weekly* 21 Jan. 398/5 We hear of the *norte*, the wind from the Equatorial belt... When the *norte* blows, many temperamental folk are apt to become unbalanced.

north. Add: **A.** *adv.* **1. b.** (Later example.)

1955 J. R. R. Tolkien *Return of King* 197 Going swiftly to lesser posts and strongholds north-away.

B. *sb.* **1. b.** *Bridge.* A person occupying a position opposite 'South'.

1926 [see *east *sb.* 4]. **1958** *Listener* 2 Oct. 541/2 North bid Three Clubs. **1965** *Ibid.* 20 May 758/2 The bidding should have made it clear to him that North was hoping to play in Two Hearts doubled. **1973** *Country Life* 21 June 1842/3 Study this deal... Dealer, North. North-South vulnerable.

2. c. (Earlier examples.)

1796 Washington *Messages & Papers* (1898) I. 217 The North, in an unrestricted intercourse with the South. **1831** J. M. Peck *Guide for Emigrants* ii. 81 The result would be more disastrous to the south and west, than the influx of foreign goods was to..the north..in 1816.

3. b. *North-of-England*, used *attrib.*, of, pertaining to, or characteristic of, the north of England.

1816 Scott *Antiq.* III. ii. 34 His father was a north-of-England gentleman. **1847** C. Brontë *Jane Eyre* I. xii. 212 A North-of-England spirit, called a 'Gytrash'; which,..haunted solitary ways. **1907** F. E. E. Bell *At Works* vi. 127 Watching football matches, a comfortless thing enough to do in a North of England winter. **1973** J. Wainwright *High-Class Kill* 221 North-of-England conformity — best-clothes-on-Sunday-speak-when-you're spoken-to.

5. north and south *Rhyming slang*, mouth; **north canoe**, a birchbark canoe once used north and west of Lake Superior, North America; **northpaw** *U.S. slang* (see quot. 1960).

1858 A. Mayhew *Paved with Gold* ii. x. 169 'I'll smash your "glass case", and damage your "north and south",' roared Bill, referring to the *face* and *mouth* of his opponent. **1928** M. C. Sharpe *Chicago May* 287/2 North and South, mouth. **1958** F. Norman *Bang to Rights* 36 Dust floating about in the air, which gets in your north and south. **1972** *Lebende Sprachen* XVII. 8/2 *North and south*, mouth. **1819** W. F. Wentzel *Let.* in L. F. R. Masson *Bourgeois de la Compagnie du Nord-Ouest* (1889) I. 134 Sir Alexander Mackenzie has suggested that one north canoe with Canadian voyageurs, and six small Indian canoes, would be a fitter outfit. **1879** H. M. Robinson *Great Fur Land* 31 The North canoe..is a light graceful vessel about thirty-six long, by four or five broad, and capable of containing eight men and three passengers. **1956** V. Fisher *Pemmican* 250 A north canoe; twenty-five foot long and from four to five feet wide, it could carry a crew of eight or nine men and their supplies, as well as three passengers. **1969** *Islander* (Victoria, B.C.) 23 Nov. 12/3 Ahead roared the Rapids of the Drowned. They gained their name after one of the Hudson's Bay Company's large north canoes capsized there with the loss of several men. **1960** Wentworth & Flexner *Dict. Amer. Slang* 358/1 *Northpaw*.., a right-handed baseball pitcher; any right-handed person. **1968** *Listener* 19 Sept. 357/1 A skilful person is 'dextrous': in its way as insulting to left-handers (they call us southpaws, though I have never heard anyone described as a northpaw) as 'white man', for someone of worthy character, is offensive to Negroes. **1972** *Daily Mail* 1 Aug. 2/6, 20 per cent of Americans are tired of grappling with things designed for northpaws.

C. *adj.* **1. d.** *North Oxford*, of, pertaining to, or characteristic of the suburban part of Oxford north of the university area, where many dons and their families live. Also as *sb.*

1935 N. Mitchison *We have been Warned* iv. 462 She had..a Sybil Dunlop moonstone on a long silver chain. A bit North Oxfordy? Well, she was North Oxford! **1950** A. Wilson *Such Darling Dodos* 79 The whimsical humour of North Oxford. *Ibid.* 97 Why you should have to drag Coleridge in, only your staunch North Oxford spirit can explain. **1973** *Country Life* 13 Sept. 720/1 The houses behind (Carolean or North Oxford Gothic?)..seem oddly familiar. **1974** F. Emery *Oxfordsh. Landscape* vii. 214 Between the rustic variety of Summertown and the old suburbs of St Giles lies the Victorian perfection of North Oxford proper.

2. d. North Circular (Road), a road passing through the northern outskirts of London.

1944 M. Laski *Love on Supertax* xi. 102 After a prolonged..journey by 'bus, Clarissa alighted way out on the North Circular Road. **1968** J. Lloyd *Death at Roman Farm* xviii. 171 You go straight through town and send the other car round the North Circular. **1974** 'A. Garve' *File on Lester* xxxi. 118 He became silent, concentrating on his driving on the busy North Circular.

4. *Comb.*, **north-facing**, facing the north; *spec.* of a window.

1952 A. G. L. Hellyer *Sanders' Encycl. Gardening* (ed. 22) 205 Position, semi-shaded or north-facing. **1973** 'I. Drummond' *Jaws of Watchdog* xii. 156 A large window. North-facing so I guess a studio.

North African, *a.* and *sb.* [f. *North Africa* the countries of northern Africa, *spec.* the region including Morocco, Algeria, Tunisia, Libya, and Egypt.] **A.** *adj.* Of or pertaining to North Africa or its inhabitants. **B.** *sb.* A native or inhabitant of North Africa.

1867 'Ouida' *Under Two Flags* II. iii. 56 The blaze and heat of the North African day. **1932** Kipling *Limits & Renewals* 321 His speech—to suit his hearers—ran From pure Parisian to gross peasant, With interludes North African If any Legionnaire were present. **1961** E. Waugh *Unconditional Surrender* ii. v. 140 Stirred by the heavy North African wine de Souza's imagination rolled into action. **1962** P. Brickhill *Deadline* vi. 88 The noise made by four North Africans on a dais. *Ibid.* 89 Belly dancing seemed to be the forte of the North African smart spots in Paris. **1972** D. Lees *Zodiac* 107 The couscous was being ladled out by a grinning North African.

North American, *sb.* and *a.* [f. *North America* the name of the northern part of the continent of America including Central America, Mexico, U.S., and Canada.] **A.** *sb.* A native or inhabitant of North America, esp. of the United States or Canada.

1766 [see North *a.* 1 c]. **1783** A. Stokes *Brit. Colonies* 144 The North Americans will refuse them their assistance. **1825** J. Neal *Bro. Jonathan* II. 1 The man of America—the Original North American..the 'Indian' as he is called. *Ibid.* III. 413 The brave North American was dead. **1974** 'E. Anthony' *Malaspiga Exit* ii. 25 You could disguise every feature except the magnificent dentistry of the North Americans.

B. *adj.* Of or pertaining to North America or its inhabitants; belonging to or characteristic of North America, living in North America, etc.

1770 J. Otis in W. Tudor *Life J. Otis* (1823) 476 My humble North American word of honour for it, my lord, these volumes will hurt neither thee, nor thy master. **1771** Franklin *Autobiogr.* (1909) 171 This was the mother of all the North American subscription libraries, now so numerous. **1775** Burke *Speech on Conciliation with America* 22 Mar. in *Wks.* II (1792) 31 The export trade to the colonies consists of three great branches. The African..the West Indian; and the North-American. **1776** (*title*) The North-American and the West-Indian Gazetteer. **1783** (*title*) The North-American Calendar. **1837** *Southern Lit. Messenger* III. 695 A declaration of the independence of the North American States. **1880** G. W. Cable *Grandissimes* iv. 23 She had..the nerve of the true North American Indian. **1916** G. B. Shaw *Androcles & Lion* p. li, Jesus himself had referred to that psalm (lxxxii) in which men who have judged unjustly and accepted the persons of the wicked (including by anticipation practically all the white inhabitants of the British Isles and the North American continent, to mention no other places) are condemned. **1923** A. L. Kroeber *Anthropol.* xiii. 356 It would be extravagant to maintain that throughout the North American continent every matrilineal tribe was culturally more advanced than every patrilineal one. **1949** *Chicago Tribune* 8 Nov. i. 12/2 The very rich head of a business is vanishing as rapidly as the North American Indian. **1952** *Oxf. Jun. Encycl.* X. 105/1 The example of the North American colonists roused other oppressed colonists to bid for freedom. **1959** M. Schlauch *Eng. Lang.* iii. 78 Words connected with the life of North American Indians also appeared after 1600. **1974** E. Ambler *Dr. Frigo* ii. 96 He was still speaking Spanish but adulterating it now with ..North American business jargon. **1975** *Country Life* 6 Nov. 1250/3 Q[*uercus*] *coccinea* is the North American scarlet oak.

north-bound, *a.* Also **northbound.** [North *adv.*] Bound for the north; travelling northwards; also, intended for such travellers, serving as a point of departure for the north. Also as *sb.*, a north-bound train.

1903 Kipling *Five Nations* 115 We gather and wait her coming—The wonderful north-bound train. **1904** W. N. Harben *Georgians* 217 The young man was at the seven-o'clock north-bound train when it stopped in the antiquated brick car-shed. **1939** G. Household *Rogue Male* 78 At the bottom of the Piccadilly escalator you turn left for the north-bound trains... I ran on to the north-bound platform. **1973** *Sci. Amer.* June 26/2 A slowing of the northbound Coastal Current. **1975** D. Beaty *Electric Train* 232 'When's the next train?' 'No more northbounds tonight.'

north country. Add: **2.** (Later examples.) Also a native of the north country of England.

1823 J. G. Lockhart *R. Dalton* II. v. v. 289 You're north-country, I believe. **1966** *Listener* 7 Apr. 501/1, I can play cockney, I can play north country. **1972** C. Drummond *Death at Bar* ii. 60 'If you borrow you make bad friends'—a flash of honest north-country peeped through the trained accent. **1974** *Times* 30 Dec. 4/7 Gracie Fields..reverts to the characteristic mixture of North Country, standard English and American overtones.

2. (Later example.)

1975 J. Mitchell *Smear Job* x. 69 Her accent was harshly Sicilian, but lurking in it was a hint of the North Country she had picked up from the soldiers.

3. (Further examples.)

1896 D. C. Murray (*title*) A capful o'nails: a North-Country story. **1933** M. Allingham *Sweet Danger* x. 129 A comfortable, homely voice with an unexpected North Country accent. **1959** R. Longrigg *Wrong Number* ii. 30 They..left in a gruff welter of north-country gratitude. **1964** V. S. Naipaul *Area of Darkness* ix. 243 The newcomer was a commercial Englishman, middle-aged, fat and red-faced. He spoke with a North Country accent.

northen, *a.* (Later U.S. examples.)

1772 D. Taitt in N. D. Mereness *Trav. Amer. Col.* (1916) 541 The Inhabitants of the Tuskigees are a remnant of Northen Indians and speak a different Language from the Creek. **1773** C. Caroll *Let.* 26 Mar. in *Maryland Hist. Mag.* (1920) XV. 58 Keep the Boy if the Northen Post be not Come in untill Monday.

norther, *sb.* Add: **1.** (Earlier and later examples.)

1827 *Western Monthly Rev.* I. 320 We were struck by a gale, that they call a norther. **1831** M. Holley *Texas* (1833) i. 19 Our voyage..is..not without hazard, on account of the Northers; as they are called. **1946** *Sun* (Baltimore) 30 Dec. 11/1 St. Mary's football squad arrived in Houston early this morning to be greeted by a Texas 'norther'. **1969** 'J. Morris' *Fever Grass* xiii. 121 A fresh 'norther' was blowing from the Gulf. **1973** *Houston* (Texas) *Chron.* 21 Oct. 1/3 (*caption*) A delightfully persistent Indian Summer lingered over Houston, luring people to parks for langorous hours of contentment before looking ahead to winter and the prospect of rain and blue northers.

2. A strong north wind blowing in other parts, esp. on the Pacific seaboard of North America.

1835 J. F. Cooper *Monikins* II. iii. 71 It may be even now questioned whether the ship would claw off..with a sending sea, and this heavy norther. **1850** in *Harper's Mag.* (1878) Jan. 279 We met a norther in coming out of the Gulf of California. **1891** *Scribner's Mag.* X. 283 The weather along the Pacific highway has been uniformly pleasant, for northers are infrequent. **1893** Kipling *Seven Seas* (1896) 28 We've slipped from Valparaiso With the Norther at our heels. **1903** —— *Five Nations* 53 That night the Norther found me—Froze and killed the plains-bred ponies.

northern, *a.* and *sb.* Add: **A.** *adj.* **1. c.** northern canoe = **north canoe.*

1820 R. Hood *To Arctic* (1974) 110 Late on the 21st Mr. Robertson, of the Hudson's Bay Company arrived, and furnished us with a guide, but desired that he might be exchanged when we met the northern canoes. **1860** S. Hancock *Narr.* (1927) 163 What is termed a northern canoe is much larger and differently shaped from those made and used by the Indians south of the Straits of Fuca. **1867** J. T. Rothrock *Flora Alaska* 434 From it the celebrated 'northern canoes' are made. These canoes, 'dug' from a single trunk and afterwards steamed into shape, will often carry four tons. **1938** *Beaver* Dec. 13/1 The northern canoe carried a crew of five or six men, apart from passengers. **1954** M. W. Campbell *Nor' Westers* 44 Each northern canoe held twenty-five ninety-pound packs instead of the sixty loaded into the Montreal canoes.

3. b. *Northern Spy*, an American variety of red-skinned, late-ripening, dessert apple. (Earlier and later examples.)

1847 J. M. Ives *New England Bk. Fruit* 46 Northern Spy.—This new native fruit, originated near Rochester, N. York. It is a fine winter apple, and is one of the most popular fruits in New York. **1850** *New England Farmer* II. 404 Northern Spy Apple. We had hoped to be able to test the qualities of this apple ourselves. **1917** D. Canfield *Understood Betsy* iii. 61 Those Northern Spies are just getting to be good about now. When they first come off the tree in October you could shoot them through an oak plank. *Ibid.* xi. 241 A basket..half full of striped red Northern Spies. **1944** *Poetry Chap Bk.* Fall 14 And fragrant windrows of crisp Northern Spies Are scattered in the tumbled twisted sheaves. **1965** Mrs. L. B. Johnson *White House Diary* 15 Jan. (1970) 221 'What are your best brands [of apple]?' asked Gromyko. Without hesitation and with honesty General McNaughton replied, 'Our best brands are Northern *Spies* and McIntosh *Reds*.'

northerner. 1. (Earlier and later examples.)

1831 J. M. Peck *Guide for Emigrants* ii. 60 Such for beauty and splendor and fragrance, the *Northerners* have never seen. **1928** I. C. Ward *Phonetics of English* i. 3 A Northerner's pronunciation of *but*..is called 'broad'. **1945** *Chicago Daily News* 1 Feb. 8/7 It's about time we long-suffering Northerners got it off our chest.

northernism (nǭ·ɹðəɹniz'm). [f. Northern *a.* and *sb.* + -ism.] Northern quality; a Northern characteristic, esp. a Northern form of expression.

1930 *Times Lit. Suppl.* 9 Jan. 28/4 Anything tainted with northernism or 'Lutheranism' is poison for Italy. **1964** *English Studies* XLV. 249 Identified..as a northernism. **1965** H. Kökeritz in Bessinger & Creed *Medieval & Linguistic Stud.* 294 Notwithstanding a few seeming northernisms..there is no vestige of his father's dialect.

northernness. Add: **2.** The quality of being from the north.

1939 C. S. Lewis *Rehabilitations* ii. 37, I knew one who could come no nearer to an explanation of Morris's charm than to repeat 'It's the Northernness—the Northernness'. **1955** E. Blishen *Roaring Boys* ii. 74 To our grey northern-ness they brought a southern gleam. **1968** G. Jones *Hist. Vikings* 2 Though they had many ties..to remind them of a shared northernness, they had but little sense of a separate Danish, Swedish, or Norwegian nationality. **1975** C. N. Manlove *Mod. Fantasy* iv. 100 Despite lifelong fascination for 'Northernness', and interests in the medieval School of Chartres and in the Renaissance, he never felt the urge to visit their sources in Scandinavia, France or Italy.

north-light. Add: Also **north light. 3.** A window, esp. in a roof, that faces north. So *north-light roof*.

1904 F. E. Kidder *Architects' & Builders' Pocket-Bk.* (ed. 14) 1304 With the 'Maze' glass, the artist may have, in all weather and in all directions, what is in effect a much-desired 'north-light'. **1919** R. Fry *Let.* 29 Nov. (1972) II. 473, I went to his tiny studio..he'd simply put a high north light in the roof of a small bedroom. **1931** *Engineering* 9 Jan. 33/2 The older bays had roofs of the saw-tooth pattern, with north lights. **1940** *Chambers's Techn. Dict.* 583/1 *North light roof*, a pitched roof of unequal slopes, of which the steeper is glazed and arranged to receive light from the north. **1958** *Times* 20 Aug. 13/6 Where only day shifts are worked the cost of installing fittings to produce the whole of the necessary illumination, instead of only those required to supplement daylight from windows or north-light roof must also be reckoned. **1961** E. E. Cummings *Let.* 22 July (1969) 272, I asked a firstrate carpenter..to make me a real—with a North Light—studio in the barn. **1972** E. Lemarchand *Cyanide with Compliments* ix. 117 Could be a studio... I can't make out if it's got a north light, though.

northman. Add: **2.** *U.S.* = Northerner 1.

1836 *Southern Lit. Messenger* II. 434 From my very heart, northman as I am, I admire and affect this good remnant of olden time. **1837** *Ibid.* III. 337 Between the Virginians and the North-men there was a wide variance.

northmost, *a.* (Later examples.)

1888 Swinburne *Armada* iii. i, Darker far than the tempests are that sweep the skies of her northmost clime. **1921** R. L. Jack (*title*) Northmost Australia.

North Sea. Add: **1. b.** *attrib. North Sea gas, oil*: raw materials discovered beneath the North Sea.

1965 *Times* 22 Sept. 10/6 The BP gas find radically alters the North Sea oil and gas search. 1967 *Guardian* 4 Jan. 3/1 Bringing North Sea gas ashore in Norfolk. 1972 *Ibid.* 8 June 9/1 The tests were carried out in households supplied by North Sea gas and those supplied by town gas. 1973 *Ibid.* 2 Mar. 15/7 North Sea oil will go a long way to reducing Britain's dependence on the volatile Middle East. 1974 *Sunday Times* (Colour Suppl.) 28 Apr. 20 It is a feature of the story of North Sea oil that men encountering its immensities can find themselves in conflict with their previous positions. 1975 *Sat. Rev.* 25 Jan. 16/3 Britain retiring into a new insularity fueled by North Sea oil.

North Star State. *U.S.* The state of Minnesota.

1862 *American Odd Fellow* I. 196/2 In the North Star State here, we rejoice in having a Grand Master, who knows no such word as fail. 1909 *World To-Day* Oct. 1108 The North Star State has been the scene of her greatest usefulness. Mrs. Potter commenced her educational work in Minneapolis. 1946 C. McWilliams *Southern Calif. Country* 173 Floats move through the streets of Long beach with captions as 'Corn is King',..'Minnesota: The North Star State'.

Norway[1]. Add: *Norway lobster* = *Dublin (Bay) prawn* (*Dublin) (earlier and later examples); *Norway pine* (see Pine *sb.*[2] 2) (earlier and later examples).

1777 T. Pennant *Brit. Zool.* (ed. 4) IV. 17 Cancer Novegicus... Norway. L[obster] with a long spiny snout. 1911, 1963 [see *Dublin (Bay) prawn* (*Dublin)]. 1720 R. Palmer *Let.* 4 Oct. in M. M. Verney *Verney Lett.* (1930) II. xxiv. 81, 36 Norway Pines and Yews for a Hedge of 38 yards long. 1784 M. Cutler *Jrnl.* 22 July in Parker & Cutler *Life & Corr. M. Cutler* (1888) I. iii. 99 We rode five miles over pitch and Norway-pine plains, with very low shrubs. 1829 J. C. Loudon *Encycl. Plants* 804 The pitch pine, *P[inus] resinosa* is generally known in its native country by the name of Norway pine; sometimes, particularly among the Canadian French, red pine. 1838 J. F. Cooper *Homeward Bound* I. xvi. 252 [He] applied his knife to try the quality of the wood, and pronounced the Norway pine of the spars to be almost equal to anything that could be found in our own southern woods. 1896 M. E. Wilkins *Madelon* 1 There were evergreens—Norway pines, spruces and hemlocks—bordering the road. 1969 T. H. Everett *Living Trees of World* 51/1 Eastern American species with paired leaves include the red pine or Norway pine (*P. resinosa*). It may at first seem to be misnamed, but its popular designation refers to the Maine village of Norway and not to the land of the Vikings. 1973 *Saint Croix Courier* (St. Stephen, New Brunswick) 26 July 1 Dories are planked with Norway (Red) Pine and clench-fastened with galvanized boat nails.

Norwegian, *a.* and *sb.* Add: **A.** *adj.* **b.** *Norwegian steam* (see quot. 1960).

1944 *Amer. Speech* XIX. 106 Norwegian steam is brute manpower,..from the tradition of the fine sailing ships. 1947 R. O. Boyer *Dark Ship* vii. 53, I heard another sailor say 'We got there by Norwegian steam' and asked him what he meant. 'By rowing,' he said. 1958 E. S. Land *Winning War with Ships* iv. 66 The power on deck for such work as weighing anchor was 'Norwegian steam'. 1960 Wentworth & Flexner *Dict. Amer. Slang* 358/1 *Norwegian steam*, manpower; muscle power. A little jocular use, esp. maritime use. 1970 *Sea Breezes* Nov. 717/2 All the hoisting and hauling is done by hand—'Norwegian steam' as we used to say years ago.

nor'-wester. Add: **1.** Esp. in the South Island of New Zealand. In full, *Canterbury nor'-wester*.

1871 Lady Barker *Christmas Cake in Four Quarters* iv. ii. 267 How many of us would be in that valley on a Christmas Day in the far future, when these trees would have struggled up against their enemy the Nor'-wester. 1933 L. G. D. Acland in *Press* (Christchurch) 2 Dec. 15/7 The ordinary man in the paddock in Canterbury recognises only three winds: east (from N.E.), sou'-wester (from S.E., S., etc., S.W., etc.), and nor'-wester. 1959 A. McLintock *Descr. Atlas N.Z.* 22 In summer the hot, dry 'Canterbury Nor'wester' is generally a most unpleasant wind. 1966 G. W. Turner *Eng. Lang. Austral. & N.Z.* viii. 163 The Canterbury *nor' wester* resembling the European *Föhn* wind..and the *three-day sou'wester* in New Zealand.

3. Also, a strong oilskin or waterproof coat worn in rough weather. In full *norwester coat*.

1689–90 J. Greenvill in W. S. Pelletreau *Early Long Island Wills* (1897) 46 My will is that my norwester Coat.. may be given to Christopher Leaming.

Norwich (nǫ·ritʃ, -idʒ). [The name of the city and county town of Norfolk.] Used *attrib.* in various collocations, as *Norwich crape* (see Crape *sb.* 1); *Norwich damask, poplin, shawl*; also *Norwich school*, an English school of painting of the early nineteenth century associated with Norwich; *Norwich stuff*, a textile fabric manufactured, or as manufactured, in Norwich; *Norwich terrier*, a small, thick-set, red or black-and-tan, rough-coated terrier with pricked ears, belonging to a breed developed in East Anglia; in North America and formerly in Britain also used for the drop-eared variety of the breed, now called the Norfolk terrier; also ellipt. *Norwich*.

1618 T. Roe *Advise for Goodes in Embassy* (1899) II. 487 Some light coullored Norwich stuffes wrought in flowers for triall, the lighter the better. 1685, 1709 [see Crape *sb.* 1]. 1741 Richardson *Pamela* IV. xiii. 75 So many Yards of Norwich Stuffs for Gowns and Coats for Girls. 1774 H. Walpole *Descr. Strawberry-Hill* 65 The room is hung with crimson Norwich damask. 1790 M. Dunsford *Hist. Mem. Tiverton* iv. 235 This year [*sc.* 1752] was introduced to Tiverton the manufacture of Norwich stuffs, camblets, tarborates, [etc.]. 1816 J. Crome *Let.* Jan. in W. F. Dickes *Norwich Sch. Painting* (1906) v. 111 You wish me to give you my opinion of your picture... How pleased I was to see so much improvement in the figures, so unlike our Norwich School; I may say they were good. 1821 *Times* 15 Aug., The ladies to wear..dark Norwich crape. *c* 1860 in A. Adburgham *Shops & Shopping* (1964) ix. 100 Paisley, Norwich and French shawls. *Ibid.*, Irish and Norwich poplins. 1877 *Encycl. Brit.* VII. 596/2 *John Crome* (1769–1821), English landscape painter, founder and chief representative of the 'Norwich School'. 1906 W. F. Dickes *Norwich Sch. Painting* 11 The Norwich School has had a powerful influence upon the Art of Great Britain. 1929 *Official Guide to City of Norwich* 25 The.. Jacquard loom for the weaving of the beautiful Norwich shawls of coloured silks. 1931 J. Lucas *Hunt & Working Terriers* xxiii. 189 In 1880 a Mr. Nichols of Wymondham, Norfolk, was breeding a small red dog which he called the Norwich terrier. 1950 '*Mercury*' *Dict. Textile Terms* 368/1 *Norwich crepe*, a cloth very similar to georgette, and made from fine silk warp and cotton weft, the crepe being produced by the weave. 1950 A. C. Smith *Dogs since 1900* xi. 179 Some of their admirers..asked permission to have them registered at the Kennel Club as Norwich Terriers. In 1932 they were admitted to registration. 1951 *Short Guide Norwich School Pictures* 3 The old masters of Holland..influenced so profoundly the Norwich School. 1958 *Times* 12 Dec. 1/7 (Advt.), Norwich Terrier for sale. 1967 R. M. R. Young *Guide to Bridewell Museum* 11 The traditional Oriental design now often called 'the Paisley pattern' was also used by the Norwich shawl makers. 1971 F. Hamilton *World Encycl. Dogs* 466 From these red East Anglian terriers..sprang the breed which was recognized..as the Norwich terrier. *Ibid.* 467 The Norwich is very hardy and adaptable. 1972 *Times* 21 June 16/5 Two large, fine views by Henry Bright, the late Norwich school landscapist. 1974 *Country Life* 28 Nov. 1652/1 The Castle Museum at Norwich has long specialized in the..Norwich School.

nose, *sb.* Add: **1. e.** In Horse-racing: the nose of a horse used as an indication of the distance between two finishing horses. Phr. *to bet* (etc.) *on the nose*: to back a horse to win (as opposed to betting for a place, or betting each way).

1908 L. Mitchell *New York Idea* i. 11 Flying Cloud slipped by the pair and won on the post by a nose in one forty nine! 1951 *Publ. Amer. Dial. Soc.* xvi. 13 Bet on the nose. 1955 *Amer. Speech* XXX. 26 Laymen.. will say that a horse won by a neck, head, or nose to describe any race which was extremely close. In this country a bet on a horse to win..is said to be on the nose. 1963 'J. Prescot' *Case for Hearing* iv. 71 Every afternoon that lad of mine is in the betting shop slapping as much as fifty quid a time on the nose. 1973 *Times* 12 Apr. 12/6 Ladbroke..assured me that I could lay £30,000 on the nose if I wished.

2. b. *fig.* (Further examples.)

1875 W. Hyde in C. F. Wingate *Views & Interviews on Journalism* 196 The 'nose for news', by which is meant unwearying alertness and insatiable hunger for something 'ahead of the other papers'. 1934 C. Lambert *Music Ho!* ii. 86 He had an astounding 'nose' for the growth of any particular movement of taste or snobbism. 1942 E. Waugh *Put out More Flags* ii. 118 One does not work in the East without acquiring a nose for a deal. 1960 *Times* 20 June 4/1 The crowds who always have a nose for personality. 1972 'J. Cassells' *Profit for Picaroon* v. 36 He was a damned good reporter..and he had a nose for a story.

c. Also, of wines.

1936 F. C. Lloyd *Art & Technique Wine* xv. 146 The bouquet, or 'nose' to use a more technical word, is very important and serves to reveal the characteristics of wines to a connoisseur. 1952 A. Lichine *Wines of France* x. 107 Its tremendous nose—*bouquet* is too delicate a word—makes it [*sc.* Chambertin] a veritable Cyrano. 1971 *Guardian* 12 Nov. 9/2 The dry white of Beaucaire.. has an aromatic 'nose' and plenty of body.

d. *on the nose* (Austral.): offensive, annoying; smelly.

1941 Baker *Dict. Austral. Slang* 49 *Nose, on the*: (said of things) disliked, offensive. 1945 T. Inglis Moore *We're going Through* 18 Withdraw! That's on the nose! 1946 K. Tennant *Lost Haven* (1947) vi. 86 'Christ! Alec,' he complained. 'This bait's a bit on the nose, ain't it?' He spat over the side as the reek of fish-heads a week old.. caught his stomach. 1953 D. Cusack *Southern Steel* 138 The beer's on the nose and the plonk'd make a willy-wagtail fight an emu. 1974 *Australian* 12 Dec. 13 She renounced her Australian citizenship and swore everlasting loyalty to the Stars and Stripes. A bit on the nose, we think.

5. (Earlier and later examples.)

1789 G. Parker *Life's Painter* xv. 151 Nose.—Snitch. 1928 E. Wallace *Gunner* xviii. 145 He was just a little thief and a nose. 1954 [see *grass *sb.*[1] 11*]. 1961 *John o' London's* 30 Nov. 610/3 Other words used for him [*sc.* an informer] include *grass, nose*, [etc.]. 1974 R. Edwards *Dixon of Dock Green* 7 He knew that CID men are allowed to drink on duty because much of their time is spent with 'noses' or informants.

6. (Later examples.)

1917 R. Fry *Let.* 23 Nov. (1972) II. 420 Millions of people..catch me on the telephone the moment I just put my nose inside the Omega. 1922 Joyce *Ulysses* 322 In Shanagolden where he daren't show his nose. 1935 J. Buchan *House of Four Winds* i. 42, I should like to put my nose inside Evallonia just to say I'd been there. 1947 M. Lowry *Under Volcano* i. 25 Geoffrey's 'nose was always in a book'. 1963 Nose wide open [see *fox *sb.* 2 c]. 1968 M. Jones *Survivor* ii. 33 It was considered anti-social to 'have your nose stuck in a book'. 1970 C. Major *Dict. Afro-Amer. Slang* 85 '*Nose wide open*' is to be in love.

7. c. *on the nose* (U.S.): accurately, precisely, to the heart of the situation; accurate, precise (esp. of time).

1937 *Printers' Ink Monthly* May 40/1 *On the button*, a program ending exactly on time... *On the nose*, see 'On the button'. 1943 *New Yorker* 30 Oct. 21/1 I'll meet you there happest twelve, but on the nose. 1944 W. C. Greet *World Words* p. v, This book has been prepared in great haste. To be readily understood, and, in radio parlance, to be on the nose, were its prime requirements. 1958 B. Holiday *Lady sings Blues* (1973) ii. 27 'You were supposed to be out weeks ago,' they told one girl. But I got out right on the nose at the end of four months. 1959 N. Mailer *Advts. for Myself* (1961) 240 Malcolm Cowley was right on the nose when he wrote that *The Deer Park* was a far more difficult book to write than *The Naked and the Dead*. 1962 P. Gregory *Like Tigress at Bay* i. 14 'That's it.' he said. 'You've hit it right on the nose.' 1972 R. H. Copperud *Dict. Contemp. & Colloq. Usage* (Eng.-Lang Inst. Amer.) 21/1 *On the nose*, right on target; exactly; accurately; on the button. 1974 Hawkey & Bingham *Wild Card* xxii. 176 The cerebroid was properly docked in the flight couch. 'Right on the nose,' Stillman said.

8. b. (Earlier example of usage with *poke*.) Conversely *to keep one's nose clean*, to behave properly, keep out of trouble (see also quot. 1909).

1850 Thackeray *Pendennis* II. xxxvi. 347 Beck! leave the room. What do *you* want poking your nose in here? 1887 *Lantern* (New Orleans) 13 Oct. 5/3 There's worse fellows than you looking for it, and if you only keep your nose clean, we'll let you have it. 1909 J. R. Ware *Passing Eng.* 162/1 *Keep your nose clean* (Army), avoid drink. 1934 J. O'Hara *Appointment in Samarra* ii. 54, I give you the sawbuck because you've just got out of the can. Keep your nose clean 1945 P. Cheyney *I'll say she Does!* i. 12 You're a guy who has gotta reputation for keepin' his nose clean, but..you're in bad with the big boy. 1959 N. Mailer *Advts. for Myself* (1961) 350 You boys on Channel Five want to keep your nose clean, now don't you? 1960 C. MacInnes *Mr. Love & Justice* 15 What we're offering you is—well, influence... How you manage there, provided you keep your nose clean, is really up to you. 1970 N. Armstrong et al. *First on Moon* i. 23 Do what people tell you, keep your nose clean and work out your academic progress. 1974 A. Ross *Bradford Business* 64 Denis Fitzgerald..a known associate of villains, but managed to keep his own nose clean.

c. Also, *to look down one's nose*.

1921 Galsworthy *To Let* iii. xi. 306 That chap Jolyon's water-colours were on view there. He went in to look down his nose at them—it might give him some faint satisfaction. 1932 *Sun* (Baltimore) 24 Oct. 8/1 It is getting more difficult for a lawyer to look down his nose at the courtroom, with consequent impairment of the prestige of the courts. 1956 A. Wilson *Anglo-Saxon Att.* i. iv. 102 When you were all little babies, I used to sing and dance all day. The English neighbours would say 'That young Mrs Middleton's quite mad', and look down their noses—so! 1973 *Times* 24 Apr. 5/4 The portrait of the famous widow, who invented the topsy-turvy logic of *remuage* and *dégorgement*..looks down her nose down the stairs at her successors in the craft.

e. *to hold one's nose*: to compress the nostrils between the fingers in order to avoid perceiving a (bad) smell. Also *fig.*

a 1592 Greene *Jas. IV* (1905) II. i. ii. 102 A stiffe docket,—hold your nose, master. 1830 Coleridge *Table-T.* 8 July (1884) 102 Son of Jacob! Thou stinkest foully. See the man in the moon! he is holding his nose at thee at that distance. 1900 *Fortn. Rev.* Jan. 74 Surely there are times when he is forced to hold his nose and shut his eyes to shut out the abominable visions he conjures up for us. 1973 *Times* 18 Sept. 18/2 Then abolish all alternatives to this public system of education, at which they hold their noses.

f. *to thumb one's nose*: to put one's thumb to one's nose and extend the fingers as a gesture of derision: to 'cock a snook'. Also *fig.* orig. *U.S.*

1903 R. Dunn *Diary* 25 July (1907) ix. 109 He thumbed his nose at us. 1929 A. C. & C. Edington *Studio Murder Myst.* iv. 37 Underlings in the studio thumbed their noses at his back. 1947 W. Motley *Knock on any Door* 119 Behind Ma's back Ang thumbed her nose at him and stuck out her tongue. 1973 J. Wainwright *High-Class Kill* 163 They are already thumbing their snotty, aristocratic noses at us.

g. *to get it up one's nose*: in P. G. Wodehouse (*a*) to become angry; (*b*) to become infatuated.

1925 Wodehouse *Carry on, Jeeves!* iii. 67 This lad seems to have chucked all the principles of a well-spent boyhood. He has got it up his nose! 1934 —— *Right ho, Jeeves* xvii. 220 So thoroughly had Gussie got it up his nose by now that it seemed to me that had he sighted me he might have become personal about even an old school friend. 1961 —— *Service with Smile* (1962) ix. 135, I have seldom seen a man who has got it so thoroughly up his nose. 1971 —— *Much Obliged, Jeeves* iii. 20 He had spoken of her..with devotion in every syllable. Plainly he had got it up his nose and didn't object to being bossed. 1973 —— *Bachelors Anonymous* viii. 92 'See what I mean?.. Got it right up his nose,' said Mr. Llewellyn. 'I have seldom seen a case where the symptoms were more clearly marked,' said Mr. Trout. 'He is taking her to dinner.'

h. *to get up someone's nose*: (see quot. 1951).

1951 Partridge *Dict. Slang* 1120/2 *Nose, get up one's*, to upset, annoy, irritate, render 'touchy'. 1975 *Daily Mail* 6 Aug. 7/1 The implication that granny was a little winning knockout with a system that couldn't be bettered ..does, I'm afraid, get rather up my nose. 1975 *Times Lit. Suppl.* 26 Sept. 1102/2 The police pulled them [*sc.* homeless alcoholics] in whenever they got up the public's nose too much.

i. *to get one's nose down* (*to*): to work arduously and concentratedly (at). Cf. Grindstone 2 b.

1962 *Times* 31 May 4/1 Getting their noses really down to business. 1966 Wodehouse *Plum Pie* i. 11 One would certainly have expected him by this time to have raised the price of a marriage licence and had the Bishop and assistant clergy getting their noses down to it.

9. e. Also, *to rub* (occas. *push*) *one's nose in it*: to remind (someone) humiliatingly of his error; to make (someone) acutely aware of (a fault, etc.).

1963 P. M. Hubbard *Flush as May* xiii. 121 I'm sorry. I've said I'm sorry... Don't rub my nose in it. 1967 'M. Hunter' *Cambridgeshire Disaster* vii. 47 It makes a change, I suppose,.. having your nose rubbed in it. 1971 D. Lees *Rainbow Conspiracy* i. 18 Using me on a hard news story would be pushing their noses in it—treating them like a branch office with printing facilities. 1972 *Times Lit. Suppl.* 3 Mar. 234/1 Discontinuity will not do on its own for a resolute dualist and Bataille wants to rub our noses in the idea of the continuous.

13. Also, the corresponding part of an aeroplane, motor vehicle, torpedo, surfboard, etc.

1899 *Royal Mag.* Jan. 251/1 In the 'nose' of the torpedo. 1899 H. G. Wells *When Sleeper Wakes* xxiv. 320 The nose of the machine jerked upward steeply. 1903 *Science Siftings* 7 Nov. 68/1 When the operator wishes to descend he pulls on a line which lowers the nose of the kite. 1906 *Strand Mag.* May 516/2 In such cases they put the nose of the machine to the opposite side. 1914 *War Illustr.* I. 406 A British Army biplane that collapsed and fell with its nose in the earth. 1942 *R.A.F. Jrnl.* 13 June 33 Always face the nose of the bomb. When I say 'nose' I mean the end which is away from the explosive charge. 1962 S. Carpenter in *Into Orbit* 57 The engineers designed the capsule so that the blunt-nose would come down first. 1962 T. Masters *Surfing made Easy* 64 *Nose*, the front of the surfboard. 1968 W. Warwick *Surfriding in N.Z.* 3/2 The nose was rounded with a slight uplift or rocker. 1971 M. Tak *Truck Talk* 110 *Nose*, the foremost part of a trailer.

14. e. A projecting part of an electric traction motor by which it is suspended from the framework of the bogie or vehicle.

1907 Parshall & Hobart *Electric Railway Engin.* x. 451 In the case of a heavy motor there is usually a nose in the frame casting which rests on a bar carried by springs on the transom. 1927 R. E. Dickinson *Electric Trains* vi. 110 On the other side of the motor case a projecting 'nose' is cast and this nose is fixed on the bogie transom with a stiff spring above and below it. 1955 E. A. Binney *Electric Traction Engin.* vii. 126 The nose end of the motor is resiliently supported on the bogie transom.

17. a. *nose-bone, -jam, -net, -peg, -pin, -tip* (later example), *weight*.

1890 W. F. Butler *Sir Charles Napier* xiii. 188 Others..have not had the nerves torn by a jagged ball passing through, breaking nose-bones and jaw-bones, and lacerating nerves. 1956 H. Gold *Man who was not with It* (1965) i. 7 Even if the sharpest nosebones worked their way into my cheeks. 1922 Joyce *Ulysses* 168 Sheepsnouts bloodypapered snivelling nosejam on sawdust. 1883 *Illustr. Sporting & Dramatic News* 6 Jan. 407/2 He calls it a 'nose net'. It consists of an ordinary bag net, just large enough to encircle the muzzle of a horse... The effect..is to make a horse keep his mouth shut and..to prevent him pulling. 1963 E. H. Edwards *Saddlery* xii. 88 A very simple device is the nose-net. ..It is put over the nose and fastened fairly tightly to the noseband and will stop the majority of tearaways. 1909 *Cent. Dict.* Suppl. II. 875/1 *Nose-peg*, a pin or stud attached to the quadrant-arm in a spinning-mule to effect an acceleration of the spindle in forming the cop. 1935 H. H. Finlayson *Red Centre* xii. 118 The nose-peg..is a hardwood cylinder expanding to a disk at one end and a smaller pointed cone at the other... The young camel is thrown, the cartilage of the left nostril pierced.., and the peg pushed through the wound. 1966 G. W. Turner *Eng. Lang. Austral. & N.Z.* iii. 39 Cook noticed the nose-pins worn by native men. 1971 *World Archaeology* III. 140 There might be an occasional stone nose-pin. 1927 W. de la Mare *Told Again* 246 The Fox then brushed himself nose-tip to stern with his brush. 1959 '*Motor*' *Man.* (ed. 36) xiii. 270 A caravan with a heavy nose weight tows more steadily than one without.

b. *nose-blowing*.

1864 J. S. Le Fanu *Uncle Silas* I. xxiv. 292 The boisterous nose-blowing that suddenly resounded from the passage. 1967 E. A. Gollschewsky in *Coast to Coast 1965–6* 87 The vehement nose-blowing..that marked her progress from room to room.

c. *nose-dropping*.

1905 J. Joyce *Let.* 19 Jan. (1966) II. 78 O, blind, snivelling, nose-dropping, calumniated Christ.

18. nose-bridge, (*a*) = Nose *sb.* 1 d; (*b*) *Archæol.* denoting a type of handle found on pottery of the Copper Age in southern Europe; **nose-candy** *U.S. slang*, a drug that is inhaled (illegally), *spec.* cocaine; **nose-clip,** a clip excluding water from the nose of a swimmer or diver; **nose door** *Aeronaut.*, a forward-facing door in the nose of an aeroplane, **nose drive,** the positioning of the engines at the front of a space rocket; so **nose-driven** *a.*; **nose drops,** a medicament intended to be administered as drops into the nose; **nose-glasses** *U.S.* (examples); **nose hangar** *Aeronaut.* (see quot. 1960); **nose-heavy** *a. Aeronaut.*, having a tendency for the nose to drop relative to the tail; hence **nose-heaviness; nose-nippers** *pl.* = Pince-nez; **nose-paint** *slang*, intoxicating liquor; also, a reddening of the nose ascribed to habitual drinking (cf. *nose-painter, -painting* s.v. Nose *sb.* 17 b); **nose paste** = *nose putty* below; **nose print,** a drawing of the facial characteristics of an animal, used as a means of identification; so **nose-printing** *vbl. sb.*; **nose putty,** a putty-like substance used in the theatre for altering the shape of the nose, etc.; **nose-ride** *v. intr.*, to stand on the nose (cf. *Nose *sb.* 13) of a surfboard (usu. as *vbl. sb.*); **nose-riders** *pl.*, spectacles; **nose suspension,** a method of supporting a traction motor from the framework of the bogie or vehicle at one end and on an axle at the other (cf. sense *14 e); so **nose-suspended** *a.*; **nose-thumbing** *vbl. sb.*, the action of thumbing one's nose (*Nose *sb.* 8 f); an instance of such behaviour; so (as a back-formation) **nose-thumb** *sb.*; **nose-to-tail** *adv.* and *a.*, (of motor vehicles) travelling, or placed, behind one another and very close together; **nose-trick,** the inadvertent inhalation or expulsion of liquid through the nose when drinking; **nose wheel** *Aeronaut.*, a wheel under the nose of an aircraft; **nose wipe** *sb.* = *nose-wiper*.

1923 A. Huxley *Antic Hay* x. 156 Shell rims with gold ear-pieces and gold nose-bridge. 1939 V. G. Childe *Dawn Europ. Civilization* (ed. 3) xiv. 245 In the pottery [of Sardinia before the nuragic age] we might distinguish: ..carinated cups and other vessels with nose-bridge handles, which persist into the nuragic age. 1935 A. J. Pollock *Underworld Speaks* 81/1 *Nose candy*, cocaine. 1935 C. F. Coe *G-Man* ii. 25 I'll lay off the booze an' you lay off the organizin', the nose-candy and the stick-ups. 1960 *Time* 25 Jan. 88/2 Cocaine..is put into crystal-line form. This enables users to sniff it ('nose candy'). 1974 *Globe & Mail* (Toronto) 28 Sept. 33/4 The movie omitted the morphine and left the cocaine because nose candy is the trendy drug. 1959 *Elizabethan* Apr. 9/2 Following the orders of the sergeant I had..fitted the nose clip in place [on an under-water breathing apparatus]. 1971 A. Diment *Think Inc.* vii. 123, I bit on the rubber mouth-piece, put on the nose-clip and flipped down my face mask. 1960 Nose door [see *beaver-tail* s.v. *Beaver[1] 6]. 1969 *Jane's Freight Containers 1968–69* 454/1 Loading is via a vizor-type, straight-in, nose-door with full-width integral ramps. 1937 *Discovery* Sept. 270/2 First he [*sc.* Goddard] directed his attention to the so-called 'nose-drive' construction. 1947 W. Ley *Rockets & Space Travel* (1948) v. 134 The gases were to be ejected through a system of nozzles at the top of the rocket; the nozzles were to pull the rocket upward, instead of pushing it upward as planned in the original Oberth Rocket... This system, known as 'nose drive'.., offered a great number of advantages. The rocket did not have to be constructed as sturdily nor did it require an elaborate steering mechanism. 1952 E. Burgess *Rocket Propulsion* v. 131 It has been stated that nose-driven rockets..are inherently more stable than those in which the motor is situated at the extreme rear. This is another common fallacy, for providing the thrust line of the motor passes through the centre of gravity of the rocket, the actual position of the motor cannot affect stability. 1942 T. Sollmann *Man. Pharmacol.* (ed. 6) 165 'Nose drops', *i.e.*, solutions in liquid petrolatum for instillation into the nares, should not contain more than 1 per cent of eucalyptol or camphor. 1970 *Women's Household* July 12/1 When he needed nose drops badly he would come over to one of us and stick up his nose. 1890 *Cent. Dict.* V. 4020/2 Nose glasses. 1901 Ade *Forty Mod. Fables* 22 He said 'Whom,' and wore Nose Glasses. 1929 D. Runyon in *Hearst's International* Oct. 64/1 To look at Judge Henry G. Blake, with his..nose glasses. 1971 *Lebende Sprachen* XVI. 11/2 US nose glasses —BE/US pince-nez. 1948 *Jrnl. R. Aeronaut. Soc.* LII. 573 In the civil field the 'nose hangar', such as that used successfully by KLM for the Constellation last winter, is a realistic effort to provide shelter where it is needed, instead of enclosing space at random. 1960 G. Blanchet *Search in North* v. 66 Nose hangars were built—sheds in which the front of a plane could be sheltered and work done on its engines. 1970 R. & J. Paterson *Cranberry Portage* xiv. 87 A great deal of time was spent warming aircraft engines with roaring fire-pots inside the canvas nose-hangars. 1919 A. Klemin *Text-bk. Aeronaut. Engin.* xv. 178 The down stream from the propellers..is said to increase the safety from the point of view of longitudinal balance, giving tail heaviness with power, and nose heaviness without power. 1930 R. Duncan *Stunt Flying* iii. 26 Nose-heaviness, or tail-heaviness, can be corrected by adjusting the horizontal stabilizer. 1959 *Times* 8 Sept. 13/6 There is no feeling of nose-heaviness, and the steering does not have any tricks on corners. 1914 S. L. Walkden *How to understand Aeroplanes* ii. 5 This 'front-heavy,' or 'nose-heavy' machine..is devoid of a self-righting effect. 1917 *Flying* I. 217/1 A stable aeroplane has its centre of gravity well forward, and normally the centre of pressure is behind the centre of gravity. Without a fixed tail plane it would therefore be nose heavy. 1945 R. von Mises *Theory of Flight* xvii. 501 If the state of motion changes or the loads are shifted, the airplane will become slightly nose- or tail-heavy so that the pilot has to operate the elevator. 1895 J. Davidson *Old Aberdeenshire Ministers* 26 The *Aberdeen Journal*, which he read aloud..in a loud monotone, nasalised by the light grip of a large pair of nose-nippers worn low. 1900 Conrad *Lord Jim* v. 40 He saw the old man lift his head from some writing so sharp that his nose-nippers fell off. *a* 1913 F. Rolfe *Desire & Pursuit of Whole* (1934) vii. 61 Smirking female with the thinnest of pinched lips and nose-nippers. 1880 A. A. Hayes *New Colorado* (1881) xi. 158 We saw..a sign, in which a name which I have never encountered elsewhere was given to stimulating beverages. This sign was '*Nose-paint* and *Lunch*'. 1901 F. E. Taylor *Folk-Speech S. Lancs.*, *Nose-paint*, a jocular term for alcoholic drink. 1922 Joyce *Ulysses* 615 A strong suspicion of nosepaint about the nasal appendage. 1968 *Amer. Speech* XLIII. 303 He [*sc.* the cowman] drinks..*nose paint* instead of 'whiskey'. 1951 N. Marsh *Opening Night* vii. 156 One cardboard box containing false hair, rouge, substance labelled 'nose paste'. 1961 Bowman & Ball *Theatre Lang.* 233 *Nose paste*, a plastic substance used to alter the appearance of an actor's nose, chin, etc. 1934 *Sun* (Baltimore) 15 Sept. 15/2 The 'nose print' of a dog is as distinctive as the finger print of a human being. 1952 *Ibid.* 8 Dec. 10/4 Three dairy scientists of the South Dakota Agricultural Experiment Station at Brookings, S.D., have worked out a system of nose prints by means of which it is possible to identify one individual cow from a million others. 1970 B. Knox *Children of Mist* iii. 56 Sometimes breeders..try to work a switch of animal. So we keep one nose-print on file in special cases..ready for comparison 1973 *Times* 26 Nov. 17/5 Four investigators followed the trail of the animals, to establish 'contact' with them and where possible, to draw 'noseprints'—simple drawings of nostril shape and facial wrinkles—to identify the particular animals and their approximate stages of development. 1934 *Sun* (Baltimore) 15 Sept. 15/2 A plan for nation-wide identification of dogs by 'nose printing' to eliminate 'dognapping'. 1960 A. Christie *Adventure of Xmas Pudding* 223 Why did I feel..I was talking to..an actor playing a part!.. What did I see..the beaked nose (faked with that useful substance, nose putty), [etc.]. 1969 K. Vonnegut *Slaughterhouse-Five* v. 88 Rosewater was a big man, but not very powerful. He looked as though he might be made out of nose putty. 1965 Farrelly & McGregor *This Surfing Life* 138 *Nose-ride*, to ride on the nose of the surfboard. 1875 E. H. Dering *Sherborne* II. xviii. 53 Sir Thomas..put on a pair of those glasses which are popularly known as nose-riders. 1962 *Austral. Women's Weekly* Suppl. 24 Oct. 3/3 *Nose-riding*, standing right at front of the board while riding a wave. 1971 *Studies in English* (Univ. Cape Town) Feb. 25 Until the end of the nose-riding era, the run of Cape Town surfers identified with the Californian scene as portrayed in *Surfer*. 1927 R. E. Dickinson *Electric Trains* vi. 111 Fig. 48 shows a bogie with two nose-suspended motors in place. 1948 D. W. & M. Hinde *Electric & Diesel-Electric Locomotives* ii. 25 During the past past few years, however, nose-suspended motors mounted on bogies have become standard practice on all American diesel-electric locomotives. 1894 K. Hedges *Amer. Electr. Street Railways* vii. 75 In the case of the G.E. 800 type when the side suspension is used, the whole of the weight is taken off the axle, whereas by the older method half the weight only was on the cross bar, resting on springs, and the remainder on the axle. One method is known as the End or Nose Suspension, the other as the Side Bar Suspension. 1927 R. E. Dickinson *Electric Trains* vi. 111 For multiple unit trains the nose suspension is practically universal. 1948 D. W. & M. Hinde *Electric & Diesel-Electric Locomotives* ii. 24 Where hammer-blow on the track and axle-loading are not limiting factors and provided that there is one motor per driven axle, nose-suspension is the simplest form of drive obtainable. 1963 *Guardian* 11 Feb. 2/6 Their medical officer of health ..is leaving his post because of the council's continued nose-thumb at the Clean Air Act. 1959 I. & P. Opie *Lore & Lang. Schoolch.* xiv. 317 That peculiar form of recognition variously known as 'the five-finger salute', 'nose thumbing',.. 'cocking a snook', or 'taking a sight' used, thirty years ago, to be demonstrated by every child in the country. 1970 *Globe & Mail* (Toronto) 26 Sept. 23/5 A nose-thumbing gesture comes as a blessed relief in a movie so painfully earnest. 1959 *Manch. Guardian* 20 July 2/5 Four L-drivers nose to tail on a busy road. 1960 *Guardian* 7 June 6/1 The accustomed queues of nose-to-tail traffic on main roads. 1963 *Times* 11 June 5/4 Yet above the garage cars sit nose-to-tail on free space within the park, and scour Mayfair for vacant meters. 1974 C. Fremlin *By Horror Haunted* 48 The nose-to-tail crawl along the motorway. 1954 P. Frankau *Wreath for Enemy* ii. i. 58 It was the 'Prendergast' that made me do what is vulgarly called the nose-trick with my lemonade. 1972 P. Dickinson *Lizard in Cup* ix. 132 Pibble almost did the nose-trick with the dung-smelling local brandy. 1934 *Flight* 6 Dec. 1301 A castor nose wheel allows the fullest use to be made of the wheel brakes of the Hammond Model Y. 1940 *Illustr. London News* CXCVII. 315 (*caption*) The 'Boston' has.. an undercarriage of the tricycle type—two rear wheels retracting into the rear of the engine nacelles and the nosewheel retracting rearwards and upwards into the fuselage. 1974 *Daily Tel.* 5 Oct. 1 A Belgian airliner.. pitched on to its nose on the main runway at Southend airport when the nosewheel collapsed shortly before takeoff. 1919 W. Deeping *Second Youth* ii. 13 Yer dirty little wretch yer; ain't yer got a nose-wipe?

nose, *v.* Add: **6. c.** (Later examples.)

1926 E. F. Spanner *Naviators* i. 9 The car nosed its way ahead on bottom gear, and at a snail's pace. 1937 *Discovery* Feb. 38/2 We nosed our way through the reeds. 1973 P. Moyes *Curious Affair* i. 14 The station wagon nosed its way along the narrow road.

d. To direct the nose of (a motor vehicle, etc.) in a certain direction.

1954 'N. Shute' *Slide Rule* iv. 89 The elevator coxswain nosed her [*sc.* an airship] upwards to about a thousand feet. 1972 D. Delman *Week to Kill* 139, I nosed the car out of town and on to 118, where I zapped it into high.

8. b. (Earlier and later *fig.* examples.)

1879 'Mark Twain' *Let.* 21 Jan. (1920) I. 187 The detectives were nosing around after Stewart's loud remains. 1887 *Lantern* (New Orleans) 3 Dec. 2/3 We

nosed around to try and find out why. 1917 Wodehouse *Uneasy Money* iv. 43 There's no harm in my nosing round, is there? Be a good chap and give me the address. 1925 —— *Carry On, Jeeves!* i. 28 He began to nose about. He pulled out drawer after drawer. 1955 A. Huxley *Let.* 25 Sept. (1969) 766 While in Guildford I read, or rather nosed about in, Penfield's book on Epilepsy. 1958 P. Kemp *No Colours or Crest* iii. 36 We nosed around the Islands, sometimes less than a mile off shore, searching for indications of our quarry.

c. (Later example.) Also with *among*.

1936 R. Lehmann *Weather in Streets* ii. 189, I thought of her nosing in my room for signs. 1941 A. Curnow in Chapman & Bennett *N.Z. Verse* (1956) 148, I am the nor'west air nosing among the pines.

d. To inform; †to turn king's evidence. *Criminals' slang*.

1811 [see Twist *v.* 9 c]. 1822 *J. Mackcoull Mem. Life* 112 *Nosed*,..watched and informed against. 1846 tr. *E. Sue's Myst. of Paris* cli. 743/1 Gros Boiteux..has already wanted to *escarper* him, (make him a stiff 'un—kill him,) because he has *mangé* (nosed, informed upon some one). 1923 E. Wallace *Missing Million* xix. 156 When a copper comes to one of the 'boys' for expert advice, it means he wants him to 'nose'. 1930 —— *White Face* x. 147 You come down 'ere an' expect us to 'nose' for you, and everybody in the court knows we're 'nosing'.

9. Also *transf.* with *ahead*, to go into the lead by a small margin.

1960 *Times* 1 July 18/1 So Miss Truman had nosed ahead at last. 1971 Wodehouse *Much Obliged, Jeeves* xiii. 134 If the McCorkadale nosed ahead of him in the voting, Florence would in all probability hand him the pink slip.

11. nose down *Aeronaut.*: to direct the nose of an aircraft downwards, to produce or undergo a (downward) steepening of the flight path.

1916 H. Barber *Aeroplane Speaks* 85 If a sharp turn necessitates banking beyond that angle, he must 'nose-down'. *Ibid.* 88 The C.G. being forward of the C.P... causes the aeroplane to nose-down. 1938 V. W. Pagé *Airplane Servicing Man.* xxiv. 818 It is considered advisable, if the engine should stop, that the plane will nose down automatically, instead of tending to stall. 1958 R. D. Blacker *Basic Aeronaut. Sci.* xi. 195/1 The vertical portion of the lift was not as great as the weight of the airplane and it nosed down, losing altitude. 1974 J. Montgomerie *Implosion* xii. 86 We crossed the coast, nosing down over pewter sea... Runway rose towards us.

12. Of an aircraft: to fall *over* on its 'nose'. Of a surfboard: to plunge downward nose first.

1928 V. W. Pagé *Mod. Aircraft* xii. 523 Always pick as smooth and level a piece of ground as possible when making a landing, as, if the ground is very soft or if there are hummocks or ditches, the machine is very likely to 'nose over'. 1953 Berg *Dict. New Words* 115/2 *Nose over*. .. *Flying*. To fall nose forward. 1963 *Surfing Yearbk.* 42/2 *Nosing*, when the nose of the surfboard goes under water while riding a wave. 1965 J. Pollard *Surfrider* ii. 20 Don't let your board 'nose'. This is what happens when the front of the surf board digs in.

nosing *vbl. sb.*: also *Comb.*, as **nosing motion** (see quot. 1940).

1883 H. E. Walmsley *Cotton Spinning* 30 The terminal velocity of the spindles may be increased in the same ratio, as their diameter decreases. This object is attained by Platt's automatic nosing motion. 1884 R. Marsden *Cotton Spinning* ix. 264 The last invention..., which seems to answer all needs, is the automatic nosing motion brought out by Messrs. Platts, and which consists of a scroll placed upon the end of the winding-on drum. 1932 W. Scott-Taggart *Cotton Machinery Sketches* (ed. 4) p. xviii, Diagram of Cop, etc., illustrating Principle of Nosing Motion. 1940 *Chambers's Techn. Dict.* 583/2 *Nosing motion* (Cotton Spinning), a motion on the mule spinning frame which, as the diameter lessens, increases the speed of the tapering spindle on which a cop is being wound.

nose-bag. Add: **1. b.** Also, the practice among holiday-makers of taking their own food or refreshments with them. Also *attrib.*, as *nose-bag crowd*.

1908 *Daily Chron.* 4 Aug. 3/4 Neither was it, as one of Messrs. Lyons's managers observed with appreciation, a 'nose-bag' crowd. 1909 *Ibid.* 7 June 5/2 The 'nose-bag' grows and flourishes.

3. Food, a meal. Also allusively. Phr. *to put* (occas. *get*) *on the nose-bag*, to eat. *slang*.

1874 Hotten *Slang Dict.* 239 To 'put on the nose-bag' is to eat hurriedly, or to eat while continuing at work. 1886 F. T. Elworthy *West Somerset Word-Bk.* 521 Well! hon I zeed zo many o' they there whit-neckangkecher fullers comin', I thinks to mysul, there's a bit of a nose-bag a-gwain on in there. 1898 J. D. Brayshaw *Slum Silhouettes* 155 Come in, we'll jist 'ave five minutes wiv the nose-bag. 1921 R. Lardner *Big Town* v. 201 We couldn't stop to put on the nose bag at the Graham. 1925 Wodehouse *Carry On, Jeeves!* vi. 145 Biffy's man came in with the nose-bags and we sat down to lunch. 1926 *Amer. Speech* I. 652/1 *Nose-bag*, lunch handed out in paper bag. 1930 D. L. Sayers *Strong Poison* xi. 137 Thanks awfully, I've had my morning nosebag. 1962 *New Statesman* 21 Dec. 897/3 The precise time the family get the nose-bag on. 1973 Wodehouse *Bachelors Anonymous* xiv. 185, I must rush. I'm putting on the nosebag with a popsy.

4. A gas-mask. *colloq.*

1915 D. O. Barnett *Let.* 11 May 135 Every one was ready and had their nose-bags on, and the gas had no effect whatever. 1940 *Everybody's Weekly* 2 Mar. 4/1 Londoners call their masks 'Dicky-birds', 'Canaries' and 'Nose-bags'.

nose-bagger. [f. Nose-bag + -er[1].] = Nose-bag 1 b.

1931 *Morning Post* 17 Oct. 12/5 The friendly little South Coast town, where only the 'nose-baggers'..are frowned upon.

nose-bleed. 2. (Earlier, later, and *fig.* examples.)

1848 *Asmodeus* (N.Y.) 73 (Th.), What's the best cure for nose-bleed, doctor? 1914 G. S. Gordon *Let.* 11 Aug. (1943) 59 Germany will get the nose-bleed of her life. 1953 'J. R. Macdonald' *Imaginary Blonde* in *Manhunt* (1964) XII. 138/1 It was a little scuffle. One of the guests suffered a nosebleed. 1960 Wodehouse *Jeeves in Offing* xv. 161 Half way through the second act..Catsmeat Potter-Pirbright, who was playing Lord Fancourt Babberley, left the stage abruptly to attend to an unforeseen nose bleed. 1972 'L. Egan' *Paper Chase* (1973) x. 160 Just a nosebleed... Somebody banged me on the nose.

no·se-cap. [f. Nose *sb.* + Cap *sb.*[1]] **1.** (See Nose *sb.* 18.)

2. *Aeronaut.* **a.** (See quot. 1950.)

1909 Kipling *Actions & Reactions* 112 From her North Atlantic Winter nose-cap (worn bright as diamond with boring through uncounted leagues of hail, snow, and ice) to the inset of her three built-out propeller-shafts is some two hundred and forty feet. 1950 *Gloss. Aeronaut. Terms* (*B.S.I.*) I. 49 *Bow cap* (*nose cap*), a structure forming the extreme forward end of the envelope or hull.

b. (See quots.)

1919 Blackburn & Newby *All about Aircraft* 106 *Nose cap*, a conical cap placed over the propeller boss to reduce head resistance. 1949 *Gloss. Aeronaut. Terms* (*B.S.I.*) ii. 19 *Nose cap*, a boss or hub fairing filled co-axially and rotating with the propeller. It does not extend aft of the front of the blade roots.

3. The metal cap on the nose of a shell or bomb which contains the device for setting the time fuse.

1917 A. G. Empey *Over Top* 301 *Nosecap*, that part of a shell which unscrews and contains the device and scale for setting the time fuse. 1919 W. Deeping *Second Youth* xxix. 244 St. Roman was not a nice place that night, with 'dud Archies' and nose-caps falling into it, as well as bombs. 1942 'R. Crompton' *William carries On* iii. 65 Rich in nose-caps and time-fuses and casings.

4. A cover protecting the nose.

1973 *Times Lit. Suppl.* 18 May 544/2 The middle-aged on the beaches clearly felt no self-consciousness in wearing little paper nose-caps as protection from the sun.

nose cone. Also **nose-cone, nosecone.** [f. Nose *sb.*] A conical nose-cap; *spec.* the cone-shaped front part of a rocket, which is designed to withstand the severe heating caused by atmospheric friction and generally contains any instruments that may be carried.

1949 *Jrnl. Brit. Interplanetary Soc.* VIII. 233 A diagram of the rocket is shown in Fig. 1. The equipment space is within a thin ogival-shaped aluminium nose-cone. 1956 *Spaceflight* I. 26/1 At this stage..the ballistic nose-cone (which protects the satellite package from aerodynamic heating) can be thrown off. 1963 *Guardian* 30 Mar. 1/5 The plane was struck by lightning .. A hole was made in the fibre nosecone. 1964 J. L. Nayler *Dict. Astronautics* 177 The nose cone may become the satellite itself..or, after shielding the satellite.., be separated by an explosive cartridge. 1974 A. Price *Other Paths* ii. ii. 128 They were trying to take the brass nose-cone off one of our howitzer shells.

no·se count. [f. the phr. *to count noses* (Nose *sb.* 6 d).] An enumeration, esp. of votes; a decision by majority vote. Hence **no·se-counting** *vbl. sb.*, assessment by numbers.

1938 *Life* 6 June 22/3 Meantime in London peace-yearning Prime Minister Chamberlain again tried to quarantine the Spanish war by proposing a nose-count and withdrawal of foreign troops. 1951 *True Police Cases* (U.S.) Nov. 42 At nose-counting time in the..hoosegow. 1958 *Listener* 30 Oct. 708/1 Was that nose-counting success due only to some adventitious notion..that *Salome* is salacious? 1960 *Times* 14 Sept. 12/6 There was much talk just then of the coming 'Nose Count' over the U.S.A. which was one way of making their census sound more personal and pointed. 1966 *Economist* 29 Oct. 450/1 The prospects of an effective tightening of Rhodesian sanctions must depend far more on the major powers' real readiness to apply a tighter squeeze than on the nominal nose-count in the Security Council. 1967 M. McLuhan *Medium is Massage* 22 Nose-counting, a cherished part of the eighteenth-century fragmentation process, has rapidly become a cumbersome and ineffectual form of social assessment.

no·se-dive, *sb.* Also **nose dive, nosedive.** [f. Nose *sb.* + Dive *sb.*] **1.** *Aeronaut.* A sudden or rapid descent by an aircraft nose first. Also *transf.*

1912 *Flight* 31 Aug. 787/1 The machine at once started a spiral nose-dive. 1914 Hamel & Turner *Flying* iii. 57 M. Adolphe Pégoud introduced side-slips, tail-dives, and nose-dives into his exhibition repertory of flying. 1917 'I. Hay' *Carrying On* i. 17 Next moment she [*sc.* the aeroplane] lurched again, and then took a 'nose-dive' straight into the British trenches. 1922 *Encycl. Brit.* XXX. 21/2 The aeroplane may..drop one wing and pass into a steep spiral glide known as a 'spinning nose-dive'. *a* 1930 D. H. Lawrence *Last Poems* (1932) 15 The dolphins leap ..and flip! they go! with the nose-dive of sheer delight. 1932 Auden *Orators* ii. 52 *Nose-dive*—Nightmare to nerves And needed by no one And dash toward death. 1952 R. S. Porteous in *Coast to Coast 1951–1952* 146 If this tub rolls over or takes a sudden nose-dive to the bottom. 1955 H. Klein *Winged Courier* viii. 52 It looked as if the 'Pioneer' was crashing to its doom in an uncontrolled nose-dive. 1975 *Aeroplane Monthly* Nov. 572 When a down-draught was encountered or the control column was pushed forward..this resulted in a nose-dive to earth.

2. *fig.*

1920 R. Macaulay *Potterism* vi. iv. 241 If you chuck the *Fact* you take away its last chance. It'll do a nose-dive now! 1946 *Sun* (Baltimore) 8 Nov. 21 An unexplained nose dive in surplus sales occurred in the July-August-September quarter. 1973 *Guardian* 28 May 6/6 After a really splendid first week, the Festival took a nosedive from which it never recovered.

no·se-dive, *v.* Also **nosedive.** [f. prec. sb., or Dive *v.*] **1. a.** *intr.* To perform a nose-dive.

1915 *Sphere* 24 July 94/2 Its engines stopped, and it nose-dived to a level of 2,000 ft. 1920 *Chambers's Jrnl.* Mar. 208/1 Checking any tendency [of a motor-boat] to nosedive. 1923 *Daily Mail* 27 Mar. 9 His aeroplane nose-dived from a height of about 150 feet and crashed on to the aerodrome. 1930 E. Blunden *Poems* 42 A hundred feet he nose-dives. 1930 *Daily Express* 6 Oct. 3/3 Then with all the lights on the airship nose dived. 1930 J. S. Huxley *Bird-Watching & Bird Behaviour* (1949) ii. 35 They [*sc.* birds] must be protected against side-slipping and nose-diving. 1968 W. Warwick *Surfriding in N.Z.* 2 He then ran to the front of the board so that it nosedived.

b. *trans.* To put into a nose-dive.

1919 Pippard & Pritchard *Aeroplane Structures* vi. 55 There is a moment upon the wings tending to nose dive the aeroplane still further. 1928 *Daily Tel.* 16 Oct. 17/5, I have nose-dived this machine at 250 miles an hour.

2. *fig.* Esp. to drop or decrease abruptly.

1920 R. Macaulay *Potterism* vi. iv. 241, I can't stop it [*sc.* a newspaper's collapsing]. But I'm jolly well not going to nose-dive with it. I'm clearing out. 1926 *Spectator* 13 Mar. 492/1 We will nose-dive straight into the middle of Mr Cobham's book. 1954 *Sun* (Baltimore) 16 Dec. 24/1 Business..has nosedived since the expressway was opened to intercity traffic. 1958 S. Hyland *Who goes Hang?* xviii. 78 'Is that all you wanted to ask me?' He said it falsetto. 'Well, as a matter of fact, no.' 'Good.' His voice nose-dived. 1969 *Daily Tel.* 23 Jan. 4/1 Allied Investments nosedived again yesterday. The market pushed the shares down another 2s to 3s 6d.

Hence **no·se-diving** *vbl. sb.* and *ppl. a.*

1917 E. W. Walters *Heroic Airmen* xii. 102 He was compelled to resort to nose-diving. 1931 *Handbk. Aeronautics* (R. Aeronaut. Soc.) iv. 288 The fuselage must have a load factor the same as for the main planes in nose diving. 1958 *Engineering* 7 Mar. 295/1 Nose diving under hard braking is no more than on more conventional vehicles. 1970 T. Hughes *Crow* 39 The nosediving aircraft concludes with a boom.

nose-do·wn, *a.* and *sb.* *Aeronaut.* [f. Nose *sb.* + Down *adv.* or the phr. *to nose down* (*Nose *v.* 11).] **A.** *adj.* With the nose directed downwards. Also as quasi-*adv.*

1916 H. Barber *Aeroplane Speaks* 87 [An inclinometer] will indicate a nose-down position by increase in airspeed. *Ibid.* 113 The aeroplane may have a tendency to fly 'nose-down'. 1933 *Jrnl. R. Aeronaut. Soc.* XXXVII. 67 She..goes into a second dive, being 1,160 ft. below her starting height at $t = 80$, and 1,440 ft. below at $t = 100$, when she is 13·5° nose down. 1965 C. N. Van Deventer *Introd. Gen. Aeronaut.* v. 88 If the engine fails the airplane will automatically assume a slightly nose-down position ready for a glide.

B. *sb.* A descent.

1942 *Tee Emm* (Air Ministry) II. 94 That steep nose-down which is often so dangerous if the sea is met sooner than expected. 1944 J. R. Hoyt *Safety after Solo* xii. 217 It is time to start the nose down toward point 3.

no-see-em (nōᵘsī-ĕm). *N. Amer.* Also **no-see-um.** [Corruption of *no see them.*] A name used for several small, blood-sucking insects, esp. biting midges of the family Ceratopogonidæ.

1848 Thoreau *Maine Woods* 2 In the summer myriads of black flies, mosquitoes, and midges, or, as the Indians call them, 'no-see-ems', make travelling in the woods almost impossible. 1902 W. D. Hulbert *Forest Neighbors* (1903) 129 The mosquitoes and black-flies and no-see-'ems had bitten him until his skin was covered with blotches. 1903 S. E. White *Forest* 154 The midge, again, or punkie, or 'no-see-'um', just as you please, swarms down upon you suddenly. 1934 A. Christie *Parker Pyne Investigates* 198 The no-see-ums are biting good and hard. 1939 K. Pinkerton *Wilderness Wife* xvi. 166 No-see-ums are after dark prowlers. 1964 *Atlantic Advocate* July 67/1 Near the bottom we struck swamp ground, a paradise for no-see-'ems, those minute flies whose burning bites are inversely related to their size. 1970 *Observer* (Colour Suppl.) 19 Apr. 59/1 Down on the banks [of a Canadian river] there were awful blood-sucking insects called 'no-see-um', giant mosquitoes. 1975 *New Yorker* 24 Feb. 82/3 No-see-ums are so small they go right through the screening of the tent. They home on flesh.

nose-end. [Nose *sb.* 17 a.] **a.** The tip of the human nose. **b.** The end of a nose in any of the transferred senses.

1611, etc. [see Nose *sb.* 17 a]. 1960 S. Plath *Colossus* (1967) 65 The nose-end that twitches. 1961 *Times* 16 Dec. 11/5 Make sure..that mud does not enter the nose-ends [of a gun].

Hence **nose-ender. a.** (See Nose *sb.* 18.) **b.** (See quots.)

1854 [in Dict. s.v. Nose *sb.* 18]. 1929 F. C. Bowen *Sea Slang* 96 *Nose ender*, a strong head wind. 1946 J. Irving *Royal Navalese* 123 *Nose-ender*, a head wind and sea; regular Destroyer Weather. 1962 W. Granville *Dict. Sailors' Slang* 82/1 *Nose-ender*, wind blowing from dead ahead, literally into one's nose.

nose-flute. Add to def.: Later used also in North America and Europe. (Later examples.)

1962 *New Yorker* 21 July 68/2 They are the manzello and the stritch, early forbears of the saxophone; a pocket-sized nose flute; the..melodica, a kind of accordion that is blown, not pumped. 1963 *Economist* 23 Nov. 775/2 Invented [jazz] instruments like the stritch and a nose-flute. 1969 *Listener* 3 Apr. 470/3 He has a large gong, a kind of..battery-run accordion, various whistles, and a nose flute.

nose-hole. 1. (Later examples.)

1922 Joyce *Ulysses* 51 His leprous nosehole snoring to the sun. 1950 R. Moore *Candlemas Bay* 242 You know how Candy looks when she's mad, like death, them black eyes and her noseholes thin.

nosema (nosī·mă). [mod.L. (C. W. von Nägeli 1857, in *Amtlicher Bericht über die Versammlung Deutscher Naturforscher* XXXIII. 133), f. Gr. νόσημα disease.] A microsporidian protozoan parasite of the genus so called, esp. *Nosema apis*, identified in 1909 as the cause of an infectious dysentery affecting bees; also, the disease itself. Also *attrib.*

1911 Fantham & Porter in *Proc. Zool. Soc.* 625 That *Nosema apis* was fatal to bees and allied Hymenoptera had been shown..by feeding healthy hive-bees, mason-bees, and wasps with honey infected with *Nosema* spores. 1912 *Ann. Trop. Med. & Parasitol.* VI. 149 Bees infected with *Nosema* seem unable to preserve their spotless cleanliness. *Ibid.* 159 Wasps have been observed carrying away bees dead of *Nosema* and then feeding their larvae on the corpses. 1914 *Bull. U.S. Dept. Agric.* no. 92, 6 The name 'Nosema disease', which the writer [*sc.* G. F. White] suggests as the common name for this disease, is, it will be observed, only a translation of the German name used by Zander. *Ibid.* One is justified in at least drawing the conclusion that Nosema infection in a colony tends to weaken the colony. 1937 W. Herrod-Hempsall *Bee-Keeping* II. xxx. 1427 Nosema Disease is caused by a microscopical animal parasite (*Nosema apis*) infesting the food canal of the bee. *Ibid.* 1429 Nosema infected bees do not, as a rule, lose the power of flight. 1955 *Sci. News Let.* 1 Oct. 214/3 Antibiotics..will give successful control over the destructive diseases, American and European foul-broods, that attack larval bees, and nosema, a killer of adults. 1973 *Times* 19 Apr. 27/5 It is against these very diseases [of bees], notably the two 'foul brood' types, but not forgetting acarine and nosema, that the Ministry is acting.

no·se-pick. [Pick *sb.*[1] 5.] An instrument for clearing the nose of mucus, etc. **no·se-picker** [Picker[1]], a person who picks his nose (with his fingers); **no·se-picking** *vbl. sb.* and *ppl. a.*

1825 *Kaleidoscope* 11 Oct. 114/3 Appended to it, by a silver chain, was..an ivory nose-pick. 1959 I. & P. Opie *Lore & Lang. Schoolch.* 47 To a nose-picker they chant: Friday, pie-day, Keep your nose tidy. 1960 Koestler *Lotus & Robot* i. i. 62 Only the nose-picking disciple kept up his activities. 1975 J. Aiken *Voices in Empty House* iv. 116 The nose-picking in itself made me feel sick... *Nobody* in polite society *ever* picked their nose.

nose-piece. Add: Also **nose piece. 3. a.** *Aeronaut.* (See quot.)

1918 W. E. Dommett *Dict. Aircraft* 32 *Nose piece*, conical metal piece attached to the propeller boss, its contour being continuous with the engine cowling.

† **b.** *Astronautics.* = *Nose cone. *Obs.*

1946 H. Harper *Dawn of Space Age* iii. viii. 113 During the initial thrusting period, while the space-vessel is ascending from the earth, the heat resisting carapace—or nose-piece..will obscure any view forward.

nose-ring. Add: **2.** (Earlier and later examples.)

1819 Keats *Cap & Bells* (1848) II. 227 His turban wreath'd of gold,.. Mustachios, ear-ring, nose-ring, and his sabre keen. 1964 S. M. Shdeek *Windswept & Other Stories* (1969) 14 Earrings and nose-rings twinkled.

Hence **no·se-ringed** *ppl. a.*

1925 E., O., & S. Sitwell *Poor Young People* 7 Like my black Nose-ringed lumbering bear.

no·se-up, *a.* *Aeronaut.* [f. Nose *sb.* + Up *adv.*[2]] With the nose directed upwards.

1933 *Jrnl. R. Aeronaut. Soc.* XXXVII. 60 The lighter air[ship]..takes 20 seconds longer than the neutral ship to return to the horizontal, beyond which it only proceeds to 1° nose up. 1971 *Flying* Apr. 35/3 All you have to do is open the throttles, accelerate to that speed, rotate to a 10- or 12-degree nose-up attitude, and blast on out of the aerodrome. 1975 *Flight Internat.* 2 Oct. 506/2, I.. adjusted the attitude..because it was more markedly nose-up than expected.

nosey, *a.* Add: Also **nosy. 3.** *slang.* Inquisitive, esp. objectionably so; curious.

1882 F. W. P. Jago *Anc. Lang. & Dial. Cornwall* 226 *Nosey*, impertinent, intrusive. 1910 H. G. Wells *Hist. Mr. Polly* vi. 163 I'm not such a blooming Geezer..as not to be able to sell goods a bit. One has to be nosy over one's buying, of course. 1928 *Daily Express* 11 Sept. 7 Marylebone man: Being nosey, I goes to 'ave a look. Magistrate: Being what? Clerk: Nosey; meaning curious. 1957 M. Spark *Comforters* iv. 91 She saw Laurence examining Eleanor's cigarette case in his nosey way. 1973 P. Evans *Bodyguard Man* xv. 104 'A nosey man...' 'Very nosey, very smart, very perceptive.'

b. *Nosey Parker, nosey parker*: an inquisitive person. Hence *Nosey-Parkering, nosey-parkering* ppl. a. and vbl. sb. and (as a back-formation) *nosey-park* v. intr.; *nosey-parkerdom, Nosey-Parkerism, nosey-parkery*, the display of inquisitive behaviour, the exercise of oppressive questioning; *nosey-parkerishness*, inquisitive tendencies.

1907 *Picture post card* (London View Co. Ltd.) (*caption*) The adventures of Nosey Parker. 1912 C. Mackenzie *Carnival* xxi. 217 'I saw you go off with a fellah.' 'What of it, Mr. Nosy Parker?' 1915 Wodehouse *Something Fresh* v. 163 'But Nosey Parker is what *I* call him,' she said. 'He minds everybody's business as well as his own.' 1925 W. Deeping *Sorrell & Son* xxx. 302 A rodent, a nasty, acute little man of the Nosey Parker genus. 1929 H. A. Vachell *Virgin* xviii. 280 I'm a pestering nosey-parkering, shilly-shallying sort of an idiot, eh? 1930 J. B. Priestley *Angel Pavement* iv. 156 That's what takes your time, my boy—doing your bit of nosy-parkering. 1932 H. Williams in *Hansard Commons* CCLXII. 1359/2 What I call the modern spirit of 'nosey-parkerism' in legislation and administration. 1937 N. Marsh *Vintage Murder* viii. 86 Shut up, Gordon... Don't nosy-park. 1939 —— *Overture to Death* xi. 118 Asking questions..out of.. nosey-parkerishness. 1947 T. H. White *Mistress Masham's Repose* xx. 150 'Good-bye.' 'And none of your Nosey-Parkering.' 1958 E. Hyams *Taking it Easy* ii. iii. 195 Is Bachelor still connected with military intelligence, or security, or whatever fancy name they have for official nosey-parkering? 1961 *Economist* 4 Nov. 426/3 This kind of officious nosey-parkery gives the British immigration authorities a bad name. 1966 C. Mackenzie *My Life & Times* V. 157 The original 'nosy parker'..was one who played Peeping Tom to love-making couples in Hyde Park. 1969 *Daily Tel.* 25 Apr. 18 Goodness knows what all this massive exercise in nosey parkerdom is costing in terms of printing, manpower, administration, etc. 1971 L. Lamb *Worse than Death* iii. 29 Mrs Marshman nosey-parked..in Mrs Marble's cupboard. 1973 *Times* 17 Feb. 14/8 As ducks are 'parkers', it is easier to understand why exploratory and curious ducklings became 'nosy parkers'. 1974 'D. Craig' *Whose Little Girl are You?* i. 19 All nosey parkers in this street. 1974 J. Cooper *Women & Super Women* 23 The social security ladies come nosy-parkering round.

B. *sb.* An inquisitive person, a 'Nosey Parker'.

1937 N. Marsh *Vintage Murder* xiv. 155 He may be a bit of a nosy, but he doesn't look like a murderer. 1942 Berrey & Van den Bark *Amer. Thes. Slang* § 399/1 *Meddler or inquisitive person*,..nosy. 1975 H. R. F. Keating *Remarkable Case* ii. 25 Tomorrow she would be out there in time, before all the noseys come out.

noseyness (nōᵘ·zinės). Also **nosiness.** [f. Nosey *a.* + -ness.] Impertinent curiosity.

1919 F. Hurst *Humoresque* 14 Haven't I learned it to you often enough a slummer must pay for her nosiness? 1941 *Penguin New Writing* X. 10, I do not think the men read these letters out of pure noseyness. 1952 J. Steinbeck *East of Eden* xviii 179 This isn't nosiness. This is the law. 1973 *Daily Tel.* 5 Oct. 17/3 Noseyness was the main motive; we always spent a lot of time speculating about other people's marriages. 1975 L. Deighton *Yesterday's Spy* viii. 60 Curiosity—even nosiness—is not yet against the law.

nosh (nǫʃ), *sb. colloq.* [Yiddish; cf. next word.] **a.** A restaurant, a snack-bar. More usually *nosh bar, -house*.

1917 R. Fry *Let.* 11 May (1972) II. 411 Come with me to Kettners' nosh at 8.0. 1959 C. MacInnes *Absolute Beginners* 154 After a quick bite at a Nosh, and two strong black coffees, I felt up to the ordeal. 1969 I. Drummond *Man with Tiny Head* i. 21 We're going to your nosh-house. 1970 *New Scientist* 2 Apr. 5/2 What dishes will they have at the epicurean nosh-houses up west? 1973 *Jewish Chron.* 2 Feb. 22/4 The quasi-kosher-culture of the nosh bar. 1974 J. Gardner *Corner Men* viii. 67 Wanted to set up a couple of class nosh houses in the Smoke, here.

b. Food, a meal. Also *nosh-up*, a (good) meal.

The more usual sense in the U.K.

1963 *Daily Mail* 15 May 4/1 Why is it that whenever I read anything about freedom from hunger it's accompanied by pictures of people having a nosh-up? 1964 *Cherwell* 4 Nov. 12 At the unofficial opening..two old ladies were dragged in off the street and given free nosh and free coats. 1968 *Punch* 14 Feb. 220 I've always found Chinese nosh both cheap and filling. 1968 A. Diment *Great Spy Race* ii. 19, I like him enough to buy him some alcohol and the occasional cheap nosh. 1970 A. Draper *Swansong for Rare Bird* ix. 79 Like most birds she didn't want to lose out on a nosh-up. 1972 C. Drummond *Death at Bar* vi. 159 Burglars go for plain, healthy English nosh.

c. A snack eaten between meals; a titbit. Chiefly *U.S.*

1965 *N.Y. Times* 9 Apr. 40 Advertising copy will stress that the company makes everything from 'soup to nosh'. (A nosh is a snack.) 1968 L. Rosten *Joys of Yiddish* 267 Many delicatessen counters display plates with small slices of salami, or pieces of halvah, with a legend affixed to a toothpick: 'Have a nosh.'

nosh (nǫʃ), *v. colloq.* [Yiddish; cf. G. *naschen* to nibble, eat on the sly.] **1. a.** To nibble, to eat a snack between meals. Chiefly *U.S.*

1957 *Observer* 3 Nov. 8/5 One eats breakfast or lunch, but one noshes in between. *Ibid.*, 'Women don't know how to nosh,' he continued, 'except chocolate and sweets.' 1958 I. Brown *Words in Our Time* 76 Women, the salesman complained, only nosh chocolates and the true nosher is a connoisseur of cheeses at once sharp and ripe. 1968 L. Rosten *Joys of Yiddish* 267 To *nosh* is to 'have a little bite to eat before dinner is ready', or to 'have a little something between meals'. 1970 *Time* 12 Oct. 42 The politician, equipped with a trowel and the Fixed Smile, gobs mortar on a cornerstone, or noshes his way along the campaign trail. 1973 *Impetus* (Toronto) June 44/2 By nightfall, when mosquitoes as large as vampire bats were noshing on my exposed flesh, I had hoed a row and a half.

b. More generally, to eat; occas., to drink.

The more usual sense in the U.K.

1962 R. Cook *Crust on its Uppers* ii. 34, I finally bought a quart of grappa... Mrs. Marengo had noshed her half down. 1965 G. Melly *Owning Up* xiv. 170 After our huge meal, we were forced to refuse although, as Mick said, 'I'd have noshed the lot if I could have done.' 1972 *Annabel* Jan. 8/3 There I sat in my steaming bath, noshing my favourite food. 1972 C. Drummond *Death at Bar* v. 122 The Sergeant..morosely noshed the veal-and-ham pie.

2. To practise fellatio (with). *coarse slang.*

1965 *Listener* 18 Nov. 803/1 Such typical modernisms as *snog, nosh, dildo*, [etc.]. 1968 [see *gamahuche *v.*]. 1972 B. Rodgers *Queens' Vernac.* 142 *Nosh*.., to suck cock.

Hence **no·shable** *a.*; **no·shing** *vbl. sb.*

1957 *Observer* 3 Nov. 8/5 Noshing—derived from the German verb *naschen*—is to sample desirable food voluptuously, surreptitiously. 1965 J. P. Carstairs *Concrete Kimono* i. 12 Too much noshin' and drinkin' the night before. 1965 W. Young *Eros Denied* xiv. 137 *Gamming*, from the French *gamahucher*, or *blowing*, or *plating*, or *noshing*, from the Yiddish *nosh*, to nibble, or eat between meals. 1966 M. Waddell *Otley* xi. 114 A particularly noshable repast—chop and peas and potatoes. 1966 *Crescendo* Nov. 10/1 This saga of happy noshing.

nosher (nǫ·ʃəɹ). *colloq.* [f. *nosh *v.* + -er[1].] A person who samples food before buying it; one given to eating snacks; a customer at a restaurant.

1957 *Observer* 3 Nov. 8/5 He jerked his head towards a round-bellied, cheerful man..who was biting large segments of Gruyère held lovingly in his hand. 'A nosher,' said the shopkeeper sentimentally. 1958 I. Brown *Words in Our Time* 76 The salesman's much-liked noshers paid up properly as they nosed and nuzzled and chewed their way from one of his cates to another. 1969 *Sunday Times* 23 Mar. 28 Hot meal vending machines in the lobby for late night noshers. 1974 *Ibid.* 20 Jan. 24/7 Gourmet foods to salivate the palates of jaded British noshers.

noshery (nǫ·ʃĕri). *colloq.* [f. *nosh *v.* + *-ery 2 b.] A restaurant; a snack-bar.

1963 R. I. McDavid *Mencken's Amer. Lang.* v. 261 Miami Beach and similar outposts of civilization are full of *nosheries*... A *Chinese Noshery* was opened in the posh Georgetown section of the District of Columbia, *c.* 1955. 1970 A. Draper *Swansong for Rare Bird* ix. 78 Let's skip your noshery and I'll take you out for some grub. 1972 K. O'Hara *Company of St George* xiii. 118 The place I'm thinking of for lunch..has the reputation of a very superior noshery. 1975 *Spectator* 11 Jan. 36/1 A chain of nosheries have just sent out a drinks guide that shows sherry in that abominable receptacle known as an 'Elgin' or 'schooner'.

noshi (nǫ·ʃi). Also **nosi.** [Jap.] A Japanese token of esteem, originally a piece of dried awabi or more recently a specially folded piece of paper, forming part of the wrapping of a Japanese gift.

1855 R. Hildreth *Japan* xli. 434 *Nosi*, a species of edible sea-weed, of which small pieces are attached to every congratulatory present. 1891 A. M. Bacon *Japanese Girls & Women* i. 2 Tied with a peculiar red and white string, in which is inserted the *noshi*, or bit of dried fish daintily folded in a piece of coloured paper, which is an indispensable accompaniment of every present. 1954 J. M. Morris *Wise Bamboo* iv. 61 It was wrapped as a Japanese wedding gift, tied with gold and white cords in the traditional knot and with a paper *noshi*, the Japanese sign of a gift, on the wrapping. 1971 R. Harbin *More Origami* 39 A simple Noshi which is a good luck item given with anything.

no show, *sb.* orig. *U.S.* Also **noshow,** (with hyphen) **no-show.** [f. No *a.* + Show *v.*] A person who reserves a place on a train, boat, or esp. an aircraft, and fails to claim or cancel it. Also *transf.* and *attrib.*

1941 *Collier's* 27 Sept. 67 He's what the Airlines call a 'no show'. 1946 *Sun* (Baltimore) 3 Sept. 6 (Advt.), The traveling public has long been aware of the so-called 'no show'. He is the person who reserves airline space, buys a ticket and then neglects to cancel his reservation when he decides not to go. 1948 *Time* 10 May 89/3 Last week ship lines were still accepting tentative bookings—but only to replace last-minute no-shows. 1949 *Birmingham* (Alabama) *News* 17 Feb. 40/3 Many passengers who were denied seats could have occupied those left empty by 'noshows'. The airlines are in the red largely because of 'noshow' passengers. **1959** *Economist* 17 Jan. 248/1 To be sure of a seat or a sleeping berth, the 'no show' traveller books on several different services and sits back to decide at leisure which booking to take up. **1961** *Flight* LXXX. 489/1 IATA announce that transatlantic operators, in order to alleviate the no-show problem, are to introduce a reconfirmation rule. **1963** *Times* 8 Feb. 19/6 Booking and cancellation charges and 'no-show' charges. **1972** R. K. Smith *Ransom* I. 25 The phone rang... Another 'no show', the Cable boy. Wasn't that on Ben Carter's route? **1973** *Daily Colonist* (Victoria, B.C.) 7 July 9/2 No-shows, passengers who simply fail to turn up for flights they've booked, have long been a major headache for airlines. **1974** *Index-Jrnl.* (Greenwood, S. Carolina) 19 Apr. 6/5 The regulatory agency said there was an increase in the number of no-shows, fans who purchase tickets but don't attend the game.

no side. *Rugby.* The (announcement of the) conclusion of a game.
1857 T. Hughes *Tom Brown's School Days* I. v. 125 Five o'clock strikes. 'No side' is called, and the first day of the School-house match is over. **1874** *Rugby Union Football Ann. 1874–75* 39 Glasgow were again compelled to touch the ball down before the call of 'No-side', when the match was declared drawn in favour of Edinburgh. **1882** [see Side *sb.*[1] 20 b]. **1904** Wodehouse *Gold Bat* xi. 126 The whistle blew for no-side. **1931** *Times* 16 Feb. 5/1 They had looked a beaten team when, a quarter of an hour from 'no side', England at last managed to score a try. **1957** *Times* 14 Nov. 17/1 When no-side arrived the Australians had had all that they wanted. **1968** *Listener* 8 Aug. 189/1 The Lions took the lead ten minutes from no side, at which point the referee began awarding penalty kicks at goal to the Transvaal side.

nosiness, var. *noseyness.

no sir: see *nossir.

no siree (nōu sǝrī·). *U.S. colloq.* Also **no sirree.** [f. No *adv.*[3] + Sirree.] No indeed; certainly not. Cf. Sirree.
1848 [see Sirree]. **1851** *Knickerbocker* XXXVIII. 496 'Can you take two to Whitlockville by five o'clock?'..'No, Siree!' **1861** [see Sirree]. **1883** *Wheelman* Apr. 2/2 No, sir-e-e, you can't have any of my boats. **1936** J. Dos Passos *Big Money* (1937) 83 No more of that stuff, nosiree. **1973** J. Di Mona *Last Man at Arlington* (1974) 49 The senator wouldn't protect him. No siree.

no-sky (nōu,skai·), *adj. phr.* [f. No *a.* 6 b + Sky *sb.*[1]] *no-sky line*: a line in a room behind which no sky is visible from table height.
1927 *Illumination Res. Techn. Paper* (Dept. Sci. & Industr. Res.) No. 7. 12 This experience of adequacy and inadequacy in front of and behind the 'no-sky' line appears to be more or less independent of weather, so long as no sun is shining into the room. **1961** J. W. T. Walsh *Sci. Daylight* xi. 255 The first step in preparing a daylight plan is, usually, to plot the no-sky line.

no smoking, *vbl. sb.* [No *a.* 4.] Usu. *attrib.*, = *non-smoking *a.*; also a formula used on a notice, sign, etc., indicating that smoking is not permitted.
1944 J. Gunther *D Day* i. 13, I wanted to smoke, but fell asleep before the No Smoking sign was switched off. **1952** T. J. Mulvey *These are your Sons* v. 108 The light over the cabin door flashed red. no smoking. fasten your seat belts—the glass panel warned. **1955** G. Willans *Fasten Your Lapstraps!* i. 17 Inside the airport there are more notices. 'No Entry', 'No Smoking', 'Staff Only'. **1961** *John o' London's* 1 June 608/3 One doesn't always notice the no-smoking sign. **1970** *New Yorker* 16 May 46/3 The no-smoking lights go out and the loud-speaker confirms that the passengers may now unfasten their seatbelts and smoke. **1971** P. Audemars *Stolen like Magic Away* i. 15 The No Smoking notice was prominently displayed. **1973** J. Ashford *Double Run* vi. 44 He reached..for a pack of cigarettes. 'This is a no-smoking compartment,' said the woman.

noso-. Add: **no·sophile** *rare* [-phil, -phile], a person who is morbidly attracted by sickness or disease.
1895 tr. *M. Nordau's Degeneration* v. i. 539 Sadists, 'bestials', nosophiles, and necrophiles, etc., find legal opportunities to gratify their inclinations. **1905** *Smart Set* Sept. 113/2 Names of Satanic painters from Hell-Fire Breughel to Arnold Böcklin..passed through the halls of this nosophile's memory.

nossir (nǒusǝ·ɹ). Chiefly *U.S. colloq.* Also **no sir.** [Corruption of *no sir*.] A formula of emphatic denial or refusal: certainly not. Cf. *no siree.
1856 *Knickerbocker* XLVII. 544 *Examiner*: Has it effected that object? *Student*: No, Sir-r-r! I don't think it has. **1930** I. Low *His Master's Voice* ii. 15 Most of his replies to questions were in the nature of 'Yessir!' 'Nossir!' or their present-day Russian equivalent. **1932** Mrs. P. Campbell *Let.* 29 Mar. in *B. Shaw & Mrs. Campbell* (1952) 298 Please don't think..that I am such a foolish creature as to look upon these letters as bouquets to *me* 'No, Sir'—as the Americans say, with the accent on the 'sir'. **1939** A. Huxley *After Many a Summer* I. iv. 43 She really cared for her old Uncle Jo, and cared for him, what was more, not merely as an old uncle—no, *sir*. **1955** Pohl & Kornbluth *Space Merchants* xv. 156 'I want to look at your books.' He shook his head emphatically. 'Nossir. Only the old man himself gets to see the books.' **1968** E. McGirr *Lead-lined Coffin* ii. 51 Joe Silverman don't like his neck being breathed down. Nossir. **1970** R. Crawford *Kiss Boss Goodbye* II. ii. 64 'Nobody gave you an answer?' 'Nossir,' Dell said. **1973** *Listener* 19 Apr. 501/3 In Texas, do you think they're going to inquire about the hanging of the venison..? No, sir. They wonder if there's any shepherd's pie.

nostalgia. Add: (Earlier example.)
1770 J. Banks *Jrnl.* in J. Cook *Jrnls.* (1955) I. 409 The greatest part of them [*sc.* the ship's company] were now pretty far gone with the longing for home which the Physicians have gone so far as to esteem a disease under the name of Nostalgia.
2. *transf.* Regret or sorrowful longing *for* the conditions of a past age; regretful or wistful memory or recall of an earlier time.
1920 D. H. Lawrence *Lost Girl* xv. 344 The terror, the agony, the nostalgia of the heathen past was a constant torture to her mediumistic soul. **1928** A. Waugh *Nor Many Waters* vi. 231 He pictures with a sense of nostalgia, too acute almost to be endured, all that marriage to Marian would have meant. **1933** D. Garnett *Pocahontas* xx. 234 Seeing all these things again filled her heart with that violent sentimental nostalgia..felt by the very young about the very recent past. **1943** Koestler in *Tribune* 26 Nov. 13/1 Even the names of Paris underground stations..become the nostalgia-imbued stimuli of conditioned reflexes. **1945** Auden *For Time Being* 37 We and They are united in the candid glare of the same commercial hope by day, and the soft refulgence of the same erotic nostalgia by night. **1951** L. P. Hartley *My Fellow Devils* xxxiii. 339 The faults for which she had been obliged to sack him no longer counted:..she was free to dwell with some nostalgia on his virtues. **1957** *Listener* 3 Oct. 512/1 When grown-ups become passionately interested in them [*sc.* children's books] some kind of nostalgia is involved. **1957** B. & C. Evans *Dict. Contemp. Amer. Usage* 322/2 Now a vogue word, *nostalgia* has come to mean any vague yearning, especially for the past and especially..when tinged with tenderness and sadness. **1959** *Observer* 8 Feb. 7/5 Nostalgia for one's childhood does not necessarily mean that the childhood was a happy one. **1971** *Sunday Times* 30 May 32/2 Nostalgia's all right, But it's not what it was.

nostalgic, *a.* Add: **3. a.** That evokes a wistful and sentimental yearning for the past. **b.** Feeling or indulging in nostalgia.
1944 *Evening News* 11 Dec. 1/3 Britain, that nostalgic lover of vanished thrones. **1950** A. L. Rowse *England of Elizabeth* v. 184 That nostalgic figure, the Town Crier, was becoming familiar. **1952** *N.Y. Herald Tribune* 28 Aug. 16/7 The Empire Theater, fifty nine years old and 'one of the fine old nostalgic houses' will be torn down next spring. **1958** B. Nichols *Sweet & Twenties* viii. 112 The charming thing snapped open, releasing a cloud of dust and a host of memories. The very sound of it was nostalgic. **1959** *Observer* 18 Jan. 19/1 It had a sudden sharp nostalgic appeal for me because I immediately recognised in one of the chief testifiers a former assistant master at my preparatory school. **1965** *Church Times* 18 June 6/4 The departure on Friday afternoon of the last train to be drawn out of Paddington Station..by a steam locomotive was..a nostalgic occasion.

nostalgically (nǫstæ·ldʒikăli), *adv.* [f. Nostalgic *a.*: see -ically.] In a nostalgic way; in evocation of the past.
1928 Galsworthy *Swan Song* III. xi. 298 A long time he sat there, nostalgically bemused, strangely unwilling to move. **1937** A. Huxley *Ends & Means* i. 2 The poor and downtrodden have always dreamed nostalgically of a man ideally well-fed, free, happy and unoppressed. **1945** *John o' London's Weekly* 24 Aug. 216 Its style of binding is nostalgically akin to that of those A.A. handbooks which one used to find (sometimes) in seaside hotels. **1957** A. E. Coppard *It's Me, O Lord!* iii. 37 There is something nostalgically primitive about the woodwind instruments. **1958** *Listener* 20 Nov. 842/2 Mesens..uses these elements lyrically and nostalgically. **1975** J. Meldrum *Semonov Impulse* iii. 57 You're going nostalgically north?.. See the boys?

‖ **nostalgie de la boue** (nǫstalʒi dǝ la bū). [Fr., lit. = yearning for mud (Émile Augier *Le Mariage d'Olympe* (1855) I. i).] A desire for degradation and depravity.
1897 G. B. Shaw *Our Theatres in Nineties* (1932) III. 168 To him those habits were 'morality'; and what was counter to them was 'nostalgie de la boue'. **1905** H. A. Vachell *Hill* vi. 121 She..said that I suffered from what the French call *la nostalgie de la boue*; that means, you know, the homesickness for the gutter. **1920** G. Dickins *Hellenistic Sculpture* ii. 27 The Hellenistic sculptors..suffered as much as any modern decadent from 'la nostalgie de la boue'. **1928** D. H. Lawrence *Lady Chatterley* xix. 358 You're one of those half-insane, perverted women who must run after depravity, the *nostalgie de la boue*. **1957** C. MacInnes *City of Spades* II. xii. 181 It's the crude animal type that attracts you... It's simply another form of *nostalgie de la boue*. **1965** *Listener* 9 Sept. 391/2 A cultural *nostalgie de la boue* for the nonsense of the 'thirties is surely perverse. **1969** R. Lowell *Notebook 1967–68* 105 The Muse shouts vacation in my ear: *Nostalgie de la boue* that shelters ape And protozoa from the rights of man. **1971** *Southerly* XXXI. 11 There may be a suggestion of *nostalgie de la boue* in his choice of sexual partners.
Similarly *nostalgie de la banlieue* [Fr., = suburbs], *des adieux* [Fr., = farewells], *du divin* [Fr., = divine], *du pavé, trottoir* [Fr., = pavement], etc.
1933 D. L. Sayers *Murder must Advertise* iii. 41 There is a *nostalgie de la banilieu* [*sic*] as well as *de la boue*. **1941** Nostalgie des adieux [see *faute de mieux]. **1944** *Horizon* IX. 13 My reaction to everything not urban has been either one of acute melancholy or an urgent *nostalgie du pavé*. **1944** M. Laski *Love on Supertax* ix. 88 'Every man has an urge to go Mayfairing once in his life.'..'*Nostalgie du trottoir*,' agreed Tatiana. **1948** W. Stevens *Let.* 6 May (1967) 596 He speaks of the nostalgie du divin, (which is obviously epidemic in Dublin). **1972** *Times* 27 Sept. 5/3 The archetypal early twentieth-century laboratory scented evocatively with *Nostalgie de la Bunsen Burner*.

nostoma·niac. *rare.* [f. Nostomania: see -ac.] A person affected with nostomania.
1913 R. W. Service *Rhymes of Rolling Stone* 50 The Nostomaniac.

nostos (nǫ·stǫs). Also **Nostos.** Pl. **nostoi.** [Gr. νόστος a return home.] A homecoming, applied *spec.* to the homeward journeys of Odysseus and the other heroes of Troy. Also, the story of such a homecoming or return, esp. as the conclusion of a literary work.
In quots. 1883, 1901, and 1962[1] *Nost(o)i* is the title of a lost poem of the Epic Cycle dealing with the return of the Greek heroes from the Trojan War.
1883 D. B. Munro in *Jrnl. Hellenic Studies* IV. 319 A passage of Pausanias (x. 28, 7), mentions, as the poems which contain descriptions of the infernal regions, the *Odyssey*, the *Minyas*, and the *Nostoi*. **1901** —— *Homer's Odyssey Bks XIII–XXIV* 380 We may regard the *Nosti* as a tragic *Odyssey*. **1920** J. Joyce *Let.* 12 July (1957) I. 143 A great part of the Nostos or close was written several years ago and the style is quite plain. **1924** T. W. Allen *Homer: Origins & Transmission* 333, I have found a reference to their Nostos in Plutarch. **1957** N. Frye *Anat. of Criticism* iii. 159 It is rare, in literature as in life, to find even a domesticated animal peacefully living through its full span of life... The exceptions, such as Odysseus' dog, are appropriate to the theme of *nostos* or full close of a cyclical movement. **1962** M. Bowra in Wace & Stubbings *Compan. to Homer* iii. 40 The tradition that he composed the *Iliad* and the *Odyssey* becomes less impressive when he is credited by different authorities with the *Thebais, Epigonoi, Cypria,* and *Nostoi*. **1962** Thomas & Stubbings in *Ibid.* ix. 292 Agamemnon, on his *nostos*, was carried away as he approached Malea. **1968** C. R. Beye *Iliad, Odyssey & Epic Tradition* v. 161 The story of homecoming had a name: *nostos*. The several *nostoi* are a leitmotiv throughout the *Odyssey* which is over-all a *nostos*, being the return of Odysseus. There is a hint in the *Odyssey* that epics of *nostoi* were currently fashionable.

Nostraine, obs. var. *Nasrani.

Nostratic (nǫstræ·tik), *a.* Also **Nostratian.** [ad. G. *nostratisch*, f. L. *nostrās, -ātis* of our country: see -ic.] (See quots.)
1931 J. W. Spargo tr. *Pedersen's Linguistic Science in Nineteenth Century* viii. 338 As a comprehensive designation for the families of languages which are related to Indo-European, we may employ the expression *Nostratian Languages* (from Latin *nostrās* 'our countryman'). **1966** B. Collinder in Birnbaum & Puhvel *Anc. Indo-Europ. Dial.* 199 Holger Pedersen used to speak about a 'nostratic' family of languages, comprising Indo-European, Semitic-Egyptian-Hamitic, Uralic, Altaic, Yukagir, and Eskimo. **1966** M. Pei *Gloss. Linguistic Terminol.* 136 *Japhetic*, a hypothetical language family claimed to include North Caucasian, South Caucasian, Sumerian, Elamite, Asianic, Basque, Etruscan, etc... *Synonyms*: Alarodian, Nostratic. **1973** *Jrnl. R. Asiatic Soc.* 46 The term Nostratic is defined by the words in brackets following the title of the book, 'Semito-Hamitic, Kartvelian, Indo-European, Uralian, Dravidian, Altaic'.

nostril (nǫ·stril), *v.* [f. the sb.] To look like or function as a nostril; to inhale or exhale through one's nose. So **no·strilled** *ppl. a.*, having nostrils, freq. of a particular kind.
The *transf.* use in quot. 1942 is unusual.
1909 *Athenæum* 31 July 125/1 The characteristically Irish 'nostrilled' portraits of the four Evangelists. **1939** E. Hemingway *Fifth Column* 131 His outstretched, wide-nostrilled muzzle. **1942** A. L. Rowse *Cornish Childhood* (1943) 36 The blue summer sea curling round the ships of those emigrant miners, the water nostrilling the stem. **1971** A. Hunter *Gently at Gallop* v. 49 He was smoking a small, sooty briar... He nostrilled a couple of wisps of smoke.

nosy, var. Nosey *a.* (in Dict. and Suppl.).

not, *adv.* and *sb.* Add: **I.** *adv.* **1. c.** With ellipsis of dependent clause after certain verbs, as *do, have, know, say not*. colloq.
1906 Kipling *Puck of Pook's Hill* 147 'Did *you* have a governess, then?' 'Did we not? A Greek too.' **1918** C. Mackenzie *Early Life Sylvia Scarlett* II. vii. 450 '..Do you remember a man called Leopold Hansberg?' 'Do I not?' Sylvia exclaimed. **1936** M. Allingham *Flowers for Judge* iv. 74 'I suppose they've been questioning you..?' 'Have they not!' Mike spoke explosively. **1973** C. Aird *His Burial Too* ii. 28 Your father always puts it in the book... I've never known him not.
2. b. Freq. with *that*.
1939 'N. Blake' *Smiler with Knife* i. 13 You never knew when..Nigel [would] be dragged into some queer criminal tangle. Not that they really needed the substantial fees he charged. **1967** *Listener* 7 Dec. 762/3 We should recall our initial experience of a creative personality as strong as Mahler's... Not that Koechlin is a Mahler, but at best he is an original.
4. Esp. with imperative force; freq. in colloq. phr. *not to worry*, do not worry.

1872 Geo. Eliot *Middlem.* II. xxxvii. 258 And he objects to a secretary: please not to mention that again. **1958** *Daily Mail* 24 July 6/5 Not to worry. By the time he ..had finished with me..I'd be doing long division. **1965** L. Meynell *Double Fault* I. iii. 31 'We'll send it for you.' 'Not to bother. I'm going down to the country this evening.' **1967** *Spectator* 11 Aug. 160/1 In short, to borrow one of Mr. William's own favoured colloquialisms, not to worry. **1970** V. Canning *Great Affair* xiv. 269 He gave me a big grin and said, 'Not to be overcome, son.'
7. c. *not half* (see Half *adv.* 3 in Dict. and Suppl.).
d. *not all that*, not exceptionally (so).
1964 T. White tr. *Simenon's Maigret & Saturday Caller* iv. 70 'Were they heavy?'.. 'Quite heavy, but not all that.' **1970** *Guardian* 24 Jan. 1/5 Asked how he assessed the capacity of the present relief system of the Federal authorities to cope with widespread starvation, Lord Hunt said: 'Now I don't feel all that confidence in the Nigerian Government's ability to do this.' *Ibid.* 28 Jan. 8/3 Without her voice, Callas is not all that impressive an actress. **1974** *Sat. Rev. World* (U.S.) 19 Oct. 56/2 Several years ago..I asked him, 'How was the picture?' He..said, 'Not all that good.' The expression 'not all that good', as far as I know, was fairly new at the time.
10. c. (Further example.) (See also *bad *a.* 1 c.) Also as *sb.*, **not-bad.**
1843 Thoreau *Let.* 24 Jan. in *Corr.* (1958) 77 He is, at any rate, one of the not-bad. **1871** Geo. Eliot *Middlem.* I. I. ix. 128 She had got nothing from him more graphic about the Lowick cottages than that they were 'not bad'. **1973** A. Ross *Dunfermline Affair* 41 A not-bad job, Farrow. Not bad at all. **1975** *Times* 20 Sept. 8/5 Brian Armstrong's *Bags of Swank* consisted of three not-bad scripts about his National Service.
II. 13*. As *sb.* or *adj. Computers.* (Usu. written in capitals.) A Boolean function of one variable that has the value unity if the variable is zero, and *vice versa*. Usu. *attrib.*
1950 W. W. Stifler *High-Speed Computing Devices* xiii. 270 This device, called a half adder, will consist of *and*, *or*, and *not* circuits..arranged so as to carry out the operations exemplified in Table 13-1. **1955** R. K. Richards *Arithmetic Operations in Digital Computers* ii. 29 A fundamental concept which is found in Boolean algebra and which has no counterpart in ordinary algebra is the 'not' function, as indicated by a line over a symbol. **1957** Goode & Machol *System. Engin.* xxv. 394 There is.. evidence that certain synapses are inhibitory, i.e. not gates. **1969** P. B. Jordain *Condensed Computer Encycl.* 341 NOT is a unary operation since it has only one argument, and it is one of the most fundamental logic operators. **1970** O. Dopping *Computers & Data Processing* i. 26 All logical connections between two-valued variables can be expressed by means of the three functions NOT, AND, and OR. **1972** *IEEE Trans. Computers* XXI. 153/2 When digital computers are synthesized with gates other than AND, OR, and NOT gates, it is difficult to design optimal ..networks directly from the Boolean expressions.
III. 14. d. (Further example.)
1867 Mill *Exam. Hamilton's Philos.* (ed. 3) xiii. 289 A complete idea of a closed figure, and of the boundary which incloses it—the outline separating object from not-object.
15. d. Freq. with *so* or *too* and an adj. forming combs. used as *sb.*, *adv.*, or *adj.*
1748 Richardson *Clarissa* II. xiii. 72, I [say]..she is too wise; that is to say..Not so young as she has been. **1851** Lytton (*title*) Not so bad as we seem. **1866** Mrs. Gaskell *Wives & Daughters* I. xxi. 235 It was also evident to her that Osborne was not too happy at home. **1870** Geo. Eliot *Let.* 15 Nov. (1956) V. 120 We..cannot yet decide whether we should..have a modest little refuge in the not-too-distant country. **1931** J. S. Huxley *What dare I Think?* vi. 190 Man believing himself the inhabitant of the Universe's central globe,..looking forward to a not-too-distant end of this terrestrial home, cannot well have the same religion as man knowing himself descended by slow evolution from the brutes. **1935** *Discovery* Sept. 262/1 The not-so-distant days of the pirates. **1941** M. Allingham *Traitor's Purse* xx. 229 'I say, it's pretty serious, isn't it?' He nodded. 'Not so hot. Time's short.' **1943** G. Ade *Let.* 5 Feb. (1973) 238 The not-so-good-news in regard to the soy bean crop does not come as a surprise. **1945** [see *atomic *a.* 2 f]. **1945** *Sun* (Baltimore) 29 Oct. 4 A British rubber inspection committee sent out to the plantations of Malaya..reported today that conditions were 'not too bad'. **1951** S. Spender *World within World* iii. 143 One of their not-too-distant neighbours was Lytton Strachey. **1952** A. G. L. Hellyer *Sanders' Encycl. Gardening* (ed. 22) 270 Soil, cool, slightly moist, and a not-too-hot position in the rock garden. **1955** *Radio Times* 22 Apr. 5/1 Where the making of home entertainment is not just a pleasant memory of the not-so-young. **1957** *Times Lit. Suppl.* 13 Dec. 762/5 The not-so-rich lady in the subway solemnly knitting a double-necked jersey. **1960** *Guardian* 20 May 4/5 Access is limited for the not-so-young, the not-so-robust, and those who like to amble. **1962** R. H. Smythe *Anatomy of Dog Breeding* iv. 77 Bred from ancestors of not-so-long-ago. **1965** *Listener* 27 May 766/2 They accepted all the good things..and the not-so-good things, with fatalistic acquiescence. **1971** *Good Motoring* Sept. 3 This incredible phenomenon is the Giant's Causeway, known to most people by pictorial sight but, oddly, not too many know its location. **1972** E. Lemarchand *Cyanide with Compliments* i. 3 Olivia watched the launch filling up. The not-so-young predominated. **1974** *Sat. Rev. World* (U.S.) 19 Oct. 41/1 All of this is seen through the eyes of a not-too-bright country boy.

notability. 1. c. (Earlier example.)
1832 J. S. Mill *Lett.* (1910) I. 33 There is need that the march of mind should raise up new spiritual notabilities; for it seems as though all the old ones with one accord were departing out of the world together. In a few days or weeks the world has lost the three greatest men in it in their several departments—Goethe, Bentham, and Cuvier.

notal (nōu·tăl), *a.*[2] [f. Note *sb.*[2] + -al.] Of, pertaining to, or employing notes.
1884 *Encycl. Brit.* XVII. 85/1 The treachery of tradition is exemplified in the loss of the rules for this once generally understood practice of notal inflexion. **1924** Lucas & Bonar tr. *Knapp's State Theory of Money* ii. 113 There are metalloplatic and papyroplatic notal money systems, with or without hylodromy.

notam (nōu·tăm). [f. initial letters of *no*tice *to* *a*ir*m*en.] A (warning) notice to pilots of aircraft.
1946 F. Hamann *Air Words* 38/1 *Notam*, notice to airmen. **1955** R. J. Schwartz *Compl. Dict. Abbrev.* 128/2 *Notam*, International Notices to Airmen. **1963** *Times* 25 Jan. 4/4 He added that until two months ago 'notams'—notices warning aviators and mariners to keep out of the way of Nato exercises—were also issued and had nearly a worldwide circulation. **1967** *Times* 16 May 10/3 All went well until the second Spanish 'notam' (notice to airmen), a week ago.

NOT AND (nǫt ænd). *Computers.* [f. Not *adv.* + And *conj.*] = *NAND.
1960 [see *NAND]. **1962** Simpson & Richards *Physical Princ. Junction Transistors* xvi. 402 Parallel *p-n-p* inverters form a 'not and' gate, for which all must be one to maintain a zero output.

notaphily (nōutæ·fili). [f. Note *sb.*[2] 18 + -phily.] The collecting of bank-notes as a hobby. Hence **nota·philic** *a.*; **nota·philism**; **nota·philist.**
1970 *Daily Tel.* 7 Nov. 15/2 Stanley Gibbons has created a new word, 'Notaphilly' [*sic*], which derives from Latin and Greek meaning 'lover of notes'... A banknote is a notaphilic item and the collector is a notaphilist. (Say it like 'philatelist'.) **1970** *Guardian* 2 Dec. 11/2 The charm of bank notes..is their history... Notaphily is the name of the hobby. **1972** *Daily Tel.* 28 Oct. 29/4 Colin Narbeth compares the position of today's notaphilist with that of Edward Stanley Gibbons..when he bought a kit bag containing thousands of Cape Triangulars for £5. **1973** *Times* 13 Dec. 17/8 Collectors have coined the ugly name notaphilists to describe themselves. About twenty of the highest-denomination notaphilists came to the auction. **1976** *Times* 27 Jan. 17/4 Biafran £1 notes..found their way on to the notaphilic market.

notarikon (notæ·rikǫn). *Jewish Lit.* Also **notaricon.** [late Gr. νοταρικόν, f. L. *notārius* shorthand-writer, Notary *sb.*] In cabalistic phraseology, the art of making a new word from letters taken from the beginning, middle, or end of the words in a sentence.
1911 'Sepharial' *Kabala of Numbers* I. iii. 31 *Notaricon*.—Selection was made of certain letters according to the rules of the art, these letters being taken from the beginning, middle, or end of the words in a sentence, so as to produce a single word from their combination. **1941** G. G. Scholem *Major Trends Jewish Mysticism* iii. 90 Techniques..popularly supposed to represent the heart and core of Kabbalism, such as *Gematria*.., *Notarikon*, or interpretation of the letters of a word as abbreviations of whole sentences. **1951** *John o' London's Weekly* LX. 716/2 Hempe..is a Tudor notarikon... Hempe is formed of the initial letters of the Tudor sovereigns Henry, Edward, Mary, Philip and Elizabeth.

notarize (nōu·tăraiz), *v.* [f. Notar(y *sb.* + -ize.] *trans.* To have (a document) legalized by a notary. Usu. in **no·tarized** *ppl. a.*
1935 *Sun* (Baltimore) 22 Nov. 10/1 In a series of notarized cases a prominent journal recently stated that in the installation of industrial machinery totaling $8,000,000 the savings effected by these installations were approximately $55,000 per plant per year. **1939** W. Faulkner *Wild Palms* 19 There was no sworn notarised statement attached. **1960** *Times* 2 Aug. 7/4 Boys under 21 and girls under 18 must be accompanied by their parents to obtain a marriage licence, or else present a notarized letter of consent. **1972** *Scholarly Publishing* III. 182 Sometimes.. he has practically to sign his life away to get his meagre five dollars or so. In extreme cases the publisher once even had to get the form notarized.

not at all. = *don't mention it* (see Mention *v.* 1 c).
1936 Wodehouse *Laughing Gas* ii. 19, I was not-at-alling and shoving the handkerchief up my sleeve again. **1963** 'E. McBain' *Ten plus One* vii. 88 'Well, thank you,' Carella said. 'Not at all. My pleasure,' Richardson answered. **1973** J. Burrows *Like an Evening Gone* i. 13 She ..pretty well knocked him flying... 'So sorry.' 'Not at all.'

notate (notē·t), *v.* [f. as Notate *a.*] **a.** To set down in musical notation. **b.** To record, note. So **nota·ted** *ppl. a.*; **nota·ting** *vbl. sb.*
1922 W. J. Locke *Tale of Triona* xii. 128 He could play, ..by ear—knowledge of notated music he disclaimed. **1965** *Listener* 16 Dec. 990/2, I would say it would help any composer to conduct if he had the gift, because, in the first place, you learn how to notate your music more precisely, so that a performing musician will know exactly what you mean. **1971** S. Cavell *World Viewed* ii. 17 Such troubles in notating so obvious a fact. **1974** *Times Lit. Suppl.* 13 Dec. 1411/2 To sing through the variants of even a single song—for example, of the beautiful 'Seeds of Love', the first which [Cecil] Sharp notated—is an enlargement of the spirit... Once notated versions of the songs exist, are the rhythmic quirks ironed out?

not at home, *adv. phr.* Not prepared to receive visitors or accessible to callers. Also as *sb.*
1829 [see At home 1 b]. **1874** Hardy *Far from Madding Crowd* I. ix. 123 'I can't see him in this state. Whatever shall I do?' Not-at-homes were hardly naturalized in Weatherbury farm-houses, so Liddy suggested—'Say you're a fright with dust, and can't come down.' **1923** *Mrs. Beeton's Bk. Househ. Managem.* i. 35 When visitors present themselves the servant charged with the duty of opening the door will..answer without hesitation if the family are 'not at home', or 'engaged'.

notatin (notē·tin). *Biochem.* [f. L. *notāt-us*, pa. pple. of *notāre* to mark, Note *v.*[2] + -in[1].] A flavoprotein produced by the mould *Penicillium notatum* which is an enzyme that catalyses the direct oxidation of glucose to gluconic acid and hydrogen peroxide and is used in the detection and estimation of glucose; = *penatin, *penicillin A, B.
1942 C. E. Coulthard et al. in *Nature* 28 Nov. 634/2 When a selected strain of *Penicillium notatum* Westling is grown on a Czapek-Dox medium an anti-bacterial substance is produced which differs in the majority of its properties from penicillin. This anti-bacterial substance was originally named 'Penicillin A', but in order to avoid confusion with penicillin we now prefer the name 'notatin'. **1949** N. G. Heatley in H. W. Florey et al. *Antibiotics* I. iii. 136 For the assay of 'notatin' (known also as 'penicillin B' or 'penatin') the medium must contain glucose, for this so-called antibiotic is an enzyme..and exerts its effect purely through the hydrogen peroxide which is formed. **1954** A. White et al. *Princ. Biochem.* xii. 264 A curious example of an antibiotic isolated from the mold *Penicillium notatum* is the substance notatin, which was later found to be the enzyme glucose oxidase containing flavin adenine dinucleotide. **1963** R. Bentley in P. D. Boyer et al. *Enzymes* (ed. 2) VII. xxiv. 568 The following materials, originally described as antibiotics, are actually enzymes similar to, or identical with, the glucose oxidase of Müller: notatin (originally penicillin A).., penatin.., and penicillin B. *Ibid.* 569 Much of the work described in this chapter has been carried out with purified preparations of notatin, and an unqualified reference to glucose oxidase in the rest of this chapter will indicate notatin.

notation. 2. (Later examples.)
1929 J. Gallishaw *Twenty Probl. Fiction Writer* 228 The wise writer depends upon recorded observations, and makes notations. **1932** L. C. Douglas *Forgive our Trespasses* (1937) vii. 128 The papers submitted on Tuesday were, in the main, satisfactory. They would be returned, with notations, to their makers, at the close of the hour.

notational, *a.* Add: (Further examples.)
1955 A. N. Prior *Formal Logic* 293 Introducing 'classes' as merely a notational device. **1964** P. K. Bock in J. A. Fishman *Readings Sociol. of Lang.* (1968) 217 The structural description of any one situation occurring in a human community must, necessarily, refer to other situations; therefore, a notational system was devised to facilitate cross-referencing. **1964** E. Bach *Introd. Transformational Gram.* vii. 145 A certain amount of facility in handling symbols and practice in manipulating the notational conventions is all that is needed. **1965** Hughes & Londey *Elem. of Formal Logic* xxxvi. 249 We can..reduce the welter of bracketing..by the following notational convention. **1965** *Language* XLI. 485 Using the same notational devices as before. **1970** J. D. McCawley in P. L. Garvin *Cognition* x. 231 A system of semantic representation that is along the lines of the notational systems used in symbolic logic. **1973** J. Hintikka *Logic, Language-Games & Information* xi. 252 A formula which is like one of the constituents or a-constituents we have defined..will be called its *notational variant*. We shall call two constituents or a-constituents *different* only if they are not notational variants of each other.
b. *spec.* Of or pertaining to musical notation.
1925 P. A. Scholes *Second Bk. Gramophone Record* 162 Under the notational description of 6/8 we get sometimes *two groups* in a bar, sometimes *three groups*.
Hence **nota·tionally** *adv.*
1896 *Musical Herald* 1 Feb. 43/1 With regard to *ba*, Mr. McNaught admitted that, notationally, there was something to be said in favour of abolishing the name, but, educationally, it was better to keep it. **1940** W. V. Quine *Math. Logic* 33 A mode which consists notationally of compounding the statements by means of 'implies' and the two pairs of quotation marks.

nota·tionist. [f. Notation + -ist.] One who uses or advocates a particular style of musical notation.
1896 *Musical Herald* 1 Feb. 41/2 Every singer should be a two-notationist. **1897** *Ibid.* 1 June 188/2 There is no reason why..Sol-faists and staff notationists should not come to a common understanding and agreement. **1928** *Musical Times* July 619 All those brought up on the Staff would benefit by a grounding in Tonic Sol-fa... The dual notationist is, in fact, the best-equipped musician.

not-being, *sb.* **1.** (Later examples.)
1907 J. R. Illingworth *Doctrine of Trinity* i. 8 We cannot possibly conceive a passage from not-being to being, as the Greeks phrased it,..except through the operation of some energy which is already actual,..and adequate. **1917** D. H. Lawrence *Look! We have come Through!* 147 It wounds me to death with my own not-being. **1933** C. Dawson *Enquiries Relig. & Culture* III. ii. 194 The element of nothingness or not-being which is inherent in the world of sensible experience.

notch, *sb.* Add: **1. b.** (Later examples.)
Quot. 1970 perh. belongs to 1 a.

1929 W. Faulkner *Sound & Fury* 399 Luster took still another notch in himself and gave the impervious Queenie a cut with the switch. **1947** *Sun* (Baltimore) 15 May 2/8 The notch, instead of $67, is $38.50, and it stops at $265.52. **1958** *Sunday Times* (Johannesburg) 21 Sept. 4/9 Each of the Transvaal's 13,000 teachers will have their pay raised by at least one notch on October 1. **1958** *Cape Times* 29 Oct. 15/5 Applicants must have auctioneering experience... The notch for appointment will depend on previous experience. **1970** E. McGirr *Death pays Wages* vii. 146 He turned up the central heating a notch.
d. A slit in the ground made to take the roots of a seedling tree.
1891 W. Schlich *Man. Forestry* II. ii. 126 An enlarged notch may be produced by swaying it [*sc.* the notching spade] to and fro. **1934** *Forestry* VIII. 21 A deep vertical notch..with adequate firming, is the best method of planting. **1970** H. L. Edlin *Collins Guide to Tree Planting & Cultivation* vii. 107 In its simplest form, the notch is just a slit cut into the ground, into which the tree's roots are inserted.
e. An incision made in a twig to stimulate the growth of a bud lower down the twig.
1916 tr. K. Koopmann in L. H. Bailey *Pruning-Manual* v. 127 Notches are made on twigs of one year's growth or more, to influence a particular bud in various ways. **1974** R. Grounds *Practical Pruning* iii. 28 Cut a notch just above the buds that you wish to form branches during April.
4. b. *spec.* An opening extending above the water level in a surface placed across a stream like a weir; also, the surface itself.
1789 [in Dict.]. **1845** *Encycl. Metrop.* III. 238/1 Theoretically the quantity discharged through a rectangular notch, which reaches to the surface, is two-thirds of what would issue through an equal orifice placed at the whole depth below the level of the fluid. **1907** W. C. Unwin *Treat. Hydraulics* v. 96 Notches for measuring purposes are weirs fitted with a plate in which an open notch is formed through which the water passes. **1908** A. H. Gibson *Hydraulics* v. 137 (*heading*) Flow over notches and weirs. **1914** W. M. Wallace *Hydraulics* xii. 188 Where it is possible to provide an artificial section for the stream the gauge notch is used, of which there are three standard forms, viz., (1) rectangular, (2) circular, (3) triangular. **1959** [see *flume *sb.* 3 a]. **1974** J. A. Fox *Introd. Engin. Fluid Mech.* iii. 99 In the case of orifices and rectangular notches the coefficient of discharge varies with both the Reynolds number and the value of *l/h*.
6. notch-back, a back on a motor car that extends approximately horizontally from the bottom of the rear window so as to make a distinct angle with it; a car having such a back; also *transf.*; opp. *fast back* (*b*) s.v. *fast *a.* 11; **notch-bar test** *Engin.* = *notched-bar test* (*notched *ppl. a.* 1); **notch-brittleness** *Engin.*, susceptibility to fracture at a notch when a sudden load is applied; hence (as a back-formation) **notch-brittle** *a.*; **notch effect** *Engin.*, the increase in the susceptibility of a specimen to fracture caused by the presence of a notch; **notch factor** *Engin.* (see quot. 1968); **notch filter** *Electronics*, a filter that attenuates signals within a very narrow band of frequencies; **notch-house** *slang*, a brothel (cf. Nautch *sb.* 2); **notch-ladder** (see quot.); **notch-planting** = *notching *vbl. sb.* 3; **notch-sensitive** *a. Engin.*, characterized by a high notch sensitivity; **notch-sensitiveness** or **-sensitivity** *Engin.* (see quot. 1970); **notch-toughness** *Engin.*, the opposite of *notch-brittleness*; *spec.* the result (in units of energy) of a notched-bar test on a specimen.
[**1959** *Motor* 28 Oct. 447/2 The Special Continental saloon..was notable for the abandonment of the sloping tail in favour of a notched back treatment.] **1965** J. Lawlor *How to talk Car* 76 *Notch-back*, body design with a separate distinct rear deck. The term is used to distinguish conventional styling from fast-back design. **1967** *Wall St. Jrnl.* 24 Apr. 6/3 The Charger will drop its fast-back roofline in 1968 in favor of a 'notchback' in which the roofline slopes more abruptly to the rear fender. **1971** *Flying* (N.Y.) Apr. 40/1 In 1962, the Cessna 182 got the Omni-Vision treatment, with a notchback after fuselage and auto-style rear window. **1972** *Practical Motorist* Oct. 69/2 In 1961..a Capri based on the notch-back Ford Classic was introduced. **1957** *Financial Times Ann. Rev. Brit. Industry* 61/4 To predict the service behaviour of the steel..the information derived from small notch bar tests is being amplified. **1966** *McGraw-Hill Encycl. Sci. & Technol.* VIII. 272/2 Notch-bar tests are usually made in either of two convenient arrangements, in both of which the specimen is broken by a freely swinging pendulum. **1958** A. D. Merriman *Dict. Metallurgy* 212/1 In notch brittle materials the notch or crack is propagated with great rapidity under sudden loading conditions. **1963** Alexander & Brewer *Manuf. Prop. of Materials* i. 16 There is a fairly well-defined threshold value of temperature below which the steel is notch-brittle and above which it exhibits relatively ductile behaviour. **1929** *Jrnl. Iron & Steel Inst.* CXX. 514 The authors first discuss two diagrams representing the ratios of the notch-brittleness obtained by means of Mesnager test-pieces and large Charpy test-pieces for a large number of steels. **1965** Matsuo & Inoue tr. *T. Yokobori's Strength, Fracture & Fatigue of Materials* vii. 162 The classical theory of Griffith explains notch brittleness as the increase of the plastic constraint factor *qe* arising from fibrous cracks induced in the early stages of the process. **1925** M. A. Grossmann tr. *Heyn's Physical Metallogr.* v. 299 With decreasing *b*, hence with increasing notch effect, the values for q and ϵ_e decrease. **1970** C. C. Osgood *Fatigue Design* iii. 104 The scale of notch sensitivity for a material then varies between no notch effect, $q = 0$, and full theoretical notch effect at $q = 1$. **1939** *Jrnl. R. Aeronaut. Soc.* XLIII. 728 The notch-factor was plotted against the notch-depth, and the resulting curve indicated that, as the notch-depth increases, the notch-factor increases up to a critical value of the notch-depth, and thereafter decreases. **1968** F. A. d'Isa *Mechanics of Metals* vii. 326 The actual effectiveness of stress concentration on fatigue strength is measured by the fatigue notch factor, K_f, which is defined as the ratio of the fatigue strength of a specimen with no stress concentration to the fatigue strength at the same number of cycles with stress concentration for the same conditions. [**1950** *Electronics* July 75/3 (*heading*) Notching filters.] **1962** *Electronic Technol.* XXXIX. 332/1 At v.h.f. it is desirable to construct notch filters from coaxial elements. **1974** *Physics Bull.* Mar. 107/3 A fully tunable notch filter for removing line frequency pick-up. **1931** *Amer. Mercury* Nov. 353/2 *Notch house*, a house of prostitution. **1956** H. Gold *Man who was not with It* (1965) xxx. 277 Nancy ran a notch-house for travelers who loved to see things. **1902** *Chambers's Jrnl.* Oct. 657/2 A notable feature of these smaller mines is the notch-ladder system of conveying the ore from the interior to the pit-head. Two masts, notched like bear-poles, form the means of ascent and descent for a more or less continuous chain of *peons*. **1953** H. L. Edlin *Forester's Handbk.* ix. 145 Notch planting..is the simplest and cheapest method, by which the majority of forest trees are planted today. **1970** —— *Collins Guide to Tree Planting & Cultivation* vii. 107 (*heading*) Notch planting. **1946** *Metallurgia* XXXIII. 250/1 A forthright statement of which material is the more notch sensitive seems impossible. **1956** M. C. Smith *Princ. Physical Metallurgy* x. 377 Because a notch creates locally a condition of triaxial tension, all metals are notch-sensitive. **1967** Lee & Neville *Handbk. Epoxy Resins* iv. 9 Cyclization, together with randomly occurring areas of low bond density for other reasons.., will lead to internal flaws within the polymer network to create notch-sensitive defects. **1934** *Jrnl. Iron & Steel Inst.* CXXX. 653 The investigation of notch sensitiveness, the use of the index of notch sensitiveness..and the connection between the index and the chemical composition..are discussed. **1934** *Jrnl. Res. Nat. Bureau of Standards* (U.S.) XIII. 535 Even for annealed copper, notch sensitivity evidently is equal to that of steels with tensile strength five times as great. **1970** C. C. Osgood *Fatigue Design* iii. 104 The ratio between the apparent increase in local stress [due to the presence of a notch or other stress concentrator] in fatigue and the increase predicted by the elastic theory of stress concentration has been defined..as notch sensitivity. **1926** *Jrnl. Iron & Steel Inst.* CXIII. 624 This method of testing enables it to be judged whether a material has been brought to its best condition in regard to notch-toughness. **1972** E. N. Simons *Testing of Metals* v. 85 A component designed for a hot climate may not give the same degree of notch toughness in a cold one if the temperature of the latter lies below the transition temperature.

notch, *v.* **3.** (Later *fig.* examples.)
1911 *Chambers's Jrnl.* Oct. 702/1 A speed of one hundred miles an hour has been notched on more than one occasion. **1963** *Sunday Express* 6 Jan. 23/5 Her earnings now are well above the £10,000 she was notching up a few years ago. **1969** 'D. Rutherford' *Gilt-Edged Cockpit* i. 17 In about three minutes Mascot would have notched up the victory they so desperately needed. **1973** *Daily Tel.* 23 Apr. 5/4 When..'Andre Previn's Music Night' notched up audiences of six million last year, he was understandably delighted and encouraged. **1974** *Nature* 29 Mar. 373/3 Albright and Wilson managed to notch up an increase in profits.

notched, *ppl. a.* Add: **1.** *notched-bar test*: any of several impact tests (as the Izod or the Charpy test) in which the energy absorbed in breaking a notched specimen in a single blow is measured.
1918 *Proc. Inst. Automobile Engin.* XII. 250 Untreated bright drawn mild steel bar..usually gives a low result on the notched bar test. **1966** *McGraw-Hill Encycl. Sci. & Technol.* VIII. 272/2 Notched-bar tests are made to estimate the ductility that may be expected in the presence of defects in structures.

notcher. Add: Also, an instrument for making notches.
1879 *Organ Voicing & Tuning* 8 Tools for voicing and tuning... 1. The *notchers* of various sizes.

notching, *vbl. sb.* Add: **3.** A method of planting seedling trees in which a slit is made in the earth to take the roots of the plant. Also *attrib.*
1847 J. Brown *Forester* iii. 85 He [*sc.* the superintendent] must examine almost each tree as it is put into the ground, whether it may be done by the pitting or notching system. **1851** *Ibid.* (ed. 2) iii. 237 The method of planting termed notching, or slitting, is done with the common spade or planting mattock. **1891** W. Schlich *Man. Forestry* II. ii. 126 Notching..differs from planting with a peg in the shape of the planting hole, which is that of a notch. *Ibid.*, The notching spade is wedge-shaped. **1930** *Forestry* IV. 18 Notching. The idea behind this type of planting is to..make an opening quickly which is of sufficient size to take in the roots of small plants. **1953** H. L. Edlin *Forester's Handbk.* ix. 145 A proper straight notching spade..has no 'crank' between handle and blade. **1967** C. E. Hart *Practical Forestry* (ed. 2) iv. 63 The most usual method [of planting] is notching with a spade or a mattock.
4. A type of pruning in which twigs are partially cut to stimulate the growth of a bud lower down the twig.
1913 J. C. Newsham *Propagation & Pruning* 23 Notching is used to force the development of particular buds, the notch being made in the bark and young wood immediately above the selected bud. **1916** L. H. Bailey *Pruning-Manual* v. 127 Notching into the wood above a bud tends to produce strong growth from that bud. **1972** G. E. Brown *Pruning of Trees* ii. 24 'Notching'..is carried out during the dormant season, when a notch is cut about 10 to 15 mm. above the bud, completely removing a wedge-shaped piece. **1974** R. Grounds *Practical Pruning* iii. 29 The lowest branches resulting from the first year's notching are pulled down and tied at the horizontal.

notchy, *a.* Add: **2.** Of a manual gear-changing mechanism: difficult to use because the lever has to be moved accurately (as if into a narrow notch).
1967 *Autocar* 28 Dec. 20/2 The gear change is rather notchy, which is somewhat typical. **1972** *Drive* Spring 150/2 Short, precise movements..are features of the four-speed gearbox, though the lever has a slightly notchy feeling. **1975** *Daily Tel.* 16 Apr. 12/6 The short gear lever is a trifle notchy in operation although I always find this preferable to vague sloppiness.

note, *sb.*[2] Add: **5. c.** In perfumery, one of the basic components of the fragrance of a perfume which give it its character.
1905 F. W. Burbidge *Bk. of Scented Garden* 34 Dr Piesse goes so far as to say that one false note amongst odours will destroy the whole harmony of the chord, just as in music or in colour. **1945** E. Sagarin *Science & Art of Perfumery* xii. 145 The odors were like sounds..and a scale could be created going from the first or lowest note, the heavy smell, to the last or highest note, the sharp smell. **1954** A. J. Krajkeman tr. *Jellinek's Pract. Mod. Perfumery* iv. 180 Honey-like odours, combining a sweet-floral with an animal note. **1960** A. Ellis *Essence of Beauty* xii. 142 The characteristic a discerning and discriminating user looks for in a perfume..is what the perfumer calls '*the soft odour note*'. **1970** *Daily Tel.* 16 Dec. 11/6 The top notes are the lightest and most volatile. **1973** E. Maple *Magic of Perfume* vi. 51 There has long existed in perfumers' language a system of 'musical notation' in which the constituents of perfume are accorded 'top notes', 'middle notes' and 'bass notes'.
10. *note of interrogation*: see Interrogation 2 b in Dict. and Suppl.
21. (sense 1) *note-head, -value*; (sense 2) *note-singing; -spinning* vbl. sb. and ppl. adj.; (sense 13) *note-block, -pad, -taker* (later examples), *-taking*; (sense 16) *note-writing* (earlier example); also as *adj.*; **note-broker** *U.S.* a broker who deals in promissory notes and bills of exchange; **note-cluster,** several neighbouring notes played simultaneously; **notehead paper,** business notepaper having a printed heading; **note-holder,** a holder of notes (sense 17) issued by a business company or the like for temporary financing; **note-layer** (sense 18), a petty thief who operates a short-change swindle; **note-row, -series** = *tone-row* (*tone *sb.* 11).
1927 R. A. Freeman *Certain Dr. Thorndyke* II. xiv. 207 Jotting down on a note-block a few brief memoranda. **1870** W. W. Fowler *10 Yrs. Wall St.* 226 This man..is an English Jew, who has gone into the business of a note broker. **1929** *Encycl. Brit.* IV. 233/1 Bill-brokers are practically unknown in the United States; their general analogue is the note-broker. **1934** C. Lambert *Music Ho!* v. 330 Van Dieren's attitude towards harmony is more indicative of future developments than the 'note clusters' of Henry Cowell. **1965** *New Statesman* 24 Sept. 458/3 A younger composer like Peter Sculthorpe attempts to enter the heart of the Australian experience by way of a static technique of cumulative ostinati and note-clusters that is more Asiatic than occidental. **1946** H. Foss in A. L. Bacharach *Brit. Mus. of our Time* iv. 78 Warlock wrote with exquisite precision:..the shape of the note-heads, the uprights, the binds and ties, were of enormous interest to him. **1974** G. Read *Music Notation* v. 63 The note-head..is somewhat oval in shape, and is either open (or 'white'..) or closed ('black'..). **1909** *Westm. Gaz.* 20 May 7/2 The..notehead paper of a London firm of stock, share, and bond dealers. **1927** *Daily Tel.* 21 June 2/3 Shareholders were prepared for unfavourable figures by the necessity for an arrangement regarding..the rights of the note-holders. **1968** *Globe & Mail* (Toronto) 13 Jan. B. 1/2 Fraudulent financial statements and a false prospectus were means used..to defraud 8,500 noteholders of Prudential Finance Corp. Ltd. of $20-million. **1928** Note-layer [see *creep *sb.* 1 d]. **1938** D. Castle *Do Your Own Time* viii. 83 They boast of having been knock-off gees, gow peddlers,..note layers, and torpedo men. **1950** H. E. Goldin *Dict. Amer. Underworld Lingo* 146/1 The note-layer usually works with an accomplice. **1922** A. Bennett *Lilian* I. iv. 41 She repeated the number, even writing it on her note pad. **1961** *Encounter* Apr. 29/2 He asked me for one of my notepads. **1975** D. Bagley *Snow Tiger* xi. 95 Smithers consulted his note-pad. **1955** *Oxf. Compan. Mus.* (ed. 9) 698/1 Note-row... This is a rigid method of composition introduced by Schönberg... All the twelve notes of the octave are employed in every composition, and all the notes are treated in such a way as to enjoy an equal footing. **1961** *Listener* 12 Oct. 578/3 He spoke of his contrapuntal technique, of note rows, their combinations and inversions. **1947** *Penguin Music Mag.* Dec. 20 'Twelve-tone technique', in which the twelve notes of the chromatic scale are used in an order known as a 'note-series', which remains the same throughout the work. **1962** *Times* 19 Oct. 18/6 The American group admit to working from a note-series. **1896** *Musical Herald* 1 Feb. 41/1 Thousands of teachers waste time in note-singing practice. **1908** E. M. Sneyd-Kynnersley *H.M.I.* xxiv. 288 The clergy encouraged note-singing for the sake of their choirs. **1946** H. Foss in A. L. Bacharach *Brit.*

Music of our Time iv. 67 He [*sc.* 'Peter Warlock'] was no devotee of note-spinning, of blowing up frog-ideas into bull-like proportions. **1961** *Listener* 20 Apr. 717/3 The ideas in the other two movements are of an altogether lower standard and there is some tiresome note-spinning. **1970** *Times* (Sat. Suppl.) 18 Apr. p. iii/5, I confess to finding it rather like late, not very strongly motivated, note-spinning Schumann. **1935** *Essays & Studies* XX. 123 All these matters can be conscientiously taught to the plodding note-taker and be redelivered at examinations. **1961** *Sunday Express* 29 Jan. 9/6 Police note-takers filled four foolscap pages. **1954** J. Masters *Bhowani Junction* II. xii. 107 They didn't want me for any more note-taking. **1955** H. Roth *Sleeper* xi. 92 The stilted feeling that note-taking causes in some interviewees. **1975** *Times* 5 July 2/5 Jurors..were often required to perform fantastic feats of attention and memory..without the aid of note-taking. **1915** *Musical Quarterly* I. 191 An Indian [*sc.* North American Indian] can give short note-values corresponding to eighth or sixteenth notes with perfect distinctness. **1917** H. James *Middle Years* iv. 46 Whereas the smartness and newness beyond the sea supposedly disavowed the low, they did so but thinly and vainly, falling markedly short of the high; which the little boxed and boiled Albany attained to some effect of,..just by having its so thoroughly appreciable note-value in a scheme of manners. **1944** W. Apel *Harvard Dict. Mus.* 497/2 The illustration shows the note values with their American terminology. **1955** G. Abraham in H. van Thal *Fanfare for E. Newman* ii. 10 The first two bars were originally written as four, in double note-values. **1963** *Listener* 17 Jan. 141/2 It [*sc.* the basic five-note figure] is repeated eight times in succession, with progressively lengthening note-values. **1814** Jane Austen *Mansfield Park* II. xiii. 291 Quite unpractised in such sort of note-writing. **1967** G. Kelly in *Coast to Coast 1965–66* 101 He became the victim of their gum-chewing, note-writing, hair-combing inattention.

note, *v.*[2] **6. a.** (Later example.)
1897 *Musical Herald* 1 June 189/1 They organise vocal music competitions, but they have no sight-singing, and no noting music by ear.

notelet. Add: **3.** A folded card or sheet of paper on which a note may be written, having a picture or design on the face of the first leaf.
1955 *Stationery Trade Ref. Bk.* 71/1 Macniven & Cameron Ltd...manufacturers of social and gift stationery. 189/1 Notelets Products. Macniven & Cameron Ltd. **1971** *Countryman* Autumn 205/2 (Advt.), S.A.E. brings our list of Christmas gifts, cards, notelets, calendars for country lovers. **1972** *Stationery Trade Rev.* Aug. 39/2 Many of the Waverley gift stationery boxes also includes [*sic*] Notelets.

note-shaver. (Earlier and later examples.)
1816 *Massachusetts Spy* 4 Sept. (Th.), We have too many note-shavers; too many gentlemen. **1905** D. G. Phillips *Plum Tree* 11 But my clients were poor, and poor pay, and slow pay. Nobody was doing well but the note-shavers. **1911** R. D. Saunders *Col. Todhunter* viii. 113 Old Eph Tucker was a note-shaver long before he was a politician, and he's got note-shavin' in his blood bigger'n a mule. **1942** Berrey & Van den Bark *Amer. Thes. Slang* ix. 520 *Shaver, note shaver*, a discounter of notes at an exorbitant rate.

no·te-shaving. *U.S. slang.* [f. prec.] The profession of a note-shaver; the making of an excessive profit on the discounting of notes.
1828 *Yankee* (Portland, Maine) I. 52/1 [By] the system of note-shaving that prevails here..the industrious and active are held in a state of bondage to the more wealthy and more lazy. **1855** P. T. Barnum *Life* 138 Had I termed the deed an extortion or note-shaving..the verdict might have been different—but I had called the act 'usury'. **1902** W. N. Harben *Abner Daniel* 38 He began to utilize this capital in 'note shaving', and other methods of turning over money for a handsome profit. **1911** [see *note-shaver].

‖ **notes inégales** (nǫtsinegal), *pl. Mus.* [Fr., lit. 'unequal notes'.] In Baroque music, notes performed by convention in a rhythm different from that shown in the score.
1927 *Grove's Dict. Mus.* (ed. 3) II. 708/2 To avoid all uncertainty, authors frequently make use of the expression *notes égales* or *notes inégales*. **1954** *Ibid.* (ed. 5) IV. 479/2 In common 4-4 time semi-quavers or quavers but not crotchets or minims are eligible as 'notes inégales'. **1965** *Times* 9 July 16 Chief among these 'errors' is the convention of *notes inégales*, which decreed that with certain specified or understood exceptions, pairs of stepwise quavers (or, in some time-signatures, semiquavers) were to be played unevenly, the first note rather longer than the second. **1970** *Sat. Rev.* 31 Oct. 55 Mr. Weaver shows a commendable awareness of performance practice, even though scholars are not unanimous—namely the *notes inégales* in the finale of No. 3 where he adjusts the left hand to the triplet meter in the right. **1972** *Times* 22 Nov. 11/5 A sound knowledge of the more ticklish stylistic points to be observed, *notes inégales*, for instance in music from Couperin's Ninth Suite. **1975** *Times Lit. Suppl.* 9 May 503/4 (Advt.), Notes on ornamentation, bowing, *notes inégales*, and other performance problems.

‖ **note verbale** (nǫt vęrbal). [Fr., lit. 'verbal note'.] An unsigned diplomatic note, of the nature of a memorandum, which is written in the third person.
1855 E. Grenville Murray *Embassies & Foreign Courts* xvii. 264 Certain diplomatic worthies invented a double-faced document called a *note verbale*, or summary of conversation. These notes are unsigned. Their object is merely to refresh the diplomatic memory... They are supposed..to combine all the advantages of writing and conversation. **1911** *Encycl. Brit.* XIX. 823/2 The so-called *notes verbales* are unsigned, and are merely of the nature of memoranda (of conversations, &c.). **1917** E. Satow *Guide to Diplomatic Practice* vii. 75 As not many *Notes Verbales* are to be met with in print, it seems worth while to reproduce it here. **1939** H. Nicolson *Diplomacy* x. 246 *Note verbale*. This is a type of communication which is less formal than a signed Note and more formal than a memorandum. It is unsigned, but it is customary that it should contain at the end some conventional expression of courtesy. It is, in fact, merely the addition of this polite tag which differentiates it from the mémoire. **1955** *Times* 20 May 8/1 Among the papers now issued by the secretariat is a *note verbale* from the permanent delegation of the United Kingdom to the United Nations. **1970** R. G. Feltham *Diplomatic Handbk.* iv. 42 The Note Verbale is not common, but it is used to clarify or confirm points raised in a previous conversation or to list items which one could not expect to be recalled precisely. **1973** I. M. Sinclair *Vienna Convention on Law of Treaties* ii. 36 The conference decided otherwise, no doubt influenced by the growing practice of constituting treaties by an exchange of unsigned *notes verbales*.

not-go (nǫ·tgo), *a. Engin.* [f. Not *adv.* + Go *v.*] Designating (part of) a gauge so made that it will not enter, or will not admit, an object whose dimensions are within a designated limit.
1917 *Proc. Inst. Mech. Engin.* Jan. 58 Those gauges that were considered right might be put into one lot.. the tests being made solely by 'go' and 'not go' check gauges. **1951** [see *go *a.* 1]. **1964** S. Crawford *Basic Engin. Processes* xiv. 296 The 'Not-Go' end is of such a size that it will not enter the hole when the hole is up to the highest limit.

'nother, colloq. var. Another *a.*, *pron.*
1934 D. Mackail *Summer Leaves* x. 341 'How long have you got here?'..''Nother month, I expect.' **1935** in Z. N. Hurston *Mules & Men* (1970) 20 People told one 'nother that God was talking in de mountains. **1972** 'L. Egan' *Paper Chase* (1973) xii. 195 'Nother little bit of the thing just occurred to me. **1973** H. McCloy *Change of Heart* iv. 39 'Nother cup of coffee, gramps?

nothing, *sb.* and *adv.* Add: **A.** *sb.* **I. 1. f.** As *adj.* or *interj.* Not at all; in no respect. *colloq.* (orig. *U.S.*).
1883 G. W. Peck *Mirth for Million* 325 'You are pretty rough on the old man..after he has..given you nice presents.' 'Nice presents nothin. All I got was a "Come to Jesus" Christmas card.' **1888** [see *nope *adv.*]. **1899** A. Nicholas *Idyl of Wabash* 175 'My account—nothing!' was her scornful ejaculation. **1899** B. Tarkington *Gentleman from Indiana* i. 10 'But you only wait—' The editor smiled sadly. 'Wait nothing. Don't threaten, man.' **1911** H. Quick *Yellowstone Nights* xi. 288 Stop nothing! Federal injunction won't do it. **1922** M. B. Houston *Witch-man* xviii. 238 'He could have found it, of course.'.. 'Found it, nothing. I saw other things he'd taken.' **1925** Wodehouse *Sam the Sudden* xiii. 93 'Two million smackers it's going to get him,' retorted Dolly. 'Two million smackers nothing! The stuff's hidden in a place where he'd never think of looking in two million years.' **1946** K. Tennant *Lost Haven* (1968) ii. 43 'How about the spooks?'.. 'Spooks nothing.' **1966** A. E. Lindop *I start Counting* xxii. 275 Grandad said, poor little mite nothing. The man that gave her a lift told the police she'd done her best to seduce him. **1969** R. Rendell *Best Man to Die* xii. 117 'Did you wait for him?' 'Wait, nothing!' said Cullam hotly. 'Why would I?' **1972** D. Lees *Zodiac* 46 'Francs?' 'Francs nothing—pounds.' **1974** T. Barling *Shooter Man* iii. 23 'It just slipped out.' 'Slipped nothing. You couldn't resist.'
g. *as slick as nothing at all*: very promptly or quickly, 'in the twinkling of an eye'. *rare.*
1884 'Mark Twain' *Huck. Finn* xl. 410 Done it just as slick as nothing at all.
h. *with* (or *having*) *nothing on*, wearing no clothes, undressed, naked.
1719 Defoe *Robinson Crusoe* 62, I stripp'd..having nothing on but a Chequer'd Shirt, and a Pair of Linnen Drawers. **1908** Kipling *Let.* in C. E. Carrington *Rudyard Kipling* (1955) xvi. 399, I cannot help blushing when I am rung-up by women—with nothing on but spectacles and a bath-towel. **1971** E. Paul *Reluctant Cloak & Dagger Man* xi. 137 We always swam here with nothing on.
i. *nothing doing*: see *do *v.* 34 c.
j. *like nothing on earth*: strange, ugly, wretched, etc., in a superlative degree.
1923 A. Christie *Murder on Links* xxvi. 286 She looked like nothing on God's earth. **1927** W. E. Collinson *Contemp. Eng.* 117 To look or feel like nothing on earth (*very bad*). **1974** M. G. Eberhart *Danger Money* (1975) iv. 39 'What's he like?' 'Nothing on earth... I wouldn't trust him with a nickel.'
k. *there is nothing* (*much*) *in it*: there is no important feature of interest or value in something; there is no significant difference between two things, etc.
1927 *Observer* 18 Dec. 19/3 The first round there was nothing much in it. In the second round Angus..punched Mansfield round the ring. **1950** Partridge *Dict. Clichés* (ed. 4) 156 *Nothing in it*, esp., *there's nothing*... (there is no appreciable—or important—difference): c. 20.
l. *to have nothing on* (someone): to be no match for (someone). See *have *v.* 14 h.
3. c. As *adj.* in trivial use: of no account, insignificant, meaningless, insipid, dull; (of a dress, etc.) discreet, elegantly unobtrusive.
1961 *Time* (Atlantic ed.) 18 Aug. 60 All these beautiful people with nothing faces. **1964** 'E. McBain' *Axe* vi. 118 This is a *nothing* game, you dig? A buck, two bucks a time, that's all. **1965** *Vogue* Aug. 43/2 Little 'nothing' sweaters and shirts for wearing with suits. **1967** W. Murray *Sweet Ride* iv. 46 It was a nothing place, just a few booths and a counter. **1969** P. Kavanagh *Such Men are Dangerous* (1971) ii. 30 The characters..were all hung up on trivia, little nothing problems in their careers and marriages. **1971** *Sunday Times* (Colour Suppl.) 23 May 53/1 A girl in one of those 'nothing' dresses with the Quant signature written all over it. **1972** P. Dickinson *Lizard in Cup* vi. 97 It's a nothing thing, like I said... But drugs aren't a nothing thing, no.
6. e. (Earlier and later examples.)
1835 J. F. Cooper *Monikins* III. iv. 93 In this happy land, there was no registration, no passports, 'no nothin' —as Mr. Poke pointedly expressed it. **1905** Kipling *Actions & Reactions* (1909) 8 'No roads, no nothing!' said Sophie. **1948** H. L. Mencken *Amer. Lang.* Suppl. II. ix. 392 There may not be no nothing. **1968** *Washington Post* 21 Sept. A.12/1 His [*sc.* Wallace's] appeal is to racial animosity, no-nothing policies.
II. 8. b. (Further examples.) Also used to typify theorizing which attempts to reduce, simplify, or explain concepts in such a way that they seem to accord with the theory propounded; so *the nothing-but*; also *nothing-but-ism, nothing-buttery.*
c **1838** W. H. Murray in M. R. Booth *Eng. Plays of 19th Cent.* (1973) IV. 160 Mark me: no amendments, no conferences—I'll have '*the* bill, the *whole* bill, and *nothing* but the bill'. **1884** *Encycl. Brit.* XVII. 701/2 Witnesses are sworn: 'The evidence you shall give..shall be the truth, the whole truth, and nothing but the truth. So', &c. **1923** R. H. Thouless *Introd. Psychol. Relig.* x. 129 The essential requirement of this theory is that it should be shown that religion contains nothing but elements of this kind, and this is exactly what Mr Schroeder makes no attempt at all to prove. **1934** J. G. Brandon *One-Minute Murder* iv. 32 As far as that poor devil's concerned..it's accident and nothing but. **1935** *Mind* XLIV. 91 Jung's formulation..is the antithesis of '*nothing-but*'-ism. **1937** A. Huxley *Ends & Means* xiv. 257 All who advance theories of mind containing the word 'nothing but', tend to involve themselves in this kind of contradiction. The very fact that they formulate theories which they believe to have general validity..constitutes in itself a sufficient denial of the validity of 'nothing-but' judgments concerning the nature of the mind. **1951** M. Lowry *Let.* 25 Aug. (1967) 252 You might call it pseudo-Freud and the philosophy of the 'nothingbut'. **1961** *Mind* LXX. 100 There is much else in the literary idiom of nature-philosophy: *nothing-buttery*, for example, always part of the minor symptomatology of the bogus. 'Love is.. nothing more, and nothing less, than [etc.]'. **1973** *Black World* June 30 A poet ain't nothin' but a bird.
10. a. (Later examples.)
1885 'M. Rutherford' *M. Rutherford's Deliverance* iv. 65 She had learned that she was nothing specially to him. **1947** E. O'Neill *Iceman Cometh* I. 79 He's nothing to you —or to me, either. *Ibid.* II. 105 The good old Cause means nothing to you any more.
c. *there's nothing to it*: it is very easy to do; there is no difficulty involved.
1934 E. O'Neill *Ah, Wilderness!* I. 21 There (*with a grin*). I know there's nothing to it, anyway. **1951** H. Wouk *Caine Mutiny* viii. 77 There's nothing to it, really, except making damn sure none of your watch-standers sit down or fall asleep standing up. **1953** F. Stark *Coast of Incense* iv. 232, I am puzzled when asked what makes my style, for there is nothing to it except a natural ear for cadence and the wish to get the meaning right. **1963** 'S. Woods' *Taste of Fears* i. 13 'There's nothing to it if you're quick, or so I'm told,' he added. **1971** D. Eden *Afternoon Walk* vii. 83 'You used to automatic drive?' 'Yes.' 'Then there's nothing to it.' **1974** *Country Life* 14 Feb. 322/1, I mentioned that I was spending the following night at a Japanese inn. He assured me that there was nothing to it.
15*. *nothing to write* (or *cable, wire*) *home about*, denoting something that is unworthy of comment, unremarkable or mediocre. *slang* (orig. *Forces*').
1917 W. Muir *Observations of Orderly* 227 Miserable conditions.., bad accommodation, doubtful food.. these, in the lingo of our now much-travelled and stoical troops, are 'nothing to write home about'. **1937** Auden & MacNeice *Lett. from Iceland* iv. 38 Fare from Hull to anywhere in Iceland, £4 10s. plus 5 kr. a day for food. The latter is nothing to write home about but eatable. **1942** Berrey & Van den Bark *Amer. Thes. Slang* i. 33 Fair to middling,..nothing to..shout about, nothing to wire *or* write home about.
16. b. *nothing-saying.*
a **1817** Jane Austen *Persuasion* (1818) IV. viii. 160 After a period of nothing-saying amongst the party.
c. (Further examples.)
1906 *All-Story Mag.* (U.S.) Aug. 593/1 The nothing-doingness of things in general outside the office. **1924** R. Graves *Mock Beggar Hall* 16 A formless lumpish, nothing-in-particular. **1924** D. H. Lawrence *England, My England* 98 They passed an agreeable, casual, nothing-in-particular evening.

nothofagus (nǫþofē·gǫs). [mod.L. (K. L. Blume *Museum Botanicum Lugduno-Batavum* (1850) I. 307). f. Gr. νόθος false + φηγός beech tree.] An evergreen or deciduous tree of the genus so called, belonging to the family Fagaceæ and native to Australasia and South America; also called southern beech.
1914 W. J. Bean *Trees & Shrubs Hardy in Brit. Isles* II. 100 Of the Nothofagus group of beeches, this [*sc. N. Moorei*] has the largest leaves of any cultivated out-of-doors in this country. **1956** *Handbk. Hardwoods* (Forest Prod. Res. Lab.) 76 *Nothofagus* is the beech of the south-

ern hemisphere. 1961 *Times* 18 July 11/7 The dark and light greens of the nothofagus forest. 1974 R. L. Fox *Variations on Garden* 138 Foresters are beginning to realize the merits of a cousin of the beech tree called nothofagus... These southern beeches grow wild in New Zealand and South America... Their leaves are small, like a small beech or hornbeam, and their autumn colouring is most remarkable: each leaf changes individually to its own shade of orange, red, brown or plain green.

nothomorph (nǫ·þomǭɹf). *Bot.* [f. Gr. νόθος cross-bred + μορφή form.] A plant produced by hybridization.
1939 R. Melville in *Proc. Linn. Soc.* CLI. 158, I therefore propose the term nothomorph (nothomorphus = hybrid form) for all hybrid forms of sexual origin, whether F_1 segregates or back-crosses. 1949 *Jrnl. R. Hort. Soc.* LXXIV. 43 This original seedling and its vegetatively produced offspring may..be the product of hybridization, i.e. a nothomorph. 1953 *Rep. Proc. 7th Internat. Bot. Congr.* 1950 xii. 544/2 These forms are recognized as nothomorphs; when desirable they may be designated by an epithet preceded by the binary name of the group and the term nothomorph. 1963 Davis & Heywood *Princ. Angiosperm Taxonomy* xiv. 482 The term nothomorph can be used to distinguish different derivatives of hybridisation between the same species.

nothosaur (nǫ·þosǭɹ). *Palæont.* [f. mod.L. name of suborder *Nothosauria*, f. generic name Nothosaurus (Georg, Graf von Münster 1834, in *Neues Jahrb. f. Min.* 525).] An extinct marine reptile belonging to the suborder Nothosauria, known from Triassic fossil remains in Europe.
1933 A. S. Romer *Vertebr. Paleont.* vii. 153 A nothosaur in which this pterygoid union had not been quite completed would make an ideal plesiosaur ancestor. 1962 *New Scientist* 5 July 34/2 Some of the nothosaurs seem to have been very lightly built, others larger and more sturdy. They all had fairly long, flexible necks and long jaws with many pointed teeth. They must have caught their food in the shallow waters..paddling along with their stubby webbed feet. 1971 *Nature* 15 Jan. 172/1 In a great many aquatic and marine reptiles, past and present, ichthyosaurs, placodonts, nothosaurs,..the external nares are just in front of the eyes.

no-thought: see *no *a.* 5 e.

no-throw. [f. No *a.* + Throw *sb.*²] In various games or sports, a throw disallowed because it does not comply with the rules.
1959 *Times* 14 Sept. 3/2 Ellis..won the hammer..with his second throw.., thereafter having four no-throws. 1964 A. Wykes *Gambling* iii. 62 The bettors place their money on whether the coins will land as two heads or two tails; a head and a tail is a 'no-throw'. 1967 Ward & Watts *Athletics* xii. 115 *Major fault* 9: point [of javelin] not coming down—no mark is made, and hence there is a 'No throw'.

not-I, not-me. [f. Not *adv.* 14 d + I, Me *pers. prons.*] That from which the subjective or personal is excluded.
1846 J. D. Morell *Hist. View Philos.* I. 59 In the same manner as the *me* implies the notion of a *not-me* from which it is distinguished..so the notion of the limited and the finite implies the correlative one of the unlimited and the infinite. 1854 A. G. Henderson tr. *Cousin's Philos. of Kant* vii. 179 The me..could only become cognizant of the not-me by means of the faculties it possesses. 1895 tr. *M. Nordau's Degeneration* iii. i. 245 It has..jumped to the conclusion that the 'I' has actually no knowledge of a 'not-I', of an external world. *Ibid.*, These wise men repeated, in a tone of conviction, the doctrine of the non-existence of the 'not-I'. 1917 D. H. Lawrence *Look! We have come Through!* 147, I suppose ultimately, she is all beyond me, She is all not-me, ultimately. 1950 A. Huxley *Themes & Variations* i. 114 Not the hyper-organic 'I', still less the divine not-I, which transcends the ego and is its ground. 1953 S. Spender *Creative Element* ii. 54 He [sc. Rimbaud] was attempting to remove the barrier which divides subject from object, the 'I' from the 'not I'. 1953 H. L. Sullivan *Interpersonal Theory of Psychiatry* x. 162 The personification of not-me is..very emphatically encountered by people who are having a severe schizophrenic episode. 1965 *Listener* 4 Nov. 691/2 We are accustomed to distinguish so sharply between 'I' and 'not-I' and between 'now' and 'not-now' that any blurring of the dichotomy may seem unnatural to us. 1974 A. Plant in M. Fordham et al. *Technique in Jungian Analysis* ii. 215 The analyst can be of use by letting the patient experience him as a 'not-me' possession.

notice, *sb.* Add: **1. c.** (Earlier example of *at —— minutes' notice*.)
1839 Dickens *Let.* 25 July (1965) I. 569 There is always a bed for you at five minutes' notice.
2. b. (Earlier examples.)
1765 Earl of Malmesbury *Let.* 16 Sept. (1870) I. 129 It is 150 florins, or fourteen guineas, a-year; but I am to try it first, and may, at any time after, quit it by giving six weeks' notice. 1836 Dickens *Let.* 5 Nov. (1965) I. 191, I have deemed it right to beg you to accept my notice from to-day.
c. An announcement read to a church congregation (freq. *pl.*).
1855 F. Procter *Hist. Bk. Com. Prayer* ii. iii. 322 The correct interpretation concerning notices to be given in church. a 1870 'Mark Twain' *Let.* in C. Clemens *My Father* (1931) i. 11 The local minister had read sixteen 'notices' of Sunday-school and Bible-class and church and sewing-society and other meetings. 1967 *Alternative Services (Second Series): An Order for Holy Communion* 4 Banns of Marriage and other notices may then be published, if they have not been published before the service.
8. a. Also, a review of a play or any public entertainment. (Earlier and later examples.)
1835 Dickens *Let.* 23 Nov. (1965) I. 97 If I take a cab and put off writing my notice 'till we return, I can easily manage it. I shall come straight from the Theatre. 1952 Granville *Dict. Theatrical Terms* 123 *Notice*, the newspaper critique following a first-night performance. 1959 N. Marsh *False Scent* (1960) ii. 76 'She's playing Eliza Doolittle,' Gantry remarked. 'Of course. Nice notices,' Marchant murmured.
9. b. notice paper, a parliamentary paper giving the current day's proceedings.
1844 T. E. May *Law, Privileges Parliament* viii. 166 Any member [of the House of Commons] may propose a question... But in order to give the House due notice of his intention, he is required to state the form of his motion on a previous day, and to have it entered in the Order Book or Notice Paper. 1884 E. W. Hamilton *Diary* 3 Apr. (1972) II. 588 After the unprecedent[ed] number of questions had been disposed of (no less than 73 being on the Notice paper).. Sir S. Northcote moved the adjournment of the House. 1956 P. & G. Ford *Guide to Parliamentary Papers* I. i. 3 The White Paper (Notice Paper)..contains certain portions of the Blue Paper and relates to the current day's sittings.

notice, *v.* Add: **3. b.** colloq. phr. *not so as you'd notice*: not to a noticeable degree.
1937 A. Christie *Murder in Mews* iii. 172 'He was fond of you?' 'Not so that you'd notice it..he rather resented my existence.' 1938 N. Marsh *Artists in Crime* (1954) xvi. 216 Garcia's not innocent, dear, not so's you'd notice it. 1966 'N. Blake' *Morning after Death* xiii. 198 'Was Chester interested?' 'Not so as you'd notice.' 1970 W. J. Burley *To kill a Cat* xi. 189 'Any luck?' 'Not so's you'd notice.'
c. *intr.* To be seen, to show, to be noticeable.
1961 Y. Olsson *On Syntax Eng. Verb* vii. 177, I have mended the hole now. I don't think it notices.
7. To write a review or 'notice' of a book, play, etc.
1854 *Punch* 15 July 20/1 The reporter who 'noticed' the diplomatists. 1859 G. H. Lewes *Let.* 5 Feb. in *Geo. Eliot Lett.* (1954) III. 10 Perhaps also you will send the 'Times' should that 'publication' notice the carpenter [sc. *Adam Bede*].

noticeability. Add: **b.** The quality, state, or fact of being noticeable.
1926 Fowler *Mod. Eng. Usage* 639 The reader will perhaps conclude that its noticeability is not a grace.

noticeable, *a.* 2. (Earlier example.)
1809 *Mem. Amer. Acad. Arts & Sci.* III. 248 The moon's limb exhibited very little of that rough or serrated appearance, which was so noticeable in 1806.

noticing, *ppl. a.* (Later examples.)
1903 E. Wharton *Sanctuary* ii. i. 79 You know she's an uncommonly noticing person, and little things tell with her. 1905 J. C. Lincoln *Partners of Tide* ii. 20 Bradley, being what his late 'Uncle Solon' had called a 'noticin'' boy', remembered Captain Titcomb's hint. 1940 R. Postgate *Verdict of Twelve* i. 12 Father was not 'noticing'; Mother was, and what's more would twist your arm till you screamed if you sulked and wouldn't answer. 1959 A. Christie *Cat among Pigeons* xx. 205 She's not what I'd call a noticing kind of child.

no-time. [f. No *a.* + Time *sb.*] A time which does not exist (*poet.*); a very short time (*U.S. local*).
a 1918 W. Owen *Poems* (1931) 69 And in the happy no-time of his sleeping Death took him by the heart. 1942 W. Faulkner *Go Down, Moses* 224 After a no-time he returned. 1943 D. Gascoyne *Poems 1937–1942* 22 Black was the No-time at the heart Of Time (the frameless mirror's back). 1963 [see *no-place *sb.*].

notion. Add: **7.** (Earlier example of *to take a notion*.)
1776 J. Verrieul *Let.* 17 Apr. in F. Chase *Hist. Dartmouth Coll.* (1891) I. vi. 347 There was Ten Regt ordert to march for Newyoark, and I toock a notion to go with them.
9. b. (Earlier and later examples.) Now *spec.* in haberdashery.
1803 F. Asbury *Jrnl.* (1821) III. 106 How would it tell to the South, that priests were among the notions of Yankee traffick? 1913 F. H. Burnett *T. Tembarom* xx. 264 The young lady from the notion counter (those wonderful shops!). 1964 *McCall's Sewing* ii. 22 While making a list of the fabrics and trims needed, check the 'Notions' section to see what notions are needed. *Ibid.* 30/2 *Notions*, all dressmaking supplies that are used in the construction of a garment: thread, zippers, tape, buttons, etc. 1965 'L. Egan' *Detective's Due* (1966) iv. 39 Varallo and Katz were led past Notions, Gift Cards and Cameras, to a counter in the center of the store.
c. *notion-peddler* (or *-pedlar*), *-peddling*, *-seller*.
1932 L. C. Douglas *Forgive our Trespasses* (1937) i. 1 The steel-bowed spectacles that had been her mother's, had of a notion peddler for two dozen eggs & a pound of butter. 1809 'D. Knickerbocker' *Hist. N.Y.* I. iv. iii. 214 He swore that he would have nothing more to do with such a squatting, bundling, guessing,..notion-peddling crew. 1839 *Chemung* (N.Y.) *Democrat* 17 Apr. (Th.), A 'notion seller' was offering Yankee clocks, &c.

notional, *a.* Add: **1. c.** *Economics.* Of a figure, profit, etc.: speculative, hypothetical; for the purposes of a particular interpretation or theory.
1958 *Spectator* 8 Aug. 204/3 The profit attributable to Iraq is the notional one which the oil companies regard as economic. 1960 *Economist* 15 Oct. 278/1 Costs per ton of storage, mainly notional interest charges, were put at roughly £5 for lead and zinc. 1964 *Financial Times* 3 Mar. 15/6 The formula for calculating this standard price has been drawn up so that it would leave the Corporation with a notional profit of £1·8 m. before tax. 1972 *Accountant* 17 Aug. 197/1 A company..will be able to obtain relief..on a notional figure.
4. b. (Earlier example.)
1791 *Gazette of U.S.* (N.Y.) 9 Feb. (Th.), If a man is a little odd in his ways, his friends say he is a notional creature, or full of notions... Love is the most notional passion.
6. *Gram.* Used (orig. by Jespersen) in relation to the semantic content, real or apparent, of grammatical forms and categories; of a word: carrying full meaning, not merely grammatical; of a verb: principal, main, not auxiliary.
1924 O. Jespersen *Philos. Gram.* iii. 56 Our examples of gender and sex will make it clear that the relations between the syntactic and notional categories will often present a similar kind of network to that noticed between formal and syntactic categories. 1933 —— *Syst. Gram.* 20 In 'he happened to fall' the notional subject is a nexus 'he..to fall'. 1942 Partridge *Usage & Abusage* 148/2 Jespersen's *The Philosophy of Grammar*..urges us to consider all words..in their three aspects, form, syntactic function, and 'natural or logical meaning'; this third aspect he calls 'notional'. 1957 R. W. Zandvoort *Handbk. Eng. Gram.* i. v. 64 The opposite of 'auxiliary' is 'notional verb', 'principal verb', or 'verb of full meaning'. 1966 M. Pei *Gloss. Linguistic Terminol.* 182 *Notional word*, a word that carries a full meaning ('he *has* luck' vs. 'he has gone'). 1971 P. J. Lucas in *Archivum Linguisticum* II. 19 There is a considerable degree of overlap (in Capgrave's *Chronicle*) between the notional signs on the one hand, and, between the grammatical signs on the other, but there is no overlap between primarily notional and primarily grammatical signs.

not-life. [Not *adv.* 14 d.] Inanimate matter, esp. as contrasted with that which contains life. So **not-living** *a.* and *sb.*
1869 T. H. Huxley in *Fortn. Rev.* 1 Feb. 140 The assumption of the existence in the living matter of a something which has no representative or correlative in the not living matter which gave rise to it. 1895 [see Not *adv.* 14 d]. 1943 J. S. Huxley *Evolutionary Ethics* i. 7 The scientific study of change, of becoming, of the production of novelty, whether of life from not-life, of a baby from an ovum. 1953 Massingham & Hyams *Prophecy of Famine* iii. 57 The harmony..between the living creature..and the not-living habitat itself. *Ibid.* viii. 169 A short-circuiting of the life-cycle in terms of not-life. 1964 Gould & Kolb *Dict. Social Sci.* 607/1 Birth rituals.. signifying the separation of the infant from the world of the dead (or not-living) and his aggregation to that of the living.

not-me: see *not-I.

not-ness. *rare.* [f. Not *adv.* + -ness.] The quality or state of not accepting something; something negative.
1933 Dylan Thomas *Let.* Sept. (1966) 21 Your 'not-ness' alone is worth all the superlatives at my command. 1945 H. Miller *Air-Conditioned Nightmare* 12 It was something negative, some not-ness of some kind or other.

notornis. Insert in etym. [mod.L. (R. Owen 1848, in *Proc. Zool. Soc.* 2).] Substitute for def.: A flightless New Zealand bird of the genus so called, which includes the single species *Notornis mantelli*, belonging to the family Rallidæ, and distinguished by blue-green plumage and pink bill and legs; = Moho¹, *takahe. (Later examples.)
The bird was believed to be extinct until Dr. G. B. Orbell rediscovered it in a mountain valley near Lake Te Anau in November 1948.
1882 *Trans. N.Z. Inst.* XIV. 253 It has always been acknowledged that Notornis is a degenerate rail. 1930 W. R. B. Oliver *N.Z. Birds* 350 Suspecting that it was a Notornis, Connor took it to the station where he carefully skinned it, preserving both the skin and the bones. 1950 *Discovery* July 217/2 The news of the discovery of *Notornis* was announced on the following day—November 21 [1948]. 1962 J. Frame *Edge of Alphabet* xii. 74 There's a valley of the notornis, the flightless bird. 1970 *Notornis* XVII. 67 This encounter [with a Takahe] was approximately 12 km from the closed Notornis area.

no-touch (nōᵘtʌ·tʃ), *a.* *Med.* [f. No *a.* + Touch *sb.*] Applied to a method of dressing wounds in an environment that is not aseptic in which no one is allowed to touch either the wound or its dressings.
1944 *Med. Res. Council Special Rep. Ser.* No. 249. 24 'No-touch' technique, and dry instruments are used throughout. 1950 *Lancet* 18 Feb. 294/1 A no-touch technique was used for all wounds; this was not a theatre technique, but comprised the exclusive use of sterile instruments and dressing swabs. 1961 R. Hare *Bacteriol. & Immunity* v. 74 For most wounds, a method of dressing known as the 'no touch' (or 'non-touch') technique is the method of choice.

not-out, *a.* Add: (Earlier and later examples.) Also as *sb.*: (*a*) A batsman who is 'not out'; (*b*) a 'not out' innings. Also *transf.*
1860 *Baily's Monthly Mag.* Sept. 426 We were most pleased with the two 'not-out' innings of 15 and 12 played by Mr. C. D. Marsham. 1869 *Ibid.* July 22 On the Tuesday the two not outs resumed their innings. 1898 *Field* 11 June 873/1 At the outset Troup and Townsend, the not-outs, were very slow. 1900 W. J. Ford *Cricketer on Cricket* 140 Last year..in 56 innings he had an average of over 30, with only one not-out to help him. 1960 *Times* 14 June 16/2 With the aid of a number of not-outs Statham's batting average for England..is more than 30. 1974 *Daily Tel.* 12 June 34/4 Graves added a not-out 55 at a run a minute.

not-quite, *a.* and *adv.* [f. Not *adv.* 10 + Quite *adv.*] **A.** *adj.* Almost, not wholly; *spec.* (*a*) not wholly committed or involved; (*b*) not wholly acceptable or respectable. Also as *sb.*
1920 D. H. Lawrence *Let.* 28 May (1962) I. 632 There is always a kind of half-measure, half-length, 'not quite' feeling about. 1948 L. MacNeice *Holes in Sky* 39 He spelled out True and Good, With their interleaving of half-truths and not-quites. 1955 N. Marsh *Scales of Justice* xii. 236 Kitty, over-painted, knowledgeable, fantastically 'Not-quite'. 1961 M. Beadle *These Ruins are Inhabited* (1963) vii. 96 We awoke on Sunday to not-quite-fog. 1962 *Punch* 15 Aug. 241/3 Marriage by special licence will be Not Quite. 1968 'Han Suyin' *Birdless Summer* 281 The 'half-caste'. The Anglo-Indian, the chi-chi, the not-quite.
B. *adv.* Nearly, almost.
1940 W. Faulkner *Hamlet* i. i. 20 The other [face] pumping up and down with metronome-like regularity to the wheel's not-quite-musical complaint. 1944 Auden *Sea & Mirror* in *For Time Being* iii. 30 The notable absence of the slightest shiver or not-quite-inhibited sneeze. 1975 M. Babson *There must be Some Mistake* xiii. 93 Karen and Jill exchanged glances of not-quite-mock despair.

no trump(s), *phr.* *Bridge.* Also **no-trump, notrump.** [f. No *a.* 1, 6 + Trump *sb.*²] **1.** *ellipt.* A call which provides for the playing of a hand without a trump suit; the play at bridge without trumps; also with preceding number (as *three no trumps*), referring to the number of tricks (above six). Also *fig.*
1899 A. Dunn *Bridge* 19 The dealer should declare 'no trumps' when he holds four aces. 1902 *Encycl. Brit.* XXVI. 370/2 With an established black suit of 5 or 6 cards the dealer should declare no-trump if he has another suit protected. 1904 J. B. Elwell *Adv. Bridge* 236 It is the rule at 'no-trump' to return partner's suit with your highest card. 1910 *Blackw. Mag.* Dec. 809/2 Nine times out of ten it is No Trumps, but sometimes the class element creeps in. 1929 M. C. Work *Compl. Contract Bridge* ii. 9 He may overcall a No Trump with a suit-bid, or overcall a suit-bid with another suit or No Trump. 1933 [see *bid *v.* 3 c]. 1952 Phillips & Reese *Bridge with Mr. Playbetter* iv. 20 Mrs. Rougenoir responded One No Trump. This was a good bid on her part. 1965 *Listener* 20 May 758/2 West has a minimum strong No Trump in terms of high cards. 1969 A. Truscott *Gt. Bridge Scandal* xx. 237 Schapiro opened this hand with a non-vulnerable weak notrump. 1974 *Country Life* 16 May 1239/3 You have to find a way to make Three No Trumps.
2. *attrib.* Made or played without a trump suit; of or pertaining to such a call.
1902 *Encycl. Brit.* XXVI. 369/2 If in a no-trump hand the partners conjointly hold 3 aces, they score 30 for honours. *Ibid.*, Each trick above 6 counts..12 in a no-trump declaration. 1906 W. Dalton *'Saturday' Bridge* 41 This is an undoubted No Trump call for the dealer. 1931 G. F. Hervey *Headlights on Contract Bridge* viii. 69 When the dealer has made an initial No-Trump bid and second hand has passed, the third hand has three courses open to him. 1973 Reese & Dormer *Compl. Bk. Bridge* v. 71 The partner of the notrump bidder..is able to judge the combined assets at once and can often select the final contract with his first call. 1975 *Way to Play* 90 In 'no trump' games, the four aces are honors. *Ibid.*, A no trump call ranks above all suits.

no-tru·mper. [f. prec. + -er¹.] A no-trump call, or a hand on which a no-trump bid is or can be made; a person who makes a no-trump call.
1899 A. Dunn *Bridge* 29 As the dealer's hand is not worth a single trick, a light 'no-trumper' means absolute ruin. 1906 W. Dalton *'Saturday' Bridge* 42 Both of the hands quoted above are sound No Trumpers. 1929 M. C. Work *Compl. Contract Bridge* ii. 12 In deciding whether to advance the No Trump..the no trumper's partner uses the same count.

not-seeing, *vbl. sb.* *Literary.* [f. Not *adv.* and *sb.* 14 a + Seeing *vbl. sb.*] The state of not seeing.
a 1930 D. H. Lawrence *Last Poems* (1932) 250 In not-looking, and in not-seeing Comes a new strength. 1932 W. Faulkner *Light in August* (1933) viii. 177 In the notseeing and the hardknowing as though in a cave he seemed to see a diminishing row of suavely shaped urns. 1963 *Listener* 10 Jan. 71/2 To the not-seeing of the flat surface of Renaissance painting Abstract Art opposes the not-seeing of figures or patterns.

not-self. Add: (Further examples.) So **not-se·lfness**.
1872 *Dublin Rev.* July 145 A philosophy that shall confirm the existence of an independent Not-Self. 1901 J. McTaggart *Stud. Hegelian Cosmol.* ix. 277 But I mean that the characteristic which experience possesses of being not-self—its 'not-selfness', if the barbarism is permissible,—will always remain as an external and alien element. 1927 J. S. Huxley *Relig. without Revelation* viii. 285 The objective outer world and the subjective un-self-organised parts of the mind are usually interwoven in what is felt as 'not-self'. 1949 D. L. Sayers tr. *Dante's Divine Comedy* I. 68 That experience of the Not-self—which, by arousing his adoring love, has become for him the God-bearing image. 1958 *Times Lit. Suppl.* 10 Oct. 581/1 Reality for him belongs exclusively to the Atman, the one impersonal Self or not-Self with which we can, and should, identify ourselves by shedding all that is personal, desirous and therefore fatally separatist in ourselves. 1960 C. Day Lewis *Buried Day* i. 25 His proper concern is with the object to be created—an object which, however much of himself goes into it, must end up as not-self.

not-the·reness. [f. Not *adv.* + There *adv.* + -ness.] The state of being absent or preoccupied.
1934 A. Huxley *Beyond Mexique Bay* 261 There is something profoundly horrifying in this immense, indefinite not-thereness of the Mexican scene. 1950 —— *Themes & Variations* 86 Rushing from one soirée to another in a strange state of alienation and not-thereness.

Nottingham (nǫ·tiŋəm; also *locally* nǫ·tiŋgəm). The name of a city in the Midlands of England, used *attrib.* to denote: **a.** Any of a variety of articles made or found there; something originating there.
1708 in *Bagford Ballads* (1877) I. ii. 389 With Nottingham Ale At every Meal. 1748 M. W. Montagu *Let.* 18 Aug. (1966) II. 410 If you would order me six dozen of Nottingham ale..it would be very acceptable. 1851 J. Pycroft *Cricket Field* vii. 151 Practise each kind of cut..and the Nottingham forward cut, with left leg over. 1866 'Capt. Crawley' *Cricket* 29 If the ball is far-pitched, carry the left foot across the front of the wicket..and hit the ball on to the ground in a direction between point and mid-wicket. This is the 'Forward Cut', or 'Nottingham Drive'. 1869 J. C. Patteson *Let.* 24 Nov. in C. M. Yonge *Life J. C. Patteson* (1874) II. xi. 392 Nottingham drill, good towelling, huckaback, &c., ought to be worthwhile to send out. 1931 *Star* 8 May 8/3 She [sc. the housewife] would be even more annoyed if we were to prove to her that she had been burning Nottingham Top Hard for years to her satisfaction. 1963 *Homes & Gardens* May 81/3 The pink Nottingham brickwork for the outside walls. 1973 *Country Life* 29 Nov. 1798/2 The pale creamy-grey of the Nottingham stone of which the house is built.
b. Nottingham catchfly, a white-flowered, perennial herb, *Silene nutans*, of the family Caryophyllaceæ, distinguished by the soft hair on its leaves and stems and the stickiness of the upper part of the plant.
[1690 J. Ray *Synopsis Methodica Stirpium Britannicarum* 140 *Lychnis sylvestris alba*... Wild white Catchfly. On the walls of Nottingham-Castle, and thereabout; shewn us first by Tho. Willsel.] 1762 W. Hudson *Flora Anglica* 165 Nottingham catchfly. Habitat in pratis montosis. On the walls of Nottingham-castle, and thereabout. 1852 A. Pratt *Wild Flowers* I. 158 No less fragrant are the flowers of one variety of the Nottingham Catchfly. 1960 *Oxf. Bk. Wild Flowers* 72/2 Nottingham Catchfly is a rare plant of dry, stony places and sea cliffs in England and Wales. 1963 W. Blunt *Of Flowers & Village* 106 Fresh seed of the Nottingham catchfly (*Silene nutans*) germinates better in the dark.
c. A type of stoneware produced at Nottingham until the end of the eighteenth century.
1855 E. Acton *Mod. Cookery* (rev. ed.) xii. 243 Fill a brown upright Nottingham jar with alternate layers of mutton..potatoes, and..onions. 1957 Mankowitz & Haggar *Conc. Encycl. Eng. Pott. & Porc.* 167/1 Examples of Nottingham stoneware may be seen in the Victoria and Albert Museum. 1960 H. Hayward *Antique Collecting* 200/2 *Nottingham earthenware*, pottery made in Nottingham from the 13th cent. and its production continued along characteristic lines until about 1800. The manufactories making Nottingham wares are unidentified. 1973 *Times* 25 Aug. 12/4 The Nottingham bear jugs and owl jugs, weirdly modelled, are particularly sought after.
d. A type of machine-made flat lace originally produced at Nottingham (also *N. net*).
1859 Geo. Eliot *Adam Bede* I. ix. 182 Hetty's dreams were all of luxuries: to sit in a carpeted parlour..to have Nottingham lace round the top of her gown. 1881 C. C. Harrison *Woman's Handiwork* iii. 187 Nottingham lace is now sold in excellent block patterns. *Ibid.* 188 Edgings of the common Nottingham antique lace at eight to ten cents a yard. 1919 T. Wright *Romance of Lace Pillow* xv. 213 The low prices at which machine lace could be sold caused great consternation among the Bucks workers. 'Nottingham Net' was followed by 'Urling's Figured Imitations'. 1921 *Daily Colonist* (Victoria, B.C.) 5 Oct. 7/7, 50 Pairs of Nottingham Lace Curtains at $1.95 a Pair. 1939 O. Lancaster *Homes Sweet Homes* 60 Small tight-shut windows, the light from which is further dimmed by a barrage of Nottingham lace. 1963 *Times* 5 June 17/2 The future of Byard Manufacturing, the Nottingham lace and textile manufacturers, was plunged into even greater obscurity yesterday by the announcement of yet another offer for the company. 1967 *Guardian* 3 July 4 Some of us still think of machine-made lace as 'Nottingham' and equate it with soot-soiled curtains in back streets. 1969 *New Yorker* 29 Nov. 55/1 Her denigratory glance travelled from the plush tablecloth to the Nottingham-lace curtains.
e. *Nottingham white* (see quot. 1969).
1883 J. W. Mollett *Illustr. Dict. Art & Archeol.* 227/2 *Nottingham white*, white lead. 1897 *Sears, Roebuck Catal.* 360/3 Colors for Artists... New Blue, Nottingham White, Olive Lake [etc.]. 1934 H. Hiler *Notes on Technique of Painting* ii. 91 White lead..also called flake-white,..Nottingham white, [etc.]. 1969 R. Mayer *Dict. Art Terms & Techniques* 263/2 *Nottingham white*, an obsolete name for white lead, used in England during the 19th century.
f. A style of fishing; also, a type of reel (see quots.).
1900 A. E. T. Watson *Young Sportsman* 22 The principle of the Nottingham men was the travelling tackle, or 'long corking'... The special tackle for this is a long and supple rod, upright rings, a Nottingham reel without check or complications..and quill or cork float. 1939 'G. Orwell' *Coming up for Air* ii. iv. 87, I could give you all the details about gut-substitute and gimp and Limerick hooks..and Nottingham reels. 1963 *Newnes Encycl. Angling* 156/1 *Nottingham reel*, a reel made in wood instead of the usual brass. *Ibid.* 156/2 The tackle used in Nottingham-style fishing is heavier than that generally adopted in the Sheffield style.
g. Nottingham (one club) system *Bridge*, a system originating at Nottingham.
1954 M. Burns (*title*) The Nottingham club system of contract bridge. 1959 *Listener* 8 Jan. 84/2 In this town our convention is the Nottingham One Club. 1959 Reese & Dormer *Bridge Player's Dict.* 154 Of the many one-club systems that were popular in the early days of contract, one of the few that have survived in Britain and America is the Nottingham system. 1959 *Listener* 31 Dec. 1178/2 The Skegness pair were convincing, using the Nottingham One Club system. 1964 *Official Encycl. Bridge* 385/1 *Nottingham Club*, a system popular in the English Midlands.

Nouba, var. *Nuba.

nougatine (nūgatī·n). [f. Nougat + -ine¹.] A form of nougat freq. covered with chocolate.
1894 E. Skuse *Compl. Confectioner* 58 Nougatines. 4 lbs. Ground Almonds. 7 lbs. Castor Sugar. 1897 *Sears, Roebuck Catal.* 7/2 Almond Nougatines. 1916 *Daily Colonist* (Victoria, B.C.) 21 July 14/5 (Advt.), Chocolate Fig Nougatines. c 1938 *Fortnum & Mason Price List* 8/1 Chocolate Nougatine Batons—per lb. 5/-.

'nough (nʌf). An aphetic form of Enough; phr. *'nough said*: used to emphasize and conclude a statement. Cf. *nuff.
1839 *Philadelphia Gaz.* 12 Nov. 2/1 N.S.M.J., 'nough said 'mong gentlemen. 1903 K. D. Wiggin *Rebecca* ii. 24 L.D.M. was talented 'nough to get Reely's money, but M.D.L. would 'a' ben practical 'nough to have kep' it. 1934 *Amer. Speech* IX. 319/2 *'Nough said*. Implies the ending of all discussion by acceptance of proposition or challenge. 1944 C. Himes *Cotton gonna kill me Yet* in *Black on Black* (1973) 197 Ain't got sense 'nough to be mad. 1973 *Amer. Speech* 1970 XLV. 76 Sure 'nough is.

nought, *sb.*, *a.*, and *adv.* Add: **A.** *sb.* **2. d.** A score of no points in a game.
1862 *Lillywhite's Cricket Scores* I. 354 It is certainly curious that Beldham should have made two noughts in this contest. 1876 *Haygarth's Cricket Scores* V. 165 Thirty-three noughts were obtained in the match.
B. *adj.* **4.** *at nought feet*: in *Aeronaut.*, very close to the ground, just above the ground.
1945 *Aeroplane Spotter* 20 Sept. 218/1 Other noteworthy flying displays were given by the Messenger I, showing off its manœuvrability at nought feet. 1949 F. Maclean *Eastern Approaches* iii. vi. 372 A large three-engined German flying boat..scared the life out of our cook by skimming past our house at what is known by the R.A.F. as 'nought feet'. 1960 *Observer* 17 Jan. 9/6 If there were an emergency at 'nought feet' the airmen could not afford the precious seconds needed to jettison the cockpit canopy normally.

noughting, *vbl. sb.* (Later example.)
1926 A. E. Taylor *Plato* ix. 225 The 'noughting' and remaking of the soul is the greatest business of life.

‖ **nouille** (nūiyᵊ). [Fr.] = Noodle *sb.*² Usu. in *pl.* Also (in *sing.*) *attrib.*, as *nouille paste*.
1845 E. Acton *Mod. Cookery* i. 6 (*heading*) To make nouilles;..Wet, with the yolks of four eggs, as much fine, dry, and sifted flour as will make them into a firm but very smooth paste. *Ibid.* 13 Make into nouille paste the yolks of four fresh eggs. 1938 *Times Lit. Suppl.* 9 July p. iii/1 Excursions to Soho to buy the authentic *nouilles* and the right kind of cheeses. 1960 *Good Housek. Cookery Bk.* (rev. ed.) 266/1 (*heading*) Noodles (Nouilles).

noumenal, *a.*, etc. Add: Now also with pronunc. (nū-).

noumenalism (nū·mĕnăliz'm, nau-). [f. Noumenal *a.* + -ism.] = Noumenism.
1902 *Encycl. Brit.* XXX. 679/2 Fechner regarded every composite body as the appearance of a spirit;..This noumenalism would not do for Lewes. 1925 J. E. Turner *Theory Direct Realism* 22 Noumenalism, meaning by this term that the character and existence of the real physical world are essentially different from and independent of the existence and attributes of sensed content.

noumenalist (nū·mĕnălist, nau-). [f. as prec. + -ist.] A believer in noumenalism. Also *attrib.* or as *adj.*
1904 G. S. Fullerton *Syst. Metaphysics* v. 88 The hypostatized abstractions of the noumenalist and the neo-Kantian. 1925 J. E. Turner *Theory Direct Realism* 8 The term 'Direct' is intended to imply further the complete absence of any representative or noumenalist factors in the process and object of perception.

noumenon. Add: Now also with pronunc. (nū·mĕnǫn). (Earlier and later examples.)
1796 F. A. Nitsch *Gen. View Kant's Princ. concerning Man* 118 The conception we have of the world of Noumena, contains no knowledge of that world, but is a mere conception of demarkation [i.e. *Grenzbegriff*, or limiting concept]. 1967 *Listener* 27 July 123/3 It was a revelation, a vision of the noumenon..and I fear that—for quite a long time—we will glory in the sensuous bliss of it all and become uncritical of the human content.

noun. Add: **1. c.** *noun-adjunct*, *-complement*, *-compound*, *-equivalent*, *-modifier*, *phrase*, *-stem*; *noun-like* adj.; *noun–adjective a.*, of or pertaining to the relationship between a noun and an adjective.
1963 *Times Lit. Suppl.* 22 Mar. 193/4 The English noun-adjective relationship..as in 'church'—'ecclesiastical'. 1962 H. A. Gleason in Householder & Saporta *Probl. Lexicogr.* 93 Among the nouns is a considerable subclass including, *United States*,..*Hague*,... These are..always preceded by *the* except in noun-adjunct position. 1964 C. Barber *Ling. Change Present-Day Eng.* v. 121 The tendency of *economy* used as noun-adjunct to develop the meaning of 'large'... The largest packet is frequently called 'economy size'. 1963 F. T. Visser *Hist. Syntax Eng. Lang.* I. iv. 624 The adoption of numerous French verbs which were construed with *à* before a noun-complement. 1966 *English Studies* XLVII. 51 Concerning the noun-complement the author discusses two points. 1914 L. Bloomfield in C. F. Hockett *Leonard Bloomfield Anthol.* (1970) 67 Any one who reads Brugmann's section on noun-compounds..will be impressed by the endless deviations..of composition-stems from independent words. 1965 J. E. Cross in *English Studies* XLVI. 108 Such noun-compounds having *wulf* as the second element are used simply of warriors to express the idea that they are anxious to fight and kill. 1935 *Jrnl. Eng. & Gmc. Philol.* XXXIV. 416 All that the primaries..have in common is their noun character; it would be simpler and clearer to call them nouns and noun-equivalents. 1954 Pei & Gaynor *Dict. Ling.* 149 *Noun-equivalent*, a word (pronoun, participle, adjective) or group of words used in the sense and function of a noun. 1963 F. T. Visser *Hist. Syntax Eng. Lang.* I. iv. 410 The direct object might be defined as the (pro)noun or noun-equivalent not preceded by a preposition. 1935 G. K. Zipf *Psycho-Biol. of Lang.* (1936) v. 31 Not all languages make, say, noun-like and verb-like distinctions. 1958 C. F. Hockett *Course Mod. Ling.* xxvi. 222 A few stems which show no inflection show syntactical behavior so nounlike that we class them as nouns. 1971 D. Crystal *Ling.* 92 We may have isolated a few noun-like words. 1955 Quirk & Wrenn *Old Eng. Gram.* iii. 68 (*heading*) Noun modifiers and pronouns. *Ibid.* iv. 109 An example of a common noun-modifier is *and* —, which has the force of 'opposite' or 'corresponding to'. 1958 W. N. Francis *Struct. Amer. Eng.* vi. 298 Nouns appear very frequently as heads of structures of modification. The modifiers in such structures may belong to any of the four parts of speech... The most common noun-modifier is the adjective. 1964 C. Barber *Ling. Change Present-Day Eng.* vii. 147 In the phrase *an old grey stone wall*,..we have the headword *wall* and as noun-modifiers the adjectives *old* and *grey*. 1951 *Mind* LX. 425 A noun-phrase used predicatively. 1965 N. Chomsky *Aspects of Theory of Syntax* ii. 63 *Frighten the boy* is a Verb Phrase..consisting of the Verb..*frighten* and the Noun Phrase..*the boy*. 1973 *Amer. Speech* 1970 XLV. 133 The grammatical transformation..operates so as to convert the noun phrase that follows the verb into the derived subject. 1935 G. K. Zipf *Psycho-Biol. of Lang.* (1936) iv. 166 The number of different verb-stems and noun-stems which enter into compounds far surpasses the number of available prefixes which may be used in compounds. 1957 R. W. Zandvoort *Handbk. Eng. Gram.* ix. i. 278 Sometimes combinations of genitive + noun and of noun-stem + noun exist side by side: *a large schoolboy hand*. 1973 *Archivum Linguisticum* IV. 37 The premiss that a noun-stem has an inherent gender.
3. (Later example.)
1930 W. Empson *Seven Types Ambiguity* viii. 300 Here we have the English language..given particular meaning by noun-adjectives in apposition.

nounal, *a.* Add: (Later examples.) Also, of an author's style: containing many nouns.
1952 *Scrutiny* XVIII. iv. 306 That same test on Milton generally bears out her observation that Milton is predominantly nounal and adjectival in the sense of described scenery rather than re-enacted experience. 1961 R. B. Long *Sentence & its Parts* xiii. 292 The demonstratives have very considerable nounal use. *This is fun*... *Who's that?*
nounally *adv.* (Later example.)
1961 R. B. Long *Sentence & its Parts* xiii. 292 But demonstratives used nounally of people can be emotional also. *That's a good fellow*.

nou·nism. *rare.* [f. Noun + -ism.] In an author's style: a marked preference for nouns.
1904 G. S. Hall *Adolescence* II. xvi. 467 Adjectivism, adverbism, and nounism, or marked disposition to multiply one or more of the above classes of words.

nou·nize, *v.* *rare.* [f. Noun + -ize.] *trans.* To make into a noun.
1871 J. Earle *Philol. Eng. Tongue* 190, 2 *Henry IV*, iv. i. 71 *there* (nounized).

nou·nness. [f. Noun + -ness.] The quality or nature of a noun.
1971 D. Crystal *Linguistics* 124 We might arrive at an explicit notion of nounness. 1971 R. Fowler in *Archivum Linguisticum* II. 143 A Pronoun..represents features proper to N (gender, noun-ness, [etc.]).

‖ **nounou** (nūnū). [Fr.] A child's name for a nurse; a wet nurse.
1894 G. du Maurier *Trilby* II. vi. 208 In the formal dusty gardens were the same piopious and zouzous still walking with the same nounous. *Ibid.*, Nounou——a wet-nurse with a pretty ribboned cap and long streamers. 1900 *Glasgow Herald* 17 July 4 He was in the charge of a nou-nou (wet nurse). 1930 *Observer* 23 Feb. 12/1 The 'nou-nous' which it was naturally the chief object of the 'piou-pious' of that period to dazzle. 1951 M. Sharp *Lise Lillywhite* i. 9 A sea-sick nounou.

nou·ny, *a.* *rare.* [f. Noun + -y¹.] Having or using many nouns; having the nature of a noun.
1926 Fowler *Mod. Eng. Usage* 654 It is as an unfailing sign of a nouny abstract style that a cluster of *-ion* words is chiefly to be dreaded.
So **nou·niness**, the state or quality of a noun.
1973 *Ann. N.Y. Acad. Sci.* CCXI. 170 Degrees of adjectiveness, degrees of nouniness.

nourishable, *a.* Restrict † *Obs.* to sense 1 and add: **2.** (Later example.)
1876 G. Meredith *Lett.* (1912) I. 269 The dear heart of him so frankly nourishable by flattery that [etc.].

nourishment. Add: **4.** *spec.* The treatment of leather with some substance to keep it soft or pliant.
1897 C. T. Davis *Manuf. Leather* (ed. 2) xlii. 596 For the nourishment of fine glacé leather, yolk of eggs is..used.

nous. 2. (Later examples.)
1927 F. B. Young *Portrait of Clare* 509 'Upon my soul, Clare,' Aunt Cathie declared, 'I thought you had more nous.' 1928 Galsworthy *Swan Song* i. ii. 12 They've got no more nous than a tom-cat. 1930 R. Campbell *Adamastor* 26 Had Creswell, Smuts or Hertzog half his nous, There would be far more goats on the Karroo And far less in the Senate and the House. 1945 R. Hargreaves *Enemy at Gate* 291 Nothing compensated for ignorance or lack of nous in a leader. 1946 [see *common *sb.*¹ 16]. 1956 [see *backwoods b]. 1959 [see *can *sb.*¹ 1 f]. 1972 *Daily Tel.* 8 Dec. 14/6, I do know how easy it would be for anyone with a camera and a little nous to film 'The Breakdown of Life' in Britain. 1973 *Times* 22 Feb. 25/1 If we had had a bit of nous we'd have probably discovered this earlier. 1975 *Daily Tel.* 29 Jan. 17 The City, extraordinary as it may sound, has very limited political nous.

‖ **nous autres** (nūzōtr'). [Fr., lit. 'we others'.] The personal pronoun *we*, somewhat stressed or emphasized.
1860 C. Lever *Day's Ride* in *All Year Round* III. 470/1 We were, in fact, henceforth 'nous autres'. 1870 J. C. Patteson *Let.* 9 Dec. in C. M. Yonge *Life J. C. Patteson* (1874) II. xii. 472 We fail to understand him, *nous autres* I mean, outside the sanctuary. 1911 E. M. Clowes *On Wallaby* v. 140 It gave us a nasty lumpy feeling in our throats—*nous autres* who had nobody to come and see us.

noust (naust, nūst). *Sc. dial.* Also † **newst, noost,** † **nowst.** [ad. Norw. dial. *naust(r)*, *nøst*, f. ON. *naust* boat-shed, dock.] 'The place in which a boat is hauled up, gen. a scooped-out trench at the edge of a beach surrounded by a shallow wall of stones, a boat-stance in gen.' (*Sc. Nat. Dict.*) Also *fig.*
The pronunc. (naust) is used in Orkney, (nūst) in Shetland.
1613 *Court Bk. Orkney & Shetland* (1967) 76 William Ewinsone..found her lying..beneath the boat nowst in Wytfoord. a 1693 T. Brown *Diary* (1898) 63 Ther wes a great boat blowen owt of the Newst at the Air. 1869 J. T. Reid *Art Rambles* 41 Down to the boat-noust the trio hirpled. 1894 L. J. Nicolson *Songs of Thule* 79 My fecht is owre wi' wind an' wave Da Noost is noo da quiet grave. 1922 *Glasgow Herald* 22 July 86 Her bow could be seen in the 'noust'. *Ibid.*, In the afternoon at ebb tide they went down to take the boats out of the 'nousts'. 1931 J. Nicolson *Tales* 55 When a boat was taken from its 'noost', and put into the water, the bow had to be turned 'sun-gaets'. 1956 C. M. Costie *Benjie's Bodle* 7 Jamie Peace sitting at the noust sorting his creels. 1971 G. M. Brown *Fishermen with Ploughs* 69 He coughed his way to the noust And launched the *Belle*.

‖ **nous verrons** (nū vęroṅ). [Fr.] We shall see; it remains to be seen.
1764 H. Walpole *Lett.* (1857) IV. 262 *Nous verrons*—the temptation [to go to Paris] is strong, but [etc.]. 1840 Dickens *Let.* ?9 Sept. (1969) II. 125, I have opened the second volume with Kit; and I saw..on looking out at sea an affecting thing that I can do with him bye and bye. *Nous verrons*. 1861 G. Meredith *Let.* 17 May (1970) I. 81, I think it will be my best book as yet... *Nous verrons*. 1889 E. C. Dowson *Let.* 21 Feb. (1967) 39, I don't feel much like matinizing at the moment *mais nous verrons*. 1951 A. Huxley *Let.* 23 Dec. (1969) 639, I may collaborate on a film about Gandhi..but feel a little hesitant, in view of the precarious condition of my eyes... *Nous verrons*.

nouveau art (nūvo ārt) = *art nouveau.
1911 G. B. Shaw *Lett. to Granville Barker* 11 Jan. (1956) 170 Olivier's new poet-earthquake [*sic*] palace of reinforced concrete is a masterpiece of *nouveau* art. a 1946 J. M. Keynes 'Dr Melchior' in *Two Memoirs* (1949) 35 The raised pattern of the *nouveau art* wall-paper. 1947 E. Taylor *View of Harbour* (1954) xi. 159 A picture of Our Lord carrying a *nouveau*-art lantern.

‖ **nouveau pauvre** (nūvo pōvr'). Pl. **nouveaux pauvres.** Also fem. **nouvelle pauvre** (*rare*). [Fr., after next.] A person who has recently become poor. Also (with hyphen) *attrib.* or as *adj.*, newly impoverished.
1965 *Punch* 27 Oct. 619/1 Maigret, holidaying with his wife in a small Normandy fishing port, is summoned to the home of a morally rotten, nouveau-pauvre family where a maid has died of arsenic poisoning after swallowing her mistress's sleeping draught. 1970 *Time* 3 Aug. 49 One of every four Americans 65 or over lives at or below 'the poverty line'. Some of these 5,000,000 old people were poor to begin with, but most are bewildered and bitter *nouveaux pauvres*, their savings and fixed incomes devoured by spiraling property taxes and other forms of inflation. 1975 *Country Life* 13 Feb. 413/1 Nouveau-pauvre? Could you do with an extra bedroom? 1975 *Harpers & Queen* May 101 Crisis for coiffeurs... Some of the *nouvelles pauvres* are clamouring for Brillo-pad mouse, the new colour and the new texture.

‖ **nouveau riche** (nūvo rīʃ). Pl. **nouveaux riches.** Also fem. **nouvelle riche** (*rare*). [Fr.] One who has recently attained to wealth; usu. with connotation of ostentation or vulgar show. Also *attrib.* or as *adj.*
1813 M. Edgeworth *Let.* 6 Apr. (1971) 15 Larry the footboy and Mrs. Rafferty's dinner are nothing to what has been seen at the dinners of les nouveaux riches at Liverpool and Manchester. 1828 Lytton *Pelham* I. xxiii. 193 You never pass by the white and modern mansion of a *nouveau riche*. 1838 W. H. Prescott *Hist. Reign Ferdinand & Isabella* III. ii. xxvi. 560 This same *nouveau riche* used to serve gold dust, says Herrera, instead of salt, at his entertainments. 1853 C. M. Yonge *Heir of Redclyffe* II. xvii. 278 His manner became so dry and repellent that visitors went away moralizing on the absurdity of *nouveaux riches* taking so much state on them. 1863 'Ouida' *Strathmore* (1865) I. vi. 96 She was a *nouvelle riche*, and brought him money. 1899 E. Wharton *Greater Inclination* 187 She had none of the *nouveau riche* prudery which classes poverty with the nude in art. 1936 F. Clune *Roaming round Darling* vi. 58 The hovels they occupy are made of mud, except a few palaces of their *nouveaux riches*. 1957 A. E. Coppard *It's Me, O Lord!* iii. 37 Modern upstarts such as the piano are as inelegant as *nouveaux riches* in a cathedral. 1957 *New Yorker* 2 Nov. 76/1 The elder members of the two families..are..rather crudely drawn caricatures of the *nouveau riche*. 1959 G. Savage *Antique Collector's Handbk.* 238 The industrializing process which started in the 1750s had gained considerable ground by the end of the century, and the Napoleonic Wars, by creating a *nouveau riche* class, succeeded in debasing taste to an extremely low level. 1966 R. Ellison in A. Chapman *New Black Voices* (1972) 404 Some of the *nouveau riche* oilmen who were members of the club. 1972 T. Aronson *Queen Victoria & Bonapartes* ix. 111 The society of the Second Empire—the courtesans, the financiers, the foreign adventurers, the *nouveaux riches*.

‖ **nouveau roman** (nūvo romaṅ). [Fr., lit. new novel.] 'A type of novel developed chiefly in France in the 1960's by such writers as Alain Robbe-Grillet, Michel Butor, Marguerite Duras, and Claude Mauriac, characterized by lack of moral, social, or psychological comment and by precise descriptions that suggest the mental state of the person experiencing or seeing them.' (C. L. Barnhart et al. *Dict. New Eng.* (1973).)
1961 *Listener* 24 Aug. 289/1 *The Key*..reads very like something young and French: it has the soberness of the *nouveau roman*. 1962 *Times* 13 Dec. 14/3 The *nouveau roman* is already a bit *vieux jeu* in France. 1965 *Harper's Mag.* July 112 The characters are many and the story jumps from one to another, often (in the manner of the *nouveau roman*) with no names other than 'he' or 'the boy' to tell you whose episode it is. 1965 *New Society* 12 Aug. 26/3 The author obviously owes a big debt both to Beckett and to the French exponents of the *nouveau roman*, with their insistent emphasis on the depiction of physical detail and visual objectivity. 1967 *Listener* 23 Mar. 391/3 One of the most successful recent attempts, for instance, is to be found in a novel by one of Robbe-Grillet's fellow-practitioners of the *nouveau roman*, Michel Butor's *Passing Time*, which combines a narrative giving us the hero's present experience and a diary containing his experiences of several months before. 1974 *Times Lit. Suppl.* 25 Jan. 69/2 The sources of Mr Gordon's off-the-peg technique are fairly clear: some Kafka; the Burroughs scissors; but mostly the *nouveau roman*. The novel, so this modish dogma asserts, is a 'vision of things', and the universe no more than the sum of the author's sensations.

‖ **nouvelle** (nūvęl). [Fr.] A short piece of fictitious narrative, freq. one dealing with a single situation or a single aspect of a character or characters.
1680 H. Savile *Let.* 24 Feb. in W. D. Cooper *Savile Corr.* (1858) 140 Disposed to those kind of books you mention of *nouvell's* and other *entretiens* of folly and levity. [1717 M. W. Montagu *Let.* 1 Jan. (1965) I. 293 Would you have me write novelles like the Countess of D'Aunois?] 1887 *Athenæum* 1 Jan. 10/2 M. de Maupassant's 'Petite Roque', a collection of *nouvelles* written with his usual cleverness. 1917 G. Saintsbury *Hist. Fr. Novel* I. iv. 88 The faults of long-windedness, of otiose padding, of unnecessary episodes, etc., are almost mechanically or mathematically impossible in the *nouvelle*. 1935 S. Spender *Destructive Element* 48 Many other of the stories of this period, especially what he [sc. Henry James] called the 'nouvelles', as distinct from the short stories which are more in the nature of the anecdote. 1959 *Times Lit. Suppl.* 27 Mar. 182/1 *Mademoiselle B—* is a *nouvelle* in a recognizable style (it reads as if it had been intelligently translated from the French) and a familiar convention. 1975 *Times* 5 Apr. 5/7 What we learnt from Henry James to call a *nouvelle*, which I take to be a fictional narrative longer than a long short-story and shorter than a short novel.

nouvelles. (Later example.)
1894 E. Dowson *Let.* 30 July (1967) 307 Write and give me your nouvelles.

‖ **nouvelle vague** (nūvęl vag). [Fr., f. *nouvelle* (fem.) new + *vague* wave.] A new movement or trend; *spec.* one in film-making originating in France in the late 1950s; also applied to other arts; also *attrib.*
1959 *Times* 4 Sept. 5/4 It is a film made by one of the old guard rather than by a member of the *nouvelle vague*. 1960 *Times* 1 Nov. 16/4 The film is in no important sense *nouvelle vague*. 1962 *John o' London's* 19 Apr. 371/2 If no clique, trend, *avant garde* or *nouvelle vague* has yet arisen to claim the creative heritage of social change, perhaps it is because the upheaval is still with us. 1962 *Movie* Sept. 34/1 *Paris Nous Appartient* is arguably the most important Nouvelle Vague film to date. 1965 *Philos. Rev.* LXXIV. 53 The *nouvelle vague* of mathematics teachers. 1973 'E. McBain' *Let's hear It* xiv. 209 Teddy normally enjoyed films, except when she was submitted to the excesses of a sadistic *nouvelle vague* camera. 1974 *Times* 26 Apr. 16/3 In terms of chronology the celebrated *nouvelle vague* of the late Fifties must by now be reckoned the old guard.

nova. 2. Add: In mod. use, a star that suddenly increases in brightness by several magnitudes and then, after a period of maximum brightness lasting from a few days to several years, decreases to its former brightness over a much longer period. (Further examples.)
Now distinguished from a *supernova.
1927 H. N. Russell et al. *Astronomy* II. xxii. 777 The brightest stars ever recorded have been novæ. Nova B Cassiopeiæ, known as 'Tycho's star', which appeared in November, 1572, was for some days as bright as Venus at her best (visible in the daytime), and then gradually waned. 1939 *Nature* 15 July 122/2 Some of these extra-galactic novæ have a brightness equal to nearly 100 million suns... In 1934 Baade and Zwicky suggested that these extremely bright objects were not ordinary novæ, but presented a class by themselves—the class of super-novæ. 1955 *Sci. News Let.* 9 July 31/1 Because of a sudden break-down of the normally stable energy conditions in its interior, a nova abruptly flares up to thousands of times its normal brightness, so that a star never before visible from the earth bursts into brilliance. 1956 R. A. Lyttleton *Mod. Universe* v. 167 Every year on the average about a score of stars within our own galaxy undergo this catastrophic development and become what are termed novae. 1968 *Times* 2 Dec. 17/2 Most of the X-ray sources are located in the Milky Way and some of them have been identified with old novae—stars that flared brightly a long time ago and are now fading.

nováčekite (novā·tʃĕkait). *Min.* Also **novacekite.** [f. the name of Radim *Nováček* (see quot. 1951) + -ite¹.] A hydrated oxide and arsenate of magnesium and uranium, $Mg(UO_2)_2(AsO_4)_2.12H_2O$, which is a member of the autunite group and is found as yellow, tetragonal crystals.
1951 C. Frondel in *Amer. Mineralogist* XXXVI. 682 In the course of examination of a small suite of secondary uranium minerals from Schneeberg, Saxony,..two specimens were found of a mineral that has proved to be a new hydrated magnesium uranyl arsenate member of the autunite group. The name novacekite is proposed for this species after the Czech mineralogist Radim Nováček (1905–1942), who made important contributions to the knowledge of the mineralogy of uranium. 1956 *Ibid.* XLI. 152 (*heading*) Novacekite from the Wichita Mountains, Oklahoma. 1965 G. J. Williams *Econ. Geol. N.Z.* xiii. 206/2 The novacekite-like secondary uranium mineral is consistent with the presence of arsenopyrite. 1971 *Mineral. Abstr.* XXII. 226/2 Localities are mentioned for ..nováčekite, especially in NE Brazil.

Novachord (nōᵘ·vākǭɹd). Also **novachord.** [f. L. *nova* new + Chord *sb.*²] Name of an electronic musical instrument of the organ family.
1940 *Chambers's Techn. Dict.* 584/1 *Novachord*, an electronic musical instrument using a single keyboard, sustaining and swell pedals. 1944 W. Apel *Harvard Dict. Mus.* 237/1 The *Novachord*, a six-octave, single-manual instrument which..resembles the spinet in form, employs a purely electronic tone-generating system. 1952 B. Ulanov *Hist. Jazz in Amer.* (1958) xxiii. 324 He made Shearing-like records with rhythm section and Novachord. 1955 L. Feather *Encycl. Jazz* (1956) 111 LPs: Bruns. 58048 (trio & Terry Gibbs);..Dec. 5297 (novachord, w. Hampton). 1970 *Oxf. Compan. Mus.* (ed. 10) 325/1 Akin to oscillating valve organs is the *Novachord*, a domestic six-octave keyboard instrument giving fourteen different tonal qualities.

novákite (novā·kait). *Min.* Also **novakite.** [ad. G. *novákit* (Johan & Hak 1959, in *Chem. der Erde* XX. 49), f. the name of Jiří *Novák*, 20th-cent. Czechoslovakian mineralogist: see -ite¹.] A tetragonal arsenide of copper, Cu_4As_3, found at Černy Důl, Czechoslovakia, which is steel-grey on fresh fracture but becomes almost black on exposure to air.
1959 *Amer. Mineralogist* XLIV. 1321 From the powder pattern, novakite is tetragonal..and..pseudocubic. 1961 *Mineral. Abstr.* XV. 292/2 Novakite occurs (with chalcocite and löllingite) as impregnations in carbonate rocks in Cu–Co–As ores. 1970 *Amer. Mineralogist* LV. 1084 (*caption*) Polished surface of kutinaite..intergrown with novákite..and calcite.

Nova Scotian (nōᵘvă skōᵘ·ʃiăn), *sb.* and *a.* **A.** *sb.* A native or inhabitant of the province of Nova Scotia in Eastern Canada.
1867 B. Murdoch *Hist. Nova Scotia* III. xxxix. 536 Captain Stewart..assured them he was a Nova Scotian in temperament. 1879 *Rep. of Nova Scotia Hist. Soc.* I. 28 As Nova Scotians, the duty of the moment..is to garner up the materials of Provincial History before they perish. 1913 B. Wilson *Nova Scotia* (ed. 2) iii. 29 Cherishing of the memory of their worthy forerunners is perhaps the most marked characteristic of Nova Scotians today. 1968 *Globe & Mail* (Toronto) 13 Feb. 3/2 A Nova Scotian, he is the only declared candidate from the Atlantic provinces. 1971 E. Jones in J. Spencer *Eng. Lang. W. Afr.* 67 The Sierra Leone settlement consisted of the following groups of people: The Black Poor, the Maroons and Nova Scotians, and the West African recaptives.
B. *adj.* Of or pertaining to Nova Scotia.
1866 B. Murdoch *Hist. Nova Scotia* II. p. iii, By almost imperceptible degrees the people acquire habits, sentiments and pursuits suited to the land in which they live.. and thus the Nova Scotian character is gradually developed. 1869 *Bradshaw's Railway Manual* XXI. 390 The Nova Scotian Government secured the passage of an Act to authorize the extension of the trunk line. 1902 W. James *Var. Relig. Exper.* viii. 173 The autobiography of Henry Alline, the Nova Scotian evangelist. 1969 in Halpert & Story *Christmas Mumming in Newfoundland* 24 South coast Newfoundlanders..manned a substantial part of the Nova Scotian..deep-sea fishing fleets.

novel, *sb.* Add: **3. b.** (Earlier example.)
1639 J. S. *Clidamas* pref. sign. A2 *recto*, Here I present you with this little Novel..which though in it selfe it be nothing, yet..may prove something.
5, 5. b. *novel-form*, *-hero*, *-length*, *-reader* (later example), *-reading* (earlier example), *-review*, *-reviewer*, *-reviewing*.
1903 A. Bennett *Truth about Author* xii. 150, I was almost bound to pander to the vulgar taste..in my short stories, but I had sworn solemnly that I would keep the novel-form unsullied. 1967 A. Burgess *Novel Now* ii. 27 In *The Waves*, whose poetic prose now reads very awkwardly, we seem to get as far away from the novel-form as possible. 1838 J. S. Mill in *Westm. Rev.* XXVIII. 439 He ..fulfils with propriety the essential functions of a novel-hero. 1948 F. R. Leavis *Great Tradition* iii. 138 He wrote ..other 'American' classics. Not to speak of short-stories and things of less than novel-length. 1972 E. Routley *Puritan Pleasures of Detective Story* vi. 72 [Austin] Freeman..was not stylist enough to sustain the technique through a story of full novel-length. 1921 P. Lubbock *Craft of Fiction* (1926) iii. 41 There is nothing more familiar to a novel-reader of today than the difficulty of discovering what the novel in his hand is about. 1789 A. Seward *Let.* 17 Aug. (1811) II. 319 The contemptible rage for novel-reading, is a pernicious and deplorably prevalent taste. 1947 'G. Orwell' in *Tribune* 31 Jan. 8/1 Raymond Postgate, who was then editor, had asked me to do the novel reviews from time to time. 1951 *Observer* 8 Apr. 7/4 Novel-reviewers receive letters from readers asking why the books we choose to notice are all..'sad, bad and mad'. 1893 E. Dowson *Let.* c 20 Sept. (1967) 292 The opinion of the average novel-reviewing Le Gallienish animal.

novelette. Add: (Earlier and later examples.) Now freq. applied to a short romantic or sentimental novel of inferior quality.
1814 J. C. Dunlop *Hist. Fiction* II. vii. 127 The endless variety of tales, or *Novelettes*,..which form so popular and so extensive a branch of Italian literature. 1914 G. B. Shaw *Misalliance* 66 'You want to be the hero of a romance and to get into the papers... A son revenges his mother's shame. Villain weltering in his gore...' *The Man* 'Oh, rot! do you think I read novelettes?' 1919 W. S. Maugham *Moon & Sixpence* xix. 161 If I am rhetorical it is because Stroeve was rhetorical. (Do we not know that man in moments of emotion expresses himself naturally in the terms of a novelette?) 1962 *Woman's Own* 3 Nov. 14/3 A ghost-lover..is a mixture of novelettes, and film stars, and pop singers, and a girl's inner yearnings as to what she thinks she wants in a man. 1965 R. Paulson *Novelette before 1900* p. v, Novelette..designates a distinctive, yet certainly vague, genre; a narrative in prose that is longer than a short story and shorter than a novel. .. It lacks the diffuseness as well as the length of a novel, and it covers more ground as well as more space than its modern offspring, the short story. 1967 A. Burgess *Novel Now* i. 16 We're unwilling to dignify books of, say, fifty thousand words and under with the title of novel, preferring to use the Italian term *novella* ('novelette' disparages not only length but content).

novele·ttish, *a.* Also **noveletteish.** [f. Novelette + -ish¹.] Pertaining to or characteristic of a novelette.
1904 J. C. Snaith *Broke of Covenden* xv. 222 Our dear stiff-backed..simple-minded, penny-noveletteish feudal baron. 1921 *Sat. Westm. Gaz.* 10 Sept. 15/2 The blue-eyed make-up of the novelettish débutante. 1940 *Illustr. London News* CXCVII. 98 What an incredible creature is Maxim de Winter himself, with his novelettish name and his novelettish riches and mansion. 1963 *Listener* 31 Jan. 217/1 The abortive seduction scene, with the couple's novelettish dialogue..was unfortunate enough. 1972 *Daily Tel.* 16 Nov. 10/4 The writing is uneven, often degenerating into the novelettish.

novelettist. (Later examples.)
1898 G. B. Shaw *Our Theatres in Nineties* (1932) III. 327 Mr Ogilvie is no mere twaddling novelettist. 1904 Beerbohm *Around Theatres* (1953) 313 The plot of it was just that which even our worst novelettists have outgrown. 1955 *Times* 28 July 3/5 She was the novelettist of the repressions of a vanished social world.

novelize, *v.* Add: **2. b.** *intr.* To write novels. *rare.*
1889 G. B. Shaw *Let.* 31 Aug. (1965) 221 Some time ago I tried novelizing again. 1961 *Times Lit. Suppl.* 27 Jan. 62/2 Dr. Johnson would certainly not have wanted to bring women novelizing into his analogy about dogs standing on their hind legs—even women novelizing at sixteen.

‖ **novella** (nove·lă). [It.] A short narrative (as the stories of Boccaccio's *Decameron*), = Novel *sb.* 3; a short novel or long short-story.
1902 W. D. Howells *Literature & Life* 116 Few modern fictions of the novel's dimensions..have the beauty of form many a novella embodies. 1911 *Encycl. Brit.* XIX. 834/2 After Bandello the decline of the Italian *novella* is evident. 1934 *Times Lit. Suppl.* 6 Dec. 868/1 A brilliant *novella* of life in a remote Welsh valley. 1957 *Ibid.* 30 Aug. 517/3 Why is any story of more than fifty pages now called 'a short novel' or 'novella'—have publishers invented a new literary category? 1958 *Ibid.* 15 Aug. 458/2 The novella is a simpler medieval form brought to perfection in the *Decameron*. Essentially the novella is an anecdote about people you know, or know of. 1959 *John o' London's* 17 Dec. 350/1 H. E. Bates has made the *novella*, which is more generally referred to as the 'long-short story' a form of fiction which is very much his own. 1974 *Times* 23 May 10/8 There are two novellas here, both written entirely in dialogue.

novelty. Add: **1. d.** An often useless, but decorative or amusing, object which relies for its appeal on the newness of its design. Hence **novelty shop.**
1901 *Daily Colonist* (Victoria, B.C.) 19 Oct. 4/6 (Advt.), Sterling Silver Novelties. An English manufacturer's range of samples: no two pieces alike. Puff Boxes, Tooth Brush Boxes, Vases, Cigarette Cases, Match Boxes, Napkin Rings. 1911 *Woman's Home Companion* Apr. 28/3 This idea can be carried out to any extent by having quantities of things which are more or less novelties to sell. 1933 *Planning* I. xvi. 14 The climax to this orgy of designs is reached at Christmas-time when the shops are filled with 'novelties' that no customer would think of wanting for himself. 1972 *Guardian* 23 Dec. 1/1 A cracker-making contract..has been withdrawn because blue jokes were found with the paper hats and novelties. 1973 'I. Drummond' *Jaws of Watchdog* xi. 145 The only retailer who handled the macabre Belgian masks had a novelty shop in High Holborn.
4. *novelty-hunter* (so *-hunting*), *-value*; **novelty number, song** (see quot. 1952).
1926 Fowler *Mod. Eng. Usage* 387/2 If each novelty-hunter struck out a line for himself, we could be content to register novelty-hunting as a useful outward sign of inward dullness, and leave such writers carefully alone. 1938 I. Goldberg *Wonder of Words* xviii. 372 We all know the novelty-hunter who uses his neologisms because he wishes to stun his reader. 1940 *Scrutiny* VIII. 397, I am thinking mostly of course of the hack of Tin Pan Alley and the ubiquitous and cynically named 'novelty' number. 1952 B. Ulanov *Hist. Jazz in Amer.* (1958) 352 *Novelty song*: a song that depends on some obvious contrivance for its appeal, such as a reorganized nursery rhyme..or an infectious sort of gibberish. 1955 L. Feather *Encycl. Jazz* (1956) 79 Novelty songs such as *Ol' Man Mose* and *Brother Bill* began to edge out the jazz material in his repertoire. 1959 *Times* (Suppl. Britain's Food) 9 Mar. p. vii/6 Novelty-value has its own appeal. 1968 'J. Christopher' *Pendulum* v. 42 The novelty value won't last.
b. Of fabrics, etc.: see quot. 1968.
1945 M. D. Potter *Fiber to Fabric* iii. 51 Novelty yarns produce the attractive nubby effects seen in tweeds. 1950 *'Mercury' Dict. Textile Terms* 368/2 *Novelty suitings*, a name applied originally to plain homespun weaves with rough, irregular fillings of different colours, but now referring to all weaves, especially brocaded or jacquard effects. 1968 J. Ironside *Fashion Alphabet* 94 *Novelty*, when used of fabrics this indicates that the material is made from more than one basic fibre and may be in an unusual weave. 1974 *Guardian* 26 Mar. 16/1, 12 leading French cloth manufacturers will be showing..jerseys, shirtings, novelty silks.

Novemberish, *a.* (Later examples.)
1939 *War Illustr.* 9 Dec. p. ii/2 We are approaching the shortest day, and the weather has been thoroughly Novemberish, in London especially. a 1945 E. R. Eddison *Mezentian Gate* (1958) xxxvii. 193 Even in that Novemberish raw weather of her years, some strength of lost youth, some glory,..lived on.

Novial (nōᵘ·viăl). [See quot. 1928¹.] Name of an artificial language created by Otto Jespersen in 1928 for use as an international auxiliary language. Hence **No·vialist**, a speaker of Novial; an advocate of Novial.

1928 O. JESPERSEN *International Language* I. 52 As such a scheme must have a name, I have called it NOVIAL = Nov(new) I.A.L., where I.A.L. stands for International Auxiliary Language. *Ibid.* II. 67 In Novial..we simplify the spelling in all words containing double letters in the national languages, from which the words are taken. *Ibid.* 137 N[ovial] therefore adopts both suffixes, but Novialists need not commit to memory which words take one and which the other ending. 1930 *S.P.E. Tract* XXXIV. 462 Novial will be suspiciously easy; keen Novialists will rattle it off together in the family circle. 1943 F. BODMER *Loom of Language* xi. 471 Novial departs from English usage in one particular. The dictionary form does the work of our past participle in compound past tenses. 1947 H. JACOB *Planned Auxiliary Lang.* iv. 83 Jespersen's Novial has many features which are advocated by the naturalistic school of interlinguists, the use of auxiliary verbs, the revised system of spelling, a certain number of flexional elements in indirect derivation, the abolition of so-called pleonastic endings, the plural form in -*s*. 1973 *Trans. Philol. Soc. 1972* 16 This work extended over many years and led eventually to the invention of his own auxiliary language, Novial... He came to think, later on, that he could have put that time to better use.

novice. Add: **3. d.** Applied to animals entered in a competitive event which have not previously (or before a specified date) won other than very minor prizes; also a competition restricted to such animals.

1903 *Forest & Stream* 21 Feb. 151 (Cent. Suppl.), Novice dogs was a large class, 28 in all. 1909 *Daily Chron.* 19 June 7/6 'Canterbury Belle' headed the list in a fine show of novice hacks. 1962 D. FRANCIS *Dead Cert* iii. 26, I was riding him in novice hurdle races. 1969 [see *CHASE *sb.*[1] 1 f]. 1975 *Country Life* 13 Nov. 1281/2 Brown Lad was a novice chaser last year... He is an absolutely top-class horse, and he stays three miles readily.

‖ **novillada** (novilʸā·da). [Sp.] A bullfight in which three-year-old bulls are fought by novice matadors.

1897 *Encycl. Sport* I. 152/2 Bulls for a *novillada* or second-rate bull-fight cost far less. 1932 E. HEMINGWAY *Death in Afternoon* ii. 23 At a novillada the spectator may see the mistakes of the bullfighters, and the penalties that these mistakes carry. 1955 K. TYNAN *Bull Fever* ii. 14 A matador's career..nowadays rarely exceeds ten years. Manolo González, who fought his first novillada in 1946 and retired rich in 1952, is a fair example. 1964 G. ERIK *Corrida* v. 21 He can become a novice matador..fighting three year old bulls with *picadors*. Such fights are known as '*corridas de novillos-toros*' or '*novilladas*'. 1967 MCCORMICK & MASCAREÑAS *Compl. Aficionado* iii. 78 They are walking back from a novillada.

‖ **novillero** (novilʸē·ro). [Sp.] An apprentice matador who has fought only in *novilladas*.

1921 L. FITZ-BARNARD *Fighting Sports* III. 166 They enter the troupe of some matador and serve an apprenticeship..; and as they improve become *novilleros*. 1932 E. HEMINGWAY *Death in Afternoon* xiii. 152 The novilleros you see killed are all victims of economics. 1955 K. TYNAN *Bull Fever* ii. 27 He took his doctorate at Seville in the spring of 1948, after a riotously successful career as novillero. 1969 *Telegraph* (Brisbane) 13 Jan. 8/1 This year could see him leaving the ranks of the novillero..and moving into the arenas of big bulls and big money. 1970 *Times* 4 May 8/7 He was the leading novillero in Spain in mid-season last year, but three bad gorings put him out of action.

‖ **novillo** (novi·lʸo). [Sp.] A young bull; *spec.* a fighting bull not more than three years old.

1838 *Quarterly Rev.* LXII. 398 A calf, a 'novillo'—which true bull-fighters place on a par with a cow. 1846 R. FORD *Gatherings from Spain* xxi. 292 The villagers..amuse themselves with baiting *novillos*, or bull-youngsters—calves of one year old. 1925 E. HEMINGWAY *The Undefeated* in *This Quarter* II. 205 You can..kill two novillos... Whatever stuff they've got in the corrals. 1967 MCCORMICK & MASCAREÑAS *Compl. Aficionado* ii. 32 The neophyte.. while gaining experience, faces not toros (four-year-olds and up) but novillos (three-year-olds).

‖ **novio** (nǭ·vio). [Sp.] A boy friend, lover.

1920 *Chambers's Jrnl.* June 359/1 She has a novio, a sweetheart. 1929 J. LANGDON-DAVIES *Dancing Catalans* iii. 63 She had a 'novio', a bad young fisherman. 1950 G. BRENAN *Face of Spain* xii. 249 They trip gaily along with their *novio* by their side.

novobiocin (novobəi·ǫsin). *Pharm.* [prob. f. as next + BIO- + *-MY)CIN.] A weakly acidic tetracyclic phenol, $C_{31}H_{36}N_2O_{11}$, that is produced by certain actinomycetes of the genus *Streptomyces* and is used in treating infections with bacteria resistant to less toxic antibiotics; also, the sodium or calcium salt of this phenol, in which forms it is usually administered.

1956 *Antibiotics & Chemotherapy* VI. 195 It has been announced that by mutual agreement of The Upjohn Co. and of Merck & Co., Inc., the generic name of novobiocin would be used for the antibiotic previously called streptonivicin and cathomycin. 1961 *Lancet* 23 Sept. 688/2 Kanamycin..was changed to a combination of erythromycin..with novobiocin 500 mg. six-hourly. 1967 *Martindale's Extra Pharmacopoeia* (ed. 25) 990/1 The usefulness of novobiocin is limited by the frequent occurrence of side-effects. *Ibid.*, Novobiocin sodium is usually administered by mouth. 1972 *Jrnl. Pharmacy & Pharmacol.* XXIV. 972 Novobiocin..is used infrequently because micro-organisms rapidly develop resistance to it.

novocain (nǭᵘ·vǫkēⁱn). *Pharm.* Also **novocaine**, and with capital initial. [f. *novo-*, comb. form of L. *novus* new + CO)CAIN(E.] A proprietary name for procaine.

1905 *Trade Marks Jrnl.* 22 Nov. 1450 Novocain... Chemical substances prepared for use in medicine and pharmacy. Farbwerke vorm. Meister Lucius & Brüning, Hoechst a/Main, Germany; manufacturers. 1910 *Practitioner* Feb. 255 For regional anaesthesia novocain has given good results. 1926 *Glasgow Herald* 25 Aug. 10 Mrs Mary Agnes Brown..died..following injections of cocaine hydrochlorate, which had been mistakenly administered instead of novocaine solution. 1952 MORIN & SMITH tr. *Herzog's Annapurna* xvi. 232 It meant injecting novocaine into the nerve ganglion. 1961 C. MCCULLERS *Clock without Hands* iii. 55 Poke, Doc's brother, just drew the tooth for me—with novocain and antibiotics. 1972 'G. BLACK' *Bitter Tea* (1973) ix. 139, I sat like a man coming from dental extractions under Novocaine, given a time of numb reprieve before the pain flared.

Novocastrian (nǭᵘvokæ·striăn), *sb.* and *a.* Also with lower-case initial. [f. L. *novo-* reduced form of *novum* new + *castr(um* castle + -IAN.] **A.** *sb.* A native or inhabitant of Newcastle upon Tyne. **B.** *adj.* Of or pertaining to Newcastle or its inhabitants.

1888 L. A. SMITH *Music of Waters* 123 They must have been scarcely as sensitive and refined as one would like to imagine the ancestors of the..Novocastrians. 1915 E. CORRI *30 Yrs. Boxing Ref.* 228, I had no friends in Newcastle when I arrived, but made many during that short visit, coming away with a tremendously high opinion of Novocastrian hospitality. 1949 H. L. HONEYMAN *Northumberland* I. vi. 104 Ralph Gardiner, a renegade Novocastrian and pupil of its Grammar School, settled at Chirton near Shields in 1650. 1959 E. L. MASCALL *Pi in High* 26 What coals to Novocastrians are, To Generals what caviare, To Hecubas what divers he's, Were mushrooms to Ozonides. 1969 C. GEESON *Northumberland & Durham Word Bk.* 4 Wilfred Whitten, the distinguished original editor of *John O' London's Weekly*, spoke of his native Novocastrian tongue as 'that abominable dialect'. 1973 I. CARR in B. S. Johnson *All Bull* 107 He quizzed me about distinguished novocastrian citizens. 1974 *Times* 29 Apr. 5/4 Mr T. Dan Smith will be remembered as the man who changed the face of Newcastle upon Tyne... The pace and scale of the change have..frightened Novocastrians.

novolak (nǭᵘ·volæk). Also **novolac.** [f. as prec. + (alteration of) LAC(QUER, LACKER *sb.*] Any of a range of soluble, fusible resins formed by condensing formaldehyde with phenol using an acid catalyst, which are used extensively in varnishes.

1909 L. H. BAEKELAND in *Jrnl. Industr. & Engin. Chem.* Aug. 547/2 In order to simplify matters, I propose to call this substance *Novolak*. 1944 H. R. FLECK *Plastics: Sci. & Technol.* ix. 203 Such properties as the durability and hardness of the Novolac type of resin were obviously very desirable properties to have in a varnish. 1970 E. PARKES et al. in K. Strauss *Appl. Sci. in Casting of Metals* ix. 359 The novolac produced in this way, when cooled down, is a hard glass-like solid and may be ground for use in the manufacture of precoated sand by the hot-coating technique. 1973 *Materials & Technol.* VI. viii. 590 The prime use of moulding powders based on novolak resins is for the production of articles and components in the electrical engineering field.

‖ **novus homo** (nǭᵘ·vŏs hǫ·mo). Pl. **novi homines.** [L., lit. 'new man'.] Orig. used in ancient Rome of the first man in a family to rise to a curule office; hence, a man who has recently risen to a position of importance from insignificance; an upstart.

1589 T. SMITH *Commonw. of Eng.* I. xx. 36 Those which were *noui homines*, were more allowed, for their vertues new and newly shewen, than the old smell of auncient race. 1764 SMOLLETT *Let.* 2 July in *Trav.* (1766) I. xvii. 280 Of these, three or four families are really respectable: the rest are *novi homines*, sprung from Bourgeois, who have saved a little money by their different occupations and raised themselves to the rank of noblesse by purchase. 1824 J. S. MILL in *Westm. Rev.* II. 391 The military leaders, being *novi homines*, were the great opponents of the aristocracy. 1956 A. TOYNBEE *Historian's Approach to Relig.* xvi. 211 All of which had to draw their diplomats from among *novi homines*. 1960 *Times* 17 Sept. 7/2 The competition of radical politicians—the *novi homines* of Africa—has weakened it [*sc.* the tribal system].

novy (nǭᵘ·vi). Also **Novy.** Shortened colloq. form of Nova Scotia, used ellipt.: (*a*) a type of local boat; (*b*) a person from Nova Scotia.

1885 *Bull. U.S. Fish Comm.* V. 8 The boats used in the ordinary fishing [along the Labrador coast] are of two kinds; those called 'novies' or Nova Scotia boats, being long and narrow. 1897 KIPLING *Captains Courageous* vi. 138 'Git aout, you Novy!' To call a Gloucester man a Nova Scotian is not well received. 1942 BERREY & VAN DEN BARK *Amer. Thes. Slang* § 385/5 *Novy*, a Nova Scotian. 1970 J. F. LEAVITT *Wake of Coasters* xvii. 165 In deep water parlance, Nova Scotia and New Brunswick vessels were usually spoken of as 'Bluenoses', but along the coast they were more frequently called 'Novies'.

now, *adv.*, *conj.*, *sb.*[1], and *a.* Add: **I.** *adv.* **1. c.** Phr. *now-it-can-be-told*, used *attrib.* to designate a book, story, etc., which reveals previously classified or unknown facts.

1932 *N.Y. Times Book Rev.* 7 Feb. 15/3 Colonel Reeve's book belongs in the now-it-can-be-told class—and he has a lot to tell. 1948 Q. WRIGHT in J. Towster *Political Power in the U.S.S.R.* p. ix, Correspondents..have produced books of the 'now-it-can-be-told' variety.

d. In colloq. expression of the type —— *now and* —— *later.*

1965 N. FREELING *Criminal Conversation* I. vii. 44 Van der Valk the tally-boy; live now and pay later. 1970 *Guardian* 18 Mar. 19/2 Profit now and pay later. 1973 *P.O. Telephone Directory* (§ 101 London area A–D) 4/1 Telephone credit cards. Talk now—pay later.

IV. 16. Revived in adjectival use: modern, fashionable, up-to-date, 'with it'.

1963 *New Yorker* 8 June 72 A black crepe dress..a now-and-future shaping of pebbly acetate-and-rayon crepe. 1967 *Time* 7 Apr. 20 The more mature of the unmarried in the Now Generation say that, far from promoting promiscuity, the pills impose a sense of responsibility. 1967 *Listener* 2 Nov. 564/2 'Drag', of course, is very now (as the copy-writers are saying). 1968 *Sat. Rev.* (U.S.) 28 Dec. 18 *Bullitt*, I find, is completely typical of the 'now' look in American movies—a swift-moving, constantly shifting surface that suggests rather than reveals depths. 1970 G. GREER *Female Eunuch* 255 Even a poet as *now* as Dylan has two kinds of female character in his imagery. 1972 *Sat. Rev.* (U.S.) 27 May 18/2 Dig the now scene. 1973 E. BULLINS *Theme is Blackness* 167 Everybody in our integrated circle of mod people is with it, man. We're the Now Crowd.

17. a. *now-forgotten.*

1785 A. SEWARD *Let.* 30 Mar. (1811) I. 53 Mr. Warton demonstrates, that the general plan of L'Allegro, Il Penseroso, was suggested to Milton by a now-forgotten work of one Burton. 1951 M. MCLUHAN *Mech. Bride* (1967) 132/2 Mere attacks on salesmanship are confusing..when these now-forgotten assumptions are missed.

nowackiite (novā·ki,əit). *Min.* [f. the name of Werner *Nowacki* (b. 1909), Swiss crystallographer + -ITE[1].] A sulphide of copper, zinc, and arsenic, $Cu_6Zn_3As_4S_{12}$, found as grey or black rhombohedral crystals at Lengenbach, Switzerland.

1965 MARUMO & BURRI in *Chimia* XIX. 501/2 Our mineral, for which we—together with Mr. J. Imhof—suggest the name *nowackiite*, in honour of Professor Werner Nowacki (Bern). 1967 *Zeitschr. für Kristallogr.* CXXIV. 354 The structure determination of nowackiite was undertaken to determine the coordination around the As atoms in the mineral, and to clarify the structural relationship to the zincblende type.

no waiting. Used (with hyphen) *attrib.* and *absol.* to designate an area, place, etc., where vehicles may not be parked.

1959 *Times* 8 Dec. 5/1 In streets where there is unilateral no-waiting, the signs only indicate the side of the road on which waiting is not allowed. 1963 M. LEVINSON *Taxi!* v. 66 A taxi-driver..must wait for..a passenger if he is asked to do so..although this law..is cancelled out by the no-waiting restrictions. 1969 *Morning Star* 18 Nov. 3 Mr Callaghan said the inquiries were about a car causing obstruction in a no-waiting area of Brixton. 1969 *Highway Code* 43 Along the edge of the Carriageway. No waiting.. at times shown on nearby plates or on entry signs to controlled parking zones. 1970 O. NORTON *Dead on Prediction* ii. 22 They had painted yellow no-waiting lines all the way.

now-a-nights. Delete *rare*[−1] and add earlier and later examples.

1841 THOREAU *Let.* 21 July in *Corr.* (1958) 45 Now-a-nights I go on to the hill. 1920 M. BEERBOHM *And Even Now* 120 The Golden Drugget is not outspread nowanights across the high dark coast-road between Rapallo and Zoagli. 1939 JOYCE *Finnegans Wake* 15 And still nowanights and by nights of yore do all bold floras..say only: Cull me ere I wilt to thee!

noway, *adv.* Add: In colloq. use (usu. written as two words): it is impossible, it can't be done.

1968–70 *Current Slang* (Univ. S. Dakota) III–IV. 86 Can we get out of the test? No Way! 1970 J. G. VERMANDEL *Dine with Devil* vii. 36 'No way I can do it any faster than that'... But Peter Angel was shaking his head. 'No way, sorry.' 1973 *Observer* 16 Dec. 40/1 A letter was dispatched to the Foreign Office asking HMG to foot the bill for any damage done—no way. 1973 *Chicago Sun-Times* 29 Dec. 25/1, I say Fred Astaire played a drunk in the made-for-TV movie 'The Over-the-Hill Gang Rides Again'. My brother says, 'No way.' Who is right? 1974 *Nature* 15 Feb. 420/2 Suppose then that an ingenious scribe..decided to use the very compound which wasn't used again for that purpose until about 1920? No way, says McCrone. 1975 *New Yorker* 20 Jan. 29/1 He said he wouldn't start up a gang today—no way.

nowhere, *adv.* Add: **4. a.** (Earlier and later examples of *transf.* use.)

1831 MACAULAY *Essay on Boswell's Life Johnson* in *Edin. Rev.* Sept. 16 Boswell is the first of biographers.. and the rest nowhere. 1928 C. A. NICHOLSON *Hell & Duchess* vi. 108 Don't imagine you have a fortune there. A hundred francs goes nowhere these days.

5. *sb.* **b.** A remote or inaccessible place; freq. in colloq. phr. *the middle of nowhere.*

1908 *Dialect Notes* III. 312 *Forty miles from nowhere*, far from any civilized or settled section. 1951 E. COXHEAD *One Green Bottle* vii. 182 My uncle's farm is on the road to nowhere... They often don't see a new face for months on end. 1960 *Times* 21 Nov. (Canada Suppl.) p. xi/2 Hydro-Quebec is starting to move far up the Manicouagan, in the middle of nowhere. 1963 A. LUBBOCK *Austral. Round-*

about 30, I got going again pretty quickly as I didn't want to be caught by the storm in the middle of nowhere. 1967 MRS. L. B. JOHNSON *White House Diary* 7 July (1970) 544 The country was sparsely populated and it was surprising to come upon such an enormous church, out in nowhere. 1967 E. COUSINS *Death in Quiet Place* i. 10 'I never heard of Boling Green.' 'You wouldn't, old boy. Fag-end of nowhere, down a two-mile lane from the second-class road.'

c. A dull person, place, or thing. Passing into *adj.*: insignificant, unsatisfactory, dull; non-existent. In most contexts *slang*.

1940 L. MACNEICE *Last Ditch* 14 The fiere and there and nowhere birds. 1948 L. SPITZER *Linguistics & Literary History* i. 18 The priestess Bacbuc (whose ambiguous response: '*Trinc!*' is just a nowhere word). 1953 W. BURROUGHS *Junkie* x. 110 The others [*sc.* patients] were a beat, nowhere bunch of people. The type psychiatrists like. 1956 B. HOLIDAY *Lady sings Blues* (1958) viii. 82 A Rolls is built for pleasure... But it's nowhere for highballing a hundred and fifty miles to make a gig. 1959 'F. NEWTON' *Jazz Scene* xii. 220 The hipster classifies..an undesirable state as *nowhere*. 1959 *Esquire* Nov. 70 *Nowhere*, the absolute of nothing. Example: That guy is nowhere. 1966 *Melody Maker* 7 May 5/2 We all thought it was the most nowhere record we'd made. 1970 *Globe & Mail* (Toronto) 26 Sept. 12/2 He wants to spread this physical act as a sign..and then..see work expanding in the nowhere parish to merge with the national scene. 1974 *Ibid.* 14 Sept. 27/4 What you'll remember is the casual dreariness of the nowhere towns and the faded dreams of the guys who never managed to get out of them.

nowhereness. Delete *nonce-word* and add later examples.

1928 D. H. LAWRENCE *Let.* 15 Dec. (1932) 766 A ghastly slummy nowhereness—but France seems all like that. 1929 —— *Pansies* 105 We can but howl the lugubrious howl of idiots, The howl of the utterly lost Howling their nowhereness.

nowheres (nǭᵘ·hwēᵊɹz), *adv.* *U.S. dial.* = NOWHERE *adv.*

1884 'MARK TWAIN' *Huck. Finn* xli. 415, I hain't been nowheres. 1909 *Dialect Notes* III. 353 *Nowheres*, nowhere.

no-win, *a.* [f. NO *a.* + WIN *sb.*[1]] Of a contest or struggle: that cannot be won.

1962 *Economist* 30 June 1310/1 He recommended an agreement with the Russians—a 'no-win' approach, in fact, 'an accommodation with tyranny'. 1966 *Ibid.* 5 Nov. 565/1 Both attack Washington for a 'no-win war policy' in Vietnam. 1973 *Daily Colonist* (Victoria, B.C.) 24 July 5/3 So the trail of broken fixes has put Nixon in a no-win position.

nowness. Delete *rare* and add: (Later examples.) Also, the quality of taking place in the present time.

1891 [see -NESS *suff.* 2 a]. 1926 *Spectator* 22 May 871/1 We enter upon a consideration of 'thenness' and nowness. 1928 A. S. EDDINGTON *Nature of Physical World* iii. 49 From this point of view the 'nowness' of an event is like a shadow cast by it into space. 1962 *Punch* 1 Aug. 153/2 Once the viewer has tasted..This-is-actually-happening, or what we might..describe as Nowness, he's going to waste no time in demanding it. 1973 *Sci. Amer.* Apr. 8/2 Nowness and Permanence. Thanks to the former, it is possible for an art historian to give at least an approximate date for its making.

nowst, var. *NOUST.

nowt (s.v. NOUGHT *sb.*, *a.*, and *adv.* In Dict.). Add later examples.

1913 [see *BLEEDER 3]. 1963 *Times* 11 Mar. 3/6 The verdict was just, anyway, for Yorkshire took their chances like men who seldom get 'owt for nowt', and were better disciplined on a day that demanded it. 1975 *Daily Tel.* 13 Dec. 9/1 There's nowt else possible in this weather.

noxa (nǫ·ksă). *Med.* Pl. **noxæ.** [L., = 'hurt, damage'.] Anything harmful to the body.

1894 GOULD *Dict. Med.* 877/2 *Noxa*, an injurious principle; especially a pathogenic microorganism, or other *materies morbi*. 1960 BLOOD & HENDERSON *Vet. Med.* I. iii. 44 The embryo..does not react to noxa in the same way as the fœtus since no inflammation or leucocytosis develops. 1966 *Ann. N.Y. Acad. Sci.* CXXV. 884 What became obviously and inevitably clear..was the extreme 'communicability' of a purely symbolic 'noxa' in working with patients in the valley of the shadow of death. 1971 *Nature* 2 July 18/2 Aspirin does not lessen pain in skin, elicited by noxious substances, but it does lessen pain in other sites, when elicited by similar noxae.

noxious, *a.* Add: **3. noxious weed,** a weed growing on neglected land, esp., in Australia and New Zealand, one considered harmful to animals, which may be the subject of regulations governing attempts to control it.

1913 F. M. BAILEY *Comprehensive Catal. Queensland Plants* 259 Many of the species [of *Calotis*] are noxious weeds, but it would be impossible to get rid of them by Act of Parliament. 1923 R. MACAULAY *Told by an Idiot* III. 256 As to Ireland, a bill was passed to reduce her docks, thistles, and noxious weeds: no other bill. 1966 G. W. TURNER *Eng. Lang. Austral. & N.Z.* vii. 142 Inspectors call [on New Zealand farmers] to make sure.. that noxious weeds, weeds officially proscribed, are controlled. 1971 *Guardian* 12 Aug. 1/5 She is not opposed to the killing..of clumps of noxious weeds like docks or thistles. 1973 *Stand. Encycl. S. Afr.* VIII. 103/2 A 'wild flower' is defined as a plant indigenous to the Republic of South Africa, except noxious weeds.

noy (noi), *sb.*[2] *Physics.* [repr. the pronunc. of the first part of *noise*.] A unit of perceived noisiness, defined so that the number of noys is proportional to the noisiness of a sound, and one noy is equal to the noisiness of a sound of specified bandwidth and intensity (see quot. 1959).

1959 K. D. KRYTER in *Jrnl. Acoustical Soc. Amer.* XXXI. 1424/1 The following..steps were taken in arriving at the procedure suggested for the calculation of the perceived noise kind of a sound: Step 1. First, the word 'noy' was coined for the units on the scale of perceived noisiness. The numerical value of 1 was assigned to the perceived noisiness of the band from 910–1090 cps of random noise at a sound pressure level of 40 db *re* 0·0002 μ bar. 1963 JERRARD & MCNEILL *Dict. Sci. Units* 96 The noisiness of a jet aircraft taking off is about 110 noys. 1970 R. D. FORD *Introd. Acoustics* vii. 138 The relationship between PN dB and noys is exactly the same as that between phons and sones. 1971 B. J. SMITH *Acoustics* ii. 29 Find the total noisiness N from $N = N_{max} + 0{\cdot}3(\Sigma N - N_{max})$ where N_{max} = highest noy value and ΣN = sum of the noy values in all octave bands.

noyade. Add: (Further examples.) Also *fig.*

1899 E. WHARTON *Greater Inclination* iv. 126 He and she..were bound together in a *noyade* of passion that left them resisting yet clinging as they went down. 1955 AUDEN *Shield of Achilles* iii. 76 Some autumn night of delations and noyades.

noyau. Add: (Earlier example.)

1787 W. DYOTT *Diary* (1907) I. 38 We dined at half-past three and drank pretty freely till eight, when we had coffee, and after noyau, etc.

b. A type of sweetmeat related to nougat.

1913 C. MACKENZIE *Sinister St.* I. I. vii. 103 Richer boys emerged from the tuckshop, sucking gelatines and satin pralines and chocolate creams and raspberry noyau. 1963 —— *My Life & Times* II. 135 A small boy tried to make up his mind whether he would invest a penny in a bar of raspberry, or greengage, or apricot noyau... Noyau was a second-best to nougat.

2. *transf.* A nucleus (of people).

1965 *Listener* 27 May 792/1 They picked out, from the far from 'little' *noyau* of young, hopeful writers who had gathered around them, any Lucien de Rubempré..who came their way. 1966 R. ARDREY *Territorial Imperative* (1967) v. 167, I have taken from the French ethologist Jean-Jacques Petter the term '*noyau*' as a label for the society of inward antagonism... It has seemed wise..to get..away..from all those English words like 'community' or 'society' which inevitably bear connotations of co-operation. *Noyau*—meaning, roughly, a nucleus—is correct in that it implies a primitive evolutionary step towards societies characterized by mutual aid.

nozzer (nǫ·zəɹ). *Naut. slang.* [Orig. uncertain: ? f. colloq. repr. of *No, sir*; but see quot. 1962.] A new recruit, a novice sailor.

1943 'TAFFRAIL' *White Ensigns* ii. 32 To make matters worse the 'nozzers', or novices, usually had their soap and water purloined by those who had been longer on board and knew the ropes. 1962 GRANVILLE *Dict. Sailors' Slang* 82/2 *Nozzer*, new entry boy at the training establishment, HMS *Ganges*... These boys have been called *nozzers*..since the establishment opened, because the petty officer in charge was nicknamed 'Nosey'.

nozzle, *sb.* **2. b.** Add to def.: Also to those in other engines, e.g. turbines (for directing the fluid on to the rotor), internal-combustion engines (for injecting fuel into the carburettor or the combustion chamber), and jet engines (for increasing the speed of the ejected fluid). (Further examples.)

1906 W. H. S. GARNETT *Turbines* I. v. 56 When it is required to reduce the power more rapidly..it is usual to deflect the nozzle so that the jet misses the buckets. 1912 *Motor Manual* (ed. 14) i. 10 When the suction of the engine occurs petrol issues through the nozzle, and an in-rush of air..passes upwards, carrying the sprayed petrol with it, and thus forming the explosive mixture. 1946 [see *AFTER-BURNING *vbl. sb.* 2]. 1949 G. P. SUTTON *Rocket Propulsion Elem.* i. 3 When a rocket unit operates, a chemical reaction occurs which generates high temperature, high pressure gases; these..are ejected through a nozzle. 1950 SKROTZKI & VOPAT *Steam & Gas Turbines* ix. 371 Mechanical spray nozzles for injecting fuel oil into combustion chambers. 1962 D. SLAYTON in *Into Orbit* 22 The job of keeping the capsule in its correct attitude during flight is taken care of by small jet nozzles fastened on the outside of the spacecraft. 1966 *Aviation Week & Space Technol.* 6 Dec. 24/1 A rocket engine with a conventional bell nozzle. 1971 B. SCHARF *Engin. & its Lang.* xiv. 207 In impulse turbines, the high pressure steam ..expands through nozzles or guide blades shaped to form nozzles.

n-p-n (e:n,pī:e·n). *Electronics.* Also **npn,** and in capitals. Designating a semiconductor device in which a *p*-type region is sandwiched between two *n*-type regions.

1951 W. SHOCKLEY et al. in *Physical Rev.* LXXXIII. 151/1 The *p-n-p* transistor has been discussed previously.. In this article we shall consider..the *n-p-n* transistor. 1962 SIMPSON & RICHARDS *Physical Princ. Junction Transistors* iii. 46 In the manufacture of *n-p-n* grown-junction transistors, the *p*-type section is grown for a very short time to give a thin base layer. 1967 *Electronics* 6 Mar. 4/2 Transistors Types: PNP and NPN. 1969 J. J. SPARKES *Transistor Switching* v. 126 Using *npn* transistors, pulsed *R–S* flip-flops using three basic types of pulse steering are available.

‖ **nritta** (n,ri·tă). Also **nrtta.** [Skr. *nṛitta* dance.] A type of 'pure', abstract Indian dance (see quots. 1967).

1917 COOMARASWAMY & DUGGIRALA tr. *Mirror of Gesture* 14 Nṛtta is here dismissed with a merely negative definition, as the object of the Abhinaya Darpaṇa is to explain how to express by gesture definite themes. 1953 F. BOWERS *Dance in India* 24 Sanskrit has no single word for dance in our sense... *Nritta*..refers to dance pure and simple, unadulterated by meaning, interpretation, gesture, or language. 1967 SINGHA & MASSEY *Indian Dances* 24 Nritta is the rhythmic movement of the body in dance... It visualizes and reproduces music and rhythm by means of abstract gestures of the body and hands... Nritta..is concerned solely with rhythmic movement in dancing and is therefore loosely termed 'pure dance'. 1967 CHUJOY & MANCHESTER *Dance Encycl.* 457/2 The repertoire of the Bharata Natya dancer includes nrtta, nrtya, and nautch. Nrtta is an abstract dance of pure lyricism; nrtya is an expressive dance, pantomimic in content; nautch is a combination of song and dance. 1971 *Femina* (Bombay) 30 Apr. 49/1 Kuchipudi is the regional variation of Bharata Natyam. Both share classical Carnatic music and the Nritta (pure dance) and Nritya (expressional dance) aspects.

‖ **nritya** (n,ri·tyă). Also **nrtya.** [Skr. *nṛitya* dance, play.] A type of Indian dance through which ideas or emotions are expressed.

1875 MONIER WILLIAMS *Indian Wisdom* xv. 463 The root *naṭ* and the nouns *nāṭya* and *nāṭaka*, which are now applied to dramatic acting, are probably mere corruptions of *nṛit*, 'to dance', *nṛitya*, 'dancing'. 1917 COOMARASWAMY & DUGGIRALA tr. *Mirror of Gesture* 14 Nṛtya should be seen by a royal audience in the court of kings. 1953 F. BOWERS *Dance in India* 25 *Nritya*..refers to expression, interpretation, and gesticulation with meaning. In other words, dance when it conveys sense and ideas. 1967 SINGHA & MASSEY *Indian Dances* 24 Nritya is that element of the dance which 'suggests *ras* (sentiment) and *bhava* (mood)'. Both..are conveyed through facial expressions and appropriate gestures... The object of both natya and nritya..is to depict ideas, themes, moods and sentiments. 1967, 1971 [see *NRITTA].

Nsima, var. *NZIMA.

nth: see N I. 4 (in Dict. and Suppl.).

n-tuple: see *-TUPLE.

n-type (e·ntəip), *a.* *Physics.* [f. *N* (repr. *negative*) + TYPE *sb.*[1]] Applied to (a region in) a semiconductor in which electrical conduction is due chiefly to the movement of electrons. Opp. *P-TYPE *a.*

1946 J. A. BECKER et al. in *Trans. Amer. Inst. Electr. Engin.* LXV. 714/1 The theoretical and experimental physicists have established that there are two types of electronic semiconductors which can be called *N* and *P* type, depending on whether the carriers are negative electrons or are equivalent to positive 'holes' in the filled energy band. *Ibid.* 715/1 *N*-type oxides, such as ZnO... *P*-type oxides, such as NiO. 1948 TORREY & WHITMAN *Crystal Rectifiers* iii. 49 A semiconductor that conducts principally by electrons in the nearly empty band is said to be an '*n*-type' semiconductor. 1962 SIMPSON & RICHARDS *Physical Princ. Junction Transistors* i. 8 If this type of impurity semiconduction is predominant the material is said to be an *n*-type extrinsic semiconductor, or simply of *n*-type. 1967 *Electronics* 6 Mar. 60/2 Light-sensitive diodes are produced by diffusing islands of boron into a substrate of n-type silicon.

‖ **nu** (nŭ), *int.* [Yiddish, f. Russ. *nu* well, well now.] An exclamation variously used to express interrogation, surprise, emphasis, doubt, etc.

1892 I. ZANGWILL *Childr. Ghetto* I. I. ii. 52 '*Nu*, Pesach, another glass of rum,' said Mr. Belcovitch genially. 1945 A. KOBER *Parm Me* 134 '*Nu*,' she said, 'is lust by you the tunk, you can't give a person a hello when you coming home?' 1967 C. POTOK *Chosen* vii. 139 'Nu,' Reb Saunders said, smiling, 'how should you not know that?' 1968 L. ROSTEN *Joys of Yiddish* 268 *Nu* is so very Yiddish an interjection that it has become the one word which can identify a Jew. 1970 L. M. FEINSILVER *Taste of Yiddish* ii. 155 Growing impatience makes for Nu-nu? and nu-nu-nu? 1971 D. MEIRING *Wall of Glass* xvii. 147 *Nu?* thought Geyra, So what?

nuance (nū,ăns), *v.* [f. the sb.] *trans.* To give nuances to. Hence **nu·ancing** *vbl. sb.*

1897 W. ARCHER *Theatr. World 1896* 94 Nor the elocutionary skill to give variety to a long speech, *nuancing* it, if I may say so, by means of his voice alone. 1959 M. SCHLAUCH *Eng. Lang. in Mod. Times* II. iii. 53 Some tendencies toward such nuancing may be detected in the late medieval distribution of *thou* and *ye* for the singular. 1973 P. A. ALLUM *Politics & Society in Post-War Naples* vi. 191 We can nuance this conclusion and reveal that in some communes the relation of the first to the fourth candidate was 2:1..; in other communes the difference between first and fourth was less than 1,000 votes out of 5,000. 1975 *Gramophone* Oct. 652/1 Take the Wagner piece.., and note the impassioned recitative, nuanced as by the human voice.

‖ **nuancé** (nü,ānse). [Fr.] = *NUANCED *ppl. a.*

1963 *Times Lit. Suppl.* 1 Feb. 71/1 Life..was extremely *nuancé*. *Ibid.* 3 May 318/5 His answer is extremely *nuancé*. 1973 P. A. ALLUM *Politics & Society in Post-War Naples* vi. 181 Macciocchi's judgement of Amendola is more *nuancé*.

nuanced (nū,ānst), *ppl. a.* [f. *NUANCE *v.* + -ED[1].] Possessing or exhibiting delicate gradations in tone, expression, etc.

1920 [see *BOSTONIAN *a.*]. 1965 *Listener* 21 Oct. 613/1 Marxists often distinguish between 'Marxism' and 'vulgar Marxism'... The former is subtle, profound, nuanced,.. and dialectical. 1969 N. FREELING *Tsing-Boum* xi. 71 An immensely nuanced and confused double-think. 1971 *Daily Tel.* 23 Mar. 16/5 He brought a wealth of finely nuanced expression to the eloquently fashioned vocal line. 1972 *Language* XLVIII. 457 In a study devoted to discriminations among adverbs, one would expect a nuanced feeling for them, and to a large extent this is borne out.

nub, *sb.*[1] **3.** Delete *U.S.* and add earlier and later examples.

1834 S. SMITH *Sel. Lett. J. Downing* 205 That's pretty much the *nub* of the business. 1933 WODEHOUSE *Heavy Weather* viii. 166 'The problem..is, How the hell is one to get it away from the blighter?' 'Quite.' 'That is, as you might say, the nub?' 1963 *Guardian* 11 Mar. 13/6 Speaking on the election of vice-presidents, he came to the nub of the unrest among active athletes. 1974 *Times* 22 Oct. 14/6 The nub of the judges' difficulty lay in Lord Widgery's reference to the claim..that the judge had no authority in law to give a direction binding on the press.

Nuba (nū·bă). Also **Nouba.** [See NUBIAN *a.* and *sb.*] The name of a group of peoples of southern Kordofan in the Sudan; a member of these peoples. Also *attrib.* or as *adj.*

1827 J. CONDER *Mod. Traveller: Egypt, Nubia, & Abyssinia* II. 242 The male Noubas, in Egypt as well as in Arabia, are preferred to all others for labour. 1884 *Encycl. Brit.* XVII. 316/2 Certainly none of the chief native races in Soudan..Maba in Wadai, Nuba in the Nile valley..can be considered as of pure Negro descent. 1910 *Ibid.* II. 51/2 The Nuba and Nuer [of the Upper Nile] worship the bull. 1911 *Ibid.* XV. 907/2 The Nubas have their own language, though the inhabitants of each hill have usually a different dialect... In the northern hills [of Kordofan] are communities of black people... They speak Arabic and are called Nuba Arabs. 1911 J. G. FRAZER *Golden Bough: Magic Art* (ed. 3) I. iii. 122 When a Nuba of north-eastern Africa goes to El Obeid for the first time, he tells his wife not to wash or oil herself. 1936 *Discovery* June 170/1 The Nubians (not to be confused with the Nuba of Kordofan). 1949 J. S. TRIMINGHAM *Islam in Sudan* i. 3 Here live the settled 'Arab' population of Kordofān, the Baqqāra cattle-breeding Arabs, the pastoral negroid Shilluk, and the pagan Nūba in the hills. 1962 OLIVER & FAGE *Short Hist. Afr.* iv. 47 The kingdom of the Nuba who are above upper Egypt. 1966 C. SWEENEY *Scurrying Bush* iii. 46 The Nuba of Kordofan..say that it was the baboon that stole the *Kako's* (hyrax's) tail. 1970 *Man, Myth & Magic* v. 138/3 Victorious Nuba wrestlers are awarded branches which are burnt; the wrestler covers himself with the ashes before the next fight. 1972 J. C. FARIS in Cunnison & James *Ess. Sudan Ethnogr.* i. 3 The Southeastern Nuba have suffered locally at the hands of a series of raiders and slavers for a long time.

nubbin. Add: (Earlier and later examples.) Also *transf.* and *fig.*, esp. something small or something that remains when the main part is worn away.

1692 in *Maryland Hist. Mag.* (1918) XIII. 209 Jones saw him buy one beaver skin..for thirty ears and nubbins of corn. 1838 B. DRAKE *Tales & Sketches* 150 A handfull of salt and a few nubbins of corn. 1857 *Harper's Mag.* Feb. 399/2 They served me, at the 'American', with a little hard *nubbin* of steak. 1904 H. F. DAY *Kin O'Ktaadn* 93 She'd squizzle all to nubbins a speech an hour long. 1910 'O. HENRY' *Strictly Business* vi. 76 A red nubbin of corn. 1915 A. C. LAWSON in *Bull. Dept. Geol. Univ. Calif.* IX. III. 38 Broad alluvial embankments..rise by a gentle slope to the summits of the ranges, where there are residual rock crests, or 'nubbins'. 1931 *Times Educ. Suppl.* 11 July 275/4 One may find 'nubbins'—hard, undeveloped, useless things—the result of non-fertilization. 1945 B. MACDONALD *Egg & I* (1946) I. iii. 45 Many of the trees bore nothing or merely two or three wizened nubbins. 1954 G. I. M. SWYER *Reprod. & Sex* App. I. 249 Shortly before puberty..a small nubbin of tissue appears beneath the areola, converting it to a cone. 1954 DYLAN THOMAS *Quite Early One Morning* I. 56 The Telecinema..blobs and nubbins and rubbery squirls receding. 1969 D. BAGLEY *Spoilers* i. 10 The lipstick was worn right down... Another lipstick worn to a nubbin.

nubbly, *a.* Add: Also *fig.*

1958 *Economist* 8 Nov. 487/1 Labour had seemed in danger of becoming obsessed with the nubbly problems of publicity. 1963 *Times Lit. Suppl.* 8 Feb. 96/5 A nubbly and idiosyncratic style.

3. Of cloth or fabric: rough-textured. Also as *sb.*

1935 *Amer. Speech* X. 192/1 A dress of *nubby* crepe may have *nubbly* knots. 1958 J. D. MACDONALD *Executioners* (1959) v. 72 She wore a green nubbly suit. 1964 *McCall's Sewing* i. 8/1 A pebbly crêpe will not noticeably add pounds, but a very rough, nubbly wool tweed may have a decided effect. 1972 *Daily Tel.* 20 Nov. 15 Janice is shopping..for new fabrics..and her mood is for nubblies, for lots of texture. 1973 J. BURROWS *Like an Evening Gone* iii. 39 A jacket of the same nubbly tweed material as the victim's skirt.

nubby, *a.* Add: Also, = *NUBBLY *a.* 3.

1935 [see *NUBBLY *a.* 3]. 1938 *Times* 19 Jan. 17/2 The nubby type jacket to the waistline is worn with vivid coloured woollen frocks. 1967 *Boston Sunday Globe* 23 Apr. 24 (Advt.), An elegant new drapery in a nubby solid and open weave design. 1974 *Union* (S. Carolina) *Daily Times* 19 Apr. 5/3 Also, when washing acrylic sweaters, socks and other wash-and-wear things turn them inside out before washing and they do not get nubby and I think they wear longer.

nubile. Add: **3.** Of women: sexually attractive.

1973 *Times* 8 Mar. 3/1 Some of the slimmest and most nubile girls in London, animadverted on by the Chancellor in his Budget statement, were on parade in London yesterday beneath the chandelier and haze of perfume of Christian Dior's thrice-repeated display of spring fashion. 1973 N. MAILER in *Atlantic Monthly* Aug. 52/2 A woman so sensitive and alive, so nubile as flesh and so evanescent as a wisp of vapour. 1975 *New Society* 17 July 154/2 Most women of my acquaintance..prefer the nubile bodies and erotic accessories of the female nudes. 1975 T. HEALD *Deadline* ii. 22 Waiting by the lift doors was a nubile blonde.

nubk. Substitute for def.: A spiny, evergreen shrub or tree of the genus *Zizyphus*, esp. *Z. spina-Christi*, belonging to the family Rhamnaceæ, native to north Africa and south-western Asia and bearing edible fruit. Cf. *ILB. (Further examples.)

1958 L. DURRELL *Mountolive* xvi. 304 He made his way ..to the great *nubk* tree standing up starkly in its clearing. 1960 —— *Clea* I. iii. 57 The 'nubk' forms the great circular palisade of trees which encircles the Moslem Paradise.

nucellar (niuse·lăɹ), *a.* *Bot.* [f. NUCELL(US + -AR[1].] Of, pertaining to, or involving the nucellus; derived from or produced in the nucellus.

1904 *Jrnl. R. Microsc. Soc.* 665 Two series of vascular bundles run in the ovule, and it is proved that the inner series, frequently described as 'nucellar', belong to the soft inner layer of the integument. 1949 DARLINGTON & MATHER *Elem. Genetics* 406 *Nucellar embryony*, a form of apomixis where the embryo arises directly from the nucellus. 1953 K. ESAU *Plant Anat.* xviii. 555 The nucellar epidermis is sometimes highly resistant and may proliferate into a nucellar cap with relatively thick walls. 1972 *Theoret. & Appl. Genetics* XLII. 314 (*heading*) Differentiation between nucellar and zygotic citrus seedlings by leaf shape.

nuck, var. *KNUCK.

nucleant (niū·kliănt), *a.* and *sb.* [f. NUCLE(US *sb.* + -ANT[1].] **A.** *adj.* Forming a nucleus. *rare*[−1].

1953 S. BECKETT *Watt* 142 In his own vitals, nucleant, he knew them clasped.

B. *sb.* A particle that initiates nucleation; = *NUCLEUS *sb.* 9.

1968 *Jrnl. Appl. Meteorol.* VII. 858/1 The relatively high base temperatures of the convective clouds makes [*sic*] it rather unlikely that their precipitation can be increased by the injection of ice nucleants. 1969 G. F. BOLLING in *Solidification* (Amer. Soc. Metals) (1971) xi. 364 The effective nucleants seem to be those active in the melt, in the sense of having been freshly formed in the melt.

nuclear, *a.* (and *sb.*). Add: **1. c.** Central, cardinal; *spec.* in *Linguistics* and *Phonetics*, being or constituting a linguistic or phonetic nucleus. Hence as *sb.*, a nuclear word, sound, etc. Cf. *NUCLEUS *sb.* 12.

1912 *Housemaster's Lett.* 91 You will forgive me if I tell you what I consider the nuclear fault underlying all this writing. 1940 H. G. WELLS *All Aboard for Ararat* ii. 82, I must look round to find those nuclear men who are needed to carry through the next revolution. 1941 *Language* XVII. 224 In the structure of the syllable vowels are nuclear, consonants marginal. 1949 E. A. NIDA *Morphol.* (ed. 2) 84 A nuclear structure consists of or contains the nucleus... In the word *formal* the nuclear element is *form-* and the peripheral element *-al*. *Ibid.* 118 There are a number of morphological classes represented by the nuclear immediate constituent. For example, there are single-morpheme nuclears, e.g. count, poet.., and multiple-morpheme nuclears, e.g. waiter, hunter. 1952 A. COHEN *Phonemes of English* 39 In general, independent nuclear words are first examined, i.e. words which do not carry morphological characteristics in the shape of phonemes and which do not show 'outer sandhi' (= external open juncture). 1962 A. C. GIMSON *Introd. Pronunc. Eng.* x. 245 In the sense that the nuclear syllable stands out from amongst its neighbours (both accented and unaccented syllables), the nucleus and its situation may be said to have a special contrastive function. 1966 G. N. LEECH *Eng. in Advertising* v. 48 The italicised portion represents a nuclear syllable. 1971 *Language* XLVII. 586 Linguists have long also recognized the existence of stress originating from phrasal structure: Chomsky and Halle call this the Nuclear Stress Rule. 1973 *Archivum Linguisticum* IV. 25 The different varieties of the same nuclear tone.

d. *Psychoanalysis.* Central to the development of the sexual components of the ego; pertaining to or being the emotional nucleus of a neurosis, esp. the Oedipus complex.

[1913 C. G. JUNG in *17th Internat. Congr. Med.* XII. 67 The unconscious existence of manifold phantasies, which have their final root in the infantile past and turn around the so-called 'Kern-complex', or nucleus-complex, which may be qualified in male individuals as the Œdipus-complex and in females as the Electra-complex.] 1916 B. M. HINKLE tr. *Jung's Psychol. of Unconscious* II. iv. 195 Taken at the roots in the case of our patients, the 'nuclear complex' (Freud) reveals itself as the incest problem. 1925 A. & J. STRACHEY tr. *Freud's Case of Obsessional Neurosis* in *Coll. Papers* III. 345 The formation of that complex which deserves to be called the nuclear complex of the neuroses. It is the complex which comprises the child's earliest impulses, alike tender and hostile, towards its parents and brothers and sisters, after its curiosity has been awakened. 1939 E. GLOVER *Psycho-analysis* x. 74 Findings of this order give rise to the tenet that the Oedipus complex is the nuclear complex of the neurosis. 1968 —— *Birth of Ego* i. 11 My views on the theoretical concept of the ego, and in particular on the 'nuclear theory' of its development. *Ibid.* 17 If one were prepared not to stick too slavishly to the idea of a *fixed* nuclear system one could put the psychic situation of the Oedipus complex in clearer perspective.

e. *nuclear family* (Sociology): a term for the basic family unit or group, consisting normally of father, mother, and offspring.

1949 G. P. MURDOCK *Social Structure* i. 1 The first and most basic, called herewith the nuclear family, consists typically of a married man and woman with their offspring. 1963 A. HERON *Towards Quaker View of Sex* 56 This taboo is of social origin, designed to protect the basic unit of society—the 'nuclear' family—from disintegration. 1966 D. JENKINS *Educated Society* iii. 98 The extended, as distinct from the nuclear, family. 1971 *Guardian* 24 Mar. 11/6 If they get rid of the nuclear family there won't be grannies any more than there will be widows. 1973 *Times Lit. Suppl.* 29 June 736/1 This is a loose network of ordinary nuclear families.

2. For 'a nucleus' read: a nucleus or nuclei, esp. atomic nuclei; also, with or by (atomic) nuclei. (Further examples.)

1914 *Engineering* 20 Nov. 607/2 A point raised by Professor Rutherford concerning the effective nuclear charge. 1920 L. DONCASTER *Introd. Study Cytol.* ii. 17 It is from the chromatin that the bodies called chromosomes, which play a great part in nuclear division, are formed. 1929 *Physical Rev.* XXXIV. 1501 The new quantum number is associated with the nucleus intrinsically in the same way that *s* is associated with the electron, and because of its similar properties might be interpreted as a nuclear spin moment. 1933 *Discovery* Apr. 106/2 With this generator, it is hoped to produce currents of the order of 1 milliampere at 5 to 10 million volts, and to insert a large vacuum tube between the sphere and ground for nuclear bombardment. 1934 *Proc. Nat. Acad. Sci.* XX. 470 When certain substances are bombarded with deutons, many and varied nuclear reactions take place. 1935 *Discovery* May 150/1 The interpretation of atomic, molecular, and nuclear radiation. 1936 [see *CYCLOTRON]. 1945 H. D. SMYTH *Gen. Acct. Devel. Atomic Energy Mil. Purposes* vi. 59 The pile was first operated..on December 2, 1942... This was the first time that human beings ever initiated a self-maintaining nuclear chain reaction. 1958 *Times Rev. Industry* Aug. 7/1 An article..on the nuclear bombardment of reactor metals..is evidence of the valuable work now being done. 1961 G. R. CHOPPIN *Exper. Nuclear Chem.* i. 4 Nuclear transformations are on the order of 10^5 to 10^6 more energetic than chemical reactions on an individual molecular or atomic basis. *Ibid.* iii. 27 Nuclear scattering is a more important factor for electrons than it is for heavy particles. 1963 *Oxf. Univ. Gaz.* 9 May 1183/2 The University has established a Professorship of Nuclear Structure in the Department of Nuclear Physics. 1966 C. R. TOTTLE *Sci. Engin. Materials* x. 235 The term nuclear cross-section is used to define the effective area that the target nucleus presents to the neutron as it moves through the lattice, on a statistical or probability basis. 1969 *Times* 20 Feb. 17/5 A single type of nuclear reaction in the sun is thought to bathe each square centimetre of the earth's surface with a flux of 10 million neutrinos every second. 1970 AMBROSE & EASTY *Cell Biol.* v. 161 During cell division in both plant and animal cells the nuclear membrane disappears.

b. *nuclear energy* or *power*: = *atomic energy* (*ATOMIC *a.* and *sb.* A. 2 d). So *nuclear-powered* adj.

In quot. 1930, merely 'energy possessed by a nucleus'. [1926 D. H. LAWRENCE *Plumed Serpent* vii. 130 She was attracted, almost fascinated by the strange *nuclear* power of the men in the circle.] 1930 *Nature* 20 Dec. 953/2 When a nucleus passes from an excited state into a state of lower energy, two different processes may occur: either a γ-quantum is emitted or one of the extra-nuclear electrons [is] thrown out of the atom..; in this case we speak of internal conversion of nuclear energy. 1941 in M. Gowing *Britain & Atomic Energy* (1964) 431 The results..indicate that it should be possible to develop a nuclear energy machine on these lines. 1945 H. D. SMYTH *Gen. Acct. Devel. Atomic Energy Mil. Purposes* xiii. 135 The possible uses of nuclear energy are not all destructive. *Ibid.*, There is no immediate prospect of running cars with nuclear power. 1945 [see sense *3 b]. 1948 *Time* 5 July 44/1 Since atomic fuel would have over two million times as much energy as gasoline, a 'nuclear-powered' plane could fly on and on. 1951 *Jane's Fighting Ships 1951–52* 6 The contract for the first nuclear-powered submarine has been awarded to the Electric Boat Co. 1958 *Daily Express* 11 Mar. 7/1 He was speaking on the hazards which arrive from the peaceful use of nuclear energy. 1962 H. D. BUSH *Atomic & Nucl. Physics* vii. 137 For an isotope to be considered as a source of nuclear power, not only must fission be induced by neutrons of moderate energies but also the cross section must be reasonably high. 1970 *Daily Tel.* 30 June 4/8 A nuclear-powered cardiac pacemaker has been placed in an unidentified female patient in the second such operation in France.

c. Applied to research and fields of study concerned with the atomic nucleus (as *nuclear chemistry, physics*) and to specialists in those fields (as *nuclear physicist*).

1933 *Discovery* Jan. 31/2 Lord Rutherford of Nelson, to whom so much of what is best in modern nuclear physics is due. 1934 *Times* 12 Dec. 19/3 The new field of nuclear chemistry was opening up with great rapidity. 1935 *Discovery* Oct. 291/1 The nuclear chemist, Professor Aston prophesied, 'will transmute and synthesise atoms as his elder brother had done molecules'. 1936 N. FEATHER *Introd. Nucl. Physics* i. 2 The astronomer and modern physicist..discard inconvenient powers of ten for ease in forming mental pictures... But they invite danger

whenever they forget what they have discarded. To such danger the nuclear physicist is particularly liable. 1945 H. D. SMYTH *Gen. Acct. Devel. Atomic Energy Mil. Purposes* 247 The first man-made atomic explosion, the outstanding achievement of nuclear science, was achieved at 5:30 a.m. of that day. 1947 *Sci. News* V. 158 The most recent development of all is the production of new elements and isotopes by bombardment of elements with neutrons or other particles, but this so-called nuclear chemistry..is really a branch of physics. 1948 *Nucleonics* June 2 Nuclear engineering, concerned with the design, construction, and operation of nuclear fission reactors. 1956 A. H. COMPTON *Atomic Quest* 5 The use of the cyclotron was of epochal importance in nuclear physics. *Ibid.* 7 Nuclear research as a subject for wartime study. 1956 *N.Y. Times* 23 July 6/5 The coming years will bring to mankind limitless ways in which this new nuclear science can advance human welfare. 1964 M. GOWING *Britain & Atomic Energy* i. 33 The sobriety of the notes and articles concealed an intense excitement amongst the nuclear scientists. *Ibid.* x. 282 Ideas whose problems were to preoccupy the nuclear engineers for many years to come. 1969 *Times* 25 Jan. 17/5 Gamma rays are a familiar form of radiation to nuclear physicists. 1970 *New Scientist* 29 Oct. 230/1 About five years ago, the University of London decided to set up the first and still the only department in the United Kingdom awarding a first degree in nuclear engineering. 1973 'D. HALLIDAY' *Dolly & Starry Bird* xviii. 278 He had offered to help track down some rather worrying leaks from our brave nuclear physicist boys.

3. Pertaining to or employing nuclear energy. (In senses a and b opp. *CONVENTIONAL *a.* 4 c.)

a. Employing nuclear energy as a source of propulsive power or electricity.

1945 *Engineering Jrnl.* XXVIII. 757/1 A large stationary power installation might be used for heat and motive power in the Arctic or Antarctic regions..where the difficulty of transporting other fuels..outweigh[s] the disadvantages and difficulties of operating and maintaining a nuclear power plant. 1955 *Tribune* 22 Apr. 4/2 Nuclear power stations are designed to be safe. 1956 *Jrnl. Brit. Interplanetary Soc.* XV. 235 The range between the energy concentrations of the typical chemical combustion rockets with moderate exhaust velocities and those of not yet realized true nuclear rockets. 1957 *Newsweek* 12 Aug. 65/2 Nuclear-power plants capable of driving jets, rockets, or even space ships tremendous distances are being developed. 1957 *Jane's Fighting Ships 1957–58* 7 Advances in nuclear propulsion enable submarines to remain submerged indefinitely. *Ibid.* 50 It was officially stated in the 1957–58 Navy estimates that progress is being made with the design of the nuclear submarine 'Dreadnought'. 1959 *Daily Tel.* 24 Feb. 11/8 The day has not yet arrived when nuclear ships can prove commercially much more attractive to ship owners than ships using conventional fuel. 1960 *Aeroplane* XCVIII. 772/2 During its development the first nuclear aircraft will inevitably have alternative chemical propulsion. 1961 *Daily Tel.* 4 Oct. 21/5 The United States plans to start flight tests of nuclear rocket engines by 1967. 1968 *New Scientist* 18 Jan. 147/2 There are still no plans for a British nuclear ship, and now the Japanese are coming into the field. 1968 *Brit. Med. Bull.* XXIV. 260/2 It offers the possibility of biological monitoring of workers such as nuclear-power-station employees who are exposed to unusual environmental hazards. 1974 L. DEIGHTON *Spy Story* xii. 118 He pointed down at the War Table... Ferdy had wiped out nuclear subs.

b. Of a weapon: deriving its destructive power from the rapid, uncontrolled release of nuclear energy.

1945 *Engineering Jrnl.* XXVIII. 752/1 In view of the source of the energy, the current terms 'atomic bomb' and 'atomic power' might well be replaced by the more exact terms 'nuclear bomb' and 'nuclear power'. 1948 *Nuclear Science Abstracts* 30 Sept. 265 Fourth, nuclear weapons have not reached their maximum size in the present type bomb. 1954 *Commonweal* 1 Oct. 621/2 It has been suggested that the nuclear warhead may be small enough to be fitted to an air-to-ground rocket. *Ibid.* 10 Dec. 279/2 The vocabulary of the nuclear age began to grow:.. 'nuclear device' vs. a deliverable weapon; [etc.]. 1955 [see *DETERRENT *sb.*]. 1956 *Time* 25 June 34/3 Hopping off on his inspection of nuclear-weapons testing grounds at Eniwetok and Bikini. 1956 *Newsweek* 3 Sept. 17/2 A nuclear device can mean either hydrogen or atomic. 1957 *Observer* 28 July 6/4 To keep the British nuclear deterrent up to date on its present scale in relation to the Soviet defence will cost more and more each year. 1958 *Times Lit. Suppl.* 15 Aug. p. xxxviii/3 After all, scientists put their skill in making nuclear bombs at the service of those who believe in things; so why cannot artists act correspondingly? 1958 in *Ann. Reg. 1958* (1959) 519 The United States Government shall provide nuclear warheads for the missiles transferred to the United Kingdom Government. 1965 H. KAHN *On Escalation* vi. 101 The U.S. should be willing to adopt the concept that the only purpose of nuclear weapons is to negate nuclear weapons... It should not try to get any 'positive' benefits from its nuclear weaponry. 1973 C. BONINGTON *Next Horizon* xxi. 283 The argument seemed settled, but then Don came on the air with the effect of a small nuclear weapon.

c. Of, pertaining to, possessing, or employing nuclear weapons; *nuclear club*, the nations that possess nuclear weapons; *to go nuclear*, to acquire nuclear weapons.

1954 *Commonweal* 30 Apr. 83 (*heading*) Nuclear war: a false dilemma. 1954 *Newsweek* 8 Nov. 30/1 Talk and thought in government circles about nuclear weapons still seems to be geared largely to the period during which we alone had the capacity to wage nuclear warfare. 1956 *Foreign Policy Bull.* 1 Jan. 59 (*heading*) Nuclear tests: psychological defeat for West. 1957 *Observer* 28 July 6/3 It does not take a very elaborate calculation to realise that our contribution to the strategic Anglo-American nuclear striking force must be very small. 1957 *Christian Science Monitor* 15 Aug. 1 *Nuclear club*, those nations which possess nuclear weapons. The only current members are the United States, Britain, and the Soviet Union. 1958 *Ann. Reg. 1957* 166 The risk of leaving no alternative save nuclear retaliation in the event of war. *Ibid.* 347 The resolution urged that the United Nations and the 'nuclear' Powers should immediately suspend all such tests. 1958 *New Statesman* 22 Feb. 218/1 The response to last Monday's inaugural meetings of the Campaign for Nuclear Disarmament suggests that it is becoming a focus for a real movement of opinion on this issue. 1958 *Economist* 22 Mar. 1006/1 The starting point of most current speeches made by the nuclear disarmers is that a hydrogen-bomb war would be an unspeakably terrible thing. 1958 *Spectator* 15 Aug. 211/2 It seems impossible to convince them that a nuclear war, should it come, will not be like that. 1959 *Daily Tel.* 27 Feb. 10 There was real anxiety lest what we should need for non-nuclear wars had been excessively cut down in favour of what is needed to deal with nuclear threats. *Ibid.* 23 Mar. 18 The rate of descent from the stratosphere of radio-active debris, including strontium 90, from nuclear explosions. 1960 *Spectator* 26 Aug. 303 The American nuclear umbrella is a myth. 1963 *Daily Tel.* 1 Feb. 22/2 The Soviet Union might threaten us with nuclear attack. *Ibid.*, A sort of last-ditch argument now being used by Conservatives under the heading of 'nuclear blackmail'. 1964 *Ann. Reg. 1963* 133 The nuclear test ban treaty..symbolized their mutual determination to secure an abatement of the arms race. *Ibid.* 138 The treaty..did at least symbolize a new and welcome flexibility of outlook on the part of the leading nuclear Powers. 1964 M. GOWING *Britain & Atomic Energy* xiii. 346 Sir John Anderson's forebodings about a nuclear arms race between competing power blocs were not the result of his concern over the specific problem of the French. 1964 J. H. ROTHSCHILD *Tomorrow's Weapons* iii. 24 Nuclear warfare would include atomic (fission) bombs, shells, and mines, and hydrogen (fusion) bombs and missile warheads. 1965 H. KAHN *On Escalation* vi. 94 Alain Enthoven has described the nuclear threshold as follows: 'In efforts to limit violence, there is..a recognizable, qualitative distinction that both combatants can recognize and agree upon if they want to.' *Ibid.* 98 As Soviet nuclear capability has grown, the Soviets have..become less aggressive. *Ibid.* 297 A nuclear stalemate..exists when the balance of central war forces is such that neither side is capable of making a disarming first strike. 1967 *Listener* 11 May 607/2 The decision of certain powers to go nuclear would have the effect of making endemic regional conflicts totally insoluble. 1970 [see *GO *v.* B. 44 a]. 1973 *Times* 22 Feb. 16/4 He was attacking the nuclear disarmers or some such, who had for a time been predicting that we would all be blown up or poisoned by the Bomb. *Ibid.* 9 Nov. 20/1 The nuclear umbrella is now demonstrably in shreds. 1974 *Sci. Amer.* July 46/1 India became the sixth member of the nuclear club on May 18 by carrying out what it called a 'peaceful nuclear explosion experiment using an implosion device'.

d. = *ATOMIC *a.* 2 f.

1954 *Commonweal* 10 Dec. 279/2 With that tragic event the facts of the nuclear age were spread out for all to see. 1960 KOESTLER *Lotus & Robot* 11 The first generation of the Nuclear Age seems to have found a like solace in Zen.

4. Special collocations: **nuclear atom,** (*a*) the concept of the atom as having the charges of one sign surrounding those of the opposite sign, which are regarded as concentrated in a much smaller central volume; (*b*) an elementary constituent of an atomic nucleus (*nonce-use*); **nuclear battery,** an electric battery that utilizes the separation of positive and negative charges accompanying radioactivity; **nuclear emulsion,** a fine-grained photographic emulsion specially designed for recording the tracks of sub-atomic particles in it; **nuclear force,** a force that acts between nucleons; now *spec.* the strong interaction; **nuclear fuel** = *FUEL *sb.* 2* d; so *nuclear-fuelled* adj.; **nuclear isomer** (see *ISOMER 2); **nuclear magnetic resonance,** magnetic resonance (see *RESONANCE) exhibited by atomic nuclei; **nuclear magneton** (see *MAGNETON); **nuclear medicine,** the branch of medicine concerned with the use of radioactive substances in research, diagnosis, and treatment; **nuclear pile** = **nuclear reactor* (see also *PILE *sb.*[3]); **nuclear reactor,** an apparatus or structure in which fissile material can be made to undergo a controlled, self-sustaining nuclear reaction with the consequent release of energy; **nuclear sap** *Biol.* (see *SAP *sb.*[1]). Also *NUCLEAR FISSION, *NUCLEAR FUSION.

1922 A. D. UDDEN tr. *Bohr's Theory of Spectra* III. i. 61 The conception of atomic structure which will form the basis of all the following remarks is the so-called nuclear atom according to which an atom is assumed to consist of a nucleus surrounded by a number of electrons whose distances from one another and from the nucleus are very large compared to the dimensions of the particles themselves. 1936 *Discovery* Jan. 31/1 The quantum mechanical theory of the nuclear atom. 1938 Nuclear atom [see *ALPHA 3 e]. 1961 POWELL & CRASEMANN *Quantum Mech.* i. 14 A nuclear atom had already been considered mathematically by Nagaoka in 1904, but it was Rutherford's analysis that established this concept as an experimental fact. 1970 G. K. WOODGATE *Elem. Atomic Struct.* i. 1 Bohr's theory of hydrogen, based on Rutherford's nuclear atom and incorporating the ideas of Planck, was the famous starting point for atomic structure. 1955 *Jrnl. Brit. Interplanetary Soc.* XIV. 85 The nuclear battery, thermo-couples, photo-electric effects, etc., present further possibilities. 1962 SIMPSON & RICHARDS *Physical Princ. Junction Transistors* iv. 74 'Nuclear batteries' in which the radiation consists of fast electrons from a β-emitter have also been built. 1949 H. YAGODA *Radioactive Measurements with Nucl. Emulsions* i. 6 These emulsions, intended for the registration of alpha-particle, proton, and fission-fragment tracks, are conveniently referred to as Nuclear Emulsions. 1958 K. M. HORNSBY tr. *Glafkides' Photogr. Chem.* I. xxiii. 409 Nuclear emulsions..have a very high concentration of silver bromide (80%), fine grain (0·1 to 0·5 μ), and are coated in very thick layers. 1966 *McGraw-Hill Encycl. Sci. & Technol.* IX. 584/2 Nuclear emulsion plates are important particle detectors for research in high-energy nuclear physics. 1935 *Sci. Abstr.* A. XXXIII. 2 (*heading*) Constitution of elementary particles and nuclear forces. 1955 FRIEDMAN & WEISSKOPF in W. Pauli *Niels Bohr* 144 Our present difficulties in understanding the saturation of nuclear forces on the basis of the free nucleon–nucleon interaction speak in favour of some change in the internucleon potential when the nucleons are closely packed. 1972 *Sci. Amer.* Oct. 100/2 Although the nuclear force acts between all nucleons, whether they are protons or neutrons, it must overcome the disruptive influence of the electrical repulsion between the positive charges of the protons in the nucleus. 1946 *Scientific & Techn. Aspects of Control of Atomic Energy* (U.N. Dept. Public Information) i. 8 The nuclear fuel provided by nature is uranium-235. 1948 Nuclear fuel [see *FUEL *sb.* 2* d]. 1970 *Daily Tel.* 4 Nov. 5 Russia is prepared to supply West Germany with enriched uranium as nuclear fuel for peaceful purposes. 1968 *Times* 18 Oct. 16/7 The element [*sc.* gadolinium] can be used in the design of nuclear-fuelled batteries. 1942 *Physica* IX. 591 (*heading*) Negative result of an attempt to observe nuclear magnetic resonance in solids. 1970 G. K. WOODGATE *Elem. Atomic Struct.* ix. 174 Some of the most accurate values of nuclear moments have been obtained directly by nuclear magnetic resonance. 1952 (*periodical title*) The American journal of roentgenology, radium therapy and nuclear medicine. 1975 *Univ. of London Bull.* Feb. 5/2 This University stands alone in the United Kingdom in having supported since 1961, in one of its medical schools, an academic department devoted solely to nuclear medicine. 1955 *Gloss. Terms Radiology* (*B.S.I.*) 17 Nuclear pile. 1962 F. I. ORDWAY et al. *Basic Astronautics* xii. 497 The effects in general would be the same as those experienced by victims of an atomic bomb blast or an accident in a nuclear pile. 1964 M. GOWING *Britain & Atomic Energy* x. 284 Now the doctors..also appreciated the great usefulness of nuclear piles for medical and biological purposes. 1945 *Sci. Amer.* Nov. 285/3 The radiation emitted by present nuclear reactors requires..heavy shielding. 1948 *Electronic Engin.* XX. 148 An atomic pile, or nuclear reactor, consists usually of an assembly of uranium metal in a pile of graphite. 1955 *Times* 16 July 6/4 The first atomic stations of the Central Electricity Authority will have two nuclear reactors each, together providing a net output of electricity of 100 to 200 megawatts. 1966 J. BETJEMAN *High & Low* 31 No nuclear reactors Bulge hideous on the downs.

B. *ellipt.* as *sb.* **a.** A nuclear weapon. Also *Comb.*, as *nuclear-armed, -free, -tipped* adjs.

1957 *Time* 15 Apr. 29/1 The British decision to convert almost completely to nuclear-armed missiles had deep meaning for all of the world. 1958 *New Statesman* 5 Apr. 423/1 The SPD are demanding in effect no nuclear weapons for the Bundeswehr, none for foreign troops stationed on German territory, and a nuclear-free zone in central Europe. 1959 *Economist* 14 Feb. 615/2 A nuclear-tipped anti-aircraft missile. 1962 *Listener* 29 Mar. 541/2 What difference does the presence of nuclear weapons make to the strategy and tactics of conventional war?.. What is it that prevents either side from going to nuclears on the battlefield? 1964 *Ann. Reg. 1963* 73 The A.L.P. supported moves to establish a 'nuclear-free zone'. 1964 *Economist* 7 Mar. 892/1 Armed with nuclear-tipped missiles. 1972 *Daily Tel.* 2 May 16 The current Nato strategy is one of flexible response, implying that both sides will seek to put off firing nuclears until the last possible moment. 1975 *Nature* 27 Mar. 281/2 It is widely believed that the CIA failed to retrieve the two chief items it wanted—the nuclear-tipped missiles and code machines.

b. A nuclear-powered submarine.

1969 *New Scientist* 28 Aug. 420/2 One can imagine an enemy submarine lying in wait..to pick up the trail of one of our patrolling 'nuclears'. 1974 'M. HEBDEN' *Pride of Dolphins* I. v. 50 'Would you say Nanjizel was a good submariner?' 'He'd done his stint in nuclears.'

nuclear fission. [f. NUCLEAR *a.* + FISSION.] **1.** *Biol.* The division of a cell nucleus (cf. FISSION 2).

1889 *Jrnl. R. Microsc. Soc.* 728 (*heading*) Phenomena of indirect nuclear fission in inverting epithelia. 1960 L. PICKEN *Organization of Cells* vii. 295 The formation of a secondary [myotube] is prepared by nuclear fission, leading to the formation of a second linear series of nuclei.

2. *Nuclear Sci.* = *FISSION 4.

1939 *Physical Rev.* LV. 418/2 As pointed out by Meitner and Frisch, the recent discovery..of the appearance of a radioactive barium isotope as the product of such transmutations offers evidence of a new type of nuclear reaction in which the nucleus divides into two nuclei of smaller charges and masses with release of an energy of more than a hundred million electron volts. The direct proof of the occurrence of this so-called nuclear fission was given by Frisch. 1945 H. D. SMYTH *Gen. Acct. Devel. Atomic Energy Mil. Purposes* i. 1 Such a conversion is observed in the phenomenon of nuclear fission of uranium. 1958 J. BETJEMAN *Coll. Poems* 230 Where Hodge sits down beside his wife And talks of Marx and nuclear fission With all a rustic's intuition. 1958 *Listener* 19 June 1005/1 Reactors operating by nuclear fission. 1971 *Sci. Amer.* June 24/3 Nuclear-fusion fuels such as deuterium and tritium liberate more energy per nucleon (neutron or proton) than nuclear-fission fuels. Unfortunately these hydrogen isotopes only burn efficiently at high temperature.

nuclear fusion. [f. NUCLEAR *a.* + FUSION.] **1.** *Biol.* The fusion of cell nuclei.

1900 *Jrnl. R. Microsc. Soc.* 73 The generation in which the nuclear fusion occurs is probably in all cases degenerate. **1970** *Protoplasma* LXX. 292 Once the outer cell surfaces have joined and two nuclei are in a common cytoplasm, nuclear fusion nearly always seems to follow.

2. *Nuclear Sci.* = *FUSION 3 d.

1952 [see *FUSION 3 d]. **1959** *Ann. Reg. 1958* 137 The Secretary-General expressed the hope that progress would be made on..the potentiality of nuclear fusion to solve the world's energy problems. **1969** *New Scientist* 25 Sept. 640/1 The *Tokamak* machines..have formed the basis of the main Soviet research effort into achieving controlled nuclear fusion during the past few years. **1971** [see *NUCLEAR FISSION 2].

nuclearist (niū·kliărist). [f. *NUCLEAR *a.* 3 + -IST.] One who advocates the possession or use of nuclear weapons; a nation that possesses nuclear weapons. (See also quot. 1952.) So **nu·clearism** (see quot. 1970[1]).

1952 *Time* 26 May 73 In Venice last week, 13 Italian painters who call themselves 'spatialists' and 'nuclearists' gave their answer with an exhibit 'inspired by the atomic bomb'. The canvases were almost as explosive as the bomb itself: furious fireballs of bright colors and bold contrasts. **1962** *Times Lit. Suppl.* 7 Dec. 948/3 In the great debate on nuclear disarmament..the advantages of sober logic and clear thought have usually lain with the anti-unilateralists, sometimes called the nuclearists. **1964** *Listener* 6 Feb. 220/2 The inclusion of a number of uncommitted countries..has..made of the Commonwealth a useful bridge,..a witness that even neutralists and a nuclearist can work closely together. **1969** *Bull. Atomic Sci.* June 39/2 The weapons..become grotesque technological deities for a debased religion of nuclearism. **1970** *Atlantic Monthly* Oct. 106 The most extreme state of contemporary deformation is..'nuclearism'. By this term I mean to suggest the passionate embrace of nuclear weapons as a solution to our anxieties..and as a means of restoring a lost sense of immortality. *Ibid.*, This deity is seen as an all-powerful force,..and the nuclear believer, or nuclearist, allies himself to that force.

nuclearize (niū·kliăraiz), *v.* [f. *NUCLEAR *a.* 3 + -IZE.] **1.** *trans.* To supply or equip (a nation) with nuclear weapons. So **nu:clearization.**

1960 *New Left Rev.* Nov.–Dec. 4/1 A generalised discontent with the prospects of a nuclearised Germany. *Ibid.* 8/1 The NEC is against the nuclearisation of West Germany. **1967** *Times* 20 Jan. 11/7 The difficulty of predicting the resolution of crises in advance, from the blockade of Berlin to the nuclearization of Cuba.

2. To render (a family, etc.) nuclear in character (see *NUCLEAR *a.* 1 e). *rare.*

1972 P. LASLETT *Household & Family in Past Time* 55 The extent to which the domestic group has been 'nuclearised', that is, has lost extension and multiplicity.

nuclearly (niū·kliăɹli), *adv.* [f. NUCLEAR *a.* + -LY[2].] In a nuclear manner (in quot. 1959, with nuclear weapons).

1959 *Listener* 17 Dec. 1062/1 The conventional forces, whether or not nuclearly armed, should be..reduced in the Soviet Union. **1966** J. E. BUSE in C. E. Bazell *In Memory of J. R. Firth* 56 Statives..also occur nuclearly in the verbal phrase.

nuclease (niū·kliēz, -s). *Biochem.* [a. G. *nuclease*, f. *nucl-* (in *nucleoproteïd* nucleoprotein): see *-ASE.

The Ger. word was coined by Emmerich and Löw 1899, in *Zeitschr. für Hygiene* XXXI. 10, but in a different sense (see quot. 1902); the present meaning originated with L. Iwanoff 1903, in *Zeitschr. für physiol. Chem.* XXXIX. 43.]

Any enzyme that catalyses the hydrolysis of polynucleotides or oligonucleotides into smaller units.

1902 VAUGHAN & NOVY *Cellular Toxins* (ed. 4) ix. 175 It is proposed that bacteriolytic enzymes be given the general name of nucleases, because they digest the nucleoproteids of the bacterial cells. *Ibid.*, In their second paper Emmerich and Löw detail the methods which they have employed in the preparation of their nucleases and immuneproteids. **1903** *Jrnl. Chem. Soc.* LXXXIV. II. 679 Various fungi..decompose nucleic acid with the liberation of phosphoric acid and purine bases; this appears to be a ferment action... The name *nuclease* is suggested for the enzyme responsible for the action. **1911** *Jrnl. Biol. Chem.* IX. 129 The term nuclease is usually understood to designate a ferment..through whose agency nucleic acid is decomposed with the liberation of purine bases. **1949** H. W. FLOREY et al. *Antibiotics* I. i. 20 The active material was supposed to be an enzyme—it was even stated to be a nuclease. **1954** A. WHITE et al. *Princ. Biochem.* xx. 582 The intestinal mucosa is also believed to form nucleases which aid in the digestion of the high molecular weight nucleic acids and polynucleotides. **1970** R. W. MCGILVERY *Biochem.* xx. 481 The effect of the lysosomal nucleases is to cleave the molecule [of nucleic acid] into smaller polynucleotide fragments, with the liberation of only a few free nucleotides.

nucleate, *a.* Add: (Examples.)

1962 R. P. LEVINE *Genetics* ix. 115 Gene recombination among the sexually reproducing nucleate organisms can occur as a result of both independent assortment and crossing over. **1972** *Nature* 28 Jan. 211/2 Cultures..of nucleate and enucleate sea urchin egg halves.

b. Applied to a kind of boiling process in which streams of bubbles rise from specific sites on a hot surface in the liquid and are recondensed in the surrounding liquid.

1938 *Trans. Amer. Inst. Chem. Engin.* XXXIII. 449 'Nucleate boiling' is what is ordinarily seen when a pan of water boils upon a stove. **1948** *Trans. Amer. Soc. Mech. Engin.* LXX. 372/1 The boiling was nucleate in the sense that the bubbles originated at favored spots on the metal surface. **1975** *Nature* 27 Mar. 322/2 This [*sc.* pulsation boiling] then progressively changes to nucleate boiling from the front of the sphere to the back.

nucleate, *v.* Add: **1. b.** To form nuclei in; to act as or provide a nucleus for.

1952 *Industr. & Engin. Chem.* June 1273/2 Such fluctuations [in local density] are occurring continuously but it is only under very special, almost critical, conditions that they become of sufficient magnitude to nucleate the phase for a transition to a more stable state. **1961** J. W. MULLIN *Crystallization* v. 109 Ethyl acetanilide can nucleate methyl acetanilide. **1969** D. K. ALLEN *Metall.* vii. 179 (*caption*) Bainite is nucleated by a ferrite crystal. **1972** *Physics Bull.* Nov. 656/1 They do predict static fatigue (since for example the probability of nucleating a crack of critical size increases with time). **1973** *Nature* 23 Nov. 212/2 Freezing nuclei are defined as particles capable of nucleating ice in supercooled water.

2. Add after 'nucleus' in both parts of def.: or nuclei. (Further examples.)

1959 *Engineering* 30 Jan. 152/2 Many of the domains appeared to 'nucleate' at crystalline imperfections. **1961** J. W. MULLIN *Crystallization* v. 104 A saturated solution cannot nucleate spontaneously. **1969** P. G. SHEWMON *Transform. in Metals* vi. 210 If ferrite containing virtually no carbon..is to nucleate in a given region of austenite containing much more carbon, several changes must occur. **1974** *Sci. Amer.* Dec. 94/2 Inclusions can nucleate, multiply and grow dendritically just as the primary metallic phase does.

nucleating *ppl. a.* (further examples); also as *vbl. sb.*

1948 *Jrnl. Colloid Sci.* III. 569 The rate of nucleation in the water drops on the metal plate is much larger than that in the water drops in a cloud, probably because of the nucleating effect of the surface of the plate and chance impurities. **1961** *New Scientist* 28 Sept. 813/1 Nucleation is associated with the similarity in spacing between the lattices of the ice crystal and of the nucleating substance. **1969** G. F. BOLLING in *Solidification* (Amer. Soc. Metals) (1971) xi. 364 (*heading*) Inoculating, nucleating and alloying. **1974** F. D. RICHARDSON *Physical Chem. of Melts in Metallurgy* II. xii. 459 Small unfilled crevices in containers can act as nucleating sites for the evolution of gases from metals at very small supersaturation pressures.

nucleate (niū·kliĕit), *sb.* *Biochem.* [f. NUCLE(IC *a.* + -ATE[4].] Any salt of a nucleic acid.

1907 *Jrnl. Chem. Soc.* XCII. I. 266 In the estimation of purine bases, much smaller yields of guanine are obtained when the copper nucleate is hydrolysed instead of the free acid. **1952** *Jrnl. Biol. Chem.* CXCVIII. 85 They attributed this discrepancy to the formation of ion pairs between sodium ions and the phosphate residues of the nucleate.

nucleated, *a.* **2.** Add to def.: esp. of buildings in villages. (Further examples.)

1942 *Rep. Comm. Land Utilisation in Rural Areas* 11 in *Parl. Papers 1941–2* (Cmd. 6378) IV. 421 In other parts, especially where the community was organised under the feudal or manorial systems, nucleated settlements or villages are the rule. **1954** M. BERESFORD *Lost Villages* vii. 233 One such district is the plain of Lancashire. Here, at first blush, we might seem to have nothing but a countryside of compact, nucleated villages. **1970** N. CHADWICK *Celts* v. 125 Houses..were grouped together in small nucleated settlements. *Ibid.*, Most settlement sites, whether nucleated or individual, were enclosed.

nucleation. Delete *rare*[-1] and add to def.: esp. by the aggregation of molecules into a new phase within a medium. **b.** The formation of something, esp. a crystal, droplet, or bubble, on or into a nucleus.

1902 *Science* 31 Jan. 177/1 After nucleation the first dense fogs were vaguely annular during the first.. exhaustions. **1906** S. S. LAURIE *Synthetica* I. iii. 39 A nucleation or involution or articulation or specific determination of Universal Being whereby an independent centre of actuality is constituted *in rerum natura*. **1906** H. T. BARNES *Ice Formation* iv. 106 An appropriate term, nucleation, might be applied to the formation of ice-crystals throughout the volume of the water, on nuclei supplied by fine particles of sand. **1933** H. G. WELLS *Shape of Things to Come* II. §3. 147 What we now call social nucleation was failing; the grouping of human beings in families and working communities was not going on. *Ibid.* 158 This chaotic nucleation of human beings about gangs and organizations for frankly criminal purposes. **1941** DOAN & MAHLA *Princ. Physical Metallurgy* (ed. 2) vi. 224 In the transformation of a solid solution, such as austenite, the rate of nucleation is limited further by the rates of diffusion of the atoms forming the precipitate. **1950** *Engineering* 13 Jan. 36/2 If the gas concentration is fairly low and the nucleation of bubbles difficult. **1966** C. R. TOTTLE *Sci. Engin. Materials* vii. 167 In b.c.c. metals, some of which do show ductile failure, the nucleation of cavities occurs at piled-up dislocation sites. **1967** A. H. COTTRELL *Introd. Metallurgy* xii. 153 Many metallurgical processes occur by nucleation and growth, e.g. the formation of CO bubbles in a steel-making bath. *Ibid.* 154 When the probability of forming a nucleus is the same everywhere, the system is in a state suitable for homogeneous nucleation.

nucleator (niū·kli,ĕitəɹ). [f. NUCLEAT(E *v.* + -OR.] A substance which provides nuclei.

1903 *Nature* 3 Dec. 103/1 Phosphorus as a nucleator suddenly bursts forth into maximum activity at about 13°. **1961** *New Scientist* 28 Sept. 813/1 Steroids can act as effective ice nucleators. **1965** *Ibid.* 4 Nov. 341/1 Glass-crystalline materials that are made from special glassy compositions containing crystal nucleators.

nucleic, *a.* Add pronuncs. (niukli·ik, -ē·ik, niū·kli,ik) and substitute for def. of *nucleic acid*: Any of the naturally occurring polynucleotides present in most cells (chiefly in the chromosomes and ribosomes), which either store genetic information or translate this into the structure of proteins; they fall into two distinct classes, deoxyribonucleic acid (DNA) and ribonucleic acid (RNA), each of which consists of long unbranched molecules of very high molecular weight and usu. occurs in combination with protein (nucleoprotein). [prob. tr. G. *nucleinsäure* (R. Altmann 1889, in *Arch. für Anat. u. Physiol.* (*Physiol. Abth.*) 524).]

1892 *Jrnl. Chem. Soc.* LXII. 224 The preparation of nucleic acid from nuclein by Altmann..gives a further means of distinguishing between the two groups. **1908** [see *glycoprotein* s.v. *GLYCO-]. **1942** *Endeavour* I. 104/1 Nucleic acid renders a further invaluable service—it makes the chromosomes visible. **1958** *Listener* 31 July 165/1 The advances of organic chemistry in the last 100 years, advances culminating in the nucleic acid story, lately so much discussed. **1960** L. PICKEN *Organization of Cells* iv. 126 The Watson and Crick helical model of nucleic acid suggests that accurate replication of a unique serial order of nucleotides might readily occur if the two helical chains could be partially uncoiled, so that each might serve as a mould for a new chain. **1972** *Sci. Amer.* June 41/3 The only other compounds that are as important in living systems are the nucleic acids, which are the repository of genetic information and direct the synthesis of proteins.

nucleo-. Add: **nu:cleo-cytopla·smic** *a.*, existing or taking place between the nucleus and the cytoplasm; relating the nucleus to the cytoplasm (with respect to some property); **nucleoge·nesis**, the formation of nuclei; *spec.* = *nucleosynthesis* below; **nucleohi·stone** (also formerly **-hi·ston**) *Biochem.*, a nucleoprotein in which the protein component is a histone; **nucleopro·tamine** *Biochem.*, any nucleoprotein in which the protein component is a protamine; † **nucleoproteid** *Biochem.* = *NUCLEOPROTEIN; **nucleosy·nthesis** *Astr.*, the cosmic formation of atoms more complicated than the hydrogen atom; hence **nucleosynthe·tic** *a.*

1905 *Publ. Carnegie Inst. Washington* No. 37. 66 In order that the nucleo-cytoplasmic equilibrium may be maintained, it [*sc.* the ascus] must be provided with an excess of nuclear material as compared with the other cells of the ascogenous hyphæ and the ascogonium. **1956** *Nature* 4 Feb. 236/2 The mean nucleo-cytoplasmic ratio was then measured by the method of Chalkley. **1968** H. HARRIS *Nucleus & Cytoplasm* p. vii, The object of this book..is to provide an introduction to some of the salient problems in the field of nucleo-cytoplasmic relationships. **1952** *Industr. & Engin. Chem.* June 1276/1 One of the more practical results of the theory of nucleogenesis is the explanation and the guidance..it has offered in preparing monodisperse colloids. **1955** *Nature* 16 July 130/2 If the assumptions..are valid,..all the radioactive elements with half-lives short compared to 4×10^8 yr. would have decayed in the time interval between nucleogenesis and the formation of the Earth. **1974** *Physics Bull.* Oct. 464/3 Nucleogenesis in stars. **1895** *Jrnl. Chem. Soc.* LXVIII. II. 52 The active agent in [blood] coagulation is regarded as a nucleo-albuminous substance, named nucleo-histon. **1914** *Chem. Abstr.* VIII. 714 There is no reason to assume that nucleohistones contain any other but the genuine nucleic acid. **1964** G. H. HAGGIS et al. *Introd. Molecular Biol.* ix. 236 Nucleic acids carry a large net negative charge, and the protein and nucleic acid components of nucleoprotamines and nucleohistones are held together largely by electrostatic forces. **1929** *Chem. Abstr.* XXIII. 4724 (*heading*) The behavior of nucleoprotamine and its components in animal metabolism. **1956** *Nature* 31 Mar. 603/1 The X-ray data indicate that the molecule [of DNA] has two grooves of unequal depths, and that in the nucleoprotamines the polypeptide chains lie inside these grooves. **1971** D. M. P. PHILLIPS *Histones & Nucleohistones* iii. 121 A conformation similar to the extended-chain form of protamine in nucleoprotamine could be present. **1886** *Jrnl. Chem. Soc.* L. 1051 The author [*sc.* E. Merck] calls 'nucleo-proteids' substances which, when boiled with water under pressure or treated with acids, alkalis, or ferments, are resolved into nuclein and albumin. **1894** Nucleoproteid [in Dict.]. **1914** M. DRUMMOND tr. *Haberlandt's Physiol. Plant Anat.* viii. 415 The reserve proteins include the various globulins, vitellins..and albumoses, also certain nucleoproteids. **1960** WASSERBURG, FOWLER, & HOYLE in *Physical Rev. Lett.* IV. 113/1 There is considerable uncertainty as to the exact time dependence of stellar evolution and nucleosynthesis in the Galaxy. **1960** FOWLER & HOYLE in *Ann. Physics* X. 281 Type I supernovae were considered to be the only events rapid enough for nucleosynthesis. [*Note*] We use the term *nucleosynthesis* rather than *nucleogenesis* for good reason. We refer..to the synthesis of the elements beyond hydrogen from..the proton, and the neutron... We reserve genesis for the creation of matter-energy. **1963** E. ANDERS in Middlehurst & Kuiper *Moon, Meteorites & Comets* xiii. 458 A second model, involving continuous nucleosynthesis throughout the Galaxy, must be considered. **1965** *Ann. Rev. Astron. & Astrophysics* III. 227 Reynold's discovery was first interpreted as indicating a surprisingly short time-interval between the end of nucleosynthesis and the formation of the solar system. **1971** *Nature* 3 Sept. 39/2 It is now clear that while this effect is operating, a large amount of nucleosynthesis takes place early in the life of a galaxy. **1960** *Physical Rev. Lett.* IV. 113/1 The nucleosynthetic processes during which iodine was made were instantaneous. **1973** *Physics Bull.* Nov. 652/1 This high percentage of helium apparently could not have been produced by the kind of nucleosynthetic processes currently occurring in stars.

nucleoid, *a.* Add: (Further example.)

1889 *Jrnl. R. Microsc. Soc.* 429 (*heading*) Nucleus or nucleoid bodies of schizomycetes.

B. *sb.* † **a.** [a. G. *nucleoïd* (M. Lavdowsky 1893, in *Zeitschr. für wiss. Mikrosk.* X. 8).] (See quots.) *Obs.*

Arnold (*Virchows Arch. für path. Anat. u. Physiol.* (1896) CXLV. 22) took over the term from Lavdowsky.

1905 GOULD *Dict. New Med. Terms* 383/2 *Nucleoid*, a term used by Arnold to designate the substance in the red corpuscles formed from the original nucleus. It is finely granular or fibrillar and is surrounded by a substance which Arnold calls paraplasm. **1913** O. C. GRUNER *Biol. Blood-Cells* 361 *Nucleoid*. Syn.: *nuclear rests, inclusion-body*. (*a*) A precipitation effect of hæmoglobin; (*b*) if basic, a special appearance of the basophile cell membrane. **1928** E. B. KRUMBHAAR in E. V. Cowdry *Special Cytol.* I. x. 307 In addition to the several bodies just mentioned, various artefacts can be produced [in erythrocytes] by standing, poor fixation, non-isotonic solutions, etc., termed 'nucleoids'.

b. [a. G. *nucleoid* (coined independently in this sense by G. Piekarski 1937, in *Arch. für Mikrobiol.* VIII. 438).] An organelle in bacteria and viruses functionally analogous to the cell nucleus of higher organisms.

1938 *Biol. Abstr.* XII. 1407/2 The author's [*sc.* G. Piekarski's] studies seem to show that the nucleoids which he observed in bacteria and sarcina are equivalent to cell nuclei. **1965** *Bacteriol. Rev.* XXIX. 277/1 The genetic material in bacterial cells forms structures which are called nucleoids... Although from the very beginning there was no doubt that these structures are nuclei with respect to their function.., their simple architecture and morphological appearance, which distinguish them from the type of nucleus present in higher organisms..make a special term desirable. **1970** *Nature* 31 Oct. 410/2 The RNA dependent activity resides in the nucleoid. **1971** *Ibid.* 30 Apr. 568 (*caption*) Cell from culture of *B[acillus] subtilis*... Two chromosomes appear as tightly packed 'nucleoids'. **1972** *Sci. Amer.* Jan. 29/2 The virion of the Rous sarcoma virus consists of a lipid-containing envelope.., an inner membrane and a nucleoid, or core, that contains the viral RNA and certain proteins.

nucleoid, var. *NUCLOID.

nucleolar, *a.* Add: Also with pronunc. (niūkli,ōu·lăɹ).

nucleolonema (niū:kliolonī·mă). *Cytology.* Pl. **-nemas, -nemata.** [a. Sp. *nucleolonema* (Estable & Sotelo 1951, in *Publicaciones Inst. de Invest. de Ciencias Biol.* I. 105), f. NUCLEOLO- + Gr. νῆμα thread.] (See quot. 1968.)

Quot. 1951 is from the authors' English translation of the summary of their Sp. paper.

[**1951** ESTABLE & SOTELO in *Publicaciones Instituto de Investigación de Ciencias Biol.* I. 123 Every true.. nucleolus is made up of two different parts... The first one is always the most important, and on account of its filamentous structure we shall call it nucleoloneme. *Ibid.*, No cell having a nucleus lacks a nucleoloneme.] **1952** —— in *Stain Technol.* XXVII. 307 In a recent paper the authors have shown filamentous structures within the nucleolus of all cells. This structure was named the nucleolonema. **1955** *Jrnl. Biophysical & Biochem. Cytol.* I. 185 (*heading*) Note on nucleolonemata in human cultured cells. **1968** R. RIEGER et al. *Gloss. Genetics & Cytogenetics* 316 As far as the ultrastructure of the n[ucleolus] is concerned, the early work was interpreted as indicating the presence of a coiled filament called the 'nucleolonema' (diameter 90–180 nm). Later this nucleolonema was found to consist of coarse granules of the size of ribosomes (diameter about 150 Å). **1973** *Jrnl. Invertebrate Path.* XXII. 405/2 The nucleolus was compact, without nucleolonema, and was completely surrounded by an electron-dense material, apparently chromatin.

nucleolus. Add: also with pronunc. (niūkli,ōu·lŏs).

† **nucleon**[1] (niū·kli̜ǫn). *Biochem. Obs.* [a. G. *nucleon* (M. Siegfried 1895, in *Ber. d. Deut. Chem. Ges.* XXVIII. 518), f. *nucle-ïn* (now *nuklein*) NUCLEIN + *pept-on* PEPTONE.] (See quots.)

1895 *Jrnl. Chem. Soc.* LXVIII. I. 314 The term *nucleon* is suggested for compounds, such as phosphorcarnic acid, which are allied to the nucleins but contain peptone instead of albumin. **1905** W. H. HOWELL *Text-bk. Physiol.* ii. 60 The discoverer of nucleon has attributed to it a very great physiological importance, as a source of energy to the muscle, and as an efficient means of transportation of iron, calcium, [etc.]... It must be stated, however, that there still remains some doubt as to the chemical individuality of the nucleon or the nucleons. **1921** *Physiol. Abstr.* VI. 477 Nucleon is a mixture of which the chief constituent is denatured protein.

nucleon[2] (niū·kli̜ǫn). *Nuclear Physics.* [f. NUCLE(US *sb.* + *-ON[1].] † **a.** = *PROTON 2. *Obs. rare.*

1923 D. LL. HAMMICK tr. *Perrin's Atoms* (ed. 2) 223 If we represent the hydrogen nucleus or positive protoatom by *h* and the corpuscle or negative protoatom by *β* we may say that all matter is made up of protoatoms *h* and *β*... The negative protoatom may be called the corpuscle and the positive protoatom the nucleon. [*Note*] Suggested by P. Auger. *Proton* has also been proposed.

b. A proton or neutron; a sub-atomic particle of which these may be regarded as two distinct states, differing in the third component of isospin. [Orig. formed as *nuclon* (see quot. 1941).]

1939 F. J. BELINFANTE *Theory of Heavy Quanta* (Leiden Univ., thesis) 40 The interaction of the heavy quanta with the heavy particles (the proton-neutron, or 'nuclon', as we shall call it briefly). **1940** PAULI & BELINFANTE in *Physica* VII. 179 The particle that is a proton in its charged state and a neutron in its neutral state, we have called a nuclon. **1940** C. MØLLER in *Physical Rev.* LVIII. 1118/1 The strength of the couplings between the nuclons and the 'vector' and 'pseudoscalar' meson fields. [*Note*] Following the proposal of Belinfante..we use the word nuclon as a common name for the nuclear constituents, the protons and neutrons. **1941** —— in *Ibid.* LIX. 323/2 Following the original proposal of Belinfante, the writer has..used the word 'nuclon' as a common notation for..neutrons and protons. In the meantime, however, it has been pointed out to me that, since the root of the word nucleus is 'nucle', the notation 'nucleon' would from a philological point of view be more appropriate. **1952** *Sci. News* XXIII. 28 It seems that each nucleon (a general name for either proton or neutron) cannot interact with all the other particles in the nucleus, but only with its neighbours. **1968** M. S. LIVINGSTON *Particle Physics* v. 97 With both protons bombarding a high-Z target, to create a nucleon–antinucleon pair.., the threshold kinetic energy must be 5·4 GeV. **1971** *New Scientist* 17 June 695/2 The questions arise why cosmic-ray nucleons or X-rays or ultraviolet photons should be produced in the centres of Seyfert galaxies.

nucleonic (niūkli̜ǫ·nik), *a.* [Partly f. prec. + -IC, partly a back-formation from next.] Of or pertaining to the nucleon or nucleonics.

1946 *Proc. Amer. Philos. Soc.* XC. 42/2 A second class of transformation..comprises the nucleonic changes which occur in the atmosphere under cosmic-ray bombardment. **1947** *Nucleonics* Sept. 1/1 Nucleonic physics..has come to be understood as the science of those changes in which there occurs a rearrangement, but no change, in the total number of nucleons present. Radioactivity, nuclear disintegration, nuclear fission, and nuclear synthesis are the processes comprehended by this science. **1947** *Electronics* Dec. 84/1 An important tool of nucleonic research is the mass spectrograph. **1955** J. A. WHEELER in W. Pauli *Niels Bohr* 163 Nuclear division calls forth a uniquely drastic kind of nucleonic re-arrangement. **1957** *Financial Times Ann. Rev. Brit. Industry* 35/5 Industrial applications..developed in conjunction with the nucleonic instrument manufacturers. **1974** *Physics Bull.* Dec. 579/3 We have been aware for many years of the inadequacy of the neutrons and protons only approach for the simplest systems because there we have had good enough nucleonic wavefunctions to show up the deficiency of the nucleons only approach.

nucleonics (niūkli̜ǫ·niks), *sb. pl.* (const. as *sing.*). [Blend of *NUCLEON[2] + *ELECTRONICS.] The branch of science and technology concerned with nucleons and the atomic nucleus, esp. with the practical applications of nuclear phenomena and associated techniques.

1945 *Nature* 10 Nov. 549/1 If science is to be controlled on a national level, the Association [of Oak Ridge Scientists] believes that the inevitable armament competition will prevent science, especially 'nucleonics', from ever being free again. **1946** Z. JEFFRIES in *Chem. & Engin. News* 25 Jan. 186 'Nucleonics' is the generic name used to some extent within the Atomic Energy Project during the war and its use is now gaining in popularity... Since released atomic energy and certain other phenomena [are] derived from the nuclei of atoms and thus from the nucleons, it seems appropriate that the generic name should be 'nucleonics'. **1950** GLASSTONE *Sourcebk. Atomic Energy* iv. 94 The use of the word 'nucleonics'.. was proposed by Z. Jeffries in July 1944. **1952** *Electronic Engin.* XXIV. 533 The newer science of nucleonics is gradually leaving the nuclear physics research laboratory and finding applications aimed at producing greater productivity of our industrial processes. **1955** J. A. WHEELER in W. Pauli *Niels Bohr* 164 Looking back, one sees in nucleonics as in so many other parts of physics how much progress can be made on the basis of simple principles. **1959** *Times* 2 Jan. 2 Developments will range from high voltage transmission systems to nucleonics and computer design. **1966** *New Scientist* 15 Dec. 623/1 Nucleonics—Soldering in place the printed circuit of a geiger counter kit, designed to patent specification of the UK Atomic Energy Authority.

nucleophile (niū·kliŏfail). *Chem.* [f. NUCLEO- + -PHILE.] A nucleophilic reagent.

1953 C. K. INGOLD *Struct. & Mech. Org. Chem.* v. 201 All bases are nucleophiles. **1959** CRAM & HAMMOND *Org. Chem.* xvii. 395 Nucleophiles such as RS^-, which owe much of their reactivity to high polarizability, tend to bring about substitution rather than elimination. **1971** *Nature* 5 Nov. 42/1 Any agent which..can therefore react with nucleophiles in the cell, is a potential carcinogen.

nucleophilic (niūkliŏfi·lik), *a.* *Chem.* [f. NUCLEO- + *-PHILIC.] **a.** Having an affinity for atomic nuclei, and so reacting at an electron-deficient bond or atom in a substrate; anionoid.

1933 C. K. INGOLD in *Jrnl. Chem. Soc.* 1121 The terms *electrophilic* (electron-seeking) and *nucleophilic* (nucleus-seeking) are suggested in place of the adjectives anionoid and cationoid introduced by Lapworth. **1946** *Nature* 20 July 94/1 Those bimolecular and unimolecular substitutions in which a nucleophilic reagent displaces halogen as halide ion from an alkyl halide. **1968** R. O. C. NORMAN *Princ. Org. Synthesis* iv. 127 A thiol anion, RS^-, although less basic than its oxygen analogue, RO^-, is more strongly nucleophilic.

b. Of a reaction: brought about by a nucleophilic reagent.

1935 HUGHES & INGOLD in *Jrnl. Chem. Soc.* 245 According as this [*sc.* the substituting agent] is nucleophilic..or electrophilic.., the reaction may be termed a 'nucleophilic' or 'electrophilic' substitution. **1946** *Nature* 20 July 94/1 That the polar effect is not absent in general from bimolecular nucleophilic substitutions is consistent with the result of exchanging the electron-releasing methyl substituents for electron-attracting groups. **1969** T. C. THORSTENSEN *Pract. Leather Technol.* vi. 91 Windus and Showell have reported the explanation of mechanism of unhairing as a nucleophilic displacement.

Hence **nucleophi·lically** *adv.*, after the manner of a nucleophile; **nu:cleophili·city**, nucleophilic character.

1953 C. K. INGOLD *Struct. & Mech. Org. Chem.* v. 201 Basicity..is a special manifestation of nucleophilic character, or nucleophilicity, that is, affinity for atomic nuclei in general. **1970** J. E. GEARIN in A. Burger *Medicinal Chem.* (ed. 3) II. xlvii. 1302/2 This group must be capable of attacking nucleophilically the carbon atom of the ester group of acetylcholine. **1971** C. J. GRAY *Enzyme-catalysed Reactions* i. 21 The nucleophilicity of the small fluoride ion is higher in non-hydroxyl solvents where the solvation is less effective.

nucleoprotein (niūkliŏprōu·tīn). *Biochem.* [a. G. *nucleoprotein* (E. Merck 1885, in *Patentschrift 35,724*), which was formerly rendered in Eng. as *nucleoproteid* (s.v. *NUCLEO-): see PROTEIN.] A combination of a protein and a nucleic acid such as occurs in living organisms.

1907 *Practitioner* Oct. 588 From the nuclein and nucleoprotein, the purin bodies, are produced, such as hypoxanthin, xanthin, adenin, and uric acid. **1938** *Ann. Reg. 1937* 352 Tobacco mosaic virus is a nucleoprotein of special character. **1944** *Adv. Protein Chem.* I. 258 The simplest and perhaps best available definition of a nucleoprotein would be to designate as such any protein with which nucleic acid is associated. **1959** *Times* 12 June 15/6 Both genes..and viruses are nucleo-proteins. **1970** *Sci. Amer.* Feb. 102 The constituents of the chromosomes—deoxyribonucleic acid (DNA) and nucleoproteins—do not absorb visible light readily.

nucleoside (niū·kliŏsaid). *Biochem.* [ad. G. *nucleosid* (Levene & Jacobs 1909, in *Ber. d. Deut. Chem. Ges.* XLII. 2475), f. after *glucosid* GLUCOSIDE: see NUCLEO-.] Any compound in which a sugar (usu. ribose or deoxyribose) is linked glycosidically to a purine or pyrimidine base; *spec.* such a compound derived from a nucleic acid by hydrolysis. (See quot. 1973.)

1911 [see *NUCLEOTIDE]. **1911** [see *ADENOSINE]. **1931** LEVENE & BASS *Nucleic Acids* vi. 126 The name 'nucleoside' was assigned to the substances of this group for the reason that..they contain sugar in a glucosidic union, and..the substances linked to the sugars are nuclein bases. **1946** *Nature* 24 Aug. 275/2 Miss Mejbaum has found.. that the pentose contained in pyrimidine nucleotides (uridylic acid and cydidylic [*read* cytidylic] acid) and nucleosides (uridine) are not determined by her method. **1969** *New Scientist* 10 July 65/2 A nucleoside like adenosine or inosine. **1973** [see *NUCLEOTIDE].

Hence **nu·cleosidase** [*-ASE], any enzyme which catalyses the hydrolysis of a nucleoside into its constituent base and sugar, or the reaction between a nucleoside and phosphate to yield a base and a sugar phosphate.

1911 LEVENE & MEDIGRECEANU in *Jrnl. Biol. Chem.* IX. 396 Under the influence of mineral acids nucleosides are readily hydrolyzed into their components. The same cleavage can be brought about by the action of enzymes present in the plasma of most of the organs tested in that direction... Since this reaction is brought about by an enzyme different from the other nucleolytic enzymes, it may be proper to refer to it under the name of *Nucleosidase*. **1935** *Biochem. Jrnl.* XXIX. 1100 If adenine nucleosidase is present it must therefore be only in relatively small amounts. **1955** CHARGAFF & DAVIDSON *Nucleic Acids* I. xv. 600 The enzymic cleavage of N-glycoside bonds of nucleic acid derivatives was first observed in nucleosides, and the term 'nucleosidases'..is still used. In the current literature this name tends to be replaced by the designations 'nucleoside phosphorylases' and 'nucleoside hydrolases'.

nucleotide (niū·kliŏtaid). *Biochem.* [ad. G. *nucleotid* (Levene & Mandel 1908, in *Ber. d. Deut. Chem. Ges.* XLI. 1907): see NUCLEO- and

*-IDE.] Any compound in which a phosphate group is linked to the sugar of a nucleoside; *spec.* any of the compounds of this type obtained by the partial hydrolysis of a nucleic acid, which are the individual monomers of which such acids are composed. (See quot. 1973.)

1908 *Jrnl. Chem. Soc.* XCIV. I. 587 It is suggested that the nucleic acids are composed of simpler complexes, the nucleotides, each formed of phosphoric acid, a carbohydrate, and a base. **1911** *Jrnl. Biol. Chem.* IX. 66 It was demonstrated that the molecule of the complex nucleic acids is composed of nucleotides and these of phosphoric acid, carbohydrate and base linked to one another in the order here given... It is possible to detach from the complex either phosphoric acid alone, giving rise to a nucleoside, or a complex of carbohydrate and base. **1937** *Nature* 30 Oct. 745/2 The co-ferments now described as phosphopyridine nucleotides, the nature of which has been at last worked out. **1952** *Sci. News* XXIV. 33 The complete nucleic acid molecules are generally considered as being built up of nucleotides, linked through the hydroxyl (OH) groups of the sugar and the acid groups of the phosphoric acid. **1957** *Times* 1 Nov. 10/3 The award to Sir Alexander Todd..of the Nobel prize for chemistry ..for his work on nucleotides and nucleotide enzymes. **1968** *Observer* (Colour Suppl.) 10 Mar. 15/1 The DNA molecule..is a two-stranded helix. Each strand is built up of smaller molecules called nucleotides. *Ibid.* 15/2 There are only four kinds of nucleotide, each with a different chemical base. These bases are adenine, thymine, guanine and cytosine. **1971** J. Z. YOUNG *Introd. Study Man* iii. 47 The symbols of the genetic code..are the nucleotides arranged in the helical DNA molecules. **1973** HENDERSON & PATERSON *Nucleotide Metabolism* p. xiv, The terms 'nucleoside' and 'nucleotide' in the strictest sense refer to N-glycosides and phosphorylated N-glycosides, respectively, derived from nucleic acids. The term is now used, however, in several broader ways. Thus, adenosine triphosphate (ATP) is not derived from nucleic acids, but is quite legitimately a nucleotide through its relation to adenosine monophosphate (AMP), which is so derived. Other N-ribosides, such as nicotinamide mononucleotide (NMN), are called nucleotides only by extension and analogy, and nicotinamide-adenine dinucleotide (NAD+), nicotinamide-adenine dinucleotide phosphate (NADP+), etc., are called dinucleotides only by a similar process. Flavin mononucleotide (FMN) is a step still further removed, as it contains ribitol instead of ribose, and flavin-adenine dinucleotide similarly extends the meaning of dinucleotide. N-glycosides such as orotidylate (OMP) and adenylosuccinate are called nucleotides through their close relationship to the 'true' nucleotides. *Ibid.* i. 10 In animal cells, a few nucleotides with sugars other than ribose and deoxyribose are known.

Hence **nu·cleotidase** [*-ASE], any enzyme which catalyses the hydrolysis of a nucleotide to a nucleoside and phosphate.

1911 LEVENE & MEDIGRECEANU in *Jrnl. Biol. Chem.* IX. 395 The enzymes performing this cleavage [of nucleotides to phosphoric acid and carbohydrate-base complex] may be referred to as Nucleotidases. **1932** *Jrnl. Biol. Chem.* XCVI. 462 The optimal activity of the nucleotidase is at a pH > 11. **1970** R. W. MCGILVERY *Biochem.* xx. 481 The phosphate group of nucleotides is liberated by hydrolysis, catalyzed by nucleotidases.

nucleus, *sb.* Add: **I. 1. b.** *Astr.* A more condensed, usu. brighter, central part of a galaxy or nebula.

1784 W. HERSCHEL in *Phil. Trans. R. Soc.* LXXIV. 442, I have seen double and treble nebulæ, variously arranged; large ones with small, seeming attendants;.. others of the cometic shape, with a seeming nucleus in the center. **1849** J. F. W. HERSCHEL *Outlines Astron.* xvii. 601 The nebula in Andromeda is visible to the naked eye... Mr. G. P. Bond, assistant at the observatory of Cambridge, U.S., describes and figures it as..very suddenly condensed at the nucleus almost to the semblance of a star. **1898** A. M. CLERKE et al. *Concise Knowledge Astron.* IV. vi. 534 The photograph shows both nuclei of the nebula to be stellar. **1955** *Sci. Amer.* May 48/3 Because of heavy dust clouds this nucleus [of the Milky Way] has not been observed visually or photographically; it was discovered by radio astronomy. **1970** *Sci. Jrnl.* Feb. 61/1 The emission line spectrum of a Seyfert nucleus is rather like that from typical hot clouds of gas such as the Orion Nebula.

3. b. (Further *attrib.* examples.)

1905 *Fortn. Rev.* 2 Jan. 15 The officers and men withdrawn..from distant squadrons will be utilised as nucleus crews in the Reserve ships. **1914** C. F. TWENEY *Dict. Naval & Mil. Terms* 163 *Nucleus crew*, the essential members of a ship's crew, such as petty officers, gunners, etc., the crew being raised to full strength in case of mobilisation. **1926** in Fowler *Mod. Eng. Usage* 713/2 Ships with nucleus crews were not so efficient as ships fully manned. **1965** M. MORSE *Unattached* i. 57 The mood of each rehearsal..depended very much on the attitude of the nucleus group.

II. 6. c. Substitute for def.: Any discrete mass of grey matter in the central nervous system. (Earlier and later examples.)

The term is used in numerous English and mod.L. combs. distinguishing the various different nuclei.

1828 J. QUAIN *Elem. Anat.* x. 622 If a vertical section be made of one of the lobes of the cerebellum, in such a way as that two-thirds of its breadth shall lie external to the incision, an oval nucleus of grey substance (*corpus dentatum*, vel *rhomboideum*) will be exposed. **1856** *Ibid.* (ed. 6) II. 493 Another division passes directly up, its fibres embracing the olivary nucleus. **1875** *Encycl. Brit.* I. 878/1 The upper mass of grey matter projects into the lateral ventricle, and is called the intra-ventricular portion or nucleus caudatus. **1899** [see *HYPOTHALAMUS]. **1968** PASSMORE & ROBSON *Compan. Med. Stud.* I. xix. 12/1 Masses of grey matter are called nuclei, some of which, e.g. the basal nuclei, are large and embedded in the depths of the brain. **1972** *Sci. Amer.* Dec. 73/1 A sensory pathway consists of peripheral sense organs and several clusters of nerve cells called nuclei. Sensory information is processed in several stages, so that each nucleus receives input from the preceding one, processes the input and sends an output to the next nucleus.

7. *Biol.* A cell organelle present in most of the cells of all organisms except the most primitive, usu. as a single subspherical structure, and consisting (except when undergoing division) of a membrane enclosing a ground substance (the nuclear sap) in which lie the chromosomes, one or more nucleoli, etc., and functioning as the repository of genetic information and as the director of metabolic and synthetic activity of the cell. Hence, by extension, applied by some writers to an organelle in some of the more primitive organisms, esp. bacteria, which is analogous in function but structurally simpler (cf. *NUCLEOID *a.* B. b).

The definition is intended to embrace and replace those of senses 5 d and 6 a. The distinction implied in the Dict. between the nuclei of plant and animal cells was artificial and modern knowledge has confirmed their essential similarity.

1831, etc. [in Dict., senses 5 d and 6 a]. **1889** [see *NUCLEOID *a.*]. **1962** *Brit. Med. Bull.* XVIII. 31/1 (*heading*) Morphology of the bacterial nucleus. **1965** [see *NUCLEOID *sb.* b]. **1968** PASSMORE & ROBSON *Compan. Med. Stud.* I. xxvi. 2/1 The mature red [blood] cell..appears as a biconcave disc with no nucleus. **1970** AMBROSE & EASTY *Cell Biol.* v. 163 This may also explain how acridine dyes penetrate the nucleus without staining the cytoplasm.

8. *Chem.* An arrangement of atoms, esp. a ring structure, characteristic of a number of organic compounds.

1845 W. GREGORY *Outl. Chem.: Org. Chem.* 512 Laurent considers benzole as in some measure the fundamental compound, or nucleus, and calls it phene. **1886** E. F. SMITH tr. *V. von Richter's Chem. Carbon Compounds* 465 The azo-group, N=N, decomposes, each nitrogen atom remaining attached as NH_2 to a benzene nucleus. **1932** *Jrnl. Chem. Soc.* 1132 Table III summarises the effect of varying the nature of the side-chain halogen in the compounds with an unsubstituted phenyl nucleus. **1951** I. L. FINAR *Org. Chem.* x. 185 If *one* acetic acid nucleus is blocked off, the fragment required is ethyl chloroacetate. **1971** *Nature* 7 May 25/1 Many psychotomimetic substances possess an indole nucleus.

9. A particle on which crystals, droplets, or bubbles can form in a fluid.

1857 [in Dict., sense 3 a]. **1886** *Proc. R. Soc. Edin.* XIII. 79 If this were the case, no nucleus would be *absolutely* requisite for the formation either of liquid from vapour or of vapour from liquid. **1906** [see *NUCLEATION]. **1939** *Q. Jrnl. R. Meteorol. Soc.* LXV. 411 If supersaturation is attained the fogs may thicken considerably owing to the deposition of water on the sea-salt nuclei. **1952** *Industr. & Engin. Chem.* June 1273/1 A nucleus in water-fog formation consists of about 80 water molecules. **1957** G. E. HUTCHINSON *Treat. Limnol.* I. ix. 583 In a free volume of water containing no minute masses of gas that can act as nuclei, an enormous excess tension is needed for bubble formation. **1967** [see *NUCLEATION].

10. A small group of bees, including a queen, used as the foundation of a new colony.

1886 F. R. CHESHIRE *Bees & Bee-Keeping* II. vii. 306 These small nuclei will sustain themselves, in average seasons. **1915** E. F. PHILLIPS *Beekeeping* iii. 39 A mere handful of bees (perhaps 200) may constitute a small colony (usually called a nucleus). **1952** H. MACE *Bee-Keeper's Handbk.* xxxiii. 154 Nucleus hives can be purchased or made, either to hold a single nucleus up to four combs, or two or more separate nuclei, separated by partitions. **1963** F. G. SMITH *Beekeeping* vii. 60 A nucleus is a very small colony of bees. It consists of a queen and up to four or six frames of brood and food, well covered with bees.

11. *Physics.* The positively charged central constituent of the atom, comprising nearly all its mass but occupying only a very small part of its volume and now known to be composed of protons and neutrons.

In Rutherford's 1911 paper called merely a 'central charge'. In the examples in the first paragraph *nucleus* is used for various speculative notions concerning the atom.

[**1844** FARADAY in *Phil. Mag.* XXIV. 141 If, in the ordinary view of atoms, we call the particle of matter away from the powers *a*, and the system of powers or forces in and around it *m*, then in Boscovitch's theory *a* disappears, or is a mere mathematical point... To my mind..the *a* or nucleus vanishes, and the substance consists of the powers or *m*. **1851** W. J. M. RANKINE in *Phil. Mag.* I. 443 The fundamental suppositions of the hypothesis of molecular vortices are the following:—*First.* That each atom of matter consists of a nucleus or central physical point enveloped by an elastic atmosphere, which is retained around it by attraction. *Ibid.*, If an indefinitely extended vibrating medium..consist of a system of atomic nuclei. **1900** *Rep. Brit. Assoc. Adv. Sci. 1900* 619 The material atom must be some kind of permanent nucleus that retains around itself an æthereal field of physical influence. *Ibid.*, If..the distances at which they [*sc.* atoms] are kept apart are large compared with the diameters of the atomic nuclei. **1903** O. LODGE *Mod. Views on Matter* 5 If the charge of electricity usually associated with a single monad atom of matter were concentrated on to a spherical nucleus one hundred-thousandth of an atom's dimension in diameter, it would thereby possess a mass about one-thousandth of that of the lightest atom known... Such a hypothetical concentrated unit of electricity it has become customary to call an 'electron'.]

1912 E. RUTHERFORD in *Phil. Mag.* XXIV. 461 In a previous paper [sc. *Phil. Mag.* (1911) XXI. 669] I have given reasons for believing that the atom consists of a positively charged nucleus of very small dimensions, surrounded by a distribution of electrons in rapid motion, possibly of rings of electrons rotating in one plane. **1919** *Conquest* I. 1. 36/2 An atom is a sort of solar system in miniature, and comprises a central core or nucleus..and a number of particles, called corpuscles, circulating round the nucleus. **1942** J. D. STRANATHAN *'Particles' of Mod. Physics* xi. 417 The nucleus must be very small, not larger than 10^{-12} cm. **1955** C. G. DARWIN in W. Pauli *Niels Bohr* 6 In 1911 he [*sc.* Rutherford] tried the idea of a heavy central electric charge repelling the α-particle—it was I think several months before it was called the nucleus—and at once the whole theory of the nuclear atom emerged. **1962** H. D. BUSH *Atomic & Nuclear Physics* iii. 62 The hypothesis that every nucleus consisted of protons and neutrons, was first suggested by Heisenberg (1932).

12. a. *Phonetics.* The syllable of a word (spoken in isolation) that bears the primary accent; in an utterance, the syllable or syllables given particular emphasis.

1922 H. E. PALMER *English Intonation* ii. 7 Each Tone-Group contains a Nucleus, which is the stressed syllable of the most prominent word in the Tone-Group. The nucleus corresponds to what is usually called sentence-stress. *Ibid.* 8 In Southern English there are four characteristic Nucleus Tones. **1924** —— *Gram. Spoken Eng.* 6 If..tonetic transcription is used, the syllable ['] will be replaced by the appropriate nucleus-symbol. *Ibid.* 14 Instead of the rise taking place in the nucleus-syllable itself, it is distributed over the nucleus and tail. **1941** *Language* XVII. 224 The present study..will deal with junctures, stresses, and consonants only in summary, and then devote itself to the syllabic nuclei. **1942** *English Studies* XXIV. 157 In combinations of noun + adjective (like French master, black bird) the marking of the nucleus tone (i.e. the point at which the pitch begins to rise or fall) is an excellent device to bring home to students the two meanings such combinations may express. **1962** A. C. GIMSON *Introd. Pronunc. Eng.* x. 244 The primary accent (or accents) in a sentence is shown by initiating a change of pitch direction, with the nucleus (falling, rising, or a combination of the two) on the appropriate syllable of the word (or words) on which attention is particularly to be concentrated. **1962** S. STUBELIUS in F. Behre *Contrib. Eng. Syntax* 151 The main features of intonation, particularly whether the sentences studied had a falling or a rising end intonation (nucleus). **1973** *Archivum Linguisticum* IV. 21 Tonality, tonicity, and tone, which refer to the number of tone-groups, the location of the tonic, or nucleus, and the choice of tone used, respectively.

b. *Linguistics.* The main word or words in a combination, phrase, or sentence; also = *KERNEL *sb.*[1] 8 b.

1934 [see *DETERMINANT *a.* and *sb.* B. 2 c]. **1949** E. A. NIDA *Morphol.* (ed. 2) 83 The nucleus of a morphological construction consists of (1) a root or (2) a combination of roots... The nonnucleus is made up of nonroots. In the construction *boyishness* the element *boy* is the nucleus and *-ishness* constitutes the nonnucleus. **1961** R. B. LONG *Sentence & its Parts* i. 20 Isolates sometimes take adjunct modifiers, much as nucleuses do. *Ibid.* 497 *Nucleuses*, minimal sequences made up of subjects, predicators, and complements, or of such of these as occur... In *come in!* the nucleus is made up of the predicator *come* and the complement *in*. **1961** Y. OLSSON *Syntax Eng. Verb* iv. 77 The relationship between a group of nuclei (like *London*, etc.) and equivalent members of the paradigms based on them (like *London-er*, etc.) give rise to an expectancy which may or may not be met in the same way. **1968** J. LYONS *Introd. Theoretical Linguistics* viii. 334 The subject and the predicate together form the nucleus of the sentence.

nuclide (niū·klaid). *Nuclear Physics.* [f. NUCL(EUS *sb.* + *-ide* (f. Gr. εἶδος form, kind).] A particular kind of atom, as defined by the number of protons and the number of neutrons in the nucleus.

Synonymous with *ISOTOPE in its broader sense, which the introduction of *nuclide* was intended to discourage in favour of the original stricter meaning of that word.

1947 T. P. KOHMAN in *Amer. Jrnl. Physics* XV. 356/2 There is at present no word in the English language to express the concept of a particular species of atom, differing from all others in the constitution of its nucleus... *Nuclear species* and the German *Kernsorte*..refer to nuclei rather than to atoms... In recent years the word *isotope* has come into use for this purpose, less by design than by default... Evidently a new word is required, and *nuclide* is proposed... The new word and its derivatives should be used in such expressions as 'stable nuclides' ..and 'nuclidic weight'. **1955** R. D. EVANS *Atomic Nucleus* xvi. 522 Because it is the only naturally occurring nuclide which undergoes fission by slow neutrons, U^{235} has attained international fame even in lay circles. **1957** *Technology* June 139/3 β and γ ionization chambers for secondary standardisation of radioactive nuclides have been designed. **1961** G. R. CHOPPIN *Exper. Nuclear Chem.* 215 Na^{22}, Na^{23}, and Na^{24} are all isotopes of sodium... Na^{22} and C^{14} are nuclides but are not isotopic to each other. **1967** *Guardian* 17 May 3/2 There was less Strontium-90 and Caesium-137 in milk last year... Levels of both nuclides were lower than at any time since 1962. **1971** *Daily Colonist* (Victoria, B.C.) 9 July 9/3 The nuclear explosive was designed to reduce the amount of residual tritium, a radioactive nuclide.

Hence **nuclidic** (-i·dik) *a.*

1947 [see above]. **1955** R. D. EVANS *Atomic Nucleus* xvi. 519 In order to identify the new artificial nuclides, it

was logical and essential to utilize the usual nuclidic notation, for example, $_{83}Bi^{212}$. **1962** *Nature* 19 May 621/2 Modern methods make possible much more exact measurements of nuclidic masses. *Ibid.*, The atomic physicist is usually concerned with individual nuclidic species—the differences in their nuclear characteristics and the nuclear, as opposed to chemical, changes they undergo. **1973** J. YARWOOD *Atomic & Nuclear Physics* xv. 429 A useful method of correlating data about the nuclides is to plot a nuclidic chart of *N*, the number of neutrons in the nucleus, against *Z*, the atomic number.

nucloid (niū·kloid). Also **nucleoid.** [f. NUCLEUS *sb.* + -OID.] (See quot. 1962.)

1908 *Westm. Gaz.* 31 July 1/3 (*heading*) 'Nucleoids' in naval manœuvres. *Ibid.* 2/2 The balance is still all on the side of the 'nucleoid' as against the newly commissioned ship. **1917** 'TAFFRAIL' *Sub* vii. 176 About June, came the summer manœuvres, when the crews of all the 'nucloids' were brought up to full strength. **1962** W. GRANVILLE *Dict. Sailors' Slang* 82/2 *Nucloid*, ship of the Reserve Fleet in peacetime, carrying a nucleus crew.

nuclon: see *NUCLEON[2] b.

nude, *a.* and *sb.* Add: **A.** *adj.* **3.** (Further examples.)

1926 E. O'NEILL *Great God Brown* I. iii. 46, I am thy shorn, bald, nude sheep! **1928** *Oxford Poetry* 3 The pale, nude flowers That I Picked for you. **1952** E. O'NEILL *Moon for Misbegotten* I. 38 It is full of nude rocks. **1964** D. VARADAY *Gara-Yaka* xii. 102 (*caption*) Vultures like baubles on a nude tree.

b. (Earlier and later examples.)

1845 *Punch* VIII. 247 A regret that Etty should content himself with merely painting from *'the nude Academy model'*. **1974** *Publishers Weekly* 26 Aug. 250/3 A novel about a nude model who longs for true love.

c. Of a revue, show, photograph, etc.: involving or portraying nude, or lightly clad, figures (usu. female).

In quot. 1870 the actresses referred to were clothed in flesh-coloured tights.

1870 O. LOGAN *Before Footlights & Behind Scenes* xv. 128 Bringing upon stage that hideous disgrace known as the 'nude drama', which took its rise with the flimsy absurdity called the 'Black Crook'. **1947** *Amer. Speech* XXII. 171 America enthusiastically adopted the word.. *nude drama*. **1957** J. OSBORNE *Entertainer* xiii. 86 Nude tableau, behind first act gauze. **1959** *Sunday Times* 16 Aug. 10/4 It is the later nude revue days of Collins's Music Hall. **1959** *Listener* 15 Jan. 132/3 The night-clubs in Calvin's city put on nude shows. *Ibid.* 6 Aug. 195/2 There are packed houses at nude revues. **1972** *Daily Tel.* 27 Nov. 17/6 An Italian magazine published nine pages of nude photographs of her last week.

4. As a colour, esp. of stockings, flesh-coloured.

1922 *Daily Mail* 18 Dec. 2 (Advt.), Ladies' Hose... Black, white,.. taupe, navy, nude, and all shades. **1926** GALSWORTHY *Silver Spoon* II. xii. 216 'Blue-stockings.' 'No, sir; they nearly all wear "nude".' **1931** M. DE LA ROCHE *Finch's Fortune* ix. 156 She had on.. 'nude' stockings. **1947** *Sun* (Baltimore) 10 Sept. 5 Nude and white in sizes 33 to 40. Corset Shop, third floor. **1973** *Philadelphia Inquirer* (Today Suppl.) 14 Oct. 17/3 (Advt.), Choose black, brown, navy or nude calfskin.

5. *Med.* Of a mouse: homozygous for a mutant gene which produces apparent hairlessness and (in most cases) a grossly hypoplastic thymus gland.

1966 S. P. FLANAGAN in *Genetical Res.* VIII. 295 [The hairless mutant was found by Dr. N. R. Grist of the Virus Laboratory, Ruchill Hospital, Glasgow... The name 'nude', symbol *nu*, has been adopted.] *Ibid.* 308 The majority of nude mice die of general body weakness within 2 weeks. **1974** *Nature* 20 Sept. 184/2 Nude mice have spread through the immunological world at a remarkable pace... The thymus abnormality leads to a marked deficiency in thymus-derived (T) lymphocytes and nude mice are rapidly replacing thymectomised mice as models of T cell deprivation and as a source of relatively pure B lymphocytes for *in vitro* studies. **1975** *Ibid.* 13 Mar. 140/2 (*heading*) Nude mice with normal thymus.

B. *sb.* **3.** *Med.* A nude mouse.

1968 *Nature* 27 Jan. 371/1 (*heading*) Section of liver of an adult homozygote nude. **1974** *Ibid.* 20 Sept. 184/2 The T cell deficiency in nudes make them highly susceptible to infection (but apparently not to cancer) and they survive poorly in conventional animal houses.

nudey, var. *NUDIE.

nudge, *sb.* Add: Also *fig.*

1922 JOYCE *Ulysses* 631 Hynes wrote it with a nudge from Corny. **1950** *Sun* (Baltimore) 20 June 21/4 Didn't you like the comeback of Cardinal Manager Eddie Dyer when given a slight nudge over his club's loss of three straight. **1953** A. HUXLEY *Let.* 19 July (1969) 679 Your publishers..would send me proofs as soon as they became available. I have heard nothing... Would you be kind enough to give them a little nudge?

nudge, *v.* Add: **1.** *fig.* (Further examples.) Also *transf.*

1877 G. H. LEWES *Let.* 27 Feb. in *Geo. Eliot's Lett.* (1956) VI. 345 He might still be induced to resume that idea if you thought fit to nudge his elbow a little. **1922** JOYCE *Ulysses* 63 Nudging the door open with his knee he carried the tray in. **1925** H. V. MORTON *Heart of London* (1926) 75 Watch the way a press of omnibuses.. will edge and nudge a way with a mere inch between their mudguards. **1936** DYLAN THOMAS *Twenty-Five Poems* 9 Half The dear, daft time I take to nudge the sentence. **1971** *Times* 28 Aug. 10/1 A smarter, cleaner place undergoing the upheaval of architectural surgery, being nudged by bulldozers and demolition gangs from one century into the next. **1973** *Guardian* 10 Mar. 1/7 Governments would use their reserves..and nudge the price if it tended to drift out of line. **1975** *Physics Bull.* Apr. 162/1 The Science Research Council's attempts to nudge university research into areas more relevant to social and economic needs are meeting some resistance.

2. Also const. *up*, to move up by pushing.

1940 DYLAN THOMAS *Portrait of Artist as Young Dog* 7 The bar was full; two fat women in bright dresses sat near the door, one with a small, dark child on her knee; they saw Uncle Jim and nudged up on the bench.

Hence **nu·dging** *pres. pple.* and *ppl. a.*, approaching, nearing, close to (used e.g. of someone's age).

1949 E. E. CUMMINGS *Let.* 23 Aug. (1969) 193 Now et comment The quote Oxford unquote Press registers alarm nudging horror. **1961** *Sunday Times* 30 Apr. 13/6 Nudging forty..Secombe knows where his ultimate responsibility lies. **1969** D. BARRON *Man who was There* ii. 30 Hughes was tough and nudging fifty. **1971** D. LEES *Rainbow Conspiracy* i. 17 The Manchester circulation is nudging the one and a half million a day mark. **1972** A. MACVICAR *Golden Venus Affair* i. 9, I was big, nudging six feet two. **1974** *Country Life* 25 Apr. 1012/2 A fuel gauge nudging zero.

nudger (nʌ·dʒəɹ). [f. NUDGE *v.* + -ER[1].] **a.** One who nudges another or others.

1910 H. G. WELLS *Hist. Mr. Polly* vi. 177 A sprinkling of girls in gay hats from Miriam's place of business appeared in church, great nudgers all of them. *Ibid.* 178 A murmer from the nudgers announced the arrival of the bridal party. **1960** S. PLATH *Colossus* 35 Nudgers and shovers In spite of ourselves Our kind multiplies.

b. *dial.* A hat.

1903 *Eng. Dial. Dict.* IV. 309/2 *Nudger*,..a hat. **1966** *Sun* 10 June 4/2 He was the only one wearing a bowler, called a 'nudger'.

nudging, *vbl. sb.* (Earlier example.)

1854 THOREAU *Walden* 97 We are not awakened by our Genius, but by the mechanical nudgings of some servitor.

nudie (niū·di). Also **nudey.** [f. NUDE *a.* and *sb.* + -IE.] A nude show; a nude person; a film, photograph, or magazine featuring nudity. Also *attrib.* or as *adj.*

1935 *Amer. Mercury* June 230/1 *Nudie*, nude show. **1939** *Amer. Speech* XIV. 4 A 'nudie' (nudist picture)..is 'cinemerotic'. **1942** BERREY & VAN DEN BARK *Amer. Thes. Slang* §509/24 *'Strip-tease'*,..nudie. *Ibid.* §583/31 *'Strip-teaser'*,..nudie. *Ibid.* §590/9 *Burlesque show*,..nudie. *Ibid.* §608/1 Bare skinema, nudie, nudie pic,..a nudist picture. **1964** *New Statesman* 21 Feb. 306/3 Trying to break away from nudies and naughties. **1967** *Punch* 4 Oct. 506/2 He had..submitted Furd's photograph to several nudie magazines. **1968** *Wall St. Jrnl.* 24 Apr., Russ Meyer is king of the 'nudies'. But his influence on American moviemaking is not inconsiderable. **1970** K. PLATT *Pushbutton Butterfly* xiv. 161, I riffled the chromos. Nudies of Janet Sanders. **1971** R. PETRIE *Thorne in Flesh* vi. 84 Desmond Morris's 'Naked Ape'. Another case of a nudey dust-jacket selling a serious book. **1971** *Petticoat* 17 July 7/3 The nudie stills appear in sex magazines all over the world. **1972** *Guardian* 16 Feb. 13/1 Male nudie pin-ups. **1974** P. GZOWSKI *This Country* 197/1 That the nudies have such phenomenal mass-market circulations says something about our sexuality.

nu·dified, *a. rare.* [See NUDI- and -FY.] Made or become bare.

1883 H. JAMES *Let.* 11 Jan. in R. B. Perry *Tht. & Char. of W. James* (1935) I. xxii. 388 This morning I went out to poor *nudified* and staring Cambridge [Massachusetts].

nudism (niū·diz'm). [f. NUDE *a.* and *sb.* + -ISM.] The cult and practice of going unclothed.

1929 *Time* 1 July 23/1 Made in Germany, imported to France, is the cult of Nudism, a mulligan stew of vegetarianism, physical culture and pagan worship. **1931** F. & M. MERRILL *Among Nudists* xv. 221 The Doctors Durville..insist that France is not ready for nudism as is Germany. *Ibid.* xvi. 233 In spite of their [*sc.* the French bourgeoisie's] modesty on the subject, nudism is growing in France. **1935** *Punch* 19 June 721/1 'A real tent... Think of the saving. Hotel bills, nothing. Meals, practically nothing. Clothes, nothing whatever.' 'Pamela,' I said imploringly, 'not Nudism.' **1944** B. MALINOWSKI *Sci. Theory of Culture* v. 44 If we were to examine from this point of view any movement, such as ..fundamentalism or nudism..we would see that in one and all we can register a certain agreement on the statement of a common purpose as between the members of the movement. **1973** *Guardian* 28 June 6/1 The principles of nudism.

nudist (niū·dist). [f. NUDE *a.* and *sb.* + -IST.] An adherent of the cult of the nude; a person who advocates or practises going unclothed. Also *attrib.*

1929 *Time* 1 July 23/1 Much publicity has been given the Nudist colony on an island in the Seine near Paris. *Ibid.*, A U.S. parallel would be if elegant Editor Frank Crowninshield of *Vanity Fair* should suddenly appear as a vegetable-eating, hairy-chested Nudist. **1931** *John o' London's* 8 Aug. 620/2 The other members of the nudist colony were..entirely normal people. *Ibid.* 620/3 The nudists of France are pursued by the police, by the clergy, by the wit of Parisian cartoonists. **1932** *Daily Express* 28 June 1/7 Nudist camps have sprung up all over England in the past fortnight, in which happy families disport themselves in the warm sunshine. **1938** L. MACNEICE *I crossed Minch* I. vi. 85 The Devil's business fell as flat As a nudist camp or an opera hat. **1960** *Guardian* 6 Jan. 1/7 One father..regularly went off and joined a nudist colony. **1966** J. BALL *Cool Cottontail* iii. 22 Have you ever been in a nudist park before, doctor? **1973** G. MITCHELL *Murder of Busy Lizzie* xiii. 146 'Did you know..that there is a nudist colony on the island?'.. 'If nudists excite you, you're welcome to them.'

‖**nudnik** (nu·dnik). *U.S.* Also **nudnick.** [Yiddish *nudnik*, f. Russ. *núdnyĭ* tedious, boring; see *-NIK.] Someone who pesters, nags, or irritates; a bore. Also *attrib.*

1947 *New Republic* 14 Apr. 42 The patrons of New York's Ruban Bleu are as boorish a collection of *nudnicks* as ever assembled in a public place. **1949** *Amer. Fabrics* No. 9. 108 *Nudnick*, a bothersome customer. **1950** *Commentary* 10 Dec. 558/2 It makes no difference to me if these *nudniks* happen to be atheists. **1955** T. STURGEON in E. Crispin *Best SF Two* (1956) 141 You are a nowhere type, a *nudnick* type, *nothing*! **1961** *John o' London's* 28 Sept. 345/2 What a pair of nudniks they are. **1964** W. MARKFIELD *To Early Grave* (1965) i. 12 'Worrier. Pest. Nudnick.' She lashed out at him. **1968** P. DURST *Badge of Infamy* v. 39 Nudnik is a kind of insulting endearment—a sort of lovable nitwit. **1968** L. ROSTEN *Joys of Yiddish* 265 A *nudnik* is someone who *nudzhes* or pesters. **1972** *New York* 8 May 70/1 Too many of our nudnik moviegoers..dread the prospect of sharing their pleasures with the plain folks.

‖**nuée ardente** (nüe ardɑ̃t). *Geol.* Pl. **nuées ardentes.** [Fr., lit. 'burning cloud'.

Introduced by A. Lacroix 1903, in *Compt. Rend.* CXXXVI. 874. In *La Montagne Pelée et ses Eruptions* (1904) 170 Lacroix says that he has since realised that the expression had earlier been used by the inhabitants of San Jorge in the Azores. He also says that by *ardent* he implies *brulant* 'burning' rather than *incandescent* 'glowing'. However the expression *nuée ardente* is usually rendered into English as 'glowing cloud' rather than 'burning cloud'.]

A hot, dense cloud of ash and fragmented lava suspended in a mass of gas, which typically is ejected laterally from the side of the dome of certain volcanoes (as Mount Pelée) and flows downhill at great speed like an avalanche. Also ellipt. **nuée.**

1904 A. HEILPRIN *Tower of Pelée* iv. 47 Professor Lacroix..refers to a number of discharges of the *nuée ardente* breaking out laterally from the base of the obelisk surmounting the crater-cone. **1912** *Amer. Jrnl. Sci.* CLXXXIV. 413 The highly viscous lavas of andesitic and trachytic nature might explode subaerially..into gas and divided solid material, causing such effects as the 'Nuées Ardentes' of Mt. Pelée. **1935** *Publ. Carnegie Inst. Washington* No. 458. 85 They witnessed the series of three rapidly succeeding *nuées* of July 9, 1902. *Ibid.* 89 The ejections followed one another at such brief intervals as to form an unbroken procession of *nuées ardentes*. **1966** *Earth-Sci. Rev.* I. 158 Large acid sheets might have a nuée origin. **1969** *Nature* 29 Nov. 864/1 Natural terrestrial examples of fluidized systems include certain quicksands ..and the volcanic eruptions known as *nuées ardentes*. **1969** BENNISON & WRIGHT *Geol. Hist. Brit. Isles* v. 94 The ignimbrites..are the product of nuées ardentes—the fiery cloud of lava and gas which occurs when lavas are silica rich and viscous. **1972** G. A. MACDONALD *Volcanoes* viii. 146 The cloud was the conspicuous feature and led Lacroix to give the name 'nuée ardente' (glowing cloud)... The name 'glowing cloud' seems, therefore, to give too much emphasis to a relatively minor feature, and many volcanologists today prefer to call the phenomenon a glowing avalanche.

Nuer (nū·əɹ). [Native name.] Name of an African people living in the south-eastern area of the Sudan; a member of this people; also, their language. Also *attrib.* or as *adj.*

1861 J. PETHERICK *Egypt, Soudan & Central Africa* xx. 362 We..made fast under two villages, which we found were inhabited by the Nouaer. **1873** E. E. FREWER tr. *Schweinfurth's Heart of Africa* I. iii. 118 We made a stop, and did some bartering with the Nueir... Here, in the heart of the Nueir population, in a district called Nyeng, we fixed our quarters. **1894** A. LEFÈVRE *Race & Lang.* v. 162 From the lakes of the Upper Nile to the Atlantic, the Negro dialects, properly so called, prevail: (1.) The Dinka group (Bari, Bongo, Chillouk, Nouer, etc.), the poorest of all, hardly issued from the monosyllabic stage. **1923** C. H. STIGAND (*title*) A Nuer-English vocabulary. **1932** C. G. & B. Z. SELIGMAN *Pagan Tribes Nilotic Sudan* i. 13 The Nuer..are tall, long-headed, very dark-skinned, and woolly-haired. *Ibid.* vi. 222 Every Nuer has a name given to him at birth by his father or other member of his family. **1940** E. E. EVANS-PRITCHARD *Nuer* iv. 184 Arab merchants..generally speak Nuer well. **1949** tr. *L. Homburger's Negro-African Langs.* iv. 89 In Nuer *ko* is used for 'we' distinct from *nè* 'I and thou' and *ne* 'we and you'. **1955** J. H. GREENBERG *Studies in African Linguistic Classification* 65 It is.. probable that the *ti* plural demonstrative and relative of Nuer..is cognate with the *ti* possessive and relative of Bari. **1964** E. A. NIDA *Toward Sci. Transl.* iii. 51 The Nuers of the Sudan have a very highly specialized vocabulary relating to cattle.

nuff (nʌf). *colloq.* (orig. *U.S.*). Also **nuf, 'nuff.** [colloq. and dial. abbrev. of ENOUGH *a.*, *sb.*, and *adv.*; cf. *'NOUGH.] **1. a. nuff said,** an indication that nothing more need be said on a particular topic. Also *nuf(f) ced*, *nuf(f) sed*, abbrev. *N.C.*, *N.S.*

1840 *Ninawah* (Peru, Ill.) *Gaz.* 20 June 2/3 'N.S. (Nuff said,)' whispered Mr. Fox. **1841** *Spirit of Times* 30 Oct.

409/1 'N.S., nuf sed', and up went *the soap*. **1873** Hotten *Slang Dict.* 235 *N.C.*, 'enough said', being the initials of *Nuf ced*. A certain theatrical manager spells, it is said, in this style. **1882** *Sydney Slang Dict.* 6/2 *N.C.*, '*Nuff Ced*', enough said. Thea. origin. **1892** J. C. Duval *Young Explorers* 151 'Nuf ced,' says Bill, 'you jess take care of your own har, and I'll see arter mine.' **1909** J. R. Ware *Passing Eng.* 184 *Nuf ced* (*From America*). Contraction of 'enough said'—absurdly spelt. Warning to say no more. Used in Liverpool chiefly. **1912** *Pedagogical Seminary* XIX. 97 [Expressions of] Negation and denial..'nuff said. **1942** N. Balchin *Darkness falls from Air* xiv. 234 'All right,' he said. "Nuff said.' **1958** *Spectator* 30 May 677/2 Mr. Randall is one of the newer members; 'nuff said. **1965** M. Spark *Mandelbaum Gate* vi. 164, I hope the food..is not unwholesome. How well I remember those weeks following your return from Spain... 'Nuff said'! **1971** J. Aitken *Nightly Deadshade* v. 49 'He and Steinherz knew one another at university before they were here.' 'Nuff said, I suppose.

b. **sure 'nuff, sho' nuff,** sure enough; often used as an intensive or to give emphasis to a point. (This sense chiefly *U.S.*, esp. in representations of Black English.)

1880 J. C. Harris *Uncle Remus* (1881) vi. 31 Den Brer Fox grit his toof sho' nuff; he did, en he look mighty dumpy. **1887** C. W. Chesnutt in *Atlantic Monthly* Aug. 257/1 Sho nuff, it rain de nex' day. **1898** J. D. Brayshaw *Slum Silhouettes* 150 An' sure 'nuff the nex' night arter, bang-bang comes a posty's knock at the door. **1921** E. O'Neill *Emperor Jones* (1953) ii. 168 I'se done up sho' 'nuff. *Ibid.* 170 I'se gone lost de place sho' 'nuff. **1935** Z. N. Hurston *Mules & Men* I. i. 26 Ole Massa begin to figger dat John musta seen somethin' sho nuff because John never had disobeyed him before. **1942** S. Kennedy in B. A. Botkin *Treas. S. Folklore* (1949) III. ii. 511 Sho nuff there was the preacher's buggy. **1971** J. D. Carr *Deadly Hall* xvii. 200 He's sho' nuff in good shape and ought to thank you.

2. Various other uses (see quots.).

1890 Barrère & Leland *Dict. Slang* II. 91/2 *Nuff* (soldiers): to have one's *nuff*, means to have had more drink than is good for one, *i.e.*, enough. **1923** J. Manchon *Le Slang* 208 *Nuff*,..abr. *enough*; *to have one's nuff*,.. être plein, ivre. **1967** C. Himes *Black on Black* (1973) 133 Go on, baby, you can be back in an hour with 'nuff bread so we can scoff. **1972** T. Kenrick *Tough One to Lose* xxi. 167 'Nuff outa you guys,' the big man called.

Nuffield (nʊ·fīld). The name of William Richard Morris, 1st Viscount *Nuffield* (1877–1963), founder of Morris Motors Ltd., used *attrib.* in **Nuffield Foundation,** a charitable trust set up by him to finance various schemes and organizations; hence *spec.* in *Educ.*, used *attrib.* and *absol.* in connection with the teaching methods and syllabuses advocated by the Nuffield Foundation since 1964, designed esp. for primary school level in mathematics, science, and French with the object of stimulating interest and individual thinking.

1943 *Times* 13 Feb. 4/4 Lord Nuffield has intimated his intention of founding forthwith a charitable trust, which will be known as 'The Nuffield Foundation'. **1946** *Nuffield Foundation First Report* 9 Early in 1943 Lord Nuffield established the Nuffield Foundation as a charitable trust and endowed it with Ordinary Stock Units in Morris Motors Ltd., to the value of £10,000,000. **1947** *Rep. Survey Comm. Old People* i. 1 The trustees of the Nuffield Foundation. **1952** C. P. Blacker *Eugenics* 307 Some voluntary organizations, such as those embodied in the Nuffield Foundation, have performed semi-official functions, such as the regional hospital surveys conducted for the Ministry of Health during the war. **1964** *Times Educ. Suppl.* 21 Aug. 249/4 If we do get new syllabuses for our nationwide school examinations they will probably be Nuffield ones. *Ibid.* 20 Nov. 932/5 Members of the Nuffield teams will act as examiners. **1971** *Where* Nov. 332/1 The sort of problem, taken from the Nuffield materials, young secondary children would soon get to work on. **1972** E. Elias *Learning & Playing* v. 81 Your child's teacher may well be using the Nuffield 'guides', a series of booklets stressing..*how* to teach. *Ibid.* vii. 128 Like Nuffield maths..Nuffield science is being introduced into more and more junior schools. *Ibid.* 129 Nuffield French has spread to schools all over the country. **1975** *Physics Bull.* Dec. 532/3 When a school starts taking Nuffield (other than Nuffield physical science) it never appears to give up doing so.

nuffieldite (nʊ·fīldəit). *Min.* [f. the name of E. W. *Nuffield* (b. 1914), Canadian geologist + -ite[1].] A sulphide of copper, lead, and bismuth, $Cu_4Pb_{10}Bi_{10}S_{27}$, found as steel-grey orthorhombic crystals at Lime Creek, British Columbia, Canada.

1968 P. W. Kingston in *Canad. Mineralogist* IX. 439 A second new sulphosalt mineral was discovered by R. M. Thompson... The new species is named nuffieldite in honour of E. W. Nuffield, who has made outstanding contributions to the understanding of many of the less well-known sulphosalt minerals. **1969** *Ibid.* X. 93 Because of the close similarity in physical properties of neyite, cosalite, aikinite, and nuffieldite, it was necessary to make numerous *x*-ray powder photographs to confirm the identity of the material used for analysis.

nuffin (nʊ·fin). Also **nuffink.** [Repr. a colloq. or dial. pronunc. of Nothing.] = Nothing *sb.* and *adv.* in var. senses.

1877 Geo. Eliot *Let.* 23 Nov. (1956) VI. 426 We have never seen Saturn's ring, nor Jupiter's moons—'nor nuffin'. **1897** [see *drift *sb.* 9]. **1898** J. D. Brayshaw *Slum Silhouettes* 1 If anyfink's wrong, yer won't get nuffink aht o' me. **1910** H. G. Wells *Hist. Mr. Polly* ix. 316, I don't stick at nuffin. **1922** 'R. Crompton' *More William* (1924) i. 24, I can't do nuffink with the mincing machine gone. **1951** 'J. Wyndham' *Day of Triffids* (1956) ii. 46 On its [*sc.* the triffid's] origins the Russians, true to type, lay low and said nuffin. **1957** *Economist* 28 Sept. 1002/1 It is..very nearly impossible for an Opposition to lie low and say nuffin at a time like this. **1958** J. Townsend *Young Devils* ii. 20 'Don't learn nuffin' in this school,' sneered another boy. **1971** G. Sims *Deadhand* II. ii. 88 Albert Chevalier's adage: 'Wot's the good of hanyfink?—Why—nuffink!' **1974** 'A. Gilbert' *Nice Little Killing* ix. 119 We don't know nuffin about the dear departed.

nugæ. Add: *spec.* in phr. **nugæ difficiles** *Philos.*, matters of trifling importance over which a disproportionate amount of time may be taken owing to their difficulty.

1710 Berkeley *Princ. Human Knowl.* 170 We may perhaps..look on all Inquiries about Numbers, only as so many *difficiles nugæ*, so far as they are not subservient to practise. **1867** Mill *Inaug. Addr.* 39, I am often tempted to ask the favourites of nature and fortune, whether all the serious and important work of the world is done, that their time and energy can be spared for these *nugae difficiles*? **1890** W. James *Princ. Psychol.* II. xxviii. 662 Such painstaking attempts..to prove all necessary judgments to be analytic..seem accordingly but *nugae difficiles*, and little better than wastes of ink and paper.

nugget, *sb.* Add: **1.** *transf.* (Earlier example.)

a **1859** T. De Quincey *Coll. Writings* (1889) I. 412 The secret truth, that rarest of all 'nuggets'.

3. Substitute for def.: A small, compact, stocky animal or person; also, a runt. (Further examples.)

1919 W. H. Downing *Digger Dial.* 36 *Nugget*, a short soldier. **1941** Baker *Dict. Austral. Slang* 50 *Nugget*, a small, weedy horse or other animal. (2) A small stocky man.

nugget(t)y, *a.* Add: **3.** Used also of other animals, and of people. Also *N.Z.*

1874 C. De Boos *Congewoi Correspondence* 141 He's just oner them short, square-built, nuggetty kinder fellers. **1887** *Daily News* 9 Apr. 5/4 The typical volunteer is gradually framing himself into a medium-sized, somewhat thickset, sturdy style of fellow, with a good chest measurement, firm on the pins, with the stamp on him of strength and hardihood—the sort of man..the Australians style 'nuggetty'. **1926** 'J. Doone' *Timely Tips for New Australians* (Gloss.) *Nuggetty*, short and sturdy. **1941** I. L. Idriess *Great Boomerang* vii. 52 Scowler was a square-faced, nuggety man, with a set scowl under a shock of hair. **1946** F. Sargeson *That Summer* 66 He wasn't a rangy specimen like me, no, he was nuggety. **1953** O. E. Middleton in C. K. Stead *N.Z. Short Stories* (1966) 187 Belle was..more nuggetty, but still powerful. She was short-haired..but had less white on the throat. **1969** O. White *Under Iron Rainbow* 107 He was a nuggety little bloke and moved as if he could handle himself in a blue. **1971** *Sunday Times* (Johannesburg) 28 Mar. 21/1 The nuggety young Wits University scrumhalf will be in the Springbok team to tour Australia. **1971** *N.Z. Listener* 19 Apr. 56/5 One was a nuggetty bloke in a sou'-wester, oilskin slicker, and bowyangs.

nuisance. Add: **2. f.** In phr. *to commit a nuisance*.

1863 *Harper's Mag.* Dec. 24/1 Commit no Nuisance. *a* **1922** T. S. Eliot *Waste Land Drafts* (1971) 5 We'd just gone up the alley, a fly cop came along, Looking for trouble; committing a nuisance, he said. **1942** Partridge *Usage & Abusage* 122/1 Euphemism may be obtained by using an extremely vague phrase, as in *commit a nuisance*.

3. *attrib.* and *Comb.*, as *nuisance action, aspect, candidate, tactics, tax*; **nuisance ground** *Canad.*, a rubbish dump; **nuisance raid,** a wartime bomb attack intended only to inconvenience or disrupt the enemy; also **nuisance-bombing, -raider; nuisance value,** the value or importance of a person or thing arising from a capacity to be a nuisance.

1964 C. Chaplin *Autobiogr.* xxix. 499 It's nothing but a nuisance action, Charlie; all the same I want you to keep from being served a summons. **1941** *Scrutiny* X. 80 There is a certain savage heightening of his nuisance-aspect towards the end. **1940** *Times Weekly* 27 Nov. 6 The change in the enemy's air tactics from general nuisance bombing to concentrated attacks on more restricted targets. **1968** *Globe & Mail* (Toronto) 13 Feb. 7/6 The only apple carts that Mr. Trudeau could not upset are those of..the three nuisance candidates. **1889** in *Herald Mag.* (Calgary) (1961) 11 Jan. 8/6 A well-grounded complaint has been made..in reference to dumping of filth in to the old cellar holes along the main road..instead of taking it to the nuisance ground. **1970** J. H. Gray *Boy from Winnipeg* 72 He had hauling contracts with the city that included carrying the manure from the city stables to the nuisance grounds. **1972** *Maclean's Mag.* (Toronto) Dec. 23/3 The town dump was known as the *nuisance grounds*, a phrase fraught with weird connotations, as though the effluvia of our lives was beneath contempt but at the same time was subtly threatening to the determined and sometimes hysterical propriety of our ways. **1942** *Hutchinson's Pict. Hist. of War* 18 Mar.–9 June 110 Some of the raids into enemy territory are merely nuisance raids to upset enemy nerves. **1943** *R.A.F. Jrnl.* Aug. 34 Cases of..nuisance raids by single rodents had been reported. **1944** *Ann. Reg. 1943* 341 The Germans..began a series of nuisance raids. **1944** *Daily Tel.* 11 July 1 German radio early to-day reported several nuisance raiders over the Berlin area. **1952** *Ann. Reg. 1951* 325 Government workers, forbidden..to strike, resorted to go-slow, work to rule, and other nuisance tactics. **1933** *Sun* (Baltimore) 4 July 6/6 (*heading*) Nuisance taxes help careerists. *Ibid.*, Authorities on taxation recently persuaded the State Legislature to abolish this tax as a nuisance, although its nuisance value was the chief reason for its existence. **1937** *Harper's Mag.* Jan. 846 If Hitler has a high nuisance value to France and England. **1958** A. Hocking *Epitaph for Nurse* v. 82 To add to her other nuisance-values, Geraldine complained and made trouble from morning to night. **1961** P. W. Brooks *Mod. Airliner* i. 22 'Nuisance value' of the aircraft to the general public on the ground (defined as the external overall noise level at climbing power at 650 ft.). **1972** *Guardian* 5 Jan. 4/6 Muskie's aides dismiss them airily as being no threat at all except for the nuisance value they may have in primaries.

‖**nuit blanche** (nwi blɑ̃ʃ). [Fr., lit. 'white night'.] A sleepless night. Also *fig.*

1853 C. Brontë *Villette* III. xli. 296 Leaving the radiant park and well-lit Haute-Ville (still well-lit, for this it seems was to be a 'nuit blanche' in Villette), I sought the dim lower quarter. **1904** W. Stevens *Let.* 14 Feb. (1967) 69, I do *not* want my dreams—my castles, my hunts, my *nuits blanches*. **1923** D. H. Lawrence *Ladybird* 81 She had gone to sleep from the *nuit blanche* of her days. **1964** L. Meynell *More Deadly than Male* xiv. 244 He lay awake for the rest of the night. A *nuit blanche*.

Nuits (St. George(s (nwi sæn ʒǫ̈rʒ). [The name of the capital town of the Côte de Nuits, in the department of Côte d'Or, France.] A red wine of Burgundy, produced in the district of Nuits St. Georges.

1841 Thackeray in *Fraser's Mag.* June 715/1 We had ..A bottle of nuits with the beef. **1875** [see *Beaune]. **1914** G. K. Chesterton *Flying Inn* xvii. 196 A strictly vegetarian beverage which bears the noble and starry name of Nuits. **1946** G. Millar *Horned Pigeon* xvi. 236 We found Eugène and his wife celebrating..with a bottle of Nuits St. Georges. **1966** B. E. Wallace *Murder in Touraine* vi. 56 Rocky swirled his Nuits St. George slowly round the glass. **1967** A. Lichine *Encycl. Wines* 383/2 Bouquet is also a Nuits characteristic, and the wines are sometimes quite pungent.

Nujol (niū·dʒǫl). The proprietary name of a paraffin oil used as an emulsifying agent in pharmacy and for making mulls in infra-red spectroscopy.

1916 *Trade Marks Jrnl.* 16 Feb. 153 Nujol... Medicinal and curative preparations..all being for human use. Standard Oil Company.., Bayonne, Hudson County, New Jersey, United States of America; manufacturers. **1923** W. Clayton *Theory of Emulsions* ii. 30 If a solution of KI be added to an emulsion of 'Nujol' in sodium oleate solution, prismatic colours are secured. **1943, 1948** [see *mull *v.*[1] 1 b]. **1956** [see *mull *sb.*[1] 2]. **1971** Rosenblatt & Davis *Course in Org. Chem.* iii. 71 Nujol is only useful in regions not involving carbon–hydrogen vibrations. **1975** D. H. Burrin in Williams & Wilson *Biologist's Guide to Princ. & Techniques Pract. Biochem.* v. 151 A Nujol mull, consisting of a fine paste of sample in liquid paraffin may be prepared.

nuke (niūk, U.S. nūk), *sb.*[2] *slang* (chiefly *U.S.*). Abbrev. of 'nuclear bomb, weapon,' etc.

1959 *N.Y. Times Mag.* 1 Feb. 46/3 Soon there may be 5-inch nuclear shells and portable Davy Crockett 'nukes' for the infantryman. **1960** *Time* 4 July 52/1 But the nuclear submarines—called 'nukes'—can cruise underwater for weeks at top speed. **1969** *Life* 29 Sept. 23/2 Once communities vied for nuclear power plants ('nukes') as passports to prosperity. **1971** J. Ball *First Team* (1972) xv. 224 'What is nukes?' the man asked. 'Nuclear specialists; no one else can handle this stuff.' **1973** *Publishers' Weekly* 14 May 44/1 They hijack a liner at sea and sink it with a baby nuke... He is given the job of detonating the big nuke.

Hence as *v. trans.*, to use nuclear weapons against; **nu·king** *vbl. sb.*

1967 *Look* 11 July 25, I remember in Saigon how disturbed General Westmoreland was after talking to a group of American editors..who told him they favored 'nuking' (A-bombing) China. **1970** *New Yorker* 4 July 52 We have to get ready to nuke 'em to kingdom come. **1972** *S. China Morning Post* (Hong Kong) 29 Sept. 2/3 Dr. Strangelove nuked the Russians on his own authority. **1972** *Japan Times Weekly* 23 Dec. 4/2, I asked how he could be sure that the Soviet Union would nuke us if we nuked China. **1973** *Daily Tel.* (Colour Suppl.) 13 Apr. 40/1 'Nuked', for those unfamiliar with modern war-parlance, means to let off a nuclear bomb.

null, *sb.*[1] Delete †*Obs. rare* and add: **b.** *Cryptography.* (See quot. 1961.)

1915 J. Buchan *39 Steps* iii. 72 It was a numerical cypher, and by an elaborate system of experiments I had pretty well discovered what were the nulls and stops. **1961** Shulman & Weintraub *Gloss. Cryptography*, *Null cipher*, a form of concealment in which most of the letters used in a text are nulls. *Nulls*, meaningless letters or numbers used in cryptograms for various purposes. **1968** 'S. Jay' *Sleepers can Kill* iv. 47 A very simple cypher, Mr. Connor, but it can of course be broken by frequency analysis... Don't forget to reverse and arrange in five letter groups, of course, with nulls to make the numbers up. **1972** *Sci. Amer.* Nov. 114/2 If this is intercepted and a translation demanded of the sender, he strikes out the symbols of the true text, explaining that they are what

cryptographers today call 'nulls', meaningless symbols inserted only to make the cipher harder to break.

c. *Linguistics.* (See quots.)

The *attrib.* example properly belongs in sense 4 of the *adj.*

1964 E. Bach *Introd. Transformational Gram.* v. 110 Since Y and Z (as usual) include the possibility of null, the transformation will have the desired effect of producing all permutations of the terminal elements. **1968** *Amer. Speech* XLIII. 277 In treating *few* in 'few people' and *much* in 'much money' as pre-article forms, Roberts assumes the existence of an unarticulated determiner he calls *null*. **1968** R. T. Harms *Introd. Phonological Theory* v. 44 If two obstruents occur initially, the first one is deleted. (Instead of 'null', the symbol 'ϕ' is sometimes used.) **1972** R. A. Palmatier *Gloss. Eng. Transformational Gram.* 111 *Null*, lacking status as a morpheme or formative—that is, having no overt representation on any level of the grammar... *Null string*, an empty string—that is, no string at all.

null, *a.* Add: Also **nul. 4.** (Further examples.) Also more widely, esp. in *Math.*, with the sense: having, being, or associated with the value zero. *null space*, a space composed of all quantities that are transformed into zero by some given transformation.

1835 W. R. Hamilton *Conjugate Functions* 23 The transition may be said to be null, or a null step, as producing no real alteration in the moment from which it is made. *Ibid.* 31 The standard or zero-moment A itself may be denoted by the complex symbol 0 + A, because it may be conceived as generated from itself by applying the null step 0. *Ibid.* 35 The null cardinal (or number none). **1884** A. Buchheim in *Phil. Mag.* XVIII. 459 A linear space of (α−1) dimensions will be called an α-point... A matrix A of order *n* is considered as operating on the coordinates of the points of an *n*-point... If A is of nullity α, the equation $A(x_1 \dots x_n) = 0$ is satisfied by all the points of a certain α-point, and conversely. I call this α-point the null space of A. **1885** *Proc. London Math. Soc.* XVI. 78 A point *x*, such that $\phi x = 0$, is called a null-point of [the matrix] ϕ. **1898** A. N. Whitehead *Universal Algebra* I. II. ii. 25 Since in combination with any other element the null element 0 disappears, the symbolism may be rendered more convenient by writing $-a$ for $0-a$. **1922** E. H. Neville *Prolegomena to Analytical Geom.* v. v. 303 An ordinary circle of radius zero is called a nul circle. **1926** H. Bagchi *Course of Geom. Anal.* v. 263 The theorem..prescribes a null value for a determinant having two coincident rows or columns. **1930** J. W. Young *Projective Geom.* ix. 158 If *A* and *B* coincide, the vector *AB* is called the null vector and is denoted by 0. **1941** Birkhoff & MacLane *Surv. Mod. Algebra* x. 268 The null-space of a linear transformation *T* is the set of all vectors ξ such that $\xi T = O$. The null-space of a matrix *A* is the set of all row matrices *X* which satisfy the homogeneous linear equations $XA = O$. **1954** H. Griffin *Elem. Theory of Numbers* i. 8 A null element is an integer that divides only itself. *Ibid.*, Zero is the null element of the rational integers. **1956** R. H. Atkin *Math. & Wave Mech.* iii. 76 As in all algebras we find it convenient to introduce two identity operators. These are called the unit operator (idem factor) and the nul operator (zero). **1961** Seifert & Brown *Ballistic Missile & Space Vehicle Systems* xvii. 404 They [*sc.* gyroscopes]..operate as angular error detectors over a very small range around their reference or null positions. **1962** C. Wexler *Analytic Geom.* v. 141 The plane cuts the cone merely in a point (the vertex), resulting in a null circle or a null ellipse. **1964** E. Bach *Introd. Transformational Gram.* ii. 14 Among the elements assumed to be available for each grammar..there is one which plays a special role, namely, the null (unit or identity) element. It functions in the mathematical system underlying the representations of a grammatical theory as does zero in ordinary addition, or the digit 1 in ordinary multiplication. **1971** L. T. Agger *Introd. Electricity* xviii. 326 Balance occurs with null, or zero, deflection of the pointer. **1971** C. W. Curtis in Powell & Higman *Finite Simple Groups* iii. 171 *f* vanishes on the nullspace of α. **1971** J. Z. Young *Introd. Study Man* xvi. 201 This ingeniously establishes the change in growth power by a nul method.

b. Mech. and Geom. *null point* [tr. G. *nullpunkt* (A. F. Möbius *Lehrb. d. Statik* (1837) I. vi. 144)], the point of intersection of all the null lines of a given wrench that lie in a given plane; *null line*, a line about which the moment of a given wrench is zero; *null plane* [tr. G. *nullebene* (A. F. Möbius, *loc. cit.*)], the plane containing all the null lines of a given wrench that pass through a given point; *null system*, a system of null points and their corresponding null planes.

1903 C. M. Jessop *Treat. Line Complex* iii. 45 The distinction between the two spaces Σ and Σ′ has now disappeared and we have a (1, 1) correspondence between the points and planes of space in which each point lies in its corresponding plane... Such a correspondence is called a null-system. The corresponding points and planes are called 'null-points' and 'null-planes'. **1911** *Encycl. Brit.* XVII. 967/1 In the 'Null-System' of A. F. Möbius (1790–1868), a line such that the moment of a given wrench about it is zero is called a null-line. **1942** Synge & Griffith *Princ. Mech.* x. 301 A rigid body is acted on by a force **F** at *O* and a couple **G**. *P* is an assigned point, with position vector **r** relative to *O*. Show that there is a single infinity of lines through *P* about which the force system has no moment; show that these lines lie in a plane... (The lines are called null lines, and the plane a null plane.) **1964** C. E. Springer *Geom. & Anal. Projective Spaces* x. 273 In a polar correlation with a skew-symmetric matrix, every point is incident with its corresponding plane. This type of correlation is called a null system, a designation used in the theory of statics as developed by Möbius.

c. *Math.* and *Logic.* Of a class or set: having no members. Of a propositional function or relation: always false; having the null class as its range.

1903 B. Russell *Princ. Math.* ii. 22 A propositional function is said to be null when it is false for all values of *x*; and the class of *x*'s satisfying the function is called the null-class, being in fact a class of no terms. **1906** W. H. & G. C. Young *Theory of Sets of Points* 288 The null-set..contains no point. **1932** Lewis & Langford *Symbolic Logic* vii. 208 The universal function is expressible as *p* v ∼*p*. The null-function is the negative of this. **1941** O. Helmer tr. *Tarski's Introd. Logic* v. 90 We have..in the calculus of relations two special relations, the universal relation V and the null relation Λ, the first of which holds between any two individuals, and the second between none. **1948** Ambrose & Lazerowitz *Fund. Symbolic Logic* x. 215 To say that a class is null, or empty, is the same as saying that there are no values satisfying its defining function. **1955** [see *domain *sb.* 4 f]. **1956** A. Church *Introd. Math. Logic* (rev. ed.) I. 31 If the range is given,..there is only one null class. But, e.g., the range of the null class associated with the form sin *x* = 2 and the range of the null class associated with the form *e* < 0 are not the same... We shall speak respectively of the 'null class of real numbers' and of the 'null class of positive real numbers'. **1966** S. Beer *Decision & Control* vi. 107 This does not matter; it simply means that the complementary set has no elements—and this is called a null set.

d. *Physics.* Existing between or joining points in space-time between which the interval is zero; *null cone* = *light cone* s.v. *light *sb.* 16.

1928 *Phil. Mag.* V. 242 A null-surface is defined as an envelope of null-cones,..while the characteristic lines on a null-surface prove to be geodesic null-lines. *Ibid.*, The history of a light wave in space-time is a null-surface. **1942** Synge & Griffith *Princ. Mech.* xvi. 475 A null line represents the history of a flash of light traveling along the axis *Ox* of a Galilean frame. **1959** J. Aharoni *Special Theory of Relativity* i. 25 An interval *ds* between two events can be either real, zero, or imaginary... In the first case the interval is called space-like, in the second case a null-interval, and in the third case time-like. **1968** T. C. Bradbury *Theoret. Mech.* xiii. 589 If the four-vector joins *O* and *C*, then $r^2 - c^2t^2 = r'^2 - c^2t'^2 = 0$. Such a vector is a null vector. Events separated by null vectors can be joined by light signals; in fact, the light cone is frequently called the null cone. **1972** J. Ehlers et al. in L. O'Raifeartaigh *Gen. Relativity* iv. 75 In a normal hyperbolic Riemannian manifold, the distinction between the interior and exterior of the null cone N_p of a point *p* can locally be defined by means of the sign of the world function $\Omega(p, q)$, for *q* near *p*.

e. Statistics. *null hypothesis*, a hypothesis that is the subject of a significance test, esp. the hypothesis that there is no difference between specified populations (any apparent difference being due to sampling or experimental error).

1935 R. A. Fisher *Design of Experiments* ii. 19 The two classes of results which are distinguished by our test of significance are..those which show a significant discrepancy from a certain hypothesis; namely, in this case, the hypothesis that the judgments given are in no way influenced by the order in which the ingredients have been added; and..results which show no significant discrepancy from this hypothesis. This hypothesis..is again characteristic of all experimentation... We may speak of this hypothesis as the 'null hypothesis', and it should be noted that the null hypothesis is never proved or established, but is possibly disproved, in the course of experimentation. **1973** *Jrnl. Genetic Psychol.* CXXIII. 86 The specific null hypotheses tested were as follows: 1. There were no differences in hostile press (fear of failure) score decreases between self-reinforcement, group therapy, and control groups. 2. Hostile press scores do not decrease significantly during the self-reinforcement condition. [Etc.] **1975** *Nature* 20 Feb. 607/1 The evidence ..for positive selection directing the evolution of early vertebrate haemoglobin is supported by a χ^2 test, performed to test the null hypothesis that sites where mutations produced acquired crucial functions did not evolve at a different rate from all other sites.

5. Passing into *sb.*: A condition of no signal; also, a direction in which no radiation is detected or emitted. Freq. *attrib.*

1926 *Bell Syst. Techn. Jrnl.* V. 295 The number of lobes is, of course, equal to the number of null directions. **1931** *Proc. IRE* XXXI. 1426 The nulls and maxima.. have been plotted in Fig. 16 for an element length of four wavelengths. *Ibid.* 1427 Figs. 15 and 16..give the null points. These are seen to be 0, 30, and 90 degrees in Fig. 15. **1931** Moyer & Wostrel *Radio Handbk.* xii. 601 A direct-current galvanometer, protected by a suitable shunt, is used as a null indicator. **1958** Condon & Odishaw *Handbk. Physics* iv. v. 66/2 Null determinations in d-c measurements are almost invariably made with a D'Arsonval-type galvanometer. **1962** *Newnes Conc. Encycl. Electr. Engin.* 94/2 The precision of a bridge measurement depends on the sensitivity of the null detector. **1968** M. Woodhouse *Rock Baby* xv. 150, I set up the loop of the D.F. set and plugged in the headset. When the signal came..I got a clear null a little west of due south. **1973** J. Hulbert *All About Navigating* vii. 96 It is much easier to pick out the softest point than the loudest point and so a null is used for direction finding. **1975** *Gramophone* Oct. 734/3 The many nulls and lobes beyond these angles are not too important, and are due to the interaction of the radiator with its housing and the corners of the cabinet.

null, *v.*[1] Add: **3.** (Further examples.) Also with *out*.

1957 *Electronics* 1 Mar. 163 The receiver uses a shielded-loop antenna to null out main powerline noise. *Ibid.*, The atmospherics..are greater in amplitude than most of the peaks of the background hash after nulling the main source of powerline noise. **1971** *Sci. Amer.* Aug. 65/2 The harmonic signal is electronically processed and sent through the feedback winding to 'null out', or cancel, the ambient field within the sensor.

null, *v.*[2] Add: Hence **nu·lling** *vbl. sb.*, the making of knurls; knurled work.

1851 C. Cist *Sk. Cincinnati* xiii. 245 All kinds of turning used by cabinet makers, including nulling of every pattern, furnished at the shortest notice. **1914** Eberlein & McClure *Pract. Bk. Period Furnit.* 63 Nulling, made up chiefly of beading, cabling and hollows, is often used to ornament the bulbous legs of Jacobean furniture. **1934** J. Gloag *Eng. Furnit.* v. 86 The edges of tables and desks were scalloped with nulling, heavily gilded. **1936** C. H. Hayward *Eng. Period Furnit.* ii. 33 The carved flutes partly filled with nullings are a feature that was used considerably in Elizabethan times. **1966** M. M. Pegler *Dict. Interior Design* 307 *Nulling*, a Jacobean wood-carving technique which produced an effect similar to repoussé or chased metalwork. The patterns were created by a series of small projections or recessions from the surface (like a boss or bead) of the wood.

nullah. Add: Also **nala.** (Further examples.)

1920 *Blackw. Mag.* Jan. 117/1 You feel him..open out again to his stride, or drop out into a sudden nala without a peck. **1933** *Discovery* June 204/1 The nalas and river beds were parched, awaiting the flood. **1963** T. Tullett *Inside Interpol* xiv. 184 After twenty minutes of careful tracking it led him to a *nala* (ditch) where he found the impression of the sole of a heavy boot. **1964** R. Perry *World of Tiger* i. 10 The tigers he shot were living in nullahs close to villages. **1972** I. Baker *Grave Doubt* vii. 90 Could it be because it was England and not some god-forsaken Afghan nullah?

nullification. 2. b. (Earlier examples.)

1798 T. Jefferson *Writings* (1905) XVII. 386 Where powers are assumed which have not been delegated, a nullification of the act is the rightful remedy. **1830** *Massachusetts Spy* 22 Sept. (Th.), Nullification nullified.

nullify, *v.* **1.** (Earlier absol. example.)

1832 J. P. Kennedy *Swallow Barn* I. xviii. 186 I'd be glad to know if we couldn't nullify.

nulliplex (nʊ·lipleks), *a. Genetics.* [f. L. *nulli-*, *nullus* no + *-plex* as in Simplex, Duplex *adjs.* etc.] Of a polyploid individual: having the dominant allele of any particular gene not represented.

1921 A. F. Blakeslee in *Amer. Naturalist* LV. 257 A Poinsettia plant may, to speak in terms of the dominant factor, be considered nulliplex with no dominant genes, or simplex, duplex or triplex with, respectively, 1, 2, or 3 dominant factors. **1929** *Jrnl. Genetics* XXI. 141 From previous results it has been assumed that 14/26 is triplex for *Y* and heterozygous for *I*, and that 'White Star' is nulliplex for *Y* and heterozygous for *I*. **1963** Lewis & John *Chromosome Marker* iv. iii. 327 On selfing, a tetraploid of this kind [*sc.* AAaa] will give not three types of offspring but five, namely AAAA (quadruplex), AAAa (triplex), AAaa (duplex), Aaaa (simplex) and aaaa (nulliplex). *Ibid.* 328 Thus, with chromatid segregation a triplex can produce nulliplex offspring.

‖**nulli secundus** (nʊ·li sĭkʊ·ndʊ̆s), *adj. phr.* Also (applied to things) **nulli secundum.** [L.] Second to none.

1869 S. R. Hole *Bk. about Roses* iv. 55 If Mr. Shirley Hibberd..can grow good Roses within four miles of the General Post-Office..it is quite certain that he would be *nulli secundus* with the full advantage of situation and soil. **1935** G. K. Zipf *Psycho-Biol. Lang.* (1966) 75 The conventional Sanskrit alphabet is an amazingly accurate phonetic alphabet, practically *nulli secundum*. **1963** *Brewer's Dict. Phr. & Fable* 649/2 *Nulli secundus*,.. the motto of the Coldstream Guards, which regiment is hence sometimes spoken of as the *Nulli Secundus Club*.

nullisome (nʊ·lisōᵘm). *Cytology.* [f. L. *nulli-*, *nullus* no + *-some[4].] A pair of homologous chromosomes lacking from a diploid chromosome complement; also, a nullisomic individual.

1944 *Genetics* XXIX. 233 Speltoidy..has been shown.. frequently to be due to the effects of a particular monosome or its corresponding nullisome. *Ibid.* 234 Seventeen of the possible 21 different nullisomes in *Triticum vulgare* have now been obtained. **1968** [see next]. **1973** G. S. Khush *Cytogenetics Aneuploids* ix. 221 In wheat, a successful method was the test of the ability of a particular tetrasome to compensate for a particular nullisome.

nullisomic (nʊlisōᵘ·mik), *a.* (*sb.*) *Cytology.* Also † **nullosomic.** [f. as prec. + *-somic*, after *monosomic *a.* (*sb.*), etc.] Having or being a diploid chromosome complement in which one or more than one pair of homologous chromosomes is lacking. Hence as *sb.*, a nullisomic individual.

1932 *Genetics* XVII. 694 In the hybrid population the nullosomics were readily recognized. **1939** *Ibid.* XXIV. 510 Neither of the nullosomic plants set seeds. **1941** *Ibid.* XXVI. 167 (*heading*) Nullisomics in *Triticum vulgare*. **1944** *Ibid.* XXIX. 233 Some of these have been nullisomic types. **1968** R. Rieger et al. *Gloss. Genetics & Cytogenetics* 319 Normally, nullisomics (or nullisomes) cannot

survive in diploids. **1973** G. S. Khush *Cytogenetics Aneuploids* i. 3 Kihara (1924) and Winge (1924) were the first to report the nullisomic plants of wheat. *Ibid.* viii. 186 Some nullisomics of *Avena* may have normal viability.

nullity. Add: **6.** *Math.* The number of columns of a matrix minus its rank; the dimension of the null space of a matrix or linear transformation (equal to the dimension of its domain minus that of its range).

1884 J. J. Sylvester in *Amer. Jrnl. Math.* VI. 274 The absolute zero for matrices of any order is the matrix all of whose elements are zero. It possesses so far as regards multiplication..the distinguishing property of the zero, viz. that when entering into composition with any other matrix ..the product..is itself over again... This is the highest degree of nullity which any matrix can possess, and (regarded as an integer) will be called ω, the order of the matrix... In general.., if all the minors of order $\omega - i + 1$ vanish, but the minors of order $\omega - i$ do not all vanish, the nullity will be said to be *i*. **1941** Birkhoff & MacLane *Survey Mod. Algebra* x. 268 Rank + nullity = Dimension of Domain. **1972** A. G. Howson *Handbk. Terms Algebra & Anal.* ix. 47 In the case of a homogeneous system, $Ax = 0$, the inverse image of 0, i.e. the set of solutions, is Ker(*t*)..which is a subspace of F^n..of dimension $n - r$, a number known as the nullity of *t* (or *A*).

nullness (nʊ·lnĕs). [f. Null *a.* + -ness.] The property or state of being null.

1949 E. Bowen *Heat of Day* xvii. 301 The coming of winter to a stop had been most felt in the absolute nullness of night. **1959** *Proc. R. Soc.* A. CCLI. 521 The wave-like character will appear in the intrinsically *null* properties of these metrics... The nullness will become apparent from the physical properties of the metrics. **1967** *Punch* 22 Nov. 795/2 This feeling of organic injury is caused by..an utter nullness of non-taste, a denatured, defused..triviality.

nullo. Add: **2.** A type of bridge in which the object is to lose rather than gain tricks, or one in which tricks gained count against a player. Also *attrib.*

1914 *London Opinion* 7 Feb. 231/1 The new 'nullo' call at auction bridge is catching on... The Editor of the *Strand Magazine* intends publishing an article about 'nullos' in the March number. **1929** M. C. Work *Compl. Contract Bridge* 241 *Nullos*, an unauthorized, and now obsolete, form of the game in which points were scored for losing instead of winning tricks. **1945** A. A. Ostrow *Compl. Card Player* 620 *Null* (*Nullo*): this is a declaration in which the bidder contracts not to win a single trick. **1972** H. Phillips *Pop. Bk. Card Games* 373 Nullos is a form of Bridge for two which I have long recommended. *Ibid.* 375 Nullos with Contract scoring is an improvement on the original game, where the scoring and conditions are based on Auction.

‖**nullus** (nʊ·lʊ̆s), *sb. rare*[−1]. [L.] No one, nobody.

1929 D. H. Lawrence *Pansies* 100 My whole consciousness is cliché And I am null; I exist as an organism And a nullus.

nully (nʊ·li). *slang.* [? f. Null *a.*; cf. *nullion* (Sc. Nat. Dict.) a stupid fellow.] A fool, a stupid person.

1973 Boyd & Parkes *Dark Number* iv. 45 You'll be a right nully if you don't mistrust all *polis* on sight. *Ibid.* vii. 78 He's a nully, a cipher. He contributes nothing. **1973** R. Parkes *Guardians* ii. 48 He's a sick, junked-up, pathetic old nully.

num, var. *yum. Cf. *nummy-nummy.

1899 R. Whiteing *No. 5 John St.* ix. 86 Her handkerchief perfumes the whole room..to which Covey's frequent 'Num! num! num!' calls embarrassing attention. **1924** *Dialect Notes* V. 274 *Num, num, nummy, num* (humming or joy). **1961** W. Sansom *Last Hours of Sandra Lee* vi. 112 Mouths ooed and nummed noises of appreciation.

numb, *a.* Add: **1. a.** Also *transf.*

1874 [see *dobber[2]]. **1892** P. H. Emerson *Son of Fens* xxxii. 349 My old head fare as numb as a beetle. **1958** [see *feel *sb.* 5]. **1970** H. E. Roberts *Third Ear* 10/2 *Numb*, dumb; stupid.

numbat (nʊ·mbæt). *Austral.* [Aboriginal name.] The banded anteater, *Myrmecobius fasciatus*, a small, rare marsupial native to south-western Australia.

1923 F. Wood Jones *Mammals of South Australia* I. 123 To the aboriginals..it is known as the Numbat, and this name will be adopted here. **1942** C. Barrett *On Wallaby* iii. 38, I trailed the numbat, in South-western Australia, taking the first photograph..of that little-known animal. **1962** *Times* 22 Nov. 24/2 The numbat, once thought to be extinct, cannot be kept in captivity. **1965** *Sunday Mail Mag.* (Brisbane) 4 Apr. 14/1 The numbat, or banded anteater, of South-west Australia..is about the size of a large rat with a coat of coarse, reddish brown fur... It also has a long bushy tail, something like the fox's brush. On the upper part of its body it has white or cream bands. *Ibid.* 14/2 Ants are ripped out of the wood with a flick of the numbat's long tongue... When eating the numbat makes a noise which sounds like 'Tut-tut'. If interrupted it makes a sound more like 'Chur-rr!'.

number, *sb.* Add: **3. e.** *U.S. slang.* Usu. in *pl.* An illegal form of gambling in which bets are taken on the occurrence of numbers in a lottery or in the financial columns of a newspaper. Esp. in phr. *to play the numbers*. Also called **numbers game, racket.** Cf. Policy *sb.*[2] 1 c. Freq. *attrib.* and *Comb.*, as *number(s)-man*; **numbers drop,** a session of such betting; **number(s)-runner,** one who collects the bets of those playing the numbers.

1897 Ade *Pink Marsh* 170 She tell Belle 'at she heah I like gin an' roll'e bones an' play numbehs. **1926** C. Van Vechten *Nigger Heaven* 286 *Numbers*, a gambling game highly popular in contemporary Harlem. The winning numbers each day are derived from the New York Clearing House bank exchanges and balances..published in the newspapers. **1934** *Sun* (Baltimore) 25 Aug. 1/2 Hawkins..identified himself as a 'pay-off man' in the 'numbers business'. **1935** *Time* 21 Jan. 45/1 In Danville, Va., operators of a 'numbers' game were bankrupt. **1949** E. E. Blanche *You can't Win* 70 The 'numbers' racket is known by different names in various sections of the country—The Numbers Policy, Clearing House, Butter and Eggs, and the Bug. **1950** H. E. Goldin *Dict. Amer. Underworld Lingo* 146 Number-man, anyone engaged in the policy numbers racket. **1958** S. Ellin *Eighth Circle* (1959) II. v. 65 He was saying something about the numbers game... The *bolita*. **1958** R. Ellison in A. Dundes *Mother Wit* (1973) 63/1 As a numbers runner he is a bringer of manna and a worker of miracles. **1959** *Times Lit. Suppl.* 16 Jan. 29/3 The 'numbers' do for Harlem what the pools do for Notting Hill Gate—and for that matter Knightsbridge as well—provide unsustaining nourishment for dreams. **1959** *Listener* 28 May 924/2, I wonder how many people now remember that prominent feature of American life in the 'thirties—the numbers racket. **1964** O. Harrington in J. H. Clarke *Harlem* 90 One of the local numbers runners dug my cartoon and.. nobody covers as much Harlem territory as the numbers man. **1965** 'Malcolm X' *Autobiogr.* 52 Betting my dollar a day on the numbers. **1968** P. Oliver *Screening Blues* iv. 134 The policy writers and numbers runners who took the bet by a rapid code of signals in the street or at the 'numbers drop' would urge them to play other numbers and at higher stakes. **1970** L. Meriwether *Daddy was Number Runner* 21 Mother played the numbers like everyone else in Harlem but she was scared about Daddy being a number runner. **1973** *Black Panther* 22 Sept. 8/3 A panel of prominent lawyers recommended last week that the District of Columbia legalize 'the numbers' racket. **1975** *New Yorker* 29 Sept. 54/3 She had met Delgado while she was selling numbers on the streets of the lower East Side.

f. *Austral.* and *N.Z.* Elementary arithmetic taught to children in primary school.

1922 *N.Z. Education Gaz.* 1 Dec. 137/1 Miss Caldwell has published a book entitled 'The Simplicity of Number', a copy of which, along with the apparatus, can..be obtained from her by teachers. **1963** B. Pearson *Coal Flat* iv. 63 You'd best make sure of his reading and his number and see if he's good enough for this class.

4. a. (Further examples.)

1898 G. B. Shaw *Philanderer* in *Plays Unpleasant* III. 135 *Julia*... What is Dr Paramore's number in Savile Row? *Charteris*. Seventy-nine. **1908** E. F. Benson *Blotting Bk.* i. 16, I saw one policeman trying to take my number. **1973** R. Lewis *Blood Money* iv. 47 'Could it be the number of the hire-car he used?'..'It's a Leeds number.'

b. (*a*) (Further examples.) In later use also *transf.* in phr. *to make one's number*, to report one's arrival, to report for duty, to pay a duty or courtesy call, to make oneself known, to make oneself acquainted. *colloq.*

1861 J. Lamont *Seasons with Sea-Horses* xviii. 293 We found that the 'Anna Louise' had only made her number twelve hours before us. **1897** P. E. Stevenson *Deep-Water Voyage* 29 We made our number, where from, where bound, and 'all well' to the steamer, which hoisted her pennant immediately. **1927** B. M. Chambers *Salt Junk* xxx. 256 Almost every ship on her way to and from South America makes her number to the island [Fernando Noronha]. **1937** C. S. Forester *Happy Return* xxiv. 281 The *Lydia* made her number, and the sound of the salutes began to roll slowly round the bay. **1942** Partridge in *New Statesman* 1 Aug. 75/1 'To *make one's number*' is still slang; it may be used absolutely, as in 'As soon as I join my unit I must make my number at Brigade', or in reference to a person, as in 'I must lose no time in making my number with one of the Staff Officers at Division'—in short, to *contact* him... By the way, one 'makes one's number' with one's *opposite number*, a phrase taken over from civil life. **1945** 'N. Shute' *Most Secret* ix. 211 Captain (D.) was there to see them off; I made my number with him as representing V.A.C.O. and we stood chatting for a time. **1951** H. Jordan *Islander* II. vii. 92 Jim brought his ketch..alongside the *Islander* soon after breakfast and made his number to the master. **1955** E. Waugh *Officers & Gentlemen* I. vi. 56 You go ahead and make your number with your CO. **1958** M. Dickens *Man Overboard* xii. 192 Ben saw himself on Speech Day, making his number with mothers in garden-party hats. **1963** P. McCutchan *Man from Moscow* ix. 87 On arrival in Moscow..Shaw made his number at the W.I.O.C.A. office. **1965** B. Sweet-Escott *Baker Street Irregular* viii. 224 We turned back and made our number with the navy there. **1974** D. Seaman *Bomb that could Lip-Read* xviii. 177 'Will you go to the conference site today?' 'Might as well make my number with the R.U.C.'

d. With reference to a lottery number, or some other number by or with which one may be identified, as an army number, esp. in fig. phr. *one's number is* (or *has gone*) *up*, one is doomed (to die), one's time is come, one is 'done for'; *one's number is on* (something, esp. a bullet or shell), one is doomed (to die, or to a particular destiny). Cf. *name *sb.* 1 g.

1806 C. Lamb *Let.* 25 Jan. in *Works* (1870) II. 89 Though this is a lottery to which none but G. Barnett would choose to trust his all, there is no harm just to call in at Despair's office for a friend, and see if *his* number is come up. **1899** C. Rook *Hooligan Nights* iv. 56 You couldn't tallygraft to Billy no more. His number's up awright, wiv no error. **1914** *London Opinion* XL. 231/2 The late Patsey Cadogan, who left £100,000 when his number went up. **1915** 'Bartimeus' *Tall Ship* i. 11, I think our number's up, old thing. **1922** Wodehouse *Girl on Boat* xi. 181 Fate had dealt him a knock-out blow; his number was up. **1925** Fraser & Gibbons *Soldier & Sailor Words* 163 *Name* (or *number*) *on, to have one's*, said of a bullet that hit a man; *i.e.*, that it was destined for him. **1929** *Mercury Story Bk.* 98 It was about midday that I first realised that his number was up. **1937** V. Bartlett *This is my Life* xi. 188 The Director-General said that he would nevertheless like me to broadcast a short talk under my own name... My number was up. **1965** Brophy & Partridge *Long Trail* 154 *Number on*. A fatalistic but consolatory superstition insisted that no man need fear any bullet or shell, however close it came, unless it *had his regimental number* (or his name and number) engraved on it. **1965** F. Sargeson *Memoirs of Peon* vi. 138 She was forgiven for insisting upon her husband's undertaking the labour which had unfortunately sent his number up. **1966** *Listener* 23 June 923/2 The endless stream of cars and lorries swept on, only momentarily slowed down when the number of one of them came up... It was the arbitrariness of accident. **1974** C. Fremlin *By Horror Haunted* 15 I'm as safe here as..any where..if it's got your number on it, you'll get it, no matter *where* you are! **1975** J. Aiken *Voices in Empty House* xviii. 331 He'd got leukaemia. He knew his number was up.

e. In *fig.* use in phr. *to get* (*take, have*) *someone's number*, to make a correct appraisal of someone's character, motives or intentions, to size someone up.

1853 Dickens *Bleak Ho.* lvii. 550 Whenever a person proclaims to you 'In worldly matters I'm a child,'..that person is only a crying off from being held accountable, and..you have got that person's number, and it's Number One. **1889** 'Mark Twain' *Yankee* xxxiv. 405 Let him go, for the present; I took his number, so to speak. *Ibid.* xxxv. 414 That was the sort of master we had. I took *his* number. **1912** C. Mathewson *Pitching in a Pinch* i. 4 'I've got your number now, Matty!' he shouted at me as he drew up at second base. **1920** W. Hard *Raymond Robins' Own Story* 190 To hurt Bolshevism you need at least to get its number. **1921** R. D. Paine *Comr. Rolling Ocean* viii. 129 Do you remember the day before when he made that crack at you in front of Miss Crozier? I had his number right then. **1934** J. M. Cain *Postman always rings Twice* ii. 15 She knew what I meant, and she knew I had her number. **1939** I. Baird *Waste Heritage* vi. 74 'Never mind who I am,' Matt said, 'I got your number anyway.' **1956** W. Graham *Sleeping Partner* xiii. 111, I was trying to think of a verse all last night... I can't think who wrote it, but he rather got my number. **1970** G. Jackson *Let.* 29 May in *Soledad Brother* (1971) 265 Big Brother. He is rather transparent. I have his number. **1975** *Times Lit. Suppl.* 21 Mar. 332/4 Field-Marshal Lord Montgomery..had [Augustus] John's number right away. 'Who is this chap?' he demanded to know. 'He drinks, he's dirty, and I know there are women in the background!'

f. A number assigned to a particular telephone (or group of telephones) which corresponds to the terminals of its line at the exchange and in modern systems is dialled by a caller in order to establish a connection with it and cause it to ring; *number-unobtainable signal* or *tone*, a sound indicating to a caller that the number dialled is unobtainable for a reason other than its being engaged; similarly *number-engaged signal*; *wrong number*, a number obtained other than the one required by the caller.

1879 *Times* 8 Sept. 12/1 The person at No. 2 calls the attention of the attendant at the exchange by means of an electric bell. At the same moment a shutter on the switchboard falls and discloses the number of the applicant. *Ibid.*, So with any other numbers; they can be instantly connected or disconnected. **1884** *List of Subscribers* (London & Globe Telephone Co.) 4 Take telephones from hooks and speak at once, giving number of subscriber wanted. **1891** [see *engaged *ppl. a.* 3]. **1911** W. J. Locke *Glory of Clementina Wing* 324 She..took up the telephone and gave a number. **1930** Wodehouse *Very Good, Jeeves!* vi. 169 A woman has tossed my heart lightly away, but what of it?.. The voice of Love seemed to call to me, but it was a wrong number. **1930** *Gloss. Terms Telegraphs & Telephones* (*B.S.I.*) 47/1 Number-unobtainable tone. **1932** E. Bowen *To North* vi. 63 Markie, too well advised to encounter Cecilia over the wire, soon traced Emmeline to her number at Woburn Place. **1942** A. Christie *Body in Library* i. 12 Miss Marple's telephone rang... 'It must be,' Miss Marple decided, 'a wrong number.' **1959** H. Hobson *Mission House Murder* xiv. 92 The phone..has been giving the number-engaged signal for over half an hour. **1965** Mrs. L. B. Johnson *White House Diary* 3 June (1970) 283, I tried to reach him, or rather his wife, to no avail. The number didn't answer. **1969** 'D. Rutherford' *Gilt-Edged Cockpit* viii. 148 He listened..to the high whine of the 'number unobtainable' signal, knowing that she had left the receiver off. **1972** H. MacInnes *Message from Málaga* xii. 183 The telephone will ring... I shall... apologize for speaking to a wrong number. **1972** 'W. Haggard' *Protectors* iii. 38 Phone me at once... You know the number.

g. *to lose one's number*, to make a gaffe, to lose stock. *rare.*

a **1936** Kipling *Something of Myself* (1937) iv. 86 He produced a bottle of real Tokay, which I tasted, and lost

my number badly by saying that it reminded me of some medicinal wine.

5. (Further examples.)

1930 E. H. Young *Miss Mole* ix. 79 She sometimes saw No. 16 [*sc.* the person living at that house] trundling towards the back gate, she sometimes heard him calling in the cats at night. **1938** S. Beckett *Murphy* v. 95 A staple recreation..had been to wait at Walham Green for a nice number eleven [bus] and take it through the evening rush to Liverpool Street and back. **1970** Y. Carter *Mr. Campion's Falcon* xii. 91 A shabby row of houses..[with] Queen Anne porches... Number Seven.. had an intricate semi-circular fanlight.

c. *number ones*, ellipt. form of 'No. 1 dress', 'No. 1 suit' (also used), a best dress uniform worn esp. in the Navy.

1829 F. Marryat *F. Mildmay* II. 132 Each was dressed out in our No. 1 suits, in most exact and unquizzable uniform. **1914** 'Bartimeus' *Naval Occasions* xviii. 157 The 'Rig of the Day' was 'Number Ones'. **1947** *Landfall* I. 287 Hughes unpacked his kit to find his number ones badly crushed, and cursing, he went in search of an iron. **1950** A. P. Herbert *Independent Member* lxi. 359 The queer-looking spectacled P.O., 'sculling about' in his No. 1's astern of the Field-Marshal. **1955** [see *doll *v.*[2]]. **1968** J. Lock *Lady Policeman* ix. 78 The PCs in their quaint, high-buttoned 'number one' uniforms look as though they have stepped out of a jaded print. **1972** *Police Rev.* 17 Nov. 1489/2 (*caption*) Probably the last Policemen ever to wear their 'number ones', the ceremonial dress of the Force which is probably being phased out at the end of the year.

d. colloq. *number one*, the finest quality, the best obtainable. As *attrib.* or *adj. phr.*, first-rate, 'tip-top'; leading, principal.

1839 *Spirit of Times* 29 June 195/1 He is the sole owner of the estate upon which the [race] track is located, and will, no doubt, do all he can to make it 'a number one' concern. **1843** F. Marryat *Trav. M. Violet* II. xi. 231 After having drained half-a-dozen cups of 'stiff, true, downright Yankee No. 1', we all of us took our blankets. **1846** *Swell's Night Guide* 40 This *sanctum-sanctorum* is.. the number one of cribberies. **1848** J. T. Fields *Let.* 15 Aug. in *R. W. Griswold's Corresp.* (1898) 242, I have some beautiful poems by me by Mrs. Barnes... They are No. 1, full of passionate feeling and eminently worthy of a place. **1855** *Trans. Mich. Agric. Soc.* VI. 495 Wheat first-rate, peas, also, oats number one. **1871** E. Eggleston *Hoosier Schoolmaster* (1872) xv. 125 Seems to me it would be number one to have God help you. **1872** —— *End of World* xi. 78 This walk seems the shortest, when I'm in superfine, number-one comp'ny. *c* **1882** in R. Pearsall *Worm in Bud* (1969) ii. 43 Awfully rollicking, fearfully frollicking, Number one Masher of all. **1897** G. B. Shaw *Let.* 16 Apr. in *Ellen Terry & Shaw* (1931) 186, I have all the British rights of Arms and all but eleven No. 1 towns for You Never Can Tell. **1904** —— *Let.* 9 Sept. in *Ibid.* 414 The tour is on the cheapest scale.. and the towns by no means all Number Ones. **1933** P. Godfrey *Back-Stage* xvi. 199 The No. 1 dates, such as Birmingham, Manchester, Liverpool, Glasgow, and Edinburgh, get the best companies. *Ibid.* 200 Actors on the No. 1's are the best paid. **1942** E. Paul *Narrow St.* xxiv. 212 The sluggish public began to scent the No. 1 scandal of the century. **1943** Koestler *Arrival & Departure* iv. 148, I could even point to a number of similarities between your No. 1 and our No. 1. **1944** *Living off Land* iv. 62 (*heading*) Panic as Enemy No. 1. **1955** W. Gaddis *Recognitions* III. ii. 752 Look, what did Schmuck's number-one boy want over there, when you stopped and talked to them. **1957** A. Grimble *Return to Islands* iii. 58 He turned out to be a number-one boat-builder. **1968** *Globe & Mail* (Toronto) 3 Feb. 51/3 One small folded letter on blue paper bearing a 12-pence stamp will be the number one feature. **1969** C. Burke *God is Beautiful, Man* (1970) 80 And it don't really make no difference if you're one or two. 'Cause with God, you're always number one. **1971** *Flying* Apr. 57/1 Pilot briefing is the number-one item in the present FSS system. **1974** *Plain Dealer* (Cleveland, Ohio) 26 Oct. 3-D/1 Officials here expect a crowd of only 35,000 for the No. 1 team in the nation.

e. *Number ten*, also *No. 10*, in full No. 10 Downing Street, the London residence of the Prime Minister. Hence used allusively to denote the influence or opinions of the Prime Minister. Also *transf.*, as quot. 1972.

1880 *Leisure Hour* 383/2 Doubtless 'oblivious forgetfulness' would occur to any one who, having been created a peer of Parliament by a late occupant of No. 10, should happen to be seen by the fallen Minister. **1905** [see *No., N°.]. **1934** *Punch* 28 Mar. 345/1 The Muse at No. 10. 'It is rare to find a Prime Minister who is also a poet.'—*Press*. **1939** *Ibid.* 13 Sept. 284/2 The mystery deepened when the man did not enter No. 10 at all. **1958** L. Durrell *Mountolive* iv. 86 Even in the rain there was the usual little cluster of tourists and loungers outside the gates of Number Ten. **1961** I. Jefferies *It wasn't Me!* iii. 36 If academic opinion differs from that at No. 10 we go our own way. **1969** 'W. Haggard' *Doubtful Disciple* i. 1 The summons to Number Ten had knocked him flat. Under-Secretaries weren't called to the Prime Minister's house. **1972** *Guardian* 14 June 12/1 Now that Mr Suto is on the brink of retirement..his wife feels she can lift the curtain on life at Japan's No. 10. **1974** *Daily Mail* 9/3 It was this which ..put Edward Heath into Number 10.

f. *number one*, a children's word or euphemism for 'urine'; similarly *number two* for 'fæces'.

1902 Farmer & Henley *Slang* V. 75/2 *Number one*,.. (nursery). Urination; also a chamber-pot. *Ibid.*, *Number two*,..(nursery). Evacuation. **1923** J. Manchon *Le Slang* 212 *I want to do number one*, je veux faire pipi. **1937** A. S. Neill *That Dreadful School* vii. 118 Our juniors have an interest in the Old English word for faeces. They use it a lot—the ones from polite homes do, I mean homes that talk of No. 2 and 'going to the House of Commons' (how appropriate a name!). **1938** I. Goldberg *Wonder of Words* vi. 108 The child is early taught to refer to his needs as 'number one' and 'number two'. **1949** F. Sargeson *I saw in my Dream* 15 You felt sick and told mother, and she felt your forehead and asked how long it was since you did number two. **1959** I. & P. Opie *Lore & Lang. Schoolch.* vi. 96 Dirty kangaroo, Sitting on the dustbin Doing his 'Number Two'. **1963** G. Greene *Sense of Reality* 47 'I want to do number one.' I blurted out... He called to Maria, 'The boy wants to piss. Fetch him the golden po.' **1967** A. Wilson *No Laughing Matter* II. 70 This little ginger [kitten] is going to do a number one if we're not careful. **1971** M. McCarthy *Birds of America* 145 When I had done Number Two, you always washed them out yourself before sending them to the diaper service.

g. *number two*: colloq. phr. (freq. attrib.), a provincial town (in contexts one not noted for its appreciation of the theatre); also, a person second in importance or rank to a head of a department, etc., a second in command.

1908 G. B. Shaw *Let.* 11 Aug. in *Lett. to G. Barker* (1956) 134 You might let her begin on a number two tour of it [*sc.* a play]. **1920** —— *Let.* 22 Dec. in *B. Shaw & Mrs. Campbell* (1952) 215 You yourself have held up the six big cities and kept poor Macdona wandering in the number twos. **1934** R. Ferguson *Celebrated Sequels* 180 My elocution..*has* been admired in such Number Two towns as Wigan. **1952** 'M. Innes' *Private View* xiv. 214 This fellow, Cadover, is your husband's Number Two? **1968** M. Woodhouse *Rock Baby* vii. 76 I'll introduce you to your Number Two... You'll need some help. **1970** *Guardian Weekly* 25 Apr. 17 Russia feels the understandable necessity to catch up in the arms race... It just does not pay to be number two. **1973** W. Fairchild *Swiss Arrangement* xiii. 173 'Lisa Kestler was in charge of the whole operation—right?' 'Right..Gray was her number two.' **1975** S. Johnson *Urbane Guerilla* II. 103 Usually we don't bother with the no. 2 man.

h. Naval. *number one*, a first lieutenant, esp. one who is second in command to the captain of a ship. Freq. as a form of address.

1909 J. R. Ware *Passing English* 184/2 *Number one*.., strictly naval for first lieutenant. **1916** 'Taffrail' *Stand By!* 120 'I'm sorry for him,' said No. One, lifting his glass with a grin. **1948** Partridge *Dict. Forces' Slang* 129 *Number one*. The sergeant in charge of a gun. (Army.) (2) The First Lieutenant. (Ward-room).

i. Forces' slang. *number nine*, an aperient pill freq. prescribed as a cure-all for minor illnesses or doubtful symptoms (see quot. 1925).

1916 *Anzac Book* 110 And should my health appear to fail And appetite grow fine, My doctor hands me—not a bill, But just a Number 9. **1925** Fraser & Gibbons *Soldier & Sailor Words* 211 *Number nine*... The popular name for the Service aperient dose or pill. From its listed number, No. 9, in the Field Hospital Case of drugs. Being the Medical Officer's stock remedy in case of doubtful ailments, or suspected malingering, in the war the expression 'A No. 9' came to be used in all kinds of applications, more or less in jest. **1926** N. Lucas *London & its Criminals* xv. 181 They have one medicine in prison for all ills—from toothache to broken limbs, this is known to all old-timers as 'white mixture' and it corresponds to the 'number nines' of the Army. **1930** Brophy & Partridge *Songs & Slang Brit. Soldier* 161 The regimental Medical Officer.. invariably gave him a standardized purgative pill, known as Number 9, and marked him down as *M.D.* i.e. medicine and duty.

6. e. An item in a programme of musical entertainment. Also, more loosely, any song or tune. Cf. senses 6 c and d in Dict.

1885 G. B. Shaw *How to become Musical Critic* (1960) 80 To tap their feet and wag their heads to the seductive swing of his numbers. **1900** E. E. Peake *Darlingtons* i. 2 After a rattling number by the band, a brief address by the Mayor, and another rattling number by the band, a neatly dressed, handsome man..advanced to the front of the platform. **1913** *Confessions of Dancing Girl* vi. 109 We had worked all the variety halls and cafés..and we had no novel 'numbers'. **1920** Wodehouse *Jill the Reckless* (1922) xi. 161 He's put over any amount of shows which would have flopped like dogs without him to stage the numbers. **1927** T. S. Eliot in *Newton's Seneca* (Tudor Translations) p. x, But the characters in a play of Seneca behave more like members of a minstrel troupe..rising in turn each to do his 'number'. **1933** P. Godfrey *Back-Stage* xiv. 173 A 'number' is any song or musical item in the programme, so named because every such item is numbered in rotation in the musical director's copy, and is always referred to by its cypher. **1948** *Penguin Music Mag.* Feb. 25 The B.B.C. could start..by putting some kind of check on the manner and matter of their inane songs—'numbers' I think they call them. **1958** *Publ. Amer. Dial. Soc.* xxx. 41 Words deeply engrained in the speech of the jazzman... Any tune is a 'number'. **1962** *Movie* Oct. 36/2 The garnish of musical numbers. **1973** J. Wainwright *Pride of Pigs* 175 It was a fine intro to a fine number. The style was traditional jazz.

f. *colloq.* A person or thing, esp. (i) an article of apparel.

1894 Somerville & 'Ross' *Real Charlotte* I. iii. 22 The shop windows..had progressed..to straw hats, tennis shoes, and coloured Summer Numbers. *c* **1900** in M. Johnson *Amer. Advertising* (1960), Indian panama horse hat. Last year our sales on this number were enormous, showing that this hat is no longer a fad. **1935** *Ladies' Home Jrnl.* Apr. 19/3 Deedee had swathed herself in an afternoon number and was happily emptying the last of my..perfume down her front. **1953** M. Steen *Anna Fitzalan* viii. 211 Petula Wimbleby's solution turned out to be an exquisite but throat-high 'little number' redeemed by lumps of jade. **1959** P. Bull *I know Face* ix. 149 The camel-hair number suffered most, as the majority of my friends wished to wear it. **1969** *Daily Tel.* 17 Jan. 17 Two of Mattli's best numbers were in impeccable white: a coat..and a wool suit.

(ii) A person, frequently with qualifying word; more usually, a woman. Cf. *article *sb.* 14 b.

1919 *Dialect Notes* V. 70 *Hot one, hot number*, used as a term of disgust. 'You're a hot one I must say.' New Mexico. **1924** H. C. Witwer in *Cosmopolitan* Apr. 70/1 Oh, she's beautiful enough!.. She's a snappy number with the skin you love to touch. **1936** L. C. Douglas *White Banners* xvi. 343 She's an odd number... I rather fancy she wears a hair shirt herself. **1938** R. C. Sherwood *Idiot's Delight* (ed. 2) 12 Bebe is a hard, harsh little number who shimmies. **1944** Auden *For Time Being* (1945) 10 And every gorgeous number may Be laid by anyone. **1950** R. Chandler *Trouble is my Business* i. 8 A girl. A red-headed number with bedroom eyes. **1955** W. Gaddis *Recognitions* II. vii. 627 Have you seen a little blond number named Adeline? **1960** C. Watson *Bump in Night* xii. 122 The fellow was rather a dull number when you get down to a straight life story. **1968** J. Sangster *Touchfeather* ii. 17, I make do with three [men]..my home number is just a nice guy who sells motor cars.

(iii) An occupation, job, assignment.

1948 Partridge *Dict. Forces' Slang* 129 *Number, quiet*, an easy job at sea or ashore. **1959** *N.Z. Listener* 24 July 5/1 A navigator's yeoman who had the cushy number of rubbing out old minefields and putting in new ones. **1968** *Listener* 19 Sept. 370/2 Transferred to what was described as a 'cushy number' with the Commandos. **1975** J. Wainwright *Square Dance* 187 He silently congratulated himself. It was a soft number, sitting here.

g. = *denier[3] 4. Cf. *count *sb.*[1] 2 b.

1923 G. G. Denny *Fabrics* I. 31 Yarn count—a number given to yarn indicating its fineness, based upon number of yards per pound, more correctly called 'yarn number'. **1927** M. H. Avram *Rayon Industry* 516 There are many systems by which the 'number', 'size', or 'count' of yarns is expressed. *Ibid.*, The number..is..the weight of a standard unit skein or hank. **1928** V. Hottenroth *Artificial Silk* ix. 160 Before the silk is ready for sale or for treatment in the dye works, it must be sorted according to quality and number (that is, thickness of thread). **1931** D. L. Pellatt *Viscose Rayon Production* xi. 97 For 150-denier yarn..the number has risen from 18 to 21, 24, 27.

19. *attrib.* and *Comb.*, as (sense 1) *number-word, -work*; (sense 3) *number continuum, series, system*; (sense 6 a) *number book, -carrier, man*; (sense 6 d) *number opera*; **number-average** *Chem.*, an average of some parameter of the molecules of a mixture calculated as an arithmetic mean with each individual molecule contributing equally, regardless of size; **number board,** a board on which numbers are displayed; **number-cloth,** the cloth bearing a horse's number in a race; **number-cruncher** *colloq.*, a machine (or occas. a person) with the capacity for performing arithmetical operations of great complexity or length; so **number-crunching; number-form,** the shapes into which series of numbers are formed in the mental imagery of some people; **number line,** a graduated line used to illustrate simple numerical concepts and operations; **number-plate,** a plate bearing a number, esp. that on a registered vehicle; **number six,** (*a*) *U.S. colloq.*, a household medicinal remedy, so called from its place on the pharmaceutical list of its inventor, Samuel Thomson; (*b*) a curl having the shape of the figure six which is dressed on to the forehead; cf. *figure-six* adj. (Figure *sb.* 26); **number six nose,** a large fleshy nose, supposed to be similar in shape to the figure six; **number theory,** the branch of mathematics dealing with the properties and relationships of numbers, esp. the positive integers (cf. Theory[1] 4 c); so **number-theoretic, -theoretical** *adjs.*; **number-theoretically** *adv.*; **number theorist.**

1935 Kraemer & Lansing in *Jrnl. Physical Chem.* XXXIX. 165 For heterogeneous materials, different methods for determining molecular weights give different 'average' values. Thus, it may be shown that freezing point, osmotic pressure, and end-group methods, when applied properly to an ideal mixture, result in an average value defined by the expression $M_n = 1/\Sigma(f_i/M_i)$ where f_i is the fractional weight of the constituent of molecular weight M_i in the mixture, and the summation is to be applied to all constituents present. This average may be designated as a 'number-average molecular weight'. **1955** *Jrnl. Polymer Sci.* XVII. 263 Number average degrees of polymerization are used to calculate the rates of initiation and transfer in vinyl polymerizations, and the extent of reaction in polycondensations. **1974** Allen & Patrick *Kinetics & Mechanisms Polymerization Reactions* vii. 419 Until recent years determination of number-average molecular weights was a most imprecise measurement. **1938** G. H. Sewell *Amateur Film-Making* v. 54 That [*sc.* numbering each shot] is done by exposing before each shot a Number Board. **1961** K. Reisz *Technique Film Editing* (ed. 9) 281 Number board, board momentarily held before the camera and photographed at the beginning of a take, recording the title of the film, number of the take and scene, in order to facilitate identification for the editor. **1969** D. Francis *Enquiry* I. i. 14 The Oxford Stewards had been elected for social reasons only..one of them couldn't read a number board at five paces. **1960** G. A. Glaister *Gloss. Bk.* 278 *Number books*, books published serially... Each part consisted of two or more sheets stitched together within blue-paper covers. **1963** *English Studies* XLIV. 149 But the kind of serialisation.. (often called 'number books' or 'subscription books') was really the issue and sale of a book in separate fascicules.. so that the purchaser could..collect the entire work and have it..bound if he so wished. **1919** M. Beer *Hist. Brit. Socialism* I. ii. ii. 108 He was successively a number-carrier, street bookseller, and editor of a democratic periodical. **1924** E. Wallace *Educated Evans* vi. 131 Catskin was the one horse..that Educated Evans would have recognized without colours and number-cloth. **1975** D. Francis *High Stakes* i. 5 People..carrying out saddles and number cloths for the next steeplechase. **1903** Number-continuum [see *continuum]. **1966** *New Scientist* 29 Sept. 729/1 The Flowers report recommended the setting up of some 'regional centres' each with a large 'number-cruncher' to take the bulk-computing load off more local machines. **1971** *Ibid.* 3 June 572/3 Tools ranging from the slide rule to the number-cruncher. **1971** A. Sampson *New Anat. Britain* v. xxvi. 497 Kenneth Keith, a brusque number-cruncher who had come into banking from accountancy. **1971** *Sci. Amer.* Aug. 100 (Advt.), Here's a calculator that speaks your language. You can customize its keyboard, memory size, display, programs and peripherals to suit your number-crunching tasks. **1883** F. Galton *Inquiries into Human Faculty* 124 The character of the forms under which historical dates are visualised contrast strongly with the ordinary Number-Forms. **1903** G. M. Stratton *Experimental Psychol.* xiii. 253 A peculiarity of this same number-form ..is that with the higher numbers the person changes his point of view. **1936** *Brit. Jrnl. Educ. Psychol.* VI. 60 The main object of the investigation..was to find if the presence of number forms is correlated to any significant degree with arithmetical ability. **1963** *Listener* 28 Mar. 547/1 Some people, whenever they think about numbers, picture them in a spatial arrangement... The experiences are called number forms. **1964** E. J. Swenson *Teaching Arithm. to Children* v. 99/1 When number lines are introduced to children, they should come in as a representation of a problem situation. **1968** Murphy & Kempf *New Math. made Simple* ii. 36 Since addition and subtraction are inverse operations, we expect subtraction to be associated with moving to the left on a number line. **1866** J. Blackwood *Let.* 21 Dec. in *Geo. Eliot's Lett.* (1956) IV. 321 The 'Number Men', i.e. men who sell the weekly and monthly publications in large numbers. **1947** A. Einstein *Music in Romantic Era* x. 117 Even a musician so retrospective as Louis Spohr could not help abandoning the number-opera towards the end of his career. **1958** *Listener* 24 July 141/3 'Die Zaubergeige', for example, is a 'number opera' whose folkish tunes stem from popular Bavarian art. **1869** *Good Words* 1 Mar. 170/2 The white porcelain number-plates upon the doors. **1901** *Motor-Car World* Apr. 74/1 We greatly fear that the number-plate is coming. **1911** *Chambers's Jrnl.* Dec. 831/1 White light to illumine the number-plate. **1973** P. Evans *Bodyguard Man* viii. 67 He..scooped up a handful of dirt from the roadside and rubbed it on to the rear number-plate until it became nearly illegible. **1975** *Drive* New Year 98/1 Numberplate collecting has always been a popular hobby in the USA. **1890** W. James *Princ. Psychol.* II. xxviii. 653 Little by little in our minds the number-series is formed. **1822** S. Thomson *Narr. Life* 63, I began with him by giving medicine to correct and strengthen the system; bathed the wound with my rhumatic drops, or No. 6. **1842** C. M. Kirkland *Forest Life* I. 71 We stick to thoroughwort,—balmony,—soot tea,—'number six',—and the like. **1909** *Bull. of Lloyd Library of Botany, Pharmacy & Materia Medica* 15 Thomson's *Compound Tincture of Myrrh and Capsicum* became celebrated as 'Number 6'. **1923** J. Manchon *Le Slang* 271 *Number sixes*,..des accroche-cœur. **1966** J. S. Cox *Illustr. Dict. Hairdressing & Wigmaking* 104/1 *Number sixes*, curls dressed onto the forehead. **1923** G. B. Shaw *Matter with Ireland* (1962) 260 Whereas my Irish nationality was formerly a valuable asset to me in England, I am now expected to apologize for it by men with wooly heads or number six noses. **1932** —— *Adventures of Black Girl* 18 A dark man with wavy black hair, and a number six nose. **1924** R. M. Ogden tr. *Koffka's Growth of Mind* v. 332 Max Wertheimer has investigated the kind of ideas employed by men who do not possess our developed number-system, in tasks where we would use numbers. **1914** *Q. Jrnl. Pure & Applied Math.* XLV. 373 We begin by listing the analogues of the algebraic invariants and then supplement these with the necessary invariants peculiar to the number-theoretic case. **1966** J. H. Cadwell *Topics Recr. Math.* xii. 133 The well-known ratio 22/7 shows that rational approximations have their uses. They are of considerable importance from a number-theoretic standpoint. **1968** A. M. Tropper tr. *H. Meschkowski's Introd. Mod. Math.* viii. 178 There are still many unsolved number theoretical problems. **1950** *Math. Tables & Other Aids to Computation* IV. 110 Something number-theoretically significant may be occurring. **1929** *Bull. Amer. Math. Soc.* XXXV. 779 There is no number-theorist who has not heard of 'Farey's series'. **1971** *Sci. Amer.* June 56/2 The pattern of occurrence of Mersenne numbers, Mersenne primes and perfect numbers continues to mystify number theorists. [**1798** A. M. Le Gendre (*title*) Essai sur la théorie des nombres. **1811** P. Barlow (*title*) An elementary investigation of the theory of numbers.] **1912** *Bull. Amer. Math. Soc.* XVIII. 335 The theory of determinants ..in the nineteenth century came to permeate all branches of number theory, algebra, [etc.]. **1939** Uspensky & Heaslet *Elem. Number Theory* p. v, Instruction in elementary number theory is given in an ever-increasing number of American universities and colleges. **1948** O. Ore *Number Theory & its Hist.* v. 76 The study of these laws in the distribution of the primes falls in the field of analytic number theory. This particular domain of number theory..is considered to be technically one of the most difficult fields of mathematics. **1964** M. McLuhan *Understanding Media* (1967) xxiii. 370 Arithmetic in grade three or nine, when taught in terms of number theory, symbolic logic, and cultural history, ceases to be mere practice in problems. **1924** R. M. Ogden tr. *Koffka's Growth of Mind* v. 334 Many peoples use other number-words in counting than the ones they use in naming sums. **1911** S. S. Colvin *Learning Process* iii. 51 Much of it might function equally well for the reading habit, or the number-work habit. **1962** *Listener* 15 Mar. 469/2 Earlier attempts to teach 'number work' are premature and cannot lead beyond the learning of meaningless rules.

numbered, *ppl. a.* Add: **2.** *Comb.*, as **numbered account**, an account at a bank, esp. a Swiss bank, which is identified only by a number and does not bear the owner's name.
1963 'Han Suyin' *Four Faces* 31 He..had money salted away in a numbered account in Switzerland. **1968** J. Blackburn *Young Man from Lima* vi. 61 Numbered accounts in Swiss banks tell no tales. **1972** W. Davis *Money Talks* 136 Italian businessmen, who appreciate this more than most, often enter Switzerland with suitcases full of banknotes and open numbered accounts on the spot.

numbhead (nʌ·mhed). *U.S. colloq.* Also **numhead.** [f. Numb *a.* + Head *sb.*, after Numskull.] = Numskull. So **nu·mbheaded** *a.*
a 1852 F. M. Whitcher *Widow Bedott Papers* (1856) x. 98 The old coot was so awful numbheaded I couldent beat anythin' into him. **1876** *Rep. Vermont Board Agric.* III. 624 The opinion too generally prevails that almost any numhead will do for a farm laborer. **1960** Wentworth & Flexner *Dict. Amer. Slang* 359/2 *Numb-head*, a stupid or dull person.

numerable, *a.* Add: Hence also **numerabi·lity**, the quality of being numerable.
1943 [see *abstractable *a.*].

numeracy (niū·mĕrăsi). [f. *numerate *a.*, after *literacy*.] The quality or state of being numerate; ability with or knowledge of numbers.
1959 *15 to 18: Rep. Cent. Advisory Council for Educ. (Eng.)* (Ministry of Educ.) I. xxv. 270 When we say that a scientist is 'illiterate', we mean that he is not well enough read to be able to communicate effectively with those who have had a literary education. When we say that a historian or a linguist is 'innumerate' we mean that he cannot even begin to understand what scientists and mathematicians are talking about... It is perhaps possible to distinguish two different aspects of numeracy that should concern the Sixth Former. **1960** *English* XIII. 44 A certain lack of 'numeracy' on the part of those trained in the Arts can make them a little purblind to the implications of figures such as these. **1960** *Rep. Proc. Conf. Univ. U.K.* 23 If scientific barbarians are to be given a veneer of literacy, and literary barbarians a veneer of numeracy, I suggest the proper apparatus for it is a lot of deep armchairs in an open access library. **1966** *Economist* 22 Jan. 310/2 The need for numeracy today is enormous. Business requires..people who..have grasped the principles of reducing a chaos of information to some kind of order. **1970** *Sci. Jrnl.* Feb. 73/2 The scientist does, however, possess the advantage of numeracy and can usually acquire quite easily the statistical and theoretical background to modern management techniques. **1972** *Daily Tel.* 22 Jan. 2/5 The plan must be welcomed for introducing pre-school children to reading, writing and numeracy.

numeraire (niū·mĕrɛ̄ᵊɹ). *Econ.* Also **numéraire.** [ad. F. *numéraire*.] The function of money as a measure of value or unit of account; a standard for currency exchange rates.
1964 Gould & Kolb *Dict. Soc. Sci.* 532/2 In a primitive society two knives might exchange for one calf. Often one good is singled out to serve as a *numeraire* or common measure for other goods, thus precious metals serve both as a good wanted for particular purposes and as a unit of value and exchange. **1965** J. L. Hanson *Dict. Econ. & Commerce* 302/1 *Numeraire*, a term sometimes used of the function of money as a measure of value. **1971** *New Yorker* 23 Oct. 118 The Bretton Woods agreement.. established the dollar as the *numéraire*, or measuring rod, against which the value of other currencies was set. **1972** *National Westminster Bank Quarterly Rev.* May 21 Three points constitute the core of the Barber proposals [at the last IMF meetings]: a) That SDRs 'become the numeraire in terms of which parities are expressed and in relation to which currencies are revalued or devalued'. **1972** *Times* 27 Sept. 21/2 The use of Special Drawing Rights rather than the dollar as the effective numeraire in the new system is widely acceptable. **1973** *Times* 26 Jan. 17/6 It is the European currencies which have been adjusted against the dollar—the numeraire in the exchange system.

numerate (niū·mĕrĕt), *pa. pple.* and *ppl. a.* Restrict †*Obs.* to sense in Dict. and add: As *adj.* [f. L. *numerus* number + -ate², after *literate*.] Acquainted with the basic principles of mathematics and science.
1959 *15 to 18: Rep. Cent. Advisory Council for Educ. (Eng.)* (Ministry of Educ.) I. xxv. 269 Little is done to make science specialists more 'literate' than they were when they left the Fifth Form and nothing to make arts specialists more 'numerate', if we may coin a word to represent the mirror image of literacy. **1960** [see *art *sb.* 7]. **1966** A. Battersby *Math. in Management* i. 21 The aim of a good Sixth Form should be to send out into the world men and women who are both literate and numerate. **1967** *Times Rev. Industry* Mar. 103/2 (Advt.), Lecturer..in Management Science... Applicants should have a good honours degree..in a 'numerate' subject (e.g., engineering, mathematics, mathematical economics, physics, statistics). **1967** C. Berners-Lee in Wills & Yearsley *Handbk. Management Technol.* 1 It has become a commonplace of the day that we must all be numerate as well as literate. *Ibid.* 8 The ability to communicate easily with a computer is one which is greatly prized by all numerate professionals in the larger commercial and industrial organizations and in the universities. **1971** *Country Life* 20 May 1264/2 It has been my impression.. that..children are much less literate and numerate than ever before.

numeric, *a.* Add: **2.** = Numerical *a.* (and *sb.*) 1, esp. 1 e.
1949 E. C. Berkeley *Giant Brains* iv. 54 It will be reasonable to use numeric codes 0 to 9 in each column.. because..numeric codes can be sorted faster than alphabetic codes. **1955** [see *alpha-numeric *a.*]. **1967** D. Wilson in Wills & Yearsley *Handbk. Management Technol.* 45 In addition to the usual numeric and alphabetic characters, a range of special characters can normally be included. **1972** *Computer Jrnl.* XV. 209/1 A monotonically increasing numeric value is associated with each partition or level of the tree. **1973** *Sci. Amer.* June 65/2 Instruments that make rapid, high-precision measurements demand numeric readouts, since analogue presentations cannot display the results in any simple way.

numerical, *a.* (and *sb.*). Add: **1. g.** Special collocations: *numerical analysis*: the branch of mathematics that deals with the development and use of numerical methods for solving problems (usu. ones too complicated for analytic methods); so *numerical analyst*; *numerical aperture*: a measure of the resolving power of a microscope objective (for a given wavelength), equal to the product of the refractive index of the medium in front of it and the sine of the semi-angle of the cone it subtends at the centre of the object; *numerical control*: the use of numerical data stored on tape or punched cards to control automatically the movement and operations of machine tools and work-pieces.
[**1930** J. B. Scarborough (*title*) Numerical mathematical analysis.] **1946** *Ann. Computation Lab. Harvard Univ.* I. 338 (*heading*) Bibliography of numerical analysis. **1947** *Sci. News Let.* 27 Sept. 201/2 The National Bureau of Standards' new Institute of Numerical Analysis will feature high-speed electronic calculators. **1952** D. R. Hartree *Numerical Anal.* i. 2 The methods of numerical analysis..have that degree of generality which entitles them to be considered part of mathematics. **1974** *Encycl. Brit. Macropædia* XIII. 390/1 The numerical solution of ordinary differential equations is one of the most important branches of numerical analysis because many physical problems lead to ordinary differential equations that cannot be solved analytically. **1956** F. B. Hildebrand *Introd. Numerical Anal.* i. 1 Generally the numerical analyst does not strive for exactness. **1878** *Jrnl. R. Microsc. Soc.* I. 19 The inner scale shows the air-angle, the outer scale the 'numerical aperture'. *Ibid.* 21 This quantity *a*, which Professor Abbe calls 'numerical aperture', gives an absolute definition of aperture..by which lenses of every kind are directly comparable. **1952** R. W. Ditchburn *Light* viii. 253 With a dry lens, the maximum value of the numerical aperture is 1·0, and values up to 0·95 have been obtained. Oil-immersion objectives with numerical apertures up to 1·65 have been constructed. *Ibid.*, Higher magnifications have been obtained using electron microscopes... The effective wavelength may be of the order of 0·1 Å., but the numerical aperture of models so far available is very low, being only of the order of 0·01. **1969** S. G. & H. Lipson *Optical Physics* ix. 278 The limit of resolution of the microscope.. is equal to λ/numerical aperture. **1952** *Final Rep. Construction & Operation of Numerically Controlled Milling Machine* (Mass. Inst. Technol. Servomechanisms Lab.) 30 July 1 (AD 22241). Summary 1 A new technique of automatic machine-tool control..promises to simplify some of the manufacturing problems of medium and small run production often encountered in the aircraft industry. Called numerical control, this technique has been applied by M.I.T...to a milling machine controlled by a numerical code punched on paper tape. **1955** *IRE Trans. Industr. Electronics* II. 3 (*heading*) Numerical control of machine tools. *Ibid.* 5/1 The numerical control for this turret punch press uses the Remington-Rand 45 column tabulating card. **1966** *Economist* 2 July 56/1 What the two companies are claiming is that..numerical control is now suitable for mass production. *Ibid.* 57/1 There are 25 British companies producing machine tools for numerical control. **1971** H. C. Town *Design & Construction Machine Tools* vii. 159 With numerical control the instructions for the various phases of a cycle are transmitted by punched or magnetic tape inserted in a reader unit.

‖ **numéro** (nümero). [Fr.] Number. Also *transf.*
Quots. 1944 and 1958 reflect the Fr. sense 'une personne bizarre'.
1944 Auden *For Time Being* (1945) 118 George, you old numero, How did you get in the Army? **1958** L. Durrell *Mountolive* v. 107 Well, you can see what a *numéro* he is. **1961** *House & Garden* May 20/2 The chair he has designed... This heavy *numéro* seems very different from the essays in lightness with which Saarinen's name has hitherto been associated. **1973** *Sunday Times* 26 Aug. Katharine Schofield opens the second act with a *numéro* of great panache.

numerology (niūmĕrǫ·lŏdʒi). [f. L. *numer(us* number + -ology.] Divination by numbers; the study of the esoteric meaning of numbers.
1911 'Sepharial' *Kabala of Numbers* I. iii. 25 A general consideration of the principles involved in the science of numerology. **1926** K. Adams *Numerology Up-to-Date* i. 11 In the science of Numerology each letter of the alphabet has a given number and each number has its special significance. **1935** *Punch* 21 Aug. 207/1 Next we come to numerology. One of the most vital factors in your life is

the Science of Numbers. **1937** M. Covarrubias *Island of Bali* (1972) ix. 312 For special occasions an offering, to be effective, must conform to certain specifications based on the influences that rule the day; the calendar, the cardinal directions, numerology and so forth. **1960** Auden *Homage to Clio* 30 The burly slave of ritual and a martyr To Numerology. **1962** *Times* 23 Feb. 5/1 Numerology is ..a term of superiority or near-abuse applied by physicists to those of their colleagues, who, by playing with numbers hope to arrive at physical understanding. **1971** *Times Lit. Suppl.* 1 Oct. 1179/1 Numerology has even had some impact on poetry in its 'concrete' forms.
Hence **numerolo·gical** *a.*, **numerolo·gically** *adv.*; **numero·logist**, one who studies or is expert in numerology.
1923 C. W. Cheasley *Numerology* (rev. ed.) vi. 67 The final digit of the whole name—the Expression..has the greatest influence of all... When Numerologists use the term 'a number 2 person', 'a number 6 person', etc., it is this Expression number that should be implied. **1926** K. Adams *Numerology Up-to-Date* ix. 79 What are the numerological properties of 1? *Ibid.* viii. 74 Cities, numerologically speaking, are a most fascinating study. **1932** *Discovery* June 204/1 It is not only the number seven that possesses a peculiar significance for numerologists. **1952** G. Sarton *Hist. Sci.* I. xvii. 439 The regular solids.. must each have some definite meaning... Plato bethought himself of the four elements; the fifth solid would then represent the whole universe. This patching up of the theory, plus the finding of a meaning for the superfluous solid, is typical of the analogies invented by numerologists and other mystical mathematicians. **1972** *Sci. Amer.* Feb. 101/1 Mel Stover, a Winnipeg numerologist, has pointed out that in Roman numerals 9 changes to 11 when turned over, but 9 in the binary system—1001—doesn't change at all. **1973** *Nature* 12 Oct. 313/1 Several dependent and independent numerological relations have been proposed. Dirac noticed that the ratio between the age of the Universe and the natural unit of time..is approximately equal to the ratio between the electrostatic and the gravitational force acting between two electrons. **1974** *Sci. Amer.* June 118/3 Anything numerologically interesting about Nixon and Watergate?

‖ **numero uno** (nū·mero ū·no). [It.] 'Number one', the best or most important (person). (Cf. *number *sb.* 5 d.)
1973 *Publishers Weekly* 25 June 71/1 Fred Carr, *numero uno* in the mutual fund game. **1973** *Philadelphia Inquirer* (Today Suppl.) 14 Oct. 24/1 (*caption*) 'Fat Rob'.. may be *numero uno* in his neighborhood, but he doesn't have anything that compares with the paneled office of Reuben Maldonado, president of the Royal Javelins in New York's South Bronx. **1975** *Publishers Weekly* 14 Apr. 55/1 The sex goddess of the moment, who stays aloof from filmland's numero uno stud only to be misused by an intellectual French director.

‖ **numerus clausus** (niū·mĕrŏs klau·sŏs). [L., lit. 'closed, or restricted, number'.] A fixed maximum number of entrants admissible to an academic institution.
1925 *Nation* (N.Y.) 8 Apr. 374 The *numerus clausus*, driven out of Russia, still keeps eager Jewish youth from the learning they crave in Poland, Hungary, and Rumania. **1959** D. D. Runes *Conc. Dict. Judaism* 178 *Numerus clausus* (Latin), limited number of Jews admitted to schools of higher education; still operative in Soviet Union. **1960** *Encounter* June 46 Owing to the *numerus clausus* imposed in Russian universities, a very large number of the students present were Jews, compelled to come West in order to graduate. **1963** *Higher Educ.* (Cmnd. 2154) v. 38 Shortage of accommodation may effectively impose a *numerus clausus* in some disciplines.

numhead, var. *numbhead.

Numidian (niūmi·diăn), *sb.* and *a.* [f. L. *Numidia* the former name of a country in North Africa + -an.] **a.** *sb.* A native or inhabitant of Numidia. **b.** *adj.* Of or belonging to Numidia. *spec.* **Numidian crane**, a grey and black crane, *Anthropoides virgo*, found in southern Europe, north Africa, and parts of Asia; = Demoiselle 2 a.
1600 Holland tr. *Livy's Romane Hist.* xxix. 731 Then the kingdome fell by descent, according to the custome and manner of the Numidians, unto Desalces the late kings brother. *Ibid.* xxx. 752 They would enlarge and set at libertie all the Numidian captives that lay in prison at Rome. **1614** A. Gorges tr. *Lucans Pharsalia* iv. 156 The vagrant fierce Numidæans. **1697** Dryden *Æneid* iv. 57 And fierce Numidians there your Frontiers bound. **1757** A. Butler *Lives Saints* III. 3 Certain false Numidian zealots. *Ibid.* 4 Seventy bishops, chiefly Numidians. **1766** [see Demoiselle 2 a]. **1836** N. Isaacs *Trav. E. Afr.* II. xvii. 324 The wild sort [of birds] are easily obtained... The Crane (Numidion [*sic*] or Demoiselle), Baleari or Crowned Crane..are common. **c 1876** E. Dickinson *Poems* (1955) III. 956 The Butterfly's Numidian Gown With spots of Burnish roasted on Is proof against the Sun. **1893** A. Newton *Dict. Birds* 111 In Europe, besides the G[rus] *communis* already mentioned, we have as an inhabitant that which is generally known as the Numidian Crane or Demoiselle..distinguished from every other by its long white ear-tufts. **1905** T. Hodgkin in L. Creighton *Life & Lett. T. Hodgkin* (1917) xi. 229, I wish I had come to this Numidian land when I was younger... Our Numidians long held Rome at bay. **1906** Kipling *Puck of Pook's Hill* 147 'Was your nurse a—a Romaness too?' 'No, a Numidian... A dear, fat, brown thing.' **1921** *Edin. Rev.* July 105 A Numidian, riding bareback and stirrupless, is throwing a lasso at a wild ass. **1957** *Encycl. Brit.* XVI. 615/1 Numidians were divided into two great tribes, the Massyli on the east, and the Massaesyli on the west... At the end of the [second Punic] war,..the Numidian kingdom surrounded Carthage except towards the sea.

numinal, *a.* For † *Obs. rare*⁻¹ read *rare* and add later example.
1927 *Contemp. Rev.* Mar. 352 Rudolf Otto in his *Idea of the Holy*, accords to music a place upon the 'wholly other' or 'numinal' of religion.

‖ **numinosum** (niūminōᵘ·zŏm). [ad. G. *numinose* the numinous.] = *numinous *sb.*
1938 C. G. Jung *Psychol. & Relig.* i. 4 Religion, as the Latin word denotes, is a careful and scrupulous observation of what Rudolf Otto aptly termed the 'numinosum'. **1950** *Brit. Jrnl. Psychol.* XL. 235 Modern psychology, with its exploration of complexes, has touched upon a taboo laid by ego-consciousness on the psychic 'numinosum' and 'tremendum'. **1973** J. Singer *Boundaries of Soul* v. 123 Many religious practices and performances seem to be carried out for the sole purpose of calling forth the power of the *numinosum* at will by invocation, incantation, sacrifice [etc.].

numinous, *a.* Delete † and add: Revived in senses: of or pertaining to a numen; divine, spiritual, revealing or suggesting the presence of a god; inspiring awe and reverence. Also *absol.* or as *sb.*
1864 R. S. Hawker *Quest of Sangraal* 17 An Orient Cruse, Fulfill'd, and running o'er, with Numynous Light. **1923** J. W. Harvey tr. *R. Otto's Idea of Holy* ii. 6 For this purpose I adopt a word coined from the Latin *numen*. *Omen* has given us *ominous*, and there is no reason why from *numen* we should not similarly form a word '*numinous*'. I shall speak..of a unique 'numinous' category of value and of a definitely 'numinous' state of mind. *Ibid.* iii. 11 The numinous is thus felt as objective and outside the self. **1934** W. Temple *Nature, Man & God* i. 23 What Otto speaks of as the '*Mysterium tremendum*', the quality in the object of religion which he describes as 'Numinous', is just that before which we do not reason but bow. **1941** G. G. Scholem *Major Trends in Jewish Mysticism* 59 The key-word of the numinous, the *Kedushah*, the trishagion from Isaiah vi, 3, in which the ecstasy of the mystic culminates: holy, holy, holy is the Lord of Hosts. **1951** J. L. Adams tr. *Tillich's Protestant Era* p. xxxix, Protestants..often are unaware of the numinous power inherent in genuine symbols. **1957** *Times Lit. Suppl.* 11 Oct. 602/5 He now urges us..to transfer finally our more numinous speculations from the altar to the earth, to the observable universe and to all its children. **1962** *Listener* 17 May, Simple English exchanges that caught the suburban yet numinous quality of the original French. **1967** G. Steiner *Lang. & Silence* 62, I want to draw attention to..the recurrent acknowledgement by poets..that music *is* the deeper, more numinous code, that language.. aspires to the condition of music. **1969** *New Scientist* 17 Apr. 114/1 There is a growing revolt of young people against what they call materialism..because they want to take seriously the experience of the inner life—in particular, experience of the 'numinous' and of 'immortal longings'. **1972** *Times Lit. Suppl.* 31 Mar. 365/4 Homer reveals the world of gods as well as the world of men, both in epic verse, and this difference from later, more numinous writers has many consequences. **1972** S. W. Sykes in Cox & Dyson *20th-Cent. Mind* II. vi. 154 [Rudolf] Otto defined numinous as the non-rational mystery behind religion, which is both awesome and fascinating. It is, he asserted, the permanent and essential feature of all religion, including Christianity.
Hence **numino·sity**, **nu·minousness**, the condition or state of being numinous.
1932 R. A. Knox *Broadcast Minds* iv. 70 What is the proper object around which the idea of 'numinousness' ought to cling? **1936** *Essays & Stud.* XXI. 132 The poet's own religious fervour, mysticism, or (if you will) poetic 'numinosity'. **1951** *Theology* LIV. 233 Somehow he manages again and again to take from the Gospel its fascination, its haunting power over the imagination, its numinosity. **1962** *Listener* 29 Nov. 940/1 Nelly, the hag, the witch, the personification of Corvey's incipient madness, lost some of her numinousness in being acted out. **1963** *Times* 29 May 11/5 If the Church is to be preserved from degenerating into a nondescript club of nebulous numinosity. **1969** E. C. Whitmont *Symbolic Quest* vii. 126 The energy which is withdrawn from the external world remains focussed exclusively upon the unconscious primitive image with its archaic numinosity.

nummion (nū·miǫn). Usu. in pl. **nummia** (-iă). [ad. Gr. νουμ(μ)ίον, dim. of νοῦμμος coin.] A Byzantine copper coin, equal to one fortieth of a *follis.
1908, **1962** [see *follis]. **1970** *Ashmolean Mus. Rep. Visitors 1969* 40 The earliest in a series of Byzantine accessions is a scarce 33-nummia coin of Justinian I. **1973** P. D. Whitting *Byzantine Coins* vii. 89 The copper nummia in their thousands were too small to have any recognizable value in themselves and could only circulate on a basis of public confidence.

nu·mmy-nu·mmy. An exclamation of pleasure; = *yummy-yummy. Cf. *num.
1929 E. Bowen *Last September* vi. 62 'Nummy-nummy,' she said, pointing out the raspberries.

num-num (numnu·m). *S. Afr.* Also **nam-nam, noem-noem.** [Afrikaans, perh. f. Hottentot or Nama name.] A spiny, evergreen shrub or small tree of the genus *Carissa*, belonging to the family Apocynaceæ and bearing fragrant white flowers; also, the edible red or purple fruit of a plant of this kind. Cf. *Natal plum* (*Natal *sb.*² 2).
1822 W. J. Burchell *Trav. Interior S. Afr.* I. x. 192 The Hottentots call this shrub 'Num'num (or Noomnoom, agreeably to English orthography), each syllable preceded by a guttural clap of the tongue. **1897** S. J. Du Toit *Rhodesia* iv. 32 Various kinds of sweet-grass and small shrubs, varied with very good large bush and trees, as..'noem noem', 'quarri', [etc.]. **1926** O. Schreiner *From Man to Man* iii. 113 The nam-nams and jasmine shrubs made a thick wall on either side. [*Note*] A shrub with a small edible berry (also 'num-num'). **1951** R. Campbell *Light on Dark Horse* ix. 127 It was the sort of scrub which reminds one of an enormous roll of vegetable barbed wire, composed chiefly of num-num and wait-a-bit thorn. **1972** Palmer & Pitman *Trees S. Afr.* III. 1899 The species [of *Carissa*] are usually known as num-nums. *Ibid.* 1903 Although the Karoo num-num is most often a bush it can grow into a small tree 3·5 m high with a round, much-branched, immensely dense and twiggy evergreen crown.

numori, var. *nomoli.

nunatak. Add: Also applied to similar mountains in other regions. (Earlier and later examples.)
1877 *Q. Jrnl. Geol. Soc.* XXXIII. 145 At Kangerdlugssuak, where the mountains are lofty and the sides of the fjords steep, there are three Nunataks..equalling the height of the neighbouring land; but above the Inland Ice of Disko..the Nunataks are lower Knolls. **1921** E. R. G. R. Evans *South with Scott* xiv. 190 Mount Hope is a nunatak of granite, about 2,800 feet in height. **1959** *Times* 9 Jan. 11/6 *Antarctic Findings*... At Shackleton Base..we lacked..solid ground. Some nunataks could be seen over 20 miles away to the east, near Vahsel Bay. **1963** 'G. Carr' *Lewker in Norway* vi. 129 The only light.. had turned snow-hummocks into far-off mountains and ricky *nunataks* into boulders.

nunc, nunk, nunks, varr. Nuncle.
1841 *Comic Almanack* Dec. 48 Come, nunks, one game at Blindman's-buff. **1876** Geo. Eliot *Dan. Der.* I. ii. xvi. 311 His uncle or 'Nunc', as Sir Hugo had taught him to say. **1884** C. M. Yonge *Armourer's Prentices* I. 142 Should you know this nunks of yours? **1915** A. Bennett *These Twain* (1916) ii. xiii. 259 'Here's Nunks!' exclaimed George. **1919** C. Orr *Glorious Things* xxiii. 279 The very night old Nunk was away.

‖ **nunchaku** (nuntʃa·ku). [Jap., f. Okinawa dialect.] Esp. in *pl.*, two hardwood sticks joined together by a strap, etc., as a defensive weapon.
1970 *Guardian Weekly* 2 May 11 The radical taste tends ..to nunchakus, which go back more than 500 years. They were..invented by Japanese peasants for self-defence when metal weapons were forbidden to all but the Samurai. **1973** *Express* (Trinidad & Tobago) 27 Apr. 31/3 The experts will give exhibitions in kobudo (weaponery) displaying their martial skill with the sai, nunchaku stick, bow, sword and daggers. **1975** *Globe & Mail* (Toronto) 17 Jan. 4/1 The proper name is nunchaku sticks. They are made of two sticks of hardwood joined together at one end by a chain, leather or rope.

‖ **nunc stans.** [L. *nunc* now + *stans* pres. pple. of *stāre* to stand.] The eternal timeless 'now' presumed, as an attribute of God, to be co-existent with Time.
1651 Hobbes *Leviathan* xlvi. 374 But they will teach us that Eternity is the Standing still of the Present Time, a *Nunc-stans* (as the Schools call it;) which neither they, nor any else understand. **1678** Cudworth *Intell. Syst.* 645 *Nunc-stans*, or a Standing Now of Eternity. **1733** A. Baxter *Enquiry Human Soul* viii. 376 The distinction of past and future vanishes with respect to such a Mind; and the expression *nunc stans* will appear to have propriety. **1854** H. L. Mansel *Lett.* (1873) 119 Augustine, and the Schoolmen after him,..speaking of Eternity as a *nunc stans*. **1896** W. Caldwell *Schopenhauer's System* viii. 393 In willing the world is at once an eternal process and an eternal stationary thing—a *nunc stans*—at the same time. **1946** J. Laird *Philos. Incursions Eng. Lit.* vi. 96 On a few occasions..Wordsworth described the specious *nunc stans* of mystical ecstasy; its apparent arrest of time itself. **1958** E. Heller *Ironic German* vi. 241 Thomas Mann was..well prepared for 'the mystery of the revolving sphere' by Schopenhauer's philosophy of the eternal *nunc stans* which resides at the centre of the illusory motion of Time. **1967** P. Merlan in *Cambr. Hist. Later Greek & Early Medieval Philos.* 100 A supreme god..lives in the *aiōn* which can also be characterized as a *nunc stans*.

nunky. Add: (Later examples.) Also **nunkie.**
1969 'A. Gilbert' *Missing from Home* vi. 86 We don't even know she had an uncle... And if it was seven o'clock ..she was running it pretty fine if she was going for a drive with Nunkie. **1973** *Times Lit. Suppl.* 30 Nov. 1472/2 She was A. J. Balfour's favourite niece... 'Nunky' died in 1930.
2. *slang.* A pawnbroker. Cf. Uncle *sb.* 3.
1921 *Daily Colonist* (Victoria, B.C.) 5 Apr. 13/5 If you happen to want a ten-dollar bill and trot round the corner to 'nunky'; why, that's really not quite done, my dear fellow. **1937** in Partridge *Dict. Slang.*

nunny bag (nʌ·ni bæg). *Newfoundland.* Also **nonny bag.** [f. Eng. dial. *noony* meal at noon + Bag *sb.*] A kind of haversack, often made of sealskin.
1842 J. B. Jukes *Excursions in Newfoundland* II. 146 Having determined to return, we hung up in the tilt a 'nunny bag' full of bread. **1895** *Dialect Notes* I. 380 *Nunny bag*, lunch bag; usually made of a piece of sealskin, and used by sealers when they go off for a day. **1919** W. T. Grenfell *Labrador Doctor* 90 Our sealers carry dry oatmeal and sugar in their 'nonny bags', which, mixed with snow, assuage their thirst and hunger as well. **1925** *Dialect Notes* V. 337 *Nonny bag*, a small knapsack to carry out on the ice; ditty bag. **1944** H. Wentworth *Amer. Dial. Dict.* 419/1 *Nunny bag*, a lunch bag, usually of sealskin. **1961** *Maclean's Mag.* 28 Jan. 47/2 He clawed through his nunny-bag till he found a bit of oatmeal.

‖ **nuoc mam** (nwǫk mām). [Vietnamese.] A spicy Vietnamese fish sauce.
1919 *Experiment Station Record* (U.S. Dept. Agric.) July 66 Analyses of nuoc-mam (a product similar to soy sauce) and other sauces are reported. **1920** *Ibid.* Oct. 458 Nuoc-mam..the Indochinese fish sauce.., has been successfully condensed for transportation and use by the Indochinese soldiers in France. **1935** M. Morphy *Recipes of All Nations* 782 A small teaspoon of nuoc-man, a typically Annamese condiment which is used in practically all their dishes. **1969** I. Kemp *Brit. G.I. in Vietnam* iv. 77 The staple Vietnamese diet of rice with fish and *nuoc mam* —'Vietnamese national sauce', they would explain, chuckling among themselves—made from rotten fish and sea water. **1970** *Harper's Mag.* July 83 Bits of rice and *nuoc-mam* are what they're used to. **1972** *Times* 25 Mar. 11 There is nothing in either Chinese or French cooking that exactly corresponds with the Vietnamese thin sauce or relish called nuóc mam formed..from the liquor of decomposing fish.

Nupe (nū·pe), *a.* and *sb.* Also 9 **Nufi, Nupé, Nyffee.** [f. the name of a former kingdom at the junction of the Niger and Benue rivers in West Africa.] **A.** *adj.* Of, pertaining to, or designating a Negro people of central Nigeria, or their language. **B.** *sb.* **a.** The Nupe people; a member of this people. **b.** The language of this people, which belongs to the Kura division of the Sudanic language family.
1829 H. Clapperton *Jrnl. 2nd Exped. to Interior of Afr.* i. 25 He hoped that we should settle the war with the Nyffee people and the Fellatah. **1841** J. F. Schön *Jrnl.* 14 Sept. (1842) 119 The Nufi Language is spoken to a considerable distance, even beyond Rabba. **1883** R. N. Cust *Sk. Mod. Lang. Afr.* I. xi. 228 At Lokója, at the confluence, Nupé is the principal language, though others are spoken. **1885** J. S. Kingsley *Standard Nat. Hist.* VI. 320 We meet the Ibo language on the lower course of the Niger, and the Nupe from the Benue up. *Ibid.* 321 The Nupe..are genuine negroes in complexion and form of face, and indisputably one of the best-formed tribes in Africa. **1888** G. T. Bettany *World's Inhabitants* vi. 614 Gando..includes the territory of the Nupe negroes. **1892** A. F. Mockler-Ferryman *Up Niger* xiii. 151 The Mohammedan Nupe lives the life of a man-about-town. **1932** J. Cary *Aissa Saved* iv. 24 The new town on the waterside, inhabited by Hausas, Nupes, Yorubas, the mixed population of a harbour. **1942** S. F. Nadel *Black Byzantium* iii. 27 The Nupe have only one word for kinship, *dengi*, which defines relationship in the widest as well as in a more restricted sense. *Ibid.* vi. 70 The people in the capital regard the Nupe spoken in the districts and on the borders of the country as an inferior brand. **1974** *African Encycl.* 375/2 The Nupe people live in valleys of the middle Niger and Kaduna rivers in Nigeria... They are famous for their work in glass, silver, bronze, and brass.

nuplex (niū·pleks). [f. Nu(clear *a.* + Com)plex *sb.*] A combined agricultural and industrial complex built around a nuclear reactor as the source of all power and providing employment for a large number of people.
1968 *Courier-Mail* (Brisbane) 27 Aug. 4/1 Mr. Seaborg [Chairman of the U.S. Atomic Energy Commission] defined the 'nuplex' as 'a giant agro-industrial complex built around nuclear reactors'. **1970** *New Scientist* 8 Oct. 89/3 One such investigation is into nuclear farms for the Middle East: half-million-acre 'nuplexes' with a reactor desalinating seawater for irrigation..and manufacturing fertilizer on the spot.

nuppence (nʌ·pĕns). *slang.* [Modelled on *tuppence*.] No money.
1886 *Longman's Mag.* VII. 551 The Americans can get our books, and do get them, and republish them and give us nothing—that awful minus quantity, nuppence! **1964** *Observer* 20 Sept. 27/7 Living on nuppence. **1973** *Times Lit. Suppl.* 30 Mar. 347/4 For the appreciation of the novel, this information matters little more than nuppence.

nuptiality. Add: **4.** The frequency or incidence of marriage within a population.
1902 *Encycl. Brit.* XXXI. 839/1 Nuptiality and Fecundity.—In connexion with the subject of natural increase may be mentioned the tendency of a people towards marriage, and the average fertility of each union. **1949** *Population Studies* II. 356 The relations between male and female nuptiality in general. **1966** E. A. Wrigley *Introd. Eng. Hist. Demogr.* iv. 111 The demographic mechanisms which produced changes in population totals—the interplay of nuptiality, fertility and mortality. **1973** *Times* 12 Dec. 2/3 Although nuptiality was higher than ever, mean family size for couples..had dropped.

Nuremberg (niū·rĕmbɜɹg). Also (erron.) -burg. [G. *Nürnberg*.] The name of a city in southern Germany. **1.** Used *attrib.* to designate a type of porcelain manufactured in Nuremberg.
1617 F. Moryson *Itinerary* III. ii. iii. 80 The Germans export..*Nurnberg* wares (so they call small wares.) **1863** W. Chaffers *Marks Pott. & Porc.* 105 *Nuremburg*, there are two plates of the XVIIIth Century in the Sèvres Museum; one in imitation of Faenza, the other an allegory of Luther. **1925** B. Rackham tr. *Hannover's Pott. & Porc.* I. iv. 355 Chinese motives..are not wanting in Nuremberg faience in the first and best period. But at an early stage they give way to plant ornament, in which a feather-like leaf..is a constantly-recurring element. **1960** R. G. Haggar *Conc. Encycl. Cont. Pott. & Porc.* 331/1 Two personalities are connected with Nuremberg *faience* in the sixteenth century, Oswald Reinhard, and Augustin Hirsvogel. **1974** *Country Life* 6 June 1411/3 The architectural detail is executed in scagliola and the walls are clad in Nuremburg tiles.
b. *Nuremberg egg*, a type of watch (see quot. 1960).
[**1895** H. L. Nelthropp *Catal. Collection of Clocks* 2 Nuremberg egg-shaped watch, metal and silver case.] **1960** H. Hayward *Antique Coll.* 201/1 *Nuremberg egg*, a misnomer applied to early South German watches. Arose from the misreading and mistranslation of 'Uhrlein' into 'Eierlein' (little clocks into little eggs). These early watches were usually drum shaped.
2. Used *attrib.* to designate a connection with the German National Socialist Party with which the city was associated, as **Nuremberg Laws**, laws promulgated in 1935 barring Jews from German citizenship and forbidding intermarriage between 'Aryans' and Jews; **Nuremberg rally**, one of the mass meetings of the German National Socialist Party which were held annually at Nuremberg from 1933 to 1938; **Nuremberg trial**(s), a series of trials of former Nazi leaders for alleged war crimes and crimes against humanity presided over by an International Military Tribunal formed from the victorious Allied Powers and held in Nuremberg in 1945–6 (also ellipt. as *Nuremberg*).
1937 Janowsky & Fagen *Internat. Aspects German Racial Policies* ii. 106 Even the official commentary on the Nuremberg laws..concluded that the German Jews were a racial (voelkische), though not a national minority. **1945** H. Nicolson *Diary* 4 June (1968) III. 64 Smuts recalled the decision..'It..shows..the value of Nuremberg. Had we shot Jodl at sight, that precious piece of evidence would never have been obtained.' **1945** *Daily Mail* 19 Nov. 2/3 All day today Lord Justice Lawrence and Mr. Justice Birkett have been discussing..the arguments for and against postponement of the Nuremberg trials. **1946** M. Belgion *Epitaph on Nuremberg* 12 The argument rested on the plea..that justice was going to be done. Of this argument the official explanation of the Nuremberg trial consisted. **1952** A. Bullock *Hitler* ii. vii. 347 The Nuremberg rallies held every year in September were masterpieces of theatrical art, with the most carefully devised effects. **1963** *Guardian* 25 Feb. 8/1 Towards the end of the Stalin era a Soviet brand of anti-Semitism had undoubtedly reached very dangerous proportions..even if it was never codified in a Soviet version of the 'Nuremberg Laws'. **1969** W. Carr *Hist. of Germany 1815–1945* xiv. 405 Hitler's violent speech at the Nuremburg party rally. **1970** *Guardian* 20 Nov. 12/6 Ever since Vietnam got big, Nuremberg has been injected into the national argument... The Nuremberg precedents demand that..American high commanders ought to be tried for war crimes. **1971** M. McCarthy *Birds of America* 110 Under the Nuremberg Laws, he would have counted as a Jew, while in Israel he would count as a Christian, since what mattered to them was your mother. **1974** *Times* 14 Feb. 16/7 Thank heavens we don't do it like the Tories. Their adulation is reminiscent of a Nuremberg rally.

Nuremberger (niū·rĕmbɜɹgəɹ). [f. *Nuremberg + -er¹.] A native or inhabitant of Nuremberg, Germany.
1673 J. Ray *Observations Journey Low-Countries* 112 Aldtdorf, a little walled Town and an University belonging to the Nurenbergers. **1894** S. Weyman *My Lady Rotha* xxxi. 349 Deserting a friend because the Nurembergers frowned upon him. **1946** H. Nicolson *Diary* 1 May (1968) III. 62 If I were a Nuremberger I should feel nothing but undying hatred for those who had destroyed my lovely city.

nurse, *sb.*¹ Add: **3. b.** Prefixed as a title to the name of one qualified to nurse, esp. in a hospital; used as a mode of address to such a person.
1702 J. Mordaunt *Let.* in E. Hamilton *Mordaunts* (1965) ii. 33, I did not doubt but that you would have great trouble in parting wth Nurse Lucas. **1791** F. Burney *Let.* 12 Sept. (1972) I. 65 Less than an hour compleated the whole business without any help excepting Nurse Whittons. **1874** [see Soul *sb.* 13 a]. **1940** A. Christie *Sad Cypress* i. i. 21 'Do you think she's really good-looking, Nurse?' Nurse Hopkins said: 'Difficult to tell what these girls really look like under their make-up!' **1955** 'A. Gilbert' *Is she Dead Too?* vi. 116 I'll be sending Nurse Wilson along. **1975** P. D. James *Black Tower* iv. 100 They made the bed together, Nurse Rainer flicking the sheets into place and neatly mitring each corner.
c. Without article in reference to the nurse in charge of a patient.
1766 [see *home-baked* (*home *sb.*¹ 14 i)]. **1914** G. B. Shaw *Misalliance* 68 Is anything the matter, John? Nurse says she heard you calling me a quarter of an hour ago. **1937** J. Betjeman *Continual Dew* 25 Nurse looked at the silent bedstead.
4. (Later examples.)
1966 *Times* 21 Apr. 16/6 Scots pines were commonly planted as nurses to oak. **1973** *Country Life* 6 Dec. 1928/1 Larch has been..used..as a nurse with hardwoods and alone.
8. *nurse-companion*, *-girl* (earlier and later examples), *-secretary*; **nurse-cell**, any cell whose function appears to be to assist another cell, esp. an ovum, in some way; **nurse cloth**, a plain-weave cotton fabric used for nurses' uniforms; **nurse-crop, -tree** = sense 4 in Dict. and Suppl.
1896 E. B. Wilson *Cell* iii. 114 In all these cases it is doubtful whether the nurse-cells are sister-cells of the egg which have sacrificed their own development for the sake of their companions, or whether they have had a distinct origin from a very early period. **1964** Bishop & Surgenor *Red Blood Cell* viii. 324 These authors present electron micrographs which depict erythroblastic islands in the bone marrow, in which a central reticulum cell (nurse cell) is surrounded by a ring of erythroblasts. **1967** *Jrnl. Cell Sci.* II. 613 Extensive nuclear fusion has been described in the trophocytes (nurse cells) in the ovaries of the milkweed bug *Oncopeltus*. **1907** *Harrods Catal.* 1407 Cotton dresses, in good quality Nurse Cloth. **1932** D. C. Minter *Mod. Needlecraft* 248/2 *Book Carrier*..Blue nurse cloth, hessian or heavy Russian crash. **1908** B. Harraden *Interplay* 210 Dr. Edgar can no doubt find you a nurse-companion. **1973** A. Christie *Postern of Fate* iii. ii. 132 She was a kind of nurse-companion with Mrs Beddingfield. **1938** *U.S. Dept. Agric. Yearbk.* 409 Wheat is a good nurse crop for the clover and grass. **1955** *Archit. Rev.* CXVII. 249 Even in hardwood country they [*sc.* conifers] are often required as nurse-crops to the deciduous trees. **1971** *Country Life* 4 Nov. 1237/1 Shall it [*sc.* seeding] be in spring or mid-August, under a nurse crop or without? **1847** C. Brontë *Jane Eyre* III. iii. 98, I will be a servant, a nurse-girl, if I can be no better. **1953** D. Lessing *Five* i. 16 There was a little black nurse-girl seated on one of the logs, under a big tree, with a white child in her arms. **1950** Nurse-secretary [see *appointment 4]. **1902** B. E. Fernow *Economics of Forestry* vii. 175 When the latter [*sc.* the young crop] has come up, the nurse trees are gradually removed to give the young seedlings the required light. **1953** H. L. Edlin *Forester's Handbk.* v. 75 Frost-tender trees should never be planted in frost-hollows without the protection of a hardier nurse-tree. **1971** E. Hyams *Capability Brown* i. v. 49 He [*sc.* Brown] used some conifers as nurse trees.

nurse, *sb.*² Add: **b.** *nurse-hound*, a name used for several dog-fish, esp. the large-spotted dog-fish, *Scyliorhinus stellaris*. (Later examples.)
1921 *Nature* 29 Dec. 585/1 The spur-dog and nurse-hound are viviparous. **1922** *Ibid.* 12 Jan. 55/2 Mr. E. Ford writes to inform us that the term 'nurse-hound' is applied at Plymouth to *Scyliorhinus stellaris*, which is not viviparous. We understand from the writer of our article that confusion has arisen from the fact that the name 'nurse-hound' is also used by fishermen in his district to refer to *Mustelus vulgaris*, which is viviparous. **1959** A. Hardy *Fish & Fisheries* ix. 179 The rough hound and the nurse hound are both spotted with dark spots..upon a lighter ground. **1967** [see *huss *sb.*]. **1969** A. Wheeler *Fishes Brit. Is. & NW. Europe* 44/1 The nurse hound is most common on rough, even rocky ground and it may be found within the algal zone close inshore, although it is most common in deeper water. **1972** [see *huss *sb.*].

nurse, *v.* Add: **1. c.** (Earlier and later examples and examples with animal as subject.)
1893 Davis & Keating *Mother & Child* xxiii. 74 A sore or cracked nipple may bleed when the infant nurses. **1938** M. K. Rawlings *Yearling* xi. 98 The fawn nuzzled her full udders and began to nurse. **1946** M. C. Self *Horseman's Encycl.* 289 It is important that the foal nurse as soon as he is strong enough to stand. **1963** M. McCarthy *Group* x. 223 If they gave him a second drink of water, he might not nurse properly when feeding time finally came. **1972** *Sci. Amer.* Dec. 18/2 After the fifth week..the kittens are able to pursue her around the cage until they can get hold of a nipple, after which the mother allows them to nurse.
3. e. (Later examples.)
1933 P. Godfrey *Back-Stage* ix. 131 A good play which does not catch on at once may sometimes be 'nursed' to genuine success. **1936** *Discovery* May 164/1 If two climbers can be 'nursed' to that height the victory [over Everest] may at last be won. **1942** *Tee Emm* (Air Ministry) II. 72 The instructor is, therefore, less able to..nurse each individual into efficiency.
5. b. (Earlier example.) Also with an injury as object.
1778 F. Burney *Evelina* I. xix. 129 She had a bad cold, and chose to nurse it. **1951** *N. Y. Times* 11 Dec., The.. pilot was safely at his home..nursing only a slight scratch on his nose.
6. d. With a drink as object: to consume slowly, holding the glass in the hand between sips.
1942 *Sun* (Baltimore) 5 Oct. 13/4 They buy several drinks in the bar, then they come in to catch the floor show, and nurse one drink along. **1962** K. Orvis *Damned & Destroyed* xiii. 88 'Don't nurse this—drink it!' I said. She gulped the straight whisky gratefully. **1964** B. Malamud *Idiots First* 70 Cronin, pretty much contented, had had one drink to her two, and he was nursing his first when she asked for a third. **1974** R. B. Parker *God save Child* (1975) xix. 134 A thin black man..was nursing a brandy glass at..the bar.

nurse-maid. Add: **b.** *slang.* (See quot.)
1943 C. H. Ward-Jackson *Piece of Cake* 44 *Nursemaid*, a long-distance fighter escort for bombers.
Hence **nu·rse-maid** *v.*, to tend and care for (a person) as a nurse-maid does her charge.

Also with non-personal object. Hence **nu·rse-maiding** *vbl. sb.*
1921 *Glasgow Herald* 19 Apr. 8 He had to be nurse-maided and chaperoned to his meals, his bath, and his bed. **1924** 'J. Sutherland' *Circle of Stars* iv. 40 'I really don't need the nurse-maiding you think I ought to have,' Gloria said impatiently. **1935** C. S. Forester *African Queen* iv. 87 That engine..was greased and cleaned and nurse-maided. **1960** *Times* 5 Apr. 14/6 The kroomen from the African mainland who nursemaided us. **1967** *Economist* 27 May 896/1 By..telling the UN force it could stop nursemaiding his country, President Nasser had scored a clear political success in the Arab world. **1973** 'A. York' *Captivator* ii. 32 Nursemaiding princesses..is not really in my line.

nursery. Add: **3. d.** An establishment for training promising young players of a particular sport.
1950 W. Hammond *Cricketers' School* viii. 80 Yorkshire does not run the usual cricket Nursery that a team like Middlesex maintains. **1954** F. C. Avis *Boxing Ref. Dict.* 76 *Nursery*, a club in which boxing talent is developed. **1961** —— *Sportsman's Gloss.* 36/1 *Nursery*, a junior club taken under the wing of a bigger club to which talented nursery players graduate. **1962** G. Scott in B. Glanville *Footballer's Compan.* 447 It seemed that lads who had been taken from the 'nursery' clubs by the slag heaps and the pits had weak heads for success.
8. a. *nursery bathroom, bedroom, -book* (earlier example), *chair, child, fender, food, -governess* (earlier example), *meal, -rhyme* (earlier and later examples), *-song, story, supper, -tale* (earlier example), *tea*; **nursery-girl**, a nursery-maid; **nursery language**, a stylized form of language used in addressing small children; **nursery word**, a non-standard word used by a child or by an adult to address a child.
1949 'J. Tey' *Brat Farrar* xii. 94 You can have the nursery bathroom all to yourself, but *do* go slow on the hot water, will you? **1941** T. S. Eliot *Dry Salvages* i. 7 His [*sc.* the river's] rhythm was present in the nursery bedroom. **1818** Keats *Let.* 23 Jan. (1958) I. 210, I was at Hunt's the other day, and he surprised me with a real authenticated Lock of *Milton's Hair*. I know you would like what I wrote thereon—so here it is—*as they say of a Sheep in a Nursery Book*. **1869** C. L. Eastlake *Hints on Household Taste* (ed. 2) viii. 191 The rush-bottomed 'nursery' chairs, of which the wood-work is stained black, with low seats and high backs..are still to be bought in the East of London. **1896** *Heal & Son Catal.* 153 Nursery Chair, low cane seat and high back. **1817** Jane Austen *Let.* 13 Mar. (1952) 484 When Caroline was sent to School some years, Miss Bell was still retained, though the others were then mere Nursery Children. **1973** *Guardian* 22 May 13/1 Nanny's pride, the nursery child, ringletted, smocked and sashed, is no more. **1907** *Yesterday's Shopping* (1969) 177/1 Nursery Fenders... Brass top, japanned diamond wire work, black bottom plate. **1913** C. Mackenzie *Sinister St.* I. i. i. 10 Round the fire was a nursery fender on which hung perpetually various cloths and clothes and blankets and sheets. **1926–7** *Army & Navy Stores Catal.* 275/2 Nursery fender, with..brass top, ⅜ in. mesh..30 in. high with 12 in. return ends. **1949** A. Christie *Crooked House* v. 29 Proper wholesome nursery food—not those queer spiced rice dishes. **1965** D. Francis *Odds Against* vi. 84 A tiny service lift used long ago to take nursery food to top floor children. **1861** C. M. Yonge *Stokesley Secret* xii. 193 She..suspected Rhoda, the little nursery-girl, who was quite a child, and had not long been in the house. **1873** —— *Pillars of House* I. v. 96 Why, you might as well turn nursery-girl at once. **1820** M. Wilmot *Let.* 4 May (1935) 60 The loss of a most valuable little french Nursery Governess. **1845** F. A. Kemble *Let.* 8 Dec. in *Rec. Later Life* (1882) III. 95 In nursery language, I *peacified* the good old lady to the best of my ability. **1925** O. Jespersen *Mankind, Nation & Individual* vii. 145 Another dialect used with regard to the person addressed is that more or less affected nursery-language which many mothers and nurses..use with small children—where 'stomach' is 'tum-tum', 'horse' is 'gee-gee', 'thank-you' is 'ta' etc. **1968** *Trans. Philol. Soc.* 107 The special structures and lexical items employed by adults when talking to young children, which we can conveniently group together under the label of *Nursery Language*. **1942** M. B. Lowndes *Let.* 15 Apr. (1971) 229 Many people..live in their country houses with relations, children, and so on. I know of one where there are three sets of nursery meals! **1953** H. Nicolson *Diary* 6 May (1968) III. 240 Dull nursery meals—beef, mutton and milk-puddings. **1832** A. Fonblanque *England under Seven Admins.* (1837) II. 304 The man of Thessaly, famed for wondrous wisdom in nursery rhymes, who, having scratched his eyes out by jumping into one hedge, jumped into another to scratch them in again. **1972** (*title*) The bedtime book of 365 nursery rhymes. **c 1820** Nursery-song [in Dict.]. **1927** W. E. Collinson *Contemporary Eng.* 9, I mention these nursery-songs. **1971** A. Mizener *Saddest Story* xxvi. 358 Ford led them in a round dance on the [Avignon] bridge to the tune of the nursery song. **1834** G. Crabbe Jr. *Life of Rev. George Crabbe* in *Poetical Works & Life of Crabbe* I. x. 304 Little tales, as nearly resembling those which had delighted his own infancy as modern systems permit..the German Nursery Stories. **1848** Thackeray *Van. Fair* xxv. 219 It was as in the old nursery-story, when the stick forgot to beat the dog. **1966** B. Ireson (*title*) The Faber book of nursery stories. **1857** C. M. Yonge *Dynevor Terrace* II. xii. 186 She often comes down after our dinner to find something for the nursery supper. **1971** J. Drummond *Farewell Party* xxi. 117, I was given a huge nursery supper by old Bertha. **1741** Richardson *Pamela* IV. lxiv. 451 You desired me to send you a little Specimen of my *Nursery Tales and Stories*, with which..I entertain.. my little Boys. **1888** Kipling *Story of Gadsbys* 4 *Miss T*... Won't you have some eggs? *Captain G*... Eggs! (*Aside.*) Oh Hades! She must have a nursery-tea at this hour. **1939** T. S. Eliot *Family Reunion* i. i. 17 Harry must often have remembered Wishwood—The nursery tea, the school holiday. **1958** M. Stewart *Nine Coaches Waiting* iii. 33 A little pantry with an electric stove for making nursery tea. **1933** L. Bloomfield *Language* ix. 157 In English almost any doubled syllable may be used, in almost any meaning, as a nursery-word. **1957** R. W. Zandvoort *Handbk. Eng. Gram.* ix. i. 287 Many of them are nursery words... *Georgy-Porgy, piggie-wiggie, tootsy-wootsies* (feet), etc.
c. In sense 3, as *nursery education*; **nursery class**, a class attached to a primary or other school for the education of children usu. between the ages of three and five years; **nursery nurse**, a person trained to care for babies and young children; so **nursery nursing**; **nursery school**, a school for children usu. between the ages of two and five years; also *attrib.* and *fig.*; so **nursery schooling**; **nursery slope** *Skiing*, a gentle slope considered most suitable for beginners; also in extended uses (see quots.).
1921 *Act* 11 & 12 *Geo. V* c. 51 § 21 Supplying.. nursery schools (which expression shall include nursery classes) for children over two or under five years of age. **1943** *Educational Reconstruction* 5 A certain number of children enter..nursery classes attached to infants' schools, at an earlier age on a voluntary basis. **1970** J. & P. Kent *Nursery Schools for All* ii. iv. 94 Nursery classes are..defined..as 'a class mainly for children who have attained the age of three years but not the age of five years', and which is part of a primary school for infants or juniors. **1938** P. E. Cusden *English Nursery School* xvii. 257 The general provision of facilities for nursery education. **1969** *Guardian* 17 Jan. 9/2 Under the new Urban Aid Scheme £3 million are promised for more nursery education. **1974** *Ibid.* 24 Jan. 13/2 Mrs Thatcher has announced a programme of expanding nursery education. **1947** A. B. Meering *Handbk. for Nursery Nurses* 1 The Nursery Nurse who prefers the care of individual children ..may become a nanny in a private family. **1967** V. C. Jones in P. J. Cunningham *Nursery Nursing* 13 Nursery nurses..care for the young child in its earliest and most impressionable years. **1972** *Guardian* 30 Aug. 11/3 The nanny or nursery nurse of today is trained in child care... Norland..describes nursery nursing as a growing profession. **1973** E. Lemarchand *Let or Hindrance* ii. 16 She was in training as a nursery nurse, and had one more year to go. **1835** D. W. Webber *Let.* in I. Butler *Eldest Brother* (1973) i. ii. 29 It was..in the year 1765 that Lord Wellesley was brought to school... It was quite a nursery school... As a kind of Preparatory School it was in great Fashion. **1891** Michaelis & Moore tr. *Froebel's Lett.* 30 He [*sc.* Froebel] thinks of christening it 'Nursery School for Little Children' or 'Self-teaching Institution'. **1918** *Act* 8 & 9 *Geo. V.* c. 39 The Board shall have regard to the adequacy of the provision of nursery schools for the area. **1958** *Economist* 24 Oct. 303/1 His [*sc.* Gaitskell's] backbenchers still belong to the nursery school of political manœuvre. **1967** O. Wynd *Walk Softly* iii. 33 A nursery-school teacher who has happily dedicated a whole life to very young minds. **1970** J. & P. Kent *Nursery Schools for All* i. i. 28 From that time the nursery school problem assumed the distinctive character which it still has today. **1974** *Times* 14 Oct. 4/1 Nursery school grants rejected by councils. Mrs Thatcher introduced a £34m programme in 1971..to make nursery schooling available to half the three to five age group by 1980. **1924** K. Furse *Ski-Running* p. vi, Every beginner should be content to devote two or three of his first days to the Nursery slopes. **1924** W. Le Queux *Crystal Claw* i. 21 She had been three times before to winter sports, and had long passed the period when she practised her 'telemarks' and 'stemmings' on the 'nursery slopes'. **1943** Hunt & Pringle *Service Slang* 47 *Nursery slopes*, the easy targets allotted to beginners on bombing tests. **1959** *Daily Mail* 14 Oct. 12/6 If you do not limber up before you go [skiing] you may find your second or third day on the nursery slopes surprisingly painful. **1972** M. Yorke *Silent Witness* ii. 12 The lifts, and even the cable-car.., had stopped... Only the two short drags on the nursery slopes were working. **1975** *Daily Tel.* 15 Jan. 13 (*caption*) An early flight [in hang-gliding] lasting about 45 seconds down a nursery slope, straight into wind.

nursey. Add: (Further examples.) Also **nursie.**
1860 G. H. Lewes *Let.* 17 Mar. in *Geo. Eliot's Lett.* (1954) III. 274, I went to see Nursie, who will 'keep house' while I am away. **1918** A. Bennett *Roll-Call* ii. i. 220 Nursey's sunshade was undiscoverable. **1968** 'D. Shannon' *Kill with Kindness* xiii. 210 He was nursie's favorite—he could coax money out of her. **1973** R. Adams *Watership Down* (ed. 2) xvii. 94 You risk the life of one of the best rabbits we've got, just to play nursey while you go wandering about like a moon-struck field-mouse.

nursing, *vbl. sb.* Add: **1. b.** The profession of a nurse (Nurse *sb.*¹ 3); the duties of a nurse.
1860 F. Nightingale *Notes on Nursing* 6, I use the word nursing for want of a better. *Ibid.* 72 She [*sc.* the writer]..honestly believes that the perfection of surgical nursing may be seen practised by the old-fashioned 'sister' of a London hospital. **1889** O'Neill & Barnett *Our Nurses* i. 2 It is commonly and justly coming to be held that nursing in all its branches is a career for educated women. **1937** E. C. Pearce (*title*) A general textbook of nursing. **1955** *Oxf. Jun. Encycl.* XI. 333/2 Florence Nightingale was the real founder of the modern profession of nursing. **1970** K. K. Guinée *Professional Nurse* i. 6 The teacher of nursing carefully selects learning experiences in the clinical area.
2. *nursing-bottle, -chair* (further examples); (sense *1 b) *nursing-training*.
1861 Mrs. Beeton *Bk. Househ. Managem.* 1041 Many kinds of nursing-bottles have been lately invented, and some mounted with India-rubber nipples. **1944** A. Seton *Dragonwyck* i. 4 She unbuttoned her bodice, snatched up the hungry baby, and settled on the low nursing chair. **1971** *Country Life* 22 July (Suppl.) 32b/2 (Advt.), A Wm. IV nursing chair of hammock shape upholstered in deeply buttoned Havana brown leather. **1914** W. Owen *Let.* 29 Oct. (1967) 291 The Nursing Training is capital for you.
b. Designating garments designed to facilitate the breast-feeding of a baby, as *nursing basque, bra, brassière, corset*.
1939 M. B. Picken *Lang. Fashion* 104/3 *Nursing basque*, basque with buttoned closings, one on each side of the front. **1969** *Sears, Roebuck Catal.* Spring/Summer 362/2 Finest nylon lace nursing bra... Easy-open clasp lets you hold baby as you open cups. **1950** Heaton & Daynes *Feeding Mothers & Babies* ii. 55 A nursing brassière.. should have a waterproof lining. **1895** *Montgomery Ward Catal.* 309/2 Dr. Strong's Tricora Nursing Corset has proved a great comfort to mothers.
c. Special Combs., as **nursing home**, (*a*) a small, private institution where the sick are cared for; also *attrib.*; (*b*) a place where certain qualities are nurtured.
1896 [in Dict., sense 2]. **1938** L. P. Smith *Unforgotten Years* vi. 149 Her barrister husband insisted..that I should be transferred without delay to what was, in his opinion, the only nursing home of reasonable thought and noble ambition—in fact, to Balliol College. **1951** [see *clinic *sb.*² 2]. **1959** T. S. Eliot *Elder Statesman* ii. 45 We've studied to avoid Anything like a nursing-home atmosphere.

nursing, *ppl. a.* Add: **1. b.** *nursing mother* (*b*) a woman who is breast-feeding her own baby.
1806 D. Wordsworth *Let.* 23 July in *Lett. William & Dorothy Wordsworth* (1969) II. 60 She thinks herself quite well, but *I* do not think she is as yet as strong as she ought to be for a nursing mother. **1897** G. Tucker *Mother, Baby & Nursery* 96 It becomes the first duty of the nursing mother to take care of herself. **1926** M. A. von Arnim *Introd. to Sally* ix. 154 Good job I ain't a nursin' mother..or the lady'd turn my milk sour. **1950** E. Pantin *Mod. Mothercraft* i. 19 The health of a nursing mother affects the quality of her milk. **1970** J. de Bairacli-Levy *Natural Rearing of Children* iii. 28 It is an error to say that a nursing mother must take milk from animals to increase her own milk. **1974** A. Huxley *Plant & Planet* xxviii. 328 Nursing mothers in the United States have so much DDT in their milk that, to quote a scientific humorist, 'in strict terms of federal law, it is illegal for them to carry their busts..across one state line to another'.

nurtural (nɜ̄·ɹtiŭrăl), *a.* [f. Nurtur(e *sb.* + -al.] Of, belonging to, or due to nurture; usually designating characteristics, etc., which can be attributed to training, environment, or the like, and are not natural or inherited.
1889 *Jrnl. Anthrop. Inst.* XIX. 78 The problem of determining purely 'racial characteristics' will be considerably simplified if we can in this way determine what may be described in contradistinction as 'nurtural characteristics'. **1922** W. R. Inge *Outspoken Ess.* 2nd Ser. 257 Professor Pearson has tabulated a long list of natural characters, and another long list of nurtural characters. **1922** *Edin. Rev.* July 47 Religion is the strongest of nurtural influences.

nurturance (nɜ̄·ɹtiŭræ̆ns). *Psychol.* [f. Nurtur(e *v.* + -ance.] Emotional and physical nourishment and care.
1938 H. A. Murray *Explorations in Personality* ii. 83 *Nurturance*,..to nourish, aid or protect... To express sympathy. To 'mother' a child. **1957** E. R. Hilgard *Introd. Psychol.* (ed. 2) vi. 134/2 Scales were designed to get at nurturance (i.e., the mother's care in feeding). **1964** Cofer & Appley *Motivation* xiv. 719 Someone else may help or sympathize with him, that is, he may receive nurturance from another. **1973** *Jrnl. Genetic Psychol.* June 185 A wide spectrum of 15 personality dimensions (e.g. need achievement, nurturance, dominance, aggression). **1974** *Nature* 9 Aug. 466/1 Warm and outgoing students (high on nurturance) also display a more favourable attitude.

nurturant (nɜ̄·ɹtiŭrănt), *a.* *Psychol.* [f. as prec. + -ant¹.] Caring or nourishing (emotionally or physically); exhibiting or pertaining to nurturance.
1938 H. A. Murray *Explorations in Personality* iii. 181 The succorance drive seeks a nurturant O and the nurturant drive seeks a succorant O. **1951** R. R. Sears in Parsons & Shils *Toward General Theory of Action* 471 The nurturant mother. **1973** *Jrnl. Genetic Psychol.* Mar. 37 Psychopaths..saw their fathers as having been less nurturant toward them and as having given them less praise. **1973** S. Fisher *Fem. Orgasm* iv. 108 Femininity is associated with being nurturant and 'nice'.

Nusranee, Nusrani: see *Nasrani.

Nusselt (nu·sĕlt). [the name of E. K. Wilhelm *Nusselt* (1882–1957), German engineer.] *Nusselt number*, a dimensionless parameter used in calculations of the heat transfer between a moving fluid and a solid, equivalent to hD/k, where h is the rate of heat loss per unit area per degree difference in temperature between the body and its surroundings, D is a characteristic length of the body, and k is the thermal conductivity of the fluid.

1933 W. H. McAdams *Heat Transmission* iv. 96 *hD/k*... Nusselt number. **1958** Condon & Odishaw *Handbk. Physics* III. ii. 35/1 For low speeds the Nusselt number is a function of the Reynolds number, the Prandtl (or Péclet) number, and the Grashof number... In the case of forced convection, for which gravitational effects are insignificant, the Nusselt number is a function only of the Reynolds number and of the Prandtl number. In natural (free) convection the Nusselt number is a function only of the Grashof number and of the Prandtl number. **1974** J. R. Welty *Engin. Heat Transfer* v. 265 Heat Transfer for Cylinders in Crossflow... At higher Reynolds numbers..the Nusselt number experiences two sudden increases, one at the separation point and one where the boundary layer undergoes a transition from laminar to turbulent flow.

nustaleek, var. *nastalik.

nut, *sb.*¹ Add: **I. 1. f.** Pl. *vulg.* The testicles. Also in various vulgar phrases.

1915 *Dialect Notes* IV. 186 *Nut*, in *pl.* testicles. **1922** Joyce *Ulysses* 467 How's the nuts?..Off side. Curiously they are on the right. Heavier I suppose. **1955** W. Gaddis *Recognitions* II. vii. 630 The lady lost her nuts, Anselm said to no one. He mumbled,—That's the world we live in, the ladies wear the nuts. **1969** B. Malcolm in A. Chapman *New Black Voices* (1972) 385 Easy way out To hate the white man..for Kicking my papa in the nuts. **1970** C. Major *Dict. Afro-Amer. Slang* 58 *Get (one's) nuts off*, sexual release, implies ejaculation more than orgasm. **1970** E. Bullins *Theme is Blackness* (1973) 167 Screwin' my best white friend's black wife makes me feel even better. Makes me get my nuts off. **1973** R. Busby *Pattern of Violence* v. 79 Russell got a boot in the nuts. **1974** J. Wainwright *Evidence I shall Give* xxi. 102 He was working his nuts off.

3. Phr. *for nuts*, in neg. contexts: at all.

1895 W. Pett Ridge *Minor Dialogues* 82 An' the eldest gal *she* thinks she can play, and, if you'll believe me, she carn't play for nuts. **1899** [in Dict.]. **1934** A. Thirkell *Wild Strawberries* xi. 237 That Miss Stevenson can't play for nuts.

6. Also const. *about*.

1920 S. Lewis *Main Street* xxiii. 280 Carrie's nuts about this Russian revolution. **1945** E. Waugh *Brideshead Revisited* I. vii. 177, I was still nuts about Rex. **1975** *New Yorker* 21 Apr. 39/1 You're nuts about me, right?

7. a. (Earlier examples.)

1846 *Swell's Night Guide* 76 Why, she's getting groggy on her pins, and if you don't pipe rumbo, she'll go prat over nut (head over heels). **1852** J. Labern *Pop. Comic Song Bk.* 76 But vun chap flung a bunch of turnips, which nearly split Dick's nut in two.

c. *Pl.* (as *adj.*): insane, crazy, 'off one's head'.

1846 *Swell's Night Guide* 75 Vhy, Owen..you knows it's no use of me being nuts, ven the donna's only nut crackers. **1914** Jackson & Hellyer *Vocab. Criminal Slang* 62 Nuts,.. As an adjective and adverb it signifies daft, mentally deranged. **1928** C. Sandburg in *Amer. Mercury* Oct. 154 There was a screw loose somewhere in him, he had a kink and he was a Crank, he was nuts and belonged in a booby hatch. **1953** [see *gee *int.*²]. **1969** I. & P. Opie *Children's Games* ii. 76 The person was..looney, nuts, a nit.

d. Phr. *to do one's* (or *the*) *nut*, to become angry, lose one's head; to be worked up about something; to be crazy.

1919 W. H. Downing *Digger Dialects* 20 *Do the nut*, lose one's head. **1936** J. Curtis *Gilt Kid* 231 The jane'd be bound to think he had done his nut. **1956** [see *choked *ppl. a.* 2]. **1957** J. Osborne *Entertainer* xiii. 86 I'm doing me nut up here. **1958** F. Norman *Bang to Rights* III. 92 The twirl would do his nut and give up. **1959** P. Bull *I know Face* xi. 199, I would be doing my 'nut' and my probable swansong for Auntie BBC. **1960** *News Chron.* 16 Feb. 6/5 Been doing his nut about little Barbara for months, he had. **1961** J. Stroud *Touch & Go* xv. 155 He's nearly done his nut over this daughter of his. **1972** J. Brown *Chancer* v. 68, I thought what Grace would say, that she'd do her nut maybe. But she didn't blink an eyelid.

e. *Pl.* Used as a derisive retort: nonsense, rubbish; I defy you. Freq. const. *to*.

1931 M. E. Gilman *Sob Sister* 267 Nuts! You'll forget Nick the minute you smell your freedom. **1934** J. O'Hara *Appointment in Samarra* iv. 86 'Nuts to you, sister,' he said. **1936** *New Yorker* 18 Jan. 20 With a hay-nonny-nonny and a nuts to you. **1938** J. Curtis *They drive by Night* ix. 109 A feather shopping bag lay on a table. He stepped across towards it. As he stepped his golf-clubs rattled. Aw, nuts to that dog barking. **1946** Wodehouse *Joy in Morning* II. xii. 91 'If you think I've got the force of character to come back with a *nolle prosequi*—' 'With a what?' 'One of Jeeves' gags. It means roughly "Nuts to you!"' **1974** D. Francis *Knock Down* ii. 25 'I'll give you a hundred.' 'Nuts.' 'A hundred and fifty.'

f. *the nuts*, an excellent person or thing. *U.S. slang*.

1932 *Amer. Speech* VII. 334 *The nuts*, denotes superlative quality. **1949** W. Stevens *Let.* 9 Sept. (1967) 647 At the Museum of Modern Art they cultivate the idea that everything is the nuts. **1955** W. Gaddis *Recognitions* II. vii. 634 Get a little cross with mirrors in it, that would be the nuts if you want to suffer your way.

8. c. A madman; a crank. *slang* (orig. *U.S.*).

1903 R. L. McCardell *Conversat. of Chorus Girl* 15 'Circus Joe'..worked the nuts on the edge of the crowd. **1914** Jackson & Hellyer *Vocab. Criminal Slang* 62 *Nut*, commonly current in all circles when the meaning is 'loco'. **1919** [see *bughouse *sb.* 1]. **1931** D. Runyon *Guys & Dolls* (1932) 213, I am commencing to think this Count Saro is some kind of a nut, and is only speaking through his hat. **1936** K. Mackenzie *Living Rough* xv. 216 We're sure a pair of nuts riding the outside over the hump this time of the year. **1960** H. Pinter *Room* 118 You're not only a nut, you're a blind nut and you can get out the way you came. **1966** T. Pynchon *Crying of Lot 49* v. 107 Why worry, she worried; Nefastis is a nut, forget it, a sincere nut. **1973** *Nation Rev.* (Melbourne) 31 Aug. 1444/1 The Worker Student Alliance, a bunch of nuts in Melbourne.

8*. *slang.* A fashionable or showy young man of affected elegance; a 'young blood', fop, or masher. Cf. Nutty *a.* 4 and *knut.

1904 in *N. & Q.* (1913) 26 July 78/1 I'm one of the nuts, one of the nibs. **1913** *Punch* 12 Feb. 115/1 Spring socks will be black and Spring ties a quiet blue. A strike of nuts is expected at any moment. **1915** Kipling in *Nash's & Pall Mall Mag.* Oct. 131/2 Winchmore, the youngest, was more on the lines of a conventional nut. **1920** W. J. Locke *House of Baltazar* xvii. 205 I've a jolly good mind to set him up regardless, like a pre-war nut—with solid silver boot-trees and the rest to correspond. **1920** R. Macaulay *Potterism* I. iv. 44 He always looked the same, calm, unruffled, tidy, the exquisite nut. **1923** *Other Lands* Oct. 3/3 The last named continue to be marks of the 'nut'.

11. a. Also *fig.*

1911 *Rep. Labour & Social Conditions in Germany* (Tariff Reform League) III. 39 When we get our nuts screwed a little tighter we shall be able to look after our own industries and mind our own business. **1973** 'J. Patrick' *Glasgow Gang Observed* x. 88 Asked why so few boys over twenty remained in the Fleet [gang], Tim replied: 'They used tae be in it but they've screwed the nut.' *Ibid.* 235 *Screw*, as in '*screw the nut*', to become sensible, to 'get wise' to oneself, to pull oneself together.

d. *fig.* in phr. (*the*) *nuts and bolts*, the practical, basic elements, or the mechanics, of a situation or thing. Freq. *attrib.*

1960 *Times* 9 Feb. 11/4 When we talk about technicians—the 'nuts-and-bolts boys'—we are all right there. **1967** *Observer* 30 Apr. 11/8 A..keen-eyed Army colonel..talks to you about 'the nuts and bolts' of the programme. **1971** *Times* 10 June 16/6 His preference was for journalism. He learnt the nuts and bolts of his profession with the Montreal *Gazette*. **1972** *Guardian* 10 June 11/7 Most of some hundreds of recommendations in the action plan and its annexe, where the nuts and bolts are, will go through. **1973** T. Allbeury *Choice of Enemies* xvii. 83 There are two kinds of security that we cover... A bit of cigarette ash on a magnetic tape could screw up a whole pay-roll..but..that's pretty well a nuts and bolts area for us. We know it inside out. **1974** *Times* 22 Feb. 14 The electors are not to be despised for this conspicuous lack of interest in the nuts and bolts of politics. **1974** *Socialist Worker* 7 Dec. 9/6 There was also a tendency to go very easy on replies to those delegates who urged the conference to adopt a fuller and finer programme rather than discuss the nuts and bolts of trying to build in the real world.

18*. *U.S. slang.* The amount of money required for a venture; overhead costs. Hence *transf.*, any sum of money.

1912 A. H. Lewis *Apaches N.Y.* 201 Every day I'm open puts me fifty dollars on th' nut. **1914** Jackson & Hellyer *Vocab. Criminal Slang* 62 *Nut*,..used by grafters whose operations involve an investment to signify an expense incurred in connection with a venture. **1933** *Sun* (Baltimore) 28 Jan. 16/4 The difficulty of 'making the nut', the term applied to accumulating the rental charge due each night to the owner of the cab. **1935** *Amer. Mercury* June 230/1 *Nut*, concession charges for booking a joint; expenses. **1936** *Amer. Speech* XI. 219 He [*sc.* the producer] decides that in order to open the show a certain amount of money will be necessary. This amount is the production *nut*. **1948** *Sun* (Baltimore) 7 Aug. 9/2 In any event the 'nut' will be close to $400,000, counting fighters, rent and promotional expenses. **1955** *Publ. Amer. Dial. Soc.* xxiv. 37 Recreation, such as gambling..and other cultural pursuits dear to the hearts of pickpockets, will cost extra... All this is counted as *nut*..and anything they make over and above this they consider income. **1956** H. Gold *Man who was not with It* (1965) xviii. 159, I was getting a nut of cash, and it felt good. **1962** J. B. Priestley *Margin Released* vi. 202 In the Thirties, when we could produce *Laburnum Grove*..for about £800..the weekly running costs—the 'get-out' or 'nut'—were round about the same figure, theatre and all. **1970** *Daily Tel.* 27 Apr. 3 New York police have their own secret slang to deal with their illegal business... 'Nut' is a cash bribe. **1972** *Publishers' Weekly* 14 Feb. 60/1 He submitted a strong script that led Fox to substitute color film and wide screen for black-and-white and the conventional small-screen ratio, and to raise the nut to $400,000.

19. a. (sense *8 c), as *nut alley*, *-doctor*, *-farm*.

1935 A. J. Pollock *Underworld Speaks* 81/2 *Nut alley*, prison insane ward. **1955** A. Huxley *Let.* 7 May (1969) 742 Next week..I go to Atlantic City to attend the Psychiatrists' Assn. meeting... I shall arrange to meet the boys on my return from the nut-doctors. *a* **1940** F. Scott Fitzgerald *Last Tycoon* (1949) i. 12 Some mystic..spouting tripe that'd land him on a nut-farm anywhere outside of California.

20. *nut-bearing*, *-questing*.

1877 L. H. Morgan *Anc. Society* (1907) I. ii. 20 In fruit and nut-bearing forests under a tropical sun, we are accustomed..to regard our progenitors as having commenced their existence. **1952** A. G. L. Hellyer *Sanders' Encycl. Gardening* (ed. 22) 130 Corylus (Cob-nut; Filbert)... Hardy deciduous nut-bearing shrubs. **1922** Joyce *Ulysses* 535 Who left his nutquesting classmates to seek our shade?

21. nut-butter, a substitute for butter obtained from the oil of nuts; **nut (milk) chocolate**, (milk) chocolate containing nuts; **nut college** *U.S. slang*, = *nut-house* (below); **nut cutlet**, a portion of meat-substitute made from nuts and various other ingredients and shaped like a cutlet; **nut factory** *U.S. slang*, = *nut-house* (below); **nut food**, food prepared from nuts; so **nut-fooder**; **nut-house** *slang*, a mental hospital; **nut-meat**, the kernel of a nut; **nut-pine**, substitute for def.: one of several species of pine producing edible seeds, native to south-western North America and the Rocky Mountains; (earlier and later examples); **nut runner**, a power tool for tightening nuts; **nut-steak**, a portion of meat-substitute made from nuts and shaped like a steak.

1907 Nut-butter [see s.v. Nucoline]. **1908** *Westm. Gaz.* 5 Aug. 2/3 Vegetarians cannot expect to be allowed to call their butter-substitute 'nut butter' when other people's butter-substitutes are called 'margarine'. **1918** C. A. Mitchell *Edible Oils & Fats* ix. 117 Deodorised coconut oil is used in the preparation of both margarine and 'nut butter'. **1961** C. Loewenfeld tr. *Bircher's Eating your Way to Health* II. iii. 246 Nut butter is a good and easily digested substitute for those who do not like, or should not have, butter. **1971** J. Hewitt *N.Y. Times Natural Foods* xv. 393 *Nut butter spread*... Place the nuts, sunflower seeds, kernels and seeds in an electric blender and blend until fine. **1975** *Times* 14 Feb. 9/7 They bake daily... They also make various nut butters. **1926–7** *Army & Navy Stores Catal.* 54/1 Chocolate..Nut (¼ lb. pkts.). **1936** 'J. Tey' *Shilling for Candles* xi. 128 'Nut or plain?' 'What?' 'The chocolate.' **1955** M. Allingham *Beckoning Lady* xvi. 228 Offering Westy half a bar of nut chocolate. **1931** *Amer. Speech* VII. 111 *Nut college*, an insane asylum. **1951** in Wentworth & Flexner *Dict. Amer. Slang* (1960) 360/2 He has been recalled by the nut college to join Napoleon..and Shakespeare, inventing paper dolls! **1908** F. A. George *Vegetarian Cookery* ix. 113 *Nut cutlets*... Make into cutlet shapes. Egg and crumb. Fry in deep fat. **1925** D. H. Lawrence *Let.* ? 17 Dec. (1962) II. 871 So Sonya will never cook us another goose, only marmite pie and nut-cutlet. **1959** 'M. Innes' *Hare sitting Up* II. vi. 112 Didn't I say something about Burgundy? Capital with nut cutlets. **1965** J. B. Priestley *Lost Empires* II. ii. 115 Mr Foster-Jones makes Health Foods... He's probably brought a case of date sandwiches and nut cutlets with him. **1973** *Times* 10 Apr. 14/1 You can no longer get nut cutlets... What you get instead now are nut rissoles. **1915** *Recruiter's Bull.* (U.S. Marine Corps., N.Y.) Oct. 15/1 It would have been impossible to have found a man any other place than a 'nut-factory' who would voluntarily have told the commanding officer that he was a deserter. **1929** J. Callahan *Man's Grim Justice* xiii. 156 They should have been in the 'nut factory'..the insane department. **1939** J. H. Chase *No Orchids for Miss Blandish* i. 33 Johnnie was a rummy... Drink had rotted him, and he was only two jumps ahead of the nut-factory. **1905** *Vegetarian Messenger* Apr. 105, I will send any readers who wish for it an address where nut-foods can be had guaranteed free of pea-nuts. **1917** N. Douglas *South Wind* x. 142 He will be an anti-vivisectionist, a nut-fooder, costume-maniac.., or a spiritualist into the bargain. **1929** *Amer. Speech* IV. 343 *Nut House*, an insane asylum. **1936** 'P. Quentin' *Puzzle for Fools* i. 6 It wasn't a sanatorium really. It was just an expensive nuthouse for people like me who had lost control. **1953** W. Burroughs *Junkie* (1972) 10, I decided I was not going to like the Army and copped out on my nut-house record... The nut-house doctors had never heard of Van Gogh. **1958** 'N. Blake' *Penknife in my Heart* iii. 42 Miriam drives you into the nut-house. **1973** 'H. Howard' *Highway to Murder* xi. 141 Supposing the plan succeeded and his wife got stuck away in a nut-house? **1974** *Radio Times* 24 Oct. 11/2 Clothing for the Government, prisons and nut-'ouses—what is it they call 'em now? **1913** A. B. Emerson *R. Fielding at Snow Camp* 102 The three boys stuck to their work..until there was a great bowl of nutmeats. **1967** *Economist* 9 Sept. 892/1 Kukui nut-meat burning gently in a shell. **1974** *Aiken* (S. Carolina) *Standard* 22 Apr. 8-A/5 The Viennese desserts called Torten are sometimes made with finely ground nutmeats without the inclusion of any flour. **1918–19** *T. Eaton & Co. Catal.* Fall & Winter 385/2 Eaton's Nut Milk Chocolate... Each bar made from fine chocolate, milk and nuts. **1932** R. Lehmann *Invitation to Waltz* III. vi. 215, I preferred to spend the afternoon on the schoolroom sofa reading *East Lynne* and eating nut-milk chocolate. **1960** *Sunday Express* 25 Dec. 13/3 When nut milk chocolate was 2d. a bar. **1845** J. C. Frémont *Rep. Expl. Exped. Rocky Mountains* 221 A pine tree..which Dr. Torrey has described as a new species, under the name of *pinus monophyllus*; in popular language, it might be called the nut pine. **1896** C. H. Shinn *Story of the Mine* 63 The nut-pine trees were soon cut down. **1949** Collingwood & Brush *Knowing your Trees* 18/1 This is one of four nut pines of the Southwest. **1969** T. H. Everett *Living Trees of World* 50/2 The piñon or Mexican stone pine (*P. cembroides*) is small and spreading, and, like its variety *P. c. edulis*, the nut pine, produces delicious edible seeds or 'nuts'. Both are natives of the southwestern United States and Mexico, the nut pine extending north to Wyoming. **1958** R. M. Barnes *Motion & Time Study* (ed. 4) xvii. 288 The multiple-spindle air-operated nut runner..is used to tighten all five wheel nuts at once. **1966** *Engineers' Digest* Dec. 97/2 Suitable for light duties and often employed with power screwdrivers and nutrunners, another type of torque limiting device ultilizes a spring-loaded steel ball. **1908** *Daily Chron.* 2 Sept. 3/4 High thinking is still nourished upon the banana and the nut-steak. **1922** Joyce *Ulysses* 163 Why do they call that thing they gave me nutsteak? Nutarians, Fruitarians. To give you the idea you are eating rumpsteak. **1966** K. Giles *Provenance of Death* iv. 103 The man..is a vegetarian... He had a nut steak.

22. Passing into *adj.* Stupid, insane (cf. senses 7, 8, and *19 a).

1919 *Sci. Amer.* 23 Aug. 184/1 Other ideas, no more revolutionary and no more absurd from the standpoint of entrenched orthodoxy, never graduate from the 'nut' class. **1922** U. Sinclair *They call me Carpenter* xix. 66, I just want to know where he got his nut ideas. **1922** S. Lewis *Babbitt* ii. 17 Ever since somebody slipped up and let you out of college..you been pulling these nut conversations about what-nots and so-on-and-so-forths. **1966** T. Pynchon *Crying of Lot 49* iii. 48 'You one of these right-wing nut outfits?' inquired the diplomatic Metzger.

nut, *v.* Add: **3.** *slang.* **a.** To think, to use one's head. Freq. const. *out*, also *up*. (Cf. Nut *sb.*¹ 7.)

1919 W. H. Downing *Digger Dialects* 36 *Nut it out*, think it out. **1925** Fraser & Gibbons *Soldier & Sailor Words* 213 *Nut out, to*, to think over. Consider. To use one's head. **1951** D. Stivens *Jimmy Brockett* 168, I did a bit of hard nutting over my plans for trotting. **1953** K. Tennant *Joyful Condemned* iv. 38 Just nut that out. **1962** A. Upfield *Will of Tribe* xix. 180, I asked him how he nutted up the idea. **1965** M. Shadbolt *Among Cinders* xiii. 112, I haven't nutted out what I'm going to say about the poultry. **1971** R. Dentry *Encounter at Kharmel* (1973) v. 81 I've been nutting the whole thing out... There's no future in it for you.

b. To butt with the head; to hit a blow on the head. Also **nu·tting** *vbl. sb.*

1937 Partridge *Dict. Slang* 575/1 *Nut*,..to punch on the head. **1963** T. & P. Morris *Pentonville* xi. 241 Few prison fights are conducted in accordance with Queensberry Rules; fists, heads (for the painful infliction of injury to the opponent's nose by 'nutting'), teeth and nails may be used at any time. **1966** D. Skirrow *It won't get you Anywhere* xiv. 61, I shot my head backwards in time to miss the nutting that was coming... The tearaway special nowadays is to hug tight, rupture his kidneys and nut him hard. **1971** J. Mandelkau *Buttons* xiii. 145 He took it off and as I was getting out of mine he nutted me in the head.

Nut, var. *Nat¹.

nutarian (nɒtɛə·riăn). [f. after *vegetarian*.] A vegetarian whose diet is based on nut products.

1914 *Chambers's Eng. Dict.* Suppl. 1281/2 *Nutarian*, one who lives on nuts.—Also adj. (From *Nut*, in imitation of *Vegetarian*.) **1922** [see *nut-steak* s.v. *Nut *sb.*¹ 21].

nutate, *v.* Delete *rare* and substitute for def.: To undergo or exhibit nutation. (Further examples of vb. and ppl. adj.)

1898 S. H. Vines *Elem. Text-bk. Bot.* iii. 211 All growing members nutate in a more or less marked manner. **1921** J. Small *Textbk. Bot.* xvii. 221 In the mature plant the top three internodes of the stem turn or nutate in a circle. **1943** R. C. Binder *Fluid Mech.* viii. 107 The nutating-disk meter or wobble-plate meter..is frequently used to meter the water supply for domestic use. **1948** C. E. Ingalls in S. N. Van Voorhis *Microwave Receivers* xv. 380 Tracking in azimuth and elevation is made possible by the use of a nutating antenna. A fixed paraboloidal reflector is combined with a dipole, which is caused to move in a small circular orbit about the focus of the reflector, to give a radiation pattern in the form of a beam that traces out a small cone. **1950** H. Goldstein *Class. Mech.* v. 168 It is not the regular precession encountered in force-free motion, for as the figure axis goes around it nods up and down..—the top nutates. **1965** Bell & Coombe tr. *Strasburger's Textbk. Bot.* 390 A nutating shoot. **1969** R. Skinner *Mech.* iv. 469 The top continues to spin about its axis of symmetry, while this axis rotates, or precesses, about a vertical axis. All the while, this axis [of symmetry] nods up and down, or nutates, as it precesses.

nutation. Add: **2. c.** Movement (as of a beam or aerial) by which an axis is made to describe a cone. (Analogous to the precession of a spinning top rather than its nutation.)

1947 *Bell Syst. Techn. Jrnl.* XXVI. 307 The axis of the beam was rotated in an orbit by 'nutation' about the mechanical axis of the antenna. **1966** *McGraw-Hill Encycl. Sci. & Technol.* XI. 204/2 Either the feed or the reflector whirls rapidly in a manner that causes the beam axis to describe a circular cone; this motion of the beam axis is called nutation.

nutational (niutēi·ʃənăl), *a.* [f. Nutation + -al.] Of or pertaining to nutation.

1881 W. D. Hay *300 Years Hence* viii. 152 The precessional and nutational movements of the earth. **1959** Van Lear & Lassen in Puckett & Ramo *Guided Missile Engin.* x. 275 If the gyro is started up with the inner gimbal initially inclined to the plane perpendicular to the outer gimbal, there is first a transitory nutational bobbing around. **1965** Bell & Coombe tr. *Strasburger's Textbk. Bot.* 361 In both kinds of movement the mechanical cause may be either a difference in growth rates on the two sides of an organ (nutational movements) or differential changes in the turgor of the cells (variational movements).

nu·tburger. Also **nutberger.** [f. Nut *sb.*¹ 1 + *burger.] A meat-substitute made from nuts, formed into a cake, and usu. served between the two halves of a toasted bun; also, a hamburger topped with nuts. Also *attrib.*

1934 M. Weseen *Dict. Amer. Slang* xix. 291 *Nutberger*, a hamburger sprinkled with nuts. **1939** A. Huxley *After Many a Summer* I. i. 6 Drive In For Nutbergers—whatever they were. **1942** *Amer. Speech* XVII. 132/2 *Nutburgers* are on sale in California. **1948** E. Waugh *Loved One* 120 Dennis..followed her to a nutburger counter. *Ibid.*, D'you know, this is the first time I've ever eaten a nutburger? **1959** *Listener* 24 Sept. 497/2 It is ready to try anything, from Mormonism to nutburgers. **1973** *Guardian* 17 Mar. 13/1 Thrifty high-protein meat substitutes,..soyabeanburgers and nutburgers and fishwiches.

nu·t-cake. [Nut *sb.*¹ 1.] **a.** *U.S.* A doughnut or fried cake. **b.** A cake containing nuts.

a **1800** *Spirit of Farmer's Museum* (1801) 235 Heap the nut-cakes, fried in butter. **1823** [in Dict. s.v. Nut *sb.*¹ 19]. **1844** *Knickerbocker* XXIV. 483 Reflection..was interrupted by the appearance..of 'nut-cakes and cider'. **1857** *Quinland* I. ii. ii. 34 By the way, Hepsy, make us some 'nut-cakes', and bring us the cider. **1873** M. Holley *My Opinions* (1891) 251 Where is the rich happy woman that wouldn't give a nutcake to a sick beggar? **1889** R. T. Cooke *Steadfast* xviii. 198 Who ever heard tell of puttin' a reason and a bit of citron into the middle of a riz nut-cake before 'twas fried? **1957** G. Mann *Bk. Cakes* 157 Nut Cake... Put into a shallow greased tin and spread the nut covering over the cake. **1966** W. I. Kaufman *Nut Cookery Bk.* 65 *Nut cake*... Fold in walnuts, stirring as little as possible... Bake in a moderate oven.

nu·t-case. *colloq.* [f. Nut *sb.*¹ (cf. *8 c) + Case *sb.*²] A crazy person; a madman.

1959 *Punch* 21 Oct. 337/2, I couldn't get anyone to talk about it openly. The way they clammed up you'd have thought I was a spy or a nut-case. **1965** A. Prior *Interrogators* iii. 28 He knew the nut-cases, the convicted sexual offenders. **1969** *Listener* 24 Apr. 586/2 You nut-case, you ought to be locked up. **1973** Boyd & Parkes *Dark Number* v. 57 They were all shams... She was a nutcase really.

nut-cracker. Add: **5. Nutcracker Man**, a nickname for the fossil hominid, *Australopithecus robustus* (or *A. boisei*), the maker of the oldest stone tools known, esp. the specimen discovered by L. S. B. and M. D. Leakey at Olduvai, Tanzania, in 1959; similar remains, including the characteristic large premolar teeth, have also been found in South Africa.

1959 *Times* 4 Sept. 8/4 He [*sc.* L. S. B. Leakey] has named the species *Zinjanthropus Boisei*... The nickname given by Dr. Leakey to the world's oldest man is 'Nutcracker Man' because of the tremendously developed teeth. **1961** *New Scientist* 26 Oct. 221 Not only is *Zinjanthropus* or Nutcracker Man 'unquestionably' human but some of his fairly distant ancestors were human as well. **1962** *Listener* 5 Apr. 589/1 Dr. Leakey's famous 'nutcracker man' Zinjanthropus (which has now been dated as having lived over 1,000,000 years ago). **1972** S. Cupitt tr. *Wendt's From Ape to Adam* iv. 228 He [*sc.* Robert Broom] found the remains of an australopithecine equipped with a particularly powerful jaw and truly nutcracker-like teeth... These 'Nutcracker men' even had a small sagittal crest on their skulls. *Ibid.* 232 At first Leakey thought that this ancient Oldowan ancestor of ours was very different from the South African 'Nutcracker man', despite his powerful back teeth. **1974** Washburn & Moore *Ape into Man* iv. 107 The huge molars..could have cracked nuts, and the Leakeys sometimes liked to call their discovery 'nutcracker man'.

nu·t-cut, *a.* *India.* [Hind. *naṭkhaṭ*.] Roguish. Also as *sb.*, a rogue, rascal.

1848 J. H. Stocqueler *Oriental Interpreter* 175/2 *Nut-cut*, roguish, mischievous. A term of reproach, good-naturedly applied in India to *vauriens*. **1901** Kipling *Kim* iv. 107 'That is a *nut-cut* (rogue),' she said. 'All police-constables are *nut-cuts*; but the police-wallahs are the worst.'

nut-grass. Substitute for def.: A small sedge of the genus *Cyperus*, esp. *C. rotundus*, whose roots form small nut-like tubers. (Earlier and later examples.)

1775 B. Romans *Conc. Nat. Hist. Florida* 129 In Carolina it [sc. *herbe au cheval*] is called nutt grass from a nutt found at its root. **1894** J. M. Coulter *Bot. W. Texas* III. 463 *Cyperus rotundus*... From the South Atlantic and Gulf States to the Texan coast... Often called 'nut grass'. **1903** 'S. Rudd' *Our New Selection* iv. 36 Nothing but burr and thistle and nut-grass grew. **1944** *Living off Land* ii. 42 The little bulbs of the onion weed (or nut grass). **1965** *Austral. Encycl.* VIII. 68/1 The latter species [sc. *Cyperus rotundus*], known as 'nut grass', is widely spread in warmer parts of the world (including Australia) and is a serious weed pest—probably the worst in areas where rice is extensively cultivated.

nuthin (nɒ·þin). Repr. of a colloq. pronunc. of Nothing *sb.*

1925 E. O'Neill *Desire under Elms* I. 69 Oceans o' trouble an' nuthin' but wuk fur reward. **1968** R. Clapperton *No News on Monday* vi. 73 Course I never said nuthin'. **1968** A. Diment *Bang Bang Birds* x. 181 Don't you cats know nuthin' about acid? **1971** *Black World* Oct. 64/2 So I don't say nuthin'. **1973** 'J. Patrick' *Glasgow Gang Observed* v. 51 Ah *hate* somewan in a company who disnae say nuthin'.

nutmeg. Add: **3.** (Example.)

1822 J. Woods *2 Yrs. Res. Eng. Prairie Illinois* 307 There are many sorts of sweet melons... I have only noticed musk, of a large size; and nutmeg, a smaller one.

5. nutmeg hickory, a species of hickory, *Carya myristicæformis*, bearing a fruit resembling a nutmeg and found in southern North America.

1810 F. A. Michaux *Hist. Arbres Forestiers de l'Amérique Septentrionale* I. 21 Nutmeg hickory nut.., nom donné par moi. **1832** D. J. Browne *Sylva Amer.* 177 This species..bears the name of Nutmeg Hickory from the resemblance of its fruits to that of the nutmeg. **1901** C. T. Mohr *Plant Life Alabama* 101 The nutmeg hickory, when full grown, resembles the shagbark hickory in its pale, shreddy bark. **1951** *Dict. Gardening* (R. Hort. Soc.) I. 404/2 Nutmeg Hickory. Tree 80 to 100 ft., shoots covered with yellowish, glossy scales... Nut ovoid, sweet, its shell hard and furrowed like a nutmeg.

nutmeggy, *a.* (Further examples.)

1928 *Daily Express* 17 Feb. 4 Luscious prunes with a creamy, nutmeggy rice pudding. **1971** *Daily Tel.* 17 Apr. 7 A large proportion of the bed could be occupied by..Cherry Pic, which has a warm nutty, almost 'nutmeggy' scent, spicy and yet sweet.

nutria. Add: Hence also, a mid-brown colour such as that of the nutria fur.

1897 *Sears, Roebuck Catal.* 234/1 Ranch Hats... Colors, light nutria or tan. **1923** *Daily Mail* 26 Mar. 6 Colours: Gold, Grey,..Nutria, Putty. **1949** *Brit. Colour Council Dict. Colours Int. Decoration* III. 19/1 *Nutria*, a colour selected in consultation with expert furriers and standardized by B.C.C. in 1934.

nutriceptor (niū·triseptǫɹ). *Immunol.* [f. *nutri-* (in *nutrient*, *nutrition*, etc., or L. *nūtrīment-um*, etc.) + Re)ceptor.] (See quot. 1926.)

1911 *Jrnl. Amer. Med. Assoc.* 7 Oct. 1210/2 According to Ehrlich's view, when a parasite becomes refractory to an immune serum it does so by developing new groups of receptors..these receptors being the same receptors as combine with the food materials, and hence called 'nutriceptors'. **1926** R. J. E. Scott *Gould's Med. Dict.* 907/2 *Nutriceptors*, Ehrlich's name for receptors which react with foodstuffs more or less exclusively, and which therefore serve the nutrition of the cell. **1932** [see *chemoceptor].

nutrient, *a.* and *sb.* Add: **B.** *sb.* (Further examples.)

1899 *Bull. Div. Veg. Physiol. & Path.* (U.S. Dept. Agric.) XVIII. 6 We can accept it as an indisputable fact that mineral matters found in plants also are real nutrients for them. **1903** H. Snyder *Chem. Plant & Animal Life* xxxvi. 344 A balanced ration is one which contains a sufficient amount of nutrients from a variety of foods to meet the requirements of the animal. **1924** *Bot. Gaz.* LXXVII. 121 (*heading*) Absorption of nutrients from subsoil in relation to crop yield. **1974** *Encycl. Brit. Macropædia* XIII. 403/2 Carbon dioxide (CO_2) and water (H_2O)..are important nutrients for all organisms. *Ibid.* 407/2 Lists of nutrients—both organic and inorganic—required by plants and animals. *Ibid.*, Essential nutrients include many amino acids, some fatty acids, many vitamins, and some minerals and trace elements.

2. Comb. *nutrient-poor*, *-rich* adjs.

1946 *Nature* 21 Sept. 421/2 In the nutrient-rich waters of the Thames type, a burst of algal growth may sometimes cease before any serious depletion of the mineral nutrient in the water has apparently taken place. **1955** *New Biol.* XVIII. 115 *N. alba* occupies a wide range of waters in the British Isles, from the oligotrophic, or nutrient-poor, peat-bottomed moorland lakes in Scotland and Ireland, to the eutrophic, or nutrient-rich fen-lodes and broads of East Anglia. **1967** *Oceanogr. & Marine Biol.* V. 108 Massive upward displacement of nutrient-rich water on to the shelf may occur a few times in a century.

nutrition. Add: Hence also **nutri·tionally** *adv.*

1890 *Cent. Dict.*, Nutritionally. **1922** *Sci. Amer.* July 42/3 A diet may furnish a sufficient amount of protein, fat, carbohydrates, salts and vitamins and yet fail to promote growth or sustain well-being unless the quality of protein is nutritionally adequate. **1949** M. Mead *Male & Female* x. 215 Foods all of which are suitable nutritionally. **1972** *Which?* Sept. 263/2 Nutritionally, there is not much difference between dairy and non-dairy [ice cream].

nutritionalist (niutri·ʃənălist). [f. Nutritional *a.* + -ist.] = next.

1956 *Nature* 24 Mar. 565/2 The importance of close co-operation between chemists, pharmacologists, biochemists and nutritionalists in this field cannot be overemphasized. **1971** *Islander* (Victoria, B.C.) 9 May 14/3 Steers a clear course through conflicting nutritionalists' claims to a reasonable assessment of the nation's dietary practices.

nutritionist (niutri·ʃənist). [f. Nutrition + -ist.] One who studies, or is knowledgeable about, food and nutrition, esp. of humans.

1926 *University of State of N.Y. Bull. to Schools* June 261/2 Dentists, dental hygienists and nutritionists when employed, are required to assist the medical inspector. **1936** *Nature* 31 Oct. 744/2 That highly trained specialist, the nutritionist—who, to be competent, must needs use the methods not only of chemistry and physics but also those of the various biological sciences. **1959** *New Biol.* XXX. 106 Animal nutritionists also have considered the possibility that traces of boron may be required in the vertebrate diet. **1969** *Daily Tel.* 21 Nov. 17/1 Until recently, nutritionists and health experts focused their attention on expectant mothers, babies and adolescents. **1973** *Nature* 27 Apr. 593/2 Few nutritionists are interested in the problems of old age.

nutshell, *v.* (Earlier example.)

1883 'Mark Twain' *Life on Mississippi* lviii. 570 The clerk nut-shelled the contrast between the former time and the present.

nutsy (nɒ·tsi), *a.* *colloq.* Also **nutsey.** [f. *nuts* (*Nut *sb.*¹ 7 c) + -y¹.] Crazy, insane.

a **1941** F. Scott Fitzgerald *Tender is Night* (rev. ed., 1953) III. viii. 175 A boy..she thought was pretty nutsey. **1942** Berrey & Van den Bark *Amer. Thes. Slang* §152/5 *Insane; crazy*,..nutsy. **1962** *Guardian* 27 Aug. 5/3 Gee, it was nutsy. **1964** W. Markfield *To Early Grave* (1965) x. 169 Take a train, you nutsy you!

Nu·tter². Also **nutter.** [f. Nut *sb.*¹ 1 + But)ter *sb.*¹] The proprietary name of a substitute for butter made from the oil of nuts; nut-butter.

1906 *Westm. Gaz.* 18 May 4/2 'Nutter', 'Nucoline', and 'Nuttene'—all representing butter made from nuts. **1909** H. G. Wells *Ann Veronica* vii. §3 Fruitarian refreshments—chestnut sandwiches buttered with nutter, and so forth. **1915** Barnett *Let.* 2 May (1915) 127 One [*sc.* a trench mortar]..fires a cylindrical thing like a *Nutter* tin. **1920** *Trade Marks Jrnl.* 19 May 971 Nutter... Fats used in cooking. Mapleton's Nut Food Company, Limited,..Liverpool; food manufacturers. **1926–7** *Army & Navy Stores Catal.* 67/1 Nutter—the ideal cooking fat. **1958** *Catal. County Stores* (Taunton) June 29 Vegetarian foods..Cooking Fat, Nutter lb. 2/1. **1974** R. B. Parker *God save Child* (1975) xi. 81 Dolly Bartlett got a package of Nutter Butter cookies from the cabinet.

nu·tter³. *slang.* [f. *Nut *sb.*¹ 8 c + -er¹.] An insane person; a violent and deranged person. Occas. used in weaker sense: an eccentric person.

1958 F. Norman *Bang to Rights* I. 36 The reason for this is to find out wether [*sic*] or not you are a nutter. **1960** *Observer* 24 July 24/7 Sally is, at first sight, one of those romantic schizoids, a near nutter. **1963** 'A. Garve' *Sea Monks* ii. 66 Reckon we'd be nutters to try it now. *Ibid.* v. 135, I reckon Chris was right, Rosie—King's a nutter. I reckon he'll go on killin' till there ain't no one left. **1963** [see *barm *sb.*³ 3]. **1965** A. Prior *Interrogators* xi. 200 A lot of 'em are nutters, I reckon. **1968** J. Lock *Lady Policeman* v. 43 The term 'nutter' was invariably used though not meant unkindly and included all types from the eccentric (a bit of a nutter) to the raving lunatic (a right nutter). **1972** R. Quilty *Tenth Session* 7 You could make out on the tipping lark—all those rich nutters.

nuttery. Add: **1.** (Later *attrib.* example.)

1932 H. Nicolson *Diary* 20 Mar. (1966) 113 What would be good..would be to put the end of the main nuttery walk at the end of a main vista.

3. *slang.* A mental hospital.

1931 'D. Stiff' *Milk & Honey Route* vi. 62 Should the sociotechnic social worker be convinced that you are not normal she will have you bound for a nuttery before sunset. **1950** H. E. Goldin *Dict. Amer. Underworld Lingo* 147/2 *Nuttery*, an institution for the criminally insane or for mentally defective delinquents.

nuttiness (nɒ·tinĕs). [f. Nutty *a.* + -ness.] The quality or state of being nutty (in various senses).

1865 R. D. Blackmore *Cradock Nowell* (1866) xv. 137 In the height of summer, [his colour was] a dappled bay; towards the autumnal equinox, a tendency to nuttiness. **1884** *Sat. Rev.* 8 Mar. 321/2 The six essays..have the 'nuttiness' of age about them. **1916** E. V. Lucas *Vermilion Box* 27 All his nuttiness has gone. You remember how his hair used to be swept right back from his forehead with lovely comb marks in it. **1926** E. O'Neill *Great God Brown* 20 And I know damn well, underneath your nuttiness, you're gone on her. **1965** H. Gold *Man who was not with It* xv. 128 It's the nuttiness of the mark and his fist in his palm. **1965** *Listener* 24 June 951/2 Scientists..regard these same assumptions as 'nuttiness from an amateur'.

nu·ttish, *a.*² [f. *Nut *sb.*¹ 8 c.] Characteristic or suggestive of a crank or a crazy person.

1909 *Punch* 24 Mar. 208/3 He indulged in a variety of eccentricities. I can imagine nothing more nuttish.

nutty, *a.* Add: **2. d. nutty slack**, coal slack in small lumps or nuts (see Nut *sb.*¹ 17). Also *fig.*

1953 *New Yorker* 31 Jan. 58/2 The low-grade small coal appetizingly known as 'nutty slack',..gives off far less appetizing fumes and dirt. **1953** *Truth* 13 Feb. 165 Durham Dilemma. We can't buy Nutty Slack. **1959** I. & P. Opie *Lore & Lang. Schoolch.* ix. 163 Stew, a not infrequent component of school dinners, is..in Croydon, 'nutty slack'. After an inapt term coined by the Ministry of Fuel (1952) for a poor quality coal, obtainable off the ration. The nuts were few and far between.

3. b. (Earlier and later examples.) Also in phr. *nutty as a fruit-cake*.

1898 S. Crane in *Cosmopolitan* Dec. 169/1 'What's the matter with that feller?' asked Martin. 'Nutty,' said the man. **1935** G. & S. Lorimer *Heart Specialist* vi. 163 'Listen, Alix, you're as nutty as a fruitcake,' I said.. 'If I were you I'd have more sense.' **1955** P. Wildeblood *Against Law* 104 He's as nutty as a fruit-cake. **1960** H. Pinter *Caretaker* III. 77 He's nutty, he's half way gone. **1963** *Daily Mirror* 6 Nov. 2/3 You have to be a real sour square not to love the nutty, noisy, happy, handsome Beatles. **1967** Wodehouse *Company for Henry* v. 84 'He doesn't strike me as unbalanced.' 'On his special subject he's as nutty as a fruit cake.' **1972** C. Weston *Poor, Poor Ophelia* (1973) xxxii. 207 Jesus, you and your nutty imagination! **1974** *Author* Spring 26 Yeats was a great poet and a fascinating critic, but if he had been hired to give a year's course of lectures on the development of English poetry his performance would have been extremely nutty.

c. Of jazz or popular music: see quots.

1955 L. Feather *Encycl. Jazz* x. 347 *Nutty*,..great, exceptional. **1959** 'F. Newton' *Jazz Scene* 290 The modern..fashion of using terms taken from mental derangement for praise (*crazy, insane, nutty*).

nyala (nyā·lă). [Native name in Tsonga and Venda languages.] A large, gregarious antelope, *Tragelaphus angasi*, or the closely related species, *T. buxtoni*, occurring in parts of southern Africa; the male is greyish-brown with several white stripes and spiral, black horns, the female is reddish-brown and hornless; = Inyala. Also *attrib.*

1899 [see *harnessed *ppl. a.* 4]. **1915** *Chambers's Jrnl.* Nov. 702/1 The horns of this antelope..approximate more to those of the nyala, one of the largest of the bushbucks. **1931** *Discovery* Feb. 61/1 It [*sc.* the nyala] is one of the few forest-frequenting antelopes, and there cannot be many hundreds left in the jungles of Zululand and Southern Nyasaland. **1947** J. Stevenson-Hamilton *Wild Life S. Afr.* xv. 109 Along the Pafuri River in the nyala bush they have become accustomed to motor traffic. **1964** *Punch* 2 Sept. 359/3 Shooting a nyala. **1975** *Country Life* 20 Feb. 444/3 A visit to..these [South African] reserves is always rewarded with views of zebra, nyala, impala, duiker, waterbuck.

nyam, *sb.* and *v.* Also **nyam-nyam.** See *yam *sb.* and *v.*

Nyanja (nya·ndʒă), *sb.* and *a.* Also **Manganja, Anyanja.** [f. Bantu *nyanja* lake + *ma-* tribal prefix, or *a-* plural prefix.] **A.** *sb.* **a.** The name of a Bantu people found in Malawi. **b.** A member of this people. **c.** The Bantu language spoken by this people. **B.** *attrib.* or as *adj.* Of or pertaining to the Nyanja people or their language.

1865 D. & C. Livingstone *Narr. Expedition Zambesi* v. 108 The Manganja generally live in villages, each of which has its own headman. *Ibid.* 123 The practice of bathing..we afterwards found to be common in other parts of the Manganja country. **1892** D. C. Scott (*title*) A cyclopaedic dictionary of the Mang'anja Language spoken in British Central Africa. **1902** H. Barnes *Nyanja-English Vocab.* p. ii, The vocabulary is primarily intended to help people to understand the Nyanja that they hear or read, and not to make up Nyanja to inflict on wandering natives. **1914** J. B. Keble in *Oxf. Survey Brit. Empire* III. x. 243 The Anyanja are a large and important group... They inhabit the western and south-western shores of Lake Nyasa and the Shiré Highlands. **1924** A. Werner in G. Lagden *Native Races of Empire* iii. 88 A Nyanja man, if addressed in Yao or Konde, would probably not understand... But a Nyanja and a Tumbuka might understand each other. **1930** A. Hetherwick *Dict. Nyanja Lang.* p. v, Not only in Nyasaland itself, but also in Northern and Southern Rhodesia..Mang'anja, or, as it is now called, Nyanja, has come to occupy the place of a lingua franca. **1966** C. G. Seligman *Races of Africa* (ed. 4) ix. 148 The spirits of dead Nyanja chiefs..are specially appealed to for rain. **1974** *Encycl. Brit. Macropædia* XI. 361/2 Nine main groups are historically associated with modern Malawi—the Chewa, Nyanja, Lomwe, Yao, Tumbuka, Sena, Tonga, Ngone, and Ngonde.

Nyassa (nəiæ·să, ny-). Also **Nyasa.** **a.** Name of a people in Malawi. **b.** A member of this people. Also *attrib.* or as *adj.*

1849 C. Pickering *Races of Man* ix. 197 The *N'yasa*, who inhabit the islands and perhaps the further shores of the Great Lake, seemed to be the most distant tribe known at Zanzibar. **1883** R. N. Cust *Sk. Mod. Lang. Afr.* II. xii. 330 Rebman..employed..a slave, whom he imagined to be a Swahili, but he overheard him speaking a totally different language, and upon inquiry he proved to be a Wa-Nyassa. **1887** A. C. Madan *Kiungani* ii. 30 We had never seen a single European in my time... They are called in the Nyasa language 'wan'tu oyela' meaning 'white man'. **1912** C. T. Domingo *Let.* 17 Mar. in Shepperson & Price *Independent African* (1958) Plate 11 (betw. pp. 158 and 159) It may take sometimes [*sic*] to possess jewels of Independance [*sic*] among *we* [*sic*] the Nyassas. Yours, lovely. For Africa. **1959** *Listener* 24 Sept. 471/1 A Nyasa is hardly thought to be a man until he has gone off to work for some years in the Johannesburg gold-mines.

nychthemeral (nikþī·mĕrăl), *a.* Also **nycthemeral.** [f. Nychthemer(on + -al.] Occurring with a variation that matches that of night and day.

1907 *Nature* 17 Jan. 287/2 The regulation of the nychthemeral cycle of temperature and its inversion in the aged. **1967** *Oceanogr. & Marine Biol.* V. 495 These nycthemeral changes of the gas tension in the different levels of water. **1974** *Nature* 13 Sept. 143/2 These animals have a nychthemeral variation of less than 2° C.

nyctinastic (niktinæ·stik), *a.* *Bot.* [a. G. *nyctinastisch* (W. Pfeffer *Pflanzenphysiologie* (ed. 2, 1904) II. xii. 476): see Nycti- and *nastic *a.*] Of the movements of flowers or leaves, caused by a regular cycle of changes in light and temperature. So **nyctina·stism**, **ny·ctinasty**, movement of this kind. Cf. Nyctitropic *a.*, Nyctitropism.

1906 A. J. Ewart tr. *Pfeffer's Physiol. Plants* III. ii. 97 Since the term 'tropism' is reserved for curvatures produced by unilateral stimuli, it becomes necessary to change the term 'nyctitropic' used by Darwin..into that of 'nyctinastic'. *Ibid.* 101 There is no reason for restricting the term nyctinastic to pronounced sleep-movements. **1921** J. Small *Textbk. Bot.* xxvi. 378 Nyctitropism or Nyctinastism includes the opening and closing of flowers.., also the rising and falling of leaves..in response to the stimulation of changes in temperature and light. **1936** J. B. Hill et al. *Botany* ix. 228 Certain leaves as well as flowers may fold up at night. These so-called 'sleep movements' of plants, brought about by the alternation of night and day, are the most common nasties and are termed nyctinasties. **1968** *New Scientist* 26 Dec. 717/1 The closing of the leaves..in fact, happens after the transfer of the plants to darkness—a so-called nyctinastic or 'sleep' movement.

nyctograph (ni·ktŏgrɑf). [f. Nycto- + -graph.] A device invented by 'Lewis Carroll' with which one can record one's ideas at night, in the dark, or when not fully awake.

1891 'L. Carroll' *Diary* 24 Sept. (1953) II. xiv. 486 To-day I conceived the idea of having a series of *squares*, cut out in card, and devising an alphabet, of which each letter could be made of lines along the edges of the squares, and dots at the corners... I shall call it 'The Typhlograph'. (24/10/91. Instead of 'typhlograph' I have adopted 'Nyctograph' at the suggestion of Warner). **1898** S. D. Collingwood *Life & Lett. L. Carroll* vii. 295 In 1891 he conceived the device..and he named it the 'Typhlograph', but, at the suggestion of one of his brother-students, this was subsequently changed into 'Nyctograph'. **1930** W. de la Mare *Eighteen-Eighties* 236 He invented..poetical acrostics and the nyctograph. **1959** R. Thomson *Psychol. Thinking* x. 198 Lewis Carroll derived so much from this source [*sc.* hypnagogic imagery] that he invented a peculiar instrument, the 'nyctograph', to enable him to jot down ideas without fully waking up.

Nylex (nəi·leks). A proprietary name of nylon.

1957 *Official Gazette* (U.S. Patent Office) 3 Sept. TM 3/2 Polymers, Inc...*Nylex* for extruded synthetic fibres, particularly adapted for use as brush bristles. **1967** E. A. Gollschewsky in *Coast to Coast 1965–66* 88 She sank into a patio chair of plaited nylex. **1969** *Guardian* 7 Jan. 7/2 For dinghy sailors, showerproof clothes in Nylex, lightweight and very strong, are..sufficient.

nylon (nəi·lǫn). Also **Nylon.** [Invented word, with *-on* suggested by *rayon*, *cotton*.

There is no evidence to support the derivations freq. given for this word in popular sources. Cf. the following quot.: **1940** *Women's Wear Daily* 9 Feb. 22 The du Pont letter, written by John W. Eckelberry, covers the general status of nylon as follows: 'The word is a generic word coined by the du Pont Co. It is not a registered name or trademark... We wish to emphasize the following additional points: First, that the letters n-y-l-o-n have absolutely no significance, etymologically or otherwise... Because the names of two textile fibers in common use—namely 'cotton' and 'rayon', end with letters 'on'..it was felt that a word ending in 'on' might be desirable. A number of words..were rejected because it was found they were not sufficiently distinct from words found in the dictionary, or in lists of classified trademarks. After much deliberation, the term 'nylon' was finally adopted.']

1. Any of the thermoplastics that are wholly synthetic polyamides with a straight-chain molecular structure, many of which are tough, lightweight, and resistant to heat and chemicals, may be produced as filaments, bristles, or sheets and as moulded objects, and are widely used for textile fabrics and industrially; *esp.* nylon 66, made from adipic acid and hexamethylenediamine.

1938 *N.Y. Times* 28 Oct. 34/3 'Nylon' is a generic name, coined by the du Pont chemists, to designate all materials defined scientifically as 'synthetic fiber-forming polymeric amides having a protein-like chemical structure; derivable from coal, air and water, or other substances, and characterized by extreme toughness and strength and the peculiar ability to be formed into fibers and into various shapes, such as bristles and sheets'. **1940** *Times* 21 Mar. 5/5 Imperial Chemical Industries, Limited, announce that progress is being made in the erection of three factories for the manufacture of nylon in England. **1942** *Industr. & Engin. Chem.* Jan. 56/2 After polymerization, the molten nylon is extruded as a ribbon onto a chilled roll. **1943** *Chem. Abstr.* XXXVII. 3947 A description of the manuf. of No. 66 Nylon (as the most important nylon). **1950** R. W. Moncrieff *Artificial Fibres* xvii. 206 '610' nylon may be preferred to '66', for use as bristles. **1955** *Sci. News Let.* 2 Apr. 217/2 The new, tempered form of nylon, named Nylon 8 by the Du Pont Co., is a liquid. It can be molded into fuel tanks, pipes, gaskets and seals. **1958** D. E. Floyd *Polyamide Resins* i. 4 Nylon-6,10 means that the diamine contained 6 carbon atoms and the dibasic acid contained 10 carbon atoms. **1963** H. R. Clauser *Encycl. Engin. Materials* 451/1 Nylons resist electrolytic corrosion, hydrolysis, fungi, bacteria and most chemicals. **1964** *Which?* Aug. 253/2 There are a number of different nylons used in textiles. *Ibid.*, Fabrics made from nylon tend to attract dirt. **1964** N. G. Clark *Mod. Org. Chem.* xvii. 361 The most famous example of this is nylon 66 (popularly known as Nylon), which is straight chain polyamide constructed from two six-carbon components. **1972** J. Wren-Lewis in Cox & Dyson *20th-Cent. Mind* II. ix. 279 In modern chemical jargon the term 'nylon' refers, not to any one specific material but to any macromolecular material made by interacting dibasic acids with diamines: such materials are sometimes also called polyamides. **1973** *Sci. Amer.* July 42/3 Today the matrix in glass-reinforced composites may be either a thermoset plastic, such as polyester,..or any of a number of thermoplastic resins, such as nylon, polyethylene or polystyrene.

2. *pl.* Nylon stockings.

1940 *Woman* (U.S.) V. ii. 68 Dunk your nylons in rich suds of neutral soap. **1948** *Daily Mail* 21 Apr. 1/2 The two cases were opened by Customs officers. Both were full of nylons and powder compacts. **1951** M. McLuhan *Mech. Bride* (1967) 33/1 Food and nylons..are consumed and promoted with moral fervor. **1957** J. Braine *Room at Top* vii. 72 High heels and nylons. **1965** *N.Y. Times* 16 May vi. 80/2 By 1964, silk and rayon stockings were almost unknown in the United States, while production of nylons had risen to 83,900,000 dozen pairs. **1966** J. Betjeman *High & Low* 67 Encase your legs in nylons, Bestride your hills with pylons, O age without a soul.

3. Fabric or cloth made from nylon yarn.

1940 *Jrnl. R. Aeronaut. Soc.* XLIV. 312 Possible use of synthetic textile Nylon as a parachute material. **1945** *Times* 6 Nov. 4/4 Arrangements are being made for the conversion into clothing and other goods of 26,000,000 yards of nylon, cotton and celanese which will become available from finished parachutes surrendered by the services as surplus. **1958** *Woman's Own* 5 Feb. 37/3 (Advt.), Nighties, blouses, slips, underwear in silk, nylon, rayon, 'Terylene', chiffon..keep their soft sheen and filmy finery. **1967** E. Short *Embroidery & Fabric Collage* iii. 81 Appliqué in more net, or in nylon, organdie, etc., would give weight and definition to the design.

II. 4. *attrib.* or as *adj.* Made or consisting of nylon.

1939 *Industrial Fibres Rev.* III. 167/1 The first considerable use of 'Nylon' yarn will be in the full-fashioned hose trade where silk is at present the raw material. **1941** *Jrnl. R. Aeronaut. Soc.* XLV. 219 Fabrics manufactured of Nylon artificial fibre, a coal derivative.., has [*sic*] been..compared with pure silk, as a possible substitute for the latter, for use in the manufacture of parachutes. **1941** *Jrnl. Amer. Med. Assoc.* 4 Oct. 1221/1 One of the principal reasons why Nylon stockings have achieved popularity is the fact that they are more 'sheer' than silk stockings. **1951** *Good Housek. Home Encycl.* 29/2 Nylon brushes are available in a variety of colours. **1958** *New Statesman* 28 June 831/1 Men who had nylon shirts and terylene suits before those fabrics got into Marks and Spencer's. **1958** L. van der Post *Lost World of Kalahari* ix. 209 We had to make our home under a nylon tarpaulin stretched taut between our Land-Rovers. **1961** *Lancet* 22 July 206/2 Nylon film as a wrapping material for sterilisation. **1968** *Bodl. Libr. Rec.* VIII. 61 It runs on four 4-inch nylon wheels fitted with roller bearings. **1973** 'E. Peters' *City of Gold & Shadows* iv. 65 In a nylon jersey house-gown..she could not possibly be anyone but Mrs. Paviour.

b. (See quots.)

1955 *Caribbean Q.* IV. ii. 103 Nylon, pronounced 'nilõ' or 'dilõ' has come to signify anything new, different, and better. The new ice house in Soufriere advertises 'nylon' ice, and nylon starch and nylon peanuts can be had, the latter being candies shaped like peanuts. **1967** Cassidy & Le Page *Dict. Jamaican Eng.* 326/2 *Nylon road*, any new very smooth asphalt-surfaced road—much smoother than the average Jamaican road. (From about 1958, FGC.)

5. General attrib.

1942 *Industr. & Engin. Chem.* Jan. 58/2 The nylon industry is only in its infancy. **1951** *Economist* 22 Sept. 686/1 Nylon output is now running at an annual rate of 100 million pounds. **1953** K. H. Inderfurth *Nylon Technol.* 11 Du Pont's third nylon plant. **1963** H. R. Clauser *Encycl. Engin. Materials* 452/1 Tubing and rod stock manufacture, plus the coating of wire and cable, are the major forms of nylon extrusion.

6. *Comb.* **a.** Instrumental, as *nylon-covered*, *-faced* adjs. **b.** Parasynthetic, as *nylon-bristled*, *-geared*, *-legged*, *-tipped* adjs.

1954 H. R. Mauersberger in *Matthews's Textile Fibers* (ed. 6) xviii. 946 Nylon-bristled toothbrushes. **1960** *Farmer & Stockbreeder* 29 Mar. 75/3 The pump..incorporates nylon-covered steel rollers to push the milk round. **1967** *Jane's Surface Skimmer Systems 1967–68* 25/1 Propulsion fans are driven by nylon-faced toothed rubber timing belts. **1961** *Listener* 5 Oct. 498/2 Nylon-geared egg beaters. **1954** J. Betjeman *Few Late Chrysanthemums* 74 And country girls with lips and nails vermilion Wait, nylon-legged, to straddle on the pillion. **1966** *Melody Maker* 23 July 10/4 His [drum] sticks are..Autocrat nylon-tipped and Japanese Star 7a.

7. Special comb.: **nylon (stocking) dermatitis**, dermatitis caused by the dye of nylon stockings; **nylon salt**, a salt formed by the reaction of hexamethylenediamine (or another diamine) with adipic acid (or another dibasic acid), which is polymerized to give nylon.

1954 H. R. Mauersberger in *Matthews's Textile Fibers* (ed. 6) xviii. 961 Any references to 'nylon dermatitis'..are distinct misnomers. **1964** *Listener* 26 Mar. 520/1 Nylon dermatitis and bunions are related to fashions of dress. **1945** *Industr. Fibres & By-Products* VII. 53/2 The nylon salt solution is..ready to be made into new nylon polymer. **1958** D. E. Floyd *Polyamide Resins* iv. 54 Nylon salts may be formed in water or aqueous alcohol solution and are crystallized from aqueous alcohol or alcohol itself. They are usually soluble in water, but insoluble in alcohol, acetone, ether, or hydrocarbons. **1964** J. G. Cook *Your Guide to Plastics* 239 Hexamethylene diamine and adipic acid are reacted together to form a salt, hexamethylene diammonium adipate, or 'nylon salt'. *Ibid.* 240 Hexamethylene diamine and sebacic acid are combined to form nylon 6:10 salt, or hexamethylene diammonium sebacate. **1947** *Jrnl. Investigative Dermatol.* IX. 207 All the subjects suffering from nylon stocking dermatitis were hypersensitive to azodyes used in the manufacture of the stockings. **1957** *Year Bk. Dermatol. & Syphilol.* 117 Nylon stocking dermatitis results from sensitivity to the dye.

nyloned (nəi·lǫnd), *a.* [f. prec. + -ed².] Clad in nylons.

1952 B. Malamud *Natural* (1963) 16 Her nyloned legs made Roy's pulses dance. **1959** *Times Lit. Suppl.* 9 Oct. 573/3 Also making the voyage are a group of American girls..normal, nyloned, streamlined, wide-eyed, cynical and destructive. **1962** *Economist* 31 Mar. 1240/1 The men are well shod, the women well nyloned. **1963** G. Freeman *Campaign* iv. 69 Lady Andover's knees emerging, pinkly nyloned, from under her black coat. **1975** I. Murdoch *Word Child* 195, I looked down, inspecting a nyloned ankle and a smart..high-heeled shoe.

nymph, *sb.* Add: **2. b.** (Further examples.)

1833 W. Tolmie *Jrnl.* 28 Mar. (1963) 133 Nymphs of the pavé numerous (in Honolulu). **1902** Farmer & Henley *Slang* V. 81 *Nymph of darkness* (or *the pavement*), ..a prostitute. **1942** Berrey & Van den Bark *Amer. Thes. Slang* §507/2 Prostitute... Nymph du pave. **1964** 'W. Haggard' *Antagonists* vii. 71 Counsellor of Embassy living with fellow-travelling nymph.... They'd do most things to muffle that one. **1965** 'S. Harvester' *Assassins Road* xvii. 180 He had been pretty sure she was a nymph. **1968** R. Stout *Father Hunt* (1969) xiii. 157 She was a nymph. She was a goddam tart.

3. b. A fishing fly made in imitation of the aquatic larval form of may-flies, insects of the order Ephemeroptera.

1910 G. E. M. Skues *Minor Tactics of Chalk Stream* iv. 32, I had tied some nymphs of appropriate colour of body. **1922** R. C. Bridgett *Dry-fly Fishing* i. 17 The place of honour [as a lure] is occupied by the artificial nymph. **1973** *Shooting Times & Country Mag.* 7 July 13/3 There were virtually no rising fish to be seen and I decided to plod on with the nymph.

6. *nymph-fishing*, *-song*; *nymph-haunted*, *-less* adjs.

1932 *Times Lit. Suppl.* 5 May 374/2 The delicate art of nymph-fishing. **1972** *Shooting Times & Country Mag.* 1 July 15/2 The Club rules permit only dry fly and nymph fishing until the end of July. **1881** O. Wilde *Poems* 69 Blue nymph-haunted seas. **1939** W. B. Yeats *On Boiler* 28 Nymph-haunted or Fury-haunted wood. **1948** C. Day Lewis *Poems 1943–47* 19 Otherwise the forest was silent: birdless; nymphless. **1930** E. Blunden *Summer's Fancy* 15 When nymph-songs echoed on the blossomed breeze.

nymph (nimf), *v.* [f. the sb., sense 3.] Of fish, esp. trout: to feed upon insect larvæ near the surface of the water. So **ny·mphing** *ppl. a.*

1963 O. Kite in C. F. Walker *Compl. Fly-fisher* iv. 140 A nymphing trout..can be seen making little movements, lifting slightly in the water from time to time. **1972** *Shooting Times & Country Mag.* 24 June 15/2 In the sheltered bay we saw the odd fish nymphing.

Nymphenburg (ni·mfənbɒɹg). The name of a former village in Bavaria, now a suburb of Munich, used *attrib.* to designate pottery manufactured there from 1761.

1863 W. Chaffers *Marks Pott. & Porc.* 183 Nymphenburg. Another form of the arms of Bavaria, also impressed on the ware. **1869** C. Schreiber *Jrnl.* 4 & 5 July (1911) I. 26 A small Nymphenburg vase mounted, £1. **1881** *Ibid.* 16 Nov. II. 374 Our only purchase was a Nymphenburg shell piece..signed and dated C.H.Z. 1771. **1882** 'Ouida' *Bimbi* 50 A little pale-faced chit of a damsel in white Nymphenburg china. **1910** *Encycl. Brit.* V. 751/1 His Nymphenburg figures are as highly esteemed as those he modelled at Höchst. **1960** *House & Garden* Oct. 112/1 A collection of Nymphenburg equestrian statuettes. **1971** L. A. Boger *Dict. World Pott. & Porc.* 247 (*caption*) Nymphenburg figure.

nymphet. Add: **b.** A nymph-like or sexually attractive young girl.

1955 V. Nabokov *Lolita* (1959) I. v. 18 Between the age limits of nine and fourteen there occur maidens who, to certain bewitched travellers, twice or many times older than they, reveal their true nature which is not human, but nymphic (that is, demoniac); and these chosen creatures I propose to designate as 'nymphets'. **1959** *Listener* 8 Jan. 63/2 A whole chorus of what the author of *Lolita* calls 'nymphets'. **1959** *Daily Mail* 31 Jan. 4/4 He is in the thick of an affair with an idealised nymphet of 20. **1963** *Spectator* 15 Feb. 199 Two nymphets visit him together, he orders them both into his bed. **1971** *Southerly* XXXI. 12 She is..at her first appearance a shameless nymphet of thirteen already *indifferent* to the number of boys who have enjoyed her favours. **1973** J. Di Mona *Last Man at Arlington* (1974) 51 Most of the 'sales executives' had turned out to be eighteen- and nineteen-year-old nymphets.

c. *attrib.* and *Comb.* in sense *b.

1959 *Spectator* 25 Sept. 406/1 Their 'baby doll' outfits had the nymphet look which has been in fashion this summer. **1960** *Spectator* 3 June 804/1 Seems to have convinced herself that..every nymphet gesture..can be repeated again and again with ever-increasing success. **1960** *Encounter* June 86/2 The hero..bails out a nymphet-loving painter. **1971** *Daily Tel.* 27 May 8/4 Mother, a randy alcoholic; father, a seedy, nymphet-chasing Peter Pan.

nympho (ni·mfo), colloq. abbrev. of Nymphomaniac *a.* and *sb.*

1935 [see *dipso]. **1954** D. Schwartz in *Avon Bk. Mod. Writing* II. 129 Some girls at school said that Phoebe was a nympho. **1958** A. Wilson *Middle Age of Mrs Eliot* 232 'That Cynthia Robertson's a bit nympho,' he announced in worldly tones. **1959** 'J. Bell' *Easy Prey* x. 106 Red-haired bitch... Distinctly nympho. **1960** 'R. East' *Kingston Black* xiv. 140 Tim Askew, degenerate turkey farmer, had pulled the trigger on the Danish nympho. **1973** 'E. Peters' *City of Gold & Shadows* xiii. 208 She was a sex-nut-case, a virgin nympho who couldn't stand being mauled but couldn't help asking for it.

nympholepsy. (Later examples.)

1955 V. Nabokov *Lolita* (1958) I. 174 The science of nympholepsy is a precise science. Actual contact would do it in one second flat. An interspace of a millimeter would do it in ten. **1974** *Times Lit. Suppl.* 8 Feb. 122/3 His congenital nympholepsy for slender girls with thin arms.

nymphology. Add: Hence **nympholo·gical** *a.* *rare.*

1953 R. Graves *Poems* 4 Confess, what elegant square or lumpish hamlet Lives free from nymphological disquiet?

nymphomania. Add: (Earlier and later examples.)

1775 E. S. Wilmot tr. de Bienville (*title*) Nymphomania, or, a dissertation concerning the furor uterinus. **1905** Moussu & Dollar *Dis. Cattle* vii. iv. 562 Nymphomania may be considered as almost invariably the result of a genital lesion. **1962** C. Allen *Textbk. Psychosexual Disorders* v. xvi. 306 Since writers..have become fascinated with the concept of nymphomania..it might be as well here to state that it is clinically very rare. **1967** L. W. van den Heever tr. *Heidrich & Renk's Dis. Mammary Glands Domestic Anim.* v. 48/1 In nymphomania the cysts should be ruptured when possible and chorionic gonadotropin..administered subcutaneously or intracystally to prevent recrudescence.

nymphomaniac *a.* and *sb.* (later examples); hence also **nymphomani·acal** *a.*

1906 J. Joyce *Let.* 13 Nov. (1966) II. 193 Cosgrave [was sure] that I would become a nymphomaniac. **1923** *Physiol. Rev.* III. 338 The writer has examined a section..of an ovary of a nymphomaniac cow. **1923** A. Huxley *Antic Hay* xx. 287 When I call my lover a nymphomaniacal dog, she runs the penknife into my arm. **1932** Gaiger & Davies *Vet. Pathol. & Bacteriol.* III. xxxv. 513 The presence of nymphomaniac cysts is always accompanied by sterility, even though the cysts are confined to one ovary. **1949** 'J. Tey' *Brat Farrar* xxiii. 208 Pawning his life to a nymphomaniacal moron. **1967** L. W. van den Heever tr. *Heidrich & Renk's Dis. Mammary Glands Domestic Anim.* iii. 34/2 Protracted nymphomaniacal estrus associated with relaxation of the pelvic ligaments, vaginal prolapse, nervousness and aggressiveness is commonly encountered. **1969** C. Allen *Textbk. Psychosexual Disorders* (ed. 2) v. xvi. 356 Nymphomaniacs who have a compulsion due to brain lesions, hormonal imbalance or other physical abnormalities. **1973** R. Hayes *Hungarian Game* viii. 59 From different sources I'd heard that she was nymphomaniacal or homosexual or frigid.

Nynorsk (niū·nǫɹsk). Also **nynorsk, Ny Norsk.** [Norw., f. *ny* new + *norsk* Norwegian.] The official name by which Landsmål is now known. Cf. *Landsmål.

1937 E. I. Haugen *Beginning Norwegian* 5 Ivar Aasen (1813–1896) conceived the idea that if one studied the most 'genuine' native dialects..one might create a form of Norwegian equivalent to what the national language would have been had Norway never been united with Denmark... He called it *landsmål*, a name by which it is still commonly known (though *nynorsk* is now official). **1952** B. Berulfsen in *Norseman* X. 187 In 1929 official action changed the names Landsmål to *Nynorsk* and Riksmål to *Bokmål*. **1966** E. I. Haugen *Lang. Conflict & Lang. Planning* vi. 257 The choice given in the proposed ballot was between 'natural Riksmål, Nynorsk, and official Bokmål'. **1971** *Computers & Humanities* V. 205 The data bank of the project, which covers both *bokmaal* and *nynorsk*, is intended to assist various types of linguistic projects in the future. **1975** *Scottish Rev.* I. 22 The 'Landsmaal', the distinctively Norwegian language that in the 70s of last century a group of writers were endeavouring to institute as the literary language of Norway as distinct from the usual Dano-Norwegian, an attempt which has developed into the 'Ny Norsk' (New Norwegian), in which a number of works are written now.

Nyon (nī·ǫ̃n). The name of a commune in Switzerland, used *attrib.* and *absol.* to designate pottery manufactured there from *c* 1780.

1869 C. Schreiber *Jrnl.* 10 May (1911) I. 2 A very fine service of Nyon, dinner and dessert. **1910** E. Dillon *Porcelain & how to collect It* xii. 181 The only distinctive mark of Nyon is a fish painted in blue under the glaze. **1960** H. Hayward *Antique Coll.* 202/1 *Nyon porcelain*, a factory was founded at Nyon, near Geneva in 1780... Current French fashions were closely adhered to and many of the wares bear floral decoration.

nyssa (ni·sā). [mod.L. (J. F. Gronovius in Linnæus *Systema Naturæ* (1735)), f. *Nysa* the name of a water nymph, in allusion to the swamp habitat of some species.] A deciduous tree of the genus so called, belonging to the family Nyssaceæ and native to North America and Asia, esp. the American species *Nyssa sylvatica*, the tupelo.

1886 G. Nicholson *Illustr. Dict. Gardening* II. 461/1 Nyssas thrive best in low, damp, moist situations, such as peat swamps. **1901** L. H. Bailey *Cycl. Amer. Hort.* III. 1109/2 Nyssas are trees or shrubs with petiolate, usually entire leaves and small flowers borne in short racemes or dense heads. **1961** *Amateur Gardening* 23 Sept. 3/3 The tulip tree..and the nyssa in the Long Walk to the Pagoda ..are among the tallest trees in the Garden [at Kew]. **1962** *Times* 24 Dec. 11/1 No exotics means..no Nyssas.

nystatin (nəi·stătin). *Pharm.* [f. the name of *New York State*, U.S.A., where it was developed + -IN^1.] A yellow antifungal substance, $C_{46}H_{77}NO_{19}$, that is produced by the growth of the bacterium *Streptomyces noursei* and is used locally in the treatment of moniliasis, esp. when caused by *Candida albicans*, and of anal and vaginal infections.

1952 *Arch. Pediatrics* LXIX. 414 Hazen et al. have shown that an antibiotic called fungicidin, Nystatin®, has inhibited coccidioides immitis spherules at 6·25 micrograms per milliliter. [*The name* Nystatin *is not in fact registered as a trade mark at the U.S. Patent Office.*] **1953** *Science* 29 May 609/2 The senior authors [*sc.* Brown & Hazen] have given the name Nystatin to their product fungicidin. It is being manufactured by E. R. Squibb and Sons under this name. **1962** E. O. Morris in Hawthorn & Leitch *Recent Adv. Food Sci.* I. 33 Yeasts appear to be particularly troublesome and it has been suggested that these organisms may be inhibited by the use of nystatin. **1974** R. N. Richards *Venereal Dis.* 104/1 The diagnosis of monilia was made and she responded well to the insertion of nystatin tablets in her vagina twice daily.

nyuck, var. *YU(C)K.

nyumyum (nyɒ·myɒ·m). *rare*$^{-1}$. = *YUM-YUM.

1922 Joyce *Ulysses* 159 O, that's nyumyum.

Nzima (n,zī·mă). Also **Nsima, Nzema.** The name of an African language spoken in Ghana.

1911 F. W. H. Migeod *Lang. W. Afr.* I. vii. 179 In Nsima (Zema) there is also a suffix, but it appears that there exists also initial change, the same as in Twi. **1945** I. C. Ward *Rep. Investigation Gold Coast Lang. Probl.* 61 Nzema..appears to be a strong admixture of two language groups, the Fante branch of the Akan language and some other group lying to the west. **1971** L. A. Boadi in J. Spencer *Eng. Lang. W. Afr.* 49 The indigenous languages [of the Gold Coast],..are still used and are, in many cases, deliberately cultivated; no less than six of them—Akan, Gã, Ewe, Nzima, Dagbani and Hausa—are regularly heard on the radio. **1975** *Archivum Linguisticum* VI. 2 In certain environments Nzema and Ahanta *k* alternates with *g*.

O

O. Add: **4*.** [orig. denoting absence: cf. O *sb.*^{1}] In *Hæmatology*, designating absence of the A and B agglutinogens of the ABO blood group system; hence (and now usu.) used to designate the blood group of individuals lacking these two agglutinogens; also, more widely, used to designate the allele involved in determining this blood group.

1926 Landsteiner & Witt in *Jrnl. Immunol.* XI. 242 It has been pointed out by one of the writers..that the isoagglutinin reactions of human blood can be possibly explained by the simple assumption of only two different agglutinogens and agglutinins. Designating these by α and β, and the agglutinogens by A and B, the following symbols are obtained for the blood groups: I—α, β; II A, β; III B, α; IV A, B—; if we include the factors A^1 and α^1 in the scheme, and if O and o signify the absence of agglutinogens or agglutinins, then the signs are: I O α, β, α^1; II A, β, and A, A^1, β; III B, α, α^1; IV A, B, o. **1927** [see *A II. 7]. **1929** L. H. Snyder *Blood Grouping* i. 3 To try to obviate the confusion still existing from a reversal of groups I and IV in the two systems, a new system of nomenclature, based on the agglutinophyllic [*sic*] capacity of the cells, has been suggested. In this system, Jansky's group IV is known as *AB*, having the two agglutinogens A and B. Group III, containing agglutinogen B, is known as group *B*, group II as group *A*, and group I, containing neither agglutinogen, as group *O*. **1948** C. C. Sturgis *Hematol.* xxiii. 825 These characteristics in the erythrocytes may be present singly or together, or they can be absent. If the absence of these isogglutinable substances is designated as O, then there are four possibilities: namely, groups, O, A, B, and AB. **1958** J. B. Miale *Lab. Med.: Hematol.* vi. 319 The O gene, when carried by both chromosomes, determines phenotype O. **1966** *Listener* 6 Oct. 493/1 Mr and Mrs H's blood was found to belong to group O, while Clive's blood was found to belong to group A2. **1968** Passmore & Robson *Compan. Med. Stud.* I. xxvi. 17/1 Every person has a pair of chromosomes each of which carries the A, B or O gene. **1969** J. H. Green *Basic Clin. Physiol.* vi. 34/2 The remainder of the population (46 per cent.) have neither A nor B on their red cells, and they are said to be Group O.

5. b. O.A.P., old age pensioner; (also formerly old age pension); O.E. (examples); also, Old Etonian; OF. (examples); OHG. (examples); ON. (examples).

1708 J. Kersey *Dict. Anglo-Britannicum* Pref., *O.F.*, Old French. **1864** R. Morris *Early Eng. Allit. Poems* p. xxxvi, The preposition *from* never occurs in the following poems; it is replaced by *fro* (Northumbrian *fra*, O.N. *fra*). **1868** —— *Old Eng. Homilies* I. p. lvi, In Gothic we find plural forms in *-a*, as *worda*, &c., which are certainly older than the O.E. forms *word*, &c. *Ibid.* 312 Cp. the O.N. *lát, læti.* **1884** O.E. [used s.v. A *adj.*^{1}]. **1884** O.H.G., O.N. [used s.v. A *adv.*]. **1887** W. W. Skeat *Princ. Eng. Etymol.* 1st Ser. x. 172 The G. *trieb* (drove) is a modern form. The O.H.G. was *dreib* or *treib*. **1891** *Ibid.* 2nd ser. iii. 43 Lat. *u*..was sometimes long, as in Lat. *nūllum*, and sometimes short, as in Lat. *mŭltum*; and was developed accordingly. Hence O.F. *nul* (nyl) and *moult* (mult). **1901** G. Frankau *Eton Echoes* 48 Or pass to hear them say with eyes askance 'The siding ass! Suppose he's some O.E.' **1912** R. W. Chambers *Widsith* 191 Kluge has pointed out that this form of the name corresponds to the O.N. *Atli*, as against the M.H.G. *Etzel*, O.H.G. *Ezzilo*. **1914** C. Mackenzie *Sinister St.* II. III. i. 516 Come and have coffee with me after hall. One or two O.E.'s are coming in, but you won't mind? **1927** *Englische Studien* 10 Nov. 81 ON. *lifr* occurs in Norwegian river-names. **1934** M. K. Pope *From Lat. to Mod. Fr.* II. xi. 172 In educated Parisian speech the denasalisation of O.F. *ã* appears to have begun in the later sixteenth century. **1936** J. Buchan *Island of Sheep* vi. 112 He wore white linen breeches, a smartly cut flannel coat, and an O.E. tie. **1940** W. O. Ross *M. E. Sermons* p. xxix, *S* appears very rarely for OE. *sc.* **1942** Partridge *Dict. Abbrev.* 70/1 *O.A.P.*, Old Age Pension(s). **1959** P. Bull *I know Face* x. 186 The O.A.P.s were very angry indeed, at not only having to witness *Waiting for Godot*, but also having to pay twelve pennies for the privilege. **1959** M. Schlauch *Eng. Lang. in Mod. Times* i. 29 The [pronoun] forms which sprang from the OE and ON datives later assumed the function of accusatives as well. **1970** B. M. H. Strang *Hist. English* iv. 274 A few native formations are calques on OF prepositions. **1972** M. L. Samuels *Linguistic Evol.* ii. 25 One of the best known of irreversible consonant-changes is that of voiceless plosives to fricatives or affricates, as in..OHG [p, t, k] > [pf, ts, kx]. **1972** E. J. Dobson *Eng. Text of Ancrene Riwle* p. cxlix, OE (ON) *ā* is normally spelt o. **1973** 'B. Graeme' *Two & Two* xiv. 139 'What about the Rexalls?' 'Apart from being O.A.P.s, I know nothing.' **1973** *Listener* 7 June 777/1 Tony sports an OE tie. **1976** *Evening Post* (Nottingham) 15 Dec. 19/7 (Advt.), Gamston Kennels (Est 1926)..Pedigree puppies ..Labradors, O.E. sheepdogs, Pekes, Poodles.

c. O.B.E., (Officer of the) Order of the British Empire; O.D.C., Order of Discalced Carmelites; O.M., Order of Merit; also by metonymy, a member of this Order; O.M.I. (see quot. 1907); O.P., Order of Preachers; O.S.A., Order of Saint Augustine; O.S.B. (examples); O.S.F.(C.), Order of Saint Francis (, Capuchin).

1798 J. Milnes *Life Challoner* 32 That zealous orthodox prelate..whose loss we at the present moment deplore, the Right Reverend Bishop Walmesley, O.S.B. *Ibid.*, Another person for whom he had deservedly the greatest respect and regard was the Rev. Pacificus Baker, O.S.F. **1839** *Dublin Rev.* May 556 English sermons were delivered by the Rev. Dr. Wiseman, F. Hughes, O.S.F., and the Rev. Messrs. McGill and Kyan. **1865** *Cath. Directory* p. lxii, Rev. F. Lawrence (Praxmarer), O.S.F.C. **1891** *Cath. Times* 6 Mar. 2/7 Very Rev. Dr Keane, O.P. **1903** *Who's Who* 760 Keppel, Hon. Sir Harry, G.C.B.; cr. 1857; O.M. **1907** *Cath. Encycl.* I. 28/1 *O.M.I.*, Oblati Mariæ Immaculatæ—Oblate Fathers of Mary Immaculate. *Ibid.*, *O.P.*, Ordo Prædicatorum—Dominicans. *Ibid.*, *O.S.A.*, Ordo (Eremitarum) Sancti Augustini—Augustinians. **1917** O.B.E. [see M.B.E. s.v. *M 5]. **1922** Joyce *Ulysses* 312 Amongst the clergy present were..the rev. P. J. Cleary, O.S.F.; the rev. L. J. Hickey, O.P.; the very rev. Fr. Nicholas, O.S.F.C.; the very rev. B. Gorman, O.D.C.;.. the rev. T. Brangan, O.S.A.;..the rev. B. R. Slattery, O.M.I.; [etc.]. **1923** *Cornh. Mag.* June 765 A Captain and an O.B.E. **1937** B. Jarrett *Eng. Dominicans* (rev. ed.) 186 Consecrated Bishop of Tiberiopolis by Pope Benedict XIII, O.P. **1955** *Times* 9 Aug. 4/7 A person of the highest character, who served with distinction in both world wars and received the O.B.E. **1955** *Essays in Crit.* V. 430 Od's life, need an O.M. swear to the truth of an epigram? **1955** 'D. Knowles' *Relig. Orders in Eng.* II. 390/1 Ashbourne, Thomas, OSA. **1957** *Oxf. Dict. Chr. Ch.* 235/1 In England and Ireland they [*sc.* the Capuchins] sign O.S.F.C. ('Ordinis Sancti Francisci Capuccinorum'). **1963** I. Wilkes *Brit. Init. & Abbrev.* 82/1 *ODC*, Order of Discalced Carmelites, 41, Kensington Church Street, London, W.8. **1969** I. & P. Opie *Children's Games* p. xvi, We have also to thank Father Damian Webb O.S.B. **1972** *Bookseller* 2 Dec. 2543 (Advt.), Sister Mary Joyce O.P.

d. O. and M., organization and methods; OAO *Forces' slang*, one and only; OAO, orbiting astronomical observatory; O.A.S., on active service; O.A.S., Organisation de l'Armée Secrète, an organization opposed to Algerian independence from France; O.A.S., Organization of American States; O.A.U., Organization of African Unity; OB, obstetrics, obstetric, or obstetrician (*U.S.*); OB, order of battle; O.B., outside broadcast; O.B.U., One Big Union; O.C., officer commanding; OCR, optical character recognition; O.C.T.U., officer cadet(s') training unit; also *Octu* (ǫ·ktu); O.D. (*U.S.*) officer of the day; olive drab; O.D., ordinary seaman; O.D., Ordnance datum; OD, organization development; O.D., o.d., outside diameter; O.D. *slang* (orig. *U.S.*), overdose; so as *v. intr.*, to take an overdose; O.D.'d, overdosed, dead of an overdose; O.D.V. *joc.*, eau-de-vie; O.E.C.D., Organization for Economic Co-operation and Development; O.E.D., Oxford English Dictionary; O.E.E.C., Organization for European Economic Co-operation; O.E.O. (*U.S.*), Office of Economic Opportunity; OGO, orbiting geophysical observatory; O.H.C., ohc., overhead camshaft; O.H.M.S., on His (or Her) Majesty's Service; O.H.V., o.h.v., overhead valve; O.K., see *O.K. *a.*, *sb.*, and *v.* (as main entry); O level, Ordinary level (of the General Certificate of Education examination); hence *O leveller*; O.N.C., Ordinary National Certificate; o.n.o., or near(est) offer; O.O.W., Officer of the Watch; O.P. (*b*) (earlier and later examples); (*d*) (examples); (*e*) observation post (also *O. Pip*); OPEC (ōu·pek), Organization of Petroleum-Exporting Countries; O.P.M., other people's money (*U.S. slang*); O.P.M., output per man; O.R., OR, operational research; O.R., other ranks; O.R.T.F., Office de Radiodiffusion-Télévision Française, formerly the state television and radio service of France; O.S., ordinary seaman; O.S., Ordnance Survey; O.S., outsize; OSHA, Occupational Safety and Health Act (or Administration) (*U.S.*); OSO, orbiting solar observatory; OSO, Ordnance Survey Office; O.S.S. (*U.S.*), Office of Strategic Services; OTB, off-track betting (*U.S.*); O.T.C., Officers' Training Corps; O.T.C., Organization for Trade Co-operation; OTC, over the counter; O.T.U., Operational Training Unit; O.U., Open University; O.U.D.S., Oxford University Dramatic Society; cf. also *OUDS; O.V.R.A. [see quot. 1961], the secret police of Fascist Italy.

1958 *Daily Mail* 3 July 4/3 Modern business techniques using 'work study' and the 'O. and M.' treatment (Organisation and Methods), can prove 'a considerable help to us in the hospitals'. **1965** *New Statesman* 7 May 707/2 An O & M survey should swiftly be initiated to decide what dead wood needs to be cut out. **1971** K. Gottschalk in B. de Ferranti *Living with Computer* v. 46 Groups concerned with efficiency in the office are sometimes called organization and methods (O & M) groups. **1936** *Nat. Geogr. Mag.* June 778/2 Or she may be the OAO—the One and Only. **1967** *Everybody's Mag.* (Austral.) 18 Jan. 36/2 In each war, a new vocabulary is created. Today, in Vietnam, Australians are again catching up on American Army slang... All would refer to a special girlfriend as their OAO—one and only. Probably, the OAO was met on skirt patrol. **1962** F. I. Ordway et al. *Basic Astronautics* iv. 119 An OAO is seen in Fig. 4.2. **1971** *McGraw-Hill Yearbk. Sci. & Technol.* 300 Although the first OAO malfunctioned, the second one (launched on Dec. 7, 1968) has..produced a wealth of important new astronomical data. **1928** Blunden *Undertones of War* 178, I remember your superscriptions, 'O.A.S.' and 'B.E.F.'. **1962** *Listener* 4 Jan. 10/2 An O.A.S. bomb. **1963** *Times* 14 Mar. 16/2 Algeria at the period when O.A.S. terrorism was at its height. **1973** C. Egleton *Seven Days to Killing* vii. 78 He was an Algerian colonist..and the French police had long been satisfied that he had never been connected with the OAS. **1949** *Ann. Organization Amer. States* I. No. 1. (title-page), Charter of the OAS. **1972** *Buenos Aires Herald* 3 Feb. 7/6 The juridical commission of the Organization of American States (OAS) has condemned the sending of British troops to Honduras. **1974** *Greenville* (S. Carolina) *News* 22 Apr. 3/5 Kissinger was asked why he had not mentioned Cuba in a speech Saturday to the Organization of American States (OAS) meeting in Atlanta. **1964** *Ann. Reg.* *1963* 322 Organisation of African Unity (O.A.U.). Established at Conference of African Heads of State at Addis Ababa, 22–26 May 1963. **1971** *Sunday Nation* (Nairobi) 11 Apr. 7/1 The announcement had little to do with any assumed prevailing trends among members of the OAU. **1944** Dorland & Miller *Med. Dict.* (ed. 20) 1005/1 *O.B.*, abbreviation for *obstetrics*. **1967** *Boston Globe* 21 May 9/3 Sitting in an office for an OB check. *Ibid.* 9/4 A two hour wait in the OB's office. **1972** *Daily Colonist* (Victoria, B.C.) 29 Feb. 2/1 My last checkup with my OB doctor revealed a fibroid tumor. **1946** Chandler & Robb *Front-Line Intelligence* xii. 137 O/B (Order of Battle) is a military science whose mission is to determine: (1) How strong the enemy is, [etc.]. **1950** *Tactics & Techniques Infantry* (U.S.) II. ii. 312 The order of battle (OB) team. **1971** *Combat Intelligence* (U.S. Dept. Army, Field Manual 30-5) vii. 7-1 Order of battle (OB) is the identification, strength, command structure, and disposition of the personnel, units, and equipment of any military force. **1975** tr. *Melchior's Sleeper Agent* (1976) III. 192 He'd sent him on to the Corps OB team, to see if there was anything in the latest Order of Battle book. **1927** *B.B.C. Handbk. 1928* 143/1 Outside Broadcast Features... Every O.B. of the simplest..nature necessitates the provision of two complete telephone line circuits..between the site of the performance and the Station Control Room. **1960** *Punch* 17 Feb. 251/1 Oh, I agree, it's [television is] splendid for sport and O.B.s. **1971** R. Busby *Deadlock* xiii. 200 You'd think he was the bloody big white chief instead of an OB technician. **1919** *Camp Worker* (Vancouver) 17 May 5/3 At Medicine Hat the Federated Railway Trades have unanimously endorsed the O.B.U. **1931** 'D. Stiff' *Milk & Honey Route* 210 *O.B.U.*, One Big Union. The ideal of the soap boxers. **1977** *Guardian Weekly* 11 Sept. 10/2 The Industrial Workers of the World... In Canada its counterpart was called the OBU, One Big Union..an attempt to prevent divisions in the labour movement by creating a single trade union. **1904** *N.Y. World Mag.* 1 May 6/5 'O.C.' is the officer in charge. **1917** F. M. Ford *Let.* 19 Feb. (1965) 84 It suits me better to write: 'O.C. Canadous will detail a fatigue party of 1 NCO & 10 men at 4:30 a.m.' **1928** T. E. Lawrence *Let.* 2 May (1938) 600 Also [you will inherit] my copyrights which now no longer include *Revolt in the Desert*: but you will be O.C. *The Seven Pillars*. **1967** G. F. Fiennes *I tried to run a Railway* iii. 25 He had been a gunner himself and had warned the O.C. of the 15 inch crew. **1966** *Computer Jrnl.* IX. 224/2 We decided to experiment to see what limitations, if any, O.C.R. would place on our stationery design. **1970** *Brit. Printer* July 57/1 The alphabet itself does not have any practical OCR use at all. **1942** E. Waugh *Put out More Flags* ii. 121 'O.C.T.U. candidates,' said the company sergeant-major. **1972** D. McLachlan *No Case for Crown* iv. 56 He reminded me sometimes of a sergeant who gave me hell in my O.C.T.U. **1929** *Papers Mich. Acad. Sci., Arts & Lett.* X. 311 *O.D.*, I, the officer of the day; II, olive drab. **1966** *Sunday Times* (Colour Suppl.) 4 Dec. 73/2 GI Jargon *OD*, officer of the day, or olive drab (both the colour and the uniforms themselves, e.g. 'I'm wearing my ODs tonight'). **1915** *Recruiter's Bull.* (U.S.) June 17/2 Two O.D. shirts you next slip in, A pair of shoes goes in between. **1975** tr. *Melchior's Sleeper Agent* (1976) III. 173 He was clad only in his OD shorts and undershirt. **1916** 'Taffrail' *Pincher Martin* i. 7 'Strumbles,' he said, ''ere's another O.D. come to join your mess.' **1962** Granville *Dict. Sailors' Slang* 83/1 *O.D.* Naval colloquialism for ordinary seaman. *OS* or *Ord* is the official abbreviation. **1926** J. Malcolm *Agric. Surveying* v. 123 The datum adopted in the Ordnance Survey of Great Britain, denoted by the letters O.D., was what was considered in 1844 to be mean sea level at Liverpool... The new datum is mean sea level at Newlyn. **1956** *Railway Mag.* Mar. 185/2 The top of the wall at the Barmouth end is 35 ft. above Ordnance Datum, dropping to 28 ft. above O.D. at the slipway. **1972** L. Alcock *By South Cadbury* ii. 25 Roughly one quarter of the hill-top, lying above four hundred and ninety feet O.D., forms a broad summit ridge. **1972** *Times* 5 June 22/6 OD. not infrequently causes strong reactions among managers. **1976** Blake & Mouton (*title*) Diary of an OD man.

1930 Walker & Crocker *Piping Handbk.* iv. 293 In sizes 14 in. and upward pipe is designated by its outside diameter (O.D.) and the wall thickness is specified. **1963** H. R. Clauser *Encycl. Engin. Materials* 120/1 Nonferrous castings are produced commercially in o.d.'s ranging from about 1 in. to 6 ft. **1967** *Electronics* 6 Mar. 15/2 (Advt.), The screw-on has an OD of only ¼ inch with a mated length of only one inch. **1960** R. G. Reisner *Jazz Titans* 162 *O.D.*, an overdose of narcotics. **1971** *Black World* Apr. 38/1 A truly brilliant Black filmmaker goes into his grave at 24..an O.D. takes him, he loses a battle of several years—the 'stuff' wins. **1972** *Telegraph* (Brisbane) 17 Oct. 70/6 A fatal dosage of drugs—*O.D.* (*successful overdose*). **1970** S. O'Callaghan *Drug Addiction in Britain* xii. 151 Diana has O.D.'d and she's dead. **1969** R. De Sola *Abbrev. Dict.* (rev. ed.) 196/1 *Od'd*, overdosed (dope addict). **1973** *Black World* Aug. 55/1 The garbage collectors found Little Prez in the alley near Six-trey, OD'd away, layin' there cool and stiff. **1839** *Picayune* (New Orleans) in *Spirit of Times* (N.Y.) 5 Oct. 368/3 Why, that in French, is nothing but O.D.V. **1886** H. Baumann *Londinismen* 124/1 *O.D.V.*,..Branntwein, Spiritus. **1965** *Acronyms & Initialisms Dict.* (Gale Research Co.) 530 *ODV*, taken from pronunciation of French eau-de-vie and used to refer to brandy. **1960** *Times* 25 Nov. 10/7 The 20 members of O.E.C.D. are the 18 full members of O.E.E.C., together with the United States and Canada. **1971** *Power Farming* Mar. 15/1 European standards, issued by the Organization for Economic Co-operation and Development (O.E.C.D.) and the European Committee of Associations of Manufacturers of Agricultural Machinery (C.E.M.A.). **1898** Morris *Austral English* p. xviii, The practice of the 'O.E.D.' has been followed in this respect. **1962** *New Yorker* 10 Mar. 132/2 The compositor..who began setting type for the O.E.D. in 1884 and was still at it when the last volume came off the presses in 1928. **1973** *Daily Tel.* 20 Oct. 11/5 The word *hoax*..at present connotes more of the mischief than of the humour mentioned in the O.E.D.'s leisurely definition. **1976** *Times* 15 Apr. 13/8 'Fanatical', in the strict OED sense of the word, is surely..appropriate..to describe those serried ranks. **1948** *News Chron.* 13 Sept. 1/2 The job they were doing had been given them by O.E.E.C. **1964** *Listener* 13 Aug. 222/2, I was nearly four years chairman of what was then called the O.E.E.C. **1965** *Economist* 17 Apr. 297/1 The testimony of..the head of the Office of Economic Opportunity, revived the emotions of last year, when Congress established the OEO. **1974** *Black Panther* 16 Mar. 4/3 Congressman Dellums is co-sponsor of a tax reform measure that will work to the same end and has fought to save the programs that were funded through the Office of Economic Opportunity (OEO). **1961** *Sat. Rev.* (U.S.) 6 May 71/3 (Advt.), Each spacecraft in the OGO series will be capable of carrying up to 50 selected scientific experiments in a single flight. **1969** *Times* 4 June 5/1 The satellite being prepared for launch tomorrow..is the sixth and last of Orbiting Geophysical Observatories (OGO). **1974** *McGraw-Hill Yearbk. Sci. & Technol.* 346/2 Figure 2 shows an altitude-density profile through the plasmasphere measured by the *Ogo* 5 satellite. **1932** O.H.C. [see O.H.V. below]. **1954** P. H. Smith *Design & Tuning of Competition Engines* iv. 57 The merits of the double o.h.c. arrangement lie mainly in the substitution of rotary for reciprocating motion right up to the valves. **1977** *Drive* Sept.–Oct. 113/1 *Ohc*, overhead camshaft. **1895** *Brewer's Dict. Phr. & Fable* (new ed.) 904/2 *O.H.M.S.*, On His (or Her) Majesty's Service. **1907** *Yesterday's Shopping* (1969) 338/2 In accordance with the provisions of the Post Office Protection Acts, Envelopes with 'O.H.M.S.' or 'On His Majesty's Service', will only be supplied to those persons who have authority to use them. **1952** L. Durrell *Spirit of Place* (1969) 115, I think one or two white lined notebooks, official Foreign Office Stationery labelled OHMS. **1972** P. Cleife *Slick & Dead* I. i. 17 Nearly all the flying I've done has been O.H.M.S. I don't think I go much on civil operations. **1932** F. J. Camm *Motor Car Upkeep* i. 15 (*caption*) Three common types of Valve Gear: Side-by-side, overhead valve (O.H.V.), operated by rocker, and overhead valves (O.H.C.), operated by camshaft. **1958** *Engineering* 28 Feb. 265/1 It is a two-door all-steel saloon of unit construction with..a flat twin air-cooled o.h.v. 600 c.c. four-stroke engine. **1968** Burdett & Ellis *Motor Vehicle Mechanics' Course* II. v. 102 The filling of the cylinder is much improved by this design, particularly where the O.H.V. is placed towards the other side of the chamber. **1949** *Joint Matric. Board Exam. for G.C.E.* *1951* 6 If a language is to be specially approved..at the O level. **1959** *Times* 3 June 8/1 O level in any subject was 'very O'. **1974** *Times* 6 July 21/3 (Advt.), Expense account—plus £25 for an 'O' level man 17–22. **1961** *Listener* 26 Oct. 659/2 The £6-odd offered (per week) by banks and post offices, etc., to sixteen-year-old 'O' levellers. **1949** *Educ. in 1948*, 44 (*caption*) in *Parl. Papers* 1948–9 (Cmd. 7724) XIV. 345 O.N.C. **1962** in H. O. Beecheno *Introd. Business Stud.* p. iii, Mr. Beecheno has written a comprehensive introduction to the commercial world. It is intended particularly for the ONC and HNC student. **1977** P. Carter *Under Goliath* xv. 79 'Our Billy got the G.C.E.,' I said. '*And* the O.N.C. He is going in for the Higher National now.' **1958** *Listener* 6 Nov. 732/1 'O.o.o' means 'one owner only', whereas 'o.n.o.' means 'or near offer'. **1973** *Country Gentlemen's Mag.* Mar. 183/1 Coffee set, thirteen pieces, £5 o.n.o. plus postage. **1977** *Drive* Sept.–Oct. 113/1 *Ono*, or nearest offer. **1923** *Man. Seamanship* (Admiralty) II. ii. 48 Any man discovering a fire..is to send a message to the O.O.W. immediately. **1958** *Spectator* 1 Aug. 169/3, I hear that the archaic and old-fashioned Officer of the Watch (briefly, OOW) is to be changed to Period Progress and Procedure Organiser and Overseer. **1790** T. Wilkinson *Mem.* II. 226 So, on their not complying with my expectations and proposals, we parted with mutual disdain, rage, and anger at O.P. and P.S. **1919** Wodehouse *My Man Jeeves* 45 Lady Malvern was a hearty..female,..measuring about six feet from the O.P. to the Prompt Side. **1933** O.P. [see *flood *sb.* 6*]. **1859** G. Simpson *Let.* 19 Nov. in *Geo. Eliot Lett.* (1954) III. 209 He says you tell him Clerical Scenes is O.P. **1921** A. Bennett *Let.* 4 Sept. (1966) I. 296 It seems to me that.. *The Old Wives' Tale* ought not to be o.p. **1916** F. M. Ford *Let.* 23 Aug. (1965) 69 George V..really was in some danger. At least he was in an O.P. that was being shelled fairly heavily. **1972** L. Lamb *Picture Frame* ii. 20 Gerry's no fool, but we don't think our o.p. has been rumbled. **1960** *Times* 15 Sept. 11/4 The conference adopted the (Iraq) proposal to establish an 'Organization of Petroleum-Exporting Countries'...The five participating states are founder-members of O.P.E.C. **1975** *Petroleum Economist* Aug. 282/1 For eighteen months the problem of OPEC's surplus oil revenues has occupied the minds of western statesmen, bankers and economists. **1916** *War Illustr.* 7 Oct. 185/1 A French 'O-Pip' in the Hills. **1919** J. Masefield *Battle of Somme* 88 Some of them were quite good trees, and we had an O. Pip in one of them (artillery observation post). **1943** Hunt & Pringle *Service Slang* 49 *O pip*, an Observation Post of the Field Artillery. ['Pip' stood for 'P' in the services' phonetic alphabet.] **1901** 'J. Flynt' *World of Graft* iv. 169 It cost me nothing to play the game, because I played it with O.P.M. (other people's money). **1969** *Time* 15 Aug. 60 No institution manages more 'O.P.M.', or Other People's Money, than Manhattan's 116-year-old United States Trust Co. **1946** J. Jewkes in *Manch. Sch. Econ. & Social Stud.* XIV. 4 Of two industries that with the higher O.P.M. is not necessarily the more efficient. *Ibid.* 5 O.P.M. may always be increased by installing new machinery but it may be the quickest way to the bankruptcy court. **1969** *New Scientist* 5 June 543 The UK with much higher R & D spending..had only a third of Japan's growth in output per man (OPM) employed. **1953** *Operational Research Q.* IV. 72 The evolution of O.R...is reflected by the number of publications. **1960** *Times* 17 Mar. 2/7 We shall require evidence of experience in either O.R. or cybernetics. **1964** T. W. McRae *Impact of Computers on Accounting* v. 118 A good number of O.R. problems can be solved by using nothing more powerful than a desk calculator. **1969** J. Argenti *Managem. Techniques* 107 Courses on OR designed for managers. **1942** Partridge *Dict. Abbrev.* 72/1 O.R., other ranks, i.e. ranks other than officers. **1947** J. Bertram *Shadow of War* VII. ii. 217 The heavy work in the camp was done by N.C.O.s and O.R.s., known as 'camp-employed'. **1967** J. Porter *Chinks in Curtain* xv. 147, I..tried to invoke the officer/O.R. relationship. **1964** *Economist* 30 May 945/2 The new [French] broadcasting service will be called ORTF, *Office de Radiodiffusion Télévision Française*. **1969** *Listener* 27 Mar. 410/2 The ORTF is never happier than with the state visit to Paris of the President of some French-speaking African statelet. **1977** *Rep. Comm. Future of Broadcasting* ii. 15 In France, years of feuding between government and broadcasters led in 1974 to the dissolution of the ORTF. **1894** *Gloss. Terms Evidence R. Comm. Labour* 60/1 in *Parl. Papers* 1893–4 (C. 7063) XXXVIII. 411, *O.S.*, an abbreviation for 'ordinary seaman'. **1908** *Army & Navy Gaz.* 7 Nov. 1066/1 The two marines, a stoker, and an O.S. caged like rats in a trap. **1914** C. F. Tweney *Dict. Naval & Mil. Terms* 164 *O.S.*, Ordinary Seaman; one who has undergone training as a ship's boy, but who is not fully qualified as a seaman. **1962** O.S. [see O.D. above]. **1962** *Punch* 10 Oct. 534/3 A good submarine forest, marked on the OS map. **1971** A. Hunter *Gently at Gallop* ii. 20 At the summit of the rise.. stood an O.S. triangulation pedestal. **1907** *Yesterday's Shopping* (1969) 768/1 O.S. Night Dresses... O.S. Chemises. **1927** *Daily Express* 3 Nov. 7 Forty inches at the hips would be O.S. A woman with 60 inches at the hips would be O.O.S., or extra outsize. **1973** *Country Gentlemen's Mag.* Mar. 184/1 For sale owing to loss of weight, full length O.S. evening dress, deep mauve velvet ..will accept £15. **1971** *New Acronyms & Initialisms* (Gale Research Co.) 56/1 *OSHA*, Occupational Safety and Health Act (1970). *OSHA*, Occupational Safety and Health Administration (Department of Labor). **1975** V. Bush (*title*) Safety in the construction industry: OSHA. **1976** G. & H. Matwes (*title*) A retailer's guide to OSHA. **1938** *Times* 2 Feb. 18/1 He could say that it would launch the O.S.O. on a new programme. **1962** *Daily Tel.* 8 Mar. 19/6 Yesterday the American National Aeronautics and Space Administration launched..the first of a new series of satellites, the orbiting solar observatories. The first was called OSO–1. **1971** *McGraw-Hill Yearbk. Sci. & Technol.* 301 Each OSO contains instruments that monitor the UV and x-ray radiation emitted by the entire solar disk. **1943** *Newsweek* 25 Jan. 26/2 OSS is the planning agency in psychological warfare for the Joint Chiefs of Staff. **1972** K. Benton *Spy in Chancery* viii. 83 We were together in Italy at the end of the war. I was in the OSS and he'd switched over to MI 6. **1964** *Horsemen's Jrnl.* Jan. 69/2 The political sponsors of the bill have figures and plans on how to 'cut-up' the tax dollar taken through O.T.B. **1971** *New Yorker* 31 July 65 Seems that the OTB computers that are linked with those at the race track developed a colic or something, and wagers at the fourteen shops around town had to be recorded manually. **1975** *Ibid.* 16 June 101/2 The OTB shops around town took in $2,442,589, of which $1,649,591 was bet on the Belmont. **1909** *Captain* XXI. p. xiv/1 Senior Divisions of the O.T.C. **1974** 'M. Innes' *Mysterious Commission* xiv. 124 He had also done rather well in what, during his public-school days, had still been called the O.T.C. **1955** *Times* 12 Aug. 8/6 The United States Congress rose without voting on President Eisenhower's proposal for entry into the O.T.C., and it cannot now be dealt with until next year. **1965** *Acronyms & Initialisms Dict.* (Gale Research Co.) 543 *OTC*, over-the-counter (Pharmacy). **1974** M. C. Gerald *Pharmacol.* ii. 20 Nonprescription (over-the-counter, OTC) sleep-facilitating products. **1976** *National Observer* (U.S.) 4 Sept. 1/3 Fourteen prescription remedies—mostly antihistamines—be sold as nonprescription, over-the-counter (OTC) drugs. **1942** *R.A.F. Jrnl.* 3 Oct. 31 At the last O.T.U. and the introduction to real aircraft. **1966** Gurnett & Kyte *Cassell's Dict. Abbrev.* 163/1 *O.T.U.*, Operational Training Unit. **1969** *Guardian* 23 July 9/3 Milton Keynes, seat of the OU from September. **1975** *Times* 28 Aug. 12/5 The OU campus seems to have a strangely insular attitude... Town and gown seldom mix. **1886** *Oxford Tatler* 15 May 35/2 No one will be surprised to hear the O.U.D.S. does not intend to put another play on the stage this term. **1959** *Oxf. Univ. Gaz.* 9 Mar. 740/2 An O.U.D.S. producer might be a clear bet for the B.B.C., a bridge and chess expert for data-processing. **1976** J. Cooper *Harriet* ii. 18 She'd seen him..in the OUDS production of *Cat on a Hot Tin Roof*. **1930** *Times* 4 Dec. 15/3 All these arrests are said to have been made by the 'O.V.R.A.', a special section of the police, dependent directly on the Ministry of the Interior. These letters are supposed in some quarters to stand for 'Organizzazione di Vigilanza Riservata sulle Associazione', whereas other persons explain the last two initials as standing for 'Repressione di Anti-Fascismo'. **1941** W. Graham *Night Journey* viii. 98 Supposing the Gestapo cut their agreement with Bonini and communicate with the O.V.R.A. **1958** *Listener* 21 Aug. 278/3 The O.V.R.A. (or Italian secret police) was by no means incompetent. **1961** C. F. Delzell *Mussolini's Enemies* i. 41 The first of these sections was the OVRA, established late in 1927 as a 'special inspectorate' with headquarters in Milan, but soon extended throughout the country. The precise meaning of the initials is still uncertain, according to the knowledgeable Guido Leto, who headed the dread agency from 1938 until 1943. [Note] Three interpretations have been suggested: 'Organizzazione di Vigilanza e Repressione dell'Antifascismo ('Organization for Vigilance and Repression of Anti-Fascism'); Organo di Vigilanza dei Reati Antistatali ('Organ of Vigilance for Anti-State Crimes'); and Opera Volontaria di Repressione Antifascista ('Voluntary Agency for Anti-Fascist Repression').

e. In *Chem. o-* (usu. italic) signifies *ortho-* (Ortho- 2 b).

1889 G. M'Gowan tr. *Bernthsen's Text-bk. Org. Chem.* xvi. 310 Thus, *o*-diamido-benzene is that one which results from the reduction of *o*-dinitro-benzene. **1926** A. Davidson *Intermediates for Dyestuffs* v. 109 *o*-Tolidine is used in making azo dyes of the same types as those derived from benzidine. **1968** R. O. C. Norman *Princ. Org. Synthesis* xi. 387 The use as a protective group is illustrated by the synthesis of *o*-nitroaniline. **1971** [see *m*- s.v. *M 5].

6. Used with reference to the shape of the letter, as *O-ring*, a gasket (usu. in the form of a ring) with a circular cross-section.

1955 J. Yarwood *High Vacuum Technique* (ed. 3) i. 60 In many ways, the best solution is the 'o' ring gasket made of rubber or, preferably, oil-resisting neoprene cord, of circular cross-section. **1959** H. Barnes *Oceanogr. & Marine Biol.* iv. 182 The sphere [*sc.* the Deep Sea Benthograph] has five openings, the largest of which is a 15-in. diameter access door closed by a cast steel plate bolted in place and sealed by two neoprene O-rings. **1971** C. M. Blow *Rubber Technol. & Manuf.* x. 443 The O-ring is very widely used, though some designers prefer the rectangular, D, or delta sections. **1975** G. Anderson *Coring* iv. 77 The pycnometer has a breech-locked lid which utilizes an 'O' ring for pressure sealing.

O, *sb.*^{1} Add: **2.** Also *Comb.*

1917 'H. H. Richardson' *Fortunes R. Mahony* III. iii. 196 He stood o-mouthed and absent-minded.

3. Name of one of several gauges of track in model railways: specifically 32 mm.; so *O gauge*. Also *OO*, 16½ mm., *OOO* 10 mm., etc.

1905 W. Ives *Something for Boys* 3 *Rails, crossings, switches*, with automatic lock action. No. O gauge, 1¼ in. **1922** *Everyday Science* Nov. 441/1 (*heading*) No. OO gauge model 'table' railways. *Ibid.*, I am pleased to see that a small gauge, i.e., 'oo' gauge, railway, is to be placed on the market shortly. **1924** H. Greenly *Model Railways* i. 3 A plan is given of Mr. H. L. Stevens' No. O gauge railway. *Ibid.* vi. 103 *No. OO Gauge 'Table' Railways.*—This standard gauge has been recently introduced by the writer..to provide for those who are limited in space to that of an ordinary dining-room table...The actual gauge is 16 mm. (⅝ in.). **1932** P. Bloomfield *Imaginary Worlds* ii. 34 Pretending..that our O gauge railway round the nursery floor is really the line taken by the 'Flying Scotsman'. **1967** C. J. Freezer *Model Railway Terminol.* 3, O. Gauge: 32 mm. Scale: 7 mm. Limited commercial support, but..in a flourishing condition. *Ibid.*, OO. Gauge: 16·5 mm. Scale: 4 mm. The most popular gauge in Britain. Fully supported commercially with ample selection of models.

-o, *suffix*^{2}. Perh. connected with O *int.* 3 and reinforced by the final syllable of abbrev. forms such as Compo, Hippo, Photo, etc. The use of the suffix is widespread in English-speaking countries but nowhere more so than in Australia (e.g. *afto*, *Arvo, *Commo, etc.).

a. Forming colloq. or slang equivalents added as a final syllable to (*a*) shortened forms of sbs., as *ag(g)ro* [f. *aggr(avation* or *aggr(ession*], *Ammo, *Beano, *Combo, *Compo^2, *Metho^1, etc.; (*b*) sbs., as *Boyo, *Bucko, *Kiddo, etc.; (*c*) adjs., as *cheapo*, *deado*.

1967 J. Burke *Till Death us do Part* ii. 31 You can buy that cheapo, cos no one wants it. **1969** *It* 13–25 June 16/2 Hustle the bread from whatever source you can... If all the above sounds like too much aggro don't..go and..run your benefit event in conjunction with an existing club. *Ibid.* 10–23 Oct. 10/1 At the moment kids are split up into different subcultural groups which have been driven by the system into a permanent state of aggro with each other. **1969** *Daily Mail* 8 Nov. 8/3 How do we get past him, man? Like he might start some agro. **1970** *Observer* 11 Jan. 28/2 Hippies and aggro-boys may look collectively and individually startling, pretty or repulsive according to tribal loyalty. **1973** A. Hunter *Gently French* iii. 29, I gets hold of the bastard and tries to pull him up. Then I sees he's bloody deado. **1973** M. Amis *Rachel Papers* 65 It wasn't day-to-day aggro, nor the drooped, guilty, somehow sexless disgruntlement I had seen overtake many relationships. **1977** 'E. Crispin' *Glimpses of Moon* xi. 215 It was possible to judge..that his aggro was strictly verbal.

b. Forming personal (chiefly occupational) sbs. from non-personal sbs., as *Bottle-o(h, *Milko, *Wino, etc.

c. Forming sbs. from adjs., as *Pinko, *Weirdo, etc.

d. As a meaningless ending in other words, as *Billy-o, *good-o, *Cheerio, *right(y)o*, etc.

oaf^2. Abbrev. of *Ofay. *U.S.*

1941 J. Smiley *Hash House Lingo* 40 *Oaf*, white person (used by negroes).

oa·fishly, *adv.* [f. Oafish *a.* + -ly^2.] In an oafish or stupid manner.

1876 F. K. Robinson *Gloss. Whitby* 134/1 *Oafishly, oafly*, absurdly; foolishly. **1908** A. S. M. Hutchinson *Once aboard Lugger* I. vii. 38 The driver becomes temporarily idiot—stands us oafishly silent. **1958** P. Kemp *No Colours or Crest* vii. 137, I stared back oafishly at him from under the brim of my hat.

oafo (ōu·fo). *slang.* [f. Oaf + *-o^2.] A lout or hooligan. Also *attrib.*

1959 C. MacInnes *Absolute Beginners* 39, I eyed the oafo. *Ibid.* 184 The oafo lot went off laughing. **1962** R. Cook *Crust on its Uppers* i. 25 The middle classes..the working classes..not to mention the oafos.

oak. Add: **8. a.** *oak-scroll, -thicket.* **c.** *oak framed* adj. **d.** *oak-pale, -trunked* adjs.

1953 E. Simon *Past Masters* I. 47 On the walls..two oak-framed prints. **1977** *Times* 15 Oct. 8/2 The house.. had..an oak-framed porch. **1922** Joyce *Ulysses* 6 His fair oakpale hair stirring slightly. **1874** G. M. Hopkins *Jrnls. & Papers* (1959) 245 A beautiful spray-off of the dead oak-scrolls. **1846–7** Thoreau *Walden* (1957) 186 Pine woods and oak-thickets. **1934** Dylan Thomas *Let.* 21 Sept. (1966) 268 No One more welcome than the oak-trunked maestro—.

9. oak barren *U.S.* (earlier example); **oak cist, coffin** (see quot. 1957); **oak flat** *U.S.*, a level expanse of ground bearing a growth of oaks; **oakleaf braid** (see quots.); **oakleaf jar** (see quot. 1960); **oak-mast** (earlier example); **oak moss,** the lichen *Evernia prunastri* or one closely related to it, often found growing on oak trees and used to produce an aromatic extract; also the extract itself; also *attrib.*; **oak-opening** (earlier and later examples); **oak-pruner** *U.S.*, a longicorn beetle *Elaphidion villosum*, the larva of which mines down the centre of hardwood twigs, causing them to snap; now usually called the twig pruner; **oak room,** an oak-panelled room; **oak towel** *slang* (see quot.); **oak wilt** *U.S.*, a disease of oaks and certain other trees produced by the fungus *Ceratocystis fagacearum*, which causes the wilting and death of foliage and eventually kills affected trees; **oak yard** *U.S.*, an enclosure in which oaks are grown.

1811 *Weekly Reg.* 12 Oct. 101/2 Our oak barrens and underwooded plains may be profitably applied to sheep. **1937** E. V. Gordon tr. *Shetelig & Falk's Scand. Archaeol.* 146 Of similar type is the other well-known form of Norse bronze-age grave, the 'oak cist', a coffin made from a thick trunk of oak, split and hollowed out. **1968** G. Jones *Hist. Vikings* I. i. 19 The tannin of the 'oak cists' of Denmark, the very flesh and fell of the wearers. **1937** E. V. Gordon tr. *Shetelig & Falk's Scand. Archaeol.* 147 The complete picture of this personal equipment is obtained from the oak coffins mentioned earlier. **1957** T. C. Lethbridge *Gogmagog* viii. 132 Occasionally in Britain and more frequently in Denmark human bodies are found buried in what are known as 'oak coffins'. These are not coffins in the ordinary sense; but are large sections of tree trunks, split lengthwise and hollowed out to contain the body. **1964** W. L. Goodman *Hist. Woodworking Tools* 10 The remarkable wooden folding stool found with an oak-coffin burial at Guldhoj in Jutland. **a 1816** B. Hawkins *Sk. Creek Country* (1848) 29 Oak flats, red and post oak,..on its left side. **1849** E. Chamberlain *Indiana Gazetteer* (ed. 3) 381 Beech and oak flats, which are adapted only to grass. **1954** *N. & Q.* June 273/1 *Oakleaf braid.*—This is the black braid supplied for hats of senior police officials and also used by St. John Ambulance. **1957** *Textile Terms & Definitions* (Textile Inst.) (ed. 3) 70 *Oakleaf braid*, a woven narrow fabric having a conventional oakleaf and acorn and border Jacquard design, now always black... It is used as a hatband for officials such as Police Inspectors. **1903** H. Wallis (*title*) Oak-leaf Jars: A fifteenth century Italian Ware showing Moresco Influence. **1960** R. G. Haggar *Conc. Encycl. Cont. Pott. & Porc.* 334/1 *Oak-leaf jars*, Tuscan fifteenth century maiolica drug pots painted with an oak-like..leaf decoration. **1849** E. Chamberlain *Indiana Gazetteer* (ed. 3) 17 Oak and beech mast is found in such quantities as to contribute largely both to feeding and fattening hogs. **1921** A. L. Smith *Lichens* x. 418 French perfumers extract an excellent perfume from *Evernia prunastri*.. known as 'Mousse des Chênes' (Oak moss), and it appears that the plants which grow on oak contain more perfume than those which live on other trees. **1921** *Times Lit. Suppl.* 25 Aug. 542/4 The oak-moss lichen is used as a basis for perfumes, the thallus on being soaked in spirit yielding a sweet and persistent odour. **1966** J. S. Cox *Illustr. Dict. Hairdressing* 104/2 Oak Moss Resin. Obtained from tree lichens, especially oak; used in perfumery. **1967** M. E. Hale *Biol. Lichens* iv. 59 These 'oak mosses' accumulated silicon, phosphorus, magnesium, iron and aluminium to a significant degree. **1975** F. Kennett *Hist. Perfume* vii. 148 The main ingredients of it [*sc.* Poudre de Chypre] are oakmoss (extracted from a species of lichen and still commonly used in perfumery, by the name *mousse de chêne*), rose-water, musk, civet, and a little sandalwood. **1830** J. M'Call in *Wisconsin State Hist. Soc. Coll.* (1892) XII. 185 From that up, on the right bank, it is oak openings. **1833** C. F. Hoffman *Winter in West* (1835) I. 142 At a sudden turning of the path, I came at once upon the 'oak openings'. **1839** C. M. Kirkland *New Home* xx. 133 The 'grubs' present a most formidable hindrance to all gardening efforts in the 'oak-openings'. **1848** J. F. Cooper *Oak Openings* I. i. 10 Giving their appellation to this particular species of native forest, under the name of 'Oak Openings'. **1882** *Econ. Geol. Illinois* II. vi. 104 There is an intermediate district occupied by oak-openings. **1970** *Daily Progress* (Charlottesville, Va.) 24 May 4/1 Trees and shrubs grew along the streams, on wooded knolls or ridges, and in occasional 'oak openings'. **1819** *Mass. Agric. Repository & Jrnl.* V. 308 From the effect of its labours, it may be called the oak pruner. **1838** *Mass. Zool. Survey Rep.* 92 The oak-pruner, so named by Prof. Peck, inhabits the white and black oaks. **1862** *Rep. Comm. Patents: Agric. 1861* (U.S. Dept. Agric.) 615 The black and white oak trees are infested with..the 'Oak-pruner'. **1899** D. Sharp in *Cambr. Nat. Hist.* VI. v. 286 *Elaphidion villosum* is called the oak-pruner in North America. **1849** Thackeray *Pendennis* I. xxiii. 213 On the other side [of the hall] the oak room. **1922** Joyce *Ulysses* 154 In the supper room or oakroom of the mansion house. **1971** D. Francis *Bonecrack* iii. 34 The account books.. are in the oak room. **1889** Farmer *Americanisms* 396/1 *Oak towel*..a stout oaken stick. There is an allusion here to 'wiping' or 'dressing one down'. **1942** *Bull. Wisconsin Agric. Exper. Stat.* No. 455. 75/1 Oak wilt, a disease now ravaging many fine Southern Wisconsin woodlots, is caused by a certain fungus. *Ibid.* 76/1 Thus far there is no way of controlling oak wilt. **1944** B. W. Henry et al. in *Phytopathology* XXXIV. 163 The present paper presents evidence on the significance, symptoms, and cause of a disease called oak wilt. **1957** J. M. Haller *Tree Care* ix. 148 The oak wilt came into prominence about seven years ago, spreading rapidly through the midwestern states. **1959** P. P. Pirone *Tree Maintenance* (ed. 3) xvi. 346 The oak wilt fungus appears to be most infectious early in the growing season. **1969** *New Scientist* 28 Aug. 430/2 They inoculate the weed oaks with the organism that causes oakwilt disease, *Ceratocystis fagacearum*. **1835** R. M. Bird *Hawks of Hawk-Hollow* II. v. 52 His father..had suddenly checked his horse at the entrance of the little oak-yard.

oaken, *a.* **4.** *oakenshaw* (further example).

1903 A. E. Housman 'The Oracles' in *Venture* I. 39 When winds were in the oakenshaws.

oa·kery. *U.S. rare.* [f. Oak + -ery.] An oak yard.

1838 C. Gilman *Recoll. Southern Matron* xxx. 213 Turning suddenly, he bounded over the fence into papa's oakery.

oa·kiness. [f. Oaky *a.*] The quality of being oaky.

1863 'G. Hamilton' *Gala-Days* 128 [In] the English Church..there is a general tone of oakiness, solid, substantial, sincere.

oak land, oak-land. Chiefly *U.S.* [Oak 8.] Land bearing a growth of oak-trees.

c **1658** in *Early Rec. Lancaster, Mass.* (1884) 271 Thare is another peice of upland..Sum part pine Land & partly oak Land. *Ibid.*, Sum part of it [is]..oake land. **1737** J. Wesley *Jrnl.* 2 Dec. (1909) I. 401 The land is of four sorts —pine-barren, oak-land, swamp, and marsh. **1751**, etc. [in Dict. s.v. Oak 8]. **1811** *Weekly Reg.* 28 Dec. 302/1 It was a piece of dry oak land. **1837** W. Jenkins *Ohio Gazetteer* 187 The soil of Franklin is, what is generally called oak land, being a mixture of clay, sand and gravel. **1849** E. Chamberlain *Indiana Gazetteer* (ed. 3) 209 The oak land is more extensive than the beech.

oaky, *a.* Add: **1.** *oaky-looking* adj.

1921 D. H. Lawrence *Sea & Sardinia* 165 Curious slim oaky-looking trees.

2. (Earlier example.)

a **1816** B. Hawkins *Sk. Creek Country* (1848) 62 The good land spreads out for four or five miles on both sides of the creek, with oaky woods.

-oan (ŏan, ōu·ăn), *suffix*. *Min.* [f. *-o* (in Ferro-, -ous c, as against Ferri-, -ic 1 b) + *-1)an 2.] Used like *-ian 2, but denoting a lower valency than that suffix (see quot.).

1930 [see *-ian 2].

oar, *sb.* Add: **5. a.** (Further examples.) Also, *to get, shove*, etc., *one's oar in*. The primary sense is 'to interfere, to be (or become) meddlesome'.

1630 R. Brathwait *Eng. Gentleman* i. 11 Youth.. putting his oare in every mans boat. **1731** C. Coffey *Devil to Pay* I. ii. 12, I will govern my own House without your putting in an Oar. **1890** Barrère & Leland *Dict. Slang* II. 92/2 '*To shove in an oar*', to intermeddle, or give an opinion unasked. **1908** J. H. Shinn *Pioneers Arkansas* xxxii. 258 The idea is always to do the other fellow before he does you, and if he does get his oar in first, come back with remark called for brevity, 'The Retort Courteous'. **1916** 'Taffrail' *Pincher Martin* vii. 109 'It ain't fit an' proper fur gals o' your age to go abart unpertected like.'.. 'And who asked you to put your oar in, Mister Billings?' **1946** K. Tennant *Lost Haven* (1947) xiv. 224 He would probably go on talking all the morning, if a bloke didn't put in his oar. *Ibid.* 226 Look here, Dip. You're shoving in your oar, so I'll just tell you what I'm up against. **1968** J. R. Ackerly *My Father & Myself* ix. 87 One who preferred to stand outside of life and observe it not (as he would have phrased it) to 'put one's oar in'. **1977** *Film & Television Technician* Mar. 5/1 The more that the workers stick their oar into the administration of actual film production the more they will weaken their strength in negotiation for the improvement of wages and terms of employment. **1978** J. Anderson *Angel of Death* i. 6 Senior police officers couldn't go around saying things like that... He'd better get his oar in first.

6. oar-fish (examples); **oar-lop** now *obs.*, a lop-eared rabbit with its ears sticking out at right angles to its head; (the form *oar-lap* appears to be an error arising from the misprint in the source of quot. 1868).

1860 J. Richardson *Yarrell's Hist. Brit. Fishes* (2nd Suppl.) 27 (*heading*) Banks's oar-fish. **1880** A. C. L. G. Günther *Introd. Study Fishes* II. 522 They [*sc. Regalecus* species] are frequently called..'Oar-fishes', from their two ventral fins, which have a dilatation at their extremity not unlike the blade of an oar. **1925** J. T. Jenkins *Fishes Brit. Isles* 120 The name Oar-fish is derived from the presence of the two pelvic fins, which are..something like the blade of an oar. **1959** A. Hardy *Open Sea* II. iv. 76 The oar-fish *Regalecus glesne*.. is world-wide in distribution... It is the largest of the so-called ribbon-fish and..it looks almost like one's idea of the mythical sea-serpent. **1972** *N.Z. News* 31 May 3/2 Two small boys rowing their boat in the Otago Harbour.. bumped into a comparatively rare oar fish, 10 feet long and weighing about 100 pounds... Only about a dozen of the species, Regalecus pacificus, have been washed up on the New Zealand coast in the past 100 years. Oar fish are a deep-water variety. **1854** 'E. S. Delamer' *Pigeons & Rabbits* 136 The Oar-lop is the next stage of deflection, when the ears extend horizontally outwards on each side, forming a line that is more or less straight, giving the idea of a pair of oars which a waterman is resting out of the water. **1868** [see Lop *sb.*^{7}]. **1872** C. Rayson *Rabbits* III. xiii. 71 The 'oar-lop'..is not fit for exhibition purposes. **1912** G. A. Townsend *Pract. Rabbit Keeping* vii. 62 If both ears stood out from the head at right angles, the rabbit was an 'Oar Lop'. **1933** F. L. Washburn *Rabbit Bk.* (ed. 2) ii. 55 The term 'oar-lop' was used to express the condition where the ears are horizontal.

oarlock. Add: Also *attrib.* in *oarlock seat*.

1874 J. W. Long *Amer. Wild-Fowl Shooting* 85 Both may row, if two sets of oarlock-seats are provided.

oarsman. (Earlier example.)

1811 *Weekly Reg.* I. XIV. 245/1 They certainly possess great dexterity as oarsmen.

oar-weed, now the usual spelling of Ore-weed; = *Laminaria. (Later examples.)

1917 *Chambers's Jrnl.* July 473/1 The 'oar weed' variety of seaweed..contains considerable supplies of potash. **1922** Joyce *Ulysses* 45 He climbed over the sedge and eely oarweeds. **1954** *New Biol.* XVII. 102 The broad oar-weeds, the Laminarias,..occupy the shore at low-water mark of spring tides and at greater depth. **1971** C. L. Duddington *Beginner's Guide to Seaweeds* iii. 47 The oarweeds (genus *Laminaria*) are large seaweeds that grow in the sublittoral zone, from just below low-tide mark down to a depth of about fifteen feet. The oarweeds are perennial plants. *Ibid.*, The oarweeds have a varied history of usefulness, first in the old kelp burning industry ..and later as a source of alginic acid and alginates.

oat, *sb.* Add: **4. d.** *to feel one's oats*, to be lively; to feel important, to display one's self-importance. *colloq.* (orig. *U.S.*).

1831 *Boston Even. Transcript* 22 Dec. 1/1 Whether the pony felt his oats,..He took a frightful canter. **1833** A. Lawrence *Diary & Corr.* (1855) 126 We both 'feel our oats' and our youth. **1843** T. C. Haliburton *Attaché* 1st Ser. II. 157 You know that, and you feel your oats, too, as well as any one. **1869** P. T. Barnum *Struggles & Triumphs* i. 33 My father..installed me as clerk in this country store. Of course I 'felt my oats'. **1897** C. M. Flandrau *Harvard Episodes* 85, I suppose he was feeling his oats when he captained his class eleven. **1959** *Listener* 5 Nov. 770/1 The new influences and pressures within a colony that was 'feeling its oats'. **1971** D. Lees *Rainbow Conspiracy* i. 17 The Manchester circulation is nudging the one and a half million a day mark and they are beginning to feel their oats.

e. *off one's oats* colloq., off one's food.

1890 Kipling in *Lippincott's Monthly Mag.* Aug. 254 I'm a bit restless and off my oats. **1898** *Bulletin* (Sydney) 17 Dec. (Red Page), The horse is a power in Australia, and a few choice expressions spring from horses..*out of collar* aptly describes out of work; *off his oats*, sickness or a state of offishness. **1930** Wodehouse *Very Good, Jeeves!* iv. 98 The poor kid, who's quite off her oats about him. **1949** D. M. Davin *Roads from Home* II. ii. 99 What's the matter, John? Off your oats this morning? **1977** J. Fleming *Every Inch a Lady* I. i. 5 It's not like to put me off me oats..but it's been a nasty day.

f. *one's oats*, sexual gratification. *slang*.

1923 J. Manchon *Le Slang* 209 *To have one's oats*, faire des bêtises avec une femme, courir la gueuse. **1941** Baker *Austral. Slang* 50 *Oats from (a woman)*, *get one's*, to coit with a woman. **1961** X. Herbert *Soldiers' Women* 265 There's nothing makes a hot-shot sheik like that so mad as being asked to pay for his oats. **1965** W. Dick *Bunch of Ratbags* 188, I was kissing her excitedly and passionately. You're doin' O.K., Cookie, you're gonna get your oats tonight for sure, I thought to myself. **1968** A. Diment *Bang Bang Birds* v. 65 Despite her lovely body it was her face that had me hooked... I like to watch something pretty..when collecting my oats. **1976** P. Hill *Hunters* vii. 90 She wouldn't let you have your oats... You wanted to go to bed with her..she wouldn't have it. **1978** J. Wainwright *Jury People* xxxvi. 108 This wife he was lumbered with. Okay—he loved her... But, even *he* wanted his oats, occasionally. He was human.

5*. oats and chaff *Rhyming slang*, a footpath.

1857 'Ducange Anglicus' *Vulgar Tongue* 14 *Oats and chaff*,..footpath. **1935** A. J. Pollock *Underworld Speaks* 82/1 *Oats and chaff*, footpath.

6. *oat-bag, -straw* (earlier examples).

1851 A. O. Hall *Manhattaner* 5 It was a modest commercial plain..with..bits of machinery, and ploughs, and oat bags, and hay bales. *a* **1650** D. Calderwood *Hist.*

Kirk (1842–9) VI. 27 A scheaffe of oat straw was sold for fourtie shillings in Edinburgh. **1850** *Rep. Comm. Patents: Agric. 1849* (U.S. Dept. Agric.) 380 Getting no other food in winter but a scanty supply of oat-straw. **1859** A. Cary *Pict. Country Life* i. 7 [He] lay..with a bundle of oat-straw for his pillow.

b. oat burner *N. Amer. colloq.*, a horse; **oat cell** *Med.*, a small oval cell with little cytoplasm and an oval, densely staining nucleus which is characteristic of a type of carcinoma of the lung (formerly regarded as a sarcoma); freq. *attrib.* in *oat-cell carcinoma, tumour*, etc.; hence **oat-celled** *a.*, containing such cells; **oat(s opera** = *OATER².

1941 *Sun* (Baltimore) 21 July 11/4 There isn't a galloper in the lot who can say 'I'm the boss', so your milkman's oat burner might do just as well as any of 'em. **1952** *Daily News* (N.Y.) 20 Aug. C 11/4 When the time comes..that even an oat-burner must sport a tax stamp on its stem or stern. **1973** B. Broadfoot *Ten Lost Years* v. 50 Them oatburners never broke down. **1903** W. S. L. Barlow *Elem. Path. Anat. & Histol.* I. ix. 190 (*heading*) Oat cell Sarcoma. **1926** *Jrnl. Path. & Bacteriol.* XXIX. 244 In obvious carcinomata of the lung 'oat cells' have been found in addition to the more readily recognisable carcinoma cells. **1956** Mayer & Maier *Pulmonary Carcinoma* iv. 96 Among anaplastic tumors belong the 'oat cell' carcinomas, called 'reserve cell' by some. **1957** A. I. Spriggs *Cytol. of Effusions* vi. 24 One of the most characteristic types of malignant cell is the oat-cell, so named after its appearance in histological sections. **1966** Wright & Symmers *Systemic Path.* I. x. 418/1 The finding of tubules in 'oat-cell' tumours..should not affect the histological diagnosis. **1972** *Brit. Jrnl. Dis. Chest* LXVI. 164 Oat cell carcinomas have a more sinister prognosis. **1926** *Jrnl. Path. & Bacteriol.* XXIX. 244 The so-called 'oat-celled Sarcoma' of the posterior mediastinum is a medullary carcinoma of the bronchi. **1948** R. A. Willis *Path. Tumours* xix. 369 'Oat-celled' or spindle-celled structure ..is common in bronchial carcinoma. **1942** Berrey & Van den Bark *Amer. Thes. Slang* § 608/9 *Western picture*, ..oats opera. **1947** *Richmond* (Virginia) *Times-Dispatch* 2 May 10/1 Roy Rogers, Gene Autry and other oat opera stars.

oat, *v.* (Earlier and later examples.)

1732 B. Lynde *Diary* 9 May (1880) 26 Next morning.. dined at Hampton;..thence to Greenland, where oated, and for 2 horses and drink, 2s. **1741** *Ibid.* 27 Oct. 121 Breakfasted and oated our 3 horses, at Deacon Tucker's. **1855** P. T. Barnum *Life* 70 Old 'Bob' was duly oated and watered.

oater² (ōu·təɹ). Chiefly *U.S.* = **horse opera*.

1951 Green & Laurie *Show Biz* 570/1 *Oater*, Western film. **1961** A. Berkman *Singers' Gloss. Show Business* 64 *Oater*, Western film. **1969** M. Pei *Words in Sheep's Clothing* (1970) iii. 22 'Western'..along with its synonyms, 'horse opera', 'oater'. **1975** *Radio Times* 10 Jan. 10/1 Borden (*Red River*) Chase wrote this expansive oater [sc. *Lone Star*] for Clark Gable.

oath, *sb.* Add: **1.** *under oath*, on or upon oath.

1712 [see Under *prep.* 14 b]. **1851** R. Glisan *Jrnl. Army Life* (1874) viii. 83 He was then under oath not to drink for six months.

c. *my* (*colonial*, etc.) *oath*, a mild expletive or exclamation: yes! of course! upon my word! *Austral.* and *N.Z. slang.*

1859 H. Kingsley *Recoll. G. Hamlyn* II. vi. 94 'You're not fit company for any man except the hangman,' said Tom,..'Oh my — — (colonial oath!)' said the other; 'oh my — — cabbage tree!' **1895** J. Kirby *Old Times in Bush* x. 143 The snake..made a hoop of hisself, and then, my oath, he did go. **1896** H. Lawson *While Billy Boils* 203 (*title*) His colonial oath. *Ibid.*, 'My oath!' he replied... 'My blooming oath!' **1899** *Bulletin* (Sydney) 25 Feb. (Red Page), I don't mean the variety whose conversation consists of expectoration and 'Yer' or 'My — oath'. **1916** J. B. Cooper *Coo-oo-ee* i. 11 Ain't that like a woman with a man? My oath it is! I know 'em. **1925** H. H. Cook *Far Flung* 12 'Perhaps ye'll square up?'..'My oath!' cried the debtor. **1928** 'Brent of Bin Bin' *Up Country* xv. 255 'My — Colonial oath!' echoed Erroll. **1941** *Coast to Coast 1941* 45 'Who said I was going to give you a quid anyway?' 'You did.' 'My oath I didn't.' **1946** E. G. Webber *Johnny Enzed in Italy* 36 'All this us der merry laugh gives, no?' I said. 'My oath!' said the Bloke. 'My colonial oath!' **1952** M. Tripp *Faith is Windsock* i. 7 'Pay twenties and twenty-ones.' 'My oath!' Bergen threw his cards (a king and a six) face down on the table. **1969** *Melbourne Truth* 12 July 3/4 Whitton, with his ear close to the receiver, listened to the reply. It was: 'My bloody oath I will.' **1974** N. Marsh *Black as he's Painted* iii. 73 'And that's when your headache really sets in, is it, Fred?' 'My oath! Well, take a look at it.' **1977** J. Wainwright *Do Nothin'* xi. 185 My oath—those couple of hours were some session.

5. *oath-breaker* (later example), *-keeper*, *-taking* (later examples); *oath-bound* adj. (later example); **oath-helper** = Compurgator 1 b; hence *oath-helping* vbl. sb.

1939 W. B. Yeats *Last Poems* 8 That all are oath-bound men. *a* **1973** J. R. R. Tolkien *Silmarillion* (1977) ix. 83 For so sworn, good or evil, an oath may not be broken, and it shall pursue oathkeeper and oathbreaker to the world's end. **1891** Oath-helper [used s.v. Compurgator 1 b]. **1902** *Encycl. Brit.* XXVIII. 331/1 One of the two litigants must prove his case, by his body in battle,.. or by an oath with oath-helpers. **1943** F. M. Stenton *Anglo-Saxon Eng.* ix. 312 It is probable that in the earliest time a man's 'oath-helpers' had been chosen exclusively from among his kin. *Ibid.* 313 Ine of Wessex orders that every person accused of homicide, whatever his status, must include at least one man of high rank among his oath-helpers. **1970** Foote & Wilson *Viking Achievement* xi. 375 After the man accused had made his statement on oath, then each of his oath-helpers swore that they believed his oath to be honest. **1973** A. Harding *Law Courts Med. Eng.* 25 The Normans let the parties fight an actual physical battle, but the Anglo-Saxons preferred trial by ordeal or by oath-helping. *a* **1973** Oath-keeper [see *oath-breaker* above]. **1960** *News Chron.* 9 July 1/3 The Mau Mau oath-taking ceremonies. **1961** *Guardian* 29 Mar. 2/1 Recent reports of Mau-Mau oath-taking in Kenya.

oathing, *vbl. sb.* (Later examples.)

1961 *Guardian* 12 May 5/5, I don't attribute the oathing to the party. *Ibid.* 7 June 11/4 The oathing ceremonies at Meru last month when more than a thousand people took an undisclosed oath. **1969** *Daily Tel.* 9 Sept. 20/6 To a Christian, secret oathing is repugnant and unacceptable. *Ibid.*, Secret oathing ceremonies.

oatmeal. Add: **2*.** A greyish-fawn colour resembling that of oatmeal. Also *attrib.* or as *adj.*

1927 *Daily Express* 2 May 7 Colours: Beige, Oatmeal, [etc.]. **1935** A. Christie *Three Act Tragedy* iii. ix. 188 The room..had walls of a rather drab oatmeal colour. **1939–40** *Army & Navy Stores Catal.* 636/2 All wool flannels. ..Oatmeal Shade. **1951** 'C. Carnac' *It's her own Funeral* i. 10 The dark panelling made a good background to her oatmeal suit. **1951** *Catal. of Exhibits, South Bank Exhib., Festival of Britain* 58/2 Woman's oatmeal glacé gloves **1960** *House & Garden* May 61/2 Settee..covered in oatmeal tweed. **1961** *Guardian* 1 Feb. 7/5 Colours were pale green, oatmeal, navy. **1963** *Times* 27 Feb. 12/5 Contrasting cravats of silk or organza set off soft oatmeal or sand tones, pale *tilleul* greens, navy, black, off-white. **1970** [see *kingfisher 3]. **1973** 'R. MacLeod' *Burial in Portugal* iii. 69 An oatmeal sports coat.

3. *oatmeal bread, stout*; **oatmeal mush** *U.S.*, porridge made with oatmeal; **oatmeal soap**, soap containing oatmeal as a mild abrasive.

1943 A. Simon *Conc. Encycl. Gastron.* IV. 83/1 Oatmeal bread. **1974** R. B. Parker *God save Child* (1975) xii. 86 He put a plate of sliced turkey..and a loaf of oatmeal bread on the table. **1883** H. P. Spofford in *Harper's Mag.* Aug. 465/1 You've been the means of starving me.. on oatmeal mush. **1903** S. Clapin *New Dict. Amer.* 291 Oatmeal-mush. **1890–91** *T. Eaton & Co. Catal.* Fall & Winter 42 Colgate's soaps—oatmeal, [etc.]. **1897** *Sears, Roebuck Catal.* 19/1 (*heading*) 'Oat Meal' Toilet Soap. *c* **1938** *Fortnum & Mason Price List* 54/1 Soaps... Violet Oatmeal. **1977** *Honey* Nov. 55/2 Friction cleaning is most effective on blackheads... You can use Boots No 7. very mild oatmeal and lavender soap 89p. **1926–7** *Army & Navy Stores Catal.* 64/1 Oatmeal Stout (recommended for its tonic properties) doz. 9/-. **1976** J. B. Hilton *Gamekeeper's Gallows* iii. 26 If he wins, he always takes it [*sc.* a rabbit] a bottle of oatmeal stout home.

Oaxacan (o,æksā·kăn), *a.* [f. *Oaxaca* (see below) + -an.] Of or pertaining to the southern Mexican state of Oaxaca.

1934 A. Huxley *Beyond Mexique Bay* 263 Our Oaxacan friend, Don Manuel, invited us one evening to join this endless procession. **1977** *Dædalus* Summer 88 This early Oaxacan elite was distinctly less sophisticated than the Olmec aristocracy of the neighboring Mexican Gulf Coast country.

ob., abbrev. of L. *obiit*, died: add examples.

1890 Barrère & Leland *Dict. Slang* II. 93/1 *Ob* (Winchester College), for *obit*. **1952** D. Balsdon *Freshman's Folly* iii. 141 The highly complexioned countenance of the Reverend Nathaniel Blunt, S.T.P., ob. 1782. **1978** C. Jones et al. *Study of Liturgy* ii. i. vii. 57 *The Apostolic Tradition*, ascribed by many scholars to Hippolytus, the schismatic bishop of Rome (ob. A.D. 236, 237?).

ob (ǫb). A representation of a pronunciation of the word Of *prep.*, supposed to occur esp. in the speech of American Blacks.

1839 *Bentley's Misc.* VI. 263 He said de Queen ob Sheba wab a dark lady, may be bery dark. **1846** *Negro Singer's Own Bk.* 5, I am glad ob it, for my part. *Ibid.* 23 One day just at de set ob sun,..When de work war did an done. *Ibid.* 29, I tell you ob a scrape I had wid a gal. **1848** S. C. Foster *Old Uncle Ned* (*song*) 3 He had no wool on de top ob de head. **1851** —— *Ring de Banjo* (*song*) 4 Den come again, Susanna, By de gas-light ob de moon; We'll tum de old piano When de banjo's out ob tune. **1882** *Judge* (N.Y.) 30 Dec. 10/3 Money am de root ob all ebil. **1891** C. Townsend *Negro Minstrels* 22 He..drank a gallon ob tangle foot. **1893** M. A. Owen *Voodoo Tales* 168 Hit am er powdeh mek outen de same hef' ob snails an' lizuhds. **1893** K. Mackay *Out Back* (ed. 2) iii. viii. 288 'Any fella longa tribe ob my sister, Queen Victoria, friend longa mine,' replied the black-fellow. **1895** Banks & Smiley in A. Dundes *Mother Wit* (1973) 256/2 Three bottles sittin' in de road, one ob dem full. **1950** R. Ames in *Ibid.* 492/2 In de fork ob de branch. **1974** R. B. Parker *Godwulf Manuscript* vii. 55 Ah is a member ob de press, baby.

oba (ǫ·bă). *W. Afr.* Also **obba.** [Yoruba.] The title of the ruler of the ancient West African kingdom of Benin, now part of Nigeria, whose power used to be absolute. Now revived as the title of a local chief.

1903 H. L. Roth *Great Benin* xi. 118 The Obba kindly promised to send a messenger round with orders. **1906** R. E. Dennett *At Back of Black Man's Mind* xvii. 175 The Oba's throne or chair was placed on a platform of mud three steps above the ordinary level of the ground. **1926** P. A. Talbot *Peoples S. Nigeria* I. iv. 155 The land of Southern Nigeria, including the kingdom of Benin, was discovered by Ruy de Sequeira. The Portuguese were presented by the Obba with..carved wooden platters. **1936** J. U. Egharevba *Short Hist. Benin* i. 8 The oracle declared..the senior wife of the Oba, to be the cause. **1957** W. M. Hailey *Afr. Survey* (rev. ed.) ii. 35 The Yoruba kingdoms in Nigeria were..highly centralized...The rulers, generally known as Obas, were traditionally sacred persons. **1967** W. Soyinka *Kongi's Harvest* 4 When an Oba stops the procession And squats on the wayside, It's on an urgent matter. **1972** R. N. Henderson *King in Every Man* i. ii. 43 The first king or oba, Eweka I, whose regalia came from the Yoruba city of Ife. **1975** *Times* 16 July 77/5 A massive Benin bronze head of an oba. **1976** *Nigerian Herald* 20 July 4/5 The Deji suggested that a law should be made by the federal military government to leave the selection and appointment of obas and chiefs to those who were versed in the custom culture of the community concerned as a means to curb chieftaincy wranglings in the country. **1976** *Sunday Times* (Lagos) 1 Aug. 24/3 It is the turn of this ruling house to provide an Oba for the town.

Obanian (obē̆i·niăn), *a. Archæol.* [f. the name of the Scottish burgh of *Oban*, Strathclyde (formerly Argyllshire).] Applied to a culture of the mesolithic period for which most evidence is found in the neighbourhood of Oban. Also as *sb.*, the Obanian culture or a person living in this culture.

1942 H. L. Movius *Irish Stone Age* ii. iii. 180 With the changing environment certain new forms were evolved to meet the new needs, but lack of flint or other easily worked rocks hampered cultural progress. It is proposed to call this culture the *Obanian*, after the type site in Argyllshire. *Ibid.*, The Obanian was discovered in 1879 by Symington Grieve (1883) at Caisteal-nan-Gillean, a large kitchen-midden 150 feet in diameter on the Island of Oronsay, Argyllshire. *Ibid.*, At Caisteal-nan-Gillean a large series of stone, bone and antler implements was found, typical of the Obanian culture. **1957** V. G. Childe *Dawn European Civilization* (ed. 6) i. 5 On the south-west coast of Scotland..the industry..is distinctive enough to be regarded as a new culture, 'the Obanian', not certainly descended from the French Azilian. **1958** *Proc. Soc. Antiquaries Scot.* LXXXIX. 91 (*heading*) Notes on the Obanian with special reference to antler- and bone-work. **1959** *Chambers's Encycl.* V. 452/1 The contents of the middens show that the Obanians hunted red and roe deer [etc.]. **1963** *Field Archaeol.* (Ordnance Survey) (ed. 4) 24 The Obanians are found on the west coast and in the islands. **1963** E. S. Wood *Collins Field Guide Archaeol.* iv. 52 The old description of the Obanian culture, based on hunting, and on fish and molluscs, as a movement of the mesolithic people called Azilian, from south-west Europe, no longer holds. **1970** Bray & Trump *Dict. Archaeol.* 165/1 *Obanian culture*...The sites are rock-shelters, and also shell middens on post-glacial raised beaches. The way of life was adapted to coastal conditions. Flintwork is scarce, but diagnostic tools include barbed spears and stone limpet-picks.

obba, var. *oba.

obbligato, *a.* Add: (Example of *transf.* use.)

1861 Geo. Eliot *Let.* 13 Apr. (1954) III. 405 Don't think about reading Silas Marner, just because it is come out. I hate obligato reading and obligato talk about my books.

B. *sb.* Also *transf.*

1888 Kipling *Departmental Ditties* (ed. 3) 58 A Wagner *obbligato, scherzo*, double-hand *staccato*, Played..by the clacking tonga-bar. **1921** G. Hopkins *City in Foreground* i. 15 He took the easiest way, which was to supply a gentle *obbligato* to the inspired melody of his companion's speech. **1955** *Times* 21 May 6/5 Lady Megan..was wreathed in smiles and her chuckle was a frequent obbligato. **1972** A. Roudybush *Sybaritic Death* (1974) ix. 82 Her complaints..formed a sort of obbligato to their drive.

obbo (ǫ·bo). *slang.* Also **obo.** [Abbrev. of Observation: cf. *-o².] Observation, esp. in police work. In military use *ellipt.* for *observation balloon.*

1925 Fraser & Gibbons *Soldier & Sailor Words* 212 *Obbo*, observation balloon. **1933** C. E. Leach *On Top of Underworld* 9, I am still 'on the trail', like you, and my thoughts hark back to the long weary days and nights of 'obbo' (observation) and shadowing. **1940** Partridge *Slang* 192 An observation balloon is an *obbo* or a *sausage*. **1968** Busby & Holtham *Main Line Kill* vi. 68 Now I got a fix on the place I got to do some obo first. Get the lie of the land. **1972** 'B. Graeme' *Tomorrow's Yesterday* xi. 117 We're keeping a man, suspected of robbery..under obbo. **1973** D. Lees *Rape of Quiet Town* iv. 68, I went to keep obbo for a bit.

obclude, var. Occlude *v.*

1894 *Proc. Zool. Soc.* 434 Only about half the iris is visible, and even some part of the lens is obcluded.

obduct, *v.* Restrict † *Obs.* to sense in Dict. and add: **2.** *Geol.* To cause to undergo obduction. Hence **obdu·cted** *ppl. a.*

1971 R. G. Coleman in *Jrnl. Geophysical Res.* LXXVI. 1216/2 Parts of the oceanic crust have been overthrust (obducted) onto thin continental edges. **1974** *Nature* 1 Mar. 38/2 The ophiolites of west Newfoundland and the Burlington Peninsula represent obducted oceanic lithosphere. *Ibid.* 20 Sept. 259/1 With the development of actual collision between continental blocks a period of tectogenesis and orogenesis (uplift) results in F_3 flexuring of the subducted lithosphere, slivers of which are consequently overthrust (obducted) on to the 'active' continental block.

obduction. Restrict † *Obs.* to sense in Dict. and add: **2.** *Geol.* The movement of a litho-

spheric plate sideways and upwards over the margin of an adjacent plate.

1971 R. G. Coleman in *Jrnl. Geophysical Res.* LXXVI. 1216/2 According to Davies.., the ophiolites represent a slab of oceanic crust and mantle emplaced in Cretaceous or Eocene time by overthrusting (obduction) oceanic crust onto the continental crust. **1972** *Rep. 24th Internat. Geol. Congr.* iii. 409 The obduction zone is..an upthrust of the oceanic crust and mantle..mostly of a marginal sea (small ocean basin). **1972** *Nature* 31 Mar. 222/2 The Alpine orogeny culminated with the collision between the continental plates of Eurasia and Africa and was probably preceded by subduction or obduction of the Tethyan plate along the European continental margin. **1975** *Ibid.* 20 Feb. 615/2 Ophiolite emplacement is integrally associated with plate subduction, either by the accretion of oceanic crust on to the wall of the upper plate (in oceanic trenches) or by the bodily thrusting (obduction) of the oceanic crust on to a continental margin.

obeah. Add to etym.: 'Also the base of Twi ɔ-bayifó, witch, wizard, sorcerer (more literally sorcery-man, "obeah-man", since *-fó* means *person*).' (Cassidy & Le Page *Dict. Jamaican Eng.* 1967).

2. Delete 'and formerly' and read 'a form of which survives in the West Indies and neighbouring countries'. (Earlier and later examples.)

1760 *Jrnls. Assembly Jamaica* 16 Dec. (1798) V. 245/1 The engrossed bill to remedy the evils arising from irregular assemblies of slaves..and for preventing the practice of *obeah*. **1930** J. J. Williams *Hebrewisms W. Afr.* 17 The word Obeah itself is really the Ashanti Obayifo, a witch or rather more properly,..a wizard, being derived from bayi, sorcery. **1934** J. Rhys *Voy. in Dark* iii. iv. 192 Anne Chewett used to say that it's haunted and obeah—she had been in gaol for obeah. **1934** *Times Lit. Suppl.* 19 July 502/4 On the subject of obia (which a Jamaican negro will never mention) they are quite unreserved. **1954** *Caribbean Q.* III. i. 5 Obeah has been prohibited by law, and is always spoken of with a laugh; but the laughter is..never sneering. **1957** *Times Lit. Suppl.* 2 May 237/3 You have also to reckon with *obeah* (in the shape of a mongrel puppy) with the other candidate Preacher, who wickedly distorts the slogan 'Vote Harbans or Die'. **1963** G. J. McCall in A. Dundes *Mother Wit* (1973) 420/1 'Hoodoo'..corresponding to *vodun* ('voodoo') and *obeah* in Haiti.

3. (Later examples.)

1917 *Chambers's Jrnl.* Apr. 248/1 No treasure was found, and the Obeahman had disappeared just when he was most wanted to counsel and guide. **1934** J. Rhys *Voy. in Dark* iii. iv. 192 Obeah-women who dig up dead people. **1970** J. Brown *Un-Melting Pot* vii. 99 Though St Lucia is 92 per cent Catholic,..obeah beliefs and practices still pervade its whole fabric of life. **1973** *Sunday Express* (Trinidad & Tobago) 1 Apr. (Suppl.) 12/2 Sometimes they would visit Rattan, the famed obeah man..and ask him to cast a good spell on them. **1974** *Practitioner* Dec. 848 An Obeahman or Obeahwoman may also be consulted about social matters, including a child's education.

obeah *v.* (later example); **obeahism** (later examples).

1902 *Chambers's Jrnl.* Feb. 82/1 The vitality of *obeahism* is surprising. **1949** *Caribbean Q.* I. ii. 45 Mr. Waugh manages to feature an immigrant London crooner ..and an obeahed French colonial official, in his treatment of..St. Lucia, Dominica, and Martinique. **1972** *Guardian* 6 Dec. 13/1 Obeahism, the Jamaican form of voodooism..that originated in Haiti.

obeche (obī·tʃī). [Bini name in Nigeria.] A large West African tree, *Triplochiton scleroxylon*, of the family Sterculiaceæ, found in lowland forests; also its light-coloured timber. Also *attrib.*

1908 H. N. Thompson in *Kew Bull.* 195 'Satinwood', 'Obeche'... At the Liverpool Market..it was classified as Satinwood... It is a very common tree. **1934** *Jrnl. R. Aeronaut Soc.* XXXVIII. 56 One of the most remarkable instances of clean fracture is in the West African Obeche. .. This wood is relatively new to commerce, and is being used..for motor-body work. **1956** *Handbk. Hardwoods* (Forest Prod. Res. Lab.) 171 Obeche grows to a height of 150 ft... Obeche is nearly white to pale straw in colour with no clear distinction between sapwood and heartwood. **1958** *Archit. Rev.* CXXIV. 41/1 The wall is of a natural obeche timber. **1965** W. Soyinka *Road* 27 And high class timber kid. High class. Golden walnut. Obeche. Ironwood. **1971** *Country Life* 1 Apr. 731/2 My obeche wood decoys have been admired a great deal in the five or six years since I made them. **1972** *Timber Trades Jrnl.* 13 May 47/2 Among the species displayed by Lathams were two hardwoods—guarea and obeche—which were comparatively unknown at the time [*sc.* 1925].

obedience. Add: **5. obedience class, test, trial**, a competition designed to test a dog's obedience; **obedience training**, the process of teaching a dog to obey orders; hence (as a back-formation) **obedience-train** *v.*

1930 E. C. Ash *Practical Dog Bk.* ii. 21 Obedience classes are held at Cruft's Show, and are always an occasion of considerable interest. **1936** J. Z. Rine *Dog Owner's Manual* xii. 195 We see these blue ribbon dogs carrying off prizes also in obedience tests. **1961** J. Holmes (*title*) Obedience training for dogs. **1971** 'L. Egan' *Malicious Mischief* (1972) i. 15 'If your dog had been obedience-trained he would not have been stolen so easily... An obedience-trained dog is impossible to steal or poison.'.. 'Obedience training? What does that mean?' **1945** C. L. B. Hubbard *Observer's Bk. Dogs* 105 The breed [*sc.* the Monkey Terrier] is surprisingly intelligent and..should do well in Obedience Trials. **1971** 'L. Egan' *Malicious Mischief* (1972) iii. 40 Most of the big bench shows have obedience trials.

obedient, *a.* (*sb.*) Add: **5. obedient plant** = *physostegia.

1948 F. Perry *Herbaceous Border* v. 103 *P*[*hysostegia*] *virginiana*, sometimes known as the Obedient Plant because the individual sage-like blossoms on the flower spikes may be moved from side to side and remain as placed. **1971** J. Raven *Botanist's Garden* xi. 201 A curious plant, of better value perhaps for the entertainment it affords children than for beauty, called the obedient plant or *Physostegia virginiana*.

obeisantly, *adv.* (Later example.)

1902 *Westm. Gaz.* 2 July 2/3, I..came in turn Of him myself obeisantly to learn.

obelia (obī·liă). [mod.L. (Peron & Lesueur 1809, in *Ann. du Muséum d'Hist. Nat.* XIV. 355), f. Gr. ὀβελ-ός + -ia¹.] A marine colony-forming coelenterate of the genus so called, belonging to the class Hydrozoa.

1868 T. Hincks *Hist. Brit. Hydroid Zoophytes* I. 148 Some of his [*sc.* Edward Forbes's] species are only various stages of one and the same *Obelia*. **1897** Parker & Haswell *Text-bk. Zool.* I. iv. 118 Obelia is a common zoophyte occurring in the form of a delicate, whitish or light brown, almost fur-like growth on the wooden piles of piers and wharfs. **1927** Haldane & Huxley *Animal Biol.* ix. 179 If an ordinary hydroid polyp like Obelia be kept in the laboratory, the hydranths, or separate organized individuals of the colony, will (unless the water is well aerated artificially) show a curious series of changes. **1974** A. Silverstein *Biol. Sciences* 178/2 *Obelia*, a typical colonial coelenterate, has a number of specialized feeding polyps and reproductive polyps, growing in a branchlike arrangement.

obelion. Substitute for etym.: [ad. F. *obélion* (P. Broca 1875, in *Bull. de la Soc. d'Anthrop. de Paris* X. 356), f. Gr. ὀβελιαῖος sagittal (given in Broca's paper as ὀβελαῖος): see *-ion².] Hence also **obe·lial** *a.*

1890 H. Allen *Clin. Study Skull* 52 The parietal foramina lie on the sides and serve as guides to this the obelial portion.

obeliscoid (ǫbĭli·skoid), *a.* [f. Obelisk + -oid.] Resembling an obelisk in form; obelisk-shaped; obeliscal.

1877 W. R. Cooper *Short Hist. Egypt. Obelisks* v. 25 An obeliscoid monolith originally erected by Osirtesen. **1901** A. J. Evans in *Jrnl. Hellenic Stud.* XXI. 173 The obeliskoid pillar of the Cretan ring.

obelisk, *sb.* (*a.*) **B.** as *adj.* For † *Obs.* read 'rare', and add later example.

1922 Joyce *Ulysses* 45 Their pushedback chairs, my obelisk valise, around a board of abandoned platters.

obeophone (ōu·biofōun). [first element uncertain + *-phone.] A type of orchestrina (Orchestrina a) (see quot. 1927). (*Disused.*)

1927 H. E. Wortham *O. Browning* xiii. 234 For the performance of chamber music he possessed a number of *orchestrine di camera*, familiarly known as 'obeophones', which represented the wood-wind, or even supplemented the strings. **1940** V. Woolf *Roger Fry* ii. 49 The host himself pedalled away at the obeophone.

oberek (obe·rėk). [Polish.] A lively Polish dance in triple time, related to the mazurka.

1938 *Oxf. Compan. Mus.* 624/2 *Oberek*, a type of Polish dance. **1952** H. Wolska *Dances of Poland* 33 Oberek. *Region*—Mazovia... *Character*—Gay and vigorous. *Formation*—Couple Dance. **1958** [see *Krakowiak]. **1976** *Times* 23 July 11/3 The Mazowsze Song and Dance Company from Poland..whirl through oberek and mazurka, polka and Krakowiak.

obertas (obēə·ɹtăs). Also **obertass.** [Polish.] = *oberek.

1889 Grove *Dict. Mus.* IV. 733/1 *Obertas*, this is described in the 'Encyklopedyja Powszechna'..as the most popular of Polish national dances. **1895** L. Grove et al. *Dancing* vii. 234 The *Obertas*, one of the most popular of national dances, is a variation of the *Mazur*. **1938** *Oxf. Compan. Mus.* 270/1 It [*sc.* a drabant] began with a solemn march and then changed to an Obertass. **1944** W. Apel *Harvard Dict. Mus.* 500/1 Chopin's Mazurka op. 56, no. 2 is in the character of an obertas. **1954** *Grove's Dict. Mus.* (ed. 5) V. 641/2 The Obertas or Oberek.

obey, *v.* Add: **1. f.** Naut. phr. *obey orders, if* (*though*) *you break owners*, obey orders, even when they are wrong.

1840 R. H. Dana *Two Yrs. before Mast* xvii. 92 It almost broke our poor darky's heart when he heard that Bess [a pet pig] was to be taken ashore... 'Obey orders, if you break owners!' said he..and lent a hand to get her over the side. **1849** H. Melville *Redburn* I. vi. 57 The motto is, 'Obey orders, though you break owners'. **1915** J. E. Patterson *Epistles from Deep Seas* xiv. 300 There was the unwritten shipboard law: 'Obey orders, even if you break owners.' **1924** R. Clements *Gipsy of Horn* iii. 50 What could be sounder than 'Obey orders, if you break owners'—meaning, do as you're told, even if you know it's wrong.

obeyance. Delete † *Obs.* and add later examples.

1921 C. E. Mulford *Bar-20 Three* x. 118 The obeyance of the order might possibly be accepted by the crowd as grounds for justification. **1939** Joyce *Finnegans Wake* iii. 540 Obeyance from the townsmen spills felixity by the toun. **1950** *Tablet* 9 Dec. 504/1 Erecting pointed arches in blind obeyance of mechanical efficiency.

‖ **obi³** (ōu·bi). *W. Afr.* [Igbo.] In Nigeria, a native hut.

1931 *Discovery* May 154/1 The more important natives have what is called an *obi* house, which is practically a shrine to the family gods and ancestors. **1937** C. K. Meek *Law & Authority in Nigerian Tribe* iii. 62 It [*sc.* a wooden pillar] is fixed inside the householder's *obi* (entrance hut) facing outwards. **1958** C. Achebe *Things fall Apart* ix. 69 'Where do you sleep with your wife, in your *obi* or in her own hut?' asked the medicine-man. **1962** —— in F. Ademola *Reflections* 24 Their grand-father..was waiting in his Obi when his grand-children arrived.

‖ **obi⁴** (ōu·bi). *W. Afr.* Also **Obi.** [Ibo: see quot. 1958.] A king of the Onitsha people of Nigeria.

1937 C. K. Meek *Law & Authority in Nigerian Tribe* x. 219 At Onitsha it was a capital offence for any one to have sexual relations with a wife of the Obi. **1958** J. S. Coleman *Nigeria* i. 28 The *obi* (an Ibo term—of likely Yoruba origin—for king or chief) was appointed by the Oba of Benin. **1973** *Times Lit. Suppl.* 9 Mar. 258/3 It is these sons who today are trying to rediscover..their ancient customs and, in particular, those relating to the office of King (Obi) of Onitsha.

Obie (ōu·bi). *U.S. Theatr.* [repr. pronunc. of *OB*, colloq. abbrev. *off-Broadway *a.* and *sb.*] One of a number of annual awards for off-Broadway experimental theatre productions. Also *attrib.*

1967 *National Observer* (U.S.) 10 Apr. 20/1 Last year several 'Obies', the prize theater awards normally reserved for the stalwarts of Off-Broadway, were given to OOB [*sc.* off-off-Broadway] veterans, among them actor Kevin O'Connor and playwright Sam Shepard. **1970** *Time* 12 Jan. 37 Meanwhile he was acting (six Broadway shows, 25 off-Broadway), collecting two Obies for off-Broadway performances, [etc.]. **1972** *Village Voice* (N.Y.) 1 June 54/2 Sharon Thie's play with its Open Theatre performers, including Sharon Gans (Obie winner for her performance as November). **1973** *Black World* Apr. 20/2 The next Proscenium production of Black significance was Derek Walcott's Obie Award-winning *Dream On Monkey Mountain.*

obit, *sb.* **1. b.** Delete *arch.* and add: (Later examples.) In mod. *colloq.* (esp. *journalists'*) use usu. regarded as an abbrev. of Obituary *sb.*

1874 *Athenæum* 12 Sept. 353 The sub-editor of a New York daily newspaper wrote to me begging me to send him the proper materials for the construction of an obit. He said it was the custom of his journal to keep obits in readiness. **1899** C. Plummer *Two Saxon Chrons. Parallel* II. p. lxiv, Then comes a period, 893–958, during which E and F are almost barren, containing only a few obits, [etc.]. **1901** Farmer & Henley *Slang* (1902) V. 85/1 *Obit*..(journalists') an obituary notice. **1935** *Atlantic Monthly* Jan. 43 (*title*) Obit for E. Harris. **1953** *N.Y. Times* 9 Aug. 8/3 This is not the obit page. **1957** D. Bethurum *Homilies of Wulfstan* 64 Wulfstan's obit is marked in MS Hatton 113, in an Ely calendar, and in the E and F *Chronicle*, 1023. **1964** W. R. Nash *How Newspapers Work* iv. 74 The 'obits'..are revised at regular intervals. **1972** *Daily Tel.* (Colour Suppl.) 11 Aug. 5/1 The obituarists polish their obits. **1975** B. Meggs *Matter of Paradise* (1976) iv. iv. 96 Doc had been given a very nice obit coverage on page 42.

obiter, *adj.* and *sb.* Add: (Later examples.) As *sb.* = *obiter dictum*.

1927 *Daily Tel.* 19 July 9/2 Lord Justice Scrutton recalled a recent obiter by Mr. Justice Eve to the effect that [etc.]. **1957** G. Schwarzenberger *Internat. Law* (ed. 3) I. xxiv. 437 The Commission's observations..were strictly *obiter dicta*. As, however, the reformulation of Article 18 ..followed the line taken in this Award, the view expressed by the Commission, though *obiter*, deserves not to pass unnoticed. **1959** 'W. Haggard' *Venetian Blind* ii. 24 Mr Justice Downderry refused an injunction. He did more. He made it very clear that his remarks were *obiter*, but he was exquisitely acidulous.

obitual, *a.* and *sb.* Add: **A.** *adj.* (Further U.S. example.)

1893 *Nation* (N.Y.) 30 Nov. 406/3 Obitual days constitute an important, distinctive, and ever recurrent feature in the proceedings of our national Legislature.

obituarian (obi:tiu,ēə·riăn). *U.S.* [f. Obituary + -ian.] = Obituarist.

1909 in *Cent. Dict.* Suppl. II. 882/3 There is one characteristic story to be told about Robert Louis Stevenson which his obituarians have missed, probably because they knew nothing about it. **1971** *Sat. Rev.* (U.S.) 6 Nov. 44/3 Alden Whitman is the obituarian for The New York Times.

obituarist. Add: (Later examples.)

1905 M. Beerbohm *Around Theatres* (1953) 399 The obituarists seem hardly to do justice to the intensely interesting personality of Irving in private life. **1930** A. Huxley *Brief Candles* 4 'Metaphysically and artistically a cretin.' 'The obituarist doesn't seem to be of your opinion.' **1961** P. Fleming *Bayonets to Lhasa* 295 'The

man' wrote one of his obituarists 'was greater than his message.' **1972** [see *obit *sb.* 1 b]. **1974** *Punch* 3 Apr. 550/3 He joins the staff of this magazine as assistant obituarist. **1978** P. Sutcliffe *Oxf. Univ. Press* iii. iii. 82 When he died in 1904 *The Times* obituarist attributed his failure to write in later life to the fact that he was unmethodical.

obituarize *v.*, also *trans.*

1969 *Observer* 21 Dec. 28/2 Stand by for a barrage of TV programmes obituarising the Sixties. **1972** *Times Lit. Suppl.* 31 Mar. 352/2 Evelyn Waugh, obituarizing Duggan, wrote [etc.].

obituary, *sb.* and *a.* Add: **A.** *sb.* **1.** (Later example.)

1952 *Latin Liturg. MSS.* (Exhib. Guide, Bodl. Libr.) 35 *Obituary* (and martyrology) Secular; cathedral of St. Ethelbert, Hereford; mid-14th century.

obituarily *adv.*: delete *rare*⁻⁰ and add examples.

1889 G. B. Shaw *London Music 1888–89* (1937) 61 Madame Ilma de Murska is dead; and an ungrateful world is describing her obituarily as a person remarkable for a compass that extended to F in alt. **1902** *Westm. Gaz.* 1 Aug. 1/3 Whatever may happen 'obituarily' or otherwise ..to the organisation and officials making these 'distributions', they have not ceased. **1974** *Daily Tel.* (Colour Suppl.) 22 Feb. 7/3 Perhaps the generation of the Seventies, will be willing to accept the notion of dressing the deceased (obituarily) in the same garments he wore in life, gravy stains and all.

object, *sb.* Add: **3. c.** *object of art* = *objet d'art* (*objet 3). Also *object of art and virtu* (see sense *3d and cf. *objet 6).

1862 E. Hall *Diary* 5 June in O. A. Sherrard *Two Victorian Girls* (1966) 294 Went..to a private view of objects of art at South Kensington..the Wedgwood ware exquisite. **1863** Mrs. Gaskell *Dark Night's Work* x. 180 The beautiful pictures and other objects of art in the house. **1879** C. Schreiber *Jrnl.* 6 Dec. (1911) II. 250 As an object of art, it is vile, but in an antiquarian point of view, most curious. **1894** G. du Maurier *Trilby* I. i. 66 I've brought you these objects of art and virtu to make the peace with you. **1918** A. Bennett *Roll Call* i. viii. 152 The perfunctory accents in which she had catalogued her objects of art. **1925** Conrad *Suspense* ii. ii. 96 Cosmo looked at it with appreciation, as if it had been an object of art. **1973** W. Just *Congressman who loved Flaubert* 119 The columnist treated Caroline as a singular object of art, a serene and delicate event.

d. *object of virtu* (see Virtu, vertu 1 c). (Further examples.)

1914 A. Huxley *Let.* 13 May (1969) 59, I hope things at Eastbourne are more or less settled now... I gather that 27 will fairly burst with bits of objects of vertu and utility. **1970** S. J. Perelman *Baby, it's Cold Inside* 54 The objects of virtu rifled from a hundred auction rooms. **1971** *Daily Tel.* 11 May 10/5 The sale, of miniatures and objects of vertu brought £15,919. **1974** *Ibid.* 9 May 6/8 The sale of miniatures, gold watches, enamels and objects of vertu, totalled £37,294.

5. *the object of the exercise*: see *exercise *sb.* 8 h.

b. *no object*, not a thing aimed at or regarded as important to obtain. Freq. also used of distance, expense, etc., not taken into account or forming no obstacle.

See C. T. Onions in *S.P.E. Tract* (1930) xxxvi. 531–4. **1782** *Morning Herald* 20 May 4/2 (Advt.), A Gentlewoman..wishes to superintend the family of a single Gentleman or Lady..and salary will be no object. **1796** *Deb. Congress U.S.* 7 Apr. (1849) 878/2 Enjoying..unexampled prosperity,..the expense of completing the frigates could be no object to the country. **1800** *Morning Herald* 4 Jan. 4/2 (Advt.), Wanted, in Chatham-place or New Bridge-street, a roomy convenient House... Rent no object, if the house is agreeable. **1855** *Poultry Chron.* III. 67/2 Where every convenience is obtainable, and expense no object. *a* **1864** R. S. Surtees *Mr. F. Romford's Hounds* (1865) i. 4 Money! Money would be no object to him! He'd give anything for a good horse! **1871** *English Mechanic* 20 Jan. 431/1 The colour of the solder is no object, as the joint will be hidden. **1873** J. H. Beadle *Undevel. West* xxxv. 762 With one team to each family (time being no object to such people) it cost them nothing to move. **1886** *Encycl. Brit.* XX. 228/1 Only those travel who travel by necessity, or to whom money is no object. **1891** Mrs. J. H. Riddell *Mad Tour* 3 The time when distance was, as the advertisements say, 'no object'. **1909** 'O. Henry' *Roads of Destiny* iv. 62 She..gave her *a la carte* to fit me out—money no object. **1916** G. B. Shaw *Pygmalion* i. 116 I'm going home in a taxi... Eightpence aint no object to me, Charlie. **1926** A. Bennett *Lord Raingo* ii. lxxii. 326 'I'm thinking of going back to town now, sir.. Unless of course you'd like me to stay.' 'No object in staying,' Sam murmured, as if in disgust. **1930** *London Mercury* Nov. 45 Distance being no object..scenes in Siam can be.. transmitted. **1956** B. Holiday *Lady sings Blues* (1973) xiv. 121 We looked for the best private sanatorium around. Money was no object. **1964** R. Braddon *Year Angry Rabbit* xii. 109 Each side, having already bankrupted itself to finance this war, declared that money was no object.

7. (Examples of *object clause*.) Also *object-case* = Accusative *a.* 1; *object complement*, a word, usu. a noun or adjective, which complements the object of a verb, expressing the state or condition of the object at the time of, or resulting from, the action; cf. **objective complement*; *object-pronoun*, a pronoun, esp. a relative pronoun, which is the object of a verb or which introduces an object clause.

1870 W. W. Goodwin *Elem. Greek Gram.* iii. 167 Object clauses depending on verbs signifying *to strive for, to care for, to effect*, regularly take the future indicative after both primary and secondary tenses. **1885** Object clause [used s.v. Ask *v.* 5 a]. **1904** C. T. Onions *Adv. Eng. Syntax* 92 The large majority of verbs took the Accusative as Object, and thus there was a tendency for the Accusative to become the universal Object-case. **1906** G. R. Carpenter *Eng. Gram.* ii. 25 An object complement is a noun or adjective completing the meaning of a transitive verb. **1927** E. A. Sonnenschein *Soul of Grammar* 10 There is no ambiguity in sentences like the following, though the object-case has the same *form* as the subject-case: 'the lion beat the unicorn'. **1957** R. W. Zandvoort *Handbk. Eng. Gram.* iii. vi. 165 *What*..may introduce a subject clause, an object clause. **1960** T. F. Mustanoja *M.E. Syntax* I. 205 Non-expression of the object-pronoun in a relative clause has not been attested in OE. **1963** F. T. Visser *Hist. Syntax Eng. Lang.* I. iv. 550 Since these added adjectives or nouns do not affect the meaning of the verb, but are merely adjuncts to the object the term 'object complement' or 'objective complement' ..seems preferable to the appellation 'predicative adjunct'. **1964** *English Studies* XLV. 386 The rapidly developing 'periphrastic genitive', in which object-case pronouns *h-r* and *h-m* (after *of*) contrasted. **1966** *Ibid.* XLVII. 55 Of special importance is the absence of the relative object-pronoun (e.g. *The spirits I have raised* abandon me), one of the most frequent idioms in coll. English. *Ibid.* 253 A third argument against calling the *that*-clause an object clause is that it lacks the noun characteristic of forming prepositional adjuncts.

10. *object-directed* adj. (so *object-directedness*); **object chart**, a chart for use in object lessons; **object choice** *Psychol.*, something external to the ego chosen as a desirable object; **object-lesson** (earlier and later examples of *fig.* use); **object libido** *Psycho-anal.*, that part of psychic energy which is directed to objects other than the ego; **object love**, love for something external to the ego or self; **object program** *Computers*, a program into which some other program is translated by an assembler or compiler; cf. *object language 3; **object-relation, -relationship** *Psychol.*, a relationship felt, or the emotional energy directed, by the self or ego towards a chosen object; also *attrib.*; **object-system**, the system of teaching by object-lessons; **object-teaching** (examples); **object-white** *Billiards*, the white object-ball; **object word**, a word which designates an object or material thing; *spec.* in the theories of Bertrand Russell, a word the meaning of which can be learnt independently of the rest of the linguistic system; **object-world**, the world external to the self, apprehended through the objects in it.

1872 *Rep. Indian Affairs 1871* (U.S.) 306 A new and original series of 'object charts' gotten up expressly for the Indians of Oregon by myself. **1920** *Internat. Jrnl. Psycho-Anal.* I. 137 Such motivation of the homosexual object-choice must be by no means uncommon. **1948** M. Klein *Contrib. Psycho-Anal.* 121 His homosexual object choice at the narcissistic level. **1965** P. L. Giovacchini in B.L. Greene *Psychother. Marital Disharmony* 43 The spouse, representing a heterosexual object choice, would ideally be associated with ego transactions. **1960** W. V. Quine *Word & Object* vii. 239 Some of us are carried away by the object-directed pattern of our thinking. **1973** *Jrnl. Genetic Psychol.* CXXII. 264 Six categories of object-directed behavior. **1963** A. Kenny *Action, Emotion & Will* 195 How..can Brentano say that object-directedness is peculiar to psychological phenomena? **1881** 'Mark Twain' *Prince & Pauper* xii. 115 In the times of which we are writing, the Bridge furnished 'object lessons' in English history. **1936** N. Streatfeild *Ballet Shoes* xi. 180 It was an object lesson she might remember always. **1977** *Early Music* July 401/3 Saul Novak gives an object-lesson in linear analysis for early music. **1920** *Internat. Jrnl. Psycho-Anal.* I. 170 Paraphrenia differs from the psychoneuroses in that the object-libido is re-converted into ego-libido. **1933** A. A. Brill in S. Lorand *Psycho-Anal. Today* 108 We thus distinguish an ego libido and an object libido. **1955** J. Strachey et al. tr. *Freud's Compl. Psychol. Wks.* XVIII. 257 The transformation of object-libido into narcissism necessarily carried along with it a certain degree of desexualization. **1964** H. Hartmann *Ess. Ego Psychol.* x. 185 It has sometimes been said..that the loss of object libido destroys the repressions. **1918** E. Jones *Papers on Psycho-Anal.* (ed. 2) xviii. 332 The transference..like every 'object love', has its deepest root in the repressed parent-complex. **1924** J. Riviere et al. tr. *Freud's Coll. Papers* II. 83 The sexual instinct passes on from auto-erotism to object-love. **1938** *Times Lit. Suppl.* 26 Feb. 132/4 The stage of 'object-love', when the moral superego takes charge and the ego is no longer coercive but submissive. **1968** A. H. Modell (*title*) Object love and reality. **1959** M. H. Wrubel *Primer of Programming* vi. 128 At the end of the assembly or translation phase.., the programmer is presented with a machine-language deck [of cards] called the 'object program'. **1970** O. Dopping *Computers & Data Processing* xix. 304 In the most common systems for automatic coding, the translation from source program to object program is done in a separate computer run, called the compilation run. **1968** L. Breger in J. Marmor *Mod. Psychoanal.* 47 The concept of 'object relation' (that is, the idea that there is a fixed quantity of energy in a more or less closed system, so that if libido is 'invested' in one object it is 'unavailable' for other purposes). **1968** R. & G. Blanck *Marriage & Pers. Devel.* vi. 70 Persons on the need-gratifying level of object relations can change partners..readily. **1970** *Jrnl. Analytical Psychol.* XV. i. 7 True object relations are the goal of our efforts towards insight into our emotions. **1926** *Brit. Jrnl. Med. Psychol.* VI. 292 The hetero-sexual stage, which is the most complete form of allo-erotic object relationship. **1946** *Internat. Jrnl. Psycho-Anal.* XXVII. 31/2 From the point of view of object-relationship psychology, explicit pleasure-seeking represents a deterioration of behaviour. **1974** K. Lambert in M. Fordham et al. *Technique in Jungian Analysis* iii. 312 This annuls the 'object relationship' between the patient and the analyst. **1869** C. L. Brace *New West* vi. 75 The improvement which we have sought so much to bring before the public in New York..—the 'Object System'—has already been adopted here. **1878** *Harper's Mag.* Mar. 607/2 This school is too large for strictly Kindergarten Teaching; but the 'object system'..was the one adopted. **1860** H. Barnard (*title*) Object teaching and oral lessons on social science and common things. **1945** C. V. Good *Dict. Educ.* 411/2 *Teaching, object*, a method of elementary-school teaching derived from the work of Pestalozzi in Europe and introduced into the United States at the Westfield, Massachusetts, State Teachers' College in 1848 and at Oswego, New York, in 1861. **1904** Mannock & Mussabini *Billiards Expounded* I. iii. 97 To enable the object-ball to go on to the baulk cushion and return up by the object-white. **1907** *Westm. Gaz.* 19 June 7/2 He got the red ball against the top cushion,..and..the object-white against the side cushion. **1914** L. Bloomfield *Introd. Study Lang.* iv. 111 The explicit predication of quality or action is impossible for languages in which every word expresses an object... In these languages the sentence consists of one or more object-words. **1940** B. Russell *Inquiry into Meaning & Truth* 80 'Object words' are defined, logically, as words having meaning in isolation, and, psychologically, as words which have been learnt without its being necessary to have previously learnt any other words. **1953** *Mind* LXII. 10 Harm has been done by the well-meaning distinction between object words and logical words... An object word such as 'table'..has meaning in isolation from other words. **1954** E. Branth tr. *H. Spang-Hanssen's Recent Theories Nature of Lang. Sign* in *Travaux du Cercle Ling. de Copenhague* IX. 75 Even in the presence of an object, an object-word will only have an extremely vague 'meaning' if it did not—through its oppositions—concentrate its meaning on a particular 'property' in the object. **1964** E. A. Nida *Toward Sci. Transl.* iv. 62 Basically there are four principal functional classes of lexical symbols: object words, event words, abstracts, and relationals. **1880** G. M. Hopkins *Sermons & Devotional Writings* (1959) ii. i. 127 Part of this world of objects, this object-world, is also part of the very self in question. **1904** W. James *Coll. Ess. & Rev.* (1920) 446 There is..no account of the fact (which I assume the writers to believe in) that different subjects share a common object-world. **1934** W. Temple *Nature, Man & God* i. vi. 135 That discussion will be primarily concerned neither with the inner life of mind, conceived as separate from environment, nor with the object-world which mind apprehends and contemplates, but with the interrelation of these two. **1948** M. Klein *Contrib. Psycho-Anal.* 313 By being internalized, people, things, situations and happenings..cannot be verified by the means of perception which are available in connection with the tangible and palpable object-world. **1964** Gould & Kolb *Dict. Social Sci.* 313/2 Our understanding and its object-world.

object, *v.* Add: **4. b.** Also with direct speech. (Later examples.)

1972 D. Bloodworth *Any Number can Play* xvi. 154 'But it would have been..a pointed piece of skin,' objected Green. **1974** *Listener* 3 Oct. 423/2 Mr Johnston objected: 'But we already have..a shop stewards' movement.'

objectant. Delete † *Obs. rare* and add: In recent use as a legal term in the U.S.

1972 *N.Y. Law Jrnl.* 22 Aug. 6/4 The proponent or the objectant seeks an examination of the adverse party before trial. **1973** *Ibid.* 19 July 13/5 In view of the failure of the objectant to proceed with this matter, his objections are denied.

objectifiable (ǫbdʒe·ktifəi:ăb'l), *a.* [f. Objectify *v.* + -able.] That is capable of being objectified.

1925 C. D. Broad *Mind & its Place* vi. 306 To be 'epistemologically objectifiable' means to be capable of corresponding to the epistemological object of some referential situation.

objectification. (Further examples.)

1883 Haldane & Kemp tr. *Schopenhauer's World as Will & Idea* I. ii. 121 (*heading*) The objectification of the will. **1927** A. N. Whitehead *Symbolism* (1928) i. 30 Thus 'objectification' itself is an abstraction. **1931** W. R. B. Gibson tr. *Husserl's Ideas* ii. ii. 123 Thanks to this objectification we find facing us in the natural setting.. not natural things merely, but values and practical objects of every kind. **1963** J. Macquarrie *Twentieth-Century Relig. Thought* xii. 203 Berdyaev goes further..in his antipathy to objectification, which he identifies with the fall of man. **1971** E. B. Ashton tr. *Jaspers's Philos.* III. i. 8 Metaphysical objectifications of transcendence.

objectified, *ppl. a.*, **objectifying**, *vbl. sb.* (Later examples.)

1927 A. N. Whitehead *Symbolism* (1928) i. 30 No actual thing is 'objectified' in its 'formal' completeness. **1931** W. R. B. Gibson tr. *Husserl's Ideas* ii. ii. 122 The intentional object first becomes an apprehended object through a distinctively 'objectifying' turn of thought. **1940** S. C. Pepper in P. A. Schilpp *Philos. Santayana* 229 The expression of moral and political greatness, however, is the satisfaction of interests quite different from objectified pleasure. **1977** R. Williams *Marxism & Lit.* ii. ii. 86 Here society is the objectified (unconscious and unwilled) general process.

objectifier (ǫbdʒe·ktifəi:əɹ). [f. Objectify *v.* + -er¹.] That which objectifies or makes objective.

1956 J. D. Mabbott in H. D. Lewis *Contemp. Brit. Philos.* (ser. 3) 307 Morality has to be sustained by conviction and conviction is a great objectifier.

objection. Add: **1. a., b.** *spec.* in *Horse Racing.*

1898 *Encycl. Sport* II. 228/1 All disputes, objections, and appeals referred to or brought before the Stewards of the Jockey Club for their decision, shall be decided by the three Stewards. **1930** *Daily Express* 6 Oct. 17/6 An objection to Kippit Lore on behalf of Ferry Maid was overruled. **1977** 'J. Le Carré' *Hon. Schoolboy* vii. 165 Objection!.. Where's the Stewards..? That horse was pulled!

objective, *a.* and *sb.* Add: **A.** *adj.* **3. a.** (Earlier and later examples.)

1838 J. S. Mill in *London & Westm. Rev.* Aug. 496 An essentially *objective* people, like those of Northern and Central Italy. **1967** H. Arendt *Orig. Totalitarianism* (new ed.) xii. 423 The Jews in Nazi Germany or the descendants of the former ruling classes in Soviet Russia were not really suspected of any hostile action; they had been declared 'objective' enemies of the regime in accordance with its ideology.

7. b. *objective complement* = *object complement* (*object *sb.* 7).

1870 C. P. Mason *Eng. Gram.* (ed. 14) 127 When the verb is transitive, and in the active voice, the complement of the predicate stands in the attributive relation to the object of the verb; as, 'He dyed the cloth red.'.. This kind of complement may be termed the *Objective Complement*, inasmuch as it is closely connected with the object of the verb. **1897** Clarke & Muller *Class Bk. Eng. Gram.* 219 The name *Objective Complement* would be applied by some Grammarians [to 'captain' in the sentence 'They elected *James captain*']. **1945** M. M. Bryant *Functional Eng. Gram.* xii. 132 In 'They made him chairman'..the second complement is called an objective complement. *Ibid.*, In the sentence 'The frost turned the leaves red', the objective complement is of a somewhat different nature, since here the word *red* is an adjective rather than a noun. **1963** [see *object *sb.* 7].

B. *sb.* **5.** *attrib.*, as **objective function**, in linear programming, the function that it is desired to maximize or minimize.

1949 *Econometrica* XVII. 207 The optimum feasible program is that feasible program which maximizes a specified linear objective function. **1958** Riley & Gass *Linear Programming* i. 5 The linear-programming problem has a linear function of the variables to aid in choosing a solution to the problem. This linear combination of the variables, called the objective function, must be optimized by the selected solution. **1969** D. C. Hague *Managerial Econ.* i. i. 16 We shall find that the aim of the business is set out in what is called an objective function—a name which indicates its purpose exactly.

objective correlative. [f. Objective *a.* 2 b + Correlative *sb.* 1.] Term applied by T. S. Eliot to the technique in art of representing or evoking a particular emotion by the presentation of physical symbols of it, as surroundings, situations, sets of objects, etc., which become indicative of that emotion and are associated with it.

[**1850** W. Allston *Lect. Art, & Poems* 16 No possible modification in the degrees or proportion of these elements [*sc.* air, earth, heat, water] can change the specific form of a plant... So, too, is the external world to the mind; which needs, also, as the condition of its manifestation, its objective correlative.] **1919** T. S. Eliot in *Athenæum* 26 Sept. 941/1 The only way of expressing emotion in the form of art is by finding an 'objective correlative'; in other words, a set of objects, a situation, a chain of events which shall be the formula of that *particular* emotion; such that when the external facts, which must terminate in sensory experience, are given, the emotion is immediately evoked. **1946** *Sewanee Rev.* LIV. 301 In that Eliot proposes the objective correlative, 'he accepts with the vast majority of his contemporaries the modern dogma that the artist is primarily concerned with emotion'. **1947** T. S. Eliot *Milton* 7 Two or three phrases of my coinage—like 'objective correlative'—which have had a success in the world astonishing to their author. **1951** H. Kenner *Poetry E. Pound* viii. 66 It is easy to see why the objective correlative, the image as sensory equivalent of an emotion, the Aristotelian equation of a poem with an action, and Eliot's claim that emotions the poet has never experienced will serve his turn as well as familiar ones, should seem in such eyes impossibly muddle-headed. **1955** D. Davie *Articulate Energy* vii. 86 The use of the objective correlative, the invention of a fable or an 'unreal' landscape. **1957** Wimsatt & Brooks *Lit. Crit.* 676 A realization that Winters' conception of poetry, like Eliot's, is ultimately 'dramatic' need not impugn the useful distinction between motive (the reason for an emotion) and objective correlative (the symbol of an emotion). **1962** T. P. Dunning in Davis & Wrenn *Eng. & Medieval Stud.* 168 The pagan element of the story is superbly handled by Chaucer as an essential feature of his *inventum*, or objective correlative. **1963** N. Frye *Romanticism Reconsidered* 3 Some belief of which Romantic poetry is supposed to form the objective correlative. **1964** J. B. Leishman *Rilke's New Poems* 9 An essentially expounding poet..might still be continuously engaged in a search for ever new 'objective correlatives' for old and unchanged convictions. **1966** K. Amis *Anti-Death League* ii. 176 This communicability..can work via an outward symbol or artefact, so that state-of-mind produces object which in turn produces state-of-mind. There are obvious analogies here with aesthetic theory, in particular with Eliot's notion of the objective correlative. **1971** *Times Lit. Suppl.* 31 Dec. 1629/4 In all these poems the objective correlative was utterly unsatisfactory from any partisan political point of view.

objectively, *adv.* Add: **3.** (Earlier examples.)

1796 [see Subjectively *adv.* 4]. **1798** A. F. M. Willich *Elem. Critical Philos.* 6 Our knowledge is called objectively true, in so far as objects must be perceived by every other being, in the same manner in which we represent them to ourselves.

objectivism. Add: (Earlier and later examples.)

1854 Geo. Eliot tr. *Feuerbach's Essence Christianity* xxi. 203 Belief in revelation is the culminating point of religious objectivism. *Ibid.* xxiii. 224 Religious objectivism has two passives, two modes in which God is thought. **1954** *Mind* LXIII. 265 'Objectivism'..is so ambiguous that there are very few ethical theories which cannot claim to be in some sense 'objectivist'. **1966** tr. *Kotarbinski's Gnosiology* ii. 68 The gap within idealist views between objectivism and subjectivism. **1972** Piccone & Hansen tr. *Paci's Function of Sci.* ii. 27 According to Husserl the struggle between objectivism and transcendentalism gives us the 'meaning' of the history of the modern spirit.

b. A derogatory term used in communist theory to describe an objective attitude towards existing conditions and acceptance of working within their limitations, as opposed to the concern for changing them according to revolutionary theories. So **objectivist.**

1933 T. B. H. Brameld *Philos. Approach to Communism* xi. 156 An objectivism which tends to deny the significance of conscious men. **1957** R. N. C. Hunt *Guide to Communist Jargon* 99 Objectivism is thus the opposite of 'party-mindedness', and is a serious offence. **1960** tr. *Lenin's Coll. Wks.* I. 401 The objectivist speaks of 'insurmountable historical tendencies'; the materialist speaks of the class which 'directs' the given economic system. **1972** tr. *Marx's On Hist. Materialism* 708 He [*sc.* Lenin] exposed their attempts to devoid Marxism of its revolutionary content and showed that their views were based on bourgeois objectivism which justified capitalism and glossed over class contradictions.

objectivity. (Later examples.)

1950 W. H. Barber tr. *Jolivet's Introd. Kierkegaard* iii. ii. 98 'Objectivity', whose idolatrous worship has been propagated by the modern nationalists, might equally well be defined as 'positivity'. **1957** R. May et al. *Existence* i. 25 He [*sc.* Kierkegaard] was convinced not only that the goal of 'pure objectivity' is impossible but that even if it were possible it would be undesirable. **1966** K. Hartmann *Sartre's Ontology* i. 7 Thus, the phenomena would cease to be moments of myself and would take on 'objectivity'. **1969** R. Strachan tr. *Malet's Thought R. Bultmann* i. 19 In its objectivity the other is only an extension of myself.

objectivization (ǫbdʒe:ktivəizēi·ʃən). [n. of action from Objectivize *v.*: see -ation.] The action or condition of making or becoming objective; an instance of this; something external in which an idea, principle, etc., is expressed concretely.

1929 [see *expressionism]. **1944** S. Putnam tr. *E. da Cunha's Rebellion in Backlands* ii. 144 It [*sc.* Canudos] was the objectivization of a tremendous insanity. **1965** D. A. Lowrie tr. *Berdyaev's Christian Existentialism* ii. 33 The objectivization of the criterion of truth usually means that that is transferred to another's consciousness and conscience. **1969** W. Glen-Doepel tr. *Metz's Theol. of World* iii. 126 The de-privatization and de-existentialization, or, to put it positively, the 'new objectivization' of the Christian message. **1970** R. J. Hollingdale tr. *Schopenhauer's Ess. & Aphorisms* 59 The will's objectivization.

objectivize, *v.* Add: Hence **obje·ctivized** *ppl. a.*

1965 D. A. Lowrie tr. *Berdyaev's Christian Existentialism* ii. 34 Objectivized knowledge in all its levels is withdrawn from the existential subject, that is from man.

object language. [Object *sb.* + Language *sb.*] **1.** A language described or analysed in terms of another language (known as the metalanguage).

1935 *Mind* XLIV. 501 Formal syntax requires a language of which it is the syntax. This language is called the 'object-language' (*Objektsprache*) in distinction from the syntax language. **1937** A. Smeaton tr. *Carnap's Logical Syntax of Lang.* 4 We are concerned with two languages: in the first place with the language which is the object of our investigation—we shall call this the *object-language*—and, secondly, with the language in which we speak *about* the syntactical forms of the object-language. **1946** C. Morris *Signs, Lang. & Behavior* 179 So if we talked about French in English, French would be the object language and English the metalanguage. **1947** A. J. Ayer *Thinking & Meaning* 26, I am using English in a twofold aspect, as an object-language and as a meta-language in which I speak about the object-language. **1947** R. Carnap *Meaning & Necessity* i. 4 In order to speak *about* any *object language*..we need a *metalanguage*. **1965** B. Mates *Elem. Logic* ii. 36 In the case of a Greek grammar written in English, Greek is the object-language and English is the metalanguage. **1977** J. Lyons *Semantics* I. i. 10 We may say that the language being described is the object-language and the language which is used to make the descriptive statements is the metalanguage.

2. In the theories of Bertrand Russell: a language consisting only of words having meaning in isolation.

1940 B. Russell *Inquiry into Meaning & Truth* iv. 63 There must..be a language of lowest type... I shall call this sometimes the 'object-language', sometimes the 'primary language'. **1954** E. Branth tr. *H. Spang-Hanssen's Recent Theories Nature of Lang. Sign* in *Travaux du Cercle Ling. de Copenhague* IX. 76 Object-words form an object-language, which is the primary languagei n a language hierarchy. **1963** J. Lyons *Structural Semantics* iv. 54, I do not accept..that there is a basic 'object-language', the elements of which are learnt in isolation and to which the rest of the vocabulary can be reduced.

3. *Computers.* A language into which a program is translated by an assembler or compiler. Cf. *object program* (*object *sb.* 10).

1961 Leeds & Weinberg *Computer Programming Fund.* ii. 49 Just as in language translation, a type of dictionary is employed in translating from the source language to the object language. **1970** P. M. Sherman *Techniques Computer Programming* vi. 109 The result of the translation process is an object program in an object language or machine language.

objectly (ǫ·bdʒėktli), *adv.* [f. Object *sb.* + -ly².] Objectively.

1925 *Blackw. Mag.* Dec. 786/1 He saw himself objectly as a felon with the mark of Cain.

o·bjectness. [f. Object *sb.* + -ness.] The state or quality of being objective or perceptible.

1933 *Mind* XLII. 510 This second relation..is a relation neither to the material object nor to 'objectness'. **1973** J. Elsom *Erotic Theatre* ix. 179 The laws concerning static nakedness exactly illustrated the idea of 'objectness'. **1975** *Russ. Rev.* (N.Y.) XXXIV. 272 In futurism, in cubism, space is almost always cultivated, but its form, being connected with objectness, does not convey, even to the imagination, the presence of world space; its space is limited by the space which is shared by the things on the earth.

‖ **objet** (ǫbʒe). [Fr., = object.] **1.** An object which is displayed as an ornament.

1857 A. Mathews *Tea-Table Talk* I. 98 Every part of it [*sc.* the house] abounded in pretty things—*objets*, as they are sometimes called, which her visitors were strictly forbidden to touch. **1960** *Good Housekeeping* Feb. 60/2 For some people a collection is worthwhile only if the pieces are rare... For others, the pleasure lies, quite simply, in possessing a number of more ordinary *objets* which intrigue, amuse or otherwise delight the owner. **1970** V. C. Clinton-Baddeley *No Case for Police* vii. 136 All those lamentable *objets* on the window-sills. **1973** *Listener* 6 Sept. 319/3 The sights and textures of bourgeois existence: the plush, the antimacassars, the *objets* in the home and in the shops.

2. One who is the object of (someone's) attentions or affection. Cf. Object *sb.* 4. *rare.*

1847 Thackeray *Van. Fair* (1848) xv. 132 Find out who is the *objet*, Briggs. I'll set him up in a shop. **1877** L. W. M. Lockhart *Mine is Thine* (1878) II. xviii. 33 He ..protested,..against being 'swindled' into further association with the *objet aimé* for the present.

3. *objet d'art* (ǫbʒe dār), a small artistic object; a curio; a precious and finely worked ornament. Also *transf.* Also *objet d'art et de vertu* (see sense 6 below and Virtu, vertu 1 c in Dict.). Cf. *object of art* (*and virtu*) (*object *sb.* 3 c).

1865 'Ouida' *Strathmore* II. xx. 236 Cachemires, sables, flowers, objets d'art, were scattered over it. **1866** Mrs. Gaskell *Wives & Daughters* II. x. 98 The various little tables, loaded with 'objets d'art' (as Mrs. Gibson delighted to call them) with which the drawing room was crowded. **1872** C. Schreiber *Jrnl.* 27 Mar. (1911) I. 145 We were surprised..to find a house containing 3 rooms entirely hung with pictures, in which were a few unimportant objets d'art. **1883** E. W. Hamilton *Diary* 29 Jan. (1972) II. 395 Mentmore..certainly is a gorgeous place, full of *objets d'art* and choice articles of furniture. **1913** W. Lawrence *Let.* 14 Oct. in T. E. Lawrence *Lett.* (1938) 159 Great collectors of objets d'art. **1921** M. Corelli *Secret Power* xxii. 267 The ceremony..has become a mere show of dressed-up manikins and womenkins, many of the latter being mere *objets d'art*,—stands for the display of millinery. **1931** *N. & Q.* 5 Sept. 175/1, I should be much obliged for any notes of the mention of platinum—as *objet d'art*, precious possession, gift, jewel or the like—in literature. **1939** O. Lancaster *Homes Sweet Homes* 38 *Objets d'art et de vertu* had been collected by rich men since the end of the seventeenth century. **1940** N. Marsh *Surfeit of Lampreys* (1941) v. 72 The jewellery and *objets-d'art* idea seemed a capital one. **1955** *Times* 27 May 14/3, I have seen him start from his chair and literally explode from the room, scattering priceless *objets d'art* to right and left. **1965** A. Nicol *Truly Married Woman* 17 He turned over the pages with the hopeless air of a connoisseur examining an objet d'art. **1973** G. Sims *Hunters Point* xiii. 119 Buchanan knew nothing of fine wines or *objets d'art* but he considered himself a connoisseur of loneliness. **1974** K. Clark *Another Part of Wood* vi. 257 At Portland Place they [*sc.* the author's children] were hidden away at the top of the house and brought down to be exhibited as *objets d'art* to our friends.

4. *objet de luxe* (see *de luxe *adj. phr.*), an especially fine or sumptuous article of value, a luxury article.

1881 C. C. Harrison *Woman's Handiwork* iii. 149 This was chiefly for the chambers of royal palaces—screens, then as now, being *objets de luxe* in the literal sense. **1934** *Burlington Mag.* Sept. 125/2 Cabinets and boxes covered with tortoise-shell veneer..are among the many 'objets de luxe' brought from India to Europe and now found in old collections. **1941** [see *BÆTYL]. **1956** K. Clark *Nude* i. 8 A few *objets de luxe*, like the Veroli Casket. **1976** *Times Lit. Suppl.* 12 Mar. 290/2 To trace the changing fortunes of Ruritanian royalty, through the dispersal of their Fabergé objets de luxe.

5. *objet trouvé* (lit. 'found object', pl. *objets trouvés*), an object found or picked up at random and presented as a rarity or a work of art. Also *transf.* and *fig.*

1937 Auden & MacNeice *Lett. from Iceland* xvii. 243 The Surrealists shall have J. A. Smith as an Objet Trouvé in disguise. **1940** Graves & Hodge *Long Week-End* xx. 352 The similarity between '*objets trouvés*'..and the neo-Victorian knick-knack collecting habit. **1956** M. Laski in *Pick of Today's Short Stories* VII. 121 *Collages* were countered by *montages, objets trouvés* by *objets faits*. **1959** P. & L. Murray *Dict. Art & Artists* 112 *Found object* (*objet trouvé*), in Surrealist theory an object of any kind, such as a shell found on a walk, can be a work of art; and such 'Found Objects' have been exhibited. **1962** *Listener* 6 Sept. 350/2 He was far ahead of his time. In the eighteen-nineties he was working on *objets trouvés*, in his case coals from his fire. **1967** L. Deighton *Expensive Place* v. 34 An *objet trouvé* is a piece of driftwood or a fine stone—it's something in which an artist has found and seen otherwise unnoticed beauty. **1970** *New Yorker* 17 Oct. 159/2 Mr. Berio allows that he has treated the Mahler movement as an *objet trouvé*. **1973** *Guardian* 17 Oct. 7/8 The City's food... Among *objets trouvés* were a screw in a cheese and tomato roll, a brass rivet in a Chelsea bun. **1974** *Sunday Times* 14 July 28/2 He [*sc.* Lord Goodman] plonked himself down, a volunteer *objet-trouvé*, and was given a studiously informal treatment.

6. *objet de vertu* (or *virtu*), a spurious transl. into French of *object of virtu* (see VIRTU, VERTU 1 c), after *objet d'art* above. (The required meaning of *vertu* does not exist in French.) Also in phr. *objet d'art et de vertu*.

1939 [see sense 3 above]. **1947** M. McCarthy in *Partisan Rev.* XIV. 63 Lady Windermere's fan becomes an *objet de virtu* as powerful as Desdemona's handkerchief. **1954** 'N. Blake' *Whisper in Gloom* ii. x. 139 Mr. Borch was in search of information rather than *objets de vertu*. **1961** *Connoisseur* Dec. p. xlvi, *Objets de Vertu* and fine Works of Art. **1974** *Times* 7 Dec. 14/7 His main interests are *objets de vertu* and antique weapons.

‖ **oblast** (ǫ·blast). [Russ.] A second-order administrative subdivision in Imperial Russia and the U.S.S.R.; a Russian province or region. Also *attrib.* Cf. *KRAY.

1886 *Encycl. Brit.* XXI. 69/2 *Oblasts*, or provinces. **1911** *Ibid.* XXIII. 875/2 For purposes of provincial administration Russia is divided into 78 governments (*guberniya*), 18 provinces (*oblast*) and 1 district (*okrug*). **1934** Webster s.v. *Soviet*, They [*sc.* the soviets]..send deputies to the higher soviet congresses: *volosts* (rural district),..*oblasts* (regional), and the congresses of the constituent republics. **1936** *Nature* 23 May 843/1 They [*sc.* the soviets] in their turn send delegates to the rayons, above which are the oblasts. **1938**, etc. [see *KRAY]. **1951** *Ann. Reg. 1950* 202 The unsatisfactory condition of agriculture in the *oblast* was attributed to organizational deficiencies... N. S. Patolichev..had spent the previous two years in the comparatively obscure post of *oblast* party Secretary. **1955** *Times* 23 Aug. 6/3 Every credit is given to those oblast and individual state and collective farms which fulfil..the plan. **1964** *New Statesman* 6 Mar. 353/1 Interference at the district level by agricultural departments of the town or village council has been superseded by 'guidance' at the *oblast* (regional) level. **1964** *Economist* 13 June 1264/2 The first oil was struck in the Tyumen oblast only in 1960. **1976** *Survey* Spring 57 He embarked on a successful career in party administration, attaining the rank of *oblast* First Secretary.

‖ **oblietjie** (ǫbli·kʸi). *S. Afr.* Also **oubl(i)etje**. [Afrikaans, f. F. *oublie* wafer (see OBLEY) + *-tjie* (dim. ending).] A type of wafer-thin tea-cake.

1890 *Hilda's Where is it? of Recipes* (ed. 2) 243 See also 'obletjes', Scones and Cakes, Puffs and Sandwiches. **1904** *Ibid.* (new ed.) 153 (Pettman), *Obletjes* (or *Oubliès*). An old-fashioned recipe for tea cakes brought to the Cape by the French refugees. **1912** *Northern News* 27 Aug. (Pettman), The one word I feel sure of is *oublietje*, that delicious, crisp, wafer-like pastry to be invariably found at bazaars in the districts settled by the Huguenots. **1947** *Cape Times* 26 Apr. 14/4 The old South African confectionery oblietjies, which the King so much enjoyed. **1947** L. G. Green *Tavern of Seas* viii. 65 She also served the rolled wafer tea cakes called oblietjies, made with cinnamon and white wine—a Huguenot contribution to Cape cookery. **1950** M. Masson *Birds of Passage* vii. 69 The cook they employed..would turn out a number of delicious local confections which included wafer thin teacakes, known as oublietjes.

obligate, *v.* **3. a.** and **5.** Add: In later use chiefly *dial.* and *U.S. colloq.*

1888 'C. E. Craddock' *Despot of Broomsedge Cove* 146 The parson..was 'obligated' to go down to the Settlement. **1900** S. R. Crockett *Little Anna Mark* xl. 340 When she came to New Milns she was obligated to go to the Scots kirk with Sir James. **1898** J. MacManus *Bend of Road* 73 I'll be happy to obligate ye. **1919** F. Hurst *Humoresque* 226 She thought maybe..I'd go over to her place for Wednesday-night supper for a change. You know how a girl like Clara gets to feeling obligated. **1955** C. McCullers in *Mademoiselle* Nov. 134/1 'I can't stay but just a minute,' John said. 'I'm obligated to sell those tickets. I have to eat and run.' **1963** *PMLA* LXXVIII. iv. ii. 11/1 Many of them felt obligated to turn out teaching materials as a kind of *quid pro quo* for their $350 stipend. **1970** N. Armstrong et al. *First on Moon* ix. 212 The foreman..has to side with his mechanics because he is obligated to a schedule. **1975** *N.Y. Times* 29 Nov. 26/1 President Ford is obligated early next month to report to Congress on the 'progress' of negotiations looking toward a Cyprus settlement.

obligately (ǫ·bligĕtli), *adv.* *Biol.* [f. OBLIGATE *ppl. a.* + -LY[2].] Out of necessity, because restricted to such a mode of life or such environmental conditions.

1952 [see *SAPROBE]. **1955** *New Biol.* XVIII. 54 These nitrogen-fixing bacteria were found to differ strikingly in that *Clostridium* is obligately anaerobic, i.e. able to grow only in the absence of oxygen. **1967** *Oceanogr. & Marine Biol.* V. 194 An obligately psychrophilic marine bacterium. **1971** [see *HALOPHILIC *a.*]. **1974** *Nature* 18 Oct. 574/2 Whether a symbiotic association is mutualistic or obligately parasitic.

obligative, *a.* Add: **2.** *Gram.* Of a verb form, mood, etc.: implying obligation. Hence as *sb.*

1877 W. D. Whitney *Essent. Eng. Gram.* v. 122 With *must* and *ought* (*to*) we make forms which may be called *obligative*, 'implying obligation': thus, *I must give, I ought to give*. **1968** J. Lyons *Introd. Theoret. Linguistics* vii. 309 The distinction between the 'obligative' and the 'inferential' sense associated with the auxiliary verb *must* in English is neutralized in a non-part sentence like *He must come regularly*. *Ibid.*, There is a further distinction within the 'obligative' in English, which has to do with the acceptance or fulfilment of the obligation. **1974** W. P. Lehmann *Proto-Indo-Europ. Syntax* iv. 105 The PIE subjunctive may resemble in meaning an obligative. *Ibid.* 131 In time the obligative meaning of the subjunctive came to be subsidiary to its function of indicating subordination.

obligatorily, *adv.* (Later examples.)

1942 Partridge *Usage & Abusage* 346/2 *Vari-coloured* and *variegated* are, the first obligatorily, the second preferably, to be used of or in reference to colour. **1961** *Amer. Speech* XXXVI. iii. 163 Postnominal modifiers can be shifted out beyond the noun obligatorily. **1975** T. P. Whitney tr. *Solzhenitsyn's Gulag Archipel.* II. iii. i. 15 Camps for forced labor were obligatorily created.

oblige, *v.* Add: **6. d.** *intr.* and *trans.* To act as a charwoman (for); to provide with domestic help. *euphem.*

1933 D. C. Peel *Life's Enchanted Cup* xix. 259 The mother took in washing and went out to 'oblige' and earned roughly 22s. a week and some of her food. **1937** E. Garnett *Family from One End St.* i. 13 She occasionally did odd work to 'oblige' Mrs. Theobald, the Vicar's wife. **1958** J. Cannan *And be a Villain* iii. 51 I'm not in service. I oblige by the hour. **1963** A. Lubbock *Austral. Roundabout* 165 Twice a week a lady came to 'oblige' in the house. **1972** 'A. Armstrong' *One Jump Ahead* i. 8 A bachelor who..paid well and wasn't too fussy..was a far better proposition than some others she had 'obliged'.

obliger. Add: **2. b.** One who obliges (sense *6 d); a charwoman. *euphem.*

1959 T. Girtin *Unnatural Break* 107, I thought that it was a bit odd the tea-things not being cleared away. I mean that looks as if they hadn't got an obliger. **1960** P. Coleridge *Running Footsteps* 63, I was agreeably surprised to be confronted by an elderly obliger, trailing a floor-mop.

oblique, *a.* (*sb.*) Add: **B.** *sb.* **3.** = *oblique case* (see OBLIQUE *a.* 5).

1695 Wheeler *Royal Gram. Reformed* vii. 26 *Qui* standing alone as a personal Relative is Englished *who* in the Nominatives, and *whom* in the Obliques. **1939** *Language* XV. 81 The obliques regularly have -ās. **1961** R. B. Long *Sentence & its Parts* 494 Obliques such as the *insist* of *she insists on paying*.

4. *Photogr.* A photograph taken at an oblique angle.

1942 *R.A.F. Jrnl.* 27 June 10 The second photograph is an earlier oblique of a similar type of *Sperrbrecher*. **1955** E. Waugh *Officers & Gentlemen* ii. i. 158 Guy, have you still got those obliques of 'Badger'?.. Bung 'em back to GHQ. **1958** C. B. Smith *Evidence in Camera* i. 19 The other two [cameras]—one on either side—were at an angle to take obliques. **1970** M. Kelly *Spinifex* ii. 48 Oosterman bracing the Newman cine-camera, F.245 aimed for obliques.

5. An oblique line, *spec.* a sloping virgule.

1961 in Webster. **1965** W. S. Allen *Vox Latina* 9 Phonemic symbols..are conventionally set between obliques, e.g. /t/. **1973** A. H. Sommerstein *Sound Pattern Anc. Greek* i. 7, I follow the practice of Chomsky and Halle, who say..that they use obliques 'for representations in which the features are functioning as classificatory devices'.

obli·teratingly, *adv.* [f. OBLITERATING *ppl. a.* + -LY[2].] In an obliterating manner; so as to obliterate.

1904 H. G. Wells *Food of Gods* i. iii. 56 He scarcely remembers the leap he must have made..so obliteratingly hot and swift did his impressions rush upon him.

obliteration. Add: **1.** Also *attrib.*, as *obliteration bombing*: heavy bombing intended to destroy a target completely.

1943 *Spectator* 24 Sept. 289/2 (*heading*) 'Obliteration' bombing. *Ibid.* 8 Oct. 337/2 The question of 'obliteration bombing'..raises more difficulties than Mr. Johnstone envisages. **1944** *Sat. Rev. Lit.* (U.S.) 8 Apr. 14/1 Vera Brittain, has become the spearhead of a movement.. directed against the mass air attacks of enemy cities, which she describes as 'obliteration bombings'. **1945** L. Mumford *City Devel.* (1946) 173 This failure would still be serious..even if obliteration bombing had not been practised.

obliterative, *a.* Add: **2.** *Zool.* **obliterative coloration, shading** = *COUNTERSHADING.

1909 [see *COUNTERSHADING]. **1926** W. P. Pycraft *Camouflage in Nature* vi. 66 This type of coloration he [*sc.* A. H. Thayer] designated 'obliterative coloration'. **1940** H. B. Cott *Adaptive Coloration in Animals* i. iii. 39 The appearance of obliterative shading can be..reproduced by means of patterns..rather than by the more usual method of graded ground colour. **1964** A. L. Thomson *New Dict. Birds* 139/2 Graded coloration, ranging from darkest on the back to lightest on the under parts, neutralises relief and thus renders the solid body as an apparently flat surface. Obliterative shading forms a basis for the coloration of nearly all cryptic birds, whether or not they carry a super-imposed pattern.

obliterator (ǫbli·tĕreitǫɹ). [f. OBLITERATE *v.* + -OR[1].] One who or a thing which obliterates.

1895 Hardy *Jude* i. i. 7 In place of it a tall new building ..had been erected..by a certain obliterator of historic records. **1900** *Pall Mall Gaz.* 18 Apr. 3/2 Fire was an obliterator of evil deeds more sure than any other.

oblivia·lity. *rare.* [f. OBLIVIAL *a.* + -ITY.] Liability to be forgotten.

1905 E. F. Benson *Image in Sand* i. 5 You certainly did not [meet him], or you would remember. Mr. Henderson has absolutely none of the quality of obliviality.

oblivion. 2. (Later example.)

1912 Galsworthy *Inn of Tranquility* 128 Hand-wrought bronze sconces and a band of metal bordering, all blackened with oblivion.

oblivious, *a.* **1.** ¶ Read: Unaware or unconscious of. Const. *of* or (esp.) *to*. Add further examples. No longer regarded as *erron.*

1926 W. de la Mare *Connoisseur* 173 Above them, as if entirely oblivious to their ranting, a glazed King Edward VII stared stolidly out of a Christmas lithograph. **1960** C. Day Lewis *Buried Day* v. 84, I stayed indoors all day for several days, oblivious to the damp heat of Falmouth.., re-living battles I had never fought. **1970** *Daily Tel.* (Colour Suppl.) 17 Apr. 54/3 For a man who has lived here all his life Makinen is oddly oblivious to the city's history.

obliviscible (ǫblivi·sib'l), *a.* [f. L. *oblīvīscī* to forget + -IBLE.] Able or likely to be forgotten.

1905 *N.Y. Times* 12 Aug. ii. 526/4 The sonnets he [*sc.* Swinburne] wrote about those poets, so obliviscible, excepting by himself.

Oblomov, Oblomoff (ǫ·blomǫf). The name of a character in Ivan Goncharov's novel 'Oblomov' (1855), represented as inactive, weak-willed, and procrastinating: used allusively. Hence **Oblomovism** (ǫblōu·moviz'm, ǫ·blomǫviz'm), conduct resembling that of Oblomov; sluggishness, inertia.

1902 *Encycl. Brit.* XXIX. 29/2 Dobroluboff said of it, '..something of Oblomoff is to be found in every one of us'. Peesareff..declared that 'Oblomoffism'..'is an illness fostered by the nature of the Slavonic character and the life of Russian society.' **1924** *Psyche* V. 55 This type of introversion..is known in Russia as 'Oblomovism'. **1925** I. A. Richards *Princ. Lit. Crit.* vii. 52 Most people in the same day are Bonaparte and Oblomov by turns. **1942** *Penguin New Writing* XII. 62 Thompson had been sinking towards semi-starvation, I to the insidious Oblomovism of the country. **1957** A. G. Meyer *Leninism* ix. 214 Oblomovism—the behavior of Oblomov, pathetic hero of Goncharov's novel of the same name, who prefers to contemplate and discuss the universe, including his own predicament, instead of taking an active part in solving his problems and participating in life. **1970** *Harper's Mag.* Oct. 44 Friends hint there may be a touch of Oblomovism combined with whatever real ills his aging flesh may be heir to.

obnoxity (ǫbnǫ·ksĭti). [f. as OBNOXIOUS *a.* + -ITY.] An obnoxious, objectionable, or offensive person or thing; an object of aversion.

1924 Lawrence & Skinner *Boy in Bush* xx. 282 The parlour was the coolest place for the meat. Easu shifted the red obnoxity, wire cover and all, to the top of a cupboard. *c* **1925** D. H. Lawrence *Virgin & Gipsy* (1930) iii. 38 That widow of a knighted doctor, a harmless person indeed, had become an obnoxity in their lives.

obnubilated, *ppl. a.* (Earlier and later examples.)

1658 [see *ADIAPHANOUS *a.*]. **1939** E. Pound *Let.* 7 Nov. (1971) 330, I loathe and always have loathed Indian art... Obnubilated, short curves, muddle, jungle, etc.

‖ **obo** (ōu·bo). [Native word.] In Mongolia, a ritual cairn of stones.

Quot. 1923 may be a different word.

1923 G. Collins *Valley of Eyes Unseen* 58 The *obo* consists of slabs of slate..inscribed all over with Tibetan characters. **1934** H. Haslund *Tents in Mongolia* vi. 59 We passed by two colossal heaps of stones, which the caravan men called *obos*. *Ibid.*, The two obo were..

erected at the point where the caravan route crosses the boundary between Inner and Outer Mongolia. **1936** P. Fleming *News from Tartary* v. 134 There were also prayer flags and a good many *obos*, which are cairns of stones with a wide range of superstitious significance. **1970** C. R. Bawden in L. Ligeti *Mongolian Studies* 65 The worship of the mountains and *obos* (ritual cairns).

obo, var. *OBBO.

oboe. Add: Now usu. with pronunc. (ōu·bo).

3. (With capital initial.) The name of a radar system for guiding military aircraft in which two ground stations interrogate a transponder in the aircraft to identify it and determine its position, which information is then transmitted to the aircraft. Also *attrib.* and *Comb.*

1945 *Daily Mirror* 15 Aug. 4/2 Next came 'Gee', the bombing beam which guided our radar-equipped bombers on to their targets, and the even more accurate 'Oboe'. **1946** *R.A.F. Jrnl.* May 169 'Oboe'-controlled Mosquito aircraft were assigned to the marking of targets. **1947** Crowther & Whiddington *Science at War* i. 59 The Oboe pathfinders started later and were faster... The pathfinder was under Oboe control, while approaching the target, for about ten minutes. **1974** *Encycl. Brit. Macropædia* XV. 371/1 The extreme accuracy of Oboe, H, or Shoran was not needed for guiding a plane between airfields on a friendly mission.

oboe (ōu·bo), *v. rare.* [f. the sb.] *trans.* To sound in the tone of an oboe.

1923 A. Huxley *Antic Hay* i. 8 Like an oboe, Mr. Pelvey intoned: 'The Lord be with you.'..those words, good Lord! that Mr. Pelvey was oboeing out of existence.

‖ **oboe da caccia** (ō·boe dā ka·tʃiă). Also **oboe di caccia**. Pl. **oboi**. [It., lit. 'hunting oboe'.] An old form of oboe, a fifth lower in pitch than the ordinary instrument. Cf. TENOROON.

1876 Stainer & Barrett *Dict. Mus. Terms* 317/2 The oboe d'amore, which was also called *oboe luonga*, produced a delicate and sweet tone, while the *oboe da caccia* corresponded to the tenoroon oboe, or corno inglese. **1880** Grove *Dict. Mus.* II. 489/1 Two important movements.. in Haydn's Stabat mater are scored for two oboi di caccia obligati. **1938** *Oxf. Compan. Mus.* 627/1 Bach does not make much sole use of the normal hautboy... The oboe d'amore and the oboe da caccia were the instruments to which he gave a solo position in his cantatas, &c. **1963** *Listener* 21 Mar. 532/2 No oboi da caccia in *Qui tollis*.

‖ **oboe d'amore** (ō·boe damō·re). Pl. **oboes, oboi**. [It., lit. 'oboe of love'.] An old form of oboe with a pitch a minor third below that of the ordinary oboe.

1876 [see prec.]. **1885** G. B. Shaw *How to become Mus. Critic* (1960) 62 The renovation of the obsolete *oboe d'amore* (love-hautboy)..proved very successful. **1930** [see *FLÛTE-À-BEC]. **1962** *Listener* 27 Dec. 1109/2 Oboe d'amore and two bassoons. **1976** *Gramophone* July 199/3 Scored only for strings and two oboi d'amore, it has some of the most remarkable writing in all Bach's church cantatas.

‖ **obosom** (obōu·sǫm). 9 b-. Pl. **abosom**. [Ashanti.] In the religious system of the Ashanti peoples of Ghana, a general name for any of the many gods inferior to the Supreme Being.

1853 B. Cruickshank *Eighteen Yrs. on Gold Coast* II. vi. 129 From the Souman, or idol of individuals, we come to the Boossum of a family or town, which frequently has no material representation. This..does not so much represent the god of an individual as a family god, or, more universally still, the god of a people. **1887** A. B. Ellis *Tshi-Speaking Peoples of Gold Coast* ii. 18 They [*sc.* the local deities] are very numerous... The general name for these deities is *Bohsŭm*. **1916** R. S. Rattray *Ashanti Proverbs* i. 30 A *suman* would seem to derive its power from the *abosom*, just as the *obosom* in turn gains its own from *Onyankǫpǫn*. **1923** —— *Ashanti* ii. 54 The great *obosom* (god) of all Ashanti is the Tano River, from which are derived countless of 'his children' as lesser *abosom*... Tano is considered as the 'son of the Supreme God'. **1960** M. J. Field *Search for Security* i. ii. 47 The Supreme God (called *Onyame* in Ashanti) is aloof... The *abosom* are the active, approachable executives of the Supreme God. The *abosom* are popularly referred to as 'the little gods', but each one is treated, in daily practice, as though he were omnipotent, omniscient and omnipresent. **1970** P. Oliver *Savannah Syncopators* 30 The occasions which I was privileged to attend, culminating in a cult or 'fetish' dance for a local *obosom*, or protective spirit, were simple and moving.

oboy. *U.S.* = *BOY *sb.*[1] 9.

1963 T. Pynchon *V.* xvi. 429 'Oboy, oboy,' said Fat Clyde wearily. **1966** —— *Crying of Lot 49* vi. 181 Oboy.. they'd call her names, proclaim her..as a redistributionist and pinko.

O'Brienism (obrai·əniz'm). [f. the name of William *O'Brien*, Irish patriot (1852–1928) + -ISM.] The conduct or anti-union policy of William O'Brien, esp. in the British Parliament about 1900 and 1901. **O'Bri·enite**, a supporter of William O'Brien.

1889 *Globe & Traveller* 18 Feb. 1/2 (*heading*) O'Brienism. **1900** *Westm. Gaz.* 22 Oct. 2/2 Mr. Redmond accepts ..the new situation caused by the triumph of the O'Brienites and the defeat of the Healyites. *Ibid.* 24 Dec. 3/1 O'Brienism is an equivalent term for constitutional anarchy. **1911** *Q. Rev.* July 241 The realists come from the north, east, and south, the strongholds of Unionism and O'Brienism.

obruchevite (ǫbrū·tʃĕvait). *Min.* [ad. Russ. *obruchevit*, f. the name of V. A. *Obruchev* (1863–1956), Russian geologist: see -ITE[1].] A mineral containing appreciable proportions of yttrium and uranium that was orig. regarded as a member of the pyrochlore group (see quot. 1977).

1955 *Internat. Conf. Peaceful Uses Atomic Energy* (United Nations) *U.S.S.R.: Sci. & Techn. Exhib.* 9 These classes of minerals include..obruchevite. **1959** *Mineral. Abstr.* XIV. 54/1 In 1945 E. I. Nefedov discovered in a pegmatite vein in the Alakuzti region (N.W. Karelia) a peculiar metamict tantalo-niobate, which he classed with ellsworthite. In 1949 A. A. Beus after a study of this mineral called it obruchevite in honour of V. A. Obruchev. ..Obruchevite replaces albite and infills cracks in quartz. It occurs in association with garnet and other minerals as nests and irregular masses up to 5 cm. in diameter. **1966** Z. Lerman tr. *Vlasov's Geochem. & Mineral. Rare Elements* II. 509 The mode of occurrence of obruchevite and its mineral paragenesis show that it forms during the latest stages of the replacement process. **1977** *Amer. Mineralogist* LXII. 407/1 [Report of the Subcommittee on Pyrochlore Nomenclature of the IMA Commission on New Minerals and Mineral Names.] Synonyms, doubtful and discredited names, and species not belonging to the pyrochlore group... *Obruchevite* (Kalita, 1957) is a name later shown to have been given to two different species (Gorzhevskaya and Sidorenko, 1969). One of these, brown obruchevite, after heating to 700°C, crystallized to the samiresite S phase... The other, black obruchevite, was subsequently renamed yttropyrochlore (Kupriyanova, 1970, unpublished). The Soviet Union's Commission of New Minerals (KNM) and Mineralogical Terminology have recommended the name *yttropyrochlore* replace this type of *obruchevite*.

obs, slang abbrev. OBSERVATION. Cf. *OBBO.

1943 *R.A.F. Jrnl.* Aug. 32 The Met. Officer has sent his obs through each hour in the usual way. **1970** O. Norton *Dead on Prediction* v. 99 Hurry up. I'm keeping obs.

obscene, *a.* **1.** Delete 'Now somewhat *arch.*' and substitute 'Now restored to general use'. (Later examples.)

1875 'Mark Twain' *Sk. New & Old* 148 The obscene Tumble-Bug. **1915** A. Huxley *Let.* 1 Feb. (1969) 65 Practically speaking, Sligger and myself are the only two possible people left alive in Oxford; even the few possibilities of last term have now vanished, leaving only a sort of obscene riff-raff. **1923** A. Bennett *Riceyman Steps* i. x. 44 The three-story houses (with areas and basements) were all alike... The fronts of the door-steps were green with vegetable growth... The areas, except one or two, were obscene. The Square..was merely decrepit, foul and slatternly. **1936** 'N. Blake' *Thou Shell of Death* i. 7 The shop windows, too, are piled with that diversity of obscene knick-knacks which nothing but the spirit of universal goodwill could surely tolerate. **1974** 'J. Graham' *Bloody Passage* i. 19 Vietnam was the most obscene episode of the century. **1974** *Greenville* (S. Carolina) *News* 23 Apr. 1/8 Energy officials have already predicted that first-quarter oil profits will be 'embarrassingly high' or 'whoppers'. Sen. Henry Jackson, D-Wash., has said they'll be 'almost obscene'. **1974** *Observer* 1 Sept. 1/7 The result..was another defeat for the image of football. In six hours of running skirmishes, 55 people were arrested; but what was particularly obscene was the mindlessness of the 3,000 or 4,000 youths who took part. **1977** *Time* 19 Dec. 19/2 Something in the very robustness of Germany's economy seemed to the terrorists and their sympathizers profoundly obscene.

2. (Later examples.)

1959 *Act* 7 & 8 *Eliz. II* c. 66 § 1 An article shall be deemed to be obscene if its effect..is, if taken as a whole, such as to tend to deprave and corrupt persons who are likely..to read, see or hear the matter contained or embodied in it. **1964** *Daily Tel.* 11 Dec. 26/2 Appeal Court judges ruled..that not only sex, but drug addiction, made a book obscene and depraved. **1972** *N.Y. Law Jrnl.* 31 Oct. 14/9 U.S.A. v. Various Articles of Obscene Merchandise.

obscenely, *adv.* **a.** Delete (*arch.*) and add later examples.

1922 F. Scott Fitzgerald *Beautiful & Damned* ii. i. 157 This cowardice sprang out, became almost obscenely evident, then faded. **1964** R. Church *Voyage Home* viii. 178 In those uncrowded years..England was less obscenely populated. **1974** *Times* 9 Apr. 14/3 Broadmoor's forbidding buildings are obscenely overcrowded.

obscenity. Add: **a.** (Later examples.)

1971 J. Trevelyan in *Mind & Mental Health* Winter 6/1 My own inclination is to apply the word 'pornography' to written or visual material concerned solely with sex, and the word 'obscenity' as a more general term covering pornography and also other things, especially violence. This view was held by D. H. Lawrence who suggested that obscenity was a matter of personal opinion, whereas pornography was something specific; he defined it as making sex dirty for money. **1972** *Police Rev.* 1 Dec. 1557/1 Mr. Chuter Ede..asked the Commissioner whether the Yard's obscenity squad—as it was then—were judges of literature and art?

b. Delete *Obs.* or *arch.* (Later examples.)

1940 *Times* 19 Apr. 7/2 What one American newspaper stigmatizes as the 'obscenity' of the attack upon Norway. **1970** *Times* 21 Mar. 3/1 The obscenity of racial hatred has been vigorously propagated over the last few years. **1975** *New Yorker* 25 Aug. 31/3 We had said 'so long' to the U.S.A., had bade farewell to..the obscenity of American life in our time. **1977** 'E. McBain' *Long Time no See* xiii. 217 One-third of all the homicides committed in this city involved a victim and a murderer who didn't even *know* each other... Perfect strangers..locked in the ultimate intimate obscenity.

obscurancy. For *rare*$^{-1}$ read *rare*. (Later example.)

1970 *Nature* 1 Nov. 448/2 Throughout these early years, *Nature* offered its 'leadership' support against obscurancy and obstruction in high places.

obscura·ntic, *a.* [f. OBSCURANT *sb.* and *a.* + -IC.] Opposed to enquiry or enlightenment. So **obscura·nticism** = OBSCURANTISM. Also **obscuranti·stic** *a.*

1926 *Contemp. Rev.* Nov. 661 The book..is full of warnings which sometimes are obvious and sometimes obscurantic. **1927** *Ibid.* Feb. 208 It would not be a work of truth or of love, but of well-meaning though mischievous obscuranticism. **1934** *Amer. Speech* IX. 278/1 The moralists..are in reality ensuring that the minds of the young will develop into the same welter of obscurantistic obsessions as their own. **1941** F. Matthiessen *Amer. Renaissance* xiv. iv. 653 Where Hawthorne's criticism runs no risk of being obscurantistic is in his portrait of Hollingsworth.

obscuration. 1. (Further transf. examples.)

1904 *Rep. Joint Comm. Phonetic Eng. Alphabet* (U.S.) ii. 12 In unstressed syllables the sounds undergo a change which, in the lack of a better name, may be called 'obscuration'. The quality and extent of this obscuration vary somewhat with the style of the discourse, the idiosyncrasy of the speaker and the nature of the neighboring consonants. **1935** J. S. Kenyon *Amer. Pronunc.* (ed. 6) 101 Not all of these pairs of stressed and stressless vowels represent the same historical stage of obscuration of the unaccented vowel. **1962** A. C. Gimson *Introd. Pronunc. Eng.* vii. 142 As a general rule, weak accent in OE led to the obscuration of short vowels and the shortening of long vowels.

obscure, *a.* Add: **4. b.** *spec.* of a vowel sound, weak and centralized; reduced. (Further examples.)

[**1568** T. Smith *De Recta et Emendata Linguæ Anglicæ Scriptione Dialogus* 14 Si Galli suum habent fœmininum obscurum, siue fuscum *e*, quod in fine dictionis positum, propè nihil sonat, auditur tamen, & apud illos est frequentissimum, nostræ linguæ prorsus incognitum. **1653** J. Wallis *Grammatica Linguæ Anglicanæ* i. 6 Eodem loci..formatur Gallorum *e* fœmininum; sono nempe obscuro. *Ibid.* 7 Ibidem etiam... Sonatur ŏ, vel ŭ, obscurum.] **1665** O. Price *Vocal Organ* B3ᵛ The ,*e*, is twofold. 1. Clear; as in let... 2. Obscure, onely when *e* is short before *r*, as in her, liberty, brother, father, merchant. *Ibid.* B4 The short û, and obscure ŏ, are formed in the throat, yet narrower then è. **1695** *Writing Scholar's Compan.* x. 36, o, is obscure, like (oo) or short (u)..before (m) as, *come* [etc.]. **1892** W. W. Skeat *Primer Eng. Etymol.* ii. 25 In the A. S. *Dūn-stān*..the *ā* has been shortened, and is now obscure. **1904** *Rep. Joint Comm. Phonetic Eng. Alphabet* (U.S.) iii. 30 The obscure [ə] is the goal to which the most of the other vowels tend when not supported by the stress. **1909** O. Jespersen *Mod. Eng. Gram.* I. ix. 249 Portuguese short *a* is an obscure vowel. **1924** J. S. Kenyon *Amer. Pronunc.* 107 Naturalness..is gained only by observing the normal relation between strest and unstrest syllables, distinct and obscure vowels. **1967** J. D. O'Connor *Better Eng. Pronunc.* v. 106 In initial position, as in..*attempt*,..*account*,..*observe*, you must again keep it very short and very obscure.

obscure, *v.* **1.** Delete †.

c. Add to def.: *spec.* to articulate (a vowel) in a weaker, more centralized position. (Earlier and later examples.)

a **1637** Jonson *Eng. Gram.* (1640) i. iii. 36 Where it [*sc.* e] endeth a last Syllable,..it either soundeth flat... Or, it passeth away obscur'd, like the faint *i*. as in these, *Written..divel*, &c. *a* **1790** B. Franklin *Autobiogr.* in *Writings* (1905) I. 358, I found his voice distinct till I came near Front Street, when some noise in that street obscur'd it. **1924** J. S. Kenyon *Amer. Pronunc.* 108 The student should rid himself of a common misconception; namely, that the obscuring of certain consonants and vowels owing to lack of stress on syllables or words is the result of a corruption of good English. **1934** *S.P.E. Tract* xxxix. 621 'Short *e*'. This is rarely raised to [i], and never, except before *-r*, obscured to [ə]. **1934** S. Robertson *Devel. Mod. Eng.* (1936) vii. 230 Obscuring and loss of formerly distinctive vowel sounds, and dropping of consonants in phonetically difficult combinations.., are both to be found in Old English. **1935** J. S. Kenyon *Amer. Pronunc.* (ed. 6) 90 The vowels of unaccented syllables have gradually become obscured to a sound quite different in resonance, or quality, from what they had formerly been. **1962** A. C. Gimson *Introd. Pronunc. Eng.* vi. 80 Vowels under weak accent are increasingly obscured to [ə] or [ɪ], or are elided.

obscured, *ppl. a.* Add: **2.** *Phonetics.* Of a vowel sound: having a neutral, centralized articulation; weakened; reduced.

1925 G. P. Krapp *Eng. Lang. in Amer.* II. 250 Difficulty was expressed in disposing of this unstressed and obscured vowel. **1934** M. K. Pope *From Latin to Mod. French* v. 119 The obscured neutral vowel in use in Modern English is buccal and central, the one in Modern French is a slightly rounded front sound. **1935** J. S. Kenyon *Amer. Pronunc.* (ed. 6) 90 The same spelling is kept for the obscured vowel that was used to spell it before it became obscured. **1962** A. C. Gimson *Introd. Pronunc. Eng.* vii. 120 As the great

variety of spellings indicates, /ə/ may represent the reduced (obscured, 'schwa') form of any vowel or diphthong in an unaccented position.

obscu·ringly, *adv.* [-LY[2].] In an obscuring manner; so as to obscure.

1902 *New Liberal Rev.* Aug. 317 The Celtic fringes hang obscuringly over our eyes, as fringes do under befeathered hats in the Old Kent Road.

obscurist (ǫbskiūə·rist). [f. OBSCURE *a.* or *v.* + -IST.] = OBSCURANTIST *sb.*

1925 *Chambers's Jrnl.* Mar. 196/1 He is no faddist or eccentric, no obscurist of any kind, but one who catches at charms in human life and paints them.

‖ **obscurum per obscurius** (ǫbskiūə·rǝm pǝɹ ǫbskiūə·riǝs), *phr.* [Late L., lit. the unclear (explained) by means of the more unclear.] An unclear argument or proposition (expressed) in terms of one that is even less clear; such an explanation. (Cf. *IGNOTUM PER IGNOTIUS.)

[**1616** W. Clerk *Withals's Dict. Eng. & Lat.* (rev. ed.) 574 *Obscurum per obscurius*. I am as wise as I was before.] **1892** C. A. M. Fennell *Stanford Dict.* 580/1 *Obscūrum per obscūrius*, phr.: Late Lat.: the obscure by the more obscure. **1949** K. Davis *Human Society* viii. 202 At its best it was an explanation *obscurum per obscurius*. **1952** G. Sarton *Hist. Sci.* I. viii. 200 Herodotus..was already combining Pythagorean ideas with Egyptian, Orphic, and Bacchic ones, and he mixed up the story of Pythagoras with that of Zalmoxis, thus explaining *obscurum per obscurius*. **1959** *Listener* 8 Jan. 58/1 Alexander's attempt to describe certain other relations in nature on the analogy of this may seem like an attempt to explain *obscurum per obscurius*.

obsequent (ǫ·bsĭkwĕnt, ǫbsī·kwĕnt), *a.*[2] *Geomorphol.* [f. OB- + *-sequent* in CONSEQUENT, SUBSEQUENT *adjs.*] Of a stream, stream valley, or drainage pattern: having a course or character opposite to that of a consequent stream, stream valley, or drainage pattern, i.e. against the direction of dip of the strata. Hence as *sb.*, an obsequent stream.

1895 W. M. Davis in *Geogr. Jrnl.* V. 134 Its escarpment face sheds short, back-flowing streams into the longitudinal subsequent valley that has been developed along the weak underlying stratum; and, even at the risk of multiplying terms unduly, I would suggest that these streams be called obsequent, as their direction is opposed to that of the initial consequent streams. *Ibid.* 145 Such obsequents are represented by the Ousel and Ivel farther east. **1902** H. J. Mackinder *Britain & Brit. Seas* 121 The Little Ouse of East Anglia is also an obsequent. **1954** W. D. Thornbury *Princ. Geomorphol.* v. 114 Obsequent valleys drain in a direction opposite to the original consequent valleys. **1968** R. W. Fairbridge *Encycl. Geomorphol.* 288/2 (*caption*) Evolution of subsequent (plus obsequent and resequent) drainage system. *Ibid.* 1187/1 Obsequents were originally defined as streams having a direction opposite to that of the consequent streams in their vicinity. Usually, however, the term is interpreted to mean merely a stream flowing against the direction of dip.

b. Of a fault-line scarp or a related feature: having (as a result of differential erosion) a relief the reverse of that originally produced by the faulting.

1913 W. M. Davis in *Bull. Geol. Soc. Amer.* XXIV. 198 If it descends toward the relatively uplifted side, as must be the case when weaker rocks occur on that side, it may be called an obsequent scarp. **1954** W. D. Thornbury *Princ. Geomorphol.* x. 260 If..the erosional topography is opposite to the original fault-produced topography, the mountain blocks would be obsequent tilt block mountains, the basins would be obsequent tilt block valleys, and the bounding scarps would be obsequent fault-line scarps. **1970** R. J. Small *Study of Landforms* iii. 103 An 'obsequent' fault-line scarp faces in the opposite direction to the original fault-scarp.

observabi·lity. [f. OBSERVABLE *a.* + -ITY.] The quality of being observable; observable character or state; capacity for being seen.

1934 *Philos. Rev.* XLIII. 137 The simplest case of verifiability—observability at will. **1944** *Mind* LIII. 220 The new positivist principle..does not say anything at all about the observability or non-observability of the facts asserted or denied in P. **1956** E. H. Hutten *Lang. Mod. Physics* iv. 154 A numerical limit to the observability of microscopic events. **1968** R. A. Lyttleton *Mysteries Solar Syst.* iv. 120 It may happen that two periods of observability occur as different parts of its path on opposite sides of perihelion bring the comet into the night sky. **1974** *Nature* 8 Feb. 352/2 It has been argued that the observability of the scattered waves is closely related to the presence of a caustic. **1974** M. Hesse *Struct. Sci. Inference* i. 9 Most accounts of the observation language were dependent on circular definitions of observability and its cognates.

observable, *sb.* **2.** Delete † *Obs. rare* and add: Something that can be perceived more or less directly; something that is knowable through the senses. (Later examples.)

1954 A. J. Ayer *Philos. Ess.* i. 9 It may be left open what situations are to count as being observable; whether, for example, we are to treat such objects as electrons as being directly accessible to observation, or only such common-sense objects as chairs and tables, or only sense-data. Whatever decision may be taken..we are..likely to be left with some descriptive expressions which do not signify observables. **1959** *Listener* 1 Oct. 520/1 According to Dingle, the observables of a science must be potentially observable 'by all normal people'. **1963** J. Lyons *Structural Semantics* i. 1 We must reject any theory of semantics the terms of which neither refer to observables nor are reducible to observables. **1964** *Philos.* XXXIX. 262 If one supposes observables to be ontologically fundamental then models appear gratuitous. **1965** P. Caws *Philos. of Sci.* viii. 52 The concepts which science borrows from the knowledge and experience of ordinary men..yield principally substantive and adjectival constructs directly linked to perception. Constructs having this direct link will be called observables. **1968** J. J. C. Smart *Between Sci. & Philos.* v. 143 The instrumentalist..would agree with the operationist in holding that in science no statements are made about entities other than macroscopic observables.

b. *Physics.* A quantity that can (in principle) be measured.

1930 P. A. M. Dirac *Princ. Quantum Mech.* ii. 25 In quantum mechanics it is more convenient to deal with something that refers to one particular time instead of to all times, analogous to the value of a classical variable at a particular instant of time. We shall call such a quantity an observable. **1955** W. Pauli *Niels Bohr* 38 All physical observables are represented by Hermitian operators. **1966** C. G. Hempel *Philos. of Nat. Sci.* vi. 74 These wavelengths are not observables in the ordinary sense of the word. **1974** H. Clark *First Course Quantum Mech.* iii. 54 In quantum mechanics..not all the observables of a system can be measured simultaneously. If some observable is measured, this act of measurement may disturb the system and change the value of some other observable... The disturbance in a classical system..can in principle be allowed for exactly but not in quantum physics. **1975** *Nature* 31 Jan. 315/1 Body temperature, enzyme activity, leaf movement, neural firing, mitotic index or other convenient observables typically persist in regular up and down periods of about 24-h, even in ostensibly constant conditions.

observant, *a.* Add: **1. b.** Acting in accordance with the precepts of behaviour associated with a particular religion, esp. Judaism.

1902 *Daily Chron.* 2 Oct. 7/1 To-day observant Jews throughout the world celebrate the commencement of their New Year. *Ibid.* Even the less observant..hasten to the Synagogue to-day to listen to the mystic sound of the Ram's Horn trumpets. **1905** *Westm. Gaz.* 25 Mar. 3/2 Someone will be suggesting [giving up] linen collars next—in which case the really Lenten-observant man will look like nothing so much as a burglar. **1972** C. Potok *My Name is Asher Lev* i. 36 The stores that were run by observant Jews were all closed on Shabbos. **1975** *Times Lit. Suppl.* 21 Nov. 1392/5 An observant Jew..declined to join a prominent Berlin literary club because he would eat only kosher food.

Observantine. Add: Also *attrib.* or as *adj.*

a **1773** A. Butler *Lives Saints* (1779) IV. 208 He [*sc.* St. James of Sclavonia] embraced with great fervour the humble and penitential state of a lay-brother among the Observantin Franciscan friars at Bitecto, a small town nine miles from Bari. **1930** F. J. Eble tr. *Grisar's Martin Luther* ii. 51 The vicar..jeopardized the canonical and disciplinary autonomy of the Observantine monasteries entrusted to his care. **1932** *Times Lit. Suppl.* 9 June 425/2 A fifteenth-century bard who joined the order at the time when the Observantine reform was making great headway in Ireland.

observation. 2. Delete † *Obs.* and add later example.

1911 W. J. Locke *Glory of Clementina Wing* xxii. 277 The daily calls to inquire after her health and happiness had grown to be a sacred observation.

10. *observation balloon, deck, gallery, platform, post, room, terrace, vehicle, window*; *observation-based* adj; (in Philos.) *observation basis, report, term* (see quots.); **observation-car** (earlier and later examples); **observation officer** = *OBSERVER 4 b; **observation-sentence** *Philos.*, in a scientific theory, a sentence that reports, or directly relates to, observed phenomena (opp. *theoretical sentence*); **observation-statement**, an observation-sentence, or the utterance of one; **observation ward** (see quot. 1927).

1909 *London Mag.* Sept. 15/2 He made numerous ascents in captive observation balloons. **1917** F. Stark *Let.* 16 Oct. (1974) I. 49 We had a great time.., going to call on a Major..who looks after the observation balloons in this sector. **1950** *Gloss. Aeronaut. Terms* (*B.S.I.*) i. 46 *Observation balloon*, a balloon fitted with a basket or car to carry passengers. **1965** *Language* XLI. 212 The value of observation-based description. **1965** P. Caws *Philos. of Sci.* xxiv. 182 The observation basis must consist of carefully formulated protocol sentences. **1872** *Harper's Mag.* May 876/1 You look out of the open 'observation car' as you sweep down from a height of 7000 feet. **1880** 'E. Leathes' *Actor Abroad* 177 An observation car, which is roomy, comfortable, and roofless is attached to the end of the train on leaving Sacramento. **1936** Wodehouse *Laughing Gas* ii. 17 These observation cars, in case you don't know, are where the guard's van is on an English train. **1973** *Guardian* 17 Mar. 14/4 The 'Rheingold' from Basle to Amsterdam..has a vista-dome observation car. **1951** A. C. Clarke *Sands of Mars* iii. 21 He..hurried out to the observation deck, wondering what happened to Earth overnight. **1976** E. P. Benson *Bulls of Ronda* vii. 51 Rafael Durán stood on the observation deck of the airport, watching the plane. **1951** Observation gallery [see *LINER *sb.*[2] 8 c]. **1904** *Daily Chron.* 22 June 9/3 The aim of the Japs..was deadly true, and the observation officers were able to see, through their field-glasses, men falling in every direction. **1922** Observation officer [see *flying officer* s.v. *FLYING *vbl. sb.* 3]. **1974** *Times* 23 Jan. 16/8 [Mr Arthur Harold Stevens] won the MC in 1916 for great courage under fire as an observation officer. **1906** F. Lynde *Quickening* 29 At the rear of the string of Pullmans was a private car, with a deep observation platform. **1943** J. S. Huxley *TVA* xv. 127 The operation building for the navigation lock..contains control machinery.., and observation platform. **1957** D. Robins *Noble One* x. 103 She could imagine him climbing up the ladders to the observation platforms on the tree tops. **1909** *Westm. Gaz.* 17 Sept. 3/1 The way of this little bird is to sit on its observation post. **1914** *Daily Express* 28 Sept. 4/5 The damage to the cathedral was the inevitable result of the French using the cathedral as an observation post. **1937** Koestler *Spanish Testament* ii. 276 At eight the prisoners came out into the courtyard again, and I took up my observation post. **1959** *Listener* 26 Mar. 553/1 In the extreme, as in modern totalitarianism, all 'observation posts' are available only to the duly qualified. **1974** K. Royce *Trap Spider* vii. 124, I could see the entrance... I had as good an observation post as any. **1964** *Amer. Philos. Q.* I. 251/1 The observation report can be a singular statement that contradicts the law. **1974** M. Hesse *Struct. Sci. Inference* i. 35 The 'meaning' of observation reports is 'theory-laden'. **1970** *Guardian* 14 Feb. 8/2 Observation rooms for looking down into the studios. **1936** R. Carnap in *Philos. of Sci.* III. 429 My testing of any sentence..refers back ultimately to my own observation-sentences. **1940** A. J. Ayer *Found. Empirical Knowl.* ii. 86 Some observation-sentence should be derivable. **1964** I. Scheffler *Anat. Inquiry* 135 Their vocabulary was appropriate, enabling derivation from observation-sentences. **1970** W. V. Quine *Philos. of Logic* i. 5 Usually observation sentences are..individually responsive to observation. This is what distinguishes observation sentences from theoretical sentences. **1946** A. J. Ayer *Lang., Truth & Logic* (ed. 2) 11 My principle..I shall restate here..using the phrase 'observation-statement' in place of 'experiential proposition', to designate a statement 'which records an actual or possible observation'. **1951** *Mind* LX. 19 Such observation statements as 'The hydrogen-oxygen mixture in this glass vessel when ignited changed into water vapour'. **1961** E. Nagel *Struct. of Sci.* xi. 348 Singular statements that either formulate the outcome of observations..or describe the overt procedures..we shall call..'observation statements'. **1966** *Philos.* XLI. 260 Various views..were held about observation-statements. **1965** P. Caws *Philos. of Sci.* xi. 78 Terms which name percepts we have agreed to call observation terms, since they refer to what is directly observed. **1974** M. Hesse *Struct. Sci. Inference* i. 10 The allegedly clear and distinct character of the observation terms. **1968** *N.Y. City* (Michelin Tire Corp.) 129 An observation terrace..offers a splendid view of the airport. **1972** *Police Rev.* 10 Nov. 1444/1 They merge into almost any background—and for this reason they are the colours selected for Police observation vehicles. **1927** W. E. Collinson *Contemp. Eng.* 58 If there is doubt as to the presence of the disease in the patient when in hospital, he may be put in an observation ward. **1961** C. Cockburn *View from West* i. 7 This ward..had..been built..probably as an observation ward for children. **1897** Kipling *Capt. Cour.* ix. 193 The secretary and typewriter sat together..by the plate-glass observation window at the rear end. **1974** P. Dickinson *Poison Oracle* i. 7 Wesley Morris stared at Dinah [*sc.* an ape] through the observation window.

observationalism. Delete *rare* and add: Also, the doctrine that observation, rather than theory, is the basis of science. Hence **observa·tionalist** *a.* (*sb.*), adhering to observationalism; practising observational as opp. to theoretical work; also, one who adheres to observationalism. Also **observationa·lity**, closeness to the level of observation; non-theoreticality.

1951 *Mind* LX. 44 The inductive account of scientific method, which is an alternative way of stating observationalism, postulated..the cautious generalization which must not go beyond the data. *Ibid.* 46 A consequence of observationalist presuppositions. **1960** W. V. Quine *Word & Object* 42 We may speak of degrees of observationality. **1966** *Philos.* XLI. 360 Opposition to dogmatic observationalism should not develop into dogmatic anti-observationalism. **1973** *Sci. Amer.* May 12/3 Sclater..writes that McKenzie 'is the theoretician and I am the observationalist'.

observationally, *adv.* Add: (Further examples.) Also, with regard to observation.

1930 A. S. Eddington *Rotation of Galaxy* 13 The effect on the apparent angular motion..remains always on the verge of what is detectable observationally. **1964** *Philos. Rev.* LXXIII. 203 We cannot observationally verify the speaker's claims. **1976** *Nature* 26 Feb. 628/2 Observationally one requires better high resolution, far-infrared and molecular maps and a near-infrared search for the dust-enshrouded protostars in the H II regions before the vastly complex physics and mechanics of these birth-places of stars can be understood fully.

observe, *sb.* Add: † **3.** *Sc.* A division of a sermon.

1833 W. L. Mackenzie *Sk. Canada & U.S.* 8, I went to hear Doctor McLeod, a steadfast Presbyterian of the old school. There..the discourse is divided and subdivided into heads and observes in true covenanting fashion.

observer. Add: **3. c.** A person who observes without participating: *spec.* (*a*) one who attends a conference, display, etc., to note the

proceedings; (*b*) one posted to an area of conflict to note events, supervise a cease-fire, etc.; also *attrib.*, as *observer force, group*.

1925 A. Toynbee *Survey Internat. Affairs 1920–23* 10 Several meetings were attended by an American observer. **1949** *Ann. Reg. 1948* 292 The U.N. Balkans Commission.. sent observers to watch the fighting there. **1958** *Observer* 10 Aug. 4/6 Any measures.., in addition to the original observer group, which would serve to ensure the territorial integrity and political independence of Lebanon. **1962** *Rep. Comm. Broadcasting 1960* 188 in *Parl. Papers 1961–2* (Cmnd. 1753) IX. 259 The three advertising organisations which now appoint one representative and one observer to the Committee formerly appointed two representatives each. **1970** *Guardian* 13 Jan. 9/8 The French and the Pope have already called for a new observer force. **1977** *Times* 9 Dec. 1/3 Sir David attended the inquest in Pretoria as an independent observer at the invitation of the Association of Law Societies of South Africa.

4. b. One whose duty it is to make military observations; *esp.* (*a*) a member of an artillery group trained to identify the target; (*b*) a person trained to notice and identify aircraft, or carried in an aeroplane († or other aircraft) to note the enemy's position, etc.; *spec.* as a rank in an air force; also *attrib.*, as *observer company, corps, flight, officer*.

[**1854** C. Tomlinson *Cycl. Useful Arts* I. 16/1 Scarcely had the observer reached the height of 3,000 feet, than he observed..a thin vapour.] **1870** tr. *F. Marion's Wonderful Balloon Ascents* iii. iv. 215 The soldiers of the enemy, all who saw the observer watching them.., came to the idea that they could do nothing without being seen. **1903** *Heavy Artill. Training* 36 If the target is not visible from the guns or ground quite close to them,..two observers are required. **1906** *Strand Mag.* May 515/1 The first of these was Mr. E. T. Fetch, accompanied by Mr. Krarup as observer, on a twelve-horse-power single cylinder Packard car. **1910** *Blackw. Mag.* Feb. 209/1 The service of an aeroplane, especially if..it could carry a military observer as well as the pilot, would be invaluable. **1914** *Field Artill. Training* ix. 325 The observer, having located the position of the target and conveyed the information to the artillery commander..receives from him the signal 'Observe for line'. **1914** *Daily Express* 28 Aug. 2/6 (*heading*) Observer flights. **1916** H. Barber *Aeroplane Speaks* 50 Quickly the Observer climbs into his seat in front of the Pilot. **1928** C. F. S. Gamble *Story N. Sea Air Station* xiii. 226 During this year [1916] the rank of Observer Officer was created. **1932** *Flight* 22 July 677/2 It is very necessary to see, at least once a year, how the Observer Corps is functioning. **1940** *War Illustr.* 5 Jan. p. ii/3 Parker.. was a Group Commander in the Observer Companies of the Royal Defence Corps, which was founded in 1916. **1954** P. K. Kemp *Fleet Air Arm* 52 During 1916, the Admiralty opened a special school for training observers and instituted a new rank of Observer Officer. **1977** J. Cleary *High Road to China* i. 26 The Bristol Fighter..carried two guns, one fired straight ahead by the pilot and the other able to be fired in a complete circle by the observer.

c. Used *attrib.* with reference to the effect of subjective factors on the accuracy or veracity of scientific observations.

1959 *Times Lit. Suppl.* 31 July 441/2 Proper consideration of evidence, allowance for observer-error, the clarification of language as a communicative medium—these.. are the barriers which stand between us and the irrational abyss. **1970** G. Greer *Female Eunuch* 90 It takes another psychiatrist to explain to her the function of observer bias. **1971** *Brit. Med. Bull.* XXVII. 6/1 Observer variability has also been shown to be a significant factor in errors in blood-pressure readings. *Ibid.* 8/1 There was little evidence to suggest variation between observers or observer bias.

obsession. Add: **3.** (Later examples.)

1901 G. B. Shaw *Three Plays for Puritans* Pref. p. xvii, The English novelist, like the starving tramp who can think of nothing but his hunger, seems to be unable to escape from the obsession of sex. **1908** Yeats & Gregory *Unicorn from Stars* ii. 56 There is another kind of inspiration, or rather an obsession or possession. **1916** G. B. Shaw *Androcles & Lion* Pref. p. lxxviii, The mass of mankind..are concerned almost to obsession with sex. **1922** Joyce *Ulysses* 130 The Roman, like the Englishman who follows in his footsteps, brought to every new shore on which he set his foot..only his cloacal obsession.

b. *Psychol.* An idea or image that repeatedly intrudes upon the mind of a person against his will and is usually distressing (in psycho-analytic theory attributed to the subconscious effect of a repressed emotion or experience).

1901 C. A. Mercier *Psychol.* 368 Obsessions are extremely varied in character. **1913** E. Jones *Papers on Psycho-Anal.* v. 126 Twenty years ago Janet separated obsessions and phobias under the title of 'psychasthenia'. **1924** J. Riviere et al. tr. *Freud's Coll. Papers* I. vii. 129 Two components are found in every obsession: (1) an idea that forces itself upon the patient: (2) an associated emotional state. **1958** M. Argyle *Relig. Behaviour* xii. 165 The similarities and differences between rituals and obsessions. **1976** Smythies & Corbett *Psychiatry* vi. 82 If we see some particularly horrible disaster, most people find it very hard to get it 'out of their minds'. An image of horror keeps coming back unwanted and the person has to struggle to get it out of consciousness. More trivial examples are tunes that keep running through one's head or fears that one might forget one's lines in a play... These are examples of mild obsessions—unwanted thoughts and images that crowd into consciousness in spite of all attempts to keep them out.

obsessional, *a.* Add: **2. a.** Characterized by or caused by an obsession (sense 3 or *3 b).

1909 *Cent. Dict.* Suppl., *Obsessional*.., pertaining to or of the nature of obsession. **1913** E. Jones *Papers on Psycho-Anal.* v. 126 Krafft-Ebing maintained the independence of obsessional states. **1940** *Mind* XLIX. 370 Thus with obsessional duties, 'I must go and make sure I have turned out the lights, turned off the taps,..made sure of security'. **1952** V. Gollancz *My Dear Timothy* i. 11 Breathless delight in ease and comfort, and an obsessional drive not merely to work but..'to work extra'. **1954** M. Fortes in E. E. Evans-Pritchard *Inst. Primitive Society* vii. 89 There are the individuals with obsessional and paranoiac fears whose fantasies about themselves.. sound like morbid caricatures of primitive beliefs. **1973** R. Lewis *Of Singular Purpose* vi. 139 There was only one word to describe Paul Mercereau's interest in Van Rijk. It was obsessional.

b. *obsessional neurosis*: in psychoanalytic theory, a psychoneurosis in which obsessional thoughts resulting from a regression of the libido lead to neurotic or compulsive behaviour. So *obsessional neurotic*, a person suffering from an obsessional neurosis.

1918 E. Jones *Papers on Psycho-Anal.* (ed. 2) xxx. 515 A detailed study of the obsessional neurosis. **1924** J. Riviere et al. tr. *Freud's Coll. Papers* I. viii. 153 Hysteria's close association with the female sex and..the preference of the male for the obsessional neurosis. **1926** [see *ISOLATION 2 b]. **1939** E. Glover *Psycho-Anal.* ii. x. 78 The obsessional neurotic, like the hysteric, recognizes the irrationality of his symptoms. **1942** [see *DE-EMOTIONALIZE *v.*]. **1968** C. Rycroft *Crit. Dict. Psychoanal.* 103 According to classical theory, the psychopathology of obsessional neurosis centres round regression to the anal-sadistic stage.

obse·ssional, *sb.* [f. the adj.] Someone whose personality is dominated by an obsession.

1928 [see *DIDDUMS]. **1945** *Brit. Jrnl. Psychol.* XXXV. 41 As a group, the obsessionals are distinctly more intelligent than the others. **1963** *Listener* 14 Mar. 474/3 The *soi-disant* realist was really just a neurotic old obsessional with an eye for detail. **1972** E. K. Ledermann *Existential Neurosis* ii. 13 Obsessionals and reactive depressives showed high sedation thresholds. **1978** P. Porter *Cost of Seriousness* 43 Dangerous modes In all weather when obsessionals walk To a favourite spur above the land.

obsessionally (ǫbse·ʃənăli), *adv.* [f. OBSESSIONAL *a.* + -LY[2].] In an obsessional manner.

1961 J. Dawson *Ha-Ha* vi. 126, I noticed the cracks between each board, and followed each one, uneasily, obsessionally, till it ran into the wall. **1961** *John o' London's* 12 Oct. 407/2 He *cares* for the theatre; passionately, obsessionally. **1966** P. Green tr. *Escarpit's Novel Computer* viii. 102 Fermigier never suspected a thing, though he was an obsessionally jealous man by nature. **1973** A. Roy *Sable Night* xvi. 167 The two men..had their attention almost obsessionally occupied.

obsessionist (ǫbse·ʃənist). [f. OBSESSION + -IST.] One who is obsessed, or subject to obsession, by a fixed idea.

1921 *Glasgow Herald* 24 June 8/4 The canards of the anti-waste obsessionists. **1928** *Daily Express* 6 Dec. 10/5, I once sat in a train for five hours opposite an obsessionist, who played chess with himself on a miniature board.

obsessive (ǫbse·siv), *a.* [f. OBSESS *v.* 3 + -IVE.] Of or pertaining to obsession; liable to obsess; obsessing. So as *sb.*, a person characterized by obsessive behaviour.

1911 I. H. Coriat *Abnormal Psychol.* ii. vi. 281 One of the most common of these obsessive states is what is known as the obsession of self-consciousness. **1918** A. E. Davis *Hypnotism* i. 46 A frequent cause of..obsessive thoughts, is the feeling of shame or regret arising out of bad habits. **1926** J. I. Suttie tr. *Ferenczi's Further Contrib. Theory & Technique Psycho-Anal.* xviii. 232 The individual suffering from obsessive ideas is really substituting thought for action. **1966** *Listener* 17 Mar. 414/1 Of the medical obsessives I particularly liked Maurice Denham as Bloomfield Bonington. **1969** N. W. Pirie *Food Resources* iii. 104 In 1939 many of us argued that protein supplies would be the main problem in feeding Britain in war-time. We were stigmatized as obsessives. **1973** T. & R. Millon *Abnormal Behav.* (1974) iii. xiii. 296 Obsessive thoughts are intrusive ideas which the person cannot block from consciousness. **1975** R. Hill *April Shroud* xii. 154 'I've got some rushes here. You want to see them?' Like all obsessives, he could not doubt the answer.

b. *obsessive-compulsive* adj. Psychol., applied to a disorder in which an obsession results in the compulsion to perform repeatedly meaningless acts; also *fig.*; so as *sb.*, a person whose compulsive behaviour is due to obsessive thoughts.

1927 Henderson & Gillespie *Text-bk. Psychiatry* xiv. 399 'Obsessive-compulsive' state is applied to a..condition in which the preoccupation issues in motor acts of an apparently trifling or meaningless kind. **1941** S. H. Kraines *Therapy Neuroses & Psychoses* i. 26 Hysteria and obsessive-compulsive neurosis in which states symbolic symptoms are more common. **1960** *Cambr. Rev.* 7 May 509/1 Of 100 severe reactions among students..the schizophrenic were the most frequent (34)... There were ..7 obsessive-compulsives. **1965** *New Statesman* 24 Sept. 426/1 [Ronald] Reagan has been associated with the obsessive-compulsive faction of the Republican right. **1970** *Jrnl. Gen. Psychol.* LXXXII. 175 Schizophrenics, obsessive-compulsives, and depressed persons have a high level of health anxiety.

obsessively (ǫbse·sivli), *adv.* [f. *OBSESSIVE *a.* + -LY[2].] In an obsessive manner; insistently, beyond reason.

1949 M. Mead *Male & Female* xv. 296 The picture can be obsessively elaborated. **1969** *Sunday Times* 2 Nov. 51/3 Expressions of a mind almost obsessively in pursuit of unification. **1971** *Physics Bull.* June 332/3 Each side.. can select instances which demonstrate its point and.. can obsessively ignore others which demonstrate the opposite. **1977** P. Dickinson *Walking Dead* iv. i. 252 The Captain continued to flick obsessively at the wheel.

obse·ssiveness. [f. as prec. + -NESS.] The quality of being obsessive.

1961 in Webster. **1966** *Listener* 8 Sept. 366/3 Mr Storey portrayed very accurately that brand of obsessiveness which salesmen call enthusiasm.

obsidianite (ǫbsi·diănait). *Geol.* [f. OBSIDIAN + -ITE[1].] = *TEKTITE.

1898 R. H. Walcott in *Proc. R. Soc. Victoria* XI. 23 As long as this uncertainty exists, some other name would be more appropriate, and I suggest and will refer to them in this paper as 'obsidianites'. **1909** [see *AUSTRALITE]. **1933** *Nature* 28 Jan. 117/1 Small, curiously shaped pieces of.. glass have long been known from certain regions, and have been called moldavites from the Moldau River in Bohemia..; australites or obsidianites from Australia; [etc.]. **1963** J. A. O'Keefe *Tektites* i. 1 Earlier writers referred to tektites as obsidianites.

obsolesce, *v.* For *rare*$^{-1}$ read *rare*. Add later examples.

1934 G. B. Shaw *On Rocks* 160 The lists of crimes and penalties will obsolesce like the doctors' lists of diseases and medicines. **1950** *N.Y. Times* 12 Nov. 96/2 (Advt.), New New a thousand times New (we'd rather die than obsolesce). **1975** *Amer. Speech 1972* XLVII. 261 *Exolinguistics* has never become a standard term in either linguistics or anthropology, but it has persisted in both periodicals and books for twenty years, and it refuses to obsolesce.

obsolescence. Add: **1. b.** *spec.* of machinery or similar assets. Also *attrib.* See also *planned obsolescence*.

1887 Garcke & Fells *Factory Accounts* v. 93 This increased production may be due to certain parts being of a more permanent type than others and added to stock with less risk of obsolescence. **1913** R. J. Porters *Pitman's Dict. Book-Keeping* 330 Any loss sustained through obsolescence is charged to Profit and Loss. **1930** *Economist* 25 Jan. 163/2 In steelworks the question of obsolescence and the need for replacement of plant depend, partly, on the present condition of the plant and partly [etc.]. **1930** Hutchinson & Lovell *Short Dict. Legal Terms* 87 *Obsolescence*. Where machinery is lessened in value not by mere usage or lapse of time, but by the fact that improved machinery is being brought into use, then this lessened value is termed obsolescence as distinguished from depreciation. **1957** Clark & Gottfried *University Dict. Business & Finance* 245/1 *Obsolescence*, with respect to an asset, the loss in value brought about through improvements in technology, changes in public taste, or a falling off in demand. **1959** Jowitt *Dict. Eng. Law* II. 1257/2 *Obsolescence allowance*, a charge allowed in determining profits for taxation purposes [etc.]. **1962** [see *built-in* (*b*) s.v. *BUILT *ppl. a.* 1 b]. **1966** *New Statesman* 28 Jan. 140/2 We have set up an obsolescence reserve of £1,500,000.

obsole·scing, *ppl. a.* [f. OBSOLESCE *v.* + -ING[2].] That is becoming obsolete.

1916 E. V. Lucas *Cloud & Silver* 71 The Mayor..still clung to the steadily obsolescing topper. **1953** *Sun* (Baltimore) 23 Oct. 2/7 This sort of conversion [of gun turrets] is, in fact, under way, with other heavy 'obsolescing' types.

obsolete, *v.* Delete 'Now *rare*' and add later (chiefly N.Amer.) examples.

1944 *Amer. Speech* XIX. 234/1 A number of new laws and regulations..will almost certainly obsolete anything we might have given you during the past couple of months. **1945** in *Amer. Speech* (1948) XXIII. 70/1 The pressure of today's events has obsoleted many filing systems. **1960** [see *PRE-TEEN *a.*]. **1961** *Flight* 3 Feb. 154/3, I have read Sir Percy Hunting's paper... It is competent, commendable stuff, marred only by the inclusion of..some Americanisms... For example: obsoleting, envisioned, peppercorn rentals, fiscal, transportation. **1967** *Technology Week* 23 Jan. 100/1 (Advt.), The new gas turbine engine that's out to obsolete the diesel. **1972** *Fortune* Jan. 52/3 Increasing emphasis upon service and repair of existing equipment, rather than mindlessly obsoleting it. **1975** *Sci. Amer.* Feb. 39/1 (Advt.), Our pre-coated sheet failed to obsolete the glass TLC plate. **1975** *Listener* 11 Dec. 788/2 (American speaker) News reporting on the hour or the half-hour—when only part of the earlier news had been obsoleted—meant a great deal of repetition. **1977** *Islander* (Victoria, B.C.) 1 May 13/1 Wire is being obsoleted in communication today.

obstacle, *sb.* Add: **3. obstacle course** (see quot. 1961); also *transf.* and *fig.*

1961 Webster s.v., *Obstacle course n*, a military training area filled with obstacles (as hurdles, fences, walls, ditches) that must be surmounted. **1973** *Black Panther* 31 Mar. 14/3 Could you describe some of the more known programs used in 'boot camp' such as the obstacle course and the firing range? **1976** J. Wainwright *Who goes Next?* 145 As an obstacle course the yard was a nothing. **1977** *New Yorker* 27 June 26/1 Then he picked two miracles attributed..to Neumann..and started them over what Father Litz called the Vatican obstacle course.

obstinancy. (Later example.)
1894 B. Thomson *Diversions of Prime Minister* xiii. 213 The steadfastness of their followers was obstinancy under the lash of persecution.

obstreperous, *a.*, **obstreperously,** *adv.* Add: Further examples of illiterate forms. Quot. 1922 is elliptical, for 'obstreperous mouth'.
1824 J. Wight *Mornings at Bow St.* 255 They were forthwith conveyed to the watch-house, and there they conducted themselves so 'obstropolously', that the constable of the night found it necessary to have them put down below. 1922 Joyce *Ulysses* 420 Hark! Shut your obstropolos.

obstruct, *v.* Add: **2. b.** *Cricket.* **obstructing the field** (formerly, **the ball**): an expression used to denote the manner of dismissal of a batsman who, in the umpire's opinion, wilfully hindered a fieldsman or interfered with the ball in order to avoid being caught, stumped, etc.
1877 C. Box *Eng. Game Cricket* 456 Obstructing the Ball.—A man may be given out, but seldom is on such an account. 1905 *Laws of Cricket* § 26 The striker is out..if under pretence of running, or otherwise, either of the batsmen wilfully prevent a ball from being caught;—'Obstructing the field'. 1952 A. A. Lilley *Twenty-Four Years Cricket* v. 57 As the ball struck it [sc. the bat] and fell to the ground, I pushed it further along with my bat, and Wheeler..gave me out. The point upon which he gave his decision was for obstructing the field.

c. In various games: to impede (a player) in a manner which constitutes an offence.
1895 H. F. Battersby *Hockey* vii. 131 A player shall not run in between his opponent and the ball so as to obstruct him. 1953 *Association Football* ('Know the Game' Series) 33/1 If an opponent is obstructing a player, the player may charge him. 1969 *F.A. Guide to Laws of Game* 144 [illegible] with an [illegible] free kick [illegible] opponent [illegible] to prevent him reaching it.

obstruction. Add: **2. b.** *spec.* in *Law.*
1908 *Encycl. Laws Eng.* (ed. 2) X. 116 *Obstruction.* This term is used in law mainly in two senses: (1) Interference with public or private rights or easements, particularly of light, way, navigation, or watercourse; (2) interference with officers of justice in the execution of their duty. 1910 *Daily Chron.* 18 Sept. 1/7 The conviction of two men for obstructing the police. After the evidence of two witnesses denying the statements made by the policemen when the obstruction charge was being considered, [etc.]. 1933 P. MacDonald *Mystery of Dead Police* v. 41 My car's outside Savarin's... There's a bobby by it. He wants me for obstruction.

c. *spec.* as an offence in various games.
1923 H. E. Haslam *How to play Hockey* xiii. 81 Turning on the ball incurs a penalty on the score of obstruction. 1935 *Encycl. Sports* 521/2 Charging and Obstruction... A player who is not running for the ball must not charge or obstruct an opponent not holding the ball. 1950 *Men's Hockey* ('Know the Game' Series) 18/1 A penalty bully is awarded..[for] a bad case of obstruction in front of goal. 1953 *Association Football* ('Know the Game' Series) 33/1 Should the obstruction take the form of a personal foul.. then the foul is penalised by a direct free kick. 1969 *F.A. Guide to Laws of Game* 197 Indirect free-kick for obstruction. 1974 *Rules of Game* 163/2 A direct free kick..is awarded for the following intentional fouls..: charging from behind (unless the opponent is guilty of obstruction).

5. obstruction light *Aeronaut.* (see quot. 1960).
1934 *Jrnl. R. Aeronaut. Soc.* XXXVIII. 728 The Air Ministry specified the number and type of obstruction lights which had to be fitted to each mast. 1960 *Guide Civil Land Aerodrome Lighting* (*B.S.I.*) 9 *Obstruction light*, a light indicating the presence of an obstruction.

obstru·ctionary, *a. rare.* [f. Obstruction + -ary[1].] Tending or disposed to obstruct.
1934 W. de la Mare *Froward Child* 13 He was not always too anxious as to what kind of self was being so obstructionary. 1975 *New Yorker* 27 Oct. 77/1 There were rules, not of Herr Kautz's making, that guaranteed us a coeducational clubroom and occasional dances, but he was always around in a silently obstructionary manner, operating the turntable, serving the punch, watching.

obstructionism. Delete *rare*—[1] and add later examples.
1941 W. S. Churchill *Second World War* (1950) III. 752 Please report to me any signs of obstructionism. 1955 *Times* 9 Aug. 11/4 Obstructionism is a game two can play. What is required to solve the difficulty is sincerity on both sides. 1964 S. Brittan *Treasury under Tories* ii. 61 He would not dream of indulging in the narrow-minded obstructionism of some of his predecessors. 1977 *Time* 5 Sept. 35/1 O'Neill made it clear to Georgia's crusty John Flynt, chairman of the ethics committee, that neither he nor Jaworski would tolerate any obstructionism.

obstructionist. (Later examples.)
1934 C. Lambert *Music Ho!* iv. 242 They were.. obstructionist rather than constructionist. 1965 [see *filibuster *sb.* 4]. 1966 M. R. D. Foot *SOE in France* ii. 11 Petty obstructionists of this kind lay about SOE's path all its life, and some have pursued it since its winding-up. 1975 *New Yorker* 26 May 94/2 The late nineteen-forties, when another President got himself returned to office by blaming nearly everything that was wrong with the country on an obstructionist Congress controlled by the opposing party.

obstruent, *a.* and *sb.* Add: **A.** *adj.* (Later examples.)
1945 R. Hargreaves *Enemy at Gate* 27 Cold, flabby, capricious, obstruent, and quite femininely vindictive, he was a creature of almost supernal selfishness, timidity and irresolution. 1973 J. Wainwright *Pride of Pigs* 70 The object of the exercise was to demolish any obstruent bushel likely to get in the way of *his* particular light.

B. *sb.* **c.** *Phonetics.* Also *erron.* **obstruant.** A fricative or plosive speech sound. Also *attrib.* and *Comb.*
1942 *Language* XVIII. 13 The first member of a cluster of two obstruents (stop or spirant) is voiceless. 1952 W. P. Lehmann *Proto-Indo-European Phonol.* ii. 7 We then arrive at three classes of phonemes: 1. those which may not function as syllabics will be called obstruents. 1955 C. F. Hockett *Man. Phonol.* 97 A number of obstruent systems include no symmetric set at all. 1956 J. Whatmough *Language* iii. 36 Sounds which are partially or completely stopped at some point—between the larynx and the lips (these are known as obstruents), e.g. *p:k*, or *f:χ*. 1962 *Word* XVIII. 312 Post-vocalic consonants cluster in exactly the opposite direction, right to left, with privilege of occurrence of more than one consonant from an obstruent group, as many as three lenes or four fortes, but never a lenis after a fortis. 1963 Ervin & Miller in J. A. Fishman *Readings Sociol. of Lang.* (1968) 71 Vowel distinctions are learned first. The order of acquisition for the remaining features is: (*a*) vowels *vs.* consonants; (*b*) sonorants *vs.* articulated obstruants [etc.]. 1965 N. Chomsky *Aspects of Theory of Syntax* iv. 168 If the second consonant is a liquid, the first must be an obstruent. 1969 *Language* XLV. 248 We may symbolize the elements involved as C (any obstruent), R (any 'resonant' or semivowel) and V (any vowel). 1970 *Canad. Jrnl. Linguistics* XV. 122 Two examples [of persistent rule] are the well known devoicing of the final obstruants in German..and the loss of final /n/ in Livonian. 1973 A. H. Sommerstein *Sound Pattern Anc. Greek* ii. 26 The derivation of [n] from /nt/ is by Obstruent Dropping. 1975 *Language* LI. 528, I observed that the alternation between distinctively paired obstruents such as [p-b, t-d, k-g] had to be stated as a morphophonemic regularity. 1977 *Trans. Philol. Soc. 1975* 4 It is clear that a certain class of obstruent-final stems has the property of having inflected forms where the stem-final obstruents differ in voicing from those in the simplex forms.

obtainability (ǫbtēi·năbi·lĭti). [f. Obtainable *a.*: see -ity.] The quality or state of being obtainable; capability of being obtained.
1932 H. H. Price *Perception* vii. 177 The existence or obtainability of the other sense-data. 1933 *Mind* XLII. 294, I fail to distinguish these merely entertained and probably false (in any case unverifiable) propositions as to the present and simultaneous actuality of the, as yet, unsensed sense, from that believed and probably true one as to their successive obtainability. 1971 *Analysis* XXXII. 55 The obtainability of the conditional.

obtrusively, *adv.* (Earlier example.)
1817 Coleridge *Biog. Lit.* I. x. 191, I have seen gross intolerance shewn in support of tolerance; sectarian antipathy most obtrusively displayed.

obtrusiveness. (Earlier example.)
a 1817 Jane Austen *Persuasion* (1818) IV. x. 213 He stood, as opposed to Captain Wentworth, in all his own unwelcome obtrusiveness.

obtundent, *a.* and *sb.* (Later examples.)
1898 H. H. Burchard *Text-bk. Dental Path. & Therapeutics* vii. 538 Under the head of obtundents are included those agents which are applied locally to benumb the terminals of sensory nerves. 1908 J. D. Patterson in C. N. Johnson *Text-bk. Operative Dentistry* xxviii. 460 Obtundents. For the purpose of obtunding, many preparations have been advocated and many methods advised. 1930 W. H. O. McGehee *Text-bk. Operative Dentistry* xxiv. 700 An Obtundent Paste devised by Doctor J. Lewis Blass..used with splendid results. 1961 MacDougall & Nixon *Guide to Dental Therapeutics* viii. 115 When drugs are applied locally to relieve pain, they are referred to as local anaesthetics or obtundents.

Ob-Ugrian (ǫb,ū·griăn). [f. *Ob*, the name of a Siberian river + Ugrian *a.* and *sb.*] A Finno-Ugric linguistic group of Siberia related to Hungarian. So **Ob-Ugric** *a.* and *sb.*
1933 L. Bloomfield *Language* iv. 68 *Ob-Ugrian*, consisting of *Ostyak*..and *Vogule*. 1954 Pei & Gaynor *Dict. Linguistics* 151 *Ob-Ugrian*, a language group (consisting of Ostyak and Vogul) which with Hungarian constitutes the Ugric branch of the Finno-Ugric (or Uralic) subfamily of the Ural-Altaic family of languages. 1955 B. Collinder *Fenno-Ugric Vocab.* p. xi, Vogul, Ostyak, and Hungarian are called Ugric languages; vg and os are called Ob-Ugric. *Ibid.* p. xiv, I made up my mind to quote even such words as occur only in Ugric and Permian, nay, even in Ob-Ugric and Permian. 1975 G. F. Cushing tr. *Hajdu's Finno-Ugrian Lang. & Peoples* iii. 119 Ob-Ugrian is the collective name for the languages and peoples known as Vogul (Man'si) and Ostyak (Chanti).

obverse, *a.* and *sb.* Add: **A.** *adj.* **4.** *Logic.* Of a proposition: obtained from another proposition by the process of obversion.
1870 A. Bain *Logic* I. 110 To each of the four Propositional Forms..there is an obverse form. 1917 J. Welton *Groundwork of Logic* v. 79 In no case is there any loss in the range of application in the obverse proposition.

B. *sb.* **3.** (Earlier example.)
1870 A. Bain *Logic* I. 110 No men are gods. The obverse is..all men are no-gods.

obversion. 2. (Earlier example.)
1870 A. Bain *Logic* I. 110 In affirming one thing, we must be prepared to deny the opposite: 'the road is level', 'it is not inclined', are not two facts, but the same fact from its other side. This process is named *Obversion.*

obvert, *v.* **3.** (Earlier example.)
1870 A. Bain *Logic* I. 110 Obvert the predicate, and prefix the sign of negation.

obverted, *ppl. a.* (Further example.)
1870 A. Bain *Logic* I. 115 These obverted forms are Particular Affirmatives, and are therefore converted simply.

obvertend. (Earlier example.)
1886 P. K. Ray *Text-bk. Deductive Logic* (ed. 2) III. ii. 129 The given proposition may be called the *Obvertend.*

obviation. Add: **2.** The use, in some American Indian languages, of the obviative (see next).
1927 L. Bloomfield in *Internat. Jrnl. Amer. Linguistics* IV. 191 Although obviation is confined to animate nouns.., verbs have obviative forms also for inanimate actors. 1974 P. H. Voorhis *Introd. Kickapoo Lang.* 45 A noun and a pronoun agreeing with it in gender, number, and obviation are used to form statements of the identity of the noun and pronoun. 1976 *Language* LII. 519 Obviation is analysed as a dimension of contrast within the 3rd person.

obviative (ǫ·bviătiv, -ēitiv). [ad. F. *obviatif*, f. Obviate *v.* + -ive.] A grammatical category of the Algonquian family of N. Amer. Indian languages that marks a third person as subordinate to another, animate, third person in a given context. Also, a similar category in certain other languages. Also *attrib.* or as *adj.*
[1866 J. A. Cuoq *Études Philologiques sur Quelques Langues Sauvages de L'Amérique* I. ii. 43 *De L'Obviatif.* Quand dans une phrase, se *rencontrent* deux 3èmes personnes de première classe, l'une *sujet* et l'autre régime de la phrase, la personne-régime se met à *l'obviatif.* *Ibid.*, Si le nom de la personne *dominante* se trouve exprimé, on le met à l'obviatif simple.] 1877 *Trans. Amer. Philol. Assoc. 1876* VII. 150 When two nouns (or a noun and a pronoun) in the third person are introduced in the same sentence, one as a subject and the other as the object of a verb, the latter takes the *obviative*—or second third-personal-form. 1899 *Proc. Amer. Philos. Soc.* XXXVIII. 186 In *pōhĕgūnūl* and the following word, we have the ending *-ūl* of the obviative, or accus[ative] of the third person, which appears in all the Algic idioms. 1922 L. Bloomfield in *Amer. Jrnl. Philol.* XLIII. 276 He arrived at no clear statement of such features as the 'obviative' (the peculiar subsidiary third person of Algonquian grammar). 1927 —— in *Amer. Speech* II. 438/2 In inflection, Menomini, like the other Algonquian languages, has an *obviative* form for subsidiary third persons. 1958 *Archivum Linguisticum* X. ii. 171 Distinction..of Proximate and Obviative, a common Algonquian distinction. 1965 *Canad. Jrnl. Linguistics* X. 80 The use of the obviative in Kutenai. 1976 *Language* LII. 519 The 3rd person in Cree includes two dimensions of contrast: proximate/obviative and animate/inanimate.

obvious, *a.* Add: **4. c.** quasi-sb., *the obvious*: something which is obvious; plain or manifest inferences, remarks, details, facts, etc.
1903 K. D. Wiggin *Rebecca* (1904) i. 10 Their steadfast gaze..had the effect of looking directly through the obvious to something beyond. 1919 M. K. Bradby *Psycho-Anal.* xiii. 175 The work of the artist who consciously and deliberately descends to the obvious..is uninteresting. 1976 G. McDonald *Confess, Fletch* (1977) iii. 19 You do the obvious and stop in at the first singles bar you come to.

oca. Add: Also **oka.** (Further examples.)
1885 W. Miller tr. *Vilmorin-Andrieux's Veget. Garden* 355 The Oka-plant is easily propagated from the tubers. 1950 *N.Z. Jrnl. Agric.* May 471/3 The oka plant.., a native of Peru,..produces tubers which can be used as vegetables. 1966 *New Scientist* 9 June 643/1 Quinoa, papa-lisa, oka and isaño are plants [in Bolivia] which produce crops with higher nutritive value and protein content than wheat, maize and rice. 1972 Y. Lovelock *Vegetable Bk.* 221 The young leaves and flowers of the Spanish-American *oca*..are used in soups and as a pot-herb, and the flowers serve as a kind of vinegar substitute in salads. It is cultivated, however, for the sake of its egg-sized tubers, which are an important staple in Mexico and the Andean states.

Occam's razor: see Razor *sb.* 1 b in Dict. and Suppl.

occasion, *sb.*[1] Add: **1.** *to rise to the occasion*: see Rise *v.* 15 b.

occasional, *a.* (*sb.*) Add: **2. b.** (Further examples.)
1829 H. C. Robinson *Diary* 13 Aug. (1967) 102 [Goethe] remarked..that *occasional* poems are among the best poems when the poet takes care to retain all the spirit of the occasion. 1834 G. Crabbe jun. in *Poet. Wks.*

G. Crabbe I. i. 15 His father took in a periodical work.. which contained, at the end of each number, a sheet of 'occasional poetry'. 1891 F. Locker-Lampson *Lyra Elegantiarum* (rev. ed.) p. x, In his [sc. the Editor's] judgment Occasional Verse should be short, graceful, refined, and fanciful, not seldom distinguished by chastened sentiment, and often playful. 1932 Chadwick & Kershaw *Growth of Lit.* I. 25 Both poems [sc. *Deor* and *Widsith*]..contain passages describing what purport to be personal experiences of the authors, and which—at least in the case of Deor—approximate to 'occasional' poetry. 1965 M. Spark *Mandelbaum Gate* ii. 18 ..had before the war published a volume of his own occasional verses. 1976 S. Hynes *Auden Generation* v. 140 Lehmann's *The Noise of History* and Spender's *Vienna*..are occasional: like..'Easter 1916' and Auden's elegy on Yeats—they mythologize history.

c. (Earlier and further examples.)
1749 J. Cleland *Mem. Woman Pleasure* II. 91 Mrs. Cole had procur'd my spark and me an occasional bed-room to ourselves. 1771 Smollett *Humph. Cl.* III. 25 At night, half a dozen occasional beds are ranged.. along the wall. 1857 Dickens & Collins *Lazy Tour* i. in *Househ. Words* 3 Oct. 316/1 A little round occasional table in a window.

B. *sb.* **2.** (Earlier and later examples.)
1867 Mark Twain *Let. to Publishers* [illegible] I am on the N.Y. Tribune staff here as an 'occasional', among other things. 1908 *Westm. Gaz.* 10 Dec. 6/1 (*heading*) Oxford University v. Oxford Occasionals. 1977 'J. Le Carré' *Hon. Schoolboy* v. 112 He never got round to blowing [sc. betraying] the Occasionals...the Occasionals were filed.

occidental, *a.* Add: **2.** Also, of, belonging to, or characteristic of, the western United States.
1809 E. A. Kendall *Travels Northern Parts U.S.* II. 28 Among the natural forest-trees, are the button-wood or occidental plane, the spruce-fir and the locust-tree. 1846 *Knickerbocker* XXVIII. 471 [illegible] an occidental correspondent, 'had carried the knife for a long time.' 1933 E. C. Jaeger *Calif. Deserts* v. 57 The occidental harvester (*Pogonomyrmex occidentalis*) is a large, reddish ant building conspicuous mounds of pebbles.

B. *sb.* **2.** An artificial language, based chiefly on the Romance languages, invented by E. de Wahl (1867–1948), an Estonian, in 1922.
1926 *Encycl. Brit.* III. 906/2 Mr. E. de Wahl.. finally produced Occidental, 'comprehensible at first sight to 10,000,000 educated Europeans without preliminary study'. To write or speak it is less easy than to read it, variety being clearly bought by the immense cost of irregularities. 1928 O. Jespersen *Internat. Lang.* I. 2, I have read articles and received letters, chiefly in Ido, but also in Esperanto and Occidental, written from very different countries. 1934 S. Robertson *Devel. Mod. Eng.* (1936) iv. 89 Jespersen now feels that there are enough points of similarity among the leading projects looking toward an international language—including Esperanto, ..Occidental, and his own creation, Novial—to justify the hope that a single adequate International Auxiliary Language will some day emerge. 1949 M. Pei *Story of Language* [illegible] The twentieth century has continued the tradition [of creating artificial languages], with..Occidental.., Monding, and a host of others.

occidentalist. Add: **c.** One who advocates or uses the artificial language called Occidental.
1946 H. Jacob *On Choice of Common Lang.* iii. 40 The Occidentalists were the most critical.

occipito-. Add: **occipito-a·ngular** (see quot. 1890); **occipito-ante·rior,** denoting that form of vertex presentation in child-birth in which the occiput is directed away from the sacrum of the mother; **occipito-poste·rior,** denoting that form of vertex presentation in child-birth in which the occiput is directed towards the sacrum of the mother.
1890 *Cent. Dict.*, *Occipito-angular*, pertaining to or common to the occipital bone and the angular convolution. 1898 Occipito-angular [see *corticipetal *a.*]. 1831 C. D. Meigs tr. *Velpeau's Elem. Tocol.* [illegible] The English accoucheurs..bestow the title of natural labour only upon the occipito-anterior position, while those upon the occipito-posterior position belongs to the class of preternatural labour. 1974 Passmore & Robson *Compan. Med. Stud.* III. 11/1 (*caption*) Vertex presentation, left occipito-anterior position, well flexed. 1831 Occipito-posterior [see occipito-anterior above]. 1971 *Brit. Med. Bull.* XXVII. 49/2 Impaired performance was associated with toxaemia, occipitoposterior presentation and delivery in an ambulance.

Occitan (ǫ·ksităn, ‖ ǫksitan). [Fr.] **a.** Old Provençal, the *langue d'Oc.* **b.** Modern Provençal; *spec.* (see quot. 1974). Also *attrib.* Also **Occitanian** (ǫksitē·niăn) *sb.* and *a.*
1940 M. Belgion tr. *de Rougemont's Passion & Society* ii. vi. 99 The whole of the Occitanian, Petrarchian, and Dantesque lyric has but a single theme—love. 1957 *Archivum Linguisticum* IX. ii. 107 [illegible] in French, the adjoining fringe of Occitanian, and Franco-Provençal. 1960 [see *Languedocian *a.* and *sb.*]. 1961 P. Green tr. *Oldenbourg's Massacre at Montségur* i. 18 Languedoc itself, the area where the Occitan dialect was spoken, was not a major Power. 1964 G. Price in *Archivum Linguisticum* XVI. 34 In the twelfth and thirteenth centuries, Occitan—or Provençal, as it is more usually known—was one of the major European literary languages. *Ibid.*, The whole Occitan-speaking area, from the Alps to the Pyrenees. 1970 *Times Lit. Suppl.* 8 Jan. 40/2 Occitanian or Provençal (raised again to the status of a literary language by the efforts of the Félibrige). 1971 P. Wolff *Western Languages* v. 147 The years around 1100 also show the sudden appearance of lyrical poetry in Occitan, first made known in written form by the Duke troubadour William of Aquitaine (1071–1127). 1974 W. B. Lockwood *Panorama Indo-European Lang.* 33 Provençal is sometimes termed Occitan. 1974 R. A. Hall *External Hist. Romance Lang.* 26 Modern Provençal has had two main standard varieties: i. That of the lower Rhône valley and the French Riviera.. often called 'Mistralien'... ii. 'Occitan', a somewhat different variety with its base in the usage of Languedoc. Much, but not all, of the difference between 'Mistralien' and 'Occitan' lies in orthographical details.

occlude, *v.* Add: **2.** Also, to cover or hide; *spec.* to cover (an eye) so as to block its sight. (Further examples.)
1909 H. G. Wells *Tono-Bungay* I. ii. 74 In the middle was the brown coffin end..half occluded by the vicar's Oxford hood. 1921 *Amer. Jrnl. Ophthalm.* IV. 179/1 The choice of the eye to be covered is usually determined by finding out which eye the patient uses for pointing or aiming at a distant object, and occluding the other, or if one eye is defective, by occluding that. 1932 *Ibid.* XV. 321/1 The nondominant eye is occluded by having its spectacle lens replaced..by a black patch. 1965 Hirsch & Wick *Vision of Children* (1964) vii. 223 For suppression or for amblyopia with central (foveal) fixation, the preferred eye is usually occluded (direct occlusion). 1975 *Nature* 6 Feb. 406/1 During early childhood amblyopia can be cured by simple treatments such as by occluding the good eye for a period.

3. *intr. Dentistry.* Of a tooth: to come into contact with a tooth or teeth of the other set.
1888 E. S. Talbot *Irregularities of Teeth* x. 64 When the first permanent molars in both jaws have erupted so that they occlude, this will prevent forward progression. *Ibid.* 65 The anterior teeth do not occlude, and when the jaws are closed quite a space is observed. 1913 A. Hopewell-Smith *Introd. Dental Anat. & Physiol.* xi. 233 The teeth in Man do not..occlude by means of their cusps, but by a perfect system of interdigitation. 1974 Gregor & Hirschfeld *Minor Tooth Movement* (ed. 3) iii. 41 The mesiobuccal cusp of the maxillary first molar occludes with the buccal groove of the mandibular first molar in occlusion.

4. *trans. Chem.* To carry (a substance) out of solution by occlusion (sense *5); to trap within growing crystals of a precipitate.
1920 *Chem. Abstr.* XIV. 3030 Ca and Mg compds. are occluded in the ppts. of $Fe(OH)_3$ by NH_4OH. 1929 L. P. Hammett *Solutions of Electrolytes* i. iii. 40 Cadmium sulfide carries down or occludes barium sulfide, although the latter is a very soluble substance. 1930 W. T. Hall *Textb. Quantitative Analysis* xi. 141 If the barium sulfate adsorbs or occludes ferric sulfate, the latter will lose SO_3 upon ignition. 1960 Kolthoff & Sandell *Textbk. Quantitative Inorg. Analysis* (rev. ed.) viii. 121 If the incorporated material fits in the crystal lattice of the precipitate (host crystal), it is occluded in the form of mixed crystals or a solid solution. 1952 Diehl & Smith *Quantitative Analysis* ii. 23 Precipitates that occlude the mother liquor seriously under such conditions that little foreign material..is present in the solution.

5. *intr. Meteorol.* Of a front or frontal system: to undergo occlusion (sense *7).
1940 [illegible] *Weather Forecasting* [illegible] There is a further supply of warm air available to provide a warm sector in the cyclone which passes through a life-history, and in due course occludes. 1955 W. J. Saucier *Princ. Meteorol. Analysis* ix. 270/1 The front occludes further, and meantime the cyclonic circulation becomes larger and more symmetric. 1969 S. Petterssen *Introd. Meteorol.* (ed. 3) xii. 212 During the further development, the cold front overtakes the warm front and the system is said to occlude.

occluded, *ppl. a.*, Add: *spec.* in *Meteorol.*, applied to a front or frontal system in which a cold front has caught up with a warm front.
1922 Bjerknes & Solberg in *Geofysiske Publikationer* III. i. 4 The remaining part of the warm sector then [illegible] disappears from the surface, so that the cyclone on the ground only consists of cold air.... For this type we have chosen the name 'occluded cyclone'. 1934 D. Brunt *Physical & Dynamical Meteorol.* xviii. 324 When a depression has become occluded, the cold front trails behind the depression. 1941 B. Haurwitz *Dynamic Meteorol.* xv. 315 In this situation the occluded cyclone can increase its kinetic energy again. 1955 [see *occlusion 7]. 1970 F. W. Cole *Introd. Meteorol.* xii. 166 If an occluded front is well advanced in development, the warm-front portion may be lifted to great heights.

occluder (ǫklū·dəɹ). *Ophthalm.* [f. Occlud(e *v.* + -er[1].] Any device designed to occlude an eye.
1930 *Brit. Jrnl. Ophthalm.* XIV. 520 Many varieties of occluder..have been devised. 1949 S. Duke-Elder *Text-bk. Ophthalm.* IV. [illegible] For older children who can be trusted not to peep round it, a simple occluder of metal..or rubber..may be attached to the spectacles. [illegible] These should be replaced by strapping on a tie-on occluder during the night. 1974 *Nature* 11 Oct. 566/2 One eye was centred on a projection perimeter (the other eye was always covered with a plasticine occluder).

occlusal (ǫklū·săl), *a. Dentistry.* [f. L. *occlūs-*, ppl. stem of *occlūd-ere* to Occlude + -al.] Of, pertaining to, or involved in occlusion (sense *3); *spec.* applied to the surface of a tooth that comes into contact with a tooth in the other jaw in occlusion.
1897 S. H. Guilford in E. C. Kirk *Amer. Text-bk. Operative Dentistry* vi. 142 Cavities upon the occlusal surface are very accessible and in full view. *Ibid.* 149 The occlusal surface of an upper first or second molar presents two points liable to decay. 1939 L. Hirschfeld *Toothbrush* i. i (*caption*) Mandible of Australian aboriginal, showing advanced occlusal wear. 1951 J. M. Schweitzer *Oral Rehabilitation* xxxi. 623 The question of whether natural teeth should have flat occlusal surfaces [illegible] occlusal surfaces that have cusps. 1974 P. E. Dawson (*title*) Evaluation, diagnosis, and treatment of occlusal problems.
Hence **occlu·sally** *adv.*, from or on an occlusal surface.
1912 B. E. Lischer *Princ. & Methods Orthodontics* xiii. 202 An arm extended from the alignment wire..is made from an ordinary tube hook and prevented from dropping occlusally, or being forced gingivally, by flattening the alignment wire with a file along its lingual surface. 1946 Tylman & Peyton *Acrylics* xxi. 408/2 Inlay wax is then added occlusally to the thin coating. [illegible] the occlusal plane is determined, and finally the occlusal surfaces and contour are carved in the wax. 1969 *Gloss. Terms Dentistry* (*B.S.I.*) 72 *Occlusally approaching clasp*, a clasp approaching from the occlusal side of the survey line and terminating in the infra area.

occlusion. Add: **3.** *Dentistry.* The position assumed by the two sets of teeth relative to each other when the mouth is closed; the state of having the jaws closed and the teeth in contact.
1880 N. W. Kingsley *Treat. Oral Deformities* xxi. 325 In extreme pain (except in cases where the patient is suffering from periodontitis, when the occlusion of the jaws intensifies the suffering), the teeth are brought together with great force. 1904 V. H. Jackson *Orthodontia & Orthopedia of Face* 196 The opening of the bite in any manner with apparatus, if continued for a considerable time, is likely to prove detrimental to the occlusion. 1962 Blake & Trott *Periodontology* xiv. 256 In the European it is usual to find that during lateral movements to either side molar and pre-molar teeth of one side remain in occlusion, while contact is completely lost on the other side. 1974 Harty & Roberts *Restorative Procedures Practising Dentist* xii. 167 It is very important that the occlusion is studied and corrected before any fixed or removable prosthesis is made.

4. *Phonetics.* The act of closing, or the period of closure of, the breath passage during the articulation of an orally-released consonant, or of the mouth passage during the articulation of a nasal consonant.
1906 *Mod. Lang. Rev.* I. iv. 346 'Occlusion', was, if I remember aright, used by Wilkins in the seventeenth century, but this hardly justifies the employment of such a pedantic term instead of 'stoppage', 'closure', or 'stop'. 1926 *Germanic Rev.* I. i. 56 Here, 'occlusion' in the glottis does not mean glottal stop, but the partial occlusion in the articulation of voice sounds. 1932 J. S. Kenyon *Amer. Pronunc.* (ed. 6) 53 This initial contact for any sound that has contact is called the closure, or occlusion, and the end of the contact is called the opening, or release. 1943 *Amer. Speech* XVIII. 39 In pronouncing such phrases as 'Hello, Pete!'..a British speaker ordinarily articulates the vowel of the final syllable, lifts his tongue to make the closure for the final consonant [illegible], holds it there from eight to twelve hundredths of a second (occlusion), and then breaks the closure with a sharp downward snap of the tongue which produces a clear 'pop' (explosion). 1966 B. [illegible] *Phonol. Analysis Present-Day Standard Eng.* (rev. ed.) ii. 11 In producing *t*, the tip of the tongue forms an occlusion against the alveolar ridge.

5. *Chem.* A kind of co-precipitation (see quots.); the trapping of foreign material by growing crystals of a precipitate.
1920 *Chem. Abstr.* XIV. 3030 (*heading*) The occlusion of lime and magnesia by ferric oxide. 1929 L. P. Hammett *Solutions of Electrolytes* i. iii. 40 Analytical separations are..never quite as satisfactory..because of the existence of occlusion. 1932 [see *co-precipitation]. 1942 W. Rieman et al. *Quantitative Analysis* (ed. 2) xvi. 294 The phenomenon [sc. coprecipitation] has also been called contamination, inclusion, occlusion, and adsorption. The last two terms should not be used as synonyms for coprecipitation because they denote certain mechanisms that apply to some but not all cases of coprecipitation. 1947 *Jrnl. Chem. Education* XXIV. 597/1 The term occlusion has been used to refer to contamination of a precipitate by the incorporation into the body of the particle [illegible] of foreign substances whether or not the latter give rise to mixed crystal formation (solid solution)...; it has been used to include both solid solution phenomena and contamination of the precipitate by adsorption...; it has been used to indicate contamination by basic salts and other substances which are not present in solid solution...; and I have heard a noted analytical chemist reserve the term to indicate the contamination of a precipitate by mechanically trapped solute or solvent ions or molecules. No doubt one could locate other senses in which occlusion has been used; the term is almost omnipresent. 1966 E. M. Rattenbury *Introd. Titrimetric & Gravimetric Anal.* vii. 146 Sometimes water or mother liquor is imprisoned by occlusion. 1969 E. S. Gilbreath *Elem. Quantitative Chem.* x. 138 Occlusion is a form of coprecipitation in which impurities are enclosed..within the lattice structure of a crystalline precipitate. *Ibid.*, Contamination by occlusion..can produce serious errors in gravimetric analysis.

6. *Ophthalm.* The covering of an eye so as to block its sight.
1920 *Brit. Jrnl. Ophthalm.* IV. 146 (*heading*) The influence of prolonged monocular occlusion in revealing errors of the muscle balance. 1973 *Nature* 26 Jan. 288/2 The technique of monocular occlusion has long been used by ophthalmologists in the therapeutic treatment of such conditions as strabismus (squint) amd amblyopia.

7. *Meteorol.* The overtaking of the warm front of a depression by the cold front, so that the body of warm air between them is forced upwards off the earth's surface by two wedges of cold air; also, the occluded front so formed.
1922 Bjerknes & Solberg in *Geofysiske Publikationer* III. i. 6 No boundary surface results when both branches meet during the occlusion of the cyclone. *Ibid.*, The occlusion then assumes the character of a cold front with a rather narrow rain zone. 1934 D. Brunt *Physical & Dynamical Meteorol.* xviii. 311 The occlusion begins at the centre, where the cold front has a shorter path to cover before overtaking the warm front. *Ibid.*, There will still be a warm sector in the upper air for some time after occlusion has taken place at the surface. 1941 B. Haurwitz *Dynamic Meteorol.* xv. 314 The part of the front where the cold front has caught up is called the occlusion. 1944 Hewson & Longley *Meteorol. Theoret. & Appl.* xvi. 277 The rate of occlusion in several types of frontal depressions may be discussed qualitatively. 1955 W. J. Saucier *Princ. Meteorol. Analysis* ix. 270/1 The cold front overtakes the warm front, resulting in an occluded front (or occlusion). 1974 *Encycl. Brit. Macropædia* V. 394/2 There still is a belt of thick cloud and rain along the line of the old occlusion.

occlusive, *a.* Add: (Further examples.) Also, characterized by occlusion.
1961 *Lancet* 22 July 192/1 Of three techniques for treating the stump [of the umbilicus]—standard non-occlusive spirit technique, antibiotic or antiseptic non-occlusive technique, and occlusive technique with or without antibiotics—they showed the last to be the most reliable and the most effective. 1972 *Where* May/June 135/2 The sheath, the various occlusive caps, even chemical contraceptives do provide some barrier to the spread of germs from one person to another. 1974 Passmore & Robson *Compan. Med. Stud.* III. xvii. 16/2 The most important cause of occlusive arterial disease is atherosclerosis.

B. *sb. Phonetics.* A consonant sound produced with stoppage of breath by the organs of speech; a stop with suppression of the explosive sound.
1902 [see *affricate *v.*]. 1943 *Amer. Speech* XVIII. 39 The only audible expression of the final consonant..is the cutting-off of the preceding vowel sound by the raising of the tongue to make the closure... The explosive stop here becomes what may most conveniently be termed a simple occlusive. 1976 *Archivum Linguisticum* VII. 183 He cannot add *H_3 and a third series of guttural occlusives to the number of fallen consonantal phonemes he has postulated.

occluso- (ǫklū·so), comb. form of *occlusal *a.* or Occlusion (sense *3), as in **occluso-cervi·cal, occlusogi·ngival** *adjs.*
1923 Dorland *Med. Dict.* (ed. 12) 755/1 Occlusocervical. 1962 J. E. H. Fowler *Heinemann Mod. Dict. Dental Students* 182/1 *Occlusocervical*, relating to the occlusal surface and the neck of a tooth. 1940 J. Osborne *Dental Mech.* xii. 136 The brush should be run in an occluso-gingival direction. 1963 C. R. Cowell et al. *Inlays, Crowns, & Bridges* iii. 14 Measuring the occlusogingival depth on the side of the tooth.

occult, *a.* (*sb.*) Add: **1. c.** *Med.* Of a disease or bodily process: †inexplicable, obscure; unaccompanied by any readily discernible signs or symptoms. *occult blood*, blood abnormally present in some material (esp. fæces) but too scanty to be readily recognized; so *occult bleeding* [tr. G. *okkulte*(*magen*)-*blutung* (I. Boas 1901, in *Deutsch. med. Wochenschr.* 16 May 315/2)].
[1690 S. Blanckaert *Lexicon Novum Medicum* 450 *Occulti morbi*... Angl. Hiden disease.] 1809 R. Carmichael *Ess. Effects of Carbonate upon Cancer* iv. 292 Three of these lucky people had occult cancers of the breast, and a fourth had an occult cancer of the lip. 1820 R. Hooper *Lexicon Medicum* (ed. 4) 623/2 *Occult quality*, a term that has been much used by writers that had not clear ideas of what they undertook to explain; and which served therefore only for a cover to their ignorance. 1854 W. E. Swaine tr. *Rokitausky's Man. Path. Anat.* I. ix. 260 This process takes place either in the depths of the growth,..as so-termed occult cancer; or upon the free surface of the body or of a mucous cavity, as so-called apert or open cancer. 1872 D. C. Black *On Functional Dis. Renal Organs* ii. 108 The condition to which Dr. Barnes has applied the term occult menstruation... Owing to the imperforate condition of the hymen, the menstrual flux accumulates in the *cul de sac*..formed by the upper portion of the vagina. 1904 *Progressive Med.* IV. 24 Occult blood is constantly found in cancer of the gastrointestinal tract. *Ibid.*, Occult bleeding in the cases of ulcer was most frequently observed when the patient had recently complained of pain in the stomach. 1905 G. R. Butler *Diagnostics of Internal Med.* (ed. 2) I. xii. 147 'Occult' blood is a term applied to blood in the fæces (or stomach contents) which is in such small quantity that it can not be recognised by the eye, or is so altered as not to be identified by the microscope. 1936 E. N. Chamberlain *Symptoms & Signs Clin. Med.* xii. 387 The most important chemical test applicable in the clinic room is that for occult blood. 1961 R. D. Baker *Essent. Path.* xvii. 458 Occult carcinoma of the prostate is a frequent finding at autopsy in persons who have died from other causes. 1962 *Lancet* 1 Dec. 1145/1 Orthotoluidine sampling of these motions was strongly positive for occult blood. 1973 *Q. Jrnl. Med.* XLII. 125 (*heading*) Occult rickets and osteomalacia amongst the Asian immigrant population. 1973 *Lancet* 4 Aug. 262/1 Occult bleeding in the alimentary tract.

occult, *v.* Add: **a.** (Later examples.)
1957 A. MacNab *Bulls of Iberia* vi. 61 In both the *gaonera* and the *mariposa* the man's body is fully exposed in front of the bull's face, and not occulted by the cloth lure. 1977 'E. Crispin' *Glimpses of Moon* v. 76 The platform for the legs competition..had a wooden screen designed to occult the competititors from the groin upwards.

c. *intr.* Of a lighthouse light: to be cut off from view as part of its cycle of light and dark.
1880, 1892 [implied in Occulting *ppl. a.*]. 1902 *Encycl. Brit.* XXVI. 464/1 The light occults every ten seconds,.. the occultations being actuated by a double valve arrangement. 1964 J. Masters *Trial at Monomoy* i. 15 Close on the north the lighthouse tower edged into the view... That light occulted every four seconds.
Hence **occu·lter,** an apparatus for occulting lights.
1902 *Encycl. Brit.* XXX. 256/1 This light shows, instead of one prolonged flash at intervals of one minute, as would be produced by the apparatus in the absence of a gas occulter, a group of short flashes.

occultist. Add: Also *attrib.*
1902 *Encycl. Brit.* XXX. 275/1 Cabalistic, occultist, Indian, and modern spiritualistic ideas and formulas. 1977 R. L. Wolff *Gains & Losses* iv. 316 Charles Maurice Davies['s]..book *The Great Secret* (1896)..was published anonymously. His occultist leanings were, however, well known.

occupance. Delete *rare* and add: *spec.* in *Geogr.*, the inhabiting and modification of an area by man. (Further examples.)
1932 *Sci. Monthly* XXXV. 266/1 France..is strikingly an aggregation of '*pays*', each of which has its distinctive character stamped upon every aspect of the natural environment and upon every form of human occupance thereof. 1934 *Ann. Assoc. Amer. Geogr.* XXIV. 81 Occupance is..defined as the process of occupying or living in an area, and the transformations of the initial landscape which result. *Ibid.*, Occupance refers not only to..economic activities but also to other activities..such as the construction of buildings, roads, and so on. The word includes also the results as well as the process of living in an area. 1974 F. Emery *Oxfordsh. Landscape* i. 35 People who could cultivate the land with digging-sticks in order to grow cereals, covering the ground..with a 'slash-and-burn' or shifting occupance of the kind we still meet today in certain forest environments.

occupancy. Add: **3. b.** *Teleph.* The proportion of the time during which a circuit or device is handling calls.
1933 *Post Office Electr. Engineers' Jrnl.* XXVI. 269 (*caption*) Probabilities *P* for a delay exceeding *t* holding times when there are *n* switches with occupancy *a*. 1960 R. Syski *Introd. Congestion Theory Teleph. Syst.* i. 10 The traffic carried (or handled) refers to the changing patterns of occupancy conditions in the group. *Ibid.* vi. 365 The limit in (4.6) exists provided the traffic intensity ('occupancy')..is less than unity.

c. The proportion of accommodation occupied or used.
1974 *Economist* 7 Sept. 51 The 'France' has had a 77 per cent occupancy rate on the North Atlantic run and a 91 per cent rate on winter cruises. 1975 *Daily Tel.* 11 Dec. 9/6 If airlines did not overbook, planes would go out 40 per cent. full instead of the 70 per cent. occupancy needed to give a profit. 1977 *Daily Tel.* 2 Dec. 10/7 Hotels in large towns outside London experienced a similar increased occupancy this year. 1977 *Time* 19 Dec. 61/3 The 27,400-acre complex..sports three Disney hotels, with an occupancy rate of about 97%.

‖**occupatio** (ǫkiŭpēi·ʃio, -pā·tio). *Rhet.* [L. (see Occupation).] = Preterition 3.
1586 [see Paralipsis]. 1928 C. S. Baldwin *Medieval Rhetoric & Poetic* x. 296 The rehearsal of all the conventionally appropriate *loci* of description at the funeral of Arcite (A2919–2966) sounds to modern ears impatient, if not sarcastic. But, after all, the whole long passage is the 'colour' *occupatio* (præteritio). The shorter *occupatio* in the Squire's Tale (F63–75) suggests sarcasm less by itself than in its connection with lines 32–40 and 401–408. 1933 F. N. Robinson *Compl. Wks. Chaucer* 772/2 It should be added that the rhetorical figure here employed—the refusal to describe or narrate, technically known as 'occupatio'—is very common with Chaucer. 1957 C. Muscatine *Chaucer & French Tradition* vi. 177 Though Chaucer omits a great deal of the tale originally told by Boccaccio in the *Tescida*, he frequently resorts to the rhetorical device of *occupatio* to summarize in detail events or descriptions in such a way as to shorten the story without lessening weight and impressiveness. 1968 J. A. W. Bennett *Chaucer's Bk. of Fame* ii. 94 But the didactic bird is not to be thwarted utterly, and launches.. into an *occupatio* summarizing the lore in Ovid's *Fasti* that would have enabled the poet to identify such constellations as Lyra, Gemini, and the Pleiades. 1975 *Times Lit. Suppl.* 28 Feb. 213/4 'Do not describe the total horror, the full degree of despair that rushes into your head, your brain and emotions...', and so on. The device used to be known as *occupatio.*

occupation. Add: **1. b.** *spec.*, by Germany and her allies during the war of 1939–45; usu., the period during which a country was held by German, etc., troops, or the state of being held by such troops.
1940 A. Huxley *Let.* 7 July (1969) 455 Jehanne was out of Paris during the occupation. 1957 *Times Lit. Suppl.* 27 Dec. 783/3 The particular man whose life has been conditioned by the hot North African sun and the cold chill of the Occupation. 1972 *Guardian* 9 Sept. 12/4 The programme is..divided into three parts... There is life under the occupation..as drawn from experience in occupied Europe.

7. Also, in military uses, as *occupation army, forces, troops;* in archæological use, as *occupation floor, layer, level, scatter, site.* Also **occupation centre,** an establishment where occupational therapy is practised or where the mentally handicapped are trained or employed; **occupation disease,** an occupational disease; **occupation neurosis** *Med.*, a painful and disabling spasm affecting muscles used more than normally because of the person's occupation; **occupation number** *Physics*, the number of particles in a system that are in any given state.
1918 E. S. Farrow *Dict. Mil. Terms* 414 Occupation Army, an army that remains in possession of a newly conquered country, retaining it as a kind of hostage, until peace is signed and the war indemnity paid. 1976 T. Allbeury *Only Good German* xiv. 100 A goon on each staircase and the lift doors padlocked... It's like an occupation army. 1940 Frazer & Stallybrass *Text-bk. Public Health* (ed. 10) xix. 440 If the home conditions are good and the defective's condition is suitable he may attend at an occupation centre where simple occupational training can be given. 1958 [see *day-hospital* s.v. *day *sb.* 23 a]. 1965 Peters & Kinnaird *Health Services Admin.* vii. 255 The [Education] Authorities establish training and occupation centres for the lower grades of the mentally handicapped who cannot take education in ordinary educational subjects. 1900 Dorland *Med. Dict.* 209/1 Occupation-disease. 1901 *Brit. Med. Jrnl.* 17 Aug. 405/1 (*heading*) The medical profession and the control of occupation diseases. 1930 F. B. Young *Jim Redlake* III. iii. 326 Overcrowding, short commons, adulterated food, occupation diseases—they're all just words in a newspaper. 1959 J. D. Clark *Prehist. S. Afr.* plate 4 (*caption*) Handaxes, cleavers, and waste flakes of evolved Chelles-Acheul culture on occupation floors. 1971 W. Tucker *This Witch* ii. 20 The native beer was vile... The occupation forces did their drinking elsewhere. 1953 R. J. C. Atkinson *Field Archaeol.* (ed. 2) i. 39 The chief use of such detectors is in the excavation of graves and occupation-layers in which metal objects may be expected to occur. 1935 *Discovery* Nov. 343/1 Further excavations..have brought to light older occupation-levels and hearth-marks. 1888 W. R. Gowers *Man. Dis. Nervous Syst.* II. v. 656 The term 'occupation neuroses', adopted from the German ('Beschaftigungs-neurosen'), is a convenient designation for a group of maladies in which certain symptoms are excited by the attempt to perform some often-repeated muscular action, commonly one that is involved in the occupation of the sufferer... The most frequent symptom is spasm. 1911 *Jrnl. Nerv. & Mental Dis.* XXXVIII. 107 An occupation neurosis is literally a fatigue cramp, and is characterized by spasms of muscles concerned in special movements, and brought on whenever these special movements, such as writing, are attempted. 1958 Sykes & Bell tr. *Landau & Lifshitz' Quantum Mech.* ix. 215 Let us seek to construct a mathematical formalism in which the occupation numbers..of the states (and not the co-ordinates of the particles) play the part of independent variables. 1974 G. Reece tr. *Hund's Hist. Quantum Theory* xiii. 180 Jordan was thus entitled to express his hope of a quantum wave theory of matter in which the numbers of particles would be the occupation numbers N_r of the discrete quantum wave states. 1954 S. Piggott *Neolithic Cultures* ix. 271 The occupation-scatter of small sherds and flints. 1939 *Oxoniensa* IV. 6 The region between Oxford and Northampton is notoriously, but probably deceptively, barren of occupation-sites. 1948 *N.Y. Jrnl. American* (Sunday Mail ed.) 9 May 1/7 Police and U.S. occupation troops are prepared for bloodshed. 1975 R. L. Duncan *Dragons at Gate* (1976) iv. 32 On the day of the surrender, before the Occupation troops could arrive, Takaeshi..set out into Tokyo Bay, and blew himself up.

occupational, *a.* (s.v. Occupation). Add: (Further examples.) *occupational disease*, (a) disease to which a particular occupation renders a person especially liable; also *joc.*; *occupational hazard, risk*, a risk accepted as part of a particular occupation; also *joc.*; *occupational therapy*, an activity, mental or physical, prescribed as an aid to recovery from disease or injury or for mental patients; so *occupational therapist*, someone skilled in supervising or trained to supervise such activity.
1901 *Brit. Med. Jrnl.* 17 Aug. 412/2 A number of valuable papers upon occupational diseases. 1915 G. E. Barton in *Trained Nurse & Hospital Rev.* Mar. 138 (*heading*) Occupational therapy. 1919 —— (*title*) Teaching the sick; a manual of occupational therapy and re-education. 1919 J. L. Garvin *Econ. Found. Peace* 328 These occupational federations acted independently of each other. 1922 *Jrnl. Mental Sci.* LXVIII. 192 The *personnel* comprises a chief and five assistant trained occupational therapists. 1923 *Ibid.* LXIX. 126 The occupational centre or 'curative workshop' serves more or less as a proving ground. 1926 B. Webb *My Apprenticeship* iii. 129 Personal vanity..was an 'occupational disease' of entertaining and being entertained. 1934 H. C. Warren *Dict. Psychol.* 184/2 *Occupational hierarchy*, the serial arrangement of occupational groups according to average intelligence. 1937 *Discovery* May 144/1 The excavation..showed a mound that had already been abandoned by 1500 B.C., and 20 occupational levels below it. 1944 M. Laski *Love on Supertax* iv. 52 You're insured against occupational risks. 1949 W. L. Warner in M. Fortes *Social Struct.* 4 Interviews..are taken from informants..from several occupational groups. 1951 R.

Firth *Elem. Social Organiz.* iv. 138 A worker's choice of employment is guided by his wife's attitude to the conditions of his work, its cleanliness, occupational risks, or security. 1952 'Vigilans' *Chamber of Horrors* 94 *Occupational hazard*, a risk necessarily run in one's work. 1956 S. Gibbons *Here be Dragons* xvii. 235, I don't think I shall marry if I'm asked. There's too much occupational risk. 1956 B. Goolden *At Foot of Hills* ix. 202 Are you the young man who wants to see over the occupational therapy department? 1959 *Listener* 15 Oct. 614/1 Exile has been regarded as an occupational hazard for poets in particular ever since Plato denied them rights of citizenship in his republic. *Ibid.* 10 Dec. 1041/3 As a museum man himself, Mr. Baxandall will be aware of.. that occupational disease of his profession which might be called 'collector's greediness'. 1959 B. Wootton *Social Sci. & Social Path.* ii. 48 Whether, on balance, prisoners are likely to exalt or to debase their occupational level is anybody's guess. 1971 *Times* 6 Sept. 7/8 The Department of Employment has occupational guidance services throughout the country. 1973 *Times* 13 Jan. 19/7 Occupational pension scheme is one established by an employer for the benefit of his employees, and wholly paid for by him and (frequently, but not necessarily) his employees. *Ibid.* 1 June 24/1 (Advt.), For further information please contact the Head Occupational Therapist. 1974 *Encycl. Brit. Micropædia* VII. 470/2 Occupational therapy provides not only training in daily living activities but also aids that make eating, dressing, and toilet less fatiguing for the sick or elderly. 1974 Passmore & Robson *Compan. Med. Stud.* III. 76/1 Asbestos has become an increasingly important occupational hazard. *Ibid.*, This led to a false sense of security about occupational disease in coal mining until 1920.
Hence **occupa·tionally** *adv.*
1952 C. P. Blacker *Eugenics* vii. 141 That there existed a group of occupationally unstable persons who are in continuous receipt of public assistance had been well recognized since the time of Charles Booth. 1971 D. Crystal *Linguistics* 132 Do they occur in certain occupationally restricted uses of language only, such as journalese? 1973 *Daily Tel.* 18 Aug. 8/1 The Council, sitting in Star Chamber, publicly ordered anyone not occupationally resident in London to leave the City forthwith.

occupa·tionalism. [-ism.] Occupational character or conduct; professionalism.
1927 E. Barker *National Character* iv. 96 It is true that a new and qualifying factor has been added to the national temper by the growth of occupationalism.

occupied, *ppl. a.* Add: Esp. of countries held by Germany and her allies during the war of 1939–45; of parts of countries under military occupation.
1940 R. W. B. Clarke *Britain's Blockade* 7 The division of France into 'occupied' and 'non-occupied' territory is nothing more than a device to relieve the Germans of the administrative difficulties created by the millions of refugees, [etc.]. 1940 A. Huxley *Let.* 9 Oct. (1969) 459 There is practically no communication between occupied France and the USA. 1941 *Times Lit. Suppl.* 18 Oct. 514/1 (*heading*) In occupied Belgium. 1965 M. Spark *Mandelbaum Gate* vii. 280 'So far as I know she's still in Israel—' Joe Ramdez clapped his hands over his ears... 'Occupied Palestine,' Freddy said with deference. 1973 L. Snelling *Heresy* I. i. 8 In Paris, capital of Occupied France, in the abnormally hot summer of nineteen-forty.

occupying, *ppl. a.* Add: **occupying power,** a state whose army occupies (part of) a foreign country; used *spec.* with ref. to the occupation of Germany after the war of 1939–45.
1946 *Ann. Reg. 1945* 452/1 (Index) Germany,..Legislation by occupying powers. 1959 *Observer* 22 Mar. 1/2 Nuclear weapons..would be in the hands of the occupying Powers and never pass into German control. 1965 A. J. P. Taylor *Eng. Hist. 1914–45* xvi. 594 The Allied leaders.. reached agreement of a kind over the reparations which each of the occupying Powers could exact from its zone of Germany. 1975 J. Cleary *Safe House* i. 26 The need to show the Occupying Powers that..they were not going to favour Nazis and ex-Nazis.

occurrence. Add: **5.** *attrib.*, as **occurrence(s book,** a record of events kept at a police station, drawn from the diaries of police officers.
1929 J. Moylan *Scotland Yard* vi. 138 Occurrence Books are kept at all stations, so that there may be a complete daily record of all occurrences, etc., at the station or within the area assigned to it. 1955 M. Gilbert *Sky High* v. 72 Everything that a policeman hears, sees and does goes down in the Occurrences Book. 1966 L. Southworth *Felon in Disguise* xi. 158 The Inspector was about to complete his Occurrence Book entries. 1972 M. Gilbert *Body of Girl* xi. 103 It would have been in the Occurrence Book... Where are the 'O' books kept?

occurrency. Restrict †*Obs.* to sense in Dict. and add: **b.** = Occurrence 2.
1935 G. K. Zipf *Psycho-Biol. of Lang.* (1936) 289 Relative frequency of occurrency of meaning. 1967 *Oceanogr. & Marine Biol.* V. 283 Most of the clusters defined by joint occurrency of species have a given geographical distribution.

ocean, *sb.* (*a.*) Add: **3. a.** Also *pl.* Lots *of.* (Further examples.)
1840 *Spirit of Times* 25 Apr. 85/3 The leader of this predatory band had oceans of money which he looked to when he sat down, and then crammed his greasy wallet back into his pocket. 1886 H. Baumann *Londinismen* 123/2 He's got oceans o' money. 1926 *Amer. Mercury* Dec. 465/1 She is a flaming flamboyant blonde with oceans of stuff. 1952 M. Laski *Village* ii. 36 Poor People's children..had oceans of pocket-money because Poor People didn't understand the value of money.

b. Phr. *ocean of being.*
1652 N. Culverwel *Worth of Souls* in *Lt. Nature* [11]. 201 All beings they are within the souls Horizon... It can take in the several drops of Being, and it can take in much of the Ocean of Being. 1690 Locke *Essay Hum. Und.* I. i. 3 We let loose our thoughts into the vast ocean of Being. 1931 G. F. Stout *Mind & Matter* 14 Knowledge of this type..leaves us adrift on the ocean of being, with oars indeed, but without rudder or compass.

4. a. *ocean-blue, -deep, -flood, -floor* (misprinted *flood* in Dict.), *front, -green, -song*; **b.** *oceanflowing* adj.; objective, as *ocean-cleaving, -dividing* adjs.
1842 W. C. Bryant *Child's Funeral* in *Fountain* 62 Flowers of the morning—red, or ocean-blue. 1936 *Times* 6 Jan. 11/3 It is in a number of good colours, including ocean-blue. *a* 1926 R. Campbell *Golden Shower* in *Coll. Poems* (1957) II. 21 The ocean-cleaving whale. 1926 J. Pedersen *Israel* I. ii. 463 He is overwhelmed by the surges of death..and desires to be pulled up from the *tᵉhōmōth* of the earth, its ocean-deep. 1966 'J. Hackston' *Father clears Out* 43 His diving-dress would enable a diver to go down to the very depths of the ocean-deep. 1954 W. Faulkner *Fable* 232 The mutual rage and fear of the three ocean-dividing nations themselves. *a* 1957 R. Campbell tr. *F. Garcia Lorca's He died of Love* in *Coll. Poems* (1960) III. 78 The ocean-flood of perjured oaths Was thundering. 1820 Shelley *Ode to Liberty* v, in *Prometh. Unb.* 211 The ocean-floors Pave it. 1884 [in Dict.]. 1968 *Times* 3 Oct. 13/3 The ocean floor is spreading out from the Mid-Atlantic Ridge at a rate of between one and three centimetres a year. 1974 L. Deighton *Spy Story* xix. 202 The ship sank to the ocean floor. 1922 Joyce *Ulysses* 656 Confluent oceanflowing rivers with their tributaries and transoceanic currents. 1934 Webster, Ocean front. 1963 *New Yorker* 8 June 104 Your own ocean-front cottages. 1975 R. L. Simon *Wild Turkey* (1976) xxii. 159 We headed..onto the street by the pier... We were alone on the ocean front. 1922 Joyce *Ulysses* 253 By bronze, by gold, in oceangreen of shadow. 1976 *Yorkshire Even. Press* 9 Dec. 20/2 (Advt.), 1973 L. 144 saloon de luxe, ocean green, black trim, two bar, wing mirrors, push button radio. 1851 H. Melville *Moby Dick* II. ii. 7 His casual stopping-places and ocean-inns, so to speak. 1939 T. S. Eliot *Family Reunion* I. i. 41 These ocean liners With all their swimming baths and gymnasiums. 1964 M. McLuhan *Understanding Media* II. x. 94 People used to say that an ocean liner might as well be a hotel in a big city. 1974 W. Garner *Big Enough Wreath* 154 Like a power-boat brought under the bows of an ocean liner. 1934 Webster, Oceanside. 1962 J. D. MacDonald *Girl, Gold Watch & Everything* vi. 71 These [rooms] interconnect so this whole oceanside can be turned into a big suite. 1975 *Sat. Rev.* (U.S.) 3 May 62/3 Immaculate oceanside apartments. 1922 Joyce *Ulysses* 261 Lips that..hummed..the oceansong.

c. ocean greyhound (later examples); **oceanless** *a.* (*nonce-wd.*) devoid of or lacking an ocean; **ocean-line** (only Melville) = *ocean-lane*; **ocean pipe-fish,** a pipe-fish, *Entelurus æquoreus*, found in oceanic waters of north-western Europe; = *snake pipe-fish* (Snake *sb.* 11 b); **ocean spray** *N. Amer.*, a shrub of western North America, *Holodiscus discolor*, of the family Rosaceæ, sometimes included in the genus *Spiræa*, and distinguished by curving branches bearing large panicles of small white flowers; **ocean wave,** rhyming slang for 'shave'.
1913 F. H. Burnett *T. Tembarom* xl. 519 An ocean greyhound had landed the pair at the dock. 1967 *Economist* 23 Sept. 1109/1 Ironically, the airlines which once had only speed to offer against the one-time ocean greyhounds, can now anticipate the lounges, cinemas, etc., which until the jumbo jets get going remain one of the few prerogatives of the ocean liners. 1941 T. S. Eliot *Dry Salvages* ii. 9 We cannot think of a time that is oceanless. 1851 H. Melville *Moby Dick* II. ii. 5 The sperm whales ..mostly swim in *veins*, as they are called; continuing their way along a given ocean-line with..undeviating exactitude. 1865 J. Couch *Hist. Fishes Brit. Islands* IV. 358, I have possessed a male of the acknowledged Ocean Pipefish which in length measured twenty-six inches. 1925 J. T. Jenkins *Fishes Brit. Isles* 372/2 Ocean pipefish. 1906 *Contrib. U.S. Nat. Herbarium* XI. 330 Schizonotus discolor... Ocean spray. 1940 *Oregon: End of Trail* 20 In the spring and early summer..sweet syringa, ocean spray, and Douglas spirea form streamside thickets of riotous blossom. 1971 *Islander* (Victoria, B.C.) 28 Mar. 12/2 A rod made from a seasoned spirea (ocean spray), a common shrub around Victoria. 1928 M. C. Sharpe *Chicago May* 287/2 *Ocean waves* [sic], shave. 1934 *John o' London's Weekly* 9 June 353/1, I 'as my ocean wave an' when I've got my mince-pie properly open I goes down the apples and pears.
Also **oceaner,** (*b*) an ocean-going vessel.
1879 W. Whitman *Specimen Days* (1882–3) 136 The proud, steady, noiseless cleaving of the grand oceaner down the bay.

Oceana (ōuʃi,ē·nă). The name of Harrington's ideal state, applied by J. A. Froude to the British Maritime Empire.
[1656 J. Harrington (*title*) The Common-Wealth of Oceana.] 1886 J. A. Froude *Oceana* 395 If Oceana is to be hereafter governed by a federal parliament, such a parliament will grow when the time is ripe for it. 1899 J. Milne *Romance of Pro-Consul* ix. 87 He was being set to the straightening-out of some twist in Oceana, to the healing of a sore which threatened one of her limbs.

oceanarium (ōuʃiănēə·riəm). orig. *U.S.* Pl. -ia. [f. Ocean *sb.* (*a.*) + *-arium*, after Aquarium.] An establishment having a pool in which large sea-creatures can be kept and observed, esp. for public entertainment.
1944 E. B. Marks *They All had Glamour* 440 The aquarium of other days is now slated to be the oceanarium. 1955 *Sci. News Let.* 2 Apr. 211 (*caption*) A puzzled fisherman hauled the strange fish in in his shad net off the northern coast of Florida and it was identified by scientists at the oceanarium at Marineland, Fla. 1962 *Times* 15 Feb. 15/5 This is the usual training manœuvre for any type of animal in circuses and in oceanaria. 1972 *Village Voice* (N.Y.) 1 June 21/5 In exploiting the killer whale for the amusement of the crowds, the oceanarium people exploded two myths about Shamu's sisters and brothers of the sea. 1977 *Time* 4 July 28/3 In San Diego, 55 miles from Disneyland, is Sea World, the best-planned, best-stocked oceanarium in the U.S.

oceanaut (ōu·ʃiănǭt). [f. Ocean *sb.* (*a.*) + *-naut*, after *aquanaut.] One who lives for a period at the bottom of the sea in an underwater 'house'.
1962 *Daily Tel.* 7 Sept. 1/5 This cylinder will be pierced in the middle by a large chimney, through which our two oceanauts will pass. 1967 *Ibid.* 30 Jan. 10/6 Cousteau looks forward to training oceanauts for completely new tasks like servicing and even laying deep underwater pipe lines for oil companies, as well as carrying out research on the almost unknown marine life found at such depths.

oceanic, *a.* Add: **1. b.** Pertaining to or inhabiting those regions of the open sea beyond the edge of a continental shelf.
1877 H. N. Moseley in *Q. Jrnl. Microsc. Sci.* XVII. 24 The present oceanic form is placed in the genus Stylochus on account of the position of the tentacles. 1902 *Encycl. Brit.* XXXIII. 936/2 The majority of the oceanic epiplankton appears to be stenothermal. 1909 [see *holoplankton* s.v. *holo-]. 1942 H. U. Sverdrup et al. *Oceans* viii. 278 The oceanic province has an upper lighted zone and a lower dark zone. 1953 E. Palmer tr. *Ekman's Zoogeogr. Sea* xiv. 312 The coastal organisms..are termed neritic and they are contrasted with the open-sea organisms; the latter are often called simply oceanic, but this term is less exact. 1974 Lucas & Critch *Life in Oceans* i. 24 The pelagic division is divided into the region inshore of the continental edge, known as the 'neritic province', and the remainder, called the 'oceanic province'. In the oceanic province some aspects of the environment may change with level.

2. a. (Later example.)
1977 *Language* LIII. 391 One must at all times bear in mind, however, that this [sc. the philosophy of history] is a field of oceanic proportions.

b. *oceanic feeling*: a phrase used in a communication with Freud by R. Rolland (1866–1944), French writer and philosopher, in describing the longing for something vast and eternal of which he and others felt aware, and which he suggested as a possible source of religious feelings, but Freud interpreted as probably the nostalgia of the psyche for the ego-completeness of infancy; so *oceanic longing.*
1930 J. Riviere tr. *Freud's Civilization & its Discontents* i. 8 It is a feeling which he would like to call a sensation of 'eternity', a feeling as of something limitless, unbounded, something 'oceanic'... I cannot discover this 'oceanic' feeling in myself. *Ibid.* 21 The 'oceanic' feeling, which I suppose seeks to reinstate limitless narcissism... I can imagine that the oceanic feeling could become connected with religion later on. 1944 O. Fenichel in *Psychoanal. Rev.* XXXI. 145 The masochist behaves masochistically because he has an oceanic longing for being united with a greater unity. 1949 Koestler *Insight & Outlook* xiii. 194 The general occurrence of the oceanic feeling, of the tendency towards cosmic self-transcendence, is a fact. 1967 *Philos. Rev.* LXXVI. 207 Correlates..may be lacking when we come to the experiencing of oceanic feelings. 1971 P. Balogh *Freud* x. 109 The feeling described by Romain Rolland of being mystically identified with the universe Freud called an oceanic feeling.

3. (Later examples.)
1937 *Discovery* Oct. 303/2 The origins of the Oceanic peoples and the remarkable affinities of culture between them and the Naga tribes of Assam. 1971 L. A. Boger *Dict. World Pott. & Porc.* 248/1 In a general sense Oceanic pottery is primarily a woman's craft. 1974 S. Marcus *Minding the Store* (1975) xiv. 288 They then proceeded to duplicate the same systematic approach to pre-Columbian terracottas and Oceanic sculpture.

4. Of a climate: influenced by proximity to a large body of water and hence having a relatively small diurnal and annual range of temperature and relatively great precipitation.
1877 A. Geikie *Elem. Lessons Physical Geogr.* v. 351 An insular or oceanic climate is one where the difference between summer and winter temperature is reduced to a minimum, and where there is a copious supply of moisture from the large water-surface. 1902 *Encycl. Brit.* XXX. 710/2 A globe whose surface is dotted with land and water, so uniformly intermixed that there can be no chance for the existence of distinct areas of continental and oceanic climates. 1922 W. G. Kendrew *Climates of Continents* xxix. 215 The east of the British Isles has a continental rather than oceanic rainfall régime. 1957 G. E. Hutchinson *Treat. Limnol.* I. vii. 443 The distinction between oceanic and continental climatic regimes. 1974 *Encycl. Brit. Macropædia* IV. 726/1 The differences

between oceanic (or maritime) and continental climates, though operative in all latitudes, are most apparent in middle latitudes.

oceanicity (ōuʃiǎni·sīti). [f. OCEANIC *a.* + -ITY.] The extent to which a particular climate is oceanic; the state of being or having an oceanic climate.

1946 *Proc. Prehist. Soc.* XII. 3 Conditions of lesser oceanicity such as are encountered for instance in the great central plain of Ireland. **1950** *Jrnl. Ecol.* XXXVIII. 335 A. Mann (1929) suggested using as a measure of the degree of oceanicity of climates an index of hygrothermy. **1950** F. E. ZEUNER *Dating Past* (ed. 2) 394 The Theory relies on periods of increased and decreased oceanicity of the climate coupled with favourable physiographic conditions. **1974** *Nature* 17 May 211/1 The term 'oceanicity' traditionally has conveyed to ecologists the impression of a climate with a low amplitude fluctuation in seasonal temperature and a high annual precipitation.

oceanite (ōu·ʃiǎnəit). *Petrogr.* [ad. F. *océanite* (A. Lacroix *Minéral. de Madagascar* (1923) III. iii. 49), f. *océan* OCEAN *sb.* (*a.*), from the abundance of the rock on islands in the Pacific + *-ite* -ITE[1].] Any of various basalts which are very rich in olivine.

[**1923** *Mineral. Abstr.* II. 146 New rock-names are doréite, kivite, and océanite.] **1926** G. W. TYRRELL *Princ. Petrol.* vi. 131 (*caption*) Oceanite (ultrabasic olivine-rich 'basalt'). **1967** HESS & POLDERVAART *Basalts* I. 201 An increase in olivine and clinopyroxene in the alkali basalts (at the expense of plagioclase) leads to the development of melanocratic lavas (alkaline picrite-basalts, 'oceanites', and ankaramites). **1971** *Nature* 8 Oct. 406/2 The area includes both aa and pahoehoe lava flows which are olivine basalt, basalt, or oceanite in composition.

oceanity (ōuʃi,æ·nīti). [ad. G. *oceanität* (W. Zenker *Der thermische Aufbau d. Klimate* (1895) II. 122): see OCEAN *sb.* (*a.*) + -ITY.] = *OCEANICITY.

1922 [see *CONTINENTALITY]. **1924** *Geogr. Jrnl.* LXIV. 43 A station near the coast will thus show marked 'continentality' or 'oceanity' of climate, according as the prevailing wind is from the land or from the sea. **1944** V. CONRAD *Methods in Climatol.* xvii. 195 It is always better to speak of continentality than of 'oceanity'. **1959** R. E. HUSCHKE *Gloss. Meteorol.* 402 *Oceanity*, same as oceanicity.

oceanization (ōuʃiǎnəizēi·ʃən). *Geol.* [f. OCEAN *sb.* (*a.*) + -IZATION.] The conversion of continental crust into the much thinner and petrologically distinct oceanic crust.

1960 *Jrnl. Geophysical Res.* LXV. 4128/2 We must do our best to find an explanation of this 'oceanization' of the earth's crust. **1961** *Jrnl. Geol.* LXIX. 653/1 The statement that the 'sial' may be transformed into oceanic crust which leads to the 'oceanization' of a certain part of the earth's surface seems to be quite reasonable. **1973** *Nature* 22 June 434/1 Subsidence being attributed variously to crustal warping associated with orogenic compression, subcrustal erosion, oceanization (and that very recently). **1975** *Ibid.* 6 Feb. 396/2 The Soviet tectonician Belousov has gone so far as to invoke extensive 'oceanisation' of continental crust to account for the ocean basins.

oceanography. Delete '*c* 1880' in etym. and substitute for def.: The science dealing with the physical and biological properties and phenomena of the ocean. Cf. OCEANOLOGY in Dict. and Suppl. (Earlier and later examples.)

1859 M. F. MAURY *Physical Geogr. Sea* (rev. ed.) xx. 330 'Text-book of Oceanography for the Use of the Imperial Naval Academy', by Dr. August Jilek, Vienna, 1857. **1931** H. B. BIGELOW *Oceanogr.* i. 8 The other [factor] is the growth of an economic demand that oceanography afford practical assistance to the sea fisheries. **1955** *Sci. News Let.* 27 Aug. 142/1 Explorers of the ocean depths have been turning up so many new peaks, ridges, basins, seamounts and other underwater landmarks that naming them all has become a major problem in the flourishing science of oceanography. **1970** D. A. ROSS *Introd. Oceanogr.* i. 3 Most oceanographers have divided oceanography into four main parts: (1) chemical oceanography; (2) biological oceanography; (3) physical oceanography; and (4) marine geology and geophysics. **1977** *Times Lit. Suppl.* 18 Nov. 1348/3 The only basic research programmes in which he was closely involved were..for high-energy physics, materials research and oceanography.

oceanological (ōuʃiǎnǒlǫ·dʒikăl), *a.* [f. OCEANOLOGY: see -LOGICAL.] Of or pertaining to oceanology.

1969 *New Scientist* 20 Feb. 382/1 A final element in oceanological development..is the promised White Paper. **1970** R. BARTON *Oceanology Today* vii. 187 Cooperation in oceanological endeavor is not confined to agreement between nations as to who owns what. **1975** *Nature* 2 Oct. 353/1 This will include atmospheric sounding, the investigation of deep-level ice cores.., oceanological investigation in the Drake Straits, [etc.].

oceanologist (ōuʃiǎnǫ·lǒdʒist). [f. as prec. + -IST.] An expert or specialist in oceanology.

1954 *Deep-Sea Research* II. 86 Any oceanologist could conceive an ideal plan, but when the problem of money is introduced, restrictions are needed. **1972** *Daily Tel.* 22 Mar. 7 (*caption*) The Russian research ship Professor Subov, 6,681 tons, which is making a three-day visit to London... The crew and team of oceanologists and meteorologists spent the day sightseeing.

oceanology. Delete *rare*—[0] and add: In mod. use freq. given a broader meaning than *oceanography* (see quots. 1969[1], 1973). (Examples.)

1898 *Ann. Rep. Board of Regents Smithsonian Inst. 1896* 295 This brings us to the equally important question of oceanology, which should comprise a complete knowledge not only of the surface currents in the Arctic seas, but also surface and deep-sea temperatures, [etc.]. **1955** *Deep-Sea Research* II. 247 The use of the words 'Oceanology' and 'Oceanologist' was well and truly pondered in days past. Today they are little used save in Russia. **1956** *Ibid.* IV. 68 In 1949 the R/V *Vitjaz* of the Institute of Oceanology of the Academy of Sciences of the U.S.S.R. started its investigations. **1969** *Nature* 1 Mar. 804/2 A brief explanation is required of the word 'oceanology'... In the early days it was often used synonymously with 'oceanography' to mean marine science... It is now often used to mean marine technology, and although it has no precise definition, is usually taken to exclude conventional naval architecture. [**1969** *Ibid.* 13 Dec. 1049/2 Much of what passes for oceanography is nothing more than the straightforward use of the sea for such things as fishing and transport.] **1970** R. BARTON *Oceanology Today* i. 12 The project becomes even more valuable when looked at in terms of the lessons..that will be learned across the whole field of oceanology, from offshore oil exploration to physiological research in diving. **1973** HICKLING & BROWN *Seas & Oceans* p. viii, Sea or ocean science is nowadays called *oceanography*, or by an even newer word, *oceanology*, coined in the 1960s. Some like to differentiate between the use of these terms, using *oceanography* to describe the more academic aspects of ocean science, and *oceanology* to describe modern ocean technology, including the political, legal, financial and economic aspects of the seas and oceans.

ochlophobia (ǫklǫfōu·biă). [f. Gr. ὄχλο-ς crowd + -PHOBIA: see OCHLOPHOBIST.] A morbid aversion to crowds. Hence **ochlopho·bic** *a.*

1894 GOULD *Dict. Med.* 884/2 Ochlophobia. **1929** C. CONNOLLY *Let.* Nov. in *Romantic Friendship* (1975) 330, I have had ochlophobia lately. **1976** R. PERRY *One Good Death* x. 157 [He] preferred to work in greater privacy. It could even be that he was ochlophobic like Toller, the Austrian killer.

ochlospecies (ǫ·klospī:sīz, spī:ʃīz). *Taxonomy.* [f. as prec. + SPECIES.] (See quot. 1967.)

1962 F. WHITE in D. Nichols *Taxon. & Geogr.* 79 Such species cannot be satisfactorily subdivided and could conveniently be distinguished from monotypic and polytypic species by calling them *ochlo*-species. **1967** *Gardens' Bull.* XXII. 6 White has coined the useful and appropriate term ochlospecies for species showing a complex reticulate pattern of variation. **1972** G. T. PRANCE in *Flora Neotropica* IX. 109 *Licania heteromorpha* is a good example of an ochlospecies.

ochone (oχōu·n), *int.* (*sb.*) Also **och hone.** Repr. an Anglo-Irish form of OHONE *int.* (*sb.*). Also as *v. intr.*

1829 G. GRIFFIN *Collegians* III. xxxiii. 54 I'm ashamed o' myself, to be always..moaning and ochoning, among the neighbours. *c* **1850** in R. Ward *Penguin Bk. Austral. Ballads* (1964) 52 The trees grew so thick I couldn't find it, ochone,..So bothered and lost was poor Paddy Malone. **1855** [see OHONE *int.* (*sb.*) b]. **1861** TROLLOPE *Tales of All Countries* (ser. 1) 67, I could plainly hear poor Larry's head strike against the stone floor. 'Ochone, ochone!' he cried at the top of his voice. **1884** D. BOUCICAULT *Shaughraun* 20/1 Och hone!—my darlin' boy, it will be a grand day for you, but your poor ould mother will be left alone..och-o-o-hone! **1913** W. B. YEATS *Countess Cathleen* in *Poems* 68 Ochone! The treasure room is broken in. **1919** G. B. SHAW *O'Flaherty V.C.* in *Heartbreak House* 186 Ochone! ochone! my son's turned agen me. Oh, whatll I do at all at all? **1920** F. SCOTT FITZGERALD *This Side of Paradise* (1921) *A Lament for a Foster Son*... Ochone He is gone from me the son of my mind. **1939** JOYCE *Finnegans Wake* 277 His sevencoloured's soot (Ochone! Ochonal!). **1977** *Irish Democrat* Mar. 6/3 When Sarsfield sailed away I wept as I heard the wild ochone.

Ochrana, var. *OKHRANA.

ochre, *sb.* Add: **4.** *ochre-yellow* (examples); **ochre-grave** (see quot.); also *attrib.*

1928 PEAKE & FLEURE *Steppe & Sown* 20 In the early type of kurgan are found skeletons..buried in a contracted position, the bones covered with red ochre. These..are now known as the ochre-graves. *Ibid.* 26 The ochre-grave folk. **1957** V. G. CHILDE *Dawn Europ. Civilization* (ed. 6) ix. 149 (*heading*) The Ochre Grave Cultures of the Pontic Steppes. *Ibid.* 157 The beginnings of the Ochre Grave culture should go back well into the third millenium. **1899** Ochre yellow [see *hæmochromatosis* s.v. *HÆMO-, HEMO-]. **1952** A. G. L. HELLYER *Sanders' Encycl. Gardening* (ed. 22) 121 *Dayana*, ochre-yellow, lip white and brown.

ochro, var. OKRO, OKRA. (Examples.)

1952 S. SELVON *Brighter Sun* v. 82 Neighbour ask me bringam dry ochro seed for she to plant. **1953** G. LAMMING *In Castle of my Skin* xiv. 274 Slicing the ochroes was ..painstaking. The prickly surface irritated the skin, and the slices fell off on the hand in a slimy mess. **1971** *Advocate-News* (Barbados) 17 Sept. (Guyana Suppl.) p. ii/2 Now the Liliputians have completed planting of ground provisions, pumpkins, ochroes, blackeye peas, celery, bora and such permanent crops as breadfruit, citrus and breadnut.

ochronosis. Substitute for entry s.v. OCHRO-: (ōukrǒnōu·sis). *Path.* [mod.L. (R. Virchow 1866, in *Arch. f. path. Anat. u. Physiol.* XXXVII. 218), f. Gr. ὠχρός pale yellow + νόσος disease + -OSIS.] An abnormal brown pigmentation of tissue, notably cartilage, esp. when a symptom of the metabolic disorder alkaptonuria and resulting from the accumulation of polymerized homogentisic acid.

1867 *Half-Yearly Abstr. Med. Sci.* XLV. 28 Virchow proposes to call this affection ochronosis. **1876** [in Dict. s.v. OCHRO-]. **1906** *Lancet* 6 Jan. 26/1 The urine of his patient.., who had the blackening of cartilages which constitutes ochronosis, was brown when passed. **1948** W. A. D. ANDERSON *Path.* iv. 88 Phenol poisoning, due to absorption of phenol from surgical dressings, has been reported to cause a pigmentation similar to ochronosis. **1962** C. A. CLARKE *Genetics for Clinician* xiii. 156 Alkaptonuria... In addition to the urinary abnormality..there is pigmentation of cartilage and ligaments (ochronosis). **1970** *Nature* 21 Nov. 770/1 Adults with alcaptonuria develop a destructive arthritis secondary to the deposition of a melanin-like pigment in their connective tissues, especially skin, tendon and cartilage (ochronosis).

Hence **ochrono·tic** *a.* [-OTIC].

1922 *Arch. Internal Med.* XXIX. 737 Hanseman observed diffuse and granular ochronotic pigment in the tissues. **1961** *Arthritis & Rheumatism* IV. 137 (*heading*) Studies on the pathogenesis of ochronotic arthropathy.

ochrous, *a.* (Earlier examples of U.S. form.)

1806 WEBSTER, *Ocherous*,..like or containing ocher. **1828** —— s.v. *Ocherous matter*, an ocherous color. **1869** *Rep. Comm. Agric. 1868* (U.S. Dept. Agric.) 427 The pasture..hardening in some such manner as 'hard-pan' forms in ocherous soil.

ocker (ǫ·kəɹ), *sb.*[2] *Austral. slang.* Also **Ocker.** [The name of a character in a series of Australian television sketches by Ron Frazer: used earlier as a colloquial variant of names like *Oscar* and *O'Connor*.] A rough, uncultivated, or aggressively boorish Australian. Also *attrib.* or as *adj.*

[**1959**: see *NAUGHTY *sb.* 2.] **1971** G. JOHNSTON *Cartload of Clay* 71 The big man would be a good player, a vigorous clubman, a hearty participant in the companionship of the club bar. He was a type Julian had sometimes talked to him about, what the boy called an 'Ocker'. **1973** *Sunday Sun* (Brisbane) 1 July 21/5 His cigarette commercials were the next step in his career. And once again the ocker image worked wonders. **1973** *Nation Rev.* (Melbourne) 24–30 Aug. 1430/4 (Advt.), Sydney Femme, 27, bored by ozzie ockers and oedipal neurotics, desires to develop dynamic dalliance with..male human beings. **1974** *Sydney Morning Herald* 24 Apr. 6 That image, of the RSL itself as a sabre-rattling élitist organisation with an over-privileged influence on governments, and of RSL members themselves as beer-swilling, 'pokey-playing' Ockers, has, executives believe, faded if not totally evaporated. **1975** *TV Times* (Austral.) 12 Apr. 10 The cult of the ocker is sweeping Australia. **1976** *Telegraph* (Brisbane) 30 July 42/1 It is no use telling Australians to wake up; it is not in the ocker character.

Hence **o·ckerdom,** ockers collectively. Also **o·ckerism,** over-assertive boorishness, uncultivated behaviour.

1974 P. PORTER in *Australian* 5 Oct. 13 The new Australian boorishness is known as Ockerism, from a slob-like character called Ocker in a television series, the embodiment of oafish, blinkered self-satisfaction. **1975** M. HARRIS in *Ibid.* 18 Jan. 18 The resurgence of an aggressive Australian ockerdom was coincident with the first election of the Whitlam Government and the discovery of a 'new nationalism'. **1977** *Sunday Times* 23 Jan. 30/8 It is seen as a defeat for the spirit of ockerism (that is, being blatantly Australian).

Ockham's razor: see *RAZOR *sb.* 1 b.

o'clock. Used, in various contexts specifying direction, esp. with reference to the target in shooting, to give bearings corresponding to the positions of the numerals on a clock face, from the standpoint of one facing twelve o'clock. (Cf. quot. 1797 s.v. CLOCK *sb.*[1] 4.)

1904 W. WINANS *Hints on Revolver Shooting* iv. 33 When the hind sight comes to the level of your eyes.., the front sight will be seen through the middle of the 'U' pointed at the bottom of the bull's-eye, the top of the front sight just touching it at 'six o'clock'. **1913** A. G. FULTON *Notes on Rifle Shooting* 15 If the bull can be seen without much difficulty a 6 o'clock aim is probably the best. **1918** E. S. FARROW *Dict. Mil. Terms* 415 *O'clock*, a term employed to indicate, by means of the divisions on the dial of the clock, the location of a hit on the target or the direction from which the wind may be blowing, as a 7 o'clock, 4 or 5 o'clock wind. **1932** J. A. BARLOW *Elements of Rifle Shooting* iv. 47 Allowances for a constant strength of wind blowing from a varying o'clock, vary according to the length of the perpendicular drawn from the hour on the clock face to the 6–12 o'clock diameter. **1943** C. H. WARD-JACKSON *Piece of Cake* 44 *O'clock*. Used thus, 'Ten bandits at 6 o'clock' means ten enemy aircraft immediately below the recipient of the message. **1948** A. M. TAYLOR *Lang. World War II* (rev. ed.) 141 *O'clock*, term used largely by fliers in designating directions, e.g. 'twelve o'clock' indicates straight ahead; 'three o'clock' directly to the right, etc. **1967** B. KNOX *Blacklight* ii. 48 The lamp continued. 'They say we have it at about our ten o'clock sir.' **1970** M. KELLY *Spinifex* vii. 119 'Black rock. Eight o'clock. One hundred metres!' He'd instinc-

tively ranged artillery in another war. **1972** *National Observer* (U.S.) 27 May 19/1 He says the hands should grasp the wheel at 3 and 9 o'clock. **1973** *Country Life* 25 Oct. 1292/1 From its position at twelve o'clock, the dog begins the critical 'lift'.

ocote (ǒko·te). [Amer. Sp., f. Nahuatl *ocotl* torch.] A resinous Mexican pine, *Pinus montezumæ*, or its wood; also = PITCH-PINE. Also *attrib.*

[**1858** G. GORDON *Pinetum* 211 It [sc. *Pinus teocote*] is the 'de'ocote' or 'Pino de'ocote' (candle wood) of the Mexicans. **1861** E. B. TYLOR *Anahuac* 336/2 *Ocote* (Aztec, *ocotl*), a pine-tree, pine-torch.] **1908** H. GADOW *Through Southern Mexico* iii. 50 The forest consists mainly of the long-leaved *Pinus liophylla* and the 'ocote' or *Pinus montezumae*, which much resembles our Scotch fir. **1926** D. H. LAWRENCE *Plumed Serp.* vii. 129 The boy.. sat apart watching the two ocote torches. **1927** —— *Mornings in Mexico* 10 There is a resinous smell of ocote wood. **1938** *Amer. Speech* XIII. 113 The name [*ocotillo*] is a derivative of Mexican-Spanish *ocote*, a word also applied to pitch-pines or the wood they yield.

ocotillo (ōukotī·yo). *U.S.* Also **ocetilla, ocochilla, ocotilla.** [Amer. Sp., dim. of *OCOTE.] A spiny shrub, *Fouquiera splendens*, of the family Fouquieraceæ, native to the south-western United States and Mexico and bearing narrow, inconspicuous leaves and panicles of red flowers. Also *attrib.*

1856 in *Dict. Americanisms*, Aside from the grass, there is a shrub called the..*Ocetilla*. **1864** *Harper's Mag.* Nov. 697/2 Passing through some dense thickets of mesquit and ocochilla, the struggling family found themselves at the foot of a rocky bluff. **1883** *Ibid.* Mar. 491/1 The houses consist of a frame-work of cottonwood or ocotilla wattles. *Ibid.* 502/2 The ocotilla is simply a wattle of sticks.. waiting to be cut down and turned into palings. **1893** *Nation* (N.Y.) 7 Sept. 169/3 Walking-sticks made of the porous ocotillo cactus. **1915** L. H. BAILEY *Stand. Cycl. Hort.* III. 1271/1 Ocotillo..is a conspicuous object in the deserts from Texas westward,..the rod-like stiff canes looking like lifeless sticks in dry weather and in its season crowned with masses of showy bloom. **1948** *Natural Hist.* Apr. 181 The Ocotillo is frankly red, adding its flaming tips to every dry stick that looked dead a week ago. **1966** MRS. L. B. JOHNSON *White House Diary* 3 Apr. (1970) 381 The ocotillo..looks like about a dozen long coach whips stuck into the ground, spraying out in a weeping fashion. **1976** *San Antonio* (Texas) *Express-News* 27 Nov. 1B/5 'You know the *ocotillo*,' Ben said, pointing out the cactus that looks like five or six spiny, 10-foot pieces of rope snaking up into the air.

-ocracy. Add: (Further examples as *sb.*) Now used generically to denote any form of government or domination to which a word in *-ocracy* can be applied.

1831 C. C. F. GREVILLE *Mem.* (1874) II. xiii. 112 It has elicited a strong Conservative demonstration, and proved that out of the rabble-ocracy (for everything is in *ocracy* now) his power is anything but unlimited. **1894** G. B. SHAW in *Fortn. Rev.* Apr. 489 Social-Democracy, like all other '-ocracies', will have a great deal more trouble with its idle and worthless members than with its able ones. **1928** —— *Intelligent Woman's Guide Socialism* xliii. 166 If it be still necessary to call the rich an ocracy of any kind, they must be called a plutocracy. **1963** F. W. FREY in L. W. Pye *Communications & Political Devel.* xvii. 299 Movement towards 'democracy'..or whatever one's preferred..'ocracy' happens to be.

octa-. Add: **b. octade·canol,** any of the alcohols with the formula $C_{18}H_{37}OH$; spec. *normal* (or *n-*) *octadecanol*, a crystalline primary alcohol which is present in whale and dolphin oils, also called stearyl alcohol; **octape·ptide,** any peptide composed of eight amino-acid residues.

1914 *Jrnl. Chem. Soc.* CV. 2233 The acetates of the dextrorotatory carbinols from ethylbutylcarbinol (γ-heptanol) on to ethylpentadecylcarbinol (γ-octadecanol) are all of negative rotation. **1949** E. CHAIN in H. W. Florey et al. *Antibiotics* II. xvii. 709 Antifoam reagent (3 per cent. octadecanol in lard oil). **1970** *Chem. & Physics Lipids* IV. 246 The major long-chain alcohols found in heart and brain tissues are hexadecanol, octadecanol and octadecenol. **1931** *Chem. Rev.* VIII. 390 Curtius..found that when his triglycylglycine ester was heated to 100°C. *in vacuo* it was converted into an insoluble infusible material having the composition $—NHCH_2CO—$. He assigned to this material the structure of a cyclic octapeptide, but a much more highly polymeric open chain structure seems more probable. **1957** *Science* 3 May 886/1 (*heading*) Synthesis and pharmacology of the octapeptide angiotonin. **1972** *Lancet* 17 June 1320/1 The superb work of Wieland and his school established that the toxins are either cyclic heptapeptides or octapeptides.

octahedrally (ǫktăhī·drăli), *adv.* [f. OCTAHEDRAL, OCTO- *a.* + -LY[2].] In an octahedral shape or arrangement.

1872 *Geol. Mag.* IX. 356 The octahedrally crystallized magnetic particles do not contain any traces of nickel. **1947** *Q. Rev. Chem. Soc.* I. 253 Half of the metal ions are surrounded, approximately octahedrally, by four oxygen atoms of AsO_4 (or PO_4) ions and by two OH ions. **1970** *Nature* 11 Apr. 141/2 It is a reasonable guess that the molecules around the ion are disposed octahedrally.

octahedrite. Add: **2.** An iron meteorite which shows a Widmanstätten structure on etching due to intimate intergrowths of plates of kamacite with narrow borders of tænite oriented parallel to the faces of a regular octahedron.

1916 *Mineral. Mag.* XVIII. 42 (*table*) Nickel-poor octahedrites. **1944** C. PALACHE et al. *Dana's Syst. Min.* (ed. 7) I. 120 A similar structureless aspect can be obtained in an octahedrite by prolonged heating at about 900°. **1974** *Encycl. Brit. Micropædia* VII. 477/2 Besides nickel-iron, major minerals in octahedrites are troilite and schreibersite.

octal (ǫ·ktăl), *a.* and *sb.* [f. OCT- + -AL.] **A.** *adj.* **1.** *Electronics.* Applied to valve bases and plugs (and the corresponding sockets) having a standard circular arrangement of eight pins with a central moulded key for determining the orientation.

1936 *Wireless World* 20 Nov. 557 (*heading*) 8-pin base (octal) connections. **1943** C. L. BOLTZ *Basic Radio* x. 165 Then from America came the international octal base, with eight pins symmetrically arranged in a circle. **1961** *Engineering* 14 July 42/4 The unit is constructed as a plug-in module..connecting with associated circuits through a standard octal plug. **1976** *Gramophone* Feb. 1403/1 At the left is an octal valve socket designed for use with a matching transformer when a moving coil cartridge is used.

2. *Math.* and *Computers.* Pertaining to or being a system of numerical notation that employs 8 rather than 10 as the base.

1948 *Math. Tables & Other Aids to Computation* III. 295 Binary numbers are set into the keyboard in the octal (radix 8) notation. **1962** T. C. BARTEE et al. *Theory & Design Digital Machines* 307 In the octal system the sequence 624 represents the sum $6 \times 8^2 + 2 \times 8^1 + 4 \times 8^0$ which is equal to the sequence 404 in the decimal system. **1963** *Rep. Comm. Inquiry Decimal Currency* 4 in *Parl. Papers 1962–3* (Cmnd. 2145) XI. 195 If..we have in mind electronic rather than human computers, a case can be made for counting in eights—an octal system—because this can be related more easily to the binary system used by such computers. **1973** *Nature* 27 Apr. 620/1 The octal system is widely used in listing machine instruction programs for small computers, where groups of six octal digits are written in place of sixteen to eighteen binary digits.

B. *sb.* The octal system; rarely *pl.*

1948 *Math. Tables & Other Aids to Computation* III. 295 The conversions octal-binary and binary-octal are very elementary. **1961** EVANS & PERRY *Programming & Coding* iii. 89 The two representations ·101010101 in binary, and ·525 in octal, are different representations of the same magnitude. **1961** *Times* 28 Dec. 9/5 (*heading*) Thinking in octals. Advantages over decimals. **1967** KLERER & KORN *Digital Computer User's Handbk.* I. ii. 39 Since $2^3 = 8^1$, a binary series of digits may be converted to octal by grouping the binary digits into three to left or right of the decimal point. Thus the binary integer $(1001011011)_2 = (1\ 001\ 011\ 011)_2 = (1133)_8$. **1977** D. BAGLEY *Enemy* xxxi. 253 This gadget is working in octal instead of decimal... Many computers work in octal internally.

o:ctame·thylcy:clote:trasilo·xane. *Chem.* [f. OCTA- + METHYL + CYCLO- + TETRA- + *SILOXANE.] A colourless oily liquid that is a cyclic tetramer, $[(CH_3)_2SiO]_4$, and an intermediate in the manufacture of many silicones.

1946 M. J. HUNTER et al. in *Jrnl. Amer. Chem. Soc.* LXVIII. 361/1 When a few drops of concentrated sulfuric acid are added to a quantity, say 10 c.c., of octamethylcyclotetrasiloxane, $[(CH_3)_2SiO]_4$, there is no marked change in appearance or temperature. **1962** BARRY & BECK in Stone & Graham *Inorg. Polymers* v. 216 The slow addition of dimethyldichlorosilane to excess water at 15 to 20°C yields a mixture consisting of 0·5% hexamethylcyclotrisiloxane (D_3), 42·0% octamethylcyclotetrasiloxane (D_4), 6·7% decamethylcyclopentasiloxane (D_5), [etc.]. **1970** *Encycl. Polymer Sci. & Technol.* XII. 503 Almost all strong inorganic alkalis and most quaternary ammonium hydroxides and phosphonium hydroxides will polymerize octamethylcyclotetrasiloxane to a high-molecular-weight polymer.

octane. Add: **1. b.** *ellipt.* for *octane number*, esp. with prefixed numeral, e.g. *105 octane*, or adj., e.g. *high octane*.

1932, etc. [see *high-octane* s.v. *HIGH *a.* 22 a]. **1935** *Oil & Gas Jrnl.* 4 July 16/1 Until this spring, 87 octane gasoline was the highest quality used for commercial take-offs. **1954** [see *ISOOCTANE b]. **1960** J. B. HILL in V. B. Guthrie *Petroleum Products Handbk.* iv. 20 A particular gasoline may have a Research octane of 88 and a Motor octane of 78. **1966** *McGraw-Hill Encycl. Sci. & Technol.* III. 309/2 High octane gasoline burns in the same manner and with the same flame velocity as low octane gasoline. **1973** J. LEASOR *Host of Extras* vii. 136 Adrenalin surged round my body like 105 octane petrol.

2. *attrib.* and *Comb.*, as (sense *1 b) *octane-booster, -quality*; *octane-scented* adj., **octane number, rating,** a number indicating the anti-knock properties of a motor or aviation fuel, equal (for numbers below 100) to the percentage by volume of isooctane in an isooctane/normal heptane mixture of equivalent performance (see also quot. 1966).

1971 *New Scientist* 5 Aug. 323/1 Those spokesmen of the oil industry who have been arguing that alkyl lead (added to petrol as an octane-booster..) is not a health hazard. **1931** *Automobile Engineer* Aug. 308/3 Lead tetra-ethyl is the only other practical alternative to cracked spirits as a means of raising the anti-knock value of motor spirit to an octane number of 70 or more (this now being the prime consideration of a premium gasoline or petrol). **1933** *Flight* 4 May 408/1 The British Air Ministry is, rightly or wrongly, taking its time over considering the matter of octane number. **1966** *McGraw-Hill Encycl. Sci. & Technol.* IX. 273/1 Fuels with octane numbers above 100 are usually rated by finding the number of milliliters of tetraethyllead required per gallon of isooctane to give the same resistance to detonation as the fuel sample. **1972** PATTERSON & HENEIN *Emissions from Combustion Engines* ii. 63 The octane number of the fuels on the market is about 94 for regular gasoline and 100 for premium gasoline. **1959** R. J. HENGSTEBECK *Petroleum Processing* i. 3 The octane quality of a gasoline may be rated differently by different engines. **1940** A. W. JUDGE *Aircraft Engines* I. v. 114 If it was found that a mixture of 78 per cent. of iso-octane and 22 per cent. of heptane matched the particular fuel under test in 'knocking' qualities the fuel in question would be said to have an octane rating of 78. **1959** R. J. HENGSTEBECK *Petroleum Processing* i. 3 Octane ratings above 100 are proportional to their performance potentials. **1972** *Daily Tel.* 15 Mar. 11/7 Lead has been used in growing quantities to raise the octane rating of petrol, allowing modern engines to make more efficient use of the fuel. **1946** *R.A.F. Jrnl.* May 151 Newly-arrived from the octane-scented atmosphere of Station life and full of inspiration.

octanoic (ǫktănōu·ik), *a.* *Chem.* [f. OCTAN(E + *-OIC.] = CAPRYLIC *a.*

1909 *Jrnl. Chem. Soc.* XCVI. II. 1378/1 (Index), Octanoic acid, ε-hydroxy-, and its lactone. **1940** T. P. HILDITCH *Chem. Constitution Nat. Fats* iii. 99 In the milk fats of sheep and goats.., the proportions of octanoic and decanoic (caprylic and capric) acids are much greater and approach, or even exceed, that of butyric acid in cow milk fats. **1972** *Materials & Technol.* V. ix. 244 $CH_3(CH_2)_6COOH$, octanoic acid, was first described in 1844 by Lerch who isolated it from butter.

Hence **octano·ate,** the anion, or an ester or salt, of octanoic acid; **o·ctanoyl,** the radical $CH_3(CH_2)_6CO—$ present in octanoic acid; also called *caprylyl*.

1945 *Jrnl. Biol. Chem.* CLXI. 416 Of the tested fatty acids octanoate was the most reactive under these conditions. **1949** *Union Internat. Chim. Pure et Appl., Compt. Rend.* XV. 145 Rule 58.3... Octanoyl. Replacing 'capryloyl' and 'caprylyl'. **1954** A. WHITE et al. *Princ. Biochem.* xviii. 482 The reaction sequence..by which octanoate might give rise to acetoacetate. **1973** *Agric. & Biol. Chem.* (Tokyo) XXXVII. 2713/2 During this acylation process with *n*-octanoyl chloride, esterification of two OH groups of 2 moles of Thr contained in a colistin nonapeptide molecule was possible. **1974** *Acta Crystallogr.* XXX. B. 2913/2 Copper(II) octanoate was prepared by adding copper carbonate to an excess of solution of octanoic acid in ethanol.

octanol (ǫ·ktănǫl). *Chem.* [f. as prec. + -OL.] Any saturated aliphatic monohydric alcohol containing eight carbon atoms, $C_8H_{17}OH$, of which many isomeric forms exist; *esp.* the primary straight-chain alcohol (*n-* or 1-octanol), or the secondary alcohol, $CH_3(CH_2)_5CH(OH)CH_3$ (2-octanol), both of which are colourless odoriferous liquids used chiefly in the perfume and plastic industries; octyl alcohol.

1905 *Jrnl. Chem. Soc.* LXXXVIII. I. 573 δ-Octanol is a mobile liquid with a pleasant odour which boils at 71° under 10 mm. pressure. **1932** *Physical Rev.* XL. 830 The series consists of twenty-one octanols which differ from one another only in the relative positions of a methyl and hydroxyl group along the chain. **1947** KIRK & OTHMER *Encycl. Chem. Technol.* I. 317, 1-Octanol..is a colorless liquid with a penetrating odor. *Ibid.*, The technical grades of 2-octanol are used as solvents for many resins and in some protective formulations to improve flow and leveling. **1950** I. MELLAN *Industr. Solvents* (ed. 2) xi. 511 Commercial *n*-octanol is derived synthetically from coconut oil. **1975** J. M. TEDDER et al. *Basic Org. Chem.* V. xii. 450 Long chain fatty alcohols (mainly octanol and decanol) are used to inhibit terminal bud growth and hence stimulate branching in many plant species.

octantal, *a.* (s.v. OCTANT). Add: **b.** Vanishing once in each octant of the compass.

1928 L. S. PALMER *Wireless Princ. & Pract.* xii. 472 Any error of this nature occurring twice in each quadrant is termed an octantal error. **1954** *Electronic Engin.* XXVI. 39/1 Marconi's have developed a unique fixed (non-rotating) aerial with low octantal error.

octaploid, var. *OCTOPLOID *a.* (*sb.*).

octapole, var. *OCTUPOLE.

Octateuch. Add: (Later examples.) Also, a manuscript or edition of the Octateuch.

1967 R. L. S. BRUCE-MITFORD *Art of Codex Amiatinus* 9 They may be read as follows: OCT·LIB LEG.. (i.e. eight books of the Octateuch and the Law..). **1976** *Amer. N. & Q.* XV. 61/2 The Octateuch of Vatopédi, going back to the thirteenth century and considered one of the most valuable of the five known octateuchs. **1977** D. CUPITT in J. Hick *Myth of God Incarnate* vii. 143 A miniature in the Smyrna Octateuch.

octavalent (ǫktăvēi·lĕnt), *a.* *Chem.* Also (now *rare*) **octo-.** [f. OCTA-, OCTO- + L. *valent-em*, pres. pple. of *valēre* to be worth.] Characterized by a valency of eight.

1880 Octovalent [see PER-[1] 5 b]. **1909** J. N. FRIEND *Theory of Valency* xvii. 117 Osmium apparently functions as an octavalent atom in the tetroxide OsO_4, as does ruthenium in RuO_4. **1950** N. V. SIDGWICK *Chem. Elements* II. 1458 There are a fair number of octovalent osmium compounds. **1973** S. E. LIVINGSTONE in J. C. Bailar et al. *Comprehensive Inorg. Chem.* III. xliii. 1209 The octavalent state [of osmium] is more stable than that of ruthenium.

Hence **octava·lency,** the property of being octavalent.

1925 *Ann. Rep. Progr. Chem.* XXI. 54 These facts are held to confirm the octavalency of osmium.

octave, *sb.* (*a.*) Add: **3. f.** (Further examples.)

1923 GLAZEBROOK *Dict. Appl. Physics* IV. 891/2 The continuous spectrum was thus extended to thirty-nine times the wave-length of sodium yellow, or five octaves into the infra-red. **1960** [see *INFRA-RED *a.* and *sb.* B]. **1961** BERKNER & ODISHAW *Sci. in Space* i. 3 Even if we add the radio-frequency window to the narrow light-wave window, the sum gives astronomy about twenty octaves.. of the electromagnetic spectrum with which to investigate the universe from the Earth.

4. b. *law of octaves* (earlier and later examples).

1865 J. A. R. NEWLANDS in *Chem. News* 18 Aug. 83/2 This peculiar relationship I propose to provisionally term the 'Law of Octaves'. **1932** J. N. FRIEND *Text-bk. Physical Chem.* I. xvi. 330 According to Newlands' Law of Octaves, when the elements are arranged in the order of increasing atomic weights, the first and last of any eight consecutive elements possess similar properties. **1965** D. ABBOTT *Inorg. Chem.* i. 31 In 1864 Newlands put forward his Law of Octaves, in which he likened the classification of elements to the musical scale.

8. octave key, a key on a wind-instrument used to produce a note an octave higher than the note that is being fingered.

1880 GROVE *Dict. Mus.* II. 487/1 In more modern instruments [*sc.* oboes] a second octave-key has been introduced..which is usually lifted on reaching A above the stave. **1911** *Encycl. Brit.* XXIV. 274/1 The first 15 semitones are obtained by opening successive keys, the rest of the compass by means of octave keys enabling the performer to sound the harmonic octave of the fundamental scale. **1957** A. C. BAINES *Woodwind Instrum.* iv. 106 With full automatic octave keys, the G♯ key, when pressed, holds down ring III.

octavic (ǫktēi·vik), *a.* and *sb.* *Math.* [f. L. *octāv-us* eighth + -IC.] **A.** *adj.* Of the eighth order or degree. **B.** *sb.* An octavic polynomial or curve.

1854 *Cambr. & Dublin Math. Jrnl.* IX. 94 The biquadratic and octavic function of x, y, $(x, y)^4$, $(x, y)^8$. **1879** *Amer. Jrnl. Math.* II. 251 For the first four orders there is but one such block, for the quintic and the sextic two, for the seventhic five, for the octavic three, and for the 9ic and 10ic four. **1897** *Nature* 11 Nov. 47/1 More general forms of octavic curves with six double points.

octet, octette. Add: **3. b.** A stable group of eight electrons in an electron shell of an atom.

1919 I. LANGMUIR in *Jrnl. Amer. Chem. Soc.* XLI. 888 After the very stable pairs..the next most stable arrangement of electrons is the group of 8 such as forms the outside layer in atoms of neon and argon. We shall call this stable group of 8 electrons the 'octet'. **1927** N. V. SIDGWICK *Electronic Theory of Valency* x. 175 The trivalent elements, having incomplete octets, strive to complete them by co-ordination. **1965** D. ABBOTT *Inorg. Chem.* ii. 70 Some atoms can remain stable when surrounded by more than an octet of electrons (they expand their octets).

c. *Nuclear Physics.* A multiplet (sense *b) of eight sub-atomic particles.

1961 in GELL-MANN & NE'EMAN *Eightfold Way* (1964) i. 13 The most important prediction is..that the eight baryons should all have the same spin and parity and that the pseudoscalar and vector mesons should form 'octets', with possible additional 'singlets'. **1962** *Physical Rev.* CXXV. 1068/2 The baryons, as well as the mesons, can form octets and singlets. **1963** [see *NONET 2]. **1968** M. S. LIVINGSTON *Particle Physics* xii. 213 Here [*sc.* in SU(3) theory] the basic octet of states is identified with the nucleon, the Σ-particle triplet, the Λ^0 particle, and the Ξ-particle doublet.

octic (ǫ·ktik), *a.* and *sb.* *Math.* [f. L. *octo* eight + -IC.] = *OCTAVIC *a.* and *sb.*

1877 *Encycl. Brit.* VI. 726/1 φ is not expressible as the square root of an octic function of θ. **1916** G. A. MILLER et al. *Theory & Applic. Finite Groups* i. 4 The Octic Group. There are eight movements of a plane which transform into itself a square situated in this plane. **1938** *Duke Math. Jrnl.* IV. 285 Normal octics with the group *G*(2, 2, 2). **1978** *Amer. Math. Monthly* LXXXV. 470 (*heading*) Rational octic and higher reciprocity laws.

‖ **octli** (ōu·ktli). [Mex. Sp.] = PULQUE.

1845 *Encycl. Metrop.* XVI. 528/1 The most favourite spirit of the Mexicans is *pulque* or *octli*, a thick ropy juice which flows from a wounded agave or maguey. **1883** *Encycl. Brit.* XVI. 213/1 The juice extracted by tapping the great aloe before flowering was fermented into an intoxicating drink about the strength of beer, *octli*, by the Spaniards called *pulque*. **1914** T. A. JOYCE *Mexican Archaeol.* ii. 43 This awakening after heavy sleep..connected the octli gods with the waxing and waning of vegetation and the moon. **1938** *Times Lit. Suppl.* 5 Mar. 157/1 Other mistakes relate to the preparation of the fermented drink 'octli'.

octo-. Add: **o·ctode** *Radio* [*-ODE[2]], any thermionic valve with eight electrodes, some of which have been used as frequency changers in superheterodyne receivers; **octofoil** *a.* (later example); **octora·diant** *a.* = *octoradial* adj.; **o·ctose** [a. G. *octose* (E. Fischer 1890, in *Ber. d. Deut. Chem. Ges.* XXIII. 934): see -OSE[2]], any monosaccharide having eight carbon atoms in the molecule.

1934 A. W. HASLETT *Radio round World* 191 There are many more complicated types of valve, ranging up to the octode which has eight different components instead of only three. **1943** *Electronic Engin.* XV. 339 Table A shows the more important frequency changers..and it may be noted that three of these, the pentode, the octode and the heptode mixer, owe some of their success to their high anode impedance. **1961** *Amat. Radio Handbk.* (ed. 3) ii. 58/2 When provided with an additional grid used as a suppressor it [*sc.* the heptode] was referred to as an octode. **1958** *Times* 11 Dec. 12/4 An octofoil plain salver. **1911** BEERBOHM *Zuleika D.* xviii. 271 He affixed to his breast the octoradiant star, so much larger and more lustrous than any actual star in heaven. **1890** Octose [see *heptose* s.v. *HEPTA-]. **1931** R. J. WILLIAMS *Introd. Biochem.* iii. 26 Aldoheptoses, octoses, nonoses and a decose have been made synthetically. **1962** D. J. BELL in Florkin & Mason *Comparative Biochem.* III. vii. 297 The first natural octose, D-glycero-D-mannoctulose..has recently been isolated from an aqueous extract of the Californian avocado.

October. Add: **3. October Revolution,** name given to the revolution in which the Provisional Government in Russia was overthrown by the Bolsheviks in 1917 on 25 October Old Style (7 November New Style). Also *transf.*

1917 *Times* 13 Dec. 8/4 The October revolution, having broken the power of the capitalists and landlords,..set up a 'Government' of People's Commissioners. **1925** P. A. SOROKIN *Sociol. of Revolution* xvii. 390 Since the latter party included the workmen, the enormous mass of soldiers.., and the peasants..the October revolution was becoming inevitable. **1932** M. EASTMAN tr. *Trotsky's Hist. Russ. Revolution* I. 17 In the historic conditions which formed Russia..we ought to be able to find the premises both of the February revolution and of the October revolution which replaced it. **1952** E. H. CARR *Bolshevik Rev.* II. xvi. 71 Even before the October revolution conditions in Petrograd..were particularly acute. **1965** *Guardian* 14 Oct. 10/2 Russian visitors..might.. conclude that our October Revolution is still to come. **1965** B. PEARCE tr. *Preobrazhensky's New Economics* 77 It would be no exaggeration to say that the most interesting and exciting question since the October revolution of 1917 ..is the question of what the Soviet system is. **1974** tr. *Sniečkus's Soviet Lithuania* 56 The October Revolution opened up a new era in mankind's social progress.

Octoberist, -brist. Restrict *nonce-wd.* to sense in Dict. and add: **2.** (Chiefly in form *Octobrist*.) **a.** [Russ. *oktyabrist*.] In Russian politics, a member of the League of the 17 October 1905 Old Style (30 October New Style), formed in response to the Imperial Constitutional Manifesto of the same date. Also *attrib.* or as *adj.*

1906 *Daily Chron.* 10 Apr. 3/6 Count Peter Heyden, of the Octoberist party. **1906** *Westm. Gaz.* 29 Aug. 2/2 One of the chief questions in Russia has been whether the Octobrists would join the Reactionaries or make terms with the Constitutional Democrats. **1912** D. M. WALLACE *Russia* (rev. ed.) xxxix. 738 The weak point in the present Assembly is that the Octobrists—the moderate party which accepts the famous October manifesto in its natural sense, and wishes to co-operate with the Government in legislative work—do not possess an absolute majority. **1958** *Times Lit. Suppl.* 10 Jan. 14/3 The majority of the *zemstva* liberals found themselves on the side of the more left-wing Kadets, and only the minority among the Octobrists. *a* **1967** A. RANSOME *Autobiogr.* (1976) xxii. 195 Pares was intimate with Gutchkov of the Octobrists.

b. [Russ. *oktyabryónok*.] A member of a Russian communist organization founded in 1925 for young people below the normal age of the 'Pioneers' (see *PIONEER *sb.* 3c).

1929 S. N. HARPER *Civic Training in Soviet Russia* iv. 66 A widening of the scope of the movement also came in 1925... The Pioneers were given the task of working among their younger brothers and sisters and of organizing them into groups of 'Little Octobrists', in honour of the Bolshevik revolution of October 1917. **1935** S. & B. WEBB *Soviet Communism* I. v. 401 The Pioneers were given the task of bringing their younger brothers and sisters, as young as eight years old, into groups of Little Octobrists. By 1926 the two junior organisations had over two million members (1,800,000 Pioneers and 250,000 Octobrists). **1959** D. W. TREADGOLD *20th Cent. Russia* III. xviii. 288 The Little Octobrists for children eight to eleven years of age, the Pioneers..and the Komsomol..were together designed to produce adults who accepted the fundamental ideological commitments and values of the Party proper. **1960** A. KASSOF in C. E. Black *Transformation Russ. Society* v. 485 The Octobrists..includes members from seven through nine years of age. *Ibid.* 496 Of particular significance was the recent decision to re-establish the Octobrists for children younger than Pioneer age. **1974** T. P. WHITNEY tr. *Solzhenitsyn's Gulag Archipel.* I. I. viii. 333 It was as just such a well-nourished little imp that our Octobrist child—Law—began to grow.

octonary, *a.* Add: *spec.* = *OCTAL *a.* 1. (Further examples.)

1891 *Bull. N.Y. Math. Soc.* I. 5 The distinction between numbers of the form $4n+1$ and those of the form $4n+3$ is of great importance in the theory of numbers, and in the octonary system it would be obvious at a glance to which of these classes a given uneven number belongs. **1957** D. D. MCCRACKEN *Digital Computer Programming* iii. 33 The simple reason for the use of octal (also sometimes called octonary) numbers is that the conversion from binary to octal can be carried out mentally. **1963** L. SCHULTZ *Digital Processing* i. 22 Requiring eight symbols, the octal (sometimes, 'octonary') system proves to be a convenient means of expressing a binary value in a number approximately one-third the length of the binary number.

octopamine (ǫktōu·pămīn). *Biochem.* [f. OCTOP(US + AMINE, the compound having been first identified in salivary extracts from an octopus.] A weakly sympathomimetic amine, $HO·C_6H_4·CHOH·CH_2NH_2$, which under the action of monoamine oxidase inhibitors may accumulate in nerves in place of the closely related noradrenaline, thereby inhibiting the transmission of nerve impulses and causing vasoconstriction; 1-(3-hydroxyphenyl)-2-aminoethanol.

1948 V. ERSPAMER in *Acta Pharmacol.* IV. 245 It is suggested that the parent substance of the adrenaline-like principle in the salivary extracts should be called 'Octopamine'. **1952** *Nature* 1 Mar. 376/1 Having identified octopamine as norsynephrine, it was obvious to expect that hydroxyoctopamine could be identified as..noradrenaline. This assumption was confirmed experimentally. **1970** PASSMORE & ROBSON *Compan. Med. Stud.* II. xv. 32/2 Octopamine is a much weaker adrenergic receptor activator than noradrenaline and replacement by it of a portion of the noradrenaline released by nerve stimulation could result in a diminished response of the effector tissue. **1972** FLORKIN & BRICTEUX-GRÉGOIRE in Florkin & Scheer *Chem. Zool.* VII. x. 323 Octopamine may result from the decarboxylation of *p*-hydroxyphenylserine.., but it is known that octopamine may also result from the oxidation of tyramine... Octopamine itself can be oxidized into..noradrenaline.

octo·pian, *a.* [f. OCTOP(US + -IAN.] Suggestive of an octopus; = OCTOPEAN *a.* Also **octo·pic** [-IC], **o·ctopine** [-INE[1]], *adjs.*

1914 CHESTERTON *Flying Inn* 248 The Captain prepared to swing himself on to one of the octopine branches [of a tree]. **1922** C. E. MONTAGUE *Disenchantment* i. 11 He had..struck..a crate, from which some octopian beast.. had reached out at him. **1968** *Punch* 6 Nov. 667/3 The sight of a Breton fisherman pulling the suckers off a live octopic leg.

octoploid (ǫ·ktǒploid), *a.* (*sb.*) *Biol.* Also **octa-.** [f. OCTO- + *-PLOID.] (Made up of somatic cells) containing eight sets of chromosomes. Also as *sb.*, an octoploid organism.

1925 C. C. HURST *Experiments in Genetics* xxxviii. 542 By successive losses of septets the decaploid species would give rise to an octoploid species, the octoploid to a hexaploid species, the hexaploid to a tetraploid species. **1931** *Genetics* XVI. 462 The octoploid may be considered as having a replication of eight basically similar genoms. **1943** *Hereditas* XXIX. 193 (*heading*) Notes on octaploid *Solanum punæ* plant. **1961** *Lancet* 26 Aug. 488/1 The finding of 8 sex-chromatin bodies in an octaploid XXXY cell can be predicted from the above formula. **1973** *Nature* 11 May 87/2 Triploids.., artificial tetraploids and octoploids were all found to synthesize orientin isomers. **1974** *Sci. Amer.* Aug. 73/2 If a hexaploid wheat (*T. aestivum*) is crossed with rye, the result is an octoploid triticale.

Hence **o·ctoploidy,** the state or condition of being octoploid.

1934 *Gen. Program 3rd Pittsburgh Meeting Amer. Assoc. Adv. Sci.* 34 Octoploidy and diploidy in *Miastor americana*. **1948** *Jrnl. Heredity* XXXIX. 42/1 Instead of abrupt doublings of the chromosome number as in mitosis, there is therefore a gradual change from diploidy to tetraploidy, from tetraploidy to octoploidy and so on. **1970** AMBROSE & EASTY *Cell Biol.* 496 (Index), Octoploidy.

octopole, var. *OCTUPOLE.

octopush (ǫ·ktǒpuʃ). [f. OCTO(PUS + PUSH *sb.*[1]] A game of underwater hockey (see quots.). Hence **o·ctopusher,** one who plays octopush.

1970 *Times* 18 Feb. 11 Octopush..is a new form of underwater hockey... The game is played by teams of six. .. The object of the game is to propel or shovel the puck.. along the bottom of the pool and into the opponents' gull [*sc.* goal]. **1971** *Observer* 23 May 19/2 (*caption*) Octopush.. the name of the game for skindivers. Players try to push a lead puck (called a squid) through the other team's goal (gulley). **1973** *Sunday Mail* (Brisbane) 7 Jan. 2/3 Octopush..is played like underwater hockey. The teams, in flippers and face masks, but no oxygen tanks, play in a pool six feet deep. Instead of hockey sticks they use pushers,—shaped something like a rake with a blade where the prongs should be. **1973** *Telegraph* (Brisbane) 7 Mar. 3/2 Not mermaids but octopushers..are members of the Safari Women's team which plays the new game of octopush—diving to push a lead weight along the bottom of a swimming pool into the opposing team's goal.

octupole (ǫ·ktupōul). *Physics.* Also **octa-, octo-.** [f. *octu-* (in OCTUPLE *a.* (*sb.*), OCTUPLET: cf. *QUADRUPOLE), OCTA-, OCTO- + POLE *sb.*[2]] A multipole of order $l = 3$. Freq. *attrib.* or as *adj.* Cf. *MULTIPOLE *sb.*

1929 P. DEBYE *Polar Molecules* ii. 26 The next configuration, constructed by a displacement of the quadrupole in an arbitrary direction, will give the octupole. **1950** D. HALLIDAY *Introd. Nucl. Physics* ii. 57 Gamma-emission

processes can be classified as to multipole order; we speak, for example, of dipole, quadrupole, octopole, and still higher-order transitions. **1954** *Physical Rev.* XCIV. 1799 (*heading*) Hyperfine structure of In[115]. Evidence of a nuclear octupole moment. **1961** A. ABRAGAM *Princ. Nucl. Magnetism* vi. 170 The existence of magnetic octopoles has been established by atomic beam methods. **1970** G. K. WOODGATE *Elem. Atomic Struct.* ix. 166 The only multipole (2^k-pole) moments which do not vanish are: magnetic dipole ($k=1$), electric quadrupole ($k=2$), magnetic octupole ($k=3$), electric hexadecapole ($k=4$), etc. **1971** *Nature* 27 Aug. 609/1 The magnetic dipole and octapole and electric quadrupole moments all appear to vanish as $(Z_c+1)^{-3}$ as Z_c, the redshift of light coming from the centre of the disk, goes to infinity.

oculism (ǫ·kiŭliz'm). *rare.* [f. L. *ocul-us* eye + -ISM; after OCULIST.] The business of an oculist; knowledge of defects of vision, diseases of the eye, etc., and the remedies.

1909 W. BOOTH in H. Begbie *Life W. Booth* (1920) II. 433 The gentleman..was a doctor and..he knew something of oculism.

oculist. Add: **3.** *Comb.* **oculist-stamp** (also *oculist's stamp*), the more usual name among antiquaries for *medicine seal, stamp* (MEDICINE *sb.*[1] 6 a).

1778 *Gentl. Mag.* XLVIII. 509 An inscription on an oculist's stamp. **1851** *Monthly Jrnl. Med. Sci.* XII. 42 Above sixty Roman oculist-stamps have been now discovered in different parts of western Europe. **1886** *Guide Exhib. Galleries Brit. Mus.* 200 Roman Implements, such as steelyards and their weights, oculists' stamps, locks and keys. **1954** R. SUTCLIFF *Eagle of Ninth* viii. 87 An oculist's stamp is a talisman to carry a man safely where a Legion could not go.

oculo-. Add: **oculo-agra·vic** *a.* [f. *oculogravic with insertion of A- 14], applied to an illusion of an apparent upward movement of objects in the visual field that is experienced when the effective force acting on a person is reduced; **oculogra·vic** *a.* [L. *grav-is* heavy], applied to an illusion of apparent tilting that is experienced when a person undergoes an acceleration that causes the effective force acting on him to change direction; **oculogy·ral** *a.* [Gr. γῦρ-os ring, circle], applied to an illusion of apparent rotation that is experienced during or just after rotational accelerations of the body; **oculogy·ric** *a.* [Gr. γῦρ-os ring, circle], relating to or involving the turning of the eyeball in its socket; *oculogyric crisis*, an attack involving the involuntary movement of the eyeball to an exaggerated position, usu. with the gaze directed upwards, and the maintenance of this position for a period.

1958 GERATHEWOHL & STALLINGS in *Jrnl. Aviation Med.* XXIX. 504 We predicted an apparent motion under conditions of reduced gravity which would be opposite in direction to the one observed at increased accelerative force. This hypothetical phenomenon, which may be observed best in the zero-gravity state, is called the oculo-agravic illusion. **1961** H. G. ARMSTRONG *Aerospace Med.* xv. 232/1 The authors attribute the oculoagravic illusion to an otolith response and noted that the direction of apparent movement of the image was the opposite of that expected from the previous work..on the oculogravic illusion. **1968** R. A. WEALE *From Sight to Light* vi. 117 The oculo-agravic illusion has been studied by means of the apparent movement of an after-image. **1947** A. GRAYBIEL et al. in *Jrnl. Exper. Psychol.* XXXVII. 170 The oculo-gravic illusion refers to the apparent displacement of an object in space which may be observed when the sensory receptors in the otolith organs are stimulated by an accelerative force which forms a resultant vector with the force of gravity. **1968** R. A. WEALE *From Sight to Light* vi. 116 The oculogravic phenomenon was accompanied by a downward turn of the eyes as acceleration increased and if no fixation light was provided. **1946** GRAYBIEL & HUPP in *Jrnl. Aviation Med.* XVII. 3/1 If..visual cues are reduced by darkness, relatively weak stimulation of the labyrinth may cause strong illusions of apparent motion which may persist after all other sensations of rotation have disappeared. To this visual phenomenon, produced in this manner, we have applied the term 'oculo-gyral illusion'. **1953** R. A. MCFARLAND *Human Factors Air Transportation* iv. 192/1 The oculogyral illusion has its origin in the stimulation of the vestibular mechanism rather than in the eye alone. **1968** R. A. WEALE *From Sight to Light* vi. 115 The astronaut Glenn reported on the oculogyral effect that he experienced in orbit. **1922** STEDMAN *Med. Dict.* (ed. 7) 690 *Oculogyric*, ophthalmogyric, oculomotor. **1927** *Jrnl. Neurol. & Psychopath.* VIII. 27 Other features may..be present, such as oculogyric crises or respiratory disorders. **1954** S. DUKE-ELDER *Parsons' Dis. Eye* (ed. 12) xxvii. 461 (*caption*) The cerebral ocular motor connections... OGA, oculo-gyric area; OGT, oculo-gyric tract. **1973** DUKE-ELDER & WYBAR in S. Duke-Elder *Syst. Ophthalm.* VI. xii. 846 The most typical spasm of vertical movements is seen in oculogyric crises, a striking phenomenon wherein spasmodic deviations of the eyes occur in any direction but usually upwards and less frequently downwards, lasting from a few seconds to some hours.

odalisque. Add: (Further examples.) Also *transf.* and *fig.*

1798 [see *domestic slave* s.v. *DOMESTIC *a.* 2 a]. **1834** R. H. BARHAM *Let.* 26 June (1870) I. v. 239 My seraglio of twelve elderly odalisques certainly does, now and then, furnish me with a job in the way of composing differences which will occasionally arise even in their well-regulated minds. **1903** G. B. SHAW *Let.* 8 May (1972) II. 322 What you want is a repertory of plays which you can carry on your own shoulders, and in which you cannot come into competition with the young odalisques of the west end. *a* **1915** JOYCE *Giacomo Joyce* (1968) 14 She leans back against the pillowed wall: odalisque-featured in the luxurious obscurity. **1926** A. BENNETT *Lord Raingo* I. v. 21 Withal she was no odalisque. She tried to improve herself, to make herself interesting to him. **1938** *Times* 18 Feb. 19/1 Evening gowns include such opposite styles as the odalisque type, with a transparent dancing skirt in black over a gown in gold lamé. **1967** *Listener* 16 Mar. 366/1 Is the creation of a cubist odalisque 'of consequence', and the devoutly humble production of an ikon not?

odd, *a.* (*sb.*) and *adv.* Add: **A.** *adj.* **2. b.** (Further examples.)

1955 J. A. WHEELER in W. Pauli *Niels Bohr* 166 The spontaneous fission rates of nuclei of odd mass number are observed to be slower than the rates for the corresponding even nuclei. **1966** *Mathematical Rev.* XXXI. 36/1 Matrices *M* of even order behave somewhat differently from those of odd order. **1970** O. DOPPING *Computers & Data Processing* ii. 50 The most common form of redundancy check is the parity check, in which the value of a check bit is determined by the parity (odd or even) of the number of ones in the unit to be checked.

f. *Physics.* Having odd parity.

1930 *Physical Rev.* XXXVI. 617 In the case of homonuclear molecules..any electron state may be either 'odd' or even. An electron state of a homonuclear molecule is said to be odd if the electronic factor of ψ changes sign on reflection in the midpoint of the line joining the nuclei, even, if it does not change sign. [*Note*] The use of 'odd' and 'even' in this sense was introduced by Hund... The words..are applied by Kronig, in a sense different from that used by Hund, to the *complete* ψ function of any molecule, homonuclear or heteronuclear. **1940** *Ibid.* LVIII. 104/1 The nearest one could come..would be to assume for the correct wave function a linear combination of two wave functions... These would correspond, respectively, to an even and odd state of the core. **1961** *Encycl. Dict. Physics* II. 786/2 Homonuclear molecules.. can have even (*g*) orbitals (*g* for 'gerade') and odd (*u*) orbitals (*u* for 'ungerade') which are, respectively, symmetric and antisymmetric with respect to interchange of the nuclei.

4. d. Add to def.: or *odds*.

1930 *Times* 25 Mar. 24/1 The balance-sheet shows a loan from the bankers of the company as at December 31 of £118,413 odds.

4*. *Math.* Of a function of one variable: having the property that changing the sign of the argument changes the sign, but not the magnitude, of the function (i.e. $f(-x) = -f(x)$).

1886 A. G. GREENHILL *Differential & Integral Calculus* ii. 80 fx is an odd function, so that $f(-x) = -fx$. **1899** F. G. TAYLOR *Introd. Differential & Integral Calculus* xi. 106 A function of x is said to be even when it is not altered..on changing x into $-x$; it is said to be odd when it is altered in sign, but not in magnitude. **1946** *Ann. Computation Lab. Harvard Univ.* I. 17 This feature is especially valuable when dealing with the interpolation of odd functions. **1969** A. M. HOWATSON *Princ. Appl. Electricity* iv. 74 Some waveforms..can be represented either by an even or by an odd function according to the choice of origin of θ. Others can be even but not odd..or odd but not even.

8. a. (Later examples.)

1930 *Daily Tel.* 1 Dec. 9/3 The 'odd' heavy tweed skirt that is worn in the English country can be left behind. **1949** D. M. DAVIN *Roads from Home* III. i. 210 He'd still be able to save the odd quid. **1959** *Times Lit. Suppl.* 7 Aug. p. ix/4 An endearing wizard liable to irritability and the odd fit of despair. **1973** G. MOFFAT *Deviant Death* i. 18 [He] could have made scarcely enough money from book reviews and the odd feature on country life to pay his petrol bill.

d. (Earlier and later examples.) *odd job* (earlier and later examples), *odd-jobber*, *odd-job man* (later and *fig.* examples); also *odd-jobs man*; *odd-job* v. intr., to do odd jobs; *odd-jobbing* vbl. sb.; *odd-timer*, *odd work*.

1743 W. ELLIS *Mod. Husbandman* Oct. xxii. 149 The Odd-man's Wages is from fifty Shillings, to four Pounds a Year. *c* **1770** in de Vries & Fryer *Venus Unmasked* (1967) 33 Miss E. P.... R.. has not received her stated allowance; is therefore obliged, in order to keep up appearances, to do *odd jobs*. **1798** J. WOODFORDE *Diary* 24 Nov. (1931) V. 148 Paid John Buck, Blacksmith, for divers odd Jobbs for the Year 1798 the Sum of 2.14.10. **1853** MRS. GASKELL *Ruth* II. iii. 64 Just try for a day to think of all the odd jobs as to be done well. **1859** DICKENS *T. Two Cities* III. ix. 206 A gentleman like yourself wot I've had the honour of odd jobbing till I'm grey at it. **1860** —— *Gt. Expect.* (1861) I. vii. 90, I was..odd-boy about the forge. **1861** MRS. BEETON *Bk. Househ. Managem.* 964 Where a single footman, or odd man, is the only male servant then..he is required to make himself generally useful. **1863** *All Year Round* 11 July 472/2 Either can rest occasionally by employing an 'odd man', of whom there are several..ready to do 'odd' work. **1897** E. L. VOYNICH *Gadfly* 190, I lived by odd-jobbing for the blacks on the sugar plantations. **1905** Odd-man [see *DOORMAN 2]. **1909** J. R. WARE *Passing Eng.* 186/1 *Odd job man*, modified description of the Shyster, who professes to do anything and only does his employer. **1915** R. BROOKE *Let.* 20 Apr. (1968) 681 Ian Hamilton..asked me if I'd like to be attached to his staff as a sort of 'galloper' and odd-jobber —'A.D.C.'. **1916** H. G. WELLS *Mr. Britling* I. i. 28 They become villa parasites and odd-job men. **1925** A. S. M. HUTCHINSON *One Increasing Purpose* I. xviii. 113 She's an odd-timer on Miss Marr's typist staff. **1933–4** WITTGENSTEIN *Blue & Brown Bks.* (1958) 44 We are tempted to describe the use of important 'odd-job' words as though they were words with regular functions. **1938** D. SMITH *Dear Octopus* I. 22 Just a sort of receptionist and general odd-jobber. **1940** L. MACNEICE *Last Ditch* 25 He was not able to read or write, He did odd jobs on gentlemen's places, Cutting the hedge or hoeing the drive. **1944** F. CLUNE *Red Heart* 32 He taught, biked, and odd-jobbed his way around Great Britain for a year. **1948** in D. M. Davin *N.Z. Short Stories* (1953) 195 The odd-jobs man who was also the cowman and gardener at the homestead. **1973** A. CHRISTIE *Postern of Fate* II. ii. 69 I've done odd jobbing, you know. **1977** *Private Eye* 4 Mar. 20/2 (Advt.), Carpentry, cleaning, electrics, decorating, gardening, handy persons, odd jobbers.

9. b. *odd bod* (freq. in *pl.*) [*BOD].

1955 [see *BOD]. **1966** *Courier-Mail* (Brisbane) 15 Sept. 1/7 A family of 12 self-styled 'odd bods' is immigrating to Spain because Australia is 'too conventional'. **1976** 'J. ROFFMAN' *Why Someone had to Die* 78 Anyone would, except you who have an inborn bias toward the odd bods in society.

D. *Comb.* **1. a.** *odd-shaped* (later examples).

1921 W. DE LA MARE *Crossings* 54 He..carries an odd-shaped fiddle. **1967** KARCH & BUBER *Offset Processes* ii. 34 Products include paper and board too large for conventional presses (over 80 inches), odd-shaped objects, posters, [etc.]. **1977** P. D. JAMES *Death of Expert Witness* III. 169 The gathering of those odd-shaped pieces of information..which, in the end, would click together to form a picture.

b. *odd-looking* (later examples), *-sounding* (later examples).

1876 'GEO. ELIOT' *Dan. Der.* III. v. xxxviii. 135 Spectators would be likely to think of him as an odd-looking Jew. **1976** J. B. HILTON *Gamekeeper's Gallows* ii. 23 Beresford was curious to know what this odd-looking character was carrying. **1853** MRS. GASKELL *Cranford* vii. 131 There she sat, as stately and composed as though we had never heard that odd-sounding cough. **1942** PARTRIDGE *Usage & Abusage* (1947) 15/1 *Admittable* is rare and odd-sounding for *admissible*.

c. the adj. with a sb., forming an attrib. phr., as *odd-number, -order, -parity*.

1922 F. F. POTTER *Teaching of Arithmetic* xvii. 325 The simple odd-number series. **1957** L. FOX *Numerical Solution Two-Point Bound. Probl.* vi. 141 This too is connected with the use of central differences for even-order equations, but of mean central differences for odd-order equations. **1962** *Gloss. Terms Automatic Data Processing* (*B.S.I.*) 33 When the numbers of ones (or zeros) is required to be odd the check is called an odd parity check, and when even an even parity check. **1967** *Bodl. Libr. Rec.* VIII. 3 The second drawback is the use of an odd-parity code, with seven-hole punching necessary for a plain tape-feed.

2. odd-and-end (earlier example); **odd-even** *a.* *Nuclear Physics*, (*a*) of or pertaining to nuclei of odd mass number and those of even mass number; (*b*) applied to nuclei containing an odd number of protons and an even number of neutrons; **odd-gaits, -gates** *local* [GATE *sb.*[2] 9 b], odd, strange, out of the way; **odd man out,** (*b*) a person or thing differing from all others of a group in some respect; **odd-odd** *a.* *Nuclear Physics*, (*a*) of or pertaining to nuclei of odd mass number only; (*b*) applied to nuclei containing an odd number of protons and an odd number of neutrons.

1846 J. BROWN *Lett.* (1912) 90, I have no continuity and thoroughness of thought,..and my style, if style it can be called, is the piebaldest, oddandendest. **1936** *Physical Rev.* XLIX. 897/1 The transitions are of the odd-even type. **1949** GAMOW & CRITCHFIELD *Theory of Atomic Nucleus* iv. 88 $\delta_A = 0$ for even-even nuclei, $= 1$ for odd-even nuclei, $= 2$ for odd-odd nuclei. **1955** J. A. WHEELER in W. Pauli *Niels Bohr* 166 Odd-even differences in spontaneous fission. The spontaneous fission rates of nuclei of odd mass number are observed to be slower than the rates for the corresponding even nuclei. **1970** G. K. WOODGATE *Elem. Atomic Struct.* ix. 195 The tentative explanation of this phenomenon, called odd-even staggering, is beyond the scope of this discussion. **1906** KIPLING *Puck of Pook's Hill* 263 Won'erful odd-gates place—Romney Marsh. **1957** H. HALL *Parish's Dict. Sussex Dial.* 88/2 *Oddgaits*, extraordinary. 'It's an odd-gaits sort of place.' **1928** F. W. CROFTS *Sea Mystery* xvi. 263 Mrs. Berlyn as hostess would reasonably be the odd man out when the change was made from snooker to billiards. **1935** MRS. BELLOC LOWNDES *Let.* 10 Aug. (1971) 129 Rex Whistler..is rather 'odd man out', though he is so quiet and unassuming that everyone likes him. **1945** F. L. GREEN (*title*) Odd man out. **1958** *Times* 29 Nov. 9/1 Here..is the fallen foe [*sc.* the pike]; not merely our foe, but..the odd man out in Nature and in the sporting Canon. **1969** in Halpert & Story *Christmas Mumming in Newfoundland* 33 These settlements are 'odd men out'; at odds with outside economic forces. **1970** *Guardian* 19 Nov. 11/4 The products..would look odd men out in a formal setting. **1973** *Listener* 15 Nov. 658/1 Eight of the present nine members of the EEC voted in favour, the odd man out being..Denmark. **1936** *Physical Rev.* XLIX. 897/1 The transition is of an odd-odd or even-even type. **1937** *Ibid.* LI. 951/2 The condition..for the instability of odd-odd nuclei should read $g + 2g_0 < g_m$. **1956** *Proc. Physical Soc.* A. LXIX. 635 (*heading*) The magnetic moments of odd-odd nuclei.

oddball (ǫ·dbǭl). *colloq.* (orig. *U.S.*). Also **odd-ball, odd ball.** [f. ODD *a.* + BALL *sb.*[1]] An eccentric person; a person of unconventional views or habits. Also *attrib.* or as *adj.*, eccentric, exceptional, peculiar.

1948 *Amer. Speech* XXIII. 221 *Odd Ball*..connoted that the individual's strangeness was not the result of prison camp experiences. 1953 Berrey & Van den Bark *Amer. Thes. Slang* (1954) § 411/2 Peculiar or eccentric person..oddball, odd-fellow, odd stick. *Ibid.* § 454/3 Social dud or bore; 'drip'..oddball, ooly-drooly, pain. 1956 W. H. Whyte *Organization Man* (1957) ix. 123 There are exceptions, but one must be a very odd ball to be one. 1956 Wallis & Blair *Thunder Above* (1959) xvi. 158 A phlegmatic odd-ball in a world of make-believe. 1957 *Time* 2 Sept. 23/2 Officialdom in Saigon thought that the threat of rebellions by the country's fanatic, odd-ball religious groups was ended at last. 1959 *Washington Post* 19 Oct. B. 10/6 Life aboard a submarine has its oddball moments. 1968 *Canad. Antiques Collector* Dec. 10/1 The classic tradition was too strong at the end of the 18th c. for any but obvious odd balls to find a 'better way' of building their houses. 1969 *Daily Mail* 15 Jan. 5/4 If that burn-time was not correct we could have gone into all sort of odd-ball orbits. 1971 C. Fick *Danziger Transcript* (1973) 41 They had been oddball friends since Harvard. 1973 M. Truman *Harry S. Truman* xvi. 331 Earlier in 1946 an oddball broke into the National Gallery and cut a hole in Dad's portrait. 1974 *Peace News* 15 Mar. 10 It's always been very much an odd ball way of doing it. 1974 D. Sears *Lark in Clear Air* ii. 27 The oddest of a family where odd-balls were not the exception.

Oddfellow, Odd-fellow. (Earlier and later examples.)
1795 in R. Humphreys *Mem. J. Decastro* (1824) 247 As many Grands and Brothers of the Odd Fellows, Bucks, Masonic and other Lodges. 1800 *Sporting Mag.* XVI. 59/1 A Member of the Odd Fellows. 1951 McWhiney & Simkins in A. Dundes *Mother Wit* (1973) 594/2 This mummery manifests itself in numerous other secret organizations such as the Odd Fellows, the Masons, and the Knights of Columbus. 1970 *New Yorker* 14 Nov. 175/1 Members of the Independent Order of Odd Fellows stuck to their symbol of an axe left cleaving into a realistic log. 1977 P. D. James *Death of Expert Witness* i. 16 A ticket for the local Oddfellows' hop.

Oddfellowship. (Later U.S. examples.)
1907 C. E. Jackson *Poet Wildey* 11 It is the Independent Order of Odd Fellows, or if you like the term better, American Odd Fellowship. 1972 *Fairbanks* (Alaska) *Daily News-Miner* 3 Nov. 7/7 Odd Fellowship..is a benevolent, fraternal order banded together for mutual benefit.

odditorium (ǫditōᵊ·riŏm). [f. Oddit(y + -orium.] A shop or venue for the display or retailing of oddities. Also *transf.*
1914 R. & E. Shackleton *Four on Tour* 182 We happened upon an 'odditorium', a delightful name adopted by a very shabby shop in a very narrow lane [in Kingston-upon-Thames] where we found some very attractive bits of old silver and china. 1933 *Sun* (Baltimore) 16 Dec. 10/6 Mr. Miller..manages the Ripley Believe It or Not 'odditorium' at Horticulture Hall, which contains thirty-nine oddities. 1940 *Time* 26 Feb. 42/2 Sanka Coffee's Tuesday-night odditorium of the air.

oddlings (ǫ·dliŋz), *sb. pl.* [f. Odd *a.* + -ling¹.] = Oddments *sb. pl.*
1854– in *Eng. Dial. Dict.* 1900 *Windsor Mag.* XI. 354 A hundred odd bits, that's all—but they are a manufacturer's oddlings.

odd lot (ǫ·d lǫt). [f. Odd *a.* + Lot *sb.*] **1.** An incomplete set or random mixture (of articles of commerce).
1897 *Sears, Roebuck Catal.* 173/1 The pants are not the size to fit the coats and vests. We then throw them into what we call our 'Odd Lots', or mixed suits. 1931 C. Maughan *Markets of London* 105 Parcels of five bags or less are known as 'odd lots' and are sold separately at the end of the auctions, with the exception of pea-berries, which are sold irrespective of the size of the lots.
2. *U.S. Stock Exchange.* Used *attrib.* of transactions involving numbers of shares smaller than is normally dealt in. Hence **odd-lotter.**
1929 *Times* 1 Nov. 22/5 The list generally moved up under heavy accumulation of standard issues, together with reports of a tremendous volume of odd-lot bargain buying. 1939 C. O. Hardy (*title*) Odd-lot trading on the New York Stock Exchange. 1968 *Economist* 28 May 94/1 The unit trust investor is not quite an odd-lotter—the American term for a small investor. 1970 *Washington Post* 30 Sept. D. 10/3 Odd lot transactions by principal dealers. 1976 *National Observer* (U.S.) 28 Feb. 8/4 But odd-lot figures..suggest that small investors were, on balance, selling in November and December of 1972 and in January of 1973. *Ibid.*, There is some statistical evidence that odd-lotters have been wrong far too often in the past.

odds, *sb.* Add: **2. c.** *What's the odds?* (earlier and later examples); *it makes no odds* (later examples).
1826 *Sessions Papers* 11 Dec. 86/1, I asked Jackson whose they were—he said, 'What odds; they are mine.' 1840 Thackeray *Shabby Genteel Story* ix, in *Fraser's Mag.* Oct. 410/1 Suppose I do die,..what's the odds? Caroline doesn't care for me. 1855 Geo. Eliot in *Ibid.* June 699/1 'What's the odds, so long as one can sleep?' is your *formule de la vie.* 1890 Kipling *Soldiers Three* 79 'Wot's the odds as long as you're 'appy?' said Ortheris. 1923 D. H. Lawrence *Kangaroo* xvi. 345 That sense of sardonic tolerance, endurance. 'What's the odds, boys?' 1967 N. Freeling *Strike Out* 84 You do neo-expressionism, or..neo-whatever-you-like, it makes no odds,..you'd still be up against the machine. 1973 'M. Underwood' *Reward for Defector* xxii. 158 What's the difficulty?.. Not that it makes any odds.

4. d. Phr. *to ask* (or † *beg*) *no odds*: to desire no advantage; to seek no favour. *U.S.*
1806 *Baltimore Even. Post* 5 Mar. 2/2 (Th.), No odds he begs Of any beast that walks upon four legs. 1834 *Vermont Free Press* 7 June (Th.), A Varmounter never uses a dog... Give him a gun, and he asks no odds. 1894 *Congress. Rec.* 29 May 5447/1 South Dakota asks no odds of any State of the Union.
e. *over the odds*, past the limit, above a generally agreed rate.
1922 'Sapper' *Black Gang* xviii. 306, I admit it seems a bit over the odds, but every word I've told you is gospel. 1930 —— *Finger of Fate* 103, I admit..that to be called a damned Englishman by Pedro Gonsalvez is a bit over the odds. 1972 *Which?* Feb. 50/1 You could even pay more than the list prices. We found some of the tools being sold for perhaps a pound or so over the odds. 1972 *Listener* 28 Dec. 897/2 Sir Michael Swann, the new Chairman of the BBC's Board of Governors..felt that Alf Garnett went 'over the odds' occasionally. 1974 'D. Craig' *Dead Liberty* xxxi. 187 His..means of pushing his conscience underground was by paying a little over the odds.
5. b. Phr. *to shout the odds* (see quot. 1925).
1925 Fraser & Gibbons *Soldier & Sailor Words* 257 *To shout the odds*, to talk too much: to brag: to grumble. 1958 F. Norman *Bang to Rights* i. 10 He was still shouting the odds about this blag which was..nothing but a dirty great romance. 1960 L. Cooper *Accomplices* ii. i. 76 There are always a few bloody fools who shout the odds about British justice and fair trials. 1967 *Sunday Times* 15 Oct. 9 For years he's shouted the odds about the Scouse way of life. 1973 'J. Patrick' *Glasgow Gang Observed* xv. 131 He still shouts the odds fae the windae when there's a ba'le [*sc.* battle] oan. 'Get *right* intae it, Tim,' he says.
6. b. *by all odds*, by far. *U.S.*
1866 W. D. Howells *Venetian Life* 50 By all odds, the loungers at Florian's were the most interesting. 1951 J. P. Marquand *Melville Goodwin* x. 163 Lee said that McClellan was the best Union general he fought against by all odds.
7. b. *odds and sods* (orig. Services' slang): see quots. 1930 and 1948; now in gen. use as a variant of *odds and ends.*
1930 Brophy & Partridge *Songs & Slang 1914–1918* 143 *Odds and Sods*, 'details' attached to Battalion Headquarters for miscellaneous offices: batmen, sanitary men, professional footballers and boxers on nominal duties, etc. *a* 1935 T. E. Lawrence *Mint* (1955) ii. ix. 125 Ten minutes late for dinner. Odds and sods to eat. 1941 *Argus* (Melbourne) *Week-End Mag.* 15 Nov. 1/4 *Odds-and-sods*, men not attached to a particular unit. 1944 *Penguin New Writing* XX. 44 The section was mixed up with the Police and one or two odds and sods as usual. 1948 Partridge *Dict. Forces Slang* 130 *Odds and sods*, men on miscellaneous duties. Men not classified. Members of other denominations than Roman Catholic, Church of England and Presbyterian. (Army.) (2) In the Navy, the rank and file—the *hoi polloi*—of the lower-deck. 1950 G. Wilson *Brave Company* xiii. 193 Add three-inch mortar, four-point-two inch mortar, Vickers and all the odds and sods. 1955 E. Waugh *Officers & Gentlemen* i. vi. 64 They left me behind with the other odds and sods. 1971 B. W. Aldiss *Soldier Erect* 205 We've got precious little strike-power..—the Assam Regiment,..a few odds and sods of the Burma Regiment, and our friends and allies of the Nepalese army! 1975 *Time Out* 26 Sept. 57/4 Although Tolkien's planned preface to the poems was never realised, his son Christopher has created one mostly out of his father's odds and sods—a radio talk plus notes. 1976 A. Hill *Summer's End* iv. 61 Beyond these were oblong-shaped archway bricks, chimney-cowlings, roof-tiles and other odds and sods I couldn't put names to.
8. odds-on *a.* (later and *fig.* examples); also (occas. without hyphen) as *adv.* and as *sb.*, the state of betting when odds are less than 1:1; an odds-on favourite.
1917 [see *Aussie *sb.* and *a.* 2]. 1923 Wodehouse *Inimitable Jeeves* xiii. 153 'Something's gone wrong with the favourite.' 'Which is the favourite, sir?' 'Mr. Heppenstall. He's gone to odds on.' 1926 E. Wallace *More Educated Evans* x. 222 Six successive odds-on chances. 1928 *Daily Mail* 31 July 11/4 The favourite.. started at heavy odds-on. 1941 Baker *Dict. Austral. Slang* 50 *An odds-on*, an odds-on favourite. 1945 J. B. Blair in *Coast to Coast 1944* 136 It seemed odds on that Wang would stop a bullet. 1955 *Times* 11 May 5/1 Lions Love..was made odds-on for the race for two-year-old fillies. 1962 D. Francis *Dead Cert* i. 7 Admiral, the odds-on certainty who had lost his first race for two years. 1972 *Guardian* 5 Dec. 13/4 In one area near Poona there have been 22 drought years in the past 30... There, as in other places, it is a two to one odds-on probability. 1973 [see *nap *v.*⁴]. 1976 H. Tracy *Death in Reserve* xix. 147 'Odds on they've gone to France.' said Phoenix. 'He'd have enough fuel.'

odds, *v.* Add: **2.** *slang.* To elude, to evade.
1958 F. Norman *Bang to Rights* ii. 52, I used to go to church every Sunday only because I couldn't odds it. 1970 G. F. Newman *Sir, You Bastard* iv. 124, I can't odds being mixed up in crime.

oddside (ǫ·dsəid). *Founding.* [f. Odd *a.* + Side *sb.*¹]. A temporary cope, usu. made of sand or plaster of Paris, in which part of a pattern is bedded while the final mould is made of the upper, exposed portion.
1836 W. H. Smith *Birmingham* iii. 33 The patterns still lie on the sand of the 'odd side', as it is termed. 1903 *Work* 11 July 364/1 It is necessary, when the mould is commenced, either to block up the half of the flask to a level with the parting line on the pattern, or to sink the same into a bed of sand as far as its greatest diameter.
This bed of sand is known as a match or 'oddside', and it has importance as a time-saver. 1928 W. Rawlinson *Mod. Foundry Operations* xiii. 177 The use of oddsides eliminates the necessity of making a joint in the moulding operation, as is the case when moulding with a loose pattern. 1960 R. Lister *Decorative Cast Ironwork* ii. 45 When the plaster has set, the sections are removed and placed aside, and the bottom section placed in another oddside. 1972 P. R. Beeley *Foundry Technol.* viii. 351 Where a flat parting plane intersects the pattern, moulding is simplified by providing a split pattern and dispensing with the oddside.

ode. Add: **1.** *spec.* A short Old English poem, esp. *The Battle of Brunanburh.*
1770 T. Percy tr. *Mallet's Northern Antiquities* II. 196 Compare the Anglo-Saxon Ode on Athelstan's Victory, preserved in the Saxon Chronicle. 1798 S. Henshall *Saxon & Eng. Lang.* 39 The first specimen we shall exhibit is the conclusion of a Saxon Ode on a Victory of King Athelstan's. 1807 S. Turner *Hist. Anglo-Saxons* (ed. 2) II. 323 It will be sufficient to add to the already copious specimens of the Anglo-Saxon poetry, the following Ode, which is appended to the menology. 1847 F. Madden *Laȝamon's Brut* I. p. xxiii, No one can read his descriptions of battles and scenes of strife without being reminded of the Ode on Æthelstan's victory at Brunanburh. 1871 H. Sweet in W. C. Hazlitt *Warton's Hist. Eng. Poetry* II. 7 There are only two poems of any merit to which we can assign with any certainty a southern origin. These are the ode on the battle of Brunanburh, and the narrative of the battle of Maldon, which were, no doubt, composed immediately after the events they record.
3. *ode-metre.*
1901 *Academy* 14 Dec. 585/2 That so-called 'irregular' ode-metre which they [*sc.* Patmore and Henley] use in common.

-ode², formative element repr. Gr. ὁδός way, path, first used in *anode, cathode, electrode*, and later in the names of thermionic valves with a specified number of electrodes, as *diode, triode, tetrode*, etc. (with two, three, four electrodes); (certain of these latter—e.g. *diode, hexode*—were orig. coined directly from the Gr. to describe forms of multiplex telegraphy).

odeon (ǫu·diǫn). [ad. Gr. ᾠδεῖον a building for musical performances.] **1.** Also **odeion** (ǫdəi·ǫn). = Odeum.
1902 Anderson & Spiers *Archit. Greece & Rome* vii. 117 No Greek example exists of the Odeon. *Ibid.* 280 *Odeon*, a circular building in which rehearsals and musical contests took place in Greece. 1908 W. R. Lethaby *Greek Buildings* iv. 196 The rotundas at Olympia..must have been covered by low cones. So must the Odeon and the Skias at Athens. 1909 A. Marquand *Greek Archit.* vi. 355 The Odeion (ᾠδεῖον), or music hall, was designed for musical contests and rehearsals of plays. 1961 L. Mumford *City in Hist.* Note to plate 12, What modern town.. can point to..as many handsome examples of bath, theater, odeon? 1977 *Jrnl. R. Soc. Arts* CXXXV. 485/2 The Department of Antiquities..is now excavating an Odeon.
2. (With capital initial.) The name of any of numerous cinemas in a chain built by Oscar Deutsch or his company in the 1930s. Hence in wider use, any cinema, esp. of similar lavish architectural style. Also *attrib.* and as *adj.*, designating this style.
1937 Klingender & Legg *Money behind Screen* 46 *Odeon Theatres, Ltd.*—Is a private company registered in October, 1933 with £100 capital... In November, 1936, the Odeon group of companies was increased by Cinema Ground Rents & Properties, Ltd... The object of the concern is stated to be primarily that of acquiring new sites for Odeon cinemas. 1946 J. P. Mayer *Sociol. of Film* iv. 51 Between August 1944 and June 1945 I spent approximately twenty Saturday mornings in the Odeon and Gaumont British Children's Cinema Clubs. 1952 A. Wood *Mr. Rank* v. 87 People on holiday..would go into an Odeon because it would remind them of the Odeon at home. *Ibid.* 88 Beginning modestly.., Deutsch by 1937 had raised the total number of Odeons to over 200... His cinemas were all built to sumptuous standards, with the aim of providing something more luxurious than the rest. .. The Odeon cinemas became almost a symbol of the British way of life in the 1930s. 1954 J. Betjeman *Few Late Chrysanthemums* 40 An Odeon flashes fire Where stood their villa. 1960 C. MacInnes *Mr. Love & Justice* 224 The two girls carried on..joint excursions to their respective Odeons. 1964 *New Statesman* 10 Apr. 576/3 The whole occasion, with its foamy seats, Odeon hues and ice-cream girls, is organised to resemble film-going as closely as possible. 1970 Collins & Dodd *Perishers Bk.* 8 4 Maisie..still finds Marlon attractive in a repulsive sort of way, especially when he wears his Odeon Admiral's uniform. 1972 H. C. Rae *Shooting Gallery* ii. 243 It was the cinema, an Odeon in the heart of Glasgow, big and modern and comfortable. 1977 *Times* 12 Feb. 7/3 M. Barrier designed his own restaurant under the influence of *fin-de-siècle*, early Odeon, and fifties tubular.

odiferous. Delete 'obs.' and add later examples.
1851 V. Lush *Jrnl.* 8 Aug. (1971) 83 It looked marvellously strange to see so rough a looking man taking up bottle after bottle with all the air of a connoisseur, and at last after much deliberation selecting those which seemed to his olfactory nerves the most odiferous. 1977 *Fanfare* (Toronto) 6 July 3/1 German barbers worked busily for years on odiferous brews to suppress the dangerous shade.

Odissi (odi·si). *India.* [f. the place-name *Orissa*, Skr. *oḍra*.] A style of Indian dance, believed to be one of the oldest in India, originating in the eastern state of Orissa. Also called *Orissi.
1965 E. Bhavnani *Dance in India* vi. 49 The classical dance as practised in Orissa, a State on the mid-north-eastern coast of India, is known as Orissi or Odissi. 1967 Singha & Massey *Indian Dances* xxv. 203 Odissi did not escape a decline any more than the other dances of India. .. Odissi draws upon several ancient texts in Sanskrit and Oriya. *Ibid.* 205 Although Odissi recitals are nowadays given on the stage, they are nevertheless essentially a form of worship in which the dancer performs an act of adoration. 1971 *Femina* (Bombay) 30 Apr. 29/1 A student of Hindustani classical music, she knows many Oriya songs and is in demand in Bombay to accompany young Odissi dancers. 1972 *Daily Tel.* 26 Apr. 14/4 Her Odissi section could hardly have been more different. In this quintessentially feminine style, she showed the ancient thrice-bent sculptural positions as well as a wide variety of spins.

odium. Add: **e.** (Earlier and later examples of *odium theologicum.*) Also, by imitation, *odium academicum* (academic), *archæologicum* (archæological), *biologicum* (biological), *ethicum* (ethical), *philologicum* (philological), *philosophicum* (philosophical), *scholasticum* (scholarly).
[1673 J. Flavel *Fountain of Life* xxx. 414 Strigelius desired to die, that he might be freed *ab implacabilibus odiis theologorum*, from the implacable strifes of contending Divines.] 1734 'Philalethes Cantabrigiensis' *Geometry No Friend to Infidelity* 13 This is the very method which the *Odium Theologicum*, the intemperate zeal of Divines has always pursued. *a* 1866 J. Grote *Exam. Utilitarian Philos.* (1870) 9 The 'odium ethicum' is even more unreasonable than the 'odium theologicum'. 1946 *Mind* LV. 98, I am accused of indulging in 'quixotic denunciations' of Spinoza, and of being actuated by a spirit of *odium philosophicum* against his memory. 1959 *Listener* 10 Sept. 405/2 The *odium biologicum* which it now seems scarcely possible for the acknowledged biologist wholly to avoid. *Ibid.* 29 Oct. 743/3 A kind of *odium archaeologicum* has been added to an *odium theologicum.* 1962 *Ibid.* 10 May 821/1 The lay public is now familiar with the wars of the learned... With Professor Dumond we are involved in something more than *odium academicum.* 1963 *Times* 14 Feb. 15/1 The question, then, arises whether the college to which he belongs and the cause of education in Oxford are to be sacrificed to the *odium theologicum* of a few infatuated dignitaries. 1970 *Language* XLVI. 246 The public will not change, and we will continue to suffer from the odium philologicum unless we can find a way of establishing more friendly communication. 1973 *Times Lit. Suppl.* 30 Mar. 347/4 Does not conceal her *odium scholasticum* at being anticipated by Firbank's bibliographer and first biographer. 1974 *Times* 21 Mar. (Art & Antiques Suppl.) p. iii/2 No punches are pulled in the book reviews, which are often filled with *odium scholasticum*, as one expert dissects the researches of another. *Ibid.* 9 Nov. 13/6 The *odium theologicum* that often characterises academic debate.

odometer. Add: Now *esp.* one in a motor vehicle. (Further examples.)
1913 *Collier's* 11 Jan. ii. 57/1 The Corbin-Brown Speedometer records speed accurately... Its hand is steady; its odometer absolutely dependable. 1962 *Which? Car Suppl.* Jan. 28/2 We measured the accuracy of speedometer, mileage recorder (odometer) and fuel level gauge. 1968 *Chicago Tribune* 9 July i. 12/1 He went out the next week-end in his sportscar and, using the auto's odometer, plotted running courses from his home. 1969 B. Weil *Dossier IX* v. 34 He checked the odometer which he had zeroed at the Porte de Versailles. Three more kilometres, then a right turn. 1972 *Drive* Spring 84 Cruising speed was restricted to 55 mph in deference to the 90,000 miles already clocked up on the broken odometer.

ödometer: see *œdometer.

odontology. (Later examples.)
1911 *Chambers's Jrnl.* June 375/1 The British Medical Association has recently founded a Section of Odontology. 1966 G. Gustafson (*title*) Forensic odontology. 1974 *Encycl. Brit. Macropædia* XI. 814/1 Other specialties on which the courts frequently require expert opinion include forensic odontology.

odorant, *a.* Restrict 'Now *rare*' to sense in Dict. and add: **B.** *sb.* Also **odourant.** An odoriferous substance; *spec.* one used to impart a desired odour to a product. Also *attrib.*
1944 R. W. Moncrieff *Chemical Senses* viii. 166 Reactions to odorants are usually mild and undefined, a sniff, a drawing away, or..disgust. 1952 Kirk & Othmer *Encycl. Chem. Technol.* VIII. 142 Biacetyl has been used as an odorant for butter, cream, milk, margarine.., cheese, coffee, confectionery.., ice cream mixes, and tobacco. 1971 *Nature* 24 Sept. 231/1 Theories of olfaction are thus legion; all attempt to answer the central question as to what properties of the odourant molecules are responsible for the wide variety of odour quality. 1972 *Materials & Technol.* V. xiv. 526 In microcellular soling mixes,.. odorants which confer the odour of leather to the soles are frequently incorporated.

† odoriferent (ǫudǒri·fěrěnt). *Obs. rare*⁻¹. [f. med.L. *odōriferens* (see Odoriferant *a.*).] = *odorant *sb.*
1858 G. A. Sala *Journey Due North* viii. 130 These boots have a peculiar..odour..of myrrh, frankincense, sandalwood, benzoin, and other odoriferents.

odorimeter (ǫudǒri·mǐtəɹ). [f. L. *odor, odōri-*, Odour + -meter.] An instrument for measuring the intensity of odours.
1898 *Amer. Jrnl. Psychol.* X. 85 We are indebted to Dr. Zwaardemaker for the words 'olfactometry' and 'olfactometer'.., 'odorimetry' and 'odorimeter'. 1966 *New Scientist* 29 Dec. 726/3 The odorimeter is well adapted for plant control, for the regular injections of plasticizer can be made to coincide with the repeated inspection of batches being deodorized.
Hence **odorime·tric** *a.*, of or pertaining to odorimetry; **odori·metry,** the measurement of the intensity of odours.
1898 *Amer. Jrnl. Psychol.* X. 85 Odorimetry is a 'side issue' of olfactometry. 1917 *Jrnl. Exper. Psychol.* II. 437 The preceding odorimetric method..enables one to stimulate the nose continuously for any desired time by means of an air current containing a known proportion of odorous molecules. 1922 *Perfumery & Essent. Oil Rec.* XIII. 5/2 Odorimetry..enables us to compare this value with fortuitous smells. 1935 *Bull. Neurol. Inst. N.Y.* IV. 15 (*heading*) Theoretical considerations, significance for quantitative and qualitative odorimetry and olfactometry. 1972 *Biol. Abstr.* LIV. 349/2 (*heading*) Odorimetric appraisal of polymers used in the construction of sealed chambers.

odoriphore (ǫu-dǒrifōᵊɹ). Also **odorophore.** [ad. G. *odoriphor* (H. Zwaardemaker *Die Physiol. des Geruchs* (1895) xiv. 240), f. as prec.: see -phore.] = *osmophore 1.
1919 *Physiol. Abstr.* IV. 13 Adsorption is..but the first step [in olfaction], on which follows the solution of the odorous substance and its odorophore in the lipoids of the epithelial cell. 1922 *Ibid.* VII. 174 The natural groups of odours..are characterised by the presence, number, size, and arrangement of odoriphore groups. 1944 R. W. Moncrieff *Chemical Senses* xii. 306 Zwaardemaker considered that odour was dependent on the presence of odoriphores, special chemical groups, each of which was responsible for a special odour.
Hence **odoripho·ric** *a.*
1944 R. W. Moncrieff *Chemical Senses* xii. 307 Odoriphoric groups. 1968 W. McCartney *Olfaction & Odours* 169 That odorous energy is molecular-vibratory..seemed to him [*sc.* A. Heyninx] the sole admissible view and on it he reached the following conclusions:..4. The specific vibrations of the 'odoriphoric' groups of the odorous molecules cause the olfactory mucous membrane..to vibrate.

odorivector (ǫu-dǒrivektəɹ). [ad. F. *odorivecteur* (A. Heyninx *Essai d'Olfactique physiol.* (1919) i. 2), f. as prec. + F. *vecteur* Vector.] Any substance the molecules of which stimulate the olfactory system.
1926 *Nature* 24 Apr. 591/2 The quality of the sensation depends on the molecular structure of the odorivector. 1964 *Ann. N.Y. Acad. Sci.* CXVI. 431 The final result of interaction of the odorivector molecules with the olfactory area is a modification in the electrical activity of the nerve cells which communicate with the sensors. 1967 J. W. Johnston in T. Hayashi *Olfaction & Taste II.* 46 It is problematical whether there are pure odorivectors that stimulate one type of sensor unit.

odour. Add: **4. b.** (Earlier and later examples.) Also without qualifying adj.
1835 Dickens *Let.* 16 Dec. (1965) I. 106 As the Tories are the principal party here, *I* am in no very good odour in the town. 1954 N. Mitford *Mme de Pompadour* xviii. 230 In 1760 St. Germain fell into bad odour with the police and Choiseul sent him packing. 1977 *Bulletin* (Sydney) 22 Jan. 33/1 There is no doubt that the Daoud affair, which has brought France into such odour abroad including a call for the boycott of French goods in the United States, was the result of conflicting interests between French diplomats and French security services.
6. *odour-reek; odour-free, -laden, -proof* adjs.; **odour-blindness,** an inability to perceive a particular smell or range of smells.
1931 *Eugenical News* July 106/1 Some were thus 'blind' to fragrance in the red flowers though perceiving fragrance in the pink while others were just reversed in their odor 'blindness'. 1973 *Nature* 23 Mar. 271/2 Three types of odour-blindness or specific anosmia have been studied. 1955 *Jrnl. Appl. Physiol.* VIII. 341/2 The odorant.. introduced through an ultramicroburette was vaporized into the odor-free test room by an atomizing jet of odor-free air. 1962 W. D. Hislop in H. W. Chatfield *Sci. Surface Coatings* xviii. 531 Use can be made of catalytic oxidation to treat the exhaust and render it odour-free. 1885 W. B. Yeats in *Dublin Univ. Rev.* Apr. 58/1 Dreaming in their soft odour-laden sleep. 1900 —— *Shadowy Waters* 44 Where time is drowned in odour-laden winds. 1934 Webster, Odorproof. 1950 *Amer. Jrnl. Psychol.* LXIII. 433 The need for rigorous experimental controls for smell soon became apparent. A large odor-proof globe..was first thought of. 1951 M. McLuhan *Mech. Bride* (1967) 62/1 A special caste of robots, who would care for the victims of such necessities in germ- and odor-proof laboratories. 1960 *Farmer & Stockbreeder* 1 Mar. 55/3 The odour-proof film minimizes the risk of contamination. 1932 W. Faulkner *Light in August* viii. 187 The odorreek of all anonymous men.

odourant, var. *odorant *sb.*

odourless. Add: Hence **o·dourlessness,** the condition of being without odour.
1890 *Retrospect Med.* CII. 149 The odourlessness of Aristol..renders the drug most useful in the treatment of soft chancre. 1901 *Westm. Gaz.* 24 Dec. 9/1 We would again lay stress on the odourlessness; it is a whale, not a smelt. 1963 W. Summer *Methods Air Deodorization* iii. 101 The simultaneous olfaction of two selected and matched odours will cause the psychological effect of odourlessness. 1968 [see *osmoceptor].

odum (odū·m). [Native name in Ghana.] = *iroko.
1920 *Nature* 29 July 692/1 Another valuable wood is that labelled Odum..which has also been exported as Iroko, sometimes falsely termed African teak... It possesses none of the qualities of teak with the exception of a superficial resemblance in colour. 1959 A. Abbs *Ashanti Boy* i. 34 The thick, odum stick he kept beside his bed. *Ibid.* 38 Low easy-chairs made of local mahogany or odum wood. 1962 *Listener* 6 Sept. 359/2 The big odum tree near the school park [in Ghana].

œcology, œcological, now written **ecology, ecological.** Cf. *ecological, ecology. (Examples of **œcological.**) So **œco·logist** (obs. var. *ecologist).
1893 J. S. Burdon-Sanderson in *Nature* 14 Sept. 465/1 Whether with the œcologist we regard the organism in relation to the world, or with the physiologist as a wonderful complex of vital energies, the two branches have this in common. 1899 *Natural Sci.* July 11 One of the most important oecological studies which has yet appeared in the United States. 1909 E. Warming *Oecol. Plants* p. v, I have given my views on oecological classification in a more comprehensive and detailed manner.

‖ **œcumene** (īkiu·meni). Also **ecumene, oik(o)umene.** [a. Gr. οἰκουμένη (see Œcumenic *a.*).] The inhabited world as known to the ancient Greeks; the Greeks and their neighbours regarded in the context of development in human society. Also *transf.*, the inhabited world (or a part of it) as known to or embraced by a later civilization.
1911 E. C. Semple *Influences Geogr. Environment* vi. 171 Humanity's area of distribution and historical movement we call the Oikoumene. 1926 M. T. Bingham tr. *P. Vidal de la Blache's Princ. Human Geogr.* 18 Ocean solitudes long divided inhabited countries (oikoumenes). *Ibid.* ii. 49 Recent conquests by the *oikoumene.* 1941 *Antiquity* XV. 6 By reinstating the archaeological evidence the continuum of the oikumene, made explicit in the medieval travellers' narratives, could be displayed as an enlargement of one already implied in the Bronze Age by the 24th century B.C. 1946 V. Ehrenberg *Aspects Anc. World* iii. 32 Greeks and Romans alike regarded the inhabited earth, the *oecumene*, as an area round the Mediterranean. 1953 *Ann. Assoc. Amer. Geogr.* XLIII. 92 Distribution of people in its..global scale, involves dividing the land portions of the earth into permanently inhabited as compared with uninhabited, or temporarily inhabited, parts. The terms ecumene and non-ecumene have been employed to represent these two major subdivisions. 1956 A. Toynbee *Historian's Approach to Relig.* 214 The political partition of the *Oikoumenê* into sixty or seventy self-governing parochial states. 1962 B. Lewis in Lewis & Holt *Historians of Middle East* xvi. 182 It [*sc.* the common stock of Muslim universal history] appears chiefly in the general introductory matter, leading up to the establishment of the Islamic oecumene. 1967 *Economist* 18 Nov. 742/2 Should the political scientist make change itself his theme? And if so, in an interdependent world, can the boundaries of his study be less than the *oikumene* itself? 1967 H. R. Friis *Pacific Basin* ii. 20 Eratosthenes..was like other philosophers before him concerned with the extent of the inhabited world—the *oikoumene.* 1972 D. M. Nicol *Last Centuries of Byzantium* v. 78 The paterfamilias, the head of the Christian family or *oikoumene*, was the emperor in Constantinople.

œcumenic, -ical, -icalism, -icality, -ically, -icity, see *ecumenic, -ical, etc.

Œdipal (ī·dipăl), *a.* *Psychol.* [f. Œdip(us + -al.] Characterized by an Oedipus complex; of or pertaining to the desire felt for a parent of the opposite sex.
1939 H. V. Dicks *Clin. Stud. Psychopathol.* v. 79 The analysis was devoted to working out the Oedipal, strongly sexualized, love for him. 1950 E. Frenkel-Brunswik in T. Wiesengrund-Adorno et al. *Authoritarian Personality* ix. 316 The problem of homosexuality relates to the different ways of failure in resolving the Oedipal conflict. 1960 *New Left Rev.* Nov.–Dec. 69/1 They interpreted the assassination in Oedipal terms, Trotsky being ..the hated father-substitute. 1967 T. Stoppard *Rosencrantz & Guildenstern* ii. 59 He at last confronts his mother and in a scene of provocative ambiguity (*a somewhat oedipal embrace*) begs her to repent. 1972 *Jrnl. Social Psychol.* LXXXVI. 157 The less dramatic resolution of the Oedipal complex experienced by females causes them to have a weaker superego than males. 1975 C. Dennis *Somebody just grabbed Annie!* 132 There was nothing Oedipal in his sleeping habits. He was just able to relax more easily in his old room.

Œdipean, *a.* (Later examples.)
1876 *Spectator* 9 Sept. 1131/2 The Œdipean trilogy of Sophocles. 1972 P. H. Kocher *Master of Middle-Earth* (1973) iii. 43 In striving to avert a danger he thinks he sees lying ahead he may take the very measures which are necessary to bring it about. All finite knowledge about the future is cursed by this Oedipean paradox.

Œdipode·an, *a.* [f. Gr. Οἰδιπόδειος of Oedipus + -an.] **a.** Of or pertaining to Oedipus or his family. **b.** = *Œdipal *a.*

1947 E. F. Watling *Sophocles' Theban Plays* 13 Each of the three plays deals with a situation in the Oedipodean family history. 1959 S. Neill *Genuinely Human Existence* x. 239 The Laian situation precedes the Oedipodean situation. 1968 H. J. Schonfield *Those Incredible Christians* xiii. 202 Athenagoras similarly speaks of the heathen charges of 'atheism, Thyestean feasts and Oedipodean intercourse'.

Œdipus. Add: **2.** *Psychol.* Used *attrib.* in *Œdipus legend, love object, phantasy, phase, situation, theory*, with reference to the psychoanalytic interpretation given to Sophocles' play *Œdipus Tyrannus* (in which Oedipus unknowingly kills his father and marries his mother) by Freud, and seen by him to exemplify the desires felt for the parent of the opposite sex by a child at an early stage of sexual development.
1910 *Amer. Jrnl. Psychol.* XXI. 97 The illustration of the attitude of son to parent is..transpicuous in the Œdipus legend. 1913 E. Jones *Papers on Psycho-Anal.* xviii. 381 (*heading*) Note on the Œdipus Saving Phantasy. 1926 J. I. Suttie tr. *Ferenczi's Further Contrib. Theory & Technique Psycho-Anal.* ix. 112 The patient..was here realizing sexual intercourse displaced 'from below upwards' (from the 'Oedipus phantasy'). 1943 H. Read *Educ. through Art* vi. 179 It is the Oedipus situation which gives rise to those idealistic tendencies in humanity which we know as religion, morality, custom, etc. 1957 N. Frye *Anat. Crit.* iii. 181 A kind of comic Oedipus situation in which the hero replaces his father as a lover. 1963 A. Heron *Towards Quaker View of Sex* 59 That acceptance which would permit the Oedipus phase to be 'worked through'. 1964 M. Argyle *Psychol. & Social Probl.* i. 19 The Oedipus theory has been found not to be universal. 1972 H. L. Muslin in S. C. Post *Moral Values & Superego Concept* v. 109 Guilt as a result of the introjection of the Oedipus love objects.
b. Œdipus complex, the name given by Freud to the complex of emotions which he found were aroused in a child by its subconscious sexual desire for the parent of the opposite sex, which, if not resolved naturally, may lead to repression, guilt feelings, and an inability to form normal emotional or sexual relationships. Cf. *Electra.
1910 E. Jones in *Amer. Jrnl. Psychol.* XXI. 72 (*heading*) The Œdipus-complex as an explanation of Hamlet's mystery. 1920 S. Freud in *Internat. Jrnl. Psycho-Anal.* I. 133 The girl had quietly passed through the normal stage of the feminine Oedipus complex, and had..begun to replace her father by a brother slightly older than herself. 1925 G. B. Shaw *Let.* 22 Feb. in J. Barrymore *Confessions of Actor* (1926) 122 [In 'Hamlet'] You..offer ..a demonstration of that very modern discovery called the Œdipus complex, thereby adding a really incestuous motive on Hamlet's part. 1927 [see *Electra]. 1942 Koestler *Yogi & Commissar* (1945) i. i. 9 He believes that all the pests of humanity, including constipation and the Oedipus complex, can and will be cured by Revolution. 1951 M. McLuhan *Mech. Bride* (1967) 76/2 Did I give my baby an Oedipus complex? 1965 T. S. Szasz *Ethics of Psychoanal.* i. iii. 50 According to classical analysis, he [*sc.* the psychoanalyst] teaches the patient about his early family situation, the Oedipus complex..transference, and resistance. 1969 *Listener* 17 Apr. 543/3 'I have to tell you, madam,' the psychiatrist said, 'that your son is suffering from an Oedipus complex.' 'What does it matter,' she replied, 'so long as he loves his mum?' 1971 J. Z. Young *Introd. Study Man* i. 4 Psychoanalysis..has revealed some startling features of our lives, but 'man is more than his Oedipus complex'. 1973 *Jrnl. Genetic Psychol.* CXXII. 156 The basic dynamics of the feminine 'Oedipus' complex are already fairly well known.
c. Œdipus effect, a term derived from that part of the legend of Oedipus which evidences the self-fulfilling nature of a prophecy or prediction (see quot. 1957).
1957 K. R. Popper *Poverty of Historicism* i. 13, I suggest the name '*Oedipus effect*' for the influence of the prediction upon the predicted event..whether this influence tends to bring about the predicted event, or whether it tends to prevent it. 1961 A. Flew *Hume's Philos. of Belief* vii. 152 Hume is here in his own way taking note of the particular sort of feedback which has recently been given the mnemonically useful label *Oedipus effect.* 1972 I. C. Jarvie *Concepts & Society* iv. 124 By our belief in them [*sc.* class beliefs] and an acute Oedipus effect, they verify themselves. 1974 *Brit. Jrnl. Pol. Sci.* IV. 255 It is precisely this research perspective which in a dialectical fashion lends strength to an objection based on the Oedipus effect.

œdometer (īdǫ·mĭtəɹ). Also (*rare*) **ödometer.** [f. Gr. οἴδημα swelling, swollen condition + -ometer.] A device for measuring the swelling of a gel when water is absorbed or the compressibility of soil.
1915 W. W. Taylor *Chem. Colloids* xi. 158 Reinke used the oedometer, by means of which he measured the pressure of imbibition, to determine the velocity of the process. 1926 C. Terzaghi in *Colloid Symposium Monograph* IV. 73 My own tests performed with coarse grained sands, by using a method which in every respect corresponds to the Ödometer method for measuring the swelling pressure of gels. 1938 *Nature* 24 Sept. 584/2 From the results of tests made with the Terzaghi oedometer together with calculation of the distribution of pressure from the building, it is possible to estimate the amount of settlement to be expected. 1969 Lambe & Whitman *Soil Mech.* ix. 116 In the oedometer test, stress is applied to the soil specimen along the vertical axis, while strain in the horizontal directions is prevented.

‖**œil-de-perdrix** (öidəpḗrdrī). [Fr., lit. 'partridge-eye'.] **1.** In French pottery and porcelain, a design of dotted circles, usu. on a coloured background, first used *circa* 1760 on Sèvres porcelain.
1865 F. B. Palliser *Hist. Lace* iv. 70 The lovely diapered ground recalls the may-flower of the Dresden and the oeil-de-perdrix of the Sèvres china of that time. 1925 W. W. Worster tr. *E. Hannover's Pott. & Porc.* III. 276 Next to the monochrome grounds, a favourite method was ..that of dotting the porcelain over with some little simple design repeated without a break in the manner of textiles. One of these was the famous *œil-de-perdrix* pattern, consisting of tiny circles or double circles dotted in green or blue and reserved in white on a monochrome or dotted ground. 1953 F. Tilley *Litchfield's Pott. & Porc.* (ed. 6) vi. 234 The names of some other decorations [on Sèvres work] occur in various catalogues and inventories: thus the *œil de perdrix*, the well-known partridge eye-pattern. 1971 *Canad. Antiques Collector* Feb. 16/2 The rim has a light blue ground upon which is the oeil-de-perdrix pattern leaving three oval reserves.
b. A similar design used as a ground in lace-making (see quots.).
1891 A. S. Cole *Suppl. Descr. Catal. Specimens Lace S. Kensington Mus.* 23 The pattern consists of tape-like stems..with a coarse ground of 'oeil de perdrix' devices or irregular little hexagons. *Ibid.* 31 The scheme of pattern consists of a series of oval compartments.., on an oeil-de-perdrix ground. 1899 A. M. Sharp *Point & Pillow Lace* vi. 148 This Lace [*sc.* Flemish Mechlin] is sometimes grounded with an ornamental 'réseau', instead of one in the usual hexagonal shape, called 'Fond de neige' or 'Œil de perdrix'. 1953 M. Powys *Lace & Lace-Making* xi. 207 (*heading*) Fond de Neige, so-called Oeil de Perdrix. An elaborate ground, often used as a filling. 1960 H. Hayward *Antique Coll.* 178/2 Other grounds were sometimes used such as the *œil de perdrix.*
2. Applied to wines which are pink or pale red in colour. Hence *ellipt.* as *sb.*
1872 E. Braddon *Life in India* viii. 305 If he required ..anything from Holloway's pills to œil de perdrix champagne..he was compelled to lay in such a stock as would carry him through. 1920 G. Saintsbury *Notes on Cellar-Bk.* v. 80 The true *œil-de-perdrix* tint is not..'synthetically' attainable. *Ibid.* xii. 188 A glass, filled with *œil-de-perdrix* wine. 1959 W. James *Word-Bk. Wine* 135 *Œil de perdrix.* A partridge-eye wine is a pink wine; the term is applied to pink champagnes, light-coloured sparkling burgundies, and to some wines made from a mixture of red and white grapes. 1966 *Country Life* 6 Oct. 820/1 In 1784 'L' Oeil de Perdrix Celleroy', presumably what we would now call pink champagne, fetched 48s. as against 38s. for ordinary Celleroy. 1971 *Times* 29 Mar. (Switzerland Suppl.) p. iii/6 From Neuchâtel..comes the best-known rosé, Oeil de Perdrix.

œllacherite (öla-kərəit). *Min.* [f. the name of J. *Œllacher* (b. 1804), Hungarian chemist + -ite¹.] A white or pink variety of muscovite containing a significant proportion of barium.
1867 J. D. Dana in *Amer. Jrnl. Sci.* XCIV. 256 (*table*) Œllacherite. [*Note*] Œllacher's margarite from Pfitschthal [Tyrol]. 1933 *Amer. Mineralogist* XVIII. 30 A massive pink mineral with obscure cleavage..appeared under the microscope to have a micaceous texture and had about the optical characters of muscovite... The varietal name oellacherite is given by Dana to this barium-muscovite. 1949 *Chem. Abstr.* XLIII. 6120 The discovery of a previously unknown mineral in the mountains of Kara-Tau in Kazakhstan is reported... The chem. compn. corresponds to œllacherite in which a part of the Al_2O_3 is isomorphically displaced by V_2O_3. 1962 W. A. Deer et al. *Rock-Forming Min.* III. 14 It is probable that roscoelite and oellacherite are distinct chemical and structural species and that there is not a complete series between each of them and muscovite.

œno-. Add: **œ·nocyte** (also † **-cyth**) *Ent.* [ad. G. *oenocyth* (H. R. von Wielowiejski 1886, in *Zeitschr. f. wissensch. Zool.* XLIII. 515): see -cyte], a large, probably secretory, wine-coloured cell of insects that is often grouped into glands and is produced in the epidermis, either remaining there or migrating elsewhere (esp. to the fat body); **œnocy·toid** *Ent.* [ad. F. *œnocytoïde* (E. Poyarkoff 1910, in *Arch. d'Anat. microsc.* XII. 344)], a large, round, non-phagocytic cell, similar in appearance to an œnocyte, that occurs in the hæmolymph of some insects; **œnophil** (ī·nofil), **œnophile** (ī·nofəil) [Gr. -φιλος loving] = *œnophilist*; so **œnophi·lic** *a.*, wine-loving.
1886 *Jrnl. R. Microsc. Soc.* 964 The second kind of cell is, in consequence of its colour, called the 'œnocyth'; these were found arranged in groups, or were very small, or formed rows, or plexiform plates or larger complexes or plates. 1891 *Ibid.* 587 Prof. V. Graber discusses the complex tissue found in the body-cavity of most insects. It includes..the yellow 'œnocytes', which Wielowiejski finds to be usually arranged in segmental groups. 1970 *Nature* 7 Nov. 581/1 Oenocytes, sometimes called abdominal endocrine glands, are the cells most probably interacting with the prothoracic glands in the control of moulting. 1925 A. D. Imms *Gen. Textbk. Entomol.* i. 125 The leucocytes exist in several forms and four types..are recognized by Hollande..as being present in most insects... These are—(1) Proleucocytes... (2) Phagocytes. ..(3) Granular leucocytes... (4) Œnocytoids. 1969 R. F. Chapman *Insects* xxxiii. 676 Oenocytoids are found in Coleoptera, Lepidoptera and some Diptera and Heteroptera. 1974 *Nature* 29 Nov. 391/2 Five cell types were identified in the haemolymph of *G[alleria] mellonella*; prohaemocytes, plasmatocytes, granular cells, spherule cells and oenocytoids. 1930 *New Statesman* 28 June 366/1 Professor Saintsbury, an oenophile who is free from the snobbish contempt his kind affect for whisky. 1961 C. Willock *Death in Covert* iii. 68 This man, who described himself as an oenophil, believed that wine was the only fit alcoholic drink. 1962 *Punch* 6 June 869/2 For not-too-adventurous oenophiles I commend Ice Peach Wine. 1969 *New Scientist* 13 Feb. 357/2 Scientific oenophiles would not need the sometimes drastic oversimplification of scientific matters. 1976 *Times* 5 May 16/8 British Transport Hotels invited distinguished oenophiles to sip and sniff their way through a representative selection of English wines. 1957 *Times Lit. Suppl.* 20 Dec. 778/2 Merely a salon volume designed for presentation to oenophilic friends.

Oerlikon (ō·ɹlikǫn). Also **oerlikon.** [Name of the suburb of Zurich where the guns are manufactured.] The proprietary name used by Werkzeugmaschinenfabrik Oerlikon Buhrle & Co. for guns and fittings manufactured by them; usu. *ellipt.* for an Oerlikon light anti-aircraft cannon.
1944 *R.A.F. Jrnl.* Aug. 272 (*caption*) Battleship..; carries..a varying number of Oerlikons.., sometimes sixty. 1949 S. P. Llewellyn *Troopships* 31 Their forward and after decks cluttered with..pom-poms, oerlikons, and improvised wash-houses, the troopships cleared Wellington heads. 1970 *Trade Marks Jrnl.* 7 Oct. 1667/2 Oerlikon... Guns and parts and fittings therefor.. ammunition for cannon; apparatus..for launching rockets; and rockets (missiles). Dieter Buhrle and Charlotte Buhrle-Schalk, trading as Werkzeugmaschinenfabrik Oerlikon Buhrle & Co.,..Zurich,..Switzerland; manufacturers and merchants. 1973 D. Lees *Rape of Quiet Town* x. 165 The frigate..sprayed the sub with Oerlikon fire.

oersted (ǭ·ɹsted). *Physics.* [f. the name of H. C. *Oersted* (1777–1851), Danish physicist.] **† a.** (Written **oerstedt.**) (See quots.) *Obs.*
1879 J. D. Everett *Units & Physical Constants* xi. 139 The practical unit of current is the current due to an electromotive force of 1 volt working through a resistance of 1 ohm. It is called a current of 1 weber per second. Its theoretical value is 1 weber per second = 1/10 of the C.G.S. unit of current. In the 'Testing Instructions of the Indian Telegraph Department' it is called the Oerstedt. 1885 W. Williams *Man. Telegr.* 206 The unit current..has already been described as the Ampère or Oerstedt.
† b. The electromagnetic unit of reluctance in the C.G.S. system, defined as one gilbert per maxwell. (In the International System of Units replaced by the reciprocal henry, equal to approximately $1{\cdot}26 \times 10^{-8}$ gilberts per maxwell.) *Obs.*
1893 [see *gilbert²]. 1896 M. Maclean *Physical Units* 92 The committee on Notation of the Chamber of Delegates of the International Electrical Congress of 1893 in Chicago recommended that—unit field intensity, unit magnetic induction, and unit magnetising force be each called a Gauss; unit flux of magnetic force a Weber; unit magneto-motive force a Gilbert; and unit reluctance or magnetic resistance an Oersted. 1903 *Electr. World & Engin.* XLI. 1010/2 For practical work..the magnetic reluctance of a cubic centimeter of all non-magnetic materials..is the same as that of an air-pump vacuum. This unit of reluctance is called the 'oersted'. 1931 S. R. Williams *Magn. Phenomena* i. 46 The oersted of reluctance corresponds to the ohm of resistance.
c. The electromagnetic unit of magnetic field strength in the C.G.S. system, equal to the field strength produced at a distance of one centimetre by a unit magnetic pole or a thin straight wire carrying half an e.m.u. (i.e. five amp.) of current; one maxwell per square centimetre. Cf. Gauss in Dict. and Suppl. (In the International System of Units the oersted is replaced by the ampere per metre, equal to $4\pi \times 10^{-3}$ (approximately 0·0126) oersted.)
1930 *Engineering* 25 July 113/2 At the meeting of the International Electrotechnical Commission..it was.. recommended that the names of maxwell, gauss, oersted and gilbert should be applied to the C.G.S. units for magnetic flux, flux density, magnetic field intensity and magneto-motive force respectively. 1938 R. W. Lawson tr. *Hevesy & Paneth's Man. Radioactivity* (ed. 2) ii. 26 By means of deflexion experiments in very strong magnetic fields (23,000 oersteds) a fine structure of the α-particles from thorium C..has been detected. 1959 *Sci. News* LI. 17 For instance, at a temperature of 0·01° K a magnetic field of the order of 100,000 oersted is needed in order to achieve the necessary nuclear alignment for the subsequent cooling by demagnetization. 1962 [see *gauss]. 1963 Jerrard & McNeill *Dict. Sci. Units* 97 The earth's magnetic field has a magnetizing force of the order 0·2 to 0·3 oersted. 1974 *Nature* 16 Aug. 535/2 Few, if any [geomagneticians] have really cared whether their magnetometers have measured *B* or *H*; the gauss, the c.g.s. unit of *B*, has by usage become totally interchangeable with the oersted, the c.g.s. unit of *H*. 1976 *Ibid.* 26 Feb. 652/1 All 36 specimens were submitted to magnetic study, including alternating field treatment up to 850 oersted.

œsophago-. Add: Also **esophago-, œso:-phagoga·stric** *a.*, pertaining to both the œsophagus and the stomach.
1954 *Amer. Jrnl. Surg.* LXXXVIII. 330/2 The procedure could be considered for esophagogastric resection

for gastric and esophageal varices secondary to portal hypertension and cirrhosis. 1971 *Nature* 9 July 117/1 The stomach was isolated between ligatures at the oesophago-gastric and gastro-duodenal junctions.

œstradiol (īs-, estrădəi·ǫl). *Biochem.* Also (*U.S.*) **estr-.** [f. *œstra(ne + di-² + -ol.] The most active known naturally occurring œstrogenic hormone in mammals, $C_{18}H_{22}(OH)_2$, which is formed in the ovarian follicles, controls the growth of the female sexual organs and some functions of the uterus, and is used, e.g., in treating the menopausal syndrome and conditions associated with hypoplasia of the genital tract.
1934 K. David et al. in *Biochem. Jrnl.* XXVIII. 1366 The increased oestrogenic potency of these substances is associated with an increased effect on the seminal vesicles. Oestradiol is about twice as oestrogenic as oestrone. 1951 A. Grollman *Pharmacol. & Therapeutics* xxvi. 579 Estradiol is also marketed in the form of its benzoic acid ester in which form its activity is greatly enhanced. 1955 *Sci. News Let.* 17 Sept. 179/3 Drs. Emil Witsch and C. Y. Chang..told..that they were able to change male Xenopus toads into egg-laying females..under the influence of a female hormone, estradiol. 1968 *Times* 4 Oct. 9/2 Injections of male or female sex hormones, testosterone or oestradiol, prevent the shedding of old antlers.

œstral (ī·s-, e·străl), *a.* Also (*U.S.*) **estral.** [f. Œstr(us + -al.] = Œstrous *a.*
1883 *Brit. Med. Jrnl.* 10 Mar. 446/2 The 'œstral' products had never found vent from the uterus. *Ibid.*, The lower animals do certainly have an 'œstral' discharge. 1972 *Biol. Abstr.* LIV. 6236/1 (*heading*) Change in the estral cycle of white rats..exposed to the..effect of gasoline..vapors.

œstrane (ī·s-, e·strēⁱn). *Biochem.* Also (*U.S.*) **estr-.** [f. *œstr(in + -ane.] (See quot. 1933.)
1933 N. K. Adam et al. in *Nature* 5 Aug. 205/2 The parent saturated hydrocarbon of the œstrin group, $C_{18}H_{30}$, containing the sterol skeleton with one methyl group but without the side chain, may be termed 'œstrane'. 1947 H. Selye *Textbk. Endocrinol.* i. 71/2 A side-chain at C_{17} does not necessarily decrease folliculoid activity in the estrane series and..it may even increase it. 1973 *Biochemistry* (Easton, Pa.) XII. 2221/1 All neutral and charged derivatives of estrane and androstane inhibit hydrolysis of *bovine* albumin.

œstrin (ī·s-, e·strin). *Biochem.* Also (*U.S.*) **estr-.** [f. Œstr(us + -in¹.] Any of various preparations of œstrogenic material; an œstrogen, *spec.* œstrone.
1926 Parkes & Bellerby in *Jrnl. Physiol.* LXI. 573 We suggest that 'oestrin' or some such term..should be applied to the oestrus-producing principle of the ovary. 1933 [see prec.]. 1935 B. Harrow in Harrow & Sherwin *Textbk. Biochem.* xxx. 756 The true theelin or estrin, as well as several related compounds, has been isolated in the chemically pure form. 1938 *Ann. Reg.* 1937 348 Oestrin, styryl blue, the acenaphthene derivatives and squalene can act as efficient evocators in embryological induction. 1970 W. B. Yapp *Introd. Animal Physiol.* (ed. 3) viii. 293 Collectively all are called oestrins or oestrogens; they are steroids.

œstriol (ī·s-, e·stri,ǫl). *Biochem.* Also (*U.S.*) **estr-.** [f. *œs(trane + tri- + -ol.] An œstrogenic hormone, $C_{18}H_{21}(OH)_3$, which is one of the metabolic products of œstradiol and is produced in the placenta during pregnancy.
1933 N. K. Adam et al. in *Nature* 5 Aug. 206/1 The full chemical names of two important derivatives..are 3,16,17-trihydroxy 1,3,5-œstratriene, and 3-hydroxy 17-keto 1,3,5-œstratriene. These names, though suitable for chemical literature, are cumbrous for general and biological use. We suggest as common names, for use where the chemical nature of the principle has been identified, 'œstriol' for the trihydroxy compound, and 'œstrone' for the keto-hydroxy compound. 1950 L. Levin in S. Soskin *Progr. Clin. Endocrinol.* xi. 410 The separation of estriol from estrone on the basis of the somewhat greater acid dissociation of estriol. 1958 *Sci. News* XLVII. 88 Oestrone has also been obtained from palm kernel extracts, and oestriol from female willow flower. 1968 *New Scientist* 22 Feb. 399/3 Measurement of the mother's oestriol excretion is a guide to the well-being of both the foetus and the placenta that nourishes it.

œstrogen (ī·s-, e·strǒdʒěn). *Med.* Also (*U.S.*) **estro-.** [f. Œstr(us + -o + -gen.] Any natural or synthetic substance which can produce or maintain the secondary female sex characteristics in a mammal and which initiates certain bodily changes associated with the menstrual or œstrous cycle.
1927 *Official Gaz.* (U.S. Patent Office) 22 Nov. 788/1 Parke, Davis & Company, Detroit, Michigan... *Estrogen.* .. Extract of gland tissue for use in the treatment of various types of ovarian disfunctions. Claims use since May 25, 1927. 1928 *Endocrinology* XII. 151 (*heading*) Estrogen, a new sex hormone. 1936 *Jrnl. Amer. Med. Assoc.* 10 Oct. 1223/2 'Estrogen' is a registered trade mark belonging to Parke, Davis and Company... This firm has commendably agreed to relinquish its proprietary rights in the name on its adoption by the Council as a generic term. 1943 *Endeavour* II. 29/2 This soon led..to the synthesis of artificial oestrogens of extraordinary potency. 1951 A. Grollman *Pharmacol. & Therapeutics* xxvi. 583 Diethylstilbestrol and other estrogens exert an ameliorative effect in carcinoma of the prostate and in certain cases of mammary carcinoma occuring in post-menopausal women. 1960 *Which?* Feb. 24/1 Oestrogens, originally, seem to have been incorporated into face creams without any direct *evidence* that they could affect the skin. 1968 Passmore & Robson *Compan. Med. Stud.* I. xxxvii. 11/1 The principal ovarian hormones..fall into three broad functional categories, oestrogens, progestagens and androgens. 1968 *Times* 28 Nov. 14/4 Most oral contraceptive pills contain both a progestogen and an oestrogen—another type of ovarian hormone which is thought to be responsible for most of the side effects, such as changes of weight. 1974 *Nature* 12 Apr. 616/2 The role of oestrogens in male reproduction remains an enigma.
b. *attrib.* and *Comb.*, as *œstrogen effect, excretion, ointment, pill, therapy; œstrogen-induced, -secreting, -treated* adjs.
1957 *Times* 2 Dec. (Agric. Suppl.) p. vi/4 A source of greater fear to those familiar with oestrogen effects is the inhalation of dust from a concentrated premix. 1972 *Endocrinology* XCI. 1273 (*heading*) Early estrogen effects on lipid metabolism in the rat uterus. 1946 *Nature* 24 Aug. 276/2 The significance of androgen and oestrogen excretion in the urine in relation to ageing. 1972 *Brit. Vet. Jrnl.* CXXVIII. 565 Urinary oestrogen excretion has been studied during the oestrus cycle in a variety of domestic animals. 1956 *Nature* 10 Mar. 478/2 Progesterone ..prevents oestrogen-induced abdominal fibroids. 1960 *Which?* Feb. 24/2 In aged women..it was claimed that treatment with oestrogen ointment had an effect. 1972 F. Warner *Maquettes* 44 Oestrogen-pills on prescription. 1948 L. Martin *Clin. Endocrinol.* ix. 191 Œstrogen-secreting ovarian tumours composed of cells resembling those of the membrane granulosa of the Graafian follicle. *Ibid.* 170 Prolonged œstrogen therapy causes hypertrophy of the uterus and endometrium and enlargement of the breasts. 1974 *Jrnl. Amer. Med. Assoc.* 4 Feb. 522/1 Patients with preexisting hyperlipemia should probably not receive estrogen therapy. 1957 *A.M.A. Arch. Path.* LXIV. 595/2 Stratified squamous epithelium may be produced in the uterus of estrogen-treated rats by two separate..methods.

œstrogenic (īs-, estrǒdʒe·nik), *a.* Also (*U.S.*) **estro-.** [f. prec. + -ic.] Of the nature of, or having the actions of, an œstrogen.
1930 *Endocrinology* XIV. 101 Estrogenic material has been obtained from follicular fluid. 1938 *Ann. Reg.* 1937 348 Knowledge [was] advanced of the relation between molecular structure and the oestrogenic activity of the sex hormones. 1951 A. Grollman *Pharmacol. & Therapeutics* xxvi. 579 The naturally occurring estrogenic substances have been displaced in therapeutics to a large extent by a number of cheaper synthetic compounds. 1957 *Times* 2 Dec. (Agric. Suppl.) p. ii/7 Recently, fresh developments have taken place in the use of oestrogenic hormones such as stilboestrol and hexoestrol for beef production both in the United States and in this country. 1965 Lee & Knowles *Animal Hormones* iii. 52 Whether oestrogenic substances are secreted by the testes [of reptiles] has yet to be clarified.
Hence **œstroge·nically,** *adv.*, as regards œstrogenic properties; **œstrogeni·city,** œstrogenic property.
1930 *Endocrinology* XIV. 389 A relationship between estrogenicity and the portion of a plant used as a source of extract. 1935 *Jrnl. Biol. Chem.* CXII. 425 In the commercial extraction of theelin and theelol from human pregnancy urine, an acidic, nitrogenous, crystalline substance may be separated from the crude extract. Preliminary Allen-Doisy tests showed this compound to be estrogenically inert. 1966 *Internat. Jrnl. Fertility* XI. 399 Both norethynodrel and norethisterone are active as a result of their estrogenicity. 1968 A. White et al. *Princ. Biochem.* (ed. 4) xliv. 948 They are estrogenically active when given orally.

œstrogenized (ī·strǒdʒěnəizd), *a.* *Med.* Also (chiefly *U.S.*) **estr-.** [f. *œstrogen + -ize + -ed¹.] Treated with œstrogen. Also **œ:strogeniza·tion,** the action or result of treating with œstrogen.
1946 *Nature* 20 July 96/1 The mobilization of lipochrome..in the serum of œstrogenized or actively laying fowl must..be familiar to all who have prepared phospholipid extracts from the sera of such birds. 1961 *Proc. 4th Nat. Cancer Conf.* (U.S.) 42/1 The lymphoid neoplasms that follow chronic estrogenization in some strains of mice. 1974 *Brain Res.* LXXX. 152 (*heading*) Inhibitory effect of neonatal estrogenization on the incorporation of [³H]lysine in the Purkinje cells of the adult male and female rat. 1975 *Nature* 5 June 498/1 The oestrogenised pituitary glands, without bromocryptine, stored less and released more prolactin.

œstrone (ī·s-, e·strǒun). *Biochem.* Also (*U.S.*) **estr-.** [f. *œstr(ane + -one.] An œstrogenic hormone, $C_{18}H_{22}O_2$, obtained from the urine of pregnant women and mares with actions and uses similar to those of œstradiol.
1933 [see *œstriol]. 1951 A. Grollman *Pharmacol. & Therapeutics* xxvi. 579 Several preparations of amorphous concentrates of pregnancy urine or placental tissue, which consist principally of estrone sulfate, are prepared commercially for oral use. 1958 *Sci. News* XLVII. 87 Oestrone was the first naturally occurring oestrogen to be isolated in pure crystalline form. 1968 R. O. C. Norman *Princ. Org. Synthesis* xix. 651 There are four asymmetric carbon atoms in oestrone, so that there are eight enantiomorphic pairs of stereoisomers, all of which have been synthesized.

œstrous, *a.* Add: Also (*U.S.*) **estrous.** Add to def.: in heat; *œstrous cycle*, the cyclic series of changes preceding, including, and following œstrus that takes place in most female mammals and involves esp. the reproductive and endocrine systems. (Earlier and later examples.)
1900 *Q. Jrnl. Microsc. Sci.* XLIV. 60 The sexual season of all mammals is evidenced by a series of phenomena which constitute, in the absence of the male, one œstrous cycle (monœstrous mammals) or a series of œstrous cycles (polyœstrous mammals). 1904 *Phil. Trans. R. Soc.* B. CXCVI. 47 (*heading*) The œstrous cycle and the formation of the corpus luteum in the sheep. *Ibid.* 73 (*heading*) Number of follicles discharging at an œstrous period. 1920 *Lancet* 2 Oct. 727/1 Under well-regulated food conditions the œstrous cycle in the guinea-pig is almost uniformly 16–17 days in duration. 1963 A. Heron *Towards Quaker View of Sex* 54 In most mammals, the oestrous cycle of the female, hormone-controlled, is an important factor. 1972 *Science* 5 May 519/1 Estrous rhythms and ovulation pattern had been reversed by substituting darkness for daylight. 1975 *Sci. Amer.* May 57/1 He [*sc.* a young male lion]..will mate if he encounters an unattended estrous female.

œstrual, *a.* Delete *rare*—⁰ and add examples. **œstruation** (examples).
1859 J. C. Dalton *Treat. Human Physiol.* ii. v. 471 The ripening and discharge of the egg are accompanied by a peculiar condition of the entire system, known as the 'rutting' condition, or 'œstruation'. 1883 *Brit. Med. Jrnl.* 3 Mar. 398/1 Observation of wild animals..shows that, after a definite period of quiescence, œstruation, or the 'rut', invariably recurs at epochs which strictly conform to some multiple of weeks. 1924 *Amer. Jrnl. Physiol.* LXVIII. 294 (*heading*) The effect of pubescence, oestruation and menopause on the voluntary activity in the albino rat. *Ibid.* 303 The oestrual cycles became more and more irregular..as the animals approached the end of their sexual life. 1936 *Ibid.* CXVI. 5 Oestrual behavior in cats represents a specific pattern of response. 1961 *Nature* 16 Dec. 1043 (*heading*) Variations in the nuclear deoxyribonucleic acid content in the adrenal cortex of the female white rat during the œstrual cycle.

œstrum. 2. b. (Later examples.)
1878 G. Fleming *Text-bk. Vet. Obstetr.* 55 The rutting, heat, œstrum, or venereal œstrum of animals is analogous to 'menstruation' in woman. 1900 [see *oestrus 2 b]. 1925 [see *acceptance 1 b]. 1939 *Brit. Birds* XXXII. 251 The 'false œstrum' which temporarily affects so many birds during this period. 1963 Jubb & Kennedy *Path. Domestic Animals* I. i. 11/1 Oestrum may not occur, or be irregular.

œstrus. Add: Also (*U.S.*) **estrus. 2. b.** In mod. use: The rut or heat (Heat *sb.* 13) of female mammals, or the period during which this lasts.
1900 *Q. Jrnl. Microsc. Sci.* XLIV. 6 Œstrus may be a brief period and exist for only a few hours. *Ibid.*, The period of œstrus is referred to by various writers as 'Brunst', 'rut', 'heat', 'season', 'brim', or 'œstrum'. *Ibid.* 17 The duration of the œstri. 1904 *Phil. Trans. R. Soc.* B. CXCVI. 67 Œstrus was noted in a Black-faced ewe. 1963 A. Heron *Towards Quaker View of Sex* 54 In the baboon.. females in oestrus will mate with other males if their overlord's attention is temporarily distracted. 1974 *Encycl. Brit. Macropædia* VIII. 1083/1 Estrogens are substances that evoke the cyclical onset of heat, or estrus, during which the animal is sexually active and receptive to the male. 1974 *Country Life* 12 Dec. 1853/1 There was mention of what is termed 'synchronised oestrus'. By the use of drugs cows are induced to come in season at a prearranged time.

‖**œuf en cocotte** (ǒf aṅ kokǫ·t). Also **œuf cocotte.** Pl. **œufs (en) cocotte(s).** [Fr., lit. 'egg in cocotte': see *cocotte 2.] An egg cooked in a cocotte, *spec.* in *pl.*, a French dish of eggs baked in butter in individual cocottes or ramekins and served thus.
1900 *French Cookery for Eng. Homes* 142 Eggs with Sweet Herbs (*Oeufs en cocottes aux fines hèrbes* [sic]). Colour in 1½ oz. of butter, some chopped onions, mushrooms, and parsley... Pour over each egg a teaspoonful of good raw cream. 1909 C. H. Senn *New Cent. Cookery Bk.* (ed. 3) xxx. 690 *Oeufs en Cocottes.* 6 to 8 eggs, a little butter, 1 tablespoonful..parsley, 6 to 8 dessertspoonfuls cream... Dish up and serve in the pipkin pans. 1960 E. David *French Provincial Cooking* 190, I always thought *œufs en cocotte* were soft. 1962 G. Z. Stone *Althea* (1964) 70 Eggs should be either very soft or thoroughly cooked,.. that was what was the matter with *oeufs en cocotte.* 1963 I. Fleming *On H.M. Secret Service* ii. 27 In England he lived on grilled soles, oeufs cocotte and cold roast beef. 1966 *Observer* (Colour Suppl.) 25 Sept. 45/3 Perhaps the perfect egg dish is *oeufs en cocotte.* 1974 *She* Jan. 83/2 Dined at home. *Oeufs en cocotte* and grouse.

‖**œuvre** (ǒvr). [Fr., = work.] A work of art, music, or literature; the corpus of work produced by an artist, composer, or writer, considered as a whole. Cf. Chef d'œuvre in Dict. and Suppl. Hence *œuvre-catalogue*, a catalogue of an artist's complete work; *œuvre de vulgarisation*, a work conveying knowledge of an academic or esoteric subject for the popular taste.
1875 Geo. Eliot *Let.* 10 Oct. in J. W. Cross *George Eliot's Life* (1885) III. xvii. 264 Our unprinted matter, our *œuvres inédites*, were safe. 1889 E. Dowson *Let.* 31 July (1967) 97 If I can only shake off Cursitor St I will go to the oeuvre like one oclock. 1890 —— *Let.* 10 or 11 June (1967) 153 The bêtise of a public which boycotts his oeuvre & buys Chomondley's pretty little ineptiae. 1898 C. Phillips *Later Work of Titian* iii. 66 The technical execu-

tion of these canvases, the treatment of landscape in the former would lead the writer to place them some years farther on still in the *œuvre* of the master. 1917 R. Fry in *Burlington Mag.* Aug. 61/2 A better opportunity for a general study of Cézanne's *oeuvre* than any other book. 1934 T. S. Eliot *Eliz. Ess.* 136 Even without an *œuvre*, some dramatists can effect a satisfying unity and significance of pattern in single plays. 1938 W. S. Maugham *Summing Up* 184 A body of work, an *œuvre*, is the result of long-continued and resolute effort. 1958 *Listener* 2 Oct. 540/1 The spreading abroad of his [*sc.* Palestrina's] bulky *oeuvre* gained considerably from the attentions of the romantics. 1959 *Ibid.* 20 Aug. 289/1 Were they in fact *oeuvres de vulgarisation* or attempts to stimulate craftsmen to new formal conquests? 1962 *Economist* 14 Apr. 141/3 An *oeuvre de vulgarisation* to bring his fifteen volume history..before a wider public. 1965 *New Statesman* 16 July 92/1 Only four oil-paintings before 1913 are listed in his oeuvre-catalogue. 1974 *Listener* 17 Jan. 66/2 A television serial which should incorporate the Palliser oeuvre entire. 1976 *Publishers Weekly* 15 Mar. 57/2 The author tells his engaging tale with disarming simplicity and the illustrator's pictures do justice to the *oeuvre*. 1976 *Times* 16 July 12/1 Raymond Lister at Cambridge..is currently working on an oeuvre catalogue for [Samuel] Palmer. 1978 *Times Lit. Suppl.* 25 Aug. 950/2 His oeuvre is the perfect place of exercise for..jocular pedantry.

of, *prep.* Add: **4. c.** *N. Amer.* and *dial.* In expressing the time: from or before (a specified hour); = To *prep.* 6 b. Also *ellipt.*

1817 T. Dean *Jrnl.* 31 May in *Indiana Hist. Soc. Publ.* (1918) VI. 278 At 15 minutes of 10 A.M. Paul Dick arrived. 1857 M. J. Holmes *Meadow Brook* v. 64 Five minutes of nine, and round the corner at the foot of the hill appeared a group of children. 1912 in *Sc. Nat. Dict.* (1965) VI. 466/3 It is a quarter of twelve. 1972 H. MacInnes *Message from Málaga* xix. 264, I have to leave by a quarter of four—at the latest. 1973 D. MacKenzie *Postscript to Dead Let.* 22 It was a quarter-of-seven. 1976 I. Levin *Boys from Brazil* iv. 135 'Can I go with you?' 'Sure... I leave at five of.'

43. b. (Further examples.) Freq. in phrases expressing surprise at something or someone unexpected.

1848 A. Jameson in G. Macpherson *Mem. Life A. Jameson* (1878) ix. 254, I ran to Ireland, of all places in the world. 1896 *Month* May 135 'I should like it of all things,' said his sister. 1906 W. Churchill *Coniston* I. xiv. 178 'Well, of all people, Cynthia Wetherell!' he cried. c 1921 D. H. Lawrence *Mr. Noon* i. in *Mod. Lover* (1934) 175 'Well, of all the idle scawd-rags!' he said as he entered with the tray. 1926 D. O. Stewart *Mr. & Mrs. Haddock in Paris* x. 266 'Maybe you would like to take a little rest—perhaps at a nice, comfortable movie.' 'Well, of all things, Will Haddock!' 1953 A. Huxley *Let.* 17 Aug. (1969) 683 It is to appear serially—of all places—in *Esquire*—which is at present engaged in serving God and Mammon. 1956 S. Beckett *Waiting for Godot* I. 40 (*He fumbles in his pockets.*)..What have I done with my spray? (*He fumbles.*) Well, of all the—— (*he looks up, consternation on his features. Faintly.*) I can't find my pulverizer! 1958 W. Plomer *At Home* ii. 38 He actually filled a glass with (of all things, on a black November afternoon) ginger beer.

d. (Later examples.) Also used (with repetition of the sb.) to denote a person who shows strong (*spec.* national) characteristics.

1895 G. A. Sala *Life* (ed. 2) I. vi. 64 Morris Barnett..was a remarkably clever man—a Hebrew of the Hebrews, with a pronounced musical faculty. 1976 *Times* 15 May 14/7 Tevic the Milkman is a Jew of the Jews in character and behaviour, yet his problems are not peculiarly Jewish. 1976 *Lancet* 13 Nov. 1094/2 A Scot of Scots, he was born in 1912 and was educated at St Andrews and Edinburgh Universities.

47. d. Typical or characteristic of (a particular period).

1923 H. Simpson (*title of play*) A man of his time. 1940 M. K. Tippett (*title of oratorio*) A child of our time. 1965 *Listener* 11 Nov. 760/1 It is a highly original picture, but like any work of art it has roots. It is absolutely of its time: all the enthusiasms of those last pre-war years are contained in it at boiling point. 1968 *Ibid.* 4 July 9/2 The 'Four' were very much of the 1890s.

52. b. (Later examples.)

1867 'E. Kirke' *On Border* iii. 67, I don't forget..how you worked of nights. 1891 Barrie *Little Minister* III. xl. 156 So long as women sit up of nights listening for a footstep. 1897 J. L. Allen *Choir Invisible* xvii. 252 You have holiday of Saturdays. I have not, you see.

of (ǫv, əv), joc. (being erroneous in Received Standard) or dial. var. Have *v.* (representing the unstressed pronunc. of *have*, esp. in such phrases as *could have, might have, must have,* and *would have*).

1837 W. Tayler *Diary* 10 May in J. Burnett *Useful Toil* (1974) II. 181 Soposing seven hundred and sixty [servants] to of advertised and the same number not to of advertised. 1844 *Southern Lit. Messenger* X. 486/2, I never would of married in the world, ef I couldn't of got jist exactly suited. 1854 M. J. Holmes *Tempest & Sunshine* viii. 115, I don't see why in the old Harry he couldn't of lived. 1913 C. Mackenzie *Sinister St.* I. I. iv. 60 Mrs. Frith used to talk about 'people as gave theirselves airs which they had no business to of done.' 1916 [see *fair *sb.*[2] 1 c]. 1924 A. S. Neill *Dominie's Five* ii. 31 If Neill had been 'ere, 'e could of told us. 1931 E. Linklater *Juan in Amer.* II. iv. 79 There's no beer-racket in Paris or Rome, and even if there had been I wouldn't of tried to muscle in on it. *Ibid.* III. iii. 226 If I hadn't of worked nights on a correspondence course I'd of been firing a furnace still. 1945 A. Kober *Parm Me* 26 I'm certainly glad to of made your acquaintance. 1946 K. Tennant *Lost Haven* (1947) i. 19 There might have been a time when I thought Kelly was too hard on the kids, and ..I might of been glad when he went off with that bloody moll. *Ibid.* xiii. 202 She must of forgot. 1957 [see *blower[1] 3 e]. 1959 [see *dozer[2]]. 1977 *New Yorker* 6 June 56/3 Sometimes I get to thinking that I could of raised up four girls and worn out a couple of saddles.

ofay (ǭu·fēi). *U.S. slang* (chiefly used by Blacks). [Orig. unknown: the balance of probability is that it is a word of African origin (but the precise attribution in quot. 1932 lacks foundation). The suggestion that it is Pig Latin for *foe* seems no more than an implausible guess.] In the use of American Blacks: an offensive term for 'a white person'. Also *attrib.* or as *adj.* Cf. *fay *sb.*[4]

1925 *Inter-State Tattler* 6 Mar. 8 We hear that 'Booker Red' has three ofays on his staff. 1926 [see *dicty *a.*]. 1927 *Amer. Mercury* Dec. 392 Ugly people there are, certainly, but the percentage of beautiful folk is unquestionably larger than among the ofay brethren. 1932 *Africa* V. 506 The last of the five words contributed indirectly by Ibibio to the English language is *Offay*... The root of the word appears to come from the Ibibio *Afia*, white or light-coloured. Hence in Harlem *Offay* means any light-coloured person and therefore a European. 1936 Mencken *Amer. Lang.* (ed. 4) v. 214 The word *ofay*, which may have come from the French *au fait* (signifying mastery), is in general use in the Negro press of the United States to designate a white person. 1940 *New Republic* 7 Oct. 472 When he goes downtown to this civil-service office, they take him for ofay... One of them young ofay chicks what clerk down there, she even tries to flirt with him. 1946 *Variety* 23 Oct. 114 Boys start out with a Negro spiritual, with their inflection and harmony making it racially authentic, although they're ofay. 1952 M. Steen *Phoenix Rising* iii. 70 'Dey run him out.' 'What for?' Karl sounded bored. 'Got after Harry's ofay chick—' 1956 B. Holiday *Lady sings Blues* (1973) v. 52 Most of the ofays, the white people, who came to Harlem those nights were looking for atmosphere. *Ibid.* ix. 89 'What will people think?' is a big deal in ofay circles. 1968 M. Richler *Cocksure* iv. 25 Was she amused by his dilemma? His ofay dilemma. 1971 *Black World* Apr. 62, I was attendant in the ofay ladies lounge. 1971 B. Malamud *Tenants* 48 Who are those cats, brothers or ofays? *Ibid.* 74 The black replies. 'No ofay motherfucker can put himself in *my* place.' 1977 *Amer. Speech* 1975 L. 89 That this word [*sc.* Yoruba *oje*] could have been brought to the United States by slaves is altogether possible... Thus *ofay* may be taken as a word said for self-protection in times of threat, which was then transferred to the source of threat, and so came to mean 'whiteman'.

off, *adv., prep., adj.,* and *sb.*[1] Add: **A.** *adv.* **1. b.** (Further examples.)

1629, etc. [see Buy *v.* 7 a]. 1902 G. H. Lorimer *Lett. Merchant* xiv. 203 By the time the real weather comes along everybody has guessed wrong and knocked the market off a cent or two. 1934 G. B. Shaw *On Rocks* I. 25 You have to buy him off with a scrap of dole. 1971 *Nature* 30 Apr. 604/1 Halfway through the book, the reasons why the rapid development of fluidics in the early 1960s has tended to fall off recently become clear.

e. *they're off* (occas. *they are off*): a colloq. phrase indicating that a race has started; *to be off and running*: to be making good progress.

1833 *Mirror of Lit.* 27 July 59/2 *They are off!* 'No, no'—cries one jockey whose horse turned his tail to the others. 1846 'Sylvanus' *Pedestrian & Other Reminisc.* xxiv. 227 The horses are paraded, the flag is dropped—'they're off!' is repeated by twenty thousand tongues. 1872 B. Jerrold *London* viii. 74 *Clear the course!*... A flutter goes through the sea of heads on the Grand Stand... *They're off!* 1928 E. O'Neill *Strange Interlude* VIII. 288 They're off!.. Navy and Washington are leading—Gordon's third. 1967 *Boston Herald* 1 Apr. 1/1 Although he has not announced it officially, Wallace appears to be off and running for the presidency. 1970 Wilson & Michaels tr. *Charrière's Papillon* vi. 211 We were off and running.

f. *ellipt.*, = off one's head (Head *sb.* 34): (somewhat) crazy. *colloq.* or *dial.*

1866 W. Gregor *Dial. Banffshire* in *Trans. Philol. Soc.* 215 *Aff, to be*, to be deranged; as, 'He wiz lang jummlet; bit he's *aff* athegeethir noo.' 1887 *Lantern* (New Orleans) 9 Apr. 3/2 Humor him as he was a little off. 1904 W. H. Smith *Promoters* i. 8 I've sometimes thought you were a trifle visionary, but I never considered you seriously off. 1927 W. E. Collinson *Contemp. Eng.* 116 Mental debility finds adequate expression in a whole series: he's not all there, a bit off (the top), he's off his chump. 1975 B. Wood *Killing Gift* II. i. 47 He was an old man, after all; perhaps he was just a little off... 'No, gentlemen, I'm not senile.'

g. In bad condition; wrong, abnormal, odd; *spec.* (*a*) of a horse or athlete: not in good condition or form; off-form; (*b*) of food: stale, sour, contaminated; (*c*) of social behaviour: unacceptable; ill-mannered; esp. in phr. (*it's*) *a bit off*.

Not always clearly distinguishable from sense 2 c.

1846 *Spirit of Times* 18 Apr. 91/1 He had endurance and speed enough to make a good race in any crowd, when 'all right', but then, he [*sc.* a horse] was liable to be oftener 'off' than otherwise. 1868 H. Woodruff *Trotting Horse* xxxvi. 300 When a trotter wins with great ease,..it is assumed, not that the loser was 'off', but that the winner is greatly superior. 1902 G. H. Lorimer *Lett. Merchant* xvi. 231, I may be off in sizing this thing up, because it's a little out of my line. 1916 'Taffrail' *Pincher Martin* ix. 153 'S'pose I'd best be makin' a move, though,' he added ruefully. 'Bit orf, I calls it!' 1922 J. Cannan *Misty Valley* 58 Audrey had taken them sailing, but she had upset the boat, and they had drifted down stream, clinging to the sail, saying that it was a bit off. Whereafter they would go down to the pier and knit jumpers. 1941 E. Bowen *Look at Roses* 68 'Your caller sounded to me a bit off.' 'Oh, Mrs. Massey's had bad news. She..didn't feel well.' 1951 C. Armstrong *Black-Eyed Stranger* (1952) xiv. 117 Ambielli's got principles. They are a little off, slightly out of whack, you know. 1952 R. Finlayson *Schooner came to Atia* 140 Meat that tasted 'off'. 1953 K. Amis *Lucky Jim* v. 56 It was rather rude, all the same. I could see Mrs. Neddy thought it was a bit off. 1960 'A. Burgess' *Right to Answer* ii. 23 That tomato juice is a bit off. Been in the tin too long. 1966 'W. Cooper' *Mem. New Man* II. v. 155 It *is* just a teentsy-weentsy bit off, isn't it, darling, not to let you know he was coming. 1974 *Listener* 24 Jan. 102/1 Something was a little off with the mechanism, so the feeder-belt chewed up the baggage in transit.

3. d. *Theatr.* = *off stage *adv.* Also *transf.* and *fig.*

1774 [see Stage *sb.* 5]. 1775 T. Campbell *Diary* 4 Mar. (1947) 44 The players..stood their ground for a long time—but were at length hissed off. *Ibid.*, Mr. Vernon attempted to speak, but he wd not be heared—still the cry was off, off. 1805 T. Dibdin in G. Colman *John Bull* p. i, To whom, thus midway placed, I say, be kind, John Bull before, Oh, spare John Bull behind (*pointing off.*). 1836 Dickens *Sk. Boz* 1st Ser. II. 255 'But you must take care you don't knock a wing down...' 'I shall fall with my head "off", and then I can't do any harm.' c 1863 T. Taylor *Ticket-of-Leave Man* I. 7 The Bellevue Tea Gardens,..ornamental orchestra and concert room... Music heard off. 1909 'I. Hay' *Man's Man* vi. 83 Portentous trampings 'off' announced the return of the glee-party. 1923 *Referee* 12 Aug. 3/3 The leopardess..was heard to roar a good deal 'off',..making her only appearance [on the stage]—in a cage. 1924, etc. [see *noise *sb.* 3 c]. 1954 T. S. Eliot *Confid. Clerk* I. 27 *Lady Elizabeth Mulhammer's voice off*: Just open that case, I want something out of it. 1965 *New Statesman* 7 May 739/1 The elaborate framework of posh voices off intoning fragments of letters.

e. Of an item of food: deleted from the menu; not available. *colloq.*

[1870 D. J. Kirwan *Palace & Hovel* x. 154 On 'off' days they have soup and thick gruel for breakfast.] 1902 Farmer & Henley *Slang* V. 89/2 'Chops is hoff' = 'there are no more chops to-day'. 1933 [see *bawl *v.* 3 c]. 1953 'M. Innes' *Christmas at Candleshoe* ii. 24 The celerity with which the less unpalatable dishes are prone to be 'off' in English hotels. 1966 N. Freeling *Dresden Green* II. 123 Sorry sir, said the waitress in the teashop, the pudding's *off*. We can do you a nice ice-cream though. 1974 D. Chantler *Man who Followed* II. 60 Tell the waiters..the Tutti-Frutti is off.

4. d. Away or free from one's work, school, service, etc.

1861 J. O'Neil *Diary* 1 Apr. in J. Burnett *Useful Toil* (1974) I. 78 At Low Moor there is a great many off. There is above a hundred looms standing. 1882 J. D. McCabe *New York* 384 Then begins five hours' patrol..after which he is 'off'. 1883, etc. [see *day off* s.v. *day *sb.* 19]. 1885, etc. [see *night *sb.* 5 a]. 1916 B. Ruck *Girls at his Billet* xviii. 242, I am sure your auntie..would be quite agreeable to letting us have the afternoon off for the ceremony. 1940 F. Sargeson *Man & Wife* (1944) 21 Of course Sally wasn't off for long. And they gave her a rise. 1970 'D. Halliday' *Dolly & Cookie Bird* v. 74, I was..bedding the avocados in lettuce. Anne-Marie was off but Helmuth did the last stages. *Ibid.* viii. 128 It was Anne-Marie's afternoon off.

6. b. Of stocks, shares, etc.: lower in value or price (by a specified amount or numbered points).

1929 *Times* 30 Oct. 14/1 Duke Power 'opened' at 130 off 39½ points, Newhaven Railway at 90 off 18. 1931 *Daily Express* 21 Sept. 14/4 Japanese bonds were only slightly off. 1964 *Financial Times* 12 Mar. 21/5 Belfast Ropework were 6d off to 35s 6d. 1977 *Times* 19 Nov. 17/5 By the close the *FT* Index was just 0·5 off at 480·5.

c. with ellipsis of *taken*. Of retail commodities: reduced in price by a specified amount.

1965 *New Statesman* 9 Apr. 562/1 Don't be put off by the fact that both packets also say '3d off'. 1966 'J. Ashford' *Consider Evidence* i. 5 Tomato soup was being sold at threepence a tin off. 1968 'J. Le Carré' *Small Town in Germany* v. 71 Don't matter what you fancy: radios, dish-washers, cars; he'll get you a bit off, like.

7. b. (Further examples.) Also *colloq.*

1824 J. Wight *Mornings at Bow St.* 21 Two young men ..were charged by a watchman with having 'bother'd him on his *bate*,' and refused to 'go along off of it when he *tould* 'em.' 1843 T. C. Haliburton *Attaché* 1st Ser. II. xii. 210 The groom has stole her oats, forgot to give her water, and let her make a supper sometimes off of her nasty, mouldy, filthy beddin'. 1884 'Mark Twain' *Huck. Finn* vi. 32 I'd borrow two or three dollars off of the judge for him. a 1922 T. S. Eliot *Waste Land Drafts* (1971) 5 The reputation the place gets, off of a few barflies. 1962 F. Norman *Guntz* i. 15, I got hold of this very very old typewriter off of a friend of mine. *Ibid.* iii. 24 After his secretary had picked him off of the floor he got on the blower to his accounts department. 1965 T. Parker *Five Women* i. 45 They'll thieve off-of anyone and jump in bed with anyone. 1974 J. Stubbs *Painted Face* xxiii. 284 Get off of me, will you, sir?

13. Used with a preceding numeral to represent a quantity in production or manufacture, esp. *one off* (see *one 29* b).

1934, etc. [see *one 29 b]. 1947 Crowther & Whiddington *Science at War* 49 Manufacturers found it very difficult to give up mass production, in order to make the 200 or so sets 'off'. 1970 *Cabinet Maker & Retail Furnisher* 30 Oct. 205/2 Without barrier coats mould breakdown can start after 60 units off. 1973 *Physics Bull.* Apr. 238/2 (Advt.), Kienzle printers. 6 off, surplus to manufacturing requirements.

B. *prep.* **5. b.** (Further examples.) Also, having lost interest in; averse to; *off form*: in bad form; *off one's game*: see *game *sb.* 6 f.

Transfer quot. 1894 in Dict. to sense 1 b.

1807 Scott *Let.* 14 Mar. (1932) I. 359 This principle that the pursuers are entitled to have their time compensated when they were *bona fide* off work. 1853 Mrs. Gaskell *Cranford* iv. 74, I had a note to say her mistress was 'very low and sadly off her food'. 1889 E. Dowson *Let.* 16 Nov. (1967) 117 You are perhaps right in being 'off' Gortsachoff's though the cooking is less deleterious than Pinolis. 1909 W. B. Yeats *Let.* 10 Dec. (1954) 544 She seemed as eager as ever about the play. I had thought she was off it. 1910 *Ibid.* 8 Jan. 546, I have not touched the long play but will come to it fresh from being so long off it. 1912 C. Mathewson *Pitching in a Pinch* vii. 142 The Chicago pitchers were away off form in the series. 1913 C. Mackenzie *Sinister St.* I. II. xviii. 452 He said he was 'off girls' at the moment. 1929 E. Bowen *Last Sept.* xxiv. 308 'What about their bungalow?' 'Oh, that was just an idea; they are quite off it.' 1973 'H. Carmichael' *Too Late for Tears* xiii. 153 I'm off my food, that's all.

9. In combs. with *the* and a sb., used *attrib.* or as *adj.*, as **off-the-course,** occurring away from a race-course; **off-the-face,** of a hat: not covering or shading any part of the face; **off-the-job,** (*a*) done or happening away from one's work; (*b*) unemployed; **off-the-rack** = **off-the-peg*; **off-the-road,** located, operated, or occurring away from roads; **off-the-shelf,** obtained from stock; ready-made; also *fig.* Also *off-the-cuff, off-the-map, off-the-peg, off-the-shoulder*; see these sbs.

1961 *Times* 2 June 22/2 She makes a tremendous off-the-course bet. 1908 *Sears, Roebuck Catal.* 1036/2 Pretty 'off the face' hat for children. 1953 *News Chron.* 2 June 1/4 The Queen Mother, in a white feathered gown and off-the-face white hat. 1962 *Guardian* 3 Oct. 8/2 Off-the-job training of men over 25 for semi-skilled work. 1967 *Time* 21 July 51 Some sort of new company-financed plan enabling an off-the-job worker to maintain 'his normal living standard' for up to a year. 1970 *Times* 9 Feb. 19 The off-the-job study is essential because it removes the pressures and distractions of the shop-floor. 1970 *Capital Times* (Madison, Wisconsin) 21 Feb. (Green Sheet Sect.) 4/2 An off-the-rack mod suit. 1974 H. Waugh *Parrish for Defence* (1975) xxxvii. 173 The dress wasn't made for it [*sc.* her figure]. That was the trouble with off-the-rack clothes. 1975 'A. Thackeray' *One Way Ticket* I. 24 He wore off-the-rack clothes. 1962 *Guardian* 4 Oct. 14/2 Off-the-road training grounds for learner drivers. 1973 *Country Life* 18 Oct. 1172/2 Another type of off-the-road transport is the all-terrain vehicle, or ATV. 1966 *Electronics* 17 Oct. 38 The fact that its roll in a normal ocean is only 1° means that off-the-shelf, land-rated equipment can be used. 1971 *Engineering* Apr. 88 (Advt.), No chance of errors and 'off-the-shelf' service.

C. *adj.* **2. a.** Also *off lead, leader, ox* (also *fig.*, a clumsy or stubborn person), *wheeler*.

1807 *Balance* (Hudson, N.Y.) 25 Aug. 267 (Th.), We behold a clumsy, awkward off ox trying the tricks of a kitten. 1823 J. F. Cooper *Pioneers* I. iv. 51, I knew just the spot where to touch the off-leader. 1848 J. R. Lowell *Biglow Papers* 1st Ser. 90 Ez to the answerin' o' questions, I'm an off ox at bein' druv. 1887 E. Custer *Tenting on Plains* xii. 354 The old reliability of a mule-team is the off-wheeler. 1890 Kipling *Barrack-Room Ballads* (1892) 35 Two's off-lead 'e answered to the name o' *Snarleyow*. 1910 J. Hart *Vigilante Girl* x. 140 An iron 'jockey-stick' ran from the near leader's hames to the off-leader's bit. 1915 *Dialect Notes* IV. 209 My grandfather was always an off-ox. 1933 *Daily Progress* (Charlottesville, Va.) 1 Apr. 4/3, I don't know him from Adam's (or God's) off-ox.

b. (Earlier and later examples.) Also of a ball or hit on this side, or a batsman who hits the ball in this direction; **off-cutter,** a cutter (see *cutter *sb.*[1] 5 b (*b*)) that turns from the off side.

1773 in G. B. Buckley *Fresh Light on 18th Cent. Cricket* (1935) 61 Having run a considerable number of notches from off-strokes. 1816 W. Lambert *Instr. & Rules Cricket* 31 Wide Bowling..should be directed at the off side of the wicket. 1836 *New Sporting Mag.* July 195 Crossing the leg over at off-balls is another rule that I should like to see more men adopt. *Ibid.*, The mode of handling the bat, for 'On hitting' must be similar to what is requisite for 'Off hitting'. 1836 E. Jesse *Angler's Rambles* 298 His off-hits between point and slip, were the admiration of the club. 1851 J. Pycroft *Cricket Field* vii. 151 A good off-hitter should send the ball according to its pitch, not to one point only, but to three or four. 1854 *Ibid.* (ed. 2) 117 A bat brought forward from the centre stump to a ball Off or to leg, must..form an angle sufficient to make Off or On hits. 1885 *J. Lillywhite's Cricketers' Compan.* 91 Mr. W. H. Woodhouse..a brilliant off-hitter. 1888 A. G. Steel in Steel & Lyttelton *Cricket* iii. 183 Left-handed batsmen are notoriously strong and powerful in their off hitting. 1895 H. G. Hutchinson *P. Steele* i. 28 This off-ball Peter..drove..so hard..that [etc.]. 1900 P. F. Warner *Cricket in Many Climes* 182 Blanckenberg..made some very good off-side strokes. 1904 F. C. Holland *Cricket* 9 Nearly all off strokes can be classed under one of the four principal hits—the off drive, the cut drive, the square cut, and the late cut. 1955 *Times* 5 July 4/2 Singh..was making his off-cutters seem more sinister than one feels they really were. 1956 R. Alston *Test Commentary* xiii. 114 Benaud packed the off-side. 1966 [see *cutter *sb.*[1] 5 b].

c. *off-verse* [tr. G. *Abvers*]: the second half-line of a line of Old English verse. Cf. *on-verse* (*on *a.* 1 b).

1935 K. Malone in *ELH* II. 291 The chief function of alliteration in OE poetry is that of binding together the two halves (i.e. the on- and off-verses) of the so-called long line. 1953 [see *ictus 1]. 1953 *Speculum* XXVIII. 451 In the on-verse position *Béowulf* 2795 has *Wuldorcyninge* and in the off-verse position *eorþ-, héah- þéodcyninges*, also *Fris-cyninge*, and *sǣ-cyninga*. 1963 F. P. Magoun in Brown & Foote *Early Eng. & Norse Stud.* 134 These verses tend to be used as off-verses. 1970 *Rev. Eng. Stud.* XXI. 134 The Anglo-Saxon poets apparently preferred beginning new clauses with the off-verse.

4. (Earlier and further examples.) Also said of a shorter period of time to denote an interval of leisure or relaxation, and of a day, night, etc., when one does not feel fit or when one's performance is not up to the usual standard. Also *transf.*

1826 F. Reynolds *Life & Times* I. iv. 151 On Mrs. Siddons's nights, Mr. Harris (being sure of an overflow from Drury-lane,) only put up his weakest bills, reserving the strongest for his *off nights*. 1843 *Knickerbocker* XXII. 325 After an 'off night' when I was allowed to stay in town. 1875 *All Year Round* 3 Apr. 23/1 That estimable lady.. had arranged those meetings on the quiet off-evenings. 1876 'Mark Twain' *Old Times Mississippi* 25 The 'off-watch' was just turning in, and I heard some brutal laughter from them. 1880 *Inter-Ocean* (Chicago) 28 May 8/3 Peters had an 'off' day. 1905 G. B. Shaw *Let.* 3 Jan. (1972) II. 487, I had quiet literary offnights at the New Shakespear Society under F. J. Furnival, and breezy literary offnights at the Browning Society. 1908 *Sketch* 11 Sept. 340/2 Rhodes had an 'off-day' and could do nothing. 1913 A. G. Bradley *Other Days* v. 161 On the rare off-days [we] raided such crows and magpies' nests as we could find. 1928 *Weekly Dispatch* 18 Mar. 2/4 He'd never have picked up that skill in the Lagos Lagoon in the few off-watches he'd get from a branch-boat. 1929 *Star* 21 Aug. 5/2 It was certainly Elder's off-night. He was not feeling in the best of form. 1930 F. E. Baily *It won't do any Harm* xiii. 260 You happened to ask me when I was in what I call an off-moment. 1932 E. E. Reynolds *Nansen* i. 9. There was plenty of fun as well as hard work. One account of an off-hour is worth extracting as illustrating Nansen's geniality in whatever company he might find himself. 1946 R. Lehmann *Gipsy's Baby* 28 She was having a bit of an off-day, unfortunately. 1959 *Manch. Guardian* 24 July 4/2 Tolstoy or Balzac in an off-moment might be almost anyone. 1959 T. Griffith *Waist-High Culture* (1960) 169 Such doubts arise in us all, at least on our off days. 1969 *Guardian* 21 Aug. 3/6 On off-days he tends his private camellias and on off-nights listens to Mozart and The Three B's. 1974 *Melody Maker* 13 Apr. 48 Everyone has their off nights and so the one who's feeling better helps the other one. 1976 *National Observer* (U.S.) 17 Jan. 19/2 Dr. Moskowitz is passionate about reading, theater, music, and writing, which is her newest off-hours activity.

b. *off year*, in the U.S., a year in which there is a Congressional election but no Presidential election. Also *attrib.*

1873 B. A. Hinsdale *Let.* 17 Oct. in *Garfield-Hinsdale Lett.* (1949) 247 About one half is lost because it is the 'off year' in politics. 1906 *N.Y. Even. Post* 5 Nov. 4 In this off-year election. 1950 *Manch. Guardian Weekly* 24 Aug. 15 A full-dress State election in an 'off-year', that is a year when there is no Presidential election. 1972 *Times* 27 Dec. 5/7 Soon after he succeeded to the Presidency the Republicans, in the off-year elections of 1946, won both Houses of Congress.

6. Corresponding to or producing the state (of an electrical device) of being off (Off *adv.* 4 b).

1899 J. Pigg *Railway 'Block' Signalling* vii. 363 In the 'off' position of the signal arm the switch makes such contacts as passes a 'holding down' current from the battery at the local station. 1935 D. L. Sayers *Gaudy Night* xvi. 327 The switch..stood in the 'Off' position, and she struck it down. 1960 *Practical Wireless* XXXVI. 425/2 A two-pole toggle switch..could be used in the S1 position thus isolating the mains completely from the equipment when the switch is in the 'off' position. 1975 L. Deighton *Yesterday's Spy* vii. 52 Schlegel bashed the 'off' button and the music ended with a loud click.

7. *Physiol.* Of, pertaining to, or exhibiting the electrical activity that occurs briefly in some optic nerve fibres in vertebrates when illumination of the retina ceases.

1903 F. Gotch in *Jrnl. Physiol.* XXIX. 393 The third or terminal portion is a second rise due to the sudden change from light to darkness; this I propose for brevity to term the OFF effect. *Ibid.* 401 The rate of development of the OFF change. *Ibid.* 403 In no instance is the OFF delay greater than the ON. 1934 *Ibid.* LXXXI. 26 The increase in P III and its rapid return at 'off' account for the increased off-effect of the light-adapted eye. 1941 S. H. Bartley *Vision* xii. 287 The off-response first appears when the light flash is very short. 1948 *Jrnl. Physiol.* CVII. 57 Of the 164 elements, 16% were pure on-elements, 5% pure off-elements, and 79% on-off-elements. *Ibid.*, The on-sensitivity to blue (*B*-on) was taken as the fundamental sensitivity... For those elements which showed no on-effect the off-sensitivity to blue (*B*-off) was used. 1972 H. Tamar *Princ. Sensory Physiol.* iv. 153/2 Some 5 per cent of all color-responsive ganglion cells seem really to have receptive fields with centers which are either 'on' or 'off' to one color.

D. *sb.* **3.** (Earlier and later examples.) Hence **off-drive** *v. trans.*, to drive (a ball) towards the off; also with the bowler as object; so *off-driver, off-driving*; **off-theory,** a theory that favours concentrating the fielders on the off side and bowling the ball at or outside the off stump.

1836 *New Sporting Mag.* XI. 193 Thus if the bat be brought forward in a straight line to meet the ball moving in the same line, the ball will be struck directly to the bowler. It is on this principle, that more to the on or off, so will it be returned, and according as it is bowled, more to his left or right. 1847 [see *break *v.* 32 b]. 1867 *Australasian* 2 Feb. 140/2 He made a very good off drive for four. 1883 W. L. Murdoch in *Longman's Mag.* Jan. 292 At the present time, when bowlers place their men on the off side and bowl on what I might term the 'off theory', batsmen should be very careful what ball they hit at. 1884 *J. Lillywhite's Cricketers' Ann.* 5 Leslie's 144 was an example of that player's very best style, his..off-driving being masterly. 1893 R. Daft *Kings of Cricket* vi. 103 C. G. Lane..could, I think, 'off drive' Jackson better than any other player of the day. 1904 F. C. Holland *Cricket* 10 Ernest Hayes of Surrey is as hard an off-driver as any. 1927 G. A. Terrill *Out in Glare* v. 95 Clement played his first ball defensively;..off-drove the next for three. 1955 *Times* 13 July 8/5 Off the third ball of what might have been the penultimate over Waite off-drove Tyson over the top of Lock. 1960 H. S. Altham in A. Ross *Cricketer's Compan.* 292 The off-theory was being to some extent abandoned. 1963 *Times* 13 Feb. 4/3 A slightly mistimed off-drive led to a fine catch by Knight, running around in front of the screen. 1975 *Cricketer* May 14/2 Max Walker showed his 'leg-cutter' that snapped back from the off. 1977 *Guardian* 3 Jan. 11/6 Greig began by off driving Chandra for four.

5. The start of a race (cf. sense A. 1 e above); also *transf.*, the start, the beginning; departure; a signal to start or depart. *colloq.*

1959 *Times* 14 Sept. 3/1 Matthews broke once and on the second 'off' knocked down the first hurdle. 1963 L. Meynell *Virgin Luck* vii. 174 The price shortened and just before the 'off' I noticed it was being offered at nine to two. 1966 J. Porter *Sour Cream* xiv. 180 It was too late. The students nearest to him..thought this was the off. They began to move forward. 1968 'H. Calvin' *Miranda must Die* iii. 34 'Time for off,' he said... He was gone. 1973 T. Allbeury *Choice of Enemies* xiii. 62 Jock..waited for someone to give him the off, and James said, 'O.K. Jock, you just give us the general picture.' 1973 'I. Drummond' *Jaws of Watchdog* xviii. 240 How long before the off will they try to put the poison in? 1978 *Lancashire Life* Apr. 50 (*caption*) Tangle-wrangle: Stan Lyons waits on the slipway for the 'off', while helpers sort-out the lines from his harness.

off, *v.* Add: **2.** (Earlier and later examples.) Cf. Off *adv.* 9. Also *to off it*, to depart; also (*slang*), to die.

1889 T. E. Brown *Manx Witch* 18 'And will you go linkin with me?' says Jack... 'I'm thinkin I'd better,' says Nessy... And offs with him. 1890 *Punch* 28 June 310/2 He found out after they'd off'd it that they didn't own a white mouse among 'em! 1930 J. Buchan *Castle Gay* iv. 72 He has probably offed it abroad. 1930 Brophy & Partridge *Songs & Slang 1914–18* 143 *Off it*, to die. 1965 *Listener* 27 May 797/2 He ups and offs from wife, job, kids.

6. To kill.

Recorded chiefly in Black English contexts in the U.S.

1968 N. Giovanni *Revolutionary Tale* in *Negro Digest* June 77/1 The only way we can ever justify offing a brother is if we have already offed twenty whiteys. 1970 *Time* 11 May 29 At the swamp, Alex was offed... Warren shot him first, Lonnie hit him a second time. 1970 *Time* 2 Nov. 25/1 The Panthers' rhetoric is inflammatory and irresponsible, and it is impossible to say how many people take their 'off the pig' injunctions seriously. 1971 *Black Scholar* Sept. 40/2 If they caught a bitch or dude fuckin' around with the honkies, they offed 'em. 1974 R. B. Parker *Godwulf Manuscript* (U.S. ed.) iii. 18 There were various recommendations about pigs being offed scrawled on the sidewalk. 1977 *Time* 31 Jan. 52/1 There is a contract out on him and he will be offed as sure as next morning's sunrise.

off-, *prefix.* Add: **4. b.** With sbs. used *attrib.* or as *adj.*, as **off-course,** situated or taking place away from a race-course; **off-design,** that is not allowed for or expected; of or pertaining to such circumstances; **off-farm,** produced or sold away from a farm; **off-gold,** of a nation: not using the gold standard; **off-highway,** of or pertaining to travel that is not on highways; **off-ice,** of or pertaining to ice-hockey players when they are not engaged in a game; **off-pitch** *Mus.*, of a note: not of the correct pitch; **off-road,** used or taking place away from roads; **off-site,** occurring away from a site; removed from a site; **off-track,** (*a*) that is off one's intended route; (*b*) done away from a race-track (*spec.* of betting). See also main entries: Off-colour etc. in Dict., *off-air etc. in Suppl.

1963 *Times* 7 Mar. 3/2 Those who bet on racecourses and have off-course businesses as well must include the net profit of the racecourse business when determining their category. 1964 A. Wykes *Gambling* viii. 195 In France, bookmaking is illegal; and the law against off-course, non-totalizator gambling is strictly enforced. 1977 *Times* 2 Apr. 14/2 If it [*sc.* the Tote] had a monopoly of off-course betting..it could provide..£35m a year for racing. 1962 S. L. Bragg *Rocket Engines* ii. 34 Its performance at off-design pressure ratios is better than that of a conventional nozzle. 1972 *Lebende Sprachen* XVII. 134/1 Motors designed for high altitude use are frequently tested under off-design conditions. 1962 *Economist* 24 Mar. 1098/2 Target prices for off-farm dairy products ..are pitched at a much lower level. 1963 *Times* 11 Jan. 9/3 In agriculture the slow progress of both total output and off-farm sales has meant that the income of the collective farm economy has not risen fast enough. 1935 *Economist* 20 Apr. 900/2 The bulk of their trade is done with off-gold countries, and in commodities whose prices are determined by conditions in those off-gold countries. 1961 *Engineering* 10 Nov. 626 The problem of off-highway mobility. 1968 *Globe Mag.* (Toronto) 17 Feb. 5 There are now lucrative possibilities in many parts of North America for Leafs who once sold their off-ice services, personal appearances and so on very cheap. 1945 Off-pitch [see *blue *a.* 3 d]. 1961 *Engineering* 10 Nov. 628 Spheres of off-road and industrial haulage. 1968 *Economist* 21 Sept. 82/2 The new engine..is more suited..to off-road vehicles (dump trucks and earth movers of all sorts). 1973 *Observer* (Colour Suppl.) 2 Sept. 51/3 Off-road racing started in California about seven years ago with the development of beach buggies and other fun cars. 1975 *Courier-Mail* (Brisbane) 11 Apr. 16/2 Off-road and recreational vehicles are becoming more and more popular. 1969 *Ibid.* 22 Feb. 14/2 Anybody involved in a fight is automatically 'off site' on the next plane. In other words all parties involved lose not only their jobs but their accommodation and have to find a new town. 1970 *Guardian* 22 Aug. 2/4 Underground nuclear explosions in Nevada..would have negligible off-site effects. 1956 Wallis & Blair *Thunder Above* (1959) i. 7 He could have deviated up into Scotland for an off-track landing at Prestwick. 1964 A. Wykes *Gambling* 340 Whereas the on-track betting turnover was the equivalent of $64,680,000, off-track betting brought in $65,240,000. The only legal way to make an off-track bet is through the Totalization Agency Board. 1970 *Globe & Mail* (Toronto) 26 Sept. B1/1 Very little of the money bet through off-track messenger betting shops reaches the pari-mutuel machines. 1976 *Times* 20 Mar. 14/6 Many off-track activities including a *concours d'elegance* of classic road cars.

5. Prefixed to the names of colours to denote colours that are almost the same as the colour specified, as *off-black, -green, -yellow*; also *off-white *sb.* and *a.*

1927, etc. [see *off-white *sb.* and *a.*]. 1930 *Daily Express* 8 Sept. 5 One of the new off-black shades.., a sort of unripe blackberry colour, is used for the third model. 1958 J. Betjeman *Coll. Poems* 250 The walls are alternately painted Off-yellow and festival mauve. 1960 *Guardian* 9 Dec. 8/5 A feeling for the gentler blues and off-greens. 1978 *Detroit Free Press* 2 Apr. 10D (Advt.), Colors included..nocturne (off-black).

offa (ǫ·fă). Repr. U.S. colloq. pronunc. of *off of* (see Off *adv.* 7 b in Dict. and Suppl.).

1935 Z. N. Hurston *Mules & Men* II. i. 233 Ah keeps her offa me too. 1954 W. Tucker *Wild Talent* (1955) iv. 43 Roll your tail offa there and come on! 1973 *Black World* Jan. 65/2 She just couldn't take her eyes offa me. 1976 G. V. Higgins *Judgement D. Hunter* xiii. 128 Get offa the pot, willya?

off-ai·r, *a.* and *adv.* [*off- 4 b, Off *prep.* + Air *sb.*[1]] **A.** *adj.* **a.** Operating on a closed circuit; not involving the broadcasting of programmes.

1961 *Times* 2 June 14/6 (Advt.), The ship of the year.. relies on Marconi Marine..for closed circuit television. A completely co-ordinated internal and 'off-air' television system—..providing passengers with closed circuit telecine and live television programmes.

b. Involving the transmission of programmes by broadcasting.

1971 *Morning Star* 10 Dec. 4/1 An attempt to prevent the allocation to ITV of Britain's last off-air television channel. 1975 *ITV Evidence to Annan Committee* i. 30 The cable companies are eager..to deliver programmes into homes by cable rather than from off-air transmission by the IBA and the BBC. 1977 *Rep. Comm. Future of Broadcasting* (Cmnd. 6753) xiv. 218 The Cable Television Association..saw no reason why those who currently contribute to off-air television should not also participate in programme making for cable television.

c. Carried out or done from received broadcast transmissions.

1973 *B.B.C. Handbk. 1974* 45/2 The off-air recording of educational programmes by educational institutions is another problem. 1980 *Amat. Photographer* 15 Mar. 119 (*caption*) Transferring film to video tape... Tuner/timer (for 'off-air' recording).

B. *adv.* From broadcast transmissions.

1974 *Author* Winter 149 A video cassette player/recorder ..will..record off-air (you can record BBC 1 while watching ITV or BBC 2). 1975 *ITV Evidence to Annan Committee* i. 31 The cable companies, if they are to offer anything more than is already available off-air, will be forced to provide a menu of feature films and big sporting occasions. 1977 *Gramophone* June 124/1 The BBC's Matrix H tapes (including one recorded 'off-air') were as good a compromise as I have heard.

off-a·xis, *a.* and *adv.* [f. *off- 4 b, Off *prep.* + Axis[1].] (Situated) away from an axis.

1939 *Sci. Amer.* Aug. 120/2 The 'off-axis' type..is exceedingly useful in practice because, with this arrangement. the photographic plate or film may be placed outside the light beam. 1962 A. Nisbett *Technique Sound Studio* ii. 36 Place the second person just the same amount off-axis in the other direction. 1971 J. H. Richardson *Optical Microscopy for Materials Sci.* i. 47 Even though a lens may be spherically corrected for objects situated on the microscope axis, spherical aberration may exist for off-axis objects.

off base, off-base, *phr.* (*adv.*) and *a.* [Off *prep.*, *off- 4 b.] **1.** Unawares, off one's guard, by surprise; mistaken. Cf. *base *sb.*[1] 15 d. *U.S.*

1936 J. Steinbeck *In Dubious Battle* vii. 109 If they can catch us off base, they'll bounce us. 1947 *Time* 20 Oct. 11/1 Your Latin American department was off base in its comparison of the Portillo Hotel in Chile with our famous Sun Valley. 1948 *Daily Ardmoreite* (Ardmore, Okla.) 26 May 6/4 There are more men caught off base at cocktail parties than ball games. 1955 W. C. Gault *Ring around Rosa* vii. 88 You're way off base, Bobby, and I think this kind of talk is in poor taste. 1971 B. Malamud *Tenants* 141, I don't feel lovely. I feel off-base, off-key, dissatisfied. 1974 *Publishers Weekly* 18 Mar. 43/1 Off base with his moralizing, an innocent in economics, overshadowed by the Kennedys.

2. Situated elsewhere than on a military base (Base *sb.*[1] 16 and *16 b).

1962 *Economist* 1 Dec. 921/2, 136 families live on the base, 220 in an off-base village. 1967 *Ibid.* 30 Sept. 1196/3 The defence establishment has often been a pioneer in ending the colour bar, both at its own installations and 'off-base'. *Ibid.*, To fight off-base discrimination against Negro Soldiers. 1977 R. Gadney *Champagne Marxist* xiv. 92 'I'd like to speak with you.' 'Fine.' 'Off-base?' 'Of course, if that's what you want.'

off beam, off-beam, *phr., a.,* and *adv. colloq.* Also **off the beam.** [f. Off *prep.*, *off- 4 b + *beam *sb.*[1] 24 c.] On the wrong track; wrong, mistaken.

1951 'J. Tey' *Daughter of Time* ix. 121 'He's away off the beam. Away off.' 'I suspected as much. Let us have the facts.' 1958 *Woman* 20 Sept. 4/4 Funny how we all get off-beam ideas of each other, isn't it? 1960 *Spectator* 20 May 729 My thinking, which was quite off-beam, convinced me at that time that the man was a charlatan. 1973 *Guardian* 15 Feb. 14/4 Mr Patrick is far from off-beam in tracing connections between environment and the behaviour he describes. 1977 *Time Out* 17–23 June 10/3 'A Question of Ulster' was not only way off beam for explanation, it was also five years ago.

off-bearer. (Earlier and further examples.)

1866 *5th Rep. Children's Employment Commission* 142/1 in *Parl. Papers* XXIV. 1 The walk flatter is often a young woman, and the off-bearer a young man. 1887 A. A. Brown *Lumbering on the Cumberland* 47 Caleb occupied the position of honor, and acted as 'sawyer'. Abe was 'setter', and Eph was 'off-bearer'. 1939 D. Hartley *Made in England* v. 170 In one shed there seemed to be about six workers, and their names indicated their share of the job; 'moulder', 'temperer', 'off-bearer', [etc.].

o·ff-beat, *sb.* and *a.* [f. Off *prep.*, *off- 4 b + Beat *sb.*[1]] **A.** *sb.* *Mus.* An unaccented beat; = Upbeat *sb.* 1; *spec.* the second or fourth beat in common time; a heavy accent on these beats.

1928 *Grove's Dict. Mus.* (ed. 3) V. 424/2 It is therefore in the nature of things that much music should begin on the offbeat. *Ibid.*, It is convenient to confine 'upbeat' to the last of the bar, leaving 'offbeat' for any other note than the first. 1946 A. Hutchings in A. L. Bacharach *Brit. Music* xvi. 205 A rhythmic figure..beginning on an off-beat so that its entry in another part will be effective and beget movement. 1958 C. Wilford in P. Gammond *Decca Bk. Jazz* ii. 31 The pianist's left hand marks the beat firmly and unvaryingly, usually also with sharply staccato off-beats. 1973 J. Wainwright *Pride of Pigs* 175 The washboard player tapped the off-beats with the tips of his thimbles.

B. *adj.* **1.** *Mus.* Of, pertaining to, or comprising off-beats; having a marked rhythm on the off-beats. Also *transf.*

1927 *Melody Maker* Sept. 926/2 In a thirty-two bar passage the rhythm might appear thus: Six bars 'off-beat', two bars 'charleston', six bars 'four in a bar', [etc.]. 1936 *Metronome* Feb. 61/2 *Off beat cymbal*, sock cymbal. 1952 J. Steinbeck *East of Eden* xvi. 151 Samuel Hamilton rode back home... Doxology's loud off-beat hoofsteps silenced the night people until after he had passed. 1959 D. Cooke *Lang. Mus.* v. 263 Another 'oom-pah' bass without the 'oom' (the cheapest of jazz-rhythms, emphasized by off-beat side-drum taps). 1970 P. Oliver *Savannah Syncopators* 15 Off-beat phrasing of melodic accents.

2. Unusual, unconventional; strange, eccentric. *colloq.* (orig. *U.S.*).

Occas. with even stress.

1938 D. Baker *Young Man with Horn* I. iv. 35 He tried ..to teach him to sweep with a utilitarian slant, all the strokes going in the same direction in such a way that.. you inevitably have a pile of whatever it is, right there in front of your broom... But..Smoke would go right back into his off-beat swishing. 1951 *N.Y. Herald-Tribune* 13 Dec. 34 The author..has a highly individual wit, a fey and off-beat slant on the process of making people laugh. 1958 S. Ellin *Eighth Circle* (1959) II. xiii. 140 One of those types who are a little offbeat when they talk to you. 1958 *Economist* 13 Dec. 993/1 The most important postwar development has been the enormous popularity of the 'off-beat' greeting card, which was first introduced seven or eight years ago. 1959 *Observer* 5 Apr. 11/4 It is the off-beat things, the eccentricities, that help give salerooms their perpetual appeal and surprise. 1960 P. Carswell (*title*) Offbeat spirituality. 1963 [see *far-out *a.* b]. 1967 *Daily Tel.* 16 Feb. 15/5 Suzanne Hywel, 23, looks off-beat but in fact has a yearning for the country that she left behind in Kent. 1976 *Times* 9 Aug. 10/1 One of London's oldest adult education institutions ran an off-beat course for reviewing the arts.

Hence **off-beatness,** the quality of being 'off-beat'; unconventionality.

1960 *Guardian* 11 Oct. 7/2 There is no deliberate off-beatness about her. 1962 J. M. Bernstein tr. *Levi's Two-Fold Night* i. 7 There is..a hint of 'off-beatness' in their conformism, which is the city's tradition.

off-board, *a.* *U.S.* [f. *off- 4 b + *board *sb.* 8 c.] Of stocks, bonds, etc.: dealt in or sold elsewhere than at the stock exchange; = *over the counter *adv.* (*a.*) a.

1950 in Webster Add. 1968 *Economist* 6 Jan. 56/2 It is suspected in Wall Street that the agency's wily staff goes out of its way to prop up those offboard markets mainly to keep 'sassy' Wall Streeters off balance. 1977 *N.Y. Times* 14 Nov. 53/4 Most brokerage and investment banking firms, however, agree that the eventual elimination of what are called 'off-board trading restrictions' is inevitable.

off-break, *sb.* [Off- 3.] **1.** The act or result of breaking off; a schism.

1866 W. Gregor *Dial. Banffshire* 7 in *Trans. Philol. Soc.* The Free-kirk's an aff-brack fae the Aul' Kirk. 1892 [see Off- 3]. 1934 *Discovery* Mar. 60/1 The line offering the most advantageous terrain must be carefully observed, glacier off-breaks noted, where and how often breaks occur.

2. *Cricket.* A ball bowled in such a way that, on pitching, it changes direction towards the leg side; such a change of direction. Also *attrib.*

1888 A. G. Steel in Steel & Lyttelton *Cricket* iii. 117 The next spin or twist on the ball..is the rotary motion from left to right. This, in cricket phraseology, is termed the 'off' break. 1903 G. L. Jessop in H. G. Hutchinson *Cricket* v. 134. On a bad wicket and with an off-break bowler the position of short leg is indispensable. 1927 G. A. Terrill *Out in Glare* iv. 74 If the spin meant the usual off-break, the ball was straighter than Verlenden had intended. It would break right away to leg. 1955, 1963 [see *Chinaman 4].

o·ff-break, *v.* [Off- 1.] **1.** *trans.* To break off, rescind. *Sc. rare*[—1].

1872 M. MacLennan *Peasant Life* (Ser. 2) 47 She winna be ony speckillation tae the pairish by offbraikin' the banns.

2. *intr.* *Cricket.* Of a ball: to break towards the leg side. Usu. as **o·ff-brea:king** *ppl. a.*

1904 *Westm. Gaz.* 10 Aug. 3/1 Mr. Bosanquet..dismissed three batsmen lbw with his 'off-breaking leg-break'. 1907 *Ibid.* 18 July 4/2 The off-breaking leg-break of the Bosanquet school has exercised a great influence over South African cricket.

off-Broadway (ǫ:fbrǭ·dwēi), *a.* and *sb.* orig. *U.S.* [f. Off *prep.* + *Broadway.] **A.** *adj.* Of a theatre, play, or performer: located in, appearing in, or associated with an area of New York other than the Broadway theatrical area; esp. with reference to experimental productions. Also *transf.*

1953 *Plays & Players* Dec. 24/3 There are still two other plays to consider... The former began life at an off-Broadway theatre and after some revisions was brought to the Vanderbilt on October 14. 1954 *New Yorker* 9 Oct. 38/1 Making a new translation of 'The Wild Duck' for an off-Broadway production. 1959 Cordell & Matson (*title*) The off-Broadway theatre. 1965 *Listener* 20 May 738/2 The Theatre of Drama and Comedy—a sort of off-Broadway Muscovite theatre. 1970 [see *fit-up]. 1971 B. Malamud *Tenants* 31 The black said his chick was an Off-Broadway actress.

B. *sb.* Off-Broadway theatres or productions collectively.

1958 *Observer* 14 Dec. 14/4 The small downtown playhouses generically known as 'off-Broadway'. 1960 L. Kauffmann *Waldo* (1962) vi. 69 What I say is, off-Broadway is going downhill. 1966 *Saturday Night* (Toronto) Oct. 25 Off-Broadway, basing itself in and around the [Greenwich] Village area, began to rock the American theatre with Albee's plays and the latest works from Europe. 1973 *Black World* Apr. 28/2 Many theater people went to Harlem and off-Broadway in hopes of improving their careers.

Hence **off-off-Broadway,** a collective term for theatres or theatrical productions away from 'off-Broadway', esp. those that are experimental or avant-garde; also *attrib.* and *ellipt.* as *off-off*.

1967 *National Observer* (U.S.) 10 Apr. 20/1 It had to happen and it did. The small, experimental theater circuit known as 'Off-Off Broadway'—more simply, just plain OOB—is changing. 1967 *Sat. Rev.* (U.S.) 10 June 20 These off-off Broadway playwrights are emerging in a period when no producer can expect to present their unconventional works except at a financial loss. 1967 *Listener* 31 Aug. 266/2 Off-Broadway and off-off Broadway are for those who genuinely love theatre. 1968 *Guardian* 8 Oct. 5/3 Off-off Broadway..never takes place in a 'real' theatre; instead, Off-off Broadway uses cafés, rooms, lofts, churches. 1971 'T. Coe' *Jade in Aries* (1973) vi. 63 He's had a couple of things produced. In coffee houses, you know. Off-Off-Broadway. 1973 *Guardian* 1 Mar. 20/3 Pierre..produces an off off-Broadway version of Peter Pan where all the characters are played by actors over sixty. 1975 *Times* 26 Apr. 7/6 The Off-Off-Broadway scene has changed... Off-Off has turned into either grant-oriented classical or imported productions, or showcases for Broadway.

off camera, off-camera, *phr.* and *adv.* [f. Off *prep.* + Camera.] Outside the range of a film or television camera; when not being filmed or televised.

1960 *News Chron.* 15 July 4/5 He is certainly the most telegenic personality of our time—yet off camera he is shy. 1970 *Guardian* 14 Nov. 8/3 Off camera Garbo..would display a charming naïveté very different from the sophistication of the rôles she played. 1973 L. Snelling *Heresy* I. iii. 20 If Frank thinks I'm going to let that pretentious little prick..keep me off-camera..he's crazy. 1976 *Scotsman* 25 Nov. 15/4 Information on..cuts is passed to Rippon by the director at a moment when she is off-camera.

off-campus, *a.* and *adv.* orig. *U.S.* [f. *off- 4 b, Off *prep.* + *campus.] Outside a campus; existing or available away from a campus. Also *ellipt.* as *sb.*

1951 M. McCarthy in *Holiday* May 162/2 Today, 'off-campus' for the students is mainly represented by Alumnae House. *Ibid.*, There is no compulsion on the part of the college that off-campus social life should be conducted under these auspices. 1965 *Times Lit. Suppl.* 25 Nov. 1057/3 The available off-campus communities were relatively few. 1969 *Listener* 8 May 632/1 Amnesty is another favourite word, and it means that student rioters shall not be bound by the laws that apply off-campus. 1972 *Sat. Rev.* (U.S.) 4 Mar. 23/3 Off-campus housing is tight and expensive. 1975 *New Yorker* 21 Apr. 8/3 It's no wonder that this wood- and brick-walled beer-and-hamburger haunt serves as an extracurricular, off-campus student union. 1978 *N.Y. Times* 30 Mar. A1/6 Most of the off-campus programs operated in Westchester have been set up in the last decade.

off centre, off-centre, *phr.* (*adv.*) and *a.* Also (*U.S.*) **-center.** [f. Off *prep.*, *off- 4 b + Centre *sb.* and *a.*] Not quite coinciding with a central position, awry; wrong, mistaken; eccentric.

1929 D. Hammett *Dain Curse* (1930) xii. 127 He had gone off center, was a dangerous maniac. 1932 *Kansas City* (Missouri) *Times* 30 May 12 This dear old country of ours can't be so terribly off center. 1934 Webster, *Off-center, Philately*, of a stamp, having margins of unequal width. 1947 J. Steinbeck *Wayward Bus* xviii. 291 The scar on his lip made the smile off-center. 1952 L. A. G. Strong *Darling Tom* 12 A round mahogany table with a doily a little off-centre. 1958 *Observer* 25 May 16/7 Miss Anna Kavan's weird, off-centre art. 1960 *Times* 23 June 15/2 The need to write..cannot be assuaged without tugging the character of its victim a little off-centre. 1971 *Listener* 25 Nov. 714/3 An out-of-date idea hung onto by slightly off-centre intellectuals. 1975 J. Wainwright *Square Dance* 196 The arrival of the Watfords had thrown little Thelma more than slightly off-centre.

off-centre, *v.* Also (*U.S.*) **-center.** [f. prec.] *trans.* To place or position off centre. So **off-centred** *ppl.a.*, not central; eccentric; **off-centring** *vbl. sb.*, displacement from a centre.

1947 R. H. Müller in J. S. Hall *Radar Aids to Navigation* vii. 273 A point 70 miles from the radar site has been designated by the intersection of two straight lines. This appears on the scope, with an off-centering of about 1·5 radii. 1947 Miller & McLaughlin in *Ibid.* ix. 351 The off-center PPI is, unfortunately, rather badly distorted as soon as it is off-centered by more than one tube diameter. 1958 *Proc. Inst. Electr. Engin.* CV. B. Suppl. No. 8. 355/1 Few radars are capable of off-centring the display by any appreciable amount. 1973 *Sci. Amer.* Dec. 87/2 When Copernicus introduced his heliocentric arrangement, the earth became one of the family of planets revolving in off-centered circles around the sun.

off colour, off-colour, *phr.* and *a.* Add: Also (*U.S.*) **off color. 1.** Also, not in good health, slightly unwell. (Earlier and further examples.)

1876 B. Harte *Gabriel Conroy* II. IV. iv. 90 Mr Hamlin had not been well, or, as he more happily expressed it, had been 'off colour'. 1898 G. Giffen *With Bat & Ball* xi. 195 The devil with which the ball..seemed to rise from the pitch..made him a nasty bowler when the wicket was off-colour. 1898 G. B. Shaw *Mrs. Warren's Profession* III. 200 *Frank*: Off colour? *Rev. S.*: (repudiating the expression) No sir: unwell this morning. 1931 A. J. Cronin *Hatter's Castle* I. iv. 80, I haven't been myself at all these last few days—quite off colour. 1955 *Times* 10 Aug. 3/3 Hampshire if slightly off colour at the end, duly won at Portsmouth yesterday. 1966 'J. Hackston' *Father clears Out* 86 He did not sit in the shed for just a while, as a man does when he suddenly feels off-colour. 1974 A. Fowles *Pastime* ii. 12 'Where's Christine?' he said. 'Over her mum's. Her mum's off colour. She's staying..till she picks up.'

2. (Earlier example.)

1860 A. de Barrera *Gems & Jewels* 164 If the manufactured diamond is found to contain a flaw, or what is technically termed 'off-color', its value is proportionately diminished.

3. Of questionable taste, disreputable, improper, indecent; esp. of jokes; risqué; hinting at obscenity; = Dirty *a.* 2.

1875 J. G. Holland *Sevenoaks* in *Scribner's Monthly* Mar. 582/1 Everybody invited her, and yet every body, without any definite reason, considered her a little 'off color'. 1883 *National Police Gaz.* (U.S.) 17 Mar. 3/1 A few choice specimens of the off color morals and hypocritical manners of the stage. 1915 *Sat. Even. Post* 23 Jan. 27/3 It is almost inevitable that sooner or later some one would be moved to tell an off-color story. 1932 *Kansas City* (Missouri) *Times* 25 Mar. 21 It seemed a bit strange for a minister to be so devoted a reader of such a (then) decidedly off-color publication. 1952 [see *fest]. 1954 Koestler *Invis. Writing* 58 It was still possible among intimate friends to pass on a joke that was politically off colour. 1958 *Spectator* 20 June 803/2 Landladies with their slangy off-colour jokes. 1972 'G. Black' *Bitter Tea* (1973) iv. 56 He had never played an off-colour commercial trick on me, possibly because I had never given him the chance.

off-coloured, *a.* (Further examples.)

1904 *Daily Chron.* 18 May 3/4 These were the 'off-coloured', the half-castes, the outcome of white supremacy in a black country. 1913 C. Pettman *Africanderisms* 317 *Mêlées*, the off-coloured diamonds from two carats down. 1936 *Times Lit. Suppl.* 12 Sept. 717/2 A rabble of races, white, black and off-coloured, thus inhabits the Southern Africa of to-day.

offcome. Add: **5.** An outsider, stranger; one who is not a native of the district. *north. dial.*

1859 J. S. Bigg *Alfred Staunton* i. 6 'Is this Morecambe Bay?'..'Eye! eye! Morkim Bay ye offcomes ca' t'; but we ca' 't t' Sands!' 1899 H. S. Cowper *Hawkshead* v. 298 The Burtons were 'offcomes', for the name does not occur in the older register. 1946 M. Lane *Tale of Beatrix Potter* v. 97 The north-country preference for distinguishing newcomers for at least a generation as 'off-comes', which can be..translated as 'rubbishing foreigners'. 1961 *Guardian* 18 Dec. 6/2 Farm workers cannot compete for cottages with..weekending 'off-comes' out of Lancashire.

So **o·ffcomed** *ppl.a.*, coming or having come from outside; **o·ffcomer,** an outsider.

1882 A. B. Taylor *Westmoreland Sketches* 27 Ise nivver fergit them two off-cumt chaps singin. 1895 *Leeds Mercury* (Weekly Suppl.) 28 Sept. 3/8 Yond's a off-comed un. 1898 B. Kirkby *Lakeland Words* 109 *Off-comer*, a stranger in the sense of not having been born in the locality. Ther's nin seea mich good i' some o' ther off-comers. 1958 *Listener* 26 June 1054/2 The educated people ..are almost entirely strangers, off-comers. 1971 *Country Life* 9 Sept. 630/2 'Ah'll wager 'twas yon.' 'Why?' 'He's an off-comer.' 1972 'G. North' *Sgt. Cluff rings True* i. 10 Give them three years: they'll cotton on to you... In the old days you were still an off-comed-un after twenty.

offcut. Add: **1.** (Further examples.) *esp.* an odd or waste piece left after the sawing of timber.

1960 'N. Shute' *Trustee from Toolroom* 284 The offcuts were turned into pulpwood for newsprint. 1967 *Times Rev. Industry* July 44/2 On the building site..a litter of broken and unbroken bricks, torn cement bags,..and timber offcuts is commonplace. 1973 *Listener* 30 Aug. 291/1 There was another piece of filming about canals which looked like the off-cuts from some more ambitious documentary.

off-diagonal (ǭfdəi,æ·gŏnăl), *a.* (*sb.*) *Math.* [f. *off- 4 b, Off *prep.* + Diagonal *sb.*] Applied to an element of a square matrix that is not on the principal diagonal (from top left to bottom right). Also as *sb.*, such an element.

1940 *Physical Rev.* LVIII. 108/2 The only approximation made..is the neglect of the off-diagonal elements. 1957 L. Fox *Numerical Solution Two-Point Boundary Probl.* iii. 59 In *U* the off-diagonal terms are identical with those in the corresponding positions of the original matrix *A*. 1972 *Jrnl. Social Psychol.* LXXXVI. 198 *All* possible comparisons are set out here—i.e., including the congruences in the off-diagonals. *Ibid.*, The off-diagonal entries are the congruences between one factor in the one study and every other factor in the other study.

off-drive: see Off *sb.* 3 in Dict. and Suppl.

off duty, off-duty, *phr.* (*adv.*) and *a.* [f. Off *prep.*, *off- 4 b + Duty.] Of a person: not engaged or occupied with one's normal work; of things, actions, etc.: of, pertaining to, or suggestive of this state. Hence *ellipt.* as *sb.*, time spent off duty.

1852 [see Duty 5 e]. 1903 'G. Thorne' *When it was Dark* III. v. 352 The ship..wore a somewhat neglected, 'off-duty' aspect. 1904 *Daily Chron.* 23 Dec. 4/4 The off-duty policeman. 1937 D. L. Sayers *Busman's Honeymoon* vii. 153, I like to do a bit o' reading in my off-duty. 1946 *Nature* 9 Nov. 646/2 To provide facilities for adults to educate themselves in their off-duty hours. 1962 R. P. Jhabvala *Get Ready for Battle* iv. 173 Joginder yawned, the luxurious off-duty yawn of a hardworking man. 1968 K. O'Hara *Bird-Cage* xxii. 180 Maureen took her off-duty in the morning. 1973 W. J. Burley *Death in Salubrious Place* i. 14 As on other off-duty Sundays, at twelve o'clock Matthew went into the Seymour Arms.

offen (ǫ·fən), *prep. dial.* (also *U.S.*). Also **affn, offan, off'n.** [var. of *off of* (see Off *adv.* 7), by substitution of *on* for *o'*: see O *prep.*[1] and *prep.*[2]] Off from; off.

1824 *Blackw. Mag.* Oct. 457/1 'Set down that bottle,' quoth I, wiping the saw-dust aff n't with my hand. 1871 E. Eggleston *Hoosier Schoolmaster* xxxi. 216 He told his wife that the master had jist knocked the hind-sights offen that air young lawyer from Lewisburg. 1871 S. Lanier *Wks.* (1945) I. 23 He picked all the rocks off'n the groun'. *Ibid.* 195, I got down off'n my load. 1903 'O. Henry' in *Ainslee's* Feb. 60/2 He'd jest light off'n his bronco and hunt a place to camp. 1922 E. O'Neill *Hairy Ape* (1923) v. 45 Clean, ain't it? Yuh could eat a fried egg offen it. 1930 *Life & Work* Feb. 63 Lottie says A maun stop letting ye borrow offan me. 1938 M. K. Rawlings *Yearling* i. 14 You knock them plates offen the table..and you'll see who's riled. 1940 W. Faulkner *Hamlet* I. ii. 43 She was taking Ab's breakfast offen the stove, onto two plates. 1944 C. Himes *Black on Black* (1973) 198 That made me mad, them sendin' that jalopy for me. But I was so high off'n them dreams I let it pass. 1977 J. Cleary *Vortex* v. 121 Who took the gun off'n her, then?

Offenbachian (ǫ·fənbā:kiăn), *a.* [f. *Offenbach* + -ian.] Of, pertaining to, or characteristic of the French composer Jacques Offenbach (1819–80), or his music.

1869 *Porcupine* 24 July 152/3 The Offenbachian piece which is now drawing all London to the theatre of the Brahams. 1888 G. B. Shaw *London Music 1888–89* (1937) 36 Its absurdly Offenbachian finale to the first act. 1937 *Scrutiny* Dec. 286 The peculiar poised vivacity of Offenbachian *opéra bouffe* is partly explained by Shaw. 1970 *Daily Tel.* 21 May 16/3 Poulenc's one-act opera..is an amusing piece in his most Offenbachian vein.

offence, offense, *sb.* Add: **5. c.** Colloq. phr. *no offence*: do not take offence; no offence is meant or taken.

1829 G. Griffin *Collegians* II. xvii. 37 'Is poor Dalton really dead?' 'He is, sir. I have already said it.' 'No offence my boy. I only asked, because if he is..it is a sign that he never will die again.' 1855 Mrs. Gaskell *North & South* II. iii. 32 I'd rather think yo' a fool than a knave. No offence, I hope, sir. 1973 R. Busby *Pattern of Violence* ii. 24 Be better when I'm out of this piss hole—no offence, gents.

9. (See quot. 1961.) *N. Amer.*

1928 G. H. Ruth *Babe Ruth's Own Bk. Baseball* ii. 19 A game of baseball is like a battle... It's a battle of defense against offense and the best organization wins. 1961 J. S. Salak *Dict. Amer. Sports* 303 *Offense*, the team, or a player of the team, on the attack, being at bat (as in baseball) or in possession of the ball (as in football). 1969 *Internat. Herald Tribune* 6 Nov. 13/4 The Leafs, with Dave Keon and Murray Oliver leading the offense.., whipped Oakland 5–2. 1970 *Toronto Daily Star* 24 Sept. 18/4 Knechtel can play offence if somebody gets hurt. 1974 *State* (Columbia, S. Carolina) 3 Mar 3-D/4 The Bears had to look elsewhere for their offense, and it came in the person of Pat Edwards, who led the charge with 16. 1976 *Washington Post* 19 Apr. D1/6 The chief defect of the Washington Bullets has been made painfully obvious in their last two playoff games with the Cleveland Cavaliers—the team is suffering from a sick offense.

offenceful, *a.* For *rare*[—1] read *rare*, and add further example.

1970 V. Nabokov in *New Yorker* 23 May 44 With each year..That separation seems more offenseful. And the offense more absurd.

offender. See also *first offender* s.v. *first *a.* (*sb.*) and *adv.* C. 2; *old offender* s.v. *old *a.* (*adv., sb.*[1]) D. 4.

offensive, *a.* (*sb.*) Add: **7.** In sports, of or pertaining to the offence (*offence *sb.* 9). *N. Amer.*

1912 C. Mathewson *Pitching in a Pinch* vi. 124 Offensive coaching means the handling of base runners, and requires quick and accurate judgment. 1928 G. H. Ruth *Babe Ruth's Own Bk. Baseball* ii. 23 The change from defensive to offensive play came gradually. 1969 *Eugene* (Oregon) *Register-Guard* 3 Dec. 1D/2 They picked Oregon's Bob Moore as the outstanding sophomore offensive back, and he more than lived up to expectations. 1970 *Toronto Daily Star* 24 Sept. 18/6 Argos had to keep an extra offensive lineman right from the start. 1974 *Anderson* (S. Carolina) *Independent* 18 Apr. 6C/1 In 1972, Heard was voted most valuable defensive player and in '73 he was picked most valuable offensive player. 1978 *Detroit Free Press* 2 Apr. 6E/2 He mentions the offensive linemen simply because they caught such hell for past Lions failures.

B. *sb.* **2.** *fig.* Forceful action or movement directed towards a particular end; a sustained campaign or effort; esp. in *peace offensive*.

1918 in S. Sassoon *Siegfried's Journey* (1945) vii. 72 There are indications that the enemies' peace offensive is creating the danger which is its object. 1919 G. B. Shaw *Peace Conf. Hints* ii. 18 Even when Germany capitulated they [*sc.* the Jingos] were still under such a terror of peace that they called her collapse 'a peace offensive'. *Ibid.* 29 There was only one really valid word in England about peace; and that was that those who preached it were the enemies of their country. Peace proposals were called peace offensives. 1939 *War Illustr.* 21 Oct. 192/1 Mr. Chamberlain stated in the House of Commons that nothing in the German 'peace offensive' could modify the attitude which Great Britain had felt it right to take. 1943 J. D. Wilson *Fortunes of Falstaff* i. 1 An excursus on Falstaff published in 1927 is, for instance, one of the more powerful offensives in the perennial campaign which Professor Stoll wages against the romantic school of Shakespearian criticism. 1952 *Ann. Reg. 1951* 321 China.. publicly supported the various manifestations of the Soviet 'peace offensive'. 1970 R. Lowell *Notebk.* 189 You mean our National Peace Offensive?

offer, *sb.* Add: **1. b.** (Further examples.)

1847 A. Brontë *Agnes Grey* xiv. 219 The conceited wretch chose to interpret my amiability of temper his own way, and at length..he actually—made me an offer! 1971 G. Mitchell *Lament for Leto* iv. 121 Ronald Dick.. certainly would make me an offer if I were free. 1976 *Scottish Rev.* Spring 6 She had plenty of flames and several guid offers.

d. (Further examples.) In *Comm.* also, the fact of being offered for sale at a low price, as sales promotion.

1966 *Listener* 9 June 830/1 The cheaper and nastier Hollywood series, which are always on offer. 1967 *Ibid.* 1 June 704/1 Purchasing the most sophisticated weapons we have on offer. 1971 *Woman's Own* 27 Mar. 21 Next week..bargain vanity case offer.

2. c. An opportunity or 'opening'. *dial.* or *colloq.*

1831 S. Lover *Legends & Stories of Ireland* 9 The first offer afther I make her as good as new. 1877 *Coursing Calendar Autumn 1876* 302 Napoleon went past Countess in the race to the hare, and..never gave his antagonist an offer. 1925 *Dialect Notes* V. 337 *Offer*. a chance (at seals).

offer, *v.* **3. b.** Delete †*Obs.* and add later example.

1939 C. Morley *Kitty Foyle* 328, I offered him to go in the bathroom to wash.

g. *refl.* To present (oneself) to a person for acceptance or refusal; to put (oneself) forward, *spec.* as a suitor.

1765 H. WALPOLE *Castle of Otranto* i. 18 In short, Isabella, since I cannot give you my son [in marriage], I offer you myself. **1893** M. E. MANN *In Summer Shade* II. xi. 28, I have this evening offered myself to Mary Burne, and she has accepted me. **1903** *Eng. Dial. Dict.* IV. 332/2 He did nothing but offer himself for her for so long as she lived. **1930** G. B. SHAW *Apple Cart* II. 75 It is my intention to offer myself to the Royal Borough of Windsor as a candidate at the forthcoming General Election. **1978** I. MURDOCH *The Sea, The Sea* 436 Charles, darling, tell me... When you came here today were you going to offer yourself to me?

† **h.** *intr.* To stand as a candidate for office. *Obs.*

1766 J. WEDGWOOD *Let.* 4 June (1965) 40 Some of our friends suspected a Candidate would offer who lived at too great a distance from the centre of the business. **1803** W. R. DAVIE *Let.* 20 Aug. in J. Steele *Papers* (1924) I. 405 The Gentlemen who prevailed upon me 'to offer' as they call it, consisted principally of the moderate men of both parties. **1835** A. B. LONGSTREET *Georgia Scenes* 234 Then lowering his voice to a confidential but distinctly audible tone, 'what you offering for?' continued he.

i. *trans.* With spoken words as object: to say tentatively or helpfully.

1881 M. CROMMELIN *Miss Daisy Dimity* I. ii. 32 'There are two hens to be set with Brahma eggs this morning, and a brood of young Cochins coming out,' offered Polly hesitatingly. **1894** 'R. ANDOM' *We Three & Troddles* iv. 21 'A coffee-mill,' suggested Wilks. 'Or a sewing machine,' I offered. **1973** J. ROSSITER *Manipulators* v. 51 'Perhaps,' Bradley offered helpfully, 'you've been name-calling somebody. And they didn't like it.' **1974** 'E. LATHEN' *Sweet & Low* v. 52 'Just like civil war in Nigeria,' offered Charlie sagely.

j. *Telephony.* To direct (a call) *to* a piece of apparatus.

1950 J. ATKINSON *Herbert & Procter's Telephony* (new ed.) II. ii. 33/2 It is readily possible to read off the traffic offered to any particular contact for any value of total traffic. **1960** R. SYSKI *Introd. Congestion Theory Teleph. Syst.* v. 194 The *N* sources originate calls which are offered to *R* channels. **1960** *Post Office Electr. Engineers' Jrnl.* LIII. 76/2 This form of control will facilitate the provision of automatic alternative routing, which will permit traffic to be offered to a direct route and then, if all circuits are engaged, to overflow to the transit network.

7. b. To put (a part of a structure, etc.) in place to see how it looks or whether it fits properly; to hold up or display (something) to test its appearance or correctness. Usu. const. *up* (occas. *on*). orig. *dial.*

1854 A. E. BAKER *Gloss. Northamptonshire Words* II. 73 One of his workmen said, 'Shall I offer up, or offer on, that frame, to see if it will fit the picture?' **1887** PARISH & SHAW *Dict. Kentish Dial.* 110, I once heard a master paper-hanger say to his assistant, when a customer was inspecting some wall-papers, 'Just offer this paper up for the lady to see.' **1903** *Eng. Dial. Dict.* IV. 332/2, I will offer the shrubs before planting them. **1952** GRANVILLE *Dict. Theatr. Terms* 125 *Offer up*, to show the producer the position of a picture or an ornament for approval before fixing it permanently, particularly mirrors which reflect the stage lighting. (2) Carpenters *offer up* doorways to fit into the door-frames, in fact they offer up anything before it is approved. The term is used by carpenters outside the theatre and is peculiar to their trade.

offerable, *a.* (Further example.)

1917 E. POUND *Let.* 26 Aug. (1971) 120 The minimum offerable arrangement would be six articles a year at 10 dollars each.

offered, *ppl. a.* Add: **1. b.** In the sense of *OFFER *v.* 3 j.

1960 R. SYSKI *Introd. Congestion Theory Teleph. Syst.* v. 194 If an offered call cannot be served immediately because at the instant of its arrival no free channel is available, the source which originated it nevertheless continues to demand service. **1974** COX & REUDINK in W. C. Jakes *Microwave Mobile Communic.* vii. 550 Sometimes estimates of offered traffic are made directly instead of estimating attempt rates and holding times separately.

offeree (ǫfərī·). [f. OFFER *v.* + -EE[1].] A person to whom something is or has been offered. Also *attrib.*

1952 *All England Law Reports* I. 1092 Provided the offer is not one which for some reason the offeree is entitled to refuse. **1967** *Economist* 20 May 754/1 The now-familiar rise in the price of an offeree company's shares before the announcement of a bid. **1972** *Real Estate Rev.* Winter 56/2 The statutes define such offerings in terms of the number of buyers, or the number of offerees, or the total number of stockholders or security holders after completion of the offering, or combinations thereof.

offering, *vbl. sb.* Add: **2. c.** Something offered to the public for entertainment, patronage, purchase, etc.; *spec.* a theatrical production.

Quot. 1820 has connotations of sense 2 a in Dict.

1820 *Offering of Sunday-School Teacher* p. iv, The Book is really what its title imports,—'The Offering of a Sunday-School Teacher' etc.; and it is equally adapted to the Sunday Schools of every denomination of Christians. **1834** *Offering* p. i, The increased demand for works of this description, has induced the Editor of '*The Offering*' to usher into the world another of the class of books which, of all others, has met with the largest share of public patronage; and it is with peculiar pleasure that he again presents himself before his friends in the capacity of a compiler or gatherer. **1848** *Sporting Life* 29 Apr. 103/2 The Easter offerings at this house [*sc.* the Strand Theatre] are *Woman's Faith* and a new burlesque extravaganza. **1901** *Munsey's Mag.* July 587/1 There were so many offerings which critics and first night audiences liked,.. which the paying public regarded with indifference. **1903** *Boston Even. Transcript* 29 Aug. 8/2 On Saturday next the Transcript will print an unusually attractive line of real estate offerings. **1932** *New Yorker* 11 June 46/2 If Mme. Sylva can summon so many listeners for subsequent offerings, her company should thrive.

3. (Examples corresp. to *OFFER *v.* 3 j.)

1938 HERBERT & PROCTER *Telephony* (ed. 2) II. ix. 371 The trunk offering selector and the trunk offering final selector together cater for a 4-digit numbering scheme. **1950** J. ATKINSON *Herbert & Procter's Telephony* (new ed.) II. xxv. 778/1 The telephonist dials a special number over the trunk offering switching train to the exchange concerned. **1964** K. H. BRINKMANN tr. *Trautmann's Design Automatic Teleph. Exch.* II. 63 An offering subgroup comprises the offering trunks which carry the traffic to such a subgroup.

offeror, var. OFFERER; *spec.* one who offers something for sale, esp. shares.

1930 A. PALMER *Company Secretarial Pract.* vi. 46 Any document by which the offer for sale is made shall..be deemed to be a prospectus issued by the company, but without prejudice to the liability of the offerors in respect of the offer. **1955** *Times* 18 May 3/8 There was no binding contract until notice of the acceptance was received by the offeror. **1972** *Mod. Law Rev.* XXXV. 74 The offer was to remain open until May 14, 1970, with the usual right being reserved to the offeror to extend the time during which the offeree shareholders could tender their acceptances.

offertorial, *a.* For *rare*[−1] read *rare* and add earlier example.

1856 *Tracts on Increase of Episcopate in Eng. & Wales* II. 9 To meet the continual demands for Church extension by stated Offertorial collections.

off flavour, off-flavour. [Cf. *OFF *adv.* 1 g.] A stale, rancid, or unnatural flavour in food.

1947 *Richmond* (Va.) *News-Leader* 3 Apr. 36/1 There are several good reasons for what they term 'off-flavors' in milk. **1950** *N.Z. Jrnl. Agric.* Mar. 265/3 White clover, subterranean clover, red clover, silage, lucerne, and chou moellier all produce an off flavour [in the cows' milk]. **1952** *Chambers's Jrnl.* June 340/2 They found that meat put under this apparatus loses its red colour, goes brown, and has a definite off-flavour. **1958** [see *ANTI-OXIDANT]. **1971** *Guardian* 19 May 8/1 Yogurts..are made with a living culture... The process is strictly controlled throughout so that off-flavours cannot develop.

off form, off-form, *phr.* and *a.* [f. OFF *prep.* + FORM *sb.*] In poor condition; 'out of form' (see FORM *sb.* 16).

1912 C. MATHEWSON *Pitching in a Pinch* 142 The Chicago pitchers were away off form in the series. **1961** *Times* 20 Feb. 6/2 Mr. Ferber sounded a little off-form. **1965** *Listener* 24 June 930/1 There is scarcely a topic in the history of science that has attracted more great men to it—yet found them so sadly off-form—than the question of whether his intelligence is inherited or due to his upbringing. **1972** H. A. WILLIAMS *True Resurrection* ii. 35 They don't say, 'He's not as good as he was,' but just, 'He's off-form today.' **1976** *Listener* 26 Feb. 250/2 The musicians may be tired, off-form.

off-gau·ge, *a.* [f. OFF *prep.* + GAUGE, GAGE *sb.*] Of steel strip: having a thickness outside the permitted tolerance. Freq. *absol.*

1940 *Sheet Metal Industries* XIV. 611/1 It is believed that the predominating cause for off gauge while at running speed is the variation in the incoming strip. **1947** *Jrnl. Iron & Steel Inst.* CLVI. 398/2 The phenomenon undoubtedly contributes to 'off-gauge' strip in wide-strip mills and other mills rolling thin-gauge material, during acceleration and deceleration. **1953** *Engineering* 9 Jan. 33/1 As the measuring head cannot be mounted close to the roll gap, there is a delay between the rolling of off-gauge strip and its indication. **1962** C. W. STARLING *Theory & Pract. Flat Rolling* ix. 149 Probably the most common cause of rejections in cold rolled sheet is 'off gauge', and as the tendency is to decrease gauge tolerances, accurate gauge control is of the utmost importance.

o·ff-glide. *Phonetics.* [f. OFF *a.* + GLIDE *sb.*] A glide that terminates the articulation of a speech-sound, when the vocal organs either return to a neutral position or adopt a position anticipating the formation of the next sound. Cf. *ON-GLIDE. Hence **off-gliding** *ppl. a.*

1879 [see BREATH 10]. **1888** [see *ON-GLIDE]. **1927** J. J. HOGAN *Eng. Lang. in Ireland* 71 An important difference between I. [*sc.* Irish] and E. [*sc.* English] is the greater value of the off-glides of the Irish consonants. **1934** [see *ON-GLIDE]. **1954** F. G. CASSIDY *Robertson's Devel. Mod. Eng.* (ed. 2) v. 104 An on-glide is one preceding the vowel (or consonant) and therefore gliding on toward it; an off-glide follows the vowel. *Ibid.*, Diphthongs develop from the addition of either on-gliding or off-gliding elements. **1964** E. J. A. HENDERSON in D. Abercrombie et al. *Daniel Jones* 418 There is frequently still sufficient pressure of air behind it for a weak off-glide to be audible. **1972** *Language* XLVIII. 865 The backness and roundness values of the following off-gliding vowels.

off-grain, *adv.* and *a.* [f. OFF *prep.*, *OFF- 4 b + GRAIN *sb.*[1]] Of a fabric: against the direction of the threads; having a grain that is not straight.

1964 *McCall's Sewing* iii. 38/2 Finishing or printing processes will often pull them [*sc.* fabrics] off-grain. In some cases a permanent finish may lock the threads in the off-grain position and the fabric can never be straightened. *Ibid.* vii. 97/1 There are several ways to straighten off-grain fabrics. **1968** J. IRONSIDE *Fashion Alphabet* 88 Cheap fabrics are often printed 'off-grain' and one is faced with the decision of either having the grain right (so that the garment hangs properly) or the pattern straight!

off-hand, offhand, *adv.* and *adj. phr.* Add: **A.** *adv.* **2.** Delete *rare* and add earlier and later examples. *U.S.*

1833 *Sk. & Eccentr. D. Crockett* ix. 119 Forty yards off-hand, or sixty with a rest, is the distance generally chosen for a shooting match. **1970** *Amer. Speech 1968* XLIII. 217 The shot could be made with a single-shot muzzle-loading rifle or with a *two-shoot gun*, fired either *off-hand* or *with a rest*.

3. *to farm off-hand*, to own or hold a farm without residing on it. *dial.*

1879 *Norfolk Archaeol.* VIII. 171 A farmer having an occupation apart from his homestead is said to farm it off-hand. **1898** *Longman's Mag.* Sept. 408 The land had been farmed 'off-hand', that is to say, the tenant did not live on the farm, but put in a working bailiff.

B. *adj.* **2. b.** *Mining.* (See quots.)

1888 W. E. NICHOLSON *Gloss. Terms Coal Trade*, *Offhand-men*, a term applied to all colliery workmen except hewers and putters. **1921** *Dict. Occup. Terms* (1927) §047 *Odd worker, off hand man, wage man*, general terms for men or boys employed above or below ground and paid by the day. **1926** [see *face-worker* s.v. *FACE *sb.* 27].

3. Of a shot: fired from a gun held in the hand without other support. *U.S.*

1856 R. GLISAN *Jrnl. Army Life* (1874) xxiv. 328, I surprised everybody by killing the duck at an off-hand shot.

4. Of a farm: owned or held by a person who does not reside there. *dial.*

1873 F. T. CHEVALLIER *Let.* 6 May in Thirsk & Imray *Suffolk Farming in Nineteenth Cent.* (1958) 106 On the off-hand farm I shall have a good opportunity of seeing what will be required during the ensuing year. **1880** R. S. CHARNOCK *Gloss. Essex Dial.* 33 Some who hold farms in different parishes call those farms where they do not reside 'off-hand farms'. **1898** *Longman's Mag.* Sept. 408 The labourer in charge of an off-hand farm. **1960** G. E. EVANS *Horse in Furrow* vii. 99 The whole capital he had sunk in the farms (the *Bransons Land* referred to was an *off-hand* holding) was not giving him the return it would have done if invested elsewhere. *Ibid.* xvi. 209 We generally spent two days at a farm, and perhaps another day at an 'off-hand' farm belonging to it.

5. *Engin.* Carried out with the workpiece held in the hand. Of a machine: intended for off-hand operations.

1931 E. P. VAN LEUVEN *Cold Metal Working* iv. 65 We refer to grinding as off-hand, semiprecision or precision grinding, depending on the type of machine and the degree of accuracy required. **1961** L. E. DOYLE et al. *Manuf. Processes* xxviii. 643 Nonprecision grinding, common forms of which are snagging and off-hand grinding, is done primarily to remove stock that cannot be taken off as conveniently by other methods. **1966** G. H. THOMAS *Metalwork Technol.* xiii. 150 'Off-hand' grinders..are of two types—bench models and pedestal (or floor) models. **1969** C. R. SHOTBOLT *Workshop Technol.* I. vi. 120 The author does not believe in any expensive cutting tool being ground by hand on an off-hand grinding machine.

6. Designating glass-ware made by hand, without a mould; also denoting such a process.

1941 C. J. PHILLIPS *Glass* viii. 156 The glass employed in 'offhand' glass blowing is usually melted in pots. **1949** P. DAVIS *Devel. Amer. Glass Industry* iv. 49 The process of blowing in a mold was closely similar to the off-hand method. **1967** C. GASKIN *Edge of Glass* vi. 144 He..took up his empty wineglass. 'If you were making this by the "off-hand" method—that's entirely by hand—you'd start with heating your batch.' **1970** *Canad. Antiques Collector* July–Aug. 14/2 You can feel the intrinsic beauty of an off-hand glass chain.

off-handed, *a.* Add: **b.** *Mining.* = prec. B. 2 b. *dial.*

1846 W. E. BROCKETT *J. T. Brockett's Gloss. North Country Words* (ed. 3) II. 59 All workmen about a coal-pit are said to be *off-handed* who are not engaged in the business of hewing and putting the coal. **1906** *Daily Chron.* 16 Oct. 5/2 The 'off-handed' men..dispersed into the four seams of the pit.

off-handedly, *adv.* (Earlier and later examples.)

1886 *19th Cent.* Oct. 541 The newspaper moralisers speak off-handedly of the skilled workman earning his two or three pounds a week. **1905** G. B. SHAW *Let.* 3 Jan. (1972) II. 485 At my third meeting I was asked to take the chair. I consented as offhandedly as if I were the Speaker of the House of Commons. **1973** J. WAINWRIGHT *Pride of Pigs* 123 He could have felt sorry for Fuller..to be systematically, off-handedly squashed.

offhandish (ǫfhæ·ndiʃ), *a.* [f. OFF-HAND, OFFHAND *adj. phr.* + -ISH[1].] Somewhat off-hand; off-handed. Hence **offha·ndishly** *adv.*, off-handedly.

1886 in H. Baumann *Londinismen* 124/2. **1926** E. M. ROBERTS *Time of Man* i. 28 The brown colt came from the other side of the enclosure, nibbling offhandishly at the wilted grass and edging always a little nearer. **1952** *Ethics* LXIII. 66/1 Commitments acknowledged only in an offhandish way by Moore.

office, *sb.* Add: **8. c.** Also followed by defining phrase, as *Office of Works.*

1880 E. W. HAMILTON *Diary* 11 Sept. (1972) I. 49 His [*sc.* W. P. Adam's] place will have to be filled up at the Office of Works. **1935** *Discovery* Aug. 217/1 Mr J. H. Markham, F.R.I.B.A., of H.M. Office of Works, has made a notable addition to London's public monuments. **1936** *Ibid.* July 199/1 A change in the Office of Works which archaeologists cannot but view with regret.

e. *Aeronaut. slang.* The cockpit of an aircraft.

1917 'CONTACT' *Airman's Outings* 123, I strapped our baggage, some new gramophone records, and myself into the observer's office. *Ibid.* 161, I withdraw into 'the office', otherwise the observer's cockpit. **1918** *Blackw. Mag.* Oct. 526/2 'Wouf!'—a deafening crash, and the old bus shakes violently as I put my head into the office. **1934** V. M. YEATES *Winged Victory* iv. 34 He put his head in the office and flew by the instruments. **1941** [see *GREENHOUSE 3]. **1942** 'B. J. ELLAN' *Spitfire!* p. x, The cockpit is called the office. **1966** *New Statesman* 13 May 687/2 'Up in the office they too knew it.' 'The office? You mean the flight deck?' 'Just that. No more. No less. The office.'

9. b. Also in *pl.*; *spec.* in phr. *usual office*(*s*.

1938 N. MARSH *Artists in Crime* vi. 84, I imagine it was to pay a visit to the usual offices. **1948** J. CANNAN *Little I Understood* ix. 124 Mildred had been too shy when Adam, indicating a door, had said, '"The usual offices"..' to open the door and look in. **1951** N. MARSH *Opening Night* ix. 220, I went to the usual office at the end of the passage. **1955** N. FITZGERALD *House is Falling* xi. 188 He was having the usual offices in his house duplicated. **1957** J. BRAINE *Room at Top* i. 13 The bathroom's to the right and the usual offices next to it. **1959** W. GOLDING *Free Fall* ii. 36 There are the usual offices indoors now. **1963** *Gloss. Gen. Building Terms* (*B.S.I.*) 10 '*Offices*' (deprecated), service rooms and W.C.s.

12. a. *office boy* (earlier and later examples; also in extended use), *building*, *chair* (later example), *copy* (various senses: earlier and later examples), *desk*, *door*, *duty* (later example), *equipment*, *expenses*, *furniture*, *girl* (earlier and later examples), *job*, *politics*, *routine*, *stool* (further examples), *work* (earlier and later examples); **b.** (sense 4) *office-holder* (earlier and later examples), *-holding* (further example; also as ppl. adj.), *-hunter* (earlier and later examples), *-hunting*, *-mongering*, *-seeker* (earlier and further examples); *office-seeking* adj. (earlier example; also as vbl. sb.); (sense 8) *office-cleaner* (later examples), *-keeper* (later examples), *worker*; *office-bound* adj.

1961 *Times* 12 Oct. 16/2 The bliss of an office-bound youngster. **1972** *Daily Tel.* 24 Apr. 25/1 'I couldn't bear being office-bound from nine to five each day' is the cry of many a sixth-former. **1846** *Knickerbocker* XXVII. 457 No songs for you, my sad street-sweeper!..Nor for you, melancholy office-boy! **1865** A. J. MUNBY *Diary* 15 June in D. Hudson *Munby* (1972) 209 She kept on writing, in a hand like an office-boy's. **1914** G. B. SHAW *Misalliance* p. lxv, An office boy of fifteen is often more of a man than a university student of twenty. *Ibid.* 7, I said, 'Make him the Office Boy.' **1944** J. S. HUXLEY *On Living in Revolution* 118 Quoting from a recent address of David Lilienthal... 'An overcentralized administration is always characterized by the fact that its field officers tend to become messengers and office boys.' **1973** A. BEHREND *Samarai Affair* i. 15 His first job had been that of office boy. **1975** *Times* 13 Oct. 13/1 Mr [Ian] Smith..has to demonstrate to his hard-liners that he is not Pretoria's office-boy. **1840** *Niles' Reg.* 23 May 182/1 The Free Trader office building has been crushed in and much shattered. **1924** R. GRAVES *Mock Beggar Hall* 62, I was aware that during the war Mock-Beggar Hall had been used as a Government office-building. **1942** *London Replanned* (R. Academy) 25/1 The Surrey bank of the River is developed with Embankment gardens and office buildings. **1874** 'H. CHURTON' *Toinette* xl. 404 The old surgeon laid down his pen..and turned his office-chair round toward his visitor. **1944** *Times* 7 June 2/2 Mr. E. Granville.. asked the Secretary to the Treasury if he would arrange that..the members of the Government Minor and Manipulative Grades Association of Office Cleaners were referred to as such, and not as charwomen or charladies, which term..was resented. **1971** J. AITKEN *Nightly Deadshade* vii. 79 Here I am, on the spot after the office cleaners have cleared off. **1974** *Times* 4 Jan. 12/3 There are practically no Hutu left in government or the civil service—not even at chauffeur or office-cleaner level. **1789** J. MORGAN *Essays Law of Evidence* I. 87 Of Office Copies. **1836** S. F. AUSTIN *Let.* 22 Nov. in *Ann. Rep. Amer. Hist. Assoc. 1907* (1908) II. 142 It contains your commission, a letter of credence to the secretary of the United States, and office copies of them. **1928** F. M. FORD *Let.* 16 Apr. (1965) 178, I don't know if you have..an office copy that you would care to sell. **1946** J. IRVING *Royal Navalese* 126 *Office copy, the*, the other half (of the drink in hand). **1957** CLARK & GOTTFRIED *University Dict. Business & Finance* 246/1 *Office copy*, in general, a copy or transcript of any document..retained for office use. In law, a copy of a document made by an officer of the court or other public officer. **1881** *Rep. Indian Affairs* (U.S.) 151 The articles manufactured by the carpenters..were as follows..one office-desk [etc.]. **1907** G. B. SHAW *John Bull's Other Island* I. 3 Against the right hand wall is a filing cabinet,..and, nearer, a tall office desk and stool for one person. **1954** T. S. ELIOT *Confid. Clerk* II. 69 It's an office desk. Sir Claude got it for me...You see, I shall do a good deal of my work here. **1715** *Boston News-Let.* 11 Apr. 2/2 A fair Alphabetical List..hung up at the Office Door, would soon resolve any Person. **1863** 'E. KIRKE' *My Southern Friends* xxiii. 235 A short rap came at the office door. *a* **1885** G. B. MCCLELLAN *Own Story* (1887) xxxii. 534 He said that he was so much occupied with office-duty that it was impossible for him to leave. **1942** D. POWELL *Time to be Born* (1943) ii. 44 The profits..had been.. put back into the business, new office equipment, printing, one thing and another. **1962** D. FRANCIS *Dead Cert* x. 115 It had once been an elegant room and even the office equipment could not entirely spoil its proportions. **1972** *Guardian* 7 June 9/6 There is really no reason why office equipment shouldn't be good looking enough to go into private houses too. **1869** *Bradshaw's Railway Manual* XXI. 161 *Expended*..Office expenses, &c—£438. **1887** 'MARK TWAIN' *Lett. to Publishers* (1967) 233 On first 10,000, we deduct $5,000 office expenses and $1750 for author. **1972** P. GRIFFIN *A-Z Office Guide* 86 They [*sc.* IOUs] should not be allowed to build up to a point when cash becomes short for office expenses. **1903** G. B. SHAW *Man & Superman* II. 66 If you were to marry the son of an English manufacturer of awffice furniture, your friends would consider it a misalliance. **1911** *Daily Colonist* (Victoria, B.C.) 4 Apr. 4/4 (Advt.), We are selling lots of office furniture these days. **1974** N. FREELING *Dressing of Diamond* 188 A dislike of metal office furniture. **1863** A. D. WHITNEY *Faith Gartney's Girlhood* xi. 97 Faith looked up, and remembered the poor office girl of three years since. **1972** J. MCCLURE *Caterpillar Cop* xii. 193 Ye Olde Englishe Tea Shoppe..was crowded with office girls, buying roast beef sandwiches with luncheon vouchers. **1818** H. B. FEARON *Sk. Amer.* 143 Those dangerous abuses in government, introduced by office-holders, which..threaten..to become inveterate. **1957** P. WORSLEY *Trumpet shall Sound* 269 The tendency of the office-holder to merge his personal interests with those of his office. **1970** R. LOWELL *Notebk.* 56 The communist committed to his commune, Artist and office-holder to a claque of less Than fifty souls. **1835** D. CROCKETT *Acct. Col. Crockett's Tour* 106 The office-holding gentry..will meet with their match in an indignant people. **1857** [see *office-seeking*]. **1936** *Discovery* Feb. 63/1 The burden of compulsory office-holding ruined the well-to-do. **1957** V. W. TURNER *Schism & Continuity in Afr. Society* iv. 93 Social Drama I illustrates the conflict that may arise ..when only a few men remain in the senior, office holding generation in a village. **1806** *Deb. Congress U.S.* 24 Feb. (1852) 506/2 It would be a struggle between office-hunters and the people. **1845** W. L. MACKENZIE *Lives Butler & Hoyt* 75 General Spicer was a keen office-hunter. **1824** *Niles' Reg.* 20 Mar. 37/2 (*heading*) Office hunting. **1889** FARMER *Americanisms* 397 Office-hunting is quite a business with the thousand-and-one 'hangers-on' to the skirts of political parties. **1923** H. CRANE *Let.* 26 Oct. (1965) 153 My mind is divided between them and an office job. **1937** M. HILLIS *Orchids on your Budget* (1938) iv. 68 We ourselves have run our one-woman ménage both with and without an office job. **1834** *Chambers's Edin. Jrnl.* III. 229/1 Has the office-keeper acquainted you with the particulars I require? **1938** *Times* 16 Feb. 8/7 A short time ago a telegram was delivered at my office.. after I had left for the day. The office keeper..had it re-telegraphed to my home address. *a* **1919** T. ROOSEVELT in Ld. Charnwood *Theodore Roosevelt* (1923) 250 These men have a gift at office-mongering, just as other men have a peculiar knack in picking pockets. **1917** H. GRANT *Two Sides of Atlantic* 45 This is known in the States as 'office-politics'. **1961** 'J. WYNDHAM' *Consider her Ways* 216 Office politics, very likely... Many a young man's gifts are stunted by them. **1907** G. B. SHAW *Major Barbara* III. 258 He could learn the office routine without understanding the business. **1911** W. OWEN *Let.* 25 Apr. (1967) 70, I am not too young to..turn to Office Routine, Customs, Revenues. **1925** H. CRANE *Let.* 7 May (1965) 204 A change from office routine for awhile. **1813** *Deb. Congress U.S.* 6 Jan. (1853) 582/2 It would augment the office-seekers, who, with the friends of the Administration, were continually haunting the Executive. **1845** *Knickerbocker* XXV. 374 A Friend writing from Washington..gives us this pleasant sketch of a 'Sucker' office-seeker. **1882–3** W. WHITMAN *Specimen Days* 259 The members..were..the meanest kind of bawling and blowing office-holders, office-seekers. **1977** *Listener* 11 Aug. 163/1 In 1881..James A. Garfield..was assassinated by a disappointed office-seeker from his own party ranks. **1857** W. R. ALGER *Genius & Posture of Amer.* 4 Office-holding partisans, office-seeking demagogues. **1860** H. GREELEY *Overland Journey* 68 If he will work right ahead,..keeping clear of speculation and office-seeking, he can hardly fail to do well. **1884** F. M. CRAWFORD *Amer. Politician* I. iv. 76 We are sick with the foul disease of office seeking. **1907** G. B. SHAW *John Bull's Other Island* I. 13 He seats himself on the office stool, and tilts it back. **1953** J. WAIN *Hurry on Down* v. 91 You couldn't rightly say whether a fella was a workman or an office stool percher or a manager. *a* **1678** J. WESTLEY in E. Calamy *Continuation of Acct. of Ministers Ejected* (1727) I. 441 They are not a People that are fit Subjects, for me to exercise Office-work among them. **1849** DICKENS *Dav. Copp.* (1850) xvi. 167, I am not doing office-work... I am improving my legal knowledge. **1886** C. M. YONGE *Chantry House* II. xiv. 136 He had spent an entire day on his hands and knees..—the office-work, as we declared. **1956** 'C. BLACKSTOCK' *Dewey Death* iv. 83, I cannot see why all the office work should be held up. **1936** *Discovery* May 146/2 The lowest value of natural illumination which an office worker requires. **1956** A. H. COMPTON *Atomic Quest* 333 Mechanics and office workers and laborers of many kinds. **1973** 'E. MCBAIN' *Hail to Chief* viii. 136 The homeward rush of office workers had already begun.

c. Special Combs., as **office block**, a block (sense *14 f) containing offices; also *attrib.* and *fig.*; **office hours**, (*a*) the hours of work at an office; (*b*) a disciplinary session *U.S. Forces' slang*; **office hymn** (see quot. 1938); **office junior**, the youngest or newest member of the staff of an office; **office party**, a party held for members of the staff of an office; **office piano** *slang*, a typewriter; **office wife**, a business man's female secretary.

1942 *London Replanned* (R. Academy) 26/2 The large octagonal building prominent in the drawing..is a suggested office block with garden court or car park. **1951** *Ann. Reg. 1950* 406 The architectural standard of most of these buildings..was very poor; most of them..resembled pre-war commercial office blocks at their most tasteless. **1963** *Listener* 10 Jan. 71/1 Harry Bertoia's beautiful but boring silvery puffball of wire is the apotheosis of what I once heard called 'office-block art'—the triumph of taste and craftsmanship over feeling. **1967** B. PATTEN *Little Johnny's Confession* 47 Maud, is that you I see Alone among the office blocks? **1972** M. GILBERT *Body of Girl* xxv. 211 If there is a covering party, I guess it'll be in the office block opposite. **1802** D. RAWN *Let.* 29 Oct. in J. Steele *Papers* (1924) I. 326 He receives in addition thereto, 300 Dollars for services *termed extra*, but wholly performed during the usual Office hour. **1841** THACKERAY *Gt. Hoggarty Diamond* (1849) ix. 100 Gus Hoskins and I, who hunted after office hours in couples. **1852** [in Dict., sense 12 a]. **1898** J. H. PARKER *Hist. Gatling Gun Detachment* 23, I don't want to hear anything about it... If you want to see me about this subject, come to me in office hours. **1903** G. B. SHAW *Man & Superman* III. 80 It is the custom ..always to put off business until to-morrow. In fact, you have arrived out of office hours. **1922** *Marine Corps Gaz.* June 212 One morning after *Office Hours* the C.O. was sitting at his desk grumbling to himself and holding his head in his hands. **1933** *Leatherneck* Apr. 14 No 'office hours' were held during the month of January for any 'A' Company Marines. **1967** A. DUBUS *Lieutenant* 41 He committed an offense, he was brought in to office hours. **1972** 'H. HOWARD' *Nice Day for Funeral* iii. 41 We only meet outside office hours. I never impose on social relationships. **1907** *New Office Hymn Bk.* II. p. v, The Office Hymns are the Hymns in the Divine Office. **1931** *N. & Q.* 19 Sept. 216/2 We are told that the best place for the Office Hymn both at Matins and Evensong is immediately before the Psalms. **1938** *Oxf. Compan. Mus.* 629/1 *Office hymn*, a liturgical hymn appointed for the Office, or Service of the day... The Office Hymns of the Roman Breviary were not transferred to the English Prayer Book. **1959** J. C. DENYER *Office Managem.* xi. 83 For office juniors, the appropriate official to approach is the Juvenile Employment Officer. **1970** J. COOPER *How to survive from Nine to Five* 81 The office junior has used hair lacquer under her arms instead of deodorant and is walking round like a penguin. **1974** R. GADNEY *Something Worth Fighting For* ix. 64 A young man, neatly dressed, the obvious clerk or what is sometimes known as an office junior. **1955** W. GADDIS *Recognitions* I. iii. 101 Who made the first one? Will somebody tell me that? said The Boss at an office party. **1967** E. MCGIRR *Here lies my Wife* iii. 105 Kellerman had been in New Orleans the previous Christmas and so had missed the..office party. **1972** G. BROMLEY *In Absence of Body* xii. 142 'You seem to have a lot of office parties.' 'Yes. People leaving or getting married, or entertaining clients.' **1942** BERREY & VAN DEN BARK *Amer. Thes. Slang* § 75/38 *Typewriter*, mill, office piano. **1945** L. SHELLY *Jive Talk Dict.* 29/2 *Office piano*, typewriter. **1970** C. MAJOR *Dict. Afro-Amer. Slang* 87 *Office piano*,..a typewriter. **1942** BERREY & VAN DEN BARK *Amer. Thes. Slang* § 542/19 *Girl Friday*, *office wife*, a female secretary. **1952** G. W. BRACE *Spire* (1953) xxvii. 268 I've been a pretty faithful office wife to him, and though he has never invited me to share a bed..he does hate to part with me. **1955** H. KURNITZ *Invasion of Privacy* (1956) vii. 52, I know all about American business executives and their secretaries. Office wives, isn't that what you call them? **1972** C. WESTON *Poor, Poor Ophelia* (1973) viii. 41 The secretary smiled sweetly... 'I'll do that, Mr. Farr. Get a good night's sleep now. 'Bye.' My office wife, he thought sourly.

office, *v.* Restrict † *Obs.* to senses 1–4 in Dict. and add: **5.** (Further examples.) Hence **o·fficing** *vbl. sb.*

1841 *Swell's Night Guide* (Gloss.) *Office*, giving warning. **1846** *Ibid.* 58 She eased him of his fawney,..officed her cullies, they pasted his nibs, and scarpered rumbo. **1859** G. W. MATSELL *Vocabulum* 60 *Officing*, signalizing; a preconcerted signal by a confederate. **1914** JACKSON & HELLYER *Vocab. Criminal Slang* 63 *Office*, noun,..a signal;..a warning... Used also as a verb in the same sense. **1926** J. BLACK *You can't Win* xiii. 182 Sanc closed the door..and 'officed' me to follow him out. **1949** PARTRIDGE *Dict. Underworld* 479/1 *Officing*, a preconcerted signal. **1955** *Publ. Amer. Dial. Soc.* XXIV. 73 The tool *offices* that they will *clip* him as he enters the ramp... The tool *offices* for a *left bridge* and a *left prat*, and the *frame* closes.

6. *intr.* To have or work in an office (sense 8 a); to share an office *with* someone. *U.S.*

1892 *Nation* (N.Y.) 21 Apr. 303/2 An attorney officing in the same building. **1917** *Dialect Notes* IV. 347 *Office with*,..to share an office with. **1936** *Atlantic Monthly* July (Contributors' Col., 4) A local newspaper has just carried two want ads containing this wording:—'Chance for public accountant to office with lawyer.' 'Chance for high grade realtor to office with lawyer.' **1973** *N.Y. Times* 11 Aug. 10/1 Mr. Mardian spoke of a man who 'officed in that same agency'.

officeful (ǫ·fisful). Also **office-full.** [f. OFFICE *sb.* + -FUL.] That amount or number of anything which would fill an office.

1963 'W. HAGGARD' *High Wire* vi. 63 He had an officeful of paper. **1966** 'E. PETERS' *Piper on Mountain* ii. 26 Put him among an office-full of civil servants, and you could lose him in a moment. **1976** J. WAINWRIGHT *Who goes Next?* 66 'You'll have witnesses, of course, sir?'.. 'A whole office-full, officer.'

office-man. Restrict †*Obs.* and *Sc.* to sense in Dict. and add: **2.** A man who works in an office; *spec.* a detective who remains at headquarters.

1904 'No. 1500' *Life in Sing Sing* xiii. 256/2 *Office Man*, headquarters detective. **1908** J. M. SULLIVAN *Criminal Slang* 17 *Office man*, headquarters detective. **1921** *Daily Colonist* (Victoria, B.C.) 12 Oct. 16/2 (Advt.), Wanted—position as office man, watchman, warehouse, or place of trust. **1949** PARTRIDGE *Dict. Underworld* 479/1 *Office man*, a headquarters detective.

officer, *sb.* Add: **3. b.** Used as a mode of address to a police officer.

1899 J. S. CLOUSTON *Lunatic at Large* II. v. 140 Keep your eye on that man, officer,..and put your plain-clothes' men on his track. **1926** GALSWORTHY *Silver Spoon* I. vii. 50 'Pardon me, officer,' he said, 'but where is Wren Street?' **1934** D. L. SAYERS *Nine Tailors* 143 It is said..that the plain bobby considers 'officer' a more complimentary form of address than 'my man', or even 'constable'. **1946** E. O'NEILL *Iceman Cometh* (1947) IV. 211 She knows I was insane. You've got me all wrong, Officer. I want to go to the Chair... God, you're a dumb dick! **1965** M. ALLINGHAM *Mind Readers* iii. 44 'Officer!' said the voice.., 'I wish to give this lady in charge.' **1976** [see *OFFICEFUL].

4. Also used in the air force, and in women's branches of the forces. Also *attrib.*

1943 C. H. WARD-JACKSON *Piece of Cake* 4 The Service experiences of all sorts of people—pilots,..old R.N.A.S. and R.F.C. officers, padres,..and the rest. **1948** PARTRIDGE *Dict. Forces' Slang* 148 R.A.F. officer term. **1952** *Oxf. Jun. Encycl.* X. 493 Members of the three [women's] services..are administered by their own officers. **1959** *Chambers's Encycl.* I. 184/2 If aircraft were used in attack and defence,..it could be argued that they were manned ..by officers and men who could claim to be members of a service which belonged neither to the navy nor to the army. **1973** K. GILES *File on Death* iii. 63 Miss Sloper.. had worked as an officer in the Second World War.

b. Phr. (*an*) *officer and* (*a*) *gentleman*, applied to a person embodying the civilized qualities expected of both, freq. used ironically, also (occas. with hyphens) *attrib.* or as *adj. phr.* Hence *officer-and-gentlemanly* adj.

1845 MRS. GASKELL *North & South* (1855) I. xiv. 164, I will bear with all proper patience everything that one officer and gentleman can take from another. **1871** *Porcupine* 29 July 275/3 They want their purchase, their officer-and-gentleman hobby, their..agreeable club of an army left undisturbed. **1888** KIPLING *Plain Tales from Hills* 123 Golightly spent..that summer trying to get the Corporal..tried by Court-Martial for arresting an 'officer and a gentleman'. **1926** —— *Debits & Credits* 334 Ignatius is one of the subtlest intellects we have, and an officer and a gentleman to boot. **1946** E. O'NEILL *Iceman Cometh* (1947) I. 48, I give you my word of honour as an officer and a gentleman, you shall be paid tomorrow. **1962** I. MURDOCH *Unofficial Rose* xxxi. 304 He was paying the penalty..for being an officer and a gentleman. **1966** A. PRIOR *Operators* iii. 28 Oh, coming the officer and gentleman touch, was he? **1969** K. GILES *Death cracks Bottle* x. 110 We will just be very nice to the police in an officer-and-gentlemanly way. **1971** 'H. CALVIN' *Poison Chasers* ix. 135 ''Evening, Sergeant,' I said. Hard, officer-and-gentleman tone. **1974** 'J. GRAHAM' *Bloody Passage* i. 11 There wasn't much I could do except put my head on the block like an officer and a gentleman.

c. *Officers Training Corps*, an organization set up in schools and universities for the preliminary training of boys and young men who may later become officers in the armed forces.

1907 *Interim Rep. War Office Comm. Provision of Officers* 10 (*heading*) in *Parl. Papers* (Cd. 3294) XLIX. 549 Proposals respecting the Officers Training Corps at Universities. **1908** *Oxford Univ. Officers Training Corps* (Misc. Paper) 1 The present Oxford University Volunteer Corps is about to be transformed into a unit of 'The Officers Training Corps'. **1925** *Officers Training Corps Gaz.* Nov. 1/1 Though our first number is devoted to the University of London Contingent,..it is proposed to include all University units of the Officers Training Corps. **1957** *Encycl. Brit.* XXII. 392/2 The Officers' Training corps (O.T.C.) was set up in 1909 under the Haldane scheme... The junior division..consisted of boys in public secondary schools..; the senior division of university contingents... After World War II the training corps was replaced by the Combined Cadet force.

6. *officer cadet* (also *fig.*), *-caste*, *-class* (also *attrib.* or as *adj.*), *-type*. Also *officer material*: see *MATERIAL *sb.* 6*.

1925 *Officers Training Corps Gaz.* Nov. 1/1 Our object.. is to foster a spirit of Unity and Co-operation among Officer Cadets. **1955** T. H. PEAR *Eng. Social Differences* iii. 101 When..the functions of Sandhurst were modified, 'officer'-cadets succeeded 'gentleman'-cadets. **1962** L. DEIGHTON *Ipcress File* xiii. 74 [He] reminded me of those N.C.O.'s who drilled officer cadets. **1973** 'S. HARVESTER' *Corner of Playground* I. iv. 38 Officer cadets due to receive their commissions at the passing-out parade. **1976** *Listener* 5 Feb. 139/1 We were a grateful generation... The fittest had survived to become the officer-cadets of the intellect. **1937** 'G. ORWELL' in *New English Weekly* 29 July 308/2 The Popular Army..modelled as far as possible on an ordinary bourgeois army, with a privileged officer-caste, immense differences of pay, etc., etc. **1936** 'N. BLAKE' *Thou Shell of Death* v. 78 His army training had given him a possibly misplaced belief in the superior wisdom of what he would never have thought of calling 'the officer class'. **1950** G. GREENE *Third Man* ii. 25 'Be quiet, can't you, sir,' my driver said. He had an exaggerated sense of officer-class. **1954** 'N. BLAKE' *Whisper in Gloom* II. xiii. 173 He treated his church wardens with a certain officer-class brusqueness. **1958** P. SHORE in N. Mackenzie *Conviction* 28 The managers are often described as an officer class and this is..an apt analogy. **1968** A. LASKI *Keeper* i. 10 Ralph's turning to Colin and saying, in the same frankly brutal officer-class manner [etc.]. **1942** E. WAUGH *Put out More Flags* i. 64 'What do you think is the right type of officer?' 'The officer-type.'.. 'Now three-quarters of your officer-type live in towns'.

officering, *vbl. sb.* (Later examples.)

1890 *Century Mag.* Dec. 207 The American system of officering..was superior to that of the English. **1907** *Daily Chron.* 15 Feb. 7/4 The preponderance of the Japanese forces;..their energetic and capable officering. **1933** BELLOC *Charles I* 231 The army was still quite unfitted to meet the better training, the larger numbers and the superior officering of the enemy. **1977** *Listener* 16 June 779/4 It was..to the manning of the Empire, or rather the officering of it, that the best products of the classical 'classical education' were destined.

officery, *a.* [f. OFFICER *sb.* + -Y[1].] Resembling an officer; having the character or nature of an officer.

1905 H. G. WELLS *Kipps* III. i. 289 Saw a lot of young officery fellers coming along.

officese (ǫfisī·z). [f. OFFIC(E *sb.* + -ESE.] = *COMMERCIALESE.

1935 A. P. HERBERT in *Punch* 19 June 730/2 It is just the misplaced effort to 'write like a book', to be elegant and flowery, that yields the sickliest growths of 'officese' and Jungle English—all these 'inst.'s' and 'ult.'s' and 'in regard to's' and 'favours' and 'representatives'. **1942** PARTRIDGE *Usage & Abusage* 218/2 Commercialese or Business English (or, as A. P. Herbert calls it, *Officese*). **1960** —— *Charm of Words* i. 22 Commercialese..has also been called officese.

official, *a.* Add: **3. b.** *official secrets*, information the disclosure of which outside official circles would constitute a breach of national security; so *Official Secrets Act*.

1889 *Act* 52 & 53 *Vict.* c. 52 § 10 This Act may be cited as the Official Secrets Act, 1889. **1911** *Encycl. Brit.* XXIV. 571/1 By the Official Secrets Act 1889 it was made a misdemeanour for an official to communicate any information or documents concerning the military or naval affairs of Her Majesty, to any person to whom it ought not to be communicated. **1931** 'G. TREVOR' *Murder at School* vi. 127 'I think once again I must plead the Official Secrets' Acts,' he answered, jocularly. **1931** *Economist* 28 Nov. 1001/1 The editor of a German Radical newspaper has been sentenced to 18 months' imprisonment on the ground that, in the course of an article on civil aviation, he revealed official secrets, whose disclosure endangered national security. **1966** A. FIRTH *Tall, Balding, Thirty-Five* v. 61 He pushed forward the buff paper. It was a shortened version of the form of submission to the Official Secrets Act of 1929. **1973** *Guardian* 29 June 14/3 Today the House of Commons debates the Franks report on the Official Secrets Acts. **1976** *Howard Jrnl.* XV. I. 24 The workings of the Official Secrets Act and the requirement of 'submission for prior approval' before a prison official can make a public statement do not encourage the development of penological expertise.

5. (Further examples.)

1957 J. PASSMORE *100 Yrs. Philos.* i. 28 The 'Scottish school'..lingered on..in the United States, where it became a sort of 'official philosophy' in the less adventurous Colleges. **1959** *Chambers's Encycl.* XII. 745 English and Afrikaans are treated on a footing of equality as official languages... Every child shall learn the second official language. **1970** *Cape Times* 28 Oct. 22/6 (Advt.), A knowledge of both official languages is required. **1976** *Scotsman* 27 Dec. 5/1 At Celtic Park, the official attendance was 47,000. **1977** *Belfast Tel.* 24 Jan. 9/4 Sinn Fein.. has failed to gain any significant support at the polls, mainly because it has been regarded as the 'front organisation' for the official IRA.

officialese. Delete *nonce-word* and add further examples.

1924 P. MACDONALD *Rasp* vi. 84 My—what's the officialese for it?—'suppression of the truth' gave Boyd clue number one. **1932** *New Statesman* 2 Jan. 7/1 Drink to me only with thine i's, and I will cross the t's, But leave no kiss within this cup of crabb'd officialese. **1941** *Manch. Guardian Weekly* 14 Mar. 214/3 Let us be fair to the Civil Servant: the official is not the only dealer in officialese. **1953** E. SIMON *Past Masters* IV. ii. 220 One was left to infer between the lines of officialese. **1960** S. FOOT *Emergency Exit* xiii. 110 Constitutional English is such a splendid language..but this was constitutional officialese. **1973** *Listener* 15 Feb. 221/3 The vocabulary Don Haworth's characters use..embodies chunks of jolting officialese. **1975** B. GARFIELD *Hopscotch* iii. 50 'You've read the backgrounding.' It was phrased in dry officialese.

officialize, *v.* Add: Hence **officializa·tion**, the rendering official in form or character.

1907 *Daily Chron.* 9 Nov. 4/4 One fails to detect a craving for any such officialisation.

officially, *adv.* Add: Also, for official purposes; in official or public statements, reports, etc. (but not in actuality).

1938 E. AMBLER *Cause for Alarm* x. 163 The price per machine will..be higher... Officially, this fact will be accounted for by the modifications. Actually, those modifications are purely nominal. **1964** L. DEIGHTON *Funeral in Berlin* vii. 51 'Does anyone have phones going across..Berlin?' 'Officially one. It connects the Russian Command..with the Allied Command.'.. 'Unofficially?' 'There *have* to be lines.' **1976** 'B. SHELBY' *Great Pebble Affair* I. 30 Officially, I had been renting my apartment for three months before I even saw it.

officina. Add: **b.** Phr. *officina gentium*, a country or area from the inhabitants of which several nations develop; also *officina gentis*, the place of origin of a nation or people.

[*c* **550** JORDANES *Getica* (1882) iv. 25 Scandza..quasi officina gentium aut certe velut vagina nationum.] **1821** DE QUINCEY *Confess.* (1822) 169 Southern Asia is.. the great *officina gentium*. **1832** *Edin. Rev.* LV. 499 The New Englanders have been the *officina gentis* to the American people. **1877** D. M. WALLACE *Russia* (ed. 2) II. xxiv. 106 As Scandinavia was formerly called *officina gentium*—a workshop in which new nations were made. **1904** W. P. KER *Dark Ages* iii. 131 [Jordanes] has a lofty conception of the destiny and fortunes of the Gothic race, and his account of the origin of the warlike nations in the Northern island, Scanzia, *officina gentium*, corresponds in prose to the epic genealogies of the poets. **1961** L. F. BROSNAHAN *Sounds of Language* ix. 195 The *officina gentium* which was Scandinavia at this period.

offing. Add: **3.** *in the offing*: nearby, at hand, in prospect, likely to happen in the near future; (in quot. 1779, exceptionally, in the distant future).

1779 J. WEDGWOOD *Let.* 30 May (1965) 234, I hope soon to say as far as 30 inches, perhaps ultimately up to 36 inches by 24, but that is at present in the offing. **1914** T. DREISER *Titan* xvii. 139 The possibility of another woman equally or possibly better suited to him was looming in the offing. **1942** D. POWELL *Time to be Born* (1943) x. 239 Somehow it didn't seem like real love without a husband in the offing. **1949** N. MITFORD *Love in Cold Climate* I. v. 57 That look of concentration which comes over French faces when a meal is in the offing. **1970** G. F. NEWMAN *Sir, You Bastard* viii. 241 Number six could be afforded if he got promoted, but promotion wasn't in the offing.

offish, *a.* (Earlier and later examples.)

1834 C. A. DAVIS *Lett. J. Downing* 75 Others are a little offish. *a* **1963** S. PLATH *Crossing Water* (1971) 31 She stopped fitting me so closely and seemed offish.

off-island, *adv.* (*phr.*), *sb.*, and *a.* [f. OFF *prep.*, *OFF- 4 b + ISLAND *sb.*] **A.** *adv.* (*phr.*) Away from an island; spec. *U.S.*, away from the island of Nantucket.

1917 *Dialect Notes* IV. 335 *Off island*, elsewhere than on the island (Nantucket). 'What would I want to go *off island* for?' **1971** *N.Y. Times* 27 June 3 One islander was heard to remark recently that he never carried more than 30 cents in his pocket unless he planned to go off-island.

B. *sb.* An island off the shore of a larger or central island.

1969 *Sunday Times* 23 Feb. 63 The off-islands may seem to be just across the nautical street but the journey can still be an experience on a rough day. **1973** W. J. BURLEY *Death in Salubrious Place* i. 14 The pleasure boats were on their various ways to the off-islands. **1973** *Publishers Weekly* 19 Nov. 61/1 Letty Ward lives with her parents on Innish, an Irish off-island. **1976** *London Calling* Apr. 2/1 The Isles of Scilly... The four inhabited 'off-islands', as they are always called, are Tresco,..St Martin's, St Agnes and Bryher.

C. *adj.* Visiting or temporarily residing on an island.

1965 *New Yorker* 13 Feb. 42 This upset afforded small comfort to off-island Republicans here..because while the Guamanians are American citizens, they cannot vote in our Presidential elections.

So **off-islander**, (*a*) a visitor or temporary resident on an island; spec. *U.S.*, in Nantucket; (*b*) an inhabitant of an off-shore island.

1882 J. G. AUSTIN (*title*) Nantucket scraps: being the experiences of an off-islander, in season and out of season, among a passing people. **1935** [see *NANTUCKETER]. **1939** S. CHAMBERLAIN *Nantucket* 4 Artists and summer visitors have discovered its allure, and many fortunate 'off islanders' now live in old Nantucket houses. **1961** D. M. DOUGLASS *Saba's Treasure* (1963) vi. 98 Off-islanders like other harbors better. **1967** *Daily Tel.* 17 May 21/1 Mr. John Knott, Conservative MP for St. Ives, Cornwall, is to meet representatives of the Duchy of Cornwall, the Scillies Council, the Steamship Co. and 'off-islanders' on Saturday. **1968** *Time* 26 July 67 Many 'off-islanders', the regular summer residents, are concerned lest their historic hideaway lose its charm.

off key, off-key, *adv.* (*phr.*) and *a.* [f. OFF *prep.*, *OFF- 4 b + KEY *sb.*[1]] Without tonal organization; out of tune; also *fig.*, wrong(ly), inappropriate(ly).

1929 M. LIEF *Hangover* 235 Eulalia Duncan sang so off-key last night that she had great difficulty moving her vowels. **1943** R. CHANDLER *Lady in Lake* xxxi. 171 There's something a little off key about everything you do. **1952** *Sci. Amer.* May 65 Friends..who sing everything in a monotone or in the same off-key pattern. **1953** *N.Y. Times* 29 Jan. 25/2 Miss Davis' performance in this scene is foolishly false—a travesty of sexy movie acting, illogical and wholly off-key. **1965** G. MCINNES *Road to Gundagai* iii. 53 A faintly off-key piano. **1971** [see *OFF BASE *phr.* (*adv.*) and *a.*]. **1973** J. WAINWRIGHT *Devil you Don't* 18 He had a peculiar, off-key voice.

offlap (ǫ·flæp). *Geol.* Also **off-lap.** [f. OFF *adv.* + LAP *v.*[2], after OVERLAP *v.* 3.] A progressive diminution in the lateral extent of conformable strata in passing upwards from older to younger strata, so that each stratum leaves a portion of the underlying one exposed; a set of strata exhibiting this.

1913 A. W. GRABAU *Princ. Stratigr.* xviii. 734 (*caption*) Diagram illustrating regressive overlap (off-lap) and the formation of a sandstone of emergence..into which the shore-ends of the successive members of the retreatal series..grade. **1948** *Jrnl. Geol.* LVI. 147/1 Marine blanket sands represent the horizontal welding of many parallel prismatic shore lines during a long, continuous period of over-lap or off-lap. **1969** BENNISON & WRIGHT *Geol. Hist. Brit. Isles* i. 14 The offlaps of marine sedimentation are accompanied, in this case, by lateral and vertical gradation into beds of continental facies. **1975** *Nature* 10 Jan. 107/1 The stratigraphic relationship of the Devensian

Tills of eastern England is therefore one of offlap..and not of overlap.

So **o·fflapping** *ppl. a.*

1906 A. W. GRABAU in *Bull. Geol. Soc. Amer.* XVII. 615 Examples of regressively overlapping or, better, off-lapping formations are frequently met. **1921** —— *Textbk. Geol.* I. xvii. 559 If, after a retreatal movement of the sea and the formation of an offlapping series.., a transgressive movement with overlapping of formations should follow, the sandstone bed of emergence would be in part reworked. **1974** *Nature* 15 Nov. 200/1 The sealevel curve is based on the series of transgression-regression cycles identified from successions of offlapping coral reefs.

off licence, off-licence. [OFF *a.* 5.] A shop or other establishment where alcoholic liquors are sold for consumption off the premises; also, a licence permitting such sales. Also *attrib.* Hence **off-licensed** *ppl. a.*; **off-licensee.**

1891, 1897 [see OFF *a.* 5]. **1907** *Daily Chron.* 16 Apr. 3/6 The number of off-license [*sic*] premises at the beginning of 1906 was 25,281. **1959** J. BRAINE *Vodi* iv. 55 It was an off-licence, very cool and smelling both earthy and antiseptic, with overtones of stale beer and tobacco. **1961** E. A. POWDRILL *Vocab. Land Planning* iii. 42 To double the population is to increase considerably the range of shops that are required, for instance, cafes, off-licences, adies' shops, etc. **1971** *Oxford Times* 15 Oct. 9 (Advt.), From our off-licensed branches. Guinness—small cans. 9p. **1975** J. AIKEN *Voices in Empty House* iii. 105 The old lady tottering..to the off-licence on the corner. **1978** *Morecambe Guardian* 14 Mar. 4/1 (Advt.), We cordially invite all Hoteliers Restaurateurs Off-Licensees, etc. to visit our new Wines & Spirits department.

off limits, off-limits, *adv.* (*phr.*) and *a.* orig. *U.S.* [f. OFF *prep.* + *pl.* of LIMIT *sb.*] Of an area, place, etc.: outside the limits within which a particular group or class of people must remain; not to be frequented or patronized, esp. by military personnel; out of bounds. Also *transf.* and *fig.*

1952 R. CUTFORTH *Korean Reporter* xv. 137 Over the door an official army notice barked in iron-mouthed print: Strictly Off Limits. **1959** *Amer. Speech* XXXIV. 155 As long as the infatuation lasts, the man is said to be *attached* and strictly *off limits* to other girls. **1960** *Encounter* Feb. 38/1 The Negroes do not blame the residents but know that..the white G.I.'s in effect declare a town off-limits. **1968** A. DIMENT *Bang Bang Birds* v. 58 We ain't going to take them out because they're brothels. If it was just that we'd mark them Off Limits to service personnel. **1968** *Economist* 5 Oct. 46/3 The federation's leaders have declared the UAW-Teamsters group to be off-limits, describing it as 'a dual labour organisation, rival to the AFL-CIO'. **1971** 'S. RANSOME' *Trap 6* (1972) xiii. 139, I didn't intrude into the investigators' office. So far as I'm concerned that's off limits. **1973** A. DUNDES *Mother Wit* 259 White females were off-limits to Negro males. **1975** *N.Y. Times* 15 Sept. 9/1 In this central highlands town, which houses the legion's Corsican command post, three bars are off limits.

off-line (stress variable), *a.* and *adv.* [f. OFF *prep.* + LINE *sb.*[2]] **A.** *adj.* **1.** Not situated or performed on a railway or by rail.

1926 HUEBNER & JOHNSON *Railroad Freight Service* xvii. 338 Many railroads maintain 'off-line' offices or agencies at important industrial and commercial centers which they reach via connecting lines. **1973** E. RATH *Container Systems* vii. 175 Bimodal containerization uses truck trailers or other highway vehicles or special containers... This service uses the rail network as the main artery and as prime source of railroad revenue. The off-line service is given as an accommodation to the freight market. **1973** *Sci. Amer.* Oct. 24/3 The achievement is made possible by off-line loading and automatic, positive control of all motions of the vehicle.

2. *Computers.* Not on-line; other than on-line.

1950 W. W. STIFLER *High-Speed Computing Devices* ii. 7 For other applications, off-line operation, involving automatic transcription of data in a form suitable for later introduction to the machine, may be tolerated. **1957** D. D. MCCRACKEN *Digital Computer Programming* xii. 157 The reel of tape may then be removed from the auxiliary ('off-line') tape unit, placed on a tape unit which is connected to the computer ('on-line'), and read in at high speeds. **1959** [see *IN-LINE *a.* 3 c]. **1967** C. BERNERS-LEE in Wills & Yearsley *Handbk. Managem. Technol.* 16 Off-line equipment is, typically, a keyboard device preparing a paper tape to be read into a computer later. An on-line keyboard, on the other hand, is one wired directly into the computer. **1969** *Sunday Times* (Colour Suppl.) 16 Feb. 22/3 The clearance and updating must be done when the computer is..'off-line', i.e. is uncoupled from any systems that the computer itself is controlling. **1972** *Accountant* 26 Oct. 530/2 Off-line operation at Hendon is through a Varian 620i computer and dual tape unit, transferring data to tape for processing later on the 370/145.

B. *adv. Computers.* With a delay between the production of data and its processing; not under direct computer control.

1950 W. W. STIFLER *High-Speed Computing Devices* ii. 7 Some teletype machines operate on line. Their operators are in instantaneous communication. Other teletype machines are operated off line, through the intervention of punched paper tape. **1959** J. K. BRIGDEN in E. M. Grabbe et al. *Handbk. Automation, Computation, & Control* II. xx. 4 Obviously all the equipment used for on-line purposes can also be used off-line. **1964** T. W. MCRAE *Impact of Computers on Accounting* i. 17 The tape is then.. transferred to a smaller computer and printed off-line on the output printer of that computer. **1967** D. WILSON in Wills & Yearsley *Handbk. Managem. Technol.* 43 Because removable media systems can store inactive data off-line, the total capacity of these systems..is virtually unlimited. **1970** *Computers & Humanities* V. 36 Although the plotter, when run 'off line', is relatively cheap, it is slow. **1971** *Sci. Amer.* Sept. 17/1 (Advt.), Until recently, the best solution was to use the complex mathematics of the Fourier transform and program a computer to do the complex signal analysis computations off-line.

off-load, *v.* Add: 'Also with stress on second syllable' and for *S. Africa* read 'orig. *S. Afr.*'. (Further examples.) Also *transf.* and *fig.*, to discard, get rid of, relieve oneself of (a person or thing). Hence **off-loa·ding** *vbl. sb.* and *ppl. a.*

1850 R. G. CUMMING *Five Yrs. Hunter's Life S. Afr.* II. xx. 82 Having off-loaded my waggon, I handed it over to Mr. Arnott, the resident blacksmith, to undergo repairs. **1903** KIPLING *Five Nations* 208 Ubique means 'Off-load your guns'—at midnight in the rain! **1916** *Chambers's Jrnl.* June 373/2 At railhead the train is off-loaded into the motor-lorries. **1942** W. S. CHURCHILL *Second World War* (1951) IV. I. ix. 146 It will..be necessary for the *Indomitable* squadrons to be off-loaded in Ceylon. **1949** *Sun* (Baltimore) 21 Feb. 4/3 The machinery was off-loaded and shipped on to Nottingham by railway. **1950** *Daily Progress* (Charlottesville, Va.) 30 Jan. 1/5 Salvage team personnel have completed offloading the Big Mo's ammunition and all beach gear has been placed. **1952** C. DAY LEWIS tr. *Virgil's Aeneid* v. 113 They enrolled the women for the colony, off-loaded the men who wanted To stay there. **1957** *Economist* 2 Nov. 429/1 The vehicle industry finds that it has to offload more of its rising production of cars and lorries on to the home market. **1958** *Times* 19 Mar. 14/2 Bales were off-loaded. *Ibid.*, Cotton was stolen from the Liverpool Docks in the off-loading of 12 different ships. **1968** M. WOODHOUSE *Rock Baby* v. 43 A Director who has to offload one of his staff and is embarrassed. **1971** C. BONINGTON *Annapurna South Face* iii. 30 This period being taken up with off-loading, Customs clearance and the journey across India. **1972** *Daily Tel.* 1 Apr. 3/2 It arrived at Heathrow as mishandled luggage, having been wrongly off-loaded in Rome from a flight from Australia to London. **1977** J. MCCLURE *Sunday Hangman* ii. 17 [He] side-stepped into the shadow of an off-loading Coke truck.

off microphone, off-microphone, *adv.* (*phr.*) and *a.* [f. OFF *prep.* + MICROPHONE.] Away from a microphone; that is distant from, or not facing, a microphone. Also colloq. abbrev. **off-mike.**

1937 *Printers' Ink Monthly* May 39/3 *Off mike*, a position away from a performer or performer's position away from the microphone. **1940** A. OBOLER *Fourteen Radio Plays* 201 Pa (*off mike*). Hangin' around with loafers in the gymnasiums! *Ibid.* 257 Dialogue or effect is off-microphone, in other words, not at full volume in relation to other speeches or effects. **1958** *Listener* 13 Nov. 799/1 Those who may have wondered what goes on in the studios before the red light comes on were given an amusing picture of off-mike life in Mr. Robert Kemp's comedy, 'Young Mother Hubbard'. **1959** *Ibid.* 30 Apr. 774/1 The story..forced Mr. Naughton to move his magnifying glass from the street and to take it, as it were, off-mike. **1962** *Times* 3 Apr. 15/2 Goonery is not the most evident attribute of his off-microphone personality. **1962** A. NISBETT *Technique Sound Studio* ii. 40 The pages should be turned well off-microphone. *Ibid.* 48 Off-microphone voices and spot effects could be blended with on-microphone voices. **1973** *Washington Post* 5 Apr. B. 2 Something electric goes on between the comedienne and her husband... There are off-mike private jokes. **1977** *Rolling Stone* 19 May 89/1 Coltrane is occasionally off-mike.

off'n, var. *OFFEN *prep.*

off-off, off-off-Broadway: see *OFF-BROADWAY *a.* and *sb.*

off-peak, *a.* [f. *OFF- 4 b + PEAK *sb.*[2]] That is not at the maximum; of or pertaining to a period of less than busiest use, consumption, business, etc.

1920 A. C. PIGOU *Econ. of Welfare* II. xv. 261 If it is impracticable to charge differential rates directly as between peak and off-peak service, this may sometimes be attempted indirectly by differentiation between continuous and intermittent services. **1924** *World Power* I. 4 (*heading*) Electric heating by off-peak load. *Ibid.*, The continental technical press..gives particulars of successful applications of off-peak power for night uses at low tariffs. **1930** *Engineering* 11 Apr. 493/2 Such systems are generally arranged for supply by off-peak current. **1933** *Discovery* Apr. 111/2 Power is sold in large quantities at a very cheap rate so long as it is used during 'off-peak' hours. **1958** *Oxf. Mag.* 6 Mar. 336/2 Televised religious discussions attract over a million viewers even at off-peak times. **1971** *Times* 2 Oct. 13/2 The new off-peak fares should actually increase carryings. **1973** *Times* 21 Nov. 19/5 To prohibit the use of off-peak electric storage heating will not help the overall fuel position. **1976** P. R. WHITE *Planning for Public Transport* v. 113 Operating costs incurred for specified service levels are related to the ratio of peak to off-peak demand.

offprint, *v.* Delete *rare* and add: Also with stress on first syllable. (Further examples.)

1951 *Catalogues MS. Collections Brit. Mus.* (verso rear cover) Offprinted from the Journal of Documentation Volume 7. **1952** M. MCCARTHY *Groves of Academe* (1953) i. 8 Maynard Hoar, author of a pamphlet, 'The Witch Hunt in Our Universities' (off-printed from the *American Scholar* and mailed out gratis by the bushel to a legion of 'prominent educators').

off-put. Add: (Further examples.) Also, one who puts off, procrastinates, or wastes time.

1866 S. GILPIN *Songs & Ballads of Cumberland* 57 It was just for an off-put. **1893–4** R. O. HESLOP *Northumb. Words* II. 509 *Off-put*, delay, one that puts off time. 'Ye'r jeest an off-put o' time, laddy.' **1923** G. WATSON *Roxburghshire Word-Bk.* 223 It was duist an offpit! A ken 'im owre weel!

Hence **off-put** *v. trans.*, to put off; to disconcert; to repel; **off-putting** *vbl. sb.* (further examples); **off-putting** *ppl. a.* (further examples); *spec.* creating an unfavourable impression, causing displeasure; **off-pu·ttingly** *adv.*, in an off-putting manner; so as to disconcert or repel.

1828 MRS. B. HALL *Let.* 25 June in *Aristocratic Journey* (1931) 295 Forgetting that we had exchanged dilatory off-putting habits of the South for the anticipation in appointments of the North. **1833** *Chambers's Edin. Jrnl.* II. 234/1 Weel, mistress,..this off-putting will do nae langer. **1866** W. GREGOR *Dial. Banffshire* 6 *Aff-pittan*, the act of procrastination. **1893–4** R. O. HESLOP *Northumb. Words* II. 509 *Offputtin*, putting off, procrastinating, dallying. 'He's varry offputtin.' **1935** 'A. BRIDGE' *Illyrian Spring* xvii. 221 Your face isn't in the least off-putting, except when you're cross. **1951** WALLACE & BAGNALL-OAKELEY *Norfolk* vi. 66 The approach from the station is very off-putting:..Victorian and later ribbon-development soon started along the station road. **1955** M. ALLINGHAM *Beckoning Lady* xii. 169 Very off-putting, homely finances. **1960** M. SPARK *Bachelors* viii. 120 'Shut up about Ronald,' Tim said. 'It's jolly off-putting.' **1961** *Times* 4 Nov. 9/4 There is something..off-puttingly impersonal about a typed letter. **1969** D. FRANCIS *Enquiry* ix. 122 Composed, cool, off-puttingly gracious, she looked ..flawless. **1970** *Guardian Weekly* 7 Nov. 15 The peculiarity of a faith that can..be so offput by the female of any species that not even a cow is allowed to pasteurise here. **1973** *Nature* 16 Mar. 212/3 The appearance of the book is a little off-putting. **1974** *Country Life* 18 July 174/1 All this may sound offputtingly sporty. **1977** *Verbatim* Dec. 10/1 In case the number of pages is off-putting, it should be mentioned that the page size is 8-1/2″ × 11″.

off-ramp. *U.S.* [OFF- 4.] A sloping one-way road leading off a main highway.

1954 HEWES & OGLESBY *Highway Engin.* viii. 160 Capacity of off-ramps is affected by the number of through vehicles using the right-hand expressway lane. **1966** M. & G. GORDON *Undercover Cat prowls Again* v. 40 He took the Hollywood Freeway... Turning right at the Balboa off-ramp, he passed a city golf-course. **1968** 'R. MACDONALD' *Instant Enemy* xxx. 192, I took the first off-ramp and drove to the Langstons' neighborhood. **1973** W. MCCARTHY *Detail* iii. 247 Stuart looked for an off ramp... The next turnoff was ten miles ahead.

offretite (ǫ·frĕtait). *Min.* Also † **offrétite.** [ad. F. *offrétite* (F. Gonnard 1890, in *Compt. Rend.* CXI. 1002), f. the name of Albert *Offret* (b. 1857), French mineralogist: see -ITE[1].] A hydrated aluminosilicate of potassium and calcium, which belongs to the zeolite group and is found as colourless hexagonal crystals at Mont Semiol, France.

1891 *Jrnl. Chem. Soc.* LX. 407 Offrétite is a zeolite which occurs in very small quantity, associated with christianite and chabasite, in the basalt of Mount Simionse [*sic*], near Montbrison. **1892** E. S. DANA *Dana's Syst. Min.* (ed. 6) 1043 Offrétite. **1894** *Amer. Jrnl. Sci.* CXLVIII. 188 Offrétite. **1967** *Nature* 3 June 1005/2 Offretite and erionite are evidently two distinct but closely related minerals, which can intergrow. **1970** *Trans. Faraday Soc.* LXVI. 1616 Synthetic unfaulted offretite can provide a wide pore molecular sieve which can sorb molecules up to ~7·1 Å in critical dimension... This zeolite appears to have possibilities as a versatile molecular sieve. **1975** *Nature* 28 Aug. 718/2 Offretite is a rare natural zeolite whose synthetic equivalent, zeolite O, is commercially important as a molecular sieve.

off-rhyme. [OFF- 4.] A partial or near rhyme.

1938 L. MACNEICE *Mod. Poetry* vii. 131 One can use internal rhymes, off-rhymes, bad rhymes, 'pararhymes'. **1944** —— *Christopher Columbus* 8 Your significant variations of rhythm, your internal rhymes, your off-rhymes and assonances and technical surprises, will get in the composer's way. **1956** E. WILSON *Piece of My Mind* (1957) vii. 124 Many kinds of liberties are countenanced—in the way of off-rhymes and irregular rhythms—in the writing of modern poetry. **1972** *Computers & Humanities* VII. 19 Taylor is not as unsystematic as might have been thought in his use of off-rhymes. **1972** J. WAIN in Cox & Dyson *20th-Cent. Mind* I. xi. 405 These half-rhymes of Owen's (para-rhymes, off-rhymes, even meta-rhymes, as they have been variously dubbed). **1977** *N.Y. Rev. Bks.* 9 June 29/4 Internal rhymes (trunks, sunk), off-rhymes (salt, silt), a lyrical vocabulary (sweetness, enchanted)—all these things are here, but none of them is obtrusive.

o·ff-saddle, *sb.* [f. the vb.] A break or rest in a journey during which horses are unsaddled.

1900 *Pall Mall Gaz.* 4 Jan. 1/3 The Cape horse..can canter along steadily all day under a burning sun, with an occasional off-saddle. **1908** *Daily Chron.* 10 Nov. 3/5, I had him by the hip at 'off-saddle' time.

OFFSADDLE

offsaddle, off-saddle, *v.* For *S. Africa* read: 'Chiefly *S. Afr.*' (Add earlier and later examples.) Hence **off-saddling** *vbl. sb.*

1837 F. Owen *Diary* (1926) 78 We off-saddled and sat on the ground on the outside of the fence. 1850 R. G. Cumming *Five Yrs. Hunter's Life S. Afr.* I. 119 Accordingly we off-saddled, and in a few minutes I was once more asleep. *Ibid.* 129 Having off-saddled our horses, we knee-haltered them. 1906 Rider Haggard *Benita* ix. 120 Directions as to their herding, and the off-saddling of the horses. 1915 Kipling *New Army* 22 The batteries off-saddled in silence. 1939 [see *knee-halter *v.*]. 1974 B. Mather *White Dacoit* ii. 22 They halted and off-saddled and picqueted the horses half a mile downstream.

off-sale. [Off *a.* 5.] The sale of alcoholic liquors for consumption elsewhere than at the place of sale.

1899 [see Off *a.* 5]. 1933 *Sun* (Baltimore) 17 Apr. 4/4 The local act also holds up issuance of an 'off sale' of beer in packages by Dressman Bros., grocers. 1963 *Times* 23 Apr. 20/3 Sales of bottles to take home ('Off-Sales') have increased very considerably, and there has been a reduction of sales by the glass. 1970 *Times* 18 Aug. 20 M. & B. wants to stop off-sales in favour of the 23 Wine Sellers off-licences it has opened in Birmingham. 1977 'J. Fraser' *Hearts Ease* ix. 104, I see you're licensed for off-sales... Does that mean you have any wines?

offscouring. 2. b. (Further examples.)

1835 J. P. Kennedy *Horse-Shoe Robinson* I. xiv. 180 Why, you off-scouring,..it is enough to make Old Scratch laugh, to hear you talk about conscience! 1871 *Scribner's Monthly* II. 546 Every Protestant is counted but the off-scouring of decent society. 1928 H. W. Shoemaker in *Publ. Pennsylvania Folk-Lore Soc.* I. iv. 10 That dressy offscouring will not come back for her... I've known him and his kind too well for more than fifty years. 1972 F. W. Lindsay *Cariboo Dream* 13/2 Thugs, gamblers and the off-scourings of the world.

off screen, off-screen, *adv.* (*phr.*) and *a.* [f. Off *prep.* + Screen *sb.*[1]] Not appearing on a film or television screen; = *off camera *phr.* and *adv.* Also *fig.*

1935 *Motion Picture* Nov. 15/1 In addition, of course, you are the perfect lady, off-screen. 1953 K. Reisz *Technique Film Editing* i. 62 A reaction shot is shown and held while the character off screen is speaking his lines. 1958 *New Statesman* 9 Aug. 168/1 There were, of course, a few minor, off-screen, political controversies. 1962 *Guardian* 14 Aug. 5/3 This disarmingly frank and articulate off-screen comedian. 1973 'E. McBain' *Let's hear It* xiv. 209 The actors made important plot points offscreen, their voices floating in over the picture. 1974 *Times* 10 Apr. 13/8 The off-screen audience, real or canned, goes into mad paroxysms.

off season, off-season. See Off *a.* 4. Also, a time of year other than the busiest or most popular time for a particular activity. Also *attrib.* or as *adj.*

1848, 1868 [see Off *a.* 4]. 1897 [see *high *a.* 4 a]. 1902 *Chambers's Jrnl.* July 433/2 The collection was not one of those which bore an historic name..and the sale took place in the off-season. 1905 A. Bennett *Let.* 5 Apr. (1966) I. 61, I have a great notion of books being issued in off-seasons. 1908 *Daily Chron.* 21 Apr. 3/5 A great part of Miss Elwin's off-season work consists in meeting girls on their way to and from school. 1930 *B.B.C. Year-bk.* 1931 25 During the summer period, normally regarded as an 'off season', the average number of new licences taken out each month..was in excess of 20,000. 1957 L. F. R. Williams *State of Israel* 44 The coaches..are primarily intended for tourists, but..carry domestic traffic in the off-season. 1959 [see *budget *sb.* 4]. 1971 W. J. Burley *Guilt Edged* iii. 59 The newcomers..disturbed the off-season peace of the Treen Hotel. 1976 *Washington Post* 19 Apr. D 1/6 With the series clinched, the Caps looked as though they were running their motors down for the offseason that begins when they fly off for a week's vacation in Hawaii.

offset, *sb.* Add: **2.** (Earlier and later examples.)

1629 J. Parkinson *Parad.* xi. 114 The root is..compassed with a number of small rootes, or of of-sets round about it. 1925 A. J. MacSelf *Bulb Gardening* xiv. 215 Almost all bulbs and corms throw what are termed offsets, which are first attached to the basal wing of the bulb, drawing nourishment from the parent bulb, but later developing an independent root system. 1970 *Sunday Tel.* 3 May 19/2 The striped pineapple plant..which I have known having to be guarded at Chelsea Show for fear of the offsets that slowly form the new plants getting missed.

3. e. *Naut.* A current flowing outwards from the shore. Also *attrib.*

1902 *Daily Chron.* 30 Aug. 5/6 There was..what maritime men call an offset at the time Holbein was swimming. *Ibid.* 2 Sept. 5/5 He had got the benefit of a good off-set current under him.

5. (Earlier and later examples.)

1769 *Conn. Col. Rec.* (1885) XIII. 207 A petition..setting forth that the petitioner and petitionee have executions against each other now in the hands of Ezekiel Williams,..upon which the petitioner prays for an off-set of the same. 1932 *Daily Progress* (Charlottesville, Va.) 13 Aug. 1/3 Mr. Minor's action is what is known as an offset. He is asking the difference between $5,000 and the amount he alleges is due to him, or slightly more than $21,000. 1973 *Times* 11 June 14/1 Instead of being 50p better off, she gained only 15p a week. The High Court ruled that this automatic offset, which was standard commission procedure, was unlawful.

7. b. (Further examples.)

1856 *Porter's Spirit of Times* 18 Oct. 106/3 The hearth-stone, garden, 'offsets',..and a thousand delicious memories..swarm before us now. 1873 J. H. Beadle *Undevel. West* xxvi. 555 He then walked along a flat offset five or six feet below the house. 1878 —— *Western Wilds* x. 157 About half way up the cliff is a small offset, where grows a beautiful pine.

10. b. Used, freq. *attrib.*, of a printing process whereby an image is first impressed on a rubber-covered cylinder and thence transferred to the paper; also used *attrib.* of paper suitable for use in this process.

1906 *Brit. Printer* Oct. 283/2 The offset process—that is, working first on the blanket and offsetting from this on to the paper fed into the machine—is as the usual method. 1909 *Brit. Pat.* 25,446 4 Nov., For printing in lithographic machines by the so-called offset process, the lithographic design used as the basis of printing is generally the opposite or the reverse of that which would be used in ordinary direct lithographic printing... By our invention we provide a press..for conveniently reversing designs or pulling proofs from reversed designs, such works being carried out by the offset method. 1918 *Pall Mall Gaz.* 29 June 8/5 A Litho Offset Press. 1921 *Dict. Occup. Terms* (1927) § 526 *Offset-litho operator*, a photographer..who obtains a negative, not reversed in position, with or without using a screen. 1926 *Encycl. Brit.* III. 220/2 By the offset method any paper, whether smooth- or rough-surfaced, may be used... Offset seems to be particularly adapted to colour-work. 1928 *Penrose Ann.* 111 The faulty soluble coating in so many classes of so-called offset papers, engenders printing troubles. 1929 *Horse* I. 61 The process employed in reproducing this notable painting is that known as offset. 1931 *Times Lit. Suppl.* 25 June p. viii/3 The rollicking offset-lithographs of Elsa Moeschlin. 1937 *Discovery* Oct. 297 An offset—a print made in a non-greasy powdered chalk. 1958 J. R. Biggs *Woodcuts* 91 Offset cartridge, that is, the paper made for printing offset lithography, has a kindly surface, and a firm feel which is suitable for some subjects. 1958 *Times Lit. Suppl.* 4 Apr. 188/3 Dr. Willoughby's book, printed by offset from type-written script. 1967 Karch & Buber *Offset Processes* Pref., Although growth for the entire [printing] industry has been great, offset-lithography has grown at twice the rate of the industry as a whole. *Ibid.* ii. 12 Letterpress printed halftones..can often be..rescreened for printing by the offset-lithographic process. 1970 [see *flexography]. 1976 *Sci. Amer.* Jan. 135/1 This work is unappealing typographically, dominated by crowded typewritten copy in grayish offset.

11. offset purchase *Econ.*, a purchase made abroad by agreement to counter-balance revenues spent in the buying country by the selling country; **offset well** (see quot. 1971); also *ellipt.* **offset.**

1966 *Economist* 20 Aug. 731/2 The Germans continue to be slow in making their military 'offset' purchases to counter the dollars which are drained away by the cost of keeping American troops in Germany. a 1974 R. Crossman *Diaries* (1975) I. 550 There was an appalling incident in the House when he was caught out about our attempt to pay for the new American planes with offset purchases. 1922 *Petroleum Gloss.* (Pan Amer. Petroleum & Transport Co.) 32 *Offset well*, a well drilled opposite to a well on an adjoining property. 1924 *Amer. Review of Reviews* Mar. 231 Adjacent to one of the California reserves were oil lands that were under active development, with wells near enough to the boundary lines of the naval territory to cause a draining away of a certain amount of oil that lay beneath the navy's property. Under such circumstances, it is customary to sink what are known as 'offset wells'. 1932 *World Petroleum* III. 467/1 Production in the light oil zone at the discovery well in the center of the field..has since been considerably augmented by the completion of offset wells. 1971 Williams & Meyers *Oil & Gas Terms* (ed. 3) 294 *Offset well*, a well drilled on one tract of land to prevent the drainage of oil or gas to an adjoining tract of land, on which a well is being drilled or is already in production. 1974 R. D. Grace in P. L. Moore et al. *Drilling Practices Manual* 349 Deviation to 10 degrees was recorded in the north offset.

offset, *v.* Delete 'Chiefly *U.S.*' and add: **1.** Also, to set off against part of something else. (Further examples.)

1930 G. B. Shaw *Androcles & Lion* Pref. 58 The triumphant solution of the first by our inventors and chemists has been offset by the disastrous failure of our rulers to solve the other. 1936 *Discovery* Sept. 278/1 To have a large family, it is essential for the parents to start at an early stage, and this is offset by the tendency to postpone marriage. 1968 *Lebende Sprachen* XIII. 51/2 The customer can offset claims on the bank only against liabilities in the same currency. 1975 E. F. L. Brech *Princ. & Pract. Managem.* (ed. 3) v. x. 813 Companies were required to deduct tax at the standard rate from all dividends paid... The only exception to this rule was if the company had itself received a dividend under deduction of tax from some other UK company... The tax on the latter could then be offset and only the net balance was remittable.

off-set (ǫ·fset), *ppl. a.* [f. Off *adv.* + Set *ppl. a.*] Set at an angle; placed off-centre or out of line.

1950 H. L. Lorimer *Homer & Monuments* v. 169 It lacks the off-set rim characteristic of the hoplite shield. 1950 *N.Z. Jrnl. Agric.* June 538/1 Equipment was further improved by acquiring sets of both tandem and off-set discs. 1960 *Farmer & Stockbreeder* 29 Mar. 76/3 This new 43in-cut offset harvester, known as the Taarup Tiger, weighs only 9½cwt.

off-set, *adv.* [f. Off- 2 + *set *sb.*] Out of range of the cameras in a film or television set or studio.

1948 L. Levy *Music for Movies* ii. 13 When we had decided on a piece of music to accompany any one scene, the band would play off set while the actor went through his actions or lines in time with the music. 1968 M. Woodhouse *Rock Baby* vii. 64 In the next shot he'd gone off-set, the camera had moved in.

offsetting, *vbl. sb.* and *ppl. a.* (Further examples.)

1857 Dickens *Perils Eng. Prisoners* iii, in *Househ. Words* Extra Christmas No., 7 Dec. 30/2 The off-settings and point-currents of the stream. 1908 *Daily Chron.* 24 Aug. 9/7, I was swimming against a strong offsetting tide. 1946 E. Hodgins *Mr. Blandings builds his Dream House* (1947) xvi. 208 The Blandings saw their house..in all its gleaming whiteness, the delicate waving leaves..the perfect offsetting foils. 1970 P. Oliver *Savannah Syncopators* 28 The offsetting of the rhythms of the different drums against the gongs and against each other, sets up an exhilarating tension. 1975 *Times* 10 July 4/8 Mr Callaghan..sees the release of Mr Hills as unconditional. No offsetting arrangement will be discussed.

off shore, off-shore. Add: **B.** *adj.* **b.** (Further examples.) Also, away from the mainland. Esp. in *off-shore island*, an island close to the mainland; *spec.* (*a*) any of a number of small islands off the coast of China, in the Formosa Strait; (*b*) Great Britain, jocularly regarded as an 'off-shore island' of Europe; hence *off-shore islander.*

1921 *Daily Colonist* (Victoria, B.C.) 5 Apr. 7/2 The seas were breaking so high over the bar here yesterday and today that some off-shore shipping was compelled to remain outside. 1946 *Sun* (Baltimore) 30 Jan. 11/3 Drastic measures should be taken to complete whatever trade negotiations and shipping are necessary to hasten receipt of raw sugars and off-shore refined sugars. 1958 *New Statesman* 30 Aug. 241/2 It is impossible to say whether it is intended merely to put a stop to the various patrol activities which Chiang's forces mount from the Quemoy and Matsu groups of islands; whether it is a prelude to the occupation of these off-shore islands, or whether, possibly, we are witnessing the first stage in the invasion of Formosa itself. 1959 M. Laski (*title of play*) The offshore island. 1963 *Listener* 14 Feb. 310/2 Reflect (as often) on the advantages of being an off-shore islander living in an intellectual fog where there is no black and white. 1966 *Economist* 8 Jan. 119/3 British Petroleum is busily finding itself another offshore rig to replace the defunct 'Sea Gem'. 1968 *Listener* 29 Feb. 264/1 The Empire had gone: England was an offshore island in competition with the giants of the world. 1972 *Guardian* 16 Aug. 12 (*heading*) View from an offshore island. Linda Christmas on the Isle of Man. 1972 P. Johnson *Offshore Islanders* 79 William I's work in rebuilding the Old English monarchy was..continued by an Angevin who became a thorough offshore islander in his turn. 1973 *Guardian* 17 Jan. 10/1 The purpose of Mr Peter Walker's new Offshore Supplies Office is..to bring more work and wealth to Britain. *Ibid.* 11 Opportunities for British industry from offshore oil and gas developments. 1973 C. Callow *Power from Sea* iv. 89 All the big off-shore pipelines have been supplied by foreign groups. 1974 E. Ambler *Dr. Frigo* i. 45 Off-shore oil? Is that what they were looking for? 1975 *Petroleum Rev.* XXIX. 397/3 With the North Sea rapidly becoming the area for the most intensive offshore diving activity in the world, the growing pains of this..profession are felt..in the UK. 1975 *Weekend Mag.* (Montreal) 15 Nov. 7/3 Jamaicans are luckier than most offshore workers. Because their government has a tax convention with Canada, they are exempt from paying income tax and Canada pension.

c. Of, pertaining to, or designating goods purchased with American dollars by and from countries other than the United States (see quots.). Also, designating the dollars used in this way.

1948 *Economist* 8 May 768/1 The 16 nations will be provided with 'off-shore' dollars for buying from Germany. 1949 *Times* 10 Sept. 5/7 Off-shore purchases, the name given to supplies to countries in receipt of Marshall aid, which are financed by Marshall aid but which do not come from the United States itself. 1952 *Economist* 13 Dec. 759 An off-shore purchase, as its name implies, is the buying of goods from countries other than the United States with dollars supplied by the United States. 1953 *Ann. Reg.* 1952 220 'Off-shore' orders for military equipment were placed with Italian industry. 1960 *New Left Rev.* July-Aug. 45/2 The buoyancy of the Japanese economy..has been..guaranteed by massive off-shore purchases of military equipment.

d. *offshore funds* (see quot. 1972).

1972 *Observer* 8 Oct. 18/1 Offshore funds, investment funds similar to unit trusts but registered abroad, usually in countries with advantageous tax situations. 1977 *Times* 29 Nov. 22/5 (*heading*) Authorized Units, Insurance & Offshore Funds.

off side, off-side, *phr.* Add: **b.** Also with stress on second syllable. (Further examples.) Also *quasi-sb.* An occasion on which a player is, or becomes, off side.

1887 in B. James *England v Scotland* (1969) ii. 50 Dewhurst shot the ball in, and Cobbold helped it past Macauley, an appeal for offside by Dewhurst being disallowed. 1899 A. H. Quinn *Pennsylvania Stories* 24 Claims of 'off-side' were freely made and repudiated by the captains. 1925 *Kansas City* (Missouri) *Star* 22 Nov. 16/4 The ball was called back, and Kansas was penalized five yards for off-side. 1972 G. Green *Great Moments in Sport: Soccer* i. 24 Menti..twice put the ball in the England net, but each time the 'goal' was disallowed for offside. 1976 *Wymondham & Attleborough Express* 19 Nov. 24/4 Hempnall were not to be denied and hit back immediately, Borroughes beating the offside trap to convert a fine pass from Tweedale. 1976 *Norwich Mercury* 10 Dec. 8/5 However, their well-drilled off-side trap prevented Thurton from ever really getting to grips with the bone-hard ground and getting any rhythm into their game.

off-side, offside, *v. Austral. colloq.* [Back-formation from *off-sider.] *intr.* To act as an off-sider or assistant. So **off-siding** *vbl. sb.*

1883 in M. Durack *Kings in Grass Castles* (1959) xxvi. 256, I have put up a yard on Galway since Uncle Jerry left—Pumpkin and Kangaroo offsiding. 1917 R. D. Barton *Reminisc. Austral. Pioneer* vii. 93, I met a blackfellow..who was offsiding for the horsedriver, and was called Archie. 1930 V. Palmer *Passage* i. v. 41 A boy like Hughie might as well climb into his coffin right away as go off-siding for old Kunkel. 1936 I. L. Idriess *Cattle King* iii. 18 I'll get you a job offsidin' for me. 1953 A. Upfield *Murder must Wait* iv. 30 He's up at Mitford, and has asked me to send you up to off-side for him. 1960 —— *Bony & Kelly Gang* 167 Bony was asked to offside for Joe Flanagan, the settlement's electrician.

o·ff-sider, offsider. [f. Off side *phr.* + -er[1].] **1.** An animal on the off side (see Off *a.* 2 a) of a team (see also quot. 1898); hence, an assistant, *spec.* a bullock-driver's assistant; a companion, deputy, partner. orig. and chiefly *Austral.* and *N.Z. colloq.*

1880 H. C. Kendall *Songs from Mountains* 41 And, as to a team, over gully and hill He can travel with twelve on the breadth of a quill, And boss the unlucky 'offsides'. 1898 *Bulletin* (Sydney) 17 Dec. (Red page), In draught work the laziest or worst horse is on the off-side, under the whip; thus it is an inferior man or thing that is a *bit off, off color*, or *an offsider*. 1903 'T. Collins' *Such is Life* 46 Jam him agen the off-sider, so's he can't shift. 1905 A. B. Paterson *Old Bush Songs* 6 They say there's no delay To get an off-sider For the old bullock dray. *Ibid.* 8 An offsider is a bullock-driver's assistant. 1924 'R. Daly' *Outpost* xx. 191 A high and mighty gentleman is Mr. Whelan nowadays! Seems to think I'm only an off-sider, and that miners are the only people who know anything about the country. 1929 W. Smyth *Girl from Mason Creek* 206 The Maori boy who acted as the cook's offsider. 1931 V. Palmer *Separate Lives* 244 Shorty's advent as offsider did nothing to help matters. He and the bullock-driver had long arguments on the way. 1934 T. Wood *Cobbers* 19 Friends with everybody though, and an offsider of the Governor, too. 1942 'M. Innes' *Daffodil Affair* ii. 46 Too right, Mr Wine. You couldn't have a better off-sider than Ron. 1952 A. Grimble *Pattern of Islands* 243 There was nothing for it but to leave me hanging around at Ocean Island as offsider to anyone the High Commissioner might send up to take charge. 1962 *John o' London's* 19 July 64/2 Jill Dawson takes a job as an offsider—'more than a secretary and different'. 1973 *Nation Rev.* (Melbourne) 31 Aug. 1451/1 'Would you wait outside,' 'Salisbury' now said to his offsider. 'I'd like a word with matron in private.' 1977 *National Times* (Austral.) 17 Jan. 5/5 He now works as a truckie's offsider.

2. In certain games: a player who is off side. *rare.*

1927 *Daily Tel.* 21 Feb. 14/1 Nor did the referee have an eagle eye for the offsider.

off spin, o·ff-spi·n. *Cricket.* [Off *a.* 2 b.] Spin that causes the ball to turn from the off side towards the leg side; bowling with this spin. Also *attrib.* Hence **off-spinner,** (*a*) a bowler who causes the ball to turn in this way; (*b*) a ball bowled with this spin; **off-spinning** *ppl. a.*

[illegible]

off stage, off-stage, *adv.* (*phr.*) and *a.* [f. Off *prep.*, *off- 4 b + Stage *sb.*] Away from the stage; that is not appearing or occurring on a stage (see also quot. 1952). Also *transf.* and *fig.*

[illegible] from what is really going on. 1952 Granville *Dict. Theatr. Terms* 125 *On-stage.* Like off-stage, this is a somatic adjective. The on-stage arm is the one nearest the centre of the stage when an artiste faces the audience. Thus, in position *stage left*, the right will be the on-stage arm, and the left the offstage one... The same applies to furniture and properties so positioned. 'Bring the settee a little more off-stage' (i.e. towards the side). 1959 J. Wain *Travelling Woman* viii. 110 The philosopher emptied one final shovelful of ashes into the bucket, carried it off to some unknown destination offstage, came back. *Ibid.* ix. 128 You didn't know me. I was just a figure off-stage somewhere. 1961 J. McCabe *Mr. Laurel & Mr. Hardy* (1962) i. 31 If any of America's millions had seen the off-stage Charles Spencer Chaplin of 1910, they might well [etc.]. 1967 M. Argyle *Psychol. Interpersonal Behaviour* vii. 125 People do not work at their image-projection all the time; there is a difference between being 'on-stage' and 'off-stage'; in the former, people feel under observation and are very concerned about the image they are projecting. 1975 *New Yorker* 21 Apr. 103/3 The climax is the obligatory offstage pistol shot.

off-street, *a.* [f. *off- 4 b + Street *sb.*] That is not on a street; esp. of parking facilities: located away from main streets. Also, that does not take place on a street.

1929 *Amer. City* Sept. 133/1 Off-street loading facilities ..reducing traffic congestion. 1937 *Sun* (Baltimore) 4 Dec. 8/1 Provision for off-street parking..should help solve the parking problem in that vicinity. 1947 *Evening Sun* (Baltimore) 12 Sept. 2/5 Fringe parking on the perimeter of the business district would help relieve traffic congestion whereas providing more off-street space for automobiles in the heart of a city would actually aggravate the problem. 1958 *Times* 1 Aug. 5/5 The provision of authorized long-term street parking places in the vicinity of railway stations where there are insufficient off-street facilities. 1969 [see *feed *v.* 7 b]. 1977 *Stornoway Gaz.* 27 Aug. 1/5 The Council say the proposals contain inadequate provision for the off-street servicing of the premises.

offtake. Add: **1.** (Further examples.) Also, the removal of oil from a reservoir or supply. Also *attrib.*

1955 *Times* 12 May 17/2 Producers have not been keen recently to tender metal for the stockpile because of the high rate of commercial off-take. *Ibid.* 25 June 9/1 A reduced offtake at home has ensued on top of a persistent decline in export sales. 1960 *Farmer & Stockbreeder* 19 Jan. 5/1 Only a moderate offtake is reported for English barley but offers remain generally small. 1969 *Hindu* (Madras) 3 Aug. 1/5 With the opening of free market shops, the off-take from the existing ration shops would go down. 1973 'D. Jordan' *Nile Green* iv. 23 We might work out some sort of guarantee based on the projected offtake of their new oil discoveries in the Western Desert. 1974 *Information Handbk. 1974–5* (Shell Internat. Petroleum Co.) 73 In Spain the Amposta field..started production in February 1973 and at the end of the year had averaged 16 800 b/d: owing to offtake difficulties, well below the scheduled 30 000 b/d. 1975 *Petroleum Rev.* XXIX. 317/1 Links with local supply systems are effected at small offtake stations, where the pipework and associated control valves appear above ground.

off target, off-target, *adv.* (*phr.*) and *a.* [f. Off *prep.*, *off- 4 b + Target *sb.*[1]] So as to fail to reach a target; that misses a target (freq. *fig.*).

1959 J. L. Austin *Sense & Sensibilia* (1962) ix. 89 The second example is ineffective, off-target, in a rather similar way. 1960 V. Jenkins *Lions Down Under* 112 His kicking, for once, was off-target. 1967 *N.Y. Herald Tribune* (Internat. ed.) 11–12 Feb. 10/6 The 'Daily Mail' carried Philip's entire speech with the headline 'Never Before a Royal Speech as Outspoken as This' and said in an editorial it thought he was way off target. 1977 R. E. Harrington *Quintain* vi. 54 Though it was off-target, and glancing, the blow rocked him back.

off-the-peg: see *peg *sb.*[1] 1 e.

off time, off-time, *sb.* [Off *a.* 4.] A time when one is off duty or free from work; a holiday; a time when normal activity is reduced or suspended.

1866 Dickens *Mugby Junction* in *All Year Round* Extra Christmas No. 10 Dec. 6/1 The answer to his inquiry, 'Where's Lamps?' was..that it was his off-time. 1894 Mrs. H. Ward *Marcella* II. iii. iv. 323 In hospital..every hour was full, and there were always orders to follow. And the 'off' times were no trouble—I never did anything else but walk up and down the Embankment..or go to the National Gallery. 1902 *Westm. Gaz.* 26 Aug. 8/1 The York Meeting..marks the close of the 'off' time, which has been experienced since the Goodwood gathering. 1936 W. R. Titterton *G. K. Chesterton* ii. iii. 139, I used the stage there at off times to advertise the plays that we ran. 1940 C. Day Lewis tr. *Georgics of Virgil* i. 24 Winter's an off-time for farmers. 1956 'J. Wyndham' *Seeds of Time* 216 In his off times he occupied himself..in teaching Lellie. 1975 *BP Shield Internat.* May 1/3 We are insisting on their having their homes within 70 miles of Aberdeen because, should there be an emergency, then they can be easily called in during their off-time to deal with it.

off-time, *a. U.S. slang.* [f. *off 4 b + Time *sb.*] Out of time; badly timed (see also quots.).

1942 Berrey & Van den Bark *Amer. Thes. Slang* § 582 *Off-time jive*, a sorry excuse. 1946 Mezzrow & Wolfe *Really Blues* xiii. 237 If somebody passed a remark that wasn't in line, he'd start singing and beating on the offtime cat. *Ibid.* 377 *Offtime*, out of harmony, old-fashioned, corny, offensive. 1973 C. Himes *Black on Black* 188, I can stop in front of a joint where the juke-box's playing and cut a step of off-time boogie.

off-white, *sb.* and *a.* [*off- 5.] **A.** *sb.* A colour very close to white, usu. with grey or yellow tinge. **B.** *adj.* Almost white. Also *fig.*, not standard; not socially acceptable; *off-white collar*: of a worker or occupation: not manual but not quite 'white collar'.

1927 *Daily Mirror* 10 Dec. 16/1 Jumper suits in white, yellow, or what the Paris dressmakers call 'off whites' will see you through. 1931 *Daily Express* 18 Mar. 5/3 Fashionable colours are all off-white shades such as palest blues, greys, pinks, and greyish-greens. 1937 F. Stark *Baghdad Sketches* 187 A voluminous nightdress very much 'off white'. 1951 H. Nicolson *Diary* 29 Aug. (1968) 208, I go to the B.B.C. to listen to recordings of King George's broadcasts. His voice is so like the present King's...it is what the B.B.C. would call 'off-white', meaning slightly cockney. 1954 M. Procter *Hell is City* iii. ix. 120 The interrogation room..had white-tiled walls and an off-white ceiling. 1962 *Guardian* 13 Apr. 1/7 Off-white-collar workers..are showing an unpatriotic reluctance to subsidise richer men's meals. 1962 A. G. Frank in *Monthly Rev.* (N.Y.) Nov. 384 There is movement into white or off-white collar jobs. 1962 M. Allingham *China Governess* (1963) iii. 51, I can be as 'off-white' as I like. I've no code to live up to. 1965 —— *Mind Readers* ix. 94 'You're telling me!' The voice could produce an off-white accent. 1969 Y. Carter *Mr. Campion's Farthing* ii. 9 A very slight off-white accent... A suggestion of the lower orders. 1971 *Nature* 15 Jan. 175/2 It yielded about 200 mg of pure wildfire toxin as a fluffy, off-white powder.

off-wring, *v. poet. rare*—1. [Off- 1.] *trans.* To wring or wrench off.

a 1889 G. M. Hopkins *Poems* (1918) 90 His twiny books Fast he opens, last he offwrings Till walk the world he can with bare his feet.

oficina. (Later examples.)

1942 *Econ. Geol.* XXXVII. 198 Fig. 9 shows an oficina that has derived its nitrate from one of the small 'islands' or 'peninsulas' rising above the playa. 1966 Pohl & Zepp *Latin Amer.* vii. 120 Most of the *oficinas* are in the north, where groundwater is available.

Oflag (ǫ·flag). [G., abbrev. of *offizier(s)lager* officers' camp.] In Nazi Germany: a prison-camp for captured enemy officers.

1941 [see *Ilag]. 1945 *News Chron.* 20 Apr. 1/6 Unlike the guards of the stalags and oflags, they never wait to give themselves up. 1958 P. Kemp *No Colours or Crest* xii. 265 Her husband..had been a prisoner of the Germans since 1939 in an Oflag in Posen. 1969 *Listener* 10 Apr. 503/3 Wing-Commander Wassler (who for practical purposes is still in his wartime Oflag). 1974 *Times* 12 Oct. 12/7 He found himself in Oflag IHA/H, a prisoner-of-war camp in Spannenburg.

oft, *adv.*, *a.* **A.** *adv.* **c.** (Further examples.)

1906 P. E. More *Shelburne Ess.* (4th ser.) 198 There are single lines here and there, such as the oft-quoted 'White arms out in the breakers tirelessly tossing', which have a magical power of evoking an image or the memory of subtle sounds and odors. 1922 Joyce *Ulysses* 269 Her first merciful lovesoft oftloved word. 1954 O. Nash *Face is Familiar* (rev. ed.) 115 The oft-quoted remark of the prominent and respectable dignitary. 1976 M. Butterworth *Remains to be Seen* i. 11 The wary air of an oft-disappointed augur reading entrails.

often, *adv.* and *a.* Add: Now also with pronunc. (ǫ·ftən). **A.** *adv.* **1. b.** In colloq. phrases: *(as) often as not, more often than not*, at least half the time; frequently; *every so often*: see *every *a.* 1 f; *once too often*: see *once *adv.* 8 d.

1911 G. B. Shaw *Getting Married* Pref. 149 Such a transaction..is as often as not the inauguration of a lifelong squabble. 1919 E. O'Neill *In Zone* in *Moon of Caribbees* (1923) 16 All them German spies they been catchin' in England has been livin' there for ten, often as not twenty years. 1960 *Observer* 25 Dec. 7/6 The driver often as not wore chauffeur's livery. 1962 M. Drabble *Summer Bird-Cage* xi. 185, I..use a small red plastic colander, and everything eels into the sink as often as not. 1977 F. Ross *Dead Runner* i. 64 More often than not the only successful outcome of a dead-run operation was the runner's grave.

oftening (ǫ·f'niŋ), *vbl. sb. rare*—1. [f. Often *adv.* and *a.* + -ing[1].] Frequent repetition.

a 1889 G. M. Hopkins *Jrnls. & Papers* (1959) 289 *Repetition, oftening, over-and-overing, aftering of the inscape* must take place in order to detach it to the mind.

oftenness. (Later example.)

1977 'E. Crispin' *Glimpses of Moon* ii. 27 The Pisser's [sc. a pylon's] noise..had actually intensified both in volume and in oftenness.

† og, ogg (ǫg). *Obs. Austral.* & *N.Z. slang.* [Cf. Hog *sb.*[1] 8.] A shilling.

1937 Partridge *Dict. Slang* 580/2 *Ogg* or *og*, a shilling. ..A corruption of *hog* (a shilling). 1945 Baker *Austral. Lang.* v. 109, 1s...*og* and *rogue* (probably a clipping of the English rhyming slang *rogue and villain*, a shilling). 1946 *Penguin New Writing* XXVIII. 125 Three quid and seven og.

‖ ogbanje (ǫgbæ·nʒi). [Ibo.] (See quots.) Also *attrib.*

1958 C. Achebe *Things fall Apart* ix. 68 The child was an *ogbanje*, one of those wicked children who, when they died, entered their mothers' wombs to be born again. *Ibid.* 68 Okonkwo had called in another medicine-man who was famous in the clan for his great knowledge about ogbanje children. 1976 *CRC Jrnl.* July 18/1 His daughter Aku-nna is so thin and delicate that it is even suggested that she might be an 'ogbanje', a 'living dead', a child only on loan to this world who will be called back to the other world while still young.

ogee. In etym. delete sentence beginning 'The use of *ogee* as the name of a moulding and curve is exclusively English': see C. Brunel in *Romania* (1960) LXXXI. 293. Add: **3. a. ogee wing** *Aeronaut.*, a wing whose outline is an ogee (used on some supersonic aircraft).

1960 *Aeroplane* XCIX. 791/1 The ogee wing is one of the highly swept delta wings with very low aspect ratio which are being proposed for supersonic airliners. Slender wings of this type have subsonic leading-edges and supersonic trailing edges. 1970 *New Scientist* 23 Apr. 172/2 The HS 133 is an ogee wing (Concorde shape).

Ogen (ōu·gen). Also with lower-case initial. [Place-name (see below).] The name of a kibbutz in Israel, used *attrib.* in **Ogen melon** to designate a small melon with pale orange flesh and brownish-orange skin ribbed with green, belonging to a variety first developed there.

1967 *Guardian* 14 June 6/7 The tiny Ogen melon... Paler-fleshed than its French antecedent, the Charentais, it has the same voluptuous flavour. 1969 *Oxf. Bk. Food Plants* 118/2 'Ogen Melon'. The name is derived from a kibbutz in Israel where it was bred and whence it has been exported during the last decade. It is said to belong to the cantaloupe group of varieties. 1973 *Observer* (Colour Suppl.) 1 July 39/1 The small ogen melon, orange ribbed with green..was first grown by Nathan Fuchs at the Ogen kibbutz, north of Tel Aviv. 1976 *Listener* 12 Feb. 171/3 This one place on the Dead Sea..is now the source of such fruits as the ogen melons we enjoy in London in midwinter.

-ogen (ŏdʒen), a suffix f. -o + -gen, q.v. In *Biochem.* appended to the names, usu. terminating in *-in*, of biologically active compounds, esp. proteins, to form the names of their inactive precursors: as *chymotrypsinogen, Fibrinogen, Pepsinogen, Renninogen.

1961 *Rep. Comm. Enzymes Internat. Union Biochem.* ix. 49 The names of enzyme precursors should no longer be formed by the use of the suffix '-ogen'; the prefix 'pre-' should be used instead.

ogg, var. *og.

oggin (ǫ·gin). *Naut. slang.* [A corruption of *hog-wash.*] The sea.

1946 J. Irving *Royal Navalese* 126 *Oggin*, the sea; the Drink; the Ditch. The term descends to us aitch-less from an earlier abuse of the sea as the Hog-Wash. 1949 W. Granville *Sea Slang Twentieth Cent.* 100 *Floggin' the 'oggin*, sailing the high seas. (Naval lower-deck.) 'Oggin is a form of 'ogwash. *Ibid.* 168 *'oggin*, the sea or ditch. A perversion of 'ogwash (hogwash). 1973 D. Lees *Rape of Quiet Town* x. 165 No one told the two gunners that the sub was about to crash-dive and they had to run like hell to avoid being left behind in the oggin.

Oghuz (ogu·z). [ad. Turk. *oğuz* (which is also the name of a legendary Turkish hero).] **1.** Also **Ghuzz** (via Arabic), **Oghus, Oğuz.** One of various Turkic tribes, now more usually included among the Turkmen, who originally inhabited Siberia and, later, Transoxiana but who crossed the Oxus in the 11th century and invaded Persia, Syria, and Asia Minor; also, a member of one of these tribes. Also *attrib.* or as *adj.*

1843 *Penny Cycl.* XXV. 395/2 For many centuries the Oghuzes were perpetually at war with the Persians. 1845 *Encycl. Metrop.* XXV. 868/2 They [sc. the Uz-beks] are called Ghuzz by the Arabs. 1854 G. Larpent in J. Porter *Turkey* I. 155 The Turks divided themselves into the Uigurs or Eastern Turks..and into the Oghus or Western Turks. 1888 *Encycl. Brit.* XXIII. 660/2 The old name Ghuzz, originally, as it seems, the Turkish Oghuz (an eponymous hero of whom Turkish chronicles tell many fables) was wholly superseded by the new name Turkman. 1899 Skrine & Ross *Heart of Asia* i. xix. 142 The Ghuz laid waste the whole of Khorāsān. 1947 Auden *Age of Anxiety* (1948) v. 103 The Ghuzz, the Guptas, the gloomy Krimchaks. 1953 [see *Kipchak]. 1965 H. M. Smyser in Bessinger & Creed *Medieval & Linguistic Stud.* 93 Ibn Fadlān's descriptions of..Oghuz (Ghuzz Turks)..are fascinating. 1972 [see *Kipchak]. 1974 G. Lewis tr. *Bk. of Dede Korkut* 10 It is known that the term 'Oghuz' was gradually supplanted among the Turks themselves by *Türkmen*, 'Turcoman', from the mid tenth century on... The Turcomans were those Turks, mostly but not exclusively Oghuz, who had embraced Islam and begun to lead a more sedentary life than their forefathers. *Ibid.*, The stock epithet of the Oghuz ladies is 'white-faced'.

2. The southern division of the Turkic languages. Usu. *attrib.* or as *adj.*

1959 J. Benzing in *Philologiae Turcicae Fundamenta* I. 2 In what follows a survey is given of the Turkic languages, in which the individual languages and dialects are set together as they seem to me, on grounds of linguistic history (phonetic and especially grammatic) to be especially closely connected... b) Southern Turkic (the Oghuz group). Here belong: 1. Osmanli... 2. Azarbaijani... 3. Turkmen. 1974 *Encycl. Brit. Micropædia* X. 198/1 Turkmen is a member of the southwestern, or Oghuz, division of the Turkic languages.

So **Oghu·zian** *a.* and *sb.*

1603 R. Knolles *Gen. Hist. Turkes* 128 For although he [sc. Othoman] were a Turke borne, yet was he not of the Selzuccian family, as were the rest, but of another house and tribe, and therefore not of them fauoured or thought to have so good right vnto any of the late Sultans prouinces or territories, as had they who being of his house and holpen with the prescription of time, enuied at the sudden rising of this Oguzian Turke, being vnto them as it were a meere stranger. 1621 Heylin *Microcosmus* 312 Solyman the chiefe of the Oguzian family and Prince of Machan, flying the fury of the Tartars, was drowned in Euphrates. 1880 A. H. Sayce *Introd. Sci. of Lang.* II. ix. 263 The one-eyed giant..reappears among the Turkish Oghuzians. *Ibid.* 264 In the Oghuzian version the story is amplified.

ogival, *a.* (*sb.*). Add: **2.** Having the shape of an ogee.

1962 *Flight Internat.* LXXXI. 269/1 One of the two Fairey FD.2 research aircraft owned by the Ministry of Aviation is being converted..to flight-test an ogival wing. 1965 *Times* 11 Sept. 7/6 Instead of being an ordinary, straight-sided delta, the Concord wing is ogival. Each wing starts as a sharp point.., spreads outwards and then curves in again. 1976 *Jane's Pocket Bk. Res. & Exper. Aircraft* 19 Special design features: Slender ogival delta wings.

ogive. Add: **4. b.** (Something having) the profile of an ogive, esp. the head of a projectile.

1904 *Sci. Amer.* 16 Jan. 44/1 It [sc. an airship] is cylindrical in form, with an ogive nose and a nearly hemispherical stern. 1947 L. E. Simon *German Research in World War II* 115 They studied the way in which the ogive (the tapering head of the projectile) broke up. 1950 E. A. Bonney *Engin. Supersonic Aerodyn.* v. 183 For a given base diameter, therefore, the drag of an ogive will be less than that of a cone of the same length. 1966 D. Stinton *Anat. Aeroplane* vi. 89 The shapes of such aircraft are determined by the need to produce favourable interactions between the relatively high-pressure regions behind shock waves and the adjacent airframe surfaces. The simplest example is shown in Fig. 6.9, in which an ogive, shedding a complete ogival Mach-cone, is split longitudinally and fitted with wings.

4*. *Statistics.* A graph in which each ordinate represents the frequency with which a variate has a value less than or equal to that indicated by the corresponding abscissa, which for many unimodal frequency distributions has the form of an ogee. (In earlier use the ordinates and abscissas were interchanged.)

1875 F. Galton in *Phil. Mag.* XLIX. 35 When the objects are marshalled in the order of their magnitude along a level base at equal distances apart, a line drawn freely through the tops of the ordinates which represent their several magnitudes will form a curve of double curvature... Such a curve is called, in the phraseology of architects, an 'ogive'. *Ibid.* 46 A law of frequency of error founded on a binominal ogive. 1930 R. Pearl *Introd. Med. Biometry & Statistics* vi. 119 The ogive and integral forms of plotting a frequency distribution are fundamentally the same. The only difference is that in the case of the ogive frequencies are plotted along the abscissal axis, and in the integral along the *y* axis as usual. 1937 Yule & Kendall *Introd. Theory of Statistics* (ed. 11) viii. 150 The values of the percentiles may be used to draw what is known as Galton's ogive curve. 1939 J. F. Kenney *Math. of Statistics* (1940) I. ii. 26 The graphs of cumulative frequencies are called ogives. 1962 A. Battersby *Guide to Stock Control* iii. 31 This curve has been derived from sales figures with a Normal distribution... It is called the 'ogive' of the Normal curve. *Ibid.*, For Normal probability paper, the scale is drawn so as to turn the Normal ogive into a straight line.

4.** *Geol.* A stripe or band of dark material stretching from side to side across the surface of a glacier, usu. arched in the direction of flow and arranged in a parallel series of similar bands.

1947 *Jrnl. Glaciology* I. 16 Then there is the question of 'Ogives' or pressure arches. 1949 *Ibid.* I. 327 Vareschi's latest and most important researches were made upon various systems of banding which appear in the tongue of the Great Aletsch Glacier, in particular what are locally called *Ogiven*. These are curved bands visible on the glacier surface, often in regular longitudinal series... The ogive itself is generally darker block ice, whilst between one ogive and the next is paler and higher *Buckel* ice. 1951 *Ibid.* I. 498 The study of the formation of ogives which in English-speaking countries are referred to as Forbes's Bands. 1974 *Encycl. Brit. Macropædia* IX. 183/1 In plan view, the ogives are invariably distorted into arcs or curves convex downglacier; hence the name ogive.

Oglala (ǫglā·lā). Also **Ogalalla, Ogallalla(h), Oglalla.** [Native word.] The chief tribe of the Teton Sioux Indians; a member of this tribe; also, their language. Also *attrib.*

1837 *Missionary Herald* (Boston, Mass.) XXXIII. 369/2 Came to the village of the Ogallallah Indians, consisting of more than two thousand persons. 1838 S. Parker *Jrnl. Exploring Tour beyond Rocky Mts.* 63 They were Ogallallahs, headed by eight of their chiefs. 1857 *Porter's Spirit of Times* 21 Mar. 34/1 They proved to be a hunting-party of Ogallallah Sioux. 1867 *Harper's Weekly* 5 Oct. 629/2 The Commission..on August 19 held a council on the steamer..with the head chiefs of the Ogiallas. 1876 H. T. Williams *Pacific Tourist* 47/1 Red Cloud is chief of the Ogalalla Sioux. 1897 C. King *Warrior Gap* (1898) 31 With him rode Baptiste,..whose mother was an Ogallala squaw. 1921 F. Hebert *40 Years Prospecting* 6 It was there that I saw Pawnee Killer, a big Ogalala Chief. 1933 [see *Dakota *sb.* 2]. 1947 B. A. De Voto *Across Wide Missouri* 123 When Sublette & Campbell brought the Oglalas to the Platte in 1834 [etc.]. *Ibid.* 191 Fort Laramie..would draw all the wandering trappers,..the Oglala Sioux, and finally the United States Army. 1973 *Freedomways* XIII. 81 There is much emphasis on the dance and its symbolic meaning to the Oglala. 1974 *Black Panther* 19 Jan. 3/3 Both Mr. Means, an Oglala Sioux, and Mr. Banks, a Chippewa, have wove the right to make opening statements to the jury once they are seated.

ogling, *vbl. sb.* Add: **b. ogling-glass** *U.S. humorous*, a monocle.

1843 *Knickerbocker* XXII. 111 There he was promenading,..an ogling-glass lifted to his eye.

o'goblin: see *goblin[2].

Ogopogo (ōu-gǫpōu·go). *Canad.* [Fanciful, said to be from a British music-hall song by C. Clark (see quot. 1974).] The name of a water monster alleged to live in Okanagan Lake, British Columbia. Also *transf.*

No contemporary (1924) copy of the song referred to in quot. 1974 has been traced.

1926 *Province* (Vancouver) 24 Aug. 7 (*heading*) 'Ogopogo' now official name of the famous Okanagan sea serpent. 1927 *Ibid.* 25 Sept. 24 Alberta claims to have an Ogopogo all its own. 1933 *Sun* (Baltimore) 24 June 14/1 He blamed the 'ogopogo', famed lake monster. 1936 A. F. Cross *Cross Roads* 71 Ogopogo's head is slightly reminiscent of Henry VIII, he has a torso like an accordion, and a tail like a shillelagh. Ogopogo is, of course, a celibate, because he lost his wife in the Carboniferous Age, and after the customary dinosaurian 1,000-year period of mourning, was once more seen about again. 1955 *Daily Progress* (Charlottesville, Va.) 8 Feb. 12/6 The dispute arose when Ogopogo's name appeared on a new bridge across the South Saskatchewan. 1962 G. MacEwan *Blazing Old Cattle Trail* vii. 46 The way was described as rough and mountainous but no Ogopogo lake serpents were reported. 1974 P. Costello *In Search of Lake Monsters* x. 222 The old Indian name for the animal was Naitaka, or Nha'a'itk; the settlers name for it is now Ogopogo, which is far more recent in origin. It is not, as some seem to have thought, an Indian word. Back in 1924 the following music hall hit from London was sung one night in Vernon: His mother was an earwig, His father was a whale. A little bit of head, and hardly any tail, And Ogopogo was his name.

‖ o-goshi, ogoshi (ōu,gǫ·ʃi). [Jap., f. *o(u)* big, major + *koshi* the waist or hips.] A hip throw in Judo.

1954 E. Dominy *Teach yourself Judo* 190 Ogoshi,..Floating Hip Throw. 1957 Takagaki & Sharp *Techniques Judo* ii. iv. 47 O-goshi: major hip throw... Throw by a pulling and twisting motion over your right hip. 1966 *Daily Tel.* 15 Nov. 21/4 A ballet dancer..broke an arm..while trying an O Goshi throw.

Ogpu (ǫ·gpu). Also **O.G.P.U.** [f. the initials of the Russ. *Ob"edinënnoe Gosudárstvennoe Politicheskoe Upravlénie* United State Political Directorate.] An organization for investigating and combating counter-revolutionary activities in Soviet Russia, which superseded the *Cheka and the G.P.U. (*G. III. 1) in 1923 and was replaced by the N.K.V.D. (*N. II. 1) in 1934.

1923 *Times* 24 Nov. 13/5 By special decree the Soviet Government has created a new Supreme International Cheka (State Political Department), which is called 'Ogpu' (Unified State Political Department). 1927 *Daily Tel.* 7 June 10/2 A report from Leningrad states that the Ogpu (Cheka) has officially announced the execution without trial of 'all the active members of a band of incendiaries'. 1927 *Glasgow Herald* 13 June 12/3 The fertile imagination of the Ogpu, or 'State Political Department'. 1927 *Daily News* 17 Nov. 7/4 Trotzky is under arrest..and is being watched by officials of the O.G.P.U. (Secret Police). 1928 *Sunday Times* 8 Jan. 11/1 The day has gone by when Ogpu sought to conceal itself. 1939 *Ann. Reg.* 1938 211 The new chief of the O.G.P.U., one Beria, appointed only in November, is stated to have dismissed 470 of the highest officers. 1940 [see *Gestapo]. 1940 *War Illustr.* 12 Jan. 608/1 Reported that Russians have brought up picked Ogpu troops into Karelian Isthmus. 1949 [see *Cheka]. 1958 C. Cockburn *Crossing Line* vi. 97 Dedicated men..with pistols in their hands ready to blaze away in case this OGPU desperado should get up to any tricks. 1972 T. Wittlin *Commissar* (1973) xxxiv. 248 For almost ten years an intensive chase after Agabekov was conducted by the men of the OGPU in Paris.

ogrillon. For *nonce-wd.* read *rare*—1 and substitute for etymol.: [a. F. *ogrillon*, f. *ogre* Ogre + suff. *-illon* in *moinillon, négrillon, oisillon*, etc.]

ohelo (ohē·lo). [Hawaiian.] The red or yellow fruit of a blueberry native to Hawaii, *Vaccinium reticulatum*. Also *attrib.*

1843 J. J. Jarves *Hist. Hawaiian Islands* i. 9 The banana, yam, sweet potato,..ohelo (a berry),..are indigenous and plentiful. 1888 W. Hillebrand *Flora Hawaiian Islands* 272 The shining fleshy berry, the 'Ohelo',..is the principal food of the wild mountain goose. 1977 *Time* 17 Oct. 41/1 Native Hawaiians have long attempted to placate the fire goddess Pele by dropping offerings—ohelo berries, liquor and, once upon a time, an occasional human—into the crater of the 4,090-ft. volcano Kilauea.

ohia (ohī·ā). Also **ohia lehua.** [Hawaiian.] = *lehua. Also *attrib.*

1824 C. S. Stewart *Jrnl.* 15 June in *Jrnl. Residence Sandwich Islands* (1828) xi. 305 The only trees and plants known to us, which we saw..were the koa, *acacia*, a large and beautiful tree of dark, hard wood, the Ohia, *eugenia malaccensis*, bearing a beautiful tufted, crimson flower, and a fruit called by foreigners, the native apple. 1825 [see *kukui]. 1866 'Mark Twain' *Lett. from Hawaii* (1967) xi. 99 Shady groves of forest trees..and, handsomest of all, the ohia, with its feathery tufts of splendid vermilion-tinted blossoms. 1888 [see *lehua]. 1917 *Nature* 20 Sept. 57/2 The ohia..also called ohia lehua and lehua, resembles..our white oak, but bears beautiful clusters of scarlet flowers with long, protruding stamens. 1937 D. & H. Teilhet *Feather Cloak Murders* x. 185 A few gnarled ohia trees..grew between the twisted rocks. 1970 S. Carlquist *Hawaii* xvi. 301 The ohia lehua lends a somber air to wet forests because of its dark green foliage, dark gray bark, and often gnarled branches. Roots are formed on the sides of trunks of ohia trees... Ohia forest can be very tall, exceeding fifty feet... The tallest ohias are on the island of Hawaii.

Ohian (ohəi·ăn), *sb.* and *a.* = *Ohioan *sb.* and *a.*

1819 in J. Flint *Lett. from Amer.* (1822) 128 The Ohian is in many cases growing up to manhood. 1836 C. R. Gilman *Life on Lakes* I. 54 He..is the very man..who should be called Buck Eye and not Ohian. 1963 *Guardian* 1 May 10/7 The average Ohian. 1977 *Financial Times* 18 Oct. 17/1 Ohians can with some justification claim to be a boilerhouse of U.S. manufacturing industry. *Ibid.* 17/7 Even those Ohian politicians who rose to national prominence rarely put their stamp on the country's affairs.

Ohio (ohəi·o), the name of a North American river, a tributary of the Mississippi, and one of the United States, used *attrib.* in *Ohio bluebell, buck-eye, sandstone*; **Ohio fever, spread** (see quots.).

1842 C. M. Kirkland *Forest Life* II. xxxix. 142 A beautiful perennial, here called the Ohio bluebell, a far larger plant than the one we know by that name. 1810 F. A. Michaux *Hist. Arbres Forestiers de l'Amérique Septentrionale* I. 38 Ohio buck eye.., nom donné par moi. 1832 D. J. Browne *Sylva Amer.* 227 It is called Buckeye by the inhabitants,..but..we have denominated it Ohio Buckeye, because it is more abundant on the banks of this river. 1948 *Chicago Sun-Times* 20 Apr. 32/2 The Ohio buckeye..is the first of all the big trees to burst forth with leaves. 1816 in T. W. Robinson *Hist. Morrill* (1944) 155 The 'Ohio Fever' took away many of our best farmers. 1831 T. Buttrick *Voyages* 57 The 'Ohio fever' became a well known expression for this desire to move West. 1835 *Knickerbocker* V. 274 Such..as some fifteen years since happened to reside in any part of New-England where what was called the 'Ohio fever' prevailed. 1881 *Harper's Mag.* Apr. 711/1 Lime stone..and gray Ohio sandstone are much used in construction. 1971 M. Tak *Truck Talk* 111 *Ohio spread*, a trailer with a separation of eight feet between the axles.

Ohioan (ohəi·oăn), *sb.* and *a.* [f. *Ohio + -an.] **A.** *sb.* A native or inhabitant of Ohio in the United States. **B.** *adj.* Of or pertaining to Ohio.

1818 E. P. Fordham *Pers. Narr. Trav.* (1906) viii. 165, I do not choose the risk of being insulted by any vulgar Ohioans. c 1848 W. Whitman in *Amer. Speech* (1961) XXXVI. 297 Ohioans [are called] *Buckeyes*. 1906 A. B. Hulbert in B. A. Botkin *Treas. S. Folklore* (1949) iii. i. 421 When exactly one half had passed on a hog gallop the Ohioan leaped down. 1927 H. Crane *Let.* 7 May (1965) 296 Under the benign influence of Ohioan pollens during the months of June, July, Sept. & October. 1929 G. L. Eskew in B. A. Botkin *Treas. S. Folklore* (1949) i. i. 13 Daniel Decatur Emmett, an Ohioan by birth, a traveling minstrel and showman by persuasion, made a song about 'Dixie, the Land of Cotton'. 1948 [see *Nebraskan *sb.*]. 1973 *Time* 25 June 11/1 A fellow Ohioan and Wooster alumnus. 1978 *Amer. Speech* LIII. 42 A wave of Ohioans in the middle nineteenth century inundated northern Indiana and a large area of Illinois.

ohm[2]. Add: (Earlier and later examples.) Now incorporated in the International System of Units, and defined as the resistance that exists between two points when a potential difference between them of one volt produces a current of one ampere.

The word *ohma* was orig. proposed for the practical unit of 'tension' (voltage), along with *volt* for the unit of resistance (see quot. 1861, which also appears in *Electrician* (1861) 9 Nov. 4 and is from the paper read at the Sept. 1861 meeting of the British Association). The origin of the change in application of the word has not been traced in print.

1861 Clark & Bright in L. Clark *Exper. Investigation Laws Propagation Electr. Current* 49 Let us derive terms from the names of some of our most eminent philosophers... We shall then have the following table: A.—Tension. 1 Daniell's Element = 1 Ohma, or unit of tension... B.—Quantity. 1 Ohma, by 1 metre square at 1 millimetre [*sic*] distance = 1 Farad, or unit of quantity... D.—Resistance. 1 Farad per second = 1 Volt, or unit of resistance. *Ibid.*, The ohma, or unit of tension, is practically a very convenient one for all battery purposes. 1865 *Proc. R. Soc.* XIV. 159 It is proposed that the new standard [of resistance] shall not be..described as so many metre/seconds, but that it shall receive a distinctive name, such as the B.A. unit, or, as Mr. Latimer Clark suggests, the 'Ohmad'. 1867 W. H. Preece in *Phil. Mag.* XXXIII. 397, I beg to suggest..to those physicists and electricians who have adopted the British-Association unit for electrical measurements, that in place of expressing this unit, as is variously done at present, by B.A. unit, Ohmad, Ohm, 10^7 metre/second, &c., it would be very convenient to apopt some universal symbol analogous to that used for degrees (°). The Greek letter ω appears to me very convenient. *Ibid.*, The conductivity of the Atlantic cable would be given by 7524ω, and its insulation by 2349Ω per knot, which may still be read Ohm and Megohm. 1867 R. S. Culley *Handbk. Pract. Telegr.* (ed. 2) ii. 30 The unit [of resistance] adopted by the electricians of this country is that determined by a Committee of the British Association, and is called the 'ohm'..., and sometimes the 'B.A. unit'. 1943 F. E. Terman *Radio Engineers' Handbk.* ii. 40 Wire-wound variable resistors of these types are available up to about 100,000 ohms, and are capable of dissipating up to about 15 watts. 1973 D. Aldous in *Pye Bk. of Audio* vii. 75/2 Most transistor amplifiers offer a wide range of output impedances, usually from 4 to 16 ohms.

2. *attrib.* and *Comb.*, as **ohm-centimetre,** the unit of electrical resistivity in the C.G.S. system, being the resistivity of a substance of which a centimetre cube has a resistance of one ohm; **ohm-metre,** the corresponding unit in the International System of Units, equal to 100 ohm-centimetres.

1920 W. T. Maccall *Continuous Current Electr. Engin.* (ed. 2) iii. 18 The resistance of a conductor 1 cm. long and 1 sq. cm. cross-section..is called the specific resistance per centimetre cube... A better name is ohm-centimetre (or microhm-cm.). 1957 H. Cotton *Electr. Technol.* (ed. 7) iii. 67 It is..logical to express ρ as so many ohms-centimetre..or ohm-metre [*sic*]... For example, for copper at normal room temperature $\rho = 1{\cdot}72 \times 10^{-6}\ \Omega\text{-cm} = 1{\cdot}72 \times 10^{-8}\ \Omega\text{-m}$. 1974 *Encycl. Brit. Micropædia* VIII. 524/1 If lengths are measured in centimetres, resistivity may be expressed in units of ohm-centimetre. *Ibid.*, The resistivity of an exceedingly good electrical conductor, such as hard-drawn copper, at 20° C (68° F) is $1{\cdot}77 \times 10^{-8}$ ohm-metre, $1{\cdot}77 \times 10^{-6}$ ohm-centimetre, or 10·7 ohm-circular mils per foot.

ohmic, *a.* (s.v. Ohm[2]). Substitute for def.: That obeys Ohm's law, or exhibits behaviour consistent with it. (Add further examples.)

1904 *Electrician* Nov. 150/1 The energy dissipation due to ohmic resistance in the metal. 1926 R. W. Lawson tr. *Hevesy & Paneth's Man. Radioactivity* i. 14 This proportionality between current and potential..only maintains for the initial part of the curve (the region of Ohmic current, cf. Fig. 3). 1949 *Bell Syst. Techn. Jrnl.* XXVIII. 471 (*heading*) Semi-conductor with two *p-n* junctions and ohmic metal contacts. 1967 *Electronics* 6 Mar. 136/2 Ohmic heating in the device induces diffusions that form p-n junctions.

Hence **oh·mically** *adv.*, by, or as a result of, ohmic resistance.

1919 *Radio Rev.* Dec. 144 A set of triodes arranged in an ohmically-coupled cascade. 1968 C. G. Kuper *Introd. Theory Superconductivity* i. 10 *Changes* in the field produce eddy currents, but they will decay Ohmically.

Ohm's law. [see Ohm[2].] **1.** *Electr.* The law that the strength of a constant electric current in a circuit is proportional to the electromotive force divided by the resistance of the circuit, and that the potential difference between any two points of it is proportional to the resistance between them. (The name was formerly used of some related principles also.)

Propounded by Ohm in *Die galvanische Kette, mathematisch bearbeitet* (1827).

1850 *Phil. Mag.* XXXVII. 463 (*heading*) On a deduction of Ohm's laws, in connexion with the theory of electro-statics. 1863 *Rep. Brit. Assoc. Adv. Sci. 1862* i. 127 The first relation is a direct consequence of Ohm's Law. 1867 *Phil. Mag.* XXXIII. 321 (*heading*) On one of Ohm's laws relating to an insulated circuit. 1910 *Encycl. Brit.* IX. 182/2 This result..obtained by Cavendish in January 1781, that the current varies in direct proportion to the electromotive force, was really an anticipation of the fundamental law of electric flow, discovered independently by G. S. Ohm in 1827, and since known as Ohm's Law. 1921 W. S. Ibbetson *Motor & Dynamo Control* iv. 131 Generally speaking, A.C. circuits do not obey Ohm's law. 1973 J. W. Gardner *New Frontiers in Electricity* vii. 138 Superconductors constitute a very dramatic violation of Ohm's law.

2. *Acoustics.* The law that a complex musical sound is heard as the sum of a number of distinct pure tones into which the sound can be analysed by Fourier's theorem.

Propounded by Ohm in *Ann. d. Physik* (1843) LIX. 513, and called *Ohm's Regel* by Helmholtz in *Die Lehre von den Tonempfindungen* (1863) ii. 54.

1875 A. J. Ellis tr. *Helmholtz' On Sensations of Tone* ii. 51 Every motion of the air..which corresponds to a composite mass of musical tones, is, according to Ohm's law, capable of being analysed into a sum of simple pendular vibrations, and to each such single simple vibration corresponds a simple tone, sensible to the ear. 1924 Wilkinson & Gray *Mechanism of Cochlea* i. 14 Ohm's Law is a generalisation from the results of experiment. 1974 S. E. Gerber *Introductory Hearing Sci.* vi. 134 Ohm's law has generated over a century of research, largely aimed at demonstrating its limitations. *Ibid.*, When tones of relatively low intensity which differ considerably in frequency are presented together, they are perceived as two distinct tones, in accordance with Ohm's law.

oho, *int.* (Further examples.)

c 1874 D. Boucicault *Shaughraun* in M. R. Booth *Eng. Plays of 19th Cent.* (1969) II. 224 Oho! if that's Robert Ffolliott, I'd like to know who's this? 1898 G. B. Shaw *Arms & Man* iii. 56 Four telegrams——a week old. (He opens one). Oho! Bad news!.. My father's dead. 1933 E. O'Neill *Ah, Wilderness!* iii. i. 93 'Oho,' they cried, 'the world is wide, But fettered limbs go lame!' 1976 P. Dickinson *King & Joker* ii. 27 Oho. So something had happened.

oh, oh (ǫ, ǫ), *int.* Also **ohoh.** [See O *int.* (*sb.*, *v.*) and Oh *int.* (*sb.*)] An exclamation of alarm or dismay in response to adverse circumstances.

1944 E. S. Gardner *Case of Black-Eyed Blonde* xvi. 156 Two police cars were closing in on them.. 'Oh, oh!' Della said under her breath. 1947 M. Lowry *Under Volcano* v. 141 'Oh...Oh.' The Consul groaned aloud... It came to him he was supposed to be getting ready to go to Tomalín. 1958 J. Morgan *Expense Account* i. 11 They were on the parkway, and Pete was thinking about the children, when he snapped his fingers and said, 'Oh, oh!' He had forgotten to buy them anything in Chicago. 1960 A. West *Trend is Up* vi. 259 Oh, oh, he thought, Silky is going to pass out before this night is through, that's for sure. 1962 J. F. Powers *Morte d'Urban* iii. 58 'Oh, oh, I was afraid of that,' Wilf said.

oh-so (ōuˈsōu), *adv.* Also **oh, so.** [f. Oh *int.* + So *adv.* III.] Prefixed as an intensive (usu. with hyphen) to adjectives or adverbs, with the sense 'ever so', 'extremely' (usu. with sarcastic or ironical overtones).

1922 *Sketch* 29 Mar. 513/3 A big grey felt hat, which looked, oh, so Spanish! 1952 M. Laski *Village* ii. 33 Her sweet but, oh, so uninteresting face. 1960 J. Betjeman *Summoned by Bells* vii. 66 That mawkish and oh-so-melodious book Holds one great truth. 1965 *Listener* 27 May 797/2 The book is worth a shelf-load of those fashionable intellectualities that oh-so-knowingly chart out the spirallings of psychotic zombies, incapable of feelings, incapable of contacts, their spiritual telephone-wires all cut. 1966 J. Porter *Sour Cream* xiii. 166 It just never entered Azatov's curly head that his oh-so-upright wife was cuckolding him every time his back was turned. 1972 J. Gores *Dead Skip* xiv. 97 The mailboxes were set against the oh-so-rustic redwood slat fence. 1973 *Radio Times* 20 Dec. 18/2 Very suave thriller with Laurence Harvey being oh-so-smooth. 1977 *Gay News* 7–20 Apr. 22/1 Most people do feel oh-so-slightly apprehensive about meeting him.

oh yeah (ōu yēə), *int. colloq.* (orig. *U.S.*). Also **O yeah.** [f. Oh *int.* + *yeah.] An exclamation or interrogative suggesting incredulity, disbelief, or scepticism; 'really?', 'is that so?'. Also as quasi-*adj.*

1930 *Forum* Dec. 376/2 Only recently, that cultural masterpiece *O yeah!* sounded its death rattle. 1931 Loeb & Schenker *Please Stand By* xii. 138 'Oh, are you William Wishtell,' the girl from the Globe asked... 'I've been dying to meet you.' 'Oh, yeah? Well, here I am in the flesh.' 1933 F. Scott Fitzgerald *Let.* 19 Oct. (1964) 237 No exclamatory 'At last, the long awaited, etc.' This merely creates the 'Oh yeah' mood in people. 1933 P. Godfrey *Back-Stage* xv. 194 They smiled indulgently, and greeted me with such remarks as 'I should worry' and 'Oh, yeah!' 1934 *Daily Mirror* 28 June 13/1 (*heading*) 'Oh Yeah!'..Observe that when you say, 'Oh, yes,' it doesn't mean at all the same thing as 'Oh, yeah,' when..you say *that*. 'Oh, yeah' means, really, 'Oh, *no*'; or 'you think you know about that do you, you guy, but I don't think you do, I don't.' 1936 *Variety* 24 June 47/3 Oh, Yeah? Agency men claim they can guess within $250 of any quoted price on a picture name. 1937 H. G. Wells *Star Begotten* vii. 120 Confronted with an idea the American says 'Oh yeah!' or 'Sez you'. 1943 *Amer. Speech* XVIII. 256 A representative group of Americanisms which have wide currency in Australia:..oh yeah. 1966 D. Francis *Flying Finish* vi. 74 'You look out, pal, you mustn't go around admitting that sort of thing...' 'Oh yeah?' I said, laughing.

Oi, oi (oi), repr. dial. or vulgar pronunc. I *pers. pron.*

1901 M. Franklin *My Brilliant Career* xvi. 132 Sure O'i can't plaze yez anyhows. 1930 J. B. Priestley *Angel Pavement* iv. 188 Now, where was Oi? Losing mi plice, wasn't Oi?.. Ow, Oi know... Whoi dew the Catholics eat Fish on Froiday? 1939 Joyce *Finnegans Wake* iii. 551 Oi polled ye many but my fews were chousen. 1962 M. Green *Art of Coarse Sailing* ii. 20 Oi couldn't 'elp overhearing... If your friend's wife's 'aving 'er confinement she oughter come ashore... A Broads boat ain't no place for a babby to be born in. Shall Oi phone for the doctor? 1975 J. Drummond *Slowly the Poison* i. 36 Oi washed down the cobbles. 1976 *Guardian* 6 Aug. 9 (Advt.), An' we're much 'bliged ta Whitbread, thassall oi can say.

oi, *int.*[1]: see *oy, oi *int.*

oi (oi), *int.*[2], var. Hoy *int.* (*sb.*[2])

1962 Jackson & Marsden *Educ. & Working Class* i. iii. 57 Father said, 'Oi, you two—you're not doing anything. Get some paint and paint under there!'

-oic, ending of the names of organic acids containing a carboxyl group, esp. in place of a methyl group. [App. first used in Caproic *a.*]

1971 *Nomencl. Org. Chem.* (I.U.P.A.C.) (ed. 2) C. 182 Carboxyl groups COOH replacing CH_3 at the end of the main chain of an alicyclic hydrocarbon are denoted by adding '-oic acid' or '-dioic acid' to the name of this hydrocarbon.

oick, oik (oik). *slang.* [Etym. obscure.] Depreciatory schoolboy word for a member of another school; an unpopular or disliked fellow-pupil. Also *gen.*, an obnoxious or unpleasant person; in weakened senses, a 'nit-wit', a 'clot'. Hence **oi·kish** *a.*, unpleasant, crude; **oi·ckman** (see quot. 1925).
1925 *Dict. Bootham Slang*, Hoick,..spit. Oick,..to spit; abbreviated form of 'ockman'. *Oickman*,..labourer, shopkeeper, etc.; also a disparaging term. **1933** A. G. Macdonell *England, their England* vi. 95 Those privately educated oicks are a pretty grisly set of oicks. Grocers' sons and oicks and what not. **1935** 'N. Blake' *Question of Proof* x. 189 Smithers is such an oick. **1940** M. Marples *Public School Slang* 31 Oik, hoik: very widely used and of some age; at Cheltenham (1897) it meant simply a working man, but at Christ's Hospital (1885) it implied someone who spoke Cockney, and at Bootham (1925) someone who spoke with a Yorkshire accent. **1940** M. Dickens *Mariana* iv. 109 The old Oik mentioned it over a couple of whiskeys. **1946** G. Hackforth-Jones *Sixteen Bells* 260 Come to think of it he must have been a bit of an oik when he worked at Bullingham & Messer. That crack about long hair was well merited. **1957** F. King *Widow* I. v. 63 He and Cooper had fought a battle with three 'oiks'—this was apparently school slang for the boys of the town. **1958** B. Goolden *Ships of Youth* vii. 162, I only need my cap on back to front to look the complete oick. **1959** W. Camp *Ruling Passion* xvi. 126 Who's that incredibly uncouth and oikish man? **1966** 'K. Nicholson' *Hook, Line & Sinker* viii. 95 So glad you got here before the oicks. **1968** *Melody Maker* 30 Nov. 24/5 Old Stinks from the third stream said: 'I say you oik, the Beach Boys latest is fab gear.' **1975** *Listener* 16 Jan. 83/1 The rigmarole about the flat was patent set-dressing, just to impress us oiks. **1975** *Times* 7 Aug. 7/7 His [*sc.* Oswald Mosley's] angels, a gang of gullible and bloodthirsty oiks ..would come pretty far down the roster of hell's legions.

oidiomycosis (o,i:diomaikōᵘ·sis). [f. Oidi(um + -o + Mycosis.] Infection of an animal or person with a fungus formerly classified in the genus *Oidium*, esp. the thrush fungus *Oidium* (now *Candida*) *albicans*. Now usu. called *candidiasis*. Cf. *moniliasis.
1901 [see *blastomycosis* s.v. *blasto-]. **1917** *Boston Med. & Surg. Jrnl.* CLXXVI. 771 (*heading*) Systemic oidiomycosis: with manifestations in the central nervous system. **1933** *Jrnl. Agric. Res.* XLVI. 169 Van Heelsbergen calls the disease oidiomycosis, and describes it as an affection of the mucous membrane of the mouth, fauces, esophagus, crop, stomach, and small intestine of birds, mammals, and man. **1938** H. T. Karsner *Human Path.* (ed. 5) x. 297 (*heading*) Blastomycosis or oidiomycosis. **1951** T. G. Hungerford *Diseases of Poultry* (ed. 2) viii. 259 Fungus infection of the digestive tract is also known as thrush, moniliasis, oidiomycosis,..and sour crop. **1975** Arnall & Keymer *Bird Diseases* viii. 139 Candidiasis, Moniliasis, Oidiomycosis, Sour Crop or Thrush (*Candida albicans* infection). Turkeys, parrots, game-birds and pigeons are mainly affected.

oik: see *oick.

oik(o)umene, var. *œcumene.

oil, *sb.*^{1} Add: **1. a.** (Further examples in sense 'mineral oil, petroleum'.) Cf. *petroleum.
1860 *Chem. News* 6 Oct. 204/2 The wells yield, by pumping, from ten to twenty-five barrels per day of the crude oil. **1862** *Ibid.* 20 Sept. 149/2 It is believed that the United States and Canada possess natural supplies of petroleum to furnish the rest of the world, for ages to come, with sufficient quantities of oil to yield all the artificial light required, and perhaps much of the fuel also. **1907** V. B. Lewes *Liquid & Gaseous Fuels* iv. 85 In these early wells the oil had to be pumped, but in 1861 a well drilled to a depth of 460 feet yielded oil at such pressure that it rose to the surface and overflowed. **1930** C. T. Brunner *Probl. of Oil* p. iii, To-day the industrial importance of oil is incalculable.. Transport interests.. depend mainly on oil..as the source of their power, while even the roads..are surfaced with a derivative of oil, bitumen. **1964** J. P. Getty *My Life & Fortunes* ii. 23 Today, oil is big business, probably the biggest of all businesses. Without oil, there would be—there could be—no civilization as we know it. **1976** *Daily Tel.* 17 June 6/5 Converting more power stations from oil to coal would push up electricity charges.
c. *midnight oil*: see Midnight *sb.* 5.
3. a. (Later examples.) Also with stronger implications: nonsense, falsehood.
1917 *Amer. Mag.* Nov. 39/2 'Why dearie!' I remarks, kissin' her; 'You know I—'. 'Easy with the oil!' she cuts me off. **1924** [see *bushwa, -wah]. **1926** Maines & Grant *Wise-Crack Dict.* 14/2 *Throwing the oil*, telling glib falsehoods. **1940** M. Marples *Public School Slang* 130 At Winchester..oil = an evasion. **1940** Wodehouse *Eggs, Beans & Crumpets* 168 Coo to him, and give him the old oil. **1954** —— *Jeeves & Feudal Spirit* i. 7 It was imperative that they be given the old oil, because she was in the middle of a very tricky business deal with the male half of the sketch and at such times every little helps.
f. For *U.S.* read 'orig. *U.S.*' and add earlier and later examples.
1862 *Amer. Ann. Cycl. 1861* 580/1 The oil, when first struck, has..been known to burst forth with great violence. **1930** 'Sapper' *Finger of Fate* 180 The general consensus of opinion was that if his cricket was up to the rest of his form, Bob had struck oil. **1936** W. S. Maugham *Cosmopolitans* 266 He'd struck oil a year or two ago and now he's got all the money in the world. **1973** N. Graham *Murder in Dark Room* xiii. 94 You stopped with Scherz. I went a little further back and struck oil. **1975** *Times* 7 Oct. 5/2 When oil is struck..the oilman needs samples for laboratory analysis.
h. *oil and vinegar*: *lit.* used together as condiments; *fig.* said of two elements or factors which do not agree or blend together, or of any two incongruous constituents, with reference to the incompatible characters of oil and vinegar when mixed.
1629 J. Parkinson *Parad.* ii. xxxvi. 503 The first shootes or heads of Asparagus..being boyled tender, and eaten with..oyle and vinegar. **1747** H. Glasse *Art of Cookery* i. 11 The French eat Oil and Vinegar with it [*sc.* broccoli]. **1777** R. Potter tr. *Æschylus' Agamemnon* in *Tragedies* 232 Pour thou oil In the same vase and vinegar, in vain Wou'dst thou persuade th' unsocial streams to mix. **1820** Keats *Let.* June (1931) II. 537 Men get such different habits that they become as oil and vinegar to one another. **1845** Thackeray *Legend of Rhine* ix, in *George Cruikshank's Table Bk.* Sept. 194 Oil and vinegar, which he took with cucumber to his salmon. **1910** *Blackw. Mag.* Oct. 562/2 We might as well try to blend vinegar and oil, as mix together these two elements in one chamber. **1930** A. P. Herbert *Water Gipsies* x. 120 'Why shouldn't our class marry his class?' 'It's oil and vinegar. They don't mix.' **1977** D. Clark *Gimmel Flask* iii. 58 Your double oil and vinegar bottle.
i. Money; *spec.* money given in order to bribe or corrupt; a bribe. *U.S. slang.*
1903 A. H. Lewis *Boss* 121 The sooner we get th' oil, th' sooner we'll begin to light up. **1935** *Detective Fiction Weekly* 31 Aug. 118/1 She didn't take care of her protection directly, that is, she didn't slip the oil to the cops herself. **1970** C. Major *Dict. Afro-Amer. Slang* 87 *Oil*, graft, pay-off to authorities.
j. *Austral.* and *N.Z. slang.* Information, news, the true facts, esp. in phr. *dinkum oil* (see *dinkum *a.*).
1916, etc. [see *dinkum *a.*]. **1919** W. H. Downing *Digger Dial.* 36 *Oil*—News; information. **1930** *Bulletin* (Sydney) 1 Jan. 50/1 On a prospect..Old bloke what died gave me the dinkum oil. **1930** L. W. Lower *Here's Luck* x. 84 We get the dinkum oil off him. **1944** J. H. Fullarton *Troop Target* ii. 18 'What's the oil, Noel?' 'Yes, spill it.' **1946** F. I. Cooze *Ten Bob Each Way* 22 I'll give you the oil according to Hoyle. **1946** J. D. Woods in *Coast to Coast 1945* 33 You'd better play a hand or two ..and get the oil about the place. **1948** V. Palmer *Golconda* xvi. 133 If anything were afoot, he told himself, Mahony would be sure to have the real oil about it, and he himself had a right to any inside information that was going. **1965** *Telegraph* (Brisbane) 5 July 8 The good oil, the drum, the griff.
4. b. Delete '*colloq.*' and 'Chiefly in *pl.*' and add earlier and later examples in *sing.*
1852 W. H. Oxberry in W. Davidge *Footlight Flashes* (1866) xii. 110 An original painting of my father, by Drummond, and a little oil, by W. Beverly. **1912** W. Owen *Let.* 26 Jan. (1967) 111, I herewith send a representation of my outward man; I[t] does not please me; nothing will, unless it were an oil by Sargent. **1938** W. T. Walsh *Philip II* xxviii. 573 One of the artists he employed was..Domenico Theotocópuli, whom he engaged to do a large oil of the martyrdom of St. Maurice. **1967** N. Freeling *Strike Out* 27 Over the chimney-piece was a large oil of three splendid horses. **1977** D. MacKenzie *Raven & Kamikaze* iii. 39 A blackframed oil of a Labrador.
6. a. *oil-broker, change, company, dilution, -feed, filter, immersion, impregnation, magnate, -patch, priming, reserve, revenue, -room, sheikh, storage, supply, tannage, -valve*; *oil bomb, bunker, -canakin, -car* (examples), *-cell* (example), *-drum, -ladle, -pan, shell, -sump, -tank* (examples); *oil depot, district, industry, platform, refinery, -region* (earlier example), *sheikdom, show* [Show *sb.*^{1} 5c], *state, terminal, -works*; *oil gas* (later example); *oil-cooker, -heater, -lamp* (earlier and later examples), *-stove* (earlier and later examples); *oil picture* (earlier example), *portrait, sketch, -work.* **b.** *oil-bearing* (later examples), *-producing* (later examples), *-retaining* adjs.; *oil-burning* ppl. a. and vbl. sb.; *oil-cracking, -drilling, -raising, -sinking, -throwing* vbl. sbs.; *oil-catcher, -cooler, -distributor* (examples), *-feeder, gusher, separator.* **c.** *oil cooling, -firing, quenching, tanning*; *oil-bathed, -bound, -cooled, -fed* (examples), *-filled, -fired* (examples), *-foul, -immersed, -impregnated, -lit* (examples), *-mixed, operated, -primed, -proof, -quenched, -related, -rich, -sleeked, -stained, tanned* adjs.; *oil-harden* vb. Also parasynthetic, as *oil-bunkered, -engined, -tanked* adjs.
1932 *World Today* LIX. 262/2 The camshaft and valvegear as a whole are oil-bathed. **1863** *Jrnl. Franklin Inst.* LXXV. 271 The out-croppings of the lowest members of the Oil-Bearing Strata. **1946** *Nature* 28 Dec. 932/1 In the search for similar oil-bearing structures, geophysical surveys have been extended over wide areas. **1977** *Times* 9 Sept. 7/3 Mao Tse-tung..wanted Peking's parks to grow fruit and oil-bearing plants. **1918** E. S. Farrow *Dict. Mil. Terms* 417 *Oil bomb*, in trench warfare, a large oil drum containing oil and a quantity of high explosive, which dissipates the burning oil in all directions. **1947** *Illustr. London News* 25 Jan. 117/1 The magnificent hammer-beam roof of Westminster Hall..was extensively damaged by an oil bomb in 1941. **1945** *Archit. Rev.* XCVII. 42 Apart from the roof the remainder of the external steel finish is oil-bound paint. **1963** *House & Garden* Feb. 77/2 Oil-bound distempers, in which the binding material is an emulsion of oil or varnish, are more correctly called water paints. **1863** *National Almanac & Ann. Rec.* 687/2 A leading Liverpool oil-broker. **1977** *Yellow Pages Classified Telephone Directory: London (North)* 250/2 Oil Brokers. **1958** *Engineering* 28 Mar. 395/3 The closing of the Canal resulted in a sharp increase in the prices of oil bunkers. **1909** *Times Lit. Suppl.* 3 June 205/2 A plutocrat..who could quell the North Sea with oil-bunkered Dreadnoughts. **1886** *Marine Engineer* VII. 283/2 The oil-burning apparatus has been fitted. **1898** *Railway Mag.* Sept. 246/1 If the atmospheric conditions of the tunnels serve as an excuse..why not use oil-burning locomotives? **1920** E. C. Bowden-Smith *Oil Firing for Kitchen Ranges* iv. 89 With regard to filtering the oil, this depends a great deal on the oil-burning system and the class of burner. **1924** *Domestic Engineering* XLIV. 191/2 The development of oil burning for land purposes has been retarded by the fluctuations in the price of oil. **1960** G. J. Gollin in W. F. B. Shaw *Domestic Heating* viii. 134 (*caption*) A complete oil-burning boiler installation. **1961** V. C. Miles *Domestic Vapouriser Burner Pract.* vii. 83 The vapourising type burner is being used in increasing quantities for the conversion of solid fuel boilers to oil burning. **1843** L. M. Child *Lett. from N.Y.* xl. 273 Children are driving hither and yon, one with a..band-box, or oil-canakin. **1876** J. S. Ingram *Centenn. Exposition* 336 The oil..was loaded by gravity upon oil cars. **1897** Kipling *Day's Work* (1898) 222 There were oil-cars, and hay-cars and stock-cars full of lowing beasts. **1850** *Rep. Comm. Patents 1849* (U.S.) 331 The complete hanger or pillow-block, with or without the oil-catcher. **1884** *Rep. Comm. Agric.* (U.S. Dept. Agric.) 363 When the cellular structure of the rind has completely developed, and the oil-cells have begun to fill. **1959** *Motor Manual* (ed. 36) x. 239 A second oil-change should be made before many hundreds of miles have been run. **1976** H. MacInnes *Death Reel* xii. 105, I am putting my car in for an oil change. **1827** J. S. Mill in *Westm. Rev.* VII. 177 A prohibition of gas-lights might be called..protection to the oil-companies. **1951** in M. McLuhan *Mech. Bride* (1967) 116/2 It's an oil company spending more money to make a better motor oil. **1974** E. Ambler *Dr. Frigo* i. 53 One oil company would be bad enough. A consortium of five..must be quite oppressive. **1932** E. Bowen *To North* xxiv. 231 She went into the scullery; here the oil-cooker was potent; she opened the window. **1977** J. Thomson *Case Closed* vi. 81 A leanto scullery..that contained the sink and an oil-cooker. **1904** A. F. Berry in M. Maclean *Mod. Electr. Pract.* II. i. vii. 79 Those [manufacturers] who use a shell-type oil-cooled construction of transformer..keep the temperature of part of the coils as nearly as possible at the temperature of the oil by spreading out the coils. **1962** *Science Survey* III. 89 (*caption*) The magnet weighs 750 tons and is energised by oil-cooled copper windings. **1973** R. W. Sillars *Electr. Insulating Materials* x. 205 Oil-cooled power equipment..requires a medium which is fluid at all climatic and operating temperatures. **1940** *Chambers's Techn. Dict.* 590/1 *Oil cooler*, a small air-cooled radiator, used in aircraft and racing cars, for cooling the lubricant after its return from the engine and before delivery to the oil tank. **1911** Bohle & Robertson *Transformers* iv. 61 Another disadvantage of oil-cooling is the fact that if a fault occurs necessitating the withdrawal of the oil and the removal of the transformer it is frequently necessary to rewind the coils. **1970** J. Shepherd et al. *Higher Electr. Engin.* (ed. 2) ix. 278 For larger transformers oil cooling is needed, especially where high voltages are in use. **1929** *Times* 31 May 9/3 Notable advances are being made in the technology of oil-cracking processes. **1954** *Encounter* Sept. 34/1 Workmen in an oil-cracking plant in Oklahoma ..got angry, because, in a collective bargaining session, the management had referred to them as semi-skilled. **1863** Oil depot [see *oil derrick*, sense *6 e]. **1970** W. G. Roberts *Quest for Oil* xiv. 141 Even the small barge which plies up and down a big river to supply a local oil depot will have to have a crew. **1949** *Gloss. Aeronaut. Terms (B.S.I.)* ii. 6 *Oil-dilution system*, a system by which the oil can be diluted to assist cold starting. **1889** *Century Mag.* Mar. 714/2 For pilot-boats oil-distributers [*sic*] are valuable when boarding vessels in breaking seas. **1909** *Daily Chron.* 17 Sept. 1/3 M. Blériot..was really thinking of the oil-distributor and the gauge showing the consumption of petrol. **1862** *Sci. Amer.* 22 Feb. 122/1 This oil district is peculiar in many respects. **1910** *Chambers's Jrnl.* Nov. 752/1 The apparatus has..demonstrated its value to the oil-district. **1937** M. Huxley *Let.* 13 Oct. in A. Huxley *Lett.* (1969) 427 We have seen so much, including..oil drilling. **1974** *Evening News* (Edinburgh) 10 Apr. 13/7 The next generation of British oil-drilling experts will be trained and produced in Scotland. **1909** *Daily Chron.* 24 July 6/3 Two small oildrums will be fixed beneath the plank. **1975** *Times* 22 July 14/2 The white Anglo-Saxon bass oil drum player. **1913** *Chambers's Jrnl.* Jan. 31/1 Oil-engined ships are..being built. **1924** *Times Trade & Engin. Suppl.* 29 Nov. 250/3 The large oil-engined liner. **1870** A. S. Evans *Our Sister Republic* ii. 50 The watchmen..with muskets in their hands, and great oil-fed lanterns by their sides. **1886** *Chambers's Jrnl.* 16 Jan. 47/2 A vessel..propelled entirely by oil-fed furnaces. **1905** *Westm. Gaz.* 15 Feb. 8/2 It has a horizontal engine,..forced oil-feed, automatic carburation, [etc.]. **1900** Conrad *Ld. Jim* vi. 74 He set the log for me; he..put a drop of oil in it too. There was the oil-feeder where he left it nearby. **1904** A. F. Berry in M. Maclean *Mod. Electr. Pract.* II. i. vii. 78 If a fire breaks out in the oil-filled tank itself, the latter may be run out into the air. **1930** *Engineering* 24 Jan. 100/3 Various special designs, including the 132-kv. single-core oil-filled cables, which..are shortly to make their appearance in this country. **1957** W. J. John *Mod. Electr. Engin.* I. iv. 126/1 In order to overcome the fire risk in oil-filled equipment, some attention has been paid to the use of chlorinated diphenyls. **1907** *Yesterday's Shopping* (1969) 713/3 Oil Filters... For filtering oil that has been used and become dirty, thus rendering it available for use again. **1925** *Morris Owner's Manual* ii. 28 Unscrew the large plug at the bottom of the sump, when the oil filter, which is

[illegible] retaining universal coupling. **1962** *B.S.I. News* Feb. 37 Bronze oil-retaining brushes and thrust washers for aircraft. **1975** P. Somerville-Large *Couch of Earth* x. 184 We should be prepared to forego a week's oil revenues. **1977** *Sunday Times* 20 Nov. 53/4 Part of the oil revenue will have to be used to tackle some of Britain's deep-seated industrial problems. **1959** *Daily Tel.* 13 Mar. 1/6 The oil-rich sheikhdoms of the Persian Gulf are still under British protection. **1975** N. Luard *Robespierre Serial* iv. 16 The profile might have fitted any oil-rich Arab. **1877**

[illegible] quot. 1943); **oil-butt**, a butt (Butt *sb.*^{2}) containing oil; also *fig.* (see quot. 1937); †**oil-can**, also (*slang*), a German trench-mortar shell (*obs.*); **oil circuit-breaker**, an oil-break circuit breaker; **oil coal**, coal from which oil is obtained; **oil-cup** (earlier U.S. example); **oil-derrick** (examples); also *fig.*; **oil-drop**, (*b*) a

[illegible] unimportant oil circuit-breakers are operated by a hand lever directly upon the breaker itself. **1964** E. A. Reeves *Installation & Maint. Industr. Switchgear* iii. 32 Medium-voltage oil circuit-breakers may be either incorporated in a cubicle or fitted on the outside of a metal-clad unit. **1873** C. Robinson *New South Wales* 52 Deposits of brown cannel oil coals and oil shales. **1850** *Rep. Comm. Patents 1849* (U.S.) 233 The combination of the tight oil cup with the axle. **1863** *Boston Herald* 16 Aug. 3/3 You see, in close proximity on every side, oil depots, oil refineries, oil derricks. **1902** 'Mark Twain' *Speeches* (1910) 367 That

long, lank cadaver, old oil-derrick out of a job. **1948** *Ada* (Okla.) *Even. News* 2 July 1/3 The work scheduled for [illegible]

Petroleum Economist Sept. 349/2 Morocco is preparing to exploit its large oil shale deposits at Timahdit in the [illegible]

molecules at air–water or oil–water interfaces must be briefly described at this point. **1971** Oil–water [see *oil pool* above].

oil, *v.* Add: **1. a.** (Later example.)
1851 H. Melville *Moby Dick* (U.S. ed.) xxv. 124 A king's head is solemnly oiled at his coronation, even as a head of salad.
b. (Further examples.) Also, to rub (a person) with oil as a protection against sunburn. Also *refl.*
1876 Geo. Eliot *Dan. Der.* II. iv. xxxiii. 332 There's a bad style of humbug, but there is also a good style—one that oils the wheels and makes progress possible. **1896** C. M. Sheldon *His Brother's Keeper* ii. 39 Have you been greasing your boots with it?.. Half a pint wouldn't oil more than one of 'em. **1909** *Daily Chron.* 6 Sept. 3/3 Her craze for the 'psychic'..oils the wheels of the plot. **1941** A. Christie *Evil under Sun* vi. 107, I oiled myself and sunbathed. **1945** E. Waugh *Brideshead Revisited* i. ii. 43 Anthony had helped oil fading beauties on sub-tropical sands. **1972** *Mainichi Daily News* (Japan) 6 Nov. 13/1 Advertisement [*sic*] is a powerful factor in boosting the economy in free countries and it is the strongest means of oiling the wheel of manufacturer-consumer relations. **1972** 'G. Black' *Bitter Tea* (1973) ix. 137 A cousin came over and oiled Sally's back. **1973** A. Holden *Girl on Beach* 51 She changed into her swim-suit, oiled herself all over. **1976** D. Francis *In Frame* xvi. 232 Our passage had been oiled by telexes from above. When we arrived..we found ourselves whisked into a private room. **1977** 'A. Stuart' *Snap Judgement* 178 He set the deal up... He was oiling the wheels for when Brigitte arrived with the secrets.
c. To cover the surface of (water) with a film of oil in order to kill mosquito larvæ.
1921 M. Watson *Prevention of Malaria* (ed. 2) xvii. 190 When a clear pool containing the ordinary floating alga is 'oiled', the alga dies. **1952** P. F. Russell *Malaria* 133 Water recently oiled is unfit for bathing. **1959** A. A. Sandosham *Malariology* vi. 252 When oiling a ravine the seepages in which the ravine stream begins should get most attention.
2. Also, *to oil the knocker*: to bribe or tip a doorman. *slang.*
1870 *Brewer's Dict. Phr. & Fable* 632/1 *To oil the knocker*, to fee the porter. The expression is from Racine, *On n'entre point chez lui sans graisser le marteau* (No one enters *his* house without oiling the knocker)—'*Les Plaideurs*'. **1901** Farmer & Henley *Slang* V. 93/1 *To oil the knocker*, to fee the porter. **1968** *Gloss. Brit. Argot* (Paramount Pictures), *Oil the knocker*, tip the porter or caretaker.
3. (Later example.)
1923 *Man. Seamanship* (Admiralty) II. 80 The pumps in the oiler should be started gradually, attempts should not be made to oil individual tanks too rapidly.
b. *intr.* To take in a supply of oil.
1914 H. H. Fyfe *Real Mexico* 201 Some day vessels will call here..to 'oil' just as they now 'coal'. **1922** *Glasgow Herald* 21 Oct. 11 After that the Renown only stopped to oil.
5. *colloq.* To move or go in a quiet or stealthy manner; (const. *in*) to enter, (*fig.*) to interfere; (const. *out*) to depart, (*fig.*) to extricate oneself. Also const. other advbs.
1925 Wodehouse *Carry on, Jeeves!* vi. 139 As man to man, do you want to oil out of this thing? **1929** —— *Mr. Mulliner Speaking* i. 28 It would be a simple task to oil in, insert the soap, and buzz back undetected. **1930** —— *Very Good, Jeeves!* iv. 110, I..oiled round to where Jeeves awaited me. **1945** 'A. Gilbert' *Don't open Door* xix. 172 As soon as he was alone he'd oil out and they could think what they pleased. **1958** —— *Death against Clock* 119 She might oil in, and..clinch a bargain on the spot. **1963** —— *Ring for Noose* x. 119 He deserves to lose his licence, oiling off and leaving you on your tod. **1968** G. Mitchell *Three Quick & Five Dead* i. 35 Do you think the girl was pestering for marriage, but that James wanted to oil out? **1977** 'J. Le Carré' *Hon. Schoolboy* xxii. 527 That twerp Enderby is oiling through the back door.
6. *intr.* *to oil up*: to clog up with oil.
[**1925**: see *oiled-up* s.v. *oiled *ppl. a.* 5.] **1960** [see *foul *v.*^{1} 1b]. **1975** *Country Life* 5 June 1470/1 In traffic..plugs oil up. And pedals are hard pressed to keep the engine alive.

oil-bag. c. (Later examples.)
a **1889** in *Century Mag.* Mar. 710/1 [I] placed two oil-bags, filled with linseed oil, over the bows. **1961** F. H. Burgess *Dict. Sailing* 152 *Oil bag*, any container that permits oil to drip out slowly to help quell the sea.

oi·l bath. Also oil-bath. [f. Oil *sb.*^{1} + Bath *sb.*^{1}]. **1.** A receptacle containing oil for cooling, heating, lubricating, or insulating apparatus immersed in it, or for other purposes.
1838 [in Dict. s.v. Oil *sb.*^{1} 6 a]. **1885** *Marine Engineer* 1 Sept. 151/2 The crank shaft at the lowest point of its revolution constantly touches the surface of an oil bath in the closed motion chamber. **1927** Mellanby & Cooper tr. *Kieser's Materials & Design Turbo-Gen. Plant* 52 The sudden drop in temperature is obtained by quenching the wheels in an oil bath. **1968** K. Ball *B.M.C. Autobk.* 4 198/2 *Oil-bath*...In air filters, a separate oil sump for wetting the wire-mesh element. **1970** —— *Fiat 600* 164/2 Oil bath Reservoir which lubricates parts by immersion. **1971** *Power Farming* Mar. 57/3 It forms an oil-bath for the drive, the oil being circulated by the gears to the outboard end and returned through piping.
2. *India.* (See quots.)
1909 *Westm. Gaz.* 28 Apr. 5/2 The term oil-bath is only a misnomer, inasmuch as the bath is not taken in oil. It is applied to a process involved in besmearing the body with

oil and then bathing in a river or tank to remove the oil from the body. **1967** Singha & Massey *Indian Dancer* ix. 96 First, the entire body is given an 'oil-bath', that is, oil is rubbed into it.

oi·l burner. 1. A device in which oil is atomized and burned to produce heat.
1900 *Engineer* 22 June 651/3 A new muffle furnace with two oil burners of the luminous lamp type, but arranged to give a blue flame. **1946** K. Steiner *Fuels & Fuel Burners* xiv. 279 Domestic oil burners usually operate intermittently.., starting when heat is required and stopping when the heat demand is satisfied. **1946** E. Hodgins *Mr. Blandings builds his Dream House* (1947) ii. xv. 204 The oil burner, starting up in answer to the call of its thermostat. **1970** D. Kut *Warm Air Heating* xx. 338 The grade of oil fuel best suited for a particular installation depends on the type of oil burner and on the hourly through-put of oil.
2. a. A vehicle or ship burning oil as fuel.
1902 *Westm. Gaz.* 29 Sept. 5/2 Tank engines were the first to be fitted as oil-burners, and now some of the newest main line locomotives are to be fitted. **1911** *Daily Colonist* (Victoria, B.C.) 20 Apr. 14/3 The steamer will be practically a duplicate of the Princess Adelaide but will be constructed as an oil-burner. **1923** R. D. Paine *Comrades of Rolling Ocean* ix. 160 That ship of his..is an oil-burner. **1942** Berrey & Van den Bark *Amer. Thes. Slang* § 766/6 *Oil burner*, a diesel-powered truck. **1971** M. Tak *Truck Talk* 111 *Oil burner*, (1) a diesel truck, as opposed to a truck that runs on gasoline.
b. *slang.* A vehicle which, because of its poor condition, uses up great quantities of lubricating oil.
1938 *Amer. Speech* XIII. 307/2 *Oil burner*, a bus which uses excessive amounts of oil. **1975** B. Garfield *Hopscotch* xvii. 166 Even in an old oil-burner he could have gone three times as far in a day's drive if he'd wanted to.

oilcake. (Earlier and later examples.)
1743 W. Ellis *Mod. Husbandman* June iv. 36 Dressing Ground with Lime, Chalk,..Oil Cake Powder, Malt Dust [etc]. **1942** *Sun* (Baltimore) 26 Nov. 8/2 The ships, operating under a safe conduct agreement between Sweden and belligerent nations, were laden with grain, oilcake and piece goods. **1971** *Post* (S. Afr., Cape ed.) (Suppl.) 9 May 10/1 (*caption*) Easy oil cake..easy and cheap to make. **1975** *Nature* 13 Feb. 488/3 The importance of the poppy crop lies primarily in the seed, as the principle [*sic*] source.., after extraction, of oil-cake for cattle feed.

oilcloth. Add: **1. b.** (Earlier and later examples.)
1796 H. Walpole *Let.* 20 Nov. (1905) XV. 430, I mentioned *carpets* made from Mr. Lysons's mosaic pavements; I ought to have said *oil-cloths*, which cost a great deal less. **1903** Kipling *Five Nations* 114 Then the oil-cloth with its numbers, as a banner fluttered free. **1904** G. Stratton-Porter *Freckles* 72 Freckles..covered the [book-]case with oil-cloth. **1974** *Country Life* 3 Jan. 33/3 Fortunately preserved under later oilcloth is decoration in the upper hall and several of the bedrooms.
2. (Later examples.)
1917 W. Owen *Let.* 8 Aug. (1967) 481 Thus I need at once... The Oil-cloth raincoat. **1954** M. Sharp *Gipsy in Parlour* xiii. 133 Oilcloth-covered tables and bentwood chairs. **1974** R. B. Parker *Godwulf Manuscript* xii. 96 The kitchen with its oilcloth-covered table.

oi·l-clothed, *a.* [f. Oilcloth + -ed^2.] Laid or covered with oilcloth. So **oi·lclothy** *a.*, suggestive of or resembling oilcloth.
1894 M. Dyan *All in Man's Keeping* II. x. 185 There came..the tap of light heels on the oil-clothed landing. **1915** Galsworthy *First & Last* in *Cosmopolitan* June 12/2 He was in a gas-lighted passage, with an oil-clothed floor. **1923** U. L. Silberrad *Lett. J. Armiter* ix. 202 Everything was slippery and oilclothy. **1974** L. Thomas *Tropic of Ruislip* ii. 41 She..followed Andrew into The Bombardier Café where he was sitting at one of the tables covered with decorated oilcloth depicting badges of Britain's fighting forces... She and her grandfather sat at the next oilclothed table staring directly at him.

oi·ldom. [f. Oil *sb.*^{1} + -dom.] The petroleum-producing districts of a country; petroleum producers and marketers regarded as a group.
1865 J. H. A. Bone *Petroleum & Petroleum Wells* 42 If he would see anything at all of oildom he must make the passage, unpleasant as it may be. **1880** *Harper's Mag.* Dec. 63 Huge shops..send..iron tanks into the oil regions—to hold the surplus of Oildom. **1904** I. M. Tarbell *Hist. Standard Oil Co.* I. 71 The rise [in freight rates] which had been threatening had come... At the news all oildom rushed into the streets. **1926** J. Ise *U.S. Oil Policy* xiii. 135 The new order immediately precipitated a fresh quarrel in oildom.

oiled, *ppl. a.* Add: **1.** (Later examples.) Cf. *to oil the wheels* (Oil *v.* 1 b) and *oil *v.* 1 b.
1932 D. L. Sayers *Have his Carcase* xx. 263 It was.. surprising to find the identification..going like oiled clockwork. **1933** *Brit. Birds* XXVII. 45 Very little has been heard of 'oiled' birds during the two years. **1971** 'D. Halliday' *Dolly & Doctor Bird* i. 2 Lying oiled in the sun. **1972** *Daily Tel.* 9 Mar. 9/2 Oiled birds, mostly razorbills and guillemots, are coming ashore in north Cornwall. **1973** J. Burrows *Like an Evening Gone* i. 8 Greta was capable. ..Her unlikely scheme proceeded as on oiled wheels. **1976** M. Millar *Ask for me Tomorrow* (1977) xvii. 135 It was the only house on Camino de la Cima, an oiled dirt road.
c. (Further examples.)
1808 M. Wilmot *Russ. Jrnls.* (1934) 316 He has a suite of apartments in the Tartar taste with oil'd paper instead of Glass windows. **1957** *Textile Terms & Definitions* (Textile Inst.) (ed. 3) 70 *Oiled wool*, unscoured or undyed knitting wool or wool dyed before spinning and containing added oil not subsequently removed. **1962** L. Deighton *Ipcress File* i. 15 His oiled teak desk. **1974** D. Ramsay *No Cause to Kill* 1. 28 The carpeting was a basket weave..effective against oiled walnut, the only wood present. **1977** J. R. L. Anderson *Death in City* viii. 119 A heavyweight oiled-wool pullover.
3. *slang.* Drunk; if unmodified (by *well*, etc.) only mildly drunk, tipsy.
1737 *Pennsylvania Gaz.* 6–13 Jan. 2 He's Oil'd. **1899** A. M. Binstead *Gal's Gossip* 169 He was certainly well 'oiled'. **1916** E. V. Lucas *Vermilion Box* 141 He was, as the slang phrase has it, 'oiled'; which is a condition of alcoholic comfort well on this side of inebriety. **1926** E. Wallace *More Educated Evans* ix. 218 He'll come out in a minute, oiled to the world. **1948** Wodehouse *Spring Fever* xvi. 162 Augustus Robb, if not actually plastered, was beyond a question oiled. **1976** J. Wainwright *Who goes Next?* 76 When they're nicely oiled, one of 'em sometimes trots around here, and buys something.
4. Executed in oil-colours.
1903 Ld. R. Gower *Rec. & Reminisc.* 246 To make an oiled copy of the framed 'Three Maries', by Carracci.
5. *Comb.*, as **oiled-down**, smoothed or plastered down with (hair) oil; **oiled-up**, fouled or choked with oil.
1907 M. A. von Arnim *Fräulein Schmidt & Mr. Anstruther* xiii. 49 Her long respectable face and oiled-down hair. **1956** H. Gold *Man who was not with It* (1965) xxviii. 264 A yellow-fleshed, oiled-down, slicked-up head. **1925** *Morris Owner's Manual* 81 Sooty or oiled-up plugs will cause erratic running, loss of power and..increased petrol consumption.

oiler. Add: **4.** For *U.S. colloq.* read *colloq.* (orig. *U.S.*) and add to def.: and/or trousers. (Further examples.)
1899 'Q' *Ship of Stars* ix. 69 Taffy wore a suit of oilers of which he was mighty proud. **1924** R. J. Flaherty *My Eskimo Friends* i. i. 6 Old Puggie, donning oilers and sou'wester,..settled down to the helm. **1969** *Islander* (Victoria, B.C.) 22 June 2/2 You can always spot Dolfie in his favorite open boat, because he almost invariably wears yellow oilers.
6. b. A naval vessel carrying oil for the use of other ships.
1916 'Taffrail' *Pincher Martin* xiv. 255 Perhaps they would be going alongside an oiler at dead of night to replenish their fuel, and the wind would get on the wrong bow. **1923** [see *oiling *vbl. sb.* 3]. **1943** *Times* 8 Dec. 3/3 A flash was followed by a sheet of flame and the oiler was burning fiercely when our aircraft left. **1960** 'N. Shute' *Trustee from Toolroom* viii. 215 She was a Fleet oiler that had discharged her cargo at Christmas Island. **1973** *Times* 15 Mar. 5/4 Sightings included a Soviet F class diesel electric submarine together with..an oiler. **1974** *Union* (S. Carolina) *Daily Times* 22 Apr. 8/1 The carrier Kitty Hawk, several escorting destroyers and an oiler were expected to sail through the Strait of Malacca today.
c. A vehicle transporting oil.
1948 Mencken *Amer. Lang.* Suppl. II. 714 A tank-car is a car or oiler. **1975** I. K. Martin *Regan & Manhattan File* 54 That truck..was a General Mills sixteen-wheel oiler.
7. a. A vessel using fuel-oil.
1911 J. Barten *Compl. Nautical Pocket Dict.* 138/2 *Oiler*, Dampfschiff mit Oelfeuerung. **1915** *Chambers's Jrnl.* Oct. 664/1 The word 'oiler', like 'steamer', for oil-driven ships..is coming into use.
b. An oil-engine.
1926 *Chambers's Jrnl.* Dec. 843/1 She is a cold-starter, two stroke 'oiler'.
c. A vehicle using oil as its fuel.
1935 *Economist* 7 Dec. 1136/2 As far as goods traffic is concerned, the growth of the Diesel lorry or 'oiler' has been at the expense of the 'steamer' as much as of the petrol vehicle.
8. *U.S. slang.* A Mexican. (*derogatory*).
1907 S. E. White *Arizona Nights* i. iv. 82 A few oilers livin' near had water holes in the foothills. *Ibid.* iii. ii. 282 We're livin' like a lot of Oilers.

oi·l field. Also oilfield, oil-field. [f. Oil *sb.*^{1} + Field *sb.*] An area or a tract of country underlain by oil-bearing strata, the oil usu. being present in amounts that justify commercial exploitation and occupying a number of distinct pools. Also *attrib.* Cf. *oil province* s.v. *oil *sb.*^{1} 6e.
1894 [in Dict. s.v. Oil *sb.*^{1} 6 e]. **1912** Tower & Roberts *Petroleum* i. 11 In most cases there is a comparatively large area which can be called an 'oil field'... Within the main field there are usually several distinct areas, or 'pools', from which the chief production comes. **1926** E. R. Lilley *Oil Industry* iii. 21 The expression 'oil field' may be used to indicate an individual area of small extent such as the Salt Creek field of Wyoming.., or a group of such areas as the Lima-Indiana field.., or even an area of as wide extent as the Mid-Continent field which includes parts of Kansas, Oklahoma, Texas, Arkansas, and Louisiana. **1930** C. T. Brunner *Probl. of Oil* i. 11 The bulk of the American output still comes from the mid-Continental and Californian oilfields. **1964** *Oceanogr. & Marine Biol.* II. 136 There is a decline of sulphate with depth in the interstitial water of cores, due to the activity of sulphate-reducing bacteria and the interstitial water eventually comes to resemble oil field brine. **1973** *Scotsman* 21 Feb. 1/4 Another oilfield has been discovered in the North Sea, just inside the British sector and about 170 miles east-north-east of Aberdeen. **1975** *Offshore Engineer* Oct. 61 (Advt.), We will be fully equipped for on the spot repair and maintenance work of all oil field tools.

oiling, *vbl. sb.* Add: **1.** Also, oil pollution; the discharge of oil from a ship.
1949 *Brit. Birds* XLII. 378 No signs of oiling were visible. **1970** *Nature* 9 May 573/2 A recent 'routine' winter oiling along the east coast of Scotland exacted a known toll of about 8,000 seabirds.
b. *spec.* The covering of the surface of water with oil. Cf. *oil *v.* 1 c.
1887 [see sense 1 in Dict.]. **1910** R. Ross et al. *Prevention of Malaria* vi. 270 (*heading*) Oiling. **1927** P. Hehir *Malaria in India* 313 Anopheles larvæ succumb to oiling more readily than Culex larvæ. **1949** F. C. Bishopp in M. F. Boyd *Malariology* II. lxi. 1340/1 Malaria cases have also been markedly reduced on estates where oiling has been practiced regularly. **1966** P. H. Manson-Bahr *Manson's Trop. Dis.* (ed. 16) iv. 70 Oiling kills mosquito larvæ probably in several ways.
3. The taking of oil on board, esp. for fuel.
1906 *Westm. Gaz.* 27 Feb. 2/1 The difficulties in the way of 'oiling', to coin a word to correspond to 'coaling', at foreign ports. **1923** *Man. Seamanship* (Admiralty) II. 79 (*heading*) Oiling from an oiler.
4. The action of *to oil out* (cf. Oil *v.* 1 b).
1859 [see Oil *v.* 1 b]. **1962** R. G. Haggar *Dict. Art Terms* 231/2 *Oiling out*, in oil painting, the process of bringing out passages which have sunk in; a drying oil is applied.., the excess is removed and the surface gently polished.

oilless, *a.* (Later examples.)
1922 [see *graphited *a.*]. **1940** *Chambers's Techn. Dict.* 591/1 *Oil-less circuit-breaker*, a circuit-breaker which does not use oil either as the quenching medium or for insulation purposes. **1972** H. Osborne *Pay-Day* i. 19 Panjeh had always been a poor relation on the Coast, an oil-less sheikhdom. **1974** *Economist* 21 Dec. 11/2 France.. has come back in from its oilless cold to work with its allies.

oilman. Add: **c.** A person in the oil industry; an employee of an oil company.
1865 *Atlantic Monthly* XV. 388 She was still considering her probable success in finding an oil-man to take her down the Creek. **1880** *Harper's Mag.* Dec. 65 There are engines and boilers and pumps to be built for the oil men. **1912** Tower & Roberts *Petroleum* v. 83 The flowing or gushing wells are always regarded as lucky strikes among oilmen. **1951** T. Sterling *House without Door* i. 6 She knew what Texas oil men looked like. **1972** *Times* 27 Sept. 21/7 The oilmen look longingly at the areas north of the 62nd Parallel. **1973** A. Price *October Men* viii. 113 The oilman Ian Howard, just back from a year..in Saudi Arabia. **1975** *Lamp* (Exxon Corporation) Winter 11/1 When they began to commit sizable sums to North Sea oil development three and four years ago, oilmen expected the investment to be a costly one.

oil-nut. b, c. (Earlier and later examples.)
1694 *Town Rec. Topsfield, Mass.* (1917) I. 86/2 From thence on a straight line to an oylenut tree which is Isaac Burtons tree marked. **1778** J. Carver *Trav. N.-Amer.* 500 The Butter or Oilnut... The tree grows in meadows, where the soil is rich and warm. **1813** H. Muhlenberg *Catal. Plantarum Americæ Septentrionalis* 96 Oil nut, (*Pyrularia* or *Hamiltonia oleifera*). **1832** D. J. Browne *Sylva Amer.* 173 In Massachusetts, New Hampshire and Vermont it [*sc.* the butternut] bears the name of Oil Nut. **1912** E. Seton *Forester's Manual* 41 White Walnut, Oil Nut, or Butternut (*Juglans cinerea*). **1933** J. K. Small *Man. S.E. Flora* 1250 P[yrularia] pubera... Buffalo-nut. Oil-nut. Mountain-coconut.

oilskinned, *a.* (Earlier example.)
1857 C. Kingsley *Two Yrs. Ago* I. iii. 80 Oil-skinned coast-guardsmen.

oilstone, *sb.* (Later examples.)
1853 [see *abrasive B. *sb.*]. **1885** [see *Arkansas 1]. **1947** J. C. Rich *Materials & Methods Sculpture* x. 300 Fine oilstones are generally used for sharpening flat chisels. **1966** A. W. Lewis *Gloss. Woodworking Terms* 63 *Oilstone slips* or *slipstones*, oilstones of a variety of shapes used to sharpen gouges, [etc.].

oi·l well. Also oil-well, oilwell. [f. Oil *sb.*^{1} + Well *sb.*^{1}] A shaft sunk to obtain oil or from which oil is obtained. Freq. *attrib.*
1847 L. Collins *Hist. Sk. Kentucky* 249 The American Oil well is situated three miles above Burksville. **1861** *Chem. News* 28 Sept. 164/2 He states that the yield from the wells has greatly decreased, and that the oil well region, for practical working, is of very limited extent. **1903** G. H. Montague *Rise & Progress Standard Oil Co.* i. 4 With the success of Drake's oil-well at Titusville, Pennsylvania, in 1859, refiners had been released from the necessity of distilling coal into petroleum before refining petroleum into kerosene. **1946** *Mod. Petroleum Technol.* (Inst. Petroleum) 86 There are two main systems employed in drilling oil-wells: (1) the rotary or mud-flush system, and (2) the cable-tool or percussion system. **1946** *Nature* 20 July 84/1 The oil industry uses large quantities of nitroglycerine explosives..in the so-called oil-well 'shooting', where the explosives are employed to shatter the underground formation and thus open up fissures through which the oil may flow freely to the well. **1976** *Globe & Mail* (Toronto) 4 Feb. 7/4 The study made conclusions similar to those of Dr. Pimlott, saying existing technology could not cope with an oilwell blowout in the Beaufort Sea.

oily, *a.* (*adv.*). Add: **2. c. oily wad,** (*a*) a torpedo boat burning fuel-oil (*disused*); (*b*) a seaman without a special skill (see also quot. 1961).
1925 Fraser & Gibbons *Soldier & Sailor Words* 213 *Oily wads*, a Navy nickname for a class of oil burning

torpedo boat destroyers. 1929 F. C. Bowen *Sea Slang* 97 *Oily Wads.* The name occasionally applied to seamen in the Navy who do not specialise in anything, from the amount of time they have to spend cleaning brass-work with oily wads. 1931 'Taffrail' *Endless Story* xxii. 344 Numbered from 1 to 36, they were generally known in the Service as the 'oily wads'. 1932 Kipling *Limits & Renewals* 199 Some oily-wad of a *Bulleana* struck up about not having got his proper bird. 1961 F. H. Burgess *Dict. Sailing* 152 *Oily wad,* a seaman with no ambition. 1963 *Times* 13 June 17/1 His first command, which he held from May 1908, to January 1910, was torpedo boat no. 14 in the Home Fleet, one of the first oil-burning ships in the Navy, known to those serving in them as 'oily wads'.

D. *sb.* An oilskin garment. Chiefly in *pl.*

1893–4 R. O. Heslop *Northumb. Words* II. 510 *Oily,* an oilskin coat. 1898 G. A. Rushton in W. A. Morgan *'House' on Sport* I. 73 Still wind and rain the next day—but we..putting on our oilies went ashore and tramped for miles. 1933 E. A. Robertson *Ordinary Families* iv. 69 Soaked to the skin in spite of their oilies, Sootie and Ronald came down into the cabin. 1959 'A. Fraser' *High Tension* x. 106, I..ran upstairs to put on a thick jersey... That and an oily would do. 1973 J. R. L. Anderson *Death on Rocks* vii. 127 His own oilies were in the club.

oink (oiŋk), *v.* [Echoic.] **a.** *intr.* Of a pig: to utter its characteristic sound. **b.** *transf.* To make a similar sound; to imitate this sound; to grunt like a pig. Also as *sb.*

1969 *New Yorker* 11 Oct. 55/2 (*caption*) I'm warning you—don't start oinking. 1971 *It* 2–16 June 11/4 Seale called the fab philosopher 'a moral coward who oinks like a pig'. 1971 E. Bullins *Hungered One* 143 The pitter-patter of the returning dog's feet came from the road, and the cricket music and an occasional pig's oink and a drowsy duck quacked at the dark. 1972 J. Wambaugh *Blue Knight* (1973) iv. 55 One young guy..leaned back in his chair and made a couple of oinks and said, 'I smell pig.' 1973 *Times* 11 Apr. 13/5 The spasmodic oink-oink of chalk on cue. 1977 C. McFadden *Serial* (1978) v. 16/2 They oinked at him, in concert, just about every time he opened his mouth.

ointment. *a fly in the ointment*: see *fly *sb.*[1] 1 e.

‖ **oiran** (oi·răn). [Jap.] A Japanese courtesan of high standing. Also *collect.*

1871 A. B. Mitford *Tales Old Japan* I. 67 They are employed to wait upon the *Oiran,* or fashionable courtesans. 1904 R. J. Farrer *Garden of Asia* xix. 181 Of all glories in Japan, the richest is that of the Oiran, or established beauty of the Yoshiwara. 1970 J. Kirkup *Japan behind Fan* 90 The boys..who are indistinguishable from the *oiran,* the geisha and other women. 1972 *Mainichi Daily News* (Japan) 6 Nov. 20/2 (Advt.), 'Oiran'—the glittering Yoshiwara courtesans—recreate the ceremonies attendant on greeting an honored patron.

‖ **Oireachtas** (e·rəχtəs). [Ir., = assembly, gathering, convocation.] **1.** The assembly and festival held annually by the Gaelic League of Ireland, a gathering similar in concept to the Welsh eisteddfod and Scottish mod. Also *attrib.*

1902 W. B. Yeats *Let.* 20 Jan. (1954) 364 Royalties..I shall ask A. P. Watt to send..to the Sec. of Gaelic League, Dublin. Do you think I should specify the purpose? Say an Oireachtas prize. 1910 *Encycl. Brit.* V. 616/1 In 1898 it was decided to hold a festival called the *Oireachtas* ('hosting, gathering') on the lines of the Welsh *Eisteddfod.* 1911 W. H. G. Flood *Story of Bagpipe* xx. 164 The *Mod* is somewhat analogous to the *Oireachtas* in Ireland. 1913 *Irishman* Feb. 4/1 The keynote of this Oireachtas must be that Irish shall be plainly and unmistakeably the dominant language. 1971 *Daily Tel.* 5 Aug. 10/3 Prominent among the tents on the field for the first time this year is one representing Eire's equivalent of the Eisteddfod—the Oireachtas, to be held in October.

2. The legislature or parliament of the Republic of Ireland, consisting of the president and two houses, a house of representatives (Dáil Éireann) the members of which are elected by proportional representation, and a partially nominated senate (Seanad Éireann).

In quot. 1947 *An t-Oireachtas* is the regular Irish form that includes the definite article.

1922 *Daily Mail* 4 Dec. 9 The Provisional Government will be out of office by Wednesday, and the Oireachtas, as the Free State Parliament will be known, will come into being, consisting of Seanad Eireann (Upper House), comprised of 60 senators, 30 of whom will be nominated by the Government, and the remainder by the elected deputies of the people, who will sit as a Lower House, under the name of Dail Eireann. 1923 W. B. Yeats *Senate Speeches* (1961) 53 A joint committee of both houses to consider how suitable accommodations for Oireachtas may be obtained. 1930 G. B. Shaw *John Bull's Other Island* Pref., in *Works* XI. 72 The Northern Parliament will not merge into the Oireachtas. 1938 *Ireland: Citizen's Manual* 27 The Oireachtas may not enact a law repugnant to the Constitution nor declare acts to be unlawful after their commission. 1947 S. Malone *Notes Procedure Oireachtas* p. vii, The National Parliament (An tOireachtas) consisting of the President and two Houses. 1969 Henig & Pinder *European Pol. Parties* xi. 447 To win an election was, and is, to win the right to a virtual monopoly of legislative and policy initiative and the power to manage and control the *Oireachtas* (Parliament). 1975 *Irish Times* 9 May 8/1 Mr. Lynch said that there had been a tradition among the staff of the Oireachtas that if there was to be any easement of Standing Orders, that easement would be given in favour of the Opposition. 1977 *Cork Examiner* 6 June 2/8 Ever since it has been bandied about both in accusation and refutation on the subsequent occasions of the passage of an electoral Bill through the Oireachtas.

oiticica (oitisi·kă). [Pg., f. Tupi *oitisica.*] A name used for several tropical South American trees, esp. *Licania rigida,* of the family Chrysobalanaceæ, whose crushed seeds yield an oil used in paints and varnishes. Also *attrib.*

1918 E. R. Bolton in *Analyst* XLIII. 251 (*title*) Oiticica oil—a new drying oil. *Ibid.,* During the early part of 1917 we examined a new oilseed sent apparently for the first time from Brazil, bearing the native name of Oiticica or Oilizika. 1925 Morrell & Wood *Chem. Drying Oils* ii. 67 The oil is extracted from the kernels of oiticica (a name applied to several species of *Moquilea* and *Conepia,* belonging to the *Rosaceæ*). 1931 B. Miall tr. *Guenther's Naturalist in Brazil* vii. 124 The handsomest tree of the Sertão is the Oiticica. A sturdy trunk, divided into many branches, lifts itself from roots which run high over the ground. From the dense green roof of foliage the older leaves hang like silver tassels. 1944 H. G. Kirschenbauer *Fats & Oils* vi. 88 Oiticica seed oil owes its outstanding drying properties to the presence of licanic acid. 1944 S. Putnam tr. *E. da Cunha's Rebellion in Backlands* ii. 123 An old abandoned shack overgrown with oiticica boughs. 1951 R. Mayer *Artist's Handbk.* iii. 112 Oiticica Oil is a Brazilian product. 1967 *Times* 27 Apr. 27/3 The rather bewildering news yesterday that the Oiticica crop in the State of Ceara in South America has been almost a complete failure may well have sent some commodity men hurrying for their dictionaries.

Ojibwa(y (ōudʒi·bwē̆). Numerous varr. [Ojibwa, based on a root meaning 'puckered' (see quot. 1824); *Chippewa is a corrupted form of *Ojibwa(y.*] **a.** A member of an Algonquian people of North American Indians, inhabiting the lands around Lake Superior and, in more recent times, certain adjacent areas from Saskatchewan to Lake Ontario. **b.** An Algonquian language spoken by this people. Also *attrib.* or as *adj.*

'*Chippewa* and *Ojibwa(y* are the same word. Of these the former was the common English form until well into the nineteenth century... *Chippewa* is now the common spelling in the U.S. and *Ojibwa(y* in Canada. Hence these forms tend to be used for somewhat different local groups, but the usage is not consistent. Ethnologists, and especially linguists, tend to use *Ojibwa(y* for all the groups in question.' (Dr. I. Goddard)

1700 in *Documents Colonial Hist. New-York* (1854) IV. 749 Upon the sides of [Lake Huron]..live several Nations, vizt. the Christinos, the Ochipoy [etc.]. 1783 in *Mass. Hist. Soc. Coll.* (1809) X. 123 Chactaws 600... Upichweys 3000. 1824 W. H. Keating *Narr. Expedition St. Peter's River* II. 151 The term Chippewa, which is generally applied to this nation, is derived from that of O'chepe'wag, which..signifies plaited shoes, from the fashion among those Indians of puckering their moccasins. 1835 C. F. Hoffman *Winter in West* II. 15 The Chippewa, or Ojibboai..is generally considered the *court language* of our North-western tribes. 1853 Dickens *Noble Savage* in *Househ. Words* 11 June 337/1 Mr. Catlin.. with his Ojibbeway Indians. 1855 Longfellow *Hiawatha* I. 13 From the great lakes of the Northland, From the land of the Ojibways. 1872 W. F. Butler *Great Lone Land* viii. 110 Little ones..jabbered the smallest amount of English or French, and a great deal of Ojibbeway, or Cree, or Assineboine. 1903 Chesterton *R. Browning* i. 7 If his great-aunt had been a Red Indian, should we not have said that only in the Ojibways and the Blackfeet do we find the Browning fantasticality combined with the Browning stoicism? 1916 [see *Montagnais *sb.* and *a.*]. 1921 E. Sapir *Language* 53 In many, as in Italian or Swedish or Ojibwa, long consonants are recognized as distinct from short ones. 1937 R. H. Lowie *Hist. Ethnol. Theory* ix. 133 Foremost among his earlier students was the part-Fox William Jones, who transcribed a superb series of Fox and Ojibwa texts. 1968 *Globe & Mail* (Toronto) 5 Feb. 6/3 A. E. Bigwin, an Ojibway who is a Toronto school principal, states [etc.]. 1972 W. B. Lockwood *Panorama Indo-Europ. Lang.* vii. 117 The biggest languages in this [*sc.* the Algonquian] family are Chippewa (USA) or Ojibwa (Canada) with 35,000 speakers, [etc.]. 1974 *Sat. Rev. World* (U.S.) 2 Nov. 23/2 Armed Ojibwa militants had occupied a 14-acre park..in the resort town of Kenora, Ontario.

‖ **ojime** (ōu·dʒime). [Jap., f. *o* string + *shime* fastening, fastener.] A bead or bead-like object, often very elaborate, used in Japan as a sliding fastening device on the strings of a bag or pouch, or of an inro.

1889 M. B. Huish *Japan & its Art* xii. 167 Japanese Art metal-work..consists of the following branches:-.. 4. Articles for personal use, notably pipes.., beads (*ojime*), [etc.]. 1902 F. Brinkley *Oriental Series: Japan* VII. v. 173 There is reason to think that the *ōjime* was the first highly ornate appendage of both the *inro* and the *kinchaku.* 1960 *Times* 2 Jan. 9/4 The inro consisted of interlocking compartments that..were opened and closed by means of sliding beads termed ojimes. 1972 *Times* 15 June 21/2 Each [inro] was complete with a coral ojime and ivory netsuke.

O.K. (ōukē̆i·), *a.*, *sb.*, and *v. colloq.* (orig. *U.S.*). Also **OK, o.k., ok.** [App. f. the initial letters of *oll* (or *orl*) *korrect,* jocular alteration or colloq. pronunc. of 'all correct': see A. W. Read in *Amer. Speech* XXXVIII (1963), XXXIX (1964), etc.

From the detailed evidence provided by A. W. Read it seems clear that *O.K.* first appeared as a jocular alteration of the initial letters of *all correct* (i.e. orl korrect) in 1839, and that in 1840 it was used as an election slogan for 'Old Kinderhook' (see sense A b). Thence by stages it made its way into general use. Other suggestions, e.g. that *O.K.* represents the Choctaw *oke* 'it is', or French *au quai,* or that it derives from a word in the West African language Wolof via slaves in the southern States of America, all lack any form of acceptable documentation.]

A. *adj.* **a.** Chiefly in predicative use or as *int.*: all correct, all right; satisfactory, good; well, in good health; also in phr. *O.K. by* (someone): acceptable to (that person); freq. used as an exclamation expressing agreement: 'yes', 'certainly', 'all right'; also appended to a statement or declaration as a strong form of challenge or appeal in which affirmation or agreement is expected. Also as *adv.*

1839 C. G. Greene in *Boston Morning Post* 23 Mar. 2/2 He..would have the 'contribution box', et ceteras, *o.k.*—all correct—and cause the corks to fly, like *sparks,* upward. 1839 *Salem Gaz.* 12 Apr. 2/3 The house was O.K. at the last concert, and did credit to the musical taste of the young ladies and gents. 1839 *Evening Tattler* (N.Y.) 2 Sept. 2/2 These 'wise men from the East'..are right..to play at bowls with us as long as we are willing to set ourselves up, like skittles, to be knocked down for their amusement and emolument. O K! all correct! 1839 *Boston Even. Transcript* 11 Oct. 2/3 Our Bank Directors have not thought it worth their while to call a meeting, even for consultation, on the subject. It is O.K. (*all correct*) in this quarter. 1839 *Philadelphia Gaz.* 12 Nov. 2/1 Yes—that's good—O.K.—I.S.B.D. [*sc.* it shall be done]. 1840 *Morning Herald* (N.Y.) 30 Mar. 2/1 A few years ago, some person accused Amos Kendall to General Jackson of being no better than he should be. 'Let me examine the papers,' said the old hero... The General did so and found every thing right. 'Tie up them papers,' said the General... 'Mark on them, "O.K.",' continued the General. O.K. was marked upon them. 'By the eternal', said the good old General.., 'Amos is *Ole Kurrek* (all correct) and no mistake.' *Ibid.* 21 Apr. 2/4 The Brigadier ..reviewed his Brigade..and pronounced every thing O.K. 1840 *Boston* (Mass.) *Daily Times* 15 Dec. 2/3 What is't that ails the people, Joe? They're in a kurious way, For every where I chance to go, There's nothing but O.K. 1847, etc. [see O 5 d]. 1853 F. Townsend *Fun & Earnest* 14 To the earnest inquiries of another, he simply responded, O.K. 1864 *Boy's Own Mag.* Nov. 450/1 No thought of taking the trouble to find out whether the order was O.K., or 'orl korrect', as Sir William Curtis phrased it. 1865 W. H. Russell *Atlantic Telegraph* 61 The communication with shore continued to improve, and was, in the language of telegraphers, O.K. 1866 *N. & Q.* 18 Aug. 128/2 The following telegram has been received from Mr. R. A. Glass..'O.K.', (all correct). 1874 E. S. Ward *Trotty's Wedding Tour* xiii. 133 We had an O.K. time till we went to bed. 1886 *Lantern* (New Orleans) 29 Sept. 3/2 Favetto umpired the game all O.K. 1894 C. H. Donovan *With Wilson in Matabeleland* xi. 253 As our American friends would say, we were still 'O.K.' 1918 [see Wet *a.* 4 b]. 1922 D. H. Lawrence *England, my England* 101 At first Joe thought the job O.K. 1922 J. Reith *Diary* 14 Dec. (1975) ii. 128 He said..that if things went OK I should get a rise soon. 1932 H. Nicolson *Public Faces* i. 8 'O.K.,' he had said, 'I'll remind old Peabottle.'.. The expression 'O.K.' was not one which should be used in the Foreign Office, and least of all by an Assistant Private Secretary. 1937 D. L. Sayers *Busman's Honeymoon* viii. 148 'I say, Mr. Superintendent, are you going to want me any more? I've got to get back to Town.' 'That's O.K. We've got your address.' 1939 *Times* 24 Oct. 4/5 'O.K.' is an abbreviation of the expression 'Orl korrec'—all correct. It is English, I think Cockney—not an Americanism. I was born in the sixties and remember it when I was a boy. 1940 'N. Blake' *Malice in Wonderland* i. ii. 20 Anything that was efficiently organised was O.K. by Paul Perry. 1941 *Coast to Coast 1941* 224 He'd have reckoned it was O.K. to have gone or to have done what I did. 1957 J. Montgomery *Twenties* xviii. 262 By mid-1929, when sound films had spread across Britain, there was hardly a town or village without some child who was saying 'O.K.' when previously he would have said 'Yes'. 1968 *Encounter* Sept. 22/1 Direct transliterations from Yiddish or 'Yinglish' versions thereof... O.K. by me. 1973 *Railway World* Apr. 172/2 It seems OK now to refer to that bit of the Mae Khlaung as the 'River Kwai'. 1973 [see the verb, below]. 1976 *Publishers Weekly* 12 Jan. 52/2 The older dog asks if Pepper will allow him to go along for awhile and Pepper says ok. 1976 *Punch* 11 Feb. (recto front cover), Harold rules—OK? 1976 *Spectator* 15 May 3/1 George Davis was released by the Home Secretary, to the delight of headline writers, and the groans of others. The *Spectator* is bored by that line OK? 1976 *Sunday Times* 16 May 42 When George Davis stepped out of Parkhurst prison last week few headline writers could resist the temptation. *George Davis is free OK?* (the Sun), *George Davis is free—but is it OK?* (the Guardian). 1976 *Observer* 13 June 1/5 He added belligerently: 'I don't want to answer no more questions, OK? No disrespect to the court.' 1976 *Sunday Express* 4 July 6/3 He kept going on and on: '..there are certain standards to be maintained in first-class compartments.'.. And when he left the train..he gave..a look which said: 'First Class Rules—O.K.?' 1977 *Times* 26 Apr. 8/4 The popular graffiti—*Rules-OK,* which originated amongst the Glasgow razor gangs of the thirties. 1977 *Zigzag* June 31/1 We could have had a great album, rather than an OK album.

b. Used as a slogan by the Democrats in the American election campaign of 1840, influenced by the initials of *Old Kinderhook,* a nickname for Martin Van Buren (1782–1862), the Democratic candidate for the presidency, who was born at Kinderhook in New York

State. *O.K. Club,* a Democratic club of New York City in 1840. *Obs.* exc. *Hist.*

1840 *Democratic Republ. New Era* 23 Mar. 3/2 (Advt.), The Democratic O.K. Club are hereby ordered to meet at the House of Jacob Colvin. 1840 *Newark Daily Advertiser* 28 Mar. 2/4 The *war cry* of the locofocos was O.K., the two letters paraded at the head of an inflammatory article in the New Era of the morning. 'Down with the whigs, boys, O.K.' was the shout of these poor, deluded men. 1840 *National Intelligencer* (Washington) 7 Apr. 1/2 Already the Locofocos have got out their banners and procession, and 'the Butt-enders' and 'Point-enders' are marching at night through our streets, led by the so-called 'O.K.' club, which is just now a cant phrase in Tammany. 1840 *Democratic Republ. New Era* 27 May 2/6 We acknowledge the receipt of a very pretty gold Pin,..having upon it the (to the 'Whigs') very frightful letters O.K., significant of the birth-place of Martin Van Buren, old Kinderhook, as also the rallying word of the Democracy of the late election, 'all correct'... Those who wear them should bear in mind that it will require their most strenuous exertions..to make all things O.K. 1948 Partridge *World of Words* (ed. 3) 175 O.K. was in 1840 the watchword of the O.K. Club, that Democratic Club of New York City which took its name from *Old Kinderhook,* nickname of Martin van Buren (1782–1862), born at Kinderhook in New York State and in 1836–40 the President of the United States.

c. Socially or culturally acceptable; correct; fashionable, modish; having or showing prestige, high-class.

1869 *Henry De Marson's New Singer's Jrnl.* xxxv. 246 The Stilton, sir, the cheese, the O.K. thing to do, On Sunday afternoon, is to toddle in the Zoo. 1899 R. Whiteing *No. 5 John St.* xxiii. 233 She objected to the parting of the ass's mane as 'too O.K. for a moke'. 1950 S. Potter *Some Notes Lifemanship* i. 30 The word 'diathesis'..is now on the O.K. list for conversationmen. *Ibid.* v. 78 Just as there are O.K.-words in conversationship, so there are *O.K.-people to mention* in Newstatesmanship. 1957 *Observer* 22 Sept. 5/4 She left her campaign to save the theatre in the elegant hands of the Piccadilly and St. James's Association Ltd.—a very O.K. organisation of local shopkeepers and business, who like to keep the district nice. 1958 *Spectator* 19 Sept. 360/3 Mr. Macmillan ended his letter by saying we must treat this crisis 'calmly and constructively'. Both these are very OK words just now. 1960 *New Left Rev.* May–June 55/1 To give up his clerk's post in favour of a much better paid (but socially less OK) job in a factory. 1963 *Listener* 17 Jan. 140/3 In an eminently 'Third Programme' talk..he drew a comparison between this opera and his recent *King Priam* which was bedevilled by O.K. names and words. 1973 *Times Lit. Suppl.* 8 June 650/5 Handy quotations from such OK literary luminaries as Macaulay, Nietzsche, Strindberg, [etc.]

B. *sb.* **a.** A member of the O.K. Club (see A. b, above). *Obs.*

1840 *Morning Herald* (N.Y.) 30 Mar. 2/1 The O.K.'s are now the most original and learned locofoco club of the day. *Ibid.* 4 Apr. 2/1 All the clubs of Buttenders, O.K.'s, N.C.'s, [etc.]. 1840 *Boston Even. Transcript* 15 Apr. 2/1 The tail of the Democratic party, the roarers, buttenders, ringtails, O.K.'s (flat burglary this latter title) and indomitables.

b. The letters 'O.K.', esp. as written on a document or the like, to express approval of its contents; an endorsement, approval, or authorization.

1841 'Dow, Jr.' *Short Patent Sermons* 106 Fortitude.. infuses new life into his soul, while Hope adds an O.K. to his condition. 1896 *Congress. Rec.* 5 Mar. 2507/1 The deputy marshall..would send word to the prosecuting attorney asking for an 'O.K.' 1901 Merwin & Webster *Calumet 'K'* xiv. 273 A formal permit..signed by Porter himself, and bearing the O.K. of the general manager. 1910 S. E. White in *Sunset* Sept. 311/1 The high office added his OK to the others. 1925 H. Crane *Let.* 21 Oct. (1965) 218 My questionnaire..had won an OK sign in the upper right corner. 1930 *Liberty* 11 Oct. 30/3 Rube copped a sneak on the joint to find out if it was ready. In twenty minutes he gives us the O.K. 1956 *Rev. Eng. Stud.* VII. 440 It is Pound who is to give the O.K. to the gods (not to God). 1961 L. Mumford *City in Hist.* xvii. 535 The fifth vice-president whose name or O.K. sets the final seal of responsibility upon an action.

C. *v. trans.* To endorse by marking with the letters 'O.K.'; to approve, agree to, sanction, pass.

1888 *Missouri Republican* 25 Jan. 10/4 The expression, 'Please O.K. and hurry return of my account,' is grammatically correct. The noun account is governed by the preposition of, and is also the object of the active transitive verb O.K. 1891 *Congress. Rec.* 13 Feb. 2635/2 If those who were to go into the clerical service of the Government were to be 'O.K.'d' by any one except the Civil Service Commission. 1898 H. E. Hamblen *Gen. Manager's Story* 82 He hunted the hook over until he found the 227's report signed, Grinnell, O.K'd., and signed by the man who had done the work. 1921 R. S. Woodworth *Psychol.* xix. 505 Not that Freud would OK our account of dreams up to this point. Far from it. 1923 Galsworthy *Captures* 192 He finished pencilling, O.K.'d the sheets,..and went back to his room. 1932 E. Wilson *Devil take Hindmost* xxi. 224 The company submits plans to us and we O.K. them... We've O.K.'d Boulder City. 1942 E. Paul *Narrow St.* xxvii. 238 Of course, he [*sc.* Petain] had not counted on having the decrees he signed dictated by a German Führer, or at least O.K.'d when their hearts beat exactly as one. 1973 P. Dickinson *Gift* v. 76 'OK, OK,' said Mr Venn soothingly... But you know quite well head office wouldn't OK it... I'm not going to risk it.' 'OK,' said Mr Palozzi. 1976 *Columbus* (Montana) *News* 27 May 3/4 Smith's report..was not officially OKed by the War Department for release until July 9.

oka, var. Oca in Dict. and Suppl.

okapi (okā·pi). [Native name.] A rare ungulate mammal, *Okapia johnstoni,* of the family Giraffidæ, about the size of a horse and reddish-brown in colour, with horizontal white stripes on the legs; native to forested regions of the Congo, where it was discovered in 1900 by Sir Harry Johnston (1858–1927), the English explorer.

1900 H. Johnston in *Proc. Zool. Soc.* 775, I found the Bambuba natives dwelling alongside the dwarfs called it 'Okapi'. 1901 *Chambers's Jrnl.* July 493/1 The native name for this strange beast, which is quite inoffensive, is the *okapi.* 1930 *Punch* 24 Sept. 337/2 A photographer has succeeded in getting a 'close-up' of the shy okapi. 1958 *Listener* 23 Jan. 154/1 The Pygmies are excellent trappers and they are paid to catch various kinds of beasts including the rare okapi, the short-necked forest giraffe, which are sold to zoos. 1960 M. Spark *Ballad of Peckham Rye* iii. 35 'You look to me like an Okapi,' he said. 'A what?' 'An Okapi is a rare beast from the Congo. It looks a little like a deer, but it tries to be a giraffe.' 1974 Mochi & Carter *Hoofed Mammals of World* i. 7 The only living relative of the giraffe is the okapi, a forest animal living in the great rain forest of the Congo.

okay (ōukē̆i·), *a.*, *sb.*, and *v. colloq.* (orig. *U.S.*). Also **okeh, okey.** [Repr. pronunc. of *O.K. *a.*, *sb.*, and *v.*] **A.** *adj.*

a. = *O.K. *adj.* a. Also as *adv.*

1919 Mencken *Amer. Lang.* 161 Dr. Woodrow Wilson is said..to use *okeh* in endorsing government papers. 1929 J. P. McEvoy *Hollywood Girl* ix. 147 *Jimmy* (*dashing out door*): I'll kill the son of a—*Girl* (*going back to kitchenette*): Okay, big boy. 1932 *Sunday Express* 3 July 9/6, I had given my hand to the comedian and heard him say: 'Hold on, baby... Hold on. It's okay. You're going to be fine.' 1934 N. Sainsbury *Gridiron Grit* xii. 142 Okay by me. 1936 J. B. Priestley *They walk in City* viii. 237 The short one took the letter... He nodded. 'Righty-o. Seems okay.' 1939 E. B. Wilson *My Memoir* 174 Approval was designated by 'Okeh, W.W.' on the margin of a paper. Someone asked why he [*sc.* Woodrow Wilson] did not use the 'O.K.' 'Because it is wrong,' Mr. Wilson said. He suggested that the inquirer look up 'okeh' in a dictionary. This he did, discovering that it is a Choctaw word meaning 'It is so'. 1944 'N. Shute' *Pastoral* i. 7 Okay. I'll tell the boys. 1945 E. Waugh *Brideshead Revisited* 304 'Don't let on to anyone that we've made a nonsense of the morning.' 'Okey, Ryder.' 1953 J. Y. Cousteau *Silent World* 91 Tailliez came up and reported, 'Everything okay. They're playing chess.' 1966 D. M. Thomas in *Listener* 17 Feb. 247/1 Okay, my starsick beauty!.. Where would you like to go? 1972 *Publishers Weekly* 6 Mar. 31/2 He says most of that magazine's ex-staffers have done 'okay' in re-adjusting. 1974 'E. Lathen' *Sweet & Low* xix. 185 He got through a third degree okay.

b. = *O.K. *adj.* c.

1958 *Observer* 7 Sept. 13/4 Being in fusion is the really okay thing now. 1966 *New Statesman* 11 Mar. 348/1 The writer's ideas are rooted deep in the soil of experience and have not been processed into pet ideas and okay-words before they have ripened. 1970 G. Greer *Female Eunuch* 159 His secretary had..moved out of Haight Ashbury when it ceased to be okay to live there. 1974 *Listener* 16 May 642/1 The current psychiatric okay-word is the cautious and benign 'disturbed'. 1976 *New Society* 1 Jan. 5/1 In spite of the levelling effect of package holidays, skiing is still an okay sport. A survey..showed that British skiers were more likely to be upper middle class ABs living in the prosperous southern counties.

B. *sb.* = *O.K. *sb.* b.

1925 Ade *Let.* 28 Mar. (1973) 104 If, while he was putting his okeh on this material, he privately disapproved of it and was sending word back east that the material was not what you wanted, of course he was putting me in a tough position. 1925 *Dollar Mag.* Dec. 207 To find new and more vivid forms of expression..in the hope that they will, in time, receive the okeh of the reading public. 1931 *North Amer. Rev.* Jan. 15/2 During the last two years Raskob has either put his okay on every major move that Jouett Shouse has made, or else suggested it himself in the first place. 1973 *Freedomways* XIII. 18 Nothing goes down without his okay. 1974 *Columbia* (S. Carolina) *Record* 25 Apr. 6-D/1 In the more than 14 years since the dam was first proposed, the estimated $75 million cost has more than doubled. Whether it will get an *okay* on environmental grounds has not been determined.

C. *v. trans.* = *O.K. *v.*

1930 *Amer. Speech* VI. 119 Parachute company stock sale okehed. 1938 *Times* 18 Jan. 13/4 The proposal to call this haunt of pleasure the 'Non-stop Journal Kino' was taken to the Supreme Court before being—as the delighted proprietor probably put it—okayed. 1945 H. I. Phillips *Private Purkey's Private Peace* vii. 40 When they okayed me at that abduction center 'for the duration' I thought it meant just for the duration of the war. 1947 *People* 22 June 5/3 Micky and ex-light-weight champion Dave Crowley did the same fight 25 times before the final take was okayed. 1958 S. Ellin *Eighth Circle* (1959) i. i. 18 Every place where you can okay it, you put down O.K. and your initials. 1968 *Listener* 5 Dec. 771/1 Okayed by Western governments, the Prague festival enjoyed a substantial dollar bonus in the form of the Illinois State University Jazz Band. 1974 *Times* 22 Feb. 19/4 Scripts Limited comes in after the screenwriter has been commissioned by a film company, has written his first draft and had it okayed.

Okazaki (ōukă·zā·ki). *Biology.* The name of Reiji *Okazaki* (1930–75), Japanese molecular biologist, used *attrib.* to designate fragments formed during the replication of chromosomal DNA, first described by Okazaki et al. in 1968 (*Proc. Nat. Acad. Sci.* LIX. 598).

1969 *Proc. Nat. Acad. Sci.* LXIV. 1065 Some of the Okazaki fragments are present in a single-stranded state in cell lysates prepared under nondenaturing conditions. 1971 *Jrnl. Molecular Biol.* LVII. 351 The assembly of small fragments of newly synthesized DNA—the so-called Okazaki fragments..—to form a larger structure. 1975 *Nature* 4 Sept. 76/3 The synthesis of new DNA strands during the process of DNA duplication seems to occur in rather short sections (now known to everyone as 'Okazaki pieces').

oke (ōuk), *a. colloq.* (orig. *U.S.*). [Abbrev. of pronunc. of *O.K. *a.*, *sb.*, and *v.*] = *O.K. *adj.* a.

1929 D. Hammett *Dain Curse* (1930) v. 45 Try not breathing so hard. Everything will probably be oke. 1934 A. Woollcott *Let.* 31 Dec. (1946) 95 Dear Lady Cavendish..gets along famously with her mother-in-law, the Duchess of Devonshire. Has the Duchess saying 'Oke' already. 1933 Dylan Thomas *Let.* Oct. (1966) 38 Laleham arrangement, though in the air, is oke by me, and if there is any one expression worse than 'sez you' this is it. 1935 *Spectator* 15 Feb. 257/1 A child replied 'oke', to something I said. 1944 L. A. G. Strong *Director* 93 'How was it, Votty?' 'Oke.' 1951 Green & Laurie *Show Biz* 570/1 *Oke,* OK. 1960 S. H. Courtier *Gently dust Corpse* xiv. 196 He's oke now. Get some brandy.

okeh, okey, varr. *okay *a.*, *sb.*, and *v.*

okey-doke (ōu:kidōu·k), *a. colloq.* (orig. *U.S.*). Also **oakie-doke, okay-doke, okey-doke, okey-dokey, okie-dokie, okle-dokle.** [Redupl. *O.K. *a.*] = *O.K. *adj.* a.

1932 *Amer. Speech* VII. 334 *Okey-dokey,* O.K. 1934 M. H. Weseen *Dict. Amer. Slang* xiii. 190 *Okey doke,* O.K. *Ibid.* xxi. 373 *Okay doke,* satisfactory; agreeable; all right. 1935 *Evening Sun* (Baltimore) 7 Feb. 7/1 An attorney asked Carl Bush, witness, to answer a 'yes or no' question. 'Oakie-Doke,' replied Bush. 1936 M. Harrison *All Trees were Green* 305 Captain Sarsfield said: 'Everything okey-doke up at the mansion?' 1936 D. Powell *Turn, Magic Wheel* i. 36 He saw that tiresome red-faced fellow.., the man who knew everybody and said 'okey-dokie' to everything. 1944 M. Laski *Love on Supertax* iv. 53 Things are okey-doke for a lot of people now. 1947 B. Schulberg *Harder they Fall* vii. 121 'Hey, Killer, tell Jack to pick me up in front of the door right away.' 'Okle-dokle,' the Killer said. 'Where we goin'?' 1958 C. MacInnes *City of Spades* ii. iii. 95 One Guinness stout, right, I thank you, okey-doke, here it is. 1968 C. Brown in A. Dundes *Mother Wit* (1973) 236/1 There are certain classic soul terms... Among the classical expressions are 'solid', 'cool',..'okee doke', [etc.]. 1977 J. Fleming *Every Inch a Lady* i. i. 7 Light out, silence, everything okey dokey, she goes off to bed. 1977 *New Yorker* 16 May 33/1 Now, if for any reason you lose your paper clips, you're in big trouble. Okeydoke.

Okhrana (ǫχrā·nă). Also **Ochrana.** [a. Russ. *okhrána,* lit. 'guarding, protection'.] An organization of political police set up in 1881 in tsarist Russia after the assassination of Alexander II to maintain the security of the state and suppress revolutionary activities, and disbanded in 1917. Cf. *Cheka.

1899 P. Kropotkin *Mem. of Revolutionist* II. vi. 246 A secret league for the protection of the Tsar [Alexander III] was started... This league still exists in a more official shape, under the name of Okhrána (Protection). 1906 *Cosmopolitan* Dec. 237/2 The exact number, local distribution, and cost of the 'okhrana'..are known to no single official. 1920 *Contemp. Rev.* June 861 The Soviet authorities were confronted with the task..of abrogating individual laws permeated by the spirit of the Tsarist Okhrana. 1928 *Illustr. Hist. Russ. Rev.* I. 65 The confidential reports of the police and the 'Okhrana' furnish convincing evidence of the political nature of the unrest in the capital. 1930 A. T. Vassilyev (*title*) The Ochrana: the Russian secret police. 1949 I. Deutscher *Stalin* ii. 36 *Okhrana,* or the Third Department, was the political police set up in 1881, after the assassination of Alexander II. 1974 T. P. Whitney tr. *Solzhenitsyn's Gulag Archipelago* I. i. ii. 67 Section 13, presumably long since out of date, had to do with service in the Tsarist secret police—the Okhrana.

Okie (ōu·ki). *colloq.* [f. *Ok*lahoma, one of the United States + -ie (see -y[6]).] A migrant agricultural worker, *spec.* one from Oklahoma forced to leave his farm during the depression of the 1930s. Also, a native or inhabitant of Oklahoma. Also *attrib.* Cf. *Oklahoman.

1938 *Forum & Century* Jan. 12 About a fifth of [the migratory workers in California] are Okies. 1939 J. Steinbeck *Grapes of Wrath* xviii. 280 Okie use' ta mean you was from Oklahoma. Now it means you're a dirty son-of-a-bitch. 1941 S. Longstreet *Last Man around World* 356 The hillbilly, the Okie..and people who once shook hands with Warren G. Harding..fill the land. [illegible] *Daily Ardmoreite* (Ardmore, Okla.) 11 July 21/5 Soon they have less reason to be offended at being called 'Okies' than residents of other states have for their nicknames. 1957 J. Kerouac *On Road* (1958) 167 This was an Okie from Bakersfield, California. 1964 *Amer. Folk Music Occasional* i. 87 The songs of the 'Okies', those modern Forty-niners from the depleted farm lands of the Southwest. 1964 E. A. Nida *Toward Sci. Transl.* viii. 180 Procházka ..describes the problem of reproducing in Czech the Okie speech used by some of Steinbeck's characters. 1970 J. Hansen *Fadeout* (1972) viii. 67 I'm a dirty, ignorant Okie to him. 1975 *New Society* 19 June 705/1 Oklahoma got most attention, but migrants drifted from other places across the Great Plains and rural south... As time passed nearly all migrants came to be called Okies once they had reached California. 1978 *Chicago* June 40/2 Glenn Allen Smith's new play about the zany misadventures of an Okie hero-type is not quite zany enough to carry its weight.

‖ **okimono** (ōukimōu·no). [Jap., = 'put thing', f. *oku* to put + *mono* thing.] A standing ornament or figure, esp. one put in a guest room of a house.

1886 W. Anderson *Pict. Arts Japan* iii. xi. 112 The ornament pure and simple, the *Okimono* of the Japanese, was..made by artists in metal from a very early period... The first of the modern *Okimono* school appeared to have been a woman named Kamé or Kamé-jo. 1890 *Artistic Japan* V. 348 It is important to distinguish between netsukés, articles made for a special purpose..and okimonos ..ornaments never intended either for use or wear. 1911 J. F. Blacker *ABC of Jap. Art* vi. 116 Elephant's tusks.. were used chiefly for *okimonos*—alcove ornaments. 1916 Joly & Tomita *Jap. Art & Handicraft* 198 (*caption*) 22 *Okimono,* figure of Fukurokujiu with staff and *tama.* 1961 *Times* 7 Nov. 24/5 (Advt.), A collection of Japanese colour prints, netsuke, and works of art..including netsuke carved in wood and ivory, okimono and ivory carvings. 1962 F. A. Turk *Jap. Objets d'Art* ii. 81 From the late figure-group Netsuke,..arose the *okimono* (i.e. a carving to stand in an alcove). 1975 *Times* 8 May 16/3 A mid-nineteenth century wood figure of a demon..is an okimono rather than a true netsuke, too elaborate and delicate to use as a button.

Okinawan (ōukinā·wăn, ǫk-), *sb.* and *a.* [f. the place-name *Okinawa,* f. Jap. *okinawa,* lit. 'rope on the sea': see -an.] **A.** *sb.* A native or inhabitant of the Okinawa Islands, esp. of Okinawa, the largest of the Ryukyu (Nansei) group south-west of Japan; also, the dialect of Japanese spoken there. **B.** *adj.* Of or pertaining to Okinawa or the Okinawa Islands, to its people or its language.

1944 *Civil Affairs Handbk.: Ryukyu Islands* (U.S. Navy Dept.) xiv. 44 These phonetic differences impart a characteristic accent to Japanese as spoken by Okinawans. 1945 *N.Y. Times* 3 Apr. 3/1 Hundreds of kimono-clad Okinawans who fled to the hills with the first shells of the American pre-invasion bombardment are now streaming into our lines. 1945 *Ibid.* 5 Apr. 3/1 Dozens of enterprising GI's..saddled small, shaggy-maned Okinawan ponies with their gear. 1947 *Sci. Monthly* Mar. 235/2 Since the dawn of recorded history the Okinawans, although nominally independent, were influenced by both China and Japan and at times paid allegiance and tribute to both. At one time Okinawan sailing vessels carried on widespread commerce with the Asiatic mainland and the islands of the Western and Southwestern Pacific. 1955 C. J. Glacken *Great Loochoo* 3 The most significant characteristic of Okinawan culture is the family system. *Ibid.* viii. 166 This fish is caught with the aid of a home-made triangular wire device (*yamaguchi* in Okinawan). 1960 B. Leach *Potter in Japan* iii. 72 About seventy relatives, Okinawans and members of the Japanese Craft Society assembled. *Ibid.* 73 Two Okinawan scholars spoke next. 1964 *Listener* 24 Sept. 473/1 Brando..has appeared in comedy before—as the Okinawan Sakini in *The Teahouse of the August Moon.* 1966 P. S. Buck *People of Japan* (1968) xiv. 176 Okinawans wish once again to become part of Japan. 1973 C. L. Hogg *Okinawa* 13 Take the very name of the island... Okinawans have always called it Uchina, but no one else has ever paid the slightest attention to what Okinawans call their island... Okinawa, the name bestowed on Uchina by the Japanese, may be translated 'rope in the offing'. 1973 *Guardian* 24 May 4/4 More than half of Okinawans dislike Americans, because of their offensive behaviour. *Ibid.* 4/6 Two Okinawan women waiting at a bus stop were killed by a drunken American driver.

Oklahoma (ōuklahōu·mă). The name of a State in the south-west of the United States (see next), used *attrib.* and *absol.* to designate a kind of rummy orig. played in Oklahoma.

1945 A. A. Ostrow *Compl. Card Player* 578 *Oklahoma Rummy...* The cards rank in sequence as in standard rummy, but ace is high only and never low. Deuces are wild. 1948 [see *canasta]. 1949 J. Scarne *On Cards* (1955) ix. 63 *Oklahoma Gin.* In this variation the twenty-first card..determines the maximum number of points in unmatched cards with which a player may knock. *Ibid.* 64 Oklahoma usually incorporates the Spades Double feature. *Ibid.* xiii. 163 *Oklahoma Rummy.* A variation of Fortune Rummy. Played for years throughout the Middle West. 1973 G. F. Hervey *Hamlyn Illustr. Bk. Card Games* 110 Oklahoma may be played by any number of players from two to five, but is generally considered best when played by three.

Oklahoman (ōuklahōu·măn). [f. *Oklahoma,* one of the United States, f. Choctaw *okla* nation, people + *homma* red: see -an.] A native or inhabitant of the State of Oklahoma.

In quot. 1894 as the name of a newspaper.

1894 (*title*) Evening Oklahoman (Oklahoma City, Okla.). 1901 *Outlook* (N.Y.) 2 Feb. 280/1 Many of the old Oklahomans who are selling out expect to secure new homes at the opening. 1934 L. Gannett *Sweet Land* iv. 40 The Texans held that 'Pretty Boy' was no Oklahoman at all. Some held that he was a genuine West Texan. 1948 *Durant* (Okla.) *Daily Democrat* 4 July 2/5 Every Oklahoman appreciates an agency which seeks to protect our lives and property. 1959 C. Ogburn *Marauders* (1960) ii. 53 John P. McElmurray, a very tall, very spare, taciturn, profane Oklahoman. 1973 *Black Panther* 1 Sept. 16/1 'Those animals,' declared Mrs. Johnson, a typical racist Oklahoman.

O.K.-ness (ōukē̆i·nės). [f. *O.K. *a.*, *sb.*, and *v.* + -ness.] The fact or quality of being O.K.; acceptability.

1935 E. Gill *Let.* 31 Jan. (1947) 321, I rejoice..to have your assurance as to orthodoxy, decency & general o.k.ness of the..'article'. 1950 S. Potter *Some Notes Lifemanship* 76 The absolute O.K.-ness of French literature..cannot be too much emphasised. 1962 M. Drabble *Summer Bird-Cage* iv. 55, I carried an aura of vicarious theatrical OK-ness. 1969 *Punch* 19 Feb. 256/2 The rest of us..should be able to talk quite fluently about the varying OK-ness of split, one-piece, or cracked-edge construction.

okoume (okū·me). Also **okoumé.** [Native name.] = *Gaboon, gaboon.

1922 C. T. Campion tr. *Schweitzer's On Edge of Primeval Forest* vi. 108 The chief sorts [of wood] dealt in are mahogany,..and okoume (*Aucoumea klaineana*), the so-called false mahogany. 1933 —— tr. *Schweitzer's My Life & Thought* xiii. 163 The trade in okoume wood..was just beginning to flourish in the Ogowe district. 1947 J. C. Rich *Materials & Methods Sculpture* x. 293 *Okoume* is another name for Gaboon Mahogany. The wood is soft and light-weight and is a light reddish-brown color. 1956 *Handbk. Hardwoods* (Forest Prod. Res. Lab.) 97 Gaboon is known as okoumé in France. The tree grows mainly in French Equatorial Africa.

okra, now the usual spelling of Okro. (Further examples.)

1878 N. A. Donnelley *Lakeside Cook Bk.* 4/2 (*recipe*) Okra Gumbo. 1923 H. C. Thompson *Vegetable Crops* xxvii. 454 Okra is a tender plant and grows best in hot weather. 1949 *Caribbean Q.* I. i. 33 The little yellow cups of witchery known as Okra flowers. 1963 J. Kirkup *Tropic Temper* 49 The chicken curry..is particularly good, with its rich, really hot sauce containing a curious, ribbed, tasteless vegetable called okras. 1964 E. Huxley *Back Street New Worlds* xiv. 141 The Shepherd's Bush market has a shop devoted wholly to West African foods ..like gbure and okra, tete and apan. 1967 *Guardian* 8 Dec. 6/4 Fresh bhindi (which some people know as ladies' fingers, others as okra). 1972 Y. Lovelock *Vegetable Bk.* 147 Perhaps the best known member [of the hibiscus family] is okra.., of African origin and said to have been cultivated in ancient Egypt. It is now grown widely in Africa, the Mediterranean region and the Americas. 1973 *News & Courier* (Charleston, S. Carolina) 4 Nov. 3-E/1 A lunch of sandwiches and okra soup will be served from noon to 1 p.m.

‖ **okrug** (ǫ·krug). [a. Russ. and Bulg. *ókrug.*] In Russia and Bulgaria, a territorial division for administrative and other purposes. Also *attrib.* Cf. *oblast.

1886 *Encycl. Brit.* XXI. 69/2 Area and population of the Russian Empire... *Okrugs,* or *otdyels* (territories) under military government. 1902 *Ibid.* XXVI. 448/2 The country [*sc.* Bulgaria] is divided into twenty-two departments (*okrŭg,* pl. *okrŭzi*), each administered by a prefect. 1935 B. W. Maxwell *Soviet State* i. i. 30 Russia, before the Revolution, was divided for purposes of administration into seventy-eight governments (*guberniya*), twenty-one regions (*oblast*), and one circuit (*okrug*). 1958 D. J. R. Scott *Russ. Polit. Institutions* ii. 74 The new administrative *okrugs*..completed in 1952, were the product of a new acute phase of the persistent concern over effective supervision of districts by higher authorities... Shortly after the death of Stalin..the *okrugs* began to be abolished again, and none of them is now in existence. 1971 J. S. Reshetar *Soviet Polity* vii. 257 The least of the ethnic autonomous administrative units is the 'national area' (*okrug*). They have been established for the numerically small peoples of the Soviet Far North and Far East who inhabit large and sparsely populated areas. 1976 *Survey* Spring 65 There are [in the USSR] 14 union republic central committees, 10 *okrug* committees..and 4,243 city and *raion* committees.

okta (ǫ·ktă). *Meteorol.* [f. Octa- by alteration.] (See quot. 1950.)

1950 *Meteorol. Gloss.* (Meteorol. Office) (ed. 3) Amendment List No. 4. 7 *Okta,* unit, equal to area of one eighth of the sky, used in specifying cloud amount. 1957 *Weather Map* (Meteorol. Office) (ed. 4) iv. 35 From January 1, 1949, cloud amount has been observed and reported in oktas or eighths of sky covered, code figure 0 representing a clear sky, 1 representing 1 okta (eighth) of sky covered or less, but not zero, 2 representing 2 oktas (eighths), and so on..8 representing 8 oktas (eighths) (sky completely covered), and 9 representing sky obscured or cloud amount cannot be estimated. 1960 *Handbk. Aviation Meteorol.* (Meteorol. Office) xii. 178 The number of oktas of sky covered by any particular cloud layer is estimated as if no other clouds were present. 1961 C. E. Wallington *Meteorol. for Glider Pilots* iii. 38 Cloud amount is usually reported in eighths of the sky covered... Sometimes the internationally convened word 'oktas' is substituted for 'eighths'. 1977 'J. Le Carré' *Hon. Schoolboy* xviii. 455 Cloud is anticipated at six to seven okta... One okta is one eighth of sky area covered.

ol (ǫl). *Chem.* [a. G. *ol* (A. Werner 1907, in *Ber. d. Deut. Chem. Ges.* XL. 2113), f. the suffix *-ol* -ol.] Used *attrib.* and in *Comb.* to designate a complex containing a hydroxyl group of which the oxygen atom is bonded to two metal atoms; also applied to the group itself.

1907 *Chem. Abstr.* I. 2537 There is no simple or normal hydroxyl group in these compounds because they will not add HX to form an aquo salt. Hence they should not be designated 'hydroxy' compounds, but may be designated 'ol' compounds, to signify that the hydroxyl group is in the complex radical. 1929 J. A. Wilson *Chem. & Leather* 29 Bjerrum succeeded in preparing an ol-compound with a nucleus containing 12 chromium atoms. 1931 *Jrnl. Physical Chem.* XXXV. 46 With the conversion of hydroxo groups to ol groups to oxo groups there results increasing resistance to the action of acids. *Ibid.* 47 The authors hazard the guess that the oxolation of ol complexes would result in a loss of the reaction with neutral salts. 1962 J. R. Leach tr. *Grinberg's Introd. Chem. Complex Compounds* ix. 266 The acid-base equilibria can.. be complicated by the formation of polynuclear complexes with 'ol' bridges, or with 'oxo' bridges. 1974 D. Nicholls *Inorganic Complexes* iii. 26 This species with the hydroxo or *ol* bridges can react in the presence of added base.

ol' (ōul). Also **ol.** A representation of a *colloq.* and Black English pronunc. of Old *a.* Cf. *ole[2].

1894 A. Morrison *Tales of Mean Streets* 224 'That's all right, ol' cock,' roared Bill Napper. 1901 W. N. Harben *Westerfelt* iv. 44 How are you, ol' hoss... Glad to see you. 1915 C. Johnson *Battleground Adventures* liv. 418 Holt met the ol' man comin' from the barn as hard as he could run. *Ibid.* 420, I got a little ol' box to sit on. 1938 M. K. Rawlings *Yearling* iv. 34 He'll rip them leaves offen the stems and cram 'em in his ugly ol' mouth like a person. 1944 C. Himes *Black on Black* (1973) 202 Just a li'l ol' knot at the bottom. 1971 *Jamaican Weekly Gleaner* 3 Nov. 32/2 'Sure hate to be back at that same old school.' 'Same ol' tests.' 1973 S. Henderson *Understanding New Black Poetry* 21 Thus in the spirituals we have..Tell ol Pharaoh to let my people go. 1977 *Washington Post* 7 Dec. B1/1 Some people will be carried away by NBC's Bette Midler special, 'Ol' Red Hair is Back', at 10 o'clock tonight on Channel 4.

-ol. 3. Add: see *-ole, which is now the usual form.

-ola, *suffix.* Chiefly *U.S.* Prob. derived from the second element of Pianola and now found esp. in commercial use as a suffix to form nouns, as *Moviola, *payola, *Victrola, etc.

See *Amer. Speech* (1961) XXXVI. 104–116.

olation (ǫlē̆i·ʃən). *Chem.* and *Tanning.* [f. *ol + -ation.

Though often attributed to Bjerrum and to Stiasny the word has not been found in their publications.]

Conversion of a complex into an ol form; linking of metal atoms by hydroxyl ligands. Used esp. with reference to chromium compounds used in tanning.

1931 *Jrnl. Physical Chem.* XXXV. 27 Bjerrum..postulated that perhaps the hydroxo groups in the complex ions were becoming more firmly bound to form larger complexes... Bjerrum called this process 'olation'. 1931 *Chem. Abstr.* XXV. 2325 The basicity of the Cr salt in a one-bath Cr liquor is greater than that of the liquor as a whole because of 'olation'; the complex ol compds. react with the free acid only very slowly. 1941 D. Woodroffe *Fund. Leather Sci.* viii. 107 This condensation is termed 'olation' and the larger complex compound obtained is described as an olated chromium salt. 1948 *Progress in Leather Sci. 1920–45* II. xxvi. 524 Stiasny..termed this process 'olation' and postulated that hydroxy groups react immediately with acid whereas 'ol-groups' only react after deolation. 1969 T. C. Thorstensen *Pract. Leather Technol.* viii. 120 At higher temperatures there is greater fixation of the chrome tanning compound by the hide protein and greater olation of the chrome complexes. 1974 D. Nicholls *Inorganic Complexes* iii. 26 The hydroxo–aquo complexes may combine together in a process known as *olation.*

Hence (as back-formations) **ola·te** *v. intr.*, to form an ol group or compound; **ola·ted** *ppl. a.*

1931 *Jrnl. Physical Chem.* XXXV. 45 The hydroxo groups then 'olated', resulting in the formation of large aggregates, eventually reaching colloidal size. *Ibid.,* It would be expected..that the greater the degree of olation, the more sluggish would be the reaction of the complex olated ion. 1936 E. W. Merry *Chrome Tanning Process* i. 9 Chrome alum solutions prepared hot are green and contain strongly olated chromium compounds. *Ibid.,* It is not possible to forecast what will be the actual basicity of the chrome salts..or the extent to which they will olate. 1941 [see above]. 1945 McLaughlin & Theis *Chem. Leather Manuf.* xiv. 416 A chrome compound is said to be 'olated' when one or more of its hydroxyl groups is held between two chromium atoms.

Olbers' paradox (ǫ·lbəɹz). *Astr.* [named after H. W. M. *Olbers* (1758–1840), German astronomer, who propounded it in *Astron. Jahrb.* (1826) 110.] The paradox that if stars were distributed evenly (in sufficient numbers) throughout an infinite static universe, the sky should be as bright at night as in the daytime, owing to the fact that whilst the apparent brightness of individual stars decreases with distance the number of stars increases in the same proportion.

1952 H. Bondi *Cosmol.* iii. 23 Olbers' paradox does not arise in a static universe in which, roughly speaking, the stars did not start to radiate until some moment which can be determined..to have been between 10^8 and 10^{10} years ago. 1969 Rosser & McCulloch *Relativity & High Energy Physics* vi. 131 Olbers' paradox can be resolved by the fact that the distant stars..are going away from the earth at high speeds (the expanding universe).

old, *a.* (*adv.*, *sb.*[1]) Add: **1. a.** *old folk(s)* (later examples), *old one, old un,* an elderly person, *esp.* one's father or mother; *any old*: see *any *a.* and *pron.* 1 e.

1836 Dickens *Pickw.* (1837) xx. 204 'It's the old 'un.' 'Old one,' said Mr. Pickwick. 'What old one?' 'My father, Sir,' replied Mr. Weller. 1854 C. M. Yonge *Heartsease* I. ii. xiv. 341 He is the great pride of the old folks at Worthbourne. 1868 *Haileyburian* I. 4/2 The Present won the toss, and completely 'penned' the 'old 'uns'. 1915 N. L. McClung *In Times like These* xix. 130 Did you ever visit an old folks' home and notice the different spirit shown by the men and women there? 1921 G. B. Shaw *Back to Methuselah* v. 253 There! What have you to say to that, old one? 1968 M. Bragg *Without City Wall* xxvi. 237 In the Women's Institute the streamers, well hung.. were still taut, after the Old Folks' Tea. 1976 *National Observer* (U.S.) 21 Feb. 2/4 Many more old folks would pay an additional charge for the protection in any one year than would enjoy its benefits against large doctor and hospital bills. And that's not the best way to harvest votes among the elderly.

c. *old bag* (see *bag *sb.* 16*); *old boy,* used of an old man (see also as main entry in Suppl.); *old geezer* (see Geezer in Dict. and Suppl.); *old girl*: see Girl *sb.* 2 a in Dict. and Suppl.; *old pot,* one's father (chiefly *Austral.*); *old trot* (see Trot *sb.*[2]); *old trout* [perh. var. of Trot *sb.*[2] infl. by Trout *sb.*[1] 4], *colloq.* applied to an old woman.

1893 G. B. Shaw *Widowers' Houses* ii. iv. 43 He wont have any news to break, poor old boy: she's read all the letters already. 1916 C. J. Dennis *Songs Sentimental Bloke* 124 *The old pot,* the male parent (from 'Rhyming Slang', the 'old pot and pan'—the 'old man'). 1930 G. B. Shaw *Apple Cart* i. 3 When they found him he was melancholy mad, poor old boy; and he never got over it. *a* 1938 C. J. Dennis in *Penguin Bk. Austral. Ballads* (1964) 234 Oh, w'erefore art you Romeo, young sir? Chuck yer ole pot, an' change yer moniker! 1938 N. Marsh *Artists in Crime* ix. 128 Miss Troy thought I wasn't good enough to come here, even if my old pot did keep a bottle store. *Ibid.* 129 'What about Mr. Pilgrim?' 'Aw, he's different... I get on with him good-oh, even if his old pot is one of these lords. Him and me's cobbers.' 1956 B. Goolden *Singing & Gold* viii. 172 Drives old ladies about the town and that... Wouldn't suit me though. Don't see yours truly rushing to carry the old trouts' shopping baskets. 1958 J. Cannan *And be a Villain* vi. 140 The old trout wasn't exactly throwing her money about, he'd say. 1964 R. Braddon *Year Angry Rabbit* x. 90 Too high and mighty for his old trout of a mum nowadays is young Gary. 1972 N. Marsh *Tied up in Tinsel* xviii. 197 'Er old pot was killed saving the colonel's life. 1974 L. Thomas *Tropic of Ruislip* ii. 42 You could scare the old boy and he'll spill his tea. He spills things all the time.

3. c. *any old*: see *any *a.* and *pron.* 1 e.

5. b. *old coon.*

1835 A. B. Longstreet *Georgia Scenes* 216 To be sure I will, my old coon—take it—take it, and welcome. 1862 *Punch* 1 Feb. 42/2, I guess them saucy Britishers Won't easy get to leeward Of such an all-fired smart old 'coon As William H. Seward. 1877 Bartlett *Dict. Amer.* (ed. 4) 436 'He's an old coon,' is said of one who is very shrewd; often applied to a political manager.

7. b. *old pal*: an old friend, freq. with reference to association or collusion in business, *spec.* in phr. *old pal's act* (and variants), favour or cooperation based on prior acquaintance.

a 1966 M. Allingham *Cargo of Eagles* (1968) ii. 26 The Old Pal's Act isn't confined to you public school types above stairs now. 1972 E. Grierson *Confessions of Country Magistrate* i. 7 What is this mysterious process by which the man in the street is suddenly transformed into the magistrate on the bench? How, if not by the Old Pals' Act or the Signs of the Zodiac, is the miracle accomplished? 1973 *Times* 23 May 2/3 All these favours given by the Post Office on the old pals network. 1975 T. Heald *Deadline* ii. 18 The old pals act will operate as far as the press is concerned.

d. Phr. (*as*) *old as the hills,* exceedingly old; perh. in allusion to Job xv. 7 'Art thou the first man that was born? or wast thou made before the hills?'.

1819 *Metropolis* I. iii. 58, I thought he was going to make a die of it! Why he's as old as the Hills. 1820 Scott *Monastery* ix. 251 If you were as good a priest as the Pope, and as old as the hills to boot, you shall not carry away Mary's book without her leave. 1821 Byron *Let.* 1 Oct. in *Works* (1901) V. 385 The Pulci Style, which the fools in England think was invented by Whistlecraft—it is as old as the hills in Italy. 1849 Dickens *Dav. Copp.* (1850) xv. 156 All the angles and corners, and carvings and mouldings, and quaint little panes of glass, and quainter little windows, though as old as the hills, were as pure as any snow that ever fell upon the hills. 1898 [see sense 7 a in Dict.]. 1914 'Bartimeus' *Naval Occasions* ix. 66 'Sides, she's as old as the hills. 1937 A. Huxley *Ends & Means* iv. 25 A violent revolution cannot achieve anything except the inevitable results of violence, which are as old as the hills. 1954 B. & R. North tr. *Duverger's Pol. Parties* ii. i. 255 Dictatorship is as old as the hills. 1956 A. Wilson *Anglo-Saxon Att.* i. i. 14 Fifty-five must seem as old as the hills to a girl like you.

e. *the old story* (and variants), a familiar tale or excuse (usu. with a connotation of implausibility).

1700 [see Story *sb.*[1] 6 a]. 1859 Geo. Eliot *Adam Bede* I. iv. 75 'What! father's forgot the coffin?' 'Ay, lad, th' old tale; but I shall get it done.' 1898 J. D. Brayshaw *Slum Silhouettes* 28 'What brought 'em to that?' Oh, the old story—liftin' their little finger. 1919 R. Frost *Let.* 4 Jan. (1964) 80 Pelle was good reading. But none of it was any news. Not a phrase but was old story. 1938 E. Ambler *Cause for Alarm* xi. 184 Too much or too little—empty stomachs or overfed ones—the old, old story.

8. a. *old bean*: see *bean *sb.* 6 e; *old boy*: see also Boy *sb.*[1] 5; *old dear*: usu. of a woman (cf. Dear *a.*[1] and *sb.*[2] B); *old fellow* (U.S.): an overseer or 'boss'; *old fruit*: see *fruit *sb.* 2 e; *old girl*: see Girl *sb.* 2 a in Dict. and Suppl.; *old hen* (see Hen *sb.* 5 a); *old horse* (cf. Horse *sb.* 4); *old hoss*: see Horse *sb.* 4, *hoss 2; *old lady*: a girl or woman, esp. one's wife or mother; also *transf.* of a man whose behaviour resembles that of an old lady; *old man* (earlier example); *old son* (cf. Son *sb.* 3 b); *old sport*: see *sport *sb.*[1]; *old thing*: see *thing *sb.*[2]; *old top* (cf. Top *sb.*[2]).

1880 [see Dear *a.*[1] and *sb.*[2] B]. 1955 J. Thomas *No Banners* ii. 23 Remember the old dear at La Souterraine who fed us on bread and ham and cheese and a bottle of wine? 1901 S. E. White *Blazed Trail* xxvii. 187 He was intensely loyal to his 'Old Fellows' [= 'bosses' of lumber camps]. 1906 Wodehouse *Love among Chickens* v. 63 Garney, old horse, you're a marvel. You think of everything. We'll buckle to right away. 1924 —— *Ukridge* i. 12 It's a wearing life, laddie. A wearing life, old horse. 1942 *R.A.F. Jrnl.* 3 Oct. 11 'Well, old horse,' I thought, 'You're going to be disappointed.' 1960 A. Christie *Adventure of Christmas Pudding* 209 He said with a remarkable lack of medical decorum: 'That you, Poirot, old horse?' 1976 'A. Hall' *Kobra Manifesto* ii. 24, I wish someone had told me, old horse. 1836 Dickens *Let.* 21 Mar. (1965) I. 141 Let me have particular word how your rheumatism is, old lady. 1859 Mrs. Gaskell *Lett.* (1966) 545 You must not send us any more work to do, old lady, for Caroline is slow, & there is a great deal to do. 1871 E. Eggleston *Hoosier Schoolmaster* (1872) xvii. 134 Here's the old lady and Shocky. 1914 Conrad *Chance* ii. i. 244 The old lady's first-rate, sir, thank you. 1932 D. L. Sayers *Have his Carcase* xii. 152 'There, there, Mother,' muttered Henry... 'Bit of a staggerer for the old lady, this.' 1938 H. Nicolson *Diary* 10 Nov. (1966) 378 This memorandum was not at all liked by the old ladies of the Executive. 1967 C. Himes *Black on Black* (1973) 133 A man called T-bone Smith sat..looking at television with his old lady, Tang. 1976 *New Yorker* 17 May 34/2 It is a sign that you wish to share your old lady. 1870 'Mark Twain' *Let.* 22 Mar. (1917) I. 172, I can make the money without lecturing. Therefore, old man, count me out. 1916 'Taffrail' *Pincher Martin* viii. 142 'Ow are we, ole son? Feelin' a bit squeamish?' 1951 S. Spender *World within World* ii. 66 Do you know, old son, this is the first time you've ever talked with me that I haven't been completely bored? 1974 L. Deighton *Spy Story* xx. 218 You're doing well, old son. 1974 N. Freeling *Dressing of Diamond* 41 Right then, old son. 1912 *Collier's* 28 Sept. 19/1 'Tough luck, old top,' he muttered. 1915 Wodehouse *Something Fresh* ii. 24, I say, Dickie, old top, I want to see you about something devilish important. 1932 A. J. Worrall *Eng. Idioms* 56, I say, old top! Do you like these?

c. In trivial use with connotations of familiarity and in jocular and mildly disparaging senses of persons and things.

Quot. 1898 under sense 6 in Dict. belongs here.

1905 *Smart Set* Sept. 117/2 No one else is going to run off with your old car. 1913 F. H. Burnett *T. Tembarom* xxxiv. 438 Whatever happens, you are both fixed all right... Whatever old thing happens. *a* 1917 E. A. Mackintosh *War, the Liberator* (1918) 91, I always wondered If our old barrage could Be half as bloody good As the Staff said it would. 1938 M. K. Rawlings *Yearling* i. 3 'There'll come a little old drizzly rain before nightfall,' he thought. 1942 Z. N. Hurston in A. Dundes *Mother Wit* (1973) 28/2, I had done..cooked you a great big old cake. 1945 *Tee Emm* (Air Ministry) V. 38 Getting the 'general impression'..to register in the old brain-box. 1949 B. A. Botkin *Treas. S. Folklore* p. xx, It is a land where the word 'old'—the Old South,..Old Man So-and-So, little old this-and-that—are terms of affection and pride. 1965 J. Bingham *Fragment of Fear* iv. 49 'I have been successful.'.. 'Good old you!' 1971 P. O'Donnell *Impossible Virgin* i. 12 When you look in poor old Tina's tum there's just a grotty old mish-mash of bits and pieces. 1971 D. Francis *Bonecrack* viii. 101 'They didn't take my advice.' 'Silly old them.'

9. b. Also *the old boy* (see Boy *sb.*[1] 6 in Dict. and Suppl.).

11. a. See also *old light(s* Light *sb.* 6d), *old school* (School *sb.* 5b), *old tenor* (Tenor *sb.*[1] 1c).

c. Prefixed to the name of a language, to denote an early period in its history, or the earliest of several periods, preceding that usu. called *middle* (see Middle *a.* 4b), as in *Old English* (see English *sb.* 1 b in Dict. and Suppl.), *Old Norse* (see Norse *sb.* 3), *Old Prussian* (see *Old Prussian *sb.* b). Abbrev. O (see O 5b in Dict. and Suppl.).

d. Prefixed to a sb. or adj. to denote a former member of a particular institution or society, esp. a public school.

1824 M. R. Mitford *Our Village* I. 147 One meets with an old Etonian, who retains his boyish love for that game [*sc.* cricket]. 1848 C. H. Newmarch *Recollections of Rugby* i. 1 'Oh! mihi præteritos referat si Jupiter annos,' is an exclamation, in which, remembering..a duck hunt at Swift's, every old Rugbæan will, I hope, most heartily concur. 1857 *Manx Sun* 4 July in *Geo. Eliot Lett.* (1954) II. 337 The writer is a gentleman of our own acquaintance, an old Cantab. 1870 *Wellingtonian* May 152 The above Match proved a very exciting one..inasmuch as it was only won by two wickets by the old Wellingtonians. 1892 (*title*) Eton of old..by an Old Colleger. 1902 S. A. Barnett in H. Barnett *Canon Barnett* (1918) II. xxxiv. The 'Old Northeyites' has kept the educational side well to the front. *Ibid.* 71 The 'Old Dalgleishers'—whose special feature is the Easter expedition—enjoyed it for the eighth year in succession. 1914 'I. Hay' *Knight on Wheels* (ed. 2) xviii. 172 Each happened to be wearing an Old Studleian tie, so common ground was established at once. 1920 Galsworthy *In Chancery* i. iv. 42 He went out to dinner alone—an old Malburian [*sic*] dinner. 1936 G. M. Young *Victorian England* xiv. 96 The Old Giggleswickian was not yet a named variety. 1964 C. Mackenzie *My Life & Times* III. 52 Henry Cruft had been at Eton and then..sent to Shrewsbury. He still considered himself an Old Etonian. *Ibid.* 126 Cyril Bailey..was an Old Pauline who had left before I went to St Paul's in 1894. 1967 V. Canning *Python Project* iii. 46 He..straightened his Old Etonian tie. 1970 P. Dickinson *Seals* i. 11 His Old Etonian tie was knotted round a starched white collar. 1975 *Listener* 6 Feb. 164/1 The son of an Army officer and an Old Harrovian.

12. a. *Old Adam* (later, allusive examples).

1888 Kipling *Wee Willie Winkie* (1889) 6 The idea that he shared a great secret..kept Wee Willie Winkie ..virtuous for three weeks. Then the Old Adam broke out, and he made..a 'camp-fire' at the bottom of the garden. 1976 *Listener* 26 Feb. 232/1 The best way to keep evil and the old Adam down was to flog the children.

b. *the old country* (earlier and later examples); also applied to a (person's) country of origin other than Great Britain, *spec.* (occas. in *pl.*) to the countries of Europe, the 'Old World'; hence *old-countryman*; *Old Commonwealth*: Canada, Australia, and New Zealand (cf. *New Commonwealth); *the Old Dominion*: see *dominion 2 b; *Old England* (further examples); *Old South,* the southern states of the U.S. before the civil war of 1861–5; *Old World* (earlier example).

c 1596 Donne *Poems* (1912) I. 76 If you from spoyle of th'old worlds farthest end. 1782 'J. H. St. John de Crèvecœur' *Lett. from Amer. Farmer* i. 3 A person who hath been to Paris, and who hath seen so many things up and down the old countries. 1796 F. Baily *Jrnl. Tour N. Amer.* 25 Dec. (1856) 172 The scenery..so very different from what we had been used to in the old country. 1828 *Amer. Q. Rev.* IV. 211 Even the illiterate of our country will distinguish an Englishman by his pronunciation, and will designate him as an 'old countryman'. 1840 *Southern Lit. Messenger* VI. 241/1 More of Old England is left in the hearts of the Old Dominion than in all the states beside, save [etc.]. 1848 Bartlett *Dict. Amer.* 239 *Old Countryman,* a native of England, Scotland, Ireland, or Wales. 1873 *Harper's Mag.* July Never in her most boastful days did the old South, under her cherished system of slave labor, produce better crops. 1884 *Boston Jrnl.* 30 Dec. 2/4 Our goods are crossing the water to keep alive old England. 1898 J. D. Brayshaw *Slum Silhouettes* 8 Loudly declaiming..about the injustice done to 'the ould counthry', and forcibly giving vent to his views upon 'Home Rule'. 1927 M. M. Bennett *Christison* xiv. 133 In 1877, twenty-five years after he had sailed from Liverpool for Victoria, Christison left Australia to visit the Old Country. 1947 E. A. McCourt *Flaming Hour* vi. 32 In the old country..there would be spinach, brussels sprouts, artichokes. 1950 W. L. James in A. Dundes *Mother Wit* (1973) 431/2 It was those cries which Negroes made famous in the Old South. 1965 *New Society* 26 Aug. 18/1 The 'old' Commonwealth consists of Canada, Australia and New Zealand; the 'new' Commonwealth includes all remaining Commonwealth countries. 1966 B. Deal *Fancy's Knell* (1967) ii. 26 Bill was Old South and Mildred wasn't. 1973 *Guardian* 26 Jan. 1/2 The rules as drafted would lead to unacceptable treatment of people from the Old Commonwealth. 1973 *Sunday Bulletin* (Philadelphia) (Discovery Suppl.) 14 Oct. 27/2 People referred to Europe (or any nation therein) as 'the old country'. 1975 *Listener* 29 May 692/2 We now see not just the African Commonwealth, but also the Old Commonwealth and the Asian Commonwealth, beginning to make their own direct links with the Community. 1975 R. Price *Our Man in Camelot* i. 25 There was much more of the Old South in Shirley's voice.

c. *Old Christmas* (*Day*) (later examples); also *Old Christmas Eve*; *Old Lady Day, Old Michaelmas* (*Day*) (earlier example), *Old Midsummer.*

1783 W. Owen *New Bk. of Fairs* 14 Friday before Old Michaelmas, meeting by custom for horned cattle. *Ibid.* 65 Monday before Old Midsummer July 5, for sheep and horned cattle. *Ibid.* 70 Monday before Old Lady Day, for broad and narrow cloths, and leather. 1931 *Sun* (Baltimore) 7 Jan. 7/2 In the church calendar the day is known as Epiphany and in England as Twelfth Night, but among the Colonists it was known as 'Little Christmas' or 'Old Christmas', deriving the name from the fact that when the calendar was changed centuries ago an error was made. 1935 *Evening Sun* (Baltimore) 5 Jan. 18/3 On Old Christmas Eve, tomorrow night, daffodils, hops and elders are supposed to shoot mysterious sprouts through snow or frozen ground. 1948 *Richmond* (Virginia) *Times-Dispatch* 8 Jan. 26/1 In Rodanthe, N.C., and probably in some other remote places, 'Old Christmas' is observed on January 5. 1956 *Sun* (Baltimore) 5 Jan. 3/2 Why residents of the Outer Banks celebrate Epiphany or Old Christmas, as well as December 25..is something lost in antiquity. 1969 in Halpert & Story *Christmas Mumming in Newfoundland* 176 From before Christmas till Old Christmas Day each Twelfth Day, they held high carnival.

d. *old days* (or *times*): past times; freq. in phr. *good* (or *bad*) *old days* (or *times*).

1828 *Oscotian* (ed. 2) I. 1 However glorious those '*good old times*' may have been, they still were destitute of one very important advantage. 1856 Geo. Eliot in *Westm. Rev.* X. 55 The aristocratic dilettantism which attempts to restore the 'good old times' by a sort of idyllic masquerading. 1898 G. B. Shaw *Mrs. Warren's Profession* ii.

197 Suppose we were both as poor as you were in those wretched old days. **1906** *Nature* 3 May (Suppl.) p. vii/2 In writing of times that are past and gone, while still within our recollection, we have all to be on our guard against a popular illusion as to the 'good old days'. **1911** G. B. Shaw *Getting Married* 261, I felt that I had left the follies and puerilities of the old days behind me for ever. **1932** H. E. Williams in N. Hodgins *Some Canadian Essays* 225 While museums exemplify the distances that we have travelled since 'the good old days', they are not the best place in which to extract the old-time flavour. **1935** *Discovery* Jan. 29/2 Even in the bad old days, however, there are some things on which Canada may well pride herself. **1950** E. H. Gombrich *Story of Art* xxv. 379 There was one thing to be said for the 'good old days'—no artist need ask himself why he had come into the world at all. **1958** A. Huxley *Brave New World Revisited* (1959) 27 In the bad old days children with considerable, or even with slight, hereditary defects rarely survived. **1973** *Archivum Linguisticum* IV. 90 Associated with 'the old days', that is with the Rana regime.

e. *old ice*: in polar regions, ice formed before the most recent winter; similarly *old snow* (see quots. 1952, 1966).

1856 E. K. Kane *Arctic Explorations* I. xii. 128 Fissures ..were beginning to break in every direction through the young ice... I therefore made for the old ice to seaward. **1885** *Encycl. Brit.* XIX. 328/1 Old ice is believed to become thicker in a second winter, and even to attain a thickness of 10 feet. **1935** *Handbk. Weather, Currents & Ice* (Meteorol. Office) vii. 102 The Arctic peak consists of old ice, which due to rafting and hummocking forms massive fields. **1952** *Jrnl. Glaciology* II. 150 The definition of *firn*, adopted by the Eidg. Institut für Schnee- und Lawinenforschung, and included in the latest 'Draft on an International Snow Classification' suggested by the Committee on Snow Classification of International Association of Scientific Hydrology, is as follows: 'old snow which has outlasted one summer at least (transformed into a dense heavy material as a result of frequent melting and freezing)'. **1966** T. Armstrong et al. *Illustr. Gloss. Snow & Ice* 30 *Old snow*, deposited snow whose transformation into *firn* is so far advanced that the original form of the ice crystals can no longer be recognized.

f. *old quantum theory*: see *Quantum theory; *Old Red Sandstone*: see Sandstone.

g. Of a coinage: designating a former monetary unit that has been replaced by a new one with the same name (see *New *a.* 4).

The French *old franc* was replaced in 1960; in Britain new' decimal currency was introduced in 1971.

1959 *Times* 10 Nov. 10/6 The new 'heavy franc', which is worth 100 old francs, is to become legal currency on January 1 next year. **1965** R. Ferguson *Woman with Secret* x. 76 She left me 40,000 francs. Old francs. **1969** *Times* 21 July (Decimal Currency Suppl.) p. 1/5 Below 5p the only old coin which will be an exact equivalent of the new will be sixpence (2½p). *Ibid.* 6 The ½p coin being smaller than the old farthing..will be universally unpopular. **1972** D. Lees *Zodiac* 46 That's over four thousand dollars..more than two million old francs. **1974** L. Thomas *Tropic of Ruislip* ix. 178 'Blimey,' she said. There's one of the old pennies in here. That's not yours, is it?' 'No,' he answered. 'I cashed all my old ones in.' **1976** *Listener* 8 Apr. 430/2 Gary was born on 5 April 1966... In those pre-decimal days, you could buy a loaf of bread for 15 old pennies.

C. *sb.*[1] **3. b.** *sing.* A type of ale noted for its

At that age..old fogeydom already lays his hand on most of us, little as we may expect it. *a* **1877** in Bartlett *Dict. Amer.* (1877) (ed. 4) 437 He's slow and rather *old-fogyish*. **1855** *Richmond* (Virginia) *Whig* 15 Mar. 1/1 Endorsed thus by two 'old liners', he was most cordially received. **1903** *N.Y. Even. Post* 31 Oct. 5 The old-liners quietly backbite him for taking up a 'fanatic' like Johnson. **1908** R. W. Chambers *Firing Line* xxix. 493, I didn't expect any cordiality..but..they classed us with the old-liners. **1925** A. Huxley *Those Barren Leaves* I. ii. 14 One of those large, handsome, old-masterish women. **1961** *Listener* 16 Nov. 822/3 There are no 'properties', no old-masterish bits and pieces, to keep the thing going. **1937** *Burlington Mag.* Mar. 137/1 The same contempt for 'old masterishness' and its devotees. **1968** S. Hynes *Edwardian Turn of Mind* ix. 317 A taste in painting that was neither Old Masterly..nor academic. **1911** H. S. Harrison *Queed* xxii. 276, I think old-soldierism is the meanest profession the Lord ever suffered to thrive.

3. *old-home, -issue, -line* (examples), *Line State* (Maryland), *-master, -standard, -wave, -year*.

1928 Blunden *Undertones of War* xvii. 177 Flinging old-home repartee at your pal passing by. **1959** *Word* XV. 147 Langer gives a considerable number of examples of old-home expressions which were lost and replaced. **1879** A. W. Tourgée *Fool's Errand* xvii. 87 Robert..was..an 'old-issue free nigger' (freed before the war). **1899** C. W. Chesnutt *Wife of his Youth* 214 Wright came of an 'old issue' free colored family, in which though negro blood was present in an attenuated strain, a line of free ancestry could be traced beyond the Revolutionary War. **1856** *Congress. Globe* 9 Jan. 180/3 Have they offered us one of my colleagues, an old-line Whig? **1908** R. W. Chambers *Firing Line* xxi. 353 I'm in an old-line institution. **1928** F. Scott Fitzgerald *Let.* 1 Feb. (1964) 383, I rode.. with the president of a very prominent club, not my own, a Princetonian of the rather old-line, conservative, very gentlemanly type. **1949** *Sun* (Baltimore) 7 Sept. 12/4 Mr. Taft's mental and moral force has been a reservoir of strength to the old-line men. **1958** *Spectator* 20 June 791/3 In spite of his reputation as an old-line Stalinist, Suslov supported Khrushchev. **1962** R. Tyre *Douglas in Saskatchewan* v. 78 The Socialists had high hopes of winning the 1934 election but the farmers were not quite ready yet to abandon their traditional support of the old line parties. **1973** *Deb. Senate Canada* 28 Mar. 2710/1 The two old-line parties are afraid of treading on the toes of the financial institutions. **1872** Schele de Vere *Americanisms* xii. 660 Maryland bears the proud title of *Old-line State* from the *Old-Line* regiments which she contributed to the Continental Army in the War of the Revolution—the only State that had regular troops of 'the line'. **1948** Mencken *Amer. Lang.* Suppl. II. 604 *Maryland Free State*..has overshadowed all the old nicknames..including *Old Line State* and *Terrapin State*. **1950** D. Gascoyne *Vagrant* 56 Though that's what this old-master lute-master opines. **1959** *Times* 19 Mar. 4/3 Still-life studies combining an old-master flavour with a slightly surrealist inclination. **1838** J. F. Cooper *Home as Found* I. x. 163 That is the First Presbyterian, or the old standard [church]; a very good house. **1962** *Listener* 30 Aug. 315/1 This reaction has not come from Old Wave film makers. **1967** *Observer* 26 Feb. 21/3 All the addicts were middle-aged...typical old-wave addicts. But what of the new ones? **1897** R. M. Stuart *In Simpkinsville* i. 14 They got him to come to the old year party one year, jest for the fun of it.

4. old bach (see *Bach *sb.* 1); **Old Baptist, Old Christian (church)** *U.S.*, names of religious denominations; **Old Bill** *slang*, the police force; a policeman; **old-clothes-man, -shop,** (later examples); **old contemptible**: see *Contemptible *a.* 4; **old crock** (see *Crock *sb.*[3] 4); **old firm,** a group of friends or associates (cf. *Firm *sb.*[1] 2 c); **old gang** *colloq.*, a group or clique of friends or colleagues, esp. politicians, accustomed to supporting each other in matters of business or policy; **Old Glory** *U.S.*, the 'Stars and Stripes'; **old guard** (see Guard *sb.* 9 b in Dict. and Suppl.); **old hand,** (*b*) (earlier and later examples); **old home week** *U.S.* (see quot. 1904); **Old Kingdom,** a name given collectively (*a*) to the Third, Fourth, Fifth, and Sixth Dynasties, which ruled Egypt from the 27th to the 22nd century B.C.; (*b*) to a period of Hittite history extending from the 18th to the 16th century B.C.; **old master** *U.S.*, the former master of a (Negro) slave; also **old mistress** (see also *Ole[2]); **old money,** old-established wealth; **old offender,** an habitual criminal; **old religion,** a religion or belief which is replaced or ousted by another, *spec.* (*a*) a pre-Christian religion; paganism; (*b*) witchcraft; (*c*) Roman Catholicism; **old-rich,** those whose wealth is old-established, opp. *new rich* (*New *a.* 8 d); also *attrib.* or as *adj.*; **Old Ritualist** [tr. Russ. *staroobryádets*], = *Old Believer; **old rope** *slang* (orig. *Naval*), rank tobacco; **old settler** [Settler 2 in Dict. and Suppl.], one of the earliest settlers in a community; **old ship** *Naval slang*, an old shipmate; **old sledge** (later examples); **old soldier** *sb. U.S. slang*, the remaining part of a smoked cigar or chewed quid; also, an empty liquor bottle (Webster 1909); **old-spelling,** the unstandardized early spelling of English; **old squaw,** the long-tailed duck, *Clangula hyemalis*; = Long-tail 1 b (in Dict. and Suppl.), Old wife 2; also *attrib.*; **old-standing** *a.* (further examples); **old style** *Typog.*, one of a group of type-faces first produced in the 19th century and modelled on the 18th-century old-face fount cast at the Caslon foundry; also *attrib.*; cf. *Caslon, *old-face; **old sweat** *slang*, an old soldier; **old thing** *Austral.* (see quots.); **old Thirteen** *U.S.*, the original thirteen American colonies, which declared their independence in 1776; also, the original 'Stars and Stripes', a flag with thirteen stars and thirteen stripes; **Old Tom** (earlier and later examples); **old witch,** a children's game; **old witch-grass,** a North American panic grass, *Panicum capillare*.

1845 A. Wiley in *Indiana Mag. Hist.* (1927) XXIII. 18, I see nothing awaiting the 'old Baptist' churches but utter annihilation. **1889** P. Butler *Personal Recoll.* 252 'Hardshell' Baptists..wish to be known as Old Baptists, or United Baptists. **1958** F. Norman *Bang to Rights* 138 Two Old Bill's came up to me and told me they had a warrant for my arrest. **1967** *Guardian* 14 Mar. 8/6 He observed a couple of men supping nearby who looked suspiciously like plainclothes men. Coulson asked the landlord. 'Oh no,' he said, 'they're drinking pints. Old Bills only drink halves.' **1970** G. F. Newman *Sir, You Bastard* viii. 272 Giving Old Bill a bung was still an offence, and there would have been no consideration for the information. **1973** K. Royce *Spider Underground* i. 19 It's me he's out to fix by bringing Old Bill about my ears. **1976** *New Statesman* 12 Mar. 322/3 If they were caught at it when the Old Bill (police) staged one of their frequent raids then we would all be up on a charge of 'maintaining a disorderly house'. **1849** E. Chamberlain *Indiana Gazetteer* (ed. 3) 175 Presbyterians, Methodists, United Brethren, Christian,..Old Christian, (or new Light) and Baptists. **1834** *Chambers's Edin. Jrnl.* III. 141/1, I feel convinced that these old-clothes-men only address persons of gentlemanly appearance. **1968** *N.Y. City* (Michelin Tire Corp.) 82 The dark smoke-filled bars which alternate with old-clothes dealers along the street..shopkeepers.. tailors and old-clothes men. **1851** A. O. Hall *Manhattaner* 6 Groups of old clo' shops, gaudily set forth with parti-colored handkerchiefs. **1930** A. P. Herbert *Water Gipsies* vii. 72 Five shillings each way... Don't desert the Old Firm! **1935** D. L. Sayers *Gaudy Night* iv. 64 If you ever want me, you will find the Old Firm at the usual stand. **1975** 'D. Jordan' *Black Account* xiii. 66 'Always happy to help the Old Firm,' he declared. **1885** J. Chamberlain in J. R. Ware *Passing Eng.* (1909) 187/1 In deference to his [*sc.* Lord Randolph Churchill's] opinion, there will no doubt be a clearance out of some of those whom the Fourth Party is in the habit of politely designating as the 'Old Gang'. **1889** *Judge* (N.Y.) XV. 368/1 Yankee Doodle came to town Astride his thoroughbred. He met the old gang going out With Grover at their head. **1891** G. B. Shaw *How to become Mus. Critic* (1960) 195 Mr Chappell has at last awakened to the fact that his stock players were becoming what vestry politicians call an old gang. **1900** —— *Let.* 4 Mar. (1972) II. 150 In excited times nominations are apt to be made freely; and what happens then is that though the old gang is pretty safe, the other seats go anyhow. **1901** *Punch* 3 Apr. 250/2 There is so much favoritism that only the Old Gang and Rank Outsiders get chosen. **1916** Mrs. Belloc Lowndes *Let.* 8 Dec. (1971) 78 Violet Markham thinks L.G. will last out a good while but that all 'the Old Gang' as people are beginning to call them, will gradually crystallise into a Peace Party. **1933** Wyndham Lewis *Old Gang & New Gang* 9 Every morning when Mr. Everyman opens his newspaper he reads about the doings of the 'Old Gang'—or rather about their non-doing and non-caring. **1934** G. G. Coulton in *S.P.E. Tract* XLIII. 103 Within six years he [*sc.* Hart] had gathered elveen colleagues whom, at a much later prize-giving, he affectionately described as 'the Old Gang', and of whom Fowler was one. **1940** H. Nicolson *Let.* 6 June (1967) 94 There is a growing feeling against what is called 'the old gang'... The men who have come back from the front feel that Kingsley Wood and Inskip let them down and must go. **1961** R. Hoggart *W. H. Auden: a Selection* 24 A feeling that 'the old gang' were always hopelessly out of touch and wrong. **1964** C. Barber *Ling. Change Present-Day Eng.* ii. 26 A rejection of Old Gang politics,..general resentment at the Establishment. **1862** W. Driver in *Salem* (Mass.) *Reg.* 10 Mar. 2, I carried my flag, 'Old Glory', as we have been used to call it, to the Capitol, presented it to the Ohio 6th. **1930** J. Dos Passos *42nd Parallel* II. 153 They wrapped me in the Stars and Stripes and brought me home on a frigate to be buried..I was wrapped in Old Glory. **1973** *Sat. Rev. World* (U.S.) 4 Dec. 16/2 The right to substitute the peace symbol for the stars in Old Glory. **1975** *Times* 15 Apr. 6/7 Come what may, the American Ambassador to Saigon..will be rescued by the Marines, with the 'Old Glory' flag rolled in his arms. **1845** C. Griffith *Present State of Port Philip* 76 The old hands are men, who, having been formerly convicts..have become free by the expiration of their sentences. **1857** R. B. Paul *Lett. from Canterbury, N.Z.* ii. 26 Only enter the dwelling of the roughest 'old hand' among us, and you will meet with.. much kindness. **1911** C. E. W. Bean *'Dreadnought' of Darling* xxxii. 283 Lots of these fellows near Bourke were 'old hands'. Some of them were decent, good fellows, and the rest—well, they were horrible—the blackest, unmitigated rascals, fearing neither God nor the devil, men who would stick at nothing. **1946** K. Tennant *Lost Heaven* (1947) 1 On one side is Limeburners', where the 'old hands' used to pound oyster shells for lime. **1904** *Boston Herald* 2 Aug. 6 In..Massachusetts this first week in August is being observed as Old Home Week, and preparations have been made for welcoming back..visitors who return to their native, or earlier, home to renew acquaintance with former scenes and companions. **1949** T. Rattigan *Harlequinade* 63 What with Mums in front and babies in the wings it's not so much a dress rehearsal

as old home week. **1973** 'I. Drummond' *Jaws of Watchdog* i. 11 He and Jenny embraced warmly. 'Old home week,' said the Princess sourly. **1905** J. H. Breasted *Egypt through Stereoscope* 22 With the accession of the 3rd Dynasty..we see Egypt rising into her first great period of power and prosperity, which we call the Old Kingdom. *Ibid.* 143 The king who made the sphinx must have dismantled some Old Kingdom mastabas to clear the rock. **1910** *Encycl. Brit.* IX. 39/1 A valuable stele from Sakkara of the beginning of the Old Kingdom was presented to the Ashmolean Museum at Oxford in 1683. **1928** C. Dawson *Age of Gods* vii. 151 We must be prepared to allow for a margin of error of more than 800 years in dealing with.. the Old Kingdom of Egypt. **1938** E. M. Sanford *Mediterranean World in Anc. Times* i. 55 The foundation of the united Old Kingdom of the Hittites was delayed by the rivalry of individual states. **1952** O. Gurney *Hittites* i. 25 Telipinus is usually regarded as the last king of the Old Kingdom. **1961** A. Gardiner *Egypt of Pharaohs* I. iv. 60 Our evidence for the Old Kingdom is purely archaeological. **1973** R. J. Williams in D. J. Wiseman *Peoples Old Testament Times* iv. 83 Great advances were made in medicine, which reached heights in the Old Kingdom never surpassed in ancient Egypt. **1974** *Encycl. Brit. Macropædia* VI. 465/1 There is good reason to regard this king [*sc.* Khasekhemui] as the founder of the Old Kingdom. **1872** S. Powers *Afoot & Alone* 61 Negroes everywhere..seemed to think they were not free unless they left the old master. **1892** 'Mark Twain' *Amer. Claimant* viii. 81 When a bell ring..en old marster tell me to—. **1949** B. A. Botkin *Treas. S. Folklore* I. iii. 58 The chief protagonist and antagonist of master-slave folklore are Old Massa and Old John. **1859** D. D. Emmett *Dixie's Land* (song) (1960), Old missus marry Will de Weaber... Here's a health to de next old missus. **1874** 'Mark Twain' in *Atlantic Monthly* Nov. 592/1 Well, bymeby my ole mistis say she's broke. **1963** *Times* 25 Feb. (Canada Suppl.) p. xv/3 Having noted that the Canadian rich, particularly the old-money rich, tend to dwell among exquisite eighteenth-century chattels, the poorer folk aspire to antiques themselves. **1966** 'D. Shannon' *With a Vengeance* i. 22 There's a lot o' money—kind of substantial *old* money, you know—her husband was a banker. **1967** L. J. Braun *Cat who ate Danish Modern* ii. 18 People with Old Money always avoid publicity on their real estate. **1817** *2nd Rep. Comm. State of Police of Metrop.* 329 in *Parl. Papers* VII. 321 The greater part of these Juvenile Offenders,..are mixed indiscriminately with old offenders of all ages. **1890** W. Booth *In Darkest Eng.* II. v. 177 C.M. Old offender, and penal servitude case. **1970** P. Laurie *Scotland Yard* x. 263 If he is an old offender, the threat failed to deter him. **1848** W. D. Cooley tr. *Erman's Trav. in Siberia* II. xii. 306 The Bugoi of the Buraets of the old religion, maintain that they know..how to deal with certain mischievous spirits. **1934** A. Huxley *Beyond Mexique Bay* 159 The old religion came..boldly out into the open in 1745. **1964** *Listener* 12 Mar. 445/3, I am glad to see that the witch-religion is becoming so respectable... Jean Morris.. now proposes..that the Templars were of the 'old religion'. **1967** D. Pinner *Ritual* x. 106 If the Old Religion possessed this village, I probably would be too frightened to be anything but a warlock voyeur. **1972** P. Dennison in N. Tiptaft *Religion in Birmingham* 140 Wherever a local squire remained Catholic there was a good chance for the survival of a small pocket of the old religion in his territory. **1973** J. Wainwright *High-Class Kill* 77 Some of 'em stumble against witchcraft. Sorcery. Demonology. The so-called 'old religion'. **1975** *Country Life* 6 Feb. 318/2 In Queen Mary's reign he was, as a reliable adherent of the old religion, put on the Council of the Welsh Marches. **1927** *Public Opinion* 18 Feb. 149/1 These mistakes..seem folly to an old-rich man. *Ibid.* 149/2 The old-rich know these things well enough, but the new-rich never discover them. **1976** T. Allbeury *Only Good German* xiv. 101 The kind of places that the old rich go to rather than the new rich. **1885** A. J. C. Hare *Studies in Russia* vi. 301 In later times the schismatics have divided into the Stároobriádtsi, or Old Ritualists..and the Bezpopoftsi, or priestless people. **1911** Old Ritualist [see *Old Believer]. **1954** G. Vernadsky *Hist. Russia* (ed. 4) v. 132 Outstanding among the leaders of the Old Ritualists, as the anti-Nikonians eventually became known, was the archpriest Avvakum. *Ibid.* viii. 180 The movement of the Old Ritualists by 1800 ceased to be a unit and broke into several separate sects. **1974** R. Pipes *Russia under Old Regime* ix. 236 Russian dissenters are customarily divided in two basic groups: the Old Believers, known to themselves as 'Old Ritualists' (*Staroobriadtsy*) and to the official church as 'Splitters' (*Raskol'niki*), and the Sectarians. **1943** Hunt & Pringle *Service Slang* 48 *Old rope*, any tobacco which offends the nostrils of those present, and especially the finer varieties such as Egyptian. **1946** J. Irving *Royal Navalese* 127 *Old Rope*, any offensive smelling tobacco. **1744** *Colonial Rec. Georgia* (1906) VI. 117 Thomas Ellis has been an old Settler in the Colony. **1815** Old settler [see Settler 2 b]. **1854** R. B. Paul *Some Acct. Canterbury Settlemt.* 5 Having now resided more than two years in the Canterbury Settlement..[I] may almost call [myself] an 'old settler'. **1964** S. M. Miller in I. L. Horowitz *New Sociol.* 293 'Old-settler' Protestant recruits largely from farm and rural areas. **1927** *Daily Express* 11 Oct. 3/4 He gave a vivid description of waiting for the train at Charing Cross, then he met an 'old ship', and they went to have a drink. **1948** Partridge *Dict. Forces' Slang* 131 *Old Ship*, a former messmate (Navy). **1884** 'C. E. Craddock' *In Tennessee Mts.* II. 82 The mingled charms of Old Sledge and apple-jack had occasioned comment. **1950** R. P. Warren *World Enough & Time* iii. 101 The groups of men who played 'Old Sledge' and 'Brag' on the sidewalk. **1834** W. A. Caruthers *Kentuckian in N.Y.* I. 12, I smokes the old sodgers what the gentlemen throws on the bar-room floor. **1869** 'Mark Twain' in *Buffalo Express* 4 Sept. 1/1 A wooden box of sand, sprinkled with cigar stubs and 'old soldiers'. *a* **1877** in Bartlett *Dict. Amer.* (1877) (ed. 4) 438 Ladies who swab our sidewalks,.. And.. Haul off old soldiers lying there at rest. **1936** *Amer. Speech* XI. 304/1 *Old soldier*, 'a partly-smoked cigar'... I have heard this more frequently as *dead soldier*, applied to empty beer or whiskey bottles but not to cigar butts. **1927** R. B. McKerrow *Introd. Bibliogr.* III. i. 246 The composition rates for old spelling texts are some 10 per cent above the normal rate. **1960** *Studies in Bibliogr.* XIII. 49 (*heading*) The rationale of old-spelling editions of the plays of Shakespeare and his contemporaries. **1969** *N.Y. Rev. Bks.* 30 Jan. 32/2 In the first place the words aren't old-spelling [in an edition of Shakespeare]. **1838** J. J. Audubon *Ornith. Biogr.* IV. 105 They have various appellations, among others those of 'old wives' and 'old squaws'. **1884** etc. [see Squaw *sb.* (and *a.*) 4]. **1892** B. Torrey *Foot-Path Way* 41 The cliffs..offer an excellent position from which to sweep the bay in search of loons, old-squaws, and other sea-fowl. **1963** *Kingston* (Ontario) *Whig-Standard* 8 Feb. 11/2 Large numbers of Old Squaw ducks were sighted during the survey. **1971** *Islander* (Victoria, B.C.) 10 Oct. 13/1 Oldsquaw is one of the few ducks with two complete annual plumage changes. **1975** *Globe & Mail* (Toronto) 9 Dec. 11/5 Almost all the affected ducks are oldsquaws, long-tailed Arctic ducks that winter by the thousand on the Great Lakes. **1927** A. Bennett *Let.* 14 Apr. (1966) I. 365 The Beaverbrook papers, which have my stuff at 1/6d a word under an old-standing arrangement. **1962** A. Sorsby in A. Pirie *Lens Metabolism Rel. Cataract* 298 The characteristic sub-epithelial opacities seen in this affection are an oldstanding observation. *c* **1869** (*title*) Specimens of old style types (Miller & Richard). **1884** Bigmore & Wyman *Bibliogr. Printing* II. 42 Perceiving the tendency to go back to a former taste in printing, this foundry [*sc.* Miller & Richard], about 1850, commenced to cut a series of what they termed 'old-style founts', the success of which has been unexampled in the annals of type-founding. **1913** J. H. Quinn *Library Cataloguing* xv. 222 'Old-style' in type does not mean old-fashioned, but the more artistic and readable type modelled on the lettering of the early printers, principally those of the Italian presses. **1966** H. Williamson *Methods Bk. Design* (ed. 2) viii. 101 In 1852 Miller and Richard..led the way to a new development by issuing specimens of a regularized old face which they named Old Style. The new class of old style types, of which this was the first, reverted to gradual shading and to oblique top-serifs, but retained vertical stress. **1919** *Athenæum* 8 Aug. 727/2 A 'gasper' is a cheap cigarette, an 'old sweat' an old soldier. **1924** A. J. Small *Frozen Gold* i. 38 You're a levelheaded old sweat, I know, or you wouldn't be carrying the button. *a* **1935** [see *Easy *a.* 13 b]. **1955** J. Thomas *No Banners* ix. 80 These were followed by two lank British privates, old sweats of the Regular Army. **1973** *Guardian* 16 Mar. 12/5, I speak feelingly as an 'old sweat' who served in Ireland in an earlier time of 'troubles'. **1848** H. W. Haygarth *Recoll. Bush Life Austral.* i. 6 The Traveller's entertainment is confined to the 'old thing', as it is contemptuously called, that is to say, beef and 'damper'. **1945** Baker *Austral. Lang.* iv. 80 It was what W. W. Dobie called the *muttonous* diet of the outback that produced the expression *the Old Thing* for a meal of mutton and damper. **1845** *Southern Lit. Messenger* XI. 584/2 Charleston..[was] the chief commercial city of the 'Old Thirteen'. **1854** B. F. Taylor *Jan. & June* 68 The 'Old Thirteen' were blazing bright—There were *only* thirteen then! **1904** *Hartford* (Connecticut) *Courant* 30 Aug. 10 We want to see the Old Thirteen draw closer and closer together. **1821** P. Egan *Real Life in London* I. ix. 187 *Old Tom*—It is customary in public-houses and gin-shops in London and its vicinity to exhibit a cask inscribed with large letters—OLD TOM, intended to indicate the best gin in the house. **1971** R. Dentry *Encounter at Kharmel* iv. 75 A bottle of Old Tom and two hot-glasses. **1881** *Harper's Mag.* Jan. 184/2 The young solks..played at 'prisoner's base' or 'old witch by the wayside'. **1906** *Dialect Notes* III. 148 *Old witch*,..an outdoor game. The players circle around one of their number, the old witch, to whom the following is addressed: 'Chickamy, chickamy, crany-crow... What time is it, old witch?' **1859** W. Darlington *Amer. Weeds* 403 Old-witch grass... Sandy pastures, cultivated grounds; throughout the United States. **1894** J. M. Coulter *Bot. W. Texas* III. 508 Old witch grass... Annual... In cultivated land everywhere.

Old Aca·demy. Also old Academy. [Academy 2.] The school of philosophy founded by Plato in the fourth century B.C., as distinguished from schools founded by later Heads of the Academy, the *Middle Academy* of Arcesilaus, and the *New Academy of Carneades. So **Old Acade·mic** *sb.* and *a.*

1702 S. Parker tr. *Tully's Five Bks. of De Finibus* v. 281, I am..well pleas'd that you've fallen upon so seasonable a Subject of Discourse, my Cousin being ambitious of a right Notion of that Hypothesis which the Old Academicks and Peripateticks you speak of, propagated about Moral Ends. **1744** W. Guthrie tr. *Cicero's Academical Treatises* in *Morals of Cicero* 304 If it were not too much Trouble, I should desire to hear anew, from your Mouth, both that and the whole System of the old Academy. *Ibid.* 316 The old Academics did not hold that all Virtue consisted in Reason. **1811** W. Ward *Acct. Writings, Relig. & Manners Hindoos* I. 357 This was the doctrine taught by Anaxagoras, and after him by Plato, and the whole Old Academy. **1874** J. S. Reid *Cicero's Academica* p. xviii, It is a positive duty to discuss all aspects of every question, after the example of the Old Academy and Aristotle. **1876** Alleyne & Goodwin tr. *Zeller's Plato & Older Acad.* xvi. 617 The Old Academy had even then, in many of its members, departed very far from genuine Platonism. **1885** J. S. Reid *Cicero's Academica* (new ed.) 15 Varro evidently means that Cicero, having in earlier works copied the *writings* of 'Old Academic' philosophers, is about to draw on the literary stores of the New Academy. **1908** *Encycl. Relig. & Ethics* I. 59/2 The 'Old Academy' carried on the discussion of the problems which Plato had raised in his oral teaching. **1937** *Oxf. Compan. Classical Lit.* 2/1 The Academy under these leaders was known as the Old Academy. **1950** F. Mayer *Hist. Ancient & Medieval Philos.* xiv. 198 Conventionally, the Platonic school is divided into three periods. The first is the period of the Old Academy, which lasted from 347 B.C. to 250 B.C. **1974** A. A. Long *Hellenistic Philos.* v. 224 Antiochus saw a very sharp distinction between the Old and the New Academy.

old age pension. [Old *a.* 2.] A pension paid in certain countries by the state or, less frequently, by a private institution, to persons who have reached a specified age and are eligible for such assistance; also *ellipt.*, as *old age*. Used *attrib.* with *act, scheme*, etc. So **old age pensioner,** one who receives an old age pension.

1879 *19th Cent.* VI. 903 When the great clubs are able to mature the scheme which they are already entertaining for the payment of old age pensions. **1890** *Chambers's Jrnl.* 8 Feb. 88/1 To qualify..the worker must contribute ..for an old-age pension for fourteen hundred and ten weeks. **1892** *Q. Rev.* Apr. 507 Old age pensions commence at seventy years of age. **1895** W. S. Cottew *Scheme for Old Age Pension Fund* 1/2 After all working and other expenses paid, interest, etc., to form an Old Age Pension Fund, and as a further addition to the fund, there shall be an Old Age Pension Rate, not to exceed one penny in the £. **1906** *Chambers's Jrnl.* 10 Mar. 239/1 There are two possible systems for an old-age pension scheme for this country. **1908** *Act 8 Edw. VII* c. 40 § 1 The sums required for the payment of old age pensions under this Act shall be paid out of moneys provided by Parliament. **1909** *Westm. Gaz.* 4 Jan. 3/3 We..believe that the Old-Age Pensions Act was almost universally intended to create a new social stratum. **1910** G. B. Shaw *Let.* 24 July (1972) II. 936 A list..which includes a choice collection of old-age-pensioners. **1931** *Times Lit. Suppl.* 4 June 441/2 The inspiration served its avowed purpose in providing him with 'an old-age pension'. **1936** *Act 26 Geo. V & 1 Edw. VIII* c. 31 § 4 Not more than one old age pension..shall be payable to any one person. **1945** A. Huxley *Let.* 30 Jan. (1969) 514 A nation like England..by 1975, will probably have declined to under forty millions, one quarter of whom will be drawing old age pensions. **1951** *Bull. Nat. Old People's Welfare Comm.* Feb. 10 In one London Borough on a given day each week old-age pensioners may attend the Municipal Baths at a cost of 1d. per bath. **1959** *Daily Tel.* 9 Apr. 1/7 Old age pensioners had much to gain from stability of prices. **1967** K. Giles *Death & Mr. Prettyman* viii. 160 'Not the old-age?'... 'The old-age for tax evaders? You're joking!' **1973** A. Christie *Postern of Fate* III. 127, I must go and talk to some old age pensioners at their club. **1976** *Sunday Tel.* 30 May 1/6 Sixteen old-age pensioners were injured..when a coach from Dundee crashed over an embankment.

Old Belie·ver. [tr. Russ. *starovér.*] A member of that section of the Russian Orthodox Church which refused to accept the liturgical reforms of the patriarch Nikon (1605–1681). Also called Raskolnik.

1814 R. Pinkerton tr. *Platon's Present State Greek Ch. in Russia* 293 On the death of the first leaders of the modern Raskolniks, some of the sects resolved to admit runaway priests into their communion, and to acknowledge their ordination on condition of their becoming Old Believers. **1879** L. B. Lang tr. *Rambaud's Hist. Russia* I. xxii. 433 She had arrested certain 'old believers'. **1911** *Encycl. Brit.* XIX. 692/1 This ruthlessness goes far to explain the unappeasable hatred with which the 'Old Ritualists' and the 'Old Believers', as they now began to be called, ever afterwards regarded Nikon and all his works. **1921** F. C. Conybeare *Russ. Dissenters* I. iii. 140 Court Kolovrat, the protector of the Old believers in Austria. **1931** *Times Lit. Suppl.* 12 Feb. 105/3 In Kiev, where he made the acquaintance of the community of Old Believers, who upheld the schism in the Orthodox Church. **1932** M. Eastman tr. *Trotsky's Hist. Russ. Revolution* I. i. 27 The struggle against the state church did not go farther than the creation of peasant sects, the faction of the Old Believers being the most powerful among them. **1957** *Oxf. Dict. Chr. Ch.* 960/2 His [*sc.* Nikon's] liturgical reforms were sanctioned..and the formation of the schismatical sect of the Old Believers followed. **1959** *Listener* 19 Feb. 338/3, I would put the number of practising Orthodox Christians (including Old Believers) [in the U.S.S.R.] somewhat lower than the figure..suggested by Mr. Kolarz. **1962** K. S. Latourette *Christianity in Revolutionary Age* IV. xix. 508 In 1957 the Old Believers were known to have a church in Moscow. **1963** V. Nabokov *Gift* ii. 122 In 1862, sixty Russian Old-Believers with their wives and children lived for half a year in these parts. **1963** N. V. Riasanovsky *Hist. Russia* xix. 221 Shchapov and numerous others have stressed the social composition of the Old Believers and the social and economic reasons for their rebellion. **1973** *Guardian* 5 Mar. 2/7 The Soviet magazine Science and Religion..was criticising small groups of Old Believers.. in Tuvinskaya province.

Old Bill. The name of a cartoon character created during the war of 1914–18 by the British cartoonist Bruce Bairnsfather (1888–1959) and portrayed as a grumbling veteran soldier with a large moustache. Freq. in allusive and *transf.* use.

1915 *Bystander* III. 4 Again, 'Old Bill' and 'Our Bert' and 'Alf', seriously comical and comically serious, fill the pages with their humour. **1925** Fraser & Gibbons *Soldier & Sailor Words* 213 *Old Bill*, a veteran. Any old Soldier; in particular one with a heavy, drooping moustache. (From Captain Bairnsfather's celebrated creation 'Old Bill'). **1930** *Daily Express* 6 Oct. 4/4 An enormous mouth fringed all around with stiff hairy bristles, just like an 'Old Bill' moustache. **1933** B. Bairnsfather *Laughing through Orient* i. 15 Old Bill who, for many years, has been so closely entwined with my existence. **1939** H. Hodge *Cab, Sir?* 54 Here comes Old Bill himself. **1942** P. V. Bradshaw *They make us Smile* 9 The creation of Old Bill was never deliberate. Bill somehow created himself. **1946** *R.A.F. Jrnl.* May 163 During the last war, Bruce Bairnsfather created 'Old Bill', a lovable grumbler, typifying the foot-slogger of the British Army in Flanders. **1973** *Times* 2 Nov. 13/7 The Old Bill moustache starts twitching.

John Quincy Adams, the 'Old Man Eloquent', expressed very happily what we now..believe. **1901** S. E. White *Claim Jumpers* i. 4 He's been pestering the old man to send him West. Old man doesn't approve. **1914** 'Bartimeus' *Naval Occasions* xix. 171 Not bad work,..bagging your Old Man's ship. **1932** [see *kid *v.*⁴]. **1946** R. Allen *Home Made Banners* xiii. 163 My old man says Quebec or no Quebec they'll have to send the Zombies over. **1974** 'J. Le Carré' *Tinker, Tailor* vi. 47 She was a sight better qualified than her old man.

b. (Earlier and later examples.)

1835 N. Ames *Old Sailor's Yarns* 53 The commander of a merchantman, although perhaps under twenty years of age, is invariably called the 'old man', by all hands on board. **1840** R. H. Dana *Two Yrs. before Mast* xxxi. 374 The 'old man'..was determined to carry sail till the last minute. **1845** *Knickerbocker* XXVI. 206 I've known the Old Man come on deck at midnight. **1873** 'Mark Twain' & Warner *Gilded Age* iv. 44 The 'old man' was the captain—he is always so, on steamboats and ships. **1897** Kipling *Capt. Cour.* vii. 143 The *Jennie Cushman*..cut clean in half—graound up an' trompled on at that! Not a quarter of a mile away. Dad's got the old man. **1916** 'Taffrail' *Pincher Martin* v. 68 Having a sherry-and-bitters with 'the old man'. **1924** 'P. Blundell' *Confessions of Seaman* ii. 22 You'd better come along and see the 'old man' now. He's just off ashore. **1958** N. Marsh *Singing in Shrouds* (1959) ix. 184 Did you ever know such a *bloody* Old Man! **1968** *Daily Tel.* 14 Aug. 15/4 'It was just like a furnace,' said Mr. Martin Jones, deckhand, of Slough, Bucks. 'The old man was grand.'

b*. Hence applied in the other Services to a commanding officer.

1830 J. P. Martin *Narr. Adventures Rev. Soldier* viii. 190 They and some others of the men..were about to have some fun with 'the old man', as they generally called the Captain. **1890** Kipling *Life's Handicap* (1891) 41 An' whin I'm let off in ord'ly-room through some thrick of the tongue an' a ready answer an' the ould man's mercy, is ut smilin' I feel? **1917** A. G. Empey *Over Top* 311 '*The Old Man*', captain of a company. He is called 'the old man', because generally his age is about twenty-eight. **1942** *R.A.F. Jrnl.* 3 Oct. 24 It was preposterous to think of the Old Man on a bicycle... The idea of the Old Man riding a bicycle set us back a long way. **1948** Partridge *Dict. Forces' Slang* 131 *Old Man*, the Commanding Officer. The Air Force gets it from the Army, which gets it from the Navy, which gets it from the Merchant Service. **1967** *Everybody's Mag.* (Austral.) 18 Jan. 36/2 Today, in Vietnam, Australians are again catching up on American Army Slang... The Company Commanding Officer is a CCO or the Old Man—even if he's all of 23. **1977** 'D. MacNeil' *Wolf in Fold* xiii. 137 The Old Man had commanded longer than most lieutenant-colonels.

d. *old man of the mountain(s)* [tr. Arab. *šaiḵ-al-jibal*], (*a*) name applied to Hasan ibu-al-Sabbah, founder of the Assassins (see Assassin 1) and his successors; (*b*) applied allusively to other political murderers, and *fig.* to persons of ruthless ambition; (*c*) a rock formation resembling the face of an old man.

1579 J. Frampton tr. *Marco Polo's Travels* xvii. 27 That way,..could not be travelled to Crerima for the crueltie of the king of that countrie,..from whome fewe coulde scape, but eyther were robbed or slayne.. And for this cause manye kings did paye him tribute, and hys name is as muche to saye, as the olde man of the mountayne. **1625** Purchas *Pilgrimes* III. i. iv. 72 Hauing spoken of the Countrey, the old man of the Mountayne shall bee spoken of, of whom Marco heard much from many. His name was Aloadine, and was a Mahumetan... Aloadine had certaine Youthes from twelve to twentie yeares of age,..other Lords and his Enemies were slaine by these his Assasines. **1773** W. Jones *Hist. Life Nader Shah* p. xiii, It may be worth while to remark in this place, that the Old man of the mountain, who is mentioned in our accounts of the Crusades, was no other than a Prince of the Ismaelian family. **1777** J. Richardson *Dict. Persian, Arabic & Eng.* p. xvi/1 He was stabbed by a Batanist, one of the subjects of the Old Man of the Mountain; whilst he was reading a petition which the assassin had presented. **1792** H. Walpole *Let.* 4 Sept. (1905) XV. 138 A whole senate has assumed the accursed dignity of the 'Old Man of the Mountain', and spawned a legion of assassins. **1818** W. Marsden tr. *Trav. Marco Polo* I. xxi. 112 (*heading*) Of the old man of the mountain; of his palace and gardens; of his capture and his death. *Ibid.* 114 There was no person however powerful, who having become exposed to the enmity of the Old man of the mountain, could escape assassination. **1837** H. Martineau *Society in Amer.* I. ii. 220 Our party..was..struck with the romance of the domestic history of the old man of the mountain, as the guide is called. **1871** *N.Y. Herald* 6 Sept. 6/6 It seems as if 'the Old Man of the Mountains' [*sc.* Brigham Young] meant to fight every step of the federal government for the supremacy of Utah. **1888** Kipling *In Black & White* (1889) 67 He might have been the original Old Man of the Mountains. **1905** H. W. C. Davis *Eng. under Normans & Angevins* xi. 308 An Arab writer lays the blame on Saladin, affirming that he had offered the chief of the Assassins, the Old Man of the Mountain, a heavy bribe. **1934** A. Huxley *Beyond Mexique Bay* 80 The lessons of Loyola and the Old Man of the Mountain. **1936** J. Buchan *Island of Sheep* vi. 118 Desperadoes who had crushed their lives were in-spanned in Castor's sense... like the servants of the Old Man of the Mountain in the Crusades. **1939** G. B. Pickwell *Deserts* 48/2 The devil's garden is well named: with boulders and 'barrels' and 'Old Men of the Mountain' it is a grotesque feature of the land of sun and wind and freakish rain. **1957** *Encycl. Brit.* II. 554/1 Before long perfervid imaginations detected the hand of the Old Man of the Mountain in political murders and attempts even in Europe. **1965** J. Fleming *Nothing is Number* i. ii. 21 You are great *assassins*,..the word itself is your very own, it comes from the Arabic *hashshash*, dating from the Crusades when your old sheik, Old Man of the Mountains, sent out his Moslem fanatics to kill the Christian leaders. They filled themselves with *hashish* to get themselves in the right mood.

e. *old man of the sea*: in the story of Sinbad the Sailor in the *Arabian Nights*, the sea-god who forced Sinbad to carry him on his shoulders for many days and nights until he was thwarted by being made so drunk that he toppled off. Hence, allusively, a person of whose company one may not easily be rid; a heavy and encumbering burden, esp. in *fig.* use. Also *attrib.*

1712 tr. *Arabian Nights' Entertainments* (ed. 2) III. lxxxiv. 57 You fell say they into the Hands of the old Man of the Sea, and are the first that ever escap'd strangling by him. **1809** W. Scott *Let.* 7 Aug. in J. G. Lockhart *Scott* (1837) II. vii. 252 About three years ago I accepted the office I hold in the Court of Session, the revenue to accrue to me only on the death of the old incumbent. But my friend has since taken out a new lease of life... Such odious deceivers are these invalids. Mine reminds me of Sindbad's Old Man of the Sea, and will certainly throttle me if I can't somehow dismount him. **1850** C. M. Yonge *Henrietta's Wish* viii. 112 Uncle Roger has got hold of him, and he is as bad as the old man of the sea. **1856** G. Meredith *Let.* 15 Dec. (1970) I. 28 The *Dulness* is something frightful, and hangs on my shoulders like Sinbad's old Man of the Sea. **1874** M. Clarke *His Natural Life* (1875) I. 9 The old-man-of-the-sea burden of parsimony and avarice which he had voluntarily taken upon him was not to be shaken off. **1899** *Strand Mag.* Mar. 308/1 When a man once gets a Cavalanci and plays to it, it sticks to him like the Old Man of the Sea. **1904** G. Stratton-Porter *Freckles* iii. 54 Again Freckles' 'old man of the sea' sat sullen and heavy on his shoulders and weighed him down until his step lagged and his heart ached. **1927** *Times* 22 July 15/4 The bad habit into which we slip almost unconsciously fixes itself about our necks as firmly as any Old Man of the Sea. **1947** M. Lowry *Under Volcano* ix. 281 What could she do under the weight of such a heritage? How could she rid herself of this old man of the sea? **1957** C. F. MacIntyre *Stephane Mallarmé: Selected Poems* p. viii, Mallarmé has been like the Old Man of the Sea, like the Biblical poor. There's no getting rid of him! **1965** N. Freeling *Criminal Conversation* ii. xx. 186, I am tired. I find you like the Old Man of the Sea. **1971** A. Price *Alamut Ambush* xiv. 169 'He's a man who likes to use others to do his own work. He likes to ride on other people's backs.' The Old Man of the Sea, thought Roskill.

f. *Theatr.* An actor playing the role of an old man, *esp.* one who specializes in such roles. Also, the role itself. Cf. *old woman 1 c.

1747 T. Whincop *Scanderbeg* 243/1 An Actor of great Humour in low Comedy, especially in the Parts of Old Men. **1762** J. Love *Let.* 5 July in D. Garrick *Private Corr.* (1831) I. 144 His feeble old men, which he has only tried one season, will increase your hopes. **1775** T. Holcroft *Let.* 1 June in *Mem.* (1816) I. ii. iv. 236, I have succeeded best in low comedy and old men. **1794** C. Mathews *Let.* 3 Aug. in A. Mathews *Mem. Charles Mathews* (1838) I. 101 He is a very respectable performer in a general line, but mostly 'old men'. **1849** *Theatrical Mirror* 24 Sept. 31 Mr. Basil Baker..is engaged for the first old men at Drury Lane. **1901** C. Morris *Life on Stage* vii. 39 A company was generally made up of a leading man.., first old man, second old man, heavy man, first comedian, [etc.]. **1957** *Oxf. Compan. Theatre* (ed. 2) 772/1 The old stock company was formed of a group of actors each of whom undertook some special line of business... The Old Man played Sir Anthony Absolute and Sir Peter Teazle, and was a person of consequence.

g. A person set in authority over others: a master, overseer, or foreman; a superintendent or senior official; a 'boss'.

1837 *Southern Lit. Messenger* III. 86, I say, darkie, the old man keeps good liquor, and plenty of belly timber, don't he? **1844** *Knickerbocker* XXIII. 83 The 'old man' himself came to the door, and looking down at his apprentice, shook his head sorrowfully. **1845** E. J. Wakefield *Adventure N.Z.* I. xi. 331 Tommy Evans, the 'old man' who headed the principal station, started..to get on board. **1887** C. B. George *40 Yrs. on Rail* ix. 167 They feel that if they can only lay it before the 'old man' it will be properly dealt with. **1913** C. E. Mulford *Coming of Cassidy* xii. 197 'Is there any chance to get a job here?' he asked anxiously. 'You'll have to quiz th' Old Man.' **1921** H. G. Wells *Grisly Folk* in *Story-Teller* Apr. 14/1 There was no Old Man who was lord and master and father of this particular crowd. **1935** A. J. Pollock *Underworld Speaks* 82/2 *Old man*, the big underworld boss; boss politician. **1949** W. Hertrich *Huntington Bot. Gardens* 32, I declined to accept it, but suggested that he deduct this amount 'from the Old Man's bill'. **1958** 'Castle' & 'Hailey' *Flight into Danger* x. 137 Is that you, Dave? Harry. Surprise for you—the Old Man is on the line. **1967** E. & M. A. Radford *No Reason for Murder* xii. 80 The Old Man is the traditional police name for a Chief Constable. **1974** 'P. B. Yuill' *Bornless Keeper* xiii. 119 Has the old man been on? He'll be wanting to ask your old mates at the Yard for help. *Ibid.*, The old man wants to hear a progress report.

h. Substituted familiarly for 'old Mr.—'. orig. and chiefly *U.S.*

1843 'R. Carlton' *New Purchase* I. 92 It ain't more nor a mile to ole-man Sturgisses. **1859** Bartlett *Dict. Amer.* (ed. 2) 301 In the South and West, instead of saying..'Old Mr. Smith', it is customary to say, 'Old man Smith'. **1862** R. R. Butler *Let.* 8 Jan. in *Congress. Globe* (4 Mar. 1868) 1664/1, I send a few lines to you by old man Jesse Price. **1902** A. D. McFaul *Ike Glidden* xvii. 126 There is old man Spencer who had always been poor. **1930** *Chicago Daily Maroon* 28 Oct. 1/3 Old Man Stagg spoke a few words in commending..the students for their show of enthusiasm. **1949** [see *old *a.* 8 c]. **1961** 'E. Lathen' *Banking on Death* (1962) xv. 123 Old man Michaels didn't like him.

i. *slang.* The penis.

1902 Farmer & Henley *Slang* V. 99/1 *Old man*,..the penis. **1968** R. Lait *Chance to Kill* xxii. 139 There was David getting out of bed in his shirt, his old man hanging out. **1971** B. W. Aldiss *Soldier Erect* 23 She had been opening up her legs before the reprise. Those glorious mobile buttocks... I felt my old man perking up again at the memory.

j. *fig.* Applied to things; spec. *Old Man River*, the Mississippi (see also quot. 1932).

1910 W. M. Raine *Bucky O'Connor* iii. 37 When Old Man Trouble comes knocking at the door. **1919** C. P. Thompson *Cocktails* 252 Why, being officially booked to meet Old Man Death on ground, I had kept the appointment in the air. **1927** Kern & Hammerstein (*song-title*) Old man river. **1932** *Sun* (Baltimore) 24 Sept. 16/3 Old Man River Sinking... The north branch of the Susquehanna river is lower than it was 112 years ago. **1933** *Lit. Digest* 12 Aug. 28/2 (*heading*) Who Owns Old Man River? **1649** *Natural Hist.* Nov. 427/3 At last they have succeeded in vaulting the natural barriers between the Great Lakes and Old Man River. **1976** B. Bova *Multiple Man* v. 55 St. Louis is a dull town... Old Man River is wide and sluggish.

4. b. *Austral.* and *N.Z. slang.* Used *attrib.* to denote the largeness or significance of the thing specified; freq. of animals (see also sense 4 a in Dict.).

1834 G. Bennett *Wanderings New South Wales* I. xv. 286 Many persons when alone are afraid to face a large 'old man' Kangaroo. **1845** R. Howitt *Impressions Australia Felix* 233, I stared at a man one day for saying that a certain allotment of land was 'an old man allotment': he meant a large allotment, the old-man kangaroo being the largest kangaroo. **1866** R. Henning *Let.* 18 July (1966) 226 Spring, a very fine kangaroo dog we have here, killed ..an old-man kangaroo about five feet high. **1888** D. Macdonald *Gum Boughs* 7 Who that has ridden across the Old Man Plain. **1902** Kipling *Just So Stories* 87 Still ran Kangaroo—Old Man Kangaroo... He ran till his hind legs ached. **1906** E. Dyson *Fact'ry 'Ands* xv. 199 Two 'underd ole-man rats that 'ad bin glued on t'Bunyip in mortil combat. **1930** J. Devanny *Bushman Burke* i. ii. 17 [He] had once taken an Old Man pig with a slasher. **1934** A. Russell *Tramp-Royal in Wild Austral.* xxix. 190 An 'Old Man' sand storm. Lashed up and hurried along by a forty-mile-an-hour gale..an inferno of swishing sand and gravel..in Central Australia. **1936** I. L. Idriess *Cattle King* xii. 107 The river-bed is indicated by wide flats, mostly lignum bush, by big old-man coolabahs, and big old gums in places. **1941** —— *Great Boomerang* xi. 82 Fifteen years may pass before an old man flood brings a miracle to the land. **1945** Baker *Austral. Lang.* xiv. 244 Especially heavy gales of this type are often called *old man southerlies* or *old man busters*. **1947** J. Stevenson-Hamilton *Wild Life S. Afr.* xxxi. 259, I was the owner of a large boarhound which killed a great many 'old men' baboons. **1953** A. Upfield *Murder must Wait* xviii. 157 An old-man red-gum growing close to the track. **1965** [see *dogger⁴]. **1972** P. Newton *Sheep Thief* ii. 17 The homestead..was fringed with a towering belt of real old-man pines. **1973** 'D. Shannon' *No Holiday for Crime* (1974) 78 You're going to be like Old Man Kangaroo, my girl. As per Mr. Kipling. Very truly sought after.

5. (Later examples.)

1920 E. Thomas *Collected Poems* 97 Old Man, or Lad's-love—in the name there's nothing To one that knows not Lad's-love, or Old Man, The hoar-green feathery herb. **1973** F. A. Boddy *Foliage Plants* iv. 62 Old world charm and sentimentality can be further satisfied with the grey, feathery, aromatic leaves of *Artemisia abrotanum*, commonly called southernwood, lad's love or old man.

7. old man salt-bush, an Australian shrub, *Atriplex nummularia*, of the family Chenopodiaceæ, used as food for sheep in dry areas.

1880 W. A. Dixon in *Jrnl. & Proc. R. Soc. New South Wales* XIV. 140 The order in which the salt-bushes proper are considered to stand from a grazier's point of view are 1st, *A. numularia* [*sic*], or old man salt-bush. **1903** 'T. Collins' *Such is Life* 16 He disappeared in the timber and old-man salt-bush. **1933** *Bulletin* (Sydney) 14 June 25/1 Old-man saltbush is a rapid grower and gives more fodder in drought-time than any other tree. **1954** B. Miles *Stars my Blanket* x. 58 A great valley, the floor..thickly covered with Old Man Saltbush. **1965** *Austral. Encycl.* VII. 541/2 The round-leaved *Atriplex nummularia* (old-man or cabbage saltbush) is one of the tallest species and may reach 10 feet in height.

8. old man's beard (earlier and later examples).

1742 W. Ellis *Mod. Husbandman* June vi. 67 In this Month [*sc.* June], be sure to cut..what we in Hertfordshire call the Old-Man's Beard. **1952** *Cape Argus Mag.* 30 Aug. 2/4 That soft yellowy-green parasite that festoons itself so theatrically over the tops of the trees, giving the forests that appearance of hoary old age, is known as 'old-man's-beard'. **1965** E. Richardson *Living Island* 10 Many standing trees are dead hosts to tattered Old Man's Beard and other lichens. **1972** R. & R. Wright *Cariboo Mileposts* 86 Old man's beard lichen, the small green plant often seen in trees. **1974** *Islander* (Victoria, B.C.) 3 Nov. 7/3 Across the path and up into the old man's beard hanging from the pines.

old-ma·nnish, *a.* [f. Old man + -ish.¹] Characteristic or suggestive of an old man.

1865 J. A. Symonds *Let.* 15 Apr. (1967) I. 535 Ughtred grows even more noisy, patronizing, old mannish, & goodnatured than he used to be. **1898** *Tit-Bits* 30 Apr. 85/1 We are all more or less acquainted with the precocious child—the 'old-mannish' boy or the 'old-womanish' girl. **1927** W. Deeping *Kitty* vii. 91 His affection for that corner of the City of Westminster grew more deep and old-mannish.

old master. *Art.* See Master *sb.*¹ 15 in Dict. and Suppl.

Oldowan (ǫ·ldowăn), *a.* Also **Olduwan.** [f. the name of the *Oldoway* (or Olduvai) Gorge, Tanzania + -an.] Belonging to an African culture of the early Pleistocene period, characterized by primitive stone tools.

1934 L. S. B. Leakey *Adam's Ancestors* v. 104 In the Kanam deposits we find examples of a culture which has been given the name of Oldowan. This name is derived from Oldoway, the site where this culture was first recognized. **1964** K. P. Oakley *Frameworks for dating Fossil Man* iv. 172 The oldest known artifacts in the world are the Oldowan pebble-tools which occur in.. South Africa, East Africa and North Africa. **1972** [see *nut-cracker 5]. **1973** B. J. Williams *Evolution & Human Origins* xi. 189/2 Even among Olduwan materials (pre-handaxe) there are stone balls that have been rounded and smoothed to a far greater degree than can be explained by any possible functional requirement.

Old Prussian, *sb.* and *a.* [f. Old *a.* + Prussian *a.* and *sb.*] **A.** *sb.* **a.** A member of a medieval people, related to the Lithuanians, who inhabited the shores of the Baltic sea east of the Vistula. **b.** The West Baltic language of this people, which ceased to be spoken in the 17th century. See also *Baltic A. *adj.* 2. **B.** *adj.* Of or pertaining to this people or their language.

1872 [see Lettic *a.* (*sb.*)]. **1891** [see *Baltic *a.* 2]. **1917** *Encycl. Relig. & Ethics* IX. 487/1 Both the Este and the Old Prussians drank mare's milk and mead. *Ibid.* 488/1 References to Old Prussian religion occur in Lives of St. Adalbert. **1922** [see *Balto-]. **1933** L. Bloomfield *Language* i. 13 A similar relation, though less close, was found to exist between the Baltic languages (Lithuanian, Lettish, and Old Prussian) and the Slavic. **1946** T. G. Chase *Story of Lithuania* i. 3 The Old Prussians..were annihilated by the Teutonic Knights. **1951** A. Spekke *Hist. Latvia* vi. 133 Christianus seems to have booked some success among the Old Prussian aristocracy. **1974** *Encycl. Brit. Micropædia* VII. 514/3 Old Prussian preserves many archaic Baltic features that do not occur in the related East Baltic languages.

old régime: see Régime 2 b.

old rose. [f. Old *a.* + Rose *sb.* and *a.*] **a.** A shrub rose belonging to a species long in cultivation or a variety grown before the development of the hybrid tea rose about 1890, generally bearing fragrant, less formal flowers during a single mid-summer period.

1885 'E. V. B.' *Ros Rosarum* p. xx, In my own garden I gather together and fondly nurture every Old Rose that can be found. **1899** G. Jekyll *Wood & Garden* vii. 78, I have also learnt from cottage gardens how pretty are some of the old Roses grown as standards. **1936** E. A. Bunyard *Old Garden Roses* p. xi, The Old Roses are restrained and never garish. **1955** G. S. Thomas *Old Shrub Roses* ii. 31 Many of the most shapely and sumptuous of our old roses were raised during the nineteenth century.

b. A shade of deep pink. Also *attrib.* or as *adj.*

1893 *Ladies' Home Jrnl.* Jan. 29/2 Old-rose and black.. is a specially fashionable combination. **1897** *Sears, Roebuck Catal.* 255/1, 36-inch all wool albatross, colors cream, pink, old rose, nile green. **1922** W. J. Locke *Tale of Triona* xxi. 241 Her mother's room, with the old rose curtains and Chippendale. **1923** W. de la Mare *Riddle* 290 The chest was empty, except that it was lined with silk of old-rose. **1932** G. Atherton *Adventures of Novelist* 349, I had brought with me an old-rose rug; I had the walls papered to match, and found an old-rose silken cover for the bed. **1948** 'J. Tey' *Franchise Affair* xi. 115 At the Alençon—cream paint and old-rose couches against the walls. **1973** *Harrods Christmas Catal.* 50 Cocktail bar with mahogany finish, and leather panels in old rose..£130.

old school. [f. Old *a.* + School *sb.*¹ 5.] **a.** A group of people or a section of society noted for its conservative views or principles; members of a profession or a political party who adhere to its traditional views or methods. Freq. *attrib.* or as *adj.* Also, *in the old school*: according to traditional or old-fashioned methods.

1798, etc. [See School *sb.*¹ 5 b]. **1806** T. G. Fessenden *Democracy Unveiled* (ed. 3) ii. 61 These bring grave old-school reflections. **1808** H. More *Lett.* (1925) 188 It was said more than twenty years ago, that I was the only one of the old school who strongly relished Cowper. **1815** *Niles' Reg.* IX. 120/2 The federal and 'old school' democratic candidate for congress. **1818** M. Edgeworth *Let.* 19 Sept. (1971) 101 Lord Bathurst is..an *agreeable* diplomatist..dry faced—of the old school. *c* **1830** [see Old *a.* 12 a]. **1838** J. F. Cooper *Eve Effingham* II. iii. 81, I could just get a look of our clergyman's wig; for he was an old school man. **1842** F. A. Kemble *Let.* 2 Oct. in *Rec. Later Life* (1882) II. 268 Some old-school Whigs, sound politicians, and great friends of mine. **1911** G. B. Shaw *Doctor's Dilemma* i. 23 Did I hear from the fireside armchair the bow-wow of the old school defending its drugs? *a* **1963** S. Plath *Crossing Water* (1971) 60 Bowing and truckling like an old-school oriental. **1972** 'E. Peters' *Death to Landlords!* i. 20, I was trained in the old school.. and by hard work I built up the business. **1973** E. McGirr *Bardel's Murder* ii. 45 He saw Captain Joyningstowe doing the old-school act with a couple of stern dowagers. **1974** *Times* 18 Apr. 19/5 The Old School certainly accept the view that the rise in the price of oil is deflationary.

b. Used *attrib.* to designate conservative or traditional religious views, as *old school Baptist, old school Church, old school Presbyterian.*

1816, etc. [see *new school a]. **1875** *Richmond* (Virginia) *Daily Whig* 3 Sept. 2/3 He should never have terminated his affiliation with the Old School Church. **1873** 'Mark Twain' & Warner *Gilded Age* vii. 80 Grandmother..was an Old-School Baptist. **1878** J. H. Beadle *Western Wilds* xii. 183 The Old School Baptisses never went nigh the Methodis' meetin' house. **1898** I. H. Harper *Life S. B. Anthony* I. 218, I recommend that you form an acquaintance..with some well-settled Old-School-Presbyterian clergyman. **1933** *Sun* (Baltimore) 21 Oct. 6/8 More than one thousand members of the Old School Baptist Churches are meeting in a three-day session at Little Creek Church. **1949** *Pacific Northwest Q.* Apr. 124 In the period before the Civil War [they] generally preferred to be called Old School Baptists. **1961** K. S. Latourette *Christianity in Revolutionary Age* III. vii. 166 We have seen the separation of the New School from the Old School Presbyterians.

old school tie. [f. Old *a.* + School *sb.*¹ 1 + Tie *sb.* 4.] A tie of characteristic pattern worn by former members of a particular school, esp. an English public school; used *transf.* and *fig.* to denote the wearer of such a tie and the behaviour and attitudes usually associated with it, esp. conservatism and group loyalty. Also *attrib.* or as *adj.* Hence **old school tie·ism.**

1932 Kipling *Limits & Renewals* 86, I was thinking over the moral significance of Old School ties and the British social fabric. **1936** S. Smith *Novel on Yellow Paper* 140 Cynthia was..an old-school-tie acquaintance of mine. **1939** G. Household *Rogue Male* 40 The only class-conscious people are..the suburban old-school-tie brigade and their wives. **1942** A. Christie *Body in Library* iii. 35 The Inspector was tacitly accusing him of favouring his own class—of shielding an 'old school tie'. **1943** J. B. Priestley *Daylight on Saturday* ix. 58 Now this Old School Tie stuff..is only a protest against inefficiency and nothing else. **1944** G. B. Shaw *Everybody's Pol. What's What?* i. 4 To the Old School Ties the dictators seem ignorant uneducated rebels. **1949** R. Chandler *Let.* 13 May in *R. Chandler Speaking* (1966) 139, I don't want to be revoltingly old-school-tie, but it does seem to me that a line has to be drawn. **1957** *Numbers* Mar. 17, I just met an old school-tie in the cloakroom. **1958** S. Hyland *Who goes Hang?* xi. 54 He was an Etonian..and he made his jokes unsmilingly as a concession to the old school tie convention. **1960** T. Hughes *Lupercal* 45 As soon Let the old school tie be rent Off their necks. **1965** *Listener* 22 July 125/1 One thing I have learned is to distrust a lot of the familiar chat about old school ties. **1969** *Ibid.* 9 Jan. 41/1 He wore an old-school tie and an air of off-hand insolence which he thought of as easy and confident. **1973** 'S. Harvester' *Corner of Playground* i. v. 55 Their being boys from adjoining villages, the African version of old school tieism.

Oldspeak (ǭu·ldspīk). [f. Old *a.* + Speak *v.*] The name used for Standard English, as opposed to the artificial language *Newspeak, in G. Orwell's novel *Nineteen Eighty-Four*, applied to normal English usage, *spec.* as distinct from technical or propagandist language. Hence **Oldspeaker**, one who uses Oldspeak.

1949 'G. Orwell' *Nineteen Eighty-Four* i. 54 You haven't a real appreciation of Newspeak... Even when you write it you're still thinking in Oldspeak. *Ibid.* 299 It was expected that Newspeak would have finally superseded Oldspeak (or Standard English, as we should call it) by about the year 2050. **1960** *Encounter* Nov. 10/1 The substitution of 'Newspeak' for 'Oldspeak' (or present-day English) is designed to effect nothing less than the destruction of human reason by linguistic means. **1974** *Daily Tel.* (Colour Suppl.) 20 Sept. 10/3 Sometimes they say that 'real' freedom, being quite different from the crude concept mistakenly used by Oldspeakers, can exist only in a socialist state.

oldster. Add: **1.** (Earlier example.)

1818 'A. Burton' *Adventures J. Newcome* ii. 77 An Oldster with a Gunter's scale Bestow'd his blows as fast as hail.

2. (Later examples.) Also *attrib.*

1938 [see *married *sb.*]. **1942** *Sun* (Baltimore) 10 Oct. 10/3 The oldsters, travelling salesmen and the like, have learned to tote their own satchel. **1957** *New Yorker* 21 Sept. 33/3 Last week, our interest in nimble oldsters led us to the twenty-fifth floor of the Whitehall Building. **1964** *Wall St. Jrnl.* 5 Feb. 16 'The youngsters are chafing at the bit and aren't willing to wait and see how the civil rights bill shapes up,' he adds, 'and we oldsters can't hold back any longer.' **1973** M. Amis *Rachel Papers* 112 What was more, the producers could afford only middle-aged actors and actresses. I shifted in my seat as the camera inexpertly focused on a parade of oldster genitals. **1974** *Anderson* (S. Carolina) *Independent* 18 Apr. 4A/2 By getting married, two oldsters in a Portland, Ore., nursing home had their Social Security cut from $412 a month to $309.40 a month because the regulations assume two can live as cheaply as one.

old-time, *a.* Add: **2. a.** Pertaining to or characteristic of an earlier or former time.

1870 'Mark Twain' in *Buffalo Express* 1 Jan. 2/6 Conrad's color came back to his cheeks and his old-time vivacity to his eyes. **1936** F. Clune *Roaming round Darling* xiv. 120 Whitney, old-time driver for Cobb and Co.'s coaches (later a grazier), planted here a score of fig-trees. **1975** *Nature* 29 May 360/2 What they have is a lot of old-time researchers who are accustomed to pursuing their own interests.

b. In ballroom dancing, applied to styles of dance and music fashionable in the nineteenth and early twentieth centuries. Also in form *old(e) tyme.* Also *absol.* as *sb.*

1887 E. Scott *Grace & Folly* iv. 64 It may not be uninteresting to enquire a little into the nature of some of the old-time dances. **1929** *Radio Times* 8 Nov. 389/2 Other records were... *Old Time Favourites*, the London Orchestra. **1933** Auden *Dance of Death* 11 Select partners for an old-time waltz. **1947** J. R. Gillespie (*title*) Old tyme dancing. **1950** A. Wilson *Such Darling Dodos* 116 Derek's crazy to take up dancing in a big way again. He adores all this old-time dancing. **1952** [see *drag *v.* 1 e]. **1960** D. Potter *Glittering Coffin* iii. 43 The primary school..is opened only for occasional old-time dances. **1967** O. Norton *Now lying Dead* iii. 54 Tuesdays he stays at home because that's Her night for her Old Time. **1974** *Radio Times* 11 Apr. 52/3 9.2 Time for Old Time in Radio 2 Ballroom. **1975** R. Butler *Where all Girls are Sweeter* vi. 75 'It's kind of romantic, really.'..'Like Old Tyme dancing on BBC radio, you mean?'

old-timer, for 'chiefly *U.S.*' read 'orig. *U.S.*' and add later examples.

1888 *New Princeton Rev.* Jan. 122 Most of us 'old-timers'..are poor now. **1910** [see *fest]. **1922** [see *animal A. 6]. **1928** *Daily Mail* 25 July 2/3 Many types are represented. There is..the old-timer, who knows more about Oxford than the inhabitants of the city themselves. **1928** D. H. Lawrence *Woman who rode Away* 60 But he was an old-timer miner. **1929** A. Woollcott in *New Yorker* 4 May 44/2 The big walnut tree that was an old-timer even in her day. **1939** *Sun* (Baltimore) 4 Apr. 12/7 Remarks about this type of vessel seem to have struck a responsive chord in the breasts of several old-timers. **1942** 'M. Innes' *Daffodil Affairs* ii. 46 You ought to meet some of the old-timers there, Mr Wine. **1962** *Coast to Coast 1961–62* 46 'Well thanks for the welcome, old-timer,'..Marlett said. **1966** *Listener* 10 Mar. 344/2, I am not sure..that some old-timers might not welcome a return to the traditional style of the Promenade Concerts. **1973** R. L. Simon *Big Fix* (1974) xviii. 146, I raised the hoe... 'Sorry old timer,' I said and brought it down on the back of his neck. **1978** *Jrnl. R. Soc. Arts* CXXVI. 194/1 This may seem familiar stuff to some of you old-timers.

o·ld-timey, *a.* Also **old-timy.** [f. Old-time *a.* + -y¹.] Old-fashioned in character; (nostalgically or sentimentally) recalling the past.

1850 A. J. Downing in *Horticulturist* V. 265 The terraced garden, too, is quaint and 'old-timey'. **1879** F. R. Stockton *Rudder Grange* xvii. 206 Things that were apparently so 'old-timey'..that David Dutton did not care to take them with him. **1892** Kipling & Balestier *Naulahka* ix. 94 The venerable..institution of matrimony is still in use here... The 'Doll's House' glanced right off this blessed old-timey country. **1921** O. W. Holmes in *Holmes–Laski Lett.* (1953) I. 372, I must have mentioned Bryce's two nights with us I think—very pleasant and old timey. **1935** 'L. Ford' *Burn Forever* 28 They're real old-timey over at the Curriers. **1936** J. dos Passos *Big Money* 221 Look how oldtimy the street looks. **1971** E. Bullins in W. King *Black Short Story Anthol.* (1972) 62 Every year in August the Mary's Shore colored community gave an ole timey camp meetin'. **1973** —— *Theme is Blackness* 103 Well, man, I don't want to hear nothin' 'bout some fat ole-timey black bitch that I'm supposed to be like. **1974** *Columbia* (S. Carolina) *Record* 24 Apr. 14-B/4 Beds have old timey coverlets and primitive barn markings form popular wall plaques. **1975** *Publishers Weekly* 17 Nov. 98/3 Large format and illustration with old-timey black-and-white drawings.

old town. [f. Old *a.* + Town *sb.* 4.] The older part of a city or town contained within its modern limits. Also *attrib.* Hence **old-towner**, an inhabitant of an old town.

1752 G. Elliot *Proposals Publ. Works Edin.* 32 In these cities, what is called the *new town*, consists of spacious streets and large buildings..while the *old town*..is more crouded than before these late additions were made. **1797** Lady Newdigate *Let.* 16 July in A. E. Newdigate *Newdegate Cheverels* (1898) xiii. 154, I am told that yᵉ Steyne & everything beyond the Old Town has been built within yᵉ last 30 years. **1842** Queen Victoria *Jrnl.* 3 Sept. in D. Duff *Victoria in Highlands* (1968) 32 We set off..for Edinburgh... The procession moved through the Old Town up the High Street. **1924** 'P. Blundell' *Confessions of Seaman* xi. 147 There is not, of course, any real 'old town' in Hamburg. Most of the city was burnt down in 1842. **1966** G. Lyall *Shooting Script* xxvi. 208 They are staying at the Colombo, on the beach front near the old town. Jiminez will control the old town, whatever happens. **1968** M. Tripp *One is One* ii. 15 The beautiful Old Town in Annecy where ancient houses were backed by wooden galleries. **1973** *Guardian* 30 May 7/1 The fears of the old-towners are certainly understandable. The population has declined. Runcorn new town..has been growing. **1973** R. Parkes *Guardians* viii. 137 The granite masses of old-town Helsinki.

old woman. Add: **1. a.** (Further examples of disparaging use.)

1852 C. M. Yonge *Two Guardians* xv. 294 What does she do but let me go muddling on with that old woman Wells! **1867** Trollope *Claverings* I. xi. 141 Who is it says so? A parcel of old women. **1876** G. M. Hopkins *Let.* 23 Sept. (1956) 142 The Pope, it is well known, is a very fine looking man but there are some smutty smirking old-

woman presentments of him. **1911** *Chambers's Jrnl.* 46/1 The new commanding officer was, however, of the genus known in the service as 'old woman', and the regiment suffered accordingly. **1918** E. Pound *Pavannes & Divisions* 39 But surely the worst of your old-women are the male ones. **1953** E. Simon *Past Masters* ii. 74 Macphail is an old wooman [*sic*]... He thought it his duty to let me know. **1975** B. Wood *Killing Gift* 177 He didn't want Marvin on guard against him... 'Give it some time, Marvin... Maybe I'm being a bit of an old woman.'

b. (Examples.)

a **1775** J. Boucher *Gloss. Archaic & Provinc. Words* (1832–3) p. l/1, Could my *old woman*, whilst I labour'd thus, At night reward me with a *smouch*, or buss. **1834** W. G. Simms *Guy Rivers* II. 97 The old woman, by whom we mean..to indicate the spouse of the wayfarer, and mother of the two youths, was busied about the fire. **1839** C. M. Kirkland *New Home* xv. 96 If my old woman was to stick up that fashion, I'd keep the house so blue she couldn't see to snuff the candle. **1869** Mrs. Stowe *Oldtown Folks* xxxvii. 481 The old woman is just as choice of her boys as if [etc.]. **1916** 'Taffrail' *Pincher Martin* xviii. 337 'Ow's Hemmeline an' Mrs Fig—yer ole woman? **1926** I. M. Peacocke *His Kid Brother* xiv. 216 His wife... a small round dumpling of a woman with rosy cheeks, whom the policeman addressed as 'Old woman'. **1976** J. O'Connor *Eleventh Commandment* xi. 143 If you went home and found someone indoors with your old woman, what would you do?

c. *Theatr.* An actress playing the role of an old woman, esp. one who specializes in such roles. Cf. *old man 1 f.

1838 A. Mathews *Mem. Charles Mathews* I. 101 Mrs. Davenport,..the inimitable 'Old Woman' of Covent Garden Theatre, having succeeded Mrs. Webb in that line soon after this period. **1901** C. Morris *Life on Stage* vii. 40 Then came the leading lady, the first old woman (who was sometimes the heavy woman)..and the ladies of the ballet. **1957** *Oxf. Compan. Theatre* (ed. 2) 772/1 The Old Woman took Juliet's Nurse.

d. old woman's tooth, a simple kind of wooden router plane used by cabinet makers.

1846 C. Holtzapffel *Turning & Mech. Manipulation* II. xxiii. 487 This plane..is generally called 'the old woman's tooth'. **1907** E. Rowe *Pract. Wood-Carving* 7 The router, very similar to the tool called by the joiner an 'old woman's tooth', may occasionally be used. **1969** E. H. Pinto *Treen* 389 Plate 417, *F*, is an 'old woman's tooth', a router plane of French walnut, probably 16th- or 17th-century.

Hence **old-wo·manishness**, behaviour characteristic of an old woman.

1941 B. Schulberg *What makes Sammy Run?* xi. 266 An old-womanishness that's won him the reputation of best-loved producer. **1977** *Listener* 28 Apr. 535/2 His mother's..puritanism and old-womanishness..seemed to hover over Owen's editorial shoulder.

old-world, *a.* Add: **2. a.** (Examples.)

1877 *Harper's Mag.* Dec. 91/2 This was..a beautiful garden kept in old-world order by a Scotch gardener. **1931** E. F. Benson *Mapp & Lucia* i. 19 Seven bedrooms, four sitting-rooms h. & c., and an old-world garden. **1965** *Canad. Jrnl. Linguistics* X. 97 Consolidation with various Old-World stocks (Uralic, Indo-European) has been sought. **1967** *Boston Sunday Globe* 23 Apr. 23 (Advt.), Old world wood finish and antique polished bronze [lamp].. antique empire gold and black candelabra.. **1977** *N.Y. Rev. Bks.* 13 Oct. 14/1 For most Americans political assassination was an Old World phenomenon of bomb-throwing Bolsheviks and Balkan fanatics.

b. Old World monkey, a catarrhine monkey belonging to the superfamily Cercopithecoidea or the family Cercopithecidæ, which includes the monkeys of Africa and Asia.

1863 H. W. Bates *Naturalist on River Amazons* II. v. 326 The Marmosets, have thirty-two teeth, like the Old World monkeys and man. **1894** H. O. Forbes *Hand-bk. Primates* I. 252 The family *Cercopithecidæ* includes all the Old World Monkeys except the Anthropoid or true Apes, and Man. **1936** E. G. Boulenger *Apes & Monkeys* vi. 120 The old-world monkeys are very widely distributed. **1968** *Times* 15 Nov. 8/6 Two fossil monkeys of the primate group known as the Old World monkeys, or Cercopithecoidea. Living representatives of the group include baboons, mandrills and macaques. **1974** S. I. Rosen *Introd. Primates* vi. 86 The Old World monkeys are biologically closer to man than the New World primates.

Hence **old-wo·rldliness** = *old-worldism.*

1934 *Archit. Rev.* LXXV. 178 The patron public was becoming patina-conscious, aware of texture, age-effects, old-worldliness, the charm of mellowness.

‖**ole¹** (ǭu·le). [Sp. *ole* Andalusian dance.] The name of a Spanish folk-dance, which is accompanied by castanets and singing.

1846 R. Ford *Gatherings from Spain* xxiii. 328 The dance..is called the *Ole* by Spaniards, the *Romalis* by their gipsies... The whole person..trembles like an aspen leaf. **1950** L. Armstrong *Dances of Spain* I. 9 *Olé*, Gaditanian (Cadiz) folk dance of apparently great antiquity. A solo with castanets. **1964** W. G. Raffé *Dict. Dance* 360 *Ole* (Spain), a woman's solo dance with castanets... The dance is accompanied by rapid vocal acrobatics..to the syllables 'ay' or 'olé' (from which the dance may take its name).

ole², dial. and Black English var. of Old *a.* Cf. *ol'.

1844 (*song-title*) Ole Bull and Ole Dan Tucker. **1874** [see *old mistress* s.v. *old *a.* (*adv.*, *sb.*¹) D. 4]. **1880** J. C. Harris *Uncle Remus* i. 18, I speck de ole 'oman en de chilluns kin sorter scramble roun' en git up sump'n fer ter stay yo' stummuck. **1907** G. B. Shaw *Major Barbara* ii. 217 Youre ony a jumped-up, jerked-off, orspittle-turned-out incurable of an ole workin man. **1935** Z. N. Hurston *Mules & Men* i. iv. 101 It laid dere for thousands of years, then Ole Missus said to Ole Massa: 'Go pick up dat box.' **1936** M. Mitchell *Gone with Wind* iv. 63 'Mammy gettin' ole,' said Dilcey, with a calmness that would have enraged Mammy. **1950** R. Ames in A. Dundes *Mother Wit* (1973) 488/1 The old time darky's.. love for 'ole marse' and 'ole mist'ess'. **1962** J. D. Salinger *Franny & Zooey* 85 That's the spirit... That's putting the ole foot down. **1971** *Jamaican Weekly Gleaner* 3 Nov. 18/2 Long debt better dan ole grudge. **1972** G. Beare *Bee Sting Deal* i. 12 'I work here,'..'But surely not at 5.30 a.m., ole gel.' **1973** [see *limer²]. **1973** *Black World* May 13/1 They succinctly and repeatedly lay the flaming blame at Ole Massa's front door. **1974** *Sunday Guardian* (Port-of-Spain) 28 July 12/2, I took a bus to Martin's Bay, and there was laughter and ole talking all the way. Down on the wharf I find gangs of limers giving jokes and ole talkin. **1977** *Time* 29 Aug. 31/1 'My God,' sighed Bell, 'I still wish we could get ole Frank Johnson to take it.'

‖**olé** (ǭuˡle·), *int.* [Sp. *olé* bravo.] Bravo. Also as *sb.*, a cry of olé. Usu. assoc. with Spanish music and dance and with bullfighting.

1922 J. Hergesheimer *Bright Shawl* 55 An uproar of applause rose from the theatre, a confusion of cries, of Olé! Olé! Anda! Anda! Chiquella! **1940** E. Hemingway *For whom Bell Tolls* v. 60 'Olé!' someone said. 'Go on, gipsy!' **1962** J. Stewart tr. *Cousseau's Death of Miss Cunningham* 44 The unheard *Olés* of the crowd..in Caracas. **1966** R. E. Pickering *Himself Again* vi. 42, I.. picked up the little glass, and drained it. 'Olé,' said Charley. **1967** J. Potter *Foul Play* i. 17 Freda and Basil had launched themselves into a perilous exhibition tango, to encouraging olés from the surrounding circle of boozers. **1973** *Sat. Rev.* (U.S.) 25 Sept. 29/1 The *plaza de toros* was packed; and..before I could even see the ring, I heard the *olés.*

-ole, *suffix* [partly f. L. *oleum* oil, partly a var. -ol], used to form chemical names. **1.** In the names of compounds containing five-membered, unsaturated rings with at least one hetero-atom, e.g. *carbazole, indazole, indole, pyrrole, thiazole, triazole.*

Although applicable to older words like *pyrrole* (1835), this systematic use of *-ole* is of more recent date.

1928 *Jrnl. Amer. Chem. Soc.* I. 3078 For five-membered rings proposals made by Widman, Bouveault and others have found their way into use. The ending *-ol* or *-ole* appears in such names as pyrrole, imidazole and even dioxole (which is non-nitrogenous). It conflicts with the ending *-ol* for alcohols and phenols (hence the modified spelling *-ole*), but its use to denote a five-membered ring is well known. **1971** *Nomencl. Org. Chem.* (I.U.P.A.C.) (ed. 3) B. 53 (*table*) No. of members in the ring..5. Rings containing nitrogen: unsaturation..-ole... Rings containing no nitrogen: unsaturation..-ole.

2. Unsystematically, in the names of aromatic ethers, e.g. *anisole, phenetole, safrole*, which were regarded as being derived from carboxylic acids in the same way as 'benzole' (benzene) is from benzoic acid.

1852 *Rep. Brit. Assoc. Adv. Sci. 1851* i. 136 The use of the analogous termination *ole* for those [bodies] formed by the abstraction of 2 atoms of carbonic acid from the same [*sc.* vegetable acids] may be apt to cause some ambiguity. Thus we use the terms benz*ole*, phen*ole*, and anis*ole*, as being derived respectively from the benzoic, salicylic, and anisic acids.

oleaginous, *a.* Add: **2.** (Later examples.)

1922 Joyce *Ulysses* 406 The scent, the smile.., the dark eyes and oleaginous address. **1945** R. Hargreaves *Enemy at Gate* 31 The shifty, oleaginous, power-greedy place-man Olympius. **1966** *Listener* 18 Aug. 244/3 Many of the other works have a rather nasty oleaginous inconsistency of texture. **1973** *Daily Tel.* 13 Oct. 11/2 A piped programme that included an oleaginous instrumental version of part of a Chopin nocturne. **1976** *National Observer* (U.S.) 17 July 5/1 For years, the conferences have been sponsored and paid for by big companies and their oleaginous lobbyists.

oleaginosity: delete † and add later example; also **olea·ginously** *adv.*

1912 L. J. Vance *Destroying Angel* x. 119 Three doors, in one of which a rotund Chinaman beamed oleaginously. **1912** W. Deeping *Sincerity* xxxii. 247 His hands were fat, his neck full of red creases, his manner towards women oleaginously gallant. **1939** Joyce *Finnegans Wake* 54 Whileas oleaginosity of ancestralolosis sgocciolated down the both pendencies of his mutsohito liptails.

oleander. (Later examples.)

1880 G. W. Cable *Grandissimes* xxvi. 193 Their long, over-arched avenues of oleander. **1915** H. H. Thomas *Greenhouse* vi. 57 Oleanders require the protection of a cool greenhouse in winter. **1956** *Railway Mag.* Mar. 165/1 A fertile valley, with magnificent oleanders. **1962** *Coast to Coast 1961–62* 13 In the shade of the giant oleander bush sat Brett's half-Persian she-cat, licking and preening herself with a too obvious absorption.

b. oleander hawk(-moth), a large moth, *Daphnis nerii*, of the family Sphingidæ whose caterpillars feed on oleander or periwinkle leaves.

1843 Humphreys & Westwood *Brit. Moths* I. 21 The Oleander Hawk-moth..measures about $4\frac{1}{2}$ inches in the expansion of its fore wings, of which the ground colour is an olive-green. **1907** R. South *Moths Brit. Isles* 1st Ser. 45 (*heading*) The Oleander Hawk-moth. **1955** E. B. Ford *Moths* i. 6 The other form [of green colouring]..is found.. in some Sphingidae such as the Oleander Hawk. **1972** *Shooting Times & Country Mag.* 24 June 20/3 When they flew they reminded me of one of the hawk moths, the Oleander perhaps, or the Elephant Hawk moth.

oleandomycin (ǭu:liˌændomai·sin). *Pharm.* [f. *oleand-rose*, a sugar of which a residue is present in the oleandomycin molecule (f. Oleandr(ine + -ose²) + -o + *-mycin.] (The phosphate of) a macrolide, $C_{35}H_{61}NO_{12}$, produced by a strain of the bacterium *Streptomyces antibioticus*, which is active against a wide range of Gram-positive bacteria and has been used in treating staphylococcal enteritis and skin infections.

1956 B. A. Sobin et al. *U.S. Pat.* 2,757,123 31 July, This invention relates to a new and useful antibiotic called oleandomycin. **1967** [see *erythromycin]. **1968** [see *macrolide]. **1974** D. Perlman in W. O. Foye *Princ. Med. Chem.* xxxi. 742/1 Oleandomycin, a macrolide bacterial antibiotic closely related to erythromycin, has been losing value as a therapeutic agent for infections caused by gram-positive bacteria because of high incidence of side effects.

olearia (ǭuli,ē̆ə·riă). [mod.L. (C. Moench *Suppl. Methodus Plantas Horti Botanici et Agri Marburgensis* (1802) 254), f. the name of Johann Gottfried *Olearius* (1635–1711), German theologian and horticulturist.] An evergreen shrub or tree of the Australasian genus so called, belonging to the family Compositæ and bearing clusters of white, yellow, or mauve flowers; = *daisy-tree* (*daisy *sb.* 7).

1839 G. Don *Sweet's Hortus Britannicus* (ed. 3) 344 (*heading*) *Olearia* DC. [English name] Olearia. **1852** *Curtis's Bot. Mag.* LXXVIII. 4638 Mr. Gunn's Olearia ..is another interesting plant of Van Diemen's Land, which braves the cold of England. **1868** *Trans. N.Z. Inst.* I. iii. 4 Along the shore there is a profusion of shrub Veronicas and Olearias. **1882** T. H. Potts *Out in Open* 64 Soon to be lost beneath a covering screen of *veronicas, olearias*, gnarled and knotted *griselinias*. **1918** *Chambers's Jrnl.* Apr. 221/1 The trees, especially the rata and the olearia, grow for nearly half their length along the ground. **1959** *Listener* 22 Jan. 174/1 He sometimes tears at the bark of the ceanolthus and olearias. **1971** *Homes & Gardens* Sept. 127/2 Many olearias have grey under-sides to the foliage.

olefine. Add: Now usu. spelt **olefin**, in accordance with the practice of reserving -ine⁵ for the naming of basic substances and heterocyclic compounds containing nitrogen. (Further examples.)

1923 T. G. Phillips *Fundamentals Org. & Biol. Chem.* ix. 131 Similar series [of compounds] may be derived from the olefin hydrocarbons. **1949** R. F. Goldstein *Petroleum Chemicals Industry* vi. 124 The higher olefins are present in all fractions of the petrol range of thermally and catalytically cracked oils. **1968** R. O. C. Norman *Princ. Org. Synthesis* viii. 262 This discussion indicates the difficulties necessarily present in effecting the dimerization of an olefin.

Hence **olefi·nic** *a.*, of, having the nature of, or characteristic of an olefin; applied *spec.* to a double bond between two carbon atoms such as is characteristic of olefins (cf. *ethylenic *a.*).

1898 J. Wade *Introd. Study Org. Chem.* lvi. 344 On gentle oxidation it [*sc.* linaloöl] yields an olefinic ketone. **1923** *Daily Mail* 15 Feb. 3 Those olefinic and empyreumatic substances which result from the distillation of coal or oil. **1968** R. O. C. Norman *Princ. Org. Synthesis* vii. 248 Although carbanions..do not react with simple olefins, they do so if the olefinic double bond is conjugated to a group of —*M* type. **1974** *Jrnl. Amer. Chem. Soc.* XCVI. 7934 (*heading*) Stereochemistry of olefinic cyclization.

olenellid (olɛne·lid), *sb.* and *a.* [a. mod.L. family name *Olenellidæ*, f. generic name *Olenell(us* (E. Billings *Palæozoic Fossils* (1861) I. 11) + -id³.] A trilobite belonging to the family Olenellidæ; of or pertaining to this family.

1892 Peach & Horne in *Q. Jrnl. Geol. Soc.* XLVIII. 239 Among the *disjecta membra* of Olenellid trilobites from the dark shales..are certain portions of the carapaces of a much larger species. *Ibid.* 240 One word of speculation as to the systematic position of the Olenellids may be admissible here. **1937** *Jrnl. Paleont.* XI. 577/1 Olenellids present a most interesting association of primitive and specialized characters. *Ibid.* 578/2 The olenellid shell is for its size excessively thin. **1969** Bennison & Wright *Geol. Hist. Brit. Isles* iv. 87 The fauna of the Lower Cambrian chiefly comprises Olenellid trilobites. **1973** P. Tasch *Paleobiol. Invertebr.* xi. 518/1 The so-called abortive segments in..other olenellids might be evidence of the type of asexual budding that occurs in certain primitive polychaete worms.

oleo, *a.* Add: **3.** [f. Oleo-.] Applied to (a system containing) a telescopic strut, used esp. in aircraft undercarriages, which absorbs shocks by means of a hollow piston into which oil is forced through a small orifice on compression of the strut (see also quot. 1965).

1916 N. J. Gill *Flyer's Guide* iii. 39 The commonest form of shock absorber is the rubber type, but Oleo gear are [*sic*] now also used to a considerable extent. **1920** *Flight* XII. 14/1 The undercarriage is at present of the usual Vee type with stream-line steel tube struts and rubber shock absorbers, but later an oleo undercarriage

will be fitted. **1935** C. G. Burge *Compl. Bk. Aviation* 596/1 The piston must return to the original position as quickly as possible after each impact, and to secure this the oleo leg frequently incorporates a compressed-air chamber. **1965** C. N. Van Deventer *Introd. Gen. Aeronaut.* vii. 149/2 The more usual method of absorbing the impact of landing is the through the use of oleo struts. Of the two general types, one has a moving piston..and the other has a fixed piston and uses a spring instead of air. **1977** D. Beaty *Excellency* i. 17 The searchlight began fingering the fuselage. The port oleo leg blossomed a brilliant silver.

B. *sb. Aeronaut.* An oleo strut or leg.

1929 F. H. & H. F. Colvin *Aircraft Handbk.* (ed. 4) iii. 64 To prepare oleo for use. 1. Pull piston out of cylinder and remove auxiliary piston. 2. Fill tube up to 6 inches from top with G.E. transformer oil No. 6. 3. Insert auxiliary piston with bolt head down. 4. Insert main piston. **1931** *Handbk. Aeronautics* (R. Aeronaut. Soc.) iii. 190 (*table*) 2 Large oleos... 2 Small oleos... 2 Front struts... 2 Rear struts. **1958** H. G. Conway *Landing Gear Design* ix. 184 The combination of tyre and oleo will thus have an efficiency well below 100 per cent. **1977** D. Beaty *Excellency* xx. 223 He felt the main wheels shake on their oleos.

oleo² (ǭu·lio). *rare.* Abbreviation of Oleograph.

1921 Galsworthy *Captures* (1923) 57 Taking up the oleos, he turned his back on the photographs. **1932** *Daily Express* 25 June 9/3 A little room hung with lace and oleos.

oleo-. Add: **a. oleophi·lic** *a.* [*-philic], having an affinity for oils or oily materials; readily absorbing oil; **oleopho·bic** *a.* [-phobic], tending to repel, or not to absorb, oils or oily materials; **oleo-pneuma·tic** *a.*, applied to a device or system which absorbs shocks by a combination of forcing oil through an orifice and compression of air or another gas; **oleotho·rax** *Med.* [ad. F. *oléothorax* (A. Bernou 1922, in *Bull. de l'Acad. de Méd.* LXXXVII. 457)], a method of treatment, now disused, in which oil is introduced into the pleural cavity; introduction of oil in this manner.

1957 *Brit. Jrnl. Appl. Physics* VIII. Suppl. 6. S23/2 For oils containing oleic acid, the behaviour is different. Zinc and copper behave initially as the other metals in becoming more oleophobic on continued contact with oil and water but this eventually gives place to a reverse trend to increasingly oleophilic behaviour. **1967** [see *hydrophobic *a.* (*sb.*) 2 a]. **1970** *Sci. Jrnl.* Feb. 21/3 Sawdust treated with appropriate silicones is water repellent but strongly oleophilic and will soak up many times its weight of oil. **1946** *Jrnl. Colloid Sci.* I. 513 The..conclusion was that the observed phenomenon was due to the adsorption of eicosyl alcohol from solution upon the interior walls of the flask, to form a film possessing the property of being oleophobic to (unwetted by) the oil solution. **1967** E. Chambers *Photolitho-Offset* xiv. 210 Others were unsuitable because of extreme low sensitivity to light, or the inherent oleophobic (ink rejecting) character of the tanned coatings. **1909** *Aëronaut. Jrnl.* Apr. 64/1 The front wheel is fitted with an oleo-pneumatic brake for safety purposes. **1930** *Engineering* 16 May 647/1 (*heading*) Oleo-pneumatic shock absorber for aeroplanes. **1960** *Times* 30 Aug. 4/5 All four wheels..are independently sprung by means of wishbones and Girling oleo-pneumatic spring and damper units. **1922** *Jrnl. Amer. Med. Assoc.* 24 June 1996/2 The normal pleura would not bear this direct treatment, but in such pathologic conditions the oleothorax answered the desired purpose. **1938** *Brit. Encycl. Med. Pract.* IX. 309 Oleothorax is a therapeutic method consisting in the introduction of a sterile oil or an oil containing an antiseptic into the pleural cavity. **1962** H. Spencer *Path. Lung* xii. 384 They [*sc.* liquid paraffin granulomas in the lung] have also been reported following oleothorax... This form of pulmonary oil granuloma has now largely disappeared following the cessation of this form of therapy.

oleograph. Add: (Earlier and later examples.) Also **oleolithograph.** Hence **oleogra·phically** *adv.*; **oleo·graphist.**

1873 *Young Englishwoman* Sept. 466/1 Oleographs are now very cheap, and, if well chosen, are most pleasing. **1897** W. Archer *Theatr. 'World' 1896* 173 Seems..incredible even in Germany, with the sacred *eikon* of the Kaiser oleographically lowering over the scene. **1907** R. Brooke *Let.* 23 Aug. (1968) 100 The book plate of one Frederick Leighton—the notable oleographist, I take it? **1922** Joyce *Ulysses* 698 Artistic oleograph on inner face of door. **1928** D. L. Sayers *Lord Peter views Body* 176 Lord Peter sat down on a red velvet arm-chair, fixed his eyes on a gilt-framed oleograph, and became wrapped in contemplation. **1939** J. Cary *Mr. Johnson* 45 He is the happy husband adoring and adored in a perpetual and rather solemn dignity like an oleolithograph of the Royal Family. **1974** N. Freeling *Dressing of Diamond* 57 It was only in books that kidnappers had a handy stammer or a gold tooth, and a flat with a large coloured oleograph on the wall of President Kennedy, or the Pope, or Marilyn Monroe.

olericulture. For *rare*⁻⁰ read *rare* and add later example.

1966 *McGraw-Hill Encycl. Sci. & Technol.* XIV. 286/1 The term olericulture, referring to vegetable production, is used occasionally.

oleum (ǭu·liǒm). Pl. **oleums.** [L., = 'oil'.] An oily liquid which is produced by dissolving sulphur trioxide in concentrated sulphuric acid in the contact process and is used in sulphonation and nitration processes; fuming sulphuric acid.

1905 Blount & Bloxam *Chem. for Engineers & Manuf.* (ed. 2) II. i. 23 Ordinary vitriol is still called in Germany English sulphuric acid, in contradistinction to the fuming acid or 'oleum'. **1949** P. W. Vittum tr. *Fierz-David & Blangey's Fund. Proc. Dye Chem.* 83 When sulfonation is done with oleum, usually only the SO_3 is used up. **1954** *Thorpe's Dict. Appl. Chem.* (ed. 4) XI. 296/1 (*heading*) Boiling-points of oleums. **1973** *Times* 20 Jan. 3/1 It [*sc.* a tanker] was carrying a maximum load of 20 tons of oleum, a corrosive acid said to react violently when in contact with a small quantity of water.

O-level, see *O 5 d.

oleyl (ǭu·lī,il, -əil). *Chem.* [f. Ole(ic *a.* + -yl.] The radical $CH_3(CH_2)_7CH{=}CH(CH_2)_7CH_2$—, which has the same carbon atom skeleton as oleic acid; **oleyl alcohol**, an oily liquid, $C_{18}H_{35}OH$, which occurs in fish oils and is used in the manufacture of surface-active agents; *cis*-9-octadecen-1-ol.

1903 *Jrnl. Chem. Soc.* LXXXIV. i. 730 Ethyl oleate gave oleyl alcohol, $C_{18}H_{36}O$, as a colourless liquid. **1951** Kirk & Othmer *Encycl. Chem. Technol.* VI. 271 The reaction of acetylene with oleyl alcohol gives vinyl oleyl ether, which is polymerizable and copolymerizable with other vinyl monomers. **1972** *Materials & Technol.* V. ix. 262 Oleyl alcohol can be epoxidised with peracids to yield polyhydroxy compounds. **1972** *Arch. Biochem & Biophysics* CL. 199/2 Oleyl CoA inhibits the adenine nucleotide transport system..of rat liver mitochondria.

olfactometer (ǫlfæktǫ·mĭtəɹ). [f. as next + -ometer.] **a.** An instrument for measuring the sensitivity of a subject to odours or the intensity of odours.

1889 H. Zwaardemaker in *Lancet* 29 June 1301/1, I have lately constructed a small instrument which, I think, deserves the name of 'olfactometer'. Its component parts consist of two tubes fitting into each other. The outer one is lined with scented material, and made to glide up and down over the inner one, of which one end remains free and is bent to fit the nostril. **1922** G. H. Parker *Smell, Taste & Allied Senses* iii. 51 Two single olfactometers may be combined so that one current carrying an odorous material..may be introduced into one nostril and another carrying a second odorous substance..into the other nostril. **1969** J. R. Hughes et al. in C. Pfaffmann *Olfaction & Taste* 173 Stimuli have been presented by a portable two-channel olfactometer. **1969** *Daily Tel.* 9 Sept. 20/8 He has developed devices, known as olfactometers, which at present measure only the intensity of odours but do not differentiate between them.

b. A device for investigating the responses of animals to odours.

1926 *Jrnl. Econ. Entomol.* XIX. 552 Adult potato beetles were collected in potato patches and were tried in the insect olfactometer. **1969** *New Scientist* 17 Apr. 134/2 Some initial experiments using an olfactometer suggested that the sense of smell did not play any part in bringing the male and female spiders together.

olfactometry (ǫlfæktǫ·mĕtri). [f. L. *olfact-*, ppl. stem of *olfacere* to smell (see Olfacient) + -o + -metry.] The measurement of the sensitivity of the sense of smell, the use of an olfactometer; also, = *odorimetry.

1898 *Amer. Jrnl. Psychol.* X. 85 Olfactometry is that branch of psychophysics which is concerned with the measurement of the keenness of smell. **1898** [see *odorimeter]. **1935** [see *odorimetry]. **1967** J. W. Johnston in T. Hayashi *Olfaction & Taste II* 48 The characterization of a simple blend as a nonresinous woody odor by means of olfactometry and G-L chromatography has shown the way to the elusive correlation of the physical parameter, odorant intensity, and the psychological dimension, subjective (odorant) strength. **1968** W. McCartney *Olfaction & Odours* 151 When olfactometry is being practised, great care must be taken to avoid fatigue (adaptation) of the perceiving organ.

Hence **olfactome·tric** *a.*, of olfactometry.

1898 *Amer. Jrnl. Psychol.* X. 101 (*heading*) Control in Zwaardemaker's olfactometric method of the factors which determine the intensity of the stimulus. **1963** *Nature* 20 Apr. 272/2 As the syntheses were completed, the compounds were tested.., using advanced olfactometric techniques, for intensity and quality of odour. **1973** *Biol. Abstr.* LV. 2568/1 An olfactometric method is proposed for estimating the attractant properties of chemical compounds for insects.

olfactorily, *adv.* (s.v. Olfactory *a.*). (Further examples.)

1944 R. W. Moncrieff *Chemical Senses* ix. 187 Structurally, the esters bear the same relation to the acids as the ethers do to the alcohols... Olfactorily, however, there is a difference. **1969** *Psychol. Bull.* LXXI. 60/1 Olfactorily acting and oral pheromones.

olfactorium (ǫlfæktōᵊ·riǒm). Pl. **-ia.** [f. as next + *-orium.] A large odour-proof enclosure in which olfactory experiments are conducted.

1950 D. Foster et al. in *Amer. Jrnl. Psychol.* LXIII. 433 The design was..made cubical and the apparatus finally devised and described here was called the olfactorium. **1968** W. McCartney *Olfaction & Odours* 148 The 'olfactoria' (capacity 500 cubic feet, for example) that have recently been constructed in the U.S.A...are essentially the same as was Fischer and Penzoldt's test room.

olfactronics (ǫlfæktrǫ·niks), *sb. pl.* (const. as *sing.*). [f. L. *olfact-*, ppl. stem of *olfacere* to smell (see Olfacient) + *elec)tronics.] The detection, analysis, and measurement of vapours by means of instruments.

1964 *Aviation Week & Space Technol.* 16 Nov. 46/1 'Olfactronics.' Chicago—Combination of chemical analysis and electronic techniques appears to offer many interesting possibilities..beyond the detection of hidden explosives on aircraft, according to Dr. Andrew Dravnieks of the Illinois Institute of Technology. **1967** *N.Y. Times* 16 July iv. 8 Eventually olfactronics may be used to guard bank vaults against burglars. **1970** *Jrnl. Reproductive Med.* Apr. 69 (*heading*) Changes in vaginal odors... A study in applied olfactronics.

Hence **olfactro·nic** *a.*, of or pertaining to olfactronics; **olfactro·nically** *adv.*, by olfactronic means.

1966 *New Scientist* 15 Sept. 623/2 A human carries an olfactronic 'image' of his recent environment, provided that this environment contained some olfactronically distinguishable features. **1969** *Product Engineering* 10 Mar. 11 Acoustic or vibration sensors, heat-detection devices, and even olfactronic detectors might prove more effective. **1970** *Jrnl. Reproductive Med.* Apr. 70/1 The atmosphere around the study object..must be known and examined olfactronically.

oligæmia (ǫligī·miă). *Med.* Also (*U.S.*) **oligemia.** [mod.L., ad. F. *oligaimie* (A. N. Gendrin *Traité philos. de Méd. pract.* (1838) I. i. 37), ad. Gr. ὀλιγαιμία (Arist.), f. ὀλίγος (see Oligo-) + αἷμα blood.] = *hypovolæmia* s.v. *hypo-; now distinguished from *anæmia*, which is applied to a diminished concentration of blood cells.

1843 T. Watson *Lect. Princ. & Pract. Physic* I. iv. 49 The blood is scanty and poor—what Andral calls (though with questionable propriety) anæmia. Oligæmia is the cacophonous but more exact name assigned to it by Gendrin; but poverty of blood is the ordinary English phrase for it, and the best of the three. **1861** T. H. Tanner *Man. Practice Med.* (ed. 4) i. i. 7 Anæmia.—Deficiency of blood (poverty of blood, spanæmia, hydræmia, or oligæmia) arises generally..where there has been deprivation. **1866** [in Dict. s.v. Oligo-]. **1942** M. M. Wintrobe *Clin. Hematol.* vii. 243 Anemia, which..refers to the concentration of oxygen-carrying substance in a certain volume of blood, is to be distinguished from oligemia, which signifies a reduction of the total amount of blood in the body. **1974** Passmore & Robson *Compan. Med. Stud.* III. iii. 2/1 Rarely, oligaemia with shock arises from loss of plasma proteins in the urine in the course of the nephrotic syndrome.

Hence **oligæ·mic** *a.*

1848 C. D. Meigs *Females & their Diseases* xxxii. 421 Your thin, pale oligæmic patient cannot take up enough oxygen out of the air she breathes, to make her *strong.* **1947** *Amer. Heart Jrnl.* XXXIII. 645 The cardiodynamic alterations are characteristic of progressive and terminal stages of shock associated with low blood volumes; that is, oligemic shock. **1965** *Proc. Soc. Exper. Biol. & Med.* CXIX. 884/1 Experimental animals were maintained until death or sacrifice at an oligemic hypotension of about 30 mm Hg.

oligarchization (ǫligāɹkəizē̆i·ʃən). [f. Oligarch(y + -ization.] Movement towards oligarchy.

1966 *Social Forces* XLIV. 328/1 Oligarchization may be defined as the concentration of power, in the Weberian sense, in the hands of a minority of the organization's members. **1974** tr. *Wertheim's Evolution & Revolution* 405 The Continued Revolution; China and the Iron Law of 'Oligarchization'.

oligo-. Add: **oligodynamic,** add: [ad. G. *oligodynamisch* (C. von Nägeli 1893, in *Neue Denkschriften d. allgem. schweiz. Ges. f. d. ges. Naturwiss.* XXXIII. i. ii. 8)]; substitute for def.: effected or exerted by minute quantities of metallic ions in solution; acting or being active at very low concentrations; (further examples); **oligoe·ster** *Chem.*, an oligomer in which adjacent monomeric units are linked together by an ester grouping, —CO·O—; **oligoha·line** *Oceanogr.* [ad. G. *oligohalin* (H. C. Redeke 1922, in *Bijdragen tot de Dierkunde* (Amsterdam) XXII. 330), f. Gr. ἅλ-ος of salt], characterized by salinity in the range immediately above that of 'fresh' water (a division of the *mixohaline category); **oligole·ctic** *a.* [Gr. λεκτ-ός chosen], of bees, gathering pollen from only a few closely related plants; **oligomenorrhæa**, now written **oligomenorrhœa** or (*U.S.*) **-rrhea**; (further example); hence **o:ligomenorrhœ·al** **-rrhœic** *adjs.*; **oligomy·cin** *Pharm.* [*-mycin], (any of) a group of antifungal antibiotics produced by the bacterium *Streptomyces diastatochromogenes*, which inhibit certain mitochondrial phosphorylation reactions; **oligonu·cleotide** *Biochem.* [ad G. *oligo-nucleotid* (F. G. Fischer et al. 1941, in *Jrnl. f. prakt. Chem.* CLVIII. 81)], any polynucleotide whose molecules are made up of a relatively small number of nucleotides; **oligope·ptide** *Biochem.* [ad. G. *oligopeptid* (Helferich & Grünert 1940, in *Naturwissenschaften* 28 June 411/2)], any peptide whose molecules are composed of a relatively small number of amino-acid residues; **oligo·phagous** *a. Zool.* [-PHAGOUS], of insects, feeding on a limited number of plants; so **oligo·phagy** (-ǫ·fădʒi); **oligophre·nia** [Gr. φρήν mind], feeble-mindedness; **o:ligopod** *a. Ent.* [ad. It. *oligopodo* (A. Berlese 1913, in *Redia* IX. 128) f. Gr. πούς, ποδ- foot], of an insect larva, having well-developed thoracic limbs; **oligosa·ccharide** *Biochem.* [ad. G. *oligosaccharid* (B. Helferich et al. 1930, in *Ber. d. Deut. Chem. Ges.* LXIII. 991)], any carbohydrate whose molecules are composed of a relatively small number of monosaccharide residues; **o:ligosapro·bic** *a. Ecology* [ad. G. *oligosaprob* (Kolkwitz & Marsson 1902, in *Mitteilungen aus der K. Prüfungsanstalt f. Wasserversorgung und Abwässerbeseitigung* I. 46): see *SAPROBE], of, being, or inhabiting an aquatic environment that is rich in dissolved oxygen and (relatively) free from decayed organic matter; so **oligosa·probe**, an oligosaprobic organism; **oligospe·rmia** *Med.* [Gr. σπέρμα seed], orig. the condition in which the amount of semen secreted is reduced; now usu. = *oligozoospermia* below; hence **oligospe·rmic** *a.*; **oligotro·pic** *a. Zool.* [ad. G. *oligotrope* (E. Loew 1884, in *Jahrb. Bot. Gartens Berlin* III. 256): see *-TROPIC], of bees, collecting nectar from only a few kinds of flower; **o:ligozoospe·rmia** *Med.* [*zoosperm* s.v. Zoo-], the condition of having the number of spermatozoa in the semen sufficiently reduced to affect fertility; **oliguria** (earlier and later examples); so **oligu·ric** *a.*, of, pertaining to, or involving oliguria; also as *sb.*, one who suffers from oliguria.

1941 *Jrnl. Marine Res.* IV. 186 Containers made of copper, zinc, tin or nickel alloys are not suitable for the collection of samples of sea water for bacteriological analysis due to the inimical oligodynamic action of the metals. **1965** B. E. Freeman tr. *Vandel's Biospeleol.* xix. 337 The Thiobacteria can synthesise oligodynamic substances (nicotinic acid..pyridoxine, vitamin B_{12}). **1973** *Times* 25 Apr. 19/5 The ability of minute amounts of these metals to exert a lethal effect upon micro-organisms is referred to as oligodynamic action. **1957** *Makromolekulare Chemie* XXIII. 31 Linear oligoesters of terephthalic acid and glycol form three polymer-homologous series: ester-diols, ester-dicarboxylic acids, and ester-hydroxyacids. **1968** *Encycl. Polymer Sci. & Technol.* IX. 491 For the synthesis of oligoesters and amides, the carboxyl group is usually activated by transformation into acid chloride or azide, as well as to a mixed anhydride or active ester. **1951** *Rep. Comm. Treatise Marine Ecol.* (National Research Council, U.S.) xi. 30 As originally proposed by Redeke, this classification was related to chlorinity rather than to the total salinity... This scheme, as set up by Redeke..is best presented in tabular form:..Cl, o/oo... Brackish water. 0·1–1·0. Oligohaline. **1971** *Nature* 24 Sept. 281/1 Next, there is a conglomerate which contains brackish oligohaline faeces fauna. **1925** C. Robertson in *Ecology* VI. 413, I have used the term oligolectic for a bee collecting pollen from a species, genus of family, where the relationship of the flowers seems to determine the preference, and polylectic for one collecting pollen from unrelated plants. **1972** *Science* 12 May 601/2 There are very few oligolectic bees. **1973** Proctor & Yeo *Pollination of Flowers* v. 151 Bees that visit only one or a few species of flowers for food are described as oligotropic, while those showing a similar restriction for pollen supplies are called oligolectic. **1974** Passmore & Robson *Compan. Med. Stud.* III. xxviii. 6/1 Although there are important exceptions, in general those conditions which cause primary amenorrhoea are congenital, while those causing secondary amenorrhoea or oligomenorrhoea are acquired. **1963** Oligomenorrheal [see *POLYMENORRHEAL *a.*]. **1955** *Obstetr. & Gynecol.* V. 661 Five of the 7 patients were amenorrheic at the time of the first visit, while 2 were oligomenorrheic with menses occurring once to three times a year. **1977** *Lancet* 15 Oct. 809/1 Bromocriptine restores normal gonadal function in some amenorrhoeic or oligomenorrhoeic patients who have normal serum-prolactin levels. **1954** R. M. Smith et al. in *Antibiotics & Chemotherapy* IV. 962 The purpose of this paper is to report a presumably new antifungal antibiotic, oligomycin. **1958** *Jrnl. Amer. Chem. Soc.* LXXX. 6093/1 The oligomycins are active against the fungi producing oak-wilt and Dutch elm diseases. **1964** *Ann. Rev. Biochem.* XXXIII. 737 In loosely coupled mitochondria, oligomycin abolishes phosphorylation, and the available evidence is consistent with the conclusion that oligomycin inhibits ATP formation from all three coupling sites of the respiratory chain. **1969** J. Dekker in D. C. Torgeson *Fungicides* II. xiii. 621 Oligomycin..is an antifungal antibiotic active against various fungal plant pathogens. The results, however, have not been good enough to warrant practical applications. **1942** *Chem. Abstr* XXXVI. 785 From a study of dialysis coeffs., the oligonucleotide has a mol. wt. 3·5–5·3 times that of a mononucleotide. **1961** *Lancet* 12 Aug. 377/2 Many oligonucleotides and polynucleotides stimulate the growth of protozoans. **1971** *Nature* 28 May 217/2 There were several oligonucleotides missing from the former RNA. **1941** *Chem. Abstr.* XXXV. 78 (*heading*) N-methanesulfonyl derivatives of amino acids and oligopeptides. **1955** *Nature* 9 July 72/2 The importance of the amino-acids and oligopeptides and the complexity of their mixtures often encountered stimulate a demand for more powerful analytical tools. **1968** *New Scientist* 22 Aug. 402/2 The first section, that on the posterior pituitary hormones, is mainly concerned with the fact that these oligopeptides do not occur free in the nerve cells. **1920** C. T. Brues in *Amer. Naturalist* LIV. 317 A distinction is made between vegetarian species with a single food-plant (Monophagous), those with several definitely fixed ones (Oligophagous) and those with quite indiscriminate food habits (Polyphagous). **1946** —— *Insect Dietary* iii. 143 (*heading*) Polyphagy, oligophagy and monophagy. **1969** R. F. Chapman *Insects* ii. 27 Other insects..feed on only a limited range of plants. They are called oligophagous. *Ibid.* 28 Oligophagy may also arise in this way [*sc.* the presence of particular chemical stimulants in certain plants]. **1899** *Allbutt's Syst. Med.* VIII. 196 *Oligophrenia*, enfeeblements of cerebral (psychical) development, with a parallel enfeeblement in the evolution of personality. **1932** *Brit. Jrnl. Psychol.* XXIII. 21 The oligophrenia, or 'small wittedness' as Continental writers call the condition, is due to an insufficiently developed brain. **1972** *Encycl. Psychol.* II. 345/2 *Oligophrenia*, synonymous with the term *amentia*..mental subnormality. [**1925** A. D. Imms *Gen. Textbk. Entomol.* ii. 179 In the oligopod phase the embryo has reached an advanced condition.] **1934** R. A. Wardle *Folsom's Entomol.* (ed. 4) iii. 173 A less active type of oligopod larva..has a cylindrical fleshy body. **1957** T. W. Kirkpatrick *Insect Life in Tropics* iv. 64 Oligopod larvae usually have well-developed thoracic legs but no abdominal feet and are typical of most beetles and Neuroptera. **1969** R. F. Chapman *Insects* 399 The least modified [larval form] with respect to the adult is the oligopod larva. **1930** *Chem. Abstr.* XXIV. 3762 The name *oligosaccharides* is suggested for the simpler cryst. sugars (intermediate between the monoses and the polysaccharides) which form 2 or more monoses on hydrolysis. **1957** *Sci. News* XLV. 87 Human breast milk..contains a number of oligosaccharides. **1968** Oligosaccharide [see *MONOSACCHARIDE]. **1925** *Bull. Illinois Nat. Hist. Survey* XV. 441 The part of a stream lying between the mesosaprobic lower limit and that of the cleanest zone normal to rivers has been called by Kolkwitz and Marsson oligosaprobic. **1933** *Water Pollution Res. Technical Paper* (D.S.I.R.) No. 3. 134 The classification of the organisms into poly-, meso-, and oligo-saprobic..is now generally in use in defining the ecological status of aquatic organisms. It is based on the conditions which result when sewage or similar polluting liquids flow into small streams or into series of lagoons; in such cases a characteristic succession of biological associations is found, beginning with the poly-saprobes living in the crude liquid and ending with the oligo-saprobes in the region where self-purification is complete. **1950** *Folia Limnologica Scand.* V. 77 The oligosaprobic zone is regarded chemically as the zone in which oxidation (mineralisation) is nearly completed. **1970** Oligo-saprobic [see *mesosaprobic* adj. s.v. *MESO-]. **1848** Dunglison *Dict. Med. Sci.* (ed. 7) 599/2 *Oligospermia*, paucity of spermatic secretion. **1897** White & Martin *Genito-Urinary Surg.* xxviii. 1027 Oligospermia, or a diminution in the quantity of semen ejaculated, may be due to deficiency in quantity or absence of any of the constituent parts of this fluid. **1936** H. Bailey *Dis. Testicle* xviii. 148 The causes of male sterility are:.. 2. Oligospermia; spermatozoa are few and inactive. **1944** R. S. Hotchkiss *Fertility in Men* ix. 185 Defective spermatogenesis results in the failure to supply the full complement of spermatozoa to the ejaculate. This varies in degree from complete atrophy of the seminiferous tubule to a reduced number of spermatozoa in the seminal discharge. The former condition produces azoospermia, while the latter causes oligospermia. **1974** Passmore & Robson *Compan. Med. Stud.* III. xxviii. 21/1 A diminution of fertility can be demonstrated when sperm density falls below 20 M/ml, and values below this level on repeated counts constitute oligospermia. **1892** *Syd. Soc. Lex.*, Oligospermic. **1971** *Nature* 19 Feb. 534/1 In veterinary and medical practice artificial insemination facilitates the use of incapacitated or oligospermic males. **1899** C. Robertson in *Bot. Gaz.* XXVIII. 29 The difference between a monotropic and an oligotropic bee may depend merely upon the accident that only one species occurs in the neighbourhood. **1919** J. H. Lovell *Flower & Bee* 106 When a species of bee restricts its visits..to a few allied kinds of flowers [it is termed] an oligotropic bee. **1946** C. T. Brues *Insect Dietary* iii. 107 Some more specialized bees and other insects restrict their visits to a much more circumscribed assortment of plant species. With these oligotropic kinds the advantages to the plant are obviously greater. **1892** F. P. Foster *Med. Dict.* IV. 2447/1 *Oligozoospermia*, of De Sinety, a variety of sterility in the male in which the spermatozoids are diminished in number and activity. **1897** White & Martin *Genito-Urinary Surg.* xxviii. 1027 Oligozoöspermia indicates a condition in which the semen ejaculated contains few spermatozoa. **1962** *Lancet* 27 Jan. 218/1 There was nothing to suggest oligozoospermia due to external causes. **1876** Dunglison *Dict. Med. Sci.* (rev. ed.) 721/2 Oliguria. **1961** *Lancet* 22 July 187/2 The œdema, oliguria, albuminuria, and absence of an impressive cardiac murmur may result in the condition being mistaken for acute nephritis. **1907** *Amer. Jrnl. Med. Sci.* CXXXIV. 77 The oliguric urine..is of higher specific gravity and contains less indican. **1918** *Endocrinology* II. 95 In man it is common to observe diuresis, especially in oligurics. **1961** *Lancet* 16 Sept. 632/1 In mushroom poisoning hæmodialysis is obviously essential where..severe oliguric renal failure occurs.

oligodendrocyte (ǫligode·ndrosəit). *Histology.* [f. *OLIGODENDRO(GLIA + -CYTE.] A kind of neuroglial cell similar to an astrocyte (but with fewer processes), found characteristically round nerve cells in the central nervous system and concerned with the maintenance of myelin.

1932 [see *OLIGODENDROGLIA]. **1966** Wright & Symmers *Systemic Path.* II. xxxiv. 1243/2 The oligodendrocyte greatly outnumbers all other cells in the brain. **1974** J. A. G. Rhodin *Histol.* xv. 316/1 The primary function of the oligodendrocytes is the formation of the myelin sheath.

oligodendroglia (ǫ:ligǫdendrǫglai·ă). *Histology.* [ad. Sp. *oligodendroglia* (P. del Rio-Hortega 1921, in *Bol. de la Real Soc. Española de Hist. Nat.* XXI. 63): see Oligo-, Dendro-, and Neuro)glia.] Oligodendrocytes. Usu. const. as *pl.*

1924 *Brain* XLVII. 430 (*heading*) Oligodendroglia and its relation to classical neuroglia. **1932** W. Penfield *Cytol. & Cellular Path. Nervous Syst.* II. ix. 437 Oligodendrocytes or oligodendroglia were first demonstrated by Robertson (1899, 1900 *a*) under the name of mesoglia by means of his platinum method. Del Rio-Hortega (1921 *b*) independently re-discovered the cells by means of a more reliable and complete staining method. He called them oligodendroglia..because, as compared with astrocytes or the astroglia, the branches of these cells were small and apparently few and he recognized them as one type of neuroglia. **1951** O. Larsell *Anat. Nervous Syst.* (ed. 2) iv. 127 The elements of oligodendroglia are related to the astrocytes but are smaller and have smaller nuclei. **1962** [see *MICROGLIA]. **1974** Bergman & Afifi *Atlas Microsc. Anat.* vi. 111 Oligodendroglia are smaller than astrocytes and have a denser nucleus and cytoplasm. **1974** J. A. G. Rhodin *Histol.* xv. 314/2 Oligodendroglia or oligodendrocytes are small, angular cells of neural ectodermal origin.

Hence **o:ligodendrogli·al** *a.*

1929 *Jrnl. Path. & Bacteriol.* XXXII. 736 An oligodendroglial cell contains a spherical nucleus which is smaller than the nucleus of a neuroglial cell. **1972** *Science* 19 May 801/3 These vacuoles..were found..in axons, astrocytes, and oligodendroglial cells.

oligodendroglioma (ǫ:ligodendrǫglai,ǭu·mă). *Path.* [f. *OLIGODENDROGLI(A + *-OMA.] A tumour derived from oligodendroglia.

1926 Bailey & Cushing *Classification Tumors of Glioma Group* 54 With few exceptions the true brain tumors may be classified under fourteen major categories as follows:..Medullo-epithelioma..Oligodendroglioma..Neuroblastoma. **1929** *Jrnl. Path. & Bacteriol.* XXXII. 735 (*heading*) Oligodendrogliomas of the brain. **1967** *Nursing Times* 27 Jan. 108/1 Some of these, such as the oligodendroglioma, grow very slowly and are compatible with normal life in many cases for 10 years or more. **1978** *Jrnl. R. Soc. Med.* LXXI. 419 Oligodendroglioma, a tumour which, in *in vitro* studies..has been shown to have a high oxygen consumption.

oligohydramnios (ǫ:ligohəidræ·mniǫs). *Obstetrics.* [f. Oligo- + Hydr(o- + Amnios.] A deficiency in the amount of amniotic fluid.

1889 [see *POLYHYDRAMNIOS]. **1928** W. G. Lee *Childbirth* xv. 227 A 'dry uterus' and an oligohydramnios..often result in either a constriction ring or a uterus closely molded to the fetal mass. **1962** D. E. Reid *Textb. Obstetr.* xxiv. 630/2 In oligohydramnios, rather than the normal 1,000–2,000 ml., only 100–200 ml. is found. **1972** E. D. Morris in C. J. Dewhurst *Integrated Obstetr. & Gynaecol.* xxii. 381 Oligohydramnios is evident clinically in two particular circumstances: in association with a small-for-dates baby, and with severe renal anomalies of the foetus.

oligomer (ǫli·gǫməɹ). *Chem.* [f. Oligo- + *-mer.] Any polymer whose molecules consist of relatively few repeating units.

1952 van der Want & Staverman in *Rec. Trav. Chim. Pays-Bas* LXXI. 379 In the course of an investigation into the chemical and physical properties of polymers of different molecular weight it appeared desirable to us to have available condensation products of ε- amino-caproic acid: $H[NH(CH_2)_5CO]_nOH$ with low and exactly specified values of *n* ('linear oligomers') e.g. *n*=2, 3, 4 and 5. **1967** [see *leucoanthocyanidin* s.v. *LEUCO-]. **1969** *Nature* 13 Sept. 1125/2 Table 1 shows the frequencies of monomers, dimers and higher oligomers of mitochondrial DNA. **1975** *Ibid.* 6 Mar. 83/2 The extent to which each of the characteristic oligomers (pentanucleotides and larger) of *Escherichia coli* 16S rRNA is conserved across phylogenetic lines.

Hence **oligome·ric** *a.*

1957 *Makromolekulare Chemie* XXIII. 32 The van der Wyk-rule, which permits calculation of the melting points of straight-chain parafins, can be applied to the oligomeric esters of terephthalic acid and glycol after introduction of suitable constants. **1970** *Nature* 13 June 1004/2 A series of oligomeric substrates, up to twelve glucose units in length and labelled at the end-group, was prepared. **1974** *Sci. Amer.* Nov. 64/1 The J chain is present in oligomeric forms, and the dimer, when found in secretions such as saliva and tears, is bonded to yet another polypeptide.

oligomerization (ǫli:gǫmĕraizē·ʃən). *Chem.* [f. prec. + -ization.] The formation or production of an oligomer from a monomer.

1958 *Jrnl. Chem. Soc.* 3563 (*heading*) The cationic oligomerisation of the stilbenes. **1972** *Jrnl. Amer. Chem. Soc.* XCIV. 6968 (*heading*) The mechanism of the oligomerization of hydrogen cyanide and its possible role in the origins of life. **1974** *Encycl. Brit. Macropædia* XIII. 716/2 Aluminum alkyls also cause dimerization, oligomerization, or polymerization of olefins.

So **oli·gomerize** *v. trans.*, to form an oligomer of.

1967 *Nature* 29 Apr. 480/1 Hydrogen cyanide could be oligomerized to the dimer, trimer, tetramer, etc., to form the parent skeletons of glycine, alanine and aspartic acid, and other organic compounds. **1972** *Proc. Nat. Acad. Sci.*

LXIX. 3389/1 Linear double-stranded molecules of simian virus (SV)40 DNA..are oligomerized by either ligase.

oligomictic (ǫligomi·ktik), *a.* [f. Oligo- + Gr. μικτ-ός mixed + -ic.] **1.** *Petrol.* [ad. Russ. *oligomiktovyĭ* (M. S. Shvetsov *Petrografiya Osadochnykh Porod* (1934) viii. 155).] (See quot. 1935.)

1935 *Jrnl. Sedimentary Petrology* V. 106/2 Rocks consisting of one to two dominant minerals are termed oligomictic and those composed of several minerals polymictic... The book is written in Russian... The review is based on a typewritten summary in English. **1949** F. J. Pettijohn *Sedimentary Rocks* xiv. 438 Schwetzoff..noted that oligomictic rocks are the characteristic deposits of epicontinental seas and are found rarely in geosynclinal depressions, whereas polymictic rocks are characteristic of geosynclinal regions. **1959** W. W. Moorhouse *Study of Rocks in Thin Section* xix. 337 The typical oligomictic conglomerate is composed predominantly of quartz. **1971** *Nature* 28 May 247/1 Structureless to planar cross-stratified, sheet-like bodies of oligomictic conglomerates and subarkoses are interbedded.

2. *Limnology.* Applied to a lake that exhibits a stable thermal stratification and only rarely undergoes an overturn.

1956 Hutchinson & Löffler in *Proc. Nat. Acad. Sci.* XLII. 84 Although at low altitudes in the humid tropics small temperature gradients can maintain stable stratifications, no stable stratification develops at the low temperatures of high altitudes, where the density difference per degree centigrade is very small. The lack of seasonal variation, that permits almost perennial stratification at low altitudes in equatorial latitudes, thus permits perennial circulation at high altitudes in the same latitudes. We propose for these two types of equatorial lake the terms oligomictic and polymictic, respectively. **1968** R. W. Fairbridge *Encycl. Geomorphol.* 617/1 The lake water body is stratified, thus oligomictic.

oligopoly (ǫligǫ·pǫli). [f. Oligo- + Gr. πωλ-εῖν to sell, after Monopoly.] A state of limited competition when a market is shared by a small number of producers or sellers.

1895 J. H. Lupton *Utopia of Sir Thomas More* i. 55/2 More makes an antithesis between *monopolium* and *oligopolium*. We have 'monopoly' but not 'oligopoly' (the sale by a few), and so cannot preserve the point of the sentence. **1933** E. H. Chamberlin *Theory Monopol. Competition* i. 8 The theory of value..has been treated..with particular reference to the problem of two sellers, or 'duopoly', and we may extend this terminology, adding 'oligopoly' for a few sellers. **1954** *Wall St. Jrnl.* 13 Dec. 3/3 'Oligopoly'—monopoly power in the hands of two or more companies. **1957** *Economist* 7 Sept. 769/2 The small but bustling textile industry is now giving a miniature demonstration of the oligopoly stage of capitalism as the big firms begin to mop up the smaller ones. **1959** [see *DUOPOLY b]. **1967** J. K. Galbraith *New Industrial State* xvi. 180 Under the cognomen of oligopoly it is assumed in its price-making to have some of the powers of a monopoly and some of the restraints of competition. **1970** *Daily Tel.* 23 Jan. 18 Deciding when monopoly or oligopoly is or is not..'against the public interest'. **1974** M. Barratt Brown *Econ. of Imperialism* iii. 52 The driving force behind the extension of commodity production under capitalism is the competition of capitalists, even under conditions of oligopoly. **1977** *Dædalus* Fall 92 Oligopolies will tend to have excess physical capacity.

Hence **oligo·polist**; **oligo:polis·tic** *a.*

1939 J. A. Schumpeter *Business Cycles* I. ii. 60 The general statement that oligopolistic prices are indeterminate would be misleading. **1947** *Jrnl. Pol. Econ.* LV. 432/1 Observed price rigidities in oligopolistic industries. **1958** *Jrnl. of Business* (Chicago Univ.) XXXI. 198/1 Frequently, in the case of older products, some measure of oligopolistic uncertainty exists. **1959** *Economist* 28 Mar. 1175/1 Increases between 1953 and 1957 came about entirely from prices in the oligopolistic industries. **1959** De Chazeau & Kahn *Integration & Competition in Petroleum Industry* xvii. 443 Having achieved the touchstone of crude oil production control, the leading firms and practically the industry as a whole behave like self-conscious oligopolists. **1967** *Spectator* 28 July 113/2 The big cleaning material manufacturers (I mean, of course, the soap and washing powder oligopolists) have all been making efforts to grab a slice of this easy market. **1971** K. Hopkins *Hong Kong* 212 Oligopolistic industries dominated by large corporations do not exist in Hong Kong. **1975** *New Law Jrnl.* 11 Sept. 892/1 The provisions of article 86 of the EEC treaty are designed to proscribe monopolistic, oligopolistic or, in certain respects, a monopsonic position being exploited at the expense of other producers or of consumers in the European Common Market. **1977** *Dædalus* Fall 92 The world is not a two-person zero-sum game, and no oligopolist expects competitors to hold their prices or quantities fixed in response to his moves.

oligopsony (ǫligǫ·psǫni). *Econ.* [f. Oligo- + Gr. ὀψων-εῖν to buy provisions; after *Monopsony; cf. Opsony.] A marketing state in which only a small number of buyers exists for a product; also *attrib.* Hence **oligo·psonist**, **oligo:psoni·stic** *a.*

1943 E. R. Walker *From Econ. Theory to Policy* iv. 61 It is surely only a matter of time before [market situation] No. 23 is christened 'oligopsony'. **1949** W. Fellner *Competition among Few* i. 11 The problem is that of oligopoly, oligopsony, and bilateral monopoly, and, of course, also of markets which are oligopsonistic on the demand side and oligopolistic on the supply side... The oligopsonist..attempts to select a definite price to be paid for the materials and services he buys. **1961** N. F. Keiser *Introductory Econ.* v. 77 At one time the major cigarette manufacturers were oligopsonists in their purchases of tobacco. **1965** D. Greenwald et al. *McGraw-Hill Dict. Mod. Econ.* 358 *Oligopsony*, a market structure with relatively few buyers... An oligopsonistic situation may lead to express or tacit collusion among the sellers. **1965** Hailstones & Dodd *Econ.* (ed. 5) xi. 229 *Oligopsony* exists when a few buyers dominate the market... The author of a college textbook faces an oligopsonistic market in the publication of his manuscript. **1972** Hunt & Sherman *Econ.* xviii. 275 Some large firms have extra market power as large buyers of commodities (technically, oligopsony power). **1972** J. Winkler *On Marketing Planning* xiii. 205 Companies operating in oligopsonistic markets find that great power flows to their salesmen.

oligotrophic (ǫligotrǭu·fik), *a. Ecol.* [ad. G. *oligotroph* (A. Thienemann *Die Binnengewässer Mitteleuropas* (1925) iv. 198 (*-trophie* sb.), 200 (*-troph*)), f. Gr. ὀλίγος small, little + τροφή nourishment: see -ic.] Relatively poor in plant nutrients and (in the case of a lake) containing abundant oxygen in the deeper parts.

[**1928** *Proc. Linn. Soc.* CXL. 100 The typical oligotroph lakes are deep, with submerged beaches narrow or absent, inconsiderable or no littoral vegetation, and an indistinct littoral animation.] **1931** R. N. Chapman *Animal Ecol.* xvi. 304 The oligotrophic type of lake is rich in oxygen even to the bottom. It owes its characteristic partly to a geologic formation which permits relatively little inwash of organic material..; and partly to biotic conditions which do not favour rapid decomposition with the consequent oxygen consumption. **1943** G. K. Fraser *Peat Deposits of Scotland* I. 3 The remains of plants nourished on rich soils (technically termed eutrophic soils)..will be able to support a greater bacterial population than those of plants grown on poor or impoverished soils (oligotrophic soils). *Ibid.* 9 Acid ground waters are usually oligotrophic and can support only short herbage such as smaller sedges. **1955** *New Biol.* XVIII. 115 *N. alba* occupies a wide range of waters in the British Isles, from the oligotrophic, or nutrient-poor, peat-bottomed moorland lakes in Scotland and Ireland, to the eutrophic, or nutrient-rich, fen-lodes and broads of East Anglia. **1972** J. G. Cruickshank *Soil Geogr.* vi. 186 Also acid and infertile is oligotrophic peat.

So **o:ligotrophy**, an oligotrophic condition.

1928 *Proc. Linn. Soc.* CXL. 109 Even if the natural course is from oligotrophy to eutrophy, the opposite process may also be found. **1957** G. E. Hutchinson *Treat. Limnol.* I. ix. 644 Hutchinson set limits [on oxygen loss] of 0.017 mg. $cm.^{-2}$ day^{-1} for oligotrophy. **1967** [see *EUTROPHY]. **1973** P. A. Colinvaux *Introd. Ecol.* xviii. 258 It is the change from oligotrophy to eutrophy, simulating as it does the natural aging of a lake, which gives rise to the idea that polluted lakes are dying or actually dead.

olingo (oli·ŋo). [Amer. Sp., f. native name.] A small, nocturnal mammal of the genus *Bassaricyon*, belonging to the family Procyonidæ, native to forest regions of Central and South America, and distinguished from the kinkajou by a straight tail which is not prehensile.

1920 E. A. Goldman in *Smithsonian Misc. Coll.* LXIX. No. 5. 155 (*heading*) Bassaricyon gabbii gabbii Allen. Bushy-tailed Olingo. **1964** E. S. Crandall *Managem. Wild Mammals in Captivity* 314 Closely allied to the kinkajou and apparently often confused with it is the olingo (*Bassaricyon gabbii*). *Ibid.* 315 The olingo is a grayer brown, over-all, than most kinkajous, with pale gray face and noticeably longer and more pointed muzzle. Its most obvious character is the long tail, which is very faintly ringed, somewhat bushy, and non-prehensile. **1965** D. Morris *Mammals* 178 Allen's Olingo bears a strong resemblance to the Kinkajou... Where the two species do differ, the Olingo is always the more primitive. **1975** *City Press* 4 Sept. 16/3 Who would suspect that the City has anything to do with 18,985 rats, 1,110 baboons, 10 anteaters and an olingo?

olio. Add: **2. b.** Further examples; *spec.* a variety act or show; also *attrib.*

1809 S. Breck *Recoll.* (1877) App. 271 We..rode round to Mr. Brent's,..with whose family we took tea, and afterward accompanied them to an olio concert. **1928** *Amer. Speech* IV. 68 Behind these *drops*,..are the *oleos*, or *act-curtains*. These..are used for small vaudeville acts... Such acts..are termed *oleo acts*, or *acts in one*. **1951** Green & Laurie *Show Biz* 570/1 *Olio*, scenery, in front of which an act, generally a 'sidewalk comedy' team performs; also specialties performed between acts in burlesque. **1956** M. Stearns *Story of Jazz* (1957) xi. 116 The second part, or olio (a word derived from the Spanish *olla*, meaning mixture), consisted of a series of solo acts that later evolved into variety or vaudeville. **1961** Bowman & Ball *Theatre Lang.* 237 *Olio*; *oleo*... A scene consisting of a specialty act..played..while another scene is being set. Also as *olio act* (or *scene*)... A backing, drop, or tableau curtain for a front scene... Also as *olio drop*... A medley of songs, dances, comic sketches and the like. **1961** A. Berkman *Singers' Gloss. Show Business* 64 *Oleo*, miscellaneous Vaudeville or Variety Acts presented between the acts or during the intermissions of a play. **1964** *Punch* 2 Dec. 852/1 The audience is invited to stay..for the Olio,..which here means an Aftershow. **1976** *Publishers Weekly* 9 Feb. 96/3 It is a mixture of self-indulgent prose, sickening violence and unbelievable happenings. The whole olio is, clearly, a bid for Bicentennial attention.

olisbos (ǫ·lizbǫs). [ad. Gr. ὄλισβος] = *dildo[1].

1887 [see *GODEMICHE]. **1941** G. R. Scott *Phallic Worship* x. 178 Aristophanes speaks of the use, by Milesian females, of an *olisbos* made of leather. **1955** V. Nabokov *Lolita* i. xxii. 126, I had blazed in her face an olisbos-like flashlight. **1967** Crosland & Daventry tr. *de Becker's Other Face Love* plate 13 (*caption*) Two women together using a double olisbos.

olistolith (ǫli·stolip). *Geol.* [f. Gr. ὀλισθ- stem of ὀλίσθημα slip, slide + -lith.] One of the discrete bodies contained within the matrix of an olistostrome.

1956 [see next]. **1974** *Nature* 26 Apr. 745/1 It is possible that these domes represent olistoliths transported over a long distance from the Turkish shelf.

olistostrome (ǫli·stostrǫʊm). *Geol.* [f. as prec. + Gr. στρῶμα bed.] A sedimentary deposit composed of a heterogeneous mixture of materials and formed by the sliding or slumping of semi-fluid sediment.

1956 G. Flores in *Proc. 4th World Petroleum Congr.* 1955 I. 122/2 By Olistostromes we define those sedimentary deposits occurring within normal geologic sequences that are sufficiently continuous to be mappable, and that are characterized by lithologically and/or petrographically heterogeneous materials, more or less intimately admixed, that were accumulated as a semi-fluid body. In any Olistostrome we distinguish a 'binder' or 'matrix' represented by prevalently pelitic, heterogeneous material... The name '*Olistolith*'..is applied to the masses included as individual elements within the binder. **1972** *Nature* 8 Dec. 328/2 The mixture is now floating in a matrix of middle Miocene olistostrome.

oliva. 1. (Later example.)

1973 A. H. Whiteford *N. Amer. Indian Arts* 133 Trade routes from the Gulf of California and Mexico brought..oliva shells into the Southwest.

olive, *sb.*[1] Add: **1. b. American olive** (examples).

1785 H. Marshall *Arbustrum Amer.* 98 American Olive tree..grows naturally in Carolina and Florida, and is a beautiful ever-green tree. **1866** *Land We Love* (Charlotte, N. Carolina) May 78 American Olive..is a very fine evergreen, producing clusters of small white flowers. **1901** C. T. Mohr *Plant Life Alabama* 14 Their banks adorned with evergreen andromedas, American olive,.. sweet bay, and azaleas.

10. A greenish-brown moth, *Zenobia* (or *Ipomorpha*) *obtusa*, of the family Noctuidæ, found in Europe and northern Asia.

1832 J. Rennie *Conspectus Butterflies & Moths Brit.* 83 The Olive..feeds on the poplar. **1908** R. South *Moths Brit. Isles* 2nd Ser. 9 The Olive..is somewhat similar in general appearance to the last mentioned [*sc.* the Double Kidney]. **1974** B. Goater *Butterflies & Moths of Hampshire* 376 The Olive..widespread but associated with *Populus* species.

11. A mayfly with an olive-coloured body belonging to the genus *Baetis*, which includes species with transparent wings, or the genus *Ephemerella*, esp. *E. ignita*, which has blue wings.

1889 F. M. Halford *Dry-Fly Fishing* ix. 206 The blue-winged olive..is known to modern entomologists as *Ephemerella ignita*. **1911** —— *Mod. Devel. Dry Fly* iii. 18, I think the dark olive is, as a rule, not so well taken by the fish as the common olive. **1949** A. C. Williams *Dict. Trout Flies* II. 267 Whereas other insects are seasonal, the olive is more or less always with us. **1971** *Country Life* 21 Oct. 1084/1 Often there is a good hatch of olives in the morning or afternoon—sometimes both—which usually brings a response not only from the grayling but from the trout.

b. An artificial fly made in imitation of an insect of this type.

1895 *Montgomery Ward Catal.* Spring & Summer 495/2 Bass Flies, consisting of the following styles:.. Oak, Olive, Montreal, Professor, [etc.]. **1907** *Yesterday's Shopping* 674/2 Special Irish Salmon Flies... Golden Olive with Blue and Jay Shoulder. **1911** F. M. Halford *Mod. Devel. Dry Fly* iii. 18 No. 7 of the series of patterns is the olive dun male. **1921** G. E. M. Skues *Way of Trout with Fly* ii. 109 The floating subimago I tried to imitate with a darkish variety of a stock pattern of olive. **1938** W. C. Platts *Mod. Trout Fishing* vii. 68 Among the wet flies in use may be mentioned..various Olives. **1968** C. F. Walker *Art of Chalk Stream Fishing* xvii. 147 My own choice would be the Rough Olive, a most successful fly.

C. a. *olive crop, culture, -grove* (later examples), *industry, spray*. **d.** *olive-backed* (earlier and later examples), *-skinned* adjs.; also with reference to the shape of an olive, as *olive-shaped* adj. **e. olive-back**, substitute for def.: a North American forest thrush, *Hylocichla ustulata*, also known as Swainson's thrush; (earlier and later examples); **olive-backed thrush** = *olive-back*, *olive thrush*; **olive-berry** (later example); **olive crescent**, a pale greenish-brown European moth, *Trisateles emortualis*, of the family Noctuidæ; **olive drab**, of a brownish green colour, used *spec.* of the colour of U.S. Army uniform; also *ellipt.*; **olive-fly** (examples); also *olive fruit fly*; **olive thrush** = *olive-back*; **olive whistler**, an Australian bird, *Pachycephala olivacea*.

1845 S. Judd *Margaret* i. xvi. 143 The olive-backs trolled and chanted among the trees. **1892** B. Torrey *Foot-Path Way* 19 The olive backs began to make themselves heard. **1945** *Mass. Audubon Soc. Bull.* Mar. 43 Two thrushes of annual interest to students are the migrant Olive-back and the Gray-cheek. **1844** J. E. DeKay *Zool. N.Y.* ii. 74 The Olive-backed Thrush is closely allied to the [hermit thrush]. **1892** B. Torrey *Foot-Path Way* 99 Two birds dashed by me—a blackpoll warbler in hot pursuit of an olive-backed thrush. **1946** T. M. Stanwell-Fletcher *Driftwood Valley* 187 We spend the long bright evenings out on the lake, listening to the chorus of olive-backed thrushes. **1958** E. T. Gilliard *Living Birds of World* 336/2 The Olive-backed Thrush..winters south to Argentina. **1869** Mrs. Stowe *Oldtown Folks* xvi. 176, I guess our olive-berries are pretty well beaten off now. **1832** J. Rennie *Conspectus Butterflies & Moths Brit.* 146 The Olive Crescent..resembles the Clay-Fan-Foot. **1908** R. South *Moths Brit. Isles* 2nd Ser. 88 The Olive Crescent..is exceedingly rare in England. **1974** B. Goater *Butterflies & Moths of Hampshire* 411 Olive Crescent... One taken in bright sunlight..in late July, 1939. **1884** *Encycl. Brit.* XVII. 762/1 Apart from occasional damage by weather or organic foes, the olive crop is somewhat precarious even with the most careful cultivation. **1977** J. Aiken *Last Movement* ii. 42 Local staff, who came and went when the orange or olive crop demanded their attention. **1893** K. A. Sanborn *Truthful Woman in S. California* xii. 155 Olive culture is just now the fad. **1957** *Encycl. Brit.* XVI. 774/1 Specialized olive culture is an important industry on hillsides throughout Greece. **1897** *Sears, Roebuck Catal.* 21/3 One gallon of this paint will cover (two coats) over 300 square feet of surface. Always order by color number as well as catalogue number... 214 Olive Drab. **1908** *Sears, Roebuck Catal.* 71/2 *Colors of..house paint*... Nile Green.. Olive Drab..Cream. **1917** A. Woollcott *Let.* Oct. (1946) 28, I was afflicted because I had signed, have been in olive drab for three months—been away from America for almost three months. **1942** —— in *Reader's Digest* Nov. 23/2 Wherefore, as he [*sc.* Irving Berlin] toils away at something for the boys in olive-drab to sing out with real emotion, he has only to listen to the bugle notes for a motif. **1948** W. J. Stokoe *Caterpillars Brit. Moths* I. 274 The caterpillar..is slaty-brown, inclining to olive-drab above. **1970** N. Armstrong et al. *First on Moon* iii. 66 Check the olive-drab lap and shoulder strapping for each man. **1886** R. C. Haldane *Subtropical Cultivations* 183 *Musca oleæ* (the olive-fly) lays its eggs in the young fruit, and is a most destructive insect. **1972** Swan & Papp *Common Insects N. Amer.* 627 The Olive Fruit Fly, *Docus oleae*, is a serious pest of olives in the Mediterranean area. **1878** O. Wilde *Ravenna* 6 Dark olive-groves and noble forest-pines. **1959** *Times* 29 Sept. 12/6 Vineyards and olive groves. **1893** K. A. Sanborn *Truthful Woman in S. California* xii. 155 Pomona is head-quarters for the olive industry. **1968** *Encycl. Brit.* XVI. 937/1 In South America and Australia, development of a commercial olive industry is..recent. **1908** *Practitioner* Sept. 360 The sounds which will best aid are those..having interchangeable olive-shaped metallic heads. **1930** T. S. Eliot tr. *St.-J. Perse's Anabasis* 65 He who fashions a leather tunic, wooden shoes and olive-shaped buttons. **1904** W. H. Hudson *Green Mansions* 4 The nervous olive-skinned Hispano-American of the tropics. **1970** H. M. Davy *Caring for your Appearance* iii. 35 Some of your friends may have a very pale skin throughout all seasons... Some others, with the very darkest colouring, we may describe as olive skinned. **1864** J. R. Lowell *Fireside Trav.* 222 Climbing the sides of the nearer Monticelli in a gray belt of olive-spray. **1957** *Encycl. Brit.* XVI. 773/2 The wild olive spray of the Olympic victor. **1904** S. E. White *Silent Places* i. 4 The white-throats and olive thrushes called in a language hardly less intelligible. **1911** J. A. Leach *Austral. Bird Bk.* 152 Olive Whistler, Olivaceous Thick-head... Olive brown; head dark-gray... Liquid, whistling note. **1933** *Bulletin* (Sydney) 20 Sept. 21/1 A curious example of vocal variation among birds..is found in the 'olive whistler'. **1944** A. Russell *Bush Ways* xxii. 105 A recluse of the open scrubs of the dry south-eastern interior, as the olive whistler is to the mountain mists of the coast ranges. **1965** *Austral. Encycl.* IX. 292/1 The olive whistler, of eastern Australia and Tasmania, is probably one of the sweetest singers among the birds of Australia.

olive-branch. 1. b. (Further examples.)

1856 R. Glisan *Jrnl. Army Life* (1874) xxiv. 324 The troops..moved up Rogue River..with the olive branch in one hand, and the sword in the other. **1936** Wodehouse *Laughing Gas* v. 60 He read his *National Geographic Magazine*. I read mine. And for some minutes matters proceeded along these lines. Then I thought to myself: 'Oh, well, dash it,' and decided to extend the olive branch. **1975** B. Garfield *Hopscotch* xxvii. 283 Abandoning the manuscript..could be an olive branch: Kendig's..assurance he was quitting.

olive-oil. Add: Also attrib.

1895 *Montgomery Ward Catal.* Spring & Summer 109/2 A pure, natural color olive oil soap made of the finest selected oil imported for this purpose. **1957** *Encycl. Brit.* XVI. 774/2 California olive oil output is normally less than 1% that of Spain. **1970** 'D. Halliday' *Dolly & Cookie Bird* ii. 10 They have..vineyards and..three olive-oil factories. **1977** P. Way *Super-Celeste* 52 The Mafia godfathers could..declare they were olive oil merchants.

2. A jocular mispronunciation of *au revoir*.

1906 *Dialect Notes* III. 148 *Olive oil, n. phr.*, au revoir. Facetious. **1909** J. R. Ware *Passing Eng.* 187/2 *Olive oil* (Music Hall, 1884). English pronunciation of *au revoir*. **1933** Partridge *Slang To-day & Yesterday* iii. iii. 206 For 'good-bye', the boys at Dulwich already in 1906 used.. *olive oil* (au revoir). **1960** Wentworth & Flexner *Dict. Amer. Slang* 364/2 *Olive oil*, good-bye.

Oliver[3] (ǫ·livəɹ). *slang.* Also **oliver.** [A male Christian name, perh. alluding to *Oliver* Cromwell (1599–1658), leader of the Parliamentary troops in the Civil War.] The moon.

1781 [see Whiddle *v.*]. **1834** W. H. Ainsworth *Rookwood* II. v. 360 Now Oliver puts his black nightcap on, And every star its glim is hiding. *Ibid.* III. v. 284 Oliver whiddles!—who cares—who cares If down upon us he peers and stares? Mind him who will? with his great white face, Boldly *I'll* ride by his glim to the chase. **1870** R. F. Burton *Vikram & Vampire* v. 171 But, look sharp, mind old Oliver, or the lamb-skin man will have the pull of us. **1882** *Sydney Slang Dict.* 6/2 *Oliver*, the moon. 'When Oliver looks pale,' when the moon is waning. **1895** *New Rev.* July 7 'There's a moon out.' 'The better for us to pick 'em off, Dan,' I returned, laughing at him. 'What—Oliver? damn Oliver!' said Zacchary. 'Let's push forward and come to quarters.' **1935** E. Weekley *Something about Words* vi. 107 *Oliver*, thieves' cant for the moon, is not in the Oxford Dictionary.

Oliver[4] (ǫ·livəɹ). [The name of William *Oliver* (1695–1764), a physician of Bath, who invented the recipe.] = *Bath Oliver* (s.v. *Bath *sb.*[2] 2 a). Also *attrib.*

1853 E. M. Sewell *Experience of Life* ii. 18 The Oliver biscuits, in the small, deep, old china dessert plates, were to my belief then never bought at any shop in Carsdale. *Ibid.* vii. 64 She was sitting..at a table..on which stood ..a china basket with Oliver biscuits.

Olivetti (ǫlive·ti). [Name of the manufacturers.] The proprietary name of a range of typewriters.

1949 *Trade Marks Jrnl.* 23 Mar. 255/1 Olivetti...Typewriters. Ing. C. Olivetti & C., Società per Azioni .., Ivrea, Province of Turin, Italy; machine manufacturers. **1966** Auden *About House* 18 The Olivetti portable, The dictionaries (the very Best money can buy). **1968** *Listener* 28 Nov. 736/1 You can still go for a walk down Gloucester Crescent and below the chatter of a dozen Olivettis catch the soft murmur of extra-marital associations. **1977** G. Fisher *Villain of Piece* i. 3, I..glared at my Olivetti. There was a sheet of copy paper in it.

olivescent (ǫlive·sĕnt), *a.* [f. Olive + -escent.] Of colour: bordering on or slightly olive.

1900 *Proc. Zool. Soc.* 506 *Kirontisa whiteheadi*... Upper-side deep olivescent brown. Fore wing with two paler olivescent marks within.

olive-yard. (Later examples.)

1880 J. H. Shorthouse *John Inglesant* xxxvii. 523 He had come into the cool pastures and olive yards. **1911** *Encycl. Brit.* XX. 87/1 Many olive-yards now exist in Upper Egypt.

olivine. c. (Examples.)

1895 A. Harker *Petrol.* 115 Numerous olivine-diabases are associated with the Carboniferous strata of the Midlands. *Ibid.* 68 The Tertiary gabbros of the western islands of Scotland..are in general olivine-gabbros. **1900** [see *KENTALLENITE]. **1936** J. S. Flett in Wilson & Knox *Geol. Orkney* xvii. 180 They are the only olivine-basalt dykes that have been discovered in the Orkneys. **1956** W. Edwards in D. L. Linton *Sheffield* 15 Boreholes in the southeastern part of the coalfield have revealed thick, sill-like beds of olivine-dolerite and analcime-dolerite. **1965** E. L. P. Mercy in G. Y. Craig *Geol. of Scotland* vii. 243/2 Rhythmic banding in the olivine-gabbros and troctolites has been described.

olland, var. *old land.

Ollendorffian (ǫlendǭ·ɹfiăn), *a.* Also **Ollendorfian.** [f. the name of Heinrich Gottfried *Ollendorff*, German educator and grammarian (1803–65) + -ian.] In the stilted language of foreign phrase-books.

[**1876** C. M. Yonge *Womankind* vi. 40 German [learnt] by the Ollendorf method.] [**1886** F. M. Crawford in H. Norman *Broken Shaft* 18 The simple but instructive dialogues of Herr Ollendorff.] **1892** *New Rev.* Feb. 252 But the characters have as much individuality as the Ollendorffian prattlers about the gardener. **1900** *Sketch* 21 Feb. 191/2 She persisted in firing off Ollendorfian French at the waiters. **1905** *Westm. Gaz.* 8 July 2/3 'Views' which we snapshot, be they views of mountains or of men, and whether we use Kodak or Ollendorffian chit-chat to produce them. **1918** L. Huxley *Life J. D. Hooker* II. xliii. 327 He started them also, and very successfully, with colloquial Latin from an Ollendorffian French handbook. **1934** H. G. Wells *Exper. Autobiogr.* II. ix. 766 A sort of Ollendorfian French.

ollunt, var. *old land.

olm (ōulm). [G.] = Proteus 3 b.

1905 A. Sedgwick *Student's Text-bk. Zool.* II. x. 307 *Proteus* Laur., the olm, 3 fingers, 2 toes, eyes hidden, Carniola subterranean waters. **1926** J. S. Huxley *Ess. Pop. Sci.* viii. 93 Three or four other species of animals, such as Proteus, the blind 'Olm' of Carniola, and Necturus, which..are not known in a land-form at all. **1955** W. Ley *Salamanders* i. 3 The olm's rather restricted habitat..is a mountainous area in southern Europe, at the northern end of the Balkan peninsula. *Ibid.* 4 The olm is blind, having no use for eyes in the cold dark caves where it normally lives. **1965** B. E. Freeman tr. *Vandel's Biospeleol.* iii. 22 It was *Proteus* which firstly, because of its size, attracted the attention of man, and [was] also the first to receive a vernacular name, the olm.

Olmec (ǫ·lmek). [ad. Nahuatl *Olmecatl*, pl. *Olmeca* lit. 'inhabitants of the rubber country'.] **1.** Also **Olmeca.** A native American people or peoples inhabiting the coast of southern Veracruz and western Tabasco during the 15th and 16th centuries, to where they probably migrated during the 12th century from the Mexican altiplano.

1787 C. Cullen tr. *Clavigero's Hist. Mexico* I. ii. 103 The Olmecas and the Xicallancas, whether one nation, or two distinct nations, but constantly allied and connected together, were so ancient in the country of Anahuac, that many authors account them prior to the Toltecas. **1883** P. J. J. Valentini in *Proc. Amer. Antiquarian Soc.* II. 193 (*heading*) The Olmecas and the Tultecas: a study in early Mexican ethnology and history. **1914** T. A. Joyce *Mexican Archaeol.* v. 125 Though bows were found among the Olmec and Huaxtec, they must have been of quite late introduction. **1931** G. Mason *Columbus came Late* xi. 238 The Olmeca people that inhabited the damp country of southern Vera Cruz and western Tabasco. **1947** M. Covarrubias *Mexico South* iv. 129 The builders of the cities of El Tajín and Teotihuacán..were the Olmecs proper. **1964** C. Gibson *Aztecs under Spanish Rule* 9 The most important migrant peoples pertinent to Valley [of Mexico] history in the Post-Classic or late pre-colonial period were the Olmeca, Xicalanca, Tolteca, Chichimeca, Teochichimeca, Otomi. .. The first five of these were historically extinct, absorbed, or expelled by the time of the arrival of Europeans.

2. A prehistoric people inhabiting the same area *c* 1200–100 BC. Also *attrib.* or as *adj.*; *spec.* designating the culture of this people or its characteristic artistic style, also found elsewhere in southern Mexico. So **O·lmecan, O·lmecoid** *adjs.*

[**1927** H. Beyer in *El Mexico Antiguo* II. 306 (*caption*) Idolo Olmego di Piedra Verdosa.] **1929** *Indian Notes* (Mus. Amer. Indian, Heye Foundation) VI. 280 Beyer published a picture of what he calls 'an Olmecan idol' formerly in his possession and now in a private collection. *Ibid.* 285 This peculiar type of mask may be safely assigned to the ancient Olmecan culture, which apparently had its center in the San Andrés Tuxtla area around Lake Catemaco. **1932** *Natural Hist.* XXXII. 519/2 There is often described in the traditions a highly civilized people called the Olmec, who lived anciently as far north as Tlaxcala, but were later dispersed to southern Vera Cruz, Chiapas, southern Puebla, and eastern Oaxaca. They were famed for their work in jade and turquoise, and were credited with being the chief users of rubber in Central America. *Ibid.* 520/2 Perhaps investigations in the Olmec area would clarify the much discussed relationship between the Mexicans and the Mayas. **1943** M. W. Stirling *Stone Monuments S. Mexico* 7 Two years later he released a study..of the giant head in which he emphasized its 'Ethiopian' features, features which have since been identified with the style of art called Olmec. *Ibid.* 54 The niche..represents the 'Olmec' open-jaguar-mouth motive. **1960** *Times* 6 Oct. 4/6 The Veracruz region, which is rich in..treasures of the Olmec..coastal cultures of ancient Mexico. **1962** G. Ashe *Land to West* viii. 222 A clay Fire-god of the prehistoric Olmecs on the Gulf coast. **1965** M. D. Coe in R. Wauchope *Handbk. Middle Amer. Indians* III. ii. 738 (*heading*) The Olmec style and its distributions. *Ibid.* 765 As one moves away from the core region and also into later time periods, many objects are encountered which are more or less Olmecoid, but these are not Olmec in our meaning. **1966** *Listener* 29 Dec. 957/1 The most ancient civilization of Mesoamerica is that of the Olmecs. **1967** L. Deuel *Conquistadors without Swords* xxi. 287 (*caption*) Substructure of Pyramid E-VII with its Olmecoid stucco masks. **1973** *New Yorker* 24 Mar. 108/2 The Olmec civilization was not identified as such until the nineteen-forties. **1973** *Black World* July 13/1 The huge stone heads of Olmec deities, exhibited an unmistakably African physiognomy. **1973** *Times* 15 Oct. 11/8 The Olmec Forerunners, the archaic predecessors of the Maya though probably not themselves Maya. **1977** *Sci. Amer.* Mar. 116/1 The Olmec, one of the earliest of the complex societies in the region, built major ceremonial centres on the low-lying coastal plain of the Gulf of Mexico; examples are San Lorenzo and La Venta. *Ibid.*, At the same time the Olmec zone of cultural influence and Olmec trade extended into much of the high plateau.

ological. (Earlier example.)

1854 Dickens *Hard T.* i. xv. 120, I hope you may now turn all your ological studies to good account.

-ology, ology, *suffix* and quasi-*sb.* Add: **2.** Also 'ology. (Further examples.)

1854 Dickens *Hard T.* ii. ix. 236 Ologies of all kinds, from morning to night. If there is any Ology left..that has not been worn to rags in this house..I hope I shall never hear its name. **1886** Kipling *Departmental Ditties* (ed. 2) 57 And after—ask the Yusufzaies What comes of all our 'ologies. **1927** G. D. H. & M. Cole *Murder at Crome House* iv. 42 'I don't think she's interested in anything except some 'ology—' 'Psychology, I think, sir,' said Johnson. **1972** C. L. Cooper in W. King *Black Short Story Anthol.* 237 With Famat's help,..I came to grasp an ology that was an intermixture of Ax, orthodox Islam, and theory. **1976** A. Price *War Game* i. ix. 172 One's doing a thesis on geology now, and the other's writing a book on meteorology. Ology is about the only thing they have in common.

ololiuqui (ōuloliū·ki). Also **ololiuhqui.** [ad. Amer. Sp. *ololiuque*, f. Nahuatl *ololiuhqui* 'one that covers'.] A Mexican climbing plant, *Rivea corymbosa* (*Ipomœa sidifolia*), of the family Convolvulaceæ; also, the narcotic drug prepared from its seeds. Also *attrib.*

1915 W. E. Safford in *Jrnl. Heredity* VI. 311 It is very strange that Mexican botanists living in the country of the Ololiuhqui have not solved the mystery of its identity. **1926** *Chambers's Jrnl.* Aug. 513/2 [He] bemused himself with the seeds of the ololiuhqui. **1941** R. E. Schultes (*title*) A contribution to our knowledge of *Rivea corymbosa*, the narcotic ololiuqui of the Aztecs. **1954** A. Huxley *Let.* 2 Mar. (1969) 699 Ololiuqu(i) is used by the Mexican and Cuban witch doctors to increase ESP faculties and relieve disease. **1962** —— *Island* xi. 171 He began to talk about the indoles recently isolated from the ololiuqui seeds that had been brought in from Mexico last year. **1966** *New Scientist* 21 Apr. 156/1 In Mexico, the seeds of climbing convulvulus plants known as 'morning glory' are made into a drug called 'ololiuqui'. **1973** *Sci. Amer.* Oct. 130/2 The seeds of certain Mexican morning glories,..the source of a magic Aztec potion, ololiuqui, used to 'communicate with the gods and receive the secret things'.

Olonetsian (ǫlone·tsiăn). Also **Olonecian, Olonetzian.** [f. *Olonets*, the name of a town and former government in N.W. Russia.] A dialect of Karelian spoken to the north east of Lake Ladoga. Also **Olo·nets.**

1932 W. L. Graff *Lang.* xi. 405 Finno-Ugric..dialects. These are..: Carelian, Olonetsian, [etc.]. **1933, 1939** [see *LUDIAN]. **1946–47** *Slavonic & E. European Rev.* XXV. 436 Olonecian (Aunus), heard on the eastern shores of Lake Ladoga.., and the Ingrian..of Ingria.., are dialectal varieties of Carelian. **1954** Pei & Gaynor *Dict. Linguistics* 154 *Olonetzian*, a member of the Finnish group of the Finno-Ugric (or Uralic) sub-family of the Ural-Altaic family of languages. **1957** B. Collinder *Survey Uralic Lang.* p. v, The dialects spoken in the Karelian-Finnish Soviet Republic are called Karelian (kr, in the north) and Olonets (ol, in the south). **1965** —— *Introd. Uralic Lang.* i. 11 Olonets (Olonetsian) is a variant of Karelian, spoken in the former province of Olonets.. and strongly influenced by Russian. **1975** G. F. Cushing tr. *Hajdu's Finno-Ugrian Lang. & Peoples* 199 Tver Carelian is spoken in the Upper Volga region, Aunus or Olonets to the north-east of Lake Ladoga.

oloroso (ǫlōrǭu·so). [Sp. *oloroso* fragrant.] A type of dry or medium sherry; a glass of such sherry.

1876 H. Vizetelly *Facts about Sherry* v. 46 The olorosos ..are deeper in colour than the amontillados. **1888** [see *FINO]. **1908** C. E. Hawker *Chats about Wine* vii. 105 There are several varieties of Sherry..the 'fino'..and the 'Oloroso'. **1920** G. Saintsbury *Notes on Cellar-Bk.* ii. 20 One might jangle a long time on Montillas and Olorosos. **1935** 'R. Hull' *Keep it Quiet* iv. 34 It seemed best..to let them try the proposed new Oloroso... Both..pronounced the new sherry to be disgusting. **1955** W. Gaddis *Recognitions* i. i. 59 He seemed prepared to sit over that dark oloroso sherry all evening. **1961** [see *AMONTILLADO a]. **1966** *Harper's Bazaar* Mar. 105 Oloroso, the darker, fuller style of sherry which, although dry in the natural state, is almost invariably sweetened for export. **1972** *Times* 20 Oct. 4/6 (Advt.), A very old, rich oloroso cream sherry. **1974** *Times* 23 Feb. 13/5 In Jerez, the producers like to start the day with a glass of natural dry oloroso. **1976** *Daily Tel.* 26 Oct. 17 [Yesterday] it [*sc.* the High Court] gave the sherry producers of the Jerez district of Spain what they had sued to obtain in 1970—exclusive right in Britain to use the descriptions *amontillado*, *oloroso* and *fino*.

olpe (ǫ·lpe). *Greek Antiq.* [ad. Gr. ὄλπη leather oil-flask, ewer, wine-jug.] **a.** A leather flask for oil or some other liquid. **b.** A kind of jug with a pear-shaped body and a handle.

1883 J. W. Mollett *Illustr. Dict. Art & Archæol.* 233/2 *Olpē*..a kind of *aryballos* with a curved handle, but no spout (originally a leather oil-flask). **1937** *Times Lit. Suppl.* 6 Feb. 93/1 An interesting eye olpe. **1961** *Oxf. Univ. Gaz.* 10 Mar. 832/1 A small late (?) Corinthian olpe with waves, dots and wavy lines. **1969** R. Mayer *Dict. Art Terms & Techniques* 269/1 *Olpe*, in ancient Greek pottery, a short-necked jug tapering to a foot... The word also denoted a leather flask used for carrying liquids. **1974** Savage & Newman *Illustr. Dict. Ceramics* 207 *Olpe* (Greek), a type of wine-jug resembling an *oenochoë*.

oly-cook, oly-koek. Add: Also **oliekoek, olycoek, -coke.** (Later examples.)

1881 *Harper's Mag.* Mar. 533/1 His favorite city has surpassed all others in..olie koeks, and New Year cookies. **1889** R. T. Cooke *Steadfast* vi. 78 Refreshing him with hot flip, oly koek, or Indian preserves. **1895** *Dialect Notes* I. 387 In the Dutch-settled districts the word *olykoeks*, which Washington Irving has made classic, is used for some of the varieties [of doughnut]. **1947** R. Berolzheimer et al. *U.S. Regional Cook Bk.* 138 The doughnut originated in Holland where it was called 'olie koeken'.

Olympia (oli·mpiă). The name of a town at the southern end of Puget Sound, the capital of the state of Washington, used *absol.* or *attrib.* in **Olympia oyster** to designate a small oyster, *Ostrea lurida*, native to the region.

[**1887** G. B. Goode *Fisheries U.S.: Geogr. Rev.* 626 No fishing is done at Olympia, the harbor being nearly bare at low water and lined with oysters. The shipment of these oysters to San Francisco is the only fishing industry of the town. The first shipment of these oysters was made two or three years ago.] **1908** *Nat. Geogr. Mag.* Mar. 225/1 New beds will have to be planted, and it will be five years before the so-called 'Olympia oyster' will again be on the market. **1911** *Encycl. Brit.* XX. 97/1 Olympia oysters are widely known in the Pacific coast region; they are obtained chiefly from Oyster Bay, Skookum Bay, North Bay and South Bay, all near Olympia. **1953** R. Froman *One Million Islands* 180 The tiny Olympia oysters also have their own unique flavor. **1957** M. McCarthy *Memories Catholic Girlhood* viii. 203 Tiny Olympia oysters... Olympia oyster cocktail. **1961** *Spectator* 8 Dec. 879/1 American ingredients and American cooking..have always to me seemed most mysterious... Olympia pan roast (olympia is an oyster) and Green Goddess dressing..turn out to be entirely local [i.e. West Coast] inventions. **1965** M. Tracy *Shellfish Cookery* v. 99 Olympia oysters are tiny West Coast oysters... They are ambrosial, and fabulously expensive even near their home waters.

Olympiad. Add: **1. b.** A quadrennial celebration of the ancient Olympic Games.

a **1490** J. Skelton tr. *Diodorus Siculus' Bibliotheca Historica* (1956) I. 382 Whiche maner of fayttis, thus ordeyned by his former instytucion, were callyd Olympiades. **1614** Ralegh *Hist. World* i. ii. xxiii. 576 These Olympiads... To tell the great solemnitie of them, and with what exceeding great concourse of all Greece they were celebrated, I hold it a superfluous labour. **1728** I. Newton *Chronol. Anc. Kingdoms Amended* i. 47 This Breviary seems to have contained nothing more than a short account of the Victors in every Olympiad. **1852** G. Grote *Hist. Greece* X. 439 They revenged themselves by pronouncing the 104th Olympiad to be no Olympiad at all. **1913** F. A. M. Webster *Olympian Field Events* i. i. 7 We have certain proof that it [*sc.* javelin throwing] was a part of the Pentathlon in the Ancient Olympiads. **1960** A. R. Burn *Lyric Age Greece* ix. 177 He [*sc.* Pheidon of Argos] was said to have marched west to Olympia and presided over the Games of the eighth Olympiad (748). **1977** *Jrnl. Hellenic Stud.* XCVII. 16 This no doubt partly accounts for the well-known string of Spartan victories in running events at the early Olympiads.

2. A (quadrennial) celebration of the modern Olympic Games revived in 1896. Hence, an occurrence of other competitions held on a regular basis. Also *fig.*

1907 *Westm. Gaz.* 1 Aug. 10/3 When the last Olympiad was held at St. Louis, U.S.A., in 1904, it was decided to hold the next in Rome. **1923** *Glasgow Herald* 26 Mar. 11/4 But the Oxford supporters were clamant in their championing, and by the time the final preparations were made the element of discord, like the seed of the apple of the Olympiads, had entered into the multitudes. **1935** *Encycl. Sports* 440/2 The next Olympiad, as the games came to be called, was held at Stockholm [in 1912]. **1957** W. Perelman tr. *Flohr's Twelfth Chess Tournament of Nations* 5 The first Olympiad was held in London in 1927, ..when 16 countries took part. **1959** *Listener* 22 Oct. 706/3 There was no reason why a British team [of Bridge players] should not be successful in the first World Olympiad next spring. **1964** —— 29 Oct. 695/1 Next week the chess olympiad begins in Tel Aviv, Israel... Unlike the Olympic Games, this is a team event, held every two years. **1967** —— 17 Aug. 213/2 Sixth-form specialisation is indeed producing diminishing returns: is the flight from maths and science compensated for by the winning of a mathematical Olympiad? **1972** *Daily Tel.* 21 June 14/5 Britain scored 42 victory points out of 60 during Monday's play in the World Bridge Olympiad. *Ibid.* 30 Aug. 9/3 Miss Pauline Dukelow, 16, paralysed from the age of three, is to receive the Andrew and Booth Courage award for the 21st Olympiad of the Paralysed at Heidelberg this month. **1976** *Radio Times* 15 May 37/1 In the following six Olympiads the Soviet Union have won over 500 medals. **1977** *Daily Times* (Lagos) 25 Feb. 17/4 This great cultural Olympiad was a gigantic step towards pan-Africanism and the unity of the black man.

Olympian, *sb.* Add: **1. b.** A competitor in the modern Olympic Games.

1976 *Billings* (Montana) *Gaz.* 27 June 7-F/6 The games are part of a 12-game tour for the *Olympians*, coached by North Carolina's Dean Smith, in preparation for the Montreal Games in July. **1977** *Time* 13 June 21/1 To support themselves during the rigors of year-round training, many Olympians have accepted deals from manufacturers and fees for appearing in track and field meets, hiding their earnings from Olympic, Amateur Athletic Union and international sports federations officials.

Olympic. Add: **A.** *adj.* **2. b.** Of or pertaining to the modern *Olympic Games*, which were revived as a quadrennial international athletic meeting at Athens in 1896 and have been held in various places since then. Also *transf.*

1896 *Scribner's Mag.* Apr. 453/1 The revival of the Olympic Games. Restoring the Stadion at Athens. **1908** *Westm. Gaz.* 31 Mar. 8/3 There is now some confusion with regard to the International Olympic games. **1914** R. Brooke *Let.* 5 Dec. (1968) 638 A New Zealand youth who was fighting in Mexico, heard the news in August, walked 300 miles to the coast, got a boat, and turned up here. He is also an Olympic Swimmer: and *knows* the *South Seas*. **1936** J. Buchan *Island of Sheep* iv. 60 It was not the Peter that you knew in the War, but Peter ten years younger, with no grey in his beard, and as trim and light and hard as an Olympic athlete. **1960** *Times* 5 Sept. 4/6 Hill..was representing a Germany united for Olympic purposes. **1962** *Listener* 1 Feb. 234/1 A match [at Bridge] between Italy and the U.S.A. in the Olympic event two years ago. **1964** D. M. Kunzle in G. C. Kunzle *Parallel Bars* 14 It was the whole conception of modern Olympic Gymnastics which was under fire. **1965** V. Canning *Whip Hand* xii. 134 He's also first class with foils and sabre, Wimbledon standard tennis, Olympic standard swimming, and a double-first Oxford. **1968** Mrs. L. B. Johnson *White House Diary* 10 Mar. (1970) 637 Lyndon..can break an Olympic record for getting dressed... We were out the door in about eight minutes. **1972** G. Lyall *Blame the Dead* i. 7 He scuttled away, carrying the pencil and cartridge case in front like a little Olympic torch. **1972** *Guardian* 19 Aug. 1/4 It is very important that he doesn't bring any infection that could affect other athletes in the Olympic Village. **1974** *Times* 21 Jan. 4/3 The boys are encouraged to join in games with them, swim in the pool of Olympic dimensions and just talk to them during breaks between intelligence tests and interviews. **1976** *Scotsman* 20 Nov. (Weekend Suppl.) 2/6 She used 30 cameramen and shot more than a million feet of film glorifying the Olympic ideal.

c. *Comb.*, with reference to the modern Olympic Games, as *Olympic-size, -sized, -style* adjs.

1966 J. Ball *Cool Cottontail* (1967) 17 A beautifully decked Olympic-size swimming pool. **1970** 'E. Lathen' *Pick up Sticks* (1971) iv. 36 The heart of Fiord Haven, where Havenites will enjoy an Olympic-size indoor swimming pool. **1969** *Guardian* 16 June 2/6 The 600-acre military 'city', 40 miles from Saigon..contains..an Olympic-sized swimming pool. **1966** *Times* 28 Feb. (Canada Suppl.) p. xiv/5 Canada's 1967 Pan-American Games, which includes Olympic-style amateur wrestling.

B. *sb.* Add to def.: Used esp. of the revived Olympic Games. Also *transf.*

1928 in *Funk's Stand. Dict.* **1948** Kieran & Daley *Story of Olympic Games* (rev. ed.) xiii. 327 The 'surest thing' in the Olympics never looked surer. **1948** *Official Rep. Olympic Games* (Brit. Olympic Assoc.) 41/1 No boxing competition of comparable size had been attempted anywhere in the world since the Berlin Olympics of 1936. **1949** W. M. Hugill in *Phoenix* (Toronto) III. 31 One school of thought has repudiated any connection between culture and the modern Olympics. **1951** *European Bridge Rev.* June 5/1 The 1951 Australian Jubilee Year World Olympic for contract bridge pairs. **1974** *Encycl. Brit. Macropædia* II. 274/1 The development of the modern Olympics into the pre-eminent athletic events in the world. **1976** *New Scientist* 28 Oct. 203/2 Any competitor in the gravitational Olympics must be able to pass the three classical tests of relativity that Einstein proposed in 1916.

‖**om** (ǭm), *int.* Also **o'm.** [Skr.] In Hinduism and Buddhism, an utterance of assent used in prayer and meditation. Also as *sb.*, an instance or example of this assent, and as *v. intr.*, to utter the assent. (See also quot. 1917.)

1788 W. Jones in *Asiatick Researches* I. 262, I am inclined, indeed, to believe, that not only *Crishna* or *Vishnu*, but even *Brahma'* and *Siva*, when united, and expressed by the mystical word *o'm*, were designed by the first idolaters to represent the Solar fire. **1810** E. Moor *Hindu Pantheon* 410 The character, that, if uttered, would yield the sound of *o'm*; or, being triliteral, better, perhaps, written *aum*. *Ibid.* 412 A suppression of breath is thus explained by an ancient legislator to imply the following meditation: '*Om*! earth! sky! heaven!' **1899** R. Whiteing *No. 5 John St.* xv. 153 The Brahmin leads us gently to the outermost courts of Nirvana. In spite of the outward symbol, we look within, and there we may hope to find *om*. **1917** A. B. Keith in *Encycl. Relig. & Ethics* IX. 490/2 The first evidence of this important position of the word is to be found in the *Aitareya Brāhmaṇa* (v. 32) in which it is declared that *om* is the world of heaven and the sun, and where it is resolved into the three letters *a*, *u*, and *m*. **1937** M. Covarrubias *Island of Bali* (1972) ix. 318 Typical Brahmana are the speculations about the sacred syllable *ong*, the Om of India. **1956** E. Wood *Yoga Dict.* 113/1 *Om*, the greatest *mantra*..or word of power. *Ibid.* 114/1 *Om* is recited by devout Hindus at the beginning of all prayers, hymns and words of worship or aspiration. **1971** *Illustr. Weekly India* 4 Apr. 31/2 The note greets me, '*Hare Krishna.* Please accept my *Om* prayers for your eternal well being. I would like to see you for a few minutes.' **1972** P. Holroyde *Indian Music* ii. 58 The sound '*om*'.., pronounced 'a-u-m', rings out like the tolling of a cathedral bell. **1976** B. Jackson *Flameout* i. 17 He..took deep yogic breaths, loudly hummed a long, resonant '*om—om—om—*'... 'It's sure good to hear you *om*-ing.'

-oma, terminal element repr. Gr. -ωμα, in which ω repr. ω (or ο) in the parent word (usu. a vb.) and -μα is a Gr. suffix forming neut. sbs., exemplified in Eng. words adopted from the Gr. such as Carcinoma, Coloboma, Derma, Diploma, Eczema, Glaucoma, Phyma, Plasma, Sarcoma, Trachoma, and in words on Gr. analogy such as Lipoma. In *Bot.* -*oma* has usu. been anglicized to *-ome. In *Med.* the examples of *sarcoma* (17th c. in English) and *carcinoma* (18th c.) have been taken as types on which to base new names of neoplasms and other localized swellings, -*oma* († -*ome*) being used as a suffix denoting 'tumour, growth' (cf. also Gr. ὄγκωμα swelling): e.g. Fibroma († Fibrome), *cementoma, *oligodendroglioma, *tuberculoma († Tuberculome).

Omaha (ǭu·mahă). Also † **Maha,** † **Omawhaw.** Pl. **Omaha, Omahas.** [ad. Omaha *umonhon* upstream people.] A Siouan people in north-eastern Nebraska, or their language; a member of this people. Also *attrib.* or as *adj.*

1804 W. Clark in Lewis & Clark *Orig. Jrnls. Lewis & Clark Expedition* (1904) I. 34 As we were pushing off this morning two Canoos Loaded with fur &c came to from the Mahas nation. *Ibid.* 124 This Village was built by a Indian Chief of the Maha nation. **1814** H. M. Brackenridge *Views Louisiana* i. vi. 76 Mahas, (or Oo-ma-ha) Reside on the Maha creek. **1823** E. James *Acct. Expedition Rocky Mts.* I. 190 Several of the Pawnee *caches*..had been broken open and robbed of their corn by the Omawhaws. **1839** *Boston Weekly Mag.* 12 Jan. 145/2 The Omaha village was one of the most beautiful that

OMANHENE

can be imagined. 1854 W. G. SIMMS *Southward Ho!* 406 The Pawnees and the Omahas were neighboring but hostile nations. 1900 G. B. GRINNELL *Indians of To-day* 12 He killed one more Omaha. 1920 [see *AVUNCULATE]. 1936 F. B. STREETER *Prairie Trails* iv. 190 He went with the Omaha Indians on a buffalo hunt. 1957 [see *KANSA]. 1964, 1968 [see *CROW *a.* 2]. 1972 W. B. LOCKWOOD *Panorama Indo-Europ. Lang.* vii. 117 The plains between the Mississippi and the Missouri were occupied by the Siouan family [of languages], of which the most significant today are Dakota (45,000) and Omaha (10,000). 1974 *Encycl. Brit. Micropædia* VII. 528/3 *Omaha*, North American Indian people of the Dhegiha branch of the Siouan language stock. *Ibid.*, Omaha social organization was elaborate, with a class system of chiefs, priests, physicians and commoners.

omanhene (ōu·manhene). [Ashanti f. *oman* council + *-hene* combining form of *ohene* chief.] Among the Ashanti people of West Africa, a paramount chief of a state or district, under whom are the lesser chiefs of villages.

1909 MOORE & GUGGISBERG *We Two in W. Africa* xi. 146, I met my first *omanhin* at Tarkwa—'king' he would have been called some years ago but his title is not officially recognised now, the head chief of a tribe being known by the native equivalent of *omanhin*. 1923 R. S. RATTRAY *Ashanti* xxii. 264, I am greatly indebted to an Ashanti chief, Osai Bonsu, *Omanhene* of Mampon, for permitting his drummer to drum this complete history into a phonograph. 1955 D. E. APTER *Gold Coast in Transition* iv. 94 The chief became known as the *omanhene* or head of council, among the Ashanti. 1959 A. ABBS *Ashanti Boy* i. 17 That was nothing strange to a conscientious Omanhene to whom the family in its widest sense meant 'the tribe'. 1961 *Guardian* 10 Nov. 1/1 Chiefs and their retinues from all over the country..Nenes, Omanhenes, Nanas, Niis, Togbes, and..[the] Fiaga of Peki. 1962 C. G. BAËTA *Prophetism in Ghana* iii. 61 The head of the Church is referred to as *Omanhene* or 'Paramount Chief'.

Omani (omā·ni), *sb.* and *a.* Also 9 **Omanee, Omany.** [Arab., f. *Oman* name of a coastal region in the south-east of the Arabian peninsula + *-i* adj. suffix.] **A.** *sb.* A native or inhabitant of Oman. **B.** *adj.* Also in form *Oman* (the place-name used *attrib.*). Of or pertaining to Oman or its inhabitants.

c 1819 F. WARDEN in *Sel. Rec. Bombay Govt.* (1856) No. 24. 433 It would give the Omanees sufficient time to unite and assemble for their general defence. 1819 —— in *Ibid.* 44 It is probable that the Oman Chiefs of the Hinavi tribe will..return to their former relations with the Imaum. 1838 J. R. WELLSTED *Trav. Arabia* I. ii. 16 In consequence of the difference in their faith, the Omán Arabs and Persians seldom intermarry. *Ibid.* xix. 292 The Omány in all ages is celebrated in the songs of the Arabs as the fleetest. 1865 W. G. PALGRAVE *Narr. Journey through Arabia* II. xv. 256 The principal family whose chiefs headed the original settlement was, according to 'Omânee tradition, that of the Ya'aribah. *Ibid.* 262 These new hostilities on the part of Islam suggested to the sectarian 'Omânees the expediency of new measures. 1871 G. P. BADGER *Hist. of Imâms & Seyyids of Omân* p. xvii, A battle was fought in which the 'Omânis were defeated with great loss. *Ibid.* p. xxxvii, In order to test the pluck of the 'Omâny sovereign..he sent him a viciously restive horse. 1928 A. T. WILSON *Persian Gulf* xii. 174 The inhabitants virtually deserted the island, by this means bringing the Omani occupation to an end. A year or two later..the Omanis seized certain islands off the Persian coast. 1931 E. WAUGH *Remote People* 165 Instead of the cultured, rather decadent aristocracy of the Oman Arabs, we have given them [*sc.* the Zanzibar Islanders] a caste of just, soap-loving young men with public school blazers. 1973 *Nat. Geographic* Feb. 209/1 Omanis danced in the streets for days. 1973 *Times* 15 Feb. 7/1 He saw in Muscat an Omani whom he recognized as having served as a political commissar in..Dhufar in 1971. 1973 *Black Panther* 4 Aug. 11/2 The struggle of the Oman people has..had wide success in the building of institutions to insure the people's survival. 1974 *Times* 18 Jan. 15/4 The 'Red Line' is the name given by the Omani guerrillas to the road linking the town of Salalah.. to the Omani capital Muscat.

Omaresque (ōumăre·sk), *a. rare.* [f. as next + -ESQUE.] Suggestive of Omar Khayyam or his poetry.

1892 *Academy* 5 Nov. 404/1 In shorter measures there is often an Omaresque effect of thought.

Omarian (omā·riăn), *a.* and *sb.* [f. the name of the Persian mathematician and epigrammatist, Ghiyāthuddīn Abulfath 'Omar bin Ibrāhīm al-Khayyāmī (*c* 1100) + -IAN.] **A.** *adj.* Of or pertaining to Omar Khayyam or his poetry; having the style or character of his poetry. **B.** *sb.* A student or admirer of Omar Khayyam; a member of the Omar Khayyam Club.

1898 *Daily News* 31 Jan. 6/3 To the devout Omarian a reproduction in black-and-white of this early MS. will carry something of the fragrance [etc.]. 1898 *Westm. Gaz.* 22 Feb. 3/4 The next service demanded of Omarian scholarship is an edition in the original Persian. 1901 *Ibid.* 8 Jan. 2/3 Marie's needs are almost Omarian in their simplicity. 1907 *Daily Chron.* 13 Feb. 3/3 Serious Omarians..are willing to do more for their master than merely turn down an empty glass. 1934 A. J. A. SYMONS *Quest for Corvo* xi. 154 An American Omarian..contributed an Introduction.

Hence **Oma·rianism, O·marism,** admiration or imitation of Omar Khayyam; the doctrines or cult of Omar Khayyam. So **O·marite** = *OMARIAN *a.*

1897 *Daily Chron.* 9 Dec. 7/2 All more or less imbued with the spirit of what is called 'Omarianism', and all.. decorously convivial. 1898 J. H. MCCARTHY in *Westm. Gaz.* 22 Feb. 3/3 The protest against what may be called Omarism. 1900 *Academy* 21 July 55/2 Mr. Fawcett called Omarism a fad. *Ibid.*, The Omarite message was interpreted: 'Get drunk as often as you can, and stay so long as you can, for there's nothing in life half so profitable.' 1918 *Naval Intelligence* xxxix. 180 Stanza after stanza, which he reeled off in an ecstatic and maudlin manner with his eyes half closed and his head wagging—all Omarites do this, you may have noticed.

Omayyad, var. *UMAYYAD *a.* and *sb.*

‖**ombré** (ǫ̃nbre), *sb.* and *a.* Also **ombre.** [Fr., pa. pple. of *ombrer* to shade.] **A.** *sb.* A fabric woven or dyed in a series of colour tones graduating from light to dark and usu. producing a striped effect. Also, such an effect or design.

1895 *Funk's Stand. Dict.* II. 1227/2 *Ombré*.., a cheap grade of silk prints. 1921 A. GANSWINDT *Dyeing Silk* 147 *Ombres.* Shaded effects on hank silk can be obtained by binding the hanks firmly on rods, using only a small amount of dyestuff in the bath [etc.]. 1930 R. CUTHILL tr. *Schober's Silk* iv. 271 *Ombré*, a fabric in a number of colours, one fading off into the other, this effect being obtained by the use of yarns printed before weaving. 1966 A. J. BLISS *Dict. Foreign Words & Phrases Current Eng.* 264 *Ombré*.., (a fabric) woven with variably dyed yarns so as to produce a shaded effect. 1975 *Country Life* 30 Oct. 1192/2 Degradé, a fabric that is shaded from light to dark and sometimes called ombré in the dress side of the trade.

B. *adj.* Shaded from light to dark in tints of one or more colours.

1918 S. KLINE *Man. Processes Winding* iii. 61 Ombré warps. 1928 *Daily Express* 27 Feb. 5/5 Rainbow effects have been produced by the use of what are known as ombre yarns, in which cotton and artificial silk of different colours are admixed. 1963 *Times* 24 Jan. 12/4 He varies the plain colours with *ombré* effect and brilliant spots. 1969 *Daily Tel.* 21 Apr. 12/6 Other leather is speckly and *ombre* with tiny splatters of black on grey highlighted with *art nouveau* silver buckles. 1974 D. RAMSAY *No Cause to Kill* i. 9 She was tempted by a blue silk *ombré* plaid.

‖**ombre chevalier** (ǫ̃nbrə ʃəvalie). Also **omble chevalier.** [Fr.] A freshwater race of the char, *Salvelinus alpinus*, of the family Salmonidæ, found in certain French and Swiss lakes, esp. the Lake of Geneva.

1884 G. B. GOODE *Fisheries U.S.: Nat. Hist. Aquatic Animals* 501 The Saibling of Bavaria and Austria is one and the same thing with the 'Ombre Chevalier' of France and Switzerland,.. the 'Char' of England and Scotland. 1905 D. S. JORDAN *Guide to Study of Fishes* II. iv. 108 The only really well-authenticated species of charr in European waters is the red charr, sälbling, or ombre chevalier. 1940 A. SIMON *Conc. Encycl. Gastron.* II. 67/1 *Omble Chevalier*... The best fish of the Swiss and Savoy lakes; it never leaves the lakes for the running waters of rivers. 1960 E. DAVID *French Provincial Cooking* 38 Some *ombres-chevaliers* from the lac du Bourget, cooked and left to get cold in white wine. 1968 V. CANNING *Melting Man* iii. 58 Tonight I shall be in France, eating *omble chevalier*, straight from the lake. 1971 *Times* 29 Mar. (Switzerland Suppl.) p. iii/5 The *ombre chevalier*, a type of salmon trout, especially from the Lake of Geneva.

‖**ombrelle** (ǫ̃nbrę·l). *poet.* [Fr.] A parasol or sunshade.

1925 E. SITWELL *Troy Park* 75 What hotels Hide their bustles and their gay ombrelles. 1942 —— *Street Songs* 13 All that I knew of shade Was the cloud, my ombrelle of rustling grey Sharp silk.

‖**ombrellino** (ǫmbrĕlī·no). [It.] (See quot. 1957.) Also, a parasol or sunshade.

1847 F. W. FABER *Let.* in R. Chapman *Father Faber* (1961) viii. 163 It was a dark still night, and the bell and lights and singing and the flashing *ombrellino* had a most touching effect among the trees. We deposited our Lord in His own tabernacle after receiving His Benediction. 1949 D. ATTWATER *Catholic Encycl. Dict.* (ed. 2) 354/1 *Ombrellino* (It., a little umbrella). A small flat canopy with one staff borne over the Blessed Sacrament when it is carried from its own altar to another for exposition, etc., and, in some countries, when it is taken to the sick. 1957 *Oxf. Dict. Chr. Ch.* 983/2 *Ombrellino*, in the W. Church the small umbrella-like canopy of white silk carried over the Blessed Sacrament when it is moved informally from one place to another. 1964 *Punch* 26 Feb. 326/3, I came round lying on the sand under an *ombrellino*.

‖**ombres chinoises** (ǫ̃nbrə ʃinwāz). [Fr., Chinese shadows.] = GALANTY SHOW.

1785 LADY NEWDIGATE *Let.* 25 May in A. E. Newdigate-Newdegate *Cheverels* (1898) iv. 61 We talk of having another representation of our *Ombres Chinoises* on Friday. 1802 *Monthly Mag.* Sept. 132/1 Besides the great theatre, there are..several smaller ones for..different exhibitions, such as ombres Chinoises. 1948 G. SPEAIGHT in M. Batchelder *Puppet Theatre Handbk.* p. xx, 'Ombres chinoises' were extremely popular in France and England during the eighteenth century, but later fell from favour. 1950 H. W. WHANSLAW *Shadow Play* vi. 48 Tronchet describes the theatre of Francois-Dominique Seraphin, in the Palais Royale,.. where he exhibited his famous *Ombres Chinoises—Le Spectacle des Enfants de France.* 1960 O. BLACKHAM *Shadow Puppets* 98 In March 1779 Astley re-opened his Amphitheatre near Westminster Bridge, and here the *ombres chinoises* continued to be advertised right up to the summer of 1790. 1974 *Encycl. Brit. Micropædia* VII. 530/2 *Ombres chinoises*.., European version of the Chinese shadow puppet show, introduced in Europe in the mid-18th century by returning travellers.

ombro-. Add: **ombro·genous** *a.*, of moorland or marsh, needing a high rainfall for its development; **ombro·philous** *a.* [a. G. *ombrophile* (J. Wiesner 1893, in *Sitzungsber. Akad. Wiss. Wien* Abth. I. CII. 503)], of a plant, able to flourish in conditions of excessive moisture; so **ombro·phily; ombro·phobous** *a.* [a. G. *ombrophobe* (J. Wiesner 1893, *loc. cit.*)], of a plant, not well adapted to very wet conditions; so **ombro·phoby.**

1939 A. G. TANSLEY *Brit. Islands & their Vegetation* xxxv. 718 Blanket moss or bog—ombrogenous. 1946 *Proc. Prehist. Soc.* XII. 3 The blanket-bog is termed an 'ombrogenous' mire, to indicate the fact that its existence is directly determined by the rainfall and evaporation to which it is subject. 1952 P. W. RICHARDS *Trop. Rain Forest* ix. 215 The vegetation of the ombrogenous moors is tall evergreen forest. 1975 J. R. ETHERINGTON *Environment & Plant Ecol.* iii. 91 Some other peat soils are topogenous rather than ombrogenous. 1895 *Jrnl. R. Microsc. Soc.* 194 Plants which, on the one hand are uninjured (ombrophilous) and on the other hand are injured (ombrophobous) by excessive rainfall. *Ibid.*, Some species..growing in moist shady situations, are nevertheless ombrophobous. *Ibid.*, Xerophilous plants..are hardly ever ombrophilous. 1903 W. R. FISHER tr. *Schimper's Plant-Geogr.* I. i. 2 Xerophytes perish after two or three days of continuous rain; they are rain-avoiding, ombrophobous, whereas hygrophytes are, as a rule, ombrophilous... Ombrophilous foliage is capable of being wetted, ombrophobous foliage is unwettable. *Ibid.* III. i. 225 Reference may here be made to Wiesner's investigations regarding the ombrophily and ombrophoby of tropical vegetation. 1897 *Jrnl. R. Microsc. Soc.* 412 Ombrophoby of flowers.—By this term Prof. A. Hansgirg designates the phenomena of curvature by which many flowers protect themselves against injury from long-continued rain or other exposure to moisture.

ombú (ombu·). Also **ombu.** [Amer. Sp., f. Guarani *umbú*.] An evergreen tree, *Phytolacca dioica*, of the family Phytolaccaceæ, native to temperate regions of South America. Also *attrib.*

1871 R. O. CUNNINGHAM *Notes Nat. Hist. Strait of Magellan* xiv. 466 A variety of cottages, each shaded by the umbrageous foliage of the large Ombu. 1878 E. CLARK *Visit S. Amer.* ix. 123 The gnarled ombus and willows, and poplars and peach trees, affording the welcome shade. 1902 W. H. HUDSON *El Ombú* i. 1 In all this district..you will not find a tree as big as this ombú, standing solitary, where there is no house. *Ibid.* 2 They say..that those who sit much in the ombú shade become crazed. 1923 *Chambers's Jrnl.* Dec. 828/2 A huge, gnarled tree, tremendously thick and spreading like an *ombú*. 1931 B. MIALL tr. *Guenther's Naturalist in Brazil* iv. 79 In the Argentine I have often admired the monumental character of one of the native trees, the Ombú. The roots rise from the ground like great brown bladders, lifting the trunk into the air, and from the trunk the branches spring like a tracery. 1941 E. NASH *I Liked Life I Lived* vi. 70 The humble dwelling was called the house of the twenty-five ombú trees. 1961 G. DURRELL *Whispering Land* i. 30 Small, neat *estancias*, gleaming white in the shade of huge, carunculated *ombú* trees, that stood massively and grimly on their enormous squat trunks. 1969 T. H. EVERETT *Living Trees of World* 144/1 The ombú becomes 60 feet tall with a branch spread of more than 100 feet. From the bottom of its very thick trunk it develops extraordinary irregular outgrowths that look like fantastic and bulky roots. *Ibid.*, The elliptic or ovate leaves of the ombú are evergreen and male and female flowers are on separate trees; the fruits are small and berry-like. 1975 *New Yorker* 14 Apr. 36/1 They give shade from the bare sun to man and beast, and men mark their way on the endless plains by remembering this or that ombu tree.

Ombudsman (ǫ·mbudzmæn). Also with small initial. [Sw. (see below) f. *ombud* commissioner, agent, repr. ON. *umboð* charge, commission, *umboðsmaðr* commissary, manager.] An official appointed to investigate complaints by individuals against maladministration by public authorities; *spec.* in U.K., the Parliamentary Commissioner for Administration. (Corresp. to Sw. *justitieombudsmannen.*) Also *attrib.* and in extended and *fig.* uses.

In Sweden, an *ombudsman* is a deputy of a group, particularly a trade union or a business concern, appointed to handle the legal affairs of the group and protect its interests generally. With the definite suffix *-en* it normally denotes a particular office. The office of *justitieombudsmannen* (abbrev. *JO*), which was instituted in Sweden in 1809, is the one which is referred to in the following quotations, and it is this office which was established in New Zealand in 1962. In Sweden the office of *militieombudsmannen* (abbrev. *MO*) was also introduced in 1809 as a parliamentary commissioner appointed to supervise and enforce the observance of laws and statutes concerned with national defence. In 1968 the offices of

justitieombudsmannen and *militieombudsmannen* were replaced by four independent officers under the joint name of *justitieombudsmännen*. The word *pressombudsman* connotes a public relations officer, and in its definite form *pressombudsmannen* an office instituted in 1969 to supervise and enforce ethical standards of the press.

The office of *ombudsman* was introduced in Finland in 1919, in Denmark in 1954, and in Norway in 1962.

[1911 *Encycl. Brit.* XXVI. 195/2 By revisers elected annually the Riksdag controls the finances of the kingdom, and by an official (*justitieombudsman*) elected in the same way the administration of justice is controlled; he can indict any functionary of the state who has abused his power. 1914 J. GUINCHARD *Sweden* iii. 200 The Riksdag has yet another form of control over the High Court of Justice and the Supreme Administrative Court in that the 'Justitieombudsman' may in certain cases arraign a member of the Court before the Court of Impeachment. 1958 S. HURWITZ in *Public Law* Autumn 236 A precedent existed in Sweden, where a 'Justitieombudsman' and a 'Militieombudsman', appointed by Parliament, were introduced in 1809 and in 1915 respectively.] 1959 *Listener* 16 July 89/1 Sweden has been running the Ombudsman system for 150 years, and Denmark has a very active Ombudsman. 1963 *Times* 23 Apr. 13/6 The perfunctoriness with which the Government recently rejected the conclusions of the unofficial inquiry conducted by Justice into the Scandinavian *ombudsman* procedure was disappointing. 1966 *N.Z. News* 28 Sept. 3/2 Britain's Ombudsman, Sir Edmund Compton, is now in New Zealand consulting New Zealand's Ombudsman, Sir Guy Powles, about the workings of his office and the way in which he deals with complaints. 1966 S. OAKLEY *Story of Sweden* xiv. 164 The private citizen was protected from the bureaucracy by a so-called *ombudsman*, elected by the Estates to hear and investigate complaints against abuses of power by public servants. 1969 *Daily Tel.* 13 Dec. 1/5 Complaints from hospital patients and staff may shortly be dealt with by a special type of Ombudsman. 1970 MORRIS & HAWKINS *Honest Politician's Guide to Crime Control* 100 We do not doubt that the ombudsman system would work well here. 1970 *Harper's Mag.* Dec. 59/1 He had come to protest to the newspaper, his only ombudsman that day. 1971 *N.Y. Post* 15 Nov. 46 She was everywhere, doing everything—columnist, lecturer,..ombudsman for every injustice. 1972 *Daily Tel.* 23 Feb. 2 The Health Service is to have its own Ombudsman or Health Service Commissioner, who will be responsible for carrying out independent investigations into patients' complaints. 1973 *Daily Californian* 1 Feb. 2/1 A state ombudsman's office, sort of an official wailing wall. 1974 M. GILBERT *Flash Point* i. 12 He had approached..his own member of parliament..the Ombudsman and..the press. 1975 *Local Council Rev.* Autumn 95 A layman's guide to the complaints machinery set up under the 1974 Local Government Act. Explains what the ombudsmen can and cannot do, how complaints should be put to them. 1976 *Daily Tel.* 2 Dec. 2/5 He complained to the Department of Education and eventually to Baroness Serota, the Ombudsman for the area, who found Surrey education committee guilty of maladministration. 1978 *Times* 7 Mar. 4/8 The appointment of local ombudsmen has had a considerable impact on local authority procedures, in the view of the authors of the first critical appraisal of their work.

Hence *ombuds-committee*; **o·mbudsmanry,** the profession or practice of being an ombudsman; **o·mbudsmanship,** the office or function of an ombudsman; **o·mbudswoman,** a female ombudsman.

1961 *Observer* 12 Nov. 40/8, I suppose I have become a sort of ombudswoman on turnstiles. 1964 *Economist* 7 Mar. 875/2 A sort of Ombudscommittee for people who feel cheated. 1970 *Daily Tel.* 6 Aug. 2/4 A resolution calling for an inquiry into promotion procedures and for the setting up of an arbitration 'ombuds-committee' has been tabled. 1971 *New Society* 25 Mar. 489/3 It is obvious that local ombudsmanry on a national scale would mean setting up a new light industry. 1966 *Times* 5 July 13 The whole field of ombudsmanship and the scrutiny of administrative practice and legislative principle..at present goes largely unmarked. 1971 *New Scientist* 3 June 597/2 The loquacious ombudsmanship of Bernard Braden. 1965 *Manch. Guardian Weekly* 21 Oct. 6 The only answer is for more councillors to see themselves as Ombudsmen and Ombudswomen. 1972 *New Yorker* 30 Sept. 6/3 Mabel Mercer, ombudswoman to all who have ever played at love. 1973 *Maclean's Mag.* June 45/1 She had previously been ombudswoman for the Status of Women Council [etc.].

omdah, omdeh. Also **omda.** [ad. Arabic *ʿumdah* column, support, trustworthy authority, village-chief, f. root *ʿmd* to support.] The headman of a village in Arab countries.

1907 *Daily Chron.* 20 Aug. 3/7 Village omdehs to be elected by the whole mass of the villagers. 1922 *Q. Rev.* Apr. 428 Omdehs and others who may be tempted to revert to the old ways would do well to reflect. 1924 *Glasgow Herald* 23 Dec. 7 The numerous 'omdas', or village headmen, who were dismissed during the Zaghlulist régime. 1926 *Blackw. Mag.* Apr. 409/1, I was reluctantly compelled to accept the hospitality of the Omdah..for lunch. 1928 *Observer* 11 Mar. 19/3 The Omdas—representatives of the central authority in the villages. 1976 *Times* 31 July 10/2, I used to go from village to village, discussing their affairs with the omdas and shaikhs.

-ome, anglicized form of *-OMA (partly through influence of G. *-om*, F. *-ome*), occurring chiefly in *Bot.* in terms such as CAULOME, *HADROME, PHYLLOME, RHIZOME, and usu. signifying a structure or group of cells forming a normal part of the anatomy, in contrast with the abnormality implied by *-oma* (cf. *MYCETOME, an organ in insects, MYCETOMA, a fungal skin disease). It also occurs in a few obs. forms of words now written *-oma*, e.g. FIBROME, TUBERCULOME.

omee (ōu·mi). *slang.* Also **omer, omie.** [Corruption of It. *uomo*, man.] A man, esp. a landlord or itinerant actor.

1859 HOTTEN *Dict. Slang* 70 *Omee*, a master or landlord. 1893 P. H. EMERSON *Signor Lippo* xiii. 42 When I got back the cullies said, 'Well, cully, how did you get on with the omer?' 1928 *Sunday Express* 14 Oct. 5 Man [in slang of busking] is 'omey'. 1937 N. MARSH *Vintage Murder* vii. 82 'A lot of omies the others were then.'.. 'Ted means they were bad actors doing worse shows in one-eyed towns up and down the provinces.' 1962 L. KNIGHT (*title*) A proper circus omie.

omega. Add: **1.** (Earlier examples.)

c 1400 MANDEVILLE *Trav.* (1725) iii. 25 What Lettres thei ben..with the Names..α Alpha, β Betha..ω Omega. 1640 S. DAINES *Orthoepia Anglicana* (1908) 11 *Oa*, sounds generally after the Greeke *Omega*. *Ibid.* 12 *Oo* in *Poore* imitates in sound the Greek *Omega*.

2. a. Also *from Alpha to Omega*: from beginning to end; from top to toe.

1929 D. G. MACKAIL *How Amusing!* 307, I was a gentleman..from alpha to omega. 1978 D. QUINN *Fear of God* II. 161 Vast explosions of silence..marching..from Alpha to Omega and..beyond the infinite.

b. Omega point, in the work of P. Teilhard de Chardin (1881–1955) a hypothesized point of convergence, absorption, or transformation which is the divine end, or God, towards which the forces of evolution are moving (see quot. 1964).

1959 B. WALL tr. *Teilhard de Chardin's Phenomenon of Man* ii. 57 He goes on to add these words which my readers would do well to recall when I come to unveil (with all due reservations and corrections) the perspective of the 'Omega point'. *Ibid.* IV. ii. 259 Accordingly its enormous layers, followed in the right direction, must somewhere ahead become involuted to a point which we might call *Omega*, which fuses and consumes them integrally in itself. 1964 N. DENNY tr. *Teilhard de Chardin's Future of Man* vi. 122 Ahead of, or rather in the heart of, a universe prolonged along its axis of complexity, there exists a divine centre of convergence... In order to stress its synthesising and personalising function, let us call it the *point Omega*. 1965 *Listener* 3 June 817/2 The idea has something in common with Teilhard de Chardin's concept of the 'Omega Point'. 1977 *Time* 28 Feb. 45/1 Teilhard believed man would eventually transcend his individualism and converge at the 'Omega Point' with the Omega—God.

2*. *Nuclear Physics.* † **a.** (Written Ω.) A former designation of certain hyperons. *Obs.*

The decay schemes in quot. 1953 are those of the sigma (Σ) particles, whilst the mass specified in quot. 1954[2] (and elsewhere) is that of the xi minus. In quot. 1954[1] a lambda particle is referred to.

1953 *Nuovo Cimento* X. 1741 Two decay schemes have been proposed by analogy with the decay of $V_1{}^0$-particles. Adopting the nomenclature proposed at the Bagnères Conference these are: (*a*) $\Omega_n{}^{\pm} \rightarrow n + \pi^{\pm} + Q$, (*b*) $\Omega_p{}^{+} \rightarrow p + \pi^0 + Q$. 1954 *Sci. News* XXXI. 62 Ω^0-particle. Mass, 2,184 ± 7 [= 1116 MeV]. Lifetime, $(3{\cdot}3 \pm 1) \times 10^{-10}$ seconds... Decay scheme: $\Omega^0 \rightarrow p^+ + \pi^-$. 1954 *Physical Rev.* XCVI. 543/1 The known hyperons, Λ^-, Ω^-, have masses equivalent to 1200 and 1320 Mev, respectively.

b. In full *omega meson.* A neutral meson with zero hypercharge, zero isospin, unit spin, and negative parity that is observed as a resonance (as when protons and antiprotons of sufficient energy collide), has a mass of 784 MeV (1534 times that of the electron), and on decaying usu. produces a positive, a negative, and a neutral pion. Freq. written as ω.

1961 B. C. MAGLIĆ et al. in *Physical Rev. Lett.* VII. 178/1 The existence of a heavy neutral meson with $T = 0$ and $J = 1^-$ was predicted by Nambu... Such a particle is also expected in the vector meson theory of Sakurai and.. according to the unitary symmetry theory; and for other reasons. We will refer to it as ω. *Ibid.* 181/1 We conclude that the data fit the qualitative criteria for an axial vector matrix element (ω meson). 1961 *New Scientist* 5 Oct. 48/1 Four physicists at the Lawrence Radiation Laboratory, Berkeley, have made observations..that appear to indicate unequivocally the presence, in a proton-antiproton reaction, of the omega-meson. 1971 *Sci. News* 10 Apr. 250 It appears that electromagnetic forces are mediated to leptons directly by photons and to hadrons by phi, rho or omega. 1971 *Sci. Amer.* July 100/2 Three vector mesons with zero strangeness are currently known: the rho, the omega and the phi. 1974 *Nature* 6 Dec. 438/2 The omega meson, which has all the same normal quantum numbers as the phi,..decays very rapidly to three pions.

c. *omega minus* (or *particle*): a negatively charged hyperon having hypercharge of −2, zero isospin, a spin of 3/2, positive parity, and a mass of 1672 MeV (3272 times that of the electron), and decaying via the weak interaction into either a xi particle and a pion or a lambda particle and a kaon. Freq. written as Ω^-.

1962 M. GELL-MANN in *Proc. Internat. Conf. High-Energy Physics* 805/2 Starting with the resonance at 1238 MeV, we may conjecture that the $Y_1{}^*$, at 1385 MeV and the Ξ^* at 1535 MeV might belong to this supermultiplet... If $J = 3/2^+$ is really right for these two cases, then our speculation might have some value and we should look for the last particle, called, say, Ω^- with $S = -3$, $I = 0$. At 1685 MeV, it would be metastable and should decay by the weak interactions into $K^- + \Lambda$, $\pi^- + \Xi^0$, or $\pi^0 + \Xi^-$. 1964 *Physical Rev. Lett.* XII. 204/1 The multitude of resonances which have been discovered recently..can be arranged as a decuplet with one member still missing... This particle (which we shall call Ω^-, following Gell-Mann) is predicted to be a negatively charged isotopic singlet with strangeness minus three. *Ibid.* 206/1 In view of the properties..established for particle 3, we feel justified in identifying it with the sought-for Ω^-. 1964 *Daily Tel.* 21 Feb. 25/2 Dr. Maurice Goldhaber, director of Brookhaven, said yesterday: 'The discovery of Omega Minus forms the capstone in a building which was so far held together only by the bold imagination of Dr. Gell-Mann and Dr. Ne'eman.' 1964 *Listener* 30 Apr. 711/1 The new theory [*sc.* that of unitary symmetry] has made one striking prediction which was subsequently found to be correct: the observation of the negatively charged Omega particle. 1972 *Daily Colonist* (Victoria, B.C.) 24 Feb. 5/2 The physicists hope to make the first observation of 'quarks'..by studying the activity of a rare and elusive sub-atomic particle called the omega-minus. 1974 FRAUENFELDER & HENLEY *Subatomic Physics* i. 6 The negative kaon..collides with a proton and produces a positive kaon, a neutral kaon, and an omega minus. The Ω^- decays into a Ξ^0 and a π^-.

3. Used *attrib.* (with capital initial) to designate a style of interior decoration and design associated with the Omega Workshops, a short-lived undertaking begun in London by the art critic Roger Fry in 1913 and influenced by the work of William Morris.

1922 D. H. LAWRENCE *Aaron's Rod* iii. 30 Into this reticence pieces of futurism, Omega cushions and Van-Gogh-like pictures exploded their colours. 1970 *Oxf. Compan. Art* 790/2 The Omega artists believed that the creative joy of the artist and craftsman should go into the making of articles for everyday use. 1973 *Guardian* 22 Jan. 8/1 Drawings and paintings of the Bloomsbury artists still crowd the walls... The dining-room's Omega chairs attend its round painted table.

omegatron (ōu·mĭgătrǫn). *Physics.* [f. OMEGA (ω being the symbol of angular frequency) + *-TRON.] A mass spectrometer that employs the principle of the cyclotron to identify and measure gases at very low pressures, a radio-frequency electric field being applied at right angles to a magnetic field so that charged particles having a certain charge-to-mass ratio impinge on a collecting electrode.

1949 J. A. HIPPLE et al. in *Physical Rev.* LXXVI. 1878/1 Since this device measures ω, it is suggested that it be called the omegatron. 1966 W. SUMMER tr. *E. von Angerer & H. Ebert's Physical Lab. Handbk.* x. 208 The omegatron..is useful as a gauge down to 10^{-12} torr. *Ibid.*, The ion-collecting efficiency of an omegatron is high and operation is satisfactory in the range 10^{-5} to 10^{-9} torr. Below 10^{-7} torr resolution up to $m = 30$ is very good... Resolution is less complete for $30 < m < 100$, but ions present can be identified. 1975 *Nature* 6 Feb. 408/2 Partial pressures were measured with an omegatron radio-frequency mass spectrometer which formed part of the UHV section.

omelet, omelette, *sb.* Add: **c.** *attrib.*, as *omelette (frying)-pan.*

1846 *Jewish Manual, or Pract. Information Jewish & Mod. Cookery* v. 99 A small omelette frying-pan is necessary for cooking it [*sc.* the omelette] well. 1879 A. D. WHITNEY *Just How* 292 Finish beating and mixing the omelette, setting on the omelette-pan when almost ready. 1948 *Good Housek. Cookery Bk.* II. 369 To season a new omelet pan. Heat the pan slowly, then melt a knob of butter in it and rub it well in with a piece of soft paper. 1977 T. HEALD *Just Desserts* vii. 174 Gabrielle won't go anywhere without her favourite omelette pan.

d. omelette (aux) fines herbes, a savoury omelette flavoured with herbs; **omelette soufflée,** an omelette made by folding the separately beaten egg whites into the mixture.

1845 E. ACTON *Mod. Cookery* xix. 489 Seasoned with minced herbs,..[it] is then called an '*Omlette aux fines herbes*'. 1846 [see *FINES HERBES]. 1928 A. CHRISTIE *Mystery of Blue Train* xix. 150 The Comte de la Roche had just finished *déjeuner*, consisting of an *omelette fines herbes*. 1845 E. ACTON *Mod. Cookery* xix. 491 (*heading*) An omlette soufflée. 1930 W. S. MAUGHAM *Cakes & Ale* xii. 145 No one could make a better *omelette soufflée* than she. 1968 *Radio Times* 28 Nov. 25/1 Colourful Cookery.. Puffed Onion Tart..Omelette Soufflée. 1975 HUME & DOWNES *Cordon Bleu Desserts* ii. 64 Omelet Soufflé with Strawberries... Omelet Soufflé 'en Surprise'. 1977 D. RAMSAY *You can't call it Murder* ii. 108 During lunch.. they disposed of..the most subtle of *omelettes fines herbes*.

o·melet, o·melette, *v.* [f. the sb.] *trans.* To make into an omelet. Also *transf.*

1872 E. EGGLESTON *End of World* xxiii. 155 The eggs.. were not poached, they were not scrambled, they were not omeletted. 1908 *Westm. Gaz.* 6 Oct. 3/1 (*caption*), I don't want to be omeleted!

omen, *sb.* Add: **c.** *omen-animal.*

1902 *Man* II. 61 The chapters on the omen-animals and the cult of skulls are of special value.

omeno·logy. [f. OMEN *sb.* + -OLOGY.] The study or science of omens.

1904 J. HASTINGS *Dict. Bible* V. 559/2 Such..occurrences as the lunar eclipse, would serve as a basis for lunar omenology.

omentopexy (ome·ntopeksi). *Surg.* [f. OMENT(UM + -O + *-PEXY.] Any operation in which the omentum is sutured to another structure, e.g. the abdominal wall.

1905 *Jrnl. Amer. Med. Assoc.* 25 Nov. 1700/2 Omentopexy was undertaken to provide collateral circulation for the portal vein. 1957 J. G. ALLEN et al. *Surgery* xxxiii. 795/2 Omentopexy was introduced by Morison, Talma and others about 1900. 1962 *Lancet* 13 Jan. 66/2 Subsequently 1 (case 12) had an omentopexy, which controlled hæmorrhage for two years.

omer, var. *OMEE.

‖**omertà** (omerta·). [dial. form of It. *umiltà* humility, with reference to the Mafia code which enjoins submission of the group to the leader as well as silence on all Mafia concerns.] Refusal to give evidence by those concerned in the activities of the Mafia.

1909 *Evening Sun* (N.Y.) 13 May 8/1 There is..the belief that it is unmanly to tell anything about a fellow countryman which could get him into trouble. It is called 'Omerta' in the Sicilian tongue, which means manliness. 1963 R. I. MCDAVID *Mencken's Amer. Lang.* 720 *Omertà*, the very strict code of the Mafia. 1965 *Times Lit. Suppl.* 25 Nov. 1058/4 He could call on a good many witnesses more bound to *omertà* than to the truth. 1965 J. WAINWRIGHT *Death in Sleeping City* II. vii. 129 They [*sc.* the Mafia] have a law... It's called the Omerta. It's an unwritten law—a code of conduct, really. 1968 *Listener* 29 Feb. 268/3 An island [Sardinia] where omerta is stronger than democracy. 1969 *Sunday Truth* (Brisbane) 30 Nov. 25/2 He had broken the highest law of the Mafia—omerta, or silence. 1970 G. GREER *Female Eunuch* 222 *Vendetta* and *omertà*..are not significant until the familial, regional community is threatened by political authority. 1977 *Time* 12 Sept. 43/1 The protection program was formally established after passage of the Organized Crime Control Act of 1970 to hasten the breakdown of *omertà*, the underworld code of silence.

Omeyyad, var. *UMAYYAD *a.* and *sb.*

‖**omi** (ōu·mi). Also, with prefixed *ō-* 'great'. [Jap.] In early imperial Japan, a high-ranking administrative official claiming imperial ancestry (cf. *MURAJI); a title of members of a family upon which such an office was bestowed.

1901 [see *MURAJI]. 1931 G. B. SANSOM *Japan* I. ii. 37 We have the *ō-omi*, or great ministers, who were appointed from among the heads of clans closely related to the imperial family. *Ibid.*, The *omi* and *muraji* of lesser standing. 1964, 1970 [see *MURAJI].

omicron (ǫ·mikrǫn, ōumai·krǫn). [Gr. ὂ μικρόν, lit. 'little O'.] The fifteenth letter (Ο, ο) of the Greek alphabet, originally having the value of short *o*.

c 1400 MANDEVILLE *Trav.* (1725) iii. 25 What Lettres thei ben..with the Names..α Alpha, β Betha,..ο Omicron ..χ Chi, [etc.]. 1631 R. HARRIS *Arraignment Whole Creature* xiii. 208 The whole Globe..cannot fill this little triangulary heart: so many Omicrons, cannot fill one little Delta. 1727–41 [see OMEGA 1]. 1893 E. M. THOMPSON *Handbk. Gr. & Lat. Palæogr.* x. 135 The very small size of *theta* and *omikron*, may also be noticed. 1947 *Jrnl. Investigative Dermatol.* IX. 215 Where *-oma* occurs as the ending in words *properly* formed from the Greek, the stem always ends in omicron or omega (as a lengthening of omicron). *Ibid.* 213 Many such Greek verb stems, but by no means all, end in *o* represented by omicron when the pronunciation is short (ŏ) or by omega (ω) when it is long (ō). 1959 A. G. WOODHEAD *Study of Greek Inscriptions* ii. 18 This differentiation, like that between *omicron* and *omega*, gradually spread to the rest of the Greek world.

omie, var. *OMEE.

omissible, *a.* Add: Hence **omissibi·lity.**

1961 in WEBSTER. 1966 *Amer. Speech* XLI. 182 Certain self-explanatory terms have been examined with a view to their omissibility from *NID 3* (Webster's Third New International Dictionary). 1971 D. CRYSTAL *Linguistics* 202 The omissibility of *here tomorrow*. 1971 T. F. MITCHELL in *Archivum Linguisticum* II. 43 The omissibility of a following noun.

‖**om mani padme hum** (ǫm mani padme hŭm), *int.* [Skr., lit. 'Hail! Jewel in the Lotus!' See *OM *int.* and quot. 1848 below.] In Tibetan Buddhism, a mantra or mystic formula intoned in prayer and meditation. Also as *sb. phr.*

1774 G. BOGLE *Narr. Mission Tibet* (1876) iii. 29 Some old women..were counting their beads and repeating their *Om mani padmi hums*! 1836 B. H. HODGSON in *Jrnl. Asiatic Soc. of Bengal* V. 88 The celebrated *Shadakshari Mantra*, or six-lettered invocation of him, viz. *Om! Mane padme hom!* of which so many corrupt versions and more corrupt interpretations have appeared. 1848 J. D. HOOKER *Himalayan Jrnls.* (1854) I. x. 229, I..returned down the 'via sacra', a steep paved path flanked by..low stone dykes, into which were let rows of stone slabs, inscribed with the sacred 'Om Mani Padmi om'.—'Hail to him of the lotus and jewel'; an invocation of Sakkya, who is usually represented holding a lotus flower with a jewel in it. 1863 E. SCHLAGINTWEIT *Buddhism in Tibet* viii. 84 Amitābha then blessed Padmapāni Bōdhisattva by laying his hands upon him, when, by virtue of this benediction, he brought forth the prayer 'Om mani padme hum'. 1895 L. A. WADDELL *Buddhism of Tibet* vi. 148 The commonest mystic formula in Lāmaism, the 'Om-ma-ṇi pad-me Hū-ṃ,'—which literally means '*Om!* The Jewel in the Lotus! *Hūṃ!*'—is addressed to the Bodhisat Padmapāni who is represented like Buddha seated or standing within a lotus-flower. He is the patron-god of Tibet. 1901 KIPLING *Kim* ii. 47 In the pauses of their talk they could hear the low droning—'*Om mane pudme hum! Om mane pudme hum!*'—and the thick click of the wooden rosary beads. *Ibid.* v. 111 He clicked the beads, and began the '*Om mane pudme hum*' of his devotion. 1924 A. HUXLEY *Little Mexican* 16 A mystic formula, a kind of *Om mani padme hum*. 1970 J. BLOFELD *Way of Power* i. i. 38 Old women twirling their prayer-wheels and intoning *Om Mani Padme Hum* while riding in buses.

† **ommateum** (ǫmătī·ŭm). *Zool. Obs.* [mod. L., f. Gr. ὄμμα, ὀμματ- eye.] (See quots.) Hence **ommate·al** *a.*

1883 LANKESTER & BOURNE in *Q. Jrnl. Microsc. Sci.* XXIII. 182 This enlarged portion of the hypodermis is, in fact, the soft or living tissue of the eye [of the Scorpion], and may be distinguished from the lens in front of it by a special name. We propose to call it the ommatēum'. The ommateum and the lens together form the eye. *Ibid.*, A well-marked 'basement membrane', which in the region of the ommateum may be called the eye-capsule, or, better, the 'ommateal capsule'. 1884 [see next]. 1898 A. S. PACKARD *Text-bk. Entomol.* 250 (*heading*) The compound or facetted eye (ommateum).

ommatidium. Add: (Earlier and later examples.)

1884 J. CARRIÈRE in *Q. Jrnl. Microsc. Sci.* XXIV. 674 The whole set of eye-units (ommatidia) of Musca vomitoria are enclosed in a chitinous capsule. [*Note*] The term 'ommateum' was introduced by Lankester in his memoir on the eyes of Scorpions to signify the entire soft parts of the non-segregate (unicorneal) eye of Arachnida and Hexapoda... The similar term 'ommatidium' is introduced in this paper to signify the units consisting each of a retinula and a vitrella, together with their sheath of pigment cells, into which the ommateum of the multicorneal (polymeniscous) eye of Arthropods, is segregated. 1925 A. D. IMMS *Gen. Textbk. Entomol.* I. 79 The distinctness of vision depends partly upon the number and size of the ommatidia. 1932 METCALF & FLINT *Fund. Insect Life* iv. 110 It is believed that each ommatidium does not form an image of the whole object, but only preserves the intensity, pattern, and color of the light coming from the particular small part of the object that is in line with its long axis. Indeed, the several ommatidia, or 'tubes', are usually so isolated from each other by pigment, that no light can pass from one to the other. 1973 *Sci. Amer.* Dec. 35/3 The compound eye of *Drosophila* is a remarkable structure consisting of about 800 ommatidia: unit eyes containing eight receptor cells each.

ommati·dial *a.* (examples).

1890 *Anatomischer Anzeiger* V. 356 This spine-bearing cornea is soon shed and a faceted one formed, each facet.. often containing in the centre the remnants of an ommatidial spine. 1925 A. D. IMMS *Gen. Textbk. Entomol.* I. 76 The hypodermis between the ommatidial pillars becomes transformed into the secondary pigment cells. 1975 *Nature* 10 Apr. 522/2 The axons from seven retinula cells of one ommatidium interweave with those of their four neighbouring facets in such a way that four bundles of three fibres, heterogeneous with regard to their ommatidial origin, plus one single axon are produced.

ommatin (ǫ·mătin). *Biochem.* Also **-ine.** [a. G. *ommatin* (E. Becker 1939, in *Biol. Zentralbl.* LIX. 622), f. Gr. ὄμμα, ὀμματ- eye: see -IN[1].] Any of the group of ommochromes characterized by weaker colours, less stability to alkalis, and lower molecular weights as compared with the ommins.

1940 *Biol. Abstr.* XIV. 1238/1 The 2 eye pigments [of *Ephestia*] are typical of those found among insects. The dark red or brown ommines occur in Lepidoptera, Hymenoptera, Coleoptera, Hemiptera Heteroptera, and in the nematocerous Diptera. The more weakly colored ommatines are found in the cyclorrhaphous Diptera, Odonata, and Orthoptera. 1965 [see *OMMOCHROME]. 1969 R. F. CHAPMAN *Insects* vii. 109 Tyndall blues are rare in insects, but the blue of dragonflies is produced in this way, the dark background being provided by a brown-violet ommatin. 1975 D. G. COCHRAN in Candy & Kilby *Insect Biochem.* iii. 237 Storage of large quantities of 3-hydroxykynurenine occurs in vesicles or dilations of the endoplasmic reticulum of tubule cells [of *Drosophila*], and, at least in certain mutants, it is converted into ommatins and ommins.

Ommiad(e, var. *UMAYYAD *a.* and *sb.*

ommin (ǫ·min). *Biochem.* Also **-ine.** [a. G. *ommin* (E. Becker 1939, in *Biol. Zentralbl.* LIX. 611), f. Gr. ὄμμ-α eye: see -IN[1].] Any of the group of ommochromes characterized by stronger colours, greater stability to alkalis, and higher molecular weights as compared with the ommatins.

1940, etc. [see *OMMATIN]. 1964 T. H. GOLDSMITH in M. Rockstein *Physiol. Insecta* I. x. 425 The principal ommin is widely distributed in arthropods and cephalopods. 1965 [see *OMMOCHROME].

ommochrome (ǫ·mŏkrōum). *Biochem.* [ad. G. *ommochrom* (E. Becker 1942, in *Zeitschr. f. indukt. Abstammungs- und Vererbungslehre* LXXX. 179), f. Gr. ὄμμ-α eye + χρῶμ-α colour: see -O.] Any of a group of insect pigments derived by condensation reactions from kynurenine and giving yellow, red, and brown body colours and commonly also found in the accessory cells of the eyes of insects.

1945 *Biol. Abstr.* XIX. 450/1 The distribution of this new group of pigments, the ommochromes, among arthropods has been examined. 1965 B. E. FREEMAN tr. *Vandel's Biospeleol.* xxv. 405 In the arthropods, melanines are replaced by ommochromes (ommines and ommatines). 1965 V. B. WIGGLESWORTH *Princ. Insect Physiol.* (ed. 6) xiii. 556 The ommochromes fall into two groups: the non-dialysable 'ommines', with large molecules, and the dialysable 'ommatines' with small molecules. 1970 *Nature* 26 Dec. 1336/2 The integumental pigments of the isopod Crustacea are ommochromes. 1974 *Ibid.* 30 Aug. 799/2 An early effect of ecdysoids at metamorphosis in some caterpillars is conversion of tryptophane into red ommochrome pigments.

omni-. Add: **o·mni-antenna,** an omnidirectional antenna; **omnici·pient** *a.* = *omnipercipient*; **omnifo·cal** *a. Ophthalm.*, designating a lens whose power changes continuously from top to bottom; also as *sb.*, such a lens; **omnifu·tuant, -fu·tuent** *adjs.* [L. *futuere* to have sexual relations with], practising or tolerant of both homosexual and heterosexual activity; **omnila·teral** *a.*, facing all directions; representing all points of view; so **omnila·terally** *adv.*; **omnipercipient** *a.* (later example); **omnipo·llent** *a.*, all-powerful; **omnipurpose** *a.*, serving all purposes; **o·mnirange** *Aeronaut.*, (part of) a navigation system in which short-range omnidirectional VHF transmitters serve as radio beacons; **omnise·ntient** *a.*, having universal feeling or sensation; **omnisu·bjugant** *a.* [cf. SUBJUGE *v.*], subjugating everything or everyone; **omnitemporal** (earlier and later examples); so **omnite·mporally** *adv.*; **omni-tooled** *a.*, possessing many tools; **omnivica·rious** *a.*, taking the place of (anything).

1974 *New Scientist* 24 Jan. 191 (*caption*) Omni-antenna. 1976 *CB Mag.* June 110 (Advt.), Beam antennas vs. omni-antenna range. 1899 BEERBOHM *More* 162 Omnicipient in material, the master of many styles. 1962 *Jrnl. Amer. Med. Assoc.* 19 May 595 (*heading*) The omnifocal lens for presbyopia. 1962 *Arch. Ophthalm.* LXVIII. 777/1 (*heading*) Use of the omnifocal. *Ibid.*, Omnifocals are used binocularly but are effective monocularly in cases where only one eye can be used. 1965 *Maclean's Mag.* 20 Feb. 1 An optical company in Ohio offers to solve this problem with an 'omnifocal' lens which has power that's gradually increased from top to bottom with no blurred area or transition zone. 1974 *Year Bk. Ophthalm.* 38 Three types of lens included..Varilux, Zoom and Omnifocal. 1929 A. HUXLEY *Do What you Will* 132 The ancient Greeks were evidently, in Sir Richard Burton's expressive phrase, 'omnifutuent'. 1966 *Listener* 24 Mar. 445/1 Stephen learns to accept himself as a homosexual only by entering a society which is innocently omnifutuant. 1967 *Ibid.* 30 Mar. 433/1 Anthony Burgess..to whom, among others, I owe such words as *omnifutuant* and *futuancy*. 1936 *Times Lit. Suppl.* 2 May 378/2 The present eight hundred pages set forth the science of omnilateral aristology. 1953 *Essays in Crit.* III. II. 374 He [*sc.* Chaucer] sees life steadily, and if he is not omnilateral, he is manysided. 1936 *Times Lit. Suppl.* 2 May 378/2 Of old, man was omnilaterally oriented. 1932 H. H. PRICE *Perception* vii. 202 This could only be avoided if we had been omnipercipient. 1922 JOYCE *Ulysses* 377 The certain sign of omnipollent nature's incorrupted benefaction. 1961 Omnipurpose [see *OMNICOMPETENT *a.*]. 1947 *Electronics* Oct. 95/2 There are also voice channels on both the runway localizer and the omnirange, which are used generally for traffic control and weather information. 1951 *Gloss. Aeronaut. Terms* (*B.S.I.*) III. 27 *V.H.F. Omni-range*, a short-range, very-high-frequency, omni-directional beacon which provides an indication in the aircraft of the bearing of the beacon, or left-right track indication. 1959 K. HENNEY *Radio Engin. Handbk.* (ed. 5) xxv. 27 As of June 30, 1955, there were 410 omniranges in operation. 1966 D. FRANCIS *Flying Finish* xvii. 199 The V. O. R.—Very high frequency Omni-range—by which one navigated from one radio beacon to the next. 1932 H. H. PRICE *Perception* iv. 72 If we were *omnisentient* beings,.. able to sense all at once all the sense-data which can ever be sensed by every sentient human or non-human. 1911 BEERBOHM *Zuleika D.* ii. 23 But would she ever meet whom, looking up to him, she could love—she, the omnisubjugant? 1956 P. FLEMING *My Aunt's Rhinoceros* 141 After the war the bureaucrats no longer held their omnisubjugant trump. 1883 B. F. WESTCOTT *Historic Faith* xi. 144 The 'eternal' does not in essence express the infinite extension of time but the absence of time, not the omni-temporal but the supra-temporal. 1970 P. A. BERTOCCI *Person God Is* xii. 223 It is..my concern to press the question on both Advaitin and Visishtadvaitin: Why not reconceive the perfection of God so that good and evil, truth and error, progress and decay, can affect the qualitative manner in which God experiences himself and the world, and in a way consistent with his omnitemporal unity and continuity? 1961 E. NAGEL *Struct. of Sci.* iv. 70 Suppose there are (omnitemporally) no physical objects that do not attract each other. 1964 *Philos. Rev.* LXXIII. 486 For something to qualify as being true omnitemporally. 1851 H. MELVILLE *Moby Dick* III. xxi.

146 This omni-tooled, open-and-shut carpenter, was,.. no mere machine of an automaton. 1967 V. NABOKOV *Speak, Memory* (ed. 2) ii. 42 The game in use was the regular 'draw poker', with, occasionally, the additional tingle of jackpots and an omnivicarious joker.

omni (ǫ·mni). Abbrev. *omnirange.*

1949 *Proc. IRE* XXXVII. 832/1 The flag alarm of the course-deviation indicator is actuated by the omni receiver. 1950 *Time* 31 July 28/2 Each omni sends out a radio signal that is different for each direction from the station.

omnibus. Add: **A.** *sb.* ¶. *Omnibi*, representing a spurious 'plural' (in quot. 1889 genitive singular) form, occurs occasionally.

1840 W. HOWITT *Visits to Remarkable Places* (ser. 1) 200 Trains of omnibuses, or omnibi, are flying down to the Broomielaw every hour. 1889 E. DOWSON *Let.* 23 June (1967) 85, I trust you arrived chez toi—in all sobriety last night & accomplished the de[s]census Av—I should say omnibi with discretion. 1969 *Times* 18 Jan. 20/3 The portmanteau term 'reprint' evades definition. It covers series, 'evergreens', omnibi, disinterments, defrostings, definitive editions, [etc.].

2. a. (Earlier example.)

1844 C. G. F. GORE *Quid pro Quo* (ed. 3) 81 What if I.. swell the 'Bravos' of the Omnibus?

b. Short for *omnibus book* (see sense *B. 3).

The word 'omnibus' was also used in the sense 'omnibus journal', i.e. a newspaper comprising a variety of items, in two early 19th-century publications, *The National Omnibus; and General Advertiser* (1831), and *The Lancashire Omnibus, a Journal of Literature and Amusement* (1832).

1930 *Writer* Jan. 74/2 One of the recent omnibuses contained selected short stories. 1931 'J. GROVE' (*title*) The omnibus of romance. 1937 'A. ARMSTRONG' (*title*) The laughter omnibus. 1976 R. USBORNE *Wodehouse at Work* (rev. ed.) iii. 92 The Preface that Wodehouse wrote for the 1974 omnibus *World of Psmith*. 1978 *Bookseller* 8 Apr. 2196/3 When..Charles Pick..dreamed up those jumbo Heinemann Octopus omnibuses, it seemed..only a matter of time before Collins put some..authors into similar vehicles.

5. *omnibus-driver, -office, -riding, sleigh, -ticket, traffic* (earlier examples); **omnibus man** (earlier example).

1843 POE *Mystery of Marie Roget* in *Ladies' Compan.* (N.Y.) Feb. 166/1 The omnibus-driver, Valence. 1870 'F. FERN' *Ginger-Snaps* 304 This honored name, shouted from lungs that would not have disgraced an omnibus-driver. 1857 *Christian Misc.* July 219/2 We know no class of men in this country who undergo a more severe life of toil..than the omnibus-men and cabmen of London. 1854 M. CUMMINS *Lamplighter* xviii. 112 You know the way from the omnibus-office. 1844 *Knickerbocker* XXIV. 91 His opinions against the omnibus-riding of so many of our idle citizens. 1860 *Boston Auditor's Ann. Rep.* 1859–60 323 One covered omnibus sleigh. 1852 E. E. HALE *If, Yes, & Perhaps* (1868) 3 This [sum].. would buy the omnibus tickets. 1869 *Engineering* 26 Nov. 348/1 With some slight modifications of detail, Mr. Wright's doors might..be advantageously applied to many railway carriages employed in working metropolitan, or as it is often called, 'omnibus' traffic. 1883 E. W. HAMILTON *Diary* 15 Dec. (1972) II. 525 Trains worked by electricity admit of any amount of subdivision, which is no small consideration as it facilitates *omnibus* traffic and dispenses with the necessity of constructing the permanent way as strong as it now is.

B. *adj.* **1.** (Earlier and further examples.)

1842 *Congress. Globe* 27th Congress 2 Sess. App. 661/1 These two articles..were caught in the omnibus, or dragnet section, which is placed in the rear of the bill. 1850 *Ibid.* 31st Congress 1 Sess. App. 524/1, I am opposed to all omnibus bills, and all amalgamation projects. 1889 *Echo* 16 Nov. 2/3 Each man pays an 'omnibus' contribution of a shilling a week for benefits. 1968 *Globe & Mail* (Toronto) 3 Feb. 40/2 All organizations said, however, that changes in existing regulations are necessary and many of the Criminal Code amendments contained in an omnibus bill presented by Justice Minister Pierre Trudeau are good ones. 1972 *N.Y. Law Jrnl.* 10 Oct. 18/9 Defendant's omnibus motion is disposed of as hereinafter indicated. 1974 *Times* 30 Aug. 15/2 The plan..is to work up some omnibus convention which most will sign.

2. *omnibus box* (later examples).

1902 *Chambers's Jrnl.* Dec. 823/2 On this level, where it touched the stage, we had an 'omnibus box', exactly after the pattern of the proverbial one at Her Majesty's, and occupied usually by the same distinguished noblemen. 1922 A. HADDON *Green Room Gossip* iv. 95 The passing of the green room reminds me of another effete institution of the English theatre—the omnibus box.

3. omnibus book, volume, etc., a book or volume having a large and occas. varied content, *spec.* one containing several reprinted works by a single author or works of the same kind and published at a price designed to place it within the reach of a wide public; **omnibus letter,** a letter intended for more than one recipient; **omnibus-sized** *a.*, of the size of an omnibus book; **omnibus ticket,** one admitting a number of persons.

1929 *Daily Tel.* 1 Jan. 6/2 It is a day of what the publishers call 'omnibus books', meaning works which carry many and varied passengers. 1933 *Mind* XLII. 525 Hume's omnibus letter addressed to Dr. Hugh Blair, and through Blair to Dr. Jardine. 1931 *Times Lit. Suppl.* 19 Nov. 918/1 This second instalment of his short stories, an almost omnibus-sized book. 1868 *Rep. Iowa Agric. Soc.* 1867 408 Some..tender hearted friends would take in their settlement [= family] and then proceed to some hole ..in the fence and hand his 'omnibus ticket' to some other parent. 1928 *Times Lit. Suppl.* 12 July 514/4 The 'omnibus' volume of 'Great Short Stories of Detection, Mystery and Horror',..runs to some 1,250 pages. 1928 *Observer* 22 July 8 The four novels together make one of the most desirable of 'omnibus' volumes. 1928 *Publishers' Circular* 14 July 39/2 (*heading*) The Omnibus Wells.

omnibus, *v.* Delete *nonce-wd.* and add: (*a*) (earlier example, in *absol.* use); (*c*) to convey by omnibus; (*d*) to publish an omnibus edition of (an author).

1833 W. C. MACREADY *Diary* 7 Nov. (1912) I. 76, I omnibused down to Drury Lane. 1863 'G. HAMILTON' *Gala-Days* 121 We were quickly omnibused to the relics of Donegana. 1933 *Times Lit. Suppl.* 5 Jan. 1/2 For the author the possibility of becoming popular enough in his lifetime to be omnibused or to omnibus himself with profit may be looked on as a new prize in the race for fame.

omnicompetent (ǫmnikǫ·mpĭtĕnt), *a.* [f. OMNI- + COMPETENT *a.*] Competent to deal with everything; *spec.* possessing jurisdiction or authority to act in all matters. Hence **omnico·mpetence.**

1827, 1889 Omnicompetent [in Dict. s.v. OMNI-]. 1900 Omnicompetence [in Dict. s.v. OMNI-]. 1926 R. H. TAWNEY *Relig. & Rise of Capitalism* ii. 124 Recent political theory has been prolific in criticisms of the omnicompetent state. 1937 *Times Lit. Suppl.* 1 May 320 His omnicompetence dealt with the history of ideas as swiftly as with the measurement of the hide. 1943 C. S. LEWIS *Abolition of Man* iii. 30 The man-moulders of the new age will be armed with the powers of an omnicompetent state and an irresistible scientific technique. 1952 V. A. DEMANT *Relig. & Decline of Capitalism* ii. 57 The principle of the omnicompetent state and the myth of the self-sufficient individual. 1961 *Economist* 30 Dec. 1276/1 A sort of ghostly middleman between omnipurpose boroughs and omnicompetent Government departments. 1967 *Punch* 11 Oct. 541 She comes from a famous wine family..so she is omnicompetent, bullying her *maître de chai*, [etc.]. 1971 *Mod. Law Rev.* XXXVI. 611 Parliament as an institution must be sovereign or omnicompetent at any given moment of time. 1973 *Times Lit. Suppl.* 28 Dec. 1588/2 The computers are no longer considered even by their supporters to be omnicompetent. 1975 *Times* 25 July 13/5 Deference to professional omnicompetence is, of course, gratifying.

o:mnidire·ctional, *a.* Also **omni-directional.** [f. OMNI- + DIRECTIONAL *a.*] Of equal sensitivity or power in all directions (usu., all horizontal directions).

1927 *Daily Tel.* 22 Mar. 13 The Marconi Company quoted £66,153 for the Beam system and £29,163 for a system of omni-directional communication. 1932 *Times Educ. Suppl.* 31 Dec. p. i/2 The other short-wave transmitter worked on 15,140 kc/s, using an omni-directional aerial. 1947 BUCK & PIERCE in J. S. Hall *Radar Aids to Navigation* ii. 51 The uhf omnidirectional beacons..are based primarily on the low-frequency system used for marine navigation. 1952 E. A. LAPORT *Radio Antenna Engin.* ii. 94 Experience has been gained with vertical radiators at many hundreds of broadcast stations, each employing one or more for omnidirectional or directive radiation. 1959 H. HOBSON *Mission House Murder* xxi. 140 They..fixed an omni-directional, remotely controlled microphone which transmitted on a closed circuit. 1969 R. B. FULLER *Operating Man. Spaceship Earth* vi. 88 Einstein formulated his famous equation E = M (matter's mass, explained in the terms of C²—speed of an omni-directional (radiant) surface wave's expansion, unfettered, in a vacuum). 1976 *CB Mag.* June 7/1 (Advt.), Electrical design of the long, 64 wavelength vertical radiating element plus full size radials guarantee..highest signal radiation efficiency in the desired vertical plane and omnidirectional, full coverage transmitting and receiving pattern.

Hence **o:mnidirectiona·lity,** the property of being omnidirectional; **o:mnidire·ctionally** *adv.*, in all directions.

1950 *Jrnl. R. Aeronaut. Soc.* LIV. 279/2 By comparing the phase of the rotating radiation pattern with that of a reference signal radiated omni-directionally, the bearing of the aircraft from the beacon is determined. 1966 L. V. BLAKE *Antennas* vii. 330 In unstabilized-spacecraft applications three-dimensional omnidirectionality may be desired. 1972 *Science* 20 Oct. 273/1 In order to achieve omnidirectionality, a second array is envisaged, located at right angles to the first. 1976 *Wireless World* Mar. 44/3 Real antennas do not radiate omnidirectionally, but concentrate the power into a directional beam.

omni·ficence. [f. OMNIFIC *a.* + -ENCE.] The fact or quality of being omnific, or of making or doing everything.

1881 RUSKIN *Our Fathers have told Us* I. ii. 62 Unwearied in protective friendship, in meekly dextrous omnificence, in latent tutorship. 1941 *Mind* L. 297 He therefore devotes the rest of the chapter to divine 'omnificence', *i.e.* the doctrine that all that is in fact done is done by God.

omnificent, *a.* (Earlier and later examples.)

1677 LOCKE *Jrnl.* 8 Feb. in *Essay Draft A* (1936) 84 They who out of a great care not to admitt unintelligible things deny or question an eternall omnificent spirit run them selves into a greater difficulty by makeing an eternall and intelligent matter. 1929 R. BRIDGES *Testament of Beauty* iv. 177 Joyful obedience With reverence to'ard the omnificent Creator and First Cause.

omniscient, *a.* Add: **omnisci·entist,** one who purports to be, or is alleged to be, omniscient.

1932 R. A. KNOX *Broadcast Minds* ii. 20 (*heading*) The omniscientists. *Ibid.* 21 We are all omniscientists now, at least in ambition... It only remains that we should pride ourselves on knowing something about everything. 1948 A. O'RAHILLY *Relig. & Sci.* iv. 32 We stand..opposed to the usurpations and truculent dogmatism of the omniscientists.

omnopon (ǫ·mnŏpǫn). *Pharm.* Also **Omnopon.** [f. L. *omn-is* all + OP(IUM *sb.* + *-on*, arbitrary ending.] A proprietary name in the U.K. for a mixture of the hydrochlorides of the opium alkaloids. Cf. *PANTOPON.

1909 *Trade Marks Jrnl.* 1 Dec. 1962 Omnopon... All goods included in class 3 [*i.e.* chemical substances prepared for use in medicine and pharmacy]. The firm of F. Hoffman La Roche & Co.,..Bale, Switzerland; chemical manufacturers. 1910 *Lancet* 15 Oct. 1169/1 Pantopon is known in Great Britain as 'Omnopon'. 1922 C. T. CAMPION tr. *Schweitzer's On Edge of Primeval Forest* v. 92 He is given an injection of omnipon [*sic*]. 1957 I. MAGILL in Gillies & Millard *Princ. & Art Plastic Surg.* I. iii. 64 For premedication Omnopon and scopolamine are preferred, as their action is synergistic. 1970 *Sunday Mail* (Brisbane) 15 Mar. 4 Police took possession of a huge quantity of cannabis, opium, L.S.D., omnopon..and cocaine. 1970 PASSMORE & ROBSON *Compan. Med. Stud.* II. v. 48/2 It is widely believed that there is no therapeutic advantage in using mixed alkaloid preparations from opium, such as papaveretum (omnopon).

omphalo-. Add: **omphali·tis** [-ITIS], inflammation of the umbilicus; **omphalo·scopy** [Gr. -σκοπία], contemplation of the navel.

1876 DUNGLISON *Dict. Med. Sci.* (rev. ed.) 722/2 Omphalitis. 1897 *Trans. Amer. Pediatric Soc.* IX. 208 Of the remaining cases..one..was due to pyæmia following omphalitis in the newly-born. 1974 PASSMORE & ROBSON *Compan. Med. Stud.* III. II. xlv. 22/1 The appearance [in the newborn] of non-specific signs of infection, whether or not there is liver enlargement, jaundice or evidence of omphalitis, is always an indication for culture of the umbilicus and the blood. 1931 T. H. PEAR *Voice & Personality* iv. 35 The psychologist, unless mental omphaloscopy contents him, must go and fetch his material. 1960 *Times* 30 Nov. 7/2 In particular he made fun of 'omphaloscopy'—gazing at the navel.

omphaloid, *a.* (Later example.)

1942 *Oxoniensia* VII. 47 Sandy, smoothed, dull black, well finished. Omphaloid base of medium-sized bowl.

omul (ōu·mul). [a. Russ. *omul'*.] A fish of the salmon family, *Coregonus autumnalis*, found in Lake Baikal and regions bordering the Arctic Ocean.

1884 *Encycl. Brit.* XVII. 605/2 Only two species of fish are of any importance [in Novaya Zembla]—the goltzy (*Salmo alpinus*) in the western rivers, and the omul (*Salmo omul*) in the eastern. 1955 *Bull. Amer. Mus. Nat. Hist.* CVI. 279/2 The salmon-herring, or omul, has a spawning run of 1500 kilometers (930 miles) in the Yenesei [*sic*] River. 1969 *Nature* 13 Sept. 1091/2 Lake Baikal is particularly noted for its reserves of omul'. 1970 G. V. NIKOLSKY et al. in Lindsey & Woods *Biol. Coregonid Fishes* 257 The omul is an anadromous fish, which lives in waters of higher salinity than other coregonid fishes... Lake Baikal omul forms a separate subspecies. 1976 'S. HARVESTER' *Siberian Road* i. 18 The Cossacks..went after sturgeon and omul, a white-fish.

on, *prep.* Add: **I. 1. d.** Also in reference to a means of communication, as *on the air* (see *AIR *sb.*[1] 1 c), *on the telephone*, etc. Hence, broadcast on a specified channel, frequency, or wavelength.

1917 R. FRY *Let.* 23 Nov. (1972) II. 420 London's just awful for me because of the millions of people that catch me on the telephone. *c* 1928 *Ibid.* 632 I've just been talking for the second time today with Vanessa on the 'phone. 1929 *Ibid.* 5 Feb. 636, I have still two 'talks' hanging over me—one at the Athenaeum Club and one on the wireless. 1944 *Sun* (Baltimore) 7 Jan. 8/3 Common usage is unerringly correct: one does not hear a program in or even by or through the radio. He hears it on the radio. 1966 *Listener* 11 Aug. 190 Viscount Dilhorne and Lord Shawcross, Q.C., interviewed by Robin Day on B.B.C.-1. *Ibid.* 8 Sept. 363/3 Earlier, on B.B.C.-2, Rozhdestvensky was back again with the Moscow Radio Orchestra and the young pianist Nicolai Petrov. 1969 D. E. WESTLAKE *Up your Banners* xviii. 124 The beautician hollered from the living room, 'Leona, come quick! You're on the TV!' 1973 P. O'DONNELL *Silver Mistress* iv. 73 Get on the phone and book three seats..to Hong Kong. 1974 *Oxford Mail* 1 May 4/7 James Dalton, the Queen's College organist, can be heard on Radio Three on Monday. 1977 *Custom Car* Nov. 11/3 Anyone with that kind of money contact Richard on Berkhampstead 71619.

i. Indicating a musical instrument which is being played, = UPON *prep.* 25. Of a musician: playing (a specified musical instrument) *colloq.* (orig. *U.S.*).

c 1386 CHAUCER *Miller's Tale* (1885, Harl. MS. 7334) 3214 And al aboue þer lay a gay Sawtrye On which he made a nightes melodye. *c* 1400 [see PLAY *v.* 26 a]. *a* 1529 [see LUTE *sb.*[1] 1 a]. 1767, 1768 [see PIANOFORTE]. 1842, 1903 [see PERFORM *v.* 7 c]. 1926 WHITEMAN & MCBRIDE *Jazz* xii. 241 Gus Miller..was wonderful on the clarinet and saxophone. 1934 *Down Beat* Aug. 4/2 Oscar Eiler remains on cello... Hunter Kahler replaced George Frewit on piano... Milt. Chalifoux is still on drums, as is Ralph Mazza on guitar and violin. 1955 SHAPIRO & HENTOFF *Hear me talkin' to Ya* xiii. 206 Our three pieces including Sonny, Sterling Conway on banjorine, and myself. *Ibid.* 251 We went into the Little Club—Gil Rodin, myself, and Benny Goodman on saxes, Glenn

Miller on trombone..; and Ben Pollack, of course, on drums. **1962** *Sunday Times* (Colour Suppl.) 10 June 3 A powerful and accomplished saxophonist—mainly on alto, originally rather too closely modelled on Charlie Parker. **1971** *Radio Times* 11 Nov. 3/4 Neither Cat Anderson on trumpet nor Lawrence Brown on trombone were able to make this tour.

j. *Math.* (Defined or expressed) in terms of (the elements of).

1934 *Ann. Math.* XXXV. 119 By an abstract ε p-simplex σ_p on *A* is meant a set of *p*+1 points of *A* whose sum is of diameter <ε. **1937** R. D. CARMICHAEL *Introd. Theory Groups of Finite Order* viii. 240 Show how to form the most general group of linear homogeneous transformations on a given set of variables. **1953** BIRKHOFF & MACLANE *Survey Mod. Algebra* (rev. ed.) vi. 119 A function defined on the elements of the set S_1 with values in T. **1968** E. T. COPSON *Metric Spaces* i. 11 Suppose we have a relation ~ defined on a set *E*. **1971** *Sci. Amer.* Aug. 94/3 The objects..will be real-valued functions *f* defined on some set S. In other words, *f* is a rule that assigns to each point *s* belonging to the set S a real number *f*(*s*) belonging to the real-number set. **1971** *Nature* 31 Dec. 527/1 We define a pairwise dissimilarity matrix *d* on a set of objects *P* as a real-valued function on pairs of elements of *P* such that $d(a, b) \geq 0$ for all *a*, *b* in *P* [etc.].

k. Addicted to or regularly taking (a drug or drugs). Cf. *ON *adv.* 1 b. *colloq.* (orig. *U.S.*).

1936 *Amer. Speech* XI. 124/2 *On drugs*, addicted. **1955** SHAPIRO & HENTOFF *Hear me talkin' to Ya* xxi. 333 The habit is a false crutch. Don't get on the H. *Ibid.* 335, I don't think all musicians are on junk by any means. **1971** *Lancet* 30 Oct. 985/1 A woman aged 49 received lithium carbonate 900 mg. daily... She was also on tranylcypromine 30 mg. daily. **1972** M. J. BOSSE *Incident at Naha* i. 60 Linda..asked what I was on these days. 'Pot, and not much of that,' I told her. **1972** D. SELMAN *Sudden Death* iv. 96, I suppose he didn't *truly* rape me... I mean I'm on the pill and everything. **1973** BOYD & PARKES *Dark Number* vii. 82 Julia went through a pretty bad time.. after the accident—on the pills, seeing psychiatrists and what have you. **1976** P. HILL *Hunters* v. 59 He dropped his mouth to hers... 'Are you on the pill?' he asked.

l. Drinking (alcoholic liquor) in large quantities or to excess. *on the booze*: see *BOOZE *sb.* 2 b; *on it* (Austral. colloq.): drinking heavily, 'hitting the bottle'.

1938 E. LOWE *Salute to Freedom* 38 He knew how drink affected Brand, and he muttered to his wife, 'He's on it proper to day mother.' **1951** E. LAMBERT *Twenty Thousand Thieves* ix. 140 'They reckon Groggy's on it again,' observed Tommy. **1955** P. WHITE *Tree of Man* (1956) 141 'It is him,' she said finally. 'It is that bastard. He is on it again.' **1962** A. SEYMOUR *One Day of Year* 12 And how long was you on it before the pubs shut? *a* **1966** 'M. NA GOPALEEN' *Best of Myles* (1968) 291 Easy seen you were on the beer last night. **1966** 'J. HACKSTON' *Father clears Out* 104 When he was on it, and wanted..another drink.. he never had to press the bell, but pressed the button with a bullet. **1967** H. STOREY in *Coast to Coast 1965–66* 203 One of the neighbours had ribbed him about 'being on the bottle'. **1976** *Daily Mirror* 18 Mar. 9/3 Watch that daily tipple, ladies. You could end up on the bottle.

6. a. Also used redundantly in *on tomorrow*, *on yesterday*, etc. *dial.* and *U.S.*

1829 *Virginia Lit. Museum* 30 Dec. 459/2 *On.* As '*on* tomorrow'; a mere expletive. Common. **1848** *Southern Lit. Messenger* XIV. 636/2 'On' yesterday, (another Southern emendation of the Queen's English, which is funny enough,) I was so unfortunate [etc.]. **1852** *N.Y. Tribune* 9 Jan. 6/1 It was the intention to send in the Treasury Report..on yesterday. **1880** W. H. PATTERSON *Gloss. Words Antrim & Down* 74 *On*..is sometimes prefixed to the words to-morrow and yesterday, thus—'I'll do it on to-morrow.' **1914** *Dialect Notes* IV. 160 On yesterday; on last week. **1922** H. C. LODGE in *Congress. Rec.* 27 Dec. 942, I took occasion to ask the Secretary of State on yesterday. **1922** JOYCE *Ulysses* 645 Lionel's air in *Martha*, *M'appari*, which..he heard, or overheard, to be more accurate, on yesterday. **1944** H. W. HORWILL *Dict. Mod. Amer. Usage* (ed. 2) 214/2 The expression *on yesterday* is an example of the Am. use of *on* where it would be considered superfluous in Eng. **1977** *Irish Press* 29 Sept. 2/1 (Advt.), Removal of remains to St. Bridget's Church, Kilcurry on today (Thursday) at 6.30 o'clock.

d. *on time*: see also TIME *sb.* 43.

10. c. *on it*: (*a*) *U.S. slang*, ready for, or skilled in, something; (*b*) *dial.*, preceded by an adverb or adjective: in a particular condition or situation, usu. one that is distressing.

1865 *Harper's Mag.* May 694/1 She's tolerable peert—the old 'oman is. Oh, she's on it, you bet. **1866** 'MARK TWAIN' in *Daily Union* (Sacramento) 22 May 3/4 In San Francisco sometimes, if you offend a man, he proposes to take his coat off, and inquires, 'Are you on it?' **1880** A. A. HAYES *New Colorado* (1881) v. 77 You bet he could cook. He was just *on it.* **1886** R. E. G. COLE *Gloss. Words S.-W. Lincs.* 103 Such phrases as 'Sorely on it', 'Sadly on it', for Sorely off, Sadly off; 'Two or three days ago I was strangely on it.' **1889** E. PEACOCK *Gloss. Words Manley & Corringham, Lincolnshire* (ed. 2) 886 He's sorely on it yit, 'cause his wife's runn'd awaay fra him. **1890** BARRÈRE & LELAND *Dict. Slang* II. 102/1 *On it* (American). This eccentric expression meant originally that a man was decidedly engaged in anything. It implied determination. 'I'm on it,' I understand it. **1946** F. SARGESON *That Summer* 96 He looked pretty crook on it.

II. 20. d. Of a joke, laugh, etc.: against or at the expense of (someone).

1866 *Harper's Mag.* July 271/2 There may be a joke about it; but if there is, it is on the Colonel, for he told me so. **1901** *Munsey's Mag.* XXV. 711/2 'It was Lanse—Lanse all the time,' he exploded. 'Oh, wasn't that one on me!' **1906** *Nation* (N.Y.) 6 Dec. 478 The people rejoiced that the laugh was on those whom they consider their natural enemies. **1933** [see *B.O.* s.v. *B. III]. **1936** N. COWARD *To-night at 8.30* 27 The joke is on us... We've never even been lovers. **1939** L. M. MONTGOMERY *Anne of Ingleside* vi. 33 The joke is on us... And a nice laugh he will have on me. **1967** *Listener* 11 May 634/1 The trouble is we enjoy a good laugh, especially if the joke's on us. **1977** P. HARCOURT *At High Risk* vi. 196 The joke was on me, and it left a sour taste.

e. Indicating a person, etc., who is to pay the bill, esp. for a treat of any kind. *colloq.*

1871 *Republican Rev.* (Albuquerque, New Mexico) 29 July 2/4 After the first round they said it was 'on me'. **1889**, etc. [see *HOUSE *sb.*[1] 2 c]. **1902** C. J. C. HYNE *Mr. Horrocks, Purser* 78 And now come and have a bit of cheap lunch. We'll consider we've tossed for it, and it's on me. **1919** 'I. HAY' *Last Million* vii. 85 'This is on us,' Al Thompson hastened to add. **1938** L. MACNEICE *Earth Compels* 22 Five minutes spent at a bar Watching the fish coming in, as you parry and shrug This is on me or this is on me. **1973** 'P. REID' *Harris in Wonderland* iii. 26 Arbuthnot wanted to drink... I explained my financial predicament but he waved it aside. 'On me, man.'

f. To the disadvantage or detriment of (a person); so as to affect or disturb. *colloq.*

1880 W. H. PATTERSON *Gloss. Words Antrim & Down* 74 'Don't break it on me,' *i.e.* don't break that thing of mine. **1892** E. LAWLESS *Grania* I. II. iv. 184 It was the devil's own abuse he got from his wife..for letting her fine spring chickens be drowned on her, which she had been months upon months of rearing. **1907**, etc. [see *DIE *v.*[1] 1 e]. **1907** J. M. SYNGE *Tinker's Wedding* II. 26 There she is waking up on us, and I thinking we'd have the job done before she'd know of it at all. **1946** K. TENNANT *Lost Haven* (1947) vii. 103, I never knew such a man for pickles——he must have eaten a bottle on me just over tea and breakfast. **1955** R. P. HOBSON *Nothing too Good for Cowboy* vii. 63 My lead cows bunched up on me and refused to face the storm. **1963, 1966** [see *GO *v.* 44 a]. **1971** M. SMITH *Gypsy in Amber* (1975) ii. 17 You're turning into a dilettante on me. **1974** M. BUTTERWORTH *Man in Sopwith Camel* iii. 34 He's passed out on me... Had some kind of seizure.

21. b. *to have a down on*: see *DOWN *sb.*[3] 5; *to have nothing* (or *something*) *on* (a person): see *HAVE *v.* 14 h.

III. 26. Restrict †*Obs.* to uses in Dict. and add: **b.** Used *colloq.* (chiefly *Austral.*) in locative senses, where 'at' would normally occur in standard use.

1853 *Bendigo* (Victoria) *Advertiser* 9 Dec. 1/1 We have for many months vainly endeavoured to procure suitable materials for publishing a Newspaper on Bendigo, to be devoted Exclusively to the Mining Interests. **1860** *Mining Surveyors' Rep.* (Mining Dept., Victoria) Aug. 198 This will be one of the richest claims on Ballarat. **1892** P. H. EMERSON *Son of Fens* xxv. 248 You grind my old Beccy [*sc.* scythe]; you're a better hand on it than I am. **1901** M. FRANKLIN *My Brilliant Career* iii. 17 Why, on Bruggabrong the women never had to do no outside work. **1966** BAKER *Austral. Lang.* (ed. 2) xvi. 348 A gold-seeker was never *at* a goldfield, but always *on* it. **1968** *Listener* 29 Aug. 267/2 The only reason she was on a bus stop at that un-Christian hour is that if she didn't catch that particular bus she would miss the only connection that would get her to the house in time to do a day's work. **1974** *Publishers Weekly* 5 Aug. 8/2 He began to put it all together, writing in the evenings and on weekends.

27. (Further examples.)

1829 D. JERROLD *Black-Eyed Susan* (1855) I. vi. 27 We found..three pilots' telescopes. This is one on 'em! **1848** THACKERAY *Van. Fair* li. 463 We're three on us—it's no use bolting. **1898** J. D. BRAYSHAW *Slum Silhouettes* 221 There was 'undreds on 'em, men, women, an' kids, an' most on 'em seemed ter be Total Obstinate Sons o' the Phoenix. **1916** 'TAFFRAIL' *Pincher Martin* xi. 204 'Ere's another on 'em! **1931** M. ALLINGHAM *Look to Lady* xvi. 170 Don't take no notice on 'im... 'E's as right as ever 'e was. **1937** —— *Case of Late Pig* xiv. 98, I don't know what he'll think on us—two on 'em instead of one. **1973** R. PARKES *Guardians* ix. 165 No doubt on it, mate. **1974** 'S. WOODS' *Done to Death* 41 Nobody as I knows on. .. It was nothing, really, to get hold on.

27*. In *dial.*, *N. Amer.*, and casual contexts, used where 'in' would normally occur in standard use.

1762 BOSWELL *London Jrnl.* 13 Dec. (1950) 82, I feel a surprising change to the better on myself since I came to London. **1787** J. ELPHINSTON *Propriety* II. II. 93 Dhe Scotch can see no incongruity in meeting a person *on* dhe street,..hwaraz dhe Inglish meet a person *in* dhe street. **1892** E. G. VINCENT *Newfoundland to Cochin China* iii. 45 The City Hall in this street, or 'on', as the Canadians would say. **1900** *Times* 6 Jan. 14/5 The genuine Dorset native always says, 'I see'd it on the paper', or 'I read it on the paper'. **1924** R. MASSON *Use & Abuse of Eng.* (ed. 4) iii. 42, I will likely know a great difference on her. **1938** J. STUART *Beyond Dark Hills* vi. 139 That took lifting and skill on flipping steel. *Ibid.* x. 396 They took 14 stitches on the other fellow. **1956** A. J. LERNER *My Fair Lady* (1958) I. viii. 86 Let the time go by, I won't care if I Can be here on the street where you live. **1972** *Scholarly Publishing* III. 181 My work as assistant editor is only a sideline. Since I do it largely on my own time, I avoid extra work and correspondence. **1972** *Time* 17 Apr. 37/2 An agreement..to help finance what is now called the Barsky Unit on the grounds of the Cho Ray Hospital. **1974** H. L. FOSTER *Ribbin'* iii. 109 Terror is waiting on line at 6:30 in the morning..for the Brooklyn Paramount to open for Alan Freed's rock and roll revue. **1974** *Melody Maker* 4 May 21/1 Hospitalized not long ago with a blood clot on his leg, he entered Beth Israel hospital again last week. **1977** *Irish Press* 29 Sept. 13/1 Michael Keane..is back on the Kerry team for Sunday's All-Ireland Under 21 Football final.

on, *adv.* (*a.*, *sb.*) Add: **1. b.** *to be on*: to be addicted to, or regularly taking, a drug or drugs; to be under the influence of drugs. Cf. *ON *prep.* 1 k. *U.S. slang.*

1938 *Amer. Speech* XIII. 188/2 *To be on*, to be addicted or actively indulging the [drugs] habit. **1955** W. GADDIS *Recognitions* I. v. 197 She's high right now, can't you see it? She's been on for three days. **1956** B. HOLIDAY *Lady sings Blues* (1973) xiv. 121 When I was on, I was on and nobody gave me any trouble. No cops, no treasury agents, nobody. I got into trouble when I tried to get off. **1968** *Sun Mag.* (Baltimore) 13 Oct. 19/3 When I took sets, I'd do it on the way to school... I'd always be on when I had to read in speech class.

11. b. Colloq. phrases: *to be on about*: to keep talking about, to harp on, to speak or write about (a subject); *to be on at*: to nag or scold (a person); *to go on about*: see *GO *v.* 84 g; *to go on at*: see GO *v.* 84 g; *to keep* (or *go*) *on and on*: to persist in speaking or questioning, to nag *at* a person.

1909 *Westm. Gaz.* 22 Sept. 8/2 Yesterday morning complainant was 'on' to him again about his religion. **1916** *Kelso Chron.* 24 Mar. 3 He was on aboot wa's that had ears. **1936** R. LEHMANN *Weather in Streets* viii. 348 Marda's always asking me why I don't get a divorce... Last year she was always on about it. **1938** N. MARSH *Artists in Crime* x. 145, I told him it would upset me but he went on and on. **1941** BAKER *Dict. Austral. Slang* 51 *On at* (*a person*), *to be*, to scold, reprove, nag at a person. **1952** A. BARON *With Hope, Farewell* 94 Well, now the second one's on at him to get married. **1958** N. F. SIMPSON *Resounding Tinkle* II. in *Observer Plays* 267 Fred was on at me to go up but I had my old coat on. **1966** *Listener* 22 Dec. 939/1 A reflection that brings us back to atmosphere and art forms, which I was on about some weeks ago. **1973** 'C. AIRD' *His Burial Too* iv. 40 The garage key that Ada Turvey was on about..that's still in the lock. **1974** G. BUTLER *Coffin for Canary* ii. 35 'You looked terrible yesterday,' he said. He kept on and on. He always did. **1974** 'J. MELVILLE' *Nun's Castle* ix. 214 He kept on and on at me till he got it out. I wouldn't have told him otherwise. **1975** K. BARCLAY tr. *P. Orum's Whipping-Boy* xxiii. 161 'He kept on and on at me.'.. 'I didn't do it.'.. 'Why can't you let me alone?'

13. b. (Further examples.) Also, of an event: arranged; going to happen. Of food or the like: placed on the stove, etc., to cook; cooking. Also *to have nothing on*: to have no engagements, business, etc.

1748, etc. [see PUT *v.*[1] 46 k]. *c* **1825** J. WALKER *Factory Lad* I. iii. 8 Interior of Allen's House—Fire place, with saucepan on. **1841** DICKENS *Barnaby Rudge* xvii. 31 Hurrah! Polly put the ket-tle on, we'll all have tea. **1841** MRS. GASKELL *Let.* 23 Dec. (1966) 46 Yesterday this plan seemed quite given up—today..it's *on* again—if all goes on well. **1883** W. AITKEN *Lays of Line* 135 The fire's black oot, and the parrich no on. **1908** R. W. CHAMBERS *Firing Line* iv. 46 If you and Virginia have nothing better on I'll dine with you tonight. **1915** R. FRY *Let.* 27 Feb. (1972) II. 383 I'm pushed away into any place and only wanted when there's nothing better on. **1938** D. RUNYON *Furthermore* vii. 129, I have nothing on of importance at the moment. **1955** M. ALLINGHAM *Beckoning Lady* ii. 21 It's still on, is it, the party? **1955** E. COXHEAD *Figure in Mist* ii. 69 'I left the potatoes on,' she muttered, and fled into the house. **1972** J. GILL *Tenant* I. iii. 25 I've nearly done, just putting the rice on. **1973** J. BURROWS *Like an Evening Gone* i. 11 Miss Limb, amazingly, had fallen in with Greta's plans. 'So it's on.' **1973** 'E. MCBAIN' *Hail to Chief* iv. 55 The television set was on, but the volume control was apparently broken. **1974** M. INGATE *Sound of Weir* xvii. 150 I'd better be getting the dinner on. We've only got sausages. **1974** *Times* 14 Mar. 16/3 It must have been the only diplomatic function on in London because many hundreds of the most notable people..were there.

c. Freq. in colloq. phr. *you are* (or *you're*) *on*: the bet or bargain is agreed.

1933 *Sun* (Baltimore) 25 Apr. 18/7 'I'll bet you a lobster dinner and all the champagne we can drink, if legal, that the dam will be filled by February 1,' said Mr. Smith... 'You're on,' said Mr. Crozier. **1961** J. SEYMOUR *Fat of Land* i. 19 'I'll let you have the two cottages..for ten pounds a year.' 'You are on,' I said. **1967** J. BURKE *Till Death us do Part* x. 153 'I'll give it [*sc.* smoking] up if you do.'.. 'All right,' snapped Alf. 'All right, you're on.' **1969** Y. CARTER *Mr. Campion's Farthing* xvi. 156 'Just to seal the bargain,' he said... 'You're on.' **1974** L. DEIGHTON *Spy Story* xviii. 194 'A quid,' I said. 'You're on,' said Ferdy.

d. *to be on*: to be in favour of, or willing to be a party to, something. *colloq.*

1888 'R. BOLDREWOOD' *Robbery under Arms* I. xi. 132 'What shall we do, Jim?..go or not?'... 'I'm half a mind to tell Warrigal to..say we're not on.' *Ibid.* 138 Now we're here what's the play called, and when does the curtain rise? We're on. **1890** —— *Miner's Right* II. xiii. 23 'I'm on,' answered Joe, a ray of humour irradiating his honest countenance. **1898** 'H. S. MERRIMAN' *Roden's Corner* xiv. 145 If there is going to be fight..I'm on. **1916** 'TAFFRAIL' *Pincher Martin* xiv. 270 'What about a glass of sherry to celebrate the auspicious occasion?' 'I'm on, Peter,..but I really think it's up to me to pay for it.' **1923** WODEHOUSE *Inimitable Jeeves* xiv. 161 This jamboree is slated for Monday week. The question is, Are we on? **1939** K. TENNANT *Foveaux* IV. 350 'Are you on?' Herb asked impatiently as he began to remove his coat. 'I'll give it a go.' **1969** V. GIELGUD *Necessary End* i. 15 I'm on—if you want to play the equivalent of Twenty Questions.

e. In a state of knowledge or awareness regarding a person, state of affairs, etc. Cf. *ON TO *prep.* 2. *U.S. slang.*

1885 *Santa Fé Weekly New Mexican* 9 July 2/2 He hoped to sell the cavalry a large lot of supplies, but Major Van Horn was 'on'. **1900** ADE *Fables in Slang* 68 The

Preacher didn't know what all This meant,..but you can rest easy that the Pew-Holders were On in a minute. **1902** [see *NEXT *a.* 13 c]. **1909** [see *DOPE *sb.* 3 a]. **1934** J. M. CAIN *Postman always rings Twice* i. 10, I saw he was on, and quit talking about the guy in the Cadillac. **1973** R. STOUT *Please pass Guilt* (1974) xviii. 164 Wolfe, turning and seeing Saul, was on as quick as I had been. He said.. 'What?'

f. In negative contexts: acceptable, allowable; possible, likely; esp. in phr. *it's* (*just*) *not on.* *colloq.*

1935 in Partridge *Dict. Slang* (1937) 587/2 The majority of amateur [snooker] players..wildly attempt shots that are not 'on'. **1957** *Economist* 19 Oct. 200/1 This proposal is no longer 'on'. **1958** M. PUGH *Wilderness of Monkeys* 38 'I say, that's a bit much,' the military man said. 'Not on.' **1963** P. H. JOHNSON *Night & Silence* xxviii. 208 He could not conceivably go on believing in Matthew's guilt. It was just not on. **1963** *Times* 29 May 10/5 How can a ship fight effectively if a third of the crew is Portuguese, a third Belgian, and a third Danish? The thing is just not on. **1968** R. V. BESTE *Repeat Instructions* xvi. 174 'I'd like to dump the last two together next time.' 'That's not on,' King told him tersely. '..Stick to procedure.' **1973** 'M. INNES' *Appleby's Answer* ii. 23 How, I repeat, is a fellow to come by a cathedral? It just isn't on. **1975** *Guardian* 20 Jan. 4/3 Reductions in the standard of living were not on.

g. Of an item of food: on the menu; available. Cf. *OFF *adv.* 3 e. *colloq.*

1949 'M. INNES' *Journeying Boy* xx. 248 Champ is off and eggs are on. **1953** R. FULLER *Second Curtain* ix. 131, I always have the curry when it's on. **1963** 'L. BRUCE' *Crack of Doom* vi. 53 Are you going to have lunch in the hotel? They've got sheeps' hearts on today.

h. *it was* (or *is*, etc.) *on* (*for young and old*): a description of complete disorder, a free-for-all. *Austral. colloq.*

1951 E. LAMBERT *Twenty Thousand Thieves* (1952) xvii. 258 Peter Dimmock bounded between the tents leaping into the air at every few paces and whooping: 'It's on! It's on for young and old!' **1955** J. MORRISON *Black Cargo* 77 A day come when some of our blokes in Sydney just put their coats on and walked off the job. It was on then for young and old. **1969** W. DICK *Naked Prodigal* 50 Just before closing time a brawl started when some bloke walking by spilt beer on Ackie so Ackie's young brother King hit him and the bloke's mate stepped in so Ackie hit him—and then it was on. **1971** D. MARTIN *Hughie* (1972) xi. 106 He almost forgot about it until the evening of Sunday when the party was due and when, in Harry's words, it was on for young and old.

B. *adj.* **1.** (Earlier and later examples.) Also of a ball or hit on this side.

1833 J. NYREN *Young Cricketer's Tutor* 34 The best way to play a ball, bowled as wide as your legs on the on side. **1836** [see *OFF *a.* 2 b]. **1836** *New Sporting Mag.* July 196 On-balls have a greater tendency to turn in towards the wicket. **1854** J. PYCROFT *Cricket Field* (ed. 2) vii. 117 An angle sufficient to make Off or On hits. **1898** K. S. RANJITSINHJI *With Stoddart's Team* (ed. 4) i. 32 For..excellence of 'on side' play he can be compared with the best players. *Ibid.* vi. 112 Iredale..got out to Hearne in attempting a huge 'on' hit. **1903** *Westm. Gaz.* 23 June 3/1 In all those on-side strokes..Fry is a master. **1904** F. C. HOLLAND *Cricket* 16 The on strokes are not so often used as the off strokes. **1909** *Westm. Gaz.* 17 Apr. 16/2 A good back and on-side player ..may confidently expect to do well under these conditions. **1959** *Times* 11 June 3/5 Bold strokes, particularly on the onside.

b. *on-verse* [tr. G. *Anvers*]: the first half-line of a line of Old English verse. Also *attrib.* Cf. *off-verse* (*OFF *a.* 2 c).

1935, 1953 [see *OFF *a.* 2 c]. **1953** [see *ICTUS 1]. **1970** *Jrnl. Eng. & Gmc. Philol.* LXIX. 438 Rime forces all final lifts in on-verse or off-verse into prominence.

3. Corresponding to the state (of an electrical device) of being on (ON *adv.* 13 b).

1899 J. PIGG *Railway 'Block' Signalling* vii. 364 When the arm is in the 'on' position, the mercury connects the two plates and completes the circuit. **1924** *Wireless World* 19 Mar. 772/2 This sort of switch is called a double-pole single-way switch, because it controls two circuits with one operation and has only one 'on' position. **1962** SIMPSON & RICHARDS *Physical Princ. Junction Transistors* vii. 142 Because the voltage across the transistor is a fraction of a volt losses in the 'on' state are small. **1962** E. G. DAVIES in G. A. T. Burdett *Automatic Control Handbk.* iii. 2 The isolator is hand operated and provided with distinct ON and OFF positions. **1967** *Electronics* 6 Mar. 133/1 With the modified circuit the relay's on-time is independent of input pulse characteristics.

4. *Physiol.* Of, pertaining to, or exhibiting the electrical activity that occurs briefly in some optic nerve fibres in vertebrates upon the commencement of illumination of the retina.

1903 F. GOTCH in *Jrnl. Physiol.* XXIX. 393 The first portion is the rise due to the sudden illumination; this I propose for brevity to term the ON effect. *Ibid.* 394 The upward movement of the image of the mercury indicates a positive (+) change in the eyeball in the ON, continuous and OFF reactions. *Ibid.* 398 The development of the ON change under these two conditions. **1937** *Brit. Jrnl. Psychol.* XXVII. 302 In the human eye the 'on effect' may occupy a time of the order of 0·1 sec. **1941** S. H. BARTLEY *Vision* xii. 286 The on-response is, in the main, duplex. **1948, 1972** [see *OFF *a.* 7].

C. *sb.*[1] (Earlier and later examples.) **on-theory**, a theory that favours concentrating the fielders on the on side and bowling the ball at or outside the on stump.

1862 *Baily's Monthly Mag.* Aug. 87 A on-drive from Jackson for 5. **1896** *Badminton Mag.* Sept. 280 A few bowlers have an 'on-theory'. **1900** W. J. FORD *Cricketer on Cricket* x. 118 George Giffen..could bowl 'off-theory' or 'on-theory'..with equal skill. **1963** *Times* 11 June 4/6 With a powerful on-drive Barker took his score to 49 and off the next ball completed a fine half century.

‖ **on** (ǫn), *sb.*[2] [Jap.] In feudal or prewar Japan, the sense of deep gratitude with an obligation or duty of service towards often highly formalized favours, as towards one's parents, teachers, lords, or the Emperor.

1946 R. BENEDICT *Chrysanthemum & Sword* (1947) v. 99 The word for 'obligations' which covers a person's indebtedness from greatest to least is *on*... *On* is in all its uses a load, an indebtedness, a burden, which one carries as best one may. *Ibid.* 101 *On* is always used in this sense of limitless devotion when it is used of one's first and greatest indebtedness, one's 'Imperial *on*'. This is one's debt to the Emperor, which one should receive with unfathomable gratitude... Every kamikaze pilot of a suicide plane was, they said, repaying his Imperial *on.* **1964** I. FLEMING *You only live Twice* iv. 56 He's acquired an ON with regard to me. That's an obligation—almost as important in the Japanese way of life as 'face'. When you have an ON, you're not very happy until you've discharged it honourably.

on (ǫn), *v.* [f. the adv.] **1.** *intr.* To go on; to move forward. Cf. ON *adv.* 9 b. *U.S. dial.*

1840 C. F. HOFFMAN *Greyslaer* II. II. x. 27 I'll see the eend of it. So with that, I ups and ons.

2. *to on with*: to place or put on. Cf. ON *adv.* 4 b. *dial.*

1843 'R. CARLTON' *New Purchase* I. xix. 170 She bethought as how she'd render off her fat; and so she ons with the grate pot. **1899** DICKINSON & PREVOST *Gloss. Words Cumberland* 231/2 Ah on's wi' my cwoat an' off teh wark. **1960** *Forfar Dispatch* 28 Jan. 8/5, I ons w'ee porridge pot.

on-, *prefix*[1]. Add: **4. onsurge**, an onward surge; **on-sweep** (earlier example).

1960 J. FINGLETON *Four Chukkas to Austral.* 16 His presence allowed the English bowlers to recover from O'Neill's onsurge. **1963** *Economist* 1 June 872/2 The real onsurge into the consumer durables revolution. **1866** *Dublin Rev.* Jan. 170 The rights of property alone.. formed the basis of resistance to the onsweep of revolution.

b. With sbs. used *attrib.* or as *adj.*, as **on-air**, while broadcasting; **on-axis**, situated or occurring on the axis; **on-course**, situated or taking place at a race-course; **on-demand**, done or available when demanded; **on-duty**, engaged or occupied with one's normal work; **on-farm**, occurring or used at a farm; **on-form**, that is in good form or condition; **on-street**, that is on a street; esp. of parking facilities; **on-track**, of betting: done at the race-track.

1972 *Guardian* 11 July 14/5 Parents shared what they thought about the programmes..by responding to on-air appeals to complete a questionaire. **1976** *Listener* 25 Mar. 362/1 The on-air behaviour of certain national and local disc-jockeys. **1962** A. NISBETT *Technique Sound Studio* 248 *Dead side* (of microphone), the angle within which the response of a microphone is low compared with the on-axis response. **1971** R. J. COLLIER et al. *Optical Holography* ii. 54 On-axis observation of either image is disturbed by the out-of-focus light from the other. **1964** A. WYKES *Gambling* viii. 195 These offices are in fact merely extensions of the on-course totalizators. **1973** *Times* 12 Apr. 12/4 The Tote, which does most of the on-course betting, earned a meagre profit of £100,000 last year. **1962** *Times* 9 Apr. (British Oxygen Co. Suppl.) p. v, Ready supply and on-demand delivery make sure your production goes smoothly. **1971** *Flying* (N.Y.) Apr. S5/3 Probably the most revolutionary innovation within this new system will be an on-demand capability. **1970** K. PLATT *Pushbutton Butterfly* (1971) xiv. 155 The on-duty cop outside the room. **1974** 'S. WOODS' *Done to Death* 184 His voice had..that wooden, on duty tone. **1969** *Times* 20 Jan. 2/1 There must be more integration with on-farm testing to select the bulls for central testing. **1970** *Daily Tel.* 26 Oct. 8/3 The economics of on-farm compounding with a mobile mill and mixer look fairly good. **1965** *Universe* 11 June 14/5 Mick Norman..last week..hit a convincing 70 against an on-form Sussex side. **1968** *Melody Maker* 23 Nov. 22/5 An on-form Phil Seamen is still one of the most exciting things to catch in a London jazz club or pub. **1959** *Manch. Guardian* 26 May 8 There might be strict prohibition of 'on-street parking' but 'not necessarily' of loading and unloading. **1973** D. WESTHEIMER *Going Public* iv. 64 A busy thoroughfare with no on-street parking. **1964** On-track [see *off-track* s.v. *OFF- 4 b].

-on, *suffix*[1]. [f. the ending of ION (and *anion*, *cation*).] **1.** *Physics.* **a.** Used (first in *ELECTRON[2]) to form the names of sub-atomic particles, as *hyperon*, *meson*, *neutron*, *proton*.

The example of *electron* gave rise to a few particle names in *-TRON, q.v.; but *proton*, chronologically the second word of this group, has proved the dominant model.

b. Used to form the names of quanta, as *graviton*, *phonon*.

2. Used, esp. in molecular biology, to form the names of some entities conceived of as units, as *codon*, *muton*, *operon*, *pedon*.

-on, *suffix*[2]. [f. Gr. -ον, neut. of -ος, nom. masc. sing. ending of many adjs.] The ending of the names of the noble gases other than helium, as *argon* (the earliest named), *radon*.

on and off, *adv. phr.* (*sb.*). Add: **b.** (Further examples.) Now usu. with hyphens.

1904 *Westm. Gaz.* 13 Jan. 2/3 The buyer resented this on-and-off policy. **1936** *Discovery* July 222/1 His left hand works an 'on-and-off' key, sounding the note when it is pressed and killing it when released. **1965** T. CAPOTE *In Cold Blood* (1966) i. 4 She had been an on-and-off psychiatric patient the last half-dozen years. **1974** *Country Life* 21 Feb. 394/3 Grazed..on a rotational or 'on and off' system. **1977** *Time* 19 Sept. 30/1 The signing of a Panama Canal treaty that was initialed last month after 13 years of on-and-off efforts through the Administrations of four U.S. Presidents.

onanism. Add: (Further examples.) Also = **coitus interruptus*. Also *fig.*

1892 A. K. GARDNER *Conjugal Relationships* vii. 96 This man put into practice, to calm the fears of his wife,.. the best calculated refinements of conjugal onanism. **1900** L. B. SPERRY *Confidential Talks with Husband & Wife* x. 150 Another scheme for the prevention of conception is to withdraw the penis from the vagina just before the ejaculation of semen. This method is appropriately called 'Onanism'. **1939** R. CAMPBELL *Flowering Rifle* III. 96 Who sponsor Onanism and Divorce And let the birthrate flounder on its course. **1952** *Encycl. Sexual Knowl.* (ed. 2) I. 78 It is therefore not correct to attribute the practice of masturbation to Onan, though the word onanism is to-day widely accepted in this latter sense and it would be pedantic not to recognize this term. **1967** G. STEINER *Lang. & Silence* 98 The recent university experiment in which faculty wives agreed to practise onanism in front of the researchers' cameras. **1977** P. ROONEY in D. Marcus *Best Irish Short Stories* II. 141 The onanism of the language, a phrase masturbated without hope of final clear expression.

onanist, onanistic *a.* (examples).

1855 W. WHITMAN *Leaves of Grass* 70 The sick-gray faces of onanists. **1867** ROBERTSON & RUTHERFORD tr. *Griesinger's Mental Path. & Therapeutics* II. iii. 173 That hidden strife betwixt shame, repentance, good intentions, and irritation, which imperiously impels to the act, we consider, after not a little acquaintance with onanists, to be by far more important than the primary, direct physical effect. **1926** tr. *M. von Gruber's Hygiene of Sex* vii. 120 The disturbances, which the physician so frequently finds in onanists are the same as those found in persons who indulge in extreme excesses of intercourse. **1934** F. SCOTT FITZGERALD *Tender is Night* II. ix. 199 There was something wooden and onanistic about her. **1961** D. HOLBROOK *Eng. for Maturity* iv. 45 Our essential cultural experiences are not celebrative, they are, rather, onanistic. **1962** *Listener* 20 Sept. 438/2 Quilty's fundamental childishness, his self-regarding, onanistic posing sets the tone of the whole film. **1966** *Ibid.* 22 Dec. 937/3 An uneasy childhood..Wykehamist Schooldays from the bullying horrors of which he escapes into an onanistic fantasy-world. **1973** *Daily Tel.* 23 Mar. 15/1 The onanistic aspect of these brief encounters in public places.

Onazote (ǫ·nāzōᵘt). Also **onazote**. A proprietary name for a type of rubber which has been expanded to a cellular condition by causing it to absorb a neutral gas under pressure during vulcanization and which is used for making life-belts and floats.

1920 *Trade Marks Jrnl.* 29 Dec. 2470 Onazote... Raw, or partly-prepared, india-rubber, balata and gutta-percha for use in manufacture. Charles Lancaster Marshall,..London,..manufacturer. **1940** *Jrnl. R. Aeronaut. Soc.* XLIV. 30 Onazote has a cellular structure in which each air bubble is completely separated from its neighbours. **1960** E. L. DELMAR-MORGAN *Cruising Yacht Equipment & Navigation* ix. 108 Though cork or Onazote lifejackets are well established and approved by the M.O.T. they are bulky and not very easily stowed in a yacht. **1960** *House & Garden* Aug. 46/4 Red and white lifebelt, onazote, £3 11s.

on board, *adv. phr.* and *a.* [f. ON *prep.* + BOARD *sb.*] **A.** *adv. phr.* See BOARD *sb.* 12 c and 14 in Dict. and Suppl.

B. *adj.* (Written *on-board*, *onboard*.) That is on board a ship, aircraft, spacecraft, or the like.

1966 *Electronics* 17 Oct. 35 In an operational launch vehicle, the signal would go to an onboard computer that would determine whether the problem was serious enough to shut down the rocket and eject the astronauts. **1967** *Technology Week* 20 Feb. 12/1 Persons who may discover the capsule are being advised that the on-board radiation source is sealed. **1969** *Jane's Freight Containers 1968–69* 433 An officially approved on-board cargo loader for Boeing 707 and 727 aircraft. **1972** *Nature* 24 Nov. 222/1 The synchronization was continuously monitored during the flight by checking the on-board clock one pulse per second against the station clock one pulse per second. **1974** HAWKEY & BINGHAM *Wild Card* xv. 128, I have to.. stand by to override the on-board computer in the event of a failure.

on-camera, *adv.* (*phr.*) and *a.* [f. ON *prep.* + CAMERA.] Within the range of a film or television camera; that is being filmed or televised.

1962 J. D. MACDONALD *Girl* xiii. 193 We are down here ..to do ten tropical commercials... We are surrounded by the Loyal Ones, on and off camera. Shrewd agency minds. **1971** R. PARKES *Line of Fire* ii. 26 To give chase, play the heroine. Maybe get interviewed on-camera afterwards. **1972** *Village Voice* (N.Y.) 1 June 26/3 Primo was an on-camera reporter for KDKA in Pittsburgh, went from there to Philly, then became director of local news here. **1974** *Keowee Courier* (Walhalla, S. Carolina) 24 Apr. 3/4 Peter Falk, star of the NBC-TV series 'Columbo', will serve as on-camera host. **1975** *Listener* 28 Aug. 258/2 Sheikh Mujib..has refused all on-camera interviews.

once, *adv.* Add: **A.** Forms. γ. 8— *dial.* and *U.S.* spellings **oncest, oncet, oncst, onct, onecest, onst.**

1789 WEBSTER *Dissertations Eng. Lang.* 111 In the middle states also, many people [say]..*oncet* and *twicet*. This gross impropriety [has]..prevalence among a class of very well educated people; particularly in Philadelphia and Baltimore. **1840** C. F. HOFFMAN *Greyslaer* II. III. xiv. 255, I ups rifle at onct, and hand on trigger to cut the string with a bullet. **1847** E. BRONTË *Wuthering Heights* I. xiii. 322 Couldn't ye uh said soa, at onst? **1851** MAYNE REID *Scalp-Hunters* I. xxi. 291 He *may* shoot well; he did onecest on a time—plum centre. **1867** A. D. RICHARDSON *Beyond Mississippi* xi. 135 Even some graduates of leading universities habitually use 'oncet' and 'twicet'. **1875** W. D. PARISH *Dict. Sussex Dial.* 105, I dunno but what you'd best shun him out of the fore-door at oncest. **1878** H. ALGER *Joe's Luck* in *Street & Smith's N.Y. Weekly* 8 Apr. 2/5, I kin whip my weight in wild cats, am a match for a dozen Indians to onst, and kin tackle a lion without flinching. **1883** H. D. RAWNSLEY in *Trans. Wordsworth Soc.* VI. 164 He niver oncst said owt. Ye're well aware if he'd been fond of children he 'ud 'a spoke. **1898** J. D. BRAYSHAW *Slum Silhouettes* 1 We was born to it, an' never expec's nuffink better; but 'e's been a real toff onct, Satan 'as. **1906** E. DYSON *Fact'ry 'Ands* viii. 95 What led me on t' wish t' be er gentleman onst more. **1909** J. MASEFIELD *Tragedy of Nan* I. 14 Why weren't I told to onst? **1913** C. E. MULFORD *Coming of Cassidy* vii. 117, I saw you onct an' I wondered if I was right. **1921** [see *LAMP *v.*[1] 4]. **1922** E. O'NEILL *Hairy Ape* (1923) i. 16 But aw say, come up for air onct in a while, can't yuh? *Ibid.* v. 45, I useter go to choich onct—sure—when I was a kid. **1932** V. RANDOLPH in B. A. Botkin *Treas. S. Folklore* (1949) III. i. 453 We seen a feller in town oncet a-wearin' a coat made out'n a piedy horse-hide. **1934** C. CARMER in *Ibid.* III. ii. 506 Chillun..think twict befo' yuh speak onct. **1967** in A. Dundes *Mother Wit* (1973) 270/2 Love me and hug me onct..again.

δ. *dial.* and *U.S.* **wance, wancet, wanst, wonst, wunst.**

1840 *Crockett Almanac* 2 Davy Crockett got skeered wunst. *Ibid.* 14, I wonst had an old flame. **1890** KIPLING *Soldiers Three* 12 Wanst upon a time, as the childerbooks say, I was a recruity. **1898** J. D. BRAYSHAW *Slum Silhouettes* 20 The poor bhoy shall be a gintleman for wance in his life. **1904** E. NESBIT *Phoenix & Carpet* v. 94, I see at wunst 'e was wuth 'is weight in flimsies. **1923** 'B. M. BOWER' *Parowan Bonanza* v. 52 Beans,..wancet they've been wrinkled wit' rain water and dried agin. **1972** L. LAMB *Picture Frame* xv. 139 Seen him wunst, I reckon. Spoke pleasant, like. **1977** *Transatlantic Rev.* LX. 152 'Lissn,' says Davey, 'Ah've been ower baurs duzzins a times an Ah've nivir wance get feart anuff tae jum paff.'

B. 3. Delete *Obs.* and add: Now *U.S. dial.*

1903 S. CLAPIN *New Dict. Amer.* 294 *Once*,..in parts of Pennsylvania settled by Germans, used as an expletive: 'Sit down *once*,' i.e. once for all. **1917** *Dialect Notes* IV. 338 'Come here *once*'..among German settlers. **1948** *Amer. Speech* XXIII. 109 Give me the knife once. **1953** *Ibid.* XXVIII. 246 Will you hand me that hammer once?

4. *once upon a time* (earlier example); also as *sb. phr.* (sometimes hyphened) and *attrib.*

1595 G. PEELE *Old Wives' Tale* sig. B1v, Once vppon a time there was a King or a Lord, or a Duke. **1876** R. E. FRANCILLON in *Gentl. Mag.* Oct. 423 There is all the difference between 'Daniel Deronda' and 'The Mill on the Floss' that lies between Now and Once upon a Time. **1927** S. SOUTHWOLD (*title*) Once upon a time stories. **1944** BLUNDEN *Cricket Country* iv. 43 So runs this once-upon-a-time in my memory. **1959** *Listener* 22 Jan. 164/2 The horse-drawn chaises of once upon a time. **1974** J. WAINWRIGHT *Hard Hit* 173 The talk between two middle-aged has-beens about once-upon-a-time days.

7. (Further examples.)

1911 KIPLING *Years Between* (1919) 7 Our ears still carry the sound Of our once Imperial seas. **1931** A. HUXLEY *Cicadas* 44 The pause and once-more fury of the gale. **1939** DYLAN THOMAS *Map of Love* 6 These once-blind eyes have breathed a wind of visions, The cauldron's root through this once-rindless hand. **1943** D. GASCOYNE *Poems 1937–42* 52 The once-met And long remembered faces. **1946** *Nature* 20 July 86/1 The once-popular 'tiger nut'. **1949** S. SPENDER *Edge of Being* 56 That past greatness and that once-willed Future Beyond the storm. **1951** W. DE LA MARE *Winged Chariot* 38 Once-green skeleton leaf. *Ibid.* 39 The angelic hierarchies Dome with their glory the once-empty skies. **1974** *Country Life* 21 Mar. 643/1 Once-popular composers..drop out of favour. **1977** J. CLEARY *Vortex* v. 135 The once-beautiful eyes, already dark with death.

b. *spec.* **once-fired** *a.*, of pottery: subjected only once to the process of firing. Also in Combs. with advs.: **once-off** *a.*, happening only once; hence as *sb.*; cf. *one off* s.v. *ONE *numeral a.* B. 29* b; **once-only** *a.*, occurring only once; **once-through** *a.*, being or employing water that enters a system, flows once through it, and then leaves it.

1952 V. ELEY *Monk at Potter's Wheel* 19 *Once fired*, an expression relating to pottery that has been glazed and fired without having received a previous firing. Mediaeval English pottery was fired in this way. **1960** H. POWELL *Beginner's Bk. Pott.* II. iv. 42 If you wish to make, decorate, and glaze in one operation, this is known as once fired. **1970** *Gloss. Industrial Furnace Terms* (*B.S.I.*) 19 *Once-fired kiln*, a kiln in which the body and the glaze thereon are fired at one and the same time, instead of in two separate firings. **1965** *Math. in Biol. & Med.* (Med. Res. Council) I. 8 But in a 'once-off' job where a standard program cannot be used..these advantages are lost. **1969** *Guardian* 1 Mar. 1/3 Steam turbines are built to a specification, often on a once-off basis or in pairs. **1973** *Times* 19 Jan. 12/1, 50 different juvenile weeklies, aside from the 'once-offs' that appear from time to time. **1976** *Gramophone* Oct. 670/3, I suppose that is the fault of once-off recordings. **1960** *Sunday Express* 28 Feb. 12/8 It is, alas, a once-only gratuity. **1963** *Listener* 7 Feb. 263/2 A 'once only' talker can never be as good as a hardened professional. **1965** M. FRAYN *Tin Men* xxiii. 125 The open-endedness of a once-only job would require a computer so complex..that it would be cheaper to use a human being. **1971** A. PRICE *Alamut Ambush* ix. 115 They wouldn't like doing it... But for a once-only job they might stretch a point. **1940** *Chambers's Techn. Dict.* 87/1 *Benson boiler*, a high-pressure boiler of the once-through type in which water is pumped successively through the various elements of the heating surface. **1946** J. N. WILLIAMS *Steam Generation* vi. 126 Forced circulation boilers may be divided into two classes according to whether the water is in continual circulation or is pumped through at one end of the heating surface and, making a single or 'once through' passage, leaves at the other end in the form of steam. **1978** *Environmental Conservation: Chemicals* (Shell Internat. Petroleum Co.) 2 Where sufficient surface water—lakes or rivers, for example—is available, waste heat from chemical plants has traditionally been discharged in the form of once-through (as opposed to re-used) cooling water.

II. 8. c. *once for all* (further examples); now usu. *once and for all*; also as *sb.*, hence *once-for-allness*; *once in a way* (examples); *once in* (or irreg. *and*) *a while* (earlier and later examples); *once in a blue moon*, rarely, exceptionally.

1781 J. WITHERSPOON in *Pennsylvania Jrnl.* 23 May 1/3 He will *once in a while*, i.e. *sometimes*, get drunk. [Used in] the middle states. **1869** MRS. STOWE *Oldtown Folks* x. 116 If he could come down here once and a while after work-hours. **1869** W. C. HAZLITT *Eng. Proverbs* 305 Once in a blue moon. **1876** Once in a blue moon [see *blue moon* s.v. BLUE *a.* 13]. **1891** J. M. DIXON *Dict. Idiomatic Eng. Phrases* 230 *Once in a way*, sometimes; at long intervals; on rare occasions. **1895** MORRIS & WYATT tr. *Beowulf* x. 23 E'en that in mind had I.., that for once and for all the will of your people would I set me to work. **1922** G. R. S. MEAD in *Quest* XIII. 490 For the Jewish eschatologist it was a once for all event he expected, whereas for such men as the Stoical thinkers it was a perpetual recurrence. **1928** E. O'NEILL *Strange Interlude* II. 76 Well, then, a little truth for once in a way! **1934** G. B. SHAW *On Rocks* I. 197, I really think, father, you might for once in a way take some slight interest in the family. **1949** *Scottish Jrnl. Theol.* II. 86 A radical misunderstanding of the New Testament teaching about eschatological once-for-allness and eschatological continuity. *Ibid.* 87 The only primitive wholeness that the Reformed Churches recognise is the once-and-for-all *wholeness* of Jesus Christ in whom God and man are at one. **1951** AUDEN *Nones* (1952) 14 The once-for-all that is not seen nor said. **1955** *Times* 26 Aug. 7/4 And when, by some mischance, once in a blue moon, the bell does ring, how startled we are. **1957** *Ann. Reg.* *1956* 137 The 1956 Budget..had..included a 'once-for-all' item of $961,000 for the 'Atoms for Peace' conference at Geneva. **1960** *Times* 12 Aug. 13/6 But contractile tissue as such seems to have been a once-for-all invention. **1960** V. NABOKOV *Invitation to Beheading* xvi. 159 Every once in a while he would jerk his flabby cheeks and his chin. **1963** *Times* 22 Feb. 16/7 Some absolute, once-and-for-all answer had to be found. **1970** T. LUPTON *Managem. & Social Sci.* (ed. 2) iv. 101 Pose questions like this and then try to think up once-for-all answers. **1972** *Police Rev.* 17 Nov. 1487/3 A once-and-for-all deduction from pension. **1975** J. B. HARLEY *O.S. Maps* i. 1 Before World War II.. there was a tendency to regard the published large scale map as something of a 'once-and-for-all' record of fact. **1976** *National Skat & Sheepshead Q.* Mar. 18 How many of you readers 'blow' the big hand—the one that appears once in a blue moon? **1977** *Times* 12 Feb. 7/3 Try not to let the once-for-allness of the occasion tempt you to eat the whole menu.

d. *once in a lifetime*, such as occurs only once in a person's life; freq. (with hyphens) *attrib.* and often used hyperbolically; *once too often*, of a thing said or done: once more than necessary or tolerable; usu. implying unpleasant repercussions.

[**1854** C. PATMORE *Angel in House: Betrothal* VIII. ii. 110 Love wakes men, once a life-time each.] **1908** YEATS & GREGORY *Unicorn from Stars* III. 122 There is a fiery moment, perhaps once in a lifetime, and in that moment we see the only thing that matters. **1915** H. T. WEBSTER *Our Boyhood Thrills* 7 (*heading*) The thrill that comes once in a lifetime. **1921** G. B. SHAW *Back to Methuselah* III. 131 Havnt you said that once too often already this morning? **1929** J. B. PRIESTLEY *Good Companions* II. v. 369 Don't be a scoffer... I've known people to scoff at these things once too often. **1932** R. ALDINGTON *Soft Answers* 122 They had simply got drunk once too often and lost their money. **1934** E. O'NEILL *Days without End* II. 68 But I warned him he'd humiliate me once too often —and he did! **1962** M. SUMMERTON *Nightingale at Noon* (1963) xii. 174 Fate presented him with a once-in-a-lifetime opportunity to get rid of her. **1962** *Times* 20 June 14/6 That once-in-a-lifetime occasion. **1973** *Mad Mag.* Oct. 8/1 Now cool it and let me really ham up this once-in-a-lifetime role! **1975** P. MOYES *Black Widower* xiii. 155 I'm afraid this is a once-in-a-lifetime trip for us. **1977** *Washington Post* 18 May C 6 (Advt.), Singer. Once-in-a-lifetime sale.

e. Phr. *once over lightly*; also (hyphenated) as *sb.* and *attrib. phr.* (chiefly *U.S.*).

1941 *Time* 12 May 55/1 Her pretty posturing, pouts, stunned, exotic stares are meaningless when she tries to do them once over lightly. **1960** WENTWORTH & FLEXNER *Dict. Amer. Slang* 364/2 *Once over lightly*, cursorily; quickly; temporarily. **1961** S. ARNE in WEBSTER s.v. *n.*, Had given political problems the once-over-lightly. **1961** S. HYMAN in *Ibid.* s.v. *adj.*, Gets the once-over-lightly treatment. **1967** M. KENYON *Whole Hog* i. 16 The young man raised the egg... Was it sunny-side-up..or once-over-lightly?

III. 9. b. *for once in your* (or *his*, *my*, etc.) *life*, on this single occasion in your (etc.) life.

1801 M. EDGEWORTH *Belinda* I. iv. 144 She has succeeded for once in her life. **1846** [see *LIFE *sb.* 3 c]. **1857** TROLLOPE *Barchester T.* III. iv. 81 And so the signora resolved..to do a good natured act for once in her life. **1859** *Blackw. Mag.* Aug. 224/2 For once in my life I agreed with my wife. **1862** W. COLLINS *No Name* I. I. ix. 115 Magdalen was caught, for once in her life, at the end of all her resources. **1881** H. JAMES *Portrait of Lady* III. xii. 189 Be a little wicked, feel a little wicked, for once in your life! **1964** 'S. WOODS' *Trusted like Fox* viii. 79 Mr. Justice Conroy..for once in his life owned himself puzzled.

c. (Later examples of *the once*.)

1887 T. DARLINGTON *Folk-Speech S. Cheshire* 282 'A thing for the once'..is an unusual or unprecedented thing. **1924** A. D. SEDGWICK *Little French Girl* I. viii. 74 'He came twice afterwards.'.. 'I didn't know that. I thought it was only the once.' **1967** N. FREELING *Strike Out* 89 [He] thought he'd live for ever... He came the once for a checkup.

o·nce-born, *a.* [ONCE *adv.* 7.] Designating or pertaining to someone whose attitude to life has retained a child-like simplicity and straightforwardness. (See also quot. 1942.) Cf. TWICE-BORN *a.* 3.

1849 [see TWICE-BORN *a.* 3]. **1902** W. JAMES *Var. Relig. Exper.* iv. 82 Another good expression of the 'once-born' type of consciousness, developing straight and natural, with no element of morbid compunction or crisis, is contained in the answer of Dr..Hale. **1942** BERREY & VAN DEN BARK *Amer. Thes. Slang* § 325/5 *Once-born*, not regenerated or not believing in regeneration.

onceness (wɒ·nsnės). [f. ONCE *adv.* + -NESS.] The fact or quality of happening only once, or all at once.

1866 R. & S. REDGRAVE *Century of Painters of Eng. School* I. iv. 108 [George] Barret's pictures are painted with the firm pencil and vigorous once-ness which characterize the works of the best painters of his time. **1917** E. POUND *Let.* 18 Apr. (1971) 109 H. Monroe seems to think that if her Chicago widows and spinsters will only shell out she can turn her gang of free-versers into geniuses all of a onceness. **1948** *Scottish Jrnl. Theol.* I. 141 In the Old Testament, he maintains, there is only one 'onceness' which is of theological significance, and that is the onceness of Christ. **1951** *Ibid.* IV. 3 That is to say, history is the field of relative uniqueness, but never of absolute once-ness.

once-over (wɒ·ns,ōᵘvəɹ). *colloq.* (orig. *U.S.*). [f. ONCE *adv.* + OVER *prep.*] A glance; a rapid inspection (often with an implication of cursoriness); *to give the once-over*, to make a rapid assessment of; to give (a person) an appraising or inviting glance; to search someone (for weapons).

1915 *Recruiter's Bulletin* (U.S.) Jan. 7/2 After giving an applicant the 'once-over' the other day I told him that I was sorry but I could not take him. **1915** *Dialect Notes* IV. 234 *Once-over*,..a glance. **1916** *Daily Colonist* (Victoria, B.C.) 27 July 6/3 Individuals whose presence on the deserted thoroughfares around the midnight hour aroused suspicions to be given what, in police parlance, is known as the 'once over'. **1926** [see *JERRY *a.*[2]]. **1927** *Daily Express* 31 Oct. 4/3 You require a couple of days, as distinct from the traditional American 'once over', to see Chelsea Old Church. **1938** E. BOWEN *Death of Heart* II. ii. 190 Daphne gave the rest of the cakes a rather scornful once-over. **1948** M. ALLINGHAM *More Work for Undertaker* xix. 219 They had entered without invitation and were giving it what Luke called 'the old once-over'. **1949** *Sun* (Baltimore) 26 Aug. 10/3 Whenever travelers enter Baltimore from abroad by ship or air, they are given the onceover by Cap'n Jenks. **1951** J. D. SALINGER *Catcher in Rye* x. 85 They probably thought I was too young to give anybody the once-over..—you'd've thought I wanted to *marry* them or something. **1953** E. TAYLOR *Sleeping Beauty* ix. 152 Len was curious about Betty..and wished to give her..what he called the once-over. **1954** *Encounter* Feb. 26/2, I could feel his eyes on me even more sharply than before. He was, so to speak, giving me a professional once-over. **1972** J. PHILIPS *Vanishing Senator* (1973) III. iv. 171 Give Mr. Styles a quick once-over, Francine, but try to remember you're looking for a gun and not trying to excite him sexually. **1977** *New Yorker* 4 July 22/1 He gave his display of perfect strawberries the once-over.

oncer (wɒ·nsəɹ). *colloq.* Also **once-er**. [f. ONCE *adv.* + -ER[1].] **1.** One who, or that which, does a particular thing only once; a thing that occurs only once; formerly *spec.* one who attends church once on a Sunday.

1892 *Review of Reviews* V. 41/2 He [*sc.* Gladstone] has a poor opinion of those whom he humorously terms 'once-ers'. **1909** *Daily Chron.* 22 Apr. 4/7 A minister regretted an increasing disposition on the part of the people to become 'oncers'. **1917** *Dialect Notes* IV. 327 *Oncer*,..he who (or that which) does a thing but once, esp. a church member who attends service but once on Sunday. **1938** *Amer. Speech* XIII. 35 For an extremely rare word..there may be only one slip, a *oncer*. **1944** AUDEN *For Time Being*

(1945) 11 Could he but once see Nature as In truth she is for ever, What oncer would not fall in love? **1972** B. RODGERS *Queens' Vernacular* 143 *Oncer*, somebody who does it once, and never again with the same person; Mr. fuck 'em and forget 'em. **1973** *Shooting Times & Country Mag.* 7 July 13/3 Eventually a fish did move under an overhanging alder, but it was a 'oncer' and I had grave doubts about it rising again.

2. *slang.* A one pound note.

1931 *Police Jrnl.* IV. 502 They spieled [*sc.* played] first for stakes of a sprazey [*sc.* sixpence],..increasing it to half a tosh,..and eventually to a oncer. **1936** [see *GEE *v.*[2] 1 c]. **1953** K. TENNANT *Joyful Condemned* iii. 27 Sure he took your oncer. But..here's another quid in place of it. **1968** L. DEIGHTON *Only when I Larf* vi. 80 He'd pay in used oncers, no cheques. **1970** A. ROSS *Manchester Thing* 122, I got my wallet out and let him look at a pound note... I laid a second oncer on top of the other. **1978** M. KENYON *Deep Pocket* vii. 82 They gave you an 'undred quid in oncers to see things their way.

oncest, oncet, see *ONCE *adv.* A. γ.

onchocercal (ǫŋkosɜ̄·ɹkăl), *a.* [f. as next + -AL.] Belonging to the genus *Onchocerca* (see next); caused by worms of this genus.

1934 R. P. STRONG et al. *Onchocerciasis* I. ii. 6 Calderon ..published a monograph on the onchocercal *Filaria*. **1974** A. W. WOODRUFF *Med. in Tropics* xiv. 237/1 In the early stages of massive onchocercal infection in African adults a dark purplish-brown plaque-like skin reaction.. may..be seen.

onchocerciasis (ǫ:ŋkosɜ̄ɹsəi·ăsis, -sɜ̄ɹkəi·ăsis). *Path.* Also **onco-**. [f. mod.L. *Onchocerca* (f. Gr. ὄγκος barb + κέρκος tail), name of a genus of parasitic filarioid worms + *-IASIS.] Infestation with or a disease caused by filarioid worms of the genus *Onchocerca*; in man *spec.* that caused by *O. volvulus* and transmitted by biting flies of the genus *Simulium*, common in tropical Africa and parts of Central America and marked by characteristic lesions of subcutaneous tissue and the eyes, often with blindness.

1911 R. T. LEIPER in *Rep. Local Govt. Board on Public Health & Med. Subjects* (Ministry of Health) No. 45. 6 The condition here termed Onchocerciasis is met with in frozen beef imported from Australia. **1911** STEDMAN *Med. Dict.* 604/2 *Onchocerciasis*, infestation of the ox with a species of *Onchocerca*. **1912** *Jrnl. Trop. Med. & Hygiene* XV. 232/1 (*heading*) Some notes and suggestions in connection with the etiology of bovine onchocerciasis. **1934** R. P. STRONG et al. *Onchocerciasis* I. iv. 19 The limitation of the centers of onchocerciasis in Guatemala to certain zones of territory..is very striking. **1967** A. W. JONES *Introd. Parasitol.* xiv. 184 In onchocerciasis of horses (and rarely of cattle), adult *Onchocerca* live in the tendons and ligaments of the back and neck region. **1971** *Nature* 21 May 151/3 In tropical and sub-tropical man-made lakes..explosive outgrowths of water weeds after filling had led to a very significant increase in the incidence of diseases carried by insects, such as malaria and oncocerciasis. **1974** *Daily Tel.* 9 Sept. 3/8 With him was.. a tropical eye disease expert from Swindon, who during the journey is to lead an investigation into onchocerciasis, a strange river blindness endemic in Central Africa. **1975** *Hansard Lords* 27 June 1706 To ask Her Majesty's Government what was the reason for the delay in presenting to Parliament the Onchocerciasis Fund Agreement 1974... Baroness Gaitskell: May I ask my noble friend to pronounce the name of the Fund? Lord Goronwy-Roberts: As if it were spelled 'Onkoserkiasis'.

onchocercosis (ǫ:ŋkosɜ̄ɹkōᵘ·sis). *Path.* [f. as prec. + -OSIS.] = prec. Hence **o:nchocerco·tic** *a.* (*sb.*)

1918 P. LUNA in *Amer. Jrnl. Ophthalm.* I. 122/2 The visual disturbances which I have met with among the carriers of this filaria..and which I shall call onchocercosis. *Ibid.* 125/2 The lesions..appear in the onchocercotic, all of whom suffer disturbances of vision. **1931** *Ibid.* XIV. 518 (*heading*) Onchocercosis in Mexico. *Ibid.* 518/1 In.. Guatemala the onchocercotic infection had been demonstrated..next to the international boundary. **1972** *Biol. Abstr.* LIV. 5491/1 In none of these cases was the infestation localized to the orbital cavity as in other filarial infections (Loa-loa, onchocercosis).

onchosphere, var. *ONCOSPHERE.

onco-. Add: **oncofœ·tal** (*U.S.* **-fetal**) *a. Med.*, occurring in tumours and in the fœtus but not in the adult; **o·ncolite** *Petrol.* [-LITE], a small, rounded body found in sedimentary rocks that is composed of incomplete concentric layers of calcareous material and is believed to be of algal origin; hence **oncoli·tic** *a.*; **oncolo·gic** *a.* (chiefly *U.S.*) = *oncological* adj. s.v. ONCO-; **onco·logist**, an expert or specialist in oncology, now esp. one concerned with treatment by means of drugs rather than surgery or radiotherapy; **oncology** (examples).

1972 P. ALEXANDER in *Nature* 21 Jan. 137/2 Table 1 is an attempt to classify the different types of onco-foetal 'antigens' (OFAs). It seems necessary to coin this phrase because the more elegant description of carcino-embryonic antigen has been pre-empted to describe one class of these compounds. **1975** *Ibid.* 25 Dec. 734/2 In man the two best known oncofoetal antigens are the α-foetoprotein (AFP)..and the carcinoembryonic antigen (CEA) of the human digestive system. **1933** R. B. YOUNG in *Trans. Geol. Soc. S. Afr.* XXXV. 32 Professor Julius Pia, in Hirmer's *Handbuch der Paläobotanik*, places the stromatolites among the *Schizophyceae* and separates them into two groups, the *Stromatolithi* and the *Oncolithi*, the latter embracing the forms..that occur as separate or individual bodies. In this paper the term 'oncolite' will be employed in the same sense, but without any implication that the bodies..differ from the rest of the stromatolites in any respect other than that generally they have formed round detached nuclei. *Ibid.* 34 When they were for a time completely stationary, the oncolitic growth became continuous, spreading over the surface of the bed to form thin but fairly extensive stromatolitic layers. **1967** *Jrnl. Sedimentary Petrology* XXXVII. 1163/2 A carbonate rock of which oncolites form a significant part may be termed oncolitic. **1972** H. BLATT et al. *Origin Sedimentary Rocks* xii. 422 Oncolites..up to 3 in. in diameter, have been found. **1974** *Nature* 22 Feb. 522/2 *Facies E.* Concentrically stacked spheroid stromatolites (oncolites). **1906** *1st Ann. Rep. Amer. Oncologic Hospital, Philadelphia* 3 The name of the Corporation shall be The American Oncologic Hospital. **1952** A. NETTLESHIP *Basic Princ. Cancer Pract.* xix. 389 The need for a clinically useful oncologic science is clear. **1925** H. GILFORD *Tumors & Cancers* xxvi. 574 Oncologists in general still adhere to older methods of operative treatment, supplemented, it may be, by actinic and other remedies. **1968** BETHELL & BURG tr. *Solzhenitsyn's Cancer Ward* (1969) I. vii. 96 On top of this a society of oncologists had been started recently. **1971** *Lancet* 21 Aug. 419/2 Although the oncologist's clinical experience with malignant disease is often limited, his pharmacological background has enabled him to make a substantial contribution to the subject. **1857** R. G. MAYNE *Expos. Lex. Med. Sci.* (1860) 810 Oncology. **1915** F. L. HOFFMAN *Mortality from Cancer* i. 1 There are few more interesting subjects for statistical analysis than cancer, or what is, perhaps, more appropriately termed the science of oncology, which comprehends tumors of all kinds, whether malignant or benign or ill-defined. **1968** BETHELL & BURG tr. *Solzhenitsyn's Cancer Ward* (1969) I. i. 3 According to the arrangement with the head doctor of the oncology clinic, the matron was supposed to wait for them at two o'clock in the afternoon. **1971** *Nature* 19 Feb. 517/3 The Imperial Cancer Research Fund (ICRF) last week announced the establishment of.. a medical oncology unit.

oncocerciasis, var. *ONCHOCERCIASIS.

oncogene (ǫ·ŋkōdʒīn). *Biol.* [f. ONCO- + *GENE.] A gene in a virus particle held to be responsible for transforming a host cell into a tumour cell.

1969 HUEBNER & TODARO in *Proc. Nat. Acad. Sci.* LXIV. 1087 It is postulated that the viral information (the virogene), including that portion responsible for transforming a normal cell into a tumor cell (the oncogene), is most commonly transmitted from animal to progeny animal and from cell to progeny cell in a covert form. **1970** *New Scientist* 3 Sept. 465/2 Occasionally, an 'oncogene' of the virus becomes switched on, and the host cell is transformed into a cancer cell. **1971** *Nature* 6 Aug. 373/2 Attempts to analyse genetically the oncogenes in adenovirus genomes, by the isolation of temperature sensitive mutants, seem most promising. **1975** *Ibid.* 7 Aug. 498/1 The oncogene theory contends that type C viruses possess oncogenic information and that malignancy is the result of activation of this genetic information.

oncogenesis (ǫŋkōdʒe·nėsis). *Biol.* [f. ONCO- + -GENESIS.] The formation or production of a tumour or tumours.

1932 in Dorland & Miller *Med. Dict.* (ed. 16) 891/1. **1944** *Jrnl. Exper. Med.* LXXX. 122 In these cases [of rapid growth of papillomas and carcinomatoids] actual oncogenesis must have been exceedingly swift, not the drawn out process it is generally held to be. **1961** *Proc. 4th Nat. Cancer Conf.* (U.S.) 41 (*heading*) Hormones and experimental oncogenesis; mammary and mammotropic tumors. **1977** *Proc. R. Soc. Med.* LXX. 393/2 It would seem plausible to think that the susceptibility to light-induced mutations..may favour viral oncogenesis in the presence of appropriate viral elements.

So **o·ncogen**, an agent that causes oncogenesis.

1969 *Jrnl. Biol. Chem.* CCXLIV. 4075/2 Present evidence does not favor 3-hydroxyuric acid as a proximate oncogen. **1971** *Nature* 5 Feb. 418/2 Hamster cells exposed to a dilute concentration of highly oncogenic compounds or to larger doses of weak oncogens can enter and pass through a complete mitotic cycle in spite of repair synthesis already in progress.

oncogenic (ǫŋkōdʒe·nik), *a.* *Biol.* [f. ONCO- + *-GENIC.] Causing the development of a tumour or tumours; of or pertaining to this effect.

1949 *New Gould Med. Dict.* 696/2 Oncogenic. **1959** *Jrnl. Nat. Cancer Inst.* (U.S.) XXIII. 277 (*heading*) The oncogenic spectrum of two 'pure' strains of avian leukosis. **1970** *New Scientist* 3 Sept. 464/2 The discovery, and confirmation, of an enzyme carried by oncogenic RNA viruses. **1971** *Nature* 29 Jan. 296/2 Both the oncogenic and protovirus hypotheses postulate that tumour viruses are escaped cellular genetic elements.

So **oncogeni·city**, the property of being oncogenic.

1944 *Jrnl. Exper. Med.* LXXX. 119 Manifestly the real criterion is the time required to render the first cell neoplastic; for oncogenicity properly speaking is contained in that act. **1967** *Cancer Res.* XXVII. 929/2 No distinction can now be made between the oncogenicities of the 7-*N*-oxide derivatives of guanine and xanthine. **1971** *Nature* 5 Feb. 416/2 The data indicate a link between oncogenicity of a compound and its capacity to provoke DNA repair synthesis.

oncolysis (ǫŋkǫ·lisis). *Biol.* [f. ONCO- + *-LYSIS.] The absorption or destruction of a tumour. So **oncoly·tic** *a.*, of, pertaining to, or causing oncolysis.

1928 STEDMAN *Med. Dict.* (ed. 10) 712/2 Oncolysis. *Ibid.*, Oncolytic. **1933** *Brit. Jrnl. Ophthalm.* XVII. 754 He finds that in these animals the oncolytic power is *nil*, and thence concludes that the eye is specially susceptible to attack by new growths. **1952** *Ann. N.Y. Acad. Sci.* LIV. 945 (*heading*) Viruses with oncolytic properties and their adaptation to tumors. *Ibid.*, Oncolysis, by viral action, has not been studied very thoroughly until recently. **1972** *Nature* 30 June 486/2 (*heading*) Oncolytic viruses. **1972** *Nature New Biol.* 5 July 8/1 No oncolysis occurred when subcutaneous tumours were treated with BEV.

on-coming, *a.* Add: Also **oncoming**. **2.** Ready to be sociable; friendly, welcoming, forthcoming, sympathetic.

1925 C. P. SLATER *Marget Pow* 183 He doesna seem to be very oncomin', for he's never been in the house yet. **1938** E. BOWEN *Death of Heart* II. iv. 248 If I were a more oncoming sort of fellow I should offer you a penny, and so on. **1942** N. BALCHIN *Darkness falls from Air* vi. 109, I think we might possibly get this through... The Secretary seemed quite oncoming. **1953** M. HOPKIRK *Queen over Water* v. 102 Some people were less oncoming. Ibid. x. 222 At Cambridge..he was welcomed effusively, but other places were less oncoming. **1967** N. MARSH *Death at Dolphin* v. 133 I'm not all that oncoming, even here.

oncornavirus (ǫŋkǭ·ɹnăvəiᵊrŭs). *Virology.* [f. ONCO- + *PICO)RNAVIRUS.] = *LEUKOVIRUS.

1970 R. C. NOWINSKI et al. in *Virology* XLII. 1152/1 We..propose the term 'oncornaviruses' to represent the oncogenic RNA viruses (following the style 'picornaviruses' for small RNA viruses). **1974** FRAENKEL-CONRAT & WAGNER *Comprehensive Virology* I. 34 Oncornavirus group. Not yet officially accepted term for leukoviruses or RNA tumor viruses, all equally imperfect names. Medium large (100 nm diameter) enveloped and ether-sensitive particles; studded in regular manner with projections and containing a dense eccentric nucleoid. **1975** *Nature* 3 Apr. 457/2 In several animal species, C-type oncogenic RNA viruses (oncornaviruses) cause malignant tumours of mesodermal origin, such as leukaemia.

Hence **onco·rnaviral** *a.*

1975 *Nature* 3 Apr. 458/1 Cultures were screened for the presence of different cytoplasmic C-type oncornaviral antigens.

oncosine (ǫ·ŋkōsīn). *Min.* Also **onko-, -in.** [ad. G. *onkosin* (F. von Kobell 1834, in *Jrnl. f. prakt. Chem.* II. 296), f. Gr. ὄγκωσις swelling (from its behaviour when heated in a blow-pipe flame): see -INE[5].] An aluminosilicate of potassium, other alkali metals, and magnesium, which is a variety of, or perhaps a mixture containing, muscovite.

1854 J. D. DANA *Syst. Min.* (ed. 4) II. 504 Onkosin... In roundish pieces, having an apple-green color; sometimes grayish or brownish. **1868** *Ibid.* (ed. 5) 480 Oncosin ..occurs in roundish masses imbedded in dolomite with mica, at Passecken near Tamsweg, in Salzburg. **1923** *Mineral. Abstr.* II. 112 Muscovite, the variety onkosine is plentiful in the salbands of the lodes. **1939** *Geol. Förening. Stockholm Förhand.* LX. 622 The vein material in the pollucite, where late muscovite occurs in the somewhat unusual form of deep mauve-coloured, compact, cryptocrystalline masses, mineralogically referred to the subspecies oncosine. **1962** *Mineral. Abstr.* XV. 393/2 Pegmatites with giant crystals of amblygonite, spodumene, and oncosine and aggregates of montebrasite.

oncosphere (ǫ·ŋkosfīᵊɹ). *Zool.* Also **oncho-.** [f. ONCO- + SPHERE *sb.*] An embryonic form of certain tapeworms. Also *attrib.*

1906 P. FALCKE tr. *Braun's Anim. Parasites Man* 202 The embryo enclosed within the embryonal shell, the oncosphere, is of spheroid or ovoid form, and is distinguished by the possession of three pairs of hooklets. **1929** H. A. BAYLIS *Man. Helminthology* 65 It [*sc.* the tapeworm embryo] is frequently referred to as a 'hexacanth embryo' or 'onchosphere'. **1949** *New Biol.* VII. 114 A ripe segment, ready to fall off the end of the worm [*sc.* the pork tapeworm]..contains 30,000–40,000 eggs, each already developed into a little six-hooked embryo (onchosphere) and protected by a shell. **1962** J. D. SMYTH *Introd. Animal Parasitol.* xx. 229 In the taenioids, the embryo develops to the oncosphere stage in the uterus of mature proglottids. **1969** A. M. DUNN *Vet. Helminthology* 112/2 The embryo is called the onchosphere. **1973** T. C. CHENG *Gen. Parasitol.* xiv. 485 The oncosphere..remains passive in the eggshell..until the embryo is ingested by a vertebrate or invertebrate intermediate host.

oncost. Add: **2.** (In general use.) Overhead expenses or costs. Also *attrib.*

1912 J. G. WILLIAMSON *Counting-House & Factory Organisation* 65 Oncost expenditure, such as wages of foremen, labourers, and general works supplies, etc...is dealt with in the same manner. *Ibid.* 71 Establishment Charges or Oncost is every expense in the Profit and Loss Account other than the prime cost of Productive work. **1924** J. STAMP *Stud. Current Probl. Finance & Govt.* 18 Such a tabulation..would enable us to..test its ratios of oncost and various kinds of unit efficiency. **1931** H. E. COLESWORTHY *Pract. Directorship* xii. 120 Overhead or running charges, such as salaries, office expenses, selling expenses, and so on, are termed 'oncost'. *Ibid.*, It is sometimes customary to divide oncost into two classes—'works oncost' and 'office oncost'. **1970** *Money Which?* Mar. 25/2 Another method is to take the figure arrived at by the first method, but add to it what is known as manufacturing 'on-cost'. This is a proportion of your manufacturing overheads, such as the rent, rates, lighting and heating of your factory or workshop. **1973** *Accountant* 23 Mar. 383/2 The problem becomes even more obscure when we look at oncost loadings—obviously the administrative backing is an essential part of running any business.

oncotic (ǫŋkǫ·tik), *a. Physiol.* [f. Onco- + -otic.] Applied to the osmotic pressure exerted by a colloid, esp. plasma protein.

1935 C. J. Wiggers *Physiol.* ii. 802 Starling demonstrated that colloids in a sol state exert a small osmotic pressure, but this varies in uncontrollable fashion owing to the fact that colloidal molecules or aggregates called micellae vary considerably. This has recently been called oncotic pressure. **1977** *Proc. R. Soc. Med.* LXX. 693/1 A reduction in oncotic pressure due to hypoproteinaemia.

ondatra. Add to etym.: [Adopted as a generic name in H. F. Link *Beyträge zur Naturgeschichte* (1795) I. ii. 52]. Substitute for def.: The North American musk-rat, *Ondatra zibethicus*; = Musk-rat 1. (Later example.)

1867 *Amer. Naturalist* I. 400 The Musk-rat, or Ondatra (*Fiber zibethicus*), so extensively diffused over North America.

‖ **Ondes Martenot** (ǫ̃nd mārtǝno). Also **Ondes, Ondes Musicales, Ondium Martenot.** [After F. *ondes musicales*, lit. musical waves; named by and after *Maurice Martenot* (b. 1898), its inventor.] An electronic keyboard musical instrument, capable of producing only one note at a time.

1936 E. S. Bensinger tr. *R. London's Film Music* iv. 177 Before all others, let us recommend the *Ondium Martenot*, the apparatus derived from Theremin's ether-wave music, but with the sound-scale anchored on a keyboard, so that no sound-fluctuations are possible. **1937** *Nature* 6 Feb. 215/1 The author's brief treatment of the comparatively new electro-musical instruments, the *Ondium Martenot*, the trautonium and the Neo-Bechstein piano, is also good. **1940** C. Sachs *Hist. Mus. Instrum.* (1942) 448 The most important monophonic instruments are Maurice Martenot's *Ondes musicales* (1928) [etc.]. **1954** *Grove's Dict. Mus.* (ed. 5) V. 591/1 Martenot..is the inventor of a radio-electric instrument called Ondes Musicales (more usually called Ondes Martenot by composers who score for it), which he first brought out in 1928. *Ibid.* 592/1 In 1937 he organized concerts with a team of eight Ondes. **1957** Manvell & Huntley *Technique Film Music* ii. 37 Ondes Martenot: an electronic instrument of great versatility, first introduced in 1928. It combines the principle of the Thérémin with a keyboard, adding additional devices for vibrato and glissando effects. **1977** *Times Lit. Suppl.* 21 Mar. 277/1 As a music student he earned a living by playing the ondes martenot in pit orchestras.

‖ **ondol** (ǫ·ndǫl). [Korean, ad. Chinese *wēn* hot, warm + *tǔ* funnel, smoke tube.] A form of domestic heating by means of a flue running underfloor from a fire or furnace, commonly used in Korea. Also *attrib.*

1964 R. Rutt *Korean Works & Days* ii. 17 The famous ondol or hot floor, built by making flues under a floor of stone or mud. **1969** *Korean Folklore & Classics* I. 8 He found the newly-laid ondol floor..pierced by a sharp drill in a thousand places... 'This ondol cost me a lot of money.' **1970** *Korea: its People & Culture* vii. 224/1 The people have held on to their traditional, radiant floor heating system, the ondol. **1974** P. M. Bartz *S. Korea* 95/2 About six o'clock the inhabitants get up, re-stoke their ondol furnaces, and start charcoal fires for breakfast.

on-drive, *v.* Add: (Further examples.)

1948 *Morning Post* 7 June 16/4 Holmes attacked the bowlers after a quiet start, twice on-driving Astill for 6. **1956** *Times* 19 May 5/1 In the next over Atkinson beautifully on-drove Johnson for his 13 [illegible].

b. *absol.* or *intr.* To drive the ball to the on. Hence **on-driving** *vbl. sb.*

1930 *Morning Post* 7 Aug. 13/1 Bryan on-drove and hooked most effectively. **1961** *Times* 21 Aug. 3/3 His cutting and on-driving were a delight. **1969** *Times* 13 Feb. 4/3 Pulling and on-driving with remarkable acumen, he dispatched six successive balls for a six and five fours.

o·n-driving, *a.* [On *adv.*] That drives on.

1884 A. de Vere *Poetical Wks.* II. 435 And ever as she sang, the on-driving snow Choked the sweet strain. **1927** *Chambers's Jrnl.* Jan. 39/1 Because there was a check, there arose long on-driving shouts from the huntsmen.

one, *numeral a., pron.*, etc. Add: **B. I. 1. c.** *like one o'clock* (further examples); also, splendidly, excellently; readily, enthusiastically.

1852 Dickens *Bleak Ho.* (1853) xx. 200 He has seen him through the shop-door, sitting in his back premises, sleeping 'like one o'clock'. **1889** E. Dowson *Let.* 31 July (1967) 97 If I can only shake off Cursitor St I will go to the oeuvre like one oclock. **1901** M. Franklin *My Brilliant Career* xix. 161 He had a taste for literature, and we got on together like one o'clock. **1924** Galsworthy *White Monkey* iii. xv. 321 Anything about the meeting, sir? Your speech must read like one o'clock! **1970** V. C. Clinton-Baddeley *No Case for Police* viii. 179 It's going to rain like one o'clock. **1973** *Guardian* 27 Oct. 11/6 Hedgehogs drink beer like one o'clock.

d. With ellipsis of *glass* or *drink*; *one for the road*, a final drink before departure. See also *quick one* (*quick *a.* 25 b). *colloq.*

1925 R. J. B. Sellar *Sporting Yarns* 165 'Did I have one over the regulation number last night?' 'Not at all..you were perfectly all right.' **1925, 1928** [see *eight *sb.* 4]. **1934** Wodehouse *Right Ho, Jeeves* xi. 126, I..put my feet up, sipping the mixture with carefree enjoyment, rather like Cæsar having one in his tent the day he overcame the Nervii. **1937** D. & H. Teilhet *Feather Cloak Murders* i. 20 You run off to bed like a good fellow. You've had one too many. **1943** J. Mercer (*song-title*) One for my baby (and one more for the road). **1948** 'E. Crispin' *Buried for Pleasure* vi. 47 How about one for the road? **1959** G. Greene *Complaisant Lover* i. i. 20 One for the road. I insist. While I call a taxi. **1968** J. Sangster *Touchfeather* xiii. 140 Didn't mean to be crude. Must have had one too many. **1972** J. Blackburn *For Fear of Little Men* xi. 119 'What about giving me one for the road, my dear.' He gulped down the remains of the sherry. **1976** *South Notts Echo* 16 Dec. 5/4 If you are driving do not have one for the road.

† **e.** Ellipt. for 'one horse' (to pull a carriage, etc.). Cf. Four *a.* 2 c. *Obs.*

1777 P. Thicknesse *Year's Journey* II. lv. 185 If you can find me out a sensible valetudinarian..who will travel as we do..in a landau and one. **1785** Cowper *Task* i. 5 Two citizens who take the air Close pack'd and smiling in a chaise and one.

f. A one-pound note or a one-dollar bill.

1846 *Illinois State Register* (Springfield) 2 Oct. 2/6 Independent of the older issues, and such as are described in the Detectors, Ones, on the Banks of 'Broome county' and 'Whitestown'..have made their appearance. **1948** *Savings & Loan News* Mar. 18/2 My billfold had a $10 bill in it, not ten ones. **1966** O. Norton *School of Liars* iii. 55 'Do you want this in ones, Mrs. Hetherington?' 'In ten-shillingses, dear boy.' **1967** 'A. Gilbert' *Visitor* iii. 45, I counted the notes, which took a ridiculously long time as they were mostly in ones. **1970** M. Kenyon *100,000 Welcomes* iii. 18 He counted out seven one-pound notes and a five..and selected three ones. **1976** J. Wainwright *Walther P. 38* 24 Drysdale started with five fives, followed by five ones, then he paused...he counted out five more singles.

g. One point or position on a scale, order, or the like; esp. in phr. *go up* (or *down*) *one*, expressing commendation (or disapprobation). *colloq.*

1909 J. R. Ware *Passing Eng.* 142/2 *Go down one*, to be vanquished. *Ibid.* 143/2 *Go up one*, applause. Derived from the school class—the scholar going one nearer the top as he goes up one. **1967** E. Lemarchand *Death of Old Girl* v. 59 'I was thinking maybe..the blood on that made the mark.' 'So was I,' said Pollard. 'Go up one.'

4. b. *murder one*: see *murder *sb.* 5.

5. b. *one in* (a specified number): designating a gradient in which the height increases or decreases by one foot (or other measure) vertically for the specified number of feet, etc., horizontally; also *ellipt.* as *sb.*

1830 M. Edgeworth *Let.* 18 Oct. (1971) 419 The inclined plane the rise of which was one in 36. **1869,** etc. [see *in *prep.* 4]. **1910** Kipling *Divers. Creatures* (1917) 322 It was all of a one in three gradient. **1968** N. Tranter *Cable from Kabul* iii. 37 Down at the foot of a one-in-three hill, I found myself in some sort of village. **1971** G. Household *Doom's Caravan* ii. 44 Its original builders had no objection to a slope of one in four. **1976** J. Wainwright *Bastard* i. 11, I slither and skid the car up the one-in-six.

6. d. See also Number *sb.* 5 b.

II. 7. c. Used as a more emphatic substitute for the indefinite article: (*a*) with adjs. in sense 'a very ——', 'an extremely ——'; (*b*) with sbs., esp. *hell* (see *hell *sb.* 4 d, *helluva). *colloq.*

1828 *Punch & Judy* i. i. 77 Toby, you're one nasty cross dog: get away with you! **1911** J. London *Let.* 7 Apr. (1966) 343 Let me tell you that you have given me one hell of a time. **1920** [see *hell *sb.* 4 d]. **1925** T. Dreiser *Amer. Trag.* (1926) i. xii. 82 He went out in the kitchen and blacked up an' put on a waiter's apron and coat and then comes back and serves us. That's one funny boy. **1934** [see *helluva]. **1948** E. Pound *Pisan Cantos* (1949) lxxviii. 66 Steele that is one awful name. **1967** 'T. Wells' *Dead by Light of Moon* (1968) xi. 111, I wondered what Mai Farmer was doing. She was one striking girl. **1967** [see *helluva]. **1972** A. Price *Col. Butler's Wolf* xii. 132 The last two, three weeks he was one worried young man. **1973** J. Di Mona *Last Man at Arlington* (1974) ii. xvi. 100 'Tell everyone I'm not Cuban,' said Medwick, hoping to get a rise out of the driver. But none came. This was one serious boy. **1976** *Publishers Weekly* 9 Feb. 85 (Advt.), Come spring, this [forthcoming book] is going to be One Hot Number.

IV. 15. b. *one of those*: a homosexual; *one of us*: a member of our group; *spec.* (*a*) a harlot (*obs.*); (*b*) a homosexual. *colloq.*; *one of these days*: see Day *sb.* 7 b; *one of those days*: see *day *sb.* 7 b.

1785 Grose *Dict. Vulgar T.*, *One of us*, one of my cousins, a woman of the town, a harlot. **1915** Conrad *Victory* i. ii. 9 Morrison was 'one of us'. He was owner and master of the *Capricorn*, trading brig, and was understood to be doing well with her. **1933** [see *nancy[3]]. **1956** [see *camp *a.* and *sb.*[5]]. **1961** Partridge *Dict. Slang* Suppl. 1207/1 *One of us, he's*, he is a homosexual. **1968** J. R. Ackerley *My Father & Myself* xvi. 185, I divined that he was homosexual, or as we put it, 'one of us'. **1976** *Times* 27 May 16/4 It would go a long way towards helping..to understand..if others would stop saying 'New Commonwealth' when they mean something like 'coffee-coloured' and 'Old Commonwealth' when they mean..'One of us'. **1977** *Gay News* 24 Mar. 18/2 Her husband..probably fits none of the stereotypes whereby she would normally identify 'one of those'.

c. *a one*: a person who is remarkable, extraordinary, outrageous, impudent, etc.; esp. in phr. *you are a one*; (*a*) *one for*: a person who likes, admires, practises, supports, etc. (something) to an outstanding degree; a devotee, champion, or admirer of (something); (*a*) *one to*: the sort of person who would (do a particular thing). *colloq.*

1880 C. M. Yonge *Bye-Words* 303 Tittering, and now and then, 'O Miss Annie, don't, pray!' 'O Miss Annie, you are a one!' **1888** —— *Our New Mistress* i. 3 Her daughters ..all married, except Lady Mary, who was always such a one for schools and poor people. **1894** S. Baring-Gould *Queen of Love* II. vi. 59, I am not one to fly in the face of Providence. **1906** E. Dyson *Fact'ry 'Ands* iii. 29 'Oh, Mr. Ellis, you are a one!' she said. **1927,** etc. [see *great *a.* 16 a]. **1932** N. Royde-Smith *Incredible Tale* 91 She was a one for football. **1934** N. Marsh *Man lay Dead* vii. 126 'The left-hand print on the stair knob is Mr. Wilde's,' said Bailey. 'Is it?' answered Alleyn without enthusiasm. 'Aren't you a one?' **1935** G. Heyer *Death in Stocks* iii. 22 Constable Dickenson had warned the Inspector that she was not one to talk. **1948** 'G. Orwell' *Let.* 10 July in *Coll. Ess.* (1968) IV. 438 Farm life seems to suit him, though I am pretty sure he is one for machines rather than animals. **1966** J. B. Priestley *Salt is Leaving* viii. 96 You're a bit of a one, aren't you, Dr Salt? **1973** J. Thomson *Death Cap* vi. 86 He's never been one for the women. I think he's a bit afraid of them.

17. c. *one thing*: something acceptable or satisfactory, contrasted with *another* (*thing*) that is unacceptable or unsatisfactory.

a **1678** H. Scougal *Life of God* (1726) 392, I do not condemn all cheerfulness and freedom, nor the innocent exercises of wit: but it is one thing to make use of these now and then when they come in our way, and another to search and haunt after them. **1735** Berkeley *Defence Free-Thinking in Math.* xxxvii. 44 It is one thing when a Doctrine is placed in various lights: and another, when the principles and notions are shifted. **1828** Scott *Chron. Canongate* 2nd Ser. I. viii. 244 It is one thing to employ the revenues of the Church..in the suitable and dutiful reception of your royal Majesty, and another to have it wrenched from us by the hands of rude and violent men. **1904** H. James *Golden Bowl* (1905) v. 66 It was one thing to have met the girl casually at Mrs. Assingham's and another to arrange with her thus for a morning practically as private as their old mornings in Rome and practically not less intimate.

d. Ellipt. for 'one or the other'. *U.S. dial.*

1895 *Dialect Notes* I. 373 *One* seems to be superfluous or else 'or the other' is omitted. 'I will see you or send word, one.' **1926** E. M. Roberts *Time of Man* (1927) vii. 257, I met a parcel of travelers that owned a bear could read or tell fortunes—one, I forget which. *Ibid.* viii. 298 It was the road overseer's fault.., or the magistrate's, one. **1937** *Scribner's Mag.* Apr. 22/2 He's making it [*sc.* liquor] on my farm or your farm, one. **1938** M. K. Rawlings *Yearling* xv. 169 Now do things go wrong again, you or Buck, one ,ride back for me. So long.

V. 21. (Further examples specially meaning the speaker himself.)

1956 R. Henriques *Red over Green* iii. 60 He meant nothing... One can't even remember his face. **1959** E. H. Clements *High Tension* ii. 19 'Do you often have your fan-mail in person?'..'Not often. One isn't in the telephone book'.

23. b. *spec.* A story or anecdote; a joke; a lie. *colloq.*

1813, etc. [see *good *a.* 1 g]. **1925** Wodehouse *Carry On, Jeeves!* x. 254 Story? Story?.. I wonder if you've heard the one about the stockbroker and the chorus-girl? **1926** D. L. Sayers *Clouds of Witness* xiii. 240 Mr. Parker endured five stories with commendable patience, and then suddenly broke down. 'Hurray!' said Wimsey... 'I'll spare you the really outrageous one about the young housewife and the traveller in bicycle-pumps.' **1931** J. Betjeman *Mount Zion* 22 Each learning how to be a sinner And tell 'a good one' after dinner. **1961** 'F. O'Brien' *Hard Life* x. 71, I will tell you a funny one, Father, Mr Collopy said. **1967** Wodehouse *Company for Henry* x. 175 The low comedians of his musical comedy days who had called him 'laddie' and begged him to stop them if he had heard this one. **1977** *Listener* 24 Nov. 674/2 'Have you heard the one about the Queen Mother?' We had not heard it, and it was very funny.

VIII. Phrases.

29. c. *all one*: see also All *adv.* 5 b in Dict. and Suppl.

29*. With following adverb.

a. one down: one point behind one's opponent in a game; inferior in one respect; disadvantaged; also (with hyphen) *attrib.* or as *adj.* Hence **one-downmanship**, the art or practice of being 'one down'; **one-downness**, the fact or state of being 'one down'. Cf. sense 29* c below.

1907 [see *down *adv.* 14 b]. **1952** S. Potter *One-Upmanship* ii. ii. 32 To increase the one-downness, bring in the washing-the-hands gambit immediately after touching hands with Patient. **1961** *Times* 8 Mar. 17 (*heading*) Handy guide to art of onedownmanship. *Ibid.* 22 Mar. 16/3 It is the Negroes who are educated, who 'talk posh', who go to university; the native English who

are one-down, with less money and less culture. **1964** M. Argyle *Psychol. & Social Probl.* iii. 36 Stephen Potter has given an amusing list of techniques for making others feel 'one-down', together with counterploys for dealing with such methods when used by others. **1967** *Punch* 4 Oct. 514/3 If he were to check his facts would he not find that on the contrary the majority of Fleet Street was indulging in a form of one-down-manship towards the British public? **1976** N. Postman *Crazy Talk* 44 He will naturally be one-down in the situation, a 'child' to the government agent's 'adult'.

b. one off: a single example of a manufactured product; something not repeated; a prototype. Freq. (with hyphen) *attrib.* or as *adj.* Also *transf.* and *fig.* Cf. *off *adv.* 13, *once-off* adj. s.v. *once *adv.* B. 7 b.

1934 *Proc. Inst. Brit. Foundrymen* XXVI. 552 A splendid one-off pattern can be swept up in very little time. **1935** *Jrnl. R. Aeronaut. Soc.* XXXIX. 41 One off per machine does not give us much opportunity for reducing production costs. **1947** *Ibid.* LI. 308/1 With the lofting technique it is possible to cut down the time required to produce a prototype aircraft for..it is possible to reproduce full-scale layouts directly on to the material to be worked..thus cutting out what was originally the factor which absorbed the most production time in the freehand manufacture of 'one offs'. **1954** *Archit. Rev.* CXVI. 411/2 Hills built the first part of Cheshunt as a 'one off' job, with no guarantees of further business, though of course it was intended to be the first of a line. **1955** *Ibid.* CXVII. 226/2 None of the motor-cars illustrated is a standardized mass-produced model; all are expensive, specialized, handicraft one-offs which can justly be compared to the Parthenon because, like it, they are unique works of handmade art. **1958** *Listener* 25 Sept. 458/2 Both the estates of the speculative 'rush' builders and the architectural one-offs are unable to keep pace with the demand [for new houses]. **1961** *Times* 3 Oct. (Computer Suppl.) p. v/3 The centres are..even able to do a 'one-off' job, such as eliminating a production bottleneck, very cheaply. **1965** R. B. Oram *Cargo Handling* iv. 70 Tailor made, or 'one-off', machines, may give great satisfaction. **1968** *Sunday Times* 29 Sept. 25 Jenkins has already made a crude stab at a wealth tax with his special charge on investment incomes... But this was a one-off effort. **1970** *Times* 28 Mar. 21 All these relationships involve money and are on a continuing basis rather than a one-off purchase. **1973** *Daily Tel.* 22 Oct. 12/4 When Barry Took's *Grub Street* (BBC-2) was screened as a one-off..I rashly predicted that it could make a series. **1974** F. Warner *Meeting Ends* ii. i. 35 But we find it much harder to shake a man off afterwards, and anyway, I don't like those 'one off' dates. I need companionship, an outing, warmth. **1976** *Scottish Rev.* Spring 33 For the most part they could only produce an endless stream of one-off building prototypes. **1977** *Hot Car* Oct. 97/1 There seems to be a good deal of misunderstanding about the way the Type Approval Regulations apply to 'one-offs' or cars built by private individuals.

c. one up: scoring one point more than an opponent; ahead of another person; (fig.) maintaining a psychological advantage; also (with hyphen) *attrib.* or as *adj.* Hence **one-upmanship**, the art or practice of being 'one up'; so **one-up** *v. trans.*, to do better than (someone); **one-upman**, an exponent of one-upmanship; **one-upness, -uppance**, the fact or state of being 'one up'.

1919 [see Up *adv.*[1] 13 e]. **1924** Wodehouse *Leave it to Psmith* i. 30 Which would make her pretty chirpy, as well as putting you one up. **1929** J. B. Priestley *Good Companions* ii. vii. 449 He can give old Omar himself points in not believing in anything, for he has cut out the book of verse, most of the loaf, and the Houri stuff, and just sticks to the jug, though he has added a clay pipe and is one up on Omar there. **1952** S. Potter (*title*) One-upmanship. *Ibid.* i. ii. 26 The establishment of one-up relations between doctor and doctor and doctor and patient and vice versa. *Ibid.* v. 64 The basic gambit is of course the achievement of the state of one-upness on the rest of the public. **1957** *Economist* 26 Oct. 295/1 This piece of applied relativity..may go down in the annals of international one-upmanship as the sputnik ploy. **1959** N. N. Holland *First Mod. Comedies* 38 This was a perhaps pardonable attempt to retain 'one-upness' in the large eyes of a rather nasty little genius. **1959** *Times Lit. Suppl.* 6 Nov. 650/4 They are one-upmen, seen from the receiving end, and they give Mr. Gibb endless opportunities for recording sillier aspects of the contemporary social scene. **1960** *News Chron.* 14 Apr. 3/1 Will Granada deny there was an element of one-upness in its satisfaction? **1961** S. Price *Just for Record* v. 37 Stephen Potter was a square in nappies compared to these one-up graduates. **1963** *Canada Month* Mar. 10/1 John Wintermeyer..one-upped the socialists by endorsing the Saskatchewan plan. **1964** 'C. E. Maine' *Never let Up* xvii. 172 It's a kind of one-upmanship. You thought you were smart, but he had to prove that he was even smarter. **1966** *Listener* 27 Oct. 622/1 Virginia Woolf can still show herself to be one up, in her literary judgements, on most current criticism. **1967** *Maclean's Mag.* Oct. 46 Another trap the psychiatrist must strive to avoid is the mistake of being one-up. The psychiatrist is in the perfect position to be the one-up man. **1969** D. S. Davis *Where Dark Streets Go* (1970) xviii. 162 You one-upped us there, Father. We came dead-end in a housing development. **1970** 'Jenner' & Segal *Men & Marriage* ii. 45 Marrying a doctor still gives a girl a bit of one-uppance amongst the neighbours. **1973** J. Wainwright *Touch of Malice* 87 Smithson's one-up-man-ship ploy of keeping a senior police officer waiting. **1975** *Times* 14 Mar. 14/5 There are one-upmanship entries. A power cruiser 'built for royalty' is offered. **1976** *Listener* 28 Oct. 544/4 The objects of human vituperation..seem to be pretty well limited to people who are one up on us, and other people's pleasures. **1977** *Time* 24 Jan. 37/1 His sweet, sporting spirit as he sits trying to absorb his defeat while graciously applauding a trickster's win is something with which any weekend athlete who has been one-upped by an allegedly friendly opponent can identify.

29.** Misc. phrases.

a. *one and the same*: used as a more emphatic form of 'the same'. Cf. L. *unus et idem*.

1869 *Bradshaw's Railway Manual* XXI. 365 This modification has..the effect of comprising in one and the same network the two lines from Paris to Lyons. **1942** H. L. Mencken *Newspaper Days* (1941) xvi. 253 His father had been, at one and the same time, a Confederate general, a French nobleman, and a graduate of both Oxford and Cambridge. **1960** C. P. Snow *Affair* v. xxxix. 364 You'd obviously got to raise the cast about Nightingale and give them an escape-route at one and the same time. **1973** D. Aaron *Unwritten War* iv. 21. 167 Abolitionism or Black Republicanism, to the South Carolinian, one and the same thing. **1976** G. Butler *Vesey Inheritance* iv. 117, I wonder..whether the King and Mr Koenig could be one and the same person?

b. *one man, one vote*: a slogan advocating that every adult man (or adult person) should have a vote; also formerly, that each voter should have only one vote; also *attrib.*

1884 A. [illegible] *Hist. of Reform* ii. 19 'One man, one vote', a cry which may have had a novel sound to some in 1883 was one of Cartwright's political principles. **1889** W. E. Gladstone in *Times* 13 June 7/2 The important measure which is specially prominent in Wales..is the measure—one man, one vote. **1891** [see sense 1 a in Dict.]. **1907** H. Lawson in Murdoch & Drake-Brockman *Austral. Short Stories* (1951) 73 The One-man One-Vote Bill was passed. **1964** *Punch* 15 July 74/3 To ensure that one-man one-vote democracy is swiftly introduced. **1977** 'G. Black' *Time for Pirates* ii. 32 The government..had declared martial law, suspending the constitution..So much for one man, one vote,' Russell said. **1975** D. Bagley *Snow Tiger* xii. 115 Not so democratic as to be a one man, one vote system.

c. *one hand for oneself and one for the ship*: a nautical proverb referring to the practice of holding on to a rope, etc., with one hand while working with the other hand; also in similar phrases (see quots.).

1812 *Port Folio* (Philadelphia) 1812 VII. 130 Always keep one hand for the owners, and one for yourself. **1902** B. Lubbock *Round the Horn* 38 The old rule on a yard is, 'One hand for yourself and one for the ship'. **1906** R. Clements *Gipsy of Horn* iii. 60 One hand for yourself and one for the owners. **1938** F. A. Worsley *First Voy. in Square-Rigged Ship* 119 One hand for the Queen and one for yourself. **1968** L. Morton *Long Wake* ii. 60, I did not know then the old adage 'one hand for oneself and one hand for the company.'

d. *one and only*: one's sweetheart; one's only child or love; also (with hyphens) as *attrib. phr.*, unique, unrepeatable.

1906 E. Dyson *Fact'ry 'Ands* ii. 4 She's er little bokker..I'm 'er one 'n' only. **1932** D. Carr *Mad Hatter Mystery* iv. 64 He'd met some girl at a dance who was the absolute One and Only. **1963** *Times* 13 May 11/3 Artur Schnabel thought that such a one-and-only performance was obtainable. **1966** *Harper's Bazaar* Sept. 64 A coat so versatile it could be the treasured one-and-only in your life. **1967** J. Hamilton *Man with Brown Paper Face* xi. 85 Daddy wasn't too happy about his one-and-only's choice of companion. **1967** J. Wainwright *Worms must Wait* xxviii. 201 He had the truculence ready for what he knew was going to be a one-and-only chance. **1975** J. McClure *Snake* iii. 259 [illegible] **1977** J. van de Wetering *Japanese Corpse* (1978) xvii. 158 [illegible] Kikuji Nagai had been her one and only.

e. *one for the* (*end*) *book*: a notable, extraordinary, or incredible event, action, saying, etc. *U.S. colloq.*

1923 H. C. Witwer *Fighting Blood* (1923) 170 Gents, this was one for the book! **1946** *Amer. Speech* XXI. 69/1 When a friend approaches with an anecdote which is strange or incredible, he often prefaces it with the remark, 'Here's one for the book'. *Ibid.*, At racetracks where pari-mutuel betting machines are not used..it was customary for book-makers to line up in a designated area... If a bettor asked unusually high odds, the bookie might comment, 'Here's one for the book', implying that no one but a green newcomer..would accept those odds. **1955** *Publ. Amer. Dial. Soc.* xxiv. 175 [illegible] one for the end book, or a story that is hard to believe.

f. (*just*) *one of those things*: something inevitable or inexplicable; a fact or happening that one cannot do anything about. *colloq.*

1934 J. O'Hara *Appointment in Samarra* ii. 25 No, it was just one of those things. **1935** C. Porter (*song-title*) Just one of those things. **1939** [illegible] Comedian Durante: 'Aw, it's just one of those things.' [illegible] 'It's just one of those things.' **1941** C. Morgan *Empty Room* i. 46 'What is it, Carey?' She smiled. 'Nothing. One of those things.' **1951** 'J. Wyndham' *Day of Triffids* ii. 28 My inability to make any column of figures reach the same total twice caused me to be something of a mystery as well as a disappointment to him [*sc.* my father]. Still, there it was: just one of those things. **1955** A. Huxley *Let.* 16 Dec. (1969) 778 Her daughter is going to have a baby—husband twenty-one and still at college, daughter supporting the household for the moment. Which is one of those things. **1971** *Daily Tel.* 19 Nov. 3/1, I know Mr Butler is a bit choked about it, but it's just one of those things. **1974** M. Babson *Stalking Lamb* xviii. 136 The price was too high..to be shrugged off as 'just one of those things'.

g. *one-of-a-kind* attrib. phr., (*a*) of only one kind; (*b*) unique.

1961 *Times* 25 Apr. 4/2 The one-of-a-kind series for racing catamarans organized last year. **1962** *New Yorker* 1 June 72 Among the one-of-a-kind mannerly materials are Paisley cotton prints. **1973** *Publishers Weekly* 23 July 66/3 A one-of-a-kind book that merits a place on the political science shelf. **1975** *New Yorker* 21 Apr. 127/3 *Children of Paradise* (1945)—A one-of-a-kind film. **1977** *Rolling Stone* 24 Mar. 48/4 Fleetwood Mac had this one-of-a-kind stage charm. They were gregarious, charming and cheeky onstage. *Very* cheeky.

30. d. *in one*. (*g*) At one stroke or attempt; esp. *to get it in one*: to succeed at the first attempt. Cf. sense (*e*) in Dict., and *hole in one* s.v. *hole *sb.* 4 a. *colloq.*

1938 J. Parris *St. Michael comes to Shepherd's Bush* 22 As a matter of fact, that's just what I am. You've got there in one. **1949** 'A. Bridge' *Frontier Passage* vi. 91 'In fact, our old friend the Hidden Hand in Biarritz runs the sabotage as well as the front—that the idea?' [illegible] enquired. 'Got it in one!' **1972** W. Garner *Ditto, Brother Rat!* xv. 106 Got it in one, old son. **1975** 'C. Aird' *Slight Mourning* iii. 36 'What we are checking on is whether someone tried to kill him...' 'Got it in one, Sloan.'

IX. 32. a. *one-child, -class, -clause* (examples), *-colour, -crop, -culture, -day* (later examples), *-deck, -digit, -drink, -electron, -family, -level, -light, -line, -member, -parent, -particle, -party, -person, -reel, -room, -sex, -star, -step, -storey* (also *-story*) (earlier, later, and *fig.* examples), *-string, -tap, -term, -volume, -word.*

1905 *Daily Chron.* 18 Nov. 6/3 It is desired to secure such a reform in the law as will bring one-child cases within the sphere of inspection. [illegible] tion can begin with a tautology, as in the following one-line derivation. **1965** Hughes & Londey *Elem. Formal Logic* xxix. 218 We shall usually write one-line proofs in this abbreviated form. **1884** E. W. Hamilton *Diary* 29 Oct. (1972) II. 720 The Tory scheme leans to the one-member principle. **1924** O. Jespersen *Philos. Gram.* 306 An old-fashioned grammarian will feel a certain repugnance to this theory of one-member sentences. **1963** J. Lyons *Structural Semantics* ii. 22 Some of the distributional classes will, of course, be one-member classes; but the majority will not. **1967** D. H. Monro *Empiricism & Ethics* xvi. 201 It is not irrational to treat an individual as a one-member class. **1969** *Times* 7 Nov. 15/7 A

committee to consider the problems of one-parent families in society. **1974** *Evening News* (Edinburgh) 12 Apr. 7/5 An Easter holiday play scheme for children from some of Edinburgh's one-parent families ended today on the slopes of Arthur's Seat. **1976** *Times* 21 May 1/6 That benefit..is paid to fewer than half the one-parent families in Britain. **1955** W. Pauli *Niels Bohr* 32 This method is essentially based on the assumption that the theory for free particles (without interactions) holds for the so-called one particle states. **1937** H. Tingsten *Political Behavior* v. 216 In certain American so-called one party states one can hardly speak of an election campaign. **1950** 'G. Orwell' *Shooting an Elephant* 156 The appearance of one-party régimes based on police terrorism, faked plebiscites, etc. **1964** T. B. Bottomore *Elites & Society* v. 95 The possible or probable concomitants of this kind of one-party rule, dictatorship and loss of personal liberty, persecution and widespread suffering. **1971** *Guardian* 11 Dec. 10/4 India..was a benign one-party state. **1975** *Times* 11 Apr. 6/5 Independent observers had little doubt that this would be the beginning of a one-party system in Portugal. **1956** J. M. Mogey *Family & Neighbourhood* 14 One-person households. **1966** J. Tunstall *Old & Alone* xiv. 281 'Under-occupation' defined as one-person household in 4+ rooms or two-person household in 5+ rooms. **1977** G. Scott *Hot Pursuit* ii. 16 It's a one-person flat. **1920** I. P. Gore in *Stage Year Bk.* 53 Such tit-bits as..a one-reel comedy founded on the rollicking antics of a malignant tumour. **1961** Getlein & Gardiner *Movies, Morals & Art* iv. 51 *The Great Train Robbery* is a one-reel film. **1897** One-room [in Dict.]. **1934** *Archit. Rev.* LXXV. 41 (*heading*) The one-room flat. **1972** C. Weston *Poor, Poor Ophelia* (1973) vi. 30 A hippie joint, you think? Anyhow, a bunch of one-room pads. **1949** M. Mead *Male & Female* xviii. 368 A one-sex world would be an imperfect world. **1966** P. Willmott *Adolescent Boys E. London* vii. 128 Younger boys more often belong to one-sex clubs. **1908** *Daily Chron.* 4 Nov. 3/3 In the meadows we did roam; And in the one-star night returned Together home. **1961** *Guardian* 24 Mar. 21/4 One-star restaurants, rather slightingly dismissed by M. Michelin as 'a good restaurant for its class'. **1975** tr. *Melchior's Sleeper Agent* (1976) iii. 168 The French general..was scheduled to tap four officers for the Legion of Honor, two one-star generals and two bird colonels. **1977** 'R. Rostand' *Killing in Rome* i. 4 A small one-star hotel. **1964** D. B. Fry in D. Abercrombie et al. *Daniel Jones* 64 He answered correctly all the items involving a one-step difference between A and B. **1964** *English Studies* XLV. 383 The difference between a normal one-step process, beginning with Scandinavian forms and limited to names, and the multi-step process. **1833** B. Silliman *Man. Sugar Cane* 64 The bagasse houses at Demerara are high one story buildings. **1858** [see One-horse *a.* 2]. **1970** J. Hansen *Fadeout* (1972) i. 3 The house was one-story, rambling, sided with cedar shakes. **1938** *Oxf. Compan. Mus.* 591/1 The one-string principle is also applied to the *Tromba Marina*. **1970** *Islander* (Victoria, B.C.) 8 Feb. 10/1 Occasionally from some nearby window you caught the strains of the [Chinese] one-string fiddle blending with the sing-song street cries of the vendors of lottery tickets. **1976** Ld. Home *Way Wind Blows* vii. 112 Some three hundred of them were gathered round, and he began to teach them, accompanied by a one-string banjo. **1952** A. Cohen *Phonemes of Eng.* 29 These two sounds (one-tap and fricative *r*) are in no way opposed. **1845** *Congress. Globe* 28th Congress 2 Sess. 122/2 The North.. never had had any but one-term presidents, democratic or federal. **1961** Y. Olsson *On Syntax Eng. Verb* ii. 34 A two-term sub-system commutable with the one-term sub-system. **1966** *Philos. Rev.* LXXV. 406 Plato confuses relations with one-term predicates. **1862** Mrs. Gaskell *Let.* 30 Sept. (1966) 698, I am going to publish a one-volume story in 'All the Year Round', where..it will occupy from ten to twelve numbers. **1961** R. B. Long *Sentence & its Parts* xix. 414 There is no possibility of doing it justice in a one-volume grammar. **1977** *Listener* 17 Nov. 651/2 The *Times* one-volume edition [of the Pentagon Papers]. **1924** R. M. Ogden tr. *Koffka's Growth of Mind* v. 320 Single words have been spoken as one-word sentences. **1956** 'H. MacDiarmid' *Stony Limits* 141 The great one-word metaphors of the Enneads. **1960** W. V. Quine *Word & Object* i. 10 'Red' as a one-word sentence usually needs a question for its elicitation. **1963** F. T. Visser *Hist. Syntax Eng. Lang.* I. iv. 603 When a noun is the object of both a merged verb and a one-word verb.

d. *one-handled, -leafed, -leaved, -membered, -minded, -pointed* (so *-pointedness*), *-seeded* (earlier and later examples), *-storied* (earlier example), *-volumed, -winged, -worded.*

1922 Joyce *Ulysses* 146 Settle down on their striped petticoats, peering up at the statue of the onehandled adulterer. *Ibid.* 564 Steel shark stone onehandled Nelson. **1952** A. G. L. Hellyer *Sanders' Encycl. Gardening* (ed. 22) 463 *Sophronitis*... Pseudo-bulbs usually small, one-leafed, stout, with one or few terminal flowers. **1875** *Amer. Naturalist* IX. 17 The singular one-leaved ash, *Fraxinus anomala.* **1946** Dylan Thomas *Deaths & Entrances* 29 Under the one leaved trees ran a scarecrow of snow. **1972** *Hilliers' Man. Trees & Shrubs* 507 [pinus cembroides] monophylla... 'One-leaved Nut Pine'. An unusual variety in which the..leaves occur singly. **1884** E. W. Hamilton *Diary* 25 July (1972) II. 659 Mr. G... rather favours one-membered constituencies. **1877** A. Dobson *Proverbs in Porcelain* 99 We, bound with him in common care, One-minded, celibate. **1941** L. MacNeice *Poetry of Yeats* x. 218 If Lawrence is..an eclectic, he is.. a one-minded one. **1811** One-pointed [see *blue grass 2]. **1958** *Listener* 11 Sept. 374/2 They receive an incredibly tough kind of training, which..produces 75 per cent. completely one-pointed fanatics. **1960** J. Hewitt *Yoga* xi. 153 If the mind takes one thought and holds it, one-pointed and still, time is erased, it ceases—psychologically—to exist. **1923** *Contemp. Rev.* Feb. 223 He has an innate tendency to 'onepointedness'—as it is sometimes called—to concentration on unity. **1941** A. Huxley *Grey Eminence* v. 120 Complete consistency comes only with complete one-pointedness, complete absorption in ultimate reality. **1960** J. Hewitt *Yoga* ix. 135 In..another method to achieve withdrawal and onepointedness, the meditator imagines that he has a diamond in each ear, [etc.]. **1796** W. Withering *Brit. Plants* (ed. 3) I. 69 Monosperma, one-seeded. **1846** D. J. Browne *Trees Amer.* 215 *Gleditschia monosperma*, the One-seeded Gleditschia. **1964** E. J. H. Corner *Life of Plants* xii. 208 Some normally one-seeded fruit, as acorns, avocado, palm fruits, or grass 'seed', will show considerable difference in the size of the true seed. **1821** W. Wirt *Let.* 29 Aug. in J. P. Kennedy *Mem. W. Wirt* (1849) II. 132 It is a small, red, hip-roofed, one-storied old house. **1880** Geo. Eliot *Let.* 19 Apr. (1956) VII. 261, I prefer Muxon's one-volumed edition of Wordsworth to any selection. **1909** *Daily Chron.* 26 July 1/1 Away it went over the cliff, that monstrous one-winged bird. **1957** One-worded [see *many-worded* adj. s.v. *many *a.* 6 c].

e. *one-rater, -roomer.*

1896 *Rudder* VII. 245/1 Next year will see a one-rater craze; or, correctly speaking, an epidemic of 20-foot racing-length yachts. **1924** D. H. Lawrence *Let.* 16 May (1962) II. 789 There's a two-room cabin where Mabel can come when she likes, and a one-roomer for Brett.

33. one-bar *a.*, of an electric fire: having only one heating element; **one-base hit** *Baseball*, a hit that enables the batter to reach the first base; also *one-baser*; **one-book** *a.*, of an author: having written only one book, or only one good book; **one-catch-all** *dial.*, a children's outdoor chasing game; **one-cross** *a.*, denoting a type of tin-plate (see quot. 1890); **one-design** *a. Naut.*, designating a yacht built from a standard design, or a class of such ships which are almost identical; also *absol.* as *sb.*; hence **one-designer**, such a yacht; **one-dimensional** *a.*, having, or pertaining to, a single dimension; hence *one-dimensionality*; **one-directional** *a.*, having, or pertaining to, a single direction; **one-egg** *a.*, (*a*) characterized by a single egg; (*b*) = *monozygotic *a.*; **one-for-one** *a.*, denoting a situation, arrangement, etc., in which one thing corresponds to, or is issued or exchanged for, each of a set of things; **one-inch** *a.*, measuring, or done at a distance of, one inch; *spec.* of a map: having a scale of one inch to the mile; also *ellipt.* as *sb.*; **one-liner** (chiefly *U.S.*), (*a*) a headline consisting of only one line of print; (*b*) a very short joke or witty remark; **one-lunger**, (*a*) a person with only one lung; (*b*) *slang*, an engine with a single cylinder; a vehicle or boat driven by such an engine; also *attrib.*; **one-man** *a.* (earlier and later examples); also in *comb.*; (*b*) loving, obedient, or attached to one man only; **one-man band,** a man who plays several musical instruments simultaneously; also *transf.* and *fig.*; **one-man show,** a show, entertainment etc., consisting of, or done by, one man only; *spec.* an exhibition of the work of one artist; **one–many** *a.*, applied to a correspondence or relation such that each member of one set is associated with or related to two or more members of a second set; **one-night** *a.*, lasting, residing, or used for a single night; **one-nighter,** (*a*) a person who stays at a place for a single night; (*b*) orig. *U.S.* = *one-night stand*; **one-night stand** [Stand *sb.*[1] 2 e] orig. *U.S.*, a single performance of a play, show, or the like at a particular place; esp. a performance given by a touring company, band, etc.; a town, theatre, etc., where such a performance or performances take place; also *transf.* and *fig.*, *spec.* a casual sexual encounter; **one-old-cat** *U.S.*, a form of baseball in which a batter runs to one base and home again, remaining as batter until the player who puts him out succeeds him; **one-one** *a.* = *one-to-one adj. (see below); **one-one,** at Cambridge University, a degree in the first section of the first class; **one-on-one** *a.* (*U.S. slang*), designating or pertaining to a situation in which two opponents or the like come into conflict; **one-over-one** *a.*, in Bridge, denoting a bid of one in a suit, made in response to a preceding bid of one in a suit; also *ellipt.* as *sb.*; **one-piece** *a.*, made or designed in a single piece; consisting of a single piece; esp. of clothing: comprising a single garment; **one-pip** *Mil. slang* (see quot. 1919); also *one-pipper*; **one-place** *a. Logic*, of an assertion, etc., in which only one thing is postulated or involved; **one-plus-one** *a. Computers*, applied to (the use of) an instruction that contains the address of an operand and that of the next instruction to be performed; **one-pole** *a.*, (*a*) (see quot. 1892); (*b*) consisting of a single pole; (*c*) (see quot. 1940); **one-pounder,** (*a*) a gun that fires one-pound shells; (*b*) a one-pound note; **one-reeler,** a film lasting for one reel, usu. for ten minutes or less; **one-ring circus,** a small circus containing only one ring; also *transf.* and *fig.*; **one-stop** *a.* (orig. *U.S.*), denoting a shop or the like that can supply all a customer's needs within a particular range of goods or services; **one-stress** *a.*, of a line of Old English verse, having only one stress; **one-suiter** *U.S.*, a suitcase designed to hold one suit; **one-tail** *a. Statistics* = next; **one-tailed** *a. Statistics*, applied to a test that tests for deviation from the null hypothesis in one direction only; cf. *two-tailed* adj.; **one-time** *a.* (earlier and later examples); (*b*) pertaining to a single occasion; done or used only once; = *one-shot *a.*; **one time** *adv.* (*colloq.* and *dial.*), (*a*) simultaneously, at the same time; (*b*) on one occasion, once; (*c*) at once, immediately; **one-time cipher, system,** etc., a cipher in which the cipher representation of the alphabet is changed at random for each letter of the message, generating a key as long as the message; **one-time pad,** a pad of keys for a one-time cipher each page of which is destroyed after being used once, so that each message is sent using a different key; **one-to-many** *a.* = *one-many* adj. (see above); **one-to-one** *a.*, applied to a correspondence or relation such that each member of one set is associated with one member of a second set, and *vice versa*; also *adv.*, as, or by means of, a one-to-one relation; **one-track** *a.*, of a person's mind: that is concentrated on, or capable of, only one line of thought or action; obsessional; also *transf.*; **one-trip** *a.*, of a bottle or other container: that is used only once; **one-two,** (*b*) *Boxing*, two punches in quick succession with alternate hands; (*c*) *Football, Hockey*, etc., an interchange of the ball between two players; (*d*) also *transf.* and *fig.*; also *attrib.*; **one-up, one-down** *a.*, designating a house consisting of one main room upstairs and one downstairs; **one-valued** *a.*, having one value (for each component); chiefly *Math.*, = *single-valued *a.*; **one-while** *a.* or *adv.*, see also While *sb.* 6 b in Dict. and Suppl.; **one-woman** *a.*, of, pertaining to, or by one woman only; *spec.* loving, obedient to, or attached to one woman only; **one-world** *a.*, of, pertaining to, or holding the view that there is only one world, or that the world's inhabitants are or should be united; hence *one-worlder*.

1962 L. Deighton *Ipcress File* ix. 54 Dalby stood..in front of a puny one-bar electric fire. **1972** J. McClure *Caterpillar Cop* 147 The gigantic fireplace..had a one-bar electric fire poised for winter in its grate. **1909** *Collier's* 5 June 11/1 The batter..would..score only a one-base hit, perhaps, instead of the home run. **1937** *Amer. Speech* XII. 244/1 *Bingle* is generic for a hit, but also indicates a one-base hit or single. **1880** *Chicago Tribune* 12 May 8/5 Clapp..was brought in by Anson's one-baser. **1949** *Los Angeles Times* 13 Mar. 25/8 Unser led off with a one-baser. **1887** *Graphic* 2 Apr. 355/1 That not uncommon literary phenomenon, the one-book man, whose endeavour to repeat a happy accident is the most imprudent thing he can do. **1890** One-book [see sense 32 in Dict.]. **1970** *Daily Tel.* 26 Sept. 8/6 The next book,.. certainly a better novel than the second, enjoyed more success but the feeling began to grow that Remarque was a one-book author. **1854** One catch all [see *cowardy *a.*]. **1876** J. Burroughs *Winter Sunshine* VIII. i. 210, I could not only walk upon the grass, but..play 'one catch all' with children, boys, dogs, or sheep upon it. **1898** A. B. Gomme *Traditional Games* II. 25 One Catch-all. The words 'Cowardy, cowardy custard' are repeated by children playing at this game when they advance towards the one who is selected to catch them. [**1818** S. Parkes *Let.* 20 Feb. in P. W. Flower *Hist. Trade in Tin* (1880) vii. 92 The following table will show the different sizes of tin plate which are made in Great Britain, and the marks by which each kind is known in commerce... Common No. 1 [size] 13¾ × 10..CI... Cross No. 1 13¾ × 10..XI.] **1890** *Cent. Dict.*, *One-cross*, a term applied to tin-plate..having the thickness of No. 30 Birmingham wire-gage, and having an average weight of 0·5 lb. per sheet. **1897** F. C. Moore *How to build Home* viii. 120 He is to furnish all tin cellar heating-pipes of best (one cross) tin. **1902** *Encycl. Brit.* XXXIII. 906/2 What are called one-design, or restricted classes [of yachts] have latterly become popular. **1904** *Rudder* Nov. 609/2 A one-design boat is one of a fleet built from the same plans. **1928** *Daily Tel.* 11 Sept. 15/6 The East Coast one-design class to the number of nine, started at 10.30 a.m. to sail a course of a dozen sea miles. **1933** E. A. Robertson *Ordinary Families* v. 76 My dinghy's in for the Orwell one-designs. *Ibid.*, The one-designs will be single-handed. **1949** *Sun* (Baltimore) 27 Aug. 8/8 Little Penguin Class dinghies, one of the most popular of the smaller one-design racing sailboats. **1928** *Daily Tel.* 11 Sept. 15/6 The second place on this occasion went to an Essex one-designer. **1883** One dimensional [see Dimensional *a.* 2]. **1909** W. M. Urban *Valuation* iii. 57 All these differences are reducible to differences in intensity and duration of a one-dimensional continuum, pleasantness-unpleasantness. **1936** V. A. Demant

Christian Polity ix. 153 History was seen as a one-dimensional continuum. **1958** M. Kennedy *Outlaws on Parnassus* ix. 141 Daniel..is as one-dimensional to her as Klesmer was. **1964** *Philos. Rev.* LXXIII. 497 A proposition is a linear or one-dimensional structure. **1970** G. K. Woodgate *Elem. Atomic Struct.* ii. 19 Equation (2.50) is in the form of a one-dimensional equation of motion. **1975** *Nature* 22 May 279/2 Fairly recently however there has been a sharp switch of attention to 'one-dimensional materials' (composed of parallel long chain molecules). **1951** S. F. Nadel *Found. Social Anthropol.* v. 90 In the order of groupings we find no exactly equivalent instances of 'one-dimensionality'. **1976** *Sci. Amer.* Dec. 105/2 These purely genetic studies were followed by cytological and biochemical work showing that the one-dimensionality of linkage maps was associated with the linear arrangement of the genes along the chromosome. **1937** *Mind* XLVI. 87 There is a one-directional character to events, an irreversibility in their order. **1950** Auden *Enchafèd Flood* (1951) ii. 65 The determination to live in one-directional historical time rather than in cyclical natural time. **1953** N. Tinbergen *Herring Gull's World* xvi. 132 In the one-egg phase the bird often stands a few feet away from the nest. **1959** *Listener* 29 Oct. 729/1 One-egg twins, being genetically identical, have exactly the same blood and tissue antigens. **1976** *Times* 23 Nov. 15/4 Siamese twins..are always one-egg twins. **1955** *Times* 1 July 15/1 The capital as doubled last year by the one-for-one capitalization issue. **1962** W. Nowottny *Lang. Poets Use* i. 4 The equivalence is more complex than a simple one-for-one relation. **1975** *Listener* 6 Feb. 162/1 Arms are to be replaced only on a one-for-one basis. **1886** T. P. White *Ordnance Survey of U.K.* vi. 102 On the one-inch map, also,..are shown the foot-paths as cross-cuts between roads. **1913** M. I. Newbigin *Ordnance Survey Maps* ii. 17 The scale of the 1-inch map is too large for any rapid form of locomotion. **1929** W. E. Collinson *Spoken Eng.* 88 I've got a one-inch ordnance [survey]. **1948** *N.Y. Jrnl. American* (Sunday Mail ed.) 9 May 1/1 The Court packing plan was defeated by a one-inch punch. **1960** *Farmer & Stockbreeder* 1 Mar. 62/1 The one-inch-to-the-mile survey sheet of Carmarthenshire. **1974** G. Moffat *Corpse Road* vii. 108 'Have you got that one-inch?' he asked. Barber brought him the map. **1904** 'Mark Twain' in *Harper's Weekly* 2 Jan. 18/1 There were headings—one-liners and two-liners—and that was good. **1969** *Harper's Mag.* May 85/2 McCarthy had a one-liner for everyone in Washington, and the reporters who found favor were those who learned to leer and feed straight lines. **1975** *New Yorker* 19 May 23/3 Gail Parent's novel is in the form of a fat girl's jokey suicide note, full of one-liners. **1976** *Times Lit. Suppl.* 11 June 688/5 His dear cousins collapse in mirth at Berry's one-liners and monologues. **1908** S. Ford *Side-Stepping with Shorty* 90 Then me and Sadie in her bubble, towin' the busted one-lunger behind. **1911** H. Quick *Yellowstone Nights* v. 124 The Old Man..was a one-lunger. **1943** 'T. Dudley-Gordon' *Coastal Command* xiii. 125 Angus, the local boatman, came alongside in his elderly 'one-lunger' motor-boat and took us ashore. **1963** Bird & Hutton-Stott *Veteran Motor Car* 15 The 'Varsity' model, and a few of the old one-lungers. **1976** *Islander* (Victoria, B.C.) 11 Apr. 10/1 This engine, a one-lunger, as it was called, drove the Scud..at a speed of six knots. **1842** *Congress. Globe* 27th Congress 2 Sess. App. 812/3 Those whose clamors are so unceasing against what they are pleased to call the 'one-man power'. **1844** *Mechanics' Mag.* XLI. 370 A common road one-man carriage. **1929** D. H. Lawrence *Lovely Lady* (1933) 99 I'm afraid Virginia is a one-man woman. **1939** L. M. Montgomery *Anne of Ingleside* xxiv. 161 There are dogs like that—one-man dogs. **1951** J. C. Fennessy *Sonnet in Bottle* vii. i. 244 Goebbels is a one-man monkey—he doesn't like anybody but me. **1956** *Railway Mag.* May 301/2 The current practice appears to be dictated..by the economics and safety of one-man operation. **1967** *Economist* 9 Dec. 1031/2 Mr Frank Cousins is facing the Government once more—this time in an attempt to keep some thousands of unnecessary bus conductors riding around on routes that could become profitable with one-man bus operation. **1975** A. Hunter *Gently with Love* xi. 32 Anne has been a one-man girl ever since she met Earle. **1976** *Dumfries & Galloway Standard* 25 Dec. 16/3 The way to make economies without too much cutting down of services is to.. bring in one-man-operated buses. **1931** (*record-title*) The one-man band. **1938** Partridge *Dict. Slang* Suppl. 1018/2 *One-man band*, a person that takes rather too much on himself..; slightly ob., as is *l'homme orchestre* supplying the origin. **1958** *Listener* 23 Oct. 663/2 The versatility.. was capitally sustained. The 'one-man band' was never 'off-beat'. **1962** *Sunday Times* (Colour Suppl.) 14 Oct. 9/1 There should also be remembered the great one-man bands of the museum world: men like Sandberg of the Stedelijk Museum in Amsterdam. **1965** *Listener* 10 June 877/1 That odd one-man-band, the three-hole pipe and tabor, which is now almost entirely confined to the border lands of France and Spain. **1974** N. Bentley *Inside Information* vi. 57 We're two mechanics short and the accountant's on holiday, so I'm a one-man band at the moment. **1977** *Listener* 7 Apr. 447/1 A young antiquarian ..did a one-man-band act..making lonesome horn noises with his mouth to accompany his own piano solo. **1896** G. B. Shaw *Our Theatres in Nineties* (1932) II. 287 The real objection to Cibber's version is that it is what we call a 'one man show'. **1905** *Today* 15 Mar. 211/1 One of the young artists..is now having his first 'one-man-show' in London. **1943** F. Scully *Rogue's Gallery* 132 He lectured on the drama at Columbia, and even took up painting in his middle age staging several one-man shows though he never had an art lesson in his life. **1955** *Ann. Reg. 1954* 373 The Beaux Arts Gallery..held..the first one-man show..of John Bratby. **1962** *Listener* 6 Sept. 358/3 Uncle Kweku..started singing 'Happy birthday' in English... It..became a one-man show. **1976** 'Z. Stone' *Modigliani Scandal* i. ii. 25 'Usher's one-man show has had it.' 'I'm afraid so..it won't harm him all that much. His talent will tell in the long run.' **1910,** etc. One-many [see *many-one* adj. s.v. *many *a.* 6 c]. **1945** R. G. Collingwood *Idea of Nature* i. ii. 71 The Platonic form is not a 'logical universal', and the things, in the natural world..to which it stands in a one-many relation are not instances but approximations of it. **1964** E. Bach *Introd. Transformational Gram.* iii. 35 Rules which replace a single symbol by one or more symbols (one-many rules). **1972** *Lect. R. Inst. Philos.* V. 77 Berkeley regarded the relation between the self and its ideas as a necessary one-many relation. **1900** H. Lawson *On Track* 124 But for the one-night lodgers..I was pretty comfortable there. **1915** T. S. Eliot *Prufrock* (1917) 9 Restless nights in one-night cheap hotels. **1943** D. Gascoyne *Poems 1937–1942* 46 Dozens of one-night rooms. **1976** *New Yorker* 15 Nov. 56/3 A fund-raising one-night gala performance. **1923** U. L. Silberrad *Lett. J. Armiter* iii. 62 The people of the house follow a sort of 'sheep and goat' plan, keeping us separate; we, the maiden-lady-some-stay visitors, sit at the upper end of the table, the one-nighters at the other end. **1937** *Amer. Speech* XII. 184/2 *One nighter*, an engagement to play for a single night. **1959** H. Hobson *Mission House Murder* iii. 22 Johnny hasn't quite become a national star yet, but he pulls in a stack of lolly doing one-nighters, mostly in the provinces. **1973** *Guardian* 28 June 15/5 Bloated, badly dressed, he was doing one-nighters with a zombie rhythm section. **1977** 'L. Egan' *Blind Search* iv. 55 We don't get so many one-nighters like we used to. **1880** D. K. Ranous *Diary of Daly Débutante* (1910) 189 This coming week..is to be what they call 'one-night stands'. **1883** *National Police Gaz.* (U.S.) 8 Dec. 3/3 One night stands are not going to be subject, if they can help it, to the experiments of one troupe of queer fakirs after another. **1896,** etc. [see Stand *sb.*[1] 2 e]. **1904** G. V. Hobart *Jim Hickey* i. 13 I'm too delicate for this one-night stand gag. I'm going to New York and build a theatre. **1912** Wodehouse *Prince & Betty* ii. 28 What's the use of a Republic in a place like this? For a little bit of a one-night stand like this you want something picturesque, something that'll advertise the place. **1916** G. B. Shaw *Let.* 14 May in *B. Shaw & Mrs. Campbell* (1952) 186, I told you not to do those one-night stands. **1937** *New Republic* 24 Nov. 70/2 The band plays a one-night stand in some town near. **1956** B. Holiday *Lady sings Blues* (1973) i. 3 When he went on the road with that band it was the beginning of the end of our life as a family. Baltimore got to be just another one-night stand. **1959** 'F. Newton' *Jazz Scene* xi. 185 One of the worst kinds of professional life, that of the touring artist, often passing through a succession of one-night stands. **1963** A. Heron *Towards Quaker View of Sex* iii. 23 These affairs may still be very promiscuous—'one night stands'—or mainly emotional. **1972** J. Wilson *Hide & Seek* viii. 138 Nearly everyone else he knew had had at least a few casual affairs or one-night stands. **1860** *Harper's Mag.* July 195/1 Mrs. Tyler Todd caught the toss, like a skilful player at 'one old cat', on the edge of her.. bonnet. **1929** *Sun* (Baltimore) 27 Mar. 10/3 Supervised play has taken the place of 'one old cat', and hockey has replaced shinney. **1949** *Chicago Daily News* 6 July 14/7 Juvenile pirates had their hang-outs and..one-old-cat and high-button-shoe football thrived. **1974** *Amer. Speech* 1971 XLVI. 84 Ball, puck, and tin-can games..one old cat baseball. **1903** B. Russell *Princ. Math.* xi. 113 This requires that there should be some one-one relation whose domain is the one class and whose converse domain is the other class. **1922,** etc. One–one [see *many–many* adj. s.v. *many *a.* 6 c]. **1950** C. M. Bowra *Romantic Imagination* 33 For him [*sc.* William Blake] allegory in the good sense is not the kind of 'one–one correspondence' which we find in *Pilgrim's Progress*. **1965** Patterson & Rutherford *Elem. Abstr. Algebra* i. 3 If..*f* is a mapping of S_1 into S_2 such that $f(x_1) = f(x_2) \Rightarrow x_1 = x_2$, then *f* is called a one–one mapping or a one–one correspondence. **1972** *Lect. R. Inst. Philos.* V. 80 If there were ideas..in this sense 'simple' they would stand only in a one–one relation to minds. **1924** *Granta* 25 Apr. 361/2 Last but not least he took a 'one one' in the French Tripos last year. **1968** K. Martin *Editor* i. 3, I had taken a One-one in my Tripos at Magdalene. **1967** *Technology Week* 20 Feb. 3/1 In the one-on-one, relatively 'simple' intercepts run during the 1962–63 test series, the 'old' *Nike-Zeus* scored on 10 of 14 attempted live ICBM intercepts. **1972** J. Gores *Dead Skip* (1973) i. 7 He had started as a field agent..three years before, when he had realized he wasn't going to be middleweight champ of the world after all; it was the only profession he knew which could give him the same one-on-one excitement he'd found in the ring. **1974** 'E. Lathen' *Sweet & Low* xi. 113 He was not in a one-on-one confrontation. There was a goodly array of..small fry present. **1932** D. Burnstine *Four Horsemen's One over One Method of Contract Bidding* i. 1 The One-Over-One system of bidding has achieved its present fame because of its use by players who have won the majority of contract tournaments. *Ibid.* 3 The One-Over-One offers leeway in arriving at the correct contract with ease. **1934** *Amer. Speech* IX. 10/1 There are..several varieties of one-over-one bids. **1959** *Listener* 19 Mar. 530/1 Many completely minimum hands..could be hamstrung by a simple one-over-one response on the first round. **1880** G. A. Sala *Amer. Revisited* (1882) II. 13 Slop-shops, or 'one-piece stores' overflowing with guernseys, pea jackets, sou'-wester hats. **1895** *Montgomery Ward Catal.* 501/2 Bathing suits. One-piece suits. **1912** *Woman's Weekly* 25 May 100/3 (*caption*) A One-Piece Dress, One-Piece Petticoat, One-Piece Drawers, and One-Piece Bodice. **1930** *Engineering* 7 Mar. 309/3 The housing and arm being also a one-piece casting. **1972** J. Mosedale *Football* ii. 21 The Van Buren uniform included a one-piece fiber crown replacing the sewn leather helmets of the pre-1940s. **1973** 'D. Halliday' *Dolly & Starry Bird* xiv. 203 A gorgeous one-piece black bathing suit. **1974** F. Warner *Meeting Ends* i. i. 2 Wrasse..in modern one-piece bathing costume. **1919** W. H. Downing *Digger Dial.* 37 *One-pip*, Second Lieutenant. **1937** Partridge *Dict. Slang* 589/1 One-pipper. **1956** D. M. Davin *Sullen Bell* 181 Whatever young one-pipper it was could get a night's leave. **1974** G. M. Fraser *McAuslan in Rough* 17 Keith was a mere pink-cheeked one-pipper of twenty years, whereas I had reached the grizzled maturity of twenty-one and my second star. **1938** *Jrnl. Symbolic Logic* III. ii. 83 Chapter 2 supplements the propositional calculus..with the Boolean algebra of one-place predicates. **1947** H. Reichenbach *Elem. Symbolic Logic* § 17. 83 The term 'property'..is usually applied only to one-place situational functions. **1967** S. C. Kleene *Math. Logic* § 27. 145 More essential use is made of the predicate calculus with quantification of one-place predicates in Example 19. **1974** *Canad. Jrnl. Linguistics* XIX. 151 But what I want to focus on here is the claim that *easy* is a one-place predicate with a sentential subject, while *eager* has a sentential object. **1959** J. W. Carr in E. M. Grabbe et al. *Handbk. Automation, Computation, & Control* II. ii. 58 In the one-plus-one addressing procedure, each instruction has a basic single-address format, but also includes a second address to be used to designate the location of the next instruction to be performed. **1969** P. B. Jordain *Condensed Computer Encycl.* 351 The one-plus-one address instruction has only the power (or flexibility) of a one-address instruction, because only one operand reference is included. **1892** J. A. Ewing *Magn. Induction in Iron* ii. 40 We may distinguish this as the 'one-pole' method, seeing that the deflection of the magnetometer is mainly caused by one of the bar's poles. **1932** 'N. Shute' *Lonely Road* i. 15 From the set of her one pole mast she might have been a Thames bawley of about fifty tons. **1940** M. Miller *Harbor of Sun* xxviii. 311 When a fisherman describes the size of a school as 'one-pole tuna', 'two-pole tuna' and sometimes 'three-pole tuna', he is but saying that the size of the tuna in that special school required one, two or three men to a team for hauling in each fish. **1845** C. M. Kirkland *Western Clearings* (1846) 27 Some scattered grains of coarse powder from near the touch-hole of the one-pounder that was fired all day by the opposition. **1893** 'Mark Twain' in *Century Mag.* Jan. 339/2 They find they've given a tramp a million-pound bill when they thought it was a one-pounder. **1916** 'B. M. Bower' *Phantom Herd* v. 69 We've made quite a haul since you left. A bunch of one-reelers. **1976** *Listener* 23–30 Dec. 833/1 When you were making the one-reelers, did each have a lengthy script? **1922** Wodehouse *Clicking of Cuthbert* vi. 142 No human being could play golf against a one-ring circus like that without blowing up. **1972** *Village Voice* (N.Y.) 1 June 40/3 An opera house that is not a cultural force is only a one-ring circus made up of vocal acrobats who use music as a trampoline. **1934** *Amer. Speech* IX. 112/2 Plenty of one-stop service stations for washing, minor repairs, lubrication, etc. **1962** *Economist* 5 May 452/1 Commercial banks which are able to offer complete 'one stop' banking service—including current accounts, consumer loans and so on. **1971** *E. Afr. Jrnl.* Mar. 34 (Advt.), You will find Text Book Centre a one-stop warehouse for all your educational requirements. **1978** *Oxford Consumer* Mar. 5/1 The store will specialise in the provision of food lines at economic prices and will be backed up by a sufficient range of household goods to enable the shopping public to derive the maximum convenience from a 'one stop' shopping trip. **1958** A. J. Bliss *Metre of Beowulf* 62 We must, in fact, recognize the possibility of one-stress verses. Sievers himself in later life envisaged such one-stress verses; Pope, too, makes one-stress verses a mainstay of his new theory. **1965** *Eng. Stud.* XLVI. 419 As examples of light, one-stress verses he gives: hu ða æðelingas. **1961** Webster, One-suiter. **1971** 'O. Bleeck' *Thief who painted Sunlight* (1972) xx. 181 He was carrying something that looked like a one-suiter. **1947** C. Eisenhart et al. *Sel. Techniques Statistical Anal.* 459/2 (Index), One-sided or one-tail tests of statistical hypotheses. **1954** *Brit. Jrnl. Psychol.* XLV. 174 The difference between these means is..only just significant at the 5% level, using the one-tail test. **1950** M. H. Quenouille *Introd. Statistics* v. 98 Here, since we are interested in deviations in one direction only, probabilities calculated using χ^2 must be halved. We are then said to be using a one-tailed test. **1969** R. H. Kolstoe *Introd. Statistics Behav. Sci.* x. 203 In a few situations the research worker, *before collecting his data*, decides that he is interested in one specific directional hypothesis... In this case a one-tailed test of significance would be indicated. **1971** B. Erricker *Advanced Gen. Statistics* xiv. 206 If we were only interested in whether the tensile strengths of the components of the first manufacturer are greater than those from the second we would only need a one-tailed test. **1850** W. Howitt *Year-bk. Country* vi. 179 Old Lodge, we salute thee for thy venerable antiquity; but we owe thee no respect as the one-time resort of the boasted virgin queen! **1870** *Appleton's Jrnl.* 5 Feb. 161/2 Then you would have one brought after the other, unless accompanied by the request, 'all at the same time', or, in their [*sc.* native Liberians'] own language, '*go fetch 'em come; both two; one time*'! **1873** C. J. G. Rampini *Lett. from Jamaica* 177 Man can't smoke an' whistle one time. **1886** F. T. Elworthy *West Somerset Word-bk.* 537 There used to be a public-house there one time, but he bin pulled down 'is gurt many years. **1899** C. J. C. Hyne *Further Adventures Capt. Kettle* ii. 31 He wouldn't stop for fighting-palaver. He'd be off for bush, one-time. **1924** *Time* 7 Jan. 30/2 Died. Richard Wittig, brother of Maximillian Harden, famed German publicist, onetime friend of the Kaiser, onetime Oberburgomaster of Posen. **1928** *Flynn's Weekly* 4 Feb. 436/1 Big Bill Douglas was enjoying a year's vacation from his usual haunts up at Sing Sing at the expense of the State. To his underworld associates he was doing a short bit in the Big House, or a one time loser. **1942** Z. N. Hurston in A. Dundes *Mother Wit* (1973) 225/2 A Zigaboo..asked a woman that one time. **1950** 'S. Ransome' *Deadly Miss Ashley* viii. 96 Duncan Westling's onetime confidential secretary was now the secretary of Duncan Westling's onetime mistress? **1959** *Listener* 10 Dec. 1023/1 The Soviet Union has lately placed some big orders with British industry... But we have a feeling that these are, or may be, one-time orders to enable the Soviet Union to progress towards self-sufficiency. **1967** *Ibid.* 2 Feb. 157/1 S. W. Johnson-Marshall, one-time chief architect at the Ministry of Education. **1973** *N.Y. Law Jrnl.* 27 Mar. 4/5 The tenant insisted that the only increase he was obliged to pay was a one-time increase. **1955** *Gloss. Soviet Military Terminol.* (U.S. Army Technical Manual 30–544) 228/1 One-time system. **1977** *Sci. Amer.* Aug. 120/3 It is easy to see why the one-time cipher is uncrackable even in principle. Since each symbol can be represented by any other symbol, and each choice of representation is completely random, there is no internal pattern. **1953** *N.Y. Times Mag.* 15 Mar. 62/2 The Russians are notorious for their reliance on a device known as the 'one-time-pad'. This means..that each message is sent in a completely different code. **1966** M. R. D. Foot *SOE in France* iv. 105 By now [*sc.* 1944] the British were using a much safer..

cipher: one-time pad... The agent held a pad of silk slips, each printed with columns of random letters or figures from which any message could be enciphered or deciphered; he..was supposed to tear each slip off and burn it after use. **1977** *Sci. Amer.* Aug. 120/3 If the one-time pad provides absolute secrecy, why is it not used for all secret communication? The answer is that it is too impractical. Each time it is employed a key must be sent in advance, and the key must be at least as long as the anticipated message. **1959** One-to-many [see *correspondence 1 b]. **1976** P. R. White *Planning for Public Transport* v. 112 In a one-to-many situation, passengers joining at the fixed point(s) request the driver to serve a destination, and no radio contact is necessary. **1873** *Proc. Lond. Math. Soc.* IV. 252 The equations..being supposed to establish a 'one-to-one' correspondence between the two integral spaces. **1882,** etc. One-to-one [see *correspondence 1 b]. **1903** B. Russell *Princ. Math.* xi. 113 Two classes have the same number.. when their terms can be correlated one to one, so that any one term of either corresponds to one and one only term of the other. **1931** C. Fox *Mind & its Body* iii. 62 One of the most important assumptions of psychological physiology was this one-to-one correspondence between neural processes and mental processes. **1936** J. R. Kantor *Objective Psychol. Gram.* xvii. 237 There is no clear-cut one-to-one relation between a grammatical form and a definite time point. **1963** J. Lyons *Structural Semantics* iii. 37 It is not so much that one language draws a greater or less number of semantic distinctions than another which prevents the matching of their vocabularies one-to-one. **1968** C. G. Kuper *Introd. Theory Superconductivity* xii. 193 There is a one-to-one correspondence between the states of a gas of non-interacting Bosons and those of a family of harmonic oscillators. **1973** One-to-one [see *nicotinic *a.* 2]. **1928** D. H. Lawrence *Lady Chatterley* xix. 363 They're all one-track minds nowadays. **1932** *Kansas City* (Missouri) *Times* 5 May 20 The persons with the one-track mind are the ones who usually have the most collisions. **1934** H. Nicolson *Curzon: Last Phase* 18 He has been accused of possessing a 'one-track mind', of being deficient in creative, as opposed to emotional, imagination. **1935** B. Malinowski in M. Black *Importance of Lang.* (1962) 78 The advertisements emanating from such one-track remedies. **1944** 'Brahms' & 'Simon' *Titania has Mother* viii. 68 Her son..had launched himself onto a one-track conversation. **1957** P. Frank *Seven Days to Never* ii. 48 The electronic machines..could distract a Russian missile's one-track mind. **1968** *National Observer* (U.S.) 3 June 15/1 It's not that I'm antisocial. It's just that I'm preoccupied. I have a very one-track mind. **1973** 'H. Howard' *Highway to Murder* vii. 77 I've got a one-track mind... All this started with a man getting shot and I keep thinking along those lines. **1967** *Times Rev. Industry* May 72/2 Most containers are 'one-trip' in the sense that the product is used and the container is thrown away. **1811** *Sporting Mag.* XXXVIII. 140/2 He..had no difficulty at getting at his man when he chose with a *one, two.* **1815** *Pancratia* (ed. 2) 359 He fought cautiously..and whenever he closed put in his *one two* with the greatest dexterity. **1910** G. W. E. Russell *Sketches & Snapshots* xlvii. 445 A smart one-two on his smeller effectually tapped his claret. **1931** *Times Lit. Suppl.* 22 Oct. 819/1 The flawless stance and one-two punch of Peter Jackson. **1948** *Economist* 20 Mar. 454/1 His [*sc.* Stalin's] one-two play in Czechoslovakia and Finland. **1952** A. Wilson *Hemlock & After* iii. i. 201 Even Ron was surprised that his 'old one two' was quite so compelling. **1960** *Times* 4 Oct. 13/4 'The old one-two', in the boxing slang of a more vulgarly robust age, indicated a quick follow-up with the right immediately after the left had landed, and the near synchronization of the two blows added immensely to their effect. **1970** *Times* 1 Oct. 10/3 Hinton came up from his position of centre back to play a one-two with Hutchinson and leave the wretched Christidis stranded. **1974** J. Gardner *Corner Men* xi. 100 Let Hart and Harvey work them over, then we can go in and do the old routine... The old one two. **1978** *Sunday Times* (Colour Suppl.) 28 May 34/4 *One-two*, using a colleague for an immediate return pass, to run on to. **1933** A. Salter in A. F. Brockway *Bermondsey Story* (1949) ii. 12 The house was one up, one down, with a small scullery. **1968** Busby & Holtham *Main Line Kill* vi. 68 Some of the back to back terraces of poky little one-up, one-down houses had been pulled down. **1898** W. B. Smith *Infinitesimal Analysis* i. 7 When to one value of the one variable there corresponds only one value of the other, this latter is called a one-valued or unique function of the former. **1913** *Trans. Amer. Math. Soc.* XIV. 481 None of the equivalent postulate-sets here referred to is in terms of its undefined entities one-valued ('categorical')—that is, each determines not a single algebra but a class of algebras. **1943** *Amer. Speech* XVIII. 220 The first stage of human development ..is that of the savage, prelogical mentality, with a one-valued semantics (or system of evaluations), in which, as Lucien Lévy-Bruhl has said, 'everything is everything else' by 'mystic participation'. **1968** E. T. Copson *Metric Spaces* vii. 85 A function is, by definition, one-valued. **1894** Hall Caine *Manxman* i. ix. 45 I'm a one-woman man, Kate; but loving one is giving me eyes for all. **1937** M. Hillis *Orchids on your Budget* (1938) iv. 68 We ourselves have run our one-woman ménage both with and without an office job. **1960** P. Tompkins in G. B. Shaw *To a Young Actress* 151 Thirty-five paintings for a one-woman exhibition at the Leicester Galleries. **1962** I. Murdoch *Unofficial Rose* ii. 32 He's a one-woman cat. **1974** J. Cleary *Peter's Pence* v. 156 I'm a one-woman man. **1926** A. E. Taylor *Plato* viii. 198 The epiphenomenalist is tied by his theory to a 'one-world' interpretation of human experience; morality presupposes a 'two-world' interpretation. **1947** *Collier's* 7 June 12 (*title*) Dunkirk—the one-world town. **1948** *Sun* (Baltimore) 22 June 2/1 His selection of Stassen for Taft's running mate was new and was surprising, in view of his oft-repeated denunciation of Stassen as a 'one-worlder' who would be unsafe in high office. **1958** *Spectator* 15 Aug. 228/2 'Men live and die for a flag; it is indeed the only thing for which they are willing to die in masses..' is a statement of a truth which one-worlders ignore at their peril. **1965** *Social Crediter* 31 July 2/2 The *active* Socialists, Communists, and One-Worlders.

one, *v.* **1.** (Later example.)
1921 B. Williamson *Supernat. Mysticism* v. 45 The human race was so oned with Adam that all sinned in him.

one-act, *a.* [One *numeral a.* 32 a.] Denoting a short play or other production consisting of a single act. Hence as *sb.*, such a play. So **one-a·cter,** a one-act play; also *fig.*
1888 *Playgoers' Mag.* Feb. 45 The story of Carton's self-sacrifice would form a touching little one-act play. **1895** One-acter [in Dict. s.v. One *a.* 32]. **1905** *Athenæum* 7 Oct. 477/3 The one-act trifle which serves as *lever de rideau.* **1912** E. Nesbit *Let.* in D. L. Moore *E. Nesbit* (1933) xv. 269, I have had a one-act play accepted by a London manager. **1927** J. Pollock (*title*) Twelve one-acters. **1940** G. Marx *Let.* 5 Sept. (1967) 25 El Capitan..has done magnificently with the Coward one-acts. **1960** A. Coren in *Introduction: Stories by New Writers* 70 The sane guy is the one who realises that life is a short one-acter. **1967** *Oxf. Compan. Theatre* (ed. 3) 223/2 The one-act play survives mainly in the productions of the amateur theatre in England and America. **1967** *Wall St. Jrnl.* 24 Apr. 18/4 The play for the evening was Lanford Wilson's 'The Madness of Lady Bright', a 45-minute one-acter. **1973** *Guardian* 23 Mar. 12/3 Tchaikovsky's seventh and last opera is a one-acter. **1976** *Scottish Rev.* Spring 17 *The Stick-Up*, a one-acter from *Fifteen Poems and a Play* (Edinburgh, 1969). **1977** L. Meynell *Hooky gets Wooden Spoon* xiii. 157 She had written a one-act play for herself.

one argument. *Logic.* [f. One *numeral a.* + Argument *sb.* 3.] The variable of a function or operator of only one variable; also *attrib.*
1941 O. Helmer tr. *Tarski's Introd. Logic* 107 In order to differentiate between two-termed and three-termed functional relations, we speak, in the first case, of functions of one variable or of functions with one argument. **1951** J. Łukasiewicz in *Proc. R. Irish Acad.* LIV. A. 27 Such values are not only constant functors of one propositional argument, as *N*, negation, for example, but also complex expressions like a functor of one argument. **1955** H. Leblanc *Introd. Deductive Logic* 191 We have studied..so-called one-argument functions like the functions square of, double of, and so on. **1957** A. N. Prior *Time & Modality* 2 A system must contain a pair of one-argument operators forming statements out of statements. **1965** Hughes & Londey *Elem. Formal Logic* iii. 19 The following are *monadic operators*, (i.e. they take one argument). **1967** H. Weber tr. J. Łukasiewicz in S. McCall *Polish Logic* iii. 47 Since this proposition is valid for all functors with one argument, it is also valid for the functor '*M*'.

one-arm (wɒ·n,ɑːm), *a.* [One *numeral a.* 32 a.] Having one arm; using only one arm; *spec.* **one-arm bandit** (orig. *U.S.*) = *one-armed bandit*; **one-arm joint** or **lunch (room)** *U.S.*, a cheap eating-house where customers sit in seats which have one arm wide enough to support plates of food, etc.; also *ellipt.* as *one-arm* sb.
1906 *Westm. Gaz.* 7 Sept. 3/2 Ordinarily I can do a one-arm press of 90 lb., but I must confess that I was almost beaten in getting hoisted some of these sheaves. **1912** M. Nicholson *Hoosier Chron.* 297 Everybody's saying 'Stop, Look, Listen!'..the white aprons in the one-arm lunch rooms say it now when you kick on the size of the buns. **1915** *N.Y. World Mag.* 9 May 14 *One arm joint*, a chair dairy lunch. **1926** *New Masses* May 9/4 Countermen in the one-arm lunches yell '*coffee-and*' not so fiercely. **1931** H. Mutschmann *Gloss. Americanisms* 43/2 *One-arm driver*, man steering auto with one arm and necking his girl with the other. **1935** J. Hargan *Gloss. Prison Lang.* 6 *One arm joint*, a cheap restaurant where one takes his food to a chair to eat. **1938** *Sun* (Baltimore) 6 Oct. 24/1 The Court of Appeals at Annapolis yesterday declared.. that the so-called 'one-arm bandit' type of slot machine is illegal. **1939** *Detective Fiction Weekly* 18 Feb. 36/2 A one-arm joint is a white-tiled place that suggests a clinic. Two long rows of armchairs line the walls. The right arm of each chair is expanded into china-topped slab. You park your food on it. **1940** J. O'Hara *Pal Joey* 57 She went with me to this one-arm where I eat. **1943** *Sun* (Baltimore) 27 Jan. 5/6 Three of the [slot] machines were described by police as the 'one-arm bandit' type. **1944** B. Hope *I Never left Home* v. 63 He flew a plane back from a mission once holding his wounded copilot in his arms. That's one-arm driving that counts. **1951** E. Kefauver *Crime in Amer.* (1952) xiii. 151 The iniquitous 'one-arm bandit' slot machines sprang up in bars and cigar stores. **1956** S. Hope *Diggers' Paradise* 156 In the exclusive clubs, with few exceptions, you may see an array of 'one-arm bandits' ranged against the walls. There fruit machines—or poker machines—are usually rigged to work with shilling discs bought from the bar steward. **1960** *Times* 2 Dec. 8/4 Six fruit machines, or 'one-arm bandits', were on display at Scotland Yard.

one-armed (wɒ·n,ɑːmd), *a.* [One *numeral a.* 32 d.] Having one arm; also *transf.*; *spec.* **one-armed bandit** (orig. *U.S.*) = *fruit machine* (see *fruit *sb.* 9).
1809 *Thespiad* 10 Every subsequent comedy would have contained a weather-beaten, one-armed sailor. **1818,** etc. [see One 32 d]. **1886** F. T. Elworthy *West Somerset Word-Bk.* 536 *One-arm'd landlord*, cant name for a pump. **1914** W. B. Yeats *Responsibilities* 26 A one-legged, one-armed, one-eyed man. **1938** *Time* 28 Feb. 33/1 Last fortnight, with her ax she demolished two more—as she called them—'one-armed bandits'. **1945** Baker *Austral. Lang.* iv. 88 *Ned Kelly* is displacing *one-armed bandit* for a poker machine. **1948** *Richmond* (Virginia) *Times-Dispatch* 8 Apr. 5/1 He was convicted of a charge of having two slot machines, familiarly known as 'one-armed bandits', in his possession. **1959** *Times* 12 Feb. 10/7 To-day the Senators were ringed with juke-boxes, pinball tables, 'one-armed bandits', and other coin-operated devices. **1971** P. Toynbee *Working Life* iv. 60 You slip your card into the slot and pull down the lever which punches the time on it. 'Talk about a one-armed bandit', someone jokes... 'Trouble is, it's never been known to pay out.' **1972** D. Francis *Smokescreen* vi. 70 There's more cars parked along the streets down there than one-armed bandits in Nevada.

one-berry. Add: **b.** *U.S.* = Checkerberry, *Indian turnip* s.v. Indian *a.* 4 b, Wintergreen.
1877 Bartlett *Dict. Amer.* (ed. 4) 319 Jack-in-the-Pulpit. (*Arisæna triphyllum*)... In Connecticut, it is called One-berry. **1892** *Jrnl. Amer. Folk-Lore* V. 100 *Gaultheria procumbens*, one-berry. **1931** M. Grieve *Mod. Herbal* II. 766 *Mitchella repens*... Partridgeberry, Checkerberry, Winter Clover, Deerberry, One-berry.

one-er, var. Oner *sb.*

one-eyed, *a.* Add: † **b.** *U.S. slang.* Dishonest. *Obs.*
1833 *Sk. & Eccentr. D. Crockett* i. 24 In the slang of the backwoods, one swore that he would never be '*one-eyed*'.
2. As a general term of disapproval or contempt: small, inferior, inadequate, unimportant; = One-horse *a.* 2, esp. of a town. *colloq.* (orig. *U.S.* or *dial.*).
1871 D. G. Rossetti *Let.* 28 Oct. (1967) III. 1021 A little hamlet called Kelmscott, the nearest town to which is Lechlade,—that being however but a 'one-eyed' town as the Yankees say. **1881** Hardy *Laodicean* III. 246, I shouldn't care for such a one-eyed benefit as that. **1887** Parish & Shaw *Dict. Kentish Dial.* 111 'That's a middlin' one-eyed place.' 'I can't make nothin' of these here one-eyed new-fashioned tunes they've took-to in church; why they've a'most done afore I can make a start.' **1937** G. Heyer *They found him Dead* i. 19, I wasn't born to this humdrum life in a one-eyed town. **1947** *E. Afr. Ann.* 1946–7 101/2 Some had said it was a grand little town; others, a one-eyed hole! **1977** *Times* 14 May 8/7 In its somewhat one-eyed way, it [*sc.* Tobago] is among the loveliest..of all the Caribbean islands.
b. Narrow in outlook; prejudiced, narrow-minded. Hence *one-eyedness.*
1863 J. Brown *Let.* Mar. (1912) 206, I do believe the man thinks he is doing God service and is honest in his way, though vain and one-eyed to ludicrosity, as you have most thoroughly and delightfully shown. **1874** Swinburne *Let.* July (1959) II. 302 With all his rhetorical power, he [*sc.* J. A. Froude] seems to me (even apart from his one-eyed prepossession and palpable special pleading) but a shallow reader of character. **1921** G. B. Shaw *Back to Methuselah* p. li, There is no reason to suspect Weismann of Sadism... It was a mere piece of one-eyedness; and it was Darwin who put out Weismann's humane and sensible eye. **1971** *Austral. Seacraft* June 4/2 It seems your correspondent is one-eyed so far as the southern part of Australia is concerned.

one-handed, *a.* Add: **B.** as *adv.* Using only one hand.
1962 J. D. MacDonald *Girl* ix. 113 You can do it okay one-handed, just push down with your thumb. **1963** *Times* 12 June 5/2 He was caught at long on, finishing his stroke one-handed. **1974** R. Adams *Shardik* v. 43 He crouched upon his knees, fumbling one-handed along the undulating tree-trunk. **1975** J. Mitchell *Smear Job* v. 37 Lonely drank one-handed.
Hence **one-ha·ndedly** *adv.*, with one hand; **one-ha·ndedness**, the state of being one-handed; (by back-formation) **one-ha·nd** *v. trans.*
1972 J. Rossiter *Rope for General Dietz* xiii. 185, I lit a cigarette one-handedly. **1973** R. Stout *Please pass Guilt* (1974) iv. 32 Jones stretched an arm and one-handed it [*sc.* the ball], and kept it. **1975** *Times* 12 Feb. 4/3 In spite of her one-handedness, [she] had followed hobbies of gardening, paper hanging, tennis. **1977** S. Coulter *Soyuz Affair* i. 9 He..onehandedly fingered a cigarette out of the packet.

one-horse, *a.* **2.** For *U.S. colloq.* read *colloq.* (orig. *U.S.*). Esp. *one-horse town*, a small or rural town; a town where nothing important or exciting happens. (Add further examples.)
1853 *Oregonian* (Portland) 19 Nov. 2/1 These *one-horse* meetings are got up by men whose capital consists in *brass.* **1855** *Knickerbocker* XLVI. 106 'In this "one-horse town",' writes a Mobile friend, 'as our New-Orleans neighbors designate it, [etc.].' **1884** *Liverpool Daily Post* 15 Oct. 5/1 The first thing it is to do is to take possession of that 'one-horsiest' of railways, from the Dock Cottages to West Kirby, and to make it into a channel of communication fit for civilised men. **1933** P. Fleming *Brazilian Adventure* iii. x. 378 Doctor Amyntas was the big noise; in this one horse town he might be said to own the horse. **1940** L. A. G. Strong *Sun on Water* 161, I was sick with I don't know what early summer passion in that one-horse store. **1969** L. Kennedy *Very Lovely People* i. 73 Their names were Homer and Arnold and they worked in a one-horse garage down a side street in Botafogo. **1973** 'D. Jordan' *Nile Green* viii. 39, I said, 'It's a one-horse town.' **1977** *Zigzag* June 23/2 I've a new song..about a girl of sixteen trying to get out of a one horse town.

Oneida (onəi·dā). *N. Amer.* [ad. Oneida *onēñyote*ʔ erected stone (the name of the main Oneida settlement at successive locations, near which, traditionally, a large syenite

boulder always appeared).] One of the five (later six) tribes of the Iroquois Confederacy of North American Indians commonly called the Five Nations (Six Nations), originally inhabiting upper New York state; a member of this tribe; their language. Also *attrib.*
1666 J. Allyn *Let.* 10 July in *Mass. Hist. Soc. Coll.* (1849) 3rd Ser. X. 63 Hereof the Mohawkes and the Oneiades have given assured notice. **1722** S. Sewall *Diary* 19 Oct. in *Ibid.* (1882) 5th Ser. VII. 311 The Messenger of the Oneidas was buried in the South Burying place. **1760** in J. W. Lydekker *Faithful Mohawks* (1938) 102 Gen! Amherst being at the Oneida Lake on the preceeding Sunday went up as far as the Oneida town. **1823, 1933** [see *Cayuga]. **1959** [see *Five Nations* s.v. *five *a.* and *sb.* C. 2]. **1965** *Canad. Jrnl. Linguistics* X. 105 The structure of Pawnee as compared with Oneida. **1969** *Observer* (Colour Suppl.) 25 May 53/3 West of them were the extraordinarily fierce Oneida. **1973** A. H. Whiteford *N. Amer. Indian Arts* 148 Lakes area silverwork probably began when the Oneida, an Iroquois tribe, moved to northeast Wisconsin.

one-ideaed, -idea'd, *a.* Also **one-idead.** Add: (Earlier example.) So **one-idea** *a.* Hence **one-idea(d)ness**, the fact or quality of being one-ideaed.
1842 *Lancet* 12 Mar. 830 The dead superstitions, and one-idead theories, of the middle ages. **1852** *Blackw. Mag.* Aug. 261/2 His absorbed one-ideadness. **1862** M. B. Chesnut *Diary* 13 Mar. (1981) 199 He is a one-idea'd man. That one idea is to fit every possible man in the ranks. **1920** H. Begbie *Life W. Booth* I. xxii. 365 It was ..this intense singleness of view, this consuming one-ideaness of soul, which made William Booth so successful. **1934** E. Bowen *Cat Jumps* 252 She..was in fact a rather one-idea girl.

oneing (wɒ·niŋ), *vbl. sb.* Also **one-ing.** [f. One *v.* + -ing[1].] Union; fusion. Also *attrib.* or as *ppl. a.*
1919 D. H. Lawrence in *Eng. Rev.* June 488 There in the sexual passion the very blood surges into communion, in the terrible sensual oneing. **1921** B. Williamson *Supernat. Mysticism* vii. 68 Sanctity is the oneing of the soul with God. **1934** *Blackfriars* Mar. 184 Mother Julian..saw it [*sc.* pain] first and foremost as the blessed instrument of our one-ing with Christ. **1958** C. Pepler *Eng. Relig. Heritage* iv. i. 224 In *The Cloud* the question of Christ's passion restoring all men to the oneing affection with God lost by Adam makes no reference to his sacrifice. *The Epistle*, on the other hand, speaks of this 'oneing' in terms of a continual offering of the sacrifice of a man's whole being.

oneiric, *a.* Delete *rare*[−1] and add further examples.
1953 *Scrutiny* XIX. 151 Their..dictation of the unconscious, oneiric delirium,..may be fragmentary..but they bear witness..to their determination to make all possible aspects of poetry incentives to life. **1963** T. Pynchon *V.* xi. 335 The green light deepened, drowning the island of Malta and the island of Fausto and Elena hopelessly deeper in its oneiric chill. **1974** *Monthly Film Bull.* (Brit. Film Inst.) Apr. 84/2 Even though his film's philosophical focus may occasionally seem blurred, the images themselves..retain throughout a haunting and oneiric quality.

oneiro-. Add: **oneirocrisy,** delete † and add later example; **oneiromancer** (later example); **oneiromancy** (later examples).
1976 *Proc. Classical Assoc.* LXXIII. 20 Intellectual constructions such as onirocrisy and physiognomony readily degenerate into the collection of *paradoxa*, upon which the temperament of the age commonly puts a religious interpretation. **1952** G. Sarton *Hist. Sci.* I. xiv. 371 There are two kinds of dreams, those of divine origin, which concern oneiromancers, and those of physiologic origin. **1931** E. Jones *On the Nightmare* ii. iii. 95 This cure for disease by Incubation—known as oneiromancy—was practised in Scotland and Ireland... Here the person slept in the skin of a sacrificed sheep, just as the worshippers of Ammon did in Thebes. **1935** J. S. Lincoln *Dream in Primitive Cultures* i. 3 At Nineveh was a collection of books on oneiromancy or dream interpretation.

one-legged, *a.* **1.** (Earlier example.)
1872 V. Lush *Jrnl.* 6 Nov. (1975) 130 In the evening Martin took Annette to the Theatre Royal to see Donato the one-legged dancer and Zuila perform on the trapeze, &c.

one-pi·pe, *a.* [f. One *numeral a.* + Pipe *sb.*[1]] **a.** Applied to a system of hot-water central heating in which radiators take water from and return it to the same pipe, which runs in a complete circuit from the boiler and back to it again.
1897 F. Dye *Hood's Pract. Treat. Warming Buildings* (ed. 3) xix. 352 It is not usual for the one-pipe system to extend to three floors above the boiler without modification. **1970** J. J. Barton *Domestic Heating* viii. 127 Two basic circuits, the 'one-pipe' and the 'two-pipe' circuit.. are commonly used in small bore heating practice.
b. Applied to a system of plumbing in which waste from sinks and the like is conveyed to the sewer by the same pipe as that from water-closets and urinals.
1910 W. P. Gerhard *Water Supply, Sewerage & Plumbing* iii. 187, I am..convinced that the one-pipe system, as I have sometimes called it, is the coming system. **1933** *Archit. Rev.* LXXIV. 54/3 The drainage and sanitary installation throughout is on the one-pipe system. **1972** T. A. Tompson *Guide Sanitary Engin. Services* vii. 228 Repetitive planning at each floor level of a multi-storey building favours localised application of single-stack principles of sanitation... This has led to the development of the modified one-pipe system.., achieving even greater economy than with the one-pipe system.

oner, *sb.* Add: **2.** (Further examples.) *spec.* One pound; one hundred pounds.
1950 *Austral. Police Jrnl.* Apr. 116 *One-r*, £1. **1962** Parker & Allerton *Courage of his Convictions* iv. i. 154 A one-er for the Guv'nor, and fifty each for me and George here, that's cut price. Two hundred all told, how's that? **1969** I. & P. Opie *Children's Games* vii. 229 When one conker breaks another into pieces so that nothing remains on the string, the winning conker becomes a 'one-er'. *Ibid.* viii. 251 He may shout a number like 'a oner', this means that the rest of the team have to jump over the person's back from the line, taking only one step. **1970** G. F. Newman *Sir, You Bastard* viii. 225 Worth a oner to you. **1974** H. R. F. Keating *Underside* xxii. 218 You'd pay me five sovereign?.. Five golden oners?

onery, onnery, o'n'ry, varr. *ornery *a.*
1849 J. J. Hooper *Night at Ugly Man's* in *Spirit of Times* 24 Nov. 471/2 We had an old one-horned cow, mighty onnery (ordinary) lookin'. **1860** M. J. Holmes *Maude* v. 63 She pronounced her 'not quite so *onery* a white woman as she at first took her to be.' **1887** E. Custer *Tenting on Plains* ix. 286 He's a good enough fellow, only he's an onery scamp of a Republican. **1905** *Dialect Notes* III. 63 The *onriest* critter. **1913** H. Kephart *Our Southern Highlanders* viii. 169 'What sort of men are they?' 'Torn down scoundrels, every one.' 'Oh, come now!' 'Yes, they are; plumb onery.' **1939** *Nat. Geogr. Mag.* Aug. 144/1 Wild sweet williams in the wood lot were much more alluring than the 'o'n'ry' weeds. **1943** *Pocahontas Times* (Marlinton, W. Va.) 4 Mar. 2/1 That onery cuss of a jay bird. **1962** W. Stegner *Wolf Willow* iii. i. 135 No Canadian steer would ever be angry or stubborn; he would be o'nery or ringy or on the prod.

one-shot, *a.* and *sb.* [One *numeral a.* 32 a.] **A.** *adj.* Achieved or done with a single shot, stroke, attempt, etc.; consisting of a single shot or try; occurring, performed, produced, used, etc., only once; single, isolated.
1907 *Westm. Gaz.* 28 Mar. 9/1 The one-shot hole..gives good play its just reward... A hole which can just, and only just, be reached from the tee by a fine driver is, therefore, an excellent hole. **1927** *Sunday Pictorial* 28 Aug. 8/4 This includes such up-to-date owner-driver features as..one-shot oiling for all other chassis points. **1948** *Sun* (Baltimore) 31 May 8/2 For this he asked a force in being fully equipped and trained. He called this 'a stop-gap, one-shot army, a plug in the dike until we rallied sufficient and effective reserves'. **1950** *N.Y. Times* 28 Dec. 3/6 A 'one-shot' insecticide system that operates when the pilot pushes a button. **1953** Pohl & Kornbluth *Space Merchants* (1955) ii. 16 Fowler Schocken was too big for one-shot accounts. What we wanted was the year-after-year reliability of a major industrial complex. **1954** K. W. Gatland *Devel. Guided Missile* (ed. 2) 29 The latter method is of chief interest for 'one-shot' rockets as the target plate burns away during running unless low specific impulse propellants are used. **1959** E. Fenwick *Long Way Down* xx. 155 It was hard to get anybody for a one-shot cleaning job. **1962** A. Nisbett *Technique Sound Studio* ii. 43 A 'one-shot' technique was used, i.e. the whole programme was taken as a continuous sequence. *Ibid.* xiii. 232 It is essential that the basic 'message' of a piece should be understood in a single hearing—for sound is basically a 'one-shot' medium. **1966** *Listener* 18 Aug. 234/1 Not enough one-shot original plays are presented on television. **1968** M. Woodhouse *Rock Baby* xvii. 163 If it had been a one-shot thing, we might have been able to do it that way... But we couldn't afford any sort of mistake. **1972** D. E. Westlake *Cops & Robbers* (1973) xvi. 251 We were pitting our one-shot plan against a normal company's normal routine. **1978** *Guardian Weekly* 7 May 15/4 Copper produces 90 per cent of Zambia's foreign exchange, and the percentage is also high for the other producing countries that have what some economists call 'one-shot economies'.
B. *sb.* An event, transaction, process, etc., that occurs only once; something that is used or intended for use only once; *esp.* a single appearance by a performer, production of a play, etc.; a story or article that has no sequel. Also *one-shotter.* orig. *U.S.*
1937 *Printers' Ink Monthly* May 40/1 *One shot*, a single program which is not one of a series. **1942** H. Haycraft *Murder for Pleasure* xi. 267 Some..magazine editors have been experimenting with novelette-length condensations ('one-shots' as they are called in the trade). **1943** *Sat. Even. Post* 20 Nov. 28/3 A one shot..is usually a charity event sponsored by a political or social organization with no professional knowledge of selling tickets. **1947** *Jrnl. Brit. Interplanetary Soc.* VI. 113 The application for which a motor is designed also has a profound effect on its design. The major variables are, magnitude and duration of thrust; fixed or variable thrust; whether for repeated use or a 'one-shot', and in the former case its total operating life. **1967** A. Arent *Gravedigger's Funeral* (1968) iv. 44 What was it going to be? A brush-off? A friendly hint that this was just a one-shotter? **1967** Wodehouse *Company for Henry* ix. 172 He..has actually sold it [*sc.* the book] as what he calls a one-shotter to a magazine. **1972** M. J. Bosse *Incident at Naha* iii. 137 'But you'd give her *my* money?' 'Sure, because you're a one-shot. I'd never have any peace if the bread came from me.'

one-sided, *a.* **2. b.** (Earlier and later examples.)
1813 H. Muhlenberg *Catal. Plantarum Americæ Septentrionalis* 49 One-sided Hawthorn. **1832** J. Lindley *Introd. Bot.* iv. 413 One-sided..; having all the parts by twists in their stalks turned one way; as the flowers of Antholyza. **1945** Step & Jackson *Wayside & Woodland Ferns* (ed. 2) 25 Wilson's Filmy-fern... Known also as the One-sided Filmy-fern.

one-sidedness. (Earlier and later examples.)
1831 J. S. Mill *Let.* 20–22 Oct. (1910) I. 11 The next thing that struck me was the extreme comprehensiveness and philosophic spirit which is in him [*sc.* Wordsworth]. By these expressions I mean the direct antithesis of what the Germans most expressively call one-sidedness. **1974** tr. *Wertheim's Evolution & Revolution* i. 20 Another factor contributing to a rather general rejection of 'evolutionism'..was connected with a certain one-sidedness in evolutionary views, as propagated about 1900. **1977** *South China Morning Post* (Hong Kong) 13 Apr. 13/6 To avoid accusations of one-sidedness and to round out the image of his subject, Dr Abrahamsen spoke to a number of Mr. Nixon's admirers.

one-step (wɒ·nstep), *sb.* Also **one step.** [f. One *numeral a.* + Step *sb.*] A ballroom dance in quick time, in which the steps resemble simple walking. Hence as *v. intr.*, to dance the one-step.
1911 *Home Chat* 7 Oct. 108/1 Camilla is just mad about the 'One-step'. **1914** V. Castle *Mod. Dancing* 44 Simply *walk* as softly and smoothly as possible, taking a step to every count of the music. This is the One Step, and this is all there is to it. **1916** H. L. Wilson *Somewhere in Red Gap* iv. 172, I caught him..in the deserted library later, while the rest was one-stepping in the..ballroom. **1921** H. S. Walpole *Young Enchanted* iii. i. 230 Bunny says I one-step better than anyone he's ever known. **1924** [see *hesitation 3]. **1938** B. Schonberg tr. *Sachs' World Hist. of Dance* vii. 445 We have shortly after 1900 the *one-step* or *turkey trot.* **1956** G. P. Kurath in A. Dundes *Mother Wit* (1973) 106/1 The one-step or turkey-trot.. was little more than a smooth walk. **1969** F. Rust *Dance in Society* x. 84 After the one-step came the fishwalk and the horsetrot—of ephemeral interest only.

one-way, *a.* [f. One *numeral a.* + Way *sb.*[1]]
† **1.** Of bread: see One *a.* 33. *Obs.*
2. Applied to a plough which can turn the furrows in one direction only. Also *ellipt.*
1884 [see One *a.* 33]. **1886** F. T. Elworthy *West Somerset Word-Bk.* 537 A two-way-zull, eens can plough vore and back in the same vore, is a handy thing like, but can't make such good work way un's can way a proper good one-way-zull. **1960** *Times* 15 Feb. 19/2 A one-way plough much lighter in weight. **1965** G. Shepherd *West of Yesterday* x. 77 We used a plough, for there was no 'one-way', as the modification of the disc plough was later named.
3. a. Leading, tending, pointing, thinking, or developing in one direction only.
1824 M. Wilmot *Let.* 5 Feb. (1935) 206 Our one way life, dearest Alicia, gives me so little to say. **1928** A. S. Eddington *Nature Physical World* 295 The notion evidently implies that something may be born into the world at the instant Here-Now, which has an influence extending throughout the future cone but no corresponding linkage to the cone of absolute past. The primary laws of physics do not provide for any such one-way linkage. **1938** L. MacNeice *Earth Compels* 61 Endurance of one-way thinking. **1951** Koestler *Age of Longing* i. 6 One-way pupils that took the light in, gave nothing out. *Ibid.* vii. 127 He put on the guarded, one-way gaze. **1953** J. S. Huxley *Evolution in Action* i. 12 All reality, in fact, *is* evolution, in the perfectly proper sense that it is a one-way process in time. **1960** Partridge *Charm of Words* 40 In a one-way dictionary, the explanations are made in the same language as that of the words defined. **1966** G. N. Leech *Eng. in Advertising* v. 48 Many intonation contrasts in English signal personal attitudes and contextual presuppositions which can scarcely apply to one-way public communication. **1973** *Australian* 17 Dec. 16/7 Your child could be studying under a Miss Brooks at school and your spouse or boy-friend may well be a one-way baby (simply a term for self-centred emotional types). **1977** *Rep. Comm. Future of Broadcasting* iii. 19 Broadcasting..is a one-way communication; viewers and listeners cannot question or express approval or disapproval.
b. *spec.* Of a ticket: entitling the holder to a journey in one direction only; 'single'. Also *fig.*
1906 *Dialect Notes* III. 148 Over three hundred negroes left Springfield, purchasing oneway tickets to many different towns. **1949** L. Hughes (*title*) One-way ticket. **1973** *Nation Rev.* (Melbourne) 31 Aug. 1464/2 One journalist..later earned himself further notoriety and a oneway ticket to Van Diemen's Land. **1976** J. Lee *Ninth Man* 258 You've bought yourself a one-way ticket to obscurity. **1977** *Times* 5 Oct. 17/8 On most days people are not having to queue to buy their one-way tickets to America at £59 a head.
c. Of a thoroughfare: along which traffic is permitted in only one direction; of traffic: passing only in one direction; also, of or pertaining to such traffic. Also *fig.*
1914 *World's Work* Aug. 302/1 Some little has already been done in the small streets off Piccadilly to *request* drivers to avoid some streets when going north and others when going south, thereby aiming at 'one-way' traffic, but there is no power to enforce the requests. *Ibid.* 304/1 Where streets are too narrow to permit of the rotary

system the difficulty can be overcome by one-way streets. **1926** *Glasgow Herald* 11 Sept. 9 A complaint has been heard from shop-keepers against the one-way system in certain streets. **1933** [see *clover-leaf* s.v. *clover *sb.* 4]. **1956** B. Holiday *Lady sings Blues* (1973) xxi. 171 This is a one-way street. If someone plants something on you and you're innocent, you have no way in the world to prove it. **1959** *Daily Tel.* 8 May 12 One-way study for London. *Ibid.*, The pros and cons of proposals for one-way traffic. **1961** [see *dream *sb.*[2] 1 b]. **1961** L. van der Post *Heart of Hunter* iii. xv. 202 The European..tends to believe that the consequences are only for the primitive and that he..is immune from them. But actually there is no one-way traffic on these eventful occasions. **1963** *Traffic in Towns* (Ministry of Transport) ii. 40/2 One-way streets and the elimination of right-hand turns have been the main features that have caught public attention. **1970** P. Laurie *Scotland Yard* iv. 96 They put up temporary one-way signs, controlled junctions. **1972** J. Gores *Dead Skip* (1973) xv. 105 Kearny entered town on one-way Howard Street. **1976** *Northumberland Gaz.* 26 Nov. 19/9 It was virtually one-way traffic in the second half as Berwick kept the visitors pinned in their own area. Heslop and Renwick added further tries and Dudgeon kicked a further three penalty goals.
d. *one-way pockets*: the pockets of a miserly person. *slang.*
1926 Maines & Grant *Wise-Crack Dict.* 11/2 *One-way pockets*, pockets of tightwad. **1961** Wodehouse *Service with Smile* (1962) ix. 143 His one-way pockets are a by-word all over England.
e. Of a window, mirror, or the like: that permits vision from one side; transparent from one side only.
a **1940** F. Scott Fitzgerald *Last Tycoon* (1949) ii. 31 Nowadays all chief executives have huge drawing rooms, but my father's was the first. It was also the first to have one-way glass in the big French windows. **1961** W. Brown *Bedeviled* 40 Obscene exhibitions viewed through peepholes made of one-way glass. **1964** F. Pohl in *Galaxy Mag.* Oct. 192/2 The cameras..that the studio people had activated for me behind every one-way mirror in the room. **1967** C. Drummond *Death at Furlong Post* iii. 27 What looked like a dirty bit of glass was a one-way window. **1972** *Jrnl. Social Psychol.* LXXXVIII. 153 Further, Ss [*sc.* subjects] were observed during the experiment through the one-way mirror for any reactions. **1975** *Times Lit. Suppl.* 7 Mar. 241/2 Two of Dizzy's aristocratic sprigs jostle at a one-way mirror watching the most high-minded statesman of all trying to reclaim a tart.
4. *Electr.* Of a switch or the like: providing only one possible path for current.
1896 W. P. Maycock *Electr. Lighting* (ed. 3) I. v. 113 Fig. 49 shows a simple or one-way switch. **1925** O. Rankin *Switches in Wireless Circuits* 56 Two ordinary 'one-way', or 'bell switches', A and B, are used to effect the usual series-parallel switching of the A.T.C. **1965** P. Honey *Planning Electricity in House* iii. 71 Wall switches are available in one-way and two-way types, the latter being used for the control of lights from two different points (e.g. on staircases).

onewhere, *adv.* (Further example.)
1887 G. M. Hopkins *Poems* (1918) 65 Each limb's barrowy brawn, his thew That onewhere curded, onewhere sucked or sank.

onflowing, *vbl. sb.* and *ppl. a.* Add: (Further examples.) Also **on-flow** *v. intr.*, to flow on.
1861 A. de Vere *Sisters* 17 No eye Finer pursued the on-flowing line: her wheel Murmur'd complacent joy. **1879** —— *Legends of Saxon Saints* ii. 23 Forward on-flowed in Apostolic might Augustine's strong discourse. **1905** R. Davey tr. *Serao's In Country of Jesus* VI. iv. 167 The rapidly on-flowing waters reflect the azure blue of the sky above. **1930** *Tablet* 26 July 107/2 The large and stately on-flowing of history.

on-glaze, *a.* (*sb.*) *Ceramics.* [f. On *prep.* + Glaze *sb.*] Of, pertaining to, or designating colour, a pattern, etc., applied on top of a glaze; = Overglaze *a.* Also as *sb.*, colour, etc., applied on top of a glaze.
1897 Sparkes & Gandy *Potters* i. 56 The class of enamel colours... The name of *overglaze* or *onglaze* is borne by the whole group. **1913** J. C. Wedgwood *Staffordshire Pott.* viii. 131 There was something crude and hard about the effect of the on-glaze printing. *Ibid.* 133 In 1770 he [*sc.* Josiah Spode] leased Banks' works in the centre of Stoke, and began making printed cream colour. This was the old 'on-glaze', or 'black' printed ware, used to guide the enameller rather than as a decoration by itself. **1957** Mankowitz & Haggar *Conc. Encycl. Eng. Pott. & Porc.* 65/2 The development of on-glaze enamelling c.1760 by the Daniel family in Hot Lane. *Ibid.* 168/2 *On-glaze*, decoration applied after the ware has been glazed and fired. **1959** *Which?* Oct. 127/1 A good deal of decorated pottery has patterns applied on top of the glaze, because this allows a much wider range of colours to be used... With this type of decoration, known as on-glaze, the colours are directly exposed to the action of the detergent. **1961** M. Jones *Potbank* xxv. 108 On-glaze decoration, working on ware that has been glazed and then fired, is more usual [than under-glaze decoration]. **1969** *Canad. Antiques Collector* Feb. 10/1 After this glost firing the onglaze enamel paintings of birds and flowers are done. **1974** *Nature* 25 Jan. 197/2 Cadmium sulphide has been shown by x-ray diffraction to be present in the onglaze decoration.

o·n-glide. *Phonetics.* [f. On *a.* + Glide *sb.*] A glide produced at the beginning of articulating a speech-sound. Cf. *off-glide. Hence **on-gliding** *ppl. a.*
1888 H. Sweet *Hist. Eng. Sounds* 10 All consonants consist of three elements, (1) the consonant itself, (2) the on-glide, and (3) the off-glide. **1919** E. Kruisinga *Handbk. Present-Day Eng.* (ed. 3) I. iv. 72 The chief difference between English and Dutch voiced stops is the difference between the on-glides of initial stops. **1934** *PMLA* XLIX. 1167 In the English pronunciation of parasitic (n) before (t) in *maintenant*, the (n) is an off-glide of the nasal vowel and an on-glide of the (t). **1934** Priebsch & Collinson *German Lang.* 385 The intermission of the breath on-glide (so noticeable in English and German). **1950** D. Jones *Phoneme* 5 The on-glide and stop of this affricate resemble those of the English t. **1954** [see *off-glide]. **1965** *Language* XLI. 478 The first allophone has a rapid palatal onglide, the second has a rapid onglide of central-vowel quality; both onglides are conditioned by the preceding consonants. **1976** *Ibid.* LII. 341 Under certain conditions of stress and sentence rhythm, syllable-final stops come to be preceded by vocalic onglides.

on-going, *sb.* **2.** Delete *rare*, and add earlier and later examples.
1860 F. W. Faber *Precious Blood* iii. 107 Not only is the continuous preservation of all things..an almost illimitable extension and ongoing of creation, but new souls of men are literally created out of nothing every moment of time. **1880** W. M. Thomson *Land & Bk.* 2 The long ongoing and outworking of the Mosaic Economy. **1962** *Listener* 11 Oct. 559/1 Experiment must be analytical and predictive in its actual on-going. **1977** *Times* 11 Apr. 5/7 The logical conclusion of a process long in the ongoing.

on-going, *a.* Add: Also **ongoing.** Also, continuing, continuous; that is in progress; current; proceeding, or developing. (Earlier and later examples.) Hence **o·n-go:ingness.**
1877 J. Blackwood *Let.* 15 Oct. in *Geo. Eliot Lett.* (1956) VI. 405 This edition..will be a steady on-going thing, a capital leading franchise in the business. **1937** [see *integration 1 b]. **1949** M. Mead *Male & Female* xvi. 336 American women have become..less willing to be merely part of some on-going operation. **1951** V. Barclay *Challenge to Darwinians* vi. 60 Bergson..held that an ongoing movement issued from that creative act. **1951** E. C. Tolman in Parsons & Shils *Toward Gen. Theory Action* iii. 352 What..would be the necessary conditions for the non-disintegration, the on-goingness, of a personality system? **1953** *Scottish Jrnl. Theol.* VI. 211 Within the time of this on-going age there would be a thousand years of political triumph for the Jews. **1954** D. Riesman *Individualism Reconsidered* (1955) xxii. 334 Sufficient time has elapsed since Freud built his system..to permit.. critical re-examination of the sort undertaken here... Its aim is..to contribute to the sociology of knowledge and to the ongoing effort. **1957** P. Lafitte *Person in Psychol.* 59 Life..is tense,..as an inescapable consequence of its being rich, varied, ongoing, and creative. **1959** P. Wheelwright *Heraclitus* iii. 38 For Heraclitus the most basic ontological fact is the ongoingness of things. **1960** in L. Pincus *Marriage* 8 The worker's capacity to learn from his on-going cases. *Ibid.* i. 16 This pattern of interaction can be understood only in the context on an on-going process. **1961** R. Kee *Refugee World* vii. 73 The refugee problem in our time is an on-going problem. **1967** G. Wills in Wills & Yearsley *Handbk. Managem. Technol.* 179 Time, in an on-going marketing situation, costs money. **1972** P. Laslett *Household & Family in Past Time* 68 The danger of mistaking a set of coincidences for an ongoing institution. **1973** *Guardian* 13 Oct. 2/3 We have an on-going military relationship which we are continuing. **1974** *Times* 1 May 6/8 The President assigned the responsibility for the on-going investigation to Mr Petersen. **1976** *Publishers Weekly* 16 Feb. 85/1 The ongoing polemic about the role of the atelier artist versus his university-based colleague. **1977** *Gay News* 24 Mar. 13/1 Dozens of other special relationships hold the potential of working well and of maintaining the symbiosis of ongoingness.

ongon (ǫ·ngon). [Russ.] In the Shamanist religion of the Buriats of Mongolia, an image of a god or spirit supposed to be endowed with the power of the force it represents; a fetish.
1901 D. Bantzaroff tr. J. Stadling in *Contemp. Rev.* LXXIX. 89 The dim idea of the immortality of the soul and a future life gave rise to the *Ongones*, the deified spirits of the ancestors. **1910** *Encycl. Relig. & Ethics* III. 12/1 The Turks of Yenisei call the ongon *tyus*, whereas among the Altaians it is named *Kurmes*. On the one hand, it is an image of God, and, on the other, God himself, a fetish possessed of his own power. The tyus, or ongon, reminds us of the rôle which among some Christian peoples is filled by the images of saints. **1936** V. A. Demant *Christian Polity* xi. 191 The Ongons of the Mongolic Buriats are effigies of dead heroes, and so are the images of Indo-China. **1950** *Funk's Stand. Dict. Folklore* II. 823/1 *Ongon*, in Buriat religion, an image embodying a god and therefore possessing the power of the god: among the Altai called *kurmes*, among certain Turks, *tyus.* **1970** *New Society* 5 Mar. 393/1 The word, 'ongon', means both a spirit and the material representation of a spirit. Drawings are made only of known spirits, each of which has particular magical powers. Since the representation *is* the spirit, the drawings themselves become magical: according to the spirit, an ongon can cure smallpox, keep young lambs healthy, give protection to fishermen and so on.

Oni (ōu·ni). Also with lower-case initial. [Yoruba.] The title given to the ruler of Ife, a large town now in the Western State of Nigeria.
1900 *Niger & Yoruba Notes* Sept. 19/1 The Oni of Ile Ife..granted..a desirable plot of ground for the Mission premises. **1904** *Jrnl. Afr. Soc.* July 472 The Oni of Ife.. has had very little political power during the present generation. **1911** *Encycl. Brit.* XXVIII. 937/1 The chief of Ife bears the title of *oni* (a term indicating spiritual supremacy). To the oni of Ife or the alafin of Oyo all the other great chiefs announce their succession. **1937** *Nigeria* xii. 4 The earliest Onis of Ife were reputed to have reigned each for as long as 200 or more years. **1974** J. R. Baker *Race* xxi. 412 Most of the Ife bronzes are in the Oni's palace. **1967** F. Willett *Ife in Hist. W. Afr. Sculpture* 22 The fact that three of the heads were crowned,..suggests that these are representations of Onis or divine kings of Ife... During this century, the Oni has appeared increasingly in public, but even the Oni Ademiluyi, who died in 1931, used to cover his mouth with a fan when he as much as took kola in public. **1976** *Daily Times* (Lagos) 8 July 11/6 Hundreds of traditional rulers including the Oni of Ife..are attending the meeting.

-onic, *suffix. Chem.* [f. -on(e + -ic, prob. after *lactonic *a.* 1.] An ending used in forming the names of acids, esp. of carboxylic acids obtained by oxidation of aldoses, as *galactonic, *gluconic, *uronic *adjs.*, etc. (Cf. also Arsonic *a.* (f. Arsonium), Phosphonic *a.* (f. Phosphonium), Sulphonic *a.* (f. Sulphone).)

onion, *sb.* Add: Forms: β. Also 9 *U.S. dial.* **ineon, ingyon.**
1825 J. Neal *Bro. Jonathan* II. 84 Ingyons are proper good, when ye're sick.
2. b. **bog onion**, substitute for def.: one of several plants with roots resembling an onion, esp. the royal fern, *Osmunda regalis*; (earlier and later examples).
1832 W. D. Williamson *Hist. State Maine* I. 120 The Brake, of which there are several varieties, the root of which is sometimes called the 'bog-onion'. **1878** W. Dickinson *Cumberland Gloss.* (ed. 2) 9/1 *Bog onion*, the *Osmunda Regalis* or flowering fern. **1892** *Jrnl. Amer. Folk-Lore* V. 104 *Arisæma triphyllum*, bog onion. Worcester Co.
5. b. *slang.* Head; esp. in phr. *off one's onion*, mad.
1890 Barrère & Leland *Dict. Slang* II. 94/2 *Off his onion* (costermongers), imbecile, cracked. **1909** H. G. Wells *Tono-Bungay* ii. ii. 176 He come home one day saying Tono-Bungay till I thought he was clean off his onion. **1922** Wodehouse *Girl on Boat* xii. 202 When.. she informed him one day that she was engaged.., he went right off his onion. **1928** *Daily Express* 11 Dec. 7/4 After four drops of beer I am properly off my onion. **1971** Wodehouse *Much Obliged, Jeeves* vi. 52 What on earth was the idea of inviting a fiend in human shape like that here?.. You must have been off your onion, old ancestor.
c. *to know one's onions*, to be experienced or knowledgeable in the subject, etc., on hand; (only P. G. Wodehouse) (*not*) *the only onion in the stew*, (not) the only person or thing to be taken into consideration.
1922 *Harper's Mag.* Mar. 530/1 Mr. Roberts knows his onions, all right. **1934** Wodehouse *Right Ho, Jeeves* vii. 75, I claim the right to have a pop at these problems.. without having everybody behave as if Jeeves was the only onion in the hash. **1952** 'E. C. R. Lorac' *Murder in Mill-Race* v. 52 If I know my onions the woman's death has been an almighty relief to the lot of them. **1956** S. Ertz *Charmed Circle* v. 86 'That old man,' he said, 'doesn't know his onions, luckily for you.' **1958** J. Cannan *And be a Villain* ix. 200 Shakespeare knew his onions, didn't he? **1958** *Times* 16 June 9/4 A man 'who knows his onions' is a man wise in the ways of the world, shrewd in affairs, a tough bargainer, by no means born yesterday. **1972** Wodehouse *Pearls, Girls & Monty Bodkin* vii. 109 She wanted to stimulate competition. By showing you you weren't the only onion in the stew she would get your attention. **1974** J. Wainwright *Evidence I shall Give* xxxii. 166 They know their onions... They are old in wisdom and experience.
6. b. *Naut.* A fraction of a knot.
1916 'Taffrail' *Pincher Martin* v. 73 We got about six and an onion out of the old bus,..and reached there about noon. **1938** F. A. Worsley *First Voy. in Square-Rigged Ship* iv. 71 The speed..was 13 knots or, as Stringer put it: 'Thirteen and an onion in the squalls.' **1958** F. H. Shaw *Seas of Memory* ii. 48 'Fifteen, sir, fifteen and an onion!' called the second mate. 'That's the way I like her to move,' said Fegan.
c. = *flaming onion* (*flaming *ppl. a.* 1 c).
1917 *Blackw. Mag.* Apr. 560/1 A line of fiery rectangles shot up... These were 'onions', the flaming rockets which the Boche keeps for..hostile aircraft. **1918** in *Amer. Speech* 1972 (1975) XLVII. 84 The airmen's pest is the 'onion', or large flaming anti-aircraft shell. **1936** 'McScotch' *Fighter Pilot* vi. 122 On heading south for the other balloon the 'onion' battery had another shot at me.
8. *onion bed* (later examples), *crop*, *-green* (also as adj.), *roll*, *salt*, *sauce*, *-seller* (later examples) (so *-selling*), *soup*, *spire*, *steeple*; *onion-loving*, *-shaped*, *-spired*, *-towered* adjs.; **onion dome**, a dome on a church, palace, etc., shaped like an onion; so **onion-domed** *a.*; **onion-fly**, for '*Anthomyia ceparum*' substitute '*Delia cepetorum*'; (earlier and later examples); **onion ring**, a circular segment of an onion; **onion set** (see quots.); **onion-skin** (*b*) (examples); also (see quot. 1879) a ballot paper of very fine paper; also *attrib.*
1857 *Quinland* I. i. xiii. 184, I spaded up the onion-bed after supper. **1874** *Rep. Vermont Board Agric.* II. 551

Raked as smooth as an onion bed. **1975** D. Green *Food & Drink from your Garden* 91 They..no longer need the special onion beds which took so many years to perfect. **1879** *Congress Rec.* 46th Congress 1 Sess. App. 120/1 The onion crop of South Carolina. **1956** R. Macaulay *Towers of Trebizond* ii. 20, I dreamed too of the Crimea, of crumbling palaces decaying among orchards by the sea, of onion domes. **1960** N. Mitford *Don't tell Alfred* xxiv. 245 The French papers were full of lines and sidelines on Russia, no photograph without its onion dome. **1973** J. M. White *Garden Game* 54 The exotic onion-dome of a church looming through the veiled whiteness. **1959** *Manch. Guardian* 26 Feb. 9/4 The Kremlin, with its three onion-domed cathedrals. **1974** *Aiken* (S. Carolina) *Standard* 22 Apr. 6-A/1 Entering through Persian onion-domed archways, guests saw the Fermata Club in Aiken transformed into a festive pavillion of purple and orange. **1840** J. & M. Loudon tr. *Köllar's Treat. Insects* ii. 159 The larva very much resembles that of the onion fly. **1966** *Punch* 6 Apr. 510/2 Sets..will grow onions..without onion fly risk. **1975** D. Green *Food & Drink from your Garden* 92 The main pest is the onion fly, which lays its eggs in May and June. **1906** S. W. Bushell *Chinese Art* II. viii. 23 The brilliant grass-greens of the Lung-ch'üan porcelain, called *ts'ung-lü*, or 'onion-green' by the Chinese. **1925** W. de la Mare *Two Tales* 71 The very ferocious onion-green dragon. **1811** Shelley *Let.* 17 May (1964) I. 76 How gets on your onion-loving Deist. **1952** M. Norton *Borrowers* iii. 24 She [*sc.* a midget] took the onion ring from Homily and slung it lightly round her shoulders. **1974** *Times* 21 Feb. 10 Thinly sliced onion rings. **1967** C. Potok *Chosen* v. 100 Lunch turned out to be a massive affair, with a thick soup, fresh rye bread, onion rolls, bagels. **1972** *New Yorker* 15 Apr. 35/3 Most of the women buy some kind of bread (a loaf of rye bread ..or a few onion rolls). **1938** E. Waugh *Scoop* iii. ii. 287 A little store of seasonings..onion salt, Bombay duck, gherkins. **1958** Onion salt [see *garlic salt*]. **1723** J. Nott *Cook's & Confectioner's Dict.* sig. X5 (*heading*) To make onion sauce. Cut..Onions into slices, put them into a Sauce-pan with some Veal-gravy,..simmer. **1787** J. Woodforde *Diary* 4 Dec. (1926) II. 356, I gave them for Dinner..a couple of Rabbitts boiled and Onion Sauce. **1877** E. S. Dallas *Kettner's Bk. of Table* 320 (*heading*) Onion sauce.—See the Soubise sauce, the Breton sauce, and the Sauce Robert. **1939** T. S. Eliot *Old Possum's Pract. Cats* 45 And when he's finished, licks his paws So's not to waste the onion sauce. **1914** W. B. Yeats *Responsibilities* 15 What th' onion-sellers thought or did. **1970** V. Canning *Great Affair* xvi. 300 Troops.. strung with hand grenades..like French onion sellers. **1915** *Daily Chron.* 23 Nov. 5/3 Onion-selling parties in England. **1886** *Harper's Mag.* Oct. 708/2 'Onion sets'.. are produced by sowing the ordinary black seed very thickly on light poor land. **1951** *Dict. Gardening* (R. Hort. Soc.) III. 1424/2 Small bulbs grown in the previous autumn and known as 'onion sets' may..be planted in spring for the raising of dry bulbs. **1975** D. & T. Hoobler *Vegetable Gardening & Cooking* 77 Growing onions from seeds takes up to four or five months, so most home gardeners buy onion 'sets', which are the baby onion bulbs, ready to be buried in early spring, 2 inches deep, 4 inches apart, in rows 12 inches apart. **1949** R. Harvey *Curtain Time* 97 And a brand-new wooden church, bright blue with a yellow onion-shaped dome. **1959** J. Braine *Vodi* vi. 87 The chapel..was a compact red-brick building with large round-headed windows, topped rather incongruously by a tower with an onion-shaped dome. **1879** C. G. Williams in *Congress. Rec.* 2 Apr. 167/2 From that time to the wee small hours of the morning onion-skin ballots went in unchallenged but not uncounted. *Ibid.* 23 June App. 120/1 The term 'onion skin' or 'tissue ballots' has obtained a generic and well-defined meaning synonymous with the 'stuffing' of ballot-boxes. **1892** *Paper & Press* July, facing p. 49 (Advt.), The Highest Grades of Typewriter Paper a Specialty. Onion Skin. Manifold Linen. **1922** *Handbk. Quality-Standard Papers* (Amer. Writing Paper Co.) 360 Onion Skin. A thin, transparent, highly glazed paper made of rag and sulphite. **1923** H. A. Maddox *Dict. Stationery* 56 *Onion skin*, an American paper trade expression..applied to very thin and crisp typewriting or bank paper, which in texture, tear and crackle has some-of the nature which characterizes the skin of an onion. **1956** S. Bellow *Seize the Day* (1957) iv. 99 He took out a substantial bundle of onion-skin papers and said, 'These are the receipts of the transactions. Duplicates.' **1970** *New Yorker* 20 June 25/2 The shredding of a quarto of onionskin stationery, to simulate the tearing up of a billion dollars. **1973** R. Thomas *If you can't be Good* (1974) xxii. 191 He handed me some folded sheets of onion skin.. I unfolded the onion-skin sheets. **1747** H. Glasse *Art of Cookery* ix. 77 An Onion Soop. **1861** Mrs. Beeton *Bk. Househ. Managem.* vi. 73 Onion Soup... 6 large onions,.. ½ pint of cream. **1942** E. Paul *Narrow St.* xxv. 223 Most of us missed our lunch but ate onion soup and sausage with sauerkraut in midafternoon. **1966** J. B. Priestley *Salt is Leaving* xiii. 179 He..opened a tin, French and good, of onion soup. **1977** P. Harcourt *At High Risk* i. 31 We settled for onion soup, a *filet* with a wine sauce, salad. **1966** *New Statesman* 17 June 893/1 The onion spires of Alaska. **1959** *Times* 25 Apr. 9/5 A baroque, onion-spired church. **1868** G. M. Hopkins *Jrnls. & Papers* (1959) 179 The churches here have those onion steeples nearly all. **1960** *Times* 11 June 11/6 A Church with Baroque 'onion' steeple. **1959** *Listener* 15 Jan. 131/1 Almost every little South Swabian and Bavarian village has its delightful onion-towered church.

oniony, *a.* (Later examples.)
1922 Joyce *Ulysses* 233 Armpits' oniony sweat. **1971** *Guardian* 5 May 9/4 Onions always smell oniony. **1975** L. Lee *I can't stay Long* 31 Garlic sprawls rank and oniony in the woods. **1977** *Times* 10 Dec. 9/6 An oniony omelette.

-onium, *suffix. Chem.* [abstracted from Ammonium.] Used in forming the names of complex cations that contain a more or less electronegative central atom, usu. bonded to a number of protons (or to other species that are regarded as substituents), as Arsonium, *carbonium, *hydrazonium, *nitronium, *oxonium, Phosphonium, *tetrachlorophosphonium*, etc.
1971 *Nomencl. Inorg. Chem.* (I.U.P.A.C.) (ed. 2) 20 Names for polyatomic cations derived by addition of more protons than required to give a neutral unit to monatomic anions, are formed by adding the ending -onium to the root name of the anion element.

onium (ōu·niǝm), *a. Chem.* Also **'onium.** [f. prec.] Applied to (compounds containing) ions of the kind named in *-onium.*
1905 *Jrnl. Chem. Soc.* LXXXVIII. i. 281 Carbon differs from other elements, which form 'onium' bases, in that it forms salts only. **1923** G. N. Lewis *Valence* ix. 108 The formation of the typical 'onium' ion is a process which differs in no essential respect from the other processes in which hydrogen or other radicals become attached to lone pairs. **1952** Kirk & Othmer *Encycl. Chem. Technol.* IX. 596 Several important classes of dyes, for example the cyanine dyes.., the azine dyes.., and the amino-substituted triphenylmethane dyes.., are onium compounds. **1953** C. K. Ingold *Struct. & Mech. Org. Chem.* v. 208 It is not necessary that the anion of the 'onium salt should be the substituting agent. **1973** J. F. Willems in R. J. Cox *Proc. Symposium Photogr. Processing Univ. Sussex* 227 Various onium compounds considerably accelerate the bleaching out of the silver.

onkosine, var. *oncosine.

onlap (ǫ·nlæp). *Geol.* [f. On *adv.* + Lap *v.*[2], after *offlap.] A progressive increase in the lateral extent of conformable strata in passing upwards from older to younger strata, so that each stratum is hidden by the one above; a set of strata exhibiting this.
1947 F. A. Melton in *Bull. Amer. Assoc. Petroleum Geologists* XXXI. 1869 The writer proposes that the simpler name *marine-onlap*, which has already been used by various authors, be substituted for the more cumbersome term used by Grabau. Marine-onlap is thus used to describe the regular progressive pinching-out of marine strata above an unconformity..in such a way that the younger beds extend farther landward than do the older beds which lie beneath... The term terrestrial-onlap can be used in connection with terrestrial formations. **1955** *Sci. Amer.* Mar. 84/2 When the sea advanced, under the simplest conditions the new deposits overlapped the older in a shoreward direction—a process called onlap. **1968** R. W. Fairbridge *Encycl. Geomorphol.* 340/1 The 'Schooley Peneplain' of the Appalachians..dips unmistakably under the mid-Tertiary transgressive onlap of the Atlantic Coastal Plain.

onlay, *v.* For † *Obs.* read '*Obs.* exc. as *ppl. adj.*' and add later examples.
1880 L. Higgins *Handbk. Embroidery* v. 54 'Onlaid appliqué' is done by cutting out the pattern in one or many coloured materials, and laying it down on an intact ground of another material. **1971** *Bodl. Libr. Rec.* VIII. 264 The covers are decorated with onlaid straw, a wide outer band of large stylized flowers..with doublures bearing similar straw onlays. **1976** *Times Lit. Suppl.* 25 June 805/4 Thomas Fassam's *An Herbarium for the Fair*, 1949, onlaid with butterflies and woodruff by a new-comer, Angela James.

onlay, *sb.* Add: (Further examples.)
1959 L. M. Harrod *Librarians' Gloss.* (ed. 2) 194 *Onlay*, a decorative panel of paper or other material glued to the cover of a book without preparing the cover to receive it. **1961** J. Carter *ABC for Bk. Collectors* (ed. 3) 139 The technique was occasionally adapted to publisher's cloth between 1840 and 1860, when the onlays were sometimes of paper. **1971** [see prec.]. **1976** *Times Lit. Suppl.* 25 June 805/1 Technical innovations since the Second World War have greatly extended the binder's decorative range. Before then decoration was limited to gold or blind-tooling and coloured onlays.
b. *Dentistry.* An occlusal rest extended so as to cover the whole occlusal surface of a tooth.
1906 J. A. Lentz *U.S. Pat.* 833,883 23 Oct., My objects are, first, to facilitate and expedite the reproduction or duplication in gold, gold alloy, or similar substance of a variety of forms, such as inlays, onlays, cusps, [etc.]. **1935** G. M. Anderson *Dewey's Pract. Orthodontia* (ed. 5) xxii. 427 If the tooth is sufficiently exposed so that one need not cut into it, an onlay may be used in conjunction with the auxiliary spring. **1973** L. Baum *Advanced Restorative Dentistry* xi. 169 Onlays are generally more acceptable than inlays in middle-aged and older patients because the design of onlay preparations provides for a casting which will bond together the remaining tooth structure.
c. **onlay graft** *Surg.*: a bone graft in which a piece of bone is fixed over a fracture.
1927 *Southern Med. Jrnl.* (Nashville, Tennessee) XX. 114/2 Of the thirty-eight bones in which the onlay graft was employed, three failed to induce osseous union. **1957** Rob & Smith *Operative Surg.* V. ix. i. 12 Fixation by on-lay graft. The technique is the same as for fixation by a metal plate except that a cortical slab graft..is used instead of the metal plate.

on-licence: see On *a.* 2.

onlie, var. Only *a.* (in Dict. and Suppl.).

on-line (stress variable), *a.*, *adv.*, and *phr.* Also **online.** [f. On *prep.* + Line *sb.*[2]] **A.** *adj.* (Usu. stressed *o·n-line.*) **1.** *Computers.* Directly connected, so that a computer receives an input from or sends an output to a peripheral device, process, etc., as soon as it is produced; carried out while so connected or under direct computer control.
1950 W. W. Stifler *High-Speed Computing Devices* ii. 7 For some applications, of which the most prominent are those in which the reduced data are used to control the process being measured, the input must be developed for on-line operation. In on-line operation the input is communicated directly..to the data-reduction device. **1957** [see *off-line *a.* 2]. **1959** [see *in-line *a.* 3 c]. **1964** T. W. McRae *Impact of Computers on Accounting* i. 17 If we are processing..a payroll,..and the output printer is directly hooked up to the computer store so that each payslip is printed immediately after it is calculated, we use the term on-line processing. **1965** *Math. in Biol. & Med.* (Med. Res. Council) VI. 295 Without time-sharing, the 'on-line' use of a fast modern machine would be unthinkably costly. **1968** *Times* 26 Oct. 4/4 It was found..by radio astronomers using the 250 ft. dish telescope connected to an on-line computer. **1971** *Computers & Humanities* V. 192 The shoebox is an automatic text-processing and retrieval system implemented for on-line operation on an ibm 360/50 computer. **1972** *Accountant* 27 Apr. 549/1 A completely integrated computer data system which..through a built-in system of analysis and recording enables managers at any level to have immediate 'on-line' access to that part of the information which is relevant.
2. Occurring or effected on the current authorized routes of an airline.
1969 *Jane's Freight Containers* 1968–69 429/2 Online and interline use [*sc.* of air cargo pallets] by JAL. **1973** E. Rath *Container Systems* x. 285 Those airlines who had purchased both the 747 and either the DC-10 or the L-1011 realized that standardization of lower-deck containers would permit them to effect on-line and interline transfers of complete containers between these different types of aircraft.
3. = *in-line *a.* 2.
1972 *Physics Bull.* Jan. 29/3 Dr K. A. Andrews.. described the progress which was being made with the problem of on-line ultrasonic testing of hot steel. **1976** *Gramophone* Aug. 354/2 Mass production calls for high speed working, sophisticated on-line testing and..a high degree of automation.
B. *adv.* (Usu. stressed *on-li·ne.*) With processing of data carried out simultaneously with its production; while connected to a computer, or under direct computer control.
1950 [see *off-line *adv.*]. **1964** *Ann. N.Y. Acad. Sci.* CXV. 654 The goal of the development has been a machine which..is fast enough for simple data-processing 'on-line' while the experiment is in progress. **1966** *Economist* 23 July 382/3 Information will be available 'on-line'..to 100 BOAC centres throughout the world. **1966** *New Scientist* 27 Oct. 161/2 All the files of the users of the system are put on-line—that is, made directly accessible to the central processor. **1968** *Amer. Documentation* Jan. 72/1 Editing will be done on-line with a display scope and keyboard. **1977** *Catalogue & Index* xlvi. 8 Those who want to work on-line will be able to work in a format designed to reflect..the way the data is stored in the computer.
C. *phr.* (Written as two words.) = *on stream *adv. phr.* a.
1968 *Daily Colonist* (Victoria, B.C.) 28 Nov. 13/6 The Skookumchuck mill recently came on line, but none of the B.C.-produced pulp destined for Japan is in the present cardboard packages. **1975** *Nature* 9 Oct. 435/3 Domestic uranium reserves will be totally committed to those nuclear reactors which are brought on line in the next 20 years.

only, *a.* Add: **2. b.** *only child*: so *only-childish* adj., characteristic or suggestive of an only child; *only-childishness*, *only-childism*, the fact or state of being an only child.
1927 *Times* 29 Dec. 7/3 They might come to speak, not of drink, but of 'only-childism', as the greatest curse of this country. **1928** *Daily Tel.* 11 Sept. 11/6 Dr. Gillespie alluded to 'Only childishness'... It had been suggested that only children were peculiarly liable to become neurotic. **1938** E. Bowen *Death of Heart* iii. ii. 341 A face at a window for no reason is a face that should have a thumb in its mouth: there is something only-childish about it. **1949** —— *Heat of Day* iii. 57 Anything that savoured of only-childishness.
5. *onlie begetter* [f. Begetter 2, quot. 1606], the sole originator; *onliest* (later examples).
1907 *Yesterday's Shopping* (1969) 1136/2 Comic and humorous songs... Ma Onliest One. **1929** H. W. Odum in A. Dundes *Mother Wit* (1973) 190 Onliest way could git him. **1937** N. Marsh *Vintage Murder* vii. 81 The Firm..was founded and built up by Mr Meyer... He was..the onlie begetter. **1969** *Australasian Post* 19 June 40/3 Isadora Duncan was..the onlie begetter of all the trends in 'free dance' which are now so familiar to us. **1971** *Black World* Oct. 62/1 The onliest time I had to say something bout it was when he was playin checkers on the stoop one time and he commenst to hummin. **1972** *Daily Tel.* 30 Mar. 6/7 Stalin's onlie begetter and mentor in murder—Lenin. **1973** *Times Lit. Suppl.* 2 Mar. 228/1 The enigmatic personality of [*Citizen*] *Kane*'s onlie begetter. **1975** *Times* 14 July 13/3 William Robson, Professor Emeritus of Public Administration, University of London, and 'the onlie begetter' of *The Political Quarterly.*

only, *adv.*, *conj.* (*prep.*) **B. 2. a.** Delete 'Now only *dial.*' and add later examples.
1899 T. Watts-Dunton *Aylwin* vii. 238 I've been a-listenin' to a v'ice as nobody can't hear on'y me. **1914**

Joyce Dubliners 44 And say what he would do to her only for her dead mother's sake. 1922 E. O'Neill Anna Christie II. 140 And only for me,..we'd be being scoffed by the fishes this minute! 1934 S. O'Casey Pound on Demand in Windfalls 195 Who else could he be, only Mr. Adams? 1939 New Yorker 13 May 23/1 Her boy friend was working his way thru the Illinois U. and didn't get to Chi only two or three times a year. 1961 W. G. Pollard Physicist & Christian (1962) 57 Yet are not we of the mid-twentieth century..just as bad off as they—only in a different way?

b. Delete † Obs. and add later examples.

1901 M. Franklin My Brilliant Career iii. 16 Only I promised to stick to the missus a while I'd scoot tomorrer. 1914 Joyce Dubliners 146 Only I'm an old man now I'd change his tune for him.

only (ǭ·nli), *sb.* [f. the adj.] **1.** Used *absol.* for 'the only chance'.

1878 J. H. Beadle Western Wilds xxvi. 417, I seed it was my first, last and only, and I sot old Sally at a gallop for that pint.

2. An only child.

1934 J. Cannan High Table ii. 17 But poor little Theodore was 'an only', said Lady Oliver. 1963 Guardian 22 Feb. 8/7 The 'only', on the other hand..envies her friends with brothers and sisters. 1975 G. Storr Chinese Egg xvii. 121 If you're an only, you're sort of a target. Everything your parents think or feel has to be worked out on you.

3. In *redupl.* form. The state of being alone. *nonce-wd.*

1946 J. B. Priestley Bright Day vi. 199 Left on my only-only today. Wife's had to dash over to Leeds.

onmun (ǫ·nmun). [Korean, ad. Chinese *yên* say(ing) + *wên* letter, language.] = *Hangul.

1948 D. Diringer Alphabet 443 The Christian missionaries, who were the first to realize that On-mun was better adapted to their use than the cumbersome Chinese characters. 1950 G. M. McCune Korea Today vi. 79 Fifteen million textbooks written in the native Onmun alphabet for use in the elementary schools. 1951 [see *Hangul].

‖onnagata (ǫnaga·ta). [Jap., f. *onna* woman + *kata* figure.] In Japanese Kabuki drama and related forms, a man who plays female roles. Commonly also called *oyama*.

1924 O. Edwards Jap. Plays & Playfellows iii. 92 Peculiar attention is given to the training and discipline of onnagata, or impersonators of female parts. 1928 F. A. Lombard Outl. Hist. Jap. Drama xi. 294 In the earlier Kabuki men had often played the part of women; but now, when it had become necessary that they should do so on all occasions, a professional class of womenfolk (onnagata) grew into prominence. 1955 A. C. Scott Kabuki Theatre of Japan viii. 169 The good onnagata must symbolize feminine qualities in a way that no actress can do. 1972 Nat. Geographic Sept. 378 Greatest of today's onnagata, or male players of female roles, Utaemon Nakamura has spent a lifetime developing the charm and grace of a leading lady.

onnery: see *ornery.

on-off, *a.* [f. On *adv.* + Off *adv.*] **1.** Of a switch or the like: that turns something on or off.

1946 Nature 12 Oct. 501/2 In the counting and control circuits, all valves are used entirely as on-off elements, not as amplitude-sensitive elements. 1958 Times Rev. Industry Oct. 92/3 Photoelectric equipment designed to perform on-off switching operations. 1960 Times 15 Mar. 18/4 The overdrive operation is therefore more complicated than the simple on-off switch. 1973 D. Francis Slay-Ride viii. 96 Behind me..stood my portable television... I..found the on-off switch, and turned the volume up to maximum.

2. = Off and on a.

1953 [see *runt 2. b]. 1962 John o' London's 19 July 66/2 Forget the hoo-ha, the on-off engagements. 1965 H. I. Ansoff Corporate Strategy (1968) vii. 108 Concern with strategy had followed an 'on-off' cycle attuned to the appearance of major strategic opportunities. 1974 J. Wainwright Evidence I shall Give xxxvii. 209 The on-off pulse of the lighthouse beam. 1976 W. Greatorex Crossover 129 Over the roast lamb Amberley had made fun of Galina's on-off vegetarianism.

onolatry (ǫnǫ·lătri). [f. Gr. ὄνο-ς ass + λατρεία -latry.] Worship of the ass. Also *fig.*

1903 Jrnl. Amer. Folk-Lore July–Sept. 203 Onolatry. Reinach, S.: Le culte de l'âne... Treats of the charges of worshipping a donkey made by the pagans against the Jews and early Christians. 1953 E. Sitwell Gardeners & Astronomers 28 The crowd's onolatries Echo that laughter.

onomasiology (ǫnǫmǣsiǫ·lŏdʒi, -ziǫ·lŏdʒi). [f. Gr. ὀνομασί-α name + -ology.] The study of the principles of nomenclature, esp. with regard to regional, social, or occupational variation. Hence **onomasiolo·gic, onomasiolo·gical** *adjs.*, **onomasio·logist.**

1931 G. Stern Meaning & Change of Meaning 2 The referents..are the basis of research in onomasiology. 1937 J. Orr tr. Iordan's Introd. Romance Ling. iii. 248 The study of a map is..a study of the nomenclature applied to such and such an object. Studies of this kind are termed onomasiological. Ibid., It should not be thought..that we a reclaiming onomasiology as a child of linguistic geography. 1954 Archivum Linguisticum VI. I. 57 Onomasiological investigation of Latin words for 'head'. 1962 Y. Malkiel in Householder & Saporta Probl. Lexicogr. 18 Array of regional or temporal counterparts of each basic entry—an arrangement sometimes called 'onomasiologic' in the Central European tradition of modern-language scholarship. 1969 Word 1967 XXIII. 578 The functional principle is useful not only in the sphere of syntax but also in the sphere of onomasiology. 1973 Archivum Linguisticum IV. 113 The onomasiological or semantic bases of the taxonomic principles. 1974 Language Sciences Aug. 28/3 Eventually the onomasiologist may be able to decide whether what we call conceptual structure is something strange of a shape we can scarcely probe by introspection. 1975 Amer. Speech 1972 XLVII. 166 These are valid onomasiological statements (that is, statements about the relationship of extra-linguistic objects to linguistic expressions), but not structural ones.

onomastic, *a.* and *sb.* Add: **B.** *sb.* **3.** *pl.* The study of the origin and formation of proper names, esp. of persons.

The *sing.* in quot. 1930 is unusual.

1930 T. S. Eliot tr. St.-J. Perse's Anabasis x. 67 The man learned in sciences, in onomastic. 1936 New Yorker 8 Feb. 54 (heading) The advance of municipal onomastics. 1957 M. Aurousseau Rendering of Geogr. Names i. 1 The scientific study of names as names, that is, of the human habit of naming things, is the science of onomastics. 1972 J. L. Dillard Black English iii. 135 The subject has not yet been investigated, but it seems possible that the West African influence on Southern onomastics has been very great indeed. 1973 Amer. Speech 1969 XLIV. 221 This collection of essays..covers the nature of language, cognition, onomastics, [etc.].

onomatology. Add: (Examples.) Hence onomatolo·gical *a.*

1919 W. De Morgan Old Madhouse 324 He therefore endeavoured to bring back the discussion from the onomatologies into which it had strayed. 1931 Times Lit. Suppl. 1 Oct. 747/2 'Onomatology', the ugly name which Mr. Ewen gives to this kind of research, is not an exact science. 1961 Brno Studies in English III. 10 Their conception of the opposition of analysis vs. synthesis is so wide as to include differences of lexical (more specifically onomatological) order.

onomatomania. [See Onomato-.] Add: Now usu. with secondary stress on first syllable. **b.** A morbid preoccupation with words; a mania for word-making.

1895 tr. M. Nordau's Degeneration III. i. 242 Trichophobia (fear of hair), onomatomania (folly of words or names), pyromania (incendiary madness). 1919 W. Osler in Proc. Classical Assoc. 28 Within the narrow compass of the primitive cell..onomatomania runs riot.

onomatopœics (ǫnǫ:mătǫpī·iks). [a. Gr. ὀνοματοποίησις the making of a name + -ic 2.] = Onomatopœsis.

1934 Times Lit. Suppl. 1 Feb. 73/2 It is obviously a tour de force in onomatopœics not quite so obvious as all the R sounds in Meredith's 'The Lark Ascending'. 1934 C. Lambert Music Ho! iv. 243 The mechanically picturesque onomatopœics of the piece [sc. Pacific 231]. 1978 Language LIV. 204 My 1969 paper on onomatopoeics in the Indian linguistic area.

onomatopoietical (ǫnǫ:mătǫpoi,e·tikăl), *a.* Also onomatopoietic. [f. Onomato- + ποιητικός creative (see Poietic) + -al.] = Onomatopœic *a.*

1709 W. King Useful Transactions in Philos. II. 29 An Onomatopoietical Formation. 1883 [see Onomatopoetic *a.*].

onomatopy. Delete † Obs. and add later examples.

a 1913 F. Rolfe Desire & Pursuit of Whole (1934) 133 'Launchchchchch' was a lovely new onomatopy for the motor-boats. 1946 Word Aug. 124 Synchronic semasiology..deals with..polysemantism, affective-value, onomatopy and congeners. 1947 Ibid. III. 9 The element of onomatopy in language is too slight to invalidate the general principle.

Onondaga (ǫnǫndā·gă). [Onondaga *onōñtā'ke* on the hill (the name of the main Onondaga settlement).] One of the five (later six) tribes of the Iroquois Confederacy of North American Indians commonly called the Five Nations (Six Nations), traditionally living near Syracuse, New York; a member of this tribe; their language. Also *attrib.* or as *adj.*

1684 in Mass. Hist. Soc. Coll. (1871) 4th Ser. IX. 187, I haue perswaded all the considerable Indians, the Maquas, Sineques, Onondages..to give up their lands. 1765 in Documents Colonial Hist. New-York (1856) VII. 719 The Onondaga Speaker Tyawarunt spoke as follows. 1823 [see *Cayuga]. 1826 J. F. Cooper Last of Mohicans I. xii. 181 The Mohawks, with their Tuscarora and Onondaga brethren. 1874 B. F. Taylor World on Wheels I. 31 The painted Senecas and the smoky Onondagas went gliding about like vanishing shadows. 1933 [see *Cayuga]. 1959 [see Five Nations s.v. *five a. and sb. C. 2]. 1971 D. Heffron Nice Fire & Some Moonpennies vii. 59 My mother is an Onondaga and my father was a Mohawk. 1974 H. Woodbury in Papers in Linguistics, Conf. Iroquoian Res. 1972 2 One way of characterizing Onondaga noun incorporation is to describe its appearance in the surface structure of this language. Ibid. 5 In Onondaga, complex sentences are subject to special rules with respect to noun incorporation. Ibid. 15 Onondaga does not have relative clauses in the same sense that English does.

onru·sh, *v. poet.* [f. On-[1] + Rush *v.*[2]] *intr.* To rush on. So **onrushing** *vbl. sb.*

1861 A. T. de Vere Sisters 71 One through deserts drear On rushing in that race distraught. 1875 W. Morris tr. Virgil's Aeneids xii. 652 Saces on his foaming steed..onrusheth to the place. 1882 A. T. de Vere Foray of Queen Meave 169, I hear the on-rushing of the car! 1887 W. Morris tr. Homer's Odyssey II. xv. 276 Grey-eyed Athene sent them a wind that blew aright through the lift on-rushing fiercely.

Onsager (unsa·gər, ǫ·nsāgəɹ). *Physics.* The name of Lars *Onsager* (1903–76), Norwegian chemist, used *attrib.* and in the possessive with reference to a theorem orig. obtained for the thermodynamics of irreversible processes, but of wide applicability in physics and biophysics, as **Onsager coefficient,** a tensor coefficient expressing the degree of interference between two irreversible processes; **Onsager('s) law** or **principle,** a statement of the reciprocal nature of the interference between two irreversible processes occurring simultaneously, *spec.* that the Onsager coefficients for each direction of flow between the two processes are equal; cf. *reciprocity theorem; **Onsager (reciprocal** or **reciprocity) relation,** a mathematical statement of the Onsager principle.

1945 Rev. Mod. Physics XVII. 343 (heading) On Onsager's principle of microscopic reversibility. 1952 Physica XVIII. 182 Verschaffelt has given an example of a linear transformation of flux and force variables which leaves the entropy production invariant and yet which destroys the Onsager relations. 1955 I. Prigogine Introd. Thermodynamics Irreversible Processes iv. 46 These Onsager reciprocity relations express that when the flux, corresponding to the irreversible process *i*, is influenced by the affinity X_k of the irreversible process *k*, then the flux *k* is also influenced by the affinity X_i through the same interference coefficient L_{ik}. 1965 W. C. Reynolds Thermodynamics xiv. 381 Some restrictions on the signs of the Onsager coefficients are provided by the requirement that the entropy-production rate be positive. 1965 Katchalsky & Curran Nonequilibrium Thermodynamics in Biophysics x. 120 Onsager's law provides a quantitative relation between the phenomena of ultrafiltration and osmotic flow. 1968 Times 31 Oct. 5/4 In the simplest terms possible, the Onsager principle or 'reciprocity theorem', as it is sometimes called, asserts that where two or more kinds of flow affect each other the equations describing them will be reciprocally related in a specific way. 1977 S. H. Chue Thermodynamics ii. 248 In the Onsager coefficients the subscript *i* refers to the flux and the subscript *j* to the driving force. 1978 B. H. Lavenda Thermodynamics Irreversible Processes ii. 29 The derivation of the Onsager reciprocal relations was based on an apparent analogy between the conditions of chemical equilibrium and the 'principle of detailed balance at equilibrium'.

‖onsen (ǫ·nsen). [Jap., ad. Chinese *wên-chüan* hot-spring.] In Japan, a thermal spring, esp. one thought to have medicinal properties; a hot-spring resort.

1933 Discovery June 189/1 The onsen, the native 'spa' or hot spring so dear to the heart of the folk of the inaka, the true countryside of this extraordinary land. 1959 R. Kirkbride Tamiko xv. 118 At the Onsen he asked for Richi. 1965 W. Swaan Jap. Lantern xvii. 197 Its onsen or hot-spring resorts more than came up to expectations.

onset, *sb.*[1] Add: **2. b.** *Phonetics.* (*a*) The movement of the speech-organs preparatory to, or at the start of, the articulation of a speech sound. (*b*) The initial part of a syllable; the consonant or consonants at the beginning of a syllable. Also *attrib.*

1933 L. Bloomfield Language vii. 118 In passing from silence to a stressed vowel, we usually make a gradual onset of the voice. 1948 J. R. Firth Papers in Linguistics 1934–51 (1957) ix. 131 These are the weak, neutral, or 'minimal' vowel, the glottal stop or 'maximum' consonant, aitch, or the pulmonic onset—all of which deserve the general name of laryngals. 1951 Trager & Smith Outl. Eng. Struct. 15 Turning to the quality of the vocalic nuclei here, we find that there is an onset in raised lower high front position. 1955 C. F. Hockett Man. Phonol. 56 Onsets are simple, consisting of one or another of some eighteen or twenty consonants.., or complex, consisting of certain clusters of some of these consonants. 1962 [see *centralization 3]. 1963 Amer. Speech XXXVIII. 57 Even if one were to agree that /h/ as an onset consonant 'is a voiceless anticipation of the following peak nucleus' [etc.]. 1966 J. C. Pope Rhythm of Beowulf (ed. 2) p. xix, These crests occur at or soon after the onset of the vowel. 1971 T. M. Lightner in W. O. Dingwall Survey Linguistic Sci. 501 Word-initial voiced consonants in English begin with voiceless onset. 1973 Canad. Jrnl. Linguistics XVIII. 115 The onset of the back-gliding diphthong is typically slightly higher and slightly backer than the onset of the front diphthong.

on shore, on-shore, *adv. phr.* (*adj.*) Add: **2.** (Further examples.)

1932 Belloc Napoleon III. 175 The..torrid day, whose heat was barely mitigated by an on-shore wind from the west. 1961 B. Fergusson Watery Maze xiv. 342 The strong onshore wind had caused the tide to rise as much as half the hour ahead of almanac time.

b. Existing or occurring on the shore or on land.

1959 Listener 5 Mar. 432/3 Glances at the on-shore life. 1973 C. Callow Power from Sea ii. 67 There is a steadily growing interest in Britain's on-shore oil potential. 1974 Daily Tel. 5 Jan. 19/2 By comparison with some of the 20 million tons a year North Sea finds it is a drop in the ocean but encouraging when measured against other onshore finds. 1975 Offshore Sept. 97/1 The discovery of oil seeps along the western shore of Cook Inlet led to the drilling of several shallow onshore wells between 1900 and 1906.

on site, on-site, *adv.* (*phr.*) and *a.* [f. On *prep.* + Site *sb.*[2]] On a particular site; occurring or situated at a site.

1959 New Statesman 17 Jan. 57/3 Russia now accepts..on-site inspection to identify suspicious phenomena. 1960 Farmer & Stockbreeder 19 Jan. (Suppl.) 36/1 Calor Propane is delivered in cylinders or to bulk storage tanks on-site. 1967 A. Battersby Network Analysis (ed. 2) xv. 271 On-site management may appear to save time and trouble through less paper-work. 1968 Sci.-Tech. News Fall 61/1 The librarian..may have to visit personally a large library..and reproduce the material on site. 1975 M. Russell Murder by Mile viii. 84 On-site moulding..had been thought preferable... 'But pre-casting might have been quicker?' Ibid. 85 He lives on site too.

on stage, on-stage, *adv.* (*phr.*) and *a.* [f. On *prep.* + Stage *sb.*] On the stage; that is appearing or occurring on a stage. Also *transf.* and *fig.*

1927 T. S. Eliot in Newton's Seneca I. p. xi, It is not at all clear whether he [sc. Hercules] destroys his family on-stage or off. 1944 New Yorker 24 June 32/1 Part of Duke's character goes well enough with the onstage Ellington who periodically throws back his head and emits a long-drawn-out 'Ah-h-h!' 1949 Theatre Arts XXXIII. 100/3 She might not have taken to sitting dangerously close to the onstage edge of the wings. 1952, etc. [see *off stage, off-stage adv. (phr.) and a.]. 1966 D. F. Galouye Lost Perception v. 52 Radcliff strode onstage, supervised a pair of attendants as they positioned the recording camera. 1975 New Yorker 5 May 51/2 Everything that he did onstage was done with an excruciating and highly theatrical intensity. 1976 Country Life 12 Feb. 346/2 The orchestra..are as well matched..as the on-stage cast. 1977 Broadcast 7 Nov. 13/2 Dickens is very skilled at keeping mechanical dolls waiting in the wings, ready to be wound up and come on-stage.

on stream (stress variable), *adv. phr.* and *a.* Also as one word (see below). [f. On *prep.* + Stream *sb.*] **A.** *adv. phr.* (Usu. as two words.) **a.** Of industrial plant and resources, etc.: in or into productive or useful operation.

1930 Refiner IX. I. 58/1 The problem which has been most annoying has been one of keeping the unit on stream. 1945 H. S. Bell Amer. Petroleum Refining (ed. 3) xvii. 262 For operating on a charge stock which produces more coke a four-case cycle may be used, with ten minutes on-stream, ten minutes for purging and valve changes, and twenty minutes for regeneration of the catalyst. 1952 Economist 6 Sept. 584/2 The Vacuum Company's refinery..is expected to come on stream this autumn. 1958 Times Rev. Industry May 52/2 This material..will be manufactured at Grangemouth where the plant..is due to come on stream in 1959. 1974 Daily Tel. 9 Feb. 15/3 Another eight large brickmaking factories are due on stream this year raising capacity to 8,000 million. 1977 N.Z. Jrnl. Agric. Jan. 5/2 The dairy came on stream early in August last year.

b. *fig.*

1965 Economist 30 Oct. 495/1 The responsiveness of Congress to the evolution of public opinion on various questions of social, economic and fiscal change had been retarded unnaturally until 1965 and then suddenly came on stream. 1972 Times 14 Mar. (Hotels Suppl.) p. i/8, London had become enormously attractive to young people all over the world..because 'le mini', Carnaby Street and the swinging set had come on stream all at more or less the same time.

B. *adj.* (Written *on-stream, onstream.*) Productive; done or occurring in the course of normal production.

1938 Proc. Amer. Petroleum Inst. XIX. III. 145/1 The catalyst activity decreases during the on-stream period of the cycle due to the accumulation of carbonaceous deposits. 1945 H. S. Bell Amer. Petroleum Refining (ed. 3) xvii. 260 A reduction of on-stream time per cycle has the same effect on a fixed-bed unit as an increase in catalyst-to-oil ratio on a continuous catalytic unit. 1971 Physics Bull. June 358/1 The main advantage of on-stream analysis is that it obviates the need for tedious sampling and separate analysis of each sample. 1975 Petroleum Rev. XXIX. 324/1 Among the advantages claimed for this system are..earlier onstream production when compared with conventional offshore field developments.

ontal (ǫ·ntăl), *a.* [f. Gr. ὄν, ὀντ- being: see Onto- + -al.] Relating to reality; comprising reality, not mere phenomena; also *absol.*

1902 J. Ward in Encycl. Brit. XXXII. 67/1 The former we may call the phenomenal, and the latter the ontal, meaning of 'aspect'. 1930 F. R. Tennant Philos. Theol. II. i. 20 Further investigation of the regularity of Nature..must wait on inquiry as to what the ontal things which underlie phenomena may be. 1935 Theology XXX. 320 A conception of experimental religion which is capable of being held in organic relation with the ontal and axiological arguments. 1970 P. Bertocci Person God Is 164 Data..provident of fresh knowledge-contact with the ontal.

Ontarian (ǫntēə·riăn), *sb.* and *a.* [f. *Ontari(o,* a lake and a province of Canada + -an.] **A.** *sb.* A native or inhabitant of Ontario. **B.** *adj.* Of or pertaining to Ontario.

1888 [see Seigneur]. 1936 Times Educ. Suppl. 18 Apr. p. iv/2 Proposals..that the Ontarian Government should acquire the house. 1967 Economist 30 Sept. p. v/1 A Maritimer earning three-quarters of the average income of all Canadians and only two-thirds of the average Ontarian, has a different perspective from a Quebecois. 1970 Globe & Mail (Toronto) 26 Sept. 5/2 After three years' striving to bring the Toronto Franco-Ontarian community together, La Chasse Galerie, a Toronto cultural organization, received official recognition from Secretary of State Gerald Pelletier yesterday. 1975 Times Lit. Suppl. 10 Oct. 1188/1 A majoritarian and imperialist view of the Canadian nation that was typically Ontarian. Ibid. 1188/5 To Ontarians especially..it seemed to point the direction of Canada's future.

o:n-the-ma·keness. *rare.* [f. phr. *on the make* (Make *sb.*[2] 8) + -ness.] The fact or state of being *on the make.*

1923 Galsworthy Captures 5 In talking with Steer one never lost consciousness of his keen 'on-the-make-ness'.

o:n-the-spo·t, *a.* [Spot *sb.*[1] 9.] Done, occurring, or located at the very place in question; observed or made by an eye-witness; immediate, instantaneous.

1886 G. M. Hopkins Let. 11 Feb. (1956) 257 Some on-the-spot account of the late riots, as witnessed by yourself or friends and informants. 1955 Astounding Sci. Fiction Dec. 8 You'll..maybe do some on-the-spot generalling. 1955 Times 5 July p. iv/1 In these 'on-the-spot' laboratories routine tests are made continuously while the machines are running. 1956 B.B.C. Handbk. 1957 90 'On-the-spot' recordings. Ibid. 121 On-the-spot reports in sound and television. 1957 [see *convectional a.]. 1960 Guardian 11 Apr. 8/1 The use of traffic wardens and on-the-spot fines. 1960 [see *Castroism]. 1972 Listener 3 Feb. 159/2 The increasing use of 'voice pieces' by on-the-spot correspondents. 1977 Herald (Melbourne) 18 Jan. 1/1 Surgeons performed on-the-spot amputations.

ontic (ǫ·ntik), *a.* [f. Gr. ὄν, ὀντ- being (see Onto-) + -ic.] Of or pertaining to knowledge of the existence or structure of being in a given entity (but see quots.). Hence **o·ntical** *a.*, **o·ntically** *adv.*

1949 W. Brock Heidegger's Existence & Being 31 One important difference between science and learning on the one hand and philosophy on the other seems to him [sc. Heidegger] to consist in the fact that every kind of scientific and scholarly knowledge was concerned with a limited set of objects, of what he called 'ontic'. 1952 Mind LXI. 131 The final outcome of ontically objective values. 1954 Scottish Jrnl. Theol. VII. 47 We have to recognise the ontic basis of our faith and obedience. 1957 M. Feagins tr. H. Kunz in P. A. Schilpp Philos. K. Jaspers II. xiii. 509 It is unavoidable to use the empirical, objectifiable data of knowledge (biological or psychological, for example) as guides to an explication of the ontic character of man as an active, experiencing and self-understanding being. 1960 W. V. Quine Word & Object iii. 120 Of the three evident advantages of 'ontic' over 'ontological', in the special sense of 'as to what there is', brevity is the least. 1962 Macquarrie & Robinson tr. Heidegger's Being & Time i. 31 Ontological inquiry is concerned primarily with Being; ontical inquiry is concerned primarily with entities and the facts about them. Ibid., The ontical inquiry of the positive sciences. 1969 A. Richardson Dict. Christian Theol. 241/2 R. Bultmann..argues that what is ontologically a human possibility, i.e. something which it is possible for men to know, is ontically actualised in Christian faith... It is much discussed today whether religious language..possesses ontic significance. 1970 J. W. Yolton Locke & Compass of Hum. Und. i. 30 The more typical passages find Locke denying any ontic sense of 'kinds'. 1975 Times Lit. Suppl. 25 July 848/2 Before you can elicit the ontic commitment of a statement, you must indulge in what R. G. Collingwood contemptuously described as the scholastic pedantry of reducing to logical form. 1976 D. E. Linge tr. Gadamer's Philos. Hermeneutics xi. 203 Anonymous intentionalities, that is, conceptual intentions in which something is intended and posited as ontically valid.

on to, onto, *prep.* Add: β (Earlier and later examples.)

1715 Duxbury (Mass.) Rec. (1893) 105 [A] place gutted away by the rain down onto Mr. Wiswells land. 1758 R. Putnam Jrnl. 3 June (1886) 62 Capt. Nixon's men..fell a tree onto some men as they were in another camp. 1788 J. May Jrnl. 30 June (1873) 75, I put powder-horn and shot-bag onto him, and a gun in his hand. 1938 L. Bemelmans Life Class II. iii. 142 Everything..can be rushed at a moment's notice onto the tables. 1954 C. S. Lewis Horse & his Boy vi. 76 He jumped down onto the rubbish. 1963 New Statesman 24 May 781/3 Russia..has heaved itself onto the plateau of the advanced industrial powers. 1973 G. Greene Honorary Consul III. iii. 132 The man was telling him to get back onto the so-called bed. 1973 [see *lead sb.[2] 1 f]. 1976 D. Heffron Crusty Crossed xxviii. 175 He was hanging onto Dot as though they both might collapse to the ground if he let go.

2. Aware of or knowledgeable about (a person, state of affairs, etc.); 'wise to' (something). Cf. *on *adv.* 13e. *colloq.* (orig. *U.S.*).

1877 Chicago Street Gaz. 20 Oct. 1/2 May Willard, why don't you take a tumble to yourself and not be trying to put on so much style around the St. Mark's Hotel, for very near all of the boys are on to you. 1888 Lantern (New Orleans) 9 July 4/3 Who is onto the rag racket. 1888 New York Mercury 21 July 3/3 A wife poisoner..ought to have for his wife a woman who is on to him, and who can meet his poison advances with a kerosene bath. 1899 A. H. Quinn Pennsylvania Stories 115 The class is about on to us, anyway, and if they find out about this.. 1911 J. C. Lincoln Cap'n Warren's Wards xii. 254 Everybody has been on to that for some time. 1919 Wodehouse Damsel in Distress xxi. 248 'So you're on to him, too?' said Billie. 'When did you get wise?' 1959 J. Osborne World of Paul Slickey I. v. 50, I can't help feeling that he's on to us... That he knows about us. 1973 G. Mitchell Murder of Busy Lizzie xiii. 151 'You didn't even tell Gavin that we may be on to something?'.. 'You may say that I have certain suspicions, but no more.'

2. *Math.* (Written onto.) Used (in place of *into*) to express the relation of a set to its image under a mapping when every element of the image set has an inverse image in the first set.

1940 C. C. MacDuffee Introd. Abstract Algebra iii. 54 If a homomorphism of *A* onto *B* exists, we write *A*∼*B*. 1962 B. H. Arnold Intuitive Concepts Elem. Topology vii. 113 Each of the sets *X* and *Y* is the set of all real numbers; $f(x) = 2x$. The transformation $f: X \to Y$ is onto. 1965 Patterson & Rutherford Elem. Abstract Algebra i. 4 We shall denote a mapping *f* of S_1 into S_2 by $f: S_1 \to S_2$... If every element *y* of S_2 is of the form $f(x)$ for some $x \in S_1$, we call *f* a mapping of S_1 onto S_2. 1971 E. C. Dade in Powell & Higman Finite Simple Groups viii. 285 An epimorphism of *A* onto a field *F*.

B. *adj.* *Math.* Used to designate a mapping of one set 'onto' another.

1942 S. Lefschetz Algebraic Topology i. 7 If the transformation is 'onto', the inverse image of the complement of a set is the complement of the inverse image of that set. 1949 —— Introd. Topology 216 (Index), Onto transformation. 1951 N. Jacobson Lect. Abstract Algebra I. 4 If α is a mapping of *S* into *T*, and β is a mapping of *T* into *S* such that $\alpha\beta = 1_S$ and $\beta\alpha = 1_T$, then α and β are 1–1, onto mappings and $\beta = \alpha^{-1}$. 1968 [see *into 2]. 1971 E. C. Dade in Powell & Higman Finite Simple Groups viii. 285 By Lemma 9.5 the map is onto.

onto-. Add: **ontotheo·logy** (see quots.); so **ontotheolo·gical** *a.*

1798 A. F. M. Willich Elem. Critical Philos. 171 Ontotheology is the cognition of a Supreme Being from bare conceptions. 1854 Geo. Eliot tr. Feuerbach's Essence Christianity ii. 38 The ens realissimum, the most real being of the old onto-theology. Ibid. 40 The onto-theological predicates are merely predicates of the understanding.

ontogenetical, *a.* For *rare*⁻⁰ read *rare* and add example.

1965 Sci. World IX. iv. 4/2, I propose to consider the processes which concern biologists as molecular, physiological, ontogenetical, historical and evolutionary.

ontogenic (ǫntodʒe·nik), *a.* [f. Ontogeny + -ic.] Pertaining to or distinguished by ontogeny; = Ontogenetic *a.*

1893 Proc. Boston Soc. Nat. Hist. XXVI. 98 The product of the evolution of an ancestor into a phylum through successive independent forms or ontogenic cycles. 1944 B. Malinowski Sci. Theory of Culture vii. 79 A new organism comes into being..starting a partly independent career of ontogenic development. 1965 B. E. Freeman tr. Vandel's Biospeleol. xxii. 367 The information available on ontogenic evolution in the Trechinae will be examined.

ontologist. (Later example.)

1856 Mill Logic (ed. 4) I. I. iii. 65 Refutation of the Ontologists from their own premises and in their own language, which he [sc. Sir Wm. Hamilton] has furnished in the first paper of his Discussions.

ontologize, *v.* Add: Hence **onto·logizing** *vbl. sb.* and *ppl. a.*

1878 S. H. Hodgson Philos. of Reflection I. 138 The 'too much' of ontologising philosophers. 1897 W. M. Urban Hist. Princ. Sufficient Reason iv. 34 The ontologising of this fundamental law is the Transcendental Ideal of the Kantian Kritik. 1940 Mind XLIX. 118 Some categories..are 'ontological or ontologising', i.e., have metaphysical import.

onwards, *adv.* **1. b.** (Later examples.)

1916 G. B. Shaw Androcles & Lion I. 9 You may be called on to appear in the Imperial Circus at any time from tomorrow onwards. 1961 New Eng. Bible Acts iii. 24 So said all the prophets, from Samuel onwards. a 1976 A. Christie Autobiogr. (1977) vii. ii. 330 From then onwards I should have first-class advice.

onychogryphosis (ǫ:nikogrifōᵘ·sis). *Med.* Also -gryposis. [f. Gr. ὀνυχο-, comb. form of ὄνυξ nail + γρύπωσις hooking of the nails.] The condition of having overgrowth, accompanied by thickening and curvature, of one or more nails (usu. of the toes).

1833 Dunglison Dict. Med. Sci. II. 104/1 Onychogryphosis, onychogryposis. 1887 Boston Med. & Surg. Jrnl. CXVII. 301/1 When the whole nail is affected, its free border has a tendency to curve downwards. It may occur in various directions, according as it is disturbed in a vertical or transverse way (onychogryphosis). 1901 Encycl. Medica VIII. 220 The subject of onychogryposis was first fully investigated by Virchow. 1972 A. Rook et al. Textbk. Dermatol. (ed. 2) II. 1662/1 Nail hypertrophy implies thickening and increase in length, whilst onychogryphosis implies curvature also.

onychomycosis (ǫ:nikomǝikōᵘ·sis). *Med.* Pl. -mycoses. [f. as prec. + Mycosis.] Fungal infection of a finger- or toe-nail, causing brittleness and discoloration.

1865 Dublin Q. Jrnl. Med. Sci. XL. 353 (heading) Two cases of onychomycosis. 1887 Buck's Handbk. Med. Sci. V. 104/2 The onychomycoses are..of only two kinds, that due to favus, and tinea tricophytina. 1954 A. C. Allen Skin xiii. 460/2 Onychomycosis is caused by many species of Trichophyton as well as by Epidermophyton floccosum, several species of Aspergillus, and Candida albicans. 1975 Daily Colonist (Victoria, B.C.) 5 Mar. 2/1 Onychomycosis (also called ringworm of the nails) is caused by a fungus infection.

onychophagist (ǫniko·fădʒist). [See Onygophagist.] One who bites his nails. So **onychopha·gia, onycho·phagy,** the habit of biting one's nails.

[1834: see Onygophagist.] 1900 Dorland Med. Dict. 456/1 Onychophagy. 1900 Daily Chron. 10 July 5/2 'Onychophagia' is far more frequent in Parisian than in provincial schools. 1903 Onychophagist [in etym. of Onygophagist]. 1907 Daily News 4 July 6/7 Dr. Didsbury..suggests that onycophagists [sic] should wear his new dental apparatus, which is fastened to the lower molars and just prevents the upper and lower teeth from meeting. 1956 D. M. Pillsbury et al. Dermatol. xlv. 1017 Onychophagia denotes biting of the free edge of the nail. 1977 Woman's Jrnl. Apr. 10 Are you an onychophagist? ..This rather long name simply means that you bite your nails.

onymously (ǫ·nimǝsli), *adv. rare.* [f. Onymous *a.* + -ly[2], after Anonymously *adv.*] With the writer's name given or attached.

1889 V. Horsley in S. Paget Sir Victor Horsley (1919) 86 He anonymously or onymously is not worth powder and shot.

o'nyong-nyong (onyǫ-ŋnyǫŋ). *Med.* Also onyongnyong. [See quot. 1960.] A mosquito-borne virus disease in East Africa, similar to dengue, which is caused by an arborvirus and carried by anopheles.

1960 A. J. Haddow et al. in Trans. R. Soc. Trop. Med. & Hygiene LIV. 517 (heading) O'nyong-nyong fever: an epidemic virus disease. Ibid., The epithet 'o'nyong-nyong' originated among the Acholi, one of the first tribes to be affected, and, being the first recorded, has been selected as the definitive name of the disease. Ibid. 518 The Map..shows the approximate dates at which o'nyong-nyong appeared in the various affected areas in Uganda. 1966 Guardian 8 Sept. 2/8 It has recently been established in Africa that a fever virus called onyongnyong is transmitted by the anopheline mosquito. 1974 A. W. Woodruff Med. in Tropics xxii. 322/1 O'nyong-nyong has caused one very major epidemic which started in Uganda in 1959 and spread to Kenya, Tanzania and Malawi involving an estimated 2 million people.

oo[2], **'oo**[2] (ū), a representation of a child-like pronunciation of *you.*

1713 Swift Jrnl. to Stella (1948) II. 644, I allow oo Six. 1900 M. Corelli Boy i. 8 Oh, Poo Sing! Does 'oo feels ill? Does 'oo feels bad? 1965 Listener 1 July 12/2 There is now an enormous gap between a tiny avant-garde and the vast mass of viewers and listeners, between people who are happy with a painting entitled 'Won't Oo Kiss Doggie?' and the few who can accept 'sculpture' made out of old motorbikes and dustbin lids.

oo[3], **'oo**[3] (ū), a representation of a colloq. (orig. Cockney) or vulgar pronunciation of *who.* So **'oom,** whom.

c 1870 A. Lloyd in W. Matthews Cockney Past & Present (1938) 91 Then left me for a feller 'oom she thought was much more grand. 1883 Kaukneigh Awlminek 7 People 'oo down't profit by experience is medder then moust loonatics. Ibid. 15 There's more then fifty thousan' pussons in Lendin oo can't write. I 'ear thet sem uv 'em is editors. 1901 G. B. Shaw Capt. Brassbound's Conversion II. 244 Oo a you orderin abaht, ih? 1903 Kipling Five Nations 199 What is the sense of 'atin' those 'Oom you are paid to kill? 1950 C. S. Forester Mr. Midshipman Hornblower 232, I was wonderin' 'oo'd come to my rescue. 1970 M. Moorcock Chinese Agent x. 71 Oo's gonna pay for all me lovely china! 1973 J. Leasor Host of Extras viii. 147 'Oo're you?' he asked belligerently.

oo[4] (o,o). Also **o-o.** [Hawaiian.] Also *oo bird.* A black and yellow bird, *Moho braccatus,* belonging to the family Meliphagidæ or honeyeaters and now believed to be extinct.

1890 S. Wilson in Ibis II. 179 Large numbers of the O-o must have been taken in old days. 1902 H. W. Henshaw Birds Hawaiian Islands 70 The brilliant shining black body feathers of the o-o were..in great demand for making cloaks. 1937 D. & H. Teilhet Feather Cloak Murders x. 181 The little Oo and Mamo birds..from which they plucked..the coloured feathers to make the cloaks. 1944 G. C. Munro Birds Hawaii 84 If it still exists no effort should be spared to save what would be the last of the famous Hawaiian oos. 1960 Guardian 3 Nov. 10/3 Robes made from the tufted feathers of the o-o bird. 1970 S. Carlquist Hawaii xi. 214 Oos were black, with tufts of yellow feathers extracted to make the yellow feather cloaks. This may well have helped to extinguish the oo. 1977 Nat. Geographic Nov. 588/1 (caption) Biologist John Sincock..of the U.S. Fish and Wildlife Service sloshes through Alakai Swamp in hope of spying the yellow-thighed oo..one of the world's rarest birds.

oo-. Add: **oogenesis** (later examples); **oogenetic** (example).

1925 R. E. Snodgrass Anat. & Physiol. Honeybee xii. 270 The process [of development] in the case of the egg cells involves oogenesis and maturation. 1960 New Biol. XXXI. 94 The process of oogenesis must not predetermine the post-fertilization history of the egg. 1974 L. B. Arey Developmental Anat. (ed. 7 rev.) iii. 31 The word 'egg' or 'ovum' is often used when referring to any stage in the course of differentiation of the female sex cell during oögenesis. 1895 D. Sharp in Cambr. Nat. Hist. V. xxii. 500 Some hypothetic rudiments they [sc. Weismann and others] consider to exist at the very earliest stage of the embryonic, or oogenetic process.

oo, *v.*: see *ooh *v.*

oocyte (ōᵘ·ŏsǝit). *Biol.* [f. Oo- + -cyte, as ad. G. *ovocyte* (now *oozyte*) (T. Boveri 1892, in Anat. Hefte Abt. II. I. 446): see Ovo-.] An egg mother-cell, which gives rise to a mature ovum by meiosis; the *primary oocyte* gives rise in meiosis I to the *secondary oocyte* and a small polar body; the secondary oocyte gives rise to the mature ovum and another polar body in meiosis II. Also (with some writers), a polar body so produced. Cf. *ovocyte.

1895 Jrnl. R. Microsc. Soc. 511 (heading) Peculiar mitosis in young oocytes of salamander. 1927 [see *ootid]. 1945 W. J. Hamilton et al. Human Embryol. ii. 12 The formation of the first polar spindle..initiates the first maturation, or reduction, division, the oocyte dividing into a larger cell, the secondary oocyte, and a much smaller cell, the first polar body. 1946 B. M. Patten Human Embryol. ii. 31 The primary oöcyte divides to form two secondary oöcytes. One of these receives little cytoplasm and is called the first polar body. 1968 Passmore & Robson Compan. Med. Stud. I. xxxvii. 10/2 All the primary oocytes so formed begin their first meiotic division before birth, but the completion of prophase is arrested until after puberty... Meiosis restarts in individual oocytes when their follicles undergo maturation in subsequent ovarian cycles... The remainder of the first meiotic division is completed by the time of ovulation, at which time a secondary oocyte is released into the tube. The second meiotic division follows immediately and..is not normally completed until the oocyte is penetrated by a spermatozoon. 1970 Ambrose & Easty Cell Biol. xii. 390 Only one ovum is formed from each oöcyte in contradistinction to the four sperm formed from each spermatocyte.

ood, var. *oud.

oodle (ū·d'l). Also -lin (in sense 1). [Of uncertain origin.] **1.** In *pl.*, large or unlimited quantities; abundance, 'heaps'. *colloq.*

1869 Overland Monthly III. 131 A Texan never has a great quantity of any thing, but he has 'scads' of it, or 'oodles', or 'dead oodles', or 'scadoodles', or 'swads'. 1887 J. C. Harris in Century Mag. Apr. 846/2 All you lack's the feathers, and we've got oodles of 'em right here. 1892 J. Barlow Irish Idylls iii. 57 A grand young pig, they'll be gettin' oodles o' money on at the fair afore Lent. 1900 Ade Fables in Slang 80 Jethro..had learned Oodles of slang up in Chicago. 1904 W. N. Harben Georgians 115 An' now you, a man with oodlin's an' oodlin's o' pore blood kin..are a helpin' at the job. 1919 H. L. Wilson Ma Pettengill iii. 78 It snowed hard. Just oodles of the most perfectly dazzling snow. 1928 Daily Sketch 7 Aug. 6/2 With oodles of 'Och Ayes', more Scots than ever bled with Wallace have flooded the office with caustic correspondence. 1929 D. G. Mackail How Amusing! 409 You wouldn't catch me coming down to the City every day, if I'd got oodles of boodle like that. 1940 O. Nash Face is Familiar 7 And he had one object all sublime, Which was to save simply oodles of time. 1957 Sunday Times 14 Apr. 13/3 Oodles for the rich, and practically nothing for the poor. 1967 She Dec. 95/1 The cover assures me that there are 'oodles of prizes', which indeed there are. 1975 New Yorker 19 May 95/2 (Advt.), Front and back yokes and oodles of shirring for fullness.

2. *Austral.* and *N.Z. slang.* Money in general. *rare.*

1941 in Baker Dict. Austral. Slang 51.

oo-er (ū,ǝ·ɹ), *int.* Also ooo-er. An exclamation expressing surprise, wonder, etc.

1912 C. Mackenzie Carnival ix. 104 'Oo-er!' cried Jenny. 'We aren't going to sleep in the dark?' 1926 C. Beaton Diary 18 Oct. in Wandering Yrs. (1961) 143 Everyone 'talked common'—the smart thing to do at the moment: 'That's a bit of all right; I don't mind if I do; oo-er!' 1933 D. L. Sayers Murder must Advertise x. 169 When told what he had missed he merely remarked 'Oo-er!' 1934 R. Ferguson Celebrated Sequels 246 Those who saunter, cry 'oo-er', bathe, knit. 1955 M. Allingham Beckoning Lady xvi. 223 'Oo-er.' Tonker bristled. 1958 S. Hyland Who goes Hang? xvi. 73 'Oo-er,' said Alec Beasley, vulgarly. 1961 Guardian 20 Jan. 7/4 The reader will either dutifully say 'Oo-er!' or..'Come off it!' 1977 J. Savage Nemesis Club vi. 76 Ian's mouth fell open. 'Ooo-er!'

oof. Add: **oo·finess,** wealth; **oo·fless** (examples); **oo·flessness,** lack of cash.

1888 Bird o' Freedom 10 Oct. 5/2 When Jack is on the spree His love may be termed free, And the tarts will oofless be Till his ship comes back. 1889 E. Dowson Let. 18 Jan. (1967) 26, I shall be very oofless tho' & must I fear be a Pinolitic pittite on the occasion. Ibid. 17 Mar. (1967) 50, I foresee great—great—ooflessness—as the result of this week—much dissipation. 1935 Wodehouse Luck of Bodkins xvii. 211 His amazing oofiness had a tendency to slip from the mind.

oof, var. Ouf, ouff *int.* in Dict. and Suppl.

1934 in Webster. 1966 L. Cohen Beautiful Losers (1970) I. 116 You wanted sock! pow! slam! ugg! oof! yulp! written in the air between you and all the world. 1976 'R. B. Dominic' Murder out of Commission iii. 27 'Oof! What a night!' he grunted.

oogamy. (Examples.)

1894 S. H. Vines Students' Text-bk. Bot. 225 Oogamy: the female organ is an oogonium. 1933 G. M. Smith Fresh-Water Algae U.S. 288 The evolution from isogamy to oögamy may be independent of any evolution in thallus structure. 1971 P. H. B. Talbot Princ. Fungal Taxon. vii. 91 Oogamy, such as occurs in some of the lower fungi, is but a specialized form of gametangial contact.

oogonial (ōᵘ,ŏgōᵘ·niăl), *a.* [f. Oogonium + -al.] Of or pertaining to an oogonium.

1902 Science 21 Mar. 457/1 The protoplasm in contact with the oogonial wall. 1938 G. M. Smith Cryptogamic Bot. I. ii. 74 An oögonial mother cell..may be terminal or intercalary in position. 1970 J. Webster Introd. Fungi 73 The oogonial initial is multinucleate.

oogonium. Add: Pl. **oogonia. 2.** *Biol.* [coined in Ger. as *ovogonium* (T. Boveri 1892, in Anat. Hefte Abt. II. I. 446): see Ovo-.] A primordial female reproductive cell that gives rise to primary oocytes by mitosis.

1895 Jrnl. R. Microsc. Soc. 511 The oogonia show a nucleus with few chromatin fragments and a very delicate, but dense linin-framework. 1920 L. Doncaster Introd. Study Cytol. v. 61 After a number of divisions..the spermatogonia and oogonia cease to divide and begin to increase considerably in size. At this stage they are called primary spermatocytes..and oocytes. 1940 G. A. Baitsell Human Biol. xii. 319 The immature egg or oögonium. 1970 Ambrose & Easty Cell Biol. x. 330 In the female the oögonia, corresponding to spermatogonia, give rise to oöcytes.

ooh (ū), *int.* Also oo, ooohh, etc. [var. Oh *int.* (*sb.*)] An exclamation of pain, surprise, wonder, disapprobation, etc. Hence as *sb.*; also **ooh-a(a)h, ooh and ah.**

1916 E. O'Neill Bound East for Cardiff in Provincetown Plays 1st Ser. 16 It hurts like hell—here... I guess my old pump's busted. Ooohh! 1919 G. B. Shaw Great Catherine iv. 152 Agh!! (She has again applied her toe). Oh! Oo! Ibid. 154 Agh! Ooh! Stop! Oh Lord! 1939 Joyce Finnegans Wake I. 149 Wee skillmustered shoul with his ooh, hoodoodoo! 1957 R. Hoggart Uses of Literacy vi. 165 A world so complex that even those who are immersed in the business of tending its more important machines can only hope to understand a little of it, is daily reduced to a local and spuriously manageable 'ooh-aah', when the paper drops on the mat. 1964 L. Deighton Funeral in Berlin xlviii. 301 There was a great 'Ooohh' and 'Aaahh' as the rocket burst. 1975 Times 19 Feb. 14/1 The oohs and ahs of a 13-year-old schoolgirl contemplating the Osmonds. 1976 New Musical Express 12 Feb. 24/2 All that mopery and Ooooh, it's so hard and lonely at the top. 1977 F. Parrish Fire in Barley ii. 19 Ooh, she's a powerful snob.

ooh (ū), *v.* Also oo. [var. Oh *v.*] *intr.* To say 'ooh'; also *trans.*, to express with the sound 'ooh'. Freq. in conjunction with *ah *v.* Also in reduplicated form *ooh-ooh.*

1953 Pohl & Kornbluth Space Merchants (1955) x. 102 Above me the respectable Costa Rican consumers oohed and ahed at the view from the prism windows. 1957 'P. Quentin' Suspicious Circ. i. 7 Monique was oohing and aahing about 'bone structure' and 'divinity of movement'. 1960 [see *ah v.]. 1961 W. Sansom Last Hours of Sandra Lee 112 Mouths ooed and nummed noises of appreciation. 1963 Times 8 May 5/6 Vociferous idiots who 'ooh' and 'ah' every time their favourites..set two feet off the stage. 1964 W. Markfield To Early Grave (1965) iv. 75 Where do you come off with..that moaning and groaning, that ooh-ing and aah-ing? 1965 J. B. Priestley Lost Empires III. x. 273 People laughed and clapped and Oo'd and Ah'd. 1971 A. Morice Death of Gay Dog xiii. 141 He..oohed and ahed his way through the column in quite the proper spirit. 1977 New Musical Express 12 Feb. 14/5 Pumping their arms in unison, pirouetting and ooh-oohing like crazy, they're very slick, very infectious and warmly humorous.

ooh-la-la (ū-lālā), *int.* Also oo-la-la, oolala, etc. [ad. F. *ô là! là!*] An exclamation of surprise, appreciation, etc. Hence as *sb.*, (*a*) the interjection 'ooh-la-la'; (*b*) *slang,* the 'naughtiness' popularly associated with the French; 'spiciness'; (*c*) an attractive or provocative girl. Also *attrib.* or as *adj.*, and as *v. intr.*

1924 Dialect Notes V. 274 Exclamations in American English (title)..oo(h):..—la la. 1940 E. A. Robertson Summer's Lease xx. 269 He went to France..believing that French girls were all 'Oo la la and snatch my garter'. 1943 Hunt & Pringle Service Slang 49 Oolala, Army French meaning O.K. or 'hot stuff'. 1950 Brooks & Warren Fund. Good Writing xii. 402 Bug-eyed young matrons oo-la-la-ing over the purchase of sheets or toothbrushes. 1952 'J. Tey' Singing Sands xiii. 215 'I like my iniquity with some ooh-la-la in it.' 'Hasn't Daphne got any ooh-la-la?' 'No. Daphne's very la-di-da.' 1952 S. J. Perelman Ill-Tempered Clavichord (1953) 72 Their silken ankles a target for the ardent glances of gendarmes twirling spiked mustaches and muttering appreciative ooh-la-las. 1959 Spectator 24 July 102/3 The ooh-la-la French maid. 1960 I. Cross Backward Sex 72 If this red-haired oo-la-la gets out of hand, I'll fix her for you. 1961 Sunday Express 7 May 17/2 The Swiss rely on precision..and the French on oo-la-la. 1970 S. J. Perelman Baby, it's Cold Inside 161 The playful slaps and oo-la-la's that rang through the valley reassured us that they were still alive. 1973 Times 10 Apr. 7/6 Those two great standbys of French fashion, quality and a little bit of ooh-la-la. 1973 A. Hunter Gently French viii. 75 He gave a dirty laugh. 'I've heard about her [sc. a Frenchwoman]. When do you get to the ooh la la?' 1975 J. F. Burke Death Trick (1976) vii. 89 'I'll have a French Seventy-five,' she said. Cathleen said, 'Oo! La-la!' and started a split of bubbly.

ooid (ōᵘ·oid). *Petrol.* [a. G. *ooid* (E. Kalkowsky 1908, in Zeitschr. f. deutsch. geol. Ges. LX. 72), f. Gr. ᾠοειδής egg-shaped.] = *oolith.

[1918 Jrnl. Geol. XXVI. 593 For an individual grain of an oolite the term 'ovulite'..appears to be preferable to Kalkowsky's 'oöid'.] 1945 Univ. Texas Publ. No. 4301. 136/1 Oöid, the individual tiny spheroid in an oölitic rock. 1949 H. W. Fairbairn Struct. Petrol. Deformed Rocks (ed. 2) iii. 49 Early structural studies of ooids include those of Albert Heim and Loretz. 1965 J. T. Greensmith Hatch & Rastall's Petrol. Sedimentary Rocks (ed. 4) viii. 199 A normal oolitic limestone or oolitic calcarenite is made up of an aggregate of spherical allochems called ooliths or ooids, usually about 1 mm. or less in diameter. 1974 Nature 15 Feb. 452/1 The formation is mainly dolomitic. ..It contains desiccation cracks,..oöids and algal stromatolites.

oojah (ū·dʒā). *slang.* Also **oojar, ujah.** [Of uncertain origin.] A substitute expression used to indicate vaguely a thing of which the speaker cannot at the moment recall the name, or which he does not care to specify precisely; a 'what-you-may-call it', gadget. So in extended forms **ooja-ka-piv** (ū·dʒăkă-piv), **(ujah-ka-piv), oojah-capiff** (ū·dʒăkăpif), **ooja-ka-pivi** (ū·dʒăkăpivi), **(ooja-ka-pivvy, oojah capivvy).** See also next word. So **oojah-cum-spiff** *a.*, all right, 'O.K.'.

1917 W. Muir Observations of Orderly xiv. 229 'Oojah', anything. 1925 Fraser & Gibbons Soldier & Sailor Words 215 Oojah (also Ooja-ka-pivi), a substitute expression for anything the name of which a speaker cannot momentarily think of, e.g. 'Pass me that h-m, h-m, oojah-ka-pivi, will you?' 1930 Wodehouse Very Good, Jeeves! i. 25 'All you have to do,' I said, 'is to carry on here for a few weeks more, and everything will be oojah-cum-spiff.' 1931 J. Van Druten London Wall II. ii. 73 There's a whole lot in the Oojah Capivvy now. 1933 Partridge Words, Words, Words! III. 192 For thingummy, Tommy says oojah, with variants oojah-ka-piv, oojah-cum-pivvy, and oojiboo. 1935 D. L. Sayers Gaudy Night viii. 178 Oh, look! your bag's opened itself wide and all the little oojahs have gone down the steps. 1941 P. Kendall Gone with Draft 118 An oojah..a gadget. 1943 Hunt & Pringle Service Slang 49 Oojah, sauce or custard. 1951 Landfall V. 89 For Pete's sake, boy, don't lose that little oojah. 1962 Sunday Times 4 Feb. 31/6 This was the catch-phrase in a music-hall song in use during the first world war... I remember the line and the tune: 'You cannot eat it, or see it, or hear it—you just ask for Ujah-ka-piv. Ibid., 'Ujah'..was used as widely and as indiscriminately as 'gimmick' and 'gadget' are used now. 1966 'L. Lane' ABZ of Scouse 78 Whur's ther oojah-capiff?, where is the hammer, spanner or whatever it might be? 1971 B. W. Aldiss Soldier Erect 94 I've seen blokes in hot countries go clean round the oojar because of the perverted practices of native women.

oojiboo (ū:dʒĭbū·). *Soldiers' slang.* [Arbitrary extension of *ooj(ah, with meaningless suffix.] = prec. So (by metathesis) **oobyjiver** (ū:bĭdʒǝi·vǝɹ).

1918 Daily Express 2 Oct. 2/5 The oojiboo may be a hammer, a saw, a spanner, but Jimmy, or anyone else, knows exactly what is wanted. Ibid., A laundry van bumped into me and carried away my oojiboo [sc. a tail lamp]. Ibid., I dropped the old oojiboo [sc. kitbag] on the platform and nipped into the refreshment-room. Wasn't gone two minutes, but d—n me if somebody hadn't won the oojiboo [stolen the kitbag]. 1925 Fraser & Gibbons Soldier & Sailor Words 215 Oojiboo, much the same as Oojah. 1933 [see *oojah]. 1963 New Society 22 Aug. 5/2 Colourful words like 'oobyjiver', meaning 'whatsis' pop up regularly.

ook (uk). *slang.* [Origin unknown.] Something slimy, sticky, or otherwise unpleasant. Hence **oo·ky** *a.*, slimy, viscous, repellent; also *fig.*

1964 S. Bellow Herzog (1965) 277 He writes poems and reads them to Mama... He looks ooky when he says them. 1969 Disch & Sladek Black Alice viii. 81 She had been ..glad..to be here, to be anywhere so long as it marked an end, so long as she could..take a shower to wash off all this brown ook. Ibid. 85 'Ain't no shower,' Clara declared flatly. 'A bath, then? I want to wash this ook off of me.' Ibid. xiv. 158 The milk was so warm and ooky it was like yogurt.

oolakan. Add: Also **eulachon, olachen, oolaghan, oolichan, ulichan.** Substitute for etym.: [f. Chinook *ūlakân.*] (Earlier and later examples.)

1834 W. F. Tolmie Jrnl. 16 Apr. (1963) 275 The canoes were laden with..dried herring spawn which they are to barter for Oolaghans. 1849 A. Ross Adventures First Settlers Oregon River vi. 97 There is a small fish resembling the smelt or herring, known by the name of ulichan, which enters the river in immense shoals, in the spring of the year. 1911 J. G. Frazer Golden Bough: Magic Art (ed. 3) I. v. 262 The Tsimshian Indians of British Columbia believe that twins..can..call the salmon and the olachen or candle-fish. 1926 B. A. McKelvie Huldowget 2 They came today for the oolichan fishing. 1953 Beaver Mar. 40/2 Oolikan, olachan, eulachon, uthlecan, hollikan and hoolican—spell it as you wish. Ibid. 43/1 Exchange value of a large box of olachen grease equalled one caribou skin.. or $1.50 in cash, a century ago. 1965 Fisherman (Vancouver) 19 Mar. 2/5 Robichaud said commercial exploitation of eulachons was banned by regulation. 1972 Evening Telegram (St. Johns, Newfoundland) 15 July 34/4 Fried and steamed clams, oysters, halibut and oolichans. 1975 H. White Raincoast Chron. (1976) 176/2 There were oolachen oil street lamps on cross beams in front of each house.

oo-la-la, oolala, varr. *ooh-la-la *int.*

oolite. Add: **4.** = *oolith.

1851 H. T. de la Beche Geol. Observer viii. 123 The little grains termed oolites, formed of concentric coatings of calcareous matter. 1907 E. H. Adye Mod. Lithology xii. 60 The oolites..formed in shallow waters are cemented together by calcareous material. 1955 E. E. Wahlstrom Petrographic Mineral. x. 335 Most oolites in clastic rocks contain a nucleus of organic matter, a fragment of shell, or a more or less rounded silicate or carbonate particle. 1961 J. H. Johnson Limestone-Building Algae 256 Algal pisolites are composed of more or less spherical masses ranging in size from that of a large oolite to spherules having a diameter as much as an inch across.

oolith (ōᵘ·ōliþ). *Petrol.* [f. Gr. ᾠόν egg + λίθος stone (see -lith).

Mod.L. *oolithus* occurs in this sense in 1721 (F. E. Brückmann Specimen Physicum Exhibens Historiam Naturalem Oolithi seu Ovariorum Piscium et Concharum in Saxa Mutatorum 5) and may be the source of both this word and Oolite (= ad. F *oölithe*).]

Each of the small rounded granules of which oolite is composed.

1788 J. H. De Magellan A. F. Cronstedt's Ess. Syst. Mineral. (ed. 2 rev.) I. 76 The Stalagmites..get a mammillary form, whilst the Stalactites acquire a conic figure: the Oolithes and Pisolites belong to the same species. 1892 Jrnl. R. Microsc. Soc. 839 These ooliths are undoubtedly the product of a lime-separating Schizophyte; and the author believes this to be the case with the greater number of the marine calcareous ooliths with a regular zoned and radial structure. 1926 G. W. Tyrrell Princ. Petrol. xiii. 227 The grains are called ooliths and the rock containing them oolite or oolitic limestone. Ooliths generally show a series of concentric coats of calcareous material in which a radiating crystalline structure can often be made out. 1938 M. Black Hatch & Rastall's Petrol. Sedimentary Rocks (ed. 3) viii. 175 A normal oolite is made up of an aggregate of spherical bodies, called ooliths, usually about 1 mm. or less in diameter, cemented by some interstitial material, usually calcite. 1971 I. G. Gass et al. Understanding Earth i. 31/2 Ooliths are formed in environments where calcite is being precipitated and there is strong and continuous wave action.

‖ oom (ōm). *S. Afr.* [Afrikaans, = Du. *oom,* G. *oheim,* OE. *ēam* Eme.] Uncle: often used as a respectful appellation when referring to or addressing an older or elderly man.

1822 W. J. Burchell Trav. S. Afr. I. xvii. 433 Old Lucas, or as he was more familiarly called, Oom Hans (Uncle Hans), now turned back with us. 1883 'R. Iron' Story Afr. Farm II. xii. 227 At the farmhouses where he stopped the 'ooms' and 'tantes' remembered clearly the spider with its four grey horses. 1885 J. Nixon Compl. Story Transvaal vi. 116 Sir Theophilus Shepstone, by direction of the High Commissioner, applied to Paul Kruger, inviting him to help with a Boer force; but 'Oom' (uncle) Paul, as he was familiarly termed, declined. 1889 H. A. Bryden Kloof & Karroo i. 42 Mr. Pieter Maynier, familiarly called by Graaff Reinetters, 'Oom Piet' (Oom, or uncle, being a term of affection in South Africa). 1913 C. Pettman Africanderisms 349 Oom Paul, the ordinary designation of the President of the late Transvaal Republic. 1923 Radio Times 28 Sept. 8/2 'Oom Jannie', as he [sc. Smuts] is known among his own people. 1951 L. G. Green Grow Lovely i. 17 That lean old man..—Oom Cappy van der Westhuysen is his name. 1971 Rand Daily Mail 27 Mar. 5/5 Mr. Sneech, still active and still running his business, is known to almost every citizen as 'Oom Harry'. 1974 State (Columbia, S. Carolina) 28 Mar. 15-B/5 Why is it that man has to remain constantly at war with himself, oom Paul?

'oom: see *oo[3].

oompah, oom-pah (ū·mpā). Also **oompa, umpah.** [Imitative.] A repetitive monotonous sound characteristic of a bass brass-instrument; hence, an instrument that makes such a sound. Also in reduplicated forms *oompah-oompah, oomp-pah-pah.* Also *attrib.* and as *v. trans.*

1877 Brooklyn Monthly Oct. 21/2 If a young lady takes her place at the piano to sing, it is your duty as a gentleman to accompany her with a very bass 'oom-pah, oom-pah, bum, bum, bum'. 1896 Scribner's Mag. July 16/2 And some, near the elephant, have set aside money sufficient for a day within sound of Seidl's orchestra, yet they prefer the oom-pah bands of rusted brass. 1904 Ade True Bills 86 He practised until he was able to crawl inside of a big Oom-pah and eat all of the Low Notes in the Blue Book. 1919 Red Cross Mag. Mar. 4/2 But the bands still bang and oom-pah. 1924 P. Rosenfeld Port of N.Y. 72 And the sheer noise..the 'rhythmic oompa of brasses'..saturate him, thrill grim, rough sardonic joy up in him. 1926 Scribner's Mag. Sept. 303/2 A genuine, imported, inimitable oompah, a horn among horns, grotesque, gigantic, inescapable. 1927 R. Hughes Paleface Leather Kid 224 That's the ole oompah-oompah born... I nachelly gotta git that oompah. 1929 W. Thurman Blacker the Berry 122 A brutal sliding trumpet and the trombone..an umpah, umpah by the bass horn..and the orchestra was playing another dance tune. 1930 I. Goldberg Tin Pan Alley 272 It was an oom-pah, oom-pah in quick tempo. 1951 Auden Nones (1952) 65 With trombone the clerk will bravely Go oompah-oompah in his minor grave. 1958 Spectator 31 Jan. 135/1 A bombastic crescendo of the utmost thematic poverty and vulgarity, complete with oom-pah off-beat chords. 1959 D. Cooke Lang. Mus. v. 257 A grotesquely galumphing 6/8 rhythm of the 'oom-pah' kind is set up, featuring the tuba and saxophone. 1961 Listener 30 Mar. 950/3 The accompaniment, an erratic oom-pah of ninth chords, over which the violin sings a hymn. 1966 Ibid. 8 Sept. 354/1 That night the clear air of New York City was heavy with the beat of drums and the oom-pah of brass bands. 1969 M. Gilbert Etruscan Net I. v. 71 They had a band..with trombones, and cornets. And a huge instrument that went oompah-oompah. 1971 Guardian 4 June 10/6 It was like watching a one-man band. You appreciate the oompah versatility but you don't want to hear him often playing. 1972 J. Wain in Cox & Dyson 20th-Cent. Mind I. xi. 395 These newer poets favoured the woodwinds of the orchestra rather than the brass section so deafeningly oompah'd by their seniors. 1974 N.Y. Times 3 Nov. 1/1 (caption) Holiday Oomp-pah-pah. 1975 New Yorker 3 Nov. 127/1 In place of oompah figures or tenths or standard offbeat chords in the left hand, he plays on-the-beat guitar chords—rump rump rump rump.

oomph (ŭmf). *slang.* (orig. *U.S.*). Also **umph, umphh, oomf.** [Of imitative origin.] Sex appeal, glamour, attractiveness; vitality, enthusiasm. Also *attrib.*, esp. *oomph girl.*

1937 Sat. Even. Post 10 Apr. 55/2 With actors, the 'it' quality has to do with their visual personality—sex appeal, magnetism, or whatever you care to call it. In back of the camera, we refer to the ingredient as 'umph'. 1937 W. Winchell in San Francisco Examiner 12 Sept. 25B Jolson's energy gave the show most of its umph and..gave the Gershwin memorial program its pace. 1939 Sun (Baltimore) 29 Apr. 10/3 The modern [girl] knows she'll make the grade if she has plenty of oomph. 1939 Life 31 July 2 Three cheers for the Oomph Girl—your, Look's and Collier's—all in one week!.. This Ann Sheridan certainly must have oomph to win the attention of the three important magazines in issues which hit the newsstands at the same time. 1939 W. C. & H. S. Pryor Let's go to Movies 52 'Look, Alice'—to the heroine—'try putting a little more 'umph' into your lovemaking.' 1942 W. B. Johnson Widening Stain 30 Lucie Coindreau, our Fanny, is the oomph-girl of the Romance Language Department. 1943 P. Cheyney You can always Duck xii. 194 She had such allure, sex appeal, oomph an' what-have-you generally. 1951 New Yorker 6 Oct. 34/1 Q—What else does it [sc. the plot of the play] lack? A—Substance, drive, authority, emotional power, and oomph. 1960 Guardian 19 May 9/3 A Lhasa belle, complete with high heels, lipstick, and 'oomph'. 1970 Daily Tel. 20 Jan. 15/2 This strictly-tailored suit has more oomph than it's ever had. 1973 Philadelphia Inquirer (Today Suppl.) 7 Oct. 41/1 If it's on plain black and white, it doesn't have the oomf. Especially to children, it just doesn't have the oomf. 1974 She Jan. 80/4 We seem to need..a fantasy element to generate enough excitement..in the children to give them the necessary oomph. 1974 San Francisco Examiner 1 May 35/1 He says he'll try to use my 'oomph' to help get BART [sc. Bay Area Rapid Transit] finished. What 'oomph' is the man talking about? Ibid. 35/2 All old World War II types will remember when Annie S. was 'the oomph girl'. 1977 Church Times 28 Oct. 9/3 This prayer may take the form of thanks—for the fact that I am alive with enough energy and oomph to my personality to hate and lust.

oonchook (ū·nʃuk, ōᵘ·n-). Also **eunchuck, owenshook, ownshuck,** etc. [ad. Ir. *óinseach,* Gael. *òinnseach* foolish woman, clown.] **†1.** *Newfoundland.* One of the men dressed as women who participated in a mummers' parade. *Obs.*

1885 W. Whittle in Evening Telegram (St. John's, Newfoundland) 21 Dec. (1962) 22 'Munn' Carter..was always a conspicuous 'fool'... Davey Foley was always the owner of a stylish rig, while his friend, Maney Murphy, appeared, I think, as an 'Owenshook'. The 'Owenshook' was always a terror to encounter, and rarely was merciful to any one who made the attack upon his wind, and woe to the man who disputed his right of giving a sound castigation for the trouble incurred. 1895 D. W. Prowse Hist. Newfoundland 402 Some were dressed as women, with long garments, known as 'eunchucks'. They were all masked, and ran after passengers with an Indian yell, and spoke in a falsetto voice. a 1930 G. J. Bond in J. R. Smallwood Bk. Newfoundland (1937) II. 259/1 Joined with these gaily bedecked figures were a smaller number of veiled men in women's clothing. They bore the appellation of Oonchooks, and were generally more persistent and punishing in their thrashing of people than their more spectacular companions. 1969 Halpert & Story Christmas Mumming in Newfoundland 49 The Newfoundland eyewitnesses draw particular attention..to both the fools who belaboured the bystanders with whips and inflated bladders, and the 'oonchooks'.

2. *Newfoundland* and *Ireland.* (Also in form **oonshick** (-ʃik).) A person who acts foolishly, a noodle. *colloq.*

1937 P. K. Devine Folklore of Newfoundland 35 Ownshook, an ignorant stupid fellow. 1955 Historic Newfoundland (Newfoundland Tourist Devel. Office) 35 Oonshick, a person of low intelligence. 1961 'F. O'Brien' Hard Life x. 83 The divil himself is in the hearts of that Corporation ownshucks. *a* 1966 'M. Na Gopaleen' *Best of Myles* (1968) 152 Begob if I used the word ownshuck you might take my meaning! 1975 *Globe & Mail* (Toronto) 1 Mar. 27/4 Moreover, all those jokes depicting Newfoundlanders as oonshicks—the vernacular for persons of low intelligence—are..ill-considered comment on the disintegration of 478 years of community loyalty and individual fortitude.

oont (ūnt). *Indian* and *Austral. colloq.* Also **unt.** [ad. Hindi (and Urdu) *ūnṭ* camel.] A camel. *Comb.* **oont-wallah,** a camel-driver.

1862 Mrs. J. B. Speid *Our Last Yrs. in India* ix. 214 The Oont-wallah or camel-man. 1892 Kipling *Barrack-room Ballads* 27 O the oont, O the oont, O the commissariat oont! With 'is silly neck a-bobbin' like a basket full o' snakes. 1894 A. G. Leonard *Camel* 101 The baggage [camel], known as 'Gamal' and 'Unt' respectively in Egypt and India. 1900 *Pall Mall Gaz.* 1 Jan. 1/3 A mule..requires more experience in handling than the bubbling oont of India. 1902 *Chambers's Jrnl.* July 431/1 To judge from the selection of pillage, some one conversant with the interior economy of the caravan was involved, and it was significant that a number of the *oont-wallahs* (camel-drivers) were missing. 1933 *Bulletin* (Sydney) 26 Apr. 33/2 Hell! what a lot of calculation had to go into piloting a couple of smelly oonts! 1945 [see *humpy sb.*[1]]. 1961 Partridge *Dict. Slang* Suppl. 1207/2 *Oont,*..a camel: Australian.

oophorectomy. Add: Hence **o:ophore·ctomize** *v. trans.* = *ovariectomize, *ovariotomize *vbs.*; **o:ophore·ctomized** *ppl. a.*

1955 *Jrnl. Amer. Med. Assoc.* 31 Dec. 1701/2 Of 38 women with metastatic breast cancer, all of whom had previously been oophorectomized, bilateral adrenalectomy produced objective remissions in 45%. 1961 *Lancet* 7 Oct. 793/1 In one case it was found that a woman who had been oophorectomised and adrenalectomised still appeared to be excreting large amounts of œstrogens. 1972 *Jrnl. Endocrinol.* LIV. 115 Oestrogens have been shown to stimulate glycolytic activity in the uteri of oophorectomized rats.

ooplasm (ōu·ǒplæz'm). *Biol.* [f. Oo- + Plasm.] The cytoplasm of an egg (see also quot. 1956).

1899 *Bot. Gaz.* XXVIII. 237 There is a stage called zonation in which the nuclei, usually in metaphase, are lined up around the ooplasm. 1939 P. Weiss *Princ. Devel.* i. 78 Only that part of the egg which consists of true oöplasm is broken up into cells, while those portions which consist mainly of yolk remain either unsegmented or cleave with considerable delay. 1956 C. H. Waddington *Princ. Embryol.* i. 16 The different regions of the cytoplasm of the egg may have specific properties, so that a particular region can only develop in one way. Such regions are spoken of as ooplasms; an older name was 'organ-forming substances'. 1974 *Acta Anat.* LXXXIX. 616 A granular or flocculent material, presumed to consist of blood proteins, in the intercellular spaces outside the oocyte, in the pinocytotic vesicles inside the oocyte, and in the ooplasm itself.

Hence **oopla·smic** *a.*

1905 *Jrnl. Exper. Zool.* II. 147 The third cleavage is equatorial... The ectoplasm is now completely segregated in the four ventral cells but the other oöplasmic substances are not as yet located in separate cells. 1925 E. B. Wilson *Cell* (ed. 3) iv. 338 The enormous increase in the cytoplasmic or oöplasmic substance during the growth of the oöcyte leads to the production of the largest known forms of cells. 1961 N. J. Berrill *Growth* xix. 483 Some insight into the ooplasmic specialization of the primitive chordate egg may be gained by a comparison among the embryonic developments in the ascidian *Styela*, the larvacean tunicate *Oikopleura*, and *Amphioxus*. 1968 F. G. Gilchrist *Survey of Embryol.* iv. 65/2 A considerable rearrangement of the materials of the cytoplasm also takes place during meiosis... The rearrangement is termed ooplasmic segregation.

oops (ūps, ups), *int.* Also **o-o-o-ps, ooops, oooops.** [A natural exclamation.] An exclamation expressing apology, dismay, or surprise, used esp. after making an obvious mistake.

1933 'R. James' *Worth Remembering* xiii. 423 Slap fighting—smiting one's opponent with the open hand—a method even Babe would scurn. Oops! Oops! Oops! 1937 L. B. Murphy *Social Behavior & Child Personality* i. i. 45 Julius picked up Gregory, carried him into the bathroom and dropped him on the floor, said 'Oops!' picked him up again, then let him walk. 1939 L. M. Montgomery *Anne of Ingleside* xliii. 334 She caught her foot in a croquet hoop... Gilbert only said 'O-o-o-ps!' and steadied her. 1939 D. Parker *Here Lies* 70 Oops, I'm sorry I joggled the bed. 1944 'P. Quentin' *Puzzle for Puppets* i. 5 Iris..said: 'Oops' as she ran into headlong collision with a Marine sergeant. 1960 V. Nabokov *Invitation to Beheading* xii. 119 She..knocked the pencil off, did not catch it in time, and said 'oops!' 1961 J. Heller *Catch-22* (1962) x. 108 'Ooops, there it goes again.' The rain began falling again. 1972 J. Burmeister *Running Scared* v. 80 If something went wrong you went Oops and called an expert. 1974 *Sunday Post* (Glasgow) 5 May 32/3 The 'keeper gave the ball the wet soap treatment—Oops, butter fingers! 1975 *Daily Mirror* 29 Apr. 16 Now, I'm all for new faces—oops, sorry, Hughie—appearing on talent shows on TV.

oo·ps-a-dai:sy, phonet. var. *upsidaisy.

1953 C. S. Forester *Hornblower & Atropos* 76 'There, baby', said the landlady. 'Daddy's going to play with you. Oops-a-daisy, then.' 1967 Partridge *Dict. Slang* Suppl. 1277/2 *Oops-a-daisy!*, a c.p. [*sc.* catch phrase] of consolation as one picks up a child that has fallen. 1970 K. Giles *Death in Church* viii. 190 'Ooops a-daisy,' said the gunman and the Inspector was hurled to the twelve-feet-high ceiling. 1974 N. Freeling *Dressing of Diamond* 173 He..smacked Colette lightly on the bottom, and said, 'Oops-a-daisy, girl: half-time.' 1976 *South Notts Echo* 16 Dec. 6/5 Only a series of oops-a-daisy mishaps, however, persuaded him to slim down in time.

oorial. [a. Punjabi *ūrial.*] Substitute for def.: A wild sheep, a subspecies of *Ovis orientalis*, having a reddish coat and long, curved horns, found in northern India and other parts of central Asia. (Earlier and later examples.)

1860 P. L. Sclater in *Proc. Zool. Soc.* XXVIII. 127 The Koch, or Oorial of the Sulimani Range, has already been well described. 1898 R. Lydekker *Wild Oxen* 172 The urial, as this sheep is termed in the Salt Range and other districts of the Punjab, is not entitled to specific separation from the sha of Astor and Ladak. 1928 V. G. Childe *Most Anc. East* ix. 201 The foot-hills to the north and west are still the haunts of urial sheep. 1969 J. Fisher et al. *Red Bk.* 168/1 The group of sheep known as Asiatic mouflon, red sheep, and urial are all forms of *Ovis orientalis*.

Oort (ūɹt). *Astr.* The name of Jan Hendrik *Oort* (b. 1900), Dutch astronomer, used *attrib.* and in the possessive to designate concepts proposed by him or arising out of his work, as **Oort('s) (comet) cloud,** a cloud of small bodies that Oort proposed orbited the sun well beyond the orbit of Pluto and acted as a cometary reservoir; **Oort('s) constant,** either of two constants in the equation relating the radial velocity of a star in the galaxy to its distance from the sun (see quot. 1977).

1941 B. J. & P. F. Bok *Milky Way* v. 106 The 'Oort' constant..measures the maximum effect [of differential rotation] at a standard distance. *Ibid.* 108 The best value of Oort's constant *A* is between five and six kilometers per second for a distance of one thousand light years. 1966 *McGraw-Hill Encycl. Sci. & Technol.* VI. 12/1 The radial velocities may be closely represented by the equation Radial velocity = *rA* sin 2*l* where *r* is the distance to the object from the Sun, *l* its galactic longitude, and *A*, known as Oort's constant, depends on many factors including the mass of the galaxy and the distance of the Sun from the galactic center. Its value is approximately 18 km/(sec) (1000 parsecs). 1968 D. C. Knight *Comets* 38 The American astronomer Fred L. Whipple recently proposed that a yet-to-be-seen comet belt, distinct from Oort's 'comet cloud', lies near the orbit of Pluto. 1976 *Nature* 29 Jan. 290/1 Some theories go even further and predict that, in addition to these comets in the Oort cloud, the remains of a primaeval comet belt may still exist at a distance of ~50 AU. 1976 *National Observer* (U.S.) 24 Apr. 6/1 Van Flandern's work also casts doubt on 'Oort's Cloud', one of the more cherished concepts of comet studies. 1977 J. Narlikar *Struct. Universe* iii. 62 Oort expressed the transverse velocity of the star in the following form $T = r(B + A \cos 2(l - l_0))$. Here *r* is the distance of the star from the Sun, *l* its galactic longitude, and l_0 is the galactic longitude of the galactic centre. *A* and *B* are constants called Oort's constants.

oosi, var. *oozi.

ootid (ōu·ǒtid). *Biol.* [f. Oo- after *spermatid.*] A haploid cell formed by the division of a secondary oocyte; by some writers restricted to the ovum, as contrasted with the polar body.

1908 F. R. Lillie *Devel. of Chick* 14 The mature ovum (oötid) and the polar bodies are the precise equivalent of the four spermatids, but..only the ovum on the female side is functional. 1927 W. Shumway *Vert. Embryol.* ii. 34 In this way four (or three, if the smaller secondary oöcyte fails to divide) oötids are produced, of which the single large cell is the ovum, while the smaller ones are known as polocytes. 1946 B. M. Patten *Human Embryol.* ii. 31 The secondary oöcyte..divides again, and in this division..the bulk of the cytoplasm goes to one of the two resulting oötids, which is then commonly called the 'matured ovum'. 1964 N. S. Cohn *Elem. Cytol.* ii. xiii. 206 The secondary oöcyte undergoes a second meiotic division, producing one large cell, the oötid, and a small second polar body. *Ibid.*, The oötid..matures into the egg or ovum. 1972 *Nature* 28 Jan. 213 It must be recalled that the sea urchin egg is an ootid whose maturation divisions ..have been completed.

ooze, *sb.*[1] Add: **2. b.** Short for *ooze leather* (see sense 4 below). Also *attrib.*

1916 *Daily Colonist* (Victoria, B.C.) 18 July 14/1 (Advt.), Ladies' 8-Inch High Laced White Ivory Ooze Boot, blind eyelets, small perforations, full Louis heel. 1922 M. B. Houston *Witch Man* vi. 78 A 'gift' volume of Shakespeare, bound in dark blue ooze.

4. ooze (or **oozed**) **leather** = *ooze-calf.*

1890 in *American Mail Order Fashions* (1961) 27 Boys' Eton Caps in ooze leather. Price, $1. 1897 *Sears, Roebuck Catal.* 324/3 A Very Good Oozed Leather Tobacco or Coin Pouch. 1928 *Publishers' Weekly* 9 June 2348 In four styles of binding..ooze leather, two colors, green or brown, $2·50. 1937 S. V. Benét *Thirteen O'Clock* 71, I could stomach Jeremy Jason, the homespun philosopher, whose small green ooze-leather booklets..produced much the same sensation in me as running a torn fingernail over heavy plush. 1960 G. A. Glaister *Gloss. Bk.* 286/2 Ooze leather, calfskins or split sheepskins prepared to give them a suede or velvet finish on the flesh side.

ooze, *sb.*[2] Add: **2.** (Earlier examples.)

1858 J. Dayman *Deep Sea Soundings* 7 The sinker was detached, and the valve..full of soft oaze. *Ibid.* 9 Between the 15th and 45th degrees of west longitude lies the deepest part of the ocean, the bottom of which is almost wholly composed of the same kind of soft mealy substance, which, for want of a better name, I have called oaze.

b. A deposit or layer of ooze on the ocean floor.

1876 *Proc. R. Soc.* XXIV. 532 In the *Globigerina*, Radiolarian, and Diatom oozes we have found..only one or two shark's teeth. 1926 G. W. Tyrrell *Princ. Petrol.* xiv. 236 The oozes which cover great areas of the ocean floor are mainly calcareous and foraminiferal. 1971 *Nature* 3 Sept. 46/1 The ophiolites in those deep-sea troughs were overlain by Upper Jurassic and Lower Cretaceous radiolarian and nannofossil oozes.

ooze (ūz), *sb.*[4] [prob. f. *ooze, oos(e)*, plur. of *oo*, Sc. form of Wool *sb.*] The nap or short fibres that project from yarn.

1892 J. Nasmith *Students' Cotton Spinning* 373 When thread is intended for lace purposes..it is passed several times through a gas flame at a high speed, so as to burn off the filaments or 'ooze' on its surface and leave it bare. 1909 *Engineer* 1 Oct. 352/1 The term 'gassing' is applied to the process of burning off the ends of fibres or 'ooze' on the different kinds of yarn.

ooze, *v.*[1] Add: **2. c.** Of persons, objects. Often with *out, up, off,* etc.

1929 D. G. Mackail *How Amusing!* 350 Whenever I came oozing along the street, he sort of edged away. 1929 Wodehouse *Mr. Mulliner Speaking* ix. 313 She had planned to lure him into the thing and then ooze off and land him with these septic kids. 1930 —— *Very Good, Jeeves!* v. 122 He oozed out, leaving me to play the sparkling host. 1935 D. L. Sayers *Gaudy Night* xi. 232 Thought I must just ooze over and pass the time of day. ?1953 [see *eel v.]. 1956 N. Marsh *Off with his Head* (1957) viii. 176, I believe I oozed off before they got going. 1963 C. D. Simak *They walked like Men* x. 57, I oozed into the place and shut the door behind me, then slid along the wall and stood there..with my back against the wall. 1966 D. Francis *Flying Finish* ii. 19 He oozed on to a bar stool, his bulk drooping around him. 1971 *Daily Tel.* 15 Sept. 9/4, 18ft 10in of gleaming black Daimler Limousine oozed up the drive and stopped outside. 1971 D. E. Westlake *I gave at the Office* (1972) 127 Decrepit people..sort of oozed out of doorways. 1977 M. Russell *Dial Death* i. iii. 27 The absence of briefcase and umbrella told her that Mr Trenchard had oozed away for the night.

3. (Further examples of *fig.* use.)

1925 E. J. P. Benn *Confessions of Capitalist* i. 21 Women over- or under-dressed, oozing money, and giving from their conversation no trace of education or of finer feeling. 1959 *Listener* 29 Jan. 228/1 The amount of charm oozed at us from the television screen. 1971 *Daily Tel.* 13 Apr. 10/7 The way he oozes bonhomie over everything from day-old chicks to old-age pensioners I find grating. 1975 B. Garfield *Death Sentence* (1976) xxix. 138 The car radio oozed wallpaper music as viscous as syrup.

‖ **oozi** (ū·zi). Also **oozie, oosi.** [ad. Burmese *ū-zi* one seated at the head of an elephant or at the prow of a boat, f. *ū* head + *si* to mount, ride on.] An elephant-driver; a mahout.

1901 G. H. Evans *Treatise on Elephants* ii. 18 Every domesticated elephant necessarily has its own particular attendant ..*oo-si* (the man who rides in front), or *mahout*. *Ibid.*, The *oo-si* should have experience of the most approved methods of fettering, catching, subduing, and approaching unruly animals. 1905 R. T. Kelly *Burma* v. 84 It is interesting to watch the elephants at work; their sagacity is remarkable, and they hardly seem to require the direction of the 'oozis' who sit astride their necks. 1930 Mitton & Yoe *Life Story of Elephant* xiii. 205 The *oozi* who had come with me from Mandalay for some reason left me, and a new man altogether took me in hand. 1960 *News Chron.* 12 July 6/1 Working with oozies (elephant drivers) who believed in magic and nats (spirits) and every form of what we call superstition, I kept a very close hold upon myself. *Ibid.*, The drinks flowed..because they knew that very soon they would be out on trek again, alone, except for their dogs, their servants, the oozies and the elephants.

oozle (ū·z'l), *v.* *Austral.* and *N.Z.* (*rare*). [f. Ooze *v.*[1] + -le.] *trans.* and *intr.* To undulate; to move slowly.

1934 T. Wood *Cobbers* xii. 153 The octopus..goggled his eyes and oozled his slimy, restless-writhing arms. 1958 *Tararua* XII. 29 The critics will no doubt note that *to oozle* and *to trickle*, which denoted much slower modes of locomotion, were comparatively little used.

oozlum bird (ū·zlǝm). [Fanciful.] A mythical or unrecognized bird (see quots.).

1899 W. T. Goodge *Hits, Skits, & Jingles* 6 It's a curious bird, the Oozlum, And a bird that's mighty wise, For it always flies tail-first to Keep the dust out of its eyes! 1951 Partridge *Dict. Slang* (ed. 4) 1126/1 *Oozlum bird*, 'a bird whose species you cannot recognise on sight', ..: Naval. 1974 P. Cave *Dirtiest Picture Postcard* xvii. 113 The fabulous oozlum bird, which flies round in ever-decreasing circles until it disappears up its own arsehole in a puff of blue smoke.

op[2] (ǫp). *Mus.* Pl. **opp, ops.** Abbrev. Opus 1 (rarely *opus v.).

In some early uses the abbreviation may be of the Italian *opera.*

1784 (*title*) A Favorite Concerto for the Harpsichord or Piano Forte with Accompanyments: Composed by Giuseppe Haydn. Op. 37. *a* 1865 Mrs. Gaskell *Let.* (1966) 817 Violet: 'Have you brought any music down Miss Gaskell?' Meta: 'No.' Violet: 'Oh—but I've brought *Op.* 7 down with me.' 1880 [see Opus 1]. 1885 W. S. Gilbert *Mikado* ii. 36 The music hall singer attends a series Of masses and fugues and 'ops' By Bach, interwoven With Spohr and Beethoven, At classical Monday Pops. 1901 *Punch* 1 May 325/2 (*caption*) Lady (referring to programme, to friend). '"Schumann, Op. 2". What's the meaning of "Op. 2"?' 'Arry (who thinks he is being addressed, and always ready to oblige with information). 'Oh, Op. 2. Second dance; second 'op, yer know. May I 'ave the pleasure?' 1921 [see *opus v.]. 1924 *Public Opinion* 12 Sept. 258/2 No longer does light and irresponsible music suffice as it did in days that are gone. The inclination of the public is towards the classical. The programmes that give the 'symphonies and ops' prove the most attractive. 1933 *Radio Times* 14 Apr. 108/2 The famous violinist, Rode, for whom Beethoven..composed this sonata for violin and pianoforte, Op. 96. 1968 *Listener* 11 July 56/2 These waltzes..differ considerably in style from the fine set of six waltzes for piano duet, Op. 22, of only a little later. 1974 *Times* 19 Oct. 9/6 Ashkenazy chooses..slow tempo for the fugues..in Op. 110. 1975 *Gramophone* July 205/1 Dvorak. Slavonic Dances, Opp. 46 and 72.

op[3] (ǫp). **1.** Colloq. abbrev. Operation. **a.** = Operation 6.

1925 W. Deeping *Sorrell & Son* xviii. 273 Motor-bus ran over her..pretty hopeless. Winter has seen her,—but thought she wouldn't stand an op. 1932 A. Christie *Peril at End House* iii. 50 Just before my op... Operation. For appendicitis. 1933 Joyce *Let.* 30 May (1966) III. 281 Dangerous for operated eye which may go blind during op. because of loss of vitreous. 1934 *Punch* 11 Apr. 397/1 No need for immediate op.; right eye untouched, but he fears left may be permanently damaged. 1953 C. Day Lewis *Italian Visit* v. 59 I'd not advise you to believe There's a slick op. to end your grief. 1964 G. L. Cohen *What's Wrong with Hospitals?* iv. 74 The probationers agreed that minor ops gave the most trouble. 1973 *Guardian* 26 May 1/5 Ops on rates. Free vasectomy operations..were available in Birmingham from yesterday. 1974 O. Manning *Rain Forest* ii. i. 139 'I was so tired the evening we reached Al-Bustan, I forgot to take that damn pill.' 'If it was as long ago as that, it *is* too late, dear, unless you want a major op.'

b. Freq. pl. = Operation 7. Also *attrib.* and *transf.*

In quots. 1941[1], 1942[1] short for *Operations Room.*

1925 Fraser & Gibbons *Soldier & Sailor Words* 215 *Ops.*, operations. 1941 *Jrnl. Aeronaut. Sci.* (Aeronaut. Rev. sect.) Jan. 32/3 The atmosphere of the room is recreated, even to the introduction of some of the new aeronautical colloquialisms, such as 'Ops' for the operations room. 1941 *Illustr. London News* CXCVIII. 434/1 The bomber pilot is aware that his squadron is scheduled for what the Service calls 'Ops' (Operations) when his batman wakes him with an early-morning cup of tea. 1942 T. Rattigan *Flare Path* i. 112, I went up to ops. at five-thirty. 1942 *R.A.F. Jrnl.* 30 May 33 The..officer..delved into his brief-case and produced the ops log. 1942 *Tee Emm* (Air Ministry) II. 89 Make certain before going out on ops. that you have your whistle. 1944 'N. Shute' *Pastoral* iii. 57 Each day we practise some new thing that we have learnt from the last op. 1949 *Radio Times* 15 July 38/3 Ann Scott,..gets into conversation..with an American girl who met a boy in the R.A.F. three months ago... 'Now he's starting Ops and he wants us to get married right away—he's like that!' 1967 O. Wynd *Walk Softly* xi. 171 'This is the ops room.' The place had one door and was about thirty feet square. 1970 *Daily Tel.* 16 June 7/1 A seasoned campaigner watched the campaign in awe. 'It is like a military op,' he said. 1973 'A. Hall' *Tango Briefing* x. 120 They'd been forced to set up the op... The decision-making had been at Prime Minister level. 1973 D. Miller *Chinese Jade Affair* xviii. 173, I looked in some awe at the maps spread out in true 'Ops Room' fashion.

2. Colloq. abbrev. Operative *sb.*, Operator. **a.** A (private) detective (see *operative *sb.* 3 b).

1926 *Clues* Nov. 162/1 *Op*, a private detective agency operator. 1927 D. Hammett in *Black Mask* May 24/1 He says in all his fifty years of gum-shoeing he's never seen such a handsome op, besides being a fashion plate and a social butterfly and the heir to millions. 1929 —— *Dain Curse* (1930) xx. 224 'Can you spare me another op?' I asked. 'MacMan is available.' 1975 J. Gores *Hammett* viii. 60 Watching the stocky two-hundred-pound op.. Hammett felt a little ill... 'You going to take over the investigation of the police department?'

b. A radio or telegraph operator (see Operator 5 in Dict. and Suppl.). (See also quot. 1970.)

1931 G. Irwin *Amer. Tramp & Underworld Slang* 138 *Op*, a telegraph operator. 1942 T. Rattigan *Flare Path* i. 94 Two bumped off—tail gunner and wireless op. 1970 *Amer. Speech* 1968 XLIII. 288 *Op*, a telephone, telegraph, or teletype operator. 1973 A. Ross *Dunfermline Affair* 13 He had been a Radio Op in the R.A.F.

op[4] (ǫp), colloq. abbrev. *optical *a.* 2 c. *op art*, = *optical art* s.v. *optical *a.* 2 c. orig. *U.S.* Also *attrib.*

1964 *Time* 23 Oct. 78/1 No less a break from abstract expressionism than pop art, op art is made tantalizing, eye-teasing, even eye-smarting by visual researchers using all the ingredients of an optometrist's nightmare. *Ibid.*, The Museum of Modern Art is planning an op show titled 'The Responsive Eye' early next year. 1964, 1965 [see *optical *a.* 2 c]. 1965 *Observer* 28 Feb. 2/6 'Op Art'—as it is known among the smart set—has taken America's contemporary art enthusiasts by storm, and made its predecessor, 'Pop Art', seem *passé*. *Ibid.* (Colour Suppl.) 23 May 23/2 Op(tical) dresses are dazzling—literally. They jar the eye with geometric billboard colour... Trad dresses are gentle and demure. Op. dresses are stark and simple. 1965 *Sun* 24 May 5/7 Then you get a blasting of Op art—that's the optical illusion art that the young artists are raving about. 1966 *Time* 28 Jan. 44 Hard on the heels of op artists, who address their work to the retina, has come a widespread number of 'kinetic' artists, who try to combine mechanics and art. 1966 *Punch* 2 Feb. 158/1 (*caption*) Or there's this op-art bowler, sir, for more formal occasions. *Ibid.* 15 June 876/3 Knee-length skirts, non-plastic fabrics, non-Op shoes; also short-haired men and neat girls in plain stockings. 1967 J. Symons *Man who killed Himself* ii. vi. 184 She was dressed now in a black and white op art dress. 1967 *Punch* 21 June 906/1 Infinite trouble is taken by op artists to produce an illusion of movement. 1967 *Spectator* 18 Aug. 194/3 With bright Op patternings, primary Pop colours and uninhibited use of synthetic materials, these designers..parallel our painters and sculptors in inventing new shapes and forms through the use of new materials. 1970 M. de Sausmarez *Bridget Riley* i. 15 Its decorative potential..has been..widely exploited commercially in 'Op' dresses, 'Op' advertising, and 'Op' packaging. 1970 W. J. Burley *To kill Cat* i. 12 A sleeveless frock in gay op-art material. 1974 *Encycl. Brit. Micropædia* VII. 545/1 The effects of Op art can be based either on perspective illusion or on chromatic tension; in painting, the dominant medium of Op art, the surface tension is usually maximized to the point at which an actual flickering is perceived by the human eye. 1974 *Listener* 24 Jan. 108/2 Artists (especially Op-artists and the like) can impose an unjustifiable authority on the observer. 1976 'Z. Stone' *Modigliani Scandal* i. iii. 35 He..looked at the desk... The grain..flowed like an op-art painting.

opacification (opæ:sifikēi·ʃən). [f. as next + -ification.] The process of rendering or becoming opaque.

1903 *Med. Rec.* (N.Y.) LXIII. 333/2 Hyperplasia, degeneration—these are results of malnutrition and the essentials of opacification [of the lens]. 1947 *Jrnl. Soc. Glass Technol.* XXXI. Abstr. Sect. 29 A boron-free enamel might be fired to a certain degree, after which opacification decreased because the TiO_2 was dissolved. 1953 *Radiology* LX. 366/1 (*caption*) The poor opacification [to X-rays] of the liver and of the portal vein is typical of cirrhosis. 1970 *Nature* 24 Oct. 363/1 Opacification of the lens began as an increase in sheen posteriorly.

opacifier (opæ·sifəi,əɹ). [f. next + -er[1].] A substance which renders something opaque.

1911 *Chem. Abstr.* V. 168 Cast Iron Enamels... SnO_2 and cryolite are the best opacifiers. 1959 *Which?* Nov. 152/1 Opacifiers (which make it look opaque and creamy) make the shampoos look more attractive. 1973 *Daily Tel.* (Colour Suppl.) 12 Oct. 72/4 The developer then gets to work on the negative and the opacifier which gradually clears to reveal the dyes, now released from the negative and visible above the white pigment. *a* 1977 *Harrison Mayer Ltd. Catal.* 23/2 Tin oxide is widely used as a glaze opacifier.

opacify (opæ·sifəi), *v.* [f. Opac(ity + -ify; cf. F. (*s'*)*opacifier.*] **a.** *trans.* To render opaque.

1940 *Chem. Abstr.* XXIV. 4268 (*heading*) Paper opacified with calcium carbonate. 1955 P. D. Trevor-Roper *Ophthalm.* xxiii. 413 Post-operative uveitis may opacify the posterior corneal layers. 1957 S. D. Gershon et al. in E. Sagarin *Cosmetics* xxiv. 615 Initially, cold-waving lotions were marketed as transparent liquids. Shortly thereafter, the advantage of marketing an opaque liquid, the milkiness connoting richness and gentleness, became evident. As a result, practically all marketed waving lotions were opacified. 1971 *Country Life* 6 May 1084/1 Milk-white glass opacified with arsenic displays a fiery opalescence if held to the light.

b. *intr.* To become opaque.

1954 S. Duke-Elder *Parson's Dis. Eye* (ed. 12) ii. 24 If either of these membranes [of the cornea] is disrupted, fluid is absorbed and the tissue opacifies. 1967 *Amer. Jrnl. Roentgenology* C. 410/2 The fluid-filled fundus of the stomach..opacifies during abdominal aortography. 1971 *Nature* 12 Mar. 120/2 Corneas rapidly opacified and swelled with the first enzyme digestion after the initial irradiation.

So **opa·cified** *ppl. a.*, **opa·cifying** *ppl. a.* and *vbl. sb.*

1914 *Chem. Abstr.* VIII. 224 The opacifying effect of the metallic oxides of Sn, Zr, Ti and Al..utilizable for white enamels. 1947 *Endeavour* VI. 117 The opacifying agent dissolves, leaving the glass transparent while molten, but causing it to become opaque upon cooling. 1954 *Amer. Jrnl. Roentgenology* LXXII. 592/1 In the later films the liver seems to be more opacified..than the spleen. 1963 *Times* 25 May 11/4 These enamels were produced by fusing what amounted to opacified glass on to wafer-thin copper. 1973 *Sci. Amer.* Oct. 128/1 A talc base will often be augmented with an opacifying pigment such as zinc oxide or titanium dioxide.

opacimeter (opæ·simītəɹ). [f. Opaci(ty + -meter.] An instrument for measuring opacity, esp. by reflection.

1919 *Chem. Abstr.* XIII. 2892 (*heading*) Opacimeter designed for the estimation of the quantity of bacteria. 1944 *Paper Trade Jrnl.* 26 Oct. 27/2 The white body consists of a white standard protected by a cover glass, as is the case in the Bausch and Lomb opacimeter. 1967 *TAPPI* Feb. 59A/1 The B & L Opacimeter reads either printing opacity (as defined by Davis) or contrast ratio (as defined by TAPPI official methods).

opacious, *a.* Delete *Obs.* and add later example (prob. used as a conscious archaism).

1953 S. Beckett *Watt* iii. 158, I was very fond of fences, of wire fences,..; not of walls, nor palisades, nor opacious hedges, no.

opacity. Add: **2.** *spec.* the ratio of the intensity of the light incident on a sample or object to that of the light transmitted by it.

1890 Hurter & Driffield in *Jrnl. Soc. Chem. Industry* 31 May 455/2 The inverse of that fraction, or $I/I_x = e^{kA}$ measures the opacity of the substance. 1926 J. W. T. Walsh *Photometry* xiii. 392 The opacity is then measured by placing the exposed area of the plate between a source of light and a photometer. 1939 *Q. Jrnl. R. Meteorol. Soc.* LXV. 417 In the case of five stations..the summer values of opacity are higher than the winter values, owing to the prevalence of sea fogs in summer. 1966 R. J. Ross *Television Film Engin.* iv. 169 The silver deposit and the opacity are so related that the logarithm of the opacity is directly proportional to the mass of silver.

b. Also used with reference to other forms of radiation. (Further examples.)

1928 [see *Lipiodol]. 1971 *Jrnl. Electron. Microsc.* XX. 124/1 The pronounced electron opacity was localized especially..on the outer membrane.

3. c. With reference to a rule in *Phonology*: the state or quality of being opaque (sense *3 c).

1971 [see *opaque *a.* 3 c]. 1975 *Trans. Philol. Soc.* 1974 113, I have myself argued..that for French the elegance of such a solution is outweighed by its disadvantages and that restructuring occurred when generalization of the loss of word-final [ə] made for opacity: the conditioning of denasalization before an intervocalic nasal was no longer discernible from surface phonetic shapes. 1977 *Language* LIII. 19 In each instance of opacity, a phonological rule which relates a large number of surface lexical items is obscured by the presence of other items in which the rule appears to fail.

opal. Add: **1. c.** The colour of an opal.

1890 O. Wilde *Pict. Dorian Gray* v, in *Lippincott's Monthly Mag.* July 41 The sky was pure opal now. 1897 *Sears, Roebuck Catal.* 212/1 Dainty colorings.. baby blue, rose pink, opal, [etc.]. 1901 [see *ash *sb.*[2] 1 e]. 1914 R. Brooke in E. Marsh *Rupert Brooke* (1918) vii. 142 Like an Italian town in silver-point..with a sea and sky of opal and pearl and faint gold around. 1966 G. W. Turner *Eng. Lang. Austral. & N.Z.* iv. 83 The sky turned into an unlucky opal.

2. (Earlier and later examples.) Also with reference to the opalescence of the glass (rather than the colour). Also *attrib.*

1885 *List of Subscribers, Classified* (United Telephone Co.) (ed. 6) 229 (Advt.), Crystal and Demi-Crystal Table Services and Ornaments..Opal, Flint, and Coloured Goods. 1949 W. A. Thorpe *Eng. Glass* (ed. 2) ix. 226 It [*sc.* a 'nine-pin' bottle] belonged in the main to Bristol and Stourbridge manufacturers, and in..white opal it lasted well into the nineteenth century. 1970 F. & L. Schuler *Glassforming* vi. 50 To the commercial decorators the term 'glass color'..means any fusible coating for glass, including both the transparent colors and the opal colors.

3. b. *spec.* Applied to an electric light bulb made of translucent white glass.

[1901 F. B. Crocker *Electr. Lighting* II. xvii. 423 Lamps are made in many colors, such as red, blue, green, amber, opal, frosted, etc., besides the ordinary clear glass bulbs.] 1904 *Electr. Rev.* 19 Feb. (Suppl.) p. ix (Advt.), The 'Ideal' half opal lamp. 1926 *Gloss. Terms Electr. Engin.* (B.S.I.) 148 *Opal lamp*, a filament lamp, the bulb of which is made of opalescent glassware so as to enlarge the source of light with a consequent reduction in surface brightness [etc.]. 1934 *Discovery* Aug. 229/2 A hole in a screen in front of an opal lamp. 1938 [see *pearl *sb.*[1] 16 b]. 1976 *Daily Tel.* (Colour Suppl.) 25 June 15/2 Opal light bulbs are primarily suitable for use where the bulb is clearly visible, for their coating hides the filament.

4. *opal-buyer, -field, -seeker; opal-black, -green, -grey, -hued* (earlier example), *-pale, -shelled* adjs.; **opal dirt** *Austral.*, the type of earth in which opal is found; **opal glass** (earlier and later examples); also *attrib.*; **opal gouger** *Austral.*, one who digs for opal; cf. *gouger c; **opal ware,** ware made of opal glass, *spec.* a type of heat-resistant opalescent ware (now, with capital initial and as one word, a proprietary name in the U.K.).

a 1963 J. Lusby in B. James *Austral. Short Stories* (1963) 225 The eyes opal-black in wrinkled slits of skin. 1911 C. E. W. Bean *'Dreadnought' of Darling* xxv. 222 The most precious colour, the opal-buyers told us, was 'fire'—the rich glow as of a red-hot horseshoe, which you find in the heart of the best opal. 1925 *Ann. Rep. Dept. Mines New South Wales 1924* 85/2 The 'Opal Dirt' is picking ground, being simply a layer of clay or sandy clay overlain by sandstone. 1963 Opal-dirt [see *bottom *v.* 5]. 1965 *Ann. Rep. Dept. Mines New South Wales 1963* 51/2 Using air compressor and jack spade on the softer opal dirt. 1902 *Chambers's Jrnl.* Aug. 496/2 There are few men on the opal-fields who do not average five pounds per week. 1866 'J. Easel' in *Queen* 11 Aug. 93/2 The opal glass, which transmits a rich and lovely iridescent light, exactly like the precious stone from which it is named. 1923 *Vogue* Oct. 47/1 A pair of feathered bird pictures and French opal glass bottles. 1956 L. M. Angus-Butterworth *Brit. Table & Ornamental Glass* iii. 10 The handle in opal glass. 1931 V. Palmer *Separate Lives* 200 The bleary-eyed opal-gougers, who spoke as if the hot winds had dried up the fountains of their speech. 1936 A. Russell *Gone Nomad* vii. 57 Lured on by the uncertainty of what the next stroke of his pick will reveal, the opal-gouger never abandons hope, until, with his funds and credit exhausted, he may hope no longer. 1955 S. Spender *Coll. Poems 1928–53* 113 The eye is carried by the choppy tide To a shore opposite of opal-green spaces. 1867 A. J. Munby *Diary* 7 June in D. Hudson *Munby* (1972) 239 All things were cool and charming, and opal-grey, in the cool sweet morning. 1881 O. Wilde *Poems*

126 Tremulous opal-hued anemones. 1946 W. de la Mare *Traveller* 25 Opal-pale..A strange and deepening lustre tinged the air. 1902 *Chambers's Jrnl.* Aug. 496/1 At that hour the opal-seeker must cease his daily toil. 1922 V. Woolf *Jacob's Room* i. 11 Out pushes an opal-shelled crab. 1894 *Montgomery Ward Catal.* 520/3 Blue Opal Ware. A new pattern, just out, made in fancy colored glass in blue opal. *Ibid.* 521/1 Opal Ware. 1929 *Encycl. Brit.* X. 412/1 In 1927 the United States had..23 manufacturers making opal ware. 1958 *Mixed Batch* (J. A. Jobling & Co.) July 47 This new glass..was a delicate pearly white; and it was given the name Opalware. 1964 *Trade Marks Jrnl.* 29 Apr. 688/2 Jobling Opalware... Tableware..made of opal glass. James A. Jobling & Company Limited,..Sunderland, manufacturers.

opaline, *a.* and *sb.* Add: **A.** *adj.* Also, resembling opal other than in colour. (Further example.)

1962 C. Frondel *Dana's Syst. Min.* (ed. 7) III. 296 Common opal. In general, opal without a play of colour. ..Includes..rock-forming opaline silica.

B. *sb.* **2.** (Later examples.) Also, translucent glass of a colour other than white.

1964 [see *make-up mirror* s.v. *make-up 6]. 1970 G. Savage *Dict. Antiques* 296/1 The manufacture of opaline was at its most popular between 1840 and 1870, after which it declined in popularity.

opalite (ōu·pǎləit). Also **Opalite.** [f. Opal + -ite[1].] Opal glass made in the form of tiles or bricks suitable for building purposes. (Formerly a proprietary name.)

1903 *Science* 13 Feb. 266/2 The feature of this building is the treatment of the interior of the cages with light-green opalite tile. 1940 *Archit. Rev.* LXXXVII. 19 The walls and counter fronts are faced with black and white opalite glass. 1975 Huntington & Mickadeit *Building Construction* (ed. 4) x. 520 Structural glass is available in the form of tile or slabs..for use as a finish on exterior and interior wall surfaces. It is known by various trade names such as *Carrara Glass, Vitrolite,* and *Opalite.*

opaque, *a.* (*sb.*) Add: **2. a.** *spec.* of glass which is not translucent.

1836 E. W. Lane *Acct. Manners & Customs Mod. Egyptians* II. 368 There is a very common kind [of bracelet]..of opaque, coloured glass, generally blue or green. 1878 A. Nesbitt *Descr. Catal. Glass Vessels S. Kensington Mus.* p. ii, It can be produced either wholly devoid of colour or tinted with any hue, and either opaque or transparent, without loss of brilliancy. 1907 E. Dillon *Glass* xvii. 291 Much opaque white glass was made in Germany.. in the first years of the eighteenth century... At South Kensington may be seen a covered beaker of this *milchglas* elaborately painted. 1926 W. Buckley *European Glass* ii. 11 The Venetians in the 14th century made an opaque red glass resembling jaspar. 1961 E. M. Elville *Collector's Dict. Glass* 145/1 Nine out of every ten glasses with opaque-twist stems will show an unknopped stem.

b. (Further examples.)

1903 Pusey & Caldwell *Pract. Application Röntgen Rays* vi. 133 It will be a great help to the radiographer if the dressings are of a material which is not opaque to the ray. 1937 I. C. C. Tchaperoff *Man. Radiol. Diagnosis* iv. 167 The emphysematous area remains transparent in the radiograph of full expiration, whereas the normal areas become more opaque on expiration. 1972 *Jrnl. Ultrastruct. Res.* XXXIX. 580 The transverse..tubules and sarcoplasmic reticulum..of *Limulus* myocardium have been examined by infusing hearts with materials which either produce electron-opaque reaction products.. or which are inherently electron opaque.

3. c. *Phonology.* Of a rule: that cannot be extrapolated from every occurrence of the phenomenon; in which not every context implies the rule.

1971 P. Kiparsky in W. O. Dingwall *Survey Linguistic Sci.* 621 Define the concept *opacity of a rule* as follows:.. A rule A→B/C—D is opaque to the extent that there are surface representations of the form (i) A in environment C—D or (ii) B in environment other than C—D... Let us refer to the converse of opacity as *transparency.* 1974 S. R. Anderson *Organization of Phonol.* xii. 209 Kiparsky gives examples..in which historical change can be seen to operate on nontransparent (or *opaque*) rules so as to make them more transparent or to eliminate them from the grammar. *Ibid.* xiii. 250 If we disregard the second possible application instead, we derive *djalum + ba: + day + be:*, which is not opaque because the long vowels are not adjacent. 1977 *Language* LIII. 18 Palatalization is opaque to the extent that there are, on the surface, some palatalized consonants not in the environment of a following *i.*

B. *sb.* **a.** (Later examples.)

1903 *Westm. Gaz.* 20 Aug. 4/1 A red batiste of voile or open canvas is more lovable than a red face-cloth or serge or tweed or linen; and, again, it is to be noted how of these red opaques the linen is better than the others. 1969 *Earth & Planetary Sci. Lett.* VII. 237/1 All the samples showed some effects—oxidation of the opaques and reddening of the silicates. 1976 *Nature* 22 Jan. 196/1 The Santiago lavas contain up to 10% phenocrysts of olivine and plagioclase in a matrix of plagioclase laths, clinopyroxene, olivine and opaques.

2. *Photogr.* **a.** A water colour or other substance for producing opaque areas on negatives, as in retouching.

1908 Schriever & Cummings *Compl. Self-Instructing Libr. Pract. Photogr.* IV. xl. 322 To make the opaque, add one ounce of No. 1 to four ounces of No. 2. 1943 *Chem. Abstr.* XXXVII. 1946 (*heading*) Removing opaque from photographic negatives. 1953 A. Sussman *Amat. Photographer's Handbk.* (ed. 4) xiv. 254 To spot a negative you can use a little India ink or you can buy a cake of opaque.

b. A print made on opaque paper.

1959 *Recomm. for Density & Contrast Range of Monochrome Films* (*B.S.I.*) 5 Prints of black-and-white photographic opaques should be made in such a way that a middle tone..will have a reflection density within the range 0·5 to 0·7. 1969 *Focal Encycl. Film & Television Techniques* 518/1 *Opaque*, term contrasting with transparency, such as a lantern slide, to denote a printed picture on opaque paper.

opaque, *v.* Add: Hence **opa·qu(e)ing** *ppl. a.* and *vbl. sb.*

1912 E. Heilmann *Brit. Pat.* 26,498, Materials which cannot be used alone as opaquing substances for enamel, yield good opaquing effects when heated so highly as to form spinels. 1913 *Chem. Abstr.* VII. 2460 Opaqueing agents for white enamels. 1967 E. Chambers *Photolitho-Offset* v. 52 Opaquing is used for spotting out pinholes, edge lines and other undesirable marking on the negatives. 1967 Karch & Buber *Offset Processes* v. 195 Paint out these holes and scratches..with an opaquing solution.

op art: see *op[4].

op. cit. (ǫp sit). Abbrev. of L. *opus citatum*, the work quoted, or *opere citato*, in the work quoted.

1883 *Nineteenth Century* Feb. 213 *Op. cit.* vol. ii pp. 200, 201. 1966 *Listener* 1 Sept. 305/1 Van Bath (*op. cit.*) has drawn attention to the relation between urban prosperity and agricultural improvement in the Netherlands. 1970 J. Needleman *New Religions* 233 J. G. Bennett, op. cit., p. 53.

Op-Ed (ǫp,e·d). *U.S.* [Abbrev. *o*pposite *ed*itorial.] In full *Op-Ed page.* A page of a newspaper, opposite the editorial page, devoted to personal comment, feature articles, etc.

1970 *Time* 10 Aug. 32 The Op-Ed page—so named because it runs opposite a newspaper's editorial page—became a journalistic tradition with the rise of the personal column. Pioneered by the Pulitzers in the old New York morning *World*, the Op-Ed provides a variety of viewpoints in dozens of major metropolitan dailies. 1970 *N.Y. Times* 21 Sept. 42/2 Through the new page opposite the Editorial Page that we inaugurate today, we hope that a contribution may be made toward stimulating new thought and provoking new discussion on public problems... The two pages together—Editorial and Op. Ed.—are designed to create an intellectual forum. 1974 *Verbatim* Dec. 6/1 The Op-Ed Page of *The New York Times.* 1977 G. V. Higgins *Dreamland* i. 8 If you don't get a regular fix of the stuff that goes on the front page, you're not going to have anything to be insightful about on the Op Ed page.

open, *sb.* Add: **4. a.** (*e*) public knowledge or view, *spec.* in phrs. *to come* (*out*) *into the open*: to reveal one's plans, acts, thoughts, etc.; *to bring* (something) (*out*) *into the open*, to bring into public notice or view.

1942 T. Bailey *Pink Camellia* v. 35 We may as well come into the open, Miss Merryman. 1965 *New Statesman* 30 Apr. 670/2 The Tory Party statement which last year brought immigration into the open as a Birmingham election issue. 1965 *Listener* 16 Sept. 399/2 The Peking *People's Daily* came out into the open, supporting Pakistan's version of events. 1976 K. Royce *Bustillo* iv. 45 They both had something to hide... It would be better out in the open.

b. (Further examples.)

1846 J. W. Webb *Altowan* I. ii. 42 All openings or natural clearings are called 'opens' by the half-breeds of the Indian country. *Ibid.* viii. 201 They cautiously entered where there seemed no indication of an open. 1880 *Encycl. Brit.* XIII. 603/1 Living in herds of from fifty to one hundred in the grassy 'opens'. 1958 *Edmonton* (Alberta) *Jrnl.* 28 July 4/3 The animal seemed to distrust the bald opens of the marsh.

5. b. *Electr.* An accidental break in the conducting path for a current.

1913 T. Croft *Amer. Electricians' Handbk.* i. 55 Open circuits in multiple wiring installations are usually readily located... The lamps on the generator side of the 'open' will..burn while those on the far side will not. 1933 F. F. Fowle *Stand. Handbk. Electr. Engineers* (ed. 6) iii. 193 Open-circuit faults or 'opens' are produced by breaks in the conductors. 1967 *Electronics* 6 Mar. 320/2 The D200's..identify shorts or opens and the polarity of diodes. 1977 *Sci. Amer.* Feb. 88/1 (Advt.), That's the beauty of the TDR, it's not limited to identifying shorts or opens—it points out any disturbances.

c. An open competition, tournament, or the like; cf. Open *a.* 14.

1926 Wodehouse *Heart of Goof* iv. 128 'After all, there is always golf.' He nodded. 'Yes... Who knows?.. The Amateur Championship—' 'The Open!' I cried... 'The American Amateur,' said Chester, flushing. 'The American Open,' I chorused. 1930 *Daily Express* 8 Sept. 11/4 Miss Brazier also has played in 'opens'. 1972 *Country Life* 7 Dec. 1600/3 Jack White who, in 1904, was the first man to break 300 in the Open. 1973 *Guardian* 28 June 13/3 The World Open is a ten-round [chess] championship open to all-comers.

d. *the Open*: the Open University. *colloq.*

1970 *Guardian* 29 Aug. 9/6 Along with Jennie Lee, Mrs Thatcher has a right to be seen as the saviour of the Open. 1972 *Ibid.* 15 June 9/5 Pupils in schools..had to wait until 21 to qualify for the Open or 23 to enter other universities as 'mature' students.

open, *a.* (*adv.*) Add: **2. c.** Of a shop, public house, etc.: accessible to use by customers (at a particular time); available for business; *they are open*: the public houses are open.

1824 E. Weeton *Let.* 8 June in *Jrnl. of Governess* (1969) II. 287 As I go to any place of Worship, fruit stalls are in the road, and confectioners shops open, as on any other day. 1836 Dickens *Sunday under Three Heads* i. 5 In streets like Holborn and Tottenham Court Road, which form the central market of a large neighbourhood, inhabited by a vast number of mechanics and poor people, a few shops are open at an early hour of the morning. 1942 'S. Campion' *Bonanza* 19 It was eleven—'they' were open. 1952 E. O'Neill *Moon for Misbegotten* iii. 152 There'll be a speak open, and some drunk laughing. *Ibid.* iv. 173 The bar at the Inn won't be open for hours. 1961 'E. Fenwick' *Friend of Mary Rose* (1962) i. 16, I was talking to Mrs. Rudd... I wanted to be sure she was open. 1965 J. Porter *Dover Three* ii. 30 Just one question, laddie... Are they open yet? 1973 A. Mann *Tiara* xiii. 118 Piccoli's will still be open. Shall I nip down and get pictures of all these types?

d. Designating a prison, borstal, or the like where the inmates are seldom or never locked up.

1946 *Rep. Commissioners of Prisons 1939–41* 47 in *Parl. Papers 1945–6* (Cmd. 6820) XIV. 281 All were now required to pass straight from the restraint of a prison wall to associated life in buildings designed as a perfectly open Borstal. 1950 *Prisons & Borstals* (Home Office) 20 The first prison camp in England was started in connection with the training prison at Wakefield in 1933... The open prison is therefore beyond the stage of experiment: it is a well-established feature of this as of many other prison systems. 1957 *Economist* 2 Nov. 397/3 Miss Size's last job was as Governor of the first open prison for women, at Askham Grange, near York, a pleasant Victorian mansion where there are neither high walls nor keys. 1964 M. Argyle *Psychol. & Social Probl.* v. 69 Open borstals had a higher success rate than closed borstals, at all levels of expected failure. 1972 P. D. James *Unsuitable Job* vii. 210 Perhaps they would send her to an open prison. Open. It was a contradiction in terms.

e. *open heart*, a practically bloodless heart that has been temporarily by-passed and cut open for examination or surgery; usu. *attrib.*

1950 *Surgery* XXVIII. 474 Recent rapid strides in the field of vascular surgery are leading inevitably to the point where a direct surgical attack on the open heart is possible. 1960 *Sci. Amer.* Feb. 84/1 The cardiac surgery group at Minnesota has now performed nearly 1,000 open-heart operations with the aid of the heart-lung machine or other methods of by-pass. 1977 *Private Eye* 1 Apr. 5/3 As he recovers from open-heart surgery (new valves) Sir Christopher Soames is confident of an early return to Parliament.

4. a. *open carriage* (earlier examples), *open car.*

1797 Lady Newdigate *Let.* 30 July in A. E. Newdigate-Newdegate *Cheverels* (1898) xiv. 200, I have never felt yᵉ Downs too hot for my open Carriage till yesterday, when I was forced to put up yᵉ Head to shade me from yᵉ Sun. 1803 M. Wilmot *Let.* 6 Aug. in *Russ. Jrnls.* (1934) i. 36 We drove about in an open Carriage, the night was lovely. 1948 M. Laski *Tory Heaven* ix. 123 An open car..ceases to be a source of pure pleasure after a certain age. 1976 *Times* 1 Mar. 13/6 Having read the correspondence for and against..seat belts, it seems that no one has put forward the case for the driver of the open car. 1977 G. V. Higgins *Dreamland* xii. 139 A small four-passenger open car with a canvas roof.

b. Of a fire: that is not enclosed in a stove or the like; also of a fireplace.

1876 [in Dict., sense 22 a]. 1886 [see *flue-curing* s.v. *flue *sb.*[3] 6]. 1894 *Country Gentlemen's Catal.* 117/1 The Nautilus dog grate... A cheerful open fire. 1926 *Daily Colonist* (Victoria, B.C.) 24 Jan. 20/1 He used to be heard singing..at night, beside his open fire. 1931 E. O'Neill *Homecoming* ii, in *Mourning becomes Electra* (1932) 51 At rear, centre, is an open fireplace. 1949 M. Laski *Little Boy Lost* x. 150 'How nice to see an open fire,' he said tritely. Madame Mercatel laughed... 'One finds open fireplaces in many old French houses.' 1965 in P. Jennings *Living Village* (1968) 122 It looks as if open coal fires are on the way out in England at last. 1976 'Trevanian' *Main* xiii. 249 He enjoys fiddling with open fires... The bark has begun to crackle and flutter with blue flame.

5. a. (Further examples.) See also *wide open* s.v. Wide *adv.* 3 b. *open jet*: see sense *22 c.

1962 *Newnes Conc. Encycl. Electr. Engin.* 728/2 The majority of switchgear manufactured up to 11 kV is of the metal-enclosed type and above this the trend is towards outdoor open type switchgear. 1968 *Gloss. Terms Offset Lithogr. Printing* (*B.S.I.*) 12 *Open arc*, an arc lamp in which the electrodes burn in free air.

c. Of a telephone line or other transmission line: above ground.

1876 [in Dict., sense 5]. 1909 *Trans. Amer. Inst. Electr. Engin.* XXVIII. 1079 Even the best cable circuit is much less efficient than an open-wire circuit. 1925 *Bell Syst. Techn. Jrnl.* IV. 524 Practically all long toll circuits were in open wire construction; that is, individual wires mounted..on poles. 1966 *McGraw-Hill Encycl. Sci. & Technol.* XIV. 48/1 Open-wire construction is used for communication or power transmission wherever practical and permitted, as in open country.

d. *Med.* Communicating with or exposed to the air; involving the deliberate exposure of an interior part of the body, esp. a fracture, so as to make it directly accessible.

1894 J. C. Da Costa *Man. Mod. Surg.* xviii. 307 Compound fracture is an open fracture, or one in which an open wound admits air to the seat of bone-injury. 1897

Stimson & Rogers *Man. Operative Surg.* (ed. 3) v. 257 (*heading*) Suture of the patella. I. Open Method. 1944 C. A. Pannett *Surg.* xliii. 541 Union of the fracture after open operation always takes longer than if a closed method has been employed. 1949 P. Kiely *Text-bk. Surg.* xxvi. 700 Injuries to the Bowel. These may be subcutaneous or open. 1969 F. T. Hoagland in S. I. Schwartz *Princ. Surg.* xlvi. 1658/2 Open reduction [of a fracture] carries the risk of local infection. 1975 *Nature* 10 Apr. 529/1 Human liver obtained by open biopsy.

e. *Med.* Of (a case of) tuberculosis: accompanied by the discharge of infectious material from the body.

1930 J. A. Myers *Tuberculosis among Children* ii. 9 The source of infection is found to be cases of open tuberculosis in the homes. 1939 *Brit. Encycl. Med. Pract.* XII. 288 The rapid decline in the number of open cases of tuberculosis..have reduced the sources of infection. 1961 *Times* 30 June 15/3 Infectious (or 'open') tuberculosis. 1974 Passmore & Robson *Compan. Med. Stud.* III. i. xiv. 2/2 Occasionally outbreaks have occurred in young tuberculin-negative adults who have been heavily exposed to open cases [of tuberculosis].

7. b. *Chem.* and *Metallurgy.* (See quot. 1938).

1938 Hume-Rothery & Raynor in *Phil. Mag.* XXVI. 130 A rough indication of the nature of a metal or alloy may be obtained by comparing the interatomic distances in the crystal with the suitably defined ionic radii of the atoms concerned. In brief, if the ionic radius is small compared with the interatomic distance, we shall have a metal of what may be called the 'open' type, whilst if the ionic radii are nearly equal to the interatomic distances we shall have a metal of the 'full' type, with many different properties. 1967 A. H. Cottrell *Introd. Metall.* xix. 322 In open metals we need consider only the valency electrons, for the electrons in the ionic shells are negligibly disturbed by the metallic binding.

8. a. Esp. in *open space*; *spec.* an area without buildings in a city or town; a small park or the like for public recreation; also *great* (or *wide*) *open spaces*: large tracts of open country.

1827 J. S. Buckingham *Trav. Mesopotamia* I. xi. 375 Attached to it are extensive stables, and a Maidan, or open space, where the horses are kept in the air. 1850 *Household Words* 3 Aug. 451/2 Suburban open spaces are being entombed in brick-and-mortar mausoleums. 1869 A. Mackenzie (*title*) The parks, open spaces and thoroughfares of London. 1896 [in Dict.]. 1910 H. G. Wells *New Machiavelli* (1911) i. iv. 131, I recall as if I had been there the wide open spaces, the ragged hillsides [of South Africa]. 1913 C. B. Purdom *Garden City* viii. 112 An open space in a city has come to mean..even a disused churchyard... In the Garden City the characteristics of the open space belong to the town as a whole. 1924 Wodehouse *Leave it to Psmith* viii. 138 You will find me somewhere out there in the great open spaces where men are men. 1942 *Ann. Reg. 1941* 286 Mr. G. E. Hatfield..bequeathed to the National Trust Marden Hall estate,..for preservation as an open space. 1943 *Our Towns* (Women's Group on Public Welfare) p. xvi, The special town conditions of overcrowding, lack of open spaces, smoke and noise. 1944 [see *closed book* s.v. *closed *ppl. a.* 3]. 1965 K. Giles *Some Beasts no More* i. 12 A phony passport, maybe, and life anew in the great open spaces. 1969 S. Coulter *Embassy* xiv. 159 The big huskies, the boys from the wide open spaces. 1971 *Daily Tel.* 13 July 2/8 The Wimbledon centre court was not an open space within the meaning of the 1936 Public Order Act. 1974 *Guardian* 28 Mar. 19/3 What do people want today—apart from open space? 1975 *Times* 8 Feb. 10/4 If you seek clean fresh air and the wide open spaces (cliche though it is, that phrase is exactly right), this is the place.

11. a. Of a brass instrument: not muted. Of a cymbal: left free to vibrate.

1926 *Melody Maker* Mar. 30 Nothing is better suited to obtain a highly successful result than the beautiful, sweet full tone of the open instrument, and I advise all artistes to try a few 'open' solos. 1927 *Ibid.* June 609/3 The cymbal must be 'open' when it is struck and must remain 'open' for practically the full length of the beat being played, only being choked out just before the next beat. 1955 Keepnews & Grauer *Pict. Hist. Jazz* xiii. 141 Cootie Williams, ..strictly an open-horn stylist until he took over Bubber Miley's chair and produced a fine, muted 'jungle' sound. 1956 B. Edwards in S. Traill *Play that Music* vi. 60 Short-damped cymbal beats or loud and frequently aimless open ones. 1967 *Crescendo* Apr. 7/2 The Les Brown trick of trumpets in tin mutes playing above open trombones. 1972 *Jazz & Blues* Nov. 11/3 Razor-sharp riffs, and sweeping, open-horn statements.

e. (Earlier example.)

1845 W. E. Jelf *Gram. Greek Lang.* I. iii. 30 When a syllable ends with a vowel it is called an open, when with a consonant, a close syllable.

f. *Electr.* Having a break in the conducting path for an electric current; esp. in *open circuit* (see sense *22 c).

1827, etc. [see **open circuit*]. 1869 *Phil. Mag.* XXVIII. 2 When the shunt is open the battery is unable to send a steady current through the voltameter. 1884 S. P. Thompson *Dynamo-Electr. Machinery* vii. 133 Dynamos.. which leave the circuits of some of the armature coils open during part of the rotation, are sometimes termed 'open-circuit' dynamos. 1901 *Chambers's Jrnl.* Sept. 617/2 If a new machine, the switches should be left open, the brushes lifted, and the machine allowed to run without load for a little time. 1962 J. P. Gregory in G. A. T. Burdett *Automatic Control Handbk.* ii. 21 With the initiating switch open, the timing capacitor..is charged to the peak value of the a.c. voltage. 1975 I. Clucas *Reed's Electr. for Deck Officers* vii. 212 The inductance in the circuit..tries to maintain the current once the switch is open.

g. Of a game of chess: developed either by gambits or by opening up the files. Cf. *close *a.* 2 c.

1856 C. Tomlinson *Chess-Player's Ann.* 75 He was a very accomplished player, and generally preferred open games, gambits, &c. 1890 R. F. Green *Chess* v. 14 An Open Game is one in which the development is effected chiefly in advance of the pawns. P to K4 as a first move on both sides, leads generally to an open game; and formerly all games begun in this way were called open—other openings being treated as close. 1917 J. du Mont tr. *Lasker's Chess Strategy* (ed. 2) iv. 43 We find an early break-up of the centre, and concurrently the opening of the Ks or Qs file for the Rooks. That is why games opened in this fashion have been classed very generally as 'open', whilst all the other openings are called 'close games'. 1936 W. Winter *Chess for Match Players* ii. 24 In the category of Open games come the vast majority of the King's side openings. 1959 H. Golombek *Mod. Opening Chess Strategy* 11 Certain openings suit certain people—some like an open type of game, others prefer a close.

h. *Med.* Applied to methods of administering anæsthetics in which the patient's respiratory tract is in communication with the air so that exhaled air is not rebreathed.

1888 D. W. Buxton *Anæsthetics* v. 78 The open method. .. A common towel is arranged so as to form a square of six folds, and enough choloroform is poured upon it to wet an area the size of a hand's palm. 1922 *Encycl. Brit.* XXX. 137/2 The induction of anaesthesia by the open method is liable to be somewhat prolonged. 1972 J. C. Snow *Anesthesia* v. 39 The Ayre T-tube system is another example of an open or semi-open system.

i. *Math.* Of a set of points: not containing any of its boundary points. Of an interval in the real line: not containing either of its end points.

1902 *Proc. Lond. Math. Soc.* XXXIV. 289 Open sets of points. 1939 M. H. A. Newman *Elem. Topology of Plane Sets of Points* ii. 25 The sum of any set of open sets is an open set. 1956 E. M. Patterson *Topology* ii. 23 An example of a set which is not open is the set defined by $0 \leq x < 1$; no ϵ-neighbourhood of $x = 0$ lies entirely in the set, for every ϵ-neighbourhood of $x = 0$ contains a point whose coordinate is negative. 1968 P. A. P. Moran *Introd. Probability Theory* iv. 185 An open interval in R_n is defined to be a set of points whose coordinates satisfy the inequalities $a_i < x_i < b_i$..where the a_i may be $-\infty$ and the b_i, $+\infty$. An open interval is clearly an open set.

j. *Logic* and *Math.* Of a statement or equation: containing at least one free variable or undetermined quantity.

1937 A. Smeaton tr. *Carnap's Logical Syntax of Lang.* i. 21 If a variable which is free in $\mathfrak{A}_1$ occurs in $\mathfrak{A}_1$, then $\mathfrak{A}_1$ is called open; otherwise it is called closed. *Ibid.*, Our classification into closed and open sentences corresponds to the usual classification into sentences and sentential functions. 1952 S. C. Kleene *Introd. Metamath.* vii. 151 Let A be a formula containing free exactly the distinct variables $x_1, \ldots, x_n$ in order of first occurrence. According as $n > 0$ or $n = 0$, we call A open or closed. 1963 W. V. Quine *Set Theory* i Imagine a sentence about something. Put a blank or variable where the thing is referred to. You have no longer a sentence..but an open sentence, so called, that may hold true of each of various things and be false of others. 1967 M. L. Tomber *Introd. Contemporary Algebra* ii. 63, $5 \cdot 3^{-1} = \frac{5}{3}$ is a solution of the open equation $3x = 5$. 1971 *Sci. Amer.* Mar. 55 The technique of long division represents a decision procedure for the predicate '*x* is divisible by *y*', where *x* and *y* can be any natural numbers. (A predicate is an open sentence: one that can be completed by assigning names to its variables.)

k. *Astr.* Of the universe: having a negative or zero radius of curvature; spatially infinite and always expanding.

1937 E. Hubble *Observational Approach Cosmol.* iii. 55 The radius [of curvature] in our universe might be positive, negative, or zero, and might be large or small... A negative curvature implies open space, an infinite universe. 1965 J. D. North *Measure of Universe* vi. 135 The logical advantages of an open model were generally thought to be fewer than those of a model with positive curvature. 1976 *New Scientist* 2 Dec. 514/1 Indirect evidence for a low-density (or 'open') Universe—one which will expand for ever—comes from the recent discovery that deuterium (heavy hydrogen) exists in interstellar space. 1978 *Daily Tel.* 27 Mar. 7/2 This question of whether the universe will prove to be 'open' or 'closed' raises an important philosophical question about the existence of God. If the cosmos is going to expand for ever,.. for countless thousands of millions of years of its future history, life anywhere will be impossible.

13. *spec.* Designating administration or government in which the public is kept well-informed and is invited to participate.

1968 R. M. Nixon in *N.Y. Times* 20 Sept. 33/2 It's time we once again had an open administration—open to ideas from the people, and open in its communication with the people—an administration of open doors, open eyes, and open minds. 1971 J. Aitken *Officially Secret* xv. 211 The absence of effective checks and controls on the activities of the contemporary Civil Service has recently led to demands for 'more open government'. 1973 *Public Administration* LI. 428 'Open government' and 'greater public participation' were becoming increasingly fashionable political slogans at this time. 1975 *Times* 11 Jan. 12/2 The need for 'open government', in which people are informed about what is being decided and have a chance to make their own suggestions. 1976 Ld. Home *Way Wind Blows* vii. 115 That this crisis was averted was in the greatest part due to the instinct for fair play and open government practised by the Tunku Abdul Rahman. 1977 *Time* (Overseas ed.) 17 Jan. 13/1 The coalition's goal, explained Dowiyogo in an interview with the *Australian*: to replace DeRobert's increasingly personal rule with 'open government—that is, to tell the people what we plan to do and why'. 1978 *Times* 17 Mar. 6/5 The Civil Service Department has refused a formal request from *The Times*, arising from the Prime Minister's open-government policy announced last July, that background material used in the preparation of the White Paper on the Civil Service published on Wednesday should be disclosed.

b. Of a place of work: in which both union and non-union workers are employed, esp. *open-shop* (cf. *closed shop*); also *attrib.* orig. *U.S.*

1896 *Typogr. Jrnl.* IX. 445 Our next efforts were directed to the Morning Leader, also an 'open' office. 1901 *World's Work* (N.Y.) July 914/2 The shop had previously been an 'open' one—that is, union and non-union men were employed without distinction. 1904 *N.Y. Even. Post* 15 Aug. 2 The Exposition is conducted along the lines of an 'open shop', by permitting the employment of both union and non-union labor. 1906 *Daily Colonist* (Victoria, B.C.) 1 Jan. 133 What is all this talk that's in the papers about the open shop? 1909 *Daily Chron.* 3 May 1/5 The strike has originated in the intention..to enforce an 'open shop' on the lake boats. 1939 *Sun* (Baltimore) 18 Apr. 22/7 As a direct result of the shutting down of union mines.., 'open-shop' mines in Garret county..were working at maximum capacity. 1964 E. H. Powell in I. L. Horowitz *New Sociol.* 333 The open-shop crusade of the 'twenties was known as the 'American plan'.

15. b. *to lay* (one) *open to*: to render (one) liable to (something), to expose (one) to.

1853 C. Brontë *Villette* I. viii. 149, I shall make blunders that will lay me open to the scorn of the most ignorant. *Ibid.* x. 183 There was something in it that pleased, but something too that brought surging up into the mind all one's foibles and weak points: all that could lay one open to a laugh. 1931 T. R. G. Lyell *Slang* 455 Judging by the people he knows and the books he reads and recommends, he certainly lays himself open to the suspicion of being one [*sc.* a Communist].

20. *open mind*, a mind accessible to all arguments or points of view, esp. in phr. *to keep an open mind.* See also *open-minded* adj. (sense 22 c in Dict. and Suppl.).

1841 [in Dict.]. 1911 G. B. Shaw *Doctor's Dilemma* i. 6 *Schutzmacher.* Oh, in my case the secret was simple enough... I'm afraid youll think it rather infra dig. *Ridgeon.* Oh, I have an open mind. What was the secret? 1914 —— *Misalliance* 29 *Lord Summerhays.* Giving the show away is a method like any other method... I should keep an open mind about it. *Johnny.* Has it ever occurred to you that a man with an open mind must be a bit of a scoundrel? 1974 'M. Allen' *Super Tour* vi. 210 We'd rather you go into all this with an open mind. 1976 *Star* (Sheffield) 29 Oct. 1/3 Senior detectives said they were keeping an open mind whether the attacker is the same man who has committed several rapes in the Barnsley area.

21. *open book*, a person or thing that can be readily understood; a person who conceals nothing; also in phr. *to read* (someone) *like an open book* (cf. Read *v.* 5 d); *open house*, welcome or hospitality for all visitors; also *attrib.* (see also House *sb.*[1] 17 b); *open letter*, a letter, esp. one written in protest against something, addressed to a particular person or persons but made public by being printed, e.g. in a newspaper.

1853 G. H. Boker *Bankrupt* iv. ii, in *Amer. Lost Plays* (1940) 105, I read your black heart like an open book. 1919 Wodehouse *Damsel in Distress* iv. 53 There's no mystery about me. I'm an open book. 1934 A. G. Street *Endless Furrow* xv. 254 Talk about old Nicholas Crawford's art and mystery in grocerin', why, that's an open book compared to farmin'. 1944 Open book [see *closed book* s.v. *closed *ppl. a.* 3]. 1973 'S. Woods' *Enter Corpse* 60 'You haven't tried to shake them off?' 'What would be the good?.. My life is an open book.' 1824 Byron *Don Juan* xvi. lxviii. 98 Though not exactly what's called 'open house'. 1836 T. Power *Impressions Amer.* II. 71 Mr. Oliver was one of a class of excellent open-house men, of which class there are specimens to be found in every part of this Union. 1907 *Westm. Gaz.* 30 Dec. 8/1 On the seventieth anniversary there was an 'open-house' reception. 1921 *Daily Colonist* (Victoria, B.C.) 23 Oct. 15/1 Victoria's three great laundries will hold 'open house' daily from 9 a.m. to 4.30 p.m. 1971 *Daily Tel.* 11 Oct. 15/2 Clarksons has an 'open house' hospitality suite at the Majestic where free liquor is flowing almost round the clock. 1878 (*title*) Open letter to the English nation from Berlin. 1890 R. L. Stevenson (*title*) Father Damien. An open letter to the Reverend Doctor Hyde of Honolulu. 1917 *Fortn. Rev.* Nov. 748 (*heading*) Problems of finance: an open letter to Lord Milner. 1966 *Listener* 6 Oct. 506/1 There are three open letters in the memorial volume. 1977 *Time* 21 Mar. 49/1 In 1972 he attacked the Black Muslims in an open letter, an act that is thought to have led to the execution of his family.

22. a. *open-cell, -class, -deck, -face, -frame, -seam, -web.*

1933 *Amer. Speech* VIII. iii. 30 *Open-cell wing*, division of the main prison building in which there were no bars on the front of the cells. 1957 *N.Z. Timber Jrnl.* Dec. 59/2 *Open-cell process*, a means of impregnating wood under pressure. The preservative is retained in the cell walls only, and the cells left empty. 1971 C. Bonington *Annapurna South Face* App. B. 243 A thin layer of open-cell foam in the sleeves. *Ibid.* 249 Open-cell foam mattresses. 1949 R. K. Merton *Social Theory* ii. iv. 136 Despite our persisting open-class-ideology, advance toward the success-goal is relatively rare..for those armed with little formal education. 1954 F. C. Avis *Boxing Reference Dict.* 79 *Open class contest*, an amateur contest consisting of four 3-minute rounds, with 1-minute intervals. 1974 *Times* 26 Oct. 12/3 P & O..the last [line] to operate two-class cruises, decided..to..revert to 'open-class' cruising. **1886** *Harper's Mag.* June 18/2 In vessels of this class it is usual to have an open-deck battery. **1906** 'H. McHugh' *Skiddoo!* v. 75 The Human Hog was invented long before the open-face street car began to stop for him. **1931** *Amer. Speech* VII. 51 Custard [in lumberjack lingo] is 'open face pie'. **1942** Berrey & Van den Bark *Amer. Thes. Slang* §91/27 *Open-face pie*, pie without an upper crust. **1946** *Sun* (Baltimore) 13 Mar. 22/7 The proposed serving of 'open-face' pies and sandwiches. **1940** *Chambers's Techn. Dict.* 595/2 *Open-frame girder*, a girder consisting of upper and lower booms connected at intervals by (usually) vertical members, and not braced by any diagonal members. **1969** *Jane's Freight Containers 1968–69* 477/2 Tilt covered open-frame containers are available. **1910** *Installation News* Jan. 4/1 The much abused open-seam conduit and the socket joint conduit. **1968** E. McCourt *Saskatchewan* iii. 36 The most fascinating by-products of open-seam mining are the miniature mountain ranges formed of the clay stripped away to expose the coal seams. **1871** T. Cargill *Strains Bridge Girders* 63 The lattice, or open web girder.

b. *open-collared, -throated, -topped.*

1945 Dylan Thomas *Let.* 30 July (1966) 277 Such lovely ladies and gentlemen.., open-collared and wild-haired in the photographers' wind. **1966** R. Ellmann *Lett. of J. Joyce* II. p. liii, The open-collared eloquence of D. H. Lawrence. **1891** 'L. Malet' *Wages of Sin* II. iv. iii. 81 The north wind blew piercing, strong and tonic... Colthurst drank it down open-throated. **1962** K. Orvis *Damned & Destroyed* 111 A loud, open-throated sports-shirt. **1904** *Westm. Gaz.* 9 Dec. 7/2 The sight should be adjustable with open-topped hood. **1964** W. L. Goodman *Hist. Woodworking Tools* 93 It is in effect an open-topped steel box.

c. open access, a system whereby users of a library have direct access to the book-shelves; also *attrib.*; **open admission** *U.S.* = **open enrolment*; **open-and-shut** *a.* (orig. *U.S.*), (*a*) simple, straightforward; esp. of a legal case in which there is no doubt as to the outcome; (*b*) of weather: characterized by alternating sunny and cloudy conditions; hence as *sb.*, (*a*) a simple or straightforward operation, case, etc.; also in phr. *as* (or *like*) *open or shut*: easily, straightforwardly; (*b*) alternately sunny and cloudy conditions; **open back** (see quot. 1923); **open bite** *Dentistry*, lack of occlusion of the front teeth when the jaw is normally closed; **Open Board** *U.S.*, an association formed in cities of the U.S. to transact dealings in options on a small scale not permitted by the local board of trade; **Open Brethren,** a branch of the Plymouth Brethren which does not practise extreme separatism; **open-cast** (earlier and later examples); also, a method of mining coal, ore, etc., by removing surface layers and working from above, not from shafts; hence as *adj.*, of, pertaining to, or designating this method of mining; also quasi-*advb.*; **open-casting,** open-cast mining; **open chain** *Chem.* (see *chain *sb.* 5 g); freq. *attrib.*; **open cheque,** (*a*) an uncrossed cheque (see Cross *v.* 7c); (*b*) a cheque for an unstated amount; also *fig.*; **open circuit,** a circuit, esp. an electric circuit, that is incomplete; freq. *attrib.*; applied to breathing apparatus in which air is exhaled into the atmosphere and so lost; hence **open-circuited** *a.*, consisting of or containing an open circuit; (as a back-formation) **open circuit** *v. trans.*; **open city,** an undefended city; *spec.* a city declared to be unfortified and undefended, and hence, under international law, exempt from enemy bombardment; **open classroom,** a classroom in which instruction is informal, individual, and free-ranging; **open community** *Ecology*, an area in which the plant cover is not dense; **open compound** *Linguistics*, a compound in which there is a space (i.e. no hyphen) between the component elements; **open cover,** marine insurance that covers all the shipments made by a person or firm without specification in advance of the details of each shipment; **open credit** *Finance*, a credit free from restrictions; **open cycle,** a cycle of operations in which a working fluid, coolant, etc., is used only once (cf. **closed cycle*); **open day,** a day when a place, e.g. a school, which is normally closed to the public is made accessible to visitors (in quot. 1892, a day kept as a holiday at Durham University); **open end** *a.* = **open-ended* adj.; *spec.* (*a*) of an investment trust (see quot. 1940); (*b*) *Spinning* (see quot. 1975); **open-ended** *a.*, having an open end; freq. *fig.*, having no predetermined limit or boundary as to time, extent, size, etc.; *spec.* of a question or test: to which the respondent frames his own answer, as opp. to selecting one or two or more pre-phrased answers; hence **open-endedness; open enrolment** *U.S.*, the unrestricted enrolment of students at schools, colleges, etc., of their choice; **open-faced** *a.*, *N. Amer.*, also of a sandwich, pie, etc.: without an upper layer of bread or pastry; **open floor** (see quot. 1932); **open go** *Austral. colloq.*, an unimpeded opportunity; a 'fair go' (see *go *sb.*[1] 4d); **open housing** *U.S.*, property that can be rented or bought without restriction on racial grounds; also *attrib.*; **open jet,** a stream of air in a wind-tunnel which is not bounded by rigid walls in the working section; **open juncture** *Linguistics*, the type of juncture (sense *2 c) found at word boundaries or marked syllable division within the word; **open line,** (*a*) a telephone line on which conversations can be overheard or intercepted by others; (*b*) used *attrib.* to denote a radio or television programme in which the public can participate by telephone; **open loop,** a control loop (*loop *sb.*[1] 4 l) without feedback, each operation or activity being affected only by those earlier in the sequence; **open market,** a market in which any buyer or seller may trade freely and where prices are determined by supply and demand; also *attrib.*; **open-minded** *a.* (examples); hence **open-mindedly** *adv.*, in an open-minded manner; **open-mindedness** (earlier and later examples); **open-neck,** a collar of a kind that leaves the neck unrestricted; *spec.* as *adj.*, of a shirt, that is worn with the collar unbuttoned, without a tie; also *fig.*; hence **open-necked** *a.*; **open occupancy** *U.S.*, occupancy of housing available to persons of any race; **open-pit** *a.* chiefly *N. Amer.*, = **open-cast* adj.; **open plan,** an architectural style allowing for no (or few) internal walls or partitions within a building, esp. an office or a school; freq. *attrib.*; hence **open-planned** *adj.*, **open-planning; open question,** a matter on which differences of opinion are legitimate; **open range,** (*a*) *N. Amer.*, a range or tract of land that is not intersected with fences; also *attrib.*; (*b*) used *attrib.* or as *adj.* = *free range* (*b*) s.v. *free *a.* D. 2; also *fig.*; **open-reel** *a.*, employing or having tape reels of such a kind that they and the tape they carry are accessible (in contrast to cassettes and cartridges); **open road,** (*a*) *U.S.*, a road that is not private; (*b*) a country road or a main road outside urban areas; (somewhat sentimentally) a road or route along which one can travel without care or hindrance; also *attrib.*; **open roof** (see quot. 1932); **open sandwich,** a sandwich without a top slice of bread; **open score** *Mus.* (see quot. 1899); **open season,** the season when hunting or fishing is allowed; also *transf.* and *fig.*, a time when something, esp. criticism, is unrestricted; **open shed** *N.Z.* (see quots.); **open shelf,** a bookshelf that is not enclosed behind a door or the like; in a library: one of any number of shelves from which the readers can take books themselves; freq. *attrib.* or as *adj.*; **open-side,** in *Rugby Football*, the side of the scrum on which the main line of the backs is ranged; opp. *blind side* (*blind *a.* and *adv.* 2 c); also *attrib.*; **open skies,** used, chiefly *attrib.*, to designate a system whereby aircraft of any nation are allowed to fly over a particular territory; *spec.* of a system whereby two or more nations permit surveillance of one another from the air; also *open sky*; **open society,** a society characterized by its flexible structure and beliefs; one having much contact with other peoples or tolerant of change or of diversity in its existing order and traditions; opp. *closed society* (*closed *ppl. a.* 3); **open stage** (see quots.); also *attrib.*; **open-stock** (*N. Amer.*), goods that are always kept in stock by a shop, etc.; esp. a crockery set for which items can be bought separately at any time; also *attrib.*; **open subroutine** *Computers*, a routine that is written, in full, directly into a program wherever it occurs; **open system,** a material system in which the total mass or energy fluctuates; an incomplete or alterable system (of ideas, doctrines, things, etc.); **open texture** *Philos.* (see quots.); hence **open-textured** *adj.*; **open-texturedness; open toe(d)** *a.*, designating a shoe that is open at the front to reveal the toes; **open-top** *a.*, having an open top; also as *sb.* (*U.S. colloq.*), a vehicle, trailer, or the like with an open top; hence **open-topped** *a.*; **open town,** (*a*) *U.S.*, a town characterized by a lack of restrictions on drinking and gambling places and the like; (*b*) = **open city*; **open tread** *a.*, of a staircase: having no risers; **open university,** a university having few if any restrictions on admission, *spec.* (with capital initials) a university founded in Great Britain in 1969 to provide courses for adult working people based on correspondence and radio and television broadcasts; also *fig.*; **open vegetation** = **open community*; **open ward,** a hospital ward designed to accommodate several patients; also *attrib.*; **open water** (chiefly *Canad.*), the melting of the ice on rivers and lakes in spring, or the time when this happens or has occurred; a stretch of water in which there is little or no ice; **open window unit** *Acoustics* (see quot. 1968[1]); **open wood, woodland** (see quot. 1889); **open woods** *N. Amer.*, a patch of woodland in which there is no undergrowth; cf. Opening *vbl. sb.* 3.

1894 *Library* VI. 344 There is absolutely no novelty about the principle of open-access. **1899** (*title*) Account of the safe-guarded open-access system in public lending libraries. **1934** *Archit. Rev.* LXXVI. 168/1 The new library has one important feature that distinguishes it from most large libraries in this country and elsewhere; I refer to the system of 'open access'. *a* **1956** A. Esdaile in D. Lodge *Brit. Mus. is falling Down* (1965) vi. 96 Free or open access can hardly be practised in so large a library as this. **1977** *Times Lit. Suppl.* 30 Dec. 1532/2 The long-standing fight about open access..is won now... Behind lay a deep-seated fear..in some librarians..of anyone from the outside world actually having the run of the shelves. **1969** *Sat. Rev.* (U.S.) 20 Dec. 54/1 (*heading*) The challenge of open admissions. Will Everyman destroy the university? **1970** *Time* 28 Sept. 36 Under its new 'open admissions' policy, CUNY [*sc.* the City University of New York] was taking such students despite their academic shortcomings. **1973** E. Taylor *Serpent under It* (1974) xii. 176 The kinds of things that stir them [*sc.* students] up these days are parietal hours and open admissions and black studies. **1976** *Times* 13 Jan. 7/4 There were three basic principles at [New York's] City University: quality education, free tuition and open admission. **1977** *Time* 28 Mar. 13/2 The country's enduring recession and unlimited open-admissions policies have turned Italian universities into what students call 'unemployment factories' or 'jobless parking lots'. **1841** *Picayune* (New Orleans) 11 Mar. 2/3 The contest between *Humming Bird* and *Maria Collier* was considered all but a 'dead open and shut game'. **1848** G. P. Burnham in F. A. Durivage *Stray Subjects* 128 That chap's snoring beat *all* the high-pressures he *ever* heerd—jest as easy as open and shet! **1890** *Dialect Notes* I. 19 The common New England maxim is 'Open and shet's a sign of wet'. **1893** *Harper's Mag.* May 975/2 The case was a dead open-and-shut one. **1904** W. H. Smith *Promoters* x. 162 It seems as if it was a dead open and shut that we've got to stay with 'em. **1930** Kipling *Limits & Renewals* (1932) 262 Like broken water, with the sun tipping it. Like Portland Race in open-and-shut weather. **1936** M. Allingham *Flowers for Judge* ix. 139 They [*sc.* the police] feel they've got an open-and-shut case. **1966** T. H. Raddall *Hangman's Beach* III. xv. 230 It was what fishermen call an open-and-shut day, with patches of black cloud, and occasional showers of rain. **1971** 'D. Shannon' *Ringer* (1972) ix. 163 This was such an open-and-shut thing. **1974** A. Morice *Death of Heavenly Twin* xiii. 138 What's the hurry if it's as open and shut as you make out? **1975** 'E. Lathen' *By Hook or by Crook* iii. 28 It's an open-and-shut case. There's no doubt. **1976** *Encounter* June 12/2 'No sweat, Mr Dennie,' said the police chief... 'It's an open and shut. The lady must have fell asleep smoking a cigarette.' **1923** H. A. Maddox *Dict. Stationery* 56 *Open back*, a bookbinding term alternatively described as hollow back, spring back, or extra... Letter-press books are either 'fast back' (in which case the leather is pasted directly on the folded sections) or 'open back' (in which case the book has a false back or is actually cased). **1961** Open back [see *loose back* s.v. *loose *a.* 9]. **1893** Smale & Colyer *Dis. & Injuries Teeth* ii. 14 Lack of anterior occlusion, or open bite may be caused in several ways—by thumb, finger, lip or tongue sucking [etc.]. **1975** W. J. B. Houston *Orthodontic Diagnosis* iv. 36 A skeletal open bite can not be satisfactorily treated by attempting to extrude the anterior teeth which have already grown as much as possible. Nor should posterior teeth be ground or extracted. **1870** J. K. Medbery *Men & Mysteries Wall St.* 16 The consolidation of the Government and the Open Boards with the old historic Stock Exchange. **1902** G. H. Lorimer *Lett. Merchant* ix. 113 If she is the daughter of old Job Dashkam, on the open Board, I should say..that she was a fine girl to let some other fellow marry. **1879** A. Miller *Brethren* v. 66 The new motto on the standard of the Open Brethren was, 'The blood of the Lamb is the union of the saints'. **1883** J. S. Teulon *Hist. & Teaching of Plymouth Brethren* ii. 18 Henceforth the Brethren parted into two hostile camps. The followers of Messrs. Müller and Craik, under the name of Open Brethren, adhered to the principles which had animated the movement in its earliest days. **1909** *Encycl. Relig. & Ethics* II. 845/2 The 'Open' Brethren..fraternize freely with other Christians. **1968** F. R. Coad *Hist. Brethren Movement* x. 159 Those who refused to apply his decree against Bethesda... Commonly called 'Open Brethren', but in this book referred to hereafter by the more accurate name of independent Brethren. **1713** in *Sc. Nat. Dict.* (1965) VI. 487/3, 33½ fadoms wrought of closs Mynding besyds what is wrought of a flagged Mynd, and oppen cast. **1789** J. Williams *Nat. Hist. Mineral Kingdom* I. ii. 272 This open-cast has been worked

to a great length upon the bearing and the old works now exhibit a horrid and frightful gulph of great length. **1802** J. Mawe *Mineral. of Derbyshire* 207 *Opencast*, when a vein is worked open from the day. **1811** [see *open-work 2]. **1903** *Copper Handbk.* III. 260 Veins are stripped and worked open-cast. **1944** *Times* 8 Apr. 2/4 The movement of opencast coal has been increased by 80,000 tons a week. **1955** *Times* 27 May 19/1 We are in process of pumping the water out of this opencast. *Ibid.* 1 July 17/3 Your company has taken a leading part in the operations of open-cast mining and has dug, during the year under review, over a quarter of a million tons of coal. **1974** *Country Life* 14 Feb. 284/2 Development and exploitation, ranging from housing estates to open-cast mining. **1886** J. Barrowman *Gloss. Scotch Mining Terms* 48 *Open-casting*, holing above the seam: working as a quarry. **1976** *Ilkeston Advertiser* 10 Dec. 11/1 Mr Alex Eadie, Parliamentary Under-Secretary of State for Energy, praised the council for its 'exceptional foresight' by co-operating with the Coal Board in planning and restoring land after opencasting. **1884** M. M. P. Muir *Treat Princ. Chem.* ii. 164 In an open chain molecule the action does not return to the carbon atom at which it started. **1928** [see *chain *sb.* 5 g]. **1968** O. R. C. Norman *Princ. Org. Synthesis* i. 23 Organic reactions frequently lead to the formation of cyclic compounds from open-chain (alicyclic) compounds. **1882** R. Bithell *Counting-House Dict.* 212 *Open cheque*, an *uncrossed* cheque, payable to Bearer or to Order on presentation. **1977** J. Wainwright *Nest of Rats* I. iii. 17 'I'll pay... Just name it.'..'An open cheque. That makes it very big.' **1827** J. Cumming *Man. Electro. Dynamics* iv. 164 In all cases of continued rotation, one of the conductors forms an open circuit. **1876** Preece & Sivewright *Telegraphy* iv. 105 Where many intermediate stations are fixed on one wire worked on the Morse principle, the closed circuit system offers considerable advantages over the open circuit system. **1893** S. R. Bottone *How to manage Dynamo* 42 This line of conductors is said to form an open circuit when there is any gap in the way; and then no current can flow. **1904** *Electrician* 13 May 139/2 If the field winding on one of the limbs of a Manchester-type dynamo is open-circuited, this limb will magnetically short-circuit the remaining limb. **1907** J. Erskine-Murray *Handbk. Wireless Telegr.* 5 Modern wireless telegraphy is, in general, open circuit telegraphy. **1927** A. E. Claxton *Performance & Design D.C. Machines* xii. 269 An open-circuited coil prevents the passage of current in its circuit. **1939** *Jrnl. Exper. Zool.* LXXXII. 420 The method employed for measuring the rate of metabolism was an open-circuit respiration system modified from the system originally developed by Haldane. **1953** [see *closed circuit* s.v. *closed *ppl. a.* 3]. **1954** *Induction Motor* (Brook Motors Ltd.) xii. 184 If one phase is open-circuited, the motor will make a humming noise when switched on. **1956** C. Evans *Kanchenjunga* ix. 92, I planned that every European going above Camp 3 should..climb with an open-circuit oxygen set. **1957** *Practical Wireless* XXXIII. 539/1 The input resistance with open-circuited output, and output resistance with short-circuited input, remain the same. **1962** J. P. Gregory in G. A. T. Burdett *Automatic Control Handbk.* ii. 7 The timing of the relay is brought about by delaying the decay of the flux in its magnetic circuit after switching off the supply to the coil, either by short-circuiting or open circuiting it. **1977** G. V. Higgins *Dreamland* vi. 59 If I don't touch an open circuit..it'll be a cave-in, or a blast. **1914** Duchess of Sutherland *Six Weeks at the War* p. xii, Unfortunately Namur is not an open city so she suffered for a short time from horrors worse than 'moral effect'. **1938** *Newsweek* 4 July 9/3 A pact restricting bombing of 'open' cities. **1944** *Ann. Reg. 1943* 57 The Government had been urged in Parliament to treat Rome as an open city. **1965** H. Kahn *On Escalation* ix. 178 Military disengagement on the open-city model. **1971** *N.Y. Times* 8 June 39 An open classroom means nothing to me unless it means that a child learns in that classroom that learning is not dependent at every level on the presence of a teacher. **1973** *Britannica Bk. of Year 1972* 732/3 *Open classroom*, a system of education in which activities involving multidisciplinary skills replace traditional subject courses. [**1909** Groom & Balfour tr. *Warming's Oecol. Plants* xxxv. 137 In some communities the soil is densely covered,..but in others the vegetable covering is so open that the colour of the soil imparts to the landscape its hue.] **1923** A. G. Tansley *Pract. Plant Ecol.* ix. 126 We get an open community in stable equilibrium with its habitat. **1929** Weaver & Clements *Plant Ecol.* vii. 142 Open communities are invaded readily. **1975** O. Rackham *Hayley Wood: Hist. & Ecol.* iii. 124 It is an open community: these plants do not form a carpet, and there is plenty of bare ground. **1961** Webster, Open compound. **1965** *Amer. Speech* XL. 41 An enterable open compound like *threshing machine*. **1884** D. Owen *Marine Insurance* (ed. 2) 56 (*heading*) Agreement to execute policies. (Off open cover.) **1895** W. Gow *Marine Insurance* xiv. 229 In every case with which the writer is acquainted the open cover is a mere document of honour. **1928** F. W. S. Poole *Marine Insurance of Goods* iv. 59 Open covers provide the merchant with continuous protection, enabling him to calculate the insurance charges for shipments ahead. **1960** *Times* 24 Oct. (Finance. Rev.) p. xiii/2 It was to meet this convenience that the 'open cover' type of policy was developed. **1903** *Pitman's Business Man's Guide* 326 *Open credit.* This is the name given to a letter of credit which contains an unconditional request to pay money to another person. **1920** J. Stephenson *Princ. & Pract. Commerc. Corr.* III. iv. 186, I..now beg to inquire whether you would be inclined to open an Account Current with me, granting me an open credit of £875. **1950** *Nucleonics* Mar. 47/1 The open cycle system is generally practical only with a plentiful and inexpensive coolant like air or water, except in the case of a rocket drive. **1957** *Gloss. Terms Nuclear Sci.* (Nat. Res. Council, U.S.) 114 *Open cycle*, cycle of operation of a heat engine in which the power fluid is used only once... Also applicable to a cooling system in which the coolant is used once and the discarded. **1958** M. J. Zucrow *Aircraft & Missile Propulsion* II. vi. 13 An open-cycle power plant is a continuous-flow prime mover using atmospheric air as the working fluid. **1971** M. M. El-Wakil *Nuclear Energy Conversion* vii. 200 In designing open cycles, the extent of induced radioactivity in the coolant and its effects on plant and surroundings should be carefully evaluated. **1973** Kettani & Hoyaux *Plasma Engin.* viii. 243 In the open cycle system, fuel is burned with the oxidiser to which -eed is added without preheating. **1892** *Durham Univ. Calendar* 6 February 22..M. Open Day. **1941** H. G. Stead *Mod. School Organisation* xv. 261 Open Days..let parents and other interested friends have an opportunity of seeing the school in action. **1953** A. K. C. Ottaway *Educ. & Soc.* vi. 117 Whenever 'open days' for visitors are possible these are appreciated by parents. **1971** *Guardian* 23 Aug. 4/1 Tours or 'open days' are being held..on sites within the Roman 'colonia' at Lincoln. *Ibid.* 12 Nov. 7/3 It was like most school open days. The women questioned the staff about how the teaching would help John's or Simon's career. **1975** B. Meyrick *Behind the Light* xi. 143 Open Days on local Trinity service vessels were an annual event. **1908** *Daily Chron.* 13 Feb. 6/2 The open-end garden-seated light car..might be trailed as a relief car. **1940** *Sun* (Baltimore) 26 Apr. 23/9 Open-end trusts are those in which shares are redeemable at their asset value at any given time. **1945** Webster Add., *Open-end*, of a contract calling for the filling by a particular contractor of all government needs for a specific product during a specified period. **1952** W. J. H. Sprott *Social Psychol.* vi. 102 The questions..may be 'open-end' questions which allow of a more elaborate answer. **1953** *Economist* 21 Feb. 500/1 An open-end trust is in all major respects similar to a flexible unit trust. It holds a portfolio of securities and cash, against which units are issued to the public, and the managers have a wide discretion in their investment policy. *Ibid.* 500/2 The principle of the open-end fund is sound. **1954** [see *closed-end* s.v. *closed *ppl. a.* 3]. **1955** T. H. Pear *Eng. Social Differences* iii. 110 Questionnaires, 'closed or open-end', might lead to interviews and written personal and private communications. **1966** *Listener* 23 June 926/1 This was what happened in the open-end discussion. **1972** *Guardian* 21 July 16/1 There was a considerable debate in financial quarters about the wisdom of open-end funds, or unit trusts, which invest in property. **1972** *Sci. Amer.* Dec. 55/1 Open-end spinning has been the subject of intense development during the past five years. Commercial frames of Czechoslovakian and Japanese design have recently become available. **1974** *Times* 22 Mar. 21/1. The demand background is there and with its lead in 'open end' and 'self twist' machines Stone should be able to exploit it whilst it lasts. **1975** A. J. Hall *Stand. Handbk. Textiles* (ed. 8) iii. 170 In an open-end spinning a sliver of, say, cotton is fed into a relatively small device.. in which the fibres under air pressure are impelled by centrifugal action and an existing air vortex on to the internal surface of a high speed (up to 40000 rev/min) rotating cylinder and are therefore carried forward with their twisting together to an exit point from which they are drawn in the form of yarn having an appropriate degree of twist. **1976** *National Observer* (U.S.) 13 Mar. 10/6 (Advt.), A managed diversified Open-end Investment Trust. **1825** Open-ended (in Dict., sense 22 b]. **1935** [see *hydraulic gradient* s.v. *hydraulic *a.* 1]. **1940** M. Mead *Male & Female* ii. 47 The problem will be whether ..those who read can keep such words as 'men', 'women', and 'children' open-ended words. **1952** *Newsweek* 26 May 39/2 This program..is bound to fail if it is looked upon as a final step. Like the original Union of our States, it should be 'open-ended'. **1952** *Sat. Even. Post* 25 Oct. 150/3 Unlike the atomic bomb, the hydrogen bomb is an 'open-ended' weapon. In theory, if you want to make it more powerful, you just shovel in more of the heavy-hydrogen mixture. **1953** E. G. Williamson in *Ann. Rev. Psychol.* IV. 344 The open-ended interviews were concerned with the manner in which an individual was currently dealing with his problem of occupational choice. **1953** K. Cobb in *Ibid.* 367 A study of visually handicapped school children, tested with an open-ended adjustment inventory printed in large print or Braille. **1957** R. K. Merton *Student-Physician* 314 Questions marked with an asterisk were initially asked in open-ended form. **1964** *Ann. Reg. 1963* 401 This at least was suggested by their answers to questions in standard intelligence tests, which tended to an alarming degree to be of the closed rather than open-ended type. **1967** M. Argyle *Psychol. Interpersonal Behaviour* ii. 39 Questions vary in the extent to which they are open or closed—an open-ended question requires a lengthy explanation rather than a choice between alternatives: the best way to get someone to talk is to ask this kind of question. **1970** D. Goldrich et al. in I. L. Horowitz *Masses in Lat. Amer.* v. 177 It seemed desirable to use mainly open-ended interviewing techniques. **1973** *Daily Tel.* 2 Mar. 2/5 Powers of punishment for contempt are open ended as no maximum prison sentence or fine is laid down. **1953** W. H. McNeill *Amer., Brit., & Russia* I. i. 63 It was almost irresistible to look at the war-time history in the light of what had come after, to search for signs and portents of the strained situation which had prevailed since 1946 or 1947, and to forget or minimize the 'open-endedness' of Allied relationships during the war years. **1973** *Amer. Speech 1969* XLIV. 289 The seeming 'open-endedness' of the set of sentences of a language presents a challenge to transformational theory. **1974** G. F. Newman *Price* vii. 218 The open-endedness of the situation worked as well for her. She was as free as Sneed to quit. **1964** *N.Y. Times* 12 Jan. E11/1 The New York system in 1960 inaugurated a policy of 'open enrollment' which permitted youngsters from designated predominantly non-white schools to apply for transfer to designated predominantly white schools. **1970** *Ibid.* 7 Jan. 42 The open enrollment policy for the city universities will be a mistake. **1974** *Florida FL Reporter* XIII. 80/3 If indeed the phrase is not to become a meaningless slogan as many people regard the term 'open enrollment' in the City University of New York. **1934** Webster, *Open-faced*.. Of pies, etc., without top crust. **1946** *Sun* (Baltimore) 26 Apr. 6/1 A spokesman for the Case-Moody Pie Corporation..said it expected to realize a one fourth savings in flour by bringing out an 'open-faced' pie in which the top crust is eliminated. **1970** *Islander* (Victoria, B.C.) 22 Feb. 14/2 This little book tells you exactly how to make.. toppings for world famous and eye-appealing Open-Faced Sandwiches. **1976** *National Observer* (U.S.) 14 Feb. 7/2 They can be fried and covered with goodies, thus becoming a pre-Columbian open-faced sandwich. **1932** T. Corkhill *Conc. Building Encycl.* 146 *Open floor*, one with exposed joists, not covered by a ceiling. **1919** W. H. Downing *Digger Dial.* 36 *Open go*, see *fair go*. **1959** H. P. Tritton *Time means Tucker* ii. 15/1 The sergeant said we were putting on a better show than the professionals. He said we could have an open go the following night. **1973** *Bulletin* (Sydney) 25 Aug. 24/1 This intimidatory behavior, the company charges, contrasts with the 'open go' policy being afforded two other major prawning operators in the region. **1966** *Guardian* 29 Aug. 7/2 The agreement provides for stronger enforcement of open-housing regulations. *Ibid.*, Labour leaders agree to support open housing. **1971** *N.Y. Times* 1 June 28 The Suburban Citizens for Open Housing, an organization pushing for integration of the outer city. **1976** *National Observer* (U.S.) 8 May 6/2 The Rev. James E. Groppi, the Roman Catholic priest who led open-housing marches in the '60s. **1932** *Jrnl. R. Aeronaut. Soc.* XXXVI. 999 In a tunnel of any sort the provision of an open jet calls for rather more power than a walled-in jet would need. **1947** A. Pope *Wind-Tunnel Testing* ii. 34 For propeller and rotor tests..the open jet offers considerable advantage. **1971** D. C. Bain et al. *Wind Tunnels* ii. 1/2 Open jet tunnels are mainly used for air flow observations and have the advantage of easy access to the model for modifications and an unobstructed view for observation and photography. **1941** Open juncture [see *juncture 2 c]. **1942** Bloch & Trager *Outl. Ling. Analysis* iii. 47 In a phonemic transcription, external open juncture is marked by leaving a space between symbols, internal open juncture by a hyphen. **1966** [see *junctural *a.*]. **1973** D. Rockey *Phonetic Lexicon* 43 Spanish permits no clusters before final pause and open juncture. **1974** *Amer. Speech 1969* XLIV. 57 The liquids occur only immediately before open juncture. **1963** 'W. Haggard' *High Wire* xiv. 151 He..booked a call to London. It would be an open line, but he didn't have access to another. **1966** 'G. Black' *You want to die, Johnny?* iii. 47 Perhaps I should tell you, even on an open line, that I've just had a summons, from Ministerial level. **1970** *New York* 16 Nov. 5/2 Her over-involvement with the Renaissance Project and the Open-Line Program present an unbalanced image. **1972** *Guardian* 25 Sept. 4 Open-line radio shows, during which the visitor sits in..a radio studio and answers questions in amplified telephone conversations with local people. **1973** 'D. Jordan' *Nile Green* xliv. 223 They had also given me an open line to London. **1947** *Jrnl. Inst. Electr. Engin.* XCIV. IIA 5/3 (*heading*) Examples of open-loop and closed-loop control. **1954, 1962** [see *loop *sb.*[1] 4 l]. **1966** *New Scientist* 30 June 830 Almost all skilled muscular activities seem to exhibit many 'open-loop', pre-programmed characteristics. An open-loop system (one that draws little or no information from the thing it governs) can always be operated more effectively than a closed-loop one, provided that its task is well defined and not subject to major disturbances. **1967** *Electronics* 6 Mar. 306/2 Typical open-loop voltage gain is greater than 200,000. **1766** Blackstone *Comm.* II. xxx. 449 Our Saxon ancestors prohibited the sale of any thing above the value of twenty pence, unless in open market. **1838** W. Bell *Dict. Law Scotl.* 627 The law of Scotland differs from that of England as to the legal effect of a sale in open market. **1870** J. K. Medbery *Men & Mysteries Wall St.* 18 The stock which has occasioned the default is sold or bought in the open market under the rule. **1930** *Economist* 22 Mar. 630/1 The failure of the Reserve Bank's buying rate to go down as fast as the open-market rate. **1933** R. McKenna *Speech Midland Bank* 10 During the period of open market buying bank deposits continued to decline, though only slightly. **1934** [see *berry *sb.*[1] 1 c]. **1961** *Wall St. Jrnl.* 24 Mar. 26 This was a departure from its open market activities in the previous four weeks, in which the System extended these operations to securities in the medium-term range, with maturities up to ten years. **1972** *Times* 26 Jan. 6/2 To have the said rent reviewed at the said time by reference to the open market rental value of the demised premises. **1828** Carlyle in *Foreign Rev.* II. 115 To open-minded, truth-seeking men, the deliberate words of an open-minded, truth-seeking man can in no case be wholly unintelligible. **1861** T. Hughes *Tom Brown at Oxf.* III. xii. 223 In fact, he is a wonderfully open-minded man for his age, if you only put things to him the right way. **1903** G. B. Shaw *Let.* 6 Mar. (1972) II. 316 You can't feel at home with anything that is strange, no matter how open-minded you may be. **1969** *Jane's Freight Containers 1968–69* 31/3 CN will be open-minded about participating. **1976** P. Donovan *Relig. Lang.* iv. 43 The words..would appear to have little relevance to testing which stems from belief and faith, or even from openminded enquiry. **1909** H. G. Wells *Tono-Bungay* II. iv. 230 'Your aunt makes Game of people,' was Marion's verdict, and, open-mindedly; 'I suppose it's all right..for her.' **1832** Carlyle in *Fraser's Mag.* V. 386/1 Boswell wrote a good Book..because of his free insight, his lively talent, above all, of his Love and childlike Open-mindedness. **1914** *Jrnl. Iron & Steel Inst.* LXXXIX. 184 Whilst he could not but admire the authors' eloquence and open-mindedness, he certainly considered that they had not proved the theory which they had set out to expound. **1972** *Science* 16 June 1209/2 Much conventional scientific training..tends to produce rigidity and avoidance of personal involvement with subject matter, rather than open-mindedness and flexibility. **1939–40** *Army & Navy Stores Catal.* 656/3 All-Wool Sweaters... V-shape open neck, plain or cable stitch. **1949** *Penguin New Writing* XXXVIII. 14 The sleeves of his open-neck shirt were rolled up. **1971** D. Bagley *Freedom Trap* vii. 152 She looked too damned fetching in stretch pants, open-neck shirt and short jacket. **1976** *Daily Mirror* 16 July 13/1 An elaborately casual outfit of trousers and jacket, in man-made fibre and worn with an open-neck shirt. **1959** E. H. Clements *High Tension* v. 83 His step-cousin's [neck] rose, long and boyish-looking, from an open-necked shirt. **1973** P. Moyes *Curious Affair of Third Dog* iii. 32 A slim, fair-haired girl in corduroy trousers and an open-necked shirt. **1975** N. Luard *Robespierre Serial* xv. 131 The burly men in open-necked shirts. **1976** *Listener* 10 June 751/1 Australia?..A free-and-easy, no-nonsense, open-necked continent. **1953** *Open Occupancy in Public Housing* (U.S. Housing & Home Finance Agency) i. 3/1 When a shifting is made from a policy of enforced segregation to one of open occupancy, clear-cut policy..is found to be mandatory. **1966** *Economist* 28 May 962/2 The landlords' association, which Core blames in part for the persistent

refusal of the City Council to pass an open-occupancy law. **1968** M. Harrington *Toward Democratic Left* x. 289 And study after study documents a correlation between high educational attainment and libertarian views on civil liberties, capital punishment, open occupancy—and the war in Vietnam. **1913** *Amer. Year Bk. 1912* 421/2 The eight-hour day in Arizona..has been extended to all open-cut workings and open-pit workings. **1959** *Times* 12 June 17/4 It [*sc.* British capital] has pioneered in the new technique of open-pit borate mining. **1971** *Daily Colonist* (Victoria, B.C.) 19 Feb. 11/3 H. M. Wright..says open-pit mines in British Columbia are 'a beautiful sight'. **1975** *New Yorker* 3 Mar. 74/2 Copper was hardly worth taking out of the ground, and times were hard in Bisbee. In 1951, Phelps Dodge began a new open-pit mine—the Lavender Pit. **1938** *Archit. Rev.* LXXXIII. 90/2 The open plan is almost universal but perhaps less easy to work than it looks. **1954** *Ibid.* CXV. 213 A few are more experimental, influenced either by the pre-war work of Arne Jacobsen and Mogens Lassen or, more recently, by the open-plan American house. **1960** M. Spark *Ballad of Peckham Rye* iii. 38 We used to have an open-plan... So that you could see everyone in the office without the glass. **1973** 'R. MacLeod' *Burial in Portugal* ii. 41 An open-plan stairway curved..towards the upper floor. **1975** in Cox & Boyson *Black Paper 1975* 30/2 The education hierarchy.. doubt the wisdom of having an open-plan school in a difficult area. *Ibid.* 31/1 For children from a less fortunate environment open plan is disaster. **1958** *Washington Post* 16 Aug. B 10/1 Open-planned kitchens have been described by designers as 'one of the most desirable areas in the home'. **1960** *Guardian* 24 Feb. 12/5 Comparatively few people have really lived in modern, open-planned houses. **1976** *Ilkeston Advertiser* 10 Dec. 12/5 (Advt.), Front shop (ideal lounge), open-planned living kitchen. **1958** *Washington Post* 16 Aug. B 10/1 The welcome theme of this architect-designed House of The Week is 'open planning'. **1958** *Listener* 25 Sept. 459/1 The trend towards open planning favours development of this method. **1859, 1983** Open question [in Dict., sense 18]. **1972** *Guardian* 2 Feb. 7/4 A sixth-form general studies unit..was designed to promote discussion of 'open' questions, to which there may be no known or agreed answers. **1976** *Howard Jrnl.* XV. I. 17 What brought about this change is an open question. **1890** *Stock Grower & Farmer* 15 Mar. 6/3 The cow men of the open ranges will make money. **1905** *Bull. Bureau of Forestry* (U.S. Dept. Agric.) No. 62, 9 The great bulk of the western stockmen are definitely in favor of the Government control of the open range. *Ibid.* 52 Under the open-range system the honest and law-abiding cattleman was at a great disadvantage. **1958** *Spectator* 11 July 60/1 Occasionally they [*sc.* the chickens] have been a little flavourless, probably because these were not open-range birds. **1959** *Times Lit. Suppl.* 6 Nov. 639/1 After the..novels laid by American and western European novelists in their batteries or deep litter, it is pleasant to discover a Greek novel as tasty as an open range egg. **1962** W. Stegner *Wolf Willow* (1963) II. ii. 45 Last survival of the open-range cattle industry, booby prize in a belated homestead rush, this country saved each stage of the Plains frontier long past its appointed time. **1970** *High Fidelity* Nov. 77/1 Here we encounter what may well be a very significant parting of the ways between cassettes and open-reel tapes. **1971** *Gramophone* Dec. 1125/1 An 8-track cartridge machine capable of recording and reproducing and a sophisticated recorder covering open-reel, cassette and cartridge recording and reproducing. **1977** *Rolling Stone* 7 Apr. 6/1 (Advt.), A cassette, unlike its open-reel counterpart, actually becomes an integral part of your system the instant you put it in your cassette deck. **1817** E. P. Fordham *Jrnl.* 31 July in *Pers. Narr. Trav.* (1906) vi. 100 This state [*sc.* Indiana] is one vast forest, intersected by a few Blaze roads and two or three open roads. **1856** W. Whitman *Leaves of Grass* (ed. 2) 223 Afoot and light-hearted I take to the open road! **1920** E. O'Neill *Beyond Horizon* III. ii. 162 So I thought I'd try to end as I might have—if I'd had the courage to live my dream. Alone—in a ditch by the open road—watching the sun rise. *a* **1930** D. H. Lawrence *Phoenix II* (1968) 220 Some of these sonnets are very fine: they stand apart in an age of 'open road' and Empire thumping verse. **1968** L. Deighton *Only when I Larf* i. 16 I'm for the open road, the jet routes, Cannes, Nice, Monte; where the pickings are rich and the living is easy. **1975** *Country Life* 16 Oct. 1007/1 Open-road motoring is inhibited by ever-lower speed limits. **1976** *Southern Even. Echo* (Southampton) 2 Nov. 15/2 Yet on the open road a new virtue is revealed, and the VX becomes an admirable companion for covering long distance in comfort. **1932** T. Corkhill *Conc. Building Encycl.* 146 *Open roof*, one in which the principals are on view. No ceiling. **1946** *Sun* (Baltimore) 13 Aug. 2/8 Some lunch counters were serving 'the open sandwich', a single slice affair. **1959** *Listener* 15 Jan. 136/2 The simplest open sandwich or even everyday platter of roast meat. **1973** D. Bagley *Tightrope Men* xxv. 172 The open sandwiches of Scandinavia. **1899** Bridge & Sawyer *Course Harmony* ii. 8 There are two methods of writing harmony—viz., in open score and short score. In open score each voice is written on a separate staff. **1979** *Early Music* Oct. 531/1 The..music examples..are offered in open score to avoid the congestion inevitable with reduction to short score. **1896** *Outing* Sept. 596/2 The first day of September marks the beginning of the open season on pheasants, grouse, and quail in Oregon. **1914** 'High Jinks, Jr.' *Choice Slang* 16 *Open season*, time when a person may expect no mercy or protection. **1918** [see *bee[1] 5 b]. **1948** *Chesterton* (Indiana) *Tribune* 28 Oct. 6/4 A brief open season on pheasants will enhance this autumn's pleasure for Hoosier sportsmen. **1958** *Listener* 18 Sept. 416/1 The open season for tropical storms is declared in the last week in August. **1969** 'J. Morris' *Fever Grass* i. 10 He remembered those thirty years, before the island became independent and open season on its security tacitly declared. **1974** *Times* 18 Feb. 14/7 Any appearance of open season for pay could spark off an explosion. **1977** D. Anthony *Stud Game* xxviii. 189 It happened to be a year the state allowed open season on does, to thin out the herd. **1872** M. A. Barker in D. M. Davin *N.Z. Short Stories* (1953) 37 Brown and Wetherby's was an 'open shed', where any shearers that came were taken on until there were hands enough. *a* **1948** L. G. D. Acland *Early Canterbury Runs* (1951) vii. 169 Clayton was an 'open shed', that is, the shearers were not engaged beforehand, but turned up and took their chance of a pen on the advertised starting day. **1821** M. Edgeworth *Let.* 12 Dec. (1971) 290 They live in the library—open shelves—mixture of half bound and bound books. **1897** *Library Jrnl.* Jan. 44/1 The adoption of the open-shelf system. **1910** A. E. Bostwick *Amer. Publ. Library* 38 Practically all small and moderate sized American libraries are now 'open-shelf', which means that the user is allowed to go personally to the shelves and select his book. **1906** Gallaher & Stead *Compl. Rugby Footballer* xi. 145 (*heading*) Tactics—combined attack on the open side. **1960** E. S. & W. J. Higham *High Speed Rugby* vii. 71 Break to the openside and close to the scrum if the open-side flank is going straight for your fly-half. **1945** *Richmond* (Virginia) *Times Dispatch* 10 Feb. 12 (*heading*) U.S. accepts agreements on aviation pact. Will exchange 'open skies' rights. *Ibid.*, Today's action confirmed the position taken by the United States delegation in favor of 'open skies' or virtually unlimited freedom of the air. **1956** *Friends Jrnl.* 18 Feb. 103/2 The United States position is that until the scientists have solved the problem of detecting stocks, priority should be given to the Eisenhower 'open skies' plan and other measures designed to create confidence rather than to affect disarmament. **1957** *Daily Mail* 7 Oct. 6/2 The day when the Communists gave an ironic answer to President Eisenhower's plea for 'Open Skies'—by creating just that. **1965** D. D. Eisenhower *White House Yrs.* (1966) II. v. xx. 470 The Open Skies proposal was criticized by the Soviets because, they said, it covered only the territories of our two homelands and would fail to cover territory where United States forces were stationed overseas. **1973** *Times* 31 Jan. (Mediterranean Suppl.) p. iv/3 To attain their ambitious targets the Greek authorities proclaimed an 'open skies' policy for charter flights. **1944** *Sun* (Baltimore) 30 Nov. 7/2 The British apparently now have the choice of going along with the 'open sky' program or facing the development of a large bloc of nations which want something along the lines of the United States. **1955** *Ibid.* 11 Nov. 1/7 Vyacheslav M. Molotov buried President Eisenhower's 'open-sky' plan for disarmament under a five-point indictment here tonight. **1935** R. Audra et al. tr. *Bergson's Two Sources Morality & Relig.* iv. 230 Never shall we pass from the closed society to the open society, from the city to humanity, by any mere broadening out. **1940** *Mind* XLIX. 116 Bergson's distinction of 'open' and 'closed' societies, when applied to the society of all mankind, leads to the conclusion that this all-inclusive society will be 'quantitatively closed'. **1945** K. Popper (*title*) The open society and its enemies. **1954** P. Mason *Ess. Racial Tension* xvii. 119 The painful transition from a society based on status to an open society in which contract and competition play a part. **1973** *Listener* 17 May 635/1 Mill's..overriding goal of maintaining an open society..in which might be realised..the flowering of human individuality in all its diversity. **1953** R. Southern *Open Stage* 41 The name 'open stage' cannot be given merely on the grounds of there being a platform free on three sides... Those three sides must be occupied by audience. **1960** *Times* 15 July 16/4 A cynic might regard the crude..Riviera settings as a powerful argument in favour of open-stage methods. **1962** *Listener* 8 Nov. 771/3 It is I think now widely accepted that the term 'open stage' is conveniently used to include not only the three-sided open stage of which Dr Southern was primarily writing in his book, but all forms of theatre where the acting area is in the same room as the audience. **1897** *Sears, Roebuck Catal.* 681 Decorated dinner-ware. Patterns sold in open stock. **1911** *Daily Colonist* (Victoria, B.C.) 23 Apr. 4/4 We are in receipt of the latest open stock pattern in Limoges China... You can always replace a broken piece at any time. **1970** *Globe & Mail* (Toronto) 26 Sept. 13/1 (Advt.), 30% off makers' suggested retail prices on open stock dinnerware patterns. **1976** *Columbus* (Montana) *News* (Joliet Suppl.) 17 June 2/4 'Open stock' is the term for piece-by-piece sales instead of sales by the place setting category. The theory behind open stock sales is that you can buy just the pieces you need as you wish them, and you can replace items as you need. **1951** M. V. Wilkes et al. *Preparation of Programs for Electronic Digital Computer* 22 The simplest form of subroutine consists of a sequence of orders which can be incorporated as it stands in a program... This type of subroutine is called an 'open' subroutine. **1958** [see *inline *a.* 3 a]. **1969** P. B. Jordain *Condensed Computer Encycl.* 353 Typically, an open subroutine will convert floating-point numbers to fixed-point numbers or fixed-point to floating-point (four to six instructions), find the next larger integer value, or find the absolute magnitude. The advantages of open subroutine [*sic*] consist of simple usage.., faster execution, and if used sparingly, conservation of memory. **1939** F. H. MacDougall *Thermodynamics & Chem.* (ed. 3) x. 134 When a transfer of matter to or from a system is also possible, the system may be called an open system. **1962** P. Strevens *Papers in Lang.* (1965) xii. 152 The study of closed-system items is what we conventionally call 'grammar', while the open systems constitute 'lexis'. **1963** A. K. Rice *Enterprise & its Environment* iv. xx. 184 Open systems exist and can only exist by the exchange of materials with their environment... An open system can achieve a time-independent steady state. **1971** J. Z. Young *Introd. Study Man* ii. 26 Organisms are not stable systems..but 'open' systems maintained in a steady state by continual expenditure of energy. **1945** F. Waismann in *Aristotelian Soc. Suppl. Vol.* XIX. 121 The failure of the phenomenalist to translate a material object statement into terms of sense data..is due to..the 'open texture' of most of our empirical concepts... I owe this term to Mr. [W. C.] Kneale who suggested it to me as a translation of *Porosität der Begriffe*, a term coined by me in German. *Ibid.* 123 It is not possible to define a concept like gold with absolute precision, *i.e.* in such a way that every nook and cranny is blocked against entry of doubt. That is what is meant by the open texture of a concept. **1956** J. Holloway in A. Pryce-Jones *New Outl. Mod. Knowl.* 35 Waismann's term for this pervasive quality of language [*sc.* its ineradicable fluidity] was 'open texture'... The view is that the meanings of most words and expressions in common use are not precisely and exhaustively fixed, and ..that it would be very inconvenient if they were. **1965** *Amer. Philos. Q.* II. 112/2 There are *practical* reasons independent of vagueness, open-texture and the like for refusing to equate names with descriptions. **1966** *Ibid.* III. 116/2 There is a certain 'open texture' about Christian beliefs. **1950** *Mind* LIX. 159 Straightway we have a case of a vague and open textured criterion. **1965** *Amer. Philos. Q.* II. 120/1 The concepts they express are open-textured. **1974** T. E. Wilkerson *Minds, Brains & People* 7 A concept is too open-textured, so we tighten it up a little. **1966** *Jrnl. Linguistics* II. 243. The ordinary language philosopher..emphasizing the intricacy, variety and 'open-texturedness' of language. **1938** *Chatelaine* Mar. 32/3 Open-toe shoes will be more popular than ever. **1942** *Sun* (Baltimore) 12 Aug. 10/7, I don't understand window dressing, but it looks to me that some of the 'girls' displaying fur coats in Charles street windows are standing in the snow with open-toe shoes on. **1942** R. Chandler *High Window* (1943) v. 43 She wore..blue and white open-toed sandals. **1965** 'M. Neville' *Ladies in Dark* viii. 80 The shoe was none too new, open-toed and with a strap round the heel. **1973** 'R. MacLeod' *Burial in Portugal* iv. 91 She had simple open-toed sandals. **1771** Open top [in Dict., sense 22 a]. **1856** *Trans. Mich. Agric. Soc.* VII. 61 John Patton..[exhibited an] open top buggy. **1935** *Discovery* June 163/1 The tin can has gradually developed into the present day open top or 'sanitary' can, with the ends rolled on by machine, and made airtight by a rubber gasket. **1955** *Amer. Speech* XXX. 92 *Open top*, a rig with sides but no permanent top. **1972** D. E. Westlake *Bank Shot* ii. 17 Open-top cartons full of paperback books. **1974** 'J. Ross' *Burning of Billy Toober* ix. 89 A veteran open-top Bentley in racing dark-green. **1964** L. Deighton *Funeral in Berlin* xix. 111 The open-topped Mercedes that drove lazily past. **1977** L. Gordon *Eliot's Early Years* v. 99 He walked across London Bridge amidst horse-drawn carts, open-topped buses. [**1901** 'J. Flynt' *World of Graft* 11 The City Hall gang went into office on the promise that the town was to be open, an' they've kept it open.] **1915** *Amer. Mag.* Sept. 51/2 On an 'open town' platform Gill was elected mayor in March, 1910. **1938** H. Nicolson *Diary* 22 Sept. (1966) 364 It may mean surrender..in return for such quite valueless concessions as..'no bombing of open towns'. **1939** R. Campbell *Flowering Rifle* vi. 148 Keep safe his bomb-dump while our patience lasts While from its store our open towns he blasts. **1946** *Reader's Digest* July 96/2 Amarillo is the most open-town in the country. **1975** J. Gores *Hammett* vi. 43 They form a Committee to clean up San Francisco, and as chairman they take the man who's been running it as an open town for sixteen years. **1960** *Guardian* 11 Mar. 8/7 Modern architects are known to favour open-tread stairs. **1972** *Daily Tel.* 29 Nov. 24/3 A circular, opentread hardwood stairway gives on to a galleried landing **1966** *New Statesmen* 14 Oct. 548/3 The Open University..is the latest and most impressive offspring of the founders of ACE (Advisory Centre for Education). **1968** *Listener* 12 Dec. 806/1 (Advt.), January 1971 is the starting date for the transmission of the BBC radio and television programmes which form part of the Open University foundation year courses. **1969** *Radio Times* 27 Nov. 12 Originally named 'the University of the Air', the Open University offers an exciting new opportunity for adults throughout the country to study for degree qualifications through the media of integrated television, radio and specially-designed correspondence courses. **1973** *Listener* 17 May 634/1 [John Stuart] Mill.. set out to be a public thinker, a one-man Open University. **1977** *R.A.F. News* 11–24 May 2/3 Gp Capt Frank Rice (Retd)..played a major part..in setting up the Cyprus Open University scheme. **1960** N. Polunin *Introd. Plant Geogr.* xiv. 447 Cacti in the New World and cactus-like Euphorbias in the Old World frequently form a characteristic feature of the usually open vegetation [in semi-deserts]. **1971** D. W. Shimwell *Descr. & Classification of Vegetation* ii. 106 Where there is space between individuals which can be colonized.., the term open vegetation is applied. **1960** A. Huxley *Let.* 27 Dec. (1969) 901 Maxwell Jones..pioneered the Open Ward system in English mental hospitals. **1965** *Nursing Times* 5 Feb. 183/2 The staff kept her under as close observation as was possible in an open ward. **1922** *Beaver* Jan. 33/1 We remained at Mountain House until open water in the spring. *Ibid.* Sept. 9/1 When the open water came, I got one hundred and thirty-two beaver. **1930** L. Munday *Mounty's Wife* iii. 50 We had to keep him at Cumberland until he could be taken to Prince Albert at open water. **1956** H. S. M. Kemp *Northern Trader* (1957) x. 130 When open-water came..we were able to pitch-off to our private trapping grounds. **1956** *Polar Record* Jan. 8 *Open water*, a relatively large area of water free of ice. **1971** T. Boulanger *Indian Remembers* 4 The people came home after open water at Oxford House. **1900** W. C. Sabine in *Amer. Architect* LXVIII. 22/1 [Hereafter all results..will be expressed in terms of the absorbing power of open windows.] *Ibid.*, The absorbing power was found to be ·73 of open-window units. **1957** D. H. Fender *Gen. Physics & Sound* xii. 398 The area of material multiplied by its absorption coefficient measures its total absorption in 'open window units'. **1968** *Punch* 11 Sept. 364/2 An Open Window Unit (o.w.u.) is a unit of sound absorption... It is equal to the absorption by an open window of one square foot in area. **1968** R. C. Stanley *Light & Sound for Engineers* xvi. 308 If the area of the surface is expressed in square feet, then the absorption is expressed in open window units or sabins. **1889** W. Schlich *Man. Forestry* I. 9 Thin Wood, or open wood, means a wood in which the crowns of the trees do not interlace. **1926** Tansley & Chipp *Study of Vegetation* x. 210 Open woodland consists of open woodland without a closed and thickly interlaced canopy. **1790** J. Armstrong *Jrnl.* in *Ohio Archaeol. & Hist. Q.* (1911) XX. 82 A course a little to the N. of W., passing through several small prairies and open woods. **1799** J. Smith *Acct. Remarkable Occurrences* 13 About the lick was clear, open woods, and thin white-oak land. **1823** C. Vignoles *Obs. Floridas* 77 Instead of the clear open woods generally seen, masses of young pine saplings are thickly spread over the rocky ground. **1824** D. E. Burch *Let.* 30 Oct. in *Florida Hist. Q.* XIV. 105 In the open woods, especially in the pine barrens, these [*sc.* fallen trees] can always be avoided by turning out. **1939** *Canad. Hist. Rev.* XX. 282 The general results..would

seem at first sight to point to the 'groves', 'open woods', 'oak openings', parklands, or whatever name one may give them, being more pronounced on the western borders of this huge territory than on the Atlantic slope.

B. *adv.* (Later examples.)

1780 J. Woodforde *Diary* 24 Oct. (1924) I. 293 He.. spoke very open and ingenuous about it. **1921** E. O'Neill *Diff'rent* II, in *Emperor Jones* 252 Tell me all about 'em. You needn't be scared—to talk open with me.

open, *v.* Add: **2. d.** (Earlier example.)

1865 W. Howitt *Hist. Discovery in Austral.* I. xiii. 207 Measures were..instituted to construct a high road through the whole distance already gone... It was then, in modern phraseology, opened by the governor, attended by Mrs. Macquarie, and an escort on horseback.

6. c. *Electr.* To break or interrupt (an electric circuit); to put (a switch or the like) into a condition in which there is no path through it for an electric current.

In quot. 1834 the meaning is the opposite, viz. 'to create, close' (Close *v.* 10 e).

[**1834** M. Faraday in *Phil. Trans. R. Soc.* CXXIV. 429 The presence of a piece of platina touching both the zinc and the fluid to be decomposed, opens the path required for the electricity.] **1836** *Ann. Electr., Magn., & Chem.* I. 71 The shock is never produced only at the moment of opening the voltaic circuit. **1876** Preece & Sivewright *Telegraphy* iv. 103 If B wishes to communicate with A he ..opens the switch. **1924** Wedmore & Trencham *Switchgear for Electr. Power Control* xxiii. 252 A smart operator can open a medium-voltage circuit safely with a plain lever switch. **1962** *Newnes Conc. Encycl. Electr. Engin.* 735/1 The circuit-breaker contacts are held closed by springs and the contacts are opened in the event of a fault by the overcurrent in a series solenoid coil. **1975** M. Mandl *Basics of Electr.* vi. 123 When voltage is applied to the coil, the flexible section is pulled down.., opening the switch.

13. b. *Bridge.* To commence (the bidding); to offer (a particular bid).

1958 *Listener* 6 Nov. 753/1 What should West open, assuming that he is playing a Two Club system? **1964** N. Squire *Bidding at Bridge* ii. 17 When you open the bidding with a *suit* your strength may be absolutely minimum. *Ibid.* 21 You should open One Diamond, as your hand is unsuitable for you to be declarer in no-trumps. **1977** *Times* 10 Dec. 13/5 North has opened One Spade at game and 30. **1977** *Harpers & Queen* Dec. 26/2 West opened the bidding with one of those..artificial Two Diamond calls.

16. (Further examples.)

1864 E. A. Parkes *Man. Pract. Hygiene* 107 The windows should open at the top, and in case the wind has a high velocity, means should be taken to distribute it. **1871** L. W. M. Lockhart *Fair to See* I. iii. 103 'A Cameron of Aberlorna!' exclaimed the host, in a tone of unaccountable astonishment, his eyes opening wide upon Bertrand. **1893** M. E. Mann *In Summer Shade* I. x. 238 Claude's eyes opened slowly upon his brother's face. **1912** *Chambers's Jrnl.* Feb. 82/1 Suddenly the great eye of the lighthouse opened. **1952** M. Allingham *Tiger in Smoke* iv. 68 It was just when we were opening... I was just getting my keys for the spirits.

d. *Electr.* Of a circuit or device: to become open (*open *a.* 11 f); to suffer a break in its conducting path.

1836 *Ann. Electr., Magn., & Chem.* I. 71 If there is a spark..it is..feeble when compared to that seen when the circuit is opening. **1924** Wedmore & Trencham *Switchgear for Electr. Power Control* ii. 14 The circuit breaker.. is designed to open freely and quickly. **1975** I. Clucas *Reed's Electr. for Deck Officers* vii. 215 A second pair of contacts..are the first to close and the last to open... They take the full brunt of the spark when opening. *Ibid.* 224 When the main switch is closed the buzzer should sound..and the individual circuits open.

22. (Further examples.) Also, to begin to speak (occas. with the quoted words as quasi-obj.).

1851 C. Cist *Sk. Cincinnati in 1851* 296 They [*sc.* strawberries] usually open at 20 to 30 cents per quart. *c* **1871** J. Albery *Apple Blossoms* I, in *Dramatic Wks.* (1939) I. 244 Come and see me open to-night. **1926** A. Bennett *Lord Raingo* I. xiii. 63 'I quite agree with you, Clews,' Sam opened immediately. **1972** *Time* 17 Apr. 28/1 Joey and Sina, whose young daughter opened in the Broadway play *Voices* last week, soon became a part of the theater-going, nightclubbing celebrity set. **1976** B. Freemantle *November Man* iv. 47 'Jocelyn is still the big tycoon,' he opened predictably. **1977** A. Morice *Murder in Mimicry* I. i. 11 Gilbert is our new lead... His name alone ensures a sell-out for the entire run..before we even open.

23. open out. f. *trans.* and *intr.* To open the throttle of (an engine); to accelerate. *colloq.* Cf. sense 24 e below.

1906 *Punch* 19 Sept. 200/1 'Open her out!' my host had said; And on the instant word The mobile monster flew ahead Like a prodigious bird. **1918** 'Q' *Foe-Farrell* 105 There was a certain amount of outcry in the rear. But I opened-out down the slope and soon had it well astern. **1922** *Encycl. Brit.* XXX. 41/1 Such a 'light' engine would not withstand being opened out fully near the ground.

24. open up. a. Also *absol.*; *spec.* to open a door or the like.

1935 M. M. Atwater *Murder in Midsummer* xxii. 210 Why didn't you open up when I knocked? **1976** 'H. Carmichael' *False Evidence* i. 14 Someone knocked at the door... 'Open up, Miss Crawford.'

c. *trans.* and *intr.* To shear wool from (a particular area, esp. the neck, of a sheep). *Austral.* and *N.Z.*

1882 Armstrong & Campbell *Austral. Sheep Husbandry* xiv. 167 The fleece should be opened up the neck, commencing at the brisket. **1904** 'G. B. Lancaster' *Sons o' Men* 81 A big Maori was making the [shearing] pace; opening up in a scientific fashion with a clean-run cut over the ear-root. **1914** H. B. Smith *Sheep & Wool Industry Australasia* vi. 37 The machine is then driven up the front of the neck several times till the neck wool is well opened up. **1956** G. Bowen *Wool Away!* (ed. 2) iii. 32 Three short sharp blows are essential here to open up the neck for clean shearing.

d. *intr.* To talk; to speak openly; to cease to be secretive.

1921 *Sat. Even. Post* 12 Feb. 61/4 We had a drink and we had another and a couple more. Finally he opened up... It took him two hours to tell his story. **1949** B. Wolfe in A. Dundes *Mother Wit* (1973) 534 How much did the Negroes tell him when they 'opened up'? Just how far did they really open up? **1952** 'N. Shute' *Far Country* 244 It's just possible she might open up with me. *a* **1953** E. O'Neill *Hughie* (1959) 23 At first, he wouldn't open up. Not that he was cagy about gabbin' too much. But like he couldn't think of nothin' about himself worth saying. **1970** M. Braithwaite *Never sleep Three in a Bed* xvi. 197 Although he never answered—or perhaps because of it—I opened up to him completely, telling him things I'd never told anyone. **1976** J. Crosby *Nightfall* xii. 66 You're not being very helpful... You must have a few ideas. Open up!

e. *trans.* and *intr.* To open the throttle of (an engine); to accelerate. Cf. sense 23 f above.

1922 *Encycl. Brit.* XXX. 41/1 At height, however, it [*sc.* an aircraft engine] could be fully opened up, and the increased power..taken advantage of. **1926** T. E. Lawrence *Let.* 27 Sept. (1938) 500 It's my great game on a really pot-holed road to open up to 70 m.p.h. or so and feel the machine gallop. **1942** *Tee Emm* (Air Ministry) II. 95 On no account..should the engine be opened up during the final stages of ditching. **1970** K. Benton *Sole Agent* vii. 78 She's a nice car, the Chevvy. She'd do ninety if I opened her up. **1973** 'D. Halliday' *Dolly & Starry Bird* ix. 131 As soon as he's got a clear stretch of road, he'll open up and you'll lose him.

f. *intr.* To start shooting (*at* or *on* someone). Also *fig.*

1939 H. L. Ickes *Diary* 30 July (1954) II. 688 Two or three days ago John L. Lewis, before the Labor Committee of the House, opened up savagely on Garner. **1974** *Black Panther* 16 Mar. 16/4 Frelimo guerrillas opened up on the train from both sides 'creating panic among the passengers', according to the report. **1974** J. Cleary *Peter's Pence* i. 25 Someone had fired a shot at the Tans and the latter had opened up as if on a duck shoot.

open air, open-air. Add: **2.** (Earlier and later examples.) Also *transf.* and *fig.*

1830 *New Baptist Misc.* Aug. 331/1 (*heading*) Open air preaching in the villages. **1842** W. Howitt *Rural & Dom. Life Germany* xvii. 237 Those open-air concerts, walks and other amusements. **1926** C. Connolly *Let.* Jan. in *Romantic Friendship* (1975) 111 It is nice being able to sit out or go..to open air cinemas full of fireflies. **1949** Koestler *Promise & Fulfilment* III. i. 296 A people living underground must be single-minded..; but these qualities when carried over into open-air politics, become a grave handicap. **1958** *Times* 25 Oct. 10/7 His editorial influence was always exerted towards purity and strength [in music]. .. His songs..are animated by the same open-air kind of ideals. **1960** C. Day Lewis *Buried Day* x. 219 It was decided we should hold an open-air meeting. **1973** 'R. MacLeod' *Burial in Portugal* ii. 36 Brightly lit open-air cafés. **1975** J. Rathbone *Kill Cure* I. ii. 17 This hotel.. overlooked..an open air cinema.

open door. Add: **1. b.** (Later examples.)

1927 *New Republic* 21 Sept. 108/1 There is some merit in the general plea for 'most-favored nation' treatment, if only under the open-door policy. **1964** *Listener* 1 Oct. 492/2 Not only Americans, but almost everyone else thinks of the 'open-door' doctrine as American. **1974** *Times* 14 Dec. 13/5 Israel's open door policy to Jewish refugees from all parts of the world. **1976** *Listener* 9 Sept. 302/3 The Americans stood by their own open door policy—that China, just like every other part of the world except the United States, should be wide open to everybody's trade.

3. b. Designating a mental hospital in which patients are allowed the maximum freedom of movement and communication.

1958 *Spectator* 11 July 49/1 All over the country mental hospitals have been converting to the 'Open Door' system. **1969** *Daily Tel.* 1 Nov. 2 The modern vogue for open-door mental hospitals, far from improving conditions for some patients, has led to reprehensible methods of restraining them. *Ibid.*, Plain common sense ..required that such patients be managed in a secure environment and not in an open-door system. **1977** *Lancet* 18 June 1302/1 A total open-door policy is considered to be progressive; but in practice it means that mentally ill people who lack insight and are troublesome ..cannot be contained there for treatment.

opener. Add: **1. a.** Also *opener-up.*

1911 *Chambers's Jrnl.* Mar. 149/2 Carl Mauch, another German opener-up of South Africa. **1946** *Mind* LV. 102 The great philosopher is an opener-up of *new* paths for the mind of man.

b. Delete †*Obs.* and add later examples. Now *U.S. slang.*

1931 G. Irwin *Amer. Tramp & Underworld Slang* 139 *Openers*, cathartic pills. **1942** Berrey & Van den Bark *Amer. Thes. Slang* §91/5 Laxative food, *loosener*, *opener*. *Ibid.* §874/14 *Openers*, cathartic compound pills.

c. An implement or device for opening tins, cases, etc. See also **bottle-opener*, **can-opener*, *tin-opener*.

1906 *Daily Chron.* 15 Aug. 5/2 An ordinary packing-case opener had been used to force the door of the case. *Ibid.*, When the robbery was discovered the iron opener was found lying on the floor. **1912** *Chambers's Jrnl.* Feb. 144/1 The man who invented an opener for tins did well. **1942** Berrey & Van den Bark *Amer. Thes. Slang* §75/6 *Opener*.., a jimmy. *Ibid.* §259/2 Key, *opener*. **1951** T. Sterling *House without Door* x. 113 She turned the bright red handle of her opener and the squat tin spun around under the knife. **1964** G. Lyall *Most Dangerous Game* i. 13 He had a bottle in one hand and an opener in the other. **1970** *Which?* Jan. 27/2 The tin which the openers found hardest to manage was the squared one. **1977** S. Coulter *Soyuz Affair* x. 110 You have a cold beer, Jim?.. There an opener someplace?

d. The first of a series of events, etc.; *spec.* in *U.S.*, the first game in a baseball match. Also *opener-upper*. orig. *U.S.*

1941 in Wentworth & Flexner *Dict. Amer. Slang* (1960) 368/2 A patriotic opener-upper, 'Under the Double Eagle'. **1942** Berrey & Van den Bark *Amer. Thes. Slang* §9/1 *Opener*, *starter-offer*, that which comes first. *Ibid.* §9/2 *Opener*, an opening remark. *Ibid.* §590/16 First act,.. *opener*. *Ibid.* §675/5 First game of a 'double-header' [at baseball],..*opener*. **1949** *Down Beat* 11 Mar. 15 *Frost* is a simple but fairly bright arrangement with a good opener. **1962** *John o' London's* 19 Apr. 388/4 *Measure for Measure* ..is a little lacking in lustre for a Stratford opener. **1967** [see *double-header c]. **1967** *Crescendo* June 10/3 The opener, Lennie's composition 'Morning Stroll', should please all stride piano fans. **1970** *Globe & Mail* (Toronto) 28 Sept. 23/3 Queen's Golden Gails and the..Mustangs struggled through three quarters of the Ontario-Quebec Athletic Association football opener here. **1974** *Cleveland* (Ohio) *Plain Dealer* 26 Oct. 5-D/3 Vaclav Nedomansky, the hulking center iceman from Czechoslovakia who will do his skating for the Toronto Toros, in town for the Coliseum opener Sunday night. **1976** *Daily Tel.* 21 June 8 The opener of the present trilogy from Granada's 'The State of the Nation'..did little to allay worries about the unaccountable power exercised by Whitehall over Westminster. **1977** A. C. H. Smith *Jericho Gun* vi. 80 I'll tell you what's going to win the opener. You see that bay gelding? **1978** *Rugby World* Apr. 25/2 It was draftsman Bertranne's tackle, after half-an-hour in Paris, in the championship opener on January 21, that put Maxwell out of the game.

e. *Cricket.* One of the two batsmen who open an innings.

1950 R. G. Strutt *Schoolboy Cricket* vi. 81 No. 3..can get on better with his brilliant scoring strokes if the openers have taken the shine off the ball. **1959** *Listener* 19 Mar. 517/1 The breakdown of our batting, chiefly of the openers. **1974** *Observer* 9 June 24/6 The other opener, Geoff Greenidge, had just flicked Lever for two fours round his legs. **1976** Dexter & Makins *Test Kill* 83 The Australian batting had collapsed on a wicket that the England openers found tolerably easy.

f. Colloq. phr. *for* (or *as*) *openers*: to begin with; for a start.

1967 *Boston Sunday Herald* 2 Apr. (T.V. Mag.) 6/2 Joey hosts Danny Thomas for openers. **1970** K. Platt *Pushbutton Butterfly* (1971) xvi. 176 'Didn't they tell you at the plant? I quit.' That was good enough for openers. **1970** E. Tidyman *Shaft* (1971) xi. 152 Shaft decided to tear the trap apart. He killed Caroli as openers, then went plunging through the door. **1971** J. Sangster *Your Friendly Neighbourhood Death Pedlar* viii. 106 It was Walpole calling from London. 'I thought you were being hanged,' he said for openers. **1974** P. Erdman *Silver Bears* iv. 54 I'd like to ask you a few simple questions.. for openers, what's with this place here? **1976** *Word 1971* XXVII. 58 These suggestions and observations call for professional reconsideration, not only of a fundamental definition of Speech per se, but of phonology, idiolect, lexicon, and Child Language as well, just for openers.

3. *Poker.* (*pl.*) Cards on which a player can open the betting. Also *fig.*

1902 *Out West* Mar. 291 'I got openers, this pot,' says he, tapping the rifle. **1909** R. A. Wason *Happy Hawkins* 114, I didn't hold openers, an' yet if I didn't draw some cards an' see it out, I stood to lose entirely. **1920** C. E. Mulford *Johnny Nelson* ii. 25 A round or two had been played when Big Tom drew his first openers. **1946** Morehead & Mott-Smith *Penguin Hoyle* 122 If the opener cannot prove to the satisfaction of other players that he held openers, his hand is dead and cannot win the pot.

opening, *vbl. sb.* Add: **1. c.** (See quots.)

1888 C. P. Brooks *Cotton Manuf.* i. 17 *Opening* or passing the matted pieces of the bales through a series of armed beaters..separating the material into small flakes and removing the heavier impurities. **1901** T. Thornley *Cotton Spinning* I. iii. 64 *Q.* Give a statement of the objects aimed at in the operation of opening... *A.* This process first opens out the matted masses of fibres to a very fleecy, soft condition; secondly, it extracts the major portion of the impurities present in the cotton..and also much seed that has escaped the ginning process; thirdly it almost now always makes a lap. **1963** A. F. W. Coulson et al. *Man. Cotton Spinning* II. ii. vi. 129 Better opening and cleaning will be obtained if the machines can be kept working and operating on the cotton continuously in a rather small quantity.

2. d. Two pages of a book, etc., that face one another.

1906 E. Johnston *Writing & Illuminating* vi. 110 Parchment sheets should have their smooth sides so placed together that each 'opening' of the book has both its pages rough or both smooth. **1914** *Trans. Bibliogr. Soc.* XII. 239 A line of type at the top of a page, above the text, is called a 'head-line'; or, if it consists of the title of the book (or of the section of the book) on every page or every 'opening' (*i.e.*, two pages facing one another), sometimes a 'running-title' or 'running-head'. **1963** *Listener* 21 Mar. 522/1 A good example of his elaborate book-production is *The Book of Ruth*, with alternate openings in full colour and in golds and greys.

4. c. (Earlier and later examples.)

1825 P. Egan *Life of Actor* vii. 264 To get up splendid Spectacles; write openings for Pantomimes. **1894** J. A.

Cave *Jubilee Dramatic Life* (ed. 2) xix. 177 For the openings of my pantomimes I was able, as opportunity occurred, to secure the services of such inimitable burlesque performers as the Vokes family.

e. *Theatr.* The first performance of a play or entertainment; a première. *U.S.*

1855 W. B. Wood *Pers. Recoll. Stage* ix. 191 The loss we sustained was less important in a pecuniary view..than in rendering our opening still more embarrassing. **1916** *Variety* 27 Oct. 12/1 Openings here next week include Marie Tempest in 'A Lady's Name' (Plymouth); 'Sybil' (Colonial); [etc.]. **1923** H. Ruby *Let.* 16 Aug. in G. Marx *Groucho Lett.* (1967) 183 The out-of-town opening.. occurred in Fairmont, West Virginia. **1959** J. Thurber *Years with Ross* xv. 247 I'm having dinner with Aleck and he's taking me to an opening. **1977** J. Aiken *Last Movement* i. 37 'What about your opening?'..'Big success. I'll show you our press notices.'

f. The start of an art exhibition, fashion show, or the like.

1905 E. Wharton *House of Mirth* II. ix. 428 Beings without definite pursuits or permanent relations, who drifted on a languid tide of curiosity from restaurant to concert-hall, from palm-garden to music-room, from 'art exhibit' to dress-maker's opening. **1952** D. Ames *Murder, Maestro, Please* xxxv. 255 Geoffrey insists on taking me to Paris for the autumn dress openings. **1969** D. S. Davis *Where Dark Streets Go* (1970) ix. 89 There was a showing of Tchelitchew drawings at the Burns Gallery... 'Not everybody who goes to an opening signs in. Especially when it's not new work.' **1972** P. Marks *Collector's Choice* ii. 62 He never went to museums except for openings.

6. opening night, the first night of a theatrical play, entertainment, etc.; **opening-time,** (*a*) the time at which a place, esp. a public house, is opened; (*b*) the time that a device takes to open.

c **1814** (*play-title*) The opening night; or, the manager hoax'd. **1828** J. Ebers *Seven Yrs. King's Theatre* 210 The interest felt by the public in the arrival of the gran maestro, on the opening night. **1929** Opening night [see *doodah 1]. **1975** P. G. Winslow *Death of Angel* 9 The milkman finds a body dressed up for opening night. **1927** D. L. Sayers *Unnatural Death* xii. 136 Within, a cheerful bustle in the bar announced the near arrival of opening time. **1943** *Gloss. Terms Electr. Engin.* (*B.S.I.*) 66 *Opening time*, applied to a circuit-breaker: the time interval from the instant of application of the tripping power when in the closed position to the instant of separation of the arcing contacts. **1971** 'H. Calvin' *Poison Chasers* i. 7 We got back around opening time... I said to Dai, 'I'll get drinks while you're telephoning.'

opening, *ppl. a.* Add: **1. a.** (Further examples.)

1783 [see *Cheltenham 1]. **1912** *More Secret Remedies* (B.M.A.) x. 157 Do the bowels act regularly without opening medicine? **1965** A. Nicol *Truly Married Woman* 72 Magnesium sulphate for opening-medicine.

b. Esp. in *Cricket*, designating or pertaining to the batsmen who open the innings, or the bowlers who open the attack.

1929 P. G. H. Fender *Turn of Wheel* iii. 92 Hendry.. had a habit of..retreating when facing Larwood, and that would never do, especially in an opening batsman. **1952** J. H. Morgan *Glamorgan County Cricket* facing p. 64 (*caption*) Arnold Dyston, a stylish opening bat. **1955** A. E. R. Gilligan *Urn Returns* 36 The pace attack of Statham and Tyson keeps Rutherford and Sawle, the opening pair, strictly on the defensive. **1971** *Times* 15 Feb. 8/2 Their opening partnerships for the seven Test matches have averaged as much as 75. **1976** J. Snow *Cricket Rebel* 25 Tony Buss, his opening partner, was far more dangerous to face than he looked from the ringside. **1977** *Times* 29 Nov. 12/5 Geoffrey Arnold's two for 95 were the best figures by a recognized opening bowler in three Tests.

open-work. Add: **1.** (Earlier examples.)

1598 Florio *Worlde of Wordes* 460/2 *Zegandla*, such small fine net worke or open worke as gentlewomen use to make and weare vpon their heads in caules. **1782** T. Pennant *Journey Chester to London* 264 The fronts are of most elegant gothic open work. **1819** Shelley *Cenci* Pref. p. xiv, With balcony over balcony of open work.

2. (Earlier example.)

1811 J. Farey *Gen. View Agric. & Minerals Derbyshire* I. 359 The sinking was further continued, and the heaps on the sides of these *open-works*, or open-casts, increased.

opepe (opī·pi). Also **epepe.** [Yoruba.] A West African tree, *Sarcocephalus diderrichii*, of the family Rubiaceæ, or its hard yellowish-brown wood; also (see quot. 1908) occasionally used for an evergreen tree of the genus *Terminalia*.

1891 *Kew Bull.* 43 Names of Yoruba Timbers... 4. Opepe. **1908** *Ibid.* 193 An unnamed species of Terminalia..is also likely to furnish timber suitable for the home markets... The Yoruba name of the tree is 'Epepe'. **1920** *Nature* 29 July 692/1 Other heavy constructional woods [in the Empire Timber Exhibition] ..are..the gamboge-coloured Opepe..and Apa. **1936** J. D. Kennedy *Forest Flora S. Nigeria* 217 The accepted trade name for this timber in England is Opepe. **1956** *Archit. Rev.* CXX. 229 The board room [has] an opepe wood block floor. **1972** *Timber Trades Jrnl.* 13 May 40/2 Opepe, an early substitute for greenheart, was originally thought to be a bottomless pit, but..the bottom of the pit had been reached about two years ago. So they were already interested in a substitute for a substitute.

opera. Add: **3.** *horse opera*: see *horse *sb.* 27a; *soap opera*: see as main entry in Suppl.; *opéra comique* (earlier and later examples); *opéra bouffe, opera buffa* (earlier and later examples); also *attrib.* or as *adj.*; *opera-bouffer* (earlier example); *opera magica* (*rare*), opera with a fantastic or supernatural subject; *opera semiseria*, seriocomic opera; *opera seria*, serious or tragic opera, *spec.* a type of Italian opera flourishing in the 18th century, usually with a classical or mythological subject.

1801 F. Burney *Jrnls. & Lett.* 22–24 Apr. (1975) V. 267 Madᵉ d'henin made a party for us all to meet again the next day, & go to the Opera Buffa. **1870** D. J. Kirwan *Palace & Hovel* xvi. 235 Mademoiselle Helena Schneider, the opera bouffe singer. **1895** G. B. Shaw *Let.* 31 Aug. (1965) I. 553 Such are the opera bouffe depths to which I have descended. **1937** *New Statesman* 25 Dec. 1102/2 The *opéra-bouffe* Jupiter who attempts to cuckold him [*sc.* Amphitryon] by assuming his form. **1970** W. Apel *Harvard Dict. Mus.* (ed. 2) 187/2 A special type of comic opera is represented by the *opéras bouffes* of Offenbach. **1882** J. J. Jennings *Theatr. & Circus Life* 76 There is something so indescribably funny in the costumes, in the facial make-up, and all that, of the happy opera-bouffer or festive burlesquer. **1802** C. Wilmot *Let.* 31 Jan. in *Irish Peer* (1920) 39 We have been to the Opera Buffa or the Italian Opera. **1842** *Ainsworth's Mag.* II. 78 It is satisfactory to see the Italian Opera..returning to the opera buffa of Fioravanti. **1963** Auden *Dyer's Hand* 184 For Falstaff, time does not exist, since he belongs to the *opera buffa* world of play and mock action. **1965** C. Hibbert *Garibaldi & his Enemies* I. i. 13 He [*sc.* Mazzini] joined the Carbonari..but the *opera buffa* ceremony of the ritualistic initiation, in which he was required to swear allegiance to unknown leaders on a bared dagger.. struck him as absurd. **1968** S. Towneley in *New Oxf. Hist. Music* IV. xv. 837 Cavalieri divides the work into three acts. And he suggests the *intermedii* should intersperse them—a practice which a century later helped in the creation of *opera buffa*. **1744** H. Walpole *Let.* 22 July (1903) II. 40 Young Churchill has got a daughter by the Frasi; Mr. Winnington calls it the *opéra comique*; the mother is an opera girl; the grandmother was Mrs. Oldfield. **1866** G. H. Lewes *Jrnl.* 30 Dec. in *Geo. Eliot Lett.* (1956) IV. 328 A wretched opera comique—*Cartouche.* **1955** *Times* 6 May 14/7 It was doubtful whether, had it not been for Grétry, we should have had some of the striking later developments of *opéra comique*. **1976** *New Yorker* 16 Feb. 110/3 In 1870, Verdi was toying with the idea of writing an opéra comique. **1956** Auden & Kallman *Magic Flute* Pref. p. xiv, We have written the dialogue in verse, because it seemed to us the right medium for the spoken word in an *opera magica*. **1963** Auden *Dyer's Hand* 484 *Die Zauberflöte*..stylistically, an *opera magica*. **1947** A. Einstein *Mus. Romantic Era* xvi. 290 The stereotyped insipidity of the Italian *opera semiseria* since Rossini. **1959** *Times* 21 Aug. 13/2 The world première of Heimo Erbse's *Julietta*, an *opera semiseria*. **1970** W. Apel *Harvard Dict. Mus.* (ed. 2) 601/2 *Opera semiseria* is a serious opera including comic elements, e.g., Mozart's *Le Nozze di Figaro*. **1876** Stainer & Barrett *Dict. Mus. Terms* 392/2 *Seria* (*It.*), serious, grave, tragic, as, *Opera seria*, a tragic opera. **1880** Grove *Dict. Mus.* II. 513/2 The gradual development of the Opera Buffa from the Interludes which were formerly presented between the Acts of an Opera Seria, or Spoken Drama. **1892** G. B. Shaw *How to become Mus. Critic* (1960) 211 The public had been educated by Gluck to expect at least a show of seriousness in an *opera seria*. **1911** *Encycl. Brit.* XX. 124/2 *Opera seria* is classical Italian opera with secco-recitative; almost always..on a Greek or Roman subject, and,..with a happy ending... The only great classic in *opera seria* is Mozart's *Idomeneo*. **1951** M. Cooper *Russian Opera* i. 14 The Empress Catherine,..favoured the opéra comique—the opera of social comment and ideas—rather than the more exclusively musical opera seria. **1975** *New Yorker* 31 Mar. 82/1 Sometime in mid-1780, Mozart was commissioned to write the principal work, an *opera seria*, for the Munich carnival season.

4. a. *opera ballet* (later examples), *band*, *-box* (earlier and later examples), *chorus*, *company* (earlier example), *-goer* (earlier and later examples), *-going*, *hero*, *-night* (earlier example), *repertory*, *stage*, *ticket* (earlier example); *opera-going* adj. (later example).

1955 *Times* 1 July 7/4 In the production at the Louvre on Tuesday the piece was given in the form of an 'opera-ballet'. **1964** *Conc. Oxf. Dict. Opera* 23/1 So popular did the opera ballet become in Paris that Wagner had to rewrite and develop the opening scene of *Tannhäuser*..to meet the demands of the Parisians. **1798** T. Holcroft *Diary* 14 Oct. in *Mem.* (1816) III. 48 Gave young Watts the letters of recommendation for the opera band. **1814** J. Mayne *Jrnl.* 2 Sept. (1909) 30 The orchestra consists of nearly sixty performers, whereas our opera band seldom musters forty. **1811** *Times* 21 Jan. 3/2 You have permitted an anonymous paragraph to appear in your columns, reflecting on a noble young lady, and accusing her of..authorising an act of..injustice, by the improper disposal of an Opera Box... It is a gross..falsehood; neither is it true that this was the cause that induced Mr. Taylor to deprive the Portland family of their Opera box. **1828** J. Ebers *Seven Yrs. King's Theatre* iv. 82 It had been customary with my predecessors to publish, at the commencement of every season, a little book, specifying what boxes were taken for the season by the different subscribers—a sort of Opera-box directory. **1865** D. G. Rossetti *Let.* 28 June (1965) II. 560 Couldn't you work up the Opera-box design? **1840** H. Cockton *Life Valentine Vox* xi. 72 His real name is Growlaway. He's in the Opera chorus, and a regular trump he is too. **1947** A. Einstein *Mus. Romantic Era* xvi. 275 It is an Italian opera chorus; one ought not to think of Handel in connection with it. **1827** W. Clarke *Every Night Bk.* 146 There are several persons in the opera-company besides those we have mentioned. **1850** 'J. Timon' *Lorgnette* (ed. 2) i. 21 A prim clergyman, who, though he is not an Opera-goer, has yet a good ear for a fiddle. **1854** H. Morley *Jrnl.* 15 July (1866) 91 The opera-goer who enjoyed that musical farce..now finds the enjoyment of it trebled by the addition of Ronconi's..drolleries. **1947** A. Einstein *Mus. Romantic Era* xix. 358 The bourgeois opera-goer of 1880 had not only the opportunity of hearing an international repertory. **1955** P. Vincent in H. Van Thal *Fanfare for E. Newman* 177 It appears that each opera-goer has his own preference for a particular 'school' of opera. **1975** *Country Life* 13 Nov. 1312/1 For Russian audiences, memories of the book [*sc.* War and Peace] may perhaps give the opera a sort of unity which it lacks for most English opera-goers. **1876** Geo. Eliot *Dan. Der.* I. ii. xviii. 359 It was enough for them to go on in their old way, only having a grand treat of opera-going (to the gallery) when Hans came home on a visit. **1947** A. Einstein *Mus. Romantic Era* xvi. 239 In it the most intimate experience of the soul is made into an *opera* and presented to the opera-going public, to the mass. **1949** *Penguin Music Mag.* VIII. 16 The beginning of my opera-going was marked by outstanding failure. **1955** P. Vincent in H. Van Thal *Fanfare for E. Newman* 175 The first years of opera-going are among our most satisfying. **1834** J. R. Planché *Deep, Deep Sea* in *Extravaganzas* (1879) I. 164 To sing a song—As Opera heroes choose Always to do, when they've no time to lose. **1855** W. B. Wood *Pers. Recoll. Stage* v. 113 As the representative of opera heroes..he had no superior. **1776** J. Wallace in D. Garrick *Private Corr.* (1832) II. 140 She had objections to both: Saturday was Opera-night; Monday degraded her. **1898** Stainer & Barrett *Dict. Mus. Terms* (rev. ed.) 329/2 Their works are still part of every Opera repertory. **1763** D. Garrick *Let.* in R. B. Peake *Mem. Colman Family* (1841) I. 90 The famous Gabrielli pleased me much; she has a good person, is the best actress I ever saw on an opera stage, and has the most agreeable voice I ever heard. **1948** *Penguin Mus. Mag.* VII. 58 Wagner..in the end..had to sell out to the commercial opera-stage. **1755** Lady W. M. Montague in *Poems by Eminent Ladies* II. 170 There was a time (oh! that I could forget!) When opera-tickets pour'd before my feet.

b. opera-cloak (earlier example); **opera-hat** (earlier example); **opera-house** (later example).

1835 Dickens *Sk. Boz* (1836) 1st Ser. II. 47 There..was the young lady, wrapped up in a hopera-cloak. **1810** *Irish Mag.* III. 226 Strutting as gentlemen, by aid of.. silk stockings, opera hats. **1976** *New Yorker* 9 Feb. 81/1 The decision to present this production in the large opera house of the Academy instead of in the pleasant new theatre downstairs did it no good.

c. Applied to styles of women's underclothing suitable for wearing with evening dress, characterized by low tops and narrow shoulder-straps, as *opera combinations, shape, top.*

1923 *Weekly Dispatch* 18 Feb. 14 (Advt.), Pure Wool Opera Combinations... Ribbon shoulder straps. **1928** *Daily Mail* 31 July 1/1 (Advt.), Artificial silk vest..opera shape. **1921** *Daily Colonist* (Victoria, B.C.) 18 Oct. 19/1 (Advt.), Women's 'Turnbull's' Mixture Combinations, with low neck, short or no sleeves, with opera tops and ankle length. **1923** *Daily Mail* 17 Feb. 4 (Advt.), Ladies' Pure Wool Combinations,..opera tops, ribbon straps. **1968** J. Ironside *Fashion Alphabet* 72 *Vest*..sometimes with shoulder-straps, and known as 'Opera top'.

operability (ǫpĕrăbi·lĭti). [f. next + *-bility.] The state of being operable; *spec.* in *Med.*, suitability for surgical treatment.

1905 *Jrnl. Nerv. & Mental Dis.* XXXII. 481 (*heading*) Brain tumors: a study of clinical and post-mortem records bearing on their operability. **1910** *Practitioner* Jan. 84 The question of operability or otherwise is a matter in which surgeons differ considerably. **1922** *Ann. Surg.* LXXVI. 396 In estimating the final value of any operative procedure for the cure of cancer, the operability rate is the crux of the situation. **1952** *Sci. Amer.* Sept. 104/2 The gun parts could be interchanged among all 10 without affecting the guns' operability. **1972** *Brit. Jrnl. Dis. Chest* LXVI. 162 (*heading*) The value of mediastinoscopy in assessing operability in carcinoma of the lung. **1975** J. Howlett *Christmas Spy* III. vi. 96 Zürich had been correct about the operability of two-way traffic.

operable, *a.* and *sb.* For **a, b** read **A, B,** restrict † *Obs.* to sense B, and add: **A.** *adj.* **1.** (Later examples.)

1911 H. S. Harrison *Queed* xiii. 160 How could this principle be..reduced to an operable law? **1961** *Medicine in Nat. Defense* (Final Rep. Office of U.S. Asst. Secretary of Defense) vi. 69 Visual and operable training aids..were used in support of Army Medical Service mass casualty exercises.

2. *Med.* Capable of being treated by an operation.

1904 *Arch. Middlesex Hosp.* III. 163 Patients..are admitted in practically all cases for operable cancer. **1925** W. Deeping *Sorrell & Son* xxxviii. 381 'But, my dear sir, if the thing is operable—' 'No, thank you. Besides, it is only a question of a few days. If you can help me to fight the pain.' **1925** H. Gilford *Tumours & Cancers* xxvi. 580 Of operable uterine cancers as large a number can be destroyed by the [radium and X-]rays as by surgical operation. **1973** *Cancer* XXXI. 180/2 Patients with carcinoma of the cervix are admitted ..only if..the tumor is considered operable and the patient potentially curable by surgery.

operand. (Further examples.)

1956 Berkeley & Wainwright *Computers* ii. 37 The first two registers receive the operands, or numbers operated on. **1965** *Language* XLI. 377 There is also a zero (and affix) causative in which the *W* contains no new verb, but the *N* and *V* of the operand sentence are permuted: *The children sat→He seated the children.* **1969** P. B. Jordain *Condensed Computer Encycl.* 354 In the mathematical operation of division, the dividend and divisor are

two operands. *Ibid.*, Most digital computer instructions have one or more operands, each indicated by a field within the instruction. **1973** D. U. Wilde *Introd. Computing Probl.* v. 106 The *or* operation is said to be true if either or both of its operands are true; otherwise, it is false. The *and* operation is true if both operands are true; otherwise, it is false.

operant, *a.* and *sb.* Add: **A.** *adj.* **b.** *Psychol.* Involving the modification of behaviour by the reinforcing or inhibiting effect of its consequences; opp. *respondent.* Cf. *instrumental *a.* 7.

1937 [see B. 1 b below]. **1938** B. F. Skinner *Behavior of Organisms* xiii. 439 Comparing conditioning of Type S (which is largely, if not wholly, respondent) with Type R (which is apparently wholly operant). **1941** *Amer. Jrnl. Psychol.* LIV. 568 (*heading*) An automatic device for providing motivation and reinforcement in operant conditioning. **1959** *Jrnl. Exper. Analysis of Behavior* II. 57 Young and Boycott (1955) describe behavior in the octopus which is almost certainly operant in nature, i.e., not elicited, but maintained by its consequences. **1963** P. Swartz *Psychol.* x. 208/2 In contrast to respondent conditioning, in which the reinforcing agent is coupled with the conditioned stimulus, in operant learning the appearance of the reinforcer is dependent upon the occurrence of the response. **1964** Ratner & Denny *Compar. Psychol.* xi. 566 A common distinction between instrumental and operant learning, at least in the laboratory, is that the animal is free to respond at any time in the operant situation but is usually given discrete trials when the learning is labeled instrumental. **1968** *Listener* 29 Aug. 265/2 The sickest man in the hospital was selected to be..the first experiment in 'operant conditioning' therapy on a severely regressed and catatonic schizophrenic in Britain. **1970** Hinsie & Campbell *Psychiatric Dict.* (ed. 4) 150/2 It has been suggested that many forms of psychotherapy are applications of operant conditioning... The patient learns what the therapist expects or wants to hear, and he modifies his own speech and behavior accordingly. **1975** *Nature* 20 Mar. 219/1 Behaviourism..puts us in a position to alter and improve the condition of mankind if only we will apply its commanding insights..into the agency of such cardinal notions as reinforcement and operant behaviour.

B. *sb.* **1. b.** *Psychol.* An item of behaviour that is held to be not a response to a prior stimulus but something which is initially spontaneous on the part of the organism, 'operating on' or affecting the environment so as to produce consequences which may reinforce or inhibit its recurrence.

1937 B. F. Skinner in *Jrnl. Gen. Psychol.* XVI. 274 There is also a kind of response which occurs spontaneously in the absence of any stimulation with which it may be specifically correlated... It does not mean that we cannot find a stimulus that will elicit such behaviour but that none is operative at the time the behaviour is observed. It is the nature of this kind of behaviour that it should occur without an eliciting stimulus, although discriminative stimuli are practically inevitable after conditioning... I shall call such a unit an operant and the behavior in general, operant behavior. **1938** —— *Behavior of Organisms* i. 21 If the occurrence of an operant is followed by presentation of a reinforcing stimulus, the strength is increased. **1941** *Amer. Jrnl. Psychol.* LIV. 568 Aversion drives..offer certain advantages which other drives, *e.g.* hunger and thirst, do not... Reinforcement can follow immediately the emission of the operant; under hunger motivation several responses may intervene between the operant that is measured and its subsequent reënforcement. **1967** Koestler *Ghost in Machine* i. 8 Operant strength is usually measured, for technical reasons, by the 'rate of extinction'—how long the rat will persist in pressing the lever after the supply of pellets has been stopped. **1972** *Jrnl. Social Psychol.* LXXXVI. 11 Recent analyses have conceived of conforming behavior as an instrumental response directed toward the attainment of functional goals... It can be considered a kind of social operant, which should change..with changes in the characteristics of reinforcement.

o·pera:table, *a. rare.* [f. Operat(e *v.* + -able.] Capable of being operated (on); operable.

1895 *Funk's Stand. Dict.*, *Operatable*, that can be operated or worked; operable. **1932** Joyce *Let.* 12 July (1966) III. 248 The right eye was still operatable 20 months ago. Now the cataract is total.

operate, *v.* Add: **4. c.** Also *transf.*, of a gambler, criminal, etc.

1883 Sweet & Knox *On Mexican Mustang through Texas* (1884) i. 16 This high-toned and honorable desperado 'operated' in one of the inland cities of Texas two years ago. **1901** 'J. Flynt' *World of Graft* 19 The West Side grafters..who have 'operated' in Chicago. **1955** D. W. Maurer in *Publ. Amer. Dial. Soc.* XXIV. 30 There are the *lone wolves*, who are professionals, but who operate predominantly alone, without the support of a mob,..for example, jewel thieves of some types, swindlers, expert forgers. **1975** T. Allbeury *Special Collection* xi. 73, I set up a network for him dealing with industrial espionage. It operated into West Germany.

d. (Later examples.)

1889 *Harper's Mag.* Aug. 448/1 Do you think all men who are what you call operating around are like that? **1961** in Webster.

e. To function, to fulfil a function, to act. (Closely related to sense 4a.)

1931 J. T. Adams *Epic of Amer.* viii. 221 There were..a thousand boats operating regularly on the Mississippi. **1932** [see *cylinder *sb.* 6]. **1932** N. M. Butler *Looking Forward* xi. 117 Government officials operating in all parts of the country. **1971** *Gloss. Electrotechnical, Power Terms* (*B.S.I.*) 1. iii. 13 A relay operates when it completes its designed function in a specified output circuit(s). **1972** *Daily Tel.* 16 Nov. 7/1 Extra buses and Underground trains will operate on most routes. **1976** *National Observer* (U.S.) 2 Oct. 18/3 Rudderless and without a keel, the 195-foot craft operates similar to a flat-bottomed fishing boat.

7. For chiefly *U.S.* read orig. *U.S.* and add later examples.

1948 'N. Shute' *No Highway* i. 19 C.A.T.O. are operating five or six of them [*sc.* aircraft] on the Atlantic route. **1971** D. Potter *Brit. Eliz. Stamps* xii. 130 Cambrian Airways..took over the operation of some internal routes previously operated only by bea. **1974** *Anderson* (S. Carolina) *Independent* 20 Apr. 2A/1 Joe King, who operates a hardware store 13 miles south of Greenville on S.C. 25, fired four bullets into their car as they fled.

8. *Surg.* To operate on. (See sense 4 b.)

1908 *Practitioner* Sept. 423, I know of two cases of pyelitis which were operated in mistake for appendicitis. **1915** W. Owen *Let.* 1 Mar. (1967) 324 Dr Denucé (who operated Sarah Bernhardt—and Charlie). **1925** Simmons & Fishbein *Art & Pract. Medical Writing* v. 43 'Operate' means, and is generally synonymous with, 'to work': the terms nearly always may be used interchangeably. The surgeon who would hesitate to say 'I worked this patient' says, without a blush, 'I operated this patient.' **1930** *Amer. Speech* V. 289 Of those questioned 26½ per cent used 'operated him', 40 per cent used 'operated on', and 33½ per cent used 'operated upon him'.

operatic, *a.*[1] Add: **b.** as *sb. pl.* The production or performance of operas.

1907 N. Munro *Daft Days* xvi. 142 He says he could never die a Christian death if he had to listen to them at their operatics through the wall. **1920** *Punch* 10 Mar. 197/1 Operatics. It has been suggested before now that Opera might be improved if the singing were done behind the scenes and the performance on the stage were carried out in dumb show by competent actors. **1928** *Daily Express* 6 Nov. 9/3 This is a real event in London's amateur operatics.

operating, *vbl. sb.* Add: **a.** (Later example.)

1913 Roberts & Smith (*title*) Practical locomotive operating.

b. *operating altitude, box, control, costs, expenses, height, revenue*; **operating crew** (see quot.); **operating system** *Computers*, a set of programs for organizing the resources and activities of a computer.

1956 D. E. Charlwood *No Moon Tonight* 15 Where the planes were circling, climbing steadily to operating altitude. **1918** 'Q' *Foe-Farrell* 117, I..found..the operating box and the gallery, switched on the lights, and shinned down a pillar to the stalls. **1930** *Daily Express* 6 Sept. 5/5 The 'operating control' can be readily grasped from the sketch. **1913** Roberts & Smith *Pract. Locomotive Operating* 26 Operating costs. **1972** *Lebende Sprachen* XVII. 34/1 US *operating costs, operating expenses*—BE working expenses. **1975** 'D. Jordan' *Black Account* xi. 54 Condon said, 'Let's settle the totals like this, then,' and wrote... Initial operating costs and interest: $85 million. **1965** Operating crew [see *flight crew* s.v. *flight *sb.*[1] 15]. **1869** *Bradshaw's Railway Manual* XXI. 417 The operating expenses are about 51 per cent. of the gross earnings. **1909** *Daily Chron.* 25 Feb. 3/5 At the same time the operating expenses had gone down from 67·9 per cent. of gross receipts in 1907 to 54·5 in 1908. **1948** 'N. Shute' *No Highway* iv. 111 I'm prepared to shut down the inboard engines after climbing up to operating height. **1930** *Daily Express* 9 Sept. 10/2 Operating revenues show a gain of about 40 per cent. at £556,936, compared with £389,518, and while profit on sales of securities is lower, the gross earnings have expanded by nearly £80,000 to £740,582. **1961** *Computer Jrnl.* IV. 222 (*heading*) The Manchester University Atlas operating system. **1963** Gregory & Van Horn *Automatic Data-Processing Syst.* (ed. 2) xii. 477 Operating Systems are programs that increase machine operating efficiency by controlling the compilation and execution of programs, supervising input and output operations, converting data from one medium to another, testing programs to debug them, and simulating the operation of one processor on another. **1973** D. U. Wilde *Introd. Computing* ii. 43 When a computer system is under the control of a monitor or operating system, the computer operator loads programs into the card reader, mounts tape reels onto tape drives, and removes results from the printer while the monitor schedules the work flow and keeps the CPU busy.

operating, *ppl. a.* (Further example.)

1904 *Daily Chron.* 16 July 7/3 The operating company is to take over the tunnel before the end of the month.

operation. Add: **4. b.** *Psychol.* A mental activity whereby the effect of actions or ideas is logically understood or predicted, esp. with reference to the supposed stages of a child's development; also *concrete operations, formal operations* (see esp. quots. 1960, 1963). Cf. *preoperational *a.*

1930 M. Gabain tr. *Piaget's Child's Concept. Phys. Causality* iv. 301 At each stage of intellectual development we can distinguish roughly two groups of operations. **1953** Mays & Whitehead tr. *Piaget's Logic & Psychol.* ii. 8 Psychologically, operations are actions which are internalizable, reversible, and coordinated into systems characterized by laws which apply to the system as a whole. **1960** J. S. Bruner *Beyond Information Given* (1974) xxiii. 415 Concrete operations, though they are guided by the logic of classes..are means for structuring only immediately present reality. **1963** J. H. Flavell *Developmental Psychol. J. Piaget* iii. 86 The period of formal operations (11–15)... The adolescent can deal effectively not only with the reality before him..but also with the world of pure possibility. **1975** J. W. Brunk *Child & Adolesc. Devel.* vi. 252 In the period of concrete operations (approximate ages 7–11), the child becomes capable of logical thought processes that can be applied to concrete problems.

5. a. (Further examples.)

1927 P. W. Bridgman *Logic Mod. Physics* i. 5 To find the length of an object, we have to perform certain physical operations. The concept of length is therefore fixed when the operations by which length is measured are fixed. **1935** *Psychol. Rev.* XLII. 517 Only those constructs based upon operations which are public and repeatable are admitted to the body of science. **1967** *Encycl. Philos.* V. 544/1 By the end of the nineteenth century, however, scientists had accepted the view that if we cannot devise operations which would disclose whether or not space was Euclidean, then no definite geometrical properties can be assigned to space.

b. (Earlier and later examples.) Also more generally, a business activity or enterprise. Also *transf.*

1832 *Reg. Deb. Congress U.S.* 22nd Congress 2 Sess. App. 107/1 The liability to be called upon for large advances, for the above operation,..makes it absolutely necessary that the limit should be strictly attended to. **1848** W. Armstrong *Stocks* 11 We conceive that this operation [*sc.* betting] is too well understood to need any particular explanation. **1851** C. Cist *Sk. Cincinnati in 1851* 236 Such is the extent of the operations of this firm. **1889** *Harper's Mag.* Aug. 448/1 One is an operation, and the other is embezzlement. **1911** J. C. Lincoln *Cap'n Warren's Wards* xi. 178, I judged..that you were well enough acquainted with Wall Street to know that queer operations take place there. **1928** F. A. Bradford *Money* xv. 283 The open market operations of the reserve banks. **1938** J. B. Williams *Theory of Investment Value* ii. 10 The rate of operations never precedes stock prices. **1960** 'E. McBain' *Give Boys Great Big Hand* iv. 30 A photo of the bag on the front pages..might not be bad for our operation... You can't buy that sort of advertising space, now can you? **1977** 'J. Le Carré' *Hon. Schoolboy* xviii. 442 The spearhead of the operation will be handled by ourselves.

7. (Further examples.) Also, the strategic movement of troops, ships, etc.; the people concerned with such movements. Freq. *attrib.* See also *combined operation*.

1915 R. W. Campbell *Private Spud Tamson* xiii. 167 Any chapter on training must also refer to night operations, generally called Night Attacks. These operations are never popular in times of training. They interfere with social engagements. **1939** *War Pictorial* 22 Dec. 9 (*caption*) The gallery above is the Operations Section of the R.A.F. Fighter Command. **1941** N. Macmillan *Air Strategy* iv. 30 We must perforce ignore..the layout of an Operations Room which functions as the brain to the body, the body in this case being the operational aircraft. **1943** J. S. Huxley *TVA* 108 The operations crew..get their orders from the control room. **1946** *R.A.F. Jrnl.* May 170 Now, in a world at peace, operations are but a faint memory for the men. **1968** R. L. Ackoff in *Internat. Encycl. Social Sci.* XI. 291/1 British military executives turned to scientists for aid when the German air attack on Britain began... These teams of scientists were usually assigned to the executive in charge of operations. **1975** A. Beevor *Violent Brink* vii. 167 In the Operations Room.. two members of the Security Committee had advocated that the two sides should be left to fight it out.

b. Used as first element of the code-name for a military or civil campaign.

1938 'Taffrail' *Operation 'M.O.'* iii. 58 It's stuffed full of secret papers that I was going to work on at home! Operation 'M.O.' **1941** *New Statesman* 26 Apr. 443/2 In this brief workmanlike account of the Evacuation from Dunkirk, Mr. Masefield... brings back a plain exact narrative of The Operation Dynamo (as the lifting was officially called). **1945** *News Chron.* 1 June 1/4 This is the first picture to be released of Operation 'Fido'. **1946** T. Driberg in *Reynolds News* 28 Apr. 4 Note to sub-editors and others: please cooperate in killing..the most overworked of current clichés—the whimsical application to a variety of topics of the military locution 'Operation —'. **1950** *Nat. Geographic Mag.* Sept. 367/1 'Operation Link', they call the project because when completed, the 4-mile bridge will link Maryland's eastern and western shores. **1968** *Listener* 23 May 656/2 Between Tet and the 'second wave', General Westmoreland launched 'Operation Final Vietcong' to sweep the enemy from the surroundings of Saigon. But this enterprise, nicknamed 'Operation Final Solution', failed. **1973** *Guardian* 25 Jan. 2/1 The operation to return the..prisoners of war..Operation Homecoming, as the Pentagon has called it.

8. a. (Further examples.) In connection with *Computers* freq. identical with sense 5; also = *function *sb.* 3d.

1885, etc. [see *logical operation* s.v. *logical *a.* (and *sb.*) 7]. **1946** *Ann. Computation Lab. Harvard Univ.* I. 50 Since the control tapes deal with operations only, they represent the solution of a mathematical situation independent of the values of the parameters involved. **1947** *Math. Tables & Other Aids to Computation* II. 356 Provision is planned for squaring, taking the reciprocal, and the maximum or minimum of two quantities, but these operations are not yet available. **1953** A. D. & K. H. V. Booth *Automatic Digital Calculators* vi. 35 It is customary, in *all* existing computers, to have circuits which will perform the operations of addition and subtraction. *Ibid.* 37 The left shift operation is more complex. **1956** Berkeley & Wainwright *Computers* ii. 38 The arithmetical operations of a computer..include addition, counting, subtraction,..truncating, rounding off, [etc.]. **1969** P. B. Jordain *Condensed Computer Encycl.* 358 In general, an operation is a single irreducible step in the performance of a computer program... In some higher-level languages, an operation is a single higher-level step (such as finding a square root). **1972** Gross & Brainerd *Fund.*

Programming Concepts ix. 248 The first hexadecimal digit is called the operation code and specifies which operation is to be performed, such as add, subtract, print, etc.

b. In the numerical solution of simultaneous linear equations by relaxation, the process of changing the trial value of one of the unknowns in order to reduce the magnitude of one of the residuals.

[**1935** R. V. Southwell in *Proc. R. Soc.* A. CLI. 65 In the operation considered, B is held fixed and A is moved.] **1940** —— *Relaxation Methods in Engin. Sci.* i. 1 The Relaxation Method takes it farther by devising methods of systematic adjustment: we apply a series of operations, each one an indirect solution of a particularly simple kind, and in this way we 'tune up' a trial solution..until it conforms with some imposed standard of accuracy. **1957** L. Fox *Numerical Solution Two-Point Boundary Probl.* iii. 38 The computations are embodied in Tables 1 and 2, showing respectively the possible operations and details of the relaxation process. **1969** W. A. Watson et al. *Numerical Anal.* I. vi. 160 Usually the operation which reduces the magnitude of the largest residual is used at any stage of the working.

11. operation code *Computers*, a character or set of characters that when put into the operation part of an instruction specifies the operation that is performed; **operation part** *Computers*, the part of an instruction that receives the operation code; **operations research** *U.S.* = *operational research* (*operational *a.* 1 b); so *operations researcher*; **operations table** *Math.*, in the numerical solution of simultaneous linear equations by relaxation, a table showing the changes in the values of the residuals that result when each unknown in turn is increased by one.

1949 E. C. Berkley *Giant Brains* vi. 103 Division has the code 76 and multiplication the code 761, and so the difference is essentially an operation code not in the third or C field. **1972** C. B. Germain *PL/1 for IBM 360* ii. 14/2 The general format of 360 instructions is an operation code..specifying the operation to be performed followed by two addresses to specify the data, or operands, involved in the operation. **1957** D. D. McCracken *Digital Computer Programming* ii. 14 Instruction Format... The first two digits..are called the operation part, and tell the machine what to do. The next four digits are termed the address part. **1969** P. B. Jordain *Condensed Computer Encycl.* 359 The operation part is usually at the left end of the instruction, and is of fixed length. **1945** E. J. King in G. C. Marshall et al. *War Rep.* (1947) 719 The application, by qualified scientists, of the scientific method to the improvement of naval operating techniques and material, has come to be called operations research. **1956** Berkeley & Wainwright *Computers* vii. 281 Much operations research can be carried out with pencil, paper, and a desk calculating machine. **1969** J. Argenti *Managem. Techniques* 81 The same problem occurs in Operations Research, i.e. in those management techniques that depend largely upon mathematics. **1970** P. M. Morse in G. J. Kelleher *Challenge to Syst. Analysis* iii. 23 Operations research has emerged as a unified area of applied science.. designed to influence policy. **1953** *Operational Research Q.* IV. 51 The operations researcher's competence stops. **1940** R. V. Southwell *Relaxation Methods in Engin. Sci.* i. 8 Having completed the calculations we can present their results in an operations table. **1957** L. Fox *Numerical Solution Two-Point Boundary Probl.* iii. 40 The full number of figures is used for the calculation of residuals, but the coefficients can be rounded-off to convenient numbers for use in the operations table. **1969** W. A. Watson et al. *Numerical Anal.* I. vi. 159 For ease of reference during the working it is convenient to summarise the effects on the residuals of unit changes in the unknowns in a table, usually referred to as the operations table.

operational (ǫpĕrēi·ʃənăl), *a.* [f. Operation + -al.] **1. a.** Of or pertaining to operation or operations; *spec.* engaged in or connected with active military operations as distinct from being under training, in reserve, etc.

1922 *Edin. Rev.* Oct. 212 The development of..air communications..ensuring the maintenance of a large and flourishing constructional and operational aircraft industry. **1928** C. F. S. Gamble *Story N. Sea Air Station* x. 146 They were placed in various groups for disciplinary and operational purposes. **1940** *War Illustr.* 5 Jan. 568/2 Each balloon can be raised to its operational ceiling in a very few minutes. **1941** *Economist* 19 Apr. 521/2 The new status is not intended merely to prevent wastage. There are operational reasons. The range of women's duties in the Army and the Air Force is extending. **1943** B. J. Hurren *Eastern Med.* xii. 129 The air forces moved up their advanced aerodromes (technically known as Operational Landing Grounds). **1962** A. Nisbett *Technique Sound Studio* 246 The soundproof room equipped with control desk, gramophone and tape reproducers and high quality loudspeaker, which is occupied by production and operational staff. *a* **1963** J. Lusby in B. James *Austral. Short Stories* (1963) 222 They were young, quiet, and looked tired. Operational men, instructing for a 'rest'. **1963** C. W. Bary (*title*) Operational economics of electric utilities. **1964** *Ann. N.Y. Acad. Sci.* CXV. 683 The problems of achieving adequate operational flexibility..underline the need for more appropriate computer systems. **1974** M. Babson *Stalking Lamb* ii. xvii. 127 He intended to make the mews house his operational headquarters... It would be safe, quiet, unsuspected. **1977** *R.A.F. News* 11–24 May 3/4 The real test lies in our operational efficiency.

b. *operational research*: a method of mathematically based investigation for providing a quantitative basis for management decisions (orig. for military planning); abbrev. O.R., OR (*O 5d); so *operational researcher*.

1941 P. M. S. Blackett in *Advancement of Sci.* (1948) V. 27/1 The work of an Operational Research Section should be carried out at Command, Groups, Stations or Squadrons as circumstances dictate. *Ibid.* 29/1 One of the tasks of an Operational Research Section is to make possible..a numerical estimate of the merits of a change over from one device to another. **1945** *World Rev.* June 49 The Operational Research Section..is a new element in the organisation of the Royal Air Force which has been evolved during the war. **1948** *Nature* 13 Mar. 377/1 In war, operational research was applied to the use of weapons, to tactics, and to strategy. In the peace-time applications of operational research, studies are directed, for example, to the use of equipment and man-power, to operating procedures, and to the solution of those many problems faced by management..or by Government authorities. **1948** *Advancement of Sci.* IV. 320/1 The operational researcher was visualised as requiring dual ranges of knowledge; on the one hand a wide and fairly detailed knowledge of technical possibilities..; on the other..a close personal knowledge of the working conditions. **1953** *Economist* 15 Aug. 465/1 This technique is called 'operational research'; and under it, teams that may include engineers, mathematicians, statisticians, economists and sociologists combine together. **1959** *Birmingham Mail* 11 Mar. 1/5 London Transport has set up an operational research team to investigate the desirability of increasing the non-smoking accommodation in Tube coaches. **1964** M. Argyle *Psychol. & Social Probl.* xvi. 194 The kind of research done by social scientists working for organizations can be divided into trend analysis, operational research and experiments. **1967** R. Whitehead in Wills & Yearsley *Handbk. Managem. Technol.* 70 The operational researcher seeks to control the situation by making a model of the outside factors from which he can to some extent predict and form a plan of production. **1967** E. Duckworth in *Ibid.* 99 The purpose of operational research is..to assist a manager to take decisions and to help him to take fewer decisions. **1971** D. C. Hague *Managerial Econ.* (rev. ed.) 6 Managerial economics is a background subject which both the line manager and the operational researcher must understand if they are to be successful.

c. In a condition of readiness to perform some intended (esp. military) function.

1944 H. St. G. Saunders *Per Ardua* xvii. 270 Only the Martinsyde F.4, not then operational, was superior to them. **1948** 'N. Shute' *No Highway* i. 19 'Could you.. find out how many hours flying these machines have done?'.. 'They can't have done much. They've only been operational for about a month.' **1963** R. M. Hare *Freedom & Reason* iii. 46 We are to make morality again (as the military writers say) 'operational'. **1965** *New Statesman* 14 May 758/1 The new Russian weapon is clearly operational. **1974** *Nature* 22 Nov. 279/2 The Cambridge group operates the majority of operational catenary instruments.

d. Of, or pertaining to, mental operations (see *operation 4 b).

1953 Mays & Whitehead tr. *Piaget's Logic & Psychol.* p. xviii, The structures which emerge in the analysis of the operational mechanisms of thought. **1963** J. H. Flavell *Developmental Psychol. J. Piaget* v. 168 The operational systems of middle childhood have certain definable properties. **1975** M. D. Smith *Educ. Psychol.* ii. 40 When a learner becomes able to deal with things that are not actually present..and to relate images and memories to predict and control the future, then he is in the formal operational stage of thought.

2. a. *Math.* Of, involving, or employing operators.

1927 H. Jeffreys (*title*) Operational methods in mathematical physics. **1937** E. Stephens (*title*) The elementary theory of operational mathematics. **1957** L. Fox *Numerical Solution Two-Point Boundary Probl.* ii. 8 We easily produce the operational equivalents $E = e^{hD}$, $E - 1 = \Delta$... With few restrictions these operators can be manipulated according to the rules of ordinary algebra. **1973** L. E. Edwards *PL/1 for Business Applications* ii. 55 An expression may consist of a single variable or a single constant or a combination of these using operators, and is then referred to as an operational expression.

b. Electronics. *operational amplifier*: an amplifier with a very high open-loop gain and a very low output impedance that is used (usu. with negative feedback) as the basis of a circuit for performing a particular mathematical operation on an input voltage with high accuracy, the relation of output to input being effectively determined solely by the arrangement and magnitude of the other, passive, circuit elements.

1947 J. R. Ragazzini et al. in *Proc. IRE* XXXV. 444/2 As an amplifier so connected can perform the mathematical operations of arithmetic and calculus on the voltages applied to its input, it is hereafter termed an 'operational amplifier'. **1962** Simpson & Richards *Physical Princ. Junction Transistors* xiii. 309 The operational amplifier is also valuable as a linear adder. Because the input and output resistances are very low, the base voltage may be made accurately proportional to the sum of a number of separate input voltages each 'weighted' by its own summing resistor. **1963** B. Fozard *Instrumentation Nucl. Reactors* xii. 152 The basic unit of an electronic analogue computer is the operational amplifier. **1972** Vassos & Ewing *Analog & Digital Electronics* v. 139 Deviations from ideality in operational amplifiers are very often negligible. For this reason the input-output relation is independent of the characteristics of the particular amplifier. This property brings about a unique simplicity of design, which is the key to the extensive use of operational amplifiers in instrumentation and control systems.

3. Of, pertaining to, or in accordance with operationalism.

1927 P. W. Bridgman *Logic Mod. Physics* i. 8 Einstein, in thus analyzing what is involved in making a judgment of simultaneity..is actually adopting a new point of view as to what the concepts of physics should be, namely, the operational view. **1935** *Psychol. Rev.* XLII. 517 Operational doctrine makes explicit recognition of the fact that a concept, or proposition, has empirical meaning only if it stands for definite, concrete operations capable of execution by normal human beings. **1937** *Harper's Mag.* Dec. 51/1 Some deductions may still be sound, but all are suspect pending operational check in modern America. **1941** A. Huxley *Grey Eminence* iii. 48 That 'operational philosophy' which contemporary scientific thinkers have begun to apply in the natural sciences. **1952** G. Sarton *Hist. Sci.* I. xvi. 404 It [*sc.* the Platonic method] is sterile because it is unworkable, or, to use our modern terminology, it is not 'operational'. **1955** A. Huxley *Genius & Goddess* 21 'Wouldn't he have lived to eighty-seven without the pills?'.. 'We can never know how his self-medication was related to his longevity. And where there's no possible operational answer, there's no conceivable sense in the question.' **1971** *Brit. Med. Bull.* XXVII. 37/2 The operational definition of hypertension in the population should be flexible.

operationalism (ǫpĕrēi·ʃənăliz'm). [f. prec. + -ism.] A theory or system which accepts only such concepts as can be described in terms of the operations necessary to determine or prove them.

1931 *Jrnl. Philos.* XXVIII. 545 Operationalism must be understood to state that a concept has no meaning unless its definition formulates performable operations. **1941** P. Frank *Between Physics & Philos.* 5 Professor Bridgman's views have..been labelled 'operationalism', although he himself is not pleased by this name. **1950** *Mind* LIX. 571 Pragmatism and its modern offspring operationalism. **1965** N. Chomsky *Aspects of Theory of Syntax* 194 Perhaps this loss of interest in theory..was fostered by certain ideas (i.e., strict operationalism or strict verificationism) that were considered briefly in positivist philosophy of science..in the early nineteen-thirties. **1965** [see *operationally *adv.*]. **1968** M. Black *Labyrinth of Lang.* vi. 142 Operationalism may rank with Freudianism as one of the major intellectual forces in the Western world between the first two World Wars. **1972** *Language* XLVIII. 418 It would appear that the appeal of various forms of operationalism, positivism, and behaviourism,.. had a great deal to do with the emergence of a new professional identification on the part of a small group of young men.

Hence **opera·tionalist** *sb.* and *a.*

1931 *Jrnl. Philos.* XXVIII. 545 When the operationalist defines a concept in terms of operations, the meaning of the concept thus defined is not restricted to performed operations or to operations which are actually going to be performed. **1934** *Mind* XLIII. 201 This Mead wants to prove by following the operationalist argument about the concepts of physics. **1941** A. Huxley *Grey Eminence* iii. 48 Buddha was not a consistent operationalist. **1965** J. D. North *Measure of Universe* xv. 335 One of the biggest objections to the operationalist philosophy is that it appears to deny meaning even to such apparently harmless dispositional words as 'imperceptible', 'movable', and so on. *Ibid.* 336 But (runs the operationalist's argument) even where D_1 and D_2 turn out to be more or less the same and even where, as a matter of convenience, one assimilates the two concepts, in any careful or philosophical account they must always be distinguished. **1977** *Language* LIII. 170 In the period when a strictly behaviorist and operationalist philosophy dominated American linguistics, constraining the linguist to discount any data beyond the 'physical' record, Hoijer delivered a paper at the 8th International Congress of Linguists on the importance of 'Native reaction as a criterion in linguistic analysis'.

operationality (ǫ:pĕrēiʃənæ·lĭti). [f. *operational *a.* + -ity.] The property of being operational.

1972 *Computers & Humanities* VII. 81 First, how well have the available [computer] programs served the current generation..? Second, considering basic design tensions—flexibility and uniformity, elegance and operationality—what directions should future design and dissemination efforts take? **1973** *Nature* 21/28 Dec. 532/1 Piaget's insistence on 'operationality' makes his experiments indirect and complicated.

operationalizable (ǫpĕrēi:ʃənăləi·zăb'l), *a.* [f. next + -able.] Capable of being operationalized.

1975 *Cooperation & Conflict* (Oslo) X. 35/1 As it consists of transactions, the transactor profile characteristic is more easily operationalizable. **1976** *Brit. Jrnl. Sociol.* XXVII. 35 However, my goal is to develop a single, broad operationalizable interpretation from the philosophical sources and to test the contemporary usages against that standard.

operationalize (ǫpĕrēi·ʃənăləiz), *v.* [f. *operational *a.* + -ize.] *trans.* To express or determine in operational terms. Hence **opera·tionalizing** *vbl. sb.*

1954 *Jrnl. Abnormal Psychol.* XLIX. 460 Once the codability variable..had been operationalized, it remained to relate this variable to some nonlinguistic behavior. **1964** [see *codability]. **1966** Hughes & Pinney in E. L. Pinney *Compar. Politics & Polit. Theory* 67 Most of these interpretations are exercises in conceptualization, rather than efforts to operationalize the concept. **1972** *Jrnl. Social Psychol.* LXXXVII. 30 Previous studies have operationalized panic in terms of average time of group escape. **1975** *Gen. Systems* XX.

116/2 By accommodating evolution, class exploitation, and industrialism, cultural materialism helps operationalize anthropology.

Hence **o:perationaliza·tion**, the process of operationalizing.

1966 Hughes & Pinney in E. L. Pinney *Compar. Politics & Polit. Theory* 67 The problem which must be solved in order to exploit the usefulness of the concept is basically one of operationalization. **1969** P. Worsley in Ionescu & Gellner *Populism* 235 The formulation and operationalization of policy..is thus a function of many other things than simply adhesion to some kind of ideology. **1973** *Sociol. Rev.* XXI. 419 The term 'operationalisation' is usually reserved for the process of trying to turn hypotheses into testable postulates, and for the problems of selecting or creating techniques to test them. **1975** *Political Stud.* XXIII. 75 That Deutsch himself is aware of the limitations of his operationalization is indicated by his calling a 'politicized people' a 'nationality', and not inhabitants of a society featuring a large public sector.

operationally (ǫpĕrēi·ʃənăli), *adv.* [f. as prec. + -ly[2].] In terms of, or as regards, operation(s), esp. the operations required to define a concept or term (cf. *operationalism).

1927 *Jrnl. Philos.* XXIV. 663 Every concept must henceforth be defined operationally, i.e., we know what a concept means when we know what operations must be performed in order to produce an instance of that concept. **1934** *Times* 26 June (Air Suppl.) p. xvii/2 It was decided, therefore, shortly after the War ended that the third Service, the Royal Air Force, should be operationally responsible for the units..engaged in the air defence of the country. **1948** *Jrnl. Abnormal Psychol.* XLIII. 143/1 There remains the evanescent residual category of 'personality', at once too broad to be operationally useful.. and too ubiquitous to be neglected. **1951** [see *operationist *a.* and *sb.*]. **1965** J. D. North *Measure of Universe* viii. 152 The belief that any theory must be provided with a 'sound conceptual foundation' is..often closely allied to the doctrine of operationalism, coinciding with that variant form according to which *every* concept must be explicitly 'operationally defined', even if only hypothetically so. **1972** *Science* 5 May 545/3 It is operationally impossible to distinguish between selection at a single locus and selection for closely linked genes. **1975** *Daily Tel.* 30 May 2/3 Operationally, the occasion on which the drop of a complete parachute brigade would be required and feasible has for long seemed remote.

operationism (ǫpĕrēi·ʃəniz'm). [f. Operation + -ism.] = *operationalism.

1935 *Psychol. Rev.* XLII. 517 The principles of operationism provide a procedure by which the concepts of psychology can be cast in rigorous form. **1942** D. D. Runes *Dict. Philos.* 219/2 Operationism makes explicit the distinction between *formal* and *empirical* sentences. **1970** *Jrnl. Gen. Psychol.* LXXXII. 113 About 25 years ago, during the period that I may call rampant operationism, a great many psychologists were misled into believing that here at last..was the long-needed panacea for guaranteeing useful scientific terms. **1975** *New Left Rev.* Nov.–Dec. 53 Bachelard's concept of the transitive dimension of science is flawed by operationism and an unrelenting hostility to the role of the imagination in science.

Hence **opera·tionist** *a.* and *sb.* = *operationalist *sb.* and *a.*

1950 *Brit. Jrnl. Psychol.* XL. 112 An unwarranted assumption is..that a person who is frustrated in a behaviourist or operationist sense, necessarily feels frustrated. **1951** *Mind* LX. 46 The operationist school in psychology. *Ibid.* 53 All operationists are rationalistic in the sense that they maintain that unless scientists define their terms operationally they will not uncover Nature's secrets. **1956** E. H. Hutten *Lang. Mod. Physics* ii. 62 The operationist theory of meaning..is not acceptable.

operative, *a.* and *sb.* Add: **A.** *adj.* **1. a.** *spec.* in legal use, applied to those words in a document which express the intention to effect the transaction concerned.

1792 [in Dict.]. **1872** in J. Russell *Rep. Cases High Court of Chancery* V. 344 If the operative part of a deed be doubtfully expressed, there the recital may safely be referred to as a key to the intention of the parties; but where the operative part of the deed uses language which admits of no doubt, it cannot be controlled by the recital. **1925** G. C. Cheshire *Mod. Law Real Property* 601 We will now turn to the operative words of the conveyance. **1951** *Times* 27 Nov. 7/4 Something to prevent that should be put into the operative part of the treaty.

b. Of words, sentences, etc.: containing the main point or key, essential to the meaning of the whole.

1926 *Sat. Rev.* 3 July 12/1 Every English sentence has an operative word. **1954** Koestler *Invis. Writing* ii. 28 The tendency of the novel had to be 'operative', that is, didactic; each work of art must convey a social message. **1963** N. Marsh *Dead Water* (1964) ii. 43 'It was nice getting your occasional letters,' Patrick said, presently. Operative word "occasional".' **1973** O. Lancaster *Littlehampton Bequest* 24 He was known..to have trailed a pike in the Low Countries when there were those..who loudly proclaimed that in his case 'trailed' was the operative word.

c. Of political ideas or principles: (*a*) capable of being put into effect; likely to be beneficial; (*b*) (see quot. 1954).

1938 H. G. Wells *Brothers* iii. 46 Are you lot over there really giving it an operative form? That's one of my phrases, brother—*operative form*... *Competent receiver* and *operative form*; two phrases for two problems that Socialism and Communism ought to have tackled forty years ago. **1954** Koestler *Invis. Writing* xx. 224 It is called the 'operative principle'. It means that you cannot write about the strategy of Communism without having worked in a factory, or Party cell, or underground organisation.

d. In weakened sense (without reference to specific activity or production): significant, important.

1955 J. L. Austin *How to do Things with Words* (1962) i. 7 But 'operative'..is often used nowadays to mean little more than 'important'. **1977** 'D. Cory' *Bennett* ii. 74 After all, she *didn't* bring the boy on that operative Saturday.

B. *sb.* **3. b.** A detective or agent employed by a detective agency; a secret-service agent.

1905 *N.Y. Press* 23 Oct. 6/4 The word 'detective' became so offensive..that it was dropped by some successful [detective] agencies. The word chosen by the Pinkertons to take its place was 'operative'. **1930** *Sat. Even. Post* 26 July 142/2 Riding on the train with him was another operative who had spent that day following Castagara. **1934** A. Christie *Murder on Orient Express* ii. ix. 137 That's not to say he'll remember me from a crowd of other operatives. **1937** *N.Y. Times* 22 Dec. 22/5 *Operative*, a spy employed by an agency. Usually has a secret designation. An operative may be a hooked man or a professional spy. **1954** W. Tucker *Wild Talent* (1955) xiv. 184 Paul wondered if this new woman in the adjoining apartment would be a plant... Slater might be playing it doubly safe and ringing in another operative on him. **1966** J. Porter *Sour Cream* iii. 36, I thought..you might just care to make your will. I advise all my operatives to do it. **1977** J. Crosby *Company of Friends* viii. 56 Sascha looked at her, pierced with reluctant admiration. What an operative!

operator. Add: **2.** *spec.* A secret-service agent. Cf. *operative *sb.* 3b.

1966 J. Porter *Sour Cream* iii. 36, I'm sure you won't find any snags. She's a most experienced operator. *Ibid.* viii. 101 I'll know by mid-morning if Feodorov's side of the operation has been successful... He's a very experienced operator. **1977** S. Coulter *Soyuz Affair* vii. 73 He keeps a close watch on his things... He's a trained operator.

4. (Later examples.) Also now freq. with a stronger implication of speculativeness or shrewdness; one who acts in an underhand manner.

1895 *Daily News* 30 Dec. 7/5 The market declined early on large receipts, but eventually improved, due to local operators covering. **1951** [see *hole *sb.* 11]. **1955** D. W. Maurer in *Publ. Amer. Dial. Soc.* XXIV. 35 Also, unlike many if not most other types of underworld operator, he tends to take his woman on the road with him. **1959** N. Mailer *Advts. for Myself* (1961) 399 He spent years hobnobbing with gentlemanly shits and half-ass operators. **1964** A. W. Gouldner in I. L. Horowitz *New Sociol.* 209 The whole world may be seen as one of marks and operators. **1970** *New Yorker* 6 June 132/3 He is what we call nowadays an 'operator', and completely unscrupulous and unashamed. **1971** D. Potter *Brit. Eliz. Stamps* i. 15 Stamps were solemnly discussed in the financial columns,..and get-rich-quick operators joined in the *mêlée*. **1974** *Times* 4 Feb. 12/6 One almost expects him to say, with J. K. Galbraith, that modesty is a much overrated virtue, but he is far too smooth an operator to be trapped into such an admission.

5. a. (Earlier and later examples.) *spec.* One who works at the switchboard of a telephone exchange (now the usual sense). Also in *Comb.*

1847 *Commerc. Rev. of South & West* Nov. 138 Its receipt [was] acknowledged by the Montreal operator in 30 minutes. **1858** E. E. Hale *If, Yes, & Perhaps* (1868) 119 It is not the business simply of 'operators' in telegraphic dens to know this Morse alphabet. **1865** *Harper's Mag.* July 169/2 Here they gathered a new force..., a telegraph operator, and workmen. **1884** *List of Subscribers* (London & Globe Telephone Co.) 2 The Special Telephone Exchange Switchboard is so constructed, that the operators ..do not overhear the conversation between Subscribers. **1921** [see *dial *v.* 4]. **1927** Haldane & Huxley *Animal Biol.* vi. 140 The human cerebrum contains more than a thousand million nerve-cells each connected by fibres with scores or hundreds of others,..so we can get some idea of its complexity by imagining a telephone exchange in which the whole human race were acting as operators. **1972** B. F. Conners *Don't embarrass Bureau* (1973) ii. 201 Operator, I'd like to call person to person to Officer Dolan. **1976** *Norwich Mercury* 10 Dec. 10/2 Bookings for operator-connected international telephone calls over Christmas will be accepted by the Post Office from Monday. **1977** G. Markstein *Chance Awakening* xix. 55 He dialled 100. He gave the operator the number. 'What seems to be the trouble?' asked the operator.

b. *U.S.* One who is licensed to drive a motor vehicle.

1967 *Boston Sunday Herald* 7 May iii. 1/1 Several witnesses to the accident have supplied police pieces of information, which, when put together, revealed the involvement of this other car—and male operator, who stopped at this point, without knowing his car—and he —had actually been involved in the tragedy. **1972** *N.Y. Law Jrnl.* 24 Oct. 18/6 At the time involved defendant was a resident of the State of New Jersey and the holder of a New Jersey operator's license. It also appears from a Motor Vehicle Bureau exhibit..that defendant's New York license had been cancelled.

6. (Earlier examples.) Also in wider use, of companies, corporations, etc.

1838 *Niles' Reg.* 13 Oct. 112/2 Our trade..is brought nearly to a stand again, by a collision between the dealers, operators and boatmen, as regards the price of freight. **1851** C. Cist *Sk. Cincinnati in 1851* 170 The largest operators in this line [manufacturing alcohol], are Lowell Fletcher & Co. **1857** *Harper's Mag.* Sept. 459/1 The leases of the operators usually covered a 'run' upon the out-crop ..of from fifty to seventy yards. **1875** *Chicago Tribune* 30 Sept. 2/4 The operators on the Pan-Handle Railroad have been paying 2½ cents per bushel for mining over 1¼ inch screen. **1953** *Manch. Guardian Weekly* 19 Nov. 9 The State, through a new sort of BBC is to own the new system and will hire it out to commercial 'operators'. **1962** *Listener* 8 Mar. 401/1 Though they [*sc.* the nationalized industries] were to be regarded partly as commercial operators, partly as public services, the relation between these two functions was not defined in the legislation that set them up. **1972** *Lebende Sprachen* XVII. 134/1 The official body representing the majority of independent operators is the British Independent Air Transport Association. **1977** *Times* 23 Mar. 14/3 Exporters have turned increasingly to road haulage for cargoes... This traffic..has attracted a large number of operators who are not equipped for the job. **1977** *Offshore Engineer* May 42/1 The results of the four wells are being closely studied by the operator Elf Aquitaine and partners ETAP and STEG.

7. (Further examples.) Also, a sign or symbol which effects other types of operation, as logical, phonological, syntactic, etc.

1925 Bryant & Correll *Alternating-Current Circuits* iii. 74 The operator, $j = \sqrt{(-1)}$, turns the vector through 90 degrees in a counterclockwise direction each time it is used. **1936** *Jrnl. Symbolic Logic* I. 60 The formal system.. employs material implication, propositional negation, universal quantification, and operators analogous to the combinatory operators *I*, *B*, and *C*. **1937** A. Smeaton tr. *Carnap's Logical Syntax of Lang.* i. §6. 21 The expressions which occur at the beginning of the sentences above ..are called the *unlimited* universal operator, the *unlimited* existential operator, the *limited* universal operator, and the *limited* existential operator respectively. **1952** *Eng. & Gmc. Stud.* IV. 12 We may regard Modern English *stone* as the result of operating with an operator that I shall write (AS. ā > MnE. [ou]) on Anglo-Saxon *stān*. **1952** S. C. Kleene *Introd. Metamath.* iv. 73 Let us call an expression of one of these ten forms an operator. In particular, ⊃, &, ∨, ¬ are *propositional connectives*, and operators of the forms ∀x and ∃x are *quantifiers*..; these six are *logical operators*. **1957** L. Fox *Numerical Solution Two-Point Boundary Probl.* ii. 8 In the theory of finite differences we carry this farther, introducing first the operator *E*, the effect of change of position, and defined by the equation $y(x + ph) = E^p y(x)$. **1964** E. Bach *Introd. Transformational Gram.* v. 113 The concatenation operator +. **1965** Phillips & Williams *Inorg. Chem.* I. i. 12 Wave mechanics sets up, as postulates, equations of the type $H\psi = X\psi$, where ψ is the wave function, and *X* some observed property such as the energy, momentum, etc. *H* is a mathematical instruction (e.g. differentiate with respect to *x*) or set of instructions..called an operator. **1966** A. Koutsoudas *Writing Transformational Gram.* i. 6 The rules of a transformational grammar consist of three types of symbols: (1) vocabulary symbols, (2) operators, and (3) abbreviators. Vocabulary symbols are symbols used to represent syntactic classes and other linguistic units; operators, as their name indicates, symbolize certain operations; and abbreviators are devices used to conflate the listing of rules. **1968** Corlett & Tinsley *Pract. Programming* ii. 14 Numbers and variables may be combined by the arithmetic operators + − × ÷ / ↑ to form an arithmetic expression. **1969** V. J. Calderbank *Course on Programming in FORTRAN IV* iii. 28 Another very common form of logical expression..is the relational expression. This has the general form $e_1\,r\,e_2$ where e_1 and e_2 are arithmetic expressions..being compared by one of the following relational operators, *r*: .EQ. Equal to (=); .LT. Less than (<); [etc.]. **1969** P. A. M. Seuren *Operators & Nucleus* iv. 116 Our hypothesis of operators as a deep structure category in grammar thus has a philosophical pedigree leading back to both the theory of quantifiers and the logic of modalities. **1976** A. R. Lacey *Dict. Philos.* 148 A logical operator is any expression whose function is to affect in a specific way the logical properties (e.g. the entailments) of an expression or expressions to which it is attached, e.g. 'and' operates on two propositions by joining them into a whole, which has entailments neither of them has separately.

8. a. In 'Basic English', an article, particle, preposition, etc., or one of certain words used as substitutes for verbs; a 'superverb'.

1929 C. K. Ogden in *Psyche* IX. iii. 1 The number of necessary names is 400, of qualifiers (adjectives) 100, of operators, particles, etc., 100. **1930** —— *Basic English* iii. 60 In conversation, the operators are frequently shortened to more convenient forms. Thus *I will* becomes *I'll*. **1946** H. Jacob *On Choice of Common Lang.* iii. 104 Three analogical extras, *say, see,* and *send*..provide a useful link between the operators and the verb-system proper. **1966** M. Pei *Gloss. Ling. Terminol.* 188 *Operator*, one of the verbal forms, prepositions, articles, etc. (about one hundred in number) in Basic English.

b. *Linguistics.* = *form-word* s.v. *form *sb.* 22, *function word* s.v. *function *sb.* 3c.

1938 B. L. Whorf *Lang. Thought & Reality* (1956) 128 Predication..operators (words specialized for predication, otherwise lexical meaning blank ('be, become, cause, do') or vague ('make, turn, get,' etc.)). auxiliary verbs. **1957** S. Potter *Mod. Linguistics* vii. 143 Operators are.. forms like articles, prepositions, conjunctions and conjunction adverbs..which perform syntactic functions. **1967** [see *functor 2].

9. *Biol.* [tr. F. *opérateur* (Jacob & Monod 1959, in *Compt. Rend.* CCXLIX. 1284).] A segment of chromosomal DNA which is thought to control the activity of the structural gene(s) of an operon, protein synthesis occurring when it is uncombined with a repressor (or is absent altogether).

1961 *Cold Spring Harbor Symp. Quant. Biol.* XXVI. 194/1 The synthesis of messenger RNA is supposed to be a sequential and oriented process which can be initiated only at certain regions, or operators, on the DNA strands. **1969** *New Scientist* 28 Aug. 416/1 In the presence of substrate, the repressor is altered, so that it can no longer bind to the operator, and enzymes are synthesized freely. **1971** D. J. Cove *Genetics* xi. 165 A mutation in the operator gene prevents the recognition of the structural genes by the repressor. **1973** *Nature* 16 Nov. 133/1 The sites on DNA to which repressors bind are called operators.

operculate, *a.* (Further examples of botanical use.)

1879 W. Phillips tr. E. Boudier in *Grevillea* VIII. 46, I would call the first section [of Discomycetes] by the name of *Operculate Discomycetes*, or simply *Operculæ*, because in this section the opening of the asci takes place by the elevation of a little lid at its summit. **1913** *Trans. Brit. Mycol. Soc.* IV. 402 The asci in the operculate series are generally larger. **1929** *Ibid.* XIV. 275 The material for this study has been gathered in great part from the accounts which have already been given by other investigators, chiefly on operculate species. **1971** P. H. B. Talbot *Princ. Fungal Taxon.* xi. 160 The dehiscent types [of ascus], both operculate and inoperculate, become turgid and discharge their ascospores forcibly.

operculum. 2. b. The lid of the ascus or sporangium of certain fungi.

1879 W. Phillips tr. E. Boudier in *Grevillea* VIII. 46 After the examination of a considerable number of Discomycetes, I am able to call the attention of mycologists to the necessity of separating this family into two very natural sections, according as to whether the mode of dehiscence is with or without an operculum. **1887, 1888** [see *astropyle]. **1971** P. H. B. Talbot *Princ. Fungal Taxon.* xi. 160 Operculate asci have an operculum at the apex: a hinged lid-like opening.

operette (ǫpěre·t). Also ‖ **opérette** (ǫpere·t). English (with accent, French) form of Operetta. Also *transf.*

1890 E. Dowson *Let.* 19 May (1967) 149 Playing snatches of opérette—Gilbert & Sullivan, Sultan of Mocha etc. **1928** *Observer* 15 Apr. 21/2 Few operettes launched on the public nowadays can compare with the delicious music of Pongrác Kacsoh. **1935** *Discovery* July 211/2 The Estonians have some charming 'operettes', too, which..without being heavy have much more in them than our average musical comedies. **1938** N. Coward (*title*) Operette. **1961** *Times* 7 Mar. 8/4 On the borderline between *opera bouffe* and *opérette*. **1977** *Listener* 30 June 867/1 Joan Aiken's *Lost Movement*, an entirely enjoyable story, offers..a pretty and melodramatic operette.

operettist (ǫpěre·tist). [f. Operetta + -ist.] A writer or composer of operettas.

1922 *Blackw. Mag.* June 717/2 There are a mass of Hungarian operettists.

operon (ǫ·pěrǫn). *Biol.* [ad. F. *opéron* (F. Jacob et al. 1960, in *Compt. Rend.* CCL. 1729), f. *opér-* (in *opérer* to effect, work, *opération* Operation, etc.): see *-on¹.] A unit of co-ordinated gene activity which is believed to account for inducible and repressible enzymes in bacteria and hence for the regulation of protein synthesis, and is usu. conceived as a linear sequence of genetic material comprising an operator, a promoter, and one or more structural genes.

1961 Jacob & Monod in *Jrnl. Molecular Biol.* III. 344 This genetic unit of co-ordinate expression we shall call the 'operon'. **1969** *Nature* 18 Jan. 219/2 Essentially, Grodzicker and Zipser's experiment was to take strongly polar nonsense mutants of the lactose operon of *E. coli*. **1969** A. M. Campbell *Episomes* ix. 116 The genes of one operon are all transcribed onto the same messenger molecule. **1971** J. Z. Young *Introd. Study Man* iii. 59 Each operon is controlled by one or more regulator genes. **1973** B. J. Williams *Evolution & Human Origins* vi. 91/1 The model of genetic control receiving the most attention today is that of the operon.

operose, *a.* **1.** (Earlier and later examples.)

1678 Cudworth *Intell. Syst.* 884 An Operose, Cumbersom, and Moliminous Business. **1855** Geo. Eliot *Let.* 25 June (1954) II. 206 Such sentences..make a style seem operose and unwieldy. **1959** *New Scientist* 19 Nov. 983 Operose and scholarly collected editions.

ope-tide. For †*Obs.* read *arch.* and add later example.

1911 Beerbohm *Zuleika D.* iii. 28 Her soul was as a flower in its opetide. She was in love.

Ophelian (ǫfī·liăn), *a.* [f. the name of the heroine of Shakespeare's play *Hamlet*: see -an.] Resembling or characteristic of Ophelia.

1903 'Marjoribanks' *Fluff-Hunters* 101 Some time ago I felt so moody and sad that I sought out a pretty pool, with an Ophelian resolution. **1928** *Observer* 18 Mar. 15/3 One can hardly fail to hear this play's Shakespearean echoes... The Gaoler's Daughter is of pure Ophelian stock. **1929** D. H. Lawrence *Pansies* 6 Its transience, its breath, its maybe mephistophelian, maybe palely ophelian face. **1962** *John o' London's* 25 Jan. 91/4 Miss Maclaine would be well advised to drop her Ophelian aspirations.

ophelimity (ǫfěli·miti). *Econ.* [f. F. *ophélimité* (also used), ad. Gr. ὠφέλιμος useful, serviceable.] (See quots.)

1896 V. Pareto *Cours d'Économie Politique* I. 3 Nous emploierons le terme *ophélimité*, du grec ὠφέλιμος, pour exprimer le rapport de convenance qui fait qu'une chose satisfait un besoin ou un désir, légitime ou non.] **1896** *Political Sci. Q.* XI. 750 The term utility, for example, has its ambiguities; and Professor Pareto substitutes the word ophélimité, meaning capacity to satisfy any want, whether rational or irrational. **1920** A. C. Pigou *Econ. of Welfare* ii. 23 Several writers have endeavoured to get rid of the confusion..by substituting for 'utility'..some other term such, for example, as Professor Pareto's 'ophelimity'. **1935** Bongiorno & Livingston tr. *Pareto's Mind & Society* I. i. 29 In pure economics my hypothesis of 'ophelimity'..remains experimental so long as inferences from it are held subject to verification on the facts. **1966** D. Mirfin tr. *Pareto's Sociol. Writings* 99 We shall employ the term *ophelimity*..to designate the relationship of convenience which makes a thing satisfy a need or desire, whether legitimate or not... *Utility* will be required for use in its ordinary accepted sense as the property which makes a thing favourable to the development and well-being of an individual, a community or the whole human species.

opherion (ofī·riǫn). Used by T. S. Eliot, perhaps in error for Orpharion.

a **1922** T. S. Eliot *Waste Land Drafts* (1971) 99 (*title*) Song. For the opherion.

ophidian, *a.* **1.** (Later example.)

1939 T. S. Eliot *Family Reunion* I. ii. 56 The dead stone is seen to be batrachian, The aphyllous branch ophidian.

ophiolite. Add: **b.** *Geol.* Any of a group of basic and ultrabasic igneous rocks, including serpentinite and serpentinized peridotite, gabbro, and diabase, which occur associated with pillow lava and radiolarian chert in a characteristic pattern of layers in the Alps and certain other regions and are thought to have been formed as a result of the submarine eruption of oceanic crustal and upper mantle material; so **ophiolite suite** or **association**, the assemblage of ophiolites, pillow lava, and radiolarian chert.

This use originated in Ger. with the recognition of the association by G. Steinmann (*Ber. d. naturforsch. Ges. zu Freiburg i. Br.* (1906) XVI. 18–49).

1937 A. L. Du Toit *Our Wandering Continents* viii. 168 There is a wide development in Morocco, Spain and the Alpine region of the so-called greenstones or 'ophiolites'. **1963** *Spec. Papers Geol. Soc. Amer.* No. 73. 204 Throughout the Alpine–Himalayan system extensive lavas occur in close association with coarse mafic and ultramafic rocks... This assemblage is the ophiolite suite of Alpine writers. It seems to represent long pre-orogenic extrusion of ultramafic magma. **1970** J. C. Maxwell in Johnson & Smith *Megatectonics of Continents & Oceans* viii. 181 The assemblage of igneous rocks which Europeans call ophiolites seems incompatible with the dominantly sedimentary melange in which it typically occurs. **1971** I. G. Gass et al. *Understanding Earth* iv. 132/2 The former is dominantly volcanic (a thick sequence of ophiolites with subsidiary cherts, carbonates and pelites). **1972** *Sci. Amer.* May 65/3 Within the Alpine–Himalayan mountain belt are narrow zones characterized by a distinctive assemblage of rocks, known as the ophiolite suite. *Ibid.* 66/3 The Ural and Appalachian-Caledonian mountain belts..have narrow zones where ophiolites are found... This implies that the Urals, for example, were created by the collision of two continental masses and that the ophiolites were generated by sea-floor spreading at a ridge axis before the continents were brought together. **1977** A. Hallam *Planet Earth* 166 The whole assemblage of sea-floor basalts and associated upper-mantle peridotites is known as the ophiolite association.

ophiolitic *a.* (Examples.)

1909 H. B. C. & W. J. Sollas tr. *Suess's Face of Earth* IV. v. 153 The sheet of the green-rocks (Ophiolitic, Rhaetic, Vindelician sheet; Steinmann). **1911** *Geol. Mag.* Decade V. VIII. 243 The constant association of dialase (spilite), serpentine, and gabbro in the northern Alps with radiolarian cherts has led Steinmann..to regard these 'ophiolitic eruptives' as the typical volcanic rocks of abysmal depressions. **1963** D. W. & E. E. Humphries tr. *Termier's Erosion & Sedimentation* xvii. 341 Radiolarites associated with the rocks of the ophiolitic suite (of geosynclines), demonstrate the subordinate role of living organisms in comparison to that played by transported material. **1971** *Nature* 3 Sept. 46/2 One of the most significant drilling results..was the discovery of a Lower Cretaceous and Upper Jurassic pelagic sequence above an ophiolitic oceanic basement, almost identical to the coeval Alpine sequences.

ophiology. (Earlier example.)

1817 *Blackw. Mag.* May 187/1 Reserving the history of the serpent tribes for the article *Ophiology*.

ophiomorphic, *a.* (Example.)

1909 Ld. Balcarres *Ital. Sculpture* i. 16 In the Celtic art of the North..these ophiomorphic meanderings of line bewilder the eye by their complexity.

ophitic, *a.*¹ Add: By some writers restricted to textures in which augite predominates and the feldspar laths do not in general touch each other. (Earlier and later examples.)

1875 G. H. Kinahan in *Proc. R. Irish Acad.* II. 118 The passage-rock..may be called ophitic hornblende-rock or amphibolic-ophyte, according to the mineral predominating. **1881** *Jrnl. Chem. Soc.* XL. 697 The ophitic rocks of the Pyrenees are characterised by the development of microliths of triclinic felspar embedded in elongated layers of pyroxene. *Ibid.*, The labradoric ophite.. showed the passage from trachytoidal to ophitic structure. **1970** K. C. Jackson *Textbk. Lithology* v. 300 Diabasic texture... Composed of tabular plagioclase with smaller interstitial granular pyroxene crystals. Ophitic texture... Small tabular plagioclase embedded in larger anhedral pyroxene crystals. **1973** G. J. H. McCall *Meteorites* xiii. 174 (*heading*) Eucrite showing ophitic intergrowth of narrow plagioclase laths.

Hence **ophi·tically** *adv.*, with an ophitic texture.

1908 *Q. Jrnl. Geol. Soc.* LXIV. 491 [The biotite] is moulded ophitically on these minerals. **1971** *Nature* 3 Dec. 265/1 The plagioclase-phyric basalts contain abundant pyroxene..,most of which has crystallized ophitically with the small calcic plagioclase laths.

ophthalmic, *a.* and *sb.* Add: **A.** *adj.* **4.** *ophthalmic acid*, a tripeptide found in the lenses of various mammals (see quot. 1958).

1956 S. G. Waley in *Biochem. Jrnl.* LXIV. 715/1 The work..deals only with acidic peptides. One such peptide (for which the name ophthalmic acid is proposed) is particularly abundant. **1958** *Ibid.* LXVIII. 192/1 The electrophoretic mobility of ophthalmic acid, a tripeptide isolated from calf lens, shows that the glutamic acid residue is γ-linked, and hence that ophthalmic acid is γ-glutamyl-α-amino-*n*-butyrylglycine. **1962** R. van Heyningen in H. Davson *Eye* I. v. 237 Although liver contains but little ophthalmic acid, extracts of liver will synthesize ophthalmic acid from the amino-acids. **1970** J. F. R. Kuck in C. N. Graymore *Biochem. Eye* iii. 206 An interesting feature of the relationship between glutathione and ophthalmic acid is that the former is a specific co-enzyme for glyoxalase while ophthalmic acid is a potent inhibitor of this same enzyme.

ophthalmologic, *a.* Delete *rare* and add: Chiefly *U.S.* (Examples.)

1901 [see *homatropine]. **1972** Aronson & Elliott *Ocular Inflammation* xiv. 314/2 Ophthalmologic examination was requested because she complained of intermittent blurring of vision in her right eye.

ophthalmophorous, *a.* (Example.)

1896 *Natural Sci.* VIII. 340 [G. A. Boulenger] has, too, for the first time, utilized the development or want of an ophthalmophorous shelf to the second suborbital as a family character.

-opia, formative element [a. Gr. -ωπία, f. ὤψ, ὠπ- eye, face: see -ia¹.] of terms denoting visual disorders and abnormalities, as Amblyopia, Myopia, Polyopia. Occas. anglicized as **-opy** [cf. -y³].

opiate, *a.* and *sb.* Add: **B.** *sb.* **1. a.** (Further example.)

1948 *Arch. Internal Med.* LXXXII. 387 These men were..again offered the choice of either an opiate or methadon.

b. *fig.* (Further examples.) See also *opium sb.* 1 b.

1927 C. Connolly *Let.* 4 Mar. in *Romantic Friendship* (1975) 276, I find covering ground rather an opiate. **1942** *R.A.F. Jrnl.* 30 May 24 There's no more beautiful feeling than the opiate of unguarded sleep on a sunny boat deck. **1960** D. Eisenhower in W. Safire *New Lang. Politics* (1968) 309/1 Hundreds of millions behind the Iron Curtain are daily drilled in the slogan: 'There is no God, and religion is an opiate'. But not all the people within the Soviet accept this fallacy; and some day they will educate their rulers, or change them. **1968** tr. M. de Unamuno in *Bartlett's Familiar Quotations* (ed. 14) 870/1 One of those leaders of what they call the social revolution has said that religion is the opiate of the people. **1975** *Gen. Systems* XX. 109/1 The European revolutionary found religion an opiate to be abolished. **1976** C. Dexter *Last seen Wearing* xxi. 158 The heady, heavy opiate of the gambling game.

2. Any drug having similar addictive effects to those of the opium alkaloids morphine and cocaine. Freq. *attrib.*

1954 *Ann. Rev. Med.* V. 318 Another interesting field of usefulness for nalorphine is in the detection of opiate addiction. Administration of 5 mg. nalorphine to individuals addicted to morphine or methadon induces abstinence symptoms within 15 min. **1960** *Federal Register* (U.S.) 5 Aug. 7351/2 The word 'opiate'..shall mean any drug..found by the Secretary or his delegate..to have an addiction-forming or addiction-sustaining liability similar to morphine or cocaine. **1961** *Jrnl. Pharmacol. & Exper. Therap.* CXXXIII. 371/1 Codeine accounts for at least 80% of the natural opiates sold on prescription. **1970** *Nature* 14 Apr. 323/1 Methadone itself is an addictive opiate. **1974** M. C. Gerald *Pharmacol.* xiii. 242 In 1973 evidence was presented demonstrating the existence of an opiate receptor in the brain. *Ibid.* 251 Although acute opiate withdrawal is thought to be a very dangerous and often fatal process, there is no evidence to support this belief.

opihi (opī·hi). [Hawaiian.] A limpet belonging to the genus *Helcioniscus*, or its shallow shell. Also *attrib.*

1915 W. A. Bryan *Nat. Hist. Hawaii* xxxv. 466 To the old-world limpets belongs the opihi of the natives. It is a favorite food with the Hawaiians... This knee-cap or umbrella shell is roughly though evenly ribbed without and pearly white within. **1976** *National Observer* (U.S.) 10 Apr. 17/2 Legend has it that when the highly prized Opihi Shell is given to one's sweetheart, the love between

them becomes eternal. Thus, wearing the Opihi around the neck assures that the shell—and their love—is preserved forevermore!

opinion, *sb.* Add: **1. b.** *public opinion* (earlier and later examples); also *attrib.*, as *public opinion investigation, poll* [*poll *sb.*¹ 7 d], *polling, survey* (see *opinion poll, survey*, sense 9 below), etc.

1735 Bolingbroke *On Parties* (ed. 2) p. xxxi, Let them stand, or fall in the publick Opinion, according to their Merit. **1751** tr. *Rousseau's Discourse Arts & Sci.* 36 They only hate all publick opinions. **1763** Chesterfield *Duke of Newcastle* in *Lett.* (1845) II. 463 The public opinion put him below his level: for though he had no..eminent talents, he had a most indefatigable industry, a perseverance. **1769** Burke *Let.* 30 July (1844) I. 181 We must strengthen the hands of the minority within doors, by the accession of the public opinion, strongly declared to the court. **1939** G. Gallup (*title*) Public opinion in a democracy. **1952** W. J. H. Sprott *Social Psychol.* 87 It is difficult to put into exact terms what is meant by 'public opinion'... We might, of course, define 'public opinion' on any particular issue..in terms of the people who actually have an 'opinion' on that issue. **1961** L. van der Post *Heart of Hunter* I. v. 88 Tom refused to let public opinion create a sense of shame in him. **1964** Gould & Kolb *Dict. Social Sci.* 477/2 *Public opinion* is a nebulous concept; it is not the simple aggregate of the opinions of the members of a public, but depends on the society's power structure, the mass media, channels of influence, etc.

attrib.

1936 L. Denning *Coming Amer. Fascism* 299 A score of great corporations can raise ten million dollars for anti-social purposes of price-fixing or public-opinion manipulation. **1937** *Sociometry* I. i. ii. 155 (*title*) Public opinion polls. **1939** G. Gallup *Public Opinion in Democracy* 3 The development during the last few years of the public-opinion survey or unofficial poll has raised..a host of new and far-reaching questions. **1940** Graves & Hodge *Long Weekend* xxiii. 401 From the United States came public opinion investigation. *Ibid.*, Public opinion investigation was first started on a large and permanent scale by the American Dr. Gallup. **1941** J. S. Huxley *Uniqueness of Man* xi. 231 The modern scientific public opinion poll, indeed, is developing such uncanny accuracy that it is infringing upon practical politics. **1944** G. Gallup *Guide to Public Opinion Polls* lxix. 91 The chief function of public opinion polls, and their chief value, is to report the *trend* of opinion. **1958** *Listener* 20 Nov. 813/1 Public-opinion polls... Public-opinion surveys. **1963** *Rep. Comm. Inquiry Decimal Currency* 5, in *Parl. Papers* 1962–3 (Cmnd. 2145) XI. 195 We considered going to further than this by commissioning a public opinion survey. **1964** *Ann. Reg. 1963* 5 His dominance of the parliamentary party was soon apparent in the major debates of the next five months, assisted, no doubt, by Labour's soaring lead in the public opinion polls. **1967** M. Argyle *Psychol. Interpersonal Behaviour* ix. 153 Public opinion polls are more closed questions, research surveys make more use of open-ended ones. **1972** *Jrnl. Social Psychol.* LXXXVII. 136 The sealing methodology employed was discussed and suggested as a possibly new public opinion polling device.

4. Phr. *a second* (or *another*) *opinion*, the opinion of a second medical adviser. Also in *transf.* and extended uses.

1885 C. M. Yonge *Nuttie's Father* II. xvii. 201 Dr. Brownlow became very grave over the injury. He said it was a surgical case, and he should like to have another opinion. **1924** J. Buchan *Three Hostages* xvi. 237 There's no cause to worry about Peter John... But if you want another opinion, why not get it? *Ibid.* 238, I think a second opinion would please Dr. Greenslade, for he too looked rather anxious. **1954** M. Sharp *Gipsy in Parlour* xxiv. 228 So my father and Miss Jones agreed... I didn't think my father would be quite so pleased to know of this second opinion, so to speak. **1966** J. B. Priestley *Salt is Leaving* xvi. 213, I have quite a competent doctor... He's never suggested a second opinion. **1970** P. Lovesey *Wobble to Death* v. 52 Herriot sought for words to influence the doctor... 'Perhaps—another opinion. Your colleague..may see the possibility of a faster recovery.' **1971** A. Price *Alamut Ambush* x. 124 So he..wanted a second opinion on what he had seen—that made sense. **1972** J. Wainwright *Night is Time to Die* 121 You're not here *as* a solicitor... Therefore, you're entitled to call some other solicitor... You might need advice—a second opinion.

9. *opinion-former, leader, maker; opinion-forming, -tapping* ppl. adjs.; *opinion-making* ppl. adj. and vbl. sb.; **opinion poll,** the assessment of the opinion of all, or of a section of, the general public by questioning a random or representative sample; hence *opinion polling, pollster*; (see also sense 1 b above); **opinion survey** = *opinion poll.

1906 G. W. E. Russell *Social Silhouettes* xiii. 90 A Journalist of this type once said to me, with all imaginable gravity, 'I should, I confess, resent any change which interfered with my position as chief opinion-former in the neighbourhood of'—Leeds or Plymouth, or whatever was the name of his town. **1962** *Times Lit. Suppl.* 24 Aug. 633/4 To hear well over half the electors of Britain talk, or read or listen to their favourite opinion-formers, one would conclude that it would be best..if no profits were made by any business. **1967** *Economist* 22 Apr. 338/1 The opinion-formers have begun their debunking. **1977** *Private Eye* 13 May 14/2 He has a fine independence of outlook and a contemptuous disregard for whatever is smart or fashionable among opinion-formers. **1959** *Encounter* Nov. 66 Literary parties..the opinion-forming fringe of the United States. **1969** *Guardian* 28 Aug. 11/6 One section of Oxfam has argued that the organisation should abandon direct aid and devote itself wholly to an opinion-forming rôle. **1974** *Broadcast* 9 Dec. 17/1 The advertising industry has few friends in the educational and opinion-forming strata. **1949** Lazarsfeld & Stanton *Communications Res.* ii. vii. 217 The opinion-leaders are not identical with the socially prominent people in the community. **1968** *Internat. Encycl. Soc. Sci.* III. 51 The hard-core noncommissioned officers constituted a cadre of 'opinion leaders' who supported the control structure. **1975** *Times* 25 Feb. 14/1 His audience of nearly 200 were predominantly..people whom the Wessex area office staff would describe only as opinion leaders. **1952** *Time* 27 Oct. 20/1 To intellectuals and other 'opinion makers', Eisenhower was infinitely preferable to the other two. **1957** *Economist* 28 Sept. 1005/1 The reactions of..professional opinion makers were more precise, but just as personal. **1975** S. Ranganathan in H. M. Patel et al. *Say not the Struggle Nought Availeth* 297 There are limits to growth as world opinion-makers are trying to explain. **1909** *Westm. Gaz.* 16 June 1/3 The Conference at the Foreign Office..exceeded expectations. The question that out-shadowed all others at this council of 'opinion-making power' was..Imperial Defence. **1956** C. W. Hills *Power Elite* xiii. 310 The means of opinion-making..have paralleled in range and efficiency the other institutions of greater scale. **1937** Opinion poll [implied in *public opinion poll*, sense 1 b above.] **1946** *Vogue* Aug. 2/2 Try a little opinion-poll for yourself. Ask a representative sample of Englishmen the following question: [etc.]. **1951** M. McLuhan *Mech. Bride* (1967) 46/2 Opinion polls function as educational rather than fact-finding agencies. **1965** *New Statesman* 30 Apr. 670/3 The substantial pro-Labour swing indicated by the opinion polls. **1971** *Guardian* 10 July 11/8 Mr Heath..interprets every unfavourable opinion poll as a clear signal that the public is one hundred per cent behind him. **1976** *Times* 27 Feb. 14/4 It is clear from opinion polls that the very large majority of people in Scotland wish to remain part of Britain. **1963** *Economist* 7 Dec. 1094/1 Opinion-polling is still a young..art in Spain. **1970** *Times* 2 June 1/2 The increase has no statistical significance when allowance is made for the tolerance limits of opinion polling. **1977** *Times* 18 Oct. 16/4 One of the weaknesses of much opinion-polling is that..it must exclude..extended questioning. **1951** M. McLuhan *Mech. Bride* (1967) 47/1 The cabdriver tends to be the opinion-pollster hero. **1970** *Times* 19 June 1/6 (*heading*) Opinion pollsters admit to wide margin of error. **1977** *News of World* 17 Apr. 1/1 Its existence is unknown to two out of every three women in the country, according to opinion pollsters. **1939** G. Gallup *Public Opinion in Democracy* 12 If elections themselves do not impose clôture on debate, is it likely that opinion surveys will? **1958** *New Statesman* 23 Aug. 213/3 The first would entail many interviews with people by the well-tried opinion-survey methods. **1948** J. Towster *Political Power in U.S.S.R.* vii. 153 The Party conference..is primarily..an opinion-tapping and effort-mobilizing agency.

opinionation (opi:nyŏnēi·ʃən). [f. Opinion *sb.* + -ation.] The state or condition of persisting in holding a dogmatic opinion or opinions.

1925 *Inner Life* x. 184 Self-expression in the immature stages may become mere self-opinionation, with an unwillingness to heed advice or guidance. **1957** R. Hoggart *Uses of Literacy* vi. 167 It can be said, with some justice, that this is an age of 'opinionation', that though few people take the trouble thoroughly to understand any problem, a great many assume that their opinions on almost every general issue will have weight. **1960** *Times Lit. Suppl.* 27 May 339/1 What is expected of him is to..advise and comment on public affairs—..to add his mite to the flood of opinionation which is slopping over the world, obscuring the inner world of values.

opinionator. Delete †*Obs. rare* and add later examples.

1930 *New Statesman* 8 Nov. 147/1, I can only regret that Mr. West devotes so much of his space to Mr. Wells the opinionator instead of to Mr. Wells the artist. **1960** *Times Lit. Suppl.* 27 May 339/1 The opinionators..are quite ready to sign petitions and organize meetings, and to these the artist is supposed to contribute his ration of right-feeling and right-thinking. **1973** *Ibid.* 16 Mar. 294/2 A pleasure that ought never to be tasted by the professional opinionator. **1974** *Ibid.* 7 June 610/2 The age demands superstars, intellectual acrobatics, personality cults, instant opinionators.

opinionnaire (opi:nyŏnēə·ɹ). Also **opinionaire.** [f. Opinion *sb.* + *-aire*, after *questionnaire.] A series of questions designed to gauge (public) opinion on a specific issue; a questionnaire.

A word of doubtful usefulness.—Ed.

1949 R. K. Merton *Social Theory* ii. 207 Interview techniques in all their numerous variety..questionnaires, opinionnaires and attitude tests. **1955** M. Reifer *Dict. New Words* 148 *Opinionaire*, a questionnaire to be filled in by various persons, for polling public opinion. **1964** P. Meadows in I. L. Horowitz *New Sociol.* 450 Researchers, equipped with pencils, pads, and opinionnaires, sought to throw on the screen of national attention the bright beams of percentages and averages of mass opinions. **1973** *Jrnl. Genetic Psychology* Mar. 56 Compliance was defined as correspondence between the child's report on an opinionnaire and the parent's report. **1975** *Bull. Canad. Assoc. Univ. Teachers* Feb. 7/3 The report..admonishes its reader not to rely on student opinionnaires in the evaluation of university teaching.

opioid (ōu·pioid). [f. Opi(um *sb.* + -oid.] = *opiate *sb.* 2.

1957 *Pharmaceutical Jrnl.* CLXXIX. 321/1 Acheson has suggested that the morphinans and other synthetic morphine substitutes should be called opioids. **1972** R. D. Dripps et al. *Introd. Anesthesia* (ed. 4) xxv. 347 Opioids administered intravenously..may be followed by hypotension. **1974** M. C. Gerald *Pharmacol.* xiii. 237 Synthetic narcotic agents or opioids, which bear only highly subtle similarities to the structure of morphine. **1974** *Nature* 20 Dec. 708/2 Antiserum obtained after immunisation with morphine-6-hemisuccinyl-bovine serum albumin..has highest and approximately equal affinity for morphine and heroin and progressively less for opioids of decreasing structural similarity.

O. Pip: see *O 5 d.

opisthe (opi·sþi, opī·st). *Biol.* [a. F. *opisthe* (Chatton & Lwoff 1936, in *Arch. de Zool. expér. et gén.* LXXVIII. 85), f. Gr. ὄπισθεν behind.] In ciliate protozoa, the posterior of the two organisms formed by transverse fission. Cf. *proter.

1950 A. Lwoff *Probl. Morphogenesis Ciliates* xi. 74 (*caption*) The peristome of the opisthe is formed from one kinetosome. **1961** Mackinnon & Hawes *Introd. Study Protozoa* iv. 214 At fission the organism [*sc.* a ciliate] divides to produce 2 daughters, an anterior proter and a posterior opisthe. **1977** *Jrnl. Protozool.* XXIV. 23/2 There is a need for enlargement or formation of oral parts such as..enlargement of oral structures in young opisthes.

opisthion (opi·sþiǫn). *Anat.* [a. F. *opisthion* (P. Broca 1875, in *Bull. de la Soc. d'Anthrop. de Paris* X. 345), f. Gr. ὀπίσθιον hinder part.] (See quot. 1878.)

1878 R. T. H. Bartley *G. Topinard's Anthropol.* ii. ii. 234 *Opisthion*, the posterior border of the occipital foramen at the median line. **1933** *Jrnl. R. Anthrop. Inst.* LXIII. 403 The remainder, which includes the opisthion, is missing. **1971** *Nature* 20 Aug. 568/1 Of the seven available specimens of *Homo erectus*, the positions of opisthion were uncertain in Java 1 and 6 and have been estimated in the literature.

opisthoglyph (opi·sþoglif), *sb.* (*a.*) *Zool.* [a. F. *opisthoglyphe*, mod.L. *Opisthoglypha* (A. H. A. Dumeril 1853, in *Mém. Acad. Sci.* XXIII. 412) f. Opistho- + Gr. γλυφή carving.] A snake belonging to a group characterized by grooves in the upper back teeth. Also *attrib.* or as *adj.* So **opistho·glyphous** *a.*

1895 Opisthoglyphous [in Dict. s.v. Opistho-]. **1896** *Proc. Zool. Soc.* 615 The grooved teeth in the Opisthoglyphs vary in number from one to three. **1923** *Nature* 14 Apr. 579/1 A comparative study of the buccal glands and teeth of opisthoglyph snakes, and a discussion of the evolution of the order from Aglypha. **1965** R. & D. Morris *Men & Snakes* viii. 177 The aglyphs and the opisthoglyphs do not release their prey, once they have bitten it. **1968** R. D. Martin tr. *Wickler's Mimicry in Plants & Animals* xii. 112 The fangs may possess a groove that connects with the outlet of a poison gland. If the hind teeth in the upper jaws are grooved, they are called opisthoglyph. Snakes with grooved hind teeth are weakly to moderately poisonous. **1969** A. Bellairs *Life of Reptiles* I. v. 186 Each of the big teeth generally has a groove down its anterior face for carrying the venom into the prey. Snakes showing this condition are called 'back-fanged' or opisthoglyphous.

†opitulation. (Later example.)

1724 R. Sutton *Let. to Sir T. Parkyns* (1726) 4, I received such Opitulation from your Dicæology.

opium, *sb.* Add: **1. b.** (Later examples.) Freq. referring to religion, esp. in phr. *the opium of the people* (see quot. 1844); also in *transf.* and allusive uses.

[**1840** Heine *Ludwig Börne* iv. 287 Für Menschen, denen die Erde nichts mehr bietet, ward der Himmel erfunden... Heil dieser Erfindung! Heil einer Religion, die dem leidenden Menschengeschlecht in den bittern Kelch einige süsse, einschläfernde Tropfen goss, geistiges Opium. **1844** Marx 'Zur Kritik der Hegel'schen Rechts-Philosophie' in *Deutsch-Französische Jahrbücher* Feb. 72 Die Religion ist der Seufzer der bedrängten Kreatur, das Gemüth einer herzlosen Welt, wie sie der Geist geistloser Zustände ist. Sie ist das *Opium* des Volks.] **1848** *Politics for People* 27 May 58 We have used the Bible as if it was a mere special constable's handbook—an opium-dose for keeping beasts of burden patient while they were being overloaded—a mere book to keep the poor in order. **1860** Geo. Eliot *Let.* 26 Dec. in J. W. Cross *George Eliot's Life* (1885) II. xi. 283, I have faith in the working out of higher possibilities than the Catholic or any other Church has presented... The 'highest calling and election' is to *do without opium*, and live through all our pain with conscious, clear-eyed endurance. **1881** T. S. Egan tr. *Heine's Ludwig Börne* iv. 169 For these people, to whom this world had nothing more to offer, Heaven was invented... Hail to this invention! Hail to that religion which could pour a few sweet soporific drops into the bitter cup of the suffering human race, spiritual opium. **1926** H. J. Stenning tr. *Marx's Selected Ess.* 12 Religion is the moan of the oppressed creature, the sentiment of a heartless world, as it is the spirit of spiritless conditions. It is the opium of the people. **1939** G. B. Shaw *Geneva* i. 28 Karl Marx—Antichrist—said that the sweet and ennobling consolations of our faith are opium given to the poor to enable them to endure the hardships of that state of life to which it has pleased God to call them. **1951** N. Annan *Leslie Stephen* vii. 201 Kingsley might protest that religion was being used as opium for the people. **1957** tr. *M. de Unamuno's San Manuel Bueno* 39 One of those leaders of what they call the social revolution has said that religion is the opium of the people. Opium...opium... opium, yes. Let us give them opium and let them sleep and dream. **1968** *Daily Tel.* (Colour Suppl.) 13 Dec. 43/4

Drink, in other words, is becoming the opium of the people. **1971** G. Steiner *In Bluebeard's Castle* iv. 93 A good deal of classical music is, today, the opium of the good citizen. **1974** *Times Lit. Suppl.* 28 June 692/5 A revue like *Here is the News* only shows politics becoming the opium of the people.

3. b. *opium addict, cellar, cigarette, -dealer, haul, house, -pipe, -smoker* (examples), *-smoking* (later example), *-smuggler, -taking* (earlier examples). **c. opium den** (earlier and later examples); **opium dream,** a dream during an opium-induced sleep; also in extended use; **opium joint** (earlier and later examples); **opium poppy,** substitute for def.: the white poppy, *Papaver somniferum*; (examples); **opium war,** a war waged by Britain against China (1839–42) following China's attempt to prohibit the importation of opium into China; also, a later war (1856–60) against China by Britain and France.

1974 *Times* 2 Dec. 8/7 Charles was a charming opium addict and rural clergyman. **1911** O. Onions *Widdershins* 278 He took me into an opium-cellar within a stone's throw of Oxford Street. **1920** Opium cigarette [see *heroin]. **1841** J. Sturge *Let.* 30 Sept. in *Visit to U.S. in 1841* (1842) p. lxiv, Waging a murderous war to compel them to make restitution to the contraband opium dealers. **1977** 'J. Le Carré' *Hon. Schoolboy* xviii. 451 By afternoon he was airborne, and..chatting merrily to a couple of friendly opium dealers. **1882** J. D. McCabe *New York* 590 Here are the headquarters of the Mongolians, their..opium dens. **1921** *Daily Colonist* (Victoria, B.C.) 25 Mar. 6/4 Lee Fong was charged in the City Police Court yesterday with being the keeper of an opium den at 536 Cormorant Street; and also with being in unlawful possession of opium. **1969** Opium-den [see *Chinatown]. **1821** Byron *Don Juan* iv. xix. 201 This is in others a fictitious state, An opium dream of too much youth and reading. **1922** W. S. Maugham *Writer's Notebk.* (1949) 202 Singapore: Opium Dream. I saw a road lined on each side with tall poplars. **1974** J. Philips *Power Killers* (1975) i. ii. 18 What about his theory? It sounds like an opium dream. **1975** *Listener* 11 Sept. 343/2 *Who Am I Now?*, a fantasy based..on the opium dreams of George Crabbe. **1974** *Evening News* 27 June 1/7 (*headline*) £250,000 opium haul. **1888** Kipling *Plain Tales from Hills* 233 It was a *pukka*, respectable opium-house, and not one of those stifling, sweltering *chardoo-khanas*. **1966** 'A. Hall' *9th Directive* vi. 54 The opium house at the Phra Chao. **1882** H. H. Kane *Opium-Smoking* 5 The principal places, known as 'opium joints', are in Mott, Pell, and Park streets. **1926** *Daily Colonist* (Victoria, B.C.) 4 July 6/2 Charged with being inmates of an opium joint in Theatre Alley, Lee and Jim, Chinese, were each fined $15. **1890** Kipling *City of Dreadful Night* (1891) vi. 40 The lamp for the opium-pipe is the only one in the room. **1976** M. Butterworth *Festival!* i. 12 A crude photo print of a naked youth.. carrying an opium pipe. **1863** Opium poppy [see Poppy *sb.* 3]. **1880** Bentley & Trimen *Medicinal Plants* I. 18 It cannot be said that the opium poppy is known anywhere in a thoroughly wild condition. **1921** A. Huxley *Crome Yellow* xxii. 247 They passed a bed of opium poppies, dispetaled now. **1931** M. Grieve *Mod. Herbal* II. 651 The Opium Poppy..is indigenous to Asia Minor, and is cultivated largely in European and Asiatic Turkey, Persia, India and China for the production of Opium. **1975** *Times* 31 May 5/3 (*caption*) Peasant women tilling an opium poppy field in Turkey. **1860** H. Greeley *Overland Journey* 259 [The Chinaman] is..an opium-smoker. **1921** *Daily Colonist* (Victoria, B.C.) 1 Apr. 6/4 The magistrate said that the accused had admitted ownership of the opium smoking paraphernalia, and to being an opium smoker himself. **1971** *Listener* 11 Nov. 644/1 A large existing market of opium-smokers. **1938** N. Marsh *Artists in Crime* (U.S.) xiii. 202 Fox had found Malmsley's opium-smoking impedimenta. **1976** *Times* 4 Sept. 7/4 What are the pros and cons of opium smoking? **1841** J. Sturge *Let.* 30 Sept. in *Visit to U.S. in 1841* (1842) p. lxii, To take under his protection one of the most extensive opium smugglers. **1966** J. Cleary *High Commissioner* iii. 50 The Chinese opium smugglers he had met before he had gone on to the Murder Squad. **1975** M. Stern in L. M. Alcott *Behind Mask* p. xxi, William Henry Thomes..had sailed aboard an opium smuggler that plied between China and California, and was himself a mine of suggestions for authors whose thrilling romances he would publish. **1821** De Quincey *Confess.* (1822) ii. 90 The whole art and mystery of opium-taking. **1836** H. C. Robinson *Diary* 29 Aug. (1967) 161 A letter written by Coleridge..in which he gives an account of his sad habit of opium-taking. **1841** J. Sturge *Let.* 30 Sept. in *Visit to U.S. in 1841* (1842) p. lxi, Opium War with China. **1917** *Encycl. Sinica* 406/1 The action which forced the foreign merchants at Canton to deliver up their stocks of opium, which was destroyed by the Commissioner, and the subsequent events which led to war with Great Britain. This has been called the 'Opium War'. **1969** V. G. Kiernan *Lords of Human Kind* v. 148 The West resorted to force, and the Opium Wars of 1840–42 and 1856–60..inducted China into..the comity of nations. **1974** *Listener* 8 Aug. 175/1 The Opium War totally destroyed Chinese confidence.

opopanax. 2. (Earlier and later examples.)

1867 *Gardeners' Chron.* 29 June 690/1 New Perfumes.—Opoponax. **1913** [see *mignonette 1 d]. **1924** Galsworthy *White Monkey* i. vii. 55 A profiteer who dropped his aitches and reeked of opoponax.

opossum. Add: **2.** (Further examples.)

Now normally replaced in Australian usage by *possum* but still the more usual form in New Zealand.

1911 C. E. W. Bean *'Dreadnought' of Darling* xvii. 162 Australia, at one time, along with its harmless marsupial kangaroos, opossums..and the rest, had its own beasts of prey. **1911** E. M. Clowes *On Wallaby* ii. 19 The only possible chance visitor is an occasional opossum on the roof. **1944** *Living off Land* iii. 57 To improvise a water container, copy the blacks, who used the skins of opossums. **1968** *Wanganui* (N.Z.) *Chron.* 15 Nov. 10/6 (Advt.), Opossum skins. Good prices, wanted immediately for overseas contract. **1973** *Massey Ferguson Rev.* (N.Z.) Mar.-Apr. 8/3 Intensive control measures around the farms by the New Zealand Forest Service and local pest destruction boards have reduced opossum numbers with the aim that this will break the cattle infection cycle and stop Tb-tested cattle from becoming re-infected.

opo·ssuming, *vbl. sb.* [-ing¹.] Opossum-hunting.

1917 'H. H. Richardson' *Fortunes R. Mahony* iii. iv. 211 There is to be opossuming and a moonlight picnic to-night.

opotherapy (ǫpǒþe·răpi). *Med.* [f. Gr. ὀπός juice + Therapy.] = *organotherapy.

1897 *Index Medicus* XIX. 899/2 Opotherapy. **1899** *Ann. Rep. Board of Regents Smithsonian Inst. 1897–98* 696 An entire new method, designated under the name of opotherapy, or treatment by organic extracts. **1902** *Brit. Med. Jrnl.* 12 Apr. 909/1 (*heading*) Placentophagy and placental opotherapy. **1908** *Practitioner* Mar. 412 The many different substances recently utilised in opotherapy. **1915** [see *ketogenetic* adj. s.v. *keto- a].

oppo (ǫ·po). *slang* (orig. Forces'). Abbrev. of *opposite number.

1939 *Airman's Gaz.* Dec., Get an oppo to relieve you for 'break'. **1942** *Gen* 1 Sept. 13/2 A sweetheart or companion in the navy is an 'oppo'. **1948** Partridge *Dict. Forces' Slang* 133 *Oppo*,... In the Navy and Air Force; a companion..or even one's wife. **1955** P. Wildeblood *Against Law* 99 Me and my oppo was in the Royal Navy. **1961** F. H. Burgess *Dict. Sailing* 154 *Opposite number, oppo*, one of two hands who perform similar duties alternately; he may be on another ship or station. **1962** K. Dobbs *Running to Paradise* 49 My oppo, my best friend in the mess, was a former Sussex poacher. **1967** D. Reeman *Deep Silence* iii. 48 Me an' the kid is oppos, see? **1971** B. W. Aldiss *Soldier Erect* 101 He's dotty on them Wog gods, aren't you, Stubby, me old oppo? **1973** *Times* 24 May 18/4 He was Mr Justice Lawson,..and what is more he was accompanied by his oppo, Mr Justice Mais.

opponens (ǫpōu·nĕnz), *a.* *Anat.* [L., pr. pple. of *opponere* to set against.] Used, ellipt. as sb. (L. *musculus* being usu. omitted), in the names of four pairs of small muscles of the hands and feet: the *opponens pollicis*, which helps to draw the thumb across the palm; the *opponens digiti minimi*, which helps to raise the little finger when the palm is stretched out flat; the *opponens digiti minimi*, of the foot; and (seldom distinguished) the *opponens hallucis* of the foot. Cf. Opponent *a.* 3.

1797 J. Bell *Anat. Bones* ii. iii. 280 The opponens pollicis, is often called the metacarpal of the thumb. **1836–9** R. B. Todd *Cycl. Anat. & Physiol.* II. 519/2 Flexor ossis metacarpi, or opponens pollicis..of a rhomboidal form. *Ibid.* 521/1 Adductor ossis metacarpi or opponens minimi digiti. **1902** D. J. Cunningham *Text-bk. Anat.* 329 The opponens minimi digiti..arises from the anterior annular ligament and the hook of the unciform bone. **1967** G. M. Wyburn et al. *Conc. Anat.* ii. 68/2 The opponens..is inserted into the shaft of the 1st metacarpal bone. **1973** *Gray's Anat.* (ed. 35) 582/2 Part of the muscle [*sc.* the adductor hallucis] may be attached to the first metatarsal, constituting an opponens hallucis.

opportunism. Add: **1. b.** *Socialism* and *Communism.* A policy of concessions to bourgeois elements of society in the development towards socialism.

1902 *Social-Democrat* Aug. 232 Bernstein's position leads him straight to opportunism, the denial of the class war, reform-politics, classes working together for the common good.., &c. **1903** [see *Marxism¹]. **1921** in J. Degras *Communist Internat. Documents* (1956) I. 247 The parties of the Communist International will become revolutionary mass parties only if they overcome opportunism, its survivals and traditions, in their own ranks. **1930** M. J. Olgin tr. *Lenin's Conference of Foreign Sect. of R.S.-D.L.P.* in *Coll. Wks.* XVIII. 148 The collapse of the Second International is the collapse of Socialist opportunism. *Ibid.*, The crisis created by the war has exposed the real substance of opportunism, revealing it in the role of a direct aid to the bourgeoisie against the proletariat. **1934** tr. Lenin in *Lenin on Britain* iv. i. 149 To explain to the masses the inevitability and the necessity of breaking with opportunism, to educate them for revolution by a merciless struggle against opportunism,..is the only Marxian line to be followed in the world labour movement.—*Autumn* 1916. **1942** M. J. Olgin tr. *Lenin's Imperialism* in *Coll. Wks.* XIX. 194 That bond between imperialism and opportunism, which revealed itself first and most clearly in England. **1957** R. N. C. Hunt *Guide to Communist Jargon* 101 Opportunism stood for that disposition of mind which rejected Marx's revolutionary teaching, with its insistence upon the total destruction of the capitalist system, in favour of concessions made by the bourgeoisie within its framework. **1974** tr. *Sniečkus's Soviet Lithuania* 26 In its relentless struggle against reformism, opportunism and Trotskyism, it [*sc.* the Lithuanian Communist Party] became even more Bolshevik in character.

2. Opportunistic state or activity. **a.** *Med.* (See *opportunistic *a.* 3.)

1962 *Laboratory Investigation* XI. 1073/1 The concept of microbial opportunism as an important occasional factor in the etiology of infectious disease has been generally recognized and its validity has been accepted.

b. *Ecol.* (See *opportunistic *a.* 2.)

1967 G. E. Hutchinson *Treat. Limnol.* II. xxii. 366 There is reason to believe that the same sort of lognormal distribution may arise by a process of evolutionary opportunism among competitive species. **1973** P. A. Colinvaux *Introd. Ecol.* xxvii. 392 If you are small and short-lived, opportunism is probably the only satisfactory strategy for life in unstable places, but if you are big, and with a low metabolic rate,..you may just sit tight through the bad times, living on your reserves and reducing your life processes to the minimum.

opportunist. Add: **1. a.** *spec.* in Socialism and Communism, an advocate of opportunism (sense *1 b).

1902 *Fortn. Rev.* Jan. 128 The mountain of the [German Social Democratic] party joined hands with the Opportunists. **1903** *Social-Democrat* VII. 87 It is constantly being brought as a reproach against the German Party that they were more severe with the revolutionary dissenters from the party theory than they have been with the opportunists. **1909** M. Epstein tr. *Sombart's Socialism & Social Movement* ii. iii. 217 All authoritative Revisionists, Opportunists, Reformers..stand firm for the class war, and..desire the total abolition of the capitalist system, and not merely its reformation. **1919** Trotsky in J. Degras *Communist Internat. Documents* (1956) I. 41 The opportunists, who before the world war appealed to the workers to practise moderation for the sake of the gradual transition to socialism. **1930** M. J. Olgin tr. *Lenin's Conference of Foreign Sect. of R.S.-D.L.P.* in *Coll. Wks.* XVIII. 149 It would be a harmful illusion to hope to restore a real Socialist International without drawing a clear line of organisational demarcation between real Socialists and opportunists. **1934** tr. Lenin in *Lenin on Britain* iv. i. 142 The *opportunists* (social-chauvinists) are working together with the imperialist bourgeoisie..in the direction of creating an imperialist Europe on the backs of Asia and Africa;..the *opportunists* are a section of the petty bourgeoisie [etc.]... *Autumn* 1916.

b. (Examples in Socialist and Communist sense.)

1902 *Fortn. Rev.* Jan. 129 Where Marx is fatalistic, Bernstein is opportunist. **1903** *Social-Democrat* VII. 86 To turn..to the German Party, the opportunist tendency has naturally always existed. **1929** J. Fineberg tr. *Selections from Lenin* I. 177 Axelrod and Martov..have dropped into the opportunist wing of our Party... They have to repeat opportunist phrases..to seek..some kind of justification for their position. **1934** tr. Lenin in *Lenin on Britain* v. iv. 212 The victory of the revolutionary proletariat is impossible..unless the opportunist social-traitor leaders are exposed, disgraced and driven out... *April–May* 1920. **1974** tr. *Sniečkus's Soviet Lithuania* 12 A reformist opportunist trend..had developed on the basis of the petty-bourgeois nationalistic ideology. **1974** J. White tr. *Poulantzas's Fascism & Dictatorship* iv. i. 147 The 'left opportunist' elements bore a very grave responsibility in the advent of fascism.

2. *Med.* An opportunistic fungus or micro-organism (see *opportunistic *a.* 3). Also *attrib.* or as *adj.*

1937 M. Frobisher *Fund. Bacteriol.* xiv. 138 These organisms..do not initiate the rot. They are opportunists. **1967** *Jrnl. Hygiene* LXV. 575 (*heading*) Classification of *Mycobacterium avium* and related opportunist mycobacteria met in England and Wales. *Ibid.*, Overt opportunist infection. **1973** *Amer. Jrnl. Med.* LV. 862/1 As an opportunist, its ability to produce human disease depends not on its intrinsic virulence but on abnormalities of host defenses. **1976** *Lancet* 27 Nov. 1169/1 There was no evidence of opportunist infection.

3. *Ecol.* An opportunistic species (see *opportunistic *a.* 2). Also *attrib.* or as *adj.*

1970 *Lethaia* III. 70 Even in habitats of low physicochemical stress and great stability, opportunists may participate in occasional invasions of an otherwise stable fauna. **1973** P. A. Colinvaux *Introd. Ecol.* xvii. 392 It is opportunist animals..which, exposed to hazard of weather in their normal lives, are most likely to have their numbers curbed by accident of weather. *Ibid.* 393 Opportunists, being specialists at dispersion, are likely to get there first but, since the fresh bare ground may be in a place of generally stable climate, the equilibrium species will not be far behind. The opportunists enjoy the new land only briefly, after which they are eliminated by competition. **1975** *Nature* 20 Nov. 197/1 These data suggest that birch is behaving as an opportunist in the composition of the forest.

opportunistic, *a.* Add: **1.** (Further examples.)

1958 J. Baldwin in W. King *Black Short Story Anthol.* (1972) 284 Their religion was strongly mixed with an opportunistic respectability and with ambitions to better society and their own place in it. **1976** *Brit. Jrnl. Sociol.* XXVII. 89 The common man is portrayed as the innocent and helpless victim of opportunistic and self-serving politicians who must be replaced by enlightened and benign rulers. **1976** *Publishers Weekly* 15 Mar. 118/3 An opportunistic TV programmer brings him to New York and takes an exploitation piece. **1977** *Listener* 20 Oct. 515 The ebullient..Furnivall comes out of it badly (vain, unprincipled,..opportunistic, unscholarly—though no one contributed more quotations for the dictionary).

2. *Ecol.* Of a species: especially suited to unexploited or newly formed habitats and occurring in populations whose size is not determined primarily by their density, being characterized by poor competitiveness in relation to other species and an ability to

increase rapidly in numbers and to disperse readily.

1960 R. MacArthur in *Amer. Naturalist* XCIV. 33 A distinction is made between opportunistic and equilibrium species. **1974** *Jrnl. Marine Res.* XXXII. 267 *Capitella capitella* and the other relatively opportunistic species discussed may be continuously present if the environment is unpredictable or may disappear as in the case of recovery following the oil spill.

3. *Med.* Of a fungus or micro-organism: not normally pathogenic but becoming so in certain circumstances, as when the body is rendered vulnerable by other agencies. Of an infection: caused by such an organism.

[**1955** *Sci. Amer.* May 31/2 Was it not possible, they argued, that the bacteria were only the secondary cause of disease—opportunistic invaders of tissues already weakened by crumbling defenses?] **1962** *Ann. N.Y. Acad. Sci.* XCVIII. 617 (*heading*) Experiences with and diagnosis of diseases due to opportunistic fungi. **1962** *Laboratory Investigation* XI. 1073/1 Opportunistic infections by bacteria, viruses, and protozoa are known. **1970** C. W. Emmons et al. *Med. Mycol.* (ed. 2) 3 The fungi which cause systemic and subcutaneous mycoses have been called 'opportunistic fungi' to emphasize the aspect of a normally saprobic fungus which can suddenly become parasitic and pathogenic when it is introduced by inhalation or traumatic implantation into the human body... It has also been used to designate fungi which cause disease only in a patient with a concurrent disease which increases his susceptibility, or in one whose innate immunity has been otherwise impaired. **1973** *Chest* LXIII. 4/1 Even more striking is the rise of opportunistic fungal infections accompanying: transplantation, immunosuppression, heart surgery and intravenous hyperalimentation. Today one can no longer accept a culture report of 'nonpathogenic fungus isolated', for indeed there may be no truly nonpathogenic fungus.

Hence **o:pportuni·stically** *adv.*, in an opportunist manner.

1958 *Times* 27 Dec. 2/4 In one brief interlude Phillips nearly scored opportunistically at the other end. **1960** W. V. Quine *Word & Object* v. 188 We can vacillate between two, opportunistically enjoying their incompatible advantages. **1972** *Maclean's Mag.* Sept. 10/1 He will act with expediency but not often, I think, opportunistically. **1976** *Sci. Amer.* Apr. 117/2 At the same time the adults will feed opportunistically on lesser prey: frogs, crabs and small fish.

opportunity. Add: **2. c.** *equality of opportunity*: equal chance and right to seek success in one's chosen sphere regardless of social factors such as class, race, religion, and sex.

1891 *Econ. Rev.* I. 474 It will possibly, however, be contended that here the ideal is equality of Opportunity. **1920** M. Beer *Hist. Brit. Socialism* II. xiv. 295 By 'equality of opportunity' Fabian women do not necessarily mean 'similarity of opportunity'. **1920** H. G. Wells *Outl. Hist.* II. ix. xli. 754/2 A sufficient measure of social justice, to ensure health, education, and a rough equality of opportunity. **1930** W. K. Hancock *Australia* ix. 183 Equality of opportunity implies free scope for natural talent, which must create new inequalities; whereas what Australian democracy desires is equality of enjoyment. **1950** G. B. Shaw *Farfetched Fables* 67 Democratic civilization is impossible, because equality of opportunity is impossible. **1973** R. R. Palmer in P. P. Wiener *Dict. Hist. Ideas* II. 146/2 The key words are fair competition, equality of opportunity, reward for merit, and careers open to talent. **1976** R. Williams *Keywords* 102 *Equality of opportunity*..can be glossed as 'equal opportunity to become unequal'.

7. Phr. *opportunity knocks* (*but once*): an opportunity presents itself (but once).

1942 Wodehouse *Money in Bank* (1946) xv. 134 Opportunity knocks but once, and he had allowed it to knock in vain. **1946** *Calif. Folklore Q.* July 241 A number of sayings surely not peculiar to Oregon; rather, trite and obvious, but still effective:..Opportunity knocks but once. **1970** *Computers & Humanities* V. 16 From the earliest English settlements to the closing of the frontier and the advent of industrialism, opportunity knocked for young men of high and low social status at fairly regular alternate intervals. **1970** *Globe & Mail* (Toronto) 28 Sept. 29/4 (Advt.), Opportunity knocks for a fluently bilingual representative over 25 with a B.Sc. **1972** *Accountant* 21 Sept. p. xvii/1 (Advt.), Opportunity knocks for an experienced qualified man.

8. *attrib.* and *Comb.* **opportunity cost** *Econ.* (see quots.); **opportunity state,** a country which offers many opportunities for advancement.

1911 H. J. Davenport in *Amer. Econ. Rev.* Dec. 725 These displacements of possible products, these foregoings of alternative openings, these sacrifices of some second thing in the process of getting some particular thing, are perhaps best indicated under the term *opportunity cost*. To go without fish to get game..may be taken as illustrative of one of the simplest aspects of the doctrine. **1926** L. D. Edie *Econ.* iii. viii. 121 The opportunity cost is the sacrifice of foregoing some alternative utility. **1936** Ault & Eberling *Princ. & Probl. Econ.* xiii. 217 Opportunity or alternative costs play a very important part in determining the nature and direction of industrial development... The principle of opportunity cost is effective in bringing about great changes in the field of production. **1951** J. R. Winton *Dict. Econ. Terms* (ed. 3) 63 *Opportunity cost.* In economic life, decisions are constantly being made which involve a choice between alternatives... One particular alternative must be chosen, ..the 'opportunity-cost' of which is represented by the alternatives foregone. **1964** Gould & Kolb *Dict. Social Sci.* 143/2 The determination of implicit cost, as well as other managerial decisions, must be based on the concept of alternative or opportunity costs, which measures cost in terms of alternatives or opportunities that are foregone. **1965** Seldon & Pennance *Everyman's Dict. Econ.* 312 To answer the question, 'Should good agricultural land be built over?' requires a comparison of the opportunity costs for society of the various alternatives. **1971** D. C. Hague *Managerial Econ.* (rev. ed.) ii. v. 121 A very useful concept is that of 'opportunity cost'... If the businessman would have invested the money at 10 per cent interest, had he not put it into the business, then the 'opportunity cost' of investing in his own business is the 10 per cent interest he has foregone. **1974** *Guardian* 27 Aug. 14/7 Couples in the United States..are well aware of the opportunity cost of having children. **1957** Opportunity state [see *contract *v.* 2 d]. **1958** *Listener* 26 June 1068/3 The mental distance between himself and his appalling younger daughters, products of the opportunity state, is brilliantly suggested.

opposed, *ppl. a.* Add: **1. b.** *Mech.* (Having pistons) arranged in pairs moving in opposite directions along the same straight line.

1910 W. A. Tookey tr. *Mathot's Construction & Working Internal Combustion Eng.* v. 71 In Europe, the two-cylinder 'twin' engine soon displaced the type with opposed or *vis-à-vis* cylinders, because the latter gave a great deal of trouble in practical work. **1925** A. W. Judge *Automobile Engines* iii. 87 Although the balance and torque of the opposed two-cylinder engine are excellent, the overall length and the cylinder dimensions limit it to power units of about 12 to 15 H.P. (maximum) for car use. **1932** A. F. Evans *Hist. Oil Engine* ii. 71 In the 'seventies we had engines with opposed pistons in one cylinder. **1966** G. F. Allen *Brit. Rail after Beeching* iv. 115 English Electric took an intricate but compact 18-cylinder opposed-piston engine originally applied to fast, small naval craft..and with it achieved a six-axle 3,300 h.p. diesel-electric unit. **1977** *Lancashire Life* Jan. 79/1 The horizontally opposed engine is air cooled.

opposer. Add: **1. b.** (Usu. with capital initial.) One of two examiners formerly appointed to carry out at Winchester College the elections to New College, Oxford. Cf. Poser¹ 1.

Rarely used, in preference to *poser*, even before 1901 (P. Yeats-Edwards, Fellows' librarian, Winchester College, private communication 28 Jan. 1977).

1891 R. G. K. Wrench *Winchester Word-bk.* 32 *Poser*, an examiner. A very old word: also still used at Eton. *Apposer* and *opposer* are other old forms of it. **1901** Rashdall & Rait *New College* vi. 132 The Warden and Opposers are not to take bribes.

opposite, *a.*, *sb.* (*adv.*, *prep.*). Add: **A.** *adj.* **5.** *opposite number*, a person or thing similarly placed in another set, etc., to the given one; a partner, a counterpart; an opponent.

1906 Kipling *Actions & Reactions* (1909) 202 'And your Opposite Number?' Penfentenyou described him. **1915** 'Bartimeus' *Tall Ship* iv. 84 We were 'opposite numbers' at your brother's wedding. **1917** *Times* 7 May 6/2 The establishment of personal contact between Sir William Robertson and his opposite number, General Pétain. **1917** E. Wallace *Kate plus Ten* (1919) vi. 107 Pick up Mr. Pretherston and don't lose him—you may choose your own opposite number. **1927** *Observer* 24 July 4/6 The 'A.A.'s' opposite number..the Automobile Association of America..has issued a very interesting map of the United States. **1969** *Times* 13 Mar. 25/2 (*caption*) The model bison the papermakers' association got from its Polish opposite numbers. **1973** J. Rossiter *Manipulators* xiii. 134 Before leaving for B Division he should have telephoned his opposite number there, clearing his proposed incursion on to another's territory.

B. *sb.* **3.** (Later example.)

1874 Swinburne *Bothwell* iv. v. 397 The task was hard with Knox for opposite To bend the council.

C. quasi-*adv.* **3.** *to play opposite*: to have (a specified actor or actress) as one's leading man or lady. Also in similar phrases.

1926 J. Agate in *Sunday Times* 7 Feb. 6/1 'Opposite him,' as they say, was Miss Ilise Marvenga, who..made Kathie into a semblance of one of those expensive mechanical dolls with a staccato utterance. **1931** P. MacDonald *Crime Conductor* ii. iv. 178 Mary Wheelwright—England's Première Leading Lady—was not to play 'opposite' Kristania in *Harlequin's Holiday*. **1936** *Times Lit. Suppl.* 31 Oct. 894/3 He was to play 'opposite to' Anna Kenney, a very famous figure on the stage.

oppositely, *adv.* Add: **1.** (Later examples.)

c **1864** E. Dickinson *Poems* (1955) II. 695 A Suspicion.. That I am looking oppositely For the site of the Kingdom of Heaven—. **1929** R. Bridges *Testament Beauty* iii. 129 Thatt other hath the arm bent down and oppositely nerved.

2. b. (Later examples.)

1909 *Bodleian* Mar. 3/1 'You can have a dead rat,.. but not a dead artist.' But oppositely, and appositely, a dead rat has no signification. **1972** *Science* 27 Oct. 425/1 Or, oppositely, should we encourage economic growth.

opposition. Add: **4. e.** *Semantics.* The state or condition of being opposite in meaning; the relationship between antonyms.

1870 [see Antonym]. **1925** P. Radin tr. *J. Vendryès's Language* iii. iii. 218 Among animal names the same opposition is frequent. Latin had *equus* and *equa*... In English *horse* is apposed to *mare*. **1932** C. K. Ogden *Opposition* i. 8 The theory of opposition offers a new method of approach not only in the case of all those words which can best be defined in terms of their opposites..but also to *any* word whose use may give rise to controversy. **1963** J. Lyons *Structural Semantics* iv. 68 The polarity of a term is a function of its opposition to its antonym, and not vice versa. **1967** R. A. Waldron *Sense & Sense Devel.* v. 107 *Black* and *white* are often two ends of a scale but in 'people are inclined to see things in black and white'..the pair is taken as the type of all 'either/or' antonymic opposition. **1976** F. R. Palmer *Semantics* iv. 82 A quite different kind of 'opposite' is found with pairs of words which exhibit the reversal of a relationship between items... Lyons suggests the term *converseness* for these, but I am more concerned to point out their essentially relational characteristics, and would thus prefer *relational opposition*.

f. *Linguistics.* A functional, or potentially functional, contrast between partially similar linguistic elements.

[**1931** *Travaux Cercle Ling. Prague* IV. 311 *Opposition phonologique*..—Différence phonique susceptible de servir, dans une langue donnée, à la différenciation des significations intellectuelles.] **1936** *Amer. Speech* XI. 110 A 'phonological system' is defined as the ensemble of phonological oppositions proper to a given language. **1953** [see *binary *a.* k]. **1963** J. Lyons *Structural Semantics* iv. 68 In Russian or German..there is an opposition to be recognized, in word-initial and word-medial position, between the voiced and the voiceless plosives. **1970** G. C. Lepschy *Survey Struct. Ling.* iii. 58 In order to have distinctive function, speech sounds must be opposed to each other (distinction presupposes opposition). An opposition can be either *distinctive* (or *phonological*), or non-distinctive. Only sounds which may occur in the same context (i.e. that are permutable) can be in opposition. **1972** M. L. Samuels *Ling. Evol.* viii. 171 In the fifteenth and sixteenth centuries there are signs that the opposition /ɛ∼ɑ/ in class iii verbs is no longer adequate. **1978** *Jrnl. Lancs. Dial. Soc.* Jan. 20 The opposition between the reflex of ME ui/oi and ME *i* is preserved.

8. *opposition bench*(*es*) (examples), *party*, *spokesman*.

1815 N. W. Wraxall *Hist. Mem.* II. 167 The Treasury Bench, as well as the Places behind it, had been for so many years occupied by Lord North and his friends, that it became difficult to recognize them again in their new Seats, dispersed over the Opposition Benches, in great coats, frocks, and boots. **1819** Keats *Cap & Bells* in R. M. Milnes *Life, Lett., & Lit. Remains J. Keats* (1848) II. 221 I'll make the opposition-benches wince. **1976** *Hansard Commons* 7 Dec. 252 The Government..cannot accept views expressed from the Opposition Benches... I object strongly to that. *Ibid.* 8 Dec. 570, I am far more concerned about the measures that would be enacted by the rabble sitting on the Opposition Benches if they were in Government. **1795** tr. *C. P. Moritz's Trav.* 60 They tell me, that at these elections when there is a strong opposition-party, there is often bloody work. **1824** J. S. Mill in *Westm. Rev.* I. 505 The Edinburgh Review..has really exhibited the vices, which we described as likely to characterize a periodical publication attached to the Opposition party. **1930** G. B. Shaw *Apple Cart* p. xxiii, Nothing has any sense or reality in it except the vituperation of the opposition party. **1974** *Times* 12 June 2/5 Mr Heseltine, Opposition spokesman on trade, said..yesterday [etc.].

oppositionary (ǫpŏzi·ʃənări), *a.* *rare.* [f. Opposition + -ary¹.] = Oppositional *a.* 2.

1926 *Contemp. Rev.* Sept. 276 Petrograd ('Leningrad') became the centre of Zinoviev's oppositionary activity.

oppositionist, *sb.* and *adj.* (Later examples.)

1962 S. E. Finer *Man on Horseback* xii. 231 The members of the nationalist movement have been reared in an oppositionist mentality. **1963** *Cambr. Rev.* 4 May 405/2 An individual perception of reality—as distinguished from a conformist perception, whether it be 'official' or 'oppositionist'—is of value and interest. **1965** A. Nove in B. Pearce tr. *Preobrazhensky's New Economics* p. xiv, Like other former oppositionists he came to the Congress to apologize for past misdeeds and to denounce Trotsky. **1971** I. Deutscher *Marxism in our Time* (1972) vi. 113 In June 1957, exactly twenty-five years will have elapsed since I was expelled from the Party as an oppositionist. **1973** *Sunday Advocate-News* (Barbados) 16 Dec. 4/3 The Consul-General attributed the rumours to be the work of 'oppositionists' who he said are trying to embarrass the Government.

oppositious (ǫpŏzi·ʃəs), *a.* *rare.* [f. Oppos(ite *a.* + -itious².] Inclined to oppose; recalcitrant.

1923 *Blackw. Mag.* Aug. 176/2 He became oppositious on leaving truly delectable posadas to left and right.

opposive, *a.* Delete †*Obs.*⁻¹ and add later example. *rare.*

1911 W. De Morgan *Likely Story* 107 He had an opposive or lazy disposition.

oppre·ssingly, *adv.* *rare.* [-ly².] So as to oppress or be oppressive; oppressively.

1925 *Glasgow Herald* 17 May 7/1 That it [*sc.* the cost] is oppressingly large no one will deny.

oppressive, *a.* Add: **2.** Also with *of.*

1972 *Times* 10 Mar. 9/1 The art world is so international today that it is not oppressive of a world-renowned art dealer living in France to have an action against him tried in England.

opry (ǫ·pri), representation of a U.S. dial. pronunc. of Opera; also *attrib.* *Grand Ole Opry*: a concert of country music broadcast on radio from Nashville, Tennessee; the type of music performed there.

Grand Ole Opry is registered as a proprietary name in the U.S.

1914 R. Grau *Theatre of Science* 23 The local manager could not see any future in exhibiting films, so he went back to the town where he had his 'op'ry house'. 1950 *Official Gaz.* (U.S. Patent Office) 18 Apr. 723/1 WSM, Incorporated, Nashville, Tenn. Grand Ole Opry. For radio program broadcasting services. 1957 *Time* 15 Apr. 49/1 Donegan..often..sounds like *Grand Ole Opry* cornball recorded at 33⅓ r.p.m. played at 78. 1961 A. Berkman *Singers' Gloss. Show Business* 64 *Opry*,..opera. 1968 *Rolling Stone* 24 Aug. 17/3 Roy Acuff tried unsuccessfully to give the audience a taste of Grand Ole Opry. He was too corny for most people's taste. 1974 *New Yorker* 6 May 46/1 They asked me where I was going. I said, To Nashville to see the Grand Ole Opry. Friday is the last show..before the Opry moves out to Opryland U.S.A. 1976 *Time* 27 Sept. 98/1 For the mass market, cruder Southern products flooded the land: hillbilly music, gospel music, the Grand Ole Opry. 1977 *Parade* (*Washington Post*) 9 Oct. 12/3 Grand Ole Opry began in 1925 almost incidentally, when Nashville station WSM put on an hour-long program featuring a country fiddler named Uncle Jimmy Thompson. The show happened to follow a broadcast of Walter Damrosch's music appreciation hour from New York, so announcer George D. Hay started out by saying: 'For the past hour you've listened to grand opera, now you're going to hear some grand *ole* opry.'

ops. (= military operations): see *OP[3] 1 b.

opsin (ǫ·psin). *Biochem.* [Back-formation from *rhodopsin* s.v. Rhodo-.] A protein liberated from rhodopsin by the action of light.

1951 G. Wald in *Science* 15 Mar. 287/2 Rhodopsin and porphyropsin are carotenoid-proteins—proteins bearing carotenoid prosthenic groups to which they owe their color and sensitivity to light... The protein probably varies from one animal to another; it may be called *opsin*, and named for the animal of origin. 1956 *Nature* 28 Jan. 174/1 When opsin is in excess.., the synthesis of rhodopsin removes neo-*b* retinene almost entirely from solution. 1970 R. W. McGilvery *Biochem.* xxvi. 644 The interaction of the conjugated hydrocarbon chain with the opsins creates the particular absorption spectrum of the visual pigments, and therefore the spectral sensitivity of the eye.

‖ **opsit** (ǫ·psit), *v. S. Afr.* [f. Du. *opzitten* to sit up.] (See quot. 1955.)

1887 Rider Haggard *Jess* viii. 72 How often do you 'opsit' (sit up at night) with Uncle Croft's pretty girl, eh? 1899 —— *Swallow* i. 6 After we had 'opsited' together several times according to our customs, and burnt many very long candles, we were married. 1900 H. A. Bryden *From Veldt Camp Fires* 195 Tobias meant to make a bit of a splash today,..although he was not prepared for the solemnity of an 'opsitting' (that all-night form of courtship, dear to the heart of the Boer). 1913 C. Pettman *Africanderisms* 351 *Opsit*.., in Cape Dutch this word is descriptive of the peculiar method of courting which in earlier days was in vogue among the Dutch farming population. 1939 S. Cloete *Watch for Dawn* 377 Why, if you wanted to court her, could you not opsit like a Burger in the sit-Kamer with a candle between you? 1955 W. Robertson *Blue Wagon* xix. 181 'In my young days I had to *opsit* for weeks.' He referred to the Boer custom of two young people sitting up by the light of a candle after the elders had gone to bed. When the candle burned out it was time for the young man to go, and if the girl did not favour him she would produce a candle-end instead of a long one as a hint she preferred his going to his company.

opsonic (ǫpsǫ·nik), *a. Bacteriology.* [f. as next + -ic.] Of or pertaining to opsonins; produced by or involving opsonins.

1903 [see next]. 1906 *Practitioner* Dec. 750 A doubling of the opsonic index means that the quantity of opsonin present has been increased in a far greater proportion. 1911 G. B. Shaw *Doctor's Dilemma* Pref. p. xxxviii, A few doctors have now learnt the danger of inoculating without any reference to the patient's 'opsonic index' at the moment of inoculation. 1929 Topley & Wilson *Princ. Bacteriol. & Immunity* I. vi. 175 Normal serum loses almost all its opsonic action when diluted 15 times with saline. 1950 C. J. Witton *Microbiol.* xxii. 297 Opsonic test. If the patient's white blood cells show unusual ability to engulf the bacteria, it indicates the presence of specific opsonins. 1969 H. I. Winner *Microbiol. in Mod. Nursing* xiii. 154 Yet another serological test, not much used today, is that to determine the opsonic index of a serum. 1972 *Pediatrics* XLIX. 225 (*heading*) Fatal familial Leiner's disease: a deficiency of the opsonic activity of serum complement.

opsonin (ǫ·psŏnin). *Bacteriology.* Formerly also **-ine.** [f. L. *obs-*, *opsōn-āre* to buy provisions, cater (f. Gr. ὀψωνεῖν) + -in[1].] A substance (usu. an antibody) in blood serum which combines with bacteria or other foreign cells and renders them more susceptible to phagocytosis.

1903 Wright & Douglas in *Proc. R. Soc.* LXXII. 366 We may speak of this as an 'opsonic' effect (opsono—I cater for; I prepare victuals for), and we may employ the term 'opsonins' to designate the elements in the body fluids which produce this effect. 1904 [see *saturation experiment* s.v. Saturation 5]. 1906 *Practitioner* Dec. 750 We know..that the presence of opsonins is necessary for phagocytosis. 1911 G. B. Shaw *Doctor's Dilemma* 10 Opsonin is what you butter the disease germs with to make your white corpuscles eat them. 1937 R. W. Fairbrother *Text-bk. Med. Bacteriol.* ix. 116 There seems to be little doubt that opsonins are quite distinct from complement. 1950 [see prec.]. 1970 W. H. Parker *Health & Dis. in Farm Animals* ix. 116 Polymorphs.. will consume many times more bacteria of a particular species if the right opsonin is present, i.e. if the animal concerned is immunised against that species.

Hence **o·psonist** (*nonce-word*), an advocate of the therapeutic use of opsonins; **opsoniza·tion**, the process of opsonizing; **o·psonize** *v. trans.*, to render (a bacterium or other particle) more susceptible to phagocytosis.

1906 G. B. Shaw *Let.* 18 Nov. (1972) II. 661 A vaccine opsinises [*sic*] your disease germs—to opsinize = à rendre friande—so that the white blood corpuscles pitch into them with an appetite. 1907 *Practitioner* Apr. 581 The hungry phagocyte is supposed to have its diet daintily opsonized for it. 1907 *Science* 13 Sept. 346/1 An alkalinity ..prevented opsonization. 1911 G. B. Shaw *Doctor's Dilemma* Pref. p. xci, Add to the newly triumphant homeopathist and the opsonist that other remarkable innovator, the Swedish masseur. 1936 *Med. Rec.* (N.Y.) CXLIII. 16/1 It seems wise..to divide the reactions occurring with antigen and antibody into three groups: A. Protective. These include neutralization.., lysis.., killing..and opsonization. 1970 Harris & Sinkovics *Immunol. Malignant Dis.* i. 6 Antibody may have opsonized the antigen before it reaches the lymph node. 1973 *Sci. Amer.* Nov. 64/3 Antibody and complement render bacteria susceptible to phagocytosis, a process called opsonization.

opsonocytophagic (ǫ:psǫ̆nosəitofæ·dʒik), *a. Bacteriology.* [f. *opson(in + -o + Cyto- + Gr. -φαγία eating (sb.) + -ic.] = *opsonic *a.*

1910 Glynn & Cox in *Jrnl. Path. & Bacteriol.* XIV. 92 We have expressed the result of comparing the combined action of leucocytes plus serum of two persons by the term Opsono-cytophagic Index. Such an index denotes.. the total phagocytic power of any blood. 1940 Schaub & Foley *Methods Diagnostic Bacteriol.* viii. 219 The opsonocytophagic test in brucellosis determines the immunity status of the individual toward Brucella. 1950 C. J. Witton *Microbiol.* xxii. 297 This figure is called the opsonic index, or opsonocytophagic index. *Ibid.* xxv. 338 The presence of antibrucella antibodies..may be tested by agglutination tests and by opsonocytophagic tests. 1969 Wilson & Mizer *Microbiol. in Nursing Pract.* xix. 447/2 Opsonins also develop, as recognized by tests that demonstrate increased phagocytosis of brucellae in the presence of the patient's serum (opsonocytophagic test).

opster (ǫ·pstəɹ). *U.S.* [f. *op[4] + -ster.] A practitioner of op art. *Temporary.*

1965 *Sat. Rev.* (U.S.) 29 May 29/3 It would seem that two older artists have been curiously misunderstood—Albers by the 'opsters' and Duchamp by the 'popsters'. 1967 *N.Y. Times* 18 Mar. 25 The intricate light-and-shadow wall reliefs of the opster Ben Cunningham.

opt, *v.* Add: **2.** Phr. *to opt out of*: to choose not to do or participate in (something); also *absol.* So, **opter-out**, one who opts out (of something). Hence phr. *to opt into*: to choose to do or participate in something; also *absol.* as *to opt in*. See also *opting *vbl. sb.*

1922 [see *opting vbl. sb.]. 1951 *Ann. Reg. 1950* 82 It became clear that many of the older and larger schools.. were likely to opt out [of the Education Act]. 1966 'A. Hall' *9th Directive* i. 12 A man wants to opt out; they have to give him an incentive that will make him opt in again. 1966 *Maclean's Mag.* 20 Aug. 43 Morgan has no cures for society so he opts out, even if his Mum does call him a class traitor. 1968 *Globe & Mail* (Toronto) 17 Feb. 1/1 Pierre Trudeau opted into the Liberal leadership race yesterday. 1968 *Listener* 20 June 790/3 But the problem is compounded if one section of the affluent majority now decides to avert its eyes from the whole sordid business of material betterment. The opters-out may not realise it, but they will make it harder to remove the remaining pockets of real poverty. 1970 *New Scientist* 30 Apr. 226/1 It is clearly impossible for the UK to opt out of the microelectronics race. 1971 'G. Black' *Time for Pirates* v. 85, I looked at the princely opter-out conscious of how much I liked him. 1973 C. Bonington *Next Horizon* xii. 174, I did not intend to go to the top, having already opted out on grounds of risk. 1976 *Country Life* 26 Feb. 496/3 Her group of rather dim students..are all determined pre-adolescents. Neither drop-outs, nor opter-outs, they have simply refused to opt in. 1977 *Church Times* 29 Apr. 2/4 The OCU said that mercy-killing extended to cases in which there was no dissent from the victim was even more permissive than the recommendations of voluntary euthanasia Bills—'in these at least the victim has to 'opt in' to be killed. Should these recommendations become law it will not easily be possible to 'opt out' of being killed.'

Optacon (ǫ·ptăkǫn). Also **optacon.** [f. Op(tical *a.* + Ta(ctile *a.* + Con(verter.] A device for enabling a blind person to recognize printed characters by touch, the fingertip being placed on an array of many tiny rods certain of which vibrate in accordance with the pattern of light from the characters.

1970 *N.Y. Times* 24 July 33/2 Another technique—that of turning symbols and letters into touch—has also become a practical reality recently because of a lightweight, portable transistor device developed by Prof. John Linvill..and Prof. James Bliss... This device, called the Optacon, consists of a tactile stimulator and an optical scanner, connected together electronically. 1973 *Times* 26 May 16/5 Mr Richard Dufton, director of research for St Dunstan's, said..that the Optacon imposed a severe learning task because the reader's right hand had to track the camera accurately across the page, picking up the print letter by letter. 1975 *Telegraph* (Brisbane) 18 May 22/3 The optacon is being used successfully in Washington, U.S., by a blind radio announcer.

optant (ǫ·ptănt). [G. and Da. *optant*, f. L. *optant-*, *optans*, pr. pple. of *optāre* to choose.] A person who, when the territory of which he is a citizen changes its sovereignty, has a choice between retaining his former citizenship, and accepting a new one.

1914 W. R. Prior *North Sleswick under Prussian Rule* 9 Nearly 40,000 of the Sleswick Danes had become optants... The peril to which their optant relatives and neighbours were exposed. 1927 *Daily Tel.* 8 Mar. 11/5 This arbitral tribunal pronounced in favour of the Hungarian optants. 1930 *New Statesman* 10 May 138/2 Both sides have made considerable concessions, especially with regard to the difficult problem of the dispossessed Hungarian optants. 1937 V. Bartlett *This is my Life* x. 143 The Hungarian optants—Hungarians in Transylvania who had the right to choose between Hungarian and Roumanian nationality after the territory had been transferred to Roumanian rule.

optation. Add: (Later example.)

1922 E. Pound *Let.* 4 May (1971) 177 As you have been so explicit in yr. optation of undisturbed solitude I hesitate to offer to prolong my sojourn in Italy.

optic, *sb.* Add: **2. b.** *pl.* The optical components of an instrument or apparatus.

1942 J. Mitchell *Ilford Man. Photogr.* xiv. 297 (*heading*) Optics of the condenser enlarger. 1948 *Rev. Sci. Instruments* XIX. 153/2 Now that better optics and energy detectors are more generally available..polarization work will undoubtedly increase. 1962 *Analyt. Chem.* XXXIV. 242/2 A double-beam infrared microspectrophotometer employing a double-beam-in-time system and reflecting optics having 8× magnification. 1972 *Physics Bull.* Mar. 155/2 It is easy to arrange the x ray optics so that the diffracted beams from a selected set of parallel crystal planes form an image of the crystal.

5. (Properly with capital initial.) The proprietary name of a device fastened to the neck of a bottle for measuring out spirits, etc.; also *optic measure*.

1926 *Trade Marks Jrnl.* 22 Sept. 2158 *Optic*, an apparatus included in Class 8, for delivering a measured quantity of Spirits or other Liquids. Gaskell & Chambers Limited. 1953 *Word for Word: Encycl. Beer* (Whitbread & Co.) 27/2 *Optic*, a measuring and dispensing device widely used for spirits. It is usually inserted into the neck of an inverted bottle. 1967 *Guardian* 18 Feb. 3/1 Overnight, with the drop in temperature, a small amount of the alcohol in optics is lost... Now the Licensed Victuallers Association has told the landlords to take the bottles down each morning and allow the liquor in the optics to flow back into the bottle. 1968 'P. Barrington' *Accessory to Murder* i. 16 Joe, the landlord, was surprised to see him...'Going to a sale somewhere, then?' he asked genially, holding a small glass under the optic measure. 1968 'A. Haig' *Sign on for Tokyo* 58 Joe could see his face reflected in the glass behind the bar, between the spirits bottles on their optic measures. 1970 V. Canning *Great Affair* xii. 215 A row of shining optics under the whisky and gin bottles. 1974 P. Cave *Dirtiest Picture Postcard* xiii. 85 She swung at her empty glass, sending it flying across the smooth bar-top to smash against a row of optics.

optical, *a.* Add: **1.** (Further examples.) Also *fig.*

c 1806 D. Wordsworth *Jrnl.* (1941) I. 253 Right before us..were several small single trees..but some optical delusion had detached them from the land on which they stood, and they had the appearance of..little vessels sailing along the coast of it. 1859 *Rep. Brit. Assoc. Adv. Sci. 1858* II. 14 (*heading*) On an apparatus for exhibiting optical illusions of spectral phenomena. 1911 *Rep. Labour & Social Conditions in Germany* (Tariff Reform League) III. 193 It may have been an optical illusion, but it certainly did seem to me that Germany was in a state of abounding prosperity. 1922 Joyce *Ulysses* 370 Looks like a phantom ship. No. Wait. Trees are they? An optical illusion. Mirage. 1937 K. Blixen *Out of Africa* v. 386 Between.. the mellow English landscape and the African mountain ridge, ran the path of his life: it is an optical illusion that it seemed to..swerve—the surroundings swerved. 1961 C. Greenberg *Art & Culture* 77 Flatness may now monopolize everything, but it is a flatness become so ambiguous and expanded as to turn into illusion itself—at least an optical if not, properly speaking, a pictorial illusion. 1971 D. Francis *Bonecrack* xiii. 169 Don't be tempted by the optical illusion that the winning post is much nearer than it really is. 1976 *Times* 20 Feb. 14/4 But it will be said, surely the growth of public spending has..been arrested... Not, alas, so. This is an optical illusion.

2. a. (Further examples.) Also used with specific reference to visible light as contrasted with other electromagnetic radiation: *spec.* operating in or employing the visible part of the spectrum.

optical density, (*b*) the logarithm to the base 10 of the opacity.

1891 W. Abney in *Jrnl. Soc. Chem. Industry* 31 Jan. 20/1 From the colour of the negatives..the photographic and optical densities were very nearly alike. 1953 J. W. T. Walsh *Photometry* (ed. 2) v. 161 The reciprocal of the transmission factor of a plate is sometimes termed the opacity of the plate and the logarithm to base 10 of the opacity is often known as the optical density of the plate. 1958, etc. [see *maser]. 1960 [see *laser[1] 1]. 1966 S. D. Rockoff in G. D. Whedon et al. *Progress in Devel. of Methods in Bone Densitometry* 7/1 The mean optical density of the bone, obtained from densitometric scanning of the radiographic image of the bone is expressed in terms of thickness of calibration wedge material which gives the same optical density. 1967 *Listener* 27 Apr. 544/2 Before the radio telescope made its impact..the optical

telescopes were able to photograph galaxies. 1970 *Sci. Jrnl.* Mar. 14 The first machine to automate completely one of the important processes of optical astronomy. 1974 *McGraw-Hill Yearbk. Sci. & Technol.* 11/2 The VLA will allow scientists to see, study, and map the radio sky at wavelengths of 1 cm or greater, with detail even greater than that possible for earthbound optical telescopes.

b. Used with reference to electromagnetic radiation other than light, and to beams of particles analogous to light: relating to the transmission of such radiation.

1933 [see *electron-optical* adj. s.v. *electron[3] 2 b]. 1938 Maloff & Epstein *Electron Optics in Television* i. 38 The electron optical problem of the electron microscope is to obtain electrostatic or magnetostatic lenses of short focal lengths capable of producing high magnifications with low image distortions. 1944 R. A. Sawyer *Exper. Spectroscopy* xi. 277 Infrared spectroradiometers differ from those used in the ultraviolet, or visible, regions chiefly because of the optical characteristics of the infrared region. 1964 *Jrnl. Optical Soc. Amer.* LIV. 15/2 A recent study of optical properties in the extreme ultraviolet.

c. Designating a form of abstract art and visual decoration in which optical effects are used to provide illusions of movement in the patterns produced. Chiefly as *optical art, artist, painting*, etc. Abbrev. *op[4]. orig. *U.S.*

1964 *Time* 23 Oct. 78/1 Preying and playing on the fallibility in vision is the new movement of 'optical art' that has sprung up across the Western world. 1964 *Life* 11 Dec. 133 Op-art is short for 'optical art', a paradoxical movement dedicated to the practice of fascinating deceptions. 1965 *Reporter* (N.Y.) 14 Jan. 46/3 They [*sc.* Israeli artists] do have the beginnings of Pop Art and Op Art (Optical Art). 1965 *Listener* 4 Feb. 196/3 To pass..to the American optical painter Edward Avedisian, and thence to the British Patrick Caulfield, is to realize the impossibility of trying to apply consistent standards to modern art. 1965 *Observer* 28 Feb. 2/6 The occasion was the private viewing of the most important show of the New York art season—an exhibition called 'The Responsive Eye', which has gathered together, for the first time, a comprehensive international collection of optical art. 1965 *New Scientist* 20 May 491 Leaving aside the question of the contemporary fashion for 'optical' painting, Vasarely's work shows—indeed it largely created—two important trends in abstract art today. 1968 N. Weston *Kaleidoscope of Mod. Art* xv. 220 The public has taken to Optical art in a big way, and it has spread rapidly from the art galleries to the Press. 1969 *Time* 7 Feb. 4 The Manhattan optical artist [*sc.* Josef Levi] has devised several new dizzying exercises with illuminated shadow boxes superimposed on black and white perforated metal screens. 1969 *New Yorker* 1 Nov. 12/3 Optical paintings by a leading Latin-American artist. 1970 C. Barrett *Op Art* i. 7/1 The evolution of abstract art was a preliminary step towards the development of optical painting. 1973 J. Lancaster *Introducing Op Art* ii. 28 Optical art is a method of painting concerning the interaction between illusion and picture plane, between understanding and seeing.

4. (Earlier examples.)

1610 H. Wotton *Let.* 13 Mar. in L. P. Smith *Life & Lett. Sir H. Wotton* (1907) I. 486 The Mathematical Professor at Padua, who by help of an optical instrument.. hath discovered four new planets. a 1666 Evelyn *Diary* an. 1641 (1955) II. 64 Those spotts in the Moone, attributed to the seas there &c according to our new Philosophy & the Phænomenas by optical Glasses.

5. *optical bleach, brightener, white*: a substance applied to textiles which produces a whitening effect by absorbing ultraviolet light and re-emitting it as blue light; so *optical bleaching, brightening, whitening*.

1947 *Jrnl. Textile Inst.* XXXVIII. A 521 'Optical bleaching' is defined as the physical alteration of a white fabric in such a way as to make it reflect, in addition to white light, a certain amount of blue light, and thereby producing a much 'bluer' and also brighter white than before. 1948 *Amer. Dyestuff Reporter* XXXVII. 432/3 Whereas 'bluings' improve the appearance of white products by absorbing red light, and 'bleaches' improve their appearance by destroying the yellow color, 'optical bleaches' operate on the entirely different principle of transforming ultra-violet and violet radiation into visible blue light. 1959 *Which?* Sept. 107/2 If a detergent with optical white is used on a white fabric too generously or too often, the fabric can get a blue or mauve tint which is not particularly pleasing. 1961 Cohen & Linton *Chem. & Textiles for Laundry Industry* v. 79 Optical bleaches will give blue by light emission and consequently improve brightness both psychologically and actually—that is, a 'whiter' white is obtained. 1964 Kirk & Othmer *Encycl. Chem. Technol.* (ed. 2) III. 739 The principle of optical bleaching was described in 1929 by Krais, but the industrial use of optical brightening began about ten years later. 1964 *Ibid.* XIX. 13 The majority of optical brighteners that are commercially available are based on stilbene derivatives. 1971 A. K. Sarkar *Fluorescent Whitening Agents* i. 1 Fluorescent whitening agents are known under various names, e.g. optical whitening agents, optical bleaching agents or optical bleaches, fluorescent bleaching agents, whiteners, brighteners, etc. 1974 *Encycl. Brit. Macropædia* XVI. 916/2 Now an integral part of all washing powders, optical brighteners are dyestuffs absorbed by textile fibres from solution but not subsequently removed in rinsing. *Ibid.*, The chemical structures of optical brightening agents are complicated.

6. Special collocations: **optical activity**, the ability of a substance to produce optical rotation; **optical axis, centre**: see sense 2 in Dict.; **optical bench**, a straight, rigid bar, usu. graduated, along which supports for lenses, light sources, and the like can be slid and to which they can be clamped; **optical character reader**, a device which performs optical character recognition and produces coded signals corresponding to the characters identified; **optical character recognition**, identification of printed characters using photoelectric devices; **optical comparator**, an instrument for facilitating comparisons of two objects by projecting shadows or transparencies of them on to a screen; **optical fibre**, a fibre that will act as a light guide in fibre optics; **optical flat** (see *flat *sb.*[3] 2 f); **optical glass**, glass of specially high homogeneity manufactured for use in optical components (see also sense 4); **optical isomer**, each of two isomeric compounds whose molecules are enantiomorphs and which are distinguishable by their equal but opposite optical rotations; so **optical isomerism**; † **optical length** = *optical path* below; **optical model** *Nuclear Physics*, a model of the atomic nucleus in which it is treated as having a potential well with an additional negative imaginary component, so that its behaviour with respect to incident particles is somewhat analogous to that of a partially absorbing body with respect to incident light waves; **optical path**, the distance which in a vacuum would contain the same number of wavelengths as the actual path followed by a ray of light, equal to the product of the actual path-length and the refractive index of the medium if the latter is homogeneous; **optical printer** = *projection printer* s.v. *projection 10; **optical pumping** [tr. F. *pompage optique* (A. Kastler 1950, in *Jrnl. de Physique et la Radium* XI. 257/2)], the production of an inversion in the population of certain energy levels in the atoms of a gas by the absorption of optical (visible) resonance radiation of suitable polarization; **optical pyrometer**, a device for measuring the temperature of an incandescent body by comparing its brightness with that of a heated filament in the instrument; **optical rotation**, the rotation of the plane of polarization of plane-polarized light by a substance through which it passes; *spec.* = *specific rotation; **optical scanner** = *optical character reader* above; **optical scanning**, scanning in which the light reflected or transmitted by the area being scanned is detected, esp. as used in optical character recognition; **optical sound** *Cinemat.*, sound recorded by optical (photographic) means on a film.

1877 *Chem. News* 23 Nov. 230/1 The optical activity disappears in those derivatives of active bodies, by the formation of which the so-called asymmetry [*sic*] of the carbon atoms ceases. 1883 Optical activity [in Dict., sense 2]. 1967 K. B. Krauskopf *Introd. Geochem.* xi. 295 If a natural material containing carbon compounds can be shown to possess optical activity, the conclusion seems inescapable that living organisms played a rôle in its formation. 1883 R. T. Glazebrook *Physical Optics* v. 113 The experiment is usually made on an optical bench. 1974 *Sci. Amer.* Apr. 28/3 An optical laboratory needs a large optical bench that is typically 20 feet long. 1962 *Proc. Symposium Optical Character Recognition* I. 133 The IBM 1418 Optical Character Reader..provides an example of the maximum tolerance for several of these variables that can be obtained in today's commercially available character readers. 1968 *Amer. Documentation* Jan. 74/2 Typed pages are transferred to magnetic tape by an optical character reader. 1974 *Encycl. Brit. Macropædia* XIV. 892/1 The U.S. Post Office has had an alphanumeric optical character reader in operation on live mail since 1965. 1962 *Proc. Symposium Optical Character Recognition* I. 93 A research and development program was initiated..to create a wide-tolerance optical character recognition system. 1970 O. Dopping *Computers & Data Processing* iii. 64 In optical character recognition (OCR), the reader responds to the darkness of the ink, just as the human eye does. 1935 O. W. Boston *Engin. Shop Pract.* II. ix. 452 An optical comparator or Optimeter is a gage used for comparing the size of various parts with that of master gages. 1967 *Economist* 14 Oct. 4 (Advt. suppl. following p. 176), The image of the gauge, magnified to 50 times its size, is projected on the screen of the Optical Comparator so that the finest details of the thread can be examined and checked. 1976 M. Maguire *Scratchproof* iii. 40 Perhaps the boys at forensic had made a gaff. Perhaps the optical comparator was malfunctioning. 1970 *Sci. Jrnl.* Dec. 68 Nobody has yet produced glass which meets all the challenging requirements of optical fibres. 1974 *McGraw-Hill Yearbk. Sci. & Technol.* 27/1 Optical fibers that act as 'light pipes' have been around for some years, serving in a variety of ways—in medical instruments, in photocopying machines, in automobile instrument panels. 1840 *Mem. R. Astron. Soc.* XI. 165 (*heading*) On the optical glass prepared by the late Dr. Ritchie. 1879 *Encycl. Brit.* X. 665/1 Optical glass is of two principal kinds—flint and crown. 1922 L. Bell *Telescope* iii. 60 The fundamental difference between the making of optical glass and the ordinary commercial varieties lies in the individual treatment of each charge necessary to secure uniformity and regularity. 1973 D. G. Holloway *Physical Prop. Glass* iv. 107 The characteristics of optical glasses are customarily represented by quoting the refractive index for one of the sodium D lines and the constringence. 1892 Optical isomer [see *enantiomorph]. 1974 *Encycl. Brit. Macropædia* V. 138/2 Optical isomers differ from the other isomeric coordination compounds in that their physical and chemical properties are identical. 1894 *Jrnl. Chem. Soc.* LXVI. I. 422 (*heading*) Optical isomerism of closed chain compounds. 1968 R. O. C. Norman *Princ. Org. Synthesis* v. 156 Optical isomerism in biphenyls is possible because the conformation..which possesses a plane of symmetry is strained with respect to non-coplanar conformations. 1894 *Phil. Mag.* XXXVII. 515 The time it takes for an impulse at G to pass completely through F will be that required by light to go over a space equal to the difference in optical length of the extreme rays GBF and GAF. 1934 W. H. A. Fincham *Optics* ii. 26, *nl* is termed the optical length of a path *l* in medium of refractive index *n*. 1952 Le Levier & Saxon in *Physical Rev.* LXXXVII. 40/1 There is an energy range for which the problem of nucleon-nuclei scattering is somewhat analogous to the scattering of light by a conducting glass sphere... Such a medium is conventionally described by introducing a complex index of refraction... We have investigated an optical model for nucleon-nuclei scattering in which a complex square well potential is used, this being equivalent to a complex index of refraction. 1963 P. E. Hodgson *Optical Model Elast. Scattering* i. 3 The initial development of the optical model was due to Fernbach, Serber, and Taylor (1949). 1970 I. E. McCarthy *Nuclear Reactions* i. iii. 70 A mechanistic description of a non-elastic reaction..involves an optical model description of the system while it is in the entrance channel and another one while it is in the exit channel. The transition between these channels is described by a model which includes details of nuclear structure. 1893 *Phil. Mag.* XXXV. 471 The phenomenon is due to the interference of two parallel gratings... Their distance, which is virtually constant, is the optical path, 2*ne*, *e* being the thickness, and *n* the index of the gelatine. 1923 Glazebrook *Dict. Applied Physics* IV. 216/1 The differences of optical paths are a direct measure of the distortion from true spherical form of the waves emerging from a lens system. 1957 G. E. Hutchinson *Treat. Limnol.* I. vi. 391 Poole and Atkins computed *b*, the mean optical path per meter, as 1·19. 1944 *Jrnl. Soc. Motion Pict. Engin.* XLII. 204 The use of the optical printer to enhance the value of the modern motion picture is demanded increasingly by studios with foresight enough to give a free hand to the man in charge of the optical department. 1953 L. J. Wheeler *Princ. Cinematogr.* v. 144 Figure 61 shows in outline the layout of the intermittent or 'stop' optical printer. 1974 L. Lipton *Independent Filmmaking* i. 39 Optical printers are used to make dissolves, special effects, frame line corrections, freeze-frame printing and so on. The optical printer is also ideal for the preparation of master printing material printed from original camera film. 1952 *Physical Rev.* LXXXV. 1051/2 The net result is an optical 'pumping' as Kastler suggests, from $m = -\frac{1}{2}$ to $m = +\frac{1}{2}$, or a tendency in the direction of nuclear orientation. This is in competition with a disorienting tendency caused by collisions. 1959 [see *multilevel a.]. 1970 G. K. Woodgate *Elem. Atomic Struct.* ix. 191 Recently..the methods of optical pumping, double resonance, and level-crossing spectroscopy have begun to provide data on the hyperfine structure of excited states. 1901 G. K. Burgess tr. *Le Chatelier & Boudouard's High-Temperature Measurements* viii. 155 The optical pyrometer, by reason of the uncertainty of emissive powers.., cannot give as accurate results as other pyrometric methods. 1922 Glazebrook *Dict. Appl. Physics* I. 649/1 It is possible to calibrate optical pyrometers by direct observations of freezing- or melting-points. 1958 Bussard & DeLauer *Nuclear Rocket Propulsion* viii. 303 The optical pyrometer has been built around the frequency-response characteristics of the human eye. 1895 C. S. Palmer tr. *Nernst's Theoret. Chem.* II. v. 288 The connection with constitution is shown in no other physical property so clearly as it is in this of optical rotation. 1929 R. A. Gortner *Outl. Biochem.* xxi. 473 Naturally-occurring tartaric acid is the *d* form, and its purity is usually determined by the optical rotation of a solution of the acid. 1967 *Oceanogr. & Marine Biol.* V. 190 Suzuki and Suzuki..have demonstrated that ovalbumin has an optical rotation of −27·6° in the native state. 1962 *Proc. Symposium Optical Character Recognition* I. 16 (*caption*) An example of the actual printed output which optical scanners are required to read. 1975 *McGraw-Hill Yearbk. Sci. & Technol.* 201/1 An optical scanner..reads a symbol printed on the product as it is passed over a slit in the checkout counter. 1958 J. Moir *High Quality Sound Reproduction* xviii. 540 Optical scanning of the sound track takes place while the film is held in contact with the drum and rotating with it. 1971 *Computers & Humanities* V. 282 Optical scanning of printed texts (avoiding the need for keypunching the material). [1933 B. Brown *Amat. Talking Pict.* vii. 146 The optical sound system of the R.C.A. portable is shown diagrammatically.] 1960 *McGraw-Hill Encycl. Sci. & Technol.* III. 124/2 The magnetic sound track is..26 frames ahead [of the picture] on 16-mm film, the same as optical sound. 1970 *New Yorker* 29 Aug. 22/1 Here we have to use optical sound—printed down the side of the film.

B. *sb.* **1.** *Cinematogr.* and *Television.* (See quots.)

1953 K. Reisz *Technique Film Editing* 281 *Optical*, any device carried out by the optical department of a laboratory requiring the use of the optical printer, e.g., dissolve, fade, wipe. 1959 Halas & Manvell *Technique Film Animation* 340 *Opticals*, mixes, fades, wipes, etc., which are made by the processing laboratories on an optical printing machine after the scenes have been photographed, instead of in the camera during photography. 1959 *Punch* 28 Oct. 366/2 Many amateurs are clearly of the opinion that this [*sc.* the subject matter] is of minor importance, their chief concern being to demonstrate their skill at devising star-burst wipes and other trick opticals. 1969 C. O. Raspor in W. R. R. Park *Plastics Film Technol.* iv. 94 It has become customary to distinguish three properties of films which are generally categorized as 'opticals': gloss, haze, and transparency. 1970 A. Fowles *Dupe Negative* xiv. 192 'I've got four hundred feet of 35 mm. ECO original here,' I said, 'how long will it take to strike a master positive?'.. 'You can have it by five this evening,' he said, 'but wouldn't you rather have an optical?' 1974 L. Lipton *Independent Filmmaking* vi. 273 When printed with an optical printer, the image can have the same orientation as it had on the camera film. If we used a contact printer to make the fades and dissolves, and then cut them into the master, we'd wind up with release prints with flopped opticals.

2. An example of optical art.

1966 *New Statesman* 5 Aug. 208/2 A sizzling red-blue optical by Ellsworth Kelly.

optician. Add: **2.** *spec.* One who makes up and dispenses spectacles and corrective lenses (sometimes also testing the eyes and providing a prescription).

1892 *Keystone* (Philadelphia) June 578/2 Some persons are unkind..enough, after the optician has spent a great deal of time testing their eyes, to ascertain from him the proper number of the glasses they need, and then..go around the corner to some street peddler of spectacles. 1897 *Sears, Roebuck Catal.* 462/1 Opticians make a practice of imposing on their customer..and give as an excuse for the prices charged, 'that the lenses were ground to order'. 1912 L. Laurance *Visual Optics* p. vii, I have endeavoured to cover..all that is essential for the sight-testing optician. *Ibid.*, No apologies are needed for mentioning some indications of pathological conditions, since a person with defective sight may go to the optician when he should go to the oculist. 1928 S. Duke-Elder *Practice of Refraction* xxiii. 333 The ophthalmologist has no more valuable and essential asset than a reliable optician with whom to co-operate. 1951 R. L. Stimson *Ophthalmic Dispensing* vi. 140 An ophthalmic dispenser's First Commandment is, an optician shall never express an opinion pertaining to a patient's vision or eye health. 1969 *Which?* Feb. 43/1 Ophthalmologist. He is a qualified doctor who has had a further training in eye disease... He will give you a prescription for your spectacles... He will not make up the spectacles... Ophthalmic Opticians are opticians who do sight tests... After giving you a prescription for your spectacles, an Ophthalmic Optician will usually expect to make them up herself. *Ibid.*, A Dispensing Optician does not do sight tests but makes up your spectacles to your prescription, wherever you have got it. 1975 *Nature* 8 May 151/1 It has sometimes been noted by opticians involved in the fitting of contact lenses that the sensitivity of the cornea seems to vary, depending on whether the patient has blue or brown eyes.

opticity (ǫpti·sĭti). [ad. F. *opticité* optical quality: see Optic *a.* and *sb.* and -ity.] In the brewing and food industries, the degree of optical activity of a solution, as a measure of its concentration.

1900 *Jrnl. Federated Inst. Brewing* VI. 219 The pyridine solutions of osazones are..very suitable for determinations of opticity. 1924 J. Grant *Chem. of Breadmaking* (ed. 4) xiii. 197 Two instruments in common use for the determination of the opticity of sugars and other optically active carbohydrates are the Laurent..and the Schmidt-Haensch.

optico-. Add: **o:pticokine·tic** *a.* = *opto-kinetic *a.*

1950 F. H. Adler *Physiol. Eye* x. 372 Opticokinetic nystagmus is induced in a subject using wide alternating stripes on a rotating drum. 1965 F. W. Newell *Ophthalm.* xxvi. 442/2 Opticokinetic nystagmus arises from looking at constantly moving objects, such as telegraph poles, from a moving automobile or train.

optimal, *a.* Delete *Biol. rare* and add further examples. Add to def.: most satisfactory.

1900 A. L. Loeb tr. *J. Loeb's Compar. Physiol. Brain* xv. 223 The greatest happiness in life can be obtained only if all the instincts—that of workmanship included—can be maintained at a certain optimal intensity. 1935 Adams & Zener tr. *Lewin's Dynamic Theory of Personality* iii. 110 Optimal environmental conditions..vary considerably with different individuals. 1935 *Mind* XLIV. 359 There might be *optimal* sense-data, such as the one which a man senses when he observes a penny head on, from a short distance, in a good light, with a normal eye, etc. 1956 Amos & Birkinshaw *Television Engin.* II. xv. 222 The constants of a feedback circuit can be proportioned to give an optimally-flat response, i.e., one with an equation having no terms in frequency. Such feedback is termed optimal and gives the widest possible frequency coverage. 1961 *Atlantic Monthly* Apr. 42/2 Students.. have often told me that it doesn't pay to be too interested in anything, because then one is tempted to spend too much time on it, at the expense of that optimal distribution of effort which will produce the best grades. 1970 *Sci. Jrnl.* Jan. 25/1 A controller..automatically adjusts the speed and feedrates of the drill so as to maintain optimal performance as the drill wears. 1971 D. C. Hague *Managerial Econ.* (rev. ed.) i. 11 An optimal decision is the one which comes as close as possible to achieving a given objective.

Hence **o·ptimally** *adv.*, in the best or most advantageous way.

1933 *Proc. R. Soc.* B. CXII. 505 The animal's tissues were not optimally hydrated. 1950 D. Halliday *Introd. Nucl. Physics* ix. 340 Phase-defocused atoms are not optimally accelerated and do not contribute to the useful beam. 1956 [see above]. 1972 *Sci. Amer.* Sept. 148/3 The time of physicians is not always optimally employed. 1976 *Lancet* 13 Nov. 1039/1 In the absence of bacteria the nitrosation reaction proceeds optimally at an acid pH.

optimality (ǫptimæ·lĭti). [f. Optimal *a.* + -ity.] The state or quality of being optimal.

1944 Von Neumann & Morgenstern *Theory of Games* iii. 162 (*heading*) Mistakes and their consequences. Permanent optimality. 1961 W. J. Baumol *Econ. Theory & Operations Analysis* ix. 184 Our condition..is a geometric representation of the following basic optimality rule. 1964 Gould & Kolb *Dict. Social Sci.* 573/2 The rational consumer of formal economic theory maximizes his expected utility, and the rational entrepreneur maximizes his expected profit. If a distinction is wanted between this very strict species of rationality and more general forms, the former may be termed optimality, the latter adaptiveness or functionality. 1970 P. A. Samuelson *Economics* (ed. 8) iv. xxxii. 609 Graduate treatises call this a case of 'Pareto optimality', named after Pareto's work at the turn of the century. 1971 *Nature* 31 Dec. 527/2 Jardine and Sibson showed that the single-link method is the only hierarchic cluster method which satisfies certain invariance and optimality conditions. 1974 *Times Lit. Suppl.* 22 Feb. 172/3 A model of the city which achieves Pareto optimality (i e, a condition in which no one can get richer through moving without making someone else poorer).

optimalization (ǫ:ptimăləizēi·ʃən). [f. Optimal *a.* + -ization.] = *optimization.

1965 *Economist* 31 July 446/3 A mechanistic commentary on optimalisation theory and resource allocation. 1971 *Nature* 15 Jan. 150/3 The council pulled its punch, and decided to 'limit itself to searching for optimalization within the possibilities existing in the present constitutional situation'. 1975 *Gen. Systems* XX. 108/2 Allen Johnson..found that legends of agricultural affinity well befitted ecological optimalization in northeastern Brazil. 1977 *Language* LIII. 327 We have seen that 'optimalization of phonotactic structure' (i.e. the *C + *h* constraint) wins out over 'paradigm coherence' (i.e. the invariable prefixation of pronouns) in Hua.

optimate, *sb.* **1.** (Later examples.)

1954 I. Murdoch *Under Net* xvii. 234 The editor was calling on the *optimates* to exercise..strong measures. 1966 Auden *About House* 20 As Nietzsche said they would, the *plebs* have got steadily Denser, the *optimates* Quicker still on the uptake.

optimific (ǫptimi·fik), *a. Philos.* [f. L. *optim-us* best + -fic.] Producing the maximum good consequences.

1930 W. D. Ross *Right & Good* 34 No one *means* by 'right' just 'productive of the best possible consequences', or 'optimific'. 1933 *Mind* XLII. 181 The 'Maximalist' Theory (this designation suggests quantity better than 'Optimific'). 1940 —— XLIX. 230 The first half of Universalistic Ethical Hedonism, to wit the theory that being optimific is the one and only right-making characteristic.

optimization (s.v. Optimize *v.*). Add to def.: The action or process of rendering optimal; the state or condition of being optimal. (Further examples.) Freq. *attrib.*

1951 Parsons & Shils *Toward Gen. Theory Action* II. ii. 123 An important superordinate problem concerning mechanisms which depend entirely on the learning-performance distinction (when this is taken as relevant to the over-all problem of the system—the optimization of gratification). 1959 *Times Rev. Industry* June 39/2 Optimization problems (e.g. minimization of costs or maximization of profits). 1961 *Aeroplane* CI. 725/1 These desiderata were then used in what the Americans delight in calling 'optimization' studies, to find the best of some 300 helicopter configurations. 1966 S. Beer *Decision & Control* x. 211 Given this framework, the process of reaching the best decision is known as optimization. 1968 *Brit. Med. Bull.* XXIV. 242/1 This resulted in a major advance known as optimization of treatment..which in essence is for the computer to give the radiotherapist a number of alternative plans, and for the radiotherapist to decide which he considers to be the most suitable. 1974 Cooper & Steinberg *Methods & Appl. Linear Programming* i. 1 In an optimization problem we seek values of the..variables which do not violate the several constraints imposed on them, but which lead to an optimal (maximal or minimal) value of the function which is to be optimized. 1976 *Daily Times* (Lagos) 22 Sept. 3/3 The socio-economic development of any nation depended on the optimisation and maximisation of land use.

optimize, *v.* Add: **2. b.** To render optimal.

1946 [see *beam *sb.*[1] 24]. 1958 *New Scientist* 17 July 410/1 They could optimise the designs to be submitted.. for the first three nuclear power stations to be built in Britain. 1965 *Economist* 15 May 788 The shape of the aircraft, i.e. its geometry, can be optimised for what ever it happens to be doing. 1966 T. Lupton *Managem. & Social Sci.* ii. 44 To optimize any one of these elements does not necessarily result in a set of conditions optimal for the system as a whole. 1974 [see prec.]. 1975 *Nature* 10 Apr. 498/2 If the expected signal wavelength is known there are specialised methods which can optimise the sensitivity of such an experiment. 197[illegible] *N.Y. Times* 30 Mar. C 11/[illegible] As a small businessman, you never really understand the importance of optimizing time and effort.

3. *intr.* To become optimal.

1971 *Nature* 23 July 251/1 The eight radii model quoted cannot optimize to the two layer model above. 1972 *Physics Bull.* Feb. 91/3 The bonding of p electrons in the broadside on (π) arrangement optimizes at shorter distances than that in the end on (σ) position.

optimum, *sb.* (*a.*) Delete ‖ and *Biol.* and add to def.: The best, or the most favourable or advantageous, condition, situation, etc.

1955 Schultz & Cleaves *Geol. in Engin.* xvii. 416 It can be seen from the shape of the curves near optimum that moisture content is extremely critical in the compaction of soils. 1970 O. Dopping *Computers & Data Processing* xxii. 340 In many planning programs..time limitations make it impossible to reach the optimum and provide proof that this has been reached. 1971 *Marriage, Divorce & Church* iii. 50 In principle the State need not encourage more children when population growth appears to be reaching, or to have passed, an optimum.

B. *adj.* (Further examples.) Cf. *optimal *a.*

1926 Rideal & Taylor *Catalysis* (ed. 2) v. 113 Most of the statements concerning promoter action by added substances have no information which would indicate what the optimum concentration of promoter is for the given reaction. 1929 S. Leslie *Anglo-Catholic* iii. 34 How shall we strike the so-called optimum density, which is best for both the health and the soul of London, and keep the population there? 1930 *Economist* 1 Nov. (Russ. Suppl.) 13/1 The question of the optimum size of such a farm is not quite settled. 1935 *Planning* 18 June 8 It is a problem of maintaining the numbers and balance of population in such a manner as to enable optimum social and economic activity. 1949 E. W. Kimbark *Electr. Transmission* xiii. 279 Another matching device is used at the sending end of the line..in order to obtain optimum power output from the transmitter. 'Optimum' signifies a compromise between maximum output and other considerations. 1950 G. B. Shaw *Farfetched Fables* Pref. 98 While the time lag lasts the future remains threatening. The problem of optimum wealth distribution..will not yield to the well-intentioned Utopian amateurs. 1954 H. M. Corley *Successful Commerc. Chem. Devel.* ix. 152 The question of the optimum production rate for a new plant is not easy to answer. 1967 E. Duckworth in Wills & Yearsley *Handbk. Managem. Technol.* 110 When the optimum order quantities have been decided, problems may occur in scheduling these through factories in the optimum manner. 1975 *New Yorker* 21 Apr. 92/2 These drivers are deviant only in that they depart from ideal or optimum behavior.

o·pting, *vbl. sb.* [f. Opt *v.* + -ing[1].] The action of the verb Opt. Freq. const. *out*. So **opting-in** *vbl. sb.*, **opting-out** *vbl. sb.*

1922 *Glasgow Herald* 13 Apr. 6 The opting of Australia out of the chain..does..impair the symmetry of the Imperial chain plan. 1958 [see *apolitical *a.*]. 1966 *Guardian* 17 Oct. 1/5 An opting out of the Church's duty. 1969 *Daily Tel.* 19 Aug. 20/1 On the crucial question of when doctors should be allowed to use the hearts of accident victims, the weight of opinion is that an 'opting in' principle should apply—only people who have signed a legal document stating that this can be done should be used as heart donors. 1970 *Globe & Mail* (Toronto) 28 Sept. 1/6 The provincial cabinet will study a proposal from Dr. Robillard for an entirely new opting-out scheme for the plan. 1973 D. Aaron *Unwritten War* III. viii. 125 Howells neither extenuated nor deprecated his opting out in his reminiscences. 1977 *Jrnl. R. Soc. Arts* CXXV. 639/2 'Alienation'..too often..is used as an alibi for inaction, lack of imagination or a sort of 'opting out' (another dangerous expression).

option, *sb.* **1.** Add to second part of def.: an alternative, a choice. (Further examples.) *soft option*: a choice which entails no difficult or strenuous actions or decisions; also (with hyphen) *attrib.*

1923 *Granta* 2 Mar. 315/1 It follows that our Tripos must be difficult; that we have little use for 'duds', for Tutors who misconceive it as being a soft option. 1953 R. Lehmann *Echoing Grove* II. 42 The lifelong consequences of a choice that, once made, is made to be adhered to with no soft option. 1957 M. K. Joseph *I'll soldier no More* (1958) xiii. 238 Odd bods from various HQs and soft-option types with vague jobs. 1967 *Technology Week* 23 Jan. 11/1 (Advt.), Everything is modular—memory, input/output processors, peripherals, central processor options, software. 1967 *Listener* 6 July 5/2 There is a tendency for many prospective students to regard social studies as a soft option. 1969 *Guardian* 14 May 1/2 Mr. Callaghan..is understood to have reserved his decision—or, in the current jargon, 'kept his options open'. 1971 J. B. Carroll et al. *Word Frequency Bk.* p. xv, This decision was guided by the results of a pilot test undertaken to try out various procedural options for the eventual work on the AHI Corpus. 1972 D. Delman *Week to Kill* 139 This cut down my options. 1973 'M. Innes' *Appleby's Answer* xxi. 183 You've been most fiendishly clever... You've kept your options open. 1976 K. Thackeray *Crownbird* v. 94 Gould had tremendous self-assurance, the ability to cope when all the options were running out. 1977 *Times* 22 Apr. 18/8 The British electorate have a powerful instinct for the soft option and a quiet life.

b. *spec.* in *Amer. Football*, a play in which a quarter- or half-back chooses whether to pass or to run with the ball; also *attrib.*

1954 *Sun* (Baltimore) 25 Nov. 15/4 We couldn't pass enough from it and our quarterbacks couldn't take the pounding on the option play as a steady diet. 1966 Rote & Winter *Lang. Pro Football* III. 127/2 *Option*, play where ball carrier has choice of running or passing; option pass play. 1974 *Cleveland* (Ohio) *Plain Dealer* 13 Oct. 1-C/3 He baffled the Badgers with the option run, gaining 146 yards and scoring on runs of 11 and six yards. 1976 *Webster's Sports Dict.* 291/2 When a halfback is running the option play, it is commonly called the *halfback option*.

2. b. Alternative; esp. in phr. *with* (or *without*) *the option* (*of a fine*).

1901 *Chambers's Jrnl.* Sept. 582/2 A third [conviction] should result in imprisonment without the option of a fine. 1903 'T. Collins' *Such is Life* (1944) iii. 106 Yet he has thoughts that glow, and words that burn, albeit with such sulphurous fumes that, when uttered in a public place, they frequently render him liable to fourteen days without the option. 1908 *Daily Chron.* 16 Sept. 4/6 A Suffragette who has been offered the option of a fine. 1914 E. Pankhurst *My Own Story* 71 They were given the option of a fine. 1925 Wodehouse *Carry on, Jeeves!* vii. 159 He will serve a sentence of thirty days in the Second Division without the option of a fine.

4. (Further examples.)

1857 *Hunt's Merchant's Mag.* XXXVII. 134 A purchase of stocks at the brokers' board, buyers' option, thirty,

sixty, or ninety days, can call for the stock any day within that time... He pays interest at the rate of 6 per cent up to the time he calls. *Ibid.*, Seller's option. This gives the seller the option to deliver any time within the time of his contract, or at its maturity... The buyer..pays interest up to delivery. 1909 D. Lloyd George in *Hansard Commons* 29 Apr. 519 'Option notes' will be charged at similar rates, calculated upon the value of the securities to which the option relates. 1928 *Daily Mail* 25 July 18/5 None of the shares of the Company are under option. 1976 *Listener* 29 Apr. 533/3 You had bought the option for a book that they hadn't yet written. 1976 L. St. Clair *Fortune in Death* iii. 25 'And the stocks you are interested in?' 'Just one. Aglia Petroleum. Thought I might pick up some calls on it.' '"Options" is our term.' 1977 *Gay News* 7–20 Apr. 22/1 WH Allen, who have the option on it here, are wary of libel suits. 1977 A. Morice *Scared to Death* i. 7 One condition of this guarantee is an option on each of the plays.

6. *option market, note, plan, price, time*; also *options exchange*.

1865 *Shareholders' Guardian* 8 Nov. 847/2 If at the expiration of the 'option' time the price be the same as the 'option' price, the person who paid the money has the right to buy, sell, or neither, as he thinks proper. 1909 Option note [see prec. sense]. 1930 *Daily Express* 30 July 10/3 (*heading*) The option market. 1961 'E. Lathen' *Banking on Death* (1962) xv. 123 He had 10 per cent of the stock already as part of an option plan. 1973 *N.Y. Law Jrnl.* 4 Sept. 7/5 (Advt.), The Chicago Board Options Exchange opened in late April.

b. (gen.) option mortgage (see quots.).

1966 *New Statesman* 4 Mar. 312/1 People will be free in future to choose between taking an ordinary mortgage and an 'option' mortgage. 1971 *Reader's Digest Family Guide to Law* 61/1 The Option Mortgage Scheme..gives the borrower the choice of foregoing tax relief in return for a Government subsidy which reduces the interest on his mortgage loan. *Ibid.* 61/2 The Option Mortgage Scheme was devised to lower the cost of home ownership to people who do not pay enough tax to obtain full advantage from tax relief on mortgage repayments. 1975 in R. Crossman *Diaries* I. 343 The Minister had announced the new option-mortgage scheme, providing subsidies on mortgages whatever the ruling interest rate if mortgagors would forgo tax relief at the standard rate. 1976 *Star* (Sheffield) 3 Dec. 6/4 Our mortgage is an Option Mortgage, therefore we get no tax relief.

o·ption, *v.* Chiefly *U.S.* [f. the sb.] *trans.* To buy or sell under option; also, to have an option on.

1934 in Webster. 1947 *Sun* (Baltimore) 3 Apr. 20/1 It was necessary for the Flock to purchase him inasmuch as Cleveland could not option him out again. 1966 E. V. Rickenbacker *Rickenbacker* (1968) viii. 127, I also optioned the land around the company for future expansion. 1968 R. Lockridge *Murder in False Face* (1969) v. 68 A friend of his had had a musical optioned a dozen times. 'Lived on options for years,' he said. 1973 *Publishers Weekly* 26 Feb. 121/2 She has written a first novel and had it optioned for films. 1975 *Bookseller* 11 Oct. 2038/3 With [the book] *Saladin* already optioned by the film makers for 150,000 dollars. 1977 *Ottawa Citizen* 8 Feb. 17/2 Irwin Meyer and Stephen R. Friedman..heard the score for Annie. They liked it; they optioned it.

optional, *a.* Add: **1.** (Further examples.)

1934 *Language* X. 120 Occasionally one finds free variants, that is, non-conditional or optional variants. 1964 E. Palmer tr. *Martinet's Elem. Gen. Linguistics* iii. 68 In the case of the actor who 'rolls' his r's on the stage but uses the 'throaty' pronunciation elsewhere, we may rather speak of 'optional' variants. 1971 *Good Motoring* Sept. 18/2 Rubber mats are standard; carpets are an optional extra at £10·88. 1972 Berman & Szamosi in *Language* XLVIII. 313 In the above examples, the phenomenon of 'optional' stress placement does not seem to correlate with significant differences in meaning or sense.

2. (Later examples.)

1930 [see *artificial *a.* 3 b]. 1962 *Gloss. Terms Automatic Data Processing* (*B.S.I.*) 39 *Optional stop instruction*, an instruction which includes the possibility of stopping the operation of the program immediately before or after the instruction is obeyed, there being some means of permitting or inhibiting this facility as required. 1967 *Technology Week* 23 Jan. 11/1 (Advt.), Input/output is managed independently by one built-in and five optional I/O processors, with up to 160 automatic I/O channels. 1970 D. Dodge *Hatchetman* i. 17 'Underwriters..are being tapped for U.S. dollar payoffs under an airtight optional-money clause.' 'What's an optional-money clause?'.. 'Most major insurers in international trade write it... You put a clause in your insurance policy giving you an option to take payment in another money of your choice.'

o·ptional, *sb.* orig. *U.S.* [f. the adj.] An optional subject or course of study; a group of students constituting a class devoted to an optional subject.

1855 *Songs Biennial Jubilee Class of '57* (Yale Univ.) 3/2 For optionals will come our way. 1857 *Yale Lit. Mag.* XXII. 291 What was never known before, since the establishment of optionals, the number pursuing the study of Hebrew is nine. 1900 *Dialect Notes* II. 47 *Optional*. 1. An optional course selected by a student in addition to his regular work. 2. A student who elects only optional or special courses. 1930 *Times Educ. Suppl.* 12 Apr. 167/3 The advanced mathematics paper set as an 'optional' by London University. 1934 *Times Lit. Suppl.* 8 Mar. 162/2 *Optional*.—An optional subject of study. One example, of 1857, is given from Yale. For over 40 years at St. Paul's School the word has meant the class studying such a subject—e.g., the Spanish optional.

optiona·lity. [f. Optional *a.* + -ity.] Optional quality; opportunity or freedom of choice.

1880 *Scotsman* 24 Jan. 6/3 How much optionality there may be in an option which is allowed to opt only in one direction may yet be a question for the learned. 1972 *Language* XLVIII. 337 (*heading*) Stress Optionality and the Global Alternative. *Ibid.* 340 The second case of optionality concerns questions. 1975 *Amer. Speech 1973* XLVIII. 39 Even so, the number of implicational relationships remains huge, suggesting..that at least some optionalities are '*not* learned by the child, but [are] predictable from exactly the kinds of substantive constraints..[and] general functional considerations..as those discussed' in Kiparsky's article and suggested to him on quite independent grounds. 1975 *Language* LI. 1015 Generative grammar has no way of formalizing optionality.

o·ptionalize, *v.* [f. Optional *a.* + -ize.] *trans.* To make optional.

1921 *Proc. Classical Assoc.* XVIII. 43 Scotland, where the disastrous results of optionalising Greek in the Universities..have had time to manifest themselves.

o·ptionless, *a. poet.* [-less.] Without choice, without an option.

1908 Hardy *Dynasts* III. I. i. 328 The hunger for embranglement That gnaws this man, has left us optionless, And hailed us recklessly to horrid war.

optoacoustic (ǫ:pto,ăkū·stik), *a.* Also **opto-acoustic.** [f. Opto- + Acoustic *a.*] Involving or being the effect whereby a light beam periodically interrupted at an audio frequency produces an audible sound when made to irradiate an enclosed body of gas.

[1959 *Science Progress* XLVII. 459 (*heading*) The optic-acoustic effect in gases.] 1971 *Jrnl. Appl. Physics* XLII. 2934/1 This technique uses the optoacoustic effect which was discovered by Bell, Tyndall, and Röntgen. 1976 *Nature* 19 Aug. 681/2 Opto-acoustic detection was discovered by Bell nearly a century ago, but was recently revived as a means of monitoring low concentrations of pollutant molecules.

optochin (ǫ·ptotʃin). Also **optoquine.** *Pharm.* [ad. mod.L. *optochinum* (coined in Ger. by M. Goldschmidt 1913, in *Klin. Monatsbl. f. Augenheilkunde* LI. II. 449), f. Opto- (from its use in treating eye infections) + G. *chin-in* Quinine.] = *ethylhydrocuprein*(e s.v. *ethyl.

1914 *Chem. Abstr.* VIII. 3692 (*heading*) Chemotherapy of the pneumococci infections of the eye, particularly of the ulcus serpens by means of optochin ointment (ethylhydrocuprein). 1925 [see *ethylhydrocuprein*(e s.v. *ethyl]. 1957 J. H. Burn *Princ. Therapeutics* xxvii. 225 When by 1918 it was agreed that, clinically, optochin was dangerous, hope of treating bacterial infections seemed to fade away. 1969 *Acta Path. et Microbiol. Scand.* XLVII. 315 The optochin test is more reliable than the bile test in differentiating between pneumococci and streptococci. 1970 *Jrnl. Gen. Microbiol.* LXI. 138 Assays were made.. for the single transformant, to erythromycin, streptomycin, and optochin resistances.

optoelectronic (ǫ:pto,ĭlektrǫ·nik), *a.* Also **opto-electronic.** [f. Opto- + *electronic *a.*] Involving or pertaining to the interconversion or interaction of light and electronic signals.

1955 *Proc. IRE* XLIII. 1906/1 The opto-electronic characteristics of electroluminescent and photoconductive transducers make them suitable for..devices and functional networks capable of light amplification, light switching, and light storage. 1959 *RCA Rev.* XX. 742 The simplest class of optoelectronic devices consists of converters. One of the earliest optoelectronic converters is the photovoltaic cell. 1969 *Sci. Jrnl.* Jan. 72/1 The integration of the semiconductor lamp into a microcircuit is a logical step towards opto-electronic applications. 1976 *Pract. Electronics* Oct. 838 (Advt.), This catalogue..offers items from advanced opto-electronic components to humble (but essential) washers.

optoelectronics (ǫ:pto,ĭlektrǫ·niks). *sb. pl.* (const. as *sing.*). Also **opto-electronics.** [f. Opto- + *electronics.] The study and application of optoelectronic effects.

1959 *RCA Rev.* XX. 742 Solid-state optoelectronics concerns the use and control of numerous relations among optical and electronic phenomena in solids. 1963 T. E. Bray in D. K. Pollock et al. *Optical Processing of Information* xvi. 216 Much of the interest in optoelectronics stems from its low cost potential. 1968 *Brit. Universities Ann.* 20 Courses in Man-Machine Systems Engineering and Opto-Electronics. 1973 *Electronics & Power* 5 Apr. 117/4 The use of an optoelectronics transmitter as a means of transmitting quantity and price data from self-service petrol pumps to garage cash desks. 1976 *Pract. Electronics* Oct. 803/3 With over 150 pages the catalogue is broken down into seven sections; Transistors; Integrated Circuits; Diodes and Rectifiers; Opto Electronics; Resistors; Capacitors and Accessories.

optokinetic (ǫptokine·tic, -kəine·tik), *a.* [f. Opto- + Kinetic *a.*] Pertaining to or designating a form of nystagmus produced by attempting to fixate objects which are rapidly traversing the visual field; also, more widely, = *optomotor *a.*

1925 H. W. Stenvers in *Acta Otolaryngologica* VIII. 545 (*heading*) On the optic (opto-kinetic, opto-motorial) nystagmus. 1947 *Jrnl. Neurol., Neurosurg. & Psychiatry* X. 110/1 When the patient watched a revolving drum the direction of which was reversed, the optokinetic nystagmus obtained to the left side was greater than that obtained to the right. *Ibid.* 116/2 The optokinetic responses ..suggest an epileptogenic zone in the left temporal zone. 1966 *New Scientist* 2 June 598/1 Many animals move their eyes or their head if they see something move, and this 'optokinetic' response is often useful as an indication.. that an animal has perceived a movement. 1974 *Nature* 8 Feb. 403/1 Newborn infants showed optokinetic nystagmus to a 4 cycles per degree pattern. 1975 *Ibid.* 22 May 330/2 If the visual surround of a walking animal is rotated, the animal turns in a characteristic way. With arthropods, this optomotor (optokinetic) response has usually been described in terms of the angular velocity of the body or eyestalks.

Hence **optokine·tically** *adv.*

1959 *Experientia* XV. 443/2 The question..seems to be whether the nystagmic movements observed..are optokinetically released by the movements of images along the photosensitive surface of the eye. 1965 *Proc. R. Soc. B.* CLXI. 243 The seeing eye is brought artificially instead of optokinetically to the same position.

optometry (s.v. Optometer). Add: (Examples); also, the occupation concerned with the measurement of the refractive power of the eyes and the prescription of corrective lenses.

1886 C. M. Culver tr. *Landolt's Refraction & Accomm. of Eye* iii. 259 One important matter to consider is the point of the retina to be chosen in optometry. 1903 *Optical Jrnl.* Nov. 658 Kindly let me know..if there is any law in New York State governing 'optometry'. 1923 Glazebrook *Dict. Appl. Physics* IV. 287/1 Optometry is a term sometimes applied to all ocular methods of estimating the refraction of the eye. We confine the application here to that generally accepted in Europe, *i.e.* to instruments where an adjustment of lenses is made by the patient in order to obtain the image of an object. 1948 H. W. Hofstetter (*title*) Optometry: professional, economic and legal aspects. 1971 *Optometry Today* (Amer. Optometric Assoc.) 15 One such program, which the profession of optometry has encouraged..is the Model Reporting Area for Blindness.

Hence **optome·tric, -ical** *adjs.*, of or pertaining to optometry; **optome·trically** *adv.*, by means of optometry; **opto·metrist** (chiefly *U.S.*), one who practises optometry; an ophthalmic optician.

1864 W. D. Moore tr. *Donders' Anomalies Accomm. & Refraction of Eye* i. 71 It was extremely important to see how far these results of measurement and calculation agreed with those of the simple optometrical investigation in the same persons. *Ibid.* (*heading*) Comparison of R and P, deduced from measurements on the eye and optometrically determined. 1886 C. M. Culver tr. *Landolt's Refraction & Accomm. of Eye* iii. 252 To determine this degree of the ametropia, we may use all the above-mentioned optometric methods. 1903 *Optical Jrnl.* Oct. 558/2 The word 'optometrist'..coined by Mr. Eberhardt [President of the American Association of Opticians], is a popular one, and likely to be adopted by the American Association at its Milwaukee Convention in 1904. 1923 Glazebrook *Dict. Appl. Physics* IV. 287/2 This optometric system is the best if one can rely upon the accurate judgment of comparison by the patient. 1945 *Amer. Jrnl. Ophthalm.* XXVIII. 669/1 It seems obvious that optometrists will continue to perform a large percentage of the refractions in America and will always be interested in matters that concern ophthalmologists also. 1957 A. Huxley *Let.* 12 Jan. (1969) 815 The medical and optometrical lobby has gone to work in the various state legislatures. *Ibid.*, What is happening in the US is that optometrists—a breed superior to opticians, in as much as they get a long training, but not fool [*sic*] blown oculists.. are steadily adopting..Batesian procedures. 1965 *Amer. Jrnl. Optometry* XLII. 50 Optometry has taken on a brighter public image in those areas where optometric educational institutions exist. 1971 'D. Shannon' *Murder with Love* (1972) v. 79 Nearly every tenant..was a professional of some kind: doctors, dentists, optometrists. 1976 *Casper* (Wyoming) *Star-Tribune* 29 June 6/1 Members voted to assist Dr. James Lane, an optometrist, with the Vision Day program on July 19. Children who will attend first grade in the fall are given a free eye examination to rule out serious problems with their vision.

optomotor (ǫ·ptŏmōutəɹ), *a.* [f. Opto- + Motor *a.*] Pertaining to or characterized by turning of the eyes or body in response to the visual perception of a moving object.

1926 *Brain* XLIX. 333 This nystagmus has been termed 'railway nystagmus', 'optomotor nystagmus' or simply 'optic nystagmus'. 1932 *Arch. Neurol. & Psychiatry* (Chicago) XXVIII. 1024 Interruption of the optomotor pathway at any point in its course would disturb the ocular adaptation to a certain direction of movement of the visual field. 1961 H. Schöne in T. H. Waterman *Physiol. Crustacea* II. xiii. 492 The fine adjustment of turning movements in locomotion is checked by optomotor mechanisms. 1971 *Nature* 9 July 128/1 In the classical optomotor experiment an animal is placed inside a rotating striped drum. 1975 [see *optokinetic *a.*].

optophone (ǫ·ptŏfōun). Also **Optophone.** [ad. G. *optophon* (E. E. F. d'Albe 1912, in *Physikal. Zeitschr.* XIII. 942/2), f. *opto-* Opto- + Gr. φωνή voice, sound.] An instrument designed to enable blind persons to read, in which a photoelectric cell is employed to scan a text and produce electrical signals that are converted into audible ones corresponding to the different characters.

1913 E. E. F. D'Albe in *Electrician* 24 Oct. 103/1 The reading optophone consists essentially of a selenium preparation illuminated by a line of light broken up into dots. 1923 *Glasgow Herald* 3 Oct. 6 Messrs Barr and Stroud..by the invention and manufacture of their 'Optophone' have supplied the blind with a practical means of reading almost any printed type. Every letter sounds a tiny musical-motive up in the treble region. 1960 *Daily Tel.* 18 Oct. 15/1 She used an optophone, an experimental instrument developed from an invention of the late Dr Fournier d'Albe. 1973 *Nature* 27 Apr. 591/1 A device for converting letters into auditory signals, the 'Optophone' of E. F. d'Albe, allowed a trained blind person to read ordinary print.., but never met the ultimate criterion of success because reading..was slow, and learning to do it was very difficult.

optoquine, var. *optochin.

optotype. Add: [first formed as mod.L. *optotypus* (H. Snellen 1875, in *Klin. Monatsbl. f. Augenheilkunde* XIII. 479).] (Examples.)

[1886 C. M. Culver tr. *Landolt's Refraction & Accomm. of Eye* iii. 229 These tables of type figures, 'optotypi', as Snellen calls them, are what we use in optometry.] 1905 *Trans. Amer. Ophthalm. Soc.* X. 648 Attempts have been made..to construct special optotypes adapted to the limited capacity and observing powers of young children and illiterates. 1963 *Arch. Ophthalm.* LXX. 113/2 Amblyopic eyes required more light to read optotypes than did normal eyes. 1970 A. H. Keeney *Ocular Exam.* iii. 29/1 Letters and numbers in the patient's own language most closely relate to his daily seeing requirements and therefore are the most practical optotypes.

o·pt-out, *sb.* [f. *to opt out* (*opt *v.* 2).] A radio or television programme broadcast by a regional station for local consumption (in preference to one distributed nationally). Also *attrib.* or as *adj.*

1962 *B.B.C. Handbk.* 27 The regions have concentrated on providing themselves with facilities for putting out television programmes which have either been fed into the national programme or have provided special programmes for local consumption on an 'opt out' basis. *Ibid.* 38, I think it would probably be convenient and sensible to place most of the regional programmes, the opt-outs as we call them, in one channel. 1964 *Listener* 31 Dec. 1049 An 'opt-out' programme is one broadcast only on a regional wavelength. 1970 *Daily Tel.* 9 Nov. 6/5 At present in Birmingham there is one weekly opt-out time slot. 1972 P. Black *Biggest Aspidistra* I. ii. 24 The regional listeners preferred London to their local station, though..you could sometimes tune in to the local opt-out as well as the net-worked programme. 1975 *Listener* 6 Feb. 166/3 The BBC in the North-East has to divide its ration of opt-out time. 1977 *Private Eye* 1 Apr. 4/2 Moreover, a mere 250,000 viewers watch *Tonight* at the best of times and it's only a regional opt-out programme anyway.

opus. Add: **3.** Applied to slighter productions, compositions, etc.

1957 J. D. Salinger *Zooey* in *New Yorker* XXXIII. 93/1 The most courageous goddam offbeat television opus you ever read. 1959 P. Bull *I know Face* ix. 147 'B' films and other ghastly opuses. 1967 *Crescendo* May 8/2 'When Lights Are Low' in the old Benny Carter opus—one of my favourites. 1967 *Telegraph* (Brisbane) 30 June 12 Nine young couples are determined to go ahead with New York's latest sex air opus—a 'wed-in'. 1976 *Publishers Weekly* 15 Mar. 49/3 A spooky chiller of a first novel, this will have readers waiting impatiently for the next Ryder Brady opus to come along.

opus (ǫ·pŭs, ō·pəs), *v.* [f. the sb.] *trans.* To include and number among the works of a composer of music. Abbreviated *Op.*

1900 W. A. Ellis *Life Wagner* I. 376 This negligence in 'opus-ing' his musical works. 1921 A. B. Smith in *Music & Let.* II. 364 A large class of composers..write pieces solely for the pleasure of opusing them. *Ibid.*, Every piece of his [Gurlitt] is Op.-ed.

‖ **opus alexandrinum** (ǫ·pŭs ælegzændrī·nŭm). Also with capital initials. [med.L., lit. 'Alexandrian work'.] A type of pavement mosaic work consisting of coloured stone, glass, and semiprecious stones arranged in intricate geometric patterns. It was much used in Byzantium in the 9th century and later found in Italy.

1852 *Murray's Handbk. N. Italy* (ed. 4) 546/1 A mosaic..of that kind which is called 'opus Alexandrinum'. 1854 Wyatt & Waring *Byzantine & Romanesque Court in Crystal Palace* 58 The *Opus Alexandrinum*—we may describe it generally as tessellated marble-work..an arrangement of small cubes, usually of porphyry or serpentine. 1875 F. E. Hulme *Princ. Ornamental Art* i. 10 The *Opus Alexandrinum* was a marble tessellation generally composed of porphyry..and serpentine..arranged into geometric patterns that were cut into the white marble slabs that composed the groundwork of the pavement. 1897 J. Ward *Historic Ornament* vii. 345 Another kind of mosaic used in pavement is that known as *opus Alexandrinum*. 1904 L. F. Day *Ornament* v. 122 But for the economic instinct prompting men always to find use for a waste product, nothing like Opus Alexandrinum might ever have been done. 1955 *Times* 18 May 12/4 The sacristy lost the whole of its contents, but still retains part of the ancient pavement of *opus alexandrinum*. 1974 *Encycl. Brit. Micropædia* I. 228/3 In the 12th century several variations of *opus Alexandrinum* evolved at local centres in Italy, including the well-known Cosmati work..of Rome.

‖ **opus anglicanum** (ǫ·pŭs æŋglikā·nŏm). Also with capital initials. [med.L., lit. 'English work'; see Anglican *a.* and *sb.*] The name given to the fine pictorial embroidery produced in England in the Middle Ages, esp. between *c* 1250 and *c* 1350, and used esp. for ecclesiastical vestments.

[1277–81 in A. G. I. Christie *Eng. Medieval Embroidery* (1938) 2 Unum pretiosissimum pluviale ad ymagines Sanctorum contextum de opere Anglicano. *c* 1840 Lady Wilton *Art of Needlework* vii. 66 So celebrated was the English work, the Opus Anglicum, that other nations eagerly desired to possess it.] 1848 C. H. Hartshorne *Eng. Medieval Embroidery* 11 English embroidery has consistently enough been called the *opus Anglicanum*, from being a manufacture extensively and skilfully pursued in our own country. 1870 D. Rock *Textile Fabrics* 281 This invaluable and matchless specimen of the far-famed 'Opus Anglicanum', or English needlework. 1909 L. de Farcy in *Embroidery* VI. 168/2, I hope that..soon I may applaud their [*sc.* that of the English] success in the revival of their *opus Anglicanum*. 1922 *Daily Tel.* 12 June 20/6 (Advt.), The collection of XV Century vestments from Whalley Abbey, including some magnificent examples of 'Opus Anglicanum', the property of the Right Hon. Lord O'Hagan. 1936 *Burlington Mag.* Oct. 182/1 The rich collection of *Opus Anglicanum* at the Victoria and Albert Museum. 1954 M. Rickert *Painting in Brit.: Middle Ages* v. 128 The fineness of quality which is to give to English embroidery, the so-called *opus anglicanum*, its wide reputation during the late thirteenth and early fourteenth centuries. 1960 D. M. Wilson *Anglo-Saxons* v. 155 In the twelfth century, embroideries of 'English work' (*opus anglicanum*) were to become famous throughout Europe. 1964 tr. *A. Geijer's Textile Treasures of Uppsala Cathedral* 23 Technical similarities to recognized examples of *opus anglicanum*, the famous English art of the High Middle Ages, have caused some scholars..to attribute this work to England. 1974 *Encycl. Brit. Micropædia* VII. 557/1 Opus anglicanum has..survived all over Europe wherever historic vestments are treasured.

‖ **opus anglicum** (ǫ·pŭs æ·ŋglikŏm). Now *rare.* [med.L., lit. 'English work'; see Anglic *a.* and cf. prec.] A type of manuscript illumination regarded as characteristically English (see quot. 1860).

1860 F. Delamotte *Primer of Art of Illumination* 11 When the graceful and luxuriant curves of foliage begin to steal into the pages of the MS. they are to be found also forming the capital of the column,..a style of illumination generally known as the *opus Anglicum*. 1901 J. W. Bradley *Hist. Introd. Coll. Illum. Lett. & Borders V. & A. Mus.* vi. 81 So excellent is the work and so famous did it become that it was considered on the Continent as typical of our national art and received the appellation of 'opus Anglicum'.

‖ **opus araneum** (ǫ·pŭs ærā·nĭŏm). [med.L., lit. 'spider's work'.] Darned netting; a type of delicate embroidery done on a ground of net. Also called *spiderwork* (Spider *sb.* 10).

1865 F. B. Palliser *Hist. Lace* ii. 17 Distinct from all these geometric combinations was the Lacis of the sixteenth century, done on a network ground (réseau), identical with the *opus araneum* or spider-work of continental writers. 1870 D. Rock *Textile Fabrics* 160 This is a good specimen of a kind of cutwork weaving, or 'opus araneum', for which Lombardy..earned such a reputation at one time. 1882 [see *spiderwork* s.v. Spider *sb.* 10]. 1882 Caulfeild & Saward *Dict. Needlework* 233/1 During the Middle Ages this Network was called Opus Araneum, Ouvrages Masches, Punto a Maglia, Lacis, and Point Conté. 1909 F. N. Jackson *Hist. Hand-Made Lace* 185 *Opus araneum*, Spider Work. The ancient name for Cluny Guipure Lace and Darned Netting.

‖ **opus consutum** (ǫ·pŭs kǫnsū·tŏm). [med.L., lit. 'work sewn together'.] = Appliqué *sb.*

1880 D. Rock *Textile Fabrics* p. cii, Our old English *opus consutum*, or, cut work, in French, 'appliqué', is a term of rather wide meaning, as it takes in several sorts of decorative accompaniments to needlework. 1882 Caulfeild & Saward *Dict. Needlework* 7/1 This work was anciently known as *Opus Consutum* or cut work. 1899 E. T. Masters *Bk. of Stitches* i. 1 The average woman needs no interest to know that she is working an *appliqué* panel..she is executing the classical *opus consutum*. 1972 *Country Life* 6 Jan. 14/1 Pieced work, or the sewing together of different materials for ornamentation—the *opus consutum* of the convents—was a common form of fabric adornment in the Middle Ages.

‖ **opus Dei** (ǫ·pŭs dē·ī). [med.L. (attributed to St. Benedict).] **1.** *Eccl.* The work of God, *spec.* the Divine Office, or liturgical worship in general, seen as man's primary duty to God.

[*c* 530–540 *Rule of St. Benedict* (1952) xliii. 102 Ad horam divini Officii mox auditum fuerit signum, relictis omnibus quaelibet fuerint in manibus, summa cum festinatione curratur... Ergo nihil operi Dei praeponatur.] 1887 F. C. Doyle *Teaching of St. Benedict* xix. 141 An 'opus Dei'—namely, the 'care of souls', which, according to the teaching of theologians, is even more the 'work of God' than psalmody. 1896 F. A. Gasquet in C. F. De Montalembert *Monks of West* I. p. xiii, The Divine Office was the daily service and formal homage rendered to the Divine Majesty. This, the *opus Dei*, was the crown of the whole structure of the monastic edifice. 1907 *Cath. Encycl.* II. 469/2 This public worship of God, the *opus Dei*, was to form the chief work of his monks. 1921 J. McCann tr. *Delatte's Commentary on Rule of St. Benedict* viii. 133 Our Holy Father and other ancient writers are well inspired when they call the liturgy in its totality the *Opus Dei* (Work of God). 1929 R. Eaton *Benedictines of Colwich* iv. 47 The nuns were desolate of spiritual help..their one consolation being to continue the 'Opus Dei' as best they could in the dismantled chapel. 1939 J. Chapman *St. Benedict & Sixth Cent.* v. 86 St. Gregory the Great follows St. Benedict in the habitual use of *opus Dei*. 1969 R. Godden *In this House of Brede* ii. 71 'Everything we do..,' said Dame Clare, 'our work, our reading, our private prayer, even our meals..are simply pauses, meant to prepare ourselves for real work, the Opus Dei.' 1977 *Church Times* 5 Aug. 10/4 The complete Opus Dei—Mattins, Solemn Eucharist and Evensong—is sung at Edington Priory Church..from Sunday evening, August 21, until the following Sunday morning by cathedral and collegiate choristers.

2. (With capital initials.) The name of a Roman Catholic organization of laymen and priests founded in Spain in 1928 with the purpose of re-establishing Christian ideals in society through the implementation of them in the lives of its members. So **Opusdei·sta**, a member of this organization.

1954 V. S. Pritchett *Spanish Temper* v. 99 The infiltrations of the members of Opus Dei who work, exactly in communist fashion, to frustrate professional groups. 1960 *Spectator* 25 Nov. 805 The organisation..is known as *Opus Dei*; the Jesuits call it 'The White Freemasonry'. 1961 *Ibid.* 9 June 830 A group of Opus Dei economists within the Spanish Government. [illegible] xv. 432 Ullastres and Navarro Rubio were known to be members of a religious society, Opus Dei, men pledged by a solemn vow to the dedication to God of all their professional talents. [illegible] Hostile to the Government: some monarchists, liberals, Christian Democrats, Communists, anarchists, the Opus Dei. 1970 J. W. D. Trythall *Franco* xii. 226 Members of Opus Dei, like freemasons, form the sort of semi-secret, loosely organised body that everybody who is not a member regards as a conspiracy. 1973 [illegible] *& Opposition* VIII. 72 The Opus Dei was founded as a religious organisation in 1928 by Father Escrivá de Balaguer in Spain. 1974 *Encycl. Brit. Micropædia* VII. 557/2 The two economic ministers were reinstated as Opusdeistas, and [illegible] other members of Opus Dei have held ministerial posts and other positions in government.

‖ **opus filatorium** (ǫ·pŭs filatō·rĭŏm). [med.L., lit. 'work of threads'.] An early name for darned netting or spiderwork. Cf. *opus araneum above.

1882 Caulfeild & Saward *Dict. Needlework* 233/1 In this lace [*sc.* Guipure d'Art]..we have the modern revival of the Opus Filatorium, or Darned Netting, or Spiderwork, so much used in the fourteenth century. 1883 J. W. Mollett *Illustr. Dict. Art & Archæol.* 233/2 *Opus filatorium*, a kind of embroidery, 14th century; modern 'filet brodé'.

‖ **opus sectile** (ǫ·pŭs se·ktile). [L., lit. 'cut work'.] A form of floor decoration dating from Roman times and made up of pieces shaped individually to fit the pattern or design, in which respect it differs from mosaic which is an arrangement of regularly shaped pieces.

1852 *Murray's Handbk. N. Italy* (ed. 4) 491/1 This Florentine mosaic seems to be the 'opus sectile' of the Romans. 1854 Wyatt & Waring *Byzantine & Romanesque Court in Crystal Palace* 59 'Opera di commesso'..that is, a mosaic formed by slices of marble, arranged somewhat on the principle of the ancient 'opus sectile', the projections of one piece being so cut as to enter into the recesses of another. 1932 E. W. Anthony *Hist. Mosaics* iii. 18 The work which the Romans designated as *opus sectile* does not come under the head of mosaic, although it has sometimes been thus classified. 1948 *Antiquity* XXII. 187 *Opus sectile* floors at the major sites of Cyrene and Tolmeita. [illegible] These existed..a technique of decoration somewhat akin to mosaic, the Saracenic *opus sectile*. 1971 P. Fischer *Mosaic* 80 Earthenware pieces are specially cut to the individual shapes required by the design..rather in the manner of the second type of Roman *opus sectile*. 1976 *Sci. Amer.* Jan. 116/2 This kind of pavement is known as *opus sectile* (the Latin phrase for 'cut work'), and the example in our house is the first Byzantine *opus sectile* floor found in a private house in Carthage that can be confidently dated.

‖ **opus signinum** (ǫ·pŭs signī·nŏm). [L., *opus* work + *signinum* Signian, of or pertaining to Signia.] A flooring material used by the Romans and consisting of broken tiles and other fragments mixed with lime mortar, being named after Signia (modern Segni), a town in Latium which was famous for its tiles.

References to *opus signinum* occur in the works of Columella, Vitruvius, and Pliny: see Lewis and Short *Lat. Dict.* s.v. Signinus. A description of the making of *opus signinum* is found in Vitruvius *De Architectura* VII. vi. §14.

1745 *Columella's Husbandry* I. vi. 34 They, it seems, contrived a plaister or flooring made with bruised tiles, or sheards of earthen vessels, and lime, tempered together. With this composition they made very durable floors, &c. and this they called *opus Signinum*. 1899 R. Glazier *Man. Hist. Ornament* 75 *Opus Signinum*, small pieces of tile. 1937 *Discovery* July 208/1 A [Roman] floor of bits of stone and tile bound together with cement (*opus signinum*). 1967 *Antiquaries Jrnl.* XLVII. 269 The two northern cells retained their floors of *opus signinum* which, in both cells, had been once renewed. 1971 P. Fischer *Mosaic* 45 The humblest mosaic-type floor decoration, *opus signinum* (named after Signia, a town in Latium), is a levelled surface made up from odd fragments of stone or pottery of different colours, set at random in lime mortar rather like raisins in a cake. 1974 *Encycl. Brit. Micropædia* VII. 557/3 *Opus signinum* was the principal form of pavement in Roman houses from the 1st century BC to about the 2nd century AD when it was rapidly replaced in main rooms by patterned pavement mosaics of the tessellated variety.

Oquassa. Substitute for def.: The name of one of the Rangeley Lakes in western Maine, used *attrib.* or *absol.* to designate a small trout, the blue-back trout, *Salvelinus oquassa*, found in these lakes. (Examples.)

1883 J. J. Lalor *Cycl. Pol. Sci.* II. 215/2 The oquassa, or blue-backed trout (*Salvelinus oquassa*) of the lakes of Maine is a noteworthy American form. 1884 *Bull. U.S. Nat. Mus.* No. 27. 457 Oquassa Trout; Blue-Back Trout. Lakes in western Maine. 1888 G. B. Goode *Amer. Fishes* 93 Tasting, eloquent,..oquassa and namaycush are among the best of them [*sc.* aboriginal American fish names]. 1903 Jordan & Evermann *Amer. Food & Game Fishes* 217 Oquassa Trout; Blueback Trout... The blueback trout is the smallest and one of the most handsome of the charrs... Formerly this fish was very abundant.

or, *conj.*² (*adv.*²) Add: **1.** *or otherwise*: see *otherwise *sb. phr., adv., adj.* A. c; *or something*: see *something *sb.* (*adj.*), and *adv.* c. In modern colloq. use *or* can introduce an emphatic repetition of a rhetorical question.

1939 P. Cheyney *Poison Ivy* vii. 122 [illegible] *Ibid.* [illegible] I blew in here a couple hours ago [illegible] an' directly I [illegible] here, did I [illegible] *Trottie True* [illegible] pride and regret, 'she may be going to marry a Lord, but can I pick 'em or can I pick 'em?'

5. *or else*: see also *else *adv.* 4 b.

7. *or* introduces questions which, in the character of an afterthought, cast doubt on a preceding assertion or on the presuppositions behind a preceding statement or question.

c 1907 W. B. Yeats *Let.* 4 July (1954) 492, I suppose [illegible] as mine, or [illegible] 112 Matilda [illegible] I'm not so [illegible] really would [illegible] think so, Sir [illegible] couldn't put up [illegible] would they, or [illegible] i. 138 They [illegible] or you'd [illegible] off. 1937 [illegible] 22/2 You [illegible] against my [illegible] ples? 1937 [illegible] these spiky [illegible] did they? [illegible] Don't let's [illegible] I want to [illegible] I did when [illegible] sure.' 1954 [illegible] didn't need [illegible] Cloy *Two-* [illegible] everyone [illegible] he? Prob- [illegible] She must [illegible] Or was [illegible] ii/7 The [illegible] and the [illegible] Or does [illegible] Or can [illegible]

II. 8. As *sb.*² or *adj.* A Boolean function of two or more variables that has the value unity if at least one of them is unity, and is otherwise zero; also called *inclusive or* (= L. *vel*); also (*exclusive or*), a function that has the value unity if at least one, but not all, of the variables are unity; (= L. *aut*). Usu. *attrib.* and in capitals, esp. designating devices for realizing this function.

[1938 C. E. Shannon in *Trans. Amer. Inst. Electr. Engin.* LVII. 713/1 There are many special types of relays and switches... The operation of all these types may be described with the words 'or', 'and', 'if', 'operated', and 'not operated'.] 1940 W. V. Quine *Math. Logic* i. 12 When 'or' is used in the inclusive sense..joint truth of the components verifies the compound. An 'or'-compound in this sense can be expressed more clearly by adding the words 'or both'. *Ibid.* 13 The exclusive use of 'or' is not frequent enough in technical developments to warrant a special name and symbol. 1947 *Proc. IRE* XXXV. 758/1 The 'or' operation is performed by a 'buffing' circuit. 1949 E. C. Berkeley *Giant Brains* ix. 149 The 'or' (as in statement 7) that is defined in the truth table is often called the inclusive 'or' and means 'and/or'. Statement 7, '1 or 2', is considered to be the same as '1 or 2 or both'. There is another 'or' in common use, often called the exclusive 'or', meaning 'or else'. 1950 [see *not *adv.* and *sb.* 13*]. 1959 *Electronic Engin.* XXXI. 591/2 The two input lines are fed into an 'or' circuit,.. which gives out a pulse whenever a pulse is received on one or both inputs. 1969 J. J. Sparkes *Transistor Switching* iv. 100 or gates in one logic system are and gates in the other. 1970 O. Dopping *Computers & Data Processing* i. 25 Several functions can be combined in one and the same equation... The order in which different operations are taken is important. The normal order is not, and, or in decreasing order of priority. 1971 J. H. Smith *Digital Logic* iv. 45 OR units are not extensively employed because they are usually constructed with components such as diodes, which have no amplification. 1972 [see *not *adv.* and *sb.* 13*].

ora². Add: **3. ora serrata** *Anat.* [L. *serrātus* Serrate *a.*], the serrated edge of the retina, just behind the ciliary body.

1839 R. Grant *Hooper's Lexicon Medicum* (ed. 7) 963/2 *Ora serrata*, the posterior serrated edge of the ciliary processes is so called. [illegible] 606 The retina..terminates behind the ciliary body in an irregular border, *ora serrata*. 1968 L. Lurance *Eye* i. 17 The sensibility of the retina to light diminishes rapidly from the macula ring to the *ora serrata*. 1973 *Brain Res.* LXII. 285 Such new cells as are added after this period are added at the margin of the *ora serrata*.

oracle, *sb.* Add: **9*.** (With capital initial.) The proprietary name of a type of teletext system. [See quot. 1976².]

1973 *IBA Technical Rev.* III. 61 Oracle—broadcasting the written word. Engineers of the IBA have recently developed and demonstrated an experimental data system, Oracle, capable of providing a continuous public information service on conventional television transmitting networks. With this system the public could receive up to 50 different 'pages' of information 'written' on their television screen, each page containing up to 880 characters, or roughly 120 words. These messages can be displayed or superimposed on the screen of a domestic television receiver without in any way affecting the reception of normal television programmes. *Ibid.* 63/1 Experimental Oracle transmissions are being made on the IBA's London television stations. *Ibid.* [illegible] requires adaptation for Oracle either by means of an independent add-on unit or,..by an internal modification. 1974 *Trade Marks Jrnl.* 17 July 1295/2 Oracle... Apparatus for the transmission or reception of television signals and apparatus for the transmission of, processing of, or visual display of alphanumeric characters and/or characters for defining parts of diagrams. Independent Broadcasting Authority.., London,..; manufacturers and merchants. 1975 *Spectator* 19 July 86/3 Within a year or two..'teletext'—the generic term for the system used by both Ceefax and Oracle—will take its place in the ordinary person's vocabulary. 1976 *Times* 19 May 1/8 At present experimental Teletext services are being broadcast by both the BBC and the Independent Broadcasting Authority, under the names Ceefax and Oracle respectively. 1976 P. R. Hutt in *IBA Technical Rev.* IX. 4/2 The author hit on the idea of the name 'oracle' one Sunday while lunching with friends. Being a classical source of advice and information the name seemed to be very apposite, and it was not long before it was made into an acronym for 'Optional Reception of Announcements by Coded Line Electronics'.

10. oracle bones, bones used in ancient China for divination (see quot. 1970).

1915 *Encycl. Relig. & Ethics* VIII. 262/1 During recent years a very interesting discovery of 'oracle bones' and tortoise-shell fragments was made in the province of Honan. 1934 K. S. Latourette *Chinese* I. ii. 40 Inscriptions on the 'oracle bones', used for divination. 1970 Bray & Trump *Dict. Archaeol.* 167/2 *Oracle bones.* Animal bones, particularly ox shoulder-blades and tortoise shells, were employed by the ancient Chinese for divination purposes. A groove was cut in the bone, after which a hot point was applied nearby, and the shape of the resulting cracks determined the answer. 1977 G. W. Hewes in D. M. Rumbaugh *Lang. Learning by Chimpanzee* i. 31 He found evidence for this theory in ancient Egyptian hieroglyphs, and in the most ancient form of Chinese writing, on the Shang oracle-bones and bronzes.

oracle, *v.* Add: **1.** (Later examples.)

1922 Joyce *Ulysses* 183 All these questions are purely academic, Russell oracled out of his shadow. 1952 C. Day Lewis tr. *Virgil's Aeneid* III. 58 This is not the land which Delian Apollo Oracled for you.

ora·culate, *v. rare.* [f. L. *ōrācul-um* (see Oracle *sb.*) + -ate³.] *trans.* and *intr.* To say or speak oracularly.

1822 E. Nathan *Langreath* II. 315, I think I behold you shaking your wise head..as you would oraculate, 'the simple Madelina little suspects' [etc.]. 1919 J. Buchan *Mr. Standfast* I. i. 32 He boomed and oraculated and the Misses Wymondham prattled. 1930 —— *Castle Gay* viii. 134 The Professor oraculated on letters, with an elephantine deference to his hearers' opinions.

oracy (ǭ·rasi). [f. L. *ōs*, *ōr-* mouth + -acy, after Literacy.] The capacity or ability to express oneself fluently in speech. Also, oral transmission of poetry, etc.

1965 A. Wilkinson *Spoken Eng.* 14 The term we suggest for general ability in the oral skills is *oracy*; one who has those skills is *orate*, one without them *inorate*. 1965 *New Society* 12 Aug. 4/2 A new qualification has been proclaimed: oracy, 'general ability in the oral skills'. The coiner of word and concept is Andrew Wilkinson, lecturer in education at Birmingham. 1967 *Daily Mirror* 16 Oct. 18 The team is making a three-year investigation into oracy ..with a £15,000 grant from the Schools Council. 1971 *Daily Tel.* 20 Apr. 14 An additional year..of the schooling ..will not..benefit them unless it has been preceded by ..a successful learning of literacy and of oracy. 1972 T. A. Shippey *Old Eng. Verse* iv. 89 Though literacy and the fixed text may have killed 'oracy' in the long run, the change need not have happened as quickly as in the present century.

‖ **oraison funèbre** (orezon fünēbr'). [Fr.] A funeral oration.

1674 H. L. Farmer *Bossuet* vii. 313 [On] the occasion of the Queen's death,..Louis XIV requested him [*sc.* Bossuet] to preach her *Oraison Funèbre*. This was in 1683. 1876 *Encycl. Brit.* IV. 70/2 In the *Oraisons Funèbres* Bossuet is unapproachable... Nowhere does his genius take such wing as at the grave's mouth. 1907 G. Meredith *Let.* 22 Dec. (1970) III. 1620 You will not care for an *oraison funèbre*. 1953 W. Stevens *Let.* 16 Nov. (1967) 802 An *oraison funèbre* is not in my line. 1957 N. Frye *Anat. Crit.* 296 The corresponding form in oratorical prose is the *oraison funèbre*, which survives in some forms of modern oratory.

oral, *a.* (*sb.*) Add: **1.** *oral history*: (the collection or study of) tape-recorded historical information concerning matters from the personal knowledge of the speaker; such a taped record; *Oral Law, oral law*: the part of Jewish religious law passed down by oral tradition before being collected in the Mishnah.

1733 tr. *B. Picart's Ceremonies & Relig. Customs* I. 46 A Man who hath made the Oral Law his principal Study..is looked upon by the Generality amongst them as a Doctor. 1797 *Encycl. Brit.* XIII. 303/2 The Mishna is divided into six parts... In the fourth..are named those by whom the oral law was received and preserved. 1879 Oral Law (in Dict.). 1907 Oral law (used s.v. Mishnah, mishna). 1962 *New Jewish Encycl.* 360/1 At present, most Jews, though they may deviate from certain of its practices, believe the Oral Law to be divinely inspired. 1971 Webster Add. [illegible] Bernard Corning *Home* v. vii. 93 The Talmud incorporates the Oral Law. The Oral Law was dictated with the written Law by God to Moses on Sinai, and..the Oral Law was received intact by the Rabbis. 1977 *Times* 16 May 7/7 Oral history..enables the historian to put the questions he wants. 1977 *Program Announcement 1977–78* (U.S. Nat. Endowment for Humanities) 11 (*caption*) Oral history taping sessions, and a more informal jam session took place as part of..an NEH Division of Research Grants project that brought together these St. Louis jazz musicians to recount their experiences... Taping sessions..produced an oral history which will be made available to music historians and centers of research through the University's Learning Resources Center. 1978 P. Thompson *Voices of Past* ii. 19 The term 'oral history' is new, like the tape recorder; and it has radical implications for the future.

b. (Of poetry, etc.) delivered or transmitted orally; of or pertaining to such poetry. Also *transf.* So **oral-formulaic** *a.*, of or pertaining to (usu. early) poetry belonging to a spoken tradition which is characterized by the use of poetic formulae as an aid to memory.

1628 [see sense 1 in Dict.]. 1767 Percy *Ess. Anc. Eng. Minstrels* (Notes) 44 He [*sc.* Asser] has however particularly recorded Alfred's fondness for the oral Anglo-Saxon poems and songs. 1774 T. Warton *Hist. Eng. Poetry* I. Diss. 1 p. l, That scalds were common in the Danish armies when they invaded England, appears from a stratagem of Alfred; who, availing himself of his skill in oral poetry and playing on the harp, entered the Danish camp habited in that character, and procured a hospitable reception. 1777 J. Brand *Observations Pop. Antiq.* p. iv, These [ceremonies]..though erazed by public Authority from the *written Word*, were committed as a venerable Deposit to the keeping of *oral Tradition*. 1892 J. Earle *Deeds of Beowulf* p. xlvi, Müllenhoff had discovered six different authors, of which the first two were oral poets, but the third had a written copy of the rudimentary work as it then existed. *Ibid.* p. xlviii, The oral Epic was simple in outline and plain in style, and therefore the contradictions, irregularities, inversions..and intolerable repetitions..can only be explained by the gradual accretion of heterogeneous elements in the process of transmission. 1898 S. A. Brooke *Eng. Lit. fr. Beginning to Norman Conquest* ii. 42 This was the origin of the early unhistoric sagas, like that of Beowulf, and such a saga was the highest form of the oral literature of the German tribes. 1906 G. P. Krapp *Andreas* 75 The poem opens with the conventional epic formula, citing the authority of oral tradition for the story. 1928 W. W. Lawrence *Beowulf & Epic Trad.* 289 A written version of *Beowulf* might conceivably have served as a guide for oral recitation. 1929 W. E. Leonard in Malone & Ruud *Stud. Eng. Philol.* 1 The intrinsic nature..of oral or chanted verse as inevitably emphasizing an organic metrical pattern. 1953 S. A. Brown in A. Dundes *Mother Wit* (1973) 40/1 The Negro was contributing..through what we call oral literature—folk literature. 1953 F. P. Magoun in *Speculum* XXVIII. 446 (*heading*) Oral-formulaic character of Anglo-Saxon narrative poetry. 1970 *Jrnl. Eng. & Gmc. Philol.* LXIX. 72 Even contemporary criticism has tended to damn with faint praise by suggesting that Old English poems were largely collections of formulae indicative of oral composition... The conclusion that the presence of formulae is an indication of oral composition has been attacked as illogical. *Ibid.* 439 It would have been possible to use [in the OE *Riming Poem*] some of the oral formulae of other elegies and add a second half-line that rimed. 1972 T. A. Shippey *Old Eng. Verse* iv. 89 The theory claims that much classical and modern poetry has been composed by an 'oral-formulaic' process. 1973 *Black World* Nov. 10/2 Black music..as a basically oral-tradition music, is lacking in the kind of documentation that..would clarify these aspects.

3. b. Involving or being sexual activity in which the genitals of one partner are stimulated by the mouth of the other; freq. in *Comb.*, as *oral-genital* adj. Cf. *cunnilingus, *fellatio.

1948 A. C. Kinsey et al. *Sexual Behavior Human Male* II. x. 373 Most prostitutes are from the lower social levels, and consequently..few of them engage freely in oral activities. 1953 —— et al. *Sexual Behavior Human Female* vii. 257 Oral stimulation of the male genitalia by the female occurs somewhat less frequently. 1958 G. S. Sprague et al. in C. Berg *Homosexuality* II. i. 213 At one end..would stand the most primitive oral pattern, fellatio. 1961 *Encounter* XVI. v. 77 His short paragraph on oral-genital techniques. 1969 'J' *Sensuous Woman* (1970) XI. 79 A few minutes of oral-genital play was a small price to pay. 1973 S. Fisher *Female Orgasm* vii. 209 Most of the women..received manual and often oral stimulation of the clitoral region. 1973 *Sunday Times* (Colour Suppl.) 11 Mar. 20 Her main discovery is that oral sex is fun. 1977 *Time Out* 17–23 June 45/3 There's more erotic charge from two seconds of Damiano's close-up oral sequences than in Ms Richmond's entire oeuvre.

4. b. Administered or taken through the mouth; involving such administration.

1957 *Amer. Jrnl. Med. Sci.* CCXXXIV. 28/1 The purpose of this study was to evaluate the efficacy of an oral alcohol-water solution of theophylline in terminating acute asthmatic attacks. 1959 *Science* 10 July 81/1 This property..led to the experimental testing of a norethynodrel-estrogen combination as an oral contraceptive. 1959 K. H. Beyer in Waife & Shapiro *Clin. Eval. New Drugs* ii. 18 Oral efficacy..is certainly a limiting factor in the acceptance of a new drug for systemic use. 1962 *Lancet* 22 Dec. 1315/2 Oral contraception is now a matter of practical politics. 1967 *Proc. Soc. Exper. Biol. & Med.* CXXIV. 483/1 The acute oral toxicity of sodium selenite in the rat does not seem to have been investigated. 1970 W. J. Burley *To kill Cat* i. 24 A sachet of oral contraceptives. 1974 Passmore & Robson *Compan. Med. Stud.* III. 1. xxviii. 34/1 A number of women..experience amenorrhoea after discontinuing oral contraception.

5. *Psychol.* In psychoanalysis, characterized by having the mouth as the main focus of infantile sexual energy and feeling.

[1910 *Amer. Jrnl. Psychol.* XXI. 316 In infantile sexuality the oral and anal-urethral erogeneous zones..as well as sadistic and masochistic impulses rule.] 1925 J. Riviere tr. *Freud's Infantile Neurosis* in *Coll. Papers* III. v. 587, I have been driven to regard as the earliest recognizable sexual organization the so-called 'cannibalistic' or 'oral' phase. 1954 R. W. Pickford *Analysis of Obsessional* iii. 68 In general the oral stages are concerned with taking in of objects and ideas, or introjection. 1972 Rosen & Gregory *Abnormal Psychol.* (ed. 2) 53/2 The child's relationship with his parents as established during the oral period.

6. *Phonetics.* Of a sound: that is articulated with the velum raised, so that there is no nasal resonance. Cf. Nasal *a.* 2. So *oral-nasal* adj.

1919 E. Kruisinga *Handbk. Present-Day Eng.* (ed. 3) I. i. i. 14 We see therefore that sounds can be produced.. with the nose-passage shut: *oral* sounds. 1924 J. S. Kenyon *Amer. Pronunc.* 36 (*heading*) Oral and nasal consonants. 1933 L. Bloomfield *Language* vi. 96 Most sounds of speech are purely *oral*; the velum is completely raised and no breath escapes through the nose. 1955 P. Strevens *Papers in Lang. & Lang. Teaching* (1965) ix. 114 In rapid speech the consonant may be omitted, leaving a nasalized vowel where Received Pronunciation would have an oral vowel followed by a nasal consonant. 1961 H. A. Gleason *Introd. Descriptive Linguistics* (ed. 2) xv. 250 If only the mouth is open, the sound is an oral resonant. 1962 *Amer. Speech* XXXVII. 228 Differences between oral and nasal vowels. 1965 *Language* XLI. 478 The oral-nasal contrast is neutralized after a nasal; such vowels are written in this paper as oral. 1970 *Ibid.* XLVI. 81 There are eight consonants in Maxakalí, comprising an oral-nasal pair at each of four points of articulation. 1975 P. Ladefoged *Course in Phonetics* i. 3 Note that the air passages that make up the vocal tract may be divided into the oral tract within the mouth and the pharynx and the nasal tract within the nose.

B. *sb.* **b.** Short for *oral examination*.

1876 G. H. Tripp *Student-Life Harvard* 18 Do something splendid on the mathematics and the 'orals', and I will wage any thing you will pass clear. 1927 W. E. Collinson *Contemp. Eng.* 124 In regard to teaching within the University the only terms worthy of notice are the use of Oral (where some universities use Viva for Viva Voce)..and tutorials. 1973 D. May *Laughter in Djakarta* xii. 194 Examiners told candidates their marks immediately at the end of the oral. 1974 B. Johnston *It's been a Lot of Fun* vi. 45 One of the dons pointed this out to me during my oral.

orality. Delete *rare*⁻¹ and add: Also, preference for or tendency to use spoken forms of language.

1946 *Hansard Commons* 17 Oct. 1055 Does the right hon. Gentleman not appreciate that it is the uncertainty about the date when written Questions will be answered which promotes orality? 1967 A. L. Lloyd *Folk Song in England* i. 25 Orality is a most important characteristic.. and we have every right to speak of the grandeurs of oral tradition. 1973 *Times Lit. Suppl.* 26 Oct. 1323/2 A synthesis of Black orality with White literacy and technocracy. *Ibid.* 1323/3 The Black man enslaved took his African orality with him.

2. In sense of *oral *a.* 5.

1934 Lewin & Zilboorg tr. *Fenichel's Outline Clin. Psychoanal.* x. 372 Our findings then are:—ambivalence, turning against the ego, orality. 1951 W. & J. McCord in *Jrnl. Mental Sci.* XCVII. 765 (*title*) The problem of 'orality' and of its origin in early childhood.

3. *Phonetics.* With reference to a sound: the quality or state of being oral (see *oral *a.* 6 above).

1949 *Word* V. 158 Nasality vs. Orality. 1952 A. Cohen *Phonemes of Eng.* ii. 36 Features that are not actually relevant in distinguishing two phonemes, e.g. alveolarity, plosion, orality..must be taken into account all the same as contributing to the existence of both phonemes. 1964 R. H. Robins *Gen. Linguistics* iv. 155 Orality and voicelessness being regarded as the absence of a feature, nasality and voice, respectively.

orally, *adv.* Add: **2.** Also in *Comb.*

1957 R. Hoggart *Uses of Literacy* iv. 86 These views usually prove to be a bundle of largely unexamined and orally-transmitted tags. 1966 C. M. Simpson *Brit. Broadside Ballad* p. ix, The orally circulating ballad of tradition. 1967 A. L. Lloyd *Folk Song in England* iii. 144 Orally-diffused amateur composition. *Ibid.* v. 368 There is the anonymous, orally-spread, firmly traditional kind of song. *Ibid.* v. 381 The text is from a broadside... The orally-transmitted versions are not so complete.

3. With the mouth, as a means of sexual stimulation. Cf. *oral *a.* 3 b.

1951 Ford & Beach *Patterns Sexual Behaviour* (1952) iii. 54 Alorese men occasionally stimulate the woman's genitals orally. 1953 A. C. Kinsey et al. *Sexual Behavior Human Female* vii. 258, 16 per cent had stimulated the male genitalia orally.

-orama, -(r)ama, *suffix.* [a. Gk. ὅραμα view, as in the second element of Cyclorama, Panorama.] As a suffix suggestive of considerable size or expanse, in commercial use to form nouns the nature of which is indicated by the first element. Cf. Diorama in Dict. and Suppl., *Cinerama.

1824 E. Weeton *Jrnl.* 17 July (1969) II. 306 Visited the Cosmorama... I had now seen many of the -ramas in London, Ignoramus' and all. 1896 E. Marriage tr. *Balzac's Old Goriot* 54 The diorama, a recent invention,.. had given rise to a mania among art students for ending every word with *rama*... 'Well, Monsieur-r-r Poiret,.. how is your health-orama?' *Ibid.* 'There is an uncommon *frozerama* outside!'... 'Why do you say *frozerama*?..it should be *frozenrama*.' 1954 *Amer. Speech* XXIX. 157 Audiorama, a display of acoustic instruments;..*striporama*, a burlesque movie. 1962 *Word Study* Dec. 6/2 An exhibition of automobiles known as Motorama,..a Launderama (coin-operated automatic washers with dryers). 1963 *Guardian* 24 Aug. 6/3, I observe from your London Letter..that a fresh verbal monstrosity is to be inflicted on the defenceless English population—something called a 'scent-a-rama'. 1973 *Advocate-News* (Barbados) 15 Dec. 6/1 With all the 'ramas' like cyclerama, brassorama, musicrama and laugharama, why can't we call this one ugly-o-rama? Doesn't it sound great? 1977 *Radio Times* 29 Oct. 20/3 *Swaporama* with Keith Chegwin.

orange, *sb.*¹, *a.* Add: **A.** *sb.* **1. c.** *Oranges and lemons* (earlier and later examples).

1873 *Young Englishwoman* Mar. 154/2 Could you.. give me the words in full of 'Oranges and Lemons'; 'I wrote a Letter to my Love'; 'Kiss in the Ring'; and any other of the old games? 1939 [see *London *bridge*]. 1969 I. & P. Opie *Children's Games* viii. 236 Players..are invited to be an 'orange' or a 'lemon' in the game of 'Oranges and Lemons'.

d. *absol.* = Orange squash, orange juice.

1950 [see *gin *sb.*³ 2 a]. 1968 T. Kinsella *Nightwalker* 45 A small jug of orange. 1972 *Guardian* 20 June 4/6 And so, back to fizzy orange and the ritual conference. 1977 N. Slater *Crossfire* iii. 60 'What can I get you?' 'Fresh orange. At least I can..set you an example.'

7. a. *orange-juice* (examples), *leaf, salad, -wood*; *orange-crate* (also fig.), *-girl* (further examples), *-man* (example); *orange bitters, cream, crush, Curaçao, gravy, sauce, squash.* **b.** *orange squeezer.*

c 1870 in H. W. Allen *Number Three St. James's St.* (1950) vii. 186/2 Orange bitters. 1877 E. S. Dallas *Kettner's Bk. of Table* 328 Parfait Amour is made of the bitter zest of limes,..syrup,..spirit of roses, and..spicy odours. It is in fact a kind of orange bitters spoilt. 1958 A. L. Simon *Dict. Wines* 121/1 *Orange bitters*, the most popular form of bitters used for flavouring cocktails and other mixed drinks. It is made from the bitter Seville orange. 1977 *Sunday Times* (Colour Suppl.) 6 June 63/2 (Advt.), Sherry..with a dash of orange bitters. 1968 'J. Le Carré' *Small Town in Germany* xiv. 229 An old lady dropped a two-Mark piece into an orange-crate. 1972 L. Anderson *Let.* 20 June in *Amer. Speech* 1972 (1975) XLVII. 38 In pilots training, we called the planes.. 'orange crates'. 1977 M. Kenyon *Rapist* ix. 107 [He] sat on an orange crate in the storeroom. 1723 J. Nott *Cook's & Confectioner's Dict.* sig. L5 *Orange Cream.* Take.. Oranges, grate the Peels into..Water; beat..Eggs.. sweeten..set it on the Fire, stir till it is as thick as Cream. 1861 Mrs. Beeton *Bk. Househ. Managem.* 736 Orange cream... 1 oz of isinglass, 6 large oranges, 1 lemon, sugar to taste, water, ½ pint of good cream... Squeeze the juice from the oranges and lemon; [etc.]. 1939–40 *Army & Navy Stores Catal.* 31/2 Kia-Ora Orange Crush—bot. 1/6. 1952 [see *crush *sb.* 4 e]. 1973 D. May *Laughter in Djakarta* iii. 54 What was almost the national drink of the Indonesian middle classes, orange crush. 1907 *Yesterday's Shopping* (1969) 100/2 Curaçao..Orange.. 3/6. 1951 E. David *French Country Cooking* 27 Grand Marnier, Mirabelle and Orange Curaçao are particularly good for soufflés and for omelettes. 1965 *House & Garden* Dec. 90/2 *Bols.* This old-established Dutch firm covers almost every liqueur—best known for Kummel, Apricot and Orange Curaçao. 1977 *Times* 9 July 10/7 A little brandy..or orange curaçao may be added. 1842 *Knickerbocker* XX. 472 The orange-girl is generally allowed to enter [an auction-store], for auctioneers are mortal, and sometimes eat oranges. 1939 G. B. Shaw *In Good King Charles's Golden Days* I. 60, I never was an orange girl; but I have the gutter in my blood all right. 1963 M. Frayn in Sissons & French *Age of Austerity* 336 The orange-girls, dressed up as replica Nell Gwyns. 1845 E. Acton *Mod. Cookery* iii. 109 *Orange Gravy, For Wild Fowl.* Boil.., in..Espagnole, half the rind of a Seville orange,..and a small strip of lemon-rind... Strain it off, add to it..port or claret. 1877 Orange gravy [see *bigarade]. 1877 E. S. Dallas *Kettner's Bk. of Table* 340 The present practice over the Continent is to stew them [*sc.* perch] in vinegar, fresh grape, orange-juice, or other sour sauce. *a* 1901 C. M. Yonge *Autobiogr.* in C. Coleridge *C. M. Yonge* (1903) iii. 85, I did not like to eat orange juice out of a pewter spoon. 1960 F. Raphael *Limits of Love* I. i. 10 Think I'll have an orange juice. 1977 J. Archer *Shall we tell President?* x. 130 'An orange juice for me. I'm watching my weight.' Doesn't he know that orange juice is the last thing to drink if you're dieting? 1838 C. Gilman *Recoll. Southern Matron* iii. 25 An orange leaf..was laid on every finger bowl. 1877 E. S. Ward *Story of Avis* 408 The splendor slept..upon the green pulses of the orange-leaves. 1880 G. W. Cable *Grandissimes* ii. 15 Perfumed ad nauseam with orange-leaf tea. 1858 *Punch* 13 Mar. 103/1 There have bawled..in his street, sweeps, orangemen, dustmen. 1845 E. Acton *Mod. Cookery* xxv. 629 *Orange Salad.* Take off the outer rinds,..from some fine China oranges; slice them thin,.. strew over them..white sifted sugar, and pour on them a glass or more of brandy. 1973 'I. Drummond' *Jaws of Watchdog* ix. 123 Elaborate spreads of cold duck and orange salad. 1867 *Common Sense Cook Bk.* 28 Orange Sauce for Game. 1977 *Vogue* Feb. 114/3 Scallops in orange sauce. 1926–7 *Army & Navy Stores Catal.* 34/1 Orange squash..Kia Ora..Schweppes'. 1936 *Discovery* June 192/1 Fruit Squashes..were analogous to the well-known orange and lemon squashes. 1975 J. McClure *Snake* iii. 42 A uniformed sergeant..was drinking orange squash..because he was on duty. 1949 M. Mead *Male & Female* xii. 247 Idly turning the pages of a catalogue that shows the best type of orange-squeezer. 1962 L. Deighton *Ipcress File* ii. 21 Stainless steel orange squeezers. 1884 G. W. Cable *Dr. Sevier* lvii. 437 He moved his orange-wood staff an inch. 1889 *Harper's Mag.* Dec. 106/2 Strangers..were detained by eager vendors of flowers and orange-wood walking-sticks. 1910 *Daily Chron.* 23 Apr. 7/3 Dilute peroxide with one-half water and apply under nails with cotton on an orangewood stick.

d. orange skin food, a type of moisturizer for the skin; **orange stick,** a short stick, usu. of orange-wood, used for manicuring the nails.

1926–7 *Army & Navy Stores Catal.* 492 An Elizabeth Arden treatment is based on..Cleansing..Toning.. Nourishing, with *Orange Skin Food* or the delicate *Velva Cream*. 1939–40 *Ibid.* 438/3 Seymour[,] Jane..Orange Skin Food—2/9. 1944 M. Laski *Love on Supertax* iii. 36 She slapped the Orange Skin Food on to her face. 1974 D. Gray *Dead Give Away* xv. 144 She now patted Elizabeth Arden's Orange Skin Food into her face. 1911 H. S. Harrison *Queed* vii. 89 Orange-stick in mouth, he went around like a museum guide. 1922 F. Courtenay *Physical Beauty* 46 You may use an orange stick..to push back the cuticle from the nails. 1966 [see *cuticle 1 d].

B. *adj.* **1. b.** *spec.* Applied to a variety of opal.

1902 *Chambers's Jrnl.* Aug. 494/1 The miners..say, one stroke of the pick may lay bare a seam of 'pin-fire' opal or break in two a rich band of 'orange'. 1971 J. S. Gunn *Opal Terminol.* 32 *Orange*, name given to opal with this distinctive colour.

2. a. *orange-fiery, -pink, vermilion* adjs. (sbs.). **b.** *orange-keyed* adj.

1922 Joyce *Ulysses* 296 The orangefiery and scarlet rays. *Ibid.* 715 Orangekeyed ware..consisting of basin, soapdish and brushtray..., pitcher and night article. 1956 D. Barnham *One Man's Window* vi. 67, I am enveloped in a world of luminous orange-pink. 1967 O. Ruhen in *Coast to Coast 1965–6* 189 The orange-pink of its desert sand. 1895 *Montgomery Ward Catal.* 253/1 Special colors... Orange Vermilion. 1951 R. Mayer *Artist's Handbk.* ii. 56 *Orange vermilion*, a variety of real vermilion.

c. orange book, a report of the Ministry of Agriculture and Fisheries dealing with marketing questions and published in orange covers; **orange grass,** (*b*) *U.S.* = *nit-weed; **orange pekoe,** a type of black tea.

1928 *Daily Express* 30 Apr. 7/4 The report is one of the Ministry's famous 'orange' books—those scientific farming pamphlets for the education of English farmers. 1932 *N. & Q.* 30 Jan. 73/2 We were glad to have a note of the reception of the Orange Books on Marketing which the Ministry of Agriculture has been putting forth. 1837 W. Darlington *Flora cestrica* (ed. 2) 324 Ground Pine. Nit-Weed. Orange-grass. 1882 E. K. Godfrey *Island of Nantucket* 36 The orange grass with its fragrance now greeting us at every turn. 1907 Orange-grass [see *nit-weed]. 1877 *Cassell's Dict. Cookery* 961/1 One pound of Moning Congo, a quarter of a pound of Assam, and a quarter of a pound of Orange Pekoe. 1911 *Encycl. Brit.* XXVI. 480/2 They [*sc.* the leaves] are now broken apart and sorted by mechanical sifters into the various grades or qualities, which are described as Orange Pekoe, [etc.]. 1960 A. E. Bender *Dict. Nutrition* 123 Orange Pekoe [is made] from the first opened leaf.

Orange, *sb.*². Add: **2.** *Orange Order.*

1940 L. MacNeice *Last Ditch* 28 A framed Certificate of admission Into the Orange Order. 1975 *Irish Times* 10 May 9/3 Mr. Thomas Passmore, grand master of the Orange Order in Belfast, said yesterday that while Britain would be simply a small member in an exclusive club if she remained in Europe, outside it she could once again earn the title of Great Britain. 1977 P. Carter *Under Goliath* i. 5 To make sure that the Protestant religion stays on top of the league in Northern Ireland, is what the Orange Order is all about.

orange-blossom. Add: **1. a.** (Later examples.)

1971 K. Wheeler *Epitaph for Mr. Wynn* (1972) xxvii. 355 Orange blossoms and murder trials don't mix.

2. A cocktail flavoured with orange juice. Also *attrib.*

1930 *Savoy Cocktail Bk.* 117 Orange Blossom Cocktail. ½ Orange Juice. ½ Dry Gin. Shake well. 1938 L. Bemelmans *Life Class* (1939) II. ii. 134 Waiters stand about with trays of cocktails, the favourite being Orange Blossoms, a mixture of gin and orange juice. 1960 B. Keaton *Wonderful World of Slapstick* (1967) 158 After taking a couple of orange blossoms, a cocktail made of orange juice and gin, Virginia got sick. 1963 I. Fleming *On H.M. Secret Service* x. 105 A sprinkling of feminine cocktails—Orange Blossoms, Daiquiris. 1965 T. Capote *In Cold Blood* (1966) iv. 223 We'd bought a bottle of ready-mix Orange Blossoms—that's Orange pop and vodka.

orange-flower. Add: **4. orange flower skin food** = *orange skin food.*

1908 *Sears, Roebuck Catal.* 798/1 Orange Flower Skin Food..acts as skin nourisher and wrinkle remover.

Orangeman. (Further examples.)

1844 Macaulay *Let.* 4 July (1977) IV. 202 If the letters were opened, it was not by any authority from the late government, but by some rascally Orangemen in the Post Office. 1894 W. B. Yeats *Let.* 16 Dec. (1954) 242, I lectured..on Fairy lore to an audience of Orangemen. 1921 *Daily Colonist* (Victoria, B.C.) 12 Mar. 2/4 Orangemen from all over the world will convene in Winnipeg in 1923, according to an announcement made today. 1975 *Irish Times* 24 May 9/4 Mr. William Douglas, Official Unionist (U.U.U.C.) Convention member for Derry said that as an Orangeman, Unionist and Loyalist he was convinced that all Ulster people who loved their country should say no to the Common Market. 1976 *Daily Record* (Glasgow) 29 Nov. 17/4 Orangeman [*name given*], from Edinburgh, will face 400 delegates from Scottish lodges at Govan next Saturday over remarks on TV about leading the Ulster Defence Association in Scotland.

orange-peel. Add: **2.** Used, usu. *attrib.*, to designate a suspended bucket or grab composed of a number of curved, pointed segments that are hinged at the top and come together to form a container.

1905 C. Prelini *Earth & Rock Excavation* x. 129 Excavator-buckets are usually either clam-shell or orange-peel buckets. 1912 C. G. Elliott *Engin. for Land Drainage* (ed. 2) xiv. 209 The orange-peel is particularly useful in building levees. 1922 Powers & Teeter *Land Drainage* xvi. 170 (*caption*) A small dry land excavating outfit with orange-peel bucket. 1959 *Micropaleontol.* V. 218/1 Two sampling devices were used, the Hayward standard orange-peel grab and a snapper sampler. 1967 *Oceanogr. & Marine Biol.* V. 527 The samples were taken with an orange-peel bottom sampler. 1975 B. Fell *Introd. Marine Biol.* iii. 21 An orange-peel grab..has four valves that appear to form the four segments of an orange when cut along its meridians.

orange-tip. For '*Euchloë Cardamines*' substitute '*Anthocharis cardamines*'. (Later examples.)

Only the male has orange tips on the wings.

1906 R. South *Butterflies Brit. Isles* 43 The Orange-tip ..has a large patch of orange colour on the outer third of its white, or creamy white, fore wings. 1930 *Times Educ. Suppl.* 4 Oct. p. iv/4 The little orange-tip flickering along the hedgeside. 1973 T. G. Howarth *South's Brit. Butterflies* 58 The Orange-tip is essentially a butterfly of lanes, hedgerows and rough fields. 1973 *Shooting Times & Country Mag.* 7 July 20/2 An orange-tip butterfly went fluttering by.

orangey, *a.* Delete *rare* and add: Also **orangy.** (Later examples.) Also, covered in orange; suggestive of oranges.

1913 H. S. Walpole *Fortitude* I. xii. 165 My fingers are all over orange... I always have an orange before dinner. .. I am orangy, but then I was late and couldn't finish it. 1919 C. Orr *Glorious Thing* xx. 245 A delicious, warm, orangy smell hung about the kitchen walls. 1966 *New Statesman* 6 May 663/1 Olivier's blue-black face and frame, Maggie Smith's pink-pallid Desdemona move for too much of the time against a dreadful orangey backcloth. 1974 N. Gordimer *Conservationist* 199 A sun as pale as last night's..moon was orangey, is stiffening the topmost leaves of that tree. 1977 C. Fremlin *Spider-Orchid* iv. 32 The orangy glow of the standard lamp.

b. *Comb.*, as *orangey-blue, -brown, -buff, -red, -yellow* adjs.

1887 A. Scholefield *Venom* III. 118 Eyes weeping, knuckles orangey-blue. 1968 H. R. F. Keating *Inspector Ghote hunts Peacock* i. 8 Its hideous orangey-brown colour, masquerading as the tan of leather. *a* 1973 'G. Ashe' *Herald of Doom* (1974) v. 49 An orangey-brown kilted suit. 1976 H. R. F. Keating *Filmi, Filmi* xiii. 123 A sportingly cut orangey-buff suit. 1977 E. W. Hildick *Loop* iv. 17 Fashionable orangey-red blush make-up. 1974 *Times* 2 May 8/7 This year's recommended colour..is a warm orangey-yellow.

orans (ǭ·rănz). = Orant. Also *attrib.*

1900 [see Orant]. 1937 *Burlington Mag.* July 25/2 Though the Trecento type, the seated Virgin with clasped hands, persists in the Quattrocento, it is the Virgin derived from the old 'orans' type, of which the Antwerp *Assumption* is the earliest Italian example. 1949 O. Demus *Mosaics Norman Sicily* 309 The Orans was seen 'spreading out her pure hands to ensure the Emperor's triumph over his enemies'.

Oraon (orā·ǫn). [Indian name, of undetermined origin.

The following are among numerous explanations of the meaning of the name: 1900 F. Hahn *Kurukh Gram.* p. iii, The Hindus, who are supposed having invented the name *Urāō* or *Orāō* for the *Kurukh* people, might have concluded that the whole nation was called by the name of this sept, *i.e. Orgorā*; this word means hawk or cunny bird, and educated Urāōs believe that the foreign designation *Orāō* or *Orā* is derived from this totemistic word. 1906 G. A. Grierson *Linguistic Survey India* IV. ii. 406 Hindūs say that the word 'Orāō' is simply the Indo-Aryan *urāū*, spendthrift, the name being an allusion to the alleged thriftless character of the people to whom it is applied. 1915 S. C. Roy *Orāons of Chōtā Nāgpur* i. 14 The name [of a monster-king] Rāwan, pronounced, as some people do, with an arrested 'O' sound at the beginning gave us the present form 'O-rāwan' or Orāon.]

(A member of) an aboriginal tribe, which calls itself *Kurukh*, dwelling in the state of Bihar in northern India; the Dravidian language of this tribe. Also *attrib.* or as *adj.*

1872 E. T. Dalton *Descriptive Ethnol. Bengal* VIII. i. 245 The Khurnkh or Orāons of Chūtiā Nāgpūr are the people best known in many parts of India as 'Dhāngars', a word that from its apparent derivation (*dang* or *dhang*, a hill) may mean any hillmen... According to the traditions I have received from the most venerable and learned of my Orāon acquaintances, the tribe has gradually migrated from the western coast of India... Orāon appears to have been assigned to them as a nickname, possibly with reference to their many migrations and proneness to roam. 1892 H. H. Risley *Tribes & Castes Bengal: Ethnogr. Gloss.* II. 138 *Orāon, Urāon, Kunokh, Kunrukh*, a Dravidian cultivating tribe of Chota Nagpur, classed on linguistic grounds as Dravidian, and supposed to be closely akin to the Mālés of the Rājmahāl hills. 1908 [see *Korku]. 1915, etc. [see *Maler *sb.* and *a.*]. 1917 *Encycl. Relig. & Ethics* IX. 501/2 The Orāons..call themselves Khuṛūkh or Kūrukh, a Dravidian term of uncertain origin, connected by some with the word *horo*, 'man', or with *kuruk*, 'a crier', or one capable of speaking, in contradistinction to the other races, whose language is not intelligible to them... This word *horo* is probably the origin of the name Orāon. 1939 L. H. Gray *Foundations of Lang.* 386 Kurukh, consisting of Kurukh or Orāon in the western part of the Bengal Presidency and the neighbouring parts of the Central Provinces, and Malto in the Rajmahal Mountains of Bengal. 1972 W. B. Lockwood *Panorama Indo-European Lang.* 224 Kurukh and Malto. The former, also termed Oraon, is used by 150,000 persons in the western ranges of the Chota Nagpur Hills, in the districts of Raigarh and Sambalpur. 1974 *Encycl. Brit. Micropædia* XIV. 563/1 Speakers of Oraon number about 1,140,000, but in urban areas, and particularly among Christians, many Oraon speak Hindi as their mother tongue.

orarion (orēᵊ·riǫn). *Eccl.* [late Gr. ὠράριον, ad. L. *orarium* Orarium.] = Orarium.

1772 J. G. King *Rites & Cerem. Gk. Ch. in Russia* 36 Plate III. represents a deacon officiating in his..Orarion which is a sort of tippet thrown over his left shoulder. 1850 J. M. Neale *Hist. Holy Eastern Ch.* I. 310 The stole was frequently called the orarion in the Western Church. 1907 A. Fortescue *Orthod. Eastern Ch.* 408 Other clerks wear a shorter sticharion and an orarion wound around them.

ora serrata: see *ora² 3.

‖ **oratio** (orēi·ʃio). [L. *ōrātiō*, *ōrātiōn-em*: see Oration *sb.*] Speech, language. Only in phrases: **oratio obliqua** [L. fem. of *obliqu-us*: see Oblique *a.* 5 b], indirect speech; **oratio recta** [L. fem. of *rect-us* straight, direct], direct speech.

1842 W. E. Jelf *Gram. Greek Lang.* II. iv. 508 The infin. and acc. follows the verb in the *oratio obliqua*, and then follows a dependent clause in which the verb stands in the *oratio recta*. 1876 Oratio obliqua, oratio recta [see Oration *sb.* 3]. 1929 R. Hughes *High Wind in Jamaica* i. 35 Then she put it into Oratio Recta, told it as a story, beginning with that magic phrase, 'Once I was in an Earthquake.' 1968 *Listener* 25 July 120/2 The stilted *oratio obliqua* of court reportage. 1962 *Times Lit. Suppl.* 26 Oct. 826/2 This involves questions of phraseology, idiom, *oratio obliqua*, and the adjustment of dialogue to the pace and mood of the narrative. 1957 R. Speaight *Life H. Belloc* ix. 175 Belloc would have been on safer ground if he had covered his quotation by making it clear that he was only giving the sense of Robespierre's words, or if he had abridged them in *oratio obliqua*.

oratist (ǫ·rătist). *rare.* [f. Orate *v.* + -ist.] One given to orating.

1887 Swinburne in *Fortn. Rev.* XLII. 173 The orotund oratist of Manhattan.

oratorial, *a.* **2.** (Later example.)

1923 *Daily Mail* 8 Aug. 8/1 The very front rank of our oratorial singers.

Oratorianism. (Earlier examples.)

1847 J. H. Newman *Let.* 31 Dec. (1962) XII. 140, I am anxious you should [try] if you have fully mastered *what* Oratorianism is. 1848 F. W. Faber in R. Chapman *Father Faber* (1961) viii. 172 There is nothing in what you say about oratorianism which takes any of us by surprise.

oratorianize, *v.* (Earlier example.)

1848 F. W. Faber in R. Chapman *Father Faber* (1961) ix. 182 It seems you have not captivated him, and he won't Oratorianize.

oratorship. (Examples.)

1592 G. Harvey *Foure Letters* (1969) iii. 17, I was supposed not vnmeet for the Oratorship of the vniuersity. 1869 R. C. Jebb *Let.* 3 Nov. (1907) 98, I was standing for the Public Oratorship.

orb, *sb.*¹ Add: **13.** *orb-like* (later examples); **orb-weaver,** (read:) a spider of the family Argiopidæ, which builds an orb-web; **orb-web,** a web formed of lines radiating from a central point, produced by a spider of the family Argiopidæ; also *attrib.*; so **orb-webbed** *a.*

1925 T. Dreiser *Amer. Trag.* I. II. xxxi. 367 Her eyes, which were now fixed on him in round orblike solemnity. 1935 W. de la Mare *Poems, 1919 to 1934* 335 The white dews drip untrembling down, From bough to bough, orb-like, unblown. 1889 H. C. McCook *Amer. Spiders* I. iii. 53 The round web of the Orbweaver probably deserves the distinction of having given the popular name cobweb to the whole spinningwork of spiders. 1941 W. S. Bristowe *Comity of Spiders* II. 244 Most orb-weavers renew their webs, other than the framework, daily. 1889 H. C. McCook *Amer. Spiders* I. iii. 53, I define an orbweb as a snare constructed of right lines radiating from a common centre. 1971 *Oxf. Bk. Invertebr.* 150/1 *Meta* is a common orb-web spinner. 1958 W. S. Bristowe *World of Spiders* xix. 256 A series of papers..is opening the door to clearer understanding of the unhesitating route pursued by the orb-webbed spiders in the course of completing their webs.

orbicular, *a.* (*sb.*) Add: **A.** *adj.* **6.** *Petrol.* Containing orbicules.

1824 H. T. De La Beche tr. *Sel. Geol. Mem. in Annales des Mines* p. ix. Diabase... Orbicular D. Spheres with concentric zones of hornblende and compact felspar in a diabase of moderately sized grains. (Orbicular granite of Corsica.) 1857 J. B. Jukes *Student's Man. Geol.* iii. 73 (*heading*) Globular diorite, orbicular greenstone, Corsican granite. 1873 *Proc. Geol. Assoc.* II. 267 The so-called 'orbicular silica'. 1954 H. Williams et al. *Petrogr.* vii. 132 A few granites have an orbicular texture. 1970 K. C. Jackson *Textbk. Lithology* v. 280 Orbicular granite... Very commonly, the orbicules show abnormal concentration of the ferromagnesium minerals.

orbicularis (ǭɹbi:kiŭlā·ris), *a.* (*sb.*) *Anat.* Pl. **orbiculares.** [L., = Orbicular *a.*] **a.** In full (*musculus*) *orbicularis oris* [L. *ōs*, *ōr-* mouth]. A muscle surrounding the lips that is partly responsible for moving the lips and the mouth.

1681 J. Browne *Compl. Treat. Muscles* 31 There is also another Constrictive Muscle, which hath gotten the name of orbicularis common to the upper lip. 1733 G. Douglas tr. *Winslow's Anat. Expos. Struct. Human Body* II. x. 136 The Semi-Orbiculares are commonly looked upon as one Muscle, surrounding both Lips, from whence it is called Orbicularis. 1797 J. Bell *Anat. Human Body* (ed. 2) I. II. i. 200 The orbicularis oris, or muscle round the mouth, is often named constrictor oris, sphincter, or osculator. 1906 *Practitioner* Dec. 726 The muscles..which..soon become implicated by the disease when cancer begins in this region [*sc.* the angle of the mouth] are: the orbicularis oris, risorius, buccinator, [etc.]. 1970 *Language* XLVI. 315 Electromyographic data were obtained from the cricothyroid and orbicularis muscles of a female speaker of American-English.

b. In full (*musculus*) *orbicularis palpebrarum* [L. *palpebra* eyelid] or (now more commonly) *oculi* [L. *oculus* eye]. A flat muscle that surrounds the orbit and occupies the upper and lower eyelids, responsible for closing the lids (voluntarily or involuntarily).

In quot. 1967 *palpebræ* is the gen. sing. rather than the more usual gen. pl. of *palpebra*.

1681 J. Browne *Compl. Treat. Muscles* 14 There is held a Dispute whether that these two Muscles be not properly one, and that *Orbicularis* being so generally received; though the one doth depress, and the other..lift up..to make a perfect close over the eye. 1733 G. Douglas tr. *Winslow's Anat. Expos. Struct. Human Body* II. x. 89 The Muscles of the Palpebræ are commonly reckoned to be two, one peculiar to the upper Eye-Lid, named Levator Palpebræ Superioris; the other common to both, called Musculus Orbicularis Palpebrarum. *Ibid.* 91 The Supercilia..may be moved..downward by the Orbiculares. 1797 J. Bell *Anat. Human Body* (ed. 2) I. II. i. 194 Orbicularis oculi, or palpebrarum, is a neat and regular muscle, surrounding the eye. 1887 J. Tomes *Syst. Dental Surg.* (ed. 3) 573 The spasm of the orbicularis was so great that the right eye could only be opened by great effort. 1902 D. J. Cunningham *Text-bk. Anat.* 376 The orbicularis palpebrarum (m. orbicularis oculi) is a transversely oval muscle surrounding and occupying the eyelids. 1967 G. M. Wyburn et al. *Conc. Anat.* v. 131/1 Incise the fibres of the orbicularis palpebræ and expose the tarsal plates. 1968 Passmore & Robson *Compan. Med. Stud.* I. xxi. 17/1 Paralysis of the orbicularis oculi abolishes the blinking response, so that the affected eye is totally unprotected from direct injury.

orbicule (ǭ·ɹbikiŭl). *Petrol.* [back-formation from Orbicular *a.*; cf. L. *orbiculus* (see Orbicle).] A spheroidal inclusion, esp. one composed of a number of concentric layers.

1931 A. Johannsen *Descr. Petrogr. Igneous Rocks* I. ii. 17 In many glassy rocks there are found certain more or less spherical bodies varying in size from pellets visible only under the microscope..to huge spheres ten feet or more in diameter. These rounded bodies are called spherulites when the constituents are arranged radially..and orbicules when these constituents are in concentric shells. 1956 L. E. Spock *Guide to Study of Rocks* iii. 26 A few gneous rocks contain orbicules. These are close spherical concentrations of amphibole arranged radially or of biotite in concentric-tangential pattern. **1970** [see *ORBICULAR *a.* 6].

orbit, *sb.* Add: **2.** (Examples referring to artificial satellites and spacecraft generally.) Also, one complete passage around the orbited body.)

1951 *Jrnl. Brit. Interplanetary Soc.* X. 219 He predicted the establishment of 'Earth satellite vehicle' rockets in orbits 25,000 miles from the Earth within about 10 years. **1951** A. C. CLARKE *Sands of Mars* vii. 82 They were now floating round Mars in a free orbit. **1962** J. GLENN in *Into Orbit* 144 Al would also give me the exact times at which the retro-rockets would have to be fired to start bringing the capsule home at the end of one, two and three orbits. **1974** *Sci. Amer.* Jan. 115/1 In its polar orbit it swings close to the North Pole, then moves south across the Equator and finally, having traversed Antarctica, returns north for the next orbit. **1977** J. SCOTT *Hot Pursuit* x. 90 Tracking stations..spotted it straight away when the orbit altered.

d. In extended use: An approximately circular or elliptical path traced by something in motion (e.g. round an atomic nucleus, in a surface wave in a liquid, or in a particle accelerator).

1827 [see *planet-wheel* s.v. *PLANET *sb.*¹ 4]. **1864** W. J. M RANKINE in *Phil. Trans. R. Soc.* CLIII. 131 The centres of the orbits of the particles in a given surface of equal pressure stand at a higher level than the same particles do when the liquid is still. **1891** *Sci. Trans. R. Dublin Soc.* IV. 599 The dominant orbit of the electron..as affected during the subsequent flight of the molecule by an apsidal perturbation. **1904** *Phil. Mag.* VII. 454 If the spectra of the elements be due to the motion of electrons revolving in circular orbits, as above supposed, several rings of orbits must exist where there are different series of spectra. **1913** BOHR in *Ibid.* XXV. 11 The forces which keep the electrons in their position—or their orbits—inside the atom. **1942** J. D. STRANATHAN *'Particles' of Mod. Physics* xi. 426 A magnetic field between two peculiarly shaped pole faces serves to guide the electron repeatedly around an orbit in this field. **1962** *Newnes Conc. Encycl. Nucl. Energy* 169/1 As the particle energy rises the radius of curvature of the path in the magnetic field increases and so the particles execute approximately spiral orbits. **1962** I. R. & M. W. WILLIAMS *Basic Nucl. Physics* i. 15 The physical significance of ψ..means that the exact position of electrons in an atomic orbit or indeed the exact location of the orbit cannot be precisely defined. **1972** M. G. GROSS *Oceanography* ix. 243 In deep water..the water parcels move in nearly stationary circular orbits... The diameter of these orbits at the surface is approximately equal to the wave height.

e. The state of being or moving in an orbit; also *fig.*; chiefly in *in, into orbit* (also with intervening qualifier).

1958 *Spectator* 22 Aug. 263/2 The US satellite now in orbit. **1959** *Economist* 21 Feb. 706/2 One observer.. describes the stock market as being 'in orbit', released from the gravitational pull of the bond market. **1959** *Daily Tel.* 21 Nov. 1/2 The United States put a satellite into orbit from Vandenberg Air Force base in California today. *Ibid.*, Recovery from orbit has never been accomplished by us or Russia. **1961** L. MUMFORD *City in Hist.* (1966) xvi. 580 Our descendants will perhaps understand our curious willingness to expend billions of dollars to shoot a sacrificial victim into planetary orbit. **1967** W. R. HINDMARSH *Atomic Spectra* ii. 9 If an ion..has, like hydrogen, a single electron in orbit about the nucleus. **1969** R. AIRTH *Snatch!* x. 99 Morland..said they were great, which sent Giorgio approximately into orbit. **1971** *Nature* 17 Sept. 160/3 A Salyut spacecraft is already in Earth orbit. **1973** R. BUSBY *Pattern of Violence* v. 81 So Charlie's an acid head... He's probably in orbit by now.

orbit (ǭ·ɹbit), *v.* [f. the sb.] **1.** *trans.* To revolve round in an orbit; to travel round.

1946 *R.A.F. Jrnl.* May 169 Orbiting the target at low level, Pathfinders' Master Bomber assessed the T.I. markers. **1949** *Jrnl. Brit. Interplanetary Soc.* VIII. 3 The way to overcome this difficulty is to have the object circle the earth at a greater distance. It might, in fact, be convenient to have it orbit the earth at a distance of 22,200 miles up. **1951** A. C. CLARKE *Sands of Mars* xi. 141 Orbiting Saturn was Titan, the largest satellite in the Solar System. **1954** *N.Y. Times* 29 Aug. 39/1 If there are satellites orbiting the earth fairly close to it, the Army Office of Ordnance Research will locate them. **1959** D. BEATY *Cone of Silence* xiv. 154 He had been slowly cruising round Mayfair in the car, orbiting huge squares. **1960** *Daily Tel.* 22 Aug. 1/6 (*caption*) Television pictures received from space by the Russians of their dogs.. as they orbited the earth in a 4½-ton space ship. **1963** *Ann. Reg. 1962* 399 Orbiting the moon, and finally landing from a parking orbit round the earth. **1973** *Sci. Amer.* Dec. 47/1 One cannot be sure that real galaxy pairs orbit each other in the parabolas or elongated ellipses demanded by our models. **1975** *Times* 11 Aug. 10/5 In 1971 and 1972, the next Mariner spacecraft to orbit the planet revealed a new..face of Mars.

2. *intr.* **a.** To move in an orbit. Const. various preps.

1951 [see *ORBITAL *a.* 2]. **1955** *Time* 14 Feb. 112/3 When Allingham sketched a sun with planets orbiting round it on a pad, he says, the visitor smiled and pointed to the fourth planet and then to his own space-suited figure. **1957** *Times* 7 Oct. 9/2 The satellite..is orbiting too high to be of maximum value for observations. **1962** F. I. ORDWAY et al. *Basic Astronautics* i. 3 Manned space capsules have orbited several hundred miles above the Earth. **1970** *Nature* 3 Oct. 11/1 American spy satellites orbiting over the Soviet Union have spotted 18 new sites. **1972** *Sci. Amer.* Nov. 105/3 Earlier investigations..showed.. how muonic atoms behave when the muons orbit within the nucleus.

b. To fly in a circle.

1952 *Sat. Even. Post* 27 Dec. 26/3 Clapp broke off and flew south to drop a flare. I orbited just north of the bridge. **1957** R. WATSON-WATT *Three Steps to Victory* 315 Dive-bombers and fighters were to orbit as required till they were joined by the slower torpedo-bombers. **1969** I. KEMP *Brit. G.I. in Vietnam* iii. 68 We had been orbiting in our helicopter for about forty-five minutes.

c. To go into orbit. Also *fig.*

1958 *Times* 30 Aug. 6/1 The Vanguard satellite which failed to orbit on May 27 probably travelled 7,500 miles into the south Atlantic. **1970** *Daily Tel.* 8 June 16/1 The company suspended dealings in March..and they should be resumed by the end of July. The shares should orbit in next to no time. **1970** *Toronto Daily Star* 24 Sept. 22/1 There is no velocity test on the British ball. So, in effect, the manufacturers could improve it to the point where it orbits.

3. *trans.* To put, send, or place in orbit.

1958 *Spectator* 14 Feb. 192/1 Soon after Explorer was orbited the air below it was filled by television stations with five-minute talks. **1961** *Listener* 20 Apr. 684/1 The news of the first man to be orbited and brought to earth. **1962** F. I. ORDWAY et al. *Basic Astronautics* ii. 25 By 1962 the two countries had orbited over a hundred objects into space. **1970** *Nature* 13 June 1011/1 France has been orbiting modest satellites with her own launcher. **1973** *Sci. Amer.* Oct. 75/3 Coronagraphs orbited in space can be constructed differently from their ground-based counterparts.

Hence **o·rbited** *ppl. a.*, **o·rbiting** *vbl. sb.* See also *ORBITING *ppl. a.*

1956 *Spaceflight* I. 6/2 Whether we can land on all of them is improbable, but orbiting trips will be made to get a closer look. **1958** *Engineering* 28 Feb. 270/1 A vehicle weighing 2,000 lb. would expend only a further 70 lb. of propellent in accelerating from an orbiting speed of 18,000 m.p.h. to the escape speed of 25,000 m.p.h. **1966** *Aviation Week & Space Technol.* 5 Dec. 22/3 The separately orbited satellite modules could be mothballed in space—fully equipped with all experiments, however—until ready for use.

orbital, *a.* Add: **1. b.** Anthrop. *orbital index* [tr. F. *indice orbitaire* (P. Broca 1875, in *Mém. de la Soc. d'Anthrop. de Paris* II. 172)], one hundred times the ratio of the height of an orbit to its width.

1879 A. DE QUATREFAGES *Human Species* xxx. 388 The smallest orbital index known is that of the old man of Cro-Magnon, which we have seen to be 61·36. **1904** W. L. H. DUCKWORTH *Stud. from Anthrop. Lab. Anat. Sch. Cambridge* xvii. 104 The mean orbital index of thirty-one skulls is 82·6: that of twenty-four males 81·2; of three females 87·4. **1955** *Chambers's Encycl.* I. 460/2 Similar conventional divisions of the orbital index..are: chamaeconchic ('low orbits') *x*–75·9, mesoconchic ('medium orbits') 76·0–84·9, and hypsiconchic ('high orbits') 85·0–*x*.

2. Delete *Astron.* and add: moving in an orbit; pertaining to such motion. (Further examples.)

1932 [see sense B below]. **1949** W. LEY *Conquest of Space* (1950) 48 Probably the manned moonship will have to be postponed until there is an orbital station. **1951** A. C. CLARKE *Exploration of Space* 47 'Orbital refuelling'..is the key to interplanetary flight. It depends simply on the fact that once a spaceship had reached circular velocity outside the atmosphere, it would continue to orbit indefinitely without the use of power. **1961** *Guardian* 6 May 1/4 The Mercury programme itself could be used as a means of sending men into substantially protracted orbital flights. **1961** *New Scientist* 27 July 203 The advent of *Midas* in orbit raises the whole question of orbital warfare. Either manned or unmanned orbital bombers are an obvious starting point for such conceptions. **1962** I. R. & M. W. WILLIAMS *Basic Nucl. Physics* i. 9 The innermost orbital electrons are the most tightly bound to the atom. **1970** G. K. WOODGATE *Elem. Atomic Struct.* iv. 55 A single-electron atom has a magnetic moment associated with the orbital motion of the electron. **1972** M. G. GROSS *Oceanography* ix. 243 (*caption*) Orbital motion and displacement of a water particle during the passage of a wave.

3. Designating a road, railway, or rail or road system encircling a large town; cf. *RING-ROAD. Also *ellipt.* as quasi-*sb.*, an orbital road.

1933 *Archit. Rev.* LXXIV. 166/2 Orbital road system around London. **1937** *Times* 13 Apr. (British Motor Suppl.) p. x/2 The plan may provide for orbital and radial roads, parkways, viaducts and tunnels, communications to aerodromes, railway stations, and docks. **1939** *N. & Q.* 1 July 1/2 The proposal to thrust an orbital road through land belonging to the National Trust. **1967** *Times Rev. Industry* Apr. 50/2 Essential features are good car parking space..and good communications. Orbital roads, motorways and similar positions are popular. **1970** *Times* 3 Feb. 2 First priority for roads, after the orbitals outside Greater London, is Ringway 2 (North and South Circular Roads). **1975** *Country Life* 16 Oct. 970 (*caption*) Near Great Warley. The London Orbital will pass across the fields in the middle distance. *Ibid.*, The London Outer Orbital Route—a far-flung bypass..running around London. **1976** *Conservation News* Nov./Dec. 7/2 Conservationists..are already presently split on their views of the proposed extension of the M25, London Outer Orbital Motorway. **1977** *Modern Railways* Dec. 459/3 This would effectively establish one of the north orbital routes discussed in the Barran Committee's *London Rail Study* three years ago.

B. *sb.* *Physics* and *Chem.* A possible pattern of electron density in space which can be realized by two electrons at the most in an atom or molecule; the wave function of a single electron corresponding to any such pattern.

1932 R. S. MULLIKEN in *Physical Rev.* XLI. 50 From here on, one-electron orbital wave functions will be referred to for brevity as orbitals. The method followed here will be to describe unshared electrons always in terms of atomic orbitals but to use molecular orbitals for shared electrons. **1956** *Nature* 11 Feb. 275/1 The resultant *g* values..will yield details on the orbitals involved in the chemical binding of the central iron atom. **1964** J. W. LINNETT *Electronic Struct. Molecules* i. 6 In quantum mechanics the orbits of the Bohr-Sommerfeld semi-classical methods are replaced by orbitals. *Ibid.* 9 Since each spatial orbital is defined by the three quantum numbers *n*, *l* and *m*, this [*sc.* the Pauli Principle] is equivalent to saying that each orbital can accommodate two electrons. **1970** *Sci. Amer.* Apr. 54/2 Each orbital is characterized by a set of 'quantum numbers', denoting various properties of the electrons in that orbital (for instance their spin, angular momentum and the probability of finding the electrons in various regions of space). **1971** J. Z. YOUNG *Introd. Study Man* ii. 26 They are atoms able to receive electrons in the orbitals of their outer shell.

orbitale (ǭɹbitē¹·li, -ā·li). [neut. of med.L. *orbitālis* cyclic, f. L. *orbita* ORBIT.] The lowest point on the lower edge of the orbit.

1920 H. H. WILDER *Lab. Man. Anthropometry* I. i. 47 *Orbitale*, the lowest point in the margin of the orbit; one of the points used in defining the Frankfort Horizontal. **1933** *Jrnl. R. Anthrop. Inst.* LXIII. 30 The orbitales..are found, with the aid of the scriber, by rotating the cranium round its auricular axis until the lowest point on the inferior margin of whichever orbit is being considered is at the same height above the drawing-board supporting the instruments as is the auricular axis. **1974** *Nature* 8 Mar. 165/1 Six bilateral and four single roentgenographic landmarks were delineated... The bilateral landmarks were: orbitale; centre of condylar shadow; [etc.].

orbiter (ǭ·ɹbitəɹ). *Astronautics.* [f. *ORBIT *v.* + -ER¹.] A spacecraft in orbit or intended to go into orbit, esp. one that does not subsequently land.

1958 C. C. ADAMS et al. *Space Flight* 140 It is not inconceivable that variations of Project Farside, or Kurt Stehling's proposal Saloon (balloon-launched satellite) may hold the key to economy-type orbiters. **1961** *Life* 3 Mar. 33 The Russians will follow their manned orbiter.. with a multi-manned moon orbiter..and space station. **1969** *Times* 4 Feb. 13/4 Maps drawn from orbiter photographs can differ by nearly two km., depending on the position assumed for the satellite when the photographs were taken. **1971** *Daily Tel.* (Colour Suppl.) 12 Nov. 12/4 While the Viking orbiter continues to survey Mars from space, the lander will start a robot examination of its surroundings. **1976** *Times* 4 Sept. 4/3 He explained that the failure occurred just after the lander separated from the orbiter and prepared to fire a crucial rocket blast that would knock it out of orbit towards Mars.

orbiting (ǭ·ɹbitiŋ), *ppl. a.* [f. *ORBIT *v.* + -ING².] That is moving in or into an orbit.

1957 *Observer* 20 Oct. 14/4 Such tricky problems as traffic regulations for orbiting satellites. **1960** *Eng. Lang. Teaching* XIV. 85 The orbiting prices also smooth the road to merger. **1962** *Newnes Conc. Encycl. Nucl. Energy* 37/2 The number *Z* of orbiting electrons equals the number of protons. **1965** *Newsweek* 20 Dec. 57 After weeks inside an orbiting laboratory, the physiological changes may be so profound that Earth becomes the alien environment. **1969** *Guardian* 17 Jan. 1 Russia yesterday created the first orbiting space station. **1971** *Nature* 23 Apr. 494/2 The orbiting Explorer 42 satellite is beginning to oust rocket and balloon borne X-ray experiments from their position of prime importance. **1976** *Field* 18 Nov. 979/1 Orbiting satellites..can tell quickly and comprehensively what is happening on earth at the present, particularly in the matter of temperatures and wind.

orbivirus (ǭ·ɹbivəɪᵊrəs). *Biol.* [f. L. *orbis* ring, circle (see quot. 1971) + VIRUS.] Any of a group of arthropod-borne RNA viruses which cause disease chiefly in higher animals and are similar to reoviruses.

1971 E. C. BORDEN et al. in *Jrnl. Gen. Virol.* XIII. 269 The authors suggest as a name for the distinctive group of viruses described, *orbiviruses* (from *orbis*, (L.), ring or circle). This name reflects the especially large, doughnut-shaped capsomeres seen on the surface of virus particles in negative contrast preparations. It is proposed as a 'genus' name, equal in hierarchy to *reoviruses*. **1974** W. K. JOKLIK in Fraenkel-Conrat & Wagner *Comprehensive Virology* II. v. 236 Since very few orbiviruses have yet been characterized with respect to the structure of their RNA or the polypeptide constitution of their capsids, and since they are serologically very diverse.., the principal criterion for admission to this genus is morphology. *Ibid.*, Although only few orbiviruses have been isolated from vertebrates, antibodies to them are widely distributed. **1976** FENNER & WHITE *Med. Virol.* (ed. 2) xxii. 407 Colorado tick fever virus is the only orbivirus so far recognized as a human pathogen.

orc. Add: **2.** Used by J. R. R. Tolkien (1892–1973) in his tales: one of an imaginary warlike people in whom are combined human and ogreish characteristics. Also *attrib.* and *Comb.*

1937 J. R. R. TOLKIEN *Hobbit* vii. 149 Before you could get round Mirkwood in the North you would be right among the slopes of the Grey Mountains, and they are simply stiff with goblins, hobgoblins, and orcs of the worst description. **1954** —— *Fellowship of Ring* 15 The last battle..was beyond living memory: the Battle of Greenfields, S.R. 1147, in which Bandobras Took routed an invasion of Orcs. *Ibid.* 23 With the help of the ring he escaped from the orc-guards at the gate and rejoined his

fellows. *Ibid.* 339 The orcs were dismayed by the fierceness of the defence. *Ibid.* 346 There was a guard of orcs crouching in the shadows behind the great door-posts towering on either side. *Ibid.* 350 The cut is not poisoned, as the wounds of orc-blades too often are. **1954** —— *Two Towers* 48 Orc-speech sounded at all times full of hate and anger. *Ibid.* 138 There the hugest Orcs were mustered, and the wild men of the Dunland fells. *Ibid.* 171 He was not so obviously orc-like as most of these were. **1955** —— *Return of King* 279 No welcome, no beer, no smoke, and a lot of rules and orc-talk instead. *Ibid.* 412 But Orcs and Trolls spoke as they would, without love of words or things; and their language was actually more degraded and filthy than I have shown it. *a* **1973** —— *Silmarillion* (1977) x. 96 And when Thingol came again to Menegnoth he learned that the Orc-host in the west was victorious.

orca (ǭ·ɹkă). [a. L. *orca* a kind of whale, adopted as a generic name by J. E. Gray in Richardson & Gray *Zool. Voy. Erebus & Terror* (1846) I. 33.] The killer whale, *Orcinus orca* (formerly *Orca gladiator*); cf. ORC, ORK I.

1866 tr. D. F. ESCHRICHT in W. H. Flower *Recent Mem. Cetacea* ii. 172 The teeth of the Orcas are of quite a different kind from those of the cachalots. **1906** *Windsor Mag.* Sept. 469/2 A school of killers or orcas had quietly come up. **1964** E. P. WALKER et al. *Mammals of World* II. 1121/1 Killer Whales; Orcas. **1977** *N.Y. Rev. Bks.* 14 July 26/1, I was called out on deck by excited voices to gaze on a pod of orcas, or killer whales, that flanked and followed our boat as if in escort.

Orcagnesque (ǭɹkānʸe·sk), *a.* [f. the nickname *Orcagna* Archangel, of Andrea di Cione (active *c* 1308–*c* 1368): see -ESQUE.] Resembling in subject-matter, style, or quality the work or manner of Andrea di Cione, Florentine painter, sculptor, and architect.

1910 W. J. LOCKE *Simon* xix. 261 Call it the Valley of the Shadow, if you like. But don't you think the attendant circumstances were rather mediæval, gargoyley, Orcagnesque? **1933** *Burlington Mag.* Oct. 173/1 He builds upon the remnants of the Orcagnesque tradition and upon Gothic sculpture. **1938** *Ibid.* Dec. 238/1 Lorenzo [Monaco]..may..have been impressed by the decorative style and the pretty colours of Jacopo [di Cione]..because of the very Sienese features which, together with the Orcagnesque basis of the forms, produce a pleasant effect.

orcelite (ǭɹse·ləit). *Min.* [ad. F. *orcélite* (also used) (S. Caillère et al. 1959, in *Compt. Rend.* CCXLIX. 1773), f. the name of Jean *Orcel* (b. 1896), French mineralogist: see -ITE¹.] An arsenide of nickel, Ni_2As, found as bronze-coloured hexagonal crystals.

1960 *Amer. Mineralogist* XLV. 753 Orcelite... The mineral has a rose bronze color, browner than that of niccolite. **1962** *Mineral. Abstr.* XV. 352/1 Orcélite.. occurs as a bronze mineral making up a large part of a vein in serpentinized harzburgite in the Trebaghi massif, New Caledonia.

orch, var. *ORK.

orchard. Add: **2.** *orchard-close, -land; orchard-circled, -fresh* adjs.

1889 W. B. YEATS *Wanderings of Oisin* 90 They will lead her home again To the orchard-circled farm. **1844** E. B. BROWNING *Lost Bower* in *Poems* II. 100 In the pleasant orchard closes, 'God bless all our gains,' say we. **1881** O. WILDE *Poems* 116 Past sombre homestead and wet orchard-close. **1922** E. K. CHAMBERS in *Poems of Today* 2nd Ser. 101, I like to think how Shakespeare..ate his pippin in his orchard close. **1971** *Countryman* Autumn 201/2 (Advt.), Direct delivery in own transport to most areas (south of and incl. Glasgow) to reach you in orchard-fresh condition. **1687** *Southampton Rec.* (1877) II. 55 One acre for his orchard land. **1903** *Daily Chron.* 4 Mar. 7/1 In the orchard-land of Normandy the privately distilled liquor is..a recognized medium of exchange. **1938** [see *grain-land* s.v. *GRAIN *sb.*¹ 18 a]. **1977** P. G. WINSLOW *Witch Hill Murder* II. 126 The Brewster land was orchard land.

orcharded, *a.* Delete *rare* and add later examples.

1791 J. BYNG *Torrington Diaries* (1935) II. 350 Most religious houses have the same kind of low sequester'd situation; and were well orcharded, and well supplied with ponds. **1968** G. JONES *Hist. Vikings* III. iii. 231 Its one perdurable asset was its rich and orcharded soil.

orcharding. **2.** (Earlier and later examples.)

1654 in C. W. Manwaring *Digest of Early Connecticut Probate Rec.* (1904) I. 155 One halfe of all my howsing, Barnes and orcharding. **1818** *Massachusetts Spy* 25 Feb. 1/2 [A farm] is well proportioned into Mowing, Pasturing, Ploughland, Woodland and Orcharding. **1863** *Rep. Maine Board Agric.* 142 One acre of orcharding on suitable soil.. will produce three times the amount in value of any other crop.

orchestra. Add: **1. c.** In modern use, a section of the auditorium of a theatre, now usually the forward part or all of the main floor. Chiefly *U.S.*

1768 STERNE *Sentimental Journey* I. 192 At the end of the orchestra..there is a small esplanade... Though you stand, as in the parterre, you pay the same price as in the orchestra. **1786** *Independent Jrnl.* (N.Y.) 5 Aug. 2 The Pit is very large, and the Theatrum and Orchestra elegant and commodious. **1872** *Chicago Tribune* 28 Mar. 5/4 The interior will contain an orchestra and three circles. **1911** *World's Work* (N.Y.) Sept. 14840/1 They were accustomed, when they went to the theatre, to pay an extra half dollar for seats in the front rows of the orchestra. **1924** D. LAWRENCE *True Story Woodrow Wilson* 117 A President.. cannot sit in the orchestra or in the balcony. **1927** *Amer. Speech* Oct. 23 In the early days of the English theatre what we know as the 'orchestra' or parquet floor of the house, was called the 'pit'. **1961** BOWMAN & BALL *Theatre Lang.* 242 *Orchestra* 1. The seating area on the main level of an auditorium.

3. *fig.* (Later examples.)

1927 CHESTERTON *Coll. Poems* 40, I salute your three violinists... They play my accompaniment; but I shall take no notice of any accompaniment; I myself am a complete orchestra. **1958** M. KENNEDY *Outlaws on Parnassus* v. 77 Writers using an orchestra of minds to tell their story for them were obliged to consider..the variety of language, as used by different minds.

4. (Further examples.) Also, **orchestra pit**, the space in front of, and below, the stage, where the orchestra plays; **orchestra seat** (*U.S.*), **stall**, a seat in a theatre in the orchestra; also, a seat in a theatre next to the orchestra and stage.

1849 *Theatrical Programme* 11 June 22 New Strand Theatre... In order to add to the convenience of the Audience, the Orchestra Stalls have been made more commodious. **1856** *Porter's Spirit of Times* 20 Dec. 262 Laura Keene's Theatre..Reserved Orchestra Seats, 75 cents... Seats in orchestra stalls, $1 each. **1872** *Chicago Jrnl.* 18 July 3/1 The house is divided into an orchestra circle, which includes the entire main floor, Mr. McVicker having decided to discard the names parquet and parquet circle, orchestra and orchestra chairs, and dress circles of first and second balcony circle. **1874** W. LENNOX *My Recoll.* II. 108, I was ensconced in a snug orchestra stall. **1895** *N.Y. Dramatic News* 19 Oct. 3/4 You wouldn't know the orchestra seats were $1,..the scale 75, 50, 25 being displayed everywhere. **1901** Orchestra stall [see STALL *sb.*¹ 5 c]. **1903** *Smart Set* IX. 57/1 There would be a modest little dinner at a quiet French restaurant..and an orchestra-chair at the Metropolitan. **1923** G. SELDES in *Vanity Fair* (N.Y.) Jan. 57/2 It [*sc.* the revue] corresponds to those de luxe railway trains which are always exactly on time..; jazz or symphony may sound from the orchestra pit, but underneath is the real tone of the revue, the steady, incorruptible purr of the dynamo. **1932** *Times Lit. Suppl.* 24 Nov. 888/3 The only other playbill reference.. is a King's Theatre bill for July 13 [1831], which advertised stalls and orchestra places at a guinea each for a concert. **1940** M. DE LA ROCHE *Whiteoak Chron.* II. vii. 241 He had got orchestra chairs for a Russian vaudeville. **1952** GRANVILLE *Dict. Theatr. Terms* 128 *Orchestra stalls*, the seats nearest the orchestra. **1956** G. DURRELL *My Family* viii. 109 Two of them were thrown into the orchestra pit before someone had the sense to lower the curtain. **1977** R. BARNARD *Death on High C's* xv. 155 The opening chords of *Rigoletto* were sounding from the orchestra pit. **1977** *Times* 24 Sept. 12/7 Two sweeping shell-like roofs..cover all but 10 rows of orchestra seats... The opera house sits on a hillside.

orchestral, *a.* Add: Also *transf.* in *orchestral effects*.

1888 KIPLING *In Black & White* 66 It [*sc.* the Religion] added an air-line postal *dak*, and orchestral effects. **1918** *Sphere* 9 Feb. 125/1 The orchestral effects of Monday night's raid.

orchestralist (ǭɹke·strălist). [f. ORCHESTRAL *a.* + -IST.] A writer of orchestral music; an orchestrator.

1899 F. J. CROWEST *Beethoven* 221 The enharmonic change in the first movement..again illustrates the wonderful resources of this king of orchestralists.

orchestrate, *v.* Add: **b.** *fig.* (Further examples.)

1956 H. WHITEHALL in *Kenyon Rev.* XVIII. 418 The traditional 'ideal' metrical patterns..have been 'orchestrated' since Marlowe. **1957** *Times Lit. Suppl.* 6 Dec. 789/2 *Nostromo*, greatest and most splendidly orchestrated of all his fictions. **1967** C. L. WRENN *Word & Symbol* 11 Swinburne's poem *Dolores*. Here there is a kind of orchestrated language which conveys a mood of meaning but no clearly describable sense. **1969** *Daily Tel.* 25 Nov. 30 Russia and America yesterday ratified the treaty banning the spread of nuclear weapons. They chose the same day by a diplomatic agreement typical of the way the two super-Powers are 'orchestrating' their moves in this front. **1974** *Guardian* 23 Jan. 2/8 The White House deployed its heavy artillery today... The counter-attack was well orchestrated. **1975** *N.Y. Times* 31 Oct. 11/6 Planning and organization were particularly striking today, when three busloads of foreign journalists were brought to the staging camps on an officially sponsored visit. The enthusiasm that greeted them was as carefully orchestrated as is the march itself. **1977** *Time* 7 Mar. 8/2 Owen helped to orchestrate the European Community's fishing agreement with the Soviet Union.

orchestration. Add: **b.** (Further examples.)

1905 *Smart Set* Sept. 113/1 Suddenly there was developed a terrific orchestration of chromatic odors. **1936** *Essays & Stud.* XXI. 150 *The Wreck of the Deutschland* has a completeness, an intellectual and emotional unity, a subtlety and variety of verbal orchestration which are unique not only in English but in the literature of the world. **1953** J. S. HUXLEY *Evolution in Action* vi. 138 The world community which we envisage and hope to bring to the birth is a variety-in-unity. In the useful phrase of the American writer, L. K. Frank, it involves an orchestration of cultures. **1956** H. WHITEHALL in *Kenyon Rev.* XVIII. 418 Yet 'orchestration' is affected by another feature of English. **1959** [see *BUILD-UP c]. **1966** *Economist* 2 Apr. 17/2 What is required is the orchestration of the western countries' common interest in reaching an arrangement with Russia. **1975** *Country Life* 16 Jan. 139/2 Braque..achieved a marvellous orchestration of colours and forms. **1975** *New Yorker* 5 May 132/2 Each night, the great orchestration of the evening news went on. **1977** *Time* 17 Oct. 22/1 A conference that collapsed because of poor orchestration was even worse than no conference at all.

c. An overcoat. *U.S. slang.*

1940 *Music Makers* May 37/3 *Orchestration*, an overcoat. **1970** in C. MAJOR *Dict. Afro-Amer. Slang.*

orchestrator (ǭ·ɹkėstrēⁱtəɹ). [f. ORCHESTRATE *v.* + -OR.] One who composes or arranges music for an orchestra, band, etc. Also *fig.*

1907 E. WALKER *Hist. Mus. Eng.* 306 As an orchestrator he [*sc.* Elgar] is among the very greatest in musical history. **1927** *Observer* 20 Nov. 14/4 Liszt, as an orchestrator, seems to overtop the other. **1954** *Grove's Dict. Mus.* (ed. 5) VII. 56/2 Although he [*sc.* Ravel] is a born orchestrator and his command of the medium is unsurpassed, hardly any of his works were originally conceived for the orchestra. **1974** *Times Lit. Suppl.* 1 Nov. 1226/3 An effort was made..to smooth away the more absurd elements of 'crusade' historiography... The orchestrator of this campaign was to be Ricardo de la Cierva, a dynamic and prolific historian. **1976** *Gramophone* May 1750/2, I had no idea what a superb orchestrator he was before I heard that record.

orchestrelle (ǭ·ɹkėstre:l). [f. ORCHESTR(A + Fr. dim. suffix *-elle*.] (See quot. 1961.)

a **1910** 'MARK TWAIN' in *Harper's Mag.* (1911) Jan. 215/1 Paine began playing on the orchestrelle Schubert's *Impromptu*. **1912** A. B. PAINE *Mark Twain* III. 1227 He added..a great Æolian Orchestrelle, with a variety of music. **1925** T. DREISER *Amer. Tragedy* (1926) I. II. xviii. 288 Fox-trots and one-steps were being supplied by an orchestrelle of considerable size. **1961** E. AMES *Daughter of House* (1963) I. vi. 95 The orchestrelle..was a kind of organ which could either be played manually, or—by switching to electrical controls—could be made to play perforated rolls like a player piano.

orchid. Add: **b.** (Further examples.)

1852 B. S. WILLIAMS (*title*) The orchid-grower's manual. **1884** *Encycl. Brit.* XVII. 818/1 Orchid-lovers have better reasons to support their fancy than had the speculative growers and barterers of tulips. **1903** *Daily Chron.* 9 Nov. 3/5 One can never tell what the orchid-seed of enterprise may blossom into. **1908** *Westm. Gaz.* 28 May 12/1 The dangers of orchid-hunting. **1909** *Chambers's Jrnl.* July 430/2 Many an orchid-hunter has sacrificed his life to his daring. **1935** N. MITCHISON *We have been Warned* I. ii. 24 Joyce's dress was long and diaphanous, orchid-coloured. **1974** *Country Life* 18 Apr. 950/4 Already my alpine house and orchid house have been shaded and humidity increased in the latter. **1974** *Encycl. Brit. Macropædia* XIII. 648/1 Discussions about orchids, whether among professional botanists or amateur orchid enthusiasts, often leave the impression that orchids are 'somehow different' from other plants. **1977** A. WILSON *Strange Ride R. Kipling* iii. 136 Singapore's orchid garden.

2. A purplish colour or tint.

1923 *Daily Mail* 15 Jan. 1 Shades are Navy, Bisque, Rust, Champagne, Orchid, Flesh, Silver, Nattier Blue, Black, Jade & Ivory. **1936** *Times* 6 Jan. 11/3 A model Court gown in orchid morganza. **1971** *Guardian* 28 Sept. 11/2 Quilted raincoat... In sand, orchid, or damson. **1975** D. RAMSAY *Descent into Dark* ii. 78 He wore orchid pyjamas of real silk.

orchidacity (ǭɹkidæ·sĭti). [f. ORCHID + -ACITY.] The quality of being 'orchidaceous' (sense 2).

1897 *Sat. Rev.* 13 Feb. 169, I have power and passion, orchidacity and flamboyancy. **1933** *Times Lit. Suppl.* 2 Nov. 747/3 He [*sc.* Kenneth Grahame].., sheltered within the sinister orchidacity of the *Yellow Book*, wrote 'pagan' papers full of the fresh air.

orchidean. Add: Also **orchidian.**

1914 C. A. MERCIER *Astrol. in Med.* 4 One eminent physician discovered..the elixir of life in orchidian extract.

orchidectomy. Add: (Earlier and later examples.) For 'the testicles' read 'one or both of the testicles'. Cf. *ORCHIECTOMY.

1870 *Austral. Med. Jrnl.* XV. 277 Dr. [D. J.] Thomas said, 'I look upon it as a case of strumous sarcocele, requiring operation, which I have taken upon myself to call Orchidectomy. Notices appearing in the daily papers of the operations to take place at this hospital, I thought, that excision of the testicle, or the word castration, would not look well.' **1947** *Nature* 4 Jan. 15/1 Cancers of the male breast yield to orchidectomy. **1967** *Med. Ann.* LXXXV. 217 In cases of male pseudohermaphroditism with testicular feminization, orchidectomy is carried out after puberty..and oestrogen therapy given. **1977** BLACK & BISHOP *Sisterhood* viii. 74 An orchidectomy involves the removal of the testicles.

Hence **orchide·ctomize** *v. trans.*, to perform orchidectomy on; to castrate; **orchide·ctomized** *ppl. a.*

1942 *Amer. Jrnl. Anat.* LXXI. 456 One castrate member..had been orchidectomized when 12 years of age. **1965** *Endocrinology* LXXVI. 1220/1 An intact group and an orchidectomized group were given ·9% saline. *Ibid.* 1222 (*heading*) Orchidectomized rat given saline.

orchidize (ǭ·ɹkidəiz), *v.* [f. ORCHID + -IZE.] *trans.* To make like an orchid. (In quots. *fig.*)

1918 A. BENNETT *Pretty Lady* xxxi. 222 In the right environment she would become another being, that was to say, the same being, but orchidised. **1922** JOYCE *Ulysses* 418 Jesified orchidised polycimical jesuit!

orchid-like (ǭ·ɹkid,ləik), *a.* [f. ORCHID + -LIKE.] Like or resembling an orchid.

1881 W. ROBINSON *Wild Garden* xii. 93 The German Irises, with their great Orchid-like blossom. **1918** C. W. BEEBE *Jungle Peace* (1919) viii. 180 And later, orchid-like, violet, butterfly peas which at first flowered among the ashes on the ground, but climbed as soon as they found support. **1923** D. H. LAWRENCE *Birds, Beasts & Flowers* 39 What would I not give To bring back the rare and orchid-like Evil-yclept Etruscan?

orchido-. Add: **o·rchidopexy.** *Surg.* [*-PEXY], fixation of a testicle, esp. of an undescended testicle in the scrotum.

1893 W. H. A. JACOBSON *Dis. Male Organs* I. ii. 83 (*heading*) Transplantation of a retained or misplaced testicle into the scrotum.—Orchidopexy. **1974** PASSMORE & ROBSON *Compan. Med. Stud.* III. I. xxvii. 8/1 The operation of choice [for imperfect descent of the testis] before puberty is orchidopexy, and after puberty orchidectomy.

orchiectomy (ǭɹki,e·ktŏmi). *Surg.* [See ORCHIDECTOMY.] = ORCHIDECTOMY in Dict. and Suppl.

1894 in GOULD *Dict. Med.* 938/1. **1948** BAILEY & LOVE *Short Pract. Surg.* (ed. 8) III. xxxiii. 602 Orchiectomy is indicated when the other testis is normal, and the cord too short to allow replacement of the maldescended organ in the scrotum. **1963** *Lancet* 5 Jan. 21/1 In four cases where the testicle was regarded as non-viable the operation was orchiectomy. Orchiopexy was carried out in the rest though it was not always thought that the organ would function. **1974** *Acta Endocrinol.* LXXVI. 237 A progressive decrease in the pituitary FSH level was observed up to 72 h after orchiectomy.

orchiopexy (ǭ·ɹki,ŏpeksi). *Surg.* [f. Gr. ὄρχι-ς testicle + -O + *-PEXY.] = *orchidopexy* s.v. *ORCHIDO-.

[**1909** F. TOREK in *N.Y. Med. Jrnl.* XC. 952/2, I had to present one of my cases of orcheopexy before this section of the Academy.] **1931** —— in *Ann. Surg.* XCIV. 97 My original spelling, orcheopexy, should therefore be changed to orchiopexy. **1938** BAILEY & LOVE *Short Pract. Surg.* (ed. 4) xxix. 544 If this treatment fails, orchiopexy should be advised. **1963** [see *ORCHIECTOMY]. **1974** *Investigative Urol.* XI. 303 Nephropexy and orchiopexy using tissue adhesives were performed on rats and rabbits.

orcinol. Add: Now the usual name for the substance (which gives a purple colour with ribose and is used in estimating nucleic acids). (Earlier and later examples.)

1880 *Jrnl. Chem. Soc.* XXXVIII. 113 (*heading*) Resorcinol and orcinol derivatives. **1921** W. T. K. BRAUNHOLTZ tr. *Moureu's Fund. Princ. Org. Chem.* vii. 372 An ammoniacal solution of orcinol undergoes oxidation in the air, giving orcein,..which is a red dyestuff. **1922** *Chem. Age* VII. 709/1 The various orders of *Variolaria*.., *Rocella* and *Lecanora* all contain orcinol in the free state, and these lichens are now used for the production of the two substitution products, litmus and archil. **1969** *Phytochem.* VIII. 2223 Orcinol (5-methylresorcinol) has been detected for the first time in higher plants, in ten species of the Ericaceae. **1973** N. C. MISHRA et al. in Niu & Segal *Role of Ribonucleic Acid* 261 DNA and RNA were determined chemically by diphenylamine and orcinol reactions respectively.

orciprenaline (ǭɹsipre·nălīn). *Pharm.* [f. ORCI(NOL + *ISO)PRENALINE.] A sympathomimetic amine, $C_6H_3(OH)_2CH(OH)CH_2$-$NHCH(CH_3)_2$, that is closely related to isoprenaline in structure and is taken (usu. as the sulphate, a white, bitter-tasting powder) for the relief of bronchitis and asthma, usu. in an inhaler or as tablets or a syrup.

1963 *Med. Digest* VIII. 95/1 Alupent (orciprenaline) is a recently introduced synthetic drug, 1-(3,5-dihydroxyphenyl)-2-isopropylamino-ethanol sulphate. **1964** *Brit. Med. Jrnl.* 18 Apr. 1017/2 Orciprenaline ('alupent'), an analogue of isoprenaline was..effective in the relief of airways obstruction..whether taken by mouth or by inhalation. **1971** *Brit. Med. Bull.* XXVII. 27/1 Death might be caused by the excessive use of aerosol inhalers containing..orciprenaline. **1974** *Times* 12 Jan. 1/3 Her pregnancy..lasted 37 weeks aided by a muscle-relaxant drug called Orciprinaline [*sic*].

order, *sb.* Add: **I. 1. c.** *Physics.* Each of a successive series of spectra formed by the interference or diffraction of light; hence, a positive number characterizing a particular spectrum or interference fringe, now recognized as equal to the number of wavelengths by which the optical paths of successive contributing rays differ.

1704 NEWTON *Opticks* II. 6 The third Circuit or Order was purple, blue, green, yellow, and red. **1722** *Phil. Trans. R. Soc.* XXXI. 244 We had here four Orders of Colours, and perhaps the beginning of a fifth, for what..I call the Purple, is a Mixture of the Purple of each of the upper Series with the Red of the next below it. **1831** [see sense 1 a]. **1874** *Phil. Mag.* XLVII. 194 In considering the influence of the number of lines (*n*) and the order of the spectrum (*m*), we will suppose that the ruling [of the diffraction grating] is accurate. **1953** SPINK & FEIGL tr. *Pinsker's Electron Diffraction* ii. 26 We obtain..$n\lambda = 2d \sin\theta$, *n* being the order of the reflection [*sc.* of electrons from a crystal]. This gives the number of whole wavelengths corresponding to the path difference for waves scattered by two neighbouring parallel planes of the direct lattice. **1967** W. H. STEEL *Interferometry* viii. 139 Observation of the interference fringes yields only the excess fraction ϵ of the order of interference, the amount by which the order exceeds some unknown integer.

II. 6. a. In the Roman Catholic Church, the orders of subdeacon, exorcist, and *ostiarius* were suppressed in 1972.

c. (Later examples of *sing.*)

1873 E. E. ESTCOURT *Question of Anglican Ordinations* i. 4 Holy Order is a Sacrament, requiring a certain matter and form. **1977** *Christian* IV. 31 There are two priesthoods... One is conferred on all, in baptism; the other on some, in the sacrament of holy order. *Ibid.* 34 The sacrament of order is a direct participation in the mystery of Pentecost.

8. *Order of Merit*; hence *Order-of-Merited* adj. (*nonce*).

1799 *Public Characters of 1799–1800* II. 164 The King of Poland..also conferred on him the honours of knighthood of the Order of Merit. **1842** T. CAMPBELL *Frederick the Great* II. xv. 71 If the order of the Black Eagle was conferred on any of the members of the order of Merit, he had to send back the latter to the king. **1880** DISRAELI *Endymion* II. xv. 150 Now you tell your master..that if he wants to strengthen the institutions of this country, the government should establish an order of merit. **1902** *Pall Mall Gaz.* XXVIII. 71/1 The King's new Order of Merit would have attracted more attention if the list had appeared alone, and not at the tail of the honours bestowed at the Coronation. **1912** G. W. E. RUSSELL *Afterthoughts* xxxix. 325 An 'Order of Merit'—as far as History goes, *the* Order of Merit—was founded by Frederick the Great in 1740; and the name was copied in turn by Hesse Cassel, Baden, Bavaria, Saxony, Oldenburg, Würtemberg, and Belgium. **1929** A. HUXLEY *Swift* in *Holy Face & Other Ess.* 64 If Swift were alive to-day, he would be the adored..the Order-of-Merited author, not of *Gulliver*..but of *A Kiss for Cinderella* and *Peter Pan*. **1959** *Chambers's Encycl.* X. 228/2 The Order of Merit (O.M.) was instituted in 1902 and is awarded to officers of the fighting services and civilians for conspicuous service. **1970** C. L. CLINE *Lett. George Meredith* I. p. xxix, In 1905 came official recognition: he became the twelfth member of the recently founded Order of Merit.

b. (Earlier example of *pl.*)

1818 KEATS *Let.* 14 Oct. (1958) I. 396 No sensation is created by Greatness but by the number of orders a Man has at his Button holes.

c. *order of the boot*: see *BOOT *sb.*³ 1 c.

10. (Further examples.) *of the same order*, said of two variables whose ratio tends to a finite number as they both tend to zero or to infinity; *to the first* (or *second*, etc.) *order*, neglecting quantities of higher order than the first (or second, etc.). Also in *Comb.* with preceding ordinal number.

1838 *Penny Cycl.* XII. 472/2 A succession of infinitely small quantities, each of which is infinitely smaller than the preceding, is said to be a series of infinitesimals of different orders. Such a series is x, x^2, x^3, &c. **1843** *Scientific Mem.* III. 172 It follows..that for an infinitely small ρ, a/ρ must itself be infinitely small; but the two values will be of the same order only if there is a finite radius of curvature. *Ibid.*, If we assume a/ρ to be of the same order as ρ^{μ},..then $a/(\rho^{1+\mu})$ represents a finite quantity varying continuously. **1880** *Encycl. Brit.* XIII. 14/1 Two infinitesimals α, β are said to be of the same order if the fraction β/α tends to a finite limit. If β/α^n tends to a finite limit, β is called an infinitesimal of the *n*th order in comparison with α. **1908** G. H. HARDY *Course Pure Math.* v. 169 We shall say that $\phi(x)$ is of the *k*th order of greatness when *x* is small if $\phi(x)/x^{-k} = x^k\phi(x)$ tends to a limit different from zero as *x* tends to 0. **1922** *Phil. Mag.* XLIII. 945 Both Sommerfeld and Epstein have obtained the value of W..by slightly different methods to the first order in F... We shall proceed to a second approximation. **1937** E. C. KEMBLE *Fund. Princ. Quantum Mech.* xi. 384 In order to get the second-order corrections to E_k and Ψ_k we differentiate Eq. (47·2) twice with respect to λ and then set λ equal to zero. *Ibid.* 386 These second-order formulas are so complicated that they are seldom used and the corrections of the third and higher order [*sic*] are still more complex. **1952** D. R. HARTREE *Numerical Anal.* ix. 192 If $a_1 \neq 0$..the number of additional correct significant figures obtained from each repetition of such a process (or..the number of repetitions required to obtain each new correct significant figure) is the same, however many figures have been obtained. Such a process is called 'first-order'. But if $a_1 = 0$, $a_2 \neq 0$..the successive errors ξ_n are ultimately related by $\xi_{n+1} = a_2\xi_n^2$... The number of correct significant figures is approximately doubled for each repetition of the iterative process... Such a process is called 'second-order'. **1962** SIMPSON & RICHARDS *Physical Princ. Junction Transistors* vi. 110 This discussion shows that h_{rb} and h_{ob} arise in such an indirect manner that they might almost be regarded as second-order effects. Their magnitudes confirm this impression. **1971** *Nature* 19 Feb. 522/2 Talk of the eclipse being an unusual strain to the Earth is idle nonsense when eclipse type conditions prevail to first order twice every lunar month.

b. *of the order of*: (*a*) *Math.* (also *of order*), having a ratio to (the quantity specified) that tends in the limit to a finite number, or that is neither a large number nor a small fraction; (*b*) *gen.* (also *in* or *on the order of*), in the region of; somewhere about.

1903 O. LODGE *Mod. Views Matter* 7 Their mass is of the order one-thousandth of the atomic mass of hydrogen. **1913** *Rep. Brit. Assoc. Adv. Sci. 1912* 398 The change of weight..should have been of the order of 1 in 10^7 per 1°C. **1927** N. V. SIDGWICK *Electronic Theory of Valency* ii. 20 The accuracy of spectroscopic measurements (of the order of one in a million). **1937** MICHELL & BELZ *Elem. Math. Analysis* I. i. 94 A number is said to be of order 10^n if its ratio to 10^n is neither large nor small. *Ibid.*, The function $f(x)$ is said to be of the order of x^n as *x* converges to zero, or to be of the *n*th order with respect to *x*, if the ratio $f(x)/x^n$ has two focal bounds of the same sign. *Ibid.* 95 The notation $f(x) = O(x^n)$ is sometimes used to express that the function $f(x)$ is of the order of x^n, that is, of the *n*th order, when *x* converges to zero. **1947** R. L. WAKEMAN *Chem. Commercial Plastics* xxvi. 786 Concentrations of catalyst in the order of 1 per cent. **1955** D. A. QUADLING *Math. Analysis* xi. 178 We may write..$|E/h^n| \leq K$, where *K* is independent of *h*. In such a case we say that *E* is 'of order h^n', or 'of the *n*th order of small quantities'. **1958** *Times* 10 Dec. 10/4 Their radioactivity was of the order of tens of millicuries. **1962** F. I. ORDWAY et al. *Basic Astronautics* x. 422 Specific impulses on the order of 3000 lb-sec/lb are possible. **1963** R. A. RANKIN *Introd. Math. Analysis* vii. 455 The statement $f(n) = O(\phi(n))$ as $n \to \infty$, or for large *n*, means that there exist real numbers $K \geq 0$ and $X \geq X_0$, which are independent of *n*, and are such that $|f(n)| \leq K\phi(n)$ for all $n > X$... The statement.. may be read as '$f(n)$ is a quantity of the order of $\phi(n)$'. **1970** *Daily Tel.* 3 Dec. 21/1 (Advt.), A salary in the order of £1,500 is envisaged. **1971** *Sci. Amer.* June 24/1 Pulses lasting on the order of a nanosecond or longer can be reliably produced. **1975** *Nature* 10 Apr. 478/1 The average flow through the gorge is of the order of 2,000–3,000 cubic metres per second.

c. *order of magnitude*: approximate number or magnitude in a scale in which equal steps correspond to a fixed multiplying factor (usu. taken as 10); a range between one power of 10 and the next; also, the order (ORDER *sb.* 10) of an infinitesimal or an infinite number. Also *attrib.* (with hyphens).

1875 *Jrnl. Anthrop. Inst.* IV. 143 The number of surnames extinguished becomes a number of the same order of magnitude as the total number at first starting in N. **1891** *Phil. Mag.* XXXII. 296 The electrochemical equivalent of gas atoms is of the same order of magnitude as that of the same atoms in electrolytes. **1903** D. A. MURRAY *First Course Infinitesimal Calculus* iii. 31 When the limiting value of the ratio *m*/*n* is a finite number, *m* and *n* are said..to be of the same order of magnitude. *Ibid.* 32 Infinite numbers, being reciprocals of infinitesimals, also have different orders of magnitude. **1909** J. P. IDDINGS *Igneous Rocks* I. vi. 193 The grains..are not all of the same order of magnitude, since one may be nine or ten times larger than another. **1937** MICHELL & BELZ *Elem. Math. Analysis* I. i. 94 Two numbers (or quantities) are of the same order of magnitude when their ratio is neither a large number nor a small fraction... It is often convenient to make a comparison of a number with a power of 10. **1941** COURANT & ROBBINS *What is Math.?* viii. 469 We shall say that b_n tends to infinity faster than a_n, or has a higher order of magnitude than a_n, if the ratio a_n/b_n (numerator and denominator of which both tend to infinity) tends to zero as *n* increases. **1968** R. A. LYTTLETON *Mysteries Solar Syst.* v. 157 The general size..would be expected to be of the order of the width of the stream. An order-of-magnitude estimate of this can be made in the following way. **1971** I. G. GASS et al. *Understanding Earth* iv. 78/2 The dynamo theory seems natural and unforced, and order-of-magnitude arguments are encouraging. **1971** *Physics Bull.* Oct. 586/1 The width of the second line of the hydrogen Balmer series, Hβ, is now considered to be a reliable measure of electron density to better than 10% over four orders of magnitude (say 10^{14}–10^{18} cm⁻³). **1974** *Sci. Amer.* June 27/1 These processes multiply the power per unit area by 14 orders of magnitude from 10^5 watts per square centimeter..to 10^{19} watts per square centimeter.

d. *Math.* (i) The number of elements in a group.

1878 A. CAYLEY in *Amer. Jrnl. Math.* I. 51 A set of symbols α, β, γ..such that the product $\alpha\beta$ of each two of them..is a symbol of the set, is a group... When the number of the symbols (or terms) is $= n$, then the group is of the *n*th order. **1941** [see next sense]. **1965** PATTERSON & RUTHERFORD *Elem. Abstr. Algebra* ii. 36 The set of all permutations of 1, 2, 3,..., *n* forms a group with respect to multiplication... It is a finite group of order *n*! and it plays an important part in the theory of finite groups.

(ii) The smallest positive integer *m* for which g^m is equal to the identity element of a group, *g* being any given element.

1897 W. BURNSIDE *Theory of Groups of Finite Order* ii. 14 Let *S* be an operation of a group of finite order *N*... If S^{m+1} is the first of the series [*sc.* S, S^2, S^3,...] which is the same as *S*,..then..$S^m = 1$... The integer *m* is called the order of the operation *S*. **1941** BIRKHOFF & MACLANE *Survey Mod. Algebra* vi. 147 Every element of a finite group *G* has as order a divisor of the order of *G*. **1968** I. D. MACDONALD *Theory of Groups* iii. 45 A periodic group is a group in which every element has finite order.

e. Each of the ranks or levels in a (non-mathematical) hierarchy in which every member save those in the lowest rank is a function of members of the next lower rank; *spec.* in *Logic* (see quot. 1908) and *Psychol.* (see quot. 1947). Freq. in *Comb.* with preceding ordinal number.

In *Math.* this sense is identical with 10 in Dict.

1908 B. RUSSELL in *Amer. Jrnl. Math.* XXX. 238 A proposition containing no apparent variable we will call an elementary proposition... Elementary propositions together with such as contain only individuals as apparent variables we will call first-order propositions... We can

thus form new propositions in which first-order propositions occur as apparent variables. These we will call second-order propositions... Thus, *e.g.*, if Epimenides asserts 'all first-order propositions affirmed by me are false', he asserts a second-order proposition. *Ibid.*, Propositions of order *n*..will be such as contain propositions of order $n-1$, but of no higher order, as apparent variables. **1929** A. W. WHITEHEAD *Process & Reality* II. ix. 285 We must provide a reason..why one 'ground' is selected rather than another... We are thus driven back to a second-order 'ground' of probability. **1936** *Mind* XLV. 170 Necessary propositions are, thus, second-order propositions, which implicitly define 'proposition' by stating the properties of anything that is a proposition. **1941** J. S. HUXLEY *Uniqueness of Man* xi. 245 Cells are first-order individuals, bodies second-order ones, and human societies (like hydroid colonies or beehives) third-order ones. **1947** L. L. THURSTONE *Multiple-Factor Analysis* xviii. 411 Factors that are obtained from the test correlations will be called first-order factors... Factors that are obtained from the correlations of the first-order factors will be called second-order factors. **1954** I. M. COPE *Symbolic Logic* 336 The hierarchy of orders prevents us from speaking about all functions or properties of a given type, permitting us to speak only about all first order functions of a given type, or all second order functions of a given type, etc. *Ibid.* 337 A proposition is of order $n+1$ if it contains a quantifier on a propositional variable of order *n* but contains no quantifier on any propositional variable of order *m* where $m \geq n$. **1961** J. B. WILSON *Reason & Morals* i. 4 Philosophers themselves are accustomed to speak of philosophical statements as being 'second-order' statements. **1971** *Sci. Amer.* Aug. 98/1 All the axioms of a complete ordered field are first-order sentences except for the completion axiom.., which talks about a property of *all* subsets. **1977** A. HALLAM *Planet Earth* 75/3 In this system, fingertip tributaries are described as first-order; when two first-order streams combine the result is a second-order stream. Two second-orders give a third-order, and so on.

f. *Chem.* The sum of the exponents of the concentrations of reactants, or the exponent of any particular reactant, in the expression for the rate of a chemical reaction. Freq. in *Comb.* with preceding ordinal number.

1902 H. C. JONES *Elem. Physical Chem.* ix. 465 Although there are only two substances, there are three molecules involved in the reaction, and we would expect it to be a reaction of the third order. **1933** E. A. MOELWYN-HUGHES *Kinetics of Reactions in Solution* vii. 219 Ionic reactions have occasionally been found to be of a higher kinetic order than is now regarded as possible. **1950** W. J. MOORE *Physical Chem.* xvii. 514 This is also a second-order reaction. It is said to be *first-order with respect to* C_2H_5Br, *first-order with respect to* $(C_2H_5)_3N$, and *second-order over-all*. **1968** R. O. C. NORMAN *Princ. Org. Synthesis* iii. 78 The decarbonylation of acetaldehyde is of non-integral order but contains both unimolecular and bimolecular steps.

g. *Physics* and *Chem.* An integer (usually 1 or 2) characterizing a change of phase of a substance, equal to the order of the lowest-order derivatives of the free energy that exhibit a discontinuity at the change. [After the similar use of G. *ordnung* introduced by P. Ehrenfest 1933 (see quot. 1933).]

1933 *Proc. Sect. Sci. Kon. Akad. Wetensch. Amsterdam* XXXVI. 152 *G* may be a function of *p* and *T* which suffers along a λ-curve (Fig. 3) a discontinuity of the second order[1], so that along that curve..$\Delta G = 0$, whereas the differential coefficients of *G* make a jump. [[1]*Note*] Cf. P. Ehrenfest. Proceedings of this meeting. [i.e. *Ibid.* 153–7 (in Ger.)]. **1946** *Nature* 28 Dec. 924/2 At low temperatures both crystalline and amorphous states [of rubber] give place to the glass-hard condition. The transition to the glassy state—the so-called second-order transition—is discussed. **1948** *Jrnl. Chem. Physics* XVI. 665 (*heading*) Note on a relation between the order of a phase transition and discontinuities in the distribution functions of molecules. **1967** A. H. COTTRELL *Introd. Metall.* xiv. 220 First-order changes such as melting and polymorphic changes of crystal structure. **1968** C. G. KUPER *Introd. Theory Superconductivity* ii. 23 The superconducting transition in the absence of a magnetic field is of second order (Ehrenfest 1933). In other words, the specific heat is discontinuous but there is no latent heat.

III. 12. b. *order of battle*, the arrangement or disposition of sections of an army or naval force; now *spec.* the organization, movements, weaponry, etc., of an enemy force; the discovery of this; a tabular record of this; also *attrib.*

1769 [see BATTLE *sb.* 12]. **1797** *Encycl. Brit.* III. 81/1 A Roman legion, ranged in order of battle, consisted of *hastati*, placed in the front; of *principes*, who were all old experienced soldiers, placed behind the former; and of *triarii*, heavy armed with large bucklers, behind the *principes*. **1889** H. R. GALL *Mod. Tactics* ii. 11 A practical and experienced soldier, seeing his enemy get under arms and form up in order of battle, will rapidly gather a lot of valuable information regarding his numbers, artillery positions [etc.]. **1924** W. C. SWEENEY *Military Intelligence* viii. 172 Enemy Order of Battle. This section is charged with maintaining the battle order of the enemy located within the area of responsibility of the commander. **1928** H. M. D. PARKER *Roman Legions* ix. 251 Arrian, in his order of battle against the Alani, shows that the legions were drawn up as a phalanx eight deep. **1934** WEBSTER s.v., *Order of battle*.., a tabular compilation by unit showing organization, commanders, movements, etc. over an extended time. **1946** CHANDLER & ROBB *Front-Line Intelligence* xii. 137 O/B (Order of battle) is a military science whose mission is to determine: (1) How strong the enemy is. (2) How he is organized. (3) What kind of weapons he has. (4) Experience of his troops. (5) Leadership of his troops. (6) Where his units are located. **1948** F. R. COWELL *Cicero & Roman Republic* ii. 36 The Greeks..were the first to invent an order of battle in which drilled men acting together as a unit.. were able to overcome unorganised enemies many times more numerous. **1950** *Tactics & Techniques Infantry* (U.S.) II. ii. 312 When a new enemy unit is identified, order of battle records will indicate its last known strength and special equipment such as tanks, armored cars, and artillery; its combat record; or any quirks of its commander. A new operation is contemplated; order of battle data will provide information as to the strength, equipment, location, mobility, and combat efficiency of the specific units the enemy can employ. *Ibid.* 313 Strategic order of battle deals with all enemy military units, regardless of location. **1966** D. G. CHANDLER *Campaigns of Napoleon* (1967) 1099 Order of battle of the Army of Italy, April 12, 1796. **1971** *Combat Intelligence* (U.S. Dept. of Army, Field Manual 30-5) vii. 7-1 In determining enemy capabilities and probable courses of action, commanders must consider order of battle intelligence together with other intelligence pertaining to the enemy, weather, and terrain. **1975** T. ALLBEURY *Special Collection* ii. 11 There were daily situation reports from both the west and east fronts including Wehrmacht orders-of-battle. **1977** S. COULTER *Soyuz Affair* v. 42 The spy..who brought you the cypher table or the enemy order of battle.

13. d. Equipment, uniform, etc., for some purpose, as *drill order, field-day order, review order; marching order*: see MARCHING *vbl. sb.* d in Dict. and Suppl. Also with sbs. descriptive of appearance, as *shirt-sleeve order*. (Orig. and chiefly *Mil.*)

1852 R. BURN *Naval & Milit. Techn. Dict. French Lang.* (ed. 2) II. 176/2 *Drill order*, tenue d'exercise, petite tenue. **1874** *Queen's Regulations Army 1873* 162 Review-order; to be worn when the Sovereign is present, for Royal escorts and guards of honour. *Ibid.* 163 Field-day-order; to be used generally for summer field-days, divisional and brigade drills,..[etc.]. *Ibid.*, Drill-order; to be used at ordinary drills and in riding-schools. **1876** Review order [see REVIEW *sb.* 3]. **1968** J. LOCK *Lady Policeman* viii. 60 My serge skirt feels heavy, my feet feel hot and sticky. Still, we are lucky to have shirt-sleeve order—the PCs haven't and look as if they are about to expire. **1973** R. HILL *Ruling Passion* II. vii. 138 The warm weather..had returned..it would be shirt-sleeve order before the day was out. **1977** 'D. MACNEIL' *Wolf in Fold* v. 49 Behind them, dressed in review order, marched the infantry of the British Army.

18. (Earlier examples.)

a **1751** in *Camden Miscellany* (1969) XXIII. 170 His vanity will make him constantly puzzling our Speaker and our Chairmen of Committees, in points of order, which in reality he will know better than they. **1781** *Parl. Reg.* 27 Nov. 46 After some debate on the point of order, respecting the right of reply, claimed by those who had made a motion.

20. *spec.* of tobacco. Cf. *CASE *sb.*¹ 5 b.

1897 M. WHITNEY in *U.S. Dept. Agric. Farmers' Bull.* No. 60. 4 'Order' or 'case' in tobacco curing means a mint condition in which the tissue will not break. **1966** *Pubn. Amer. Dial. Soc.* XLV. 18 The tobacco has to be in order before it can be properly stripped.

21. (Further example.)

1938 J. CARY *Castle Corner* 435 The sentry threw his gun to the order and shouted in one word, 'alt-oo-go dar. Pass, friend'.

IV. 23. a. (Earlier and later examples of *under orders*.) Cf. *under starter's orders* s.v. *STARTER.

1835 DICKENS *Let.* *c* 29 Dec. (1965) I. 113, I regret to say that my being under orders from The Chronicle will prevent my enjoying the pleasure of seeing you tomorrow. **1969** H. R. F. KEATING *Inspector Ghote plays Joker* iii. 55 This authoritative figure took the microphone..and made an announcement that the horses were under orders. **1977** A. C. H. SMITH *Jericho Gun* iv. 52 The PA commentary told him when they were under orders, and off.

b. *Order in Council*: an order issued by the British sovereign († or the governor of a British colony) on the advice of his or her privy council; also, an order issued by a government department under powers bestowed by Act of Parliament.

[**1674–5** (*title*) His Majesties Declaration for enforcing a late Order made in Council.] **1746** in *New Jersey Archives* (1882) 1st Ser. VI. 369 An Embargo on all Vessels in this Province for the Space of one Month unless his [*sc.* the president's] Order in Council shall be first Obtained for the Sailing of any Vessel. **1785** [see COUNCIL *sb.* 6 a]. **1809** *Ann. Reg. 1807* xii. 227/2 English commerce..was not only greatly cramped, but lay prostrated on the ground, and motionless, before a protecting and self-defensive system was interposed by our orders in council. **1867** A. TODD *On Parl. Govt. in Eng.* I. v. 287 The crown has no right, by a mere Order in Council,..to sanction a departure from the requirements of an existing law. **1892** W. R. ANSON *Law & Custom of Constitution* II. i. 47 An Order in Council is practically a resolution passed by the Queen in Council, communicated by publication or otherwise to those whom it may concern. **1911** *Encycl. Brit.* XX. 187/2 At the present day orders in council are extensively used by the various administrative departments of the government, who act on the strength of powers conferred upon them by some act of parliament. **1928** A. FITZROY *Hist. Privy Council* iv. 85 The Orders in Council in reply to Napoleon's Milan and Berlin decrees. **1961** *Halsbury's Laws Eng.* (ed. 3) XXXVI. 477 Proclamations and Orders in Council are instruments made by the Crown, the latter, by which the great majority of powers conferred on the Crown are required to be exercised, being orders expressed to be made by and with the advice of the Privy Council. **1964** *Mod. Law Rev.* XXVIII. III. 335 His Majesty could by Order in Council provide..that the registers of a particular country..should be deemed to be 'a public register'. **1973** *Trinidad & Tobago Overseas Express* 28 May 4/2 If the initiative came from Britain then the Order-in-Council method would be applied. **1977** *Gay News* 24 Mar. 1/3 Mr. Mason is expected to draw up an Order in Council.

c. *doctor's orders*: instructions from one's physician; *fig.*, any injunctions which cannot be evaded.

1841 DICKENS *Let.* 18 Jan. (1969) II. 189, I have been obliged to make up my mind—on the doctor's orders—to stay at home this evening. **1886** H. MUNBY *Let.* 12 Mar. in D. Hudson *Munby* (1972) 410 Oh the miserable & false step you took when you separated me from you, by the doctor's orders. **1932** A. CHRISTIE *Peril at End House* ix. 104 No one..will be admitted... 'Doctor's orders,' they will be told. A phrase very convenient and one not to be gainsayed. **1940** W. FAULKNER *Hamlet* 73 He returned to the gallery offering his candy about. 'Doctor's orders,' he said. 'He'll probably send me another bill now for ten cents for advising me to eat a nickel's worth of candy.' **1970** *Guardian* 26 Nov. 3/4 The absence of East German leader, Herr Ulbricht, whose official explanation of 'doctor's orders' failed to convince a Communist journalist. **1976** *Sci. Amer.* Mar. 127/2 'Doctor's orders' excuse almost any behaviour, yet they are mere advice.

d. Phr. *orders are* (also vulg. or joc. *is*) *orders*: commands must be obeyed.

1852 H. MELVILLE *Pierre* XVI. ii. 323, I am sorry, sir, but orders are orders:..I can't disobey him. **1933** 'HAY' & 'ARMSTRONG' (title of play) Orders are orders. **1939** A. RANSOME *Secret Water* i. 18 I'm awfully sorry, you people. It just can't be helped. Orders is orders. **1973** *Times* 2 June 12/3 The delicious ridiculousness of the telegram perhaps has to be explained... But orders were orders.

e. *Computers.* = *INSTRUCTION 4 c, *COMMAND *sb.* 1 d; *esp.* one in machine language or another low-level language.

1946 GOLDSTINE & VON NEUMANN in J. von Neumann *Coll. Wks.* (1961) V. 26 In performing a multiplication one usually performs about 3 or 4 associated additions or subtractions or comparisons; hence at least 4–5 orders must be given and at least that many numbers transferred—it is assumed that an order specifies only one basic operation, together with its transfers. *Ibid.*, We agree to store our orders in the same place as our numbers. **1948** [see *INPUT *sb.* 2 d]. **1958** [see *INSTRUCTION 4 c]. **1967** KLERER & KORN *Digital Computer User's Handbk.* I. i. 10 Machine-language coding uses the machine order code, which is directly interpreted by the instruction register. **1970** O. DOPPING *Computers & Data Processing* vi. 98 The detailed information sent to the input/output units from the local control units can be called orders.

24. b. (Examples with sense 'postal order'.)

1891 YEATS *Let.* Dec. (1954) 186, I had intended to return the £1 at once... Some days passed by..the order which I enclose being all the time on my table awaiting posting. **1913** W. OWEN *Let.* 16 Dec. (1967) 221, I have cashed the Order long ago.

c. *a large order* (further examples). Also, *a big order, a strong order; a tall order*: see TALL *a.* 8 d in Dict. and Suppl.

1880 TROLLOPE *Duke's Children* I. xxiv. 284 In her opinion it would be best that the Duke should..give them money enough to live upon. 'Is not that a strong order?', asked the Earl. **1907** G. B. SHAW *Major Barbara* I. 210 *Barbara.* Yes. Give us Onward, Christian Soldiers. *Lomax.* Well, thats rather a strong order to begin with, dont you know. **1919** V. WOOLF *Night & Day* xxiv. 340 Well, Greek may be rather a large order. I was thinking chiefly of English. **1923** H. G. WELLS *Men like Gods* II. ii. 174 'You mean to jump this entire Utopian planet?' said Mr. Hunter. 'Big order,' said Lord Barralonga. **1927** *Sunday Times* 6 Mar. 23/3 There is no technical necessity now for the spark system, but it would be a rather big order to ask that all ships should abolish it. **1958** 'A. BRIDGE' *Portuguese Escape* viii. 125 This is quite a large order, isn't it? Suppose you tell me a bit more.

e. *colloq.* A request for refreshments or food, e.g. in a restaurant or public house; a portion or helping of a dish or article of food or drink served in a restaurant, snack-bar, etc.

1836 DICKENS in *Bell's Life in London* 17 Jan. 1/1 'Pray give me your orders gen'lm'n—pray give me your orders'..and demands for 'goes' of gin, and 'goes' of brandy, and pints of stout, and cigars of peculiar mildness, are vociferously made. *c* **1863** T. TAYLOR in M. R. Booth *Eng. Plays of 19th Cent.* (1969) II. 90 Now then James! Jackson, take orders. Interval of ten minutes allowed for refreshment. Give your orders, gents. **1898** A. BENNETT *Man from North* v. 29 A waitress, who approached and listened condescendingly to his order. **1904** 'O. HENRY' in *N.Y. World Mag.* 27 Mar. 10/4 And all this while she [*sc.* the waitress] would be performing astounding feats with orders of pork and beans, pot roasts, [etc.]. **1905** —— in *N.Y. World* 16 July (Oregon Fair Suppl.) 3/2 The screaming of 'short orders'..and all the horrid tumult of feeding man. **1934** G. B. SHAW *Village Wooing* 120 *Z*... Will you take a string bag? *A.* Yes. *Z.* Thanks very much. Shall I put the rest of the order into it? *A.* Of course. What else do you suppose I am buying it for? **1934** *Punch* 8 Aug. 158/3 The publican wanted to call on his clients with orders. 'What orders?'..'Beer.' **1949** *Crisis* (N.Y.) Nov. 305/2 They looked like the best tasting flapjacks in the world. They went inside and had an order. **1963** V. NABOKOV *Gift* v. 302 One could already hear the energetic '*psst, psst*' of Shahmatov, who had been served the wrong order. **1973** J. SHUB *Moscow by Nightmare* xiv. 165 Two orders of stuffed vine leaves, please. **1978** K. O'HARA *Ghost of T. Penry* xiii. 112 The pub sign was swinging furiously in the wind..inside.. they were taking last orders.

f. *order to view*: a requisition from a house or estate agent to an occupier to allow a client to inspect his premises.

1911 W. J. LOCKE *Glory of Clementina Wing* xxiii. 337 A caretaker took the order-to-view given by the estate agents and conducted the party over the place. **1922** E. H. Young *Bridge Dividing* III. xi. 301 It's to let. I've got an order to view. **1940** L. MacNeice (*title of poem*) Order to view. **1967** C. Drummond *Death at Furlong Post* iv. 36 Vacant these fourteen years... There have been many orders to view. **1971** M. Tripp *Five Minutes with Stranger* II. v. 125 I'll call in personally tomorrow and get an order to view.

V. See also *10, *23, *24, etc. **25. a.** (Earlier examples.) **c.** (Later examples.)

1698 *House of Commons Jrnl.* 8 Apr. (1742–62) XII. 198/2 The House, according to the Order of the Day, resolved itself into a Committee of the whole House to consider further of Ways and Means for raising the Supply granted to his Majesty. **1729** E. Knatchbull *Parliamentary Diary* (1963) 95 The orders of the day were moved for and so this day's debate ended. **1779** *Parl. Reg.* 5 May 401 The order of the day was read for the House to resolve itself into a committee of supply. **1959** *Times* 19 Sept. 7/7 The restorers are at work: *anastylosis* is the order of the day. **1976** *Abingdon Herald* 9 Dec. 5/2 The removal of ice from the moving parts and sheets was the order of the day. The light air and bright cold conditions required a high degree of concentration.

27. in order. c. Appropriate to or befitting the occasion; suitable; called for; also, in fashion, current, correct. orig. *U.S.*

a **1861** T. Winthrop *John Brent* (1862) viii. 85 If the gent has made a remark what teches you, apologies is in order. **1878** J. H. Beadle *Western Wilds* xxv. 399 One week sufficed to conclude my business in Oregon, but before leaving a few general notes are in order. **1903** *N.Y. Times* 4 Sept. 2/3 Good byes were in order on the Erin last night. **1931** G. T. Clark *Leland Stanford* xiv. 457 It was quite in order..that when this bill was before the Senate, he should express himself upon it. **1973** 'M. Innes' *Appleby's Answer* xv. 128 A confidential and man-to-man note will be in order. **1977** N. Marsh *Last Ditch* vi. 151 Is it in order for us to ring up your father and ask him to dine?

d. *in* (or *at, on*) *short* (or *quick*) *order*: without delay, immediately, summarily. orig. *U.S.*

1834 W. G. Simms *Guy Rivers* I. 204 Be off now in a hurry, or I shall fire upon you in short order. *a* **1852** F. M. Whitcher *Widow Bedott Papers* (1856) xxv. 307 If ever you dew it agin you'll git your walkin'-ticket on short order. **1892** *Outing* Apr. 19/1, I was so thoroughly comfortable that I went to sleep in short order. *a* **1916** H. James *Ivory Tower* (1917) III. iv. 198 Your solution, is marriage to a wife at short order. **1932** N. Hodgkins *Some Canad. Ess.* 202 We had made a sailor of him in short order. **1973** E. Berckman *Victorian Album* 180 The woman checked... This doesn't mean she failed to tally, because she did in short order—and all the more savagely. **1976** *Publishers Weekly* 24 May 58/3 Linda descends on twenties London to become, in short order, a model, the toast of lords [etc.].

30. out of order: (Later examples of mechanical or electrical devices.) Also (sometimes hyphenated) *attrib.*

1926 E. O'Neill *Great God Brown* III. i. 70, I forgot to tell him something important this morning and our phone's out of order. **1928** D. L. Sayers *Unpleasantness at Bellona Club* xii. 141 The telephone cabinet..was so annoyingly labelled 'Out of Order'. **1950** T. Walsh *Nightmare in Manhattan* III. 82 A phone booth behind the news-stand—it has an out-of-order sign on it. **1971** R. Thomas *Backup Men* xxii. 190 The two elevators wore *out of order* signs. **1977** A. Scholefield *Venom* IV. 172 She had also telephoned the house..and had received an out-of-order tone.

31. *order-maker, -making* vbl. sb. and ppl. adj.; *order–disorder* adj.; **order form** (examples); **order man, orderman,** a man who takes or makes out orders; **order mark** (see quot. 1963); **order pad,** a pad (Pad *sb.*[3] 4) of order forms; **order-paper,** (*b*) in the House of Lords, a publication of questions, etc., for the remainder of the session; **order wire** *Teleph.*, a wire used to communicate verbal information about the setting up of a connection for a customer, or between operators at different manual exchanges, or between a customer and an operator in establishing a data link.

1938 *Nature* 9 Apr. 643/1 Fröhlich has tried to interpret the λ-phenomenon of liquid helium as an order-disorder transition. **1964** *Discovery* Oct. 65/2 The theory of order–disorder transformations in alloys. **1894** *Country Gentlemen's Catal.* 3 We hope..that subscribers..will use our Enquiry and Order Forms. **1929** *Radio Times* 8 Nov. 114/2 Note in Order Form below the extra saving made by ordering 500 or 1000 [cigarettes] at a time. **1972** *Accountant* 26 Oct. 504/1 The phrase 'order forms' is to be understood to mean forms which the company makes available for other persons to order goods or services from the company. **1906** W. James *Mem. & Stud.* (1911) ix. 222 Not only in the great city, but in the outlying towns, these natural ordermakers, whether amateurs or officials, came to the front immediately. **1902** —— *Var. Relig. Exper.* viii. 170 Unhappiness is apt to characterize the period of order-making and struggle. **1963** *Times* 22 Feb. The order-making machinery in the Bill. *a* **1951** A. C. Headley in Murdoch & Drake-Brockman *Austral. Short Stories* (1951) 367 It was the rent and the order man, and a new pair of shoes. **1977** *N.Z. Herald* 5 Jan. 2-11/5 (Advt.), An experienced timber orderman is required for timber yard in western suburbs. **1912** A. Brazil *New Girl at St. Chad's* vi. 99 By general custom all pencils ..or other stray possessions were put into what was known as the forfeit tray, whence their owners might reclaim them by paying the penalty of the loss of an order mark. **1963** Barnard & Lauwerys *Handbk. Brit. Educ. Terms* 141 *Order mark,* a punishment (usually confined to girls' schools) for offences of a comparatively trivial kind. **1936** L. C. Douglas *White Banners* x. 226 She pushed the order-pad and pencil towards him. **1972** M. Kaye *Lively Game of Death* (1974) i. 4 Manufacturers..whip out order pads and hope to sell enough merchandise. **1946** *May's Treat. Parliament* (ed. 14) II. xii. 245 Together with the Minutes of Proceedings is printed the Order Paper, consisting of a programme of future business so far as appointed. **1912** Thiess & Joy *Toll Telephone Pract.* xiv. 214 (*caption*) Phantom circuit used as an order wire. **1948** J. Atkinson *Herbert & Procter's Telephony* (new ed.) I. xvii. 345/1 The out-going order-wires are multiplied throughout all positions at the originating exchange. **1973** R. N. Renton *Data Telecommunication* ix. 211/1 Communication with the customer for setting up and clearing connections is effected over telephone circuits (order wires) via the normal telephone exchange.

order, *v.* Add: **6.** Also with ellipsis of *to be.* Chiefly *U.S.*

1781 J. Witherspoon in *Pennsylvania Jrnl.* 9 May 1/2 These things were ordered delivered to the army. **1794** [in Dict.] **1799** in *Essex Inst. Hist. Coll.* (1877) XIII. 61 But the wind growing faint, I ordered the signal taken in. **1873** J. H. Beadle *Undevel. West* xi. 191 My bill was introduced by Senator Williams of Oregon, read by title, and ordered printed. **1875** J. G. Holland *Sevenoaks* in *Scribner's Monthly* Sept. 599/1 He went out,..jumped into Mr. Talbot's waiting coupe, and ordered himself driven home. **1938** W. T. Walsh *Philip II* xxi. 423 The Duchess ordered ships fitted out to meet and escort him. **1972** *Sci. Amer.* July 76/1 Frederick ordered the children raised in silence, so that they would not hear one spoken word. **1976** M. Machlin *Pipeline* l. 510 Coutts ordered the ship's speed reduced to six knots. **1977** *Time* 19 Dec. 9/2 When the local military commander was ordered removed after having congratulated the throng on its patriotic singing, Lagoa angrily summoned the marchers back on the pavement.

7. b. (Earlier and later examples of *to order about.*)

1853 C. Brontë *Villette* I. iv. 74, I refused to be ordered about and thrust from him. **1942** R. G. Collingwood *New Leviathan* 201 For a man of weak or undeveloped will nothing is so pleasant as being ordered about.

c. *to order up,* in the game of euchre: to order (the suit of the card turned up by an opponent who is dealing) to be adopted as trumps; also *absol.*

1847 J. S. Robb *Streaks of Squatter Life* 129 His antagonist ordered the king up. **1878** [see *Assist *v.* 7 c]. **1950** *Hoyle's Games Modernized* (ed. 20) 88 If the non-dealer thinks his hand good enough, with the suit of the turn-up card as trumps, to make three tricks, he says..'I order it up'. **1963** G. F. Hervey *Handbk. Card Games* 184 The elder hand (non-dealer) may either order up or pass. If he orders up, the suit of the exposed card becomes the trump suit, and the dealer must take up the exposed card and discard..a card.

8. Also const. *up* and *absol.*

1763 J. Woodforde *Diary* 3 Sept. (1924) I. 31 Mrs. Bacon pressed me to drive with her, but I had ordered in Hall, and I could not. **1895** *Montgomery Ward Catal.* Spring & Summer 1 Please read remarks and rules before ordering. *Ibid.*, How to order. Commence your order similar to the sample heading on page 2. **1930** A. Bennett *Imperial Palace* II. liv. 400 The waiter wrote and vanished. When Gracie returned, Evelyn said: 'I've ordered.' **1946** D. Stivens *Courtship Uncle Henry* 197 We all drank together and ordered again. **1967** 'L. Egan' *Nameless Ones* xvi. 212 'Would you like to order, sir?' Obsequious waiter. **1973** J. Gores *Final Notice* (1974) xxvi. 166 We'll have some more sparkling burgundy and then we can order. **1976** B. Lecomber *Dead Weight* vi. 72, I..ordered up two toasted ham sandwiches. **1976** J. M. Brownjohn tr. *Kirst's Time for Payment* vi. 134 Order up, ladies and gentlemen, and don't worry about the breath test.

orderable, *a.* Add: **b.** That may be arranged in series. **c.** That may be ordered (at a snack bar, etc.).

1949 *Mind* LVIII. 194 Recently it was realised that it was not necessary to regard satisfactions as additive; all previous conclusions about economic behaviour could still be deduced if they were merely regarded as orderable. **1962** C. O. Frake in J. A. Fishman *Readings Sociol. of Lang.* (1968) 438 Some, but apparently not all, orderable items at a lunch counter are distinguished by the term *something to eat.*

order-book. Add: **c.** (Examples.)

1844 Erskine May *Law of Parl.* viii. 168 Each member ..rises and reads the notice he is desirous of giving, and afterwards..delivers it..to the second clerk assistant, who enters it in the Order Book. **1929** G. F. M. Campion *Introd. Proc. House of Commons* ii. 65 The Order Book, which is also coloured white, is issued each day before the meeting of the House. **1964** Abraham & Hawtrey *Parl. Dict.* (ed. 2) 128 The Order Book of the House of Commons is published in the afternoon of each sitting day. It contains a complete list of all the orders, notices and questions for the following and subsequent days up till the end of the session.

d. (Earlier and later examples.) Also *transf.*, the amount of orders to be fulfilled; also *fig.*

1771 J. Wedgwood *Let.* 10 Apr. (1965) 105, I had immediate recourse to the order book where I did not doubt of finding some of the Vases we had ordered unmade. **1856** *National Rev.* III. 354 Passions are contending; life is a discipline; there is a reference every moment to the directory of the discipline—the order-book of the passions. **1910** A. Bennett *Clayhanger* II. i. 161 I've shown him he's wrong by our order-book, but he wouldn't see it. **1955** *Times* 13 May 19/2 The order book at the year-end was satisfactory, comparing very favourably with the position at the end of 1953. **1958** *Times Rev. Industry* Sept. 79/1 Order-book position is patchy. **1971** *Daily Tel.* 4 Aug. 13/3 Much of the industry remains pessimistic, with many companies still facing stagnant or declining order books. **1974** *Listener* 24 Oct. 531/3 You would think there was a boom on, from our order book.

ordered, *ppl. a.* Add: **2. c.** *Math.* Of a set: having the property that there is a transitive binary relation, >, such that for any elements *a, b* of the set $a>b$, $b>a$, or $b=a$; *ordered pair*, a pair of elements (a, b) such that $(a, b) = (u, v)$ if and only if $a=u$ and $b=v$; similarly *ordered triple, n-tuple.*

1901 *Bull. Amer. Math. Soc.* VII. 225 Consider the ordered assemblage 1, 2, ..., *n'*, ..., *n*, ...*n''*, where *n'*, *n* and *n''* are definite and subject to the condition that, in (S), *n'* comes before *n* and the latter before *n''*. **1906** W. H. & G. C. Young *Theory of Sets of Points* vi. 121 A set in given order will be called an ordered set... Its components..may be distinguished as ordered components. **1941** Birkhoff & MacLane *Survey Mod. Algebra* ii. 55 There are many other ordered fields: the field of real numbers, the field..of numbers $a+b\sqrt{2}$.., and other subfields of the real number field. **1953** *Mind* LXII. 541 A notion so little abstruse as that of an ordered pair. **1963** H. J. Ryser *Combinatorial Math.* i. 5 Let *S* be a set and let $(a_1, a_2, \ldots, a_r)$ be an ordered *r*-tuple of not necessarily distinct elements of *S*. **1966** Meyer & Hanlon *Fun with New Math* vii. 90 To each point in 'the Cartesian Plane'..there corresponds a unique ordered pair of real numbers. **1968** E. T. Copson *Metric Spaces* i. 5 The set of all ordered triples (x, y, z) of real numbers.

orderedness. (Later examples.)

1935 *Jrnl. Theol. Stud.* XXXVI. 314 The belief that the world's orderedness or knowability is an expression of mind. **1974** *Sci. Amer.* Mar. 43/1 (Advt.), In liquid-crystal work, one deals with the different forms and degrees of orderedness among molecules.

orderly, *a.* and *sb.* Add: **A.** *adj.* **4.** *orderly buff* (slang) = **orderly sergeant* (*b*); (see also quot. 1948); *orderly corporal*, (*a*) a corporal who attends upon an officer to carry orders or messages; (*b*) a corporal whose turn it is to attend to the domestic affairs of his corps or regiment; *orderly dog* (slang) = **orderly corporal* (*b*); (see also quot. 1948); *orderly officer* (*b*) (examples); *orderly pig* (slang) = *orderly officer* (*b*); (see also quot. 1948); *orderly room* (earlier and later examples); *orderly sergeant* (further examples); (*b*) a sergeant whose turn it is to act as officer of the day.

1757 Loudoun & Lyman *Gen. Orders of 1757* (1899) 126 Each Corp is to have an Orderly Searj[ean]t Ready. **1802** C. James *New Mil. Dict.* s.v., *Orderly Officer.* See Officer of the Day. *Ibid.*, Orderly serjeants when they go for orders are sashed. Orderly corporals and orderly men wear their side arms. **1806** H. Burrard *Let.* 26 Mar. in *Circumstantial Rep. Charges against Duke of York* (1809) 269 Your messenger knows where to find me, as I am at this Orderly Room for two or three hours most days. **1833** *Mirror of Parliament* 11 June 2216/2 They (often from connexion or friendship with the Orderly-room clerks) got their ages entered as eighteen. **1867** J. M. Crawford *Mosby* 121 Horace Johnson of Warrenton..was appointed orderly sergeant. **1917** A. G. Empey *Over Top* 302 *Orderly-corporal*, a non-commissioned officer who takes the names of the sick every morning. **1918** E. S. Farrow *Dict. Mil. Terms* 420 *Orderly Officer*, the officer of the day, or that officer of a corps or regiment whose turn it is to supervise for the day the arrangements for food, cleanliness etc. **1925** Fraser & Gibbons *Soldier & Sailor Words* 216 *Orderly buff*, orderly sergeant. *Ibid.*, *Orderly dog*, orderly corporal. **1934** V. M. Yeates *Winged Victory* 224 Grey..was censoring the men's letters, being orderly dog for the day. **1943** C. H. Ward-Jackson *Piece of Cake* 45 *Orderly buff*, orderly corporal. *Ibid.*, *Orderly dog*, orderly sergeant. *Ibid.*, *Orderly pig*, orderly officer. **1948** Partridge *Dict. Forces' Slang* 132 *Orderly buff, dog, pig.* Strictly speaking, *dog* seems to have been officer or corporal, *buff* sergeant, and *pig* corporal. The non-regular Army, however,..used all three phrases indiscriminately. **1953** K. Tennant *Joyful Condemned* xviii. 163 David..was signed into the care of an orderly sergeant. **1964** A. Powell *Valley of Bones* i. 31 Tell the Orderly Corporal Mr Bithel is reporting sick this morning. **1971** S. Milligan *Adolf Hitler* III. 51 Oh those military meals!..Visits from orderly officers did little to help. **1974** G. M. Fraser *McAuslan in Rough* 39 When I was..doing my recruit training..there was a villainous orderly sergeant who used to get us up in the mornings. **1977** J. Tarrant *Rommel Plot* xviii. 187 He asked the operator to put him through to the Orderly Room.

5. Also *orderly box.*

1904 *Daily Chron.* 18 Mar. 6/3 They had..been seen bearing up documents and throwing them into orderly boxes.

B. *sb.* **1.** (Earlier and later examples.)

1781 *Calendar Virginia State Papers* (1875) I. 452 The orderly, his wife and negro woman at York..have never received one single shilling. **1966** [see *Batman[2]].

‖**ordinaire** (ǫrdinę̄r), *sb.* [Fr.] Short for *vin ordinaire. Also as *adj.*

1861 Thackeray *Roundabout Papers* xiv, in *Cornh. Mag.* July 123 A sound genuine ordinaire, at 18*s*. per doz. let us say. **1888** Mrs. H. Ward *R. Elsmere* II. III. xxii. 215 He ate his boiled mutton and drank his *ordinaire* like a man. **1906** *Daily Chron.* 24 July 6/5 What is 'ordinaire' as applied to wine?.. That is a local term which has existed for two centuries in France. **1920** [see *bourgeois *a.* 4]. **1936** H. G. Wells *Croquet Player* ii. 11 He..ordered another half litre of wine. Either out of ignorance or preference he was drinking red ordinaire. **1959** *Good Food Guide* 32 Spanish ordinaires at 1/6 a glass. **1960** *House & Garden* July 13/1 Two decent, slightly better than *ordinaire* wines. **1972** *Ibid.* Feb. 100/4 Roodeberg is an honest wine, distinctly better than an *ordinaire.*

ordinary, *sb.* Add: **10.** Add to def.: usu. with capital initial, in the Roman Catholic rite, those parts of a service, esp. the mass, which do not vary from day to day; *spec.*, those unvarying parts which form the mass as a musical setting (Kyrie, Gloria, Credo, Sanctus, Benedictus and Agnus Dei). Also *transf.*, of other rites. (Later examples.)

1905 Proctor & Frere *New Hist. Bk. Common Prayer* (rev. ed.) i. 12 In the Liturgy, the 'Canon' or central prayer..was the Roman canon, and in fact the rest of the invariable framework of the public service (or 'Ordinary') was that adopted from Rome. **1929** E. C. Thomas *Lay Folks' Hist. Liturgy* I. v. 22 These Nestorian Liturgies differ in the Anaphora, but the Ordinary of the Mass is the same in all. **1944** W. Apel *Harvard Dict. Mus.* 427/2 Around 500, the Mass consisted only of the chants of the Proper, alternating with lections from the Epistles, etc. Gradually, the chants of the Ordinary were introduced, probably in the following chronological order: Sanctus, Kyrie, Gloria, Agnus Dei, Credo. *Ibid.* 523/2 Other services [than the Mass]..also comprise invariable and variable portions. For instance, the Magnificat forms a part of the Ordinary of Vespers. **1974** *Daily Tel.* 20 July 7/8 Surely the way Haydn in his 70s appears tormented by sudden doubts at the word 'mortuorum' belies the confidence of the preceding 'Et expecto resurrectionem', uniquely so in any setting of the Ordinary. **1976** *Gramophone* Apr. 1653/3 It is found in ninth-century manuscripts and may well be connected with two well-known items of the Ordinary, Sanctus I and Sanctus XI.

14. a. (Later examples.)

1908 G. B. Shaw *Lett. to Granville Barker* (1956) 114 Charlotte at a farmers' ordinary at Towcester was immense. **1928** *Daily Chron.* 9 Aug. 4/4 Lord Beaconfield [*sic*] was accustomed to make some of his most important pronouncements at the farmers' ordinary at Aylesbury. **1976** N. Roberts *Face of France* xv. 157 The establishment in an English market town which still does a good farmers' ordinary.

c. (Earlier examples.)

1637 in *Essex Inst. Hist. Coll.* (1869) IX. 55 Mr. John Holgraue..hath undertaken to keep an ordinary for the entertainment of strangers. **1650** *Archives of Maryland* (1883) I. 294 Wine or other Provisions to bee expended in any Ordinaries within this Province. **1680** in *New Hampsh. Hist. Soc. Coll.* (1866) VIII. 15 What person soever..shall profane ye Lord's Day..by Dining at ordinarys in time of publique worship..shall forfeite 10s.

16. a. (Later examples.)

1909 *Times Lit. Suppl.* 20 May 185/2 Shakespeare introduces the ordinary, whether in characters or in events, only as a foil to the extraordinary. **1977** 'D. Cory' *Bennett* iv. 121 The case I'm engaged on..is rather out of the ordinary. **1978** *Atlanta Jrnl. & Constitution* 14 Jan. 23T (Advt.), Tapas. When you're fed up with the ordinary... Our European chef proudly presents over 40 delightfully different hot and cold appetizer treats.

c. One of a class of inmates in a poor-house.

1910 *Daily Chron.* 14 Jan. 8/5 The 'ordinaries' (whom we should call able-bodied) were able to roam all over the building.

17. c. (Later example.)

1964 *Financial Times* 23 Mar. 12/3 Plans are being considered to fund back indebtedness through..a 1-for-1 rights issue of 5s Ordinary at par.

18. b. (Later examples.)

1934 *Burlington Mag.* Oct. 182/2 In this very year, January 1431, the King's painter to Charles IX of France was Henry Mellein, and his painter-in-ordinary was Conrad de Vulcop. **1944** *Ibid.* Dec. 307/1 Jacques d'Arthois, painter-in-ordinary of the Forêt de Soignes, is one of the leading figures of the Brussels landscape school of the seventeenth century.

19. a. *ordinary-keeper* (earlier examples).

1645 in *Essex Inst. Hist. Coll.* (1869) IX. 136 To provide for a ordinarie keeper. **1662** *Archives of Maryland* (1883) I. 447 All Ordinary Keepers debts either upon bill or accompt..shall be allowed due.

ordinary, *a.* Add: **3. d.** Of language, usage, discourse, etc.: that most commonly found or attested, *spec.* as contrasted with logical symbolism or a specialized terminology.

1685 tr. *Arnauld & Nicole's Logic* II. x. 221 As when I say, *All Men have two Arms.* This Proposition ought to pass for true, according to ordinary use. *Ibid.* III. xv. 70 Whereas it is the method of the Schools to propound the Argument entire, and afterwards to prove the Proposition which receives the difficulty, that which is usual in ordinary discourse, is to join to doubtful propositions the Proofs that confirm 'em. **1690** Locke *Essay Hum. Und.* II. xxi. 121 Philosophy it self,..must have so much Complacency, as to be cloathed in the ordinary Fashion and Language of the Country. **1828** J. S. Mill in *Westm. Rev.* IX. 145 Arranging all these propositions in that *order,* which (so strongly does ordinary language corroborate our view of the case) is termed their *logical* order. **1843** —— *Logic* I. I. i. 25 We must begin by recognising the distinctions made by ordinary language. *Ibid.* II. IV. v. 268 These changes, by which words in ordinary use become more and more generalized. **1874** W. Wallace tr. *Hegel's Logic* 43 The deeper and philosophical meaning of truth can be partially traced even in the expressions of ordinary language. **1892** —— *Ibid.* (ed. 2) 52 The deeper and philosophical meaning of truth can be partially traced even in the ordinary usage of language. **1902** W. James *Var. Relig. Exper.* ii. 36 Trifling, sneering attitudes even towards the whole of life... It would strain the ordinary use of language too much to call such attitudes religious. **1906** B. Russell in *Mind* XV. 256 One of the objects to be aimed at in using symbols is that they should be free from the ambiguities of ordinary language. **1909** W. M. Urban *Valuation* ii. 33 Our ordinary usage, at least, makes a clear distinction between feeling and will. **1932** H. H. Price *Perception* viii. 256 We also stick to common sense and the ordinary usage of language. **1939** *Mind* XLVIII. 62 While it is true that a formal calculus frequently assists in detecting errors which are unnoticed in ordinary language, each formal calculus carries with it new sources of confusion. **1949** *Mind* LVIII. 147 The redefinitions which are implicit in philosophical paradoxes do quite often..receive a certain backing from ordinary usage. **1951** J. Holloway *Lang. & Intelligence* viii. 123 Ordinary language is the language of persons unacquainted even with the idea of conforming to a dictionary. **1957** J. Passmore *100 Yrs. Philos.* xviii. 438 Not all ex-students of Wittgenstein look with kindness on the 'ordinary language' philosophies which have latterly dominated the philosophical scene at Oxford. **1977** *Oxford Times* 9 Dec. 5/3 He is the leading exponent of ordinary language philosophy, and became a fellow of the British Academy in 1960.

5. b. *ordinary wine* (Fr. *vin ordinaire*). Cf. *ordinaire *sb.*

1814 M. Birkbeck *Notes Journey through France* 102 Such is the habitual temperance..that the inns..seldom have any liquor stronger than their ordinary wine. If you call for brandy, they are obliged to send for it to the Caffé. **1860** Dickens *Uncomm. Trav.* (1861) vii. 92, I was in the dear old France of my affections. I should have known it, without the well-remembered bottle of rough ordinary wine.

c. *Comm.* Of shares, stock, etc.: forming part of the common stock and without 'preference'; also applied to shareholders holding such stock.

1869 *Bradshaw's Railway Manual* XXI. 10 No dividend was declared in March on the ordinary stock. *Ibid.* 42 An obligation..to pay to the ordinary shareholders a dividend..at the rate of 2 per cent. **1878** [see Preference 8]. **1891** [see Share *sb.*[2] 2]. **1955** *Times* 12 May 17/7 The issue of three shares to Ordinary shareholders for every five which they hold at present. *Ibid.* 1 July 16/5 The balance-sheet shows the increase in ordinary share capital arising from this capitalization and the manner in which the reserves have been applied. **1974** *Terminol. Managem. & Financial Accountancy* (Inst. Cost & Managem. Accountants) 60 *Ordinary shares,* shares which entitle the holders to the remaining divisible profits (and, in a liquidation, the assets) remaining after prior interests (e.g. preference shareholders) have been satisfied. **1977** *Times* 1 Dec. 20/7 Earning per Ordinary Share 4.44p... Ordinary Dividend 3.4914p.

d. Of people: typical of a particular group; average; without exceptional experience or expert knowledge.

1855 Geo. Eliot *Let.* 12 May (1954) II. 201, I really think a taste for descriptive writing is the rarest of all tastes among ordinary people. **1902** G. B. Shaw *Mrs. Warren's Profession* Pref. p. xvii, The ordinary Briton thinks that if every other Briton is not under some form of tutelage..he will abuse his freedom viciously. **1903** —— *Man & Superman* III. 76 But I am well aware that the ordinary man—even the ordinary brigand, who can scarcely be called an ordinary man (Hear, hear!)—is not a philosopher. **1922** M. A. von Arnim *Enchanted April* iii. 50, I don't think references are nice things at all.. between ordinary decent women. **1952** A. Christie *Mrs McGinty's Dead* ii. 21 It's not the sort of thing that an ordinary man—or a jury—can believe. **1971** *Listener* 28 Oct. 566/1 The language of everyday speech is used in verse..because the role of the poet is an ordinary-man role. **1974** *Times* 11 Oct. 14/8 A government claiming that it wants to involve ordinary people in decision-making. **1975** T. Heald *Deadline* ii. 19 The *Globe*..made its appeal to 'the man in the street'. Leader writers were instructed..to spice their texts with frequent references to 'ordinary folk'.

e. *ordinary level,* the lowest of the three levels of the General Certificate of Education; abbrev. *O level* (*O 5 d).

1947 *Examinations Secondary Schools* (Secondary Schools Exam. Council) 8 An examination at 'Ordinary', 'Advanced' and 'Scholarship' levels should be available each year to candidates who are at least sixteen on Sept. 1st. **1959** *Listener* 29 Jan. 195/2 The arts sixth former who has perhaps 'done' a little science to 'ordinary level'. **1960–1** *Where* Winter 14/1 Ordinary (O) level is normally taken at 16 after a 5-year course. **1963** Barnard & Lauwerys *Handbk. Brit. Educ. Terms* 99 In 1951 the *School* and *Higher School Certificate* examinations in secondary schools were replaced by a General Certificate of Education..examination at three levels—ordinary, advanced, and scholarship. **1978** *Nature* 27 Apr. 784/1 Asimov carefully explains the inverse-square law of force in a manner that should be comprehensible to a pupil considered incapable of taking ordinary level school physics.

6. (Later U.S. example; cf. *ornery *a.*)

1800 *Aurora* (Philadelphia) 1 May 2/3 This ordinary drunken wretch is supposed to be the perpetrator.

ordinate, *sb.* **b.** Add: In mod. use, the distance of a point from the left-to-right (*x*) axis measured parallel to the other (*y*) axis; the *y* co-ordinate of a point. (Further examples.)

1855 I. Todhunter *Treat. Plane Co-ordinate Geom.* i. 2 *OM* is called the abscissa of the point *P*; and *ON*, or its equal *MP*, is called the ordinate of *P*. **1880** *Proc. R. Soc.* XXX. 511 The horizontal ordinates give the stress, the vertical ordinates give the elongation. **1896** *Min. Proc. Inst. Civil Engin.* CXXVI. 233 The area of the loop, with the magnetizing force as abscissas, and the magnetization as ordinates, represents the energy dissipated. **1948** *Electronic Engin.* XX. 10/1 This apparatus was constructed to produce a trace on a cathode ray tube in which the ordinates were proportional to shutter opening and the abscissa to time. **1951** R. M. Garrels *Textbk. Geol.* 470 In order to graph these changes, it is customary to begin by drawing two lines at right angles to each other. One is usually drawn horizontally (the *X* axis or abscissa), the other vertically (the *Y* axis or ordinate). **1971** Niles & Haborak *Calculus with Analytic Geom.* i. 4 The directed distance *OM* is the *x*-coordinate or abscissa of point *P* and is denoted by *x*. The directed distance *ON* is the *y*-coordinate or ordinate of point *P* and is denoted by *y*.

ordinate, *v.* Add: **5.** *Statistics* and *Ecology.* To subject to the mathematical operation of ordination (sense *1 c).

1962 *Ecol. Monogr.* XXXII. 137 (*heading*) Ordinating forest communities by means of environmental scalars and phytosociological indices. **1969** E. C. Pielou *Introd. Math. Ecol.* xx. 255 When we wish to ordinate vegetation by means of a principal components analysis, there are four decisions to make.

ordination. Add: **I. 1. c.** *Statistics* and *Ecology.* [tr. G. *ordnung*.] The arrangement of a set of points, given as in a multidimensional space, into a space of fewer dimensions with minimal distortion.

1954 D. W. Goodall in *Austral. Jrnl. Bot.* II. 323 Factor analysis does not result in a classification of vegetation in the ordinary sense, but in an arrangement of the vegetational data in a multi-dimensional space. For such an arrangement, there appears to be no word in English which one can use as an antonym to 'classification'; I would like to propose the term 'ordination'. **1969** E. C. Pielou *Introd. Math. Ecol.* xx. 256 As a method of summarizing the results of a survey, ordination has two great advantages over classification: it obviates the need for setting up arbitrary criteria for defining the classes and there is no need to assume that the classes (if there are any) are hierarchically related. **1971** Blackith & Reyment *Multivariate Morphometrics* xvi. 229 Proctor..found that ordinations of some British liverworts gave readily interpretable polarities.

ordinator. 1. (Later examples.)

1929 R. Bridges *Testament of Beauty* i. 134 And wouldst thou play Creator and Ordinator of things. **1952** G. Sarton *Hist. Sci.* I. xvi. 421 The world architect (*dēmiurges*) is not a creator but, like the *nus* of Anaxagoras, an ordinator.

ordnance. Add: **5. ordnance datum** (examples); **ordnance map** (earlier example); **ordnance survey** (earlier and later examples); also *fig.*

1833 W. Dyott *Diary* Nov. (1907) II. 170 On the 17th Captain Gosset, who had been employed last year stationed at Lichfield on the ordnance survey service called. **1839** C. Fox *Jrnl.* 8 Oct. (1972) 59 Sir H. Vivian was chuckling over the admirable Ordnance map. **1886** T. P. White *Ordnance Survey of U.K.* vi. 106 The Ordnance datum for Ireland is not the same as for Great Britain. **1889** G. W. Usill *Pract. Surveying* ix. 137 The ordnance datum of this country was determined by the ordnance authorities to be 'the approximate mean level at Liverpool'. **1934** C. Lambert *Music Ho!* 11 This book makes no attempt to be an ordnance survey of modern music. **1956** Ordnance datum [see O.D. s.v. *O 5 d]. **1972** L. Alcock *By South Cadbury* ii. 24 The summit is a little over five hundred feet above Ordnance Datum, with the hill itself standing about two hundred and fifty feet above the surrounding countryside. **1973** *Times* 13 Aug. 4/1 Mounting pressure from country walkers..may save the popular 2½ in. Ordnance Survey map.

ordoñezite (ǭɹdǭnyḗi-zəit). *Min.* [f. the name of Ezequiel *Ordoñez* (1867–1950), Mexican geologist + -ite[1].] An antimonate of zinc, $ZnSb_2O_6$, which is found as brown tetragonal crystals at Guanajuato, Mexico.

1954 *Mineral. Abstr.* XII. 303 (*heading*) Ordoñezite, zinc antimonate, a new mineral from Guanajuato, Mexico. **1955** *Amer. Mineralogist* XL. 66 Ordoñezite occurs as drusy or stalactitic masses of repeatedly twinned tetragonal crystals having a maximum size of 2 mm. **1968** I. Kostov *Mineral.* II. iv. 264 To this [byströmite] group are referred the following oxides, which are isostructural with rutile and tapiolite. ($P4_2/mnm$): byströmite.., ordoñezite.., and tripuhyite.

Ordovician, *a.* Substitute for def.: Of, pertaining to, or designating the second earliest period of the Palæozoic era, following the Cambrian and preceding the Silurian. Also *absol.*, the Ordovician period or its rocks. Add earlier and later examples.

1879 C. Lapworth in *Geol. Mag.* Decade II. VI. 14 The whole of the great Bala district where Sedgwick first worked out the physical succession among the rocks of the intermediates or so-called Upper Cambrian or Lower Silurian system..lay within the territory of the *Ordovices*; a tribe as undaunted in its resistance to the Romans as the Silures... Here, then, we have the hint for the appropriate title for the central system of the Lower Palæozoics. It should be called the Ordovician System. **1902** A. J. Jukes-Browne *Student's Handbk. Stratigr.*

Geol. viii. 118 In Ayrshire..the Ordovician has the ordinary facies of a formation accumulated at no great distance from a continental coast-line. **1955** *Times* 4 June 8/5 It is a site of great geological interest for its variety of Ordovician volcanic lavas, with intrusive igneous rocks interbedded with fossiliferous mudstones and slates. **1967** D. H. Rayner *Stratigr. Brit. Isles* iv. 80 In the British Isles the first fragmental remains of vertebrates are known from the Silurian beds, although bony plates have been found in the Ordovician of the United States.

ore[2]. Add: **1.** (Further examples.) Also applied to minerals mined for their content of non-metals.

1910 J. F. Kemp in *Jrnl. Canadian Mining Inst.* XII. 357 Sometimes..in the mining of the non-metallic substance sulphur, the output of the mine is called 'sulphur-ore', although no metal is involved at all. Yet while we may not especially controvert this usage, it cannot be said to seriously affect the general and large conception of ore as limited to the metalliferous minerals. **1913** W. Lindgren *Mineral Deposits* i. 4 The use of the term 'ore' is not quite consistent. Ordinarily it implies a metal, but the expression 'sulphur ore', meaning pyrite, is sometimes seen, and occasionally such terms as 'sapphire ore' are found. **1939** G. A. Roush *Strategic Mineral Supplies* xiv. 401 The deposits of ore, or caliche, are highly irregular. **1951** A. F. Taggart *Elem. Ore Dressing* i. 2 The miner was principally responsible for making ore of the low-grade California gravels by discovering ways to mine them that are..cheap. **1970** *Materials & Technol.* III. ii. 97 There are no fundamental differences between the treatment of metalliferous ores and other types of ore.

3. a. *ore bin, -bucket, -pass, -vein* (example). **b.** *ore-carrier, -carrying, dresser, -dressing* (further examples). **c. ore body** (examples as one word); **ore-shoot** = Shoot *sb.*[1] 7.

1935 *Economist* 8 June 1334/1 It will be necessary..to sink the shaft..below the reef and to cut stations and ore bins. **1962** R. B. Fuller *Epic Poem on Industrialization* 197 The magnificent horizontal and vertical lines of its highways,..ore bins, and skyscrapers. **1955** *Times* 12 July 15/6 Difficulties being met are the poor bearing quality of the ground..and the presence of unconsolidated footwall beds associated with considerable volumes of water in No. 1 shaft area—which is holding up the advance towards the orebody. **1971** *Wall St. Jrnl.* 19 Feb. 20/5 Bad weather during the past three weeks delayed stripping of the overburden of the Black Cub orebody. **1977** *Bulletin* (Sydney) 22 Jan. 42/1 But a mineral deposit doesn't become an orebody unless the mineral concerned can be extracted and sold at a profit. **1912** *Chambers's Jrnl.* Dec. 784/2 The men..had begun to send the ore-buckets down empty. **1936** *Atlantic Monthly* CLVII. 164/2 Our great..ore-carriers..are no exception, because their existence is due to the State's primary intervention in granting monopoly rights to the rental value of the.. ore-fields they tap. **1975** 'D. Jordan' *Black Account* viii. 46 Ore shipping studies comparing the capacity of Japanese ore carriers with the proposed berthing and loading facilities. **1909** *Westm. Gaz.* 29 Nov. 8/2 During the storm three ore-carrying steamers were beached near the entrance to the harbour. **1974** *Encycl. Brit. Micropædia* IV. 194/2 This ability of the ore dresser to modify the flotability of minerals made possible many seemingly magical separations. **1909** H. Louis *Dressing of Minerals* i. 4 It appears better to treat coal-washing and ore-dressing as one and the same subject. **1914** S. J. Truscott tr. *Beyschlag's Deposits Useful Minerals* I. 72 There should be at least sufficient iron present..to cover the costs of ore-dressing and of metallurgical treatment. **1946** *Nature* 27 July 140/1 To provide an information service dealing with publications concerning all branches of geology, mineralogy,..ore-dressing and production metallurgy. **1878** *Encycl. Brit.* XVI. 453/2 E the main lode, H permanent levels, and K ore-pass reserved amidst the rubbish (*deads*) D. **1884** J. A. Phillips *Treat. Ore Deposits* 50 As a general rule, all the ore-shoots in a given vein dip in the same direction. **1944** *Q. Jrnl. Geol. Soc.* C. 251 A mineral vein may carry several ore-shoots, separated by barren stretches. **1906** *Chambers's Jrnl.* Feb. 159/2 A few digs with the shovel laid bare the outcropping of the ore-vein.

orectic, *a.* **a.** *Philos.* Delete *rare* and add later examples. Hence **orecti·vity.**

1906 S. S. Laurie *Synthetica* I. 161 Let us rather call it Orectivity or Conation. **1947** [see *affective *a.* 7 b]. **1952** C. P. Blacker *Eugenics* 216 We may perhaps be on the threshold of a period when similar advances will be made in tests of the so-called 'orectic' functions; these comprise the qualities which make up what is commonly called character. **1970** G. Greer *Female Eunuch* 67 For, no matter which theory of the energy of personality we accept, it is inseparable from sexuality... Flügel called it orectic energy.

oregano (ǫrẹ̀gā-no, ǫre-gāno). [Sp. and Amer. Sp. var. of Origanum.] The dried leaves of wild marjoram, *Origanum vulgare*, or, esp. in North and Central America, the dried leaves of a shrub of the genus *Lippia*, esp. *L. graveolens*; both are used as seasonings for food, the latter having a stronger flavour.

1771 J. R. Forster tr. *Osbeck's Voy. to China* I. 33 *Origanum Creticum*, Spanish *Oregano*, known by the name of Spanish hops, is used to make anchovies and other meats more palatable. **1889** S. Watson in *Proc. Amer. Acad. Arts & Sci.* XXIV. 67 *Lippia* (*Zapania*) *Palmeri*... 'Origano'; with a strong sage-like odor and used as a pot-herb. **1889** *Contrib. U.S. Nat. Herbarium* V. 226 Lippia spp. Oregano. The leaves of oregano are very much used to flavor food. *Ibid.* 227 The name 'oregano' seems to be a generic term applied to the leaves thus used of several species of Lippia. **1959** *Listener* 2 Apr. 601/2 The hallway smelled of herbs, of oregano and basil. **1969** F. Rosengarten *Bk. Spices* 276 The pungent Mexican oregano, indigenous to the warmer areas of the Western Hemisphere, is of the genus *Lippia*... The milder European oregano..is of one of several species of *Origanum* native to the Mediterranean region, principally *Origanum vulgare* L. The perplexing confusion between marjoram and oregano, and between the two types of oregano, is not limited to botanists. **1972** M. J. Bosse *Incident at Naha* i. 17, I rolled some grass from the oregano jar where we keep ours and smoked the joint. **1976** *National Observer* (U.S.) 2 Oct. 11/4 Cook the meat loaf for 15 minutes, and then sprinkle the top with seasoning mixture of oregano, cinnamon, and sugar.

Oregon (ǫ·rigǫn). The name of one of the United States of America, situated on the Pacific coast, used *attrib.* to designate plants and animals found in the region, as **Oregon ash,** a species of ash, *Fraxinus oregana*, or its wood; **Oregon cedar** = *Lawson cypress* (*Lawson); **Oregon fir** = *Douglas fir* (*Douglas[1]); **Oregon grape,** an evergreen shrub, *Mahonia aquifolium*, bearing racemes of yellow flowers followed by dark berries resembling grapes; also, the berry itself; **Oregon junco,** a small black, brown, and white bunting, *Junco oreganus*; **Oregon lily,** one of the hybrid lilies produced by Jan de Graaff at the Oregon Bulb Farms; **Oregon pine** = *Oregon fir.*

1869 *Amer. Naturalist* III. 407 Oregon Ash... This first appears at the Dalles. **1969** T. H. Everett *Living Trees of World* 287/1 Other American ashes worthy of note include the Oregon ash..most important of Western species, which grows to a height of 80 feet and may have a trunk 4 feet in diameter. **1872** F. F. Victor *All over Oregon & Washington* xxvii. 279 The Oregon cedar..grows very abundantly near the coast. **1971** F. H. Titmuss *Commercial Timbers of World* (ed. 4) 94 Alternative names for the timber [of Lawson cypress] include Oregon Cedar and White Cedar. **1904** E. O. Wooton *Native Ornamental Plants of New Mexico* 15 The Douglas Spruce or Oregon Fir..and the Bull Pine..would well repay the care necessary to getting them established. **1851** *Oregon Statesman* (Oregon City) 27 June 3/1 Oregon Grape, so called, is not a grape, but resembles the grape in size and appearance. **1873** G. M. Grant *Ocean to Ocean* 283 A dark green prickly-leaved bush like English holly, called the Oregon grape. **1949** *Jrnl. N.Y. Bot. Gdn.* July 153/1 In one bed is Oregon grape and beauty-berry. **1971** *Daily Colonist* (Victoria, B.C.) 27 May 53/3 The meat is then placed in the pit and covered with kelp, seaweed, Oregon grape and other wild plants. **1974** J. E. Underhill *Wild Berries* 69 Still other people are devotees of Oregon Grape wine. **1917** T. G. Pearson *Birds Amer.* III. 47 Maybe the handsomest is the Oregon Junco..with a black head and breast sharply defined against a mahogany-brown back, white under parts, and pinkish-brown sides. **1964** A. Wetmore *Song & Garden Birds N. Amer.* 364/2 In winter flocks of Oregon juncos roam western foothills, canyons, and suburbs. **1971** *Islander* (Victoria, B.C.) 13 June 13/3 Oregon junco..come from the north during winter. **1964** *Horticulture* Dec. 49/1 (Advt.), *World Famous Oregon Lilies*, Higo Iris, Hardy Cyclamen and Exbury Azaleas. Send for free catalog. Rex Bulb Farms..Newburg, Oregon. **1967** J. de Graaff *Lilies* 31 Those [lilies]..are becoming known as the Oregon lilies because they are bred and raised in Oregon. **1845** *N. Amer. Rev.* LX. 166 One of those gigantic Oregon Pines..whose prostrate trunk Douglas found to be two hundred and fifty feet in length. **1888** *Encycl. Brit.* XXIV. 386/2 The principal timber is yellow and red fir, ordinarily known as 'Oregon pine', which constitutes the bulk of the forests. **1889** [see *Nootka fir*, s.v. *Nootka B. *adj.* 2]. **1947** R. Peattie *Sierra Nevada* 148 Douglas fir..is known in the trade as Oregon pine. **1964** *House & Garden* Oct. 95/1 Her kitchen units, in Oregon pine.

Hence **Oregone·se,** the people of Oregon; **Orego·nian** *a.*, belonging to Oregon; as *sb.*, a native or inhabitant of Oregon; **O·regonly** *adv.*, after the manner of an Oregonian.

1848 E. Bryant *California* xv. 197, I think the Oregonese had a little the advantage of us in this respect. *a* **1861** T. Winthrop *John Brent* (1862) xxii. 243 Armstrong's opinion was only my own, expressed Oregonly. **1863** *Harper's Mag.* Sept. 570/2 If one does not know the 'lingo', he will often be troubled in conversing with the Oregonians. **1873** J. H. Beadle *Undevel. West* xxxv. 762 There is a distinctively Oregonian look about all the natives and old residents. *Ibid.* xxxvi. 772 No Oregonians will eat of salmon caught above the mouth of the Willamette. **1974** *New Yorker* 25 Feb. 88/3 Oregonians, in whatever part of the state they reside, tend to be small-town-ish.

oregonite (ǫ·rigǫnəit). *Min.* [ad. G. *oregonit* (Ramdohr & Schmitt 1959, in *Neues Jahrb. f. Min. Monatshefte* 247), f. prec.: see -ite[1].] An arsenide of nickel and iron, Ni_2FeAs_2, which occurs as white hexagonal crystals with a metallic lustre in Oregon and has been made artificially.

1960 *Mineral. Abstr.* XIV. 500/1 A new nickel-iron arsenide from Josephine Creek, Josephine County, Oregon, is named *oregonite*. **1968** *Ibid.* XIX. 285/2 By heating at 470°C under a pressure of 1600 kg/cm² mixtures of arsenic ground with iron and nickel or with Fe–Ni alloys (36, 50, and 77% Ni) oregonite,..was obtained.

orenda (ore·ndā). [Iroquoian.] (See quots.).

1902 J. N. B. Hewitt in *Amer. Anthropologist* IV. 33 (*title*) Orenda and a definition of religion. *Ibid.* 37 This subsumed magic power is called..*manitowi* by the Algonquian, *pokunt* by the Shoshonean, and *orenda* by the Iroquoian tribes. And it is suggested that the Iroquoian name for the potence in question, *orenda*, be adopted to designate it. **1911** *Encycl. Brit.* XVII. 306/2 Everything in nature, and particularly all animate objects, have their *orenda*; so have gods and spirits... *Orenda* is above all the power of the medicine man. **1917** *Encycl. Relig. & Ethics* IX. 556/1 The term *orenda* is, in fact, only one of a large group of terms, members of which are found in most, if not all, Indian languages, which have the same general meaning—invisible power or energy. **1920** [see *mana]. **1947** C. S. Lewis *Miracles* xi. 100 It [*sc.* pantheism] may even be the most primitive of all religions, and the *orenda* of a savage tribe has been interpreted by some to be an 'all-pervasive spirit'.

Oreo (ǭə·rio). *U.S. slang.* Also **oreo.** [See quot. 1973.] A derogatory term for an American Black who is seen (esp. by other Blacks) as part of the white establishment.

1968–70 *Current Slang* (Univ. S. Dakota) III–IV. 89 *Oreo, n.*, a black who thinks like a white or tries to join white society. (Blacks on the outside but whites on the inside). **1969** *Harper's Mag.* Mar. 61 Trouble is Negroes been programmed by white folks to believe their products are inferior. We've developed into a generation of Oreos—black on the outside, white on the inside. **1970** H. E. Roberts *Third Ear* 11/1 *Oreo*, a black person with white-oriented attitudes. **1971** *Black World* June 31/1 Every Black man or woman who refers to his Black brother or sister by a derogatory label such as *Tom, nigger,* or *oreo*.. is deliberately walking into the enemy's trap. **1973** A. Dundes *Mother Wit* 596 One term of derision is 'oreo' for such an individual [*sc.* a Negro with 'white' mentality]. The term comes from a standard commercially prepared cookie which has two disc-shaped chocolate wafers separated by sugar cream filling. An 'oreo' is thus brown outside but white inside, hence, a Negro who has internalized white values. **1975** *Amer. Speech* 1972 XLVII. 151 Black educators are sometimes included in this hostile cosmos, for they are labeled *Oreos.*

‖**ore rotundo** (ǭə·ri rotʊ·ndo), *adv. phr.* [L. abl. of *ōs* mouth + abl. of *rotundus* round.] Lit. 'with round mouth'; with round, well-turned speech. Cf. Orotund *a.* (*sb.*).

[Horace *Ars Poetica*, Grais ingenium, Grais dedit ore rotundo Musa loqui]. **1720** Swift *Let.* 1 Dec. in *Works* (1859) II. 300/1 Is taught there to mouth it gracefully, and to swear, as he reads French, *ore rotundo*. **1845** R. Ford *Hand-bk. for Travellers Spain* I. 82 The Castilian speaks with a grave distinct pronunciation, *ore rotundo*; he enunciates every letter and syllable. **1922** Blunden *Bonadventure* 144 Replying You. Big. Stiff *ore rotundo*. **1931** M. Summers *Supernatural Omnibus* 7 In his fine stage voice *ore rotundo* he would declaim some half a dozen wilting lines and demand applause.

orf[2] (ǭɹf). [Var. of dial. *hurf*, prob. f. ON. *hrufa* crust or scab on boil; cf. dial. *reef* skin eruption, dandruff.] A virus disease of sheep, cattle, and goats, characterized by a secondary infection with the bacillus *Fusiformis necrophorus*, which causes ulcers and scabs in and around the mouth and on the feet or other parts of the body; also called scabby mouth, contagious ecthyma, or contagious pustular dermatitis.

1868 J. C. Atkinson *Gloss. Cleveland Dial.* 362 Orf... Scurf on an animal's skin. **1876** C. C. Robinson *Gloss. Mid-Yorks.* 95/1 Orf.., applied to a running sore on cattle. **1948** *Brit. Jrnl. Dermatol.* LX. 405 The virus of orf could produce lesions in man. **1959** *News Chron.* 9 July 5/5 Live sheep..have been found to be suffering from orf, a skin disease. **1972** *Country Life* 30 Mar. 812/3 The flock is then dipped, vaccinated against orf and allowed to scavenge.

orf (ǭɹf), *prep.* and *adv.* Representing a 'phonetic' spelling of a vulgar or affected pronunciation of Off *adv., prep., adj.,* and *sb.*[1]

1889 [see *green *a.* 1 j]. **1901** B. Pain *De Omnibus* i. 2 Pressintly a gint calls 'im an' orf 'e goes. **1916** [see *cobber *sb.*[2]]. **1937** N. Marsh *Vintage Murder* xx. 226 She tells 'im orf a treat. **1955** M. Allingham *Beckoning Lady* iv. 55 Just then orf come 'is 'at, and lord luvaduck! **1976** M. Butterworth *Remains to be Seen* v. 82 Do you mind awfully if I push orf? There's a point-to-point. **1978** A. Price *'44 Vintage* xxiii. 260 Everyone had scarpered—cleared orf.

orful (ǭ·ɹful), *a.* Representing a 'phonetic' spelling of an affected or emphatic pronunciation of Awful *a.*

1890 Kipling *Courting of D. Shadd* (ed. 2) 166 'Got any money?' 'On'y a little—orful little.' **1901** A. H. Rice *Mrs. Wiggs of Cabbage Patch* ix. 123 He coughs all the time... Other day he had an orful spell. **1946** A. Christie *Hollow* iv. 34 'Feel pretty bad, don't you?'.. 'Orful, I feel.'

Orford (ǭ·ɹfǝ̄ɹd). The name of the *Orford* Copper Company of New Jersey, U.S.A., used *attrib.* to designate a process it developed for separating nickel from copper by making use of the difference in the solubilities of their sulphides in molten sodium sulphide.

1895 *Mineral Industry* III. 458 The Orford Copper Company was employed some time ago by the United States Navy Department to separate copper from nickel in a large quantity of Canadian matte purchased by the department, and to deliver nickel in the form of an oxide

for use in alloying steel. In order to do this the company employed a new process, which may fairly be called the Orford process, and this has been from time to time improved until there is now produced from the Canadian mattes by a fire process metallic nickel which is from 99% to 99·3% pure. **1923** U. R. Evans *Metals & Metallic Compounds* III. 183 Copper sulphide is soluble in fused sodium sulphide; nickel sulphide is not (the principle of the Orford process). **1967** J. R. Boldt *Winning of Nickel* 276 Inco completed replacement of the Orford method by the matte separation process at the Copper Cliff smelter in 1948.

org (ǭɹg). *colloq.* abbrev. of Organization. Also in *Comb.*

1936 *Esquire* Sept. 160/3 The Joe Breen (Hays org) influence on pix. **1951** Green & Laurie *Show Biz* 570/2 *Org*, organization. **1970** *Harper's Mag.* Apr. 86 They [*sc.* White Anglo-Saxon Protestants] drew the institutions around themselves, moved to the suburbs, and became org-men. **1973** R. Hayes *Hungarian Game* xxi. 129 A private C and D org? **1976** *Times Lit. Suppl.* 5 Mar. 273/5 Systems theorists wish, unlike biologists, to 'explain' the brain..in cybernetic terms of the global properties of organizations ('orgs' as Mr Serebriakoff calls them). **1977** *Time* 6 June 48/2 Many are anti-org types and hard to count.

organ, *sb.*[1] Add: **5. d.** *Jacobson's organ:* see also *Jacobson.

8. (sense 2) *organ-music* (fig. examples), *-note* (later fig. examples), *recital, -stop* (fig: cf. *stop *sb.*[2]), *-tone* (earlier examples); *organ-toned* adj.; (sense 5) *organ regeneration, transplant, transplantation*; **organ-cactus,** substitute for def.: the giant cactus or saguaro, *Carnegiea gigantea*, found in south-western North America and so called from its resemblance to the pipes of an organ; (later example); **organ clock** (see quot. 1962); **organ specificity** *Biol.*, specificity towards a particular organ, esp. as exhibited by an antigen, so **organ-specific** *a.*

1947 *Time* 10 Mar. 18/2 The two Presidents rode between rows of organ cacti. **1956** G. H. Baillie et al. *Britten's Old Clocks & Watches* (ed. 7) 155 Musical and organ-clocks became more popular during the last half of the eighteenth century. **1962** E. Bruton *Dict. Clocks & Watches* 126 *Organ clock*, clock playing a small pipe organ every three hours. Popular in the second half of the eighteenth century. **1973** *Times* 2 Jan. (Europe Suppl.) p. xii/2 The centrepiece of the show will be an organ clock by Charles Clay. **1869** G. Meredith *Let.* 25 June (1970) I. 382 He [*sc.* Poe] gave the idea shape in a fine roll of organ music. **1934** Organ music [see *mimsey *a.*]. **1921** A. Huxley *Let.* 8 Sept. (1969) 204 Papini..one can read with much pleasure... Great sharpness and clarity and wit combined with melody and organ notes and sweeping gesture. **1942** *R.A.F. Jrnl.* 3 Oct. 26 The organ-note of four engines hoisting the flying boat into the air. **1958** Wodehouse *Cocktail Time* iii. 29 A good bishop, denouncing from the pulpit with the right organ note in his voice. **1881** *Harper's Mag.* May 814/1 Piano and organ recitals have long been fashionable. **1923** *Radio Times* 28 Sept. 9/1 3.0.—Organ recital at Steinway Hall, London. **1974** *Times* 12 Nov. 15/2 No one would be happier than I to have Sunday afternoon organ recitals once again at the Palace. **1927** Haldane & Huxley *Animal Biol.* ix. 174 Among reptilia, lizards are the only animals which possess even the power of organ-regeneration. **1972** L. V. Polezhaev *Organ Regeneration in Animals* i. 3 It is conventional to divide all animals into..those which are capable of organ regeneration and those which are not. **1936** K. Landsteiner *Specificity Serol. Reactions* iii. 64 Upon injection of the organ specific substances. **1971** J. Z. Young *Introd. Study Man* xvii. 210 Organ-specific inhibitors are known from many tissues. **1911** *Jrnl. Exper. Med.* XIV. 48 Absolutely no organ specificity is demonstrable as regards the agglutination experiments. **1968** H. Harris *Nucleus & Cytoplasm* iv. 78 It has been contended that the pattern of puffing shows organ specificity but the evidence for this does not seem to be at all conclusive. **1955** A. L. Rowse *Expansion Eliz. Eng.* 27 The Queen.. was furious... At once all the organ-stops are out. **1971** P. Worsthorne *Socialist Myth* iii. 37 The Tory Party has..a liking for the language of togetherness, and real patriotic fervour. These are the organ-stops it can pull out with genuine faith and zest. **1819** Keats *Hyperion* I, in *Lamia & other Poems* (1820) 148 She spake In solemn tenour and deep organ tone. **1894** 'Mark Twain' in *Century Mag.* Mar. 773 There was nothing weak in the deep organ tones that responded. **1895** *New Age* 12 Sept. 372/2 An organ-toned voice of prodigious depth. **1922** Joyce *Ulysses* 454 The strains of the organtoned melodeon. **1968** *Listener* 18 July 90/2 Idea-transplants are as difficult as organ-transplants: in both cases human beings have a built-in mechanism for rejection. **1970** *Memorandum Organtranspl.* (Netherlands Red Cross) § 5c Should an organ transplant be considered..it will be desirable not to have the diagnosis of the donor's death made by one physician only. **1971** *Essentials from Rep. Organtranspl.* (Netherlands Red Cross) 3 It was considered necessary, in the context of organtransplantations, to establish a precise..criterion of..death.

organal, *a.* **3.** Of or pertaining to the medieval style of part-singing known as organum. (Cf. Organum[1] in Dict. and Suppl.)

1916 Stanford & Forsyth *Hist. Mus.* 128 At this time the organal voice had..become finally fixed in its position above the plain-song. **1932** *Music & Lett.* XIII. 190 The melody has a long reciting note on b, which, as he [*sc.* Otker] says, 'has no proper organal response.' **1977** *Early Music* July 337/1 The upper voice seems rather like an ornamented organal voice.

organelle (ǭɹgăne·l). *Biol.* Formerly also **organella.** [mod.L. *organella* (see Organ *sb.*[1] and -el[2]), after earlier *organulum* (K. Möbius 1884, in *Biol. Centralbl.* IV. 392; O. Bütschli *H. G. Bronn's Klassen und Ordnungen des Thier-Reichs* (1888) I. III. 1412), dim. of Organum[1] (see -ule).] Any of various specialized structures of an individual cell, analogous to the organs of multicellular organisms.

1909 H. M. Woodcock in E. R. Lankester *Treat. Zool.* I. I. 212 A Trypanosome was merely described as possessing an unmistakable nucleus, and also a small deeply-staining element..situated at the root of the flagellum, and termed variously 'blepharoplast', centrosome, or micronucleus. It is to Schandinn that we are indebted for the revelation of the essential nuclear nature of the latter organella. **1912** E. A. Minchin *Introd. Study Protozoa* i. 1. However complex the structure and functions of the body [of Protozoa], the organs that it possesses are parts of a cell ('organellæ'), and are never made up of distinct cells. **1924** Hegner & Taliaferro *Human Protozool.* vi. 233 Division of the trophozoite is by longitudinal fission, the posterior end of the animal splitting before the anterior organelles are completely divided. **1926** L. W. Sharp *Introd. Cytol.* (ed. 2) iii. 59 It is also frequently urged that 'organelle' rather than 'organ' should be used for intracellular differentiations. **1955** *New Biol.* XIX. 27 Essentially similar organelles, cilia, cover many membranes..of animals. **1969** F. E. Round *Introd. Lower Plants* ii. 19 Within the outer cytoplasmic membrane occur all the normal organelles—chloroplast, nucleus, dictyosome(s) (Golgi body), endoplasmic reticulum, granules, and vacuoles. **1974** *Sci. Amer.* Oct. 45/1 Bacteria also have flagella, but these are an entirely different organelle.

Hence **organe·llar** *a.*

1970 *Genetics* LXVI. 305 (*heading*) Organellar damage and revision as a possible basis for intraclonal variation in *Paramecium*. **1975** *Nature* 13 Mar. 160/2 If this were the case in my wild carrot protoplasts, about 20% (the approximate proportion of organellar DNA in these cells) of the dimers would have remained after maximum excision.

organetto (ǭɹgăne·to). [It.] A small portative organ used in the Middle Ages.

1876 Stainer & Barrett *Dict. Mus. Terms* 340/2 *Organetto* (*It.*), a little organ. **1952** W. L. Sumner *Organ* iii. 58 Francesco Landini..was an excellent player on several instruments, but he preferred the portative or organetto. **1959** *Collins Mus. Encycl.* 474/1 *Organetto*.., small portative organ of the Middle Ages. **1960** D. J. Grout *Hist. Western Mus.* (1962) iv. 129 In addition to the portative organ or organetto, there were positive organs. **1976** D. Munrow *Instruments Middle Ages & Renaissance* ii. 16/1 The *organetto*, usually known today as the portative organ, was one of the most popular instruments, regularly illustrated from the thirteenth to the sixteenth centuries.

organic, *a.* Add: **4. b.** (ii) Of an element: contained in an organic compound.

1868 *Jrnl. Chem. Soc.* XXI. 87 Estimation of the carbon and nitrogen contained in the organic portion of the solid constituents (organic carbon and nitrogen). **1900** [see *nitrifying *ppl. a.*]. **1924** L. Doncaster *Introd. Study Cytol.* (ed. 2) ii. 20 Practically the whole of the organic phosphorus in the nucleus is contained in..nucleic acid. **1957** G. E. Hutchinson *Treat. Limnol.* I. xii. 735 There is no information available as to the fraction constituting the soluble or colloidal organic phosphorus of lake waters. **1972** *Limnol. & Oceanogr.* XVII. 349/2 Dissolved organic carbon.., particulate organic carbon.., and particulate nitrogen..were measured to determine the distribution of organic matter.

d. Of a fertilizer or manure: produced from natural substances, usually without the addition of chemicals.

1869 S. R. Hole *Bk. about Roses* vi. 76, I made anxious experiment of a multiplicity of manures—organic and inorganic, animal and vegetable. **1942** *Organic Farming & Gardening* I. 3/2 Compost fertilizer is a purely organic material as distinguished from mineral fertilizers (chemicals). **1952** C. E. L. Phillips *Small Garden* iii. 18 Compared with the chemical fertilizers, the organic ones are slow in action but enduring in effect. *Ibid.* 19 Of other forms of organic manure, the following are valuable. **1960** *Times* 27 Feb. 9/2 A good organic-based general fertilizer. **1975** D. Green *Food & Drink from your Garden* v. 38 There is probably something in the theory that vegetables have their quality improved by the use of organic fertilizers.

e. *organic soil*: (see quot. 1928).

1928 *Bull. Amer. Soil Survey Assoc.* IX. 33 *Organic soils*, soils composed mainly of organic material; the organic content being sufficient to dominate the soil characteristics. **1943** Millar & Turk *Fund. Soil Sci.* ii. 63 Deposits of organic soils are of common occurrence in the northern border states of Minnesota eastward. **1966** *McGraw-Hill Encycl. Sci. & Technol.* XII. 423/1 Organic soils such as peats and mucks may contain as much as 95% carbonaceous material.

f. Of farming or gardening: growing plants without the use of chemical fertilizers, pesticides, etc., adding only organic fertilizers to the soil.

1942 J. I. Rodale in *Organic Farming & Gardening* I. 3/1 What is claimed roughly for these organic methods of farming is that they increase the fertility of the soil, produce much better tasting crops,..reduce weeds, do away with the necessity of using poisonous sprays, improve the mechanical structure of the soil. **1948** *Sci. Monthly* June 482/1 Considerable success is claimed in the humid tropics with 'organic farming' where labor is plentiful..and where fertilizers are very expensive or difficult to obtain. *Ibid.*, The organic devotees are primarily interested in production. *Ibid.* 482/2 Great claims have been made for soil improvement by earthworms, usually as a special phase of 'organic' gardening. **1971** *Islander* (Victoria, B.C.) 13 June 14/2 An organic gardener uses natural mineral and organic fertilizers to build his soil. **1973** *Country Life* 6 Dec. 1986/4 Organic farming community. Six professional couples with children wish to purchase..large country residence with small-holding acreage. **1975** *Listener* 14 Aug. 203/2 The great thing about organic farming is that you..build up life in the soil, using natural organic manures such as compost and farmyard manure.

g. Of food: produced without the use of chemical fertilizers, pesticides, etc.

1972 *Daily Tel.* 12 Feb. 6/7 The organic food market is booming. *Ibid.*, 'Ninety per cent of the "organic" apple juice being sold in California is not made from organically grown apples,' said a spokesman for one of the country's biggest organic food wholesalers. **1972** R. Bloch *Night-World* (1974) xiv. 93 Past the organic-food hangouts for the health freaks. **1975** *Times* 5 Sept. 14/8 Another great interest was the growing of food organically, which resulted in a number of books on organic food..in the late 1940s and early 1950s.

5. a. (Further examples.)

1895 E. B. Titchener tr. *Külpe's Outl. Psychol.* I. ii. 140 By 'organic sensations' we mean the sensations adequately stimulated by changes in the condition of the bodily organs,—muscles, joints, etc. **1901** W. James *Let.* 10 July (1920) II. 158 What I *crave* most is some wild American country. It is a curious organic-feeling need. **1933** G. Murphy *Gen. Psychol.* viii. 124 From what has been said about hunger and thirst it seems reasonable to believe that these organic sensations depend partly upon the compounding of simple sense qualities.

d. *organic selection* (see quot. 1942).

1896 J. M. Baldwin in *Amer. Naturalist* XXX. 444 We may simply..apply the phrase, 'Organic Selection', to the organism's behavior in acquiring new modes or modifications of adaptive function. **1942** J. S. Huxley *Evolution* vi. 304 We have here a beautiful special case of the principle of organic selection,..according to which modifications repeated for a number of generations may serve as the first step in evolutionary change. **1970** T. Dobzhansky *Genetics Evol. Process* ix. 303 The term organic selection has been coined to describe the parallelism between racial genotypic and environmental phenotypic variability.

6. a. (Further examples.)

1817 Coleridge *Biog. Lit.* I. xii. 237 The fairest part of the most beautiful body will appear deformed and monstrous, if dissevered from its place in the organic whole. **1847** W. Smith tr. *Fichte's Characteristics Present Age* 94 What this organic unity of a work of Art..really is,—will be asked by no one to whom it is not already known. **1870** S. H. Hodgson *Theory of Practice* II. 166 Rome..was unequal to..incorporating into an organic whole the nations included in her empire. **1874** W. Wallace tr. *Hegel's Logic* 19 The truths of philosophy are valueless, apart from their interdependence and organic union. **1923** *Psychol. Rev.* XXX. 371 Thinking is not an isolated fact... It is the final step in an organic learning process. *a* **1943** R. G. Collingwood *Idea of Hist.* (1946) 123 Marx..conceived this unity not as an organic unity.

b. (Earlier and later examples.)

1849 *Congress. Globe* 30th Congress 1 Sess. App. 47 [The origin of a Territorial Government] is not from *such people*, but from the law of Congress, usually styled the 'organic law', establishing it. **1857** in Bartlett *Dict. Amer.* (1860) 304 The powers of the corporation of Washington are only those which are conferred by the organic law, the charter. **1963** M. Khadduri *Mod. Libya* vii. 184 Both the Tripolitanian and Fazzanese organic laws permit the amendment of any provision during the first session of the legislative assemblies by a simple majority of all the members.

c. Phr. *organic composition of capital* (Econ.): see esp. quot. 1887.

1887 Moore & Aveling tr. *Marx's Capital* II. xxv. 625 The composition of capital is to be understood in a twofold sense. On the side of value, it is determined by the proportion in which it is divided into constant capital or value of the means of production, and variable capital or value of labour-power, the sum total of wages. On the side of material, as it functions in the process of production, all capital is divided into means of production and living labour-power... I call the former the *value-composition*, the latter the *technical composition* of capital. Between the two there is a strict correlation. To express this, I call the value-composition of capital, in so far as it is determined by its technical composition and mirrors the changes of the latter, the *organic composition* of capital. Wherever I refer to the composition of capital, without further qualification, its organic composition is always understood. **1937** M. Dobb *Pol. Econ. & Capitalism* i. 14 The important simplifying assumption that the ratio of labour to capital employed in different lines of production was everywhere equal: what Marx termed equality in the 'organic composition of capital' or what later economists would have called uniformity of the 'technical coefficients'. **1966** J. Robinson *Essay Marxian Econ.* ii. 7 We can avoid ambiguity, without falsifying Marx's meaning, if we use symbols *c*, *v* and *s* only for rates per unit of time of depreciation and raw material lost, wages and profit, and speak of the organic composition of capital, not as *c/v* but as capital per man employed. **1972** G. C. Harcourt *Some Cambr. Controv. Theory Capital* 8 The assumption..is akin to that of Marx.., namely, a uniform organic composition of capital for the processes..of each technique. **1974** M. B. Brown *Econ. of Imperialism* iii. 54 With the increasing application of machinery to production the organic composition of capital would rise. **1975** *Chinese Econ. Studies* VIII. iv. 84 Capital accumulation and capital concentration inevitably increase the organic composition of capital.

B. *sb.* An organic compound. Usu. *pl.*

1953 R. E. Grim *Clay Mineral.* iv. 62 Studies of the methylation of certain organics during their adsorption by montmorillonite. **1970** *Nature* 11 July 149/1 Small samples..of the meteorite..were ground with a small chisel previously heated to a dull red to remove organics. **1974** *Sci. Amer.* May 75/1 The biological material in Dean's recipe..represents 2,000 times the amount of organics normally present in seawater.

organically, *adv.* Add: **1. e.** Without the use of chemical fertilizers, pesticides, etc.

1971 *Countryman* Autumn 203/1 (Advt.), Homely atmosphere, quality food, organically grown vegetables, log fires. **1972** *Guardian* 3 Apr. 7/3 Apple pie... Wash apples (don't peel if organically grown). **1975** *Listener* 14 Aug. 204/2 Their last crop of the season, hand-grown, organically-manured leeks.

organicism. Add: **1. b.** The doctrine that everything in nature has an organic basis or explanation; that everything in nature is part of an organic whole (in sense of ORGANIC *a.* 6 a).

1912 A. Tridon tr. *Delage & Goldsmith's Theories Evol.* 163 In that respect, organicism is the perfect antithesis of Weismannism. **1928** *Jrnl. Philos. Stud.* Jan. 39 This is the reason why modern organicism, the organic theory of nature, seems so important for modern biology. **1945** [see *HOLISM]. **1960** *Encounter* XV. ii. 73 Mr. Tate and his friends..were accused of sentimental organicism, of naïvely hoping to revive the virtues of the antique world by restoring its economic forms. **1969** *Times Lit. Suppl.* 20 Nov. 1341 Organicism..holds that some organic properties are not reducible to those of smaller parts. **1976** *Nature* 3 June 439/1 Reductionism rests on the belief that the whole can be fully explained in terms of the parts whereas organicism (biological holism) asserts that the whole cannot be fully explained in this way.

organicist, (*b*) one who holds the organic theory of nature; also *attrib.* or as *adj.* So **organici·stic** *a.*

1912 A. Tridon tr. *Delage & Goldsmith's Theories Evol.* 164 Roux and the other organicists lay special stress on the factors of individual evolution. **1928** *Jrnl. Philos. Stud.* Jan. 29 That which was common to the organicists, said Delage, was that they regarded..life, the form of the body..as resulting from the reciprocal play or struggle of all its elements. *Ibid.* 39 The organicistic schema formerly covered the living world, and now covers also the world of the non-living. **1941** J. Needham in P. A. Schilpp *Philos. A. N. Whitehead* 251 About the historical origins of the organicistic viewpoint in biology a great deal could be said. **1941** W. M. Urban in *Ibid.* 304 Bergson from whom..the organicist philosophy has got its main insights. **1954** D. Riesman *Individualism Reconsidered* vi. 401 All such 'organicistic' analogies are..dangerous. **1969** P. A. Robinson *Freudian Left* 164 He [*sc.* Marcuse] argued that the Fascist conception of the state was in fact heir to the organicist tradition in political theory. **1971** *Nature* 24 Dec. 490/1 These factors are added to a resurgence of organicist philosophy and a revulsion against Jensenism. **1974** D. L. Hull *Philos. Biol. Sci.* 125 Exchanges between the so-called mechanists and organicists, materialists and vitalists, reductionists and holists, to mention but a few of the terms used to characterize the two sides of this perennial dispute. **1976** *Times Lit. Suppl.* 15 Oct. 1301/1 Extending the organicist thinking of his 'master', the Scottish planner and regionalist Patrick Geddes.

organicity (ǭɹgǎni·siti). [f. ORGANIC *a.* + -ITY.] The quality or state of being organic.

1936 V. A. Demant *Christian Polity* ix. 161 Instead of a return to true organicity, we have Collectivism, both in the patchwork of decaying Capitalism and in Russian Communism. **1945** *Mind* LIV. 53 This fourth and last of the distinctively mental properties..might be called 'organicity' as well as 'integration'. **1970** *Jrnl. Gen. Psychol.* July 110 Concepts relating to organicity and psychosis were excluded in order to limit the population of concepts. **1977** A. Sheridan tr. *J. Lacan's Écrits* vi. 213 Freud first threw light on the evolution itself of the process, thus making it possible to illuminate its own determination, by which I mean the only organicity that is essentially relevant to this process.

organification (ǭɹgæ:nifikēɪ·ʃən). [f. ORGANI(C *a.* + -IFICATION.] Incorporation into an organic compound.

1937 *Nature* 15 May 836/1 (*heading*) Rate of 'organification' of phosphorus in animal tissues. **1966** Wright & Symmers *Systemic Path.* II. xxxi. 1099/1 Thiocarbamide and aniline derivatives..impair organification of iodine in the thyroid. **1976** *Lancet* 27 Nov. 1191/2 Congenital goitres and hypothyroidism have been caused by maternal ingestion of iodides, presumably because the iodides had blocked the organification of iodine and induced pituitary-dependent thyroid hyperplasia.

organigram (ǫɹgæ·nigræm). Also **organogram.** [f. ORGANI(ZATION + -GRAM.] = *organization chart* (*ORGANIZATION 4).

1962 A. Sampson *Anat. Brit.* xxvii. 437 (*caption*) The organogram of Shell October 1959. *Ibid.* xxx. 490 The aircraft companies, built up by brilliant pioneers like de Havilland or Sopwith, are passing painfully into an era of accountants and organograms. **1967** *Economist* 22 Apr. 392/2 A current joke is that Montedison is drawing up, not an organigram, but a 'baronigram'. **1975** A. Beevor *Violent Brink* iii. 64 Many notes had been taken..and doodles drawn. The CGS had a complicated organigram sketched in front of him. **1977** *Official Jrnl.* (Patent Office) 23 Nov. 3791 The first issue of the Official Journal of the European Patent Office will appear in December 1977. The contents will include..an organigram of the European Patent Organisation.

organism. Add: **2. b.** *Philos.* The theory that in science everything is eventually an organic part of an integrated whole.

1925 A. N. Whitehead *Sci. & Mod. World* (1926) 112 This doctrine involves the abandonment of the traditional scientific materialism, and the substitution of an alternative doctrine of organism. **1928** *Jrnl. Philos. Stud.* III. 33 He [*sc.* Lloyd Morgan] saw no reason why the term organism should not be applied to all those 'natural entities', as he called them, existing throughout the universe in emergent degrees of complexity. **1959** A. W. Levi *Philos. & Mod. World* xii. 486 The 'philosophy of organism'..suggests the synthesis of incompatibles. **1965** E. E. Harris *Foundations of Metaphys. in Sci.* xiv. 282 The appropriate philosophy for contemporary science must be..a philosophy of organism.

4. organism-environment, designating the relationship between an animal and its surroundings.

1946 C. Morris *Signs, Lang. & Behavior* iii. 84 One action rather than another..is 'required' by the organism-environment situation. **1958** *New Biol.* XXVI. 84 His [*sc.* man's] social, mental, and technological achievements do not make his 'organism-environment' relationship less important than that of other animals. **1969** *Listener* 13 Nov. 655/1 What lemmings are supposed to do when they get too many has become almost apocryphal and the simile has been used often enough to prophesy courses of human behaviour by people who have no understanding of lemmings or their environment or of the organism-environment relation.

organismic (ǭɹgǎni·zmik), *a.* [f. ORGANISM + -IC.] Of or pertaining to an organism; applied esp. to theories, etc., relating to interdependence or organic unity. Hence **organi·smically** *adv.*

1886 J. C. Burnett (*title*) Diseases of the skin, from the organismic standpoint. **1921** C. M. Child *Origin & Devel. Nervous Syst.* i. 3 The problem..is the problem of the nature and origin of the pattern which constitutes the organism as a whole, whether it consists of one cell or many, in short, the problem of organismic pattern. [*Note*] In view of the fact that the word 'organism', which implies the existence of a unity and order in the entity so designated, is universally accepted and employed, the word 'organismic' is not only biologically and etymologically justified but fills a need which is becoming more and more apparent. **1923** *Psychol. Bull.* XX. 684 (*heading*) The organismic vs. the mentalistic attitude toward the nervous system. **1934** *Jrnl. Philos. Sci.* I. 474 Having thus given, in organismic terms, a brief over-view of the nature and content of thought, we may now consider for a moment its logical aspect. **1943** C. L. Hull *Princ. Behavior* ii. 28 The basic principles of organismic behavior are to be viewed against a background of organic evolution. **1948** M. Sherif *Outl. Social Psychol.* i. ii. 29 The perfectly defensible organismic position..holds that typical reactions of the organism are not fragmentary and that the conceptual or abstract level of psychological functioning is the level of human functioning in the normal conditions of civilized life. **1951** E. E. Evans-Pritchard *Social Anthropol.* iii. 54 Professor Radcliffe-Brown has..clearly and consistently stated the functional, or organismic theory of society. **1961** Webster, Organismically. **1967** *Encycl. Philos.* V. 549/1 The term 'organismalism' was coined by the zoologist W. E. Ritter in 1919 to describe the theory that..'the organism in its totality is as essential to an explanation of its elements as its elements are to an explanation of the organism'. Subsequent writers have largely replaced 'organismal' with..'organismic' as a title for this theory. **1971** *Jrnl. Gen. Psychol.* LXXXV. 230 This finding suggests that cognitive style is organismically based. **1975** *Nature* 27 Mar. 370/2 Luria emphasises molecular aspects of biology, but not to the exclusion of organismic aspects.

organistic, *a.* Add: **b.** Of or pertaining to an organism; based on organisms.

1910 *Fabian News* XXI. 16/1 He adopts the organistic view of society—that society is a being. **1953** *Essays in Crit.* III. 429 Is it speech about individual things..or speech about their feeling of being related, a Whiteheadian organistic 'feeling'?

organistrum (ǭɹgǎni·strəm). [a. med.L. *organistrum*, f. *organum* organ.] A name for the earliest form of hurdy-gurdy (see quots. 1954, 1974.)

[*c* **1350** *Nova Legenda Anglie* (1901) II. 310 Tundalus.. uidit quasi castrum et papiliones plurimas..in quibus chordas et organa, timpana et citharas cum organistris et cymbalis canentes..audierat.] **1870** C. Engel *Descr. Catal. Mus. Instrum. S. Kensington Mus.* 38 The *organistrum* had three strings, producing three different tones, which appear to have consisted of the tonic, fifth, and octave. **1940** C. Sachs *Hist. Mus. Instruments* (1942) xiv. 272 The hurdy-gurdy then [*sc.* in the 13th century] was no longer called *organistrum*, but *symphonia* in Latin (or rather Greek), *chifonie* in French and *cinfonia* in Spanish. **1954** *Grove's Dict. Mus.* (ed. 5) IV. 416/1 In the 12th century the organistrum was about five feet in length, and two executants were required, one of them to turn the handle which moves the wheel (*rotulus*) and the other to manipulate the key-mechanism (*plectra*). There were three strings, the outer tuned in octaves, the middle one a fourth or fifth below the highest string. **1960** *Times* 10 June 4/5 The New York performers [of the *Ludus Danielis*], however, have added an orchestra consisting of a trumpet, soprano recorder, oboe, carillon, handbells, viola, hurdy-gurdy (organistrum), [etc.]. **1974** *Encycl. Brit. Macropædia* X. 437/1 The organistrum, a large medieval hurdy-gurdy operated by two players: one turned a crank rotating a wheel that rubbed against one or more strings to make them sound, while the other produced different notes by turning the key-shaped levers that stopped the strings at various points.

organistry (ǭɹgǎnistri). *rare.* [f. ORGANIST + -RY.] The post of organist.

1890 *Pall Mall Gazette* 19 July 4/1 He..held the local town hall organistry.

organity. Restrict † *Obs.* to sense in Dict. and add: **2.** An organized whole or organism.

1929 R. Bridges *Testament of Beauty* iv. 801 These perfected unify'd organities..all act in response to external stimulants.

organization. Add: **2. c.** Esp. as *social organization* in *Sociol.* and *Anthropol.*

[illegible]

d. Phr. *organization and methods* (see quot. 1968).

[illegible]

4. Special combs.: **organization centre** *Embryol.* [tr. G. *Organisationszentrum* (H. Spemann 1921, in *Arch. f. Entwicklungsmech. d. Organismen* XLVIII. 568)], a region of an embryo that acts as an inductor (*INDUCTOR 5); **organization chart,** a graphic representation of the structure of an organization showing the relationships of the positions or jobs within it; **organization man** orig. *U.S.*, a man who subordinates his individuality and his personal life to the organization he serves.

[illegible] urge for tidiness. **1958** *Economist* 4 Oct. 27/1 Are we gradually getting our equivalents of the 'organisation man', smooth, able, well-adjusted, unexceptionable—and unexceptional? **1960** Koestler *Lotus & Robot* 277, I loathe the crooners and swooners..the Organization Man and the *Reader's Digest*. **1966** N. Freeling *King of Rainy Country* 27 Canisius is just an accountant, an organisation man. A nobody. **1972** 'J. Quartermain' *Rock of Diamond* xiii. 74 He's expendable, an organisation man and a useful commodity in America.

organizational, *a.* Add: (Later examples.)

1938 I. Kuhn *Assigned to Adventure* xxx. 315 Fox itself went through three complete organizational changes in less than two years. **1960** *Guardian* 12 July 1/7 The high noon of the twentieth century 'organisational man'. **1962** A. Battersby *Guide to Stock Control* vii. 62 Such organizational problems are combined with investigations of the nervous systems of animals and the design of servo-mechanisms in the new studies called cybernetics. **1964** M. Argyle *Psychol. & Social Probl.* xiv. 171 *Organizational pathology* refers to the tendency for organizations to become ineffective in several characteristic ways. The most familiar trouble is the over-elaboration of formal rules and procedures, seen by the outsider as 'red-tape'. **1965** *New Society* 22 Apr. 14/3 Industry has its problems. Can organisational psychology help? **1969** J. Argenti *Managem. Techniques* v. 22 *All* organisations..have organisational problems. **1970** T. Lupton *Managem. & Social Sci.* (ed. 2) iii. 71 The organizational environment for maximum performance and human satisfaction. **1973** A. Dundes *Mother Wit* p. xiii, Having explained the organizational plan of the book, I should like to add a final word about the title.

Hence **organiza·tionally** *adv.*

1933 *Times Lit. Suppl.* 15 June 415/3 This is a moment for fresh, organizationally detached and sympathetically cooperative thinking. **1959** *Internat. Jrnl. Appl. Radiation & Isotopes* VI. 305/1 Even were technical feasibility successfully achieved, the firms organizationally capable of using the process and exploiting its marginal advantages are few. **1962** *B.B.C. Handbk.* 94 Organizationally, the External Services are an integral part of the BBC. **1976** *Nature* 8 July 88/3 'Organisationally,' it says, 'OTA lacks the minimum of orderly structure.'

organizator (ǭ·ɹgǎnəizēɪ:təɹ). *Embryol.* [ad. G. *organisator* organizer (given this sense by H. Spemann 1921, in *Arch. f. Entwicklungsmech. d. Organismen* XLVIII. 568).] = *ORGANIZER 2, *INDUCTOR 5.

1924 *Nature* 23 Feb. 276/2 Spemann has proved that the dorsal-lip region is a differentiator (or 'organisator' as he styled it). **1928** *Biol. Abstr.* II. 1320/2 Embryos with 1 axis are harmonically built, developing from germs in which 'organizators' lie close beside each other. **1939** E. E. Just *Biol. Cell Surface* xi. 290 By experiment it is possible to analyze the factors which set up the conditions for differentiations in a more normal or natural manner than..in experiments with transplantations involving conceptions of 'organizators' and the like.

o:rganiza·tory, *a.* [f. med.L. *organizat-*, ppl. stem of *organizāre*, see ORGANIZE *v.* + -ORY².] Of or pertaining to organization.

1921 *Public Opinion* 17 June 560/1 The merely organisatory work of delivering wooden houses and materials to France. **1926** *Ibid.* 2 July 15/3 If the question of women's professions was a theoretical and ethical problem in the beginning, it must be regarded as an intellectual and organisatory one nowadays.

organize, *v.* Add: **2. b.** (Further examples.)

1904 'Mark Twain' *$30,000 Bequest* (1906) 7 When we organize, we'll get three shares for one. **1966** *McGraw-Hill Encycl. Sci. & Technol.* II. 614/2 They may organize into complex structures such as plastids. **1976** *Spare Rib* Nov. 26/4 This is not the first time women have organised for peace.

c. *trans.* With a person as obj.: to provide for; to make (special) arrangements for. Also *refl.*

1892 'Mark Twain' *Amer. Claimant* ii. 35 I'll get you organized in no time. **1952** M. Tripp *Faith is Windsock* iv. 72 'Where's Arthur?' asked Bergen. 'Gone off with a Waaf, I think. Jake's organised too.' **1959** 'M. Cronin' *Dead & Done With* viii. 122 There were going to be official complaints reaching the office in the morning about his activities... He felt he could organize himself out of most of it. **1977** B. Pym *Quartet in Autumn* v. 46 She was the kind of person who liked to keep herself to herself and must not be organized in any way.

d. *trans.* To arrange (personally); to take responsibility for providing (something); to 'fix up'. *colloq.*

1952 M. Laski *Village* ix. 141 Martha organized a scratch meal. **1972** G. Durrell *Catch me a Colobus* ix. 179 We spent the rest of the day organizing a car to take us to Mexico City the following morning. **1976** P. Hill *Hunters* v. 55 Got a big job fer you, son... Organize some sandwiches from the pub.

4. *trans.* To acquire deviously or illicitly; to obtain cleverly (orig. *Mil. slang*). Cf. G. *organisieren* Mil. slang in same senses.

1941 *New Statesman* 30 Aug. 218/3 *Organise*, to acquire illicitly. (A new R.A.F. equivalent for the last-war word 'win', meaning to 'scrounge'). **1942** *R.A.F. Jrnl.* 16 May 12 Even the plugs in the washbasins are replaced. Why do people like to 'organize' those plugs? They just fade away..and have to be replaced; but what use they are to the lads who make them souvenirs, few know. **1957** H. Roosenburg *Walls came tumbling Down* ii. 53 The verb 'to organize' had been widely in use in the [Nazi] prisons and camps and simply meant to acquire what one needed by stealing, bamboozling, or any other means at one's disposal. *Ibid.* iii. 73 Those Frenchwomen..were busy organizing some sausages from a reluctant butcher.

organized, *ppl. a.* Add: **2. b.** Acquired deviously, illicitly, or cleverly. (Cf. *ORGANIZE *v.* 4.) *slang.*

1957 H. Roosenburg *Walls came tumbling Down* v. 127 They had moved in..with a few organized mules and removed all the stores.

4. Of or pertaining to a coordinated criminal organization directing operations on a large or widespread scale, esp. in phr. *organized crime.*

1929 J. Landesco *Organized Crime in Chicago* ii. 25 Organized crime is not, as many think, a recent phenomenon in Chicago. *Ibid.* ix. 205 Newspaper writers,.. interested in establishing the national and international ramifications of organized criminals. **1931** F. D. Pasley *Muscling In* iii. 94 Next to beer and booze, organized prostitution yielded the heaviest profits. **1941** H. Asbury *Underworld of Chicago* ix. 299 During the last few months of Mayor Harrison's final term Chicago was probably as free from organized vice as at any time in its history. **1952** Turkus & Feder *Murder, Inc.* i. 9 In all the history of crime, there has never been an example of organized lawlessness to equal the Syndicate. **1973** *Black Panther* 5 May 2/2 It is widely known that Inman is himself a kingpin in the city's organized crime and racket rings, as is Atlanta Mayor Sam Massell. **1975** *Globe & Mail* (Toronto) 3 June 1/9 He was named in police evidence before the Quebec organized crime inquiry as one of the four top lieutenants of the Godfather of organized crime in Montreal.

5. Special combs.: **organized games,** athletics or sports as organized in a school, college, etc.; **organized labour,** workers affiliated by membership in trade or labour unions.

1933 D. L. Sayers *Murder must Advertise* xviii. 304 In Brotherhood's régime of bread and circuses, organised games naturally played a large part. **1944** L. P. Hartley *Shrimp & Anemone* i. 16 The world of day-schools and organized games. **1974** *Times* 5 Jan. 10/3 At modern Oxbridge there has been a decline in the participation by undergraduates in organized games. **1885** in C. Evans *Hist. United Mine Workers of Amer.* (1918) I. 131 To organized labor..and to the generous and sympathetic public..we return our sincere and heartfelt thanks. **1924** L. Wolman *Growth of Amer. Trade Unions* 82 The number of wage earners..would not be considered by some a thoroughly fair base for measuring the achievement in size of an organized labor movement. **1926** *Brit. Worker* 10 May 2/4 The fight to maintain the workers' standard of living will be won by the united forces of organised labour. **1948** *Time* 15 Mar. 27/2 He thought of himself as the leader of all the people, not just of organized labor. **1975** *Times* 7 Jan. 12/8 The rise of both organized labour and high management around the turn of the century.

organizer. Add: **1. a.** Also, one who 'organizes' criminal activity (cf. *ORGANIZE *v.* 4).

1945 C. Burney *Dungeon Democracy* i. 19 He was an admirable 'organiser' at worst,..and succeeded in building up a private stock of those luxuries which did not come on the standard lists of the canteen. **1976** E. Ward *Hanged Man* iii. 15 Gold was now an organizer—setting up criminal work on commission.

2. *Embryol.* [tr. G. *organisator* (see *ORGANIZATOR).] = *INDUCTOR 5.

1925 H. Spemann in *Brit. Jrnl. Exper. Biol.* II. 500, I have given the name of 'organisers' to cells capable of inducing the formation of new anlagen. **1927** —— in *Proc. R. Soc.* B. CII. 177 (*heading*) Organizers in animal development. **1934** [see *EVOCATOR b]. **1946** [see *INDUCTOR 5]. **1970** A. M. Winchester *Concepts Zool.* xviii. 474/1 The transplanted dorsal lip acted as an organizer. **1975** *Nature* 13 Nov. 129/2 The organiser for the axial pattern of the whole body [of *Xenopus*] during amphibian development is a small group of cells at the dorsal lip of the blastopore.

organo-. Add: In many compounds a secondary stress may be given as *o:rgano-* or *orga:no-*, and is not indicated in the individual words listed below. **organoleptic** *a.*, also, involving the use of the senses; (examples); hence **organole·ptically,** *adv.*, as regards organoleptic properties; **organometallic** *a.* (earlier example); also *absol.*, an organometallic compound; **organosedime·ntary** *a. Geol.*, produced by or involving sedimentation as affected by living organisms.

1852 T. R. Betton tr. *Regnault's Elem. Chem.* I. 13 (*heading*) Of the different physical and organoleptic characters by which bodies are distinguished. *Ibid.*, The organoleptic characters are those impressions produced on the organs of taste, smell, and touch. **1940** *Nature* 21 Dec. 796/1. Unimpeachable organoleptic evidence exists for the statement that even under modern conditions cheese-making is not by any means a fully controlled industrial process. **1963** W. Summer *Methods Air Deodorization* iii. 228 Science has failed, so far, to conceive of an instrument which might be called an artificial nose and which would allow comparisons..between organoleptic sensations and instrumental measurements. **1970** *Daily Tel.* (Colour Suppl.) 25 Sept. 32/2 Each morning he must check the contents of the 50 vats to ensure they are suitable for bottling. His responsibilities there are organoleptic, that is, he is concerned with the qualities of the senses—sight, taste and smell. **1976** *Daily Colonist* (Victoria, B.C.) 30 May 4/7 The U.S. Food and Drug Administration would be crippled without its organoleptic analysts. **1959** *Proc. Florida State Hort. Soc.* LXXII. 145/1 Celery air-expressed from California was compared with freshly harvested Florida celery in April and May of 1957. The samples were rated organoleptically and analyzed chemically for several constituents. **1970** H. E. Nursten in A. C. Hulme *Biochem. Fruits & Products* I. x. 246 (*heading*) Organoleptically significant components of specific fruits. **1852** *Phil. Trans. R. Soc.* CXLII. 417, I have continued my researches upon the organo-metallic bodies. **1938** H. Gilman *Org. Chem.* I. iv. 463 They [*sc.* organomercurials] are the only organometallics in the first two groups which can be manipulated in water. **1965** *New Scientist* 2 Dec. 658/1 Organometallics have already amply proved their worth as catalysts. **1974** *Encycl. Brit. Macropædia* XIII. 716/1 The syntheses of many specific organometallics are dictated by the particular properties or reactions of a single compound or group of compounds. **1976** *Nature* 4 Mar. 11/1 Thus, so far as σ-bonded transition metal organometallics are concerned, the realisation of the difference between their kinetics and thermodynamic stabilities has led several research groups to design stable molecules. **1964** *Jrnl. Geol.* LXXII. 81/2 Algal stromatolites are laminated organosedimentary structures formed by the activity of algal mats in binding fine particulate sediment. **1972** *Ibid.* LXXX. 592/1 'Molar tooth' structure..most likely resulted from organosedimentary processes similar to those forming algal stromatolites.

2. *Chem.* **a.** Prefixed to the names of elements to form adjs. designating compounds in which an atom of the element is bound to an organic radical; as *organochlorine, -lead, -lithium, -magnesium, -mercury, -phosphorus, -silicon, -tin, -zinc.* These may also be used *absol.*

1961 *Jrnl. Econ. Entomol.* LIV. 636/1 The..effectiveness of six organochlorine insecticides applied to soil were determined in the field against *Hippelates* gnats. **1970** *Motor Boat & Yachting* 16 Oct. 25/2 Two serious disadvantages about the early organochlorines such as D.D.T. were that they were..concentrated in certain tissues of the bodies of successive predators. **1974** *Country Life* 26 Dec. 1984/2 The gradual decline in the use of organochlorine pesticides has allowed predatory birds.. to re-establish themselves. **1861** *Q. Jrnl. Chem. Soc.* XIII. 228 Organo-lead compounds are arranged under the types of sesquioxide and peroxide of lead. **1974** *Physics Bull.* May 180/1 Lead alkyl petrol additives provide virtually the only source of organolead compounds in the environment. **1932** *Jrnl. Amer. Chem. Soc.* LIV. 1957 It is possible to prepare many organolithium compounds by the direct interaction of lithium with an RX compound in ether or benzene. **1974** *Chem. & Pharm. Bull.* XXII. 1711 Many reports have been published so far on the reactions between organolithiums and open chain compounds. **1901** *Jrnl. Chem. Soc.* LXXX. i. 263 The organo-magnesium compound produced crystallises in colourless, flattened needles. **1968** R. O. C. Norman *Princ. Org. Synthesis* vi. 202 Organomagnesium compounds, known as Grignard reagents after their discoverer, are the most widely used of organometallic reagents. **1860** *Chem. News* 30 June 26/1 At present the representatives of the organo-mercury series all belong to the mercuric type. **1963** A. J. Hall *Textile Sci.* v. 263 Shirlan is also much used for protecting cotton against mildew attack. A number of complex organo-mercury compounds are also effective for this purpose. **1974** *Nature* 20 Sept. 236/2 The ability of organo-mercury compounds, particularly methylmercury, to concentrate in tissues of fish and other animals at levels which are toxic for human consumption, is of increasing environmental concern. **1950** G. M. Kosolapoff *Organophosphorus Compounds* i. 7 In many classes of organophosphorus compounds we find mixtures that are inseparable by fractional distillation. **1962** *Brit. Birds* LV. 431 Certain organophosphorus sprays were used on Brussels sprouts. **1971** *Homes & Gardens* Aug. 89/3 A servicing company can apply a long-lasting organo-phosphorus insecticide to pendant fittings. **1941** *Jrnl. Amer. Chem. Soc.* LXIII. 1194/1 The lack of a clear understanding of the behavior of the substituted organo-silicon compounds and the possibility of producing useful resinous polymers for them led to an investigation of some of the disubstituted compounds. **1955** Brown & Dey *India's Mineral Wealth* (ed. 3) x. 391 The resultant organosilicon chlorides are hydrolised [*sic*] to silanols which condense into the polysiloxanes or silicones. **1974** *Encycl. Brit. Macropædia* XIII. 716/1 The organosilicon halides..are prepared by a special reaction using copper as catalyst. **1866** Watts *Dict. Chem.* IV. 220 No organo-tin compounds containing only one equivalent of positive radicle, have hitherto been produced. **1960** *Times Rev. Industry* Apr. 83/3 Among the new chemicals are..organotin compounds of germicidal.. nature. **1860** *Chem. News* 30 June 26/1 Organo-zinc compounds are decomposed by water, oxide of the metal and hydride of the radical being produced. **1966** *McGraw-Hill Encycl. Sci. & Technol.* IX. 403/1 For many years organo-zinc compounds were used for synthetic purposes.

b. Prefixed to the names of various types of compounds to form sbs., indicating the presence of an organic radical in the molecule, as *organoalkali, -borane, -mercurial, -metal, -phosphate, -siloxane*; these may also be used *attrib.* Also **organocompound,** an organometallic compound.

1932 *Jrnl. Amer. Chem. Soc.* LIV. 1958 A corresponding simpler technique might be used with organoalkali compounds like phenyl-lithium. **1968** G. E. Coates et al. *Princ. Organometallic Chem.* iii. 34 A large class of charge-delocalized organoalkali compounds consists of the addition compounds between alkali metals and bi- or polynuclear aromatic hydrocarbons. **1957** Organoborane [see *HYDROBORATION]. **1971** J. D. Roberts et al. *Org. Chem.* xix. 536 The boron halides and the organoboranes (BR_3) are Lewis acids. **1976** *Nature* 15 Jan. 163/2 Chapters 4–6 .., on four-coordinate organoboranes, organodiboranes, and other polyboranes, are particularly welcome. **1866** Watts *Dict. Chem.* IV. 224 Aluminium series. — The only known organo-compounds of this series are aluminic methide, $Al'''(CH^3)^3$, and aluminic ethide, $Al'''(C^2H^5)^3$. **1974** *Nature* 13 Dec. p. x (Advt.), Each chapter provides a complete review of all aspects of the organocompound in question. **1972** *Ibid.* 25 Feb. 414/3 The dumping of substances likely to find their way into the food chains is completely prohibited—these include organohalogen and organosilicon compounds. **1977** *Offshore Engineer* May 29/1 The real villains on the 'blacklist' are certain organohalogens (for example the extremely toxic and readily absorbed printing by-product PCB), mercury and cadmium. **1866** Watts *Dict. Chem.* IV. 221 A brisk action with considerable elevation of temperature attends the formation of the organo-mercurial compound. **1938** H.

Gilman *Org. Chem.* I. iv. 463 Organomercurials are the least active organometallic compounds of the first two groups. **1961** *Times* 27 Mar. 5/3 It is comparable with the organo-mercurials against apple scab. **1971** *Nature* 23 July 222/1 Only inhaled elemental mercury vapour is comparable with the above organo-mercurials in inducing intoxication. **1866** Watts *Dict. Chem.* IV. 230 Arsenic Series.—This series..contains the first discovered organo-metal, cacodyl. **1971** *Nature* 31 Dec. 518/1 The use of metal complexes as organo-metal catalysts for the synthesis of industrially important organic intermediates and polymers. **1958** *Jrnl. Econ. Entomol.* LI. 714 (*heading*) New organophosphate insecticides developed on rational principles. **1974** M. C. Gerald *Pharmacol.* vii. 133 Medically, organophosphate compounds are used for the treatment of glaucoma. **1946** E. G. Rochow *Introd. Chem. Silicones* i. 2 The period since 1940 has seen rapid development of the organosiloxanes or silicone polymers. **1970** *Sci. Jrnl.* Feb. 21/3 Cement and plaster can be made water repellent by incorporating into the mix small quantities of organosiloxanes which have a large proportion of silicon bonded hydroxyl groups.

organogenic, *a.* (s.v. ORGANOGENY). Add: (*b*) *Petrol.* = *ORGANOGENOUS *a.* Hence **organoge·nically** *adv.*

1934 Webster, *Organogenic Petrog.*, derived from organic substances. **1949** F. J. Pettijohn *Sedimentary Rocks* x. 301 These crinoidal limestones may be termed 'organogenic conglomerates'. **1967** *Oceanogr. & Marine Biol.* V. 550 The coralligenous biocoenosis..is particularly well developed..on rocky as well as on organogenically fixed bottoms.

organogenous (ǫɹgǎnǫdʒe-nəs), *a. Petrol.* [f. ORGANO- + *-GENOUS.] Of a rock: formed from organic materials.

1881 E. Renevier in *Proc. Geol. Assoc.* VI. 426 (*table*) Organogenous. — Sedimentary by organic means. **1967** *Oceanogr. & Marine Biol.* V. 503 The soft substrata of the circalittoral zone are made up of terrigenous sediments.. and organogenous remnants.

organoid, *a.* Add: **B.** *sb. Biol.* = *ORGANELLE.

1930 Maximow & Bloom *Text-bk. Histol.* i. 5 The constituents of the cytoplasm..may be classified as the organoids and the inclusions. The organoids are structures ..which are probably endowed with the ability to divide ..in contrast to the inclusions which are passive, lifeless, temporary constituents of the cell. The organoids comprise the mitochondria, the Golgi apparatus, the centrioles, and fibrils. [*Changed to* organelle *in ed.* 7 (1957).] **1956** *Anatomical Rec.* CXXV. 481 The presence of mitochondria in smooth muscle was described by Cowdry ('34), long after these organoids had been studied in other tissues. **1957** H. S. D. Garven *Student's Histol.* i. 19 The centrosome... Not infrequently this organoid is found lying embedded in the skein of the Golgi body.

organology. Add: **4.** The study of the history of musical instruments.

1959 *Times Lit. Suppl.* 17 July 428/4 More specialized aspects of 'organology' (to use the term proper to the study of old instruments) formed the topics of the fifteen papers read by members of the Galpin Society. **1960** *Times* 18 Mar. 4/6 Organology..pursues one branch of ethnomusicology, the comparative study of instruments as they are found in the various communities. **1971** *Times Lit. Suppl.* 19 Nov. 1453/5 (Advt.), Studies in keyboard organology. **1977** *Early Music* July 405/2 Only one thing is lacking to help the student of organology: a photograph of each instrument.

organological *a.*, **organologist** (later examples, in sense *4).

1976 *Early Music* July 293/1 Munrow's emphasis on the musical use of instruments reflects a refreshing and welcome new departure in organological studies. **1976** *Jrnl. Amer. Mus. Instrument Soc.* II. 120 He must be counted among the finest organologists of our generation.

organosol (ǫɹgæ·nǒsǫl). [a. G. *organosol* (E. A. Schneider 1892, in *Ber. d. Deut. Chem. Ges.* XXV. 1164): see ORGANO- and *SOL *sb.*⁶] A dispersion in which the dispersion medium is an organic liquid; *spec.* one of particles of a synthetic resin in a liquid consisting of plasticizer and volatile components, which can be converted into a solid plastic simply by heating (cf. *PLASTISOL).

1892 *Jrnl. Chem. Soc.* LXII. 775 Organosol Ag ([in] ethyl alcohol) is formed by the dialysis of the hydrosol in absolute alcohol. **1931** E. S. Hedges *Colloids* xiv. 192 When the organosols are treated with a liquid which is soluble in the dispersion medium, but does not dissolve the disperse phase, the latter is precipitated. **1946** *Mod. Packaging* Mar. 262/2 Conversion..from solution coating to an organosol dispersion more than doubled the unit output of one modern coating plant. **1960** *Times Rev. Industry* May 23/2 The best known of the organosol coatings are the vinyl synthetic resin types. **1963** H. R. Clauser *Encycl. Engin. Materials* 454/2 Several plants are using spread coaters for the application of organosols and plastisols to strip steel to provide materials competitive with the light metals and plastics. **1972** *Materials & Technol.* V. xi. 338 Organosols are easier to spray in conventional equipment, and give thinner films with a greater range of flexibility and hardness.

organotherapy (ǭ:ɹgǎnoþe·rǎpi). *Med.* [f. ORGANO- + THERAPY.] Treatment by the administration of preparations made from animal organs, esp. glands.

1896 *Med. Times & Hosp. Gaz.* XXIV. 545/2 The belief..that diseases arising from a lack of the normal secretion of a certain gland, may at times be treated with benefit by administering the secretion of that gland from lower animals..forms the basis of modern organotherapy. **1915** [see *KETONE 2]. **1939** M. A. Goldzieher *Endocrine Glands* i. 1 Organotherapy is as old as mankind and is used by primitive peoples to-day. **1958** *Internat. Jrnl. Fertility* III. 315 It is difficult to determine the incidence of endocrine disorders in studies of sterility. It is largely for this reason that organotherapy is either neglected or applied empirically. **1968** *Guardian* 27 Apr. 7/6 Organotherapy—the use of organs, animal or human, as a form of treatment.

Hence **o:rganotherapeu·tic, -the·rapic** *adjs.*, of or pertaining to organotherapy; **o:rganotherapeu·tics** *sb. pl.* = *ORGANOTHERAPY.

1900 *Lancet* 25 Aug. 610/2 Dr. J. G. Soutar..regarded the subject of organo-therapeutics as one of great importance. **1905** *Ibid.* 19 Aug. 554/2 Tests whereby organo-therapeutic substances might be recognized. **1908** *Practitioner* Sept. 428 All forms of medical treatment of this affection, whether hygienic, dietetic, medicinal, organotherapic, or electrical in nature, are unsatisfactory. **1911** *Encycl. Brit.* XXVI. 798/2 The success which has been achieved has led to the use of many other organs in a raw or compressed form, or as extracts, in other diseases... To this method of treatment the name of organo-therapeutics or opo-therapy has been given. **1923** H. R. Harrower *Index of Organotherapy* 26 Measures that I know have been unusually effective..are not necessarily organotherapeutic. **1928** F. W. Broderick *Dental Med.* xi. 223 There may exist..certain definitely recognisable conditions in which appropriate organotherapeutic and other measures will be extremely useful. **1934** *Q. Cumulative Index Med.* XV. 934/2 Action of certain organotherapeutic preparations on coronary vessels of isolated heart of animals and of man.

organ-pipe. Add: **2. d.** = *organ-cactus* (ORGAN *sb.*¹ 8 in Dict. and Suppl.).

1854 *Colburn's United Service Jrnl.* Feb. 274 A specimen of [cactus]..which from its shape is commonly called 'the organ pipe' rose to the height of about twenty feet. **1957** J. Kerouac *On Road* (1958) IV. v. 276 We began to see the ghostly shapes of yucca cactus and organ-pipe on all sides.

3. organ-pipe cactus = sense 2 d above.

1908 W. T. Hornaday *Camp-Fires on Desert* 352 The mines are quite the northern limit of the organ-pipe cactus. **1977** *Times* 21 Apr. 16/8 Organ-pipe cacti still grow alongside..golf course greens [in Arizona].

organum¹. Add: **2.** (Later examples.) Also *attrib.*

1884 W. H. Frere *Winchester Troper* p. xxi, He [*sc.* Notker] first tried his hand with the melody known as Organa. *Ibid.* p. xxxix, The Organum became not a mere mechanical repetition of the principalis, but another part more or less independent of it. **1932** *Music & Lett.* XIII. 185 This singing in two parts..was also popularly called 'Organum'. *Ibid.* 189 The alto and bass have the melody, the others the organum. **1965** *Listener* 20 May 756/3, I specially liked the alternating plainchant and two-part polyphony in *organa* style of the Kyrie. **1977** *New Yorker* 23 May 126/3 The 'Hymn for a New Age' is an antiphonal chant given out by the children in organum fourths, accompanied by oboes and English horn.

organza (ǫɹgæ·nzǎ). [ad. F. *organsin*, It. *organzino*: see ORGANZINE.] A thin stiff transparent dress-fabric of silk or synthetic fibre.

1820 M. Edgeworth *Let.* 4 June in C. Colvin *M. Edgeworth in France & Switz.* (1979) 144 The distinguishing characteristic is a silk organza handkerchief. **1934** *Times* 22 June 17/4 The latter in checked organza in red, white and black colourings. **1956** 'R. Crompton' *Matty & Dearingroydes* x. 93 The dress was egg-shell blue organza. **1964** D. Francis *Nerve* xvi. 210 Alice front-view in spotted organza at a Hunt Ball. **1973** *Country Life* 19 Apr. 1114/2 The pink and green flowers used for the cap sleeves..are cut from the exquisite organza fabric.

orgasm. Add: Also *attrib.*

1936 H. M. & A Stone *Marriage Manual* viii. 276 Orgasm incapacity is more frequent. **1949** *Orgone Energy Bull.* Apr. 94 *Orgasm reflex*, the unitary involuntary contraction and expansion of the total organism in the acme of the sexual embrace. **1965** P. & E. Kronhausen *Sexual Response in Women* ii. i. 64 The kind of 'orgasm anxiety' to which our informant referred is very common among women. **1967** P. S. Catling *Experiment* xxxvi. 228 You might have expected higher I.Q. levels to mean ..an enhanced orgasm yield. **1968** R. Kyle *Love Lab.* xxiv. 322 With his help, she had broken through into orgasm country. **1973** S. Fisher *Female Orgasm* i. 31 The amount of time spent by the husband stimulating his wife does not correlate with her orgasm frequency.

Hence **orga·smal** *a.* = *ORGASMIC *a.*

1964 *Brit. Jrnl. Med. Psychol.* XXXVII. 63 Failure to achieve normal adult heterosexual adjustment was shown ..by frigidity, and sometimes promiscuity with orgasmal disappointment. **1969** C. Allen *Textbk. Psychosexual Disorders* (ed. 2) iv. 71 Castration seems to have little effect on the desire or orgasmal capacity of the woman.

o·rgasm, *v.* [f. the sb.] To experience a sexual orgasm.

1973 S. Fisher *Female Orgasm* vii. 207 It often takes me as much as 15 minutes of stimulation before I can orgasm. **1974** *New Direction* IV. iv. 5/3 He stroked my clitoris until I orgasmed. **1977** *Observer* 25 Sept. 23/7 Approved wisdom has it that women should orgasm from that which achieves male orgasm and reproduction —penile thrusting; not from direct clitoral stimulation.

orgasmic (ǭɹgæ·zmik), *a.* [f. ORGASM + -IC.] Of or pertaining to sexual orgasm; in a state of sexual orgasm. Also *transf.* and *fig.*

1935 R. V. Storer *Sexual Technique* xxi. 278 The orgasmic contractions of the uterus act as a kind of suction pump. **1946** M. Peake *Titus Groan* 368 What had gone wrong? The orgasmic moment [of murder] he had so long awaited was over. **1947** J. Steinbeck *Wayward Bus* xiv. 197 Back in the bus he had felt, in anticipation, a bursting, orgasmic delight of freedom. **1953** A. C. Kinsey et al. *Sexual Behavior Human Female* ix. 390 It had acquainted the girl with the nature of an orgasmic response. **1966** Masters & Johnson *Human Sexual Response* ix. 131 The female is capable of rapid return to orgasm immediately following an orgasmic experience. **1968** *New Statesman* 16 Aug. 208/1 The Trojans meet the Greeks like lovers, almost naked, agog for the dark orgasmic flutter of killing or being killed. **1969** *Daily Tel.* (Colour Suppl.) 7 Mar. 7/2 She lay naked on the floor doing the standard orgasmic heaving. **1971** *Daily Tel.* 3 Apr. 14 Television commercials are often blatantly erotic, so that even the simple act of eating a chocolate bar is turned into an orgasmic experience. **1971** 'V. X. Scott' *Surrogate Wife* 13 Women do *not* ejaculate any sort of fluid.., but they *do* have an orgasmic reaction that is physiological, not just mental. **1976** L. Deighton *Twinkle, twinkle Little Spy* xviii. 186 From Katerina came a long orgasmic whimper.

Hence **orga·smically** *adv.*

1972 D. F. Barber *Pornography & Society* iii. 93 The orgasmically satisfied man or woman is unlikely to throw a bomb. **1974** *Forum* VII. 30/2 While she is still glowing orgasmically, he should enter her.

orgasmist (ǭɹgæ·z'mist). *rare*⁻¹. [f. ORGASM + -IST.] One who delights in sexual excitement.

1938 Dylan Thomas *Let.* 6 July (1966) 205 It's a crack at young Georgians, not at New-Versers, intellectual muckpots leaning on a theory, post-surrealists and orgasmists, tit-in-the-night whistlers, [etc.].

orgastic, *a.* Add: (Later examples.)

1930 *Internat. Jrnl. Psycho-Anal.* XI. 439 In many cases the trauma of punishment falls upon children in the midst of some erotic activity, and the result may be a permanent disturbance of what Reich calls 'orgastic potency'. **1942** T. P. Wolfe tr. W. Reich in *Internat. Jrnl. Sex-Econ. & Orgone Res.* Mar. 33/2 Psychic as well as somatic disturbances are due to the *stasis* (damming-up) of energy in the organism. This stasis is due to *orgastic impotence*; only *orgastic potency*, i.e. biologically correct discharge of sexual energy, guarantees a normal energy household (sex-economy). **1963** H. I. Schneer *Asthmatic Child* vi. 78 She said that it was often impossible to have an orgastic response with her husband. **1969** P. A. Robinson *Freudian Left* 17 Orgastic potency was *defined* in economic terms; it was 'the capacity for complete discharge of all dammed-up sexual excitation through involuntary pleasurable contractions of the body'.

Hence **orga·stically** *adv.*

1941 *Internat. Jrnl. Psycho-Anal.* XXII. 215 The patient was also orgastically potent in Reich's sense of the term. **1953** W. Reich *Murder of Christ* 189 Orgastically impotent physicians in the realm of medical orgonomy will mess up the medical techniques to establish the orgonotic streaming in sick organisms or will forget them. **1973** S. Fisher *Female Orgasm* i. 32 The more a woman is capable of responding orgastically to her spouse,..the happier their marriage will be.

Orgatron (ǭ·ɹgǎtrǫn). [f. ORGA(N *sb.*¹ + *ELEC)TRON(IC *a.*] A kind of electronic organ.

Registered as a trade mark in the U.S. in 1935 but this mark was later cancelled. It was re-registered as a trade-mark in 1975.

1935 *Official Gaz.* (U.S. Patent Office) 16 July 518/2 Everett Piano Company, Chicago, Ill... Orgatron..For Musical instruments—Namely, Key-Board Instruments for Producing Pipe Organ Tones through Electronic Means. Claims use since Apr. 12, 1935. **1935** *Piano Trade Mag.* May 8 The Everett Piano Co., South Haven, Mich., this month announces its new musical instrument, the Orgatron, an electronic instrument with the tone of a fine pipe organ. **1937** *Sun* (Baltimore) 12 Nov. 19/4 The feature of the entertainment program will be a recital each afternoon and evening on 'The Orgatron', one of the latest developments of the electric organ. **1940** *Chambers's Techn. Dict.* 599/1 *Orgatron* (*Acous.*), an electronic musical instrument using the pneumatic action of a reed organ. The electrical current for the operation of radiating loudspeakers is obtained by electrostatic pick-ups operated by the motion of the languids of the air-operated reeds, using adequate amplifiers. **1959** *Collins Mus. Encycl.* 218/1 The Everett Orgatron..amplifies the vibration of harmonium reeds. **1975** *Official Gaz.* (U.S. Patent Office) 1 Apr. TM 23/1 GTR Products, Inc., Cranford, N.J.... Orgatron. For Electronic Organs.. First use Oct. 29, 1973.

orgiast. Delete *rare* and add later examples.

1939 Joyce *Finnegans Wake* 254 Orion of the Orgiasts. **1957** M. Spark *Comforters* iv. 87 'He's an orgiast on the quiet.' 'A what?' 'Goes in for the Black Mass.' **1967** *Punch* 8 Nov. 723/2 That's why orgiasts have to be slightly tight And orgiate in the middle of the night. **1975** R. H. Rimmer *Premar Experiments* (1976) ii. 194 Sorry, love bug. I'm not an orgiast.

orgiastical, *a.* Add: (Further example.)

1930 *Observer* 26 Jan. 10 The Greek Tragedy up to date includes a number of ballads, revolutionary songs, and orgiastical dances.

Hence **orgia·stically** *adv.*

1930 A. Bennett *Imperial Palace* xliii. 310 A grand climacteric of display designed orgiastically to receive the New Year into the infinite succession of years. **1934** [see

*COLOURISTIC *a.*]. **1961** J. Heller *Catch-22* (1962) xxiii. 238 Yossarian and Dunbar were busy in a far corner pawing orgiastically at four or five frolicsome girls and six bottles of red wine. **1965** *Eng. Stud.* Feb. 28 We may get nearer the truth by seeing Venus..as having fused orgiastically with Adonis.

orgone (ǭ·ɹgōᵘn). [f. ORG(ANISM, ORG(ASTIC *a.* + *-one* as in *HORMONE.] In the psycho-analytical theory of Wilhelm Reich (1897–1957), a vital energy or life force which supposedly informs the universe and can be collected and stored in an *orgone accumulator* or *box* for subsequent use in the treatment of mental and physical illnesses. Also *attrib.*

1942 T. P. Wolfe tr. *Reich's Discovery of Orgone* I. ix. 341 This energy, which is capable of charging non-conducting substance, I termed *orgone*. *Ibid.*, The orgone energy can be demonstrated visually, thermically and electroscopically in the soil, the atmosphere and in plant and animal organisms. **1942** —— tr. W. Reich in *Internat. Jrnl. Sex-Econ. & Orgone Res.* July 138/2 Our orgone therapy experiments with cancer patients consist in their sitting in an orgone accumulator. *Ibid.* 143/2 The patient left the orgone box. **1948** —— tr. *Reich's Discovery of Orgone* II. iv. 95 The orgone accumulator consists of an outer wall of organic material such as wood or celotex and an inner wall of sheet metal. **1949** *Orgone Energy Bull.* Apr. 95 *Physical orgone therapy*, application of physical orgone energy concentrated in an orgone accumulator to increase the natural bio-energetic defenses of the organism against disease. *Ibid.*, *Psychiatric orgone therapy*, mobilization of the orgone energy in the organism, i.e., the liberation of biophysical emotions from muscular and character armorings and the establishment of orgastic potency. **1952** M. McCarthy *Groves of Academe* (1953) ii. 24 A senior girl's voice, plaintive, 'Dr. Mulcahy, really, do we have to believe in orgones?' **1955** W. Gaddis *Recognitions* i. v. 194 Max was discussing orgone boxes as though he had lived in one all of his life. **1957** J. Kerouac *On Road* (1958) 152 Why don't you fellows try my orgone accumulator? **1959** N. Mailer *Advts. for Myself* (1961) 295 God who is It, who is energy, life, sex, force, the Yoga's *prana*, the Reichian's orgone. **1973** A. S. Neill *Neill! Neill! Orange Peel!* ii. 141, I could not understand Reich's theory of Orgone Energy... He had a small motor which was charged by an orgone accumulator. **1973** *Sat. Rev. World* (U.S.) 6 Nov. 4/1 Reichian orgone therapy.

orgonity (ǭɹgǫ·niti). [f. *ORGON(E + -ITY.] (See quot. 1949.) So **orgono·tic** *a.*, **orgono·tically** *adv.*

1942 T. P. Wolfe tr. *Reich's Discovery of Orgone* I. ix. 342 Many biologists..have observed the blue coloration of frogs in sexual excitation, or a bluish light emanating from flowers; we are dealing here with the biological (orgonotic) excitation of the organism. **1942** —— tr. W. Reich in *Internat. Jrnl. Sex-Econ. & Orgone Res.* Nov. 205/2 Mix on a microscopic slide biologically (i.e., orgonotically) weak blood with rot bacteria or T-bacilli. **1945** —— in *Ibid.* Apr. 19/2 Protozoa form in the organism only in the case of orgonotic weakness in the respective organs, and that they disappear with strong orgonity. **1949** *Orgone Energy Bull.* Apr. 95 *Orgonity*, the condition of containing orgone; the quantity of orgone contained. **1953** W. Reich *Murder of Christ* ii. 17 The qualities of the freely functioning orgonotic living system..bear out this mystified religious inkling of a basic truth.

orgonomy (ǭɹgǫ·nǒmi). [f. as prec. + -NOMY.] The study or investigation of 'orgone'. Hence **orgono·mic** *a.*, pertaining or relating to orgonomy; **orgo·nomist,** one who practises orgonomy.

1949 *Orgone Energy Bull.* Jan. 23 (*heading*) The First Orgonomic Conference at Orgonon, August 30 to September 3, 1948. *Ibid.*, On Sunday evening, Aug. 29, 1948, 35 physicians, educators, and laboratory workers gathered in the laboratory at Orgonon, Rangeley, Maine, for a 4-day conference in the field of orgonomy. *Ibid.* 27 Reich concluded by saying..orgonomy represented a new way of thinking and a new science. **1950** *Ibid.* Apr. 93 It happens again and again that a physician who has not finished his training in medical orgonomy, or has never even entered training, poses as a trained medical orgonomist. **1953** W. Reich *Murder of Christ* 200 Orgonomy.. is the factual comprehension of the universal 'Cosmic Orgone Energy'. **1964** *Parade* (Austral.) Mar. 47/2 Indeed they came in such numbers that the Austrian [*sc.* Wilhelm Reich] was forced to put on a staff of orgonomists—regular doctors who saw quick profits in this new branch of medicine. **1969** P. A. Robinson *Freudian Left* 59 The student who has immersed himself in Reich's early work will find the science of Orgonomy curiously familiar.

orgulous, *a.* (Add examples in modern literary use.)

1922 Joyce *Ulysses* 383 Then spoke young Stephen orgulous of mother Church that would cast him out of her bosom. **1928** V. Woolf *Orlando* i. 46 There was an orgulous credulity about him which was pleasant enough. *Ibid.* iv. 151 A covey of swans floated, orgulous, undulant, superb. **1929** Wyndham Lewis *King Spider* (1930) iv. 227 Charles, baffled here, turns his eyes elsewhere, filled with orgulous dreams. His imagination and his early successes have turned his head. **1941** Auden *New Year Let.* 187 That the orgulous spirit may while it can Conform to its temporal focus with praise. **1946** E. Linklater *Dark of Summer* 60 Coloured prints..all were bright, fantastic, orgulous—and serenely defiant of war and the cold Atlantic. **1976** M. Spark *Takeover* x. 147 This confidence ..frequently over-rides with an orgulous scorn any small blatant contradictory facts.

oribatid (ǫri·bǎtid), *sb.* and *a.* [f. mod.L. family name *Oribatidæ*, f. generic name *Oribata* (P. A. Latreille in C. N. S. Sonnini *Buffon's Hist. Nat. Insectes* (1802) III. 65), perh. f. Gr. ὀρειβάτης mountain-ranging: see -ID.³] **A.** *sb.* A small, oval, dark-coloured mite belonging to the family Oribatidæ or the order Oribatoidea, which includes non-parasitic mites with a thickened integument, giving them a leathery appearance. **B.** *adj.* Of or pertaining to this group of mites.

1875 *Encycl. Brit.* II. 276/1 The Oribatides in general live on vegetable matter. **1914** *Brit. Mus. Return* 170 in *Parl. Papers* LXXI. 193 Thirty-three Oribatid Mites from Hawaii. **1924** *Glasgow Herald* 12 Jan. 4/2 The hard-shelled 'beetle-mites', or Oribatids feed on decaying vegetable matter. **1967** M. E. Hale *Biol. Lichens* vii. 101 The lichens..are in turn infested with oribatid mites that appear to be lichenivorous. **1972** J. Balogh *Oribatid Genera of World* 9 The Oribatid mites are one of the richest soil Arthropodan groups. *Ibid.* 14 Permanent mounts with framing are wholly unsuitable for the study and conservation of Oribatids. *Ibid.* 15 The Oribatid specimen to be studied is carefully lifted from the lactic acid.

o·rielled, *a. rare.* [f. ORIEL + -ED².] Provided with oriels (sense 2).

1905 *Westm. Gaz.* 4 Nov. 6/2 Tawny sunlight works bright wizardries In orielled cloisters.

oriency. Add: Also **orience.** *poet.*

c **1865** G. M. Hopkins *Poems* (1967) 120 Once it was scarce perceivèd Lent For orience of the daffodil. *Ibid.* 139 The heightening dawn with milky orience Rounds its still-purpling centreings of cloud.

orient, *sb.* Add: **6. Orient Express,** the name of a train which ran (from 1883 to 1961) between Paris and Istanbul and other Balkan cities, via Vienna, and of its successors (see quots. 1961, 1977). Also *attrib.* in *fig.* sense, in allusion to its association with espionage and intrigue.

1883 *Times* 2 Nov. 10/1 A small folded card,..the back giving the timetable of the journey up to Constantinople, ..and the front, under the heading 'Orient Express', the direction 'M.—is requested to take his seat, in carriage —, bed No. —'. **1904** A. E. Housman *Let.* 23 Sept. (1971) 75, I *can* pay the sum they ask, but I very much object to, as Constantinople and the Orient Express are both pretty expensive. **1920** *Cornh. Mag.* Jan. 23 From Paris onwards, my three days journey was happily in the diplomatic train, the one-time Orient Express. **1925** C. Connolly *Let.* May in *Romantic Friendship* (1975) 81, I was thinking of..vanishing on the orient express. **1937** E. Ambler *Uncommon Danger* i. 24 He could see a destination board on..one of the sleeping cars—Wien, Buda-Pesth, Belgrade, Sofia, Istanbul. The Orient Express looked warm and luxurious inside. **1960** O. Manning *Great Fortune* i. 5 The day before had been spent on familiar territory, even if the Orient Express had kept to no schedule. **1961** *Guardian* 29 May 11/3 The Orient Express is no more. After 78 years of existence the train..is today [May 28] on its last journey from the Gare de l'Est station here [Paris] to Bucharest. **1965** *Observer* 16 May 9/1 Amateurish-sounding Orient Express techniques..are widely used... Master agent Lonsdale passed messages via drawing pins stuck..in a Lyons Corner House. **1977** *Daily Tel.* 2 May 6 (Advt.), Thursday, 19th May, 1977, will mark the death of a legend... On that day..*The Orient Express*..will begin its final run... On 22nd May the great train comes to rest in Istanbul's Sirkeci Station. **1977** *Ibid.* 19 May 10/8 The train, known since 1969 as the 'Direct-Orient', goes..to Belgrade, where first- and second-class coaches, including sleepers, separate. One section then heads for Athens and the other for Istanbul. After tonight there will be no through coaches to Athens or the Bosphorus... All that will be left..will be the Simplon Express from Paris Gare de Lyon to..Belgrade. The sole survivor of the original northern route, still wistfully and only symbolically labelled 'Orient Express', will run from the Gare de l'Est in Paris to Budapest and Bucharest.

orient, *v.* Add: **1. b.** (Further examples.)

1896 *Science* 3 July 11 We are now at a loss to orient the several parts of the cranium. **1926** C. E. Mulford *Cassidy's Protégé* xiii. 170 Hesitating for a moment while he oriented the report, he started toward the edge of the hill-top. **1953** E. Lynam *Mapmakers' Art* ii. 48 All early maps were oriented with the East at the top (whence our words 'to orient'). **1965** *Orienteering* ('Know the Game' series) 31 Once a map has been 'set' or oriented an orienteer should be aware of his position at all times by relating the map to the ground over which he is moving. **1972** N. J. W. Thrower *Maps & Man* vii. 102 A network of fundamental survey lines oriented predominantly in cardinal directions.

2. (Further examples.) Also, to assign or give a specific direction or tendency to.

1940 W. Faulkner *Hamlet* 223 So he held himself still .., trying to orient himself by looking back up the slope, to establish whether he was above or below the tree, to the right or left of it. **1952** G. Sarton *Hist. Sci.* I. xxii. 579 The orator's art consist[s] in shaping and orienting the passions of the people who listen to him. **1972** *Sci. Amer.* Dec. 6/2 The program is oriented toward the long-range goal of providing small power sources, automobiles included, with nonpolluting synthetic fuels. **1977** D. Bennett *Jigsaw Man* iv. 88 He had recognised the lie of the land... He was fully oriented.

4. *Chem.* **a.** *intr.* Of a substituent in a ring: to direct atoms and groups to a specified position in the ring when they enter it as substituents.

1924, 1937 [implied in *ORIENTING below]. **1949** English & Cassidy *Princ. Org. Chem.* vi. 106 Why should one group orient predominantly ortho and para, and another predominantly meta? **1971** [see *ORTHO *a.* (*adv.*) 1].

b. *trans.* To ascertain the relative positions of the substituents in (a ring or a cyclic compound).

1941 F. E. Ray *Org. Chem.* xv. 375 To prove the structure (orient the ring) of an unknown di-substituted compound. **1958** Read & Gunstone *Text-bk. Org. Chem.* xxiv. 419 After sound methods of orienting substituted benzenes had been devised it became possible to study more satisfactorily the substitution reactions concerned.

5. *trans.* To cause the molecules of (a plastic or other material) to assume a position in which their axes are parallel.

1958 W. D. Paist *Cellulosics* xi. 252 Considerable enhancement of the physical properties of many resin films has been realized on biaxially orienting the formed film. **1969** W. R. R. Park *Plastics Film Technol.* ii. 26 Virtually any thermoplastic material can be oriented.

Hence **o·rienting** *vbl. sb.* and *ppl. a.*; *spec.* in *Chem.* (cf. sense *4 a).

1727–41 [in Dict. s.v. ORIENT *v.* 1 a]. **1924** *Jrnl. Chem. Soc.* CXXV. 1377 (*heading*) The orienting influence of the thiocyano-group in aromatic compounds. **1937** F. C. Whitmore *Org. Chem.* III. 723 Benzotrichloride and other meta orienting compounds. **1962** F. I. Ordway et al. *Basic Astronautics* iv. 159 (*caption*) Magnetic field measuring devices carried by Soviet Sputniks, showing orienting apparatus. **1965** *Orienteering* ('Know the Game' Series) 32 The engraved arrow inside the compass housing points to the top of the map and the orienting lines lie parallel to the grid lines. **1977** 'A. Stuart' *Snap Judgement* 167, I did some orienting..by looking out of the [helicopter] window.

orientable (ǭə·riĕntǎb'l), *a.* [f. ORIENT *v.* + -ABLE.] Capable of being oriented; in *Math.* [tr. G. *orientierbar*], applied to a surface for which it is possible, if each point is regarded as surrounded by a small closed curve, to assign a sense (clockwise or anticlockwise) to each curve so that they are the same for all points sufficiently close together; not non-orientable; also used analogously of spaces of higher dimension.

1935 A. P. Herbert *What a Word!* iii. 85 One of our great motor-manufacturers advertises 'A very neat orientable anti-glare visor.' **1949** S. Lefschetz *Introd. Topology* ii. 82 Two orientable connected closed surfaces are homeomorphic if they have the same genus. **1952** P. Nemenyi tr. *Hilbert & Cohn-Vossen's Geom. & Imagination* vi. 306 It can be demonstrated that all two-sided surfaces are orientable. **1960** L. Picken *Organization of Cells* vii. 265 Cleveland's material suggests only that in hypermastigine flagellates the centrioles utilize all orientable material. **1965** tr. *Lietzmann's Visual Topology* 120 A one-sided surface is not orientable. **1968** A. H. Wallace *Differential Topology* vi. 79 The sphere is orientable but the projective plane is not. **1975** W. M. Boothby *Introd. Differentiable Manifolds* v. 215 A manifold *M* is orientable if and only if it has a covering..of coherently oriented coordinate neighborhoods.

Hence **o:rientabi·lity,** the property of being orientable.

1949 S. Lefschetz *Introd. Topology* ii. 76 Orientability implies that the triangles of *K* may be 'oriented' (in an intuitive sense) so that adjacent triangles always have their orientations disposed as in Fig. 37. **1956** E. M. Patterson *Topology* i. 9 The idea of orientability is derived from the physical idea of two-sidedness. **1972** *Nature* 13 Oct. 387/1 Within general relativity it is necessary to impose time orientability on the E_4 manifold such that the arrows placed on timelike world lines agree in sign.

oriental, *a.* and *sb.* Add: **A.** *adj.* **3. b. oriental poppy,** a perennial poppy, *Papaver orientale*, with large scarlet flowers, native to western Asia; (earlier and later examples).

1731 P. Miller *Gardeners Dict.* s.v. Papaver, *Papaver; Orientale*... Very rough Oriental Poppy, with a large Flower. **1963** W. Blunt *Of Flowers & Village* 139, I think the oriental poppy..is the most exciting of them all.

c. *Oriental stitch*, substitute for def.: a long straight stitch tied down with a short diagonal stitch in the centre. (Further example.)

1900 Day & Buckle *Art in Needlework* 66 Oriental-stitch, sometimes called 'Antique-stitch', is a stitch in three strokes, just as feather-stitch is a stitch in four.

d. Other Special Combs.: *Oriental carpet, rug*: a hand-knotted carpet or rug made to one of various designs in the Orient; a carpet or rug made to a similar design elsewhere; *Oriental Jew*, a Jewish person from the Middle or Far East, esp. from Yemen, Ethiopia, Iraq, or India; *Oriental Languages*, Eastern languages; these as a subject of university study; *oriental-looking* ppl. adj.; *Oriental Lowestoft*: name given to Oriental porcelain erron. thought to have been made or decorated at Lowestoft, England; see sense *B. 4.

1868 C. L. Eastlake *Hints Household Taste* 267/2 Oriental carpets. **1894** *Country Gentlemen's Catal.* 114 *Oriental carpets and rugs*. In Selected Designs and Colourings, Imported direct. **1972** *Guardian* 8 Sept. 11/5

ORIENTALIA

You'll find an Aladdin's cave crammed full of authentic, handmade Oriental and Persian carpets. **1938** R. T. Feiwel *No Ease in Zion* xxi. 298 One-fifth of Tel Aviv consists of Oriental Jews. **1961** L. Finkelstein *Jews* II. xxv. 1179 Shakespeare's *Comedy of Errors*,..appealed greatly to the imagination of the Oriental Jew. **1968** Mrs. L. B. Johnson *White House Diary* 7 Feb. (1970) 628 Some interesting excerpts: Between 60 and 65 percent of the people of Israel are 'Oriental' Jews. **1822** M. Edgeworth *Let.* 23 Jan. (1971) 334 We have just walked to see Hertford College... There are eight professors—two for classical literature—three Oriental languages, [etc.]. **1970** M. Kelly *Spinifex* i. 23, I went up to Cambridge, doing Oriental Languages. **1972** 'J. Bell' *Death of Poison-Tongue* i. 8, I have come to Polford to do Oriental languages. **1869** 'Mark Twain' *Innoc. Abr.* viii. 79 A ragged, oriental-looking negro. **1964** P. F. Anson *Bishops at Large* viii. 281 This long-bearded, oriental-looking prelate. [**1866** W. Chaffers *Marks Pott. & Porc.* (ed. 2) 317 There is such a peculiarity in the form and quality of the Lowestoft porcelain that we are surprised any one at all conversant with..collections of china, could ever mistake it for Oriental.] **1949** G. Savage *Ceramics for Collector* ii. 41 The former belongs to the 'Oriental Lowestoft' or 'East Indian China' group. **1971** L. A. Boger *Dict. World Pott. & Porc.* 67/1 It [*sc.* Chinese Lowestoft] is also called Chinese Export Porcelain and Oriental Lowestoft. **1974** Savage & Newman *Illustr. Dict. Ceramics* 208 *Oriental Lowestoft*, an erroneous term, first given currency by W. Chaffers.., for the enormous quantity of porcelain made in the 18th century in China for export to Europe. **1881** C. C. Harrison *Woman's Handiwork* III. 138 Oriental rugs are so generally used. **1931** A. U. Dilley (*title*) Oriental rugs and carpets. **1966** M. G. Eberhart *Witness at Large* (1967) vii. 95 The wide hall upstairs had faded oriental rugs placed at spots almost sure to trip anybody. **1976** J. van de Wetering *Corpse on Dike* ii. 20 A table, covered with a thick oriental rug.

B. *sb.* **4.** Denoting a variety of porcelain imported from China by European countries from *c* 1700 to *c* 1835; also known as *Oriental Lowestoft, Chinese Lowestoft, Chinese Export Porcelain.* Also *attrib.*

1863 W. Chaffers *Marks Pott. & Porc.* 134 Brameld. This mark is in red, on porcelain vases, in imitation of Oriental. **1873** C. Schreiber *Jrnl.* (1911) I. 201 A collection of choice specimens of Oriental. **1926** [see *Minton].

5. Used *ellipt.* for *Oriental carpet, pattern, rug*, etc.

1897 *Sears, Roebuck Catal.* 220/2 Extra Fine Lace Back Suspenders... A magnificent assortment of patterns. Persians, Orientals, Dresdens. **1938** I. Goldberg *Wonder of Words* v. 91 The noun *oriental* has ceased, or half-ceased, to mean a rug woven in the Orient; it has come to mean a rug of a certain design and coloring. **1969** M. G. Eberhart *Message from Hong Kong* xvii. 152 The rugs in the hall were old Orientals, worn thin too, but still glowing in reds and blues. **1972** E. Berckman *Fourth Man on Rope* i. 19 On its polished floor-boards lay a thin faded Oriental, once a very good one. **1977** C. McFadden *Serial* (1978) xxx. 67/2 Martha..began to pull Kate unsteadily across the Oriental.

Orientalia (ōᵊri,entē̆i·liă), *sb. pl.* Also **orientalia.** [mod.L., neut. pl. of L. *orientālis* oriental.] Things, esp. books, relating to or characteristic of the Orient.

1916 *Asiatic Rev.* VIII. p. iii, (*Index*) Orientalia. **1928** H. Crane *Let.* 28 Mar. (1965) 322, I enjoyed your historical notes and orientalia. **1932** *N. & Q.* 16 Jan. 35 (Advt.), *Books, prints, autographs*... No. 534. Orientalia. **1973** *Country Life* 20 Sept. (Suppl.) 73 19th Century Orientalia. **1975** *Sat. Rev.* (U.S.) 22 Mar. 57/2 Gumps, San Francisco—celebrated for *objets d'art*, orientalia, china, glass, jade.

orientalizing, *vbl. sb.* and *ppl. a.* (In Dict. s.v. Orientalize *v.*) Add: *spec.* designating a style of Greek art, or the period to which it is dated (*c* 750–*c* 650 B.C.), in which influences from the art of the Near East are discernible.

1902 *Encycl. Brit.* XXV. 574/2 From Ionia the style of vase-painting which,..may best be termed the 'orientalizing', spread to Greece proper. **1939** J. D. S. Pendlebury *Archaeol. Crete* vi. 335 Courby, in his study of such vases, divides them into three groups... Orientalizing, which he dates from 750 to 650. *Ibid.* 336 In the Orientalizing Period a number of important works of art in bronze was produced in Crete. **1948** [see *black-figure* s.v. *black *a.* 19]. **1950** H. L. Lorimer *Homer & Monuments* ii. 74 Late Geometric and Early Orientalizing graves. **1960** T. Burton-Brown *Early Mediterranean Migrations* iii. 74 There was a group which..knew the same kinds of procedure in architecture and sculpture, as the Greeks used from the Orientalizing Period. **1973** P. Green *Conc. Hist. Anc. Greece* 55 (*caption*) The domestication of mythical and other beasts was typical of 'Orientalizing' art.

orientate, *v.* Add: **3.** *trans.* (*Chem.*) = *orient *v.* 4 b.

1924 E. J. Holmyard *Outl. Org. Chem.* xix. 368 When several compounds have been orientated in this way, the constitution of other substances may be ascertained by converting them into substances of known constitution. **1926** J. Read *Text-bk. Org. Chem.* xxiv. 550 Such multitudes of benzene derivatives have been orientated that it is a comparatively simple matter to apply this method.

o·rientated, *ppl. a.* [f. Orientate *v.* + -ed[1].] **1.** = *oriented *ppl. a.* 1.

1886, 1900 [see Orientate *v.* 1].

2. = *oriented *ppl. a.* 3 (and similarly hyphenated).

1967 *Indexer* V. 162/2 We must, as far as possible, be customer- or user-orientated. **1968** *Listener* 1 Aug. 153/1 Stravinsky's Webern-orientated style. **1971** *Guardian* 18 Oct. 8/1 Polytechnics..are too big, too static, too institutional, too degree-orientated. **1972** C. Jones *Introd. Middle English* 4 Earlier works on the subject of medieval English language..have tended to be performance-orientated. **1974** *Cape Times* 1 Aug. 1/8 The highest percentage of votes appeared to be recorded at the Progressive Party-orientated..polling districts. **1975** *Daily Tel.* 26 July 11/3 His attitude..has been condemned as 'irresponsible and politically orientated'.

o·rientating, *ppl. a.* [f. as prec. + -ing[2].] That orientates or orients; *spec.* in *Chem.* (cf. *orient *v.* 4 a).

1876 *Jrnl. Chem. Soc.* XXIX. 240 The author closes this section with some remarks on the value of the 'orientating' influence exercised by various radicles. **1920** *Christian World* 19 Aug. 7/1 Upon these young men and women the lecture must have had a great orientating effect. **1921** E. Herman *Creative Prayer* 104 For that world of reality ..is..Love, and its highway—the great orientating path that gives it coherence—is Christ. **1952** *Mind* LXI. 484 The use of warning, priming or orientating signals. **1966** G. P. Ellis *Mod. Textbk. Org. Chem.* v. 101 The substituent A possesses a specific directing or orientating effect on the incoming group.

orientation. Add: **2. c.** Also, = Orientalization.

1914 G. K. Chesterton *Flying Inn* viii. 81 He also wants to drive a tunnel—between East and West—to make the British Empire more Indian; to effect what he calls the orientation of England and I call the ruin of Christendom.

4. b. *Chem.* The process of ascertaining the relative positions of the substituents in a ring.

1891 *Jrnl. Chem. Soc.* LX. 1199 The method employed by Claus and Runschke..for the orientation of 4:6-dichlorometaxylene. **1903** Walker & Mott tr. *Holleman's Text-bk. Org. Chem.* II. 473 Oxidation is another important aid in their orientation, and is employed to determine whether the substituents are attached to the same or to different rings. **1953** Astle & Shelton *Org. Chem.* xxii. 420 (*heading*) Korner's absolute method of orientation.

5. b. An introduction to some subject or particular situation; a briefing. Also *attrib.*, as (U.S.) *orientation course, program* (see quots.).

1942 in Partridge *Usage & Abusage* 226/2 *Orientation course*, American pedagoguese for an introductory, general or historical study, usually of the social sciences, designed for college freshmen or sophomores. **1953** K. Reisz *Technique Film Editing* II. viii. 140 Had we, for instance, opened the sequence with the long continuous scene of the bearded forest (as an *orientation* scene of the locale in which the tale was set) we would have no preparation to understand and appreciate its charms and mysteries. **1968** *Globe & Mail* (Toronto) 17 Feb. B6 (Advt.), Selected applicants will be offered a comprehensive orientation program in branch banking. **1968** *N.Y. Times* 23 July 41/1 Mr. Mailer was giving an 'orientation' (or was it a sophisticated party game?) for nearly 100 participants in his third film venture. **1970** *Toronto Daily Star* 24 Sept. 2/2 A student orientation program at the University of British Columbia. **1972** D. Delman *Sudden Death* (1973) iii. 73 'I needed..an insight into the way you tennis cats think.' 'In other words I've just delivered what amounts to an orientation lecture.' **1976** *Columbus* (Montana) *News* 10 June 4/2 Four prospective LABO host families met Sunday evening at the home of Mr. and Mrs. Bill Wright to go over an orientation program and participate in a Japanese style dinner. **1976** J. Crosby *Nightfall* xxxviii. 231 Hawkins had read Wittgenstein only because Theresa had. Her books were his orientation course.

6. *Chem.* The orienting effect of a substituent in a ring on other atoms or groups (see *orient *v.* 4 a).

In quot. 1890 the word could be interpreted in sense 2 b.

1890 *Jrnl. Chem. Soc.* LVIII. 484 The study of substitution phenomena, especially in the aromatic series, shows that the so-called orientation rules are dependent on the atomic or molecular weight of the atom or radicle which dominates or directs the position taken up by the substituting-group. **1946** A. A. Morton *Chem. Heterocyclic Compounds* ii. 33 Replacement reactions are unique in that the position of the entering group is largely determined by the nature of the reactant, not by any orientation by groups. **1971** J. D. Roberts et al. *Org. Chem.* xx. 574 When the two substitutents have opposed orientation effects, it is not always easy to predict what products will be obtained.

7. Special Comb.: **orientation triad** (see quot. 1962).

1953 H. Haber *Man in Space* 155 If all three components of the orientation triad are intact, the human body is fully equipped to reckon with the force of gravity, to keep its balance and to remain properly aligned relative to the vertical. **1962** F. I. Ordway et al. *Basic Astronautics* xii. 475 The center of the body's orientation system is located in the inner ear; the system, however, consists of three elements, often called the orientation triad. The first component is sight, and the second is the system of mechanoreceptors or nerve endings..that are sensitive to pressure. But the vestibular apparatus of the inner ear is the heart of the system since it contains the mechanism that senses acceleration.

orientational (ōᵊri,ĕntē̆i·ʃənăl), *a.* [f. Orientation + -al.] Pertaining to or involving orientation, esp. of variable elements in a specified context.

1952 [see *cognitive *a.*]. **1962** Corson & Lorrain *Introd. Electromagn. Fields* iii. 113 We then considered orientational polarization in which molecules with a permanent dipole moment tend to be aligned by an external field. **1968** J. Lyons *Introd. Theoret. Linguistics* vii. 275 The notion of *deixis*..is introduced to handle the 'orientational' features of language which are relative to the time and place of utterance. **1974** R. Jessop *Traditionalism, Conservatism & Brit. Pol. Culture* i. 17 Most political scientists employ 'political culture' as a mere catchword.. for all sorts of influences..that include both structural and orientational factors. **1976** *Nature* 23 Sept. 353/2 An extraordinary orientational relationship between the rhombohedral green rust..and magnetite..has been reported.

Hence **orienta·tionally** *adv.*

1975 *Nature* 31 Jan. 310/2 The molecules are positionally ordered but orientationally disordered and mobile.

oriented (ōᵊ·ri,ented), *ppl. a.* [f. Orient *v.* + -ed[1].] **1.** Having a definite or specified orientation; *spec.* in *Math.*

1918 O. Veblen *Projective Geom.* II. ix. 426 If two oriented points are similarly oriented with respect to a line *l*, they are similarly oriented with respect to a line *m* if and only if *l* and *m* do not separate the two points. **1950** D. J. Struik *Lect. Classical Differential Geom.* i. 6 The sense of increasing arc length is called the positive sense on the curve; a curve with a sense on it is called an oriented curve. **1951** C. Palache et al. *Dana's Syst. Min.* (ed. 7) II. 181 From Tsumeb..as a secondary mineral associated with smithsonite (in part as oriented growths thereon). **1960** L. Picken *Organization of Cells* vii. 265 Once we admit..that the centromeres..are 'sticky' to oriented proteins, we may have conceded all that is necessary for them to figure as centres of orientation. **1970** A. Goetz *Introd. Differential Geom.* i. 6 The space with a chosen ordered triple e_1, e_2, e_3 of independent vectors is called an oriented space. **1975** *Nature* 29 May 389/1 A shipboard palaeomagnetic reconnaissance of 295 vertically oriented basement samples.

2. Of a plastic or other material: having the constituent molecules oriented with their axes parallel to one another.

1947 R. Nauth *Chem. & Technol. Plastics* viii. 195 Table 32..indicates the qualities of 'oriented' and 'unoriented' vinylidene chloride plastics. **1950** V. Stannett *Cellulose Acetate Plastics* i. 17 If a poorly oriented fibre is loaded only those chains lying parallel to the axis take the load. **1969** L. S. Mounts in W. R. R. Park *Plastics Film Technol.* v. 139 Oriented polystyrene films have excellent clarity, sparkle, and gloss.

3. Having an emphasis, bias, or interest indicated by a preceding sb. (usu. joined by a hyphen) or advb.

1950 *Psychiatry* XIII. 181 Our language patterns..are adult-oriented. **1957** *Jrnl. Nervous & Mental Dis.* CXXV. 459/2 The 'disease-oriented' physician and his 'person-oriented' fellow commonly encounter difficulties in agreeing. **1960** R. K. Webb *Harriet Martineau* x. 295 Our psychologically oriented, relativistic age. **1968** *Globe & Mail* (Toronto) 17 Feb. B7 (Advt.), Multi-faceted and expansion oriented company with operations in Eastern Canada and United States. **1971** *Computers & Humanities* VI. 30 sims will provide a user-oriented language, making it easy for the user to define, validate, interrogate, and analyze the data. **1971** M. McCarthy *Birds of America* 269 Why are you so art-oriented, all of a sudden? **1972** *Lebende Sprachen* XVII. 134/1 Environmentally oriented research. **1974** *Times* 12 Feb. 11/5 We simply must produce managers in the future who are design oriented. **1978** *Jrnl. R. Soc. Arts* CXXVI. 744/2 To move away from performance-oriented drama towards more reflective drama.

orienteering (ōᵊri,ĕntīᵊ·riŋ). [ad. Sw. *orientering* orienteering; cf. Orient *v.*] The competitive sport of finding one's way on foot across rough country with the aid of map and compass; this sport as undertaken on horse-back or by car, canoe, etc. Also *attrib.* So **orientee·r,** one who engages in orienteering; also as *v. intr.* and *fig.*

1948 *Amer. Ski Ann.* 1949 146/1 Senior Scouts representing 15 Boy Scout Councils from New York and New Jersey met at Snow Ridge Ski Center last winter to compete in the first official Ski Orienteering race to be held in America. **1949** *Univ. Mass. Executive Bull.* 8 Dec., Dr. Bjorn Kjellstrom, from Sweden, and instigator of the new sport of Orienteering will show a new colored film on training for Cross-Country and Touring Skiing. **1954** *Time* 1 Nov. 58 Known officially as 'orienteering', the sport dates back to 1918 when the first Swedish club was formed to hold formal competitions. **1956** *Official Gaz.* (U.S. Patent Office) 10 Apr. TM92/2 Bjorn Kjellstrom, d.b.a. American Orienteering Service, New York, N.Y... *Orienteering*. For Instructing in Map and Compass Reading Through Lectures Delivered in Person With or Without Illustrations by Means of Film. Use since August 1946. **1965** *Orienteering* ('Know the Game' series) 10 The orienteer must remember that peat hags can be very rough pasture. *Ibid.* 36 Ankle support is very important in rough orienteering country. *Ibid.* 40 There are fascinating opportunities to orienteer by canoe, cycle, pony or ski. **1971** 'D. Halliday' *Dolly & Doctor Bird* v. 60 You were supposed to be treating my blisters, not orienteering all over my torso. **1971** *Sunday Times* 10 Oct. 30 Furthermore, Swedish orienteers bled their way into medical history a few years ago following an epidemic of the disease. **1973** *Whole Earth Catal.* 23/1 Within recent years, Orienteering as a sport, in the form of 'Orienteering Races' has swept Europe... Orienteering events have become regular features in many athletic and outdoor clubs. **1976** *Observer* 26 Sept. 16/3 Egil Johansen, asked to single out the most compelling attraction of orienteering, says simply: 'It is the fellowship of the other orienteers.'

orientite (ōᵊri,e·ntəit). *Min.* [f. *Orient-e*, the name of the province in Cuba where it was first found + -ite[1].] A hydrated silicate of calcium and manganese, $Ca_4Mn^{III}{}_4Si_5O_{20}\cdot 4H_2O$, found as light brown or pink orthorhombic crystals.

1921 Hewett & Shannon in *Amer. Jrnl. Sci.* CCI. 491 As the mineral is known to occur in two localities in Oriente Province, where many manganese deposits are found, and it may be widespread in the region, it is appropriate that the geographic relation be perpetuated in the name *orientite*. **1961** *Amer. Mineralogist* XLVI. 227 The largest known crystals of orientite do not exceed 1 mm. in length.

orificial, *a.* Restrict †*Obs. rare*[−1] to sense in Dict. and add: **2.** Of or pertaining to an orifice or orifices; in *Med.* used *spec.* with reference to a theory that many pathological conditions arise from irritation of the orifices of the rectum and urogenital system and can be relieved by surgery or other treatment of these areas.

1887 E. H. Pratt *Orificial Surg.* i. 14, I believe that all forms of chronic diseases have one common predisposing cause, and that cause is a nerve-waste occasioned by orificial irritation at the lower openings of the body. **1926** *Spectator* 25 Sept. 473/1 With a fine inconsistency, they belong to societies of so-called 'orificial surgery' and follow strange cults of electrical healing. **1960** [see *clitoridectomy]. **1973** *Biol. Abstr.* LV. 6287/1 This..revealed an orificial pulmonary stenosis which clinical investigation had failed to detect.

origami (ǫrigā·mi). Also **origame.** [Jap., f. *ori* fold + *kami* paper.] The Japanese art of folding paper into intricate designs. Also *attrib.*

[**1922** F. Starr in *Japan* (San Francisco) Oct. 43/1 Their book on paper-folding in schools compares favorably with any we have. It is entitled *shikaka origami dzukai*, paper-folding explained with figures.] **1956** 'R. Harbin' *Paper Magic* 14 The art of *origami* has been handed down from father to son through countless generations. **1959** R. Condon *Manchurian Candidate* (1960) ii. 56 While they had light he..amused them or startled them or flabbergasted them with the extent of his skill at *origami*. **1961** E. Kallop in S. Randlett *Art of Origami* (1963) 16 Apart from origami as an art in the sense of the individually unique, folded paper has a role in the ceremonial etiquette of Japanese life. **1963** 'R. Harbin' *Secrets of Origami* 11 If you can obtain a supply of Japanese Origami paper, so much the better. **1968** R. V. Beste *Repeat Instructions* vii. 67 You should try origame..Paper-folding. It's a Jap word. **1972** C. Fremlin *Appointment with Yesterday* xi. 83 The Origami cut-outs they'd had such a craze for over Christmas, they were on the bed too. **1973** M. Crowell *Greener Pastures* 101 There are..paper stars and origami birds.

Origenian, *a.* (Earlier and later examples.)

1661 G. Rust *Origen* 19 A perfect explication of the Trinity after the Origenian way. **1879** R. Ornsby in *Dubl. Rev.* July 64 A fourth branch of the Origenian evidence is prophecy and its fulfilment.

Origenism. (Later examples.)

1903 W. Bright *Age of Fathers* II. xxix. 54 Anastasius ..condemned Origenism in a Roman synod. **1908** L. B. Radford (*title*) Three Teachers of Alexandria: Theognostus, Pierius and Peter. A Study in the Early History of Origenism and Anti-Origenism. **1960** H. C. Graef tr. *Altaner's Patrology* v. 242 He [*sc.* Methodius] is important principally as a successful opponent of Origenism. **1966** H. Chadwick *Early Christian Thought & Classical Tradition* iv. 120 In judging the system of Origen as a whole it is important to remember that some of the most characteristic features of 'Origenism' are not his personal invention.

original, *a.* and *sb.* Add: **A.** *adj.* **4. c.** *original print*, a print made directly from a master image on wood, stone, metal, etc., which is executed by the artist himself, printed by him or under his supervision and, in recent times, usually signed by him.

1961 *What is an Original Print?* (Print Council of Amer.) 9 An *original print* is a work of graphic art, the general requirements of which are: 1. The artist alone has made the image in or upon the plate, stone, wood block or other material, for the purpose of creating a work of graphic art. 2. The impression is made directly from that original material, by the artist or pursuant to his directions. 3. The finished print is approved by the artist. *Ibid.* 12 The difference in the price commanded by an original print and a reproduction acknowledged as such is largely a reflection of the difference in their aesthetic qualities. **1965** Zigrosser & Gaehde *Guide to Collecting Original Prints* ii. 14 When we speak of an *original* print, we mean that the artist both conceived and executed it. **1970** P. Gilmore *Mod. Prints* 7 Several committees between 1960 and 1965 tried variously to define, protect, and elevate the 'original' print, always equating it with artist handwork. **1970** *Studio Internat.* June 283/3 These [assemblage] prints look just as individual as most original prints. **1972** J. Heller *Printmaking Today* (ed. 2) p. v, The 'original' print has become a significant feature of contemporary life.

B. *sb.* **3. e.** An image or impression produced during an actual photographing or recording session from which copies may subsequently be made.

1918 H. Seymour *Reprod. Sound* 16 At first, every record sold to the public was an original, or what is technically described as a 'master', but means were soon found by which copies could be secured from the master. **1949** Frayne & Wolfe *Elem. Sound Recording* xiv. 266 Making the master from the original is a process involving several operations. **1970** A Fowles *Dupe Negative* xiv. 196 This type of film doesn't have a negative... The piece of film that actually runs through the camera is called the original..from which all subsequent prints are struck. **1971** L. B. Happé *Basic Motion Picture Technol.* ix. 277 (*caption*) Duplicates from reversal originals.

4. b. *spec.* in Fashion and *haute couture*, a garment specially designed by a couture house for exhibition in a collection, or a copy of such a garment made to order. Also *Mus.*, (usu. jazz), a piece written by the performer(s).

1946 B. G. Chambers *Keys to Fashion Career* x. 86 Partner and designer of the firm of Young Originals. **1957** M. B. Picken *Fashion Dict.* 238 *Original*, a garment designed in and produced by a couture house, bearing the label of the house. It is usually a duplication made to order of the model shown in the collection. Each order is called a 'repeat' by the couture house. **1966** *Crescendo* Oct. 22/3 The Monk Quartet was playing originals—'Hackensack', 'Rhythm-A-Ning' and 'Epistrophy'. **1967** *Melody Maker* 28 Jan. 15/5 The material is a nice mixture of originals, blues and ballads. **1975** R. H. Rimmer *Premar Experiments* (1976) ii. 195 My sister, wearing Pucci originals. **1976** *Observer* 22 Feb. 32/5 (Advt.), Anna Belinda announce that until February 28th they will continue to add a further distinction to their hand-made originals in silks, velvets and Liberty prints.

originality. Add: **1.** (Earlier example.)

1742 H. Walpole *Let.* 14 July (1903) I. 256 It is one of the most engaging pictures I ever saw. I have no qualms about its originality.

3. (Earlier example.)

1742 T. Gray *Let.* 24 May (1935) I. 206 My Lady of Queensbury is come out against my Lady of Marlborough; & she has her Spirit too, & her Originality, but more of the Woman, I think, than t'other; as to the Facts it don't signify two pence, who's in the right.

‖ **orihon** (ǫ·rihǫn). [Jap., f. *ori* fold + *hon* book.] A book formed by folding a printed roll alternately backwards and forwards between the columns, and usu. fastening it with cord down one side.

1907 C. Davenport *Book* ii. 28 The Chinese and Japanese..by help of the ancient device of 'stabbing' the flattened roll along one of its sides,..produce a form called an 'Orihon', easy to consult, strong. **1910** *Encycl. Brit.* IV. 216/2 A roll [of vellum, paper, etc.] of this kind can be folded up, backwards and forwards, the bend coming in the vacant spaces between the columns of writing. When this is done it..becomes a book, and takes the Chinese and Japanese form known as *orihon*—all the writing on one side of the roll or strip of paper and all the other side blank... The earliest fastening of such books consists of a lacing with some cord or fibre run through holes stabbed right through the substance of the roll, near the edge. Now the *orihon* is complete, and it is the link between the roll and the book. **1951** S. Jennett *Making of Bks.* xi. 155 This form of book was much used throughout the East, and was known as an orihon. **1960** G. A. Glaister *Gloss. Bk.* 287 *Orihon*, a manuscript roll on which the writing was done in columns running the short way of the paper with margins between each. The roll was then folded, the margins having the effect of a closed fan. **1968** E. G. Turner *Greek Papyri* 173 It has been suggested that the codex evolved out of an intermediate form of leather or skin roll in which stitches were inserted in the spaces between every second column, the skin being folded on itself at the intervening column, so that the whole roll was folded concertina-fashion. Books of this kind (termed *orihon*) are still in everyday use in China and Japan.

orinasal, *a.* (*sb.*) Add: The form **oronasal** is now usual. (Further examples.)

1882 *Brit. Med. Jrnl.* 1 July 13/1 Metal Respirators. 1. The principal forms made by Jeffries..are three: one for the mouth only..; a second.., for the mouth, with a scarf; and a third, for the mouth and nose (ori-nasal). **1888** Martindale *Extra Pharmacopœia* (ed. 5) 163 (*heading*) Oro-nasal inhalations. **1926** Jordan & Kindred *Textbk. Embryol.* xx. 407 The oronasal groove is obliterated by the fusion of the median nasal and maxillary processes to form the upper lip. **1938** *Jrnl. Aviation Med.* IX. 184/2 Since most aviators breathe through the nose, it would be necessary to use an oronasal type of mask only in the presence of nasal obstruction. **1960** J. J. Sharry in W. L. McCracken *Partial Denture Construction* xxi. 468 The palatal repair itself may look adequate, and yet we find that there may be an oronasal perforation in the labial mucobuccal fold. **1970** *Jrnl. Physiol.* CCVI. 22P (*heading*) Oronasal distribution of inspiratory flow during various activities.

Hence **orona·sally** *adv.*, by means of the mouth and nose.

1970 *Science* 15 May 858/2 In the course of previous experiments, more than 100,000 newborn..mice received reovirus 3 oronasally.

O-ring: see *O 6.

Orisha (ori·ʃă). [Etym. obscure (see quot. 1926).] A name given to a number of native deities of Southern Nigeria. Also *attrib.*

1926 P. A. Talbot *Peoples S. Nigeria* II. iii. 29 A hierarchy of Orisha (derived perhaps from 'ri', see, and 'sha', select—or from the Edo word Oyisa). **1929** A. C. M. Burns *Hist. Nigeria* ii. 39 A number of minor deities (Orishas) who are more directly interested in mundane matters. **1937** M. Perham *Native Admin. Nigeria* II. xii. 189 The House of the Orishas or sacred images stands in the market outside the official residence of the Ataoga. *Ibid.*, The visits to Orisha-houses and consultations of oracles. **1949** G. Parrinder *West Afr. Relig.* ii. 16 The chief divinities, generally non-human spirits, often associated with natural forces (called *abosom, vodū, orisha*). **1961** J. Jahn in A. Dundes *Mother Wit* (1973) 97/1 Without the drums it was impossible to call the orishas. *Ibid.* 98/2 The procedure which in the African orisha cult evokes ecstatic immobility..produces, in the Negro churches, 'mass ecstasy'. **1974** *Afr. Encycl.* 548/3 Many Yoruba people are Christians or Muslims, but still follow the traditional religion, which has several powerful Gods ..and many less important ones called 'orisha'. **1976** *Wilson Q.* Autumn 77 The African deities (*orishas*) became identified with various Christian saints.

Orissi (ori·si). *India.* [f. *Orissa* a state of eastern India.] = *Odissi.

1960 C. Fabri in *Mārg* (Bombay) XIII. ii. 5 Orissi.. bears great resemblance not only to Bharata Natya, but also to Kuchipudi, in Andrha, which may be described as 'next door'. **1962** B. Gargi *Theatre in India* 51 Orissi dance, practised in Orissa in South-East India, is recognised by scholars as an authentic classical dance... This dance style..claims to be over two thousand years old. **1965** E. Bhavnani *Dance in India* vi. 50 The Orissi dance has its roots in devotional ritual. **1971** *Femina* (Bombay) 2 Apr. 9/1 The first part of the programme was in Bharat Natyam, the second and third in the Kuchipudi and Orissi styles of dancing. **1974** *Encycl. Brit. Macropædia* XIII. 741/1 The classical dance of Orissa, known as the *orissi* dance, has survived for more than 700 years.

-orium, *suffix.* Add: Now used, esp. in America, in many, often hybrid, formations, as *barbatorium, bobatorium, healthatorium*, etc.

1925 *Amer. Speech* I. 38/2 *Barbatorium*, a barber shop. *Bobatorium*, a place where hair is bobbed. *Healthatorium*, synonymous for *sanatorium*. *Infantorium*, a sanatorium for infants... *Motortorium*, an automobile repair shop... *Suitatorium*, a place where suits are cleaned and pressed. [etc.] **1943** *Ibid.* XVIII. 71/1 Perhaps the following forms on the *-orium* ending have not previously been noted: *furnitorium* (a furniture store), *hairitorium* (a store dealing in wigs and hair goods), and *puritorium* (a Hebrew ritual bath). **1957** *Journal des Traducteurs* II. 49 In recent decades advertisers intent on catching the public eye have freely exploited such suffixes as *-orium*..to form *hairorium, meatorium, sportorium*. **1959** *Times* 28 Oct. 13/4 A market gardener in New South Wales..keeps a lonely stall bearing the notice 'Potatorium'. *Ibid.* 31 Oct. 7/7 Lubritorium (United States) is, inevitably, not Latin. *Ibid.* 2 Nov. 13/4 Thirty years ago there was a 'pantatorium' in Cambridge (Mass.), which pressed pants.. rapidly. **1963** R. I. McDavid *Mencken's Amer. Lang.* 221 The former [*-orium*] has given *lubritorium* (the lubricating rack in a filling station), *printorium, corsetorium, hotdogatorium, parentorium* (a parent guidance center), *puritorium* (a Jewish ritual bath) and *eatatorium*.

Oriya (orī·yă), *a.* and *sb.* Also **Ooreah, Ooriya, Uriya,** etc. [f. Skr. *Odra* name of a region of India.] **A.** *adj.* **a.** Of or pertaining to Odra, an ancient region of India corresponding to the State of Orissa. **b.** Of or pertaining to the State of Orissa, which takes its name from Odra. **B.** *sb.* **a.** A native of Odra. **b.** The Indo-European language of Odra, which is spoken widely in Orissa.

1801 *Asiatick Researches* VII. 225 Utcala or Ódradésa is co-extensive with the Subá of Órésá... The language of this province, and the character in which it is written, are both called Uríya... the Bráhmens of this province use the Uríya character in writing the Sanscrit language. **1831** A. Sutton *Introd. Gram. Oriya Lang.* p. vii, The Oriyas speak every word with the bold rusticity of an English countryman... A Bengallee can scarcely be met with who speaks Oriya, but he may instantly be detected by his peculiar mode of pronunciation. **1848** J. H. Stocqueler *Oriental Interpreter* 177/2 The *Ooreahs* are, in some respects, excellent servants; they are very careful of furniture. *Ibid.* 179/1 The language of the Ooreah nation is a dialect of the Sanscrit, much resembling the Bengalee, and called the Ooreah. **1855** H. H. Wilson *Gloss. Indian Terms* p. xxii, The Telugu..meeting on the north with Uriya, on the west with Maráthí and Karnáta. **1866** G. Campbell in *Jrnl. Asiatic Soc. Bengal* XXXV. II. (Suppl.) 52 They [*sc.* the 'Bhooyas']..speak Oorya on the Ooriah borders..and Hindee farther north. *Ibid.* A resemblance in appearance to the Ooryahs, among whom high cheek bones seem to prevail. **1872** W. W. Hunter *Orissa* I. v. 171 The Uriyá language held its own for centuries, almost to the walls of Kalingapatnam itself. **1873** E. Balfour *Cycl. India* (ed. 2) IV. 266/1 A monster snake..was worshipped by the Ooryah as a deity. *Ibid.* 267/1 In Vizianagram, Telugu is only spoken in the open country, and Urya in the mountains. **1903** G. A. Grierson *Linguistic Survey India* V. II. 367 It is called Oṛiyā, Ōḍrī, or Utkalī, that is to say the language of Ōḍra or Utkala, both of which are ancient names of the country now known as Orissa. *Ibid.* 368 The Oṛiyā verbal system is at once simple and complete. **1930** R. D. Banerji *Hist. Orissa* I. x. 136 The average Oriya..is usually dwarfish in stature and brownish black in complexion. **1954** Pei & Gaynor *Dict. Linguistics* 155 *Oriya*, an Indic language (also called *Odri*) spoken in Orissa, Bihar, Bengal, the eastern regions of the Central Provinces and in the northern part of the Madras Presidency, by a total of about 13,000,000 native speakers. **1971** *Femina* (Bombay) 30 Apr. 29/1 A student of Hindustani classical music, she knows many Oriya songs. **1972** W. B. Lockwood *Panorama Indo-European Lang.* 206 Oriya is the state language of Orissa, where it is the native medium of sixteen and a half millions. *Ibid.*, The Oriyas reached their present seats after advancing from the west.

ork (ǫɹk). *slang.* (orig. and chiefly *U.S.*). Also **orch.** [abbrev. of Orchestra.] An orchestra, *spec.* a jazz or dance band.

1936 *Metronome* Feb. 61/3 Orville Knapp and ork back in town... Cecily Riggs and ork home from the Santa Rita Hotel, Tucson engagement. **1937** in *Variety* 10 Nov. 58/3 Philly Orch on Thursday (11) night will present.. 'Stars Pool'. **1950** *Down Beat* 1 Dec. 13 (*title*) Ina Ray Ork looks good on TV; plays well, too. **1959** C. MacInnes *Absolute Beginners* 136 The Dickie Hodfodder ork, led by Richard H. in person, playing away merrily. **1972** *Jrnl. Jazz* 41/4 'Weeping Willow'—recorded in London backed..by Georgie Fame, Colin Green and the Norrie Paramour Ork!

Orkney (ǭ·ɹkni). The name of a group of islands off the north coast of Scotland, applied *attrib.* to various local animals and products; **Orkney sheep**, a small feral sheep distinguished by horns curving backwards and a brown, white, or speckled fleece; **Orkney vole**, a larger subspecies of the European vole, *Microtus arvalis orcadensis*, found only in the Orkney Islands.

[**1805** G. Barry *Hist. Orkney Islands* III. 319 The Sheep ... is a peculiar breed, and, from some features in its character, seems to have sprung from the same stock with those of Iceland, the Feroes, and Shetland.] **1862** Mrs. Beeton *Bk. Househ. Managem.* 320 The Leicestershire breed [of sheep] is the best example of this lymphatic and contented animal, and the active Orkney, who is half goat in his habits, of the restless and unprofitable. **1905** J. G. Millais *Mammals Gt. Brit. & Ireland* II. 279, I have noticed that both the Orkney and the Water Voles often possess a white tip to the end of the tail. **1912** H. J. Elwes in *Scottish Naturalist* 6 Orkney Sheep.. Sheep never seem to have been so important here as in Shetland. **1926** *Daily Colonist* (Victoria, B.C.) 7 Jan. 5/3 A group of smartly designed coats of the popular Orkney cloth in half a dozen beautiful shades. **1935** *Discovery* June 168/2 All our voles, including the Orkney and Skomer voles, seem as numerous as they have ever been. **1952** L. H. Matthews *Brit. Mammals* vi. 155 In captivity the Orkney vole is noticeable for its pugnacious disposition and its readiness to bite. **1966** T. Fitzgibbon *Art Brit. Cooking* 133 Orkney cheese is a creamy Cheddar type cheese made in the Orkney Islands, but exported to England. **1971** [see *invented plait* s.v. *invented *ppl. a.* 9]. **1972** *Country Life* 3 Aug. 273/1, I shall be grateful if you can give me any information regarding Orkney chairs... As the name suggests, this is a traditional Orkney kind of furniture. **1973** *Scotsman* 21 Feb. 10/1 Chewe straws made from Orkney cheese. **1974** J. M. Dorey et al. in P. A. Jewell et al. *Island Survivors* iv. 124 The litter size of these [Soay] lambs was 1.28, which is the same as..the Orkney sheep on North Ronaldshay. **1978** *Vole No.* 6. 34/2 Orkney is distinguished in having its own special vole, the Orkney vole, larger than its mainland cousin.

Orkneyman (ǭ·ɹknimæn). = Orcadian *sb.* A native or inhabitant of the Orkney Islands.

1775 Hearne & Turnor *Jrnls.* (1934) 191, I have interfered so far as to ask what encouragement I may expect in case the Orkneymen seem'd to intenate that 12l. per annum would induce them to be active & useful. **1842** *Trans. Lit. & Hist. Soc. Quebec* IV. 1333 The animals frequenting this country (include)..the Common Hare of Canada, called Rabbits, by the Orkney men in the service of the Hudson's Bay Company. **1936** *Beaver* (Winnipeg) Dec. 4/1 In the year 1799 about five hundred and thirty persons were employed by the Hudson's Bay Company, at their fur trade posts in North America, of whom four hundred and sixteen..were Orkneymen. **1956** V. Fisher *Pemmican* 50 He had heard that..there were..a few red-faced Orkneymen, a few Moravian sisters and brothers. **1961** J. W. Anderson *Fur Trader's Story* i. 1 For nearly two hundred years Orkneymen played a prominent part in the fur trade of Canada. **1969** G. M. Brown *Orkney Tapestry* 26 Hardly a thing is known about these first Orkneymen..apart from the monuments they left behind them.

orl (ǭl), *a.* and *adv.* Representing a 'phonetic' spelling of a vulgar pronunciation of All *a.* and *adv.*

1864 [see *O.K. *a.*]. **1898** J. D. Brayshaw *Slum Silhouettes* 14, I could 'ear the plates abreakin' pretty nigh orl night. **1923** 'R. Crompton' *William Again* xiv. 240 That..looks a bit of orl right. **1939** [see *O.K. *a.*]. **1955** M. Allingham *Beckoning Lady* ii. 31 Orl right, orl right, I'll go. **1971** *Listener* 9 Sept. 341/1 'To her our own quaint phraseology 'May bien, she was a bit of orlright. **1972** *Buster & Jet* 15 Jan. 33 Orl right, you ol' Scrooge! I'm meetin' some real tightwads today, it seems!

Orleanian (ǭɹlī·ăniăn). [f. *Orlean(s* + -ian.] An inhabitant of New Orleans in the United States.

1946 *New Orleans Times-Picayune* 23 Mar. 17/4 (*heading*) Orleanian tells of Jap tortures. **1947** in B. A. Botkin *Treas. S. Folklore* (1949) III. ii. 525 Perhaps the most famous of romantic Mardi Gras stories that Orleanians tell is that one concerning the 'ghost dinners' served each Shrove Tuesday. **1948** *Highway Traveler* Dec. 18/1 Because Orleanians at one time depended upon the duel as a means of settling disputes, the extensive display of dueling pistols and swords on the second floor has special interest. **1952** B. Ulanov *Hist. Jazz in Amer.* (1958) v. 46 The dance-music instrumentation familiar to most Orleanians.

Orleanist. (Earlier and later examples.)

1834 tr. C. M. *Catherine de Villemarest Life Prince Talleyrand* II. ix. 184 The letter of the Abbé Maurice [*sc.* Talleyrand]..proves to me, that after having been an anarchist, an Orleanist, and not having been able to become a Robespierrist..he has now become a Directorist. **1976** W. Gérin *E. Gaskell* xiv. 155 Mary Anne Clarke..an Orleanist to the backbone..had the good fortune to charm Chateaubriand.

Orlon (ǭ·ɹlǫn). Also **orlon.** A proprietary name of a man-made polyacrylonitrile fibre which makes a soft, warm yarn for textiles and knitwear. Freq. *attrib.*

1948 *N.Y. Times* 25 Aug. 43/3 The DuPont Company announced today it had adopted the trade mark 'Orlon' for a synthetic textile fiber on which it has been conducting research for several years and which previously has been known as Fiber A. **1950** *Official Gaz.* (U.S. Patent Office) 17 Oct. 677/1 E. I. DuPont de Nemours and Company... *Orlon*. For yarns of synthetic fibers. Claims use since Aug. 14, 1948. **1952** *Trade Marks Jrnl.* 9 Jan. 31/1 *Orlon*... Raw or partly prepared synthetic textile fibres. E. I. DuPont De Nemours and Company.., Wilmington, State of Delaware, United States of America; manufacturers. **1952** [see *Acrilan]. **1956** A. Huxley *Let.* 17 Nov. (1969) 811 It is interesting to find cotton coming back—in orlon and nylon, to ironing, no starch. **1957** P. Wildeblood *Main Chance* 86 His smiling teeth looked as expensively synthetic as his orlon shirt and dacron suit and nylon socks. **1969** 'J. Munro' *Innocent Bystanders* vii. 98 The woman was as grizzled a grey, and in it the yellow Orlon dress had lost its glitter. **1973** *Materials & Technol.* VI. 327 Polyacrylonitrile, best known under the trade names of 'Courtelle', 'Acrilan' and 'Orlon'.

o-rmer, *n.* *Channel Islands.* [f. the sb.] To collect ormers. Chiefly as **o·rmering** *vbl. sb.*; also *attrib.*

1903 *Eng. Dial. Dict.* IV. 359/1 He's gone ormering. **1953** W. D. Hooke *Channel Islands* iv. 110 Ormering is another important amusement. A very low tide is always referred to in the Islands as 'an Ormering tide'. **1953** S. P. Mais *Channel Islands* 67 It was a full moon and a spring tide, 'ormering' time. **1953** M. Pears *Mr. Pye* x. 71 'Where are you going, Tingeluce?' 'Ormering.' 'I d-didn't know there was an ormer tide.' **1965** 'J. Christopher' *Wrinkle in Skin* vi. 77 Mathew had walked out occasionally, on an ormering tide. **1968** R. M. Lockley *Channel Islands* 74 Even hard-hearted employers have forgiven employees absent from work over a low tide when the excuse has been 'ormering'. **1976** W. Rothero *Jersey* iii. 40 The land and the sea were exploited to the utmost for the islanders' subsistence, and from them came the traditional custom: the ormering parties, held when the *oreille-de-mer* were gathered.

orming (ǭ·ɹmiŋ), *a.* *dial.* [The same word as Hawm *v.*] Ungainly and clumsy in movement, tall and awkward. (quot. 1913) (standing about) gawping and staring.

E.D.D. Hawm v.[1] cites from 'Sc., Yks., Chs., Der., Not., and Lin.'.

1903 *Eng. Dial. Dict.* IV. 359/2 *Orming*, ppl. adj... Notts. Tall and awkward. **1913** D. H. Lawrence *Love Poems* 52 Niver a baby had eyes As sulky an' ormin' as thine. **1922** —— *England, my England* 216 Mrs. Goodall.. fairly hated the sound of correct English. She *heard* and *tha'd* her prospective daughter-in-law, and said: 'I'm none so ormin' as I look, seest ta.' Fanny did not think her prospective mother-in-law looked at all orming.

ornament, *sb.* Add: **2. d.** *Mus.* A grace note; a decorative figure used to embellish a melodic line.

1664 J. Playford *Introd. Skill of Musick* (ed. 4) 1. 58 There is made now always in indifferent and confused use of those Excellent Graces and Ornaments to the good manner of Singing, which we call Trills, Grappes, [etc.]. **1779** W. Waring tr. *Rousseau's Dict. Mus.* 196 *Graces*, Ornaments... By this term are called, in the French music, certain turns and shakes in the throat, and other ornaments joined to the notes, which are in such or such a position, according to the rules prescribed by a taste in singing. **1801** Busby *Dict. Mus.* p. xxx, Of the graces in music. To these ornaments may be added the *Slur*,..the *Staccato*. **1889** G. B. Shaw *How to become Mus. Critic* (1960) 67 Her voice is not yet quite solid, and..she appended a dreadful 'ornament' to Hark, those Chimes. **1964** *Listener* 7 May 885/2 The important indications of Rameau's ornaments were either unnecessarily simplified or omitted, regrettably since ornaments are not optional embellishments in music of this period; they form an integral part of the harmony, frequently devised to prepare or resolve dissonance. **1969** *Ibid.* 26 June 904/3 The ornaments are not frequent, and only come on emotional words.

ornamental, *a.* Add: **b.** *ornamental-leaved* adj.

1870 B. S. Williams (*title*) Choice stove and greenhouse ornamental-leaved plants. **1952** A. G. L. Hellyer *Sanders' Encycl. Gardening* (ed. 22) 122 Ornamental-leaved kinds [of coleus] require to have points of their shoots pinched off in early stage of their growth to ensure dwarf or well-shaped plants.

B. *sb.* **2. 1.** A tree or shrub grown for the sake of its attractive appearance.

1903 *Pop. Sci. Monthly* Jan. 277 It could be done more easily with strawberries, or with some of the common ornamentals that do not reproduce true to seed. **1972** *House & Garden* Feb. 42/3 When planting ornamentals, plant fruit trees and bushes too.

ornamentalist. (Further examples.)

1858 C. C. Perkins *Italian Sculptors* vi. 234 After his [*sc.* Piero Giacomo Ilario's] day Mantuan sculptors were few, and are generally ornamentalists in marble or stucco. **1934** *Burl. Mag.* 12 May 208/1 Pastrycooks... First hand required. Must be first-class decorator and ornamentalist.

ornamentation. Add: **3.** *Mus.* The use of grace notes to provide embellishment of a melodic line.

1879 [see *arabesque *sb.* 5]. **1938** *Oxf. Compan. Mus.* 675/1 Ornamentation in the old Italian vocal style had become so much second nature with them that nothing he could say would induce them to keep to the copies before them. **1947** A. Einstein *Mus. Romantic Era* xvii. 310 The ornamentation of the Romantic-Oriental folk song. **1962** *Listener* 2 Aug. 189/1 Her ornamentation is always designed to stress dissonances at crucial points. **1965** *Ibid.* 29 Apr. 643/3 Should we not consider carefully whether, say, several piano chords are not as obligatory in playing Ravel as correct ornamentation in playing Bach?

orné, *a.* Add: (Earlier and later examples.) Also, decorated, ornate.

1781 J. Byng *Torrington Diaries* (1934) I. 48 The place [*sc.* a garden bower] is not sufficiently *orné*. **1781**, etc. [see *cottage *a.*]. **1864** Trollope *Small House at Allington* I. ix. 91 A solitude in the centre of a wide park is now the only site that can be recognized as eligible. No cottage must be seen, unless the cottage *ornée* of the *gardener*. **1959** N. Mitford *Blessing* I. iii. 22 If she could propose an 18th-century summerhouse, or cottage *orné* as the object of a walk, he would accompany her.

orneriness (ǭ·ɹnərinĕs). Chiefly *U.S.* [f. *ornery *a.* + -ness.] Meanness, cussedness, contrariness.

1899 B. Tarkington *Gentleman from Indiana* iv. 45 They..let loose their devilitries just for pure orneriness. **1927** W. E. James *Cow Country* 119 The bronk's orneriness had come to the top, and that pony..begin to get sort of desperate and he wanted a bucking way out. **1957** *Economist* 9 Nov. 489/2 Some of these young and innocent..sold property like Indonesia's island dissidents, some on tribal differences, like Ghana's Ashantis, and some arise out of pure political orneriness. **1959** J. Thurber *Years with Ross* v. 75 Some orneriness of mood aggravated by..peptic ulcers. **1973** W. H. Auden in *Listener* 22 Feb. 243/1 Our mother sentiments said: 'I'll drive cars for the U.S.'

ornery (ǭ·ɹnəri), *a. dial.* and *colloq.* Also **ornary.** [dial. var. of Ordinary *a.* 5.] Commonplace, of poor quality, coarse, unpleasant; low, mean, cantankerous. (Now chiefly *U.S.*)

1816 U. Brown *Jrnl.* in *Maryland Hist. Mag.* (1915) X. 369 The Land is old, completely worn out, the farming extremely ornery in general. **1861** T. Winthrop *John Brent* (1862) vii. 71 Good company betters the ornerest sort er weather. **1875** C. M. Yonge *My Young Alcides* II. ix. 261 If I refused, he should think..that I couldn't take that 'ornary' object, as he had overheard himself described that day. **1914** Galsworthy *White Monkey* II. xii. 190 A low-down thought—mean and ornery. **1937** M. P. K. Devine *Folklore of Newfoundland* 35 Ornery, ugly or plain; no good. **1938** I. Kirk *Stand in Adventure* xiv. 128 If he had not mentioned the clothes I should not have become ornery. But these clothes were a sore point with me. It was always..the ornery thing. **1939** *Sun* (Baltimore) 28 Jan. 10/3 We are forced to spend all this money solely because mankind up to now is too ornery to organize civilization on some more sensible basis. **1941** J. Faulkner *Men Working* 201 Males is the orneriest critters. **1944** T. D. Clark *Pills, Petticoats & Plows* xvi. 28 Not a thing, you ornery cuss do you've got to work. **1958** M. Sakel *Epilepsy* (1959) 40 He may, contrary to his usual behavior, become aggressive, pugnacious or 'ornery' in behavior. **1962** G. MacEwan *Blazing Old Cattle Trail* xxi. 143 There the sheep were in one of their ornery moods, determined not to get their nice Merino wool wet. **1972** *Newsweek* 10 Jan. 18/3 The public as a whole might turn ornery if some semblance of prosperity were not quick to return.

ornithischian (ǭɹnipə·skiăn), *a.* and *sb.* [f. mod.L. order name *Ornithischia* (H. G. Seeley 1887, in *Proc. R. Soc.* XLIII. 170), f. Ornith(o- + Ischium: see -ia[2], -ian.] **A.** *adj.* Of, pertaining to, or designating an ornithischian. **B.** *sb.* A herbivorous dinosaur of the order Ornithischia, which includes forms having a pelvic structure resembling that of birds.

1901 H. G. Seeley *Dragons of Air* xvii. 199 In some at least of the..Ornithischian Dinosaurs, there is no antorbital vacuity. **1933** A. S. Romer *Vertebrate Paleontol.* ix. 152 It is probable that few ornithischians were entirely bipedal in habits. **1962** E. H. Colbert *Age of Reptiles* i. 19 The ornithischian dinosaurs, in which the rod-like pubic bone of the pelvis had rotated backwardly to run parallel to the ischium,..were of late Triassic origin. **1967** *New Scientist* 16 Mar. 534/2 [In Lesotho] they also found a triangular piece of jawbone ..belonging to an elongated bird-like pelvis. **1971** *Nature* 12 Nov. 75/1 Before 1962 the genus *Geranosaurus*, which is represented by a single jaw, was the only undisputed Triassic ornithischian on record, but since that date knowledge of the earliest ornithischian dinosaurs has increased considerably. **1977** *Radio Times* 17 Dec. 45 One of the liveliest of these disputes concerns the two great groups of dinosaurs, the saurischians (with hip-bones like those of lizards) and the ornithischians (in which they were more bird-like).

ornitho-. Add: **o:rnithocopro·philous** *a.* [f. *coprophilous* (s.v. Copro-)] (see quot. 1928).

1928 H. D. Jackson *Gloss. Bot. Terms* (ed. 4) 434/1 *Ornithocoprophilous*,..applied to lichens, which benefit by the excreta of birds. **1958** J. L. Bakkman *Phytosociol. & Ecol. Cryptogamic Epiphytes* iii. 104 *Xanthoria candelaria* is highly ornithocoprophilous on rocks in Sweden. **1967** [see *nitrophilous *a.*].

ornithogalum (ǭɹnipǫ·gælə̆m). [mod.L., ad. L. *ornithogalē*, Gr. ὀρνιθόγαλον, f. ὀρνιθο- bird + γάλα milk, adopted as the name of a genus by Linnæus (*Hortus Cliffortianus* (1737) 124) and earlier botanists.] A bulbous plant of the genus so called, belonging to the family Liliaceae, native to Europe, Asia, or Africa, and usually bearing spikes or racemes of white flowers; = Ornithogal, *Star of Bethlehem* (Star *sb.*[1] 22 c).

1664 [see Ornithogal]. **1755** [see *Star of Bethlehem* s.v. Star *sb.*[1] 22 c]. **1792** *Curtis's Bot. Mag.* VI. 190 (*heading*) Golden ornithogalum. **1825** [see Ornithogal]. **1901** L. H. Bailey *Cycl. Amer. Hort.* III. 1174/2 Ornithogalums may be divided into hardy and tender groups, and each of these may be subdivided into dwarf and tall groups. **1931** M. Grieve *Mod. Herbal* II. 770/1 Only one [species] is truly native to Great Britain, the spiked Ornithogalum, *O. pyrenaicum* (Linn.), and is not common, being a local plant, found only in a few counties. **1962** F. Palmer *Plants of Cambeboo* xvii. 284 Milk-white Ornithogalums of various species grew here and there, Chinkerinchees to us all.

ornithologically, *adv.* (Examples.)

1865 *Ibis* 87 The country..is rather poor ornithologically. **1970** *Daily Tel.* 12 May 11/4 This month is about the best for the ornithologically minded. **1977** *Listener* 30 June 867/1 A superb setting, a marshy bird reserve in Southern India where..the..narrator's ornithologically outrageous mania which traps him into seeing too much, forces him into flight among sordid tourist development.

ornithologize, *v.* (Earlier example.)

1874 *Amer. Naturalist* VII. 268 At Topeka..we also tarried for ten days, devoting the time almost exclusively to ornithologizing.

ornithophile (ǭ·ɹnipofail). [f. Ornitho- + -phile.] **1.** = *Ornithophilist.

1959 *Listener* 10 July 66/1 Irresistible to ornithophiles.

2. *Bot.* A plant pollinated by birds.

1970 *Watsonia* VIII. 67 Mountain regions..are..also regions where ornithophiles from diverse plant genera congregate.

ornithophilous, *a.* (Later examples of *Bot.* use.)

1940 G. F. Scott-Elliott in *Ann. Bot.* IV. 265 (*title*) Ornithophilous flowers in South Africa. **1906** J. R. A. Davis tr. *Knuth's Handbk. Flower Pollination* I. 76 The following Cape plants are ornithophilous. **1975** *New Phytologist* LXXIV. 368 There are species [of giant lobelia] which are ornithophilous.

ornithophily. Add: **2.** *Bot.* [cf. mod.L. *Ornithophilæ*, name of a group of plants pollinated by birds (F. Delpino in H. Müller *Die Befruchtung der Blumen* (1873) 15).] Pollination by birds.

1903 W. R. Fisher tr. *Schimper's Plant-Geogr.* I. vi. 121 The ornithophily of a species of Erythrina was also established by Fritz Müller. **1970** *Watsonia* VIII. 67 The plants (75 colour photographs, mainly of birds visiting flowers) emphasize the importance of ornithophily in the U.S.A. **1975** *New Phytologist* LXXIV. 368 Ornithophily is replaceable by autogamy.

ornithopod, *a.* and *sb.* [Add to etym. after *Ornithopoda*: (O. C. Marsh 1881, in *Amer. Jrnl. Sci.* 3rd Ser. XXI. 423).] Substitute 'suborder' for 'order' in def. and add examples.

1933 A. S. Romer *Vertebrate Paleontol.* ix. 156 (*caption*) Dorsal and lateral views of the ornithopod dinosaur *Camptosaurus*. *Ibid.* 157 The armor of the back was stiffened in ornithopods by a latticework of tendons. **1975** *Nature* 23 Oct. 668/1 We describe here ornithopod dinosaur remains from the Lower Cretaceous of western North America. **1977** A. Hallam *Planet Earth* 275 Larger ornithischians, known as ornithopods, had appeared in the middle Jurassic.

ornithopter (ǭɹnipǫ·ptəɹ, ǭ·ɹnipǫptəɹ). *Aeronaut.* Also †-ptere. [ad. F. *ornithoptère* (P. Renard 1908, in *L'Aérophile* 15 Jan. 35), coined to replace *orthoptère* (*orthopter 2) because of the latter's etymological inappropriateness and its entomological meaning; see Ornitho-.] A machine designed to achieve flight by means of flapping wings.

1908 *Aeronautics* I. 86/1 *Ornithoptère* denotes a machine in which the means of sustentation and propulsion consist of beating-wings. **1909** *Flight* I. 99/2 There are many who watch most anxiously for the success of artificial flapping flight by the aid of machines which have been variously named 'orthopters' and 'ornithopters'. **1933** *Jrnl. R. Aeronaut. Soc.* XXXVII. 205 This gentleman.. resolved to employ the ornithopter type for military observation. **1957** *Listener* 19 Dec. 1021/2 Where the ornithopter..was concerned, Leonardo gave most of his attention almost obsessionally to the flapping wing ornithopter. **1960** *Observer* 17 Apr. 17/1 A Tass announcement said that 'the ornithopter—appropriately powered by a motor-cycle engine—was flown by 'for several yards' in Moscow. **1972** *Nature* 10 Nov. 173/1 A book largely concerned with a theoretical synthesis of man-powered ornithopters. **1976** *Globe & Mail* (Toronto) 4 Aug. 2/6 Mr. Newell, 64, has built five of his airplanes, called ornithopters, in the past 12 years.

ornithosaur. (Examples.)

1887 H. G. Seeley in *Phil. Trans. R. Soc.* B.CLXXVIII. 191 This osseous condition approximates to that which characterises the bones of Ornithosaurs and Birds. **1913** *Q. Jrnl. Geol. Soc.* LXIX. 372 An Ornithosaur from the Wealden Shales of Atherfield (Isle of Wight).

ornithosaurian, *a.* (*sb.*) (Examples.)

1888 R. Lydekker *Catal. Fossil Reptilia Brit. Mus.* I. 24 Considerable portion of the skeleton of a large Ornithosaurian..from the Wealden of Brook, Isle of Wight. **1901** H. G. Seeley *Dragons of Air* xvi. 177 In many ways the Ornithosaurian animals are like Birds. **1913** *Q. Jrnl. Geol. Soc.* LXIX. 372 The late Rev. W. D. Fox..discovered..many associated ornithosaurian bones.

ornithosis (ǭɹnipōᵘ·sis). [f. Ornitho- + -osis.] A disease affecting birds, certain small mammals, and man, caused by a micro-organism belonging to the genus *Chlamydia*, and producing severe, sometimes fatal, pneumonitis in man and respiratory or generalized infection in birds and other animals. Cf. Psittacosis in Dict. and Suppl. Hence **ornitho·tic** *a.*

1930 K. F. Meyer et al. in *Proc. Soc. Exper. Biol. & Med.* XLI. 573 (*title*) Complement fixation test..as an aid in recognizing latent avian infections (ornithosis). **1947** H. V. Dawson *Dis. Poultry* 288 There are, however, cases of an infection in..the pigeon, which is equally ornithotic. **1951** *Lancet* 29 Sept. 572/2 This ornithosis is primarily ..a review of present knowledge of the clinical features [etc.]..among inhabitants of the Faroe Islands. **1962** *Times* 25 Mar. 8/6 A generalized virus infection known as psittacosis or ornithosis. **1969** *Daily Tel.* 25 Mar. 13/4 much bigger danger..is the wood pigeon, which..suffers from ornithosis, a near ally to..psittacosis..which sufferers can give man a form of pneumonia. **1973** *Observer* 4 Nov. 5/3 A steadily increasing number of people..catch psittacosis or ornithosis..from birds..from imported parrot-like birds.

oro-, formative element repr. Gr. ὄρος mountain (as in *orometric* (1774), *orology* (1781)). For their use, see etym. of Orography.

orobus. Delete *Obs.* and substitute for second part of etym.: [adopted by Linnæus (*Genera Plantarum* (1737) 325) and earlier botanists as the name of a genus.] Substitute for def.: A herb of the genus so called, belonging to the family Leguminosæ and now usually included in the genus *Lathyrus*; = *bitter-vetch* (Bitter- *a.* 2). (Later examples.)

1876 tr. *van Oort's Dutch Gardener* II. cv. 155 Of the Orobus. This is a small blue purplish Flower, bears plentifully; it is a sort of Vetches. **1801** *Curtis's Bot. Mag.* XV. 521 (*heading*) Early-flowering Orobus. **1894** W. Robinson *Wild Garden* (ed. 4) xi. 53 Among the plants that are suitable for hedgerows and lanes, &c. are; May Apple, Orobus in variety, many Narcissi. **1903** *Flora & Sylva* I. 202/1 The name 'Orobus' is useful, for it expresses a race of plants..distinct from the climbers to which the name 'Lathyrus' was formerly exclusively applied. Thee Orobi..supply some of the most beautiful of spring and early summer flowers.

orocline (ǭ·rəklain). *Geol.* [f. *oro- + κλίν-ειν to bend.] (See quot. 1955.)

1955 S. W. Carey in *Papers & Proc. R. Soc. Tasmania* LXXXIX. 257 An orocline is an orogenic system which has been flexed in plan to a horse-shoe or elbow shape, the name *orocline* is proposed. **1970** Lipson & Moseley in *Cambr. Anc. Hist.* (ed. 3) I. i. 11 Carey recognizes six oroclines in the orogenic belt in Europe. **1977** *Nature* 13 Oct. 589/2 The orocline concept is incompatible with simple plate tectonics and offers no reason why arcs should form.

orocratic (ǫrokræ·tik), *a.* *Geol.* [f. *oro- + κράτ-ος strength + -ic.] Characterized by an increase in the roughness of relief as a result of crustal upheaval.

1924 W. Ramsay in *Geol. Mag.* LXI. 155 Only after the orogenic phases was the relief comparable to that of the present day, or still higher and more broken. For such a period I propose the name orocratic time. **1962** J. Wells *Physiogr. Evol. Brit.* vi. 71 The Carbo-Permian times witnessed..one of the greatest orocratic periods of the earth's history. **1964** *Times* 28 Aug. 9/4 A 'phillocratic period' is a relatively quiet one between two 'orocratic periods'.

orogen (ǭ·rǫdʒĕn). *Geol.* Also **orogene.** [a. G. *orogen* (L. Kober *Der Bau der Erde* (1921) i. 21): see *oro- and -gen.] An orogenic belt.

1923 *Bull. Geol. Soc. Amer.* XXXIV. 167 Geanticlines and orogens may remain as dry lands or may sink into the depths of the oceans. **1934** *Geogr. Jrnl.* LXXXIII. 517 The orogen on the west North American Cordillera. **1953** [see *mountain system s.v. *mountain 9 a]. **1964** L. U. de Sitter *Struct. Geol.* (ed. 2) xxxi. 424 The Pyrenees are essentially a Hercynian orogene reactivated in the Alpine Period. **1970** *Nature* 28 Nov. 838/2 The width of the orogen is anomalously great in the region of the central United States. **1975** *Ibid.* 20 Feb. 599/2 The effects are restricted to the western seaboard of Scandinavia and within overthrust nappes on the foreland edges of the Caledonian orogene.

orogenesis. (Further examples.)

1928 [see *epeirogeny]. **1970** *Nature* 9 May 498/1 Basaltic dykes have been used to separate chronologically two principal periods of orogenesis.

orogenetic (ǫrodʒĕne·tik), *a.* *Geol.* [f. Oro(genesis + -genetic.] = Orogenic *a.*

1888 J. J. H. Teall *Brit. Petrogr.* 441 *Orogenetic*, that which relates to the formation of mountains. **1925** J. Joly *Surface-Hist. Earth* i. 24 The orogenetic movements which developed the Appalachians. **1969** *Nature* 24 Jan. 356/2 The age of the crystalline [of the earth] was greater in the past.., and the suggests that orogenetic activity may have decreased. **1970** R. J. Small *Study of Landforms* iii. 89 Over much of the earth compressive forces in the crust, stemming mainly from orogenetic movements of diverse age, have produced folding of sedimentary rocks.

Hence **orogene·tically** *adv.*

1934 *Bull. Geol. Soc. Amer.* XXXIV. 158 There must have been another crustal element orogenetically active. **1949** *Ibid.* LX. 1756/2 He is..strongly convinced that the anorogenic periods were of long duration, as against the short orogenetically active phases.

orogenic, *a.* Add: *orogenic belt*, a strip of the earth's surface which has been subjected to folding or other deformation during an orogeny.

1942 M. P. Billings *Structural Geol.* v. 93 Some geologists, however, believe that entire orogenic belts are sliding off of continental-scale emergent topographic highs. **1944** A. Holmes *Princ. Physical Geol.* xviii. 378 It is essential to discriminate carefully between the geographical concept of a mountain range or system and the geological concept of an orogenic belt: the one refers to the height and relief of the land; the other to the structure of the rocks, whether the region be high, low, or submerged. **1960** *Sci. Jrnl.* Feb. 52/2 The entire orogenic belt bordering the eastern margin of the Pacific may have been caused by the sea floor spreading and the consequent underthrusting along the eastern Pacific margin.

Hence **oroge·nically** *adv.*, by orogenesis.

1935 *Geogr. Jrnl.* LXXXVI. 36 Neither in the orogenically quiet nor..remarkably well preserved at 1973 L. G. Gass et al. *Understanding Earth* xii. 231/2 The ancient distributions of sediments of the Ivory Coast were orogenically deformed some 2000 million years ago.

orogenital (ǭ:rǭdʒe·nităl), *a.* [irreg. for *origenital*; f. L. *ōs, ōr-* mouth + Genital *a.*] = *oral *a.* 3 b.

1963 A. Heron *Towards Quaker View of Sex* 16 In the U.S. and Great Britain orogenital contact seems much less common in what are often called the lower social classes. **1971** *New Society* 1 Apr. 533/1 Human sex play has a large, though tabooed, orogenital component. **1974** E. Rosser et al. *Abnormal Psychol.* (ed. 2) xv. 277/2 The technical terms for oro-genital activity are cunnilingus..and fellatio.

orogeny (s.v. Orogenesis). Add. **a.** = Orogenesis.

1962 [see *epeirogeny]. **1950** F. E. Zeuner *Dating Past* (ed. 2) xi. 349 The periods of regression and orogeny result in intensified erosion on the continent. **1972** *McGraw-Hill Yearbk. Sci. & Technol.* 303/1 Orogeny results from interactions of this global, continuously evolving system of oceanic ridges and trenches, according to the concepts of..lithosphere plate tectonics.

b. A geological period of mountain-building.

1914 *Jrnl. Geol.* XXII. 647 Throughout its range the Laramide orogeny was marked by great volcanic activity. **1940** *Geogr. Jrnl.* XCVI. 31 The conception of the orogenic cycle..gets little support from a study of the Variscan orogeny of Central Europe. **1970** *Nature* 7 Nov. 471/2 Is it generally correct to associate all previous Chinese orogenies with times of rapid sea-floor spreading?

orographic, *a.* Add: **b.** *Meteorol.* Applied to precipitation which results from moist air being forced upwards by mountains, and to the action of mountains in producing such precipitation.

1920 *Q. Jrnl. R. Meteorol. Soc.* XLI. 47 These considerations at once brought into prominence the distinction between orographic and cyclonic rainfall, which was perhaps one of the most puzzling features in British meteorology. **1938** E. G. Bilham *Climate Brit. Isles* v. 101 Thunderstorm rains..do not often produce total amounts of rainfall..of the same order of magnitude as are observed when intense cyclonic rains are augmented by orographic action. **1955** W. J. Saucier *Princ. Meteorol. Analysis* x. 241 The dominant effect..is therefore orographic lifting of the moist and turbulent air is sufficient to produce precipitation without needing convergence at the surface. **1968** *Jrnl. Appl. Meteorol.* VII. 857/1 Orographic precipitation has long been considered to be affected by large-scale and cloud seeding. **1976** B. Lecomber *Dead Weight* iii. 41 It's the same pattern on all the West Indian Islands;..on the high ground, where the orographic clouds form and double or treble the rainfall, you find the forests.

orographical, *a.* Add: **b.** *Meteorol.* = prec.

1909 H. R. Mill in *Geogr. Teacher* V. 75 Placing this established fact of the dependence of the annual rainfall on the height or the configuration of the ground, side by side with the equally established fact of the complete independence of heavy cyclonic or thunderstorm rains on terrestrial conditions, we are obliged to divide rain into two classes; one greater in intensity but shorter in duration, and, in amount, irregular, in a manner which we may call meteorological rain (in thunderstorms and cyclones), and the other weaker in intensity but greater frequency and duration, we may call orographical rain. **1921** *Geofysiske Publikasjoner* II. 10/3 The effects of orographical rain are strongly restricted by the tendency of all stable air currents to curve round the mountains horizontally. **1947** *Q. Jrnl. R. Meteorol. Soc.* LXXIII. 16 The orographical rain was well developed over North Wales and the English Lake District.

Hence **ogra·phically** *adv.*, in accordance with, or by, orography; by the action of mountains.

1873 *Q. Jrnl. Geol. Soc.* XXIX. 389 These two lakes.. are separated by a prolongation of the parallel ridges of the Schaffberg massif, so that orographically the Wolfganger See lies in a synclinal trough which may be traced along a line of lakeless valleys, and is separated from the head of the Foschelsee by a narrow ridge. **1877** *Ibid.* XXXIII. 143 The part of North Greenland here described can be geologically and orographically divided into three districts. **1902** D. G. Hogarth *Nearer East* 14 Here is a continuous parting of waters, but not, orographically, a continuous mountain range. **1947** *Q. Jrnl. R. Meteorol. Soc.* LXXIII. 13 Warm front rain is intensified orographically, but not so much as warm sector rain. **1971** *Nature* 23 Apr. 504/1 The development of ice fog in relatively still, orographically protected areas.

orography. Add: (Further examples.) Also, the orographical features of a region.

1853 E. Hitchcock *Outl. Geol. Globe* 10 A knowledge of the Hydrography of a country aids as much in determining its geology as does its Orography,—that is, a description of its mountains. **1873** *Q. Jrnl. Geol. Soc.* XXIX. 382 As the general circumstances of climate and orography are fairly similar throughout the great Alpine chain, it may reasonably be expected that any explanation suggested for the lake-basins of one district should apply also to another. **1904** *Geogr. Jrnl.* XXIII. 183 How the system of orography of Eastern Siberia accords with the orography of Asia altogether, such as it now appears from the recent explorations in Central Asia. **1923** *Q. Jrnl. R. Meteorol. Soc.* XLIX. 226 The 'normal' orographical distribution of rainfall depends of course not only on the orography of the country but on the constancy of the prevailing wind. **1955** W. J. Saucier *Princ. Meteorol. Analysis* x. 304/1 Local circulations due to convection or orography may..cause discrepancy between local winds and the flow pattern of the surroundings. **1975** *Weather* XXX. 141 Orography has the main influence in mountainous regions while the latitude effect is most important in regions with smooth orographies.

orohydrography (ǫ:rohəidrǫ·grăfi). [f. *oro- + Hydrography.] (See quot. 1967.) Hence **o:rohydrogra·phic, -gra·phical** *adjs.*

1892 *Syd. Soc. Lex.*, *Orohydrography*, a description of the water sheds of mountains. **1900** *Geogr. Jrnl.* XVI. 35 The oro-hydrographic conformation of the Andine region extending southwards of Mount Tronador is extremely complex. **1908** C. R. Enock *Peru* i. 2 A glance at the map will show the remarkable series of parallel topographical—or rather oro-hydrographical—features which exist, due to this agency. **1936** *Geogr. Jrnl.* LXXXVIII. 268 There are..maps, geological and orohydrographical, covering the whole region. **1967** E. Vollmer *Encycl. Hydraulics* 208 *Orohydrography*, the branch of hydrography which deals with the relation of mountains to drainage. **1970** *Soviet Hydrology: Selected Papers* No. 1. 19 As an example of one of the most ancient orohydrographic discrepancies we can mention the crossing of the Penzhina Range by the Talovka River.

orometric, *a.* (Later example.)

1945 *Q. Jrnl. R. Meteorol. Soc.* LXXI. 44 At the other end of the orometric scale it seems pretty evident that the British mountains which only exceptionally reach an altitude of even 4,000 feet are neither high enough nor compact enough to produce the maximum amount of orographic rainfall.

orometry (orǫ·mĕtri). *Geogr.* [f. *oro- + -metry.] The measurement of forms of relief.

1898 *Geogr. Jrnl.* XI. 205 As the geoid is treated in geodesy, he treats the *oroid* in orometry. **1972** P. Tilley tr. *Hettner's Surface Features of Land* 160 Ideas used in analytical geometry to express geometrical figures arithmetically have been copied in orometry.

oronasal: see also *orinasal *a.* (*sb.*).

oropendola (ǫrope·ndolă). [a. Sp. *oropendola* golden oriole.] Any of several birds belonging to one of the species of the family Icteridæ, found in tropical regions of south and central America. Also *attrib.*

1898 *Auk* XV. 327 (*heading*) Cassin's Oropendola. **1912** Brabourne & Chubb *Birds S. Amer.* I. 431 Swainson's Oropendola. **1955** *Sci. News Let.* 23 July 64/3 Nests of the oropendola, or 'giant oriole', are narrow bags two to three feet long, woven of straws and weeds. **1968** *Sci. Jrnl.* Nov. 11/3 Nests of oropendolas—a bird found in the Republic of Panama—containing also one or two young of the cowbird..often produce more oropendola fledglings than 'unparasitized' ones. **1970** R. Meyer de Schauensee *Guide Birds S. Amer.* 352 This family [*sc.* Icteridæ] comprises an assortment of birds ranging from forest-inhabiting oropendolas and caciques to the more familiar grackles, cowbirds and meadowlarks. **1974** *Nat. Geographic* Nov. 687 Oriole-like oropendolas were building their hanging nests in nearby trees.

Oropesa (ǫropī·ză). The name of the vessel first used to test the apparatus, used *attrib.* to designate a minesweeping device first developed during the war of 1914–18.

1939 *War Illustr.* 9 Dec. 399/2 Minesweepers work in pairs, with each unit 300 to 500 yards apart. Between them, sometimes suspended from two sets of apparatus called Oropesa floats, is drawn the sweep wire, which has a series of steel cutters. Should this come into contact with a mooring cable, the mine will rise to the surface and it can then be destroyed by gunfire. *Ibid.* 29 Dec. 538/1 *Oropesa*, a type of minesweeping float. The name originated from that of a trawler in which the newly-invented gear was first tested. **1940** *Manch. Guardian Weekly* 12 Apr. 297 A little farther on a minesweeper was being reconditioned, grim-looking with bristling gun, and oropesa sweeps. **1949** J. S. Cowie *Mines, Minelayers & Minelaying* vi. 105 The spread of an Oropesa type of sweep. **1965** K. Langmaid *Approaches are Mined!* ii. vii. 138 The introduction of..the Oropesa Float made it possible for a *single* vessel to sweep a wide area in a reasonably short time.

oro-pharyngeal, *a.* (Earlier and later examples.)

1885 *N.Y. Med. Jrnl.* XLII. 376/1, I experimented with regard to the isolation of the temperature sense in the nasal and oro-pharyngeal cavities. **1967** *Nursing Times* 8 Sept. 1196/2 An oropharyngeal airway will help, but the proper position of the attendant's fingers is far more important.

oropharynx. (Examples.)

1887 L. Browne *Throat & its Dis.* (ed. 2) iii. 53 (*heading*) Inspection of the mouth, fauces, and oro-pharynx. **1894** P. W. Williams *Dis. Upper Respiratory Tract* ii. 12 For clinical purposes the pharynx is divided into three regions, the naso-pharynx, the oro-pharynx and the laryngo-pharynx. **1935** R. S. Stevenson *Rec. Adv. Laryngol. & Otol.* ii. 38 Norman Patterson is of opinion that at the present time a combination of diathermy and radiation is usually best in the treatment of malignant disease of the oro-pharynx. **1976** *Lancet* 4 Dec. 1248/1 There was diffuse ulceration in the oropharynx.

orosomucoid (ǭrǭu·somiū·koid). *Biochem.* [f. Gr. ὀρός serum + -o + *mucoid *sb.*] A glycoprotein which forms the major component of seromucoid.

1955 R. J. Winzler in *Methods Biochem. Analysis* II. 281 The major component in human seromucoid is an electrophoretically and ultracentrifugally distinct acidic glycoprotein which has been crystallized and quite well characterized by chemical, physical, and immunological methods... It is appropriate to assign to this protein the name orosomucoid..to indicate its source and nature. **1965** *Biochim. & Biophys. Acta* CI. 336 The human and bovine orosomucoids had molecular weights of 41600 and 37300 respectively by the sedimentation diffusion method. **1972** *Res. Communications Chem. Path. & Pharmacol.* III. 663 Important increases in the level of orosomucoid accompany exudative inflammatory processes.

orotic (orǫ·tik), *a. Chem.* [ad. It. *orotico* (Biscaro & Belloni 1905, in *Ann. della Soc. Chim. di Milano* XI. 18), f. Gr. ὀρό-ς serum, whey + It. *-otico* -otic.] *orotic acid*, a colourless crystalline heterocyclic acid, $C_4HN_2(OH)_2$-COOH, which is found in milk and is a growth factor for some micro-organisms.

1905 *Jrnl. Chem. Soc.* LXXXVIII. i. 672 The occurrence of small crystals of an organic compound in the mother liquor of lactose has led to the discovery of a new acid, orotic acid, which has been found to be a normal constituent of milk. **1944** *Nature* 26 Feb. 251/2 Four out of five of the strains of Group C streptococci of various types which were tested also needed uracil or orotic acid for optimal growth. **1965** T. L. V. Ulbricht *Introd. Nucleic Acids* vi. 70 Studies with bacterial mutants showed that orotic acid..was an effective precursor of both DNA and RNA pyrimidines.

Hence **oro·tate** [-ate¹ c], the anion, or an ester or salt, of orotic acid; **oro·tidine** [*-idine d], a nucleoside containing an orotic acid residue, the phosphate of which is a precursor of pyrimidine nucleotides.

1905 *Jrnl. Chem. Soc.* LXXXVIII. i. 672 Methyl orotate,..a white, crystalline powder with a somewhat bitter taste, melts at 248–250° and dissolves in water or alcohol, forming faintly acid solutions. **1973** Henderson & Paterson *Nucleotide Metabolism* xi. 182 Under certain circumstances, the carbamyl phosphate product from the liver enzyme may be diverted..into the orotate pathway. **1951** A. M. Michelson et al. in *Proc. Nat. Acad. Sci.* XXXVII. 396 (*heading*) A new ribose nucleoside from Neurospora: 'orotidine'. **1963** A. M. Michelson *Chem. Nucleosides & Nucleotides* ii. 44 Periodate oxidation and enzymatic synthesis..and conversion into uridine show that orotidine is 3-β-D-ribofuranosylorotic acid. **1973** J. R. Bronk *Chemical Biol.* xii. 465 The resulting orotidine 5′-phosphate is then decarboxylated to give..uridine 5′-phosphate.

orotic aciduria (orǫ:tik æsidiū·riă). *Path.* Also as one word. [f. *orotic acid* (see prec.) + -uria.] A rare genetic disorder in which an enzyme deficiency impairs the metabolism of orotic acid, resulting in anæmia and excessive amounts of the acid in the blood and urine. Hence **oro:tic acidu·ric** *a.*

1959 Smith & Baker in *Jrnl. Clin. Invest.* XXXVIII. 798/1 Only recently has a 'pyrimidine disease', orotic aciduria, been described. **1961** *Ibid.* XL. 662/2 No large survey has..been undertaken to document the true incidence of the orotic aciduric defect. **1962** *Jrnl. Laboratory & Clin. Med.* LIX. 852 The family of an infant with congenital oroticaciduria excreted normal amounts of pseudouridine. **1972** F. Nour-Eldin *Haematol.* xxi. 218/2 Hereditary orotic aciduria is a rare disorder of the pyrimidine metabolism with megaloblastic anaemia, refractory to vitamin B_{12} but responding to uridine.

orotundity. (Examples.)

1909 in *Cent. Dict.* Suppl. **1922** J. M. Murry *Probl. Style* 20 Wordsworthians were there to discover the hallmark of genius on his most insignificant orotundities. **1936** 'M. Innes' *Death at President's Lodging* ix. 167 D.C. was absorbed in his narrative now: the self-consciousness, the orotundity were gone. **1960** *Spectator* 2 Sept. 347 Those thudding clichés, those meaningless orotundities. **1963** *Punch* 16 Jan. 105/3 An orotundity which isn't meant to be funny.

Oroya fever (oroi·ă fī·vəɪ). *Path.* [f. La *Oroya*, the name of a town in central Peru + Fever *sb.*¹] An acute, frequently fatal, febrile and hæmolytic disease which occurs in Peru as the first stage of infection with the bacterium *Bartonella bacilliformis*, the second, chronic, stage being verruga peruana.

1873 T. J. Hutchinson *Two Yrs. in Peru* II. xx. 61 The 'Oroya fever', as it was called, from the simple circumstance of its having occurred on this line [*sc.* the Callao–Lima–Oroya railway line] (although more than a hundred miles distant from the terminus at the little town of Oroya), caused a dreadful mortality here during the years of 1870 and 1871. **1903** *Encycl. Medica* XIII. 326 Death ensues without any appearance of eruption, constituting the grave form or Oroya fever. **1949** M. A. Jennings in H. W. Florey et al. *Antibiotics* II. xxxi. 1033 *Bartonella bacilliformis*, the organism of Carrion's disease—known also in its acute stage as Oroya fever, and in its chronic stage as verruga peruana—is highly sensitive to penicillin. **1962** Gordon & Lavoipierre *Entomol. for Students of Med.* xx. 138 In addition to being a severe biting nuisance, sandflies are responsible for the transmission to man of several forms of leishmaniasis, a sandfly fever and Oroya fever or Carrion's disease. **1974** Passmore & Robson *Compan. Med. Stud.* III. i. xii. 86/2 In those who survive Oroya fever, after an interval..the dermal lesions of verruga peruana develop.

orphan, *sb.* and *a.* Add: **A.** *sb.* **2. b.** *slang.* A discontinued model of a motor vehicle.

1942 Berrey & Van den Bark *Amer. Thes. Slang* § 81/7 *Orphan*, a discontinued make. **1948** Mencken *Amer. Lang.* Suppl. II. 724 *Orphan*, or *off-breed*, an obsolete model. **1967** W. & M. Morris *Dict. Word & Phr. Origins* II. 280 *Load, orphan, pig*,..and *iron* all designate poor cars.

3. orphan asylum (examples).

1811 *Freemason's Mag.* (Philadelphia) Nov. 97 The other proclaims its benevolent character in its name, 'The Orphan Asylum Society'. **1833** J. B. Purcell in *Catholic Hist. Rev.* (Washington) (1919) V. 241 There was a benefit at the L. [Louisville] Theatre for the Orphan-Asylum, at this time. **1921** E. O'Neill *Diff'rent* ii, in *Emperor Jones* 281 He'll go and leave all he's got to some lousey orphan asylum. **1964** D. Owen *Eng. Philanthropy* (1965) ii. v. 159 This London Orphan Asylum at Clapton, opened in 1825... Again..a movement was launched which, by the early 1840's, had established the Infant Orphan Asylum at Wanstead. **1978** E. Healey *Lady Unknown* ii. 60 Faraday's earliest letters to her..were..mostly requests for proxy votes for his protégés at the Orphan Asylum of which they were both patrons.

B. *adj.* **b.** *Orphan Annie*: see *little Orphan(t) Annie* (*little *a.* 13).

2. *Path.* **orphan virus,** any virus that is not known to be the cause of a disease.

1954 J. L. Melnick in *Amer. Jrnl. Publ. Health* XLIV. 572/1 The remainder of this report is concerned with... 3. The detection of new viruses, provisionally called 'orphan viruses' (as we know so little to what diseases they belong) from patients suspected of having nonparalytic poliomyelitis. **1955** [see *echo virus]. **1961** P. L. Carpenter *Microbiol.* ix. 110 A considerable number of viruses recently isolated from the feces of healthy as well as ill individuals are not known to produce disease... These viruses..are called orphan viruses, and those isolated from humans are known as ECHO viruses.

orphaned, *a.* Add: **3.** *slang.* Of a motor vehicle: discontinued as a model. Cf. *orphan *sb.* 2 b.

1920 F. B. Scholl *Automobile Owner's Guide* 3 Orphaned cars may run as well..as anybody could ask for, but when a company fails or discontinues to manufacture a model, the car immediately loses from one-third to one-half of its natural value.

orphenadrine (ǭɪfe·nădrīn). *Pharm.* [f. Or(tho- + *-phenadrine* (contraction of *di)phenhydramine).] A bitter-tasting ortho-methyl derivative of diphenhydramine used (as its white crystalline citrate or hydrochloride) as an antispasmodic, esp. in the treatment of Parkinsonism; $(CH_3C_6H_4)$-$CH(C_6H_5)\cdot O\cdot CH_2CH_2N(CH_3)_2$.

1957 *Jrnl. Amer. Med. Assoc.* 13 Apr. 1352/1 The new drug, orphenadrine (Disipal) hydrochloride..has proved exceptionally beneficial in the control of..symptoms of Parkinsonism, without increasing tremor. **1963** *Practitioner* Nov. 646 Two cases are recorded in which the distressing symptoms of persistent hiccup were relieved by orphenadrine citrate. **1977** *Lancet* 16 Apr. 858/2 He was put on orphenadrine but because of progressive extrapyramidal disability levodapa with benserazide ('Madopar') was prescribed.

Orphic, *a.* (*sb.*) Add: **A.** *adj.* **3.** Of, pertaining to, or characteristic of Orphism (*Orphism 2).

[**1913** G. Apollinaire *Méditations Esthétiques* vii. 25 *Le cubisme orphique* est l'autre grande tendance de la peinture moderne. C'est l'art de peindre des ensembles nouveaux avec des éléments empruntés non à la réalité visuelle, mais entièrement créés par l'artiste. **1914** A. J. Eddy *Cubists & Post-Impressionism* (1915) v. 69 *Cubism Orphique* is created entirely by the artist; it takes nothing from visual, objective realities, but is derived wholly from the painter's imagination; it is pure art.] **1950** D. Cooper tr. *Raynal's Hist. Mod. Painting* III. 52 Apollinaire used to distinguish between 'scientific Cubism', 'physical Cubism', 'instinctive Cubism' and 'Orphic Cubism'. **1959** *Listener* 19 Nov. 869/1 He [*sc.* Apollinaire] distinguished two kinds of pure Cubism—scientific and Orphic... Orphic Cubism..dealt with the universe of mind and imagination, the inner world. **1974** *Encycl. Brit. Micropædia* VII. 594/3 Apollinaire's use of the word 'Orphic' recalls both the Symbolist painters' use of the term 'Orphic Art'..and the poetry of Orpheus.

Orphism. Add: **2.** A movement within Cubism, identified by Guillaume Apollinaire (1880–1918) and pioneered by a group of French painters calling themselves *Le Section d'Or*, which emphasized the lyrical use of colour in pure abstract designs.

1914 A. J. Eddy *Cubists & Post-Impressionism* (1915) xiv. 207 Superficial Impressionism leads naturally to the painting of pure color effect—*color music, orphism, compositional* painting. **1915** *Forum* (N.Y.) Dec. 663, I have always held that Orphism and Simultaneism are merely extended Impressionism. **1959** H. Read *Conc. Hist. Mod. Painting* iii. 91 Robert Delaunay..was responsible for another deviation from orthodox Cubism... This deviation Apollinaire christened Orphism. **1959** Brown & Slater tr. *Pasternak's Safe Conduct* xi. 259 In its symbolism, that is to say, in everything in the imagery verging on orphism,..the romantic view of life is devastatingly vivid and it is incontrovertible. **1971** J. Willett in A. Bullock *20th Cent.* x. 235/1 Expressionism..was an all-absorbent force which sucked up every other new tendency since Fauvism (Cubism, Futurism and Delaunay's near-abstract Orphism). **1976** *Telegraph* (Brisbane) 20 Oct. 2/3 Apart from orphism, futurism and several other arty isms, the doc is an expert on early Australian painters.

Hence **Orphist** (less commonly **Orphiste**) *sb.* and *a.*

1914 A. J. Eddy *Cubists & Post-Impressionism* (1915) v. 60 Today we have the 'Neo-Impressionists',..the 'Futurists', the 'Orphists', [etc.]. *Ibid.* 64 A form of dramatic representation that is essentially Cubist, Futurist, and Orphist in its expression. **1915** *Blast* July 41 Less interesting..is the Orphiste movement. Delaunay is the most conspicuous Orphiste. **1959** J. Golding *Cubism* i. 39 Léger,..whose art was now becoming more obviously divergent, is also sometimes referred to as an Orphist. **1959** *Listener* 13 Aug. 253/2 After 1911..he [*sc.* Chagall] begins to place his figures among arbitrary arcs of colour which are like caricatures of *Orphist* forms. **1970** C. Barrett *Op Art* i. 9/2 Delaunay..and..other Orphists were familiar with Chevreul's theory of simultaneous contrast.

orpiment. (Later example.)

1969 R. L. S. Bruce-Mitford *Art of Codex Amiatinus* 3 The third leaf, a purple-stained folio with text written in yellow orpiment (not gold), carried on its recto and verso respectively a prologue and a table of contents.

orra, *a.* Add: *orra man* (also *orraman*) (earlier and later examples.)

1802 J. Sibbald *Chron. Scot. Poetry* IV. (Gloss.), *Orrow man*, a day labourer. *a* **1867** E. Smith *Mem. Highland Lady* (1898) ix. 161 Orraman means the jobber or Jack-of-all trades. **1937** F. Niven *Staff at Simson's* xxv. 243 Of his return the cashier was made aware by the arrival in his office of the odd-job man—the 'orra man'. **1955** in *Sc. Nat. Dict.* (1965) VI. 494/2 (Advt.), Orraman (married) or Orrawoman required for Hillhead, attend some cattle in winter. **1973** *Courier & Advertiser* (Dundee) 14 Feb. 1/3 (Advt.), Tractor-Orraman wanted for intensive livestock and cropping.

Orrefors (ǫ·rĕfǭɪz). The name of the town in Sweden where glass is manufactured, used to designate glassware produced there and the style of decoration characteristic of it.

1928 T. Palm tr. *Wettergren's Mod. Decorative Arts Sweden* 23 The relation between the figures of the Orrefors glasses and these daughters of the Renaissance and of the baroque style proves to be a result of the material..and of the technique. **1929** *Encycl. Brit.* X. facing p. 404 (*caption*), Modern cut and engraved European glass. 1. 'Fashion' vase of engraved Orrefors glass designed by Edward Hald. **1931** G. Janneau *Mod. Glass* ii. 21 The Orrefors production is not entirely limited to this style. **1935** N. Mitchison *We have been Warned* iv. 427 The Orrefors mirror and the linen sheets on the bed. **1961** E. M. Elville *Collector's Dict. Glass* 174/2 The examples ..are illustrative of the vigorous, muscular figures which have established the Orrefors tradition. **1970** *House & Garden* May 41/1 Scandinavian elegance of design..is, literally, crystallized in Orrefors glass. **1970** W. Wager *Sledgehammer* (1971) iii. 13 He put down the Orrefors pitcher.

orright (ǭrəi·t), repr. a vulgar or colloq. pronunciation of *all right* s.v. Right *a.* 15 c.

1941 M. Treadgold *We couldn't leave Dinah* iii. 58 Thomas gave a little sigh of relief. 'Orright,' he said. **1969** E. McGirr *Entry of Death* v. 89 It's orright, yer money's on. **1971** J. Wainwright *Dig Grave* 54 'Here, have another toffee, Billy.' 'Orright. Ta.' **1971** J. Ripley *Davis doesn't live Here* 18 'You wanna say summat?'..'Orright. Say it.' **1978** J. Wainwright *Jury People* xxiii. 68 'I'll do orright.' 'On social security?' 'Aye.'

orsell, os(s)el, ossil, oz(z)el, etc. varr. Norsel *sb.*

1750 in *Sc. Nat. Dict.* (1965) VI. 495/3 The Ossels..each 18 inches long..are fixed to two Mashes at one End by an Eye. **1881** *Proc. Soc. Antiquaries Scotl.* III. 150 The nets ..are attached to a strong..rope by means of thinner cords known as 'ozzels'. **1921** [see *norseller]. **1972** M. F. Wakelin *Patterns Folk Speech Brit. Isles* 18 Figure 2 shows the *sole-rope* as made fast to the net by the *ossils*. *Ibid.* 24 There are four fairly well-defined usages. *Ossil* extends, with one or two breaks, from Burnmouth as far as Lossiemouth. **1973** W. Elmer *Terminol. Fishing* ii. 54 The bait..is put on a *skewer*..or hung on an *orsle*..(a short piece of line), and the whole is weighted with stones.

orseller, var. *norseller.

ort. (Later examples.)

1913 D. H. Lawrence *Love Poems* 50 Then what art colleyfoglin' for?—I'm not havin' your orts and slarts. **1917** —— *Look! We have come Through!* 60 To me it seems the seed is just left over From the red rose-flowers' fiery transience; Just orts and slarts. **1922** Blunden *Shepherd* 44 With hungry hubbub begging crusts and orts. **1940** V. Woolf *Writer's Diary* 30 May (1953) 334 Scraps, orts and fragments. **1946** K. Tennant *Lost Haven* (1947) xi. 169 'Orts and leavings, that's what pigs eat.' The idea of a set of pigs fattening on what Bee-Bonnet left from his scanty meal amused the loungers. **1950** R. Moore *Candlemas Bay* 223 Neal took the orts out to the hens and hurled them, dish and all, over the henyard wall. *Ibid.* 224 Neal started back to the house, kicking the orts dish along the gravel walk in front of him. **1972** J. Metcalf *Going Down Slow* iv. 61 When you've eaten every last ort and scrap, would you like dessert? Coffee? Brandy? **1976** 'M. Innes' *Gay Phoenix* iv. 54 A waiter..wheeled up a trolley of elaborately bedized scraps, orts and broken meats.

ortanique (ǭɪtănī·k). [f. Or(ange *sb.*¹ + Tan(gerine *sb.* 2 + Un)ique *a.*] A citrus fruit resembling a slightly flattened orange, produced by crossing an orange and a tangerine and cultivated in the West Indies. Also *attrib.*

1937 *Times* 16 Feb. 11/2 Jamaica shows ortaniques, a cross between the usual orange and a tangerine. **1961** *Spectator* 9 June 857 The West Indian experimental ortaniques, a cross between a tangerine and an orange. **1969** 'J. Morris' *Fever Grass* viii. 72 I've got to risk turning fifty acres of certain sugar into avocadoes, ortaniques and Bombay mangoes. **1969** *Harrod's Summer Food News* 6/2 Canadian Ortanique Juice 19 oz. tin 3/3.

‖ **orteguina** (ǭɪtegī·nă). [Sp., f. the name of Domingo *Orteg(a* (born 1906), Spanish bullfighter, who practised it + *-uina*.] In bullfighting, a decorative pass made with the muleta; = *manoletina.

The more usual term is *manoletina*.

1957 A. MacNab *Bulls of Iberia* viii. 86 Domingo Ortega introduced a variant in which the man..passes his left arm behind his back and grips the lower edge of the cloth with his left hand. This was called the *orteguina* but is now generally called *manoletina* owing to its having been popularised by the late Manolete. **1959** [see *manoletina]. **1962** B. Conrad *Encycl. Bullfighting* 174/1 *Orteguina*, a cape pass attributed to Domingo Ortega in Spain... Also may refer to a *muleta* pass similar to the *manoletina*. **1967** McCormick & Mascareñas *Compl. Aficionado* iv. 147 The manoletina is another embellishment... Formerly, it was called the *orteguina*, after Domingo Ortega, who took it from the nineteenth-century repertoire and refurbished it. Manolete picked it up and further refined (i.e. vulgarized) it.

orter (ǭ·ɪtəɪ), a representation of a colloquial or vulgar pronunciation of *ought to*. Cf. *oughta, oughter.

1864 Hotten *Slang Dict.* 196 Where's the party as 'ad a orter be lookin' arter this 'ere 'oss? **1886** F. H. Burnett *Little Lord Fauntleroy* xiii. 243 'Seems like somethin' orter be done,' said Mr. Hobbs. **1897** Kipling *Capt. Cour.* iv. 88, I orter ha' warned you. **1917** E. O'Neill *Long Voyage Home* in *Smart Set* Oct. 84/2 Orter wear a muzzle, you ort! **1931** *Amer. Speech* VII. 91 Ater Maud's death he orter do better'n' he use' to. *a* **1966** M. Allingham *Cargo of Eagles* (1968) v. 73 His regular mate..wasn't there to 'elp 'im..like wot 'e orter 'ave bin.

‖ **orterde** (ǭ·ɪtēəɪdə). *Soil Sci.* [G.] (See quot. 1939.)

1930 *Bull. Wisconsin Geol. & Nat. Hist. Surv.* LXXVIIa. 86 Dark rusty brown sandy loam containing coffee-brown cemented lumps (the orterde of the Podsol). **1936** J. S. Joffe *Pedology* vi. 132 These chemical compounds are responsible for the condition of ortstein and orterde (hardpan). **1939** *U.S. Dept. Agric. Yearbk. 1938* 1173 *Ortstein*, hard irregularly cemented, dark-yellow to nearly black sandy material formed by soil-forming processes in the lower part of the solum. Similar material not firmly cemented is known as orterde. **1965** B. T. Bunting *Geogr. Soil* xiii. 152 Modal iron podsols are the true forest podzol, usually under spruce, with deep F1, 2, 3 layers, and deep orterde rather than pan.

orthaxial, *a.* (Example.)

1886 J. A. Ryder in *Rep. U.S. Comm. Fisheries 1884* 985 The word orthaxial is used to designate the archaic straight type of vertebrate axis which is not bent upwards at its posterior extremity.

orthesis (ǭɪþī·sis). *Med.* [f. Gr. ὀρθ-ός straight, right + *-esis*, after Prosthesis.] = *orthosis.

1956 R. L. Bennett in *Physical Therapy Rev.* XXXVI. 721/1 To my knowledge, the words, 'orthesis', 'orthetics', and 'orthetic devices', have not been used before... The word 'orthesis' may be defined as a medically prescribed device applied to or around a weakened bodily segment to give support and increased function. **1963** *Med. Electronics & Biol. Engin.* I. 511/1 Prehensile motion of the hand..is to be performed by an externally powered orthesis, but is not to be preprogrammed and is to be completely controlled by the patient. **1971** *Rand Daily Mail* 4 Sept. 14/1 He has developed various aids for the disabled. The most useful, perhaps, is the motorised orthesis or hand splint which can be connected to a blow-and-suck pipe or any other control system.

orthicon (ǭ·ɪþikǫn). [abbrev. of next.] A kind of television camera tube similar to the iconoscope but having a transparent target plate, so that the scanning beam can be made to strike normally the opposite side to that on which the image is projected, and employing low-speed electrons for the beam, resulting in increased sensitivity and an absence of the spurious signals associated with the iconoscope. See also *image orthicon* (*image *sb.* 8).

The term was registered in the U.S.A. as a proprietary name in 1940, but it is now a generic term in the public domain.

1939 [see next]. **1942** *Electronic Engin.* XV. 127 Because of the greater light sensitivity of the orthicon it is expected that the equipment will fill a need for lightweight equipment to be used under adverse light conditions. **1955** G. M. Glasford *Fund. Television Engin.* iv. 98 The disadvantages of the orthicon have encouraged the search for better tubes in spite of the improvement over the iconoscope in the matter of increased sensitivity. The orthicon suffers from a very poor contrast scale with a very small range of intermediate grays as compared with the iconoscope. **1968** *Brit. Med. Bull.* XXIV. 261/2 Tubes such as..orthicons integrate the light energy incident upon each part of the tube over the entire frame-scanning interval.

orthiconoscope (ǭɪþəikǫ·nǒskǒᵘp). Now *rare* or *Obs.* [f. Orth(o- + *iconoscope; see quot. 1939.] = prec.

1939 *Electronics* July 11/1 On June 7th Albert Rose and Harley Iams of the RCA Manufacturing Company Research Laboratories at Harrison, revealed..the details of a new developmental television pick-up tube which..is one of the most significant advances in television electronics since the advent of the iconoscope itself. The new tube's formal name is 'Orthiconoscope' ('Orthicon' for short) from the Greek root ortho meaning 'straight' and iconoscope for 'image-viewer'. The name derives from the fact that the curve between input light and output current is a straight line, in contrast to the similar curve of the iconoscope which is not linear. **1940** D. G. Fink *Princ. Television Engin.* iii. 111 The orthiconoscope displays no spurious signal, and there is accordingly no background shading defect to be compensated. **1953** Amos & Birkinshaw *Television Engin.* I. 81 The orthicon (orthiconoscope in full) was developed by Rose and Iams in America and used in that country immediately before the Second World War for outside broadcasts.

ortho-. Add: **1.** **o·rthocone,** the conical shell of certain fossil nautiloid cephalopods or a fossil characterized by a shell of this shape; hence **orthoco·nic** *a.*, of or pertaining to a fossil or a shell of this type; **o·rtho-cousin,** one of cousins whose related parents are of the same sex (cf. *cross-cousin* s.v. *cross- B); **ortho-di·gita** (see quot. 1939); hence **orthodi·gital** *a.*; **orthofe·rrite,** any of the compounds with the formula $AFeO_3$, where A is a trivalent metal ion (usu. a rare earth), which have an orthorhombic crystal structure and exhibit weak ferromagnetism at room temperature; **o·rthoform** *Pharm.* [after Chloroform *sb.*], methyl *m*-amino-*p*-hydroxybenzoate, $C_8H_9NO_3$, a crystalline compound with anæsthetic properties which has been used as a dusting powder for wounds and ulcers and in dentistry; **o·rthogeosy·ncline** *Geol.* [ad. G. *orthogeosynklinale* (H. Stille 1935, in *Sitzungsber. d. preuss. Akad. d. Wissensch.* (*Phys.-mat. Kl.*) 182)], a linear geosyncline between a continental and an oceanic kratogen (craton), typically comprising a miogeosyncline and an adjacent eugeosyncline; hence **o:rthogeosyncli·nal** *a.*; **o·rthogneiss** *Petrogr.* [a. G. *orthogneiss* (H. Rosenbusch *Elem. d. Gesteinlehre* (1898) 467)], gneiss derived from igneous rocks; **orthokine·sis** *Zool.* [*kinesis 2], a kinesis in which linear movement is shown; hence **orthokine·tic** *a.*; **o:rthopercu·ssion** *Med.*,

very light diagnostic percussion of the chest by means of one finger striking the knuckles of another bent at right angles and with its tip resting in an intercostal space; **orthophony** (further examples); hence **orthopho·nic** *a.*, pertaining to orthophony; reproducing sounds correctly; **orthopho·ria** *Ophthalm.* [Gr. φορός bearing], the state of perfect oculomotor balance, in which the visual axes tend towards being parallel in the absence of a fusion stimulus; hence **orthopho·ric** *a.*; **o·rthopho:to, orthopho·tograph,** an image produced optically or electronically from aerial photographs by eliminating distortions of angles and scales so as to give a result corresponding to a planimetric map; **orthopho·tomap,** a map made up from an assembly of orthophotographs on which relief has been indicated by contours or shading; **o·rthophyre** *Petrogr.* [a. F. *orthophyre* (H. Coquand 1856, in *Mém. de la Soc. d'Émulation du Dép. du Doubs* I. 64)], porphyry in which the phenocrysts are chiefly of orthoclase; **orthophy·ric** *a. Petrogr.* [ad. G. *orthophyrisch* (H. Rosenbusch *Mikrosk. Physiogr.* (ed. 2, 1887) II. 594)], (of the groundmass of porphyritic rocks) containing short, stout feldspar crystals of rectangular or quadratic cross-section; **o·rthoploid** *a. Cytology* [a. G. *orthoploid* (H. Winkler 1916, in *Zeitschr. Bot.* VIII. 422)], having a complete or balanced set of chromosomes: with most authors = *euploid *a.*; **orthopyro·xene** *Min.* [ad. G. *orthopyroxen* (E. Düll, at the suggestion of F. Rinne, in *Zeitschr. f. Krystallogr. und Mineral.* (1902) XXXVI. 654)], any orthorhombic pyroxene; **ortho·(r)rhaphous** *a. Ent.* [f. mod.L. *Orthorhapha* (later *Orthorrhapha*) (F. Brauer *Monographie der Oestriden* (1863) 33), f. Gr. ῥαφή seam], belonging or pertaining to the suborder Orthorrhapha, a group of dipterous insects in which the adult emerges from the puparium through a straight or T-shaped split; cf. *cyclorrhaphous* adj. (*cyclo-); **orthotecto·nic** *a. Geol.* [ad. G. *orthotektonik* sb. (H. Stille *Einführung in den Bau Amerikas* (1940) i. 9)] formed by, or of the nature of, a deformation which produces complicated and crowded systems of fold belts such as the Alps and is characterized by much magmatism and lateral thrusting (believed to be characteristic of orthogeosynclines); cf. *paratectonic* adj. (*b*) s.v. *para-¹ 1; **orthoto·pic** *a. Med.* and *Biol.* [Gr. τοπικ-ός of place], involving transplantation of a structure to the same site in the recipient as it occupied in the donor; also said of the transplanted structure; hence **orthoto·pically** *adv.*; **o·rthovoltage** *Med.*, a voltage (in an X-ray tube) of the size used in conventional deep therapy (200–400 kilovolts); usu. *attrib.*; **o·rthowater** = *polywater.

1900 A. Hyatt in C. R. Eastman tr. *von Zittel's Text-bk. Paleont.* I. 573 An orthocone is the young of the straight as well as many of the coiled forms [of fossil cephalopod]. **1935** Twenhofel & Shrock *Invertbr. Paleontol.* ix. 368 Early cephalopod shells were dominantly straight or but slightly coiled... Among the Nautiloidea this type of shell is known as an orthocone. **1969** Bennison & Wright *Geol. Hist. Brit. Isles* iv. 88 In the Durness sequence..the commonest fossils are gastropods..together with orthocone cephalopods. **1970** R. M. Black *Elements Palaeont.* viii. 78 *Orthoceras* and similar orthocones range in length from about 3 cm to possibly 460 cm. **1926** A. F. Foerste in *Jrnl. Sci. Lab. Denison Univ.* XXI. v. 304 (*heading*) Orthoconic genera. *Ibid.* 310 The specimen is figured as having its ventral side ribbed and fluted vertically in a manner very similar to that of *Kionoceras*, a form of ornamentation unknown in any other orthoconic triangular cephalopod. **1935** Twenhofel & Shrock *Invertbr. Paleontol.* ix. 367 (*caption*) Idealized diagram of an orthoconic cephalopod. **1974** *Nature* 8 Feb. 396/1 Glaciomarine beds..in South West Africa have yielded..an orthoconic nautiloid. **1918** J. G. Frazer *Folk-Lore in Old Testament* II. ii. vi. 98 It has become customary to call the marriageable cousins *cross-cousins*, because..the related parents are of opposite or cross sexes. There has hitherto been no special name for the unmarriageable cousins, the children of two brothers or of two sisters, but for convenience I propose to call them *ortho-cousins* to distinguish them from cross-cousins. In the case of ortho-cousins the related parents are of the same sex. **1932** [see *cross-cousin* s.v. *cross- B.]. **1937** R. Stout *Red Box* iii. 38 Ortho-cousins are..the children of two brothers or of two sisters. **1972** D. Davies *Dict. Anthropol.* 141 *Ortho-Cousin*, a term little used now,..can be a synonym for parallel cousin..or for a parallel cousin of the same unilineal descent..group as the person concerned. **1937** *Fiopian Footprints* VIII. 3/2 Let us hope that the history of orthodigita will in time be replete with names of those who will..have illuminated Podiatry. **1939** H. A. Budin in *Jrnl. Exper. Podiatry* I. 19 (*heading*) New and improved orthodigital appliances in the non-surgical correction of deformities of the..toes. *Ibid.*, In his first lecture on this subject, delivered at a local society meeting in February 1934, the writer introduced some of the appliances and the technic which he had devised... For the purpose of designation of this newer phase of therapeusis, the author..chose the term Orthodigita... Orthodigita may be defined as the amelioration or correction, by non-surgical means, of toe deformities or malalignments. **1968** Fisher & Whitney in F. Weinstein *Princ. & Pract. Podiatry* xii. 265/1 Although permanent orthodigital correction is sought in adults also,..the patient may need to wear the appliance continually in order to keep the toes in proper alignment. **1978** *Chiropodist* XXXIII. 105 Although I personally favour the qualities (for most orthodigita) of KE 20 silicone rubber, I have found that Otoform provides..an inexpensive and effective method of introducing removable silicone appliances to patients. **1956** *Jrnl. Chem. Physics* XXIV. 1239 (*heading*) Magnetic properties of a gadolinium orthoferrite, $GdFeO_3$, crystal. **1966** *McGraw-Hill Encycl. Sci. & Technol.* V. 219/2 Many of the orthoferrites are strongly ferromagnetic at liquid helium temperatures. **1971** *Sci. Amer.* June 83/2 The first magnetic materials found to have the desired properties for studying the new bubble technology were orthoferrites, a special class of ferrites with the chemical formula $RFeO_3$, where R represents yttrium or one or more rare-earth elements. Samarium terbium orthoferrite is a good example. **1897** *Lancet* 18 Sept. 738/1 It has been found by Dr. Einhorn and Dr. Heintz, of Munich, that the compound methylic ether of amidoxybenzoic acid is possessed of remarkable anæsthetic, or rather analgesic, properties when locally applied. To this substance the name of 'orthoform' has been given. **1940** F. R. Davison *Synopsis Materia Medica* xi. 383 Orthoform..has also been used in dentistry, nasal catarrh, hay fever, and in similar conditions. **1965** Faulconer & Keys *Foundations Anesthesiol.* II. iv. 806 In the light of these considerations it is understandable that the Orthoform group anesthetizes better than the corresponding benzoyl combinations. **1941** Orthogeosynclinal [see *parageosyncline* s.v. *para-¹ 1]. **1945** *Bull. Geol. Soc. Amer.* LVI. 1172 Marginal geosynclines gaining principal detritus from uplifts in orthogeosynclinal belts. **1975** *McGraw-Hill Yearbk. Sci. & Technol.* 223/1 The orthogeosynclinal belts of Stille and Kay have come to be understood as an assemblage of crustal features related to continental shelf subsidence. **1936** tr. H. Stille in *Bull. Amer. Assoc. Petroleum Geologists* XX. 853 Although the orogenic movements took place at the same time in different areas, they were strong (Alpine type) only in certain mobile belts which had developed as orthogeosynclines. **1951** *Mem. Geol. Soc. Amer.* XLVIII. 88 The later Devonian and Carboniferous are of argillite and graywacke when detritus could reach the subsiding structural basin from distant highlands of rocks laid in orthogeosynclines. **1968** Orthogeosyncline [see *foredeep* s.v. *fore- 5]. **1902** Orthogneiss [see *paragneiss* s.v. *para-¹ 1]. **1932** A. Harker *Metamorphism* xvii. 271 The more or less distinctly banded crystalline rocks which are conveniently styled orthogneisses (in contradistinction to paragneisses, which are highly metamorphosed sediments) attain in some countries a vast development. **1962** *Mineral. Abstr.* XV. 553/2 Alkaline syenitic orthogneisses form a lenticular body 5 km in length in biotite gneisses in the Cevadais area, near Ouguela, Alto Alentejo, Portugal. **1937** D. L. Gunn et al. in *Nature* 18 Dec. 1064/2 Variations in generalized, undirected, random locomotory activity..are kineses... We propose to divide kineses into (*a*) orthokineses..variations in linear velocity (previously called simply kineses), [etc.]. **1940** Fraenkel & Gunn *Orientation of Animals* ii. ii. 17 Woodlice aggregate in moist air.. by means of an ortho-kinesis. **1971** J. D. Carthy in J. E. Smith et al. *Invertebr. Panorama* xii. 251 This behaviour, consisting of changing rates of movement with different levels of stimulation, is known as an orthokinesis. **1958** —— *Introd. Behaviour Invertebr.* xii. 317 When the stones are beneath the surface..they [*sc.* chitons] do not tend to congregate by a simpleorthokinetic response. **1973** *Nature* 16 Nov. 168/1 An orthokinetic effect may serve to influence adult movements, as their rate of progress over the reefs is almost certainly affected by the ease with which they can find adequate food and shelter from daylight. **1907** *Practitioner* Apr. 530 The chief disadvantage of Goldscheider's ortho-percussion is that it requires an absolutely silent room. **1916** L. F. Barker *Monographic Med.* II. 499 In orthopercussion, the force of the blow is directed exactly perpendicular to the surface. **1966** *Lancet* 31 Dec. 1469/1 The technique of percussion (orthopercussion) of the patient's abdomen in order to recognise peritoneal irritation is shown. **1926** *Daily Colonist* (Victoria, B.C.) 2 July 6/6 (Advt.), The genuine His Master's Voice Victrola is the only true orthophonic. **1927** *Gramophone* V. 309/2 The gramophone part of it contains an improved form of orthophonic horn. **1954** Pei & Gaynor *Dict. Linguistics* 155 *Orthophonic*, relating to orthophony..; conformable with the standard or accepted rules of pronunciation. **1969** *John Edwards Mem. Foundation Q.* V. ii. 81 This valuable discographic aid lists all Victor Recordings..starting with the introduction of the electrical 'orthophonic' recording system in February, 1925. **1845** W. Russell *Orthophony* p. xiii, The term orthophony is used to designate the art of cultivating the voice. The systematic cultivation of the vocal organs,..is a branch of education for which our own language furnishes no appropriate designation. The compiler of this manual has ventured to adopt, as a term convenient for this purpose, the word *orthophony*—a modification of the corresponding French word '*orthophonie*', used to designate the art of training the vocal organs. **1954** Pei & Gaynor *Dict. Linguistics* 155 *Orthophony*, Correct pronunciation or articulation. **1886** Orthophoria [see *exophoria* s.v. *exo-]. **1907** J. H. Parsons *Dis. Eye* xxviii. 563 In cases of latent squint the position of rest is not orthophoria, with the visual axes parallel, but heterophoria, with some deviation of the axes. **1950** F. H. Alder *Physiol. Eye* x. 386 It is unfortunate that the implication is frequently made that orthophoria is the normal condition and heterophoria an abnormal one. This is not true. **1888** *Arch. Ophthalm.* XVII. 159 In the orthophoric state the eyes are able to unite images when a prism of 2° to 3° is introduced with its base up or down before one of the eyes. **1954** S. Duke-Elder *Parsons' Dis. Eye* (ed. 12) xxix. 483 Since the position of rest is usually one of slight divergence, few people are orthophoric and some degree of heterophoria is almost universal. **1965** *Photogrammetric Engin.* XXXI. 223/1 Several stereo aerial models from test areas were successfully compiled into contour maps and orthophotos. **1972** *McGraw-Hill Yearbk. Sci. & Technol.* 245 The orthophoto, like any good map, allows the engineer to lay out a proposed highway with accurate scale, direction, and curvature. **1955** *Photogrammetric Engin.* XXI. 529/2 Given an orthophotograph, the engineer, surveyor, forester, geologist,..can correlate points imaged on the orthophotograph with points observed on the ground, and..can make direct measurements on the orthophotograph to determine distances between points. **1970** J. A. Howard *Aerial Photo-Ecol.* xii. 136 No doubt orthophotographs will have a wide application in natural resource studies not requiring a stereoscopic examination. **1967** *Photogrammetric Engin.* XXXIII. 274/1 The altitude contours derived from the height measured can then be added, along with any desired annotations, to form an accurate 'orthophotomap' of the area of interest. **1974** *Geo Abstr.* G. 520 The Topographic Division of the U.S. Geological Survey produces a series of orthophoto products ranging from separate photos for in-house use..to the multicoloured orthophotomap. **1890** *Cent. Dict.*, Orthophyre. **1895** A. Harker *Petrol.* viii. 102 The most usual type of orthoclase-porphyry (orthophyre of Rosenbusch) is exemplified by dykes and sills in the Carboniferous of Thuringia. **1930** Peach & Horne *Geol. Scotl.* iv. 108 Examples of dykes of orthophyre occur on Sgonnan Mòr. **1947** E. E. Wahlstrom *Igneous Minerals & Rocks* x. 301 Orthophyre is a porphyritic trachite consisting largely of orthoclase. **1895** A. Harker *Petrol.* viii. 102 Other porphyrites have the 'orthophyric' type of groundmass (with short felspar-prisms), as in the porphyries. **1964** G. A. Joplin *Petrogr. Austral. Igneous Rocks* v. 65 Most trachytes are porphyritic with phenocrysts of anorthoclase in an orthophyric and/or trachytic groundmass. [**1920** *Svensk Bot. Tidskr.* XIV. 301, I have in my material found the following numbers: 14, 21, 28, 35 and 42. Further, in some specimens, I met with numbers not being a multiple of 7 (anorthoploid forms).] **1932** *Proc. 6th Internat. Congr. Genetics* II. 63 The ratios between orthoploid and aneuploid gametes in the cases studied were found to be 1·4:1. **1937** T. Dobzhansky *Genetics & Origin of Species* ix. 268 In a translocation heterozygote at least six classes of sex cells can be produced... Classes 1 and 2 carry normal gene complements..; 1 and 2 are termed regular or orthoploid. **1963** *Portugaliae Acta Biol.* A. VII. 8 The unfortunate terms orthoploid and anorthoploid have been used in various senses also by more recent writers. *Ibid.* 9 There is no doubt that euploid and aneuploid should be maintained and orthoploid and anorthoploid definitely dropped. **1903** *Mineral. Mag.* XIII. 374 Following Rinne, E. Düll..proposes the terms ortho-pyroxene and klinopyroxene. **1940** *Amer. Mineralogist* XXV. 282 Orthopyroxenes of plutonic igneous rocks normally show well developed diopsidic lamellae. **1963** W. A. Deer et al. *Rock-Forming Min.* II. 33 Many orthopyroxenes can be distinguished from clinopyroxenes by their characteristic pink to green pleochroism. **1970** *Science* 28 Aug. 866/2 (*heading*) Orthopyroxene-plagioclase fragments in the lunar soil from Apollo 12. **1890** *Cent. Dict.*, Orthorhaphous. **1899** D. Sharp in *Cambr. Nat. Hist.* VI. vii. 458 In the Mesozoic epoch the Order [*sc.* Diptera] is found as early as the Lias, the forms being exclusively Orthorrhaphous. **1946** *Nature* 2 Nov. 636/2 Against the orthorrhaphous Diptera, 'Gammexane' shows a high degree of activity. **1961** Orthorrhaphous [see *cyclorrhaphous* adj. s.v. *cyclo- 1]. **1956** L. U. de Sitter *Struct. Geol.* i. 16 A later uplift, separated from the last orogenic paroxysmal phase by a period of erosion, is typical of orthotectonic regions. **1969** *Earth & Planetary Sci. Lett.* VI. 189 The orthotectonic orogens forming island arcs such as Japan lie entirely within ocean basins. **1969** J. F. Dewey in M. Kay *N. Atlantic* xxiv. 309/1 Strata ranging in age from late Precambrian through early Ordovician constitute a northern orthotectonic belt, characterized by complex recumbent and commonly triclinic fold geometry and high-grade metamorphism. **1921** *Jrnl. Exper. Zool.* XXXII. 7 Limb bud placed in natural location—orthotopic transplantation. **1958** *Immunology* I. 46 The survival times of successive sets of orthotopic scale (skin) homografts revealed that increasing systematic immunity develops rapidly in stepwise fashion. **1968** *National Observer* (U.S.) 29 Jan. 4/4 Dr. Starzl performed the first orthotopic liver transplant, in which the diseased liver is removed and another implanted, in 1963. **1921** *Jrnl. Exper. Zool.* XXXII. 61 The shoulder-girdle in orthotopically grafted limbs is derived in part from the host and in part from the transplanted tissue. **1974** *Nature* 11 Oct. 553/1 When solid tissue allografts are transplanted orthotopically to alien hosts they are rejected with a characteristic tempo and vigour that depends primarily on the immunogenetic disparity between donor and host. **1976** *Ibid.* 22 Jan. 209/1 We used a standardised H-test procedure in which tail-skin grafts were orthotopically exchanged in a 'reciprocal circle' among syngeneic mice. **1967** M. E. J. Young *Radiological Physics* (ed. 2) xi. 372 Superficial therapy..60–120 kVp. Medium voltage therapy... 120–140 kVp. Deep therapy or orthovoltage therapy..200–400 kVp. Megavoltage therapy... Above 1 MV. **1972** Barnes & Rees *Conc. Textbk. Radiotherapy* iii. 63 In recent years there has been a tendency for deep X-ray therapy (orthovoltage radiation) to be largely superseded by megavoltage therapy as the advantages of the higher energy radiation become more generally appreciated. **1976** *Lancet* 6 Nov. 1031/2 All patients received induction treatment..for 4 weeks followed by 2400 rad of orthovoltage cranial irradiation plus five intrathecal injections of methotrexate. **1966** B. V. Derjaguin in *Discussions Faraday Soc.* XLII. 118 The usual state of water and certain other liquids is thermodynamically metastable... It would be convenient to call 'usual water' metawater, and the anomalous columns—orthowater. **1969** *Nature* 27 Dec. 1293/1 Water condensed into glass or quartz capillaries has unusual properties, which have been ascribed to the formation of a new polymer termed 'orthowater', 'anomalous' water or 'polywater'. **1970** *Compton Yearbk.* 176/2 The substance, variously called orthowater, anomalous water, polywater, and superwater, differs radically from ordinary water.

2. b. **orthoxy·lene,** an isomer of xylene that is a colourless mobile liquid and is now obtained from petroleum naphtha for use esp. as a source of phthalic anhydride.

1872 *Jrnl. Chem. Soc.* XXV. 893 (*heading*) Preparation of orthoxylene from liquid bromotoluene. **1968** *Economist* 14 Dec. 63/3 Both BP Chemicals and ICI have existing plants for phthalic anhydride. ICI's 20,000 ton plant at Wilton manufactures from naphthalene, an older method being replaced by the ortho-xylene method as oil supersedes coal tar as a source of raw materials for the chemical industry.

3. *Physics* and *Chem.* Of, pertaining to, or designating the form of some homonuclear diatomic molecules in which (as in orthohydrogen) the two nuclei have parallel spins (see also quot. 1940²); also more widely, characterized by the presence of parallel spins. Also as an independent word.

1927 T. Verschoyle tr. *Haas's Atomic Theory* v. 182 No spectroscopic transition between the normal para-term and the lowest (two-quantum) ortho-term is possible. **1939** J. W. T. Spinks tr. *Herzberg's Molecular Spectra* I. iii. 150 The modification with the *greater* statistical weight is usually called the ortho modification and that with the smaller weight the para modification. **1940** Glasstone *Text-bk. Physical Chem.* i. 79 From the spectrum of helium it is known that the ortho-levels have less energy than the par-levels with the same values of the quantum numbers *n* and *l*. *Ibid.* 96 Symmetrical polyatomic molecules, such as water, deuterium oxide, cyanogen and acetylene, exist in ortho- and para-forms; they behave in a manner similar to hydrogen, deuterium and nitrogen molecules, since the other atoms, viz., carbon and oxygen, have no nuclear spins. **1966** D. H. Whiffen *Spectroscopy* ix. 114 The best-known example is hydrogen where the ortho states with odd *J* have three times the degeneracy of the even *J* or para states. **1970** P. J. Wheatley *Chem. Consequences Nucl. Spin* xi. 50 The hydrogen molecule, $^1H_2$... The rotational levels with *J* odd are associated with symmetric nuclear states, that is with ortho states, and..the rotational levels with *J* even are associated with antisymmetric nuclear states, that is with para states. *Ibid.* 51 The deuterium molecule, $^2D_2$,.. Ortho-D_2 will be associated with rotational levels having even values of *J*, and para-D_2 with those having odd values of *J*. **1977** *Sci. Amer.* Oct. 66/3 The parallel quark spins combine to give each meson one unit of spin. This arrangement of spins is known in atomic physics as the 'ortho' configuration.

ortho (ǭ·ɪþo), *a.* (*adv.*) [f. Ortho-.] **1.** *Chem.* (Now usu. italicized.) Characterized by or relating to (substitution at) two adjacent carbon atoms in a benzene ring; at a position adjacent *to* some (specified) substituent in a benzene ring. Also as *adv.*

1876 *Jrnl. Chem. Soc.* I. 240 An influence is exercised by the NO_2-group in favouring the displacement of Cl, Br,.. &c., *only* when it is in the ortho (1:2) or para (1:4) position relatively to one or other of these radicles. **1903** Walker & Mott tr. *Holleman's Text-bk. Org. Chem.* II. 347, 1:2 = 1:6 substitution-products are called *ortho*-compounds. **1920** *Conquest* Apr. 260/3 The ortho acid is the one of use to the Saccharin manufacturer, the para form producing a compound of very slight sweetening power. **1924** [see *meta *a.*]. **1938** L. F. Fieser in H. Gilman *Org. Chem.* I. ii. 73 The ortho coupling and ortho bromination of phenols involve substitution at the carbon atom connected to that carrying the hydroxyl group by a double linkage. **1949** [see *orient *v.* 4 a]. **1968** G. E. Coates et al. *Princ. Organometallic Chem.* ii. 26 Contrast the effect of heat on mercuric benzoate, when substitution of the aromatic ring *ortho* to the carboxyl group occurs. **1971** J. D. Roberts et al. *Org. Chem.* xx. 573 Halogen substituents..strongly orient *ortho* and *para* through conjugation of the unshared electron pairs.

2. *Photogr.* [f. Ortho(chromatic *a.*] Orthochromatic. Also *ellipt.*, an orthochromatic plate.

1904 G. B. Shaw *Let.* 26 July (1972) II. 435, I have..a box of half-plate extra-rapid Ortho plates which I got 6 months ago. **1906** *Westm. Gaz.* 17 May 14/2 The Barnet medium ortho is about half the speed of the well-known extra-rapid Barnet ortho-plate... For the best results the Barnet ortho screen is recommended. **1921** *Glasgow Herald* 27 Apr. 9, I have said something in previous articles about ortho plates. **1956** E. Mytum *Introd. Photogr. Materials* iii. 58 If the emulsion is optically sensitized by a dye which absorbs green and yellow light, it is said to be orthochromatic, often abbreviated to 'ortho'.

3. See *ortho- 3.

orthochoanite (ǭɪþokǭu·ănəit). *Palæont.* Also Ortho-. [f. Ortho- + Gr. χοάν-η funnel + -ite¹ 2 a.] A nautiloid cephalopod having a straight, very short septal neck; also, a member of the obsolete sub-order Orthochoanites, of which such necks were characteristic.

1898 A. Hyatt in *Proc. Amer. Assoc. Adv. Sci.* XLVII. 364 The suborders..are as follows:..IV. Orthochoanites. The siphuncles small except in primitive forms and without deposits, or, if present, these are irregular and are gathered around the funnels. No endosiphuncles. The funnels are straight and the siphuncle is apt to be tubular. **1944** E. O. Ulrich et al. *Ozarkian & Canad. Cephalopods III* (Geol. Soc. Amer. Spec. Papers, No. 58) 27 Some of the Canadian nautilicones have lamellar deposits within the camerae... Apparently they are not present in the holochoanites, but they are being found in many orthochoanites and cyrtochoanites of the Paleozoic and the Triassic. **1952** R. C. Moore et al. *Invertebr. Fossils* ix. 342/1 The Cyrtochoanites have been found to include a very diverse assemblage of which some are closely related to 'Holochoanites' and others to 'Orthochoanites'.

Hence **orthochoani·tic** *a.*

1905 [see *holochoanitic *a.*]. **1944** E. O. Ulrich et al. *Ozarkian & Canad. Cephalopods III* (Geol. Soc. Amer. Spec. Papers, No. 58) 26 In all the coiled Canadian cephalopods, the siphuncle is orthochoanitic in structure, the septal necks being short and straight. **1964** C. Teichert in R. C. Moore *Treat. Invertebr. Paleont.* K. 95/2 Kuhn.. separated the straight, orthochoanitic nautiloids as a new 'suborder' termed Orthoceracea. *Ibid.* 174/2 Siphuncle large..; septal necks very short, orthochoanitic.

orthochromatic, *a.* Add: In mod. use, having a sensitivity which is more nearly uniform throughout the visible range than that of silver halide used alone, but which is relatively low in the red and high in the blue part of the spectrum (cf. *panchromatic *a.*). (Further examples.)

The emulsions, etc., orig. called 'orthochromatic' were in fact orthochromatic in the foregoing sense.

1903 A. Payne *Pract. Orthochromatic Photogr.* vi. 88 Orthochromatic plates possessing additional sensitiveness to the yellow and green rays only may be handled in a red light possessing abrupt absorption... Plates.. sensitive to the whole of the spectrum..require to be treated in darkness. Such plates are occasionally termed panchromatic. **1919** *Conquest* Nov. 24/1 Plates known as 'orthochromatic' or 'isochromatic' are sensitive to some of the greens and yellows... Nevertheless, even with these plates, the delicate greens of springtime, the yellow of the gorse, and many other colours in nature, are rendered unsatisfactorily, while the red leaves of autumn photograph as black [*sic*]. **1920** *Jrnl. Franklin Inst.* CLXXXIX. 25 In the commercial orthochromatic (sensitive to green and yellow) and panchromatic (sensitive to all colors) plates, the dyes are incorporated in the emulsion and the mixture flowed over the glass plate. **1944** *Electronic Engin.* XVI. 326/1 It is essential to use either an orthochromatic or panchromatic emulsion. **1970** *Amat. Photographer* 11 Mar. 60/1 As positive film is orthochromatic (that is, it is not sensitive to the red component of the visible spectrum), it can be handled under a red photographic safelight.

2. *Biol.* Exhibiting or characterized by the same colour as that of the stain used.

1899 *Jrnl. R. Microsc. Soc.* 379 Orthochromatic macrocytes which are without nuclei and appear late in embryonic life and in anæmia. **1930** Maximow & Bloom *Text-bk. Histol.* v. 114 In this way erythroblasts arise in which the protoplasm is purely acidophil, and stains a bright pink with the Romanowsky mixture. These cells are called orthochromatic erythroblasts or normoblasts. **1971** *Cancer Res.* XXXI. 505/1 This behavior was paralleled by a shift in the pH of the transition from orthochromatic to metachromatic staining of the nucleus after methanol fixation.

So **orthochro·matized** *ppl. a.*, **orthochro·matizing** *vbl. sb.*

1902 P. H. Mell *Biol. Lab. Methods* xii. 175 M. Monpillard says that, for scientific purposes, he prefers ready orthochromatized plates. **1903** A. Payne *Pract. Orthochromatic Photogr.* iv. 46 This process is termed orthochromatising, and plates so treated are known as orthochromatic or isochromatic plates. **1956** E. Mytum *Introd. Photogr. Materials* iii. 59 Sometimes the word *isochromatic*..was applied to the earlier orthochromatised emulsions.

† **orthodiagraphy** (ǭɪþodəi,æ·grăfi). *Med. Obs.* [ad. G. *orthodiagraphie* (F. Moritz 1900, in *Münch. med. Wochenschr.* 17 July 992/2): see Ortho-, Dia-¹, and -graphy.] A technique for producing sketches showing the exact sizes of organs by projecting the shadow formed by a narrow beam of X-rays on to a fluorescent screen perpendicular to the beam and drawing round the shadow.

1904 C. Beck *Röntgen Ray Diagnosis* iii. 44 (*heading*) Orthoradiography. **1907** *Practitioner* Apr. 524 By means of orthodiagraphy..the exact size and shape of an organ or tumour can be ascertained. **1930** D. A. Rhinehart *Roentgenographic Technique* 382/1 (Index), Orthodiagraphy of heart.

Hence **orthodi·agram,** a sketch produced by orthodiagraphy; **orthodi·agraph,** an instrument for orthodiagraphy; **orthodiagra·phic** *a.*, of or pertaining to orthodiagraphy; performed by means of orthodiagraphy; **o:rthodiagra·phically** *adv.*

1904 C. Beck *Röntgen Ray Diagnosis* iii. 44 The orthodiagraph made by Hirschmann permits direct tracing by the aid of a movable screen. *Ibid.*, The fact that the size of all skiagraphs is larger than that of the objects they represent, led to the construction of orthodiagraphic apparatus, by the aid of which the exact size of a body is determined. **1907** *Practitioner* Apr. 526 Changes in the size of the heart..can be determined by the system..of measuring orthodiagrams which have been taken direct on paper. **1930** D. A. Rhinehart *Roentgenographic Technique* xvii. 305 With certain types of fluoroscopes or with special attachments that may be fastened to other kinds, orthodiagraphic tracings of the cardiac outlines, giving nearly the exact dimensions of the heart shadow, can be made. **1931** P. Kerley *Rec. Adv. Radiol.* iv. 76 When screening the heart at 2 metres' distance, it takes only twenty seconds for the operator to map out an outline with a pencil, and this measurement is as accurate as that of an orthodiagram. **1938** K. C. Clark *Positioning in Radiogr.* xviii. 290/2 As an alternative to teleradiography,..the orthodiagraph may be used to record the size of the heart, a short anode-film distance being employed. **1938** *Jrnl. Amer. Med. Assoc.* 14 May 1718/1 Trimani found the parapericardial triangular shadow at the left lower arch of the roentgen silhouette of the heart in fifty-seven of 3,568 teleroentgenograms and 400 orthodiagrams which he studied. **1938** *Brain* LXI. 118 Measured Orthodiagraphically, the transverse diameter of his chest was 26 cm.

orthodontia (ǭɪþodǫ·ntiă). *Dentistry.* [f. as next + -ia¹.] Orthodontics.

1849 C. A. Harris *Dict. Dent. Sci.* 554/2 Orthodontia. **1850** —— *Princ. & Pract. Dental Surg.* (ed. 4) i. xi. 149 Fauchard and Bourdet were among the first who turned their attention to orthodontia. **1908** *Practitioner* Dec. 858 A case showing the importance of moving the roots of the teeth in orthodontia. **1939** S. J. Bregstein *Dentist & his Control of Pract.* v. 155 Preventive orthodontia, however, is regarded..as more commendable than corrective realignment of teeth and jaws. **1970** *Biol. Abstr.* LI. 3601/2 Surgery and orthodontia in 'retained teeth'.

orthodontic (ǭɪþodǫ·ntik), *a.* and *sb. Dentistry.* [f. Ortho- + Gr. ὀδοντ-, ὀδούς tooth + -ic.] **A.** *adj.* Serving to correct the positions of the teeth; of or pertaining to orthodontics.

1905 *Brit. Dental Jrnl.* XXVI. 993 (*heading*) Where extraction is justifiable in connection with orthodontic operations. **1939** S. J. Bregstein *Dentist & his Control of Pract.* v. 155 Speech defects, psychiatric disturbances, personality changes, rhinological pathoses all are correctable through orthodontic measures. **1962** Blake & Trott *Periodontology* xiv. 148 More permanent and more satisfactory immobilization can be obtained by fitting stainless steel orthodontic bands to each individual tooth. **1974** *Trans. European Orthodontic Soc. 1973* 403 A practitioner undertakes treatment in co-operation with the fully-trained orthodontic specialist.

B. *sb. pl.* The branch of dentistry concerned with the treatment and prevention of irregularities of the teeth and jaws. Usu. const. as *sing.*

1909 *Trans. Brit. Soc. Study Orthodontics* II. 86 The President..said that it was decided..that some change should take place in the name of the society. The council ..thought it desirable..the name of the society should be 'The British Society for the Study of Orthodontics'. He said..they had had the opinion of Dr. Murray the great philologist, of Oxford, to guide them. Mr. [C.] Schelling, as the originator of the suggested new title, here explained why he introduced the word Orthodontics. Mr. Schelling proposed..that the title of the society should be changed to the above, and..the motion was carried. **1939** L. B. Higley in Higley & Boyd *Dentistry for Children* ix. 218 Many difficult problems confronting the general dentist and orthodontist could be eliminated if preventative orthodontics were more thoroughly understood and practiced by all dentists who do work for children. **1941** J. D. McCoy *Appl. Orthodontics* (ed. 5) 6 Orthodontics has an abundant literature. **1955** G. M. Anderson *Dewey's Pract. Orthodontics* (ed. 8) ii. 89 Does orthodontics today differ from the practice of forty..years ago? **1971** *Daily Tel.* 24 Sept. 11/5 Unknown to many mothers, orthodontics (correction of irregularities in teeth) are available on the National Health Service.

Hence **orthodo·ntically** *adv.*, with regard to orthodontics.

1935 G. M. Anderson *Dewey's Pract. Orthodontia* (ed. 5) xxii. 458 Cases which though completed orthodontically were never satisfactory esthetically. **1940** M. G. Swenson *Complete Dentures* xxvi. 436 Orthodontically any movement of a tooth or teeth..will cause a change in the relationship of the inclines of the moved teeth to the remainder of the teeth. **1972** *Biol. Abstr.* LIV. 4614/2 An orthodontically interesting pair of twins.

orthodontist (ǭɪþodǫ·ntist). *Dentistry.* [f. *orthodont(ic *a.* and *sb.* + -ist.] One who practises orthodontics.

1903 *Dental Ann.* 130 Orthodontist, a newly-coined word, already considerably used in the United States to signify one who studies and practises orthodontia, or the treatment of irregularity, malposition, or malocclusion of the teeth. **1939** [see *orthodontic *sb.*]. **1969** *Daily Tel.* 7 Feb. 22 (*caption*) Mr. Lewis, an Australian orthodontist, is returning to Sydney as he is dissatisfied with working under the National Health Service.

orthodox, *a.* and *sb.* Add: **A.** *adj.* **5.** Also **Orthodox.** Of Judaism or Jews: adhering to the rabbinical interpretation of Biblical law and its traditional observances.

1853 *Jewish Chron.* 15 July 326/2 In all affairs of this kind, concessions are expected only on the orthodox side. **1858** *Manch. Guardian* 26 Mar. 3/4 He would not limit himself to a mere apology for their position, but confidently proclaim their right to the title of 'The truly orthodox Manchester Hebrew congregation'. **1876** Geo. Eliot *Dan. Der.* III. vi. xlvi. 308 He is not a strictly orthodox Jew, and is full of allowances for others. **1898** W. J. Locke *Idols* iv. 38 An elderly..Jewess of the most orthodox faith. **1904** *Jewish Encycl.* VII. 368/1 The stability and the immutability of the Law remained from the Orthodox standpoint one of the cardinal principles of Judaism. **1938** *Time & Tide* 12 Mar. 343/1 Peggy Simon, who was Jewish too, didn't have to bother about prayers.. because Peggy's father and mother weren't Orthodox. **1960** 'E. McBain' *Give Boys Great Big Hand* iv. 31 The family was..practising Orthodox Judaism. **1966** *Guardian* 28 July 6/5 The traditionalist right flank of Orthodox Judaism. **1973** *Jewish Chron.* 19 Jan. 22/2 It is ..unlikely that he would have been allowed to speak in an Orthodox pulpit in this country. **1974** *Encycl. Brit. Macropædia* X. 301/1 For many, the foreign designation 'orthodox' (used by Reform Jews for traditionalist Jews) makes little if any sense. **1973** J. Ryder *Trevayne*

(1974) xxxii. 247 It is the Hebrew Sabbath.. This house is Orthodox. **1975** *Nature* 6 Nov. 9/2 For an Orthodox woman, sexual intercourse is only permitted during a limited period each month. **1977** *Rolling Stone* 21 Apr. 72/2 Conversation among Orthodox Jews never strays far from questions of ethics, points of law, one's religious activities.

6. Applied to sleep characterized by the absence of rapid eye-movements and probably of dreams and by lesser physiological activity as compared with 'paradoxical' or REM sleep.

1967 W. P. Koella *Sleep* ii. i. 16 Subjects after being aroused from..'orthodox' sleep stages rarely recalled dreams. **1971** U. J. Jovanović *Normal Sleep in Man* ii. 75 We shall..use the Kleitman (1963) and Jouvet classification (1961, 1965, 1968), and divide up the entire polygraphic period of sleep in man into normal (orthodox) sleep and paradoxical sleep (periods of dream phases). *Ibid.* viii. 259 A phase of orthodox sleep lasts for about one-and-a-half hours.

B. *sb.* **1. c.** An orthodox Jew (see sense *A. 5).

1889 I. Zangwill in *Jewish Q. Rev.* I. 391 With the 'unintelligently orthodox', this mental attitude is generally associated with ignorance of our history and of the fluidity of ceremonial forms. **1892** I. Zangwill *Childr. Ghetto* II. 296 Now at last we poor orthodox will have a voice. **1914** *Encycl. Relig. & Ethics* VII. 608/2 The choice of method, unpleasing though it be to the orthodox, must be left to the conscience and judgment of the liberals themselves. **1927** E. O'Neill *Lazarus Laughed* ii. 35 Their former distinctions of Nazarenes and Orthodox are now entirely forgotten. *Ibid.* 42 The Nazarenes and the Orthodox separate and slink guiltily apart. **1964** E. E. Klein in W. Berkowitz *Ten Vital Jewish Issues* 48 The Conservative [*sc.* Jew] as always keeps one [*sc.* a yarmulke] in his pocket, and the Orthodox wears it on his head.

orthodoxy. Add: **d.** The orthodox practice of Judaism; the body of orthodox Jews.

1888 *Jewish Q. Rev.* I. 55 The Rabbis..would have utterly rejected the man's orthodoxy, or would have denied that his views were really what he professed them to be. **1892** *Ibid.* IV. 213 Let us..hope that Dr. Friedländer's conception is by no means Orthodoxy's last word. **1899** B. Drachman *Neo-Hebraic* … Hirsch set up that view of Judaism called in Germany 'Denkgläubigkeit', which we may translate as 'intellectual or enlightened Orthodoxy'. **1955** M. Sklare *Conservative Judaism* i. 25 It would be desirable to conclude this historical introduction with some analysis of the growth or decline of Orthodoxy, Conservatism and Reform. **1966** H. Kemelman *Saturday the Rabbi went Hungry* (1967) viii. 51 Some of the older congregants brought up in Orthodoxy. **1975** *Times* 13 Aug. 12/4 The mayor of Hackney..says: 'I never travel by car on Saturdays or Jewish festivals.'..Now the mayor aims to combine his orthodoxy with a little money-making by getting people to sponsor his walks.

orthodromic, *a.* Delete *rare* and add: Also, representing great circles as straight lines. (Examples.) Also **o·rthodrome,** a great circle or a route forming part of one.

1883 J. Pryde *Treat. Pract. Math.* 455 The arc of a great circle, which is the shortest distance between two places, is called the orthodrome. **1922** R. Keen *Direction & Position Finding by Wireless* iv. 103 A chart on which all Great Circles are represented as straight lines is known as orthodromic. **1928** L. S. Palmer *Wireless Princ. & Pract.* xii. 487 This type of projection is the orthodromic projection and the map of the district surrounding the transmitting point is termed a Gnomonic chart. **1935** *Geogr. Jrnl.* LXXXV. 466 The equation of the orthodrome (geodesic line). **1956** *McGraw-Hill Yearbk. Sci. & Technol.* VI. 267/2 The longer the distance, the greater the deviation between loxodromes and orthodromes.

2. *Physiol.* Being or involving a nerve impulse that is propagated in the normal direction.

1943 *Jrnl. Neurophysiol.* VI. 143 (*heading*) The interaction of antidromic and orthodromic volleys in a segmental spinal motor nucleus. **1954** *Jrnl. Physiol.* CXXVI. 502 (*heading*) Interaction between direct current and orthodromic stimulation. **1972** *Science* 2 June 1044/1 The prolonged inhibition of the monosynaptic excitation of motoneurons by orthodromic volleys in muscle and cutaneous afferent fibers.

Hence **orthodro·mically** *adv.*

1954 Penfield & Jasper *Epilepsy & Functional Anat. Human Brain* v. 203 Distant projection of the impulses from a focal after-discharge has been shown to be conducted only orthodromically (not antidromically) over transcortical bundles of fibers. **1976** *Nature* 4 Mar. 56/2 When the interval between the two spikes was reduced progressively, the antidromic response eventually disappeared because it collided with the direct spike travelling orthodromically along the same nerve fibre.

orthoepic, *a.* Add: (Later example); **ortho·epically** *adv.* (later example).

1969 *Computers & Humanities* III. 259 The authors aimed at producing a broad phonetic transcription similar to the phonetic transcription already established for orthoepic French. **1975** *Amer. Speech 1973* XLVIII. 113 Being perhaps abnormally tenacious orthoepically and considerably past middle age, I use [aɪ] in all these words, but scholars and laymen alike have been seen to wince at the crudity of my speech.

orthoepist. Add to def.: used esp. of those 16th- and 17th-century writers whose aim was to describe a 'correct' pronunciation of English, to reform the spelling system to make it reflect such a pronunciation more accurately, etc. (Earlier and later examples.) Also *attrib.* or as *adj.*

1791 J. Walker *Crit. Pronouncing Dict.* s.v., *Orthoepist*, ..one who is skilled in Orthoepy. **1909** O. Jespersen *Mod. Eng. Gram.* I. ix. 248 Up to quite recent times, most orthoepists have disregarded natural pronunciation. **1917** J. M. Clark *Vocab. Anglo-Irish* 18 It is a well-known fact that in Tudor English a much more open ō was pronounced than is the case today. Apart from the testimony of orthoepists, the proof is to be found in contemporary American and Irish usage. **1920** H. C. Wyld *Hist. Mod. Colloq. Eng.* iv. 113 Hitherto writers upon the history of Modern English have relied mainly upon the Orthoepists. **1927** H. E. Zachrisson *Eng. Pronunc. at Shakespeare's Time* p. x, The old-fashioned types of pronunciation were, as a rule, taught by English orthoepists. **1957** E. J. Dobson *Eng. Pronunc. 1500–1700* I. ii. 193 The tendency to regard the spelling reformers as primarily interested in teaching 'correct pronunciation'..may be due to the difficulty of finding a term to cover all the sixteenth- and seventeenth-century writers on pronunciation; 'grammarians' is clearly inexact except in a few cases, and so the term 'orthoepists' has been widely accepted. **1969** A. C. Partridge *Tudor to Augustan English* viii. 173 On their own admission, orthoepists were students of language who sought to establish the principles of correctness in speech. **1972** M. L. Samuels *Linguistic Evol.* vii. 144 For Early Modern English, there is much orthoepistic, phonetic and other evidence. **1975** *Language* LI. 74/1 It would be of great value if a presumably accurate report of upper-class speech of the period [*sc.* Early Modern English] could be carefully analysed and related to witnesses such as Hunt and the other orthoepists.

orthoepistic *a.* (later example); hence also **orthoepi·stical** *a.*

1913 E. Zachrisson *Pronunc. Eng. Vowels 1400–1700* 3 Statements in early grammars and spelling-books (the orthoepistical evidence). **1957** E. J. Dobson *Eng. Pronunc. 1500–1700* I. i. 36 There is no other orthoepistical evidence for this pronunciation. **1972** *Eng. Studies* LIII. 506 This type of pronunciation was vulgar or dialectal..but at least in some words it must have infiltrated educated speech, as is shown by its frequent appearance in the orthoepistic works. **1972** P. M. Wolfe *Linguistic Change* 18, 31 I have in general limited myself to English orthoepistical works.

orthoepy. Add: Also, the study of the relationship between pronunciation and a writing system. (Later examples.)

1913 D. Agate in H. C. O'Neill *Guide to Eng. Lang.* I. v. 74/1 To these four divisions of grammar many grammarians have added Orthoepy, which treats of pronunciation. **1957** E. J. Dobson *Eng. Pronunc. 1500–1700* I. ii. 193 In spite of his title *Orthoepia Anglicana*..what he [*sc.* Daines] sets out to teach is orthography, not orthoepy. **1969** A. C. Partridge *Tudor to Augustan English* viii. 181 Though his was not the last shot fired in the hundred years' war of English orthoepy, Cooper's *Grammar* established that the criterion of correct Standard English rests firmly on its pronunciation. **1976** *Visible Language* X. 20 Phonetization of the alphabet and other writing systems is a province of orthoepy.

orthogenesis (ǭɹþodʒe·nĭsis). *Biol.* [a. G. *orthogenesis* (W. Haacke *Gestaltung und Vererbung* (1893) ii. 31), f. Ortho- + -genesis.] A series of similar variations in successive generations, leading to evolutionary change produced by these mutations. Also **orthogene·tics** *sb. pl.*

1895 *Nature* 3 Oct. 554/2 Prof. Eimer, of Tübingen, spoke..on the subject of orthogenesis. **1897** *Jrnl. R. Microsc. Soc.* 108 The causes of orthogenesis are to be found in the action of environment upon the constitution of the organism. **1911** *Encycl. Brit.* XXVII. 912/1 Many successful series,..as they have survived, must inevitably display orthogenesis to some extent. **1930** G. R. de Beer *Embryol. & Evolution* iv. 32 This incorrigible tendency to produce larger and larger horns, to vary continuously in the same direction, has been given the name of orthogenesis. **1937** A. Huxley *Ends & Means* xiv. 261 Neither Lamarckism nor the orthogenetic theory seems to be demonstrably defective. **1962** *Nature* 18 Feb. 309/1 His [*sc.* Arthur Trueman's] contributions to the more philosophical aspects of palaeontology ranged from lineage and orthogenesis, to a reconsideration of the species concept. **1965** R. Howard tr. *de Beauvoir's Force of Circumstance* x. 500 They talked, not about birth control but about the joys of maternity, not about contraception but about orthogenesis. **1970** *Watkins* VIII. 178 A telling argument in favour of mutation pressure (orthogenesis) rather than natural selection. **1972** *Nature* 10 Aug. 375/1 A non-Darwinian orthogenesis had sealed the fate of *Megaloceros* under the oppressive weight of its own enlarging antlers.

orthogenetic (ǭɹþodʒene·tik), *a.* [f. prec.; see -genetic.] Of, pertaining to, or characterized by orthogenesis. Also *transf.* Hence **orthogene·tically** *adv.*

1895 H. Gadow in *Proc. Cambr. Philos. Soc.* X. 35 (*title*) Orthogenetic variation in the shells of Chelonia. *Ibid.* 37 Since these variations all lie in the direct line of descent.. I call this kind of atavistic variation orthogenetic. **1911** J. Ward *Realm of Ends* xvii. 383 Can we conceive this world evolving orthogenetically, as a biologist would say? **1927** Haldane & Huxley *Animal Biol.* xi. 253 The orthogenetic series can be perfectly well explained by natural selection. *Ibid.*, The extinct cephalopod molluscs ..often evolved orthogenetically into the most bizarre forms. **1930** W. R. Inge *Christian Ethics & Mod. Probl.* i. 13, I shall not maintain that the evolution of Christianity has been..orthogenetic. **1965** B. E. Freeman tr. *Vandel's Biospeleol.* xiii. 198 Biospeologists..are more interested in orthogenetic evolution which proceeds in a parallel manner, in different phyletic lines. **1973** *Jrnl. Genetic Psychol.* CXXIII. 231 In its most general sense, Werner's theory of development centers on the orthogenetic principle.

orthogonal, *a.* Add: **2.** *Math.* **a.** Of a linear transformation: preserving lengths and angles; leaving unchanged quantities of the form $x_1^2 + x_2^2 + \ldots + x_n^2$ and the inner product of any two vectors.

1859 G. Salmon *Lessons Introd. Mod. Higher Algebra* xv. 153 We shall call the orthogonal transformation is to transform simultaneously a given quadratic function, and $x^2+y^2+z^2+w^2+\&c.$, so that the latter remaining of the same form, the former may become $Ax^2+By^2+Cz^2+Dw^2+\&c.$ **1893** L. G. Weld *Short Course Theory Determinants* ix. 227 The transformation, in analytical geometry, from one set of axes to another, without changing the origin, is orthogonal. **1941** Birkhoff & MacLane *Survey Mod. Algebra* ix. 225 A linear transformation *T* is orthogonal if it preserves the absolute value of every vector ξ, so that $|\xi T| = |\xi|$. **1972** F. E. Hohn *Introd. Linear Algebra* viii. 237 An orthogonal transformation of $\mathcal{E}^n$ maps orthogonal vectors onto orthogonal vectors and nonorthogonal vectors onto nonorthogonal vectors.

b. Applied to the group of all orthogonal matrices of a given order.

1898 *Bull. Amer. Math. Soc.* IV. 196 A linear substitution *S* on the marks of a Galois Field of order p^n, will be called orthogonal if it leaves absolutely invariant $\xi_1^2 + \xi_2^2 + \ldots + \xi_m^2$. The order of the orthogonal group *G* on *m* indices in the $GF[p^n]$ is thus [etc.]. **1941** Birkhoff & MacLane *Survey Mod. Algebra* ix. 225 This subgroup of the full linear group is called the orthogonal group O_n; it is isomorphic to the group of all orthogonal transformations of the given Euclidean space. **1957** F. E. Hohn *Introd. Linear Algebra* viii. 252 Show that the set of all linear operators on $\mathcal{E}_n$ of the form $T = UX$, where *U* is orthogonal, constitute a group (the orthogonal group).

c. Of a square matrix: representing an orthogonal transformation; such that the rows (and likewise the columns) are orthonormal when considered as vectors; equal to the inverse of its transpose; (these three properties are equivalent).

1907 M. Bôcher *Introd. Higher Algebra* xi. 154 An orthogonal transformation. [*Note*] The matrix of such a transformation is called an orthogonal matrix. **1964** N. N. Hancock *Matrix Anal. Electr. Machinery* ii. 18 The value of the determinant of an orthogonal matrix is necessarily ±1, but the converse is not true.

d. Of two vectors or functions: perpendicular; having an inner product equal to zero. Of a set of vectors or functions: such that the inner product of any two is zero if and only if the two are distinct.

1913 *Proc. London Math. Soc.* XII. 297 The theory of Fourier series and of other series of orthogonal functions. **1926** E. W. Hobson *Theory of Functions of Real Variable* (ed. 2) II. x. 754 If $\{\phi_n(x)\}$ be a complete sequence of linearly independent functions for the interval (*a*, *b*), a normal orthogonal and complete system of functions $\{\psi_n(x)\}$ can be determined such that $\phi_n(x)$ is a linear function of $\psi_1(x), \psi_2(x), \ldots, \psi_n(x)$. **1941** R. V. Churchill *Fourier Series* iii. 45 The functions … are orthogonal in the interval … **1967** A. A. Goldstein *Constructive Real Analysis* iii. 112 We define an inner product space… Two functions *f* and *g* in *I*(*a*, *b*) are said to be orthogonal if $(f, g) = 0$. *Ibid.* 115 Two points *x* and *y* of a Hilbert space *H* are orthogonal if $(x, y) = 0$. Similarly, two subspaces *M* and *N* of *H* are said to be orthogonal if $[M, N] = 0$. **1968** C. G. Kuper *Introd. Theory Superconductivity* i. 3 Bardeen, Cooper and Schrieffer (1957) constructed a variational wave function for a ground state with complete electron pairing, and orthogonal functions for low-lying excited states having only a few such pairs broken.

e. *Statistics.* Of a set of variates: statistically independent. Of an experimental design: such that the variates under investigation can be treated as statistically independent.

1933 *Jrnl. Agric. Sci.* XXIII. 110 In an ordinary replicated field experiment of the randomised block or Latin square type the differences of the means of plots receiving the same treatments are taken without hesitation to be true measures of treatment differences, but this is only so because the experiment has been specially arranged so as to be orthogonal. **1950** M. H. Quenouille *Introd. Statistics* iv. 59 If..comparing a series of..measurements on people to determine the effect of age, these comparisons may be complicated by the effect of sex.. The only manner in which we can assume that sex does not enter into the comparison is to choose the same proportion of each sex in each age group. The effect of sex is then said to be 'orthogonal' to the effect of age. **1967** *Word* XXIII. 219 Another model which provides a relevant comparison to phonological distinctive features is the mathematical method of factor analysis... The various mathematical methods employed lead to the positing of a number of independent 'orthogonal' factors and each test or other set of responses is described in terms of positive or negative loadings on each factor. **1973** *Jrnl. Genetic Psychology* CXXII. 45 Implicit in the work..is the concept that creativity and intelligence are relatively orthogonal (i.e., unrelated statistically) at high levels of intelligence.

orthogonality (ǭɹþǫgǒnæ·lĭti). [f. Orthogonal *a.* + -ity.] The property of being orthogonal. Freq. *attrib.*

1892 O. Heaviside *Electr. Papers* II. 583/2 (Index), Orthogonality of electric and magnetic forces. **1930** P. A. M. Dirac *Princ. Quantum Mech.* iii. 50 The orthogonality theorem. **1933** *Jrnl. Agric. Sci.* XXIII. 108 Orthogonality is that property of the design which ensures that the different classes of effects to which the experimental material is subject shall be capable of direct and separate estimation without any entanglement. **1935** Pauling & Wilson *Introd. Quantum Mech.* 441 (*heading*) Proof of orthogonality of wave functions corresponding to different energy levels. **1964** Corson & Lorrain *Introd. Electromagn. Fields* iv. 158 This property of orthogonality of the Legendre polynomials is important in evaluating the coefficients of Eq. 4-139. **1970** G. K. Woodgate *Elem. Atomic Struct.* ii. 17 The spherical harmonics have the orthogonality property. **1972** A. W. F. Edwards *Likelihood* vi. 106 Such cases may be expected to be rather rare ..and the most important application of the concept of orthogonality is to the quadratic support surface. **1973** *Jrnl. Genetic Psychol.* CXXII. 203 The correlations between the E and N dimensions are low enough to justify the assumption of orthogonality of these two dimensions of personality.

ortho:gonaliza·tion. *Math.* [f. as prec. + -ization.] The procedure of constructing an orthogonal set of functions or vectors from ones that are linearly independent but not orthogonal.

1922 *Proc. London Math. Soc.* XXI. 97 We have now only to derive from $\{\chi_n(t)\}$ a new set $\{\phi_n(t)\}$ by the 'orthogonalisation theorem' which is due to Schmidt. **1953** Courant & Hilbert *Methods Math. Physics* I. ii. 50 The orthogonalization..normalized set possessing the property (3). **1966** G. Arfken *Math. Methods for Physicists* ix. 342 Consider two (nonparallel) vectors **A** and **B** in the *xy*-plane. We may normalize **A** to unit magnitude and then form **B**′ = *a***A** + **B** so that **B**′ is perpendicular to **A**. By normalizing **B**′ we have completed the Schmidt orthogonalization for two vectors.

So **ortho·gonalize** *v. trans.*, to render orthogonal (and, often, to normalize); **ortho·gonalized** *ppl. a.*

1930 Ruark & Urey *Atoms, Molecules & Quanta* 747 Other systems of polynomials often used in wave mechanics are obtained by orthogonalizing the system, $p^{1/2}, xp^{1/2}, x^2p^{1/2}, \ldots$ where $p(x)$ is a so-called 'weight function'. **1937** Michell & Belz *Elem. Math. Analysis* II. xvii. 914 If the set of functions $\phi_n(x)$..is orthogonalized by means of the function $p(x)$. **1939** C. H. Goulden *Methods Statistical Analysis* xii. 219 If the number of varieties is 21, the numbers would be written out as below..and we would have to use a completely orthogonalized 4×4 square..to which the varieties are assigned at random, as described above. **1948** Brown & Campbell *Princ. Servomech.* iii. 66 After the initial transformation, certain manipulations enable us to orthogonalize the mathematical forms. **1966** *McGraw-Hill Encycl. Sci. & Technol.* II. 93/2 Several techniques exist for solving the wave equation, at least for certain states, notably the Wigner-Seitz method and the orthogonalized-plane-wave method.

orthograde (ǭɹþǫgrēd), *a.* [irreg. f. Ortho- + L. *-gradus* walking: see Grade *sb.*] Holding the body upright.

1902 A. Keith in *Jrnl. Anat. & Physiol.* XXXVII. 18 He [*sc.* the author] regards the primates as divided into two very distinct groups—those which carry the axis of the body in a horizontal position—the Pronograde Primates..; and those which carry the axis of the body in an upright position—the Orthograde Primates, into which group fall the gibbon, orang, chimpanzee, gorilla, and man. *Ibid.* 19 It is now generally recognised that the anthropoids, in their natural habitat, carry their bodies in the upright or orthograde posture. **1925** J. Laird *Our Minds & their Bodies* 48 An orthograde [or erect] animal, like man. **1940** *Nature* 6 July 27/1, I regarded the Gibbon as a representative of the primates of the orthograde stock. **1973** B. J. Williams *Evol. & Human Origins* viii. 112/2 Many features of this arboreal primate indicate..a more orthograde, upright posture. *Ibid.* 114/1 Man being completely orthograde had..a recurved spine that bends back sharply in the lumbar region.

orthography. 1. b. (Earlier example.)

1588 W. Kempe *Educ. Children* sig. F 3v Orthographie ..teacheth with what letters euery syllable and word must be written, and with what points the sentence and parts thereof must be distinguished... Which expressing and skill of the hand, belongeth properly to the Arte of Painting, and not vnto Grammar.

orthohelium (stress variable). [f. Ortho- + Helium, as an antonym of the earlier *par(a)-helium*.] The form of helium whose spectrum exhibits a fine structure of triplets owing to the spins of the two orbital electrons being parallel.

1923 A. D. Udden tr. *Bohr's Theory of Spectra* iii. 86 Helium was at first assumed to be a mixture of two different gases, 'orthohelium' and 'parhelium'; but now we know that the two spectra simply mean that the binding of the second electron can occur in two different ways. *Ibid.* 87 The metastable state..is the final stage of the process giving the orthohelium spectrum. **1967** Powell & Crasemann *Quantum Mech.* xii. 458 Spectroscopically, the singlet and triplet systems are independent of one another, and have been given the names *parahelium* (singlets) and *orthohelium* (triplets). The ground state of the orthohelium system..is therefore stable with respect to optical transitions of the usual kind, and has a correspondingly long lifetime.

orthohydrogen (stress variable). Also **ortho hydrogen** and with hyphen. [f. Ortho- + Hydrogen as tr. G. *orthowasserstoff* (Bonhoeffer & Harteck 1929, in *Naturwissenschaften* XVII. 182/1), coined on the analogy of *orthohelium* (see quot. 1935).] The form of molecular hydrogen in which the two nuclei in the molecule have parallel spins, so that the spectrum exhibits a hyperfine structure of triplets; it differs slightly in physical properties from the other form (*parahydrogen) and forms 75 per cent of hydrogen in equilibrium at room temperature.

1929 *Chem. Abstr.* XXIII. 2614 (*heading*) Experiments on para- and ortho-hydrogen. **1935** A. Farkas *Orthohydrogen, Parahydrogen & Heavy Hydrogen* ii. 4 The parahydrogen molecules have antiparallel nuclear spins and even rotational quantum numbers, while the orthohydrogen molecules possess parallel nuclear spins and odd rotational quantum numbers. *Ibid.*, The expressions orthohydrogen and parahydrogen were chosen by Bonhoeffer and Harteck on analogy with the nomenclature for the helium atom (orthohelium and parhelium), but it must be emphasized that the distinction between the hydrogen modifications is based on the different orientations of the nuclear spins, while in the case of helium it depends on the orientation of the electron spins. **1968** D. J. E. Ingram *Introd. Adv. Inorg. Chem.* xiii. 368 Ordinary hydrogen gas is a tautomeric mixture of orthohydrogen and parahydrogen. **1966** D. H. Whiffen *Spectroscopy* ix. 114 In the absence of magnetic materials near-equilibrium mixtures of ortho and para hydrogen are stable for many months. **1969** H. T. Evans tr. *Hägg's Gen. & Inorg. Chem.* xviii. 452 At 0°K..only parahydrogen exists at equilibrium.

orthometric, *a.* Add: (Examples.)

1854 J. D. Dana *Syst. Min.* (ed. 4) I. 23 In Crystallography there are three axes; and orthometric forms are either at right angles with one another, producing orthometric forms, or oblique, producing clinometric forms. **1883** *Encycl. Brit.* XVI. 349/2 All crystals may be divided into 'orthometric' or erect forms and 'clinometric' or inclined forms.

2. *Surveying.* Of, pertaining to, or being a height measured by orthometry.

1909 G. L. Hosmer *Geodesy* x. 254 The United States Coast Survey has adopted the orthometric method… **1923** D. Clark *Plane & Geodetic Surveying* II. v. 103 A line of constant orthometric elevation is parallel to the mean sea level surface. **1967** Heiskanen & Moritz *Physical Geodesy* iv. 172 Orthometric heights are the natural 'heights above sea level', that is, heights above the geoid. **1974** *Encycl. Brit. Macropædia* XVII. 838/2 To correct these distortions, orthometric corrections..must be applied to long lines of levels at high altitudes that have a north-south trend.

Hence **orthometrically** *adv.*

1952 G. Bomford *Geodesy* iii. 155 If no error of observation is made, BB and aA will be measured orthometrically.

orthomolecular (ǭɹþomole·kiŭlăɹ), *a.* *Psychol.* [f. Ortho- + Molecular *a.*] (See quot. 1968.)

1968 L. Pauling in *Science* 19 Apr. 265/1, I have reached the conclusion..that another general method of treatment, which may be called orthomolecular therapy, may be found to be of great value... Orthomolecular psychiatric therapy is the treatment of mental disease by the provision of the optimum molecular environment for the mind, especially the optimum concentrations of substances normally present in the human body. **1970** [see *megavitamin* s.v. *Mega- a]. **1971** *Nature* 15 Oct. 452/2 The term 'orthomolecular psychiatry' introduced by Professor L. Pauling in 1968 has taught American psychiatrists to appreciate a principle well known to scientists. **1972** *Daily Colonist* (Victoria, B.C.) 16 July 25/3 Heavy doses of certain vitamins..correct biochemical imbalances. It's called the orthomolecular approach. **1977** *National Observer* (U.S.) 22 Jan. 11/5 Megavitamin, or orthomolecular, therapy, is sometimes used to treat mental retardation, psychoses, hyperactivity, autism, dyslexia, and other learning disorders.

orthomorphic, *a.* Restrict *rare* to sense 1 and add further examples of sense 2.

1910 [see *conformal *a.* 2]. **1937** *Jrnl. Optical Soc. Amer.* XXVII. 358/2 The only telescope with orthomorphic object plane and image plane is the supermagnifying telescope. **1938, 1957** [see *Conformal *a.* 2]. **1971** *Jrnl. Photographic Sci.* XIX. 144/1 A small error in *r*, d*r*, was used for orthomorphic copying.

Hence **orthomo·rphism,** the property of being orthomorphic.

1919 *Wireless World* May 69/2 Orthomorphism..may be possessed by many different types of graticule. **1969** *Geogr. Jrnl.* CXXXV. 381 It is well to preserve orthomorphism, if only for its help in solving great circle problems. **1975** J. B. Harley *O.S. Maps* iii. 19 The projection stretched the topography equally in all directions, rather than only in a north-south direction, and this gave it the property of conformality or orthomorphism, in which there is a minimal distortion of shape over small areas and the scale..is likewise equal in all directions at any one point.

orthonormal (ǭɹþonǭ·ɹmăl), *a.* *Math.* [f. Ortho(gonal *a.* + Normal *a.*] Both orthogonal and normalized.

1932 M. H. Stone *Linear Transformations in Hilbert Space* i. 7 Two elements *f*, *g* of 𝔥 are said to be orthogonal if (*f*, *g*) vanishes... A subset 𝔊 of 𝔥 is said to be an orthonormal set if, when *f* and *g* are elements of 𝔊,

$$(f, g) = \begin{cases} 1, & f = g \\ 0, & f \neq g \end{cases}$$

1941 R. V. Churchill *Fourier Series* iii. 35 The symbol $\{\phi_n\}$ will be used to denote an orthonormal set whose vectors are ϕ_1, ϕ_2, and ϕ_3. The simplest example of such a set is that consisting of the unit vectors along the three coordinate axes. **1965** Patterson & Rutherford *Elem. Abstr. Algebra* v. 173 It is frequently desirable to choose an orthonormal basis: that is, a basis of which each vector is of unit length..and such that any two basic vectors are orthogonal. **1968** G. Ludwig *Wave Mech.* i. iii. 33 The degree of such a complete orthonormal system can exceed the finite dimension of a Hilbert space.

Hence **orthonorma·lity,** the property of being orthonormal.

1949 L. I. Schiff *Quantum Mech.* 401 (Index), Orthonormality. **1959** G. Troup *Masers* ii. 16 This orthonormality is an expression of uniqueness of the states. **1975** *Physics Lett.* A LIII. 3 … (the C_k^α) must satisfy the orthonormality relation.

orthonormalize (ǭɹþonǭ·ɹmălaiz), *v.* *Math.* [f. prec. + -ize.] *trans.* To make (a set of vectors or functions) orthonormal by orthogonalizing them and then multiplying each by an appropriate factor. Hence **o:rthonormaliza·tion.**

1935 *Trans. Amer. Math. Soc.* XXXVII. 309 When we orthonormalize $\{T^{n-1}\chi_1\}$ we obtain $\{\phi_n\}$. **1968** D. Kripke *Introd. Anal.* iii. 57 Orthonormalize the functions 1, *x*, *x*² with respect to the inner product … **1972** A. Kyrala *Appl. Functions Complex Variable* xiii. 318 Establish the orthonormalization relations for the associated Legendre functions. **1974** Avery & Dempster *Introd. Optimization Methods* iv. 89 Rosenbrock's method with linear search does not exhibit quadratic termination, due to the orthonormalization of the search directions in the second part of each iteration and the alignment of one of them towards the minimum.

orthopædic, -pedic, *a.* Add: *orthopædic bed*, a bed in an orthopædic ward; now any one individually designed to relieve specific skeletal symptoms; *more generally*, a bed with a very firm mattress or board; also *orthopædic bedding, divan*, etc.; *orthopædic shoe*, a shoe designed to ease or correct deformities of the feet (cf. quot. 1842 s.v. Orthopædical *a.*); also *orthopædic boot, footwear*.

1943 Funsten & Calderwood *Orthopedic Nursing* iv. 90 Orthopedic beds may be made with two lines placed over the end of the bed, rather than by tucking it in at the end, to protect the patient. **1951** B. Malamud *Tenants* ii. The tall mother..wears a plain white dress, orthopedic shoes, and a blue cloche hat that hides her eyes. **1974** N. Gordimer *Conservationist* 241 She..turns a foot on the heel of one of those clogs, like orthopaedic shoes, the women are wearing these days. **1976** *P.O. Teleph. Directory: London Postal Area* June, Orthopaedic Bedding Centre. **1976** P. van Rjndt *Tetramachus Collection* (1977) ii. 11 A cripple, obliged to wear a heavy orthopedic boot on his right foot. **1977** *Evening Post* (Nottingham) 27 Jan. 19/1 (Advt.), Modern single bed with mattress £15. Single bed base £10. Single orthopaedic mattress as new £30. *Ibid.*, Orthopaedic 4ft divan complete. As new, £50. **1977** *Daily Express* 1 Feb. 19 (Advt.), OBC orthopaedic beds..look like any good quality bed.. The big difference is this: They are designed with medical help and hand-assembled in thousands of different layers, to give the maximum support to back sufferers whatever their weight or type of build. **1977** *Observer* 13 Feb. 18/6 (Advt.), Beds, soft, firm, extra firm, orthopaedic. **1977** *Times* 14 May 25/7 (Advt.), Orthopaedic footwear our speciality.

orthopantomography (ǭɹþopæntǫmǫ·grăfi). *Med.* [f. Ortho- + *Pantomography.] A modification of pantomography in which the X-rays are made to be more nearly normal to the line of the jaws, so that a radiograph can be obtained showing all the teeth and adjacent tissue in a straight line.

1959 Y. V. Paatero in *Acta Radiologica* LI. 449 Since stereoscopy has been successfully adapted to ordinary pantomography..theoretical and experimental investigations into the possibility of obtaining equally good stereoscopic effects with the new orthopantomographic method ('orthoradial pantomography' (or orthopantomography for short), were considered to be justified. **1961** *Oral Surgery* XIV. 947 (*heading*) Pantomography and orthopantomography. **1968** *Brit. Jrnl. Radiol.* XLI. 872/2 They reported that during orthopantomography the exposure at the eccentric axis was 15 R and at the skin surface it varied from 0 R to 0·9 R.

Hence **orthopanto·mogram,** a radiograph obtained by orthopantomography; **orthopanto·mograph,** an instrument for performing orthopantomography.

1959 Y. V. Paatero in *Annales Medicinae Internae Fenniae* XLVIII. Suppl. 28. 223 As no 'orthopantomograph' suitable for clinical use was yet available, the accompanying pictures were taken of a dry skull with a hand-rotated chamber with the film. **1961** — in *Acta Radiologica* LII. 452 The jaws appear flat in the orthopantomogram and not curved as in ordinary pantomograms. **1967** L. M. Ennis et al. *Dental Roentgenol.* (ed. 6) x. 283 In operation of the Orthopantomograph, the patient remains stationary while the x-ray tubehead circulates from his right side around behind his neck to the left side, while the film rotates about an axis and at the same time, revolves from the left side of the patient's face, around the front and to the right side of the face. **1971** *Brit. Dental Jrnl.* CXXX. 429/2 There are two image layers in the orthopantomograph, one on either side of the rotational

centre, the object further from the film presenting reversed images. *Ibid.* 433/2 This delicate spur of bone is not visible on the orthopantomogram.

orthopod (ǭ·ɹþopǫd). *slang.* [Alteration of Orthopaedic *a.*] An orthopædic surgeon.

1960 'R. Gordon' *Doctor in Clover* ix. 76 We were interrupted by the surgeon himself, a big, red-faced, jolly Irishman. Most orthopods are, when you come to think of it. **1966** I. Jefferies *House-Surgeon* vii. 131 We had two male beds and one female, and the orthopods had two spare beds. **1969** D. Francis *Enquiry* xii. 164, I telephoned to the orthopod who regularly patched me up after falls. **1978** *New Yorker* 13 Mar. 82 The problem now was to persuade the orthopod to go in and remove the screws.

orthopraxy. Add: **1.** (Further examples.) Also **orthopraxis.**

1951 *Jrnl. Theol. Stud.* II. 98 The complete obedience of Jesus must be taken to be a complete vindication of the Law, and therefore the champions of legal orthodoxy (and orthopraxis), such as James and Peter, are the heroes of Jewish Christianity. **1960** J. Parkes *Foundations Judaism & Christianity* vi. 297 We cannot..imagine an orthopraxy, which made a mizwah of reciting a special blessing over a fruit tree in bloom, attached to a Puritan theology which was quick to threaten Hell-fire for any slight disobedience. *Ibid.*, Historically however, rabbinic orthopraxy was lived with an entirely different background. *Ibid.* 311 There does not appear to have been any single system, nor was any particular method of choice a matter of 'orthopraxis'. **1971** *Clergy Review* LVI. 218 The orthodoxy of faith in the coming universal kingdom must constantly be made true in the ortho-praxy of creative flight forward with the world. **1976** E. Maclaren *Nature of Belief* vii. 73 No amount of impeccable ortho*doxy* is belief. Belief is ortho*praxis*, commitment to certain action.

orthopsychiatry (ǭɹþosəikəi·ătri). [f. Ortho- + Psychiatry.] A branch of psychiatry concerned especially with the prevention of mental or behavioural disorders. Hence **orthopsychia·tric** *a.*, **orthopsychi·atrist.**

1924 *Survey* (N.Y.) 15 Aug. 536/1 'Straightness of Spirit'—interpreting this title literally—is the goal toward which the recently organized Association of American Orthopsychiatrists will bend their efforts. **1930** (*title*) American journal of orthopsychiatry. A journal of human behavior (American Orthopsychiatric Association). **1956** O. Pollak *Integrating Sociol. & Psychoanal. Concepts* III. ix. 221 A field as energetic and imbued with the spirit of experimentation in practice as orthopsychiatry. **1971** E. M. Bower *Orthopsychiatry & Educ.* 17 (*heading*) The challenge to education and orthopsychiatry.

orthopter. Add: (Later example.)

1935 *Discovery* July 199/1 Another orthopter, *Ephippigera vitium* Latr...lives in the Pacific and west Mediterranean areas, and in some places in Central Europe as a 'Pontic relic'.

† **2.** *Aeronaut.* Also **-ptere.** [ad. F. *orthoptère* (de Ponton d'Amécourt 1862: see S. Stubelius *Balloon, Flying-Machine, Helicopter* (1960) 90); so called because of the 'straight' (vertical) motion of the wings.] = *ornithopter. *Obs.*

App. misinterpreted at first in Eng. as referring to a clockwork flying model. The word was superseded *c* 1909 by *ornithopter*, and for a time attempts were made (chiefly in dicts. and glossaries) to differentiate the meanings of the two words on etymological grounds. (See S. Stubelius *Balloon, Flying-Machine, Helicopter* (1960) 93–7.)

1868 *Catal. First Exhib. Aeronaut. Soc. Gt. Brit.* 11 (*heading*) Working models. 12 Orthoptere. Viscount de Ponton d'Amecourt,..Paris. **1873** J. B. Pettigrew *Animal Locomotion* 217 MM. Nadar, Pontin [*sic*] d'Amécourt, and de la Landelle have constructed clockwork models (*orthopteres*), which..raise themselves into the air. **1887** tr. *J. Verne's Clipper of Clouds* vii. 65 If the orthopter —striking like the wings of a bird—raised itself by beating the air, the helicopter raised itself by striking the air obliquely with the fins of the screw as it mounted on an inclined plane. **1906** *Sci. Amer.* 18 Aug. 115/3 'Aéronef', or 'appareil d'aviation' (aviation apparatus) means an apparatus heavier than air, of which there are several kinds, such as..(3) L'Orthoptère (orthopter) or mechanical bird, i.e., an aëronef sustained and propelled by beating wings. **1909** [see *ornithopter]. **1909** *Westm. Gaz.* 23 Mar. 4/2 The Lamplough orthopter is not dependent for its sustentation in the air on rapid motion. **1910** *Flight* II. 58/1 All types of helicopeters [*sic*] and orthopters. **1917** *Jane's All World's Aircraft* A. 10/2 *Orthopter*, an intended-to-fly machine in which the wings are flapped mechanically in a manner which the designer believes would be the right way for a bird to flap its wings if its Creator had known more about aero-dynamics.

orthopteran, *a.* (Examples.)

1900 *Proc. U.S. Nat. Museum* XXIII. 393 The Orthopteran genus *Trimerotropis*. **1956** *Nature* 10 Mar. 490/1 This is a preliminary report of an investigation in progress dealing with cytogenetics of orthopteran insects of the southern hemisphere.

orthopteroid (ǫɹþǫ·ptĕroid), *a.* and *sb.* *Ent.* [f. Orthoptera *sb. pl.* (P. A. Latreille in C. S. Sonnini *Buffon's Hist. Nat. Crustacés & Insectes* (1802) III. 267) + -oid.] **A.** *adj.* Belonging or pertaining to a group of insect orders closely related to the Orthoptera. **B.** *sb.* An insect included in this group.

1887 A. Heilprin *Geogr. & Geol. Distribution Animals* 146 The discovery..of an apparent orthopteroid (Palæoblattina) in the most nearly equivalent deposits of Calvados, France. **1889** Nicholson & Lydekker *Man. Palæont.* (ed. 3) I. 593 The Orthopteroid section of the *Palæodictyoptera* includes a group of forms representing the modern Cockroaches. **1910** *Encycl. Brit.* XIII. 432/1 Orthopteroid wing-neuration. **1942** E. O. Essig *College Entomol.* ii. 15 Frons—a single sclerite between and below the branches of the epicranial suture; carries the single frontal ocellus of orthopteroid insects. **1973** W. S. Romoser *Sci. of Entomol.* xi. 323 Like the paleopterous forms, orthopteroid insects are hemimetabolous.

orthopterous, *a.* (Later examples.)

1895 D. Sharp in *Cambr. Nat. Hist.* V. viii. 198 Three millimetres is the least length known for an Orthopterous insect. **1920** W. J. Lucas *Monogr. Brit. Orthoptera* p. v, One or two new ones [*sc.* species] may fairly be looked for, when those naturalists who investigate our orthopterous fauna have become more numerous. **1965** D. R. Ragge *Grasshoppers, Crickets & Cockroaches Brit. Isles* 2 The various types of Orthopterous insect have little in common.

orthoptic *a.* (*sb.*) Add: **3.** Substitute for def.: Employing the principles of orthoptics; of or pertaining to orthoptics. (Earlier and later examples.)

1886 C. M. Culver tr. *Landolt's Refraction & Accomm. of Eye* v. 407 We may hope to effect a cure of the strabismus by means of orthoptic treatment, with the aid of stereoscopic exercise. **1907** J. H. Parsons *Dis. Eye* xxviii. 559 If there is any evidence of some degree of binocular vision it may be advisable to attempt to cultivate this by orthoptic treatment. **1932** *Brit. Med. Jrnl.* 14 May 918/2 The Royal Westminster Ophthalmic Hospital established an orthoptic department..in January, 1930. **1968** Katzin & Wilson *Strabismus in Childhood* viii. 65 In most patients with crossed eyes orthoptic exercises play a prominent role in treatment.

B. *sb. pl.* **orthoptics** (const. as *sing.*). The treatment (esp. by means of eye exercises) of defects in the action of the eye muscles, esp. those causing defective binocular vision.

1934 *Brit. Jrnl. Ophthalm.* XVIII. 429 The first examination in Orthoptics to be held in London will take place on July 3rd and 4th, at the Royal Westminster Ophthalmic Hospital. **1957** *New Scientist* 9 May 38/2 Many opticians nowadays have specialised in orthoptics..and it is now included in the training of all ophthalmic students. **1957** A. Huxley *Let.* 12 Jan. (1969) 815 Optometrists.. are steadily adopting more and more the Batesian procedures into their system of 'orthoptics'. **1963** [see *orthoptically* below].

Hence **ortho·ptically** *adv.*, by means of or with regard to orthoptics. Also **ortho·ptist,** one who practises orthoptics.

1937 Lyle & Jackson *Pract. Orthoptics in Treatment of Squint* i. 3 The orthoptist must remember that most squinting children who are old enough to have experienced the taunts of their schoolfellows suffer considerably from self-consciousness and inferiority. **1945** *Brit. Jrnl. Ophthalm.* XXIX. 420 (*heading*) An analysis of one hundred cases of strabismus treated orthoptically. **1963** *Arch. Ophthalm.* LXX. 117/1, 177 cases of accommodative strabismus..had been discharged as orthoptically satisfactory after treatment with glasses or miotics alone, or in combination with orthoptics and surgery. **1969** H. A. Knoll in R. Kingslake *Appl. Optics* V. x. 282 The orthoptist is trained to diagnose and treat by nonmedical means binocular muscle anomalies. **1972** *Daily Tel.* (Colour Suppl.) 28 Apr. 10/1 On the staff at Ryegate..are a physiotherapist, an occupational therapist, an orthoptist (a therapist for the eyes), [etc.].

orthoroentgenography (ǭ:ɹþorǫntgěnǫ·grăfi, *etc.*). *Med.* [f. Ortho- + *roentgenography* s.v. *roentgen-.] A technique for producing radiographs showing the exact sizes of organs or bones by using a narrow beam of X rays perpendicular to the plate or film.

There are other pronuncs. of this word and its derivs. analogous to those of *roentgenography, -gram, etc.

1946 W. T. Green et al. in *Jrnl. Bone & Joint Surg.* XLIV. 60 In a study of growth by the Havard Infantile Paralysis Commission, it was found necessary to develop a method of measuring the lower extremities which would meet the following requirements... The method to be described has been designated 'orthoroentgenography'. **1972** M. O. Tachdjian *Pediatric Orthopedics* II. vii. 1486/2 The advantages of orthoroentgenography are that the true length of each bone can be measured, because magnification due to divergence of rays is eliminated by directing only perpendicular rays at the ends of the long bones.

Hence **orthoroe·ntgenogram,** a radiograph produced by orthoroentgenography; **o:rthoroentgenogra·phic** *a.*, (done by means) of orthoroentgenography.

1946 *Jrnl. Bone & Joint Surg.* XLIV. 63 The true length of each bone can be measured directly from orthoroentgenograms without computation. *Ibid.* 65 Orthoroentgenographic measurement of a dissected adult femur, 45·7 centimeters long, gave a length of 45·5 centimeters. **1951** L. A. W. Kemp *Students' Radiol. Math.* v. 62 In the second orthoroentgenographic technique, a narrow slit.. in a sheet of metal opaque to X-rays, is arranged to be vertically below the source S, so that at any instant there is only a narrow line of X-rays..across the patient, the plane containing the rays being vertical. **1972** M. O. Tachdjian *Pediatric Orthopedics* II. vii. 1486/1 The technique [*sc.* teleoroentgenography] is not satisfactory for serial mensurations and bone detail is much less than that seen on spot orthoroentgenograms.

orthoscopic, *a.* Add to def.: *spec.* of binocular vision: without the reversal of convexity and concavity produced by pseudoscopic instruments. (Earlier and later examples.)

1853 *Q. Jrnl. Microsc. Sci.* I. 305 To produce orthoscopic binocular vision, simple, not erecting eye-pieces, are required. **1881** *Jrnl. R. Microsc. Soc.* I. 204 Orthoscopic vision is always obtained, when the right half of the right pupil and the left half of the left pupil only are employed —pseudoscopic vision in the opposite case. **1937** *Jrnl. Optical Soc. Amer.* XXVII. 333/1 The imagery should be as nearly orthoscopic as possible.

orthosis (ǭɹþǭᵘ·sis). *Med.* Pl. **-oses.** [f. Gr. ὄρθωσις making straight (f. ὀρθοῦν to set straight: see -osis.)] An artificial external device, as a brace or splint, which may be powered or unpowered and which prevents or assists relative movement in the limbs or the spine.

1958 H. A. Rusk *Rehabilitation Med.* ix. 196 Above all it is necessary for doctor, orthotist, and therapist to be fully aware of the tremendous physical and emotional impact of the orthosis on the total rehabilitation of the patient. **1966** *3rd Ann. Rocky Mountain Bioengin. Symposium* 79/1 The Rancho Electric Arm is the outgrowth of seven years of experimental work in upper extremity external power orthoses. **1970** J. Kjølbye in G. Murdoch *Prosthetic & Orthotic Pract.* xi. 459 Orthoses used in conjunction with physiotherapy are of greatest use in the prevention of deformity by protecting the weaker group of muscles from the overactivity of their antagonists.

So **ortho·tic** *a.*, serving as an orthosis; of or employing an orthosis or orthoses; **ortho·tics,** the application of orthoses; **ortho·tist,** one who practises orthotics.

1951 *Jrnl. of OALMA* May 34/1 Skilled technicians now called Certified Orthotists or Prosthetists. **1955** *Artificial Limbs* May 99/1 On March 12 and 13, a two-day session dealing with prosthetic and orthotic devices brought together at the Statler Hotel in Los Angeles a record attendance of prosthetists, orthotists, and orthopedic surgeons. **1957** *Ibid.* Spring 116/1 Approximately 60 percent of the class time in orthotics was used for actual laboratory practice. **1968** *Math. Biosciences* III. 156 Essential problems in prosthetics, orthotics, remote handling, and robot design have a common theoretical background. **1970** R. D. Muckart in G. Murdoch *Prosthetic & Orthotic Pract.* xi. 481 The orthotist would thus be able to devote his time more profitably to the solution of the splintage problems of the severely disabled patient. **1975** *Observer* 8 June 1/7 A critical report of the orthotic industry..is now circulating among orthopaedic surgeons. **1976** *Alyn & Deeside Observer* 10 Dec. 22/3 (Advt.), The post also offers experience in Plaster Work and Orthotics.

orthostat (ǭ·ɹþostæt). *Archæol.* Also **-state** (-stēⁱt). [ad. Gr. ὀρθοστάτ-ης upright shaft, pillar, building stone laid with the longest edge vertical.] An upright stone or slab, either forming part of a building or set in the ground as a monument.

[**1909** A. Marquand *Greek Archit.* ii. 67 Walls of temples and other buildings were usually provided with both base and crown. The orthostatai were set off from the vertical face of the wall.., and, even when the entire wall was covered with stucco, formed a more or less visible base.] **1926** D. G. Hogarth *Kings of Hittites* ii. 26 (*caption*) Orthostats of south gateway. **1933** *Antiquity* VII. 222 The orthostates rest against pairs of jambs kept apart by sills rising to about half the height of the chambers. **1950** G. E. Daniel *Prehist. Chamber Tombs Eng. & Wales* iii. 34 Megaliths used as orthostats, i.e. set in the ground and standing upright. **1950** H. L. Lorimer *Homer & Monuments* 419 The actual remains of Geometric temples would lead us to expect a few courses of undressed stones (possibly with an outer facing of orthostats) supporting a wall of crude brick. **1970** *Encycl. Brit.* XVIII. 454/1 The practice of setting a series of stone slabs, called orthostates, at the bottom of a wall below the mud-brick upper parts became common in the Assyrian period and was seen again in Hittite architecture. **1972** Y. Yadin *Hazor* ii. vii. 72 The most important of these is a small orthostat of a lion... The lion is clearly an entrance-jamb orthostat. **1974** F. Emery *Oxfordshire Landscape* i. 36 They, or their close followers the Beaker people, also built circles of standing stones..and orthostats.

orthostatic (ǭɹþostæ·tik), *a.* [f. Ortho- + Static *a.*; in sense 1 coined as F. *orthostatique* (J. Teissier 1899, in *Semaine Medicale* 425/1).]

1. *Med.* Caused by, or resulting from, an upright posture; manifested or occurring while a person is standing up.

1902 *Med. Ann.* XX. 90 The condition..variously called intermittent or cyclical albuminuria, is perhaps best defined by the term orthostatic albuminuria, for..position is the factor which determines the appearance and disappearance of the albumin. **1927** *Physiol. Rev.* VII. 466 Disturbed vascular conditions, such as orthostatic albuminuria. **1961** *Lancet* 26 Aug. 475/1 The unusual disorder known as orthostatic hypotension of unknown origin. **1971** *New Scientist* 29 July 249/1 The cosmonauts of Soyuz 9 suffered from severe orthostatic hypotension and for several days were unable to assume the erect posture unaided.

2. *Archæol.* Set on end; constructed of stones or slabs set on end.

1912 T. E. Peet *Rough Stone Monuments* p. iv, The first and most important principle, that on which the whole of the megalithic construction may be said to be based, is the use of the orthostatic block, i.e. the block set up on its

edge. *Ibid.*, The orthostatic slabs were often deeply sunk into the ground where this consisted of earth or soft rock. **1926** D. G. Hogarth *Kings of Hittites* ii. 29 Ground-courses of orthostatic slabs. **1941** *Proc. Prehist. Soc.* VII. 1 The use of large stones for orthostatic walling. **1950** G. E. Daniel *Prehist. Chamber Tombs Eng. & Wales* iii. 34 In almost all the chambers with orthostatic walls there is also intercalary walling to fill up the spaces.

Hence **orthosta·tically** *adv.*

1950 *Chambers's Encycl.* IX. 234/1 The burial chamber or chamber tomb, a vault or chamber walled with megaliths set orthostatically side by side.

o:rthostereosco·pic, *a.* [f. Ortho- + Stereoscopic *a.*] Showing solid objects with their true proportions and perspective, *spec.* without the reversal of convexity and concavity produced in pseudoscopic instruments.

1892 *Jrnl. Quekett Microsc. Club* V. 46 If orthostereoscopic vision is required the transposition must be corrected. **1937** *Jrnl. Optical Soc. Amer.* XXVII. 339/2 In the x-ray clinic, orthostereoscopic radiographs help to reduce error in diagnoses. **1941** R. M. Allen *Photomicrogr.* v. 236 It is not essential, especially for higher-power work with single objectives, that true orthostereoscopic effects be obtained. **1966** H. Asher tr. *Valyus's Stereoscopy* ii. 94 Observation of these conditions ensures that an orthostereoscopic image is produced, i.e. one which shows a correct proportion in depth and undistorted perspective, and allows the visual fusion of the two stereograms into a single spatial percept.

Hence **orthoste·reoscope,** a binocular microscope giving orthostereoscopic images; **o:rthostereosco·pically** *adv.*, in an orthostereoscopic manner; **o:rthostereo·scopy,** the production of orthostereoscopic images.

1892 *Jrnl. Quekett Microsc. Club* V. 52 If the two prisms were joined into one, it would..make a very efficient orthostereoscope. **1928** B. J. Leggett *Theory & Pract. Radiol.* III. ix. 430 This process of ortho-stereoscopy has important practical bearings. **1937** *Jrnl. Optical Soc. Amer.* XXVII. 333/2 A Greenough type microscope, if the magnification is not too high, may be built to operate orthostereoscopically for related combinations of objective and eyepiece pairs. *Ibid.* 339/2 Accuracy in the making of contour maps by aerial stereophotography depends upon the complete achievement of orthostereoscopy. **1966** H. Asher tr. *Valyus's Stereoscopy* ix. 380 Only when all these conditions are observed will the primary image system 'see' the object orthostereoscopically. *Ibid.* 409 To reproduce a natural relief the conditions of orthostereoscopy must be observed.

orthotropic, *a.* Add: **2.** Having three mutually perpendicular planes of elastic symmetry at each point.

1943 *Q. Appl. Math.* I. 128 Another important special case is that of an orthotropic elliptic plate bent by a linear load, for which the solution is new. **1963** P. Fern tr. *Lekhnitskii's Theory Elasticity of Anisotropic Elastic Body* i. 21 Delta-wood and plywood can be considered as homogeneous and orthotropic in the first approximation. **1971** P. J. Dowling in K. C. Rockey et al. *Devel. in Bridge Design & Constr.* 557 After the war the orthotropic deck bridge was developed and this form of bridge has now become a common form of construction..where saving in weight or depth of construction are important parameters. **1975** Cusens & Pama *Bridge Deck Anal.* i. 18 For long-span steel bridges, the deck is frequently a plate with longitudinal stiffeners (stringers). This form of construction is known to steel designers as an orthotropic plate.

orthotropy. Add: **2.** The condition of being orthotropic.

1966 F. J. Plantema *Sandwich Construction* v. 118 Numerical results have been computed only for a few typical cases of orthotropy corresponding to corrugated-core sandwich plates and sandwich plates having standard glass-fabric laminate faces. **1974** R. Szilard *Theory & Anal. Plates* iii. 375 If we assume that the principal directions of orthotropy coincide with the *X* and *Y* coordinate axes, it becomes evident that four elastic constants..are required for the description of the orthotropic stress-strain relationships.

orticant (ǭ·ɹtikănt), *a.* (*sb.*) [ad. It. *orticante* stinging, Urticant *a.*] Irritating to the skin. Hence as *sb.*, an orticant agent.

1939 L. W. Marrison tr. *Sartori's War Gases* xiv. 217 The introduction of one or more sulphur atoms..between the chloroethyl groups..confers orticant properties. **1944** R. W. Moncrieff *Chem. Senses* vi. 140 In addition to the lachrymatories..there are the skin irritants or 'orticants'. *Ibid.* 141 Sym-dichloroacetone..has an 'orticant' action on the skin as well as being lachrymatory. **1965** *Economist* 27 Mar. 1361/3 Orticants are related to these [*sc.* vesicants like mustard gas] and cause itching. **1970** *Daily Tel.* (Colour Suppl.) 20 Feb. 19 Harassing agents..come in a profusion of varieties. There are..lachrymators..; orticants, which irritate the skin, and vomiting gases.

Ortolani's sign (ǭ·ɹtǒlāni). *Med.* [named after Marino *Ortolani*, Italian who described it (*La Lussazione congenita dell' Anca* 1948)).] A click which can be obtained from and is diagnostic of congenital dislocation of the hip in the newborn.

1965 Rains & Capper *Bailey & Love's Short Pract. Surg.* (ed. 13) xvii. 346 The Ortolani test takes precedence over radiography. **1974** A. Henry in R. M. Kirk et al. *Surgery* xv. 322 Diagnosis of the unstable or 'dislocatable' hip should be made in the newborn, and all midwives and doctors who participate in the care of the newborn child should be familiar with the examination of the baby's hip—Ortolani's sign.

‖**ortstein** (ǭ·ɹtʃtəin). *Soil Sci.* [G.] A hardpan, esp. one in the B horizon of a podzol that is cemented with iron and organic matter.

1906 E. W. Hilgard *Soils* x. 184 The latter class of hardpans is especially conspicuous in the case of swampy ground and damp forests, where 'moorbedpan' and reddish 'ortstein' are characteristic. **1932** G. W. Robinson *Soils* iii. 57 The most widely spread type of pan is that formed by deposition of hydrated ferric oxide, the so-called iron pan or 'ortstein'. **1972** C. B. Hunt *Geol. Soils* x. 234 Some hardpans are formed by accumulations of organic matter..as in the well cemented, lower layers (ortstein) of a Ground Water Podzol Soil.

Ortygian (ǫɹti·dʒiăn). [f. L. *Ortygi-us* (f. Gr. Ὀρτυγία (ὄρτυξ a quail) Quail-island) + -an.]

1. Of or pertaining to Ortygia, the ancient name of the island of Delos, held to be the birthplace of Apollo and Artemis.

1640 J. Gower tr. *Ovid's Fasti* v. 120 These prayers make Mercury in heaven to smile, Remembring his Ortygian cheat yerwhile. **1729** G. Adams tr. *Sophocles' Tragedies* II. 180 And his [Apollo's] Ortygian Sister likewise..Fire-bearing Diana. **1866** J. B. Rose tr. *Ovid's Metamorphoses* i. 27 A votress of the power Ortygian. **1956** E. Pound tr. *Sophocles' Women of Trachis* 12 Sylvan Artemis, torch-lit Artemis With thy Ortygian girls.

2. Of or pertaining to Ortygia, an island which forms part of the city of Syracuse in Sicily. *rare.*

1820 Shelley *Arethusa* in *Posthumous Poems* (1824) 160 And at night they sleep In the rocking deep Beneath the Ortygian shore.

Orvieto. Add: (Earlier and later examples.) Also *attrib.*

[**1673** J. Ray *Observations Journey Low-Countries* 363 Heer [*sc.* in Rome] is great variety of Wines..as Greco, Lagrime of Naples..Orvietano.] **1846** Dickens *Pictures from Italy* 159 Such wine in flasks, as the Orvieto. **1849** Thackeray *Pendennis* I. xxxv. 344 We had some Orvieto wine for dinner. **1926** P. M. Shand *Bk. Wine* viii. 225 Umbria is chiefly noted for its white wines, of which the most famous name is Orvieto. **1940** M. Healy *Stay me with Flagons* 70 Orvieto, now sweet, now dry, offers varied charms. **1967** A. Lichine *Encycl. Wines* 535 Orvieto is white, and one of Italy's most consistently delightful wines, some of it semi-sweet, some of it dry... The market for Orvieto *amabile* or *abboccato*..has fallen off... Sweet Orvieto has a particular charm and special delicacy... Dry Orvieto is usually considerably higher in alcohol. **1974** N. Marsh *Black as he's Painted* ii. 39 Bottles of dry Orvieto..and other Italian wines.

b. Used *attrib.* to designate a type of majolica ware manufactured there.

1925 B. Rackham tr. *E. Hannover's Pott. & Porc.* I. III. ii. 96 It requires a practised eye to distinguish between a genuine and a spurious Orvieto jug. **1959** G. Savage *Antique Collector's Handbk.* 70 The earliest *maiolica* was a series of wares painted in green and manganese purple at Orvieto and elsewhere, and most are loosely called 'Orvieto ware'. **1960** R. G. Haggar *Conc. Encycl. Cont. Pott. & Porc.* 338/2 *Orvieto*, a distinctive class of early Italian maiolica painted in coppergreen and manganese-purple with decorations and shapes of a distinctly 'Gothic' type. **1973** *Times* 3 Nov. 2/2 A Parke Bernet expert called in to look over what was there found the Medici bowl in a cupboard under a broken Orvieto dish.

Orwellian (ǫɹwe·liăn), *a.* [f. the name of 'George *Orwell*', the nom-de-plume of the English writer Eric Blair (1903–50) + -ian.] Characteristic or suggestive of the writings of 'George Orwell', esp. in his satirical novel *1984* which portrays a form of totalitarian state seen by him as arising naturally out of the political circumstances of his time. Hence as *sb.*, an admirer of the ideas of Orwell. So **Orwe·llianism, O·rwellism.**

1950 M. McCarthy *On Contrary* (1962) 187 A leap into the Orwellian future. **1952** *Time* 1 Sept. 19/2 Under the new rules, the Politburo and the Orgburo will be merged and the two ugly Orwellian names replaced by the stern old Latin 'Presidium'. **1958** *Times* 20 Feb. 11/3 In Orwellian language, 'imperious'. **1959** N. Mailer *Advts. for Myself* (1961) 309 Virtually perfect Orwellian ambivalences—(War is Peace, Love is Hate, Ignorance is Knowledge). **1961** [see *Huxleyan, Huxleian *a.* b]. **1963** *Observer* 3 Nov. 33/1 There is an Orwellian grimness about the naming of the Abolition of Passes Act in South Africa, which ensured that an African had to carry 27 papers combined into one booklet. **1967** *Listener* 2 Nov. 583/2 It is Orwellianism transferred to the world of the commercial, in which machines work beautifully, everybody is on a kind of holiday and wears a blazer and a redcoat number, the daily flavour of the ice-cream is announced on the morning radio, [etc.]. **1970** *Guardian* 30 Nov. 11/2 His election night insistence that the blood on his face was nothing less than the blush of victory can now be seen as a triumph of public relations.., an interesting example of the progress of Orwellism in national politics. **1971** *Ibid.* 14 Jan. 7/1 McLuhanites and Orwellians are likely to block our view of their masters' arguments. **1972** *Ibid.* 17 Feb. 14/5 The Orwellian Newspeak style. **1974** *Daily Tel.* (Colour Suppl.) 20 Sept. 27/4 One sees a future Robin Day as an Orwellian Official Moderator of the Ministry of Received Truth. **1976** *Time* 5 Apr. 22/2 In foreign affairs, says Moynihan, there is something almost Orwellian about the transformation of the word liberal to mean the opposite of what it meant a decade or so ago.

orygine (ǫ·ridʒəin), *a.* *Zool.* [f. L. *oryg-*, stem of *oryx* (see Oryx) + -ine¹.] Resembling an antelope of the genus *Oryx*, esp. belonging to an African group including the addax, gemsbuck, roan, and sable antelopes, which share certain characteristics with the oryx, such as long horns, tufted tails, and large, square teeth.

1898 *Proc. Zool. Soc.* 352 The Addax, I think, is on the whole more an orygine type than a hippotragine. **1947** J. Stevenson-Hamilton *Wild Life S. Afr.* xi. 75 (*heading*) The orygine antelopes.

oryzenin (orəi·zěnin). *Biochem.* [f. L. *orȳz-a* (Gr. ὄρυζα) rice + *-enin*, prob. after *gluten-in.] A glutelin that is the chief protein in rice.

1908 Rosenheim & Kajiura in *Jrnl. Physiol.* XXXVI. p. lv, These three proteins appear to be the only ones present in rice and we propose to call them respectively: Rice-globulin, Rice-albumin, and Oryzenin. **1926** *Jrnl. College of Agric. Hokkaido Imperial Univ.* XVI. 76 The differences between the common and the glutinous rice oryzenins are marked by their physico-chemical properties. **1952** *Chem. Abstr.* XLVI. 2747 As tested by fermenting each constituent of rice by *Aspergillus oryzae*, the odor of sake was found to be due mainly to the oryzenin. **1972** B. O. Juliano in D. F. Houston *Rice* ii. 41 Glutelin or oryzenin is the major protein fraction of brown and milled rice.

‖ **osaekomi waza** (ōsa‚ekōmi waza). *Judo.* Also **osae waza.** [Jap., lit. 'art of holding', f. *osae* to press upon or against + *komi* to be packed up + *waza* art.] (See quot. 1932.)

1932 E. J. Harrison *Art of Ju-Jitsu* v. 64 '*Osaekomi-waza*', otherwise methods of holding down one's opponent on the mats for a longer or shorter interval. *Ibid.* 65 The first trick of *osaekomi-waza*..is the so-called Locking of the Upper Four Quarters. **1941** M. Feldenkrais *Judo* 173 Immobilizing or holding down (osae-waza). **1956** K. Tomiki *Judo* iii. 90 Practice in *katame-waza* (art of grappling) had better be based on that in *osae-waza* (art of holding, or hold-down). **1962** LeBell & Coughran *Handbk. Judo* vi. 103 (*heading*) Osaekomiwaza—hold on there! Hold down techniques. **1970** A. P. Harrington *Judo Guide* i. 31 Methods of holding the opponent with his back largely on the ground with one or both arms under restrictive control are known as Hold-downs (Osaekomi-waza).

Osage (ōᵘ·sēⁱdʒ), *sb.* and *a.* [ad. Osage self-designation *wazhazhe*.] **A.** *sb.* A member of a Siouan Indian people coming originally from the Osage river valley, Missouri. Also, their language. **B.** *adj.* Of or pertaining to this people.

1698 tr. *Hennepin's New Discovery* I. 141 Several Savages of the Nations of the Osages, Cikaga, and Akansa, came to see us. **1722** D. Coxe *Descr. Carolana* 16 The Yellow [River] is called the River of the Massorites, from a great Nation inhabiting in many Towns near its juncture with the River of the Osages. **1804** [see *fix *v.* 14 b]. **1832** [see *heap *sb.* 4 d]. **1835** W. Irving *Tour on Prairies* i. 7 He spoke a Babylonish jargon of mingled French-English, and Osage. **1906** *Indian Affairs: Laws & Treaties* (U.S.) (1913) III. 253 All lands belonging to the Osage tribe of Indians..shall be divided among the members of said tribe. **1931** C. Turner tr. *von Schmidt-Pauli's We Indians* xiii. 132 The long-drawn-out and piercing war-cry of the Osages. **1933** L. Bloomfield *Language* iv. 72 The Siouan family includes..Dakota,..Omaha, Osage, [etc.]. **1945** J. J. Mathews *Talking to Moon* 87 She..said in Osage to the chief, 'I want to tell my son about the way we did things.' **1973** A. H. Whiteford *N. Amer. Indian Arts* 90 Oto and Osage also wove hat and neck bands of horsehair. **1974** *Encycl. Brit. Micropædia* VII. 602/3 The discovery of oil on the Osage reservation in the late 19th century.. made the Osage a uniquely prosperous people.

Osage orange. [f. prec. + Orange *sb.*¹] = *maclura, a tree native to Arkansas and neighbouring regions formerly occupied by the Osage tribe; also, the fruit of this tree. Also *attrib.*

1817 [see Orange *sb.*¹ 3 a]. **1838** H. W. Ellsworth *Valley Upper Wabash* v. 52 These fences, whose tops are covered with a luxuriant growth of the wild locust hawthorn, or Osage Orange. **1846** D. J. Browne *Trees Amer.* 465 The Osage Orange-tree. **1857** *Trans. Illinois Agric. Soc.* II. 23 The practicability of successfully cultivating the Osage Orange plant into a protective hedge. **1859** [see Orange *sb.*¹ 3 a]. **1891** *Harper's Mag.* Sept. 579/2 There was a hedge of Osage Orange on one side of the yard. **1940** E. Fergusson *Our Southwest* ix. 157 Mrs. Hayden found a stately adobe house with water piped in, an osage orange hedge around an orchard. **1970** *New Yorker* 28 Feb. 41/2 When we were little..we could bowl the Osage oranges down. **1974** A. Dillard *Pilgrim at Tinker Creek* vi. 100 Beneath the overarching limbs of tulip, walnut, and Osage orange, I see the creek pour down.

Osagyefo (ōᵘsagye·fo). [Fante]. Redeemer: a name given to Kwame Nkrumah (1909–72), first prime minister of Ghana (1952–60), president of the Republic (1960–66).

1961 *Guardian* 29 Sept. 10/1 The Osagyefo (or Redeemer, as he is called by his press). **1964** *Economist* 14 Mar. 983/3 The Osagyefo offered a careful welcome. **1965** *Ibid.* 24 Apr. 389/1 Everyone's suspicions of the Osagyefo (it means redeemer) were discussed. **1975** H. R. Isaacs in H. M. Patel et al. *Say not the Struggle Nought Availeth* 265 He found it impossible to follow Nkrumah on the way to becoming the Osagyefo, the Messiah, the Savior of the People.

osazone (ōu·săzōun). *Chem.* [f. the suffix *-osazone*, ad. G. *-osazon* (E. Fischer 1884, in *Ber. d. Deut. Chem. Ges.* XVII. 580): see -ose[2], Azo-, and -one.] Any of the yellow crystalline solids whose molecules contain two adjacent hydrazone groups, which are obtained by treating compounds containing the groups —CO·CHOH— or —CO·CO— with phenylhydrazine and are used for characterizing sugars.

1888 *Jrnl. Chem. Soc.* LIV. 1267 The osazones of all saccharoses dissolve in cold, fuming hydrochloric acid with a dark-red coloration. **1938** M. L. Wolfrom in H. Gilman *Org. Chem.* II. xvi. 1404 The three fundamental procedures used by Fischer..in his great feat of elucidating the configuration of the sugars were osazone formation, oxidation to *meso* acids.., and the methods for building up or degrading the members of the sugar series. **1972** J. W. Suttie *Introd. Biochem.* iii. 28 Sugars which differ in configuration only at carbons 1 or 2, such as glucose, mannose, and fructose, will give the same osazone.

Osborne (ǫ·zbǭəɹn). The name of a former royal residence on the Isle of Wight used *attrib.* and *absol.* to designate a type of sweetish plain biscuit.

Said by the manufacturers (private communication 14 March 1977) to have been first made in 1860. It was originally intended to name them after Queen Victoria, but she asked that the biscuit should be named after her favourite home—Osborne—on the Isle of Wight.

1876 *Official Guide Cunard Steamship Co.* 158 (Advt.), Huntley & Palmers..List of Biscuits..Osborne Biscuits. **1888** *Mrs. Beeton's Bk. Househ. Managem.* ii. 31/2 Biscuits..Osborne..1*s.* 6*d.* per tin. **1911** A. Bennett *Hilda Lessways* vi. vi. 402, I should like another biscuit. But I don't want the Osbornes—the others. **1926** W. de la Mare *Connoisseur* 49 My stranger in the tea shop had been refreshing himself with Osborne biscuits. **1938** S. Beckett *Murphy* v. 96 The biscuits..were the same as always, a Ginger, an Osborne, a Digestive, a Petit Beurre. **1960** R. Collier *House called Memory* ii. 28 My mother's first memory..was of sitting on the steps..and being given some Osborne biscuits. **1972** D. Bloodworth *Any Number can Play* x. 77 The gibbon..grabbed three Osborne biscuits from a tin on the low, marble table.

osbornite (ǫ·zbǭəɹnəit). *Min.* [f. the name of George *Osborn-e* (see quot. 1870) + -ite[1].] Titanium nitride, TiN, found as small yellow octahedra in oldhamite in the meteorite which fell at Bustee, India.

1870 N. S. Maskelyne in *Phil. Trans. R. Soc.* CLX. 198 This microscopic mineral I wish to name Osbornite in honour of Mr. Osborne, and in order to commemorate the important service that gentleman rendered to science in preserving and transmitting to London in its entirety the stone which his zeal saved at the time of its fall, and in recording all he could collect about the circumstances associated with that fall. **1941** *Mineral. Mag.* XXVI. 36 The minute octahedra of osbornite are easily visible with a lens in the meteorite itself. **1962** B. Mason *Meteorites* x. 156 The only nitrogen mineral recorded from meteorites is osbornite, TiN, a rare accessory known only from Bustee (an enstatite achondrite).

Oscan (ǫ·skăn), *a.* and *sb.* Also 6–8 **Oscian.** [f. L. *Osc-us* (pl. *Osci*) Oscan + -an.] **A.** *adj.* **a.** Of or pertaining to the Osci (also called Opsci, Opici), a pre-Sabellian people centred on Campania in southern Italy. **b.** Of or pertaining to the Italic language called Oscan (see sense b of the sb.).

1598 R. Grenewey tr. *Tacitus' Annals* iv. 95 The Oscian play, a light sport pleasing the peoples humor. **1600** Holland tr. *Livy's Romane Hist.* x. 365 Such as were well seene in the Osciane toung. **1649** J. Ogilby tr. *Virgil's Aeneid* vii. 24 Saticulus with them And Oscian bands. **1728** T. Gordon tr. *Tacitus' Annals* iv, in *Works* I. 164 The Oscan Farce, formerly only the contemptible delight of the vulgar. **1731** J. Trapp tr. *Virgil's Aeneid* vii, in *Works* III. 92 And the Oscian Band. **1797** G. Baker tr. *Livy's Hist. Rome* II. x. 356 He..sent persons, who understood the Oscan language. **1845** [see Umbrian *a.* 1]. **1948** D. Diringer *Alphabet* II. ix. 503 The Oscan script..was an offshoot of the Etruscan alphabet in its southern Campano-Etruscan sub-species. **1958** E. Pulgram *Tongues of Italy* xvii. 229, I do not believe in the existence of Umbrian, Oscan, and Latinian types of dialects outside of Italy. **1958** E. Badian *Foreign Clientelae* ix. 195 Of other patrons of Marius we know of the Herennii,..themselves of Oscan origin. **1969** E. T. Salmon *Roman Colonization* v. 88 Atella and Calatia, also joined Hannibal, as did a number of the Oscan and Greek allies of Rome in southern Italy. **1974** *Encycl. Brit. Macropædia* XIV. 789/2 The Oscan village of Pompeii..soon came under the influence of the cultured Greeks..across the bay.

B. *sb.* **a.** A member of the Osci. **b.** A name given to the Italic dialects of central and southern Italy, used by the Sabellian peoples who displaced or absorbed the Osci. See *Osco-Umbrian *a.* and *sb.*

1753 C. Pitt tr. *Virgil's Aeneid* vii, in *Works* III. 355 The rough Saticulan and Oscan stood. **1813** J. C. Eustace *Tour through Italy* II. x. 258 Oscan was not unknown even in the age of Cicero and Augustus. **1835** [see *Ligurian *sb.*]. **1882** [see Umbrian *sb.* 2]. **1897** R. S. Conway *Italic Dial.* II. 472 In some few cases in Oscan the final -*s* of the Nom. and Gen. is still wanting. **1934** S. Robertson *Devel. Mod. Eng.* (1936) ii. 21 Of the latter [*sc.* non-Latin Italic dialects], the Oscan and the Umbrian..are to be distinguished. **1939** [see *Marsian *sb.* and *a.*]. **1948** D. Diringer *Alphabet* II. ix. 501 The Oscans..or Osci, in Greek, Oskoi or Opikoi, from Opsci or Opici..were Italic tribes who inhabited southern Italy in the second half of the first millennium B.C. **1969** H. V. Morton *Traveller in S. Italy* vii. 272 We encountered a workman, still apparently speaking Oscan. **1969** E. T. Salmon *Roman Colonization* ii. 44 In 354, Rome signed an alliance with the Samnites, an Oscan-speaking people of the south. *Ibid.* ix. 149 Sulla's colonies in the north and south had helped to bring the use of Etruscan and Oscan to an end. **1974** *Encycl. Brit. Macropædia* XIV. 789/2 It seems certain that Pompeii, Herculaneum, and Stabiae were first settled by the Oscans. **1976** *Archivum Linguisticum* VII. 62 Latin shares a number of lexical items with Oscan.

Oscar[1] (ǫ·skăɹ). Also with small initial. *Austral.* and *N.Z.* [Rhyming slang on *Oscar Asche* (an Australian actor, 1871–1936).] Cash, money.

1919 W. H. Downing *Digger Dial.* 36 *Oscar*, money. **1931** W. Hatfield *Sheepmates* xix. 161 Sit in, some o' yous that aint flyblown—.., an' their IOU's is good, if there's no real Oscar about the joint. **1942** L. Mann *Go-Getter* ii. 16 Get the oscar off Tom soon's I see him. He's honest. **1945** E. G. Webber *Johnny Enzed in Middle East* 20 'Well, me lending you my balance to get you out of the cart,' said the bloke, 'and them sending me the oscar so that you can pay it back.' **1949** *Newsweek* 31 Oct. 60/2 He would have been glad to buy me a pail of suds if he'd had any Oscar. **1959** D. Niland *Big Smoke* i. 21 If you'd been fighting all those blokes in the ring you'd have more oscar in your kick now than the Prime Minister himself. **1969** [see *dinkum B. *adj.*].

Oscar[2]. [Said to be an arbitrary use of the Christian name of *Oscar* Pierce, 20th-c. American wheat and fruit grower (see below).] One of the statuettes awarded by the Academy of Motion Picture Arts and Sciences, Hollywood, U.S.A., for excellence in film acting, directing, etc. These awards have been made annually since 1928. So *Oscar-winner, -winning*.

A former secretary of the Academy of Motion Picture Arts and Sciences is said to have remarked in 1931 that the statuette reminded her of her 'Uncle Oscar', namely Oscar Pierce.

1936 *Time* 16 Mar. 56/2 Neither Director Ford nor Screenwriter Nichols appeared to claim their prizes—small gold statuettes which Hollywood calls 'Oscars'. **1949** [see *Emmy]. **1958** *Punch* 25 June 838/3 A quiet scene, with the camera in close-up, every word counting, and the actors playing for an Oscar. **1962** *Times* 30 Oct. 16/1 The Oscar-winning Fellini pictures. **1968** B. Foster *Changing Eng. Lang.* i. 45 The Oscar statuette is..gold-plated. **1974** *Times* 1 Mar. 13/4 Luis Buñuel's eminence as..1973 Oscar-winner. **1975** C. Nesbitt *Little Love & Good Company* xix. 246 If I were to start thanking the Lunts for all the gifts..that I have received from them I should sound like an Oscar-winning starlet. **1976** *Southern Even. Echo* (Southampton) 6 Nov. (Advt. Suppl.) 5/1 Ed Begley (he got an Oscar) as a vicious town boss whose daughter (Shirley Knight) gets ruined by Newman.

b. In *transf.* use applied to any award for an outstanding performance or achievement.

1941 *Time* 2 June 82/2 That these trials..did not keep Producer Gabriel Pascal from turning out a polished and distinguished product is a transcendent Oscar in the one-time cavalryman's lap. **1947** *Sun* (Baltimore) 1 July 7/6 René Clair's 'Silence Est D'Or' (Silence Is Golden) won the grand prize of Brussels' world film competition today and was awarded the Belgian 'Oscar', a small statuette of Brussels' patron saint, St. Michael. **1949** *Natural Hist.* Nov. 417/3 If there are any 'Oscars' to be awarded in the world of animal acting, the vote of many naturalists will.. go to the hognose snake. **1954** *Economist* 29 May 739/2 The first British 'Oscar' for a company report, represented by a pair of silver wall sconces, has been given to Thomas W. Ward. **1959** *Times Lit. Suppl.* 13 Nov. 662/3 Once a year it publishes an annual which is in effect a kind of collection of Oscars for design in these fields. **1963** *Guardian* 8 Feb. 8/2 An ensemble for spring..which won an export Oscar last year. **1971** *Daily Tel.* 18 Dec. 10 A local artist is being commissioned to make an 'Oscar' which will be presented annually to the publishing house which..has made the most distinctive contribution to literature. **1973** J. Wainwright *Pride of Pigs* 215 An Oscar-winning performance, lad. **1978** *Guardian* 27 Feb. 16/5 QPR had been given their first goal when an Oscar-winning dive in the penalty area by Shanks brought an unbelievable penalty decision from the referee.

oscillate, *v.* Add: **1. c.** *Electronics.* Of a circuit or device: to cause oscillations in an electric current flowing in it.

1917 R. D. Bangay *Elem. Princ. Wireless Telegr.* I. 84 The methods employed for causing an aerial to oscillate, and thus radiate electric waves, fall under two headings. **1928** *Times* 23 Mar. 20/1 If too much retroaction is employed the circuit resistance may become negative, when the whole system will begin to oscillate. **1948** A. L. Albert *Radio Fund.* x. 370 If some of the output signal voltage is fed back into the control-grid in the proper manner, a vacuum tube will oscillate. **1971** L. T. Agger *Introd. Electr.* xxiv. 432 A circuit containing inductance and capacitance will oscillate at its own natural frequency, provided it is left undisturbed after the oscillation has been started.

d. Of a radio or (*transf.*) its user: to transmit radio waves owing to faulty operation.

1921 *Wireless World* 29 Oct. 481/1 For a set to 'react' or 'oscillate' it is necessary that the main inductance and the reaction coil should be so connected that the direction of windings bear a certain relation to one another. **1926** *Westm. Gaz.* 3 Feb. 3/1 Listeners-in who oscillate may find themselves deprived of their licences. **1933** 'R. Stranger' *Elem. Wireless* xxxvi. 181 A good way of telling when a receiver is oscillating is to have a milliammeter connected in the anode circuit of the detector... Without the milliammeter you may oscillate and not even know it. **1943** C. L. Boltz *Basic Radio* xiv. 222 An interesting effect is observable if the tuning condenser is used when the receiver is oscillating.

3. (Further examples.)

1883 *Phil. Trans. R. Soc.* CLXXIII. 663 The ring (with wire circuit open) was oscillated backwards and forwards. **1905** Preece & Sivewright *Telegraphy* (new ed.) 462 The spark gap which oscillates the energy.

4. *intr. Math.* To increase and decrease alternately as successive terms are taken (in the case of a series) or as the variable tends to infinity (in the case of a function).

1898 Harkness & Morley *Introd. Theory Analytic Functions* viii. 102 In the series $(1-1)+(1-1)+(1-1)+...$, each term is 0 and the limit is 0, but the series $1-1+1-1+...$ oscillates. **1940** C. A. Stewart *Adv. Calculus* i. 7 $n^2\cos n\pi, n+(-1)^n n^2$ oscillate infinitely. **1973** D. G. Ball *Introd. Real Anal.* iii. 43 A sequence which has no limit at all is said to oscillate.

oscillating, *ppl. a.* Add: **2.** *Math.* Of a series or function (see *oscillate *v.* 4).

1898 Harkness & Morley *Introd. Theory Analytic Functions* viii. 102 Most English text-books regard oscillating series as not divergent. **1973** D. G. Ball *Introd. Real Anal.* iii. 59 Consider the sequence $2\frac{1}{2}, -1\frac{1}{2}, 2\frac{1}{4}, -1\frac{1}{4}, 2\frac{1}{8}, -1\frac{1}{8}, 2\frac{1}{16}, -1\frac{1}{16},...$ This sequence is bounded and oscillating.

3. *Electronics.* Of an electric current or the like: undergoing rapid periodic reversals in direction. Of a circuit or device: characterized by such a current (cf. *oscillate *v.* 1 c).

Oscillating as used of a current implies a much higher frequency than *alternating* and also a different origin (see *oscillator 2 a).

1906 A. E. Kennelly *Wireless Telegr.* viii. 94 (*heading*) Energy of electric oscillations, or oscillating currents, set up in a vertical receiver. **1920** E. W. Stone *Elem. Radiotelegr.* xi. 44 If two oscillating circuits containing inductance and capacity be..coupled together, they act very differently than when they are allowed to oscillate by themselves. **1948** A. L. Albert *Radio Fund.* x. 386 Variations in the plate output circuit will not be reflected back into the control-grid oscillating circuit. **1974** 'I. Drummond' *Power of Bug* xi. 166 The tiny transmitter.. emitted a high, oscillating signal on ultra short wave.

oscillation. Add: **2. b.** *Psychol.* Fluctuation of attention or mental efficiency.

1895 *Amer. Jrnl. Psychol.* VII. 84 *Schwankungen der Aufsmerksamkeit*, oscillations or fluctuations of the attention. **1927** C. S. Spearman *Abilities of Man* xix. 319 We may now conveniently turn to a phenomenon which may be described as oscillation of cognitive efficiency, and which probably has an intimate connection with..fatigue. *Ibid.* 326 Here in oscillation, then, we have come upon a new single and universal factor, a third in addition to *g* and perseveration. **1943** C. L. Hull *Princ. Behavior* xvii. 306 One may reasonably conjecture that it was produced by some factor in the experimental situation other than the primitive oscillation tendency. **1950** P. E. Vernon *Struct. Human Abilities* viii. 89 An Oscillation factor in rate of fluctuation of reversible perspective figures.

3. *Electronics.* A rapid alternation in the direction of flow of a current; the state of a circuit in which this is occurring; also, a rapidly varying electromagnetic field produced by such a current.

1853 tr. H. Helmholtz in *Sci. Mem. Trans. Foreign Acad.* (*Nat. Philos.*) 143 It is easy to explain this law if we assume that the discharge of a battery is not a simple motion of the electricity in one direction, but a backward and forward motion between the coatings, in oscillations which become continually smaller until the entire *vis viva* is destroyed by the sum of the resistances. **1885** *Electrician* 18 Dec. 106/1 It shows that if the resistance be reduced to nothing, whilst the coefficient of self-induction of the circuit is finite..the oscillations continue for ever undiminished in strength. **1911** *Encycl. Brit.* XXVI. 532/2 The transmitting antenna wire is alternately charged to a high potential and discharged with the production of high frequency oscillations in it. **1932** E. V. Appleton *Thermionic Vacuum Tubes* 102 Since the applied anode potential is negative no anode current flows unless oscillations are present. **1951** A. Sheingold *Fund. Radio Communication* xii. 254 The desirable operational properties of an oscillator may include the property of being self-starting, the ability to maintain continuous oscillation, [etc.]. **1959** R. L. Shrader *Electronic Communication* xii. 307 If it is desired to prevent a radio circuit from going into oscillation, it may be necessary to introduce degeneration in it, or neutralize it. **1975** Barron & Joyce *Electricity* xv. 450 The electrical oscillations of a capacitor-inductor loop can be used to transmit oscillations of radio frequency.

4. *Math.* **a.** The difference between the greatest and the least values of a function in any given interval.

1893 Harkness & Morley *Treat. Theory of Functions* ii. 49 If $f(x)$ be a discontinuous function of x, which is always finite between a and b, its values within an interval δ will have an upper and a lower limit. The difference between these upper and lower limits is named the oscillation of the function. **1937** Michell & Belz *Elem. Math. Analysis* I. i. 73 If the focal bounds are unequal, the sequence is said to be divergent and to have a finite oscillation. *Ibid.* 91 The functions $\sin x$, $(\sin x)/x$. The former function has a finite oscillation -1 to $+1$ at $x=+\infty$. **1973** G. Klambauer *Real Analysis* iii. 75 The oscillation is a monotone mapping in the sense that if J_1 and J_2 are bounded open intervals and $J_1 \subset J_2$, then $\omega(f; J_1) \leqslant \omega(f; J_2)$.

b. Variation consisting of alternate increase and decrease (cf. *oscillate *v.* 4).

1908 T. J. I'A. Bromwich *Introd. Theory Infinite Series* ii. 22 If all the terms..of the series are positive, the sequence..steadily increases; and so..the series Σa_n must be either convergent or divergent; that is, oscillation is impossible. **1968** Fox & Mayers *Computing Methods for Scientists & Engineers* vii. 129 A characteristic of a good approximation is a type of oscillation property, effectively produced by the presence of $n+2$ alternating maxima and minima of the error.

5. *attrib.*, as † *oscillation circuit* (now called *oscillator circuit*), *detector*, † *valve*; **oscillation constant** (see quot. 1940); † **oscillation transformer** = *jigger *sb.*[1] 5 p.

1906 A. F. Collins *Man. Wireless Telegr.* ii. 32 The resistance of an oscillation circuit is practically negligible. **1923** —— *Everybody's Wireless Bk.* xii. 198 The resistance of an oscillation circuit, whether it is a closed circuit or the aerial, must be made as low as possible in order to let the high frequency currents oscillate in it freely. **1908** C. C. F. Monckton *Radio-Telegr.* ii. 36 For different circuits, as long as the oscillation constants are the same the natural periods of vibration are the same. **1940** *Chambers's Techn. Dict.* 601/1 *Oscillation constant*, the square root of the product of the inductance (in henries) and the capacitance (in farads) of a resonant circuit. **1908** J. A. Fleming *Elem. Man. Radiotelegr.* ix. 324 It is necessary to employ in the receiving circuit an oscillation detector which is..not merely affected by oscillations, but affected to some extent proportionately to their amplitude. **1906** Oscillation transformer [see *jigger *sb.*[1] 5 p]. **1923** A. F. Collins *Everybody's Wireless Bk.* viii. 130 There are two distinct types of tuning coils used for continuous wave sending sets, and these are (1) the helix, or close-coupled coil, and (2) the oscillation transformer, or loose-coupled coil. The former gives much better results. **1906** *Proc. Physical Soc.* XX. 177 (*heading*) The construction and use of oscillation valves for rectifying high-frequency electric currents. **1908** J. A. Fleming *Elem. Man. Radiotelegr.* vi. 205 A very simple but effective form of oscillation valve was invented by the author in 1904.

oscillator. **2. a.** Substitute for def.: an apparatus for generating oscillatory electric currents by non-mechanical means. (Add earlier and later examples.)

1889 *Electrician* 9 Aug. 359/1 A practical application is to the theory of a Hertzian oscillator. **1907, 1908** [see *Hertzian *a.*]. **1933** *Geogr. Jrnl.* LXXXII. 327 In the shoal range the sound producer is a Type 399 oscillator mounted inside the ship. **1951** A. Sheingold *Fund. Radio Communication* xii. 263 Special oscillators, viz., the klystron and the magnetron, have been designed for operation in the very-high-frequency range. **1962** A. Nisbett *Technique Sound Studio* xii. 202 An oscillator is a valid 'concrete' sound source. **1970** B. Zeines *Electronic Communication Syst.* viii. 327 Essentially, a feedback oscillator is an amplifier deriving its input signal from its own output. **1977** N. Freeling *Gadget* v. 222 The oscillator is a multi-vibrator affair using a couple of integrated circuit chips.

b. A person who causes or allows a radio to transmit radio waves (cf. *oscillate *v.* 1 d).

1927 *Daily Tel.* 1 Mar. 6 Cases where engineers..have succeeded in tracking down oscillators. **1927** *Glasgow Herald* 20 Sept. 8/2 The running-to-earth of offending oscillators.

3. Something that moves to and fro.

1911 *Sci. Abstr.* A. XIV. 402 In the theory referred to [of 1901] the emission of the elements of energy is assumed to be excited by the continuous steady absorption of energy from the radiation incident upon one of the assumed oscillators. This the author [*sc.* Planck] now replaces by the assumption that the emission and absorption are entirely independent. **1936** P. M. Morse *Vibration & Sound* ii. 27 Every driven oscillator is mass controlled in the frequency range well above its natural frequency ν_0, is resistance controlled near ν_0.., and is stiffness controlled for frequencies much smaller than ν_0. **1970** G. K. Woodgate *Elem. Atomic Struct.* iii. 45 In trying to make a model of a one-electron atom out of three classical oscillators (one for each direction of polarization) we do not take account of the fact that the atom can emit many frequencies..whereas the classical oscillators emit only one frequency. **1973** *Country Life* 19 July 151/2 The best oscillators have a device which enables the degree of oscillation to be varied from a wide to a narrow throw.

4. *attrib.*, as *oscillator circuit*.

1931 Moyer & Wostrel *Radio Handbk.* vii. 390 (*heading*) Typical oscillator circuits. **1970** B. Zeines *Electronic Communications Syst.* viii. 326 Oscillator circuits are commonly used in communication systems.

oscillatory, *a.* Add: Now also with pronunc. (ǭsi·lătəri). **2.** *Electronics.* Of a current: = *oscillating *ppl. a.* 3. Of a circuit or device: capable of sustaining oscillations (*oscillation 3).

1853 *Phil. Mag.* V. 399 It is probable that many remarkable phenomena which have been observed in connexion with electrical discharges are due to the oscillatory character which we have found to be possessed when the condition..$C < 4A/k^2$ is fulfilled. **1878** *Jrnl. Soc. Telegr. Engin.* VII. 319 The discharge is oscillatory if R is less than 40,000 ohms, and continuous if it is greater than that amount. **1905** *Electrician* Feb. 614/1 The effect on the rate of radiation, of varying the length of that portion of the air wire included in the coupled oscillatory circuit. **1948** Slurzberg & Osterheld *Essent. Radio* x. 449 The essential parts of a vacuum-tube oscillator are (1) the oscillatory or tank circuit,..(2) a vacuum-tube amplifier, (3) a feedback circuit. **1971** L. T. Agger *Introd. Electr.* 442 Calculate the natural frequency of an oscillatory circuit in which the inductance is 0·1 H and the capacitance is 10 μF. **1973** J. Yarwood *Electr. & Magnetism* xi. 428 The oscillatory current obtained by the discharge of a capacitor through an inductance decays in amplitude with a logarithmic decrement of $RT/2L$.

3. *Math.* = *oscillating *ppl. a.* 2.

1908 T. J. I'A. Bromwich *Introd. Theory Infinite Series* i. 16 Since an oscillatory sequence always contains at least two convergent sub-sequences (those giving the extreme limits),..an oscillatory series can always be made to converge by grouping the terms in brackets. **1968** H. M. Lieberstein *Course in Numerical Anal.* (1969) vii. 126 The error term is oscillatory and unbounded... In computation with small *h* one will see for small *n* very good answers being produced, but eventually as *n*..becomes larger, a 'flowering' will take place as the oscillatory error term takes over.

oscillogram (ǭsi·lǒgræm). [f. as prec.: see -gram.] A record obtained by means of an oscillograph.

1903 *Jrnl. Inst. Electr. Engin.* XXXII. 43 (*heading*) A study of the phenomenon of resonance in electric circuits by the aid of oscillograms. **1917** G. D. Shepardson *Telephone Apparatus* iii. 32 As illustrative of the original oscillograms Fig. 7 shows the record of the sound *d* in *day*. **1938** *Amer. Speech* XIII. 69/1 Simultaneous oscillograms of accented vowels taken from above and below the vocal cords reveal no measurable difference in time for the beginning of the voice vibrations. **1950** *Ann. Reg. 1949* 428 There had been brought back..more than 400 oscillograms from explosions in depths between 300 and 3,500 fathoms. **1976** *Word 1971* XXVII. 57 No matter how phonemically obvious certain vocalized continua may appear to be, the differing articulations producing these phonemic samenesses will not produce phonetically same oscillograms or sound spectrograms, the only reliable measuring artifacts in such assessments. **1976** *Gramophone* Mar. 1530/3 The accompanying oscillograms (Fig. 4) show the distortion at 40Hz for power levels of 1 and 5 watts.

oscillograph (ǭsi·lǒgrɑf), *sb.* [f. L. *oscill-āre* to swing + -o + -graph.] **1.** An instrument for detecting and measuring the motion of a ship or of the sea.

1874 E. Bertin *Notes on Waves & Rolling* 118 We may ..hope for favourable results from the employment of the oscillograph with two pendulums which Mr. Froude has invented; the French Admiralty has caused a similar instrument to be constructed. **1896** *Trans. Inst. Naval Archit.* XXXVII. 322 The graphic tracing of the movements of the sea with the help of the 'oscillograph' was undertaken. **1904** C. H. Peabody *Naval Archit.* 344 Investigations of the rolling of ships in quiet water and among waves have been made..by aid of instruments known as oscillographs which have slow and quick pendulums, and registering devices.

2. [ad. F. *oscillographe* (A. Blondel 1893, in *Compt. Rend.* CXVI. 502).] Any instrument for displaying as a continuous curve the form of a varying voltage (e.g. that associated with an oscillatory or alternating current, or one derived from a bodily or non-electrical phenomenon); properly restricted to recording instruments, but freq. also used (esp. formerly) to denote the cathode-ray oscilloscope even without the camera needed for recording the display.

1893 A. Blondel in *Electrician* 17 Mar. 571/1 The object of this communication is to describe some new galvanometric apparatus or oscillographs allowing one to determine by direct observation..the periodic curves of alternating currents. **1910** G. W. Pierce *Princ. Wireless Telegr.* xviii. 181 The necessary sensitiveness..was finally obtained with a Braun's cathode tube oscillograph. **1913** *Electrician* 7 Nov. 172/1 For many purposes, particularly in the teaching of alternating currents to elementary students and in the elucidation of certain problems in the higher branches of alternating-current engineering, the kathode-ray oscillograph is convenient and practical. **1925** *Lit. Digest* 11 July 25/1 The stethoscope makes the heart-beat audible, and the oscillograph gives graphic presentation of its action. **1927**, etc. [see *oscilloscope d]. **1932** *News Chron.* 23 Sept. 10/6 In one of the wooden huts at Slough I watched lightning flashes from thousands of miles away being recorded on the glass face of an oscillograph some 9in. in diameter. **1933** E. W. Golding *Electr. Measurements* xv. 526 In some forms of cathode-ray oscillograph special provision is made for photographing the wave-form under observation. **1968** M. Woodhouse *Rock Baby* ix. 93 The single hooded eye of an oscillograph peered from the shadows. **1974** *Sci. Amer.* Mar. 94/3 In the electromechanical oscillograph..a tiny mirror is attached to the moving coil of a galvanometer, which oscillates in sympathy with the applied voltage.

3. = *oscillogram.

1936 *Discovery* June 197/2 Some excellent oscillographs are shown..of damped and undamped waves. **1957** *Electronics* 1 May 163/2 The same tube inserted in a tapered S-band waveguide circuit produces an oscillograph as shown in Fig. 1B. **1957** *New Biol.* XXIII. 35 Plate 11..shows some oscillographs (records made with the oscilloscope) of the songs of two species of insects. **1975** *Nature* 5 June 514/3 The photographic and line illustrations are adequate, but the oscillographs and sonograms have reproduced poorly.

oscillograph (ǭsi·lǒgrɑf), *v.* [f. prec. sb.] *trans.* To record or display by means of an oscillograph.

1910 G. W. Pierce *Princ. Wireless Telegr.* xviii. 182 The drum must be driven synchronously with the alternating current which is being oscillographed. **1926** *Physical Rev.* XXVIII. 554 The signal is detected, amplified, and oscillographed.

oscillographic (ǭsi:logræ·fik), *a.* [f. *oscillograph *sb.* + -ic.] Of, pertaining to, or employing an oscillograph.

1908 J. Erskine-Murray tr. *Ruhmer's Wireless Teleph.* ii. xiii. 145 Two oscillographic records from a singing arc. **1934** *Amer. Speech* XIV. 311/2 From an oscillographic investigation the author concludes that loudness depends on frequency. **1957** Manvell & Huntley *Technique Film Music* iii. 171 Soon afterwards Scholpo and Rimsky-Korsakoff began the oscillographic analysis of natural sounds. **1972** C. N. Herrick *Instruments & Measurements for Electronics* xv. 356 (*heading*) Oscillographic camera equipment.

Hence **osci:llogra·phically** *adv.*, by means of or as an oscillograph. Also **oscillo·graphy**, the use of oscillographs.

1925 *Jrnl. Inst. Electr. Engin.* LXIII. 1091/1 After Braun (1897) had produced the first two-dimensional electron-jet instrument.., Zenneck (1899) proceeded to work it oscillographically. **1931** *Proc. Physical Soc.* XLIII. 502 (*heading*) A time base for the cathode-ray oscillography of irregularly recurring phenomena. **1938** *Physical Rev.* LIV. 34/2 Average heights of oscillographically recorded pulses. **1964** *Times Rev. Industry* Apr. 5/3 (Advt.), The workshop..will feature demonstrations of industrial and scientific applications of instant photography including: Photomicrography, Oscillography, [etc.]. **1974** *Nature* 8 Nov. 122/1 Each utterance was analysed oscillographically.

oscillometer (ǫsilǫ·mītəɹ). [f. as prec. + -meter.] **1.** A gyroscopic form of oscillograph (sense *1).

1899 *Sci. Amer.* 29 July 71/2 An interesting gyroscopic device termed the 'oscillometer', has been put on the market by a Milanese firm.

2. *Med.* [ad. F. *oscillomètre* (V. Pachon 1909, in *Compt. Rend. hebd. des Séances et Mém. de la Soc. de Biol.* 735).] An instrument for indicating or recording the magnitude and rhythm of the pulse.

1910 *Brit. Med. Jrnl.* 3 Dec. 1765/2 With the oscillometer..the disappearance or reappearance of the pulse is indicated by the needle. **1934** C. J. Wiggers *Physiol. Health & Dis.* (1935) xl. 606 Some blood pressure instruments are equipped with a form of oscillometer by means of which the magnitude of the pressure fluctuations created within the bag can be read or recorded. **1974** Passmore & Robson *Compan. Med. Stud.* III. i. xvii. 3/2 The oscillometer measures the amplitude of pulsation at different levels of the limb and provides an index of pulsatile blood flow.

Hence **osci·llometric** *a.*; **oscillo·metry**, the use of an oscillometer.

1933 *Amer. Heart Jrnl.* VIII. 398 Oscillometric studies carried out on 214 patients revealed that the maximal oscillometric phase..did not occur at the point of disappearance of the systolic tone. **1934** Webster, Oscillometry. **1961** *Neurology* XI. 25/2 Carotid oscillometry is easy to manage and is harmless as blood pressure determination. **1974** J. D. Maynard in R. M. Kirk et al. *Surgery* xi. 234 Oscillometry, skin temperature measurements, and plethysmography at present are only research investigations. **1974** J. R. Marshall in Lichtiger & Moya *Introd. Pract. Anesthesia* ix. 93 The oscillometric method of measuring blood pressure requires a wide cuff containing two narrow inflatable bags.

oscilloscope (ǭsi·lǒskōup). [f. as prec. + -scope.] † **a.** (See quot.) *Obs.* exc. as in d.

1909 Webster, *Oscilloscope*, an instrument for showing visually the changes in a varying current; an oscillograph.

† **b.** (See quots.) *Obs.*

1915 R. Knox *Radiogr.* i. 53 The best method of detecting reverse current is by the use of an oscilloscope tube... Two aluminium wires, separated by a small gap, are enclosed in an oblong glass tube, and the wire connected with the negative pole becomes, when the current passes, surrounded by a violet fluorescence, but if each wire is alternately negative and positive both wires become fluorescent and the length of the fluorescent band indicates the intensity of the current. **1926** *Gloss. Terms Electr. Engin.* (Brit. Engin. Stand. Assoc.) 206 *Oscilloscope*, an auxiliary discharge tube in which the length of the negative glow affords an indication of the amount of current passing.

† **c.** An apparatus in which the principle of the stroboscope is employed to render visible irregularities in the motion of rapidly rotating or oscillating machinery. *Obs.*

1922 *Glasgow Herald* 10 Oct. 6 The Elverson oscilloscope ..is an apparatus for slowing down to the eye..any high-speed machinery. **1924** A. J. H. Elverson in *Jrnl. Sci. Instrum.* I. 116 The 'Elverson Oscilloscope' is an opto-mechanical device which has been designed to facilitate the examination of the behaviour of high speed movements *under working conditions*. This is effected by presenting, in rapid succession, accurate pictures of the movement at predetermined intervals of angular displacement in such a way that, owing to persistence of vision, the eye of an observer sees a true representation of the movement at a speed which is a fraction of the true speed. **1927** *Trans. Amer. Inst. Electr. Engin.* May 550/1 There was a device produced in England a few years ago by a Mr. Elverson, designed for visualizing the movements of rapidly oscillating mechanisms. It is purely mechanical and optical in its nature and is now generally known as the Elverson oscilloscope.

d. More fully cathode-ray oscilloscope. An electronic instrument in which a moving spot on the screen of a cathode-ray tube represents by its position the relationship between two variables, usu. a steady or varying signal voltage (vertically) and time (horizontally), and which is capable of displaying a periodic variation in voltage as a stationary trace.

1927 Bedell & Reich in *Trans. Amer. Inst. Electr. Engin.* May 546 These limitations..may be removed and the field of usefulness of the cathode-ray oscillograph so widened that it becomes practically a new instrument. As the instrument developed for this purpose..is primarily intended for visual observation, we have given it the name 'oscilloscope'. Permanent record may be obtained.. by a photograph in the usual way. On the other hand, an *oscillograph* of the Blondel or Duddell type, both in name and in fact, is primarily for graphical record. **1936** Karapetoff & Dennison *Electr. Lab. Exper.* xix. 451 The need exists for an oscillograph possessing negligible inertia. Such an instrument is the cathode-ray oscillograph, or oscilloscope. **1953** A. C. Clarke *Prelude to Space* xxiii. 121 Clifton seemed to be hypnotised by a cathode-ray oscilloscope, the screen of which was filled with fantastic geometrical figures, continually shifting. **1967** G. F. Fiennes *I tried to run a Railway* viii. 99 One weekend the oscilloscope at one depot, Liara, found seven axles flawed. **1972** C. N. Herrick *Instruments & Measurements for Electronics* xv. 350 A conventional oscilloscope can be converted into an oscillograph by mounting a camera in front of its crt screen. **1976** *Word 1971* XXVII. 522 A small computer generated displays of letter-like elements on an oscilloscope screen.

Hence osci:llosco·pic *a.*, osci:llosco·pically *adv.*

1949 *Jrnl. Appl. Physics* XX. 1105/1 In all of the oscilloscopic records at 200 volts there are 10 examples of the ends of transient opens. **1961** *Trans. Symposium Electrode Processes 1959* 186 All measurements were made oscilloscopically. **1962** *Trans. Faraday Soc.* LVIII. 389 The potential changes were generally recorded oscilloscopically. **1965** *Math. in Biol. & Med.* (Med. Res. Council) iv. 136 Few, if any, systematic studies have been made to assess an experimenter's information-handling capacity as applied to his ability to view oscilloscopic traces or examine film records.

Oscotian (ǫskōu·ʃən), *sb.* [f. *Oscot(t)* + -ian.] A member or former member of the Roman Catholic college and seminary called St. Mary's at Oscott, near Birmingham. Also as *adj.*

The date of the original foundation, at Old Oscott, was 1794, and the college was transferred to New Oscott in 1838.

1828 (*title*) The Oscotian or Literary Gazette of St. Mary's..edited by the students of Oscott College. **1828** *Oscotian* I. 45, I met a body of youngsters escorted by a gowned superior: these I concluded to be Oscotians. **1853** J. Morris in J. H. Pollen *Life J. Morris* (1896) iv. 88 An Oscotian, thoughtful, gentlemanly, having a turn for philosophy. **1860** F. C. Husenbeth *Life Mgr. Weedall* iii. 43 Every Oscotian knows the favourite Oscott game of 'Bandy'. **1908** *Catholic Times* 22 May 8/2 No man was ever more of an Oscotian than he. **1956** *Oscotian* 83 Oscott has now acquired a tape-recorder... The 'sounds off' have not yet managed to convince any Oscotian that he is a budding Peter Ustinov. **1968** *Ibid.* 42 Allowing for the time it takes Oscotian cricketers to warm up—..we managed about ten days sporadic cricket.

Osco-Umbrian (ǫ·skoˌʊ·mbriăn), *a.* and *sb.* Also **Oscan-Umbrian.** [f. L. *Osc-us* *Oscan *a.* and *sb.* + -o + Umbrian *sb.* and *a.*] **A.** *adj.* Of or pertaining to a group of Italic languages including Oscan and Umbrian and related dialects. **B.** *sb.* This language-group. Also, a member of the peoples who spoke languages of this group.

1894 J. Rhys in *Trans. Philol. Soc. 1891–4* 117 The Romans used *qu* just as the ancient Irish did..for the Osco-Umbrian dialects replaced *qu* by *p*. *Ibid.* 119 The Siculo-Latin race had already settled down when the Osco-Umbrians arrived. **1895** C. D. Buck (*title*) The Oscan-Umbrian verb-system. *Ibid.* 135 The plural and passive forms have developed independently and on different lines in Oscan-Umbrian and in Latin. **1897** R. S. Conway *Italic Dial.* II. 469 (*heading*) Accidence of the Osco-Umbrian dialects. **1904** C. D. Buck *Gram. Oscan & Umbrian* p. iv, This grammar is called a Grammar of Oscan and Umbrian, not of the Oscan-Umbrian dialects, for it does not pretend to treat systematically the minor dialects included under the name Oscan-Umbrian. **1939** [see *Latino-Faliscan* s.v. *Latino-]. **1948** D. Diringer *Alphabet* ii. ix. 501 'Italic' is mainly used..to indicate the Osco-Umbrian sub-branch of the Italic branch of the Indo-European family. **1958** E. Pulgram *Tongues of Italy* xvii. 228 Beeler..explains the agreements of Latin with Oscan-Umbrian by their existing in close vicinity over several..centuries. *Ibid.* xviii. 232 The chronologies suggested for all these invasions are hopelessly..irreconcilable, going as low as..2300 B.C. for the Osco-Umbrians. **1971** *Archivum Linguisticum* II. 99 The *sē*-stem is..an imperfect subjunctive in Latin as well as in Osco-Umbrian. *Ibid.* 100 The complex system of Latin and Osco-Umbrian subjunctives..is also an innovation of the Italic group. **1974** A. Watson *Legal Transplants* iv. 26 It is not absolutely certain that *poena* is a direct borrowing from Greek: it is possible, though unlikely, that it was borrowed at second hand, first passing through some intermediate language like Osco-Umbrian. **1976** *Archivum Linguisticum* VII. 60 Comparison of the above types with their Oscan-Umbrian counterparts reveals separate historical origins.

-ose[2]. Add to def.: Now extended to carbohydrates which are not isomers of glucose, saccharose, or cellulose, as *arabinose, rhamnose, ribose, xylose*, etc., and to classes of sugars, as *aldose, furanose, hexose, pentose, pyranose*, etc.

-ose[3], a suffix corresponding to -osis, used to form the names of fungal diseases of plants, as *erinose.

Osgood–Schlatter (ǫ·zgud ʃlæ·təɹ). *Med.* The names of Robert Bayley *Osgood* (1873–1956), U.S. surgeon, and Carl *Schlatter* (1864–1934), Swiss surgeon, used *attrib.* and in the possessive to denote a disease described independently by them in 1903, viz. epiphysitis of the tibial tubercle.

[**1909** *Edin. Med. Jrnl.* II. 249 The name of Schlatter's Disease has been given to this lesion since it was described by that surgeon in 1903, though Müller had published an account of it in 1887.] **1912** *Amer. Jrnl. Orthopedic Surg.* IX. 317 Some have suggested Osgood-Schlatter disease, but it is my own opinion that the condition was recognized..many years previous to the work of these two gentlemen. **1921** *Ann. Surg.* LXXIII. 77 This condition is usually referred to as Osgood-Schlatter's Disease, but should really be classed as an injury and not a true disease. **1932** W. Mercer *Orthopædic Surg.* ix. 32 The true Osgood-Schlatter disease shows..characteristic bony changes which stamp it as a definite disease entity. **1961** R. D. Baker *Essent. Path.* xxi. 559 Names of men designate the disease in its various locations. For example, Osgood-Schlatter's disease is idiopathic aseptic necrosis of the tibial tubercle.

‖ **oshibori** (ǭ·ʃibōəri). [Jap., f. *o-* deferential prefix + *shibori* that which has been wrung out.] A towel which has been wrung out, usually in hot, but sometimes in cold, water; used in Japan to wash the hands and face before a meal. Also *oshibori towel*, etc.

1959 R. Kirkbride *Tamiko* vii. 47 They..wiped their hands upon first the hot and then the cold oshibori brought to them in bamboo baskets. **1963** *New Yorker* 22 June 14 A fragrant o-shibori hot towel to refresh you. **1970** *Guardian* 12 Dec. 6/6 The little thoughtfulness of the oshibori hot towel to begin a meal. **1974** *New Yorker* 22 Apr. 131 (Advt.), Hot *oshibori* towels. Or, how to freshen up without getting up.

-oside (osəid), *suffix. Chem.* [f. -ose[2] + -ide, after Glucoside, *glycoside.] Used to form the names of glycosides and classes of glycosides, as *furanoside, ganglioside, glucuronoside, pyranoside*, etc.

Osirian, *a.* Add: (Later examples.) Hence **Osi·rism**, the cult or ritual associated with Osiris.

1906 W. M. F. Petrie *Relig. Anc. Egypt* v. 38 The earliest phase of Osirism that we can identify is in portions of the Book of the Dead. **1968** *New Larousse Encycl. Mythol.* (new ed.) 19/1 Isis, in the Osirian myth, represents the rich plains of Egypt. **1971** E. Iversen in J. R. Harris *Legacy of Egypt* (ed. 2) vii. 183 Annius of Viterbo ..used Diodorus' version of the myth as a background to his curious efforts to establish a heroic genealogy for his papal patron Alexander VI Borgia, identifying the bull of the papal coat of arms with the Osirian Apis, and making the Pope a descendant of the god himself. **1972** P. M. Fraser *Ptolemaic Alexandria* I. v. 256 This identification [of Sarapis with the powers of the Underworld], which also proceeds from the Osirian character of Sarapis, is not attested by dedications.

Osirify (osəiə·rifəi), *v.* [f. the name *Osiris* (see Osirian *a.*): see -fy.] *trans.* To identify (a man or god) with Osiris. Hence **Osi:rifica·tion**, identification with Osiris.

1890 *Cent. Dict.*, Osirify. **1906** W. M. F. Petrie *Relig. Anc. Egypt* iv. 23 The most renowned was the *Hapi* or Apis bull of Memphis..who was Osirified and became the Osir-hapi. **1912** *Encycl. Relig. & Ethics* V. 238/1 In the earliest royal monuments the dance of men in the festival of Osirification of the King is represented.

Oslo (ǫ·zlo). The name of the capital of Norway, used *attrib.* in phr. *Oslo breakfast*: a type of meal for children planned to supply nutritional deficiencies in their diets and introduced into Norwegian schools by Dr. Carl Schiotz. So *Oslo meal*, etc.

1937 *Bull. Health Organisation* (League of Nations) VI. 197 Certain schools [in Norway] introduced the 'Oslo breakfast', whilst other schools continued the older practice of giving a hot lunch. The quality of this hot meal was less good than that of the 'Oslo breakfast', which consists of protective foods (milk, Kneipp rusks and Kneipp bread, cheese, butter, orange, apple or raw carrot). **1938** *Rep. School Medical Officer* (*London County Council*) 47 Considerable attention has of late been directed to the method of giving school meals to necessitous school children and others in the schools of Norway and other northern countries. This method is generally referred to as 'the Oslo breakfast'... It has attracted the attention of dieticians in other countries..and suggestions have been made that it should be introduced into London schools in substitution for the present provision of hot mid-day meals. *Ibid.* 49 It may be held that the Oslo meal would have the same effect if given at midday as if given first thing in the morning. **1949** D. Macardle *Children of Europe* vii. 123 The Oslo breakfast was planned by Professor Schiotz to give the child calories..vitamins..and to..strengthen their jaws by chewing. It consists of.. pasteurized milk, two biscuits made of wholemeal wheaten flour,..margarine..cheese made of goat's milk,..rye bread.., a raw carrot.., or half a banana..orange or.. apple. **1952** J. J. Morel *Progressive Catering* I. ii. 34 An Oslo breakfast, or a health dinner as it is known in England. **1958** R. Stow *To Islands* i. 19 Feed him on..free milk and Oslo lunches. **1962** *Guardian* 26 Oct. 5/2 'The Oslo breakfast'..consisted of..milk, half an orange, wholemeal bread, and goat's milk cheese.

osmanthus (ǫzmæ·nþǫs). [mod.L. (J. de Loureiro *Flora Cochinchinensis* (1790) 28), f. Gr. ὀσμή scent + ἄνθος flower.] An evergreen shrub of the genus so called, belonging to the family Oleaceæ, usually native to eastern Asia or North America, and bearing clusters of small white or cream, usually fragrant, flowers.

1877 *Gardeners' Chron.* 24 Feb. 239/2 (*heading*) Note on Osmanthus. **1912** *Curtis's Bot. Mag.* CXXXVIII. 8459 The *Osmanthus* which is here figured is one of the most pleasing of new evergreen shrubs. **1976** *Country Life* 26 Feb. 478/1, I am advocating..lighter, altogether less domineering shrubs like..phormiums, variegated osmanthus, [etc.].

osmate. Substitute for def.: A salt or ester of osmic acid (H_2OsO_4) in which osmium has an oxidation state of 6, and in the case of alkali metals having the formula $M_2[OsO_2(OH)_4]$ and obtainable by the action of a reducing agent such as alcohol on an alkaline solution of osmium tetroxide. (Add examples.)

1852 G. Fownes *Man. Elem. Chem.* (ed. 4) II. 368 Osmate of potassa is produced when the metal is fused with nitre. **1868** J. P. Cooke *First Princ. Chem. Philos.* (1870) xix. 420 Osmic anhydride..is unknown, but potassic osmate..can easily be obtained. **1909** L. Kahlenberg *Outl. Chem.* xxxi. 525 The chlorides $OsCl_2$, Os_2Cl_6, and $OsCl_4$ are known, as are also osmates like $K_2OsO_4.2H_2O$. **1950** N. V. Sidgwick *Chem. Elements* II. 1501 The tetroxide reacts with potassium or caesium hydroxide in methyl alcohol to give the tetramethyl osmate $(MO)_2Os^{VI}(O{\cdot}CH_3)_4$. **1973** S. E. Livingstone in J. C. Bailar et al. *Comprehensive Inorg. Chem.* III. xliii. 1230 Potassium osmate, originally formulated as $K_2OsO_4{\cdot}2H_2O$, has the octahedral structure $K_2[OsO_2(OH)_4]$.

osmatic (ǫsmæ·tik), *a.* [ad. F. *osmatique* (P. Broca 1878, in *Revue d'Anthropologie* VII. 397), f. Gr. ὀσμή smell + -atic.] Having well-developed olfactory organs and a good sense of smell; cf. *macrosmatic *a.*, *microsmatic *a.* So **o·smatism**, the degree of development of the olfactory organs.

1890 W. Turner in *Jrnl. Anat. & Physiol.* XXV. 106 He [*sc.* Broca] has classified the Mammalia, in relation to the magnitude of their olfactory apparatus, into two groups: osmatic mammals, which possess a well-developed rhinencephalon with a keen sense of smell, and anosmatic mammals, in which the rhinencephalon and olfactory sense are either feeble or not developed at all. **1903** *Amer. Anthropologist* V. 638 The related doctrines that the olfactory organs are large in osmatic, small or absent in anosmatic animals. **1903** *Trans. Linn. Soc.* (*Zool.*) VIII. 369 The size of the hippocampal formation does not seem to vary directly..with the degree of osmatism.

osmiate. Add to def.: = *perosmate. Now *Obs.* (Earlier and later examples.)

The passage in ed. 8 (1895) of *C. Bloxam's Chem.* corresponding to quot. 1890 occurs s.v. *perosmate.

1844 *Phil. Mag.* XXIV. 394 A solution of osmiate of potash. **1852** [see *osmite]. **1854** *Chem. Gaz.* 1 July 242 The air..which is still saturated with vapour of osmic acid, passes into a solution of potash, and finally to the aspirator; the osmiate of potash thus produced is treated with a few drops of alcohol, and collected in the form of crystallized osmite of potash. **1890** Thomson & Bloxam *C. Bloxam's Chem.* (ed. 7) 428 By dissolving osmic anhydride in potash and adding alcohol, the latter is oxidised at the expense of the potassium osmiate, and rose-coloured octahedral crystals of potassium osmite ($K_2OsO_4.2Aq$) are obtained.

b. = *osmate.

1905 Gooch & Walker *Outl. Inorg. Chem.* II. xviii. 490 By fusing osmium compounds with potassium hydroxide and potassium nitrate, potassium osmiate, K_2OsO_4, is formed. **1962** P. J. & B. Durrant *Introd. Adv. Inorg. Chem.* xxiv. 1033 All three elements in the Group, iron, ruthenium, and osmium, are present in the oxidation state VI in the ferrates, the ruthenates, and the osmiates; the general formula is K_2MO_4.

osmic, *a.* Add: *osmic acid*, (*b*) the acid H_2OsO_4, known chiefly in the form of its salts (osmates).

1879 Roscoe & Schorlemmer *Treat. Chem.* II. II. 458 In addition to these the salts of osmic acid, H_2OsO_4, are known, but neither the acid nor the corresponding oxide, OsO_3, have been prepared. **1936** J. W. Mellor *Comprehensive Treat. Inorg. & Theoret. Chem.* XV. lxxii. 705 H. Moraht and C. Wischin obtained a black substance by the action of heat on a mixture of potassium osmate and nitric acid... [They] found that when dried over phosphorus pentoxide, in vacuo, its composition corresponds with the hydrate, $OsO_3.H_2O$, or osmic acid, H_2OsO_4. O. Ruff and K. Bornemann could not confirm the analysis. **1950** N. V. Sidgwick *Chem. Elements* II. 1499 Hexavalent osmium... There are two binary compounds..and a considerable number of complex salts, all of which are in some sense derivatives of osmic acid H_2OsO_4.

osmic (ǫ·zmik), *a.*[2] [f. Gr. ὀσμή smell, odour + -ic.] Of or pertaining to odours or the sense of smell. Hence **o·smically** *adv.* Cf. *osmics *sb. pl.*

1938 G. M. Dyson in *Chem. & Industry* XVI. 647/1 When the mucus linings are inflamed and covered with thickened mucus..the ability to smell will be considerably diminished..by reason of the osmic sensory processes being cut off from access to the air. *Ibid.* 648/1 Certain chemical groups and configurations..lead to the development of intra-molecular frequencies capable of affecting the osmic sensory processes. These are referred to, subsequently, as the osmic frequencies. **1964** *Ann. N.Y. Acad. Sci.* CXVI. 557 If this can be confirmed, it will help to define a lower limit to the range of 'osmic frequencies'. **1966** *Nature* 5 Feb. 551/1 The osmic properties that have been identified are the non-specific ones. They do not correlate with particular odours such as those of rose..or peppermint. The fact is that the osmically specific properties of odorous substances have still not been securely identified.

osmicate (ǫ·zmikĕit), *v. Biol.* [f. Osmic *a.* + -ate[3].] *trans.* To stain or treat with osmic acid (osmium tetroxide). So **o·smicated** *ppl. a.*

1914 Stedman *Med. Dict.* (ed. 3) 658/1 *Osmicate*, to stain with osmic acid. **1943** *Proc. Nat. Acad. Sci.* XXIX. 228 The classical Golgi apparatus of the fixed and osmicated cell. **1971** *Nature* 2 Apr. 334/2 Random pieces of grossly normal thyroid tissue..were diced.., fixed in 1·5% glutaraldehyde..post-osmicated, dehydrated and embedded in 'Araldite 502'. *Ibid.* 17 Sept. 199/1 Teased preparations of osmicated nerves from paralysed rabbits revealed focal, segmental myelin loss characteristic of EAN.

So **osmica·tion**, treatment with osmic acid.

1899 *Jrnl. Morphol.* XV. Suppl. 73 A curious effect of osmication was noted in some peripherally lying cells.. in a preparation of *Spelerpes ruber*, which had been fixed in Flemming's chromo-aceto-osmic mixture. **1928** *Biol. Rev.* III. 337 Nassonow..has made a..study of *Paramecium* and some other protozoa by methods of fixation (osmication) and by study *in vivo*. **1934** L. W. Sharp *Introd. Cytol.* (ed. 3) iv. 69 After osmication or silver impregnation they may appear like hollow structures with a blackened periphery. **1972** B. M. Wagner in I. Mandl *Collagenase* II. 68 Fixation in glutaraldehyde in the presence of 3% basic fuchsin, followed by osmication, resulted in increased stainability of collagen with phosphotungstic acid (PTA).

osmics (ǫ·zmiks), *sb. pl.* (const. as *sing.*). [f. Gr. ὀσμή smell, odour: see -ic 2.] The branch of science concerned with odours and the sense of smell.

1922 J. H. Kenneth *Osmics* 3 Osmics..is a convenient term to connote that area of the field of science which is concerned with smell... Briefly, osmics is the science of the stimuli, organs, and the sense of smell. **1965** *Cold Spring Harbor Symp. Quant. Biol.* XXX. 635/1 It must surely be conceded that the science of osmics has come of age with the recognition of its modulating psychophysical principle as being related to the sizes and shapes of volatile molecules.

osmio-. Add: Also used in other words, as *osmiophilic *a.*

osmiophilic (ǫzmiofi·lik), *a. Biol.* [f. Osmio- + *-philic.] Having an affinity for, or staining readily with, osmium tetroxide.

1927 *Biol. Bull.* LIII. 182, I am..of the opinion that these fatty granules of Guilliermond..have nothing to do with the osmiophilic platelets here described. **1942** *Nature* 10 Jan. 52/1 Granules..which at some stages produce osmiophilic substances. **1971** *Ibid.* 25 June 535/1 In OsO_4-fixed ultrathin sections of the costo-chondral junctions of 1 month old guinea-pigs, Bonucci has found osmiophilic bodies with mean diameters of 500–2,500 Å.

osmiridium. Add: Now usu. distinguished from *iridosmine* (see quots. 1968, 1973). (Further examples.) [For references to G. *osmiridium* (1828, 1831), *osm-iridium* (1824), and *osmium-iridium* (1821) see *Mineral. Mag.* (1963) XXXIII. 716.]

1938 *Mineral. Abstr.* VII. 162 Osmiridium (nevyanskite) from Bolshaya Victorievka mine, Kusuetsky Alatau, West Siberia..gave Os 40·3, Ir 41·6, Ru 4·2, Pt 1·1, Au 1·8, Fe 8·5. **1963** [see *nevyanskite]. **1966** Phillips & Williams *Inorg. Chem.* II. xxxiv. 609 The platinum metals occur native in the alloy osmiridium. **1968** I. Kostov *Mineral.* II. i. 90 Iridosmine (Ir, Os) with Os > 35%, and osmiridium (Os, Ir) with Os < 35%; hexagonal... Occurs as small triangular or hexagonal plates. **1973** *Canad. Mineralogist* XII. 105/1 Iridosmine occurs with osmiridium in two grains (samples 10120, gr. 2 and M12339, gr. 1). *Ibid.* 110 An excellent historical review on the nomenclature of natural Os-Ir alloys was made by Hey (1963)... He suggested that the most suitable nomenclature..be the following: For the *cubic* alloys: *osmiridium* with Os < 32 at. %. For the *hexagonal* alloys: *iridosmine* with 32 < Os < 80 at. %. Native *osmium* for Os > 80 at. %... Our proposals for alloys in the Os-Ir-Ru system are that:.. e) *Iridosmine* of Hey (1963) be redefined as *hexagonal* (Os, Ir) alloys with no single other element > 10 at. % of total, and where Os < 80 at.% of (Os+Ir)... f) *Osmiridium* of Hey (1963) be redefined as *cubic* (Ir, Os) alloys with no single other element > 10 at. % of total, and where Ir < 80 at. % of (Ir+Os).

osmite. Add: = *osmate. Now *Obs.* [first formed as F. *osmite* (E. Frémy 1844, in *Jrnl. de Pharm. et de Chim.* V. 189).]

1844 *Phil. Mag.* XXIV. 394 The liquor..deposits a crystalline powder of osmite of potash. **1852** tr. *Regnault's Elem. Chem.* II. III. 352 Osmite of potassa is obtained by pouring a few drops of alcohol into a solution of osmiate of potassa. *Ibid.*, No osmite of ammonia is known. **1891** W. Ramsay *Syst. Inorg. Chem.* xxix. 483 Potassium osmite, $K_2OsO_4.2H_2O$, is prepared by dissolving the tetroxide in potassium hydroxide, and adding alcohol.

osmium. Add: **b.** Special comb.: **osmium lamp**, a filament lamp in which the filament is made of osmium; **osmium tetroxide**, OsO_4, a poisonous, pale yellow solid that has a distinctive pungent and harmful vapour and is used in solution as a biological stain, esp. for lipids, and a fixative; also called *osmic acid*.

1907 *Westm. Gaz.* 16 Feb. 14/2 The osmium lamp..was expensive to start with, and could be used only in the pendent position. **1952** H. Hewitt *Mod. Lighting Technique* iv. 34 Whilst further developments such as the Nernst lamp, the osmium lamp and the tantalum lamp were all of technical interest, it was not until the adoption of the tungsten filament by Coolidge in 1909 that there was any considerable advance towards more efficient electric lamps. **1876** *Encycl. Brit.* V. 537/2 Osmium tetroxide is reduced at red heat. **1920** *Jrnl. R. Microsc. Soc.* 133 The Golgi apparatus has the following reactions: ..2. Black in Kopsch's or Mann-Kopsch's osmium tetroxide methods. **1954** H. W. Deane in R. O. Greep *Histol.* iii. 43 Many lipid substances will also blacken with osmic acid (osmium tetroxide). In most instances..this blackening apparently depends on the fact that unsaturated fatty acids reduce the colorless osmic acid to black osmium dioxide. However, many nonfatty protoplasmic constituents..may also be blackened with osmic acid. **1968** J. March *Adv. Org. Chem.* xv. 616 There are many reagents which add two OH groups to a double bond. OsO_4 and alkaline $KMnO_4$ give cis addition, from the less hindered side of the double bond. Osmium tetroxide adds rather slowly, but almost quantitatively. **1974** *Nature* 18 Jan. 145/1 White Leghorn chicken embryos..were fixed in glutaraldehyde and osmium tetroxide.

osmo-[1]. Add: Also used to repr. Osmium (cf. Osmio-), as in **osmophi·lic** *a.*[1] = *osmiophilic *a.*

1961 in Webster. **1961** *Lancet* 16 Sept. 656/1 The appearance of osmophilic densities in the zones of lamellar discontinuity at the nodes of Ranvier. **1972** *Jrnl. Electron Microsc.* XXI. 85/1 In order to check the nature of the osmophilic granules..in enlarged axons, the distribution, origin and properties in the area postrema were studied in a morphological comparison..with neurosecretory granules in the hypothalamus.

osmo-[2]. Add: **osmometer**[2], also, an instrument for the measurement of osmotic pressures; (further examples); **osmometry**[2], **-metric** *a.*[2] (examples); hence **osmome·trically** *adv.*; **osmophi·lic** *a.*[2] *Biol.* [ad. G. *osmophil* (A. A. von Richter 1912, in *Mycolog. Centralbl.* I. 74)] tolerating or thriving in a medium which exerts a high osmotic pressure; so **o·smophile**, an osmophilic organism; **o·smoreceptor** *Biol.*, any sensory organ which reacts to changes in osmotic pressure (concentration) in the body fluids.

1903 M. H. Fischer tr. *Cohen's Physical Chem.* ix. 139 Pfeffer measured the osmotic pressure of sugar solutions of various concentrations with a mercury manometer, and obtained with such an osmometer the following results. **1974** Tombs & Peacocke *Osmotic Pressure Biol. Macromolecules* iii. 86 Claesson and Jacobsson..have made an osmometer with a very precise optical method for determining the difference in height of two menisci. **1976** *Nature* 12 Aug. 578/1 The total osmolality of the fluid was obtained with a Clifton nanolitre osmometer. **1913** *Chem. Abstr.* IV. 298 In a series of expts. in an 'osmometric vessel'..the following mol. wts. are obtained by balancing the pressure of the salt against a sugar soln. on the other side of the membrane. **1964** J. Eliassaf tr. *Rafikov's Determination Molecular Weights* vi. 169 The osmometric measurement of molecular weight is based on the fact that the osmotic pressure..is proportional to the number..of gram-molecules of dissolved material in a definite volume of solution. **1943** *Jrnl. Physical Chem.* XLVII. 69 The molecular weight of a carefully fractionated sample determined osmometrically checks the value obtained for the same polymer by means of the ultracentrifuge. **1964** J. Eliassaf tr. *Rafikov's Determination Molecular Weights* vi. 223 It was found that for molecular weights of less than 75,000 a difference begins to appear between the molecular weight determined osmometrically and the molecular weight computed by the Mark-Houwink equation. **1913** *Chem. Abstr.* VII. 297 (*heading*) Osmometry of saline solutions and the theory of Arrhenius. **1973** *Nature* 27 Apr. p. xv (Advt.), It also shows how osmometry, ultracentrifugation, light scattering,..and gel filtration are used to analyze polydisperse systems. **1961** P. L. Carpenter *Microbiol.* xiii. 201/1 Microorganisms that have become adapted to high osmotic pressure are called osmophiles. **1969** L. do Carmo-Sousa in Rose & Harrison *Yeasts* I. iii. 88 She also suggested the possibility of finding obligate osmophiles..in Antarctic soils which have a high content of soluble salts. **1920** F. W. Tanner tr. *A. Guilliermond's Yeasts* iv. 120 The maximum concentration for spore formation in a yeast depends upon the species. For an osmophilic species like *Zygosaccharomyces Mandshuricus* the concentration is high. **1960** L. E. Hawker et al. *Introd. Biol. Micro-Organisms* xvi. 380 Sugar concentrations of 50 to 70 per cent effectively prevent the growth of most micro-organisms... A few osmophilic yeasts and bacteria may grow slowly. **1972** *Sci. Amer.* Apr. 95/2 Because of the high osmotic pressure of honey, they are yeasts of the type called osmophilic, meaning that they live or thrive in a medium that has a high osmotic pressure. **1946** E. B. Verney in *Lancet* 30 Nov. 782/1 The osmoreceptors, wherever they may be, do not accommodate during short-period exposure to a rise in the osmotic pressure of the carotid plasma produced by NaCl. **1947** —— in *Proc. R. Soc.* B. CXXXV. 68 It becomes justified, therefore, to introduce the term 'osmoreceptors' as descriptive of the autonomic receptive elements with which the neurohypophysis is functionally linked, and through whose activation the pituitary antidiuretic substance is released. **1970** A. J. Vander et al. *Human Physiol.* xii. 354/2 Receptors must exist which are sensitive to extracellular osmolarity. These osmoreceptors are located in the hypothalamus. **1973** *Nature* 14 Dec. 383/1 The osmoreceptors which control the salt glands in marine birds are located in or near the heart.

osmoceptor (ǫ·zmoseptǭɹ). *Physiol.* [a. G. *osmoceptor* (L. Ruzicka 1920, in *Chem.-Zeitung* XLIV. 94/2), f. Gr. ὀσμο- Osmo-[1] + G. *re*)*ceptor* (now *rezeptor*) Receptor.] A sensory receptor for the sense of smell.

1944 R. W. Moncrieff *Chem. Senses* xii. 314 (*table*) Author... Ruzicka... Date... 1920... General class... Chemical... Salient features... Osmophore and osmoceptor. **1952** Pirenne & Abbott tr. *H. Piéron's Sensations* II. v. 105 Such a correspondence between osmoceptors and osmophores may be important in the explanation of the qualitative differentiation between smells. **1968** W. McCartney *Olfaction & Odours* 159 If a substance reacts with the primary but not with the secondary osmoceptors, odourlessness (fatigue) follows saturation of the primary osmoceptors.

osmol (ǫ·zmǫl). Also **osmole** (-mōul). [Blend of Osmotic *a.* and *mole *sb.*[7]] A thousand milliosmols.

1942 J. L. Gamble *Chem. Anat., Physiol. & Path. Extracellular Fluid* (ed. 4) Notes to chart 17-B, 0·64 osmoles per liter. Solute concentration. **1956** A. C. Guyton *Textbk. Med. Physiol.* xxvi. 302/2 If each molecule ionizes into two ions each of which is osmotically active, then 1 mol of solute equals 2 osmols. **1964** L. Martin *Clin. Endocrinol.* (ed. 4) i. 51 In *normal* subjects they found that the initial serum osmolality was 285±4·4 m-osmoles/kg. of water. **1971** W. S. Beck *Human Design* vii. 207/2 The numerical unit in which osmotic pressure is expressed is the osmol. We speak of osmols or milliosmols per liter.

osmolal (ǫzmōu·lăl), *a.* [Blend of Osmotic *a.* and *molal *a.*] Of the concentration of a solution: expressed as an osmolality.

1939 [see *milliosmol* s.v. *milli-]. **1971** W. S. Beck *Human Design* x. 348/2 The osmotic concentration measured by freezing point depression is properly termed an *osmolal* concentration—milliosmols of solute per kilogram of solvent—in contrast to an osmolar concentration —milliosmols of solute per liter of solution. The two values differ only slightly in ordinary dilute solutions, however.

Hence **osmola·lity**, the number of osmotically effective dissolved particles per unit quantity of a solution, esp. when expressed as (milli)osmols per kilogramme of solvent. Cf. *osmolarity.

1959 *Q. Jrnl. Med.* LII. 237 The blood was centrifuged, and the plasma osmolality (m-osmoles per kg.) estimated from the freezing-point depression. **1963** *Jrnl. Amer. Med. Assoc.* 31 Aug. 699/1 Normal subjects are characterized by an initial serum osmolality between 273 and 293 milliosmoles (mOsm) per kilogram. **1968** Passmore & Robson *Compan. Med. Stud.* I. v. 29/1 Sweat..is essentially a weak solution of sodium chloride. The osmolality lies between 100 and 200 mOsm/l. **1972** [see *milliosmole* s.v. *milli-]. **1974** *Nature* 12 Apr. 605/1 Controls of the osmotic pressure of the solutions of inorganic ions..showed that their osmolalities were practically independent of temperature. **1976** *Lancet* 25 Dec. 1414/1 A man of 61 was admitted to the neurology unit in stupor, which proved to be hyponatræmic (serum osmolality 236 mos-mol/kg).

osmolar (ǫzmōu·lăɹ), *a.* [Blend of Osmotic *a.* and *molar *a.*[3]] Of the concentration of a solution: expressed as an osmolarity.

1942 J. L. Gamble *Chem. Anat., Physiol. & Path. Extracellular Fluid* (ed. 4) Notes to chart 17–B, Multiplying osmolar concentration by cubic centimeters of urine produces a measurement of total solutes as milliosmoles. **1944** *Proc. Amer. Philos. Soc.* LXXXVIII. 152/1 If we divide degrees of freezing point depression by 1·86 we obtain osmolar concentration. Multiplying osmolar concentration by cubic centimeters of urine, defines the total output of solutes as milliosmoles. **1963** *Jrnl. Amer. Med. Assoc.* 31 Aug. 700/2 The serum osmolar concentration of 25 normal subjects..was 285±4·4..milliosmoles (mOsm) per kilogram of water. **1971** [see *osmolal *a.*].

Hence **osmola·rity**, the number of osmotically effective dissolved particles per unit quantity of a solution, esp. when expressed as (milli)osmols per litre of solution. Cf. *OSMOLALITY.

1953 *Lancet* 12 Sept. 540/2 An assumed plasma osmolarity of 310 milliosmols per litre. **1962** J. H. KINOSHITA et al. in A. Pirie *Lens Metabolism Rel. Cataract* 406 The final glucose concentration..was 5 µmoles/ml. The total osmolarity was calculated as 307 µosmoles/ml. **1965** *New Scientist* 24 June 868/1 The most important functions of the kidney are to keep constant the volume, osmolarity and composition of the fluid which surrounds the cells of the body. **1973** *Jrnl. Biol. Chem.* CCXLVIII. 4172/1 Solutions of low osmolarity.

osmophore (ǫ·zmofǭəɹ). [f. OSMO-[1] + -PHORE; in sense 1 a back-formation from *OSMOPHORIC *a.*] **1.** A chemical group whose presence in the molecules of a substance causes it to have a smell.

1919 *Perfumery & Essent. Oil Rec.* 21 May 105/1 Both Rupe and Majewski and Cohn point out that one osmophore can often replace another without distinctly changing the odour. **1944** R. W. MONCRIEFF *Chem. Senses* ix. 185 The ether group is only a weak osmophore and is easily overpowered by other features of the molecule. **1963** W. SUMMER *Methods Air Deodorization* i. 66 One and the same osmophore appearing with different molecules usually causes different odours.

2. [ad. It. *osmoforo* (G. Arcangeli 1883, in *Nuovo Giornale Bot. Ital.* XV. 75).] A scent gland found in the flowers of certain plants belonging to the families Orchidaceæ, Araceæ, Aristolochiaceæ, and Asclepiadaceæ.

1966 S. VOGEL in *Proc. 5th World Orchid Conf.* 254/1 We could find such glands, called osmophores, in many different orchidaceous groups... A genuine scent organ or osmophore..may be defined as a glandular, multicellular and clearly differentiated tissue within the floral region, which is well exposed to the atmosphere. **1967** *New Scientist* 22 June 725/2 Several orchids and a few other flowers have..developed glands (osmophores) producing scent substances (terpenes) in liquid form. **1974** C. L. WITHNER et al. in C. L. Withner *Orchids* vi. 305 The scent tissue is ultimately organized into scent glands, or osmophores.

So **osmopho·ric** *a.* [ad. G. *osmophor* (Rupe & Majewski 1900, in *Ber. d. Deut. Chem. Ges.* XXXIII. 3402)], of, pertaining to, or being an osmophore (in either sense).

1901 *Jrnl. Chem. Soc.* LXXX. I. 103 (*heading*) Osmophoric groups. **1922** G. H. PARKER *Smell, Taste, & Allied Senses* iii. 79 Osmophoric groups are such as the hydroxyl, aldehyde, keton, ester, nitro, and nitril groups. None of these..is associated with a particular odor, but any one may be the occasion of odor, if it occupies an appropriate place on a benzene ring. **1968** W. MCCARTNEY *Olfaction & Odours* 133 These investigations suggest that the quality of the odour of a compound depends largely on the steric structure and is modified..by the presence of osmophoric groups. **1974** C. L. WITHNER et al. in C. L. Withner *Orchids: Sci. Stud.* vi. 305 The richness of the osmophoric cells in reserve materials relates to their production of fragrant terpene oils.

osmoregulation (ǫ:zmoregiŭlēi·ʃən). *Physiol.* [a. G. *osmoregulation* (used, prob. for the first time, by R. Höber 1906, in *Physik. Chem. der Zelle und der Gewebe* (ed. 2) ii. 31): see OSMO-[2] and REGULATION.] The maintenance of a more or less constant osmotic pressure in the body fluids of an organism.

1931 *Q. Cumulative Index Med.* IX. 870/2 (*heading*). Disturbances of osmoregulation in experimental uremia. **1932** *Sci. Rep. Tôhoku Imperial Univ.* VII. 229 (*heading*) On the osmoregulation of the blood of several marine and fresh water molluscs. **1964** *Oceanogr. & Marine Biol.* II 306 Osmo-regulation is known in all bony fishes. **1971** *Nature* 16 Apr. 469/2 The role of the amphibian urinary bladder in osmoregulation is well documented.

Hence **osmore·gulate** *v. intr.*, to maintain the osmotic pressure of the body fluids at a constant level; **osmore·gulating** *vbl. sb.*; **osmore·gulator**, an organ or part of the body concerned in osmoregulation; an organism capable of osmoregulation; **o:smoregula·tory** *a.* [ad. G. *osmoregulatorisch* (Höber *loc. cit.*)], of, pertaining to, or effecting osmoregulation.

1911 STEDMAN *Med. Dict.* 616/2 *Osmoregulatory*, influencing the degree and rapidity of osmosis. **1927** *Biol. Abstr.* I. 238/2 Parallel experiments with sucrose instead of urea showed that the more rapid recovery from plasmolysis in etherized solutions was due to more rapid penetration of urea, not to any osmoregulatory or ether stimulus to the production of new cell solutes. **1935** *Biol. Rev.* X. 357 The excretory organs of numerous freshwater animals act as osmoregulators. **1958** *Jrnl. Exper. Biol.* XXXV. 234 The very great osmo-regulating ability of *Artemia* has been described. *Ibid.* 241 Animals whose branchial epithelium has been damaged by a brief exposure to saturated $KMnO_4$ solution have lost the ability to osmo-regulate. **1959** SOUTHWOOD & LESTON *Land & Water Bugs Brit. Isles* xiv. 395 Its osmo-regulatory mechanism enables the bug to live in waters with from 5 to about 18 parts per thousand of salt in solution. **1960** *Biol. Abstr.* XXXV. 2729/2 The degree of activity of their osmoregulating mechanisms. **1963** R. P. DALES *Annelids* v. 109 In view of the lack of evidence that the nephridia do more than act as drains as far as nitrogenous wastes are concerned, we can but incline to the view that they are primarily osmoregulators. **1964** *Oceanogr. & Marine Biol.* II. 307 Holeurysaline osmo-regulators can regulate in salinities ranging from that of pure fresh water to that of full strength sea water or higher. **1969** *New Scientist* 30 Jan. 243/1 Herring embryos could osmoregulate even before chloride cells had developed. **1970** *Nature* 24 Oct. 378/1 A primary osmoregulatory function of prolactin is the reduction of extrarenal sodium outflux.

osmosis. Add: This, rather than *osmose*, is now the usual term.

2. *fig.* Any process by which something is acquired by absorption.

1900 [in Dict., sense 1]. **1930** E. POUND *XXX Cantos* xxix. 137 Languor has cried unto languor about the marshmallow-roast (Let us speak of the osmosis of persons). **1968** *Times Lit. Suppl.* 26 Sept. 1079/1 A director born (like Godard) in 1930 is in a position to know, by a sort of unconscious osmosis, more than a director born in 1898 (like René Clair) can hope to learn. however conscientiously he may try. **1970** *Author* LXXXI. 113 It is not a question how much you teach them but how much they learn, perhaps largely by osmosis. **1977** P. D. JAMES *Death of Expert Witness* III. 128 News percolated through a village community by a process of verbal osmosis.

osmotic, *a.* Add: *osmotic pressure*, the excess pressure that must be applied to a solution to prevent the entry into it of pure solvent when they are separated by a semipermeable membrane, or the excess pressure that develops in the solution when osmosis is allowed to occur in such circumstances; *osmotic shock*, rupture of a cell following a sudden drop in the osmotic pressure of the surrounding liquid, owing to the inflow of liquid that occurs.

1888 *Jrnl. Chem. Soc.* LIV. 778 (*heading*) Osmotic pressure in the analogy between solutions and gases. **1950** T. F. ANDERSON in *Jrnl. Appl. Physics* XXI. 70/1 The similar viruses, T2, T4, and T6 which appear in the electron microscope to have membranes surrounding the internal structures of the heads can be disintegrated by what might be termed 'osmotic shock'... Presumably, the virus heads swell when the osmotic pressure is suddenly reduced, and actually burst if the reduction is sufficiently large and sudden. **1970** AMBROSE & EASTY *Cell Biol.* ii. 83 A solution containing one gramme-molecule of non-ionizable solute in 22·4 litres exerts an osmotic pressure of 1 atmosphere at 0°C. The use of a delicate manometer by Adair and Adair enabled them to determine the molecular weight of proteins by comparing their osmotic properties with those of known solutes. **1973** D. A. ANDERSON *Introd. Microbiol.* x. 110/1 The cells of many bacteria ..are likely to burst when placed in distilled water. This method (osmotic shock) is often used to release components from inside the cell for biochemical analysis. **1973** R. KRUEGER et al. *Introd. Microbiol.* v. 201/2 The cytoplasmic water contains a tremendous variety of small organic and inorganic molecules and numerous ions and soluble enzymes... Gram-positive bacteria have an osmotic pressure of 22 atm (atmospheres).

2. In *fig.* senses. Cf. *OSMOSIS 2.

1952 W. D. JACOBS *William Barnes* i. 10 There is also the strong and numberless clan which utilizes all the latinic iridescence at its command..to rejoice that the language had such osmotic good fortune. **1965** *Economist* 24 Apr. 451/2 In the osmotic way these things happen, virtually all of them [*sc.* workers] were absorbed by other local industries. **1975** B. GARFIELD *Hopscotch* xxii. 236 The joy she took from flying..in some profound osmotic way.. had communicated itself to him.

osmotically, *adv.* Add: Also *fig.*

1974 *Times Lit. Suppl.* 20 Dec. 1437/5 A reminder of how much royal legend is osmotically absorbed by even the most reluctant reader of royal biography. **1976** *Times* 20 Jan. 11/6 The chic of the actresses was absorbed, osmotically, into his ranges [of clothes].

osmunda (ǫzmʊ·ndă). [med.L. (see OSMUND[2]), adopted as a generic name by J. Petiver in *Musei Petiveriani centuria VI & VII* (1699) 53.] A fern of the genus so called, esp. the royal fern, *Osmunda regalis*.

1789 E. DARWIN *Bot. Garden* II. 11 The fair Osmunda seeks the silent dell. **1818** [see *cinnamon fern*]. **1858** J. A. SYMONDS *Let.* 6 June (1967) I. 144 Is the Osmunda yet in flower? **1865** M. EYRE *Lady's Walks S. of France* xxv. 272 The *Osmunda* grew at the other end of the lake. **1894** W. BROCKBANK in W. Robinson *Wild Garden* (ed. 4) xiii. 143 The brooklet was..fringed with marsh plants.. together with Osmundas, Hart's-tongues, and other Ferns. **1974** R. GROUNDS *Ferns* v. 57 This method [*sc.* container-growing] is ideal if one wants to use a strong-growing osmunda, for example, as a specimen plant.

2. In full, *osmunda fibre*. A fibre generally made from the roots of *Osmunda cinnamomea* or *O. claytoniana*, used as a potting medium for orchids.

1910 *Gardeners' Chron.* 5 Nov. 329/3 He [*sc.* H. G. Alexander] decided that Osmunda fibre was the best material procurable. **1932** *Bull. Amer. Orchid Soc.* Dec. 87/2 Osmunda fibre taken from swampy land should not be used. **1942** LAURIE & RIES *Floriculture* xix. 460 Osmunda peat makes the best medium for this group [of epiphytic orchids]. **1942** C. L. WITHNER *Orchids: Sci. Survey* viii. 349 Its [*sc.* tree fern's] water-retention and drying capacities are very similar to osmunda. **1951** *Dict. Gardening* (*R. Hort. Soc.*) III. 1440/2 Osmunda is at present the best material for general use [in potting orchids] but in fact several other fibres are in use to eke out and modify Osmunda fibre. **1959** *Listener* 17 Dec. 1094/3 A piece of osmunda fibre, or wire netting filled with moss and fastened round a cane, can be driven into the pot and kept moist with spraying. **1966** T. B. MORRIS *Orchids with Murder* i. 7 The pleasantly familiar smell of damp loam and sphagnum and osmunda fibre. **1974** C. L. WITHNER *Orchids: Sci. Stud.* iii. 154 Osmunda fiber plus water provided adequate fertility for orchids.

osmundine (ǫ·zmʊndīn). *N. Amer.* [f. *OSMUNDA + -INE[4].] = *OSMUNDA 2.

1932 *Bull. Amer. Orchid Soc.* Sept. 52/2 The best compost to use for growing Cattleyas should consist of good fibrous brown fern root, osmundine as it is called. **1934** R. STOUT *Fer-de-Lance* vii. 101 Supplies—pots, sand, sphagnum..osmundine.

Osnaburg. Delete †*Obs.* and add later examples. Also with small initial.

1917 J. HERGESHEIMER *Three Black Pennys* (1918) 38 Tobacco and shoes, ozenbrigs and molasses and rum. **1938** M. K. RAWLINGS *Yearling* xi. 110 Beyond the utensils were the dress goods; calico and Osnaburg, denim and shoddy, domestic and homespun. **1949** *Caribbean Q.* I. I. 12 Every October cloth was issued, at the rate of seven yards of osnaburgs. **1959** POTTER & CORBMAN *Fiber to Fabric* (ed. 3) ix. 153 When made of waste mixed with low-grade cotton it is known as part-waste osnaburg.

b. *attrib.*, made of Osnaburg linen.

1681 *Rec. Court. of New Castle on Delaware* (1904) 493 Twoo Remnants of Osnabriggs Linnen. **1758** in *Essex Inst. Hist. Coll.* (1874) XII. 145 Others very much soaked in their Osombrige Tents. **1774** in *Maryland Hist. Mag.* (1911) VI. 41 John Johnson..had on..a pair of leather breeches and osnabrig trousers. **1813** J. TAYLOR *Arator* 137 A regular supply of a winter's coat,..two oznaburg shirts, a good hat and blanket. **1841** *Southern Lit. Messenger* VII. 775/2 Our slaves in the South-West are annually supplied with two cotton Oznaburg shirts. **1863** 'E. KIRKE' *My Southern Friends* vii. 99 The thin Osnaburg gown.

osone (ōᵘ·sōᵘn). *Chem.* [ad. G. *oson* (E. Fischer 1889, in *Ber. d. Deut. Chem. Ges.* XXII. 87), f. -*ose* -OSE[2] + -*on* -ONE.] Any compound containing two adjacent carbonyl groups, obtained by hydrolysing an osazone.

1889 [see *glucosone* s.v. *GLUCO-]. **1938** PRESCOTT & RIDGE *Org. Chem.* xx. 353 The osone, which may be isolated as the lead compound, is reduced with zinc dust and acetic acid to a ketose. **1957** J. W. GREEN in W. Pigman *Carbohydrates* vi. 334 The osones exist only as amorphous or sirupy materials.

‖ **o-soto-gari** (ǭsōᵘtogāri). Also osotogari, **O-.** [Jap., f. *o* grand + *soto* outside + *gari* f. *kari* to mow, to reap.] The name of a throw in Judo.

1941 M. FELDENKRAIS *Judo* 45 Pull the opponent's sleeve and attack by the first leg throw(*O-Soto-Gari*). **1956** E. J. HARRISON tr. *H. Aida's Kodokan Judo* iii. 58 Should you anticipate your opponent's attempted Osotogari then just at the moment when he is contemplating the reaping action with his right leg swiftly pull him near you to counter. **1963** P. BUTLER *Judo Complete* iii. 59 The *osotogari* is my own favourite throw and many variations are possible upon the basic method. **1970** A. P. HARRINGTON *Judo Guide to Black Belt* i. 28 Now..try the O-soto-gari against your opponent.

Ossetian (ǫsī·ʃăn), *sb.* and *a.* Also **Ossetan** (-ī·tăn), **Osset(e)** (ǫ·set, -īt), **Ossetic** (ǫse·tik). [f. Russ. *osetin*, f. Georgian *os*, *oset'i* Ossetia (place-name) + -IAN.] **A.** *sb.* **a.** A member of a people of the Central Caucasus, inhabiting North Ossetia (the North Ossetian Autonomous Soviet Socialist Republic) and South Ossetia (an Autonomous Oblast of the Georgian Soviet Socialist Republic). **b.** The language of this people, one of the Eastern Iranian group.

1814 [see *CHECHEN]. **1841** [see *MEDIAN *sb.*[2] 2]. **1869** C. ENGEL *Examples of Art Workmanship* 4 The Ossetes, or Ossetines, are an Indo-Germanic race dwelling in Central Caucasus. They call themselves Irôn; Oseti is the name given to them by the Georgians and other neighbouring nations. **1888** J. WRIGHT tr. *Brugmann's Elem. Compar. Gram. Indo-Gmc. Lang.* I. 5 Ossetian (spoken in the neighbourhood of the Caucasus). **1902** [see *INGUSH]. **1913** [see *ALAN[2]]. **1925** P. RADIN tr. *Vendryès's Language* ii. 38 This phenomenon [*sc.* sound-shifting] is found in other languages besides the Germanic: in Armenian, for example, and in Ossetic. **1933** C. D. BUCK *Compar. Gram. Greek & Latin* 8 Modern Iranian is represented by..the isolated Ossetan in the Caucasus. **1933** L. BLOOMFIELD *Language* iv. 62 An isolated offshoot [of Iranian], far to the west is Ossete, in the Caucasus, spoken by some 225,000 persons. **1933** *N. & Q.* CLXIV. 192/1 Ossetian is a modern Iranian vernacular, spoken by the Ossetes, a tribe in central Caucasus. **1944** G. A. NEBOLSINE tr. *Vernadsky's Hist. Russia* (rev. ed.) i. 16 The Caucasian Alans were called *As* or *Os*—a name which their descendants, the Ossetians, still bear. **1959** B. GEIGER et al. *Peoples & Lang. Caucasus* 47 Ossetian has the status of a literary language. **1964** R. H. ROBINS *Gen. Linguistics* viii. 328 Georgian, Armenian, and Ossetic, languages spoken in contiguous regions, but belonging to different families, Armenian and Ossetic being I-E languages, and Georgian one of the members of the Caucasian family. **1965** G. Y. SHEVELOV *Prehist. of Slavic* 615 Ossetian, the language which to a certain degree continues the Ir[anian] dialects north of the Black Sea, is known only in its modern form. **1970** D. A. RUSTOW in *Camb. Hist. Islam* I. IV. vi. 691 The neighbouring Circassians, Chechens and Ossetians proved almost equally tenacious. **1971** L. ZGUSTA et al

Man. Lexicogr. vii. 300 The glosses would probably be given in Ossetic, the dictionary being determined for the Ossetes.

B. *adj.* Of or pertaining to this people or their language.

1877 A. H. KEANE tr. *Hovelacque's Sci. of Lang.* v. 207 The Ossetian declension is fuller than the Persian. **1910** *Encycl. Brit.* V. 552/2 The Mamison Pass, over which runs the Ossetian military road (made passable for vehicles in 1889)..lies at an altitude of 9270 ft. **1931** *Times Lit. Suppl.* 2 June 398/1 Joseph..Jugashvili (otherwise Koba, otherwise Stalin) was the son of a Georgian father and an Ossetian mother. **1933** R. G. KENT *Old Persian* (ed. 2) 7 The Ossetic dialects, in the general region of the Caucasus; derived from the Scythian of Southern Russia. **1962** D. M. LANG *Mod. Hist. Georgia* i. 9 The Ossetian Military Road runs northward from Kutaisi towards the Mamison Pass. *Ibid.* ii. 49 The Ossete mountaineers and the villagers of Mtiuleti were forced to toil without payment. **1974** *Country Life* 24 Jan. 146/4 An Ossete fold-ballad, adapted by the poet Kosta Khetagurov (the Caucasian equivalent of Robert Burns), begins: The fox has been whetting her teeth for the badger.

‖ **ossia** (ǫsī·ǎ, ǫ·sīǎ), *conj.* [It., f. *o sia*, or maybe.] In musical directions: or rather, or alternatively.

1876 STAINER & BARRETT *Dict. Mus. Terms* 340/2 *Ossia*..or else, as *ossia più facile*, or else in this more easy way. **1959** *Collins Music Encycl.* 476/2 Ossia, 'or'. Used to indicate an alternative passage..

ossiculectomy (ǫsi:kiŭle·ktŏmi). *Surg.* [f. L. *ossicul-um*, dim. of *os*, *ossi-* bone + *-ECTOMY.] Excision of the auditory ossicles.

1899 W. MILLIGAN in *Jrnl. Laryngol.* XIV. 1 An uncomplicated otorrhoea which has resisted all forms of treatment for six months is certainly a case for ossiculectomy. **1959** G. E. SHAMBAUGH *Surg. Ear* xix. 515 (*heading*) Partial ossiculectomy.

ossified, *ppl. a.* Add: Also *fig.*

1901 *Yale Fun* 55/1 'Did you hear about the row over in Peabody Museum?' 'What was the trouble?' 'A lot of the exhibits got ossified.' **1932** R. LEWIS *Babbitt* xxii. 337 Oh, but wasn't I. Some ossified! **1934** WEBSTER s.v., *Ossified*..2. Figuratively, fixed, hardened; set in a conventional form; ultraconservative. **1953** E. O'NEILL *Long Day's Journey* (1956) iv. 156 What's the matter with the Old Man tonight? Must be ossified to forget he left the light on. **1961** in WEBSTER s.v., Bitterly criticized the organization for being ossified. **1976** *Times* 5 Nov. 14/8 That lack of any spirit of initiative..that is normally the hallmark..of ossified totalitarianisms.

‖ **osso bucco** (ǫ·so bu·ko). Also **osso buco**. [It., lit. 'marrow-bone'.] Shin of veal stewed in wine with vegetables.

1935 M. MORPHY *Recipes of All Nations* 144 *L'osso buco* (veal stew). This is one of the best-known Italian recipes. **1962** *Guardian* 22 Apr. 8/7 Florentine Veal Stew is reminiscent of the Osso Bucco beloved of all Italy... **1969** A. WILSON *Old Men at Zoo* ii. 97 She..had arranged with Grazia my favourite meal—osso bucco, a light red chianti, zabaglione. **1969** E. HUMPHREYS *Gift* ii. 1. 109, I ordered osso buco and half a litre of Chianti. **1966** *Daily Tel.* 9 Nov. 13/4 There is a super snack bar..where you can fortify yourself with Osso Bucco, Beef Stroganoff or roast lamb. **1974** *Times* 2 Nov. 11/5 Prepared beef olives and osso bucco.

ostatki (ǫste·tki), var. *ASTATKI.

1913 V. B. LEWES *Oil Fuel* 71 The oil remaining in the retort, called 'Ostatki' (the Russian distillers) and 'Residuum' in America, is used for fuel.

ostectomy (ǫste·ktŏmi). *Surg.* Also †**osteectomy** (ǫsti,e·ktŏmi). [f. OST(EO- or OSTE(O- + *-ECTOMY.] (See quots.)

1894 GOULD *Dict. Med.* 946/1 *Osteectomy*, excision of a portion of bony tissue. **1900** DORLAND *Med. Dict.* 465/1 *Ostectomy, osteectomy*, the excision of a bone. **1966** *Gloss. Terms Dentistry* (*B.S.I.*) 36 *Ostectomy*, the division of a bone at two points with removal of the intervening portion of bone. This procedure is, in effect, similar to a resection..but the term 'ostectomy' is reserved for operations designed to correct some structural deformity of a congenital, developmental, or acquired nature whereas 'resection' applies to the removal of diseased bone. **1977** *Proc. R. Soc. Med.* LXX. 432/1 Bell & Dann (1973) found two changes in the surgically repositioned bone and only minor changes in incisor overbite and overjet in 25 patients following anterior maxillary ostectomies and lower labial segmental procedures.

ostension. Add: **1. b.** Logic. = *ostensive definition* (see next).

1950 W. V. QUINE in *Jrnl. Philos.* XLVII. 619 The ostensions which introduce a singular term. **1960** —— *Word & Object* 141 Or the ostension of [the] word in the language..essentially degenerative with no suggestion of the natives' ... *Ibid.* ostension on the part of the natives. **1963** J. LYONS *Structural Semantics* iv. 54, I accept that ostension plays a necessary part in the normal process of learning a language. **1968** —— *Introd. Theoret. Linguistics* ix. 409 The difficulty of explaining the meaning of any word without using others to limit and make more explicit the 'scope' of 'ostension'.

ostensive, *a.* Add: **2. b.** *ostensive definition* (Philos.), the explanation of a word by pointing at or otherwise indicating, or by presenting, one or more objects to which it applies.

1921 W. E. JOHNSON *Logic* I. vi. 94 We may now introduce the technical term 'ostensive' which will suggest as its opposite the familiar term 'intensive'... Imposing a name in the act of indicating, presenting or introducing the object to which the name is to apply,..this it is that constitutes ostensive definition. **1949** A. J. AYER *Found. Empirical Knowl.* ii. 88 This is effected by the method of ostensive definition. **1950** R. ROBINSON *Definition* ii. 15 'Ostensive definition' is the name of a method, the method that makes use of pointing or physical introduction. **1953** I. M. COPI *Introd. Logic* iv. 108 An ostensive definition refers to the examples by means of pointing or some other gesture. **1961** *Amer. Speech* ... **1968** J. LYONS *Introd. Theoret. Linguistics* ix. 409 Ostensive definition... **1963** ... Ostensive definition being the meaning of 'dog' ostensively.

Hence **oste·nsiveness**, the state or quality of being ostensive.

1933 *Mind* XLII. 190 (*heading*) Tests for ostensiveness. **1971** T. F. MITCHELL in *Archivum Linguisticum* II. 40 Ostensive meaning. 'Specification' as a category of linguistic experience may encompass such varied grammatical classes as articles, ordinal numerals, and deictics... It is perhaps particularly to the area of deixis that 'ostensiveness' belongs.

ostensively. Add: **c.** By pointing, gesture, or presentation.

1921 W. E. JOHNSON *Logic* I. vi. 94 A simple adjective name—such as red—cannot be defined analytically but only ostensively. **1948** B. RUSSELL *Human Knowl.* 85 How children learn the word 'dog' ostensively. **1953** W. V. QUINE *From Logical Point of View* iv. 78 Once a fund of ostensively acquired term is at hand there is no difficulty in explaining additional terms discursively.

osteo-. Add: **osteoarthri·tic** *a.*, of, pertaining to, or affected by osteoarthritis; **osteoarthritis**, substitute for def.: degeneration of the joints of the body, which occurs to a greater or lesser extent from the third decade of life, is manifested as pain, discomfort, and stiffness in the joints, and results from progressive deterioration of articular cartilage until finally bone is rubbing directly against bone; **o:steo-arthro-pathy** [*arthropathy* s.v. ARTHRO-], any disease which affects both the bones and the joints; *spec.* a syndrome (pulmonary osteoarthropathy) marked by broadening and thickening of the fingers, painful swollen joints, and enlarged distal ends of long bones, and seen chiefly as a complication of various chest diseases; hence **o:steoarthropa·thic** *a.*; **osteoarthro·sis** [-OSIS] = *osteoarthritis*; hence **osteoarthro·tic** *a.*; **o·steocyte** [-CYTE], an osteoblast that has ceased its bone-forming activity and is enclosed within a lacuna in the bone matrix; **o:steodontokera·tic** *a.* *Anthrop.* [ODONTO- + KERATO- + -IC], of a culture based on the use of bone, tooth, and horn implements; **osteola·thyrism** *Med.* [LATHYRISM], an experimental skeletal disease of animals produced by the ingestion of seeds of some plants of the genus *Lathyrus* or certain chemicals; **osteo·lysis** [*-LYSIS], the pathological destruction or disappearance of bone tissue; so **osteoly·tic** *a.*, causing or characterized by osteolysis.

1902 *Amer. Jrnl. Med. Sci.* CXXIV. 808 The frequency with which osteo-arthritic changes are found in Paget's disease has not received the attention which they invite. **1962** *Lancet* 8 Dec. 1233/1 Mr. Philip Newman and Mr. Harry Piggott described a ten-year follow-up of osteoarthritic hips. **1963** JUBB & KENNEDY *Path. Domestic Animals* I. ... **1890** *Lancet* 12 Apr. 794 ... Osteoarthritis is a non-inflammatory disorder of movable joints characterized by deterioration and abrasion of articular cartilage, and also by formation of new bone at the joint surfaces. **1903** *Med. Rec.* (N.Y.) 21 Feb. 312/1 Walter Brent reports a case which shows the intimate relations which exist between nerve lesions and osteoarthropathic changes. **1927** HOLLANDER & MCCARTY *Arthritis* (ed. 8) iv. 1009 Osteoarthritis ... the commonest form of joint disease... **1892** *Brit. Med. Jrnl.* ... (*heading*) Three cases of 'hypertrophic pulmonary osteo-arthropathy', with remarks. **1901** *Encycl. Medica* IX. ... In leprosy osteo-arthropathies have been described by Heiberg which have many of the characteristics of the osteo-arthropathies of tabes. **1958** *Jrnl. Bone & Joint Surg.* XL. B. 538 (*heading*) Familial osteoarthropathy of the fingers. **1974** J. D. MAYNARD in R. M. Kirk et al. *Surgery* x. 216 Polyarthritis and pulmonary osteoarthropathy..are sinister clinical findings. **1932** W. BOYD *Text-bk. Path.* xxxii. 898 Osteoarthrosis.—This is commonly called osteoarthritis, but as the condition is essentially degenerative with no suggestion of inflammation it would appear preferable to speak of osteoarthrosis. **1970** *New Scientist* 4 June 487/1 One, rheumatoid arthritis, is an inflammatory condition of unknown cause, starting in the synovial lining... The second, osteoarthrosis, is essentially a disorder of cartilage. **1974** PASSMORE & ROBSON *Compan. Med. Stud.* III. I. xxv. 37/1 Osteoarthrosis (osteoarthritis) is a common disease of diarthrodial joints in both men and animals. **1964** W. S. C. COPEMAN *Textbk. Rheumatic Dis.* (ed. 3) xiii. 276 Restriction of the use of a joint may protect it from developing osteo-arthrotic changes. **1974** PASSMORE & ROBSON *Compan. Med. Stud.* III. I. xxv. 37/1 There are biochemical differences between senescent and osteoarthritic cartilage. **1943** *Q. Jrnl. Exper. Physiol.* XXXII. 9 Superficially placed osteocytes. **1965** M. C. HALL *Locomotor Syst.: Funct. Histol.* vi. 87 Once it has formed the matrix around itself this cell, the osteoblast, becomes an inhabitant of its own secretions. Its function changes from a bone forming cell to a bone maintaining cell and it is then known as an osteocyte. **1957** R. A. DART in *Transvaal Mus. Mem.* No. 10. 1 The purpose of this paper..is to show that..the essential culture of *Australopithecus prometheus* was osteodontokeratic... This long name indicating literally 'bone-tooth-horn' may appear unduly ponderous. **1959** J. W. KITCHING (*title*) Bone, tooth & horn tools of palaeolithic man: an account of the osteodontokeratic discoveries in Pin Hole Cave, Derbyshire. **1967** *New Scientist* 27 Apr. 202/1 This has been published by Dr Dart under the jaw-cracking title of the osteodontokeratic (literally bone, tooth and horn) culture of the Australopithecines. **1957** H. SELYE in *Revue Canad. de Biol.* XVI. 1 An apparently quite unrelated skeletal disease, 'osteolathyrism', can be induced experimentally in laboratory animals by feeding them the seeds of other types of Lathyrus plants, especially *L. odoratus*. The active principle of the latter is aminopropionitrile. **1971** *Sci. Amer.* June 51/1 One form of this disease, called osteolathyrism, can be produced experimentally in animals by administering aminonitriles and related compounds. In osteolathyrism the inhibition of cross-links in elastin and collagen brings about structural abnormalities in the connective tissues, particularly those of blood vessels and bone. **1859** S. WILKS *Lect. Path. Anat.* I. 34 There is a third form of cancer,..to which Lobstein has given the name of osteolytic cancer, or osteolysis. **1875** —— & MOXON *Ibid.* (ed. 2) 63 These formations appear to be of the same nature as those first described by Lobstein. **1926** *Surg., Gynecol. & Obstetr.* XLIII. 308/2 There is regression of bone (osteolysis). **1969** B. S. EPSTEIN *Spine* (ed. 3) xi. 692/2 As a result of the destructive process of the vertebrae..osteolytic lesions of the spine or of marrow with Gaucher's cells ... **1974** PASSMORE & ROBSON *Compan. Med. Stud.* XXXII. 840 (*caption*) Osteolytic osteogenic sarcoma in the femur of a child. **1974** PASSMORE & ROBSON *Compan. Med. Stud.* III. I. xxvi. 31/1 Giant cell tumour (osteoclastoma)... Commonly a thin shell of bone covers the lesion, which is osteolytic.

osteoclastoma (ǫ:stiŏklæstō·mă). *Path.* Pl. -omas, -omata. [f. *osteoclast* s.v. OSTEO- + *-OMA.] A giant-cell tumour of bone characterized by the presence of numerous osteoclast-like cells; orig. applied to a variety of mostly benign tumours, but now restricted to a type that is often malignant.

1926 *Jrnl. Path. & Bacteriol.* XXIX. 399 (*heading*) A case of osteoclastoma (myeloid sarcoma, benign giant-cell tumour) with pulmonary metastases. **1931** *Brit. Jrnl. Surg.* XIX. 242 Generalized osteitis fibrosa with multiple osteoclastomata. **1948** [see *MYELOMA]. **1961** *Lancet* 30 Sept. 753/2 Osteoclastomas respond well to irradiation. **1974** PASSMORE & ROBSON *Compan. Med. Stud.* III. I. xxvi. 31/1 Giant cell tumour (osteoclastoma).. The nature of the cells is not definitely known, but the possibility that the giant cells are osteoclasts has led to the alternative name, osteoclastoma.

osteoderm (ǫ·stiodɜːm). *Zool.* [Back-formation from *osteodermal* adj. s.v. OSTEO-.] A bony plate in the skin, esp. in reptiles.

1898 H. GADOW *Classification of Vertebrata* 27 Body scaly without osteoderms. **1931** *Proc. Zool. Soc.* I. 208 Exquisite examples of true dermal bones are those ossifications 'within the skin' which in Amphibia and Reptiles are now generally called osteoderms. **1960** A. BELLAIRS *Life of Reptiles* II. vii. 194 Many lizards have an armour of bony scutes or osteoderms in the dermal layers of the skin.

osteodystrophy (ǫstiodi·strŏfi). *Med.* Also as mod.L. **-dystrophia**. [f. (in Ger.) as mod.L. *osteodystrophia* (v. Mikulicz 1905, in *Verhandl. d. Ges. deutsch. Naturforscher und Ärzte* LXXVI. II. Med. Abt. 108), f. OSTEO- + *dystrophia* s.v. DYS- : see *-Y[3].] Any of several disorders affecting the whole skeleton in which there is defective bone development owing to a badly balanced diet or faulty metabolism; *spec.* one in which there is increased resorption of bone and its replacement by fibrous and poorly mineralized tissue, producing skeletal pain and brittle bones, which are often enlarged and deformed in the young; it occurs in animals, esp. horses, as a result of too high a ratio of dietary phosphorus to calcium (*osteodystrophia fibrosa*) [coined in Ger. by T. Stenholm in *Pathologisch-anat. Studien über die Osteodystrophia Fibrosa* (1924) 90]), and in man in association with chronic renal insufficiency and hyperparathyroidism (*renal osteodystrophy*).

1930 *Jrnl. Exper. Med.* LII. 669 (*heading*) Experimental fibrous osteodystrophy (ostitis fibrosa) in hyperparathyroid dogs. *Ibid.* 690 The hypostotic-porotic form of osteodystrophia fibrosa. **1932** W. BOYD *Text-bk. Path.* xxxi. 883 Among these..osteitis fibrosa, osteitis deformans, osteomalacia, rickets,..hereditary chondrodysplasia, and marble bones may be mentioned. As they are all diseases of the growth of bone they may be considered together under the heading of the osteodystrophies. **1953** *Jrnl. Path. & Bacteriol.* LXV. 302 In severe renal osteodystrophy the

local upheaval in bone formation may result in the development of lesions resembling either osteopetrosis or Paget's disease. **1963** JUBB & KENNEDY *Path. Domestic Animals* I. i. 8/1 The classical osteodystrophies are rickets, osteomalacia, and osteodystrophia fibrosa. *Ibid.* 23/2 (*heading*) The osteodystrophy of fluorine poisoning. **1966** WRIGHT & SYMMERS *Systemic Path.* II. xxxvii. 1324/4 Renal osteodystrophy is..one of the commonest metabolic diseases of bone. The bone changes consist of fibrous replacement of bone..together with rickets or osteomalacia. **1970** A. R. JENNINGS *Animal Path.* xii. 233 Equine osteodystrophia fibrosa arises if horses are fed on a diet containing large quantities of bran which is rich in phosphorus. The same situation is sometimes seen in pigs... In dogs, however, the prime cause is renal insufficiency with secondary hyperparathyroidism. **1974** PASSMORE & ROBSON *Compan. Med. Stud.* III. I. xxvi. 30/2 The clinical features of renal osteodystrophy are most marked in children, in whom there is growth retardation.

So **o:steodystro·phic** *a.*

1935 *Physiol. Abstr.* IX. 529 Osteodystrophic factors in different animals. **1960** *Proc. Zool. Soc.* CXXXIV. 307 Osteodystrophic conditions have been recognized in New World monkeys for many years.

osteofibrosis (ǫstiofaibrō·sis). *Vet. Sci.* [f. OSTEO- + *FIBROSIS; perh. ad. F. *ostéofibrose* (Achard & Thiers 1925, in *Gaz. des Hôpitaux civils et militaires* XCVIII. 917/1).] Osteodystrophia fibrosa of animals.

1936 *Vet. Rec.* XLVIII. 1400/1 The third osteodystrophic disease in animals is one which has only come to be recognised within recent years. It is osteodystrophia fibrosa, also known as osteofibrosis. Although by no means uncommon..it has most often been described as osteomalacia or osteoporosis, conditions which we now know to have a different pathology. **1961** J. O. L. KING *Vet. Dietetics* viii. 99 Osteofibrosis is a bone disease, commonly known as 'big head' or 'miller's disease', which is found in horses fed on rations which are very low in calcium, especially where there is an excess of phosphorus.

Hence **o:steofibro·tic** *a.*

1938 J. R. GREIG et al. in *Hutyra's Special Path. & Therapeutics* (ed. 4) III. 224 Osteofibrotic pigs with thickening of the cranial bones (snuffles).

osteogenesis. (Further examples.)

1950 A. W. HAM *Histol.* xvi. 192/2 The formation of bone is usually spoken of as ossification or osteogenesis. **1963** JUBB & KENNEDY *Path. Domestic Animals* I. i. 9/2 Alkaline phosphatase..is very active in osteogenesis.

b. *osteogenesis imperfecta* [mod.L. (W. Vrolik *Tabulæ Illustrandam Embryogenesin Hominis et Mammalium* (1849) tab. 91)], an inherited disease characterized by extreme weakness of the bones, which break frequently even before birth.

1903 *Amer. Jrnl. Med. Sci.* CXXV. 762 The disease called by Vrolik and by Stilling osteogenesis imperfecta, which previously in all probability was classed under foetal rickets, is a definite, intrauterine process, the chief characteristics of which are great brittleness and softness of the bones, numerous fractures, and resulting deformities, due to as yet obscure disturbances of myelogenic and periosteal bone formation. **1923** *Brit. Jrnl. Surg.* XI. 737 Osteogenesis imperfecta is the name given to a disease which is characterized by a congenital defect in the evolution of the osteoblast, and recognised clinically by defective ossification of the cranium and a multiplicity of fractures resulting from trivial causes. **1974** PASSMORE & ROBSON *Compan. Med. Stud.* III. II. xlv. 47/2 Osteogenesis imperfecta congenita is easily distinguished from achondroplasia but unnaturally short limbs, the result of prenatal fractures, create a superficial resemblance.

osteogenetic, -genic, -genous *adjs.* (Examples.)

Osteogenic is the usual adj.; *osteogenous* is rare.

1867 *Quain's Elem. Anat.* (ed. 7) I. p. cv, This soft transparent matter, which becomes ossified, may..be distinguished by the name of 'osteogenic substance', as proposed by H. Müller, or simply of 'osteogen'. **1874** A. E. J. BARKER tr. *Frey's Histol. & Histochem. of Man* 258 It will be proper to designate..the similarity of the osteogenetic process to that in other parts of the system. **1905** J. S. FERGUSON *Normal Histol.* xi. 177 The osteogenous tissue of this layer, containing osteoblasts, osteoclasts, and developing blood vessels, grows into the cartilage. **1938** M. SINCLAIR *Fractures* xv. 180 Certain forms of ununited fractures may therefore be stimulated into union by the presence of actual osteogenetic tissue in a bone graft. **1947** *Radiology* XLIX. 310/1 Similar osteogenic sarcomas were described in persons poisoned with radium for eight to ten years. **1951** S. GILDEN tr. *Lacroix's Organization Bones* xv. 204 Autoplastic grafts of marrow are osteogenic in adult animals as well as in young animals. **1975** *Nature* 29 May 373/3 It was felt that statistical analysis could be based only on that bone tumour type for which solid prognostic data exist, agreed by all to be classical osteogenic sarcoma.

osteoid, *a.* Add: *spec.* consisting of or being the uncalcified amorphous matrix that is the organic constituent of bone. (Earlier and additional examples.)

osteoid sarcoma, a characteristic kind of benign tumour of bone, small and usu. painful.

1840 C. WEST tr. *Müller's Nature of Cancer* I. ii. i. 136 The osteoid tumor of the bones..is a growth composed entirely of osseous substance. **1859** *Proc. R. Soc.* IX. 661 It seems to follow that the peculiar distribution of real osseous tissue and of the 'osteoid' structure, as the osseous tissue without [bone-]corpuscles may be called, has a deeper signification. **1859** S. WILKS *Lect. Path.* *Anat.* I. 31 There is encephaloid and scirrhous cancer of bone,..if the latter is wholly ossified, we have osteoid cancer. **1875** —— & MOXON *Ibid.* (ed. 2) 54 The microscope shows a structure which can best be compared to those plates of osteoid cartilage which are so common on the spinal pia mater after middle life, i.e. it closely resembles bone which has been deposited. **1916** E. H. KETTLE *Path. Tumours* ii. 96 A tumour may form osteoid tissue composed entirely of ossified cartilage... In every respect except that there is no deposition of calcium salts in the matrix. **1932** W. BOYD *Text-bk. Path.* xxxii. 898 These ... the abnormal bone ... **1935** H. L. JAFFE in *Arch. Surg.* XXXI. 724 One feels forced to conclude that one is dealing here with a benign bone neoplasm the distinctiveness of which has not hitherto been recognized and which I am designating 'osteoid-osteoma'. **1950** A. W. HAM *Histol.* xvi. 190/2 Under normal conditions..newly formed bone exists as either uncalcified or osteoid state for only a transitory period. **1966** WRIGHT & SYMMERS *Systemic Path.* I. xxxvii. 1400 An osteoid osteoma takes the form of a rounded mass of gritty, reddish-grey tissue... Histologically..the lesion consists of vascular osteoblastic tissue, containing irregular trabeculae of osteoid. **1970** *Nature* ... by W. A. BERESFORD *Lect. Notes Histol.* vii. 49 The osteoid seam is a very poorly mineralized zone of matrix, 1-3 µ wide, seen with light microscopy between the true bone and the active osteoblasts.

B. *sb.* † **a.** A kind of malignant tumour composed of osteoid tissue. [The original sense, after G. *osteoid sb.* (J. Müller *Ueber den feinern Bau und die Formen der krankhaften Geschwülste* (1838) 44), and that in quot. 1847–9 in Dict. (given more fully below).] *Obs.*

1847–9 R. B. TODD *Cycl. Anat. & Physiol.* IV. I. 135/2 Osteoid.—Under the names of osteoid or ossifying fungus, tumour, Müller describes a growth..composed of a greyish white, vascular, nodulated substance, of the consistence of fibro-cartilage. **1854** W. E. SWAINE tr. *Rokitansky's Man. Path. Anat.* I. ix. 185 This series [of new growths]..separates into the osteoid, and into the bony concretion. *Ibid.* 185 Müller's osteoid the osteoid..is the cancer ...

b. Osteoid tissue, uncalcified bone. Also *attrib.*

1934 *Vet. Jrnl.* XC. 157 They were not considered to be genuinely rachitic, because the pathognomic osteoid formation was absent, or the amount of osteoid was too considerable... **1953** *Bull. Johns Hopkins Hosp.* LXXII. 236 Osteoid in the normal adult individual is usually absent or very scanty. **1963** JUBB & KENNEDY *Path. Domestic Animals* I. i. 9/2 Osteoid consists of fibrillar protein (collagen) in a non-fibrillar medium which is probably largely of mucopolysaccharides. **1972** H. L. JAFFE *Metabolic Dis. Bones* xv. 387 It is the presence of abundant osteoid that characterizes the histologic picture of both rickets and osteomalacia. **1972** *Science* 2 June 1023/1 Light microscopic observations indicated no obvious differences between treated and control cultures with respect to proliferation of boneforming cells or degree of osteoid formation.

osteologically, *adv.* (Examples.)

1891 *Proc. Zool. Soc.* 196 It will at once be seen that, osteologically, *Stactolaema* is quite different from any of our other pigeons. **1946** F. E. ZEUNER *Dating Past* vii. 289 This form is osteologically quite distinct from *Lynx pardina* Temminck.

osteon (ǫ·stiǫn). *Histology.* Also **osteone**. [a. G. *osteon* (W. Biedermann 1914, in H. Winterstein *Handb. d. vergleich. Physiol.* III. I. 1150), f. Gr. ὀστέον bone.] = *Haversian system*.

1928 MOORE & KEY tr. *Leriche & Policard's Normal & Path. Physiol. Bone* i. 14 Recently, several histologists have tried to make the Haversian systems the structural units of bone, and have given them the name of osteones. This nosologism is useless. **1958** *Jrnl. Bone & Joint Surg.* XL. A. 463 Haversian systems or osteons course primarily longitudinally in the long bones. **1968** PASSMORE & ROBSON *Compan. Med. Stud.* I. xvi. 12/1 The osteons branch and interweave one with another. **1971** *Nature* 30 July 335/1 The osteons were orientated along the length of the bone and perpendicularly to the surface of the frontal bone of the skull.

osteopetrosis (ǫstiopetrō·sis). *Path.* [f. OSTEO- + L. *petra*, Gr. πέτρα rock + -OSIS.] A rare hereditary disease in man in which there is excessive formation of dense trabecular bone with resulting brittleness; also, any similar disease of animals.

1926 H. C. KARSHNER in *Amer. Jrnl. Roentgenol.* XVI. 405/1 Osteopetrosis is an hereditary disease. *Ibid.* 405/2 The term osteopetrosis (stony bones) is chosen because in one word it describes the primary pathological condition, bone petrification. **1947**, etc. [see *marble bone* s.v. *MARBLE *sb.* 9]. **1948** *Jrnl. Exper. Med.* LXXXVIII. 579 (*heading*) Hereditary osteopetrosis of the rabbit. **1974** PASSMORE & ROBSON *Compan. Med. Stud.* III. I. xxvi. 29/2 (*caption*) In osteopetrosis the bone texture is uniformly dense and no medullary cavity is present.

Hence **osteopetro·tic** *a.*

1951 *Jrnl. Bone & Joint Surg.* XXXIII. A. 937 The transverse lines in the metaphyses of osteopetrotic bones are fractures in various stages of healing. **1965** *Ibid.* XLVII. A. 1365 (*heading*) Avian osteopetrotic bone.

osteoporotic, *a.* (s.v. OSTEOPOROSIS). (Examples.)

1910 *Brit. Jrnl. Dental Sci.* LIII. 481 The bone itself presenting a fibrous and osteoporotic appearance. **1970** *Sci. Jrnl.* Aug. 72/3 All individuals, whether with normal or osteoporotic, therefore lose a small amount of bone mineral at night which is made up during the day.

Osterizer (ǫ·stəraizəɹ). Also with lower-case initial. [f. the name *Oster* (see quot. 1949) + -IZE + -ER[1].] The proprietary name of a type of electric food mixer.

1949 *Official Gaz.* (U.S. Patent Office) 5 Apr. 39 John Oster Manufacturing Company..Osterizer. For electric food mixers. **1967** E. B. NICKERSON *Kayaks to Arctic* xv. 142 Back in the days before celluloid tape, the Osterizer was virtually the only plastic known. **1977** *Rolling Stone* 30 June 73/2 Her kitchen is very white—walls, doors, floors, white appliances, Braun coffee grinder and Osterizer, white salt and pepper shakers, [etc.].

Ostiak, var. *OSTYAK.

ostial (ǫ·stiăl), *a.* *Anat.* [f. OSTI(UM + -AL).] Of, pertaining to, or having an ostium or ostia; of the nature of an ostium (in an insect's heart).

1900 MIALL & HAMMOND *Struct. & Life Hist. Harlequin Fly* 76 All the valves found in the insect are of two Chironomus, whether cellular, ostial, or aortic, appear to be derived from the semicircular muscle-cells. **1910** *Practitioner* Jan. 37 The ostial end of the tube dilates to allow of the passage of the mole. **1969** R. F. CHAPMAN *Insects* xxxii. 661 (*caption*) Incurrent ostial valves in the larva of *Chaoborus* at different phases of the heartbeat. **1969** *Jrnl. Thoracic & Cardiovasc. Surg.* LVII. 792/2 At autopsy all showed varying degrees of coronary ostial sclerosis with thrombosis.

ostiate (ǫ·stiˌēit), *a.* *rare.* [f. OSTI(UM + -ATE[2].] = *OSTIAL *a.*

1897 *Natural Sci.* Apr. 266 The parapodial jaws and the ostiate heart cannot be supposed to have been both developed independently in each group of arthropods.

‖ **ostinato** (ǫstinā·to), *a.* and *sb.* *Mus.* [It., obstinate, persistent.] **A.** *adj.* Recurring, frequently repeated. **B.** *sb.* Pl. ostinati, ostinatos. A musical figure which recurs unchanged and at the same pitch. Also *transf.* Cf. *basso ostinato* s.v. *BASSO.

1876 STAINER & BARRETT *Dict. Mus. Terms* 340/2 *Ostinato* (It.) lit. obstinate, used in the sense of 'frequently repeated', as *basso ostinato*, a ground-bass. **1928** *Daily Express* 27 Aug. 5/1 At the Queen's Hall there are three principal themes, the *ostinato* on page three obtruding itself against a version of the second theme. **1934** C. LAMBERT *Music Ho!* II. 126 We find the juxtaposition of short lyrical phrases..with ostinatos of extreme and deliberate bareness. **1948** G. ABRAHAM in L. A. Bacharach *Brit. Music* iii. 62 Neurittis..had exaggerated his [*sc.* Holst's] mannerism of ostinato bass-figures through the quotation which they could be indicated by repeat-signs. **1947** A. EINSTEIN *Mus. Romantic Era* xx. 349 The combination of *doloroso* and *agitato*, of *cantabile* and rhythmic *ostinati*, is typical. **1959** R. FULLER *Ruined Boys* II. ix. 131 He became aware of the noises of summer—of insects, larks, leaves—that provide the normally unidentified ostinato that nevertheless enriches the obvious themes of colour, sun and cloud. **1971** *Times Lit. Suppl.* 1 Oct. 1180/1 In *Erwartung*, the way becomes an ostinato figure that marks the beginning of Schoenbergian serialism. **1972** *Listener* 8 ... **1974** *Early Music* II. 227 A series of fanfare-like ostinati alternating between the lower two voices. **1975** *Country Life* 2 Oct. 848/1 Mime's forge is equipped with a... mechanical hammer which superimposes its own ostinato on Wagner's music. **1976** *Gramophone* May 1766/3 The only genuine mistake is one in the Ravel..where the bars before fig. 129, in the 'Action rituelle', where the tam-tam suddenly sticks out of the percussion ostinato of which it forms a part.

ostium. Add: **2.** Also, a slit-like valve in an insect's heart. Also in combs. with mod.L. adjs. (Earlier and later examples.)

1828 J. QUAIN *Elem. Anat.* viii. 537 The fimbriated border presents a fissure or opening, (ostium abdominale) which is directed backwards and downwards... Fallopian tube... **1874** *Jrnl. Anat.* ... **1883** *Trans. Linn. Soc.* ... (ostium internum)... **1915** F. W. LEWIS ... **1925** A. D. IMMS *Gen. Textbk. Entomol.* ... **1947** C. KEMPER *Jrnl. Anat.* ... **1956** ... **1969** R. F. CHAPMAN *Insects* xxxii. 661 The anterior and posterior lips of each ostium are reflexed into the heart so that they form a valve permitting the flow of blood into the heart.. but preventing its outward passage. *Ibid.* 663 There are unpaired excurrent ostia in the heart of Plecoptera and Embioptera.

Ostmark (ǫ·stmɑːk). [G., f. *Ost* east + *mark* MARK *sb.*[2] 2c.] The name sometimes given in western countries to the currency of the German Democratic Republic, usu. to distinguish it from the West German Mark.

1948 *Times* 9 July 5/6 In the Soviet sector [of Berlin],.. only the *Ostmark*, the temporary currency of the Soviet

zone, circulates legally. *Ibid.*, There was no intention that the *Deutschemark* [*sic*] should rival the *Ostmark* as a currency in the western sectors. [**1950** *Britannica Bk. of Year* 300/2 According to a report of the Deutsche Notenbank the note circulation in Eastern Germany was estimated in Feb. 1949 at Deutsche Mark (Ost) 4,112 million.] **1959** *Times* 18 Feb. 14/4 West German demand for Ostmarks is relatively weak. *Ibid.* 14/5 The 'real' value of the Ostmark is about ·50 west mark. **1972** R. W. LAST tr. *Freund's From Cold War to Ostpolitik* 28 The commandants of the western sectors declared that the Soviet orders relating to the introduction of the Ostmark into Greater Berlin were null and void.

ostomy (ǫ·stŏmi). orig. *U.S.* [f. *COL)OSTOMY and similar words.] An operation such as colostomy or ileostomy that involves making a permanent artificial opening in the body.

1957 *Ileostomy Q.* July 62/2 (Advt.), 'Ostomy' Skin Cream... United Surgical Supply Co. **1961** *Ibid.* Summer 65/2 Mr. Reynolds... doesn't look on his 'ostomy' as a handicap or hindrance to work or play. **1963** (*title of periodical*) Ostomy quarterly. **1964** *Hospitals* XXXVIII. 88 At present 35 ostomy clubs..in this country [*sc.* the U.S.A.] provide a means for persons with ileostomy, colostomy, ureterostomy, and ileal bladder to meet together and learn more about the management of their condition. *Ibid.* 90 Diet need not be a major problem to those with an ostomy. **1975** *Globe & Mail* (Toronto) 22 Aug. 25/8 There are three main types of ostomy: urostomy or urinary diversion, mostly due to cancer or birth defects; ileostomy or diversion of the small intestine due mostly to ulcerative colitis and Crohn's disease; and colostomy or diversion of the large intestine, due mostly to cancer and birth defects.

Hence **o·stomate** [cf. -ATE[1]], a person who has had an ostomy.

1966 *Ostomy Q.* Spring 15/2 What can be more rewarding than the feeling you have after visiting a new ostomate and knowing you have helped them? **1973** *Daily Colonist* (Victoria, B.C.) 2 Mar. 17/7 The ostomate faces a life of carrying the plastic bag outside the body. **1975** *Globe & Mail* (Toronto) 22 Aug. 25/8 Whether or not you are one of the 100,000 or more ostomates in Canada you should be interested in the current convention of the United Ostomy Association going on in Toronto. **1977** *Telegraph* (Brisbane) 2 Aug. 2/1 Ostomates are people who have had operations to by-pass bowel or urinary tracts, and who have to wear bags attached to their body to collect their body wastes.

Ostpolitik (ǫ·stpǫliti:k). [G., f. *Ost* east + *Politik* policy.] German policy towards Eastern Europe, associated mainly with the Federal Republic of Germany's cultivation of good relations with the Communist block during the 1960s, but applied also, by extension, to the policies of other western countries regarding the East as a whole.

1961 T. PRITTIE *Germany Divided* vi. 155 They will scarcely overlook Hitler's statement,..'The goal of *Ostpolitik* is to open up an area of settlement for one hundred million Germans.' **1967** *Economist* 6 May 558/3 Herr Kiesinger..promised that the government would not pursue its Ostpolitik 'behind the backs of the expellees'. **1968** *Ann. Reg.* 1967 253 Immediately the East German Government, supported by Moscow, took steps to hamper further progress by Bonn's *Ostpolitik*. **1970** *Atlantic Monthly* July 26 In the west, the big change was Willy Brandt's narrow victory in the West German elections last October, and the formation of a new Bonn coalition government dominated by the Social Democrats, prepared to abandon the rigidities of the Adenauer foreign policy of the last twenty years and embark on an entirely new and dynamic course of Ostpolitik. **1971** *Times Lit. Suppl.* 15 Oct. 1246/2 The politicians of Bonn are rather unhappy at the widespread use of the term 'Ostpolitik' by their Western allies. In the history of twentieth-century Germany, this term has signified a whole range of activities, from Hindenburg's humiliation of Russia at Brest-Litovsk to the East-West balancing act of Rapallo, and back to the domination of the East by Schacht's financial diplomacy and Hitler's armies. **1971** *New Yorker* 23 Oct. 156 Nixon as a risk-taker is something of a surprise... But his *Ostpolitik* is daring. It is a repudiation of his entire past. **1972** R. W. LAST tr. *Freund's From Cold War to Ostpolitik* 75 Brandt knows better than anyone else that Berlin is the real test of the new Ostpolitik.

ostracod. Now the usual spelling of OSTRACODE *a.* and *sb.* (but see quots. 1953 and 1974[2]). Substitute for etym.: [a. mod.L. *Ostracoda* and Fr. *ostracode* (P. A. Latreille *Genera Crustaceorum et Insectorum* (1806) I. 17), formerly *Ostrachoda, ostrachode* (P. A. Latreille in C. S. Sonnini *Buffon's Hist. Nat. Crustacés & Insectes* (1802) II. 361), f. Gr. ὀστρακώδης testaceous.] (Later examples.)

1935 TWENHOFEL & SHROCK *Invertebr. Paleontol.* x. 436 Ostracods are small, bivalved Crustacea which are found inhabiting all waters but are most abundant in marine habitats. **1953** *Ibid.* (ed. 2) xiii. 548 Ostracodes are minute, lentil-shaped crustaceans having a bivalve carapace that completely encloses the indistinctly segmented body. **1956** [see *MYSID]. **1957** *New Biol.* XXIV. 70 Some of the ostracods..possess haemoglobin. **1965** B. E. FREEMAN tr. *Vandel's Biospeleol.* ix. 112 Hypogeous ostracods are known from Europe, North America and Japan. **1969** BENNISON & WRIGHT *Geol. Hist. Brit. Isles* viii. 164 Ostracod shales were laid down contemporaneously in deeper waters in the Torquay area. **1974** A. DILLARD *Pilgrim at Tinker Creek* viii. 132 An ostracod, a common fresh-water crustacean of the sort I crunch on by the thousands every time I set foot in Tinker Creek. **1974** *Smithsonian Contrib. Earth Sci.* No. 13. 22/1 Constituents were counted and grouped in the following classes: terrigenous (mica and other, including quartz), bioclastic remains (pelagic forams, benthonic forams, pteropods, and other), plant fragments, and other (including ostracode valves, sponge spicules, unidentifiable fragments, etc.).

ostracoderm (ǫ·străkodɜːm), *sb.* and *a.* [a. mod.L. sub-class name *Ostracodermi* (E. D. Cope 1889, in *Amer. Naturalist* XXIII. 852), f. Gr. ὀστρακόδερμος hard-shelled.] **A.** *sb.* A small, primitive, fossil fish belonging to the group formerly designated the sub-class Ostracodermi. **B.** *adj.* Of or pertaining to this group of fossils.

1891 A. S. WOODWARD *Catal. Fossil Fishes in Brit. Mus.* (*Nat. Hist.*) II. p. xvii, The Arachnid theory is based upon a complete misapprehension of the most fundamental points in Ostracoderm skeletal anatomy. **1898** [see OSTRACO-]. **1933** A. S. ROMER *Vertebr. Paleont.* ii. 24 Almost all are covered with various types of armor, a feature to which the name 'ostracoderms' ('shell-skinned') is due. **1935** *Amer. Jrnl. Sci.* CCXXIX. 323 (*title*) The ostracoderm genus *Dartmuthia* Patten. **1968** A. S. ROMER *Procession of Life* viii. 148 By that time [*sc.* early Devonian] higher fish types descended from the ostracoderm stock were already evolving. **1969** BENNISON & WRIGHT *Geol. Hist. Brit. Isles* viii. 182 The jawless Ostracoderms with dorsal shield and granular armour..had made their appearance in the Upper Ludlow Beds.

ostrakon (ǫ·străkǫn). Also **ostracon**. Pl. **ostraka, -ca.** [ad. Gr. ὄστρακον potsherd.] A sherd of pottery or (more rarely) limestone used in antiquity as a surface for writing or inscribing, often, at Athens and in other Greek cities, to cast a vote (see OSTRACISM 1), or as a common writing material. Used (freq. in *pl.*) of archæological finds of this kind in the Middle East.

1883 *Proc. Soc. Biblical Archaeol.* V. 84 The British Museum has lately acquired..a considerable number of ostraka or potsherds discovered at Elephantine, Thebes, and other places. *Ibid.* 119 Two ostraka or slices of limestone formed for the purpose, inscribed with hieratic inscriptions. **1900** *Athenæum* 23 June 783/1 The study of Greek ostraca is a comparatively new one. **1921** G. A. F. KNIGHT *Nile & Jordan* 251 The name Bata has been recovered in a hieratic ostrakon. **1934** *Discovery* Apr. 90/2 The smaller finds include..actual *ostraca* from the voting of 483 B.C. when 'Aristides the Just' was banished. **1952** G. SARTON *Hist. Sci.* I. iv. 114 Two late mathematical papyri,..as well as Coptic ostraca from Wādī Sarga (near Asyūt)..contain unmistakable examples of Egyptian computation. **1960** [see *FIND *sb.* 4]. **1968** V. EHRENBERG *From Solon to Socrates* Notes 414 The number of ostraca found by archaeologists gives little evidence as to the actual results of the voting. **1972** *Times* 18 May (Egypt Suppl.) p. iv/4 During the excavation of the temple terrace, many fragments of stone stelae and *ostraca* (inscriptions on potsherds) were found bearing dedications to Isis. **1978** *N.Y. Times Bk. Rev.* 21 May 39/1 Demotic is known today from papyri and ostraca—broken bits of pot that have been scribbled on.

ostrich[1]. Add: **1. c.** Short for 'ostrich skin'.

1939 R. STOUT *Some Buried Caesar* xiii. 161 The brown ostrich card-case, gold-tooled. **1973** J. DRUMMOND *Bang! Bang! You're Dead!* iii. 6 She walked from the room, carrying the jewel-case and a matching ostrich purse.

2. a. (Further examples.)

1598 J. MARSTON *Pigmalion* Sat. I. 34 Fie that his Ostridge stomack should digest His Ostridge feather. **1844** E. B. BROWNING *Let.* 11 Jan. (1954) 212 But the squeamishness of this Age,..this Ostrich age..which exposes its own eggs, and then hides its head in the sand,..is really to me quite monstrous. **1952** DYLAN THOMAS *Let.* 6 Nov. (1966) 380 These ostrich griefs were always with me. **1976** *Listener* 6 May 585/2 The typical ostrich-Briton of today.

b. *ostrich-like* (later examples), *-skin*; **ostrich-egg cup**, a decorated cup made from an ostrich-egg; **ostrich-farm** (earlier and later examples); **ostrich-farming** (later examples).

1937 *Burlington Mag.* Apr. p. xxiv/2 The Leipzig *Ostrich Egg Cup* called thus because of the egg being decorated with the figures of birds. **1960** H. HAYWARD *Antique Coll.* 205/2 *Ostrich egg cup*, sometime supposed to be the eggs of griffins or phoenixes, ostrich eggs were often mounted as cups (and occasionally made into flasks) in the 16th cent. and later. Many surviving specimens are German. **1876** C. M. YONGE *Three Brides* II. i. 10 He has been acting as manager on an ostrich farm. **1926** *Daily Colonist* (Victoria, B.C.) 18 July 21/5 The White and Gold Room at Buckingham Palace, in which ladies sit in rows before passing into the Throne Room to curtsy to Their Majesties, is irreverently referred to on court-nights by junior members of the household as 'the ostrich farm'. **1927** CHESTERTON *Coll. Poems* 180 Old Noah he had an ostrich farm and fowls on the largest scale. **1974** *Encycl. Brit. Micropædia* VII. 618/1 This demand [for plumes] led to the establishment of ostrich farms in South Africa, the southern U.S., Australia, and elsewhere. **1902** *Chambers's Jrnl.* Jan. 53/2 His unconscious host,..prosed on concerning himself chiefly with.. the future of ostrich-farming. **1957** *Encycl. Brit.* XVI. 959/2 Ostrich farming is carried on in Cape Colony, Egypt, Algeria, the French Riviera, Southern U.S. and elsewhere. **1944** *Sci. Jrnl. R. Coll. Sci.* XIV. 64 Profound influences are continually at work causing changes in the general constitution of man, and no amount of ostrich-like behaviour will prevent their action. **1966** *Guardian* 30 July 3/5 Britain is more ostrichlike in its approach to the problem than America. **1976** *Times* 5 Apr. 3/2 It would be an ostrich-like attitude on the part of the executive if this chance was turned down. **1926–7** *Army & Navy Stores Catal.* 84/3 Ostrich skin Cigar Case..each 8/6. **1971** P. DRISCOLL *White Lie Assignment* ii. 17 Transferring the cigarettes..into an ostrich-skin case. **1976** J. MCCLURE *Rogue Eagle* xiii. 222 He was loading his pipe from an ostrich-skin pouch.

ostrich-feather. Add: **1.** (Later examples.) Also *fig.*

1932 D. GASCOYNE *Roman Balcony* 12 The ostrich feathers Of the waves That flap against the shore. **1957** M. B. PICKEN *Fashion Dict.* 238/2 *Ostrich feathers*, wing and tail feathers from an ostrich. Often several feathers are carefully glued together as one to give abundant appearance. Dyed in all colors and beautifully curled. **1977** *Listener* 15 Dec. 794/1 Before I graduated to ostrich feathers..I had absorbed..[a] vital principle..that dramatic privilege exists mainly not to be made use of.

Comb. (Later examples.)

1884 *List of Subscribers* (London & Globe Telephone Co.) Henry, C. S. & Co... Ostrich Feather Merchants, 12, Jewin Crescent. **1908** *Westm. Gaz.* 27 June 12/1 This was worn with a hat and ostrich-feather ruffle of a very charming soft grey tone. **1966** J. LAVER *Victoriana* 131 Towards the end of the [19th] century large ostrich-feather fans became fashionable.

ostrichism. (In Dict. s.v. OSTRICH[1].) Delete (*nonce-wd.*) and add later examples in allusive and extended uses.

1944 J. S. HUXLEY *On Living in Revolution* 3 The fact that a world war existed and the ostrichism of our reactions to it were most obvious in the case of Spain. **1945** R. HARGREAVES *Enemy at Gate* 285 A departure into Maginot Line ostrichism which had ended..with a vital chain of defence works being 'turned'. **1958** *New Statesman* 1 Mar. 260/2 Geoffrey Dawson's calculated ostrichism towards the Fascist dictators..during the Thirties. **1960** *Spectator* 15 July 106 A new wave of ostrichism in regard to defence is sweeping the country.

ostrobogulous (ǫstrobǫ·giŭləs). *slang.* [Etym. obscure; see quot. 1973.] A word associated with the writer Victor B. Neuburg (1883–1940), used with various shades of meaning to describe the bizarre, unusual, or interesting (see quots.). Hence **ostrobo·gulatory** *a.*; **ostrobogula·tion, ostrobogulo·sity,** *sbs.*

1951 A. CALDER-MARSHALL *Magic of My Youth* i. 31 'Ostrobogulous' was Vickybird's favourite word. It stood for anything from the bawdy to the slightly off-colour. Any *double entendre* that might otherwise be missed by his audience was prefaced by, 'if you will pardon the ostrobogulosity.' **1952** A. GRAVES *Ostrobogulous Pigs* 7 Once upon a time there were..five ostrobogulous pigs... flipperty filthy grubby muddy little pigs. *Ibid.* 10, I can no longer endure this ostrobogulatory behaviour. *Ibid.* 11 'I can no longer endure the odorous and obnoxious ostrobogulations of those creatures,' said Angelina Boghurst-Fisher. **1963** *Sunday Times* 29 Dec. 10/2 (*heading*), An ostrobogulous year for the toy men... 10/6 Minnie King works full time for Ostrobogulous Toys... 'It's a word, they say, used by children and means mischievous but gorgeous'. **1965** J. O. FULLER *Magical Dilemma of V. Neuburg* I. iv. 58 Some of the entries were not printed because they were ostrobogulous. This was a favourite word of Vicky's. It was used in the place of indecent or pornographic, and had the advantage..that it implied no moral attitude. *Ibid.* 59 He would speak of an ostrobogulous tale... He took Morton's opinion as to its ostrobogulosity. **1968** *Times Lit. Suppl.* 24 Oct. 1203 There has been no developing tradition of the whimsical and ostrobogulous (to use the word Victor Neuburg applied to this work) merely because there have been no similar collections. **1972** *Ibid.* 30 June 757/4 His career, fabulous, prestigious, sordid, sinister, and in the word of Victor Neuburg ostrobogulous. **1973** *Ibid.* 27 July 871/1 It was sick, dirty, or more precisely, 'ostrobogulous', which according to Victor Neuburg..meant etymologically full of (Latin, *ulus*) rich (Greek, *ostro*) dirt (schoolboy, *bog*).

Ostrogoth. Add: (Earlier example.)

1605 J. SYLVESTER tr. *Du Bartas's Devine Weekes* 456 Normans, Allains, Ostrogothes.

c. as *adj.* = OSTROGOTHIC *a.*

1920 H. G. WELLS *Outl. Hist.* 350/1 The adventurous wanderings that ended at last in the Ostrogoth Kingdom in Italy. **1954** ARNOLD-BAKER & DENT *Everyman's Dict. Dates* 275/1 Theodoric..became a figure of German legend, and founded an Ostrogoth kingdom in Italy *c.* 500.

Ostwald (ǫ·stwăld). The name of Wilhelm *Ostwald* (1853–1932), Russian chemist, used *attrib.* and in the possessive to denote apparatus invented and principles enunciated by him, as **Ostwald('s) dilution law**, the law that for dilute solutions of a binary electrolyte the square of the degree of dissociation of the solute, multiplied by its concentration, and divided by one minus the degree of dissociation, is a constant for the solute; **Ostwald pycnometer**, a pycnometer in the form of a bulb joined at the top to a horizontal capillary tube and at the bottom to a U-tube bearing a graduation at the level of the capillary; **Ostwald('s) viscometer**, an instrument in which the viscosity of a liquid is determined by the time taken for a measured volume to pass through a capillary bore in a U-tube.

[1899 J. Walker *Introd. Physical Chem.* xxi. 227 Certain empirical relations have..been found connecting the degree of dissociation and the dilution, and these have a form similar to that of Ostwald's dilution formula.] 1902 H. C. Jones *Elem. Physical Chem.* vii. 354 (*heading*) Testing the Ostwald dilution law. 1930 C. W. Davies *Conductivity of Solutions* i. 12 Ostwald's dilution law breaks down completely when it is applied to solutions of the common salts. 1973 A. W. Adamson *Textbk. Physical Chem.* xii. 538 Weak electrolytes obey the Ostwald dilution law well, and their behavior is thus determined primarily by a dissociation equilibrium. 1910 A. Ewell *Text-bk. Physical Chem.* 25 An Ostwald pyknometer is very convenient as a weighing pipette. 1962 Perry & Koenig in A. Pirie *Lens Metabolism Rel. Cataract* 306 For viscosity measurements, Ostwald viscometers were used to measure flow rates at 25°C. Densities were measured with Ostwald pycnometers. 1924 *Abstr. Bacteriol.* VIII. 2 Direct inoculation of the sterilized gelatin culture media in Ostwald's viscometer was tried for a period of 3 months. 1967 Margerison & East *Introd. Polymer Chem.* ii. 103 The usual method employed to measure the solution and solvent viscosities utilizes an Ostwald viscometer.

Ostyak (ǫstia·k). Also Ostiac, Ostiack, Ostiak, etc. [Russ. *ostyák*.] **a.** (A member of) a Finno-Ugric people, also called *Khantý*, living in the Ob River basin in Western Siberia. **b.** The language of this people, belonging to the *Ob-Ugrian group. Also *attrib.* or as *adj.*

1722 tr. *Muller's Manners & Customs of Ostiaks* in tr. *F. C. Weber's Present State of Russia* II. 56 The Ostiacks and Samoieds often venture over those high Rocks into the Country, where they kill Elks and Rain-Deer. 1757 J. Dyer *Fleece* iv. 143 Land of the lazy Ostiacs, thin dispers'd. 1841 *Penny Cycl.* XXI. 467/1 South of the Samoyedes are the Ostiaks, who occupy both banks of the river Obi. 1859 G. W. Dasent *Pop. Tales from Norse* p. lxvii, The Ostjaks, a tribe akin to the Lapps. 1870 J. Lubbock *Origin of Civilisation* iii. 96 The Ostiaks regard it as a crime to marry a woman of the same family or even of the same name. 1880 A. H. Sayce *Introd. Sci. of Lang.* II. viii. 204 It is difficult..to distinguish the Ostiak forms ..from the persons of the Sanskrit verb. 1889 T. Duka *Essay Ugor Lang.* 32 Klaproth..makes mention of five Finn dialects, but describes..the Ostjak around the river Ob, and the Magyar. 1911 J. G. Frazer *Golden Bough: Magic Art* (ed. 3) II. ix. 11 The Ostyaks and Woguls, two peoples of the Finnish-Ugrian stock in Siberia. 1933, etc. [see *Ob-Ugrian]. 1938 *N. & Q.* 23 Apr. 291/1 The Ostyaks of the Tobolsk region have a spirit of death whom they call *xeina*. 1944 [see *Nenets]. 1959 *Chambers's Encycl.* XI. 432/2 The Uralics, so termed by Bunak, are perhaps best represented by the western Siberian Voguls and Khuntu or Ostyaks, between the Urals and the Ob basin, in reality a single people who call themselves 'Mansi'. 1966 T. Burrow *Sanskrit Lang.* (ed. 2) 23 Vogul and Ostyak are now found to the East of the Urals, but are considered to have moved there from the West. 1971 [see *Mordvin]. 1972 W. B. Lockwood *Panorama Indo-European Lang.* 152 East of the Zyryene-speaking area, in Siberia, lie Ostyak and Vogul which with Hungarian form the Ugric division of Uralic. 1974 T. P. Whitney tr. *Solzhenitsyn's Gulag Archipel.* I. i. i. 6 From Karger they took his archive of the Yenisei Ostyaks. 1975 G. F. Cushing tr. *Hajdu's Finno-Ugrian Lang. & Peoples* iii. 123 Anthropologically the Voguls and Ostyaks..are classed by anthropologists as Europo-Sibirid or Uralic.

‖**osu** (ǫ·su). *W. Afr.* [Igbo.] An outcast, an 'untouchable'.

1958 C. Achebe *Things fall Apart* xviii. 140 These outcasts, or *osu*, seeing that the new religion welcomed twins and such abominations, thought that it was possible that they would also be received. 1960 —— *No Longer at Ease* vii. 71 'I am an *osu*,' she wept... 'So you see we cannot get married,' she said, quite firmly, almost gaily—a terrible kind of gaiety. *Ibid.* 75 Obi knew better than anyone else that his family would violently oppose the idea of marrying an *osu*. *Ibid.* xiv. 133 Our fathers in their darkness and ignorance called an innocent man an *osu*, a thing given to idols, and thereafter he became an outcast, and his children, and his children's children for ever. *Ibid.* 134 Obi repeated his points again. What made an *osu* different from other men and women? Nothing but the ignorance of their forefathers. 1960 *Spectator* 21 Oct. 616 He falls in love with Clara, an *osu* or 'untouchable'. 1973 *Black World* June 39/2 This..explains the violent objection of the clansmen to Obi marrying an *osu* (a cult-slave; for this reason ostracized from the normal life of the village).

Oswego (ǫzwī·go). [The name of a river and a town in the northern part of the state of New York.] **1. a.** = *Oswego bass.

1857 [see *Lunge *sb.*³].

b. A proprietary name for a type of cornflour.

1881 J. T. Gill *Compl. Bread, Cake & Cracker Baker* i. v. 75 Maizena..is maize deprived of all its albuminoid or flesh-forming constituents..and is, therefore, simply pure starch... Corn flour and Oswego are only other names for the same substance. 1907 *Official Gaz.* (U.S. Patent Office) 29 Jan. 1758/2 National Starch Co., Jersey City, N.J. and Oswego, N.Y. Filed Dec. 17, 1906. Used ten years. Oswego. Particular description of goods.—Corn-starch. 1911 *Encycl. Brit.* XVII. 449/1 When deprived of the gluten it [*sc.* maize] constitutes oswego, maizena or corn flour.

c. = *Oswego biscuit.

c 1900 in A. Davis *Package & Print* (1967) (*frontispiece*) Peek, Frean & Co. Biscuits. Oswego. 1907 *Yesterday's Shopping* (1969) 8 Biscuits..Oswego..

2. *attrib.* **Oswego bass** = *large-mouth (bass)* s.v. Large *a.* 15 a in Dict. and Suppl.; **Oswego biscuit, cake,** a biscuit or cake made with **Oswego flour** (= sense *1 b); **Oswego tea,** a herbaceous perennial plant, *Monarda didyma*, belonging to the family Labiatæ and native to eastern North America; its leaves were formerly used to make a medicinal tea.

1758 C. Rea *Jrnl.* 20 July in *Essex Inst. Hist. Coll.* (1881) XVIII. 112 The Lake affords plenty of a Fish call'd Oswego Bass. 1840 J. F. Cooper *Pathfinder* I. ix. 130 Even the Major himself..will sometimes swear that an oat-meal cake is better fare than the Oswego bass. 1884 [see *large-mouth (bass)]. 1965 A. J. McClane *Standard Fishing Encycl.* 473/1 The largemouth bass is regionally known as green bass, green trout, Oswego bass, and black bass. 1936 Chesterton *Autobiogr.* 324 Do let me offer you an oswego biscuit. 1963 C. Mackenzie *My Life & Times* II. 94 However well Huntley and Palmer may have made Oswego and Osborne biscuits..they could not reproduce the authentic flavour of real Petit Beurres. 1907 M. I. Rivers *Tips for Tea* vii. 50 Oswego cakes... Cream the butter and sugar together until they are white; add the eggs one at a time, and stir in lightly the oswego. 1949 A. R. Daniel *Baker's Dict.* Oswego flour. A particular type of flour obtained from maize, but not just pure cornflour. 1752 P. Miller *Gardeners Dict.* (ed. 6) s.v. *Monarda*, The inhabitants drink an infusion of this Herb as Tea, and call it Ozweega Tea. 1759 —— (ed. 7) s.v. *Monarda*, It [*sc.* Monarda with Flowers collected in Heads] is commonly called Oswego Tea, by which title it was brought to England. 1760 [see Tea *sb.* 6]. 1789 [see *Monarda]. 1850 S. F. Cooper *Rural Hours* 117 Humming-birds..are partial to the bee larkspur also, with the wild bergamot or Oswego tea. 1947 *Nat. Geogr. Mag.* July 62/1 Among the many useful plants they found ..were two of this same Mint Family, Oswego Tea (*Monarda didyma*), and Wild Bergamot (*Monarda fistulosa*). 1954 C. Hylander *Macmillan Wild Flower Bk.* 344 Oswego Tea, also known as Bee-balm, has a natural range from New York to Michigan..; it is frequently cultivated as an ornamental. 1970 B. Miles *Bluebells & Bittersweet* vii. 113/3 *M[onarda] didyma* (Oswego-tea) is usually red, raising round heads of lipped flowers 4 feet in midsummer.

Otaheitean (ōtāhī·tiăn), *a.* and *sb.* *Obs.* Also Otahitean, Otaheite, Otaheitan. [f. *Otaheite* early name of the Pacific island of Tahiti.] = *Tahitian *a.* and *sb.*

Cf. Otaheite apple in Dict.

1773 W. Wales *Jrnl.* 31 Aug. in Cook *Jrnls.* (1961) II. 796 With regard to the Personal Beauties of the Otahitean Ladies, I believe it would be most prudent to remain entirely silent. 1792 W. Bligh *Voy. to South Sea* vi. 81 Among people so free from ostentation as the Otaheiteans ..the strictness with which the punctilios of rank are observed, is surprising. *Ibid.* xii. 147 In a small vocabulary, that I made..only four words, out of twenty-four, differed from the Otaheite. 1793 F. Burney *Jrnl.* 3 May (1972) II. 104, I saw nothing of this *Tio*:—I accept your otaheité epithet, & like it much. 1799 *Sporting Mag.* XIV. 203/1 Our Otaheitean girl, who was tolerably fair, and had a comely person. 1817 T. Cogan *Ethical Questions* v. 233 The good-natured Otaheite *feels* it to be an obligation of hospitality, to present his wife or daughter to a stranger. 1819 Shelley *Let.* 3 Nov. (1964) II. 140 A North American Indian, or an Otaheitan. 1851 *Illustr. Catal. Gt. Exhib.* iv. 980/1 All the above-mentioned sugars are the produce of the Otaheite or Tahiti cane. 1861 R. Bentley *Man. Bot.* 674 The starch known as Tacca starch, Tahiti Arrow-root, or Otaheite Salep.

otavite (otă-vəit). *Min.* [ad. G. *otavit* (O. Schneider 1906, in *Centralbl. f. Mineral., Geol. und Paläont.* 389), f. *Otavi*, name of a town in northern South West Africa: see -ite¹.] Naturally occurring cadmium carbonate, $CdCO_3$, occurring as crusts of minute rhombohedral crystals usu. white or greyish white in colour.

1906 *Jrnl. Chem. Soc.* XC. ii. 620 A new mineral which is named otavite, after the locality. 1943 *Mineral. Abstr.* VIII. 366 Otavite, previously described as a basic carbonate of cadmium from Tsumeb, South-West Africa, forms thin crusts of minute rhombohedra often in parallel growth on smithsonite. X-ray powder photographs agree with Zachariasen's (1928) determination for artificial $CdCO_3$. 1967 *Soviet Physics: Crystallogr.* XII. 417 Artificial otavite, $CdCO_3$, crystals were obtained, which were activated by Co^{+2} so that EPR spectra of the magnetic Co^{+2} ion were observed at varying orientations of the external magnetic field at 4·2°K.

other, *adj. pron.* (*sb.*). Add: **A.** *adj.* **2. c.** *the other half*: (*a*) the other half of the world; people of a different class or those enjoying a different (usu. more affluent) way of life, *spec.* in phr. *how the other half live(s)*; (*b*) orig. *Naval slang*: a second drink; a drink bought in return for another.

(*a*) [1532 Rabelais *Pantagruel* (1547) II. xxxi. 206 La moitié du monde ne sçayt comment l'aultre vit. 1607 J. Hall *Holy Observations* xvii. 26 One halfe of the world knowes not how the other liues.] 1640 G. Herbert *Outlandish Proverbs* No. 907, in *Witts Recreations* sig. D7ᵛ Halfe the world knowes not how the other halfe lies [*sic*]. 1830 Marryat *King's Own* I. x. 141 It is an old proverb that 'one half the world do not know *how* the other half live.' Add to it, nor *where* they live. 1890 J. A. Riis *How Other Half Lives* 1 Long ago it was said that 'one half of the world does not know how the other half lives'. *Ibid.* 3 The sufferings and sins of the 'other half'..are but..a just punishment upon the community that gave it no other choice. 1945 N. L. McClung *Stream runs Fast* xiii. 106 We were only amateurs but we did find out a few things about how the 'other half' lived. 1965 E. O'Brien *Aug. is Wicked Month* xiii. 141 'Why not, see how the other half live...' she said... They would have the pool and servants to wait on them. 1968 J. Sangster *Touchfeather* xiii. 140 He said if I was ever in Los Angeles to look him up. Glad I did. Talk about how the other half lives! 1970 'D. Shannon' *Unexpected Death* (1971) ix. 141 'My God,' said Higgins. 'How the other half lives.' 1975 *Times* 15 Jan. 15/4 In the interests of national unity, may I support your plea to the other half (and if they do not know they are in it, that is half the trouble) to accept that the shocking conditions of the London comprehensives as reported by you are really good for their children, and for society.

(*b*) 1922 W. S. Maugham *On Chinese Screen* lii. 211 No sooner was your glass empty than he was prompt with the China phrase: 'Ready for the other half?' 1931 C. Lithgow *Simple Sailor* xv. 184 You won't have the other half? Sure? 1936 'G. Orwell' *Keep Aspidistra Flying* v. 117 Drink up!..It's time we had the other half of that. 1965 R. Jeffries *Dead against Lawyers* vii. 69 You'll have the other half, Inspector? Two whiskies under the belt are better than one. 1966 A. Prior *Operators* ii. 16 'The other half please, George.' 'Yessir, Mr. Barclay.'.. The barman turned to him. 1975 E. Berckman *Indecent Exposure* viii. 101 'Have to be shoving off now, sorry—.' 'The other half,' Dennison objected. 'What were you drinking, whisky—?'

d. *the other side*: (*a*) the world to come, the world beyond the grave, esp. as inhabited by the spirits of the dead; (*b*) *Austral.* and *N.Z. slang* (see quots.); (*c*) an opponent or an opposing side; one regarded as such.

(*a*) 1684 Bunyan *Pilgrim's Progress* II. 220 So he passed over, and the Trumpets sounded for him on the other side. 1819 [see Side *sb.*¹ 12a *fig.*]. 1926 A. Conan Doyle *Hist. Spiritualism* I. viii. 187 The sharp detail which we receive from the Other Side is incompatible with any vague grandiose idea of the sort. 1941 Auden *New Year Let.* 38 It is Utopian to be dead, For only on the Other Side Are Absolutes all satisfied. 1945 A. Huxley *Time must have Stop* vi. 66 Are they still obese on the other side? I'd like to ask next time you have a séance. 1960 M. Spark *Bachelors* ii. 22 When Patrick's under the control I shouldn't think he could help saying what comes to him from the other side. 1973 *Listener* 8 Mar. 306/2 Max [Aitken]..is devoted to the memory of his father [*sc.* Beaverbrook] and..allows his father to edit the paper from the Other Side. 1974 'D. Shannon' *Crime File* xii. 185 They went to Katie May Blaine's funeral... 'You don't want to worry,' she told Mrs. Blaine simply. 'She's being looked after, the other side.'

(*b*) 1855 W. Howitt *Land, Labour & Gold* ii. 362 Scenery precisely like hundreds of miles which I have seen 'on the other side', as they call Victoria, and as the Victorians call Van Diemen's Land. 1884 A. Cox *Recollections* 125, I ax your pardon, zur, but were you ever at the Yan Yean works over the other side? *a* 1948 L. G. D. Acland *Early Canterbury Runs* (1951) 389 *Other side*,..Australia. 1963 X. Herbert *Disturbing Element* 2 My parents..were what were called T'othersiders, meaning people who had come to West Australia from the other side of the continent.

(*c*) 1916 'Taffrail' *Pincher Martin* xiv. 259 Their expeditions to that region known as 'the other side', for the express purpose of discomforting the Hun. 1939 'N. Blake' *Smiler with Knife* iii. 57 They've youth and independence and courage. They're England. And you know what the other side says—'Woman is for the recreation of the warrior'. 1966 I. Asimov *Fantastic Voyage* i. 19 I've met him several times at scientific conferences on the other side. 1967 B. Norman *Matter of Mandrake* xxv. 211 There was a change of plan... The Other Side was becoming too worried. 1972 *Sat. Rev. Society* (U.S.) Dec. 33/2 The way is far more open..to similar wars of 'aggression' or 'national liberation' or whatever the Vietnam War has been. The 'other side' can now engage in such activities with the understanding that the United States will be very reluctant to intervene. 1976 B. Freemantle *November Man* iv. 45 Hugo will know it was an attempt on his life... If he runs to the other side, everything is going to be easy for us.

e. *the other place*: one place regarded from the point of view of or with reference to another place; *euphem.*, Hell (as opp. Heaven). Also in depreciatory senses; *spec.* Oxford as regarded in Cambridge (and vice versa).

1841 F. A. Kemble *Let.* 29 Dec. in *Rec. Later Life* (1882) II. 156, I conclude that letters will occasionally come *to* heaven, and always be written in—the other place. 1874 'Mark Twain' *Gilded Age* xi. 108 Washington was alternately in paradise or the other place just as it happened that Louise was gracious to him or seemingly indifferent. 1880 Trollope *Duke's Children* I. xx. 245 Shall I go to heaven for doing that?.. Or mayn't I rather go to the other place? 1920 'O. Douglas' *Penny Plain* i. 13, I wouldn't much care to go to heaven myself, for all my friends are in..the Other Place. 1944 A. Thirkell *Headmistress* ix. 204 Sir Hosea Weaver..a Cambridge man..had..taken the highest kind of degree that the other place can give in Political Economy. 1958 *Listener* 14 Aug. 232/1 Cambridge has always tried to be more typical and less exotic than the other place. 1967 V. Gielgud *Conduct of Member* ii. 15 There were Oxford men who persisted in speaking of Cambridge as 'the other place'. 1970 *Guardian* 22 Apr. 24/2 In the *other place*, it had been the rule to have sponsors since 1688 and Stormont had the same rule. 1972 'M. Innes' *Open House* xi. 101 There being neither youth or age, sir, in the 'evenly mansions—no, nor in the other place either. 1973 *Deb. Senate S. Afr.* 17 May 2807, I am thinking..of the important task of industrial decentralization about which so much has been said in this House and in the Other Place. 1973 'M. Yorke' *Grave Matters* v. ii. 84, I don't know Oxford at all well... I know the other place better. Isn't that what you call it? 1974 *Oxford Times* 5 July 1/4 (*heading*) By punt to the other place. 1976 *Gramophone* Oct. 552/2 In the old days, 'the other place' never seemed to me to compete at all, but it is very different now with several excellent offerings from Magdalen under Bernard Rose.

f. *the other thing* (colloq.): the contrary, opposite, or reverse; something quite different. *euphem.* sexual activity; the penis. Phr. *to do the other thing*: to do as one pleases (usu. as an expression of contemptuous dismissal).

1846 *Swell's Night Guide* 89 The wealthy voluptuary cannot choose but be gratified, as far as feasting, drinking, and the other thing goes. 1848 Trollope *Kellys & O'Kellys* I. vii. 172 They'd ax him to come and see his sister married, and av' he didn't like it, he might do the other thing. 1885 C. M. Yonge *Nuttie's Father* II. xix. 224 It's the sort of thing that one only laughs at because otherwise one would have to do the other thing! 1913 A. Bennett *Regent* i. vi. 165 You mean you won't!.. Well, you can do the other thing! 1922 Joyce *Ulysses* 359 Besides there was absolution so long as you didn't do the other thing before being married. 1923 E. P. Mathers tr. *Mardrus's Bk. of Thousand Nights & One Night* VII. 55 His heart is hard, his other thing is soft. 1929 J. van Druten *Young Woodley* xii. 241 'You don't believe me?' 'I do not.'.. 'Then you must do the other thing.' 1953 H. Clevely *Public Enemy* xxii. 165, I couldn't have a better home.., and anybody who doesn't like it can do the other thing. 1977 'D. Cory' *Bennett* iv. 127 The C.D.I. wouldn't like it, no. But, then, he could always do the other thing.

g. *the other man* (and varr.): a man with whom a woman already in an amatory relationship forms a new attachment; a lover. Similarly *the other woman*: a woman with whom a man forms a new attachment in such circumstances; a mistress.

Quot. 1867 may not have the overtones of the other examples.

1855 Browning *Men & Women* 88 Why must I..Put any kiss of pardon on thy brow? Why need the other women know so much? 1867 Trollope *Last Chron. Barset* I. xxxviii. 331 A woman, when she is jealous, is apt to attribute to the other woman with whom her jealousy is concerned, both weakness and timidity. 1886 Kipling *The Other Man* in *Plain Tales from Hills* (1888) 80 They married her when she..had given all her poor little heart to another man... We will call him the Other Man. 1909 F. Barclay *Rosary* xxiii. 242 The 'other man' is always a problem. 1912 T. Dreiser *Financier* xxxvii. 418 Curiously, the other woman did not seem so vastly important—that is, who she was. 1920 *Ladies' Home Jrnl.* Apr. 36 The cast includes Thomas Meighan as the husband, Gloria Swanson as the wife he changed, and Bebe Daniels as the other woman. 1927 E. Glyn *'It'* xiv. 137 What if being in the corner should make Ava go for help to the *other man*? 1935 *Mademoiselle* Aug. 3/2 Mr. Montgomery is the erring husband and Mr. Tone the 'other man'. 1946 G. Millar *Horned Pigeon* xxi. 360 She told me that her engagement had been broken, in fact her fiancé..was already married to 'the other woman'. 1953 K. Tennant *Joyful Condemned* xxxix. 391 'Who's the other guy?' 'There isn't any other guy.' 1966 'S. Ransome' *Hidden Hour* ii. 20 She had been here before. With the 'other man'? 1973 G. Moffat *Deviant Death* i. 15, I had to sit in the corner and keep quiet. I knew how the other woman feels at the posh family funeral of her lover. 1975 *Daily Mirror* 29 Apr. 9/5 They married in 1967 after Miss Smith was named as the 'other woman' by Mr. Stephens' first wife.

h. *the other end*: the person (or his location) with whom one is communicating by telephone.

1941 B. Schulberg *What makes Sammy Run?* vi. 121 Julian managed to get Sammy on the other end of a telephone. 1974 R. B. Parker *Godwulf Manuscript* iii. 17 The phone rang... The girl's voice at the other end was thick and very slow. 1978 T. Allbeury *Lantern Network* xi. 160 The fruity voice on the other end of the line.

5. f. *other ranks*: in the armed forces, non-commissioned officers and ordinary soldiers, seamen, etc. Occas. in *sing.*, a member of the other ranks. Also *transf.* and (in form *other-rank*) *attrib.* Cf. Rank *sb.*¹ 5 b.

1925 Fraser & Gibbons *Soldier & Sailor Words* 216 *Other ranks*, the usual official designation for N.C.O.'s and privates in orders, etc., as distinguished from Commissioned Officers. 1926 F. M. Ford *Man could stand Up* ii. ii. 106 There were all these inscrutable beings; the Other Ranks, a brownish mass, spreading underground, like clay strata in the gravel. 1929 T. E. Lawrence *Let.* 18 Apr. (1938) 652 M.B. is an amateur of the R.A.F., like me: but he doesn't know the other ranks in it, and won't like their dirt and brutality. 1931 —— *Let.* 20 Aug. (1938) 733 A book written by an 'other rank' would not mention the officers. 1946 *R.A.F. Jrnl.* May 150 The Sussex Square Club and Hostel..has been opened for male other ranks. 1959 *Encounter* July 85/1 Before the war the young officer really believed that the social circumstances of his upbringing and birth entitled him to give orders to 'other ranks'; and the 'other ranks' were on the whole quite satisfied. 1960 J. MacLaren-Ross *Until Day she Dies* vi. 94 There was a mob of other ranks sitting around on their kitbags. 1960 A. Waugh *Foxglove Saga* vi. 107 The other ranks of the Pigs were mostly recruited from the criminal and the stupid. 1965 G. McInnes *Road to Gundagai* ii. 26 The promenade deck was abruptly curtailed by a wooden grille..with the sign 'Other Ranks Only. No Entry. Military Police.' 1966 *Times* 9 July 9/6 The Army have given the expression 'other ranks' its marching orders. Commands have been told the term 'soldier' is preferred. 1968 R. West *Sh. Vietnam* ii. 46 Many marine other-ranks have college degrees. 1971 S. Hill *Strange Meeting* iii. 169, B Company has lost 2 officers and 3 wounded, and about 30 of other ranks. 1973 *Listener* 5 July 22/1 They, like their other-rank colleagues, knew well enough what it was all about. 1974 K. Royce *Trap Spider* i. 8 'Good man,' he said, as if he'd invited an 'other rank' to the officers' mess for a special treat. 1976 *Daily Tel.* 20 July 4/3 The withdrawal of British regular units from Oman will not affect..the secondment of British officers and other ranks of all three services to the Sultan's forces.

B. 10. as *sb.* Sexual activity; sexual intercourse. Also occas., homosexual practices. *slang.*

1922 Joyce *Ulysses* 358 They would be just good friends like a big brother and sister without all that other. *Ibid.* 429 Bit light in the head. Monthly or effect of the other. 1928 D. H. Lawrence *Lady Chatterley* xiv. 241 She loved me to talk to her and kiss her... But the other, she just didn't want. 1936 J. Curtis *Gilt Kid* 135 'Doing half I was.' 'What for? The other?' 'Yes.' The pansy simpered. *Ibid.*, He gets a stretch for screwing and another stretch on top of that for the other. 1969 F. Norman *Banana Boy* 127, I..usually managed to get Mary behind a haystack for a 'bit of the other'. 1974 *Spectator* 22 June 764/2 I've got to be noticed by any guy who's on the prowl away from home and looking for a bit of the other.

D. 1. *other-dimensional* (of or from another dimension); *other-mindedness*.

1926 *Public Opinion* 30 Apr. 436/3 The habit of..other-mindedness. 1940 J. Betjeman *Old Lights for New Chancels* 50 Coffee and Ulysses, Tennyson, Joyce, Alpha-minded and other dimensional, Freud or Calvary? Take your choice. 1961 *John o' London's* 25 May 592/2 His humour is so naturally other-dimensional.

2. *other-centred* (centred in others); *other-directed* Sociol. (applied to persons whose behaviour and goals are directed by standards they feel acceptable to others, esp. some kind of peer group; cf. *inner-directed* (*Inner *a.* (*sb.*²) 1 n), *tradition-directed*; hence as *sb.*; also *other-directedness, -direction*); *other-regard* (regard for others); *other-regarding* (further examples); *other-regardingness* (further examples).

1925 *Inner Life* (Ser. 2) 219 Love of the large room is characteristic of souls that are other-centred. 1950 D. Riesman et al. *Lonely Crowd* i. 9 The society of incipient population decline develops in its typical members a social character whose conformity is insured by their tendency to be sensitized to the expectations and preferences of others. These I shall term other-directed people. *Ibid.* 20 It is also my impression that the conditions I believe responsible for other-direction are affecting increasing numbers of people in the metropolitan centers of the advanced industrial countries. *Ibid.* 23 What is common to all other-directeds is that their contemporaries are the source of direction for the individual. 1953 *Brit. Jrnl. Psychol.* XLIV. 187 From having been in the nineteenth century inner-directed..most Americans have now become 'other-directed', taking standards and guidance mainly from their contemporaries at each stage in their lives. 1957 V. Packard *Hidden Persuaders* xvi. 168 The increasing desire of Americans to make a good impression on their peer group, as a part of the trend to other-directedness. 1959, etc. [see *Inner *a.* (*sb.*²) 1 n]. 1966 D. Jenkins *Educated Society* i. 27 Dangers..of excessive 'other-directedness' and the production of 'organization men'. 1973 *Listener* 17 May 635/1 [J. S.] Mill's fears were those of post-war American sociologists like David Riesman, who saw American society as full of 'other-directed' men. 1975 R. H. Rimmer *Premar Experiments* (1976) i. 48 The deep joy and satisfaction he discovers in other-directedness and his love for people-involvement. 1938 *Times Lit. Suppl.* 8 Oct. 635/2 When we come..to the consideration of the two major types of Sentiment, Self-regard and Other-regard (or Love), we are on firmer ground. 1923 J. S. Huxley *Ess. Biologist* vii. 273 The instincts that are self-regarding and those that are other-regarding. 1947 *Mind* LVI. 277 If the common good is our principle, we should expect a preference to be given to other-regarding ends over self-regarding ones. 1952 V. Gollancz *My Dear Timothy* xx. 274 At the bottom of it..was an element of something other-regarding. A sense of public service should have taken its place. 1969 *Listener* 6 Feb. 164/1 And protest in our society seems to be of these two kinds also—self-regarding and other-regarding—in rather comparable proportions. 1958 *Times Lit. Suppl.* 31 Jan. 54/5 Reason, objectivity, tolerance, charity, other-regardingness—these are not natural gifts of men.

otherliness (ʌ·ðəɹlinės). *rare.* [f. Other *a.* + -ness; cf. -ly¹.] The quality of being different or apart in some way.

1949 Koestler *Promise & Fulfilment* ii. i. 194 That eery odour of otherliness, of vagrancy and jugglery which surrounds Mr. Abramowitz. 1967 E. Grierson *Crime of one's Own* x. 86 It was another fault of the romantic spirit to imagine that..agents..[had] some special quality of 'otherliness', whereas their very profession must demand the opposite. 1976 A. Koestler in D. Villiers *Next Year in Jerusalem* 101 Do I really consider myself a member of a chosen race... If not, what right have I to go on.. inflicting on my children the stigma of otherliness?

otherwise, *sb. phr.*, *adv.*, *adj.* Add: **A. c.** Phr. *or* (occas. *and*) *otherwise*, following a noun, adjective, adverb, or verb, to signify a corresponding word of opposite or different meaning.

1886 *Rep. Brit. Assoc. Adv. Sci. 1885* 872 The index number..is 1 or 100, according to the use or otherwise of the decimal point. 1892 [see *playing week* s.v. Playing *vbl. sb.* 2]. 1895 *Pall Mall Mag.* Jan. 35 The most amusing feature of the case was the conflict of professional evidence as to the merits, or otherwise, of Mr. Whistler's paintings. 1910 *Practitioner* Jan. 84 The question of operability or otherwise is a matter in which surgeons differ considerably. 1911 E. C. Worden *Nitrocellulose Industry* II. xiv. 697 These enamels may be closely imitated..by taking a given pattern, enlarging it pantographically or otherwise, [etc.]. 1922 C. Mackenzie *Altar Steps* vii. 56 Mrs. Lidderdale's dread..was that her son would acquire a West country burr, and it was considered more prudent, economically and otherwise, to let him go on learning with his grandfather and herself. 1966 *Listener* 22 Sept. 427/2, I do not question the eruption at Santorin,..but the supposed connection of the underwater survey with the historicity or otherwise of the Atlantis myth. 1972 W. A. Pantin *Oxf. Life* iv. 52 Professor Southern gave us some stimulating reflections about the aims, development, and achievements (or otherwise) of the Honour School of Modern History. 1972 *Times Lit. Suppl.* 13 Oct. 1233/1 The circumstances of the publication of the Penguin Books version of D. H. Lawrence's *Lady Chatterley's Lover* in November 1960 led to much public discussion of the desirability or otherwise of printing these long-banned sexual words. 1973 *Oxford Times* 30 Nov. 10, 12,000 Cowley workers enjoyed (or otherwise) an enforced holiday because of a strike by plant attendants at the car assembly factory.

other world, other-world, *sb.* and *a.* Also o·therworld. Add: **1.** (Earlier and later examples.) Also (in sense c) more *gen.*, a range of experiences conceived in imagination or fantasy as lying outside the world as normally known.

c 1200, 1611 [see World *sb.* 1 d]. 1612 W. Strachey *Trav. Virginia* (1849) i. iv. 60 The liuetenant..with his dagger, sent him to accompanye his master in the other world. 1679 [see Guarantee *sb.* 1]. 1762 Sterne *Tr. Shandy* V. xlii. 141 Baldus..entered upon the law so late in life, that every body imagined he intended to be an advocate in the other world. 1804 M. Wilmot *Let.* 29 June in *Russ. Jrnls.* (1934) i. 107 What think you..the place is like?..to let you at once into the secrets of Other Worlds, know that Kattova is very like a wooden village. 1880 G. M. Hopkins *Let.* 22 Dec. in Hopkins & Dixon *Corr.* (1935) 37 The other-world of imagination. 1887 G. B. Shaw *Short Stories, Scraps & Shavings* in *Works* (1932) VI. 101 With gho—with people from—with ladies and gentlemen from the other world. 1920 D. H. Lawrence *Women in Love* iv. 46 The whole otherworld, wet and remote, he had to himself. 1953 A. Huxley *Let.* 21 June (1969) 678 His [*sc.* the schizophrenic's] commonest experiences are of an Other World, not heavenly but infernal and purgatorial. 1960 S. Plath *Colossus* 39 These..sheets..Speak in sign language of a lost otherworld, A world we lose by merely waking up. 1968 T. Wolfe *Electric Kool-Aid Acid Test* v. 60 The whole *other world* that LSD opened your mind to. 1975 *Times Lit. Suppl.* 25 Apr. 445/3 Mr McCarthy sets his saga of Servier County in the strange otherworld of a Tennessee winter.

2. (Further examples.)

1892 J. S. Stuart-Glennie in *Proc. Internat. Folk-Lore Congr. 1891* 225 Myths which..I would name..the Sacerdotal. By these I mean especially all the Otherworld Myths. 1917 [see *Immram]. 1957 G. Ashe *King Arthur's Avalon* i. 28 A Lake Villagers' burial ground on Ynyswitrin, with a resulting Ghosts' High Noon presided over by other-world deities.

otherworldly, *a.* Add: **1.** (Further examples.)

1955 A. Huxley *Let.* 10 Jan. (1969) 720, I took mescalin yesterday, for the second time... The experience had a human content, which the earlier, solitary experience, with its Other Worldly quality..did not possess. 1957 G. Ashe *King Arthur's Avalon* iii. 107 The poem entangles Arthur in a network of other-worldly themes, a network which takes in Glastonbury and the quest for a miraculous vessel. 1972 *Where* Oct. 275/3 Other..readers..may find the results almost as other-worldly as might a non-literate home in the East End.

3. Also, of or pertaining to the world to come. Also *absol.* as *sb.*

1920 E. I. Watkin in C. Hess *God & Supernatural* 141 Nor have these souls merely desired sufferings as the unavoidable price of an other-worldly reward. 1932 C. P. Curran in F. J. Sheed *Irish Way* 134 The other-worldly man must put on the man of affairs, the monk-bishop become a politician. 1950 'G. Orwell' *Shooting Elephant* 106 The other-worldly, anti-humanist tendency of his [*sc.* Gandhi's] doctrines. 1961 *New Eng. Bible Luke* xvi. 8 For the worldly are more astute than the other-worldly in dealing with their own kind.

other-worldness (ɒ·ðəɹwɒ·ɹldnės). [f. Other world *sb.* and *a.* + -ness.] = Otherworldliness.

1915 J. London *Jacket* i. 1 These child glimpses are of other-worldness, of other-lifeness, of things that you had never seen in this particular world of your particular life. 1956 I. Bromige *Enchanted Garden* ii. ii. 87 It was an expressive face. In repose, it had an appealing wistfulness, an other-worldness. 1961 P. Dougherty *Mother Mary Potter* iii. 35 In the Portsea–Southsea district..Mary's persevering and unobtrusive other-worldness was inevitably discussed.

otitis. Add: **b.** In mod.L. collocations: *otitis externa* [tr. F. *otite externe* (J. M. G. Itard *Traité des Maladies de l'Oreille* (1821) I. i. 164)], inflammation of the external ear; *otitis interna* [tr. F. *otite interne* (loc. cit. 170)], inflammation of the inner ear († or the middle ear); = *Labyrinthitis; *otitis media*, inflammation of the middle ear.

1864 D. B. St. J. Roosa tr. *von Tröltsch's Dis. Ear* vi. 60 By Otitis interna I understand the purulent catarrh of the middle ear, or cavity of the tympanum. *Ibid.*, Otitis externa, or diffuse inflammation. 1874 J. Hinton tr. *von Tröltsch's Surg. Dis. Ear* viii. 56 (*heading*) Acute suppurative catarrh of the middle ear, or acute otitis media. 1883 J. P. Cassells tr. *Politzer's Text-bk. Dis. Ear* 711 (*heading*) Inflammation of the labyrinth (otitis interna). 1959 *Woman* 16 May 31/1 She has all the signs of a really bad acute otitis media—or infection of the middle ear. 1974 Passmore & Robson *Compan. Med. Stud.* III. ii. xxxii. 5/1 If there is no time to soften the wax in this way,

a proprietary wax solvent may be used, but these are more likely to cause otitis externa than the simpler substances. 1976 M. Machlin *Pipeline* iii. 36 Diseases that could be easily treated, such as chronic *otitis media*,..were let to run their courses for lack of doctors.

oto-. Add: ototo·xic *a.*, having a toxic effect on the ear or its nerve supply; so o:totoxi·city, the property of being ototoxic.

1951 *Trans. 10th Conf. Chemotherapy of Tuberculosis* 224 The growing list of tuberculostatic antibiotics which have an ototoxic action..suggests that still other substances derived from Streptomyces may be expected to show a similar toxicity. 1967 Busch & Lane *Chemotherapy* v. 86/2 The ototoxic effect is exerted primarily on the cochlear division of the 8th nerve. 1974 *Arch. Path.* XCVI. 304/1 Methyl mercury is a unique ototoxic agent. 1951 *Trans. 10th Conf. Chemotherapy of Tuberculosis* 224 (*heading*) Ototoxicity of hydroxystreptomycin. 1975 *Nature* 3 Jan. 45/1 Kanamycin ototoxicity yields such cochlear lesions that histological and audiometric measures are well correlated.

otolaryngology (o:tolæriŋgǫ·lŏdʒi). *Med.* [f. Oto- + *laryngology* s.v. Laryngo-.] The branch of medicine concerned with the ear and the larynx; also often used to include the nose (avoiding the cumbersome *otorhinolaryngology).

1897 *Jrnl. Laryngol., Rhinol., & Otol.* XII. 554 (*heading*) Holocaine [*printed* holoraine] in oto-laryngology. 1945 *Electronic Engin.* XVII. 555 A book which embodies the experience of twenty five years of research and teaching..in otolaryngology. 1955 *Sci. News Let.* 13 Aug. 109/1 The cat's hearing span was measured by Drs. William D. Neff and Joseph E. Hind of the University of Chicago's Laboratories of Physiological Psychology and Otolaryngology. 1973 Paparella & Shumrick *Otolaryngology* I. p. ix/1 We, the editors, felt both honored and challenged when the Saunders Company invited us to prepare a new and definitive reference source in otolaryngology. The new book was to replace the classic Jackson work, *Diseases of the Nose, Throat and Ear*. 1975 Felton & Fowler *Best, Worst & Most Unusual* 250 Dr. Eugene M. Batza of Cleveland Clinic's Department of Otolaryngology, examined a five-member rock combo and found that all of them suffered from..laryngitis.

Hence o:tolaryngolo·gic (chiefly *U.S.*), -lo·gical *adjs.*, of or pertaining to otolaryngology, or to the ear, (nose) and throat; o:tolaryngo·logist, one who specializes in otolaryngology.

1898 (*title*) Transactions of the Ophthalmologic Division of the Western Ophthalmological and Oto-Laryngological Association. 1898 *Laryngoscope* May 311 One of the most practical factors of the organization was the division into two sections, one the ophthalmologic, the other the oto-laryngologic. 1911 *Jrnl. Laryngol., Rhinol., & Otol.* XXVI. 661 The moral responsibility..rests upon oto-laryngologists as teachers. 1961 *Lancet* 30 Sept. 782/1 A meeting of the North of England Otolaryngological Society is to be held on Saturday, Oct. 7. 1964 D. A. Dolowitz *Basic Otolaryngol.* p. vii, These gains have lengthened the otolaryngologic resident training period to four years after internship. 1973 *Sci. Amer.* Sept. 92/1 Gynecologists, urologists, ophthalmologists, otolaryngologists and others each operate on their respective organs and tissues. 1974 *Encycl. Brit. Macropædia* XVII. 822/2 Otolaryngologic surgery is performed in the area of the ear, nose, and throat. *Ibid.*, Benign and malignant tumours of the upper air passages..are currently dealt with by the otolaryngological surgeons.

Otomi (ōutōmi·). Pl. Otomi, Otomies. [Sp., f. Nahuatl *Otomí*.] An Indian people inhabiting parts of central Mexico; a member of this tribe. Also, the language of the Otomi. Also *attrib.* or as *adj.*

1787 C. Cullen tr. *Clavigero's Hist. Mexico* I. ii. 105 The Mazahuas were once a part of the nation of the Otomies, as the languages of both nations are but different dialects of the same tongue. 1845 [see *Huastec]. 1877 L. H. Morgan *Anc. Society* ii. vii. 194 The confederacy was confronted by hostile..tribes..the Mechoacans on the west, the Otomies on the northwest. 1883 *Encycl. Brit.* XVI. 207/1 The languages of the Nahua nations..show no connexion of origin with the language of the Otomi tribes. 1891 D. G. Brinton *Amer. Race* 338 The Otomi presents so many sounds unfamiliar to the European ear that the attempt to represent it by our alphabets can be only remotely accurate. 1948 D. Diringer *Alphabet* vii. 124 They [*sc.* Chichimeca]..are considered by some experts as of Otomi origin. 1954 *Bible Translator* V. 61 Otomi people suffer from considerable cultural insecurity. 1964 *Language* XL. 81 A great variety of person-tense-aspect prefixes are available to the speaker of Otomi. 1972 *Ibid.* XLVIII. 847, 6b is given by Hockett for..Otomi nasal vowels. 1972 [see *Mixtec].

otorhinolaryngology (o:torəi:nolæriŋgǫ·lŏdʒi). *Med.* [f. Oto- + Rhino- + *laryngology* s.v. Laryngo-.] The branch of medicine concerned with the ear, nose, and throat. Cf. *otolaryngology, E.N.T. s.v. *E III.

1900 Dorland *Med. Dict.* 468/1 Otorhinolaryngology. 1902 *Nature* 2 Oct. 554/1 The congress will be divided into the following sections:—..oto-rhino-laryngology, [etc.]. 1962 *Lancet* 27 Jan. 195/2 No special techniques need to be learnt, as in neuro-surgery, urology, or otorhinolaryngology. 1968 W. McCartney *Olfaction & Odours* 235 Author of 'Essai d'Olfactique Physiologique' (1919) and of many papers on oto-rhino-laryngology.

Hence o:torhi:nolary:ngolo·gical *a.*; o:torhi:nolaryngo·logist, one who specializes in otorhinolaryngology.

1938 W. H. C. Romanis in Rolleston & Moncrieff *Mod. Anæsthetic Pract.* x. 185 Local anæsthesia..its use in ophthalmic or oto-rhinolaryngological operations. 1948 *Brit. Dental Jrnl.* LXXXV. 223/1 He had..no history of oto-rhino-laryngological trouble. 1960 *Times* 3 Mar. 5/2 Then there is 'otorhinolaryngologist'. No one would wish to charge large fees if he called himself an ear, nose and throat specialist. 1962 *Listener* 10 May 809/1 Oto-rhino-laryngologists, botanists or X-ray crystallographers.

otosclerosis (ōutosklĕrōu·sis, -sklĭərōu·sis). *Path.* [mod.L., ad. G. *otosclerose* (A. Politzer *Lehrb. der Ohrenheilkunde* (ed. 4, 1901) 263): see Oto- and Sclerosis.] A disease of the ear in which the normal tissue of the temporal bone is replaced by spongy bone, with the result that movement of the stapes becomes impeded and deafness ensues. Hence otosclero·tic *a.*, of, pertaining to, or affected by otosclerosis; also as *sb.*, one who suffers from otosclerosis.

1901 *Arch. Otology* XXX. 279 Otosclerosis, so frequent, distressing, and little amenable to treatment, receives 12 pages. 1904 *Jrnl. Laryngol., Rhinol., & Otol.* XIX. 518 It was especially difficult to differentiate between true otosclerosis and fixation of the stapes following tympanic inflammations. 1933 M. Yearsley *Otosclerosis* iii. 7 Körner..recorded the marriage of two otosclerotics. *Ibid.*, The marriage of an otosclerotic male with his otosclerotic niece resulted in seven otosclerotic children. 1933 *Punch* 20 Dec. 700/1 He..exhibits Swift with much pathos as the sufferer from the particularly cruel disease which now bears the ugly name of otosclerosis. 1974 I. Friedmann *Path. Ear* v. 247 Otosclerosis is very common in Indians and whites. *Ibid.*, Every tenth adult person has otosclerotic foci within his temporal bone.

‖**otriad** (ǫtrya·d). [a. Russ. *otryád* a detachment.] In Russia: a detachment, group of soldiers (see also quot. 1916).

1916 *Yorkshire Post* 23 Feb. 4/4 An Englishman who works with a volunteer ambulance or otriad, behind the Russian lines. 1919 H. S. Walpole *Secret City* i. xvii. 117 Zinaida Fyodorovna had just come back from her Otriad on the Galician front. 1933 —— *Vanessa* iv. i. 672 The Retreat had begun and with the rest of the Otriad he had been flung into the little town of O——.

Otshi-herero (ōu:tʃiherē·ro). Also Otjiherero, Otyi-. The name used by the Hereros for their language, generally called *Herero.

1859 H. Hall *Man. S. Afr. Geogr.* xviii. 73 Damara and Orampoland, or the land of the Otjiherero, present a belt of sandy country between the coast and the high inhabited table lands. 1871 J. Mackenzie *Ten Yrs. North of Orange River* 494 Otyiherero (Damara). 1880 *Encycl. Brit.* XI. 732/2 By their language (Otyiherero) the Herero belong to the great Bantu family. 1910 *Ibid.* XIII. 358/1 They call themselves Ovaherero and their language Otshi-herero. 1919 H. H. Johnston *Compar. Study Bantu & Semi-Bantu Lang.* I. 350 Oci-herero.

Ottawa (ǫ·tăwă). [a. Canad. F. *Outaouais*, f. Ojibwa tribal name *otāwā*.] A North American Indian people of the Algonquian family first encountered on the shores of Lake Huron; a member of this tribe. Also *attrib.* or as *adj.*

1687 in *Documents Colonial Hist. New-York* (1853) III. 442 A party of Sinnekes and Onnondages have plundered some French..and have also taken some Ottawa Indians prisoners. 1754 [see *Chippewa]. 1768 J. Lees *Jrnl.* Aug. (1911) 39 On the Miami River, the greatest part of the nation of the Ottawas inhabit, they..are spread about it..chiefly on the North side of Lake Huron..and towards the North West. These, the Pous and Chipewas have almost the same language, and is [*sic*] called the Ottawa-Tongue. 1833 A. Jackson in *Messages & Papers of Presidents* (1896) III. 38, I transmit..a treaty concluded between the commissioners on the part of the United States and the united nation of Chippewas, Ottawas, and Potawatamies, at Chicago. 1835 [see *Kickapoo]. 1865 [see *Algonquin, -kin *sb.* and *a*]. 1890 J. G. Frazer *Golden Bough* II. iii. 113 When men of the Bear clan in the Otawa tribe killed a bear, they made him a feast of his own flesh. 1910 F. W. Hodge *Handbk. Amer. Indians* II. 171/1 There were 197 Ottawa under the Seneca School, Okla. 1962 D. H. Hymes in J. A. Fishman *Readings Sociol. of Lang.* (1968) 129 The Ottawa believed the cries of infants to be meaningful. 1967 D. Jenness *Indians of Canada* (ed. 7) xviii. 282 The Iroquois turned their arms against the Ottawa and drove them from Georgian bay. Some fled west towards lake Superior; others took refuge with their Potawatomi kinsmen in the United States... Many of these refugees returned to..the north shore of lake Huron. 1977 *Detroit Free Press* 11 Dec. 16–C/1 'It's a hidden attack by commercial interests on native Americans,' said Mrs. Waunetta Dominic, chairwoman of the Northern Michigan Ottawa Association in Petoskey.

otter, *sb.* Add: **4. c.** A type of paravane, used esp. by merchant vessels. Later, any paravane.

[1910 *Blackw. Mag.* June 899/1 We might adapt to naval use those poaching expedients, the 'cross-line' and the 'otter'.] 1920 *Nature* 8 Jan. 487/1 The paravane or otter..proved a very effective weapon against both mines and submarines. 1920 *Rep. Brit. Assoc. Adv. Sci. 1919* 273 The Protector Paravanes, or Otters, carry a form of cutter, but no explosive charge whatever. 1954 Bradford & Quill *Gloss. Sea Terms* 138/2 The mooring of a mine coming in contact with this taut tow-line slides along to the otter where it enters a pair of cutting jaws and is cut adrift.

7. *otter-skin* (later examples); **otter-board,** (*a*) a fishing-tackle consisting of a board with several hooks attached; (*b*) = *door 4 b; **otter-man,** a fisher who uses an otter-line or otter-board; †**otter-sheep** *U.S.*, a variety of sheep with short, crooked legs; cf. Otter *sb.* 6; **otter tail** (see quot. 1932); **otter-trawl** (examples).

1901 *Field* 5 Jan. 19/2 The otter-board was only employed..upon those lakes where the trout were indifferent to the angler's flies. 1904 *Daily Chron.* 21 Nov. 5/7 He lost his otter-board and had to put a new trawl on next morning. 1936 J. Buchan *Island of Sheep* xiii. 247 Look at the trawl. It's absurd. It has no otter-boards... There's something wrong with this ship. 1971 *Daily Tel.* (Colour Suppl.) 21 May 21/3 The net flops into the water and sinks away sideways. When it is some way out it is followed by the shuddering crash of the two otter-boards, ton-weight doors, that are towed at an angle to keep the net open on the bottom. 1901 *Field* 5 Jan. 19/2 The otterman must chuckle inwardly when he sees a perspiring and jaded angler..with one or two fish in his basket. 1809 E. A. Kendall *Trav. Northern Parts U.S.* I. 309 Some of the farmers [in Connecticut] are partial to a remarkable variety of sheep, which they call the otter-sheep. 1863 H. S. Randall *Pract. Shepherd* (ed. 7) v. 42 A family of them, the Otter Sheep—so termed from their short, crooked, rickety legs. 1873 *Amer. Naturalist* VII. 742 The otter sheep..originated on the farm of Seth Wright, near Charles River, Mass. 1884 *Century Mag.* Feb. 516/1 There were also the Otter sheep, said to have originated on some island on our eastern coast. 1849 F. Parkman *Calif. & Oregon Trail* x. 144 The dandy carried a bow and arrows in an otter-skin quiver at his back. 1971 *Country Life* 28 Oct. 1128/3 Red chokers about sun-burned necks, and more than one round hat of otter-skin. 1932 Lady Howe in A. C. Smith et al. *Hounds & Dogs* viii. 69 The tail..should be very thick towards the base gradually tapering towards the tip, of medium length, should be practically free from any feathering, but should be clothed thickly all round with the Labrador's short, thick, dense coat, thus giving that peculiar rounded appearance which has been described as the 'otter' tail. 1948 B. Vesey-Fitzgerald *Bk. Dog* 1000 Otter tail: A thick tapering tail similar to that of an otter, much desired in the Labrador Retriever. 1973 P. R. A. Moxon *Gundogs* (ed. 9) iv. 71 The Labrador Retriever..is undoubtedly ideal for water, the sleek water-resisting coat and 'otter' tail seeming almost designed for the job. 1899 W. C. McIntosh *Resources of Sea* 93 The new otter-trawls capture more round than flat fishes. 1936 Russell & Yonge *Seas* (ed. 2) 275 In the case of the otter-trawl two 'otters' or 'doors' are used. To these the sides of the net's mouth are attached, and they are set at such angles that as they are drawn over the sea bottom they diverge farther and farther from the centre of the net's mouth until an equilibrium point is reached and the mouth of the net is stretched agape. 1973 W. Elmer *Terminol. Fishing* ii. 71 In many places the otter trawl has ousted the beam trawl. 1973 *Fisheries Fact Sheet* (Environment Canada Fisheries & Marine Service) No. 1. 2/3 The former two [*sc.* the trawler and dragger]..catch fish by dragging an otter-trawl or similar device. This is a large baglike arrangement of nets which captures fish as the vessel tows it through the water.

o·ttered, *ppl. a.* [f. Otter *v.* + -ed¹.] That has been fished with otter tackle. So **o·tterer; o·ttering** *vbl. sb.*

1901 *Field* 5 Jan. 19/2 An observant gamekeeper..tells me that he knows well the signs of an ottered lake. *Ibid.* 19/3 Very gradually, may be, the otterers will learn that they are ruining many fine waters by their malpractices. 1907 *Westm. Gaz.* 24 Jan. 2/1 Trout..obtained by the unsportsmanlike method of netting or 'ottering'.

otto¹. a. (Later examples.)

1908 *Westm. Gaz.* 30 Mar. 10/3 As a scent otto of violets has become increasingly popular each year. 1919 S. Kaye-Smith *Tamarisk Town* i. ii. 48 There was a drift of faint perfumes: flowers, macassar oil, otto of roses, lavender and peau d'espagne. 1939–40 *Army & Navy Stores Catal.* 431/1 Cold cream, 'Otto of Rose' jar, 1/3.

Otto³ (ǫ·to). The name of Nikolaus August *Otto* (1832–1891), German engineer, used *attrib.* to designate (*a*) the four-stroke cycle employed in most petrol and gas engines (cf. *four-stroke *a.*), idealized as adiabatic compression followed by heat addition at constant volume, adiabatic expansion, and heat rejection at constant volume; and (*b*) an engine employing this cycle.

The cycle was orig. proposed by A. Beau de Rochas in 1862, but Otto was the first to build an engine employing it (in 1876) after conceiving the idea independently.

1878 *Sci. Amer.* 30 Mar. 195/1 The new Otto horizontal gas engine..closely resembles the ordinary horizontal steam engine. 1885 W. Macgregor *Gas Engines* ii. 81 The original and classic type of the Otto engine has received improvements at the hands of both its German and English manufacturers. 1886 D. Clerk *Gas Engine* vii. 183 The indicator diagrams prove the very efficient nature of the Otto cycle. 1930 *Engineering* 7 Feb. 186/1 Theoretically, the Otto cycle..promised higher efficiencies than the constant-pressure Diesel cycle. 1966 *McGraw-Hill Encycl. Sci. & Technol.* VII. 201/1 For an Otto engine, an increase in either the air temperature or density increases the tendency of the engine to knock. 1975 *Sci. Amer.* Jan. 34/3 According to the DOT-EPA report, the

40 percent improvement by 1980 should be attainable with the present Otto-cycle (four-stroke) gasoline engine, in combination with improved transmissions, reduced weight and aerodynamic drag and improved accessories.

† **b.** Used *ellipt.* for *Otto engine*. *Obs.*

1886 D. Clerk *Gas Engine* vi. 106 The Otto is only half single acting. 1903 *Work* XXV. 18/2 Petrol car engines are of the vertical single-acting Otto type, any variations consisting chiefly of horizontal Ottos.

ottoman, *sb.*² **2.** (Later examples.)

1922 Barker & Midgley *Analysis Woven Fabrics* (ed. 2) xv. 301 *Ottoman cloth*, a dress fabric of a warp-rib structure, usually made from hard, crisp yarns. 1948 G. L. Fraser *Textiles by Britain* Gloss., *Ottoman*, dress material in either silk or rayon with a broken rib face and plain-weave back. It is produced in various sizes of rib. 1951 A. T. C. Robinson *Rayon Fabric Construction* viii. 79 A high-class Ottoman suitable for dressy coats. 1968 Mrs. L. B. Johnson *White House Diary* 9 Apr. (1970) 659, I had worn my yellow ottoman cotton with an easy skirt. 1972 M. L. Joseph *Introductory Textile Sci.* (ed. 2) xxii. 239 Many rib-weave fabrics have heavy yarns inserted as picks. Examples of this construction include poplin, faille, bengaline, and ottoman.

Ottomanism (ǫ·tŏmăniz'm). [f. Ottoman *a.* and *sb.*¹ + -ism.] The culture (or aspects of it) of the Ottoman Turks; Ottoman civilization.

1911 *Q. Rev.* July 261 On behalf of an Ottomanism honestly applied to all the Ottoman nationalities. 1930 *Times Lit. Suppl.* 7 Aug. 635/1 Her leaders had been experimenting with Ottomanism. 1930 [see *levantinize *v.*]. 1974 C. E. Dawn (*title*) From Ottomanism to Arabism.

ottomanize, *v.* (in Dict. s.v. Ottoman *a.* and *sb.*¹). (Later example.) So **O:ttomaniza·tion,** the fact or process of enforcing Ottoman ideals or Ottomanism. **O·ttomanizing** *ppl. a.*

1912 *Chambers's Jrnl.* Dec. 817/2 If the Young Turks..had tried fraternisation instead of persisting in Ottomanisation, Turkey's credit would have risen immediately. 1920 *Glasgow Herald* 17 Mar. 10 The inhabitants were anticipating enforced Ottomanisation when the Italian fleet arrived at Astypalea. *Ibid.* 15 July 4 His liberalism earned him the utter hatred of the ottomanising Committee of Union and Progress. 1936 R. C. K. Ensor *England, 1870–1914* xiii. 436 Turkey..had alienated the liberal Powers by reverting to policies of Ottomanization and massacre. *Ibid.* 463 They were united by the Ottomanizing policy of the Young Turks. 1972 D. Dakin *Unification of Greece* xii. 178 The new policy was to ottomanise the empire, to abolish the nationalist organisations, and to disarm the warring factions.

Ottonian (ǫtōu·niăn), *a.* [ad. late L. *Ottoniānus*, f. the name of Otto I (cf. Neronian *a.*): see -ian.] **a.** Of or pertaining to the East Frankish dynasty of the Holy Roman Empire founded by Otto I (912–73), which ruled from 962 to 1002. Also (*rare*) as *sb.*, a member of this dynasty. **b.** Pertaining to or characteristic of the art of this period, in part a revival of Carolingian art and extending into the 11th century.

1898 H. Fisher *Medieval Empire* I. iii. 95 The dominion of the Ottonian house was one thing in Saxony and another thing outside the Saxon borders. 1928 E. F. Jacob *Holy Roman Empire* v. 73 Yet the theory of the Empire reached its finest and truest form when the days of the Hohenstaufen were over and it was obvious that the unity of the Ottonian Reich could never be restored. 1936 A. W. Clapham *Romanesque Archit.* viii. 179 The splendours of the Ottonian revival in painting, miniature, ivory, and metal work. 1938 *Times Lit. Suppl.* 12 Nov. 732/2 A group of over seventy Carolingian and Ottonian bindings. 1939 *Archit. Rev.* LXXXVI. 130/3 It is no service to German art to maintain that it is best when purest; its real triumphs are those happy marriages of Teutonic exuberance and Latin elegance which bear their first fruit in Ottonian miniatures. 1943 *Burlington Mag.* Sept. 228/2 The style of the work is 'Ottonian', an early phase of Romanesque. 1951 A. R. Lewis *Naval Power & Trade Mediterranean, 500–1100* vi. 223 The increase of Byzantine gold coins in Germany, the luxury of the Ottonian court, the influence of Byzantine art motifs in German Romanesque architecture. 1958 *Times Lit. Suppl.* 3 Jan. 4/4 Paintings of the Rheims school in Carolingian times are shown to herald the transcendental outlook of Ottonian miniatures. 1967 *Cambr. Hist. Later Greek & Early Medieval Philos.* 587 After 950 the Ottonian dynasty were capable of re-establishing monarchical power. *Ibid.* 590 Bishop Adalbold of Utrecht..was a man of many-sided activities in his diocese, his territory and at court, representing a type not infrequent under the Ottonians. 1968 *Eng. Hist. Rev.* LXXXIII. 24 It is impossible even to suggest a figure for the size of Saxon contingents during the decades when the Ottonians acquired their *imperium*. 1970 *Oxf. Compan. Art* 799/2 The Byzantine strand, always present in Ottonian art, was particularly strong in the School of Regensburg. 1976 *Times Lit. Suppl.* 19 Nov. 1463/3 Ottonian art (as this art of the Saxon and Salian dynasties is generally called).

Otyiherero, var. *Otshi-herero.

ou (o·ū). [Hawaiian.] A green and yellow bird, *Psittirostra psittacea*, belonging to the subfamily Coerebinæ, or honeycreepers.

1887 L. Stejneger in *Proc. U.S. Nat. Mus.* X. 93 The Ou, feeds on bugs and sings on the wing. 1890 *Ibis* II. 194 The constant twittering the Ou almost invariably makes while feeding at once betrays its identity. 1903 [see *ieie]. 1944 G. C. Munro *Birds of Hawaii* 123 With its yellow head and bright green body in varying shades on different parts the ou is a beautiful bird. 1970 S. Carlquist *Hawaii* xi. 191 The ou eats fruits, seeds, leaves, and caterpillars.

ouad, var. *oued.

ouananiche. For '*Salmo salar* var.' substitute '*Salmo saler ouananiche*'. (Add earlier and later examples.)

1873 *Forest & Stream* 4 Sept. 53/2 When you have had a surfeit of fresh water salmon *ouinanish*—retrace your steps to Chicoutimi. 1907 J. G. Millais *Newfoundland* i. 2 Newfoundland is a most attractive place, with its thousands of lakes and pools; picturesque streams teeming with salmon, trout, and ouananiche. 1966 *Globe & Mail* (Toronto) 15 Jan. 29/5, I..tied into my first ouananiche with him on Trout Lake near North Bay. 1969 H. Horwood *Newfoundland* xvi. 128 All [*sc.* North Harbour, Rocky, Colinet, and Back Rivers] have exceptional fishing for sea trout, brook trout, or ouananiche.

oubaas (ōu·bās). *S. Afr.* Also **ou baas, oud baas.** [Afrikaans, f. *ou(d)* old + *baas.] Elderly head of a family; elderly man, old gentleman. Also used as a form of address and prefixed to a surname or a Christian name.

1869 T. Baines *Diary* 2 July (1946) I. 59 They recognised at once and were rather pleased with the likeness of 'Oud Baas', Mr. Hartley. 1914 L. H. Brinkman *Breath of Karroo* ii. 24 The master of the house is addressed as 'Ou baas' and a young man, 'Klein baas'. 1942 'B. Knight' *Sun climbs Slowly* iv. xxx. 286 No, it was not the *Oubaas*, she said, but someone she had not seen before. 1946 V. Pohl *Land of Distant Horizons* 50 They were so incredulous that the Hottentot exclaimed indignantly: 'Oubaas, if I have never told the truth, I am telling it now'. 1947 *Forum* (Johannesburg) 19 Apr. 1/1 The problems which faced the Oubaas this week ranged from the spreading boycott of Indian traders in the Transvaal..to the need for separate trade unions for Africans. 1952 E. H. Burrows *Overberg Outspan* vi. 161 These were the people who congregated on Sundays with the family to hear the *oubaas* read a chapter of the Bible. 1959 *Cape Times* 14 Feb. 3/4 Joe told him that he and Moyisi had hit the *oubaas* with kieries. *Ibid.* 5 June 13/5, I..recognized him as *oubaas* De Greef. 1973 *Deb. Senate S. Afr.* 17 May 2807 Old Koos..said: 'Oubaas,..I don't know why you should start so late.'

oublietje, var. *oblietjie.

ouch, *int.*¹ Add: (Earlier and later examples.) Also as *sb.*²

1838 J. C. Neal *Charcoal Sk.* 38 'Ouch!' shrieked Dabbs; 'my eye, how it hurts!' 1843 'R. Carlton' *New Purchase* I. ii. 9 The tiers becoming all vocal with 'bless my soul's'—'my goodnesses!'—and vulgar 'ouches!' 1918 Galsworthy *Five Tales* 235 Freda gurgled: 'Ouch! You *are* a beast!' 1958 R. Godden *Greengage Summer* v. 56 'Ouch!' said Joss and looked as if she would be sick again. 1972 D. Delman *Sudden Death* (1973) iv. 110 'Ouch,' she said, grinning. 'Ouchie-wowchie. Well, that tears that, doesn't it?'

Ouchterlony (ǫχ-, ǫktərlōu·ni). *Immunol.* The name of O. T. G. *Ouchterlony* (b. 1914), Swedish microbiologist and immunologist, used *attrib.* and in the possessive with reference to a standard precipitin test devised by him, which normally involves placing antigen and antibody in separate wells sunk into a layer of agar on a plate and observing the line of precipitation which develops as the substances meet after diffusing through the agar.

1952 *Methods Med. Res.* V. 368 Ouchterlony's technique deviates from the ideal particularly with respect to points (1) and (2). 1954 *Jrnl. Immunol.* LXXIII. 232/1 We undertook an experimental evaluation of Ouchterlony's test. 1955 *Ibid.* LXXV. 460 (*heading*) Interpretation of the Ouchterlony precipitin test. 1962 Luntz & Wright in A. Pirie *Lens Metabolism Rel. Cataract* 319 Serum from each patient was put on Ouchterlony plates and tested against..human lens protein. 1975 *Nature* 21 Aug. 656/1 This antiserum gave a single line on Ouchterlony gel diffusion with MAF.

‖**oud** (ūd). Also **ood, oude, ʿūd.** [ad. Arab. *ʿūd*, lit. 'wood'.] A form of lute or mandolin played principally in Arab countries.

1738 T. Shaw *Trav. Barbary & Levant* 270 They [*sc.* the Moors] have the *Rebebb*, a Violin of two Strings, which is played upon with a Bow: the A-Oude, a Bass double stringed Lute, bigger than our Viol, which is touched with a Plectrum. 1836 E. W. Lane *Acct. Manners & Customs Mod. Egyptians* II. v. 69 The *'ood* is a species of guitar, which is played with a plectrum. 1870 C. Engel *Descr. Catal. Musical Instruments S. Kensington Museum* 8 The *Oud*..was brought by the Moors to Spain, where it is still known as the *laud*. *Ibid.*, The Oud is generally provided with frets made of cords of gut. 1883 *Encycl. Brit.* XV. 70/1 The modern Egyptian ''ūd' is the direct descendant of the Arabic lute. 1931 H. G. Farmer *Stud. Oriental Mus. Instruments* (Ser. 1) viii. 92 That the *mizhar* and the *ʿūd* were distinct types of lute we know from several authorities. 1950 P. Bowles in *Penguin New Writing* XXXIX. 10 As he passed over into the unlighted district he heard a few languid notes being strummed on an oud. 1957 H. G. Farmer in *New Oxf. Hist. Music* I. 446 The greatest of all the instruments of Islamic peoples was the *ʿūd* or lute. 1960 *New Yorker* 16 July 86 A couple of Near Eastern love songs performed..on the oud, a large Egyptian stringed instrument that resembles a gourd sliced in half and that emits an urgent nasal, tinging sound. 1972 J. Wambaugh *Blue Knight* (1973) xiv. 252 He plucked and stroked those *oud* strings with the quill of an eagle feather. It's a lute-like instrument and has no frets like a guitar. 1976 D. Munrow *Instruments Middle Ages & Renaissance* 25/2 The earliest tuning employed [for the lute] is thought to have been a series of fourths, adopted from the *'ūd* and still used by Arab players today.

Ouds (ɑudz). = O.U.D.S. (*O. 5 d).

1914 C. Mackenzie *Sinister St.* II. iv. 567 Are you running the Ouds as well as The Oxford Looking-Glass? 1967 E. Coxhead *Thankless Muse* vii. 135 Presently the Ouds producer appropriated her.

‖**oud-stryder** (ōu·tstrǝi:dər). *S. Afr.* [Afrikaans, = ex-soldier.] A veteran of the South African War (1899–1902) who fought on the side of the Boer republics. Also *gen.*, an ex-soldier.

1947 *Cape Argus* 29 Mar. 1 More than 1,600 Oudstryders to-day waited to give the Royal Family their own special welcome. 1947 *Cape Times* 31 Mar. 6/5 One of the happiest features of the visit of the Royal Family to Pretoria was the sincere welcome they received from the Oudstryders. 1948 *Ibid.* 21 Sept. 4/3 Calculating the means limit of *oudstryders*. 1954 *Sunday Times* (Johannesburg) 26 Sept. 1 Many oudstryders and ex-servicemen of both World Wars will attend. 1957 *Cape Argus* 12 Apr. 6/8 Legislation..had provided for oudstryders' pensions. 1975 *Eastern Province Herald* (S. Afr.) 13 Oct. 11 The uncompromising Boer oudstryder..told the Queen that he could never forgive the British for fighting against the Boers. The Queen was all sympathy; as a Scot, she said, she understood his feelings perfectly.

‖**oued** (wed). Also **ouad.** [Fr. rendering of Arabic *wādī* Wadi, wady.] = Wadi, wady.

Usu. used only in contexts relating to those territories in N. Africa which were formerly under French control.

1854 J. R. Morell *Algeria* vii. 121 The Ouad-Foddah, or river of silver, has its rise in a high rugged mountain. 1874 *Q. Jrnl. Geol. Soc.* XXX. 117 Lignite occurs..at El Kheicha on the banks of the Oued M'Zi, a branch of the L'Aghouat river. 1883 *Encycl. Brit.* XV. 608/1 Mascara, a fortified town of Algeria..occupies two small hills separated by the Oued Toudman. 1920 *Glasgow Herald* 27 Nov. 4 For an hour we followed the course of the oued (or river). 1965 Mountjoy & Embleton *Africa* iv. 180 A large number of small oueds descend from the mountains and water this immediate piedmont zone.

ouf, ouff, *int.* Add: (Also with pronunc. uf, ŭf.) **1. a.** (Earlier and later examples.) Also, expressing a sense of alarm or annoyance.

1851 Longfellow *Jrnl.* 9 Feb. in S. Longfellow *Life H. W. Longfellow* (1886) II. 189 There was no violent discussion; so that the Count did not, so often as usual, clasp his round head with both hands and say, 'Ouf!' 1913 G. B. Shaw *Let.* 9 July in *B. Shaw & Mrs. Campbell* (1952) 130 You shut the door on me..and said 'Ouf!' when it slammed. 1951 R. Senhouse tr. *Colette's Chéri* 33 Back in Paris again—ouf!—I'll pack him off to his precious studies. 1958 L. Durrell *Balthazar* iv. 237 Ouf! I fled. 1967 A. Laski *Seven Other Years* iv. 59 'Ouf!' he said... 'This is not my dance.'

b. As an expression of resignation or relief.

1909 Mrs. H. Ward *Daphne* iii. 56 Daphne, with an 'Ouf!' of fatigue, took off her hat. 1921 G. B. Shaw *Back to Methuselah* v. 261 All. Ouf! (*A great sigh of relief.*) 1964 V. Nabokov *Defence* xiii. 220 'Ouf,' sighed Mrs. Luzhin, 'we're finally rid of them.' 1976 N. Freeling *Lake Isle* xiv. 109 'Ouf,' said Sophie, sitting down and kicking her shoes off.

2. Cf. Wuff *sb.*

ought, *sb.*² (Examples of *pl.*)

1908 A. Bennett *Human Machine* 43 You have a special apparatus within you for dealing with a universe where *oughts* are flagrantly disregarded. 1933 W. de la Mare *Lord Fish* 269 His master had told him little about his oughts. 1941 *Daily Progress* (Charlottesville, Va.) 3 July 10/7 (Advt.), Oughts for a great vacation. Bathing, boardwalk and 'loaf-life' specials. Mens & Boys Swim Trunks.

ought, *v.* Add: **IV. 8.** With periphrastic auxiliary *did*, corresponding to uses under sense 5. *dial.*, *colloq.*, and *vulgar*.

1854 C. M. Yonge *Heartsease* II. iii. ix. 236, I..told him he didn't ought to go. 1867 R. Young *Rabin Hill's Excursion to Weston-super-Mare* 12 That's jist how things did ought to be. 1876, 1888 [see next]. 1932 D. L. Sayers *Have his Carcase* xxvii. 356, I did ought to have spoke up at the time. 1942 'M. Innes' *Daffodil Affair* i. 17 And I hope that none here will say I did anything I didn't ought. For I only done my duty.

oughta, oughter (ǭ·tə), a representation of a colloq. or vulgar pronunciation of *ought to* (see Ought *v.* 5, 7c and *8). Cf. *orter.

1876 C. M. Yonge *Three Brides* II. i. 7 They ought to be ashamed of themselves, they did oughter. 1888 Rider Haggard *Col. Quaritch* III. v. 79 You are my lawful husband, and I calls on you to cease living as you didn't oughter and to take me back. 1901 M. Franklin *My Brilliant Career* iii. 16 You oughter go out more. 1926 F. M. Ford *Man could stand Up* i. ii. 34 'He hadn't ought'er done it!' He hadn't really oughter. 1935 'R. West' *Harsh Voice* ii. 151 He reckoned I oughta see London and Paris and Rome. **1943** K. Tennant *Ride on Stranger* iii. 24 She didn't oughter do it. **1945** A. Kober *Parm Me* 180 No joking, Mac, you oughta take it easy. **1959** N. Mailer *Advts. for Myself* (1961) 54 He could see the other house. It had oughta be away from the town. **1963** [see *daddy-o]. **1967** E. Grierson *Crime of one's Own* ii. 22 'I expect 'e buys 'er things.' 'So 'e oughter.' **1971** D. Heffron *Nice Fire & Some Moonpennies* i. 11 Indian huh? Well she oughta go for this then. **1974** W. Garner *Big enough Wreath* i. 8 You didn't oughter've done that, Mr Smith. You know the regs. **1976** M. Maguire *Scratchproof* iii. 45 Somebody oughta lock you away.

oughtness. Delete '*rare*' and add later examples.

1918 [see *good *a.* 14 c]. **1931** W. M. Urban *Fund. Ethics* i. 9 The forms of conduct or behavior which have this character of oughtness are then called standards or norms. **1948** A. O'Rahilly *Moral Princ.* iv. 21 On the occasion of experience we see the oughtness of certain acts, we have an intuitive appreciation of an objective moral order. **1958** R. C. Angell *Free Society & Moral Crisis* ii. 16 The peculiar power of our moral lenses is to pick out and see clearly all the elements in society that reveal 'how oughtness is organized'. **1967** D. von Hildebrand *Trojan Horse in City of God* xxv. 190 The victory of truth and value is the fulfillment of an oughtness.

Ouidaesque (wīdă,e·sk), *a.* [f. *Ouida*, the nom-de-plume of the English novelist Marie Louise de la Ramée (1839–1908) + -esque.] Characteristic or suggestive of the novels of 'Ouida'; marked by extravagance or lack of restraint.

1909 *Westm. Gaz.* 6 Apr. 4/2 It is the case of the 'Ouida-esque' young man with the big cigar of the motor shows over again. **1915** W. J. Locke *Jaffery* xxi. 287 Like the Ouidaesque hero, who could ride a Derby Winner with one hand, and stroke a University Crew to victory with the other. **1929** *Sunday Express* 20 Jan. 9 The 'dark Odyssey of Gilbert Stroud' is almost Ouidaesque. **1930** *Times Lit. Suppl.* 17 Apr. 338/3 A hero of almost Ouida-esque impressiveness. **1971** E. Mavor *Ladies of Llangollen* xii. 201 In Mr Penruddock's final Ouida-esque version, the successful elopement occurs on the night of a grand ball.

Ouija (wī·dʒa). [f. F. *oui* yes + G. *ja* yes.] A proprietary name for a board having the letters of the alphabet and other signs used for obtaining messages and answers in spiritualistic séances and in the practice of telepathy. Also (with lower-case initial) applied generally to spiritualistic spelling devices. Also *ouija-board*.

1891 *Official Gaz.* (U.S. Patent Office) 3 Feb. 510/2 Toys known as Talking Boards.—Kennard Novelty Company, Baltimore, Md... Used since July 1, 1890. The word 'Ouija'. **1895** *Montgomery Ward Catal.* 236/2 Ouija, or Egyptian Luck Board. **1895** I. M. Rittenhouse *Maud* (1939) 590 Once or twice he had referred to something a Ouija-board in Chicago had said, and how it had spelled my name in full. **1904** *Pop. Sci. Monthly* Jan. 195 The various alphabet-using forms of amateur mediumship, such as table tipping, the 'Ouija-board', and certain other devices for making our muscles leaky. **1909** H. Carrington *Physical Phenom. Spiritualism* 67 The phenomena of table-tipping, of ouija and planchette writing. **1922** O. Lodge *Raymond Revised* 45 By the use of instruments known as 'planchette' and 'ouija', often employed by beginners. **1931** *Times Lit. Suppl.* 24 Dec. 1036/4 When receiving communications Ingeborg..is all the while writing, or moving swiftly with the pointer of the ouija board. **1944** Auden *For Time Being* (1945) 114, I have prohibited the sale of crystals and ouija-boards. **1949** *Official Gaz.* (U.S. Patent Office) 20 Sept. 638/2 William Fuld, Baltimore, Md. Ouija..for Gameboard, utilizing a Planchette and sometimes known as a Talking Board. Claims use since July 1, 1890. **1968** *Trade Marks Jrnl.* 22 May 835/2 Ouija... Board games, being parlour games providing answers to questions. Parker Brothers, Inc. (a Corporation organised and existing under the laws of the State of Maine, United States of America), 190, Bridge Street, Salem, State of Massachusetts, United States of America; Manufacturers. **1968** *Daily Tel.* 10 Sept. 20/6 A television advertising campaign to promote Ouijas as games at Christmas. **1973** *Listener* 20 Sept. 386/2 Professor [John] Taylor's schedule of talks to a medium, a ouija-board operator, a scientist and a hard-line theologian. **1974** *Sci. Amer.* Jan. 108/1 Tens of thousands of young people in the U.S. (particularly in California), caught up in the current occult explosion and eager to know more about Eastern mysticism and early Chinese history, are now consulting the *I Ching* as seriously as they consult the Ouija board or the tarot cards.

ouklip (ǒu·klip). *S. Afr.* [Afrikaans, f. Afrikaans, Du. *oud* Old *a.* + *klip* (see *klip *sb.*).] A kind of lateritic conglomerate found in southern Africa.

1892 *Graham's Town (Cape Province) Jrnl.* 20 Sept. 2 A few months ago it was discovered..that large beds of Ou Klip (honeycomb gravel rock) on the farm were literally saturated with mercury. **1940** *Min. Resources Union S. Afr.* (Dept. of Mines) (ed. 3) iv. 458 There are many types of laterite or 'ouklip' in this country. The harder types of conglomerate ouklip generally form a very good basecourse for bitumen... Some types of soft conglomerate ouklip which contain an appreciable quantity of sandy soil binder generally yield satisfactory sand-clay bases... The pebble type ouklip..consists mostly of fairly hard lateritic pebbles and soil binder. **1950** *Cape Times Week-end Mag.* 8 Apr. 4/7 This was an outcrop of granite and ouklip. **1955** J. H. Wellington *S. Afr.* I. ii. x. 289 In the western highlands of Natal, Peutz recognizes a third grassveld type..in which the infertile sandy soils, underlain by 'ouklip' (i.e. pirolitic ironstone) produce a poor type of veld. **1961** M. M. Cole *S. Afr.* iv. 87 The weathering processes..result in the formation of soils comprising an A horizon of some 12 to 24 inches of friable sand.. overlying a B horizon of mottled clayey-sand containing many ferruginous concretions which in the lower part are cemented to form a hardpan called 'ouklip'.

ould (ɑuld), a representation of an Ir. pronunciation of Old *a.*

c **1675** *Purgatorium Hibernicum* (MS. 470, Nat. Libr. Ireland) 100 Singing 'Ould Rose' and 'Tory Rory'. **1803** G. Colman *John Bull* ii. ii. 17 I'm as aisy as an ould glove. **1829** G. Griffin *Collegians* I. vii. 153 O, wirra, Eily! this is the black day to your ould father. *c* **1874** D. Boucicault in M. R. Booth *Eng. Plays of 19th Cent.* (1969) II. 174 This cabin where the remains of the 'ould family', two lonely girls, live. **1898** J. D. Brayshaw *Slum Silhouettes* 8 The ould counthry. **1936** 'N. Blake' *Thou Shell of Death* iii. 46 She's a bit of an ould stick, but there's no harm in her. **1970** S. J. Perelman *Baby, it's Cold Inside* 152 From the moment I had first set foot on the Ould Sod I had yearned to pick up a typical sample of the local crafts. **1977** *Time* 12 Dec. 30/1 Overseas Chinese may not want to go back to live in the People's Republic any more than a U.S. enthusiast for Ireland wants to live on the Ould Sod.

‖ **Ouled Nail** (ū·lėd nā·il, -nail, -nēl). [Fr., f. *Ouled Naïl*, ad. Arab., lit. 'sons of Nail'.] A group of Arab peoples of Algeria; *spec.* in North African cities: an Arab professional dancing girl belonging to these peoples.

1881 A. A. Knox *New Playground* xiii. 324 Perhaps the less said the better about the dancing women, who are, if I remember right, called 'oulad nail'. They inhabit a street apart at Biskra, and come from some distant part of the desert. **1906** M. W. Hilton-Simpson *Algiers & Beyond* iv. 77 One of the most interesting features..is the *cafés maures*, in which the Ouled Nail girls dance. *Ibid.* 82 The Ouled Nails will speak to anyone... Their morality will not bear close investigation. **1914** M. D. Stott *Real Algeria* ii. 20 Khadava admitted with the most sweet of cynical smiles that it was only to English and Germans that she confessed to being an Ouled Nail—it was found to be more profitable. **1956** L. Morgan *Flute of Sand* v. 82 'The psychology of an Ouled Nail dancing girl,' a French administrator had told me, 'is very difficult to understand.' **1973** Wodehouse *Bachelors Anonymous* ix. 122 'Oh dear. Are you hurt?' she wailed, and ran to where Mr Trout was pirouetting like an Ouled Nail dancer with his hand to his mouth. **1975** 'P. Loraine' *Ask Rattlesnake* i. ii. 53 The Ouled Nails, the North African tribe which trains its women to become whores in the dissolute coastal cities.

ouma (ǒu·mā). *S. Afr.* [Afrikaans = grandmother, f. *ou* old + *ma* mother.] A name used in addressing or referring to one's grandmother or to an elderly woman.

1910 D. Fairbridge *That which hath Been* 42 A coloured person, mevrouw;—'Ou'ma Jannetje' she calls herself. **1929** P. Smith *Little Karoo* (rev. ed.) 181 Was it but three days ago..that Ou-ma had buried her son? **1937** S. Cloete *Turning Wheels* xxvii. 423 'I am sorry, Ouma,' he said, 'to find you like this.' **1952** *Cape Times* 2 Sept. 14/3 Three drunk skollies approached her and asked whether ouma had no one to take her home. **1953** J. Packer *Apes & Ivory* iii. 28 Sybella Margaretha Krige, known to her friends as 'Isie' until the Second World War when the Springbok soldiers she helped so devotedly christened her 'Ouma' (Grannie). **1971** *Cape Herald* 15 May 2/9 But how does Ouma busy herself besides sweeping?

ounce, *sb.*[1] Add: **4. b.** *ounce force*, a unit of force equal to the weight of a mass of 1 ounce, esp. under standard gravity (cf. *gramme and *gramme force).

1961 *B.S.I. News* Oct. 26/2 A similar distinction is made between..ounce (oz) and ounce-force (ozf). **1966** [see *gramme force* s.v. *gramme b].

‖**oung** (ɑuŋ), *v.* Also **aung.** [Burmese.] *trans.* In Burma of an elephant: to push, roll, or drag logs from one place to another or down a stream.

1900 M. & B. Ferrars *Burma* v. 118 (caption) Pushing the logs off the shoals (aung). **1901** G. H. Evans *Treatise on Elephants* i. 9 A well-trained tusker always commands a good price; he is so much more useful both in the yards and forests, as with his tusks he can 'oung'.., stack timber, assist in getting logs over obstacles, &c. *Ibid.* xvi. 191 It is advisable for some time after such an accident [to the loins] that the animal be loaded lightly; it is also as well not to put him to any heavy work such as *aunging* heavy timber. **1935** R. Campbell *Teak-Wallah* iv. 44 Is there any spectacle, I wonder, that can surpass in magnificence the sight of twenty or thirty elephants, all in the prime of condition, 'ounging' timber down a swollen jungle stream? *Ibid.* xiv. 208 Mounted on a tusker, I spent all day riding up and down the river, superintending the work of the 'ounging' elephants and seeing that they did not allow stacks to form. **1974** *Encycl. Brit. Macropædia* V. 971/2 In India, Sri Lanka..and Burma, their chief use now is in lumbering—they drag logs through the jungle and push floating logs around bends and off sandbanks ('aunging').

oupa (ǒu·pā). *S. Afr.* [Afrikaans = grandfather, f. *ou* old + *pa* father.] A name used in addressing or referring to one's grandfather or to an elderly man.

1920 R. Y. Stormberg *Mrs. Pieter de Bruyn* 40, I had secret misgivings that the Nooitgedacht sheep wouldn't pass the test, even though Oupa Cloete is nearly stone blind. **1934** 'N. Giles' *Ridge of White Waters* i. xiv. 165 Ou'pa Wessels, although well over seventy, insisted on accompanying the commando. **1939** S. Cloete *Watch for Dawn* 161 What is it? Tell me, oupa. **1951** P. Abrahams *Wild Conquest* 88 But Oupa Johannes didn't kill us and we didn't kill him. **1953** U. Krige *Dream & Desert* i. 13 When his father had told him of Oupa's death, Jannie had been very sad since he had loved Oupa. **1976** J. McClure *Rogue Eagle* xii. 204 Wolraad grinned... 'Where's Pa gone, oupa?' he asked.

our, *pron.* Add: **B. 1. c.** Used familiarly with a Christian name to denote a relative (esp. a child) or acquaintance of the speaker. Also, with surname, an employee of a company, and, in joc. address to a patient, the diseased or injured part of the body.

1847 A. Brontë *Agnes Grey* xi. 163, I sent our Bill to beg Maister Hatfield to be so kind as look in on me some day. **1856** Dickens *Dorrit* (1857) i. xiii. 118 Now, let's see whether there's anything else the matter, and how our ribs are? **1864** —— *Mut. Fr.* (1865) I. ii. ix. 246 Sloppy explained..that the Orphan (of whom he made mention as Our Johnny) had been ailing. **1911** F. H. Burnett *Secret Garden* xxvii. 298 He..took a golden sovereign from his pocket and gave it to 'our 'Lizabeth Ellen', who was the oldest. **1932** N. Royde-Smith *Incredible Tale* ix. 131, I sent up our Mr. Wilkinson, who has lived in Russia. **1936** 'G. Orwell' *Diary* 11 Feb. in *Coll. Ess.* (1968) I. 175 The son 'our Joe', just turned 15. **1952** 'W. Cooper' *Struggles of Albert Woods* i. ii. 23 'What do you think of it, our Albert?' his mother cried. **1968** 'J. Fraser' *Evergreen Death* v. 40 'What are you doing out here then, our Arnold?' his sister asked. **1977** G. Markstein *Chance Awakening* xxii. 70 'He's a swinger,' said Chance. 'Our Mike gets around.'

f. *our hero*: used familiarly of the hero by the writer of a work of fiction, biography, etc.

1804 J. Bisset *Crit. Ess. Young Roscius* p. x, Our little Hero caught the first theatric spark. **1854** Rawdon Brown *Let.* 6 May in M. Lutyens *Millais & Ruskins* (1967) 205 He..said that he had amused him more than anyone since Robinson Crusoe! A greater compliment could certainly not have been paid our hero. **1905** H. A. Vachell *Hill* v. 110 Much of our hero's time was spent in the company of the Duffer. **1961** *Mind* LXX. 104 And so our hero escapes from his appalling predicament: with one bound, Jack was free. **1975** *Radio Times* 29 May 13/1 Most of the books have a fair amount of physical violence which leaves Our Hero battered almost to pulp.

Ouranian (ɑuərēi·niăn), *a.* [f. Gr. οὐράνι-ος heavenly + -an.] Of or pertaining to heaven or the upper regions. (Cf. Uranian *a.*[1])

1908 G. G. A. Murray in R. R. Marett *Anthropol. & Classics* 68 A great proportion of our anthropological material is already to be found in prehistoric Crete..the stones, the beasts, the pillars, and the ouranian birds.

ouroboros, var. *uroboros.

ours, *poss. pron.* Add: *spec.* = our regiment, chiefly in phr. 'of ours'. Also in *transf.* and extended uses.

1787 W. Dyott *Diary* Nov. (1907) I. 38 The company at dinner was..Captain Gladstanes, 57th regiment; Captain Dalrymple, 42nd; Hodgson of ours, and myself. **1823** *Spirit of Public Jrnls.* M.DCCC.XXIII (1825) 1 What is't attracts the optic pow'rs Of Ensign gay, when fortune show'rs Down prospects of 'a step' in 'ours'? **1847** Dickens *Dombey* (1848) xxi. 206 Edith Skewton, Sir,.. married (at eighteen) Granger of Ours. **1847** Thackeray *Van. Fair* (1848) xxvii. 235 Run Simple (Ensign Simple, of Ours, my dear Amelia). **1877** G. M. Hopkins *Let.* 6 Jan. (1938) 94 Lancashire..from where a good many of Ours come. **1894** 'Mrs. Alexander' *Choice of Evils* II. ii. 47 We have a young fellow..in 'ours', who has just saved the Colonel's life by a lucky shot. **1922** W. Cather (*title*) One of ours. **1975** *Listener* 25 Dec. 893/3 A short, highly professional story of a competent war time spy of Ours who goes back to the game. **1977** T. Heald *Just Desserts* i. 10 You know he was one of ours?.. Provided us with information, tip-offs, odds and ends.

‖**oursin** (ūərsæ̃n). [Fr.] = Sea-urchin 1.

1928 R. Hall *Well of Loneliness* xl. 376 Many loved Prunier's..because of its galaxy of sea-monsters. A whole counter there was of incredible creatures—Oursins, black armoured and covered with prickles; Bigorneaux; serpent-like Anguilles Fumées. **1931** *Daily Express* 21 Sept. 3/4 There are..other strange fish known as oursins, literally 'little bears'. **1950** E. David *Bk. Mediterranean Food* 142 Oursins (those spiny sea-urchins cut in half from which you scoop out the coral with a piece of bread). **1966** P. V. Price *France: Food & Wine Guide* 274 There are also *oursins* (sea-urchins), which look just like prickly chestnuts, and taste rather like a snail that has been taken to the sea. **1972** *Guardian* 1 Jan. 11/4 Men bring nets full of oursins to quayside cafés. **1973** D. Miller *Chinese Jade Affair* xvii. 156 A basket of shell-fish..belons, oursins, moules, crevettes and Marennes.

Ouspenskyist (ŭspe·nski,ist). [f. the name of Peter Demianovich *Ouspensky* (1878–1947), Russian philosopher + -ist.] A follower of Ouspensky or his teaching. Also **Ouspe·nskian, Ouspe·nskyite** *adjs.*

1958 L. Durrell *Balthazar* ii. 29 Alexandria is a city of sects..Steinerites..Ouspenskyists, Adventists. **1968** T. Wolfe *Electric Kool-Aid Acid Test* iv. 54 Alpert soars in Ouspenskian loop-the-loops. **1975** M. Bradbury *History Man* v. 81 A radical Catholic priest and his Ouspenskyite mistress.

ouster[1]. Delete 'now implying a wrongful dispossession' and add: Also, eviction (from office, etc.) by judicial process or as a result of revolution or political upheaval. **b.** In lay use: dismissal, expulsion; the action of manœuvring out of (a place or position). Now chiefly *U.S.*

1961 P. Holmes *Sheppard Murder Case* ix. 82 Mr. Y, who had been asked no questions about a possible criminal record, had answered in the negative when asked if he had ever been a witness in any court. If it could be shown that Mr. Y had testified at his 1943 trial this answer could be made the basis for his ouster. **1967** K. Giles *Death in Diamonds* ix. 164 Mary Smith had to leave because of her bad influence on the other girls. She was fifteen when she got the ouster. **1968** *Telegraph* (Brisbane) 3 May 11/1 Mr. Cecil Harmsworth King lost his job as chairman of the International Publishing Corporation in a 'palace revolution' by his own directors. The ouster came three weeks after..a critical article. **1972** *Newsweek* 10 Jan. 25/3 When the court ouster came, the vets..marched out, clenched fists raised. **1973** *Listener* 20 Dec. 842/1 It is the hope..that enough damning evidence would be found to force the ouster of the President overnight—to make him resign. **1974** *Spartanburg* (S. Carolina) *Herald* 22 Apr. A1/2–3 The report..forced the resignation of Lt. Gen. David Elazar, the military chief of staff, and fueled demands for the ouster of Defense Minister Moshe Dayan. **1975** *N.Y. Times* 12 Sept. 6/1 The Communists appeared eager not to be isolated from power as a result of the recent ouster from the premiership and the High Council of the Revolution of Gen. Vasco Gonçalves, whom they had backed. **1976** *Honolulu Star-Bull.* 21 Dec. A-2/2 Hay said one problem still open in Chile concerns persons missing since Allende's ouster and death. **1977** *Time* 9 May 17/2 He was especially anxious to court the Kremlin in view of the rapid cooling of the U.S.'s interest in Ethiopia following the junta's ouster of the Emperor.

oustiti (ū·stiti). [ad. F. *ouistiti* (used in same sense) (see Wistiti).] = Outsider 4.

¶ Properly spelt *ouistiti*.

1941 G. Heyer *Envious Casca* x. 178 'Which would lead one to suppose that the murderer found the door locked, and turned the key from the outside.' 'With an oustiti,' nodded the Sergeant. **1962** 'D. Betteridge' *Package Holiday Spy Case* ii. 19 An essential item of a burglar's tool kit is..an oustiti. It resembles a long pair of pliers, but..has two semi-circular metal tongues. Inserted into a keyhole, it will grip firmly a key on the other side, and gentle but powerful wrist-work will turn it in the lock.

out, *adv.* Add: **I. 3. c.** *from this out*: henceforth, from now on; also *from here (on) out*, *from that out*. *colloq.* (chiefly *U.S.* and *Anglo-Irish*).

1867 F. A. Buck *Lett.* (1930) 214 Now, I am going to try to be a Jew from this out. **1882** W. D. Howells *Mod. Instance* in *Century Mag.* Apr. 925/1 I'll take a back seat from this out. **1899** W. B. Yeats *Let.* 28 Nov. (1954) 330, I imagine I am about the only person who belongs to the orderly world she is likely to meet from this out. She seems to be perfectly mad. **1905** H. Corkran *Lucie & I* 36 From this out I will think of you as a young *diablesse*. **1907** J. J. Horgan *Great Catholic Laymen* (ed. 2) i. 37 Napoleon was then at the height of his power. From that out his Empire began to decline. **1922** Joyce *Ulysses* 432 Mrs Marion from this out, my dear man, when you speak to me. **1941** in H. Wentworth *Amer. Dial. Dict.* (1944) 434/1 He has run the race and is fairly entitled to sit on the sidelines from here out. **1942** *Ibid.*, Bob Maslow's at the controls from here on out. **1972** A. Friedman in Cox & Dyson *20th-Cent. Mind* I. xii. 420 There are symbolic signs everywhere that Charlotte's and Adam's lives from here on out will be lives of protracted emptiness and captive anguish.

7. b. *to have it out*, add to def.: also with other objects; also, to discuss fully or reveal (a matter); to settle (a dispute or misunderstanding) with someone; (earlier and further examples).

1811 Jane Austen *Sense & Sens.* II. viii. 129 She had better have her cry out at once and have done with it. **1825** H. Wilson *Mem.* I. 77 O let us have it all out now, and have done with it. **1839** Dickens *Nickleby* xxxi. 305, I shall double-lock myself in with him and have it out before I die. **1847** J. A. Froude *Shadows of Clouds* iv. 52 The result was the advice which best harmonized with the suggestion of his own heart, to go off at once to Morlands, have it all out with Emma, and put his father's letter into Mr. Hardinge's hands. **1880** Trollope *Duke's Children* II. xi. 126 Let us have this out, Mabel, before we go. **1932** E. Bowen *To North* xxvi. 385 You must have this out with Emmeline, find how she stands with this young man and..strongly discourage the whole affair. **1959** H. Hamilton *Answer in Negative* x. 118, I was trying to decide whether I ought to leave it alone or have it out with him. **1971** *Where* Dec. 361/1 We eventually went to the LEA and saw an assistant education officer who agreed that we should meet the head and the doctor and have the whole thing out.

13. a. (Later examples.) From the 1960s frequently used as a chant (preceded by the name of a politician or by a word signifying something unwanted) by political demonstrators. Also used in written political slogans.

1968 *Times* 8 July 1/7 More than 100 Pakistani students took over the Pakistan High Commission in Lowndes Square, S.W., last night... Demonstrators lined the balcony chanting and waving placards saying: 'This building is occupied' and 'Ayub out'. **1970** B. Levin *Pendulum Years* xv. 259 The streets of Britain continued [in the 1960s] to echo to cries of 'Americans out of Vietnam!' and the ritual chanting of 'Ho-ho-ho Chi Minh!' **1970** D. Neville-Rolfe *Power without Glory* ii. 244 Even the stone-throwing rioters displayed their own brand of charm when they paraded in front of the Embassy their very home-made banners boldly declaring '*out—perfide albino*'. **1970** (*recorded from oral evidence*) Wilson *out! out!* **1973** *Times* 21 Nov. 18/4 'Anti-abortion—*out out out*; Free Abortion on Demand—*in in in*,' they chanted. **1976** *New Society* 26 Aug. 435/2 (*heading*) Hashish, out. **1977** *Woman's Own* 26 Mar. 31/3 Then you notice the blackened buildings, boarded up houses and the painted slogans in the housing estates: 'Brits out.' 'Join the IRA.'

b. *out with it*: an exhortation to a speaker to admit or assert something over which he is hesitating.

1709 [in Dict.]. **1924** Galsworthy *Forest* i. i. 11 Out with it, Mr. Farrell. **1974** 'S. Woods' *Done to Death* 35 Come on, Dick, out with it. What do you know about them?

II. 15. b. (Further examples.) So *school is out* (chiefly *U.S.*): school is at an end. See also *day out* (*day *sb.* 19), *night out* (Night *sb.* 5a in Dict. and Suppl.). *out to lunch*: see *lunch *sb.*[2] 2b.

1827 W. Taylor *Poems* (ed. 2) 91 In that whimp'ling burn when the school was out. **1843** Mrs. Stowe *Mayflower* 172 But, when 'school was out', James's spirits foamed over as naturally as a tumbler of soda-water. **1870** [see Evening *sb.*[1] 3]. **1911** in *Sc. Nat. Dict.* (1965) VI. s.v., Word went through the toon like lichtenin, for the school wis out. **1925** Wodehouse *Carry On, Jeeves!* v. 108, I have already visited some of New York's places of interest on my evening out. **1948** 'J. Tey' *Franchise Affair* ix. 94 'Shouldn't leave your car. Take it with you. .. It's Saturday.' 'Saturday?' 'School's out.' 'Oh, I see. But there's nothing in it..that's movable.' **1956** B. Holiday *Lady sings Blues* (1973) ii. 17, I finished up the fifth grade, and as soon as school was out Grandpop put one those big tags around my neck, saying who I was and where I was going. **1965** *Times Lit. Suppl.* 22 Apr. 317/1 In a recent road safety cartoon on B.B.C. television, I was surprised to hear the Americanism 'when school is out'. **1974** 'R. Tate' *Birds of Bloodied Feather* vi. 127 'I trust I'm not disturbing you?' 'What's the time?' 'Four.' 'School's out'... 'Come on in'. **1974** *Times* 8 Apr. 14/7 A retired bricklayer..spoke to me in the owlishly conspiratorial tones of one who has been cheered by his evening out.

c. (Further examples of sense 'on strike'.) Cf. sense *19b. *out there*, in the war of 1914–18: at the Western Front; in France. *colloq.*

1917 A. G. Empey *Over Top* 302 *Out there*, a term used in Blighty which means 'in France'. Conscientious objectors object to going 'out there'. **1920** W. J. Locke *House of Baltazar* xii. 150, I want to kick myself for sitting here in luxury when there's so much to be done out there. I had got my platoon—I was acting first lieutenant—like a high-class orchestra. **1929** *Papers Mich. Acad. Sci., Arts & Lett.* X. 312/2 Out there, England's equivalent for 'Over there'. **1974** *Times* 27 Feb. 14/3 The miners are still out; world prices are still rising. **1977** 'J. Le Carré' *Hon. Schoolboy* vi. 121 Pound's in the soup again... Electricians out. Railways out.

i. *to be out for*: to have one's interests or energies directed to, to be intent on (something); also *to be out to* (do something). Cf. *all out *adv. phr.* 4. orig. *U.S.*

1901 Merwin & Webster *Calumet 'K'* i. 13 They're mostly out for results up at the office. Let's see the bill for it. **1901** S. E. White *Westerners* xxix. 272 When they are out to have a good time,..they want somebody they can have their sort of fun with. **1907** Beerbohm in *Sat. Rev.* 13 Apr. 457/1 She is not 'out for' fun. She is an ardent suffragist. **1912** *Humanitarian* Oct. 76, I am sure that no person in this country, save him who is out for personal gain, wishes us to lag behind in this movement. **1913** H. Walpole *Fortitude* iii. x. 388 She's out for happiness at any cost and you're out for freedom. *Ibid.* iv. iii. 474 He was 'out' to defend his whole life. **1926** A. L. Maycock *Inquisition* v. 116 All the Inquisitors..were out to convert and reconcile, not to condemn. **1956** A. L. Rowse *Early Churchills* xii. 236 The Dutch vetoed a battle... Marlborough was exceedingly disappointed: he was out for decision. **1959** N. Mailer *Advts. for Myself* (1961) 42 They kept actin' like they was out to get him first.

19. b. (Further example.) Also, out of work, unemployed. Cf. sense 15c in Dict. and Suppl.

1878 T. Wood in J. Burnett *Useful Toil* (1974) iii. 309, I..was regarded as an enthusiast in some places for seeking work when so many were out who were known to the masters. **1890** W. Booth *In Darkest England* i. iv. 38, I would often be out of work a fortnight to three weeks at a time. Once earned £3 in a week, working day and night, but then had a fortnight out directly after. **1920** J. Ferguson in *Northern Numbers* 98 She had been 'out' since May, Her 'panto' savings now were well-nigh spent. **1935** N. Mitchison *We have been Warned* i. 74 He's a riveter. He came..when there was work going at the docks... Now he's out. **1968** J. Bingham *I Love, I Kill* viii. 95 When I told him I was 'out', he bought me a pint instead. And he gave me two tickets for the show he was in. **1973** *Listener* 29 Nov. 736/1 A British prime minister discovered as having been implicated in the same kind of depths as Nixon would be out. **1977** *Times* 6 Dec. 6/6 Because my husband's out of work he's not a man to take anything off us... Since the time he has been out they've never gone short of shoes.

c. For 'In *adv.* 5c' read 'In *adv.* 6d'. (Earlier example.) Freq. *not out* (cf. Not-out *a.* in Dict. and Suppl.); also *transf.* and *fig.*

1609 R. Armin *Hist. Two Maids of More-clacke* D2v *Tutch.* What doe you call it when the ball sir hits the stoole? *Filbon.* Why out. **1746** in J. Nyren *Young Cricketer's Tutor* (1833) 111 England, 1st Innings... Newland 18—not out. **1881** *Sportsman's Year-Bk.* 137 He..has been in 36 times, and 'not out' four times. **1906** E. Dyson *Fact'ry 'Ands* vi. 62 The..thermometer.. registered 103°, not out. **1937** Partridge *Dict. Slang* 424/2 *Not out* (96), 96 and still alive. **1955** *Times* 11 July 4/2 Waite, when he had made 61 and with the total at 314, was given not out caught behind the wicket off Lock.

d. (Further examples.)

1930 E. Wallace *Lady of Ascot* i. 13 'How long have you been out?'.. 'I don't know what you mean,' he said. 'How long have you been out of gaol?' **1934** D. L. Sayers *Nine Tailors* 279 Well, as you know, I wasn't out. I was inside again, owing to a regrettable misunderstanding. **1967** M. Proctor *Exercise Hoodwink* xiii. 91 He was a hardened criminal... The days 'out' were great days, a life of affluence and excitement. The days 'in' were the price he paid. **1976** 'B. Graeme' *Snatch* v. 56 It was Reg Abbott who got two [years], wasn't it? Reg should be out by now.

e. Unconscious; *spec.* in *Boxing*, defeated through failing to rise within the ten seconds allowed after being knocked down; so *out on one's feet*: dazed or barely conscious, although still in a standing position; *out like a light*: see *light *sb.* 5f; *out to it* (*Austral. slang*): dead drunk; fast asleep.

1898 B. J. Angle in W. A. Morgan *'House' on Sport* I. 45 A competitor stopped by a blow on the mark is as much 'out' as though rendered helpless by a hit on the point. **1901** R. Fitzsimmons *Phys. Cult. & Self-Defense* 159 Time was up. The champion was out. **1918** *War Birds* (1927) 150 She responded..by hitting him playfully over the head with an empty port bottle... It was a terrific crack and he was out for some time. **1941** Baker *Dict. Austral. Slang* 52 *Out to it*, dead drunk. **1947** 'N. Shute' *Chequer Board* 3, I..fell down..on the floor, clean out. **1946** K. Tennant *Lost Haven* (1947) xi. 171 He was properly out to it that night. We made speeches about how sorry we was to see him go. **1952** M. Allingham *Tiger in Smoke* xix. 270 You're ill... You may not know it, but you're out on your feet. **1955** E. Hillary *High Adventure* 175 For God's sake, Charles, keep an eye on John! He's out on his feet but doesn't realise it! **1963** N. Marsh *Dead Water* (1964) ix. 246 When he opened his eyes he thought with astonishment: 'I was out.' **1973** 'H. Howard' *Highway to Murder* vii. 85 He was still out cold but he began coming round just before the ambulance got there.

22. b. (Further examples.)

1936 M. Mitchell *Gone with Wind* xii. 227 He had seen no pantalets on the streets, so he imagined they were 'out'. **1954** [see *in *adv.* 6 i]. **1959** *Encounter* Dec. 16/1 It is becoming steadily easier for newspaper or television programmes to dictate what is out, what is in. **1972** *Daily Tel.* 15 Mar. 14 Creativity is 'in', while spelling, punctuation and well-formed handwriting..are 'out'.

c. Out of the question, impossible, not to be considered; unwanted, unacceptable, prohibited; out of place, irrelevant.

1936 W. Stevens *Let.* 27 Jan. (1967) 307 Any form of hell raising is simply out. **1938** *Topeka* (Kansas) *Capital* 15 June 10/1 (*headline*) Rail legislation out? **1940** 'M. Innes' *There came both Mist & Snow* ii. 26 The revolver-shooting fad to which I had been so unexpectedly introduced appeared to me childish in itself and oddly 'out' in the sort of house-party characteristic of Belrive. **1945** *Tee Emm* (Air Ministry) V. 33 Unauthorised low-flying should be out, repeat out. **1956** I. Bromige *Enchanted Garden* ii. ii. 91 Fiona left her velvet coat and tulle dress in the back of the car, tied the raincoat tightly round her waist and rolled up the sleeves... Glamour was out that evening. **1973** 'H. Howard' *Highway to Murder* i. 16 'Tell your boss to have a quiet word with the law.'.. 'No, that's out.'

23. Also, having exhausted one's supply of a particular thing; out of stock of a specified article.

1885 *List of Subscribers* (United Telephone Co.) p. xv, The hotel cellarman came up... 'Sir,' said he, 'we have had a run upon minerals, and are nearly out.' **1935** J. Steinbeck *Tortilla Flat* iii. 36 Run down and get four bottles of ginger-ale. The hotel is out. **1942** 'A. Bridge' *Frontier Passage* xi. 194 You haven't got a gasper, have you? We're out. **1972** J. McClure *Caterpillar Cop* iii. 30 Got a smoke? I'm out.

b. Used in radio communication to indicate that the speaker has finished speaking and expects no reply. *over and out*: see *over *adv.* 6b.

1950 'D. Divine' *King of Fassarai* xi. 73 He called the signaller. 'Take this down... "No signs occupation. Out."' **1955** E. Waugh *Officers & Gentlemen* i. ix. 108 He took the instrument. 'Headquarters to D Troop. Where are you? Over... You can't be... Damn. Out.' **1958** 'Castle' & 'Hailey' *Flight into Danger* i. 23 The acknowledgment came on the air. 'Flight 714. This is Winnipeg Control. Roger. Out.' **1966** D. Holbrook *Flesh Wounds* 218 'Hallo Roger Baker, Hallo Roger Baker. Able Zebra asks for hornet support. Roger Baker over.' 'Roger Baker O.K. Out.' **1971** J. Wainwright *Dig Grave* 16 At County Headquarters Wireless Operations Room they used the more powerful, country-wide air waves... 'Purple Fifteen to Control. Understood and out.' **1976** L. Dills *CB Slanguage Dict.* (rev. ed.) 51 *Out*, through transmitting.

26. b. (Earlier examples.)

c **1792** Jane Austen *Minor Wks.* (1954) 151 This mighty affair is now happily over, and my Girls *are out*. **1813** —— *Pride & Prej.* II. vi. 72 Are any of your younger sisters out, Miss Bennet? *Ibid.* 73 The younger ones out before the elder are married!

c. (Further examples.)

1872 B. Jerrold *London* xv. 127 The ginger-beer merchant..gesticulating and pattering one sultry morning... 'The Best Drink Out!' was his perpetual cry. **1973** *Times*

15 Oct. 22/8 A Triumph is still the best bike out, as a Norton represents a compromise between design criteria and production costs.

IV. 32. *out and return*. = sense 30 a.

1963 *Times* 31 May 16/2 The lengths of the out-and-return paths to the ionosphere. **1966** A. Battersby *Math. in Managem.* vi. 157 Each unit represents 1,000 tons making an out-and-return journey of 2 miles.

out, *sb.* Add: **1. b.** *slang.* (See quots.) So *three-out*, a glass holding a third of some measure of liquor.

1835 [see *gin *sb.*[2] 2 a]. **1903** *Daily Chron.* 24 June 7/2 'Two Bass's and three outs' is an order which seems to be instantly comprehended by a barmaid who distributes the contents of two bottles among three glasses. **1908** *Ibid.* 6 Feb. 4/7 In a gin palace an 'out' is a dram glass.

3. c. (Earlier examples.)

1823 M. R. Mitford *Our Village* (1824) I. 209 He.. thinks nothing of contending with both sides, the ins and the outs, secure of out-talking the whole field. **1853** F. Gale *Public School Matches* 13 The 'Outs' are pleased at the steady pace, and the 'Ins' are equally pleased with the steady batting.

d. *colloq.* An out-patient at a hospital. So *outs*, the out-patient department.

1933 Partridge *Slang To-day & Yesterday* iii. iii. 192 *Outs*, out-patient department. **1964** G. L. Cohen *What's Wrong with Hospitals?* iv. 76 Distinction between the 'ins' and the 'outs' is inevitably fading.

4. d. An attempt, undertaking; the achievement of a particular result; progress, success; usu. in phr. *to make an out*. *colloq.* and *dial.*

1843 H. Y. Webb *Diary* 4 May in *Amer. Speech* (1951) XXVI. 183/1 A man..that made half as many good resolutions as I have or made a worse out in sticking to them. **1845** M. M. Noah *Gleanings* 148 He slipped the fatal jack of diamonds from the bottom of the pack, and claimed the money then in stake... I at first thought him in jest, and laughed at him for making so bungling an out. **1853** J. G. Baldwin *Flush Times Alabama* 31, I might have made a pretty good out of it, if I had thrown myself upon the merits of my case. **1854** A. E. Baker *Gloss. Northamptonshire Words* II. 82 'He made a good out of that speculation.' 'He made a poor out of his speech.' **1893** J. Salisbury *Gloss. Words SE. Worcestershire* 27 'Making a goodish out' or 'a poorish out', are terms applied to any undertaking when successful or the reverse. **1904** W. N. Harben *Georgians* xix. 176 Warren got down on his knees then and actually tried to pray; but he made a pore out. **1938** M. K. Rawlings *Yearling* xxvi. 347, I often figger I made a sorry out of it, not encouragin' you. **1951** H. E. Giles *Harbin's Ridge* xv. 125 Let the woman of a house get sick, and it just goes to pieces. In the city, now, a man can make out very well... But in the country he makes a poor out of it when the hub of the house comes down.

e. *Baseball.* The act of getting a player out. *U.S.*

1860 in *Amer. Speech* (1947) XXII. 204/1 Three 'outs' and one 'run'. **1886** H. Chadwick *Art of Pitching & Fielding* 15 Mere speed costs more in wild pitches, and called and passed balls than it yields in outs or strikes. **1973** *N.Y. Herald Tribune Internat.* 15 June 15/4 Evans, whose only out in 17 straight appearances was a sacrifice fly, wiped out a 3-2 Pirate lead with his 14th homer. **1974** *Index-Jrnl.* (Greenwood, S. Carolina) 18 Apr. 10/1 After two outs in the second, winning pitcher Jack Davenport singled and scored when the next three batters walked.

f. A way out, means of escape; an excuse, defence, alibi. *slang* (orig. *U.S.*).

1919 R. Lardner *Real Dope* iii. 79, I am not one of the kind that are looking for an out and trying to hide behind a desk..because I am afraid to go into the trenches. **1926** J. Black *You can't Win* vi. 69 If a copper grabs you you've got an out. You ain't exactly beggin'. **1934** R. Stout *Fer-de-Lance* xvii. 287 There are times when I would welcome..an escape from life's meaner responsibilities—what Mr. Goodwin would call an out. **1953** P. Frankau *Winged Horse* i. i. 4 You like thumbing your nose at common sense... And it gives you an Out from me. **1970** G. F. Newman *Sir, You Bastard* 12 He wanted an out, a plausible story that would extricate his head from the chopping block. **1974** *New Yorker* 22 Apr. 130/2 Ardent pro-Europeans on Mr. Wilson's team would stand firm..against those ministers who will probably be for rejection of any terms and a quick out.

5. b. A defect, disadvantage, blemish. *colloq.* and *dial.* (chiefly *U.S.*).

1886 E. S. Ward *Burglars in Paradise* 48 Sound as sense! Hadn't an out about him. **1893** K. Sanborn *Truthful Woman in S. California* 69 Are there no 'outs', no defects in this Pasadena? **1917** H. Garland *Son of Middle Border* xiii. 129 Even hostling had its 'outs', especially in spring when the horses were shedding their hair. **1955** W. W. Denlinger *Compl. Boston* 167 A perfection in one part cannot make up for serious 'outs' elsewhere in the whole dog.

7. Phr. *at* (or *at the, on the*) *outs*: at variance or enmity (with someone or something). *colloq.* and *dial.*

1824 W. Carr *Horæ Momenta Cravenæ* 97 'To be at outs,' is to be at variance. **1877** E. Peacock *Gloss. Words Manley & Corringham, Lincolnshire* 185/2 They fell at outs last Brigg fair was three year, an' hev nivver hed a good wod for one another sin'. **1884** *Congress. Rec.* 23 Apr. 3326/1 His church and the Unitarians [were] very much at outs. **1915** D. H. Lawrence *Rainbow* iv. 90 She was always at outs with authority. **1917** G. B. McCutcheon *Green Fancy* 87 My daughter and I are..what you might say 'on the outs' at present. **1928** A. Waugh *Nor Many Waters* ii. 84 We were at outs pretty badly about that time. And when you're at outs it doesn't take much to send you off. **1936** M. de la Roche *Whiteoak Harvest* ix. 113 You could scarcely have done a worse thing. Renny and Alayne are at the outs. **1955** W. Gaddis *Recognitions* iii. i. 732 It's all right, don't explain. I'm on the outs with them too. **1973** J. Porter *It's Murder with Dover* vii. 67 Soon as he [sc. a cat] gets at outs with one of the guests the old devil's off upstairs making a convenience of the chap's bed.

out, *a.* Add: **2. b.** *out island*, also (often with capital initials) *spec.* any of the outlying islands of the Bahamas (see quot. 1957). Hence *out islander*. (Often hyphened.)

1875 *Encycl. Brit.* III. 238/1 The inhabitants of the out-islands were reduced to indigence and want. **1957** *Ibid.* II. 928/1 New Providence.., although not one of the larger islands, is the most important..; the others are known collectively as the Out Islands. **1971** 'D. Halliday' *Dolly & Doctor Bird* i. 6 A former minor Ambassador ..living on one of the Bahamian out-islands. **1971** *Bahamas* XXIII. iii. 20 Bahamian Out Islanders are among the world's friendliest people. **1973** *Whitaker's Almanack 1974* 775/1 There are a General Post Office in Nassau, 4 branch offices in New Providence and 109 sub-offices in the Out Islands.

7. Unfashionable; opp. *in *a.* 2.

1966 *Punch* 29 June 946/1 Nowhere have I come across a word of guidance for the 'out' crowd—the vast, non-swinging, switched-off, palateless, utterly without-it lot who dominate the community. **1969** *Daily Tel.* 24 July 17/6 They [sc. children] want to eat savoury things most of all; but there are certain 'in' sweet-stuffs and a very great many 'out' ones.

out, *v.* **1.** Delete 'now *Obs. exc. dial.*' and add later examples.

1927 H. A. Vachell *Dew of Sea* 269 I'll out 'em both, even if it breaks the contract. **1941** E. R. Eddison *Fish Dinner* vii. 103 Should a been unlorded long since, outed of all his hopes, for's misgovernment. **1942** E. Waugh *Put Out More Flags* ii. 137 It was just a question of outing those fellows in the government. Sir Joseph had seen many governments outed... He'd soon out Hitler if he were alive and a German. **1968** *Daily Mirror* 27 Aug. 7/3 No one throws things away any more. They 'out' them.

b. Delete '*Obs. exc. dial.*' and add later examples. So *out pipes* (*Naut.*), to cease smoking pipes; also as *sb.* (see quots.).

1899 *Daily News* 30 Jan. 6/4 The water flooded high the stoke-holes, outing the fires. **1900** *Black & White Budget* 1 Sept. 684/1 The times set apart for smoking are generally from noon till about 1.15, when the marine drummer beats a long roll on his drum as a signal to 'Out-pipes'. **1916** 'Taffrail' *Pincher Martin* ii. 25 At one-ten the bugle sounded 'Out pipes', and the decks were cleared up. **1950** *Publ. Amer. Dial. Soc.* xiv. 50 *Out*,..to extinguish, as a lamp, a fire. **1961** F. H. Burgess *Dict. Sailing* 154 *Out pipes*, the order to stop smoking.

c. (Further examples.) Also, to murder.

1900 G. R. Sims *In London's Heart* xlviii. 294 He glanced contemptuously at the prostrate form of his accomplice. 'Looks like I've outed him,' he said. 'Good job if I have—he'll never blab again.' **1913** E. C. Bentley *Trent's Last Case* ii. 27 The body not being robbed looks interesting, but he may have been outed by some wretched tramp. **1915** E. Corri *30 Yrs. Boxing Ref.* 221 Lewis.. promptly hit him a terrific punch on the point. 'Outed' by bluff! **1927** E. Wallace *Feathered Serpent* xviii. 229 I've heard fellers in Dartmoor say that if ever they got the chance they'd 'out' him.

d. In a ball game, esp. lawn tennis: to send (the ball) outside the court or playing area.

1865 W. S. Banks *List Provincial Words Wakefield* 3 *Ahted*, put out. 'Ahted t'first ball.' **1927** *Daily Express* 22 June 2/2 Raymond, striving for extra speed, netted and outed a succession of returns. **1928** *Ibid.* 5 July 11/4 He outed and netted two drives.

e. In cricket: to put or declare (a batsman) out. ? *Obs.*

1899 *Captain* I. 517/1 Never forget that there are other ways of outing a man besides clean bowling him. **1906** *Daily Tel.* 23 Aug. 9/7 Myers went in, but was almost immediately 'outed' under singular circumstances.

4. b. (Further examples.)

1896 G. B. Shaw *Let.* 11 Feb. (1965) I. 596 There is something fundamentally unfriendly in having a grievance and not outing with it. **1942** W. Faulkner *Go Down, Moses* 167 The negro he was shooting at outed with a dollar-and-a-half mail-order pistol..only it never went off. **1975** *Bookseller* 17 May 2540/1, I was just getting ready to say that as sometime chief solo-boy at Exeter College, Oxford, I was as good as Ernest Lough, when McCarry outed with: '*My Hear My Prayer* was very sweet.'

c. Of information, news, etc.: to become known. *dial.*

1893–4 R. O. Heslop *Northumb. Words* II. 515 It suin outed"—became commonly known. **1905** E. Phillpotts *Secret Woman* i. ix. 87 Yet it outed as she'd said 'no' to him.

out, *prep.* **1.** Delete '*Obs.* or *arch.* exc. in *from out*' and add later examples.

Not current in Received Standard in the U.K. but common in dial., and in various regions abroad. Several of the examples are U.S., Austral., and N.Z.

1926 A. G. McAdie *Man & Weather* 19 The ship would make easier weather by proceeding out the western entrance. **1958** *Otago Daily Times* 24 Feb. 5/2 He flew out the side of the cloud to warmer air. **1960** M. Spark *Bachelors* x. 150 You should of pushed him out the nest long ago. **1961** *Coast to Coast 1959–60* 34 During this time he had tried to throw everything movable out the ward-room window. **1962** *Amer. Speech* XXXVII. 269 To drive with the left arm out the window. **1967** *Southerly* XXVII. 73 She looked out the window..at all the other houses. **1968** K. Weatherly *Roo Shooter* 111 Sam was really crook, leaning out the window spewing. **1969** *Listener* 31 July 162/2 Schoenberg kept the 12 notes we ended up with but threw the hierarchy out the window. **1969** *Eugene* (Oregon) *Register-Guard* 3 Dec. 1D/1 'And,' continued Belko, 'that's when our whole plan went out the window.' **1972** D. E. Westlake *Cops & Robbers* (1973) iii. 46 He looked out the windshield. **1973** *Black World* Aug. 55/2 He slid on back out the kitchen door. **1975** *New Yorker* 29 Sept. 43/2 Mrs. Santana and her children contribute to the refuse by throwing their trash and garbage out their windows.

out-. Add: **A. I. 1.** *out-district* (examples), *school, -shed.

1849 in *Worcester* (Mass.) *City Documents* No. 1, 33 Most of the schools in the out-Districts, have been conducted.. under the new order. **1858** J. Morgan *Let.* 9 June in *Richmond-Atkinson Papers* (1960) I. 408 The sale of spirits on the two-gallon system..I consider to be the chief curse of the out districts. **1927** *Scots Observer* 8 Oct. 11/4 Back this summer from six months in the district in charge of out-schools. **1957** V. W. Turner *Schism & Continuity in Afr. Society* v. 154 Chikimbu was the problem child of the local Mission out-school. **1895** J. Roberts *Diary* 6 This led into someone's outshed.

2. *out-nurse, -porter, -pupil* (earlier example), *-sister* (later examples).

1909 *Englishwoman* Apr. 269 If she has a baby, it has to be dragged from bed and carried to some out-nurse. **1902** *Chambers's Jrnl.* Nov. 717/2 'Boots' will select for him that out-porter who will most briskly wheel his colossal pile of cases. **1927** *Daily Express* 14 July 9/2 Both men were out-porters at Snow Hill Railway Station. **1853** Mrs. Gaskell *Ruth* II. vii. 183 I'm a sort of out-pupil of yours. **1877** G. M. Hopkins *Let.* 6 Jan. (1938) 93 The eldest, Milicent, is given to Puseyism: she is what is called an out-sister of the Margaret Street Home. **1939** M. Philip *Companions of Mary Ward* I. i. 6 Her entrance.. into the Convent of Poor Clares as an out-sister.

4. *out-party* (later examples).

1949 *Manch. Guardian Weekly* 11 Aug. 3 The Opposition is an 'out' party in the brawling 18th century sense. **1965** *N.Y. Times* 18 July iv. 8 For a minority out-party, any position except 'me too' almost inevitably is going to become simple opposition. **1976** *Guardian Weekly* 26 Sept. 7/3 Whichever party does not control the White House—the 'out-party'—does not even have a leader.

6. *out-path* (earlier and later examples).

1573 T. Cartwright *Replye to Whitegift* 27 It is our partes to walke in the broade and beaten way, as it were the common caussie of the commaundement, rather then an outpathe of the example. **1897** G. MacDonald tr. Schiller in *Rampolli* 64 Could I but the outpath follow—Ah, how were my spirit blest!

II. 7. *outstress*.

1881 G. M. Hopkins *Sermons & Devotional Writings* (1959) 197 The first intention then of God outside himself or, as they say, *ad extra*, outwards, the first outstress of God's power, was Christ; and we must believe that the next was the Blessed Virgin.

B. 10. *outflooding, -hanging*.

1909 R. Kane *Sermon of Sea* xix. 306 Its eager essence is roused, directed, loosened, and flung forward in the out-flooding force of a soul's quest, in the torrent-like tide of love. **1850** W. Howitt *Year-bk. Country* ix. 313 In the lower, out-hanging towers are dungeons. **1851** H. Melville *Moby Dick* I. ii. 13, I at last came to a dim sort of outhanging light not far from the docks. **1972** D. Haston *In High Places* iii. 38 A few words of explanation and 1,500 feet of long, out-hanging, body-burning abseils.

11. *out-flung* (later examples), *out-fought* ppl. adjs.

1940 W. Faulkner *Hamlet* iii. i. 159 This time his outflung hands touched the farther bank. **1955** V. Cronin *Wise Man from West* xiii. 243 They heaved themselves up by hand and foot over outflung eaves of the plateau. **1892** Stevenson & Osbourne *Wrecker* xii. 189 Our outfought enemy [sc. a squall] only a blot upon the leeward sea.

12. Also, away from the thing named, as *out-shore*.

1947 [see *kick-back, kickback c]. **1961** *Times* 2 Aug. 4/1 Meanwhile, the outshore current looked more attractive to three other good starters.

C. I. 14. *outflood, -lean* (later example).

1920 D. H. Lawrence *Women in Love* xxiii. 349 The marvellous fulness of immediate gratification, overwhelming, out-flooding from the source of the deepest life-force. **1900** Hardy *Poems of Past & Present* (1902) 170 The land's sharp features seemed to be The Century's corpse outleant.

15. a. *outfan, outheave* vbs.

1930 R. C. Campbell *Adamastor* 87 Victory-vanned, with her feathers out-fanned, The palm tree alighting my journey delayed. **1957** —— *Coll. Poems* II. 253 The Spring with rosy spinnaker outfanned Comes curling silver fleeces through the land. **1850** W. B. Ullathorne *Remarks on Proposed Education Bill* 14 A momentum with which to outheave from the soul of youth both the principle of authority and the positive doctrines of religion together. **1908** Hardy *Dynasts* III. iii. iii. 104 Till dawn began outheaving this huge day, Pallidly—as if scared by its own bringing.

b. *out-tire* (later examples).

a **1877** Swinburne *Lesbia Brandon* (1952) xvi. 165 Her limbs shuddered now and then..as if cold or out-tired. **1905** 'Q' *Shining Ferry* i. vii. 91 And so, out-tired with their long day,..they came at nightfall..to the palace of enchantment.

16. *out-elbow* vb. (*poet.*).

1936 Dylan Thomas *Twenty-Five Poems* 9 Now that my symbols have outelbowed space.

II. 18. *out-achieve, -bat, -bowl, -break, -cook, -field, -hustle, -skate, -sprint, -wait* (later examples).

1960 V. Packard *Waste Makers* (1961) xxiv. 295 The Russians outachieved the United States in launching

earth satellites. **1970** *Time* 17 Aug. 39 Getting along with parents has never been easy in the U.S. America has almost begged for trouble by expecting children to outachieve their parents. **1773** J. Duncombe in R. Freeman *Kentish Poets* (1821) II. 364 To see the Surry cricketers Out-bat them and out-bowl. **1873** *Chicago Tribune* 4 June 1/7 The Mutuals outbatted their opponents. **1970** Outbat [see *outbowl* below]. **1773** Outbowl [see *outbat* above]. **1823** *Lady's Mag.* July 388/1 There was no doubt that Andrews could, if he chose, out-bowl Samuel Long, and out-bat Tom Coper. **1970** *Sunday Tel.* 20 Dec. 21/7 This weakened M.C.C. side..have been outbatted, outbowled and outfielded by South Australia. **1944** *Sun* (Baltimore) 15 Jan. 9/3 Outbreaking her rivals, the light-coated filly opened up a length advantage in the run to the turn. **1955** *Ibid.* 19 May 20/4 Aeschylus outbroke the opposition but could not keep pace. **1839** Dickens *Let.* 9 Sept. (1965) I. 578 A woman..who..far out-cooked the cook of Petersham! **1970** N. Armstrong et al. *First on Moon* ii. 34 Mike Collins, who had become fond of dishes like coq au vin.., definitely could outcook Lew Hartzell. **1976** *S. Wales Echo* 23 Nov., He proved it last night when he out-cooked five girls to win the South Glamorgan round of the Wales Gas schools cookery competition. **1875** *Chicago Tribune* 17 Aug. 5/6 The Browns were outbatted and outfielded. **1960** *Times* 19 May 21/2 They had been out-batted, out-bowled, and out-fielded. **1970** Outfield [see *outbowl* above]. **1961** in Webster, It is one thing to be beaten and quite another to be out-hustled. **1966** *Daily Progress* (Charlottesville, Va.) 8 June 30/2 We out-hustle them, out-position them, and in all but two games have actually out-rebounded them. **1975** *New Yorker* 7 Apr. 116/2 The other Knicks,..out-hustling the Lakers, double-teamed them all over the court. **1963** *Times* 4 Mar. 3/7 M. Schnelldorfer..had some bad falls and was outskated by his fellow Bavarian, the ebullient S. Schönmetzler. **1968** *Globe & Mail* (Toronto) 15 Jan. 19/1 Detroit Red Wings outskated, outhustled and outshot Chicago Black Hawks Saturday night. **1938** *Times* 25 July 5/1 Pender had struck two even shrewder blows for his side by out-sprinting A. Pennington twice. **1963** *Times* 11 Feb. 3/3 Snell was outsprinted in a mile. **1929** D. H. Lawrence *Pansies* 95 Still a man can be A meeting place for sun and rain, Wonder outwaiting pain As in a wintry tree. **1957** T. Hughes *Hawk in Rain* 28 Where the insects couple as they murder each other, Where the fish outwait the water. **1977** J. B. Hilton *Dead-Nettle* i. 9 She had..succeeded in out-waiting her antagonists. Her patience..had become too much for their nerves.

20. *out-game, -modern* vbs.

1940 *Sun* (Baltimore) 14 June 20/1 The Greentree filly outgamed Rosetown in a thrilling battle to the wire. **1957** *Ibid.* 1 Feb. 22/1 Careless Miss..caught Miss Erlen inside the sixteenth pole and then outgamed that filly, thanks to Brooks's superior handling. **1922** *19th Cent.* Apr. 654 The old dog could in truth out-modern the best of them. **1935** Out-modern [see sense *23 a].

21. *out-feature, -figure* (example), *-gambit, -luck, -machine, -monster, -poll* (later examples), *-rebound* vbs.

1929 R. Bridges *Testament of Beauty* i. 714 True beauty of manhood outfeatureth childish charm. **1866** Hardy *Time's Laughingstocks* (1909) 54 Intently busied with a vast array Of epithets that should outfigure thee. **1962** L. Deighton *Ipcress File* v. 34, I felt tired and out-gambited. **1916** H. Titus *I Conquered* ix. 119 The hind legs straightened, that mighty force bore on his footing—and the stone slipped. The Captain [sc. a horse] was out-lucked. **1928** *Daily Express* 9 July 13/1, I ran into Charles Kingsley there, who..was just outlucked as [*sic*] Wimbledon. 'I drew Patterson in the third round... What do you think of that for bad luck?' **1942** *Ann. Reg. 1941* 39 General Wavell had to conduct simultaneously a number of campaigns in each of which he was outnumbered and outmachined. **1955** E. Bowen *World of Love* iv. 74 See, today, how even Antonia had been out-monstered. **1968** *Economist* 11 May 21/1 A list of delegates pledged to support Mr. Kennedy at the Democratic nominating convention in August..outpolled by a margin of two to one a list pledged to Mr. Humphrey. **1973** *Guardian* 19 Apr. 4/5 In Oakland, California, incumbent Republican John Reading outpolled Black Panther party chairman Bobby Seale by more than 34,000 votes in an eight-man race for Mayor. **1976** *Time* 27 Dec. 8/2 Soares..had hinted he might resign if his party was heavily out-polled. **1966** Out-rebound [see *out-hustle* s.v. sense *18]. **1974** *State* (Columbia, S. Carolina) 8 Mar. 4-B/5 Maryland badly out-rebounded us when they beat us by 11 points at College Park. They are a great board team. **1977** *Detroit Free Press* 11 Dec. 5-D/2 This is the first time we've been outrebounded all year.

23. a. *to out-fiend fiends, outfish fish, out-modern the moderns, outmonster the monstrosities, outrainbow the rainbow*.

a **1918** W. Owen *Coll. Poems* (1963) 53 The few who rushed in the body to enter hell, And there out-fiending all its fiends and flames With superhuman inhumanities. **1930** E. Blunden in *Time & Tide* 3 Jan. 16 The new painters, with their endeavours to outmonster the monstrosities of uninspired futurists. **1935** *Amer. Speech* X. 192/2 She *out-moderns* the moderns in a frock that is made for *cocktailing*. **1956** 'H. MacDiarmid' *Stony Limits* 33 The range of the tartans outrainbowing the rainbow. **1960** T. Hughes *Lupercal* 46 Four-legged yet water-gifted, to outfish fish.

outa (ɑu·tă), a representation of a colloq. or vulgar pronunciation of Out of *prep. phr.* orig. *U.S.* Cf. *outer, *outta.

1896 [see *out-of-sight *adj. phr.* 2]. **1906** H. Green *At Actors' Boarding House* 42 Outa twelve hundred a week yuh kin pay it back an' have ten-fifty left. **1931** E. Linklater *Juan in Amer.* iii. ii. 215 My act's *different*. Outa the ordinary, see? **1952** B. Malamud *Natural* (1963) 93 Yeah..but we're outa the cellar now and who done that—the wind? **1963** [see *beezer 1]. **1966** 'J. Hackston' *Father clears Out* 51 We must've taken close on a 'undred an' fifty thousan' pounds' worth of gold outa that mine. **1970** R. D. Abrahams *Positively Black* iii. 73 The white man's rooster would then just beat the goddamned shit outa that son-a-bitch. **1973** *Black World* July 56/2 They all looked high outa it. **1976** *New Musical Express* 12 Feb. 28/6, I felt like it might sometimes have got a bit outa hand. The rhythm got lost.

out-act, *v.* (Later examples.)

1906 *Westm. Gaz.* 2 June 6/3 The best of Hamlets [is sometimes] outacted by the worst of gravediggers. **1975** D. Gray *Ride on Tiger* viii. 62 She knew she could out-act Anna any day.

outage (ɑu·tėdʒ), *sb.* orig. and chiefly *U.S.* [f. Out *adv.* + -age.] A period or state in which (esp. electrical) apparatus is not operating as a result of disconnection or failure; *spec.* a power cut.

1903 *Electr. World & Engin.* 18 Apr. 653/1 The lamp hours were 54,187; percentage of lamp outage, 6–10; globes broken, 23. **1951** *Engineering* 5 Jan. 28/2 Outages will occur for about 50 per cent. of the flashovers so that trouble due to lightning can be expected once per annum per 500 miles. **1955** *Tweed* (Ontario) *News* 14 Apr. 10/6 Defective lamps will be replaced on Friday of each week. To report outage phone 207 before Friday each week. **1958** J. G. Brown *Hydro-Electric Engin. Pract.* III. vi. 118 The outage which is to be expected with hydro plant is less than with steam turbo-generators. **1963** K. Neville in D. Knight *100 Yrs. Sci. Fiction* (1969) 73 There's an outage in the Silver Lake Area. The brakes on a bus failed and took out an overhead section. **1974** *Indian Express* 26 Dec. 1/7 An inspection of the pipelines in all the reactors in America had been ordered... It had been decided that the first unit would have an outage for refuelling from January 13. **1976** *Washington Post* 19 Apr. A8/3 The outages were caused by overheating of the fuses that sit atop the power lines' poles. **1976** *Cody* (Wyoming) *Enterprise* 23 June 16/7 Mr. Royale has lived in Wapiti Valley since 1921, has had electric power since 1947, and remarked that this was the longest power outage he could remember.

outarm (ɑutā·ɹm), *v.* [Out- 18 a.] *trans.* To exceed in possession or acquisition of weapons of war. Also *refl.*, to provide (oneself) with more arms than a competitor.

1930 H. Belloc in *G. K.'s Weekly* 25 Jan. 309/1 Those [governments] who have the less money seek for a pledge from those who have the more not to outbuild them and not to outarm themselves against their power competitors. **1950** *N.Y. Times* 31 Dec. 22/1 We shall not only out-arm our foes but 'out-sacrifice' them. **1955** V. Cronin *Wise Man from West* xiii. 243 Outnumbered and outarmed, he had recourse to a trick. **1966** *Listener* 8 Sept. 355/3 A country..which is accustomed to..out-arming everybody.

outasight (ɑu·tăsəi·t), *a.* Also **out-a-sight, outasite.** Colloq. contraction of *out-of-sight *adj. phr.*

1893 S. Crane *Maggie* xviii. 150 You're the kind of man we like, Pete. You're outa sight! **1968** *Surf International* (Austral.) I. vii. 44 Sydney's got a lot of girls, but Cape Town has an unbelievably high standard, I reckon. We check them out at lunch time—outasight! Too much! **1969** *Observer* 16 Feb. 40/6 Pendennis boogaloos, falls by cats into numbers, and lays down heavy outasight rappings. Translation: Pendennis arranges events, calls on people in the money, and produces a lot of fantastic gossip. **1971** *Oz* May 5, I gave this guy nearly £30 and he rented a room for me in some outasite neighbourhood. *Ibid.* 7 On they trotted. 'Outasite, outasite,' yelled Lee after them. **1973** *Black World* June 63 This Sistuh here sho give some out-a-sight sets.

ou·t-at-elbow(s. 1. See Elbow 4c. **2.** Of a dog (see quots.). Used predicatively (usually without hyphens) and *attrib.*

1922 F. T. Barton *How to choose a Dog* 155 The elbows turn outwards from the chest wall, so plainly seen in the Bulldog, but in other dogs to be out at elbows is a serious fault. **1943** H. N. Beilby *Staffordshire Bull Terrier* vii. 34 A dog that is out of condition..is just as unsound as one which is out-at-elbow due to incorrect placement of the shoulder-blades. **1954** C. L. B. Hubbard *Compl. Dog Breeders' Man.* 225 Out at Elbows. Having the elbow joints noticeably turned away from the body due to faulty front formation. **1975** T. Gray *Beagle* (ed. 3) 196 Out at Elbow. Elbows turned away from the chest, uneven in appearance, loose.

out-back, *adv.* Add: Also **out back, outback.** (Earlier and later examples.)

1878 'R. Boldrewood' *Ups & Downs* iii. 31 There was not a streak of crimson in the pearly dawnlight, as the whole party..rode silently along the indistinct trail which led 'out back'. **1893** H. Lawson *Coll. Verse* (1967) I. 447 Scenery outback isn't like Illawarra. **1901** M. Franklin *My Brilliant Career* iii. 12 The boys, as they attained manhood, drifted Out Back to shear, drove, or to take up land. *Ibid.* xxxiv. 289 George Melvyn had a large station Out Back. **1909** *Daily Chron.* 29 Jan. 4/7 Under the title of 'The Church Outback', Dr. G. H. Frodsham, Bishop of Northern Queensland..has published in an Australian paper a lively account of his experiences in..his enormous diocese. **1936** I. L. Idriess *Cattle King* i. 2 Fascinating stories..of the big mobs outback. **1942** C. Barrett *On Wallaby* i. 13 Out Back I have met honest sundowners. **1944** *Living off Land* iii. 49 Anyone who has been out-back..will know..how every waterhole becomes alive with frogs.

2. (As two words.) Outside at the back of a house or other building; in or into the back garden or back yard. *U.S.*

1892 'Mark Twain' *Amer. Claimant* iii. 36 There was a message, now, from out back, and Colonel Sellers went out there in answer to it. **1964** T. Wolfe in *Esquire* Feb. 97 If this wasn't such a high-class joint we would take wiseacres like you out back and beat you into jellied madrilene.

B. *adj.* Of, pertaining to, or characteristic of the Australian interior or back-country.

1900 H. Lawson *Darling River* in *Prose Wks.* (1948) 269 'The Queenslan' rains'..seem to be held responsible.. for most of the out-back trouble. **1906** A. B. Paterson (*title*) An outback marriage. **1913** W. K. Harris *Outback in Australia* i. 2 Of course, you get various opinions of Outback hospitality. **1931** *Times Lit. Suppl.* 1 Oct. 738/1 Sydney was the natural centre of outback adventurers. **1957** *Times* 11 May 7/6 In its own way the project is as brave as any bit of outback pioneering for metals or minerals. **1971** *Southerly* XXXI. 27 We've already rejected the proposition that wallabies are too shamingly outback to be possible material for poetry. **1977** A. Wilson *Strange Ride R. Kipling* iii. 157 'Men only' bars are disappearing even in outback Australia.

C. *sb.* **1.** The Australian interior or back-country.

1907 *Gentl. Mag.* July 78 These young dwellers in the Out Back have often no educational opportunities. **1911** in E. M. Clowes *On Wallaby* iv. 115 The Outback can still breed some true mates. **1920** B. Cronin *Timber Wolves* 9 In the seclusion of the outback they are at liberty to revert to grossness unspeakable. *Ibid.* ii. 40 Such men are not uncommon in the outback. **1930** *Times Lit. Suppl.* 10 July 577/2 Its travesty of the essentially peaceable 'out-back' is no worse than the scenario writer would think permissible. **1955** *Times* 6 July 17/3 The Duke of Edinburgh..cited, and discussed in detail, the scope for aircraft in the outback. **1971** *Sunday Australian* 8 Aug. 17/3 Australia's tourist future lies in the outback. **1977** *Hongkong Standard* 12 Apr. 9/6 It still has not as yet agreed to finance the cost of building a prototype solar-energy collection and storage plant in the Australian outback.

2. *transf.* Applied to other regions or countries with allusion to the Australian interior. Also *fig.*

1959 *Listener* 15 Jan. 140/3 Tramp your way through Wales's 'outback'. **1963** *Times* 20 Apr. 9/7 The recent wave of village school and mosque building in this hitherto neglected outback [in Turkey]. **1972** *Guardian* 3 Apr. 2/5 Senator Humphrey's statement came after a 14-hour swing around the Wisconsin outback. **1975** *Times* 5 Dec. 1/4 Mr Wilson's statement..provided a field day for.. skinheads of the parliamentary outback. **1976** J. van de Wetering *Tumbleweed* x. 91 Cunucu, that's the outback, the outback of Curaçao. **1978** *Globe & Mail* (Toronto) 4 Mar. 8/3 A B.C. government plan to use television and radio to bring higher learning to the outback has thrown academia into a tizzy.

Hence **ou·tbacker**, a native or inhabitant of the out-back; **ou·tbackery**, the cultivation of attitudes and values characteristic of the outback.

1913 W. K. Harris *Outback in Australia* i. 3 Another feature in the character of the Outbacker..is his honesty. **1918** R. H. Knyvett *Over There* iii. 28 In their enthusiasm the people of the capital city practically mobbed these 'outbackers', loading them..with cigarettes and candy. **1927** *Blackw. Mag.* Oct. 461 A grove of giant Tasmanian tree-ferns,..the 'old man' fern of outbackers. **1933** *Bulletin* (Sydney) 18 Jan. 11 Paddy Whelan is far from being the first outbacker to do a perish for water. **1966** T. Ronan *Once there was Bagman* 124 The phase of life, now sneered at by our pharisaical, suburban, scholarship-nurtured intelligentsia—'Outbackery', they call it—has its intervals of excellence. **1971** *Southerly* XXXI. 22 We began to dismiss the image of the explorer, the bushranger, the whole outbackery. **1971** *Bulletin* (Sydney) 14 Aug. 49 No violence, no sex, no self-conscious outbackery; the only complaint, the book is too short.

ou·t-basket. [f. Out *a.*, Out- 6 + Basket *sb.*] In an office, etc.: a basket or tray for outgoing correspondence or other documents. Cf. *out-tray.

Sometimes written as two separate words with *out* regarded adjectivally.

1940 *Amer. Speech* XV. 247 His incoming mail is put in an in-basket and his outgoing mail in an out-basket. **1944** 'N. Shute' *Pastoral* v. 115 He sealed the letter... He tossed it into the OUT basket. **1952** *Chambers's Jrnl.* Aug. 493/2 The phones stood primly, the in and out baskets were in their places, and he was in charge of it all. **1968** Mrs. L. B. Johnson *White House Diary* 20 Aug. (1970) 704 It looks as if the Russians have emptied their out-basket to us. They have answered *all* of the President's correspondence. **1970** [see *in-basket].

out-blo·ssoming, *vbl. sb.* [Out- 9.] The act of blossoming out or forth; a flowering; usu. *fig.*

1907 *Daily Chron.* 31 July 4/4 'Sunday out' has become a well-nigh universal out-blossoming. **1924** W. B. Selbie *Psychol. Relig.* 178 The religious awakening of adolescence ..is..generally an outblossoming of the whole nature into a larger and more wonderful world. **1932** M. Joynt tr. *Gougaud's Christianity in Celtic Lands* ii. 45 The wonderful out-blossoming of Christianity which distinguished Ireland in the following ages.

ou·tblowing, *vbl. sb.* [Out- 9.] A blowing out or outwards. Also **ou·tblowing** *ppl. a.*, that blows out.

1900 *Geogr. Jrnl.* XVI. 406 Blowing towards and in upon the polar regions to make good the drain caused by the surface outblowing south-easterly winds. **1909** *Daily Chron.* 31 May 4/4 An intaking and outblowing of the breath between the teeth. 1928 PEAKE & FLEURE *Steppe & Sown* 14 The borders..had acquired their characteristic loess soil..through the outblowing of the winds from the ice sheets over the loose detritus.

outboard, *a.*, *adv.* Add: **A.** *adj.* **2.** Of a motor: attached to the outside of a boat, at the stern; also of a motor-boat propelled by such an engine. Also *ellipt.* as *sb.*, such a motor or boat. Hence **outboard motor-boating, -motoring, -motorist(e).**
1909 *National Sportsman* Mar. 488/1 (Advt.), Make a motor boat of any boat in 5 minutes..with the Waterman Outboard Motors. 1914 *Yachting Monthly* XVI. 408 The demand for the outboard motor steadily increases. 1926 S. LEWIS *Mantrap* i. 14 When we don't use the out-board motor, they [sc. the Indians] do the paddling, not us. 1928 *Daily Express* 21 Apr. 10/3 The racing 'outboard' boat has given us a fascinating pastime. 1928 *Daily Tel.* 10 July 17/5 Miss Joan Spicer..is one of the best-known of 'out-board-motorists'. 1928 *Daily Mail* 25 July 17/4 The new pastime of outboard motor-boating. 1928 *Ibid.* 7 Aug. 19/6 One of the big appeals of outboard motoring is the ease with which the boat and engine can be handled. 1935 *Discovery* Mar. 77/1 The journey will be made in the semi-decked whale boat which will be fitted with a 'Sea-gull' outboard engine. *Ibid.* Aug. 225/1 There are four outboards, with a speed of twenty-five knots. 1943 J. W. DAY *Farming Adventure* iii. 41 In peace-time you would find..the creek noisy with outboards. 1959 P. CAPON *Amongst those Missing* 165 It's probably a canoe with an outboard motor. 1972 D. BLOODWORTH *Any Number can Play* xxi. 214 There's a man looking after the outboard I came in. 1973 J. LEASOR *Host of Extras* v. 66 The roar of an outboard engine splintered my dreams. 1974 *State* (Columbia, S. Carolina) 8 Mar. 4-B/1 (Advt.), Meet the 1974 Mercs. The most advanced outboards in boating. 1977 *Daily Tel.* 24 Feb. 19/7 The water was like glass as the Royal barge came in, chased by..outboards.
3. (See quot. 1956.)
1956 W. V. PAGE *Mod. Aircraft* vii. 267 The parasitic resistance..can be greatly reduced by properly streamlining the engine supports, especially those of the outboard engines as properly housing the engine mounted in the fuselage offers no particular difficulty. 1956 W. A. HEFLIN *U.S. Air Force Dict.* 365/1 *Outboard*, of aircraft components: Out toward a wing or airfoil tip; away from the fuselage or hull... *Outboard engine*, on an airplane having four or more engines, an engine farthest from the fuselage or hull.

outbreathed, *ppl. a.*[1] (Later example.)
1914 R. M. JONES *Spiritual Reformers 16th & 17th Cent.* 177 This entire manifested or out-breathed universe is, he says, the expression of the divine desire for holy sport and play.

ou·tbred, *ppl. a.* [OUT- 11.] Bred from parents not closely related.
1903 *Biometrika* II. 171 Waltzing mice must be crossed with in-bred and out-bred pure-bred albinos and in-bred and out-bred cross-bred albinos. 1955 *New Biol.* XVIII. 38 In an outbred population..things are very different. 1959 C. D. DARLINGTON *Darwin's Place in Hist.* x. 68 All the 'plasticity' (or variability) is due to the inbreeding of outbred plants and animals, or to the outbreeding of inbred plants and animals. 1977 *Times* 14 Apr. 14/4 There were consistent differences [in I.Q.] between the outbred children and the children of first cousins.

outbreed, *v.* [OUT- 14, 18.] **1.** (Stress even or on first syllable.) *trans.* and *intr.* To breed from parents not closely related.
1926 *Times* 24 Aug. 14/3 Whether plants are inbred or outbred is a matter which is left to regulate itself. 1968 *Times* 23 May 17/4 In the wild it is usually an advantage for plants to outbreed—to reproduce by fertilization with other individuals not themselves.
2. (Stress on second syllable.) *trans.* To be quicker or more prolific in breeding than.
1903 [listed s.v. OUT- 18]. 1926 BELLOC *Comp. Wells's Outl. Hist.* ii. 24/2 A slightly faster minority of swallows always outlive and outbreed their slower rivals. 1976 K. BONFIGLIOLI *Something Nasty in Woodshed* i. 10 Only the sparrow..can outbreed magpies by diddling his mate all the year round.

ou·tbreeder. [OUT- 8 + BREEDER 1.] A plant which is not self-fertile, or an animal in which breeding pairs are not closely related.
1963 E. MAYR *Animal Species & Evolution* xiv. 421 The entire breeding system of outbreeders is so organized as to accumulate and preserve genetic variation. 1963 DAVIS & HEYWOOD *Princ. Angiosperm Taxon.* xiii. 424 In many forest trees which are outbreeders, pollen may disperse for long distances. 1975 S. K. JAIN in Frankel & Hawkes *Crop Genetic Resources* ii. 22 We speak of..outbreeders, inbreeders, and apomicts.

ou·tbreeding, *vbl. sb.* [f. *OUTBREED *v.*] Breeding from parents not closely related.
1901 *Bull. U.S. Dept. Agric. Div. Veg. Physiol.* XXIX. 38 'In-and-in breeding', 'outbreeding' and other expressions relating to the close or distant relationship of parents have been prominent subjects among animal breeders. 1902 *Encycl. Brit.* XXV. 372/2 The tribes practised far more in-breeding than outbreeding. 1919 EAST & JONES *Inbreeding & Outbreeding* i. 13 Interest in the effects of inbreeding and of outbreeding is not confined to the professional biologist. 1940 *Nature* 30 Mar. 485/2 If we are to regard unisexuality simply as one of a number of outbreeding mechanisms, it is necessary to account for the fact that it is frequently found in some groups..but rare in others. 1959 C. D. DARLINGTON *Darwin's Place in Hist.* xii. 70 Outbreeding is necessary for the recombination of differences. 1964 E. J. H. CORNER *Life of Plants* xi. 197 Flowers have devices..to carry pollen from one flower to another, and, preferably, from one plant to another, to secure outbreeding. 1967 *Economist* 16 Dec. 1108/2 There is no evidence..that dairy stock are suffering from too close breeding through artificial insemination. Too much out-breeding..would be a more valid criticism. 1975 S. K. JAIN in Frankel & Hawkes *Crop Genetic Resources* ii. 22 Many specific genetic factors are known which regulate the degree of outbreeding.

outburst, *sb.* Add: **1. b.** *Astr.* A solar radio emission of great intensity and several minutes' duration which occurs in conjunction with a solar flare.
1947 C. W. ALLEN in *Monthly Notices R. Astron. Soc.* CVII. 387 Besides steady noise and bursts, one can detect, rather rarely, sudden outbursts of radio noise, which last for a few minutes, fluctuating violently, and then disappear. *Ibid.* 394 It is this correlation between flares and outbursts that shows that an outburst has a particular physical significance. 1955 *Sci. Amer.* June 42/3 The more common type of burst is a brief surge of intensity lasting only a few seconds; we call this a 'radio flash'. During a week of intense activity on the sun, there may be 100 flashes. The second type, far less frequent, is a burst lasting several minutes; this is called an 'outburst'. The outbursts..come only during a solar flare. 1971 J. S. HEY *Radio Universe* v. 104 Large flares are often accompanied by very intense outbursts of radiation on metre wavelengths lasting between about 5 and 30 min. In a classification of radio bursts according to their characteristic properties, these outbursts have been designated as Type II radio bursts. *Ibid.* 108 The microwave outbursts, often called microwave Type IV, sometimes accompany large flares and are particularly interesting because of their association with solar cosmic rays.

ou·t-camp. Chiefly *N. Amer.* [OUT- 1.] A camp at some distance from the main camp.
1844 *Knickerbocker* XXIII. 116 The Sioux..would not fail to attack, according to their custom, the out-camps. 1944 T. ONRAET *Sixty Below* ix. 96, I had a little rice and bacon in one of my out-camps. *a*1951 E. HILL in Murdoch & Drake-Brockman *Austral. Short Stories* (1951) 292 The lone little cavalcade..passed a deserted out-camp. 1970 *Islander* (Victoria, B.C.) 1 Nov. 12/2 From his main cabin, Sam had four outcamps along his trapline.

outcaste, *v.* Add: Also *reflex.*
1915 KIPLING *New Army in Training* 64 What will be the position..of the young man who has deliberately elected to outcaste himself from this all-embracing brotherhood?

ou·t-city, *a.* [OUT- 12.] Situated outside a city; suburban.
1939 J. STEINBECK *Grapes of Wrath* xiii. 118 Then the buildings grew smaller... The wrecking yards and hot-dog stands, the out-city dance halls. 1963 *Economist* 9 Nov. 571/1 Those [rents] to be paid in out-city areas.

outclass, *v.* Add: (Further examples.) Also *transf.* (with connotations of *CLASS *sb.* 5b).
1909 *Chambers's Jrnl.* Jan. 61/1 In the process of production..the Americans soon found themselves outclassed. 1911 *Ibid.* Oct. 702/2 It [sc. the aeroplane] can even outclass the telegraph upon short journeys. 1955 *Times* 20 June 13/3 The Ferrari team..was quite outclassed and the chief opposition to the German cars came from the Maseratis. 1966 M. WOODHOUSE *Tree Frog* xii. 105 He was practised in these techniques..and..I was outclassed. 1977 R. L. WOOLF *Gains & Losses* i. 35 A bumbling Church of England rector..hopelessly outclassed in argument by three chief Catholic spokesmen.

out-countenance, *v.* For *Obs.* read *Obs.* exc. in *arch.* use and add later example.
a 1945 E. R. EDDISON *Mezentian Gate* (1958) xxxix. 222 In him..burned..that same recklessness and superfluity which, when he..went on..into known instant peril of death at Middlemead, had outcountenanced the great lamp of heaven.

ou·t-country, *a.* [OUT- 1, 12.] **a.** Associated with or suggestive of the country (as opp. a town). **b.** Situated or coming from a particular country.
1943 *Time* 22 Feb. 53/2 With his clear, healthy, out-country look..baggy, tweedy clothes..Will White had strolled into his office. 1963 [see *IN-COUNTRY].

ou·t-county, *a.* [OUT- 12.] Situated, or coming from, outside a particular county.
1961 *Observer* 26 Mar. 12/6 A..version of this scheme..was shown..to representatives of the L.C.C.'s expanding 'out-county' towns. 1963 *Times* 16 Feb. 12/3 This is one of the oldest of the 15 'out-county' estates of the L.C.C. and contains altogether 9,584 dwellings. 1972 *Where?* Feb. 62/2 If the neighbouring borough is the only reasonable alternative then they are pretty well duty bound under Section 8 of the 1944 Act to approach that borough for a place, *but Section 8 in no way binds the out-county borough to offer one. Ibid.* 63/2 She could opt to go to another college outside the area, so long as she was prepared to pay the higher fees which generally apply to out-county students.

ou·t-cri:cket. [f. OUT- 3 + CRICKET *sb.*[2]] In cricket: bowling and fielding, as opposed to batting.
1884 *J. Lillywhite's Cricketers' Ann.* 65 Kent's first victory of the season, the result mainly of creditable out cricket. 1904 *Westm. Gaz.* 22 Feb. 3/1 Bad batting lost us the game, for our out-cricket in both innings was up to a very high standard. 1927 M. A. NOBLE *Those 'Ashes'* 106 Larwood's fast bowling was the feature of England's out-cricket. 1963 *Times* 25 Feb. 4/2 Between luncheon and tea, New Zealand's courage made for such splendid out-cricket that Barrington and Cowdrey were compelled to display most of their resources. 1975 *Sunday Tel.* 27 Apr. 32/3 Lancashire's out-cricket was far too positive for their total of 215 for six to be more than a pious dream for Yorkshire, whose innings closed at 137 for seven. 1977 *Guardian* 3 Jan. 11/5 India, who won the toss, were dismissed by some brilliant English outcricket for 155.

outcrier. a. (Later examples.)
1902 BEERBOHM *Around Theatres* (1924) I. 344 If..any of our outcriers harbour the delusion that a School would inculcate something more than technical tricks [etc.]. 1931 *Tablet* 23 May 673/1 Once more the outcriers have cried out before they are hurt. 1942 BEERBOHM in *Listener* 24 Sept. 389/1 One is tempted to say that the outcriers are entirely altruistic men.

ou·tcropper. [f. OUTCROP *sb.* + -ER[1].] One who takes coal from an outcropping seam or vein.
1926 *Glasgow Herald* 7 Oct. 8/3 The outcroppers are doing good business for themselves. *Ibid.* 5 Nov. 7 Prosecutions which reveal extensive damage being committed by coal outcroppers..were brought in the Airdrie J.P. Court yesterday.

outcross, *sb.* Add: (Further examples.) Also *attrib.*
1918 *Genetics* III. 475 When double-throwing Matthiola is used as egg parent in an outcross to ordinary singles, half the offspring are doubles. 1949 R. B. KELLEY *Sheep Dogs* (ed. 3) iv. 62 It is not necessary to purchase such outcross dogs. 1971 *Farmer & Stock-breeder* 16 Feb. 49/1 An unplanned out-cross with an unknown wheat took place and this brought much improved fertility. 1973 *Times* 25 Aug. 8/5 Original outcrosses were Northumberland-bred Scottish Blackfaces. 1977 *Horse & Hound* 14 Jan. 15/3 Malacate stands out as a potential outcross for the majority of high-class European mares.

outcross, *v.* (Stress even.) [f. the sb.] *trans.* To cross (an animal or plant) with one not closely related to it. Also *absol.*
1918 *Genetics* III. 437 Each [headed fly] was outcrossed separately to a fly from some non-headed stock. 1931 E. B. FORD *Mendelism & Evolution* ii. 40 If the now highly inbred stock be outcrossed to ordinary wild-type flies, it is found that the extracted recessives..have returned to the original condition. 1949 R. B. KELLEY *Sheep Dogs* (ed. 3) iv. 62 It will be necessary to periodically outcross by introducing dogs from other sources, or the danger of close-breeding is apparent.

ou·tcrossing, *vbl. sb.* [f. *OUTCROSS *v.*] The crossing of an animal or plant with one not closely related to it.
1950 *Brit. Jrnl. Psychol.* XL. 132 Vigour is restored by out-crossing. 1973 PROCTOR & YEO *Pollination of Flowers* xii. 382 It is seldom easy to establish the amount of outcrossing which takes place under natural conditions. 1973 B. J. WILLIAMS *Evolution & Human Origins* xiv. 249/2 Heterosis can occur as a result of outcrossing, or mating between breeding isolates.

outcry, *sb.* Add: **2.** Also *U.S.*
1931 *Amer. Speech* VII. 20 *Public outcry*, an auction sale. 1961 in WEBSTER. 1974 *Times & Reporter* (Chester, S. Carolina) 15-A/6 (Advt.), It is the executor's duty to sell it at public outcry... I, the undersigned Special Referee will sell at public outcry to the highest bidder..the following described real estate.

outcry, *v.* **1. a.** (Later example.)
1849 THACKERAY *Pendennis* II. xxiii. 204 She at once took side with Helen against Doctor Portman, when he outcried at the enormity of Pen's transgressions.

ou·t-cue:. *Broadcasting.* [OUT- 7.] A cue (*CUE *sb.*[2] 1c) that indicates when a particular recording or transmission is about to end.
1962 A. NISBETT *Technique Sound Studio* viii. 139 The main things to mark on the script are as follows:..(iv) Duration (or other means of indicating when the outcue is coming up). (v) Type of fade out. 1969 S. HYLAND *Top Bloody Secret* i. 81 The BBC's duty director..waited for the Presentation announcer's out-cue. 1972 D. LEES *Zodiac* 19 It was Zodiac's television experience that led him to leave on a good 'out cue'.

ou·t-curl. *Curling.* [OUT- 7.] = *OUT-TURN.
1903 [see *IN-CURL].

ou·tcurve. [OUT- 7.] **1.** *Baseball.* The bending or curving of a ball outwards (i.e. away from the batter); the course of such a ball; a ball pitched so as to curve in this way.
1881 *N.Y. Herald* 29 July 6/5 Reipslaugher,..not being used to the difficult delivery of Bond, found great difficulty in handling the in-shoots and out-curves. 1886 [see *INCURVE *sb.*]. 1897 [see *INSHOOT]. 1910 *Encycl. Brit.* III. 459/2 The commonest of these swerving deliveries, and the first invented, is the out-curve, the ball coming straight towards the batsman until almost within reach of his bat, when it suddenly swerves away from him towards

the right, if he be right-handed. 1943 *Amer. Speech* XVIII. 106 Various types of curve ball are the sinker.., the in-curve, the out-curve,..and the screwball.
2. *gen.* An outward curve or prominence.
1903 *Encycl. Brit.* XXVIII. 622/2 It is convenient to employ a specific name for a projection of a coast-line less pronounced than a peninsula, and for an inlet less pronounced than a bay or bight; outcurve and incurve may serve the turn. 1912 GALSWORTHY *Inn of Tranquillity* 68 The lovely outcurve of his sweeping line. 1958 *Sun* (Baltimore) 19 July 7-O/5 This design bows to the new round-shouldered vogue, but builds here in an 'outcurve', to avoid the drooping look apparent in many of the new styles.

ou·tcurved, *a.* [OUT- 11.] Of the bow of a violin or other stringed instrument: having a curve in the direction away from the hair.
1954 *Grove's Dict. Mus.* (ed. 5) I. 855/2 The mechanics of a bow with slight positive curve (away from the hair; out-curved) necessarily differ somewhat from those of an incurved bow. 1960 *Daily Tel.* 5 May 16/2 She had used an archaic 'outcurved' bow to play the music written by J. S. and W. F. Bach on a modern viola. 1976 *Early Music* Oct. 513/1 The baroque cello..combined with the out-curved bow, which help maintain the correct style of playing.

outdated, *ppl. a.* Add: (Later examples.) Hence **outda·tedness.**
1937 'C. CAUDWELL' *Illusion & Reality* vi. 115 The development of capitalist production remorselessly turns the craftsman into a labourer... Eventually, employed as a factory hand, he may still cherish his outdated skill by making models. 1960 *20th Cent.* Mar. 259 By his lofty and elusive outdatedness, he has outwitted the practically minded, the men of action and the men of defeat. 1967 KARCH & BUBER *Offset Processes* v. 167 Causes of fog... Outdated film. 1976 R. LEWIS *Witness my Death* i. 35 For you, it's a matter of protecting a professional reputation, threatened by outdated methods.

out-door, outdoor, *a.* (*adv.*) Add: **1.** (Earlier example.)
1748 RICHARDSON *Clarissa* III. 208 In other words, to employ itself rather in the out-door search, than in the in-door examination.
5. Special collocations: **outdoor department,** that section of a public house that sells liquor for consumption off the premises; also *ellipt.* as *sb.*; **outdoor girl,** a girl or young woman who likes an open-air life; **outdoor things,** clothes that are worn out of doors.
1958 *Times* 29 Apr. (Beer in Britain Suppl.) p. ix/6 Jug-and-Bottle... For the younger customer..a licence... Off-Sales, Outdoor Department: the modern equivalent. 1961 M. JONES *Potbank* i. 6 Every pub has its off-licence department, known as the Outdoor. 1971 R. ROBERTS *Classic Slum* vi. 94 A great deal of beer bought in jugs from the 'Outdoor Department' was drunk at home... Lower-class women..stood crushed together drinking in the 'Outdoor'. 1907 J. WEBSTER *Jerry Junior* xiii. 198 Nannie was a big wholesome outdoor girl, of a purely American type. 1947 M. LOWRY *Under Volcano* ix. 275 The Hawaiian Islands gave us this beautiful outdoor girl who is fond of swimming, golf, dancing. 1969 D. THOMAS *Honey Bk., Ready* 31/2 Outdoor girls who swear by hair spray are bound to... 1847 E. BRONTË *Wuthering Heights* II. xiv. 185 I snatched my outdoor things... 1904 E. NESBIT *Phoenix & Carpet* x. 185 Every one put on its outdoor things..and all was ready. 1960 *Don't lie to Police* xi. 187 They told me Miss Cart was there when you returned. In her outdoor things?

outdoors, *adv.* Substitute for entry: **out-doo·rs,** *adv.* (*sb.*) [OUT *prep.*] **A.** *adv.* Out of doors; in the open air.
1817 S. R. BROWN *Western Gazetteer* 113 The chimney is sure to be placed out doors. 1846, 1882 [in Dict.]. 1913 G. STRATTON-PORTER *Laddie* xiii. 404 What you learn there [sc. at school] doesn't amount to a hill of beans compared with what you can find out for yourself outdoors. 1937 C. MORTON *Bible comes Alive* vi. 149 At Jericho..the people mostly lived out doors.
b. Hence phr. *all outdoors* (*U.S. colloq.*), the whole world; everybody; freq. in comparisons alluding to the immensity of the areas out of doors.
1832 J. NEAL *Bro. Jonathan* I. 111 Stuffy feller (that bear) as ever you see'd; big as all out o' doors. 1830 S. SMITH *Life & Writings J. Downing* (1833) 64, I had a letter from him..as long as all out doors. 1844 [in Dict.]. 1846 *Quincy* (Illinois) *Whig* 17 Feb. 2/2, I was going to say the President's message—Jimmy K's statement to all out-doors, and some parts of Aabey. 1862 [in Dict.]. 1940 W. FAULKNER *Hamlet* III. 213 The hot sun of July falling through the shadeless and even curtainless windows open to all outdoors. 1948 *Chicago Tribune* 28 Mar. 6/3 Its spirit is literally 'as big as all out-doors'.
B. *sb.* = OUT-OF-DOOR, -DOORS *sb. phr.* Freq. in phr. *the great outdoors*, the 'great open spaces' (see *OPEN *a.* 8a).
1857 N. P. WILLIS *Convalescent* (1859) 121 The 'down party'..were enjoying the river from the uncommon outdoors of Mr. Grinnell's broad prairies. 1932 *Ann. Reg.* 1931 II. 47 The 'great outdoors' was represented most elaborately in 'Cimarron', which starred Richard Dix in a fine dramatisation of the opening up of Oklahoma, and 'Trader Horn', a picture of wild life in Africa. 1940 [see *CLIFF-HANGER]. 1949 *Skyline Trail* Mar. 14/2 Trail Hikers, like all lovers of the outdoors, are by nature animal lovers as well. 1968 *Globe & Mail* (Toronto) 17 Feb. 44 During the 30 years (good heavens, is it that long?) that I have been writing about the outdoors, changes have been great. 1968 *Listener* 30 May 713/1 The Victorians worked hard to get the great outdoors inside a theatre. 1977 *Times* 23 Apr. 12/7 Ideal for anyone who wants that taste of the great outdoors without straying too far from..civilization.
Hence **outdoo·rsman,** one who likes outdoor activities; **outdoo·rsy** *a.*, associated with or characteristic of the outdoors; fond of an outdoor life; hence **outdoo·rsiness.**
1952 BERG & SAMUELS *Lady on Beach* vi. 184 In my attempts to be a truly outdoorsy woman at all times I had a ludicrous crab-hunting misadventure of my own. 1958 *Globe & Mail* (Toronto) 24 July 8/6 He's kept busy coping with appetites of outdoorsmen. 1960 I. WALLACH *Absence of Cello* (1961) 105 It was hardly an outdoorsy vacation. 1961 *Guardian* 4 Mar. 4/7 These new-style outdoorsmen. 1967 E. B. NICKERSON *Kayaks to Arctic* xii. 112, I..find no personal satisfaction in..showing an obnoxious out-doorsiness in a dozen..ways. 1973 J. GORES *Final Notice* (1974) vii. 43 Larry Ballard, good-looking in a wind blown, outdoorsy way. 1974 *Publishers Weekly* 18 Mar. 42/1 A poor boy who..became a prodigious outdoorsman and mountain climber. 1976 *Ibid.* 11 Oct. 91/2 His interviews read like scenes out of a good outdoorsy novel. 1976 *Parliamentarian* July p. viii, The hazards of translation in the European Parliament have given Mr Michael Stewart an undeserved reputation as an outdoorsman of unusual dexterity. He had spoken of his job as being like 'paddling a canoe shooting rapids', but the German interpreter thought he had said '..shooting rabbits' and translated accordingly. Mr Stewart was promptly wished good hunting by two German Members. 1976 *National Observer* (U.S.) 18 Sept. 1/4 They're the sort of out-doorsy fellows we usually associate with television shows about tagging grizzlies and spying on beavers in Wyoming.

outdra·wn, *ppl. a.* [OUT- 11.] Extended, drawn out.
1905 E. F. BENSON *Image in Sand* i. 8 Bank after bank of out-drawn stops and keyboard coupled to keyboard makes the air thick with tumultuous melody.

outdrive, *v.* Add: **2. b.** *Golf.* To drive farther than; = *OVERDRIVE *v.* 5.
1906 *Daily Chron.* 22 May 9/4 MacFarlane, after being outdriven from the tee, played a perfect approach to within a yard of the hole. 1923 *Weekly Dispatch* 13 May 1 He seemed to set out not to out-drive his opponent..but to steer a straight course. 1952 B. DARWIN *James Braid* ii. 29 James in a state of bliss straightway outdrove all those who but lately had had the audacity to outdrive him.

outed, *ppl. a.* Add: (Further examples.) Also *slang*, killed.
1919 W. H. DOWNING *Digger Dialects* 37 *Outed*, killed. *a* 1945 E. R. EDDISON *Mezentian Gate* (1958) ii. 23 Put case I had fallen in with your fine design to match me to yonder outed Prince of Akkama. 1950 PARTRIDGE *Here, There & Everywhere* 73 The English synonyms for death..are less numerous than the French... *Wiped out*, whether of one's person or one's military unit;..*outed*, from boxing.

outen, *adv.*, *prep.* (*a.*) Add: **B.** *prep.* **1. b.** Out of; out from. *dial.* and *colloq.* (chiefly *U.S.* and *Sc.*).
1854 *Laird of Logan* 441, I cud get as gweed a yane onyday out'n a hedge at the road-side. 1867 'MARK TWAIN' *Celebr. Jumping Frog* 38 He'd yank a sinner outen (Hades), And lam him with the blest. 1895 *Horse Rev.* 31 Dec. 1840/3, I hearn Marse Henry cum outen de house lak he did in de days ob old. 1898 R. BLAKEBOROUGH *Wit, Folklore of North Riding* 425 Sha tummel'd outen t' winder. 1926 [see *FEISTY *a.*]. 1944 C. HIMES *Black on Black* (1973) 197 They done left me outen it altogether. 1945 in *Sc. Nat. Dict.* (1965) VI. 511/1 Ye're aye tryin tae mak an auld wumman ooten me. 1976 T. GIFFORD *Cavanaugh Quest* i. 16, I come outen the elevator and I'm heading through the delivery hallway.

outen (au·tən), *v.* *U.S. dial.* [perh. f. OUT *adv.* 22a, 23: see -EN[5].] **1.** To extinguish, put out, erase.
1916 *Dialect Notes* IV. 338 Be sure to outen the light when you go to bed. 1933 M. K. RAWLINGS *South Moon Under* 331 Outen that light. 1937 *Amer. Speech* XII. 205 The average [Pennsylvania German] speaker will employ in speech, and often in writing, all the forms ascribed above to the educated person, and in addition..such forms as 'outen the light', [etc.]. 1950 *Publ. Amer. Dial. Soc.* XIV. 50 *Outen*: *v. t.*, to extinguish, as a fire. To erase, as writing on a slate. 1961 in WEBSTER, You might outen the candles there.
2. To come out with, utter, tell.
1951 L. CRAIG *Singing Hills* viii. 68 Finally Maje said, 'Outen it, fellow. Give out what you know.' *Ibid.* xiii. 126 There are words we want to outen and we can't.

outer, *a.* (*sb.*[1]). Add: **1. b.** *Printing.* In sheet work, designating the forme containing the type pages from which the outer side of the sheet is printed and including the type page for the first page of the printed sheet.
1755 J. SMITH *Printer's Gram.* x. 262 They [sc. compositors] lay one extremity thereof against the hind side of the Fifteenth page, if it is an Inner Form; or against the hind side of the Thirteenth page if it is an Outer Form. 1808 C. STOWER *Printer's Gram.* vii. 171 A sheet in Folio. Outer Form. Inner Form. 1841, etc. [see *INNER *a.* (*sb.*[4]) 1 e].
c. *Phonetics.* Denoting an articulation in a part of the mouth nearer the lips than that designated by the term qualified by 'inner'.
1867 [see *CLOSURE 5c]. 1888 [see *INNER *a.* (*sb.*[4]) 1 g]. 1972 HARTMANN & STORK *Dict. Lang. & Ling.* 39/2 The outer closure may occur at the lips, the teeth, the alveolar ridge, the palate, velum, pharynx, or glottis or any intermediate point.
3. **Outer Circle,** the road running round the perimeter of Regent's Park, London; **outer form,** (*a*) (see sense 1 b above); (*b*) *Linguistics* (see quot. 1972); also *outer speech form*; **outer multiplication** *Math.*, the formation of an outer product; **outer product** *Math.* [tr. G. *äusseres produkt* (H. Grassmann *Die lineale Ausdehnungslehre* (1844) p. xi)], a vector product (*rare*); more commonly, a related product of two vectors or tensors that yields a tensor of higher rank than either of them; **outer space,** the region beyond the earth's atmosphere or beyond the solar system; *from outer space*, a colloq. phr. implying outlandishness and frightfulness as of creatures described in some science fiction; **outer suburb,** one of the more remote suburbs of a city or town; so *outer-suburban* adj.; **outerwear,** clothing designed to be worn outside other garments; opp. *underwear*.
1829 *Picturesque Guide to Regent's Park* 29 The outer circle or Ring consists of a fine level drive, planted with trees on each side; within this is another circle or pathway. 1867 H. LARGE *Large's Way about London* 33 *Outer Circle*, Regent's Park, N.W., from the Marble Arch along Oxford st., through Orchard st, Portman sq, Baker st, Upper Baker st and Clarence gt to the Outer circle. 1938 E. BOWEN *Death of Heart* i. 8 Cars slid lights all round the Outer Circle. 1974 *Kelly's Post Office London Directory* (ed. 175) Regent's Park.. Outer Circle. Hanover Lodge..Zoological Society's Gardens. 1972 HARTMANN & STORK *Dict. Lang. & Ling.* 113/2 The grammatical and semantic structure of a particular language is unique to that language; outer form, however susceptible its sound system (*outer form*) may be to influences from other languages. 1898 A. N. WHITEHEAD *Treat. Universal Algebra* I. iii. 207 Progressive and Regressive Multiplication are called Outer Multiplication. 1959 M. R. SPIEGEL *Schaum's Outl. Theory & Probl. Vector Analysis* viii. 169 Inner and outer multiplication of tensors is commutative and associative. 1960 H. W. TURNBULL *Theory of Determinants* xi. 215 It involves a determinantal factor (*ab*) which is an outer product of the symbolic linear sets *a*, *β*, *γ*. 1959 M. R. SPIEGEL *Schaum's Outl. Theory & Probl. Vector Analysis* viii. 169 The outer product of two tensors is a tensor whose rank is the sum of the ranks of the given tensors. This product which involves ordinary multiplication of the components of the tensor is called the outer product. 1963 J. ABRAM *Tensor Calculus* iii. 27 We could have written eqn. (3.14) as... This bracket is an anti-symmetric tensor of order two. It is known as the outer product of the two vectors... Only in three dimensions can the outer product be replaced by the vector product. 1970 BEDFORD & DWIVEDI *Vector Calculus* iv. 168 Since the outer product of two vectors is not commutative, the order of the two factors in the outer product is important. 1901 H. G. WELLS *First Men in Moon* iii. 45 After all, to go into outer space is not so very different from a polar expedition. 1935 *Discovery* Apr. 105/1 In America it was proposed to explore Mars or Venus or even the very realms of outer space. 1958 *Observer* 16 Mar. 1/2 The Soviet Union yesterday put forward a plan for banning the military use of outer space. 1961 WODEHOUSE *Ice in Bedroom* xxiii. 191 Every time I see this little horror from outer space, I want to sock him. 1964 *Ann. Reg.* 1963 31 On 25 July the treaty banning nuclear tests in the atmosphere, in outer space, and under water was signed. 1972 'H. CALVIN' *Take Two Popes* xii. 135 If your job says expend somebody, you expend... Do you think I'm an inhuman thing from outer space? 1972 *Times* 4 Dec. 8/4 Pioneer should become the first man-made object that goes past Jupiter..into outer space. 1901 Outer space [see *ASPIRING]. 1974 F. HERWOOD *Planning & Human Need* vi. 104 The rise of the ghetto must be met, at least in part, in the outer suburbs. [see *CREEPING *ppl. a.* 2b]. 1964 P. F. ANSON *Bishops at Large* vi. 202 Ethelbert Lodge..had been the home of an episcopal curia and outer-suburban country estate of the Landaffs for several years. 1928 *Daily Express* 7 May 4 (Advt.), Seven guineas and three guineas are probably the two most popular prices for outerwear garments. 1946 *Daily Tel.* 27 Mar. 5/8 From April 1 elastic can be used without restriction in the manufacture of underwear,..outerwear,..and miscellaneous articles. 1965 [see *LEISURE *sb.* 6 a and c]. 1977 *Ashes to Asher* ix. 91 They were hairy outerwear and came from socially aware suburbs. 1977 *Times* 24 Dec. 17/1 In the six months to October 31, Forminster, which makes ladies' and children's outerwear clothing, turned in a pre-tax profit of £564,000.
B. *sb.* **2.** *Electr. Engin.* In a three-wire distribution system, either of the two conductors whose potentials are respectively above and below that of the earth or neutral by equal amounts.
1900 [see *NEUTRAL *a.* and *sb.* B. 4]. 1932 R. RAWLINSON in E. Molloy *Pract. Electr. Engin.* V. 1590/2 The potential difference between the outers is still 500 volts, and..they may therefore be used for power supply, while

lighting load may be taken on a circuit between either outer and the mid-wire. 1970 H. FERRY *Electr. Supply* I. ii. 35 The cancellation of the two currents flowing in opposite directions in the middle wire..enabled the section to be safely reduced to one-half that of the outers.
3. (*a*) *pl.* Outdoor clothing. (*b*) An outer garment or the outer part of a garment.
1904 E. NESBIT *Phoenix & Carpet* vi. 111 'We'd best put on our outers in case—' 'We might rescue a traveller buried in the snow.' 1971 C. BONINGTON *Annapurna South Face* xi. 129 It needed a distinct effort of will..to force on frozen boots. I kept the felt inners inside the sleeping-bag, but the outers were too bulky and I used these as a pillow; even so, each morning they were frozen solid. T. FROST in *Ibid.* xviii. 222, I put on my..proofed nylon outers. 1976 *Horse & Hound* 3 Dec. 18/2 (Advt.), The Husky Riding Waistcoat, with its strong nylon outer, filled with polyester for thermo-insulated warmth, will keep those biting winter winds at bay.
4. *Austral. slang.* The part of a racecourse outside the enclosure; also *transf.* Hence in phr. *on the outer*: penniless; out of favour, excluded.
1924 *Truth* (Sydney) 27 Apr. 6 *Outer, on the*, to be poor; to be outside. 1926 'J. DOONE' *Timely Tips for New Australians* Gloss., *Outer*, a slang word denoting a betting ground overlooking a race-course. 1928 A. WRIGHT *Good Recovery* 157 You told me yourself that you were the cause of my being on the outer. 1953 T. A. G. HUNGERFORD *Riverslake* 174 And you're on the outer for sticking up for him? 1963 A. ROSS *Australia 63* iii. 86 Fine drizzle delayed things for half an hour, then shirts were ripped off again in the Outer, the beer cans were set up, and play proceeded. 1970 I. SOUTHALL *Bread & Honey* 54 Warren had always been on the outer, like a stray dog, always getting pushed.
5. An outer container into which one or several objects already enclosed in their own containers are packed for transport or display.
1920 J. STEPHENSON *Princ. & Pract. Commerc. Corr.* II. xii. 129 Size and description of outer. 1950 J. G. DAVIS *Dict. Dairying* 356 Packing of freshly frozen and packaged ice cream into 'outers' containing dozens of units..considerably slows the rate of hardening. 1955 *Sales Appeal* Jan.–Feb. 37/2 Display outers, cut-outs and showcards. 1967 *Times Rev. Industry* May 83/3 An instrument..provided with a carrying case was packed in a corrugated outer for transit. 1971 *Guider* Nov. 407 (Advt.), Your Net Profit per [chocolate] bar: 5p. Your Net Profit per outer (18 bars per outer): 90p.

outer, colloq. var. OUT OF *prep. phr.* Cf. *OUTA, *OUTTA.
1856 [see *GEEWHILLIKINS *int.*]. 1898 J. D. BRAYSHAW *Slum Silhouettes* 221 ''Ave a cigar,' an' 'e pulls a 'andful outer 'is skyrocket. 1926 *Opportunity* Mar. 83/2 The devil often assumes an importance entirely unbecoming to one who has been summarily 'kicked outer heaven'. 1954 M. PROCTER *Hell is City* II. iv. 53 If I could get outer 'ere I could be 'ome in less nor an hour.

outfall. 3. b. (Further example.)
1933 *Catholic Bulletin* Mar. 182 In *The Commonweal*, Padraic Colum..now finds a suitable outfall for his anti-Irish spate.

outfield, out-field, *sb.* Add: **3. a.** (Earlier and later examples.) Also *transf.* and *fig.*
1851 W. CLARK in W. Bolland *Cricket Notes* 136 If you are in the out field, and the batsman is on the alert, he will steal a run. 1868 H. CHADWICK *Game of Base Ball* 73 The Irvingtons..placed the substitute in the out-field. 1948 *Denison* (Texas) *Herald* 2 July 12/3 Have you ever played the outfield? 1971 *Times* 16 Feb. 7/6 The groundsman at the cricket ground here has been mowing the outfield. 1975 *Cricketer* May 19/1 When they batted the opening pair put on 85 in the first 20 overs, a splendid rate on an extremely wet outfield. 1976 *Billings* (Montana) *Gaz.* 16 June 1-C/3 Playing both first base and the outfield, Rudi was regarded as one of the most valuable players in the league every season from 1970–75. 1976 P. HARCOURT *Dance for Diplomats* ii. 15 We taxied around the runways..and came to a stop somewhere in the out-field. 1977 J. LE CARRÉ *Hon. Schoolboy* iii. 58 Martindale..in the Whitehall outfield lived in a state of primaeval innocence about the reality of Smiley's world.
b. (Earlier and later examples.) Also, the out-fielders collectively.
1867 *J. Lillywhite's Cricketers' Compan.* 160 [He is] a fine out-field. 1867 H. CHADWICK *Base Ball Player's Bk. Reference* 138 The Out-Field—The out-fielders are the left[,] centre and right-field positions. 1868 —— *Game of Base Ball* 17 The out-field being neither active in their movements, or sure catchers. 1910 *Blackw. Mag.* Jan. 93/2 Tyldesly..ten years ago was one of the finest out-fields imaginable. 1948 *Daily Ardmoreite* (Ardmore, Okla.) 30 Mar. 6/4 If the Boston Braves win.., they will do it without an outfield which is particularly strong defensively. 1956 *People* 13 May 13/3 Cliff is an attractive all-round cricketer and an especially good long-throwing fast-moving outfield.

out-field, *v.* *Cricket* and *Baseball.* **1.** *intr.* [f. the sb.] To field in the out-field. *rare.*
1862 *Baily's Monthly Mag.* Aug. 85 The Surrey people ..selecting..an F. Lee, a Daniel, and an E. B. Rowley to out-field.
2. *trans.* (See *OUT- 18.)

out-fielder. (Earlier and later examples.)
1868 H. CHADWICK *Game of Base Ball* 73 The Irvingtons..took an out-fielder from his regular position. 1942 *Sun* (Baltimore) 3 Apr. 18 Estel Crabtree, Cardinal outfielder, attributed his greatly improved batting in 1941 to constant work with the machine. 1972 *Village Voice* (N.Y.) 1 June 69/2 He'd lyricize an outfielder's grace and then demean baseball as a little boy's game. 1974 *Anderson* (S. Carolina) *Independent* 19 Apr. 4B/3 The A's used Allan Lewis, an outfielder with little hitting ability, as a pinch running specialist the last two World Series.

out-fielding, *vbl. sb.* (Earlier and later examples.)
1860 in *Ball Players' Chron.* (1867) 12 Dec. 3/1 The out-fielding was only so-so. 1861 *Times* 25 May 9/4 Notwithstanding the truly fine bowling and general good out-fielding, Mr. Burbidge and Griffith defended their wickets in a masterly manner. 1973 B. RICHARDS *On Cricket* x. 98 Outfielding, too, requires concentration.

ou·t-fighting, *vbl. sb.* [OUT- 2.] Fighting that is not at close quarters.
1848 *Sporting Life* 5 Feb. 297/2 At out-fighting, Bateman was decidedly the quickest and the best. 1905 *Times* 6 Mar. 4/2 It is probable that this outfighting, before the adversaries close, will be fruitful in important lessons relating to the art of naval war.

outfit, *sb.* Add: **2. a.** (Further examples.) Also, equipment of any kind; a set of articles for a particular purpose.
1867 J. F. MELINE *Two Thousand Miles on Horseback* 74 The saddler who sold me my saddle assured me it was the best outfit he had furnished for some time. Bought a hat, and was told, 'Well, Sir, I call that a good outfit.' 1869 A. K. MCCLURE *3,000 Miles through Rocky Mts.* 211 Everything is an 'outfit', from a train..to a pocket-knife. 1873 E. B. TUTTLE *Boy's Bk. Indians* 45 Friday had a beautiful set of arrows, bow and quiver, which I desired to purchase... Friday would not sell his 'outfit', as it is called, for money. 1924 T. E. LAWRENCE *Home Lett.* (1954) 359 A solo isn't as secure on a wet road as a side-car outfit. 1958 *Amer. Speech* XXXIII. 271 *Outfit*, a man's equipment.
c. A person's clothes; a set of garments.
1852 [in Dict., sense 2 a]. 1875 *Scribner's Monthly* Dec. 286/1 The comfortable dress for the mother or flannel outfit for the baby, can be sent. 1946 *Chicago Daily News* 10 Aug. 12/3 Elaborate ceremonial outfits are fashioned by the women of the tribe. 1968 J. IRONSIDE *Fashion Alphabet* 104 Accessories..really can change the whole feeling of an outfit and turn a last year's garment into something swinging or chic. 1976 C. BERMANT *Coming Home* i. vii. 105 She did not wear the same outfit two days running.
d. The apparatus used by a drug addict for taking drugs.
[1935 A. J. POLLOCK *Underworld Speaks* 84/2 *Outfitted*, to get a supply of dope.] 1951 *N.Y. Times* 14 June 22/2 'John' went into a drugstore on upper Park Avenue, asked for an 'outfit', and for 45 cents got an envelope containing a medicinal dropper and a hypodermic needle, such as used by 'mainliners' for heroin injections. 1953 W. BURROUGHS *Junkie* (1972) ix. 86, I asked him to come back to my apartment to take a shot. We went back to my room, and I got out my outfit that hadn't been used in five months. *Ibid.* xii. 121 She keeps outfits in glasses of alcohol so the junkies can fix in the joint and walk out clean. 1960 *Times Lit. Suppl.* 16 Sept. 589/4 Anyone who has snorted or used the outfit and then kicked would find it [sc. a book] of absorbing interest.
3. (Earlier examples.)
1867 *Harper's Mag.* July 137/2 Their 'outfit' (in the language of the plains this word signifies the conveyance, its contents, and the team) consisted of a Concord coach [etc.]. 1869 S. BOWLES *Our New West* viii. 163 With a mounted escort of about twenty gallant young miners.. we made up a grand 'outfit'.
b. A group of people; an organization; a business firm or concern.
1883 'MARK TWAIN' *Life on Mississippi* ii. 31 In that day, all explorers travelled with an outfit of priests. De Soto had twenty-four with him... The expeditions were often out of meat, and scant of clothes, but they always had the furniture and other requisites for the mass. 1925 *Amer. Speech* I. 149/2 The big cattle companies or 'outfits'. 1926 J. BLACK *You can't Win* iv. 37, I was left on the bench with the two drunks... The desk man pointed to us. 'What will I do with this outfit, Hayes?' 1927 [see *LAY-OUT 2 c]. 1930 D. L. SAYERS *Strong Poison* iv. 48 You must get me passed in as part of your outfit. 1935 WODEHOUSE *Blandings Castle* xii. 310 'Come and join my little outfit,' he said heartily. 'I've always room for a personal friend.' 1939 F. W. CROFTS *Fatal Venture* xi. 154 He's carrying this entire outfit on his shoulders. Only for Stott your job and mine might go phut. 1943 H. L. MENCKEN *Heathen Days* vii. 89, I was presently playing trios and quartettes with an outfit that devoted four hours of every week to the job. 1951 C. W. MILLS *White Collar* I. ii. 23 The great bulk of businesses are small outfits, which do not last long. 1958 S. ELLIN *Eighth Circle* (1959) II. i. 33 There's a couple of other agencies—Inter-American, Fleischer—pretty good outfits that might do just the job you want. 1962 *Observer* 25 Nov. 12/8 Aldermaston is 'less stuffy than Harwell' and usually beats the other nuclear outfits at rugger. 1975 *New Yorker* 21 Apr. 69/2 Investigators working not only for Retail Credit but for other large consumer-investigation outfits confirm this. 1977 J. WAINWRIGHT *Do Nothin'* v. 67 Some of the modern outfits don't have brass. Just a four-piece sax line-up.
c. *spec. Services' colloq.* A group of servicemen; a regiment, squadron, or the like.
1916 [see *anti-tank* s.v. *ANTI-[1] 4 (iii)]. 1922 C. E. MONTAGUE *Disenchantment* ii. 18 A man is wanted for Post Corporal... 'Cushiest job in the 'ole outfit!' 1930 F. A. POTTLE *Stretchers* 28 The bowlegged officer flew into a disciplinary rage and addressed the boy as follows: 'What outfit do you belong to? How long have you been in the army?' 1951 H. HASTINGS *Seagulls over Sorrento* I. i, in J. Trewin *Plays of Year* IV. 52 Incidentally, wot are you doing in an outfit like this? 1959 *Times Lit. Suppl.* 30 Jan. 57/3 The sergeant, though a war hero and the only soldier in a slack outfit, is a psychopathic bully and a pathetically lonely man. 1966 *Inland* (Inland Steel Co., Chicago) Autumn 13/1 Recruits from all over the United States are assigned to the same military outfit. 1978 A. PRICE *'44 Vintage* xix. 220 France in '40, then the Middle East... And finally Yugoslavia as a weapons adviser to a big Partisan outfit.
4. *Canad.* The name given to the fiscal year of the Hudson's Bay Company; hence, a year.
1791 in *Beaver* (1947) Dec. 11/2 The difficiency in the order for the carrot tobacco arises from the great overplus we have for the ensuing outfit. 1833 J. G. MCTAVISH *Let.* 18 Dec. in G. de T. Glazebrook *Hargrave Corr.* (1938) 124, I had thought, of getting away in one or two outfits. 1841 G. SIMPSON *London Corr.* (1973) 9 The lease..will expire..with the close of Outfit 1841/2. 1874 in *Alberta Hist. Rev.* (1956) Spring 16 The Athabasca accounts for the Outfit were closed and the Packet given over to the care of Mr. Moberly. 1913 I. COWIE *Company of Adventurers* 280 The end of each business year—called 'Outfit'—was May 31, upon which date the inventory of everything belonging to the Company at the fort was taken. 1935 in D. JENNESS *Eskimo Admin.* (1964) II. 51 Altogether sixteen deaths occurred during the outfit and only four births that I know of. 1973 W. R. SAMPSON *J. McLoughlin's Business Corr. 1847–48* p. xix, He was to be made a wintering partner with a share in the profits of the concern for Outfit 1814.
5. A person (usu. pejorative). *slang* (chiefly *U.S.*).
1867 J. F. MELINE *Two Thousand Miles on Horseback* 74 To cross the plains, or go to the mountains, every one must get an outfit; and having outfitted, you become yourself an outfit. 1924 C. E. MULFORD *Rustlers' Valley* xi. 130 You ain't believin' everythin' *this* outfit tells you, are you? 1925 *Ladies' Home Jrnl.* May 26/2 'But that young outfit will drive me wild,' protested Mrs. Denmeade. 1942 BERREY & VAN DEN BARK *Amer. Thes. Slang* § 379/2 *Person*,..outfit.

outfit, *v.* Add: **a.** (Further examples.) Also simply, to provide or supply (a person or thing) *with*.
1924 W. M. RAINE *Troubled Waters* xvi. 167, I outfitted some of the boys with guns. 1928 *T. Eaton & Co. Catal.* Spring & Summer 219/4 Two or three suits like this would outfit him for the Summer. 1935 A. SQUIRE *Sing Sing Doctor* xiv. 205 The condemned man is outfitted with new clothing. 1953 K. TENNANT *Joyful Condemned* xxxix. 381 She was smartly outfitted: white shoes, white hat.. and a black-and-white striped dress. 1971 *Sci. Amer.* Aug. 24/1 When the slim merchantman was ready for launching, its owner must have visited a ship chandler to buy the gear needed to outfit the vessel. 1972 P. H. KOCHER *Master of Middle-Earth* (1973) ii. 23 They helped to outfit the dwarf expedition when it was penniless. 1973 'D. HALLIDAY' *Dolly & Starry Bird* iv. 55 Enough polo necks to outfit the entire British Raj. 1975 *Nature* 10 July 150/3 Such cameras are intrinsically sensitive to the near ultraviolet band, and to be used for ultraviolet viewing need only be outfitted with an ultraviolet-transmitting lens and filter. 1976 *National Observer* (U.S.) 19 June 8/6 And began to practice out of the van that he and several seabee friends had outfitted.
b. (Earlier and later examples.)
1881 *N.Y. Times* 18 Dec. 4/3 To 'outfit' is to fit out for any purpose whatever. 'We outfitted at St. Paul.' 1902 S. E. WHITE *Blazed Trail* xxv. 168 It's a good place to outfit from because we can probably get freight rates direct by boat. 1919 H. L. WILSON *Ma Pettengill* iv. 129 He outfitted at the Chicago Store in Tucson, getting the best all-wool ready-made suit in Arizona. 1924 C. E. MULFORD *Rustlers' Valley* x. 115 Yestiddy was pay-day, an' if they don't outfit now, some of 'em won't have no money after to-night. 1976 H. & G. GORDON *Ordeal* xv. 108 A party of four went down the trail... Three males and a female. Ring up the Big Rock Trading Post... Get the trader up and find out if they outfitted there.
Hence **ou·tfitted** *ppl. a.*
1975 *Offshore Progress: Technol. & Costs* (Shell Internat. Petroleum Co.) 5 Several semi-submersibles delivered early in 1975 have hulls 300 ft long, and a total outfitted weight of up to 16,500 tons. 1977 *Time* 19 Dec. 31/1 Outfitted in quilted parkas, they can be seen roaming the snow-covered hills and hollows of Appalachia in search of game to keep down meat bills.

outfitting, *vbl. sb.* and *ppl. a.* (Further examples.) Also *ellipt.*
1908 *Westm. Gaz.* 24 Apr. 7/4 Until the cold weather and overcoats finally disappear there will be no improvement in outfitting. 1932 D. L. SAYERS *Have his Carcase* xxxi. 409, I went to the men's outfitting and asked for collars. 1964 A. ADBURGHAM *Shops & Shopping* xii. 130 Outfitting was the euphemism employed for both men's and women's underclothes. 1976 *Cody* (Wyoming) *Enterprise* 23 June 1/1 Many of the complaints about the grizzly bear's critical habitat center around the effect such a definition would have on the outfitting business. A confrontation between outfitting camps and grizzlies within a critical habitat area would result in the relocation of the camp rather than the grizzly. 1977 *Times* 30 Nov. 2/4 The decision by more than 1,600 angry outfitting trades at Swan Hunter's [shipbuilding] yard to continue their overtime ban.

outfla·nker. [-ER[1].] One who outflanks.
1920 *Q. Rev.* Jan. 107 As fast as Joffre created a new Army to prolong his left..so fast did the Germans cover their threatened right and seek to outflank their would-be outflankers.

outflash, *v.* **a.** (Earlier and later examples.)
1833 BROWNING *Pauline* 59 Do I not..burn to see thy calm, pure truths out-flash The brightest gleams of earth's philosophy. 1939 JOYCE *Finnegans Wake* 210 A pretty box of Pettyfib's Powder for Eileen Aruna to whiten her teeth and outflash Helen Arhone.

outfling, *v.* Add: (Further examples.) Hence **outfli·nging** *vbl. sb.*
1894 A. S. WAY tr. *Euripides' Tragedies* I. 135 What speech in thy frenzy outflingest thou? 1896 *Ibid.* II. 186 Through Hades' hall to thee I call, Day after day my cries outflinging. 1922 JOYCE *Ulysses* 415 Outflings my lord Stephen, giving the cry. 1932 S. GIBBONS *Cold Comfort Farm* v. 73 Judith thrust the words aside with a heavy movement of her hand, like the blind outflinging of a tortured beast. 1950 R. BRADBURY *Martian Chronicles* 86 If art was no more than a frustrated outflinging of desire, ..what good was life?

outflow, *v.* (Later example.)
1909 *Daily Chron.* 3 Mar. 5/7 Then outflowed a stream of facts and figures whose accumulated force swept the critics off their feet.

outflowing, *vbl. sb.* (Further examples.)
1884 H. B. MACKEY tr. *St. Francis de Sales' Treat. Love of God* VI. xii. 267 The outflowing of a soul into her God is a true ecstasy. 1932 F. J. SHEED *Irish Way* 332 Work itself was..an outflowing or a special directing of the steady fixedness of the soul in God. 1953 R. C. JOHNSON *Imprisoned Splendour* i. 31 Mystical experiences..are, an out-flowing into awareness of something from a higher level of the self even than the buddhic.

outflowing, *ppl. a.* (Later example.)
1975 *Sci. Amer.* Mar. 33/3 The compact star in its orbit plows through this outflowing wind and captures only a tiny part of its matter.

outflux. Add: In mod. use, *spec.* the outward movement of ions through a cell membrane.
1949 *Physiol. Rev.* XXIX. 132 If the high K concentration in muscle and nerve cells as compared with the K concentration of the surroundings is due to a Donnan equilibrium then the K influx and the K outflux through the cell membranes must be the same. 1974 *Nature* 11 Jan. 96/1 The..rate of sodium outflux (15 min) in turkey erythrocytes was not modified.

outfly, *v.* Add: **2.** *spec.* of aircraft and their pilots: to surpass in terms of skill or speed in flying.
1908 H. G. WELLS *War in Air* viii. 253 Light as this armament was.., it was sufficient for them to outfight as well as outfly the German monster airships. 1942 *Tee Emm* (Air Ministry) II. 94 It's no good outflying the Hun if you can't shoot him down. 1975 *Listener* 17 July 77/1 It could outfly the Focke Wulf Condor and..that sort of German bomber. 1976 *Globe & Mail* (Toronto) 28 Jan. 29/2 Capt. Stinson..flatly states that the computer can 'out-land a human pilot three to one. I'll admit it can outfly me.'

outfox (autfǫ·ks), *v.* [f. OUT- 18b + FOX *v.* 2c.] *trans.* To outdo in deception or cunning; to outwit.
1962 K. ORVIS *Damned & Destroyed* xii. 81 He finally out-foxed himself. 1965 C. D. EBY *Siege of Alcázar* (1966) ii. 54 Pozas was no fool. Perhaps he had even outfoxed them. 1965 R. SHECKLEY *Game of X* (1966) xix. 133 We have outfoxed Forster at every turn, and we shall outfox him now for the final time. 1973 J. JONES *Touch of Danger* lvi. 320, I guess you've got the drop on me... You outfoxed me. 1976 *National Observer* (U.S.) 19 June 6/5 As one Republican National Committee official put it: 'The Ford people have no right to be bitter. They were just out-foxed, outmaneuvered, and out-organized.'

out front, *advb.* and *adj. phr.* Chiefly *U.S.* [f. OUT *adv.* + FRONT *sb.*] At or to the front; in front; *spec.* (*a*) at or to the battle-front; (*b*) *Theatr.*, in front of the stage; in the auditorium; (*c*) *fig.*, in the forefront (of a political or intellectual movement); progressive. Also (with hyphen) *attrib.*
1916 'BOYD CABLE' *Action Front* 200 These average good men who had 'joined up' freely, who had longed for the end of home training and the transfer 'out Front'. 1934 J. M. CAIN *Postman always rings Twice* vi. 57 She went out front with an order, and me and the Greek sat down. 1937 C. HIMES *Black on Black* (1973) 129 This little white lain pulled up out front in a big Lincoln touring a block long. 1962 *Listener* 11 Oct. 570/1 Obsolete accommodation, backstage and out-front. 1963 AUDEN *Dyer's Hand* 186 There is no difference for Falstaff between those on stage and those out front. 1968 T. WOLFE *Electric Kool-Aid Acid Test* iv. 53 A community of intelligent, very open, out-front people—out front was a term everybody was using. 1973 *Black World* Mar. 36 His behind-the-scenes rather than vigorous outfront leadership. 1976 L. ALTHER *Kinflicks* ii. 22 'Have you seen the new fishburger franchise?' Mrs Yancy asked, pointing out the window at a red and silver building with a sign out front featuring in neon a one-legged pirate tangoing with a laughing swordfish. 1977 *New Yorker* 10 Oct. 156/3 Powell himself was said to be deeply bothered by that, and to have realized that he had been too 'out front' on the issue.

outgas (autgæ·s), *v.* [OUT- 26.] **1.** *trans.* **a.** To drive off sorbed gas or vapour from (a solid), esp. by heating in a vacuum.
1921 [implied in *OUTGASSED *ppl. a.*]. 1925 *Physical Rev.* XXVI. 658 When the surface layer of gas is removed from a fresh specimen the increase in the photo-electric current is greater than the decrease from the maximum value as the specimen is outgassed. 1953 *Electronic Engin.* XXV. 19 The cathodes were out-gassed by eddy-current heating. 1965 C. M. VAN ATTA *Vacuum Sci. & Engin.* iii. 101 After the gauge tube and elements have been thoroughly outgassed, an opposite effect becomes noticeable.
b. To release (sorbed or dissolved gas or vapour).
1971 I. G. GASS et al. *Understanding Earth* ix. 137/2 Only gradually, as volcanoes continued to outgas volatile products still trapped in the mantle, will a secondary atmosphere and ocean have replaced the primary envelope. 1974 *Nature* 31 May 438/1 If NH_3 is outgassed from the Martian crust it would be photolysed.
2. *intr.* To give off sorbed gas or vapour.
1962 [implied in *OUTGASSING *vbl. sb.*]. 1965 C. M. VAN ATTA *Vacuum Sci. & Engin.* ix. 365 Untreated metal samples outgas at the rate of about 10^{-7} torr liter/sec cm^2 after 1 hr of vacuum pumping at room temperature. 1975 *Sci. Amer.* Feb. 110/3 Any polymer surface will outgas (a 'virtual leak' into your clean volume) two orders of magnitude more than steel.
So **outga·ssed** *ppl. a.*, **outga·ssing** *vbl. sb.*
1921 *Proc. Nat. Acad. Sci.* VII. 115 The coarsely granular sample of the thoroughly outgassed material is weighed and placed in a steel pressure bomb which is then evacuated until all adsorbed gases are removed. 1925 *Physical Rev.* XXVI. 657 His thermionic measurements show that the value of *A* in the Richardson equation.. decreases from $4{\cdot}76 \times 10^{17}$ for the slightly outgassed position to $1{\cdot}7 \times 10^{24}$ for continued outgassing. 1952 *Trans. Faraday Soc.* XLVIII. 739 The vessel F was 'protected' by a series of traps in liquid air, some of which..contained carefully outgassed granulated charcoal. 1962 F. I. ORDWAY et al. *Basic Astronautics* iii. 49 Whatever outgassing that may occur from the surface would expectedly give rise to a more substantial, albeit tenuous, atmosphere. 1971 I. G. GASS et al. *Understanding Earth* ii. 43/2 This so-called 'excess' argon is probably produced by heating and outgassing of ancient, potassium-bearing rocks. 1973 *Nature* 3 Aug. 272/2 The relative abundances of these gases will depend critically on the subsequent history of the outgassed methane. 1973 B. J. WILLIAMS *Evolution & Human Origins* vii. 97/2 If these dates are correct then the evolution of life must have begun, as we would expect, almost immediately after the earth's crust melted and the outgassing of the early atmospheric gases ceased.

outgeneral, *v.* Add: **b.** *transf.* and *fig.*
1859 J. S. MILL in *Fraser's Mag.* LX. 767/1 A nation which thinks of nothing but of outwitting and outgeneralling its neighbours. 1910 J. DRISCOLL *Ringcraft* 14 He was the better boxer and the stronger man, but was outgeneralled during two-thirds of the bout. 1940 W. FAULKNER *Hamlet* III. ii. 184 At last he outgeneralled himself with his own strategy:..even his father admitted that there was nothing else about the farm for him to learn. 1973 WODEHOUSE *Bachelors Anonymous* x. 128 However confident he may be that he has outgeneralled a woman, a man likes to have reassurance on the point from a knowledgeable third party. 1976 *Billings* (Montana) *Gaz.* 26 June 1-B/6 Tanner has played better tennis —little Kirmayr out-generalled him time and again—but always in reserve was his 140 m.p.h. cannonball service.

outgiving (au·tgiviŋ), *ppl. a.* [OUT- 10.] That gives out; open-hearted, generous. Hence **outgi·vingness.**
1942 J. LEES-MILNE *Ancestral Voices* (1975) 31 K. as outgiving as ever. 1961 *Spectator* 26 May 763/2 [Brendan] Behan is still..a talker and singer for talking and singing's own sake, spontaneous and out-giving. 1963 *London Mag.* Sept. 11 Her face was round and pleasantly fleshed, her eyes cool and outgiving when she was not anguished or perturbed. 1968 *Listener* 10 Oct. 458/2 She had lost a great deal, I think, of her out-givingness in that way and the novels perhaps reflect this. 1972 *Times* 3 July 16/7 There can rarely have been such an outgiving man who was less of an extrovert.

outgo, *sb.* **4.** (Earlier example.)
1869 S. BOWLES *Our New West* i. 26 The great Salt Lake of Utah..has no visible outgo, though richly fed from various quarters.

outgo, *v.* **1.** For *Obs.* read '*Obs.* except *poetic*' and add later examples.
1899 P. H. WICKSTEED tr. *Dante's Paradiso* xiii. 161 That living Light which so outgoeth from its Source that it departeth not therefrom. 1905 *Outlook* 4 Nov. 629/1 So you, dear Frank, were last of those To whom a tender thought outgoes.

outgoing, *vbl. sb.* **2.** Delete † *Obs.* and add later *literary* examples.
1864 G. M. HOPKINS *Poems* (1967) 130 The hill Which with its lined and creased flank The outgoings of the vale does block. 1918 D. H. LAWRENCE *New Poems* 49 Each door, each mystic port Of egress from you I will seal and steep in perfect chrism..So you shall feel Ensheathed invulnerable with me, with seven Great seals upon your outgoings.

outgoing, *ppl. a.* Add: **c.** Suitable for wearing when one goes out.
1867 QUEEN VICTORIA *Let.* 19 Nov. in R. Fulford *Your Dear Letter* (1971) I have just returned from a drive with dear Marie F... I can't judge of the figure in her out-going dress. 1909 'O. HENRY' *Roads of Destiny* x. 162 Take him back..and fix him up with outgoing clothes. Unlock him at seven in the morning.
d. Extrovert, sociable, open-hearted, friendly.
1950 *Brit. Jrnl. Psychol.* XXII. 107 Out-going primary or secondary need-determined behaviour may be thwarted and aggressiveness may result. 1955 in D. Tidyman *Dummy* (1974) iii. 40 He is a rather attractive, out-going child. 1964 J. PHILIPS *Laughter Trap* (1965) III. iii. 132 She was a warm, outgoing girl. 1966 *Tablet* 22 Oct. 1183/1 The..poem, whose outgoing, solicitous concern is for the whole of man's relation to the world around him. 1973 C. BONINGTON *Next Horizon* ix. 138 He appeared outgoing, frank, and immensely enthusiastic. 1978 E. HEALEY *Lady Unknown* ii. 62 Gregarious and out-going, he brought a brilliance and radiance to every gathering.
outgoingness (further examples).
1960 G. ASHE *From Caesar to Arthur* vi. 160 For the moment we may leave him in Ireland, fitly typifying the outgoingness of the British saints. 1967 C. FREMLIN *Prisoner's Base* ii. 22 Claudia's gifts..tolerance—out-goingness—sympathy. 1971 *New Scientist* 1 Apr. 44/2 Bondi's natural eloquence and outgoingness could be usefully harnessed to his new job.

outgrabe (autgrēɪ·b), *v.* A factitious word introduced by 'Lewis Carroll' (see quot. 1855[2]). (In quot. 1903 used for 'outdo', after the style of *out-Herod*, etc.)
Quot. 1855[1] also occurs in the first verse of 'Jabberwocky' in *Through the Looking-Glass* (1871) i. 21.
1855 'L. CARROLL' *Rectory Umbrella & Mischmasch* (1932) 139 All mimsy were the borogoves; And the mome raths outgrabe. *Ibid.* 140 *Outgrabe*, past tense of the verb to *outgribe*. (It is connected with the old verb to *grike* or *shrike*, from which are derived 'shriek' and 'creak'.) 'Squeaked.' 1876 —— *Hunting of Snark* v. 50 The Beaver had counted with scrupulous care, Attending to every word: But it fairly lost heart, and outgrabe in despair, When the third repetition occurred. 1903 *Sat. Rev.* 7 Feb. 164/1 Deadmanship! wrote..Dr. Shrapnel..; and the word is fit to stir the jealous admiration of Carlyle or even Lewis Carroll. Indeed Dr. Shrapnel 'outgrabed' them both.

ou·t-group. [Cf. OUT *a.* 1.] Those people not necessarily forming a group themselves, who are excluded from or do not belong to a specific in-group; also *attrib.* Hence **out-grou·per,** an individual who does not belong to a specific in-group.
1907 [see *IN-GROUP]. 1934 K. YOUNG *Introd. Sociol.* I. i. 13 One is prejudiced against the members of the out-group. 1949 R. K. MERTON *Social Theory* II. vii. 186 The systematic condemnation of the out-grouper continues largely irrespective of what he does. 1952 M. MCCARTHY *Groves of Academe* (1953) vi. 119 Where discrimination exists, protection of the out-group is mandatory. 1967 M. ARGYLE *Psychol. Interpersonal Behaviour* iv. 71 Two people from different groups are apt to treat each other as 'outsiders', members of the out-group, and to reject one another through their failure to conform to the norms of the in-group. 1970 *Jrnl. Gen. Psychol.* Oct. 259 An error was considered..outgroup intrusion if a word paired with a dissimilar stimulus was elicited. 1970 C. T. RESTREPO in I. L. Horowitz *Masses in Lat. Amer.* xiv. 516 Violence developed the conflict with respect to the out-group and institutionalized it. 1976 *Times Lit. Suppl.* 2 Jan. 2/4 Gypsy legends..provide a charter for the in-group rather than the out-group reference of their morality.

outgrow, *v.* Add: **1.** Also *refl.*
1878 HARDY *Ret. Native* I. I. iii. 54 His mother cried for scores of hours when 'a was a boy, for fear he should outgrow himself and go for a soldier.

outguess (autge·s), *v.* [f. OUT- 18b + GUESS *v.*] *trans.* To outwit (someone) by guessing more cleverly or shrewdly.
1913 *Jrnl. Animal Behaviour* III. 90 It was clearly a case of the chick 'outguessing' the experimenter. 1921 E. O'NEILL *Emperor Jones* i. 163, I kin outguess, outrun, outfight, an' outplay de whole lot o' dem all ovah de board. 1936 M. MITCHELL *Gone with Wind* I. v. 78 Mammy sighed resignedly, beholding herself outguessed. 1956 'J. WYNDHAM' *Seeds of Time* 187, I don't know whether he outguessed me or whether he was just lucky. 1972 P. H. KOCHER *Master of Middle-Earth* iii. 44 Her function in the story is to warn them, and herself, to tend to the duty in hand and not rashly to presume that finite minds can outguess the supreme architect who plans the whole. 1975 *Publishers Weekly* 6 Jan. 52/3 Murders and cover-ups follow, with the reader trying to outguess the author all the way.

out-gun, *v.* Add: (Further examples.) Also, to surpass in gunnery; to outshoot. Also *fig.* Hence **outgu·nned** *ppl. a.*
1942 *R.A.F. Jrnl.* 13 June 12 The wolf—helpless, out-ranged and outgunned so far as the *Devonshire* was concerned. 1945 *Sun* (Baltimore) 17 Mar. 9-0/1 We're just out-tanked and out-gunned, that's all. 1960 *Observer* 28 Aug. 25/6 The Corporation has shown that it can outgun its rival. 1973 *Daily Tel.* 28 Nov. 15 The Battle of the River Plate, in which three outgunned cruisers took on the German pocket battleship Admiral Graf Spee. 1977 *Times* 5 May 1/8 Mr Callaghan, Dr Owen..and Mr Rees.. could well be outgunned by a majority of ministers.

outgush, *sb.* Add: Also *fig.*
1871 BROWNING *Balaustion's Adventure* 170 Frank outgush of the human gratitude Which saved our ship and me. 1919 [see *LA, LA].

ou·thalf, out-half. *Rugby Football.* [f. OUT- + HALF *sb.* 6 d.] = *fly-half* (*FLY *sb.*[2] 8). Cf. *outside half* s.v. *OUTSIDE *a.* 6.
Fly-half is the more usual term in the U.K.
1961 in WEBSTER. 1973 *Irish Times* 2 Mar. 3/7 At their heels was scrum-half Early.., while Spring is a cool out-half. 1976 *Sunday Tel.* 23 May 39/8 'Oh my,' said Mike Quinn, the outhalf, 'if the New Zealand centres are as quick as that what are their wings like?' 1977 *Belfast Tel.* 27 Jan. 25/8 No sooner had he settled into the side than out-half Keith Gilpin was sidelined.

outhouse. Add: **b.** A privy. Chiefly *U.S.*

1819 W. Sewall *Diary* 14 June (1930) 53/1 Near the paddles or wheels which propel the vessel forward are two outhouses, which is a very great convenience. 1832 W. Irving *Jrnl.* Nov. (1919) III. 185 Old Spanish wooden building, with piazza—out houses—French buildings, with casement. 1912 *Dialect Notes* III. 584 *Out-house*,..a privy. 1921 H. Kephart *Camping & Woodcraft* (1928) I. xii. 210 If the well is near a stable or outhouse, or if dish-water is thrown near it, let it alone. 1968 C. Helmericks *Down Wild River North* I. xviii. 286 They still clung steadfastly to the old outhouse. 1973 *N.Y. Law Jrnl.* 26 July 16/7 (Advt.), Executive hideaway, L.I. Rustic appointments throughout..attractive 2 seater outhouse.

c. At a school, a house (see House *sb.*[1] 4 c) separate from, or subsidiary to, the main or central house.

1900 Farmer *Public School Word-Bk.* 144 *Out-houses* (Charterhouse),—all the boarding houses except Sanderites, Verites, and Gownboys. The names of the eight out-houses are Girdlestonites, Lockites, [etc.]. 1933 L. A. G. Strong *Sea Wall* 187 A tall, genial boy named Adams, belonging to another of the out-houses.

outing, *vbl. sb.* Add: **3.** (Earlier example.)

1849 *Bell's Life in London* 30 Sept. 6/2, I Zingari averages, 1849. Results with bat and ball... Innings... Runs... Outings... Bowled.

4. b. An appearance in an athletic contest, race, etc.

1943 *Sun* (Baltimore) 22 Apr. 18/1 Benefitting from a previous outing at the meeting, Mrs. Ray Feinberg's Charge, ridden by Danny Scocca, closed fast. 1976 J. Snow *Cricket Rebel* 29 Any young cricketer brought into the traditional twelve for his first Test outing could expect a back page headline in the national newspapers. 1976 *Evening Times* (Glasgow) 30 Nov., Bula, although only once successful in three outings so far this season, is my confident choice to register a repeat victory at Haydock Park.

7. (Further examples.) **outing flannel** *U.S.*, a type of flannelette; also *attrib.*

1890 *Advt.* (Schairer & Millen, Ann Arbor), 15 pieces new stripe outing flannels, 15 cent quality, now 10 cents. 1897 *Sears, Roebuck Catal.* 258/3 Outing flannel. This flannel is suitable for ladies' wrappers, waists, night robes, childrens' cloaks, also men's and boys' shirts and night shirts. 1916 *Daily Colonist* (Victoria, B.C.) 9 July 3/6 (Advt.), New awning stripe outing skirts. *Ibid.* 22 July 2/1 (Advt.), Take a Camera Wherever You Go. What is more delightful than having pictures of outing parties, picnics, scenery, etc.? 1925 T. Dreiser *Amer. Tragedy* (1926) I. II. xv. 262 He was pleased by the..summery appearance he made in an outing shirt and canvas shoes. 1938 M. K. Rawlings *Yearling* v. 51 The gray fur was as soft as his mother's outing flannel nightgown. 1967 *Boston Sunday Herald* 14 May II. 17/2 Every summer its members provide rafts and guides for vacationing families, outing groups, and student expeditions.

outland, *sb.* and *a.* **A.** *sb.* **3. a.** Delete †*Obs.* and add later examples. Also *transf.*

1931 *Sun* (Baltimore) 21 Oct. 14/6 Were it [*sc.* the weasel] the size of a bear or lion..the outlands would be unsafe for man unless he carried a gun. 1961 in Webster, In the outlands, the Yankees had been strangers. 1970 R. Lowell *Notebk.* 190 Someone comes here from the outlands, Trinidad.

outlander. Delete '(Now *poetic*, or a literary revival, or a mannerism of translation.)' and add earlier and later examples.

1598 Florio *Worlde of Wordes* 265/1 Pellegrino..a stranger, an alien, an outlander, an outlandish man. 1909 W. Tuckwell *Pre-Tractarian Oxford* iii. 59 Unfortunate outlanders whose digestion of the dinner and relish of the port wine were spoiled by these animated dialectics, went away complaining that Oriel Common Room *stunk* of Logic. 1930 *Time & Tide* 20 Sept. 1167, I think, if it is not an impertinence for an outlander to make a suggestion, that England does need [etc.]. 1938 *Englische Studien* LXXII. 323 The printed page marks the outlander by the use of italics and preservation of the French accents. 1952 G. Wilson *Julien Ware* 117 But, having admitted this outlander to his home and table, Mr. Craig felt it necessary to talk for his education and profit. 1964 *Listener* 24 Sept. 453/2 Mr [Robert] Kennedy is of course Boston Irish... He lives in Virginia... He saw the beguiling prospect of a Senate seat in New York. This State now echoes..with the cry of 'carpet-bagger' and 'outlander'. 1976 *Ibid.* 15 July 38/2 He tried to go to New York to the big law firms and they turned him down—he was an outlander from California. 1976 *Verbatim* Sept. 8/a In three essays on the phonology of Georgia place-names, Goff discusses local designations, recording divided usage,..and clarifying accentuation when needed. Outlanders are given the preferred native pronunciations of *Albany* ('All' benny'), *Aragon* ('Arrow'gun' or 'Arrer'gun'), *Schley* ('Sly''), and *Taliaferro/Bolivar* (which make a perfect rhyme in Georgia).

outlandish, *a.* Add: **2.** Also, immoderate, exceeding proper limits. (Partly arising from sense 3.)

1955 *Times* 6 June 7/4 One or two [people]..will shatter the monotonous efficiency of the [refreshment drinks] machine with some outlandish demand for a highly individual brew. 1977 *Time* 10 Jan. 46/1 The outlandish cost of armaments—$25 million for an F-15 today, v. $4 million for a Phantom jet in 1970—along with the rising prices of other imports, pushed the inflation rate into the stratosphere.

3. (Earlier examples.)

1792 J. Byng *Torrington Diaries* (1936) III. 53 So in an *outlandish* place I must creep to bed and pray for summer. 1842 Dickens *Let.* 16 Apr. (1974) III. 202 The inns in these outlandish corners of the world would astonish you by their goodness.

outlaugh, *v.* **2.** (Later example.)

1908 Swinburne *Duke of Gandia* I. 32 Her..Whose eyes outlaugh the splendour of the sea.

outlaw, *sb.* Add: **3.** Also *attrib.* or quasi-*adj.*, esp. in **outlaw strike**, a withdrawal of labour without the authority of a trade union, an unofficial strike.

1903 *Wide World Mag.* Mar. 546/2 The whole Western country was scoured for the wildest and most vicious 'outlaw' bronchos that could be found. 1912 C. A. Siringo *Cowboy Detective* v. 87, I told him to trot out his outlaw horse. 1920 *Harvey's Weekly* 17 Apr. 5/2 The 'outlaw' railroad strikes..are unjustifiable. 1931 *Economist* 6 June 1215/2 Arbitration would consume a good many months at best, and would probably be followed by 'outlaw' strikes. 1937 *Times* 22 Nov. 21/3 On top of this disappointment came a fresh outbreak of 'outlaw' strikes in the motor industry. 1949 *N.Y. Times Bk. Rev.* 13 Mar. 22/4 His brother Billy was thrown by an outlaw bronco. 1977 *New Yorker* 6 June 90/2, I did bring in fifty of them outlaw steers that way once.

outlawed, *ppl. a.* Add: **b.** That has been allowed to run wild. *U.S.*

1907 C. E. Mulford *Bar-20* xx. 197 Yu has got about as much show catchin' one of them as a tenderfoot has of bustin' an outlawed cayuse.

outlawry. 1. (Further *fig.* examples.)

1924 Kellor & Hatvany (*title*) Security against war... Vol. II. Arbitration. Disarmament. Outlawry. 1946 *Rep. Internat. Control Atomic Energy* (U.S. Dept. of State) I. 4 We..studied..the factors..involved in an international inspection system supposed to determine whether the activities of individual nations constituted evasions or violations of international outlawry of atomic weapons. 1964 *New Statesman* 1 May 682/2 When Baldwin writes about homosexuals he writes with a Negro's sense of another 'outlawry'.

outlet, *sb.* Add: **1.** *spec.* a shop, a retail store; an institution disposing of the produce of a manufacturer; a market (for goods).

1919 *Brit. Manufacturer* Nov. 28/1 India..is the most important outlet for British goods. 1933 [see *consumer 2 c]. 1962 *Daily Tel.* 29 May 12/2 The Six are reluctant to guarantee 'comparable outlets' in advance. 1964 *New Society* 27 Feb. 20/2 The shopper..has a choice between self-service and counter-service, and..between the supermarket and smaller self-service... Given this rather bewildering variety of 'outlets' for groceries it is of interest to consider why any housewife should choose one shop rather than another. 1966 *Listener* 17 Mar. 376/1 One may hope that the stimulus given by National Library Week will..lead to a permanent improvement in the outlets for the distribution of reading matter. 1975 *Sci. Amer.* Oct. 122/3 Both the motor and its companion controller are available from W. W. Grainger, Inc., an electrical-supply firm with retail outlets in all major U.S. cities. 1976 *National Observer* (U.S.) 27 Nov. 8/2 Yet Moldafsky and Bird are outlet experts. Each has more than 10 years' experience in outlet shopping, and both have researched and written guide books to outlet stores.

b. (Further examples.)

1916 Joyce *Portrait of Artist* (1969) ii. 64 The ambition which he felt astir at times in the darkness of his soul sought no outlet. 1928 E. O'Neill *Strange Interlude* I. ii. 67 She's got to find normal outlets for her craving for sacrifice. 1938 'G. Graham' *Swiss Sonata* 260 You think Mlle Lemaitre and the average wife are doing the same thing; in different ways each one is seeking an outlet. 1978 E. Healey *Lady Unknown* iv. 138 Dickens..needed ..an outlet for that driving energy.

c. *Anat.* The opening of a cavity of the body formed by the skeleton; used orig. of the pelvis and later of the thorax. Opp. *inlet *sb.* 4*.

1797 J. Bell *Anat. Bones, Muscles & Joints* I. v. 140 The outlet of the pelvis is the lower circle again, composed by the arch of the pubis and by the sciatic ligaments. 1828, 1906 [see *inlet *sb.* 4*]. 1960 E. Gardner et al. *Anat.* xxix. 339/1 The thoracic cavity communicates with the abdomen by the inferior thoracic aperture, or thoracic outlet, which is closed by the diaphragm. 1974 Passmore & Robson *Compan. Med. Stud.* III. II. xliii. 2/1 There are two areas which present the greatest potential mechanical obstruction to descent of the fetus, the pelvic brim and the effective bony outlet at the level of the ischial spines.

d. = *point *sb.*[1] A. 19 e.

1892 E. A. Merrill *Electr. Lighting Specifications* 77 The building shall be wired to —— lamp outlets, —— switch outlets... At each outlet the loose wire shall be neatly coiled and the ends carefully taped. 1917 A. L. Cook *Interior Wiring* III. 213 Wherever there is an outlet, such as a lighting fixture or a switch. 1925 G. A. Willoughby *House Wiring* i. 68 Some local ordinances require the number of watts to be counted for each type of outlet in various rooms. 1958 M. Dickens *Man Overboard* iv. 54 There was only one electric outlet from which a multiple plug sent fraying wires in all directions. 1968 Paschkis & Ryder *Direct Analog Computers* xxvi. 377 A second board is provided having male plugs fitting the female outlets on the board. 1972 C. L. Cooper in W. King *Black Short Story Anthol.* 221 He set it [*sc.* a tape recorder] down next to an..easy chair, unlatched the top, and plugged the cord into one of the wall outlets.

5. (sense 1 b) *outlet mechanism*; (sense 1 d) **outlet box**, a box used to contain connections to wires where they are led out of conduits.

1906 N. Harrison *Electric Wiring* vi. 133 With all metallic conduits whether flexible or not there are employed junction and outlet boxes. 1971 W. N. Alerich *Electr. Construction Wiring* v. 92/1 An outlet box or the equivalent must be inserted at every point in the system where access to enclosed wires is necessary. 1949 Koestler *Insight & Outlook* v. 69 In the gradually emerging sense of humour, we have a further outlet mechanism.

outlier. Add: **1. c.** (Later examples.)

1939 Joyce *Finnegans Wake* 97 From his holt..the outlier, a white noelan.., led bayers the run. 1976 *Abingdon Herald* 25 Nov. 7/2 Another outlier found near Sparsholt went through Sparsholt Copse, across to the Spinneys again, and back to Westcot. 1977 *Field* 13 Jan. 52/1 Hounds found an outlier at the back of Alexton village and hunted him past the hall.

2. a. (Later examples.)

1955 A. Thom in *Jrnl. R. Statistical Soc.* A. CXVIII. 275 Many of the circles have one or more outliers, i.e. single upright stones outside the ring. *Ibid.* 283 Why not also include *Little Meg* as an outlier to the circle at *Long Meg and her Daughters*?

c. (Further examples.) Also *attrib.*

1885 R. F. Burton tr. *Arabian Nights' Entertainments* V. 177 They took leave of him and departing to the outliers of the city, flew..to their several abodes. 1926 [see *Hattic *a.*]. 1928 *Library Assoc. Rec.* Dec. 244 The Central Library has wisely recruited several of the larger public libraries..to act as outlier libraries. 1961 T. Landau *Encycl. Librarianship* (ed. 2) 259/2 Its 'Outlier' libraries, which lend their specialized books and periodicals on the N.C.L.'s request when other resources fail, now number 281. 1973 *Computers & Humanities* VII. 136 What..differentiates 'La Comtesse d'Escarbagnas', at the top of the diagram, from the remaining plays? What is special about an outlier such as 'Dom Garcie'? 1977 *Jrnl. R. Soc. Arts* CXXV. 269/2 The Library is an 'outlier' library of the British Library.

outline, *sb.* Add: **2. c.** The representation of a word in shorthand.

1850 *Phonography, or, Phonetic Shorthand* 1/2 It [*sc.* Phonography] is..as intelligible as speech itself;..and as used in verbatim reporting, a perfect and intelligible outline. 1886 *Encycl. Brit.* XXI. 839/2 By this method the number of possible readings of an unvocalized outline is greatly reduced. 1898 [see *Pitman's]. 1933 D. L. Sayers *Murder must Advertise* i. 14 He..went on dictating and my hand was so shaky I could hardly make my outlines. 1969 G. Charlesworth *Effective Teaching Techniques in Commercial Subjects* 9 The difficulty of outlines should be very gradually increased and only one point of theory at a time should be so revised. 1973 E. M. Arlett *So You want to be a Shorthand Typist* ii. 18 The pen races smoothly over the pages, filling them with neat outlines that presently will be read back with practised ease.

d. *Typogr.* A display type-face in which the letters are drawn only in outline, perhaps with added shading. Also *attrib.*

1878 *Specimens of Newspaper, Book & Ornamental Founts* (Sir Charles Reed & Sons), Grotesque Outline... Clarendon Outline... English Black Outline. 1970 W. P. Jaspert et al. *Encycl. Type Faces* (ed. 4) 243 Zephyr... An accentuated outline titling, giving a somewhat three-dimensional effect.

3. Also, a précis of a proposed article, novel, scenario, etc.

1928 E. O'Neill *Strange Interlude* I. iv. 123 Charlie's coming to bring his suggestions on my outline for Gordon's biography. 1967 H. Van Siller *Biltmore Call* 164 You know the way he worked. A short outline first, then the first draft. 1968 D. Francis *Forfeit* i. 9 And when you've thought out how you'd like to present it, send us an outline. 1969 L. Hellman *Unfinished Woman* vi. 66 It was..too hard to write a shooting script, or even an outline, about a war I did not know. 1976 D. Quinn *Limbo Connection* xi. 189 It is a good story... I just delivered a few pages of outline to my agent... So let nobody forget, I own the film rights.

6. outline plan, a draft or sketch lacking many details; **outline planning permission**, permission sought by or from an authority, for building, demolition, or industrial development; so *outline planning application*; **outline stitch** (further examples).

1972 *Guardian* 5 Aug. 11/2 Westminster City Council are sitting on outline plans to knock down..a tenement of 89 flats. 1968 *Act* 16 & 17 *Eliz. II.* c. 72 §66 'Outline planning permission' means planning permission granted, in accordance with the provisions of a development order, with the reservation for subsequent approval by the local planning authority or the Minister of matters..not particularised in the application. 1971 *Reader's Digest Family Guide to Law* 100/1 If the householder does not wish to spend a lot of money on detailed plans and drawings, he can submit brief details and ask for approval of the building work in principle. If this outline planning permission is granted, it will justify further expenditure on more detailed plans. 1973 *Guardian* 25 Apr. 8/7 A meeting has been called..to protest against an outline planning application..for underground mining in the parish. 1908 M. H. Morgan *How to Dress Doll* iii. 34 Fill in with the chain or outline stitch and work with the embroidery buttonhole stitch. 1960 B. Snook *Eng. Hist. Embroidery* 14 Couching, laid-work and outline stitch are used.

ou·t-lot, outlot. *U.S. Obs.* exc. *Hist.* [Out- 1.] A lot or piece of ground situated outside a town or other area.

1643 *Rec. Colony & Plantation New Haven* (1857) I. 94 Mris Eldreds out lotts. 1774 in *Amer. Archives* (1837) (ser. 4) I. 278 An out lot, of ten acres, contiguous to the town, shall be laid off for such as desire the same at an easy rent. 1779 [see *in-lot 2]. 1837 W. Jenkins *Ohio Gazetteer* 148

A tract of land on the east side of the town has like-wise been divided into 23 outlots of five acres each. 1886 Z. F. Smith *Hist. Kentucky* 29 [They gave] to each man a half-acre lot and a ten-acre outlot. 1948 [see *in-lot 2].

ou·t-migrant. [Out- 8.] One who leaves one country or place to settle in another. Cf. *in-migrant *sb.* and *a.*

1953 *Caribbean Q.* II. iv. 53 This year, the number of out-migrants may outstrip even 1951's record figure. 1971 *Sci. Amer.* July 18/3 The state of New York for the first time had more out-migrants than in-migrants.

ou·t-migrate, *v.* [Out- 14.] *intr.* To leave one country or place to make one's home in another. Hence **ou·tmigra·tion**. Cf. *in-migration.

1953 *Caribbean Q.* II. iv. 53 This January, twice as many net-out-migrated as in January of 1949, 1950, 1951. *Ibid.*, Out-migration grows like snowballs running down hill. 1970 S. L. Barraclough in I. L. Horowitz *Masses in Lat. Amer.* iv. 157 Without greater agricultural production and accelerated out-migration, however, incomes resulting from reform would be dissipated by population increase within a generation or so. 1971 [see *in-migration]. 1975 *N.Y. Times* 17 Nov. 24/3 The problem for cities that are victims of outmigration is that population decline correlates with at least a relative fall in average income. 1976 *Time* 27 Sept. 55/3 Meanwhile, the out-migration of young blacks and whites has been reversed. 1977 *Jrnl. R. Soc. Arts* CXXV. 551/1 High rise flats.. proved to be particularly unsuitable for families with young children, with the result that out-migration of such families was accelerated.

outmode (autmōu·d), *v.* [Out- 21; cf. F. *démoder*.] *trans.* To put out of fashion. (Chiefly in pa. pple.) So **out-moded** *ppl. a.*, no longer in fashion, out-of-date.

1903 *Academy* 17 Jan. 71/1 Jesse Berridge is a poet, not a poetess, to use a somewhat outmoded word. 1906 R. S. Hichens *Call of Blood* ii. 15 He was not wholly emancipated from la petite femme tradition, which will never be outmoded in Paris. 1906 *Westm. Gaz.* 26 May 2/3 Even the out-moded globe-trotter will find that his trot must be maddeningly slow. 1915 T. Burke *Nights in Town* 392 The poor laddie is sadly outmoded, but he doesn't know it. 1926 W. J. Locke *Stories Near & Far* 133 The joined fragments showed an old photograph of a young man, in out-moded raiment. 1927 *Daily Express* 24 Sept. 8/4 George Ade has been out-moded by Will Rogers, whose jests die as soon as they are born. 1942 *Amer. Speech* XVII. (Reprints & Monogr. No. 4) 4 The people of this region might also have preserved out-moded features of speech. 1952 R. Neill *Moon in Scorpio* xi. 96 It's new Penny—and you're outmoding all Whitehall. 1971 *Sci. Amer.* Oct. 27/1 He believes such a union would outmode the present broadcast networks. 1978 E. Healey *Lady Unknown* vi. 149 Mrs Livingstone in a queer out-moded bonnet.

outmost, *a.* Add: **1.** Also *fig.*

1866 Ruskin *Crown Wild Olive* ii. 108 The outmost and superficial spheres of knowledge.

b. (Later examples.)

1887 W. Morris tr. *Homer's Odyssey* I. I. 2 Tha far-dwellers outmost of menfolk. 1927 E. S. Rae *Hansel fae Hame* 1 The hert rugs hame fae outmaist eyens o' earth.

c. (Later example.)

1887 W. Morris tr. *Homer's Odyssey* I. ix. 166 And but little it lacked, but the outmost of the helm it lighted on.

out of, *prep. phr.* Add: **I. 5. e.** From a base in; using (a place) as a centre of operations.

1960 'E. McBain' *Give Boys Great Big Hand* xii. 146 We were going to run away together... I could always get work out of Miami. 1974 *Publishers Weekly* 24 June 56/2 Working out of Bozeman, Montana, Jack Folsom has enjoyed the help of some 60 'friends'. 1975 *Listener* 27 Nov. 712/2 Marshall McLuhan still works out of a ramshackle office in a converted coach-house on the edge of the University of Toronto. 1975 B. Garfield *Hopscotch* xv. 145 He was the District Director out of Atlanta. 1976 *Times Lit. Suppl.* 25 June 784/4 The miscellaneous radio amateurs and visionaries who worked out of shacks and garages. 1976 *Church Times* 26 Nov. 7/4 Mrs. Briant now works out of Vancouver's Christ Church Cathedral, where she has set up a ministry for shut-ins.

II. 8. e. *out of this world*: see *world *sb.*

9. (Further examples.) *out of one's head*: out of one's mind or memory (see also Head *sb.* 36 in Dict. and Suppl.).

1814 Jane Austen *Mansf. Park* I. i. 6 She could not get her poor sister and her family out of her head. 1901 'L. Malet' *Hist. R. Calmady* v. x. 469 Obviously it was impossible to go back. He must go on rather—out of sight, out of mind. 1912 F. M. Hueffer *Panel* I. iv. 109 You meant to get her out of your head. 1935 M. de la Roche *Young Renny* iv. 32 Lord, what a waist he has! Do you suppose he can put a solid meal out of sight? 1938 W. de la Mare *Memory* 40 So gaily resigned To out-of-sight being out-of-mind.

10. d. *out of it* (earlier and later examples.)

1830 M. Edgeworth *Let.* 8 Dec. (1971) 442 Poor Davies Gilbert to whom the place was in every way unsuited is well out of it. I hope he thinks so. 1880 *Punch* 25 Dec. 299/1, I was out of it, jolly clean out of it. 1904 H. James *Golden Bowl* I. I. xxi. 344 He..moved her by..taking pity ..on her just discernible depression... He guessed that she felt herself, as the slang was, out of it. 1916 Galsworthy *Sheaf* i. 15 She is simply too 'out of it' to know anything. 1955 *Times* 13 Aug. 8/7 Feeling not a little out of it, we nevertheless took tea with the ladies of the parish. 1959 *News Chron.* 10 July 3/4 Bungalow dwellers ..may well have felt out of it. 1973 R. Lewis *Of Singular Purpose* vii. 157 You're well out of it, Harry. Believe me, you're well out of it.

e. (See quots.)

1963 *Amer. Speech* XXXVIII. 174 Drunk: soused, out of it, stoned, bombed. 1967 Wentworth & Flexner *Dict. Amer. Slang* Add. 698/1 *Out of it*... 3. Not concerned with mundane things, as when under the influence of a drug, obsession, all-consuming idea, etc.; in a state of euphoria. 4. Not 'with it'; stupid; 'square'. 1973 *To our Returned Prisoners of War* (Office of U.S. Secretary of Defense) 8 *Out of it*, to be out of touch with reality when under the influence of a drug, especially hallucinogens. To lack understanding and awareness, especially in a subculture. 1973 *Black Panther* 27 Oct. 12/3 James Jenkins ..describes several inmates on 'F' block, who were once 'sharp dudes' as being 'completely out of it' following 'therapeutic sedation'.

11. b. *out of work, out-of-work*: see Work *sb.* 27 in Dict. and Suppl.

12. (Further examples.)

1858 Geo. Eliot *Scenes Clerical Life* II. 189 Furnishing sugar or vinegar to..families that found themselves unexpectedly 'out of' those indispensable commodities. 1973 'D. Halliday' *Dolly & Starry Bird* iii. 36, I hadn't a light for his cigar; Charles and I were out of matches.

13. In current use, both in the *spec.* sense, and as a *fig.* development of this.

1892 A. W. Pinero *Magistrate* I. 24 You nominated yourself for the Matrimonial Stakes. Mr. Farringdon's The Widow, by Bereavement, out of Mourning. 1924 Galsworthy *White Monkey* III. xii. 295 The room seemed to him to have been got by a concert-hall out of a station waiting-room. 1950 'P. Woodruff' *Island of Chamba* vii. 110 Their [*sc.* Muslims'] thought has the same pedigree as ours: by Greece out of Palestine. 1956 N. Marsh *Off with his Head* (1957) viii. 177 Teutonic Dancer by Subsidise out of Substituteon. 1968 *Listener* 3 Oct. 452/2 If there were a radio stud-book, it could contain some such entry as 'Any Questions? by After-Dinner Speaker out of Group Therapy'. 1973 'I. Drummond' *Jaws of Watchdog* xvi. 210 Humblebee was bred..by an imported French champion out of a mare by an Argentine quadruple-crown winner.

14. Used with *balance, *phase, *register: see the sbs.

III. *out-of-awareness* (also as *sb.*), *out-of-balance* (also as *sb.* and *vb.*), *out-of-bounds* (earlier example; also as *sb.*), *out-of-breath*, *out-of-condition*, *out-of-context*, *out-of-control* (also *ellipt.*), *out-of-focus* (later examples; also as *sb.*), *out-of-form*, *out-of-hours*, *out-of-key*, *out-of-office*, *out-of-phase*, *out-of-place* (later example), *out-of-pocket* (further and *ellipt.* examples), *out-of-school* (further examples), *out-of-season* (further examples), *out-of-sync*, *out-of-the-body*, *out-of-the-ordinary*, *out-of-the-season*, *out-of-tune* (examples; also *ellipt.*); *out-of-wedlock*.

out-of-round: see *round *sb.*; *out-of-true*: see *true *sb.*; *out-of-truth* see *truth *sb.*

Further derivatives, as *out-of-breathness*, *out-of-the-worldness*, *out-of-touchness*, *out-of-trueness*.

1965 *Canad. Jrnl. Linguistics* Fall 36 When one hears the paralanguage of a speaker, one first, rapidly and out-of-awareness, establishes the base line of the speaker. 1974 *Florida FL Reporter* XIII. 11/2 Children work on language in an out-of-awareness situation. 1921 W. S. Ibbetson *Motor & Dynamo Control* vi. 194 It would not be correct to balance an out-of-balance pulley by fixing a counter-weight on the armature core. 1932 R. Rawlinson in E. Molloy *Pract. Electr. Engin.* V. 1590/2 The function of voltage balancing on the two sides is done by a '3-wire balancer set', which is designed with regard to the maximum out of balance current which the supply company considers it necessary to legislate for. 1958 *Listener* 13 Nov. 780/1 Friction..and the slightest out-of-balance of the motor cause the axis of the gyroscope to deviate. 1967 L. Holmes *Odhams New Motor Manual* viii. 189/2 Out-of-balance can be caused by wheel damage or brake drum eccentricity. 1968 Burdett & Ellis *Motor Vehicle Mechanics' Course* II. xii. 284 Any 'out-of-balance' weight on the wheel..may give rise to a tendency to 'throw' the wheel up from the road and back down on to it once in every revolution of the wheel. 1974 Harvey & Bohlman *Stereo F.M. Radio Handbk.* v. 119 In this type of detector ..the setting of R_4 is arranged to out-of-balance the signal currents in Q_{10} and Q_{11}. 1857 T. Hughes *Tom Brown's School Days* I. ix. 219 Many of the old wild out-of-bounds habits stuck to them as firmly as ever. 1947 M. Lowry *Let.* 24 July (1967) 150 Into what roughs, out-of-bounds and quagmires we shall get ere the book be finished..one knows not. 1973 *Guardian* 21 May 22/3 The long sixteenth, a treacherous hole into the wind with an out-of-bounds lurking on left. 1939 G. Greene *19 Stories* (1947) 156 Low out-of-breath tones. 1900 E. Glyn *Visits of Elizabeth* 98 His snorts of out-of-breathness could be heard for miles. 1972 C. Fremlin *Appointment with Yesterday* xi. 85 She should have realised that her middle-aged, out-of-condition body would..rebel. 1977 F. Branston *Up & Coming Man* xiii. 133 A puffy, out-of-condition young man. 1951 *Mind* LX. 91 The unused, out-of-context sentence specifies no speaker. 1973 R. Ludlum *Matlock Paper* ii. 16 Rumor; out-of-context statements..; constructed evidence. 1961 *Daily Mail* 20 July 9/8 His out-of-control tractor plunged 20 ft. into the River Nar. 1974 H. L. Foster *Ribbin'* vii. 321 Most workers have strong feelings against intervening physically with a child's out-of-control behavior. 1974 G. Jenkins *Bridge of Magpies* vii. 104 Finally the out-of-control twisting of the boat eased. 1977 *O.D.* No. 3. 9/1 The circumstances of our interview with Schmidt were curious, weird, if not verging on the out-of-control. 1946 *Nature* 30 Nov. 786/1 Thus any error in magnification due to slight out-of-focus in one camera was compensated by the other. 1962 L. S. Sasieni *Princ. & Pract. Optical Dispensing* xii. 303 From this we see the extent of the out-of-focus region. 1966 D. G. Brandon *Mod. Techniques Metallogr.* 29 The out-of-focus image of the specimen surface. 1967 E. Chambers *Photolitho-Offset* xi. 160 Slight out-of-focus and gentle vibration of the front of the camera during exposure are 'dodges' also used. 1961 *Times* 27 Dec. 4/1 Parfitt..must now have a chance of gaining his first Test cap in place of the out-of-form M. J. K. Smith. 1977 *South China Morning Post* (Hong Kong) 22 July 18/5 Rick McCosker, Australia's out-of-form opener, boosted his chances of retaining his place for the third cricket Test against England next week with a fighting 77 against Warwickshire yesterday. 1967 *Guardian* 3 Aug. 3/7 A scheme for out-of-hours deliveries in Greater London. 1977 *Times* 9 Sept. 2/3 An out-of-hours repair service for customers. 1962 *Times* 17 Jan. 13/1 The..slightly out-of-key episode with the travelling salesman. 1976 'J. Fraser' *Who steals my Name?* vii. 89 A tiny little oddness, one of those strange out-of-key facts a policeman is trained to spot. 1961 *Times* 4 Sept. 5/1 His out-of-office activities as a dosshouse owner. 1973 'W. Haggard' *Old Masters* i. 11 An out-of-office politician. 1938 L. F. Blume *Transformer Engin.* xiv. 369 It would be of no advantage to introduce a circulating current having an out-of-phase component. 1968 C. G. Kuper *Introd. Theory Superconductivity* iv. 65 The real part and the imaginary part of the dielectric constant respectively relate the in-phase and out-of-phase parts of the displacement to the electric field. 1974 *Country Life* 26 Dec. 1997/3 A smaller telescope..would still be an out-of-place intrusion on the landscape. 1902 G. B. Shaw *Let.* 22 Oct. (1972) II. 284 Actual out-of-pocket loss. 1971 D. C. Hague *Managerial Econ.* iii. 67 It is unlikely that prices will be cut below out-of-pocket costs. 1972 G. Durrell *Catch me a Colobus* ix. 182, I said that..I would be willing to cover his out-of-pocket expenses if he'd join the expedition to help us with our work. 1973 'I. Drummond' *Jaws of Watchdog* vii. 92 He was to be paid £15,000 in used notes, for his own pay-off and his out-of-pocket. 1930 *Times Educ. Suppl.* 26 July 329/3 Organizational activities are relegated to out-of-school hours. 1959 *Times* 20 Jan. 9/3 The result is to be seen in those out-of-school activities which may land them in the juvenile courts. 1970 G. E. Evans *Where Beards wag All* xix. 217 One of Mrs. Jay's out-of-school jobs was keeping pigs. 1944 M. Laski *Love on Supertax* ii. 12 Lack of..out-of-season fruits. 1960 *Farmer & Stockbreeder* 12 Jan. 41 (Advt.), Details of the January out-of-season discounts are now available from your New Holland Dealer. 1966 *Listener* 24 Nov. 783/3 A boy, bored and neglected, at an out-of-season hotel. 1948 *Proc. IRE* XXXVI. 904/1 The effect would be to prolong excessively the out-of-sync condition whenever a discontinuity occurred in the transmitted sync signal. 1956 M. Stearns *Story of Jazz* (1957) xxiv. 301 The hectic, 'out-of-sync' short, *The Jazz Dance*, was filmed in one evening. 1967 *Listener* 21 Sept. 380/1, I rather miss the deliciously ludicrous sight of a singer trying to catch up with his own voice, like an out-of-sync film. 1977 *New Yorker* 25 July 19/3 Rumbles of thunder interspersed the lightning flashes in belated, out-of-sync fashion. 1946 G. N. M. Tyrrell *Personality of Man* vii. xxii. 199 These out-of-the-body cases are of exceptional interest. It is worth pointing out that in two such cases.. the percipients describe the process of getting out of their bodies in almost identical terms. 1969 *New Scientist* 3 July 33/2 Old wives tales..such as spontaneous telepathy, out-of-the-body experiences or poltergeists. 1931 *Times Lit. Suppl.* 15 Oct. 788/1 A disturbing..absolutely out-of-the ordinary life-story. 1974 *Times* 9 Nov. 12/5 Andorra.. has caught the imagination of many skiers looking for an out-of-the-ordinary holiday. 1955 *Times* 3 May 4/2 Permission has been given by the Rugby Union for the Coventry team to play an out-of-the-season game. 1876 H. Sidgwick *Let.* 24 Aug. in A. & E. Sidgwick *Henry Sidgwick* (1906) 323 There is a great charm in this scenery and in the feeling of out-of-the-world-ness. 1916 D. H. Lawrence *Let.* 9 Jan. (1932) 306 But come and see us here, because of the sea and the silence and peace and the out-of-the-worldness of it all. 1957 M. Stewart *Thunder on Right* iii. 38 The out-of-the-worldness of the place pressed heavily upon her. 1952 M. Lowry *Let.* May (1967) 317 The essential..points are too often clouded as a result of the technical out-of-touchness of the writing. 1960 *Guardian* 21 July 8/4 'Out-of-touchness' has produced a fervent desire for recognition. 1921 *Spectator* 26 Feb. 268/2 When you start your wall there seems by eye very little or nothing wrong with it, but when you have got it up some thirty or forty feet the out-of-trueness is appalling. 1803 H. Wynne *Diary* 6 July (1940) III. 83 A beautifull out of tune Symphony, consisting of hair-dressers, butchers, &c. opened the play. 1917 T. S. Eliot *Prufrock* 20 The voice returns like the insistent out-of-tune Of a broken violin. 1930 J. Dos Passos *42nd Parallel* I. 129 A little out of tune orchestra was playing. 1977 J. Wainwright *Day of Peppercorn Kill* 125 Somebody was whistling an out-of-tune version of *Sleepy-Time Gal.* 1961 M. Beadle *These Ruins are Inhabited* (1963) ix. 129 Some Irish girls emigrate briefly to give birth to out-of-wedlock babies. 1972 *Guardian* 22 July 5/7 A single woman experiencing an out-of-wedlock pregnancy.

out-of-date, *adj. phr.* Add: Also *absol.*

1928 *Manch. Guardian Weekly* 17 Aug. 132/1 This column..is apt to specialise in the out-of-date. 1938 *Times* 27 Aug. 3/5 There is to the comparatively unmodern and out of date something indefinably comforting in a winning score of 300.

ou:t-of-da·teness. [f. Out-of-date *adj. phr.* + -ness.] The state or condition of being out-of-date; obsoleteness.

1915 E. Carpenter *Healing of Nations* xvii. 208 Finally..one realizes the monstrosity and absurdity of the present conflict—its anachronism and out-of-dateness in the existing age of human thought and feeling. 1928 *Sunday Dispatch* 30 Dec. 10/5 Consider the rich opportunity to tell the good man of his stuffiness, his out-of-

dateness. 1930 A. Huxley *Vulgarity in Lit.* i. 3 It took several centuries to reduce Dante's guide-book to out-of-dateness. 1940 'G. Orwell' *Inside Whale* 103 They continue to be read in spite of their obvious out-of-dateness.

out-of-door, -doors. Add: **B.** *sb.* (Earlier and later examples.)

1819 Keats *Let.* 4 Feb. (1958) II. 37 One not ill enough to forget out-of-doors. 1970 D. Mathew *Courtiers of Henry VIII* III. vi. 206 Henry VIII had been a great King of the out-of-doors. 1975 R. V. Redinger *Geo. Eliot* (1976) vi. 377 The many scenes which take place in the unhampered out-of-doors.

ou:t-of-doo·rness *nonce wd.*, the state or condition of being out-of-doors.

1929 W. Deeping *Roper's Row* xxiii. 255 Hazzard liked..the play of the wind through his aggressive hair. It gave him a feeling of out-of-doorness and of freedom.

out of print, *adj. phr.* [See Out of *prep. phr.* 14, III and Print *sb.* 7 b.] Of a book: no longer available from the publisher; also used with reference to gramophone records. Also as *sb.*

Written with hyphens when used *attrib.*

1674 [see Print *sb.* 7 b]. 1800 C. Lofft in R. Bloomfield *Farmer's Boy* (ed. 3) 128 (*Appendix*), I am happy to be stopt here, by so good a cause as the urgency of the Publishers to complete a Third Edition; they informing me that the second is entirely out of print. 1895 [see Print *sb.* 7 b]. 1896 [see Out of *prep. phr.* III]. 1927 R. B. McKerrow *Introd. Bibliogr.* p. vi, I have..been repeatedly asked to reissue the pamphlet (now out of print). 1938 *New Statesman* 13 Aug. 262/1 There are two admirable lieder records. Karl Schmitt-Walker..sings.. two of the very best of Hugo Wolf's songs..the latter only available in the out-of-print Gerhardt album of Wolf. 1950 *John o' London's Weekly* 7 July 398/3 Every reader has got his own favourites among the out-of-print. 1965 *Amer. N. & Q.* Oct. 24/1 *New Variorum Edition of Shakespeare*—What is the status of in- and out-of-print volumes in the set? 1973 L. Snelling *Heresy* I. 12 Some out-of-print unknown... Why, by the way, is he unknown and out of print? 1975 *Times Lit. Suppl.* 22 Aug. 956/6 (Advt.), Any American books, new or out-of-print. 1977 *Gramophone* Sept. 448/2 (Advt.), We still have second-hand LP and 78 Departments where many unusual and out of print records may be found.

out of series, *adj. phr.* [f. Out of *prep. phr.* 8, III + Series.] (See quot. 1952.)

Written with hyphens when used *attrib.*

1952 J. Carter *ABC for Bk. Collectors* 128 Of an edition specifically limited in number, there will usually be printed some extra copies... Such copies are understood not to invalidate the certificate of limitation; and their status is sometimes indicated by the words out of series, instead of a number. 1977 *Q. Jrnl. Library of Congress* July 233/2 The copy acquired by the Library is an out-of-series, unnumbered example on Arches paper.

ou:t-of-si·ght, *adj. phr.* (*sb.*) [f. Out of *prep. phr.* 9, III + Sight *sb.*[1] 10 b.] **A.** *adj. phr.* **1.** Outside the range of sight; distant.

1876 Geo. Eliot *Dan. Der.* II. iv. xxxi. 280 She was really getting somewhat febrile in her excitement... Was it at the novelty simply, or the almost incredible fulfilment about to be given to her girlish dreams of being 'somebody'—walking through her own furlongs of corridor and under her own ceilings of an out-of-sight loftiness.

2. Excellent, incomparable, superior; delightful, exciting, surprising. *slang* (orig. *U.S.*).

1896 S. Crane *Maggie* (rev. ed.) v. 43 D'way I plunked dat blokie was outa sight. 1896 Ade *Artie* xi. 94 She looked out o' sight! Some of 'em have got their sealskins and their sparklers, but this little girl, with that new make-up and the flowers, beat the best of 'em. 1897 C. W. Chesnutt *Let.* 24 Sept. in H. M. Chesnutt *Charles Waddell Chesnutt* (1952) x. 81, I saw the Shaw Memorial and the new Public Library Building, which are 'out of sight'. 1902 J. D. Corrothers *Black Cat Club* i. 26 'Out o' sight!' yelled a dozen voices as the poem was concluded. 1927 C. Sandburg *Amer. Songbag* 279 The corn we raise is our delight, The melons, too, are out of sight. 1961 *Down Beat* 5 Jan. 23, I find some of the musicians I've encountered on the road rather ridiculous... It seems everything is 'something else' these days. Or is it 'out of sight'? 1966 *Surfer* VII. iv. 11 The waves are a perfection 10 to 15 feet and straight over. Really up tight and out of sight! 1970 *Times* 6 July 8/5 Whenever possible he liked to make a point of talking to drug users on their own ground... An action that led one girl student to remark 'for an official person he's absolutely out of sight.' 1973 *Ottawa Jrnl.* 14 July 24/2, I met this groovy dude at the bowling alley. He is out of sight.

B. *sb.* One who or that which is unseen. *rare.*

1930 Auden *Poems* 18 Better where no one feels, The out-of-sight, buried too deep for shafts.

ou:t-of-sta·te, *adj. phr.* [f. Out of *prep. phr.* 8, III + State *sb.* 31 c.] Originating from outside a state of the United States. Also *ellipt.* as *sb.*, the area outside a particular state. Hence **ou:t-of-sta·ter**, a person originating from outside a particular state.

1935 *Amer. Mercury* July 290/2, I was also faced with their prejudice against out-of-town and out-of-state teachers. 1938 *Amer. Speech* XIII. 178 The larger southern part [of Idaho]..since the days of the Oregon trail has received a greater percentage of 'out-of-staters'. 1943 *Sun* (Baltimore) 28 Apr. 28/3 Dealers in a poultry black market on the Eastern Shore, including..live-chicken buyers from out-of-State and Baltimore wholesalers. 1948 *Herald-Press* (St. Joseph, Mich.) 14 Aug. 1/1 Many a speeding out-of-state motorist is of the opinion that the St. Joseph policemen are a lot of meanies. 1964 J. Masters *Trial at Monomoy* i. 11 Bunch of crackpots, long-hairs, foreigners, out-of-staters. 1973 *N.Y. Law Jrnl.* 26 July 1/6 The law students interviewed, most of them from out-of-state, made these observations. 1973 D. Barnes *See Woman* (1974) 90 A Buick. It had out-of-state plates on it. 1974 *Columbia* (S. Carolina) *Record* 25 Apr. 18-C/5 Many out-of-staters are showing a lot of interest in purchasing real estate in the Santee-Cooper complex. 1976 *CB Mag.* June 23/1 (Advt.), Special Bonding Service—protects you from being arrested if you ever get caught speeding out-of-state without enough money to pay the fine.

out-of-the-way, *adj. phr.* Add: **B.** as *adv.* Oddly; exceptionally, extraordinarily.

1717 [in Dict., sense 4]. 1901 'M. Franklin' *My Brilliant Career* x. 80, I really believe that on that night I did not look out of the way ugly. 1928 E. M. Forster in *Life to Come* (1972) 105 He was completely sincere when he told the Trevor Donaldsons that he had had an out-of-the-way pleasant weekend. 1939 —— in *Ibid.* 118 If only he wasn't so handsome, so out-of-the-way handsome.

C. as *sb.* A remote spot, an out-of-the-way place.

1971 *Islander* (Victoria, B.C.) 7 Nov. 13/4 Colorful people who lived at these out-of-the-ways on the west.. coast.

out-of-the-wayness. (Further examples.)

1899 Kipling *From Sea to Sea* I. xii. 307 It [*sc.* Kobé, in Japan] lives among hills, but the hills are all scalped, and the general impression is of out-of-the-wayness. 1926 S. T. Warner *Lolly Willowes* II. 129 There's not such another village in Buckinghamshire for out-of-the-wayness.

out-of-time, *adj. phr.* Delete †*Obs.* and add later example.

1909 *Westm. Gaz.* 4 Sept. 13/2 He is so full of admiration for James III.—the 'Old Pretender', in common language —that he casts an out-of-time vote for him.

ou:t-of-tow·n, *adj. phr.* (*sb.*) [Out of *prep. phr.* 8, III.] **A.** *adj.* Situated, originating from, or occurring outside a town. Also, unsophisticated.

1825, 1891 [in Dict. s.v. Out of *prep. phr.* III]. 1903 R. Bedford *True Eyes* 198 The two walked down the rise on its out-of-town side. 1930 E. Waugh *Vile Bodies* x. 174 There were a great number of journalists making the best of an 'out-of-town' job. 1937 M. Hillis *Orchids on your Budget* iv. 71 When an out-of-town cousin turns up, you undoubtedly entertain her more lavishly. 1957 H. Roosenberg *Walls came tumbling Down* ix. 209 She was going..to make some out-of-town phone calls. 1959 *Manch. Guardian* 10 Aug. 3/1 Scotland's 'Theatre in the Hills' is a model for out-of-town theatres. 1971 *Daily Hampshire Gaz.* (Northampton, Mass.) 9 Nov. 1/3 Servicemen based in Northampton led groups of out-of-town based gas men to various sections of the town. 1973 J. Mann *Only Security* vii. 81 I've never seen anything so horrible as one of those out-of-town shopping centres I saw in the Midlands. 1977 *New Yorker* 9 May 35/1 Most of them [*sc.* students] were so nice and out-of-town that they were not completely comfortable with escalators and revolving doors.

B. *sb.* A person originating from outside a particular town. Also **ou:t-of-to·wner**.

a 1911 D. G. Phillips *Susan Lenox* (1917) II. vi. 162 Except for 'out-of-towners', the married men were the chief support of their profession. 1941 *Reader's Digest* May 29/1 Sometimes out-of-townerz leave no tip at all. 1958 *Manch. Guardian* 7 June 4/6 The out of towners swarm in and laugh in the wrong places. 1966 'W. Haggard' *Power House* xiii. 136 The driver..[had] mistaken him for an out-of-town and he'd cruise him through Regent's Park and..claim the full fare. 1973 J. Drummond *Bang! Bang! You're Dead!* xxv. 84 Out-of-towners or won't-works agitate at so much the hour, never mind what about. 1976 *New Yorker* 15 Nov. 162/2 It had found that many of the out-of-towners in charge of choosing convention sites..had had to be talked out of changing their plans at the last minute.

out-party: see Out- 4 in Dict. and Suppl.

outpass, *v.* Add: **1.** (Later examples.)

1928 J. H. Mozley tr. *Statius* I. 47 Mayst thou outpass the limits of old Nestor's age. 1929 R. Bridges *Testament of Beauty* IV. 150 That fadeth only as it outpasseth mortal sight.

2. (Later example.)

1930 *New Statesman* 28 June 360/1 Parties which have historically played their part..but now find themselves outpassed by newer parties.

out-patient. Add: **b.** *pl.* The out-patient department of a hospital.

1910 *Practitioner* July 87 After death..she was recognised as the woman who had previously attended at out-patients. 1968 'L. Black' *Outbreak* i. 8 There's a smallpox suspect in the Outpatients at St Swithin's. 1977 B. Pym *Quartet in Autumn* v. 47 A visit to out-patients at the hospital.

ou·t-peeping, *ppl. a.* [Out- 10.] That peeps out.

1908 A. Austin *Sacr. & Prof. Love* 72 And on out-peeping roots the sun-god shoots The shafts of his golden quiver.

outperform (autpəɹfǭ·ɹm). [f. Out- 18 + Perform *v.*] *trans.* To perform better than; to surpass in a specified activity or function.

1960 *Farmer & Stockbreeder* 9 Feb. 92 It has been outperforming other light tractors ever since it was introduced and using less fuel in doing it. 1966 *Economist* 26 Mar. 1239/2 The fund..will be expected to outperform other mutual funds. 1972 *Times* 30 Sept. 9/5 Salzburg's cast, in voice and temperament, outperform their studio rivals. 1973 M. Woodhouse *Blue Bone* xii. 122 It outperforms structural steel by a factor of ten times. 1975 *Publishers Weekly* 30 June 22/1 In the sweeping stock market recovery of 1975, book publishing companies—after years in the Wall Street doghouse—have sharply outperformed the general market. 1978 R. V. Jones *Most Secret War* xli. 390 The Mustang could outperform all the standard German day fighters.

out-place. Delete †*Obs.* and add later examples.

1838 F. Owen *Diary* 8 Jan. (1926) 94 Most of the inhabitants were absent at the out places or villages. 1911 *Chambers's Jrnl.* Apr. 221/1 It is this longing..that sends the sportsman into the out-places. 1956 W. R. Bird *Off-trail in Nova Scotia* vi. 176 But that reminds me of my scrap book. You know we in these out places keep such things.

outplace, *v.* [Out- 18, 15.] *trans.* **a.** To displace or oust. **b.** (see quot. 1970.)

1928 *Daily Express* 16 Jan. 5/3 Skirts dipping at one side will outplace in many houses the skirt dipping at the back that was so popular during the winter. 1970 *Time* 14 Sept. 83 Instead of simply bouncing a subordinate, the boss can send him to a firm that specializes in helping unwanted executives to find new jobs. The practitioners have even coined a euphemistic description for the process: 'outplacing' executives who have been 'dehired'.

outplay, *v.* (Further examples.)

1938 L. Bemelmans *Life Class* II. iii. 144 The bands always tried to outplay each other, waging a musical warfare. 1972 G. Green *Great Moments in Sport: Soccer* ii. 35 England has been outplayed..in a number of ways. 1977 J. Wainwright *Do Nothin'* v. 67 Krupa on tom-toms and James on the horn..two of the best in the business trying to out-play each other.

out-po·cketing, *vbl. sb. Biol.* [Out- 9.] The outward movement of part of a surface so as to form a pocket- or sac-like cavity.

1924 L. B. Arey *Developmental Anat.* i. 8 Circumscribed folds..produce (*a*) evaginations, or out-pocketings, and (*b*) invaginations, or in-pocketings. 1968 *Progress in Brain Res.* XXIX. 55 Portions of the outer nuclear envelope..could project into the cytoplasm and pinch off... Out-pocketings of the outer nuclear envelope are frequently seen in early neuroepithelial cells. 1971 *Nature* 16 Apr. 472/2 The glands arise as an outpocketing of the vagina.

outpoint, *v.* Add: **3.** In various sports and games, esp. boxing: to score more points than; to defeat on points. Also *transf.* and *fig.*

1903 *Westm. Gaz.* 19 Feb. 7/3 In the second [coursing] ties, Priestlaw, notwithstanding his speed, was outpointed cleverly by Handsome Creole. 1909 *Ibid.* 2 Feb. 12/2 With Aiken unable to settle down [in a billiards match] and failing to make any material use of some nice openings, he continued to be outpointed. 1909 *Ibid.* 20 Feb. 16/3 Driscoll outpointed the American featherweight champion. 1922 *Weekly Dispatch* 12 Nov. 11 In a ten-rounds boxing contest here to-night Bermondsey Billy Wells (England) outpointed Johnny Tillman. 1949 *Sun* (Baltimore) 16 July 9/6 Barfly regularly outpoints and outfoots all who sail against her. 1955 *Times* 12 May 4/3 Eddington, the coloured American, had been outpointed in Ireland. 1959 *Economist* 11 Apr. 159/1 BOAC happens to have been outpointed in the never-ending game of poker that the airlines play with traffic rights for chips. 1970 *Globe & Mail* (Toronto) 28 Sept. 21/3 Scot Ken Buchanan won the world lightweight boxing title Saturday in San Juan, Puerto Rico, when he outpointed champion Ismael Laguna of Panama under a blazing sun. 1976 *Daily Record* (Glasgow) 4 Dec. 30/5 The Baillieston man outpointed Irish champion John McLoughlin after both had been floored in an amazing second round.

outport[1]. Add: **1. b.** Chiefly in Labrador and Newfoundland, a small remote fishing village. Also *attrib.* Hence **ou·tporter**, an inhabitant of an outport.

1820 in C. R. Fay *Life & Labour in Newfoundland* (1956) viii. 138 Almost every fifth fisherman is what is termed a 'Planter', particularly in the outports of the Island. 1904 *Westm. Gaz.* 6 May 10/1 The Newfoundland outporters are hardy, courageous, boldly adventurous, simple-lived. 1907 J. G. Millais *Newfoundland* p. xv, I have tried to enter into the life of the true Newfoundlander—the man of the outports. 1949 [see *new light 3 a]. 1964 L. E. F. English *Historic Newfoundland* 6 Visit the fishing villages, the so-called outports of the Province of Newfoundland. 1966 A. R. Scammell *My Newfoundland* 17 Environment and circumstance..developed in the young outport lad initiative, and a sense of responsibility. 1973 B. Broadfoot *Ten Lost Years* xix. 214 Mallory and Derek, pretty fancy names for a couple of outporters, eh? 1974 *Nat. Geographic* Jan. 116 Most islanders cling to the seaside in isolated villages called outports. 1978 *Globe & Mail* (Toronto) 14 Jan. 8/2 Trudeau's Government.. killed a subsidy that keeps a useful ferry service going between the mainland and outer islands in the Queen Charlottes and between the Vancouver supply area and the tiny outports of the northern coast.

c. A small port, located to support the commerce of a main port.

1935 J. A. Fraser *Spain & W. Country* x. 108 It was from Seville and its little out-port Sanlucar de Barrameda ..that nearly all the early Spanish voyages of discovery went forth. 1952 F. W. Morgan *Ports & Harbours* 76 In order to prevent a loss of trade the port undertakes the development of an 'outport' nearer the sea, which can attract the larger vessels. 1956 *Sun* (Baltimore) 3 Sept. 4/5 It just wouldn't work to have a New York contract cover each of the outports.

out-posi·tion, *v.* [f. Out- 21 + Position *sb.*] *trans.* In various sports and games, to secure an advantage over (an opponent) in terms of position; to defeat in a contest for a particular position.

1928 *Sunday Express* 24 June 21 Tilden got to everything with that long, easy stride of his, and Hunter was often outpositioned by shots that he did not expect would be returned. 1960 E. S. & W. J. Higham *High Speed Rugby* xi. 131 The further he progresses, the more necessary it will become to out-position and out-think the opposition. 1966 *Daily Progress* (Charlottesville, Va.) 8 June 30/2 We out-hustle them, out-position them, and in all but two games have actually out-rebounded them.

outpost, *sb.* Add: **b.** (Further examples.)

1881 D. G. Rossetti *Ballads & Sonnets* 196 Shall my sense pierce love,—the last relay And ultimate outpost of eternity? 1917 E. V. Lucas (*title*) Outposts of mercy.

c. (Further examples.)

1776 *Battle of Brooklyn* II. i. 19 We are the remains of the out post guard. 1870 De B. R. Keim *Sheridan's Troopers* 206 This simple means is also resorted to by the troops on outpost duty. 1923 Kipling *Irish Guards in Great War* I. 226 The blockhouse..was absorbed into our outpost-line. 1946 *R.A.F. Jrnl.* May 175 In Berlin and Hamburg..and on airfields and outpost stations, the R.A.F. has taken root. *Ibid.* 176 Then, these outpost men are busy.

2. A trading settlement situated near a frontier or at a remote place in order to facilitate the commercial contacts of a larger and more centrally situated town or settlement. Also, by extension, any of various other kinds of remote settlements and institutions (see examples). Also *attrib.*

1802 in E. Coues *New Light on Early Hist. Greater Northwest* (1897) I. 204 [I] made up the assortment of goods for the outposts, equipped the summer men, clerks, etc. 1911 A. K. Chignell (*title*) An outpost in Papua. 1955 W. G. Hardy *Alberta Golden Jubilee Anthol.* 212 In Alberta, after World War I, it set up 'outpost hospitals' for the brides and families of soldier-settlers in the outlying reaches of the province. 1956 H. S. M. Kemp *Northern Trader* 17 The red [thumbtacks] represent the Company's permanent, year-round establishments; the blue ones, the winter posts and outposts. 1961 L. van der Post *Heart of Hunter* I. v. 80 Tsane had once possessed a district commissioner, but had declined into a remote police outpost. 1970 *Islander* (Victoria, B.C.) 22 Feb. 5/3 The post here [*sc.* Cambridge Bay] is what is called an Outpost where goods are landed to be drawn on by other posts that may run short.

3. The furthest territory of an empire, esp. in the phr. *outpost of* [*the British*] *Empire*, common since the end of the nineteenth century, and now occasionally used in a nostalgic or ironic sense. Also *attrib.*

1912 Kipling *Songs from Bks.* 94 There he shall blaze a nation's ways with hatchet and with brand, Till on his last-won wilderness an Empire's outposts stand. 1924 E. M. Forster *Passage to India* xx. 182 His simple words had reminded them that they were an outpost of Empire. 1924 *Granta* 25 Apr. 361/1 Attock Fort, in the Punjab... An outpost of Empire, where the Cabul and the Indus meet. 1929 J. B. Priestley *Good Companions* II. vii. 448 May she marry the outpost-of-Empire lad in the Sudan. 1934 Dylan Thomas *Let.* 25 Apr. (1966) 111 Then I shall walk back..covering up a..weakness with a look of fierce & even Outpost-of-the-Empire determination. 1937 'G. Orwell' *Road to Wigan Pier* ix. 173 In an 'outpost of Empire' like Burma the class-question appeared at first sight to have been shelved. 1970 R. Johnston *Black Camels* ii. 26 You've sure got an impressive outpost of empire here. 1971 R. Roberts *Classic Slum* vi. 81 Some families..had male members..who had soldiered in the outposts of empire.

outpromise, *v.* (Later example.)

1938 *Sun* (Baltimore) 9 July 6/1 Many more experienced statesmen have thought that they had outpromised the field, only to find themselves sadly in arrears at election time.

outpunch (autpʊ·nʃ), *v.* [f. Out- 18 + Punch *v.*[1] 3.] *trans.* In boxing, to surpass (an opponent) in punching ability.

1950 F. Mills *Twenty Years* xiii. 168 For the next few rounds.., he was gradually coming back..and for the seventh, eighth and ninth, I think he outboxed and outpunched me. 1956 —— *Forward the Light-Heavies* ix. 99 Delaney made a bad start in 1927 by being out-punched by Jim Maloney over 10 rounds at the new Garden. 1960 *Times* 26 Apr. 17/3 He was outthought, outboxed and.. outpunched by the Mexican. 1961 *Times* 6 June 5/4 J. Malcolm was outpunched and outpointed.

output, *sb.* Add: **1.** (Further examples.)

1942 *R.A.F. Jrnl.* 30 May 34 Dislocation of other factories depending on the Billancourt output. 1959 C. Singer *Short Hist. Sci. Ideas* iv. 116 This drug book [of Pliny's] is the prototype of the medical output of the next fifteen hundred years. 1965 Seldon & Pennance *Everyman's Dict. Econ.* 315 Output is normally understood to be gross output; but..the most useful definition..relates to gross output less the goods and services used in production; this amount is called net output. 1971 *Cabinet Maker & Retail Furnisher* 24 Sept. 524/1 The United Kingdom is the planned market for 80% of the output. 1972 *Value Added Tax: Gen. Guide* (H.M. Customs) 16 The goods and services he [*sc.* the taxable person] supplies are called his outputs and the tax he charges is his output tax. 1976 *Times* 21 May 4/1 *Manpower Paper No. 8*... estimated the output of graduates to 1981.

c. Energy produced by a device or system; *spec.* an electrical signal delivered by, or available from, an electronic device.

1884 S. P. Thompson *Dynamo-Electr. Machinery* vii. 113 The result is an extraordinary increase in the 'output', or, as Sir William Thomson terms it, 'activity' (i.e. amount of work done per second) of the machine. 1902, 1933 [see *input *sb.* 2 b]. 1956 *B.B.C. Handbk.* 1957 57 The equipment used in studio control cubicles for selecting and mixing the outputs of the various microphones. 1962 A. Nisbett *Technique Sound Studio* i. 17 The output of individual studios is fed in 'live', and linked together by station identification and continuity announcements. 1969 *Times* 7 Mar. 15/1 A television camera scans the object to be viewed, and output from the camera controls whether the pad vibrates. 1975 *Which?* Sept. 258/1 For the same output of heat, one of the gas fires..might land you with a bill of £26 a year, another with one of only £16. 1976 *Gramophone* June 116/3 Using the Bruel and Kjaer wave analyser to measure separate harmonics, rather than distortion factor including noise, suggested even lower figures: 0.048% at full output, 0.038% at −20dB and 0.074% at −30dB.

d. *Computers.* Data or results produced by a computer; also, the physical medium on which these are represented.

1948 *Math. Tables & Other Aids to Computation* III. 7 The 'output' or result of computation consists of numbers only. It has been proposed..to build a 'thinking' machine whose output would be orders rather than numbers. 1949 [see *input *sb.* 2 d]. 1959 E. M. McCormick *Digital Computer Printer* ix. 133 The output punch could be activated and would punch the output into the card. 1964 F. L. Westwater *Electronic Computers* vi. 105 Quite sophisticated computers use punched cards as a supplementary output. 1967 D. Wilson in Wills & Yearsley *Handbk. Managem. Technol.* iii. 44 Figure 3.1 outlines the basic steps for validating (editing), processing (sorting and calculating, etc.), and recording the output (writing the data on magnetic tape or disk, printing out, etc.). 1971 P. Harvey *Computer Sci.* vii. 102 If a decimal print out is required for visual inspection the binary output must be converted to a form suitable for operating a printer.

e. *Linguistics.* A structure resulting from the application of a lexical, grammatical, or phonological rule. Cf. *input *sb.* 2 f.

1961, etc. [see *input *sb.* 2 f]. 1968 P. M. Postal *Aspects Phonol. Theory* iii. 34 The phonetic representations are the final output of the entire set of phonological rules. 1970 *Language* XLVI. 261 The use of transformational rules applied to the output of structure-free grammars. 1971 *Archivum Linguisticum* II. 139 We can represent the output of the realization rules in the conventional way, partly by orthographic forms and partly by generalized morpheme-symbols such as -s.

2. A place where, or device through which, an output is delivered by a system, esp. an electronic device.

1933, 1946 [see *input *sb.* 3]. 1958 *Electronic Engin.* XXX. 1/2 The weakest part of a computer installation is, in general, the input and output which is usually slow compared with the speed of the machine itself. 1962 D. S. Halacy *Computers* iii. 66 (*caption*) A high-speed printer is the output of this computer. 1973 N. H. Crowhurst *Basic Audio Syst.* (1974) ix. 169 The essential ingredients of a feedback loop..are: an amplifier of some sort and feedback from output to input.

3. The action or process of supplying an output.

1947 [see *input *sb.* 4]. 1947 D. R. Hartree *Calculating Machines* 12 Functions of calculating machine components... (iii) Input (reception of data from the outside world). (iv) Output (supply of results to the outside world). (v) Transfer. (vi) Control. 1960 Gregory & Van Horn *Automatic Data-Processing Syst.* ii. 64 Output of results is the fourth stage in the flow of data. 1967 D. Wilson in Wills & Yearsley *Handbk. Managem. Technol.* iii. 46 In the early types of machine, input, computing, and output occurred serially so that large areas of expensive hardware were unused for much of the time. 1970 O. Dopping *Computers & Data Processing* xi. 154 When a computer is used for process control, the input and output is largely effected via..devices which convert analogue to digital information and vice versa.

4. *attrib.* and *Comb.* **a.** simple attributive, as *output circuit, device, impedance, punch, routine, stage, transformer, tube, unit, valve.* **b.** constituting output, as *output current, power, voltage.*

1920 H. J. Van der Bijl *Thermionic Vacuum Tube* vii. 178 Distortionless amplification is obtained if the amplified current in the output circuit is..an exact enlarged reproduction of the input current. 1973 N. H. Crowhurst *Basic Audio Syst.* (1974) vi. 118 The input circuit requires a fixed component of voltage or current to insure that the device operates at the correct combination of voltage and current in its output circuit. 1920 H. J. Van der Bijl *Thermionic Vacuum Tube* vii. 168 This would produce distortion since the output current is not an exact reproduction of the input. 1962 Simpson & Richards *Junction Transistors* ix. 219 The shift due to the rise in ambient temperature is thus relatively small and can be tolerated for peak-to-peak output-current swings of about 7mA. 1929 K. Henney *Princ. Radio* xii. 281 Output devices are used to (1) Keep d.c. current from the loud speaker winding; (2) Prevent serious loss in plate voltage; [etc.]. 1948, 1968 Output device [see *input device* s.v. *input *sb.* 5]. 1930 Moyer & Wostrel *Pract. Radio Construction & Repairing* (ed. 2) iv. 54 The plate resistance..in the case of audio-frequency amplification may be considered as the output impedance. 1962 J. H. & P. J. Reyner *Radio Communication* x. 409 (*heading*) Effect of feedback on output impedance. 1920 H. J. Van der Bijl *Thermionic Vacuum Tube* vii. 237 If the tube is used to amplify modulated high-frequency oscillations.. it must obviously be capable of giving a much larger output power. 1972 *IEEE Trans. Geoscience Electronics* X. 13/1 A CW output power of 1 W with a power efficiency of 70 percent has been achieved. 1959 Output punch [see sense *2 d]. 1962 *Gloss. Terms Automatic Data Processing* (*B.S.I.*) 96 *Automatic tape punch, output punch* [deprecated], a tape punch which automatically transcribes coded electrical signals into rows of holes in a paper tape and moves the tape as necessary. *Ibid.* 43 *Output routine*, a routine which organizes the output process of a computer, e.g. starts the output equipment, presents data to it at suitable intervals of time, and specifies format. 1926 *Wireless World* 1 Sept. 317/2 If, however, we are *not* within five miles of a B.B.C. station, the value of extra low impedance in the output stage is unnecessary. 1962 J. H. & P. J. Reyner *Radio Communication* x. 409 Voltage feedback..has the effect of reducing the effective internal impedance of the output stage. 1929 K. Henney *Princ. Radio* xii. 281 An output transformer is necessary to provide maximum energy transfer from the tube to the speaker. 1968 L. G. Sands *Easy Way to service Radio Receivers* (1973) ii. 52 Between points 1 and 3 you should get a higher resistance reading... If not, the output transformer may be grounded. 1929 J. H. Morecroft *Elem. Radio Communication* vii. 242 If the speaker is to use 100 milliwatts on the average,..the output tube should be drawing from its battery at least 10 watts. 1959 M. H. Wrubel *Primer of Programming* i. 9 In some cases the output unit can produce a printed sheet of answers directly; in other cases..the cards or tapes produced by the computer must be fed into an auxiliary machine. 1940 *Chambers's Techn. Dict.* 603/2 Output valve. 1942 *Electronic Engin.* XIV. 726 The stages in the receiver are covered in sequence from R.F. amplifiers to L.F. output valves. 1937 W. G. Dow *Fund. Engin. Electronics* 267 The useful output voltage..is the alternating component of the voltage across R_L. 1962 A. Nisbett *Technique Sound Studio* 259 Where a microphone impedance is strongly capacitative..its output voltage is fed to the grid of a valve.

output, *v.* **4.** (Further examples.)

1946 *Nature* 12 Oct. 504/1 Results are output in the form of punched cards. 1965 K. Nicol *Elem. Programming* vii. 31 A comparable output device is the graphical plotter which can directly output the results of a calculation as curves or points on a paper chart. 1972 P. B. Gove in H. D. Weinbrot *New Aspects Lexicogr.* 153 The *Seventh Collegiate*..has been completely programmed so that most of the relationships seen by the eye on a printed page can be output by the computer. 1976 *Physics Bull.* July 298/2 The most common device for inputting and outputting information is still the venerable teletype.

outputter[2].

For *Mod. Newspr.* read 1902 *Spectator* 22 Nov. 784/1.

out-quencher. For † read '*Obs.* exc. *hist.*' and add later example.

1959 L. Gross *Housewives' Guide to Antiques* viii. 84 Similar in appearance are the 'douters' or out-quenchers, which were used to extinguish candle flames...these had two discs between which the burning wick was nipped.

outrage, *sb.* Add: **3. d.** A person of strange or wild appearance, or one who is extravagant in behaviour.

1869 'Mark Twain' *Innoc. Abr.* 35 Who is that smooth-faced, animated outrage yonder in the fine clothes? 1884 —— *Huck. Finn* 236 Blamed if he warn't the horriblest-looking outrage I ever see. 1909 'O. Henry' *Roads of Destiny* 351 This old medical outrage floated down to my shack when I sent for him.

out-rai·l, *v.*[2] *nonce-wd.* [f. Out- 21 + Rail *sb.*[2] 2 b.] *trans.* To surpass in respect of a railing.

1866 Ruskin *Crown Wild Olive* p. xii, The publichouse-keeper on the other side of the way presently buys another railing, to out-rail him with.

outrake[2]. Restrict †*Obs.* to sense a in Dict. and add: **b.** (Later example.)

1976 G. Moffat *Short Time to Live* v. 49 This lady at Burblethwaite... I saw her walking up the outrake.

outrance. b. (Further examples.)

1837, 1860 [see *à l'outrance]. 1955 *Times* 19 May 10/2 Powerful Mau Mau 'generals'..are in favour of continuing the struggle *à outrance*. 1959 *Listener* 8 Oct. 589/1 The destruction of her fragile world and the war *à outrance* to keep something of it intact.

outrange, *v.* Add: **1. b.** *transf.* In certain ball games, to have a greater command of the field of play than (an opponent).

1930 *Times* 15 Mar. 6/1 At fullback, Scotland will have ..R. C. Warren, but he may be outranged by J. C. Hubbard.

3. c. *Aeronaut.* Of an aircraft: to have a greater range than (another aircraft).

1942 [see *out-gun *v.*].

outrank, *v.* Add: (Earlier and later examples.) Hence **outra·nked** *ppl. a.*

1842 *Spirit of Times* (Philad.) 1 Sept. (Th.), It won't be long before he fills the place of some one of the drones and cakes who now outrank him. 1903 *Westm. Gaz.* 26 Aug. 4/1 Barr did not outrank Wringe here as he had done at the start. 1948 K. Anthony *Lambs* i. 11 She laid some strange claim to a gentility outranking her husband's. 1973 J. Wainwright *Pride of Pigs* 179 You haven't a leg to stand on... You don't even out-rank me. 1974 S. Gulliver *Vulcan Bulletins* 21 The luckless, outranked, beautifully-spoken captain of shooting.

outré, *a.* (Further examples.)
1883 C. M. Yonge *Stray Pearls* I. iv. 42 Madame de Port Royal..is..suspected of being outrée in her devotion. 1934 C. Lambert *Music Ho!* v. 304 He will adopt a more eclectic and less outré manner. 1961 J. McCabe *Mr. Laurel & Mr. Hardy* (1962) vii. 139 Physical humour was becoming outré. 1975 L. Farago *Aftermath* xix. 320 The prosecution of criminal Nazis is coming to be widely regarded as a kind of séance in which the spirits of another era are conjured up by some *outré* hocus-pocus. 1975 *Times* 20 Sept. 6/2 Poirot..was the Englishman's notion of the comic foreigner, *outré* as everybody knew foreigners to be.

outreach, *sb.* Add: Also, the extent or length of reaching out; *spec.* the fact or extent of an organization's involvement in the community. (Further examples.) Also *attrib.*
1941 F. Matthiessen *Amer. Renaissance* iii. ii. 114 That has caught Browne's ability to take the familiar and to give it an unexpected outreach. 1950 *Theology* LIII. 417 The spiritual outreach of the body politic. 1965 R. B. Oram *Cargo Handling* i. 15 The management should have, ready for hire, cranes that may provide an outreach of 38 metres, as are to be found in Rotterdam. 1967 *Times* 27 Jan. 17/1 (*caption*) Each lantern is attached to 12ft. long outreach brackets on..tapered steel columns. 1967 *Gleaner* (Jamaica) 12 Nov. 9 Means and methods of furthering the outreach of the Jamaican church. 1972 *Evening Telegram* (St. Johns, Newfoundland) 24 June 35/5 Real preaching instead of the kind that people grow fat on; real outreach concern, whether it's over the back fence or overseas. 1974 *Times Lit. Suppl.* 18 Jan. 50/3 That was how he came to think of the Church, in its 'outreach' into the complex social whole. 1977 M. Wiles in J. Hick *Myth of God Incarnate* viii. 162 In his attitudes towards other men his life was a parable of the loving outreach of God to the world. 1978 *Amer. Libraries* IX. 67 As assistant deputy director of Buffalo and Erie County (N.Y.) Public Library, William Miles..oversees such special services as library outreach for community centers.

outreaching, *vbl. sb.* and *ppl. a.* (Further examples.)
1875 H. E. Manning *Internal Mission of Holy Ghost* xiv. 405 The symmetry and outlines of the Tree of Life, with its outreaching branches. 1902 A. T. Mahan *Retrospect & Prospect* iv. 111 This outreaching of an imperialistic arm by all the greater nations..constitutes..the motive to a closer union. 1905 *Daily Chron.* 14 Dec. 3/4 Such entire absence of out-reachings towards 'virility', 'grip', and 'tenseness'. 1972 *Science* 26 May 855/1 Certainly strong, outreaching, and complete fertility-control programs are desirable.

out-reckon, *v.* Restrict †*Obs.* to sense in Dict. and add: **b.** To overestimate. *rare.*
1898 Hardy *Wessex Poems* 154 But though, your powers outreckoning, You hold you dead and dumb.

out-relief. Add: **a.** Also *fig.*
1973 G. E. Aylmer *State's Servants* ii. 51 Ireland was still regarded as an out-relief centre for the less hopeful and worse-qualified members of the English governing class.
†**b.** *concr.* A person receiving out-door relief. *Obs.*
1904 *Westm. Gaz.* 22 Apr. 3/2 Still worse is the case of the aged 'out-relief', with his 3s. a week.

outride, *sb.* Add: **3.** In the writings of Gerard Manley Hopkins: (see quots.).
1880 G. M. Hopkins *Let.* 22 Dec. (1935) 41 By means of the 'outrides' or looped half-feet..I secure a strong effect of double rhythm, of a second movement in the verse besides the primary and essential one. *c* 1883 —— *Poems* (1918) 5 Two licences are natural to Sprung Rhythm. The one is rests, as in music... The other is *hangers* or *outrides*, that is one, two, or three slack syllables added to a foot and not counting in the nominal scanning. 1934 C. Day Lewis *Hope for Poetry* ii. 10 What Hopkins called 'outrides', unstressed syllables occasionally placed before the stressed ones at the beginning of the foot. 1973 *Studies Eng. Lit.: Eng. Number* (Tokyo) 24 In certain of the poems written in sprung rhythm..'outrides' appear to be extensively used, and these Hopkins has taken great pains to indicate... Without the help of these signals, there is little chance of the reader being able to distinguish between an 'outriding' and an ordinary foot.

outride, *v.* Add: **2. c.** To ride out of or beyond.
1903 J. L. Weston tr. *Sir Gawain at the Grail Castle* i. 15 In that one night had he outridden Britain and all that country.
4. (Example in Canad. use.)
1964 *Albertan* (Calgary) July 15/6 He's driving two chuckwagon outfits [and] will probably outride for a couple more.
5. *trans.* To keep cattle from going beyond (a tract of land) by riding along the boundaries of it. *U.S.* (? *Obs.*).
1874 J. G. McCoy *Hist. Sk. Cattle Trade* 375 He does not herd his cattle but designates certain bounds within which the employees permit the stock to range at will. This manner of holding stock is termed 'out riding' the country.

outrider. Add: **2. b.** *spec.* A fellow of New College, Oxford, accompanying the Warden on an official visitation of the estates of the college. Hence **outri·dership.**
1901 Rashdall & Rait *New College, Oxf.* 187 The outridership..was claimed by two Fellows,..who both wanted to accompany the Warden on progress. *Ibid.* 251 The Warden (or Sub-warden) accompanied by a Fellow known as 'Out-rider'..and the Steward, visit the farms on some part of the College estates. 1952 A. H. Smith *New College, Oxf.* iii. 47 An out-rider is still appointed each year to go with the warden on his summer progress around the estates.
5. (Further examples in *fig.* use.) *spec.* An escort mounted on a motor-cycle.
1851 H. Melville *Moby Dick* II. 164 Sharks also are the invariable outriders of all slave ships crossing the Atlantic. 1939 *Sun* (Baltimore) 25 Aug. 6/4 A motorcycle out rider was killed in Vermont while escorting a troop train. 1957 *Economist* 21 Sept. 912/2 The panoply of police cars, blue lights and motorised outriders with which the Chancellor chooses to move about. 1971 *Daily Tel.* 9 Aug. 11/4 We were given the full VIP treatment, including an escort of police motor-cycle outriders all the way from Boulogne to Le Touquet. 1975 T. Allbeury *Special Collection* xiii. 86 Under his chin were the telltale outriders of a dewlap. 1977 *Belfast Tel.* 22 Feb. 5/5 The Belfast councillors were driven through Dublin's streets in a convoy, with Gardai motorcycle outriders halting traffic so the entourage could travel speedily.
6. *U.S.* A mounted herdsman who prevents cattle from straying beyond a certain limit (see also quot. 1872).
1872 *Kansas Magazine* 319/2 Where the grower does not drive his own stock to market, the buying and driving is done by a class of speculators known in Texas as 'outriders'. 1874 J. G. McCoy *Hist. Sk. Cattle Trade* 348 [The] trail escapes the vigilant eye and Indian cunning and proficiency of the herdsman or outrider. 1907 S. E. White *Arizona Nights* i. vi. 117 We saw..the whole herd and the outriders and the mesas far away. 1939 P. A. Rollins *Gone Haywire* 230 Cowboys, patrolling as 'outriders' and 'line riders', had always to keep an eye on them. 1968 R. F. Adams *Western Words* (rev. ed.) 214/2 *Outrider*, a cowboy who rides about the range to keep a sharp lookout for anything that might happen to the detriment of his employer; also called *range rider*.
7. *U.S.* A mounted official who escorts race-horses to the starting post.
1947 *Sun* (Baltimore) 8 Nov. 11/4 Them outriders ought to get extra pay for steeplechases. 1961 Atkinson & Freeman *All the Way!* iv. 50 The outriders in their red hunting coats also accompany the field. Years ago there was only one..but today there are three at several of the major tracks. *Ibid.*, One outrider stands just to the outside in front of the gate, one takes a position outside the gap, three-sixteenths down the track, and the third remains behind the gate. 1968 M. T. Malloy *Racing Today* 42/1 A couple of other horses meanwhile may be running away with their jockeys hanging on for dear life, and with the track's red-coated outriders in hot pursuit.
8. *Canad.* 'In a chuckwagon race,..one of the four riders who load the wagon, direct the horses during the starting turns, and gallop with the outfit to the finish line' (*Dict. Canad.*).
1955 W. G. Hardy *Alberta Golden Jubilee Anthol.* 169 There are four outriders to each of the four outfits in every heat. When the starting-horn blows, one outrider holds back the team of horses fighting to be on its way. Another throws the stove in the rear of the chuckwagon. The remaining two pitch the flies and poles into the covered wagon. 1958 *Encycl. Canadiana* IX. 393 A new feature introduced in the 1923 [Calgary] stampede was the chuckwagon race... Each wagon is pulled by a four-horse team..and each has its outriders. 1964 *Albertan* (Calgary) 7 July 1/2 Three crack-ups..saw all drivers, outriders and horses come out unscathed.

outriding, *vbl. sb.* Add to def.: Also, *U.S.*, the work of an outrider (sense *6); a spell of executing this.
1907 C. E. Mulford *Bar-20* 6 Skinny Thompson took his turn at outriding one morning after the season's round-up. 1926 D. Branch *Cowboy & his Interpreters* 94 Groups of cowboys rode on inspection trips, 'out-ridings', to locate the scattered groups of cattle, to note the condition of grass and water [etc.]. 1968 R. F. Adams *Western Words* (rev. ed.) 214/2 *Outriding*, performing the duties of an *outrider*..; also called *range riding*.

outriding, *ppl. a.* Add: Applied to a syllable in the poetry of Gerard Manley Hopkins: see *outride *sb.* 3.
1877 G. M. Hopkins *Let.* 21 Aug. (1935) 45 There are no outriding feet in the *Deutschland*. An outriding foot is..a recognized extra-metrical effect; it is and it is not part of the metre..not being counted... Outriding feet belong to counterpointed verse. *c* 1883 —— *Poems* (1918) 5 These outriding half feet or hangers are marked by a loop underneath. 1953 W. H. Gardner in G. M. Hopkins *Sel. Poems & Prose* 224 The rhythm is sprung and outriding. 1973 [see *outride *sb.* 3].

outrigger. Add: **4. b.** An addition to a trailer to increase its carrying capacity.
1971 M. Tak *Truck Talk* 113 *Outriggers*, the short brackets that extend, if needed, from the sides of a low-boy trailer. 1973 *Amer. Speech* 1969 XLIV. 207 *Outrigger*, device used for increasing the width of a trailer.
5. b. *Aeronaut.* A supporting structure that projects outwards from the main part of an aircraft or spacecraft.
1909 *Flight* 27 Mar. 176/2 One of the most characteristic features of the machine is that derived from the appearance of the outrigger framework which carries the biplane elevator in front and the rigid biplane tail behind. 1922 *Encycl. Brit.* XXX. 20/2 In the pusher..the controlling surfaces are carried on an open frame ('outriggers') in front, at the rear, or in both positions. 1928 Chatfield & Taylor *Airplane & its Engine* xii. 235 The tail surfaces of the flying boat are usually carried on the stern of the hull, but sometimes they are supported on outriggers, or booms. 1969 *Times* 3 June [illegible] p. iii/7 The clusters were mounted on outriggers 90° apart on the ascent stage. 1976 *Aeroplane Monthly* Jan. 6/1 The new W.11 [illegible] used the same engine and transmission as the original design, but married them to a skeletal fuselage and outriggers with three-bladed rotors.
8. *outrigger canoe.*
1908 E. J. Banfield *Confessions of Beachcomber* II. i. 238 The grandfathers of the blacks of Hinchinbrook Island..have been popularly credited with the art of making out-rigger canoes, such as were common on the coast to the north. 1968 J. Severson *Great Surfing* 155 *Outrigger canoe*, a canoe employing the use of an outrigger. 1974 'M. Allen' *Super Tour* (1975) ix. 322 The [Fijian] natives..rode in outrigger canoes.

ou·tright, *v. U.S. Sports slang.* [f. Outright *adv.* (*adj.*).] *trans.* To give (a baseball player) a free transfer.
1975 *Cleveland* (Ohio) *Plain Dealer* 6 Apr. 6-C/1 It was very difficult when Joe was outrighted to Oklahoma City on Saturday. 1979 *New Yorker* 25 June 46/1 The Pirates had finally released him late in March ('outrighted' him, in baseball parlance).

outroar, *sb.* Delete *rare* and add examples.
1845 *Dublin Rev.* June 314 'God strike you, Satan' was the Reformer's outroar. 1891 G. Meredith *One of our Conq.* II. ii. 28 As it were, the towering wood-work of the cathedral organ in quake under [illegible] of its multitudinous outroar. 1955 V. Cronin *Wise Man from West* vii. 139 On the day of [illegible] eclipse..all the inhabitants of China were [illegible], to frighten away with crystals, drums and an outroar of yelling the monster that would otherwise swallow the sun.

out-room. Delete †*Obs.* and add examples in N. Amer. use. Also *attrib.*
1865 A. D. Whitney *Gayworthys* ii. 21 Gersham..ran up and down the out-room staircase. *Ibid.* 28 As she came into the out-room again. 1920 J. [illegible] *House* xii. 88 The floors were..sprinkled with white sea sand, that the 'out-rooms' [illegible] in fanciful patterns by brooms. 1971 *Islander* (Victoria, B.C.) 22 Aug. 5/2 He has a basement and an outroom full of bottles waiting to receive attention.

out-run, *sb.* Add: **3.** *spec.* The outward run of a sheepdog.
1938 [see *fetch *sb.*[1] 2]. 1955 [see *lift *sb.*[1] 5i]. 1973 *Country Life* 25 Oct. 1292/1 The collie's gathering outrun follows a wide curving natural cast..and ends behind the sheep.
4. [G. *auslauf.*] In Skiing: see quots.
1913 F. H. Harris *Dartmouth Out o' Doors* 101 The 'outrun'—the level stretch at the foot of the hill on which the jumpers check their speed. 1957 *Encycl. Brit.* XX. 749/2 He leans far forward over the points of his skis with arms outstretched, planing his body to increase his distance, lands with a slight give to his knees and speeds down the outrun. 1963 *Amer. Speech* XXXVIII. 206 *Out run*, in general, the bottom end of a ski run. In ski-jumping, the distance between the take-off and the landing point. 1974 *Encycl. Brit. Macropædia* XVI. 837/1 After the slope levels off, the jumper stops by turning on the outrun.

outscape, *sb.* Add: **2.** *rare.* [Scape *sb.*[8]] The outward appearance of a region. Cf. *inscape.
1868 G. M. Hopkins *Jrnls. & Papers* (1959) 184 In the afternoon we took the train for Paris and passed through a country of pale grey rocky hills of a strong and simple outscape.

outsco·re, *v.* [Out- 18.] *trans.* In various sports and games: to score more than; to surpass in scoring.
1958 *Times* 17 Oct. 20/1 Keenan..then settled down to outscore the Canadian consistently. 1960 E. W. Swanton *W. Indies Revisited* [illegible] Sobers, who found himself [illegible]. 1968 *Globe & Mail* (Toronto) 5 Feb. 17/2 In other games yesterday, Brampton Rocketts outscored Aurora Tigers 8–7 and league-leading Dixie Beehives edged North York Rangers 3–2. 1968 [see *face *v.* 5]. 1977 *Daily Mirror* 16 Mar. 30/4, I thought Mo outscored [illegible] five punches to one throughout most of the fight.

outseg (au·tseg), *v. U.S. colloq.* [f. Out- 18 + *seg, abbrev. of *segregationist.] *trans.* To support or advocate a more segregationist policy than (someone else).
1967 *Time* 13 Oct. 19 Governors Spiro Agnew of Maryland and Winthrop Rockefeller of Arkansas won office even though their Democratic opponents 'outsegged' them. 1970 *Manch. Guardian Weekly* 4 Apr. 14 Brewer..acquired a reputation as an effective administrator, and, most important, he has no intention of being 'Out-segged' by Wallace. 1970 M. Pei *Words in Sheep's Clothing* xvii. 166 'Segregation', a word that has given rise to such slang abbreviations as 'seg' and 'to outseg'. (Someone was once described as 'outsegging' Wallace.)

outsell, *v.* **2.** (Later examples.)
1961 Webster s.v., Nonfiction..continues to outsell fiction in the bookstores—*Publishers' Weekly.* 1968 *Globe & Mail* (Toronto) 27 Feb. 40 (Advt.), Again in 1967, Islington Plymouth outsold every metro Plymouth dealer. 1976 *National Observer* (U.S.) 10 Apr. 3/1 Washington [illegible] arrival in the [illegible]. 1977 *Times* [illegible] sold on its [illegible]

outsetting, *vbl. sb.* Add: **1.** (Later examples.)
1903 W. B. Yeats *In Seven Woods* 4 It's time to build up Emain that was burned At the outsetting of these wars. [illegible] The start was made from Blois a few days later, as strange an outsetting as ever was [illegible]
3. What is outside the self. *rare.*
1880 G. M. Hopkins *Sermons & Devotional Writings* (1959) 127 This applies to the universal mind or being too; it will have its inset and its outsetting; only that the outsetting includes all things, with all of which it is in some way..

outshoot, *sb.* Add: **4.** *Baseball.* = *outcurve 1.
1887 *Courier-Jrnl.* (Louisville, Kentucky) 5 May 6/3 He had a clever drop and out-shoot on which McQuaid failed to give him strikes. 1903 R. H. Barbour *Weatherby's Inning* 230 Then followed an out-shoot and a drop, neither of which did Joe take to. 1911 W. Patten *Bk. Baseball* 63/1 The plain horizontal outshoot, by the way, is no more in fashion. 1972 B. Shaw *Pitching* x. 86 Years ago, terms like 'drops', 'downer', 'out-shoot' were used to describe the curve ball.

outshot, *sb.* Add: **1.** (Later examples.)
1957 E. E. Evans *Irish Folk Ways* v. 69 In the traditional house of the north-west..a small bed-wing—the outshot or *cailleach*—projects out from one of the side-walls at the chimney end of the kitchen. 1961 M. W. Barley *Eng. Farmhouse & Cottage* ii. 88 Sometimes the outshot room is called 'the room below the entry'; in other cases it is a 'backed' [illegible] in East Anglia. In two cases it is an 'outend', presumably an outshot. [illegible] rubble-walled [illegible] cottage [in rural Ireland] [illegible] projection or 'outshot' for the bed—a feature of the western and northern fringes.
3. b. (Examples.)
1880 J. Dunbar *Pract. Papermaker* 13 Fines, Seconds, Thirds, Cords both good and light, Outshots, Prints, and the various qualities of Hemp and Jute Bagging. 1883 R. Haldane *Workshop Receipts* (Ser. 2) 369/1 Fines consist of fine white cottons; seconds, soiled white cottons;..outshots, good, strong, and sound rags. 1937 E. J. Labarre *Dict. Paper* 171/2 *Outshots*, a class of canvas or low white cottons found among rags.

outshout, *v.* Add: (Later examples.) Also *fig.*
1934 T. N. Wilder *Heaven's my Destination* ix. 165 The [illegible] shut up, you two! [illegible] 1962 R. P. Jhabvala [illegible] outshouted them [illegible]. 1962 *Spectator* [illegible] outshouts the rest of the cast. 1971 *Daily Tel.* (Colour Suppl.) 8 Jan. [illegible] the voices calling [illegible] in favour of [illegible]. 1976 *New Yorker* 15 Nov. [illegible] to speak [illegible] a mouthful of pebbles.

outsho·ve, *v.* [Out- 18.] *trans.* To outdo or surpass in shoving; to shove harder than.
1938 D. Runyon *Furthermore* vii. 142 Then the next thing anybody knows, the Yales outshove the Harvards, and now the game is over. [illegible] *Times* [illegible] Mar. 3/5 Often they outshoved [illegible] in the tight [illegible] Wales were unable [illegible]

out-shut, var. Outshot *sb.* 1.
1624 in S. O. Addy *Gloss. Words Sheffield* (1888) 167 [illegible] 1637 *Ibid.*, [illegible] the chiefe dwelling house [illegible] and out shuts [illegible] at the yearly rent of [illegible]. 1943 H. J. Massingham *Men of Earth* iii. 26 A [illegible] village [illegible] and even [illegible] outshuts and barns. 1945 —— *Wisdom of Fields* [illegible] half-frozen [illegible] [sc. a basketer].

outside. Add: **A.** *sb.* **3. b.** In isolated regions of Northern Canada and Alaska: the world outside these regions, esp. as an area of settlement and civilization.
1827 in *Beaver* (Winnipeg) (1927) Dec. 141 He was to bring in the last letters from outside which we could expect until next spring. 1898 *Yukon Midnight Sun* (Dawson, Yukon Territory) [illegible] men who have just arrived from the outside. 1904 J. Lynch *Three Yrs. Klondike* 54 On September 22 the last boat left for the 'outside' viâ the Lakes and Skagway. 1941 G. de M. Poncins *Kabloona* 13 Only the Arctic existed for them; and everything that lay below the Mackenzie River, was to them the remote, the virtually non-existent 'Outside'. 1968 R. M. Patterson *Finlay's River* 108 From there the survey party travelled with the horses, passing over the Wolverines in the first snow, headed for Fort St. James and 'the outside'. 1972 *Globe & Mail* (Toronto) 5 Dec. 35/1 He believes the story of Old Crow is valuable because it also shows how little people in The Outside—the rest of Canada in Yukon language—understand of the situation inside the territory.
c. *Austral.* The unsettled areas in the interior or bush.
1888 'R. Boldrewood' *Robbery under Arms* I. vii. 95 Dick Dawson came in from outside, and he said things are shocking bad; all the frontage bare already, and the water drying up. 1949 *Geogr. Mag.* Feb. 371 Rural life offers you such terms as *backblocks*, *outback* and *outside*, meaning remote, inland country. 1959 Baker *Drum* 132 *Outside*, unsettled districts in the interior or bush.
d. *slang.* The world out of prison; also the world out of the Army, civilian life.
1903 'J. Flynt' *Rise of R. Clowd* ii. 80 A boy in a Reform School with a 'plant' on the 'outside' takes a high place among his companions. 1919 D. G. Rowse *Doughboy Dope* 9 A is the Army at that stage of your young life when you were on what the Army calls 'the outside'. 1933 *Amer. Speech* VIII. iii. 26/2 'Whitey', who escaped three times from solitary confinement clear to the outside, was an acknowledged eel. 1965 C. D. B. Bryan *P. S. Wilkinson* 369, I never asked..what you did on the outside. 1972 C. Drummond *Death at Bar* ii. 56 Kath hasn't been having it so good, what with a couple of worthless sons who haven't the sense to keep on the outside.
e. In Surfing: see quots.
1963 *Surfing Yearbk.* 42/2 *Outside*, surfing area past the breaking surf. 1963 *Pix* 28 Sept. 62/3 *Outside*, the surfing area outside the breaking wave.
6. (Earlier and later examples.)
1789 J. Woodforde *Diary* 13 June (1927) III. 114 For the remaining part of our fare paid..for 1 outside 12/0. *a* 1800 in *Norfolk Fair* (1970) Nov. 31/7 This Coach from Norwich to London by Newmarket every Day Convey 8 Insides..and 6 Outsides in the most Pleasant And Agreeable Stile. 1902 *Chambers's Jrnl.* Nov. 715/1 The 'George' at Grantham is still..one of the best inns in England, as it was when these two prudent 'outsides' left the Yorkshire coach and 'turned in' there. 1914 'I. Hay' *Lighter Side School Life* vi. 152 He was called at half-past two..and by three o'clock was off as an 'outside' upon the Tally-Ho Coach.
b. = Outsider 2.
1898 [see *behind *sb.* 2 b]. 1899 *Captain* II. 186/1, I headed out to the right, [and] saw our outside get it. 1906 *Field* 13 Oct. 610/1 Their outsides showed so crude a conception of passing that [etc.]. 1927 *Observer* 21 Aug. 18/3 The team are young and play attractive football, with a clever set of outsides who combine well. 1963 *Times* 23 Feb. 3/1 Although their forwards were playing such a solid game the Westminster outsides were too slow to beat their opposite numbers.
8. Short for *outside paddle-wheel* on a river steamboat.
1876 'Mark Twain' *Tom Sawyer* 29 Come ahead on the stabboard! Stop her! Let your outside turn over slow. 1894 —— in *Century Mag.* May 19 Set her back on de outside... Come ahead on de inside.
B. *adj.* **1.** *outside* (*jaunting*) *car* (earlier examples); *outside passenger*, one who travels on the outside of a conveyance.
c 1810 W. Hickey *Mem.* (1960) 326 One of the stage-coaches, with a number of outside passengers..frightened our horse. 1815 Outside passenger [in Dict.]. 1829 [see Jaunting-car]. 1849 W. Allingham *Diary* 30 June (1907) 49 Our party took leave and mounted a back outside-car in Gloucester Street. 1849 Dickens *Dav. Copp.* (1850) v. 51 The story of my supposed appetite getting wind among the outside passengers, they were merry upon it. 1874 'G. Ramsay' *Thomas Grant* i. 23 They drove up on an outside car to the quays.
2. a. (Earlier and later examples of *outside world*.)
In some uses indistinguishable from sense 3.
a 1902 S. Butler *Way of all Flesh* (1903) lxxx. 366 Days before those in which he had begun to bruise himself against the great outside world. 1926 Galsworthy *Silver Spoon* i. iii. 22 Our trade-unionists despise the outside world. They've never seen it. 1947 E. F. Russell in Aldiss & Harrison *Decade 1940s* (1975) 168 He..looked out through the dome... The outside world slumbered. 1969 *Times* 9 Jan. 4/5 They [sc. the Hutterites]..live in almost complete isolation from their neighbours and the outside world. 1976 *Listener* 29 July 104/2 All of us [blacks] will want to see blacks on television. The outside world will be astonished at the talent we have. 1976 *Norwich Mercury* 19 Nov. 2 Mr Sims finished his comments with a reference to the outside world's contact with the school—the wide range of evening classes available, [etc.].
c. *spec.* of a water closet: situated outside the house, building, etc.
1939 M. Spring Rice *Working-Class Wives* iv. 73 Mrs. P. of Glasgow... Under the drawbacks of her house she says 'Outside lavatory, (used by six families.) Public house at close which is objectionable [etc.].' 1943 *Our Towns* (Women's Group on Public Welfare) iii. 88 The outside closets are generally much less well kept than the inside ones... The majority of houses have outside W.C.'s only. 1959 B. J. Farmer *Murder Next Year* ii. 7 The keys of the house had been left in the outside w.c. 1960 P. Hastings *Sandals for my Feet* i. i. 13 I'm tired of..the smell of damp and that filthy outside loo. 1974 P. Highsmith *Ripley's Game* ii. 11 The little brick structure, formerly an outside toilet, that served as a tool shed.
3. b. In Northern Canada and Alaska: belonging to or obtained from another part of the world, esp. from the settled or urbanized parts.
1896 C. Whitney *On Snow-Shoes to Barren Grounds* 40 Gairdner had annoyed me a great deal, and no doubt we had worried him not a little, breaking in upon the even and lethargic tenor of his monotonous life with our 'outside' (as the great world is called by the denizens of this lone land) hustling ways. 1904 J. Lynch *Three Yrs. Klondike* 141 The leader is always a small 'outside' dog, usually of the Scotch collie breed. [illegible] *Huntress* [illegible] raised outside [illegible] mission. 1958 P. Berton *Klondike Fever* [illegible] expense!.. [illegible] Outside word. 1977 *Globe & Mail* (Toronto) [illegible] By special [illegible] the dance.
6. Special collocations: **outside broadcast** (see quot. 1941); **outside broadcaster,** (*a*) one who makes or supervises an outside broadcast; (*b*) an outside contributor to broadcasting; **outside broadcasting,** the action of making an outside broadcast; **outside cabin,** a cabin with a window or porthole on the side of the ship; **outside chance,** a very unlikely chance; **outside forward,** in association football and hockey, either of the two players, called the outside left or right (see below), of the forward line; **outside half** = *fly-half* (*fly *sb.*[1] 8); **outside interest,** an interest not directly or necessarily connected with one's everyday life or interests; **outside job** *slang*, a crime committed in a house, etc., by a person not connected or associated with the household or building concerned; **outside leaf,** an outer leaf on a vegetable, esp. cabbage; **outside left, right,** in association football and hockey, a player playing on the extreme left or right of the forward line; **outside line,** a telephonic connection with an external exchange (cf. *line *sb.*[2] 1 c); **outside man** *U.S. slang*, one involved in any of various special roles in a confidence trick or robbery; **outside right** (see *outside left* above).
1927 *B.B.C. Handbk.* 1928 [illegible] *Outside broadcasts*, a broadcast item taking place at some point other than the studio. [illegible] 1941 *B.B.C. Gloss. Broadcasting Terms* 22 *Outside broadcast* (abbrev. O.B.), programme originating elsewhere than in the studio [illegible]. 1953 *News Chron.* 2 June 3/5 It is a long way from 1937, when B.B.C. television mounted its first outside broadcast from Hyde Park Corner on another Coronation day. [illegible] 1941 [illegible] *Outside broadcasting* [illegible] 1972 P. Black *Biggest Aspidistra* ii. 27 In the early days the bands played mostly in clubs and hotels, and so were part of outside broadcasting. [illegible] Outside cabin [illegible] 1966 *Guardian* 29 Oct. 5/5 Two-berth outside cabins with private shower and w.c. 1971 'A. Garve' *Late Bill Smith* [illegible] a limited number of privileged passengers; fully air-conditioned and stabilised; swimming pool and two bars; all outside cabins. 1909 *Daily Chron.* 11 Jan. 4/6 The chance that the right marriage of poetry and music should come is an outside one. 1958 R. A. Knox *Footsteps at Lock* xiv. 135 By an outside chance you might find it lying about somewhere. [illegible] *Gipsies* xx. [illegible] all out for an outside chance. 1973 D. Rutherford *Kick Start* viii. [illegible] moon stones [illegible]. [illegible] *Encycl. Sport* I. 517/1 [illegible] on the right [illegible] 1898 J. Goodall *Assoc. Football* 18 [illegible] outside forward [illegible] *Encycl. Sport* [illegible] is usually done by a half-back or by an outside forward. 1949 *Rugby League Football* ('Know the Game' Series) 3 Stand off half [illegible] (Llanelli v. [illegible]) [illegible] Outside-half [illegible] gave a superb exhibition of outside-half play. 1860 J. W. Palmer [illegible] fault of the [illegible] the cases with which I have been occupied. [illegible] Hall *Saturday Life* [illegible] and devoted [illegible] outside interests? [illegible] Rendell *Face of Trespass* [illegible] outside interest [illegible] 1925 A. Christie *Secret of Chimneys* [illegible] he was killed by someone in the house [illegible] unlatched the [illegible] like an outside job [illegible] house *Money for Nothing* [illegible] an outside job. [illegible] The police are [illegible] an 'outside job'—I mean, it wasn't a burglar. The broken open window was faked. 1972 Outside job [see *inside job* s.v. *inside *a.* e]. 1739 E. Smith *Compl. Housewife* (ed. 9) 37 Take a well-shap'd Cabbage, peel off some of the outside leaves. 1747 H. Glasse *Art of Cookery* iv. 57 Take a fine White-heart Cabbage..half boil it..take great Care not to break off any of the the outside Leaves. 1861 Mrs. Beeton *Bk. Househ. Managem.* 560 Boiled cabbage... Pick off all the dead outside leaves. 1960 *Good Househ. Cookery Bk.* (rev. ed.) 217/1 Wash the chicory and remove the outside leaves. *Ibid.* 218/1 *Leeks*... Remove the coarse outside leaves. 1900 *Football Who's Who* 134 Cassidy, Joseph, Manchester City (outside left). 1905

Gibson & Pickford *Assoc. Football* I. 161 Every one knows Alec Smith. He's the outside left of the Rangers. 1960 B. Liddell *My Soccer Story* xvi. 102 To get down to my final choice at outside-left, I vote for Peter McParland, an unorthodox type of winger with a wonderful turn of speed. 1965 *Men's Hockey* ('Know the Game' Series) (rev. ed.) 24/2 The outside right when about to pass to his left will find it more convenient to have the ball a little in front of his left foot. The outside left, however, must make a half turn to the right when passing right [etc.]. 1975 *Liverpool Echo* (Football ed.) 1 Feb. 3/1 There seems to have been some debate as to who was the Liverpool outside left before Billy Liddell. 1944 Outside line [see *line *sb.*[2] 1 c]. 1962 E. S. Gardner *Case of Blonde Bonanza* (1967) x. 117 You can't get an outside line on these phones unless they connect you. 1972 D. Bloodworth *Any Number can Play* xx. 203 Ivansong seized the telephone (which..was automatically switched to an outside line). 1926 J. Black *You can't Win* ix. 111 He..made his living serving as 'target' or outside man, for the yegg mobs that preyed on country banks. 1937 *N.Y. Times* 22 Dec. 22 *Outside man*, a spy under a cover, but not masquerading as an employe of a plant. 1938 F. D. Sharpe *Sharpe of Flying Squad* xiv. 151 She was acting as look out or 'outside man' for two expert safe breakers. 1947 *Amer. Speech* XXII. 169/1 *Outside-man*, the member of a *shell-mob* who locates promising suckers on the lot, steers them to the game, and assists in the play. 1890 C. W. Alcock *Football: Assoc. Game* 48 The outside-right should not be more than eight or ten yards beyond him. 1974 *Sunday Mail* 14 Apr. 39/1 Both goals..were scored by the outside rights.
C. *adv.* **1. b.** In Northern Canada and Alaska: in the settled or urbanized areas outside these regions; abroad.
1898 F. Russell *Explor. Far North* 80 To 'go in', by the way, is to descend the Athabasca; to return to civilization is to 'go outside'. 1923 F. Waldo *Down Mackenzie through Gt. Lone Land* 246, I had thought that life beyond the 65th parallel or so was life beyond the pale; but I was now to learn that the Arctic Circle is the inner circle, and the real outsider is—of course—the one who lives 'Outside'. 1945 R. W. Service *Ploughman of Moon* 321 If I had been Outside it would have taken me five years to save a thousand dollars. 1955 *Whitehorse* (Yukon Territory) *Star* 24 Feb. 2/1 One of the outstanding characteristics of the Yukon is the general indifference to what is going on Outside. 1970 *Islander* (Victoria, B.C.) 1 Nov. 13/3 Sam Otto spent 14 years in the Barren Lands and Northwest Territories, without going outside once.
c. *Austral.* In the interior or bush. *rare.*
1911 C. E. W. Bean *'Dreadnought' of Darling* xxxv. 317 But, be the 'inside' country never so tame and densely populated, there will always be a huge stretch of country 'outside' which cannot by any known means be closely settled.
d. *slang.* Out of prison; in civilian life.
1919 W. Lang *Sea-Lawyer's Log* ix. 108 You *got* to 'ave some bloody religion in the Navy. Now, wot church did you go to outside? 1937 *Research Stud.* (Washington State Coll., Pullman) V. 19 A boy entering this institution [*sc.* a reformatory] learns more bad habits than he would ever think of learning out side. 1961 Partridge *Dict. Slang* Suppl. 1210/1, I don't care what you were 'outside'; you're in the Andrew now, so don't forget it, or you'll be in the rattle.
e. In Surfing: see quots.
1962 T. Masters *Surfing made Easy* 65 *Outside*, out past the breaking waves, or at the furthest break. 1962 *Austral. Women's Weekly* 24 Oct. (Suppl.) 3/3 *Outside* or *out the back*, a long way out at sea, beyond the first line of breakers.
3. a. (Further examples.)
1869 *Galaxy* June 831 Don't let's get outside of more'n a bottle apiece, and that plain whiskey. 1886 [see *get *v.* 40*]. 1915 J. Webster *Dear Enemy* 174 He likes to dine outside of the family vault. 1943 G. Greene *Ministry of Fear* i. ii. 29 Murderers..are very, very seldom..gentlemen. Outside of story-books. 1975 *Nature* 20 Mar. p. xx (Advt.), These books are..distributed outside of the U.S.A. and Canada by Academic Press. 1976 *Gramophone* Nov. 903/2, I was able to spend some time in their CD-4 four-channel disc cutting room—the most important such facility outside of Japan.
b. For *U.S. colloq.* read *colloq.* (orig. *U.S.*). Add earlier and later examples.
1859 A. L. Elwyn *Gloss. Supposed Americanisms* 82 *Outside*, this word is frequently used by writers in newspapers in a sense not known to the language. In a *Ledger* of a late date, there is a phrase..'*outside* of the Secretary of War', for 'no one but that official'. 1913 R. Fry *Let.* 5 Apr. (1972) II. 367 I'm very much interested by what you said about the need of some big belief outside of art. 1968 *Listener* 22 Aug. 234/1 The only power we have is to expel a union for corruption or for following a communist or a fascist policy. Outside of that the unions pretty much take care of their own business. 1972 *New York* 8 May 62/2 Outside of a slightly annoying tendency to call all female customers 'Hon', everything about Mr. Blume inspires confidence.

ou·tside-i·nside, *a. rare.* [f. Outside *a.* + Inside *a.*] Of or pertaining to both the outside and the inside.
1930 R. Graves *Ten Poems More* 11 Neat outside-inside, neat below-above Hermaphrodising love. 1951 M. McLuhan *Mech. Bride* (1967) 35/2 It can be picture-framed by a Pond's ad for the 'special outside-inside face treatment'.

outsider. Add: **1.** (Further examples, with varying degrees of specificness.) Also *attrib.*
1897 J. McCarthy *Gladstone's Life* xxvii. 90/2 The outsider class..quarreled with Mr. Gladstone because he was always giving them a surprise. 1912 T. E. Lawrence *Let.* 10 Feb. (1938) 136 About the Jerablus seals:—I can't give you those, only the five outsiders: the Jerablus ones were bought [etc.]. 1935 *Amer. Speech* X. 271/2 *Outsiders*, buyers who ship special kinds of livestock to other markets. 1944 F. Brown in B. W. Aldiss *Introd. SF* (1964) 69 No one knew who the Outsiders were..or from what far galaxy they came. 1958 *Amer. Speech* XXXIII. 167 (*Australian Cattle Lingo*) *Outsider*,..a stray. 1974 *Nat. Geographic* Jan. 114/2 The fishermen talked shyly in the presence of an outsider from upalong. (An 'outsider' is any off-islander, including even other Canadians.) 1977 *Globe & Mail* (Toronto) 5 July 8/3 So far, he says, the inquiry has been lucky enough to be seen as a unit, not as two Yukoners—one for whites and one for Indians led by an outsider.
b. (Earlier examples.) Also *transf.*, a person who fails to gain admission to the 'ring'; a person who habitually backs outsiders in a race; *rank outsider*, (*a*) an outsider at very long odds; (*b*) a person who is considered socially inferior (cf. sense 1 c below).
1836 R. S. Surtees *Let.* in A. Mathews *Mem. Charles Mathews* (1839) IV. ix. 185 An unfortunate outsider, called Astracan. 1836 *Spirit of Times* 5 Mar. 20/1 The Brother to Maria, the Babel colt, and Taishteer, are a shade worse, owing, no doubt, to the money laid out upon Brother to Nell Gwynne. No change amongst the outsiders. 1845 *Ibid.* 31 May 158 The 'outsiders' won 'smartly' on both races, and the staunch friends of Fashion, who have backed her 'all through', have 'got hunk' and a good deal over. 1855 J. R. Planché *New Haymarket Spring Meeting* in *Extravaganzas* (1879) V. i. 94 Which are the favourites, and which outsiders? 1871 R. A. Proctor *Light Sci.* (Ser. 1) 288 The success of a rank outsider will be described as 'a misfortune to backers'. 1890 Barrère & Leland *Dict. Slang* II. 170 *Rank outsider* (common), a vulgar fellow, a cad. From a racing term applied to a horse outside the rank. 1902 Farmer & Henley *Slang* V. 116/1 *Outsider*,.. (racing), a person who fails to gain admission to the 'ring' from pecuniary or other causes. 1908 *Magnet* I. i. 8/2, I ask you if you ever saw such a rank outsider in all your natural?
c. A person who is isolated from or does not 'fit' into conventional society either through choice or on account of some social, intellectual, etc., reason. Often deprecating. *spec.* In literary criticism: the archetypal artist or intellectual seen as a person isolated from the rest of society. Also *attrib.*
1907 'I. Hay' *Pip* x. 322 'I didn't think you ought to play [golf] with him,' said Pip coolly. 'He's an utter outsider.' 1908 *Magnet* I. i. 7/1 'You rotten outsider!' said Bulstrode, in tones of concentrated rage. 'You're not fit to be at a decent school.' 1913 H. Kephart *Our Southern Highlanders* xiii. 294 A bastard is a woods-colt or an outsider. 1946 S. Gilbert tr. *Camus's L'Étranger* (*title*) The Outsider. 1956 C. Wilson *Outsider* i. 14 Many great artists have none of the characteristics of the Outsider. Shakespeare, Dante, Keats were all apparently normal and socially well-adjusted. 1957 *Times Lit. Suppl.* 25 Oct. 640/1 His [*sc.* C. Wilson's] original contribution was simply the Outsider gimmick. 1958 [see *down and out *adj. phr.*]. 1958 J. Raymond *England's on Anvil!* 40 Like Proust the Jew, Pope the Roman Catholic son of a linen-draper was an outsider. 1959 *Times Lit. Suppl.* 1 May 261/4 It throws light on two generations of 'outsider' philosophers grappling with their own sudden emergence into the world of letters and art. 1963 *Spectator* 4 Oct. 430/1 *The City of Dreadful Night*..is Outsider poetry. 1966 C. Sweeney *Scurrying Bush* xiv. 201, I remember an odd fellow when I was in Nigeria. Bit of an outsider, really, but do anything with snakes.
Hence **outsi·derdom, outsi·derhood,** the condition or state of being an outsider (in sense 1 c, above); **outsi·derish** *a.*, of the nature or character of an outsider; so **outsi·derishness; outsi·derism,** the theory or practice of being an outsider; **outsi·derliness,** the quality or fact of being an outsider (in sense 1 c); **outsi·derly** *a.*, characteristic of an outsider or of outsiderliness.
1956 C. Wilson *Outsider* viii. 216 He had accepted his 'Outsider-ishness', not as a symptom of some strange disease, but as a sign that his healthy soul was being suffocated in a world of trivial, shallow, corrupted fools. 1957 *Times Lit. Suppl.* 25 Oct. 640/4 That the seeds of outsiderliness are found in us all may well account for Mr. Wilson's success. 1958 *Ibid.* 28 Mar. 165/3 A final view of Mr Freund..might be that he is a sort of Colin Wilson without a theory of Outsiderdom, searching for the religious viewpoint that will include the complexities of modern science. 1958 *Listener* 26 June 1070/1 Genuine outsiderhood, as experienced by delinquents, psychotics, alcoholics, unmarried mothers, and a whole host of people who..find themselves the wrong side of the law. *Ibid.* 10 July 63/3 His 'outsiderism' made him enjoy shocking the professional scientists. 1959 *Times Lit. Suppl.* 23 Jan. 44/1 His account of that heredity is wry, humorous, and outsider-ish. His Jewish ancestry has its roots in Poland and Slovakia. *Ibid.* 4 Sept. 503/3 The outsiderly novel by Henri Barbusse, *L'Enfer*, which was the starting point of Mr. Wilson's first book. 1960 *Ibid.* 8 Apr. 221/1 Poetic abstraction and 'outsiderish' philosophical terminology. 1961 *Guardian* 23 June 9/6 The jigging of today's young is their alternative to 'outsiderism'. 1961 *John o' London's* 16 Nov. 548/2 The trouble with Outsiderdom as a philosophy is the squalid assortment of fellow-travellers it attracts. 1962 *Times Lit. Suppl.* 21 Sept. 710/2 Accompanying such pieces of outsiderly narcissism is a certain amount of fashionable philosophizing. 1966 *New Statesman* 8 July 61/1 An outsiderly Old Etonian whose rebellion against the ethics of his upbringing has driven him mad.

ou·tsize, *sb.* and *a.* [f. Out *a.* + Size *sb.*[1]] **A.** *sb.* **a.** A person or thing larger than the normal or the majority; esp. a ready-made garment larger than the standard sizes. Also *transf.*
b. Greater size than normal.
1845 *Ainsworth's Mag.* VII. 213 The borrowed child being ra-ther an out-size..rendered Cora's carrying him a matter of difficulty. 1883, 1894 [see Out *a.* 5]. 1902 E. Nesbit *Five Children & It* viii. 211 Robert was indeed what a draper would call an 'out-size' in boys. 1924 *Mod. Draper* II. 69 With regard to all ladies' underclothing it is necessary to keep a good assortment of outsizes. 1924 A. Christie *Man in Brown Suit* xii. 96 'I don't think he'll have any out sizes,' murmured Pagett, measuring my figure with his eye. 1933 Chesterton *St. Thomas Aquinas* i. 14 In the present case the outline is rather an outsize. The gown that could contain that colossal friar is not kept in stock. 1957 M. B. Picken *Fashion Dict.* 238/2 *Outsize*, size larger than the regular sizes; generally used for sizes larger than 46 bust. Also applied to hosiery having a top larger than average. 1970 V. Gielgud *Candle-Holders* i. iv. 34 This impression of massive outsize was only exaggerated by the sight of George Eltham standing at the top of the flight of shallow steps which linked the drive with the great front doors.
B. *adj.* Larger than the average, usual, or stock size. Also *transf.* Also **o·utsized** *ppl. a.* Hence **outsi·zeness,** the fact or quality of being outsize.
1880 *Good Words* 46/1 He was what is sometimes called an 'outsized man'..imposing in appearance. *c* 1890 in *American Mail Order Fashions* (1961) 12 Ladies' outsize plated silk hose. 1895 *Westm. Gaz.* 20 July 2/1 She was a great outsized woman. 1904 H. G. Wells *Food of Gods* i. ii. 19 He conceived a picture of coops and runs, outsize and still more outsize coops, and runs progressively larger. 1906 *Westm. Gaz.* 5 May 5/1 'An out-sized cat, I call him,' remarked the cook soon after his arrival. 1922 Joyce *Ulysses* 715 A pair of outsize ladies' drawers of India mull, cut on generous lines. 1928 *Blackw. Mag.* May 709/1 A valley of utter desolation,..with an outsize snow mountain..at either end of it. 1937 Auden & MacNeice *Lett. from Iceland* xii. 173 She wears an amazing woollen helmet with earflaps which combined with her goggles and general outsizeness makes her look like a piece of Archaic Greek sculptuary. 1952 *Manch. Guardian Weekly* 9 Oct. 3/4 The Republicans are undoubtedly united in an outsize bed where there is room for everybody except the few liberals who got the General [*sc.* Eisenhower] his nomination. 1958 *Times* 31 Oct. 3/5 Her voice is of a rich mezzo-soprano quality, not outsize, but certainly amply big enough for most lieder. 1974 A. Williams *Gentleman Traitor* v. 86 He..dressed in an outsize suit of white slub-silk. 1976 *Publishers Weekly* 26 Apr. 3/3 (Advt.), A massive, out-sized book of more than 800 pages.

outskirt, *sb.* **1.** For 'Now only in *pl.*' read 'Now usu. in *pl.*' and add further *sing.* examples.
1891 Hardy *Tess* I. xix. 245 The outskirt of the garden in which Tess found herself had been left uncultivated for some years. 1943 J. Betjeman *Eng. Cities & Small Towns* 14, I am reminded of that moving passage about a provincial suburb in Gissing's story *Fate and the Apothecary* describing, I think it must be, an outskirt of Exeter.
2. (Later example.)
a 1930 D. H. Lawrence *Last Poems* (1932) 46 Corpse-eaters They dwell in the outskirt fringes of nowhere.

outsmart (autsmā·ɹt), *v.* [f. Out- 20 + Smart *a.*] *trans.* To get the better of or overcome by superior craft or ingenuity; to prove too clever for; to outwit. Also *refl.*
1926 H. C. Witwer *Roughly Speaking* iii. 95 Young Farrell seemed to have more than recovered from Ben's terrible right hand blow, for he was now doing what Pete told me was 'outsmarting' Ben. 1948 E. Waugh *Loved One* 105 'All his stories are about the same thing—American innocence and European experience.' 'Thinks he can outsmart us, does he?' 1954 J. Steinbeck *Sweet Thursday* xxi. 131 It is such fun to outsmart a smart guy. 1957 P. Frank *Seven Days to Never* 48 It takes a machine to outsmart a machine. 1961 'B. Wells' *Day Earth caught Fire* viii. 128 Hell, we're all so bloody clever at outsmarting nature. Anything you can split I can split better. 1974 J. Heller *Something Happened* 296 'I'm going to sock you one, Daddy,' he squeals in frustration, as he feels himself outsmarted. 1975 J. F. Burke *Death Trick* (1976) xii. 156 Like all smart crooks, he outsmarted himself. 1977 *R.A.F. News* 8–21 June 4/2 While the authorities dither and disagree on a plan of action, the beast continues his rampage of death, outsmarting every move to trap him.

outspan, *sb.*[1] (Earlier examples.)
1822 W. J. Burchell *Trav. S. Afr.* I. iv. 92 These *uitspan*, or *outspan places*, are, in fact, the caravanserays of the Cape. 1844 *Colburn's United Service Mag.* May 23 *Outspan*,..place of rest, where the oxen are unyoked and turned out to graze.

ou·tspeech. *rare.* [Out- 7.] Frank or candid words; plain language or terms; outspeaking.
1919 W. de Morgan *Old Madhouse* 439 Outspeech would be the safest course as well as the easiest, with this girl. 1921 —— *Old Man's Youth* xxvii. 267, I was sorry a moment after for my own outspeech.

outspeed, *v.* (Later examples.)
1911 *Chambers's Jrnl.* Jan. 57/1 If the black, whirling maelstrom of a cyclone looms up before him, he will make a detour or even outspeed it. 1930 R. Campbell *Adamastor* 31 Bold is he..And swift—outspeeding as he runs The corposants of Leda's sons. 1962 *Times* 28 Mar. 4/2 He had outspeeded Spinks in the early rounds. 1977 'O. Jacks' *Autumn Heroes* viii. 112 It became clear that the truck would survive once it outsped the fire line.

outspread, *v.* Add: **3.** *intr.* To spread out, extend itself.
1906 *Westm. Gaz.* 26 June 2/3 Each young branch, outspreading in the sun, Reflects in shadow on the sod below. 1925 [see *asile].

outsta·ndingly, *adv.* [f. Outstanding *ppl. a.* + -ly[2].] In a notable or outstanding manner; in or to an exceptional degree; remarkably or conspicuously.
1909 *Westm. Gaz.* 18 Jan. 12/2, I don't mean to say that he is an outstandingly good putter. 1922 A. S. M. Hutchinson *This Freedom* ii. ix. 150 There was outstandingly one such day of absorption in delight..for Rosalie. 1928 *Observer* 18 Mar. 23/3 Her Wagner songs are outstandingly fine. 1961 R. J. Corsini et al. *Role-playing in Business* ix. 142 A district manager who had been recently appointed to his position after being outstandingly successful as a store manager, found that in supervising other managers he was having a difficult time.

ou·tstate, out-state, *a. U.S.* [f. Out- + State *sb.* 31 c.] **a.** Of or pertaining to a part of a state away from the largest population centre (see also quot. 1931). **b.** Coming from or living in another state; = *out-of-state *adj. phr.*
1931 *Amer. Speech* VI. 310 'Out-state' is a compound word not yet recognized by the dictionaries but frequently used by Nebraskans, Iowans, Coloradans, and Wyomingites. University students from these states reported two meanings: out in the state away from the main city, and out in the state away from the speaker's home. 1934 *Sun* (Baltimore) 2 Nov. 2/1 The vote which Mr. Picard would have to pile up there to overcome his opponent's out-State lead. 1961 in Webster, Lost the governorship because the outstate vote went against him. *Ibid.*, A gorge of unusual natural beauty which few out-state visitors see—M. W. Fishwick. 1967 *National Observer* (U.S.) 12 June 5/1 The pattern of an all-too-familiar Western movie, the one about the clever outstate cattle barons displacing the local nesters.

out-station. Add: (Further examples.) Esp. in *Austral.* and *N.Z.* use (cf. Station *sb.* 14).
1844 *Port Phillip Patriot* 11 July 1/3 (Morris), There are four out-stations with huts, hurdles..and every convenience. 1859 F. Fuller *Five Years' Residence N.Z.* viii. 157 Out-stations are started for the shepherds, who watch separate flocks, to live in. 1862 R. Henning *Let.* 29 Sept. (1966) 107 Another bedroom wherein reside any other members of the 'staff' who happen to be at home—more than half are always at the out-stations. 1911 C. E. W. Bean *'Dreadnought' of Darling* xxxv. 311 When the out-station was reached, she rang up. 1944 F. Clune *Red Heart* 64 Southwards he trudged, came to the out-station of the Darling Downs. 1947 P. Newton *Wayleggo* (1949) 154 Some high country districts are so extensive that it is necessary to have a second homestead situated in some distant part of the property. Such a place is termed an 'out-station'. 1954 G. Durrell *Three Singles to Adventure* v. 122, I prefer not to remember the ride to the outstation.
b. A subordinate branch of a business or other enterprise. Also *attrib.*
Some examples not clearly distinguishable from those in sense a.
1872 W. F. Butler *Great Lone Land* (ed. 2) xi. 161 About five miles from the mouth of Rainy River there was a small out-station of the Hudson Bay Company kept by a man named Morrisseau. 1962 *Housewife* (Ceylon) Feb. 8 For the benefit of our outstation members, who have borne with us for so long, we will be opening branch Associations in Kandy and Galle very shortly. 1968 D. Lampe *Last Ditch* xii. 135 Underground broadcasting from fixed stations is untenable. The out-station operators would have been in great danger. 1970 *Guardian* 24 Aug. 14/1 Small outstations dealing with separate processes in the manufacturing cycle have been established in towns up to 20 miles from the city. 1973 *Sunday Advocate-News* (Barbados) 21 Jan. 2/5 He first appeared in uniform on normal patrol work attached to Central Station. He next underwent a short spell of duty at 'out-stations' before returning to the Bridgetown Division office to take up clerical duties. 1975 *Times* 9 June 13/8 (Advt.), National Railway Museum..York..an outstation of the Science Museum London.

outstay, *v.* Add: **2. b.** To surpass in endurance.
1877 *Coursing Calendar Autumn 1876* 5 Laughter made the early points with Lady Don, but the latter fairly outstayed the dog. 1951 *Publ. Amer. Dial. Soc.* xvi. 48 *Outstay*, to be able to equal the speed of one's nearest competitors long enough to defeat them... *Outstay the field: phr.*, of a horse: to take the lead and hold it until the finish.

outstep, *sb.* Restrict *rare* to sense in Dict. and add: **2.** In full *outstep well*. An oil well drilled beyond an area already drilled in order to extend the productive area or ascertain its limits.
1947 *Economist* 5 July 40/1 Some of these wells were considerable outsteps to test the possibilities of outlying areas, thereby extending the limits of the proven producing area to the west and north. 1955 *Bull. Amer. Assoc. Petroleum Geologists* XXXIX. 1353 In the latter part of the year a contract National 80 B rig began drilling a program of outsteps to evaluate the oil find at Alcata-I. 1973 R. E. Chapman *Petroleum Geol.* xii. 256 More problems arise when planning an 'out-step' well, to prove an extension to the accumulation, or to define its limits.

outstretch, *sb.* **2.** (Later example.)
1918 A. Symons *Cities & Sea-Coasts* iii. 312 Grass, or any soil, was but a rare interval between a broken and distracted outstretch of grey rock.

outstride, *v.* (Further examples.)
1798 Jane Austen *Let.* 17 Nov. (1952) 27 Though she does not gain strength very rapidly, my expectations are humble enough not to outstride her improvements. 1972 *Times* 11 Sept. 3/1 The department outstrides those of many universities in equipment and reputation.

outstrip (autstri·p), *v.*[2] [f. Out- 18 + Strip *v.*[1]] *trans.* To surpass in stripping; to wear less clothing than. (With punning reference to Outstrip *v.* in Dict.)
1887 *Daily News* 29 Dec. 5/3 'Yes,' replied his cynical friend, after a glance at the young lady, 'I admit that she out-strips them all.' 1897 W. C. Hazlitt *Four Generations Lit. Family* II. 155 The abridged petticoats of the ladies proceeded to an intolerable pitch; and many tried, as Byron said, to outstrip one another. 1938 H. M. Alexander *Strip Tease* 19 'Then it was competition that was responsible for the peeling.' 'Yeah.' Garns laughs. 'They tried to outstrip each other.'

outswinger (au·tswi:ŋəɹ). [f. Out- 7 + Swing *v.*[1]] **1.** *Cricket.* A ball bowled with a swerve or swing from the leg to the off in its flight; also, the bowler of such a ball. So **ou·tswing,** the swerve or swing imparted to such a ball. Also **ou·tswi:nging** *ppl. a.*
1920 E. R. Wilson in P. F. Warner *Cricket* 67 Mr. E. W. Clark bowled 'out-swingers'. 1925 *Country Life* 18 July 93/2 To make a ball swerve in the air from the leg stump into the slips (the out-swinger, it is called). 1953 Miller & Whitington *Cricket Typhoon* 144 It was perfectly controlled late out-swing. 1955 A. Ross *Australia 55* 211 He drove the out-swinging ball. 1958 P. Richardson *Tackle Cricket this Way* iii. 50 Trueman, also, is primarily an outswinger. 1963 A. Ross *Australia 63* iii. 82 Barrington, reaching out, played an out-swinger beautifully wide of mid-off. 1968 N. Cardus in J. Arlott *Cricket: Great Bowlers* 21 He told me that he obtained his out-swing by spin. 1975 *Cricketer* May 9/1 His lifting out-swing is more difficult and his pace changes more skilful.
2. (*Assoc.*) *Football.* A pass, usu. across the mouth of the goal, in which the flight of the ball curves away from the centre of the goal.
1959 *Times* 19 Mar. 18/3 Riley lost his man and sent over an out-swinger. 1961 F. C. Avis *Sportsman's Gloss.* 37/1 *Outswinger*, a centre pass, particularly from a corner kick, that moves in its flight to goal in a slight arc away from the centre of the goal.

outta (au·tă), colloq. contraction of Out of *prep. phr.* orig. and chiefly *U.S.* Cf. *outa.
1937 C. Himes *Black on Black* (1973) 142 You keep outta dis, yellow niggah. 1958 Keith & Bergman (*song-title*) Outta my mind. *Ibid.*, And I came out-ta my spin, That dizzy spin I was in, Don't ask me, 'Where have I been?' Been out-ta my mind! 1967 *Boston Sunday Herald* 14 May (Comic Section), Outta the way, Barrey boy. 1971 B. Brandon (*title*) Outta sight, Luther. 1973 *Black World* Jan. 65/1 Tears was dripping down the Grand Worthy Matron's face, making a mess outta all her powder and rouge. 1977 *Ripped & Torn* vi. 2/2 Is this a last ditch attempt by me to make some money outta this thing?

out-take (au·t,tēik), *sb.* [f. Out- 7 + Take *sb.*] A length of film or tape rejected in editing.
1960 O. Skilbeck *ABC of Film & TV* 92 *Out takes*, takes rejected in the cutting room. 1970 *Guardian* 27 Jan. 1/3 Unused film, what in the industry are called 'the out-takes'. 1972 I. Hamilton *Thrill Machine* xxxii. 147, I got the original neg. all the out-takes, and seven prints..and the film was out of circulation. 1974 T. Chastain *Pandora's Box* (1975) xv. 149 We're going to show you..all the film we shot yesterday..including the out-takes—that is, the stuff we won't be using on the air after we've finished editing. 1977 *Zigzag* Mar. 29/3 Both the album's opener, Gene's 'Home Run King', and the traditional 'In The Pines' could have easily been outtakes from either D & C album. 1977 *Time* 24 Oct. 54/3 *Looking for Mr. Goodbar* has narrative lapses, jerky editing and confusing fantasy sequences that look like Ken Russell outtakes.

out-think, *v.* Add: **2.** (Later examples.)
1934 *Sun* (Baltimore) 10 Jan. 1/1 The farmers..outthought the railway officials. They telephoned back to Burlington and appraised the pickets there of the train's approach. 1941 Steinbeck & Ricketts *Sea of Cortez* xvi. 161 The mule..knows he can out-think a horse and he is pretty sure he can out-think a human. 1954 D. Dodge *Lights of Sharo* ii. 50 They could outrun us, outfight us, or overwhelm us, but they couldn't out-think us. 1962 L. Deighton *Ipcress File* xiii. 78 What chance did I stand..they were both out-thinking me at every move. 1972 'H. Calvin' *Take Two Popes* vi. 48 His quarry had outthought him, and fled into the unexpected area on the other side of the road.

out-thrown, *ppl. a.* (Further examples.)
1927 *Month* May 398 The sides of the outthrown headland are too steep to be rushed. 1978 *Nature* 26 Jan. 318/2 The 'grid', however, may result principally from the overlapping of outthrown debris produced by the giant basin forming collisions.

out-thrust, *sb.* Add: (Later examples.) Also *transf.* and *fig.*
1950 J. Jenks *From Ground Up* xxi. 215 It was social and economic self-reliance at home that made possible an out-thrust of such vigour, that we became for a time not only the workshop of the world, but its merchant, carrier, and banker as well. 1955 J. R. R. Tolkien *Return of King* 162 An out-thrust of the eastward hills. 1973 *Baptist Federation of Canada Prayer Calendar* 16 Aug., Pray that the out-thrust of the Gospel by individual Christians will be unobstructed by timidity, busyness and fear.

out-thrust, *v.* Delete *rare* and add later examples.
1887 W. Morris tr. *Homer's Odyssey* I. x. 89 Sheer outthrusting nesses each other hold in face. 1892 G. Meredith *Poems: Empty Purse* 40 From him are the brutal and vain, The vile, the excessive, out-thrust. 1907 *Daily Mail* 7 Dec. 6/4 He sat his horse as if he were a part of it, the reins dangling carelessly, his feet out-thrust in the huge Moorish stirrups. 1963 C. D. Simak *They walked like Men* iii. 17 He lay back in the chair, with..his long legs outthrust into the shadow underneath the radio console.

out-tray (au·t,trēi). [f. Out- 3 + Tray *sb.*[2]] In an office, etc.: a tray for outgoing and completed correspondence and other papers. Cf. *in-tray, *out-basket.
Sometimes written as two separate words with *out* regarded adjectivally.
1943 [see *in-tray]. 1943 G. Greene *Ministry of Fear* i. ii. 25 There were two trays on his desk marked In and Out, but the Out tray was empty. 1947 L. Hastings *Dragons are Extra* ix. 198 Invariably they were removed to the 'out' tray, probably with the tongs. 1959 *Punch* 30 Sept. 249/1, I looked around at the dusty chaos and saw how orderly it all could be..files here, records there, in-trays, out-trays. 1969 J. Argenti *Managem. Techniques* 247 Put this piece of paper in the Out-tray. 1976 Deakin & Willis *Johnny go Home* vi. 90 His desk is adorned with..wire In and Out trays.

out-turn. Add: (Further examples.) spec. *Econ.*, an amount or result attained, as distinct from an estimate.
1928 R. S. Troup *Silvicultural Syst.* ix. 115 Both [the farmer and the gardener] take such measures as they can to improve the quality and increase the outturn of their field or garden crops. 1930 *Economist* 5 July 19/2 The wheat harvest has begun in the south-west... The outturn promises to be about as large as last year. 1932 *Times* 29 Sept. 15/3 The Budget had been balanced on paper, but it remained to be seen what the actual result would be at the end of the financial year. The outturn proved the soundness of the balancing. 1957 *Times* 18 Nov. (Ann. Financial & Commerc. Rev.) p. xxxii/1 Barring frosts or other calamities, the out-turn delivered to the market can be reduced only by the action of growers or of some intermediary agency in withholding stocks. 1963 *Guardian* 15 Mar. 3/2 The figures for the outturn of defence expenditure in 1961–62 are provisional. 1972 *Accountant* 21 Sept. 350/1 Some indication of the outturn for the current year. 1976 *Daily Tel.* 3 Apr. 17 There is little point in the Government producing White Papers on public expenditure since the outturn has been so consistently different from official projections.
b. *gen.* A result, outcome.
1881 W. Paul *Past & Present of Aberdeenshire* 19 Rahab, spoken of in the Bible, made a bad beginning, but she had a fine out-turn,—she had a fine out-turn, she married Salmon. 1961 *Atom* Feb. 12/2 Any statement on the future cost of power involves a complicated judgement of the out-turn of many factors.
2. *Curling.* A turning motion given to a stone which causes it to curve to the left.
1890 J. Kerr *Hist. Curling* 411 No curler is..entitled to be reckoned a graduate of arts in curling until he has mastered the knowledge of the *in-turn* and the *out-turn*. 1900 —— in A. E. T. Watson *Young Sportsman* 200 The in-turn is made when the curl is to be toward the right, the out-turn when it is to the left. 1969 R. Welsh *Beginner's Guide Curling* xi. 82 The..'in-turn' and 'out-turn', are activated in different ways.

ou·t-vote, *sb. rare.* [Out- 2.] The vote of an out-voter; such votes collectively.
1790 E. Sheridan *Let.* June in T. Moore *Mem. Life R. B. Sheridan* (1825) 465, I suppose you have sent for the out-votes; but, if they are not good, what a terrible expense will that be! 1945 G. B. Grundy *55 Yrs. at Oxf.* 121 Rosebery was defeated on the out-vote.

out-voter. Add: (Further examples.) Also *transf.*
1931 *Birmingham Post* 24 Oct. 8/1 (*headline*) Aeroplane to bring up out-voters. 1965 W. R. Ward *Victorian Oxford* v. 96 Out-voters were told that a scheme designed by the heads to keep dissenters out on respectable grounds was a plan to let them in. 1967 *Economist* 6 May 575/1 Some distinguish between freeman electors and ten-pound householders, and between resident voters and out-voters.

outwall. Add: **b.** Used by Edmund Blunden in the sense 'outward appearance'.
1933 E. Blunden *Charles Lamb* v. 131 He [*sc.* Wordsworth] does express the *altitudo* of Lamb's personality and influence far more thoughtfully than a host of subsequent writers to whom Lamb's outwall with its Punch and Judy shows and all the fun of the fair has been the principal thing to report. 1937 —— in *Essays & Stud.* XXII. 60 Acquaintances who did not always separate the man from his outwall. *Ibid.*, The candour and keenness of the first period, when the outwall had not yet become necessary.

outwander, *v.* Delete *Obs. rare* and add later *literary* example. Hence **outwa·ndering**

OUTWARD

vbl. sb., a wandering out or outwards (in quot. *fig.*).
1880 H. Collins *Heaven Opened* II. xiv. 215 God does not mind the out-wanderings of our vagabond imaginations. **1922** Joyce *Ulysses* 379 On her stow he ere was living with dear wife and lovesome daughter that then over land and seafloor nine years had long outwandered.

outward, *adv.* Add: **4.** *outward-looking, -steeled, -turning* adjs.
1890 W. James *Princ. Psychol.* I. x. 296 Our considering the spiritual self at all is a reflective process, is the result of our abandoning the outward-looking point of view. **1927** A. Huxley *Proper Stud.* 52 How repulsive, how incomprehensible I find the philosophy which is the rationalization of these people's outward-looking passion for their fellows! **1975** *Times* 5 Apr. 3/3 Teachers try to capture the interests of these pupils with vocationally oriented, outward-looking courses. **1888** G. M. Hopkins *Poems* (1918) 69 The heroic breast not outward-steeled. **1930** *Times Lit. Suppl.* 30 Oct. 888/4 As in a maze, the outward-turning paths lead back to the centre. **1976** *Listener* 12 Aug. 176/3 This bodily prosperity, this outward-turning energy.

outward-bound, *a.* Add examples relating to travel other than by sea.
1777 P. Thicknesse *Year's Journey* II. xlvi. 110 My entertainment at this house, *outward-bound*, was half a second-hand roasted turkey. **1832** *Chambers's Edin. Jrnl.* I. 86/3 He would find himself on the top of one of the outward-bound coaches of the metropolis.
2. (With capital initials.) The name of a sea school founded by Kurt Hahn at Aberdovey in 1941, on the basis of which an Outward Bound Trust was formed in 1946 with the aim of establishing further residential schools for the training of boys and girls in mountaineering as well as naval and other outdoor activities. Also *transf.* Hence **Outward-Bound course, scheme, school,** etc. (see quots.).
1943 *Times* 28 June 2/4 Dr. G. M. Trevelyan..at Aberdovey on Saturday gave the name of Garibaldi to a sea-going ketch presented by the partners of Alfred Holt and Co., shipowners, to the Outward Bound Sea School. **1947** *Times* 23 Sept. 2/5 To-day the experiment begun in war-time is firmly established as a permanent undertaking operated by the Outward Bound Trust. **1950** *Times Educ. Suppl.* 16 June 474/1 Variety is the spice of the Outward Bound life. **1957** K. Hahn in D. James *Outward Bound* 10 In July 1941..Lawrence Holt, the shipowner.. secured the financial support of his firm, the Blue Funnel Line, for the foundation of the first Short-Term School at Aberdovey. The name Outward Bound Sea School was his invention. **1958** *Listener* 12 June 976/1 Sir Richard Livingstone and 'Outward Bound' enthusiasts both saw in the training of character a panacea for a world adrift. **1961** Visct. Montgomery *Path to Leadership* xi. 173 In my opinion the best character training given to boys in Britain is that provided by the Outward Bound Trust, through their sea and mountain schools... Outward Bound can take only some 4,000 boys a year. **1965** M. Morse *Unattached* v. 159 He had been sent on the Outward Bound course by his firm. **1973** C. Bonington *Next Horizon* xiv. 213 Sebastian in the mist, looking like a bedraggled outward-bound schoolboy in the mists of Wales or the Lakes. **1977** P. Theroux *Consul's File* 42 He was the local magistrate. An Outward Bound type..dead keen to go camping.

outward-bounder. Add: (Earlier example.)
1851 H. Melville *Moby Dick* II. xi. 72 The long absent ship, the outward-bounder, perhaps, has letters on board.
2. (With capital initials.) A pupil at an 'Outward Bound' school; an advocate of such schools and their methods.
1961 *Sunday Times* 19 Feb. 34/2 Knowing how to tie reef-knots,..with all the other indispensable appurtenances of the Outward Bounder. *Ibid.* 26 Feb. 39/3 We 'intelligent' talkers have to thank the 'Outward Bounders' for the liberty and freedom to be so.

ou·twash. *Geol.* [Out- 7.] Material (chiefly sand and gravel, or further away silt and clay) carried out from a glacier by melt-water and deposited beyond the terminal moraine. Freq. *attrib.*
1894 T. C. Chamberlin in *Jrnl. Geol.* II. 533 There were, however, tracts of assorted material formed by waters outflowing from the ice where no definite terminal ridging took place. Such forms may be designated *outwash* aprons in distinction from *overwash* aprons. **1905** *Ibid.* XIII. 245 One of the pronounced features of the outwash is the pitted-plain development. **1908** *Amer. Jrnl. Sci.* CLXXV. 108 The river terraces of outwash gravel. **1934** *Antiquity* VIII. 306 The retreat of the ice and the formation of the outwash plain in front of the moraine. **1947** Auden *Age of Anxiety* v. 116 And you, bright Prince,..O stiffly stand, a staid monadnock, On her peneplain; placidly graze On her outwash apron, her own steed. **1957** [see *overwash *sb.*]. **1963** D. W. & E. E. Humphries tr. *Termier's Erosion & Sedimentation* vii. 163 During the recession of the Quaternary ice sheets, detrital accumulations were left behind, either in the form of outwash fans deposited at the snouts of glaciers by subglacial streams.., or as fluvio-glacial ridges often filling in lakes (eskers). **1971** R. F. Flint *Glacial & Quaternary Geol.* vii. 187 In the downstream direction, outwash is diluted by an ever-increasing proportion of nonglacial alluvium derived through tributary streams that did not originate in glaciers. **1972** J. G. Cruickshank *Soil Geogr.* ii. 59 Outwash alluvium is usually similar in texture, but may include some fine debris where the melt water has spread over a large area.

out west (aut we·st), *sb.* and *adv. phr.* [f. Out *adv.* 2, 16 + West *adv.* 2 c.] **A.** *sb.* Orig., the territory to the west of the early American settlements; by extension, the distant West of the U.S. as regarded by inhabitants of the East (cf. West *sb.* 3 b). Also *attrib.* **B.** *adv. phr.* In or to this region. Also *transf.* in or to the western parts of Canada or Australia.
1835 C. F. Hoffman *Winter in West* II. 119 Old Kaintuck..whips all 'Out-West' in prettiness. **1848** R. W. Griswold *Passages from Corr.* (1898) 243 The 'out West Editor' would inform her..as to who he is. *Ibid.* 244 Why is it your new volume is not out West? **1857** *Lawrence* (Kansas) *Republican* 4 June 2 Any one who has spent any time in farming 'out west', will see that this is a mistake. **1887** C. B. George *40 Yrs. on Rail* 62 New York State was considered 'out West' then. **1890** S. M. St. Maur *Impressions Tenderfoot* xiii. 173 All ponies and dogs 'out West' seem shy of women. **1898** P. L. Ford *Tattle-Tales of Cupid* 205 As they say out West, it's come to stay and grow up with the country. **1944** *Living off Land* iv. 81 The country out-west is full of dead men's bones. **1952** *Manch. Guardian Weekly* 17 July 15 He worked as a section hand out west on the Union Pacific railroad. **1961** *Maclean's Mag.* 29 July 36/1 I've worked freights that have carried more passengers than the passenger runs did, especially out west in the Depression.

outwinter (au·t,wintəɹ), *v.* Also **out-winter.** [Out- 15.] *trans.* To keep (animals) in the open during the winter. So **ou·twintered** *ppl. a.*, **ou·twintering** *vbl. sb.* Cf. *in-winter *v.*
1959 *Times* 2 Nov. 21/1 Our first decision was to have our own acclimatized breeding outwintered cattle. **1960** *Farmer & Stockbreeder* 8 Mar. 109/3 Gimmer hoggs that are outwintered well in North-East Cheshire have yielded fleeces up to 11 lb. *Ibid.* 15 Mar. 81/3 An average of 35s for outwintering. **1965** Cooper & Thomas *Profitable Sheep Farming* viii. 75 Grazing obtained by outwintered ewes in a more normal season is not appreciable.

outwith, *prep.* and *adv.* For 'Chiefly *north.*; now only *Sc.*' read: 'Chiefly *north.* and *Sc.*'
A. *prep.* **1.** (Later examples.)
1927 W. D. Simpson *Historical St. Columba* 17 Of the four peoples..who struggled for mastery in what is now Scotland, the Angles alone were entirely outwith the pale of Christianity. **1947** H. Farmer *Hist. Music Scotl.* 216 In music, there was but one name, John Abell (d. 1724), and he gained his fame *outwith* Scotland, to use the appropriate Scots word. **1970** 'E. Ferrars' *Seven Sleepers* iv. 46 I'm moving into a small bungalow outwith the town. **1972** G. Henderson *Early Medieval* vi. 232 Outwith history painting the image of the crucified Christ, as an evocation of the Passion and redemption, was slow to make itself felt. **1975** C. N. Manlove *Mod. Fantasy* iv. 99 What he came to demand was 'rational' friendship, the bond of a common interest in something outwith the self. **1978** *Dumfries & Galloway Standard* 21 Oct. 1/8 Attempts might be made..to promote the greater use of existing static caravan sites outwith the July/August period.
b. (Later examples.)
1705 in *Aberdeen Jrnl. N. & Q.* (1909) II. 309/1 It shall be leisum to them..to take themselves outwith the family. **1958** *Times* 5 Dec. 8/7 Their value may also extend outwith the narrow sphere of medicine.

outworking, *vbl. sb.* Add: (Further examples.)
1881 W. M. Thomson *Land & Bk.* 2 The long ongoing and outworkings of the Mosaic Economy. **1958** *Church Times* 5 Dec. 1/3 If the outworking of *apartheid* policies meant such cruelty and callousness, then even the Slightest Smell of a *Compulsory apartheid* must be removed from our Churches. **1961** B. R. Wilson *Sects & Society* III. xiv. 288 Christadelphians do take a keen interest in political events, as out-workings of biblical prophecy. **1976** M. Wiles *What is Theology?* iv. 108 Philosophical assumptions implicit in these [Freudian] ideas and their practical outworking in therapy.
2. The work of an out-worker.
1970 G. Greer *Female Eunuch* 131 As an alternative to nursing or outworking, waitressing..is not conspicuously preferable. **1974** *Daily Tel.* 12 Jan. 2/6 There are..areas of the country where 'out-working' has been traditional, as with lace at Nottingham.

outworldish, *a.* (Later example.)
1880 G. Macdonald *Bk. Strife* 73 A strange auroral bliss, an Arctic awe, A new, outworldish joy awoke intense.

outy (au·ti). *colloq.* [f. Out *adv.* + -y[6].] The act of letting a dog or other pet out of the house, or taking it for a walk; 'walkies'. Freq. in *pl.*
1949 C. H. B. Kitchin *Cornish Fox* xi. 167 He was Mrs. Ropford's dog, and Mrs. Steele was giving him his evening 'outy'. **1962** N. Marsh *Hand in Glove* vii. 230, I was going to bed and he asked for outies. **1967** B. Whitaker *Chained Crocodile* iii. 37, I must see to Skipper first, 'e wants outies rather bad.

outyield (autyī·ld), *v.* [f. Out- 18 + Yield *v.* 8.] *trans.* To surpass in terms of yield; to produce more than.
1927 *Daily Tel.* 15 Nov. 12/1 The Danish swede crop outyields ours by about six tons per acre. **1957** *Times* 2 July (Agric. Suppl.) p. i/1 Wheat after sugar beet will almost invariably outyield wheat taken after a one-year mixed ley.

ouvala, var. *uvala.

ouvarovite, var. Uvarovite.

‖ **ouvert** (uvɛ̄r). *Ballet.* [Fr., = open.] (See quots.)
1914 *Techn. Encycl. Theory & Pract. Art of Dancing* 110 *Ouvert*,..any movement in which the legs are open sideways to right or left. **1952** Kersley & Sinclair *Dict. Ballet Terms* 72 *Ouvert(e)*,..a position of the feet in which the feet do not touch. **1968** J. Winearls *Mod. Dance* (ed. 2) iv. 103 The name given to all opening and closing movements of this type is Ouvert.

‖ **ouvreuse** (uvrøz). [Fr.] In France, a woman who 'opens' theatre boxes; an usherette in a French theatre or cinema.
1892 G. B. Shaw *Let.* 21 Apr. (1965) I. 338 It took the united strength of my three companions, the *ouvreuse*, the acting manager, the fireman, and a commissionaire to hold me down and restrain me from hurling an opera glass at her head. **1944** H. Croome *You've gone Astray* i. 11 A disgruntled old *ouvreuse* in a Paris theatre. **1968** *Guardian* 27 Apr. 9/8 An ouvreuse in a theatre or cinema. **1972** *Ibid.* 5 Feb. 3/8 An inescapable one franc for the cloakroom and another for the *ouvreuse*.

‖ **ouvrier** (uvrie). [Fr.] A workman. Also *fem.* **ouvrière,** a working woman.
1848 J. Arnold *Let.* 28 Mar. in J. Bertram *N.Z. Lett. T. Arnold* (1966) 36 France has been absorbed into the great cities, the cities into Paris, Paris into the middle and lower classes—these into the 'ouvriers', the Ouvriers into the clubs. *a* **1855** C. Brontë *Professor* (1857) II. xxiv. 205 You, a scion of Seacombe, have proved your disdain of social distinctions by taking up with an *ouvrière*. **1857** C. M. Yonge *Dynevor Terr.* I. xx. 334 She..had seen him ..preceded by a brave and faithful *ouvrier*. **1904** A. Bennett *Great Man* xxiv. 267, I come from the *ouvriers*, ..the working peoples. **1972** R. Cobb *Reactions to French Rev.* iv. 142 All that the *ouvrière* might know of the Revolution was that it appeared to be..oppressive and interfering. **1975** —— *Paris & its Provinces* v. 199 An extended family group..Marie-Anne Deloutre, *tricoteuse*; Ludovine Deloutre, *ouvrière*; and Constance Deloutre, *ouvrière*.
Hence **ouvrierism, -isme, ouvrierist** (see quots.).
1969 G. Stedman Jones in Cockburn & Blackburn *Student Power* 29 Such ouvrierism—the belief that the traditional working class has a monopoly of socialist potential—is a mystification. **1974** J. White tr. *Poulantzas's Fascism & Dictatorship* IV. i. 166 The fascist and national socialist *leaders* are extremely cautious in the use of this double-edged weapon, the 'ouvrierist' use of corporatist themes. **1976** F. Zweig *New Acquisitive Society* II. x. 134 Ouvrierism implies that the worker can never be the wrongdoer, he can be wronged but he cannot wrong others. **1977** *Foreign Affairs* LV. 805 There has been little of the *ouvriérisme* in this party that has marked the French.

ouwarovite, ouwarowite, varr. Uvarovite.

ouzo (ū·zo). Also **ouso.** [mod.Gr. οὖζο.] A Greek spirituous drink flavoured with aniseed; a glass of this.
The etymology of the Greek name οὖζο is disputed. A popular etymology derives it from the Italian designation *uso Massalia* 'for the (commercial) use of Marseilles' stamped on packages of selected silkworm cocoons exported in the 19th century via Volos from the Thessalian town of Tyrnavos. The designation came to stand for 'superior quality', which the spirit distilled as *ouzo* was thought to possess: see A. Tzartzanos in *Indogermanische Forschungen* (1932) LII. 217–20.
1898 H. N. Brailsford *Broom of War-God* 4 The Prefect had placed a chair beside him and had pledged his in *ouso*. **1935** *Chem. Abstr.* 880 (*heading*) Method for controlling the purity of alcohol used in the preparation of 'ouzo'... The beverage known as 'ouzo' in Greece has an alc. content of 35–45%. **1957** L. Durrell *Bitter Lemons* 25 The excellent *ouzo* and his general affability transformed the journey. **1957** F. King *Man on Rock* ii. 61, I devoured a whole tin of American bully-beef, washed down with ouzo. **1965** O. Manning *Friends & Heroes* xx. 202 Alan had a bottle of ouzo on his table and he started filling the glasses. **1973** D. Lang *Freaks* 8 Stavros downed another ouzo. **1975** *Daily Colonist* (Victoria, B.C.) 18 May 4/6 Constantine, late of Greece, could..take to peddling ouzo.

ovablastic (ōuvăblæ·stik), *a.* *rare*⁻¹. [f. L. *ova*, pl. of Ovum egg + Gr. βλαστικ-ός springing forth.] Making eggs burst open (in the womb).
1922 Joyce *Ulysses* 397 Mr Dixon..took on to ask Mr Mulligan himself whether his incipient ventripotence.. betokened an ovablastic gestation in the prostatic utricle.

Ovaherero (ōu·văheri³·ro) = *Herero; also a tribe of the Hereros.
1855 F. N. Kolbe in W. Holden *Hist. Natal* 436 Damara-land is inhabited by a nation divided into two principal tribes, the Ovaherero and Ovampantera. **1856** C. J. Andersson *Lake Ngami* iv. 52 The Damaras are divided into two large tribes, the *Ovaherero* and the *Ovapantieru*. **1880** *Encycl. Brit.* XI. 731/2, 110,000 are Herero (80,000 Ova Herero and 30,000 Ova-mbanderu). **1884** A. Lang *Custom & Myth* 20 The Ovahereroes in

South Africa..appease with a black sheep the spirits of the departed. **1920** *Encycl. Brit.* XIII. 358/1 Herero or Ovaherero.

oval, *a.*[1] and *sb.*[1] Add: **A.** *adj.* **2.** *Oval office* the office of the President of the United States in the White House.
[**1966** *N.Y. Times Mag.* 8 Apr. 38 (*caption*) New look—The oval office in the White House's west wing reflects increasingly the interests and personality of the man who spends upwards of seven hours of his long working day in it.] **1969** H. P. Jones *First Bk. White House* 30 In 1909, seven years after the President's offices had been moved into the West Wing, the President's Oval Office was added. **1970** B. B. Fay *Pleasure of his Company* xxiv. 159 One Saturday morning in the Oval Office I asked the President, 'Is there any truth to the rumor that you intend to dump Lyndon in '64?' **1972** W. McGivern *Caprifoil* viii. 133 This is the first time I've ever interrupted the President... Mary Donovan..walked to the door that connected her office to the Oval Office. **1973** *Times* 18 June 1/6 Mr Krogh told him that the order for the break-in came 'from the Oval Office'. [illegible]

5. a. *oval-faced* (examples).
1886 Ruskin *Præterita* I. 326 A graceful oval-faced blonde of fifteen. [illegible]

B. *sb.* **2.** (Further examples.)
[illegible]

oval (ōu·văl), *v.* [f. Oval *a.*[1] and *sb.*[1]] **a.** *trans.* To make oval, to give an oval shape to. **b.** *intr.* To move in oval-shaped curves. Hence **ovalling** *ppl. a.*
[illegible] **1922** Joyce *Ulysses* 424 The odour of the sickening weed floats towards him in slow round ovalling wreaths. **1969** 'R. Stark' *Blackbird* iv. 26 Grofield's plane ovalled between massed gray clouds and the grubby sprawl of New York City.

ovalbumen, -in. Add: In mod. use written **ovalbumin** and applied to the albumin that is the principal protein of egg-white. (Further examples.)
1905 C. E. Simon *Text-bk. Physiol. Chem.* (ed. 2) xxi. 457 According to Gautier and some of the older observers, white of egg (albumen) contains a number of different albumins, which in part seem to belong to the true albumins and in part to the globulins. [illegible] **1965** [see *ovoglobulin]. **1969** K. W. McGilvery *Biochem.* viii. 150 The principal protein of egg whites, ovalbumin, is especially susceptible to denaturation in this way.

ovality (ovæ·lĭti). [f. Oval *a.*[1] and *sb.*[1] + -ity.] = Ovalness.
[illegible] **1962** *Gloss. Terms Glass Industry* (B.S.I.) 41 Ovality, deviation of a glass article from a circular towards an elliptical cross section. [illegible]

Ovaltine (ōu·văltīn). Also with small initial. [Prob. a fanciful extension of Oval *a.*[1] 3.] The proprietary name of a powder composed principally of malt extract, milk, and eggs; a drink made from this.
1906 *Trade Marks Jrnl.* 13 June 820 Ovaltine... Alimentary products..Albert Wander,..Berne, Switzerland; manufacturer of alimentary products. **1907** *Yesterday's Shopping* (1969) 516/2 Ovaltine—tin, 1/4½. **1912** R. Brooke *Let.* 26 Jan. (1968) 354 These things go on round,..Mrs Digby on India, and Mrs Fox at Bridge, and Ovaltine. **1930** J. Cannan *No Walls of Jasper* 64 The parlourmaid..set down a cup of Ovaltine on the corner of the writing table. **1937** J. Betjeman *Continual Dew* 22 He gives his Ovaltine a stir And nibbles at a 'petit beurre'. **1940** F. Stark *Winter in Arabia* xv. 290 Qasim woke me with hot ovaltine at one-thirty. **1960** S. Plath *Colossus* 58 You fed My brother and me cookies and ovaltine. **1969** *Trade Marks Jrnl.* 19 Feb. 306/2 Ovaltine... Food preparations (not medicated) in powder or tablet form composed principally of malt, milk and eggs, the malt predominating; and rusks. A. Wander, Limited,.. London,..manufacturing chemists. *Ibid.* 29 Oct. 1802/2 Ovaltine... Preparations of milk, malt and eggs, flavoured with cocoa, the malt predominating, for use in making food beverages; biscuits (other than biscuits for animals), cakes, rusks, chocolate and non-medicated confectionery. A. Wander, Limited,..London,..manufacturing chemists. **1973** 'P. Loraine' *Voices in Empty Room* II. iv. 115 Lulu Jenkins..lay slumped up on the pillows of her large bed, sipping Ovaltine.

Ovambo (ovæ·mbo), *sb.* and *a.* Also **Ambo, Avamba, Ovampo.** [f. Bantu *ova-* pl. prefix + *ambo* man of leisure.] **A.** *sb.* **a.** A member of a Bantu people living in the northern part of South-West Africa; this people collectively. **b.** The language of the Ovambos. **B.** *adj.* Of or pertaining to the Ovambos.
1853 F. Galton *Narr. Explorer Trop. S. Afr.* vi. 179 The Ovampo were twenty-four in number with a tall enterprising-looking young man as captain. **1856** C. J. Andersson *Lake Ngami* xiv. 163 At a considerable distance to the north, there lived a nation called Ovambo. **1864** T. Baines *Explor. S.-W. Afr.* ii. 40 The Damaras.. could not even make an assegai, but bought their weapons of the Ovampo. **1884** *Encycl. Brit.* XVII. 318/2 Many..of the Bantu-speaking southern races..[including] Ovambos of the south-west coast..are also variously affected by foreign elements. **1897** J. Bryce *Impressions S. Afr.* v. 42 On the higher grounds and generally in the far northern parts [of German South West Africa], where the Ovampo tribe dwell, grass is abundant. **1902** *Encycl. Brit.* XXXII. 736/2 The Ovambo or Ambo, in the northern part of the protectorate, are agriculturists. **1909** *Daily Chron.* 28 July 4/4 Three and a-half days' work with a mere handful of Ovambo 'boys'. **1911** J. G. Frazer *Golden Bough: Magic Art* (ed. 3) I. iii. 63 The Ovambo of South-western Africa believe that some people have the power of bewitching an absent person by gazing into a vessel full of water till his image appears to them in the water. **1953** L. G. Green *Lords of Last Frontier* ii. 18 Battels deserted and lived for sixteen months among the Ovambos. *Ibid.* xvi. 235 Finnish missionaries learn Ovambo before they leave Finland. **1959** *Chambers's Encycl.* X. 277/2 The Ovambo people are the largest community of South-west Africa. **1967** *Courier-Mail* (Brisbane) 20 Apr. 18/6 It was interesting to find the tall, slim, gentle houseboy employed by my host and hostess was an Ovambo. **1973** *Daily Tel.* 21 Nov. 4/6 They were familiar with Ovambo tribal customs. **1974** *Encycl. Brit. Micropædia* I. 295/3 Ambo, also known as Ovambo or Avamba, people located in the dry grassland country of northern South West Africa and southern Angola... They speak a language of the Bantu group. **1974** *Times* 14 Oct. 7/2 Oshakati, the Ovambo capital near the Angolan border.

ovariectomy. Add: (Further examples.)
1932 S. Zuckerman *Social Life Monkeys* v. 77 If the ovaries of any mature mammal are experimentally removed (the operation of ovariectomy), all cyclic activity in the accessory reproductive organs ceases. **1958** *Sci. News* XLVII. 85 Removal of the ovary (ovariectomy) from a young hen results in the growth of cock-like spurs, a large comb, and male plumage. **1969** J. H. Green *Basic Clin. Physiol.* xviii. 104/2 Once the placenta has developed, the corpus luteum of pregnancy in the ovary is no longer essential for the maintenance of pregnancy and an ovariectomy (oophorectomy) could be carried out.
Hence **ovarie·ctomize** *v. trans.*, to deprive of one or both ovaries; **ovarie·ctomized** *ppl. a.*
1924 *Physiol. Abstr.* IX. 33 In two ovariectomised hens ..a testis was found. **1928** *Proc. Soc. Exper. Biol. & Med.* XXV. 490 Guinea pigs..on the fourth day after œstrum were ovariectomized. **1958** *Sci. News* XLVII. 85 The antlers of young stags do not develop after castration, and ovariectomized female deer may produce horns. **1974** *Nature* 5 Apr. 525/1 All rats were ovariectomised and thyroidectomised on day 1 of the experiment.

ovariole. (Later examples.)
1925 A. D. Imms *Gen. Textbk. Entomol.* 147 Each organ [*sc.* ovary] is composed of a variable number of separate egg-tubes or ovarioles which open into the oviduct. **1965** B. E. Freeman tr. *Vandel's Biospeleol.* xxii. 364 The female of *Aphaenops*..has a single ovariole on each side of the body. **1976** *Nature* 17 June 614/2 Cells which display the highest levels are those which have been assumed to be the most active and specialised, such as..the trophocytes of insect ovarioles.

ovariotomy. Add: (Earlier example.) Hence **ovario·tomize** *v. trans.* = *ovariectomize *v.*; **ovario·tomized** *ppl. a.*
1844 *Lond. & Edin. Monthly Jrnl. Med. Sci.* IV. 58 Her case..is believed only by the few who have lately come into the field as the champions of Ovariotomy. **1916** *Biol. Bull.* XXX. 293 The ovariotomized duck may or may not undergo a change in plumage, corresponding to that of the male. **1927** *Jrnl. Physiol.* LXII. 312 Does a lactating ovariotomised mouse require more œstrin to produce œstrous symptoms than a non-suckling ovariotomised mouse? To answer this question a number of mice were ovariotomised soon after parturition. **1964** *Biol. Abstr.* XLV. 7553/1 Ovariotomized rats..were starved for 20–22 hours.

ovate, *a.* Add: **1. c.** *absol.* as *sb. Archæol.*, an implement having an oval blade.
1946 F. E. Zeuner *Dating Past* ix. 283 It is clear, however, that by the end of this interglacial the Acheulian had acquired all its characteristic attributes, like ovates and the S-twist. **1956** A. L. Armstrong in D. L. Linton *Sheffield* 91 Three more hand-axes, all ovates of middle Acheulean type and refined technique, are recorded from our area. **1959** J. D. Clark *Prehist. Southern Afr.* vi. 157 The same assemblage of wood-working tools occurs—small, nearly parallel-sided picks, small flat ovates, miniature 'tea cosies', [etc.].

ovational, *a.* Add: (Later example.) Also, resembling or in the nature of an ovation.
1928 *Music & Lett.* July 235 The ovational ecstasy is not essentially connected with the musical impression.

oven, *sb.* Add: **2. c.** *transf.* A small oven-like tomb built at ground level.
1851 E. S. Wortley *Trav. U.S.* I. xxi. 237 The graves are also elevated. The dead are buried in sepulchral houses, which are termed here 'ovens'. **1879** *Cassell's Techn. Educ.* IV. 267/2 Owing to the damp nature of the ground..there are no graves in the cemeteries, the coffins with the dead being deposited in tombs or 'ovens' erected above the soil. **1921** *Chambers's Jrnl.* Aug. 511/1 There was no system in the arrangement of the 'ovens'.
d. A cremation chamber; *spec.* one of the chambers used by the Germans during the war of 1939–45 for the cremation of Jewish corpses.
[illegible]
3. *fig.* [illegible] *to have something in the oven* (and variants), to be pregnant. See also *bun *sb.*[1] 1 a, *pudding *sb.* 5 c.
[illegible]
4. *oven-dry, -hot, -ready* adjs.; **oven-bottom(ed)** *a.*, designating cake or bread baked on the bottom of the oven; **oven-cloth,** a heat resistant cloth used for handling dishes in an oven; **oven-cook** *v.*, to cook in an oven; **oven-glass,** glass ware suitable for use in an oven; **oven-glove,** [illegible]; **oven mit(t)** = *oven glove* above; **ovenproof** [Proof *a.* (*adv.*) 1 b], suitable for use in an oven; **oven timer** (see quot. 1961); **oven-to-table** *a.*, designating ovenware designed also for use at the table for serving; **ovenware,** dishes that can be used for cooking in an oven; **oven wood** (later examples).
[illegible] **1961** *Harper's Bazaar* Feb. 29/2 Two casserole dishes..are flame-proof and ovenproof. **1974** *Country Life* 5 Dec. 1735/3 Ramekins in ovenproof pottery. **1960** A. E. Bender *Dict. Nutrition* 90/2 *Oven ready*, term applied to poultry that have been plucked, neck, legs and entrails removed,..and finally sealed into heat-shrinkable Cryo-vac wrapping—ready for oven without any further handling. **1960** *Farmer & Stockbreeder* 23 Feb. 64/3 A new firm..has been formed with the aim of becoming one of the largest producers of oven-ready turkeys and ducklings in the country. **1962**

[see *fish finger* s.v. *fish *sb.*[1] 7]. **1973** *Times* 16 Nov. 4 Wholesale prices of oven-ready turkey and chicken have fallen slightly in the past week. **1961** *Which?* Oct. 250/1 One cooker..had an automatic *oven timer*. This, like the ringers had a clockwork mechanism, which turned the oven on and off after a pre-set time. **1977** *Transatlantic Rev.* LX. 87 Then he [mimes] a man shaving and showering in a flurry of interruptions: the phone, the doorbell, the oven timer. **1979** *House & Garden* Mar. 78/3 (Advt.), Oval casserole from versatile range of oven-to-table ware. **1977** *Jrnl. R. Soc. Arts* CXXV. 215/2 A wide range of ceramic items from the early beginnings to the latest oven-to-table ware. **1926–7** *Army & Navy Stores Catal.* 785 Pyrex transparent 'glass' oven ware. **1933** *Archit. Rev.* LXXIV. 26 (*caption*) Some very well shaped Vitreosil quartz ovenware. **1959** *Listener* 5 Feb. 267/1 Bake fillets of cod or haddock in an ovenware dish. **1973** *Guardian* 23 May 9/5 Prestige's second eleven price range, Skyline, for their ovenware passes on many of the blessings of their first team's design.

ovenette (ɒv'ne·t). [f. Oven *sb.* 2 + -ette.] A small or subsidiary oven.
1919 [see *kitchenette]. **1976** *Southern Even. Echo* (Southampton) 11 Nov. 7/2 (Advt.), Eye-level grill-rotisserie-ovenett [*sic*], plus roomy autotimed warming drawer.

over, *adv.* Add: **2. b.** *to be* (someone) *all over*: to be very characteristic of (that person); to be exactly what one might expect of (someone specified). Also *transf.*
1721 R. Palmer *Let.* 31 Aug. in M. M. Verney *Verney Lett.* (1930) II. xxiv. 90 [Mr. Churchill is] Vulponi all over. **1799** C. Lamb *Let.* 20 Mar. (1935) I. 153 The last stanza hath nothing striking in it, if I except the two concluding lines, which are Burns all over. **1821** Scott *Pirate* II. v. 114, I see where you would be—this is Sebastian and Dorax all over. **1852** [see *all over *advb. phr.* 1 a]. **1863** J. S. Mill *Let.* 22 Nov. (1910) I. 310 This is Spencer all over; he throws himself with a certain deliberate impetuosity into the last new theory. **1898** J. D. Brayshaw *Slum Silhouettes* 14 Ah! gal, that's married life all over—fight and agree, fight and agree! **1906** Galsworthy *Man of Property* II. xii. 257 That's Phil all over—he was always like that. **1913** A. Bennett *Regent* II. viii. 249 He's his father all over, that lad is! **1945** H. Closs *High are Mountains* 56 It was old Longshanks all over to send one off on some futile errand. **1973** A. Holden *Girl on Beach* 37, I could have killed Dick when he..said he'd asked these two men to dinner, but that's Dick all over of course, just expects me to cope.
6. a. (See also *get *v.* 66 e.)
b. Used in radio communication to indicate that the speaker has finished speaking and intends his communicator to reply. *Over and out*: used to indicate that the communication is at an end. Cf. *out *adv.* 23 b. Also *transf.*, esp. in *over to you*: it is your turn (to speak, act, etc.).
1926 J. L. Pritchard *Bk. Aeroplane* viii. 144 'Hullo, Croydon,..now passing Biggin Hill. Over!'..The final word 'Over' tells the Croydon operator that the pilot is switching his transmitting apparatus over to receiving so that he can hear what Croydon has to say. **1940** 'Gun Buster' *Return via Dunkirk* II. iv. 117 X calling Robert Eddy...I can hear you...remain on receive...over to you over. **1955, 1966** [see *out *adv.* 23 b]. **1967** 'R. Foley' *Fear of Stranger* (1968) x. 105 Over to you, pal, Kay thought in amusement. **1969** *Guardian* 22 July 11/1 Thank you Peter, thank you Paul, to name but two. Over, like, and out. **1972** N. Marsh *Tied up in Tinsel* v. 123 'Well, ta for the tip anyway. Over and out.' Alleyn hung up. **1973** G. Mitchell *Murder of Busy Lizzie* xv. 175 'But, for the moment, we are concerning ourselves with the Lovelaine family, I thought.' 'Sorry! Over to you, then.' **1974** P. Wright *Lang. Brit. Industry* xiv. 135 It [*sc.* language repetition] occurs particularly where.. actions and accompanying words amount to a drill, as in the radio operator's *Over and out*. **1976** L. Dills *CB Slanguage Dict.* (rev. ed.) 51 *Over*, through transmitting but listening.
9. b. Until a later time or period; till the next season; overnight.
1861 *Trans. Illinois Agric. Soc.* IV. 317 Old bugs live over, and produce eggs the following season. **1884** J. Hay *Bread-Winners* xi. 172, I am so glad you resolved to stay over. **1899** A. Nicholas *Idyl of Wabash* 53 We don't want to winter them steers over. **1953** N. Gordimer *Lying Days* II. xx. 168, I was going to sleep over at the house of an old friend of my mother's. **1968** J. Sangster *Touchfeather* xiv. 149 'You're staying over.' 'But I haven't brought any clothes.' **1973** 'D. Shannon' *No Holiday for Crime* (1974) vi. 91 He..put a second shirt and a razor in a briefcase in case he had to stay over.
14. Phr. *over (and done) with*: completed, finished; dispensed with.
1938 R. D. Finlayson *Brown Man's Burden* 53 The others..were glad when they were free to go, for..the salutations were over and done with. **1970** Auden in *New Yorker* 21 Feb. 118/1, I have one slight criticism.. which I will get over with at once. **1977** S. Woods *Thief or Two* 136, I thought, if I was going to do it, I'd get it over with.

over, *prep.* Add: **1. c.** (Earlier and later examples.)
1805 in *Spirit of Public Jrnls.* (1806) 96 A writer over the signature of Zanga, is another buckram expression. **1826** *New Harmony* (Indiana) *Gaz.* 22 Mar. 207/2 A writer over the signature of 'A Farmer'..states that he has been completely successful..in saving his wheat [from weevils]. **1934** H. G. Wells *Exper. Autobiogr.* II. viii. 626 Bennett ..wrote much of the little weekly paper, *Woman*, he was editing..over the signature..of 'Aunt Ellen'. **1946** *Sunshine Mag.* Apr. (front cover), New preface over the author's own signature.
7. a. (See also *all over *advb. phr.* 1 c).
d. (Further examples.)
1916 T. MacDonagh *Lit. in Ireland* 120 The characteristic qualities of the ancient Irish lyrics are those of good lyric poetry the world over. **1930** *Publishers' Weekly* 23 Aug. 675/2 There are many more like me, the country over, really anxious to feed their fanaticism. **1971** *Guardian* 14 Jan. 11/8 Policemen love one another the world over.
e. *Math.* (Defined or expressed) in terms of (the elements of); *esp.* having coefficients or co-ordinates in, or having elements with coefficients or co-ordinates in.
1932 *Trans. Amer. Math. Soc.* XXXIV. 171 (*heading*) Theory of cyclic algebras over an algebraic number field. **1938** A. A. Albert *Mod. Higher Algebra* ii. 40 The most interesting and important linear sets for our purposes are those of finite order *n* over a field $\mathcal{F}$... Their elements may be thought of as points in an *n*-dimensional space with coordinates in $\mathcal{F}$. **1965** J. J. Rotman *Theory of Groups* vi. 103 A number $a \in C$ is algebraic over *F* in case $F(a)$ is a finite-dimensional vector space over *F* (otherwise *a* is transcendental over *F*). **1972** A. G. Howson *Handbk. Terms Algebra & Analysis* xi. 55 The polynomials form a subring..called the ring of polynomials over *K*.
12. b. *over the wicket*: see *bowl *v.*[1] 4 b.
13. Add to def.: by means of (a telephone, radio communication, or the like); = *on *prep.* 1 d.
1899 [in Dict.]. **1928** Blunden *Overtones of War* iv. 43 Persons who, speaking over the field telephones, gave away any information at all..would be court-martialled. **1929** *Radio Times* 8 Nov. 387 'Pickwick', and other such novels, should be read serially over the microphone. *Ibid.*, Over the wireless a reading can be listened to without.. irrelevant disturbances. **1946** *Ibid.* 8 Feb. 3/3 His boys and girls who had been heard over All-India Radio. **1966** *Listener* 17 Nov. 725/1 He is asking you over the telephone, so you cannot point or use gestures. **1969** *N.Y. Rev. Books* 2 Jan. 5/1 In his Security Gap speech over CBS on October 25, Nixon said one of his major aims would be to 'correct its (the Pentagon's) over-centralization'.
15. b. Having recovered from (an illness, disease, or the like). Cf. Get *v.* 41 b.
1929 'S.N.D.' *Sir W. Howard, Visct. Stafford* iii. 29 He was in England, just over an illness, and straitened for lack of money in the autumn of 1646. **1942** D. Powell *Time to be Born* (1943) vi. 139 You're over it, aren't you, Vicky? That's wonderful. **1964** L. Deighton *Funeral in Berlin* xviii. 109 Finally there is not being in love and liking that—you are over it then—cured. **1975** J. Grady *Shadow of Condor* (1976) i. 18 My wife just got over the flu..she's over it now. **1977** P. Smalley *Trove* ii. 78 He had guessed about the alcoholism as soon as Daley said he had been in hospital... He probably was over it, but you never knew for sure.

over, *sb.*[2] **2. b.** *pl. Printing.* Copies printed in excess of the number ordered, to allow for wastage.
1888 C. T. Jacobi *Printers' Vocab.* 92 Overs, the 'plus' copies beyond a certain number. **1901** D. Cockerell *Bookbinding* I. ii. 36 The printers usually keep a number of 'overs' in order to make good such imperfections. **1946** J. A. Eisler in H. Whetton *Pract. Printing & Binding* xxvii. 328/2 Finding the net amount of paper entailed in the production of a job (exclusive of overs) should present few difficulties to the man familiar with ordinary paper usage. **1961** T. Landau *Encycl. Librarianship* (ed. 2) 269/2 Overs. The number of sound copies over after the printing of the net number of copies ordered.
4. Add to def.: Since 1900, an 'over' has normally consisted of 6 balls, except in Australia, and recently occas. elsewhere, where it is 8. (Earlier and later examples.) Also *attrib.* Cf. Maiden *a.* 4 b in Dict. and Suppl.
1833 *New Sporting Mag.* V. 325 The Anglesea are in the field, and Floyer bowls the over. **1921** Ld. Harris *Few Short Runs* xi. 284 Under such circumstances how we welcome the umpire's 'Last over, gentlemen.' **1955** *Times* 9 May 15/1 When one says that only one hook was aimed at Tayfield in 37 overs the reader will get some idea of the fullness of his length. Before each over he stands over his stumps and performs a kind of ritual. **1960** E. W. Swanton *W. Indies Revisited* 282 The over-rate during the First Test was higher than in any subsequent one. **1974** B. Johnston *It's been a lot of Fun* xvi. 116 'Stick to the play, Percy, and keep that sort of chat for between the overs,' said the producer. **1977** *Times* 18 Jan. 9/8 The fact that the ball had to be replaced three times,..and that the sightscreens are not easily shifted, all helped to bring down the Indian over rate..to just under 11 to the hour.
5. *Mil.* (chiefly *pl.*). A bullet, shell, or other missile that passes beyond its target. *colloq.*
1915 W. H. L. Watson *Adventures Despatch Rider* v. 66 He believes the Uhlans were North Irish Horse and the bullets 'overs'. **1928** Blunden *Undertones of War* iv. 43 A familiar place far enough from the Brickstack which we held to receive the 'overs'. **1944** A. Jacob *Traveller's War* 238 Men on the fringe of the battle area..receive the 'overs' and keep ducking flat as they hear the hissing approach of tank ammo, that has missed its mark. **1969** I. Kemp *Brit. G.I. in Vietnam* vi. 140, I..laid them behind a tree..; they should be moderately safe there from 'overs'.

over, *v.* **4.** (Further examples.)
1847 A. Brontë *Agnes Grey* xi. 163, I was sore distressed Miss Grey—thank God it's owered now. *Ibid.* xii. 185 You'll *both* stay while this shower gets owered. **1933** L. A. G. Strong *Sea Wall* II. xiv. 219 He done an operation on a woman and she never overed it. **1936** 'N. Blake' *Thou Shell of Death* xiii. 231 Master Dermot was killed in France, the year of the Easter Rising. His da never overed it. **1949** *Amer. Speech* XXIV. 111 *Over*, to recover from, as a disease or an injury.

over-. Add: **I. 8.** Also with the sense 'upon the surface so as to cover in part', as in *overpaint *sb.*, *overprint *v.* II. **a.** *over-screen, -stamp* vbs.
1906 Hardy *Dynasts* II. I. ii. 17 Draw down the curtain, then, and overscreen This too-protracted verbal fencing-scene. **1935** *Burlington Mag.* June 288/1 Over-stamping on Sheffield-made candlesticks the London date-letter 1775–6. **1963** *Times* 23 May 9/6 She would inquire about what they were suffering from 'and they would tell me nerves, flu, bronchitis, or gastritis and so on, and I just wrote out the certificate and signed "T. G. Boyle", and overstamped'. **1977** *Belfast Tel.* 22 Feb. 3/1 Your book will be overstamped to include an extra £1 a week for your first child from 4th April, 1977, and posted back to you.
c. *over-bolster, -boot, -cover, -gaiter* (further example), *-jumper, -mitt, -shirt* (earlier and later examples), *-sock, -stocking, -trousers.*
1917 D. H. Lawrence *Phoenix II* (1968) 64 A single bed, opened for the night, the white over-bolster piled back. **1939–40** *Army & Navy Stores Catal.* 607/2 Motoring Overboots, in Brown Sheepskin. **1959** *Times* 2 Oct. 14/6 Her..macintosh, rain hood, and over-boots testified to a careful preparation for the realities of the English climate. **1971** C. Bonington *Annapurna South Face* xiii. 143 Mick got ready for the next pitch, removing crampons and overboots in readiness for what was obviously going to be a hard piece of free rock climbing. **1915** F. M. Ford *Good Soldier* IV. ii. 224 Fishing-rods in green baize over-covers. **1963** *Times* 23 Feb. 11/3 Plentiful over-cover induced the deer to stay. **1908** 'O. Henry' *Voice of City* 233 It was Rosalie, in..gray walking suit, and tan oxfords with lavender overgaiters. **1975** *Times* 7 Oct. 11/4 Over-jumper with wide sleeves..and a square neck. **1971** Overmit [see *liner[3] 3]. **1971** C. Bonington *Annapurna South Face* 298, Gloves with waterproof over-mitts are standard. **1805** Lewis & Clark *Orig. Jrnls. Expedition* (1905) II. 159 The weather being warm I had left my leather over shirt and had woarn only a yellow flannin one. **1974** *Country Life* 2 May 1096/2 The man wears a cotton..striped overshirt, denim shorts and pull-on hat. **1911** Webster, Over-sock. **1929** *Footwear Organiser* Jan. 31 (*heading*) The Oversock vogue spreads throughout the country. **1971** 'D. Halliday' *Dolly & Doctor Bird* iii. 29 For golf, I have always worn..oversocks with good shoes. **1892** Kipling *Lett. of Travel* (1920) 6 The driver with red mittens on his hands, felt overstockings that come up to his knees, and, perhaps, a silvery-gray coon-skin coat on his back, walks beside me. **1852** *Harper's Mag.* Apr. 707/1 My duck over-trousers..were beginning to be rather tender in certain places. **1968** *Daily Tel.* 28 Sept. 9/4 Come rain or snow the lot would be covered by overtrousers or over-skirts and waterproof jackets. **1976** *Good Motoring* Nov. 24/1 Coats, jackets and overtrousers.
13. b. Prefixed to a plural number (or occas. a singular number used *attrib.*) to denote persons who are older than that particular age.
1940 Graves & Hodge *Long Week-End* xvii. 303 The *Evening News*..throwing open its columns to the over-forties. **1959** *Manch. Guardian* 19 Aug. 3/6 Sir Compton Mackenzie and Miss Ruby Miller, for the over-70s, sparring with youthful zest. **1960** *Guardian* 13 Apr. 6/6 There seems to be no place for the over-fifties. **1960** C. Watson *Bump in Night* ii. 25 We shall want to take a closer look.. without being trampled to death by the Over-Sixty clubs. **1972** M. J. Bosse *Incident at Naha* 47 They had their hair done in an over-thirty style. **1973** M. Amis *Rachel Papers* 22 The over-twenties, I grant you, must see it [*sc.* sex] largely as a matter of obligation, too: but obligation to the partner, not to oneself, like us. **1975** B. Meyrick *Behind Light* xiv. 183 After the boys' competitions, where I came in second in the over-twelves, came..community hymn singing. **1977** *New Wave* No. 7. 8 The only night spots, right, are an over-25's place and Mecca.
19. *over-matter* (further examples).
1928 *Daily Express* 7 Feb. 3/6 Early buyers of lingerie had all the advantages, for most of the real bargains.. belonged to ranges that..were 'overmatter' that had to be cast out of stock. **1967** *Economist* 2 Dec. p. iii/2 This one book is only a bit of overmatter from all his earlier over-writings. **1972** *Observer* 12 Mar. 16/8 Large quantities of titanium overmatter..were just thrown into the dustbin. **1977** *Oxf. Diocesan Mag.* Oct. 4/1 The carry-forward of over-matter means that no issue can be planned as 'an island, entire of itself'.
II. 23. *over-bowl* (earlier and later examples). Cf. also sense *27 a.
1844 *W. Lillywhite's Illustr. Hand-bk. Cricket* 18 Do not over bowl yourself by random bowling. **1962** *Punch* 1 Aug. 152/3 A Cowdrey who would certainly not overbowl himself.
27. a. *over-accentuate, -bowl, -commit, -complicate, -condense, -control, -cook, -deflate, -dramatize, -elaborate, -emphasize* (examples), *-enrich* (example), *-express* (examples), *-ink, -interpret, -invest, -linger* (later example), *-order, -prove* (examples), *-rank, -regulate, -rev* (trans. and intr.), *-secrete, -stress* vbs.
1885 A. Brereton *Dramatic Notes* 31 She slightly over-accentuated certain passages. **1977** *Gramophone* Jan. 1160/1 If anything the conductor over-accentuates at the expense of broader phrasing. **1962** *Times* 15 May 4/2 It must be a temptation for his captain to overbowl him. **1976** J. Snow *Cricket Rebel* 41 In his first full championship season in 1974 Andy [Roberts] was overbowled consistently throughout the summer while Hampshire tried

to retain the county title. **1964** Y. Malkiel in *Archivum Linguisticum* XVI. 15 W. J. Entwistle may have over-committed himself. **1973** *Guardian* 11 Apr. 8/7 A few [families] were found to have overcommitted themselves with hire purchase. **1966** A. Battersby *Math. in Managem.* v. 130 One can easily over-complicate a model, and the manager and mathematician must collaborate closely to decide not only what is relevant, but what is significant. **1976** H. Tracy *Death in Reserve* xiv. 113 You're over-complicating the whole thing. **1933** *Mind* XLII. 391 The actual statement of the theory is, in view of its importance, somewhat over-condensed, and ought perhaps to have been expanded. **1962** A. Nisbett *Technique Sound Studio* xiii. 232 Pace is not achieved by over-condensing vital information. **1941** *Sun* (Baltimore) 28 June 1/5 He simply overcontrolled the ship. **1904** *Daily Chron.* 30 May 8/3 So you can over-cook even a sauce. **1963** R. Carrier *Great Dishes of World* xiii. 225 Be careful not to overcook *pasta*. **1975** I. Daly in D. Marcus *Best Irish Short Stories* (1977) II. 32, I want the steak medium-rare... All you Irish overcook meat. **1977** *Harpers & Queen* Sept. 28/1 The salmon was overcooked. **1962** *Daily Tel.* 15 June 14/2 The reported trends are signs ..that the economy has been over-deflated and that confidence in future expansion needs fostering. **1974** *Times* 28 Feb. 19/2 A Labour Government would be tempted to overdeflate in the Budget in order to make itself more attractive to foreign lenders. **1955** S. Spender *Making of Poem* iv. 63 Perhaps I over-dramatize the affair. **1976** M. Butterworth *Remains to be Seen* vii. 111 The flat..he now saw as a fortress... He hoped he was over-dramatizing his situation. **1905** *Daily Chron.* 2 Sept. 3/1 Mr. Phillpotts has resisted the temptation..to over-elaborate his descriptions of natural scenery. **1933** W. E. Orchard *From Faith to Faith* ix. 206 These [dogmas] were over-elaborated during the early controversies, and have only obscured His personal power by theories about Him. **1905** *Outlook* 7 Oct. 485/1 He over-emphasises when he suggests that Hungary is a solid State and Austria but a bundle of provinces. **1926** J. S. Huxley *Ess. Pop. Sci.* 153 This..we must discount unless we are to over-emphasize the antinomy between the microcosm and the macrocosm. **1968** H. Harris *Nucleus & Cytoplasm* i. 12 It cannot be over-emphasized that actinomycin D is an extremely toxic compounds. **1852** Mill *Pol. Econ.* (ed. 3) I. II. ii. 276 Wealth which could no longer be employed in over-enriching a few. **1883** 'Mark Twain' *Life on Mississippi* 399 Terms which did not over-express the admiration with which the people viewed him. **1959** N. Mailer *Advts. for Myself* (1961) 17, I am not suited for this sort of confrontation despite..a bloody season of overexpressed personal opinions as a newspaper columnist. **1927** *Observer* 12 June 9 The literary man..is apt to over-ink his pictures of contemporary morals. **1939** T. S. Eliot *Family Reunion* II. i. 77 You overinterpret. I am sure that your mother always loved him; There was never the slightest suspicion of scandal. **1963** *Times* 9 May 16/4 She..sometimes fell into the opposite trap of overinterpreting detail. **1975** *Nature* 18 Dec. 562/3 There is a danger, however, that such a negative result could be over-interpreted as suggesting that recombinant experiments are inherently safe. **1934** Webster, Overinvest. **1958** *New Statesman* 25 Jan. 94/3 The trouble about Poland today..is that we are rather like a furniture manufacturer who has plenty of table legs but no table tops. We have over-invested in legs and now we want more capital for the tops. **1895** W. B. Yeats *Poems* 23 He has over-lingered his welcome. **1950** *Times* 20 Feb. 7/7 It was not surprising that, when steel was most scarce, the distribution scheme worked least well: firms over-ordered and accumulated stocks and there was nothing to encourage them to use as little steel as possible. **1977** D. Bennett *Jigsaw Man* v. 106 'You aren't liking your good grub.' 'I think I over-ordered.' **1912** A. Lang *Shakespeare, Bacon & Great Unknown* iv. 81 Mr. Collins, 'a violent Stratfordian', overproved his case. **1929** R. Graves *Poems* 31 Now is a sheet of paper, A not blank expectation,..A being, over-proved, A report of happiness. **1958** *New Statesman* 23 Aug. 222/3 In answering poll-questionnaires..'we tend to over-rank ourselves'. **1972** *Korea Times* 17 Nov. 2/3, I am no longer willing to remain patient with the parade of overranked non-entities whose actions reflect their own ignorance. **1938** *Sun* (Baltimore) 16 Apr. 8 Mr. Roosevelt is equally muddled in his general attitude toward trade and industry. For five years he has overregulated trade and industry. **1973** *Sci. Amer.* Sept. 165/1 Some observers believe that the pharmaceutical industry is now overregulated and that bureaucratic interference with the industry has reached such a level that the American public is being denied certain drugs available overseas. **1935** C. G. Burge *Compl. Bk. Aviation* 87/1 The control stick is pushed forward to give the diving position and the throttle eased slightly back to avoid over-revving the engine. **1978** *Daily Tel.* 26 July 3/2 Mr Wheatcroft adjusted the accelerator, but..he noticed that the coach seemed to be 'over-revving', as though the driver was not using the brakes to slow it down. **1927** Haldane & Huxley *Animal Biol.* viii. 164 If the pituitary begins to over-secrete before the epiphyses have been joined by bone to the shafts, the patient becomes a giant. **1916** T. MacDonagh *Lit. in Ireland* 66 An Irish reader would be content to pronounce the words as they come,..not overstressing 'in' and 'up'. **1933** *Mind* XLII. 238 It is also admitted that children have to be taught cleanliness (which parents often over-stress from snobbishness!). **1970** *Daily Tel.* 26 Sept. 9/7 It would however be wrong to overstress the importance of colour in lithography. **1977** *Bitumen* (Shell Internat. Petroleum Co.) 7 Knowledge of such properties has made it possible to design roads and airfield runways on sound engineering principles, ensuring that no part is over-stressed even under the heaviest loads.
b. *over-blessed, -characterized, -concentrated, -concerned, -involved, -preoccupied, -recovered, -rehearsed, -represented* (further examples), *-technicized* pa. pples.
c **1804** Wordsworth *Vaudracour & Julia* in *Misc. Poems* (1820) I. 283 His spirit sank, Surcharged, within him,—overblest to move Beneath a sun that wakes a weary world. **1918** *Nation* (N.Y.) 7 Feb. 130/1 Not by any means a leader even in a body that has not of late been overblessed with outstanding personages. **1977** D. Clark *Gimmel Flask* iii. 58 Green wasn't over-blessed with good manners. **1959** *Times* 9 Nov. 6/1 To begin with they [*sc.* the figures] are heavily over-characterized (by the dangerous means of self-description). **1957** K. G. Wittfogel *Oriental Despotism* 24 On-the-spot rains create additional dangers when they are overconcentrated. **1934** Webster s.v. Over- 6, Overconcerned. **1941** *Mind* L. 2 You are under-concerned about the cases which don't trouble you at the moment, and over-concerned about the one that is striking you at the moment. **1976** P. Hill *Hunters* vi. 67 Was he then weak..over-concerned with what others thought of him. **1965** M. Morse *Unattached* iii. 90 Even the most..skilled of workers can..become over-involved with the..situation at hand. **1975** *Times Lit. Suppl.* 13 June 669/1 Historians ..had been overpreoccupied with what was done to the slaves and had slighted what slaves had done for themselves. **1962** *Which?* (Car Suppl.) Jan. 11/2 Where there is a minus figure, it means that the braking system had over-recovered and needed less pressure than usual; this may make control difficult. **1967** D. Goch in Wills & Yearsley *Handbk. Managem. Technol.* 147 Fixed overheads, being relatively unaffected by fluctuations in the number of units produced during the period, will be either under- or over-recovered to the extent that output varies from that which was assumed to be normal when the standards were set. **1976** *Gramophone* May 1732/1 Most concerts are under-rehearsed and rely too much on the inspiration of the moment, or are over-rehearsed and so dead. **1965** J. Hajnal in Glass & Eversley *Population in Hist.* vi. 121 The deaths of young women are very probably over-represented. **1974** *Howard Jrnl.* XIV. 39 ESN schools, where West Indian youths are significantly over-represented, mainly because they have been wrongly placed there. **1953** *Mind* LXII. 424 It is a joy to go back to the beginnings of a subject which has since become over-technicised.
28. a. *over-articulate, over-clean* (examples), *-complimentary, -concise, -conscious* (example), *-controversial, -dependent, -elaborate, -fast, -friendly, -fussy, -ingenious, -insistent, -keen, -logical* (later examples), *-mellow* (later example), *-mighty* (later examples), *-nimble* (example), *-obvious, -optimistic, -patient* (example), *-pessimistic, -picturesque, -plump, -prolific, -prone* (later example), *-rapid, -ready* (earlier and later examples), *-sensational, -shy, -susceptible* adjs.
1975 *Christmas Greeting* (Rhodes House, Oxford) 6 When dons behave badly, they behave very badly: it is partly the fault of being over-articulate, though early rearing probably has something to do with it too. *c* **1806** D. Wordsworth *Jrnl.* (1941) I. 327 Two beds, with not over-clean bedclothes. **1818** 'A. Burton' *Adventures J. Newcome* I. 32 The Bed-cloaths, when by daylight seen, They did not fancy over-clean. **1867** Mill *Exam. Hamilton's Philos.* (ed. 3) p. vii, Some of the writers are.. even over-complimentary. **1940** W. Stevens *Let.* 30 Aug. (1967) 374 The trouble here is that the lines are over-concise. **1965** *Language* XLI. 142 In his desire to be complete..and informative, Kukenheim is over-concise. **1851** H. Melville *Moby Dick* I. i. 5 Whenever I begin to..be over conscious of my lungs. **1862** H. Sidgwick *Let.* 28 Jan. in A. & E. M. Sidgwick *Henry Sidgwick* (1906) 74 It seems smashing, but he loses by being over-controversial. **1975** *New Yorker* 21 Apr. 127/2 We fought the war for them and made them overdependent on air support. **1931** A. Esdaile *Student's Man. Bibliogr.* vi. 198 Incised bindings..also became over-elaborate, especially in Germany. **1934** Webster, Overfast. **1949** R. Blesh *Shining Trumpets* ii. 41 Overfast tempos did not appear in Afro-American music until very recently. **1939** R. Campbell *Flowering Rifle* vi. 142 In their own tanks they ha ve to be locked up As in a box an over-friendly pup. **1962** E. Godfrey *Retail Selling & Organization* xii. 131 Personal comments and an over-friendly manner also amount to discourtesy. **1974** N. Freeling *Dressing of Diamond* 135 Richard asked..whether I'd perhaps let myself get over-friendly with Colette. **1962** *Times* 16 Feb. 15/2 The ballet-boyish treatment of the pirate chorus is new, and inclined to look over-fussy. **1858** Bagehot *Coll. Works* (1965) II. 70 It would be over-ingenious to argue..that he had no peculiar interest in young ladies in general. **1977** A. Wilson *Strange Ride R. Kipling* iv. 208 The over-ingenious method of Kipling's narration. **1915** D. H. Lawrence *Let.* 26 Feb. (1962) I. 323, I wish you'd tell me when I am foolish and overinsistent. **1977** *N.Y. Rev. Bks.* 13 Oct. 35/2 The slightly overinsistent Ciceronianisms here draw attention to themselves. **1934** Webster s.v. Over- 6, Overkeen. **1959** I. & P. Opie *Lore & Lang. Schoolch.* x. 181 The word 'sap'..at Eton is primarily used to castigate someone who is over-keen on his work. **1977** J. Bingham *Marriage Bureau Murders* x. 130 A girl in trouble with an over-keen lover. **1920** W. R. Inge *Truth & Falsehood in Relig.* 19 Exclusive intellectualism in religion..commits us to an over-logical scheme. **1966** *Eng. Stud.* XLVII. 299 Occasionally Williams seems to make Shakespeare over-logical. **1930** Wyndham Lewis *Let.* 30 July (1963) 190 Joyce is like an over-mellow hot-house pear. **1887** W. Morris tr. *Homer's Odyssey* I. x. 188 There was one Elpenor, the youngest;.. In war not over-mighty. **1920** G. Robinson *David Urquhart* 11 The ephemeral predominance of an over-mighty subject. **1950** *Catholic Times* 17 Feb. 6/3 Some will perceive first the dangers of the Overmighty State. **1966** *Economist* 15 Jan. 170/3 Officialdom is also angry with the overmighty bishops for taking a political initiative without consulting the government. **1978** *Jrnl. R. Soc. Arts* CXXVI. 212/2 Monopolistic bodies have a tendency to become over-mighty. **1885** W. B. Yeats in *Dublin Univ. Rev.* June 111/2 Cease! no more! Thou hast an over-nimble tongue. **1925** I. A. Richards *Princ. Lit. Crit.* xxxv. 287 If to some readers parts of it appear unnecessary—either *irrelevant*, in the one case; or *over-obvious* in the other—I have nothing to add. **1951** Koestler *Age of Longing* II. iv. 243 Not to mention such over-obvious facts as the disparity in the number of divisions. **1953** *Encounter* July 48/1 If *The Tempest* is over-pessimistic and manichean, *The Magic Flute* is over-optimistic and pelagian. **1976** *Broadcast* 29 Mar. [illegible] Over-optimistic predictions of BBC income in the coming year. **1881** 'Mark Twain' *Prince & Pauper* [illegible] 137, I like not much bandying of words, being not overpatient in my nature. **1934** Webster, Overpessimistic. **1953** Over-pessimistic [see *over-optimistic* above]. **1938** L. MacNeice *Mod. Poetry* 10 His [*sc.* Housman's] hanged man, his soldiers, are over-picturesque. **1932** W. Faulkner *Light in August* xiii. 300 Hightower leans there..in the August heat, oblivious of the odor in which he lives—..that odor of overplump desiccation and stale linen as though a precursor of the tomb. **1923** D. H. Lawrence *Birds, Beasts & Flowers* 179 Those over-prolific white mice. **1976** N. Freeling *Lake Isle* viii. [illegible] His ma..was overprone, maybe, to well-meant advice about bringing up the children and such. **1848** Mill *Pol. Econ.* I. II. xi. 413 Where a labouring class..refrain from over-rapid multiplication, the cause..has always hitherto been, either actual legal restraint, or a custom of some sort [etc.]. **1964** *Ann. Reg. 1963* 263 Credit also became tighter, and full employment created the conditions for an over-rapid rise in wages and salaries. **1628** O. Feltham *Resolves* (ed. 3) xcvii. 283 And such there are, that are over-ready in the wayes of pleasing, and labour. **1859** Bagehot *Coll. Works* (1965) II. [illegible] We may seem to make unusual criticisms, and to be over-ready with depreciation or objection. **1906** *Westm. Gaz.* 24 Mar. 2/3 They may be gazing on the..over-ready-to-burst chestnut-buds. **1959** *Times* 3 Sept. [illegible] He is sometimes oversensational and has tried to stuff too much into one book. **1939** L. MacNeice *Autumn Jrnl.* xi. 46 Over-shy at times, morose, defeatist. **1934** Webster, Oversusceptible. **1966** *Eng. Stud.* XLVII. [illegible] The connection..is made by the over-susceptible sensibility of Emily.
b. *over-demanding, -pressing, -refining* ppl. adjs.
1949 M. Mead *Male & Female* iii. 74 Too much emphasis upon the assertive demanding aspects of the mother may build a female picture that is over-active, over-demanding, and threatening. **1893** *Harper's Mag.* Aug. 335/1 The finding..such a palpable motive as revenge against an overpressing and clamorous creditor tipped the balance. **1940** Dylan Thomas *Portrait of Artist as Young Dog* 122 And I never felt more a part of the remote and overpressing world, or more full of love and arrogance and pity and humility. **1855** Bagehot *Coll. Works* (1965) I. 322 The sceptical, over-refining Toryism of Hume and Montaigne.
c. *over-arranged, -controlled, -cultivated, -emotionalized, -expanded, -inflated, -inked, -mechanized, -nourished, -oiled* (examples), *-padded, -perfumed, -polished, -qualified, -schematized, -sensitized* (also as sb.), *-sophisticated, -speculated, -spiritualized, -structured, -sugared* ppl. adjs.
1924 R. H. Mottram *Spanish Farm* 148 She was after all only saying the same thing in French, when a frail, fair, over-arranged lady kept her waiting in the glove department of the Bon Marché. **1956** G. Coulter in M. T. Williams *Art of Jazz* (1960) 167 Most of these bad performances take place in an over-arranged, be-violined setting. **1977** *New Musical Express* 12 Feb. 10/5 We hated over-arranged stage acts and gimmicks constructed just to go with one particular piece of music. **1964** M. Argyle *Psychol. & Social Probl.* ix. 123 Anxiety neurosis is an extreme case of the over-controlled personality resulting from a strict upbringing. **1968** J. S. & B. Bruner in *Internat. Jrnl. Psychol.* III. 239 In early human growth, the initially well-organized systems seem to be predominantly of the automatic or 'overcontrolled' type, as with breathing, swallowing, and initial sucking. **1868** W. James *Let.* 5 Apr. in R. B. Perry *Tht. & Char. W. James* (1935) I. 268 The cool acceptance by the bloody old heathens of everything that happened around them [etc.]..would all make their society perfectly hateful to these overcultivated and vaguely sick complainers. **1965** S. L. Barraclough in I. L. Horowitz *Masses in Lat. Amer.* iv. 138 Erosion of overcultivated hillsides is prevalent. **1938** R. G. Collingwood *Princ. Art* iii. [illegible] Plato..thinks that the new art of the decadence is the art of an over-excited, over-emotionalized world. **1965** H. J. Habbakuk in Glass & Eversley *Population in Hist.* [illegible] 157 These epidemics..cannot have been a Malthusian punishment inflicted on an overexpanded population. **1934** E. Pound *Eleven New Cantos* xxxiii. 14 The meeting decided we were over-inflated. **1964** W. G. Smith *Allergy & Tissue Metabolism* i. 3 The post mortem picture shows over-inflated lungs. *a* **1974** R. Crossman *Diaries* (1975) I. 585 The economy was over-inflated now, we needed to take some of the heat out and to drive a little of the employment out of it. **1978** *New York* 3 Apr. [illegible] His reputation is overshadowed..by that of Sir Edward Elgar, on whose inflated scores a grossly overinflated revival is now being perpetrated. **1967** Karch & Buber *Offset Processes* viii. 281 A heavy flow of ink will cause over-inked copy resulting in scum. **1960** *Farmer & Stockbreeder* 1 Mar. 135 (Advt.), If you haven't got a Lundell today, you're overmechanised. **1977** *Listener* 20 Oct. 517/2 It provides..a record..of what happens when a lumpish, over-mechanised and wrongly-trained army..meets a lightweight but adept Asian guerilla force in its own country. **1931** H. Read *Meaning of Art* [illegible] 94 'Barock' with its dark and loaded sound implying well the heavy, swollen, over-nourished forms that must be urged into movement to make their impression. **1929** E. Pound tr. *Rimbaud* 13 A woman's head with brown over-oiled hair. **1938** L. MacNeice *Mod. Poetry* i. 4 Every action from this poetry, which they felt to be priggish, pontifical or merely dull and overpadded. **1857** Bagehot *Coll. Works* (1965) II. 28 An over-perfumed softness pervades the poetry of society. **1938** H. Nicholson *Let.* 21 Apr. (1966) 337 An ex-diplomatist with those overpolished manners, that *boulevard extérieur* elegance, which always faintly annoys me. **1957** Manvell & Huntley *Film Music* iv. 180 The lighting cameraman may be concerned with a range of problems, from the over-polished conversation [illegible]

of an electric organ to the number of arcs required to illuminate the Royal Albert Hall. **1968** *N.Y. Times* 3 Feb. 19/5 It is often hard to get the message across to personnel men 'who make points hiring over qualified people for less than they're worth'. **1969** *Time* 28 Mar. 41 Applications are flooding colleges across the country. The problem is how to cull the lucky few from the over-qualified many. **1977** *Times* 19 Aug. 6/3 The unemployed PhD.., the over qualified school-leaver, have already brought home..the consequences of..rising unemployment and rising qualifications. **1962** U. WEINREICH in Householder & Saporta *Probl. Lexicogr.* 35 Over-schematized though it may be, ad hoc intralinguistic considerations suggest [etc.]. **1965** *Language* XLI. 504 Rigid, overschematized synchronic analysis. **1926** E. HEMINGWAY *Sun also Rises* xiv. 154, I read the Turgenieff..in the oversensitized state of my mind after much too much brandy. **1965** *Punch* 17 Mar. 397/2 This double disability may at first sight seem hard on the over-sensitised. **1975** *New Yorker* 26 May 18/2 *Blow-up*..An over-sensitized and wildly misaccented account of the mod, mad world of London, 1966. **1918** Oversophisticated [see *film fan* s.v. *FILM *sb.* 7 c]. **1971** *Guardian Weekly* 10 Apr. 14/4 An overdeveloped oversophisticated country. **1971** D. POTTER *Brit. Eliz. Stamps* xv. 163 Prices began to rise on all sides. Only the overspeculated commemoratives failed to make progress. **1951** S. SPENDER *World within World* iii. 118 With him I escaped to some extent from the over-spiritualized, puritan, competitive atmosphere in which I had been brought up. **1959** N. MAILER *Advts. for Myself* (1961) 18, I find to my perhaps over-structured horror that I rather enjoy the high-pressured rubber of bridge. **1971** *Nature* 27 Aug. 591/1 The organization that has resulted from all the wrangles and compromises is over-structured. **1906** *Westm. Gaz.* 15 Aug. 4/2 No better corrective of their over-sugared literature, with its artistic embellishment, could be suggested than Mary Wollstonecraft's unflattering plain-dealing. **1968** *Daily Tel.* 17 Dec. 13/7 Much of this [Amontillado] sold here tends to be rather flat and over-sugared.

d. *over-banked, -muscled* adjs.

1930 *Times* 27 Mar. 21/2 The Port of Karachi..is considerably over-banked. **1966** *Economist* 18 June p. xxxiii/1 It has reduced its branches in Scotland, which is more overbanked than England. **1956** H. GOLD *Man who was not with It* (1965) ix. 74 His weary overmuscled body. **1977** *Gay News* 7–20 Apr. 23/1 No, I don't like that, over-muscled... I do like people who keep their bodies in shape though.

29. a. *over-caring, -drugging, -farming, -meddling* (example), *-packaging, -packing, -padding, -planning, -revving, -soiling* vbl. sbs.

1938 BELLOC *Sonnets & Verse* 39 Believe in none and die of over-caring. **1868** A. B. GARROD *Materia Medica* (ed. 3) 375 Much discredit has been thrown upon the whole subject of the medicinal treatment of disease by the practice of indiscriminate prescribing and over-drugging. **1946** *Nature* 23 Nov. 733/1 In his keen observation, in his reflexion and deductions, and in his dislike of over-drugging, More had all the endowments of a wise physician. **1943** J. S. HUXLEY *TVA* vi. 33 Over-farming was not the only exploitation. **1861** MILL *Repr. Govt.* iv. 82 A government..required to hold its hands from over-meddling..is not to the taste of such a people. **1972** *Computers & Humanities* VII. 81 Rarely has the sociology of knowledge provided such an obvious example of technology shaping the formulation of research conception as in the over-packaging of most social science statistical analysis. **1967** KARCH & BUBER *Offset Processes* vii. 262 Repeat until satisfactory, but avoid overpacking. **1962** *Economist* 8 Sept. 925/1 Wasteful overpadding of junior executive posts. **1974** T. P. WHITNEY tr. *Solzhenitsyn's Gulag Archipel.* I. i. x. 393 To defend *quality*..amid the general uproar about *quantity*, planning, and overplanning. **1977** *Times* 16 Feb. 15/5 The country has suffered from intensive over-planning. **1976** F. GREENLAND *Misericordia Drop* i. v. 39 After some predictable over-revving with the clutch out, they were off. **1959** *Times* 12 Jan. 11/3 It is advisable to avoid over-soiling and consequent hard rubbing.

b. *over-accentuation, -aspiration, -blame, -classification, -commitment, -consumption* (examples), *-control, -dramatization, -expansion, -expression, -influence, -insistence* (examples), *-interpretation, -involvement, -moralization, -multiplication, -nutrition, -organization, -ornamentation, -recovery, -regulation* (examples), *-reliance* (examples), *-secretion, -sophistication, -speculation* (earlier and later examples), *-stress, -tension* (examples).

1907 R. FRY *Let.* 11 Jan. (1972) I. 280, I find even in Dürer's portrait of himself here a certain over-accentuation, a self-consciousness. **1928** I. C. WARD *Phonetics of Eng.* xiii. 115 In order to cure over-aspiration, it is necessary to tell the pupil to make the contact firm and the release vigorous. **1964** CRYSTAL & QUIRK *Syst. Prosodic & Paralinguistic Features Eng.* iii. 38 Voice qualities... Over-aspiration (excessive pressure being released as compared with normal articulation) particularly noticeable on vowels, and on those consonants where there is normally little aspiration. **1874** TENNYSON *Merlin & Vivien* in *Wks.* VI. 12 Of overpraise and overblame We choose the last. **1955** *Bull. Atomic Sci.* Apr. 127/3 In other defense activities it is undoubtedly true that overclassification is the rule. **1961** *Lancet* 26 Aug. 497/1 He was..a little bit impatient with the fussy over-classification that was coming into vogue in his specialty. **1964** Y. MALKIEL in *Archivum Linguisticum* XVI. 14 Undercommitment versus overcommitment. **1934** WEBSTER, Overconsumption. **1974** *Daily Colonist* (Victoria, B.C.) 15 Nov. 1/6 The Norwegian delegation is pushing a conference resolution stating that overconsumption impairs the health of the affluent. **1941** O. E. PATTON *Aircraft Instruments* vii. 114 A follow-up mechanism is a necessary part of an automatic control, in order to prevent overcontrol, which gives rise to oscillations, or hunting, of the aircraft. **1957** J. S. BRUNER in *Psychological Rev.* Mar. 144/2 George Klein's work..suggests that, in general, people who are not able to shift categorization under gradually changing conditions of stimulation tend also to show what he describes as 'overcontrol' on other cognitive and motivational tasks. **1973** *Sociometry* XXXVI. 135 A false fire-alarm went off precisely as the stimulus-subject in a severe condition was screaming from the electrical shock, providing an extremely amusing overdramatization of an already impactful event. **1976** R. HILL *Another Death in Venice* i. i. 21 Out there was a young *mafioso*... No, that was an absurd over-dramatization. **1935** *Planning* III. LIII. 5 The legacy of war-time over-expansion, often in uneconomic locations, and of post-war over-capitalisation is a second [factor]. **1964** E. H. POWELL in I. L. Horowitz *New Sociol.* xx. 332 The drive for profit produced a disastrous fluctuation of the business cycle, with periods of over-expansion and prosperity followed by bleak times of contraction and business failure. **1899** W. JAMES *Talks to Teachers* viii. 75 This ceaseless over-tension, over-motion, and over-expression are working on us grievous national harm. **1908** *Edin. Rev.* July 71 That greatest snare of Faber's unquestionable eloquence: over-expression. **1966** 'H. MACDIARMID' *Company I've Kept* iii. 85 The Scottish public has been..debauched and distorted by English over-influence. **1934** WEBSTER, Over-insistence. **1953** K. REISZ *Technique Film Editing* ii. 114 An over-insistence on one aspect of the theme..may..divert the spectator's attention from the main theme. **1965** W. S. ALLEN *Vox Latina* Appendix B. 109 His [Erasmus'] conclusions appear to arise partly out of an over-interpretation of Marius Victorinus. **1968** F. G. LOUNSBURY in J. A. Fishman *Readings Sociol. of Lang.* 53 The above..is possibly also an overinterpretation of the facts. **1964** P. WORSLEY in I. L. Horowitz *New Sociol.* 385 Their problem would not be separateness..but over-involvement. **1976** *Times* 20 May 18/7 [Jerry] Brown..has reportedly made himself a pain in the neck by over-involvement in detail and incapacity to delegate. **1933** A. N. WHITEHEAD *Adventures of Ideas* xii. 201 The forgetfulness of this doctrine leads to an over-moralization in the view of the nature of things. **1931** J. S. HUXLEY *What dare I Think?* i. 28 Can he [*sc.* the biologist], by studying the pest in its original home, discover what are the other species that normally act as checks on its over-multiplication? **1899** Overnutrition [see *undernutrition* s.v. UNDER-[1] 10 b]. **1899** *Discovery* Apr. 98/2 While diseases associated with under-nutrition (tuberculosis, etc.) are steadily decreasing here, those associated with good nutrition—not to say over-nutrition—(diabetes, etc.) are on the increase. **1971** *New Scientist* 25 Feb. 407/1 Enormous problems of malnutrition and overnutrition remain unsolved and untackled. **1976** *Sci. Amer.* Sept. 40/2 Malnutrition may come about in one of four ways. A person..may be taking in too many calories or consuming an excess of one component or more of a reasonable diet; this condition is overnutrition. **1946** *Nature* 21 Sept. 392/2 There was no disposition on the part of the delegates to encourage the over-organisation of such interchange or movement of scientific workers. **1968** C. A. DOXIADIS *Betw. Dystopia & Utopia* 18 In 1959, Aldous Huxley..explains what disasters we should expect because of over-population, over-organization, and brain-washing. **1933** R. TUVE *Seasons & Months* iv. 188 Poets had come to take delight in over-ornamentation. **1967** D. GOCH in Wills & Yearsley *Handbk. Managem. Technol.* 148 At an output level of £510,000 there is an over-recovery of 10,000 × 1s. od. = £500. **1875** *Encycl. Brit.* II. 575/1 Appointments..were made under the purchase system... Every regimental commission had a fixed regulation price..in addition to which an over-regulation price, which sometimes even exceeded the regulation price, had sprung up. **1950** A. L. ROWSE *England of Elizabeth* iv. 113 The natural energy, inventiveness, enterprise of the people, that was..not discouraged, thwarted and stifled by over-regulation. **1976** *National Observer* (U.S.) 4 Dec. 10/2 What is really wrong with government overregulation is not that business people find it burdensome and costly, but that its principal victim is the consumer. **1833** J. S. MILL in *Monthly Repos.* VII. 663 Over-reliance on our own judgment is one thing, over-reliance on the judgment of the world when in unison with our own, is another. **1961** L. F. BROSNAHAN *Sounds of Language* iii. 50 An apparent over-reliance on the spelling as a means of identification of dental fricatives. **1948** MARTIN & HYNES *Clin. Endocrinol.* i. 20 Acromegaly... A disease due to over-secretion of the hormones of the acidophil cells of the pars anterior in adult life. **1934** C. LAMBERT *Music Ho!* i. 44 That modern craving—essentially a product of over-sophistication—for the dark and instinctive that we find in D. H. Lawrence. **1857** J. S. MILL in *Coll. Wks.* (1967) V. 502 To prevent the Bank, at times when there is a tendency to overspeculation, from encouraging that tendency. **1940** *Times* 27 Feb. 14/4 The Spanish Council of Ministers has approved the reopening of the stock exchanges at Madrid, Barcelona, and Bilbao on March 1... The reopening is subject to various restrictions aimed at curbing over-speculation. **1965** G. J. WILLIAMS *Econ. Geol. N.Z.* iii. 23/1 A period of depression resulted from over-speculation. **1923** J. S. HUXLEY *Ess. Biologist* i. 55 Whatever overstress and maladjustment the complexity of modern civilization has brought with it, [etc.]. **1971** *Daily Tel.* 26 July 11/1 The effect of overstress is cumulative, causing structural weaknesses over a period of time. **1899** Over-tension [see *over-expression* above].

c. *over-capacity, -consciousness, -dominance, -exactness* (example), *-expressiveness, -intensity* (examples), *-lusciousness, -optimism, -susceptibility* (example).

1934 WEBSTER, Overcapacity. **1960** *New Left Rev.* May–June 20/1 Under-capacity use of the railways..and over-capacity use of the roads. **1971** *New Scientist* 8 Apr. 96/1 Industry is currently suffering from some over-capacity following the cutback in the aerospace effort. **1911** J. WARD *Realm of Ends* xv. 337 *Only* to differentiate this 'Over-consciousness' from all such consciousness as we can conceive is the term 'the Unconscious'..applied to it. **1960** *Farmer & Stockbreeder* 1 Mar. 65/1 From a geneticist's point of view in-breeding deterioration is explained by a decrease in those combinations showing overdominance. **1976** *Nature* 15 July 227/2 Population geneticists have never agreed on the extent to which over-dominance of fitness—that is, the situation in which the fitness of the heterozygote exceeds the fitness of both homozygotes—is responsible for the maintenance of genetic variability in populations. *a* **1866** J. GROTE *Treat. Moral Ideals* (1876) 169 In danger of erring on the side of..over-exactness. **1976** *Gramophone* Apr. 1598/3, I found a trace of over-expressiveness in such movements as the second of the *Norwegian Melodies* and the Sarabande from the *Holberg Suite*. **1858** BAGEHOT *Coll. Works* (1965) II. 101 We endure the over-intensity..of the surrounding misery. **1899** W. JAMES *Talks to Teachers* viii. 74 Our faces, all contracted as they are with the habitual American over-intensity and anxiety of expression. **1978** *N.Y. Rev. Bks.* 23 Feb. 30/1 TB was understood, like insanity, to be a kind of one-sidedness: a failure of will or an overintensity. **1898** G. SAINTSBURY *Short Hist. Eng. Lit.* x. i. 671 A certain over-lusciousness traceable in his [*sc.* Keats's] earlier work. **1963** *Times* 25 Apr. 12/7 If the facts are as stated this is the biggest step towards the controlled release of the energy obtainable from the fusion of heavy hydrogen nuclei since the phase of early overoptimism represented in Britain by Zeta (Zero Energy Thermonuclear Apparatus). **1976** B. FREEMANTLE *November Man* ii. 14 Dennison had an aptitude for over-optimism. **1843** J. S. MILL *Let.* 21 Oct. in *Coll. Wks.* (1963) XIII. 600, I cannot charge myself with any oversusceptibility in the matter.

d. *over-ambition, -elaboration, -emphasis* (further examples), *-precision.*

1929 A. N. WHITEHEAD *Process & Reality* p. xi, Speculative philosophy and overambition. **1973** *Times* 18 Oct. 15/3 His career was blighted by over-ambition. **1901** *Chambers's Encycl.* VII. 141/2 A frequent over-elaboration of style and strainedness of wit that fatigues rather than exhilarates. **1940** *Mind* XLIX. 69 Many readers may feel that there is a danger in mathematical over-elaboration. **1974** tr. *Wertheim's Evolution & Revolution* i. 72 Phonetic spelling could not be realized in Egypt precisely because of the over-elaboration of the cumbersome hieroglyphic system. **1912** J. S. HUXLEY *Individual in Animal Kingdom* i. 24 By an over-emphasis of the species-individuality of which we are the parts, it is often said that our bodies are only 'cradles for our germ cells'. **1935** *Planning* III. XLIX. 1 Many of these arguments..are based on a more or less crude over-emphasis of certain aspects at the expense of the whole. **1949** M. MEAD *Male & Female* x. 215 Human children..will make well-balanced but individual selections [of food], compensating one day for an over-emphasis of the day before. **1926** FOWLER *Mod. Eng. Usage* 684/1 That is the logical arrangement, which..is free from any taint of over-precision. **1952** C. P. BLACKER *Eugenics* 112 It is difficult..to steer a course that shall keep clear of the mudflats of platitude on the one hand, and not come to grief against the rocks of over-precision on the other.

30. *over-often* adv. (later example); *over-cheerily, -closely, -diligently* (example), advbs.

1947 DYLAN THOMAS *Let.* 20 May (1966) 307 Did you receive the postcard, overcheerily scribbled with messages? **1909** *Westm. Gaz.* 15 Apr. 12/2 The loving parent does well not to examine overclosely into the reasons for this regret. **1690** LOCKE *Essay Hum. Und.* III. vii. 228 This part of Grammar has been, perhaps, as much neglected, as some others over-diligently cultivated. **1976** R. BARNARD *Little Local Murder* iii. 38 His sports jackets did not go over-often to the dry cleaner's.

III. 31. *overpage* adv.

1870 D. G. ROSSETTI *Let.* 29 July (1965) II. 893, I send another correction overpage. **1932** R. A. CRAM in *Newsletter Mediaeval Acad. Amer.* 15 Nov. 3 Over page is a list of a few books recently issued by these publishers. **1970** *Daily Tel.* 5 Sept. 15 Money-go-round is continued overpage.

32. *over-centre, -life-size, -ocean, -shoulder* adjs.

1975 *Sunday Times* 23 Feb. 17/4 McDonnell Douglas designed a rear-cargo door with four electrically-driven 'over-centre latches'. They were to close over spools in the aircraft body and pull the door shut against its seal. **1937** *Burlington Mag.* Mar. 133/2 The little terra-cottas are to the finished, often over-life-size, sculptures what the drawings are to the big pictures of the same period. *Ibid.*, A finely executed sketch for the over life-size bronze statue of *Innocent X*. **1955** S. SPENDER *Making of Poem* 40 Over-life-size people seen through the eyes of his childhood. **1967** E. SHORT *Embroidery & Fabric Collage* iii. 64 A round rug could be worked with one giant over-life-size sunflower as its basis. **1906** *Daily Chron.* 22 Feb. 3/2 Mr. Raleigh is at some pains to show how those over-ocean discoveries and adventures acted on the poetry and imagination of their own times. **1946** R. A. MCFARLAND *Human Factors Air Transport Design* x. 420 In the early stages of overocean flying, an extra station for the navigator has received little criticism. *Ibid.* 410 (*caption*) The flight deck of the B-314 flying boat. The photograph shows a typical layout of stations for multiple flight crews employed in long-range overocean flying. **1957** *Economist* 21 Dec. 1021/2 On the flight deck other qualified Clipper pilots (at least four are on every overocean flight) relieve him. **1955** W. GADDIS *Recognitions* II. viii. 674 Some of the guests were leaving, with over-shoulder looks of last-minute anticipation. **1959** *Daily Tel.* 22 July 7/1 Air attack 'over-shoulder' bombing... There was a spectacular demonstration of 'over-the-shoulder' bombing, when projectiles..climb thousands of feet before plunging into the sea while the delivery aircraft flies off at top speed.

over-abundance. (Later example.)

1971 *Nature* 19 Feb. 548/1 The large overabundances of Li, Be and B are probably due to spallation.

overabundant, *a.* (Later examples.)

1964 *Ann. Reg. 1963* 249 And though wine, wheat, and beef production were well below the previous year's exceptionally high figures, milk, fruit and vegetables were over-abundant. **1971** *Nature* 19 Feb. 548/1 The element F is undetected in several energy ranges, but may be over-abundant in the 50–200 meV/nucleon range.

over-achie·ver. *Psychol.* [f. OVER- 22 + ACHIEVER.] One who achieves more, as a result of environmental or personality factors, than tests based only on intelligence predict; someone who achieves more than is expected.

1953 *Jrnl. Abnormal Psychol.* XLVIII. 533/1 Ambitious students regularly achieve beyond their predicted 'aptitude' by dint of hard work; such 'overachievers' will not usually fail. *Ibid.* 534/2 Overachievers among public school boys who are most frequently and fiercely driven by ambition. **1964** M. ARGYLE *Psychol. & Social Probl.* ii. 26 There is evidence that people who produce a lot of achievement imagery tend to be over-achievers at school work. **1973** *Times* 17 Nov. 12/2 Mr Kirstein turned out to be one of nature's over-achievers. He founded a great American school of ballet. **1975** *Listener* 13 Mar. 331/2, I would never have been an over-achiever if it hadn't been for that unhappy love-affair with my mother. **1976** *Kingston* (Ontario) *Whig-Standard* 10 Jan. 6/3 It is even more boring than all those grabbers about the man of the year, the women of the year, and grannies, do-gooders, athletes, tots, and other over achievers of the year.

So **over-achie·ve** *v. trans.* and *intr.*; **over-achie·vement; overachie·ving** *vbl. sb.* and *ppl. a.*

1953 *Jrnl. Abnormal Psychol.* XLVIII. 533/2 Academic overachieving and underachieving were measured by the difference between a man's Predictive Rank List..and his attained Rank List. *Ibid.* 536/1 Academically acceptable groups who are over- and underachieving. **1961** J. S. BRUNER in *Harvard Educ. Rev.* XXXI. 26 Our tests on such children show them to be lower in analytic ability than those who are not conspicuous in overachievement. **1967** *Economist* 2 Sept. 788/2 Mr Funston has..over-achieved his goal of selling America on the delights of investing. **1968** *N.Y. Times* 2 May 58 This succinct yet passionate ballet overachieves its immediate purpose by choreographically summing up the Dumas story with a series of brilliantly visualized cinematic-style vignettes. **1971** *Time* 15 Feb. 33 When the 'morning after' rolls around, many an overachieving boozer prays for a hang-over cure. **1972** *Accountant* 6 Apr. 444/1 Where budgetary control is employed..the sales variance analyses would explain the reasons for over- or under-achievement. **1973** *Jrnl. Genetic Psychol.* CXXIII. 252 Overachieving high school males tend to rely less on RV. **1976** *Woman's Day* (N.Y.) Nov. 54/2 David continued to 'overachieve' all through high school, college and a distinguished law school.

over against: see OVER *adv.* 7 b.

overage (ōu·vərèdʒ), *sb.*[2] [f. OVER *a.* 3 + -AGE.] A surplus, an excess; an additional amount; *spec.* an actual amount (of goods, money, etc.) greater than that estimated.

1945 MENCKEN *Amer. Lang.* Suppl. I. 366 *Overage* (a bank term: the opposite of a *shortage*). **1949** *Richmond* (Virginia) *Times-Dispatch* 30 Aug. 2/2 The warehousemen agreed to the..sales plan with an 'overage schedule' which is intended to assure Danville the sale each day of 8,800 baskets of tobacco. **1957** *Britannica Bk. of Year* 512/1 *Overage*, costs in excess of estimated or contracted price. **1965** *Economist* 11 Dec. 1235/1 Those export earnings..have not been offset by earlier unforeseen *increases* in export proceeds, inelegantly christened 'overages'. **1968** *Punch* 27 Mar. 447/1 A barman who doesn't show a regular ten per cent 'overage' in favour of his employers will be dismissed because he either gives the customers too much (or too little ice) or he fiddles on his own behalf beyond the customary limits tolerated by the management. **1971** *Daily Colonist* (Victoria, B.C.) 26 Nov. 39/2 His normal welfare allowance would be only $95 a month. However..the man was eligible for overages, which would make his total monthly allowance $120.50. **1973** *Times* 10 May 25/5 A good average for overages is about 6 per cent on monthly sales, minus the minimum rent. In other words, if a store does $50,000 of business in a month and pays a minimum rent of $1,000 a month it has to pay an additional $2,000 in overages. **1975** *Budget* (Sugarcreek, Ohio) 20 Mar. 8/2 Every shortage, or overage, has to be accounted for.

over-aged, *a.* Add: **2.** *Metallurgy.* Subjected to over-ageing.

1953 *Jrnl. Inst. Metals* LXXXII. 265/2 The normal-purity alloys gave ductile fractures when tested in the overaged condition. **1967** A. H. COTTRELL *Introd. Metallurgy* xxi. 405 After plastic working, the over-aged structure gives the stronger alloy.

over-a·geing, *vbl. sb. Metallurgy.* Also **over-aging.** [OVER- 29 a.] Prolonged artificial ageing of metal so that its hardness begins to decrease.

1954 *Gloss. Terms Iron & Steel* (*B.S.I.*) 21 Overageing. **1955** *Jrnl. Inst. Metals* LXXXIII. 529/1 This work provides direct evidence for the supposition that over-ageing during fatigue has a controlling influence on the fatigue strength of precipitation-hardening aluminium alloys. **1970** *Materials & Technol.* III. i. 32 This 'over-aging' effect is due to the increased diffusion giving rise to precipitation and agglomeration into more massive particles.

overall, *sb.* Add: **2. b.** Also, close-fitting trousers worn as a part of army uniform. Also *overall trousers*.

1900 *Dress Regulations Officers of Army* 76 Units may decide to wear either white waistcoat, kamarband instead of waistcoat, white mess dress with kamarband, or white jacket with kamarband and cloth overalls or trousers. **1938** J. CARY *Castle Corner* 152 Some sticky substance on the General's chair had glued him to his seat. His overalls were ruined. **1942** E. WAUGH *Put out More Flags* iii. 177 He looked very elegant and old-fashioned in his blue patrol jacket and tight overall trousers. **1960** A. WAUGH *Foxglove Saga* xii. 223 Martin had donned the Pig's Full Dress Mess Kit..and in its smart red monkey-jacket, narrow blue overalls with a broad red stripe..he really did look most striking. **1967** *National Observer* (U.S.) 3 July 12/3 France entered the war and started uniforming the American troops: Blue regimental coats, waistcoats, breeches or overalls, and black cocked hats.

d. Now often made with a bib and strap top, and sometimes worn by themselves or with a shirt and not over trousers.

1897 *Sears, Roebuck Catal.* 178/2 Painters' white drill overalls, made with apron and shoulder straps. **1926–7** *Army & Navy Stores Catal.* 748/1 Bib and brace overall. Blue and Brown Dungaree. Price, 8/6. **1949** 'G. ORWELL' *Nineteen Eighty-Four* i. 6 A smallish, frail figure, the meagreness of his body merely emphasized by the blue overalls which were the uniform of the Party. **1969** R. T. WILCOX *Dict. Costume* (1970) 253/2 *Overalls*, loose-fitting overtrousers with a front bib held by a strap around the neck.

Hence **o·veralled** *a.*, wearing overalls.

1908 *Smart Set* June 94/1 The familiar spectacle of half-grown boys and overalled and unshaven men. **1916** C. H. STAGG *High Speed* i. 2 He saw an overalled boy jump into the air and crack his heels together. **1928** *Sunday Express* 12 Feb. 9 Plainly there was something seriously amiss with the engine... The overalled mechanic gave an impatient stamp of his foot. **1972** *Guardian* 23 Oct. 11/1 White-overalled, stethoscoped Napoleons. **1974** N. FREELING *Dressing of Diamond* 53 Two overalled characters..on the garbage-collection round.

overall, over-all, *adv.* Restrict †*Obs.* to senses in Dict. and add: **1. c.** Taking all aspects into consideration; generally.

1958 *Spectator* 6 June 753/2 Overall, the profits from trading for the year..were somewhat less than in 1956. **1959** *Wall St. Jrnl.* 30 June 2/3 'Over-all our line is up,' J. D. Bassett, Jr., of Bassett Furniture Industries,..said. **1967** *Autocar* 5 Oct. 50/3 In the up to 1,300 cc race, John Fitzpatrick..was mopping up the opposition to win the up to 1,000 cc class easily and also came fourth overall. **1974** *Cape Times* 1 Aug. 1/6 The United Party was thought to have the edge overall through having a commanding lead in the postal and special votes. **1975** *Physics Bull.* Aug. 365/3 Overall, an excellent and stimulating book. **1976** *Church Times* 16 July 6/5 'Sylvia's Lovers' is the most tantalising of Mrs. Gaskell's books: a failure overall, maybe, but embodying some exceptionally brilliant scenes. **1976** *Oxford Mail* 6 Mar. 1 All these figures are up. So is crime overall. **1976** *Nature* 25 Mar. 376/3 There are some strange omissions too, in what is overall a very comprehensive text. **1976** *Gramophone* Mar. 1525/2 The latter was virtually flat overall.

over-all, *adj. phr.* Add: Now usu. written **overall** (stressed *o·verall*) and treated as a fully developed adj. Also with wider meaning: Considered over the whole range of components, features, or aspects. (Further examples.)

1904 R. M. WALMSLEY *Mod. Pract. Electr.* IV. iv. 984 (*caption*) Overall dimensions of tramcar motors. **1927** R. T. NICHOLSON *Austin Seven Bk.* iii. 6 Overall length, 9 ft. 2 ins. Overall width, 3 ft. 10 ins. **1930** *Daily Express* 6 Sept. 3 The New B P has 'high overall volatility'. All of it evaporates quickly. **1940** *Economist* 20 July 83/1 The recent fall in the over-all rate of net interest..has been due..to higher taxation on all investment income. **1941** *B.B.C. Gloss. Broadcasting Terms* 22 *Over-all merit*, technical rating of a radio channel, expressed in numerals ranging from nought to five to represent the combined effect on reception of signal strength, fading, interference, depth of modulation, and distortion. **1956** A. L. ROWSE *Early Churchills* viii. 139 We must add an appreciation of the importance of sea-power in the conduct of over-all military operations. **1958** *Oxf. Univ. Gaz.* 7 Mar. 679/2 Approximately equal in its overall area to the Weldon Room it differs from the latter in design in that it is not a single room covering the entire space available, but has been subdivided into three. **1958** *Engineering* 11 Apr. 455/2 The estimated overall efficiency of the rectifier is 94·2 per cent. **1967** M. ARGYLE *Psychol. Interpersonal Behaviour* x. 195 In some studies there has been an overall decline in effectiveness, possibly due to unskilled trainers. **1972** P. OLYSLAGER *Handbk. Ford Cortina Mk. III* 3 The overall length is the same as on the Mk II Cortina. **1974** *Country Life* 26 Dec. 2019/1 It is the overall balance of trade that is important. **1976** *Daily Tel.* 30 June 4/6 They had taken a 'significant step forward in cooperation' aimed at seeing that the current overall recovery..does not touch off a new round of inflation.

o·ver-and-u·nder, *a.* [OVER *adv.* 1.] Designating a kind of shotgun in which the barrels are mounted not side by side as is usual but one above the other. Also *ellipt.* as *sb.*

1930 G. BURRARD *In Gunroom* 30, I have ordered a pair of Over and Under guns as I prefer the grip on this type of gun. **1961** C. WILLOCK *Death in Covert* iv. 89 Under his arm he carried a beautiful Churchill over-and-under gun. **1968** *Globe & Mail* (Toronto) 17 Feb. 44 Most of the over-and-unders have only one trigger, to be pulled once for each shot, the fastest two shots in the sport. **1973** *Country Life* 28 July 268/2 Over and under shotguns. **1973** D. LEES *Rape of Quiet Town* vii. 118 He was carrying an over-and-under that must have set him back the thick end of a thousand quid, and, behind that much gun, even plus-fours..couldn't make him look silly. **1976** *Field* 30 Dec. 1272 (Advt.), These superb over-and-unders meet the growing demand from the modern sportsman for a genuine dual purpose gun, combining the advantage of the over-and-under for clay shooting with balance and handling qualities of the game gun.

over-anxiety. (Earlier and later examples.)

1826 J. S. MILL in *Parl. Hist. & Rev. 1826* 658/1 The discredit into which the small notes had fallen through their over-anxiety to get rid of them in a hurry. **1829** —— *Autobiogr.* (1924) 302 They did not consider it very dignified to evince an over-anxiety to stand forth in defence of themselves on slight occasions. **1955** *Times* 13 July 3/4 The issue hung in the balance for the last two hours and both sides made mistakes through over-anxiety. **1976** J. WAINWRIGHT *Who goes Next?* 145 He'd have to watch his step..not to make a hash of things, because of over-anxiety.

overarching, *ppl. a.* Add: (Further examples.) Also *fig.*

1913 [see *lamp-shine* s.v. *LAMP *sb.*[1] 4 a]. **1926** J. S. HUXLEY *Ess. Pop. Sci.* 192 The great biological invention, the amnion, came into existence—an overarching membrane grown by the embryo for its own protection. **1929** V. WOOLF *Granite & Rainbow* (1958) 98 Some over-arching conception, something which we may call 'a reading of life.' **1938** E. BEVAN *Symbolism & Belief* iii. 62 The wholly separate world he sees overhead..gives, as nothing else can give, the vision of overarching immensity. **1972** *Listener* 9 Mar. 301/3 There is a hunger for sociological theory—but there is no over-arching Newtonian scheme..by which the differences can be resolved. **1976** *Brit. Jrnl. Sociol.* XXVII. 348 The 'world economy'..is a *world*-system like the world-empire—but which has no overarching political structure.

overarm, *a.* Add: **1.** (Further examples.) Also as *adv.*

1897 K. S. RANJITSINHJI *Jubilee Bk. Cricket* iii. 85 All bowling—fast, medium, or slow—may be delivered either over-arm, round-arm, or under-arm. *Ibid.* 92 Over-arm bowling..is the kind most generally adopted now. **1907** *Westm. Gaz.* 22 Aug. 20/1 The earliest over-arm bowlers were very fast, and..when Brighton Browne and Mynn were in rapid mood, nine of the ten fieldsmen were placed behind the wicket. **1934** W. J. LEWIS *Lang. Cricket* 30 *Over-arm* or *over-hand bowling*, that in which the hand is raised above the level of the shoulder in delivery, the ball being delivered with a downward swing of the arm. **1970** R. BOWEN *Cricket* viii. 127 The Demon bowler, F. R. Spofforth, who is generally credited with introducing true over-arm bowling to England. **1975** *Oxf. Compan. Sports & Games* 195/1 Virtually all bowling nowadays is over-arm. *Ibid.* 200/2 Over-arm—as distinct from the lower bowling actions (under-arm and round-arm)—had a profound effect upon the character of the game.

2. *Swimming.* Applied to a stroke in which one or both arms are lifted out of the water before being advanced; also of a swimmer, that employs an over-arm stroke.

1887 *Encycl. Brit.* XXII. 770/1 Harry Gardener..used the overhand or overarm stroke. **1893** SINCLAIR & HENRY *Swimming* iii. 79 The old-fashioned over-arm swimmer lay on the water, with his shoulder blades at right angles to the surface. **1908** [see *BACK-STROKE c]. **1912** F. SACHS *Compl. Swimmer* 133 Until the last few years a swimmer who desired to race, first endeavoured to master the over-arm side stroke. **1933** [see *BACK-STROKE c]. **1968** W. ANDERSON *Teaching Physically Handicapped to Swim* iii. 37 In the prone position he can cultivate a side stroke and an alternative over-arm stroke with bi-lateral breathing. **1975** *Oxf. Compan. Sports & Games* 1015/1 From breast-stroke came side stroke, then English overarm, or side overarm stroke.

3. *Lawn tennis.* Of a style of service, in which the racket is swung above the shoulder to hit the ball.

1929 W. E. COLLINSON *Spoken Eng.* 90 Last time some of your overarm serves were unplayable. **1978** J. SYMONS *Blackheath Poisonings* II. 111 I'm better than George, because..his horizontal service..can never really compare with an overarm service in strength.

o·vera:rm, *sb.* [OVER- 1 d.] An overhanging arm, esp. that which extends over the work-table of a milling machine.

[**1903** W. H. VAN DERVOORT *Mod. Machine Shop Tools* xxiii. 321 Suitable ties are now furnished with most makes of milling machines connecting the outer end of the over-hanging arm with the knee.] **1922** H. D. BURGHARDT *Machine Tool Operation* II. viii. 169 Braces for tying the overarm, outer arbor support and knee together. **1964** S. CRAWFORD *Basic Engin. Processes* vi. 148 The overarm is accurately located on the top face of the column, providing support and correct alignment for the cutter arbor. **1976** *Gramophone* Oct. 695/1 The British-made Collaro B610..is a record changer for up to six records, or its overarm can be removed and a stub spindle can be substituted for single record use.

o:ver-arti·culate, *v.* [OVER- 27.] *trans.* To articulate or pronounce too carefully. Hence **o:ver-arti·culated** *ppl. a.*; **o:ver-articula·tion.**

1921 H. E. PALMER *Princ. Lang. Study* 72 The teacher may have considered it his duty to over-articulate his words. **1935** A. L. JAMES *Broadcast Word* iii. 103 Is he pedantic? (i.e. over-articulating sounds); is he clerical? **1935** G. K. ZIPF *Psycho-Biol. of Language* (1936) 217 It [*sc.* the speech of the obsessed speaker] offends the auditor because it is over-articulated in meaning. *Ibid.* 218 The normal stream of speech steers between the Scylla of over-articulation and the Charybdis of under-articulation. **1975** *Time Out* 9 May 13/2 To overarticulate is a mistake because making a film is an attempt to express the unconscious.

o·ver-award. [OVER- 19.] In Australia, used *attrib.* to designate a sum paid by an employer in addition to an agreed minimum wage or salary award.

1950 A. W. FOSTER in Copland & Barback *Conflict of Expansion & Stability* (1957) ix. 677 The Court..has no power to fix maximum rates nor to impose any sanction upon an employer who pays over award rates. **1963** G. PALMER *Guide Austral. Econ. Statistics* v. 101 The wages referred to..are simply the minimum wages as prescribed in specific awards. They do not include overtime, over-award payments, bonuses, etc. **1965** *Economist* 14 Aug. 583/1 Australian arbitration awards are minima only and a major problem of the post war period has been the growth of 'over-award' payments (i.e. wage drift). **1969** *West Australian* 5 July 2/2 A sub-contractor on the S.E.C. power house site..could increase his contract price to cover an award rise of $5 a week. However, his contract would not allow him to pass on over-award rises.

overbank, *v.* Add: **2.** *Aeronaut.* [OVER- 27.] **a.** *trans.* To bank (an aircraft) too much when making a turn; also with the turn as obj.

1915 *Tech. Rep. Advisory Comm. Aeronaut. 1914–15* 307 If a turn be overbanked it will bring into play a lateral component of gravity which produces sideslip. **1919** W. G. ASTON *Aeronaut. made Easy* xvii. 160 If on the turn the machine is overbanked, it will side-slip inwards. **1936** *Discovery* Mar. 72/2 It is essential that the machine is not overbanked on a turn, since there are no ailerons to correct this.

b. *intr.* Of an aircraft: to bank too much. Also said of the pilot.

1929 F. A. SWOFFER *Learning to Fly* iv. 38 (*heading*) Why an aeroplane overbanks. **1932** D. GARNETT *Rabbit in Air* I. 17, I overbanked and didn't use enough rudder. **1952** A. Y. BRAMBLE *Air-plane Flight* xii. 181 During the turn..there is a greater tendency to overbank in a climb than in level flight.

Hence **overba·nking** *vbl. sb.*

1915 *Tech. Rep. Advisory Comm. Aeronaut. 1914–15* 307 (*heading*) Objections to extreme overbanking. **1921** *Sci. Amer.* 15 Oct. 275/3 The side-slip—a lateral movement of a plane caused by overbanking or by underbanking—is measured.

o·verbank, *sb. Aeronaut.* [OVER- 29.] The action of overbanking (*OVERBANK *v.* 2).

1919 A. W. JUDGE *Handbk. Mod. Aeronaut.* xiii. 676 Inward slip..results from an overbank, which causes the machine to turn inwards, and slip down sideways. **1955** M. ROYCE *Studies for Student Pilots* II. 94 The inner mainplane now experiences the greater A[ngle of] A[ttack] and the increment of lift it obtains tends to neutralize the overbank tendency.

over-bark. [f. OVER- 8 c + BARK *sb.*[1] 1.] Used *attrib.* to designate measurements of logs taken before the bark has been removed.

1953 H. L. EDLIN *Forester's Handbk.* xiv. 214 As a rule, logs are measured..while they still have their bark on. This is over-bark measure, and is of course greater than the volume of actual timber. **1967** SCOTT & PALMER *Hiley's Woodland Managem.* (ed. 2) ix. 131 (*caption*) Percentage of bark in the over-bark volume.

overbear, *v.* Add: **4.** *intr.* To produce too much fruit, thereby affecting the quality of it.

1863 *Horticulturalist* XVIII. 295/2 You can now point out every tree that was allowed to overbear. **1872** *Rep. Vermont Bd. Agric.* I. 118 The Bartlett and Louise Bonne de Jersey commence bearing young, and are inclined to over bear. **1901** *U.S. Dept. Agric. Yearbk. 1900* 387 It is a great mistake to allow pear trees to overbear.

overbelief. [OVER- 1 e, 18, 29 c.] **a.** A belief which determines other beliefs. **b.** A belief surviving from the past. **c.** Belief in more than is warranted by the evidence or in what cannot be verified; also, such belief beyond that which is customary among adherents of a particular faith or sect.

1897 W. JAMES *Will to Believe* p. xiii, The most interesting and valuable things about a man are his ideals and over-beliefs. **1900** J. MORLEY *Oliver Cromwell* I. iii. 51 Faith in the literal construction of the word was pushed to an excess..resembling a true superstition or over-belief. **1901** [see OVER- 18]. **1920** 'W. S. PALMER' *Christianity & Christ* 153 We have these 'over-beliefs'; and we even count men poor who are without them. **1930** *Times Lit. Suppl.* 27 Nov. 1011/1 All is well, or will be well, when the new over-beliefs dominate. **1961** M. LASKI *Ecstasy* xxviii. 295 Whatever may be the ultimate source of these beliefs, which I shall call primary overbeliefs, it is generally accepted that their expression must at least partially be a temporal, local and natural matter. **1971** E. CARPENTER *Cantuar* VI. iv. 340 He [*sc.* Archbishop Tait] was sensitive to the contemporary agnosticism of such as Tennyson, and was convinced that this could not be combated by insisting on the 'over-belief' characteristic of the protagonists of the Oxford Movement. **1973** M. PAFFARD *Inglorious Wordsworths* II. xii. 162 Their overbeliefs—their own assumptions or conclusions about the significance (if any) of the experiences they described.

overbend, *v.* **1.** Delete '(Only in *pples.*)' and add further example of sense b.

c **1886** G. M. HOPKINS *Poems* (1967) 97 Her earliest stars, earlstars, stárs principal, overbend us.

o·verbend, *sb.* [OVER- 5 b.] The curved stretch of pipe above the point of inflexion in the S-shaped length of pipeline being lowered on to the sea bed from a barge. Cf. *SAGBEND.

1969 *Preprints 1st Ann. Offshore Technol. Conf.* II. 38/1 As the lay barge proceeds into deep water the articulated stinger curves downward and the suspended pipe span acquires a distinct S-shaped curve. The upper part of this curve, called the over-bend, is supported by the stinger. **1976** *Offshore Platforms & Pipelining* 130/1 The overbend is supported by rollers on the barge and stinger.

overbid, *v.* Add: **2. c.** *trans.* and *intr.* In *Bridge* = *OVERCALL *v.* Also *fig.*

1908 R. F. FOSTER *Auction Bridge* 51 If the hand is overbid, the suit named may be a guide as to the advisability of changing to no-trumps. **1908** [see *BID *sb.* 2]. **1917** E. BERGHOLT *Royal Auction Bridge* 87 It is imperative to overbid with Two Clubs, as a warning—colloquially known as a 'rescue'. **1918** R. F. FOSTER *On Auction* (1919) 169 Overbidding a suit just because there are four honours in it is quite unnecessary. **1923** *Daily Mail* 6 Oct. 6/4 The partner of the under-bidder..cannot make any further bid unless the opponents double or overbid. **1936** A. HUXLEY *Eyeless in Gaza* iv. 35 'Poor child!' his father said to himself; and then, overbidding as it were, 'Poor motherless child!' he added deliberately. **1947** S. HARRIS *Fund. Princ. Contract Bridge* I. i. 17 If one of the adversaries did overbid him, North would almost certainly make a sacrifice bid of four Spades. **1952** I. MACLEOD *Bridge* vii. 82 Be chary of overbidding a suit bid with one No Trump. **1952** PHILLIPS & REESE *Bridge with Mr. Playbetter* xxviii. 118 Mrs. Portly raised to Two Spades; an aggressive player, at the score, might have risked a shut-out bid of Three Spades, but Mrs. Portly knew better than to overbid when playing with Hurry. **1973** *Bridge Mag.* Feb. 123/2 Let us not forget that partner might have overbid slightly in this position. **1974** *Times* 15 June 9/1 They scrape up a bid whenever they can, convinced that by overbidding him they make the declarer work harder for his contract. **1975** *Times* 5 Dec. 14/6 Some European leaders, in overbidding the Brussels game, had seriously misled the public.

Hence **o·verbidder,** one who makes an overbid (see next); **o·verbidding** *vbl. sb.*, the action of one who overbids.

1912 F. IRWIN *Fine Pts. Auction Bridge* 85 There is no fault as common in Auction as overbidding. **1929** M. C. WORK *Compl. Contract Bridge* iii. 15 This artificial system is apt to cause overbidding. **1936** E. CULBERTSON *Contract Bridge Complete* II. xxix. 345 When the overbidder is vulnerable, the extent of the overbid should be less than two tricks. **1964** *Official Encycl. Bridge* 404/1 The overbidder must not be allowed to think that he is playing with an underbidder, or worse will follow. **1974** *Country Life* 17 Jan. 98/1 It is rewarding to study hands..and see what caused a poor result... Here is the tragedy of *Overbidding*.

o·verbid, *sb. Bridge.* [f. the vb.] A bid that is higher than is justified by one's cards; also = *OVERCALL *sb.*

1917 [see *flag-flying* s.v. *FLAG *sb.*[4] 7]. **1947** S. HARRIS *Fund. Princ. Contract Bridge* I. i. 15 If one of the adversaries then makes an overbid of Diamonds or Clubs,..North can then bid his Spades. **1952** I. MACLEOD *Bridge* iv. 46 When I use the term 'bid suit' that means a suit bid as a genuine suit—it does not include cue-bids or over-bids in the opponent's suit. **1967** P. ANDERTON *Play Bridge* v. 40 The number of points lost when you go down on the overbid should represent a considerable saving on what you would lose if the opponents obtained game and rubber. **1969** *Bridge Mag.* Oct. 219/2 North made a slight overbid of three hearts, but ensured the right strain. **1972** *Ibid.* June 386/2 He bid three clubs. Scared of..the overbid of three spades or the underbid of two spades.

o·verbite. *Dentistry.* [OVER- 8(?).] The overlapping of the lower teeth, esp. the incisors, by the corresponding upper teeth; now usu. confined to overlap in a vertical direction. Cf. *OVERJET.

1887 W. G. A. BONWILL in W. F. Litch *Amer. Syst. Dentistry* II. 487 It will be found in 95 per cent. of cases that the upper teeth project over the lower, and that the depth of overbite varies as the depth of the cusps of the bicuspids are deep or shallow. **1924** T. GOODHUGH *Art of Prosthetic Dentistry* x. 246 The least amount of overbite occurs with the wisdom teeth, and the greatest amount at the incisors. **1947** E. HYAMS *William Medium* vii. 137 An immensely long upper lip rising from a prominent overbite and barely covering a number of long, yellow horse-teeth. **1971** R. M. & F. M. KEESING *New Perspectives in Cultural Anthropol.* 55 The 'overbite' in the mouths of most of today's readers is apparently an adaptation in the maturation process to eating soft foods.

over-bitter, *a.* (Examples.)

1600 R. PERSONS in *Publ. Catholic Rec. Soc.* (1906) II. 120 But the rest for that it was longe and over bitter against some particular men, I thought good to leave it out. **1927** H. CRANE *Lett.* (1965) 283 A good dig at certain people, but I think the sarcasm is over-bitter. **1943** *Pope & People* (Catholic Truth Soc.) ix. 171 There is now commonly much dispute, and sometimes over-bitter dispute, on this topic.

overblea·ch, *v.* [OVER- 27.] *trans.* To bleach excessively so that the material bleached deteriorates. Hence **overblea·ched** *ppl. a.*, **overblea·ching** *vbl. sb.*

1921 S. H. HIGGINS *Bleaching* xiii. 111 The 'copper value' standardised by Schwalbe is the most definite measure available for the diagnosis of chemical modification in celluloses, particularly by overbleaching. *Ibid.*, The copper value of strongly overbleached cottons may rise as high as 16. **1946** L. E. WISE *Wood Chem.* vi. 150 In pulp manufacture, overbleaching gives a paper with lowered strength, due to oxycellulose formation. **1950** B. E. HARTSUCH *Introd. Textile Chem.* vii. 188 If the specific viscosity of the cotton is greater than 1, the cloth has been overbleached. **1963** A. J. HALL *Textile Sci.* ii. 25 Overbleached cotton containing oxycellulose resists dyeing with direct cotton dyes, so that if this overbleaching is not even, the bleached yarn or fabric is liable to dye unevenly. **1972** L. PALLADINO *Princ. & Pract. Hairdressing* xi. 153/1 Overbleached hair when wet is almost like chewing gum.

o·ver-blouse. [OVER- 8 c.] A blouse worn over another outer garment.

1921 *Daily Colonist* (Victoria, B.C.) 2 Apr. 20/1 (Advt.), A new style Over-Blouse of Silk Crepe de Chine, in shades of white, flesh, navy and black, effectively embroidered with silk floss in contrasting shades. **1923** *Daily Mail* 13 Feb. 15 In usefulness no similar garment can compete with the over-blouse. **1960** *Times* 20 Jan. 8/4 Suits were generally accompanied by sleeveless silk overblouses. **1963** *New Yorker* 23 Nov. 189 Our double-knit 2-piece dress with slender skirt and scalloped overblouse. **1974** *Country Life* 28 Feb. 456/3 Full overblouse with..a matching full-length skirt.

overblow, *v.*[1] Add: **6.** (Further examples.) Also *intr.* for *refl.*

1938 *Oxf. Compan. Mus.* 228/1 The cornet is horrible when overblown. **1946** MEZZROW & WOLFE *Really Blues* (1957) 363 They have to overblow their instruments, fighting to be heard. **1956** M. STEARNS *Story of Jazz* (1957) xvii. 214 Benny Goodman reacted differently: 'This is the first time..that I've ever heard a tenor sax played the way it should be and not overblown.' **1976** *Early Music* Oct. 511/1 Then Jacob Denner (1732) developed an instrument, which, by means of a key, overblew into the higher register, at an interval of an octave plus a tritone (the modern clarinet overblows at an interval of a twelfth).

7. *Metallurgy.* To subject (a charge) to an excessive length of blast. Cf. OVERBLOWN *ppl. a.*[1] 3.

1869 *Chem. News* 9 Apr. 170/2 If a charge is 'overblown'—that is, if it be subjected to the action of the air for too long a period,..the steel will be found to be defective in proportion to its unskilful treatment. **1932** E. GREGSON *Metall.* ii. 32 Great experience is necessary at this point, since if the metal is 'over-blown' for only 15 seconds, steels containing a large proportion of iron oxide are obtained. **1951** G. R. BASHFORTH *Manuf. Iron & Steel* II. ii. 26 Frequently a heat that has been overblown may be dirty.

overblow (ōu·vəɹblōu), *sb. Metallurgy.* [f. prec. vb.] A period or instance of overblowing.

1879 *Jrnl. Iron & Steel Inst.* 158 One minute overblow and the phosphorus came down to ·75 per cent. **1946** *Ibid.* CLII. 12P This rather long overblow only caused a reduction of about 0·03% in the blown-metal carbon content.

overblow·ing, *vbl. sb.* [f. OVERBLOW *v.*[1] + -ING[1].] **1.** *Metallurgy.* Subjection to an excessive length of blast.

1879 *Jrnl. Iron & Steel Inst.* 121 The removal of phosphorus was assisted by slight over-blowing. **1890** W. M. WILLIAMS *Chem. Iron & Steel Making* xvi. 301 An experimentally overblown sample..in spite of overblowing, produced a high quality of mild steel after the addition of spiegeleisen. **1932** E. GREGORY *Metall.* ii. 35 Over-blowing results in the production of over-oxidised and 'wild' metal. **1949** *Jrnl. Metals* Dec. 27/2 One of the most important features of end-point control is the elimination of overblowing.

2. *Mus.* In the playing of a pipe or wind instrument: production of a harmonic or overtone instead of the fundamental note through extra force of air.

1879 *Organ Voicing & Tuning* 17 *Overblowing, or speaking the octave.* Causes:—(*a*), languid too low; (*b*), excessive wind-hole. **1898** [see OVERBLOW *v.*[1] 6]. **1938** *Oxf. Compan. Mus.* 59/1 Other wood-wind instruments..obtain their second octave by overblowing. **1954** *Grove's Dict. Mus.* (ed. 5) VI. 468/2 Overblowing greatly increases the natural compass of wind instruments. **1977** *Times* 13 Apr. 11/1 Michel Portal, a splendid clarinettist, was playing ill-toned microtones and strident chords (faked by overblowing in a manner now common in new music). **1977** *Early Music* July 351/1 A minute hole in the crook, especially on lower shawms, aids overblowing into the second octave.

overblown, *ppl. a.*[1] Add: **2.** (Later examples.)

1929 R. BRIDGES *Testament of Beauty* III. 55 The empty mind may float lightly in the full moonshine of o'erblown affluence. **1971** *New Yorker* 30 Oct. 25/2 This overblown, frolicsome Western [film].

3. (Examples.)

1879 *Jrnl. Iron & Steel Inst.* 156 The paper admitted that ferrous oxide was employed in the process, and he should wish to ask..how much of it was produced by oxidation in an overblown charge. **1946** *Ibid.* CLII. 9P A later shut-off results in overblown metal. **1951** G. R. BASHFORTH *Manuf. Iron & Steel* II. ii. 21 It is claimed that this application has reduced the likelihood of over-blown heats and considerably improved the control of quality. **1958** A. D. MERRIMAN *Dict. Metallurgy* 222/2 If the blast is allowed to continue after this, oxidation of part of the iron occurs and the charge is then overblown.

overblown, *ppl. a.*[2] (Further examples.)

1844 E. B. BROWNING *Poems* II. 121 From those over-blown faint roses, Not a leaf appeareth shed. **1916** JOYCE *Portrait of Artist* v. 260 The great overblown scarlet flowers of the tattered wallpaper. **1933** *Jrnl. R. Hort. Soc.* LVIII. 232 No useful purpose is served by leaving the flowers until they are in the overblown condition. **1960** P. GALLICO *Mrs. Harris goes to N.Y.* 178 Tired greens, dispirited cabbages and overblown sprouts. **1976** 'J. ROSS' *I know what it's like to Die* v. 39 A creamy, overblown peach blonde.

overboard, *adv.* Add: **2.** Also, excessively, beyond one's means; esp. in phr. *to go overboard*, to behave immoderately; to go too far; to display excessive enthusiasm.

1931 D. RUNYON in *Collier's* 26 Sept. 8/2 We go overboard today. We are washed out. We owe every bookmaker.., and now we are out trying to raise some scratch to pay off. *Ibid.* 9/4 We do not have anything to bet on these races, or any way of betting on them, because we are overboard with every bookmaker we know. **1945** [see *BUST *v.*[2] f]. **1951** J. P. MARQUAND *Melville Goodwin* (1952) viii. 113 Did you ever hear about General Goodwin going overboard over an American girl in Paris..? **1953** 'S. RANSOME' *Hear no Evil* (1954) xv. 140 The man went overboard in a big way morally—he made himself a thief, deserted his family. **1960** *N.Z. Listener* 30 Sept. 11/1, I cannot admire 'abstract' interpretations any more than I can go overboard about sculpture rigged up out of bicycle parts. **1968** *Wall St. Jrnl.* 28 Feb. 16/2 It is easy to go overboard on the new techniques, for all their virtues. **1971** *Jrnl. Gen. Psychol.* Jan. 153 Many psychologists..have gone overboard in hypothetical and speculative associations of this sort. **1978** *Times* 9 Jan. 8/6 Lord Allen..has been convinced of the importance of money..without going overboard in defence of it.

o·verboil, *sb. rare.* [f. the vb.] phr. *on the overboil*: in an overboiling condition, a state of ebullience.

1883 RUSKIN *Let.* 30 Oct. in *Igdrasil* (1890) June 218 And my brains always on the overboil, if I don't mind.

overbook (ōuvəɹbu·k), *v.* [f. OVER- 27 + BOOK *v.*] *trans.* To make more bookings for (a theatre, hotel, aircraft, etc.) than there are places or seats available; to book an excessive number of (customers, passengers, etc.). Also *intr.* Hence **overboo·ked** *ppl. a.*; **overboo·king** *vbl. sb.*

1903 *Daily Chron.* 10 Nov. 9/1 The booking clerks had by some oversight overbooked the theatre. **1964** G. L. COHEN *What's Wrong with Hospitals?* v. 104 A Ministry report on waiting time [at hospital clinics] pointed to the same causes of delay: over-booking and arbitrary appointment intervals. **1967** N. BUXTON *Travel '67* 674 The traveller is told..that the flight has been over-booked and that he must wait for the next plane. *Ibid.* 675 Over-booking may sometimes happen even in the most competently run organization. **1971** *Guardian* 28 July 7/8 An approach to end overbooking at some Spanish holiday hotels will begin in Madrid. **1972** J. POTTER *Going West* 77 All airlines overbooked by fourteen per cent as a matter of policy. **1973** *Daily Tel.* 5 Sept. 6/7 If we did not overbook our flights the jumbos would be flying with a mass of empty seats. *Ibid.*, We overbooked 31 passengers on a flight to Canada today. **1975** *Times* 24 June 19/5 Over-booking—the practice of selling seats twice over, followed by almost every airline in the world. **1976** *Times* 26 Jan. 19/4 The over-booked passenger, despite his protests, finds himself 'bumped' off the flight. **1978** *TV Times* 28 Jan. 65 (Advt.) Want a really carefree holiday this year? Like to stay overnight where and when you like, and not bother about over-booked hotels?

overborne, *ppl. a.* (Later examples.)

1901 G. B. SHAW *Devil's Disciple* I. 19 Uncle Titus, overborne, resumes his seat on the sofa. **1961** in WEBSTER *s.v.*, Art, industry, and commerce, so long crushed and overborne. **1977** *Navy News* Sept. 7/1, I seem to remember reading in Navy News a while ago that the Navy was so overborne with cooks that some could expect to be drafted to non-cooking billets.

overbought, *ppl. a.* [f. OVERBUY *v.*] (See quot. 1957.)

1957 CLARK & GOTTFRIED *University Dict. Business & Finance* 252/2 *Overbought.* 1. The condition of having purchased more than is needed to meet requirements. However, as used, the condition does not necessarily result from excessive buying, but may be the result of shrinking needs or resale volume. 2. In securities trade usage, a condition in which the demand for securities at existing prices has been filled, so that prices tend to drop. **1961** *Spectator* 26 May 774 When a market becomes..overbought, prices are extra sensitive to bad news. **1962** E. GODFREY *Retail Selling & Organization* xx. 199 If a buyer is overbought, he is informed of this fact immediately by stock control office. **1968** *Economist* 16 Mar. 86/2 One of the Americans' hopes would be that, after an initial soaring, the free market price of gold could come down again; partly because hoarders and speculators are now so overbought. **1974** S. MARCUS *Minding Store* iv. 61 They..would never pass a desirable garment, however overbought they might be.

overbound, *v.*[1] For *Obs. rare*[—1] read *rare* and add later example.

1956 T. DRIBERG *Guy Burgess* iii. 45 Churchill seemed a little pleased. 'My eloquence!' he said. 'Ah, yes, that..Herr Beans can rely on in full and indeed..some would say, in overbounding measure.'

over-bow, *v.* Add: **1.** Also *refl.*, to adopt too strong a bow.

1939 P. H. GORDON *New Archery* 72 Beginners should be most careful not to overbow themselves. **1976** *St. Louis* (Missouri) *Globe-Democrat* 17 Sept. 5 B A bow with a draw weight of 40 pounds or more is adequate for the taking of many deer. Many beginners 'overbow' themselves and sacrifice accuracy for unnecessary power.

over-breathe, *v.* Add: **3.** [OVER- 26.] *intr. Physiol.* = *HYPERVENTILATE *v.* a.

1928 F. W. BRODERICK *Dental Med.* xi. 221, I..advise the parent to see that the child is sitting or lying down at the time they are over-breathing. **1961** *Lancet* 5 Aug. 304/2 All divers should therefore be instructed not to overbreathe before diving.

overbreathing *vbl. sb.* (examples in the sense of *OVER-BREATHE *v.* 3).

1920 *Jrnl. Biol. Chem.* XLIII. 9 A continuation..of overbreathing and blowing off of CO_2 results in a compensatory disappearance of alkali from the blood. **1954** W. MAYER-GROSS et al. *Clin. Psychiatry* iv. 131 A prolonged period of over-breathing, causing alkalinity of the blood. **1968** *Brit. Med. Bull.* XXIV. 202/1 On overbreathing there are a few bilateral episodes of delta activity, maximal posteriorly.

over-bridge, *sb.* Add: (Earlier and later examples.) Also, a bridge across a road. Also as *adj.*, travelling or placed across bridges.

1876 [see UNDER-BRIDGE]. **1905** *Daily Chron.* 3 Feb. 6/6 Who..were the people who objected to the over-bridge trams? *Ibid.* 1 June 4/3 The Select Committee on the over-bridge tramways. **1959** *Manch. Guardian* 11 Aug. 6/5 More pedestrian overbridges and underpasses. **1962** *Engineering* 14 Dec. 775 The works will include..four overbridges. **1973** *Inverness Courier* 31 July 6/5 Because of the single track and few overbridges, this would be a relatively cheap route to electrify.

overbuild, *v.* Add: **2.** (Later examples.) Also *fig.*

1909 *Daily Chron.* 17 July 3/2 This is an enjoyable book. ..It has faults of plot—it is over-built—and of character-drawing [etc.]. **1946** *Sun* (Baltimore) 4 June 11/7 France also has some merit in his charge of 'overbuilding' players. **1946** 'R. WEST' *Train of Powder* (1955) 9 The German tendency to overbuild which has done much to get them into..recurring financial troubles. **1977** *Time* 8 Aug. 43/1 As a result of the slowdown in the growth of petroleum consumption and some reckless overbuilding by shipyards in the early 1970s, the tanker business is in the worst depression in memory.

3. (Further examples.) Also *fig.*

1865 W. WHITE *Eastern England* II. xxi. 283 To me..its especial charm is that it is not overbuilt or cockneyfied. **1870** A. T. DE VERE *May Carols* (ed. 2) 12 That, wholly over-built by grace, Nature might vanish, like some isle In great towers lost. **1893** A. JESSOPP *Studies by Recluse* (ed. 2) iii. 97 It was one of the many religious houses that started in a very ambitious way, and early overbuilt themselves. **1939** *Sun* (Baltimore) 11 Apr. 3/3 'It seems a paradox,' he added, 'that in order to revive building operations and furnish employment for a class of workmen it must be carried on in places already overbuilt and becomes the real cause of the idleness of this class of workers.' **1961** *Wall St. Jrnl.* 24 Mar. 1/1 'I don't think the lower interest rates are going to have the slightest effect on housing demand,' says Henry Bubb, president of Capitol Federal Savings & Loan Association of Topeka, 'We're just over-built.' **1977** *Sat. Rev.* (U.S.) 17 Sept. 54/2 Is little New Orleans overbuilt? Perhaps not. By early summer the Hilton had already booked one million room-nights.

over-bu·ll, *v.* [f. OVER- 27 + BULL *v.*[1] 2.] *trans.* To raise the price of (stocks, etc.) excessively. Hence **over-bu·lled** *ppl. a.*; **over-bu·llishness.**

1905 *Daily Report* 14 Oct. 2/4 Readings were heavy, having apparently been over-bulled. **1938** *New Statesman* 16 July 132/1 Happily Wall Street has gone through a corrective period and over-bullishness has given way to more restrained optimism. **1965** *Economist* 23 Oct. 436/1 The market ignores the sad fate of so many American wonder electronic stocks: over-bulled, then crashing at the first set-back.

o·ver-bump, *v.* [f. OVER- 14 + BUMP *v.*[1] 3.] In bumping-races, to catch and bump (a boat ahead of a pair of other boats that have withdrawn after a bump), thus going up three places. So **o·ver-bump** *sb.*

1905 *Daily Chron.* 2 Mar. 7/3 One boat (Corpus) over-bumped Selwyn and ascended three places. **1920** 'Two OF 'EM' *Guide Cambr. Univ. Life* 10 If three or more boats bump simultaneously the rear boat goes to the front of those boats, and is said to have made an 'over-bump'. **1930** *Magdalene Boat Club 1828–1928* v. 19 The year 1868 is notable as providing the only occasion upon which the first boat ever made an overbump. **1963** *Times* 31 May 5/3 Merton II also attempted a big task in going for an overbump on Queen's II and it was only by one foot that they failed to make it.

overburden, -burthen, *sb.* Add: (The spelling *-burthen* is no longer current.) **2. b.** The material lying over any particular point underground, esp. over a tunnel or pipeline; also, the pressure due to the weight of this material.

1948 TERZAGHI & PECK *Soil Mech. in Engin. Pract.* ii. 68 With respect to the present overburden, the clay on the right-hand side is a precompressed soft clay, and that on the left-hand side is a normally loaded soft clay. **1968** G. N. SMITH *Elem. Soil. Mech.* iv. 88 The overburden pressure at a point in a soil mass is simply the weight of the material above it. The effective overburden is the pressure from this material less the pore water pressure. **1970** *Daily Tel.* 27 June 2/5 The pipeline was laid in December 1967..but the unexpectedly heavy scouring action of the bottom waves in the shallow North Sea soon removed the two foot overburden.

c. Loose, unconsolidated material lying above bedrock.

1955 *Proc. Colorado Sci. Soc.* XVI. 102/1 *Overburden*,..a term used by geologists and engineers in several different senses. By some it is used to designate material of any nature, consolidated or unconsolidated, that overlies a deposit of useful materials, ores, or coal, especially those deposits that are mined from the surface by open cuts. As employed by others *overburden* designates only loose soil, sand, gravel, etc., that lies above the bedrock. The term should not be used without specific definition. **1969** *Civil Engineering* (U.S.) June 43/2 A more detailed soil investigation of the upper 50 ft or so of the overburden..showed that the allowable bearing capacity was only about 200 to 300 psf, much too low for the proposed bridge foundations.

overbuy, *v.* Add: **1. b.** To buy goods at a (wholesale) price beyond the means of (a competitor).

1886 HARDY *Mayor Casterbr.* II. iii. 36 We'll under-sell him, and over-buy him, and so snuff him out.

2. (Later examples.) Also, of a wholesaler or retailer: to buy goods, materials, etc., in excess of those needed; to accumulate surplus stock. See also *OVERBOUGHT *ppl. a.*

1938 *Sun* (Baltimore) 26 Feb. 18/8 The high level of delinquencies in Maryland results from several conditions—business recession, unemployment, 'overbuying'. **1950** *Times* 11 Mar. 9/7 Last year the Government overbought imported frozen fish and now fresh fish was being sold at a very low price. **1966** J. M. SHEWAN in *Proc. Internat. Symp. Food Irrad.* (Internat. Atomic Energy Agency) 496 Merchants, both wholesalers and retailers, frequently 'over-buy', particularly when supplies are plentiful and prices cheap.

overcall (ōuvəɹkǭ·l), *v. Bridge.* [OVER- 22, 26.] **a.** *trans.* and *intr.* To make a bid higher than (a previous bid or another player).

1908 R. F. FOSTER *Auction Bridge* 54 Neither should he bid a number of tricks which is more than necessary to overcall the previous declaration. **1916** F. IRWIN *Compl. Auction Player* 66 There is also a perfectly good suit with which to over-call. **1917** E. BERGHOLT *Royal Auction Bridge* 85 Here Y...is justified in over-calling with One Spade. **1918** R. F. FOSTER *On Auction* (1919) 183 The partner may be called upon to assist..when the second hand overcalls the dealer. **1929** M. C. WORK *Compl. Contract Bridge* ii. 9 He may overcall a No Trump with a suit-bid, or overcall a suit-bid with another suit or No Trump. **1974** *Country Life* 17 Jan. 98 To overcall One Club with One Spade on a modest hand is worthwhile.

b. To bid more on (one's hand) than it is worth. Also *fig.*

1927 A. H. POLLEN in 'Neon' *Gt. Delusion* p. xvi, The bright young conjurers of Kingsway have been overcalling their hands. **1930** *Time & Tide* 11 Apr. 463 His partner had been overcalling. **1934** L. H. DAWSON *Hoyle's Games Modernized* 31 Sometimes a forced bid is also a forcing bid, in that the intention is to make the opponents over call.

overcall (ōu·vəɹkǭl), *sb. Bridge.* [f. the vb.] A bid which is higher than a previous bid (see also quot. 1959). Cf. *OVERBID *sb.*

1916 F. IRWIN *Compl. Auction Player* 65 To bid against your partner, when no one else has bid, is to use the over-call. **1917** E. BERGHOLT *Royal Auction Bridge* 85 Third hand should not carry on the contest further, unless he has some additional reason which his first overcall was not sufficient to proclaim. **1959** REESE & DORMER *Bridge Player's Dict.* 159 An overcall is a bid made after an opponent has opened the bidding. **1973** *Country Life* 10 May 1331/1, I had to take West's One Spade overcall into account. He was likely to have both missing Kings.

over-care. (Further examples.)

1693 DRYDEN tr. *Persius' Satires* II. 81 The very over-care, And nauseous pomp, wou'd hinder half the Pray'r. **1833** C. LAMB *Last Essays* 103 Sunday itself—that unfortunate failure of a holyday..with my sense of its fugitiveness, and over-care to get the greatest quantity of pleasure out of it—is melted down into a week day. **1937** V. MCNABB *God's Way of Mercy* xii. 103 There is no over-care about tomorrow; just a love of God.

overcarefully, *adv.* (Examples.)

1865 DICKENS *Mut. Fr.* I. II. xiv. 296 'The sister,' said Bradley, separating his words over-carefully, and speaking as if he were repeating them from a book, 'suffers under no reproach.' **1870** —— *Edwin Drood* ii. 9 Life, for *you*, is a plum with the natural bloom on; it hasn't been over-carefully wiped off for *you*. **1939** JOYCE *Finnegans Wake* 122 Three *basia* or shorter and smaller *oscula* have been overcarefully scraped away.

overcarry, *v.* Add: **1. b.** (Later example.)

1972 M. MEAD *Blackberry Winter* xi. 148 The National Research Council had insisted on mailing my checks to me, and the next boat overcarried the mail. This meant that for six weeks I had no money.

Hence **overcarried** *ppl. a.*

1897 [see sense 1 b in Dict.]. **1903** KIPLING *Five Nations* 25 The galloping breakers stride, And their over-carried spray is a sea—a sea on the landward side.

overcast, *sb.* Add: **2.** (Further examples.) *spec.* in *Aeronautics*: cloud-cover which restricts visibility and necessitates reliance on instruments for navigation.

1938 [see *contact flying* s.v. *CONTACT sb. 6]. **1942** *Tee Emm* (Air Ministry) II. 69 In the past you scraped above the undergrowth, now, through the overcast. **1946** *Happy Landings* (Air Ministry) July 5/1 The aircraft..was seen emerging from the overcast. **1946** R. A. McFarland *Human Factors Air Transport Design* ix. 396 The flight was progressing normally at an altitude of 10,000 ft until it entered a rapidly forming thunderstorm. Shortly thereafter, parts of the plane began to fall from the overcast. **1967** S. Blanc *Rose Window* (1968) xix. 183 It was only mid-morning and a high overcast that obscured the sun had not yet burned away. **1972** B. F. Conners *Don't embarrass Bureau* (1973) i. 6 He felt the sun starting to burn through the overcast as he climbed the ladder.

5. (Earlier examples.)
c**1840** Lady Wilton *Art of Needlework* xx. 317 There is back stitch—overcast—and seam stitch. **1867** C. Austen *My Aunt Jane Austen* (1952) 7 She was a great adept at overcast and satin stitch—the peculiar delight of that day.

overcast, *v.* Add: **7.** Also in *Bookbinding* (see quot. 1956).
1880 J. W. Zaehnsdorf *Art of Bookbinding* iii. 13 Each section is then overcast or oversewn along its whole length. **1901** [see *OVERCAST *ppl. a.* 3]. **1951** L. Town *Bookbinding by Hand* xx. 244 At this point the book must be marked up for sewing, as this cannot be done after the sections have been overcast. **1956** *Bookman's Conc. Dict.* 209/1 *Overcast*, to sew leaves, especially single leaves, in bookbinding, with a long hem-stitch style of sewing; also known as Whip Stitch. **1963** B. C. Middleton *Hist. Eng. Craft Bookbinding Technique* iii. 25 The usual method..involves overcasting separately a series of groups of leaves.

overcast, *ppl. a.* Add: **3.** Also in *Bookbinding* (see *OVERCAST *v.* 7) and in *Lace-making*.
1865 F. B. Palliser *Hist. Lace* xiii. 181 It is only 'bride ordinaire';..very different from the clear 'over-cast' hexagon of the last century. **1901** D. Cockerell *Bookbinding* v. 81 To 'overcast' the first and last sections..fails in the object aimed at by merely transferring the strain to the back of the overcast sections.

overcasting, *vbl. sb.* Add: **4.** Also in *Bookbinding*: see *OVERCAST *v.* 7.
1885 W. J. E. Crane *Bookbinding for Amateurs* vi. 54 There is another way of overcasting more used in London. **1901** D. Cockerell *Bookbinding* 316 Overcasting, over-sewing the back edges of single leaves or weak sections. **1931** A. Esdaile *Student's Man. Bibliogr.* vi. 182 Overcasting, a more respectable form of stabbing, by which the folds are sewn together beyond, and enclosing, the ordinary sewing.

overcasting, *ppl. a.* (Further examples.)
1901 D. Cockerell *Bookbinding* ii. 51 The custom with binders is to overcast the backs of the leaves in sections, and to sew through the overcasting thread. **1964** *McCall's Sewing* 134/2 A row of machine-stitching close to the raw edge serves as a guide for keeping overcasting stitches even.

o:ver-centraliza·tion. [Over- 29 b.] Excessive centralization of administrative functions, leading to inefficiency. So **overce·ntralized** *a.*
1882 E. W. Hamilton *Diary* 20 June (1972) I. 290 He spoke strongly of the dreadful 'administrative chaos' and over-centralisation which existed in Dublin Castle. **1896** L. T. Hobhouse *Theory of Knowl.* 276 The over-centralisation of the imperial government. **1944** J. S. Huxley *On Living in Revolution* xi. 118 An overcentralized administration is always characterized by the fact that its field officers tend to become messengers and office boys. **1949** I. Deutscher *Stalin* vii. 241 Stalin obviously erred in the direction of over-centralization. **1962** *Punch* 7 Feb. 261/3 Jellicoe..had to fight with a machine encrusted with over-centralisation. **1971** *Time* 29 Mar. 36/3 Now the country [*sc.* Poland] is tense but quiet, as [Edward] Gierek attempts to consolidate his position and cope with an appalling economic mess caused by years of overconcentration on heavy industry, overcentralization and postponement of reforms. **1977** *Times* 7 Nov. 14/6 The [Soviet] economy is still an overcentralized shambles.

o·vercheck, *sb.*[2] [Over- 8.] On cloth: a check pattern superimposed on a pattern of smaller check.
1906 *Daily Chron.* 25 Apr. 8/5 There will be a few very neat stripes,..while the finely-traced overchecks will be much in evidence. **1923** [see *GLENURQUHART, GLEN URQUHART]. **1959** [see *HOUNDSTOOTH]. **1967** N. Freeling *Strike Out* 27 The suit was..mohair in a large complicated overcheck of fuchsia, havana and off-white. **1974** *Daily Tel.* 6 Mar. 15 Tweedy, sporty men should be in their element as checks..get bigger, plaids get bolder and overchecks more aggressive.
So **overche·cked** *ppl. a.*
1925 *Eaton's News Weekly* 2 May 10 A fine brown and blue tweed, overchecked in rust. **1960** *Harper's Bazaar* Aug. 56/2 Predominantly bright red, it [*sc.* a coat] is overchecked in yellow and black. **1969** *Times* 30 Sept. 14/5 (Advt.), Stunning dresscoat in bold overchecked..wool. **1970** *Vogue* May 21/2 A willow green tweed jacket..overchecked in brown and navy.

over-cho·sen, *a.* *Sociology.* [Over- 28 c.] Denoting those who, in a sociometric group study, are chosen by others well above the average number of times; the 'stars' of a sociometric group; also *ellipt.* as *sb.* Cf. *UNDER-CHOSEN *a.*
1943 H. H. Jennings *Leadership & Isolation* iv. 66 The test of whether or not this is true might be made by examining the choice behavior of the 'over-chosen' as compared with the 'under-chosen'. *Ibid.* 68 In Group A_{11} in which 16 percent are over-chosen..the number by whom they are chosen extends from 16 to 24 individuals, while the under-chosen who comprise 17 percent..are chosen by from zero to 3 individuals. **1956** J. Klein *Study of Groups* 91 The fact that certain people are over-chosen or under-chosen implies that members share certain standards which the over-chosen exemplify and the under-chosen fall short of. **1958** W. J. H. Sprott *Human Groups* ix. 150 On work and living—with choice..there was considerable overlap among what she calls the 'over-chosens'.

overclimb, *v.* (Later example.)
1882 'Ouida' *Bimbi* v. 149 A loggia..all overclimbed by hardy rose-trees.

overclo·sure. *Dentistry.* [Over- 29.] A condition in which the lower jaw is raised more than normal in relation to the upper jaw when put into the rest position.
1934 *Ann. Otol., Rhinol. & Laryngol.* XLIII. 7 (*caption*) Broken lines spheno-mandibular ligaments and pterygoid muscles further relaxed by marked overclosure of edentulous mouth. **1948** *Brit. Dental Jrnl.* LXXXV. 225/1 Men with severe recently acquired malocclusion (extreme voluntary overclosure after removal of their full dentures). **1975** H. Thomson *Occlusion* iv. 51 A vertical extension of mandibular displacement where the IOD [*sc.* interocclusal distance] is in excess of 4 mm. is referred to as mandibular overclosure.

overcoat. Add: (Earlier and later examples.) Also *attrib.* Also *fig.* *spec.* with ref. to means of disposing of a body; *wooden overcoat*, a coffin. *slang.*
1802 *Monthly Mag.* XIV. 325/1 [He] presented to the king of Spain a very light and thin over-coat, which rain could not penetrate. **1807** *Salmagundi* 31 Dec. 395 Observing it to be dressed in a man's hat, a cloth overcoat, and spatterdashes, I framed my apology accordingly. **1857** P. Cartwright *Autobiogr.* 201 My pistol..was in my overcoat pocket. **1860** Mayne Reid *Odd People* 136 Not contented with being tatooed, these also *paint* their bodies, by way of 'overcoat'. c**1864** E. Dickinson *Poems* (1955) II. 705 The Spirit turns away Just laying off for evidence An Overcoat of Clay. **1903** Farmer & Henley *Slang* VII. 362/2 *Wooden-overcoat* (or *-surtout*),..a coffin. **1909** *Dialect Notes* III. 389 *Wooden-overcoat*, coffin. **1940** H. W. Thompson *Body, Boots & Britches* xix. 486 Your wooden overcoat won't have any pockets. **1942** [see *MEASURE *v.* 2 c]. **1958** M. Kelly *Christmas Egg* i. 9 He put his hands in his overcoat pockets. **1969** *Times* 15 Nov. p. vi/7 Felco secateurs are exceedingly popular... Their model..has one handle with a plastic 'overcoat' which rotates as you squeeze the handles together. **1971** *Guardian* 12 Aug. 11/2 The paratroops were edgy and the one who let me through the barricade reckoned I would come out in a wooden overcoat. **1972** K. Bonfiglioli *Don't Point that Thing at Me* xvii. 147 He..had to get us to ground of his own choosing before he could fit us for cement overcoats. **1974** *Times* 14 Dec. 1/7 The Foreign Office..asked Miami police searching for Mr John Stonehouse, the missing British MP..for a special report of..a Mafia style 'concrete overcoat' known to have contained a body. **1976** *Publishers' Weekly* 19 Apr. 80/1 The neighborhood gossips about a most unlikely overcoat flasher.

overcoating. (Further examples.)
c**1900** in *American Mail Order Fashions* (1961) 25 The fabric is a regular overcoating Vicuna... A firm but soft finished overcoating in one of the new shades. **1948** *Hansard Commons* 16 Mar. Written Answers 224 We have stipulated that half of the Italian supplies should be in men's suitings and overcoatings. **1967** *Times Rev. Industry* Apr. 53/3 J. and J. Crombie is to make 96,000 yards of overcoating, worth more than £250,000, for the Soviet Union.

o·vercoatless, *a.* [f. Overcoat + -less.] Not having, or not wearing, an overcoat.
1908 L. A. Harker *His First Leave* vi. 60 An old gentleman, overcoatless, umbrellaless. **1929** J. B. Priestley *Good Companions* ii. v. 389 Her hatless and overcoatless companion. **1936** E. Sitwell *Victoria of Eng.* xii. 149 So, overcoatless, since his threadbare overcoat was in pawn, he went out into the rainy streets.

overcome, *v.* Add: **2. c.** Phr. *we shall overcome*, used as a slogan by minority groups, with allusion to the text of a Negro Gospel song.
[**1901** C. A. Tindley in Miles & Clifton *New Songs of Gospel* No. 27 (*title*) I'll overcome some day.] **1948** *People's Songs* Sept. 8 We will overcome, We will overcome, We will overcome some day. Oh Down in my heart, I do believe, We'll overcome some day. **1961** *Jet* 14 Dec. 53 That the Freedom Riders left their imprint on the prison was evident from the songs and slogans they scribbled on the walls. A favorite was the crusaders' theme *We Shall Overcome*. **1963** *N.Y. Times* 23 July 21/1 The theme song of the integration movement, 'We Shall Overcome', has had its own history of integration, passing from Negro singers to white and back again to Negro. **1968** 'Ebon' *Revolution* 15 'We shall overcome' And black Truth bombs Explode In the back, Alleys, Of Newark's Asshole. **1973** *Black World* Sept. 8/1 The Christian strains of 'We shall Overcome' gave way to the more Garrisonian 'We Shall be heard!' as the Sixties progressed.
7. (Further examples.)
1835 A. B. Longstreet *Georgia Scenes* 7, I had overcome about half the space which separated it from me. **1875** W. Morris tr. *Virgil's Aeneids* xii. 907 And e'en the hero-gathered stone..O'ercame not all the space betwixt.

o:ver-co·mpensate, *v.* Also as one word. [Over- 24.] *trans.* and *intr.* To compensate excessively for (something); spec. in *Psychol.*, to exhibit over-compensation (see next). Hence **o:ver-co·mpensated** *ppl. a.*; **o:ver-compensa·tory** *a.*
1768–74 [see Over- 24 a]. **1917** *Psychol. Bull.* XIV. 207 By psychological investigation and analysis, one may disclose the psychic phase of these compensatory and over-compensatory processes. **1934** H. C. Warren *Dict. Psychol.* 190/1 *Overcompensate*, to make more than the necessary amount of allowance or adjustment. **1937** 'M. Innes' *Hamlet, Revenge!* ii. viii. 194 One builds on the over-compensated Oedipus—Dad Advises Sonny-boy. **1947** *Partisan Rev.* XIV. 476 As a result he..over-compensates by consciously immersing himself in parochial attitudes shared by the folk. **1949** M. Mead *Male & Female* iv. 88 Their sons again grow up similarly focussed on women, similarly in need of over-compensatory ceremonial to rescue them. **1958** 'E. Crispin' *Best SF Three* 11 As to human intellect, science fiction has emphasised its inadequacy over and over again. The *genre* 'overcompensates', no doubt, in showing the man animal so often defeated, or all but defeated, by the Other Thing, and this over-compensation is the origin of the accusations of pessimism that are so often levelled against it. **1962** Simpson & Richards *Physical Princ. Junction Transistors* xii. 280 In case (b) (perfect compensation), the dashed curve coincides with the extrapolated linear sections of the transfer characteristics, and in the overcompensated case (c) it lies between the two linear sections. **1968** L. Durrell *Tunc* i. 17 We sparred gracefully in the fashion of well-educated Englishmen overcompensating. **1973** 'D. Halliday' *Dolly & Starry Bird* xviii. 277 Innes..will shortly give you a brandy because he is overcompensating. He thought you were guilty and is now ashamed. **1973** C. Mullard *Black Brit.* ii. iv. 44 In others [*sc.* classrooms] the teachers overcompensated and patronized their new pupils.

o:ver-compensa·tion. Also as one word. [Over- 29 b. Cf. G. *überkompensation*.] A term used in psychological analysis by A. Adler to denote the exaggerated striving for power, etc., that can activate someone suffering from a severe sense of inferiority; an exaggerated response of making allowance or amends for something; more than equitable compensation.
1917 *Psychol. Bull.* XIV. 207 Adler thought he had discovered 'a remarkable relationship between somatic inferiority and somatic psychic over-compensation'. **1917** Glueck & Lind tr. *Adler's Neurotic Constitution* (1921) i. i. 4 This fetal character..furnishes the increased possibility for compensation and over-compensation. **1932** *Brit. Jrnl. Psychol.* Apr. 347 Occasionally, over-compensation for this error [in colour memory] seems to be taking place. **1941** J. S. Huxley *Uniqueness of Man* xi. 228 Bias of this type has the additional danger that those who make an effort to discount it may readily swing into over-compensation—a bias of opposite sign. **1955** T. H. Pear *Eng. Social Differences* vi. 138 Theories of compensation and over-compensation for inferiority-complexes. **1969** V. H. Vroom in Lindzey & Aronson *Handbk. Social Psychol.* (ed. 2) V. 220 The evidence reported concerning overcompensation and undercompensation pertains to its short-term effects on performance. **1977** D. Clark *Gimmel Flask* vi. 106 Her over-compensation in over-spending may be due to her disappointment.

o:verco·mpound, *a.* *Electr. Engin.* [f. Over- 24 b + Compound *a.*] = next.
1931 G. C. Blalock *Princ. Electr. Engin.* ix. 95 The overcompound generator is adapted to more remote loads, since its rising characteristic will compensate for linedrop. **1972** I. L. Kosow *Electr. Machinery & Transformers* iii. 99 Most commercial compound dc dynamos..are normally supplied by the manufacturer as overcompound machines.

o:vercompou·nded, *ppl. a.* *Electr. Engin.* [f. Over- 24 b + Compound *v.* + -ed[1].] Of a dynamo: compounded in such a way that the voltage increases with load.
1892 S. P. Thompson *Dynamo-Electr. Machinery* (ed. 4) xi. 295 For such work as supplying current to an electric tramway, an over-compounded dynamo with laminated field-magnets is the best generator. **1927** A. E. Clayton *Performance & Design D.C. Machines* viii. 181 With an over-compounded machine the value of the current in the shunt winding actually increases with increasing load. **1971** L. T. Agger *Introd. Electr.* xvi. 299 If..the load has to be supplied over a considerable distance, the rising characteristic of the over-compounded generator..may be desirable.
So **overco·mpounding** *vbl. sb.*; also (as a back-formation) **overco·mpound** *v. trans.*
1892 S. P. Thompson *Dynamo-Electr. Machinery* (ed. 4) xi. 295 The same process suits for over-compounding, the excitation at full load being raised until the volts at terminals rise to the higher number of volts that will allow for the drop in the leads. **1914** A. Gray *Princ. & Pract. Electr. Engin.* xiii. 75 For railway service the generators are overcompounded so as to maintain the trolley voltage at some distance from the power house. **1937** H. E. Stafford *Troubles of Electr. Equipment* iii. 46 To over-compound a generator not provided with diverters, the only permanent way is to increase the strength of the series and interpole fields. **1966** J. Meisel *Princ. Electromechanical-Energy Conversion* ix. 394 Making $G_{af}\omega_r$ greater than $Ra + Rs$, or overcompounding the machine, gives an increasing output voltage with increasing load current. *Ibid.*, By a slight overcompounding the droop

can be taken into account and the output voltage at no load..can be made equal to the output voltage at full-load current.

o:verconso·lidated, *ppl. a.* [Over- 24 b.] Of soil or clay: consolidated to a greater degree than could have been produced by the present pressure of overburden.
1936 *Proc. Internat. Conf. Soil Mech. & Foundation Engin.* III. 52 Tests with strongly overconsolidated soils. **1957** Bishop & Henkel *Measurement of Soil Properties in Triaxial Test* 115 A series of tests on an over-consolidated clay. **1969** C. R. Scott *Introd. Soil Mech. & Foundations* vi. 139 Dense sands and over-consolidated clays increase in volume when sheared, if they are free to do so.
So **o:verconsolida·tion.**
Quot. 1960 represents a simple non-technical use.
1936 *Proc. Internat. Conf. Soil Mech. & Foundation Engin.* III. 51 In order to investigate the influence of the voids ratio on the shearing resistance, the test specimens were reconsolidated in three different ways, so that they at the start of the shearing test would be in a state of either natural consolidation, simple overconsolidation, or cyclic overconsolidation. **1960** *Farmer & Stockbreeder* 1 Mar. 94/3 There is no danger of over-consolidation with tractors in the silo. **1969** Lambe & Whitman *Soil Mech.* xxix. 452/2 The natural water content considered in relation to the liquid and plastic limits gives some idea of the degree of overconsolidation.

over-correct, *v.* Add: **1.** Also, to correct (a lens) so that there is an aberration opposite to that of the uncorrected lens. (Earlier and later examples.)
1829 H. Coddington *Treat. Reflexion & Refraction of Light* vi. 253 If a concave flint glass lens..be too powerful for a convex crown glass one when placed in contact, so as to over-correct the colour, as it is said, the achromatism may be made more perfect by separating them a little. **1885** T. Longmore *Illustr. Optical Manual* (ed. 3) ii. 43 It should be ascertained if the M[yopia] has not been over-corrected by the lenses supplied, and the eyes brought into a condition of H[ypermetropia]. **1975** M. Ruben *Contact Lens Pract.* ii. 19/2 In hypermetropia..one eye (the non-dominant) can..be overcorrected by as much as 1·5 D so as to give the best binocular acuity for distance.
2. *trans.* and *intr.* To make an excessive correction to or in (something); to correct (someone) too frequently.
1956 *Publ. Amer. Dial. Soc.* xxvi. 45 But he may still have trouble remembering which sound belongs in which words, and in his anxiety he will overcorrect and hear *sung* for *sun* or *some*. **1966** J. Derrick *Teaching Eng. to Immigrants* iii. 117 In this, as in all questions of pronunciation teaching, the greatest care and tact is necessary, so that the nervous pupil, from being over-corrected, does not become more nervous and inhibited. **1966** *Publ. Amer. Dial. Soc.* 1964 xlii. 5 To overcorrect a slide, causing rear of car to waggle like a fishtail. **1966** *Rep. Com. Inquiry Univ. Oxf.* II. 325 This calculation almost certainly over-corrects for any error. **1967** J. Rathbone *Diamonds Bid* iv. 39 The car..slewed towards the right. I over-corrected and the on-side fender hit a low wall. **1974** L. Deighton *Spy Story* xx. 215 The machine tottered into the air, swinging as the nervous pilot over-corrected. **1976** *Columbus* (Montana) *News* 10 June 4/3 The driver overcorrected, causing the vehicle to go into a skid, overturning in the median after crossing the road.

over-corre·ction. [Over- 29 b.] An excessive correction; a correction which results in error in the opposite direction; *spec.* (*a*) *Optics*, correction of a lens to such a degree as to produce the opposite aberration; (*b*) *Linguistics* = *HYPERCORRECTION.
1885 T. Longmore *Illustr. Optical Manual* (ed. 3) ii. 43 If there be no over-correction, but the concave lenses ordered are found to be only equivalent to the excess of refraction, or, in other words, to the degree of M[yopia] which they have been calculated to neutralise, [etc.]. **1947** E. H. Sturtevant *Introd. Linguistic Sci.* viii. 80 Sometimes..he has caught himself creating such forms as [dju-, tju-] for *do* and *two*. Such 'over-corrections' have been observed very often in many languages. **1957** F. & R. Lockridge *Practise to Deceive* (1959) x. 137 That was prejudice, to be corrected. But, then, there was the danger of over-correction. **1961** F. G. Cassidy *Jamaica Talk* iii. 46 All too easily this attempt leads to what is called 'over-correction', for he puts sounds into places where they do not belong in Standard. **1966** A. Battersby *Math. in Managem.* vii. 185 Fifteen weeks after the sales increased, the rate of production at the factory had surged up to an increase of 45%; forty weeks later it had swung back in an over-correction. **1975** M. Ruben *Contact Lens Pract.* ii. 19/1 In hypermetropia associated with convergence,..overcorrection with a contact lens can be just as effective as spectacle lenses.

over-cre·dulous, *a.* (Later example.)
1936 L. MacNeice tr. *Aeschylus' Agamemnon* 29 The over-credulous passion of woman expands In swift conflagration.

over-critical, *a.* (Earlier and later examples.)
1851 Mrs. Gaskell *Let.* 1 Sept. (1966) 161 She..is inclined to be over-critical & fastidious with everybody & everything. **1946** *Nature* 28 Sept. 457/1 The points of inflexion of various thermodynamical functions in the overcritical region. **1976** J. Wilson *Let's Pretend* v. 57 Lately she'd been over-critical of Margaret, finding her exasperating instead of endearing.

overcroft (ōᵘ·vəɹkrǫft). [f. Over- 8 c + Croft *sb.*[2]] In early church architecture, a series of small rooms below the roof (see quot. 1925).
1925 P. Power *Early Christian Ireland* ii. 22 On the inside the little building is roofed, or ceiled, by a barrel vault, between which and the outer stone roof there is an overcroft, or series of very small rooms, lighted by opes in the gable. **1964** B. Whelpton *Unknown Ireland* xi. 149 The thirteenth-century Church of St. Doulough with its steep stone roof and overcroft.

overcrop, *v.* Add: **II. 3.** (Later examples.) Also *transf.*
1946 *Nature* 2 Nov. 606/2 Sometimes his [*sc.* man's] responsibility is brought home to him by physical disaster,..as in the effects of deforestation, over-cropping or overgrazing. **1960** *Farmer & Stockbreeder* 9 Feb. 97/1 Many instances of 'soil sickness'..in the past would have been put down to thin or hungry soil, overcropping or bad drainage. **1974** *Times* 25 Apr. 12/6 He worries..about the over-cropping of the sea.

overcrowded, *ppl. a.* Add: (Earlier and later examples.) Hence **overcrow·dedness.**
1836 J. S. Mill in *London & Westm. Rev.* Apr. 18 In every overcrowded department there will arise a tendency among individuals..to unite their labours or their capitals. **1902** *Daily Chron.* 29 Sept. 3/3 It is here that the over-crowdedness of his pages comes most into evidence. **1911** G. B. Shaw *Doctor's Dilemma* p. lxxxiv, He is struggling for life in an overcrowded profession. **1950** T. S. Eliot *Cocktail Party* ii. 126 But they can't *all* stay there! I mean, it would make the place so over-crowded. **1960** *Encounter* Mar. 56/1 The spiritual effects of over-crowdedness. **1976** H. Nielsen *Brink of Murder* ix. 81 Small businessmen, seeking relief from overcrowded cities..gravitated seaward to create new forms of overcrowding. **1977** *Listener* 17 Mar. 339/1 A cell which..would beggar description in terms of coldness, smallness, overcrowdedness.

overcrowding, *vbl. sb.* (Earlier and later examples.)
1848 Mill *Pol. Econ.* I. ii. xiv. 472 A sufficient degree of overcrowding may depress the wages of women to a much lower minimum than those of men. **1865** J. Simon in E. R. Pike *Human Doc. Victorian Golden Age* (1967) vi. 292 Where 'overcrowding' exists in its sanitary sense, almost always it exists even more perniciously in certain moral senses. **1932** J. S. Huxley *Probl. Relative Growth* vii. 207 With overcrowding and less favourable food conditions 'degeneration' set in. **1955** M. Gilbert *Sky High* ii. 22 Every time we meet he's got a fresh reason for it. Overcrowding, parking offences, congestion of pavements. **1976** *Language* LII. 30 The unit in relation to which overcrowding must be described is the syntactic configuration, and not just the simplex S.

o·vercure, *sb.* [Over- 29.] The process or result of overcuring; overvulcanization.
1915 *Jrnl. Soc. Chem. Industry* 15 Oct. 990/1 The load at breaking point was found to increase with the time of cure to a certain point and then to collapse, the rubber becoming very brittle at an overcure. **1952** J. Delmonte *Plastics Molding* ix. 265 Translucent urea molded parts are the first to show the effects of overcure—by dullness or chalkiness in finished products. **1964** [see *OVERCURING *vbl. sb.*].

overcure (stress variable), *v.* [Over- 27.] **a.** *trans.* To cure (plastic or rubber) for longer than the optimal period; to overvulcanize. Also *absol.* **b.** *intr.* To undergo overcuring.
1916 *Jrnl. Soc. Chem. Industry* 31 Aug. 872/1 It is..common in the case of many goods for manufacturers when vulcanising to undercure rather than overcure the rubber so as to avoid the danger of over-vulcanising and consequent deterioration. **1949** B. L. Davies *Technol. Plastics* xxi. 405 The whole board is cured in the usual manner in the press, care being taken not to overcure. **1952** J. Delmonte *Plastics Molding* viii. 211 Before the advent of thorough internal heating by high frequency, thick-molded parts would overcure on outside surfaces before the interior reached molding temperature. **1972** *Materials & Technol.* V. xiv. 523 The outside of a thick rubber article has to be cured longer than the optimum curing time (overcured) in order to ensure an optimum cure within the article.
So **o·vercured** *ppl. a.*; **overcu·ring** *vbl. sb.*
1912 *Jrnl. Soc. Chem. Industry* 16 Dec. 1100/1 It is important to note that (3) was almost fully cured at 35 lb. and over-cured at 45 lb. *Ibid.* 1101/2 Beyond this point it was probable that a decrease merely indicated overcuring. **1952** J. Delmonte *Plastics Molding* ix. 265 An overcured plastics part is even more difficult to recognize than an undercured piece, primarily because most plastics are not critical in their time of cure. **1964** A. E. Juve in Alliger & Sjothun *Vulcanization of Elastomers* ii. 30 Vulcanization occurs in three stages; (1) an induction period, (2) a curing or cross-linking stage, and (3) a reversion or over-cure stage... Some systems..have little or no tendency toward reversion or increased tightness of cure as the result of overcuring.

o·vercurrent. *Electr.* [Over- 29.] A current in excess of that which is normal, safe, or allowed for.
1931 H. W. Brown *Electr. Equipment* (ed. 2) i. 19 (*heading*) Automatic protection against ground, short circuit, and overcurrent. **1971** Moore & Elonka *Electr. Syst. & Equipment for Industry* ii. 31 Direct-acting over-current trips are generally used on low-voltage circuit breakers.

overcut, *sb.* Add: **c.** *Mining.* A cut at or near roof level in a seam.
1940 *Trans. Inst. Mining Engin.* XCIX. 55 When the undercut is in coal, the importance of effective shovelling may be less, but it must always remain to give freedom of working to the machine, and it is only when an intermediate or overcut is taken that it is unnecessary. **1960** J. Sinclair *Winning Coal* vi. 170 When overcuts at a greater height are required..a hydraulic turret machine..is generally most satisfactory.
d. In electrochemical machining, the distance between the outside surface of the cathode and the side of the cut in the part being machined.
1965 *New Scientist* 5 Aug. 336/2 If the voltage is allowed to fluctuate, the size of the overcut will fluctuate, and it becomes impossible to machine parts to close tolerances. **1974** J. A. McGeough *Princ. Electrochem. Machining* v. 146 The amount of overcut can be diminished by several devices, including insulation along the external side walls of the cathode.

over-cut, *v.* **1.** [Over- 27.] *trans.* To fell too many trees in (a forest) at once, upsetting the regular supply of trees suitable for cutting.
1906 W. Schlich *Man. Forestry* (ed. 3) I. v. 75 Private owners are inclined to favour their own monetary interests to the disadvantage of future generations by overcutting ..their forests. **1913** *Q. Rev.* Oct. 446 In the case of private ownership, there is always a danger of the forests being overcut to obtain quick returns.
2. *Mining.* [Over- 1.] *trans.* (See quot. 1967.) Also *absol.*
1907 *Trans. Inst. Mining Engin.* XXXI. 387 So far as the machine is concerned, it will overcut as well as undercut. **1947** *Ibid.* CVI. 18 The first seam to be overcut with a machine designed specially for roof-forming was the Cockshead Seam in North Staffordshire. **1967** *Gloss. Mining Terms* (*B.S.I.*) viii. 20 *Overcut*, to cut by machine at or near roof level in a seam.
3. [Over- 27.] *intr.* To cut or produce a groove in a gramophone record with such amplitude as to run into an adjacent groove.
1935 H. C. Bryson *Gramophone Record* iv. 73 For sounds of constant absolute intensity over the frequency range of 30–250, all the sounds have an equal tendency to over-cut. **1962** W. R. Wellman *High Fidelity Home Music Syst.* (ed. 2) viii. 175 When a recording is made, it is possible for the recording stylus to 'overcut', or run into an adjacent groove. **1976** *Gramophone* Sept. 514/3 The audible improvements over the transformer are subtle but very real, totally abolishing my previous reservations about slight sibilance on overcut records with the SL15/Mk. II.
Hence **o:vercu·tting** *ppl. a.* and *vbl. sb.* Also **o·vercu:tter,** a machine for overcutting.
1923 *M. & C. Machine Mining* II. 283 (*caption*) A repeat order for overcutting turbine driven universal machines for heading and longwall service. **1928** M. D. Williams *Pract. Machine Mining* iv. 45 The 'Samson' chain machine when adapted for overcutting is fitted with a jacking device which permits the raising or lowering of the machine in case of variation in the position of the band to be cut. **1928** *Daily Tel.* 9 Oct. 9/7 Wasteful over-cutting, forest fire, fungi, insects, and wind combined are rapidly wiping out Canada's available trees. **1935** H. L. Bryson *Gramophone Record* iv. 83 If overcutting seems likely to occur, the electrical input of the recorder is cut down to the safety line, i.e., the amplitude is reduced. **1944** J. S. Huxley *On Living in Revolution* xi. 117 Over-cutting of the forests resulted in the closing of its [*sc.* Elma's] one big mill. **1946** *Trans. Inst. Mining Engin.* CVI. 20 (*caption*) Overcutting coal-cutter arranged to cut at 7 ft. *Ibid.* 21 An overcutter was required to cut at 4 ft. **1962** W. R. Wellman *High Fidelity Home Music Syst.* (ed. 2) viii. 175 Overcutting is prevented, during recording, by limiting the swing of the cutting stylus on low frequencies. **1966** S. D. Woodruff *Methods of Working Coal & Metal Mines* III. A. iv. 158 (*caption*) Hydraulic turret overcutter with down-curved jib for roof cutting. *Ibid.* 173 The cutting unit consists of a special long-wall coal cutter which is equipped with two horizontal jibs—one for undercutting and the other for overcutting.

overda·mped, *a.* [Over- 24.] Of a physical system: damped to a greater extent than the minimum needed to prevent oscillations. So **overda·mp** *v. trans.*
1922 Glazebrook *Dict. Appl. Physics* II. 373/1 The galvanometer system is non-oscillatory and overdamped. **1936** *Jrnl. Sci. Instruments* XIII. 101 Even with a resistance which overdamps the movement the rapidity of indication is said to compare favourably with that obtainable with pivoted instruments of a similar type. **1939** *Amateur Radio Handbk.* v. 79/2 The aerial should damp the network to flat resonance, though if it overdamps it, it may not be possible to draw any power from the anode circuit. **1962** Simpson & Richards *Physical Princ. Junction Transistors* xv. 374 If the reverse is true the response will be overdamped and the output voltage will rise exponentially. **1975** *Nature* 10 July 121/1 If large, complex systems have the property that their linear connectivity is low, then they are more likely to be stable, and if stable, are more likely to be over-damped than to oscillate.

o·verday, *a.* [Over- 32]. Designating a herring that is not freshly caught. Also *absol.*
1883 [see Over- 32]. **1889** *Tit-Bits* 17 Aug. 298 About 24 hours after capture the herring is liable to the pouring out of extravasation of blood about his gills and fins, which..bruised appearance is quaintly called in the fish trade *over-day tarts*. **1927** *Sunday Express* 23 Oct. 5/1 Some of the herrings caught to-day are what is known as 'overdays'. A herring becomes an 'overday' if not sold within twelve hours of leaving the water, and its price

drops precipitately. **1958** *New Statesman* 23 Aug. 218/1 'Overdays' herring, not completely fresh, can still make oil and meal, though 'sludge' needs fresh fish.

overdee·pen, *v.* *Geol.* [Over- 22; *overdeepening* is tr. G. *übertiefung* (A. Penck 1899, in *Verhandl. des 7en Internat. Geographen-Kongr.* (1901) II. 232).] *trans.* To deepen further, to make even deeper. So **overdee·pened** *ppl. a.*, **-dee·pening** *vbl. sb.*
1900 *Proc. Boston Soc. Nat. Hist.* XXIX. 308 Penck has suggested that glaciated valleys of the Alpine kind should be called 'overdeepened'. **1902** *Q. Jrnl. Geol. Soc.* LVIII. 703 The hanging valleys in the lower part of the Val Ticino are attributed to the overdeepening of the main valley by ice. **1905** *Jrnl. Geol.* XIII. 392 As soon as a shallow wind-blown hollow is formed, that part of the integrated drainage system which leads to the hollow will supply waste to it whenever rain falls there;..the coarser waste will accumulate, and thus the tendency of the winds to overdeepen local hollows will be..counteracted. **1913** *Bull. Geol. Soc. Amer.* XXIV. 214 Similarly, a valley glacier, while actively overdeepening its trough, might develop a more or less abrupt step in the trough floor. **1968** R. W. Fairbridge *Encycl. Geomorphol.* 327/1 Regardless of whether the overdeepening of the river mouth is primarily due to eustatic rise of sea level or tectonic subsidence, it is said to be a drowned valley. *Ibid.* 743/2 The rock floors of overdeepened troughs remain bare in parts. *Ibid.*, Overdeepening is responsible for the production of numerous lakes, some of very large size. **1970** C. A. Lewis *Glaciations Wales* ii. 25 The major through-valleys of Snowdonia were overdeepened and straightened at this time by lowering of the watersheds by ice moving radially outwards from a centre near Llyn Tegid.

overdesi·gn, *v.* [Over- 26.] *trans.* To design to a standard of reliability or safety higher than the usual or minimum standard.
1964 M. Gowing *Britain & Atomic Energy, 1939–1945* x. 278 The team in fact decided to 'over-design' the pile. **1969** *Word Study* Apr. 4/2 Rocket boosters must contain..hundreds of redundant systems to preclude failure and to provide reliability; they are overdesigned, well beyond what is necessary for normal function. The same over-design appears in much of the prose. **1970** *Nature* 18 July 218/2 Once a tower has been shown to live up to its specification, the designer likes to see by how much he has over-designed it, and about fifty per cent of the towers are eventually tested to destruction.
Hence as *sb.*, the action of overdesigning; an instance of this.
1969 [see above]. **1972** *Lebende Sprachen* XVII. 134/2 Greater reliability, without penalties caused by over-design, are expected to be achieved through use of a simulator.

o:verdetermina·tion. [f. next.] The existence of more than one cause or contributory factor; *spec.* in *Psychol.*, the expression in one symptom of two or more needs or desires.
1917 C. R. Payne tr. *Pfister's Psychoanal. Method* 143 We have often had opportunity..to show these over-determinations. **1925** A. & J. Strachey tr. *Freud's Analysis of Case of Hysteria* in *Coll. Papers* III. 73 In the world of reality..a complication of motives, an accumulation and conjunction of mental activities—in a word, over-determination—is the rule. **1940** *Mind* XLIX. 370 A doubt such as 'I can never really know what another person is feeling' may arise from more than one of these sources. This over-determination of sceptical symptoms complicates their cure. **1955** J. Strachey tr. *Freud's Infantile Neurosis* in *Compl. Psychol. Wks.* XVII. 56 The contradiction is easily resolved if we regard it as a case of overdetermination. **1967** *Philos.* XLII. 374 There is a special kind of plurality, namely overdetermination. **1970** B. Brewster tr. *Althusser & Balibar's Reading Capital* (1975) 315 The overdetermination of a contradiction is the reflection in it of its conditions of existence within the complex whole, that is, of the other contradictions in the complex whole, in other words its uneven development. **1973** S. Heath in *Screen* Spring/Summer 111 Underground cinema..has attempted to break through..the particular ideological overdetermination of the camera. **1975** *New Left Rev.* Nov.-Dec. 14 The ideological over-determination is itself a forced response, and what forces it is a material dilemma.

overdete·rmine, *v.* [Over- 24.] *trans.* To determine, account for, or cause in more than one way, or with more conditions than are necessary. So **overdete·rmined** *ppl. a.*, having more determining factors than the minimum necessary; having more than one cause; *spec.* in *Psychol.*, giving expression to more than one need or desire; **overdete·rmining** *vbl. sb.*
1879 *Encycl. Brit.* X. 377/2 The definitions which have not been mentioned are all 'nominal definitions', that is to say, they fix a name for the thing described. Many of them overdetermine a figure. **1917** C. R. Payne tr. *Pfister's Psychoanal. Method* 143 The neurotic symptom has several determining factors, at least two. Therefore, it is called over-determined. **1924** C. M. Baines tr. *Freud's Ætiology of Hysteria* in *Coll. Papers* I. x. 213 The idea chosen as the basis of a symptom will be one which various factors combine to arouse and which is stirred up from several directions simultaneously;—a state of affairs I have elsewhere tried to formulate by saying that hysterical symptoms are over-determined. **1950** *Mind* LIX. 199 My large-scale map of the small area occupied by θ will show that its display of Q is over-determined. **1959** I. Pool in Saporta & Bastian *Psycholinguistics* (1961) 308/1 The third of the overdetermining influences in the development of contingency analysis was structural linguistics. **1969** G. Stedman Jones in Cockburn & Blackburn *Student Power* 30 Mass student insurgency is *par excellence* an 'overdetermined' phenomenon. **1974** J. White tr. *Poulantzas's Fascism & Dictatorship* i. iii. 40 The economic process is overdetermined by the class struggle, which has primacy.

over-develop, *v.* Add: Hence **o:ver-deve·loped** *ppl. a.*
1974 *Environmental Conservation* I. 17/2 The kind of agricultural system that now predominates in the over-developed countries. **1974** B. Pearce tr. *Amin's Accumulation on World Scale* I. ii. 157 In the overdeveloped economies,..the tendency to (relative) underconsumption weighs heavily upon investment. **1977** G. Fisher *Villain of Piece* ii. 14 Majorca's over-developed coastline.

over-development. (Later examples.)
1974 B. Pearce tr. *Amin's Accumulation on World Scale* II. iv. 507 A favourable trade balance has beneficial effects only if saving tends to be superabundant, in a context of overdevelopment. **1976** *Survey* Summer-Autumn 156 The recognition of the planet's economic unity springs from a justifiable alarm at overdevelopment.

o:ver-differentia·tion. *Philol.* [Over- 29 b.] The unnecessary differentiation of elements in a phonemic, graphemic, or grammatical system, or in its analysis. So **o:ver-differe·ntiated** *a.* Cf. *DIFFERENTIATION 1 b.
1933 L. Bloomfield *Language* 223 Some irregular paradigms are *over-differentiated*. Thus, corresponding to a single form of an ordinary paradigm like *play* (*to play, I play, we play*), the paradigm of *be* has three forms (*to be, I am, we are*). *Ibid.* 269 Among the substantives are some pronoun-forms which, by over-differentiation, do not serve as actors: *me, us, him,* [etc.]. **1955** H. A. Gleason *Introd. Descriptive Linguistics* xvii. 174 Over-differentiation can be discovered and corrected from the record alone by rigorous procedures. **1962** *Canadian Jrnl. Linguistics* VII. 105 The choice of an under- as against an over-differentiated graphemic system must be made by comparing the practical difficulties of each in relation to reading and writing. **1965** *Language* XLI. 67 This grammar is characterized by functional overdifferentiation... I mean the distinguishing of constructions that are functionally identical. **1965** *Amer. Speech* XL. 108 Overdifferentiation of phonemes involves the imposition of phonemic distinctions from English into the phonemic system of Texas German, where they are not required.

over-dispersion (ōᵘ:vəɹdispə·ɹʃən). *Ecology.* [f. Over- 29 b + Dispersion.] A greater unevenness in the distribution of individuals than would be the case if the existence and position of each were independent of the rest, so that there is an increased proportion of the area with a large or a small concentration of individuals. Hence **o:ver-dispe·rsed** *a.*, distributed in this manner.
1936 *Jrnl. Ecol.* XXIV. 234 If in a set of counts of numbers of individuals per sample area the relative variance is greater than unity, this indicates that the dispersion is greater than would be expected on the assumption of random (Poisson) distribution... [This] first condition, over-dispersion, implies that individuals are scattered less evenly..than would be expected. *Ibid.* 250 Over-dispersion shows itself in an excess of quadrats containing no 'individuals' or a large number of 'individuals', there being a corresponding deficit in central classes. **1946** *Ecology* XXVII. 329/2 The writer [*sc.* L. C. Cole] prefers the terms 'contagious' and 'negatively contagious' to 'overdispersion' and 'underdispersion' both of which have suffered reversals of meaning in the hands of different authors. **1948** *Bot. Rev.* XIV. 226 The study of over-dispersion is of major importance in statistical ecology. *Ibid.*, Over-dispersion is the rule rather than the exception among animals. *Ibid.* 227 Cole applied this technique to the analysis of various over-dispersed populations. **1957** P. Greig-Smith *Quantitative Plant Ecol.* iii. 56 The former has been termed overdispersed (referring to the distribution curve obtained) and the latter underdispersed. Unfortunately these terms have sometimes been used in the reverse sense (referring to the pattern of individuals on the ground)..and are better avoided. **1964** V. J. Chapman *Coastal Vegetation* ii. 24 The plants are over-dispersed or aggregated into clumps.

overdo, *v.* Add: **5.** (Further examples.) Also in phr. *to overdo it*, to do too much for one's health; to overtax one's strength.
1817 M. Whalley *Let.* 23 Apr. in J. Constable *Corr.* (1962) I. 164, I trust however that your Darling is better than when you wrote, & was not overdone with company yesterday. **1853** J. Ruskin *Let.* 18 Aug. in M. Lutyens *Millais & Ruskins* (1967) 85 He overdid it last winter and now evidently stands in need of rest. **1866** Geo. Eliot *Let.* 5 June (1956) IV. 267, I..have been a new creature ever since, though a little over-done with visits from friends and attention (miserabile dictu!) to petticoats etc. **1901** M. Franklin *My Brilliant Career* v. 31 We were too overdone to make more than one-worded utterances, so waited silently in the blazing sun, closing our eyes against the dust. **1920** N. Coward *I'll leave it to You* ii. 24 You work terribly hard. I only hope you won't over-do it. **1924** A. Huxley *Let.* 3 Dec. (1969) 237 We must be careful not to make him overdo it, unless we want him laid up. **1973** N. Meyer *Target Practice* (1975) xv. 185, I worked at being normal. Perhaps I overdid it.

o·verdog. [Over- 2 b.] A superior dog; usu. *fig.*, a dominant or victorious person. (Opp. Underdog; cf. *top dog* s.v. Top *sb.*[1] 32.) Hence **o·verdoggery.**
1908 *Westm. Gaz.* 17 Oct. 5/1 But the smart terrier was an Overdog, And knew a trick worth two of that. **1938** *Richmond* (Virginia) *Times-Dispatch* 6 Apr. 8/4 It's contrary to all instinct, yet, as one looks over the world today, one has to feel sorry for the overdogs. **1957** T. Gunn *Sense of Movement* 30, I praise the overdogs from Alexander To those who would not play with Stephen Spender. **1962** T. Zinkin *Caste Today* 40 When the underdog begins to feel that he is as good as his neighbour, the result depends largely upon whether the overdog has stopped believing in his own over-doggery. **1966** *New Statesman* 28 Oct. 638/3 Underdogs abandoned by or abandoning inadequate male overdogs..revisit old haunts, look up old lovers, drift unsatisfactorily through a hostile world. **1970** R. Lowell *Notebk.* 204 Dear Mary, with her usual motherly Solicitude for the lost overdog.

overdone, *ppl. a.* (Examples in sense 'over-cooked'.)
1781 J. Woodforde *Diary* 24 Sept. (1924) I. 322 It was a shabby dinner and overdone. **1853** Dickens *Child's Hist. Eng.* II. xvi. 25 All the sandy prospect lay beneath the blazing sun burnt up like a great overdone biscuit. **1964** L. Deighton *Funeral in Berlin* xx. 117 A finger like a Lyons sausage—slightly overdone. **1975** B. Wood *Killing Gift* (1976) ii. iv. 75 The waiter..brought two plates of the overdone roast beef. **1977** P. G. Winslow *Witch Hill Murder* i. 20 She was not fond of overdone hamburger.

over-door, *sb.* (Further examples.)
1873 C. Schreiber *Jrnl.* 25 Oct. (1911) I. 236 Went on to Dirksen's, bought an over-door, also Watteau, for Ivor. **1967** A. Eeles *Canaletto* 9/2 Smith also commissioned a group of 'over-doors', obviously intended to be part of a decorative scheme.

overdo·sage. [Over- 29 b.] The administering or taking of too large a dose (of medicine, drugs, etc.).
1922 *Encycl. Brit.* XXX. 137/2 To find a method of preventing these chloroform deaths, by enquiring into the conditions of overdosage. **1929** *Irish Jrnl. Med. Sci.* Apr. 183 Such symptoms were obviously due to overdosage. **1964** L. Martin *Clin. Endocrinol.* (ed. 4) i. 12 The excessive secretion of, or therapeutic overdosage with ACTH produces the clinical picture of Cushing's syndrome. **1976** *Times* 11 Mar. 2/3 The recommendations on aspirin and similar pain killers reflect growing concern about overdosage.

overdose, *sb.* For quot. 1690 in Dict. substitute quot. 1700 below and add further examples.
1700 Locke *Essay Hum. Und.* (ed. 4) ii. xxxiii. 223 Had this happen'd to him, by an overdose of Honey, when a Child, all the same Effects would have followed. **1916** [see *COBBER *sb.*[2]]. **1931** H. Crane *Let.* 22 June (1965) 375, I have to leave most of this to your judgment of the potency and malfeasance of an overdose of tequila. **1952** M. Laski *Village* ix. 147 An overdose of Miss Beltram produced the inevitable effect..you couldn't help taking the other point of view. a**1953** E. O'Neill *Long Day's Journey* (1956) iii. 105, I hope, sometime, without meaning it, I will take an overdose. **1965** *New Statesman* 7 May 729/3 Rosetti's..wife died after only two years of marriage, of an overdose of drugs. **1971** *Black Scholar* June 53/2 There are brothers..who have been singled out for overdoses of the atrocities that we are being subjected to. **1973** *Times* 16 July 14/7 Others were either seeking immediate emotional release or were trying to produce a dramatic effect on friends or relatives—and in both such cases an overdose achieved the desired result.

overdose, *v.* Add: **3.** *intr.* To take an overdose of drugs. Hence **o·verdosing** *vbl. sb.*
1973 R. Ludlum *Matlock Paper* xxx. 261 The doctor told me that he'd prescribe heavier 'medication' but warned me not to overdose. **1974** G. McDonald *Fletch* (1976) xvii. 100 She was dead... He guessed she had overdosed. **1977** *Times* 19 Jan. 14/2 Heroin smoking..throws overdosing figures into doubt. **1977** Wood & Geasland *Twins* 21 You don't take a full bottle of an anticonvulsant if you mean to overdose on Seconal.

overdraw, *sb.* **2.** (Examples.)
1902 A. D. McFaul *Ike Glidden* xvi. 122 He was prancin'..until he got him hitched inter this new bitin' gear an' overdraw. **1905** *Springfield* (Mass.) *Weekly Republ.* 8 Sept. 5 Much has been accomplished to abolish the pernicious practice of docking horses, but it is just as important that the abuse of the overdraw check should be corrected.

overdraw, *v.* Add: **II. 6.** (Earlier and later examples.)
a**1817** Jane Austen *Northanger Abbey* (1818) II. vii. 140 Characters, which Mr Allen had been used to call unnatural and overdrawn. **1912** *Chambers's Jrnl.* June 359/1 Perhaps it will be said that the above statements are overdrawn.
7. *intr.* In card-games: to exceed the maximum permissible score by drawing too many cards. Also *refl.*
c**1805** Jane Austen *Watsons* in J. E. Austen Leigh *Mem. J. Austen* (1871) 358 Vingt-un is the game at Osborne Castle... Lord Osborne enjoys it famously... I wish you could see him over-draw himself on both his own cards. c**1863** T. Taylor in M. R. Booth *Eng. Plays of 19th Cent.* (1969) II. 150 (*Draws card.*) Thirty-four—overdrawn—confound it! Now let's see your hand. **1950** *Hoyle's Games Modernized* (ed. 20) 158 Many players habitually stand at fifteen, and if the dealer is a reckless player, with a tendency to overdraw, it may be good policy to stand upon an even smaller figure.

Hence **overdra·wer,** one who overdraws a bank account, or has an overdraft.
1906 W. De Morgan *Joseph Vance* xxxvii. 378 Among the overdrawers, C. Vance & Co. was a conspicuous instance, figuring for a good round sum among the Debtors.

overdrawn, *ppl. a.* Add: *spec.* Of tea: infused too long.
1847 A. Brontë *Agnes Grey* xii. 189 Other thoughts assisted to..impart a relish to the cup of cold, overdrawn tea. **1969** V. C. Clinton-Baddeley *Only Matter of Time* 94 Cigarettes and cups of overdrawn tea.

overdress, *sb.* **1. a.** (Later examples.)
1975 *Country Life* 6 Feb. 346/1 The chemise has now become universally known as the overdress..because..we have been pulling it over sweaters and shirts. **1976** *Ibid.* 19 Feb. 442 The beautiful, bright-green suede overdress by Jean Muir.

overdress, *v.* Add: **2.** (Later example.)
1947 *Sat. Rev. Lit.* (U.S.) 1 Mar. 23/2 There are times, too, when Mr. Miller overdresses the phrasing of his dialogue.
Hence **o:verdre·ssed** *ppl. a.*; **o:verdre·ssiness**; **o:ver-dre·ssing** *vbl. sb.*
1820 M. Edgeworth *Let.* 10 Aug. in C. Colvin *M. Edgeworth in France & Switz.* (1979) 209 Enter 3 English ladies overdressed in silks blonde and flowers! **1836** Dickens *Sunday under Three Heads* i. 2 There is a great deal of very unnecessary cant about the over-dressing of the common people. **1874** J. Brown *Let.* 4 Sept. (1907) 240 We saw the Duke with his eagle's feather, the Duchess in purple and lace, petite and overdressed. **1891** O. Wilde *Pict. Dorian Gray* i. 9 After I had been in the room about ten minutes, talking to huge overdressed dowagers and tedious Academicians, I suddenly became conscious that some one was looking at me. **1927** *Daily Express* 26 Aug. 1 What would look indecent in a London hotel becomes overdressiness here. **1932** W. W. Jacobs *Night-Watchman* iv. 680 Nothing looks worse than an over-dressed woman. **1939** Joyce *Finnegans Wake* 441 What's overdressed if underclothed? **1962** J. D. MacDonald *Girl* ix. 113, I got maybe forty [girls] stripped entire... Compared to them sixty broads, I was *over-dressed*. **1975** J. F. Burke *Death Trick* (1976) ix. 132 An overdressed old mama.

overdrink, *v.* Add: (Further examples of *intr.* use.)
1904 'J. O. Hobbes' *Vineyard* iv. 55 They over-eat and over-drink, and they try to forget what they really want. **1932** Kipling *Limits & Renewals* 79, I steadily overdrank for a fortnight out of pure hunger. **1952** H. Waugh *Last seen Wearing* (1953) 53 She..says he's a show-off and over-drinks. **1967** N. Freeling *Strike Out* 29 Typical Bavarian, overate, overdrank, no exercise. **1976** *Times* 20 Oct. 14/3 Many of us over-eat, over-drink, and over-smoke.
overdrink *sb.*, delete †*Obs.* and add later examples; **o:ver-dri·nking** *vbl. sb.*
a**1902** S. Butler *Way of all Flesh* (1903) vi. 23 Even his excellent constitution was not proof against..over-feeding and what we should now consider overdrinking. **1906** *Westm. Gaz.* 19 Sept. 2/3 The Bishop of Oxford, in the protest..against the over-drinking of healths, appears to think that little was done in the way of similar protest between the far-off days of King Ahasuerus..and those of Queen Victoria. **1907** *Ibid.* 29 Oct. 12/1 The publicans are held responsible for the over-drinking of their customers. **1924** J. J. Walsh *Catholic Church & Healing* vi. 66 Bartholomew was aware, moreover, that insanity may come from overeating as well as from the overdrinking of strong wine. **1951** T. Sterling *House without Door* vii. 89 Her hands..had that luminous, slightly puffy look which comes from overdrinking. **1974** E. Brawley *Rap* (1975) 7 And her mother..had followed him two years later of love end overdrink. **1977** *Times* 7 Sept. 7/6 A first line of recovery from over-drink.

overdrip, *v.* (Later examples.)
1897 F. Thompson *New Poems* 176 Shake the lilies till their scent Over-drip their rims. **1898** A. S. Way tr. *Tragedies of Euripides* III. 198 'The altar, overdripped with Hellene blood?'..'Blood-russet are its rims in any wise.'

o·verdrive, *sb.* [Over- 26.] **1. a.** In a motor vehicle, a speed-increasing gear which may be brought into operation in addition to the ordinary (reducing) gears, so providing a gear higher than direct drive (the usual top gear), and in some cases correspondingly raising other gears, and thereby enabling engine speed to be reduced for a given road speed.
1929 *Trans. Soc. Automotive Engineers* XXIV. 335/1 The functioning of a transmission with either an overdrive or an underdrive through a double-internal-gear set in combination with a suitable clutch is exactly equivalent to that of a two-speed axle. **1932** Elliott & Consoliver *Gasoline Automobile* (ed. 4) xviii. 479 As a general thing, the high gear [on heavy trucks and buses] is a direct drive, although in some cases the high gear ratio is an overdrive, that is, the engine crankshaft turns at a slower speed than the propeller shaft. **1938** *Times* 23 Aug. 8/5 The overdrive is engaged by merely pushing in a lever just below the instrument board. **1958** *Times* 1 Oct. 8/3 To this can be added at extra cost a Laycock de Normanville overdrive operating on second and top gear. **1959** [see *gear ratio* s.v. *GEAR *sb.* IV]. **1959** C. Campbell *Sports Car* (ed. 2) ix. 174 Some designers prefer to use the overdrive on top gear only, giving in effect a 5-speed gearbox. Others prefer to make it operative on both top and third, so that overdrive third gear is an intermediate step between direct third and direct top. **1970** *Motoring Which?* July 83/1 The 3-litre was comfortable up to about 95 mph (helped by its overdrive on 3rd and top). **1973** J. Wainwright *Devil you Don't* 5 It was a great car—a Jag. Mark II—well capable of three-figure speeds at the flick of the overdrive switch.
b. *fig.*
1962 L. Deighton *Ipcress File* i. 18 Dalby's voice trailed off as he slipped his mind into over-drive. **1967** *Times* 28 Oct. 20/2 Always the narrative is in smooth overdrive, even when it whisks you in successive paragraphs through Haiti..or the Preseli. **1969** *Times* 13 Dec. p. iv/7 The Websters knew everyone... Here, too, are Lilian Baylis and Sybil Thorndike, in overdrive as ever. **1970** G. Greer *Female Eunuch* 44 If women find that the clitoris has become the only site of their pleasure instead of acting as a kind of sexual overdrive in a more general response, they will find themselves dominated by the performance ethic. **1977** *Rolling Stone* 30 June 101/1 He shifts into overdrive on the first track, a fiery and propulsive Clarke composition entitled 'The Heat of the Battle', and never looks back for the whole album.
2. *Science Fiction.* = *HYPERDRIVE.
1953 [see *INTERSTELLAR *a.*].

overdrive, *v.* Add: **5.** *Golf.* To drive farther than (an opponent); to outdrive.
1900 *Gentl. Mag.* Feb. 126 If the opponent is a longer driver one is spared the temptation..of pressing to avoid being overdriven.
Hence **o:verdri·ving** *vbl. sb.*
1837 De Quincey *Revolt of Tartars* in *Blackw. Mag.* July 101/2 The cattle suffered greatly from over-driving. **1909** *Chambers's Jrnl.* Mar. 203/1 The electrical equipment is provided with suitable automatic devices to prevent over-driving. **1972** *Science* 16 June 1236/2 Control of the intensity is necessary to prevent overdriving by strong signals.

over-dry, *a.* Add: (Later examples.)
1891 W. Schlich *Man. Forestry* II. 32 *Dry Mould*..is formed by the decomposition of certain lichens on over-dry soil. **1959** A. H. Nissan *Textile Engin. Processes* xi. 293 It is then necessary to bring up the water content of the overdry parts by re-moistening them to a uniform and acceptable value. **1962** J. T. Marsh *Self-Smoothing Fabrics* xiii. 199 If the fabric is not sufficiently dry, then the subsequent mechanical operation is adversely affected; if the fabric is over-dry, then the durability of the finish, but not the initial effect, is adversely affected.
over-dryness (later example.)
1879 *Encycl. Brit.* X. 753/2 In extremely frosty weather ..they acquire so little moisture that then a difficulty arises from their over-dryness.

overdry, *v.* Add: **b.** (Later examples.) Hence **over-dry·ing** *vbl. sb.*
1867 K. H. Digby *Day on Muses' Hill* 146 The subsoil may be overdried. **1888** *Encycl. Brit.* XXIV. 657/1 Over-drying of wool has to be specially guarded against. *Ibid.* 657/2 Unless the wool is spread with great evenness..at points where the hot air escapes freely it may be much over-dried. **1959** A. H. Nissan *Textile Engin. Processes* xi. 293 It has already been demonstrated that overdrying the material before it is allowed to leave the drier consumes disproportionately longer time than drying it to the water content required in use. *Ibid.*, Unless there is close control, manual or automatic, on the uniformity of the final moisture content of the material in all directions, it is necessary to overdry the material so that its wettest part is within drying specification. **1962** J. T. Marsh *Self-Smoothing Fabrics* iii. 24 For many years the effect of drying has been recognised by the practical dyer in a qualitative manner, and there has been much warning as to the dangers of 'over-drying'. **1966** A. W. Lewis *Gloss. Woodworking Terms* 14 *Check*,..also called 'honeycombing'; due to over-drying too rapidly in the kiln.

overdu·b, *v.* [f. Over- 8 + *DUB *v.*[5]] *intr.* and *trans.* To impose (additional sounds) on to an existing recording. So **overdu·bbing** *vbl. sb.*
1962 *John o' London's* 16 Aug. 162/4 His particular interest lies in multi-taping, or 'overdubbing'. **1969** *Rolling Stone* 28 June 19/1 On rhythm dates I only do bass, drums, guitar, overdub the guitar, overdub whatever,..and then Al overdubs the strings, the horns, or whatever. **1970** *Times* 18 July 7 Apparently Crosby, Stills and Nash went back to the studios to overdub their vocals on 'Suite: Judy Blue Eyes', but the rest are left as originally played. **1971** *Times* 5 Jan. 9 She also prefers to use her own voice, overdubbed several times, as a backing choir. **1973** *Sci. Amer.* Apr. 2/2 (Advt.), Overdubbing has become a familiar term to every knowledgeable musician. .. To overdub properly, the artist recording on the second track has to listen to the material recorded on the first track while performing in perfect synchronization to it. **1977** *New Yorker* 8 Aug. 66/3 The singing actors..then overdubbed the lyrics.
Hence **o·verdub** *sb.*, an act or instance of imposing additional sounds on to an existing recording.
1976 in *6,000 Words* 146 The last big album it took them eight months of overdubs to produce. *Ibid.*, Vocal overdubs. **1976** *Gramophone* Aug. 353/2 By means of numerous synchronized re-recordings or 'overdubs', a typical pop recording is now built up over a period of days or weeks. **1977** McKnight & Tobler *Bob Marley* vi. 83 Wayne Perkins..is supposed to have added some overdubs.

overdue, *a.* Add: **c.** Of a library book: that has been retained by the borrower or reader longer than the period allowed; *overdue notice*, a notification sent to a reader requesting the return of an overdue book. Hence as *sb.*, an overdue book.
1890 T. Greenwood *Public Libraries* xxiii. 376 With this, as with other indicators,..books are shown in or out instantaneously,..and overdue books can be detected with little trouble. *Ibid.* 378 The borrowers' cards if arranged in a series of dated compartments, can be made to show the overdues. **1903** J. D. Brown *Man. Library Economy* xxix. 384 Overdue books could easily be detected by a register of this kind, on simply scanning the column of returns, filled up by the librarian. *Ibid.* 387 This personal form of ledger..makes the detection of overdues difficult. **1938** L. M. Harrod *Librarians' Gloss.* 109 *Overdue notice*, a request to a reader asking for the return of books which have been kept out beyond the time allowed. **1965** Smith & Baxter *College Library Administration* vii. 133 Whether the incidence of seriously overdue books is significantly reduced by charging fines is a matter for conjecture. **1966** E. V. Corbett *Introd. Librarianship* (ed. 2) i. xviii. 205 (*heading*) Fines and overdues. *Ibid.* 206 Other libraries may find that it is more economical to send out the first overdue notice not earlier than two months after the date due for return. **1970** *Times* 19 Nov. 15/6 He smiled only when he had returned them, like library books both overdue and unread, to the safety of the house. **1978** T. Allbeury *Lantern Network* xiii. 199 A library reminder for two overdue books.
d. Of a woman: not having had a menstrual period at the expected time. *colloq.*
1970 M. Tripp *Man without Friends* iii. 28 She placed her hand on mine, tenderly, 'I'm not overdue,' she said. **1972** F. Warner *Maquettes* 16 He doesn't even know I'm overdue. And he hasn't had it for a week. **1976** D. Francis *In Frame* viii. 115 She's feeling sick... I don't want a kid yet. She isn't overdue or anything.

overdye·, *v.* [Over- 8.] *trans.* To dye (something already dyed) with a second dye. Hence **overdye·ing** *vbl. sb.*
1946 Horsfall & Lawrie *Dyeing of Textile Fibres* (ed. 2) iv. 71 The question of overdyeing particularly concerns the production of shirtings or dress materials which are woven from grey unmercerised cotton, together with grey cotton dyed with fast-to-bleaching dyestuffs, and either undyed or dyed immunised cotton. These goods are piece bleached and overdyed with dyestuffs fast to light and washing in such a manner that the immunised and dyed cotton effects are unaffected. **1952** *Dyeing of Nylon Textiles* (I.C.I.) xi. 124 A two-bath process is necessary, the Vat dyestuff first being applied and oxidized and the union material then being over-dyed with the acid dyestuff.

over-early, *adv.* and *a.* (Later examples.)
1871 C. Kingsley *At Last* II. x. 49 This over-early marriage among the Coolies is a very serious evil. **1922** M. E. Christie *Henry VI* iv. 119 The young Henry..now found the responsibilities of government over-early thrust upon him.

over-easily, *adv.* (Later example.)
1977 A. Wilson *Strange Ride R. Kipling* iii. 153 Trix's view of Violet Garrard has been..over-easily accepted.

overeat, *v.* **1. a.** *intr.* Delete (Now *rare*) and add further examples. **b.** *refl.* Delete (The usual construction) and add later example.
1879 [see *MOST *a.* II. 5 e]. **1888** Nye & Riley *Railway Guide* 8 My appetite is four sizes too large for a man of my height and every little while I over-eat. **1904** [see *OVERDRINK *v.*]. **1944** L. P. Hartley *Shrimp & Anemone* iii. 33 You must see that he doesn't..over-eat himself. **1946** M. C. Self *Horseman's Encycl.* 291 A horse that from illness or some other cause is not being exercised will overeat. **1967** [see *OVERDRINK *v.*]. **1972** D. S. McLaren *Nutrition & its Disorders* viii. 158 The view, commonly held, that all obese people overeat has not been substantiated by several surveys.

over-e·ducate, *v.* [Over- 27.] *trans.* To educate to excess or for too long. So **over-e·ducated** *ppl. a.*; **o:ver-educat·edness.**
1845 J. S. Mill in *Edin. Rev.* LXXXI. 510 The fears of the patrons and managers lest the poor should be 'over-educated'. **1899** W. James *Talks to Teachers* 257 To be imprisoned or shipwrecked or forced into the army would permanently show the good of life to many an over-educated pessimist. **1922** N. Coward *Coll. Sketches & Lyrics* (1931) 24 I've over-educated myself in all the things I shouldn't have known about at all. **1935** D. L. Sayers *Gaudy Night* xvii. 364 How many women care..about anybody's intellectual integrity? Only over-educated women like us. **1970** *Time* 19 Oct. 66 Critics contend that..new students are being 'over-educated' for non-existent jobs. **1970** *Worship* Oct. 491, I distinctly recall how self-conscious I felt of my over-educatedness the moment I sat down with him. **1977** *N.Y. Rev. Bks.* 14 Apr. 43/1 (Advt.), Seattle Professional Man, 36, newly settled, witty, travelled, over-educated, gentle, 5'9", seeks attractive woman counterpart.

over-egg, *v.* [Over- 27 a.] *fig.*, in phr. *to over-egg the pudding*, to argue a point with disproportionate force; to exaggerate.
1892 *Review of Reviews* Jan. 8/2 It is possible to over-egg your pudding and M. Stambuloff is doing it. **1961** *Daily Tel.* 9 Feb. 24/3 Mr Foot seems to have been ill served by his seconder, Mr Will Griffiths.., who is reported to have made the mistake of 'over-egging the pudding'. **1976** *Times* 2 Nov. 4/5 Mr Page, though remaining confident, wonders whether some of the recent [election] forecasts may have over-egged his pudding.

over-employ·ment. [Over- 29 b.] A situation in which vacancies for jobs, esp. skilled

jobs, exceed the number of unemployed, producing a labour shortage; a state of insufficient unemployment.
1944 *Times* 9 Nov. 5/3 Sir William Beveridge believes that..'Full employment'..'should mean a floating balance of not more than about three per cent unemployed'... The dangers of inflation implicit in an attempt to secure 'over-employment' are real enough in any full employment policy. **1958** *Times* 11 June 11/5 The contraction of demand has been felt in the United States in the form of unemployment, and in Great Britain in a relaxation of the state of over-employment.

o:ver-enginee·r, *v.* [OVER- 26, 27.] *trans.* To engineer to a standard higher than is technically necessary, or to an extent greater than is technically desirable. Hence **over-enginee·ring** *vbl. sb.*
1964 M. GOWING *Britain & Atomic Energy 1939–1945* ix. 259 Certain units were very much 'over-engineered' which in itself led to complications. **1967** *Gramophone* Dec. 317/1 Walter gives the warmest reading, though..the recording is certainly over-engineered. **1971** *Flying* Apr. 89/2 (Advt.), The brakes have been purposely over-engineered by twenty percent. **1977** *Gramophone* June 113/2 Some over-engineering seemed to have taken place, however, at the expense of complete fidelity.

over-enthu·siasm. [OVER- 29 d.] An excess of enthusiasm. So **o:ver-enthusia·stic** *a.*
1927 *Melody Maker* Aug. 781/2 Some of my readers..have said that I am over-enthusiastic about Harry Richman. **1962** A. NISBETT *Technique Sound Studio* viii. 148 Over-enthusiasm has..resulted in the operator flicking the disc completely free from the turntable and skimming it across the room! **1962** A. BATTERSBY *Guide to Stock Control* viii. 77 There is also the over-enthusiastic salesman who offers to provide non-standard varieties of his company's products. **1970** *Nature* 24 Oct. 303/1 An over-enthusiasm for classifying projects of very marginal secrecy may have hampered progress. **1976** M. DRABBLE *Genius of T. Hardy* 166 The dog that destroys Gabriel Oak's sheep is over-enthusiastic, not malicious.

over-eru·ption. *Dentistry.* [OVER- 29.] Excessive extension of a tooth in the direction of the opposing teeth.
1961 H. R. B. FENN et al. *Clin. Dental Prosthetics* (ed. 2) xxiii. 628 Consideration must be given..to the extraction of the offending tooth..if the over-eruption is gross. **1975** H. THOMSON *Occlusion* viii. 130 Over-eruption of unopposed teeth..can be prevented by the muscle forces of tongue or cheek.
So **over-eru·pt** *v. intr.*, to undergo or exhibit over-eruption; **over-eru·pted** *ppl. a.*
1961 H. R. B. FENN et al. *Clin. Dental Prosthetics* (ed. 2) xxiii. 628 (*heading*) Over-erupted teeth. **1963** C. R. CROWELL et al. *Inlays, Crowns & Bridges* vii. 79 This wear may proceed to such an extent that the opposing teeth gradually overerupt. **1974** E. M. BARNETT *Pediatric Occlusal Therapy* vi. 161/1 The twin-wire arch is set so that the anterior wires lie slightly apical to where they should lie if the teeth were not overerupted.

over-estimate, *v.* (Earlier examples.)
1823 J. S. MILL in *Bermondsey Bk.* (1929) VI. 16, I may be told..that I over-estimate the effect of these motives on bad men. **1825** —— in *Westm. Rev.* Apr. 297 The habitual propensity of mankind to over-estimate advantages which they do not possess.

over-estimate, *sb.* (Earlier and later examples.)
1854 C. M. YONGE *Heartsease* I. II. xv. 357 A very good child, but spoilt..by John's over-estimate of her. **1895** *Geogr. Jrnl.* VI. 184 The liability..to an over-estimate of density [of population]. **1977** *Listener* 17 Mar. 327/1 The real membership of the Labour Party..is now 445,000..and..even this figure is probably an overestimate.

over-excitement. (Earlier example.)
1822 M. EDGEWORTH *Let.* 7 Jan. (1971) 313, I shall so arrange and *limit* our goings out..that Fannys health shall not suffer by over excitement.

over-exploi·t, *v.* [OVER- 27.] *trans.* To exploit excessively. So **o:ver-exploita·tion**; **over-exploi·ted** *ppl. a.*
1922 W. SCHLICH *Man. Forestry* (ed. 4) I. 273 The more accessible areas of the merchantable forests have been considerably over-exploited in the past. **1952** G. SARTON *Hist. Sci.* I. xi. 296 The mines were overexploited in the fifth century. By the middle of the following century only old workings were open. **1957** MANVELL & HUNTLEY *Film Music* iv. 191 Like all good ideas, they tend to be over-exploited to their own detriment. **1961** *Carolina Q.* Spring 18 It seems that only in today's Southern fiction does Tobacco Road..continue to live—but only as a weary, overexploited phantom. *Ibid.* 21 This legend..well deserves a rest after the overexploitation of the past century. **1963** *Times* 6 Feb. 8/7 U. P. Gerasimov and E. K. Fedorov, commenting on over-exploitation of fisheries, alleged that trawlers..'are ploughing the seas near African and Asian shores in search of fish'. **1964** V. J. CHAPMAN *Coastal Vegetation* vi. 137 The question has been asked whether such dunes are the result of man's negligence or over-exploitation. **1970** *New Scientist* 3 Dec. 374/1 Improved international cooperation should limit overexploitation of important [fish] stocks. **1977** *Sci. Amer.* May 26/1 The floods of 1969 in Tunisia replenished all the reservoirs that up to then had been described as overexploited.

over-expose, *v.* Add: (Further examples.) Also *absol.*
1894 B. POTTER *Jrnl.* 12 Sept. (1966) 338 Went to Berwick again with papa who had over-exposed on Monday. We got out at Tweedmouth and he photographed from the same side at first and afterwards amongst the boats. **1925** B. BEETHAM in E. F. Norton *Fight for Everest, 1924* 325 We were all inclined to under-expose in tropical Sikkim and to over-expose in arctic Tibet. **1937** AUDEN & MACNEICE *Lett. from Iceland* iv. 45 In Iceland, even if you are using a meter, there is a tendency to over-expose. *a* **1963** S. PLATH *Ariel* (1965) 46, I could draw no breath, Dead and moneyless, Overexposed, like an X-ray. **1971** *Radio Times* 23 Sept. 62 They hope that by not being over-exposed they'll be able to last longer, keep their standard higher. **1976** *Gramophone* Dec. 978/2 Nobody could say that this much maligned instrument [*sc.* the saxophone] is over-exposed now that it has faded from the pop scene.
over-exposure (further examples); *spec.* an excessive number of public appearances by an entertainer, actor, or the like.
1947 *Radiology* XLIX. 364/2 Is the peripheral blood picture as reliable an indicator of over-exposure as radiologists have considered it to be? **1969** *Guardian* 15 July 18/4 The company's general manager said that 'over-exposure' of records inhibited sales. **1971** *Ibid.* 6 Feb. 6/6 Future plans? A repeat of the 'Basil Brush Show' on the BBC, but careful avoidance of overexposure. **1974** *Times* 1 Feb. 19/6 We are all for exposure [on television], but we are against over-exposure. As anyone knows who handles films and photographs, you get unwanted results from over-exposure. **1974** *Physics Bull.* May 179/1 CO is not cumulative either in man or the environment but although moderate overexposure seems to cause no permanent damage to health, concern has been expressed about possible effects on the foetus following exposure of the pregnant mother. **1978** *Times* 5 Jan. 5/5 Perhaps renaissance is only the flip-side of over exposure.

over-exte·nd, *v.* [OVER- 27.] *trans.* **a.** To extend or reach further than (something). *rare.* **b.** To extend (a thing) too far. **c.** To take on (oneself) or impose on (another) an excessive burden of work, commitments, etc.; to attempt more than is practicable. Hence **over-exte·nded** *ppl. a.*; **over-exte·nsion.**
1937 R. ERSKINE *Stout Adventure M. Stewart* iii. 62 A culture and a civilization,..which, reckoned in the gross, outweighed and over-extended by a deal feudal, that is to say, English culture, manners and customs. **1938** *Sun* (Baltimore) 6 May 3/5 We are no longer over-extended in new construction or in capital equipment. **1962** L. DAVIDSON *Rose of Tibet* ii. 46 The arrangement was for a car to pick him up..but when..no car appeared, he realized he must have overextended Mr. Mukherjee, and took a bus instead. **1963** C. R. COWELL et al. *Inlays, Crowns & Bridges* iii. 25 The air-turbine instrument is usually less unpleasant for the patient, but there is risk of overextension. *Ibid.* vi. 64 If a temporary crown is overextended cervically the gingivae may be forced back. **1966** *Economist* 5 Mar. 898/2 Mr McNamara has defended himself vigorously, calling it 'absolutely false to say that we are over-extended.' **1968** *Blues Unlimited* Dec. 12 He never overextended himself and impressed with his command of the idiom. **1970** F. C. WEFFORT in I. L. Horowitz *Masses in Lat. Amer.* xi. 386 The sociological analysis..often considers what is merely possible, thus overextending the limits which the Brazilian historical situation allows for in the way of planned social change. **1973** *Daily Tel.* 7 Feb. 16 They have overextended themselves in the purchase of jumbo jets and do not wish to make further capital commitments at this time. **1976** *National Observer* (U.S.) 3 July 9/4 In much the same way..it's easy to overextend yourself on vacations by scheduling too much sight-seeing and other activity.

overface, *v.* Add: **1. a.** (Further examples.) Also in non-*dial.* use, *esp.* to alarm or intimidate (a person, animal, etc.) by presenting too great a task or obstacle. Hence **overfa·ced** *ppl. a.*
1926 A. BENNETT *Lord Raingo* I. xliv. 196 There she stood, over-faced and dumb and apologetic in her plain brown dress. *Ibid.* lvii. 254 He scorned them, but in their collectivity they still over-faced him. **1944** *R.A.F. Jrnl.* Aug. 261 You cannot imagine any situation which would over-face them. **1950** W. A. RILEY in C. R. Acton *Dog Ann. 1951* 94 Personally, I think it is a great mistake to over-face them [*sc.* dachshunds], they are so game and will 'have a go', but they are not killers. **1958** *Times* 13 Dec. 9/4 Undoubtedly one of the most important points to remember is not to 'overface' the pony or its enthusiastic rider. **1971** R. ROBERTS *Classic Slum* vi. 94 Lower-class women, bold enough to enter a pub but too 'overfaced' to sit,..stood crushed together drinking in the 'Outdoor'. *Ibid.* vii. 109 Social intimidation..confused and 'overfaced' the simple. **1976** *Horse & Hound* 10 Dec. 69/3 (Advt.), This horse has been brought on slowly, never being overfaced.

overfall, *sb.* Add: **1.** (Further examples.) Also *transf.* and *fig.*
1947 A. C. DOUGLAS *Gliding & Advanced Soaring* i. 32 The cloud currents, the heat turbulence, high winds and mountain overfalls, which the aeroplane pilot finds so unpleasant or even dangerous, are regarded by the sail-plane pilot as friends, not as enemies. **1961** B. FERGUSSON *Watery Maze* iv. 89 Radiating in every direction from C.O.H.Q. were heavy overfalls, as they are called in seaman's parlance, or troubled waters, in landsmen's, needing more than a modicum of oil. **1970** *Motor Boat & Yachting* 16 Oct. 35/1 We didn't bother to avoid an area marked on the large scale chart with those squiggly lines that denote dangerous overfalls. **1975** J. R. L. ANDERSON *Death in North Sea* viii. 139 There were two main tidal streams... They might run up to about two knots, with a somewhat faster rate by some overfalls off Spurn Head.
3. Delete † *Obs.* and add later examples.
1811 D. BUCHAN in K. Winter *Shananditti* (1975) ii. 24 This day's distance is estimated at eleven miles allowing seven from the island..up to the overfall. **1921** H. GUTHRIE-SMITH *Tutira* i. 5 A meandering serpentine creek..which..breaks into a series of overfalls.

over-familiar, *a.* (Later examples.)
1936 R. W. CHAPMAN *S.P.E. Tract* XLVII. 237 'Your wife' may be over familiar, if I do not know Jones very well. **1955** H. KUTTNER in Aldiss & Harrison *Decade 1950s* (1976) 173 He saw the reflection of the over-familiar scene. **1976** B. BALL *Keegan: One-Way Deal* 16 Holmyard frowned. Ross was over-familiar. There was a time and a place for such things.

over-feeding, *vbl. sb.* Add: (Further example.) Also as *ppl. a.*
1621 H. WOTTON *Let.* 26 Nov. in L. P. Smith *Life & Lett. Sir H. Wotton* (1907) II. 219 As may excuse the English, whom otherwise they think an overfeeding nation. **1902** G. B. SHAW *Let.* 18 Nov. (1972) II. 289 Who is Doctor Buzzi? Is it starvation cure, or overfeeding cure, or water cure, or faith healing, or what?

overfill, *v.* Add: **1.** (Further examples: cf. next.) Hence **overfi·lling** *vbl. sb.*
1908 B. STOUGHTON *Metall. Iron & Steel* viii. 200 With open passes, the collars cannot be made to quite touch..and the pressure may squeeze some metal between them, forming a 'fin' along the side of the piece. This is known as 'overfilling the pass'. **1953** D. J. O. BRANDT *Manuf. Iron & Steel* xxxii. 239 When the first portion of the rod enters the stands, there is no tension and the passes are overfilled but if there is tension when all the stands are rolling this will cause underfilling which will go on until the back end of the rod leaves the first stands, and the tension is relaxed, then overfilling will take place again. **1968** R. N. PARKINS *Mech. Treatm. Metals* iii. 135 The method adopted in this latter work was to determine the filling characteristic of a pass sequence, since this may be used to determine whether a given ingoing bar will under- or over-fill the pass. *Ibid.* ii. 78 Overfilling will cause the excess metal to spread into the roll joints..giving fins.

o·verfill, *sb.* *Metallurgy.* [f. the vb.] A projection on rolled metal due to the metal being too large for the aperture through which it was forced in rolling, so that the excess spread between the junction of the rolls; also, a bar or the like that is too large for the rolling it is to undergo.
1924 F. W. DENCER *Detailing & fabricating Struct. Steel* xxvii. 355 Material with overfills is not necessarily defective. **1929** *Rolling Mill Jrnl.* Jan. 16/1 If a bar when passing through a pass does not fill the groove, it is spoken of as an underfill. If it more than fills the groove and squeezes out at the sides, it is spoken of as an overfill and the material squeezing out at the sides is known as a fin or a flash. **1957** *Making, Shaping & Treating of Steel* (U.S. Steel) (ed. 7) xxx. 556/2 Overfills are broad and less sharp than fins. As a rule, overfills occur more frequently than fins and in many cases are associated on the same bar with underfills. **1964** N. WEINSTEIN tr. *Polukhin's Rolling Mill Pract.* xvii. 286 Fills and overfills obtained in rolling section steel are illustrated.

over-fish, *v.* Add: Now used esp. in reference to marine fishing grounds and the types of fish or shellfish caught there. (Later examples.) **over-fishing** *vbl. sb.* (Later examples).
1925 J. T. JENKINS *Fishes Brit. Isles* 137 All this statistical information has an important bearing on the vexed question of over-fishing [of cod]. **1941** *Sun* (Baltimore) 29 Nov. 6/3 Next year we won't have any oysters, and then, no doubt, we will be told that we overfished the beds this year. **1946** *Nature* 10 Aug. 189/2 Russell and Graham, discussing the overfishing problem, have recently emphasized that sea fisheries under present conditions have reached, if not over-reached, the limits of profitable yield. **1958** *New Statesman* 23 Aug. 218/1 Everybody who knows anything about fish..is anxious about over-fishing. **1973** *Daily Colonist* (Victoria, B.C.) 21 June 3/3 Industry spokesmen..have been pressing for action by Ottawa to halt alleged overfishing by foreign fleets. **1975** *Nature* 24 Jan. 290/3 Some [species], notably the European hake..have been heavily overfished.

overfit (ōu·vəɹfit), *a.* *Physical Geogr.* [f. OVER- 28, after *misfit.*] Pertaining to or designating a stream which, if its average flow in the past was at present-day levels, would be expected to have eroded a larger valley than it has done.
1913 *Ann. Assoc. Amer. Geogr.* III. 18 Conversely, the capturing branch, AB, and its river, BD, below the entrance of the capturing branch, being increased in volume, should exhibit for a time an overfit relation to their valley curves, in the sense of actively enlarging them. **1954** W. D. THORNBURY *Princ. Geomorphol.* vi. 156 It is difficult to cite examples of overfit rivers, or streams with floodplains too small for the size of the stream. Hence there may well be a question whether overfit streams exist. The reason..may be that a stream cannot long remain overfit, for an increase in volume will be accompanied by increased erosive power and rapid adjustment of valley size. **1968** R. W. FAIRBRIDGE *Encycl. Geomorphol.* 706/2 Davis..also recognizes overfit streams.

overflight (ōu·vəɹflait). [f. OVER- 4 + FLIGHT *sb.*[1]] The flight of an aircraft over specified territory without landing. Formerly also used of birds. Also *transf.*
1598 FLORIO *Worlde of Wordes* 383/1 *Soruólo*, an ouer-flight, a surflight. **1883** J. S. STALLYBRASS tr. *Grimm's Teutonic Mythol.* III. xxxv. 1133 Our early ages appear also to have seen a meaning in the *overflight* of certain birds. **1958** *Times* 8 Aug. 7/2 United States aircraft have now been given permission to fly over, but it was not clear yesterday whether British aircraft had permission. The question of overflights has been given much prominence. **1960** *N.Y. Times* 12 June 6E/2 The U-2 reconnaissance 'overflights' provided, by aerial photography and tape recording of Soviet radio and radar emissions, the most important intelligence gathered by the C.I.A. **1966** T. PYNCHON *Crying of Lot 49* ii. 37 The can [of hair spray] hit the floor, something broke,..propelling the can swiftly about the bathroom. She looked up,..her field of vision cut across by wild, flashing overflights of the can, whose pressure seemed inexhaustible. **1970** C. DUERDEN *Noise Abatement* v. 88 Overflights are the problem. Once they are experienced, the reaction of the general public will be the deciding factor and supersonic flights may be prohibited over this country. **1970** *New Scientist* 1 Oct. 37/2 A single supersonic overflight cracked 95 per cent of the eggs of the terns breeding on the Dry Tortugas in Florida. **1973** *Sci. Amer.* Feb. 17/3 When Francis Gary Powers was shot down in May, 1960, the U-2 overflights ended, except for certain minor incursions. **1976** *Daily Tel.* 21 Jan. 2/6 It was bewildering that Britain and France, builders of the plane [*sc.* Concorde], did not allow supersonic overflights.

overflood, *v.* Add: (Further example.) Hence **overfloo·ded** *ppl. a.*; **overfloo·ding** *vbl. sb.* and *ppl. a.*
1881 E. W. HAMILTON *Diary* 3 Nov. (1972) I. 182 The danger for the moment in Ireland seems to be the over-flooding of the Land Court. **1921** W. DE LA MARE *Veil* 47 Then silence, and o'er-flooding noon. **1955** E. POUND *Classic Anthol.* I. 16 At the over-flooded ford. **1973** *Nature* 21/28 Dec. 450/2 The whole is cut by a dyke swarm which is especially dense at the contact between the layered plutonic rocks and the other components of the formation, and was subsequently overflooded by three different basalt series.

overflow, *sb.* Add: **2. c.** *Telephony.* A situation in which more calls are directed to a group of switches or lines than they are able to handle; a call so directed. Usu. *attrib.*
1924 J. G. MITCHELL *Mech. Manual Switching* i. 27 When the traffic reaches a peak so high as to prevent the prompt dispatch of calls by the operators on duty, those lines which are not receiving service will display a continuous signal on the overflow section. **1934** G. S. BERKELEY *Traffic & Trunking Princ. Automatic Telephony* vii. 146 The overflow meter..records the number of calls lost due to the insufficiency of the plant. *Ibid.* 147 One overflow meter is provided to record the number of overflows occurring on the two levels. **1962** J. RIORDAN *Stochastic Service Syst.* iii. 39 Turn now to the size of overflow traffic, and write $P_n(t; r)$ for the stationary probability that in an interval of length *t*, whose initial point is an arbitrarily chosen point, there are *n* overflow calls from the first *r* servers. **1972** *Guardian* 19 Feb. 8/6 The BBC..recorded 4,300 overflow calls. **1974** R. N. RENTON *Internat. Telex Service* x. 383/2 Calls failing to find a free outlet in the level will hunt to the 11th bank contact, receive the appropriate service signal and operate the overflow meter. *Ibid.*, Tables of critical overflow are published.
d. *Computers.* The generation of a number having more digits than the capacity of the device holding it; also, the excess digit(s). Freq. *attrib.*
1951 *Proc. IRE* XXXIX. 275/1 *Overflow.* (1) The condition which arises when the result of an arithmetic operation exceeds the capacity of the number representation in a digital computer. (2) The carry digit arising from this condition. **1959** J. JEENEL *Programming for Digital Computers* iv. 187 The instruction..will cause program execution to proceed from 0104 to 0105 regardless of whether the overflow indicator is on or off. **1965** SWALLOW & PRICE *Elem. Computer Programming* xv. 279 The arithmetic operations of addition and subtraction are slightly more complex in the IBM 1401 than in the 141. One difference is in the handling of overflows. *Ibid.* 287 Whenever an overflow bit is carried into the zone of the high-order digit during an ADD or SUBTRACT operation, this indicator is turned on. **1969** P. B. JORDAIN *Condensed Computer Encycl.* 364 The most significant digit(s) are considered to be overflow. **1973** C. W. GEAR *Introd. Computer Sci.* vi. 241 Most systems provide monitor subroutines that can be called to specify what is to happen in the case of overflow or underflow.
4. b. *Geol.* A natural notch or channel formed by water overflowing from a lake.
1902 *Q. Jrnl. Geol. Soc.* LVIII. 481 When the watershed is very uniform in height, and the ice has at one stage actually surmounted it, then several parallel overflows may be developed out of the gutters which are trenched in the outer slope by water flowing off the ice itself. The overflows..which cut through the Northern Cleveland watershed above the village of Egton are typical of this arrangement. **1973** R. J. PRICE *Glacial & Fluvioglacial Landforms* v. 128 So long as evidence for the existence of the ice-dammed lakes does not simply consist of the meltwater channels themselves, the interpretation of such meltwater channels as overflows cannot be disputed.
5. *overflow meeting* (further examples); **overflow channel** *Geol.* = sense 4 b above; **overflow table,** a table used to accommodate extra people attending a dinner, meeting, etc.
1902 *Q. Jrnl. Geol. Soc.* LVIII. 473 The criteria by which ancient extra-morainic lakes can be recognized are mainly four:—(1) beaches; (2) deltas; (3) floor-deposits; and (4) overflow-channels. **1969** BENNISON & WRIGHT *Geol. Hist. Brit. Isles* xvi. 356 Associated water-cut channels (called overflow channels) from ice-dammed lakes generally thought to result from water flowing marginally along ice sheets may, in many cases, also be a product of subglacial drainage. **1923** *Radio Times* 28 Sept. 2/1 The voice of a public man has been..made to operate loud-speakers of overflow meetings. **1974** *Times* 20 Feb. 1/1 Two eggs were hurled at Mr Wilson as he struggled through a crowd to enter Oxford Town Hall last night to address an overflow meeting of more than 1,000 people. The eggs went wide. **1973** A. BEHREND *Samarai Affair* i. 14 There was a smaller overflow table where the members of the Pilotage Committee..liked to congregate.

overflow, *v.* Add: **I. 3. a.** (Later examples.) Also *fig.*
1890 *Forum* (N.Y.) VIII. vi. 700 The fiery lava of passion overflowing the appointed bounds. **1916** G. B. SHAW *Androcles & Lion* p. cviii, The causes which have produced this sudden clearing of the air include the transformation of many modern States, notably..the tight little Island of Britain, into empires which overflow the frontiers of all the Churches. **1973** R. J. WILLIAMS in D. J. Wiseman *Peoples of Old Testament Times* iv. 80 Fed by the yearly rains in the Abyssinian highlands, the river [*sc.* the Nile] rose steadily and overflowed its banks.

overflowing, *vbl. sb.* Add: **2.** Esp. in phr. (*full* or *to fill*) *to overflowing*: more than full, so as to overflow.
1879 R. A. STERNDALE *Afghan Knife* II. vii. 69 In the meantime fugitives kept pouring into the house, which was full to overflowing. **1898** SKEEL & BREARLEY *King Washington* (1899) xii. 75 The boys..were filled to overflowing with the excitement of the hour. **1920** H. M. PIM *Short Hist. Celtic Philos.* v. 62 They might have produced a Tartarus, and filled it to overflowing. **1961** *Washington Post* 1 June A24 The jails were filled to overflowing with political prisoners who had incurred his displeasure.

overflowingness. (Further example.)
1883 G. MEREDITH *Let.* 30 Oct. (1970) II. 718, I have been hearing from Will of your radiant overflowingness.

overfly, *v.* Add: **1. d.** Of an aircraft or its passengers: to fly over (a specified point, area, etc.). Also *absol.* Hence **o·verflying** *vbl. sb.* and *ppl. a.*
1944 L. L. SELL *Eng.-Spanish Techn. Dict.* 861 Overfly the field, to. **1946** *Happy Landings* (Air Ministry) July 7/1 Istres was planned as our first stop, but a 'mistral' put that airfield out of action and we had to overfly to Elmas (Sardinia). **1948** *Shell Aviation News* No. 120. 8/3 Following winds enabled us to overfly Marseilles, which we had intended as our first stop. **1957** *Times* 19 Sept. 8 A proposal by South Africa that she should be granted certain over-flying rights in the High Commission Territories has been accepted in principle by the British Government. **1958** 'N. SHUTE' *Rainbow & Rose* vi. 228 With the greater range of these aircraft we could overfly Canton and go direct to Fiji. **1965** *Listener* 2 Sept. 338/1 Any society will have..its quota of nuclear installations where overflying is prohibited. **1966** *New Scientist* 22 Dec. 669/2 The aircraft overflown by the supersonic liners, will 'take a fine old walloping'. **1973** *Daily Tel.* 17 Mar. 7/1 Mr Josefsson has been fighting for years to get rid of the Nato base at Keflavik, a vital link in tracking overflying Russian aircraft. **1973** 'A. HALL' *Tango Briefing* ii. 22 Did you get official overflying permission?..Did you get official permission from the Algerian government to overfly their territory? **1977** *R.A.F. News* 5–18 Jan. 2/4 By remarkable coincidence, the golf course was overflown by two Phantom aircraft on a local low-level route.
2. (Further example.)
1954 D. A. BANNERMAN *Birds Brit. Isles* III. 278 It is of interest to note that the northern birds in general arrive somewhat earlier in the south of the Iberian Peninsula and so 'over-fly' their relatives from more southerly breeding places.
4. To exceed (the maximum flying-time allowed by regulations).
1966 *Daily Tel.* 1 Nov. 12/3 One or two of the pilots had 'overflown' their regulation number of hours and had to present themselves for 'medicals'.

overfulfil, *v.* Restrict † *Obs.* to sense in Dict. and add: Also (*U.S.*) **overfulfill.** **2.** To achieve more than the mere fulfilment of (a plan, goal, etc.); to reach (a target) before the expected time. So **o:verfulfi·lment,** (*U.S.*) **-fulfillment.**
1950 *Sun* (Baltimore) 2 June 14/7 When the Russian whalers returned to their home port of Odessa on the Black Sea they had overfulfilled their production plan by 29·9 per cent. *Ibid.* 7 Aug. 10/2 If present rates [of Soviet production] are continued to December 31, a massive overfulfilment of the 1950 goals will result. **1952** KOESTLER *Arrow in Blue* 70 To omit a single one [*sc.* slogan]—say 'the strengthening of the production-offensive for the over-fulfilment of the light metal industry's revised counter-plan'..would have laid the lecturer open to the accusation [etc.]. **1964** K. G. LOCKYER *Introd. Critical Path Analysis* vii. 61 Bars to the *left* of the observation line represent under-fulfillment, whilst those to the *right* represent over-fulfillment. **1966** J. PORTER *Sour Cream* iv. 51 Even the waitresses had something to take their minds off their feet and the difficulties of over-fulfilling the latest five-year-plan. *Ibid.* v. 68 The current ten-year plan was going to be over-fulfilled. **1971** *New Society* 25 Mar. 475/2 Factories have undertaken commitments to overfulfil production targets in honour of the congress. **1975** *Nature* 16 Oct. 527/2 Soviet planners see agriculture as being essentially one more branch of industrial production, amenable to the same system of fulfilment and overfulfilment of plans. **1976** *Chinese Law & Govt.* IX. 101 We should fulfill or overfulfill the Fourth Five-Year Plan in 1975.

o·verfur. [OVER- 8.] The outer layer of an animal's fur.
1913 [see *guard hair* s.v. *GUARD *sb.* 18.] **1968** J. IRONSIDE *Fashion Alphabet* 153 *Top fur.* This refers to the guard hairs or overfur; it is stiffer, coarser and usually darker than the underfur.

overgaze, *v.* **2.** (Later example.)
1879 A. T. DE VERE *Legends Saxon Saints* 41 Meadow banks, Not yet o'er-gazed by Windsor's crested steep Or Reading's tower.

over-ge·neralize, *v.* [OVER- 27.] *intr.* To draw general conclusions from inadequate data; to argue more widely than is justified by the available evidence, by circumstances, etc. Also *trans.*, to draw an over-general conclusion from (data, circumstances, etc.). Hence **o:ver-generaliza·tion; over-ge·neralized** *ppl. a.*
1904 [listed in Dict. s.v. OVER- 27 a]. **1937** *Mind* XLVI. 243 He over-generalises. For example, he claims that implicit is superior to explicit comparison. **1947** J. G. WEIGHTMAN *On Language & Writing* 87 In trying to think with precision we oscillate between two extremes, over-generalization and over-particularization. **1956** *Nature* 10 Mar. 478/2 The properties of larval blow-fly carbohydrases do not completely exemplify Weidenhagen's hypothesis of bond specificity, the over-generalized nature of which has been criticized more recently. **1957** K. A. WITTFOGEL *Oriental Despotism* 370 Overgeneralizing the experience of a rapidly changing Western world, they naively postulated a simple, unilinear, and progressive course of societal growth. **1970** S. L. BARRACLOUGH in I. L. Horowitz *Masses in Lat. Amer.* iv. 97 To avoid falling into the more obvious errors of overgeneralization about Latin America as a whole. **1971** *Sci. Amer.* Aug. 75/3 Although Sherrington was careful not to overgeneralize from his findings, less circumspect workers soon adopted the extreme position that most animal behavior consists of reflexes. **1975** R. V. REDINGER *Geo. Eliot* (1976) v. 245 Slanted and overgeneralized as are many of its discussions, this book fulfilled a real need.

overget, *v.* Add: **4.** To prevail upon; to take possession of (a person).
1904 *Tradesman's Price-List* (Herbert Morris & Bastert) 11 Similarity of appearance so far overgets a customer as to induce him to pass us. **1928** *Sunday Dispatch* 16 Sept. 2/3 The thought to marry Fanny overgot the man, and he set out to see if it could be done. **1953** D. D. C. P. MOULD *Rock of Truth* xi. 193 When I had been a Catholic some time and perhaps begun to grasp some idea of what God was really like, then the fear of hell did overget me.

overglaze, *sb.* Add: (Earlier and later examples.) Also in *Painting.*
1880 *Harper's Mag.* Nov. 904/1 The work most familiar to us as taught in America during the last three or four years has all been on the over-glaze. **1947** J. C. RICH *Materials & Methods Sculpture* ii. 49 Overglazes can be applied in several ways, the most frequent of which are direct painting with a soft brush, and fine spraying. **1948** F. A. STAPLES *Water-Color Painting* x. 116 Now that the first tone..is dry, the second, an overglaze, is painted. **1974** SAVAGE & NEWMAN *Illustr. Dict. Ceramics* 208 Overglaze..is by means of colours termed enamels.

overglaze, *a.* (Earlier and later examples.)
1879 J. C. L. SPARKES *Handbk. Pract. Pottery Painting* 28 Oil mediums may be used for over-glaze and for under-glaze work. *Ibid.* 31 In painting on china, earthenware and all 'over-glaze' ware with enamel colours the procedure is as follows. **1881** *Harper's Mag.* May 835/1 It was..the most extensive and satisfactory exhibit of amateur overglaze decoration made up to that time. **1936** *Burlington Mag.* Oct. 145/2 Figural designs in polychrome over-glaze painting. **1970** *Ashmolean Mus. Rep. Visitors 1969* 46 Ginger jar with underglaze blue landscape decoration and later added dragons and flowers in overglaze colours. **1973** *Country Life* 7 June 1591/3 By the early 1770s..the head colourman, Constantine Smith, had invented a dark, semi-matt, over-glaze blue enamel of exceptional brilliance.

overglaze, *v.* Add: Hence **o:vergla·zing** *vbl. sb.*
1947 J. C. RICH *Materials & Methods Sculpture* ii. 49 Overglazing is a process in which color or overglazes are applied over glazes for decorative effects; they require an additional exposure to heat. *a* **1977** *Harrison Mayer Ltd. Catal.* 18/2 Over-glazing. This usually occurs on very porous or easy fired ware. Over-glazing can frequently lead to stuck ware and can cause underglaze colours to run.

overgo, *v.* Add: **4.** (Later examples.)
1917 *Ampleforth Jrnl.* Jan. 127 He resolves to emulate, perhaps even to overgo, the 'Orlando Furioso'. **1923** W. RENWICK *Spenser Selections* p. ix, Ronsard's *Franciade*..*Orlando Furioso.* Spenser would overgo both.
5. (Later example.)
1924 [see *FEATHER *sb.* 1 b].

overgoing, *vbl. sb.* (Later example in sense 'transgression'.)
1903 J. K. JEROME *Tea-Table Talk* v. 95, I was very severe upon both the shortcomings and the overgoings of man.

over-govern, *v.* Add: **2.** (Later examples.)
1938 *Sun* (Baltimore) 16 Apr. 8 For five years he [*sc.* F. D. Roosevelt] has overgoverned. It was not enough to have specific reform of specific evils, say, stock market practices. **1976** *Daily Record* (Glasgow) 29 Nov. 14/2 Another sign of distrust is shown by the fact that nine out of ten replies believe that Scotland is in danger of being over-governed.
So **over-go·verned** *ppl. a.*; **over-go·verning** *vbl. sb.*; **over-government** (further examples).
1847 J. S. MILL *Lett.* (1910) I. 131 The habitual over-governing by which power and importance are too exclusively concentrated upon the Government and its functionaries. **1848** —— *Pol. Econ.* II. v. xi. 529 The inferior capacity for political life which has hitherto characterized the over-governed countries of the Continent. **1861** —— *Repr. Govt.* iv. 83 The more popular the institutions,..the more monstrous the overgovernment exercised by all over each, and by the executive over all. **1976** *Times* 13 May 4/2 Scotland is over-governed, over-taxed and over-subsidized. **1976** *Scotsman* 15 Dec. 12/3 The Tories may have added to 'over government' by their reform of local government.

overgrazed (ōu·vəɹgrēⁱzd), *ppl. a.* [OVER- 27.] Of grassland: made susceptible to erosion by the destruction of vegetation through excessive grazing. Also *fig.*
1929 WEAVER & CLEMENTS *Plant Ecol.* vii. 142 The latter [*sc.* dominant species] are handicapped..in over-grazed mixed prairie. **1949** *Pacific Discovery* July–Aug. 2/1 In the South I have seen cattle tracks in overgrazed land become gullies ten feet deep. **1972** 'G. NORTH' *Sgt. Cluff rings True* i. 10 Thin, patchy grass, overgrazed, badly in need of lime. **1977** *Times Lit. Suppl.* 1 Apr. 402/2 He [*sc.* Christopher Isherwood] may cover ground..that no doubt will be covered until there is not a blade of grass left on it when the PhD herd have moved on from the overgrazed 1920s.

overgra·zing, *vbl. sb.* [OVER- 29 a.] Damage to vegetation, esp. grassland, by excessive grazing.
1935 *Discovery* Aug. 232/2 Overgrazing has removed most of the nutritious native grasses. **1943** J. S. HUXLEY *TVA* vi. 33 The method of row cropping allowed the top-soil to be washed away. Overgrazing followed, and now the land is completely useless. **1946** *Nature* 2 Nov. 606/2 There is also the risk of over-grazing, especially where the natural vegetation consists rather of shrubs and bushes than of turf. **1956** *Ibid.* 3 Mar. 417/1 Failure of natural regeneration in juniper forests in Somaliland is now attributed to micro-climatic changes following over-grazing. **1964** E. J. H. CORNER *Life of Plants* xv. 268 We can form no idea of the destructive power..of over-grazing where trees, for lack of rain, are disappearing. **1971** *Country Life* 12 Aug. 390/2 Examination of the intensively grazed patches showed no overgrazing such as occurs when hill sheep become concentrated on heather. **1976** *Conservation News* Sept./Oct. 4/1 Over-grazing..has led to the whole vast area becoming degraded.

overground, *sb.* Restrict † *Obs.* to sense in Dict. and add later examples (see quots.: in most modern senses used in deliberate antithesis to *underground*).
1931 *N. & Q.* 11 Apr. 267/2 There has been for some long time past a line of motor-omnibuses running from Hadley Woods and Barnet to Victoria Station on the sides of which..is the word 'Overground'. **1966** *Evening Standard* 24 Feb. 16/2 The overground is an aerial railway with completely automatic operation. *Ibid.*, Buses would run through the suburbs to the edge of the city's centre... Passengers would change on to one or other of the overgrounds, a series of independent, six-mile rail loops. **1969** *Gandalf's Garden* VI. 11/1 *Overground*, like the Underground, from which it grew, it exists in the spirits of those who are living it, in the act of seeking a deeper understanding of life, in the expression of an aspirational lifestyle working in harmony with natural and mystical laws.

overground, *a.* Add: **b.** *fig.* Overt; unconcealed; publicly acknowledged. Opp. UNDERGROUND *a.* 4.
1943 *Ann. Reg. 1942* 244 'Overground' resistance to the Germans was as strong as ever. **1961** *Times* 14 Jan. 7/7 But wherever they went the journalists were approached by 'overground' sympathizers with the independence fighters. **1970** *New York* 16 Nov. 50/2 They have been..whipped around in the over- and underground press. **1971** *Times* 15 Jan. 12/8 Now even overground publishers are jumping on the revolutionary bandwagon. **1971** *Guardian* 27 Sept. 14/6 Overground media generally treat sex, drugs, and violence in a misguided matronly tone.

o:vergrou·nd, *adv.* [OVER- 31.] Above the ground; into the open; opp. *underground.*
1930 *Sat. Even. Post* 22 Mar. 15/2 There was a rumbling as of a subway train heard over-ground. **1944** F. CLUNE *Red Heart* 6 After descending a few feet underground to have a look at the lode, I felt a desire to fly a few thousand overground to get a different angle of view. **1951** E. D. M. *St. Philomena the Wonder-Worker* (ed. 6) i. 16 The Christians could not safely perform the burial services in the presence of their heathen enemies over ground. **1963** *Times* 22 Apr. 11/1 Thought has been given to procedures by which rebels would 'come overground' and give up their weapons. **1968** *Economist* 3 Feb. 15/2 This seemed likely to mean that commercial gaming would continue underground in the pre-1960 manner..; but instead gaming popped overground into open places. **1973** *Times Lit. Suppl.* 23 Nov. 1455/4 Douglas Hayes has had an underground reputation... It is time for that reputation to appear overground.

overgrown, *ppl. a.* Add: **1.** (Later examples.) Also *fig.*
1804–20 W. BLAKE *Jerusalem* iv, in *Compl. Writings* (1972) 732 Of blood thro' all my nervous limbs; soon overgrown in roots I shall be closed from thy sight. **1907** R. BROOKE *Let.* Dec. in E. Marsh *Rupert Brooke* (1918) 26, I have already..got some faith in the real, sometimes overgrown, goodness of all men. **1938** E. AMBLER *Cause for Alarm* xvi. 273 There's an old disused road... It's overgrown by trees now. **1975** A. DILLARD *Pilgrim at Tinker Creek* xii. 213 The forested cliffs..gave way to overgrown terraces.

overgrowth. 2. (Earlier and later examples, in *lit.* sense.)
1879 *Jrnl. Chem. Soc.* XXXVI. 769 Those substances only should be considered as isomorphous which are capable of forming mixed crystals or which are capable of forming 'overgrowths' (*Ueberwachsen*), *i.e.*, when a crystal of the one is suspended in a solution of the other, the crystal increases in size, owing to the deposition on it of the substance in solution. **1896** G. B. SHAW *Let.* 14 Jan. (1965) I. 586 Those auburn tresses..concealed a grey—nay, a *white*—undergrowth, which is now an overgrowth. **1940** GLASSTONE *Text-bk. Physical Chem.* (1941) v. 342 Many instances of the production of overgrowths, e.g.,..alums, monoclinic double sulfates, etc., are known. **1971** I. G. GASS et al. *Understanding Earth* xiii. 170/2 Silica can be deposited as overgrowths around detrital quartz grains.

overhand, *adv.* and *a.* Add: **A.** *adv.* **3. b.** *Archery.* (See quots.)
1875 *Encycl. Brit.* II. 378/2 Shooting over-hand is to shoot at the mark over the bow-hand. **1939** P. H. GORDON *New Archery* 404 *Overhand*, of shooting, same as..*Forehand.*
B. *adj.* **1.** Delete † *Obs.* and add later example.
1888 KIPLING *Barrack-Room Ballads* (1892) 122 The overhand stabbing-cut silenced the yell.
2. (Earlier and later examples.)
1828 in W. Denison *Cricket: Sk. Players* (1846) 44 It is the over-hand delivery to which I principally object,..it has all the qualities of a throw, except the force. **1975** *New Yorker* 17 Nov. 154/2 The Oakland scouting report on him warned he had six pitches..all of which he could serve up from the sidearm, three-quarters, or overhand sectors, and points in between.
b. *Swimming.* Applied to a variety of the side-stroke in the performance of which one hand is raised above the water and carried forward. Also *ellipt.* as *sb.*
c **1881** 'CAPT. CRAWLEY' *Swimming* 38 There are two styles of Side Swimming, severally known as the side-stroke and the over-hand. **1886** J. FINNEY *Hints on Swimming* 5 Taking into consideration the high rate of speed attainable by its means, the overhand stroke, when exhibited by a first-class swimmer, far exceeds any other style. **1905** 'NATATOR' *Swimming* 13 Overhand stroke. To change from the ordinary side stroke to the Overhand is simple. When the upper arm has finished its stroke, advance it *above* water, beyond the head, where it again enters, with the palm of the hand facing out. **1931** J. BUCHAN *Blanket of Dark* xiv. 260 Peter's long arms in an overhand stroke devoured the waters.
c. *Lawn Tennis.* Of a stroke: made with the racket above the arm or shoulder.
1889 H. W. W. WILBERFORCE *Lawn Tennis* 30 The form of service almost universally used is the overhand service. **1900** A. E. T. WATSON *Young Sportsman* 379 For the ordinary overhand service the ball should be thrown up in line with the right ear and slightly backwards. **1911** *Encycl. Brit.* XVI. 301/2 High overhand service, by which alone any great pace can be obtained, was first perfected by the brothers Renshaw between 1880 and 1890. **1978** J. SYMONS *Blackheath Poisonings* II. 105 For..ladies an overhand service was a great waste of strength.

overhand, *v.* Delete *arch.* and add further examples. So **overhanding** *vbl. sb.*
1908 M. H. MORGAN *How to dress Doll* ii. 20 Overhanding is the real sewing of the seam... simply sewing over-and-over, close to the edge, with very small stitches. *Ibid.* v. 51 Overhand the neck and armholes with lace. **1964** *McCall's Sewing* ii. 31/1 *Overhanding*, a straight stitch used to hold finished edges together when a strong, flat, invisible seam is needed, as in table linen, undergarments, sewing on lace or patching.

over-handed, *a.* Add: **2. b.** *overhanded knot* = *overhand knot* (OVERHAND *a.* 4).
1883 *Man. Seamanship for Boys' Training Ships R. Navy* (Admiralty) (1886) 91 *Q.* How do you make a reef-knot..? *A.*..First make an over-handed knot round the foot of the sail, [etc.].

overhang, *sb.* **a.** Delete 'Chiefly *Naut.*' and add further examples.
1908 H. G. WELLS *War in Air* v. 158 The overhang of the gas-chambers intervened. **1919** T. K. HOLMES *Man from Tall Timber* ix. 101 The two women..lived alone on Paradise Knoll, just under the overhang of its crown. **1924** J. BUCHAN *Three Hostages* xxi. 306 The corrie face..seemed nothing but slabs and rotten rocks, while the few chimneys had ugly overhangs. **1940** W. FAULKNER *Hamlet* I. i. 19 Once more Varner looked down into the cold impenetrable agate eyes beneath the writhen overhang of brows. **1957** *Brit. Commonwealth Forestry Terminol.* II. 89 *Hang*, the forward lean given to the blades in a vertical frame saw so that the teeth engage with the wood in succession and not simultaneously (the *lead-in*). Syn. *Overhang.* *Ibid.* 127 *Overhang*, the forward slope given to the front of a stack of converted timber for protection against weathering. **1960** *Times* 22 Oct. 9/4 One lone perch will take up residence in a deep undercut or an overhang of bushes in much the same manner as a big trout. **1973** C. BONINGTON *Next Horizon* v. 81 The rock juts steeply above, in a series of bristling overhangs, cut by a broken crack. **1975** *New Yorker* 21 Apr. 34/2 She retreated only after high winds began to shred the plastic overhang that had been keeping the stage dry.
b. *Electr. Engin.* The part of an armature winding which projects beyond the armature core.
1915 M. WALKER *Specification & Design Dynamo-Electr. Machinery* vii. 172 (*caption*) Dimensions of the overhang of concentric coils. **1936** SAY & PINK *Performance & Design A.C. Machines* x. 193 The problem of the estimation of eddy current losses in overhang conductors is very difficult.
c. *Aeronaut.* (*a*) (The length of) the part of a wing beyond its outermost point of support; (*b*) in a biplane or multiplane, (the length of) the part of a wing that extends beyond the tip of an adjacent wing.
1915 *Flight* 9 Apr. 248/2 In plan form the main planes, of which the upper one has a slight overhang, have a pronounced taper towards the tips. **1919** PIPPARD & PRITCHARD *Aeroplane Struct.* iv. 17 The position of the top plane extending beyond the outermost interplane strut is called the extension plane or the overhang. **1928** V. W. PAGE *Mod. Aircraft* v. 180 A typical training biplane which has both a pronounced forward stagger and an overhang as well. **1933** W. MUNRO *Marine Aircraft Design* vi. 91 In older types of biplanes the upper wing used to be built with a much longer span and overhang than the lower wing. **1953** J. H. STEVENS *Shape of Aeroplane* i. 29 The weight of the overhang when on the ground was taken by wires attached to kingposts protruding from the top plane above the outer pair of interplane struts. *Ibid.*, It was common practice on early aeroplanes to make the upper wing of larger span than the lower. This extra length was called the overhang. **1977** *Aeroplane Monthly* May 274/2 The B.E.2e was..distinguished by the large overhang of the upper wing extensions.
d. In a turntable unit, the distance between the stylus point and the centre of the turntable when the pickup arm is placed so that these two points and the pivot of the pickup are in line and the turntable centre is between the other two points.
1937 *Radio Engin.* Mar. 17/1 The curves in Fig. 3 have been calculated to show the initial tracking angle which occurs with a conventional pickup arm as a function of the radius R of the playing circle, the length of the arm L, and the overhang D, which is the distance between the center of the turntable and the needle point when the needle point is in line with the centers of the base and turntable. **1945** *Electronics* Mar. 111/2 If the needle point overhang *D* is 13/16 inch..the tracking angle varies from 32 deg at 6 inches through 27 deg at 3·3 inches back to 32 deg at the 2-inch radius. **1953** G. A. BRIGGS *Sound Reproduction* (ed. 3) xxvi. 330 The calculations for best offset angle and overhang..were made on the basis of zero tracking error at the inside groove of a standard 12" record. **1958** S. KELLY in E. Molloy *High Fidelity Sound Reproduction* viii. 139 With a 7½ in. arm, this [*sc.* minimum distortion] requires an offset angle of 24½° and an overhang of 0·56 in. **1976** *Gramophone* Feb. 1406/1 The headshell has..a slot screw fitting to permit accurate setting of the effective pivot-to-stylus distance (overhang).
e. *Econ.* An excess of (estimated) expenditure over available or budgeted funds.
1953 *Sun* (Baltimore) 13 May 4/2 Humphrey pointed out that no revenue has been provided to cover the $81,000,000,000 'overhang' of appropriations which will be outstanding at the close of this fiscal year. **1954** *Britannica Bk. of Year* 638/1 Economic policy produced *Overhang*, an appropriation in excess of actual funds. **1974** *Financial Times* 19 July 22/8 The dollar overhang has disappeared, the Middle East war has ended. **1976** *Washington Post* 13 Mar. A19/3 That [*sc.* a big increase in the outflow of U.S. capital] meant a sizeable addition to the existing 'overhang' of dollars, amounting to nearly $200 billion held outside of the United States.
f. In sound recording and reproduction, the (usu. undesired) continued oscillation of a system after the cessation of the signal causing it; *spec.* that of a loudspeaker, esp. in the bass when the cone is insufficiently damped near its resonant frequency. Cf. *HANG-OVER, HANGOVER 3, quots. 1961, 1967.
1971 J. EARL *How to choose Pickups & Loudspeakers* v. 126 When the cone oscillates at resonance it vibrates quite violently unless well damped. This can cause 'overhang' effects at the bass end. **1975** G. J. KING *Audio Handbk.* ii. 34 A very low R_s in parallel with the loudspeaker inhibits overshoot and rings on transient type signal, and this is particularly desirable at the loudspeaker's bass resonance frequency where, without such damping, the cone can oscillate vigorously when triggered by a transient, an effect which is responsible for 'overhang' and 'boomy' bass. **1976** *Gramophone* Dec. 1028/3 The sound has warmth and resonance (the long overhang of the King's acoustic is expertly handled without the slightest muddying of the textures). **1977** *Ibid.* May 1774/3 Square wave response of all the cartridges showed virtually no overhang and an adequate rise time with little ringing.

overhaul, *v.* **3.** (Further *transf.* examples.)
1933 D. L. SAYERS *Hangman's Holiday* 187 He..drove on, overhauling the police car. **1955** *Times* 12 May 4/3 (*heading*) American boxers overhauled. **1976** *Horse & Hound* 3 Dec. 53/2 Their pilot..circling left-handed..doubled back to Loscar Common Plantation again and was overhauled in some kale after a good 35 min.

overhead, *adv., sb., a.* Add: **B.** *sb.* **2.** That which is above; the firmament.
1865 G. M. HOPKINS *Poems* (1967) 151 The grass was red And long, the trees were colour'd, but the o'er-head, Milky and dark, with an attuning stress Controll'd them to a grey-green temperateness. **1911** *Chambers's Jrnl.* Jan. 79/1 It forms a handy guide, philosopher, and friend to the vast unfathomable overhead. **1959** E. COLLIER *Three against Wilderness* i. 8 Live sparks rocketed up into the smoky overhead.
3. (Freq. *pl.*) Ellipt. for *overhead charges, expenses*, etc.
1914 *Automobile Topics* XXXIV. 31/2 One of the numerous fallacies of business..is the argument that the small organization is in a position to serve the customer to better advantage for the reason that its 'overheads' are small. **1915** *Lit. Digest* 21 Aug. 360/1 (Advt.), Her typewriter is standing idle and adding to 'overhead'. **1922** *Public Opinion* 29 Dec. 629/2 We are able to reduce overheads through the employment of far more automatic machinery. **1930** J. B. PRIESTLEY *Angel Pavement* i. 36 The first thing, the very first thing, we've got to do is to reduce the overheads in this business. **1954** *Encounter* Dec. 79/1 The two million families..are the enemies of the workers in the modern sectors of the French economy. Their overhead is more than the traffic will bear, and the worker feels, if he does not understand, this. **1972** *Accountant* 17 Aug. 215/1 Work in progress to be stated at cost including overhead. **1972** *Computer Jrnl.* XV. 199/1 A possible objection to the use of streams might be that the overheads associated with their structure make them excessively inefficient. **1974** *Terminol. Managem. & Financial Accountancy* (Inst. Cost and Managem. Accountants) 19 *Overhead*, the total cost of indirect materials, wages and expenses.
4. *Lawn Tennis.* An overhead stroke.
1969 *New Yorker* 14 June 46/3 His overhead is hit with his whole arm—no mere flick of the wrist. **1972** D. DELMAN *Sudden Death* vi. 145 I'll hit lobs to you so you can work the kinks out of your overhead. **1977** *Transatlantic Rev.* LX. 108 He slashes, he wheels, he whaps an easy overhead into the net.
C. *adj.* **1. a.** (Further examples.)
1917 'CONTACT' *Airman's Outings* 36, I awoke to the roar of engines, followed by an overhead drone as a party of bombers circled round until they were ready to start. **1921** *Times Lit. Suppl.* 8 Sept. 574/3 Stress is laid on the complete clearance of overhead cover [from teak plantations]. **1949** P. LATHAM in Aldiss & Harrison *Decade 1940s* (1975) 204 Stoddard turned off the overhead light. **1959** *Chambers's Encycl.* V. 97/2 With A.C. systems difficulties may arise due to the inductance of overhead lines and the capacitance of underground cables. **1976** *Billings* (Montana) *Gaz.* 16 June 10-C/4 (Advt.), Comm. Warehouse Space 300, 400, 600 and 1000 sq. ft., paved, lighted, security fence, overhead door, cold storage.
b. *overhead cam-shaft*, a cam-shaft mounted above the cylinder block of an internal-combustion engine; *overhead valve*, a valve in an internal-combustion engine which has its seat in the top of the combustion chamber, in the surface opposite the piston.
1912 R. W. A. BREWER *Motor Car Construction* iii. 42 One of the principal objections to overhead valves in the past was the difficulty of driving an overhead camshaft... With the introduction of silent chain drives for camshafts these difficulties no longer exist. **1921** A. W. JUDGE *Automobile & Aircraft Engines* viii. 316 Another reason for the better volumetric efficiency of the overhead valve lies in the fact that in the case of the side-by-side valves the charge has to pass over a larger combustion chamber area before it arrives in the cylinder. **1958** *Times* 1 July 6/6 Instead of having a side valve engine of 1,265 c.c., the new model has the overhead valve 1,390 c.c. engine. **1966** *McGraw-Hill Encycl. Sci. & Technol.* XIV. 263/2 In overhead valve engines, the cam shaft may be mounted on the cylinder head near the valves. **1974** *Country Life* 21 Nov. 1579/1 The engine with its twin overhead cam-shafts is buzzy at high revs.
3. *Lawn Tennis.* Of a stroke: made with the racket above one's head.
1904 J. P. PARET et al. *Lawn Tennis* 345 *Overhead*, with the racket above the head. **1919** C. HIERONS *Lawn Tennis* xiv. 61 In overhead volleying there is far too much pat ball. **1925** K. MCKANE *Lawn Tennis* vi. 98 The most important of all overhead strokes—the service. **1951** HARMAN & MONROE *Use your Head in Tennis* v. 45 For the overhead slice serve, take hold of your racket handle in the eastern grip.
4. Of costs or expenses: incurred in the production of a batch of articles apart from the prime cost of each (cf. ONCOST), or in the upkeep of plant and premises.
1909 J. L. NICHOLSON *Factory Organization & Costs* i. 7 The distribution of manufacturing expenses, sometimes called overhead charges, and in other instances, indirect expense or burden. **1911** F. E. WEBNER *Factory Costs* xvii. 212 Under such a system most of the usual overhead expenses become direct. **1922** *Westm. Gaz.* 8 Dec. 6/2 The overhead cost of every factory that is gas-lighted will go up. **1930** A. H. CHURCH *Overhead Expense* I. i. 1 Overhead expense in manufacturing is defined usually as consisting of the so-called 'fixed' charges (such as rent, interest, depreciation, insurance, taxes, etc.) plus all that large class of expenditure on labor and materials which cannot be charged definitely to any given job or lot of product. **1958** J. F. MAGEE *Production Planning & Inventory Control* iii. 26 Under absorption costing, the value includes not only direct costs but also allocated overhead charges (usually only factory overhead). **1970** *Encycl. Brit.* XVI. 1167/2 Overhead costs are various business expenses that cannot be readily identified with specific products or services produced or sold. **1974** *Terminol. Managem. & Financial Accountancy* (Inst. Cost and Managem. Accountants) 42 *Overhead distribution sheet*, a columnar form used for the purpose of distribution of overhead expenditure over cost centres, and for the apportionment of the accumulated expenses of service cost centres over others.

overhear, *v.* Add: **3.** Also *absol.*
1706 SWIFT *Baucis & Philemon* in *Poems* (1958) I. 92 The Strangers overheard, and said [etc.]. **1913** *Cassell's Mag.* June 2/1 Glancing over his shoulder to make certain that the nurse hadn't overheard. **1922** JOYCE *Ulysses* 114 Mr Bloom, chapfallen, drew behind a few paces so as not to overhear. **1929** E. O'NEILL *Dynamo* III. i. 132 He lowers his voice carefully as if he didn't want the dynamo to overhear. **1976** M. MILLAR *Ask for me Tomorrow* (1977) ii. 19, I couldn't talk to you freely this morning because I didn't want..that witch in his office to overhear.

overheat, *v.* Add: **2.** *intr.* To become too hot.
1902 C. S. ROLLS in A. C. Harmsworth et al. *Motors & Motor-driving* 172 How to tell when a Motor is Overheating. **1908** *Westm. Gaz.* 27 Oct. 4/1 The engine overheated twice,..but this was when the car was taken out without any water in the radiator. **1950** *Sci. News* XV. 81 It may be found that the rocket overheats in spite of this cooling. **1971** *Sci. Amer.* Aug. 108/3 The compressors developed the required low pressure but overheated after several hours of continuous use. **1974** *Country Life* 21 Mar. 654/3 The mill oversped, the stones overheated, and the mill caught fire.
3. (Usu. as *pa. pple.* or as *ppl. a.*) Of a national economy: to bring about a condition of marked inflation by placing excessive pressure on resources during a period of expansion in demand. Also **overheating** *vbl. sb.*, the inflationary consequences of such a state or condition.
1956 *Ann. Reg. 1955* 227 Heavy industry had..been 'over-heated', but there was no strain on producers of consumer goods, and retail prices in this field had hardly risen. **1962** *Daily Tel.* 16 July 10/2 Most people would accept that an economy must not become 'overheated'. **1962** *Economist* 1 Sept. 800/2 Its [*sc.* Japan's] balance of payments crises and periods of 'overheating'. **1965** *Ibid.* 16 Oct. 269/1 The money and securities markets in New York have allowed themselves in the past month a nice frenzy of fear about the economy's 'overheating'. **1971** *Observer* 14 Mar. 8/4 It was argued in the 1960s that a faster rate of growth produced so-called 'overheating'—the situation in which demand exceeds resources, whether of labour or plant. **1973** *Times* 16 May 27/3 The danger of an over-heated economy was 'reasonably small'. **1974** *Times* 23 Mar. 13/2 The second layer of trouble was the progressive overheating of the economy last year. *Ibid.* 9 Oct. 5/2 Successive governments..had overheated the economy by increasing public spending and boosting demand, consequently pushing up inflation. **1976** F. ZWEIG *New Acquisitive Society* II. ii. 94 No labour legislation can replace..general economic policy based on the discipline of the market system and avoidance of overheating of the economy.

overheating, *vbl. sb.* Add: (Further examples.) Cf. also prec. sense 3.
1961 *Family Handyman* Oct., Too light a feed..causes overheating of the tool and burning of the cutting edge. **1969** *Gloss. for Landscape Work* (*B.S.I.*) v. 13 *Overheating*, an undesirable spontaneous temperature rise, due to the action of bacteria during the decomposition of vegetable matter. **1973** J. LEASOR *Host of Extras* i. 13 Early models..beset by unlucky snags like over-heating, and gears that jumped out on the over-run. **1977** M. SOKOLINSKY tr. *R. Merle's Virility Factor* xvi. 314 The rumors had appeared spontaneously among the blacks due to the overheating caused by the tense period.

over-hit, *v.* **b.** (Further example.)
1919 C. HIERONS *Lawn Tennis* xiii. 57 The beginner should take care that he does not over-hit the ball.

over-housed, *ppl. a.* Add: (Further examples.) Hence (as a back-formation) **over-house** *v. refl.*, to have house accommodation in excess of one's requirements or means.
1863 W. M. THACKERAY *Let.* 23 Sept. in J. Brown *Lett.* (1907) 332 If I don't mistake there was a man who lived at Abbotsford [*sc.* Sir W. Scott] overhoused himself. **1921** *Spectator* 26 Feb. 261/1 Young people marrying on little and determined not to 'overhouse' themselves. **1963** *Guardian* 8 Mar. 4/4 Many of the chronic invalids were overhoused in that they were not able to make full use of their accommodation. **1970** *Times* 14 Mar. 3 In 1965 they found themselves over-housed and decided to sublet the fourth floor.

overhung, *ppl. a.* Add: **1. b.** = OVERHANGING *ppl. a.*
1923 H. G. WELLS *Men like Gods* II. iii. 206 The gully was..difficult, he thought, to ascend, but quite practicable downward. It was completely overhung.
3. (Later example.)
1928 C. F. S. GAMBLE *Story N. Sea Air Station* 8 Somewhat similar to a Blériot monoplane, except for..the 'overhung' system of mounting the engine.
4. = *hung-over* (*HUNG *ppl. a.* 4).
1964 I. FLEMING *You only live Twice* v. 60 He was considerably overhung. The hard blue eyes were veined with blood. **1974** *Times* 4 Apr. 20/6 A young man hurried in at about 10, looking overslept and overhung. **1977** K. BENTON *Red Hen Conspiracy* xiv. 115 Juan arrived..looking rather overhung.

over-hu·nt, *v.* [OVER- 27.] *trans.* To hunt (an animal) to such an extent that an excessive number is killed; to hunt (a country) too much or to depletion. Also *fig.*
1862 *Q. Rev.* III. 229 That enthusiastic temper which leads men to overhunt a beaten enemy. **1936** *Discovery* Sept. 293/1 From time to time the moose is over-hunted in some districts, but after a few years' protection they come back again. **1968** C. HELMERICKS *Down Wild River North* II. xxiv. 385 The whole North has been overtrapped and overhunted.

o:verhydra·tion. *Med.* [OVER- 29 b.] An excessive amount of water in the body or a part of it.
1943 *Jrnl. Clin. Investigation* XXII. 471 (*heading*) The relationship of dehydration and overhydration of the blood plasma to collapse in the management of artificial fever therapy. **1956** *Circulation* XIV. 1029/1 The physiologic and clinical consequences of an absolute excess of water consist of cellular overhydration, pulmonary edema..and convulsions. **1974** PASSMORE & ROBSON *Compan. Med. Stud.* III. II. xlix. 6/2 While the physiological defences against the naturally occurring hazard of dehydration are extremely effective, those against the iatrogenic hazard of overhydration are much less efficient.
So **o:verhydra·ted** *ppl. a.*
1943 *Jrnl. Clin. Investigation* XXII. 482/1 Although definitely overhydrated..the patient did not become edematous. **1974** PASSMORE & ROBSON *Compan. Med. Stud.* III. II. xlix. 7/2 Patients with acute renal failure or oliguria may readily become overhydrated since water can be lost only by extrarenal routes.

o:verinclu·sion. *Psychol.* [OVER- 29 b.] The indiscriminate inclusion of irrelevant responses to a stimulus, observed in some cases of severe mental illness. Hence **o:ver-inclu·sive** *a.*
1939 *Jrnl. Mental Sci.* LXXXV. 1019 The inability to select and to restrict, and to eliminate the less closely related elements from the conceptual structure, means that the psychological boundaries are functionally insufficient. The result of this situation is *over-inclusion.* **1942** HANFMANN & KASANIN *Conceptual Thinking in Schizophrenia* vii. 95 Some of the examples that Cameron brings as illustrating overinclusion can be also seen as examples of extreme primitivisation of thinking. **1951** CAMERON & MAGARET *Behavior Pathol.* xv. 457 The concept of overinclusion was first developed operationally, in connection with the sorting of behavior of schizophrenic patients. But its use since then has been expanded to cover a wide range of behavioral disorganization, both normal and abnormal. *Ibid.* 458 Excitement of any kind can lead to overinclusive behavior. **1974** *Psychiatric Q.* XLVIII. 109 (*title*) Investigation of factors related to stimulus overinclusion.

over-indulgence. (Earlier example.)
1754 RICHARDSON *Grandison* II. xxxvi. 354 The Lady having, by her early over-indulgence, ruined the morals of her child.

o:verinhi·bited, *a.* *Psychol.* [OVER- 28 c.] Applied to a person whose reactions, esp. in a social context, are abnormally inhibited. So **o:verinhibi·tion,** the state or fact of being overinhibited.
1899 W. JAMES *Talks to Teachers* xv. 179 Certain melancholiacs furnish the extreme example of the over-inhibited type. **1942** PARTRIDGE *Usage & Abusage* (1947) 353/2 Thanks to the Freudians, we have..heard almost too much about the over-inhibited person. **1946** HEWITT & JENKINS *Fund. Patterns Maladjustment* 31 The..syndrome pattern of overinhibited behavior is of unquestionable familiarity to clinicians and mental hygienists. **1970** *Jrnl. Gen. Psychol.* LXXXII. 162 The apparent over-inhibition by schizophrenic Ss.

o:ver-insu·rance. [OVER- 29 b.] Insurance (of goods, property, etc.) in excess of their real value.
1755 N. MAGENS *Ess. Insurances* I. 92 The Law of Spain..ordains, 'That the Insurers who signed last in Date should return the Premium, in case of an Over-Insurance.' **1802** S. MARSHALL *Treat. Insurance* I. I. iv. 118 In the case of an over-insurance..the first underwriters on the policy were formerly holden to be answerable, to the extent of the loss, and the subsequent ones discharged. **1880** *Encycl. Brit.* XIII. 185/1 When the value proved under an open policy falls short of the sum originally insured, the difference..is technically termed an over-insurance. **1965** 'W. HAGGARD' *Hard Sell* xii. 136 Suspicion, the over-insurance of risks which were seldom worth the premium. **1970** M. GREENER *Penguin Dict. Commerce* 232 *Over-insurance.* Property is over-insured if insured for more than it is worth. **1977** *Times* 5 May 4/3 If cover were provided for a longer period an element of over-insurance would lead to unnecessarily high subscription rates.

o:ver-insu·re, *v.* [OVER- 27.] *trans.* To insure for more than the real value; to insure excessively. (Chiefly as pa. pple.)
1904 [listed in Dict. s.v. OVER- 27 a]. **1910** *Times* 28 June 6/1 An old vessel, trading at a loss, over-insured. **1922** *Blackw. Mag.* Sept. 318/2 Ship and cargo are over-insured about ten times, I suppose? **1970** [see *OVER-INSURANCE]. **1976** D. FRANCIS *In Frame* iv. 68 If it was over-insured it was to allow for inflation.

o:ver-intelle·ctual, *a.* [OVER- 28 a.] Excessively intellectual; concerned too much with reason or mental processes.
1854 BAGEHOT *Coll. Works* (1965) I. 219 It had no feverish excitement, nor over-intellectual introspection. **1944** J. S. HUXLEY *On Living in Revol.* xv. 184 Over-intellectual and over-specialized.

So **o:ver-intellectualiza·tion**; **o:ver-intelle·ctualize** *v. trans.*; **o:ver-intelle·ctualized** *ppl. a.*
1924 P. C. Buck *Scope of Music* 72 Preaching the over-intellectualization of their art until the red blood has gone out of it. **1929** A. N. Whitehead *Process & Reality* 263 The interest in logic, dominating overintellectualized philosophers, has obscured the main function of propositions. **1933** —— *Adventures of Ideas* iii. 53 Even here we must not over-intellectualize the various types of human experience. **1975** R. L. Simon *Wild Turkey* (1976) xii. 86 The sauna of some over-intellectualized exhibitionists. **1975** *Verbatim* Sept. 7/1 Over-intellectualization is rampant: if the Old Believers in Russia avoided the use of the future tense, the only effect would have been to create a new future of some kind, to refer to tomorrow's planting of potatoes, with the old future relegated to taboo—no deep change in speech habits would have resulted. **1977** *Early Music* July 308/2 We must avoid the temptation to over-intellectualize the artists' procedure.

o·verjet. *Dentistry.* [Over- 13(?).] The extent to which the upper teeth, esp. the incisors, project forward in a horizontal direction beyond the corresponding lower teeth. Cf. *overbite.
1930 I. G. Nichols *Prosthetic Dentistry* vi. 149 A greater overjet or projection forward of the maxillary teeth is required in cases of deep overbite than in cases of shallow overbite. **1940** M. G. Swenson *Compl. Dentures* xxvi. 445 The incisal guidance is governed mostly by the operator's choice of the desired inclination. This inclination may be changed by the amount of overbite (vertical overbite) and overjet (horizontal overbite). **1962** Blake & Trott *Periodontology* xv. 155 In patients with abnormal overbite or overjet an habitual rest position may be established which assists the patient to form an efficient lip seal. **1977** *Proc. R. Soc. Med.* LXX. 432/1 Bell & Dann (1973) found..only small changes in incisor overbite and overjet in 25 patients following anterior maxillary ostectomies.

o·verki·ll (stress var.), *v.* orig. *U.S.* [Over- 27.] *trans.* and *intr.* To kill or destroy to a greater extent than is necessary. Also *fig.*
1946 *Sun* (Baltimore) 17 Jan. 4/5 It pointed all, or a great majority, of the guns at a single object. This method resulted in missing most of the in-coming attackers and of over-killing those which could be hit. **1958** *Lincoln* (Nebraska) *Evening Jrnl.* 8 Aug. 4/4 The argument that you do not need the power to 'overkill', if you already have H-bombs [etc.]. **1965** *New Statesman* 30 Apr. 690/3 His magnanimity towards those who ordained that Dresden should be overkilled. **1967** *Economist* 23 Dec. 1227/3 Mr Humphrey's oratory..overkilled the McCarthy ridicule of what the Senator assails as an immoral and, equally unforgiveably, an irrational war. **1968** *Punch* 27 Nov. 751 We maintained armed forces to defend a non-existent Empire and spent uselessly and prodigally in a vain attempt to keep abreast of the titans in capacity to kill and overkill. **1971** B. Callison *Plague of Sailors* iv. 157 When you intend to decimate a whole nation, why get puritanical about over-killing a few dozen more?

o·verkill, *sb.* orig. *U.S.* [Over- 29.] **a.** The capacity, esp. of nuclear weapons, to kill and effect destruction in excess of strategic requirements. Also *attrib.*
1958 *Time* 17 Mar. 25/2 A word coming more and more into Pentagon usage is 'overkill'—a blunt but descriptive term implying a power to destroy a military target not once but many times more than necessary. **1959** *Times* 18 May 7/2 The Chiefs of Staff of the Navy and the Army..told Congress..that this 'over-kill' capacity is unnecessary. **1962** *Economist* 30 June 1307/1 It is military nonsense for Britain and France to produce nuclear weapons, when the United States has an 'overkill' of those weapons coming out of both its ears. **1965** H. Kahn *On Escalation* 280 Overkill by a factor of ten or more, so that even the blind..would understand the situation. **1968** W. Ash *Ride Paper Tiger* xii. 191 There's no point in plastering a target which has already been demolished. Anyone carrying the weapons you do has to be a bit careful about the problem of overkill. **1971** *Guardian* 27 Sept. 13/3 The nuclear club reached the point of H-bomb overkill. **1976** J. Cox *On Warpath* 7/1 A mere pin-head of a man-made poison could kill everyone alive today. Military strategists talk of 'Doomsday' and 'Overkill'.
b. *transf.* and *fig.*
1965 *New Scientist* 24 June 841/1 There is only a limited number of whales in the sea and the delegates must decide between an irrational short-term overkill or long-term conservation. **1967** *New Yorker* 1 Apr. 94 Its producer..is a misguided champion of cinematic overkill: twice as large is twice as good, twice as loud is twice as convincing. **1968** *Guardian* 20 Mar. 10/1 Just how much Mr Jenkins ought to cut consumption is arguable. The world monetary crisis provides a strong psychological reason for going for 'overkill'. **1970** *Globe & Mail* (Toronto) 26 Sept. 7/5 The..social and economic consequences..ascribed to 'advertising overkill'. **1973** *Times Lit. Suppl.* 21 Dec. 1555/3 It is astonishing, in these days of critical overkill, that Peter Wolfe's little book is the only one yet written on Rebecca West. **1975** *Listener* 3 July 22/3, I have only the smallest objection to the [Wimbledon] coverage, apart from the serious danger of overkill (three hours daily on BBC 1, over six on BBC 2). **1976** *Times* 30 Jan. 17/2 What point is there in producing things if over-kill taxation means that nobody will buy them?

overland, *a.* (Further examples.) Also *ellipt.* as *sb.*
1841 *Niles' Reg.* 6 Feb. 353/2 The news from China and India we have received by the overland mail. **1848** *Alfred in India* 158 Passengers went..across the desert to Alexandria, and from thence in another steamer... This is called the 'overland route'. **1861** B. I. Hayes *Let.* 11 Feb. in *Pioneer Notes from Diaries* (1929) vii. 253 By the Overland Stage arriving here on the 8th inst., I received your valued favor of the 12th ult. **1862** Mrs. J. B. Speid *Our Last Yrs. in India* iv. 79 Her Majesty's mail! What would England..say, could they witness the bi-monthly arrival of the overland here! **1901** *Daily Colonist* (Victoria, B.C.) 24 Oct. 1/5 The northbound Southern Pacific overland express..was held up by robbers..but the robbers secured little booty. **1977** C. Allen *Raj* 20 Before the opening of the canal the fastest means of getting to and from India was by taking the overland route..from Alexandria to Suez.

overland, *v.* Add: (Further examples.) So **overla·nding** *vbl. sb.* and *ppl. a.*
1871 M. Clarke *Old Tales of Young Country* 163 'Overlanding' was a profitable and, withal, romantic occupation. Young men of spirit, wearied of the capital, and prompted by love of gain and adventure, purchased cattle and sheep in New South Wales, and drove them 'overland' to the 'New Orleans' of Colonel Torrens. **1916** *Chambers's Jrnl.* Nov. 729/1 Men live in the saddle..when 'overlanding' cattle. **1925** *Ibid.* Dec. 810/1 Overlanding drovers who were opening up new country passed by. **1933** *Bulletin* (Sydney) 16 Aug. 20/2 Overlanding from the N.T. to Queensland. **1941** I. L. Idriess *Great Boomerang* xviii. 128 Any of the great overlanding or exploring treks would have been more instructive to remember. **1965** *Sunday Truth* (Brisbane) 17 Oct. 32/4 In 1839 the entire Wills establishment..overlanded to..western Victoria. *Ibid.* 32/5 Heading an overlanding expedition of 25 men and women and 10,000 sheep Wills led his party slowly north to a destination in central Queensland. **1975** *Australasian Express* (London) 10 Oct. 24/2 Overlanding is a wonderful way of travelling.

overlander[2]. Add: **1. a.** Also *N.Z.* (Earlier and further examples.) Also *slang*, a sundowner, a tramp.
1841 G. Grey *Jrnls. Two Expeditions of Discovery* ii. 183 The Overlanders are nearly all men in the pride of youth, whose occupation is to convey large herds of stock from market to market and from colony to colony. **1852** *Lyttleton Times* (N.Z.) 27 Mar., Mr. A. Clifford has succeeded in driving about 1500 ewes from the Wairau district... Two other parties of 'over-landers' are reported to be close on his heels. **1898** Morris *Austral. Eng.* 333/1 *Overlander*... (2) A slang name for a *Sundowner*. **1907** A. Searcy *In Austral. Tropics* 125 If a crowd of overlanders and backblockers happened to be present, things would be made lively. **1933** L. G. D. Acland in *Press* (Christchurch, N.Z.) 11 Nov. 15/7 *Overlander*, an Australian word for a man driving sheep or cattle a long distance. It was sometimes used in Canterbury in the early days. **1941** Baker *Dict. Austral. Slang* 52 *Overlander*, a traveller. (2) One who makes long expeditions from one State to another with stock. (3) A settler from another State. (4) A drover. (5) A sundowner. **1967** *Woman's Day* (Austral.) 27 Feb. 10/3 Now in her early sixties, she still lives her life with all the zest of her overlander days.
b. *N. Amer.* One who moves from one part of the country to another; a migrant. *Obs.* exc. *Hist.*
1857 *Hutching's Mag.* Mar. 398/1 Reader, if you have never been an *over-lander*, I will tell you a little about camp life. **1916** A. C. Laut *Cariboo Trail* 55 Some of the Overlanders had narrowly escaped a massacre. **1950** B. Hutchinson *Fraser* 88 The most remarkable immigrants of all deserve to be remembered—the Overlanders of '62, the men..who walked to Cariboo across the Rocky Mountains. **1963** *Canad. Geogr. Jrnl.* Oct. 112/3 Among those who heard the call of 'Gold in the Cariboo!' were the Overlanders. **1968** E. Russenholt *Heart of Continent* iii. vii. 116 This summer [*sc.* 1859], three parties of 'Overlanders', some 60 in all, leave from Assiniboia for the Cariboo.
2. In general use: one who travels overland to a country which can also be reached by sea or air; one who travels a long distance overland.
1953 J. Packer *Apes & Ivory* xxiii. 240 There were many 'overlanders' after the war, when it was impossible to get a sea-passage to Southern Africa. **1960** *Guardian* 22 Nov. 7/5 Everywhere beyond Austria the overlander will attract..the attentions of the idle bystander. **1974** *Country Life* 26 Dec. 2008/1 The intrepid long-distance overlander of today.

overlap, *sb.* Add: **1. a.** Also *transf.* and *fig.*
1880 [in Dict.]. **1931** A. Keith *Place of Prejudice* 19 Head and heart are never quite separated; there is a large overlap in their fields of action. **1955** *Bull. Atomic Sci.* June 205/3 Perhaps outside help with parts of the investigation should be arranged, with some of the 'second laboratory' type of overlap providing the spur of competition. **1960** *Times* 31 Oct. 4/4 Later Glover scored a good try after Willcox had made the overlap in a set-piece movement. **1962** L. Deighton *Ipcress File* xxiv. 156 That camera went out of action..but luckily we have overlap on the camera fields. **1970** K. Ball *Fiat 600, 600D Autobook* 164/2 *Overlap*, period during which inlet and exhaust valves are open together. **1970** *Times* 1 Oct. 10/3 Hollins again at the 29th minute..joined in an overlap with Mulligan. **1974** *Country Life* 3/10 Jan. 25/3 For the Garvan Gallery the William and Mary style runs from 1685 to 1730, Queen Anne from 1715 to 1765, Chippendale from 1750 to 1790 and Federal from 1788 to 1830. The overlaps are perfectly acceptable. **1974** *Times* 19 Dec. 6/8 There is a considerable overlap between the committee's membership and that of the Council for..Arab–British understanding. **1975** *Times* 11 Jan. 12/4 Allowing for duplications and overlaps..the CIA..'spies on 100,000,000 Americans'. **1976** *Sunday Post* (Glasgow) 26 Dec. 36/2 Better goal-kicking did the trick, since the try count was three-all, but Gala could have popped the result into their Christmas stockings a lot earlier if they had not so persistently kicked away overlaps.

b. overlap fault, an overthrust fault.
1883 W. S. Gresley *Gloss. Terms Coal Mining* 180 *Overlap fault*, a peculiar kind of fault where a seam is reversed or doubled back over itself. **1886** J. Prestwich *Geol.* I. xv. 257 (*caption*) Great slide or overlap fault in the Radstock coalfield.
c. In yacht-racing, a position in which a yacht overtaking another is debarred by the rules from passing one side, or in which the yachts concerned cannot turn toward each other without risking a collision.
1898 *Encycl. Sport* II. 585/2 If in rounding any mark in a race, or any obstruction, an inside yacht has an overlap, the outside yacht or yachts must give her sufficient room. **1935** *Encycl. Sports* 760/2 Provided that the overtaking yacht makes her overlap on the side opposite to that on which the overtaken yacht then carries her main boom, the latter may luff as she pleases [etc.]. **1958** *Times* 23 Sept. 14/2 The American yacht then went smoothly forwards to establish an overlap.
d. *Phonetics.* Concurrence of the concluding sound of one phoneme with the opening of the next, as represented spectrographically; homonymy (as French /sañ/, *cent* and *sans*).
1942 *Amer. Speech* XVII. 42 There is no over-lap between the length of this vowel when full grade..and its length when reduced grade. **1964** *Language* XL. 62 It is possible to raise the question of phonemic overlap... This instance of overlapping phonetic values..need not be regarded as a violation of phonemic principles.
e. *Linguistics.* (See quot. 1948).
1948 E. A. Nida in *Language* XXIV. 431 Instances of 'overlap', i.e. forms which are in complementary distribution except at certain points where there is a contrast resulting from fluctuation of forms. **1973** *Archivum Linguisticum* IV. 12 /de/ and /bi/ and /bi/ and /na/ do have areas of overlap.
2. *Geol.* The extension of a stratum beyond or over the edges of younger underlying strata; also, the upper stratum; esp. = *onlap (a *transgressive overlap*); also = *offlap (a *regressive overlap*); less commonly = *overstep *sb.* 2.
1846 H. T. De la Beche in *Mem. Geol. Survey Gt. Brit.* I. 24 Proceeding westward..the overlap of the Old Red Sandstone becomes such that from Middleton Hall to the vicinity of Caermarthen, it rests on the lower Silurian rocks, covering up the higher Silurian beds. A. C. Ramsay in *Ibid.* 319 This..shows a tendency to a certain amount of contemporary depression of coast and sea bottom on the south to admit of an overlap during the formation of these deposits. **1857** [in Dict.]. **1876** *Q. Jrnl. Geol. Soc.* XXXII. 377 By the term *overlap* no unconformity is here meant, but the concealment of lower beds by the extension, through a progressive subsidence, of those next in time upon the old rocks. **1883** [see *overstep *sb.* 2]. **1906** A. W. Grabau in *Bull. Geol. Soc. Amer.* XVII. 569 The types of overlap of sedimentary strata may be classified as follows: A. Irregular or discontinuous overlap. B. Regular continuous or progressive overlap. 1. Marine. *a.* Transgressive. *b.* Regressive. 2. Non-marine. *c.* Fluviatile. **1913**, etc. [see *offlap]. **1947** *Bull. Amer. Assoc. Petroleum Geol.* XXXI. 1868 There is much confusion in the uses of the word 'overlap' in geological literature. In recent years it has often been used to describe..: (1) the regular and progressive pinching-out of sediments above unconformities, and (2) the regular truncation of sediments below unconformities. **1962** Read & Watson *Introd. Geol.* xii. 616 Stratigraphical traps are the result of unconformities, lensing, overlap and reflected buried hills. **1972** B. B. Brock *Global Approach to Geol.* xvii. 224 The Lower Division forms remnants of a regressive overlap left by a contracting and retreating inland basin.
3. *Railways.* The distance beyond a signal that must be unoccupied before an engine is allowed to approach it past the previous signal.
1925 Tweedie & Lascelles *Mod. Railway Signalling* x. 145 To work without overlaps it would be necessary to have some system of positive speed control which would bring a train to a stand at a stop signal independently of the driver. **1956** *Railway Mag.* Mar. 351/2 The standard British block telegraph overlap of 440 yd. beyond the home..signal can be modified by special instructions. **1969** H. R. Broadbent *Introd. Railway Braking* i. 6 The overlap beyond the stop signal is therefore a form of margin on the distance between the warning and stop signals to cover for contingencies.
4. *Computers.* The strictly simultaneous performance of two or more operations during the execution of a program.
1963 L. Schultz *Digital Processing* xi. 239 In a machine with a modular memory..the execution of a program might be significantly speeded by locating variables in one module and instructions in another, and by providing an overlap in the time needed for executing one instruction and fetching the next. **1969** P. B. Jordain *Condensed Computer Encycl.* 365 I/O overlap is a great improvement over strictly sequential operations, but it still leaves the electronics waiting for the slower main memory.

overlap, *v.* Add: **1. b.** (Later examples.)
1911 R. Brooke *Let.* 13 Dec. (1968) 325 Ka appears some days later. James for a weekend... But some (e.g. Ka and Margery) will not overlap; so there'll not be too much of a crowd. **1971** *Nature* 22 Oct. 509/2 The Roths child inquiry and the Dainton inquiry overlap only at the edges. **1977** K. O'Hara *Ghost of T. Penry* viii. 65 The generations overlapped because the Binns were an enormous family.

overla·pped, *ppl. a.* [-ed[1].] That overlaps or is overlapped (in various senses).

1839 [see *double coal* s.v. *double *a.* A. 6]. **1898** G. Saintsbury *Short Hist. Eng. Lit.* viii. ii. 498 The constant preference of overlapped or enjambed lines for the strict couplet. **1926** J. Adams *Christian Good of Scotland* viii. 126 To neglect or overlook the nobler ideals of the Church, because of its presently divided and overlapped system, is neither politic nor wise. **1962** A. Nisbett *Technique Sound Studio* 263 *Permanent joint*, cemented joint in slightly overlapped tape.

overlapping, *ppl. a.* (Further examples.)
1926–7 *Army & Navy Stores Catal.* 854/3 Golfing requisites..Harry Varden's own overlapping grip. Each 4/6. **1958** *Jrnl. Social Issues* XIV. i. 39 A major research question in changing attitudes and behavior is involved in this issue of overlapping situations. **1964** *Language* XL. 62 This instance of overlapping phonetic values therefore need not be regarded as a violation of phonemic principles. **1971** *Brit. Med. Bull.* XXVII. 7/1 According to the theory of overlapping population distribution, the population screened is comprised of a diseased and a non-diseased group, both of whom possess the attribute being measured, though with different frequencies at various test levels. **1977** *New Yorker* 10 Oct. 124/2 A certain touring professional golfer barely missed equalling the record for fewest putts in a tournament round by a man using the ordinary overlapping (as opposed to the more popular reverse overlapping) grip. **1978** A. Price '*44 Vintage* ix. 103 The flick-knife was held, for the upwards [blow]..which came in under the overlapping ribs.

overlapping, *vbl. sb.* (Further examples.)
1896 [see *coalescence 3]. **1935** *Discovery* Sept. 278/1 Room within this waveband..to accommodate several independent high-definition sound and picture channels without overlapping or interference. **1942** *Amer. Speech* XVII. 46 'Coarticulation' and 'overlapping'. **1956** Jakobson & Halle *Fund. of Lang.* ii. 14 The so-called overlapping of phonemes confirms the manifestly relational character of the distinctive features... The same sound [e] in one position implements the diffuse, and in another, the compact term of the same opposition. **1963** B. Fozard *Instrumentation Nucl. Reactors* x. 112 Overlapping, i.e. failure to resolve the pulses, may cause errors of measurement in three main ways.

over-late, *a.* and *adv.* Add: **a.** *adj.* (Later example.)
1958 T. Stanwell-Fletcher *Clear Lands* 245 The diminishing supply of native animals..despite overlate measures of restocking and conservation.
b. *adv.* (Later example.)
1875 W. Morris tr. *Virgil's Aeneids* vii. 597 And overlate the Gods thou shalt adore.

overlay, *v.* Add: **2. b.** (Examples.)
1888 C. T. Jacobi *Printer's Vocab.* 93 *Overlay*, to make ready by overlaying—the reverse of underlaying. **1894** *Amer. Dict. Printing & Bookmaking* 413/1 *Overlay*,..to secure proper effects in printing by means of graduated impression between the impressing surface and the sheet, using different thicknesses of paper. **1940** *Chambers's Techn. Dict.* 604/2 *Overlay*.., to adjust the impression surface of a machine by cutting and patching.

overlay, *sb.* Add: **2. a.** (Later examples.)
1888 Wilson & Grey *Pract. Treat. Mod. Printing Machinery* xxi. 337 The object of an overlay is not to equalise the impression, but to intensify the pressure upon the dark parts or solids, that they may be firm and bright, and to lessen the impression upon the lighter shades, in order to give them that degree of delicacy and cleanliness that would be altogether wanting if the pressure exerted were uniform. **1946** V. S. Ganderton in H. Whetton *Pract. Printing & Binding* xi. 132/2 In letterpress the gradation in tone of a half-tone plate corresponds to a gradation in resistance to pressure... The object of an interlay or overlay is to balance the variation in resistance by varying the pressure on the different areas. **1970** E. A. D. Hutchings *Survey of Printing Processes* iv. 57 Many of these operations [in make-ready] can be carried out..in the pre-press department, using..register tables..and other precision equipment such as mechanical overlay systems.
b. In offset lithographic printing: (see quot. 1974).
1967 Karch & Buber *Offset Processes* 548 *Overlay*, in offset, the transparent..covering on the copy on which directions..are placed in conjunction with the original. **1974** J. Craig *Production for Graphic Designer* 172/1 The copy for each additional color is pasted on acetate overlays, each one representing a color. *Ibid.* 190/2 Overlay. Transparent paper or film flap placed over artwork for the purpose of (1) protecting it from dirt or damage, (2) indicating instructions to the platemaker or printer, or (3) showing the breakdown of color in mechanical color separations.
3. a. (Further special senses.) Also *attrib.*, as *overlay mattress*.
1886 J. Barrowman *Gloss. Scotch Mining Terms* 48 *Overlay*, the material above the rock in a quarry. **1926–7** *Army & Navy Stores Catal.* 1076 Overlay mattresses for children's bedsteads and cots. **1930** *Daily Tel.* 9 Apr. 6/3 (Advt.), No bumps, no lumps, no sag in the 'Vi-Spring', the overlay mattress. **1949** H. M. Cautley *Norfolk Churches* 38 Another feature came into perfection, namely the overlay, with its imposed and crocketted hood-moulds on the face of tracery, giving such depth and solidity to the whole, as at Scarning..and Bedingham. **1953** *B.U. Encycl. Handbk.* (Bedding Publications Ltd.) App. 198 Upholstered overlay mattresses with spring or cellular rubber interior. **1962** A. McIntosh in Davis & Wrenn *Eng. & Medieval Stud.* 234 Just as we concluded earlier that Thornton transcribed B_1 and M_1 without very much tampering, so must we now conclude that S transcribed M_1 with sufficient fidelity for his own overlay not to have obliterated strong traces of an underlying language. **1968** *Lebende Sprachen* XIII. 104/2 *Overlay*, the action accomplished in the interfacing technique wherein linking events are meshed into a single event. **1972** H. Kurath *Studies Area Linguistics* viii. 124 The replaced language was an overlay (superstratum), as French in England or Frankish in northern France.
b. A transparent sheet bearing additional information, which is laid over a map or diagram.
1938 E. Raisz *Gen. Cartogr.* xv. 172 Transparent tracing papers are made of straw and cornstalk base and are used in map work for sketching, for copying, and for tissue overlays which indicate various colors and tints. **1952** V. Canning *House of Seven Flies* viii. 128 He put the overlay on the chart so that the cross..fell on the position of the house. **1964** G. Lyall *Most Dangerous Game* vi. 45, I did the real work using a celluloid overlay with wax-pencil marks. **1973** J. S. Keates *Cartogr. Design & Production* xx. 200/2 It is more satisfactory to produce the pattern stick-up on a separate overlay, and then combine this with the line image in a separate processing stage. **1974** *Sci. Amer.* May 126/1 The plastic overlay rotates around the pin. The track on the earth's surface over which the satellite passes during any orbit is plotted as a curved line on the overlay.
c. A layer of coloured glass added on top of clear glass in decorative glass-ware; usu. *attrib.* in *overlay glass*, *overlay paper-weight*, *overlay weight*; hence *ellipt.* for *overlay paper-weight*.
1940 E. H. Bergstrom *Old Glass Paperweights* ii. 12 In an encrusted overlay weight, the color overlays and faceting were apparently followed by a final dip into clear crystal to complete the weight. **1954** E. M. Elville *Paperweights & Other Glass Curiosities* i. 16 Also highly prized by connoisseurs are the overlay paperweights. Overlays were usually made with the millefiori mushroom in a crystal globe..given..a final casing of a colour such as red, blue or green. **1958** G. B. Hughes *Eng. Glass for Collector 1660–1860* xx. 220 Those who toured the Continent in the early nineteenth century enthusiastically adorned their dining-tables and dressing rooms on their return with specimens of colourful Bohemian work known as cased or overlay glass. **1967** Wodehouse *Company for Henry* iii. 45 'A French eighteenth-century paperweight alluded to as follows:..Clichy double overlay weight. **1968** *Canad. Antiques Collector* Aug. 9/1 We know that Burlington produced many sophisticated types of glass including overlay glass. **1973** J. Mackay *Glass Paperweights* viii. 54 The majority of these overlay weights are doubles with an inner overlay of white and an outer overlay in shades of red, green or blue.
d. *Med.* A gelling layer spread on top of a layer of cells in culture and containing an indicator of the presence or absence of some cell product.
1954 *Jrnl. Exper. Med.* XCIX. 168 The agar overlay, used to overlay the cultures after infection, consisted of 12 parts of 2·7 per cent agar (*A*), 12 parts of neutral red solution (*B*), 8 parts of fourfold Earle's saline (*C*), and 5 parts of embryo extract (*D*). **1960** *Virology* X. 377 The difference in diameter of plaques grown under agar overlay with and without viral antibodies serves as a measure of antigenic difference. **1974** *Nature* 20 Dec. 745/2 Virus pools were grown in Vero cells and titrated by plaque assay with a methylcellulose overlay as before.
4. *Dentistry.* A structure intended to improve the occlusal surface of the tooth or teeth over which it fits.
1935 G. M. Anderson *Dewey's Pract. Orthodontia* (ed. 5) xxi. 443 To stabilize the canine anchorage, the canine overlays are connected on the lingual surface with a clasp metal wire. **1954** *Brit. Dental Jrnl.* XCVII. 268/1 An upper acrylic bite overlay, sliding the mandible forward at the same time as opening the bite, gave complete relief in three days by preventing the condyles from being forced back in the fossæ. **1973** Anderson & Storer *Immediate & Replacement Dentures* (ed. 2) iv. 33 The retentive type of the overlay prosthesis should be considered where the residual ridges are almost non-existent and in cases where for reasons of speech, retching or excessive salivation, a palateless type of upper denture is necessary.
5. In betting, odds which are unjustifiably high. *U.S.*
1944 *Sun* (Baltimore) 21 Sept. 17/4 There is always a section that keeps in keen pursuit of the 'overlay', which means a horse rated, say, at 4 to 1 which closes at 10 to 1 or 12 to 1, well above his 'morning line' price. **1944** D. Runyon *Runyon à la Carte* (1946) 102 Everybody around is saying The Sky makes a terrible over-lay of the natural price in giving Brandy Bottle a G against his soul. **1955** *Amer. Speech* XXX. 27 The term *overlay* originally described the situation which existed if a bookmaker laid odds which, viewed objectively, were too high. At the present time *overlay* is used to describe odds meeting this test which appear in the pari-mutuel machine. Most often odds are considered an overlay if they are higher than the odds quoted in the so-called *morning line*.
6. *Computers.* The process of transferring a block of instructions or data to internal storage in place of what is already there, esp. in order to utilize a limited high-speed memory for those parts of a program that are in active use; also, a section of program so transferred.
1963 *Automatic Data Processing Gloss.* (U.S. Bureau of Budget) 38/1 *Overlay*, a technique for bringing routines into high-speed storage during processing, so that several routines will occupy the same storage location at different times. **1965** Swallow & Price *Elem. Computer Programming* xiv. 273 Once the new table is determined for a given run, the original table and the program for calculating the new one can be discarded, so the main program may be loaded over the previous one... This is frequently called an overlay. **1969** P. B. Jordain *Condensed Computer Encycl.* 367 When a program and its data are too large for the computer, the program must be divided into segments and so constructed that only the active segments (or overlays) need be in core. **1970** O. Dopping *Computers & Data Processing* xiv. 219 We can have a special memory area for overlay, or we can let a part of the program normally used—the main program, be overwritten. **1972** J.-L. Baer in A. F. Cardenas et al. *Computer Sci.* v. 170 If subroutines A and B will never call each other, then they can occupy the same positions in memory. If, at run-time, B is called while A is present, an overlay will be performed replacing A by B, and vice versa.

overlayer. (Later example.)
1917 *19th Cent.* Jan. 132 Faith in God and in a hereafter has been accompanied in history by an overlayer of superstition.

overlaying, *vbl. sb.* (Earlier and later examples of *Typogr.* sense.)
1839 T. C. Hansard *Treat. Printing & Type-Founding* (1841) 117 Anciently, the artist in wood contented himself with producing his lights and shades by cutting his lines..upon a plane, leaving to the printer the task of producing the required effects by a tedious process of overlaying. **1967** V. Strauss *Printing Industry* vii. 428/2 Overlaying serves two purposes: one is to level the impression and the other to provide varying pressure.

overlea·rn, *v.* [Over- 27a.] *trans.* To learn excessively; *spec.* in *Psychol.*, to learn (something) beyond the stage of initially successful performance. Hence **overlea·rned** *ppl. a.*; **overlea·rning** *vbl. sb.*
1874 [see Over- 27a]. **1918** E. C. Tolman in *Psychol. Monogr.* XXV. i. 48 The first hypothesis would assume that this longer time corresponded to relatively more just supraliminal and relatively fewer 'over-learned' associations. **1929** R. S. Woodworth *Psychol.* (1930) iii. 94 Material that has been 'over-learned', i.e., studied beyond the point where it can barely be recited without error, is forgotten more slowly. **1939** —— *Experimental Psychol.* ii. 30 Nagel went further; he pushed the original learning to the point of great overlearning. **1948** E. R. Hilgard *Theories of Learning* xii. 339 The most evident effect of overlearning is this one upon recall. It provides the explanation for the long retention of overlearned skills like swimming or bicycle riding. **1953** H. Schuell *Aphasia Theory* (1974) x. 203 Using some of the highly over-learned automaticisms of speech often produces this result quickly. **1963** W. B. Kolesnik *Educ. Psychol.* x. 248 She wants him to overlearn the material. **1972** J. L. Dillard *Black English* vii. 269 The technique consists of intensive practice to the point of overlearning.

over-length. Add: Also as quasi-*adj.*
1959 Halas & Manvell *Technique Film Animation* xviii. 170 If a film..becomes overlength through the addition of essential sound, it is usually better to consider cutting out a complete sequence. **1962** A. Nisbett *Technique Sound Studio* vii. 117 A programme..turns out to be seriously overlength. **1977** *Times Lit. Suppl.* 29 Apr. 537/1 *Middlemarch*, an over-length novel even by the generous Victorian reckoning.

over-light, *a.* (Later example.)
1908 *Daily Chron.* 21 Apr. 4/4 Now and then he was a trifle..over-light in his treatment of opponents.

overline, *v.*[1] Add: (Further example.) Hence **overli·ned** *ppl. a.*, having a line above it (usu. said of a printed character or number).
1879 *Proc. Lond. Math. Soc.* X. 21 The terms underlined and over-lined may both be omitted. **1963** *Amer. Jrnl. Physics* XXXI. 339/1 The overlined quantities are suitably taken average values of the specific heats indicated. **1967** A. Battersby *Network Analysis* (ed. 2) xii. 200 An overlined duration, e.g. $\overline{12}$, means that the present method calls for one man, but two men could be put on the job and would do it in half the time.

overling. Delete †*Obs.* and add later examples.
1917 [see *idle *a.* 8 a]. **1976** *New Yorker* 22 Nov. 109/2 In Pinter's world, servants and other underlings are always filled with menace; the ominous likelihood is that they will prove to be overlings.

overlip (ōᵘvəɹli·p), *v. literary.* [Over- 1.] *trans.* To overflow the lip or brim of.
1868 G. M. Hopkins *Jrnls. & Papers* (1959) 181 The Jumeaux and..the Breithorn, both over-lipped with heavy cowls of snow. **1872** S. Butler *Erewhon* v. 38 The clouds rolled up to the very summit of the pass, though they did not overlip it.

overlive, *v.* Add: **a.** (Further examples.) Also, to live beyond (one's income).
1749 J. Cleland *Mem. Woman Pleasure* I. 124 He was the only son of a father, who..rather over-liv'd his income. **1842** Tennyson *Poems* II. 102 Perish in thy self-contempt! Overlive it—..be happy! wherefore should I care? *a* **1877** Swinburne *Lesbia Brandon* (1952) iii. 36 His clear, wary, untameable eyes which had seen many dangers through, had overlived and overcome much trouble. **1881** S. Evans *Evans's Leicestershire Words* (new ed.) 207 *Overlive*,..to out-live; survive.
c. Also, to live too intensely, or too actively.
1921 Galsworthy *To Let* ii. ii. 126 He had only just relapsed, from having overworked, or overlived, himself again.
Hence **overliving** *vbl. sb.*
1817 Scott *Let.* 11 Aug. (1933) IV. 496 The task of maintaining a poor renderd effeminate and vicious by over wages and over-living and necessarily cast loose upon society.

overload, *sb.* Add: (Further examples.)
1937 J. Orr tr. *Iordan's Introd. Romance Linguistics* iii. 166 Semantic hypertrophy, or semantic overload, as it has been called. **1938** *Sun* (Baltimore) 11 July 2/6 The cabin behind the cockpit was crowded with..elaborate radio and navigation equipment. The plane weighs 25,000 pounds—an overload of 7,500 pounds. **1969** *Advancement of Sci.* XXVI. 72/1 Much the same applies to officials..who are responsible for the continuity of policy. They frequently live in a state of perpetual overload from which there is no obvious escape. **1974** *Gen. Systems* XIX. 62/2 A confused oldster may easily be maneuvered to sign a contract accepting the machine's uncompensible expensiveness. She thereby converts her overload into culturally-encouraged 'pathways to madness'.
b. *spec.* A current or voltage in excess of that which is normal or allowed for. Freq. *attrib.*, esp. designating devices for protecting against overloads.
1904 *Westm. Gaz.* 1 Dec. 8/1 A representative..was conducted through the mighty power-house... 'This is the biggest thing of its kind running in England. It is designed for 20,000 horse-power; at a pinch it could stand a 50 per cent. overload.' **1908** *Installation News* II. 38 There is always some novel addition to our Conduit System,..in addition to various side issues such as overload cut-outs. **1930** *Engineering* 3 Jan. 32/2 The overload trips are operated through a relay. **1962** A. Nisbett *Technique Sound Studio* 10 Striking a reasonable balance between noise levels, on the one hand, and overload or peak distortion, on the other, may take so much time and effort that the best part of the programme is lost. **1974** *Sci. Amer.* Nov. 34/1 The blackout was traced to the tripping of a circuit breaker in Ontario during a momentary overload. **1975** *Hi-Fi Answers* Feb. 38/1 The disc input has a fixed sensitivity of 2.5mV with an overload margin of 100mV. **1975** G. J. King *Audio Handbk.* v. 112 A stereo amplifier switched to the mono mode may have an overload value which differs from that in the stereo mode.

overload, *v.* (Further examples.) Also *absol.*
1962 A. Nisbett *Technique Sound Studio* 254 Frequency modulation... Limiters are not needed to avoid overloading transmitter valves. **1973** *Times* 30 Nov. 6/7 He thought some circuits were overloaded by as much as four kilowatts. **1976** G. A. Browne *Slide* (1977) 8 The woman..hoped she wouldn't overload again... Anyway, today she was prepared with eight extra fuses.
b. *intr.* To become overloaded.
1961 *Jrnl. Water Pollution Control Federation* XXXIII. 1280/1 Ice caused the aerator to overload, straining the drive belts. **1977** *Times* 16 July 5/8 The safety devices to stop them [*sc.* power lines] overloading came into action.
overloaded *ppl. a.* (further examples); **overloading** *vbl. sb.* (earlier and later examples).
1889 R. Fry *Let.* 2 Aug. (1972) I. 123 Ashbee..got hold of..*Atalanta in Calydon*, and read it with such overloading of sentiment as is usual with him. **1907** *Daily Chron.* 3 Oct. 2/2 The rapid increase in the number of companies not being admired by the somewhat overloaded bulls. **1958** *Spectator* 11 July 68/3 It is essential to retain its [*sc.* the Government's] legislative apparatus of control of borrowing. Overloading will never be prevented without it. **1962** A. Nisbett *Technique Sound Studio* iii. 67 There should be no trouble with this except where the output of a close balance on heavy brass is fed through pre-amplifiers of fixed gain; this can result in overloading and consequent distortion. **1962** R. H. Smythe *Anat. Dog Breeding* 70 These are the muscles which when well developed cause 'overloading' of the shoulders. *Ibid.* 77 What judges term, 'overloaded shoulders'. **1974** *Times* 9 Jan. 3/3 Surplus, damaged and overloaded goods were sold to people in the trade.

overlock, *v.* Add: **c.** *trans.* and *intr.* To secure (the edge of cloth) so as to strengthen it and to prevent fraying; also, to oversew (by machine). Usu. as **o·verlocking** *vbl. sb.*
1901–2 *T. Eaton & Co. Catal.* Fall & Winter 64/2 Men's fine imported natural wool night robes, made with collar attached,..overlocked seams, [etc.]. **1909** *Public Ledger* (Philadelphia) 24 June 5/2 Fishnet Lace Curtains, overlocked edge. **1921** *Dict. Occup. Terms* (1927) § 376 *Overlocking machinist*,..stitches round scolloped edge of finished lace curtain with overlocking machine. **1960** *Textile Terms & Definitions* (Textile Inst.) (ed. 4) 106 *Overlocking*,..effecting the joining of two or more pieces of fabric by means of an double or treble chain stitch... This operation is performed on overseaming machines. **1973** *Guardian* 12 Mar. 9/4 Most of the women work in..linking, cutting, and overlocking. **1976** *Leicester Trader* 24 Nov. 17/2 (Advt.), Klynton Davis require experienced employees with the following skills: lockstitching, overlocking, welting, [etc.]. **1978** *People's Friend* 13 May 24/2 (Advt.), Every detail is perfect—like the needle overlocking technique that ensures incredible strength at the seams.

overlocker (ōᵘ·vəɹlǫkəɹ). [f. *overlock *v.* c + -er[1].] (See quot. 1921.)
1921 *Dict. Occup. Terms* (1927) § 419 *Overlocker*.., guides cut edges and seams of knitted fabrics under needles of power-driven overlock machine, thereby joining them together..with an overlock stitch. **1964** *Age* (Melbourne) 15 Aug. 62 (Advt.), Experienced overlockers and plain sewing machine operators and juniors to be trained are required for our making-up rooms. **1975** *Evening Herald* (Dublin) 8 May 12/5 (Advt.), Experienced overlockers, button machine operators and flat machinists. This staff is required for our slacks factory. **1976** *Leicester Mercury* 16 July (Advt.), Leta Knitwear Ltd., outerwear and underwear manufacturers require very urgently overlockers, binders, and designer.

overlooker. 2. (Later examples.)
1963 *Times* 25 May 9/7 'But t'brig isn't t'world', a sewing-shop overlooker says over his gill of mild. **1973** *Guardian* 12 Mar. 9/1 She has worked as an overlooker at Mansfield Hosiery for five years. **1974** *Times* 5 Dec. 5/2 A mill in the Preston area [where]..the first Asian overlooker was recently promoted. **1976** T. Jeal *Until Colours Fade* ix. 111 A group of all but naked men, working under an overlooker, raking out white-hot coke.

overlord, *sb.* Add: **1. b.** *transf.* A person (occas. an animal) in a position of superiority, authority, or power; *spec.* in British politics, a member of the House of Lords given charge of one or more government ministries; hence in the politics of other countries, and in industry.
1932 S. Zuckerman *Social Life Monkeys* xiv. 228 Baboons are not promiscuous. Very few observations have been made of females having sexual relations with males other than their overlords. **1939** Joyce *Finnegan's Wake* i. 97 To ongoad and unhume the great shipping mogul and underlinen overlord. **1951** *Economist* 8 Dec. 1394/2 Departmental ministers subject to an overlord would, under arrangements of this kind, rarely have direct access to the cabinet. **1953** *Ann. Reg. 1952* 36 The Opposition had long been critical of Mr. Churchill's 'overlords'—super-departmental Ministers screened from cross-examination by their membership of 'another place'. **1954** *Economist* 31 July 355/1 General Perón..did reduce the nominal number of ministries..; but this was accompanied by the appointment of four..'overlords'..concerned with defence, economic, technical, and political affairs. **1957** *Observer* 13 Oct. 1/4 Instead of appointing a missile overlord, as is being urged, the President has referred the whole problem..to yet another committee. **1969** *Daily Tel.* 20 Nov. 3/1 Lord Beeching..is being tipped for the job of 'overlord' to run BOAC and BEA. **1970** *Guardian* 11 May 10/1 The old problem of the Overlord Minister which re-emerged from the Churchillian past after Mr Wilson's last big reshuffle. **1970** *Daily Tel.* 11 Sept. 6/5 Under the reorganisation Vauxhall will become part of a European division of General Motors... Mr L. Ralph Mason..will become the European 'overlord'. **1977** *Time* 21 Feb. 24/2 In Mexico the destruction of planted fields and the arrests of several overlords,..have led to fierce internecine battles for control of the business.
2. (With capital initial.) The code-name for the allied invasion of German-occupied Normandy in June 1944.
1943 J. Reith *Diary* 18 Sept. (1975) vi. 310 Meeting..about a ridiculous Churchill demand for a twenty-five per cent increase in Overlord (invasion of Europe) force. **1947** J. R. Deane *Strange Alliance* i. ii. 22, I emphasized our commitments to Overlord, the Mediterranean, the Pacific. **1948** C. Falls *Second World War* xxiv. 219 The title given to the plan for the invasion of the European Continent from the west, Operation Overlord, possessed a special significance. It was the over-riding operation. **1950** W. S. Churchill *Second World War* III. ii. xxxiv. 585 It was my earnest desire that the crossing of the Channel and the liberation of France (the operation then called 'Round-up', which was subsequently changed to 'Overlord') should take place in the summer of 1943. **1961** E. Waugh *Unconditional Surrender* iii. ii. 239 'Overlord', that one huge hazardous offensive operation on which, it seemed, the fate of the world depended. **1974** G. Markstein *Cooler* lxvi. 230 We've been working very hard to sell them the idea that the Pas de Calais is the objective of Overlord.

overlord, *v.* Delete *rare* and add later examples. Hence **overlo·rded, overlo·rding** *ppl. adjs.*
1910 Galsworthy *Sheaf* (1916) 132 Our dim consciousness of this serene and overlording principle of Equity. **1959** *Catholic Herald* 27 Nov. 5/3 The Ukrainians of Kiev..resisted the attempts of the Moscow Orthodox Patriarchate to overlord them. **1961** B. Vawter *Conscience of Israel* vi. 139 A fact that can help account for its economic expansion in a world overlorded by the Great King. **1966** *Economist* 24 Sept. 1234/3 They were feeling too overlorded. **1970** *Daily Tel.* 11 May 16/5 The appointment of a single executive to overlord the Dutch and British ends of the business would represent a major switch. *Ibid.* 16 Oct. 19/5 (*heading*) David Barran to 'overlord' Shell.

overly, *a.* Add: **2.** (Later example.)
1833 *Chambers's Edin. Jrnl.* 27 Apr. 97/1 Some day, your wife mentions to you, quite in what Mrs Pringle called an overly way, that she happened to meet her old friend Mrs Nicholson on the street that day.

overly, *adv.* Add: **1.** Now more widespread but still regarded as an Americanism in the U.K. (Further examples.)
1854 [see *dogan]. **1903** Kipling *Five Nations* 21 Yet, caring so, not overly we care To brace and trim for every foolish blast. **1924** B. G. Elliott *Automobile Repairing* x. 131 An overly rich mixture may be caused by the fuel nozzle valve being open too wide. **1926** J. A. Muller *Stephen Gardiner* ii. 11 The intense, overly conscientious Thomas Bilney. **1929** M. C. Work *Compl. Contract Bridge* v. 111 Overly sanguine or 'bad break' slam tries. **1942** *Sun* (Baltimore) 10 Oct. 10/2 A panting, frantic lady..is..not overly generous after she has stood for a while..shouting for an invisible porter. **1956** D. Karp *All Honourable Men* 50 She took one with an overly feline movement of her body and looked into his face as she accepted his light. **1963** *Wall St. Jrnl.* 25 Jan., As for the future, however, even the most annoyed American official seems not to be overly alarmed. **1968** *Globe & Mail* (Toronto) 17 Feb. 8/1 The Manitoba Minister of Agriculture is not overly impressed with the horsemen's woes. **1970** *Nature* 17 Jan. 213/1 Scientists are not going to be overly interested in the 'sociology of the sociology of science' if nothing is forthcoming except internal disagreement. **1970** G. F. Newman *Sir, You Bastard* i. 27 Whoever had covered them hadn't been overly careful about the tubes. **1972** *Times Lit. Suppl.* 2 June 624/4 Those overly rationalistic readers who demand to see in her work plentiful evidence of a higher or deeper 'sanity'. **1977** *Dædalus* Summer 157 This is the methodological point lying behind Popper's overly propositional thesis about the eternal falsifiability of scientific truths.

overman, *sb.* Add: **4.** [tr. G. *übermensch*.] = Superman.
1895 tr. *M. Nordau's Degeneration* iii. v. 470 The 'bullies' gratefully recognise themselves in Nietzsche's 'overman'. **1900** *Q. Rev.* July 116 In such old religion he discovers no prophecy of the man that is to be; he reaches forward to some 'overman' beyond it. **1908** H. G. Wells *War in Air* xi. 365 His mind ran to 'improving the race' and producing the Over-Man. **1915** *Lond. Q. Rev.* Jan. 59 Such a process of superabstraction would involve either an overman or a *deus ex machina*. **1928** A. Huxley *Point Counter Point* vi. 108 If you were a little less of an overman,..what good novels you'd write! **1932** [see *Nietzschean *sb.* and *a.*]. **1971** *Black Scholar* June 50/1 The 'magnificent savage', 'the mindless overman', is even dying with the deficiency. **1976** M. & G. Gordon *Ordeal* xi. 69 He would overlook Charlie's shortcomings. An overman had to... Vince had read as much of Nietzsche as necessary to learn about the overmen—the supermen—and the inferiors, the masses.

overman, *v.* Add: **1.** (Further example.)
1851 H. Melville *Moby Dick* I. xxxvii. 270 My soul is more than matched; she's overmanned—and by a madman!
Hence (in sense 2) **over-ma·nning** *vbl. sb.*
1971 *New Scientist* 9 Sept. 553/1 Output has reflected this over-manning. **1975** *Broadcast* 28 July 4/1 The ACTT does not consider overmanning..to be a central issue in the financial problems confronting ITV. **1975** *Times Lit. Suppl.* 12 Sept. 1028/4 During the winter overmanning was enforced—one labourer per fifteen acres. **1975** *Daily Tel.* 19 Sept. 3/3 (*heading*) Sir Keith urges TUC to curb overmanning. **1978** *Jrnl. R. Soc. Arts* CXXVI. 656/1 The overmanning in all the factories that I visited was very considerable.

overmark, *v.* Add: **4.** [Over- 27a.] To award too many marks to (a candidate in an examination, competition, etc.)
1947 C. S. Lewis *Miracles* xvii. 198 Some examiners tend to overmark any candidate whose opinions and character, as revealed by his work, are revolting to them. **1970** *Times* 5 Mar. 13 One judge admitted that she had overmarked Wood, for no good reason that I could discover other than sympathy for a champion in distress.

overmast, *v.* Add: (Later examples.) Hence **overma·sting** *vbl. sb.*
1902 Conrad *Youth* 40 She was certainly over-masted. **1906** —— *Mirror of Sea* xi. 58 It was a fine period in ship-building, and also..a period of over-masting. **1930** J. Masefield *Wanderer of Liverpool* 27 Mr. Potter was inclined to think that the *Wanderer* had been over-masted.

overmastering, *ppl. a.* (Later examples.)
1860 Geo. Eliot *Mill on Floss* III. 209 He had been tortured by scruples, he had fought fiercely with overmastering inclination. **1903** *Cambr. Mod. Hist.* II. xviii. 644 He [*sc.* Paul III] was not a zealot, possessed with one overmastering idea. **1915** F. S. Oliver *Ordeal by Battle* iv. 120 Nietzsche is not concerned to evolve a sovereign and omnipotent state, but a high overmastering type of man. **1942** *R.A.F. Jrnl.* 13 June 2 What a blockade by sea can do..an overmastering force of bombers can most certainly do. **1957** J. S Huxley *Relig. without Revelation* (rev. ed.) v. 110 Some have this in an overmastering degree.

overmastery, *sb.* Restrict † *Obs.* to sense in Dict. and add: **b.** Supreme authority, sovereignty.
1901 T. Shaw *Patriotism & Empire* (Young Scots Soc. Publ. No. 1) 4 We are free..to think with a lifting of the heart of the struggle for Scottish independence against the overmastery of England. **1971** J. F. Wathen *Great Sacrilege* ii. 28 God Himself..would prevent such a thing from happening, either directly or through His ordinary overmastery of all creatural actions.

over-measure, *v.* **a.** (Later example.)
1877 Tennyson *Harold* iv. iii. 119 By St. Edmund I over-measure him.

overmod, *sb.* Delete quot. *c* 1200, which illustrates a different word (see *Medium Ævum* (1952) XXI. 38–9).

o:vermodula·tion. *Electronics.* [Over- 29b.] Amplitude modulation that is so great as to result in unacceptable distortion; *spec.* modulation that causes the amplitude of the carrier to become zero for a significant part of each cycle of the modulating wave.
1927 *Exper. Wireless & Wireless Engin.* IV. 3/1 The strengths of..undesired harmonics relative to the desired fundamental decrease as the modulation ratio *a*/*a* is decreased. This is the ground for the term 'over-modulation' as a fault at the transmitter. Over-modulation shows, of course, only during the relatively loud passages. **1937** *Wireless World* 2 Dec. 563/1 Over-modulation of broadcast transmitters has been a serious problem. **1962** A. Nisbett *Technique Sound Studio* vii. 132 One of the problems which may arise when tape is re-used is a background 'chatter' from the previous recording. This can be due..to overmodulation of the previous recording. **1968** *Radio Communication Handbk.* (ed. 4) ix. 4/2 In frequency modulation there is no condition equivalent to over-modulation.

So **overmo·dulate** *v. trans.*, to subject to overmodulation; *intr.*, to cause or suffer overmodulation.
1928 L. S. Palmer *Wireless Princ. & Pract.* xi. 418 As long as the modulator be not overworked, the oscillatory current cannot be over-modulated. **1937** *Wireless World* 2 Dec. 563/1 Automatically graduated compression, or limiting,..makes it almost impossible to overload or over-modulate the transmitter. **1962** A. Nisbett *Technique Sound Studio* ii. 48 Completely dead sound is difficult to balance: the transient peaks tend to over-modulate.

overname, *v.* (Later example.)
1902 J. H. Skrine *Pastor Agnorum* 31 Twenty faces in three ranks, and, though no face is like another, we could have safely overnamed the varieties before we fronted them.

overner (ōᵘ·vəɹnəɹ). *local.* [? f. Over *adv.* + -er[1], after *northerner*, *southerner*, etc.] In the Isle of Wight, a visitor or immigrant from the mainland; = Overer *sb.*[2] Cf. *overun *a.* and *sb.*
1886 W. H. Long *Dict. Isle of Wight Dial.* 46 *Overner*, or *overun feller*, a person whose home is over the water, on the main land; not a native of the Island. West countrymen, who come to work in the Island, are always 'overun fellers'... 'I wish it had capsized they there overners, comen across.' **1951** B. Vesey-Fitzgerald in E. Molony *Portraits of Islands* 65 You may still find in the interior men who speak of people from the mainland as 'overners' (foreigners). **1965** L. Wilson *Portrait of Isle of Wight* i. 15 As well as a strong feeling of security it confers on the inhabitants a sense of identity which 'overners', as they call them, people from over the water, do not possess. **1974** *Isle of Wight County Press* 12 Oct. 19/6 Now there's myself (an Overner, surely) shedding a tear for a scene that had to go.

over-nicety. (Earlier example.)
1754 Richardson *Grandison* VI. xxvii. 165 A little over-nicety at setting out, will carry them into a road they never intended to amble in.

overnight, over night, *adv. phr.* (*sb.*, *a.*) Add: **A.** *adv. phr.* **3.** In the course of a single night; hence, rapidly, instantaneously; without any perceptible or significant passage of time.
1939 Joyce *Finnegans Wake* ii. 378 The unnamed nonirishblooder that becomes a Greenislander overnight! **1942** E. Waugh *Put out More Flags* 246 Alastair's battalion found itself overnight converted from a unit in the early stages of training into first line troops. **1955** H. Roth *Sleeper* viii. 60 Adults don't change—rarely, at any rate, and not overnight. **1955** 'A. Gilbert' *Is she Dead Too?* ii. 30 She'd been there two years when Alice Poulden died. It all seemed to happen overnight as you might say. **1957** F. & R. Lockridge *Practise to Deceive* (1959) vi. 90 Lane..was already 'the' polo player. A handicap justifying such phrasing is not acquired overnight. **1963** *Cambr. Rev.* 4 May 400/1 An article which overnight catapulted him to fame. **1966** *Word Study* Dec. 5/2 Overnight the vernacular of space became a popular idiom. **1972** P. Green *Shadow of Parthenon* 15 The classics have nearly half a century of revolutionary critical development to catch up on: the thing cannot be achieved overnight. **1972** *Observer* 12 Nov. 36/7 The great original thinker, the noble if flawed human being, the entranced and hallowed poet-sage—all these have virtually been blown away overnight. **1973** [see *ouster[1]].
B. *sb.* **2.** A stop or halt lasting for one night; also, a person who stops at a place for a single night (see quots.).; something that arrived during the night.
1959 *Times Lit. Suppl.* 9 Oct. 573/4 A highly convincing background of aviation and suburbia, dinettes, lounges (domestic, not airport), overnights and rosters. **1964** *Economist* 11 Jan. 114/2 The YHA..had a record number of..'overnights' (the total number of nights that beds were occupied). **1968** J. Lock *Lady Policeman* xiii. 117 The gaoler brings the 'overnights' from the cells. **1974** *Guardian* 23 Mar. 14/3 With two overnights in Interlaken..it sounded..soothing enough. **1976** C. Weston *Rouse Demon* (1977) i. 4 Ten minutes to spare, pounding up the stairs to the Detective Bureau. ..Time to skim through the overnights before the morning rundown.
C. *attrib.* or *adj.* **2. a.** Designating a price or value as at the end of business on a particular day. **b.** Applied to money lent or borrowed, or otherwise made available, from one day to the next.
1909 *Westm. Gaz.* 7 Aug. 12/1 Finishing on the other side very strongly, the overnight prices were well above those ruling in the 'House' at the close of yesterday afternoon. **1928** *Daily Mail* 9 Aug. 18/6 Borrowers occasionally paid up to 4½ per cent. for fresh overnight money. **1930** M. Clark *Home Trade* xxix. 233 Sometimes the loan is merely one from one day to the next—often termed 'overnight loans'; at other times it is for a short period of say seven days ('weekly loans'). **1973** *Times* 2 Feb. 14/7 New York money men have been known to quake in the knowledge that they have on deposit $1,000m of overnight money—money which can be withdrawn the next morning.
3. Of a person: transient, staying overnight; *fig.*, instantaneously esteemed or popular. Of a thing: achieved or accomplished rapidly.
1934 [see *air hostess* s.v. *air *sb.*[1] III. 4]. **1960** *News Chron.* 18 Mar. 6/6 He has become an over-night hero to countless people. **1961** *Sunday Express* 7 May 15/5 One really eye-catching picture in a top magazine, and she is an overnight star. **1974** S. Sheldon *Other Side of Midnight* iv. 86 The war, like all wars, had created overnight millionaires. **1974** Hawkey & Bingham *Wild Card* xv. 127 His novel advances in the field of biodegradable materials..had brought him an overnight reputation.
4. Special collocations, as *overnight bag, case*, a light case or holdall carried by a traveller; *overnight telegram*, a telegram intended for delivery the following morning.
1925 *New Yorker* 17 Oct. 32/2 This is simply wonderful for travelers, though, as an overnight bag, it has the disadvantage of having no room in the bag for anything except, perhaps, a French nightie. **1955** 'N. Shute' *Requiem for Wren* i. 2 He took the overnight bag from me. **1972** M. Crichton *Terminal Man* i. v. 43 She lifted a small blue overnight bag. **1934** Webster, Overnight case. **1935** *Montgomery Ward Catal.* 582 Wards finest—tray fitted overnight case. **1952** 'J. Tey' *Singing Sands* xiv. 238, I indicated my small overnight case which was lying..on the bunk. **1970** G. F. Newman *Sir, You Bastard* 265 The overnight case belonged to Sneed. **1955** *P.O. Guide* July 309 Overnight telegrams are accepted between the hours of 8 a.m. and 10 p.m. daily for delivery next day..normally by the first post. **1974** *Ibid.* Nov. 367 An Overnight telegram may be sent between 8 a.m. and 10.30 p.m. for delivery, normally by first post, the following morning... The charge is 40p for 10 words or less and 2p for each extra word.

overni·ght, *v.* [f. the adv. phr.; cf. G. *übernachten*.] *intr.* To pass the night (*at* or *in*); to lodge for the night.
[**1876** Geo. Eliot *Let.* 2 Sept. (1956) VI. 275 We saw no English or American visitors, except such as 'übernachten' there and pass on.] **1891** M. M. Dowie *Girl in Karpathians* xiii. 177 He invites us to over-night at his house. Will you go? **1962** *Daily Progress* (Charlottesville, Va.) 2 Nov. 6/6 When a stewardess 'overnights' she spends the night in the city where the flight terminates. **1965** *Harper's Bazaar* Dec. 80/1 Anyone who has over-nighted in pre-Hilton Istanbul. **1971** R. Falkirk *Chill Factor* v. 47 Her Icelandic philosophy was immensely popular..wherever Icelandair overnighted. **1971** *Advocate-News* (Barbados) 20 Mar. 1/2 At the conclusion of his South American visits the President and party will again over night in Barbados on March 31 and leave for home on April 1. **1972** *Drive* Spring 49/2 With the horses provided you could trek deep into the forest, overnighting in a bivouac. **1976** J. B. Hilton *Gamekeeper's Gallows* xviii. 182 Fletcher over-nighted at Derby.
Hence **overni·ghting** *vbl. sb.* (also *attrib.*).
1948 K. Anthony *Lambs* vi. 132 The clambering in and out of stage-coaches, the overnighting in strange taverns. **1966** *Guardian* 19 Mar. 1/3 Overnighting facilities for climbers have been cut off, and visitors generally discouraged. **1969** 'J. Morris' *Fever Grass* x. 95 D'you have a spare shirt for this big man here? He didn't plan on overnighting.

o:verni·ghter. [f. Overnight *adv. phr.* + -er[1].] **a.** = *overnight bag* (see *overnight C. 4).
1959 *Sears, Roebuck Catal.* Spring–Summer 865/1 Easier Packing... 18-inch Overniter. **1967** 'S. Marlowe' *Second Longest Night* vi. 58 Armed with two overnighters and the necessary toilet articles. **1967** K. Giles *Death in Diamonds* iv. 75 One suitcase..and a little overnighter with electric razor and stuff. **1972** M. Crichton *Terminal Man* i. v. 45 Benson placed the screwdrivers into the overnighter.
b. A person who stops at a place overnight (cf. *overnight *sb.* 2).
1961 F. & R. Lockridge *Murder has its Points* (1962) ix. 96 At the Hotel Dumont there had, at the time in issue, been twenty-three overnighters. **1977** *Daily Tel.* 7 June 1/1 The area approaching St Paul's Cathedral was also becoming filled with overnighters.

o·ver-note. [Over- 1e.] A note heard through or above other sounds; an overtone.
1917 Conrad *Shadow-Line* 204 He..burst into..a loud laugh... It was a provoking, mocking peal, with a hair-raising, screeching over-note of defiance.

over-old, *a.* (Later examples.)
1875 W. Morris tr. *Virgil's Aeneids* viii. 509 My body over-old for deeds begrudged such government. **1883** Ld. R. Gower *My Reminisc.* II. 140 Their children..have a delicate over-old look for their age.

overpaint (ōᵘ·vəɹpēⁱnt), *sb.* [Over- 8c.] A second or further layer of paint; a paint used for overpainting.
1958 *Times* 1 July 5/4 The overpaint has now been cleaned off, revealing not only the skull which Petty holds but also two other skulls engraved on the page of the book to which the left hand is pointing. **1973** *Canad. Antiques Collector* Jan.–Feb. 17/1 This cupboard, when found, was over-painted; the over-paint was carefully removed to expose the original red on the cupboard proper, and the blue-green on the cornice and decorative mouldings. **1973** F. Taubes *Painter's Dict.* 168 The evidence or absence of brushstrokes in the underpainting should be considered, for these may interfere with the overpaints.

o:verpai·nted, *ppl. a.* [Over- 8b.] That has been painted over another painted surface.
1967 J. N. Barron *Lang. of Painting* 195 The painter plans the underpainting to obtain the variety of effects and optical mixtures he wishes in combination with, or in contrast to, the subsequent overpainted layers of paint. **1973** *Guardian* 16 Mar. 13/5 The National Portrait Gallery is..showing again..the family portrait of Sir Thomas More... The overpainted nineteenth century brown background has been removed.

overpai·nting, *vbl. sb.* [Over- 8b.] The action of the verb Overpaint (sense 1); a layer of paint applied over another.
1928 *Daily Express* 20 Dec. 1/3 The explanation of the over-painting is simple. The sitter..decided to have his sheriff's robes painted over the clothes in Holbein's picture. **1935** E. Neuhaus tr. *Doerner's Materials of Artist* iv. 203 The overpainting must be applied somewhat more liquidly—at least in the first layer—than the underpainting. **1954** M. Rickert *Painting in Brit.: Middle Ages* i. 23 The overpainting in thick white and yellow pigment. **1958** tr. *K. Herberts's Artists' Techniques* 140 Overpainting must be done with great care, since the lower layers of paint remain effaceable. **1973** *Guardian* 16 Mar. 26/1 Restoration..has removed the overpainting which had made the picture look unnaturally flat.

overpa·rk, *v.* *U.S.* [Over- 27.] *intr.* To park a motor vehicle for longer than the permitted period. So **overpa·rked** *ppl. a.*, **overpa·rking** *vbl. sb.*
1938 *Daily Progress* (Charlottesville, Va.) 3 Mar. 1/8 The case of Clyde Anable, charged with overparking. **1957** *Times* (Seattle) 12 Sept. 38 The City Council today authorized the Police Department to hire ten 'meter maids', uniformed women who will patrol the streets and write tickets for overparking. **1974** *Keowee Courier* (Walhalla, S. Carolina) 24 Apr. 7/1 Gaillard also reported that the police are now issuing 'courtesy tickets' to people who overpark. These tickets provide only a friendly warning, and no fine, unless a person accumulates two in one day or three within a week. **1976** 'J. Ross' *I know what it's like to Die* xxviii. 177 His Renault car had..been there all day... It had incurred fines for overparking. **1977** 'L. Egan' *Blind Search* iv. 53 Officious meter maids checking overparked cars.

overparted, *a.* Add: (Further examples.) Also *transf.*
In quot. 1975[2] the sense is 'having a voice too strong for the part'.
1896 G. B. Shaw *How to become Mus. Critic* (1960) 240 As Siegmund the Unlucky he was quite overparted. *Ibid.* 244 He was overparted in Siegfried. **1959** *Times* 19 Nov. 16/2 Mr. William McAlpine seemed vocally over-parted as Boris. **1966** *New Statesman* 25 Mar. 437/3 Overparted Mr Constantine is persuaded to deliver himself of some grand lines, among which '*Je crois aux données immédiates de la conscience*' is probably as ludicrous as any. **1975** *Bookseller* 26 July 315/1 Seemed rather over-parted as the *Sunday Telegraph* lead reviewer. **1975** *Gramophone* Sept. 505/1 For some, Fischer-Dieskau is considerably overparted. **1977** *Listener* 5 May 592/2 The baritone and tenor soloists..rather seized the attention from the sadly overparted soprano.

overpass, *v.* Add: **3.** (Later example.)
1938 *Times* 16 Aug. 15/4 The stream..swelled uproariously. It did not anywhere overpass its deeply engraved channel, but raced helter-skelter and bank high to the road and the beach.
7. b. (Later examples.)
1905 *Daily Chron.* 24 Oct. 1 The Russian and Austrian agents in Uskub overpass their duties. *a* **1973** J. R. R. Tolkien *Silmarillion* (1977) 262 But the design of Manwe was that the Númenóreans should not..desire to overpass the limits set to their bliss.
8. (Later example.)
1872 G. M. Hopkins *Let.* 5 Mar. (1956) 118, I cannot tell how I have overpassed your birthday and only been recalled to it now too late by seeing the date March 3 on a letter.
overpassing *vbl. sb.* (further example).
1865 Mill *Auguste Comte* 14 He deemed all real knowledge of a commencement inaccessible to us, and the inquiry into it an overpassing of the essential limits of our mental faculties.

o·verpass, *sb.* orig. *U.S.* Also **over-pass.** [Over- 1d.] A raised stretch of road or railway line that passes over another road or railway line; = *fly-over 1. Also *attrib.*
1929 *Amer. City* Oct. 104/2 In certain cases where the construction of under- or over-passes cannot be avoided..my system simplifies them to an astonishing extent. **1933** [see *clover-leaf* s.v. *clover *sb.* 4]. **1938** *Sun* (Baltimore) 31 Aug. 7/3 Overpasses were built in both communities after years of agitation and numbers of serious crossing accidents. **1952** [see *expressway]. **1959** *Daily Tel.* 9 Nov. 1/1 But they refused to allow a car, scooter or even a bicycle to be pushed on these overpass roads. **1964** L. Deighton *Funeral in Berlin* li. 313 Dominating the whole scene is the gleaming stone pillar of the Cenotaph like the freshly-built leg of a new overpass. **1969** *New Scientist* 17 Apr. 105/1 A major earthquake would..cause the overpasses into the city to collapse. **1973** H. Nielsen *Severed Key* i. 11 The traffic lanes leading away from the airport were packed. Once over the overpass, Keith made a sharp right turn. **1974** *Anderson* (S. Carolina) *Independent* 23 Apr. 3A/5 A 180-foot overpass over Clinchfield railroad tracks on secondary road 126 in Spartanburg went to Dickerson, Inc., of Monroe N. C., which entered a low bid of $335,360. **1976** *National Observer* (U.S.) 5 June 1/1 As you top an overpass, your eyes are drawn to a red, white, and blue water tower on the horizon.

over-pe·dal, *v.* [Over- 27.] *intr.* and *trans.* To over-use a piano's sustaining pedal; to play using the sustaining pedal too much. Hence **over-pe·dalled** *ppl. a.*, **over-pe·dalling** *vbl. sb.*
1961 *Times* 12 Apr. 6/1 He [*sc.* the pianist] was liable to force his tone, over-pedal. **1968** *Daily Tel.* 1 Nov. 21/3 The tempestuous coda to the Ballade, for instance, was smudged through over-pedalling. **1976** *Gramophone* June

69/1 Berman himself often overpedals in an over-resonant studio so that bigger climaxes are just a confused noise. **1976** *Ibid.* Sept. 421/3 The cadenza..is over-pedalled. **1976** *Daily Tel.* 29 Nov. 11/1 More control of rubato and over-pedalling is needed, but she has all the musical and technical talent to be an important pianist.

over-persuasion. (Later example.)
a **1817** JANE AUSTEN *Persuasion* (1818) III. vii. 142 It had been the effect of over-persuasion.

overpitch, *v.* **1.** (Earlier and later examples.) Also *absol.*
1851 [see *BOWL *v.*^{1} 4 b]. **1958** D. BRADMAN *Art of Cricket* 104/1 If one's length is faulty, over-pitch rather than under-pitch that new ball. **1963** *Times* 28 May 4/5 On a perfect pitch, he played each good length ball with care, but those overpitched he punished severely and his 100 included four sixes and 10 fours.
2. (Later examples.)
1976 N. ROBERTS *Face of France* vi. 69 He is nondescript and correct, she high-coloured and over-pitched. **1977** *Church Times* 28 Jan. 6/4 The tone of much of Kingsley's writing..now seems overpitched to an almost hysterical degree.
Hence **overpi·tched** *ppl. a.*, of a ball: that is pitched too far.
1855 [see *BREAK *sb.*^{1} 5]. **1897** [in Dict.]. **1900** A. E. T. WATSON *Young Sportsman* 147 He has lunged out as far as he can reach, hoping to 'smother' a somewhat overpitched ball. **1958** D. BRADMAN *Art of Cricket* 101/2 Learn to bowl the yorker if you can but be prepared to get hit for some fours off the overpitched balls in the process.

over-play, *v.* Add: **1. b.** *fig.* To emphasize too much; to attach too great an importance to; *spec.* in phr. *to overplay one's hand*, to spoil a good case by exaggerating its value.
1930 *Times* 27 Mar. 15/3 Conditions are clearly more favourable to agreement..provided only that *Nahas Pasha* does not over-play his hand. **1933** *Sun* (Baltimore) 16 Aug. 10/7 American newspaper headline writers..'overplay' the news for which they write captions. **1952** *Essays in Crit.* II. 325 He [*sc.* Empson] thinks Tillyard and Dover Wilson..overplayed their hands in attending too exclusively to the 'official' explanations of Shakespeare's history plays. **1956** A. L. ROWSE *Early Churchills* 269 Here was the one chance of the Allies..thrown away by overplaying their hand. **1960** I. PEEBLES *Bowler's Turn* 190 Dexter over-played his luck and ran himself out. **1965** *New Statesman* 20 Apr. 673/3 One building society told me that the 'crisis' had been 'very much overplayed' and that there were already signs of the investment situation easing. **1968** *Globe & Mail* (Toronto) 13 Jan. 29/2 The problem has been overplayed, he said. The recent slump doesn't indicate a trend. **1977** F. DURBRIDGE *Passenger* iii. 146 Judy may have over-played her hand and tried to cut herself in on one of Andy's little rackets.

overplus, *sb.* (*adv.*, *a.*) Add: **A.** *sb.* **c.** (Earlier and later examples.)
1721 M. W. MONTAGU *Let.* May (1966) II. 5, I beleive [I] shall take care another time not to involve my selfe in difficulties by an overplus of Heroic Generosity. **1794** D. O'CONNELL *Let.* 22 Apr. (1972) I. 17, I could spend three months at home in my native air free from all cost; which would compensate for the overplus of travelling charges. *a* **1817** JANE AUSTEN *Two Chapters of Persuasion* (1926) 27 To..pay for the overplus of Bliss, by Headake & Fatigue. **1934** E. POUND *Eleven New Cantos* xxxvi. 28 Cometh he to be when the will From overplus Twisteth out of natural measure. **1969** *Worship* XLIII. 394 In origin, the sacred is an overplus of meaning expressed with such power that it overwhelms everyone who perceives it. This overplus is beyond analysis.

over-pole, *v.* Add: Hence **overpo·led** *ppl. a.*, **overpol·ing** *vbl. sb.*
1742 W. ELLIS *Mod. Husbandman* Aug. xx. 98 Overpoling [of hops] is worse than Under-poling. **1758, 1861** Overpoled [in Dict.]. **1890** Overpoling [in Dict., sense 2]. **1910** *Jrnl. Inst. Metals* IV. 207 In the case of 'overpoled' copper the gases were released in such a quantity as to not only neutralise the effect of shrinkage, but to elevate the surface of the ingot. *Ibid.* 230 'Poling' could be pushed further, before the stage of 'overpoling' was reached, than could be done in the case of pure copper. **1930** *Ibid.* XLIII. 121 Hence a very slight further poling beyond the point *A* will cause a marked increase in the porosity of the ingot; the surface will rise and all the defects of overpoling will appear. **1937** ARCHBUTT & PRYTHERCH *Effect of Impurities in Copper* iii. 34 Overpoled metal is not in a satisfactory condition to withstand rolling and fabrication for two reasons. **1949** J. E. GARSIDE *Process & Physical Metall.* xxii. 378 If poling has been too prolonged the cuprous oxide content is very low. This gives rise to the evolution of considerable quantities of water vapour on solidification and the metal expands in the mould, forming a ridge. Such material, too brittle for many purposes, is termed 'overpoled copper'.

over-population. (Earlier and later examples.)
1823 J. S. MILL in *Black Dwarf* XI. 754 Not only the master manufacturer but the landowner also, has an interest in over population. **1826** MALTHUS *Diary* 10 July (1966) 265 Landlord at Kenmore complained of the drought, the fall in the price of cattle, and the overpopulation of the country. **1959** A. HUXLEY *Let.* 26 Nov. (1969) 880 It may turn out to be hideously tragic when their efforts to modernize the country break down under the combined pressures of inefficiency and over-population. **1971** *Daily Tel.* (Colour Suppl.) 3 Dec. 24/3 Forrester has designed a model of the world system to try to discover the long term effects of pollution and overpopulation. **1977** *Times* 31 May 5/4 Mexico..is exporting its over-population.

overpotential (ōu·vəɹpǒtenʃăl). [OVER- 19, 29.] = *OVERVOLTAGE I, 2.
1920 *Jrnl. Amer. Chem. Soc.* XLII. 94 Overpotential varies with the nature of the electrode. **1961** A. C. WHISH in G. F. Tagg *Pract. Electr. Engin.* II. 292 For the over-potential tests the transformer is excited in the normal way on one winding to twice or more than twice the value of its rated voltage. **1974** *Encycl. Brit. Macropædia* VI. 644/1 The overpotential can be considered as logarithmically dependent on the current density.

o:verprescri·be, *v.* [OVER- 27.] *trans.* and *intr.* To prescribe an excessive amount of (a drug). Hence **o:verprescri·bing** *vbl. sb.*; **o:ver-prescri·ption.**
1953 *Times* 31 Oct. 4/3 Many doctors admitted that since the introduction of the shilling prescription they often tended to over-prescribe to save the patient coming back for an extra shilling's worth. **1967** N. LUCAS *C.I.D.* x. 138 The over-prescribing of drugs by a small number of doctors. **1968** *New Scientist* 28 Mar. 679/2 There is the agricultural merchant who bends the regulations, there is the veterinary surgeon who over-prescribes, and finally there is the unscrupulous retail pharmacist. **1969** *Observer* 9 Nov. 3/5 Doctors could exercise greater control by restraining themselves from over-prescribing drugs. **1970** *Times* 4 Nov. 6 Dr. Hindmarch blames overprescription by doctors as the main source of illicit amphetamines. **1972** *Science* 26 May 883/2 Yet physicians themselves have a sense that they, as a group, over-prescribe and overuse psychoactive drugs. **1974** M. C. GERALD *Pharmacol.* xi. 205 In a recent survey of 55 physicians in the Boston area, 37 felt that their fellow physicians were overprescribing sedatives.

overpressure. Add to def.: pressure (of a fluid) in excess of that which is normal or allowed for. (Further examples.)
1936 B. JONES *Elem. Pract. Aerodynamics* xviii. 290 The pressure gages should be watched closely, especially before take-off. Overpressure indicates a stoppage in the line. **1941** O. E. PATTON *Aircraft Instruments* x. 153 The suction and overpressure tests are given to find out what will happen to the gauge if it is subjected to pressures below atmospheric or exceeding their normal range. **1962** *Trans. Faraday Soc.* LVIII. 194 A quantitative hydrogenation of a solution of the polymer in benzene was attempted using..an over-pressure of hydrogen (*ca.* 700 mm). **1963** *Ann. Med. Internae Fenniae* LII. 212 The idea of employment of over-pressure directed into the human organism may be questioned in principle. **1977** *Sci. Amer.* Mar. 82/1 Just above the vocal folds are the two 'false' vocal folds, which are engaged when someone holds his breath with an over-pressure of air in the lungs.
b. *spec.* The difference between the (highest) instantaneous pressure at a point subjected to a shock wave and the ambient atmospheric pressure.
1955 *Communications Pure & Appl. Maths.* VIII. 340 We evaluated the *D* corresponding to the shock over-pressure of 5°. **1961** *Shell Aviation News* No. 278. 9/2 It's quite conceivable that a ground overpressure of only 0·1 lb per square foot may be decidedly unacceptable to the farmer... and 0·04 lb per square foot to those who value peace and quiet. **1967** *Guardian* 5 July 1/8 Overpressures will vary from 1 to 1·5 lb a sq. ft. which is less than the maximum Concord bang. **1975** *Sci. Amer.* July 15/3 Overpressure is proportional to the energy released by a nuclear charge and is inversely proportional to the cube of the distance from the point of explosion.

overpri·ce, *v.* [OVER- 22, 27.] *trans.* To price (something) more highly or excessively highly; to price a commodity beyond the means of (someone). Also *absol.* Hence **overpri·cing** *vbl. sb.*
1605 P. ERONDELLE *French Garden* sig. K^v, Buye for me yonder waistcoate..for if I cheapen it, they will ouer price it me by the halfe, As for you, they knowe you haue better skill in it. **1972** K. BONFIGLIOLI *Don't point that Thing at Me* i. 5 He was the second greatest art-dealer of the century: he poisoned his life trying to over-price Duveen out of the field. **1976** 'Z. STONE' *Modigliani Scandal* I. iii. 29 My view is that you have been over-priced for some time... At present few of your canvases deserve to fetch more than £325. **1977** D. CLARK *Gimmel Flask* iii. 49 The antique world offers tremendous scope for faking,..underselling, overpricing and so on.

over-priced, *a.* Add: (Later examples.) Also, of a commodity: priced too highly.
1977 T. HEALD *Just Desserts* vii. 153 Rubbery prawns with over-priced vinegary Mexican wine. **1978** M. KENYON *Deep Pocket* i. 5 The vendors of over-priced ice-cream at Marble Arch.

overprint, *sb.* Add: **2.** [OVER- 8c.] **a.** Over-printed matter, esp. on a postage stamp (see quot. 1913). **b.** The action or result of over-printing.
1876 *Let.* 6 Sept. in J. Easton *De La Rue Hist. Brit. & Foreign Postage Stamps, 1855–1901* (1958) xxiv. 710 We should be furnished with the duties which are to fall in the stamps clearly written or printed, so that we might avoid mistakes in making the overprint. **1899** *Captain* I. 421/2 The correct over-print should have been 'Z.C. de peso'. **1912** KNECHT & FOTHERGILL *Princ. & Pract. Textile Printing* VII. 319 Its darker colour will mask the paler tint of the over-print. **1913** E. B. EVANS *Stamps* (ed. 4) 58 *Overprint*, some addition to the design or inscriptions, printed or written upon a stamp which was already complete and fit for use without any such addition. **1928** *Daily Mail* 7 Aug. 18/4 On three values of this printing some sheets received the overprint upside down. **1938** KNOPF & INGERSON *Struct. Petrology* xiv. 197 Upon this earlier movement there was stamped an oblique overprint of a later deformation, now recorded in the quartz fabric. **1938** E. RAISZ *Gen. Cartography* xvi. 188 Each drawing must have register marks for perfect overprint. **1965** *Jrnl. Neurosurg.* XXII. 346 The pulses received by the scaler may be recorded by statistical overprint with a telegraphic printer. **1971** *Nature* 18 June 463/1 (*caption*) At top, 3 weeks raw data in °F with overprint of best fit cosine wave. **1973** *Daily Tel.* 23 June 25/6 The 10F on 90c exists with inverted overprint and makes about £1,500 in this condition.

over-print, *v.* Add: **2.** To print too many copies of (a book, etc.). Also *absol.*
1909 in WEBSTER. **1931** H. G. WELLS *Work, Wealth & Happiness of Mankind* (1932) ix. 353 The belligerent governments withdrew gold from internal circulation and resorted to the printing press to replace it. Each in its own measure overprinted. **1962** *Which?* Sept. 274/2 We always overprint and there are a few copies of this available if you should want it.
II. [OVER- 8.] **3.** *trans.* **a.** To print additional matter on (a surface already bearing print); to mark by a subsequent printing process. Also *transf.*
1863 in J. Easton *De La Rue Hist. Brit. & For. Postage Stamps, 1855–1901* (1958) xii. 263 Printed first from the Sixpenny plate and afterwards overprinted one penny & fourpence. **1876** *Let.* 6 Sept. in *Ibid.* xxiv. 710 As it is impossible to overprint the full sheet at one operation, and at the same time ensure the overprinted matter falling truly in the panels left blank for it, we could only overprint half the sheet at a time. **1899** *Captain* I. 187/1 The current stamps of Great Britain were overprinted with the company's name. **1912** KNECHT & FOTHERGILL *Princ. & Pract. Textile Printing* VII. 319 Dry the goods well and then—(2) Over-print them with either a cover or pad roller in steam Alizarin red or pink. **1912** *Chambers's Jrnl.* Nov. 750/2 In 1903 permission was again granted to firms to overprint the backs of stamps. **1950** *Chambers's Encycl.* II. 788/1 A white spot on a coloured ground will be obtained by first printing the fabric with a paste containing a reserve chemical and then over-printing with a thickened solution of a dye which will be destroyed or its fixation be prevented by the reserve chemical. **1967** *Q. Jrnl. Geol. Soc.* CXXIII. 274 In the north-west corner of Sheet 10/iii at Lepeidere the *A* and *B* domains come together and both are 'overprinted' and partly obliterated by *C* tectonism. **1974** *B.S.I. News* Jan. 14 BSI is prepared to accept bulk orders for copies of any one standard and have the covers overprinted to the purchaser's requirements. **1974** *Nature* 27 Sept. 296/2 Granite was part of a widespread event which overprinted any isotopic record of the early history of the gneisses in the area.
b. To print (additional matter) on a surface already bearing printing; to add by a subsequent printing process. Also *transf.*
1926 C. F. D. MARSHALL *Brit. Post Office* I. vi. 54 On the 1st of January, 1883, the 3d...and 6d...made their appearance in lilac, with the value overprinted in carmine. **1937** *Q. Jrnl. Geol. Soc.* XCIII. 583 There may be complete obliteration of the earlier fabric.., but frequently a second fabric is 'overprinted' on the earlier.. without complete loss of the latter's characteristics. **1938** E. RAISZ *Gen. Cartography* xvii. 199 The French overprinted a network of even kilometer squares upon their maps. **1975** J. B. HARLEY *O.S. Maps* ii. 21 This grid..was overprinted on War Office editions of Ordnance Survey maps.
So **o:verpri·nted** *ppl. a.*, **o:verpri·nting** *vbl. sb.*
1876 Overprinted [see sense 3 a above]. **1912** KNECHT & FOTHERGILL *Princ. & Pract. Textile Printing* VII. 320 The ground will be plain or 'patterned' according as the roller used in the second- or over-printing was a 'pad' or a cover. **1931** H. G. WELLS *Work, Wealth & Happiness of Mankind* (1932) ix. 353 The overprinting of paper money continued. **1962** *Collier's Encycl.* XVIII. 679/2 *Surcharged and overprinted*, stamps on which a new value or name has been printed; 'surcharge' is used when overprinting involves change in value. **1971** I. G. GASS et al. *Understanding Earth* ii. 51/1 Overprinted age patterns may span the entire interval between two (or more) thermal events. **1975** J. B. HARLEY *O.S. Maps* ix. 143 In addition to the coloured administrative diagrams, 1:100 000 scale base maps printed in grey are available without overprinting.

over-pri·vileged, *a.* [OVER- 28d.] Possessing or enjoying too many privileges. Also *absol.*
1934 *Word Study* Mar. 6/2 If this country has an over-privileged class, it is the railroad phoneticians. **1941** *Times Educ. Suppl.* 6 Dec. 581/1 The over-privileged classes..and in general what is crudely described as the 'old school tie' influence in Government, business, and the professions, see their privileges threatened. **1956** C. W. MILLS *Power Elite* i. 14 The moral conception of the elite, however, is not always merely an ideology of the over-privileged. **1971** P. WORSTHORNE *Socialist Myth* viii. 192 Today..the majority..is not only overprivileged politically; it is overprivileged economically. **1973** *Nature* 16 Mar. 210/2 The weak and the poor are under-privileged because we are over-privileged.

over-production. Add: (Later examples.) Also *attrib.*
1934 C. LAMBERT *Music Ho!* IV. 233 The present age is one of overproduction. **1948** G. CROWTHER *Outl. Money* (rev. ed.) v. 151 'Over-production' and a slump result. *Ibid.* 153 The 'over-production' theory, then, like the 'under-consumption' theory, is, in a sense, sometimes right, but for the wrong reasons. **1960** *Farmer & Stockbreeder* 19 Jan. (Suppl.) 50/3 Over-production of Eggs sank prices to rock-bottom in 1959. **1974** M. B. BROWN *Econ. of Imperialism* viii. 199 The general opinion held in Germany in the mid 1880s was that colonies were needed

to solve the problem of overproduction. **1974** *Country Life* 26 Dec. 2019/1 In America there has been overproduction resulting in shutdowns. **1975** *New Yorker* 21 Apr. 7/2 She is most effective as a live performer of her own material—projecting a loose, direct, unpretentious, spontaneous style that, unfortunately, is lacking on her records, which suffer from overproduction.

over-proof, *a.* (*sb.*) Add: (Later examples.) Also *fig.*
1906 *Daily Chron.* 11 May 9/2 The appeal to the Government..asking them to prohibit the importation of over-proof spirits into British territory. **1967** N. MARSH *Death at Dolphin* iv. 94 She really is..the original overproof *femme fatale*. **1973** J. WOOD *North Beat* x. 131 'Try a punch at the brew,' the sergeant suggested. Collins tried a punch. He coughed..'What is it—some of the overproof stuff?'

o:verprote·ction. *Psychol.* [OVER- 29 b.] The condition or act of protecting (someone, esp. a child) to an undue or unhealthy extent.
1930 *Smith Coll. Stud. Social Work* I. 42 Such over-protection would be increased if the child were sickly or handicapped in any way. **1938** D. M. LEVY in *Psychiatry* I. 569/2 We would thus succeed in isolating those personality factors..that would be..a result of maternal overprotection. **1949** S. A. STOUFFER et al. *Amer. Soldier* iv. 135 A theory currently of considerable interest in psychiatry seeks to trace some types of neurotic behavior to overprotection by the mother. **1964** M. ARGYLE *Psychol. & Social Probl.* ix. 123 Hysteria and psychosomatic complaints, especially in women, are related to a history of overprotection by the mother. **1970** H. EDELSTON *Foundations & Growth of Character* III. i. 96 Just as maternal deprivation can be harmful in its effects, so can maternal overprotection. **1977** A. WILSON *Strange Ride R. Kipling* vi. 275 Much of her over-protection may have come from her sense of his strain.
So **overprote·ctive** *a.*, that protects to an undue or unhealthy extent; **overprote·ctiveness**, the state or condition of being over-protective. Also **overprote·ct** *v. trans.*; **overprote·cted, -prote·cting** *ppl. adjs.*
1930 *Smith Coll. Stud. Social Work* I. 58 One or both parents over-protective, rejecting, or were overambitious, as the category indicates. **1938** D. M. LEVY in *Psychiatry* I. 569/2 How their personality traits resulting from growth in the medium of the overprotecting mother have shaped their destinies. **1949** M. MEAD *Male & Female* xv. 310 He is often..over-protective towards his son. **1957** R. B. CATTELL *Personality & Motivation* iv. 135 It is a sociological pattern, varying from family to family of 'Overprotectiveness—vs—Tough Neglect'. **1961** WEBSTER s.v. *Overprotect*, Overprotected children. **1964** M. ARGYLE *Psychol. & Social Probl.* vi. 80 Women are prone to hysteria, especially when they have had a dominating mother who has over-protected them, thus concentrating attention on bodily complaints. *Ibid.* ix. 120 Several investigators reported that schizophrenics have mothers who are over-protecting, but when control groups have been used no such difference is found. **1969** M. D. VERNON *Human Motivation* x. 161 Even over-protectiveness..tends to create anxiety. **1973** R. LEWIS *Blood Money* v. 60 She hadn't wanted him to talk to her daughter... She was over-protective. **1976** *Times* 1 Sept. 14/1 A warning against over-protecting elderly people. **1977** R. BARNARD *Blood Brotherhood* xiv. 150 Over-protected and over-driven, Philip had drifted through.. life.

o:ver-pu·blicize, *v.* [OVER- 27 a.] *trans.* To publicize too much or to excess; to give undue importance to by publicizing. So **o:ver-pu·blicized** *ppl. a.*
1939 *War Illustr.* 4 Nov. p. ii/1, I regard Lindbergh's pronouncement on the War as a piece of gratuitous impertinence... One of the most grossly over-publicised personalities of our age he expects too much if he thinks his words must carry weight just because he once flew the Atlantic. **1957** L. FEATHER *Bk. of Jazz* (1959) xv. 132 Admittedly the drummer today is over-publicized, over-featured and over-praised in proportion to the role he should play as a member of an ensemble. **1964** E. A. NIDA *Toward Sci. Transl.* xii. 252 It is unfortunate that MT (standard abbreviation for machine translation) has been over-publicized. **1965** 'W. HAGGARD' *Hard Sell* i. 1 Over-publicized world beaters which mysteriously disintegrated.

o·verpunch, *sb. Computers.* [OVER- 1 d.] A hole or hole position in the upper portion of a punched card.
1969 MAISEL & WRIGHT *Introd. Electronic Digital Computers* xiii. 325 If the overpunch is a 12, the number is positive, and if it is an 11 punch, the number is negative. **1970** O. DOPPING *Computers & Data Processing* ii. 44 The two top rows are called zone punches or overpunches. The uppermost row is called 12 or Y, the second from the top, 11 or X. In addition, the uppermost numerical row, the 0 row, is used as a third overpunch. **1973** MURRILL & SMITH *Introd. Computer Sci.* vi. 219 When programs written in programming languages such as Cobol read numerical data, the sign does not normally occupy a separate column. Instead, it is entered as an overpunch in the rightmost digit.
So **o·verpunch** *v. trans.*, to represent by means of an overpunch; **o·verpu:nching** *vbl. sb.*
1962 *Gloss. Terms Automatic Data Processing* (*B.S.I.*) 90 *Over-punching*, the use of the upper curtate to represent a digit independently of the use of the lower curtate. **1973** MURRILL & SMITH *Introd. Computer Sci.* vi. 220 Figure 6·6 also illustrates the situation when the & symbol is overpunched in the same column as the 3.

overrange (ōu·vəɹ,rēindʒ), *sb. Electr.* [OVER- 5, 13.] **a.** A signal larger or condition stronger than an instrument is designed to accept or measure. Freq. *attrib.* **b.** An extension of the nominal range of an instrument.
1941 T. J. RHODES *Industr. Instruments for Measurement & Control* iii. 73 This may result in dangerously high pressures if the instrument is subjected to temperatures much higher than those for which it was designed. Usually these overrange temperatures are accidents which can be avoided. *Ibid.* 74 Where the instrument is subject to periodic overrange..it is necessary to provide a means of overrange protection. **1974** *Nature* 15 Nov. p. ix/2 (Advt.), 4-digit electronic display read-out, with 100% over-range, i.e. up to 19999. **1976** *Physics Bull.* Mar. 131/3 This instrument also has a digital voltmeter output and meter overrange protection.

overrate, *v.* Add: **2.** *Rowing.* [OVER- 22.] To row at a faster rate than (an opponent).
1960 *Times* 4 Apr. 14/1 They [*sc.* Oxford] were still overrating Cambridge. **1961** *Times* 10 July 4/6 Lady Margaret made no mistakes in the Ladies' Plate, snatched an early lead, and, always overrating Eton, came home by a length and a third.

overreach, *sb.* Add: **1. b.** (Later examples.)
1961 B. FERGUSSON *Watery Maze* xv. 370 In Burma the Japs made their classic over-reach between March and June of 1944, when..they attempted to surround and defeat the British and Indian forces in Manipur. **1977** *Time* 10 Jan. 55/3 Felker's personal grandeur may match his managerial overreach. Since last spring he has asked for: 1) a 25% increase in his 1975 salary of $120,461, 2) the wherewithal to buy a house in Long Island's ducal Hamptons, and 3) company purchase of his super-duplex.
2. (Later examples.) Also *attrib.*
1932 J. BUCHAN *Gap in Curtain* iv. 193 Verona's mare got an overreach in a bog. **1949** 'J. TEY' *Brat Farrar* xxviii. 250 All the horses were safely back and all well except that Buster had an overreach. **1963** E. H. EDWARDS *Saddlery* xx. 151 A common injury sustained when jumping is caused by an over-reach and, in show jumpers, this often occurs low down on the heel or just above it. A rubber over-reach boot is usually the answer. **1976** *Horse & Hound* 10 Dec. 5/1 You would still have to change your clothes before riding and again when you returned, in order to remove muddy over-reach boots and turn your horse out.

overreaching, *vbl. sb.* (Further examples.)
1802 [see *DICKER *v.*]. **1933** H. BELLOC *William the Conqueror* 55 At that moment appeared the go-between who settled the whole affair, earning thereby the permanent gratitude and protection of Godwin and his sons. But it was an overreaching. They had better have kept away from such an ally! **1971** *New York Law Jrnl.* 23 Nov. 18/5 The 1955 amendment..was adopted..to protect such tenants from overreaching by landlords who authorized commercial and business leases upon tenants. **1977** A. WILSON *Strange Ride R. Kipling* iv. 202 His apprehensions of disaster brought on by overreaching and vainglory.

over-react (ōu:vəɹ,rī,æ·kt), *v.* [OVER- 27.] *intr.* To respond with excessive force or emotion to a given situation. Hence **o:ver-rea·cting** *vbl. sb.*; **o:ver-rea·ction.**
1961 L. MUMFORD *City in Hist.* i. 26 At the same time, the male over-reacted against the feminine side of his own nature. **1962** *Times* 21 Aug. 5/6 The more emotional.. person tended to over-react to stress. **1965** *Economist* 13 Nov. 705/3 Some critics do think that the United States is in danger of getting its priorities wrong, by over-reacting to the threat of China and under-reacting to the possibilities of easing the Soviet dominance of east-central Europe. **1967** C. COCKBURN *I Claud* xxxiv. 424 This 'over-reaction' to what were..reasonable queries and doubts, was psychologically very revealing. **1968** *Listener* 26 Sept. 410/1 This was the first escalation, triggered off when the militants goaded the authorities into over-reacting, closing the faculties and calling in the police. **1971** J. OSBORNE *West of Suez* I. 22, I just hope he..gets some innocent pleasure out of it, which he's entitled to without censorious philistines like me over-reacting. **1973** E.-J. BAHR *Nice Neighbourhood* x. 106 Jack cried despairingly. He always over-reacted. **1973** *Times* 16 Nov. 7/2 A number of Government members believed that the espionage charges levelled against the editors constituted an over-reaction to the articles published in the magazine. **1974** M. C. GERALD *Pharmacol.* i. 12 Whether this was an over-reaction to questionable laboratory results or a sound scientific decision destined to rescue mankind remains to be seen at a future, less emotionally charged time. **1976** *Times Lit. Suppl.* 22 Oct. 1324/4 The hostility to the medium which he encountered in the early days led Strand (like some other major photographers) to over-react, to overcompensate by being rather too serious, too ponderous, too unbending.

over-read, *v.* **2.** Delete †*Obs.* and add later example.
1925 W. DE LA MARE *Two Tales* 14 He was merely over-reading what he had read.

over-reave (ōuvəɹ,rī·v), *v.* [Etym. uncertain; perhaps f. OVER- 16 + a confusion of REEVE *v.*^{1} and WEAVE *v.*^{1}] *trans.* A term used of the metre of his poetry by G. M. Hopkins, to denote scansion continued for a complete stanza, as distinct from that confined to individual lines (see quots.). So **over-rea·ving** *vbl. sb.*
1879 G. M. HOPKINS *Lett. to R. Bridges* (1935) 86 These little graces help the 'over-reaving' of the verse at which I so much aim, make it flow in one long strain to the end of the stanza and so forth. **1880** —— *Let.* 22 Dec. in Hopkins & Dixon *Corr.* (1935) 40 In lyric verse I like sprung rhythm also to be *over-rove*, that is the scanning to run on from line to line to the end of the stanza. **1881** —— *Lett. to R. Bridges* (1935) 120 In my lyrics in sprung rhythm I am strict in overreaving the lines when the measure has four feet, so that if one line has a heavy ending the next must have a sprung head. **1973** *Studies in Eng. Lit.: Eng. Number* (Tokyo) 25 'Over-reaving', or the scanning of verse without break from line to line, is a structural necessity following upon the method of scansion Hopkins proposes for sprung rhythm.

over-record, *v.* [OVER- 27.] *trans.* **a.** To record using too large a signal, so that distortion occurs. **b.** To make too many recordings of (a work or performer). So **over-reco·rded** *ppl. a.*, **over-reco·rding** *vbl. sb.*
1961 E. N. BRADLEY *Records & Gramophone Equipment* i. 23 Inter-groove modulation is caused by over-recording, or by too small a spacing between grooves. **1976** *Gramophone* Sept. 432/2 The comparatively little-recorded *Hamlet*..is linked with the vastly over-recorded *Serenade* for strings. **1977** *Ibid.* Mar. 1467/1 They might well buy the disc for Sabicas, who is not over-recorded these days. **1977** *Rolling Stone* 24 Mar. 79/3 The expanded 'louds' would cause overrecording or tape saturation.

override, *v.* Add: **3. b.** (Further example.)
1932 *Amer. Jrnl. Physiol.* C. 116 There is a significant difference between the amounts of theelin needed to blot out the normal progestational picture as compared to the amount it takes to over-ride a rabbit unit of injected corporin.
c. To cause the operation of (an automatic device) to be suspended, esp. in favour of manual control.
1946 [implied in *overriding* ppl. adj. below]. **1949** *Gloss. Aeronaut. Terms* (*B.S.I.*) II. 11 *Boost control override*, a device to override the boost control so that a pressure higher than the normal controlled pressure can be obtained. **1967** *Instrumentation Technol.* Aug. 38/1 (*caption*) Differential pressure across column overrides temperature control. **1971** *Daily Tel.* 20 Oct. 2/2 It has four forward gears and reverse controlled by a speed-sensing governor which can be overridden by the driver using a gear lever. **1975** *Nature* 20 Mar. 193/1 An interactive computer display system using manual intervention where necessary to override automatic procedures has proved adequate.

o·verride, *sb.* [f. prec. vb.] The action or process of suspending an automatic function; a device for performing this. Freq. *attrib.*
1946 *Aircraft Engin.* XVIII. 112/1 A manual over-ride for landing operation would be necessary. **1949** [see *OVERRIDE *v.* 3 c]. **1952** A. Y. BRAMBLE *Air-Plane Flight* x. 154 There is an 'over-ride' control introduced, by which the automatic limitation of boost can be exceeded at the will of the pilot in particular circumstances. **1957** *Practical Wireless* XXXIII. 697/1 The only connections yet to be made are those coupling up the override switch to the appropriate part in the circuit. **1963** *Amat. Photographer* 7 Aug. 43/1 (Advt.), Fully automatic [camera] with manual over-ride. **1968** *Instrumentation Technol.* Aug. 53/1 Overrides can be designed to provide gradual, rather than abrupt, corrective action and they can function in both directions so that manual reset is not required. **1974** 'A. HAIG' *Peruvian Printout* 39 He pressed the override switch and the computer came instantly to life. **1976** *Offshore Platforms & Pipelining* 203/1 Mechanical overrides are provided should any of the automatic equipment fail.

overrider (ōu·vəɹ,rəi:dəɹ). Also **over-rider.** [f. OVER- 8c + RIDER 12.] One of a pair of projecting pieces attached to the bumper of a car, affording added protection to the bodywork.
1937 *Times* 18 Oct. 20/3 The bumpers have over-riders. **1959** *Times* 6 Jan. 12/1 The overriders on bumpers must not have dangerous protuberances. **1963** J. T. STORY *Something for Nothing* iii. 83 The Jaguar with its big over-riders had suffered not at all. **1963** [see *mud flap* s.v. *MUD *sb.*^{1} 5]. **1974** 'A. HAIG' *Peruvian Printout* 123 The bastard was ramming her back bumper. Thank God she had over-riders fitted. **1975** J. PIDGEON *Flame* ii. 32 The charging Chevy's [*sc.* Chevrolet's] chrome overriders gored the big red and white arrow that pointed diagonally across the carriageway and tossed it high over the roof.

overriding, *vbl. sb.* and *ppl. a.* Add: (Further examples.) *spec.* **overriding commission**, an extra or additional commission. Also **o:verri·dingly** *adv.*
1894 Over-riding commission [in Dict.]. **1906** *Westm. Gaz.* 15 Feb. 11/1 As a rule, the terms of commission, both underwriting and 'over-riding', are very literal. **1930** W. S. CHURCHILL *My Early Life* xiii. 184, I had not heard a word in Cairo of how Sir Herbert Kitchener had received the over-riding by the War Office of his wishes upon my appointment. **1930** A. PALMER *Company Secretarial Pract.* 47 Companies frequently pay 'over-riding' commission also. This is a commission paid to persons for procuring other persons who are willing to underwrite blocks of shares. **1946** R. A. MCFARLAND *Human Factors Air Transport Design* x. 148 Many pilots feel..that in the event of an emergency they must have immediate access to engine instruments. Over-riding throttle controls for the pilot may be advisable. **1956** A. H. COMPTON *Atomic Quest* II. 126 If Oppenheimer has an 'Achilles' heel, it is his overriding loyalty to his

friends. **1957** J. S. HUXLEY *Relig. without Revelation* (rev. ed.) ix. 232 If the full development of human individuals and the fulfilment of human possibilities is the over-riding aim of our evolution, then any over-population which brings malnutrition and misery, or which erodes the world's material resources.., is evil. **1968** *Lebende Sprachen* XIII. 5/1 All diverse organizations must be harnessed together into a single team directed towards the single overriding objective. **1969** W. K. ROOTS *Fund. Temperature Control* v. 118 An inexpensive overriding command system provides both the 'night shut down' and 'vacant-room shut down', features that can significantly reduce the heating cost of public buildings. **1973** A. QUINTON *Nature of Things* 378 He cannot..maintain that morality is overridingly authoritative in this sense.

over-ripeness. Add: (Later examples.) Also *fig.*
1876 G. MEREDITH *Beauch. Career* III. x. 183 Immense wealth and native obtuseness combine to disfigure us with the aspect of over-ripeness, not to say monstrosity. **1904** [see *hypermaturity* s.v. *HYPER- IV.]. **1976** *Listener* 12 Aug. 172/1 Henry James appraised Warwickshire... He felt a ripeness, he hints at an over-ripeness.

overroof, *v.* (Earlier example.)
a **1828** D. WORDSWORTH *Jrnl.* (1941) II. 272 The track ..was..over-roofed, like an outside staircase of a Castle.

over-ru·ff, *v.* [OVER- 22.] *trans.* To over-trump. Also *absol.* and (with stress on first syllable) as *sb.*, an act or instance of over-ruffing.
1813 *Hoyle's Games of Whist & Quadrille* 50 *Ruff*, and *over-ruff*, to trump a suit led, second or third hand. *c* **1890** *Up to Date Games of Cards* 37 Ruff means to trump a suit second or third hand, when you are rid of that suit: *over-ruff* means to trump above. **1906** *Westm. Gaz.* 13 Oct. 14/1 Had A held neither of these cards he would have surely led a diamond.., instead of putting his partner to an over-ruff in the spade. **1926** M. C. WORK *Auction Bridge Complete* II. ii. 331 He should lead trumps before taking the ruff and so avoid any chance of an over-ruff... Dummy's ruff of the losing Club might be over-ruffed. **1974** *Oxford Times* 12 July 10/8 North can ruff in whenever he likes, but dummy overruffs and plays a diamond. **1975** *Times* 16 Aug. 7/4 East was end-played, whether or not he over-ruffed. **1976** *Country Life* 26 Feb. 498/1 When you see the chance of an easy overruff, don't be in too much of a hurry to take it.

overrun, *sb.* Add: **2. b.** An excess of expenditure over that estimated or budgeted for.
1956 *Wall St. Jrnl.* 10 Oct. 12/3 Some of our government officials get carried away with the thought of spending $156 million plus the over-run beyond the estimate. **1960** *Times* 21 Nov. (Canada Suppl.) p. xiii/2 Among these were cases of capital overruns and operating returns poorer than expected. **1973** *Nature* 23 Mar. 224/3 If there are cost overruns on the first two missions, the third may be scrapped. **1974** *Times* 26 Oct. 15/1 Britain's own advanced gas cooled reactor programme is hopelessly compromised by massive cost overruns brought about by.. constructional delays. **1976** *Sci. Amer.* July 122/1 The total cost had been $8 million, an overrun of some 40 per cent. **1978** *Daily Tel.* 13 Apr. 21 This sum..is just under half what remains in the contingency reserve for overruns on public expenditure.
c. An excess of production.
1958 T. LANDAU *Encycl. Librarianship* 230/2 *Overrun*, copies surplus to the number ordered. **1962** J. N. WINBURNE *Dict. Agric.* 539/2 *Overrun*,..the excess amount of lumber actually sawed from logs over the estimated volume or log scale, usually expressed as a percentage of log scale. **1970** *Toronto Daily Star* 24 Sept. 27/1 (Advt.), Our huge purchase includes many carloads of the top lines of merchandise, plus close-outs, over-runs, sample bales.
3. (Examples.)
1898 J. SOUTHWARD *Mod. Printing* I. xxxiv. 210 When there is a long over-run, the matter should be placed upon a small galley, which should be turned, so that the last line rests against its head. **1902** T. L. DE VINNE *Pract. Typogr.: Correct Composition* (ed. 2) xvi. 309 Every paragraph containing an alteration that compels one or more overruns should be re-read. **1935** B. PERRY *And gladly Teach* vii. 169 When the forms were made up, there was an over-run of three lines. **1977** *New Yorker* 26 Sept. 64/3 The *Times*..ran a front-page story, with a four-column overrun on a rear page.
4. The proportional increase in bulk that occurs when butter fat is made into butter or an ice-cream mix is made into ice-cream.
1906 H. SNYDER *Dairy Chem.* vii. 71 During the process of butter making, the slight loss of fat in the skim milk and buttermilk is more than compensated for by the added water, casein, and salt in the butter. The additional butter made from a pound of butter fat is called the overrun. **1922** MOJONNIER & TROY *Techn. Control Dairy Products* xv. 443 Insufficient overrun greatly increases the cost of the ice cream, and yields a product that is immediately detected by its heavy and soggy appearance. **1958** *Sunday Times* 22 June 23/6 Overrun is the aeration or amount any given mix [for ice-cream] will swell in volume when subjected to the freezing process. **1972** *New York* 15 May 4/1 (Advt.), The best coffee ice cream in New York... Sixteen per cent butterfat, 50 per cent overrun; one pint weighs 12¾ ounces.
5. Motion of a vehicle at a speed greater than that being imparted by the engine; freq. in phr. *on the overrun*. Also *attrib.*, designating a system of braking in a towed vehicle (see quot. 1967).
1928 *Observer* 8 Jan. 21/4 The engine runs smoothly and quietly throughout most of its range. There is a certain drumming noise, rather difficult to define and trace, on the over-run, but it is comparatively trifling. **1959** *Motor Manual* (ed. 36) v. 141 This..is at a maximum when the engine is on the over-run. *Ibid.* xiii. 273 When the car brakes are applied, or the car slows down against a closed throttle, the caravan tends to overrun, thus causing the bar to move back against the spring, push back the operating lever and thus apply the caravan brakes. This is known as the 'over-run' method. **1962** *Which? Car Suppl.* Oct. 131/2 The washers..would only operate properly when the engine was on the overrun, i.e. when the foot was taken off the accelerator. **1967** *Gloss. Caravan Terms* (*B.S.I.*) 2 *Overrun braking*, a system of braking in which the caravan brakes are automatically operated by the momentum of the caravan when the towing vehicle is braked. Normally this is achieved by mounting the coupling head on a shaft moving on the drawbar and restrained by a compression spring or a damper. **1969** J. G. GILES *Gears & Transmissions* i. 23 On down gradients heavy vehicles will drive the engine... On these over-run conditions, the transmission torques are reversed. **1977** *Good Motoring* May 3/1 A trailer with over-run brakes can weigh more than the kerb weight of the car, providing a limit of 40 mph is observed.

overrun, *v.* Add: **6.** (Further examples.)
1907 A. T. RITCHIE *Let.* 4 July (1924) xii. 273 It [*sc.* Norway] is like Switzerland, but softer and bigger and not over-run. **1914** G. B. SHAW *Misalliance* p. li, We are over-run with Popes.
9. a. For 'Now *rare*' read 'Now chiefly *Mech.*'; *spec.* to rotate faster than.
1932 J. A. MOYER *Gasoline Automobiles* (ed. 4) viii. 382 The free-wheeling unit or overrunning clutch *G* is given this name because in this arrangement the driven member may overrun the driving member. **1955** W. H. CROUSE *Automotive Transmissions* v. 123 The inner race drives, the outer race is driven. Also, the outer race can over-run, or turn faster than, the inner race. **1959** *Motor Manual* (ed. 36) xiii. 274 As the caravan over-runs the car, the shaft moves backwards against the spring and operates the brake-actuating lever. **1966** *McGraw-Hill Encycl. Sci. & Technol.* IX. 455/2 The second function [of an over-drive] is to permit the output shaft to overrun the transmission shaft.
10. a. (Further examples.)
1867 *Ball Players' Chron.* 14 Nov. 4/3 He fell over Murtha, who was in his way, and overran his base. **1889** E. DOWSON *Let.* 24 Mar. (1967) 54 This appears to be an *extra* special [letter]: it is overrunning [*sic*] all limits. **1948** *News-Palladium* (Bluton Harbor, Mich.) 14 Aug. 6/3 Hazel, going down to second, overran the base as Joe Mack rifled the ball to McCoy. **1973** *Guardian* 1 Sept. 3 When he whispered to her she had overrun the schedule. **1974** M. S. EHRICH *Reincarnation of Peter Proud* xiv. 117 Daley stopped the Movieola... 'I overran it a little..I'll reverse the film.' **1977** *Times* 7 Feb. 7/2 In the opening three minutes Macdonald put Ross clean through, only for the wing half to overrun the ball after dribbling past the goalkeeper.
d. (Examples.) *spec.* in *Broadcasting*, to exceed the allotted time.
1959 *Sunday Express* 30 Aug. 17/3 Last time Borge overran by 15 minutes—and was kept on the air. **1962** A. NISBETT *Technique Sound Studio* i. 17 The announcer on duty in the continuity studio..must intervene if any contribution under-runs, over-runs, breaks down in the middle, or completely fails to materialize. **1962** *Rep. Comm. Broadcasting 1960* 95 in *Parl. Papers 1961–2* (Cmnd. 1753) IX. 259 The BBC bulletin is free to over-run when the service requires it. **1974** *Listener* 14 Feb. 209/1 Arthur Henderson, leader of the Labour rump, lost his head halfway through, thinking he was going to over-run, and ended in a gabble.
III. 12. [OVER- 27.] To run (something) excessively; *spec.* (see quot. 1899).
1899 W. P. MAYCOCK *Electr. Wiring* i. 48 If ordinary lamps..marked for 100 volts..be put on a circuit at, say, 105 volts, the light given will be increased by about 25 per cent., and the watts absorbed per candle-power diminished. This is called over-running lamps, and their life will..be short. **1926** T. T. BAKER *Wireless Pict.* v. 67 It has been found convenient to use a 4·5 volt lamp over-run by a 6, 8 or even 10-volt battery. *Ibid.* 68 When a lamp is over-run it..becomes highly incandescent instantaneously. **1938** G. H. SEWELL *Amateur Film-Making* iii. 35 The Photoflood is essentially a tungsten lamp of normal type which is 'overrun' by having a much higher pressure (voltage) of current passed through it than is normal for domestic burning. **1962** *Which?* Oct. 297/1 It is possible that many people over-run their [water-softening] units—that is, they are not aware of the moment when the water starts running hard, and go on using the hard water for a time.

overrunning, *vbl. sb.* (Further examples.)
1884 G. B. SHAW *Let.* 29 Aug. (1965) I. 95 The proofs.. ought not to be paged, as the insertion of additional slips would cause over-running. **1908** *Westm. Gaz.* 3 Apr. 12/1 Over-running in the cricket-field had..brought the doctors ..several youthful cases of a rather severe type. **1970** J. SEYMOUR *Compan. Guide E. Anglia* ix. 166 The fishermen haul the shanks by hand using the technique known as 'overrunning'—or letting the boat drive down over the shank carried by the swift tide.

overrunning, *ppl. a.* Add: **2.** *overrunning clutch*: a clutch in which the normally driven member is able to rotate faster than the normally driving member.
1921 J. A. MOYER *Gasoline Automobiles* vii. 172 The engine is prevented from driving the [starting] motor by the use of an over-running clutch, which slips when the engine tends to drive the motor. **1930** [see *FREE WHEEL, FREE-WHEEL]. **1966** *McGraw-Hill Encycl. Sci. & Technol.* III. 224/1 The driven shaft can run faster than the driving shaft with an overrunning clutch. This action permits freewheeling as the driving shaft slows down or another source of power is applied.

over-sail, *v.*^{1} Add: **3.** To sail beyond.
1851 H. MELVILLE *Moby Dick* III. xlix. 295 I've over-sailed him [*sc.* Moby Dick]... Aye, he's chasing *me* now; not I *him*.

oversail, *v.*^{2} Add: **2. a.** (Later examples.)
1960 N. SCARFE *Suffolk* 97/2 Columbine Hall..is an ancient moated manor house of beauty, standing straight out of the moat, its upper storey oversailing. **1978** A. & G. RITCHIE *Anc. Monuments Orkney* 31 The lower parts of the walls are vertical, but the upper courses oversail slightly as they rise.
c. To project beyond or overhang (a base).
1912 C. E. POWER *Eng. Mediaeval Archit.* II. 483 In the Decorated period the triple roll base..begins to rise in height, often oversailing the plinth with flat under-side. **1931** *Antiquity* V. 48 The successive courses of the inner wall begin to oversail one another. **1938** *Proc. Prehistoric Soc.* IV. 199 The lowest layer was laid horizontally and each succeeding course was laid at an angle, each stone oversailing the other.
So **oversai·ling** *ppl. a.*
1833 J. C. LOUDON *Encycl. Cottage, Farm & Villa Archit.* 227 These walls..should have what is called a. Welsh cornice (two or three oversailing (protruding) courses of brickwork). **1880** R. BLACKMORE *Mary Anerley* I. xvii. 278 Strong sunshine glared upon the over-saling [*sic*] tiles, and white buckled walls, and cracky lintels. **1954** N. PEVSNER *Essex* 291 It has very heavy timbers and brackets to carry an oversailing upper storey. **1972** J. FLEMING et al. *Penguin Dict. Archit.* (ed. 2) 209/2 *Oversailing courses*, a series of stone or brick courses, each one projecting beyond the one below it. **1976** *National Trust Newsletter* Autumn 12/1 A striking black and white timbered building with high oversailing gables.

oversalt, *v.* (Further examples.)
1816 JANE AUSTEN *Emma* II. iii. 41 They must not over-salt the leg. **1939** S. SPENDER tr. *Toller's Pastor Hall* I. 14 Julie, tell cook not to oversalt the roast.

oversa·turated, *ppl. a.* [OVER- 24 b.] Super-saturated; chiefly in *Petrol.*, applied to a rock or magma in which there is free silica (or some other specified oxide). Hence **o:versatura·tion,** the property of being oversaturated.
1913 S. J. SHAND in *Geol. Mag.* Decade V. X. 510 Any rock which contains free quartz or tridymite of magmatic origin will be termed oversaturated. *Ibid.* 513 The connexion between mineral and chemical composition need not be obscured thereby, but rather the reverse, especially if..the relative degrees of oversaturation or undersaturation were introduced..in the classification. **1946** *Nature* 19 Oct. 549/1 We would rather expect such condensations to show no more rotation than the water droplets in a fog formed from oversaturated vapour. **1947** E. E. WAHLSTROM *Igneous Minerals & Rocks* ix. 253 Rocks may be oversaturated, saturated, or undersaturated with respect to alumina or other oxides as well as with respect to silica. *Ibid.*, Free silica, which indicates oversaturation, is not present in rocks containing undersaturated minerals such as leucite and nephelite. **1974** A. D. EDGAR in H. Sørensen *Alkaline Rocks* v. i. 357/1 The genesis of oversaturated alkaline rocks. **1974** W. C. LUTH in *Ibid.* vi. vi. 506/2 Since the liquid is always silica oversaturated the crystalline assemblage must be silica undersaturated when a liquid is present.

oversay, *v.* Restrict †*Obs.* to senses in Dict. and add later example of sense b.
1874 [see UNSWEAR *v.*].
c. To exaggerate, overstate.
1900 *Scribner's Mag.* Sept. 368/2 This is oversaying it, of course, but the truth is in what I say. **1933** *G.K.'s Weekly* 21 Sept. 41/2, I assure you that if what I say runs towards superlatives it does not oversay what I still think and feel.
Hence **oversay·ing** *vbl. sb.*
1916 T. MACDONAGH *Lit. in Ireland* 46 Latin dispenses with the redundancies, the over-sayings, compressing a phrase into a verb.

overscore, *v.* Add: **b.** (Later examples.) Also *fig.*
1901 E. F. BENSON *Luck of Vails* III. xii. 115 The gentle hum of the warm afternoon came languidly in. Suddenly a fuller note began to overscore these noises in gradual crescendo. *Ibid.* xx. 233 The boon of the doctor's arrival quite overscored that sinister impression he had formed of him.
c. [OVER- 27.] To score (music) with excessively elaborate orchestration. Also *absol.*
1947 N. CARDUS *Autobiogr.* III. 263 There is a fine sensibility moving darkly in the symphonies of Arnold Bax, but he tends to let his texture become congested; he over-scores. **1947** C. GRAY *Contingencies* i. 37 The characteristic vice of overscoring is significant [in music of the Edwardian era]. **1977** *Gramophone* Feb. 1249/1 It is all too easy to dismiss his post-war orchestral works as garrulous, repetitive and over-scored.

overscraw·l, *v.* [OVER- 8.] *trans.* To scrawl over or on. Hence **overscraw·led** *ppl. a.*
1871 BROWNING *Prince Hohenstiel-Schwangau* 28 Why keep each fool's bequeathment, scratch and blur Which overscrawl and underscore the piece? **1879** G. MEREDITH *Egoist* II. xi. 220 A yet more instructive passage than the over-scrawled Seventieth, or French Section.

over-scrupulous, *a.* (Later examples.)
1908 *Westm. Gaz.* 17 Aug. 2/1 Mr. Bryan's over-scrupulous attitude towards advertisements. **1952** E. GRIERSON *Reputation for Song* xxi. 173 She was..a giver, if he'd ever seen one, and not over-scrupulous in conscience or other things.

oversea, *a.* and *adv.* Add: **A.** *adj.* **3.** (Later examples.)
1931 *Times* 17 Feb. 9/1 The competition of our rivals in the home and oversea markets. **1953** [see *DOMINION *sb.* 2 b]. **1955** P. TOWNSEND *China Phoenix* 9 He was Fukienese by birth, from a province of China from which came many oversea Chinese. **1959** *Times* 30 July 6/6 (*heading*) Council formed for oversea research. **1969** *IEEE Trans. Antennas & Propagation* XVII. 254/1 In ship-to-ship detectability studies it is important to have a method of estimating oversea radar range distribution.
B. *adv.* (Further examples.)
1903 W. B. YEATS *In Seven Woods* 15 And you are more high of heart than she For all her wanderings over-sea. **1955** *Times* 9 May 11/2 Private investment oversea. **1974** *Black World* Jan. 55/2 He have to go leave that Ford car when he go oversea from Fort Benning, and it stay in our front yard.

over-seas, *adv.* Add: **b.** quasi-*sb.* Foreign parts; abroad.
1919 *Empire Rev.*, Munition workers who have come from overseas. **1926** A. BENNETT *Lord Raingo* I. lix. 264 Every traveller from overseas was knocked silly by the spectacle. *Ibid.*, Britons whose secret conceit, compared to the ingenuous self-complacency of overseas, was as Mount Everest to Snowdon. **1966** *Listener* 8 Sept. 335/1 In the years before the war our financial income from overseas provided finance to pay for more than a third of our imports.

oversea·s, *a.* [f. the adv.] = OVERSEA *a.*; *overseas Chinese*, a native of China residing in another country.
Overseas is now more frequently used than *oversea*.
1892 KIPLING *Lett. of Travel* (1920) 47 Some day a man will bethink himself and write a book..called 'The Book of The Overseas Club'. **1905** *Daily Chron.* 29 Mar. 3/2 The political liberties of these islands were..deeply endangered by the overseas dominion..of Spain. **1908** *Westm. Gaz.* 26 June 9/3 The magnitude of the overseas possessions which we had to defend. **1912** *Chambers's Jrnl.* Nov. 754/1 In athletic prowess we are now far inferior to those overseas descendants of our race. **1918** P. S. ALLEN *Let.* 9 June (1939) 146 At Merton we are hoping to have some 'overseas' undergraduates next term. **1920** *Act 10 & 11 Geo. V* c. 29 (*title*) An Act to authorise the granting of credits and the undertaking of insurances for the purpose of re-establishing overseas trade. **1933** A. THIRKELL *High Rising* xi. 195 She'll be able to vamp the overseas students and have a splendid time. **1942** 'G. ORWELL' *Diary* 21 June in *Coll. Ess.* (1968) II. 433 The BBC simply isn't listened to overseas, a fact known to everyone concerned with overseas broadcasting. **1947** *Sun* (Baltimore) 22 Aug. 6/4 'Overseas' Chinese will participate in the elections not only as voters but as candidates as well. **1961** S. CHANDRASEKHAR *Communist China Today* viii. 155, I learned that Communist China was anxious to attract the savings of the overseas Chinese who would like to return to their homeland. **1966** [see *HOME *a.* 3 b]. **1968** D. TORR *Treason Line* 16 He was tall for an Overseas Chinese. **1970** V. CANNING *Great Affair* xii. 205 That..was in the overseas edition of *The Times*. **1972** J. BALL *Five Pieces Jade* viii. 96, I consider myself a Chinese-American, in other words an American citizen of Chinese descent. So do almost all of us. But to the Chinese in China—Taiwan or the mainland—either way, we are overseas Chinese. **1974** *Guardian* 27 Mar. 1/1 Tax to be charged on 90 per cent of overseas earnings whether remitted to UK or not. **1975** *Encounter* Feb. 43/2 My wife, phoning Chicago from London, asked, 'Is this the overseas operator?' A pitying male voice replied, 'This is *one* of them, madam.' **1977** 'J. LE CARRÉ' *Hon. Schoolboy* iv. 81 It was an overseas Chinese outfit.

overseer, *sb.* Add: **1. d.** *U.S.* A member of a board of officials which manages the affairs of a college, esp. Harvard College, Massachusetts.
1643 *New Englands First Fruits* 13 Over the Colledge are twelve Overseers chosen by the generall Court. **1812** in *Proc. Mass. Hist. Soc.* (1890) 2nd Ser. V. 176 [Harvard Commencement] The Corporation and Overseers arrived at 20 minutes past ten. **1832** W. D. WILLIAMSON *Hist. State Maine* II. 563 Its government was committed to a board of 13 Trustees, including the President, and a supervisory body of 45 Overseers. **1900** *Dialect Notes* II. 47 *Overseers, board of*, a special governing board of Harvard, chosen by the alumni from their own number. **1946** *Sat. Rev. Lit.* (U.S.) 14 Sept. 5 Then he too retired, but was promptly elected to a seat on the Board of Harvard Overseers. **1973** *Amer. Universities & Colleges* (ed. 11) 729/1 *Harvard University*... *Governing Board.* President and Fellows of Harvard College: self-perpetuating board of 7 members; life terms. Many board actions subject to consent of Board of Overseers, which is composed of 30 members elected by alumni for 6-year terms.
e. A Friend (FRIEND *sb.* 7) chosen for the pastoral supervision of the congregation to which he belongs.
1785 *Bk. of Discipline New England Yearly Meeting Soc. of Friends* 39 That each monthly meeting choose two or more sober and judicious men friends, and two or more women friends, to be overseers in each preparative-meeting, which overseers are to render account of their service to the monthly meeting at least once a quarter, and to be annually appointed or re-chosen. **1832** W. D. WILLIAMSON *Hist. State Maine* II. 699 Each society [of Quakers] has at least four Overseers, two males and two females. **1921** R. M. JONES *Later Periods Quakerism* I. iv. 131 These Overseers..were in the course of time charged with responsibility for the moral life of the membership. *Ibid.* 132 Like the Elders, they had no absolute rules to guide them, but there slowly accumulated..a body of Advices and Queries which furnished the Overseers with a pretty clear line of procedure. **1963** A. HERON *Towards Quaker View of Sex* i. 7 Particularly does this apply to elders and overseers in the Society of Friends. **1974** G. HUBBARD *Quaker by Convincement* IV. iii. 209 The responsibilities of Overseers are to encourage attendance at Meetings for Worship and for business, to exercise a care over younger members and children and those in need of assistance, [etc.].

o·ver-self. [OVER- 2 b.] The finer, stronger, or more assertive part of one's nature (see also quot. 1960).
1888 E. CLODD *Story of Creation* xi. 223 The terrible mass of wrong-doing can only be lessened and finally removed by suppression of the over-self. **1908** *Daily Chron.* 30 Apr. 3/1 It is the Shakespeare that projected his over-self into two score of masterpieces of poetry and drama that is Shakespeare for us. **1960** J. HEWITT *Yoga* 7 The Yogi believes that there is a universal Overself with whom he can make contact and identify himself in moments of higher consciousness. *Ibid.* xi. 155 To the Yogi Samadhi is the merging of the individual Soul or Self with the universal Soul or Overself.

oversell, over-sell, *v.* Add: **1. b.** To make excessive or unrealistic claims for (goods advertised or offered for sale, etc.); to give (someone) an exaggerated idea of the value or worth of something. Also *absol.*, *transf.* and *fig.*
1928 *Publishers' Weekly* 10 Nov. 1978/2 We remember— how perfectly!—the names and the publishers of books on which we were oversold last season, and had, subsequently, to send the way of all deadwood. **1957** *Technology* July 174/1 The word 'syndicate' and the syndicate method in management training has been over-sold for some time. **1960** *20th Cent.* Sept. 234 Mr Wesker's enemies dismiss him as a mere brand-name oversold by the theatrical Left. **1970** A. TOFFLER *Future Shock* (1974) xx. 463 It would be foolish to oversell the ability of science, as yet, to forecast complex events accurately. **1971** P. DICKINSON *Sleep & his Brother* iii. 56 'Why does he want to see me?' 'Aha! I fear I may have oversold you. We are his hobby, and he is not a patient man.' **1971** *Nature* 19 Nov. 118/3 The current disenchantment with science arises because science was oversold in the postwar years. **1973** E. LEMARCHAND *Let or Hindrance* xi. 132 Chap oversells himself..but he knows his way round in business. **1977** R. E. HARRINGTON *Quintain* iv. 34 'They believe they're safe.' Diamond..knew he was over-selling to Felix, and he damned himself for it.
Hence **overselling** *vbl. sb.* (later example); **oversold** *ppl. a.* (further examples).
1934 *Sun* (Baltimore) 5 Apr. 27/1 Word of this amendment..caught the wheat market in apparently an over-sold condition. **1968** L. SELLERS *Doing it in Style* 206 If a reporter writes a story too hard it may get into the paper unqueried and result in trouble. If he oversells in advance he will only infuriate the executives... Overselling is the way to trouble. **1971** *Daily Tel.* 28 Oct. 19/4 The market is in a heavily oversold condition.

oversell (ōu·vəɹsel), *sb.* [f. the vb.] Excessive or ambitious claims for, or promotion of, goods offered for sale; also *transf.* and *fig.*
1969 *Computers & Humanities* IV. 53 No doubt we are partly the victims of oversell by our IBM salesmen and computer directors, who promised us the computer would do things it is quite unsuited for. **1970** G. GREER *Female Eunuch* 13 Perhaps the sexual sell was oversell. **1974** 'G. BLACK' *Golden Cockatrice* i. 17 It was another case of oversell, like that soap powder campaign..which drove irritated women to buy the brands which didn't promise.. a ten per cent whiter wash.

over-sensitive, *a.* Add: (Further examples.) So **over-sensiti·vity.**
1953 K. REISZ *Technique Film Editing* III. xv. 266 He becomes over-sensitive to his surroundings and suspicious of casual passers-by. **1965** HOUSE & STOREY *Lett. of Dickens* I. 146 Seymour..seems to have suffered from over-sensitivity. **1977** H. INNES *Big Footprints* II. ii. 137 The over-sensitivity of a boy who has lost his natural parents. **1978** E. HEALEY *Lady Unknown* i. 43 An over-sensitive skin that broke out in a rash when she was under strain.

oversew, *v.* Add: Also *transf.*
1938 L. M. HARROD *Librarians' Gloss.* 109 *Oversewing*, the act or process of sewing over and over the leaves of a book. **1969** R. MAINGOT *Abdominal Operations* (ed. 5) I. xxv. 428/2 The duodenal stump is next oversewn, securely closed, and inverted with sutures of fine silk.

over-sexed (ōuvəɹse·kst), *a.* [OVER- 28 d.] Having sexual propensities or qualities in an excessive degree; inordinately desirous of sexual gratification. Also *fig.* Hence **over-se·xedness.**
1898 C. P. STETSON *Women & Econ.* iii. 43 The male.. is a far more normal animal than the female of his species, —far less over-sexed. *Ibid.* 44 The secretion of milk is a maternal function,—a sex function. The cow is over-sexed. **1901** G. B. SHAW *Three Plays for Puritans* p. xi, I did not find that matters were improved by the lady pretending to be 'a woman with a past', violently over-sexed. **1908** A. NOYES *William Morris* 98 A creature so gluttonously over-sexed and selfishly serpentine as Gudrun. **1923** C. MACKENZIE *Parson's Progress* viii. 95 Nobody questions the ethical value of Christianity except a few oversexed egomaniacs. **1923** *Daily Mail* 5 Feb. 5 His..terra-cotta-coloured nudes..are repulsive and over-sexed. **1937** G. FRANKAU *More of Us* x. 112 Fearful lest one more line of such o'er-sexed stuff Leave us no energy to do our next stuff. **1941** AUDEN *New Year Let.* III. 54 Hearing how circumstance has vexed A broker who is over-sexed. **1953** *Encounter* Nov. 31/1 Y. is the political equivalent of a *nymphomaniac*... This kind of neurosis..flourishes chiefly in the climate of the Left—for, generally speaking, the Left is politically over-sexed. *Ibid.* 31/2 The unfulfilled urge 'to belong' may lead to 'political oversexedness', expressing itself in blind, self-sacrificing devotion to some unholy cause. **1973** 'D. HALLIDAY' *Dolly & Starry Bird* xv. 229 You've laid your oversexed little claws on my girl. **1977** 'H. CARMICHAEL' *Grave for Two* ix. 111 To put it mildly she was inclined to be over-sexed.

over-sharpness. (Examples.)
1858 BAGEHOT *Coll. Works* (1965) II. 84 'Over-sharpness' in the student is the most unpromising symptom of the logical jurist. **1955** J. L. AUSTIN *How to do Things with Words* (1962) vii. 90 Alice-in-Wonderland over-sharpness. **1970** *Nature* 19 Dec. 1218/1 The quality of supra-threshold vision in the fovea results from a balance between optical unsharpness..and neural 'oversharpness' (lateral inhibition).

overshoot, *v.* Add: **1. d.** *trans.* and *intr.* To fly beyond (a designated landing-point) while attempting to land an aircraft. Also *transf.*
1920 *Flight* XII. 368 (*caption*) Pilot heads..for aerodrome, knowing for certain he will overshoot. **1928** *Lit. Digest* 12 May 73/1 To 'put her on hot' is to land fast, usually resulting in 'overshooting' the field. **1932** D. GARNETT *Rabbit in Air* I. 35 In my first attempt I thought I had overshot for some reason when I had undershot hopelessly. **1958** 'CASTLE' & 'HAILEY' *Flight into Danger* vi. 79, I can't guarantee at all that this plane will get down on the field. She's just as likely to pan down short or overshoot. **1973** C. BONINGTON *Next Horizon* xviii. 248 We over-shot the runway once. **1974** P. ERDMAN *Silver Bears* iii. 54 The MG suddenly swung off the road... Doc was caught by surprise and overshot. Slowly he backed up.

overshoot (ōu·vəɹʃūt), *sb.* [f. the vb.] The action or result of OVERSHOOT *v.*; *spec.* in *Electronics* (see quot. 1971).
1944 *Flight* 1 June 584/1 Uncorrected over-shoot generally means a write-off. **1945** *Electronic Industries* Sept. 214 *Overshoot*, an excessive potential attained by a portion of the main body of a pulse. **1947** R. LEE *Electronic Transformers & Circuits* ix. 249 Added to this is a slight oscillation overshoot. **1956** AMOS & BIRKINSHAW *Television Engin.* II. i. 25 The consequent disturbance of the phase relationships frequently causes the phenomenon of overshoot illustrated in Fig. 11. This is an effect which can be compared with that of inertia in mechanical systems and causes the voltage, after executing the transient, to exceed momentarily the final steady value. **1963** R. P. DALES *Annelids* viii. 156 They have found that with posterior pieces there is often an 'overshoot', the new thoracic region having more segments than it should. **1969** J. J. SPARKES *Transistor Switching* iv. 113 Overshoot or undershoot of the output waveforms will result. **1970** *New Scientist* 17 Dec. 515/2 He sees the market economy as a system with high gain and strong feedback, possessing self-regulation, but troubled by overshoots. **1971** *Gloss. Electrotechnical, Power Terms* (*B.S.I.*) III. i. 37 *Overshoot*, transient exaggeration of the magnitude of the leading or trailing edge of a steep-sided signal. **1977** *Gramophone* Oct. 744/3 The square wave showed one sharp overshoot and was then well damped.

overshot, *a.* **1.** (Further examples.)
1880 [see *HIGH *a.* 4 c]. **1904** KIPLING *Traffics & Discov.* 389 Mechanically, an overshot wheel with this head of water is about as efficient as a turbine. **1914** *Chambers's Jrnl.* Mar. 205/1 The sewage passes over a wheel of overshot or undershot type. **1968** J. ARNOLD *Shell Book of Country Crafts* xii. 175 The overshot wheel was used where there was a higher fall of water and was intended to turn by the *weight* of the descending water against the buckets or floats. **1978** J. B. HILTON *Some run Crooked* iii. 22 The Powder Mill..was a mid-nineteenth-century ruin..an overshot water-wheel, now crippled over a leaking weir.

overshot, *ppl. a.* Add: **3.** (Later examples.)
1931 T. R. G. LYELL *Slang* 668 There are innumerable synonyms applicable to different degrees of intoxication. Those most in general use are..*muddled, overcome, overshot*, [etc.]. **1942** BERREY & VAN DEN BARK *Amer. Thes. Slang* § 106/7 Drunk,..*overshot*.
7. Of a pattern or weave: that is characterized by uninterrupted lines of weft where the yarn has been made to pass over two or more warp threads before re-entering the fabric.
1952 H. J. BROWN *Hand Weaving* vi. 102 Any design that can be arranged on a formation of squares may be reproduced in overshot weaving... Many of the fine old American colonial coverlets that have been handed down as heirlooms were made in overshot design. **1965** E. TUNIS *Colonial Craftsmen* iv. 102 Such were the 'overshot' coverlets in wide variety which are prized examples of old weaving. **1970** E. REGENSTEINER *Art of Weaving* v. 77/1 In the weaver's language, patterns produced by this system are called 'overshot', because when the harnesses are raised, the weft threads skip or 'float' over the groups of warp threads.

overshroud, *v.* (Later example.)
1916 A. S. WAY tr. *Virgil's Aeneid* I. III. 113 A night of rain overshrouds The sky.

oversight, *sb.* Add: **1. a.** (Further examples.)
1931 H. J. ROSE tr. *W. Schmidt's Orig. & Growth Relig.* xvi. 275 The Supreme Being, thus exercising oversight on the doings of men, is likewise able to reward..and punish. *Ibid.* 277 His oversight of what men do and leave undone in the moral sphere. **1935** F. H. HAYWARD *Alfred the Great* xii. 81 There was no centralised mint and probably little centralised oversight. **1971** *Nature* 4 June 292/1 The need to continue investigations on a broad front to keep

OVERSIMPLIFICATION

an ecological oversight of the biogeodynamics of each metal. **1977** *Time* 21 Feb. 38/1 Congressional oversight has proliferated.

c. A survey, view. *rare.*

1889 F. E. Gretton *Memory's Harkback* 291 You have a closer and more direct oversight of the home, or Herefordshire, view.

2. a. (Later examples.)

1927 *Public Opinion* 8 Apr. 329/2 The generous-hearted demand that we accord to China the recognition due to a modern nation is sometimes made in oversight of the fundamental elements in the problem. **1959** J. L. Austin *Sense & Sensibilia* (1962) ii. 13 These are the quite common cases of misreadings, mishearings, Freudian oversights, &c. **1976** *Washington Post* 19 Apr. A4/5 The back-room view in the White House is that there should be a single oversight committee on Capitol Hill to answer the clamor for corrective action.

b. Also, a person who is passed over.

1955 T. H. Pear *Eng. Social Differences* 241 When one studies the failures among those who were selected for grammar schools and the oversights among those who were not selected, not a few mistakes..might have been avoided had the child's social environment been taken into account.

Hence **oversi·ghted** *ppl. a.*, overlooked.

1857 J. Hyde *Mormonism* (ed. 2) ix. 215 There is one oversighted contradiction that stares us in the face.

o:versimplifica·tion. [Over- 29 b.] The action or process of simplifying to excess; the result of this; a simplistic style or procedure.

1930 R. A. Fisher *Genetical Theory Nat. Selection* 42 It is a patent oversimplification to assert that the environment determines the numbers of each sort of organism that it will support. **1934** *Discovery* Dec. 339/2 The danger of over-simplification of exceedingly complex problems. **1958** 'P. Bryant' *Two Hours to Doom* 91 'Since the war a dozen countries have gone Communist.'..'I realise that. But I think it's an over-simplification of the issue.' **1968** M. S. Livingston *Particle Physics* p. vi, The author accepts responsibility for any shortcomings or oversimplifications in these descriptions. **1975** *Listener* 4 Dec. 734/3 When the cars are there, they sell... This, of course, is an over-simplification.

oversi·mplify, *v.* [Over- 27.] *trans.* To render excessively or delusively simple; to explain in simplistic terms. Also *absol.* Hence **oversi·mplified** *ppl. a.*; **oversi·mplifying** *vbl. sb.* and *ppl. a.* Also **oversi·mplifier,** one who oversimplifies.

1934 Webster, Oversimplify. **1936** *Mind* XLV. 222 Preformation..errs by over-simplifying the problem. **1940** *Amer. Speech* XV. 67 The old fallacy of the over-simplifiers, searching for 'the' cause where there usually is a complex of causes, has also bedeviled philology. **1942** *Scrutiny* X. iv. 360 The difference cannot be explained simply by saying that the comic parts of *Chuzzlewit* are good and the 'serious' or 'sentimental' parts bad, because that is an over-simplifying of the case. **1946** *Sun* (Baltimore) 11 Mar. 10/3 (*heading*) It is easy to oversimplify about the Russians. **1946** J. S. Huxley *Unesco* ii. 47 In somewhat over-simplified terms. **1953** D. F. Pocock tr. *Durkheim's Sociol. & Philos.* iii. 74 This..would be preferable in our schools to the over-simplified..explanations with which we too often deceive the curiosity of youth. **1963** *Times* 21 Jan. 9/2 These two stubborn, over-simplifying old men. **1965** C. Walsh in J. Gibb *Light on C. S. Lewis* 114 He berated Lewis as an oversimplifier. **1975** R. Browning *Emperor Julian* i. 27 An over-simplified summary of results must suffice. **1976** *Listener* 28 Oct. 550/1 Magnus Pyke..has been accused..of 'popularising' his subject, of oversimplifying.

oversize, *sb.* Add: **1.** (Further examples.)

1920 E. Sitwell *Wooden Pegasus* 88 Neutralize The overtint and oversize. **1920** *New Yorker* 12 Sept. 90/1 (Advt.), This most unusual watch,..shown here in its actual over-size.

2. That which is above a certain size.

1902 *Encycl. Brit.* XXXI. 374/1 It then goes to a screen with eleven holes to the linear inch, and yields a granular undersize and oversize. **1905** *Electrochem. Industry* Mar. 124/2 The oversize, which contains no slime whatever, is delivered directly to four Wilfley concentrating tables.

oversize, *v.*[1] Delete †*Obs.* and add later U.S. examples of both senses.

1879 'Mark Twain' *Lett. to Publishers* (1967) 114, I say $1100 instead of $1042 to cover little possible mistakes in over-sizing the plates. **1904** —— in *Harper's Weekly* 2 Jan. 19/1 The whole of that is intelligible to me ..except..[one] remark... That one oversizes my hand. Gimme five cards. **1930** *Flight* 25 Apr. (Suppl.) 460e/1 With various wheel forms, the effect of oversizing tyres and the use of smooth or safety treads. **1957** R. Lister *Decorative Wrought Ironwork* 230 *Oversizing.* Iron oxidizes while being worked, and it is therefore sometimes necessary to work it slightly larger than its intended final size, or oversize it, in order to counteract any loss by this.

o·versize, *a.* [f. the sb.] = *outsize *a.*

1909 *Cent. Dict.* Suppl., *Oversize*.., of excessive size; specifically, noting material which is too large to pass through the meshes of a given screen or size. **1924** *T. Eaton & Co. Catal.* Spring & Summer 292/2 Oversize cord tires need special cord size inner tubes. **1936** W. Faulkner *Absalom! Absalom!* vi. 195 A new oversize overall jumper coat. **1960** I. Cross *Backward Sex* i. 12 That bald head, like an over-size tennis ball, the worse for much wear. **1973** *Publishers Weekly* 25 June 16 (Advt.), A magnificently illustrated, oversize book. 10"×11¾". **1976** *Billings* (Montana) *Gaz.* 1 July 2-D/5 (Advt.), Nearly new three bedroom split entry home featuring large bedrooms, a fully equipped oversize kitchen with lots of cabinets.

overskirt. (Earlier examples.)

1870 *Harper's Bazaar* 22 Oct. 675 Over-skirts are elaborate, and show great variety in design. **1873** *Young Englishwoman* Mar. 131/2 An over-skirt of tulle looped up with a scarf sash.

overslee·ping, *vbl. sb.* [-ing[1].] The action of the verb Oversleep.

1908 *Westm. Gaz.* 31 Oct. 3/2 What with your smashings, and your over-sleepings, and burning the dinner on Sunday, and all. **1912** R. Brooke *Let.* 25 Feb. in E. Marsh *Rupert Brooke* (1918) iv. 70 My cure consists in perpetual overeating and oversleeping. **1977** J. Aiken *Last Movement* vii. 136 Over-sleeping is as bad as over-eating.

o·verslung, *a.* [Over- 1 c.] Supported above the main part, or some particular part, of an apparatus.

1960 Shepherd & Withers *Mech. Cutting & Loading of Coal* v. 78 A number of low-type cutters provide for a range of appropriate adjustable positions between the underslung and overslung position of the reversible turret. **1971** B. Scharf *Engin. & its Lang.* xvi. 228 Overhead travelling cranes consist of (1) a load girder with a roller mounted carriage at each end running on top (overslung gantry type), or along the bottom flanges of gantry rails (underslung gantry type); (2) a trolley..; and (3) a hoist. **1971** J. M. Paxton *Man. Civil Engin. Plant* (ed. 2) I. 190/2 Both rear axles are driven via double reduction hubs with an over-slung worm and wheel from a 12-speed gearbox, giving 12 forward and 3 reverse speeds.

oversold: see Oversell *v.* in Dict. and Suppl.

over-soon, *adv.* (Later example.)

1878 Hardy *Ret. Native* II. ii. vi. 10, I told him 'twas barely decent to come so oversoon; but words be wind.

over-soul, *sb.* Add: Also, = Superman.

1908 [see *beyond *adv.* and *prep.* D].

o·ver-soul, *v.* [f. the sb.] *trans.* In passive, to be ruled or dominated in respect of the soul. Hence **over-sou·ling** *vbl. sb.*

1916 'A.E.' *National Being* ii. 13 None of our modern States create in us such an impression of being spiritually oversouled by an ideal as the great States of the ancient world. **1925** D. H. Lawrence *Refl. Death Porcupine* 168 A primrose has its own peculiar primrosy identity, and all the oversouling in the world won't melt it into a Williamish oneness.

o:ver-specializa·tion. [Over- 29 b.] Too much specialization, esp. in education or evolution.

1931 J. S. Huxley *What dare I Think?* iv. 144 Over-specialization produces..individuals with scientific hypertrophy and religious atrophy. **1936** *Discovery* Oct. 328/2 The author emphasises the importance of correlation between laboratory and clinical observation and deplores the over-specialisation which has gradually crept in and checked many valuable conclusions in medical science. **1957** *Technology* Mar. 3/4 The evils of over-specialization at school are never easily wiped out. **1958** *Spectator* 22 Aug. 258/2 Two extant species [of elephant] alone survive from 352 branches. Mr. Carrington suggests that the elephant declined more from this over-specialization than from unscrupulous Romans or Asiatics. **1974** tr. *Wertheim's Evolution & Revolution* i. 75 As soon as the dead end produced by over-specialization is reached, only a forceful..breakthrough can produce a reversal of the involutionary trends.

o:ver-spe·cialize, *v.* [Over- 27.] *intr.* To specialize too much (in a particular endeavour). So **o:ver-spe·cialized** *ppl. a.*

1926 J. S. Huxley *Ess. Pop. Sci.* p. vi, Science herself is over-specialised. **1930** G. R. de Beer *Embryol. & Evolution* xiii. 95 If a race has become excessively over-specialized, even the younger stages of the ontogenies of its individuals may have lost their plasticity. **1944** J. S. Huxley *On Living in Revolution* xv. 184 To overwork and over-specialize. **1953** N. Tinbergen *Herring Gull's World* ii. 5 They [*sc.* gulls] are not over-specialised gliding fliers like the shearwaters. **1960** *Farmer & Stockbreeder* 16 Feb. (Suppl.) 8/2 The emphasis must be changed from the over-specialized Wiltshire baconer.

o:ver-spe·cify, *v.* [Over- 27.] *trans.* To specify too narrowly; to limit excessively in scope; to specify in excessive detail. So **o:ver-speci·fic** *a.*; **o:verspecifica·tion.**

1957 R. K. Merton *Social Theory* (rev. ed.) x. 389 The practical problem had been overspecified in its initial formulation. *Ibid.*, This overspecification for a time diverted our attention from salient alternatives of investigation. **1962** U. Weinreich in Householder & Saporta *Probl. Lexicogr.* 32 A definition like *triangle* 'a figure that has three sides and three angles, the sum of which is 180°' is avoided as overspecific, since *triangle* is sufficiently defined by the number of sides. **1965** *Language* XLI. 234 Using it..might risk being overspecific in this instance, but this is preferable to a less unified system of description in which several concepts are used with less general application. **1968** M. S. Livingston *Particle Physics* ix. 167 The relationships between isotopic spin, multiplicity of charge states,..and baryon number for hadrons are overspecified. **1975** N. Chomsky *Logical Struct. Linguistic Theory* vii. 200 It is at once clear that the grammar 26 provides an overspecification since the following simpler grammar corresponds to 26.

o·ver-speech. [Over- 29 b.] Loquacity; indiscretion.

1865 see Over- 29 b]. **1920** E. Pound *Umbra* 115 Arnaut loves, and ne'er will fret Love with o'er-speech. **1922** E. R. Eddison *Worm Ouroboros* iv. 48 'Keep thou thy lips from overspeech,' said the King. 'These be mysteries whereon but to think may snatch thee into peril.'

o·verspeed, *sb.* [Over- 29 b.] (An instance of) overspeeding.

1914 H. Pender *Amer. Handbk. Electr. Engineers* 1315 Over-Speeds.—All types of rotating machines shall be so constructed that they will safely withstand an over-speed of 25 per cent. **1926** J. Kirsopp *Use of Power in Colliery Working* ii. 156 The entire device prevents overspeed of any description. **1950** *Sun* (Baltimore) 12 Apr. 32/1 The plane, flying at a speed of 135 miles per hour, developed 'a structural failure due to overspeed'. **1969** *Power System Protection* (Electr. Council) II. ix. 250 The governing system which requires an actual overspeed to produce a response and take corrective action.

overspee·d, *v.* [Over- 27.] *intr.* To drive or operate faster than is permitted or allowed for. So **overspee·ding** *ppl. a.* and *vbl. sb.*

1906 *Westm. Gaz.* 24 Apr. 4/2 The police had been..engaged elsewhere to look out for over-speeding drivers. **1913** *Collier's* 11 Jan. ii. 50/2 There were but three convictions for overspeeding. **1950** W. W. Lewis *Protection Transmission Syst. against Lightning* xi. 367 It is believed that generator overspeeding and loss of load as single factors without solid faults will seldom give rise to dangerous overvoltages. **1971** M. Tak *Truck Talk* 114 *Overspeed*, to run an engine at an excessive number of revolutions per minute for the gear being used. **1972** *Lebende Sprachen* XVII. 134/2 A turbine undergoing tests at Calder Hall 'B' atomic power-station in Cumberland overspeeded last night and exploded. **1974** [see *overheat *v.* 2].

overspend, *v.* Add: **1. b.** (Later example.)

1951 L. MacNeice tr. *Goethe's Faust* 39 He backs away, gives way, the day is overspent.

2. c. (Later examples.)

1946 L. P. Hartley *Sixth Heaven* v. 107, I doubt if it's even wise to offer to pay half... You mustn't overspend yourself. **1953** E. Simon *Past Masters* iii. 159 From the outset [you] overspent... The money has all gone on inessentials. **1959** M. Summerton *Small Wilderness* xiii. 159 Money? Overspent yourselves in this place?

Hence **overspe·nding** *vbl. sb.* (in sense 2 of the vb.).

1932 *Ann. Reg. 1931* 300 The country [*sc.* the U.S.A.] regrets the over-spending of the past few years. **1963** *Times* 9 Jan. 9/2 They cover underspending less comprehensively than overspending, and without knowledge of both it is impossible to make an accurate interim assessment of trends in Government expenditure. **1976** *Times* 20 Oct. 14/3 The borrowing requirement (a euphemism for government overspending).

overspill, *sb.* Add: **a.** (Further examples.) Now usu. the movement of surplus population from a city to a less heavily populated area of the same country; this surplus population or the housing or new area occupied by it. Also *transf.*

1930 *Times* 22 Apr. 6/7 On the south lie the famous South Downs, within range of the overspill from the seaside towns. **1940** J. Buchan *Memory Hold-the-Door* vi. 145 Emigration undertaken as a reasoned policy..and not as a mere overspill of population. **1944** *Daily Tel.* 12 July 4/4 When one member objected to Mr. Morrison's use in connection with population of the word 'overspill', the Minister admitted that it was ugly, though convenient. **1946** *Nature* 13 July 39/1 Public interest has been stimulated equally by the controversies over the proposals for dealing with Manchester's overspill in a new town at Mobberley, or the even larger overspill problem of Liverpool. **1947** *Daily Mail* 22 May 1/1 We are apt to be too much concerned with the new satellites and 'overspills'. We should first reconstruct the other cities. **1955** *Times* 12 May 15/2 Since 1951 Socialist power there has been improved by an overspill of families of men and women who work in the factories and machine shops of Swindon. **1958** *Times* 3 Oct. 14/5 Although an overspill has been necessary to accommodate them all comfortably..most of the paintings look very handsome. **1959** *Economist* 3 July 42/1 Diversification may thus proceed from an overspill of strength in one department or another. **1965** A. Garner *Elidor* xv. 111 'That's what you must expect when you have overspill in a decent area,' said Mrs. Watson. 'They shouldn't be allowed to build out in the country. People aren't going to change when they move from the city.' **1972** F. Warner *Lying Figures* ii. 15 Epigyne is lit by the overspill from the two spots. **1972** *Times* 21 Dec. 4/4 *Overspill*, the planned movement of people who do not want to go to towns that do not want to have them. **1976** 'D. Halliday' *Dolly & Nanny Bird* xvi. 212 The next wave..struck us..and the two men huddled on the floor of the cockpit received the first overspill..from the lee side. **1976** T. Stoppard *Dirty Linen* 9 An overspill meeting room for House of Commons business in the tower of Big Ben.

b. *attrib.*

1945 *Ann. Reg. 1944* 63 The Bill for the purchase of so-called 'overspill' areas where those who were crowded out could be accommodated. **1946** [see sense a above]. **1952** *Economist* 21 June 799/2 No less than 28 of these [district councils] are intended under the plan to absorb 'overspill' population coming from Wolverhampton, Walsall,..and other congested towns... Over a quarter of the new houses ..would form part of 'overspill' schemes. **1958** *Spectator* 30 May 710/1 Recent 'overspill' housing policies. **1966** *New Statesman* 28 Jan. 140/2 Nor are we likely to get any transitional or 'overspill' benefit because of the restrictive provisos. **1972** *Guardian* 8 Sept. 6/5 Official overspill schemes and the movement to the suburbs..account for only half the population drift: the rest are leaving the

country. **1975** Cox & Boyson *Black Paper 1975* 30/1 The school is built on the edge of a city overspill estate. *Ibid.* 31/2 There will be difficulties in any overspill area where people are moved away from family and familiar surroundings.

overspill, *v.* Add: (Further examples.) Also, to cause (something) to spill over; *spec.* to remove (surplus population) from a city. Also *intr.*, to spill over; to overflow. Hence **overspi·lling** *ppl. a.*

1958 *Times Rev. Industry* Feb. 24/1 Some 70,000 people are to be 'overspilled' from Glasgow City..into the new towns of East Kilbride and Cumbernauld. **1961** *Times Lit. Suppl.* 27 Jan. 51/2 We overspill our savings..onto the backward nations. **1962** *Times* 7 Mar. 11/4 The process of being over-spilled in a familiar country-side..is bound to be at least slightly deterrent to the true Cockney. **1963** *Listener* 21 Mar. 516/1 The need to re-house ever more overspilling Londoners. **1970** G. F. Newman *Sir, You Bastard* vii. 196 The eighteen prisoners who finally stood charged over-spilled the dock at the committal proceedings. **1972** *Accountant* 28 Sept. 392/2 It is an illusion..for internal auditors to think that, by overspilling into management functions and systems design, they will enhance the status..of their profession. **1977** *Daily Tel.* 14 Feb. 6/7, 30,000 homes were bulldozed inside 10 years and the occupants 'overspilled' beyond the city boundaries in the new or expanded towns of Runcorn, Skelmersdale, [etc.].

overspin, *v.* Restrict †*Obs.* to sense in Dict. and add: **2.** To confuse (an opponent) with overspin (see next).

1940 G. Marx *Let.* 5 Sept. (1967) 25 He's a left-hander and slashes and cuts and overspins his opponent dizzy.

o·verspin, *sb.* [Over- 6.] In *Cricket* and other ball games: a rotating motion imparted to a ball in which the upper part turns in the direction of flight, or is struck with upward inclination. Also (in full *overspin ball*), such a delivery. Now usu. called *topspin*. So **o·verspi:nner**, a ball delivered with overspin.

1904 F. C. Holland *Cricket* 54 The over spin gives to the ball the same over-and-over motion that is seen in a ball that has been topped at golf or billiards. **1908** A. W. Myers *Compl. Lawn Tennis Player* 109 A strong forward or over-spin is thus imparted to the ball. **1925** *Country Life* 18 July 93/1 The over-spin ball is the logical outcome of the googlie, inasmuch as..the hand turns over more than in the leg-break, but not so much as in the googlie. **1927** M. A. Noble *Those 'Ashes'* 178 Hendren was bowled by a faster overspin, which he mistook for a leg break. **1930** C. V. Grimmett *Getting Wickets* iii. 57 It would..have what is called 'overspin', and, after striking the pitch, gather pace as the spin took effect. *Ibid.* 60 An overspinner will be produced if the back of the hand is outwards, and the hand pointing horizontally to the demonstrator's left. **1961** F. C. Avis *Sportsman's Gloss.* 130/1 *Overspin*, a forward rotating movement of the ball in flight, causing an acceleration off the pitch. *Ibid.* 259/1 (Lawn Tennis), *Overspin*, top spin imparted to the ball, to give power and unexpected movement off the court. **1970** H. Taylor *Golf Dict.* 150 *Overspin*, a word sometimes used for topspin.

o·verspray. [Over- 19.] Sprayed liquid that does not adhere to the object or area being sprayed.

1948 L. W. Lammiman in W. von Fischer *Paint & Varnish Technol.* xxvi. 453 Arcing the gun on a surface causes overspray to bounce obliquely off the surface. **1955** Edwards & Wray *Aluminium Paint & Powder* (ed. 3) iv. 95 The amount of paint sprayed is greatly reduced as most of the overspray is eliminated. **1972** *Materials & Technol.* V. xi. 371 Spraying is normally carried out in spraybooths..and overspray is removed by an efficient exhaust system. **1977** *Hot Car* Oct. 50/4 For a single pin stripe you must of course mask either side of the tapes to stop overspray going on to the base colour.

over-spru·ng, *ppl. a.* [Over- 27 b.] Fitted with too many or too flexible springs.

1923 *Daily Mail* 12 July 12 The saddle for my weight was over-sprung, and over pot-holes was inclined to bounce on to the frame. **1937** *Daily Herald* 16 Apr. 18/2, I have no very high opinion of the rather flashy high-powered over-sprung American car. **1962** L. Deighton *Ipcress File* xxi. 146, I..pulled the big oversprung Lincoln Continental on to the road.

oversquare (stress variable), *a.* [Over- 28.] Of an internal-combustion engine or its cylinders: having a bore greater than the stroke.

1959 *Times* 23 Sept. 5/4 The new engine is 'over-square' and has a cubic capacity of 6¼ litres. **1966** *Economist* 9 July p. xxv/1, If exaggerated, the over-square cylinder tends to have a wafer-thin combustion chamber with limited compression turbulence, and to suffer from indifferent combustion at low operating speeds. **1967** *Ibid.* 8 July p. xxix/1, Hauliers also fear wear on the fast (up to 3,300 rpm) V-engines: these are often oversquare, with neat chunky cylinders. **1971** *Motor* 16 June 26/1 The new 1971 cc engine in the 504 is a four-cylinder in-line over-square unit.

over-stai·n, *v.* [Over- 8, 27.] **1.** (In Dict. s.v. Over- 8 a.)

2. *Biol. trans.* To stain (tissue) excessively, usu. in order that certain parts may be differentiated by selective removal of some of the stain; also *absol.*

1885 C. O. Whitman *Methods of Research in Microsc. Anat.* ii. 39 If by any chance the sections are over-stained, the superfluous color may be extracted by a brief sojourn in very dilute ammonia. *Ibid.* 48 Diffuse staining may generally be avoided by first overstaining and then withdrawing the color to any desired extent by means of alcohol. **1935** Kingsbury & Johannsen *Histol. Technique* 38 In regressive staining the tissue is over-stained and the excess of stain removed by the application of a differentiator. *Ibid.* 40 The rule is to over-stain and watch the differentiation carefully with the microscope. **1971** A. Lamela *Introd. Med. Lab. Methods* xxiii. 238 The tissue is over-stained and then decolorized until the correct depth of color is obtained.

Hence **over-stai·ning** *vbl. sb.*

1885 C. O. Whitman *Methods of Research in Microsc. Anat.* ii. 42 Minot's picric-acid carmine..gives a stronger differential coloring than Ranvier's picro-carmine; but over-staining must be most carefully avoided. **1929** C. E. McClung *Handbk. Microsc. Technique* i. 23 Regressive stains..are allowed to act until overstaining is accomplished, after which the desired degree of differentiation is brought about by removal of the excess coloration.

overstand, *v.* Restrict †*Obs.* to sense 2 and add later examples of sense 1.

1888 G. M. Hopkins *Poems* (1967) 198 Fairyland; silk-beech, scrolled ash, packed sycamore, wild wychelm, hornbeam fretty overstood By. **1938** *Venerabile* Oct. 61 The lofty roof o'erstands the graceful shrine.

3. To pass over; to cross.

1949 *Sun* (Baltimore) 27 Aug. 8/8 But Colie outguessed him, for while Reckord was watching the Jersey boat, he overstood the mark. Colie whipped about and Scanty was second before the Baltimore sailor realized what had happened. **1976** *Yachts & Yachting* 20 Aug. 372/1 Colin Evans overstood the line, thus losing the much-disputed 3rd place to Geoff Tindale.

4. *to be overstanding for honours*: in the University of Oxford, to be incapable of obtaining honours in an examination because of the lapse of more than the permitted number of terms (normally twelve) since matriculation.

The phrase 'disqualified for Honours by standing' is preferred in the O.U. *Examination Decrees and Regulations.*

1933 V. Brittain *Testament of Youth* iii. x. 476 They had undertaken, since I was so excessively 'over-standing' for Honours, a special procedure on my behalf. **1965** N. Coghill in J. Gibb *Light on C. S. Lewis* 52 Had we taken two [years] we would have been overstanding for Honours.

o·verstayer. *N.Z. colloq.* [f. Overstay *v.* + -er[1].] A Polynesian or other immigrant who stays beyond the time permitted by a work permit.

1977 *N.Z. Herald* 5 Jan. 1-4/4 While expressing sympathy for the plight of overstayers, the Maori leader said the laws of the nation had to apply to everyone, regardless of race. **1977** *N.Z. Woman's Weekly* 10 Jan. 38/4 We have heard so much lately about the overstayers and while agreeing wholeheartedly that the law must be held in regard and obeyed, I have been wondering if we realize just how much we depend on some of these Island people. **1978** *Guardian Weekly* 22 Jan. 9/2 In October, 1975, the Auckland police suddenly cracked down on 'overstayers'—those Pacific Islanders who had stayed beyond the length of their work permits.

overstee·pened, *ppl. a.* *Physical Geogr.* [Over- 22.] Steepened further; *esp.* (of a valley) having a greater steepness than running water would have caused owing to the predominantly downward erosive action of a glacier. So **overstee·pening** *vbl. sb.*

1900 W. M. Davis in *Proc. Boston Soc. Nat. Hist.* XXIX. 308 In the same way, the waterfalls from the hanging valleys, the showering waste that forms the falls, and the landslides from the basal cliffs, all show that the banks of the glacial channel—the lower walls of the existing valleys—are too steep; and they may be therefore called 'oversteepened'. **1922** *Bull. U.S. Geol. Survey* No. 730. 13 In the Black Canyon..the broad floor of the valley and the steep cliffs that rise precipitously 2,000 feet at its side suggest oversteepening of the walls by ice. **1942** O. D. von Engeln *Geomorphol.* xix. 462 That side of a valley reach against which the glacier current impinges is regularly oversteepened. **1964** *Amer. Jrnl. Sci.* CCLXII. 783 Over the bar itself the steepened profile increases velocity and tends to erode the obstruction. This enhancement of erosion in the over-steepened reach and deposition in the flat area will..tend to eliminate the original bar. **1970** R. J. Small *Study of Landforms* iv. 131 Each stream profile displays several apparent knick-points or oversteepened sections coincident with the gorges.

o·versteer, *sb.* [Over- 29 b.] A tendency in a motor vehicle to increase the sharpness of the turn when made to deviate from the straight.

1936 *Proc. Inst. Automobile Engin.* XXX. 759 This proportion of the total roll couple carried by the front and rear tyres is found to have a very marked effect on steering, and has to be proportioned by trial to avoid excessive effects in the direction either of 'oversteer' or 'understeer'. **1957** *Which?* Autumn 25/2 Slight oversteer is noticeable if the boot is well loaded, but this is easily compensated by a slight increase of pressure on the rear tyres. **1972** *Sci. Amer.* Aug. 20/2 In a vehicle with oversteer the driver must turn the steering wheel away from the center of the turn as speed increases. **1977** *Hot Car* Oct. 23/2 The average driver can handle understeer more safely than oversteer.

overstee·r, *v.* [Over- 27.] *intr.* To exhibit oversteer. So **overstee·ring** *vbl. sb.* and *ppl. a.*

1936 *Proc. Inst. Automobile Engin.* XXX. 730 Parallel-motion rear springing would give excessive over-steering because of the slip angle of the rear tyres. *Ibid.*, Almost every car over-steers to a certain extent. **1948** R. Dean-Averns *Automobile Chassis Design* ii. 49 An oversteering vehicle when negotiating a bend above the certain critical speed of the vehicle for straight running will tend to run into its turn. **1959** *Motor Manual* (ed. 36) v. 147 When the slip angle is greater at the rear, the car oversteers (i.e. turns more sharply than the driver intends). **1962** *Times* 10 Apr. 6/5 Does this Volkswagen oversteer? **1962** *Which? Car Suppl.* Oct. 117/2 The car changed from having a strong understeering characteristic to the opposite oversteering one—that is, the rear wheels tended to turn the car more than the driver wanted. **1971** C. Williams *Car Conversions* ii. 29 An oversteering car..does not slow down appreciably when the steering correction is applied. **1972** S. Abbey *Bk. of Marina* xii. 110 If the rear tyre pressures are too low, the car will 'oversteer'.

overstep, *v.* Add: *to overstep the mark*: see *mark *sb.*[1] 12 e; also *ellipt.*

1931 W. Faulkner *Sanctuary* xvi. 118, I made a fire in the stove. I guess I over-stepped.

2. *Geol.* Of the upper strata of an unconformity: to extend over (underlying strata) in such a way as to form an overstep. Also *intr.* with *on to*. Chiefly *British*.

1883 J. G. Goodchild in *Geol. Mag.* Decade II. X. 227, I have found it convenient..to speak of this stratigraphical relation of unconformable beds to the various rocks immediately beneath as Overstepping. For example, I should say that the Roman Fell Beds in the neighbourhood of Melmerby overlap the Upper Old Red, while the Carboniferous formation..oversteps the older rocks there. **1937** *Q. Jrnl. Geol. Soc.* XCIII. 107 Some of the older records..seem to suggest that the Gault in East Sussex oversteps the Folkestone Sands within a short distance of their outcrop. **1938** A. K. Wells *Outl. Hist. Geol.* ii. 12 In the diagram section the Cambrian rocks overstep the Pre-Cambrian, and higher divisions overlap lower ones. **1969** [see *overstep *sb.* 2]. **1972** *Gloss. Geol.* (Amer. Geol. Inst.) 507/1 An unconformable stratum that truncates the upturned edges of the underlying older rocks is said to 'overstep' each of them in turn (except where the stratum and the underlying beds have the same strike).

So **overste·pping** *vbl. sb.*, *spec.* in *Cricket*, the action of bowling with either foot illegally positioned in relation to the creases.

1869 Mill *Subj. Women* i. 32 An overstepping of the proper bounds of authority. **1959** *Oxford Mail* 2 Feb. 8/7 Rorke..lost a lot of his fearsomeness after being rightly no-balled because of his long drag, called over-stepping in Australia. **1976** J. Snow *Cricket Rebel* 98 Rowan reports remarks I am alleged to have made..after I had been no-balled for overstepping.

overstep, *sb.* (s.v. Overstep *v.*). Add: **2.** *Geol.* In an unconformity, the structural relationship between the lowest stratum of the upper series and the truncated ends of the underlying strata when, as is often the case, these have a different dip from the upper series. Chiefly *British*.

1883 A. J. J. Browne in *Geol. Mag.* Decade II. X. 336 Both cases involve an unconformity, and..the difference between them is really this: in overlap the basement member of the upper series has a limited extension, while in overstep the basement bed has a continuous extension. .. The unconformity between the two series will generally be much greater in the case of overlap than in the case of overstep, for in the latter the beds all dip in the same direction, and the existence of an unconformity is usually only made patent by the fact of overstep. **1937** *Q. Jrnl. Geol. Soc.* XCIII. 120 In the two marginal parts of the Wealden trough..there is a definite overstep, due to the fact that while the freshwater Wealden beds were being deposited in the central subsiding area the margins of the trough were being gently uptilted. **1948** *Bull. Amer. Assoc. Petroleum Geologists* XXXII. 2297 The word *overstep*..in more than 60 years of existence..has not succeeded in gaining any recognition among American geologists and very little in England. **1969** Bennison & Wright *Geol. Hist. Brit. Isles* i. 10 In the case of any angular unconformity the lowest bed of a series of strata is seen to rest on beds of differing ages. Such a phenomenon is known as overstep, and the post-unconformity bed is said to overstep onto successively older beds.

over-stimulate, *v.* (Example of *absol.* use.)

1928 A. B. Callow *Food & Health* 24 Condiments..have the effect of stimulating gastric secretion, but..they tend to over-stimulate.

overstock, *sb.* **2.** (Later examples.)

1976 *Author* Summer 51 The remainder merchants, who prefer to be called overstock dealers. *Ibid.*, A firm called B.S.C. Remainders—founded 12 years ago as an overstock wholesaler. **1976** *Billings* (Montana) *Gaz.* 18 June 4-B (Advt.), Powell's furniture outlets buy factory overstock.

overstocked, *ppl. a.* (Earlier example.)

1844 Mill *Ess. Pol. Econ.* ii. 71 An overstocked state of the market is always temporary.

o·verstorey. [Over- 1 d.] = *over-wood.

1959 *Ecology* XL. 478 The survival and average total height growth of loblolly pine seedlings in Arkansas were best under overstory openings of 43 feet in diameter and larger. **1962** A. Fry *Ranch on Cariboo* xv. 155 A rolling country where the cattle browsed under an overstorey of

open timber. **1976** *Nature* 22 Jan. 207/2 It [*sc.* the cocoa crop] is thus in intimate contact with an overstorey of forest trees and an understorey of ground vegetation.

overstrain, *sb.* Add: *spec.* the condition of having been strained beyond the yield point.

1895 [see *fatigue *sb.* 1 d]. **1931** H. J. Tapsell *Creep of Metals* iii. 32 The raising of the yield point at air temperature of mild steel was even more marked if the stress producing the overstrain were continued for a time. **1941** H. Gilkey et al. *Materials Testing* 132 The general nature of cold-working, of whatever sort, is simply overstrain.

overstrain, *v.* Add: **3. a.** *spec.* to strain (a substance) beyond the yield point.

1899 J. A. Ewing *Strength of Materials* iii. 40 This operation was carried far enough to overstrain the piece a second time, and curve *D* then shows that a very imperfectly elastic condition has reappeared. **1931** H. J. Tapsell *Creep of Metals* iii. 33 The effect of a period of rest on iron and steel overstrained at air temperature is to produce a recovery of elasticity and an increase in hardness. **1962** R. E. Smallman *Mod. Physical Metall.* vii. 242 If a specimen which has been overstrained to remove the yield point is allowed to rest.., the yield point returns as shown.

overstrained *ppl. a.*, **overstraining** *vbl. sb.* and *ppl. a.* (further examples).

1895 *Proc. R. Soc.* LVIII. 131 The tendency to creeping is found..to be much reduced in consequence of the hardening and recovery of elasticity which the overstrained material undergoes. **1900** *Phil. Trans. R. Soc.* A. CXCIII. 15 Curve No. 2 illustrates the semi-plastic condition of the material immediately after the removal of the overstraining load. **1931** H. J. Tapsell *Creep of Metals* iii. 32 An overstrained material does not resemble its former state: it is physically a new material. **1962** R. E. Smallman *Mod. Physical Metall.* vii. 242 The absence of a yield point at the beginning of plastic flow is characteristic of a specimen in an overstrained condition. **1962** P. G. Forrest *Fatigue of Metals* vi. 191 The beneficial effect of tensile overstraining on the fatigue strength of a notched bar. **1971** *Afr. Wildlife* XXV. 51/1 During the past few years an increasing number of wild animal species has been found to be prone to the development of a disease complex for which the new name, overstraining disease, is suggested. *Ibid.*, It is characterized by muscular degeneration, paralysis especially of the hind limbs and the passage of dark red-brown urine. The course is variable and affected horses may die from acute heart failure or the accumulation of toxic amounts of excretory products, resulting from kidney damage. Overstraining disease in game..may result from various capturing techniques. **1974** *Nature* 22 Feb. 577/1 Capture myopathy (so-called overstraining disease) in wild animals has gained increasing prominence.

overstretch, *sb.* (Later examples.)

1964 *Listener* 3 Sept. 335/2 Her overstretch was made fatal by this by-passing strategy. **1974** *Daily Tel.* 7 Feb. 2/6 Reasons for this are said to be the considerable 'overstretch' in the Navy's resources caused by shortage of ships and manpower.

overstride, *v.* **2.** (Later example.)

1925 *Glasgow Herald* 5 Nov. 11/2 In conception and in achievement it [*sc.* The British Empire Exhibition] overstrode the confines of mere commercial partisanship.

overstrike, *v.* Add: **3.** [tr. F. *surfrapper*: see Over- 8.] *trans.* To strike (a coin) with a new die, imposing a second design on the original; to strike (a new design) on a coin. Hence **o·verstrike** *sb.*, an overstruck coin; **overstru·ck** *ppl. a.*

[**1884** *Encycl. Brit.* XVII. 630/2 A coin is said to be *surfrappé* when it has been struck on an older coin, of which the types are not altogether obliterated.] **1905** *Numismatic Chron.* V. 110 Supposing a sufficient number of overstruck pennies of the same type are available. *Ibid.*, A well-known instance of overstriking coins in modern times occurred in 1804, when..two million Spanish dollars..were overstruck with new dies in the Boulton presses at Soho, and issued as British currency. **1911** *Encycl. Brit.* XIX. 871/1 A coin is said to be 'over-struck' or 're-struck' when it has been struck on an older coin, of which the types are not altogether obliterated. **1914** *Brit. Mus. Return* 114 in *Parl. Papers* LXXI. 193 Another [penny] of the same reign showing the ninth type..overstruck on the seventh. **1932** *Proc. Brit. Acad.* XVIII. 212 Sextantes, with mint-marks C and MA, of the same class as certain early denarii, are commonly found in Sardinia overstruck on Sardinian bronze. We cannot assign such overstrikes to any date earlier than 237 B.C. **1936** *Proc. Prehist. Soc.* II. 144 The Whaddon Chase type coins have been heavily overstruck. **1955** C. Seltman *Greek Coins* (ed. 2) 24 The coiners in a particular mint saved themselves the trouble of preparing metal blanks, and employed instead the actual coins of some other city, heating them first in the furnace and then striking them between their own punch- and anvil-dies... Such coins, technically known as overstruck coins, [etc.]. **1970** *Ashmolean Mus. Rep. Visitors 1969* 33 Fifth century overstrikes at Rhegium and Messana.

o·verstroke. [Over- 6.] An overarm stroke in swimming.

1902 *Encycl. Brit.* XXV. 696/4 Over-stroke. **1934** E. Pound *Eleven New Cantos* xxxix. 44 Came Mara swimming with light hand lifted in overstroke. **1948** —— *Pisan Cantos* (1949) lxxvi. 44 No overstroke No dolphin faster in moving Nor the flying azure of the wing'd fish under Zoagli.

o:verstu·ffed, *ppl. a.* [Over- 8 b, 27 b, 28 c.] **1.** Of furniture: completely covered with a thick layer of stuffing.

1925 T. Dreiser *Amer. Trag.* (1926) I. ii. xxi. 302 An old, faded and somewhat decrepit overstuffed chair. **1928** S. Lewis *Man who knew Coolidge* i. 46 You sit at home in the ole over-stuffed chair. **1930** U. Parrott *Strangers may Kiss* 28 The two armchairs were guaranteed by the landlady to be overstuffed. **1931** *Times* 16 Mar. 22/6 (Advt.), Overstuffed Chesterfield suites in tapestry. **1962** M. Duckworth in C. K. Stead *N.Z. Short Stories* (1966) 362 He passed a..man seated on an overstuffed couch. **1973** W. M. Duncan *Big Timer* x. 66 There were three over-stuffed armchairs.

2. *fig.* Inflated, exaggerated; obese, fat.

1936 L. C. Douglas *White Banners* xi. 233 Hannah's fears that an overstuffed optimism might involve them all in a financial disaster were gradually allayed. **1946** E. Linklater *Private Angelo* xviii. 230 Some great overstuffed history of the world's calamities. **1972** *Daily Tel.* 15 Jan. 9/2 The vulgar antics of the overstuffed strip-tease dancers.

over-sum, *v.* Delete †*Obs.* and add later example.

1929 R. Bridges *Testament of Beauty* iv. 108 The imperativ obligation cannot be over-summ'd.

over-supply, *sb.* Add: (Later examples.) Also *attrib.*

1932 *New Yorker* 14 May 42/2 You will write things..that will get the public interested in that oversupply of pongee somebody is trying to unload. **1964** *Times Rev. Industry* Apr. 3/2 Until the oversupply position improves, there will not be enough work for Britain's shipyards. **1971** *New Scientist* 2 Sept. 510/2 Even in these days of apparent oversupply of scientists it is clear there are never enough really creative minds on the genius level to go around. **1975** *Physics Bull.* Nov. 477/1 In Britain, despite the oversupply of staff and the shortage of money, every department is well equipped. **1977** *Living with Tanker Surplus* (Shell Internat. Petroleum Co.) 2 An impressive amount of new capacity will continue to add to the over-supply.

over-supply, *v.* (Earlier and later examples.)

1865 M. Eyre *Lady's Walks S. of France* ii. 18 The markets at Bordeaux and Pau and Tarbes were *over-supplied* with fruit, butter, poultry, and eggs. **1960** *Farmer & Stockbreeder* 15 Mar. 74/3 The market becomes over-supplied.

over-sweet, *a.* (Later examples.)

1935 [see *icky, ikky *a.* and *sb.*]. **1951** S. Spender *World within World* v. 267 At the end of a lecture on the effect of gases.., I hid for half an hour in a telephone box, overwhelmed by the vision of human beings asphyxiating one another in poisonous over-sweet scents. **1976** *Times* 20 Oct. 14/7 The 2½-lb Christmas puddings..taste fine, if a bit over-sweet.

o·verswing. [Over- 29 b.] **1.** An excessive swing.

1921 *Studies* Sept. 385 The overswing of a pendulum slipping from benevolent despotism of the priest into confiscation of church property. **1926** *Amer. Speech* I. 632 *Overswing*, a common fault [in golf]. **1971** I. G. Gass et al. *Understanding Earth* x. 148/2 We can speculate that such an overswing would eventually correct itself.

2. [Over- 6.] In gymnastics, a movement in which the body swings or turns over.

1955 *Simple Gymnastics* ('Know the Game' Series) 22 *Overswing vault*—Take off both feet and spring high above the apparatus. Drop into the angle position with bent arms and head well back beyond the far edge of the apparatus. Keep the legs down while the hips fall forward. Then extend the hips and push hard with the hands. **1964** G. C. Kunzle *Parallel Bars* ii. 63 You can also do this overswing off one bar outwards. **1965** *Trampolining* ('Know the Game' Series) 43/2 A simple dismount is to walk to the end of the bed, drop into a knee bounce, place the hands on the end rail and perform an overswing on to the feet, in the same way as you would do an overswing over a box horse.

overtake, *v.* Add: **1. a.** Now esp. of a motorist: to drive a vehicle past (another vehicle travelling in the same direction). Also *absol.*, and with the vehicle as subj.

1936 J. Prioleau *Motorist's Compan.* xxix. 418 The commoner examples of dangerous driving are..cutting-in and overtaking another car travelling at any but an obviously lower speed [etc.]. **1938** M. Campbell *Key to Motoring* ix. 107 One of the greatest dangers of fog-driving lies in overtaking or in passing stationary vehicles. **1959** [see *cheesed *a.*]. **1959** E. H. Clements *High Tension* vi. 100 'The journey back was worth it... West overtook me.'..'Where did he overtake you?' 'Rannoch.' **1973** R. Hill *Ruling Passion* ii. iv. 114 A slow lorry suddenly appeared ahead... He swung out sharply to overtake. **1976** *Southern Even. Echo* (Southampton) 12 Nov. 5/2 Marshall was riding a motorcycle on dual carriageway..when a car in front pulled out to overtake.

10. *Bridge. intr.* To take with a higher card a trick already taken by one's partner. Also *trans.*, to play a higher card than (the card played by one's partner).

1904 J. B. Elwell *Advanced Bridge* 164 With no re-entry in a hand, overtaking is often the only means of making a suit. **1939** N. de V. Hart *Bridge Players' Bedside Bk.* xviii. 73 Declarer ducked in dummy, but Herr von Bludhorn overtook. *Ibid.* xxxviii. 118 Declarer played out the King, Knave, and Ten of Trumps, overtaking the Ten with dummy's Queen. **1959** Reese & Dormer *Bridge Players' Dict.* 161 South..leads the jack of spades and, depending on which suit West unguards, overtakes or not with dummy's queen of spades. **1974** *Country Life* 17 Jan. 98/3 East should have overtaken the Spade King and switched to Diamonds.

overtaking *vbl. sb.* and *ppl. a.* (further examples).

1880 R. G. Marsden *Treat. Law Collisions at Sea* vi. 187 Article 20 is express as to the duty of an overtaking sailing-ship to keep out of the way. **1928** R. Townsend *Motoring made Easy* v. 74 No attempt should be made at overtaking on a curve in the road or at a corner. **1960** [see *cutting *vbl. sb.* 9 c]. **1961** F. H. Burgess *Dict. Sailing* 155 *Overtaking light*, a fixed white light, screened so as to be visible 12 points, i.e. from right astern to 2 points abaft the beam each side, and from two miles away; carried by all ships under way. **1966** [see *fast lane* s.v. *fast *a.* 11]. **1970** *Motoring Which?* July 116/1 Five mirrors specifically designed for door-mounting (often called 'overtaking mirrors'). **1975** J. Cleary *Safe House* ii. 100 He saw the beam of the overtaking vehicle's headlamps.

overtalk, *v.* **b.** (Further example.)

1903 E. M. Forster in *Temple Bar* Dec. 682 They shook him and tried to overtalk him, but he still went on.

overtasked, *ppl. a.* (Earlier U.S. example.)

1869 'Mark Twain' *Innoc. Abr.* 289 Relief for overtasked eyes and brain from study and sightseeing.

overtaxation. (Earlier example.)

1823 J. S. Mill in *Black Dwarf* XI. 749 Over-taxation cannot lower wages.

overtaxing, *vbl. sb.* (Example.)

1877 Tennyson *Harold* i. i. 6 Nay, there be murmurs for thy brother breaks us With over-taxing.

over-the-board, *a.* [Over *prep.* 12.] Of chess competition: with the participants actually present, as opp. to correspondence play; with opponents facing each other across the chess-board.

1932 E. Lasker *Manual of Chess* iv. 228/2 His talent for over-the-board play was not considerable. **1954** H. Golombek *World Chess Championship 1954* 54 He has retired from over-the-board play and has devoted himself to writing about and teaching the game. **1974** C. H. O'D. Alexander *On Chess* iii. ix. 214 For various reasons—no chess club in the locality, lack of time, finding over-the-board play too much nervous strain—a number of people don't play it.

over the counter, *adv.* (*a.*). [Over *prep.* 12.] **a.** See Counter *sb.*[3] 4. **b.** *N. Amer.* With reference to the selling of stocks and shares: as a direct transaction, (business concluded) outside the system of a recognized stock exchange. **c.** Hence of purchase and selling generally: (transacted) directly between seller and buyer; openly, legitimately. (Cf. *under the counter* s.v. *counter *sb.*[3] 4 b.) Also *transf.* and (usu. with hyphens) as *adj. phr.*

1875, 1889 [see Counter *sb.*[3] 4]. **1921** *Mag. of Wall St.* 10 Dec. 179/2 There is another field which readers have expressed the desire to see us cover. That is the great field comprising unlisted securities, which are dealt in over-the-counter... In response to this demand, we are inaugurating, beginning with this issue, a new department to appear under the caption, 'Over-the-Counter'. *Ibid.*, It will be our effort to confine our analysis of over-the-counter stocks to as brief a space as possible. **1925** A. M. Sakolski *Princ. Investment* x. 108 'Over-the-counter' transactions (i.e., those which occur privately, whether consummated directly by negotiation between buyer and seller or through dealers and brokers) generally, in the absence of special agreements, follow the common practices of the exchanges. **1929** Willis & Bogen *Investment Banking* iii. 50 This market is referred to as the unlisted or over-the-counter market, because business is transacted..'over the counter' of the individual broker or dealer, rather than at..an exchange. **1934** *Sun* (Baltimore) 27 Apr. 2/3 The committee has heard evidence of extensive manipulation in certain New York bank stocks after their withdrawal from the New York Stock Exchange and while they were being sold 'over the counter'. **1936** Wodehouse *Laughing Gas* xv. 165 'You think this tooth could be sold?' 'Over the counter, sir, over the counter.' **1944** *Amer. N. & Q.* July 64/1 A North Carolinian illustration of this futile neologistic tendency found in high places is 'over-the-counter salesperson', for the simple..clerk. **1957** Clark & Gottfried *University Dict. Business & Finance* 253/2 The securities are traded on a face-to-face, or over-the-counter basis. In the actual operation of over-the-counter trading, a trader who specializes in a particular security arranges for all transactions, either by bringing buyers and sellers together, or by buying and selling the security for his own account. **1958** *Spectator* 11 July 58/3 The over-the-counter service of spit-roasted chicken is a development in the catering trade. *Ibid.* 60/1 The over-the-counter charge for the spit-roasted bird is 12s. 6d. **1963** *Times* 23 Apr. 20/3 Sales of bottles to take home ('Off-Sales') have increased very considerably, and there has been a reduction of sales by the glass—'Over the counter'—with a consequent reduction in profit. **1965** *McGraw-Hill Dict. Mod. Econ.* 367 In addition to common and preferred stocks, almost all U.S. government securities and municipal and corporate bonds are traded over the counter. **1968** *Globe & Mail* (Toronto) 17 Feb. B 2/1 An over-the-counter speculative issue that rose from 25 cents a share to $4.50 in the past six months. **1969** *Guardian* 25 July 6/6 If 'over the counter' pregnancy

testing is to come to Britain, the tests should be carried out by the pharmacist. **1972** *Times* 16 May (Wall Street Suppl.) p. iv/7 For 25 years he was a broker and then from 1964 to 1967 headed the National Association of Securities Dealers which regulates over-the-counter deals. **1974** M. C. Gerald *Pharmacol.* ii. 20 Nonprescription (over-the-counter, OTC) sleep-facilitating products..have capitalized on the drowsiness induced by methapyrilene. **1974** *Guardian* 22 Mar. 11/5 The retailers wanted to simplify the collection of air fuel surcharges on package holidays by including the sum on the over-the-counter invoice presented to the customer.

over the moon: see *moon *sb.* 3 b.

over-the-road, *a.* [Over *prep.* 13.] Of, pertaining to, or used in long-distance road transportation.

1945 *Sun* (Baltimore) 24 Oct. 14-0/8 Approximately 1500 members of Local 557, Freight Drivers and Helpers ..yesterday went on strike which threatens to halt about 50 per cent of trucking operations in the city and practically all over-the-road, inter-city hauling into and out of Baltimore. **1967** *Jane's Surface Skimmer Systems 1967–68* 63 Movement of shipment in cages installed on air-in-floor pallets eliminates rehandling between city pick-up trailers, terminal site, over-the-road trailers, second city terminal site and final delivery unit. **1969** *Jane's Freight Containers 1968–69* 407/3 Both the 20 ft and 40 ft units are designed to be used as over-the-road trailers. **1971** M. Tak *Truck Talk* 114 *Over-the-road driver*, a driver who hauls goods long distances. **1977** *Time* 13 June 36/2 Last week's decision ended a long and hard-fought case against..a trucking company based in Lubbock, Texas, which the Government accused of discriminating against blacks and Hispanics in deciding who would get over-the-road driving jobs.

overthought, *sb.* Add: **b.** [Over- 2 b.] Conscious thought; an explicit concept; = Over-soul.

1883 G. M. Hopkins *Let.* 14 Jan. (1938) 105 Two strains of thought..the overthought that which everybody, editors, see..which might for instance be abridged or paraphrased..the other, the underthought, conveyed chiefly in the choice of metaphors etc used and often only half realised by the poet himself. **1884** W. James in R. B. Perry *Tht. & Char. W. James* (1935) I. 583 Those who must answer this question negatively are forced to the notion of an Over-thought behind the phenomenal real.

overthrow, *sb.* Add: **1. d.** The state of being overthrown.

1903 *Daily Chron.* 12 Sept. 5/1 Half a dozen great trees were torn up by the roots, and lay in disorderly overthrow ready for the saws. **1906** F. Thompson *Ode Eng. Martyrs* in *Wks.* (1913) II. 136 Till she shall know This lesson in her overthrow: Hardest servitude has he That's jailed in arrogant liberty.

4. *Cricket* (earlier and later examples). *Baseball* (examples).

1748 in H. T. Waghorn *Dawn of Cricket* (1906) 21 To play or pay, bye balls, and overthrows to count. **1856** *Spirit of Times* 8 Nov. 165/1 Gessner [made]..three homes in succession, one of them being helped by an overthrow. **1949** *Telephone-Reg.* (McMinnville, Oregon) 4 Aug. 1/2 There is no sliding and a player cannot run on overthrows. **1955** *Times* 13 July 3/4 The Middlesex fielding was uncertain and Warwickshire were helped by indiscriminate throwing, which led to many over-throws.

5. *Archit.* A panel of decorated wrought-iron work forming the architrave of a gateway or arch.

1911 J. S. Gardner *English Ironwork 17th & 18th Centuries* 26 The base of the overthrow took the form of a latticed girder, or two bars braced together by scrolls. **1932** *Times Lit. Suppl.* 10 Nov. 835/2 Making artistic ironwork—gates simple or elaborate, with or without 'overthrows'. **1957** R. Lister *Decorative Wrought Ironwork Gt. Brit.* iii. 89 The overthrow, a composition built up from sixteen simple scrolls, reaches up symmetrically in the centre of the stretcher..terminating into a large disc bearing a garter and shield, and crested by a coronet. **1971** *Illustr. London News* Oct. 54/1 A good quality 18th century Italian Istrian marble wellhead with wrought iron overthrow. **1975** *Oxf. Compan. Decorative Arts* 481/2 The overthrows of the gates, swollen to gigantic size, plainly prefigure the Rococo.

overthrow, *v.* Add: **6.** *trans.* [Over- 27.] To throw farther than is necessary or desired; to throw too far; *spec.* in *Cricket*, to throw (the ball) beyond or wide of the wicket, so as to concede overthrows.

1833 *Field Bk.* 141/1 The batters may take the advantage of running when a ball has been over-thrown. **1862** *Chambers's Encycl.* III. 320/1 Misconception of this [distance] may lead to overthrowing the ball, or throwing it short. **1875** *Baily's Mag.* Apr. 403 A ball..overthrown, on the ground where he learnt his cricket, means the loss of four, five, or even six or seven runs.

overthrowal (ōuvəɹþrōu·āl). [f. Overthrow *v.* + -al II.] The act of overthrowing; subversion, defeat.

1916 W. J. Locke *Wonderful Year* xxiii. 333 Thus came the overthrowal of all Corinna's scheme of values. **1920** —— *House of Baltazar* xxi. 257 The ultimate object of this gathering was the overthrowal of the Government. **1941** 'R. West' *Black Lamb* II. 196 The upsetting of kings and the overthrowal of empires. **1949** *Tablet* 3 Sept. 147/1 The overthrowal of the Second Reich of the Hohenzollerns. **1968** *Progressive* Nov. 10/3 Spiro Agnew..had coined at least two new words by early October: 'overthrowal' and 'uprisal'.

overthrust, *sb.* Add: In mod. use (also *overthrust fault*), a reverse fault in which the fault plane makes a relatively small angle with the horizontal; formerly, any reverse fault. (Further examples.)

1903 A. Geikie *Text-bk. Geol.* (ed. 4) I. 690, 1. Normal Faults... 2. Reversed Faults or Overthrusts. **1944** A. Holmes *Princ. Physical Geol.* vi. 80 Reverse or Thrust Faults... When the resulting fracture is inclined at an angle between 45° and the horizontal..the corresponding fault is described as an overthrust. **1957** *Bull. Geol. Soc. Amer.* LXVIII. 168/1 The Medicine Lodge overthrust, a low-angle fault with a displacement of many miles. **1969** *Ibid.* LXXX. 953 Overthrust fault surfaces are actually undulatory rather than plane.

overthrust *v.* (examples); hence **o·verthrust** *ppl. a.*

1900 [see *crust *sb.* 13 b]. **1901** *Nature* 3 Jan. 234/1 Huge masses of country have been overfolded, fractured, and overthrust, the older being pushed over the newer. **1942** M. P. Billings *Structural Geol.* x. 184 The thrusts dip north, and the overthrust sheets have traveled northward relative to the underlying formations. **1956** W. J. Arkell *Jurassic Geol. World* ix. 225 The Jurassic and Cretaceous systems were strongly folded and overthrust in the post-Oligocene, pre-Miocene orogeny. **1968** [see *foreland 4*].

overthwart, *v.* **2.** (Later example.)

1937 *John o' London's* 5 Feb. 761/2 My parents were for ever overthwarting me, both on 'em. Always to school I had to go till I was twelve, and to church I had to go regular as clockwork.

overthwarting, *vbl. sb.* (Later example in sense 1 c of the verb.)

1942 W. Rose *Good Neighbours* iii. 20 A field was first evenly ploughed all over, after which cross ploughing—called *over-a-thurting*—often followed, severing the furrows and leaving the soil thoroughly exposed to the air.

overtilt, *v.* Delete †*Obs.* and add later example.

1905 *Westm. Gaz.* 1 Feb. 2/3 Our house tottereth To ruin; because this people with the breath Of pity would overtilt it.

overtime, *sb.*, *adv.* Add: **A.** *sb.* **a.** (Earlier and later examples.) Also, payment for work performed in excess of normal hours. *overtime ban*, industrial action in which the working of overtime is suspended.

1846 *Swell's Night Guide* 42 There are instances of the awful enemy lodging itself here, through some private failing in overtime. **1911** *Daily Colonist* (Victoria, B.C.) 16 Apr. 14/5 A conference is to be held..with regard to the demand being made by the seamen for overtime pay. **1916** 'Boyd Cable' *Doing their Bit* iv. 63 Their haggling over 8*d.* or 8½*d.* an hour pay, or Saturday half-holidays, or double overtime for Sunday. **1928** F. B. Young *My Brother Jonathan* iii. iii. 446 Joe Matthews..'picking up' six pounds a week, to say nothing of overtime, in Higgins's shell factory. **1952** *Times* 1 Dec. 2/5 (*heading*) Miners' overtime ban ended. **1968** *R. Comm. Trade Unions & Employers' Associations Res. Papers* No. 9, 82 The negotiations broke down during 1964 and the unions imposed an overtime ban and a work to rule. **1972** *Times* 15 May 17/3 Workers at another Ferranti factory are due to impose the restrictions—an overtime ban, work to rule and boycott of piecework—from today. **1973** C. D. Garratt *Masterpieces in Steam* 128 My visit to Brynlliw was during the overtime ban prior to the miners' strike. **1973** *Times* 1 Dec. 3/6 The miners' overtime ban yesterday went into its fourth week.

b. In sporting contests: extra time added in the event of a draw. *N. Amer.*

1921 *Daily Colonist* (Victoria B.C.) 15 Mar. 11/1 Overtime game goes in favor of Towers. The Towers Club..defeated the Senators..by 8 goals to 7, the game having to go into overtime to decide the match. **1946** *Richmond* (Virginia) *Times-Dispatch* 20 Mar. 17/3 His keen floor generalship..built up to a terrific climax—his 55-foot shot in the last two seconds that tied the score at 74–74 and led to the Ram's 82–79 overtime triumph. **1961** J. S. Salak *Dict. Amer. Sports* 315 *Overtime*, continuance of play after the regulation time for a contest has expired when the score is tied. **1970** *Washington Post* 30 Sept. D 1/7 The Senators..lost two overtime games to the Baltimore Orioles. **1974** *State* (Columbia, S. Carolina) 3 Mar. 2-D/5 Freshman Walter Davis banked in a 35-foot jump shot at the end of regulation time to cap an incredible North Carolina comeback Saturday as the fourth-ranked Tar Heels nipped arch rival Duke, 96–92 in overtime.

B. *adv.* Also *fig.*

1938 E. Ambler *Cause for Alarm* xvi. 262 In that moment my brain worked overtime. **1953** A. Huxley *Let.* 25 Sept. (1969) 685 They are pork manufacturers, with a farm where five thousand sows work overtime eating the garbage of the city of Long Beach and producing fifty thousand piglets per annum. **1971** *Nature* 26 Nov. 179/1 Washington's science policy machinery has been working overtime on plans that could significantly alter relationships between science and government. **1974** G. Butler *Coffin for Canary* i. 9 My tongue had worked overtime at the week-end..talking with my sister.

o·ver-ti:mer. [f. Overtime *sb.* + -er[1].] One who works overtime.

1926 S. Baldwin *On England* 99 The just and the unjust, the half-timers, the whole-timers and the over-timers.

overti·p, *v.* [f. Over- 27 + Tip *v.*[4]] *trans.* To give an excessive gratuity to (one who has been of service). Also *absol.* Hence **o:verti·pping** *vbl. sb.*

1926 E. Hemingway *Fiesta* (1927) iii. xix. 269 The waiter seemed a little offended..so I overtipped him. **1928** A. Huxley *Point Counter Point* xxi. 388 The fare was three-and-six. Philip gave the driver two half-crowns... He made a habit of over-tipping. **1938** L. MacNeice *Earth Compels* 38 And there are the men who appear to be men of sense,..For fear of opinion over-tipping in bars, For fear of thought studying stupefaction. **1941** *Penguin New Writing* VIII. 62 The guests in the guest-house, they were charmed with him too. They over-tipped him. **1965** *New Statesman* 7 May 729/3 If we had reason to believe that he once overtipped a cab-driver, it would be recorded here. **1976** E. Ward *Hanged Man* xvii. 98 He stopped the taxi..paid and overtipped. *Ibid.* xxxiii. 214 A taxi-driver collected the message for him. Anonymity is worth over-tipping.

overtone, *sb.* Add: **1.** Also, an analogous component of any non-acoustic oscillation, having a frequency that is an integral multiple of the fundamental frequency.

1922 A. D. Udden tr. *Bohr's Theory of Spectra* iii. ii. 83 This apparent difficulty is explained by the occurrence in the motion of the hydrogen atom..of harmonic components corresponding to values of *r*, which are different from 1; or using a terminology well known from acoustics, there appear overtones in the motion of the hydrogen atom. **1937** Jenkins & White *Fund. Physical Optics* xii. 283 If the charged oscillator is bound by a force which does not obey Hooke's law,..it will be capable of re-radiating not only the impressed frequency, but also various combinations of this frequency with the fundamental and overtone frequencies of the oscillator. **1973** *Physics Bull.* July 421/2 The first overtone spectrum of HBr near 4·95 μm has been obtained in this way.

2. *fig.* (Freq. in *pl.*) Applied to literature, esp. poetry: what is suggested or implied by the sound or meaning of the words. More generally, a connotation or subtle implication in thought, language, or action.

1890 W. James *Princ. Psychol.* I. ix. 258 Let us use the words psychic overtone, suffusion, or fringe, to designate the influence of a faint brain-process upon our thought. *Ibid.* 281 The total idea..is the overtone, halo, or fringe of the word, as spoken in that sentence. **1904** J. G. Huneker (*title*) Overtones, a book of temperaments. **1911** Brereton & Rothwell tr. *Bergson's Laughter* ii. 96 There would be nothing amusing in the saying, 'It serves you right, George Dandin', were it not for the comic overtones that take up and re-echo it. **1922** H. Crane *Let.* 2 Apr. (1965) 83 The very effective literary device of under-accentuation in just the right place to produce 'overtones' of overwhelming effect. **1940** *Mind* XLIX. 209 Stripped of these mental overtones, his definition is quite close to the etymological meaning of the word *infinite*. **1952** *Jrnl. Theol. Stud.* III. 64 Once Israel had again fallen under the 'yoke of the heathen', the festival of dedication would have reminded the Jews of the exploits of Judas and could easily have taken on nationalistic overtones. **1957** *Economist* 21 Sept. 953/1 Seven per cent would have been a stronger gesture of resolve, if it had not been left so late as to suggest some overtones of desperation. **1965** *Listener* 25 Nov. 869/1 There are overtones of Aldous Huxley and Norman Douglas, and echoes of Firbank. **1976** 'D. Halliday' *Dolly & Nanny Bird* xix. 258 The kidnapping and holding to ransom of an American child, in order to raise money for the self-styled Croatian Liberation Army. A simple crime with political overtones which the militia would work out for themselves.

overtone, *v.* Add: **2.** (Earlier example.)

1868 [see Tone *v.* 5].

3. To give an overtone or implication to. *rare.*

1871 G. Meredith *H. Richmond* II. xvii. 269 She threw a kindly-comical look, not overtoned, at the miniature ships on the mantelpiece, and the picture of Joseph leading Mary with her babe on the ass.

overtop, *adv.* (Later Austral. example.)

1921 'J. O'Brien' *Around Boree Log* (1937) 82 And every creek a banker ran, And dams filled overtop.

overtopped, *ppl. a.* Add: Of a small tree: growing beneath the canopy formed by larger trees and receiving no direct light. (Later examples.)

1917 *Jrnl. Forestry* XV. 74 The crown classes usually distinguished are: Dominant... Co-dominant... Intermediate... Overtopped. Trees with crowns entirely below the general forest canopy and receiving no direct light. **1948** *Ibid.* XLVI. 833/2 Seedlings overtopped but with considerable side light survive and grow just about as well as seedlings in small openings. **1976** G. W. Sharpe et al. *Introd. Forestry* (ed. 4) ix. 187 Suppressed. Pertaining to trees with small crowns that are entirely below the level of the canopy receiving no direct light from above or from the sides. Also called overtopped.

overtopping, *vbl. sb.* and *ppl. a.* (Later examples.)

1903 W. B. Yeats *In Seven Woods* 39 And he, The one over-topping man that's in the world, Keeps far away. **1959** *Times* 2 Nov. 21/1 Undertopping in no case exceeded 5 lb. to the hundredweight and overtopping with three machines did not exceed 3·4 lb. **1976** M. Green *Children of Sun* vi. 234 America's overtopping of England—England's overshadowing and diminution.

overtower, *v.* (Later examples.)

1928 E. Blom *Limitations of Music* 18 As an artist..he [*sc.* Mozart] overtowers every other figure in musical history. 1954 N. Pevsner *Essex* 385 The church is..cut off by the railway from the village and now overtowered by a modern factory.

over-train, *v.* **a.** (Later examples.)
1910 *Blackw. Mag.* Jan. 135/1, I was as hard as nails..; but was over-trained, and after a time did the walking, and even the shooting, with some loss of the keenness with which I began it. 1971 *Nature* 10 Sept. 126/2 Here rats are overtrained (by means of foot shock) to avoid entering a black box. 1976 A. Grey *Bulgarian Exclusive* v. 35 They're over-training the Olympic squad.

over-tra·p, *v.* [Over- 27.] *trans.* To trap (a region) too much in such a way as to deplete the fauna. So **over-tra·pped** *ppl. a.*
1964 P. Worsley in I. L. Horowitz *New Sociol.* 381 Over-trapped areas around the settlements. 1968 [see *over-hunt *v.*].

overtravel, *sb.* Add: **b.** Movement of part of a machine beyond the desired point; an allowance made for such travel. Freq. *attrib.*
1923 R. Grierson *Electr. Lift Equipment* xiii. 87 Overtravel spaces at both the top and the bottom of the shaft are essential, as the brake is subject to wear. 1939 R. S. Phillips *Electr. Lifts* xii. 177 The top overtravel is defined as the distance provided for the car floor to travel above the level of the top terminal landing before the car is stopped by the ultimate limit switch. 1962 E. G. Davies in G. A. T. Burdett *Automatic Control Handbk.* iii. 21 Series limit switches..are generally employed as emergency over-travel switches not called upon to operate except after the failure of the normal stopping or reversing pilot limit switch. 1977 *Economist* 3 Sept. 50/3 (Advt.), Overflow valves. Overtravel switches. Oxide magnets.
overtravel *v.* (*b*) (see quot. 1909).
1909 *Cent. Dict. Suppl.*, Overtravel... In *mech.*, to travel farther than is necessary to do a certain act or perform a definite function. 1955 *Archit. Rev.* CXVII. 143/4 At top and bottom of the well there are Limit Switches which automatically cut off the lift motor in the event of the car overtravelling in either direction.

over-traw·l, *v.* [Over- 27.] *trans.* To trawl (a fishing-ground) too much or to depletion. So **over-traw·ling** *vbl. sb.*
1913 *Q. Rev.* Apr. 444 In 1892 the Trawlers' Society..again protested against over-trawling.

o·vertrick. *Bridge.* Also with hyphen. [Over- 19.] A trick taken in excess of the number contracted for. Also *attrib.*
1921 F. Irwin *Compl. Auction Player* 15 Each over-trick is worth twice its value below the line. 1927 *Observer* 5 June 19/2 All over-trick bonuses gained by a vulnerable side..count double. 1929 M. C. Work *Compl. Contract Bridge* i. 2 A trick-score of 20 or 30 plus two over-tricks each worth 50. 1947 S. Harris *Fund. Princ. Contract Bridge* 88 This play may cost West an overtrick. 1966 *Listener* 16 June 891/3 One overtrick is a small premium with which to insure a hazardous contract. 1973 *Times* 10 Feb. 12/8 If he had kept three spades, declarer would have made an overtrick by putting him on play with a club or a heart.

overtrim, over-trim, *v.* **2.** (Earlier example.)
1816 Jane Austen *Emma* II. xvii. 328, I have the greatest dislike to the idea of being overtrimmed.

overturn, *sb.* Add: **6.** *Limnology.* The mixing or circulation of the water in a thermally stratified lake that usu. occurs once or twice each year as a result of the cooling or warming of the epilimnion.
[1898 G. C. Whipple in *Amer. Naturalist* XXXII. 27 Soon the surface and bottom layers come to have substantially the same temperature, and vertical currents extend from top to bottom. This is the 'period of spring circulation', or the 'spring overturning'.] 1911 Birge & Juday in *Bull. Wisconsin Geol. & Nat. Hist. Survey* xxii. p. xi, The stratum of circulating water becomes increasingly thinner until..the permanent summer conditions are established. Thenceforward, until the overturn, only the water of the epilimnion can have direct contact with the air. 1935 P. S. Welch *Limnology* iv. 38 During the spring and fall overturns.., when the water is of uniform density from top to bottom, return currents may extend even to the bottom of relatively deep inland lakes. 1972 *Ann. Rep. Freshwater Biol. Assoc.* XL. 42 Before the overturn, the epilimnetic population of both Grasmere and Blelham Tarn went through the whole [nitrate] reduction process. 1974 *Encycl. Brit. Macropædia* X. 607/1 Mixing due to cooling or warming processes that increase the density of surface waters sufficiently to cause them to sink results in what is termed circulation, or overturn, of lake water.

overturn, *v.* Add: **3. b.** To reverse (a judicial decision).
1969 *Morning Star* 8 Aug. 1/3 Last month, the US circuit court of appeals in Boston overturned the verdict. 1973 *Observer* 22 July 10/1 The House of Lords unanimously overturned this decision.

overturned, *ppl. a.* Add: *spec.* in *Geol.* applied to a fold or the limb of a fold that is tilted beyond the vertical (cf. Overfold *sb.*).
1861 *Amer. Jrnl. Sci.* LXXXI. 218 Sometimes it may overlie the overturned Utica formation. 1896 [see Overfold *v.* 2]. 1907, 1970 [see *foreland 4*].

overun (ōu·vərən), *a.* and *sb. local.* Also **overrun(ner.** [f. Over *adv.* or *a.* + Un².] **A.** *adj.* In the Isle of Wight, coming from the mainland; not native to the Island. **B.** *sb.* = *overner.
1881 H. & C. R. Smith *Isle of Wight Words* 48 *Overun*, coming from 'across the water', from the mainland of the county. 1886 [see *overner]. 1889 F. Cowper *Captain of Wight* x. 130 There'll have to be some overrunners asked over. *Ibid.* xiv. 176 That's what I call a hardy knight...'Tis a pity he's an overrun.

overve·ntilate, *v. Physiol.* [Over- 26.] **a.** *trans.* = *hyperventilate *v.* b.
1917 J. S. Haldane *Organism & Environment* i. 13 The very rapid and shallow breathing of a dog in hot weather does not over-ventilate its lungs. 1931 *Amer. Jrnl. Physiol.* XCVIII. 202 A dog with a low Ca of 6·05..was overventilated..with the corresponding precipitation of violent convulsions. 1963 Spalding & Smith *Clin. Pract. & Physiol. Artific. Respiration* iii. 43 A patient who has been overventilated during artificial respiration.
b. *intr.* = *hyperventilate *v.* a.
1927 *Jrnl. Physiol.* LXIV. p. xxiv, Nahun showed that there is a decrease in limb volume when a normal man over-ventilates. 1961 *Ibid.* CLVI. 240 The subject was instructed to overventilate from a 100 l. Douglas bag containing 4% carbon dioxide in air.

o:verventila·tion. *Physiol.* [Over- 26.] = *hyperventilation.
1911 *Amer. Jrnl. Physiol.* XXVIII. 387 Over-ventilation of the lungs..is at once suggested as an explanation. 1927 *Jrnl. Physiol.* LXIV. p. xxiv, The vessels of the skin play some part in maintaining the blood-pressure during over-ventilation in man. 1963 Spalding & Smith *Clin. Pract. & Physiol. Artific. Respiration* v. 76 Artificial overventilation may alter the response to CO_2 of the mechanism controlling respiration and may induce a volume-dependent type of respiration.

overview, *v.* Delete † *Obs.* and add later examples.
Quot. 1864 is in sense 1, quots. 1935, 1977 in sense 2. The use in quot. 1977 is influenced by *overview *sb.* 2.
1864 S. Warner *Old Helmet* I. xii. 247 Mrs. Powle's fair face would overview a moral desolation more hopeless and more cheerless [etc.]. 1935 E. R. Eddison *Mistress of Mistresses* xi. 197 From the sweep of eagles' wings it becometh us overview the matter, and what's just and allowable of our greatness, choose that. *Ibid.* xii. 225, I would have you, as a politic prince.., refer the whole estate you are in to your highness' deliberate overviewing again. 1977 *Canad. Jrnl. Linguistics* 1976 XXI. 1. 17 The interplay of physiological-cognitive factors with respect to early speech perception and production has been critically overviewed by Gilbert (1975).

overview, *sb.* Restrict † *Obs.* to sense in Dict. and add: **2.** orig. *U.S.* A survey, summary, or comprehensive review of facts or ideas; a concise statement or outline of a subject. Hence **o·verviewer**, one who formulates an overview.
1934 *Amer. Speech* IX. 318/2 Editors testify that the new noun *overview* is now being worked as hard by educationists as 'purposeful', 'challenge', 'objective', 'motivation', et al., have been in the past. 1944 *Mind* LIII. 276 According to the jacket..the purpose of this book is to present 'an overview of present-day philosophical trends'. 1958 *Times Lit. Suppl.* 28 Mar. 173/4 Her 'overview' of psycho-analytical thought is therefore followed by detailed expositions of the views of the principal dissidents. 1967 *Wall St. Jrnl.* 4 May 4/3 The first step is to take an 'overview' of the problem. 1969 M. Crichton *Andromeda Strain* v. 44 A scientist with a conscience, an overview, an appreciation of the significance of events. 1969 M. Scriven in N. R. Hanson *Perception & Discovery* p. vii, There is a..tension between the demands of exact historical scholarship and the more free-ranging interests of the overviewer. 1973 *Nature* 6 July 59/3 The chief drawback of this volume is that it lacks a chapter devoted to a critical overview of the comparative relevance of particular models as they relate to particular experimental ends. 1974 *Florida FL Reporter* XIII. 31/2 With presumably Ms. Hess as overviewer making some critical commentary on which of the arguments were stronger and why. 1975 A. S. Miskimin *Renaissance Chaucer* viii. 230 In an overview such as this, many questions will be begged. 1975 *Sci. Amer.* May 119/2 The first 50 pages of overview include a neat account of the logical architecture of possible machines. 1975 *Listener* 31 July 130/2 In its latest policy 'overview', the State Department ponders [etc.].

o·vervo:ltage. [f. Over- 19, 20 + Voltage², as tr. G. *überspannung* (W. A. Caspari 1899, in *Zeitschr. f. physikal. Chem.* XXX. 91).] **1.** The difference between the electrode potential for a reaction (as the liberation of a gas) in practical, irreversible conditions and the theoretical, reversible value.
1907 Whitney & Brown tr. *Le Blanc's Text-bk. Electro-Chem.* viii. 299 Not only in the case of hydrogen, but also in that of oxygen, an over-voltage which varies with the nature of the electrode..is produced by the separation of the gas. 1922 Glazebrook *Dict. Appl. Physics* II. 60/1 Platinised platinum as a cathode has the lowest overvoltage among the elements commonly available. 1965 Phillips & Williams *Inorg. Chem.* I. xi. 421 Using a mercury electrode the over-voltage of hydrogen is so great that the sodium is discharged and taken into the mercury as an amalgam.
2. A voltage in excess of that which is normal, safe, or allowed for.
1921 S. Q. Hayes *Switching Equipment for Power Control* v. 157 Relays are built to furnish protection on A.C. or D.C. circuits against overvoltage, no voltage, overload, no load, reverse load and reverse phase. 1963 B. Fozard *Instrumentation Nucl. Reactors* v. 56 It may sometimes occur that a tube must be used with an insensitive scaling equipment, in which case a high overvoltage may be necessary to give satisfactory operation.

o:vervulcaniza·tion. [Over- 29 b.] Vulcanization for longer than is necessary to achieve the maximum modulus of elasticity.
1900 W. T. Brannt *India Rubber, Gutta-Percha & Balata* iii. 111 If the articles are allowed to remain too long in the solution, over-vulcanization may take place, that is, the surface of the article becomes hard and brittle. 1967 tr. *W. Hofmann's Vulcanization & Vulcanizing Agents* i. 24 Its symptoms [*sc.* those of postvulcanization] are thus very similar to the rising characteristic of tensile stress at a given elongation which is associated with overvulcanization.

overvu·lcanize, *v.* [Over- 27.] *trans.* To subject (rubber) to overvulcanization; to make hard or brittle by vulcanizing. Hence **overvu·lcanizing** *vbl. sb.*
1911 P. Schidrowitz *Rubber* xi. 172 Bysow..found that with a total sulphur of from 3·44 per cent. to 8·91 per cent. equilibrium was not produced until the samples were over-vulcanized. 1916 [see *overcure *v.*]. 1922 *Encycl. Brit.* XXX. 35/1 In 1916 some resistance to petrol was introduced by using pure para heavily loaded with mineral matter and rather over-vulcanized. 1967 tr. *W. Hofmann's Vulcanization & Vulcanizing Agents* i. 24 When an article is wholly or partly overvulcanized (thick-sectioned articles, for example, are sometimes overvulcanized at the surface), efficient protection against ageing is always necessary.

o·ver-walker. *rare.* [-er¹.] One who walks too much or too far.
1876 R. L. Stevenson in *Cornhill Mag.* XXXIII. 685 It is here that your over-walker fails of comprehension.

overwash, *sb.* (Further examples.)
1894 [see *outwash]. 1909 *Jrnl. Geol.* XVII. 375 Locally overwash plains are conspicuous topographic features. 1957 J. K. Charlesworth *Quaternary Era* I. xxii. 442 Thus outwash may occur with or without moraines; the two types constitute respectively Chamberlin's overwash aprons and outwash plains or aprons.

o:ver-wa·ter, *a.* [Over- 32.] **a.** Performed or proceeding across water (in quot. 1900 = 'foreign'). **b.** Situated on or located over water.
1900 W. C. Murray *Week's Rambling with 'Bra Quamin'* 11 De talkin' tree is a ober water duppy tree. 1933 *Sun* (Baltimore) 7 Dec. 1/7 It was the longest over-water flight of their aerial survey tour. 1942 *R.A.F. Jrnl.* 16 May 16 It has a long range..which renders it suitable for over-water operation from shore bases. 1946 *Sun* (Baltimore) 30 July 8/2 If geophysical reports are favorable, over-water rigs will be erected later for drilling tests. 1960 *Washington Post* 20 Mar. B8/4 Air Guard spokesmen emphasized that missions..are routed over nonpopulated areas to an over-water course to the Buck Target Range in North Carolina. 1967 *Jane's Surface Skimmer Systems 1967–68* 2/1 Research and development programmes are dedicated in the main to the task of speeding up over-water transport. 1976 B. Lecomber *Dead Weight* II. vii. 82 All aero-engines immediately go into auto-rough whenever you think about them on the over-water flight.

overweight, *sb.* Add: **1.** (Later examples.)
1886 J. Barrowman *Gloss. Scotch Mining Terms* 48 *Overweight*, excess weight of disposals over outputs. 1888 W. E. Nicholson *Gloss. Terms Coal Trade Northumb. & Durham* 62 *Overweight*, the difference between the standard weight and the average weight for a fortnight when over the standard.
3. b. Applied to a person: obesity, excess of weight.
1917 *Med. Times* (N.Y.) Aug. 217/2 (*heading*) Reduction cures for overweight. 1918 J. Broadhurst *Home & Community Hygiene* xviii. 298 (*heading*) The relation of overweight to death rates. 1923 J. W. Barton *That Body of Yours* 102 There are two kinds of overweight. First.. the kind you inherit. You have always been a little overweight. Then there is the kind that is acquired at the age of 27 to 37. 1925 Mrs. Belloc Lowndes *Diary* 10 Jan. (1971) 102 Lord Northcliffe was..distressed at what he regarded as my overweight. He told me that he and his wife weighed each other constantly. 1951 I. B. Allen *Youth after Forty* vi. 53 Raw fruits do not induce overweight. 1973 M. Senechal *Guarding your Family's Health* iv. 56/1 One of the commonest of all health problems revealed by a checkup is overweight. 1975 *Times* 20 Dec. 12/5 She needs to lose weight..an overweight caused by a careless diet.
5. *Pros.* An instance of overweighting (see quot.).
1940 J. R. R. Tolkien in Clark Hall & Wrenn *Beowulf & Finnesburg Fragment* p. xxxi, An example with double overweight would be *wéllmàde wárgeàr.*

over-weight, *a.* Add: (Now usu. written as one word.) *spec.* of a person: too heavy for one's height and build; obese. Also *absol.*
1899 G. R. Shepherd in *Med. Examiner* July 211/2 From our mortality records the overweights are clearly

less desirable than the underweights. *Ibid.* 212/1 Are people who are overweight likely to live longer than those who are..underweight? 1910 A. Bryce *Laws of Life & Health* vii. 247 No 'over-weight' dies of old age or senility. 1925 C. E. Turner *Personal & Community Health* ii. 52 The tendency to become overweight is no doubt constitutional. 1935 H. Roberts *Everyman in Health & Sickness* II. iv. 152 Insurance companies find that 'overweights' are bad lives. 1941 F. Silver *Foods & Nutrition* iv. 117 A child..is considered overweight if he is 20 per cent above the average for his age and height. 1951 I. B. Allen *Youth after Forty* vi. 50 'She is an obese person' sounds right down [*sic*] unattractive—but if she is pounds overweight, that's just what she is. 1958 *Which?* Autumn 20/1 Overweight people can reduce their weight by reducing their food intake. 1974 *Times* 4 Jan. 5/3 The overweight have become America's largest, least protected minority group. 1976 'L. Black' *Healthy Way to Die* ii. 12 How often must he have heard every possible comic remark about overweight women!

overweight, *v.* Add: **3.** *Pros.* To stress as in overweighting.
1940 J. R. R. Tolkien in Clark Hall & Wrenn *Beowulf & Finnesburg Fragment* p. xxxi, The second dip of B, C and the dip of D, E may not be overweighted.

overweighted, *ppl. a.* (Later examples.)
1927 Bowley & Stamp *Nat. Income 1924* 23 When we pass from the accidental grouping in the returns, overweighted by banks, to the whole numbers according to the census, the averages are brought up again to the medians. 1948 *Mod. Philology* XLVI. 86 Besides these.. light or weak verses there are, of course, many which are extra-heavy, 'overweighted', and many which seem over-long. 1949 *Penguin Music Mag.* Feb. 85 Such a combination provides a monotonous tone colour and the bass seems turgid and overweighted.

o:verwei·ghting, *vbl. sb.* [-ing¹.] The act or fact of giving or having too much weight; overloading, overload (see also quot. 1940).
1905 *Westm. Gaz.* 1 Sept. 2/3 The frightful over-weighting of the postman because of the flood of pictorial postcards. 1914 R. M. Jones *Spiritual Reformers 16th & 17th Cent.* xv. 289 Their gravest difficulty being an over-weighting of learning which they sometimes failed to fuse with their spiritual vision. 1940 J. R. R. Tolkien in Clark Hall & Wrenn *Beowulf & Finnesburg Fragment* p. xxxi, Overweighting and extension..are a means of including certain common but slightly excessive patterns in the metre; also of adding weight to the line where required, and of packing much significant word-material into a small space... It consists in replacing the dip by a long (subordinate) stress.

overwhe·lmment. [f. Overwhelm *v.* + -ment.] = Overwhelmedness; Overwhelmingness.
1866 W. M. Thomas tr. *Hugo's Toilers of Sea* II. vii. i. 90 There is a degree of overwhelmment which abstracts the mind entirely from its fellowship with man. 1960 *Angling Times* 9 Sept. 6/2 Our river may cause love at first sight, a sudden instantaneous overwhelmment with its manifest beauty.

o·verwind (-waind), *sb. Mining.* [Over- 29.] An instance of overwinding.
1892 *Trans. Fed. Inst. Mining Engin.* I. 58 With 'The Visor' applied to prevent a fast overwind, the arrangement in headgear to prevent the engines being started the wrong way, and good detaching hooks, disastrous overwinding seems an impossibility. 1929 H. Cotton *Electr. applied to Mining* x. 531 In the case of an overwind the pressure cylinder of the brake engine is opened to exhaust. 1958 I. C. F. Statham *Coal Mining Pract.* II. iv. 347 To protect against the aggravation of an overwind by starting the engine in the wrong direction, the overwind switches are connected through directional contacts in the master controller for the motor. 1973 *Daily Tel.* 31 July 1 An 'overwind' could have caused the descending cage, which works on the same winding system, to run out of control.

overwinder (s.v. Over-wind *v.*). Add def.: A device which guards against overwinding. (Further examples).
1912 McCulloch & Futers *Winding Engines* xv. 328 On overwinders working in pits in which double or treble decking takes place, a 'fourth attachment' is added. 1935 G. Poole *Haulage & Winding* xiii. 309 If..the cage is exceeding the predetermined speed at a particular point, the speed hook engages with the winged nut moving on the screw, and the overwinder comes into action. 1958 I. C. F. Statham *Coal Mining Pract.* II. iv. 343 The Whitmore overwinder is shown in Fig. 59, from which it may be seen that there is a separate swinging overspeed arm and travelling nut for each cage or skip.

over-winter, *v.* Add: **3.** *intr.* To live through the winter: said esp. of insects and fungi.
1933 *Jrnl. R. Hort. Soc.* LVIII. 227 The larvae [of apple sawfly] overwinter in cocoons in the soil. 1946 *Nature* 28 Sept. 454/2 The most probable causes of the disease [*sc.* tomato stem and fruit rot] are spores in propagating soil and spores which have overwintered in cracks of old cases. 1953 D. A. Bannerman *Birds Brit. Isles* I. 138 Either the Greenland or Hornemann's redpoll 'over-winters' in north-east Greenland. 1959 *New Biol.* XIX. 60 Temperate species [of midge] must normally overwinter as larvae. 1960 *Times* 9 July 9/2 Black spot overwinters on pieces of rose foliage. 1967 *Times* 23 Nov. 4/8 Two species of ticks and four of fleas..overwinter on the island [*sc.* Macquarie Island]. 1972 Swan & Papp *Common Insects N. Amer.* xxii. 589 *Culex* mosquitoes overwinter as fertilized females. 1977 J. L. Harper *Population Biol. Plants* xiii. 431 The butterfly lays its eggs in batches of 20–400 on the leaves of *Aster*, the larvae feed gregariously until the third instar when they disperse, over-winter and then start feeding again as solitary feeders.
4. *trans.* To keep (animals or plants) alive through the winter.
1945 *Scythe* Mar. 10 Primitive farming with low yields and no fodder crops on which to over-winter more than a minimum of domestic cattle. 1961 *Listener* 7 Dec. 992/3 The Chabaud and Nice types [of carnation] raised from August sowings would be safer if over-wintered in frames or under cloches. 1970 *Jrnl. R. Hort. Soc.* XCV. 358 The glaucous-leaved *Eucalyptus globulus*, sown in the previous July or August and overwintered under glass. 1976 *Evening Chron.* (Newcastle) 26 Nov., In this case the plants can be potted up into fairly large pots and overwintered in a cold frame until they can be planted out when the better weather comes along.
Hence **overwi·ntered** *ppl. a.*, **overwi·ntering** *vbl. sb.* and *ppl. a.*
1901 *U.S. Dept. Agric. Yearbk. 1900* 90 The overwintering crop. 1916 J. P. Lotsy *Evolution by Means of Hybridization* vii. 90 This experiment is yet in progress with the overwintered F_1 plants. 1923 *Glasgow Herald* 17 July 7 Preparations [are to be] made for a probable overwintering there next year. 1925 *Jrnl. Agric. Res.* XXXI. 1 (*title*) Further studies on the overwintering and dissemination of cucurbit mosaic. 1936 *Amer. Bee Jrnl.* LXXVI. 452 (*title*) Influence of pollen reserves on the surviving population of over-wintered colonies. 1958 *New Biol.* XXVI. 33 A lateral bud grows out from the corm [of the bulbous buttercup] in the autumn and gives rise to an overwintering rosette of foliage leaves. 1963 *Field Archaeol.* (Ordnance Survey) (ed. 4) 120 Overwintering sites may not have had very strong defences since they made use of natural obstacles and did not look far beyond the needs of a season. 1967 *Oceanogr. & Marine Biol.* V. 417 The spawning and over-wintering concentrations of Bank herring. 1971 *Farmer & Stockbreeder* 23 Feb. 45 (Advt.), Dose your overwintered cattle with 'Helmatac' now, and remove the winter worm burden. 1973 *Times* 18 June 16/2 The overwintered wild oat can produce at least twice as much seed to cause future trouble as those that start their lives in the spring. 1974 A. Dillard *Pilgrim at Tinker Creek* viii. 135 Look at an overwintering ball of buzzing bees.

o·ver-wood. [Over- 1 d.] The layer of vegetation formed by the tallest trees in a forest. Cf. *overstorey.
1889 W. Schlich *Man. Forestry* I. iii. 202 In fertile low lands Oak appears as overwood and underwood. 1928 R. S. Troup *Silvicultural Syst.* iv. 39 Silviculture may require the overwood to be retained for seeding purposes or for the protection of the young crop. 1969 *Gloss. Landscape Work* (*B.S.I.*) v. 34 *Overwood.* 1. Large free-growing trees growing above coppice. 2. The upper layer of a two-storeyed forest.

overwork, *v.* Add: **II. 3. a.** Also *transf.* and *fig.*
c 1878 G. M. Hopkins *Loss of Eurydice* in *Poems* (1967) 72 No Atlantic squall overwrought her. 1922 *Times Lit. Suppl.* 12 Oct. 642/2 Gobineau has carefully avoided any such threadbare device as that of the missing heir so overworked by Scott. 1951 T. Sterling *House without Door* iii. 35 Wall Street—you know that phrase is overworked. 1963 P. Phillips in Sissons & French *Age of Austerity* vi. 148 There was a New Look in daffodils.. in housing..in furniture... The phrase was disastrously over-worked. 1977 *Times* 7 Apr. 20/4 The episode is characteristic (the one word he overworks in this book).
overworked *ppl. a.* (earlier example); also *transf.* and *fig.*
1854 C. M. Yonge *Heartsease* II. iii. xiii. 292 Violet.. weak, anxious, and overworked. 1959 M. Summerton *Small Wilderness* i. 5 That over-worked truism about the wife being the last to know. 1978 *Times* 14 Jan. 14/1 The phrase 'low profile'..an over-worked image almost evacuated of meaning.

o·verworld. [Over- 1 d.] **1.** The celestial or immaterial world.
1858 [see Over- 1 d]. 1905 V. McNabb *Oxf. Conf. on Faith* 33 We would then represent our agnostic as saying .. 'We are quite certain that there are natural forces and a great world of nature, of which we are a part. But we cannot get beyond this world. There may be an over-world, or an immanent world.' 1966 *Punch* 1 June 803/1 Since West Indies cricket was baptised..in 1928, it has produced players fit..to play in some overworld a representative company of cricket immortals. 1973 *Times Lit. Suppl.* 2 Mar. 249/2 Krishnamurti at thirteen found himself translated..into the theosophical over-world or under-world inhabited by the Lords Maitreya and Koot Hoomi on the astral..plane.
2. The terrestrial world, the earth, land, viewed from beneath water.
1911 Beerbohm *Zuleika D.* xix. 290 He floated up. There was air in that over-world. 1958 L. Durrell *Mountolive* i. 15 Water-tortoises and frogs and sliding fish—a whole population disturbed by this intrusion from the overworld.
3. The community of conventional, law-abiding citizens, as opposed to the 'underworld'.
1938 F. D. Sharpe *Sharpe of Flying Squad* i. 13 The difference between the Underworld and the Overworld folk is that one lot works for a living; the other 'acquires' wealth and regards toil as sin. 1950 [see *Mr. Fix* s.v. *Mr. 2 e]. 1959 *Encounter* May 29/2 The counter-society or underworld is, like the society or overworld which has expressed it from its own body, class-ridden.

overwrap, *v.* Add: (Further examples.) Hence **overwra·pping** *vbl. sb.*
1959 *Times* 12 Aug. 15/6 Rigidity is essential for a material intended for overwrapping. 1967 C. R. Oswin in F. A. Paine *Packaging Materials* xix. 262 This is particularly helpful when bulk overwrapping blocks of small packets. 1972 *Daily Tel.* (Colour Suppl.) 12 May 13/2 Overwrap anything large like poultry with mutton cloth for extra protection [in the freezer]. 1972 Briston & Neill *Packaging Managem.* ix. 180 Regenerated cellulose film, commonly known as 'Cellophane'..is the traditional material used for the overwrapping of products such as biscuits, cigarettes and chocolate.

overwrap (ōu·vəræp), *sb.* [Over- 8 c.] A flexible wrapping fitted over packaged goods.
1956 *Visible Packaging of Flour Confectionery* (British Cellophane Ltd.) 3 Pastries should not be included in packs which have an overwrap of fully-moistureproof film. 1969 L. S. Mounts in W. R. R. Park *Plastics Film Technol.* v. 122 The greater stiffness [of medium density polyethylene films] improves the machinability for overwrap applications.

overwrite, *v.* Add: **1. c.** *Computers.* To place new data in a section of memory and destroy what is already there: used with the old data or the location as obj.
1959 [implied in overwriting *vbl. sb.* below]. 1962 *Gloss. Terms Automatic Data Processing* (*B.S.I.*) 25 The previous data is said to be overwritten in this process. 1970 [see *overlay *sb.* 6]. 1972 *Computer Jrnl.* XV. 200/2 Only one field, which contains an entry stating who is allowed to overwrite the file, may be altered by the programmer. 1973 C. W. Gear *Introd. Computer Sci.* vi. 245 Storing into a data area will overwrite another piece of data.
3. (Further example.)
1933 W. E. Orchard *From Faith to Faith* v. 67 It did not seem to me to matter much whether the Book of Isaiah had one or more authors, or the Pentateuch had been overwritten by different hands much later than Moses.
4. a. (Further example.) **b.** (Earlier example.)
1815 Jane Austen *Let.* 31 Dec. (1952) 449 It encourages me..to believe that I have not yet..overwritten myself. a 1889 Browning *Let.* in F. L. Lucas *Tennyson* (1957) iii. 21, I have written too much, my dear Mr. Gosse; I have over-written; I have written myself out.
d. *trans.* (freq. in *ppl. adj.* and *vbl. sb.*). To write too elaborately or ornately; to write in a high-coloured, over-rich style.
1923 F. M. Ford *Let.* 15 Oct. (1965) 155, I have by me a story of his that I don't like *much.* I *might* print it—but it is extremely over-written and..the extravagance I least like is over-writing. 1931 *Daily Express* 31 Jan. 8/2 A fortnight ago an over-written but under-nourished play called 'Colonel Satan' broke the Haymarket record for brevity of life. 1937 C. S. Lewis *Let.* 8 Mar. (1966) 158 Have you read F. L. Lucas' *Decline and Fall of the Romantic Ideal?* Hideously over-written in parts, but well worth reading. 1947 —— in *Ess. presented to Charles Williams* p. vii, In the earlier stories..there was a good deal of over-writing, of excess in the descriptions and, in dialogue, of a false brilliance. 1968 [see *Nabokovian *a.*]. 1977 *Times* 28 Apr. 22/3 This book..is..so over-written at times it irritated me.
Hence **overwri·ting** *vbl. sb.*, **overwri·tten** *ppl. a.*
1850 J. Miley *Hist. Papal States* I. 8 This palimpsest, or multifariously over-written document, which we have thus discovered the superficies of the Papal territory to be. 1939 H. J. Massingham *Eng. Countryside* iv. 73 These are the legitimate South Downs, the over-written and exploited range. 1949 W. G. Hoskins *Midland England* i. 3 Stone that gives to not a few Midland towns and villages a beauty unsurpassed by the most over-written Cotswold places. 1956 N. R. Ker *Pastoral Care* 23 Besides additions, the text contains erasures, some with and many without overwriting... Overwriting where it occurs is mostly in the main hand. 1959 M. H. Wrubel *Primer of Programming* v. 113 A common error made by beginners is to erase program steps by storing data in the same location. An error of this kind, called 'overwriting', is easily detected.

overwrou·ghtness. *rare.* [-ness.] An overwrought condition.
1923 W. Deeping *Secret Sanct.* xxii. 229 He..poured out a glass of white wine for Stretton, sensing the man's overwroughtness, and noticing the tense mouth and the troubled eyes.

ovibos (ōu·vibǫs). [mod.L. (H. de Blainville 1816, in *Bull. Sci. Soc. Philomatique Paris* 76), f. L. *ovis* sheep + *bōs* ox, as the animal was considered to represent a type intermediate between the sheep and the ox.] A small, stocky ruminant of the monotypic genus so called, bearing long, shaggy, dark brown fur and native to Arctic regions of North America and Greenland; = Musk-ox.
1903 R. Lydekker *Mostly Mammals* 287 No objection can be taken to the prefix 'musk'..yet the English title 'ox' is in the highest degree misleading, while the technical 'Ovibos', which suggests characters intermediate between the oxen and the sheep, is equally unsatisfactory. 1921 V. Stefánsson *Friendly Arctic* 342 We found the ancient and far-decayed skull of a female ovibos. *Ibid.* 582, I shot two ovibos as all we needed out of the fifteen or twenty seen. 1925 *Chambers's Jrnl.* 14 Feb. 167/1 Here ..will roam large herds of the domesticated musk-ox or ovibos. 1929 *Encycl. Brit.* II. 306/1 With the ovibos domesticated, the potentialities of the Arctic will be greater. *Ibid.*, Ovibos beef is indistinguishable from ordinary domestic beef.

ovicide². [f. Ovi-¹ + -cide.] An agent that kills eggs, esp. those of insects. Hence **ovici·dal** *a.*²
1930 *Jrnl. South-Eastern Agric. College, Wye* XXVII. 147 Ovicides, contact insecticides and grease-banding.. having been tried. 1932 *Ibid.* XXX. 63 (*heading*) Studies on the ovicidal action of winter washes. 1961 *Lancet* 12 Aug. 371/1 AzUR given orally is an effective ovicide in the mouse, terminating early pregnancy at dosage levels which produce no detectable toxic effect in the mother. *Ibid.*, Different batches vary somewhat in activity, and their ovicidal activity and their toxicity are not necessarily related. 1967 *New Scientist* 20 Apr. 154/2 The experiments dramatically illustrate the effectiveness of juvenile hormone as an ovicide, as well as an insecticide.

ovigenetic (ōu:vidʒĕne·tik), *a.* [f. Ovi-¹ + -genetic.] = Ovogenetic *a.*
1908 *Lancet* 23 May 1495/2 The spermatogenic and ovigenetic cells of the sexual glands in higher animals.

oviposition. Add: (Later examples.) So **ovipositional** *a.*, of or pertaining to oviposition.
1931 *Jrnl. Morphol.* LI. 22 The females [of Orthoptera] usually abandon them [*sc.* the eggs] as soon as oviposition is complete. 1965 Lee & Knowles *Animal Hormones* ii. 28 There is evidence that LTH is important for oviposition in some of the fishes. 1971 *Nature* 13 Aug. 484/1 Work is in progress..making use of the parasites' ovipositional response as a bioassay for the kairomone. 1973 *Ibid.* 30 Nov. 270/1 After oviposition these juveniles [of the nematode, *Deladenus siricidicola*] escape from the egg and enter a free living phase.

ovocyte (ōu·vosəit). *Biol.* [a. G. *ovocyte*: see *oocyte.] = *oocyte.
1905 *Jrnl. Acad. Nat. Sci. Philadelphia* XIII. 9 In a young [ascidian] ovocyte..there is no trace of yolk. 1941 J. F. Nonidez *Histol. & Embryol.* 83 After synapsis the female germ cells are known as primary ovocytes. 1967 *Nature* 30 Dec. 1315/2 *Xenopus* ovocytes have hundreds of nucleoli and *Asterias* ovocytes only one.

ovoflavin (ōuvoflēi·vin). *Biochem.* [a. G. *ovoflavin* (R. Kuhn et al. 1933, in *Ber. d. Deut. Chem. Ges.* LXVI. 318): see Ovo- and *flavin 2.] Riboflavin found in egg-white.
1933 *Chem. Abstr.* XXVII. 3480 The ovoflavin crystallizes from 2N AcOH in brown-orange needles. 1943 [see *hepatoflavin* s.v. *hepato-].

ovoglobulin (ōuvoglǫ·biŭlin). *Biochem.* Also † ovi-. [ad. F. *ovoglobuline* (Corin & Berard 1889, in *Arch. de Biol.* IX. 12): see Ovo- and Globulin.] A globulin present in egg-white.
1889 *Jrnl. Chem. Soc.* LVI. 1075 Their coagulation temperatures are:—oviglobulin α at 57·5°; oviglobulin β at 67°. 1905 [see *ovalbumen, -in]. 1959 *New Biol.* XXX. 16 Although the white of the egg contains both albumin and globulin these are chemically different from the blood proteins of the same name and they are usually distinguished by means of a prefix: ovalbumin and ovoglobulin.

ovomucin (ōuvomiū·sin). *Biochem.* [f. Ovo- + Mucin.] A water-insoluble mucoprotein in egg-white. Cf. *ovomucoid.
1898 A. Eichholz in *Jrnl. Physiol.* XXIII. 163 A new constituent of egg-white—ovomucin. I have to call attention to another constituent which has so far not been definitely described as a constituent of white of egg. This substance possesses all the properties assigned generally to mucin, and differs from the mucoids in being insoluble in distilled water. 1936 *Poultry Sci.* XV. 350/2 This change, which involves a decrease in the water insoluble fraction (ovomucin)..accompanies the loss of viscosity in the egg white. 1972 *Agric. & Biol. Chem.* XXXVI. 947 The carbohydrate content of ovomucin gel (B) obtained from the eggs stored in an atmosphere of carbon dioxide was higher than that of ovomucin gel (B) obtained from the eggs stored in air.

ovomucoid (ōuvomiū·koid). *Biochem.* [ad. G. *ovomukoid* (C. Th. Mörner 1894, in *Zeitschr. f. physiol. Chem.* XVIII. 526): see Ovo- and Mucoid *a.*] A water-soluble mucoprotein in egg-white (also called *ovomucoid* α); also (*ovomucoid* β) = *ovomucin.
1894 *Jrnl. Chem. Soc.* LXVI. i. 264 Ovomucoid... This is the name given to a proteid-like substance which can be obtained from white of egg, after boiling, acidifying, and filtering to separate albumin and globulin. 1938 K. Meyer in *Cold Spring Harbor Symp. Quant. Biol.* VI. 91 A logical classification of the hexosamine-containing compounds should be based on the nature of the carbohydrate radical... Table I shows the classification which we propose...A. Mucopolysaccharides...B. Glycoproteins, containing neutral mucopolysaccharides of unknown composition. a. ovomucoid-α (formerly called ovomucoid). b. ovomucoid-β (formerly called ovomucin). c. serum mucoid, serum glycoid. d. globulins... e. pregnancy urine hormone. *Ibid.* 100 A very viscous fraction precipitating out by dilution of egg white with water has been called 'ovomucin'. In order to avoid the term 'mucin' we propose the name ovomucoid-β. 1954 A. White et al. *Princ. Biochem.* viii. 142 Well-defined soluble mucoproteins have been obtained from egg white (ovomucoid α), from serum.., and from human pregnancy urine. *Ibid.* 143 Insoluble mucoids have been obtained from egg white (ovomucoid β), chalazae, submaxillary glands, and vitreous humor. 1972 *Jrnl. Biol. Chem.* CCXLVII. 6450/2 The four proteins studied in this paper, ovalbumin, conalbumin, ovomucoid, and lysozyme,..together account for 85 to 90% of the egg white proteins.

ovonic (ovǫ·nik), *a. Electronics.* Also **Ovonic.** [f. *ov-* (in the name of Stanford R. Ovshinsky (b. 1922), the U.S. physicist and industrialist who discovered the property) + *-onic* (in *electronic*).] Pertaining to, involving, or utilizing the property of certain amorphous semiconductors of making a rapid, reversible transition from a non-conducting to a conducting state on the application of an electric field stronger than some minimum value.
1966 *Electronics* 19 Sept. 191/2 This month, Ovshinsky ..was able to let the cat out of the bag. His company.. calls the device by its trade name, Ovonic. 1968 S. R. Ovshinsky in *Proc. Electronic Components Conf.* (Washington) 313 The Ovonic Threshold Switch (OTS) and the Ovonic Memory Switch (OMS) are described. 1968 *Daily Tel.* 12 Nov. 22/6 A company founded by Mr. Ovshinsky to develop glass semi-conductors is producing 150,000 'ovonic' devices a day. These are in the form of glass film a twentieth of the thickness of a human hair. 1968 *Economist* 30 Nov. 45/1 The big electronics research laboratories have been sceptical and even a bit cross that Mr Ovshinsky has put his name on to the 'ovonic devices' and the 'Ovshinsky effect'. Bell Telephone Laboratories declared that one of its scientists presented the first paper describing glass semi-conductors six years ago. 1970 S. R. Ovshinsky in *McGraw-Hill Yearbk. Sci. & Technol.* 359/1 Since ovonic switches have a number of unique characteristics, they are applicable to a wide range of uses. 1971 R. G. Neale et al. in W. B. Riley *Electronic Computer Memory Technol.* vii. 192/2 Each cell in the memory consists of an Ovonic amorphous semiconductor device and an isolating diode in series on a silicon substrate. 1973 *IEEE Trans. Electron Devices* XX. 190/1 A very high-speed electrically alterable read-only memory using Ovonic memory cells..has been designed and tested for use in the terminal multiplexor.
Hence **ovo·nics** *sb. pl.* (const. as *sing.*) [-ic 2], the study and application of ovonic effects and devices.
1968 *Economist* 30 Nov. 45/2 There are problems on the ovonics side too. The reliability of the glass semi-conductors is still suspect. 1969 *Sci. Jrnl.* Aug. 78/3 Other fields of application for Ovonics lie in a.c. control where the bidirectionality of Ovonic switches will be of prime importance.

ovovitellin (ōuvo‚vite·lin). *Biochem.* Also **-ine.** [f. Ovo- + Vitellin.] The vitellin of egg-yolk.
1906 *Westm. Gaz.* 29 Aug. 2/2 The hæmoglobin of the blood of the chick is formed by certain cyanic ferruginous compounds which are found associated with ovovitelline. 1950 Sumner & Myrbäck *Enzymes* I. i. 20 It is likely that such food proteins as casein and ovovitellin are elaborated to serve as sources of amino acids for the suckling mammal or for the incubating embryo.

ovoviviparity. (Later examples.)
1931 H. R. Hagan in *Jrnl. Morphol.* LI. 18 Ovoviviparity. That type of viviparity in which the egg contains sufficient yolk to nourish the embryo until hatching. 1965 Lee & Knowles *Animal Hormones* iii. 62 The term ovoviviparity, in preference to viviparity, is sometimes applied to the birth of live young in reptiles, as it is claimed that the nutrition of the foetus in this class depends on the egg-yolk and not on the placenta. 1973 *Nature* 27 Apr. 617/1 Morphological adaptations to ovoviviparity would..include some means of internal fertilization.

Ovshinsky (ǫvʃi·nski). *Electronics.* The name of S. R. *Ovshinsky* (see *ovonic *a.*), used *attrib.* in the sense 'ovonic'.
1966 *Electronics* 19 Sept. 192/3 The potential versatility of the 'Ovshinsky effect' was his reason for secrecy. 1968 *Daily Tel.* 12 Nov. 22/6 Mr. Stanford Ovshinsky, 45, a scientist, from Troy, Michigan, explains in the journal the theory behind his discovery of 'The Ovshinsky Effect', which is that amorphous glasses have qualities similar to those of semi-conductors. 1971 *New Scientist* 1 Apr. 29/2 Amorphous semiconductors—the well-known 'Ovshinsky devices'—are being produced and used in the United States..in a 256-bit 'read-mostly' computer memory.

ovulate, *v.* Add to def.: Said of an organism and of a Graafian follicle. Hence of an ovum: to be discharged. (Examples.)
1888 H. N. Martin in B. C. Hirst *Syst. Obstetr.* I. 92 Women who have never menstruated have borne children: they must have ovulated without menstruating. 1928 *Q. Jrnl. Exper. Physiol.* XVIII. 197 The group of follicles which will ovulate at an œstrus period undergo nearly their whole maturation growth after the time when the œstrus stimulus becomes operative. 1959 *New Biol.* XXX. 21 The ovary at the time of ovulation looks like a bunch of grapes, but with the difference that all the grapes, or follicles, are of different size. The largest will ovulate first. 1968 D. W. Wood *Animal Physiol.* xiii. 303 Many [mammals] ovulate spontaneously as part of the oestrus cycle, but others require the stimulus of copulation before they will ovulate. Such induced ovulators include the ferret, the cat and the rabbit. 1974 *Biol. of Reproduction* X. 199/1 Amphibia are good subjects for *in vitro* studies of ovulation, for mature ovaries can be maintained in simple salt solutions at room temperature and can be induced to ovulate by the addition of crude extracts of..pituitary glands.
b. *trans.* To discharge (an ovum).
1924 *Amer. Jrnl. Physiol.* LXIX. 587 The periodic formation of new ova in the adult ovary makes it unnecessary to consider that ova in the immature ovary lie dormant until they either degenerate or are ovulated. 1934 *Science* 16 Nov. 462/1 Our observations would indicate that it is not necessary for the egg to be ovulated directly into the infundibulum. 1973 *Nature* 13 July 72/1 One species, the plains viscacha (*Lagostomus maximus*), has a bizarre reproductive habit in that it ovulates between 200 and 800 eggs from ovaries that look more like those of a fish than a mammal.
Hence **o·vulated** *ppl. a.*, **o·vulating** *vbl. sb.*
1910 F. H. A. Marshall *Physiol. Reproduction* iv. 133 The second spindle is formed at about the ovulating stage, and the second polar body is discharged into the interior of the Fallopian tube. 1940 *Record of Proc. Amer. Soc. Animal Production* 302 (*heading*) Potential fertility of artificially matured and ovulated ova in cattle. 1959 *New Biol.* XXX. 21 The ovulating hormone..causes the follicle to split along the line of the stigma, thus releasing the mature yolk from the ovary. 1970 *Sci. Jrnl.* June 94/1 This can be righted by supplying the hormone artificially and the ovulated eggs can then be fertilized. 1972 *Amer. Jrnl. Veterinary Res.* XXXIII. 1589, 25 ewes were given estradiol to induce release of endogenous ovulating hormone.

ovulation. (Earlier example.)
1848 C. D. Meigs *Females & their Diseases* xxxi. 403 The ovulation is marked by a natural return of the mensual hemorrhage.

ovulator (ōu·viulēitəɹ). *Physiol.* [f. Ovulat(e *v.* + -or.] An animal that ovulates (in a specified way).
1961 *Recent Progress Hormone Res.* XVII. 120 Both the cat and the rabbit are induced ovulators. 1968 [see *ovulate *v.* a]. 1971 J. Z. Young *Introd. Study Man* xv. 194 In the doe rabbit and domestic cat ovulation occurs only after copulation... Such animals are..said to be reflex ovulators. It has recently been shown that the rat, long considered to be a spontaneous ovulator, can be stimulated by coitus..to ovulate earlier than it would do otherwise.

ovulatory (ōuviulēi·təri), *a.* [f. as prec. + -ory².] Of or pertaining to ovulation.
1931 *Jrnl. Mammalogy* XII. 139 The sterility of the animals is due to the absence of an ovulatory cycle. 1972 *Endocrinology* XCI. 1253/1 The administration of various drugs on the afternoon of proestrus blocks the ovulatory surge of pituitary ovulating hormone. 1977 *Daily Mirror* 15 Mar. 7/4 In some cases when women come off the Pill, we can stimulate the return of periods with an ovulatory drug.

ow (au), *int.*² [A natural exclamation: cf. O *int.*, Oh *int.* and varr., Ow, ou *int.*, and Ouch *int.*¹] An exclamation expressing sudden pain.
1919 G. B. Shaw *Great Catherine* iii. 146 (*Claire twists herself loose; turns on him; and cuffs him furiously*) Yow—ow! Have mercy, Little Mother. *Ibid.* iv. 155 Ow! Youve nearly pulled my teeth out. 1926 —— *Translations & Tomfooleries* 239 *Reginald.* Oh! Oh! Oh! The crocodiles! Stop! Ow! Oh! 1969 D. E. Westlake *Up your Banners* (1970) xviii. 121 She threw another hammerlock on me. 'Ow,' I said. 1976 R. B. Parker *Promised Land* (1977) xi. 60, I..hugged her. 'Ow,' she said. I eased up a little on the hug.

owdacious (audēi·ʃəs), *a. colloq.* (orig. *U.S.*) [? A 'portmanteau' blending of Audacious *a.* and Outrageous *a.*] Impertinent, mischievous, bold. Hence **owda·ciously** *adv.*, outrageously.
1846 in Bartlett *Dict. Amer.* (1848) 243 He had a daughter Molly, that was the most enticin', heart-distressin' creature that ever made a feller get owdacious. 1847 in *Ibid.* 243, I was never so owdaciously put out with the abominable abolitionists before. 1857 C. M. Yonge *Dynevor Terr.* I. vi. 81, I wonder you aren't ashamed of yourselves, and the family in such trouble! Downright owdacious! 1947 W. de la Mare *Coll. Stories for Children* 98 Some crabbed old woman said they were owdacious, or imperent, or mischeevious.

owe, *v.* Add: **2.** *Sports.* To be under an obligation to give one's opponent in a match (a number of strokes or points) as a handicap.
1904 J. P. Paret *Lawn Tennis* 345 *Owe-fifteen* (*thirty or forty*), a term used in handicap play to indicate that one player must make one (fifteen), two (thirty), or three (forty) points in each game before he begins to score. 1908 *Daily Chron.* 24 Aug. 9/3 Mr. F. Scarf..owing one stroke, beat Mr. R. C. Oppenheimer,..(handicap 15), by 7 holes up and 5 to play.
2. b. (Later examples.)
1970 'E. Queen' *Last Woman* ii. 135 'She'll come', Newly said grimly. 'After that yarn of hers, she owes me.' 1972 D. Anthony *Blood on Harvest Moon* i. 17 'Another job.'..'I couldn't turn this one down,' I said. 'I owe the lady.'

Owen² (ōu·ən). The name of E. E. *Owen* (1915–49), Australian inventor, used *attrib.* or alone to designate a sub-machine-gun invented by him.

1958 D. P. Mellor *Role of Sci. & Industry* xv. 329 The Owen gun was an Automatic firearm of the usual recoiling breech bolt type, with a fire control member cooperating directly with the trigger. 1961 D. Dexter *New Guinea Offensives* ii. 51 All sections testing the Owen preferred it to the Tommy-gun. 1965 *Austral. Encycl.* VII. 34/1 In the field of military inventions, one of the best-known is the Owen sub-machine gun, patented in 1941 by its inventor, E. E. Owen. 1967 'E. Lindall' *Time too Soon* iii. 32 An Owen gun slung across his body. 1970 M. Kelly *Spinifex* viii. 132 The Owen gun best and only friend.

Owenism. (Earlier examples.)
1830 *Mechanic's Press* (Utica, N.Y.) 10 June 254/2 What a precious compound of almost all that is unprincipled, is here presented:—Agrarianism, Owenism. 1833 J. S. Mill in *Tait's Edin. Mag.* III. 352 This doctrine..might easily have misled a less expanded mind..into the vagaries of Spenceanism and Owenism.

Owenite. Add: (Earlier and later examples.) Also *attrib.*
1829 Southey *Sir Thomas More* I. vi. 144 But wherefore do you think that the Owenite scheme is likely to be carried into effect only by sectarian agency? 1831 E. G. Wakefield *Householders in Danger from Populace* 9/2 The desperadoes..may be divided into two classes, which I shall designate as *Huntites* and *Owenites*. *Ibid.* 10/1 The Owenites..are bent on the overthrow of all existing laws. 1836 'Bronterre' tr. *Buonarroti's Hist. Babeuf's Conspiracy* ii. 363 My readers of the Owenite or co-operative school will be forcibly reminded..of the many doubts.. addressed to Robert Owen, touching the possibility of reducing his system to practice. 1843 Mill *Logic* II. vi. ii. 485 If the Owenite stops here, he is in a position from which nothing can expel him. 1919 M. Beer *Hist. Brit. Socialism* I. ii. ii. 131 George Mudie, an Owenite and journalist. 1950 G. B. Shaw *Farfetched Fables* Pref. 81, I am not stigmatizing all Owenites, Marxists, and Darwinists as immoral. 1956 W. H. G. Armytage in D. L. Linton *Sheffield* 205 This was an Owenite centre (Robert Owen had first visited the town on 30 December 1833), where a considerable amount of adult education was undertaken. 1975 V. Cunningham *Everywhere spoken Against* vii. 186 William Taunton..was a physical-force Chartist, an Owenite Socialist, manager of Coventry's first Co-operative Store.

owl, *sb.* Add: **2. b.** *Brown Owl*, the name given to the adult leader of a Brownie Guides pack; *Tawny Owl*, a Brown Owl's assistant.
1918 R. S. S. Baden-Powell *Girl Guiding* i. ii. 17 The Brown Owl (that is, the leader of the Pack) takes her place by the toadstool. *Ibid.* 21 Each Pack is under the charge of a grown-up leader—the Brown Owl. 1921 in —— *Brownies* (ed. 2) 60 A Brownie Pack consists of not less than two Sixes..under a Brownie Guider, who is called the Brown Owl, and her assistant the Tawny Owl. 1932 [see *pack *sb.*¹ 3 d]. 1950 *Oxf. Jun. Encycl.* IX. 254/1 Brownies are divided into 'Packs' of 18–24 children, under the leadership of two adult leaders, known to the Brownies as 'Brown Owl' and 'Tawny Owl'. 1968 M. Finch *Eye with Mascara* xiv. 149 She sounded like Brown Owl chivvying her Brownies. 1973 *Brownie* 10 Jan. 7/1 Our Pack has a membership of 20 keen Brownies... Our meetings are held at Brown Owl's house... Tawny Owl is a Sister at Hetune and walks all the way to our meetings. 1977 *Guider* July 331/2 She was a Guider in this Company, a Brown Owl of the 1st Teignmouth Pack, a Sea Ranger Skipper and a District Commissioner.
7. *owl-cote, -down, -hoot, -light, -time*; parasynthetic and similative, as *owl-dark, -dusk, -dusked, -eyed* (later example), *-headed* (later examples), *-soft, -wise* adjs.; also *owl-wise* adv.; **owl bus** *N. Amer.*, a bus running during the night; **owl car** *N. Amer.*, a tramcar running during the night; **owl jug**, a porcelain jug shaped like an owl; **owl-train** (earlier and later examples); **owl trolley** = *owl car.
1947 *Sun* (Baltimore) 24 June 10/2 Operators of all other all-night busses and trolleys have been directed to connect with the owl bus, just as they did with the..owl trolley. 1975 *Washington Post* 26 Dec. A 22/3 Chances are the owl bus riders will..simply fade away.., just another segment of the population abandoned by the Metro system. 1889 Farmer *Americanisms* 405/2 *Owl-Car*, a tram-car plying late into the night. 1904 *N.Y. Even. Post* 7 May 1 The driver of an 'owl car' that rattled eastward on Spring street. 1911 *Daily Colonist* (Victoria, B.C.) 21 Apr. 12/7 An 'owl car' service has been inaugurated by the B.C.E.R. Company at Vancouver. 1947 *Sun* (Baltimore) 24 June 10/2 The No. 17 owl or all-night car has been supplanted by a No. 28 bus. 1863 'G. Hamilton' *Gala-Days* 107 For the substantial stone city..turns out to be a miserable little dirty, hutty, smutty, stagnant owl-cote. 1920 E. Sitwell *Wooden Pegasus* 41 In owl-dark garments goes the Rain. 1924 —— *Sleeping Beauty* ii. 18 Smoothing the dusky dawn's owl-down. 1928 —— *Five Poems* 18 That sang sweet country songs in owl-dusked leaves:..but time drifts owl-dusk o'er the brightest eyes. 1925 F. Scott Fitzgerald *Great Gatsby* iii. 45 A..man, with enormous owl-eyed spectacles. 1960 Auden *Homage to Clio* 55 Steatopygous, sow-dugged and owl-headed. 1968 *Listener* 11 July 59/2 A Chou Dynasty bronze bell with tiger-handle and an owl-headed drinking vessel. 1938 W. de la Mare *Memory* 11 Came owl-hoot From the thicket. 1925 B. Rackham tr. *E. Hannover's Pott. & Porc.* I. 554 The origin of the faience owls..is fully discussed..by Walter Stengel..in the *Jahrbuch für Kunstwissenschaft* for 1924, p. 26. He gives good reason for regarding these owl-jugs,..as being of Nuremberg origin. 1936 [see *double eagle 2]. 1960 R. G. Haggar *Conc. Encycl. Cont. Pott. & Porc.* 339/1 Faience owl jugs were made at Nuremberg in Germany... Specimens dating between 1540 and 1560 are recorded by Rackham. 1934 *Owl light* [see *moth-light* s.v. *moth *sb.*¹ 3]. 1936 Dylan Thomas *Twenty-Five Poems* 42 Altarwise by owl-light in the halfway-house The gentleman lay graveward with his furies. 1924 E. Sitwell *Sleeping Beauty* xvi. 59 An owl-soft shadow falling over folly. 1953 W. de la Mare *O Lovely England* 32 Owl-soft his wings. 1972 R. Adams *Watership Down* xxiii. 159 By owl-time Bigwig and his helpers had scratched out a kind of lobby inside the entrance to one of the runs leading down from the wood. 1856 *N.Y. Herald* 8 Jan. 1/2 The 'Owl Train', due at Jersey City at five o'clock yesterday morning, did not arrive until afternoon. 1876 S. & A. Warner *Gold of Chickaree* 248 Must take the morning train. It's not quite an 'owl train'—but comes along, I believe, by eight o'clock. 1910 *N.Y. Even. Post* 22 Dec. 3 The engine of the 'owl train'—for by this term the one leaving New York after midnight is called..went off the track. 1947 *Los Angeles Times* 18 Jan. 1/1 (*heading*) 7 killed and 86 hurt in wreck of owl train. 1947 Owl trolley [see *owl bus* above]. 1906 Kipling in *Tribune* 16 Jan. 4/3 Jimmy..rolled his congested eye-balls, owl-wise. 1912 W. Deeping *Sincerity* xxxviii. 276 His round, lard-coloured, mildly owl-wise face. 1939 Joyce *Finnegans Wake* i. 78 The eternals were owlwise on their side every time.
b. *owl fly* (*b*) = *owl midge; **owl midge** = *moth-fly* (Moth *sb.* 3).
1932 Riley & Johannsen *Med. Entomol.* xiv. 195 The Psychodidae, popularly known as moth flies, owl flies, sand flies, or papataci, are minute dark-coloured insects whose body and wings are densely covered with hairs. 1951 Colyer & Hammond *Flies Brit. Isles* 84 The Owl Midges or Hairy Moth-flies are easily recognisable; they may often be seen on windows, where they either run actively with a curious, jerky gait or remain perfectly still. 1962 Gordon & Lavoipierre *Entomol. for Students of Med.* xx. 131 Flies belonging to the other three subfamilies [of Psychodidae]..are known as moth flies or owl midges.

owly, *a.* Delete †*Obs.* and add later examples.
1864 O. W. Norton *Army Lett.* (1903) 203 Last night I was out all night in the rain..and I feel owly to-day. 1873 C. G. Leland *Egyptian Sketch-bk.* 33 Up started a little, dark, old, owly, goblin, night-ghoul of a creature. 1978 R. Holles *Spawn* vii. 61 The round, slightly owly features of the woman in nursing uniform.
b. owly-eyed *a.* Also (*U.S. dial.*), intoxicated.
1900 *Dial. Notes* II. 47 *Owly-eyed*, intoxicated... Wise.

own, *a.* Add: **1.** (Further examples.) Phr. *to do one's own thing*: see *thing *sb.*¹
1841 Thackeray *Gt. Hoggarty Diamond* (1849) viii. 91, I would not have taken the lord mayor's own daughter in place of Mary with a plum for her fortune. 1931 M. Allingham *Police at Funeral* xiv. 200, I wonder if you would tell me in your own words how you came to have such a wound? 1962 L. Deighton *Ipcress File* 7 'Just tell me the whole story in your own words, old chap...' I was wondering whose words he thought I might have used. 1974 G. Butler *Coffin for Canary* ii. 47 If we were every man his own Hitchcock, we wouldn't need to go and see the films.
b. *ownest* (later examples).
1907 G. B. Shaw *Major Barbara* i. 253 *Lomax*: How is my ownest today? *Ibid.* 272 My ownest, there is no danger. 1922 Joyce *Ulysses* 352 Then mayhap he would embrace her gently..and love her, his ownest girlie, for herself alone. 1939 G. B. Shaw *Geneva* iii. 53 My ownest and bestest, you are a Dame of the British Empire.
c. (Later examples.) Also, *to be one's own woman.*
c 1374 Chaucer *Troylus* ii. 750, I am myn owene woman wel at ese. 1966 J. Potts *Footsteps on Stairs* (1967) i. 14 This final encounter with Vic was a necessary part of the ritual of release. With it behind her,..she was once more her own woman. 1969 *Guardian* 17 Feb. 2/7 He freely admitted that he had learned something from all the early masters... But..he was determined to be his own man. 1972 P. Dickinson *Lizard in Cup* xi. 164 His own personal desire..to be his own man, to act and conquer outside Caesar's provinces. 1974 *Times* 2 Nov. 4/8 Mr Brown insists he is not a liberal... In truth, Mr Brown is his own man. 1975 D. Bagley *Snow Tiger* xvi. 138 There'll be no strings. I'm my own woman, I am.
2. c. *own goal*: a goal (see Goal *sb.* 3) scored against one's own side. Also *fig.* (see quots.).
1947 *Sporting Mirror* 7 Nov. 10/3 Huddersfield were extremely unlucky to go under by an own-goal score to Charlton. 1952 *Times* 27 Dec. 8/1 Yesterday the Albion, with the help of two 'own goals', won a great game. 1962 *Punch* 11 Apr. 569/4 Mal holds the record for equalising own-goals. 1976 *Guardian* 11 Aug. 10/8 Two youngsters of Provisional IRA blown up by premature explosion of own bomb while crossing peace line... described as own goals by smiling Army press officers. 1976 *Norwich Mercury* 10 Dec. 8/6 With no one taking control J. Purling eventually left M. Warman stranded with a back header that lopped just under the crossbar for an own-goal. 1977 *Observer* (Colour Suppl.) 2 Jan. 12/2 The two men who had blown themselves up—'own goals' in the army's gruesome parlance. 1978 *Guardian* 30 Nov. 1/6 The Parliamentary scene was set last week for one of those gentlemanly arrangements which allows the opposition to have its say without actually scuppering the Government. Unfortunately, the Government scored an own goal.
3. *absol.* (Further example.)
1854 Thackeray *Newcomes* I. xxiv. 233 Her teeth [were] as regular and bright as Lady Kew's own.
c. *to call (a thing) one's own*: cf. Call *v.* 17 d; *to come into one's own*: to get possession of one's rightful property; to be properly esteemed; *to get one's own back*: see *get *v.* 55 b (quots. 1910–22); *on one's own*: delete (*slang* or *colloq.*) and add later examples.
1613–1857 To call a thing one's own [in Dict. s.v. Call *v.* 17 d]. 1902 J. Milne *Epistles of Atkins* iv. 63 His one thought how to 'get his own back'. 1912 T. Dreiser *Financier* vii. 71 The ready-made shoe—machine-made to a certain extent—was just coming into its own. 1917 A. G. Empey *Over Top* 302 *On your own*, another famous or infamous phrase which means Tommy is allowed to do as he pleases. An officer generally puts Tommy 'on his own' when he gets Tommy into a dangerous position and sees no way to extricate him. 1925 D. H. Lawrence *Phoenix II* (1968) 482 At night, when the silence of the moon, and the stars, and the spaces between the stars, is the silence of me too, then I am come into my own by night. 1929 —— *Ibid.* 580 For what does goodness mean? It means, in the end, being like everybody else, and not having a soul to call your own. 1930 A. Huxley *Let.* 18 Oct. (1969) 343 I've really had very little time to call my own. 1931 *Week-End Rev.* 24 Oct. 515/1 It looks as if the music of Jean Sibelius were at last coming into its own this winter. 1936 *Discovery* July 222 Electrical instruments will not come into their own until a large repertory of music has been composed specially for them. 1946 R. Allen *Home Made Banners* x. 115 But in these last moments each of them was on his own. 1969 *Listener* 24 July 109/1 At this point, alas, the Art Nouveau comes into its own. 1976 *National Observer* (U.S.) 18 Dec. 1/1 'Nixon spurred interest in this type of prosecution, but since he left we're more or less on our own,' says a lawyer who works part time on obscenity matters.
4. a. own brand, a class of goods marked with the name of the retailer instead of the manufacturer; also *attrib.*; **own category** *Psychol.*, a type of attitude test in which the subject is asked to select suitable categories into which to grade controversial statements and thereby reveals his own emotional involvement; also *attrib.*; **own-label** *attrib.*, of merchandise marked with a label showing the name of the retailer instead of the manufacturer; occas. (without hyphen) in non-*attrib.* position; **b. own-rooted** *a.* = *own-root.*
1970 *Times* 5 Feb. (Pedigree Dogs Suppl.) p. iii/6 A range of 23, some of which were 'own brands'. 1970 *Times* 16 Feb. (Food in Britain Suppl.) p. ix (*heading*) Own brands are money-savers. *Ibid.*, The principles of own-brand groceries date back to the turn of the century when stores such as Lipton and Home & Colonial did much of their own packaging. [1953 Sherif & Hovland in *Jrnl. Abnormal Psychol.* XLVIII. 135/2 Ultimately it may provide a means of utilizing the individual's own categorization of statements as a behavioral index of his stand on an issue.] 1961 —— & —— *Social Judgment* v. 118 (*heading*) Judgment of items with individual choice of categories—'own' categories. *Ibid.* 126 If future investigations bear out the promise of our results, it may prove feasible to order the stands of individuals on a controversial social issue through their placement of relevant items within their 'own' categories. 1970 *Jrnl. Gen. Psychol.* LXXXII. 147 The basic task for the Ss was that frequently used in cognitive complexity research, the free-sorting or own-categories technique. 1972 *Jrnl. Social Psychol.* LXXXVIII. 84 In the present study, the Q-sort variant known as the own-categories technique was used to investigate some effects of redundancy and congruence on judgement scales. 1973 N. Lemon *Attitudes & their Measurement* vii. 199 The development of the own categories procedure as a method of measuring involvement. 1961 *Economist* 11 Mar. 983/2 In the grocery and provisions trade, the larger multiples were almost all engaged in some food manufacture between the wars, supplemented by agreements for 'own-label' products from other manufacturers. 1969 *Times* 13 Mar. 23/3 Tesco and Woolworths have withdrawn supplies of 'Tuf' shoes and launched 'own label' brands. 1972 *Sunday Times* 31 Dec. 63/3 Lyons is particularly a supplier of supermarket own-label items and catering foods. 1975 *Times* 16 May 8/3 'Own label'..is the trade term for shops that sell groceries made by well-known companies but with the name of the shop, not of the maker, prominent on the packet. 1977 *Daily Tel.* 14 Jan. 1/6 Sainsbury's said most of its London shops were without supplies of its own-label bread, which it was still selling at 17p, but had normal supplies of proprietary loaves selling at 19p. 1915 M. E. King *Gothic Ruin & Reconstruction* 12 Let the renascent art blunder at first, as it must if it be own-rooted and not parasitic.

own, *v.* Add: **2. c.** Of hounds: to show recognition of (the scent of the quarry).
1781 P. Beckford *Thoughts on Hunting* xx. 255 Foxes will run the roads at..times, and hounds cannot always own the scent. 1838 T. Smith *Extracts Diary of Huntsman* v. 126 Owning a scent, when hounds throw their tongues on the scent. 1893 W. C. A. Blew *Radcliffe's Noble Sci. Fox-Hunting* ix. 161 A couple or two, or a single hound, may have come across and struck upon the scent of a fox which has shifted, unseen, across a ride. The scent in the stuff is too stale for them freely to own. 1954 J. I. Lloyd *Beagling* 142 Hounds own a scent when it is strong enough for them to speak to it. 1971 G. Wheeler *Year Round* 21 Now the kale comes really alive as hounds drive through it converging on Ladybird's corner. One after another they own her line.
5. c. *to own up* (earlier and later examples.)
1853 J. A. Benton *California Pilgrim* 55 However, you 'own up', and confess. 1858 S. A. Hammett *Piney Woods Tavern* 28 I'm willin' to own up that I'm ginerally considered to rather have a gift that way myself. 1861 *Harper's Mag.* Mar. 463/2 The English have long since resigned even the name of competitors,..as far as fishing on the Grand Bank is concerned... They have quit the field, 'owned up beat'. 1951 *People* 3 June 7/1 It will be difficult to find many regular backers who could truthfully own up to a good week at Epsom. 1966 *Listener* 10 Mar. 342/1 It is the usual thing to address the class sternly and demand that the culprit should 'own up'.

owner. Add: (Examples with sense 'a race-horse owner'.) *Slang.* The captain of a warship, barge, or other boat; also of an aircraft. So **ownere·ss,** the captain's wife.

1863 *Chambers's Encycl.* V. 428/2 The income of a jockey..is often very large: £1000 has frequently been given by a grateful owner. 1898 A. E. T. Watson *Turf* v. 124 The winner of a selling race has..to be sold by auction; the owner receives no more than the entered selling price. 1903 Kipling *Traffics & Discoveries* (1904) 49 I'm goin' to deviate to the owner's comfortable cabin direct. 1914 'Bartimeus' *Naval Occasions* iii. 21 That there launch precious near fouled the mark-buoy... Their owner sailing 'er too. 1916 G. Taylor *With Scott* 213 Scott was invariably known as The Owner, a naval term always applied to the captain of a warship. 1923 *Blackw. Mag.* Apr. 445/2 The Owner and Owneress have a very jolly little cabin. 1930 in C. Allen *Raj* (1977) ix. 123 Cricket. Owners, Trainers and Jockeys, Vs. Patrons, Stewards and Officials. 1943 C. H. Ward-Jackson *Piece of Cake* 45 Owner, the Commanding Officer, the captain of an aircraft. 1971 'D. Halliday' *Dolly & Doctor Bird* xiii. 192 Johnson slept for an hour. I left the wheel to go into the owner's cabin to rouse him. 1977 D. Francis *Risk* ii. 14 Binny, Tapestry's trainer, didn't want me on the horse. 'Not in the Gold Cup,' he'd said..when the owner had proposed it.

b. appositive, as *owner-breeder, -driver, -manager, -occupant, -occupier, -operator;* also *owner-occupation, -occupiership; owner-driven, -managed, -occupied* adjs.

1937 E. Rickman *On & off Racecourse* i. 4 The most successful owner-breeders are in the game because they have a genuine love of the thoroughbred and of the sport. 1971 *Daily Tel.* 20 Oct. 17/4 Sir Humphrey, the fourth baronet, owner-breeder of Parthia, the 1959 Derby winner. 1919 *Honey Pot* I. iii. 4 (Advt.), For immediate delivery. Daimler 57-hp special..owner driven. 1960 *Amer. Speech* XXXV. 240 In truckers' language a 'gypsy' is an owner-driven truck. 1918 A. Bennett *Pretty Lady* xxxi. 214 The interior of the cab,..was ornate with toy-curtains..to indicate to the world that he was an owner-driver. 1924 *Morris Owner* Mar. 20/1 A sympathetic understanding of his car and of road-craft by an owner-driver..makes for more pleasurable motoring. 1963 *Times* 23 May 8/3 Last year there were 7,000 cabs in London, and 10,400 drivers, of whom 2,919 were owner-drivers. 1972 *Police Rev.* 17 Nov. 1484/1 In the case of haulage firms, and in particular the owner-driver..it pays to overload. 1972 *Accountant* 5 Oct. 411/3 Independent 'owner-managed' business as distinct from the large multi-national corporations. 1965 H. I. Ansoff *Corporate Strategy* iv. 62 During the high-growth phase of the electronics industry many new firms were started by owner-managers. 1967 C. Margerison in Wills & Yearsley *Handbk. Managem. Technol.* 18 While they were never entirely a law unto themselves, the owner-managers of the nineteenth century were largely role-determining actors—they were able to control their factories and affairs very much in the manner that they wished. 1970 *Globe & Mail* (Toronto) 26 Sept. B2/3 Residual lending activities of CMHC in the owner-occupant market appear to have increased. 1958 *Ann. Reg.* 1957 71 The slowing up of the property market, especially in the sale of houses for owner-occupation. 1970 *Daily Tel.* 16 Mar. 11/1 The change-over to owner-occupation really began to show about 1920, when some 20 per cent. of the land was in the hands of those who farmed it. 1952 *Time* 9 June 66 (Advt.), Here in Philadelphia, the percentage of owner-occupied homes is greater than in any other large city in America..greater than the national average. 1960 *Times* 23 May 3/5 Northern Ireland is a country of small farms, mainly owner-occupied. 1961 E. A. Powdrill *Vocab. Land Planning* iii. 44 This recognition is evolved from a study of the age and condition of buildings, densities, incidence of owner-occupied properties, and rateable values. 1972 M. Jones *Life on Dole* xi. 84 The old houses are, in general, owner-occupied. 1935 *Planning* II. xliii. 2 There are for example the approaches of State ownership at the one extreme, and of sub-division among many thousands of small owner-occupiers at the other extreme. 1958 *New Statesman* 4 Jan. 7/3 Here and there an enterprising tenant, owner-occupier or determined landlord has repaired and repainted and the contrast is startling. 1971 *Reader's Digest Family Guide to Law* 460/1 An owner-occupier..does not pay capital gains tax on the sale of his home. 1974 *Times* 8 Aug. 18/7 It is questionable whether any public interest is served by requiring owner-occupiers to let commercial and factory premises where they have ceased to use them for their own purposes. 1924 *Glasgow Herald* 28 Nov. 9 As to owner-occupiership, the figures were..encouraging. 1957 E. Bott *Family & Social Network* 161 Owner-operator of a small tobacco and sweet shop. 1971 M. Tak *Truck Talk* 114 *Owner-operator,* a trucker who both owns and drives his rig. 1976 *Woman's Day* (N.Y.) Nov. 50/2 'You can do damage if you don't replace a radiator cap..correctly,' warns Jim Gottfredsen, longtime owner-operator of Gup's West Side Service in Racine, Wisconsin.

ownership. (Further examples.)

1899 T. Veblen *Theory of Leisure Class* ii. 23 The practice of seizing women from the enemy as trophies, gave rise to a form of ownership-marriage. 1906 *Westm. Gaz.* 20 June 7/1 These were the 'ownership' voters, which were a scandal of the franchise. The speaker knows of a case where one man had sixty-seven ownership votes. *Ibid.,* As an instance of this plural voting by ownership, Wimbledon had 3,350 non-resident voters who owned property in the borough. 1910 *Ibid.* 10 Jan. 2/1 What..are the advantages which are claimed for the ownership system? 1944 W. Temple *Church looks Forward* xxii. 158 At an earlier date Ownership and Management were very closely connected. 1956 H. Gaitskell in Gould & Kolb *Dict. Social Sci.* (1964) 457/2 Nationalization..is generally understood to mean the taking over by the State of a complete industry so that it is owned by and managed and controlled for the Community, and public ownership..strictly speaking means the ownership by the community of any property whether individual or not, whether embracing the whole of an industry or only part of it. 1975 *Chinese Econ. Stud.* VIII. iv. 6 The ownership pattern refers to who owns the means of production (including means of labor, such as machines, plants, land, and objects of labor, such as raw materials).

owney-oh (ōu·ni,ōu). *joc.* Also **owneo, ownio, ownie-o, owny-oh.** [f. a popular song (1907) *Antonio & his Ice-Cream Cart.*] Phr. *on one's owney-oh,* on one's own; alone. (Cf. Own *a.* 3 c.)

1922 Joyce *Ulysses* 96 He's as bad as old Antonio. He left me on my ownio. 1956 A. Wilson *Anglo-Saxon Att.* I. iv. 117 As I see it, when you haven't anything more to give a person, well, then you're on your ownio. 1963 'A. Gilbert' *Ring for Noose* xi. 132 'On your owney-oh?' she said. 1967 J. Symons *Man who killed Himself* I. vi. 54 Soon I shall be able to go shopping without worrying, all on my owneo. 1969 F. Sargeson *Joy of Worm* iii. 75 For that matter how in Hades have I managed with the job? Solo. All on my ownie-o. 1976 'W. Trevor' *Children of Dynmouth* xi. 200 She was crying and moaning in the wind, sir, up there on her owny-oh with nobody giving a blue damn about her.

owning (ōu·niŋ), *ppl. a.* [-ing².] That owns property, plant, business interests, etc.

1904 *Electrical Investments* 7 Dec. 773/1 A set-off against any advantage the owning company may be said to secure in extra traffic by the connection. 1909 *Westm. Gaz.* 19 Jan. 2/1 Of the five owning companies three at least have other routes which are more profitable to them. 1923 M. Sadleir *Desolate Splendour* 80 Morvane and the literal appellation of its owning family.

ownsome (ōu·nsəm). [f. after Lonesome *a.*] Phr. *on one's ownsome,* alone.

1939 M. Harrison *What are we waiting For?* 130 You tucked up for bye-byes all on your little ownsome. 1948 D. Ballantyne *Cunninghams* 248 We'll call at the cottage..and dance on our ownsome. 1961 J. Maclaren-Ross *Doomsday Book* 103 I'm absolutely on my ownsome, old feller. 1967 R. Petrie *Foreign Bodies* xi. 163 Oh, snap out of it. You'll pull through on your ownsome. 1976 G. Seymour *Glory Boys* xii. 149 He's been left on his ownsome, and doesn't like it.

ownty-downty (ōu·nti,dōu·nti), *a.* Also **ownty-donty, owny-towny.** [A rhyming jingle.] A familiar or nursery extension of Own *a.*

1815 D. Humphreys *Yankey in Eng.* 19 My owny, towny, Lydy Lovett. 1871 L. M. Alcott *Little Men* v. 68 How nice it is to do it all my ownty donty self! 1882 O. W. Holmes *Let.* 18 Mar. in J. Brown *Lett.* (1912) 449 It is told, the story, without any affectation, but so lovingly that the blessed little creature becomes our own child, our 'ownty-downty', as New England nursery small talk has it.

owt (aut). Repr. dial. pronunc. Aught *sb.*² Esp. in phr. *owt for nowt,* anything for nothing.

1847 E. Brontë *Wuthering Heights* II. xviii. 344 'All well at the Heights?' I inquired of the woman. 'Eea, f'r owt Ee knaw!' she answered. 1895 J. T. Clegg *Works* I. 238 There's olez tuthri cliverdicks to smile At owt they thinken raythor eaut-o'th'-road. 1913 D. H. Lawrence *Let.* 1 Feb. (1962) I. 183, I should think you've forgotten the Yorkshire proverb, 'An' if tha does owt for nowt, do it for thysen'. 1935 'L. Luard' *Conquering Seas* 128 He's got tongue that would fair make one think owt to nowt. 1963 [see *nowt]. 1977 E. W. Hildick *Loop* xviii. 123 Owt's possible, any bloody thing.

ox. Add: **4. a.** Delete † *Obs.* and add later examples. *dumb ox:* see *dumb *a.* 7 b.

1906 E. Dyson *Fact'ry 'Ands* x. 126 You don't see 'em buckin' up, or playin' ther frivolous ox. 1922 Joyce *Ulysses* 9 Don't you play the giddy ox with me! 1923 *Brewer's Dict. Phr. & Fable* (new ed.) 809/2 *To play the giddy ox,* to act the fool generally; to behave in an irresponsible or over-hilarious manner.

ox. Add: **5. a.** *ox-chain, -gad, -goad* (later examples), *-mill, -sawmill, -sled* (earlier and later examples), *-team* (later examples), *-train, -wagon* (earlier and later examples). **b.** *ox-driver* (earlier and later examples); *ox-broad* adj.

1953 Dylan Thomas *Under Milk Wood* (1954) 28 P.C. Attila Rees, ox-broad, barge-booted, stamping out of Handcuff House in a heavy beef-red huff. 1785 G. Washington *Diaries* (1925) II. 441 [1] Oxe Chain. 1817 J. K. Paulding *Lett. from South* I. 128 Next came three men,..chained together with an ox-chain. 1842 [see *goose-yoke* s.v. *goose *sb.* 8]. 1866 *Rep. Indian Affairs* (U.S.) 292, I also repaired 20 wagons, 15 ox chains, 15 grain cradles. 1828 A. Royall *Black Bk.* II. 114 He was one of your right down flat footed ox-drivers. 1843 *Yale Lit. Mag.* VIII. 332 'Gee Bright!' shouted the stentorian voice of an ox-driver. 1916 G. B. Shaw *Androcles & Lion* I. 23 *The ox driver.* The menagerie service is the Emperor's personal retinue. 1937 K. Blixen *Out of Africa* IV. 269 One strong young animal gave..his Native ox-drivers endless trouble. 1836 *Knickerbocker* VIII. 681 His father kept a long ox-gad to whip him with. 1843 *Ibid.* XXI. 125 The ladies requested the loan of Mr. Diddlemas's ox-goad to knock down chestnut burrs. 1848 E. Bryant *California* iii. 32 The crack of the ox-goad, the 'whoa-haws',..create a most Babel-like and exciting confusion. 1916 G. B. Shaw *Androcles & Lion* I. 23 A man with an ox goad comes running through the central arch. 1826 T. Flint *Recoll.* 211 Steam-mills arose in St. Louis, and ox-mills on the principle of the..tread-mill. 1837 J. M. Peck *Gazetteer Illinois* (ed. 2) I. 33 Ox mills on the inclined plane and horse mills by draught, are common throughout ..the state. 1817 in *Trans. Illinois State Hist. Soc. 1910* (1912) 150 An inclined Wheel ox Saw Mill with two saws. 1842 in *Kansas State Hist. Soc. Coll.* (1918) XIV. 755, I made also an oxsled. 1844 *Knickerbocker* XXIII. 445 Let us ride..home on the ox-sled. 1863 H. S. Randall *Pract. Shepherd* (ed. 7) xix. 228 The old-fashioned, lively and merry scene of hauling out hay on an ox-sled. 1904 M. E. Waller *Wood-Carver of 'Lympus* 82 Uncle Shim is driving the ox-sled down the Pent Road. 1776 in *Huntington (N.Y.) Town Rec.* (1889) III. 17 Carting Genll Tryons Baggage from Huntington to Jamaica with an Ox team. 1848 E. Bryant *California* i. 14 Ox-teams seem to be esteemed as preferable. 1913 J. London *Valley of Moon* 297 The chest of drawers..had crossed the Atlantic by sailing ship and the Plains by ox team. 1974 M. Fido *R. Kipling* 77/1 Hiring labourers and ox-teams. 1849 in E. Page *Wagons West* (1930) 120 We will now push off for good and any ox train that gets ahead of us will have to travele. 1850 L. H. Garrard *Wah-to-Yah* 72 Overtaking a United States ox-train, with which I traveled and stayed all night. 1869 *Bradshaw's Railway Manual* XXI. 433 Four years ago the only way of traversing these 1,721 miles between the Missouri and the Pacific was by mail coaches, or by mule or ox trains. 1887 E. Custer *Tenting on Plains* 357 There is no picture that represents the weariness and laggard progress of life like an ox-train. 1968 E. McCourt *Saskatchewan* x. 112 Some settlers arrived..by ox train and Red River cart. 1831 J. M. Peck *Guide for Emigrants* II. 135 From twelve to fifteen large ox waggons are employed..in hauling it [sc. coal] to market. 1857 D. E. E. Braman *Information Texas* iii. 56 The ox-wagons, the 'peculiar institution' of this country, are hauling away cotton. 1878 T. J. Lucas *Camp Life & Sport S. Afr.* iii. 42 The Cape ox-waggon is quite an institution, and has been called, like the camel, the 'ship of the plains'. 1946 E. O'Neill *Iceman Cometh* (1947) I. 44, I vas so tough and strong I grab axle of ox wagon mit full load. 1960 [see *backveld]. 1971 *Sunday Express* (Johannesburg) 28 Mar. 11/1 Students to whom I spoke described the move as 'archaic and back to the ox-wagon'.

6. ox-beef (later example); **ox-chip,** a piece of dry ox-dung; **ox-frame,** a frame for holding oxen while they are being shod; also **ox-shoeing frame; ox-yard,** also, a yard where oxen are kept.

1878 *Amer. Home Cook Bk.* 5 Ox-beef, when it is young, will have a fine open grain, and a good red colour. 1857 E. Bandel *Frontier Life in Army* (1932) 178 No timber to be seen yet, and our wood is gone. We must get along on what few buffalo or ox chips we can gather. 1857 W. Chandless *Visit to Salt Lake* I. vii. 122 Some one pitched on an old camping-place studded with 'ox-chips'. 1844 *Knickerbocker* XXIII. 155 A little slab-roofed smithy... An ox-frame standing by the door, and at one side a shed. 1890 N. P. Langford *Vigilante Days* I. xxvi. 384 We sat down upon the ox-shoeing frame, and talked over the whole matter. 1885 W. Morris in *Commonweal* I. 12/1 The straw from the ox-yard is blowing about. 1910 J. Masefield *Ballads & Poems* 42 The red cock in the ox-yard crows.

b. ox-balm (examples).

1854 *Trans. Michigan Agric. Soc.* V. 130 The plants were very numerous, among which were oxbalm..and marsh grass. 1931 W. N. Clute *Common Names of Plants* 97 The ox-balm (*Collinsonia*) is merely a larger balm.

ox-. Add: **3.** Form of *oxa- before a vowel.

oxa-. Also before vowels **ox-.** Combining element in systematic chemical names used to denote the presence of an oxygen atom (regarded as replacing a —CH_2— group), as in *6-oxa-3-thiadecanenitrile, 2H-2-oxapyrene, oxirane, oxolane.*

1928 *Jrnl. Amer. Chem. Soc.* L. 3075 In order to avoid confusion with the ordinary meanings of *oxy-, thio-, azo-,* etc., it is recommended that the forms *oxa-, thia-, aza-,* (etc.) be employed to indicate the presence of hetero atoms in a ring (the *a* being dropped before a vowel). 1971 *Nomencl. Org. Chem.* (I.U.P.A.C.) (ed. 3) B. 53 (*table*) Element Oxygen.. Prefix Oxa.

oxacillin (ǫksăsi·lin). *Pharm.* [f. *is)oxa(zole* (s.v. *iso- b) + *peni(cillin.*] A semisynthetic penicillin, $C_{19}H_{18}N_3O_5NaS.H_2O$, that is used as an alternative to methicillin, having the same resistance to penicillinase and being in addition resistant to acid so that it can be taken orally; (5-methyl-3-phenyl-4-isoxazolyl)-penicillin sodium. Also called *oxacillin sodium* and *sodium oxacillin* (in the British and U.S. pharmacopœias respectively).

1962 *Proc. Mayo Clinic* XXXVII. 137, 5-Methyl-3-phenyl-4-isoxazolyl penicillin (Prostaphlin*). [*Note*] *Trade name of Bristol Laboratories, Inc... Since this paper was prepared for publication, 'oxacillin' has been adopted as the generic name of this drug. 1963 *New & Non-Official Drugs* 148 Sodium oxacillin is a semisynthetic penicillin salt for oral administration. 1967 *Martindale's Extra Pharmacopoeia* (ed. 25) 994/1 Oxacillin sodium is more resistant to destruction by the acid gastric secretion than benzylpenicillin or methicillin sodium. 1970 *Atlantic Monthly* Mar. 50 He was also given heavy doses of antibiotics, including a gram of chloramphenicol, a gram of oxacillin, [etc.].

oxal-. Add: **oxalace·tic acid,** a dicarboxylic acid, $HOOC{\cdot}CO{\cdot}CH_2{\cdot}COOH$, which crystallizes as an enol form and is produced *in vivo* by transamination from aspartic acid and in

the Krebs cycle by oxidation of malic acid; so **oxala·cetate,** the anion, or an ester or salt of, oxalacetic acid.

1891 *Jrnl. Chem. Soc.* LX. 1333 On mixing..a benzene solution of carbon oxychloride with copper oxalacetate..the copper salt takes up an appreciable quantity of chlorine. 1969 Oxalacetate [see *glyoxylate]. 1896 *Jrnl. Chem. Soc.* LXX. I. 599 Nef's ethylic ethoxyfumarate..when hydrolysed with hydrochloric acid, gives oxalacetic acid. 1939 *Ann. Reg.* 1938 375 In the biological fixation of nitrogen by root nodule bacteria the formation of aspartic acid via the oxime of oxalacetic acid was confirmed. 1972 *Arch. Biochem. & Biophysics* CLIII. 226/1 Oxalacetic acid..functions as a key substrate in metabolism as the keto form; however, the pure compound crystallizes as the *cis* enol of hydroxymaleic acid.

oxalate (ǫ·ksălēit), *v. Med.* [f. the sb.] *trans.* To add an oxalate to, esp. so as to prevent coagulation of blood.

1911 *Amer. Jrnl. Physiol.* XXIX. 204 The tissue extract was itself oxalated to remove any calcium that may have been present. 1934 *Brit. Med. Jrnl.* 7 July 10/2 Blood collected under paraffin was oxalated and centrifuged and the plasma examined spectroscopically. 1954 *Blood* IX. 610 The serum was decanted and 2·8 ml. were oxalated by adding 0·5 ml. of 0·1 M. potassium oxalate.

So **o·xalated** *ppl. a.,* containing added oxalate.

1893 *Jrnl. Path. & Bacteriol.* I. 443 (*heading*) Effect of graduated additions of calcium chloride to oxalated blood. 1946 *Nature* 16 Nov. 708/2 The prothrombin concentration in normal human oxalated plasma averaged approximately 2 mgm. per 100 ml. when expressed as protein nitrogen. 1964 W. G. Smith *Allergy & Tissue Metabolism* vi. 69 Oxalated blood samples were collected both before and for several minutes after shock.

oxalo-. Delete 'used before consonants' and add: **oxaloa·cetate** = *oxalacetate* s.v. *oxal-; **oxaloace·tic acid** = *oxalacetic acid* s.v. *oxal-; **oxalo·sis** *Path.* [-osis], a rare disorder of metabolism in which crystals and stones of calcium oxalate are deposited in the kidneys and elsewhere, often causing death during childhood as a result of renal failure; **oxalosu·ccinate,** the anion, or an ester or salt, of oxalosuccinic acid; **o:xalosucci·nic acid,** a tricarboxylic acid, $HOOC{\cdot}CO{\cdot}CH(COOH){\cdot}CH_2{\cdot}COOH$, which is an intermediate in the formation of α-ketoglutaric acid from isocitric acid in the Krebs cycle.

1943 Sumner & Somers *Chem. & Methods of Enzymes* xviii. 324 Malate, oxaloacetate, or succinate could replace fumarate in reaction (c). 1962 S. G. Waley in A. Pirie *Lens Metabolism Rel. Cataract* 356 Another γ-keto acid that undergoes enzymatic decarboxylation is oxaloacetate. 1937 *Nature* 18 Sept. 503/2 α-Ketoacids other than pyruvic, for example..oxaloacetic or phenylpyruvic acid, may equally serve as acceptors for the amino group of glutamic acid. 1940 [see *α-ketoglutarate* s.v. *keto- a]. 1968 Passmore & Robson *Compan. Med. Stud.* I. ix. 14/2 Acetyl-CoA reacts with oxaloacetic acid to produce citric acid. 1952 Ying Chou & Donohue in *Pediatrics* X. 660 (*heading*) Oxalosis. Possible 'inborn error of metabolism' with nephrolithiasis and nephrocalcinosis due to calcium oxalate as the predominating features. 1973 N. M. R. Buist et al. in Forfar & Arneil *Textbk. Pediatrics* xix. 1171/2 Treatment of oxalosis includes alkalinization of the urine, dietary restriction of calcium and a large fluid intake. 1911 *Chem. Abstr.* V. 3240 Tri-Et oxalosuccinate..is best prepared by means of EtOK. 1962 S. G. Waley in A. Pirie *Lens Metabolism Rel. Cataract* 355 In the citric acid cycle..two molecules of CO_2 are formed per turn of the cycle; one comes from oxalosuccinate, a β-keto acid. 1925 *Chem. Abstr.* XIX. 4423 (Index), Oxalosuccinic acid, triethyl ester. 1948 *Jrnl. Biochem.* CLXXIV. 144 The instability of oxalosuccinic acid makes an accurate estimation of this constant rather difficult. 1966 F. A. Robinson *Vitamin Co-Factors of Enzyme Syst.* viii. 541 In this organism [sc. *E. coli*], biotin appears to function in the conversion of oxalosuccinic acid into α-ketoglutaric acid.

oxamide. Add: [first formed as F. *oxamide* (J. Dumas 1830, in *Ann. de Chim. et de Physique* XLIV. 130).]

oxazepam (ǫksē^i·zĭpæm). *Pharm.* [f. Ox- 1 + Az(o- + *-ep(ine* (suffix designating an unsaturated seven-membered ring containing nitrogen) + Am(ide.] A tricyclic, creamy-white powder, $C_{15}H_{11}ClN_2O_2$, which is a tranquillizer given to relieve anxiety states and to control the withdrawal symptoms of alcoholism.

1964 *Jrnl. Pharmaceutical Sci.* LIII. 1181/1 Oxazepam, 7-chloro-1,3-dihydro-3-hydroxy-5-phenyl-2*H*-1,4-benzodiazepine-2-one, has been characterized pharmacologically in our laboratories as an anticonvulsant and mild central depressant and is currently under clinical investigation as an antianxiety agent. 1966 *Jrnl. Amer. Med. Assoc.* 21 Nov. 952/1 Six days after oxazepam was stopped completely, her husband reported she had been up all night, was talking irrationally, and was having visual hallucinations. 1974 *Brit. Jrnl. Clin. Pract.* XXVIII. 65/1 Oxazepam, one of the benzodiazepine group, has an anxiolytic action with very little sedative potential.

oxazole (ǫ·ksăzōul). *Chem.* [ad. G. *oxazol* (Hantzsch & Weber 1887, in *Ber. d. Deut. Chem. Ges.* XX. 3119): see Ox- 1, Azo-, and *-ole.] **a.** A weakly basic, heterocyclic compound, O·CH:N·CH:CH, which is a volatile liquid. **b.** Any of the derivatives of this compound obtained by substituting for hydrogen.

1888 *Jrnl. Chem. Soc.* LIV. 574 Oxazoles are obtained by the condensation of α-halogen-ketones with amides. 1892 [see *imidazole]. 1929 R. A. Gortner *Outl. Biochem.* xiv. 350 Polypeptides may be considered to enolize ..yielding substituted imidazoles or substituted oxazoles. 1966 *McGraw-Hill Encycl. Sci. & Technol.* IX. 461/1 Oxazole is miscible with water and organic solvents. 1968 A. Albert *Heterocyclic Chem.* (ed. 2) vi. 289 Oxazole alkaloids have been isolated from flowering plants in the Rutaceae and Graminae, and oxazolidines (their reduced analogues) occur in cabbages.

oxazolidine (ǫksăzǫ·lidīn). *Chem.* [f. prec. + *-idine.] Any of the compounds obtained by substituting for hydrogen in the hypothetical parent compound $CH_2CH_2NHCH_2O$ (which is the fully hydrogenated form of oxazole), some of which are anticonvulsants and are used in treating petit mal.

1902 *Jrnl. Chem. Soc.* LXXXII. I. 56 (*heading*) Synthesis of oxazolidines by the action of aldehydes on hydramines. 1953 *Chem. Rev.* LIII. 315 The oxazolidines are liquids or solids of basic character; their stability to hydrolysis is generally low. 1961 A. Goth *Med. Pharmacol.* xix. 229 In the clinical use of the oxazolidine derivatives, the following toxic effects have been reported: drowsiness and ataxia, photophobia, and a strange visual disturbance.

oxazolone (ǫksæ·zǫlōun). *Chem.* [f. as prec. + -one.] Any compound containing the nucleus obtained by hydrogenating one of the double bonds of oxazole and replacing a methylene group by a carbonyl group; = *azlactone.

1899 Japp & Findlay in *Jrnl. Chem. Soc.* LXXV. I. 1027 It occurred to us that, by substituting an α-hydroxy-acid for the α-keto-alcohol in the foregoing reaction, it might be possible to prepare oxazolones (ketodihydro-oxazoles). 1947 *Sci. News* IV. 70 The synthesis of penicillin G starts with a benzyl oxazolone and with penicillinamine, and attempts to recombine them. 1968 A. Albert *Heterocyclic Chem.* (ed. 2) vi. 290 Of the oxazolones, the 5-isomer..and its derivatives are the best known. 1968 R. O. C. Norman *Princ. Org. Synthesis* xviii. 604 The oxazolones or azlactones, prepared by the dehydration of *N*-acyl-α-amino-acids, are employed in Erlenmeyer's synthesis of α-amino-acids. 1975 *Nature* 13 Nov. 149/2 We have examined..the production of anti-parasite antibodies and parameters of T-cell function (the response to phytohaemagglutinin (PHA) and oxazolone).

ox-blood (ǫ·ks,blʌd). [f. Ox + Blood.] The blood of the ox; a colour resembling this; also used *attrib.* or as *adj., spec.* of a colour of opals, of porcelain, and of leather.

1705 *Whole Art of Dying* II. 53 Ox-blood Colour. First Tinge the Stuffs Yellow,..and work them till they are sufficiently beautiful, then..put into the Kettle a Tub of stale Urine, and boil it again till they take the Dye. 1707 [in Dict. s.v. Ox 5 a]. 1897 *Sears, Roebuck Catal.* 194b/3 Men's Hard Cash Lace, best Russia Calf, latest Ox Blood (dark wine) color. 1936 *Burlington Mag.* Jan. 10/2 The splendid ox-blood and peach bloom reds of the Ch'ing dynasty. 1937 D. Jones *In Parenthesis* 118 You feel the pack of the Ox-blood Kid—it's as light as the Reg'mentals—there's a whole lot of them that work it. 1941 'Brahms' & 'Simon' *No Bed for Bacon* ii. 37 It was Elizabeth of England in ivory and ox blood. 1950 H. McCloy *Through a Glass, Darkly* i. 5 A bowl of ox blood porcelain. 1950 C. Fry *Venus Observed* II. i. 34 Umber, bronze and brass, ox-blood, damson, Crimson, scalding scarlet. 1967 S. Lloyd *Lightning Ridge Bk.* 24 In the reds, miners describe colour as 'ox blood', 'pidgeon's blood', 'port wine red', and so on. 1968 *Listener* 27 June 825/1 One of the occupants..was wearing 'ski pants and ox-blood-coloured shoes'. 1968 D. Torr *Treason Line* 69 He..walked over to the Chinese vases in the window. He put one vase with an ox-blood glaze..into the window on the left. 1971 J. S. Gunn *Opal Terminol.* 32 *Ox blood,*..name given to the deepest of the red-coloured opals. 1974 'G. Black' *Golden Cockatrice* x. 163, I got another vase..imitation ox blood. 1975 T. Stoppard *Travesties* I. 27, I think to match the carnation, oxblood shot-silk cravat.

ox-bow. Add: **1.** (Further U.S. examples.)

1833 S. Smith *Life & Writings J. Downing* 106 A farmer ort to stick to his ox bows and goard sticks. 1846 R. B. Sage *Scenes Rocky Mts.* iii. 26 An extra quantity of ox bows, axle-trees..in case of accidents or breakage. 1881 *Rep. Indian Affairs* (U.S.) 398 Ox-bows, 2-inch..doz. 51.

2. b. More fully *ox-bow lake.* A curved lake left in a former meander of an adjacent river after the river has changed its course and cut through the narrow neck of the meander.

1898 W. M. Davis *Physical Geogr.* ix. 245 An abandoned meander is occupied by nearly stagnant water... In time it becomes an ox-bow lake. 1902 [see *Mortlake]. 1937 [see *cut-off *sb.* 2]. 1944 A. Holmes *Princ. Physical Geol.* x. 165 If a flood occurs when only a narrow neck of lands is left between adjoining loops, the momentum of the increased flow is likely to carry the stream across the neck... A deserted channel is left, forming an ox-bow lake which soon degenerates into a swamp. 1957 L. Eiseley *Immense Journey* 22 Fishes of this type who get themselves immured in oxygenless ponds or in cut-off oxbows buried in winter drifts. 1961 *Listener* 19 Oct. 614/1 In his lifetime the river flowed through his ox-bow. 1962 [see *Mortlake].

3. *Comb.* (in sense 1) **ox-bow key,** a key for fastening the end of an ox-bow; **ox-bow stirrup,** a stirrup resembling an ox-bow in shape; also *ellipt.*

1882 *Rep. Indian Affairs* (U.S.) 480 Ox-bow keys, 2 inch. *a* 1918 J. Bratt *Trails of Yesterday* (1921) xiii. 52 In the morning I was set to work..making ox-bow keys and fitting bows to yokes. 1907 S. E. White *Arizona Nights* I. i. 5 Uncle Jim sat placidly on his white horse, his thin knees bent to the ox-bow stirrups, smoking. 1942 Berrey & Van den Bark *Amer. Thes. Slang* §915/5 *Ox bows,*..wide old-fashioned wooden stirrups.

Oxbridge (ǫ·ksbridʒ). [Short for 'Oxford and Cambridge'.] A name used to designate the universities of Oxford and of Cambridge; the characteristics common to both, esp. as distinct from other universities in the British Isles. Also *attrib.* Cf. *Camford.

1849 Thackeray *Pendennis* I. xxix. 286 'Rough and ready, your chum seems,' the Major said. 'Somewhat different from your dandy friends at Oxbridge.' 1906 V. Woolf in Q. Bell *Virginia Woolf* (1972) I. 205 You see a pink cheeked boy whose only talk is of cricket..enter upon his first term at Oxbridge. 1907 G. W. E. Russell *Seeing & Hearing* v. 35 We ran a neck-and-neck race at the University... In those days I little thought of settling down in Oxbridge. 1912 H. G. Wells *Marriage* i. §1. 5 A ..meretricious dressing-bag of imitation morocco, which had been one of her chief financial errors at Oxbridge. 1924 E. F. Benson *David of King's* i. 5 Useless..to delude the intelligent reader into believing that it was Queen's Parade at Oxbridge or Prince's Parade at Camford. 1955 T. H. Pear *Eng. Social Differences* i. 20 He would often wish his sons to go to a public (boarding) school and to Oxbridge. 1958 *New Statesman* 30 Aug. 244/1 Whatever its merits or demerits, Oxbridge remains first choice for a majority of university applicants. 1960 Auden *Homage to Clio* 88 Oxbridge philosophers, to be cursory, Are products of a middle-class nursery. 1960 *Times* 14 Mar. 13/4 The University 'Rag'..is now more closely identified with what the jargon calls 'Redbrick' rather than 'Oxbridge', with the world of Kingsley Amis rather than that of the young Compton Mackenzie. 1964 S. Brittan *Treasury under Tories* i. 21 One characteristic that Whitehall does have in common with other élite groups is its overwhelmingly Oxbridge character. 1967 *New Scientist* 6 Apr. 39/2 Father is very likely public school and 'Oxbridge'. Son is 'redbrick'. 1973 D. Robinson *Rotten with Honour* 9 We'll have—Hale... Twenty-six, Oxbridge, degree in languages. 1975 *Sunday Tel.* 4 May 30/2 (*heading*) Fine win by Oxbridge. The Universities beat Worcestershire by 66 runs at Fenners. 1976 *Globe & Mail* (Toronto) 1 July 35/8 And as if to make the thing complete the prayer was often offered by a clergyman who read the liturgy with an Oxbridge accent even though he was a product of southern Ontario.

Hence **Oxbri·dgean, -ian** *a.* and *sb.*

1959 *Cambr. Rev.* 7 Feb. 315/1 They [sc. the Americans] are so polite, so ready to believe that the visiting Oxbridgean must find Yale or Harvard an anti-climax. 1959 *Guardian* 14 Dec. 6/4 'The mere fact,' you fretted, that such Oxbridgean institutions 'have refused the student loan money may make their actions suspect'. 1960 *Mind* LXIX. 419, I have wished I was reading one of those Oxbridgean philosophers who, had they taken the right turning early in life, would long ago have brought the filing and card-indexing systems of the British Civil Service up to a new peak of perfection. 1970 *Atlantic Monthly* May 132 As apt as an eighteenth-century Oxbridgean with the Latin tag. 1971 *New Scientist* 17 June 706 The ivy-covered, Oxbridgian atmosphere of Yale.

ox cart. [Ox 5.] A cart drawn by an ox. Also *transf.*

1749 J. Hempstead *Diary* 6 July (1901) 526 No ox Carts in these parts & very few Horse Carts. 1877 [see Ox 5]. 1918 E. S. Farrow *Dict. Mil. Terms* 427 *Ox Cart.*—A slang name for a heavy French shell which moves with a moderate velocity. 1973 *Country Life* 29 Nov. 1802/3 Continuously exploring..deeper into veldt and jungle, leading, on his Arab pony, a train of ox carts. 1975 *Times Lit. Suppl.* 5 Dec. 1462/1 Crane's noble and well-loved print 'The Triumph of Labour' with its procession of garlanded oxcarts and rejoicing workers.

Oxfam, OXFAM (ǫ·ksfæm). [Short for Oxford Committee for Famine Relief.] An organization for the distribution of food, funds, etc., in disaster areas and to poor countries.

1963 D. Mitchell in *Oxfam Ann. Rep.* 1962/63, At first the Oxford Committee for Famine Relief—Oxfam, as it is now known all over the world—was one of many similar groups with a temporary mission: the emergency relief of war suffering in Europe. 1968 *Listener* 8 Aug. 162/1 A grim, obsessed Oxfam worker, struggling to get a few scraps and leavings to millions of starving children. 1969 N. W. Pirie *Food Resources* 15 Oxfam posters used to read 'Oxfam hates hungry babies' and went on to suggest ways of getting rid of the hunger. 1974 I. Murdoch *Sacred & Profane Love Machine* 304 Her few inexpensive jewels had gone to Oxfam. 1975 J. Blackburn *Mister Brown's Bodies* i. 10 Those disgusting pictures of famine victims you see in the Oxfam shops.

Oxford. Add: **Oxford accent,** a style of pronouncing English popularly supposed to be particularly characteristic of members of the University of Oxford and (esp. before 1939) to be marked by affected utterance; **Oxford bags** [Bag *sb.* 16], a style of trousers very wide at the ankles; so **Oxford-bagged** *a.*; **Oxford blue,** a

dark shade of blue, adopted as the colour of the university; **Oxford Blues:** see Blue *sb.* 7a; **Oxford clay,** add to def.: earlier called *clunch clay* (Clunch *sb.* 6), *plastic clay* (Plastic *a.* 5b); (earlier and later examples); **Oxford cloth** (see quots.); **Oxford(shire) Down,** a sheep of the breed so called, produced by crossing Cotswold and Hampshire Down sheep and developed by Samuel Druce at Eynsham about 1830; **Oxford English,** English spoken with an Oxford accent; the speech popularly supposed to be characteristic of a member of the University of Oxford; **Oxford frame** (later examples); **Oxford grey** (see *Oxford mixture*) (examples); also, the colour of such cloth; **Oxford Group:** see *group *sb.* 3 d; **Oxford hollow,** in *Bookbinding,* a flattened paper tube inserted between the spine of the book and its cover, to strengthen the spine and allow the book to be opened flat more easily; **Oxford John,** a dish of sauced and stewed mutton with other ingredients; **Oxford marmalade,** a kind of coarse-cut marmalade originally manufactured in Oxford (registered as a trade-mark by Frank Cooper in 1908 and 1931); also *attrib.* and *fig.*; **Oxford plant** = *Oxford weed;* **Oxford punch** [Punch *sb.*³] (see quots.); **Oxford ragwort,** an annual herb, *Senecio squalidus,* belonging to the family Compositæ and bearing heads of yellow flowers (a native of southern Italy now naturalized in Britain, after having escaped from the Oxford Botanic Garden); **Oxford sausage,** a kind of sausage; also *fig.*; **Oxford scholar** *slang,* a crown; five shillings; a dollar; **Oxford shirt,** a shirt made from Oxford cloth; **Oxford shirting** = *Oxford cloth;* **Oxford trousers** = *Oxford bags;* **Oxford unit** *Pharm.,* a unit of penicillin originally adopted at the Sir William Dunn School of Pathology in the University of Oxford, being the amount which when dissolved in 1 c.c. of water gave the same inhibition as a certain partly purified standard solution; cf. **penicillin unit;* **Oxford voice** = *Oxford accent;* **Oxford weed,** the ivy-leaved toadflax, *Cymbalaria muralis.*

1904 J. K. Jerome *Tommy & Co.* v. 174 Somerville's Oxford accent is wasted here. 1924 Galsworthy *On Expression* 8 And dare we condemn cockney—a lingo whose waters, in Southern England, seem fast flooding in over the dykes of the so-called Oxford accent, and such other rural dialects as are left? 1934 *S.P.E. Tract* xxxix. 616 It might be said perhaps that the 'Oxford Accent' conveys an impression of a precise and rather foppish elegance, and of deliberate artificiality. 1940 G. Arthur *Concerning Winston Spencer Churchill* 194 Lacking the 'Oxford accent' he spoke as a Briton to Britons. 1959 J. Braine *Vodi* vi. 93 Dick assumed an Oxford accent. 'It's naht old-fashioned, dear brethren, to think of Hell in the language of fire and brimstone.' 1974 P. Dickinson *Poison Oracle* iii. 45 The Sultan's manner is very deceptive. ..His Oxford accent and his slang..are all a sort of parody of our civilisation. 1961 *Times* 18 May 16/6 Eton-cropped maidens sporting decorously with Oxford-bagged partners. 1927 *Dancing Times* Jan. 573/2 Oxford bags, plus fours, in fact, any old thing. 1933 P. Balfour *Society Racket* ii. 61 We wore high-necked jumpers and 'Oxford bags'. 1938 J. Betjeman *Oxf. Univ. Chest* i. 9 The pale-faced mechanics in Oxford bags and tweed coats, walk down the Cornmarket. 1948 H. Acton *Mem. Aesthete* vi. 119 Instead of the wasp-waisted suits with pagoda shoulders and tight trousers affected by the dandies, I wore jackets with broad lapels and broad pleated trousers. The latter got broader and broader. Eventually they were imitated elsewhere and were generally referred to as 'Oxford bags'. 1971 *Daily Tel.* 2 Aug. 9/5 By night you're either..Dietrich in a velvet blazer and Oxford bags, or Carole Lombard in a halter-neck satin top..and a long satin kilt. 1973 *Country Life* 19 Apr. 1115/1 A single-breasted striped suit for the bridegroom, which has Oxford bags and patch pockets. 1866 Mrs. H. Wood *Elster's Folly* I. viii. 191 'Strike your colours, ladies, you that sport the crimson and purple!' called out a laughing voice from one of the skiffs. 'Oxford blue wins.' 1875 *All Year Round* 20 Feb. 444/2 The hues peculiar to the best period of Sèvres were gros-blue, a dark heavy Oxford blue. 1923 Joyce *Let.* 26 Feb. (1966) III. 72 The smaller Oxford blue volume. 1959 M. Gilbert *Blood & Judgement* xvi. 176 He picked up a thin, Oxford-blue folder. 1973 R. Ludlum *Matlock Paper* xi. 101 A half-unbuttoned, oxford-blue shirt. 1818 W. Phillips *Selection of Facts Geol. Eng. & Wales* 66 In these ..are included the three strata..namely, the Forest marble, the Cornbrash limestone, and the clunch clay (Oxford Clay). 1967 D. H. Rayner *Stratigr. Brit. Isles* ix. 298 Around Peterborough, vast pits have been dug in the Oxford Clay for brick-making. 1964 *McCall's Sewing* iv. 58/2 *Oxford cloth,* plain-, basket- or twill-weave cotton, often used for shirting. It is fairly heavy cloth in which two yarns travel as one in the warp, and one filling yarn is equal in size to the two-warp yarns. 1968 J. Ironside *Fashion Alphabet* 243 *Oxford cloth,* a heavy cotton cloth used for shirts and sometimes jackets or summer suits. 1969 *Sears Catal.* Spring/Summer 24 Rajah shirt with soil release. Oxford cloth of polyester and cotton. 1976 *National Observer* (U.S.) 17 Jan. 10/3 (Advt.), Our fine quality, breathable, pure cotton oxford cloth pyjamas are unbeatable for wear all year 'round. [1849 *Jrnl. R. Agric. Soc.* X. 436 The Cotswold is a large breed of sheep, and is the stock from which the class called new Oxford is sprung.] 1859 *Ibid.* XX. 345 Amongst the 'other short-woolled sheep' exhibited, we have Hampshire Downs, Oxfordshire Downs, [etc.]. *Ibid.,* The Oxford Downs date from the year 1833..when a neat, well-made Cotswold ram was used with Hampshire ewes. 1912 R. Lydekker *Sheep* v. 106 The formation of the Oxford down was commenced about the year 1833. 1970 *Observer* (Colour Suppl.) 26 Apr. 36/1 All six Down breeds..are shortwools, Oxford Down..being heaviest... Bold-looking sheep with top-knot of wool above dark face. 1926 D. H. Lawrence *Plumed Serp.* iii. 31 An odd, detached, yet cocky little man, a true little Indian, speaking Oxford English in a rapid, low, musical voice. 1932 *S.P.E. Tract* xxxvii. (*title*) 'Oxford' English. 1952 M. Steen *Phoenix Rising* iv. 72 Americans come over and proceed to acquire what they think is Oxford English. 1969 'H. Pentecost' *Girl Watcher's Funeral* (1970) ii. ii. 100 It was Oxford English with a slight accent which I took to be French. 1975 *ITV Evidence to Annan Committee* 15 In the spoken word the traditional currency was, till recently, Oxford English. 1939 'N. Shute' *What happened to Corbetts* iii. 88 He laid her on the ornate, gilded iron bed beneath a picture of the 'Stag at Bay' and a text in a wood Oxford frame that told them 'God is love'. 1973 J. Thomson *Death Cap* xiii. 177 The pictures on the walls were..pre-Raphaelite prints in Oxford frames. 1836 W. F. Tolmie *Diary* 28 Oct. in *Jrnls.* (1963) 332, 1 pr Extra S. fine dark Oxford grey trousers. 1864 J. S. Le Fanu *Uncle Silas* II. xii. 191 An old Oxford gray surtout that showed his lank length to advantage. 1903–4 Oxford grey [see *Donegal]. 1964 S. Bellow *Herzog* 237 A stylish oxford gray summer suit. 1973 R. Hayes *Hungarian Game* xxxix. 235 Urkowitz' face was turning a shade of fine Oxford gray. 1956 H. Williamson *Methods Bk. Design* xix. 308 Another method is to fix on the spine of the section a tube of paper, or Oxford hollow, and to fix the cover to this. 1960 G. A. Glaister *Gloss. Bk.* 289/1 *Oxford hollow:* a flattened paper tube which is attached to the back of a book..so that when the book is opened the back opens up independently of the spine... The O[xford] U[niversity] P[ress], who have supplied this note, state 'presumably the word Oxford was used to describe this kind of binding because undoubtedly the old Oxford bindery was the first to use it, particularly for leather-bound Bibles. It is properly applied only to a leather-bound book and is a style which is now used by most leather binders.' 1965 L. S. Darley *Introd. Bookbinding* 30 Another way of lining the spine, after the mull has been glued in place, is by making a tubular hollow—sometimes called an Oxford hollow—a device which provides a lining for the spine, a hollow for the cover and an additional point of union between the book and its case. 1892 *Encycl. Pract. Cookery* II. 68/2 Oxford John. 1952 F. White *Good Eng. Food* ii. iv. 136 Oxford John. Take a well hung leg of mutton [etc.]. 1907 *Yesterday's Shopping* (1969) 24 Marmalade..Oxford (Frank Cooper's). 1942 C. Morley *Thorofare* lxx. 459 There was Cooper's bitter Oxford marmalade—the only Oxonian product to which Uncle Dan would grant supremacy. 1962 *Sunday Express* 25 Feb. 6/3 Wyatt's thick-cut Oxford marmalade voice. 1973 'S. Harvester' *Corner of Playground* i. viii. 74 A new jar of Oxford lime marmalade. 1856 Hawthorne *Jrnl.* 31 Aug. in *Passages from Eng. Note-Bks.* (1870) II. 150 We looked also at the outside of the wall [of New College], and Mr. Parker..showed us a weed growing..hanging plentifully downward from a shallow root. It is called the Oxford plant, being found only here, and not easily, if at all, introduced anywhere else. 1845 E. Acton *Mod. Cookery* xxvi. 637 Oxford punch... Lemons..oranges..calf's foot jelly..white wine ..French brandy..Jamaica rum [etc.]. 1877 E. S. Dallas *Kettner's Bk. of Table* 320 Oxford punch.—The great characteristic of this punch is its having a quantity of calf's foot jelly dissolved in it. 1892 *Encycl. Pract. Cookery* II. 69/1 Oxford punch. 1884 W. Miller *Dict. Eng. Names Plants* 249/2 Senecio..squalidus. Oxford Rag-wort. 1886 G. C. Druce *Flora Oxfordshire* 158 Oxford Ragwort... Very plentiful in and around Oxford, where it was first noticed by Sir Joseph Banks. Dillenius sent seeds to Linnæus but whether he gathered them from the Oxford Garden or the wall of the town no memorandum exists. 1926 *Nat. Hist. Oxford District* 72 A few brave adventitious plants may be seen on the walls, including the ubiquitous Oxford Ragwort. 1948 Prime & Deacock *Shorter Brit. Flora* 149 Oxford Ragwort..; waste places, greatly on the increase. (This plant is very common on bombed sites in London.) 1969 *Nature* 27 Sept. 1303/2 Oxford ragwort (*Senecio squalidus*) spread throughout the railway system after seeds had been dispersed from the Botanic Garden to Oxford station. 1973 Gilmour & Walters *Wild Flowers* (ed. 5) xiii. 198 The Oxford ragwort..has achieved..an astonishing recent spread. 1764 (*title*) The Oxford sausage. 1877 E. S. Dallas *Kettner's Bk. of Table* 100 The Oxford sausage is a crépinette, can be made at home. 1926 *Daily Colonist* (Victoria, B.C.) 5 Jan. 6/2 (Advt.), Fresh Made Oxford Sausage, 3 lbs. for 32 c. 1937 Partridge *Dict. Slang* 596/1 *Oxford Scholar,* five shillings (piece or sum): New Zealanders' rhyming s. on *dollar:* C. 20. Also from ca. 1870, in the S.W. of England. 1938 F. D. Sharpe *Sharpe of Flying Squad* 332 *Oxford Scholar,* dollar. 1960, 1965 Oxford scholar [see sense *2]. 1967 Oxford scholar [see *caser]. 1926–7 *Army & Navy Stores Catal.* 717/1 Oxford shirts—With stiff cuffs, soft turnover cuff or small wristband to link and button—each 12/9. 1959 *Listener* 4 June 982/1 The cloth cap and the collarless Oxford shirt. 1971 *New Yorker* 9 Oct. 29/2 (Advt.), Our own make long staple cotton oxford shirt. [1907 *Yesterday's Shopping* (1969) 742/2 Fancy cotton shirting... Oxford mat, best quality.] 1926–7 *Army & Navy Stores Catal.* 697/2 Oxford shirting (soft finish) 29 in. wide.. 1/1 Per yd. 1940 *Chambers's Techn. Dict.* 606/1 *Oxford shirting,* a plain-weave cotton fabric, generally striped, used for shirting. 1948 G. L. Fraser *Textiles by Britain* 165 Oxford shirting, plain or fancy woven striped shirting cloth. 1925 *Punch* 4 Mar. 244 (*caption*) Perils of the Dance. The terror of the Oxford trousers. 1937 J. Laver *Taste & Fashion* xvii. 241 The advent of Oxford trousers in the middle twenties. 1942 Florey & Jennings in *Brit. Jrnl. Exper. Path.* XXIII. 122 For those using the dilution method it may be stated that the 'Oxford unit' is that amount of penicillin which when dissolved in 50 ml. of meat extract broth just inhibits completely the growth of the test strain of *Staphylococcus aureus.* 1948 Wright & Montag *Textbk. Pharmacol. & Therap.* (ed. 4) xxxiii. 548 For the treatment of mild to moderately severe infections daily dosages of 80,000 to 120,000 Oxford units are sufficient. 1952 W. T. Salter *Textbk. Pharmacol.* xlix. 1084/1 The new international unit and the old Oxford unit are very close. 1920 A. Huxley *Limbo* 89 When the Military Representative spoke, he could hear again that wretched Nut's rendering of the Eton and Oxford voice. 1924 E. M. Forster *Passage to India* xxiv. 221 'We object to the presence of so many European ladies and gentlemen upon the platform,' he said in an Oxford voice. 1931 *Atlantic Monthly* Feb. 149/1 The pronunciation of the common people left its impress indelibly on the so-called best people, with a few languid drawls, terminal *ars,* clipped *gs* and feeble *hs* thrown in..., which..acquired the name of the Oxford voice. 1834 W. Baxter *Brit. Phænogamous Bot.* I. 23 This very pretty plant is a native of Italy, and is said to have been originally introduced into England by means of its seeds having been brought in some marble sculptures from that country to Oxford, where it has long established itself on the walls of the Colleges, gardens, &c. in such abundance as to have obtained the name of 'Oxford-weed'. 1976 C. Oman *Oxford Childhood* vii. 133 Mrs Pemberton was a qualified botanist and I was soon flattered by being sent up to the top of a crumbling wall..to get her specimens of Oxford Weed.

b. Used *attrib.* or as *adj.* to denote the characteristic manner, speech, behaviour, etc., of a present or former member of the University of Oxford; freq. = *Oxford accent.*

1877 H. James *Let.* 28 Feb. in R. B. Perry *Tht. & Char. W. James* (1935) I. 375, I lunched the other day with Andrew Lang to meet J. Addington Symonds,—a mild, cultured man, with the Oxford perfume. 1897 G. B. Shaw *Our Theatres in Nineties* (1932) III. 108 Stage smart speech, which, like the got-up Oxford mince and drawl of a foolish curate, is the mark of a snob. 1909 P. Gibbs *Street of Adventure* iii. 37 'Because I tell you so,' said Luttrell, with a touch of his Oxford manner. 1913 A. Lunn *Harrovians* ii. 27 Mr. Lee had neither a double chin nor an Oxford manner. 1919 J. B. Morton *Barber of Putney* ii. 24 Up and down one heard the Oxford drawl. 1922 Joyce *Ulysses* 6 He thinks you're not a gentleman... Because he comes from Oxford. You know, Dedalus, you have the real Oxford manner. *Ibid.* 431 In youth's smart blue Oxford suit with white vestslips. 1926 A. S. L. Farquharson in J. C. Wilson *Statement & Inference* I. p. xxv, 'The trouble is that one feels life is so short, *ars longa* but philosophy seems very much longer.' This is the scholar's last lesson, the clue perhaps to what is sometimes called Oxford irony. 1928 D. H. Lawrence *Woman who rode Away* 152 But, in a voice more expostulatingly Oxford than ever, he said [etc.]. *Ibid.* 153 Jimmy got up, with a bit of an Oxford wriggle, and held out his hand. 1934 *Spectator* 5 Jan. 18/2 Surely it is permissible to suggest..the Oxford Bleat by writing down the directions given me the other day as 'past a whaite house, between the water-tah and the pah station'. 1937 *N. & Q.* 12 June 418/1 What we term Oxford pronunciation, and wrongly so call it. 1938 F. D. Sharpe *Sharpe of Flying Squad* xvii. 189 A gentleman with an Oxford drawl..in darkest Hoxton. 1957 G. Avery *Warden's Niece* viii. 153 A young man..clumsily trying to propel his punt from the stern instead of the conventional Oxford position in front. 1958 B. Nichols *Sweet & Twenties* v. 74 If the worst came to the worst I could 'rely on the Oxford manner'. 1960 W. B. Gallie *New University* vi. 115 They were delighted by the fact that he so often appeared to be joking—for so they described Lindsay's elaborate Oxford irony.

2. *ellipt.* for Oxford English Dictionary, Oxford grey, marmalade, mixture, scholar, shirt, shoe, trousers, etc.

? 1890 in *Amer. Mail Order Fashions* (1961) 28/2 Women's tan Dongola Kid, square or pointed toe, fox heel Oxfords. French stay. 1890 *Illustr. London News* 24 May in L. de Vries *Victorian Advts.* (1968) 51/3 New Range of Coloured Oxfords, Cambrics, and Calicottas for Shirts and Pyjamas. 1902 Farmer & Henley *Slang* V. 119/1 *Oxford,* a crown piece. 1903 in S. Nowell-Smith *Edwardian England* (1964) facing p. 180 (Advt.), The 'Oxford'. Blacking leather or glacé kid. 1914 *Glasgow Herald* 7 Sept. 10/2 Glasgow firms manufacture..zephyrs, Oxfords, shirtings, and dress goods. 1926 *Daily Colonist* (Victoria, B.C.) 21 July 16/4 (Advt.), A 4-ply worsted wool in shades of pink... Oxford. 1929 G. Mitchell *Mystery of Butcher's Shop* xi. 120 He fell down, and tore chunks out of his Oxfords on the brambles. 1932 *New Yorker* 11 June 45 Waterproof leather oxfords or ghillies..; suede oxfords at Brooks and Rogers Peet... Golf oxfords at Spalding [etc.]. 1945 M. D. Potter *Fiber to Fabric* viii. 155 *Oxford,* a plain weave of medium and heavy weights. 1950 W. Stevens *Let.* 21 Nov. (1967) 699, I look it up either at the office, where we have a Webster, or have someone look it up for me in the State library, where there is an Oxford. 1957 V. Nabokov *Pnin* i. 8 His conservative black oxfords had cost him about as much as all the rest of his clothing. 1958 [see *beetle-crusher* s.v. *beetle *sb.*¹ 5]. 1960 'A. Burgess' *Doctor is Sick* 103 'We'll say a quid deposit, returnable on return of the hat, and a straight charge of an Oxford for the loan. Right?' 'Right.' The young man handed over his Oxford scholar. 1962 L. Deighton *Ipcress File* i. 13 He rocked on his hand-lasted Oxfords. 1964 J. Symons *End of Solomon Grundy* i. ii. 29 The routine of breakfast..Cooper's Oxford, the electric percolator. 1965 *Australasian Post* 4 Mar. 46 From 'dollar' we have the rhyming slang 'Oxford scholar', which eventually became shortened to an 'Oxford'. 1970 *Catal. L. L. Bean* (Freeport, Maine) Fall 32 Heavy duty nylon oxford outside is waterproof. 1971 D. E. Westlake *I gave at the Office* (1972) 130 Dressed in..new clothes—down to expensive black oxfords. 1972 'I. Drummond' *Frog in Moonflower* 10 The Master..spread a piece of toast with Cooper's Oxford. 1976 *New Yorker* 26 Jan. 52/3 It is like

seeing a pair of oxfords suspended from an ornate chandelier. 1978 *Spectator* (New Canaan High School, Connecticut) 66 Then I..pulled out four shirts: a turtleneck, a Lacoste 'alligator' shirt, a flannel shirt, and a wrinkled, white button-down Oxford.

b. The University of Oxford; *collect.,* the members of the University; quasi-*adj.,* belonging to or supporting the University. With specific adj., any of various school examinations conducted under the auspices of the University.

a 1697 Aubrey *Wiltshire: Topogr. Coll.* (1862) 17 At Oxford, (and I believe at Cambridge) the rod was frequently used by the Tutors and Deans. 1886 H. Baumann *Londinismen* 129/2 Are you Oxford or Cambridge? 1899 Beerbohm *More* 155, I was a modest, good-humoured boy. It is Oxford that has made me insufferable. 1916 W. Owen *Let.* Apr. (1967) 389, I hear you are applying yourself to some solid study for the J[unior] Oxford. 1930 *Times Lit. Suppl.* 25 Dec. 1103/1 There is encouraging evidence elsewhere that young Oxford is beginning to recognize that mere cleverness is poetically sterile. 1966 *Rep. Comm. Inquiry Univ. Oxford* I. 17 Members of Oxford.

Oxfordian, *a.* Substitute for def. (s.v. Oxford): Of, pertaining to, or designating a division of the Upper Jurassic in Britain lying below the Kimeridgian and above the Callovian (in continental Europe restricted to the lower part of this division). Also *absol.,* the Oxfordian stage or period. [In this sense ad. F. *oxfordien* (J. Thurmann 1830, in *Mém. de la Soc. d'Hist. nat. de Strasbourg* I. 22).] Add earlier and later examples.

1849 *Q. Jrnl. Geol. Soc.* V. 179 The clear definition of an equivalent of the Oxfordian group..had not been defined in the Southern Alps until M. von Buch demonstrated to the Italian geologists, that their 'Ammonitico rosso' was of Oxfordian age. *Ibid.* 182 The overlying stage..is therefore a good representative of the Oxfordian of the Alps. 1885, 1946 [see *Callovian *a.*]. 1967 D. H. Rayner *Stratigr. Brit. Isles* ix. 275 The base of the Upper Jurassic is sometimes taken at the base of the Oxfordian stage. 1975 A. Hallam *Jurassic Environments* ii. 16 The problem of correlation becomes much more serious from Upper Oxfordian times onwards. *Ibid.,* The increased difficulty of correlating the Upper Oxfordian is a consequence of northwards retreat of the boreal cardioceratid faunas which had ranged widely into southern Europe during the Lower Oxfordian.

2. [f. the title of Edward de Vere, Earl of *Oxford* (1550–1604) + -ian.] Used with reference to the theory that the Earl of Oxford wrote the plays attributed to Shakespeare. Also as *sb.*

1930 P. Allen *Case for Edward de Vere* 20 These poems of Chapman..are, in my judgement, enough..to prove the Oxfordian authorship of 'Shakespeare'. 1930 *Times Lit. Suppl.* 11 Sept. 712/2 Oxfordians seem to start from the basic assumption that the association of the work of Shakespeare with the Stratford player needs explanation. 1932 *Ibid.* 23 June 462/4 For the ordinary reader, and especially for the reader unfamiliar with the Oxfordian theory, Mr. Allen's book is not wholly satisfactory. 1958 *Listener* 17 July 100/3 Baconians, Oxfordians, Rutlandians, Derbyites. 1970 S. Schoenbaum *Shakespeare's Lives* vi. ix. 609 Freud read the Looney book. It converted him to the Oxfordian faith. *Ibid.* 610 The contempt felt by reputable scholars for the Oxfordians.

Oxfordish (ǫ·ksfədiʃ), *sb.* and *a.* [f. Oxford + -ish¹.] **A.** *sb.* Oxford jargon or slang. **B.** *adj.* Of, pertaining to, or suggestive of the University of Oxford.

1863 C. Reade *Hard Cash* I. 16 Ploughed is the new Oxfordish for plucked. 1921 R. Macaulay *Dangerous Ages* iv. 74 A pleasant, Oxfordish room, with the brown paper and plain green curtains of..college days. 1931 W. Holtby *Poor Caroline* v. 196 'Ah, a very opportune arrival, sergeant,' he began in his formal Oxfordish voice. 1962 J. D. Salinger *Franny & Zooey* 127, I can never bring myself to smile back at him when he's being charming and Oxfordish. He's on lend-lease or something from Oxford.

O·xfordy, *a.* [f. Oxford + -y¹.] Oxfordish, Oxonian.

1924 D. H. Lawrence in *Criterion* Oct. 27 It was his manner, his rather Oxfordy manner, more than anything else, that went beyond her. 1927 in *Lett. Gertrude Bell* I. i. 12 My sister..begged me to send Gertrude to stay with them for the winter..opining that frequenting foreign diplomatic Society might be a help for Gertrude 'to get rid of her Oxfordy manner'. *a* 1935 T. E. Lawrence *Mint* (1955) ii. iii. 108 The rough end of the hut tries to copy the accent he displays when he reads our nominal roll. It's an Oxfordy drawl. 1959 *Encounter* Aug. 74/1 This deliciously fat Oxfordy volume.

oxic (ǫ·ksik), *a.* [f. Ox(ide *sb.,* Ox(ygen + -ic.] **1.** *Soil Science.* Applied to a subsurface mineral soil horizon more than 30 cm. thick that is characterized by the virtual absence of any weatherable materials and the presence of hydrated oxides of iron and aluminium, highly insoluble minerals such as quartz, and clays of the type in which single sheets of silica tetrahedra alternate with single sheets of alumina octahedra. Cf. *Oxisol.

1960 *Soil Classification: 7th Approximation* (U.S. Dept. Agric.) v. 53/1 The concept of the oxic horizon presented here is very tentative, as it has had little testing... The oxic horizon is one from which weathering has..removed or altered a large part of the silica that is combined with iron and aluminium, but not necessarily the quartz or 1:1 lattice clays. 1970 P. M. Ahn *W. African Soils* (1974) viii. 221 The boundaries of the oxic horizon are usually gradual and diffuse, and the horizon shows no rock structure or very little. 1972 Foth & Turk *Fund. Soil Sci.* (ed. 5) x. 262 All soils with oxic horizons belong to the Oxisol order. 1976 D. Steila *Geogr. Soils* xi. 151 It is believed that the most extensive areal development of the oxic horizon may have taken place under paleo-climates of much higher rainfall.

2. [Back-formation from *anoxic *a.*] Involving, characterized by, or related to the presence of oxygen.

1970 *Acta Radiologica: Therapy, Physics, Biol.* IX. 257 The investigations..were conducted with the aim of comparing the reactions of different cell lines to oxic and anoxic roentgen radiation. 1972 *Radiation Bot.* XII. 151 The damage produced by anoxic irradiation can be enhanced by O_2 after irradiation, to give oxic damage. 1975 *Nature* 26 June 740/2 We have measured in oxic and anoxic cells the yield of strand breakage on bacteriophage λ DNA superinfecting lysogenic bacteria. 1975 *Ibid.* 4 Dec. 415/1 It is formed from phytol thermocatalytically in oxic conditions.

Hence **oxi·city,** oxic condition.

1978 *Nature* 16 Mar. 216/2 The available data suggest that the oxicity/anoxicity of the sediment and the water column affects the amount and nature of the organic matter incorporated into the sediment during deposition.

oxidant. Delete *rare* and add to def.: = Oxidizer 1 in Dict. and Suppl. (Further examples.)

1930 M. Stephenson *Bacterial Metabolism* iii. 49 The position of a system..is determined by the Eh when [the concentration of] oxidant equals [that of] reductant. 1950 *Engineering* 29 Dec. 575/3 Fuels can be burned..with fluorine when no oxygen is present at all—although a fluorine compound would still be known as the 'oxidant'. 1961 *New Scientist* 12 Oct. 110/1 Oxidants may include oxygen, air, chlorine,..and so on, and many combinations of fuels and oxidants are possible for fuel cell development. 1972 *Daily Tel.* 27 Nov. 3/2 Engineers pumped more than a million gallons of fuel and oxidant in the fuel tanks in a count-down rehearsal.

oxidase (ǫ·ksidēiz, -s). *Biochem.* Also †Oxydase in Dict. and Suppl. [ad. F. *oxydase* (G. Bertrand 1896, in *Compt. Rend.* CXXII. 1217), f. *oxyde* Oxide *sb.*: see *-ase.] Orig., any enzyme which brings about oxidation (now called an *oxidoreductase); now used only of such enzymes that react with molecular oxygen, esp. those that catalyse the transfer of hydrogen from a substrate to oxygen so as to form water or hydrogen peroxide (cf. *oxygenase).

1896 *Chem. News* 19 June 293/2 (*heading*) A new oxidase, or oxidising soluble ferment of vegetable origin. 1935 *Times Lit. Suppl.* 28 Feb. 127 Xanthine oxidase is unable to attack nucleosides. 1949 P. H. Mitchell *Textbk. Biochem.* xii. 340 Cytochrome Oxidase.—This enzyme specifically catalyzes the oxidation of cytochromes by molecular oxygen. 1956 *Sci. News* XL. 43 When fresh or withered [tea] leaf is damaged it absorbs oxygen from the air with the aid of an enzyme, an oxidase, occurring in the leaf cells. 1964 *Rep. Internat. Union Biochem. Comm. on Enzymes* in Florkin & Stotz *Comprehensive Biochem.* XIII. vi. 33 All enzymes catalysing oxido-reductions will be named 'oxidoreductases' in the systematic nomenclature... [In the trivial nomenclature] the terms 'dehydrogenase' or 'reductase' will be used much as hitherto... 'Oxidase' will be used only for cases where O_2 acts as an acceptor, and 'oxygenase' only for those cases where the O_2 molecule is directly incorporated into the substrate. 1970 R. W. McGilvery *Biochem.* xvii. 407 Aromatic rings may be opened by oxidations utilizing a complete molecule of oxygen. Examples:..the oxidation of hydroxyanthranilate by hydroxyanthranilate oxidase.

oxidation. a. (Further examples.) Add to def.: Also, the removal of hydrogen from a compound.

1866 [see *dehydrogenation]. 1900 *Jrnl. Chem. Soc.* LXXVIII. i. 301 (*heading*) Oxidation of the nature of dehydrogenation by means of ferricyanides. 1959 Cram & Hammond *Org. Chem.* vi. 73 In organic chemistry, oxidation involves the removal of hydrogen and (or) the addition of oxygen or some other hetero atom to a compound.

b. Add to def.: The partial or complete removal of an electron from an atom or molecule; an increase in the proportion of electronegative constituents in a molecule or compound. (Further examples.)

1907 Whitney & Brown tr. *Le Blanc's Text-bk. Electro-Chem.* vii. 256 There must be, in every galvanic cell, an oxidation at one electrode and a reduction at the other. 1928 H. L. Hind tr. *Schoen's Probl. Fermentation* xiv. 167 There is a change from divalent iron to iron at a higher degree of oxidation. 1968 Passmore & Robson *Compan. Med. Stud.* I. viii. 5/2 It is also characteristic of biological oxidations that they are often linked together to form a chain along which the electrons flow.

2. Special combs.: **oxidation number,** the charge (expressed in units of the negative of the electron charge) which is assigned to an atom on the assumption that the bonding in the substance or radical in which it exists is completely ionic; the average formal charge so assigned to atoms of a particular element in a compound or radical; **oxidation potential,** the electrode potential required to bring about a particular oxidation reaction at the electrode; **oxidation–reduction** = *redox; freq. *attrib.*; **oxidation state,** oxidation number; the state of having a particular oxidation number.

1948 *Jrnl. Chem. Educ.* XXV. 278/2 The O_2 molecule..is made up of two neutral atoms..; the oxidation number of the O atom is consequently zero. 1964 Cram & Hammond *Org. Chem.* (ed. 2) vi. 98 The system described is similar to the inorganic system if effective oxidation numbers, ranging from −4 (in CH_4) to +4 (CCl_4 or CO_2), are assigned to *individual carbon atoms* within organic molecules. 1968 J. March *Adv. Org. Chem.* xix. 853 Carbon in propane has an oxidation number of −2·67 and in butane of −2·5, though organic chemists seldom think of these two compounds as being in different oxidation states. 1900 *Jrnl. Chem. Soc.* LXXVIII. ii. 642 An attempt was made to measure a number of oxidation potentials. 1942 C. E. K. Mees *Theory Photogr. Process* viii. 308 When a silver salt solution is added to a reducing solution, there is an adjustment of the oxidation potentials of each to a common value intermediate between the two. 1966 C. R. Tottle *Sci. Engin. Materials* x. 225 The readiness to ionize by losing electrons in the presence of an ionizing solvent is called the oxidation potential. 1909 *Chem. Abstr.* III. 2648 It is therefore unnecessary to assume..that the e.m.f. in an oxidation-reduction cell is due to H_2 or O_2 at definite pressure in the electrode. 1951 *New Biol.* XI. 29 The over-all reaction in photosynthesis is of a type, extremely important in living organisms, known as an oxidation-reduction, in which one compound becomes oxidized at the expense of another which is reduced. 1974 *Sci. Amer.* Dec. 65/2 Chemical processes of this kind, in which electrons are transferred from one molecule to another, are called oxidation-reduction reactions. 1975 *Jrnl. Biol. Chem.* CCL. 9329 The oxidation-reduction equilibrium of the γ chains of human fetal hemoglobin. 1942 Sneed & Maynard *Gen. Inorg. Chem.* vi. 120 In many cases an element in its oxide or chloride is not reduced to the free state (by hydrogen) but to a lower oxidation state. 1973 J. C. Dick *Analytical Chem.* ii. 15 The hydrogen atom is assigned an oxidation state of +1, except in hydrides where the oxidation state is generally −1.

oxidative, *a.* Substitute for def.: Involving, pertaining to, or characterized by oxidation. (Further examples.)

1923 *Q. Jrnl. Med.* XVI. 145 (*heading*) The rate of oxidative recovery from exercise in man. 1935 C. F. & G. T. Cori in Harrow & Sherwin *Textbk. Biochem.* xx. 567 This so-called 'oxidative quotient' was found to be between 3 and 6 for all tissues examined. 1962 Dardenne & Kirsten in A. Pirie *Lens Metabolism Rel. Cataract* 419 Oxidative deaminations are of greater importance for the degradation of amino acids. 1972 *Daily Colonist* (Victoria, B.C.) 13 Feb. 27/5 It may be that the human need for Vitamin E is increasing due to the oxidative atmosphere man is creating around the world.

Hence **o·xidatively** *adv.,* by an oxidative process.

1964 *Oceanogr. & Marine Biol.* II. 152 The reineckate ion is removed with silver nitrate, and interfering substances (basic amino acids) are oxidatively deaminated by boiling with silver oxide. 1972 Cotton & Wilkinson *Adv. Inorg. Chem.* (ed. 3) xxiv. 773 Molecules that contain multiple bonds may be added oxidatively without cleavage to form new complexes which have 3-membered rings. 1976 *Lancet* 11 Dec. 1312/2 *Pseudomonas* metabolises carbohydrates oxidatively.

oxide. Add: **b.** *attrib.* and *Comb.,* as *oxide-coated* adj., *oxide coating.*

1933 J. H. Morecroft *Electron Tubes* ii. 23 In making the modern oxide-coated filament the coating is applied in the form of a white carbonate. *Ibid.,* The amount of CO_2 released from the oxide coating of an ordinary rectifier tube..., when the carbonate is being reduced to the oxide, is sufficient to fill the bulb to about 10 mm Hg pressure. 1964 F. Rosebury *Handbk. Electron Tube & Vacuum Techniques* (1965) 95 Experimental tubes often do not require their oxide-coated cathodes to have exceptionally long life. 1966 *McGraw-Hill Encycl. Sci. & Technol.* XIV. 246/2 Oxide coatings commonly have a work function of the order of 1·1 ev.

oxidize, *v.* **1.** Add to def.: More widely, to cause to undergo oxidation; to remove an electron from, completely or partly. (Further examples.)

1894 Perkin & Kipping *Org. Chem.* I. vi. 95 Alcohol is readily oxidised by chromic acid, yielding acetaldehyde. 1913 McPherson & Henderson *Course in Gen. Chem.* xxxiv. 471 Ferrous chloride is said to be oxidized to ferric chloride. 1942 Sneed & Maynard *Gen. Inorg. Chem.* xviii. 438 Chlorine oxidizes bromide ions and iodide ions to free bromine and iodine respectively. 1968 R. O. C. Norman *Princ. Org. Synthesis* iv. 143 The methyl group is oxidized when methane is converted by bromine into methyl bromide, because the electron-pair in the C—Br bond is less under the control of the carbon atom than the pair in the original C—H bond.

oxidizer. Add: **1.** *spec.* one used to support the combustion of fuel in a rocket engine or fuel cell. (Further examples.)

1950 *Sci. News* XV. 78 Liquid fluorine could be used as an 'oxidiser' with suitable fuels, but great care is needed when handling this liquid. 1952, etc. [see *FUEL *sb.* 2* c]. 1973 *Daily Tel.* 3 Aug. 34/3 Like all rocket engines they use a fuel, monomethyl hydroazine, and an oxydiser, nitrogen tetroxide.

oxidoreductase (ǫ:xidorĭdǝ·ktēiz, -s). *Biochem.* Also †oxydo-. [ad. F. *oxydo-réducase* (Battelli & Stern 1921, in *Arch. internat. de Physiol.* XVIII. 413), f. *oxyde* OXIDE *sb.* + *réduc-tion* REDUCTION: see *-ASE.] Any enzyme that catalyses oxidoreduction.

1922 [see *HYDRATASE]. 1922 [see *HYDROLASE]. 1928 *Physiol. Abstr.* XIII. 141 The oxydoreductase of yeast, unlike that from milk, acts on methyl glyoxal almost as strongly as on acetaldehyde. 1958 *Nature* 15 Feb. 452/2 Enzymes catalysing oxido-reductions would be named oxidoreductases in the systematic nomenclature, according to the scheme 'donor-acceptor reductase'. 1964 [see *OXIDASE]. 1974 *Encycl. Brit. Macropædia* VI. 897/2 Oxidoreductases and transferases account for about 50 percent of the approximately 1,000 enzymes recognized thus far.

oxidoreduction (ǫ:ksidorĭdǝ·kʃǝn). *Biochem.* [f. OXID(ATION + -O + REDUCTION.] A process in which one substance is oxidized and electrons from it reduce another substance.

1934 *Jrnl. Faculty Sci. Hokkaido Univ.* II. (Ser. 3) 18 In animal and in plant cells, oxidoreduction occurs in parallel with the formation of peroxide. 1958 [see *OXIDOREDUCTASE]. 1971 *Ital. Jrnl. Biochem.* XX. 129 Crystalline lactate dehydrogenase..catalyzes the dismutation of glyoxylate to glycolate and oxalate producing the oxido-reduction system of NAD coenzymes illustrated in Figure 1.

So **o:xidoredu·ctive** *a.*, involving oxidoreduction.

1951 WHITBY & HYNES *Med. Bacteriol.* (ed. 5) xx. 314 Others obtain energy from amino-acids, probably by paired oxido-reductive deaminations. 1971 J. Z. YOUNG *Introd. Study Man* xxvi. 376 Oxido-reductive reactions are linked with phosphorylation to produce ATP.

oximeter (ǫksi-mĭtǝɹ). *Med.* [f. OXI- + -METER.] A device for measuring the proportion of hæmoglobin in the blood which is in the oxidized form.

1942 G. A. MILLIKAN in *Rev. Sci. Instruments* XIII. 434 (*heading*) The oximeter, an instrument for measuring continuously the oxygen saturation of arterial blood in man. 1951 *Anesthesiology* XII. 549 With the oximeter it is possible to show how the anesthetist in his manipulations may induce trends which often lead to severe anoxemia. 1966 *McGraw-Hill Encycl. Sci. & Technol.* IX. 469/2 A second type of oximeter is designed to measure the oxygen saturation of blood outside the body during or shortly after withdrawal of the blood from various sites in the vascular system. 1975 *Sci. Amer.* Feb. 69/1 (Advt.), With the ear oximeter, the physician can make the measurement simply by attaching an optoelectronic device to the patient's ear and reading percent oxygen saturation on the display.

Hence **oxime·tric** *a.*, employing an oximeter; **oxi·metry**, the use of an oximeter.

1944 G. A. MILLIKAN in O. Glasser *Med. Physics* I. 900/1 (*heading*) Oximetry: continuous measurement of blood oxygen. 1948 *Federation Proc.* VII. 104/2 (*heading*) Oximetric determination of cardiac output in man. 1955 *Canad. Jrnl. Psychol.* IX. 67 (*heading*) Studies on the physiology of awareness: An oximetrically monitored controlled stress test. 1966 *McGraw-Hill Encycl. Sci. & Technol.* IX. 469/2 The physical basis of oximetry stems from the difference in absorption by oxygenated and reduced hemoglobin of red light of wavelengths in the region of 640 mμ. 1971 *Biol. Abstr.* LII. 9988 (*heading*) Cerebral venous oxygen saturation during rapid changes in the arterial blood pressure: an oximetric study in dogs.

oxine (ǫ·ksīn). *Chem.* [ad. G. *oxin* (Hahn & Vieweg 1927, in *Zeitschr. f. anal. Chem.* LXXI. 123), f. *oxychinolin*, hydroxyquinoline.] 8-Hydroxyquinoline, C_9H_7NO, a crystalline phenol which forms water-insoluble complexes with many metal ions and is used in analysis and as a deodorant and antibacterial agent.

1927 *Chem. Abstr.* XXI. 2444 Oxine itself is difficultly sol. in water..but is more sol. in AcOH. 1947 [see *HYDROXYQUINOLINE]. 1956 R. G. W. HOLLINGSHEAD *Oxine & Its Derivatives* IV. xxxiv. 1167 The growth of athlete's foot is prevented by the use of an absorbent powder, such as talc or starch, containing oxine and finely divided boric acid. 1973 T. MOHÁCSY tr. *Burger's Organic Reagents in Metal Analysis* ii. 97 Oxine complexes are usually intensely coloured in chloroform. This renders possible the spectrophotometric determination of many metal ions.

Oxisol (ǫ·ksisǫl). *Soil Science.* [f. *OXI(C *a.* + *-SOL.] A type of stable, highly weathered mineral soil found in tropical regions (see quots. 1960, 1971). Cf. *OXIC *a.* 1.

1960 *Soil Classification: 7th Approximation* (U.S. Dept. Agric.) xvi. 238/1 The Oxisols include the soils that, in recent years, have been called Latosols, and many, if not most, of those that have been called Ground-Water Laterite soils... All soils that have oxic horizons are included in the order. 1970 P. M. AHN *W. African Soils* (1974) vii. 221 The Oxisol order probably includes only the oldest and most highly weathered of West African soils. 1971 *Gloss. Soil Sci. Terms* (Soil Sci. Soc. Amer.) 26/1 *Oxisols*, mineral soils that have an oxic horizon within 2 m of the surface or plinthite as a continuous phase within 30 cm of the surface, and that do not have a spodic or argillic horizon above the oxic horizon. (An order in the USDA soil taxonomy.) 1972 [see *OXIC *a.* 1]. 1976 D. STEILA *Geogr. Soils* xi. 150 There are two main regions of Oxisol concentration: in South America surrounding the alluvial soils of the Amazon and in equatorial Africa.

Oxo (ǫ·kso). Also **oxo**. [f. OX + -O.] The proprietary name of an extract of beef, orig. a liquid, later in solid tablets, used as the basis of drinks and soups; also, the drink made from such tablets.

1899 *Trade Marks Jrnl.* 24 May 626 Oxo... Fluid beef. Liebig's Extract of Meat Company, Limited,..London,..and..Antwerp; manufacturers of Liebig Company's extract of meat, and manufacturers, shippers, and importers of South American produce. 1904 C. L. NEIL *Mod. Physical Culture* 15 Perhaps the handiest and best form in which to take this [*sc.* beef tea] is Oxo, made by Liebig's Extract of Meat Company. 1905 *Car* 24 May p. viii/1 A 4 oz. bottle of Oxo, and a few biscuits. 1907 *Yesterday's Shopping* (1969) 5/2 (*heading*) Liebig Co.'s Oxo Fluid Beef. 1914 D. O. BARNETT *Let.* 6 Nov. (1915) 4 Chocolate would be very welcome, also Oxo tablets. 1922 *Glasgow Herald* 13 Apr. 2 We shall carry..a thermos flask for Oxo. 1936 E. M. ANDERSON *Pract. Camp Cookery* 84 Brown sauce... 1 or 2 Oxo cubes. *Ibid.*, Use Oxo, relish, etc., if required, and season to taste. 1963 B. NILSON *Bk. Meat Cookery* 14/2 Pour over the Oxo stock. 1975 C. FREMLIN *Long Shadow* i. 10 Popping in with cups of tepid Oxo. 1976 A. HILL *Summer's End* i. 6 It smelled like a ton of flowers pressed into the space of an oxo-cube.

oxo(-) (ǫ·kso), *prefix* and *a.* *Chem.* [f. OX(YGEN + -O.] **1.** As a word-forming element in the names of organic compounds used to denote the presence of a carbonyl group, as in *oxodecanoic acid*, *3-oxovaleric acid*, *OXOSTEROID.

1971 *Nomencl. Org. Chem.* (I.U.P.A.C.) (ed. 3) C. 176 In the presence of a group having priority over carbonyl for citation as principal group, or when all the carbonyl groups cannot be included in the ketone functional class name, the presence of carbonyl-oxygen atoms is indicated by the prefixes 'oxo-', 'dioxo-', *etc.*

2. With a hyphen, as quasi-*adj.* without a hyphen, or joined, as one word, to a sb. **a.** Applied to an oxygen atom linking two other atoms, and to compounds containing such a grouping. Now *rare*.

1921 *Jrnl. Chem. Soc.* CXIX. 1657 This is the value of the factor for the oxo-linking, —O—. 1922 *Ibid.* CXXI. 1811 In the oxo-compound the molecule is in one straight line. 1934 *Jrnl. Amer. Chem. Soc.* LVI. 795/1 Ol compounds readily change to oxo compounds upon heating their solutions. 1962 [see *OL].

b. Applied to compounds, ions or groups containing one or more oxygen atoms bonded to another atom (*spec.* in *Org. Chem.* indicating the presence of a carbonyl group; cf. 1 above), as *oxo acid*, *anion*, *group*, *ion*.

1935 H. J. LUCAS *Org. Chem.* xxxix. 514 If the oxo group is on the end of a carbon chain, the compound is an aldehyde acid. 1958 *Jrnl. Physiol.* CXL. 154 The normal metabolism of the intestinal cells overshadows or compensates for any changes in the oxo acid concentrations in the blood. 1959 *Nomencl. Inorg. Chem.* (I.U.P.A.C.) v. 44 Most of the common acids are oxoacids, *i.e.*, they contain only oxygen atoms bound to the characteristic atom. 1965 PHILLIPS & WILLIAMS *Inorg. Chem.* I. xiv. 523 Oxyacids are sometimes called oxo-acids to stress the fact that they contain XO groups in their structures. 1969 H. T. EVANS tr. *Hägg's Gen. & Inorg. Chem.* xxi. 500 Charged oxide molecules, that is oxo ions, should also be considered as oxides. The particular descriptions of these..and the corresponding oxo acids, are given under the respective central-atom element. 1971 *Nomencl. Org. Chem.* (I.U.P.A.C.) (ed. 3) C. 120 Trivial names for oxo carboxylic acids..may be used for the acyclic component. 1972 COTTON & WILKINSON *Adv. Inorg. Chem.* (ed. 3) v. 170 The second main class of acid behavior is shown by compounds with X—OH groups; these are called oxo acids, and generally have a formula of the type H_nXO_m, for example H_3PO_4. *Ibid.* xxi. 639 Many of the most important ligands having oxygen donor atoms are oxo anions, such as NO_2^-, NO_3^-, SO_4^{2-} and ClO_4^-. *Ibid.* xiv. 408 Tetrahedral oxo ions such as PO_4^{3-}, ClO_4^-, MnO_4^-.

c. Also **OXO, Oxo.** Applied to the hydroformylation process or reaction.

1947 *Chem. & Industry* 1 Feb. 73/2 The OXO reaction is one of general applicability to compounds containing the ethylenic double bond. 1949 R. F. GOLDSTEIN *Petroleum Chem. Industry* x. 187 The Oxo reaction is exothermic, 35k. cal. being liberated per mol of olefin reacted. 1966 *McGraw-Hill Encycl. Sci. & Technol.* VI. 543/1 Cobalt catalysts are universally used for the Oxo reaction. 1969 [see *hydroformylation* s.v. *HYDRO-]. 1971 *Nature* 20 Aug. 537/1 Roden..discovered the oxo or hydroformylation reaction which generates aldehydes and alcohols from olefins, carbon monoxide, and hydrogen in the presence of cobalt catalysts.

Oxonian. Add: **B.** *sb.* Also, used of residents of Oxford who are not members of the University.

1966 *Oxford Mail* 16 Mar. 8/3 We real Oxonians do still exist.

Oxonolatry (later example).

1932 A. QUILLER-COUCH in *Fifty Years* 46 What we, the lighter-hearted, did not renounce was the charm of the place, that 'Oxonolatry', if you will, which Swinburne had scoffed at.

oxonium (ǫksōu·niǝm). *Chem.* [f. OX- 1 + *-ONIUM.] The hydroxonium ion, H_3O^+, or any derivative of this in which one or more of the hydrogen atoms are replaced by organic radicals. Usu. *attrib.*

1899 COLLIE & TICKLE in *Jrnl. Chem. Soc.* LXXV. 717 If oxygen can replace phosphorus, nitrogen, or sulphur in bases, then these oxygen compounds can be viewed as derivatives of the hypothetical base, oxonium hydroxide, OH_3OH. 1929 R. A. GORTNER *Outl. Biochem.* xxx. 621 The anthocyanin compounds readily add acids to form oxonium salts. 1959, 1966 [see *HYDRONIUM]. 1974 *Nature* 31 May 474/2 By dynamic reversal of the H_2O binding step before pseudorotation, the two oxygens in the pair of oxoniums could equilibrate with the H_2O oxygens of the medium.

oxosteroid (ǫksostīǝ·roid, -ste·roid). *Biochem.* [f. *OXO- 1 + *STEROID.] = *ketosteroid* s.v. *KETO- a.

1956 *Biochem. Jrnl.* LXII. 1P Urinary 17-oxosteroid conjugates consist of a mixture of glucuronides and sulphates. 1968 PASSMORE & ROBSON *Compan. Med. Stud.* I. xxxi. 13/1 (*caption*) 17-Oxosteroids are derived mainly from androgens and are excreted in urine principally as their 3-sulphate derivatives.

oxotremorine (ǫksotre·mǒrīn). *Pharm.* [f. *OXO- + *TREMORINE.] A crystalline compound, 1-(2-oxopyrrolidino)-4-pyrrolidino-2-butyne, $C_{12}H_{18}N_2O$, a metabolite of tremorine, which is capable of inducing the symptoms of Parkinsonism and is used in research into this disease.

1961 A. K. CHO et al. in *Biochem. & Biophys. Res. Communications* V. 276 For purposes of rough measurement, a unit was defined as the dose per Kg. which on intravenous injection in mice gave this threshold effect. It proved to be equal to approximately 15 μg. of 2-oxotremorine base. 1970 PASSMORE & ROBSON *Compan. Med. Stud.* II. v. 560/2 Another method [of inducing symptoms of Parkinsonism] involves the use of the drug tremorine.. or its active metabolite oxotremorine which induce tremor, rigidity, akinesia, analgesia and signs of peripheral cholinergic stimulation in many animal species. 1972 BRIMBLECOMBE & PINDER *Tremors* iv. 96 Leslie & Maxwell (1964) first suggested that oxotremorine is a more desirable tool in screening compounds for anti-Parkinson activity than is tremorine.

oxprenolol (ǫkspre·nǒlǫl). *Pharm.* [f. OX- 1 + *pren-* (f. *ISOPRENALINE) + reduplicated -OL (after *PROPRANOLOL).] The compound 1-(*o*-allyloxyphenoxy)-3-isopropylamino-2-propanol, $C_{15}H_{23}NO_3$, which is an adrenergic blocking agent used mainly (in the form of the white crystalline hydrochloride) in the treatment of cardiac arrhythmia, angina and hypertension.

1968 *Jrnl. Amer. Med. Assoc.* 29 July 292/1 The following nonproprietary names..have been adopted by the United States Adopted Names (USAN) Council.. Oxprenolol hydrochloride. 1969 *Jrnl. Appl. Physiol.* XXVII. 366/1 Oxprenolol, a specific beta-receptor blocking agent, diminished effort tachycardia by 15·0% and emotional tachycardia by 34·2%. 1977 *Lancet* 5 Nov. 954/1 Players receiving oxprenolol feel calmer before performing and happier with their recital.

ox-tail. *ox-tail soup* (examples).

1834 H. W. BRAND *Simpson's Cookery* v. 64 Ox-tail soup. Blanch two ox-tails, cut into pieces of a size suited to be served at table. 1841 THACKERAY *Gt. Hoggarty Diamond* (1849) xi. 142 Three silver tureens of soup: viz. mock-turtle soup, ox-tail soup, and giblet soup. 1865 C. M. YONGE *Clever Woman* II. iv. 53 The whole party were in a little den at the pastrycook's; the boys consuming mutton pies, and the ladies ox-tail soup. 1868 M. JEWRY *Warne's Model Cookery* 172/2 (*heading*) Ox-Tail Soup. 1939 JOYCE *Finnegans Wake* I. 133 Can rant as grave as oxtail soup and chat as gay as a porto flippant. 1970 SIMON & HOWE *Dict. Gastron.* 286/1 It is said the first ox-tail soup was made by a starving French nobleman during the Reign of Terror.

oxter, *sb.* (Further examples in Irish, and occas. wider, use.)

1901 G. B. SHAW *Admirable Bashville* II. 304 But many felt that Byron shewed bad taste In taking old Ned Skene upon his back, And, with Bob Mellish tucked beneath his oxter, Sprinting a hundred yards to show the crowd The perfect pink of his condition. 1914 JOYCE *Dubliners* 206 Many a good man went to the penny-a-week school with a sod of turf under his oxter. 1932 AUDEN *Orators* II. 70 The madman keeper crawls through brushwood, Axe under oxter. 1956 H. SUTHERLAND *Irish Journey* iii. 25 Each carrying a loaf under his 'oxter'. 1964 *Listener* 19 Mar. 494/3 Alan Whicker..stood..on that bubbling pitch lake of Trinidad..and let us hear a calypso from a man who'd fallen into it up to his oxters. 1977 D. BAGLEY *Enemy* ix. 63 Benson's carrying a gun in his oxter. 1978 *Jrnl. Lancs. Dial. Soc.* Jan. 15/1 [Durham] *Oxter*, armpit.

c. *Comb.* **oxter-plate** (see quot. 1904).

1885 H. PAASCH *From Keel to Truck* 46 Plate,..oxter—Tôle de voûte contre l'étambot. Achselgrube-Platte.

1904 A. C. HOLMS *Pract. Shipbuilding* I. 526 The oxter plates are those which take the sternpost, immediately below, or partly on, the transom. 1927 G. F. LEECHMAN *Theory & Pract. Steering* 51 The rotary current applies considerable pressure upon the hull in the vicinity of the oxter plate. 1948 R. DE KERCHOVE *Internat. Maritime Dict.* 511/1 *Oxter plate*, a shell plate riveted to the stern frame in way of the rudderpost head.

oxy, *a.* For *rare*⁻¹ read *rare* and add: **b.** Resembling an ox.

In quot. with pun on *Oxford*.

1922 JOYCE *Ulysses* 9 Tell that to the oxy chap downstairs and touch him for a guinea.

oxy (ǫ·ksi), *a.*² *Chem.* The prefix OXY- (in sense 2) used without a hyphen as a quasi-*adj.*

1910 *Encycl. Brit.* VI. 51/1 Oxyaldehydes and oxyketones (viz. compounds containing an oxy in addition to an aldehydic or ketonic group) undergo both condensation and oxidation when treated with phenylhydrazine. 1923 W. M. CUMMING et al. *Systematic Org. Chem.* v. 74 In the following section are discussed the more important of those condensations which give rise to oxy compounds—aldehydes, ketones, and quinones. 1962 COTTON & WILKINSON *Adv. Inorg. Chem.* v. 136 Many metal ions whose solutions are acidic may be regarded as oxy acids. 1975 *Inorg. Chem.* XIV. 1232/1 These chloroaluminate solvents are known to be effective chlorinating agents for a number of oxides and oxy anions.

oxy-. Add: **2.** Further examples of the use of *oxy-* to denote a combination or mixture of oxygen with a fuel gas, as *oxy-acetylene*, *-fuel*, *-gas*, *-propane*; all usu. *attrib.*; **oxy-helium**, a mixture of oxygen and helium, used as a breathing mixture in deep-sea diving.

1909 *Westm. Gaz.* 19 Jan. 4/2 A special weldless steel tubing brazed together by an oxy-acetylene process. 1939 L. TIBBENHAM *Welding Cast Iron* iii. 29 The temperature of an air-acetylene flame is about three-quarters that of oxy-acetylene. 1959 *Listener* 8 Oct. 583/3 Before the days of oxy-acetylene. 1961 C. WILLOCK *Death in Covert* iii. 56 A desk which had been designed by a modern sculptor whose chosen tool was the oxy-acetylene welder's torch. 1975 R. C. JAIN tr. *Castro & de Cadenet's Welding Metall.* iii. 20 The oxy-acetylene process is being largely replaced by the electric-arc methods since it has a number of disadvantages in the welding of stainless steels. 1969 *New Scientist* 8 May 284/2 The oxy-fuel burner was then lit and the charge melted with the burner operating under reducing conditions. 1951 E. G. WEST *Welding Non-Ferrous Metals* iii. 48 The oxy-gas cutting of ferrous metals, except stainless steels, depends on the rapid oxidation of the iron by the oxygen stream. 1974 *Nature* 4 Jan. 53/2 In this operation, in which oxy-gas torches have been used at the work face, temperatures are in the region of 3,000–3,500° C. 1966 A. B. CAMERON in P. Hepple *Petroleum Supply & Demand* 38 The use of oxyhelium equipment now enables them [*sc.* divers] to remain as deep as 525 ft for periods up to 30 minutes. 1969 *Physics Bull.* Feb. 51/2 There is a problem of speech communication between divers breathing oxy-helium and the men on the surface. 1976 *Offshore Engineer* Apr. 23/1 Comex physiologists first described the High Pressure Nervous Syndrome (HPNS) during oxy-helium dives as long ago as 1968. 1963 A. C. DAVIES *Sci. & Pract. Welding* (ed. 5) vi. 371 Iron and steel can be cut by the oxy-hydrogen, oxy-propane, oxy-coal gas and oxy-acetylene cutting blow-pipes with ease, speed and a cleanness of cut. 1970 *Daily Tel.* 24 Sept. 2/2 On the morning of the explosion two workmen were using oxy-propane cutting tools.

oxyanion (ǫ:ksi,æ·nǝiǫn). *Chem.* Also **oxy-anion** (with hyphen). [f. OXY- 2 + ANION.] An anion containing one or more atoms of another element each linked to one or more oxygen atoms.

1940 *Jrnl. Chem. Soc.* 131 (*heading*) The interchange of heavy oxygen between water and inorganic oxy-anions. 1952 *Chem. Rev.* L. 456 Since oxyanions have a large amount of resonance stabilization, it is necessary to overcome this stability in order that a chemical reaction take place. 1972 W. R. KNEEN et al. *Chemistry* xvii. 427 The phosphorus acids and oxyanions all contain approximately tetrahedral four-co-ordinate phosphorus.

oxy-arc (ǫ·ksi,ăɹk). Also **oxyarc.** [f. OXY- 2 + ARC.] An arc struck in an atmosphere of oxygen between a work-piece and a hollow electrode through which the oxygen is supplied. Usu. *attrib.*

1956 A. C. DAVIES *Sci. & Pract. Welding* (ed. 4) vi. 349 (*heading*) Oxyarc cutting process. 1962 J. BELL *Crime in our Time* I. 15 They..attempted to burn the safe open with oxy-arc equipment. 1974 *Petroleum Rev.* XXVIII. 676/1 Oxy-arc, at present widely used, may be replaced by the technique of arc-plasma. 1976 *Offshore Platforms & Pipelining* 155/3 This entails..cutting the pipe with an oxy-arc tool.

oxycellulose (ǫksise·liulōus). [f. OXY- 2 + CELLULOSE *sb.*] Any of various substances obtained by the oxidation of cellulose, some of which are used as gauze or lint in cases of hæmorrhage.

1893 CROSS & BEVAN in *Jrnl. Chem. Soc.* XLIII. 22 A portion of the mass [of cellulose]..yields but slowly to the action of the nitric acid, in consequence, we find, of its conversion into an oxidised derivative, to which we have provisionally given the name *oxycellulose*. 1938 *Ann. Rev. Biochem.* VII. 58 When oxycelluloses are boiled with dilute alkalis, the fluidity is virtually unaffected and there is only a slight decrease in tensile strength. 1954 KIRK & OTHMER *Encycl. Chem. Technol.* XIII. 551 Oxycellulose, when properly prepared, is a hemostatic material and is gradually absorbed by bodily fluids when it is used to pack wounds. 1963 [see *OVERBLEACHED *ppl. a.*].

oxychlorocruorin (ǫ:ksiklōǝrokrū·ǒrin). *Biochem.* [f. OXY- 2 + *chlorocruorin* s.v. CHLORO-¹.] The oxygenated form of chlorocruorin.

1870 *Jrnl. Anat. & Physiol.* IV. 129 (*caption*) Green blood of *Sabella ventilabrum*, shewing the two bands of Oxy-chlorocruorin. 1924 *Proc. Cambr. Philos. Soc.: Biol. Sci.* I. 217 The absorption bands of oxychlorocruorin and reduced chlorocruorin resemble those of oxy- and reduced haemoglobin shifted towards the red end of the spectrum. 1950 *Sci. News* XV. 89 Oxyhaemoglobin and oxychlorocruorin have each their characteristic absorption spectrum and the blood of *Serpula* shows both.

oxychromatin (ǫksikrōu·mătin). *Biol.* [a. G. *oxychromatin* (M. Heidenhain 1894, in *Arch. f. mikrosk. Anat.* XLIII. 543), f. Gr. ὀξυ- sharp, acid: see *CHROMATIN.] A supported component of chromatin characterized by a greater affinity for acid dyes and a smaller content of nucleic acid than chromosomal chromatin.

1895 *Jrnl. R. Microsc. Soc.* 159 'Oxychromatin' and 'cyanophilous granulation' are independent structural parts of the chromatin framework. 1924 E. G. CONKLIN in E. V. Cowdry *Gen. Cytol.* ix. 550 The oxychromatin which escapes into the cell body at mitoses is often more abundant than that which goes to form the chromosomes. 1948 W. ANDREW tr. *E. D. P. de Robertis's Gen. Cytol.* vii. 140 With ultracentrifugation the nucleus is deformed and the nuclear structures are stratified in the following layers..: (1) nucleolus, (2) basichromatin, (3) oxychromatin, (4) karyolymph.

oxydase. Substitute for def.: obs. var. *OXIDASE. Add earlier and later examples.

1896 *Jrnl. Chem. Soc.* LXX. II. 571 The author gives the generic term *oxydase* to laicase, tyrosinase, and other oxidising ferments of vegetable origin. 1929 R. P. WALTON tr. *Waldschmidt-Leitz's Enzyme Actions* vii. 231 In these cases, special oxydases..appear to be present.

oxydoreductase, obs. var. *OXIDOREDUCTASE.

oxygen. Add: **1.** Also *fig.*

1849 LYTTON *Caxtons* I. II. i. 45 Having thus exhausted all the oxygen of learning in that little receiver [*sc.* a preparatory school], my parents looked out for a wider range for my inspirations. 1932 *Sunday Times* 15 May 6 That cheerful noise which is the oxygen of 'society'.

b. An atom of oxygen.

1950 F. H. HATCH et al. *Petrol. Igneous Rocks* (ed. 10) I. ii. 68 The essential hydroxyl groups are included in the planes containing the 'free' oxygens in the tetrahedra. 1966 *Mineral. Mag.* XXXV. 1071 These compounds contain oxygens associated with three nearly equidistant protons forming H_3O^+ ions. 1974 *Nature* 4 Jan. 15/1 The main chain carbonyl oxygens were not resolved from the rest of the main chain, so that the orientation of the peptide bonds could not be determined directly.

3. b. *oxygen-carrier* (earlier and later examples), *lack*, *saturation*, *tension*; *oxygen-carrying* (examples), *-dependent*, *-free*, *-poor* adjs.; **oxygen bottle**, a cylinder of compressed or liquid oxygen; **oxygen debt** (see quot. 1923); **oxygen lance** (see *LANCE *sb.*¹ 6**); **oxygen mask**, a mask fitting over the nose and mouth through which oxygen or oxygen-enriched air may be supplied for breathing; **oxygen tent**, a tent-like enclosure for placing over a patient in order to provide him with an oxygen-enriched atmosphere.

1932 *Illustrierte Technische Wörterbücher* XVII. 234/1 *Sauerstoffbombe* (f.)—oxygen bottle—bombe (f) à oxygène—bombola (f) d'ossigeno. 1941 *Flight* 16 Jan. 48/1 The rigger changes the oxygen bottles and fits the starting motor to the aircraft. 1974 HAWKEY & BINGHAM *Wild Card* xix. 156 The other [trolley], an anesthesia machine, carried three yoked oxygen bottles. 1888 *Jrnl. R. Microsc. Soc.* 596 Schunck..regards chlorophyll as a respiratory pigment, but probably a carbonic acid-carrier, not an oxygen-carrier. 1972 *Cytobiologie* V. 52 This was the first application of the perfusion medium with a fluorocarbon as oxygen carrier. 1916 A. P. MATHEWS *Physiol. Chem.* xi. 496 There may be in addition a union with the hemoglobin, which will retard its oxygen-carrying capacity. 1968 *Times* 13 Nov. 16/1 The beads may be minute clumps of haemoglobin, the oxygen-carrying protein. 1923 HILL & LUPTON in *Q. Jrnl. Med.* XVI. 142 The 'oxygen debt' is defined as the total amount of oxygen used, after cessation of exercise, in recovery therefrom. 1947 T. K. CURETON et al. *Physical Fitness Appraisal* xiv. 437 The lactic acid [in the blood] increases very rapidly, with an estimated increment of 7 grams for each liter of oxygen debt. 1969 J. H. GREEN *Basic Clin. Physiol.* xi. 66/1 It is possible to run a short distance (100 yards) without breathing... After the exercise has been completed the subject breathes deeply and rapidly for the next few minutes in order to take in oxygen to 'repay' the oxygen debt. 1956 *Nature* 17 Mar. 531/1 Oxygen-dependent reactions. 1972 *Jrnl. Exper. Marine Biol. & Ecol.* IX. 217 The respiratory rate of the oxygen-dependent prosobranch, *Buccinum undatum* L., increases with decreasing salinity. 1933 *Mining & Metallurgy* XIV. 340/1 Oxygen-free high-conductivity copper..that is now being commercially offered for the first time represents a notable achievement in electro-metallurgy. 1972 *Radiation Res.* XLIX. 507 The oxygen-carrying capacity of *Busycon* and *Limulus* hemocyanins is eliminated following irradiation of the protein with cobalt-60 γ-rays in neutral, oxygen-free media. 1925 *Physiol. Rev.* V. 554 Carbon dioxide and oxygen lack produced marked increases in respiration with very little change in the reaction of the blood. 1963 R. P. DALES *Annelids* iv. 93 Among oligochaetes the abilities of *Tubifex* to withstand oxygen-lack are well known. 1920 *Abstr. Papers in Sci. Trans. & Periodicals* July 177 A description is given of the oxygen mask used by the French military aviators. 1930 E. RICE *Voy. to Purilia* i. 16 The air grew rarer... We were obliged to don the oxygen masks..so carefully and laboriously constructed for us. 1964 L. S. BRUNNER et al. *Text-bk. Medical-Surgical Nursing* xii. 189/1 Those individuals whose need for supplemental oxygen is greatest..are the very ones who are most prone to resist the application of an oxygen mask. 1975 *Times* 4 Sept. 1/8, I saw an air hostess run into the first-class compartment with an oxygen mask. 1951 M. ABERCROMBIE et al. *Dict. Biol.* 128 *Lung*... Present in early fishes before origin of Amphibia, probably as an accessory breathing organ adapted to oxygen-poor fresh waters in which earliest vertebrates probably lived. 1969 *Listener* 6 Feb. 163/1 Mars has an atmosphere which is appreciable, even though it is too thin and too oxygen-poor to support any Earth creatures. 1942 *Electronic Engin.* XIV. 724 The ratio of their output currents will remain constant for any given degree of oxygen saturation. 1969 CROFTON & DOUGLAS *Respiratory Dis.* i. 33/1 The oxygen saturation is relatively uninfluenced until the partial pressure of oxygen in alveolar gas falls to relatively low levels. 1916 A. P. MATHEWS *Physiol. Chem.* xi. 481 The arterial blood had an oxygen tension varying in different experiments from 91·6–104·4 per cent of the tension in the alveolar air. 1953 J. HUNT *Ascent of Everest* 273 The respiratory centre in the brain..responds normally not to the oxygen tension of the blood..but to the direct effect of carbon dioxide tension of the arterial blood. 1965 B. E. FREEMAN tr. *Vandel's Biospeleol.* xviii. 318 The oxygen tension of subterranean waters is variable. [1921 *Jrnl. Physiol.* LV. p. xx (*heading*) A simple oxygen bed tent and its use to [*sic*] a case of œdema and chronic ulcer of the leg.] 1925 *Sci. Amer.* Sept. 181/2 (*caption*) A portable oxygen tent for pneumonia patients. 1974 C. HILL *Scorpion* 55 Michael was lying under an oxygen tent ..the top of his head covered in bandages.

oxygenase (ǫ·ksidʒĕnēiz, -s). *Biochem.* [a. G. *oxygenase* (Chodat & Bach 1903, in *Ber. d. Deut. Chem. Ges.* XXXVI. 607): see OXYGEN and *-ASE.] Any enzyme which catalyses the incorporation of molecular oxygen into a substrate; orig. used in the narrowest sense of *OXIDASE.

1903 *Jrnl. Chem. Soc.* LXXXIV. I. 378 Most oxydases contain principles of both types, and it is proposed to retain the term peroxydase for those substances..which are not themselves oxidisers, but impart activity to, and thus destroy, peroxides, whilst the new term, 'oxygenase', is proposed for those substances..which are capable of producing hydrogen peroxide, but leave it in an inactive condition. 1920 M. W. ONSLOW in *Biochem. Jrnl.* XIV. 536 The function of this additional enzyme is to catalyse the oxidation of the catechol substance with the formation of a peroxide... It is proposed to call this second enzyme an oxygenase, a term originally used by Chodat and Bach for the portion of an oxidase which can be replaced by hydrogen peroxide. 1964 [see *OXIDASE]. 1971 M. F. MALLETTE et al. *Introd. Biochem.* viii. 282 Both oxygenases and hydroxylases use O_2 to alter substrates... Both classes of enzymes reduce O_2 but do it by incorporating oxygen atoms into organic substrates. On the other hand, oxidases reduce O_2..and do not incorporate it into organic compounds.

oxygenator. Add: **c.** *Med.* An apparatus for oxygenating the blood.

1928 *Jrnl. Physiol.* LXVI. 443 Drinker, Drinker and Lund used a modification of Hooker's oxygenator in bone marrow perfusion experiments. 1961 [see *HOLE *sb.* 7 e]. 1968 *Sci. Jrnl.* Nov. 63/2 Attempts to replace the living heart and lungs by a mechanical pump and an artificial oxygenator were unsatisfactory.

oxygenless (ǫ·ksi,dʒĕnles), *a.* [f. OXYGEN + -LESS.] Not containing oxygen.

1935 E. R. BURROUGHS *Pirates of Venus* 46 Even though she [*sc.* the planet Venus] be oxygenless. 1963 I. FLEMING *On H.M. Secret Service* ix. 98 James Bond followed her, holding his breath against the searing impact of the Arctic, oxygenless air. 1972 *Sci. Amer.* Apr. 58/2 We still have the 'fossils' of this life-style in the universal process of anaerobic (oxygenless) fermentation.

oxylith (ǫ·ksiliþ). Also **-lithe.** [ad. F. *oxylithe*, f. *oxy-* OXY- 2 + *-lithe* -LITH.] A commercial name for calcium peroxide, CaO_2, as used in breathing apparatus as a convenient source of oxygen (evolved by reaction with carbon dioxide).

1902 *Chem. News* 16 May 240/2 The recent invention of 'oxylith', due to M. Jaubert, will help to make the use of nascent oxygen still more wide in its applications. 1924 *Chambers's Jrnl.* Oct. 701/2 The helmet of this appliance.. is supplied with a substance called 'oxylithe', which gives off pure oxygen when breathed upon. 1948 R. H. DAVIS *Breathing in Irrespirable Atmospheres* vi. 206 Where 'Oxylithe' was employed as the main source of oxygen in the earlier designs, it took a couple of minutes to evolve oxygen in sufficient quantity to allow work to begin.

oxyluciferin (ǫksiliusi·fĕrin). *Biochem.* [f. OXY- 2 + *LUCIFERIN.] The oxidised form

of (a) luciferin produced by the action of luciferase. (Cf. the note s.v. *LUCIFERIN.)

1919 E. N. HARVEY in *Jrnl. Gen. Physiol.* I. 135, I suggest also that luciferin when oxidized be designated oxyluciferin. 1953 J. RAMSBOTTOM *Mushrooms & Toadstools* xiv. 163 Luciferin is oxidised by luciferase in one part of a cell producing oxyluciferin which in another part is reduced again to luciferin. 1964 *Oceanogr. & Marine Biol.* II. 351 Chromatography yields three isolated substances, two of which are active in luminescence and have been designated oxyluciferin A and oxyluciferin B.

oxymoronic (ǫ:ksimǒǝrǫ·nik), *a.* [f. OXYMORON + -IC.] Suggestive of oxymoron; incongruous, self-contradictory. So **oxymoro·nically** *adv.*

1901 BEERBOHM *Around Theatres* (1924) I. 270 A little bore, whose oxymoronically belated-premature death we hail..as a merciful release. 1970 G. GREER *Female Eunuch* 164 It would be oxymoronic to claim to be gently, reliably or sensibly in love. 1972 P. GREEN *Shadow of Parthenon* 35 'Imperial democracy', that oxymoronic contradiction in terms. 1975 *Times Lit. Suppl.* 17 Oct. 1226/5 Professor Hardy's point..is signalled by the oxymoronic 'dubious consolations' of her title.

oxymyoglobin (ǫ:ksimǝi,ǒglōu·bin). *Biochem.* [f. OXY- 2 + *MYOGLOBIN.] The oxygenated form of myoglobin.

1935 *Physiol. Abstr.* XIX. 578 The dissociation curve of oxymyoglobin is a hyperbola. 1953 FRUTON & SIMMONDS *Gen. Biochem.* vi. 167 At a given oxygen pressure oxymyoglobin is less dissociated than oxyhemoglobin, i.e. myoglobin has a higher affinity for oxygen. 1974 M. D. RANKEN in Birch & Parker *Vitamin C* ix. 122 At the surface of fresh meat, in contact with oxygen, the colour is the bright red of oxymyoglobin.

ox yoke. [OX 5.] A yoke used for draught oxen. Also *transf.*

1573, 1688 [see OX 5]. 1785 G. WASHINGTON *Diary* 16 Nov. (1925) II. 441 A good Oxe Cart—2 Oxe yokes. 1809 *Austin Papers* (1924) I. 164 One Plough, one Harrow.. and two Ox yokes. 1847 in H. Howe *Hist. Coll. Ohio* 188 Journeys..of 20 or 50 miles, for the sole purpose of having the staple of an ox-yoke mended. 1879 B. F. TAYLOR *Summer-Savory* xvii. 138 Awkward H's like a pair of leaning bar-posts with one bar, and B's like ox-yokes. 1938 M. K. RAWLINGS *Yearling* iv. 41 Penny.. went to the sink-hole with a wooden ox-yoke supporting two wooden buckets over his thin shoulders.

oxyphil, *a.* Add: Also **-phile.** For 'acids' read 'acid dyes'. (Earlier and later examples.)

1893 *Brit. Med. Jrnl.* 25 Feb. 400/1 The complete histological differentiation of a leucocyte depends upon staining it in such a manner that all the oxyphile and basophile elements are brought out in relief. 1925 H. J. CONN et al. *Biol. Stains* vii. 79 Yellowish eosin..is used in various technics for staining the oxyphile granules of cells. 1970 *Jrnl. Path.* CII. 193 One of the adenomata in this series is composed of oxyphil cells and transitional oxyphil cells when viewed by light microscopy. 1976 *Clinical Endocrinol.* V. Suppl. 377 The one oxyphil adenoma tested by Schorr *et al.* (1972) showed no response.

oxyphilic (ǫksifi·lik), *a.* *Biol.* [f. prec. + -IC.] Readily stained by acid dyes; oxyphil.

1901 C. E. SIMON *Text-bk. Physiol. Chem.* xiii. 303 The granules which are found in certain forms of leucocytes are apparently of an albuminous nature. According to their affinity for acid, basic, or neutral dyes, they are termed oxyphilic, basophilic, and neutrophilic, respectively. 1973 *Cancer* XXXI. 253/1 The endoplasmic reticulum is poorly developed in oxyphilic cells, which means that protein-synthesis must be scanty.

oxyphilous (ǫksifilǝs), *a.* *Biol.* [f. as prec. + -OUS.] = prec.

1893 *Brit. Med. Jrnl.* 25 Feb. 400/2 All the granules found in normal circulating leucocytes are oxyphilous. 1968 *Zeitschr. f. Mikrosk.-Anat. Forsch.* LXXIX. 551 Oxyphilous Welsh cells of the parathyroid glands have glandular functions.

oxyproline (ǫksiprōu·līn). *Biochem.* [ad. G. *oxyprolin* (H. Leuchs 1905, in *Ber. d. Deut. Chem. Ges.* XXXVIII. 1937): see OXY- 2 and *PROLINE.] = *hydroxyproline* s.v. *HYDROXY-.

1928 *Physiol. Abstr.* XII. 599 Tetrasulphocyandiamminechromic acid..may be used to give crystalline precipitates with proline and oxyproline. 1931 *Chem. Rev.* IX. 264 Hammarsten..has synthesized γ-oxyproline. 1935 A. K. ANDERSON *Essent. Physiol. Chem.* xvi. 177 The pyrrol rings are present in the amino acids proline and oxyproline. 1972 *Biol. Abstr.* LIV. 4493/2 As silicosis progressed the quartz and oxyproline content in the lungs increased.

oxytetracycline (ǫ:ksitetrăsǝi·klīn). *Pharm.* [f. OXY- 2 + *TETRACYCLINE.] The 5-hydroxy derivative, $C_{22}H_{24}N_2O_9$, of tetracycline (usu. administered in the form of its yellow hydrochloride), which is an antimicrobial substance produced by cultures of the bacterium *Streptomyces rimosus* and has actions and uses similar to those of other tetracyclines.

1953 *Jrnl. Amer. Med. Assoc.* 3 Jan. 46/1 Both compounds possess a common four-ringed skeleton, for which the generic term tetracycline has been proposed. They differ only in that aureomycin has a chlorine atom at ring one while terramycin has a hydroxyl group at ring three. The chemically descriptive generic names chlorotetracycline and oxytetracycline, respectively, have therefore been proposed for the two compounds. 1958 *Times* 29 Sept. 2/7 The promise of antibiotics as a means of controlling certain plant diseases has been brought a step nearer fulfilment by a recent Order which permits the use of streptomycin and oxytetracycline. 1966 *Economist* 9 Apr. 169/1 This February, Pfizer's patent monopoly on the drug, oxytetracycline, expired. 1969 CROFTON & DOUGLAS *Respiratory Dis.* xviii. 321/1 Sometimes oxytetracycline or chlortetracycline is better tolerated [than tetracycline]. 1974 R. M. KIRK et al. *Surgery* ii. 24 Oxytetracycline is given to prevent secondary infection.

oxytocin (ǫksitōu·sin). *Med.* [f. OXYTOC(IC *a.* and *sb.* + -IN¹.] A hormone present in the neurohypophysis of mammals which stimulates uterine contractions and the ejection of milk from a lactating breast and is made synthetically for use in inducing labour and controlling bleeding after delivery.

1928 *Jrnl. Amer. Chem. Soc.* Feb. 575 α-Hypophamine is being supplied under the trade name of Oxytocin..and β-Hypophamine under the trade name Vasopressin... They were first supplied for clinical use in August, 1927. 1928 *Official Gaz.* (U.S. Patent Office) 11 Sept. 258/1 Parke, Davis & Company, Detroit, Mich... *Oxytocin* for preparation for stimulating contractions of uterine and other unstriped muscular fibers. Claims use since Sept. 15, 1927. 1930 *Druggists Circular* Oct. 60/2 In 1928 Oliver Kamm succeeded in separating the two principles of the posterior lobe of the pituitary body. One of them, which acts upon the uterus, Kamm named oxytocin. 1937 *Dispensatory U.S.A.* (ed. 22) 854/1 The oxytocic principle has been called pitocin, oxytocin, [etc.]. 1965 LEE & KNOWLES *Animal Hormones* ii. 29 The second hormone, oxytocin, appears to exert its action only in mammals, but it may play some part in oviposition in the other vertebrates. 1969 *Daily Tel.* 19 Dec. 11/8 Oxytocin is used regularly to start or speed up a labour that is lagging. 1974 *Nature* 20 Dec. 630/1 Neurones in the hypothalamus synthesise the hormones vasopressin and oxytocin. 1977 'E. CRISPIN' *Glimpses of Moon* xii. 256 It read like an instruction from an obstetrician to a lady who..has been given oxytocin to hurry matters up.

Hence **oxyto·cinase** [*-ASE], an enzyme that inactivates oxytocin.

1949 *Chem. Abstr.* XLIII. 2330 (*heading*) Effect of cysteine and glutathione on the oxytocinase activity of blood plasma. 1974 PASSMORE & ROBSON *Compan. Med. Stud.* III. II. xl. 1/2 A serum oxytocinase which rapidly inactivates oxytocin appears in increasing amounts during pregnancy.

oxytonic (ǫksitǫ·nik), *a.* Also **oxyto·nical.** [f. OXYTONE *a.* and *sb.* + -IC.] Characterized by an oxytone; designating a language in which the majority of words are oxytones.

1890 *Century Dict.*, Oxytonical. 1954 PEI & GAYNOR *Dict. Linguistics* 156 *Oxytonic language*, a language in which the majority of the words bear the main accent on the last syllable. 1963 *Amer. Speech* XXXVIII. 217 Armenian, like French, is oxytonic; with rare exceptions, the stress in a word falls on the last syllable. 1972 HARTMANN & STORK *Dict. Lang. & Linguistics* 160/2 *Oxytonic language*, a language in which most words have fixed.. stress on the final syllable.

oy, oi (oi), *int.* [Yiddish.] An exclamation used by Yiddish-speakers to express dismay, grief, etc. Occas. in wider use. Also *oy vay*, *vey* [a. G. *Weh* woe] (see quot. 1968).

1892 I. ZANGWILL *Childr. Ghetto* I. xii. 270 The dispute thickened; the synagogue hummed with 'Ei's' and 'Oi's' not in concord. 1924 *Dialect Notes* V. 274 Exclamations in American English... Oy: —, — yoy. 1928 H. CRANE *Let.* 22 Feb. (1965) 317 Oy-oy-oy! I have just had my ninth snifter of Scotch. 1932 L. GOLDING *Magnolia Street* III. viii. 570 'And if you hadn't..oi! oi!' said Ada. 'It would have been awful!' 1934 WODEHOUSE *Right Ho, Jeeves* xi. 141 A story about a Scotchman, an Irishman, and a Jew... I said 'Hoots, mon,' 'Begorrah,' and 'Oy, oy.' 1939 MRS. P. CAMPBELL *Let.* 28 June in *B. Shaw & Mrs. Campbell* (1952) 332 A Jewish Mother... The Mothers dialogue consisting of: 'Oi; Oi; tch: tch.' 1959 B. KOPS *Hamlet of Stepney Green* I. 12 Children, *oy vay*, don't talk to me about children. 1963 V. NABOKOV *Gift* iii. 181 His trick of garbling Russian, in imitation of a farcical Jewish accent as when he said..'Oy, vat a mudnik!' 1968 L. ROSTEN *Joys of Yiddish* 14 Two A[lteren] K[ockern] had sat in silence on their favorite park bench for hours, lost in thought. Finally, one gave a long and languid '*Oy!*' The other replied, 'You're telling *me*?' *Ibid.* 273 *Oy* is often used as lead-off for '*oy vay!*' which means, literally, 'Oh, pain', but is used as an all-purpose ejaculation to express anything from trivial delight to abysmal woe. 1975 *New Yorker* 3 Mar. 34/3 The family gathered round for Passover. Oy, nephew! What's the world coming to. 1976 K. THACKERAY *Crownbird* v. 91 'Oy vey, but he's recovered well,' Stein thought.

‖**oyama** (oya·ma). [Jap.] = *ONNAGATA.

1963 *Guardian* 21 Jan. 9/3 The extraordinary verisimilitude achieved by the 'oyama' or actors playing female rôles. 1965 *This is Japan 1966* 51/2 He was considered to be the last of the great *oyama*, male performers of female roles on the Shimpa stage.

-oyl, ending of the names of acid radicals, formed on the names of the corresponding carboxylic acids ending in *-ic* or *-oic*; e.g. *FUMAROYL, *HEXANOYL.

1971 *Nomencl. Org. Chem.* (I.U.P.A.C.) (ed. 2) C. 185 The name of a univalent or bivalent acyl radical formed by removal of hydroxyl from all the carboxyl groups is derived from the name of the corresponding acid by changing the ending '-oic' to '-oyl'.

oyster. Add: **1. c.** *the world is my oyster*: the world offers opportunities for profit, etc.; also in extended uses.

1598 SHAKS. *Merry W.* II. ii. 2 Why then the world's mine Oyster, which I, with sword will open. 1930 J. A. WILLIAMSON *Short Hist. Brit. Expansion* (ed. 2) II. VI. ii. 183 *Laissez-faire*, with its cosmopolitan view of the world as the trader's oyster. *a* 1938 T. WOLFE *You can't go Home Again* (1940) III. xxix. 469 Drake was self-contained: the world his oyster, seas his pastures, mighty distances his wings. 1942 A. BRYANT *Yrs. of Endurance* xi. 230 He [*sc.* Napoleon] was not going to waste his incomparable genius to make the world—his oyster—safe for Barras and the plutocrats of the Luxembourg. 1949 A. MILLER *Death of Salesman* I. 39 The world is an oyster, but you don't crack it open on a mattress. 1975 *New Yorker* 26 May 66/3 A few weeks after the conference, he told American farmers that the world food market could 'be our oyster'. 1977 'J. GASH' *Judas Pair* vi. 67 The world was my oyster. My uneasy mood had vanished.

d. A reserved or uncommunicative person.

1925 M. WILTSHIRE *Thursday's Child* xi. 221, I wouldn't mind betting Jane's worrying herself sick over it; and he—goodness knows what he's doing or feeling. I never saw such an oyster. 1930 J. B. PRIESTLEY *Angel Pavement* vi. 305, I never knew anybody so close, you old oyster you!

e. A type of unmoored submarine mine detonated magnetically or acoustically as a vessel passes over it. Also *attrib.*

1947 CROWTHER & WHIDDINGTON *Science at War* iv. 177 The mine mechanism [*i.e.* of German pressure mines] consisted of a rubber air bag with an aluminium diaphragm. With a change of pressure air escapes from the bag, the diaphragm is moved, and after a time closes an electrical detonating circuit. A change of about 1/1,000th in the total pressure, equivalent to that of about ½ inch of water, exerted for about six seconds, was needed to operate the mechanism. This device, called an 'oyster', is shown in Plate XLV. 1950 A. P. HERBERT *Independent Member* lii. 308 The enemy..had new terrors by then—the 'oyster' or pressure mine. 1955 J. F. TURNER *Service most Silent* xi. 155 For the first four years of the War both the Germans and ourselves were developing the top secret mine of the War—the oyster. *Ibid.*, The Luftwaffe produced 'acoustic oysters' and the Navy 'magnetic oysters'. 1965 K. LANGMAID *Approaches are Mined!* xiv. 240 The first 'Oyster' minefields were laid by German light craft on the night of the 6th/7th of June [1944].

4*. A greyish-white colour resembling that of an oyster. Cf. *5 c.

1922 *Daily Mail* 11 Dec. 14 (Advt.), Silk hose... In black, white,..peacock, flame, oyster. 1960 *Housewife* Apr. 97 Cotton sailcloth... In a choice of three good colours—oyster, light royal blue, or black. 1978 H. MACINNES *Prelude to Terror* ii. 18 A..study in greys, from silver carpeting to pale oyster walls.

4.** One of the cross-sections of wood in an oyster veneer.

1924 G. O. WHEELER *Old Eng. Furnit.* (ed. 3) iii. 22 These 'oysters' are often in kidney-shape and the welding of a mosaic was no easy task. 1974 *Country Life* 26 Sept. (Suppl.) 60 William III oyster olivewood chest. The top with concentric rings of oysters of decreasing size.

5. a. simple *attrib.*, as *oyster shoal*; connected with the taking, breeding, keeping, selling, or eating of oysters, as *oyster-bar* (examples), *-dish*, *house*, *-merchant*, *navy*, *pirate*, *saloon*, *season*, *-shop* (examples), *-smack* (examples), *-stall*, *stand*, *supper*; made of oysters, as *oyster cocktail*, *cracker*, *-patty* (examples), *-pie* (later examples), *-sauce* (earlier examples), *soup*, *stew*, *stuffing*.

1878 R. L. STEVENSON in *London* 8 June 441/1 They were driven by a sharp fall of sleet into an Oyster Bar. 1925 E. SITWELL *Troy Park* 74 That child is the small wicked ghost Of Metropoles and oyster bars. 1972 E. HARGREAVES *Fair Green Weed* I. 12 I've been eating something called escovitched fish in an oyster bar. 1895 *Funk's Stand. Dict.*, Oyster cocktail. 1905 *Granville Centennial Cook Bk.* 121 Oyster Cocktail. *c* 1938 [see *COCKTAIL 4]. 1957 M. MCCARTHY *Memories Catholic Girlhood* viii. 203 Olympia oyster cocktail and devilled Dungeness crabs. 1975 M. ORR *Rich Girl, Poor Girl* (1977) xii. 147 The meal..began with an oyster cocktail and progressed to a cold Senegalese soup. 1873 *Kansas Mag.* III. 273/2 Our commissary department was poorly supplied.. four small oyster crackers. 1924 *Amer. Mercury* Apr. 430/1 The custom that some Baptist churches have fallen into of oyster crackers and cubes of bakers' bread in the Lord's supper is to my mind unscriptural. 1975 BYFIELD & TEDESCHI *Solemn High Murder* (1976) iii. 53 Baxter tore open a cellophane envelope of oyster crackers. 1865 GEO. ELIOT *Let.* 8 Jan. (1956) IV. 174, I am so much pleased with..the pretty oyster-dish. 1834 H. J. NOTT *Novelettes* I. 94 He can escape from the empty pageant to the substantial and homely comforts of a beefsteak or oyster house. 1949 *Fishing Gaz.* Oct. 96/2 Hampton oyster houses are George T. Elliott, M. F. Quinn, and J. S. Daly and Son. 1726 B. FRANKLIN *Jrnl. Voy. Philadelphia* in *Mem. Life & Writings* (1818) I. App. p. iii, The oyster-merchants fetch them..from other places. 1869 *Rep. Comm. Agric. 1868* (U.S. Dept. Agric.) 340 An oyster merchant of Rochelle, doing business with the growers of the adjacent islands of Oleron and Ré, will say £250,000 per annum. 1932 *Sun* (Baltimore) 19 Sept. 6/5 (*heading*) Sour note on a recent addition to the oyster navy. 1962 *Daily Progress* (Charlottesville, Va.) 21 Mar. 3 (*heading*) 'Oyster navy' gets radar. 1807 M. E. RUNDELL *New Syst. Dom. Cookery* 133 (*caption*) Oyster patties. 1843 *Ainsworth's Mag.* IV. 97 An eulogy of the excellence of Lord Marmiton's oyster patties. 1932 AUDEN in *Rev. Eng. Stud.* (1978) Aug. 285 Over oyster patties, I'll explain it all. 1953 K. TENNANT *Joyful Condemned* vi. 51 Mrs. Mike was carrying a plate of oyster patties. 1851 A. O. HALL *Manhattaner* 59 Some of them [*sc.* mosquitoes] are dainty, and associate only with fat people whose nightmares are based upon turtle steaks and oyster pies. 1976 R. CONDON *Whisper of Axe* I. xxiii. 146 They ate oyster pie and crab cakes. 1903 J. LONDON *Let.* 9 Mar. (1966) 147 When the oyster pirates..arrived, they forced the two watchmen off into the water. 1930 J. DOS PASSOS *42nd Parallel* II. 133 Oyster pirates used to shanghai young fellers. 1833 *Knickerbocker* I. 117 To be seen about taverns and oyster saloons. 1905 J. C. LINCOLN *Partners of Tide* vi. 105 The pair entered a little battered restaurant with the sign 'Atwood's Oyster Saloon' over the door. 1727 'E. DORRINGTON' *Hermit* I. 27 The boil'd Meat and Oyster-Sauce. 1798 JANE AUSTEN *Let.* 24 Oct. (1932) I. 9 We.. had some beef-steaks and a boiled fowl, but no oyster sauce. 1861 MRS. BEETON *Bk. Househ. Managem.* p. xxiv (Index), Oyster..season. 1866 'F. KIRKLAND' *Pictorial Bk. Anecdotes* 181/2 [He] traded up and down the James and York rivers, especially during the oyster season. 1977 *Harpers & Queen* Nov. 276/3 When the British oyster season is over, the clam trade continues. 1827 J. L. WILLIAMS *View W. Florida* 16 The entrance to this bay is obstructed by sand bars and oyster shoals. 1823 in *Spirit of Public Jrnls.* (1825) 8 Charged with assaulting David Tullock, *Esq.* at an oyster-shop in Brydges-street. 1841 DICKENS *Old C. Shop* xxxix. 12 Kit, walking into an oyster-shop as bold as if he lived there. 1913 MRS. P. CAMPBELL *Let.* 25 Mar. in *B. Shaw & Mrs. Campbell* (1952) 102 Many a rendez-vous at Cheesmans oyster shop. 1802 E. WYNNE *Diary* 22 Oct. (1940) III. iv. 69 Several other sailors' bodies have been thrown on shore, it was an oyster smack that was lost. 1976 *Times* 27 Aug. 17/1 A string of oyster smacks..will be competing in the Thames Oyster Smack Race. 1741 E. SMITH *Compl. Housewife* (ed. 10) 62 (*heading*) Oyster soop. 1861 MRS. BEETON *Bk. Househ. Managem.* 103 Oyster soup. 1935 M. MORPHY *Recipes of all Nations* 599 *Oyster soup*.. is one of the favourite soups in America. 1836 DICKENS *Pickw.* (1837) xxii. 227 Here's a oyster-stall to every half-dozen houses. 1922 E. SITWELL *Façade* 15 Oyster-stall notes. 1830 *Boston Even. Transcript* 29 Sept. 2/4 The oyster stands in New Orleans have been leased for..the same price as last year. 1851 A. O. HALL *Manhattaner* 7 Oyster stands, where dirty mouths and flickering tallow candles grinned ghostly satisfaction. 1977 *Times* 14 May 13/4 For £4,000 to £6,000 each..you may be able to buy two Sèvres oyster-stands. 1846 D. CORCORAN *Pickings* 128 Mrs. Smith was never known to have an oyster stew of an evening that she did not divide it with Mrs. Jones. 1973 P. A. WHITNEY *Snowfire* xiii. 254 He..brought me back a bowl of hot oyster stew and crackers. 1935 M. MORPHY *Recipes of All Nations* 611 Oyster stuffings for poultry are frequently found in old English cookery books. 1971 M. G. EBERHART *Two Little Rich Girls* (1972) ix. 105 The dinner..with its oyster stuffing for the turkey and its huge mince pie. 1741 B. LYNDE *Diary* 17 Apr. (1880) 107 Oyster supper with all the Court. 1856 H. B. STOWE *Dred* II. 221 He drinks and frolics, and has his oyster-suppers. 1949 *Missouri Hist. Rev.* Apr. 215 He is uneasy at oyster-suppers at the 'Opera House'.

b. *oyster-eating* (examples), *-opener* (examples), *shucker*; also instrumental, as *oyster-covered* adj.

1882 W. D. HAY *Brighter Britain!* I. iii. 72 Oyster-covered rocks. 1905 *Westm. Gaz.* 10 Aug. 1/3 Oyster-covered iron beams and girders. 1933 *Sun* (Baltimore) 7 Aug. 14/2 An old-fashioned Southern Maryland 'oyster eatin''. 1977 *Times* 22 Jan. 12/3 The return to oyster-eating. 1827 W. CLARKE *Every Night Bk.* 62 If the visitor make an ally of the waiter or oyster-opener, he may often have people pointed out to him there, who are rather worth seeing. 1900 W. STEVENS *Let.* 21 Oct. (1967) 47 Fishmen, and grizzly oyster-openers. 1969 E. H. PINTO *Treen* 140 A simple but effective oyster opener, in Colchester Museum, is a wooden block, hollowed out to take a large oyster; [etc.]. 1898 GOULD *Pocket Med. Dict.* (ed. 2) 234 *Oyster-shuckers' keratitis*, a form due to corneal traumatism from pieces of embedded oyster-shell. 1969 L. HELLMAN *Unfinished Woman* vi. 61 The oyster shucker.. would open oysters for my father. 1973 *Daily Colonist* (Victoria, B.C.) 1 July 9/2 Rumor has it that the professional oyster shucker eventually becomes immune to the poison and simply counts his scars.

c. *oyster eye*, *kiss*; *oyster-coloured*, *-white* (examples) adjs.; also, further examples in the sense of oyster-coloured.

1901 *Sketch* 11 Sept. 303 Her wedding-dress of oyster-white satin. 1904 *Daily Chron.* 5 May 8/4 Women are wearing bronze shoes with their golden-brown costumes, ..oyster-coloured suede with a costume of that shade. 1920 S. LEWIS *Main St.* xi. 141 An oyster-coloured blouse. 1922 JOYCE *Ulysses* 114 Oyster eyes. 1931 J. CANNAN *High Table* xi. 168 The mother of the bride wore a handsome dress of oyster satin. 1938 S. BECKETT *Murphy* 117 Oyster kisses passed between them. 1951 [see *GIBSON²]. 1952 P. ATKEY *Juniper Rock* i. 5 A short, pink man in oyster silk pyjamas. 1958 L. DURRELL *Balthazar* i. 13 Pearl ground with shadowed oyster and violet reflections. 1960 *Harper's Bazaar* Apr. 84 A short narrow dress of delustred oyster satin. 1969 [see *Irish crochet* s.v. *IRISH *a.* 2 c]. 1974 N. MARSH *Black as he's Painted* i. 20 The glossy walls were an agreeable oyster-white. 1975 G. MOFFAT *Miss Pink* iii. 48 Bridget wearing oyster lace over silk. 1976 *Southern Even. Echo* (Southampton) 11 Nov. 9/1 The bride wore an oyster-coloured empire-line dress.

d. *oyster-boat* (later examples); hence *oyster-boatman*; *oyster-cellar* (earlier examples); *oyster-farm*, *-farming* (examples); hence *oyster farmer*; **oyster-crab** (later examples); **Oyster Feast**, a traditional feast held at Colchester to mark the beginning of the oyster-fishing season; **oyster-fish** (*b*) for '*Batrachus tau*' substitute '*Opsanus tau*'; (examples); (*c*) (examples); (*d*) (see quot. 1903); **oyster fitting**

(see quot. 1940); **oyster-knife** (later examples); **oyster-like** *a.* (later examples); **oyster-man** (further examples); **oyster-piece**, a piece of oyster veneer; **oyster-plant**, (*a*) (later examples); (*b*) (examples); **oyster-scow** *U.S.*, a scow engaged in oyster-fishing; **oyster-tongs** (later examples); **oyster-veneer**, a whorled veneer obtained esp. from small boughs of trees; cf. *OYSTERING *vbl. sb.* b; also *oyster-veneered* adj., *oyster-veneering*; **oyster walnut** (see quot. 1944); freq. *attrib.*

1813 J. K. PAULDING *Lay of Scottish Fiddle* i. 18 The sailors..urg'd in dreams the gallant chase Of oyster-boats far up the bay. 1891 *Scribner's Mag.* X. 472, I will try to describe how the deck of an oyster-boat must be trimmed for its work. 1859 G. A. SALA *Twice round Clock* 251 Listen to the slang of oyster-boatmen and bargees. 1830 J. F. WATSON *Ann. Philadelphia* 220 Oyster Cellars..did not at first include gentlemen among their visiters. 1842 DICKENS *Amer. Notes* I. vi. 208 Lamps, marking the whereabouts of oyster-cellars. 1844 J. E. DEKAY *Zool. N.Y.* VI. 12 The P[innotheres] *depressum* of Say, is..the male, or we suppose the young, of the Common Oyster Crab. 1884 J. A. RYDER in G. B. Goode *Fisheries U.S.: Nat. Hist. Aquatic Animals* I. x. 744 Some Oysters were dredged up by the crew which contained some Oyster-crabs. 1902 H. L. WILSON *Spenders* 131 Now the oysters will be due—fine fat Buzzard's Bays—and oyster crabs. 1938 L. BEMELMANS *Life Class* I. iii. 49 All maîtres d'hôtel..are especially fond of little fried things.. whitebait, oyster crabs, fried scallops. 1960 C. M. YONGE *Oysters* vii. 118 An essentially parasitic crab which has been the cause of considerable damage.., especially in Delaware Bay, is the small Oyster Crab. 1940 *Sun* (Baltimore) 18 Oct. 3 Chesapeake oysters, which are now raised on 'oyster farms'. 1975 *Times* 24 Apr. 3/4 Their oyster farm is claimed to be the only one of its kind in Europe. 1946 *Nature* 26 Oct. 587/1 Many of the newly settled spat perish in the first weeks of sedentary life, and in spite of all the care of the oyster-farmers. 1953 *Sun* (Baltimore) 5 Feb. 19/5 If private leasing of such beds were allowed, they could produce enough seed to supply oyster 'farmers' their all-important seed oysters. 1977 *Harpers & Queen* Nov. 275/4 Fertilised eggs, which oyster farmers call..white, grey or black sick, depending on how ripe it is. 1943 *Sun* (Baltimore) 5 Feb. 10/1 A system of 'oyster farming' combining a free fishery with close State management. 1962 D. NICHOLS *Echinoderms* iii. 35 In the early days of oyster-farming the fishermen would drag a dredge across the beds to collect the starfish. 1974 *Country Life* 21 Nov. 1561/3 A high mercurial content in the water ..prevents oyster farming. 1888 E. L. CUTTS *Colchester* xviii. 171 The annual feast on the election of the mayor is called the Oyster Feast...people take it for granted that the feast derives its name from the bivalve for which the town is famous; but this origin of the name may be questioned. 'Oyster feasts' are common at the beginning of a new official reign in many places. 1924 F. MUIRHEAD *England* (ed. 2) 575 The opening of the oyster fishing is celebrated by an 'Oyster Feast' on Oct. 20th. 1934 A. E. HOUSMAN *Let.* 15 Sept. (1971) 360 The chief ambition of my life has long been to be invited to the Colchester Oyster Feast. 1972 R. COBB *Reactions to French Revolution* 4 He had once dressed up as a Roman Emperor for the carnival at the time of the Oyster Feast. 1855 S. F. BAIRD in *Rep. Bd. Regents Smithsonian Inst.* IX. 340 The toad-fish, or, as it is called at Beesley's point, the oyster-fish, on account of its frequenting the oyster beds, is one of the fisherman's pests. 1878 *Proc. U.S. Nat. Mus.* I. 374 *Tautoga omitis* [sic].—Oyster-fish. Rather common. 1884 G. B. GOODE *Nat. Hist. Aquatic Anim.* I. III. 251 The toad-fish, *Batrachus tau*, called also on the coast of New Jersey and in some parts of the Southern States 'Oyster-fish', is one of the most repulsive looking fishes upon our coast. 1903 T. H. BEAN *Fishes N.Y.* 598 This [*sc. Tautoga onitis*] is better known in New York as the blackfish; farther south it is styled..Moll, Will George and oyster fish. *Ibid.* 656 *Gobiosoma bosci*... Naked Goby; Mud Creeper; Oysterfish. 1940 *Sun* (Baltimore) 30 Apr. 6/3 The oyster fish, sometimes called 'devil fish' by local fishermen, has a big head and mouth with which it crushes oysters for food. 1962 K. F. LAGLER et al. *Ichthyology* xi. 361 In the oyster toad-fish (*Opsanus tau*)..blood sugar levels are raised upon intramuscular injection of corticosteroid compounds. 1892 J. W. URQUHART *Electric Ship-Lighting* vii. 227 The oyster fitting, without the guard, is much used for cabins. 1940 *Chambers's Techn. Dict.* 606/2 *Oyster-fitting*, a bulkhead fitting designed to emit light on both sides of the bulkhead or other partition upon which it is mounted. *a* 1841 W. P. HAWES *Sporting Scenes* (1842) II. 120 Oyster-knives and blood become well acquainted. 1907 *Yesterday's Shopping* (1969) 120/3 Place a stiff knife similar to an oyster knife between the rubber ring and flat metal top, and..it will easily come off. 1973 *Daily Colonist* (Victoria, B.C.) 1 July 9/1 An oyster knife..is obviously patterned on the rapier. 1937 *Daily Herald* 16 Feb. 19/6, I have never known Mr. Rinder so definitely oyster-like as when I tackled him yesterday on the line he proposed to take in the broadcast. 1975 *Country Life* 13 Sept. 1320/4 Its [*sc.* salsify's] stale fish flavour, described as oyster-like. *a* 1976 A. CHRISTIE *Autobiogr.* (1977) x. ii. 488 There was something oyster-like about Rosalind. 1753 in E. Singleton *Social N.Y. under Georges* (1902) 350, I am informed that an oysterman..may clear eight or ten shillings a day. 1853 O. S. FOWLER *Home for All* (rev. ed.) 23 Those persons who would economize, have only to order those very shells which the oyster-man has to pay to have carted from his cellar. 1955 *Times* 31 Aug. 5/1 But the oystermen of Cornwall eat the succulent grade four oyster with greater relish than its bigger brother. 1974 'A. GARVE' *File on Lester* xxxix. 143, I talked first with some oystermen along the front here. 1925 Oyster piece [see *OYSTERING *vbl. sb.* b]. 1960 *Times* 9 Dec. 18/7 The beautiful 'oyster-piece' veneers and marquetry. 1973 *Daily Tel.* (Colour Suppl.) 16 Mar. 47/2 Laburnum wood which has a dark heart and a yellow sapring when cut across the grain produces highly decorative 'oyster pieces'. 1821 W. COBBETT *Amer. Gardener* iv. 257 Salsafy, called by some oyster plant, is good in soups, or to eat like the parsnip. 1841 *Cultivator* VIII. 114 Oyster Plant, or Salsify..after boiling soft, make gravy of flour, butter, etc. and add to them, and really they are rich substitutes for oysters. 1885 E. P. ROE *Nature's Serial Story* xxvi. 186 Will your nose become *retroussé* if I ask you to aid me in planting parsnips, oyster-plant, carrots and.. onions? 1938 R. GATHORNE-HARDY *Wild Flowers in Brit.* vi. 40 Another beautiful blue flower to be found on the coasts of Scotland and northern England is the uncommon Oyster Plant, *Mertensia maritima*,..which spreads its fat leaves and drooping blue flowers over the sea-shingle. 1960 *Oxf. Bk. Wild Flowers* 170/1 Sea Lungwort or Oyster Plant (*Mertensia maritima*)..is a rare plant of stony sea-shores in the north. 1972 Y. LOVELOCK *Vegetable Bk.* 201 The roots [of salsify] also are eaten..and are said to taste like asparagus; others suppose them to taste like oysters, from which belief it gains its name vegetable oyster and oyster plant. 1976 *Billings* (Montana) *Gaz.* 1 July 2-A/4 Ten percent vegetables and fruits: Beets, carrots, onion, oyster plant (salsify) rutabagas. 1824 *Nantucket Inquirer* 26 Jan. (Th.), He wore a hat of the new oyster-scow cut. 1856 *Dollar Times* (Cincinnati, Ohio) 11 Dec. 2/5 Our river boats are palaces of paint and gilding, but a leak from the bowsprit of an oyster-scow will sink one in fifteen minutes. 1835 J. J. AUDUBON *Ornith. Biogr.* III. 608 My host carried with him..a pair of oyster-tongs. 1949 R. J. SIM *Pages from Past Rural N.J.* 74 In oyster tongs the pin is thirty-two inches or more above the heads. 1909 G. O. WHEELER *Old English Furnit.* (ed. 2) iii. 115 Sections of small walnut branches were built in veneers,..resembling..oyster-shells, and.. this particular work has come to be classed as 'oyster veneer'. 1974 *Country Life* 30 May 1538/1 The use of walnut oyster-veneers in England is common. 1914 EBERLEIN & MCCLURE *Pract. Bk. Period Furnit.* 86 When the cabinets were 'oyster' veneered, inlaid with marqueterie or lacquered. 1976 *Country Life* 27 May (Suppl.) 48d/1 (Advt.), A rare 17th century oyster-veneered walnut side table. 1916 E. W. GREGORY *Furnit. Collector* vi. 91 The well-known 'oyster' veneering is also typical of the style. 1944 C. DREPPERD *Primer of Amer. Antiques* 241/2 *Oyster Walnut*, the burl in walnut having oyster shapes and forms in it and obvious when cut on the bias. A fine veneer pattern. 1972 *Country Life* 8 June (Suppl.) 51 A small William & Mary 'oyster walnut' chest of drawers.

oyster, *v.* Add: **b.** *trans.* To feed on oysters; with *up*.

1861 T. WINTHROP *Cecil Dreeme* 156 Boys, I've got a sick man to oyster up.

c. *intr.* To shut *up*; be silent. *slang.* Cf. *CLAM *v.*⁴ 2.

1973 R. PARKES *Guardians* xii. 225 Once they got him down the station he oystered up proper. Not another word.

oystered (oi·stǝɹd), *a.* [f. OYSTER *sb.* + -ED².] **1.** Of a veneer: bearing an oyster-shaped or whorled pattern.

1914 EBERLEIN & MCCLURE *Pract. Bk. Period Furnit.* ii. 57 The middle or end of the Carolean epoch when the whorled or 'oystered' veneer made from the transverse slices of small boughs came into vogue.

2. Eaten with oysters. *Nonce-use.*

1932 DYLAN THOMAS *Let.* (1966) 4 Oh, woe..unto Mumbles and the oystered beer.

oystering, *vbl. sb.* Add: **2.** Oyster veneer or work done with this.

1914 EBERLEIN & MCCLURE *Pract. Bk. Period Furnit.* iii. 90 Marqueterie, oystering and lacquer were freely used in their decorations. 1925 PENDEREL-BRODHURST & LAYTON *Gloss. Eng. Furnit.* 117 The slices are referred to as Oyster-pieces, and the arrangement as Oystering.

oyster-shell. Add: Also *fig.*

1876 TROLLOPE *Prime Minister* IV. vi. 87 When Aristides has been much too just the oyster-shells become numerous.

b. **oyster-shell bark-louse** *N. Amer.* = *oyster-shell scale* (*a*); **oyster-shell scale** *N. Amer.*, (*a*) a scale insect, *Lepidosaphes ulmi*, which attacks many trees and shrubs; (*b*) the disease produced by this insect, characterized by small curved scales on the plant's bark; **oyster-shell veneer, veneering** = *oyster veneer*, *veneering* s.v. *OYSTER 5 d.

1868 *Rep. Comm. Agric. 1867* (U.S. Dept. Agric.) 73 The oyster shell bark louse or scale insect..is said to be exterminated by washing the tree with a mixture of two parts of soft soap, eight parts of water, and lime enough to give it the consistence of whitewash. 1877 *Rep. Vermont Board Agric.* IV. 150 Dr. Worcester has also shown me a branch covered with oyster shell bark-louse. *Ibid.*, The insect is shorter and stouter than that of the oyster shell scale. 1924 LAWRENCE & SKINNER *Boy in Bush* ii. 21 He was glad to reach the oyster-shell path running up Wellington Street. 1924 C. C. DEAM *Shrubs of Indiana* 233 Some individuals of several species are attacked and even killed both in cultivation and in the wild state by the oyster-shell scale. 1929 G. G. & F. GOULD *Period Furnit. Handbk.* xiv. 147 Veneer:..oyster-shell—in walnut, irregularly shaped oval pieces, cross sections cut from small boughs, the wood graining suggesting oyster-shells, popular in the late Charles II period. 1955 R. FASTNEDGE *Eng. Furnit. Styles* 286 This oystershell veneering, used for cabinet doors and drawer fronts, was introduced from Holland in the late seventeenth century. 1971 A. CHRISTIE *Nemesis* viii. 86 A William and Mary oyster-shell bureau. 1972 SWAN & PAPP *Common Insects N. Amer.* 163 Oystershell Scale: *lepidosaphes ulmi*... The armor is light to dark brown, shaped like a tiny oyster shell.

Ozalid (ōu·zălid). [Formed by reversing Diazo- and inserting *l*.] A proprietary name used esp. in connection with a diazotype copying process in which the light-sensitive coating of the paper contains the coupling compound as well as the diazonium salt, so that the image may be made visible by exposure to gaseous ammonia. Hence, a photocopy produced by this process.

1924 *Trade Marks Jrnl.* 16 July 1608 Ozalid... Paper.., stationery, and bookbinding. Kalle & Co., Aktien Gesellschaft.., Biebrich-on-Rhine, Germany; manufacturers and merchants. **1928** *Official Gaz.* (U.S. Patent Office) 13 Nov. 292/2 Kalle & Co. Aktiengesellschaft, Wiesbaden-Biebrich, Germany. Filed Jan. 30, 1928. *Ozalid* for light-sensitive copying and photographic papers. Claims use since Mar. 19, 1923. **1929** *Encycl. Brit.* XVII. 803/1 G. Kögel has patented..the use of diazoanhydrides..for paper that may be used for the same purpose; this process is known as 'Ozalid'. **1939** *Thorpe's Dict. Appl. Chem.* (ed. 4) III. 589/2 Ozalide [*sic*] papers, which have largely displaced blueprint paper, are based on the principle that a light-sensitive diazoanhydride may be mixed with a phenol or aromatic amine without coupling until the mixture is rendered slightly alkaline. **1941** *Official Gaz.* (U.S. Patent Office) 23 Dec. 767/2 General Aniline & Film Corporation, New York, N.Y. Filed Oct. 22, 1941. *Ozalid.* For light-sensitive diazotype paper, cloths, films, etc., for machines for developing the photoprints thus produced and parts of such machines. Claims use for light-sensitive diazo type materials since Mar. 19, 1923; and for printing and developing machines since May 1, 1936. **1944** *Trade Marks Jrnl.* 12 Apr. 167/2 Ozalid... Photographic and photocopying apparatus, instruments, and utensils. Ozalid Company Limited,.. London,..manufacturers and merchants. **1967** V. Strauss *Printing Industry* v. 267/2 Diazo papers..are also known as Ozalids and whiteprints. **1970** E. A. D. Hutchings *Survey of Printing Processes* vii. 114 Development of the intermediates is carried out in an Ozalid ammonia vapour dyeline machine. **1975** J. Butcher *Copy-editing* v. 61 The final proof is a photographic proof (usually an Ozalid).

Ozark (ōu·zāɪk). Also **Osark.** [ad. F. *aux Arcs* at the Quapaw, ult. ad. Illinois *akansea* Quapaw Indian.] The Quapaw, a North American Indian people, or perhaps a local group of this tribe.

1816 H. Ker *Trav. Western Interior U.S.* 40 We were visited by a few of the Osark tribe of Indians, who came to us in canoes... They are called by the name of a river they inhabit, on the west side of the Mississippi. **1821** T. Nuttall *Jrnl. Trav. Arkansa* vi. 81 The aborigines of this territory, now commonly called Arkansas or Quapaws and Osarks, do not..number more than about 200 warriors. **1910** F. W. Hodge *Handbk. Amer. Indians* II. 180/2 *Ozark*, a term at one time applied to a local band of Quapaw, from their residence in the Ozark mountain region of Missouri and Arkansas.

ozonation. (Further examples.)

1948 Kirk & Othmer *Encycl. Chem. Technol.* II. 426 A triozonide is formed with difficulty by ozonation of benzene. **1972** *Adv. Chem.* CXII. 1 A competition exists during ozonation of olefins between ozonolysis and epoxide formation.

ozone. Add: **c. ozone-sonde,** a radiosonde for transmitting information on the ozone content of the atmosphere; also without hyphen as one word or two.

1960 *Monograph Internat. Geodetic & Geophysical Union* No. 3, 20 During the IGY many successful balloon soundings with the ozone-sonde were obtained. **1964** *Bull. Atomic Sci.* Jan. 29 In the past two years there has been an increased emphasis on several aspects of antarctic meteorology...albedo programs, meteorological studies aboard the Eltanin, and the inclusion of vertical coverage through..ozonesondes, and gammasondes. **1969** McIntosh & Thom *Essent. Meteorol.* vii. 111 One form of ozone sonde..is that devised by A. W. Brewer. Air is bubbled through a small electrolytic cell filled with neutral potassium iodide solution.

ozoner (ōu·zōunəɹ). *U.S. slang.* [f. Ozone + -er[1].] A drive-in cinema. Also *ozoner cinema.*

1948 *Time* 26 Apr. 96/2 This week, New York City will get its first 'ozoner': a 600-car..affair on Staten Island. **1949** *Sat. Rev. Lit.* (U.S.) 11 June 4/1 There are now between 1,000 and 2,000 drive-ins in the U.S...Hollywood calls them 'ozoners'. **1962** *Punch* 24 Jan. 167/1 Virtually every picture window of the motel rooms face[s] out on the ozoner cinema.

ozonide. Add: **2. a.** [ad. G. *ozonid* (C. Harries 1904, in *Ber. d. Deut. Chem. Ges.* XXXVII. 840).] Any of the compounds containing the ring C—O—O—C—O, which are formed by the addition of ozone to olefinic double bonds and are explosive oils or amorphous solids.

1904 *Jrnl. Chem. Soc.* LXXXVI. i. 361 The ozonides are mostly highly explosive. **1929** R. A. Gortner *Outl. Biochem.* xxxi. 640 On treatment of the oleic acid ozonide with water, it decomposes into hydrogen peroxide, pelargonic acid, and azelaic acid semi-aldehyde. **1959** E. L. Mascall *Pi in the High* 28 Though many facts that art. provides *Re* substances called Ozonides. **1968** R. O. C. Norman *Princ. Org. Synthesis* xviii. 505 Lithium aluminium hydride reduces ozonides to alcohols.

b. The ion O_3^-, or a salt of this ion.

1949 *Chem. Abstr.* XLIII. 4170 The compd. can be formulated K^+,O_3^-, and termed K ozonide. **1962** P. J. & B. Durrant *Introd. Adv. Inorg. Chem.* xxi. 799 The ozonide ion, (O—O—O)⁻, is present in potassium ozonide, KO_3. **1966** *McGraw-Hill Encycl. Sci. & Technol.* XII. 409/2 Sodium also forms an ozonide, NaO_3, when ozone is passed into a solution of sodium in liquid ammonia.

ozonization. Add: **b.** Reaction with ozone, esp. in an ozonolysis process.

1906 *Jrnl. Chem. Soc.* XC. 227 The ozonisation of elaidic acid could only be carried out in chloroform solution. **1936** *Jrnl. Amer. Chem. Soc.* LVIII. 2272 The isolation of 1,2-diketones from the ozonization of disubstituted acetylenes. **1964** Roberts & Caserio *Basic Princ. Org. Chem.* vii. 192 Ozonides..may explode violently and unpredictably. Ozonizations must therefore be carried out with due caution.

ozonize, *v.* Add: **2.** (Further examples.) Also, to cause to react with ozone.

1906 *Jrnl. Chem. Soc.* XC. 227 A solution of sodium oleate was ozonised and then evaporated under reduced pressure. **1930** *Jrnl. Amer. Chem. Soc.* LII. 2550 Methyl neopentyl ketone..and trimethylacetaldehyde were produced in the ratio of approximately 3·7 to 1 when diisobutylene was ozonized and the ozonide hydrolyzed. **1968** *Adv. Chem. Ser.* LXXVII. lxii. 35 To determine what mechanistic pathways might be operating to account for the different cis-trans ratios, we have ozonized the olefin stereoisomers in the presence of the respective, necessary aldehydes. **1971** *Nature* 26 Nov. 213/1 The source [*sc.* mercury] was ozonized and mechanically spread on the Pt backing plate.

ozonolysis (ōuzǒnǫ·lĭsis). *Chem.* [f. Ozon(e + -o + *-lysis.] The cleavage of double or triple carbon–carbon bonds by reaction with ozone.

1931 *Jrnl. Amer. Chem. Soc.* LIII. 358 In order to obtain the ketone indicated from hydrocarbon number 1 it is necessary to assume that the pinacolone rearrangement suggested by Butlerow has taken place during ozonolysis. **1936** *Jrnl. Org. Chem.* I. 145 The ozonolysis of triple bonds has been studied with..six acetylene representatives. **1951** I. L. Finar *Org. Chem.* I. iv. 57 Ozonolysis..is probably the best method for determining the position of a double bond in any olefinic compound. **1971** I. G. Gass et al. *Understanding Earth* ix. 133 The ozonolysis products of the kerogens are aromatic in the Onverwacht.

Hence **ozonoly·tic** *a.*, involving ozonolysis.

1956 *Jrnl. Polymer Sci.* XXII. 213 A simple ozonolytic method has been developed which enables the natural rubber trunk chains of rubber-polymethyl methacrylate and rubber-polystyrene interpolymers to be degraded into low molecular weight fragments. **1972** *Angewandte Chemie* (Internat. Ed.) XI. 1089/2 (*heading*) Ozonolytic degradation of a catenane.

ozonosphere (ōuzōu·nosfiəɹ). *Meteorol.* [f. Ozon(e + -o + Sphere *sb.*] The region of the atmosphere where there is a significant concentration of ozone, at an altitude of 10 to 50 km. (6–30 miles); esp. the part between about 20 and 25 km (12–15 miles) where the concentration is greatest.

1933 *Jrnl. Inst. Electr. Engin.* LXXIII. 578/2 The upper part of the stratosphere is conveniently dealt with under a different name, even though there is no definite boundary between the stratosphere and the ozonosphere. **1951** *Jrnl. Brit. Interplanetary Soc.* X. 22 The composition up to the ozonosphere is then fairly well established. Above this both oxygen and nitrogen still form the major part of the atmosphere. **1963** *New Scientist* 1 Aug. 232/2 Apart from the effect that this might have in..giving us acute sunburn, there is another potentially important outcome of contaminating the ozonosphere. **1965** K. E. Spells in J. A. Gillies *Textbk. Aviation Physiol.* iii. 49 The region from about 12 to 22 miles altitude is sometimes called the ozonosphere.

P

P. Add: **I. 1.** *attrib.* Used *spec.* to designate one of the two main groups of languages which developed from Common Celtic, so called because its distinctive phonological features include the development of IE. **qᵘ* to *p*, as *P-Celtic, -division, -group,* etc.; *P-Celt*, a speaker of P-Celtic.

1891 J. Rhys in *Trans. Philol. Soc.* 1891–4 104 (*title*) The Celts and the other Aryans of the *P* and *Q* groups. *Ibid.* 111 We are entitled to conclude that the Q Celts arrived in the west before the P Celts, as they are found occupying the furthest parts of the Celtic area... The conclusion is scarcely to be avoided that the later comers, the P Celts, came as invaders and conquerors. **1892** [in Dict.]. **1913** J. M. Jones *Welsh Gram.* 1 Keltic: (*a*) the Q division, consisting of dialects in Gaul and Spain, and the Goidelic group, comprising Irish, Scotch Gaelic and Manx; (*b*) the P division, consisting of Gaulish, and the British group, comprising Welsh, Cornish and Breton. **1949** *Antiquity* XXIII. 23 By birth-place and blood, Kieran was closely associated with the P-Celtic tribe of Corcu-Loigde. *Ibid.* 27 The Ulaid (the Ulstermen of the Saga), were once P-Celts, in O'Rahilly's view. **1953** [see *Brittonic *a.* and *sb.*]. **1972** W. B. Lockwood *Panorama Indo-Europ. Lang.* 74 The term Goidelic is chiefly used to denote Irish as distinct from British or, more technically speaking, to denote Q-Celtic as opposed to P-Celtic. **1977** *Word* 1972 XXVIII. 133 One may wish to see Pictish interpreted as somewhat less different from Cumbric and the rest of insular p-Celtic than Jackson would argue.

4. *P.Z. exercise* (*R.N.*), an exercise at sea.

1905 *Trans. Inst. Naval Archit.* XLVII. ii. 305 The P.Z. exercises have been so conducted as to be deceiving. **1916** 'Taffrail' *Pincher Martin* viii. 140 Gunnery, gunnery, *toujours* gunnery—unless it was torpedo-running, steam tactics, or P.Z. Exercises—was carried on throughout the year. **1962** Granville *Dict. Sailors' Slang* 85/2 *PZ*, tactical exercise in the Fleet at sea in peacetime when the Code flags *PZ* were run up at the start of the exercise.

II. *p* = *piano* (examples); P., 'prompter side' in a theatre; cf. *P.S.* below; p, parental generation (see quot. 1902); p, p., pence, penny, in decimal currency (see *new *a.* 4); Pa, pascal; p. a., per annum (Per *prep.* I. 2); P.A., personal assistant; P.A. [see quot. 1972], a canvas climbing boot with a rubber sole strengthened with a steel plate; P.A., political agent, press agent, Press Association, programme assistant; P.A., PA, p.a., public address; P.A., power amplifier; PAL, phase alternation line (name of a colour television system); P. and O. (earlier example); PAR, precision approach radar; P.A.S., PAS, *para*-aminosalicylic acid; PAYE, P.A.Y.E., pay as you earn; PAYV, pay-as-you-view; P.B.I., Poor Bloody Infantry(man), so P.B. used with other sbs.; p.c., per cent; p.c., P.C., postcard; P.C. = Privy Councillor (examples); PC, propositional calculus (see *propositional *a.* b); also *attrib.*; PCB, pcb, printed circuit board; (see also sense II. d below); PCM, pcm, pulse code modulation; PCP, phencyclidine; p.c.u., passenger car unit; P.D., p.d., potential difference; P.D., preventive detention (also, detainee); PDI, powered descent initiation (of a spacecraft); P.D.Q., p.d.q., pretty damn quick; P.E., physical education; P.E., p.e., plastic explosive; P.E.N., PEN, Poets, Playwrights, Editors, Essayists, and Novelists; PEP (*Radio*), peak envelope power; P.E.P., PEP, Political and Economic Planning; *p/e ratio*, price-earnings ratio; PERT (orig. *U.S.*), program(me) evaluation (and) review technique (orig., program(me) evaluation research task); (= **network analysis* b, esp. as used to deal with events of uncertain duration); P.E.S.C., Public Expenditure Survey Committee; p.f. (*Mus.*), pianoforte, [It. *piano forte*] soft then loud, [It. *più forte*] more loudly; p.f.a., P.F.A., pulverized fuel ash; PFC, p.f.c. (*U.S.*), Private 1st Class, poor foolish (forlorn, etc.) civilian; PG, parental guidance (*N. Amer.*, a cinema film classification); P.G., p.g., paying guest; hence p.g. *v. intr.*, to reside as a paying guest; PGR, P.G.R. [ad. G. *p.g.R.* (O. Veraguth 1907, in *Monatsschr. f. Psychiatrie und Neurol.* XXI. 387)], psychogalvanic reflex, response; Ph.D. [L. *Philosophiæ Doctor*], Doctor of Philosophy, Doctorate of Philosophy; P.I. (*U.S. slang*), pimp; PI, p.i., private investigator; PIB, Prices and Incomes Board; PIDE [Pg. *Polícia Internacional e de Defesa do Estado*], International Police for the Defence of the State; p-j, P.J., pyjama; PK., Pk., psychokinesis, psychokinetic (see quot. 1943); P.K.I. [Indonesian *Partai Komunis Indonésia*], Indonesian Communist Party; PKU (*Med.*), phenylketonuria; P.L.A., People's Liberation Army; P.L.A., Port of London Authority; P.L.M. [Fr. *Paris–Lyon–Méditerranée*], Paris–Lyons–Mediterranean (Railway); P.L.O., PLO, Palestine Liberation Organization; PL/I, PL/1 (*Computers*), 'Programming Language One', a versatile and powerful high-level language designed to replace both Fortran and Cobol in their respective fields; PLP, P.L.P., Parliamentary Labour Party; PLSS, personal life support system; P.M., particular (*or* peculiar, *or* proper) metre; p.m. = afternoon (further examples); p.m., P.M., *post mortem*; P.M., Prime Minister; P.M.A. (*Dentistry*) (see quot. 1969); P.M.G., Postmaster General; P.M.S., pregnant mare's serum, or a gonadotrophic extract of it; PNdb, PNdB, perceived noise decibel(s) (see quot. 1959); P.N.E.U., Parents' National Educational Union; P.O., post office (examples); also, postal order; P.O.A., Probation of Offenders Act; P.O.D., pay on delivery; P.O.D. (*U.S.*), Post Office Department; P. of W., Prince of Wales; POL, petrol, oil, and lubricants; P.O.O., Post Office Order; POP, Post Office Preferred; P.O.P., POP, printing-out paper; POPOP [f. the repeated initials of Phenyl and *oxazole, the molecule consisting of five such rings joined in this order], 1,4-di[2-(5-phenyloxazolyl)]benzene, a substance used in solution as a scintillator; POUM, P.O.U.M. [Sp. *Partido Obrero de Unificación Marxista*], Workers' Party of Marxist Unity; P.O.W., Prince of Wales; P.O.W., POW, prisoner of war; p.p., *per procurationem*, by proxy (examples); *pp* or *ppp* = *pianissimo* (examples); P-P, PP, pellagra-preventive or -preventing (formerly a designation of the vitamin now called niacin); ppb, p.p.b., parts per billion; P.P.C. (further examples); hence P.P.C. *v. intr.*; PPC, P.P.C., progressive patient care; P.P.D., PPD, purified protein derivative (of tuberculin); P.P.E., PPE, Politics, Philosophy, and Economics (a course of study at Oxford University); P.P.I., p.p.i., plan position indicator; PPK, *Polizei Pistole Kriminal* [G., police criminal pistol], a type of handgun; PPLO, pleuropneumonia-like organism(s); ppm, parts per million; P.P.S., Parliamentary Private Secretary; P.P.S., P.P.P.S., (*post*) *post post scriptum*; PPU, Peace Pledge Union; P.Q., PQ, parliamentary question; P.R., photographic reconnaissance; P.R., Pre-Raphaelite; P.R., prize ring; P.R., proportional representation; P.R., public relations; P.R., PR, Puerto Rico, Puerto Rican; P.R.A. (examples); P.R.B., Pre-Raphaelite Brother(-hood); P.R.O., Public Record Office; P.R.O., public relations officer; PROM, programmable read-only memory (cf. ROM s.v. *R II. 2a); PROP, Preservation of the Rights of Prisoners; P.R.S., Performing Rights Society; P.S. = *post scriptum*, postscript (examples); P.S., 'prompter side' in a theatre; p.s.c., passed staff college; p.s.i.(a.), psi, pounds per square inch (absolute); P.S.V., p.s.v., public service vehicle; P.T., physical training; P.T., PT, purchase tax; P.T.A., Parent–Teacher Association; PT boat (*U.S.*), patrol torpedo boat; PTC, phenylthiocarbamide; P.T.I., physical training instructor; P.T.O. (examples); p.t.o., PTO, power take-off; Pty. (*Austral. commercial*), proprietary; P.U.O., pyrexia of unknown origin; P.U.S., PUS, Permanent Under-Secretary; pw, p.w., per week; P.W.D., Public Works Department; PWR, pressurized-water reactor; PX (*U.S. mil.*), Post Exchange. Also *PABA, *pH, *p-n-p, *p-type.

1740 J. Grassineau *Mus. Dict.* 173 *P*, in the Italian music, frequently signifies *piano*, which is what we called soft. **1888** Kipling *Masque of Plenty* in *Departmental Ditties* (1890) 48 (*adagio dim.*) Fillèd with praise... (*p*) Ay, paint our swarthy billions The richest of vermilions. **1957** H. Shanet *Learn to read Music* iv. 123 Between *f* and *p*, there are *mezzo forte*.., and *mezzo piano* (medium soft). **1977** G. Warfield *How to write Music Manuscript* 133 Place a 'p' under the first note and a 'pp' under the last in these two examples. **1901** G. B. Shaw *Let.* 7 Nov. in *B. Shaw & Mrs. Campbell* (1952) 14 Titheradge's determination to die parrallel [*sic*] to the float with his heels O.P. and his head P..rather spoils the picture. **1933** P. Godfrey *Back-Stage* i. 18 The amber circuits in No. 1 batten, floats, and P. and O.P. perches. **1902** W. Bateson et al. *Rep. Evolution Comm. R. Soc.* I. 160 We suggest as a convenient designation for the parental generation the letter P. In crossing, the P generation are the pure forms... Starting from any subject-individual, P_2 is the grandparental, P_3 the great-grandparental generation, and so on. **1918** Babcock & Clausen *Genetics Rel. Agric.* x. 180 (*caption*) Re-appearance of parental values in the F_2 offspring. P_1 Leaf Factor. **1952** Srb & Owen *General Genetics* ix. 164 Verify this by diagramming a sequence of crosses through F_2 where the P generation is the reciprocal of the one shown in the text. **1975** V. Grant *Genetics of Flowering Plants* i. 9 The experimental results can be summarized as follows: P round yellow × wrinkled green. **1934** Webster, *p.*,..penny. **1968** *Times* 17 Apr. 6/1 The 10p. and 5p. coins could appear in small change almost at once. **1973** *Guardian* 18 Dec. 13/3 We couldn't get away with a half p in tax evasion—but they do. **1976** 'W. Trevor' *Children of Dynmouth* iv. 95 Yesterday had officially been the last day for entries, but he'd seen no reason to turn away the man's fifty p. **1964** H. S. Hvistendahl *Engin. Units* iii. 29 In the French decree of May 3rd, 1961, the name pascal (Pa), is adopted for the N/m². **1975** *Nature* 2 Oct. 371/2 The density of vitreous silica is affected irreversibly by the application of pressures of more than 2 × 10⁹ Pa. *a* **1912** W. T. Rogers *Dict. Abbrev.* (1913) 145/2 *p.a.*, *Per annum* (For the year). **1931** *Times* 16 Mar. 22/4 (Advt.), Present rental value £300–£350 p.a. **1955** *Times* 7 July 1/5 Salary, £1,200–£1,600 p.a., plus free furnished quarters, fuel, light, water. **1942** Partridge *Dict. Abbrev.* 73/1 *P.A.*... Personal Assistant. **1943** N. Balchin *Small Back Room* i. 14 D'you think Higgins goes in for women? We might hire him a suitable P.A. **1969** D. Clark *Nobody's Perfect* ii. 61 Couldn't his P.A. have rung you when you got home? **1975** M. Sinclair *Long Time Sleeping* iii. 38 Gilbert Winter's office and the adjacent one of his long-suffering PA. **1963** [see *kletterschuh]. **1972** D. Haston *In High Places* ii. 35 Neil and I [were] ahead leaving the other two arguing about who should wear the one pair of P.A.'s. (These are special boots for hard rock-climbing, with stiff, smooth rubber soles and canvas uppers. The initials are those of their inventor, Pierre Allain, a famous French climber before the Second World War.) **1913** E. M. Forster *Let.* 1 Jan. in *Hill of Devi* (1953) 25 The Political Agent from Neemuch..brought a party... The P.A...planted himself on the State for the night. **1937** F. Stark *Baghdad Sketches* 187 [They] send messages to the P.A. **1936** *Amer. Speech* XI. 220 In terms of the theater, the *P.A.* is the Press Agent. **1958** *Spectator* 11 July 53/2 The press box was empty except for PA and *The Times*. **1915** M. MacDonagh *Diary* 6 Oct. in *London during Gt. War* (1935) II. iii. 80 My friend Howe, of the 'P.A.'. **1942** Partridge *Dict. Abbrev.* 73/1 *P.A.*... Publishers' Association. But also Press Association. **1972** D. McLachlan *No Case for Crown* iii. 45 I'll deal with the P.A.; their news editor used to work under me. **1968** *Listener* 4 Apr. 442/3 Four of these programme assistants form the nucleus of Radio Sheffield's staff... My immediate task is to look at the material..left for me the previous night by one of the other PA's. **1936** *Amer. Speech* XI. 220 In radio, a *P.A.* system is a public address system. **1953** Pohl & Kornbluth *Space Merchants* (1955) ii. 23 The PA system announced that my flight was ready. **1963** *Times Lit. Suppl.* 1 Mar. 156/1 Marlowe anticipated Whitman's barbaric yawp by setting up a national PA system of blank verse. **1964** S. Bellow *Herzog* (1965) 35 Over the p.a. system the management begged the spectators not to throw pennies. **1940** *Chambers's Techn. Dict.* 608/1 *P.A.*, power amplifier. **1971** *Melody Maker* 4 Sept. 20 The giant PA's distort their guitars out of all recognition. **1963** J. R. Davies *Understanding Television* xiii. 485 Mention must be made of the recently introduced PAL system, developed by Telefunken... The PAL system has been investigated by the European Broadcasting Union... PAL is based on the N.T.S.C. system. **1968** *Listener* 21 Nov. 687/2 It's not quite true..that PAL and SECAM are 'irreconcilable' now that at least one inventor is trying to market a cheap conversion kit. **1975** *New Scientist* 31 July 274/1 Commercial TV resolution in the US is 525 lines, against Europe's PAL standard of 625 lines. **1863** Dickens *Uncomm. Trav.* in *All Year Round* 6 June 350/2 The well-known regularity of the P. and O. Steamers. **1951** *Gloss. Aeronaut. Terms* (*B.S.I.*) iii. 21 *Final controller*, a radar controller employed in the transmission of PAR talk-down instructions to the pilot of an aircraft on the final approach to the runway, and in passing monitoring information to the pilot when using a landing aid other than PAR. **1966** *McGraw-Hill Encycl. Sci. & Technol.* X. 578/2 In common practice, the ground PAR operators call instructions to the pilot. **1946** *Lancet* 5 Jan. 15/2 Treatment with *p*-aminosalicylic acid (P.A.S.) was given in three periods with concomitant falls in temperature. **1959** J. Braine *Vodi* vi. 85 They'd tried strep. and

P.A.S. **1971** *Brit. Jrnl. Dis. Chest* Jan. p. vi (Advt.), A choice of flavoured drinks... the acceptable way of taking PAS and Isoniazid. **1944** *Times* 4 Apr. 2/2 (*heading*) PAYE begins. **1956** 'C. Blackstock' *Dewey Death* iv. 89 Miss Holmes..[was] doggedly working out the P.A.Y.E. for the thirty members of I.L.D.A. staff. **1972** *Accountant* 21 Sept. 343/1 Scare stories..about the implications of the proposed letter suffixes to employees' PAYE code numbers, have been officially denounced. **1958** *Spectator* 27 June 829/2 The need is to make the idea of PAYV much more familiar than it is. There have been many references to it from time to time in the press in the last few years, but for some reason the idea has never caught on. **1916** *B.E.F. Times* 1 Dec. f. 4/1 So here's to the lads of the P.B.I. Who live in a ditch that never is dry. *a* **1918** J. T. B. McCudden *Five Yrs. in R. Flying Corps* (1919) 134 The famous Ypres salient..was by no means regarded with friendly feelings by the Infantry—or P.B.I. as they generally call themselves. **1946** *Jrnl. R. United Service Inst.* XCI. 52 He is the 'P.B.I.' of the service on whom the final success of the scheme depends. **1949** F. Swinnerton *Doctor's Wife comes to Stay* 149 He's only the P. B. Author. **1952** *Sunday Times* 14 Dec. 7/3 Procedural remedies are being sought, mostly by back-benchers—the 'P.B.I.' of Parliament. **1968** W. Winward *Conscripts* xii. 154 The p.b.i. gets the chopper, but never the officer. **1972** *Guardian* 1 Feb. 12/2 In the trenches the PBI.. await the order to go over the top. **1976** *Times Lit. Suppl.* 30 Jan. 106/1 The Crossman interpretation of the position of the MP whom he sees as the PBI of the mass party. **1874** 'Mark Twain' *Lett. to Publishers* (1967) 80 Bliss had contracted to pay me 10 p.c. on my next book... He paid 7½ p.c. on Roughing It and 5 p.c. on Innocents Abroad. **1931** *N. & Q.* 26 Dec. 465/2 A 10 p.c. solution of oxalic acid will be useful if ink-stains be present. **1889** E. C. Dowson *Let.* 21 Feb. (1967) 39, I enclose a P.C. wh. I had just written—it is no longer necessary—but you may as well post it. **1951** R. Macaulay *Lett. to Friend* (1961) 194, I had..a nice picture p.c. from Father Pedersen from Rome. **1881** E. W. Hamilton *Diary* 22 Nov. (1972) I. 185, I told Mr. G. he ought to make May a P.C. **1973** *Whitaker's Almanack* 1974 84 The Duke of Buccleuch and Queensberry, P.C., K.T., G.C.V.O., aged 78. **1960** P. H. Nidditch *Elem. Logic Sci. & Math.* 29 In a PC of the usual, axiomatic type the only definitions are those of connectives. **1965** Hughes & Londey *Elem. Formal Logic* xv. 101 In 1910, in the first volume of *Principia Mathematica*, Whitehead and Russell presented an axiomatization of PC. *Ibid.* xxix. 211 Their validity can be determined by PC methods alone. **1973** J. J. Zeman *Modal Logic* xi. 181 One might ask if there is a modal system bearing an analogous relationship to the classical PC. *Ibid.* xii. 191 The definition of complete modalization was extended to include certain PC theorems. **1977** *Engin. Materials & Design* Aug. 9/2 Thought to be the most powerful calculator/watch combination on the market, hybrid construction is used to mount the chips on a small pcb which also carries a miniature 5 by 4 matrix keyboard. **1977** *Gramophone* Nov. 960/3 It is not usual to mount heavy components on PCBs. **1947** *Bell Syst. Techn. Jrnl.* XXVI. 395 This paper describes an experiment in transmitting speech by PCM, or pulse code modulation. **1966** *Punch* 10 Aug. 224 PCM will enable each existing pair of telephone cables to carry twelve times as many conversations as before. **1972** [see *modulation 7]. **1977** *Broadcast* 7 Nov. 10/1 Sound radio signals..are distributed in pcm multiplex form along analogue television links. **1973** PCP [see *phencyclidine]. **1977** *Time* 18 July 35/3 Disturbing increases in the use of a dangerous new street drug called PCP. **1960** J. Drake in E. Davies *Roads* iv. 85 The term 'passenger car unit' or p.c.u. is used in capacity measurements to make allowance for mixed traffic—all motor vehicles count as one unit, except heavy goods vehicles, buses, and coaches which count three. On a road having moderately high volumes of heavy traffic it is found that the p.c.u. count is 50% more than that for motor vehicles. **1966** *New Scientist* 29 Sept. 711/3 The unit of traffic he used was the 'passenger-car unit' (pcu) which is employed by the Ministry of Transport. A bus is rated at 3 pcu, for example. **1887** W. E. Ayrton *Pract. Electr.* vii. 371 An influence machine can produce a P.D. between its terminals of some hundreds of thousands of volts. **1935** J. N. Friend *Text-bk. Physical Chem.* II. vii. 297 Experimental measurement of the P.D. between two liquids presents many difficulties. **1963** A. F. Abbott *Ordinary Level Physics* xxxvii. 487 The terminal p.d. is always less than the e.m.f., and the difference..represents the p.d. required to send the current through the internal resistance of the cell. **1956** 'C. Raven' *Underworld Nights* 30 The last I heard of him he was done for pinching a shaving brush from Woolworth's and sentenced to eight years P.D. under the new act. **1959** *New Statesman* 24 Jan. 102/2 My seven hosts were all preventive detainees (PDs) serving terms of five to 14 years imprisonment, and with three or more convictions. **1973** J. Wood *North Beat* vi. 81 The thought of preventive detention appalled him. There was no remission with P.D. **1969** *New Scientist* 17 July 115/1 The critical operation is then a 'three-phase powered descent initiation' or PDI, the braking manoeuvre which begins at this low point and reduces the vehicle's velocity to zero at a height of around 7000 feet. *c* **1875** B. Woolf *Mighty Dollar* in B. H. Clark *Favorite Amer. Plays 19th Cent.* (1943) 489 That's right, you'd better step P.D.Q., pretty damn quick. **1890** *Ladies' Home Jrnl.* July 12/4 The P. D. Q. Camera. **1891** Kipling *Life's Handicap* 189 He went as his instructions advised *p.d.q.*—which means 'with speed'. **1926** [see *make *v.*[1] 65 c]. **1961** B. E. Wallace *Death packs Suitcase* v. 55 I'd come back here P.D.Q., because here I'd know my way around. **1974** 'A. Haig' *Peruvian Printout* 33 Whoever is messing about with our computers, I want him found p.d.q. **1956** J. Edmundson *P.E. Teachers' Handbk.* vi. 30 Quite effective P.E. lessons can be taken in a classroom even with.. very limited space. **1973** J. Burrows *Like an Evening Gone* iii. 40 Sporting equipment of a modest kind,..a vaulting horse and a set of P.E. mats. **1976** 'W. Trevor' *Children of Dynmouth* v. 115 Stringer, the headmaster, was rubbish; the P.E. man went after the girls. **1949** F. S. Chapman *Jungle is Neutral* ii. 19 We piled the dicky high with Tommy-guns, cases of P.E. (plastic high explosive), grenades, and an assortment of demolition and incendiary devices. **1971** P. O'Donnell *Impossible Virgin* xiii. 261 He had some fuse and plastic explosive, but... using p.e. to set off a bullet would produce the wrong sort of noise. **1923** *Times* 2 May 11/3 Mr. John Galsworthy presided last night over a company of playwrights, poets, essayists, and novelists at an international dinner given by the P.E.N. Club. **1924** G. S. Gordon *Let.* 20 Sept. (1943) 176 A private dinner in the evening with the Stockholm Pen Club (P—poets; E—editors; N—novelists). **1931** T. E. Lawrence *Let.* 13 Apr. (1938) 718 The P.E.N. suggestion is rather astonishing. **1966** 'H. MacDiarmid' *Company I've Kept* xiii. 270 Saurat rendered great service to the International P.E.N., as one of its Vice-Presidents. **1969** L. Hellman *Unfinished Woman* xiii. 195 A reception for the president of PEN, an Englishman. **1956** *Proc. IRE* XLIV. 1710/1 For a 0·5 watt SSB signal (1w PEP) there is 0·095 watt in the AM component. **1971** *Gloss. Electrotechnical, Power Terms* (*B.S.I.*) iii. vii. 23 *Peak envelope power; P.E.P.* of a radio transmitter. The power supplied to the aerial transmission line or specified artificial load by a transmitter during one radio frequency cycle at the highest crest of the modulation envelope, taken under conditions of normal operation. **1976** Perkowski & Stral *Joy of CB* vii. 70 As in AM transmissions, the peak envelope power (PEP) is still limited to 12 watts, but since the carrier is reduced or suppressed, additional power can be put into the sideband. **1933** *Planning* vi. 15 They began more than two years ago to study..the possibilities of renewing friction in..industry, agriculture, finance, the social services... The PEP budget..is raised entirely from among those interested in the work. **1941** J. S. Huxley *Uniqueness of Man* xi. 233 It [*sc.* group work] is far more necessary in social science, where various bodies, such as P.E.P., are studying how to perfect it as a research method. **1970** I. Sieff *Mem.* ix. 164, I would not want to try and write a history of PEP here if I could but I would like to say something about it. **1965** *Acronyms & Initialisms Dict.* (Gale Research Co.) 558 P/E, Price/Earnings Ratio (Relation between price of a company's stock and its annual net income). **1969** *Times* 30 Apr. 30/5 It leaves the historical p/e ratio on the ordinary shares..looking vulnerable for a newspaper company at 20.8. **1959** *Amer. Statistician* Apr. 10/1 This Program Evaluation and Review Technique (code-named PERT) is applied as a decision-making tool designed to save time in achieving end-objectives. **1960** *IRE Trans. Engin. Managem.* VII. 103/2 PERT (Program Evaluation and Review Technique) utilizes the network concept of R and D projects, and analyzes the 'time to completion' variable. **1962** [see *network analysis* s.v. *network 5]. **1964** A. Battersby *Network Analysis* ix. 134 Pollack has published a detailed description of how PERT was brought in to control the construction of the $47,000,000 Zero Gradient Synchrotron at the Argonne National Laboratory. **1969** J. Argenti *Managem. Techniques* 72 The technique known as PERT..is used when the duration of an activity is not accurately known. **1974** *Encycl. Brit. Macropædia* XIII. 600/1 Critical path method (CPM) is an optimizing procedure applicable only to certainty-type formulations of such problems. Project evaluation and review technique (PERT) is applicable to risk- as well as certainty-type formulations but does not always yield optimal solutions. *a* **1974** R. Crossman *Diaries* (1976) II. 126 Then we moved on to housing where we had a very strange situation because, after agreeing to the cuts PESC demanded, the Minister of Housing made an extraordinary Ministerial announcement virtually saying that the target of 500,000 houses a year had been abandoned. **1976** H. Wilson *Governance of Brit.* iii. 61 This meeting or 'PESC' was concerned not with detailed allocations of expenditure, as finally announced in February 1976, but with basic priorities. **1876** Stainer & Barrett *Dict. Mus. Terms* 348/2 *P.f.*, abb. of (1) Pianoforte. (2) *Piano, forte*, soft then loud. (3) *Più forte*, louder. **1938** *Oxf. Compan. Mus.* 712 *Pianoforte*,..often abbreviated pf. **1958** *Archit. Rev.* CXXIII. 326 The ground-floor walls are of cavity construction with an inner skin of insulating p.f.a. blocks and yellow bricks outside. **1970** *Sci. Jrnl.* Aug. 78/2 Marketing officers of the CEGB are today developing PFA sales for a wide range of civil engineering and building activities. **1941** *Amer. Speech* XVI. 167/2 PFC, Private 1st Class. **1947** *Ibid.* XXII. 112 References to rates and ranks are numerous. One variously caricatures P.F.C. ('private first class') as 'poor foolish civilian'... A *double P.F.C.*, however, is a corporal, since he has two chevrons on his sleeve. **1948** J. T. Appleby *Suffolk Summer* iii. 13 The third, an embittered P.F.C., did odd jobs of typing and message carrying. **1955** *Daily Progress* (Charlottesville, Va.) 24 Aug. 18/1 Pfc's with PhD's teach generals and others the fundamentals of atomic weapons. **1963** T. Pynchon *V.* i. 13, 'I would like to sing you a little song.' 'To celebrate your becoming a PFC' said Ploy... 'Pore Forlorn Civilian, We're goin to miss you so.' **1977** 'E. McBain' *Long Time no See* xii. 198 'A man named James Harris, served with the Army.'.. 'Rank?' 'Pfc.' **1972** *New Acronyms & Initialisms* (Gale Research Co.) 81/2 *PG*, parental guidance suggested (some material may not be suitable for pre-teenagers) (movie rating). **1974** *Daily Colonist* (Victoria, B.C.) 6 Oct. 27/5, 61 per cent rated the film PG (Parental Guidance) or G (General Audience). **1976** *New Yorker* 12 Jan. 70/2 Why would anybody want a PG-rated Peckinpah film? **1977** *Time* 11 Apr. 38/3 Modest, well crafted, less bloody and less bloody-minded than most TV shows, it is a PG film that any P ought to be happy to G the kids through. **1923** U. L. Silberrad *Lett. J. Armiter* ii. 49 They have made the suggestion that I should p.g. with them for the autumn and winter. **1925** F. Stark *Let.* 1 July (1974) I. 93, I am afraid I shall not be well enough to look after p.g.'s after all. **1933** M. Allingham *Sweet Danger* v. 62 We've got one P.G. already... She's been with us three years. **1959** D. Wallace *Richard & Lucy* v. 87 Terribly expensive rail fare, and they'd probably expect us to p.g. when we got there. **1972** *Times Lit. Suppl.* 28 Apr. 500/4 A decayed-gentlewoman's nice home for p gs. **1977** N. Slater *Crossfire* i. 22 I'm going to descend on you... I'd like to PG—handsome rental available. **1938** R. S. Woodworth *Exper. Psychol.* xiii. 277 The name 'psychogalvanic reflex' was introduced by Veraguth..., who made a comprehensive study of the Féré effect... In the present chapter we will call it PGR. **1949** *Brit. Jrnl. Psychol.* XL. 86 When the individuals' P.G.R. scores are obtained for a given attitude they must be expressed for each person relative to his own general P.G.R. reactivity. **1954** Woodworth & Schlosberg *Exper. Psychol.* (rev. ed.) vi. 137/2 The rapid changes in conductance have been studied extensively and suffer from too many names. The oldest is *psychogalvanic reflex* (PGR), but many dislike the implications that it is psychic or a reflex. **1962** *New Scientist* 22 Mar. 672 PGR (psychogalvanic response) records were taken for the same purpose. **1869** *Atlantic Monthly* Jan. 89/2 His cousin, the Ph.D. from Göttingen, cannot help despising a people who do not grow loud and red over Aryans and Turanians. **1903** W. James *Mem. & Stud.* (1911) 331 A Ph.D. in philosophy would prove little..as to one's ability to teach literature. *Ibid.*, He was of ultra Ph.D. quality. **1906** [see *D.Phil.* s.v. *D III. 3]. **1936** *Discovery* May 156 Julius Grant, Ph.D., M.Sc., F.I.C. **1966** J. Betjeman *High & Low* 40 Doubtless some pedant for his Ph.D. Has ascertained the facts. **1973** G. Mitchell *Murder of Busy Lizzie* xv. 174 Why should anybody want to strangle a harmless little Ph.D. like Mr Lovelaine? **1931** G. Irwin *Amer. Tramp & Underworld Slang* 144 *P.I.*, a pimp or pander, merely a euphemism by contraction. **1970** C. Major *Dict. Afro-Amer. Slang* 90 *P.I.*, pimp. **1960** *Acronyms Dict.* (Gale Research Co.) 165 PI..Private Investigator. **1970** G. F. Newman *Sir, You Bastard* vi. 170 The PI had his licence revoked. **1973** *Publishers Weekly* 13 Aug. 48/3 This is the third p.i. mystery featuring Shock and his partner. **1966** *Economist* 16 July 227/1 The Government should use the PIB only when it is ready to back the board's recommendations to the hilt. *a* **1974** R. Crossman *Diaries* (1975) I. 421 We can't afford to let the policy fail and yet quite soon we are going to face the 18-per-cent increase of army pay which the P.I.B. will almost certainly award. **1959** *Listener* 9 July 45/1 The widespread activities of the state security police, known as PIDE. **1970** *Ann. Reg.* 1969 276 On 19 November the [Portuguese] Government dissolved the PIDE and placed a similar organization under the direct control of the Ministry of the Interior. **1974** *Daily Tel.* 21 Aug. 4/7 The Portuguese Legion..was outlawed together with the PIDE/DGS. **1964** S. Bellow *Herzog* 257 Put on those p-j's now. **1967** R. De Sola *Abbrev. Dict.* (rev. ed.) 208/1 *Pj's*, peejays (pajamas). **1970** *New Yorker* 4 Oct. 122/3 (Advt.), Cotton sleep culotte rated perfect for P.J. parties. **1943** L. E. & J. B. Rhine in *Jrnl. Parapsychol.* VII. 20 This is the first of a long series of research reports describing experiments on what is called the 'psychokinetic' or 'PK' effect. The PK effect is colloquially called 'mind over matter', and means the direct influencing of a physical system by the action of a subject's effort, without any known intermediate energy or instrumentation. *Ibid.* 21 Up to the present, nothing has been published on the topic of the PK effect. **1949** *Mind* LVIII. 391 In PK the mind is supposed to cause changes in physical objects outside its own body, not by means of the nervous system and the muscular apparatus, but directly, by mere thought or 'will'. **1973** *Times* 4 Dec. 17/7 There is nothing new in PK (psychokinesis) and telepathy. **1976** *Times Lit. Suppl.* 13 Feb. 172/4 Demonstrating PK in chickens, and even in fertile eggs, which appear capable of influencing mentally an electronic randomizer controlling the switching mechanism of a lamp. **1939** J. S. Furnivall *Netherlands India* viii. 250 Semaoen, the leader of the revolutionary section, formed a Communist party (P.K.I.). **1944** B. H. M. Vlekke *Nusantara* xv. 341 On May 23, 1920, the Social Democrat Club of Semarang decided to take the name Communist Party of the Indies, in Malayan, *Perserikatan Kommunist di India* (P.K.I.). **1973** J. M. van der Kroef in R. F. Staar *Yearbk. Internat. Communist Affairs* 469 The oldest such party in Asia, the Communist Party of Indonesia (Partai Komunis Indonesia; PKI) formally came into existence on 23 May 1920 as an outgrowth of the 'Indies Social Democratic Association' founded six years previously by Dutch Marxists. **1964** *Observer* 19 July 8/6 The younger had been afflicted by the same dread disease—PKU (phenylketonuria)—which had produced retardation in her sister. **1976** *Lancet* 6 Nov. 1031/1 We have looked for differences in the mono-oxidation of phenylalanine between plasma (or serum) from normal persons and patients with P.K.U. Identification of different enzymatic activities within these two groups would be one of the first steps in the development of a test for diagnosing variants of P.K.U. **1962** E. Snow *Other Side of River* (1963) xxxix. 290 Chinese don't use the old word *ping*, or 'soldier', any more; the P.L.A. has only *chan-shih*, or 'fighters'. **1967** *Ann. Reg.* 1966 372 The Military Affairs Commission of the Central Committee appointed Mao's wife..'adviser on cultural work to the People's Liberation Army' (PLA). **1972** *Observer* 19 Nov. 8 The PLA is none the less a revolutionary army with traditions rooted in a guerrilla past. **1977** *Times* 12 Oct. (China Suppl.) p. iii/4 The People's Liberation Army, the PLA, is making little secret of its desire to modernize equipment under the leadership of Chairman Hua. **1925** *P.L.A. Monthly* Nov. 16/1 The warehouses of the P.L.A. become the Mecca of the woolbuyers of the world. *Ibid.* 22/1 Final tests preparatory to the opening of the P.L.A. automatic telephone system were carried out. **1936** *Discovery* Aug. 232/1 As for using it [*sc.* the Thames] for transport, that is left to P.L.A. tugs, brick-barges, and an occasional pleasure steamer. **1969** S. Hyland *Top Bloody Secret* i. 27 The PLA man in charge of the landing stage. **1898** W. J. Locke *Idols* xvi. 230 The great P.L.M. train carried Hugh swiftly northwards. **1919** R. Fry *Let.* 6 Oct. (1972) II. 458 The P.L.M. is really worse than the S.E.R... It's almost impossible to travel by it. **1948** W. Fortescue *Beauty for Ashes* xxxiii. 251 For the whole of my Paris visit..he was indefatigable in his attentions, finding a free Entr'aide car..to take me to my various destinations, and battling with the P.L.M. to get me a reserved seat and, if possible, a wagon-lit to take me to the South. **1972** R. Cobb *Reactions to French Revolution* iii. 87 The P.L.M... bypassed the river valley, to take in Montbard and Dijon. **1965** *Times* 23 Mar. 9/2 Ahmad Shukairy, leader of the Palestine Liberation Organization (P.L.O.) has extended his search for help to China. **1974** *Guardian* 23 Jan. 2/4 According to the official Egyptian news agency, Yasser Arafat, the PLO chairman, addressed a message disowning the PLO statement. **1974** *Jewish Chron.* 20 Dec. 11/1 The Palestinians, through the PLO, must show magnanimity and statesmanship. **1976** *Time* 27 Dec. 14/1 Although badly battered from its losing role in the Lebanese civil war, the P.L.O. remains an important force. **1965** *PL/I: Language Specifications* (IBM Form C28-6571-0) *title-page*, This manual is a description of the full facilities of PL/I to be implemented under Operating System/360. **1966** E. A. Weiss *PL/1 Converter* p. iii, Many of the limitations..of FORTRAN have been eliminated in PL/I. **1970** A. Cameron et al. *Computers & Old Eng. Concordances* 27, I myself will be very surprised if the next generation of machines will not accept Fortran programming and probably Cobol, Algol, and PL 1 programming. **1972** *Computers & Humanities* VII. 12 Work in language computation is frequently done in PL/1. **1955** R. T. McKenzie *Brit. Pol. Parties* vii. 385 As a prelude to a discussion of the contemporary structure of the PLP it is necessary to recall that the party in Parliament from its earliest years was plagued by two problems. **1963** Butler & Freeman *Brit. Pol. Facts 1900–60* ii. 94 During the war-time coalition the P.L.P. elected an Administrative Committee of twelve, with Peers' representation, all of whom were non-ministers. **1974** *Times* 22 Mar. 1/8 (*heading*) Mr Mikardo elected PLP head. **1976** H. Wilson *Governance of Britain* viii. 160 A Labour prime minister has to operate in a number of intersecting party political circles. The first is PLP meetings. **1968** *McGraw-Hill Yearbk. Sci. & Technol.* 359/1 A portable life support system (PLSS) back pack will provide breathing oxygen, suit pressurization, carbon dioxide removal, liquid cooling and oxygen temperature control, biomedical monitoring package, suit pressure and high oxygen flow sensors, and communications. **1970** N. Armstrong et al. *First on Moon* iii. 63 Complete with the PLSS, the Armstrong and Aldrin suits weighed one hundred eighty-three pounds. **1764** A. Williams *Universal Psalmodist* (ed. 2) 87 Halli-fax. Hymn 50th..P.M. *Ibid.* 91 Dalston. Psalm 122d Dr W. P.M. **1798** [see Metre *sb.*[1] 1 b]. *a* **1912** W. T. Rogers *Dict. Abbrev.* (1913) 152/2 *P.M.* (mus.), peculiar metre (of hymns). *c* **1830** in M. Johnson *Amer. Advertising, 1800–1900* (1960), Worcester, Ms. and New York Mail Stage Line..leaves Worcester, Wednesday and Saturday mornings at 3, and arrives in Norwich at 4 same p.m. *Ibid.*, The Fanny..arrives in Norwich next morning; and in Worcester, by stage, in the p.m. **1845** *Punch* VIII. 54/2 The lights along the Hampstead Road still persist in turning day into night, and burning for several hours after P.M. **1965** *New Statesman* 23 Apr. 661/3 While I am abroad..entries must reach the office Monday p.m. **1974** F. Nolan *Oshawa Project* xx. 119 Staff meeting this pm at 1500 hours. **1911** Webster, *P.M.* or *p.m.*...post mortem, or post mortem examination. **1922** *Lancet* 12 Aug. 358/1 It seems possible that death resulted from a dislocation of the neck. Were the vertebræ examined p.m.? **1928** D. L. Sayers *Unpleasantness at Bellona Club* xvi. 187 The advisability of a P.M. in all cases of sudden death. **1938** S. Beckett *Murphy* xii. 259 They carried him into the p.m. room. **1973** R. Hill *Ruling Passion* i. ii. 15 We haven't had the PM yet, but the doctor was very certain it happened last evening. **1907** W. S. Churchill *Let.* 27 Mar. in R. S. Churchill *Winston Churchill* (1969) II. Compan. I. 653 Could not you or the PM send him a 'private & personal' urging him not to fail us. **1915** Lloyd George *Family Lett.* (1973) 178 It was found impossible ..for the P.M. to hold Exchequer during the time I am occupied in organising Munitions. **1972** M. Sinclair *Norslag* ii. 15 The P.M. wants you round at the Cabinet Office. **1948** Schour & Massler in *Jrnl. Dental Res.* XXVII. 733 A quantitative method of assessing the prevalence of gingivitis in large groups of persons is proposed. Each gingival unit consisting of a papillary portion (P), a marginal portion (M), and an attached portion (A) is examined and counted separately. This method is tentatively termed the P-M-A-Index (P-papillary gingivitis; M-marginal gingivitis; A-attached gingivitis). **1962** Blake & Trott *Periodontology* iii. 27 The difficulty of assessment of gingivitis has been partly overcome by the use of the P.M.A. index. **1969** *Gloss. Terms Dentistry* (*B.S.I.*) 67 *P.M.A. index* (papillary, marginal and attached index), an epidemiological index for scoring the extent of gingival inflammation. **1890** Webster, *P.M.G.*, Postmaster-general. **1908** G. B. Shaw *Coll. Lett.* (1972) II. 803 Your letter..did not overtake me until I arrived here (Bayreuth), too late for a rejoinder to the P.M.G. **1927** [see *franking machine]. **1968** *Listener* 29 Aug. 285/2 Vague recent statements by the PMG, Mr Stonehouse. *a* **1974** R. Crossman *Diaries* (1976) II. 71 When Ted Short replaced Tony Benn as P.M.G. the Post Office was delighted. **1942** J. Hammond et al. in *Jrnl. Agric. Sci.* XXXII. 308 To avoid the constant use of cumbersome phrases, gonadotrophic extracts of the urine of pregnant women (chorionic gonadotrophin) are referred to throughout the text as 'U.P.', and extracts of the serum of pregnant mares as 'P.M.S.' **1957** *Times* 2 Dec. (Agric. Suppl.) p. vi/2 In experiments with Romney Marsh, Cheviot and Southdown sheep, P.M.S. injections have increased the number of lambs born of each ewe mated. **1970** W. H. Parker *Health & Dis. in Farm Animals* vii. 84 The method that appears most likely to succeed is a combination of progesterone and P.M.S. **1959** K. D. Kryter in *Jrnl. Acoustical Soc. Amer.* XXXI. 1425/1 The translation from perceived noisiness in noys to perceived noise level in PNdb is expressed by the equation PNdb = $(1.2I \log_{10} N)/0.03$, where *N* is the number of noys. By definition, the perceived noise level of sound 'X' (in PNdb) is the sound pressure level in db *re* 0·0002 μbar of the 910–1090 cps band of random noise that is judged by an average listener to be acceptable (or, inversely, as unacceptable) as sound 'X', under specified conditions of listening and testing. **1966** *Science* 18 Mar. 1346/3 As a practical matter, the loudness level, in phons, and the perceived noise level, in PNdb's, of a sound are usually calculated from acoustical measures of the sound rather than found by subjective judgement tests. **1971** *New Scientist* 18 Mar. 604/1 In calculating the NNI for any given point, a number of standard flight paths are assumed and the PNdB level for each type of aircraft moving along each possible flight path is then calculated, together with the number of aircraft movements. *a* **1912** W. T. Rogers *Dict. Abbrev.* (1913) 153/1 *P.N.E.U.*..., Parents' National Educational Union. **1931** *Times Educ. Suppl.* 14 Mar. 98/1 P.N.E.U. methods... The paper on English Teaching..is of especial interest to the Parents' National Educational Union. **1972** R. Asher *Talking Sense* viii. 103, I was looking at the material my daughter of thirteen is studying at the P.N.E.U. school she attends. **1824** E. Weeton *Jrnl.* May (1969) II. 280, I wished to see the General Post Office... I was close by the P.O., and could not tell which was it. **1861** Geo. Eliot *Let.* 17 May (1954) III. 415 You are at liberty to imagine a kiss from me, or else to accept a note for it payable at sight. They don't give *P.O.*'s for such payment here. **1891** A. Beardsley *Let.* 25 Dec. (1971) 31, I shall be glad of a few more copies of the November *Bee*, so enclosed PO for 1/-. **1973** *Guardian* 18 Apr. 12/3 Is not the ratio of inland to EEC-bound mail such that the PO (which is no longer G) would more than cover its costs. **1944** J. H. Bagot *Punitive Detention* i. 13 In the years 1936 to 1939 a marked tendency is disclosed for the proportions discharged P.O.A. and fined to decrease and for the proportion placed on probation to increase. **1945** *N. & Q.* 10 Mar. 106/1 'Dismissed P.O.A.' simply means that a case is dismissed under the Probation of Offenders Act, 1907... Such dismissals are of such daily occurrence in the courts that P.O.A. are initials as commonly understood as are R.A.F. or Y.M.C.A. in their respective spheres. **1859** P.O.D. [see *C.O.D.]. **1890** Webster, *P.O.D.*, Post-Office Department; pay on delivery. **1870** A. J. Munby *Diary* 13 July in D. Hudson *Munby* (1972) 288 Today the Thames Embankment was opened, *not* by the Queen, but by the P. of W. and his sister Louise. **1974** *Listener* 3 Jan. 17/2 Lord Berners..invited a smart lady to luncheon to meet 'the P of W'. She arrived agog, expecting to meet the Prince of Wales and found it was the Provost of Worcester. **1944** *Times* 7 June 6/1 The operational planning for invasion had to be interpreted at an early stage in terms of P.O.L. requirements. **1955** *Bull. Atomic Sci.* Feb. 56/3 As it was, POL (gasoline and liquid fuels) *were* excluded from Pusan and unloaded by offshore floating lines. **1977** *R.A.F. News* 22 June–5 July 11/3 Training is centred on the ground attack role, with simulated attack profiles (SAPs) being flown against..POL objectives (Petrol, Oil, Lubricants). **1856** J. A. Symonds *Let.* 28 June (1967) I. 75 Thank Papa very much for the P.[ost] O.[ffice] O.[rder] which I got cashed without difficulty. **1886** W. S. Churchill *Let.* 13 July in R. S. Churchill *Winston Churchill* (1967) I. Compan. I. iv. 123, I received the P.O.O. which you sent me and am very thankful for it. **1966** Gurnett & Kyte *Cassell's Dict. Abbrev.* 169/2 *P.O.O.*, Post Office Order. **1968** *Which?* 11 Jan. 2/2 The Post Office have told us that they are introducing a new system of envelope sizing, called Post Office Preferred (POP). Packets which are not the size the Post Office prefers will not qualify for the cheapest postal rates. **1971** D. Potter *Brit. Eliz. Stamps* ix. 95 As part of the standardisation programme, the Post Office issued POP (Post Office Preferred) sizes during the course of 1968. **1895** W. K. Burton *Man. Photogr.* viii. 126 Paper for the [gelatino-chloride] process..is sold under various names... Examples are 'Solio-artistotype', 'Artisto-platino' and 'P.O.P.' **1925** P. R. Salmon *All about Photogr.* xx. 94 There is practically no difference between the cost of a finished print on P.O.P. and one on self-toning paper. **1972** A. Tyrrell *Basics of Reprogr.* vi. 105 POP should perhaps be remembered as more nearly a reprographic than a photographic process. **1955** *Science* 9 Dec. 1139/3 A solution of 0·4-percent PPO and 0·01-percent 1,4-di(5-phenyl-2-oxazolyl)benzene (POPOP) has been used, which gives..a pulse height of 121 percent. **1975** *McGraw-Hill Yearbk. Sci. & Technol.* 261/2 The scintillator dye POPOP, heated in an oven to a vapor pressure of about 10 torrs.., when pumped with 400 kW of ultraviolet light from a nitrogen laser, has produced 30 kW of tunable laser output. **1937** F. Borkenau *Spanish Cockpit* ii. 82 The depth of the antagonism between Esquerra and PSUC on the one hand and CNT and POUM on the other becomes intelligible. **1940** N. Mitford *Pigeon Pie* iii. 48 'It's Hitler and Stalin now, don't forget the wedding bells.' Mary had gone P.O.U.M., so she grudgingly conceded this point. **1916** F. M. Ford *Let.* 23 Aug. (1965) 69 The P.O.W.—who was quite unrecognizable, was perfectly businesslike. **1966** Gurnett & Kyte *Cassell's Dict. Abbrev.* 170/1 *P.O.W.*...Prince of Wales. **1919** W. H. Downing *Digger Dial.* 39 P.O.W., prisoner of war. **1941** *War Illustr.* 31 Jan. 101/1 P.O.W. camps in Germany and Poland are shown in this map. **1953** *News Chron.* 2 June 2/3 In Pusan..Pyun Yung Tai repeated his threat that foreign troops arriving to supervise P.o.W.s after a ceasefire would have to 'fight their way into Korea'. **1957** J. Braine *Room at Top* xiii. 128 What do you think a POW gets to eat? **1973** *Black Panther* 21 July 6/3 Former POW Sergeant Robert P. Chenoweth, 25, of Portland. **1882** R. Bithell *Counting-House Dict.* 235 *P.P. Endorsements.* Endorsements by procuration—that is, *per-procuration*... The following is the usual form of a per-pro endorsement. 'Pay to the Order of Blanc & Co. 'Per Pro Shipley & Sons. 'Thos. Brown.' **1882** [see Per *prep.* I. 7]. **1922** F. von Hügel *Let.* 29 Nov. (1927) 364 Yours very sincerely F. v. Hügel (pp. S.B.). **1967** E. Lemarchand *Death of Old Girl* i. 12 All the other letters can wait... Sign the ones we've done p.p. **1724** *Short Explic. Foreign Words in Mus. Bks.* 53 The letter P is often used as an Abbreviation of the Word *piano*: and PP as an Abbreviation of the Words *piu piano*: and PPP as an Abbreviation of the Word *pianissimo*. **1966** *Listener* 2 June 815/2 Helga Pilarczyk..often ignored the composer's repeated demands for *pp* or sometimes *ppp* singing. **1925** Goldberger & Tanner in *Public Health Rep.* (U.S. Public Health Service) XL. 77 It would seem as if the heretofore unrecognized pellagra-preventive factor, to which we shall hereafter refer as factor P-P, were capable of preventing the disease with little if any cooperation from the protein factor of the diet. **1926** —— in *Ibid.* XLI. 307 If the so-called growth-promoting water-soluble vitamin of the yeast is distinct from the antineuritic and from the P-P factor, then [etc.]. **1935** *Biochem. Jrnl.* XXIX. 2830 (*heading*) The vitamin B_2 complex. Differentiation of the antiblacktongue and the 'P.-P.' factors from lactoflavin and vitamin B_4. **1942** Bicknell & Prescott *Vitamins in Med.* v. 259 Nicotinic acid was at first hailed as the PP or pellagra preventing factor, but it is now known that pellagra is a multiple deficiency disease and that lack of nicotinic acid is only one of the factors in its causation. **1967** H. A. Guthrie *Introd. Nutrition* xii. 240/2 Niacin, another water-soluble vitamin identified with the B complex, has been known as nicotinic acid and as the pellagra preventative (P–P) factor. **1956** J. Cholak in P. L. Magill et al. *Air Pollution Handbk.* xi. 10 The normal fluoride content of the atmosphere is extremely low (2 to 8 ppb). **1970** *Nature* 25 July 403/1 Locally manufactured peanut butter was highly contaminated [with aflatoxin] in almost every case, with a mean approaching 500 p.p.b. **1975** *Ibid.* 23 Oct. 632/3 Figures in terms of p.p.b. (parts per 10⁹) are as commonplace nowadays as p.p.m. once were and for some compounds analysts can measure as little as 1 part in 10¹¹. **1838** H. C. Robinson *Diary* 30 Oct. (1967) 190, I called with a p.p.c. card on Samuel Rogers. **1863** Mrs. Gaskell *Dark Night's Work* vii. 108, I don't see any reason he had to come calling and P.P.C.-ing. **1883** Kipling *Let.* 14 Aug. in C. E. Carrington *Rudyard Kipling* (1955) iv. 53, I distributed my P.P.C. cards. **1909** J. R. Ware *Passing Eng.* 191/1 *To P.P.C.* (*Soc.*, 1880 on), to quarrel and cut. **1966** Gurnett & Kyte *Cassell's Dict. Abbrev.* 170/1 *p.p.c.*, pour prendre congé, to take leave. **1960** *Brit. Med. Jrnl.* 10 Sept. 783/2 You cannot travel far in the United States hospital world nowadays before hearing the phrase 'progressive patient care' (P.P.C.)... P.P.C. is defined as 'the organization of facilities, services, and staff around the medical and nursing needs of the patient'. **1964** G. L. Cohen *What's Wrong with Hospitals?* vi. 128 The very consultants who attack P.P.C. for disrupting continuity of care champion the same idea in maternity units. **1934** *Amer. Rev. Tuberculosis* XXX. 766 Under current arrangements the Committee endorses the use of a special designation, namely 'Tuberculin, P.P.D. (Purified Protein Derivative)', by these two manufacturing houses. **1951** *Proc. R. Soc. Med.* XLIV. 1046 Even the best P.P.D. preparations on the market, of about 90% tuberculoprotein content, still contain appreciable quantities of polysaccharide and nucleic acid which are not known to be concerned in the intradermal tuberculin reaction at all. **1963** Lincoln & Sewell *Tuberculosis in Children* iii. 41 The Heaf test requires a special apparatus that makes 6 skin punctures 1 mm deep through a layer of concentrated PPD containing 100,000 TU per ml. **1955** *20th Cent.* June 584 Every university may have its department of economics or philosophy or sociology, but only Oxford —with or without sociology—has PPE. **1964** E. Waugh *Little Learning* viii. 173 A new, disreputable school named Modern Greats (now dubbed P.P.E.) was for 'publicists and politicians'. **1972** *Times Lit. Suppl.* 3 Nov. 1310/2 Edward Heath and Harold Wilson read PPE. **1945** *Electronic Engin.* XVII. 683/1 With the Battle of Britain by night must be associated..the use of G.C.I., or Ground Control Interception, which Dr. Denis Taylor developed, using the P.P.I. (Plan Position Indicator) for the first of its many applications. **1959** *New Scientist* 23 July 97/2 In the ppi system the position of the echoes with respect to the ship is built up in plan or map form on the face of a cathode ray tube. **1966** *McGraw-Hill Encycl. Sci. & Technol.* I. 163/2 PPI presentations also change with altitude and direction of approach, making identification more difficult. **1946** W. H. B. Smith *Walther Pistols* 26 The P.P.K. means *Polizeipistole Kriminal*, indicating that the arm is intended for detectives and other police not in uniform who need a smaller weapon which can be readily concealed about the person. **1948** —— *Small Arms of World* (ed. 4) 424 Characteristics of PPK Model: Length 5.8″. Barrel 3.25″. Weight 19 oz. Capacity 7. **1973** J. M. White *Garden Game* 173 The Jensen..had been thoroughly searched, but the PPK had been clipped back under the dash. **1975** J. McClure *Snake* iii. 36 Zondi..checked his PPK automatic. **1976** G. Seymour *Glory Boys* x. 121 For close protection work he favoured the PPK (Polizei Pistolen Kriminal) Walther... The PPK was a small weapon, manufactured..by..the Karl Walther factory at Ulm. **1947** *Proc. Soc. Exper. Biol. & Med.* LXIV. 165/2 The most reliable method for the identification of the P.P.L.O. is the use of stained agar preparations. *Ibid.*, The P.P.L.O. are resistant to sulfonamides and to penicillin. **1965** *Listener* 11 Mar. 372/1 Unlike viruses, PPLO can grow in the absence of living host cells. **1972** *Science* 5 May 504/1 The mycoplasmas (originally called pleuropneumonia-like organisms or PPLO) have been studied both by people who want to grow them for study because of their pathogenicity and small size and by people who want to get rid of them because they are common tissue culture contaminants. **1913** *Bull. Univ. Illinois: Water Survey Series* No. 10, 42 (*table*) Dissolved oxygen, p.p.m. **1948** *New Biol.* V. 64 At 1–5 ppm it can be used..as a means of inducing tomatoes to set without pollination. **1964** *Punch* 25 Nov. 806/2 A peregrine found dead on Lundy Island contained 78 ppm of total chlorinated hydrocarbon in its liver. **1936** H. Nicolson *Let.* 12 June (1966) 265 Then his P.P.S. took him by the arm and he left the House for ever. **1959** *Times* 23 Oct. 14/6 Mr. Barber had served as a whip before he became P.P.S. to the Prime Minister in February, 1958. **1973** O. Lancaster *Littlehampton Bequest* 94 At the last Government reshuffle he was appointed P.P.S. to the Minister of Exploitation. **1976** H. Wilson *Governance of Britain* ii. 30 Apart from this, my experience of consultation is that, in ninety per cent of the cases, the senior minister concerned recommended his own P.P.S. for promotion in his own or another department—advice I did not always follow. **1841** Dickens *Let.* 9 July (1969) II. 325 P.S. Half asleep... P.P.S. They speak Gaelic here. **1900** G. B. Shaw *Let.* 30 Dec. (1972) II. 216 PPS I have been reading 'Herod' (I never go to the theatre now). *c* **1921** E. E. Cummings *Sel. Lett.* (1969) 81 P.S. am waiting for you... P.P.S. Elaine writes your painting is awfully good... P.P.P.S. Enjoyed the Krazy [Kat] you sent B. **1967** *Listener* 21 Dec. 814/2 PS: There'll be a special Christmas edition of *Round the Horne*. PPS: The new series of *Round the Horne* starts in February. PPPS: No advertising. **1937** A. Huxley *Let.* 30 Mar. (1969) 416, I have talked to the secretary of the PPU and he agrees that it will be best to go ahead with the 6d. edition in paper. **1973** *Freedom* 1 Sept. 4/1 Convinced pacifists of the PPU type. **1948** *R.A.F. Rev.* Feb. 3 (*heading*) The path of the 'P.Q.'. **1962** *Observer* 6 May 21/5 The parliamentary question—the 'P.Q.' **1975** M. Sinclair *Long Time Sleeping* ix. 117 'Everything stops for a PQ,' he said... 'Is there a Parliamentary Question down?' Pringle asked. **1976** H. Wilson *Governance of Britain* vii. 132 Prime ministers approach the bi-weekly ordeal by questions in different frames of mind, but of two things I am sure: no prime minister looks forward to 'PQs' with anything but apprehension; every prime minister works long into the night on his answers. 1943 *Aeroplane Spotter* 3 Dec. 278 (*caption*) The De Havilland P.R. Mosquito. 1946 *R.A.F. Jrnl.* May 153 We are indebted for the photographs and specialized articles with which the P.R. people supplied us. 1958 *Times Lit. Suppl.* 24 Jan. 48/5 Another part of the secret was to have P.R. grouped together. 1851 *Art Jrnl.* July 186/2 A school..that..will continue to exist unless Mr. Ruskin and his friends the P.-R.s upset it. 1874 L. Troubridge *Life amongst Troubridges* (1966) ix. 75 Amy's present rage..is to make her room pre-Raphaelite, with a border of P.R. bulrushes all round it. 1829 P. Egan *Boxiana* 2nd Ser. II. 8 A boxer of considerable notoriety in the London P.R. 1863 'Ouida' *Held in Bondage* I. iii. 64 Heroes of the Turf and the P.R. 1966 Gurnett & Kyte *Cassell's Dict. Abbrev.* 170/2 *P.R.*, Prize Ring. 1885 H. Sidgwick in A. & E. M. Sidgwick *Henry Sidgwick* (1906) 400 The application of the principle of P.R. to University Constituencies. 1909 (*heading*) P.R. Pamphlet No. 1 (Revised March 1909). 1924 [see *labour *sb.* 2 c]. 1935 H. Finer (*title*) The case against PR. 1974 *Times* 2 Mar. 14/4 If PR is to come in Britain the crucial issue is whether it is to be the single-member or multi-member kind. 1942 Partridge *Dict. Abbrev.* 77/1 *P.R.*.. Public Relations. (The publicity department of certain Services and Ministries.) 1944 A. Jacob *Traveller's War* 200 The remains of the P.R. unit set off down the desert road. 1963 H. Kubly *Whistling Zone* (1964) II. xv. 177 Your students are giving you an excellent PR. 1977 *Time* 31 Jan. 48/3 The p.r. man behind this is the star. 1909 *Cent. Dict.* Suppl., *P.R.*, an abbreviation..(*b*) of *Porto Rico*. 1966 *Publ. Amer. Dial. Soc.* 1964 XLII. 41 Terms used exclusively for Puerto Rican include..*P.R.* 1972 D. E. Westlake *Cops & Robbers* (1973) 9 There were no customers in there; just the Puerto Rican clerk... The PR was neutral as gray paint. 1895 G. B. Shaw *Our Theatres in Nineties* (1932) I. 32 If the friend of Sir Joshua Reynolds had been Sir David Garrick, and if every successive P.R.A. had had for his officially recognized peer the leading actor of his day. 1975 *Country Life* 18 Dec. 1736/2 Sir Gerald Kelly..the first of the televised art popularisers as PRA. 1849 D. G. Rossetti *Let.* 1 Oct. (1965) I. 73 Love to our family, the P.R.B., and all. 1850 W. M. Rossetti *P.R.B. Jrnl.* in *Preraphaelite Diaries & Lett.* (1900) 283 Collins has not established a claim to the P.R.B.-hood..the connexion would not be likely to promote the intimate friendly relations necessary between all P.R.B.'s. 1852 J. Brown *Let.* (1912) 128 The other morning I saw a scene which, were I a P.R.B. and a genius, I would make immortal. 1973 *Country Life* 8 Feb. 330/1 Rossetti is strangely well suited to one aspect of today's tastes... His aims went sensationally beyond those of his fellow members of the PRB. 1892 F. W. Maitland *Let.* 6 Sept. (1965) 105, I ought to be at P.R.O. next week. 1931 *N. & Q.* 5 Dec. 408/2, I cannot find the Returns of officers' services for these regiments at the P.R.O. 1958 *New Statesman* 6 Sept. 330/2 Farewell, adieu, BM and PRO, My time is up, reluctantly I go. 1941 H. Nicolson *Diary* 8 July (1967) 177, I drive down to White's Club with Duff [Cooper] and beg him to treat the P.R.O.s this afternoon with all gentleness. They are a touchy lot. 1966 'H. MacDiarmid' *Company I've Kept* i. 22 Philip Jordan, Attlee's P.R.O., was with us. 1970 J. Tunstall *Westminster Lobby Correspondents* v. 56 The Prime Minister's scarcity value to the Lobby men allows him (a) to speak to the Lobby mainly through a P.R.O. and personally only on irregular occasions; (b) the P.R.O. still attracts a substantial daily attendance even when he says little. 1973 *Jrnl. Electrochem. Soc.* CXX. 1001 (*heading*) Reliability of NiCr 'fusible link' used in PROM's. 1977 *Sci. Amer.* Sept. 139/1 Information stored in ROM's and PROM's is nonvolatile. 1972 *Guardian* 1 Sept. 1/6 Prisoners were..vowing to stay up [on the roof] until the Home Office recognised the prisoners' union, PROP. *Ibid.* 1/8 PROP—'Preservation of the Rights of Prisoners'. 1973 D. Curtis *Dartmoor to Cambridge* xiii. 121 We decided to organise the British movement by forming a prisoners' union. That was the first step in the conception of PROP, the Preservation of the Rights of Prisoners, a movement that was destined to shake the British prison system to its core four years later. 1976 *Daily Mail* (Hull) 16 Dec. 11/4 PROP, the prisoners' rights organisation, is to hold its own public inquiry into the riot at Hull Jail. 1976 A. Miller *Inside Outside* xi. 176 Where PROP went wrong..was in inviting men to break the prison rules. 1927 *Melody Maker* Aug. 755/2, I am assuming..that the hall is not already licensed by the P.R.S. 1968 *Listener* 8 Aug. 177/3 A lot of the more superior publishers..weren't members of the PRS. 1616 T. Roe *Let.* 30 Nov. in *Embassy to Court of Gt. Mogul* (1899) II. 359 P.S.—I humbly desire your Honor to doe me the fauour to thanck Sir Thomas Smyth in my behalfe. 1757, etc. P.S. [see main entry in Dict.]. 1771 J. Wedgwood *Let.* 11 May (1965) 108 PS The letter to which this is a ps I did not like to send by the post. 1842 Dickens *Let.* 1 May (1974) III. 228 Look over leaf for the PS. 1853 Mrs. Gaskell *Cranford* ix. 163 So she ended her letter; but in a P.S. she added, she thought she might as well tell me what was the peculiar attraction to Cranford just now. 1969 *Listener* 15 May 682/3 PS. These are only hints. Please do not repeat verbatim. 1790 P.S. [see O.P. s.v. *O 5 d]. 1838 R. B. Peake *Quarter to Nine* i, in B. Webster *Acting National Drama* II. 5 Apartment; Frolick's lodgings; closet door. Enter Mrs. Jervis, P.S. 1942 Partridge *Dict. Abbrev.* 78/2 *p.s.*, prompt side (of a theatre). Also *P.S.* 1896 *Oxfordshire Light Infantry Chron.* 1895 26 Lt.-Colonels... (1) Johnstone, J., *p.s.c.* 16 Mar. 92. 1920 *Punch* 24 Mar. 225/1 Upon my first arriving on his Staff he had said to me, 'Oh, by the way, P.S.C., of course?'.. 'You have Passed Staff College, of course?' he said a little less affably. 1972 *Times* 7 Sept. 16/2 Without the magic letters 'psc' (passed staff college) after their names their chances of being promoted above major are, at best, doubtful. 1944 E. W. F. Feller *Air Compressors* xiv. 438 Figure 410 shows the piping arrangement for the 250 psi method. 1959 *Motor* 2 Sept. 75/2 This is with 5·90–13 Goodyear tubed tyres, inflated to 26 p.s.i. back and front. 1968 M. Woodhouse *Rock Baby* xviii. 183 She fitted a little pressure gauge to the drill-guide... The needle on the gauge flicked across at once, to nine p.s.i. 1975 *Offshore Engineer* Sept. 44/1 Phillips has a contract with the British Gas Corporation (BGC) to meet the required amount per day at a specified pressure which is 1,000psi. 1951 psia [see *isentropically* adv. s.v. *iso-]. 1975 *Petroleum Rev.* XXIX. 91/1 The stabilised crude, with a vapour pressure of 5–7 psia, will be piped to the Greatham site. 1932 *Motor Transport* 28 Mar. 351/3 A bus driver who was refused a p.s.v. driving licence..has appealed with success to a bench of magistrates. 1944 L. D. Kitchin *Road Transport Law* 9/2 Every p.s.v., except those first registered on or before January 1, 1932, must be capable of turning in a circle not exceeding 60ft. 1972 *Police Rev.* 1 Dec. 1577/1 A licensed p.s.v. is not being used as a p.s.v. when it is not carrying passengers. 1922 T. E. Lawrence *Let.* 1 Sept. (1938) 364 If I can get able to sleep, and to eat the food, and to go through the P.T. I'll be all right. 1938 *Times* 14 Feb. 10/4 A Half-Day Course in P.T. 1965 W. Lamb *Posture & Gesture* viii. 107 If a woman she may be respected for her vigour but is too likely to be drained of feminine attractiveness—'P.T. hag' is the profession's own term. 1973 M. Amis *Rachel Papers* 137 The Darwin-born PT instructress, on the other hand, her glossy shoulder-muscles rippling in the ninety-degree heat, threw her bulk round the court in frank virility. 1958 *People* 4 May 2/1 (Advt.), Victor £498 plus £250 7s. PT. 1963 *Which?* Mar. 71/1 [Price] excluding 10 per cent P.T. 1966 *Punch* 27 July 132/2 The Government's determination not to flinch, if necessary, from a ten per cent PT increase on musical instruments? 1925 *Kansas City* (Missouri) *Star* 4 Feb. 11/1 (*heading*) P.-T.-A. plans celebration. 1962 L. Deighton *Ipcress File* xviii. 106 The old stuff about re-treads, P.T.A. meetings, and where to go for a good divorce. 1973 J. Burrows *Like an Evening Gone* iv. 47 We're doing a P.T.A. play at the school. 1976 *Publ. Amer. Dial. Soc.* 1973 LX. 12 Mother of Informant 8. Active in church affairs and P.T.A. 1942 PT boat [see *expendable *a.*]. 1961 W. Vaughan-Thomas *Anzio* v. 67 On 28 January Clark himself, after being nearly killed on the deck of his PT boat, arrived at Anzio. 1974 *Lebende Sprachen* XIX. 38/1 US *PT boat*—BE/US *motor torpedo boat*, BEa. *E-boat.* Schnellboot. 1932 Blakeslee & Fox in *Jrnl. Heredity* XXIII. 97/1 The long name, phenyl-thio-carbamide, we are shortening to the nickname *P.T.C.*, an abbreviation which we shall use throughout this paper. *Ibid.* 98/1 To those who find P.T.C. strongly bitter, it seems incredible that any one could call it tasteless. 1965 *Punch* 10 Nov. 688/1 Tallness, or colour-blindness, or the ability to taste the substance known as PTC (phenylthiocarbamide) are inherited. 1976 PTC [see *phenylthiocarbamide* s.v. *phenyl 2 b]. 1909 *Army & Navy Gaz.* 1 May 430/3 Foil v. Foil (P.T.I's only). 1916 'Taffrail' *Pincher Martin* iii. 45 The next turn was by the P.T.I. (Physical Training Instructor). 1964 J. Hale *Grudge Fight* vi. 92 Buck Jones the P.T.I. 1968 M. Woodhouse *Rock Baby* xxi. 202 He was wearing a sweatshirt and blue drill trousers, like a grossly overweight P.T.I. 1977 *R.A.F. News* 11–24 May 2/4 There will also be a trampoline display by PTIs of the RAF. 1859 Geo. Eliot *Let.* 24 Feb. (1954) III. 24 Yours ever truly George Eliot. P.T.O. I have reopened my letter to ask you [etc.]. 1902 H. G. Wells *Let.* 14 Feb. in *A. Bennett & H. G. Wells* (1960) 76 P.T.O. 1966 Gurnett & Kyte *Cassell's Dict. Abbrev.* 172/2 *P.T.O.*, Please Turn Over. 1951 A. B. Lees *Farming Machinery* xv. 144 If p.t.o.-driven trailers of compact design could be made available to hill farmers at an economic price they might well prove to be a major factor in increasing food production from marginal land. 1967 *Jane's Surface Skimmer Systems* 1967–68 64/2 Air can be supplied from a PTO-driven blower mounted on the tractor unit. 1973 *Country Life* 28 June 1904/2 The Bearcat combined roughage and grain grinder... Available as a pto driven static or trailed machine..[it] incorporates a pre-breaker with bevel edge. 1904 *Age* (Melbourne) 20 May 1 (Advt.), Ball and Welch Pty. Ltd. 1938 *Act* (Victoria) 3 *Geo. VI* no. 4602, sect. xxvi, §5 The word 'proprietary' or the abbreviation thereof 'Pty.' shall form part of the name of a proprietary company. 1969 *Northern Territory News* (Darwin) 11 July 4/3 (Advt.), N.T. Real Estate Pty. Ltd. 1934 V. M. Yeates *Winged Victory* III. ix. 365 'I wish you'd tell me what PUO means.' 'What do you want to know that for?' But the M.O. overcame his professional love of mystery, and added: 'It stands for Pyrexia of Unknown Origin.' 1964 M. Hynes *Med. Bacteriol.* (ed. 8) xii. 192 A clinical diagnosis of typhoid is rarely possible in the early stages, and the patient will be investigated as a case of 'P.U.O.' or pyrexia of unknown origin. 1976 *Proc. R. Soc. Med.* LXIX. 557/1, I..told him what a lot of interesting medical cases had come to the hospital under that useful army diagnosis of 'PUO'—pyrexia of unknown origin. 1933 C. Mackenzie *Water on Brain* viii. 112 'The P.U.S.?' 'The Permanent Under-Secretary,' Hunter-Hunt explained. 1974 P. Gore-Booth *With Great Truth & Respect* 324 What governed the whole of my life in this final period was the circumstance that, apart from the Secretary of State, the PUS was the only person in the Office whose obligation it was to have some knowledge of everything. 1939 *London Weekly Advertiser* 7 June 6/5, 35/- P.W. Incl. 1968 *Punch* 19 June 871/1 Let's be terribly sympathetic to the BOAC pilots, skilled, resourceful and too well-mannered actually to call their £30 pw pay rise offer 'insulting', the vogue word for this sort of thing. 1975 *Irish Times* 9 May 22/3 (Advt.), Rent inclusive of heating. £10 p.w. 1976 *Burnham-on-Sea Gaz.* 20 Apr. 7/7 (Advt.), I have sacked my advertising man. I have been paying him £1.15 pw and what do I get. 1909 *Cent. Dict.* Suppl., *P.W.D.*, an abbreviation of *Public Works Department.* 1922 G. Bell *Let.* 17 July (1927) II. xxii. 645 Mr. Cooke, Major Wilson and I accepted the invitation of Sabih Bey, Minister of P.W.D., to bathe from his house in Muadhdham. 1958 G. Durrell *Encounters with Animals* iv. 168 There was the little cockney P.W.D. man who..offered to drive me a hundred-odd miles, over atrocious African roads. 1971 *Illustr. Weekly India* 11 Apr. 45/3 Those of a structural nature which the P.W.D. were able and willing to solve. 1975 O. Sela *Bengali Inheritance* ix. 76 The flat was..originally built for European PWD engineers. 1954 *Mech. Engin.* July 585/1 The Westinghouse Electric Corporation, as of July 1953, was assigned responsibility for the development and design of a pressurized light-water reactor (PWR). 1976 *New Scientist* 5 Aug. 290/2 PWRs are now better proven with a greater reliability than two years ago, and there are several large units now working in the US and West Germany. 1929 *Papers Mich. Acad. Sci., Arts & Lett.* X. 317/1 *PX*, post exchange. 1936 *Amer. Speech* XI. 62 The army took avidly to the name canteen—so avidly that after..the stores were officially re-named 'Post Exchanges', the name persisted... Even the modernistic abbreviation 'PX' does not seem to displace it. 1959 C. MacInnes *Absolute Beginners* 102 My ivy-league outfit a GI got for me last year from his PX. 1971 M. McCarthy *Birds of America* 184 If I give him a divorce, they'll take away my PX card and my QC privileges. 1975 *Publishers Weekly* 13 Jan. 59/1 Two American truck drivers get lost in Vietnam and take shelter in an abandoned supply depot that is still stocked with PX goodies like stereos, TVs and canned food.

b. *Teleph.* P = 'private' in PABX, private automatic branch exchange; PAX, private automatic exchange; PBX, private branch exchange; PMBX, private manual branch exchange.

1923 *P.O. Electr. Engineers' Jrnl.* XV. 309 On P.A.B.X.'s a jack is provided on the manual board for every line. 1976 *Computing Europe* 2 Sept. 5/2 Lines can be intercepted within buildings (particularly at PABXs or distribution boards) and even at Post Office telephone exchanges. 1923 *P.O. Electr. Engineers' Jrnl.* XV. 315 These plants are in some cases working as single P.A.X.'s. 1974 *Ibid.* Oct. 19 (Advt.), Pye Business Communications' capability in PAX and PABX telephone systems can improve the efficiency of your existing installation or provide you with a completely new system. 1976 *Eastern Even. News* (Norwich) 9 Dec. 16/2 (Advt.), Norwich Airport has a vacancy for a temporary Clerk/Telephonist (part-time) involving manual operation of a small PABX switchboard, some simple accounts and typing. 1917 G. D. Shepardson *Telephone Apparatus* xv. 253 The current through the talking subscriber's circuit experiences a considerable fall of potential due to the resistance in the trunk line between the control exchange battery and the P.B.X. board. 1940 R. Chandler *Farewell, my Lovely* xxxii. 240 A uniformed man dozed behind a pint-sized PBX set into the end of a scarred wooden counter. 1958 'P. Bryant' *Two Hours to Doom* 19 He lifted the phone to the PBX. 1976 *National Observer* (U.S.) 10 Apr. 9/2 The FCC program, which takes effect Aug. 1 for the PBX, key, main, and coin telephones. 1932 Herbert & Procter *Telephony* (ed. 2) I. xiv. 614 Lines terminating on a P.M.B.X. are connected to consecutive jacks. *Ibid.* 619 The P.M.B.X. operator withdraws the plug from the exchange line jack. 1975 *Post Office Electr. Engineers' Jrnl.* LXVIII. 61/2 The PMBX No. 4 was originally designed to meet the requirements of single-position installations of up to 160 extensions.

c. p = *pico-, as in pF, picofarad(s).

1940 *Chambers's Techn. Dict.* 631/2 *pF.*, *pf.*, abbrev. for *pico-farad.* 1958 *Engineering* 21 Feb. 228/2 The sensitivity of the Tektor Major is better than 1·0 pF under all conditions.

d. P = 'poly-' in PCB, polychlorinated biphenyl; PTFE, polytetrafluoroethylene; PVA, (*a*) polyvinyl acetate; (*b*) polyvinyl alcohol; PVC, polyvinyl chloride; PVP, polyvinylpyrrolidone. (All these often occur with full stops, and some are occas. given in lower case letters.)

1966 *New Scientist* 15 Dec. 612/3 In Sweden, PCB is known to be used in electrical insulations, hydraulic oils, high-temperature and high-pressure lubricating oils, paints, lacquers and varnishes, and as pigments in various plastics. 1971 *Observer* 12 Dec. 4/3 Levels of PCBs..are roughly 10 to 100 times higher in plankton from the open ocean than from coastal waters. 1974 J. Burton *Pollution* v. 33 Thousands of seabirds died in the Irish Sea in 1969, and it is believed that PCB was responsible. 1977 *Jrnl. R. Soc. Arts* CXXV. 240/1 Our wildlife has..been affected by the accumulation of the persistent organo-chlorine pesticides and by PCBs. 1949 *Electronic Engin.* XXI. 220/1 Two new types of seals will shortly be available, one of which is made from an entirely new material, polytetrafluorethylene (P.T.F.E.). 1962 *Which?* Aug. 255/1 There are two kinds of non-stick frying pans—those with a silicone finish and those with a plastic called polytetrafluoroethylene, or PTFE. 1973 *Materials & Technol.* VI. viii. 545 Trade names of PTFE plastics include: Teflon (USA); Fluon (UK); and Hostaflon TF (Germany). 1943 Simonds & Ellis *Handbk. Plastics* viii. 393 (*table*) Polyvinyl alcohol 'PVA' Resin. *Ibid.* 1007 (*Gloss.*) Trade name/PVA Type/Polyvinyl alcohol Typical applications/Tubing, rubber substitute (see Resistoflex). 1960 *Farmer & Stockbreeder* 22 Mar. (Suppl.) 10/2 Again, modern materials come to the rescue, and this time it is the relatively new PVA (Polyvinyl acetate) emulsion. This is a synthetic resin which, when used in emulsion form and added to sand, cement, and certain aggregates, will give a jointless, waterproof floor. 1966 A. W. Lewis *Gloss. Woodworking Terms* 38 *Polyvinyl acetate* (*pva*), emulsion glue of a white creamy consistency. It is used cold and does not require a hardener. 1969 L. S. Mounts in W. R. R. Park *Plastics Film Technol.* v. 140 Three water soluble plastic films are currently produced from polyvinyl alcohol (PVA), methyl cellulose and polyethylene oxide. 1977 *36 Home Handyman Projects* (Austral. Home Jrnl.) 74/2 Cut thin strips of very thin felt and apply a PVA glue to one side of the felt. 1941 *Electronic Engin.* XIV. 541/2 Polyvinyl chloride and copolymers. Examples—Welvic, B.X.P.'s, 'P.V.C.', Chlorovene. 1957 *Economist* 12 Oct. 161/1 Similar plastic-coated steels, all using pvc, have been offered in the United States for some time. 1971 *New Scientist* 10 June 630 The PVC-coated fabrics that have previously been the main synthetic material used in women's footwear. 1972 *Country Life* 16 Mar. 612/1 Playing around with a few old bricks..and a cowl made of wire and pvc sacks. 1975 W. G. Roberts *Quest for Oil* (rev. ed.) vii. 77 Plastics, particular examples being Polythene, polystyrene, PVC and a number of synthetic resins. 1951 *Lancet* 19 May 1096/1 A Saline solution incorporating a P.V.P. compound was prepared and used in Germany with some success as a plasma substitute during the 1939–45 war. 1959 *Jrnl. Inst. Brewing* LXV. 73/2 Beer can be chill-proofed, without risk of pasteurization haze, by adding PVP during storage in the cellar. 1966 J. A. Brydson *Plastics Materials* xiv. 286 In the field of cosmetics p.v.p. is used because of its unique property of forming loose addition compounds with skin and hair. Hair lacquers may be formulated based on 4–6% p.v.p. in ethyl alcohol. 1974 M. C. Gerald *Pharmacol.* ix 164 The plasma substitutes, dextran and polyvinylpyrrolidone, PVP.

III. 1. *P trap*, a trap consisting of a U bend the upper part of whose outlet arm is bent horizontal or nearly so.

1885 P. J. Davies *Stand. Pract. Plumbing* I. 103 Fig. 205 is the ordinary half ∽-trap, wrongly called ℧-trap. 1890 W. R. Maguire *Domestic San. Drainage* vi. 206 No. 1 is the S-trap; No. 2, half S-trap; No. 3, P-trap. 1976 R. Day *All about Plumbing* 64/1 In a ground floor w.c. it is usual to fit an S trap,..but in an upstairs floor w.c., a P trap is usually installed.

2. *p* or *P* (*Physics* and *Chem.*) = principal: orig. used to designate one of the four main series of lines in atomic spectra, but now more frequently applied to electronic orbitals, states, etc., possessing one unit of angular momentum.

1890 J. R. Rydberg in *Phil. Mag.* XXIX. 335 A few examples will suffice to show the arrangement and the use of this system. K $(D_1, 4)$ denotes the fourth line of the first diffuse series of the spectrum of potassium..Rb ($P_{1,2}$, 2) the second doublet of the principal group of Rb [etc.]. 1910 [see *F III. 1 j]. 1922, 1955 [see *D III. 3 b]. 1964 J. W. Linnett *Electronic Struct. Molecules* ii. 29 In the oxygen atom two 2*p* orbitals are half-filled..and so..bonds can be formed to two hydrogen atoms.

3. [initial letter of *primary*.] Used, chiefly in *P wave*, to denote an earthquake wave of alternate compression and rarefaction (the faster of the two main kinds of wave transmitted through the earth).

1908 C. G. Knott *Physics Earthquake Phenomena* xi. 199 Although Rebeur Paschwitz had suggested the possibility, Oldham, of the Geological Survey of India, was the first clearly to establish the existence in the complete record of two distinct phases in the Preliminary Tremors. These will be distinguished as *P* and *S*. *Ibid.* xii. 225 (*heading*) Chordal and arcual speeds of P and S phases of preliminary tremors in earth-radius per minute. 1936 V. B. Macelwane in Macelwane & Sohon *Introd. Theoret. Seismol.* I. ix. 248 When it became clear that the *P*-waves were of the condensation-rarefaction type and the S-waves of the shear type, and individual earthquakes had been observed at a sufficient number of stations, attempts were made to draw up time–distance curves for the arrival times and to correlate these with the time of occurrence of the earthquake. 1966 *McGraw-Hill Encycl. Sci. & Technol.* XII. 152/1 The seismic body waves (*P*, *S*, and composite types like *PS*) have predominant periods in the range 1–15 sec, with *P* and *S*, respectively, at the short-period and long-period end of this range. 1971 I. G. Gass et al. *Understanding Earth* iii. 54/1 The core-mantle boundary is marked by an abrupt reduction in the velocity of compressional or *P* earthquake waves.

4. On the analogy of *pH, used to denote the negative of the common logarithm of a concentration or activity expressed in moles per litre; similarly p*K*, the negative of the common logarithm of a dissociation constant; pF (see quot. 1971).

1924 N. H. Furman in H. S. Taylor *Treat. Physical Chem.* II. xiii. 828 By graphic interpolation, plotting —log *k* against pOH, we find pOH to be 5·1. 1929 H. T. S. Britton *Hydrogen Ions* iii. 43 The hydrogen-ion concentration has a *p*H value of $\frac{1}{2}pK_w$ at neutrality. 1935 R. K. Schofield in *Trans. 3rd Internat. Congr. Soil Sci.* II. 39 It has proved convenient to use a new scale, which I have called the pF scale, to express what has, in the previous pages, been called 'suction'... The symbol 'p' expresses its logarithmic character, while the symbol 'F' is intended to remind us that by defining pF as the logarithm of the height in centimetres of the water column needed to give the suction in question, we are really using the logarithm of a free energy difference measured on a gravity scale. 1946 Lutz & Chandler *Forest Soils* ix. 292 Designation of the energy relations of soil water in terms of *p*F is analogous to specification of reaction in terms of *p*H. 1965 R. G. Kazmann *Mod. Hydrol.* v. 141 The energy gradient of the soil (*p*F) works against the force of gravity. 1968 Passmore & Robson *Compan. Med. Stud.* I. vi. 5/2 In a solution containing a mixture of buffers the conjugate acid base ratio is determined at any pH by its p*K*. 1971 *Gloss. Soil Sci. Terms* (Soil Sci. Soc. Amer.) 12/2 *pF* (obsolete), the logarithm of the soil moisture tension expressed in centimeters height of a column of water. 1972 *Wastewater Engin.* (Metcalf & Eddy, Inc.) vii. 255 With pOH, which is defined as the negative logarithm of the hydroxyl-ion concentration, it can be seen..that, for water at 25°C, pH + pOH = 14. 1973 F. G. Shinskey *pH & pIon Control in Process & Waste Streams* i. 4 Increasing activity is indicated by a decreasing pIon.

5. *P-marker* (Linguistics) = *phrase-marker* (*phrase *sb.* 7).

1955 N. Chomsky *Logical Struct. Linguistic Theory* (microfilm, Mass. Inst. Technol.) x. 735 We define 'K is the P-marker of Z' as: K is the set of strings which appear as a line of one of the members of E, where E is an equivalence class of S_1-derivations of *Z*. 1963 Chomsky & Miller in R. D. Luce et al. *Handbk. Math. Psychol.* II. 301 A grammatical transformation, then, is a mapping of *P*-markers into *P*-markers. 1964 E. Bach *Introd. Transformational Gram.* iii. 39 A representation of immediate constituent structure for a string, such as is given by a labeled bracketing or labeled tree, is called a *phrase marker* (P marker). 1967 D. G. Hays *Introd. Computational Linguistics* xiii. 210 The parser would..submit, to a higher level source-language processor or to a translator, base P-markers—constituency diagrams with all transformations undone. 1976 *Language* LII. 110 A successful asymmetrical derivation rule must operate on a P-marker and not on an individual lexical entry.

6. [repr. *proton*.] *p-process* (Astr.): a process believed to occur in stars by which heavy proton-rich nuclei are formed from other nuclei, esp. in circumstances of high proton flux such as may obtain in supernovae.

1956 F. Hoyle et al. in *Science* 5 Oct. 613/3 At such temperatures (*p*, γ) reactions occur in a time of the order of 10 seconds, even on the heaviest nuclei (*p*-process). 1957 *Rev. Mod. Physics* XXIX. 617/2 It is probable that the maximum number of protons which can be added to C^{12} through the duration of the *p* process is only two. 1977 J. Narlikar *Struct. Universe* ii. 50 Apart from these two processes there is a rarer process which produces proton-rich isotopes by exposing the r-process and s-process material to a fast flux of protons or of high-energy photons. This is known as the p-process.

paan (pān). [see Pan *sb.*⁵] = Pan *sb.*⁶; the leaf of the betel pepper, *Piper betle*, used to enclose slices of betel nut (*Areca catechu*) mixed with lime; also, the masticatory formed by this mixture. Also *attrib.*

1964 *New Statesman* 3 Apr. 517/3 Eating the green leaf called paan..and the nut of the betel palm..is common in India, in Malaya and in Ceylon... The best paan leaf comes from Banaras. 1967 Singha & Massey *Indian Dances* vi. 68 The betel or *paan* as it is called, is covered with silver leaf and sprinkled with rose water. 1971 *Illustr. Weekly India* 4 Apr. 48/1 Paan chewing is an ancient Indian habit. *Ibid.* 51/2 The Mughal rulers were great connoisseurs of paan. 1974 *Times* 7 Dec. 10/8 Paans, those leaf-wrapped chews.

‖ **pa'anga** (pā,ā·ŋgā). [Native.] The monetary unit of Tonga.

1966 *Times* 21 May 8 Tonga has decided against calling its new decimal currency unit the dollar because the native word, 'tola', also means a pig's snout, the soft end of a coconut, or, in vulgar language, a mouth. The new unit, to be introduced next year, will be called 'pa'anga', which has only two alternative meanings—a coin-shaped seed and, not surprisingly, money. 1973 *Whitaker's Almanac* 1974 986/2 Pa'anga (T$) of 100 *Seniti*.

paarlmoer, var. *perlemoen.

PABA (pī:ē̆i,bī,ē̆i·, pā·bā). *Pharm.* Also **Paba, paba.** [f. the initial letters of the formative elements of the chemical name.] = **para-aminobenzoic acid*.

1944 *Jrnl. Exper. Med.* LXXIX. 337 (*heading*) Comparison of PABA production by staphylococci (method of Lewis) with the concentration of sodium sulfathiazole required to inhibit *in vitro* growth of the cultures. 1945 *Proc. Soc. Exper. Biol. & Med.* LVIII. 262/1 In addition to its activity against typhus rickettsiae, preliminary experiments have shown that paba also inhibits growth of the rickettsiae of Rocky Mountain spotted fever. 1951 A. Grollman *Pharmacol. & Therapeutics* xxvii. 606 The use of Paba has also been advocated in various 'collagen' diseases. 1963 *Times* 2 May 14/1 (Advt.), Para-aminobenzoic acid..has long been known simply as PABA which, even when pronounced 'pahbah' is clinically and aurally acceptable. 1966 *McGraw-Hill Encycl. Sci. & Technol.* XIII. 244/2 The mode of action of sulfonamides is considered to be an antimetabolite activity, dependent upon the inhibition of enzyme systems involving the essential PABA.

pablum (pæ·blŭm). Also **Pablum.** The proprietary name of a children's breakfast cereal. Also *fig.*

1932 *Official Gaz.* (U.S. Patent Office) 19 July 596/2 Mead Johnson & Company, Evansville, Indiana. Pablum. For specially prepared cereal food consisting of a mixture of wheat meal, to which have been added wheat embryo, dried yeast, powdered dehydrated alfalfa leaf and powdered beef bone prepared for human use. Claims use since June 4, 1932. 1941 *Trade Marks Jrnl.* 5 Feb. 47/2 Pablum... A food for infants and invalids. Mead Johnson & Company,..Evansville, State of Indiana, United States of America; manufacturers. 1953 Pohl & Kornbluth *Space Merchants* i. 9 Over the breakfast juices and the children's pablum..they spoke persuasively to each other about how wise and brave they had been to apply for passage in the Venus rocket. 1956 P. Frank *Forbidden Area* i. 40 He would want her to quit her job, and eventually..there would be children. She would find..her research downgraded to equations involving boiled water, evaporated milk, Karo syrup, and pablum. 1964 A. Tyler *If Morning ever Comes* xi. 160 *Joanne* never picks up. I had to scrape pablum off the damn *toaster* this morning. 1970 *Listener* 22 Oct. 538 In one week Spiro Agnew ascribed moral decay to the universities, Dr Spock, and the Presidential Commission on Campus Unrest, which, he said, had produced a report which 'was sure to be taken as more pablum for the permissivists'. 1971 W. Hillen *Blackwater River* ii. 9 After a start on pablum and eggs and milk..he developed into a strong dog. 1976 *Amer. N. & Q.* XIV. 147/2 Its [*sc.* Oscar Wilde's *The Happy Prince*] obvious allegory and its appeal to a patronizing benevolence are palatable as pablum, but hardly to be trusted to sustain Eliot's confused, but intellectual Fisher King.

Pabst (pabst). *U.S.* The name of a lager beer.

Registered in the U.S. as a proprietary name.

[1906 *Official Gaz.* (U.S. Patent Office) 26 June 2969/2 Beer. Pabst Brewing Company, Milwaukee,..Used ten years..the name 'PABST' on the upper part of the band.] 1920 *Ibid.* 15 June 579/2 Pabst Brewing Company, Milwaukee..Pabst..Beer and Malt Extracts. *Claims use* since June, 1889. 1963 R. Wellek in N. Frye *Romanticism Reconsidered* 112 'It is impossible to think seriously with words such as Classicism, Romanticism, Humanism, or Realism.'.. But of course these terms are not labels: they have a range of meaning very different from Pabst Blue Ribbon or Liebfrauenmilch. 1974 *Black World* Mar. 59/1 He opened the refrigerator. In its confines were two cans of Pabst. 1976 *Time* 27 Sept. 47/2 The room is a cacophony of the ping-pong-dingdingding of the pinball machine, the pop-fizz of another round of Pabst, [etc.]. 1977 *Time* 7 Feb. 60/1 Belting down tumblers of Jim Beam and Pabst, they compared horror stories.

pac (pæk). Also **pack.** *N. Amer.* [Of Lenape (Delaware) Indian origin.] **a.** (See quots. *a* 1877 and 1961; cf. *shoepack.) **b.** With initial apostrophe, *ellipt.* for *shoepack.

a 1877 Knight *Dict. Mech.* II. 1590/1 *Pac; Pack*, a moccasin having a sole turned up and sewed to the upper. Though now made of leather of various kinds, the *pac*, as used by the Indians of the Six Nations, for instance, was made of hide boiled in tallow and wax; or of *tawed* hide subsequently stuffed with tallow or wax. 1922 *Outing* May 68/1 Footwear, pac boots 16 inches; rubber boots. 1931 'Grey Owl' *Men of Last Frontier* 181 And here is where my hard-soled 'packs came in. 1944 *Sears, Roebuck Catal.* 345 Leather top work Pac... Not rationed... If you wear size 6½ or 7 shoe, order size 6 pac. 1961 Webster 1616/1 *Pac* also *pack*,..2. a laced heelless sheepskin or felt shoe worn inside a boot or overshoe in cold weather. 1973 B. Wright *Four Seasons North* 12 The trails are slippery in our rubber-soled pacs.

pace, *sb.*¹ Add: **III. 7. b.** *spec.*, in *Cricket*, the speed of a bowler's delivery; the velocity of a ball bowled. Also *ellipt.*, = pace-bowling.

1800 T. Boxall *Rules & Instr. Cricket* 15, I think it very proper not to bowl always the same pace. *Ibid.* 34 The striker must judge for himself what pace the bowler is bowling. 1816 W. Lambert *Instr. & Rules Cricket* 18 Never bowl faster than you can do..well, varying your pace as you may judge proper. 1900 P. F. Warner *Cricket in Many Climes* 90 In Cobb and Kelly the New York team had two capital bowlers. The former is right-hand, and varies his pace well. 1955 [see *back-lift* s.v. B*ack- B]. 1976 *0–10 Cricket Scene* (Austral.) 7/1 Many [innings] have been an amazing barrage of brutal driving, hooking and cutting against both pace and spin.

c. *pace of the wicket, ground* (further examples).

1888 R. H. Lyttelton in Steel & Lyttelton *Cricket* ii. 52 In back play, unless the ball is very short, the pace of the ground may beat a man. 1903 A. C. MacLaren in H. G. Hutchinson *Cricket* ix. 252 Too much importance ought never to be attached to the opening game, owing to those who have not previously visited Australia being wholly unaccustomed..to the fast pace of the wicket. 1955 *Times* 9 May 15/1 He spun it, too, and one can imagine how dangerous he might be on wickets with any pace in them.

8. a. Later example with *hold*.

1893 *Nat. Observer* 30 Sept. 505/1 Watts would have made a better race with La Flèche if he had pushed to the front, since she can hold a hot pace.

b. *to go the pace* (further examples); *to set the pace* (examples).

1890 *Licensed Victuallers' Gaz.* 5 Dec. 363/1 Fresh from Oxford, Arthur had been going the pace. 1892 [see *blind *a.* (and *adv.*) I i]. 1905 A. Bennett *Tales of Five Towns* I. 109 Well, you *have* been going the pace! We always knew you were a hot un, but really—. 1928 Barrie *Peter Pan* I, in *Plays* 19 Nana must go about all her duties in a most ordinary manner..; naturalness must be her passion; indeed, it should be the aim of every one in the play, for which she is now setting the pace. 1928 E. Wingfield-Stratford *Hist. Brit. Civilization* II. iii. 1090 Britain was ceasing to set the pace to her neighbours; she was beginning to show signs of flagging in the race. 1958 *Engineering* 4 Apr. 424/2 Do things before anyone else not wait to see what someone else does—set the pace and keep them hopping.

IV. 14. pace-bowling (hence *pace bowler*), **pace-change, pace-man** (cf. senses 7 b, c above).

1951 *People* 3 June 8/6 Good news to-day about Alec Bedser and Trevor Bailey, the two pace bowlers likely to be chosen to open England's attack in the First Test. 1976 *Scotsman* 24 Dec. 15/7 But whatever flaws the Australian batting may suggest, their pace bowlers, Dennis Lillee and Jeff Thomson, seem the likeliest match-winners. 1958 *Listener* 16 Oct. 604/1 England had no powerful reserves of pace-bowling strength to call on. 1951 R. Robinson in A. Ross *Cricketer's Compan.* 1960 III. 396 Not satisfied with..a wide range of pace-changes, he rings in a leg-break or a round-armer now and again. 1974 *Times* 25 Nov. 10/2 Prasanna, coming in at No. 10, edged, slashed and drove the West Indies pacemen to distraction. 1976 *0–10 Cricket Scene* (Austral.) 11/1 The deadpan wickets in England were enough to cause a speedster to retire, but Lillee's speed genius and dedication earned him 21 wickets—more than any paceman on either side.

pace, *v.* Add: **2. a.** Also *fig.*

1921 W. de la Mare *Veil* 24 She paced in pride The uncharted paths men trace in ocean's foam. 1955 *Sci. News Let.* 26 Mar. 201/1 The red maple is one of the first trees to wear its now flower-patterned spring frock. Experts use it as a milestick for pacing spring weather northward because it is one of the few trees that grows from Florida to Quebec.

5. a. Also *transf.* and *fig.*

1961 A. Berkman *Singers' Gloss. Show Business* 65 *Pacing an act*, making a line-up plan of the songs used in an act, so that the interest and enthusiasm of the audience builds up to the end. 1962 *Listener* 31 May 947/1, I had been pacing him in meetings all over the country—that is, talking, to keep the crowd..until Cook should arrive. 1967 *Technology Week* 23 Jan. 64/1 The past history of the design and construction of the vehicles and their propulsion systems..has been paced by the major developments in the materials fields. 1968 *Globe & Mail* (Toronto) 15 Jan. 18/9 The Hawks outscored the Gens 4–1 in the third period to take the victory. Dennis Giannini paced the winners with two goals. 1968 P. Oliver *Screening Blues* ii. 82 'Lining out' in which a lead singer paces a line and the congregation follows with the same line or a refrain response with a linear reply. 1969 *Times* 13 Dec. (Sat. Rev.) p. iv/4 He knows to a nicety just how to pace a book. 1973 *Internat. Herald Tribune* 15 June 15/4 In the American League, Jim Spencer drove in two runs, one with his third homer of the season, in the seventh, to pace Texas to a 4–2 triumph. 1977 *Sci. Amer.* June 138/3 It is the development of the remarkable military reconnaissance programs that has indirectly paced civilian technology in the postwar years.

b. *Med.* To make (the heart) beat at an appropriate rate by stimulating it with pulses of electricity.

1963 *Brit. Heart Jrnl.* XXV. 299 When the heart was paced by ventricular stimulation, mean left atrial pressure was higher..than when the atrium was paced at the same rate. 1973 Segel & Samet in P. Samet *Cardiac Pacing* iv. 82 The coronary sinus provides another area from which the heart can be paced.

pace (pē̆i·si, pā·ke), *prep.* [L., abl. sing. of *pax* Peace as used e.g. in phr. *pāce tuā* by your leave.] By the leave of (a person).

Used chiefly as a courteous or ironical apology for a contradiction or difference of opinion.

1863 *Fraser's Mag.* Nov. 662/1 Mendelssohn was an artist passionately devoted to his art, who (*pāce* Dr. Trench) regarded art as *virtù*. 1883 *Standard* 1 Sept. 2/2 *Pace* the late Sir George Cornewall Lewis, Mr. Scofield is right. 1911 *Chambers's Jrnl.* Nov. 720/1 The colour [of fruit]..is a tacit invitation (*pace* the gardener) to the feast. 1931 R. L. Mégroz *Joseph Conrad's Mind & Method* vii. 170 Stevenson..was regarded by the English critics who, (*pace* Mr. Olage) so dislike fine writing on fine subjects, as a master of prose in fiction. 1955 *Times* 7 July 9/6 Nor, *pace* Mr. Smith, was I for one moment defending immorality in the journalist. 1973 A. H. Sommerstein *Sound Pattern Anc. Greek* iv. 116 Indeed, pace Chomsky and Halle, we would probably want it to be impossible for mid glides to exist at all. 1974 'M. Underwood' *Pinch of Snuff* iii. 23 It was something he greatly prized—*pace* Whitby-Stansford. 1976 *Conservation News* Nov./Dec. 4/2, I find (a) incredible (*pace* Herman Kahn).

b. In Latin phr. *pace tanti viri*: by the leave or favour of so great a man.

1771 Smollett *Humph. Cl.* II. 101 Dr. Shaw..says, he has seen flakes of sulphur floating in the well.—*Pace tanti viri*; I, for my part, have never observed any thing like sulphur, either in or about the well. 1855 *Sat. Rev.* 8 Dec. 100/2 But who seeks for them in Harry Brougham's speeches, or even—*pace tanti viri* be it spoken—in Henry Lord Brougham's *Historical Sketches*? 1865 Mill *Exam. Hamilton's Philos.* xxvii. 544 Admiration, *pace tantorum virorum*, is a different thing from wonder.

paced, *a.* Add: **2.** Also *fig.*

1953 *Brit. Jrnl. Psychol.* Nov. 295 The task..was simply to touch the contact corresponding to the one lamp which was alight. In one condition (the 'paced' one) the lamp went out. *Ibid.* 296 Group II, which was paced, showed a fall-off. 1958 *Oxf. Univ. Gaz.* 27 Jan. 523/2 This year has seen the conclusion of a programme of work on 'Conditions influencing the rate of learning in paced and unpaced tasks'.

pace-maker. Add: **1. a.** Also, one of the leading runners in a race. Also *transf.*

1900 *Field* 4 Aug. 186/1 Up to this he had been one of the pacemakers, but even now he was not to be left behind. 1942 *Sun* (Baltimore) 29 Apr. 19/1 El Toreador scored by three parts of a length over Seaway, pacemaker for most of the mile and 70-yard route of the seventh event. 1951 R. & N. McWhirter *Get to your Marks!* iii. 66 Latching on to an over-awed local pace-maker in a 3-man field, Whitfield swept through. 1957 J. Peters et al. *Mod. Middle- & Long-Distance Running* xii. 55 He..missed a great opportunity of using Hansenne as a pacemaker. 1961 *Times* 11 Apr. 14/5 An athlete training with an automatic pacemaker. 1978 G. A. Sheehan *Running & Being* xii. 174 A pacemaker..had zeroed in on the perfect pace.

b. One who sets the rate of working for others or the standards to be achieved by others; a 'trend-setter'. Also *transf.*, of things.

1906 U. Sinclair *Jungle* xi. 130 They would get new pace-makers and pay them more. 1967 *Daily Tel.* 17 May 17/7 Public schools may well continue to act as pacemakers and trend-setters for this country's education as a whole. 1970 *Ibid.* 14 Feb. 9/8 Both Rameau and Gluck were pacemakers in their day, progressive composers who consciously set out to change the musical landscape of the society in which they lived. 1976 *Star* (Sheffield) 30 Nov. 14/4 Chris Guthrie's hopes of a return to face the Second Division's pacemakers Chelsea at Bramall Lane on Friday night soared today. 1977 *Times* 22 Dec. 12/5 Pacemakers to watch in 1978 are..good Victorian dining furniture, early Continental oak and walnut, and tallboys.

3. *Physiol.* **a.** That part of an animal's heart which determines the rate at which it contracts and where the contractions begin (in man and other mammals normally the sino-atrial node).

1910 *Heart* II. 39 The normal auricular complex is most closely simulated by beats excited from the neighbourhood of Keith and Flack's node. The pace-maker of the heart is therefore situated in the neighbourhood of the superior cavo-auricular junction. 1927 Haldane & Huxley *Animal Biol.* vii. 146 In a mammal the beat starts at the entrance of the great veins to the right auricle in a special piece of tissue known as the 'pace-maker' which does not contract but stimulates the neighbouring muscle. 1951 *Ann. Surg.* CXXXIV. 8/2 Pratt..stated that the pacemaker is governed, at least in part, by the chemical content of venous blood returning to the right auricle. 1961 *Lancet* 9 Sept. 574/1 After a 550-volt shock complete auriculo-ventricular dissociation with a nodal or high bundle pacemaker at a rate of 30 per minute evolved spontaneously. 1968 Passmore & Robson *Compan. Med. Stud.* I. xxviii. 10/2 If the rate of discharge of the SA node is depressed, the portion of the heart with the next highest spontaneous discharge rate becomes the pacemaker, usually the AV [*sc.* atrioventricular] node.

b. A structure which controls the rate of rhythmic activity of an organ other than the heart.

1949 Koestler *Insight & Outlook* xx. 138 A certain region near the sinus end of the heart controls the rate of beat, and similar 'pacemakers' function in the stomach, the ureters, and so forth. 1963 R. P. Dales *Annelids* vi. 132 If there are two pacemakers maintaining two muscular activities it may well be asked if one can dominate the other. 1966 S. Ochs in E. E. Selkurt *Physiol.* (ed. 2) vi. 147 A pacemaker for rhythmicity of the EEG wave was located in the thalamus. 1974 McLennan & Sandberg *Synopsis Obstetr.* (ed. 9) x. 145 Any group of excited myometrial cells anywhere in the uterus may serve in the pacemaker role and initiate electrical propagation throughout the myometrium.

c. A man-made device which supplies electrical signals to the heart, stimulating it to beat at an appropriate rate.

1951 *Ann. Surg.* CXXXIV. 9/1 Our attention has been directed to the study of the problem of cardiac arrest to determine if the use of an electrical artificial pacemaker in resuscitation would be of value. 1963 *Daily Tel.* 16 Apr. 17/1 A transistorised pacemaker little bigger than a match box and weighing only a few ounces has been devised for implanting in the abdominal wall. 1967 *New Scientist* 9 Feb. 331/1 Battery-powered pacemakers normally need their batteries replacing after no more than 2–3 years, needing a surgical operation each time. 1974 *Times* 4 Mar. 6/6 He is to undergo an operation for the installation of a pacemaker to help him overcome a heart condition which has affected him on several recent occasions.

pa·ce-ma:king. [Pace *sb.*¹] **1.** The act or practice of making or setting the pace for competitors in a race.

1896 Ade *Artie* xi. 94 Ain't I tellin' you that we done the pacemakin'? 1900 *Field* 8 Sept. 384/1 Pacemaking has long since exceeded the original function it was designed to fulfil as a preventive of waiting tactics in ordinary bicycle races. 1911 *Encycl. Brit.* XXIII. 854/1 An element unknown to sprinting enters into middle- and long-distance runs, namely that of pace-making. 1968 B. Tulloh *On Running* ix. 114 You can often help each other along by sharing the pacemaking.

2. *Med.* = *pacing *vbl. sb.* 3.

1963 *Amer. Jrnl. Cardiol.* XI. 594 (*heading*) Hemodynamic sequelae of idioventricular pacemaking in complete heart block. 1973 Scherlag & Lazzara in P. Samet *Cardiac Pacing* xiii. 309 (*heading*) Pacemakers and pacemaking in the AV junction.

pa·ce-se:tter. [Pace *sb.*¹] One who sets the pace, trend, or fashion. (Chiefly *fig.*)

1895 [see Pace *sb.*¹ 14]. 1946 *Sun* (Baltimore) 2 Aug. 14/1 Perlina, the early pacesetter, was second about two lengths in front of..White Ford. 1958 *Listener* 23 Oct. 653/3 The new middle-class society and the new pace-setters within it. 1961 *Times* 8 May 16/4 Music is limited to punctuation,..or the role of pace-setter for words and action. 1969 *Times* 22 Oct. (Ghana Suppl.) p. i/2 Ghana was the pacesetter for modern Africa when it became the first sub-Saharan black country to move from colonial status to independence. 1970 *Globe & Mail* (Toronto) 25 Sept. 31/7 Keith Alexander of Calgary continued to be the Canadian pacesetter, firing a 71 yesterday for a two-round total of 143. 1973 A. E. Wilkerson *Rights of Children* 307 The White House Conferences on Children have served since 1909 as pacesetters in child welfare. 1975 *N.Y. Times* 16 Oct. 43/7 'For all the city's problems, New York has been and will continue to be the pace-setter for a high quality of urban life in this country,' the developer said in an interview.

pa·ce-se:tting, *a.* [Pace *sb.*¹] That sets the pace, trend, or fashion. (Chiefly *fig.*)

1965 *Economist* 13 Feb. 645/2 A moratorium on other settlements while a clearly 'pace-setting' wage claim for the year is under review. 1967 *Jane's Surface Skimmer Systems* 1967–68 1/1 If the pacesetting operations by British Rail and Hover-Lloyd alone cannot silence the sceptics, they can be invited to look to the Mediterranean. 1976 M. Birmingham *Heat of Sun* iii. 24 The educational initiative which the state had taken over in founding the pace-setting Achimota school.

pacey (pē̆i·si) *a.* Also **pacy.** [f. Pace *sb.*¹ + -y¹.] Having pace or speed; fast. (*lit.* and *fig.*)

1906 J. J. Munro *Let. to F. J. Furnivall* (MS.) 25 Aug., In the practice of the day before yesterday, Cantab was perceptibly the pacier boat. 1927 *Observer* 29 May 28/4 These hitters, when once they get a real start, play havoc with pacey bowling. 1967 *Listener* 25 May 688/2 The production..is..pacey and vivid. 1968 *Daily Mirror* 27 Aug. 7/2 This is considered very pacey, which is the new word for trendy. Nobody, but nobody, says trendy any more. 1969 C. Booker *Neophiliacs* ii. 48 The whole world..had been reduced to the same grainy, pacy, ever more 'realistic' dream. 1969 *Pony* July 512/1 *Show Jumping Summer* is a 'pacey' book, in which events follow each other in quick succession. 1977 *Daily Tel.* 13 Jan. 17 (Advt.), The Celeste's low slung, pacey appearance isn't just for show. The 2 litre model has a top speed of 105 mph. 1977 *Times Lit. Suppl.* 11 Feb. 145/4 (Advt.), Pacy, turbulent story with an excellent and authentically researched diamond mining setting.

pachinko (pătʃi·ŋko). Also **pachinco.** [Jap. *pachin* onomatopœic word repr. the sound of something triggered off + *ko* dim. suffix.] A variety of pin-ball popular in Japan. Also *attrib.*

1953 *Encounter* Nov. 7/2 In Tokyo there are 5,000 registered *pachinko* halls. 1954 J. L. Morse *Unicorn Bk.* 1953 262/1 An interesting development in Japan was the popular craze for *pachinko*, a kind of poor-man's pinball game. 1964 *Listener* 8 Oct. 540/2 Pachinko is played with handfuls of ball-bearings. You drop them, one by one, into the machine, flick them round, and if they land in a winning cup, the machine coughs back fifteen ball-bearings which are bought in the first place, twenty-five at a time, for fifty yen (one shilling). If you amass enough of them, they can be exchanged for prizes. 1971 *Guardian* 11 June 11/6 Pachinco machines, dozens of them side by side in rows..are all identical. A trigger shoots off a ball which may find its way into a slot and produce a jackpot of balls. 1973 A. Broinowski *Take One Ambassador* v. 55 This [joint] next door's *pachinko*..reminds me of some of the leagues clubs at home. *Ibid.* 56 They'll [*sc.* the Japanese] ..spend their time in a useless game like this *pachinko*.

pachisi. Add: (Further examples.) Also played in other countries besides India. Also *attrib.*

The spellings *parcheesi* and *parchesi* are now usual.

1892–3 *T. Eaton & Co. Catal.* Fall & Winter 67/2 Games... Parcheesi, linen board, 50c. 1895 *Montgomery Ward Catal.* 235/3 Royal Game of Parchesi complete..including..folding board,..8 dice,..and 16 counters. *a* 1910 'O. Henry' *Rolling Stones* (1916) 37 John Tom Little Bear, in full Indian chief's costume, drew crowds away from the parchesi sociables and government ownership conversaziones. 1948 C. McCullers *Discovery of Christmas* in *Mademoiselle* Sept., Budge..was too little to count straight, to play Parcheesi, to wipe himself. 1950 O. Nash *Family Reunion* (1951) 15 They can't pass it in bridge or parchesi or backgammon. 1969 L. Kennedy *Very Lovely People* i. 35 The first night we were there two girls came over with a parcheesi board, and for the next two months we played nothing but parcheesi. 1972 *Jrnl. Social Psychol.* LXXXVII. 108 This procedure was chosen over Vinacke's parchesi game and Gamson's simulated political convention to emphasize the relative importance of the subjects [*sic*] initial resources. 1972 *Listener* 13 July 51/3, I was playing a lot of board games at home. We had Pachisi, Monopoly..Chinese chequers and what-not.

‖ **pachuco** (pătʃu·ko). [a. Mexican Sp. *pachuco* flashily dressed, vulgar.] A juvenile delinquent of Mexican-American descent, esp. in the Los Angeles area; in extended use, a derogatory term for any Mexican-American. Also *attrib.*

1943 C. Himes in *Crisis* July 200/1 *Pachuo* is a Mexican expression which originally meant 'bandit' but has degenerated by usage into a description of a juvenile delinquent... In Mexican districts in the county of Los Angeles, small bands of pachuos have organized into gangs to fight each other. 1944 *Time* 10 July 26/2 Pachuco... Mexican for zootsuiter. 1946 C. Himes *Black on Black* (1973) 256 Some pachuco kids were ganged about the juke box, talking in Mex. 1947 *Common Ground* Summer 79/1 The Pachuco dialect is a mélange composed of *Caló*, Hispanicized English, Anglicized Spanish, and words of pure invention. 1950 G. C. Barker (*title*) Pachuco: an American-Spanish argot and its social function in Tucson, Arizona. *Ibid.* (1958) I. 13 In many cities of the American Southwest there are today Mexican-American boys who are known..as pachucos. These boys..may be distinguished by certain peculiar characteristics of dress, behavior, and language. 1954 J. Steinbeck *Sweet Thursday* 11 In Los Angeles..he led a gang of pachucos. 1966 T. Pynchon *Crying of Lot 49* i. 11 Hostile Pachuco dialect, full of chingas and maricones. 1972 J. Wambaugh *Blue Knight* (1973) v. 70 *'Órale, panzón,'* he said, like a pachuco, which he put on for me. He spoke beautiful Spanish..but the barrios of El Paso Texas died hard. 1976 *Word* 1971 XXVII. 294 Pachuco, also known as *tirilí*, *tirilongo*, is used not only by felons, delinquents,..and others outside respectable society, but also by younger males throughout the Southwest as a street variety and for its slangy effect.

pachycaul (pæ·kikǫl), *sb.* (*a.*) *Bot.* [f. Pachy- + Gr. καυλ-ός stem, stalk.] A tree having a thick primary stem and few or no branches; also *attrib.* or as *adj.* Hence **pachycau·lous** *a.*; **pa·chycauly**, development of this type. Cf. *leptocaul *sb.* and *a.*

1949 E. J. H. Corner in *Ann. Bot.* XIII. 392 The pachycaulous Cycad. *Ibid.* 393 The old clumsy pachycaul with massive and slow-growing branches. 1954 *Phytomorphology* IV. 264/1 In general, six effects accompany the transition from pachycauly to leptocauly. 1964 E. J. H. Corner *Life of Plants* ix. 154 'Pachycaul' (with thick primary stem) denotes massive construction as of the rosette tree or cabbage tree. *Ibid.* 155 The pachycaul plants establish themselves by robust growth. 1967 E. A. Menninger *Fantastic Trees* 16 (*heading*) The pachycaulous trunks. *Ibid.*, One conspicuous example of this pachycaulous curiosity today in the forests of the Ivory

Coast and Nigeria is the aky tree, also called the forest papaw. Despite its enormous size..this tree's trunk is soft, porous, and spongelike, and it is generally unbranched. It is a living relic of an ancient age. 1973 A. J. Willis *Introd. Plant Ecol.* v. 55 A tendency to a pachycaul habit is seen in the ash, with its pinnate leaves and thick stubby twigs. 1974 *Kew Bull.* XXIX. 535 The pachycaul Giant Lobelias are some of the most spectacular plants of the tropical African highlands. *Ibid.* 549 In herbaceous species [of giant lobelia]..the development is similar to that of the forest pachycauls. 1974 *New Phytologist* LXXIII. 971, I propose a general hypothesis on the evolutionary trends involving pachycauly in *Senecio*. 1976 *Phil. Trans. R. Soc.* B. CCLXXIII. 359 The purpose of this account [of climbing species of *Ficus*] is to provide new evidence for the general theory of angiosperm evolution from pachycaul to leptocaul vegetation.

pachysandra (pæ:kisæ·ndrā). [mod.L. (A. Michaux *Flora Boreali-Americana* (1803) II. 177), f. Gr. παχύς thick + ἀνήρ, ἀνδρό-ς male, in reference to the thick stamens of the male flowers.] A small evergreen subshrub of the genus so called, belonging to the family Buxaceæ, native to eastern North America or eastern Asia, and bearing white or pinkish-white flowers.

1813 W. T. Aiton *Hortus Kewensis* (ed. 2) V. 260 Trailing Pachysandra. Nat[ive] of North America. Intro[duced] 1800, by Messrs. Fraser. 1818 *Curtis's Bot. Mag.* XLV. 1964 Trailing Pachysandra... Pachysandra was first described in Michaux's Flora of North-America, and received its name from the remarkable thickness of its stamens. It has very little beauty to recommend it to the flower-garden; but the curious Botanist will regard it with some interest. 1914 W. J. Bean *Trees & Shrubs Hardy in Brit. Isles* II. 118 The Pachysandras thrive in any moist soil, and do not mind shade; they make neat tufts, but are of only moderate decorative value. 1941 R. S. Walker *Lookout* 52 Pachysandra, or mountain spurge, blooms in March and April in the rich soil in Lookout Mountain woods. 1961 *Amat. Gardening* 18 Nov. 1/1 The pachysandra is..one of those borderline plants that are half shrub half herbaceous perennial. 1975 *New Yorker* 23 June 38/3 Both laughing, he supervised John's spitting out the lettuce and paper and tobacco into the pachysandra.

pachytene (pæ·kitīn). *Cytology.* [ad. F. *pachytène* (H. von Winiwarter 1900, in *Arch. de Biol.* XVII. I. 63): see Pachy- and *-tene.] The third stage of the first meiotic prophase, following zygotene, during which the paired chromosomes shorten and thicken, the two chromatids of each separate, and exchange of segments between chromatids may occur.

[1900 *Jrnl. R. Microsc. Soc.* 654 As the chromatic thread spreads itself again through the nuclear space, this duality disappears, and the thread is single, thick, and moniliform (pachytænic stage).] 1912 *Jrnl. Exper. Zool.* XIII. 378 All the threads still stain deeply and are very much thicker than in the leptotene-stage; hence these nuclei may be called the pachytene-nuclei. 1932 *Proc. 6th Internat. Congr. Genetics* I. 257 It is assumed that every chiasma represents a crossover which has occurred between two of the four chromatids at pachytene. 1965 Bell & Coombe tr. *Strasburger's Textbk. Bot.* 34 In pachytene the pairing of homologous chromosomes is completed. 1974 *Cytogenetics & Cell Genetics* XIII. 330 Breakdown of spermatogenesis at the pachytene stage of meiotic prophase was observed in most germ cells.

pacifarin (pæsi·fārin). *Med.* [f. L. *pāci-f(ic)ār-e* to make peace + -in¹.] Any biologically produced substance which, when introduced into an organism, protects it from the harmful effects of an infection without killing the pathogen.

1963 H. A. Schneider in *Proc. Amer. Philos. Soc.* CVII. 445/2 As a new and third category of ecological ectocrines, which already embraces vitamins and antibiotics, we add as the first example of its class the substance I have described above, and name the class 'pacifarins' from the Latin verb 'pacificare', to make peace, to pacify. (I wish to thank my colleague, Dr. Ludwig Edelstein, for guiding me in this choice.) The particular pacifarin, the salmonellosis pacifarin, is, we believe, addressed only to the typhoid diseases, and for other diseases there are, we postulate, other pacifarins waiting to be identified. 1967 *Daily Tel.* 10 May 14/6 The basis of the discovery is that a microscopic amount of pacifarin extracted from wheat and dried egg-white will protect mice against infection by salmonella. 1975 *Infection & Immunity* XI. 69/2 Certain bacterial products other than enterobactin are also now known to possess pacifarin activity.

pacific, *a.* and *sb.* Add: **A.** *adj.* **2. b.** = *peaceful *a.* 4.

1906 *Chambers's Jrnl.* Jan. 61/2 It is by their mastery of the policy of 'pacific penetration' that the Chinese make themselves such formidable neighbours.

4. *pacific blockade* (see quots.).

1880 *Encycl. Brit.* XIII. 194/1 The right of 'pacific blockade', *i.e.*, the blockade of ports belonging to a nation with which we profess not to be at war, has been asserted in a few doubtful instances. 1889 A. C. Boyd *Wheaton's Elem. Internat. Law* (rev. ed.) IV. i. 404 The above-mentioned proceedings against Greece and Brazil furnish instances of what is called 'pacific blockade'; the blockading power blockading the coast, or a certain portion of the coast, of the blockaded power, but declaring, at the same time, that a state of peace is maintained. 1895 T. A. Walker *Man. Public Internat. Law* II. iv. 96 Pacific blockade consists in the cutting off by one state of communication with the ports or a particular portion of coast of another, otherwise than in the case of declared war, with the object of preventing commercial relations by sea. 1935 T. A. Taracouzio *Soviet Union & Internat. Law* x. 299 Another form of redress to which nations sometimes resort, and which is yet not considered war, is pacific blockade.

B. *sb.* **2. b.** *Pacific coast* (further examples), *coaster, Northwest, seaboard, state* (examples), *style*; **Pacific slope,** (*a*) (see quot. 1902); hence *Pacific sloper*; (*b*) an escape across the Pacific Ocean to avoid arrest (chiefly *Austral.* and *N.Z. slang*); **Pacific time,** time as reckoned on the 120th meridian west of Greenwich.

1872 R. G. McClellan *Golden State* xxxi. 523 The Pacific coast..contains an area equal to one-half of the whole territory of the Republic of America. 1948 *Denison* (Texas) *Herald* 2 July 12/2 The most valuable fish is the Pacific Coast salmon. 1970 J. H. Paterson *N. Amer.* (ed. 4) xix. 287/1 Like other Pacific coast cities, it has a small steel output, based on scrap. 1883 *Harper's Mag.* Nov. 943/1 [The completion of the Union Central route has not] given the 'boost' to California that the 'Pacific coasters' so fondly dreamed of. 1889 *Wealth & Resources of Oregon & Washington* (Union Pacific Railway Co.) 3 The resources and industries of the Pacific Northwest are so varied..as to not only suggest but enforce its consideration in sections. 1938 G. Cash *I like Brit. Columbia* 100 The Provincial Library at the Parliament Buildings is quite one of the most interesting in the Pacific Northwest. 1977 *Time* 12 Dec. 61/3 They take the armchair beachcomber on a scenic tour..past the cypresses of Monterey and the great coastal forests of the Pacific Northwest to the fog-shrouded Aleutians. 1838 *Knickerbocker* June 556 Where the prairie stretches away..shall sweep the long, hissing train of cars, crowded with passengers for the Pacific seaboard. 1845 J. C. Frémont *Rep. Exploring Expedition* 274 [We were] now about to turn the back upon the Pacific slope of our continent. 1855 [in Dict.]. 1901 Hall & Osborne *Sunshine & Surf* 38 There is such a thing known in Australia, America, and Canada as 'the Pacific slope', which, being interpreted, means a hurried departure, down to these regions [*sc.* Tahiti] of gentlemen who find these countries too hot to hold them. 1902 [in Dict.]. 1902 W. S. Walker *Zealandia's Guerdon* 292 Perhaps he [*sc.* the missing man] accomplished the 'Pacific Slope'. *Ibid.* 326 He [*sc.* the detective] has packed so many 'confidence men' off to penitentiary that the others have done the 'Pacific Slope' in various directions, chiefly towards Australia. 1915 H. B. Niver *Elem. Geogr.* 167/1 By means of irrigation, the Pacific Slope has become one of the greatest fruit-growing sections of the world. 1938 R. Gilkison *Early Days in Dunedin* xiii. 133 In the 'seventies and 'eighties many fraudulent debtors, embezzlers and rich thieves escaped from New Zealand before arrest, by doing what came to be known as the 'Pacific Slope'. 1945 Baker *Austral. Lang.* xiv. 243 *Eucalyptian, the Pacific slope*,..and *tiersman*, are not so important to our language that we could not do without them. Yet we would tend to class these as standard. 1954 E. Gunther in Freeman & Martin *Pacific Northwest* (ed. 2) 16 Salmon runs occur in all streams in the Pacific slope. 1876 *Benton Democrat* (Corvallis, Oregon) 18 Aug. 2/3 (*heading*) Pacific slopers. 1883 *Harper's Mag.* Mar. 648/1 'Well,' said the Pacific sloper, 'if it's a private funeral, what do they call it a reception for?' 1820 W. Tudor *Lett. on Eastern States* 57 When the future Pacific states come to be represented in congress. 1949 *Los Angeles Times* 6 Nov. 1/8 The overall increase for Pacific States is 5,251,000 or 53·9%. 1976 *National Observer* (U.S.) 2 Oct. 1/2, 16 South Atlantic and South Central states got back $11.5 billion more than they paid in taxes. Thirteen Pacific and Mountain states came out $10.6 billion ahead. 1959 *Wall St. Jrnl.* 13 July 1/4 Increasing numbers of home builders..are experimenting with the new style, often called..'Pacific style'. 1883 *N.Y. Herald* 18 Nov. 12/3 In the United States the standards will be known as the 'Eastern', 'Central', 'Mountain' and 'Pacific' times. 1958 'Castle' & 'Hailey' *Flight into Danger* ii. 36 'How soon do you expect to land?' 'About five a.m., Pacific Time.' 1976 *National Observer* (U.S.) 7 Aug. 16/1 Broadcast times apply to Eastern and Pacific time zones.

c. Used to designate a type of steam locomotive with a 4-6-2 wheel arrangement designed to pull express passenger and freight trains; also *absol.*

1903 *Amer. Engineer & Railroad Jrnl.* Oct. 351 (*caption*) Pacific-4-6-2 Type Passenger Locomotive—Chicago, Rock Island and Pacific Railway. 1905 *Railroad Gaz.* 9 June 620/1 (*heading*) Pacific locomotive with superheater for the Erie Railroad. 1908 *Westm. Gaz.* 31 Dec. 3/1 The most interesting locomotive novelty of the year was the Great Western Company's 'Pacific' type of express engine. 1910 *Ibid.* 25 Jan. 2/1 French railways..built their first 'Pacific' not long before ours, and this season most of the 'Riviera' expresses will be horsed by these vast machines. 1938 L. M. Beebe *High Iron* iii. 97 The most important U.S.R.A. designs, for present purposes, fall into four wheel arrangements: the 4-6-2 or Pacific, the 4-8-2 or Mountain, the 2-8-2 or Mikado, and the 2-10-2 or Santa Fe types. 1972 B. C. Blanton *400,000 Miles by Rail* iii. 37/1 The last major rail trip I took with my parents was in November, 1910. Our route was over the Katy's rails to St. Louis. The *Katy Flyer* was now headed by a Pacific locomotive. *Ibid.*, The *Royal* Blue was advertised as a solid-vestibuled train... It was headed by a Pacific and carried a Pullman parlor-observation car with open platform. 1978 *Observer* 26 Mar. 2/5 'Swanage'..was built in 1950 and died in 1964. It is a 4-6-2 Pacific type.

pacification. Add: **a.** (Later examples.) *spec.* *U.S.*, a process or operation (usu. a military operation) designed to secure the peaceful cooperation of a population or an area where one's enemies are thought to be active.

1946 'G. Orwell' in *Horizon* XIII. 76 Defenceless villages are bombarded from the air, the inhabitants driven out into the countryside, the cattle machine-gunned, the huts set on fire with incendiary bullets: this is called *pacification*. 1966 *N.Y. Rev. Bks.* 3 Mar. 4/3 It would be wrong to predict a priori that President Johnson's new 'counter-insurgency' and 'pacification' programs, based on plans for economic and social development in the Southern villages, will fail as totally as did the quite similar plans sponsored by the French and later by the Diem regime. 1967 *New Yorker* 14 Oct. 55 For God's sake, Hinton! You mean all this time I've been talking about pacification you thought I meant *peace*? 1969 A. G. Frank *Latin Amer.* xxv. 401 The latest effort, for instance, is to have the Latin American military occupation forces improve their reputation in the countryside by undertaking Latin American versions of the imperialist 'pacification' program in Vietnam. 1969 *Listener* 12 June 814/2 US civilians are busy with pacification programmes to make the peasants more hostile to communism and more loyal to Saigon. 1974 *Black Panther* 16 Mar. 2/3 This prison we're in is a military camp and it has the most propagandist and pacification program in all the camps in America. 1974 *Encycl. Brit. Macropædia* XIII. 847/1 The formula that peace is the aim of war..has time and again been expressed in the paradoxical concept of pacification, which means exactly those violent actions through which an expanded area of peace shall be won and maintained.

pacificator. (Later examples.)

1907 G. B. Shaw *Let.* 10 June (1972) II. 692, I..have just created a scandal among the German pacificators by informing the leading Viennese newspaper that I consider disarmament..absurd. 1968 *Daily Tel.* (Colour Suppl.) 13 Dec. 43/4 Drink, in other words, is becoming the opium of the people: the great pacificator.

pacificism (pæsi·fisiz'm). [f. Pacific *a.* + -ism.] **a.** Rejection of war and violence as a matter of principle; = *pacifism. **b.** Advocacy of a peaceful policy; rejection of war in a particular instance.

As an ideological term *pacifism (influenced by Fr. *pacifisme*) is now the preferred form.

1910 W. James *Mem. & Stud.* (1911) xi. 283 Pacificism makes no converts from the military party. 1912 *Q. Rev.* July 203 With the old Pacificism, the Pacificism of the Quakers, of Tolstoi, and of all those who hold that war must not be tolerated..the world has long been familiar. 1916 G. G. Coulton (*title*) The main illusions of pacificism. 1920 *Q. Rev.* Oct. 396 The revolution [in Japan, 1868].. was a reaction against these centuries of pacificism. 1936 F. M. Ford *Let.* 20 Aug. (1965) 254, I am just on the verge of litigation with the Oxford University Press over modifications they have arbitrarily made in my pacificisms quasi-communism and other outrages that they have committed. 1957 A. J. P. Taylor *Trouble Makers* ii. 51 By 'pacificism' I mean the advocacy of a peaceful policy; by 'pacifism' (a word invented only in the twentieth century) the doctrine of non-resistance. 1978 J. Meyers *Katherine Mansfield* x. 131 Weekends [with Lady Ottoline Morrell] ..were characterized by high spirits and high-mindedness, pacificism, poetry and all that was ultra-modern in the arts.

pacificist (pæsi·fisist). [f. as prec. + -ist.] **a.** One who rejects war and violence as a matter of principle; = *pacifist *sb.* **b.** One who advocates a peaceful policy as the first and best resort (see prec., sense b). Also *attrib.* or as *adj.*

In sense a *pacifist (influenced by Fr. *pacifiste*) is now the preferred form.

1907 *Westm. Gaz.* 2 Apr. 2/2 We have..a picture of Germany going to war in order 'to demonstrate the futility of the dreams of the Pacificists'. 1908 *Ibid.* 4 June 5/1 It is not sufficient to simply call him 'Pacificist' to prevent him denouncing these follies anew. 1910 W. James *Mem. & Stud.* (1911) xi. 275 In my remarks, pacificist though I am, I will refuse to speak of the bestial side of the war-*régime*..and consider only the higher aspects of militaristic sentiment. 1910 [see *milk-blooded* s.v. *milk *sb.* 10]. 1912 *Q. Rev.* July 204 To make war impossible, the older Pacificists appealed to the heart and soul of man; the new Pacificists make their appeal to his pocket. *Ibid.* 217 In places he draws the usual Pacificist conclusion. 1919 J. Buchan *Mr. Standfast* i. 21 It was bad enough for anyone to have to pose as a pacificist, but for me, as strong as a bull and as sunburnt as a gipsy and not looking my forty years, it was a black disgrace. *Ibid.* 35 You were bidden..turn yourself from a successful general into a pacificist South African engineer. 1923 *Blackw. Mag.* June 822/2 These people are instinctive pacificists. 1923 R. Macaulay *Told by Idiot* III. iv. 192 Stanley was in these days a stop-the-war, pacificist Little Englander, anti-militarist, anti-Chamberlain, anti-Concentration Camp. 1965 D. A. Martin *Pacifism* v. 73 The dissenting opposition to war discussed here is *pacificist* not pacifist. The dissenters did not hold that war was always wrong but that it should be avoided wherever humanly possible. 1966 *New Statesman* 3 June 815/2 We are all..'Pacificists' —that is, we believe that war should be avoided wherever humanly possible

pacifico (pasi·fiko). [Sp.] A person of pacific or peaceful character, *spec.* a native of Cuba or the Philippines who submitted without active opposition to Spanish occupation.

1897 R. H. Davis *Cuba in War Time* 41 His [*sc.* General Weyler's] object..was to prevent the pacificos from giving help to the insurgents. 1898 *Harper's Weekly* 19 Feb. 174/2 The pacificos who are in the fields supply the food for the army [*sc.* the insurgents in Cuba, February, 1898] and are under military supervision. 1905 A. G. Robinson *Cuba & Intervention* iii. 35 These became known as the *Pacificos*. 1916 G. B. Shaw in *To-day* 13 May 38/2 One who accidentally tackled Mr Ponsonby, and, miscalculating the mettle of the true British Pacifico, had his head heartily punched for his pains.

pacifier. Add: (Further examples.)
1956 H. Gold *Man who was not with It* (1965) ii. 12 Telling *as if*, secure in morphine or other pacifiers,..we found in the show that forgotten moral thickness for which so many of us were sick. 1969 *Daily Tel.* 10 Nov. 2 By 1990 most of us..will be taking synthetic mood modifiers, pacifiers and general comforters.
2. A baby's dummy. *U.S.*
1904 F. Crissey *Tattlings Retired Politician* 367, I put away my teething ring and baby 'pacifier' several years ago. 1949 M. Mead *Male & Female* xiii. 271 The very modern pediatrician may recommend a pacifier—the same old pacifier that still lingers on the back-street, in the little..drug stores. 1960 *Encounter* Mar. 19/2 Minnie Foote's baby got held up to see and dropped his pacifier in the box. 1963 M. McCarthy *Group* xiv. 321 Norine removed the pacifier from the baby's mouth. 1976 *Billings* (Montana) *Gaz.* 4 July 5-A/3 The government has closed the books on a case that started with the choking death of a five-month-old boy and ended with the recall of more than 100,000 baby pacifiers.

pacifism (pæ·sifiz'm). [ad. F. *pacifisme* (see quot. 1902): see -ISM, *PACIFICISM.] The policy or doctrine of rejecting war and every form of violent action as means of solving disputes, esp. in international affairs; the belief in and advocacy of peaceful methods as feasible and desirable alternatives to war.
1902 *Proc. 10th Universal Peace Congr.* 74 M. Emile Arnaud... Speaking at length, in French,..said:... The negative programme of Pacifism is anti-War-ism. 1906 *Times* 30 July 5/4 It can bring its naval policy into harmony with its foreign policy and give pledges to 'pacifism'. 1915 *National Rev.* Mar. 54 The greatest war in history is now being fought in the cause of Pacifism. 1917 *Atlantic Monthly* June 745/2 To such people pacifism is a religion, an interpretation of Christianity. 1919 G. B. Shaw *Heartbreak House* p. xviii, There was only one virtue, pugnacity: only one vice, pacifism. That is an essential condition of war. 1929 Chesterton *Thing* 111 Nothing that I say here has any connection with what is commonly called pacifism. I think that our friends and brethren fell ten years ago in a just war. 1935 *Fellowship* Mar. 3/1 Pacifism does not renounce the struggle, but carries it on with the more effective weapons of non-violence. 1936 A. Huxley in G. K. Hibbert *New Pacifism* ii. 39 Humanism was once a favourable environment for pacifism. It has now become wholly inimical. 1937 P. S. Mumford *Introd. Pacifism* i. 10 Pacifism is not simply a negative policy of refusing to fight. It is a constructive policy of showing that there are more powerful and better ways of opposing your enemies. 1941 A. Huxley *Let.* 17 Nov. (1969) 470 In war time, it would seem, psychological conditions are such that the application of pacifism to politics is for all practical purposes impossible. 1945 G. C. Field *Pacifism & Conscientious Objection* 4 Pacifism..is not one single, simple creed, but a number of creeds. 1955 *Bull. Atomic Sci.* Sept. 265 If armaments are not acceptable to the pacifist, does this mean that he will submit to the aggressor and meekly resign himself to what he considers evil? The answer is emphatically no. This is to confuse pacifism with appeasement. The pacifist is definitely not a passivist. 1957 A. J. P. Taylor *Trouble Makers* ii. 51 Even Bright, who was sometimes nearer to pacifism, did not plead 'that this country should remain without adequate and scientific means of defence'. 1964 Gould & Kolb *Dict. Social Sci.* 481/2 Pacifism has never yet been adopted as official policy by any state. 1976 *Pacifist* Jan. 10/1 It is an integral part of pacifism not only to expect one's own freedom but also to allow everybody else their freedom. 1976 *Christian Cent.* 15 Sept. 753/2 Most American evangelicals have been less than enthusiastic about pacifism.

pacifist (pæ·sifist), *sb.* and *a.* [ad. F. *pacifiste*: see -IST, *PACIFICIST.] **A.** *sb.* A proponent or advocate of pacifism; one who believes in resort to peaceful alternatives to war as means of settling disputes. Also in Comb., as *pacifist-minded* adj.
1906 *Times* 30 July 5/4 The French 'Pacifists' will appeal to England's example in order to induce France also to cut down her naval programme. 1907 *Academy* 2 Mar. 204/2 In 1890 he [sc. Carducci] wrote on 'War' an ode that passionately condemned the Pacifists. 1917 *Atlantic Monthly* June 748/1 My friend is a pacifist because he dreads the tantalizing consequences upon himself of resisting aggression by violent physical methods. 1929 Chesterton *Thing* 133 One does not need to be a pacifist to think that gunpowder need hardly go on being useful on quite such a grand scale. 1930 W. S. Churchill *My Early Life* xxvi. 346, I have always been against the Pacifists during the quarrel, and against the Jingoes at its close. 1930 *Sun* (Baltimore) 30 Dec. 7/3 Most people like to have their soldiers represented as brave... But not these German pacifists and defeatists. 1936 A. Huxley *Let.* 2 Mar. (1969) 401 What the pacifist suggests is the eminently reasonable course of using intelligent generosity to begin with—rather than waiting to use it till the evil act has been committed. 1937 P. S. Mumford *Introd. Pacifism* ii. 15 The Pacifist believes: 1. That war, i.e. mass murder, as a political policy is morally wrong, and consequently will never produce good results... 2. That security for nations, ideals or personal freedom can be obtained only by non-violent resistance, [etc.]. 1948 'J. Tey' *Franchise Affair* xix. 225 The murderous rage that fills the pacifist-minded when their indignation is roused. 1961 E. S. Turner *Phoney War* xiii. 179 There was an arrogance among certain militant pacifists which..prevented them from respecting the views of those who thought freedom worth fighting for. 1969 *Listener* 10 Apr. 505/3 Perhaps the oddest aspect of dispatches from Vietnam is that pacifist-minded conscripts are so readily made available to say they don't think words like honour and patriotism make any sense out there. 1970 R. Sampson *Anarchist Basis of Pacifism* 1 My usage of the term 'pacifist' includes only those who live by the principle that they will not intentionally take human life, cost what it may. 1974 A. Price *Other Paths to Glory* i. 18 'What did your father do in the last war?'..'He worked on a farm... He was a conscientious objector—a pacifist.'
B. *adj.* Of or pertaining to pacifism; characterized by rejection of war and belief in peaceful alternatives. Also **pacifi·stic** *a.*, suggestive of or inclined to pacifism; **pacifi·stically** *adv.*
1908 *Times Lit. Suppl.* 10 Dec. 453/3 He made a speech remarkable for its string of pacifist commonplaces. 1909 *Westm. Gaz.* 1 Sept. 9/1 Prussia was led to 1806–7 by pacifist and humanitarian ideas. 1918 A. Huxley *Let.* 3 Mar. (1969) 146 The one blow at the eleventh hour was the Vice Provost's refusal to allow De La Warr's article on Labour and the War to be printed, as being too revolutionary and pacifist. 1920 W. J. Locke *House of Baltazar* xxi. 256 It contained the names of representatives of all the disgruntled and pacifist factions in England. 1927 H. D. Lasswell *Propaganda Technique in World War* iii. 62 H. G. Wells may be taken as an example of the pacifistically inclined Liberal..whose support of the War came at the cost of inner struggle. 1930 *Sun* (Baltimore) 31 Dec. 6/2 What Mr. Bouton characterizes as a peculiarly German Socialist, defeatist and pacifist attitude of 'a class apart' is far from that. 1936 A. Huxley *What are you going to do about It?* 4 It has seemed best to state the pacifist case in terms of a series of answers to common antipacifist objections. *Ibid.* 17 War..can only be prevented from breaking out if at least one government of an important sovereign state chooses to act pacifistically towards its neighbours. 1945 *Fellowship* Aug. 152/1 A fearful responsibility rests upon us who believe in the pacifist way. 1956 A. H. Compton *Atomic Quest* i. 37 The public attitude..had been strongly pacifistic. 1961 E. S. Turner *Phoney War* xix. 278 Vendors of Fascist, Communist or pacifist literature were liable to be arrested. 1973 D. Aaron *Unwritten War* III. vii. 108 Both [sc. William and Henry James] felt the need..to display a pacifistic heroism. 1976 *Pacifist* Jan. 5/2 It doesn't seem to me to be a pacifist way of overcoming things we dislike just to use our position of power (as parents or as government) to impose our will upon the people over whom we happen to have some influence.

pacing, *vbl. sb.* Add: **2.** *Cycle-Racing* and *Athletics.* The act of (tactical) pace-making, and hence of artificially increasing the speed of a competitor by allowing him to proceed in the slip-stream of a (usu. motorized) vehicle; also, the act of distributing effort carefully over a race to ensure optimum performance, esp. by utilizing the wind resistance offered by other competitors. Also *attrib.*
1895 G. L. Hillier *Cycling* (ed. 5) 342 Appended are the Rules for 'Herne Hill Pacing', which, if strictly enforced, insure fair pacing all round... No pacer is to remain on the path, unless actually pacing. 1897 *Encycl. Sport* I. 62/1 (Athletics) *Pacing*, going in front so as to quicken the speed at which the race is being run, or at which some particular competitor is running. *Ibid.* 287/2 As a natural consequence the trade appeared upon the scene, the friendly character of the assemblies gradually disappeared, things became more business-like, pacing was supplied, tandems and often larger multicycles swooped down in shoals upon the highways. 1902 *Encycl. Brit.* XXVII. 327/2 The introduction of pacing by multicycles and motors next took from cycle racing what interest was left. 1935 *Encycl. Sports* 206/2 Surprising records have been created by pacing, since motorcycles and cars became available for this purpose, with wind-shields attached to the rear to protect the cyclist from wind resistance. These wind-shields also act as a sucker and help to draw him along. 1955 R. Bannister *First Four Minutes* x. 123 This was the first of many occasions when Chris Chataway helped me with the pacing in the early stages. 1974 *Encycl. Brit. Macropædia* XVIII. 544/2 Pacing styles have changed over the years, and this is particularly evident in mile running. The early style was to start fast, relax somewhat during the middle two laps, then finish as rapidly as possible. 1974 *Sunday Mirror* 21 July 38/1 Schaer..indulged in some unfair 'pacing', tucking himself in behind..Doyle and refusing official requests to move away.
3. *Med.* Artificial stimulation of the heart so as to make it beat at an appropriate rate. Cf. *PACE *v.* 5 b.
1962 *Lancet* 29 Dec. 1369/2 We report here our experience of artificial pacing in nineteen patients, with special emphasis on the management of Stokes-Adams attacks in hospital, and the emergency control of rhythm by the electrode catheter. 1969 J. P. P. Stock *Diagnosis & Treatment of Cardiac Arrhythmias* x. 144 Paired pacing has been employed in severe heart failure to increase the force of contraction of the heart. 1975 J. Fleming in F. J. Fawcett *Cardiol.* I. 24 Electrical pacing of the heart on a short or long term basis is now firmly established as a valuable measure.

pack, *sb.*[1] Add: **1. c.** (Later examples.)
1897 *Outing* XXX. 374/1 Men..shoulder their packs of general cussedness, and..hit the trail. 1962 J. Braine *Life at Top* vi. 101 Suddenly the pack was on my shoulders again; there was no quietness in the room.
d. *Photogr.* A set of two or three plates or films sensitive to different colours which are superimposed and exposed simultaneously. Cf. *BI-PACK, *TRI-PACK.
1907 *Brit. Jrnl. Photogr.* 19 July 547/2 By interspersing ..filters with films in sets for tri-chromatic negatives..the respective exposures can be made in rapid sequence without removing the pack from the camera. 1929 *Penrose Ann.* XXXI. 41 To assert that the colour analysis of the pack is equal to that of orthodox trichromatic work would be incorrect.
e. A knapsack, rucksack, usually with a wooden frame. Chiefly *Forces'* and *N.Z.*
1916 'Boyd Cable' *Action Front* 49 The neutral ground ..was a sea of mud..littered with..packs which had been cut from or slipped from the shoulders of the wounded. 1925 Fraser & Gibbons *Soldier & Sailor Words* 218 *Pack*, the infantry knapsack. 1958 *Tararua* XII. 27 Food and gear have to be carried. Everyone in New Zealand..puts it in a *pack*. 1968 *N.Z. Listener* 15 Mar. 6/1 You women can't go carrying all that stuff. Here, Joyce, give us your pack, Joyce! 1969 *Ibid.* 21 Feb. 4/1 Hobnail boots and canvas pack..just the gear for pushing through scrub and supplejack. 1971 *Ibid.* 22 Feb. 51/2 Pack carrying is still the same old personal battle between man and gravity. 1973 *Parade Sunday Bull.* (Philadelphia) 7 Oct. 31/2 Packs: Most versatile pack is a tubular metal pack frame, contoured to the body with a waist strap that transfers the weight to the legs and hip muscles.
f. A packet or package, esp. of cigarettes. More usual in the U.S. than in the U.K.
1924 *Saucy Stories* May 54/1 Miss de Rose..reached for a pack of Strikes. 1936 *Discovery* Nov. 345/2 Ten nuts are the equivalent of one pack of Golden Bat cigarettes. 1937 J. A. Lee *Civilian into Soldier* v. 219 He emptied his pack of issue cigarettes. 1951 *N.Y. Times* 14 June 22/6 It comes in a little pack. 1958 *Listener* 19 June 1015/1 Six packs of American cigarettes. 1959 *Housewife* June 80 The fine white Table Salt in the gaily coloured packs! 1959 N. Mailer *Advts. for Myself* (1961) 218 Stoned with lush, with pot,..Milltown, coffee, and two packs a day, I was working live, and overalert, and tiring into what felt like death. 1963 *B.S.I. News* Apr. 20/1 The 'shelf-appeal' pack designed to catch the eye of the ordinary shopper. 1974 'J. Le Carré' *Tinker, Tailor* xxiii. 201 Gerstmann was a chain-smoker: Camels. I sent out for several packs of them—*packs* is the American word?
g. The container into which a parachute is packed.
1926 *Sci. Amer.* Aug. 100/1 (*caption*) This photograph.. shows the pilot parachute just emerging from the pack. 1930 C. J. V. Murphy *Parachute* 43 The jumper, with the pack strapped on his back, dived from the wing of a plane. 1940 *Aeroplane* 13 Sept. 298/2 A small pilot chute..pulls the main parachute out of its pack. 1969 D. Dwiggins *Bailout* vi. 88, I would have to jump, but first squeeze from my turret and reach my parachute pack from its rack in the fuselage. 1976 A. White *Long Silence* vii. 59 The snap when the fixed line broke open the pack, and the jerk when the pack pulled out the chute.
2. (Earlier and later examples as a measure of furs.)
1744 A. Dobbs *Acct. Countries adjoining Hudson's Bay* 39 He had four Packs of Beaver of 40 each. 1774 S. Hearne *Jrnl.* 11 Oct. (1934) 122 By the Masters account ..65 or 70 Packs or Caggs, called by them Pieces, are put on board each canoe. c 1840 D. Thompson *Narr. Explorations W. Amer. 1784–1812* (1916) iv. 417, I traded three packs of Furrs (a pack is 90 lbs). 1961 Phillips & Smurr *Fur Trade* II. 330 [He] fined him thirty packs of beaver, which was just the quantity he had.
3. c. *Rugby Football.* The forwards of a team, who form one half of the scrummage; also, the scrummage itself.
1887 M. Shearman *Athletics & Football* II. iii. 305 The chief business of the half-back then became to snap up the ball..as soon as it came away from the pack. 1900 A. E. T. Watson *Young Sportsman* 253 Form a compact scrummage with the heads down. Long and straggling packs are easily broken through. 1909 *Westm. Gaz.* 11 Dec. 20/2 Cambridge have an exceptionally fine pack, to whom they must look almost entirely for victory, their halves and three-quarters being but moderate. 1927 H. S. Walpole *Jeremy at Crale* xvi. 278 Mellon's probably the best three-quarter playing on any school side this season. But that needn't worry us. We've got a better pack than theirs. 1955 *Times* 1 Aug. 3/3 The British forwards.. were beaten time and again by the Rhodesian pack. 1960 E. S. & W. J. Higham *High-Speed Rugby* III. xii. 147 Only those who have played in the pack know what will-power it sometimes requires to stand up from a scrum in the last ten minutes and force the weary legs to run. 1972 G. Slatter *Football is Fifteen* i. 16 Tom Morrison, manager of the All Blacks, said only the forwards would know what the loss of Simpson meant to the pack. 1976 *Eastern Even. News* (Norwich) 29 Nov. 13/7 Pressure from Beccles led to a five-metre scrum where they pushed the Union pack back over the line to give Shannon a try.
d. The organizational unit of the Brownie and Wolf Cub movements.
1918 [see *BROWNIE[1] 2]. 1932 U. M. Williams *For Brownies* 111 Brown Owl is guarding the rest of the pack. 1945 'Gilcraft' *How to run a Pack* 5 The man or woman who in a weak moment has consented to run a Wolf Cub Pack. 1965 *Wolf Cub Jubilee Bk.* 31 Some Cub Packs in Canada have a real wolf's head on the top of their totem pole. 1973 *Guardian* 1 Apr. 11/3 Brown Owl said she'd understand if I wasn't quite happy in the pack.
e. In the war of 1939–45, a number of German submarines operating together.
1943 *Times* 13 Dec. 2/1 The story is told below of the defeat of a pack of U-boats in the North Atlantic. 1944 *Daily Tel.* 11 July 3 Captain Walker and his crew smashed U-boat packs lying across the Arctic and North Atlantic convoy routes. 1956 R. Braddon *Nancy Wake* ix. 96 The Bay of Biscay was to be the main target area for U-Boat packs. 1961 S. E. Ellacott *Ships under Sea* x. 100 A common practice among U-boat packs was to lie in line at one- or two-mile intervals across a shipping lane. 1978 *Jrnl. R. Soc. Arts* CXXVI. 252/2 The German submarine packs..were threatening to starve us into submission.
11. b. *Surg.* A soft pad usu. composed of several layers of gauze sewn together, used esp. for wedging organs of the body during an operation.

1916 Parker & Breckinridge *Surg. & Gynæcol. Nursing* xx. 263 At the Mayo clinic three sizes of packs are used, (1) 4 × 8 inches, (2) 5 inches by 3 yards, (3) 3 inches by 2 feet. The latter are used for packing about the gall-bladder. 1944 W. W. Babcock *Princ. & Pract. Surg.* xviii. 285 Salt packs consist of gauze soaked in 10 per cent hypertonic solution in which 5- or 10-grain tablets of sodium chloride are embedded. 1955 *Times* 15 July 11/4 The plaintiffs' cause of action was that during an operation on Mrs. Urry for the delivery of a child by lower Caesarian section, a swab or pack was left in her body. 1955 M. G. Lynch in Ochsner & DeBakey *Christopher's Minor Surg.* (ed. 7) xxi. 500/1 Gelfoam packs will often control this bleeding. 1970 H. Haxton *Surg. Techniques* vii. 45 Most bleeding can be controlled by the pressure of a pack or a finger on the right spot.
c. *Dentistry.* A substance applied in a plastic state to the gums around and between the teeth, subsequently hardening, to serve as a dressing after disease or surgery of periodontal tissue.
1923 A. W. Ward in *Jrnl. Amer. Dental Assoc.* X. 478/2 In order to avoid infection, pain, sensitiveness of the roots..I have devised a quick setting pack. This pack is mixed like cement and flowed between the teeth and all over the exposed surface. The tissues regenerate under the pack, which is allowed to remain four to six days after the operation. 1953 I. Glickman *Clin. Periodontology* xliv. 743 The mixed pack is separated into small masses. *Ibid.* 746 If a portion of the pack fractures off within three days after it was placed, the entire pack should be replaced. 1974 D. L. Allen et al. *Periodontics for Dental Hygienist* (ed. 2) x. 206 The placement of a periodontal dressing or pack following surgery is extremely important.
d. = *face-pack* s.v. *FACE *sb.* 27.
1934 M. Verni *Mod. Beauty Culture* I. v. 29/1 In many schools of beauty, the pupils are taught to sponge the face with hot water before applying the pack. 1944 R. G. Harry *Mod. Cosmeticology* (ed. 2) v. 55 The tightening effect is produced by the drying of the pack, and is enhanced by the presence of albumin and/or certain gums. 1964 Wells & Lubowe *Cosmetics & Skin* II. vii. 202 The significant mechanism operative in the use of face packs is the drying of the pack on the skin surface. 1972 *Vogue* Jan. 15/2 To transform a dry skin..use this simple pack.
13. (Earlier example.)
1866 N. Chevalier *Reminisc. Journey across South Island* (typescript) 7 The pack [was] a strong heavy old chap, the third pretty good. The fourth a flea bitten Arab mare.
13*. Slang phr. *to send to the pack* (see quot. 1916); also *to go to the pack*, to lose a (high) position, to 'go to pieces', to deteriorate. Chiefly *Austral.* and *N.Z.*
1916 C. J. Dennis *Songs Sentimental Bloke* 94 I've sent the leery bloke that bore me name Clean to the pack wivout one pearly tear. *Ibid.* 127 *To send to the pack*, to relegate to obscurity. 1919 W. H. Downing *Digger Dial.* 26 *Go to the pack*, deteriorate. 1934 T. Wood *Cobbers* xvi. 200 The country was going to the pack. 1939 Joyce *Finnegans Wake* 269 If she can't follow suit Renée goes to the pack. 1946 K. Tennant *Lost Haven* (1947) xvi. 250 Everything'ull go to the pack unless they're let go home again. 1952 D. Niland in *Coast to Coast 1951–52* 196, I can't let him go to the pack like that. 1958 G. Casey *Snowball* 118 You wait till he gets a bit older. Them abos always go t' the pack. 1963 D. Crick *Martin Place* 196 Things are goin' to the pack. If they get any shorter of work, they'll close down.
14. a. *pack-animal* (earlier example), *-dog*, *-mule* (earlier and later examples), *-pony* (examples). **c.** **pack-clouds** (*poet.*), densely massed clouds; **pack-drill,** also in phr. *no names, no pack-drill*: see *NAME *sb.* 1 h; **pack-frame,** a frame, usu. of metal, into which a knapsack or other pack is fitted for easier transport; **pack-ice** (earlier and later examples); **pack-leader,** the leader of a group of animals; **pack-peddler,** one who travels round from village to village with a pack of small items for sale; **pack-rat,** substitute for def.: the North American bushytailed woodrat, *Neotoma cinerea* (further examples); also *attrib.* and *fig.*; hence as *v. trans.*, to collect an assortment of objects, as a pack-rat does; **pack-sack,** the container into which goods comprising a pack are put, a rucksack; also *attrib.* in phr. *pack-sack citizen* (*Canad.*), a vagrant; **packshot,** in television advertising, a close-up picture of the advertised product in its wrappings; **pack-strap(s,** the strap or straps which secure a load round the forehead or shoulders of a person or to the back of a pack animal; **pack tactics,** the practice of German submarines of operating in groups; **pack-track, -trail,** a path or route suitable for a pack-train; **pack-train** (earlier and later examples).
1847 *Santa Fé* (New Mexico) *Republican* 16 Oct. 2/2 They left their wagons and took pack animals, and ten days' provisions. 1871 G. M. Hopkins *Note-bks. & Papers* (1937) 141 If you look well at big pack-clouds. 1844 *New Orleans Picayune* 18 Mar. 38/1 The only assistant they took with them was an Indian-trained pack dog. 1913 I. Cowie *Company of Adventurers* 323 Pack-ponies were also used; also pack dogs, the latter bearing frequently burdens mountains high in comparison with their size. 1933 B. Willoughby *Alaskans All* 18 We four stood clinging to the collars of our pack dogs, wondering what marvels lay beyond. 1976 T. Walker *Spatsizi* xii. 132 Travelling slowly with their pack dogs, they walked 150 miles through the mountains. 1955 E. Hillary *High Adventure* vii. 118 Her [sc. the Sherpani's] method of carrying it [sc. her load] was with a headband, and as I had no pack frame with me I had to follow suit. 1963 *Guardian* 9 Aug. 7/4 To get to Dyrfjoll was a whole day's march from the nearest road and the pair used a sledge pack-frame on the way in. 1973 [see above sense 1 e]. 1976 G. Moffat *Over Sea to Death* x. 119 The paraphernalia of [mountain] rescue: rucksacks, pack frames, radio sets. 1850 R. A. Goodsir *Arctic Voy. Baffin's Bay* 108 As long as there was a chance of procuring whales in Prince Regent's Inlet, he might have persevered..great as the risk would have been in pushing through the heavy pack-ice we had fallen in with. 1930 *Times Educ. Suppl.* 25 Jan. p. iv/1 From the air it was also observed that the great region of heavy pack-ice..gives place to waters comparatively little encumbered. 1965 *Kingston* (Ontario) *Whig-Standard* 3 Apr. 4/5 About 300,000 of these seals are killed on the pack ice every spring. 1975 *Nature* 18 Dec. 594/1 In the foraminifera-poor beds which we believe represent periods free, at least seasonally, of packice. 1902 J. H. M. Abbott *Tommy Cornstalk* 35 In work where there is a probability of being under fire..the pack-leader might be left behind. 1975 W. H. Nesbitt in M. W. Fox *Wild Canids* xxvii. 394 The female pack leader [of a group of feral dogs] often 'scouted' ahead before moving the pack. 1835 A. Underwood in *Southwestern Hist. Q.* (1928–9) XXXII. 139 In company with Messrs. Money, Gay..and William Pruit attended by a Mexican with a pack mule we took our departure. 1839 Z. Leonard *Adventures* (1904) 61 We now scattered over a considerable range of country for the purpose of hunting, leaving ten or twelve men only to bring on the pack-mules. 1909 W. R. Harris *Catholic Church in Utah* 128 We..entered a small mountain forest of pine trees in which we lost one of our pack mules. 1934 F. Stark *Valleys of Assassins* ii. 74, I..crouched with my back to the gale on the pack-mule. 1868 *Harper's Mag.* Aug. 348/2 Ten years ago a pack peddler went through the town. 1880 *Ibid.* Nov. 892/1 There was a pack peddler with smuggled shawls and laces at the door. 1944 G. Wilson *Passing Institutions* 70 We..married, and died in a small area, learning of the big outside world only through books and an occasional pack peddler or clock tinker who came in. 1870 De B. R. Keim *Sheridan's Troopers on Borders* 201 [Indians] drive the herds and pack-ponies, or else on foot lead them. 1923 J. H. Cook *50 Yrs. Old Frontier* 98 We used pack ponies on the return trip. 1936 D. McCowan *Animals Canad. Rockies* xxii. 196 From the fact that it habitually transports sundry articles from one place to another the animal [sc. the wood rat], in the West, is commonly called Pack rat. 1955 Priestley & Hawkes *Journey down Rainbow* iii. 47 A mass of bat and pack-rat droppings. 1963 *Spectator* 21 June 803/2 Obsessed with some impulse, like a packrat fear of throwing anything away. 1966 H. Marriott *Cariboo Cowboy* iii. 40, I had other visitors every so often in the shape of sharp-faced, long-tailed rats which were known as pack rats. 1970 R. Lowell *Notebk.* 22 The horrifying mortmain of Ephemera: keys, drift, sea-urchin shells, Packratted off with joy. 1970 *Publishers' Weekly* 8 June 154 A pack rat is somebody who wants to have his own information material, his own personal library or files, even if this means indulging in a little petty thieving. 1973 *Ann. N.Y. Acad. Sci.* CCXI. 308 The sporadic pack-rat collecting we all do every day by habit amidst the print, graffiti, and speech that encompass our peculiar lives. 1973 'D. Shannon' *No Holiday for Crime* (1974) vi. 94 When I came to, they were busy as packrats carting stuff out. 1851 W. Kelly *Excursion to California* I. ix. 159 We, the packers, were now busily employed making pack-sacks of a uniform size. 1920 *Rod & Gun in Canada* Nov. 715/1 A good old-time packsack. 1966 *Globe & Mail* (Toronto) 18 Jan. B5/7 [He] was a pack-sack citizen and appeared on Skid Row streets..with..caulk boots which would be later hocked for the last bottle. 1970 'E. Lathen' *Pick up Sticks* (1971) viii. 70 The packsacks under Thatcher's chair contrasted strongly to the matched sets of luggage piled everywhere. 1960 O. Skilbeck *ABC of Film & TV* 94 *Pack Shot*, the egregious scene with which most T.V. 'Commercials' conclude: a C.U. of the Sponsor's wrapped product. 1966 G. N. Leech *Eng. in Advertising* v. 42 In seven-second commercials there is little time to show anything except a title card or a 'pack shot' establishing a visual image of the product. 1969 *Focal Encycl. Film & Television Techniques* 128/1 Television advertising, for instance, makes much use of cinemacrography in the so-called 'pack shots' but these are normally filmed at a scale of less than 1:1. 1897 J. W. Tyrrell *Across Sub-Arctics of Canada* 12 Western half-breeds, trained in the use of the pack-straps as well as the paddle. *Ibid.* 70 We both took a turn at the pack-straps. 1902 S. E. White *Blazed Trail* 113 The solitary man with the packstraps across his forehead and shoulders had never seen so many [wood creatures]. 1949 P. Newton *High Country Days* iv. 38 The swags..lashed together [on a pack-horse] with the long packstraps. 1956 M. Duggan *Immanuel's Land* 53 He walked along..with the packstraps cutting into his shoulders. 1956 H. S. M. Kemp *Northern Trader* 25 Our canoemen tied their packstraps around a hundred-pound piece, piled another hundred-pound piece atop it, squatted down cross-legged while they adjusted the headband, heaved themselves up and jogged off. 1960 B. Crump *Good Keen Man* 109 The only reason my pack-straps didn't go the same way was that I noticed Harry eyeing them. 1942 *Sun* (Baltimore) 21 Feb. 2/8 The Nazi in command of the U-boat fleet, had promised to use 'pack tactics' on the Eastern Atlantic and save the largest submarines and best crews for attacks off American shores, to cripple Allied tanker strength. 1944 *Hansard Commons* 7 Mar. 1897 It might have seemed as if perhaps after all, the U-boats with their pack tactics might defeat the convoy system. 1870 *App. Jrnls. House Reps. N.Z.* D. xl. 6 It will be desirable to connect them [sc. No Name diggings] by a metalled pack-track with Marsden to the Greenstone. 1930 L. G. D. Acland *Early Canterbury Runs* (ser. 1) ix. 219 It [sc. a hut] is miles by pack-track from the nearest neighbour. 1843 in *Utah Hist. Q.* (1929) II. 116 There is little grass in the mountains and the pack trail bad. 1911 *Daily Colonist* (Victoria, B.C.) 29 Apr. 17/6 The completion of a pack trail into the valley of the Naas. 1965 *Beaver* Autumn 54/1 Along the pack trail we met trappers coming out of the bush. 1849 K. Webster *Diary* 19 June in *Gold Seekers of '49* (1917) iii. 50 It is said at Fort Kearney that the wagons passed here already this season, en route for California, number 5,400, and also three pack trains. 1862 R. C. Mayne *Four Yrs. Brit. Columbia & Vancouver I.* 148 From thence pack-trains could make Alexandria.. in 14 or 15 days. 1922 *Beaver* Nov. 64/1 The daily progress of a pack train is a single drive of ten to fifteen miles. 1965 *Beautiful Brit. Columbia* Summer 9/1 He..operated a pack train for the Hudson's Bay Company.

pack, *v.*[1] Add: **I. 1. b.** (Later examples.)
1831 *Reg. Deb. Congress U.S.* 8 Feb. 133 It is believed that, in Cincinnati alone, there were slaughtered and packed this year one hundred thousand hogs. 1852 *Trans. Mich. Agric. Soc.* III. 230 Mullet..are sometimes used as pan-fish, and are packed to a limited extent.
2. c. Now freq. without *up*. Also used in passive of a person: to have finished packing.
1906 'O. Henry' in *N.Y. World Mag.* 1 July 8/1, I am packed and was to have left for the North Woods this morning. 1907 G. B. Shaw *John Bull's Other Island* I. 18 *Doyle*: Hodson. *Hodson*..: Did you call, sir? *Doyle*: Pack for me too. I'm going to Ireland with Mr. Broadbent. 1912 R. Brooke *Old Vicarage Grantchester* (1916) 10 God! I will pack, and take a train, And get me to England once again! 1958 J. Cannan *And be a Villain* iv. 84 I'm packed, but I must..tell them I'm leaving. 1962 J. Braine *Life at Top* xxii. 248, I turned away without speaking and went upstairs to pack. 1969 G. Lyall *Venus with Pistol* xxxv. 234 We're all packed up. I don't know if I've got all your things in the right bags. 1974 J. Johnston *How Many Miles to Babylon?* 66, I have to catch the Dublin train. I..should go and pack.
d. *to pack up* (or *in*) (*intr.*), to stop working; to give up an enterprise; to surrender; to die; to cease to function; to collapse.
1925 Fraser & Gibbons *Soldier & Sailor Words* 219 *Pack-up*.., *To*, to stop (as opposed to 'carry on'). To give up. To finish. To die. 1926 E. F. Spanner *Naviators* i. 8 It was about five in the afternoon when Sir Joseph decided to pack up for the day. 1928 C. F. S. Gamble *Story N. Sea Air Station* xii. 201 To make matters worse another engine packed up, and this increased the heavy list of the ship. 1940 'Gun Buster' *Return via Dunkirk* II. xii. 174 The Belgians have packed up... They've laid down their arms. 1948 C. Day Lewis *Otterbury Incident* iv. 45 There seemed nothing to stop Toppy unless his voice packed up; so Ted and I left him. 1953 W. Burroughs *Junkie* (1972) vii. 70 He said, 'We've got to pack in. We can't last with this crowd.' 1956 'J. Wyndham' *Seeds of Time* 87 The laterals aren't firing... I mean they *won't* fire. They've packed up. 1959 'M. Innes' *Hare sitting Up* I. i. 17 The most surprising people will pack up under strain. 1962 *Economist* 3 Nov. 440/2 The Algerians seem to think that Mr Khemisti..broke off the talks as a gesture of solidarity with Cuba. But American rumours suggest that it was the State Department itself that decided to pack in. 1967 J. L. Anderson *Vinland Voyage* 90 None of us had much confidence in it [sc. our auxiliary engine] and it packed up a few days later. 1973 A. Broinowski *Take One Ambassador* i. 14 I'm Mrs Norrice..Mr Norrice passed on last year..Bert's ticker packed up. 1977 G. Fisher *Villain of Piece* iv. 40 Where is the nearest telephone?..It's packed in.
e. *to pack up* or *in* (*trans.*), to stop (doing something), to give up, finish with; freq. in phr. *to pack it up* (or *in*), to stop working, abandon an attempt, etc.; also as imp., be quiet, 'cut it out', behave yourself. *slang* and *colloq.*
1942 Berrey & Van den Bark *Amer. Thes. Slang* § 205/4 *Stop talking*; '*shut up*',..pack it up. 1943 Hunt & Pringle *Service Slang* 50 *Pack it up* or *in*, stop talking or fooling; cut it out. 1945 J. B. Priestley *Three Men in New Suits* iv. 104 'Pack it up,' she warned him. 1949 'G. Orwell' *Nineteen Eighty-Four* I. viii. 86 'Oh, pack it in!' said the third man. 1951 A. Baron *Rosie Hogarth* 126 Pack it up, Joyce. I'm telling you. 1951 'N. Shute' *Round Bend* 10 It looked as if the public were getting a bit tired of it. Sir Alan packed it up. 1953 W. Burroughs *Junkie* (1972) xii. 121 Every month or so she hires a new lover, gives him shirts and suits and wrist watches, and then packs him in when she has enough. 1959 'O. Mills' *Stairway to Murder* viii. 92, I packed up my job last week... I just told you... I packed in my job last week. 1963 N. Hilliard *Piece of Land* 43 They saw nothing... At eight o'clock they decided to pack it in..headed back to town. 1971 B. W. Aldiss *Soldier Erect* 47 'Why don't you pack in ordering us about, Wally?' I asked. 1974 J. Wilson *Hide & Seek* viii. 137 Rob Millar didn't finish work until gone eleven, and then decided he'd have to pack it in. 1974 K. Clark *Another Part of Wood* ii. 56 He had long ago 'packed it in', and spent his life sitting by the window dozing, with a volume of Pepys' Diary upside down on his knee. 1976 *News of World* 14 Mar. 11/2 He has been ordered to pack in his job and return for the final four weeks of term. 1977 *Daily Mirror* 18 Mar. 24 Hey! Pack it in! That's my missus—pack it in!
3. d. (Earlier example.)
1850 R. G. Cumming *Five Yrs. Hunter's Life S. Afr.* II. 141 The ground all round was packed flat with their spoor.
f. *Theatr.* Phr. *to pack them in*: to attract a capacity audience. Also *transf.*
1943 *N.Y. Times* 9 May II. 5/4 Harry James and his band have been helping to pack them in at the Paramount. 1970 *Guardian* 31 July 9/3 Bolton's Octagon Theatre is packing them in for Old Tyme Musical Hall. 1972 *Ibid.* 1 Dec. 11/1 She's still at it at 49; still packing 'em in, and getting the odd rave review. 1977 *New Statesman* 24 June 809/2 His rejigged Radio 4 *Today* programme is now packing the listeners in.

g. *Computers.* To compress (stored data) in a way that permits subsequent recovery; *spec.* to represent (two or more items of data) in a single word. Also *absol.*
1954 *Computers & Automation* Dec. 18/1 *Pack*, to combine several different brief fields of information into one machine word. 1959 M. H. Wrubel *Primer of Programming* viii. 189 If the data consist of only a few significant digits, two or more numbers can be 'packed' into a single 10-digit word. They will be transferred from the card to the machine as a 10-digit word, which must subsequently be 'unpacked' by an appropriate program. 1961 L. W. Hein *Introd. Electronic Data Processing* ix. 165 One of the bad situations is that of the 31-word record in a 60-word fixed record-length computer... A decision to pack means some rather complex programming. 1964 *IBM Systems Jrnl.* III. 125 Decimal digits, packed two to a byte, appear in fields of variable length (from 1 to 16 bytes). 1972 *Computer Jrnl.* XV. 199/1 A stream..might be formed by one stream function which unpacks words into bytes, followed by another one which packs them all up again. 1973 [see *PACKED *ppl. a.*[1] 1 c].
5. a. (Further examples.) Also of a group of runners in a long-distance race.
1844 in *Rep. U.S. Comm. Patents 1846* (1847) 34 It [sc. cotton] does not pack and becomes hard. 1887 A. W. Tourgée *Button's Inn* 200 It [sc. the storm] filled the road with a slippery mealy mass, which did not cling or pack. 1890–3 E. M. Taber *Stowe Notes, Lett. & Verses* (1913) 8 The snow packs so readily that I can walk without much difficulty. 1908 *Westm. Gaz.* 27 July 9/3 The failure of the British representatives..was undoubtedly due to their failure to 'pack' well.
b. (Later examples.)
1939–40 *Army & Navy Stores Catal.* 656/1 Foulard Silk [dressing gown]..of very light weight to pack small. 1946 *Mod. Lang. Notes* LXI. 444 This dress washes and irons and packs easily. 1974 *Janet Frazer Catal.* Spring & Summer 455/2 Pneumatic 'Igloo' tent... Packs away compactly.
c. Of the forwards in Rugby football: to form or take their places in the scrummage. Also const. *in* and *down*.
1874 *Rugby Union Football Ann. 1874–5* 15 A good forward will..pack in again at the back of the scrummage. 1887 M. Shearman *Athletics & Football* II. iii. 313 There is many a good scrimmager who packs quickly. 1900 A. E. T. Watson *Young Sportsman* 252 Be the first to form the scrummage and pack quickly. 1927 Wakefield & Marshall *Rugger* II. iii. 162 If..his opponents have the right of putting the ball in, he ought to pack opposite their loose-head to be ready to check their scrum-half. 1949 *Rugby League Football* ('Know the Game' Series) 31 The front row forward who packs nearest to the referee has what is known as the 'loose head'. 1968 Hudson & Dyer *Your Bk. of Rugger* v. 49 'Number 8'..packs down in between the second row, with his back parallel to the ground and his feet spread evenly apart. 1970 G. Slatter *On the Ball* vi. 139 The scrum packs down on our 25-yard line and Haigh takes the ball.
6. c. *Surg.* To fill, wedge, or cover with a pack. Also *absol. to pack off*, to wedge (an internal organ) with packs so as to keep it away from a region of interest.
1889 Caird & Cathcart *Surg. Handbk.* vii. 53 The cavity of the nostril may be packed with a long strip of lint. 1897 Stimson & Rogers *Man. Operative Surg.* (ed. 3) I. 28 Much of the hemorrhage can be stopped..by packing with sponges or pads of gauze. 1906 H. M. Davies *Man. Minor Surgery & Bandaging* (ed. 13) ii. 36 Large pieces of gauze..are very convenient for packing off the intestines..from the rest of the abdomen. 1924 R. Howard *Surg. Emergencies* iv. 74 The site of the obstruction should be isolated..by..packing the surrounding intestine off with abdominal pads. 1940 R. Maingot *Abdominal Operations* I. I. ii. 46 The little sinus that remains may be lightly curetted out and packed with gauze which has been soaked in..penicillin. 1955 M. G. Lynch in Ochsner & DeBakey *Christopher's Minor Surg.* (ed. 7) xxi. 499/2 If the hemorrhage is severe..the nose should be packed. 1972 Nealon & Grossi in P. F. Nora *Operative Surg.* i. 7/2 The wound is packed open with gauze over a simple layer of nonadherent material.
7. (Further example.)
1910 C. E. Mulford *Hopalong Cassidy* xxxviii. 242 Hall carefully packed his pipe and puffed quickly.
b. Also const. *out.*
1932 C. C. Martindale *What are Saints?* 58, I had to go straight from Westminster Cathedral..to a church in Chiswick,..packed out with people observing his [sc. St. Edward the Confessor's] feast-day. 1944 G. Texidor in D. M. Davin *N.Z. Short Stories* (1953) 301 The domain in front of the hotel was packed out with cars and lorries. 1973 *Times* 7 June 15/5 The Rolling Stones couldn't play the Ken Colyer Club one night and The Dimensions appeared instead... It was packed out. 1977 D. Clark *Gimmel Flask* vi. 101 This place is packed out for lunch.
c. *to pack them*, to hold back diarrhœa caused by nervousness; hence to be terrified. Also with explicit alternative objs. *Austral. slang.*
1951 E. Lambert *Twenty Thousand Thieves* 132 He's packing them badly. He's quite useless. 1952 T. A. G. Hungerford *Ridge & River* iii. 46, I suppose the poor cow *would* pack 'em a bit. He's on'y a kid, by the look of him. 1959 'D. Forrest' *Last Blue Sea* 69 You know something, thought Ron Fisher, you're no good. You're packing them. 1961 R. Braddon *Naked Island* 44 'Who's panicking?' 'You are, son. Fair packing 'em, y'are.' 1971 D. Ireland *Unknown Industrial Prisoner* 132 They were packing the shits when he went off his head in the control room last time.
9. a. (Earlier examples.) Also, to carry in any manner; to wear habitually; to possess. Also *absol.*
1805 W. Clark *Jrnl.* 15 Dec. in Lewis & Clark *Orig. Jrnls. Lewis & Clark Expedition* (1905) III. xxi. 280 Proced up the 1st. right hand fork 4 miles & pack the meat from the woods to the canoes. 1816 U. Brown *Jrnl.* in *Maryland Hist. Mag.* (1916) XI. 360, I let him know that I..meant to hire a horse of him to pack our provisions. 1845 J. C. Frémont *Rep. Exploring Expedition* 73 It would have been a work of great time and labor to pack our baggage across the ridge. 1863 S. Butler *First Year in Canterbury Settlement* v. 61 The back country..is inaccessible by dray, so that all stores..have to be packed in and packed out on horseback. 1874 E. Eggleston *Circuit Rider* vii. 71 My shoes hurts my feet, an' I have to pack one of 'em in my hand most of the time. 1890 N. P. Langford *Vigilante Days* II. xviii. 282 No man that ever packed a star in this city can arrest me. 1902 A. H. Lewis *Wolfville Days* v. 61 He finds this person ain't packin' no gun. 1903 *Dialect Notes* II. 323, I never did pack a watch. 1913 [see *HEFT *v.*[1] 1]. 1927 *Amer. Speech* II. 361 He packed the child home. 1930 L. G. D. Acland *Early Canterbury Runs* (ser. 1) viii. 200 George Harper.. used to pack the wool out on bullocks, three sacks on each. 1940 R. Chandler *Farewell, my Lovely* iii. 22 Don't you pack no rod? 1952 *Picture Post* 6 Dec. 37/2 A revival of flogging was in loud demand, and there was the barmaid who insisted she wouldn't feel safe until every policeman packed a gun. 1970 C. Wilson *Campbell of Yukon* ii. 9 Here the cargoes &c have to be 'packed' on the men's shoulders from water to water. 1973 *Washington Post* 5 Jan. B3/6 Actress Ali McGraw 'packs all the glamor of a worn-out sneaker'.
b. To travel with one's goods or merchandise in packs.
1842 M. Crawford *Jrnl.* (1897) 14 Some of the company preparing to pack from here. 1857 W. Chandless *Visit to Salt Lake* II. vii. 264 Waggoning through the settlements ..and thence 'packing' to California. 1903 S. E. White *Forest* ii. 15 Do not carry a coat..you will never wear it while packing. 1911 J. F. Wilson *Land Claimers* i. 1 It isn't much fun packing along that trail.
c. To be capable of delivering (a blow) with force; esp. in phr. *to pack a punch*; also *fig.* *colloq.*
1921 H. C. Witwer in *Collier's* 19 Feb. 22/3 He packed a wicked right and had stopped a lot of good men before Kid Roberts cut him short with a one-round knockout. 1922 E. O'Neill *Hairy Ape* (1923) i. 7 He packa da wallop, I tella you!..No fightin', maties. 1934 M. H. Weseen *Dict. Amer. Slang* 239 *Pack a terrific punch*, to hit hard, or to have the ability to do so. a 1938 T. Wolfe *Lett.* (1956) 45, I think my play 'The House' will 'pack a punch'. 1957 *Listener* 20 June 1008/1 An artist who packs such a violent literary punch might be expected to make use of a savage, expressionist line. 1958 Wodehouse *Cocktail Time* xvi. 137, I take it that she busted you one... These nannies pack a wicked punch. 1971 *Sunday Express* (Johannesburg) 28 Mar. 22/3 Like Kies, he packs a powerful punch, but he does not wade in like Jan. 1973 W. M. Duncan *Big Timer* xxi. 137 That Carver packed a wallop, didn't he? I should have plugged him sooner.
II. 10. b. Also *to go packing*.
1926 T. E. Lawrence *Seven Pillars* lxxxviii. 468 As both example and guilt were blatant, the others went packing into the far room while their chiefs forthwith executed sentence.
c. Const. *off*. Also, to die.
1766 Sewel & Buys *Compl. Dict. Eng. & Dutch* I. 549/3 To pack off, (to die) *Stērven*. 1914 W. Owen *Let.* 15 June (1967) 260 The alternative would be to come home immediately..and at once to pack off to some other part of the world. 1933 T. E. Lawrence *Let.* 10 Aug. (1938) 774, I would like myself and those I care for to pack off all together.

pack, *v.*[2] Restrict † *Obs.* to I and add: **II. 4.** (Later examples.) Also, to secure (a particular decision or result) by selecting or arranging the body of voters, etc.
1925 A. Toynbee *Survey Internat. Affairs 1920–23* 80 In order to prevent any possibility of 'packing' the vote, the date of residence was not fixed..as the day when the Treaty came into force, but as the day when it was signed. 1932 *Ann. Reg. 1931* II. 275 China... Together with others from the Canton faction they issued a manifesto declaring that the elections to the Convention would be a fiasco, that the Convention would be 'packed' by Chiang, and that its whole purpose was to seat him more firmly in the dictatorship. 1955 *Times* 26 May 8/3 The Minister had tried to cloak the Bill with respectability but this had to be seen against his earlier statements about 'packing' the courts and the Senate. 1965 *Mod. Law Rev.* XXVIII. 517 He was not above packing the House in order to curb such activities. 1973 *Black Panther* 14 Apr. 12/2 The Supreme Court is being systematically packed despite the defeat of Carswell and Haynesworth. 1976 *Daily Tel.* 24 Apr. 2/6 Vauxhall Labour party is to meet..to investigate allegations that the general management committee is being improperly 'packed' with members to ensure the selection of a merchant banker to succeed the present MP.
5. (Further example.)
1927 *Amer. Speech* II. 352/2 They packed the deal on the other players.

packability (pækăbi·lĭti). [f. Packable *a.*: see *-BILITY.] The capacity to be packed; *spec.* of clothes and fabrics, the ability to be packed easily into, and to travel without damage in, a suitcase.
1958 *Vogue* May 108 A traveller's dream of packability, a pleated dress..that emerged uncreased after days in a suitcase. 1963 *New Yorker* 29 June 44/1 (Advt.), All the lightness, comfort, and packability that Arnel is famous for. 1967 *Daily Tel.* 24 Oct. 19 Dress and coat ensembles of great packability. 1977 *Shooting Times & Country Mag.* 13–19 Jan. 33/1 One of the main advantages of pistols—'packability'—is lost.

packable, *a.* Delete *rare* and add further examples. Also *ellipt.* as *sb.*
1932 *Times Lit. Suppl.* 21 Jan. 40/4 So easily packable, pocketable, readable a copy of the book. 1962 *Sunday Express* 30 Dec. 16/5 Opaque and eminently packable pyjamas. 1972 *New Yorker* 29 July 16/1 (*caption*) A packable that goes to town dressed up or down! 1976 *Sci. Amer.* Feb. 123/3 In any Euclidean space an infinite set of boxes packable by a given brick has a finite subset of packable boxes that can be used for packing all the others.

package, *sb.* Add: **2. b.** *fig.* (See quots.) *slang.*
1933 *Sun* (Baltimore) 17 Aug. 8/6 The 'package', as the kidnapped victim is called, is rushed across the State line and delivered to the 'keepers'. 1935 A. J. Pollock *Underworld Speaks* 85/1 *Package*, the kidnaped victim. 1945 L. Shelly *Jive Talk Dict.* 15/2 *Package*, a girl. 1960 Wentworth & Flexner *Dict. Amer. Slang* 371/1 *Package*, an attractive, usu. small and neat, girl or young woman. 1963 *Listener* 4 Apr. 585/1 A 'package' was criminal jargon for a dead body with three or more bullets placed in a most efficient manner at the base of the skull.
3. b. *transf.* and *fig.* orig. *U.S.* A combination or collection of interdependent or related abstract entities. (Cf. Parcel *sb.* 7 b.) Hence *attrib.*, esp. (*a*) of negotiations, as *package deal*, a transaction or proposal agreed to as a whole, the less favourable items as well as the more favourable; so *package offer, proposal*, etc.; (*b*) of holidays, tours, etc., one in which all arrangements are the responsibility of agents; (*c*) of a series of acts in a vaudeville show, on television, etc. (For a considerable body of further evidence see *Amer. Speech* May 1958, 73 ff.)
Dict. Americanisms has an 1846 quot. for *package ticket* in the sense 'one entitling the holder to a specified number of journeys'.
1931 *Social Sci. Abstr.* 15838 Insurance in a 'package'... The exact size of the 'package' offered to any employee depends upon age, sex, and length of service. 1948 *Investor's Reader* 26 May 9 Often units of preferred and common shares are tied together in one big economy package which offers two for the price of one. 1951 W. F. Buckley *God & Man at Yale* 40 This report is primarily a highly controversial package of political and economic ideas. 1952 *N.Y. Herald Tribune* 9 Apr. 26/7 On this basis the [steel] industry is confronted not with a package [sc. the wage recommendations of the Wage Stabilization Board] costing 29.8 cents an hour, but one that will cost 59.6 cents. 1953 in *Amer. Speech* (1958) XXXIII. Suppl. 74 Eventually the agency offers a package—a violinist, a pianist, a singer, a quartet and perhaps a small string orchestra. 1958 *Economist* 8 Mar. 849/1 The [U.S.] Administration is thinking of proposing a simplified 'package' [for summit talks]. 1959 *Observer* 26 Apr. 6/2 The package would link the future of Berlin with a settlement of the German problem and European security. 1962 M. McLuhan *Gutenberg Galaxy* 128 (*heading*) The *camera obscura* anticipated Hollywood in turning the spectacle of the external world into a consumer commodity or package. 1965 *Melody Maker* 25 Sept. 20 Few fans will turn out every week for expensive concerts. Yet, in the past, some towns have had two or even three big packages in their area in one week. 1966 *Listener* 3 Mar. 300/2 It was the orthodox deflationary package that was to be expected in an off-year for the British balance of payments. 1967 M. McLuhan *Medium is Massage* 22 The mass audience..is..merely given packages of passive entertainment. 1968 *Globe & Mail* (Toronto) 3 Feb. B3/3 Although the Canadian union gained the greatest economic package in its 30-year history from the company the day before, hundreds of workers were reluctant to return to work. 1969 L. G. Arthur in A. E. Wilkerson *Rights of Children* (1973) 132 The great advancement of the juvenile courts is that they can be imaginative and flexible: that they can design a package of dispositions to fit the needs of a particular child. 1969 *Listener* 2 Jan. 31/2 You can arrange as part of your package to hire a car. 1970 *Daily Tel.* 3 Mar. 2/1 The Hillman Minx de Luxe model will also cease to be available, but a 'comfort' package of carpets and central console will be available for the Minx. 1972 *Guardian* 25 Mar. 1/1 Northern Ireland's Parliament at Stormont is to be prorogued for a year under the peace package of Ulster announced in the Commons yesterday. 1972 *Times* 16 Aug. 1/4 National bargaining on the unions' package claim... The whole package amounted to 40 per cent... Union officials at local level will now be able to pursue those parts of the package that have not been met. 1975 J. De Bres tr. *Mandel's Late Capitalism* xi. 358 The underdeveloped country sends its entire export package to the imperialist country. 1977 *Time* 17 Jan. 17/1 The Carter package offers something to just about everybody. 1977 'A. Stuart' *Snap Judgement* 192 The tired tanned look of tourists at the end of a two-week package.
attrib. (*a*) 1951 *N.Y. Herald Tribune* 14 Dec., President Truman in effect wrapped up the whole subject of creeping corruption in his administration..into one package policy which he laid before the nation. 1952 *Ibid.* 4 Mar. 1/8 (*headline*) State to offer city 'Package' Aid Plan. *Ibid.*, The Dewey [State] administration was reported ready tonight to offer New York City a 'package' solution for the city budget crisis. 1952 *N.Y. Times* 9 Mar. 15/1 The plan for West German sovereignty was tied up in a 'package deal' with the plan for West Germany's military and financial contribution to Western defense. 1952 *N.Y. Herald Tribune* 19 May 14/3 The ensuing battle for diplomatic advantage..seemed pretty well played out by the end of April, when the United Nations Command made up its 'package' offer of settlement. 1955 *Times* 13 May 8/4

An attempt at some sort of 'package deal', whereby a disarmament settlement is made dependent on the solution of other world problems. 1963 *Ann. Reg. 1962* 247 The Chancellor..criticized the American suggestion for a 'package deal' with Moscow. 1964 M. McLuhan *Understanding Media* (1967) II. xxxi. 334 He tends instead to accept the full image as a package deal. 1967 *Listener* 20 July 74/2 In the peculiarly modern field of mass productive factory design, the so-called package-dealer holds sway, with his offer of a complete design-and-build service. 1972 *New Building* Oct. 11/3 Fully conscious of the wastefulness of normal tendering methods, Simonbuild consider their package deal system the only sensible and economic approach for the needs of today. 1972 Wodehouse *Pearls, Girls & Monty Bodkin* iii. 41 This step-daughter who had come to him as part of the package deal when he and Grayce were joined in holy wedlock. 1973 I. M. Sinclair *Vienna Convention on Law of Treaties* v. 141 But the 'package' proposal did not envisage any stage of compulsory arbitration or judicial settlement, except in relation to the *jus cogens* articles. 1976 *Economist* 16 Oct. 73/3 It now says that what Mr Smith accepted in his September 24th broadcast was not—as he says he was told it was—a package deal that had been firmly agreed upon with Mr Kissinger by Britain and the African presidents; it was merely a set of 'proposals'.
(*b*) 1952 *N.Y. Herald Tribune* 14 Apr. 6 (Advt.), Unique, new, all-expense fishing 'package' trips. 1952 *N.Y. World-Telegram & Sun* 29 Apr. 25/1 Eastern Air Lines has mapped out an ambitious schedule of package vacations. 1958 *New Statesman* 11 Oct. 502/2 The great increase in British visitors, however, has been in the less expensive areas, and this is reflected in the range of 'package' tours offered by the travel agencies. 1960 *News Chron.* 11 Apr. 9/4 Pony trekking..is probably the most popular of all the package-holidays. 1964 M. McLuhan *Understanding Media* (1967) II. xx. 212 They can have..Berlin or Venice in a package tour. 1967 *Spectator* 1 Dec. 699/3 Cut-price package trips [are] practically keeping the international airlines going during the slack season. 1971 *Daily Tel.* 9 Jan. 18/3 Moscow..is one place where the package tourist scores heavily over the independent tourist. 1973 J. Symons *Plot against Roger Rider* II. 45 Are they going to be stranded in a foreign city?.. Reassured, the package tourists sink into their seats. 1976 A. Grey *Bulgarian Exclusive* xiv. 99 A large Russian package tour group. 1976 *Field* 18 Nov. 980/2 We read of the pilgrimages to Eton..and the crowds who flocked there—'those package tourists of medieval times'. 1977 *Irish Times* 8 June 4/1 Mr. Adrian Hopkins, of Bray Travel, which organises package holidays in the Seychelles, said yesterday that they could have full refunds if they decided not to travel.
(*c*) 1946 *Sun* (Baltimore) 18 Feb. 11/5 The boys in blue finally have come up with touring package shows of their very own. 1952 *N.Y. Jrnl.-American* 23 May 13/1 Judy Garland and Sid Luft are investing close to $100,000 in producing their own package shows to tour the country. 1956 B. Holiday *Lady sings Blues* (1973) xx. 165 When theatre dates began to slack off, Mr. Levy decided to get together a whole package show, with me as the star and fronting the band. 1958 *Amer. Speech* XXXIII. (Suppl.) 73 A package show on radio or television may involve a number of varied performances given once a week throughout the year, with a core of main performers and with others who are brought on for one or more performances, but not for all. 1959 *Tatler* 1 Apr. 34/1 Stan Getz, star of last year's 'Jazz At The Philharmonic' package show in Britain.
c. Any related group of objects that is viewed or organized as a unit.
1947 Crowther & Whiddington *Science at War* Pl. ix (*caption*) A late model; it is a 'package', with magnetron and magnet complete with rectangular wave guide out of which power pours. 1952 *New Yorker* 12 Apr. 45 (Advt.), The big, beautiful, package that comprises the restaurants of Rockefeller Center. 1958 *Observer* 12 Oct. 1/3 The objectives of this lavish experiment were..to test the interaction of the three stages..and..to achieve a trajectory that could be controlled enough to put the final package within some 50,000 miles of the moon. 1962 F. I. Ordway et al. *Basic Astronautics* iv. 128 The cosmic ray package contains two proton monitors..and devices to detect electrons. 1962 S. Carpenter in *Into Orbit* 56 Another item which is thrown off, just before re-entry, is the package of three retro-rockets which ride strapped to the heatshield during most of the mission. 1965 *IEEE Trans. Reliability* XIV. 114 (*heading*) Impact testing of plug-in circuit packages for high reliability. 1968 *Times* 10 Dec. 6/7 The orbiting observatory carries two experimental packages, both intended to make observations in the ultraviolet region of the spectrum. 1970 *Sci. Amer.* Feb. 26/1 If..one wanted to design a central processor with 1,000 logic elements, one could begin by selecting 200 integrated-circuit packages, each containing an average of five gates of the kind desired. 1972 *Computers & Humanities* VII. 82 A collection of canned programs maintained as a unit is a library. If they are closely related in content, e.g., all statistical, then the library is a package. 1976 *Offshore Platforms & Pipelining* 23/2 The pump package of Noble's Rig 27 varies slightly from the specification guide. 1976 *Physics Bull.* May 202/3 A Japanese company has recently developed a multicolour three-dimensional image video package, using 35 mm film for the storage of the holograms, which is aimed at the advertising, educational, medical and recreational markets. 1977 *Modern Boating* (Austral.) Jan. 81/2 Every boat-trailer-motor package which goes out of their doors is fully equipped with the recommended number of fire extinguishers as well as all other recommended safety gear as part of the package.
d. Misc. *attrib.* uses of sense 3 a. (All *N. Amer.*)
1821 *Deb. Congress U.S.* (1855) 16th Congress 2 Sess. App. 1526 Package sales..by the assortments of merchandise they combine, excite most interest. 1888 M. Deland *John Ward* i. 2 They were at the mercy of Phibbs, the package man, who brought their wares on his slow, creaking cart..from Mercer. 19[illegible] H. A. Smith *View from Chivo* xiii. 127 Jim Davey clos[illegible] the package store two hours early and shut off all beer sales. 1976 *Amer. Speech 1974* XLIX. 115 *Package bar*, establishment that sells packaged goods (wine, beer, and hard liquor) as well as serves liquor by the drink.

package, *v.* [f. the sb.] *trans.* To wrap up, make into a package. Also *fig.* and *absol.*
1928 M. H. Weseen *Crowell's Dict. Eng. Gram.* 445 *Package*, commercial cant as a verb. 1947 M. McCarthy *On Contrary* (1961) 11 An image of happiness as packaged by the manufacturer. 1954 N. Tomalin in *Granta* 6 Nov. 21/1 Two licensed pigeon-feed sellers who..earn ten pounds a day, have factories in Ealing packaging the peas, and are worried acutely by their income tax. 1959 *Listener* 30 July 186/3 There is no lack of playwrights who carry on the old trade of packaging sentimental hokum into a tough shell. 1967 *Boston Sunday Herald* 26 Mar. VI. 7/6 Something for everybody is the way Air France describes its 1967 Jet Away Holidays tour program, recently packaged and distributed to travel agents. 1970 *Natural Hist.* Feb. 18/3 Philadelphia is now packaging its trash, putting it into empty coal cars and then shipping the material back to the strip mines as land fill. 1972 *New Yorker* 2 Dec. 162/2 The black films are packaged, financed, and sold by whites. 1976 N. Postman *Crazy Talk* 70 Billy Graham has packaged his 'message' in a way not unlike that in which Scope packages its 'message'. 1976 *NBR Marketplace* (Wellington) III. 5/1 It brings in its products from factories overseas (US, UK, Australia, and even Brazil) and packages here.

packageable (pæ·kėdʒăb'l), *a.* [f. *PACKAGE *v.* + -ABLE.] Capable of being packaged, in the various senses of the verb.
1961 *Aeroplane* CI. 578/2 From the present generation of packageable propellents (e.g. nitric acid and a mixed amine fuel), a practical realizable specific impulse of about 245 sec. can be obtained. 1977 *Church Times* 6 May 7/1 Fundamentalism, like Maoist Marxism, is positive and highly packageable.

packaged (pæ·kėdʒd), *ppl. a.* [f. *PACKAGE *v.* + -ED[1].] Wrapped up, made into a package; pre-packed.
1933 R. Simmat *Princ. & Pract. Marketing* viii. 67 The various types of packaged product may be divided into two classifications—the liquid and the solid. 1934 S. B. Whipple *Noble Experiment* 173 The exorbitant prices for packaged spirits and wines. 1945 P. Cheyney *I'll say she Does!* ii. 41 This Confucius certainly knew his packaged goods. 1952 [see *KNOCK-DOWN *a.* 3]. 1958 *Times* 4 Aug. 9/2 The pile of débris that results from unpacking a pound of packaged biscuits has to be seen to be believed. 1964 M. McLuhan *Understanding Media* (1967) II. xx. 207 The age of the consumer of processed and packaged goods. 1969 *Jane's Freight Containers 1968–69* 184 Packaged Timber Berth.
b. *fig.*
1946 R. Blesh *Shining Trumpets* (1949) xiv. 325 The mediocre, but financially profitable slick packaged musical trash. 1947 *Richmond* (Virginia) *Times-Dispatch* 11 May 1-D/4 Scheduled for this Summer are some 14 weekly 'packaged' tours. 1951 *N.Y. Herald-Tribune* 24 Apr. 23/6 Fifteen 'packaged' lessons have been prepared. 1953 R. Chandler *Long Good-Bye* iv. 22 The goddam women will start..making up with the packaged charm. 1954 *Sun* (Baltimore) 16 Oct. 12/6 Many insurance companies offer homeowners a 'packaged' policy..at a saving over the cost of a multiplicity of policies each to cover a limited field. 1958 *Observer* 7 Dec. 15/6 Some of the units would be 'packaged' kitchens, bathrooms or bedrooms—that is to say, units with built-in furniture and fittings. 1959 *Economist* 7 Feb. 511/2 A packaged holiday, usually consisting of two weeks or ten days.. for an all-in charge that includes air ticket and hotel bill. 1959 *Listener* 23 Apr. 700/1 A foreign policy that is a packaged deal. 1961 L. MacNeice in *New Statesman* 30 June 1054/3 The weather is packaged and the spacemen In endless orbit. 1967 A. Cowan in Wills & Yearsley *Handbk. Managem. Technol.* 160 It is not possible to write a prescription which can be applied as a finite system to a business. This is a disappointment to those who hope that a neat, packaged solution can be found. 1970 *Daily Tel.* 10 Nov. 17/5 Even ghost-hunters have been offered a packaged tour of the haunted houses of England. 1975 *Times* 18 June 5/2 The packaged tourists pay homage and more solid currency to Anne Hathaway's cottage.

packager (pæ·kėdʒəɹ). [f. *PACKAGE *v.* + -ER[1].] One who packages, in the various senses of the verb.
1959 *Wall St. Jrnl.* (Eastern ed.) 14 Jan. 1/1 They're rough-and-tough 'action-adventure' shows, and all have been produced by independent packagers—TV program producers who assemble the ingredients of story, director and cast, and then put them together to make a show. 1969 *Listener* 17 Apr. 538/1 'These two are the outstanding producers and packagers of guilt in our time,' whines Alexander Portnoy of his parents. 1976 *New Yorker* 19 Apr. 31/2 Morty Mann, the packager of 'Back-Chat', who claims to have lost 'a small fortune' in the enterprise, says that the program failed because the Prince just couldn't 'relate'. 1977 W. Marshall *Thin Air* ii. 22 We'll go and see your food packager.

packaging, *vbl. sb.* (in Dict. s.v. Package.) (Further examples.) Also *fig.*
1934 *Planning* II. xxxvi. 10 They must submit samples, manufacturing formula, chemical analysis.., description of process of manufacture and packaging. 1950 *Sunday Times* 7 May 8/4 It is apparent..that in the last 12 months packaging has improved considerably. 1954 *Economist* 20 Feb. 568/1 Some of the subsidiaries do 'quite a substantial amount of trade' in packaging materials. 1957 L. F. R. Williams *State of Israel* 96 Much attention is being paid to grading, packaging and marketing. 1962 *Rep. Comm. Broadcasting 1960* 34 in *Parl. Papers 1961–2* (Cmnd. 1753) IX. 259 A trivial presentation may consist ..in an excessive interest in smart 'packaging' at the expense of the contents of the package. 1962 M. McLuhan *Gutenberg Galaxy* 268 Language and the arts would become mere packaging devices. 1976 R. Barnard *Little Local Murder* ix. 105 Slovenly shoppers who could be deceived by the most patently bogus special offer or the most obviously inflated packaging device. 1977 W. Marshall *Thin Air* ii. 24 I've seen the food packaging people... Apparently the packaging is done in the buildings.

packaway (pæ·kăwēi), *a.* [f. Pack *v.*[1] + Away *adv.*] Capable of being folded into a small space when not in use.
1957 *Archit. Rev.* CXXII. 355/3 Other garage doors to be shown... A new packaway door..which was exhibited in prototype at the last exhibition..has been modified and improved since. 1973 *Times* 4 May 15/1 Up in London's Seven Sisters Road, they are marketing the best value, child-to-adult, packaway bicycles I have seen. 1974 *Times* 5 Mar. 9/1 (*caption*) Those amorphous, ectoplasmic packaway jobs [sc. mackintoshes] the earnest tourists wear.

packed, *ppl. a.*[1] **1.** Add to def.: Of meals, packaged for transporting and eating on a picnic or in an informal manner.
1958 B. Pym *Glass of Blessings* xvi. 186 Sitting in mackintoshes, eating packed lunches..tramping home again through the rain—one can see how he would pine after Portugal. 1959 'M. Innes' *Hare sitting Up* III. i. 167 He'll be back in no time. With a nice packed lunch. 1965 E. Salter *Once upon a Tombstone* II. xvii. 152 How about a trip up the glacier? We could take a packed lunch and make a day of it. 1968 *Guardian* 30 Mar. 10/5 You can be met at Benbecula by a hired car with a packed meal in it. 1973 G. Mitchell *Murder of Busy Lizzie* iv. 49 We're going to explore the island. Do you think, Father, that we could ask for a packed lunch? 1976 *Times Lit. Suppl.* 5 Nov. 1398/4 What an outing that would have been; you would have needed a packed lunch.
b. *Med.* Applied to blood cells separated as much as possible from plasma (usu. by centrifugation); esp. in *packed-cell volume*, the proportion of a sample of blood, by volume, occupied by cells after they have been allowed to settle; cf. *hæmatocrit* s.v. *HÆMATO-, HEMATO-.
1933 *Amer. Jrnl. Med. Sci.* CLXXXV. 59 Several students of the blood have again become interested in the determination of the volume of packed red cells by means of various types of hematocrit. 1943 *Lancet* 6 Nov. [illegible] The mean corpuscular hæmoglobin concentration, which is the ratio of the hæmoglobin content of the blood (grammes per 100 c.cm.) to the volume of the packed red cells (hæmatocrit). 1955 *Ibid.* 17 Dec. 1274/2 *Investigations.*—Blood-count: hæmoglobin 4·7 g. per 100 ml.; packed-cell volume 14%; mean corpuscular hæmoglobin concentration 33%; [etc.]. 1961 *Ibid.* 26 Aug. 490/1 A thirteen-year-old boy required a 250 ml. packed-cell transfusion every four to five weeks. 1967 S. Taylor et al. *Short Textbk. Surg.* i. 11 Packed-cell transfusions are viscid and slow to flow. 1974 *Jrnl. Appl. Physiol.* XXXVII. 976/1 These data were obtained under very carefully controlled conditions (e.g., careful measurement of packed-cell volume and resistivity..). 1976 *Nature* 1–8 Jan. [illegible] 0.25 ml of packed cells were added to 2 ml of a medium..
c. *Computers.* Of, pertaining to, or being a decimal number stored with successive digits represented by successive half-bytes and the sign by the rightmost half-byte.
1964 *IBM Jrnl. Res. & Devel.* VIII. 95 (*in figure*) Packed decimal number. 1966 R. Silverstone in I. Opler *Programming the IBM System/360* xi. 126 All internal decimal arithmetic operation[s] must be performed in packed format. *Ibid.*, Since packed numbers require half as much space as zoned decimal numbers, they should be used in storing or writing intermediate files which are not to be directly printed. 1973 Murrill & Smith *Introd. Computer Sci.* vi. 219 To obtain the packed representation from the zoned (or unpacked) representation, we need only to (a) remove all zone fields and (b) transpose the sign and digit fields in the rightmost byte... Systems such as the IBM 360 have hardware instructions to 'pack' and 'unpack' decimal numbers.

packer[1]. Add: **3. a.** Also *Canad.* and *N.Z.* Also in extended use.
1859 *Brit. Colonist* (Victoria, B.C.) 6 June 2/3 The arrival of over one hundred pack mules from the Chilliwack country, where they have been wintering, and offers by the packers to take freight to Lytton City for eighteen cents, has failed to revive trade. 1874 A. Bathgate *Colonial Experiences* x. 135 The rear [of string of pack horses was] brought up by the packer on horseback, his broad-brimmed wide-a-wake hat pulled well over his weather-beaten face. 1952 H. Innes *Campbell's Kingdom* 22, I was wakened with the news that the packer was in from Come Lucky and would be leaving after lunch. I was taken out and introduced to a great ox of a man who was loading groceries into an ex-army truck. 1958 G. Terry *Hist. & Legends of Chilcotin* 7 Tom Hutchinson was a packer and worked with a pack train of 300 mules between Yale and Barkerville. 1968 R. M. Patterson *Finlay's River* 101 When they travelled with horses they had, with him, a competent packer.
b. Also *Canad.* Also, a pack-dog.
1908 A. C. Laut *Conquest Gt. Northwest* II. 270 Getting two or three of the wise old bell-mares, that are in every string of packers, at the end of a long rope, the canoemen shot across the whirl of mid-stream and got footing on the opposite shore. 1944 J. Martin *Canad. Wilderness Trapping* 13 In spring and fall when it is impossible to haul your supplies, dogs come in handy as packers.
c. One who transports goods in a pack

his back; also, in more recent use, one who carries a rucksack containing all the necessities of travelling. Chiefly *Canad.*

1873 G. M. GRANT *Ocean to Ocean* 356 We could see that continuous labour for one or two years in solitary wilderness..as surveyor, transit-man..or even packer, is a totally different thing from taking a trip across the continent. **1892** E. S. BROOKES *Frontier Life* xiii. 117, I have often watched the packers, who would carry a load of seventy-five pounds on their backs, through a rough survey line for six or seven weeks. **1921** A. HEMING *Drama of Forests* 320 Upon the first his companion placed two more packs; then, stooping beneath the weight of 240 pounds, the packers at a jog-trot set off uphill and down, over rugged rocks and fallen timber. **1968** R. M. PATTERSON *Finlay's River* 29 He was a short, stocky man—the ideal build for a packer—and it was nothing for him to pack a two hundred-pound load over a long portage. **1974** *Weekend Mag.* (Montreal) 9 Mar. 20/3 It is something peculiar to the Spanish that they look on every packer as a hippie-freak—and they don't like hippie freaks.

4. (Further example.)

1902 *Census Bull.* (U.S.) No. 216. 28 June 61/1 Types succeed each other in the packer with 3-em space between the words, until a continuous line is formed.

5. (Examples.) For '*U.S.*' read 'orig. *U.S.*' and substitute for def.: A device inserted into an annular space in an oil well (such as that between the casing and the tubing) in order to block the flow of oil and gas.

1885 *Encycl. Brit.* XVIII. 718/1 An indiarubber packer is then attached in such a manner that within it the pipe that is above it slides in that which is below it, and the rubber is forced against the sides of the drill-hole. **1904** *Dialect Notes* II. 386 When a well has sufficient gas to flow its product through a two-inch pipe, but will not make its production through the casing, a packer is placed at or near the top of the sand to compel the gas or oil to relieve itself only through the tubing. **1922** D. T. DAY *Handbk. Petroleum Industry* I. 291 The fundamental principle of all types of packers embodies the vertical compression and lateral expansion of a resilient substance ..between casing or tubing and the wall of the hole, between two strings of casing, or between tubing and casing. **1960** C. GATLIN *Petroleum Engin.* xiii. 256/1 The ratio of hole diameter to unexpanded packer diameter is kept as low as possible, and commonly ranges from 1·1 to 1·2. **1973** J. W. JENNER in Hobson & Pohl *Mod. Petroleum Technol.* (ed. 4) iv. 141 In some areas where the wells are not easily accessible..a packer is run on the tubing and set just above the pay zone. Completion fluid of high enough density to kill the well is contained in the annulus above the packer and production is via the tubing. **1977** *Sunday Times* 24 Apr. 17/4 The plug is called a packer, and it blocks off the bottom of the outer casing.

packer². (Later example.)

1905 W. O'BRIEN *Recoll.* 295 Mr. Peter O'Brien.. afterwards earned the titles of Lord O'Brien of Kilfenora and..'Pether the Packer'.

packet, *sb.* Add: **1. f.** *slang* (chiefly *Mil.*). A bullet or other missile; hence, trouble, misfortune; *to stop* (or *cop*, etc.) *a packet*, to be killed or wounded; to get into trouble; to be reprimanded.

1917 P. MACGILL *Brown Brethren* xx. 284 Wot's she doin' standin' out in the street like that?.. She'll stop a packet if she's not careful. **1925** FRASER & GIBBONS *Soldier & Sailor Words* 219 *Packet*, a bullet wound, *e.g.* it would be said of a wounded man:—He 'stopped a packet' or 'bought a packet'—*i.e.*, got hit by a bullet. Also, any trouble or unexpected bad luck. **1933** D. L. SAYERS *Murder must Advertise* vii. 120 I'm really fearfully sorry you copped that packet that was meant for me. **1946** J. IRVING *Royal Navalese* 130 *Packet*, trouble, in some form or another. 'So-and-so caught his packet on the Russian convoys'... 'Smithie caught a packet from "The Bloke".' **1948** PARTRIDGE *Dict. Forces' Slang* 135 Blimey, old Bill didn't half cop a packet from the C.O. **1958** B. HAMILTON *Too Much of Water* xi. 236, I was a bit vague as to when Swete got his packet. **1960** 'H. CARMICHAEL' *Seeds of Hate* viii. 70 Frank Mitchell copped a packet on the river bank. **1978** A. PRICE '*44 Vintage* iii. 39 We've been disbanded... The same thing's happening to the 2nd Northants, they've caught a packet too.

g. A large sum of money. *slang.*

1922 M. ARLEN *Piracy* III. viii. 214 Tarlyan and Cypress had both won a packet at *chemin de fer*. **1928** WODEHOUSE in *Strand Mag.* Aug. 114/1 'Get in on the short end,' said Aurelia earnestly, 'and you'll make a packet.' **1930** W. S. MAUGHAM *Bread-Winner* II. 76 It cost me a packet. **1955** 'E. C. R. LORAC' *Ask Policeman* vii. 101 Lived in style for years and must have spent a packet. **1959** J. FLEMING *Miss Bones* xv. 171 I've cleaned up a nice little packet..but Walpurgis still owes me quite a bit. **1966** J. BETJEMAN *High & Low* 74, I bet your racket brings you in a pretty packet. **1972** P. D. JAMES *Unsuitable Job* iii. 80 That awful cross of roses... Poor old nanny, it must have cost her a packet.

h. *Physics.* A localized disturbance of a field or medium that retains its identity as it travels; usu. = *wave packet* (*WAVE *sb.* 10).

1928 *Proc. R. Soc.* A. CXVII. 278 We may imagine that by means of Heisenberg's γ-ray microscope we have detected an electron near a hydrogen nucleus in the form of a packet like (5·4) with x_0, etc., so adjusted that on the older quantum theory the particle would describe the *n*th circular quantum orbit. **1934** *Discovery* May 125/1 Photons (quanta or packets of electro-magnetic energy) are in general more efficient in bringing about atomic changes than particles of corresponding energy. **1956** A. A. TOWNSEND *Struct. Turbulent Shear Flow* v. 102 If this is true, a packet of turbulent fluid that has been moved across the flow by large eddy motion will have a turbulent ntensity determined by its rate of energy gain over a considerable part of its previous existence as turbulent fluid. **1970** *Nature* 29 Aug. 937/2 As long as the incoming packet is spatially sharp enough, the reflected packet will manifestly carry information about the scattering mass.

2. (Further examples.)

1852 J. R. PLANCHÉ *Invisible Prince* ii. 14 Fierce whiskered gents,..Smoked bad cigars, on board the penny packets. *a* **1936** KIPLING *Something of Myself* (1937) ii. 26 Turkey..turned up, usually a day or two late, by the Irish packet, aloof, inscrutable.

3. *packet-ship* (earlier examples), *-steamer*; *packet cigarettes, goods* (examples), *mix, soup, tea* (examples); **packet rat**, a derogatory name for a seaman, *spec.* one who specialized in the short voyage across the Atlantic.

1909 *Westm. Gaz.* 5 June 11/1 The demand for packet cigarettes..has given rise to the manufacture..of special brands. **1958** *Observer* 9 Feb. 5/3 Packet goods... Loose goods. **1977** 'J. FRASER' *Hearts Ease* ix. 103 Shelves of tinned and packet goods. **1968** D. E. ALLEN *Brit. Tastes* i. 35 Housewives in the South are fonder of all the speeded-up ways of cooking... Packet mixes are used with less compunction. **1894** STEVENSON & OSBOURNE *Ebb-Tide* II. ix. 172, I fought my way, third mate, round the Cape Horn with a push of packet-rats that would have turned the devil out of hell and shut the door on him. **1906** *Daily Chron.* 11 Aug. 4/6 It is almost as far a cry from the days of the Liverpool 'packet rat' as it is from the craft of to-day to the 'coffin ships' of Plimsoll memory. **1920** *Punch* 7 Apr. 266/1 An' the blessed lights o' Liverpool a-winkin' through the rain To welcome us poor packet-rats come back to port again. **1935** J. MASEFIELD *Victorious Troy* 68 'Who in hell said "Time, too?"' Cobb asked... 'Which of your damned packet-rats said "Time, too", then?' **1967** A. L. LLOYD *Folk Song in England* iv. 296 The packet-rats sailing under the house-flags of the Black Ball, Red Star, Dramatic and Swallowtail lines. **1782** R. MORRIS *Let.* 7 Oct. in J. Jay *Corr. & Public Papers* (1891) II. 349 Joshua Barney..[is] now commanding the Packet Ship *General Washington*. **1837** A. LANGTON *Jrnl.* in *Gentlewoman Upper Canada* (1950) 9, I should strongly recommend avoiding a crowded packet-ship..or perhaps a packet-ship at all. **1962** *Which?* Jan. 20/1 For convenience in shopping, the packet soups (the least heavy and bulky) are obviously better than the tinned soups. **1974** A. ROSS *Bradford Business* 125 We stoked up hurriedly on packet soup and woody pork chops. **1865** T. P. KETTELL *Hist. Great Rebellion* xx. 246 These two vessels had been packet-steamers, running to New York. **1883** E. EGGLESTON *Hoosier School-Boy* 115 The little packet-steamer was landing at the wharf. *c* **1870** in A. Davis *Package & Print* (1967) Pl. 192 (Advt.), Niblett's farm house bread stores...Agent for the celebrated packet tea. **1907** *Yesterday's Shopping* (1969) 1 Packet Teas packed on the estates in China and India. **1931** *N. & Q.* 14 Nov. 353/2 One of the earliest distributors of packet tea was the one-time old-fashioned firm of Horniman. **1976** *Times* 28 July 1/6 Packet tea and large, sliced loaves are among the eight foods for which manufacturers have voluntarily restricted prices.

packetarian (pækĕtēə·riăn). *U.S.* [f. PACKET *sb.* 2 + *-ARIAN.] One of the crew of a packet-boat.

1882 *Harper's Mag.* July 281/1 The typical 'Jack' of the pre-propeller age—the 'packetarian', and the able seaman of the clipper-ship fleet—has..utterly vanished. **1887** S. SAMUELS *From Forecastle to Cabin* 265 The 'packetarians' came last, and they invariably found themselves reduced to the same toggery in which they boarded the ship. **1930** R. CLEMENTS *Grey Seas* 110 No 'packetarians' these days, Mr. Findlay.

†**packeteer** (pækĕtīə·ɹ). *Obs.* except *Hist.* [f. PACKET *sb.* + -EER.] **1.** *Canad.* A carrier (often an Indian) of letters and documents, esp. in the fur trade.

1784 J. THOMAS in *Publ. Hudson's Bay Rec. Soc.* (1954) XVII. 27 Tradesmen at their separate employs, Hunters, Trappers, fishermen, and pacqueteers. **1943** *Beaver* Mar. 30/1 Mail packets were operated on a time-table, just as are mail flights by aeroplanes today. 'Packeteers' were never armed.

2. = *PACKETARIAN.

1922 *Short Stories* Feb. (early issue) 141/2 *Au revoir*, Joe Pichegru, you sun-smoked son of a packeteer!

packeter (pæ·kĕtəɹ). *Canad.* [f. PACKET *sb.* + -ER¹.] = *PACKETEER 1.

1893 J. HORDEN *Forty-Two Yrs. amongst Indians & Eskimo* xv. 147 The packeters returning from Abbitibbe with the letters..to Moose. *Ibid.* xvii. 169 The 'packeters' were espied crossing the river, in snow-shoes. **1961** J. W. ANDERSON *Fur Trader's Story* iii. 24 Others again would drive dog teams..while others would be 'packeters', hauling the mail..in winter.

pack-flat (equal stress), *a.* [f. PACK *v.*¹ + FLAT *a.*, *adv.*] Capable of being made into a flat package.

1951 *Good Housek. Home Encycl.* 107/2 Glass-top coffee table with pack-flat base. **1969** E. H. PINTO *Treen* 375 The 18th-century, 'pack-flat', mahogany wig stand..is exactly like modern travelling millinery stands. **1974** *Sunday Tel.* 6 Oct. 20/1 (Advt.), Sportsmans snug will keep you warm & dry in the worst weather! Full size 54″ × 30″ × 20″. Pack flat 30″ × 6″ × 2″.

packie (pæ·ki). *N.Z. colloq.* [f. PACK *sb.*¹ + -IE.] = *PACKMAN 2 b.

1945 J. D. PASCOE in *N.Z. Geographer* I. 20 Next on the list is the 'packie'—half-cook, half-handyman, always good with horses or mules—who takes blankets and provisions into 'camp'. **1947** P. NEWTON *Wayleggo* (1949) 14 The process is repeated, the 'packie' moving on again. **1963** *Weekly News* (Auckland) 8 May 39/1 The packie happened to come in that morning for more bread. **1972** P. NEWTON *Sheep Thief* iv. 31 An old packie by the name of Paddy Roper lost two of his team there.

packing, *vbl. sb.*¹ Add: **I. 1. c.** (Earlier example.)

1861 GEO. ELIOT *Let.* 10 Dec. (1954) III. 472 As I hope the Florentine hydropathist may not be a quack as Dr. Gully at Malvern certainly is, I shall be disappointed if there is no good effect to be traced to judicious 'packing' and sitz baths.

d. The transporting of goods on pack animals.

1843 *Amer. Pioneer* II. 162 Merchandise..was principally carried on pack horses until after 1788. Packing continued to be an important business in Kentucky until 1795. *Ibid.* 215 The grain would not bear packing across the mountains; a horse could not carry more than four bushels of it. **1897** *Boston Daily Globe* (evening ed.) 4 Aug. 5/2 Prices for packing across the pass have risen. **1948** *Hungry Horse News* (Columbia Falls, Montana) 24 Sept. 8/1 Roy owns a valuable string of pack horses and does considerable packing for the forest service.

e. An extra charge added to the cost of delivered goods to cover the cost of packing them.

1901 *Pitman's Business Terms, Phr. & Abbrev.* 155 *Packing*... The *charge* made for packing. **1974** *Parker's Wholesale Catal.* (J. Parker Dutch Bulbs..Co.) Autumn 13/2 All our prices are inclusive of duty and packing.

f. The spatial arrangement of the constituent atoms of a crystalline structure relative to one another.

1917 [see *HEXAGONAL *a.* 3]. **1945** C. W. BUNN *Chem. Crystallogr.* vii. 276 The mode of packing of atoms, ions, or molecules in crystals may be regarded as controlled by two principles—the principle of close packing.., and, where ions are concerned, the tendency for an electrically charged unit to surround itself with units of opposite charge. **1966** C. R. TOTTLE *Sci. Engin. Materials* iii. 54 The closeness of packing of atoms in a crystal lattice affects thermal and mechanical properties. **1973** H. D. MEGAW *Crystal Struct.* ii. 54 Packing of ions as rigid spheres determines the coordination number (the number of anions surrounding a cation).

II. 2. b. *slang.* Food, particularly if of inferior quality.

1891 J. BENT *Criminal Life* 272 Packing,..food. **1925** FRASER & GIBBONS *Soldier & Sailor Words* 219 *Packing*, rations. Food in general. **1973** 'P. MALLOCH' *Kickback* iv. 27 'When you've had the kind of packing I've had for three years, this is a treat.' He..began to eat.

III. 3. a. *packing-house* (examples), *-paper* (examples), *-plant, -room, -shed* (examples). **c. packing box** (earlier examples); also *attrib.* = *packing case*; **packing case**, also *attrib.* used disparagingly of a type of modern architecture alleged to resemble packing-cases in its regularity and monotony; **packing density** *Computers*, the density of stored information in terms of bits per unit of storage medium; **packing fraction** *Nuclear Physics*, 10,000 times $(M-A)/A$ (or $(M-A)/M$), where M is the atomic weight of a nucleus and A is its mass number; cf. *mass defect* s.v. *MASS *sb.*² 10 d; **packing needle** (later examples); **packing station**, *spec.* an official depot where eggs are graded and packed; also (with hyphen) *attrib.*

1774 in *Mass. Hist. Soc. Coll.* (1914) LXXI. 214 To a packing Box £0. 9. 4. **1800** JANE AUSTEN *Let.* 25 Oct. (1952) 77 The charge of 3s 6d for the Packing box. **1909** *Daily Chron.* 8 July 8/3 (*heading*) No 'packing box' houses. They are not turned out by the score or the hundred all to one pattern like packing cases. **1935** *Fortnightly* Apr. 410 So we are given the packing-case building—rectangular boxes with holes punched for doors and windows. **1961** *Times* 11 Apr. 4/2 Sir Jacob Epstein apparently intended a certain reproach to the 'packing-case' type of modern architecture in the bronze group, 'Pan'. **1958** *Wescon Convention Record of IRE* IV. 49/1 The first limitation in packing density is the number of pulses per inch that can be recorded on each track, which is limited by the basic resolution of the head and tape combination. **1967** MCLACHLAN & MOLSOM *Data Processing* xi. 171 The speed of reading and writing..will depend upon the physical speed of the tape past the read/write head, and the packing density of the information written. **1927** F. W. ASTON in *Proc. R. Soc.* A. CXV. 501 The mean gain or loss of mass per proton when the nuclear packing is changed from that of oxygen to that of the atom in question..will be called the 'packing fraction' of the atom and expressed in parts per 10,000. **1938** R. W. LAWSON tr. *Hevesy & Paneth's Man. Radioactivity* (ed. 2) xix. 179 The idea of the 'packing fraction' has been introduced, by which we understand the difference between the mass of the atom and the integral part of its mass-number, divided by the mass-number. **1949** FRIEDLANDER & KENNEDY *Introd. Radiochem.* ii. 38 The mass defect Δ is the difference between the atomic mass M and the mass number A: $\Delta = M - A$... The packing fraction f is the mass defect divided by the mass number: $f = \Delta/A$. (Sometimes f is defined as Δ/M; the difference is negligible.) **1955** A. E. S. GREEN *Nuclear Physics* ii. 55 Packing fractions are positive (0·6 to 0 mMU) for the stable nuclei from 1 to 20, negative (0 to −0·8 to 0 mMU for nuclei from 20 to 170, and positive again (0 to 0·6 mMU) for the very heavy nuclei. **1968** G. M. MOSSOP *Advanced Level Atomic Physics* ix. 150 The packing fraction is the mass defect per nucleon. **1834** C. F. HOFFMAN *Winter in West* (1835) II. xxxii. 136 One of the packing-houses, built of brick, and three stories high, is more than a hundred feet long, and proportionably wide. **1901** *Chambers's Jrnl.* Mar. 208/1 Two of the largest packing houses had in their cold-storage chambers no fewer than two hundred and sixteen million eggs. **1968** *Globe & Mail* (Toronto) 5 Feb. 5/4 He was leading the campaign to organize packinghouse workers. **1977** *Time* 22 Aug. 43/1 By 14 he had quit school and started work as a janitor and a packing-house laborer. **1880** L. HIGGIN *Handbk. Embroidery* iv. 34 It should now be braced with twine by means of a packing needle. **1937** A. M. MIALL *Making Home Furnishings* x. 161 Thread your long curved packing needle with string, and with a few large stitches through the hessian secure the tops of the springs to it. **1861** D. G. ROSSETTI *Let.* Jan. (1965) II. 392, I shall have it printed on common brown packing-paper. **1939** *Army & Navy Stores Catal.* 305/2 Packing paper and cloth for export parcels. **1921** *Daily Colonist* (Victoria, B.C.) 17 Mar. 2/5 Representatives of ten national packing plant unions today pledged their support to the Amalgamated Order of Meat Cutters. *Ibid.* 9 Oct. 31/5 Fire, which apparently originated in a smokehouse last night, destroyed the packing plant of the H. F. Lewis Company. **1854** *Harper's Mag.* Mar. 456/1 The 'packing-room' is the loft of the gin-house. **1900** H. LAWSON *On Track* 94 One day I went downstairs to the packing-room and saw a lot of phosphorus in jars of water. **1960** J. BETJEMAN *Summoned by Bells* ii. 12 Bang through the packing-room! **1901** *Chambers's Jrnl.* Feb. 99/1 An expert to accompany the fruit from the orchard, through the packing-shed, on to the port of shipment. **1946** R. TENNANT *Lost Haven* (1947) vii. 100 Jack Starbrace had fallen over backwards into the packing shed. **1930** E. BROWN *Brit. Poultry Husbandry* 347 Packing stations. **1938** L. PEARCE-GERVIS *Compl. Poultry Keeper & Farmer* v. 150 Each grade has its own particular colour..and contains the registered number of the packing station. **1960** *Farmer & Stock-breeder* 1 Mar. 55/3 Many of us..are not at all happy about current packing-station prices.

packing, *ppl. a.* Add: **2.** As the second element in adj. combs.: habitually carrying, esp. of a weapon, as *pistol-packing*, etc. (See *PACK *v.*¹ 9 a.)

1936 E. AMBLER *Dark Frontier* xi. 174 How do you suppose we're going to stop a mob of eight dagger-packing Greeks? **1943**, etc. [see *pistol-packing* ppl. adj. s.v. *PISTOL *sb.* 2]. **1959** *Times Lit. Suppl.* 13 Nov. 664/4 He is forced into carrying a Luger-packing German entomologist on a hunt for a rare and mysterious beetle.

packman. Add: **2. a.** *N. Amer.* One who transports goods by means of pack-animals or in a pack on his own back.

1828 in *Kansas Hist. Q.* (1936) Aug. 251, I & the two pack-men returned to the creek with six horses and all the baggage. **1847** *Ex. Doc. 31st U.S. Congress 1 Sess. House* (1849) No. 5. III. 627 Two of Judge Burt's packmen arrived today for provisions to take to the parties south. **1908** W. R. NURSEY *Story Isaac Brock* vi. 45 Brock would watch these packmen as, thus handicapped with a load weighing from two to five hundred pounds, they set out across the rough portage.

b. *N.Z.* A sheep-station handyman whose principal duties are conveying goods by pack-animal from camp to camp and cooking; hence also **packman-cook.**

1933 L. G. D. ACLAND in *Press* (Christchurch, N.Z.) 11 Nov. 15/7 *Packer, packman*, one who loads the packhorses and leads or drives them from camp to camp. He also cooks for the musterers. **1961** B. CRUMP *Hang on a Minute* 76 The packman-cook [on the sheep-station] was a muttering-to-himself old man called Joe. **1963** *Weekly News* (Auckland) 31 July 37/1 Packman-cook can be a tough job in some places. **1972** P. NEWTON *Sheep Thief* vii. 51 'Come and get it or I'll chuck it out.' The packman's rude yell rang through the hut and all hands stirred.

packsaddle, *v.* [f. the sb.] *trans.* To convey on a packsaddle.

1912 *Red Mag.* Mar. 508/1 They had a burro on another ledge of the estate, which packsaddled things in from where the stage dropped them.

pact, *v.* For *Obs.* read *rare* and add later example of sense b.

1940 *Economist* 6 Jan. 10/2 To-day Germany has pacted with Communism, Italy is neutral, the West fights Germany.

‖**pacta sunt servanda** (pæ·kta sunt sɜ̄ɪvæ·nda), *phr.* [L., lit. 'agreements must be kept': cf. Cicero *De Officiis* III. xcii pacta et promissa semperne servanda sint; *Digesta Iustiniani* II. xiv ideo servandum erit pactum conventum.] The principle, esp. in international law, that agreements are binding and inviolable.

1855 R. PHILLIMORE *Commentaries upon Internat. Law* II. v. vi. 56 *Pacta sunt servanda* is the pervading maxim of International, as it was of Roman jurisprudence. **1925** E. SATOW in *Cambr. Hist. Jrnl.* I. 295 (*heading*) *Pacta sunt servanda* or International Guarantee. **1939** E. H. CARR *Twenty Years' Crisis* xi. 232 War writers..have attempted to treat the rule *pacta sunt servanda* not merely as a fundamental rule of international law, but as the cornerstone of international society. **1945** J. L. KUNZ in *Amer. Jrnl. Internat. Law* XXXIX. 197 *Pacta sunt servanda* means the institution, by general international law, of a special procedure—the treaty procedure—for the creation of international norms... Valid treaty norms must be kept, but they can, by appropriate procedures, be revised. *Pacta sunt servanda* means the inviolability, not the unchangeability, of treaties. **1958** *Reports of Judgments* (Internat. Court of Justice) 121 The maxim *pacta sunt servanda* is of special significance in considering this contention of the Government of Sweden. **1962** *Times* 27 June 8/6 The ordinary rule was that contracts were to be enforced—*pacta sunt servanda*. **1973** I. M. SINCLAIR *Vienna Convention on Law of Treaties* iii. 53 Article 26 of the Convention reproduces, in lapidary language, the basic principle *pacta sunt servanda*, designated by the Commission as 'the fundamental principle of the law of treaties'.

paction, *sb.* (Further examples.)

1856 BOUVIER *Law Dict.* (ed. 6) II. 277 *Pactions*. International Law. When contracts between nations are to be performed by a single act, and their execution is at an end at once, they are not called treaties, but agreements, conventions or pactions. **1883** *Wharton's Law Lexicon* (ed. 7) 586/1 *Paction*, a bargain or covenant. **1964** *Mod. Law Rev.* XXVII. 314 In *Learmonth* v. *Sinclair's Trustees*..Moncrieff L.J.-C. said: ..Now custom, whether general or local, when it is effectual operates not by implied paction or contract but by law.

pacu. Also **paku.** Substitute for def.: A large, vegetarian, freshwater fish, *Colossoma nigripinnis*, belonging to the family Characidæ and native to the northern parts of South America. (Later examples.)

1938 A. H. VERRILL *Strange Fish & their Stories* 218/1 Pacu. **1959** P. CAPON *Amongst those Missing* 181 Not too bad, I guess. A paku, and two of those fish sort of like bream. **1962** D. W. TUCKER tr. *Sterba's Freshwater Fishes of World* 112 (*caption*) The Pacu is a rarely-imported species from northern South America. **1974** H. MACINNES *Climb to Lost World* vi. 86 In the river there was plenty of Pakuweed, so called by the Indians because the Paku fish is supposed to feed on it. **1977** D. J. COFFEY *Encycl. Aquarium Fish* 89/1 *Colossoma nigripinnis*. Pacu. 71 cm. (28 in.) This fish comes from the Amazon... Body colour is silver. The dorsal and anal fins are red in colour.

pacy, var. *PACEY *a.*

pad, *sb.*² Add: **1. b.** *Austral. spec.* A track made by bullocks, cattle, camels, etc. Cf. *cattle-pad* s.v. *CATTLE 9.

1911 C. E. W. BEAN '*Dreadnought*' *of Darling* xxiii. 207 The white track was the pad made by the feet of many camels. **1934** A. RUSSELL *Tramp-Royal in Wild Australia* ii. 20 The pad was winding and rocky and scarcely discernible. One sometimes wondered if it really was a pad, so little did it look like one. **1941** I. L. IDRIESS *Great Boomerang* ix. 67 You'll see the pad leadin' back to the horse paddock. **1954** B. MILES *Stars my Blanket* xix. 138 In places we were able to follow a wandering bullock-pad. **1966** 'J. HACKSTON' *Father clears Out* 19 We almost wore a series of pads through the bush trailing it. **1968** K. WEATHERLY *Roo Shooter* 37 They strode down the cattle pad at a fast walk.

pad, *sb.*³ Add: **I. 1. b.** A bed; hence, a lodging, a place to sleep; one's residence. Also, a room frequented by narcotic (esp. marijuana) users. *slang* (orig. *U.S.*).

1718 C. HITCHING *Regulator* 19 The names of the Flash Words now in vogue amongst Thieves... The Padd, *alias* Bed [etc.]. **1846** *Swell's Night Guide* 67 The only question she asks is, 'vot pad do you vont?' **1914** JACKSON & HELLYER *Vocab. Criminal Slang* 64 Pad..., a bed; a place to sleep. **1938** *New Yorker* 12 Mar. 36/3 Pads where semi-conscious smokers are robbed of their money are creeper joints. **1956** 'E. MCBAIN' *Cop Hater* (1958) ix. 79 'If Ordiz is a junkie, what's he doing on Whore Street?' 'He's blind in some broad's pad.' **1959** [see *beatnik* s.v. *BEAT GENERATION]. **1959** N. MAILER *Advts. for Myself* (1961) 346, I went with my wife and my friend..to a cold-water pad, south of the village. **1961** *Spectator* 25 Aug. 266 In this half world of cats, pads and hipsters there is a residual morality. **1965** 'MALCOLM X' *Autobiogr.* 57 Cats' pads, where with the lights and the juke down mellow, everybody blew gage and juiced back and jumped. **1967** *Boston Sunday Herald* 26 Mar. IV. 1/3 'Four out of five times when we go into a pad where we have been told there is pot, we find it', says the detective. **1967** N. LUCAS *C.I.D.* x. 134 A 'pad' is a bed—in a flat, a house, a bed-sitter or even in a shack. **1973** 'D. SHANNON' *No Holiday for Crime* (1974) vi. 89 She's got a pad over on Nadeau Street. **1974** K. MILLETT *Flying* I. 26 The usual university ghetto pad, an old house gone hip. **1977** *Time* 17 Jan. 8/3 They later searched the apartments of several employees, as well as Starckmann's swank pad in Neuilly.

c. A padded cell: cf. PADDED *ppl. a.*²

1938 S. BECKETT *Murphy* 167 The padded cells, known to the wittier as the 'quiet rooms', 'rubber rooms' or, in a notable clip 'pads'. **1964** G. L. COHEN *What's Wrong with Hospitals?* vii. 147 The side-rooms are in fact 'pads' remaining from the pre-tranquillizer regime. **1965** *New Statesman* 23 July 119/2 Quondam padded cells, now pink and chintzy, are given to old reliables who could leave hospital if they had anywhere to go. 'Yes, we do keep two 'pads' for isolation purposes,' barked the burly N.C.O.

3. e. A strip of rubber (etc.) material fitted in the road which when depressed by traffic operates road signals. Cf. *detector-pad.

1933 H. WATSON *Street Traffic Flow* ix. 165 Electric contact pads or strips, called 'detectors' laid in the carriageway, and actuated by vehicles. **1935** *Times* 3 Dec. 11/4 The installation of pedestrian control signals, to be operated by traffic pads. *Ibid.*, The Hendon Council, however, did not consider that the provision of pad-operated pedestrian signals would serve the desired purpose on such a road. **1960** H. MANZONI in E. Davies *Roads* vii. 176 The signals are actuated by vehicles passing over a detector pad consisting of two hollow rubber treads.

f. = *launching pad* s.v. *LAUNCHING *vbl. sb.* b.

1949 *Gloss. Guided Missile Terms* (GM 51/8) (Res. & Devel. Board, U.S. Dept. Defense) 75 *Pad*, a permanent or semipermanent base constructed to support a missile-launching device. **1953** *Air Univ. Q. Rev.* Fall 32 (*caption*) To withstand the pressures and intense heat of the exhaust blast during take-off, the 100-foot-square pad must be two-and-one-half feet thick. **1958** *Times* 1 Mar. 6/3 The missile had been on its pad for days before firing. **1964** MRS. L. B. JOHNSON *White House Diary* 24 Mar. (1970) 101 Saturn I will be used to send an unmanned Apollo spacecraft into orbit—in fact, one is on the pad..now for launching later this spring. **1971** *Sci. Amer.* Sept. 229/3 In 1961..the Russians had just 14 big liquid-fuel SS-6's on open pads. **1973** *Times* 17 Oct. 1/4 Our Egyptian rockets ..are now on their pads ready to be launched by the single order to press a button.

g. A take-off or landing point for a helicopter.

1960 *Washington Post* 1 Dec. D13/1 The mushrooming uses to which the 100-by-100 foot concrete helicopter 'pad' at the Pentagon is being put today. **1972** L. HANCOCK *There's a Seal in my Sleeping Bag* ix. 238 We were walking along the boardwalk past the helicopter pad. **1974** *BP Shield Internat.* Oct. 18/2 With a slight bump and shudder, we landed on the helicopter pad. **1976** 'L. BLACK' *Healthy Way to Die* ii. 17 The circling helicopter was descending towards the pad. **1977** *Time* 31 Jan. 15/1 On the drive from a helicopter pad to his office, his swift-moving convoy was guarded by three select commando battalions.

h. = *cow-pad* (*COW *sb.*¹ 8).

1971 *New Scientist* 1 July 36/2 The dung is also the incubation medium of many helminth parasites of stock, the eggs of which are passed in the pads. **1973** *Nature* 30 Nov. 271/1 Such dung pads soon dry into a hard cake. **1974** *Sci. Amer.* Apr. 101/2 On the average 12 dung pads are dropped by a single adult bovine every day. **1976** *Australasian Express* 3 Sept. 2/1 Dung pads are being eaten at a rapid rate in the northern half of Australia.

III. 13*. A padding machine. Freq. *attrib.*

1935 *Textile World* LXXXV. 1860/2 Formation of the patches is a danger signal, however, and—as is true of all danger signals in the pad dyehouse—should be taken heed of at once if serious trouble is to be avoided. **1951** *Jrnl. Soc. Dyers & Colourists* LXVII. 508/1 In dyeing practice, the nearest approach to printing technique was the so-called pigment-padding process, where the vat dye was applied on the pad as a suspension of unreduced dye. **1955** *Ibid.* LXXI. 896/1 Difficulties may arise from this increase [in width], e.g. from wrinkles passing the pad nip. *Ibid.* 900/1 Does the preferential uptake of water by the dry cloth entering the pad box cause a slight change in the concentration of direct dye and give a length of cloth off shade at the beginning of a run? **1961** COCKETT & HILTON *Dyeing of Cellulosic Fibres* xi. 358 The essential parts of a pad are a nip of two or more rollers, a trough, and means of applying pressure to the nip. **1966** *Encycl. Polymer Sci. & Technol.* V. 238 The machine consists of two or three squeeze rolls mounted over a shallow trough, or 'pad box', provided with guide rolls for the cloth.

13.** *Electr.* A resistance network inserted into a transmission line to attenuate all frequencies equally by a known amount.

1931 *Electronics* Feb. 508/1 The term 'pad' as commonly employed in connection with audio frequency circuits, refers to an attenuation device used to reduce the power at a point in a circuit by some desired value... Regarded as an electric circuit, a pad consists of a one-section artificial line whose elements are pure resistances. **1951** W. J. CREAMER *Communication Networks & Lines* v. 46 Non-symmetrical pads may be employed for the purpose of matching impedances, but there will be a minimum loss below which it is not possible to go without getting into the difficulty of a negative resistance element. **1959** K. HENNEY *Radio Engin. Handbk.* (ed. 5) xxi. 22 The minimum attenuation setting of a ladder pad normally corresponds to its insertion loss, which amounts to approximately 2·5 db. **1967** D. H. HAMSHER *Communication Syst. Engin. Handbk.* vi. 31 Two-wire trunk circuits may contain a switchable 2-db pad.

13*.** *U.S. slang.* (See quots.)

1970 *Daily Tel.* 27 Apr. 3 New York police have their own secret slang to deal with illegal business... A 'pad' is an establishment that provides police with regular payoffs. **1971** *N.Y. Times* 19 Oct. 47 The gamblers of the city paid off the policemen on a regular monthly basis after they had been placed on what is called 'the pad'. **1971** *Guardian* 28 Oct. 13/6 [He] was thrilled with becoming a plainclothesman because..'he was now on the pad'. The pad is the regular sum paid to officers for ignoring illegal activities. **1971** *Time* 1 Nov. 23 When a cop was transferred to a new post, the pad from his old station kept up for another two months. **1973** M. TRUMAN *Harry S. Truman* iii. 72 In Kansas City there was a tradition of carrying one or two thousand city employees 'on the pad' without requiring them to show up for work.

IV. 14. *pad-foot* (examples), *-mark* (later examples); **pad eye** *Engin.* (see quot. 1909); **pad mangle**, a padding machine; **pad money** *U.S. slang* (see quots.); **pad-play** (examples); **pad room** *U.S.*, in a theatre, a waiting-room for performers; **pad-steam**, used *attrib.* and *absol.* to denote a process in which fabric is first padded and then steamed; **pad stitch** (see quot. 1968); also (with hyphen) as *vb.*; so **pad stitching; padstone** (see quots.).

1909 *Cent. Dict. Suppl.*, *Pad-eye*, in ship-building, a flat rectangular piece of metal with an eye or ring projecting edgewise from its surface, the whole forming one solid piece. It is attached to the surface on which it is placed by screws or rivets through the flat part. **1972** L. M. HARRIS *Introd. Deepwater Floating Drilling Operations* vi. 76 The principal use of buoys in floating-drilling operations is to mark anchors... The unit should have strong padeyes on top and bottom for attaching handling lines and pendants. **1976** *Offshore Platforms & Pipelining* 239/3 For added strength, cables are run from each corner of the frame to the existing pad eyes on the mass anchor. **1905** F. S. ROBINSON *Eng. Furnit.* xii. 181 The legs of these tables are somewhat too straight to be classed as cabriole, and have pad feet. **1955** R. FASTNEDGE *Eng. Furnit. Styles* 286 *Pad foot*, resembling the club foot but set on a disk. **1974** *Country Life* 7 Mar. (Suppl.) 40 A fine quality George II red walnut Armchair in original condition having cabriole legs and pad feet. **1955** *Jrnl. Soc. Dyers & Colourists* LXXI. 777/2 On leaving the pad-mangle, the goods should pass directly into a Mather & Platt ager. **1966** R. C. CHEETHAM *Dyeing Fibre Blends* i. 59 Piece-dyeing is usually carried out on a pad-mangle, with the least possible duration in the wet state. **1926** T. E. LAWRENCE *Seven Pillars* lxvi. 344 We..marched south..seeing tracks of gazelle..with, in one spot, stale padmarks of leopard. **1973** *Times* 14 Aug. 14/5 Only pad marks of mink were found. **1904** 'No. 1500' *Life in Sing Sing* 256/2 Pad money, money for lodgings. **1927** *Dial. Notes* V. 457 *Pad money*.., money for a night's lodging or for admission into an opium den. **1920** E. R. WILSON in P. F. Warner *Cricket* 94 [They] used to bowl round the wicket in order to get batsmen l.b.w. with their off break, which, if bowled over the wicket, permitted 'pad play'. **1956** N. CARDUS *Close of Play* 15 Hobbs and Sutcliffe twice frustrated an Australian attack on 'sticky' pitches—on one of the most vicious of all, at Melbourne, largely by pad-play. **1960** J. FINGLETON *Four Chukkas to Australia* 39 His reliance on so much pad-play had ruined his stroke-play. **1927** K. NICHOLSON *Barker* 149 *Pad room*, waiting room for performers. **1931** *Amer. Mercury* Nov. 353/2 *Pad room*, a dressing tent. **1945** *Textile World* Jan. 84 (*heading*) Pad-steam dyeing reaches volume production. **1961** COCKETT & HILTON *Dyeing of Cellulosic Fibres* xi. 361 The several forms of pad-steam..give greater production because of their continuous character. **1966** *Encycl. Polymer Sci. & Technol.* V. 240 The Du Pont pad-steam process employs steam under controlled conditions to bring about dye fixation. **1924** W. D. F. VINCENT et al. *Cutters' Pract. Guide Cutting & Making Body Coats* 30/3 Whether or not you pad-stitch the collar, or stitch together by machine, give it firmness as well as shape. **1964** *McCall's Sewing* ix. 126/2 *Pad stitch*, this stitch is very similar to diagonal basting stitch. **1968** J. IRONSIDE *Fashion Alphabet* 84 *Pad-stitch*, 1. In Tailoring: A diagonal basting stitch used to hold interlining or canvas to fabric. 2. In Embroidery: Filling stitches over which fancy stitches are worked; used to give bulk. **1964** *McCall's Sewing* xiv. 254/1 End pad-stitching at seam line. **1963** *Gloss. Gen. Building Terms* (*B.S.I.*) 19 *Padstone*, a block of stone or concrete built into a wall to distribute the pressure from a concentrated load. **1964** J. S. SCOTT *Dict. Building* 222 *Padstone*, a stone or concrete pad in a wall.

pad, *v.*¹ Add: **1. c.** (Later examples.)

1904 [see *BOOT *sb.*² 1 c]. **1916** J. B. COOPER *Coo-oo-ee* x. 130 A dog with you breaks the lonesomeness of the bush, I know. We've padded the hoof together. **1966** 'J. HACKSTON' *Father clears Out* 108 When the people..missed the coach, and had to pad the hoof into the town.., they'd take the short cut. **1970** *N.Z. Listener* 21 Dec. 8/4 We pondered on the day years ago when we were padding the hoof ourselves.

2. b. Also of a person, and reduplicated *pad-pad.*

1899 C. J. C. HYNE *Further Adventures Capt. Kettle* v. 84 Naked feet pad-padded quickly up over the dust and grass. **1926** A. BENNETT *Lord Raingo* I. ii. 7 A nice thing, that with five servants in the place, and him a millionaire, he should be reduced to padding about in his socks! **1966** 'J. HACKSTON' *Father clears Out* 84 Father was changing as we padded along. **1975** M. BRADBURY *History Man* vii. 126 Felicity pads at Howard's side down the long bright passage.

pad, *v.*² Add: **2. b.** To extend or increase (an official list, expense account, claim for payment, etc.) with unauthorized or fraudulent items.

1913 *Maclean's Mag.* Mar. 104 (*heading*) Padding the expense account. *Ibid.* 105/2 To pad this account by magnifying the cost of hotel accommodation, meals and railway fare, was most distasteful to him. **1921** *Daily Colonist* (Victoria, B.C.) 23 Oct. 1/2 John W. Duncan, charged with padding a cheque.., was found not guilty. **1928** *Observer* 15 Apr. 12 They claim that the list of members..was heavily 'padded' by the inclusion of persons without their knowledge and consent. **1967** *Boston Sunday Herald* 26 Mar. IV. 3/1 Beating the system can be done by padding bills, adding a dollar here, two dollars there, for medical items, clothing and so forth. **1968** *Economist* 14 Dec. 27/2 A report by a Granada television team in Britain showed that the [voting] lists were padded. **1976** 'M. ALBRAND' *Taste of Terror* xi. 67 Forster never paid but preferred to be sent a bill in spite of the fact that he knew it was padded.

3. Substitute for def.: To impregnate *with* a liquid or paste by squeezing between rollers, the substance applied being either on one of the rollers or in a bath preceding them. (Further examples.)

1897 C. T. DAVIS *Manuf. Leather* (ed. 2) 324 They first pad the leather with a solution of alizarine rendered slightly alkaline with ammonia. **1927** C. E. MULLIN *Acetate Silk* (1928) xxx. 375 In applying the developed colors on cotton, the fiber is usually first padded in the naphthol bath and then the color developed in a second bath. **1933** *Chem. Abstr.* XXVII. 2045 Fabrics having a cellulose acetate pile are printed or padded with a soln. of regenerated cellulose in caustic alkali to obtain a local or over-all deposition of cellulose. **1972** D. HILDEBRAND in K. Venkataraman *Chem. Synthetic Dyes* VI. iii. 431 The fabric which has been padded with dyestuff and anti-crease agent is stored in the presence of mineral acid as catalyst for 1–3 hours.

pada (pā·dā). Also **padam.** [Skr., = foot.] **a.** An Indian lyrical poem set to music. Also *attrib.*

1880 F. S. GROWSE *Mathurā* (ed. 2) 212 He is said to have written a Hindi poem called the Dasratna, together with a few short *Sākhis* and *Padas* in the same language. **1898** B. A. PINGLE *Indian Mus.* (ed. 2) ii. 64 The Dhrupada is a later form of the original Padas or Bhajanas (prayer or praise). **1914** [see *BHAJAN]. **1957** O. GOSVAMI *Story Indian Mus.* xx. 212 The *Padam* is sung in the same way as the *Kriti*. **1967** SINGHA & MASSEY *Indian Dances* ii. 43 Padas are love lyrics which cover every conceivable aspect of love from the mystic and divine to the earthly and profane. **1968** *Jrnl. Mus. Acad. Madras* XXXIX. 102 Annamacharya says that the songs of one ignorant of the proper style and pattern of *pada*-composition..would be worthless. **1972** P. HOLROYDE *Indian Mus.* 275 Padams demand a very dramatized technique of mime gesture and facial expression. **1974** B. C. DEVA *Indian Mus.* v. 76 The songs of Bhadrachala Ramadas..are referred to as pada-s.

b. A group of words forming a section of Sanskrit verse; a mode of reciting this.

1887 M. MONIER-WILLIAMS *Brāhmanism & Hindūism* (ed. 3) xv. 409 These Vedic texts may be recited according to any one or more of the five different Pāthas, or modes of recitation, called Samhitā, Pada, Krama, Jaṭā, and Ghana. **1916** A. A. MACDONELL *Vedic Gram. for Students* 449 A hemistich of two or more Pādas is treated as a unit. **1917** —— *Vedic Reader for Students* p. xiii, The earliest expedient of this kind was the formation of the Pada or 'word' text, in which all the words of the Saṃhitā text are separated and given in their original form as unaffected by the rules of Sandhi. **1965** *Language* XLI. 11 Three instances are accounted for by the pāda *devān devayatē yaja*. **1971** *Language* XLVII. 59 The first three pādas are hyposyllabic as transmitted.

padanda, var. *PEDANDA.

‖**padang** (pa·daŋ). [Malay.] An open grassy space; a field, esp. a playing-field; also, scrub vegetation.

1915 *Blackw. Mag.* Dec. 793 Two others were caught near the Europe Hotel, sitting on the 'padang' by the sea. **1927** R. J. H. SIDNEY *In Brit. Malaya To-Day* 45 We are standing on my large verandah overlooking the school *padang*. **1933** L. AINSWORTH *Confessions Planter in Malaya* 74 They had been given leave to gather on our Padang (a large field in the centre of all our coolie lines). *Ibid.* 182 The clubhouse is set in very pleasant surroundings, with a large padang, or playing-field, attached. **1952** P. W. RICHARDS *Trop. Rain Forest* ix. 211 The vegetation was the so-called padang, a kind of heath-like scrub or poor forest. **1964** K. G. TREGONNING *Hist. Mod. Malaya* x. 226 E. W. Birch..was one of those whose delight it was to see a *padang*, a shady playing field, laid out in every kampong of the State. **1972** *Malay Mail* (Kuala Lumpur) 25 May 4/5 The pasar malam..will be held at Dato Kramat padang from Wednesday. **1977** *Borneo Bull.* 7 May 4-A/2 Red Cross Day will be celebrated here tomorrow (Sunday) with a parade on the town padang followed by a ceremony in the Youth Centre.

padauk, var. PADOUK (in Dict. and Suppl.).

padded, *ppl. a.*² For 'treated with a mordant in calico-printing' substitute: Impregnated throughout with a dye or the like by padding (see *PAD *v.*² 3). (Further examples.) *padded cell* (examples); also *fig.* and *transf.*; also (with hyphen) *attrib.*; *padded room* (see quot. 1976); *padded shoulders*, the shoulders of a suit, etc., padded to give the appearance of breadth; also *fig.*; *padded soap.*

1891 S. P. SADTLER *Handbk. Industr. Org. Chem.* ii. 61 Soaps made in this way retain all the glycerine..and belong to the class known as 'filled' or 'padded' soaps. **1935** H. G. WELLS *Things to Come* 14 The fact that in the future various light apparatus such as a portable radio, electric torch, notebook, will have to be carried on the person and that this will probably necessitate a widening of those broadly padded shoulders which are already necessary in the costume of contemporary men because of their wallets and fountain pens. *a* **1936** KIPLING *Something of Myself* (1937) vii. 176 A 'comfortable nursery' proved to be a dark padded cell at the end of a discreet passage! **1959** N. N. HOLLAND *First Mod. Comedies* 39 'Dapperwit' implies a comparison of clothes and wit, as though Dapperwit's pretensions to wit were a kind of padded shoulders to cover his actually feeble intelligence. **1961** COCKETT & HILTON *Dyeing of Cellulosic Fibres* xi. 361 Where vat dyes are concerned, development of the padded goods with alkaline reducing agent may be carried out discontinuously on jigs. **1962** A. NISBETT *Technique Sound Studio* ii. 48 Of the various possible sound-deadening systems, it is best to try to avoid those which give a padded-cell effect. **1963** WODEHOUSE *Stiff Upper Lip, Jeeves* xxii. 167 Stiffy, who is pure padded cell from the foundations up, was planning to marry the Rev. H. P. Pinker. **1972** D. HILDEBRAND in K. Venkataraman *Chem. Synthetic Dyes* VI. iii. 424 If the substantivity of the dye is insufficient, the dye will migrate when the padded goods are dried under inhomogeneous drying conditions. **1973** *Jewish Chron.* 19 Jan. 14/5 To give readers some idea of what passes for dialogue in the padded cell of the drama department, here is the kind of exchange in which Mr Mackie specialises. **1976** *Amer. Speech* 1973 XLVIII. 207 An uncontrollable patient may be sent to the *quiet room* or *QR*, formerly called the *padded room*. **1977** *Time* 3 Jan. 56/3 You will recall that the last we saw of Inspector Clouseau he had succeeded in..driving his immediate superior, Chief Inspector Dreyfus, completely, totally, padded-cell mad.

padder, *sb.*² Add: **2.** A padding machine.

1927 C. E. MULLIN *Acetate Silk* (1928) xxviii. 361 It is usually best to dye the acetate silk first and then cross-dye the cotton either in the padder or jig. **1955** *Jrnl. Soc. Dyers & Colourists* LXXI. 894/2 In order to get as good impregnation as possible before the fabric reaches the nip of the padder, this small volume is combined with a comparatively long path through the padder trough. **1971** E. I. VALKO in H. Mark et al. *Chem. Aftertreatment of Textiles* iii. 105 Impregnation of fabric is frequently carried out by first immersing the fabric and then pulling it between the rolls of a padder.

3. *Electronics.* Also **padder capacitor, condenser.** A usu. adjustable capacitor connected in series in a tuned circuit in order to improve the tracking with another tuned circuit at low frequencies when the tuning of the two circuits is ganged (as in a superheterodyne). Cf. *TRIMMER.

1936 J. H. REYNER *Testing Radio Sets* (ed. 3) xiii. 190 (*in figure*) Padder. **1939** H. J. HICKS *Princ. & Pract. Radio Servicing* ii. 21 C10, C11, C12 and C13 are the trimmer and padder condensers employed in the oscillator circuit to make it track with the r.f. tuning condenser. **1950** A. MARCUS *Radio Servicing* xi. 453 In addition to the series padder, a small trimmer capacitor is usually connected in parallel with the main tuning capacitor. **1971** A. MARGOLIS *Mod. Radio Repair Techniques* vi. 82 To maintain the tracking difference, a smaller oscillator coil and padder capacitor is added to the circuit.

padding, *vbl. sb.*² Add: **1.** (Further examples, corresponding to *PAD *v.*² 2 b and PAD *v.*² 3 in Dict. and Suppl.)

1954 *Jrnl. Soc. Dyers & Colourists* LXX. 383/2 The polymer emulsion..is then suitably diluted for application to the fabric, and applied by padding and drying. **1964** 'E. LATHEN' *Accounting for Murder* (1965) x. 94 It would take more than a little juggling of expense accounts to explain the situation... Somebody must have gone into a panic about some minor padding. **1973** *Times* 31 July 1/8 Several delegations have submitted a variety of inflated expense statements ranging from high living to outright padding of the bills.

attrib. **1912** J. HÜBNER *Bleaching & Dyeing Veg. Fibrous Materials* xvi. 371 The padding machine may be used for..impregnating with the aniline solution in the dyeing of Aniline Black. **1935** *Chem. Abstr.* XXIX. 4179 Dyeing in concentrated dye bath... The mech. and operating details of padding mangles of various manuf. are described. **1963** A. J. HALL *Textile Sci.* iv. 193 The textile material..is impregnated evenly with the dye liquor with the aid of a so-called padding mangle. **1973** *Materials & Technol.* VI. vii. 461 The most popular type of padding machine has two or three rollers giving one or two immersions in the padding solution.

3. *Electronics.* The use of a padder; *padding capacitor* or *condenser* = *PADDER *sb.*² 3.

1935 A. T. WITTS *Superheterodyne Receiver* iv. 37 Condenser C_3, the padding condenser, has a capacity that is large in comparison with the tuning condenser C_2, with the result that at the lower settings of the latter the padding condenser has very little effect. **1936** J. H. REYNER *Testing Radio Sets* (ed. 3) xiii. 190 The alternative method, that of padding and trimming, is usually adopted. *Ibid.* 191 We..have three variables, namely, the oscillator inductance, the parallel trimming condenser, and the series padding condenser, and three frequencies which are to be 'spot on'. **1946** C. A. QUARRINGTON *Mod. Pract. Radio & Television* I. xviii. 140 The [condenser] vanes are shaped to give the necessary effect when working on the highest frequency band incorporated in the receiver, and the lower frequency band or bands are corrected by means of padding. *Ibid.* 141 The series condenser..is called the padding condenser and is usually..of the pre-set type..; the small parallel capacity..is called the trimming condenser, and is invariably of the pre-set type. **1962** J. H. & P. J. REYNER *Radio Communication* ix. 371 The padding and trimming capacitances must..be altered for each wave range and are usually..changed over by the switching which alters the coils.

4. padding stitch = *pad stitch.*

1913 M. E. WILKINSON *Embroidery Stitches* 123 *Padding stitch*, close Satin stitch worked over a raised or padded grounding. **1955** E. A. MANSFIELD *Clothing Construction* xv. 323/2 To shorten the fold-line so that the collar will lie close to the neck,..put in two rows of small padding stitches along the fold line.

paddle, *sb.*¹ Add: **III. 7. d.** In *Leather-making* (see quot.).

1885 C. T. DAVIS *Manuf. Leather* xviii. 356 The motion of the wheels causes the stock to move up in front, pass under the wheels, and down on the concave bottom to the back of the vat, and thus by means of the paddles, and the constant changing position of the stock a thorough and gentle agitation is maintained.

8. b. (Later examples.)

1970 *Wall St. Jrnl.* 16 June 1 The 'board of education', a paddle applied smartly to the backside, may be making a comeback in the classroom. **1977** *New Yorker* 30 May 27/1 One of the boys..was said to have been struck more than twenty times with a paddle.

c. A short-handled bat with a broad, flat blade, used in various ball games.

[**1925**: implied in *paddle tennis*]. **1935** MASON & MITCHELL *Active Games & Contests* xxi. 388 Paddle Ball... This is an excellent game played with a paddle-tennis ball and paddle. **1949** P. B. BARRINGER *Natural Bent* xviii. 127 To play two-hole cat, four boys, two bases, a ball, and two bats were needed. These bats were sometimes called paddles. **1974** E. TIDYMAN *Dummy* vi. 80 He..accepted a challenge to play table tennis..offering the doctor instruction on the proper way to hold his paddle. **1975** *Oxf. Compan. Sports & Games* 745/1 Bigger and heavier paddles are used, and the ball may be played into court off the walls. **1976** *Webster's Sports Dict.* 299/2 The table tennis paddle is slightly smaller than the paddleball paddle.

9*. *Astronautics.* A paddle-shaped array of solar cells projecting from a spacecraft.

1959 *Listener* 13 Aug. 247/1 The four 'paddles' recharge the satellite's batteries by converting sunlight into electricity. **1966** *Electronics* 17 Oct. 36 The two solar paddles each extend 19 feet from the Agena and together provide 15 kilowatts. **1972** *Nature* 17 Mar. 90/1 The solar paddles and antenna were also reported to be working satisfactorily.

IV. 10. *paddle-boat* (examples), *punt, steamer* (earlier example), *tug*; (sense 4) *paddle-foot.*

1874 J. W. LONG *Amer. Wild-Fowl Shooting* 79 Now the building of a paddle-boat is not so simple an undertaking as many of my readers may suppose. **1891** *Scribner's Mag.* X. 13 She was a paddle-boat, built of wood, and was 207 feet long. **1938** M. K. RAWLINGS *Yearling* xx. 261 The remaining bears were scrambling across the swamp like paddle boats, churning the water behind them. **1954** J. R. R. TOLKIEN *Fellowship of Ring* II. ix. 399 It wasn't a log, for it had paddle-feet. **1909** *Yachting Monthly* Dec. 93/2 The 'paddle punt' is about 14 ft. long, strongly built on the Deal model. **1918** N. DUNCAN *Battles Royal* IV. ii. 242 Old Elihu Maul, with a hook and line, had fished the Boiling Pot in civil weather from a paddle-punt. **1970** E. J. MARCH *Inshore Craft Gt. Brit.* II. ii. 82 The second class [*sc.* punts]..cost £12. Generally known as 'paddle punts', they were used for inshore fishing. **1886** *Outing* VIII. 26/1 The Ripple, [a] paddle steamer of the river steamer type. **1923** *Man. Seamanship* (Admiralty) II. viii. 152 For long tows at sea the screw tug is the most efficient, as owing to the propellers being totally submerged they are not affected by the sea to the same extent as paddle tugs. **1930** J. MASEFIELD *Wanderer of Liverpool* 15 The Paddle-tug Wrestler arrived at an hour ere flood. **1955** *Times* 6 June 6/6 Experience has shown that paddle tugs are more efficient than screw-driven tugs for work in confined basins because of their great manoeuvrability and power.

11. paddle ball, a game played with a light ball and wooden bat in a four-walled handball court; **paddle board**, (*b*) a wooden board for supporting a person in water, esp. when surfing; **paddle-box** (earlier and later examples); also *transf.*; **paddle foot** *U.S. slang*, (*a*) an infantryman; (*b*) a member of an airforce ground crew; (see also sense 10 above); †**paddle-plane** *Aeronaut.* = *cyclogiro, -gyro* s.v. *CYCLO-; **paddle tennis**, a type of tennis played in a small court with a sponge-rubber ball and wooden or plastic bat; **paddle-vat** = *paddle-tumbler*; **paddlewheeler**, a paddle steamer.

1935 Paddle ball [see sense 8 c above]. **1962** *Times* 14 Nov. 3/6 Games of..paddle-ball (Rugby fives with table tennis bats). **1973** *Daily Colonist* (Victoria, B.C.) 20 May 2/2 Our daughter..gets blisters on her soles after a hard game of tennis or paddle ball. **1975** *Listener* 9 Oct. 484/1 A *cinquantaine sportive* concern to keep fit through paddle ball at the Downtown Club. **1967** J. SEVERSON *Great Surfing* Gloss., *Paddleboard*, a square-sided, hollow surf-craft usually constructed of plywood. **1968** *Surfer Mag.* Jan. 56/1 He surfed a hollow paddle board that he made at home. **1974** 'R. B. DOMINIC' *Epitaph for Lobbyist* xvi. 139 He was going to find time for healthful exercise—paddle-board, or a few turns in the pool. **1833** *Chambers's Edin. Jrnl.* 1 June 140/2 The captain now takes his station on the paddle-box. **1908** G. B. SHAW *Let.* 31 Dec. (1972) II. 823 Charlotte wrecked it [*sc.* the car] the first day. The professional kept her in countenance by knocking off the paddle-box against the gate. *a* **1936** KIPLING *Something of Myself* (1937) iv. 102, I saw..a woman crouching on the paddle-box of a crowded boat. **1976** P. LOVESEY *Swing, swing Together* xxxvii. 175 A coat of white paint on the paddle-box, lifeboats and funnel. **1946** *Amer. Speech* XXI. 34/2 *Paddlefeet*.., Infantrymen. **1948** MENCKEN *Amer. Lang.* Suppl. II. 727 The airmen..use many derisory terms in speaking of themselves..*e.g.*,.. *paddlefoot*..for a member of the ground crew. **1950** *Life* 2 Jan. 98/2 Murray was a paddlefoot in Europe. **1957** *New Yorker* 23 Nov. 67/3 A paddlefoot mess officer in North Africa. **1960** WENTWORTH & FLEXNER *Dict. Amer. Slang* 372/1 *Paddlefoot*.., an infantry soldier. **1933** *Flight* 2 Feb. 107/2 Our Berlin correspondent indicates that the Rohrbach 'paddle plane' has the circumferential speed of the paddles approximately equal to the top speed. **1950** *Gloss. Aeronaut. Terms* (*B.S.I.*) I. 30 *Cyclogyro* (paddle-plane). **1925** *Playground* Mar. 710/1 He secured permission from Park Commissioner Francis B. Gallatin to mark several paddle tennis courts in Washington Square Park. **1944** F. G. MENKE *Encycl. Sports* (rev. ed.) 490 Frank P. Beal..originated Paddle Tennis in 1924 to provide children with a game that would teach them the rudiments of tennis. **1972** *N.Y. Times* 3 Nov. 22/4 (Advt.), A new recreational facility featuring tennis, paddletennis..and a barbeque and picnic area with charming pavilion has been completed. **1903** L. A. FLEMMING *Pract. Tanning* 23 Sheepskins are also very satisfactorily tanned with one-bath chrome liquors in paddle-vats. **1924** H. A. TRIPP *Shoalwater & Fairway* vii. 128 That Belle steamer was a mine-sweeper in the War, and jolly useful they found the old excursion paddle-wheelers. **1970** S. TRUEMAN *Intimate Hist. New Brunswick* vi. 83 He built the *Reindeer*, a paddlewheeler that easily outraced bigger and fancier river boats. **1976** *National Observer* (U.S.) 13 Mar. 8/1 (Advt.), Return to heartland America aboard the legendary paddlewheeler Delta Queen, or the luxurious new Mississippi Queen.

paddle, *sb.*³ Add: **2.** Usu. in form **paidle.** (Earlier example.) Also *attrib.*

1879 *Cases Court of Session, Scotl.* (ser. 4) VI. 1324 Nets ..often have also a barrel-shaped trap or paidle attached to them. **1882** *Ibid.* IX. 186 The respondents..earn part of their living by fishing on the shores of the Solway by means of small stake-nets, locally called 'paidle-nets'.

paddle, *sb.*⁴ Restrict †*rare* to sense in Dict. and add: **2.** (Also *Sc.* **paiddle, paidle.**) An act of paddling in mud or shallow water.

1866 W. GREGOR *Dial. Banffshire* 121 The twa bairns keepit a paidle..in the lint-cobble, catchin' wattir-horse. **1880** LONGMUIR & DONALDSON *Jamieson's Etym. Dict. Scottish Lang.* (rev. ed.) III. 430/1 We..had a gran' paidle in the saut watter. **1896** A. M. BISSET *Poets Linlithgowshire* 188 But woe to the imp that..damm'd up the burn for a paiddle or wade. **1942** 'N. SHUTE' *Pied Piper* v. 106 Wouldn't you like to take your shoes off and have a paddle, then? **1976** *Morecambe Guardian* 7 Dec. 25/7 They decided to go for a swim and walked into the sea first for a paddle.

paddle, *sb.*⁵ Add: *Comb.* **paddle-over** [after WALK-OVER], an easy victory in a boat race.

1906 *Westm. Gaz.* 4 July 5/1 Little more than a paddle-over for the Cambridge men.

paddle, *v.*¹ Add: **II. 4.** (Later examples.)

1908 H. G. WELLS *War in Air* iii. 72 Then he got up, paddled about, rearranged the ballast bags on the floor,.. and turned over the maps on the locker. **1970** F. DURBRIDGE *P. Temple & Harkdale Robbery* v. 51 Tam Coley paddled cheerfully out [of the room] with a nod to Paul.

5. *Comb.* **paddle-pond**, a pond in which children may paddle.

1930 *Time & Tide* 14 Feb. 195/2 He saw that these spaces were..empty, and he resolved that some..of them should be filled; hence the goal-posts and paddle-ponds.

paddle, *v.*² Add: **2. b.** (Earlier and later examples.)

1828 J. HALL *Lett from West* 261 It seems that they were not so well skilled in navigation as the *Lady of the Lake*, who 'paddled her own canoe' very dexterously. **1834** W. G. SIMMS *Guy Rivers* II. 225 He guessed therefore, best haul off, and each..'must paddle his own canoe'. **1887** *Harper's Mag.* Mar. 547/1 They couldn't see how he was to paddle his canoe all alone by himself. **1924** [see *BUNCH *v.*¹ 1 c]. **1924** M. KENNEDY *Constant Nymph* xvii. 232 Why can't she leave the fellow to paddle his own canoe? **1949** *Time* 4 July 25/2 They seem more interested in paddling their own canoes than shaping a strong third force that would be the best weapon against the communism they all hate.

3. (Later examples.)

1919 L. F. CODY *Memories Buffalo Bill* 31, I had started from the porch to paddle every one of them [*sc.* the children]. **1976** 'D. HALLIDAY' *Dolly & Nanny Bird* iv. 48 The first thing a Maggie Bee nurse does in any British household is to ask the mother if she minds if the offspring get paddled from time to time.

4. To use a paddle, in various special senses of the sb.: (*a*) to stir or mix (molten ore) with a paddle; (*b*) to wash or dye (leather) by means of a paddle; (*c*) to stir (the lye in soap-making) with a paddle.

1873 E. SPON *Workshop Receipts* (ser. 1) 382/2 The paddling should be continued until a ring drawn with the spatula may be recognized. **1874** J. A. PHILLIPS *Elem. Metallurgy* 544 The pot-skimmings..are now thrown into the furnace and well paddled with the charge. **1909** H. G. BENNETT *Manuf. Leather* 171 When a quick and even colouring is desired..the goods may be paddled in the first liquors.

5. To use (something) like a paddle.

1929 W. DEEPING *Roper's Row* xxxii. 363 He spread his table napkin, and finding the soup too hot, paddled his spoon in it.

pa·ddle-fish. [f. PADDLE *sb.*¹ + FISH *sb.*¹] A large freshwater fish of the family Polyodontidæ, which includes the two genera *Polyodon* and *Psephurus*, characterized by a projection resembling a paddle attached to the upper part of its head.

[**1686** tr. *Relation Invasion Florida* xxiv. 121 We caught another sort of fish also, called *Pexe-palla*, the *Palat-fish*; the head of it is covered with a kind of an elbow-hood, the upper point whereof is shaped like a Palet or Lingel.] **1807, 1892** [in Dict. s.v. PADDLE *sb.*¹ 11]. **1908** *Century Mag.* July 457/1 In Louisiana it [*sc. Polyodon spatula*] is known as billfish, billdom, and paddlefish. **1948** *Sat. Rev. Lit.* (U.S.) 15 May 26/2 They were assailed by questions about the Paddlefish, the Brindled Stonecat, or the Tessellated Darter. **1962** K. F. LAGLER et al. *Ichthyol.* iv. 110 Examples of partly scaled fishes include the paddlefish (*Polyodon*)..that inhabits streams in Central North America and has also a near-relative, the freshwater swordbill, *Psephurus*, in China. **1976** *Billings* (Montana) *Gaz.* 16 June 1-C/1 The paddlefish is a living fossil like the Coelacanth..and is found only in two places on earth, the Mississippi River and its tributaries and the Yangtze drainage in China.

2. = *oar-fish* (s.v. OAR *sb.* 6 in Dict. and Suppl.).

1953 *Sun* (Baltimore) 18 Sept. 11/5 One of the world's rarities, an oarfish, or paddle fish, believed to be the origin of sailors' tales of sea serpents, has been caught off Sydney Heads.

paddler¹. Add: **2.** *pl.* A child's waterproof knickers or overall.

1928 *Weekly Dispatch* 27 May 15/7 All-black bathing suits. Besides suits, there are the much needed rubber paddlers, caps, and shoes.

paddle-wheel. Add: **3.** A device shaped like the wheel of a paddle-boat, used in a game of chance.

1926 ADE *Let.* 26 Oct. (1973) 114 We had games as follows: one roulette, one hazard,..one wheel of fortune for cash, one paddle wheel for fancy baskets, dolls, boxes of candy etc. **1935** *Sun* (Baltimore) 22 July 7/4 Operation of paddle wheels, bingo devices and other alleged games of chance would no longer be tolerated. **1939** *Ibid.* 24 Apr. 18/1 Prince George's county..secured passage of a measure that would legalize not only pinball games, but bingo, paddle wheels and other similar devices. **1961** J. SCARNE *Compl. Guide Gambling* xix. 459 *Paddle wheel or raffle wheel*, a carnival wheel each of whose numbered sections contain one, two, or three numbers. Most such wheels have a counter laydown raffle chart on which bets are placed.

4. *attrib.* †**paddle-wheel aeroplane** or **aircraft** = *cyclogiro, -gyro* s.v. *CYCLO-.

1935 *Technical Rep. Aeronaut. Res. Committee 1933–34* I. 8 We have also considered a number of proposals for the construction of..paddlewheel aeroplanes. **1939** *Jrnl. R. Aeronaut. Soc.* XLIII. 756 Unusual aircraft such as flapping wing aircraft, paddle wheel aircraft, tail-first aeroplanes.

paddling *vbl. sb.*¹ (s.v. PADDLE *v.*¹ in Dict.) Add: *Comb.* **paddling pool.**

1932 T. SHARP *Town & Countryside* x. 204 These playgrounds should be equipped with swings,..and a sand-pit and perhaps a paddling pool. **1958** *Listener* 6 Nov. 727/1 Children's paddling pools. **1972** *Guardian* 1 Aug. 15/6 Emma went to the crèche-playgroup where there were swings, slides, books, a paddling-pool, a Wendy-house.

paddling, *vbl. sb.*² (s.v. PADDLE *v.*² in Dict.) (Further examples.)

1874 J. A. PHILLIPS *Elem. Metallurgy* 542 The alternate raking and paddling of the charge is continued at regular intervals. **1888** L. A. SMITH *Music of Waters* 325 The following is a specimen of the paddling-songs, which really form the principal water-music of the Tonga Islands. **1953** P. PROVENCHER *I live in Woods* i. 6 The crews commenced a lively paddling song on quitting the shore.

paddock, *sb.*² **1. c.** Substitute for def.: In Australia and New Zealand, any field or piece of land enclosed by fencing, irrespective of size or land use. (Further examples.)

1822 J. DIXON *Narr. Voy. New South Wales* 58, I saw a few paddocks of clover and English grasses, in as good condition as I have seen the same fields in England. **1847** A. HARRIS *Settlers & Convicts* xvi. 279 In ten months' time from their occupying the farm, there was one of the sides of a paddock fence put up. **1873** TROLLOPE *Austral. & N.Z.* II. xxiii. 368 Vast paddocks containing perhaps 20,000 acres each. **1911** C. E. W. BEAN '*Dreadnought*' *of Darling* i. 7 The Western Division is inhabited—indeed it is all fenced into paddocks. **1916** J. B. COOPER *Coo-oo-ee* ix. 115 Sandy was in his pumpkin paddock ('pumpkins were the things to feed pigs on'). **1924** H. T. GIBSON *That Gibbie Galoot* xv. 56 By the way, you Colonials call a field a paddock, or more often a paddeck. **1930** L. G. D. ACLAND *Early Canterbury Runs* (ser. 1) i. 7 In the early sixties..expense and the scarcity of water prevented much sub-division into paddocks on most of the plains stations. **1937** 'W. HATFIELD' *I find Australia* iv. 59 The 'horse-paddock' near the homestead was eight miles by eight, and that wasn't a big 'paddock'. **1957** *N.Z. Listener* 22 Nov. 4/3 'Creek' and 'paddock' are New Zealandisms, because they mean something quite different in the English of England. It is of some significance that Katherine Mansfield uses both words only in their New Zealand sense. **1962** J. FRAME *Edge of Alphabet* xv. 81 With the gate into the field (they call it paddock) shut. **1963** B. PEARSON *Coal Flat* xvii. 310 In the lush paddocks on the river flat, a few prosperous farmers ran sheep. **1968** K. WEATHERLY *Roo Shooter* 7 In the forty-five-thousand-acre paddock at the bottom of the rocky hill country the rains have scoured great washouts in the slopes. **1977** *National Times* (Austral.) 17 Jan. 16/4 (Advt.), Lot 7: Homestead Block: 166.05 ha (410 acres) Freehold, gently undulating country running down to river, subdivided into 9 paddocks.

3. (Further examples.)

1862 *Otago: Goldfields & Resources* 34 Sod walls..are largely used in making dams and 'paddocks'. **1874** A. BATHGATE *Colonial Experiences* viii. 93 This process is carried on for months, the tail-race being prolonged into the space from which the ground has been washed away, until a larger hole or 'paddock' is taken out, with precipitous sides, varying in height from a couple of feet to two hundred or more.

4. *paddock fence, gate* (later examples), *sheet*; **paddock-grazing**, in dairy farming, a method of pasture management developed by the French farmer, André Voisin, in which several fields are used in rotation; hence, as a back-formation, **paddock-graze** *v. trans.*

1864 R. HENNING *Let.* 4 Mar. (1966) 156 The 'Station Creek'..came down a roaring river..and swept down with it the whole of the paddock fence which crossed its bed, though it was built in that part of entire trunks of trees. **1908** E. J. BANFIELD *Confessions of Beachcomber* II. iii. 326 The tribe cut off the iron bracing from the paddock gates. **1911** C. E. W. BEAN '*Dreadnought*' *of Darling* i. 11 The coach had stopped at a paddock gate. **1969** 'J. ASHFORD' *Prisoner at Bar* v. 37 A herd of Frisians were paddock-grazing the nearest fields. **1960** *Farmer & Stockbreeder* 16 Feb. 77/3 It was no use introducing strip- or paddock-grazing unless this was accompanied by a marked increase in stocking. **1962** K. N. RUSSELL *Fishwick's Dairy Farming* (ed. 3) II. 228 This..is the gospel according to Voisin, whose system of rotational paddock grazing Mr. Pearson has recently adopted. **1970** C. S. BARNARD et al. *Milk Production* xv. 240 Paddock grazing is an alternative to strip grazing which obviates the daily task of moving electric fences. Semi-permanent fences are erected to create paddocks of a size and number that enable fresh grazing to be offered every 1–7 days. **1970** R. JEFFRIES *Dead Man's Bluff* ix. 85 The bloke who invented paddock grazing knew a thing or two. **1971** *Power Farming* Mar. 40/2 Paddock grazing is the modern way to really efficient grassland management. It enables grass to be accurately rationed ensuring that each cow has the correct intake of highly nutritious herbage, resulting in increased milk yield per acre and improved milk quality. **1975** *Country Life* 26 June 1702/3 Modern paddock-grazing owes much to the work of ICI..pioneers of the one-day, 21-paddock, two-sward system. **1963** E. H. Edwards *Saddlery* xxi. 160 The everyday exercise sheets..are the same shape as a paddock sheet, but usually a few inches larger. **1977** *Horse & Hound* 10 June 28/1 (Advt.), Coloured Rollers, For day rugs or paddock sheets.

paddock, *v.* Add: **1.** (Further examples.)

1873 Trollope *Austral. & N.Z.* II. xii. 214 The sheep are all 'paddocked',—that is, kept in by fences,—so that shepherding is unnecessary. **1930** L. G. D. Acland *Early Canterbury Runs* (ser. 1) v. 111 He paddocked the sheep one night at the Rangitata Bridge. *Ibid.*, Meaning to have breakfast at the hotel where his sheep were paddocked. **1941** Baker *N.Z. Slang* v. 40 *To paddock land*, to put up fences; *to paddock stock*, to put stock into a paddock.

2. Also, To excavate washdirt in shallow ground (see Paddock *sb.*[2] 3 a); occas. const. *out*.

1860 *National Mag.* VIII. 307/1 Those who have seen Chinamen at work 'paddocking' in the worn-out alluvial gold-diggings of Australia, can speak for their steady, untiring industry. **1863** V. Pyke in *App. Jrnls. House Reps. N.Z.* D. vi. 18 Many..who held river claims worked very successfully by wing dams, consisting of bags of sand laid into the stream, so as to cut off a portion of its bed, which, being drained by pumping, was paddocked out and passed through the cradle.

paddy, *sb.*[1] Add: **1.** Also, the rice plant. Now freq. written *padi*. (Further examples.)

1893 F. A. Swettenham *About Perak* 41 The country for miles round Parit Buntar has been converted from jungle into fields of sugarcane and padi. **1894** [see *nasi]. **1900** C. O. Blagden in W. W. Skeat *Malay Magic* iii. 58 In the inland villages it is regarded as a great crime to use the sickle (*sabit*) for cutting the *padi*. **1931** *Economist* 19 Dec. 1168/2 Thousands of acres of paddy are being planted in isolated plots that were merely abandoned swamps. **1943** *Sun* (Baltimore) 10 June 12/2 Unmilled or rough rice, growing or cut, is known as 'paddy'. **1966** S. M. Sadeek *Windswept & Other Stories* (1969) 2 The strong winds whistled their..tunes..through the whispering sugarcanes and the sheets and sheets of shimmering padi —green, or golden—under the tropic sun. **1969** J. M. Gullick *Malaysia* ii. 47 'Padi' is the term for unhusked rice and is used to denote the rice plant. **1971** R. Russell tr. *Ahmad's Shore & Wave* i. 9 All around is flat country.. and dotted about the plains are muddy bluish pools which from the air look like big pieces of blue glass set in the fields of green waving paddy. **1972** M. Sheppard *Taman Indera* 163 The grey water bottles are left to dry for two or three days and are then fired in a shallow trench, using coconut fibre, coconut shells and padi husks.

b. = *paddy-field*.

1948 *Amer. Speech* XXIII. 229/2 *Paddy*, a rice field. **1972** *Sci. Amer.* May 23/1 The entire immediate area had been a rice paddy, but during the years when no cultivation had occurred, the rice had been replaced by a very tall reed. **1974** *Encycl. Brit. Micropædia* VII. 668/3 *Paddy*, a wet field used for growing rice. **1974** *Indonesian Observer* 26 July 1/3 The President was informed that the irrigation project whose construction work was begun 1936 but stopped until 1971 will be capable of irrigating 50,330 hectares of paddies. **1974** *Nat. Geographic* Aug. 252 A tree-lined road cut through harvested grainfields and paddies resting under a crystal-blue sky.

3. *paddy-field* (further examples).

1937 *Discovery* Jan. 7/2 Spacious padi fields. **1971** *Illustr. Weekly India* 25 Apr. 42/1 The large tracts of golden paddy fields blended with the molten gold of Niger flowers. **1977** *Borneo Bull.* 7 May 1/5 Four people.. were electrocuted in a padi field at Kampong Keriam, near Tutong, last month when an overhead power cable collapsed on to the field.

Paddy, *sb.*[2] Add: **1.** Also used as a form of address, often felt to be derog., for an Irishman.

1907 G. B. Shaw *John Bull's Other Island* III. 75 *Hodson*... Dont you be taken in by my ole man, Paddy. *Matthew*... Paddy yourself! How dar you call me Paddy? **1916** 'Taffrail' *Pincher Martin* ii. 29 'Stop yer bloomin' noise, Paddy!'.. And Pincher suffered no further inconvenience at the hands of Peter Flannagan.

c. The proprietary name of an Irish whiskey; a drink of this. Also (sometimes with lower-case initial) Irish whiskey generally.

1925 *Trade Marks Jrnl.* 23 Dec. 2827 Paddy... Whisky. Cork Distilleries Company, Limited,..Cork, Ireland; distillers. **1971** J. Aiken *Nightly Deadshade* vii. 77 Milly is drinking port..and O'Grady, double Paddys. **1974** D. Seaman *Bomb that could Lip-Read* vii. 51 Will you gentlemen join me in a drink now?.. Three Paddies, then, is it? **1975** *New Yorker* 25 Aug. 40/3 Did she put Irish whiskey in your glass, that Paddy junk? We asked for Scotch. **1976** N. Freeling *Lake Isle* xx. 140 The drop of paddy's fearfully dear here. I've no opinion of the stuff the supermarket calls Scotch.

4. (Further examples.)

1929 J. Owen *Shepherd & Child* i. 14 Tristina went—and that without pulling the door behind her 'in a paddy', as she would have done if the order had come from Miss Trellis. **1933** [see *Irish *sb.* 5]. **1959** 'O. Mills' *Stairway to Murder* v. 56 It was my awful temper. I used to get into the biggest paddies when I was a kiddie. **1959** I. & P. Opie *Lore & Lang. Schoolch.* x. 178 They taunt the person:..'Don't get in a paddy.' **1975** J. Cowley *Mandrake Root* (1976) xvi. 280 You're a pigheaded Stilwell... Got a real paddy when you let go.

7. In Black English, a white person; also *attrib.* or as *adj.*

1946 C. Himes *Black on Black* (1973) 256 'Hey, don't spit in the sink where you wash the glasses,' some paddy down the bar said. **1962** [see *boot *sb.*[3] 1 e]. **1966** *Sat. Rev.* (U.S.) 15 Oct. 74/2 Man, how I hate Paddies (white people)! **1967** *Trans-Action* Apr. 6/1 This field worker.. had run with 'Paddy' (white), 'Chicano' (Mexican), and 'Blood' (Negro) sets since the age of twelve. **1970** R. D. Abrahams *Positively Black* i. 8 The black became beautiful and the white became nothing but a honky and a paddy. **1972** J. Wambaugh *Blue Knight* (1973) xiv. 240, I spotted a paddy hustler taking a guy up the back stairs. *Ibid.* 241 Paddy hustling was always a Negro flimflam and that's where the name came from, but lately I've seen white hustlers using this scam on other paddies. **1973** D. Barnes *See the Woman* (1974) 68 'What are you?' Grear said to West. 'This paddy's interpreter?' *Ibid.* 70 Biggest paddy whore in Normandie Avenue. **1976** —— *Yesterday is Dead* (1977) II. 199 Hollister..found he was the only white face in the place... 'You know..I'm the only paddy in here.'

8. *Railway* and *Colliery slang*. (See quots.)

1965 H. Sheppard *Dict. Railway Slang* 8 *Paddy*, colliery train from mine to railhead. **1971** D. J. Smith *Discovering Railwayana* x. 58 *Paddy*, train conveying coal from the pithead to distant sidings. **1977** *Guardian Weekly* 4 Dec. 19/4 Once out of the cage, there was a quarter of a mile to walk over pit railway sleepers, dodging heavy equipment, to the 'paddy' whose proper name is 'the endless rope haulage manrider'.

9. In *Combs.* of **Paddy** or **Paddy's**: **Paddy Doyle** *Services' slang*, confinement in the cells, esp. in phr. *to do*, or *doing, Paddy Doyle*; **Paddy's hurricane** *Naval slang*, a flat calm; **Paddy's lantern** *colloq.*, the moon; **Paddy('s) lucerne** *Austral.*, a local name for the tropical evergreen shrub, *Sida rhombifolia*, of the family Malvaceæ, a pest in parts of Australia, although cultivated elsewhere for the fibre it yields; **paddy mail** = sense 8 above; **paddy wagon** *slang* (orig. *U.S.*), a police van; occas., a police car; **Paddy Wester** *slang*, an inefficient or inexperienced seaman; (see also quots.).

1919 Paddy Doyle [see C.B. s.v. *C III. 3]. **1932** E. Weekley *Words & Names* xii. 174 Doing '*Paddy Doyle*' as a euphemism for doing 'time' in the cells. **1948** Partridge *Dict. Forces' Slang* 136 *Paddy Doyle*, a lower-deck term for 'detention cell'—singular or collective. *a* **1865** Smyth *Sailor's Word-bk.* (1867) 514 *Paddy's hurricane*, not wind enough to float the pennant. **1891** H. Patterson *Illustr. Naut. Dict.* 132 *Paddy's hurricane*, when there is little or no wind, so that the pennant hangs down alongside the mast. **1897** 'F. B. Williams' *On Many Seas* 43 We came on deck to find a 'Paddy's Hurricane'—a calm. **1903** A. Sonnichsen *Deep Sea Vagabonds* vii. 114 The winds here never blew at all, or, after the manner of Paddy's hurricane, up and down. **1958** J. G. R. Bisset *Sail Ho!* v. 48 A dead calm was known as 'Paddy's hurricane'. **1933** P. A. Eaddy *Hull Down* v. 104 Work round the deck and up aloft is a hundred times easier when 'Paddy's Lantern' is hung out. **1937** Partridge *Dict. Slang* 600/2 *Paddy's lantern*, the moon: nautical... Prob. after parish-lantern. **1898** Morris *Austral Eng.* 195/2 Hemp, Queensland,..name given to the common tropical weed *Sida rhombifolia*... Called also Paddy Lucerne. **1926** 'J. Doone' *Timely Tips for New Australians* (Gloss.), Paddy's lucerne.—A prevalent type of weed. **1965** *Austral. Encycl.* VIII. 127/2 Paddy's lucerne is so tough and difficult to eradicate that it is reckoned one of the most formidable weed-pests in warmer parts of the Commonwealth. **1945** *Penguin New Writing* XXIII. 85 Colliers were drawn toward it from all the surrounding parts. His mate, Ron Loss, came in by the paddy-mail. **1976** *Star* (Sheffield) 29 Nov. 1/2 A man died today when ten miners were thrown from a paddy mail which crashed at a pit on the outskirts of Barnsley... The paddy mail—a train for carrying miners underground—struck a wooden roof support which had become dislodged, and was derailed. **1930** *Chicago Tribune* 26 Mar. 3/6 He was informed by the pink faced lockup keeper that all Chicago's 'paddy waggons' are motor driven. **1932** J. T. Farrell *Young Lonigan* vi. 259 First thing you know they'll have you in a jam, and you'll be riding in the paddy wagon. **1946** C. Himes *Black on Black* (1973) 260 The police..held all four of us there waiting for the ambulance and the paddy wagon. **1964** M. Banton *Policeman in Community* iii. 51, 2 Patrolmen in a wagon ('Paddy-wagon' or 'Black Maria'). **1967** *N.Z. Listener* 20 Jan. 3/4 A policeman is my guide, and a paddy-wagon my carriage, for a late evening trip around the town's night-spots. **1972** H. C. Rae *Shooting Gallery* iv. 258 The Paddy-wagon, customarily on duty at the cul-de-sac beyond the junction. **1973** *Sunday Mail* (Brisbane) 18 Mar. 25/6 A Police paddy wagon..cruised by. **1974** *Times* 21 Sept. 14/2 Police, dogs, ambulances..a gigantic paddy-wagon. **1927** F. H. Shaw *Knocking Around* xiii. 125 He was not an actual Paddy-Wester, but he had sailed shipmates with many of them. **1929** F. C. Bowen *Sea Slang* 100 *Paddy Wester*, a fake seaman with a dead man's discharge, after a notorious boarding-house keeper in Liverpool who shipped thousands of green men as A.B.'s for a consideration. **1937** Partridge *Dict. Slang* 600/1 *Paddy Wester*; occ. *paddywester*. A bogus seaman carrying a dead man's discharge-papers; a very incompetent or dissolute seaman. **1938** W. E. Dexter *Rope-Yarns* 125 They had a pack of fake seamen sailing on dead men's discharges—a crew of 'Paddy Westers'.

paddy (pæ·di), *a.* [f. Pad *sb.*[3] + -y[1].] Having pads; cushion-like; soft, mild; also, 'comfortable', placidly self-satisfied.

Not in common use. The contextual sense is not entirely clear in some of the examples.

1865 C. M. Yonge *Clever Woman* II. iii. 38 A pair of plump, paddy-looking old friends. **1873** —— *Pillars of House* II. xix. 156 The paddy good-natured face in bed. **1958** L. M. Boston *Chimneys of Green Knowe* 132 He was woken by Orlando's whiskery face poking him in the ear, and a paddy foot on his eyelid. **1962** N. Marsh *Hand in Glove* ii. 53 The impressive things about Sergeant Raikes were his size and his mildness... He said: 'Good afternoon, miss,' in a loud but paddy voice.

paddywhack, -wack. Add: **3.** Also in occas. use in *Austral.* and *N.Z.* (Examples.)

1898 B. Kirkby *Lakeland Words* 111 Ah gev yon beggar paddy-whack fer his sauce, an' he'll nut fergit it in a hurry. **1923** G. Watson *Roxburghshire Word-bk.* 227 *Paddy-whack*.., a stroke or blow; a whack or whacking. **1924** *Truth* (Sydney) 27 Apr. 6 *Paddywhack*, a beating. **1965** F. Sargeson *Mem. Peon* iv. 75 'Of course Michael is not going to be unsociable,' she announced. 'I'll have to give him a paddy-whack if he is.'

padenda, var. *pedanda.

padge (pædʒ). *dial.* Also **pudge.** [Cf. Pudge[1].] The barn owl, *Tyto alba*, which has white plumage flecked with brown or grey. Also *attrib.*

1848 A. B. Evans *Leicestershire Words* 65 Padge-owl, the common owl. **1881** *Ibid.* (new ed.) 208 Padge..the common barn-owl. **1885** C. Swainson *Provincial Names & Folklore Brit. Birds* 126 Familiar names [of the barn owl]... Padge, Pudge, or Pudge owl (Leicestershire). **1937** Auden in Auden & MacNeice *Lett. from Iceland* viii. 103, I'll never grant a more than passing beauty To pudge or pilewort, petty-chap or pooty.

padi, var. Paddy *sb.*[1] in Dict. and Suppl.

‖ **padkos** (pa·tkɔs). *S. Afr.* Also **padkost,** erron. **pat-koss.** [Afrikaans, f. *pad* road + *kos* (Du. *kost*) food.] Food for the journey; provisions.

1848 R. Gray *Jrnl.* 16 Dec. (1849) 95 Having got careless as to our 'pat-cop' [*sic*] as we approached home, we fared but badly, and finished our meal by a draught of not the clearest water in the world. **1850** —— *Jrnl.* 6 Nov. (1851) 173, I was not allowed to depart without a good supply of pat-koss, and other comforts provided by the kindness of the parishioners. **1878** *Cape Monthly Mag.* Nov. 273 A dish of 'sesaties' and a couple of loaves..often formed part of the 'padkos'. **1895** in *Funk's Stand. Dict.*, Padkost. **1950** *Cape Times* 19 Sept. 14/1 With apples, biscuits and fish and chips as padkos, Mr. C. J. Kirstein..arrived..from Cape Town..in 12 hours, 45 minutes. **1957** —— 3 Apr. 9 (*heading*) 'Padkos' Passengers. *Ibid.*, One thing the South African Railways have always had to contend with is South Africa's habit of taking along 'padkos' on a journey. **1971** *Sunday Mail Mag.* (Brisbane) 24 Oct. 11/2 No Afrikaner ventures forth on a journey of more than a few miles without his padkos ('food for the road'), and our new friends were seemingly equipped for a fortnight's safari.

padouk. Also **padauk.** Substitute for def.: A large deciduous or evergreen tree of the genus *Pterocarpus*, belonging to the family Leguminosæ, esp. *P. soyauxii* of West Africa, *P. dalbergioides* of the Andaman Islands, and *P. macrocarpus* of Burma and Thailand; also, the reddish hardwood produced by these trees. Also *attrib.* (Earlier and later examples.)

1839 H. Malcom *Trav. South-Eastern Asia* I. II. 189 The Pa-douk, or Mahogany.., is plenty in the upper provinces... It grows very large, and is mostly of the branched or knotty kind. **1858** C. T. Winter *Six Months Brit. Burmah* xv. 115 The pa-douk..a highly ornamental evergreen tree with bright yellow papilionaceous flowers, which are very fragrant, exudes a gum. **1908** W. R. Fisher *Schlich's Man. Forestry* (ed. 2) V. 590 Many foreign woods are used for piano-cases—mahogany, American walnut and maple, padauk, satin wood, etc., ebony for keys, and Florida-cedar for the hammers. **1928** *Observer* 25 Mar. 13/2 Counter-tops at the Bank of England are made of Andaman padouk. **1930** *Times Lit. Suppl.* 5 June 483/3 He would be a tiro indeed who could mistake the fierce red and lively grain of padouk, or the dull heavy texture of sabicu for mahogany. **1956** *Handbk. Hardwoods* (Forest Prod. Res. Lab.) 182 Padouk wood is variable in texture. **1967** G. Sims *Last Best Friend* vii. 61 The weight of the Padouk door might have been the first sign to the perceptive caller that this was a rather special house. **1973** *Times* 1 Feb. 19/3 In the furniture section padouk wood display cabinets..were selling in the £200–£500 range.

Padovan (pæ·dǒvăn), *sb.* and *a.* [f. It. *Padova* (see Paduan *a.* and *sb.*) + -an.] **A.** *sb.* = Paduan *sb.* 1. **B.** *adj.* = Paduan *a.*

1973 M. West *Salamander* i. 42 That girl in the Ferrari, she's a Venetian, a Veronese, a Padovan. **1978** *Sci. Amer.* June 26/1 Eighteen years a professor in Padua, he had published only two books, one an instruction manual.., the other a witty polemic against a Padovan student who had sought to rip off that very instruction book!

pædeia (paidai·ă). Also **paideia.** [a. Gr. παιδεία child-rearing, education.] In ancient Greek society: education or upbringing; more gen., a society's culture; the sum of physical and intellectual achievement to which the human body and mind can aspire. Also *transf.*

[**1875** F. Hueffer tr. *Guhl & Koner's Life of Greeks & Romans* 196 The education proper of the boy (παιδεία) became a more public one, while the girl was brought up by the mother at home.] **1939** G. Highet tr. *Jaeger's*

Paideia I. II. iii. 283 The age of Sophocles saw the beginnings of an intellectual movement... This was the movement mentioned in our introductory chapter: it was *paideia*, education, or rather culture, in the narrower sense. The word paideia, which at its first appearance meant 'childrearing', and which in the fourth century, the Hellenistic, and the Imperial Roman ages constantly extended its connotation, was now for the first time connected with the highest aretê possible to man: it was used to denote the sum-total of all ideal perfections of mind and body. **1962** *Listener* 30 Aug. 323/2 The Lycurgan training for public service enriched Greek 'paedeia'. **1967** *Ibid.* 17 Aug. 201/3 Marx is..built into my intellectual experience, what the Greeks would have called my *paideia*. **1977** G. W. H. Lampe *God as Spirit* ii. 49 Wisdom is a holy spirit of *paideia*, which, in opposition to materialistic (Epicurean) culture, is the disciplined observance of the Law.

pæderast. (Later examples.)

1925 R. Fry *Let.* 7 Sept. (1972) II. 581 We had a long talk on the tyranny of the Paederasts and Sapphists. **1935** E. E. Cummings *Let.* 2 Jan. (1969) 131 Scientists are of course pederasts, as we neither know nor care; & unnaturally enough this natural history museum is a temple or cathedral of the scientific spirit. **1963** A. Heron *Towards Quaker View of Sex* 69 Socially the paederast is the most isolated of homosexuals. **1969** *Listener* 14 Aug. 205/3 A divorced woman on the throne of the House of Windsor would be a pretty big feather in the cap of that bunch of rootless intellectuals, alien Jews and international pederasts who call themselves the Labour Party. **1971** P. Quennell *Marcel Proust* 11 The sense of his own separateness, as a paederast who loved women,..and a sick man..intensified his gift of observation.

pædiatric (pīdi,æ·trik), *a.* *Med.* Also **pediatric.** [see Pædo-, pedo- and Iatric *a.*] Of, pertaining to, or dealing with pædiatrics or the diseases of children.

1880 A. Jacobi in *Trans. Amer. Med. Assoc.* XXXI. 709 (*heading*) Address on the claims of paediatric medicine. **1894** [in Dict. s.v. Pædo-, pedo-]. **1927** W. P. Lucas *Mod. Pract. Pediatrics* iii. 20 From the moment a child is conceived it is a pediatric problem. **1963** *Times* 16 May 13/3 For all these reasons it is not surprising that opinion among paediatricians and paediatric nurses is divided. **1965** *Math. in Biol. & Med.* (Med. Res. Council) v. 228 In a well known text book of paediatric surgery there is a photograph of nine children with tracheo-oesophageal fistula, successfully treated. **1973** D. Morley *Paediatric Priorities in Developing World* i. 1 Doctors with the relevant paediatric training can organize a service which will prevent more than one-half of the deaths in infancy and early childhood without awaiting any great change in environment.

Hence **pædia·trically** *adv.*

1949 M. Mead *Male & Female* ix. 192 If her suckling of the child [is] replaced by a formula pediatrically prescribed—then also we may find very serious disturbances in maternal attitudes.

pædiatrician (pīdi,ătri·ʃăn). *Med.* Also **ped-.** [f. prec. + -ician.] A specialist or expert in pædiatrics.

1903 *Med. Rec.* (N.Y.) LXIII. 513/2 Dr. L. Emmett Holt said he thought all pediatricians would agree that most of the cases which had given trouble in diagnosis were those in which there was a prolonged fever. **1932** *Lancet* 12 Nov. 1072/1 (*heading*) Congress of German pædiatricians in Vienna. **1959** *Times Lit. Suppl.* 13 Mar. 148/1 The authors, a pediatrician and a psychiatrist, show a broad understanding of the sociological scope of adolescents' difficulties. **1971** C. G. Parsons in S. M. Bates *Pract. Paediatric Nursing* p. ix, Paediatricians have always depended on nurses to help children to settle into hospital..and to take the mother's place when she has to be separated from her baby. **1976** *Daily Tel.* 1 Mar. 2/1 Women who become pregnant despite being fitted with intra-uterine contraceptive devices..may risk giving birth to seriously malformed babies, a consultant paediatrician has warned.

pædiatrics (pīdi,æ·triks), *sb. pl.* (const. as *sing.*). *Med.* Also **ped-.** [f. as prec.: see -ic 2.] The branch of medical science dealing with the study of childhood and the diseases of children.

1884 (*periodical title*) Archives of pediatrics. **1924** *Glasgow Herald* 10 June 6 The new foundation is the Sampson-Gemmell Chair of Medical Pædiatrics at the Royal Hospital for Sick Children. **1946** *Nature* 24 Aug. 277/2 Geriatrics must also come to occupy a part not less important in medicine than pediatrics to-day. **1956** Ld. Amulree in A. Pryce-Jones *New Outl. Mod. Knowl.* 213 Infant and child welfare services are available in most civilized countries, and pediatrics and the care of children is becoming more and more a preventive service. **1975** *Physics Bull.* Oct. 458/1 (*caption*) In the field of paediatrics the x ray examination of infants and young children is a diagnostic procedure that is on the increase.

pædiatrist (pīdai·ătrist, pīdi,æ·trist). *Med.* Also **ped-.** [f. as prec. + -ist.] = *pædiatrician. Also *attrib.* or as *adj.*

1897 [in Dict. s.v. Pediatrist]. **1928** O. Wilkinson *Strabismus* x. 143 The advice and insistent counsel of the family physician and the pediatrist. **1977** *Chicago Tribune* 2 Oct. XII. 12/2 (Advt.), 1 deluxe optometric or pediatrist suite.

pædogamy (pīdǫ·gămi). *Biol.* Also (chiefly *U.S.*) **pedogamy.** [ad. G. *pädogamie, paedogamie* (M. Lühe 1902, in *Schriften d. Physikal.-ökonom. Gesellsch. Konigsberg, Sitzung biol. Sektion* XLIII. 5), f. Pædo- + *-gamy.] In certain protozoans, reproduction by the fusion of gametes derived from the same parent cell (see also quot. 1953).

1910 G. N. Calkins *Protozool.* iv. 146 (*heading*) Fertilization by endogamy (pedogamy, Prowazek). **1953** R. P. Hall *Protozool.* ii. 80 Pedogamy appears to be an unusual type of syngamy in which the two gametes are not more than one or two cell-generations removed from a single gametocyte. **1965** Poljanskij & Chejsin *Dogiel's Gen. Protozool.* (ed. 2) vii. 322 In certain cases sister individuals, originating from a single nucleus of one and the same mother cell, act as gametes. Phenomena of this type are known as paedogamy.

Hence **pædo·gamous** *a.*, of or pertaining to this type of reproduction.

1926 G. N. Calkins *Biol. Protozoa* xi. 510 Fertilizations have been described as exogamous, endogamous, autogamous, or pædogamous. **1940** L. H. Hyman *Invertebrates* I. iii. 161 The fertilization [of Myxosporidia] is seen to be a paedogamous autogamy.

pædogenesis (pīdodʒe·nĭsis). *Biol.* Also (chiefly *U.S.*) **pedo-.** [mod.L., coined in Ger. (K. von Baer 1866, in *Bull. Acad. Imp. Sci. St.-Pétersbourg* IX. 96), f. Pædo- + Genesis.] Reproduction by larval or immature forms of animals, esp. certain insects; cf. *neoteny b. Hence **pædogene·tic** *a.*, pertaining to or characterized by pædogenesis.

1871 W. S. Dallas tr. O. von Grimm in *Ann. & Mag. Nat. Hist.* VIII. 32, I had before me an insect [*sc.* a species of *Chironomus*] which is subject to what Von Baer calls pædogenesis. *Ibid.* 36 Different animals may be subject to pædogenesis at different stages of development. **1888, 1889** [in Dict. s.v. Pædo-, pedo-]. **1891** F. V. Theobald *Acct. Brit. Flies* 42 The ovaries [of flies of the family Cecidomyidæ] become fully developed and bud off eggs; [*Note*] = Pædogenesis (i.e., the production of ova by the immature animal, and is in the insecta always parthenogenetic). **1895** D. Sharp in *Cambr. Nat. Hist.* V. iv. 142 A very rare kind of parthenogenesis, called paedogenesis, has been found to exist in two or three species of Diptera. **1920** H. Reinheimer *Symbiosis* II. viii. 158 The 'wages' of a prolonged transgression against the law of Symbiosis is thus indeed death—in the shape of diathesis, dissolution, and of a kind of Paedogenesis—precocious sexuality. **1951** Colyer & Hammond *Flies Brit. Isles* iii. 71 Parthenogenetic reproduction by immature stages is known as paedogenesis. *Ibid.* iv. 83 Those [*Miastor* larvae] which are paedogenetic have no 'breast-bones', while those which will pupate and produce normal flies possess this organ. **1964** R. M. & J. W. Fox *Introd. Compar. Entomol.* vii. 243 Neoteny (pedogenesis) involves the precocious maturity of the ovary so that young are produced by a mother who has not reached the imaginal instar.

pædomorphic (pīdomǭ·ɹfik), *a.* Also **ped-.** [f. Pædo- + Morphic *a.*] **1.** *Biol.* Exhibiting pædomorphism or pædomorphosis.

1891 *Proc. Acad. Nat. Sci. Philadelphia* 209 It might be expected that pedomorphic varieties closely resemble each other when the same disposition is exhibited in closely allied species. **1922** *Jrnl. Linn. Soc.* (*Zool.*) XXXV. 97 Crinoids are as 'pædomorphic' as any Perennibranchiate Amphibian. **1957** L. Eiseley *Immense Journey* 119 The pedomorphic features of man—his almost hairless body, his helpless childhood, his surprisingly developed brain. **1959** J. D. Clark *Prehist. S. Afr.* iv. 90 The Boskop type exhibits a continuation of certain pedomorphic (infantile) characteristics into the adult state, as does the Bushman. **1965** B. E. Freeman tr. *Vandel's Biospeleol.* xi. 165 This feature may be considered in addition to that of small size, as a paedomorphic character. **1970** G. Greer *Female Eunuch* 31 This..is an observation which is frequently made about the whole female body, that it is infantilized or pedomorphic.

2. (After *anthropomorphic*.) Having (or attributing to other objects) the form or characteristics of a child.

1903 H. G. Wells in *Fortn. Rev.* Jan. 184 He will look out on the world with anthropomorphic (or rather with pædomorphic) eyes. **1907** H. Ellis in *19th Cent.* May 767 The Child..imagines a colossal magician, of anthropomorphic (if not paidomorphic) nature.

pædomorphism (pīdomǭ·ɹfiz'm). *Biol.* Also **ped-.** [f. Pædo- + *-morphism.] The retention of juvenile characteristics in certain adult mammals.

1891 *Proc. Acad. Nat. Sci. Philadelphia* 208 Dr. Harrison Allen spoke of the disposition occasionally exhibited in adult mammals, for the proportions of different parts of the body to remain as they were in the immature individuals... Dr. Allen proposed for this peculiarity the term pedomorphism. **1931** *Amer. Jrnl. Physical Anthropol.* XVI. 203 (*title*) Pedomorphism in the pre-Bushman skull. **1970** G. Greer *Female Eunuch* 333 The pedomorphism of women has always been remarked upon. **1973** B. J. Williams *Evolution & Human Origins* vii. 98/1 Paedomorphism is said to occur when certain developmental processes are retarded in such a way that important features of the larval or infantile form are maintained into sexual maturity.

pædomorphosis (pīdomǭ·ɹfōsis, -mǭɹfōᵘ·sis). *Biol.* [f. Pædo- + Morphosis.] Phylogenetic change indicated by the retention of juvenile characteristics in the adult form.

1922 W. Garstang in *Jrnl. Linnean Soc.* (*Zool.*) XXXV. 100 In other articles I propose to deal with the origin and significance of larval forms, and to draw attention to some further examples of the influence of larval characters upon adult organisation, to which I apply the term 'Pædomorphosis'. **1932** J. S. Huxley *Probl. Relative Growth* vii. 240 (*caption*) Diagram to illustrate positive and negative mutations in rate-factors, leading to recapitulation and paedomorphosis, respectively. **1965** B. E. Freeman tr. *Vandel's Biospeleol.* ix. 134 Partial neoteny, a phenomenon to which Garstang's term paedomorphosis may be more exactly applied. **1970** *Times Lit. Suppl.* 4 June 619/4 The most serious lacuna in the book is the absence of any mention of the part which the well-known principle of retardation of development, or paedomorphosis, has played in the evolution of man. **1971** [see *neoteny]. **1974** G. L. Stebbins *Flowering Plants* xi. 249/2 The evolutionary mechanism by which this [*sc.* the evolution of herbs and subshrubs into shrubs or trees] could happen involves the principle of paedomorphosis.

pædophile (pī·dofəil), *sb.* and *a.* Also **pedo-.** [ad. Gr. παιδόφιλ-ος loving children (cf. -phil, -phile).] **A.** *sb.* A person with pædophilia. **B.** *adj.* = *pædophiliac, -philic *adjs.*

1951 *Group Psychotherapy* IV. 166 (*heading*) Psychodramatic treatment of a pedophile. **1954** *Jrnl. Projective Techniques* XVIII. 352/1 This sexualized view of a late middle-aged female, by a 26-year-old subject, reflects the strikingly immature confusion of sexual and maternal figures found in the pedophile group. **1975** *Sunday People* 1 June 2/6 Many paedophiles who read the article will have heard of P.A.L. for the first time and will be anxious that the organisation survives to continue this service. **1976** *Publishers Weekly* 23 Aug. 75/2 Hilary is nine... She's at the mercy of the old man she calls the Devil, actually a pathetic pedophile. **1977** *Sunday Times* 30 Jan. 41/2 The paedophile authors he discusses include the diarist Kilvert, Lewis Carroll..and J. M. Barrie.

pædophilia (pīdofi·liă). Also **pedo-,** † **paido-.** [f. Pædo-, pedo- + *-philia, or f. as prec. + -ia[1].] An abnormal, esp. sexual, love of young children.

1906 H. Ellis *Stud. Psychol. Sex* V. i. 11 Paidophilia or the love of children..may be included under this head [*sc.* abnormality]. **1926** *Med. Jrnl. & Rec.* CXXIV. 161/1 One must keep clearly in mind in dealing with pedophilia the distinction between that mediating homosexuality, and the much more pure perversion which is our subject. **1952** E. A. Gutheil tr. *Stekel's Patterns of Psychosexual Infantilism* i. 62 Some eager lady friends of the mistress of the house who expand their friendship to include the younger male generation, may be suspected of pedophilia. **1962** *Listener* 20 Sept. 438/1 The film certainly is not..a study in paedophilia, of a middle-aged professor's grotesque passion for a twelve-year-old girl. **1963** A. Heron *Towards Quaker View of Sex* 69 Paedophilia is not always homosexual. **1973** *Times Lit. Suppl.* 8 June 647/2 He was driven to a pedophilia in which he played the role of both parents to the children of his fancy.

Hence **pædophi·liac, -phi·lic** *adjs.*, pertaining to or characterized by pædophilia; also as *sb.*, a pædophilic person.

1927 *Psychoanal. Review* XIV. 191 It is only in rare cases that one encounters an individual who has pedophilic predilections and at the same time is suffering from venereal disease. *Ibid.*, Krafft-Ebing..in his attempt at psychological explanation falls back on 'a morbid disposition only' on the part of the pedophalic [*sic*] as the motivating factor. **1951** *Group Psychotherapy* IV. 170 He then insisted he had never had the slightest amount of pedophiliac desire, and that his crime was a total mystery to him. **1954** *Jrnl. Projective Techniques* XVIII. 348/1 The rapists probably do differ from the pedophiles, however, on the variable of aggression, the majority of the pedophilic acts having been of a passive and seductive nature. **1960** *Spectator* 8 July 69 The..survey..shows the paedophiliac to be a type altogether distinct from the adult-seeking homosexual. **1963** A. Heron *Towards Quaker View of Sex* 69 A variety of early experiences and inadequacies of upbringing..make the paedophilic especially sympathetic towards..the state of childhood and immaturity. **1974** J. Bancroft *Deviant Sexual Behaviour* vi. 157 Paedophiliac offenders frequently have personalities in which self-deception and deception of others is marked. **1976** *Publishers Weekly* 26 Apr. 52/1 He contacted fellow pedophiliacs and through them was able to sample many kinds of young girls.

Paelignian (peli·gniăn, pəi-), *sb.* and *a.* Also **Pelignian.** [f. L. *P(a)eligni* + -an.] **A.** *sb.* **a.** A member of an Oscan-Umbrian people centred on Corfinium in southern Italy. **b.** The language of this people. **B.** *adj.* Of or pertaining to this people or their language.

1600, etc. [see *Marrucinian *sb.* and *a.*]. **1853** C. Merivale *Fall of Roman Republic* iii. 77 The Samnites and Pelignians reclaimed 4,000 of their own countrymen who had thus established themselves in the Latin town of Fregellæ. **1862** W. P. Dickson tr. *Mommsen's Hist. Rome* II. xi. 332 When..the Samnites and Pælignians applied to the senate for a reduction of their contingents, their request was based on the ground that during recent years 4000 Samnite and Pælignian families had migrated to the Latin colony of Fregellæ. **1909** W. E. Heitland *Roman Republic* II. xliii. 434 For the revolted Allies some centre was necessary as the headquarters of the confederate government. This was found at Corfinium in the Paelignian country on the eastern side of the Apennine. **1933, 1939** [see *Marsian *sb.* and *a.*]. **1939** R. Syme *Roman Revolution* xiv. 193 The fierce Marsians and Paelignians had long and bitter memories. **1967** E. T. Salmon *Samnium & Samnites* iv. 177 Ovid.. had his own Paelignian homeland in mind.

paella (pa,e·lă, pəi,e·lă). [Cat. *paella*, f. OFr. *paele* (mod. *poêle*), f. L. *patella* pan, dish.] A Spanish dish of rice with chicken, seafood, vegetables, etc., cooked and served in a large shallow pan. Also *fig.*

1892 *Encycl. Pract. Cookery* II. 84/2 *Paela*, a favourite Spanish dish containing the usual oil and garlic. **1926** B. Reynolds *Cocktail Continentale* 66 You lunch at Antiqua Casa Botin. Crab soup. Paella a la Valencian, a mixture of chicken, fish, meat, rice, vegetables, snails and clams. **1939** R. Campbell *Flowering Rifle* II. 52 Dead Charlies climbing on each other's backs To make a huge paella of the plains. **1955** J. Thomas *No Banners* viii. 69 An appetizing *paella* and a goblet of *vino tinto* for each man. **1960** 'W. Haggard' *Closed Circuit* x. 121 The food was excellent: he could give her the *paella* which he knew she loved, and now she was eating it with relish..her mouth still full of rice and chicken. **1965** *Punch* 24 Feb. 289/1 The locals mostly speak with a macaroni Italian accent, the Dons with some kind of *paella* Spanish. **1970** 'D. Halliday' *Dolly & Cookie Bird* vii. 99 Paella with all the right things in it, squids and octopuses and chicken and lobster-tails and paprika and sherry and peas and onion and pimento and pork, all done with saffron rice. **1972** *Village Voice* (N.Y.) 1 June 24/3 The underworld provided the paella for the detective stories, including the celebrated 'Pinktoes'. **1976** *National Observer* (U.S.) 23 Oct. 19/3 The restaurant serves up an imaginative assortment of house specialities, many European inspired including..the Spanish Paella Valenciana.

‖ **paepae** (pai·pai). Also 9 **pi-pi.** Pl. **paepae.** [Native name.] **a.** An elevated stone platform on which Polynesian houses were often built. **b.** A paved area in front of some Polynesian buildings. **c.** A type of raft.

1846 H. Melville *Typee* xxiv. 219 Like all the other edifices of any note, it was raised upon a small pi-pi of stones. **1919** *Century Mag.* Aug. 446/1, I sat..on the palm-shaded *paepae* of my cabin above the blue lagoon. **1923** R. Linton *Material Culture of Marquesas Islands* 272 Many of the *paepae* on hill slopes were simple terraces with a trench at the rear to carry off water. *Ibid.* 273 The perfect *paepae* contains stones of three sorts. **1927** P. H. Buck *Material Culture of Cook Islands* i. 39 In the well-preserved house site..the cobbled *paepae* terrace was a foot lower than the house terrace. **1930** —— *Samoan Material Culture* 56 When a high *paepae* platform was made, most of the posts did not reach the ground level. **1958** T. Heyerdahl *Aku-Aku* x. 334, I was the man who had travelled to Raroia with my friends on a *pae-pae*. **1968** N. A. Rowe in T. Heyerdahl *Sea Routes to Polynesia* 205, I had wondered for many years what this could mean but I see now..that it was a reference to a raft or *pae-pae*. *Pae-pae* can also mean stone platform: hence the confusion. **1974** T. Heyerdahl *Fatu-Hira* ii. 88 Here we stumbled upon human vestiges..mostly overgrown terrace walls and stone platforms, *pae-pae*, where native huts had once stood. *Ibid.* iv. 155 Some *paepae* had been declared *tabu* by ancient medicine men and often contained burials and old artifacts.

paff (pæf), *int.* Also **paf.** [Imit.] An expression of contempt. Also used to represent the sound of a blow.

1851 Longfellow *Golden Legend* v. 242 These beggars ..Lamed and maimed, and fed upon chaff, Chaunting their wonderful piff and paff. **1897** *Pall Mall Gaz.* 28 Sept. 2/3 The combatants used their fists only... Paf! paf! one for you, and paf! paf! for your opponent. **1910** [see *crash *v.* 6 a]. **1922** Joyce *Ulysses* 40 What offence laid fire to their brains? Paff!

pagani·stic, *a.* [f. Pagan *sb.* and *a.* + -istic.] = Paganish *a.* 2.

1933 Dylan Thomas *Let.* (1966) 71 But the more paganistic..one becomes, the less one feels the desire to write. **1938** *Sun* (Baltimore) 14 Nov. 2/5 He will ignore the madman Hitler and his cripple-minded Goebbels with their paganistic philosophy. **1948** L. Spitzer *Linguistics & Lit. Hist.* v. 194 A comparison of Claudel's Christian ode with the paganistic ode of Ronsard.

page, *sb.*[1] Add: **I. 5. c.** (Examples in sense 'an attendant upon a legislative body'.)

1840 *Boston Even. Transcript* 18 Feb. 2/1 A page took them to the Clerk—the Clerk handed them to the Speaker. **1878** B. Harte *Man on Beach* 104 Obtaining political influence through caucuses, I became at last page in the Senate. **1949** *Time* 27 June 61/1 The Capitol Page School ..is attended by the House's 49 page boys, the Senate's 21, the Supreme Court's four, and a few more Capitol-employed boys.

d. (Further examples.)

1955 *Times* 8 July 10/4 She was attended by two pages, James Mostyn and Viscount Quenington, two child bridesmaids, Harriet and Sarah Duckworth, and five older bridesmaids. **1973** 'M. Yorke' *Grave Matters* III. i. 54 Maybe he'll bring the girl to see us... Then you can start planning Andrew's page's outfit.

II. 9. *page-boy* (further examples); (*b*) used *attrib.* and *absol.* to designate a woman's hairstyle in which the hair is worn in a long bob with the ends turned under and hanging on the shoulders; (*c*) = *pager *sb.*[3]

1874 A. J. Munby *Diary* 29 June in D. Hudson *Munby* (1972) 368 'Goodbye, William!' she said to the page boy who opened the hall door for us. **1939** R. Chandler *Big Sleep* i. 13 Her hair was a fine tawny wave cut much shorter than the current fashion of pageboy tresses curled in at the bottom. **1951** 'A. Garve' *Murder in Moscow* i. 27 She still wore her fair hair in a fringe with a page-boy bob. **1961** L. P. Hartley *Two for River* 53 'Mr. Lenthall, please, Mr. Lenthall, please,' intoned a page-boy in a high-pitched nasal sing-song. **1971** *Guardian* 27 July 9/2 Hair is in long page-boy bobs with hair slides. **1973** *Times* 17 Jan. 4/5 The Post Office yesterday launched its radio 'pageboy', a tiny 'bleeper' which can be set off by a telephone call. **1975** H. McCutcheon *Instrument of Vengeance* vi. 94 Her hair was..cut in pageboy style. **1976** G. McDonald *Confess, Fletch* (1977) xxvii. 125 Her hair was a perfect black, shining pageboy.

page, *sb.*[2] Add: **1.** Also, a complete leaf of a book, etc.

1728 Swift *Poetical Works* (1967) 346 Tim set the Volume on a Table, Read over here and there a Fable, And found, as he the pages twirl'd, The Monkey, who had seen the World. **1786** Burns *Poems, chiefly in Scottish Dial.* 71 Now moths deform in shapeless tatters Their unknown pages. **1819** Byron *Don Juan* I. xcv. 50 By the wind Even as the page is rustled while we look, So by the poesy of his own mind Over the mystic leaf his soul was shook. **1895** Conrad *Almayer's Folly* 254 Books open with torn pages bestrewed the floor. **1919** G. B. Shaw *Great Catherine* 115 A play that will leave the reader as ignorant of Russian history as he may be now before he has turned the page. **1925** F. Scott Fitzgerald *Great Gatsby* iii. 55 Knew when to stop, too—didn't cut the pages. **1950** W. Stevens *Auroras of Autumn* 99 On the pedestal, an ambitious page dog-eared. **1969** I. Murdoch *Bruno's Dream* 27 He..folded the page into a paper dart. **1974** 'J. Le Carré' *Tinker, Tailor* I. xiii. 113 The pages had been excised with a razor blade.

d. That which is (actually or notionally) written, printed, etc., on a page. Cf. sense 2 a.

1805 W. Blake *Let.* 19 Jan. (1966) 857 The first page of the Poem was beautifully executed. *c* **1862** E. Dickinson *Poems* (1955) I. 376 Tell him the page I didn't write. **1903** G. B. Shaw *Revolutionist's Handbk.* viii, in *Man & Superman* 210 Whilst these pages are being written an English judge has sentenced a forger to twenty years penal servitude. **1951** L. Hughes *Montage of Dream Deferred* 39 Up to my room, sit down, and write this page. **1966** *McGraw-Hill Encycl. Sci. & Technol.* V. 177/1 Railroads employ high-speed facsimile (several pages per minute) for the transmission of waybills. **1975** *Times* 24 Sept. 2/8 Using the Keypad, the user would call up the Viewdata service and select the 'pages' of information to be displayed on the television screen. **1978** *Broadcast* 6 Mar. 10/2 Viewdata's standard page of 24 lines of text requires 480 lines of actual picture information.

e. *Computers.* A division of the main store of a computer consisting of a certain number of 'words' (commonly a few thousand); also, a corresponding amount of data or part of a program.

1948 *Ann. Computation Lab. Harvard Univ.* XVI. 46 A 'page' number marks a section of 'blocks' in much the same fashion as the page of a book would contain several lines of data, while the block number identifies the line of data. **1962** *IRE Trans. Electronic Computers* XI. 226/2 The main core store is..partitioned into blocks..which for identification purposes are called pages. **1970** O. Dopping *Computers & Data Processing* ix. 124 In systems with extensive core swapping it is advantageous to have the program subdivided into 'pages' of a prescribed size, e.g. 1,000 words, and consider primary storage to be subdivided into blocks, where each block has room for exactly one page.

3. *page-size*; *page-long* adj.; **page charge,** a fee of so much per page requested from an organization when a learned journal publishes a paper by one of its members; **page-galley,** (*a*) a galley containing enough type to print a page; (*b*) a galley proof on which the type has been divided into pages and numbered; **page printer,** a printer (sense 2) whose output is in the form of printed or typed pages; so **page-printing** *ppl. a.* and *vbl. sb.*; **page-proof** (examples); **page reference,** a reference to a specific page or group of pages in a book or periodical; **page-turner,** (*a*) a mechanical device for turning the pages of a book (so **page-turning** *ppl. a.*); (*b*) *fig.*, a very enjoyable or readable book; (*c*) one who turns the pages of a musician's score, usu. during a performance.

1966 *Jrnl. Amer. Chem. Soc.* 5 Feb. 8A/2 A page charge is assessed to cover in part the cost of publication. Payment is expected but is not a condition for publication. **1968** J. M. Ziman *Public Knowl.* vi. 117, I am informed that the decision of the referees on the paper is quite independent of whether this 'page charge' is honoured. **1969** *Physics Bull.* Jan. 23/1 Page charges were first introduced by the AIP in 1930 for *The Physical Review*..; for *The Physical Review*, for example, they currently amount to $60 a page. The charges, which are not mandatory, are paid by authors' institutions and are honoured by the majority of institutions... These are used to offset those items of expenditure classified as 'input' production costs; subscription prices are kept relatively low and are applied towards 'output' production costs. **1970** *Nature* 29 Aug. 892/1 The actual pressure behind page charges is the desire of each editor to keep the selling price of his journal within the means of the individual subscriber. **1927** R. B. McKerrow *Introd. Bibliogr.* I. vi. 63 The page 'galley'... The compositors of the Elizabethan period normally finished a page of work at a time. **1964** F. Bowers *Bibliogr. & Textual Crit.* III. i. 65 An example of the routine practice of the trade would be the transfer of lines of type from the stick directly into the Elizabethan page galley instead of the long or slip galley of later times. **1971** *Library* XXVI. 297 The earliest hint of a 'long' galley, as opposed to a 'square' or page galley, seems to appear in an 'oral testimony'..which claims that the news galley of 1770 contained at least 132 lines of matter in long primer between eighteen and twenty ems wide. This suggests a galley with dimensions of about 20 in. × 4 in. **1918** E. Pound *Let.* 4 June (1971) 136 You have got all the points I noted in the page-galleys, so I was right in not cabling about them. [**1975** J. Butcher *Copy-Editing* xii. 213 The printer may be asked to provide page-on-galley proofs. **1976** *Gloss. Documentation Terms* (*B.S.I.*) 46 *Page-on-galley proof*, a single-stage alternative to galley proof and page proof.] **1930** *Times Lit. Suppl.* 30 Oct. 886/2 The page-long questions and answers of the *cause célèbre*. **1961** *Times* 31 Aug. 11/1 These convoluted evocations, often in page-long parentheses. **1899** *Electrical Engineer* (N.Y.) 2 Mar. 249/2 There is..no simple page printer having such speed and such perfect control over page and line as is here secured. **1948** *Ann. Computation Lab. Harvard Univ.* XVI. 69 The output devices of the machine are page printers and tape punches. **1967** Burkhard & Clare in D. H. Hamsher *Communication Syst. Engin. Handbk.* ii. 30 Until recent years page printers were developed almost exclusively for teletype-writer use and were almost entirely mechanical. With a trend toward higher speeds and toward data applications many of the mechanisms are being replaced by electronic circuits. **1895** *Jrnl. U.S. Artillery* Oct. 593 (*heading*) Experimental use of the Essick page printing telegraph for transmitting information in seacoast artillery firing 1895. **1959** J. W. Freebody *Telegr.* ii. 55/2 This machine, known as a teleprinter, printed the messages on a paper tape. In 1931, a page printing machine was introduced. **1881** W. Whitman 24 Aug. in *Daybooks & Notebooks* (1978) I. 256 The first batch of page-proofs of the new volume, to-day. **1901** T. L. de Vinne *Pract. Typogr.: Correct Composition* xvi. 301 Page proofs seriously add to the expense of the work when the author makes much alteration. **1934, 1951** [see *galley *sb.* 5 b]. **1951** W. Stevens *Let.* 30 July (1967) 724 On receipt of page proofs [I] will give them prompt attention. **1975** J. Butcher *Copy-Editing* v. 59 It is usually cheaper to proceed straight to page proof. **1925** *Manual of Style* (Univ. Chicago Press) (ed. 8) 195 Unfilled page references must be queried. **1953** R. L. Collison *Indexes & Indexing* I. 68 Page references should be carefully stated. **1971** *Nature* 30 Apr. 602/1 The index has suffered greatly from the speed of production, being incomplete both in subject matter and the page references given. **1975** J. Butcher *Copy-Editing* x. 184 Form of text reference. The author's name, date of publication and page reference (if one is needed) are given in parentheses. **1929** H. Crane *Let.* 30 Aug. (1965) 344, I think we ought to change our plan regarding page size and use. **1946** *Nature* 24 Aug. 267/1 The members of the Institute hope that British periodicals which adopted reduced page-sizes as a war-time measure will as soon as possible revert to full size. **1969** *Daily Tel.* 11 Aug. 18/3 The page-turner, made of plywood, plastic toy gears and commonplace lamp batteries, operates with the user's suck or blow a miniature electrical device. **1974** *Publishers Weekly* 27 May 5/1 (Advt.), What happens next makes *Eagle in the Sky* a moving, exciting, and ultimately joyous page-turner. **1976** *Gramophone* Mar. 1482/1 As host of the first runthrough by the composer..of the *Elègie*, and page-turner at the English premieres of the two string sonatas, I may be forgiven for having a nostalgia for this music. **1976** *Washington Post* 19 Apr. C5/1 The last time I saw her she was up on that stage without an orchestra; just herself, the piano player, and the page turner. **1976** *Publishers Weekly* 19 Apr. 81/2 Like the other crime novels from the British author, this is a real page turner. **1969** *Daily Tel.* 11 Aug. 18/2 I've been using this page-turning gear without trouble for 18 months.

page, *v.*[1] Add: **c.** To send for, search for, or communicate with (a person) by means of a page; to have the name of (a person) called out by a page. Also in extended use (of various electrical or electronic devices). orig. *U.S.* So **pa·ging** *ppl. a.* and *vbl. sb.*[2]

1904 L. Bell *At Home with Jardines* 65 The name of Jardine was paged through the corridors and billiard-room and café. **1904** *Sun* (N.Y.) 21 Aug. 5 A bell boy is called. 'Here, page Mr. Smith, Room 186', the clerk will say. The process of 'paging' Mr. Smith consists of calling out his name in the dining and other public rooms of the hotel. **1916** H. L. Wilson *Somewhere in Red Gap* ix. 368 A.. mining promoter from Arizona..has himself paged by the boys about twenty times a day so folks will know how important he is. **1923** *Daily Mail* 31 July 6/5 The telephone operator..turned to me. 'Stay around awhile,' she instructed. 'I'll "page" you when I'm through.' **1936** H. F. Olsen in *RCA Rev.* I. I. 58 (*heading*) General announce and paging systems. *Ibid.* 59 For certain types of general announce, paging, and sound distributing installations,..the intensity level required is relatively low. **1938** Wodehouse *Code of Woosters* xiii. 283 Jeeves, go and page Mr. Spode. Tell him I want him to come and put a bit of stuffing into my alibi. **1959** A. Sexton in *Hudson Rev.* Spring 80 Out in the hall The intercom pages you. **1960** *IRE Trans. Vehicular Communications* Dec. 48 (*heading*) Personal radio paging in the VHF band. **1970** *Railway Mag.* Oct. 579/1 Post Office staff at Waterloo have been issued with two-way speech radio paging equipment supplied by Modern Telephones Limited to enable a central control point to speedily contact personnel handling letter bags on the platforms and concourse. **1971** *Daily Tel.* (Colour Suppl.) 22 Oct. 9/2 It works on the principle of the short-range radio paging systems used in factories and large-office complexes. **1973** *Times* 21 May (Telecommunications Suppl.) p. v/2 In connexion with pocket paging systems, dialling a code on the telephone will signal the appropriate paging receiver. The person paged will then dial a reply code on his nearest telephone. **1976** *New Yorker* 26 Jan. 54/2 We'd better have him paged. **1977** *Time* 2 May 49/1 A portable paging device about the size of a cigarette pack, the beeper is a mini-radio receiver that puts the person carrying it on instant call from office, home or anywhere else.

page, *v.*[2] Add: **3.** *intr.* To look *through* the pages of a book.

1943 *Amer. Speech* XVIII. 138 The following notes, taken as I paged through the book at random. **1966** E. Palmer *Plains of Camdeboo* xviii. 291 Paging through the books is an experience for every Palmer of every generation, for a single entry can recall a drama..forgotten for many years. **1970** *Physics Bull.* Sept. 401/1 The selection of a metal..for use in the body is not a simple matter which can be accomplished by paging through a handbook.

pageant, *sb.* Add: **2.** For *Obs.* read '*Obs.* exc. *Hist.*' and add later examples.

1825 T. Sharp *Diss. Pageants Coventry* 20 It is evident that the 'scaffolds' were placed upon wheels, and moved

with the Pageant, to which it probably was attached. **1954** *Oxf. Jun. Encycl.* XII. 261/1 Each guild was usually responsible for one play, which was performed on a special two-decker waggon called a *Pageant*. **1960** Beckson & Ganz *Reader's Guide to Lit. Terms* (1961) 153 The pageant was built on wheels. **1974** S. J. Kahrl *Trad. Medieval Eng. Drama* ii. 36 What distinguishes the pageants referred to in this procession from the stages otherwise described is that the structures for the Jesse tree, for St John and St Edward, the four Cardinal Virtues, and the censing angels..are all built around permanent architectural features.

5. b. A spectacular representation (usually in the form of a procession) of scenes or events belonging to the past history of a place.

1883 D. Cook *On Stage* I. x. 219 In the pantomime season, or whenever any great pageant or spectacle is to be produced, these plots are of prodigious extent. **1905** *To-day* 7 June 180/2 The inhabitants are preparing a pageant. **1908** *Westm. Gaz.* 1 Oct. 2/3 On the sixth of these [days]..there will be presented a historical pageant. **1939** W. Ward *Theatre for Children* xiii. 248 Many playgrounds end their season with a festival or pageant in which every group has some part. **1970** Burton & Lane *New Directions* iii. 78 A great many pageants have been so gruesome—Merrie Englande with rain—the form has earned itself a bad reputation. **1977** K. O'Hara *Ghost of T. Penry* xvii. 173 It was that charity pageant the old mistress put on up Kelletts one year. She needed a robe for an archbishop..when they were acting a crowning.

6. b. *pageant cart, drama, -master* (later examples), *stage, vehicle, wagon*; *pageant-like* adj.

1974 S. J. Kahrl *Trad. Medieval Eng. Drama* ii. 39 Such structures occupy space, as they do in pictures of continental pageant carts. **1975** P. Happé *Eng. Mystery Plays* 27 The civic records of York and Chester show that the plays were performed on pageant carts. **1953** *Travel* Apr. 36/2 This spectacular pageant-drama is a civic non-profit enterprise of the twin cities of Hemet and San Jacinto in Riverside County, California. **1974** *Encycl. Brit. Macropædia* XIII. 862/1 The pageant dramas of the West have tended to be largely open-air performances given in front of mass audiences. **1933** R. Tuve *Seasons & Months* i. 41 A pageant-like processional march of familiar figures. **1973** M. Amis *Rachel Papers* 215 I've included a break-down of one of your more pageant-like essays. **1937** Auden *Spain* 11 Tomorrow the hour of the pageant-master and the musician. **1963** *Times* 17 Apr. 13/3 Miss Gwen Lally, O.B.E., pageant master, play producer, and lecturer, died on Sunday. **1977** *Daily Tel.* 23 June 18 Pageantmaster to Hammersmith & Fulham Silver Jubilee Committee. **1974** S. J. Kahrl *Trad. Medieval Eng. Drama* ii. 34 The most usual event which called for the erection of pageant stages in the street was a royal entry. **1825** T. Sharp *Diss. Pageants Coventry* 18 The different Companies Accounts..refer to the Pageant vehicles. **1932** T. W. Stevens *Theatre* vii. 61 The pageant wagon, interesting in itself, was a sterile device. **1958** A. C. Cawley *Wakefield Pageants* p. xxvi, A pageant-wagon with the superstructure of a ship, would have provided a strong scenic attraction for the *Processus Noe*. **1974** S. J. Kahrl *Trad. Medieval Eng. Drama* ii. 37 Nor, when we come to visualize the early pageant wagons as opposed to the fixed pageant stages, are we helped by the nature of the surviving evidence.

pageanteer. Delete †*Obs.* and add: **b.** One who takes part in a pageant (sense *5 b).

1910 *Daily Chron.* 11 Apr. 1/7 The pageanteers must be enjoying themselves all the time. **1927** *Daily Express* 15 July 2/4 The pageanteers—3,000 of them—assembled on the green and sang 'Land of Hope and Glory'.

pageantry. Add: **1.** (Later example.)

1908 *Daily Chron.* 21 July 4/6 The pageantry brings the classes together.

2. Also *fig.*

1909 E. Pound *Personae* 53 Slow-moving pageantry of hours.

paged, *a.* Add: **a.** (Further examples.)

1868 Geo. Eliot *Let.* 29 Apr. (1956) IV. 433 As I shall not see these paged sheets again, will you charitably assure me that the alterations are safely made? **1869** D. G. Rossetti *Let.* 21 Aug. (1965) II. 715 Replace it at the end of the first section of sonnets—not as paged. **1930** A. E. Housman *Let.* 29 June (1971) 297 In the paged proofs..I find that the printers have inserted..a comma.

c. *Computers.* Divided into pages (*page *sb.*[2] 1 e).

1966 *IEEE Trans. Electronic Computers* XV. 857/1 The relationship between hardware and software in a paged and segmented system environment greatly affects system performance. **1973** *Nature* 6 Apr. 361/2 Among the features that made it [*sc.* Atlas] one of the most advanced machines in the world at the time were a number of facilities that have since become standard such as its permanent master programme and paged store.

pager (pēi·dʒəɹ), *sb.*[2] [f. Page *sb.*[2] + -er[1].] Following a numeral (usu. with hyphen): a book, newspaper or the like having the number of pages indicated by the numeral.

1966 'H. B. Taylor' *Triumvirate* i. 9 He'd gone all out on school supply ads..and turned out a thirty-two pager. **1973** *Times* 5 Sept. 18/7 The Fabian pamphlet, a 50-pager which should be out in a fortnight. **1977** *Times* 12 Sept. 12/6 Pamphlets on..how to get the most out of your phone service. This last is a 14-pager containing 59 pieces of information.

pager (pēi·dʒəɹ), *sb.*[3] [f. Page *v.*[1] + -er[1].] A radio device that emits a sound when activated by a telephone call, used to contact a person carrying it.

1968 *Guardian* 10 Apr. 7/3 There are already..in this country devices called radio pagers. You carry in your pocket the pager, which is linked by radio connection to your telephone. When the telephone rings, the pager blips, and you can answer the call by speaking into the pager. As things stand the pager is illegal. **1973** *Sci. Amer.* Aug. 57 The receiving pager..has a range of 50 miles. It emits a tone when the person carrying it is wanted on the telephone. **1977** *Time* 2 May 49/1 Customers can either rent the pocket pagers..or buy them outright.

Paget (pæ·dʒĕt). *Path.* The name of Sir James *Paget* (1814–99), English surgeon, used in the possessive to designate: **a.** A disease (also called *osteitis deformans*) that affects chiefly the elderly and is often symptomless, being characterized by the localized alteration of tissue in one or more bones (most often in the spine, skull, or pelvis), which become thickened and may undergo fracture or bending.

1877 *Guy's Hosp. Rep.* XXII. 337 (*heading*) Osteoporosis or Paget's osteitis deformans. **1889** *Index-Catal. Library Surg.-General's Office, U.S. Army* X. 353/2 (*heading*) Paget's disease of the bones. *See* osteitis deformans. **1939** *Times* 18 May 4/6 Dr. Alexander Wilson said that Mr. Desnos had Paget's disease—the bones of his skull were brittle and at least 50 per cent. softer than a normal skull. **1963** *Lancet* 5 Jan. 34/2 In Paget's disease stones are formed as a result of increased excretion of calcium due to increased turnover of bone. **1966** Wright & Symmers *Systemic Path.* II. 1390/2 Paget's disease is a precancerous condition. **1974** Passmore & Robson *Compan. Med. Stud.* III. xxvi. 17/2 Paget's disease is rarely a painful disorder in the absence of complications such as fracture, tumour or arthritis.

b. A reddish eczematous condition of the skin associated with cancer of the underlying tissue and usually occurring in the female nipple; freq. called *Paget's disease of the nipple*; so *Paget* († or *Paget's*) *cell*, a large cell with clear cytoplasm and hyperchromatic nuclei found in the epidermis of the affected area.

1880 *Brit. Med. Jrnl.* 24 Jan. 128/1 As attention was first called to it by Sir James Paget, he [*sc.* J. E. Erichsen] would suggest for it the name 'Paget's disease' of the nipple. **1910** *Jrnl. Cutaneous Dis.* XXVIII. 383 The epidermis..contained many of the large round, so-called Paget cells. **1917** *Lancet* 7 Apr. 519/2 The degenerate deeper layers of the epidermis..contain the large clear cells known as Paget's cells. **1923** *Brit. Jrnl. Surg.* XI. 317 Carcinoma in the breast, with which Paget's disease of the nipple is usually associated, is a primary carcinoma of the breast epithelium. **1954** *Jrnl. Obstetr. & Gynæcol.* LXI. 758 (*heading*) Histochemical characterization of the specific cells in Paget's disease of the vulva. **1966** Wright & Symmers *Systemic Path.* I. xxviii. 995/2 Paget's disease of the breast..accounts for between 2 and 3 per cent of all mammary carcinomas. *Ibid.* 996/2 Patients with Paget's disease are still very likely to be treated for supposed dermatitis of the nipple. **1974** Passmore & Robson *Compan. Med. Stud.* III. xxx. 9/2 Some believe that the Paget cells represent intra-epidermal spread of this [*sc.* cancer].

paging, *vbl. sb.*[1] Add: **2.** *Computers.* Division (of storage) into pages (*page *sb.*[2] 1 e); the transfer of pages between the central store and an auxiliary store.

1966 *IEEE Trans. Electronic Computers* XV. 855/1 The computer addressing techniques known as paging and segmentation will be familiar. **1970** O. Dopping *Computers & Data Processing* ix. 124 Paging systems and hardware relocation..are comparatively new inventions. **1972** *IEEE Trans. Computers* XXI. 1053/1 Under demand paging, a single program's execution and its resulting page swapping can be overlapped.

paging, *vbl. sb.*[2]: see *page *v.*[1] c.

pagoda. 5. pagoda sleeve (earlier and later examples).

1872 *Queen* 3 Feb. 71/2 Pagoda sleeves..embroidered all over. *Ibid.* 1 June 391/3 Pagoda sleeves with rich lace ruffles beneath. **1873** *Young Englishwoman* Apr. 183/1 The casaque..has a large pagoda sleeve. **1952** C. W. Cunnington *Eng. Women's Clothing* ii. 33 A 'pagoda sleeve' (long and narrow, opening some five or six inches above the wrist with small white undersleeves).

‖ **pagri** (pa·gri). Var. Puggree. Also *pagri-cloth*.

1901 [see Puggree 2]. **1930** *Aberdeen Press & Jrnl.* 22 Apr. 5/2 He has no British officers and no uniform except a distinguishing kind of pagri (head-dress). **1930** *Punch* 1 Oct. 392/2 Mr Thompson should not allow this bee to find a permanent home in his *pagri*. **1934** [see *drill *sb.*[4]]. **1974** 'B. Mather' *White Dacoit* 18 Sowars straightened tunics and pagris. **1978** 'M. M. Kaye' *Far Pavilions* vi. 98 She slept soundly..tied to him by a length of *pagri* (turban) cloth that prevented her from falling.

-pagus, *suffix*, f. Gr. πάγος that which is fixed (f. πηγνύναι to fasten), used to form the names of different kinds of Siamese twins according to their site of attachment; as *thoracopagus* s.v. Thoraco-.

pah, pa, *sb.* Add: (Further examples.)

The spelling *pa* is now the more usual.

1858 *App. Jrnls. House Reps. N.Z.* E. iv. 4 They seem ..at present inveterate in their adherence to their Native habits, and to their residence in Pas. **1863** T. Moser *Mahoe Leaves* 14 A pah is strictly a *fortified village*, but..a collection of huts forming a native settlement is generally called a pah now-a-days. **1880** J. C. Crawford *Recoll. Trav. N.Z. & Austral.* 28 A large pa (village) was visible at Pitone. **1900** H. Lawson *Over Sliprails* 82 Things, according to pa gossip, had gone wrong with her from the date of the tragedy. *Ibid.* 88 A poor pa outcast who had negro blood in her veins. **1905** J. M. Thomson *Bush Boys N.Z.* iii. 45 In every Maori 'pah' or village, there is one specially large 'whare', or house. **1911** W. H. Koebel *In Maoriland Bush* xii. 170 The old Maori did not appear to have lost caste among the brethren of his own race who still dwelt in their quaintly carved Pahs. **1920** *Outlook* (N.Z.) 29 Nov. 18/3 Her parish is roadless, deep ravines and steep hills separate pa from pa. **1938** R. Finlayson *Brown Man's Burden* 52 The old pa where many of the chieftain's ancestors and relatives were laid to rest on the topmost parapet. **1958** S. Ashton-Warner *Spinster* 72 When she got to the pa they cried in the Meeting House. **1959** G. Slatter *Gun in my Hand* 99 Unpainted whares at the pa with cardboard at the windows and new cars parked on the lawn. **1963** D. Adsett *Magpie Sings* 109 Perhaps they would go and live in the pa and get big and fat like lots of other Maori women.

‖ **paha** (pā·hā). Pl. **paha.** [Malay.] In Malaysia, a unit of weight used esp. for gold and equal to ¼ tahil, equivalent to ⅓ oz. (9·4 grammes-weight).

1839 T. J. Newbold *Pol. & Statistical Acct. Straits of Malacca* II. xiv. 236 A fine of ten tahils and one paha is to be exacted. **1947** R. O. Winstedt *Malays* 42 If the ordinary person pay five *paha* of gold in consideration of his low birth, then the marriage can take place.

Pahari (pahā·rī), *sb.* and *a.* Also **paharia, Pahaṛiya.** [Hind. *pahāṛī* (language) of the mountains, f. *pahāṛ* mountain.] **A.** *sb.* **a.** An Indo-Iranian language group to which belong the languages spoken in the lower ranges of the Himalayas from Nepal to Chamba. **b.** (Also **pahareen.**) A native or inhabitant of this region. **B.** *adj.* Of or pertaining to this region, its inhabitants, or the languages spoken by them.

The name was originally applied to one of the languages or dialects which formed the group (so quots. 1857, 1886).

1811, etc. [see *Maler *sb.* and *a.*]. **1857** *Jrnl. Asiatic Soc. Bengal* XXVI. 317 The languages included in..these two papers are..Dahi or Darhi..Dênwár..Pahi or Pahari, [etc.]. **1876** [see *Kashmir 1]. **1884** [see *Malto *sb.*]. **1886** Kipling *Plain Tales from Hills* (1888) 1 The Kotgarh Chaplain christened her Elizabeth, and 'Lispeth' is the Hill or *pahari* way of pronouncing it. **1887** R. N. Cust *Linguistic & Oriental Ess.* (ser. 2) iii. 73 It is..more convenient..to treat Hindi as the unit, and then allow full room to its magnificent dialects, such as..Bagri, Pahári, [etc.]. **1901** Kipling *Kim* iv. 93 'Huh! It is only a *pahari*' (a hillman), said Kim over his shoulder. 'Since when have the hill-asses owned all Hindustan?' *Ibid.* 107 A *pahareen*—a hillwoman of Dalhousie, my mother. **1905** *Daily Chron.* 9 Sept. 3/1 This particular district has been inhabited, since the British occupation, by two distinct races, the Santals and Paharias... The Paharias are a mountain race. **1908** T. G. Bailey *Lang. N. Himalayas* III. p. i, Of these all, except Lāhulī, belong to what is at present called the Western Pahāṛī language of the Northern Group of the Sanskritic Aryan Family. **1916** G. A. Grierson *Linguistic Survey India* IX. IV. 1 The Pahāṛī languages fall into three main groups. *Ibid.* 2 The mass of the Aryan-speaking population of the Himalayan tract in which Pahāṛī is spoken belongs, in the West, to the Kanēt and, in the East, to the Khas caste. **1923** A. Turnbull *Nepali Gram. & Vocab.* (ed. 3) 1 They also frequently refer to the language as 'Pahāṛī', or 'Pahāṛiyā', though, strictly speaking, it is only one of the many forms of Pahāṛī. **1943** R. Godden *Rungli-Rungliot* 37 Monbad speaks Hindustani and Paharia. **1950** *Encycl. Brit.* XVIII. 773/2 The hill dialects, known as Pahari, are akin to the language spoken in Rajputana. **1971** *Hindustan Times* (New Delhi) 7 Apr. 7/3 Mr Khandalavala's books include Pahari Miniature Painting, an exhaustive survey of painting in the Punjab hills. **1974** *Encycl. Brit. Micropædia* VII. 676/3 *Pahari painting*, miniature painting and book illustration that developed in the independent states of the Himalayan foothills in India. **1974** *Listener* 31 Jan. 136/3 The shops are run by Sikhs: the Paharis, the hill people, are not really interested.

‖ **pahit** (pa·hit). [Malay, = bitter.] In full *gin pahit*. Gin and bitters.

1914 G. Frankau *Tid'apa* (1915) ii. 11 He had shouted for *pahit* and for *stinger* till the hot, strong bane of them swept In flame to each brain-cell's tinder. **1923** W. S. Maugham in *Nash's & Pall Mall Mag.* Apr. 32/1 The Irishman ordered a dry martini for her and a gin pahit for himself. **1932** —— *Book-Bag* 21 Shall you be ready for a gin pahit in ten minutes? **1961** Conyn & Marten *Bali Ballet Murder* iii. 41 *Gin pahit* as an aperitif. **1963** J. Kirkup *Tropic Temper* 213 Patrons are requested to reserve their tables for dinner before ordering their pahits. **1965** O. A. Mendelsohn *Dict. Drink* 251 *Pahit*, Malay Club term for long gin drink. **1968** *Punch* 12 June 852/2 There was the euphoria of foreseeing ourselves in Maugham fiction—white-tuxedoed dinner-parties, the fans turning on the ceiling, gin *pahits* on the veranda, humid adultery behind the jalousies.

Pahlavi, *a.* and *sb.* Add: (Earlier and later examples.) Also **Pehlavi, Pehlevee, Pehlevian.**

1773 W. Jones tr. *Astarabadi's Hist. Nader Shah* App. 159 Barzuien learned the Indian tongue, and, having.. procured a copy of the book, translated it into the Pehlevian dialect: about an hundred and forty years after, his work was turned from Pehlevi into Arabick. **1777** J. Richardson *Dict. Persian, Arabic, & Eng.* p. iv/2 The idiom of Farsistan (Persia Proper)..had an extensive range over the most civilized of the lower districts: whilst the Pehlavi prevailed chiefly around the Mazenderan or Caspian Sea. **1789** [see Zend 1]. **1815** M. Elphinstone *Acct. Kingdom of Caubul* II. iv. 191 Some of these Zend and Pehlevee words are, however, common to the Shanscrit. **1950** R. G. Kent *Old Persian* 6/2 Middle Iranian includes the Iranian dialects as they appear from about 300 B.C. to about 900 A.D. They are in general called Pahlavi. *Ibid.* 7/1 Arsacid Pahlavi was the official language of the Arsacid dynasty of Parthia, which ruled from 250 B.C. to 226 A.D... The Sasanian or South-west Pahlavi was the official language of the Sasanian dynasty, which ruled from 226 A.D. until..652. **1968** K. Jahn *Rypka's Hist. Iranian Lit.* 34 There is a great wealth of Middle Persian book-literature, usually known as Pahlavi literature. The language is also called book-Pahlavi. **1972** W. B. Lockwood *Panorama Indo-European Lang.* 235 Middle Persian or Pahlavi, as the language of this literature is also called.

‖**pahoehoe** (pāhōᵘ·ihōᵘ,i). *Geol.* Also †**paho-ihoi.** [Hawaiian.] A form of solidified lava that is undulating or billowy in form and has a shiny appearance. Cf. *Aa².

1859 [see *Aa²]. **1864** R. Anderson *Hawaiian Islands* 142 The broken lava is piled ten or fifteen feet above the smooth, hard *pahoihoi*. **1869** *Q. Jrnl. Geol. Soc.* XXV. 434 From this a stream of the smooth satin-like lava called 'pahoehoe' in Hawaii flowed for a few hours. **1972** *Islander* (Victoria, B.C.) 24 Sept. 4/2 One is pahoehoe—a taffy-like lava that has hardened into folds and creases that give it a smooth, ropy look, like frosting that has spilled over the top of a cake. **1975** *Nature* 29 May 387/1 The eruptions of this period include two..which produced pahoehoe instead of the more normal aa flows of Etna.

paiche (pai·ʃe). [Amer. Sp.] = *Arapaima, *Pirarucú.

1961 E. S. Herald *Living Fishes of World* 99/1 The largest species in the family [*sc.* Osteoglossidæ] is the fabulous *Arapaima gigas*, called paiche in Peru, pirarucú in Brazil, and arapaima in British Guiana. **1962** N. Maxwell *Witch-Doctor's Apprentice* xx. 251 At each stop, the block of salt fish grew higher. It was all *paiche*, a giant fish often attaining a weight of more than two hundred pounds, which is found only in the Amazon River system. **1976** 'A. Hall' *Kobra Manifesto* xiii. 174 We were eating *paiche* with *farinha*.

paid, *ppl. a.* Add: **2.** (Further examples.)

1961 Webster s.v., A good job and a paid vacation. **1967** *Listener* 30 Nov. 694/2 The shortening of hours of work, improvements in housing standards, paid holidays, the prohibition of child labour,..should all, on the face of it, have helped to intensify family cohesiveness. **1978** *Ibid.* 19 Jan. 78/3 After the Second World War, paid holidays became common.

4. a. *paid for*, esp., an instruction given to a dog that has 'earned' a reward; *paid-up* (earlier and further examples).

1848 Mill *Pol. Econ.* II. v. ix. 463 A known and large amount of paid-up capital. **1854** J. C. Maitland *Cat & Dog* 8 To please Lily, I learned to sit patiently watching the most tempting buttered crust on the ground under my nose, when she said 'Trust, Captain!' never dreaming of touching it till she gave the word of command, 'Now it is paid for;' when I ate it in a genteel and deliberate manner. **1874** 'Mark Twain' *Gilded Age* xlviii. 435 What the insurance companies call the 'endowment', or the 'paid-up' plan, by which a policy is secured after a certain time without further payment. **1913** [see *Extended *ppl. a.* 2 c]. **1922** 'R. Crompton' *Just—William* xii. 238 They taught him to sit up and almost taught him 'Trust' and 'Paid For'. **1934** 'R. Hull' *Murder of my Aunt* ii. 107 It seemed eternity to me, and I suppose to So-so too, waiting for the glad sound of 'Paid for'. **1970** J. Fleming *Young Man, I think you're Dying* ix. 127 The manner of dogs put on *Trust* for a biscuit, watching the biscuit until the words: *Paid for!*

b. In phr. *paid-up member*, a member of a club, society, etc., who has paid a subscription.

1959 H. Hobson *Mission House Murder* xxvi. 176 A fully paid-up member of Medina, Soho's equivalent of the old Mafia. **1960** *News Chron.* 1 July 6/7 Prince Philip always..[visits] the club when he attends Cowes Week—he's a fully paid-up member. **1970** 'H. Carmichael' *Remote Control* ii. 24 He was a paid-up member of the Kennel Club. **1976** *Conservation News* Nov./Dec. 12 That issue contained the application for non-voting tickets for paid-up members not attending as delegates.

paidle, var. Paddle *sb.*³ in Dict. and Suppl.

paido-: see Pædo-, pedo-.

paigeite (pēⁱ·dʒəit). *Min.* [f. the name of Sidney *Paige* (1880–1968), U.S. geologist + -ite¹.] A borate of iron and magnesium, $(Fe^{II},Mg)_2Fe^{III}BO_5$ with more ferrous iron than magnesium, which is found as black orthorhombic crystals.

1908 Knopf & Schaller in *Amer. Jrnl. Sci.* CLXXV. 324 The other mineral, for which we propose the name *paigeite*, in honor of Mr. Sidney Paige of the Geological Survey, was found at two localities, at Brooks Mountain [Alaska] in loose blocks, and at Ear Mountain, 40 miles to the northeast, *in situ*. **1954** [see *Hulsite]. **1961** *Doklady Earth Sci.* CXXXIV. 1004 (*heading*) On the discovery of warwickite and paigeite in Precambrian dolomitic marbles of North Korea.

‖**pai-hua** (pai,hwā). Also **bai hua, báihuà.** [Chinese *báihuà*, f. *bái* white, clear, plain + *huà* language, speech.] The standard written form of modern Chinese, based on the northern dialects, esp. that of Peking; the vernacular literary style (opp. *Wenyen). Also *attrib.* Cf. *Putonghua.

1923 B. Karlgren *Sound & Symbol in Chinese* iii. 37 Some modern newspapers have tried to introduce *pai hua* 'white language', i.e. vulgar style, colloquial, at least in one or two columns, but with no great success. **1932** O. M. Green in *Asiatic Rev.* XXVIII. 114 The Literary Revolution of 1917 to 1919..secured the adoption of the Pai Hua, the most widely spoken language in China, for all literary purposes. Not only newspapers and magazines but many standard works are now printed in the Pai Hua. **1936** N. Wales in E. Snow *Living China* 336 The healthy parvenu *pai-hua*, 'plain speech', literature of the people in the spoken language, ashamed of itself and despised and outcast by the *wen-yen literati*. **1937** E. Snow *Red Star over China* i. 44 These Shensi hill people have a dialect of their own..but they understand *pai-hua*, or mandarin Chinese. **1950** J. De Francis *Nationalism & Lang. Reform in China* i. 7 The *paihua* or colloquial style was, roughly speaking, speech reduced to writing in the ideographic script. **1968** P. Kratochvíl *Chinese Lang. Today* v. 163 Perhaps the most influential factor..was the outburst of writing in the new style which followed the rejection of *wényán*. This style called *the new báihuà*, or only *báihuà* (usually translated as 'vernacular'; the etymology of the Chinese term which means something like 'plain language' is not quite certain), grew partly out of the tradition of popular old *báihuà* writing mainly represented by the great medieval novels.

pai kau (pai kau). Also **pai kow, pie-gow.** [Cantonese, f. *p'ai* tablet + *kau* nine.] A Chinese gambling game played with dominoes.

1906 [see *Chuck-a-luck]. **1969** R. C. Bell *Board & Table Games* II. vi. 102 Pai Kow is played with a full set of Chinese dominoes. **1972** *Guardian* 29 May 16/2 Mr Liu ..wants..to attract his countrymen away from Mah Jong and Pai Kau—a sort of dominoes played like poker. **1978** *Daily Tel.* 13 July 1/7 The legalising of exotic games such as fan tan and pai kau in casinos.

‖**pailou** (pai·lōᵘ). [Chinese, f. *p'ai* tablet + *lou* tower.] An elaborate Chinese commemorative or ornamental gateway.

1836 J. F. Davis *Chinese* II. xviii. 321 The emperor occasionally orders a *pae-low* to be erected at the public expense. **1887** *Chinese Times* 1 Oct. 785/2 The homes of the dead, with their *p'ailou* and carved images of lions, sheep, &c. **1923** *Blackw. Mag.* Jan. 101/2 In front of one of them [*sc.* temples] stands a white stone *pailou*, which shines in the sunlight like a flamboyant Stonehenge trilith. **1947** *Archit. Rev.* CII. 13 The p'ailou is a commemorative gateway, set up either to mark a famous place or to serve as a monument to the dead and having three or five openings with double or triple lintels and tiled roofs. **1948** R. Alley *Gung Ho* 28 Under the great pailou framing the main street. **1958** W. Willetts *Chinese Art* II. viii. 736 This class of monumental stone arch or gateway... Commemorative or triumphal in function... *P'ai-lou* are almost always built of stone, in close imitation of a wooden prototype and with carpentry technique. They have one, three, or five openings. **1977** *N.Y. Rev. Bks.* 26 May 21/2 The destruction of Peking started in the 1950s, when all the pailous that spanned the main thoroughfares of the old city were eliminated.

pain, *sb.*¹ Add: **3. d.** *pain in the neck* (colloq.) (also simply *pain*), an annoying or tiresome person or thing; also, in same sense (but *vulg.*), *pain in the arse*. Also, *to give* (someone) *a pain* (*in the neck* or *arse*), to be annoying or tiresome (to someone).

1908 R. E. Knowles *Web of Time* xiv. 144 'There's naethin' like the guid auld oatmeal.' 'You Scotch folks give me a pain,' broke in David. **1912** *Maclean's Mag.* Nov. 68/1 Bill, you give me a pain. **1924** Wodehouse *Leave it to Psmith* ix. 188 He got there first, damn him! Wouldn't that give you a pain in the neck! **1933** E. B. White *Let.* Mar. (1976) 112 All through the campaign I thought Mr. R. was something of a pain. **1934** [see *Ass]. **1937** [see *Dinkum *a.*]. **1941** W. A. Percy *Lanterns on Levee* 77, I was a sickly youngster..a frail problem child, a pain in the neck. **1951** 'A. Garve' *Murder in Moscow* x. 102 What do we really know of Mullett, except that he was a pain in the neck to everybody? **1958** *Spectator* 7 Feb. 175/3 The Liu was almost as big a pain in the neck as the previous night's Preziosilla. **1967** W. Soyinka *Kongi's Harvest* 14 Your uncle is a pain in the neck. **1970** *Times* 7 Jan. 7/7 Anthony Quinn.. plays a wise, noble, feckless, life-loving Greek dispenser of advice, lay preacher and general pain in the neck. **1972** D. Ramsay *Little Murder Music* 8 Hey, Jack, how does his royal pain in the ass intend to take the *Scherzo*? **1972** *Times* 13 Sept. 7/5 He represents about 1 per cent of the blokes... He is a pain. **1973** 'E. McBain' *Hail to Chief* i. 6 Homicide cops..were pains in the ass to detectives actually..trying to solve murder cases. **1975** *New Yorker* 21 Apr. 103/3 She is a pain, and, unconsciously, the source of many of the troubles that follow. *Ibid.* 17 Nov. 125/1 Fiction, in whatever form, about real people is more often than not a pain, and sometimes downright pernicious. **1976** A. White *Long Silence* xiv. 123 Lieutenant Otto Andersen..was a consistent pain in the arse to Colonel Birkenkamp, constantly reporting his men for slovenly dress. **1977** *Rolling Stone* 7 Apr. 12/2 It was an increasing pain in the ass to do the same material each night.

7. a. pain point *Physiol.* = *pain spot*; **pain spot** *Physiol.*, a small spot on the surface of the skin that is sensitive to pain; **pain-threshold**, the upper limit of tolerance to pain.

1897 tr. *T. Ribot's Psychol. of Emotions* 27 Goldscheider..admits pain-points (points sensible to pain), but not a specific organ for pain nor special nerves to transmit it. **1954** S. Rothman *Physiol. & Biochem. Skin* v. 136/2 There is no doubt that itching is produced with great ease when the stimulus is weak and repetitive and when several pain points are stimulated simultaneously. **1888** W. Stirling tr. *Landois's Text-bk. Human Physiol.* (ed. 3) xiv. 831 The pain-spots can be isolated by means of a needle, or electrically. **1927** Haldane & Huxley *Animal Biol.* v. 126 Stimulation of a single pain-spot will only cause movement after a long time or never. **1973** C. P. Swanson *Nat. Hist. Man* viii. 237/2 (*caption*) Pain spots on normal hands of two individuals. **1902** Pain-threshold [see *Misery 8]. **1969** 'I. Drummond' *Man with Tiny Head* xvi. 182 Sandro guessed that his pain-threshold would be high and that he would give nothing away.

b. *pain-shot* adj.

1911 Beerbohm *Zuleika D.* iv. 44 He was gazing at the girl with pain-shot eyes.

c. *pain-giving, -killing, -relieving* adjs.

1889 'Mark Twain' *Connecticut Yankee* xviii. 215 The executioner..was a good, pains-taking and pain-giving official. **1890** W. James *Princ. Psychol.* II. xxi. 306 Locke expressly makes the *pleasure*- or *pain-giving* quality to be the ultimate human criterion of anything's reality. **1934** R. Bodley *Japanese Omelette* iv. 30 The geta..supposed to fit the sole of a Japanese foot..are, to my mind, the most paingiving form of footwear ever devised. **1964** J. J. Walsh *Understanding Paraplegia* iv. 24 In many cases, the so-called 'harmless' pain-killing drugs are not sufficiently strong to stop the pain. **1974** 'J. Graham' *Bloody Passage* ix. 125 The pain-killing injection had helped. **1908** *Practitioner* Dec. 850 The experiment was absolute proof of the pain-relieving quality of congestion. **1935** *Discovery* Aug. 226/2 One very fortunate property which such a generator appears to possess is its pain-relieving virtue. **1966** *Lancet* 31 Dec. 1436/1 Digitalis, quinidine,.. and pain-relieving drugs were given when indicated to both groups.

d. pain-killer (earlier and later examples); **pain-proof** *a.*, having immunity from pain.

1853 *La Crosse* (Wisconsin) *Democrat* 7 June 2/4 Ayer's Cherry Pectoral, Perry Davis' Pain Killer. **1855** I. C. Pray *Mem. J. G. Bennett* 200 The many pain-killers invented have diminished largely the amount of human suffering. **1863** W. B. Cheadle *Jrnl. Trip across Canada* (1931) 102 Milton rubs his face with pain-killer. **1932** J. Steinbeck *Pastures of Heaven* x. 236 Pat could hear the sizzle of mentholatum and painkiller gushing from containers and boiling into the fire. **1959** [see *benzocaine* s.v. *Benzo-]. **1973** 'R. MacLeod' *Burial in Portugal* iii. 68 Finding the painkiller tablets, he swallowed a couple. **1977** A. Morice *Murder in Mimicry* I. iv. 33 In those days you had to get by with ordinary painkillers which were about as effective as a slug of brandy to a man having his leg amputated. **1903** 'Mark Twain' in *North Amer. Rev.* Jan. 3 No C[hristian] S[cience] family would consider itself..pain-proof without an Annex.

pain, *sb.*² **2.** *pain perdu* (later examples.)

1723 [see *Cream *sb.*¹ 7 b]. **1941** W. A. Percy *Lanterns on Levee* 11 Oh, the poor little boys who never put a lump of butter into steaming butter-bread (spoon-bread is the same thing) or lolled their tongues over pain-perdu. **1961** T. Henrot *Belgium* 189 *Pain-perdu*—rusk softened in warm, sweetened milk then browned in butter. **1972** *Guardian* 28 Jan. 9/5 Pain Perdu or Gilded Crusts..are very popular with children.

pain, *v.* **3.** (Later example.)

1885 W. Pater *Marius* II. 213 Christ, paining in him, set forth a copy to the rest.

painedly (pēⁱ·ndli, pēⁱ·nĕdli), *adv.* [f. Pained *ppl. a.* + -ly².] In a pained manner.

1921 D. H. Lawrence *England, my England* (1922) 270 Mr. Enderby looked up painedly. **1926** Fowler *Mod. Eng. Usage* 418/2 *Painedly*. A bad form.

painfully, *adv.* Add: **4.** *fig.* Excessively, to an alarming degree.

1900 [see *Effortful *a.*]. **1909** A. Lancastre Saldanha *Recoll.* xi. 151 Sir Robert Peel was painfully shy with strangers. **1941** N. Coward *Australia Visited* i. 2 An R.A.F. plane which seemed to me almost painfully small. **1961** *Flying* (N.Y.) Feb. 33/1 At this point it should be painfully obvious that cities, being 'soft', and the people within them are ideally suited to destruction by nuclear weapons.

pa-in-law. *colloq.* = Father-in-law.

1886 H. Baumann *Londinismen* 130/1 *Pa-in-law*,.. father-in-law. **1949** E. Coxhead *Wind in West* xii. 289, I couldn't throw away the chance of poking my fingers into pa-in-law's pie. **1952** M. Allingham *Tiger in Smoke* xiv. 213 Are you worrying about..my future pa-in-law?

pai·nstakingness. [f. Painstaking *a.* + -ness.] The fact or habit of taking pains; assiduous effort.

1927 *Sunday Express* 19 June 19/3 The sportingness of owners, the painstakingness of trainers, and the brilliance of jockeys. **1936** *Times Lit. Suppl.* 2 May 366/3 His dicta are analysed by the author with the same painstakingness.

paint, *sb.* Add: **2. e.** Phr. *as smart* (*pretty*, etc.) *as paint*: superlatively smart, pretty, etc.

1850 F. E. Smedley *Frank Fairleigh* xli. 340 Why, Oaklands, man, you are looking as fresh as paint; getting sound again, wind and limb, eh? **1883** R. L. Stevenson *Treas. Isl.* viii. 65 You're a lad, you are, but you're as smart as paint. I see that when you first came in. **1905** *Tatler* 22 Feb. 306/3 Half the English [Rugby] side were played out twenty minutes before the end while the Irish-

men were still as fresh as paint. **1918** A. Quiller-Couch *Foe-Farrell* 176 He stared..across at the grouped rustic buildings, all as pretty as paint. **1930** H. A. Bryden *Enchantments of Field* 187 After all, your hounds may be as handsome as paint, but if they fail you in nose, cry and hunting-power they are worse than useless. **1963** N. Marsh *Dead Water* (1964) iii. 58 Miss Emily arrived at noon on Monday. She had stayed overnight in Dorset and was as fresh as paint. **1975** J. I. M. Stewart *Young Pattullo* ii. 40 Not always wholly agreeable, perhaps, but as clever as paint.

5*. *sb.* and *a.* = Pinto *a.* and *sb.* (Chiefly *U.S.*)

1848 Bartlett *Dict. Amer.* 243 In some of the Southern States, a horse or other animal which is spotted, is called a *paint*. **1869** *Overland Monthly* III. 126 A black-and-white-paint horse, fifteen hands high. **1909** 'O. Henry' *Roads of Destiny* vi. 96 Sam Kildrake's old paint hoss that killed hisself over-drinkin' on a hot day. **1948** *Sun* (Baltimore) 18 June 15/4 The Appaloosa somewhat resembles the paints, pintos and calicos, so popular with the plains Indians. **1955** W. Foster-Harris *Look of Old West* viii. 226 An animal violently splotched with different colors was called a paint. **1975** J. Hansen *Trouble Maker* i. 2 She led out a little paint mare. *Ibid.* xii. 125 The sorrel followed the paint. **1976** *Billings* (Montana) *Gaz.* 2 July 11-C/6 (Advt.), 8 yr old paint gelding. Child broke. Possible show horse.

6. *paint-drum, -job, -oil* (earlier example), *-rag, -shop, -stoving, -work* (further examples); *paint-remover* (earlier and later examples), *-stripper, -thinner*; *paint-dappled, -daubed, -speckled* adjs.; **paint box** (earlier examples); **paint-brush** (*b*) = *Indian paint-brush* (*Indian A. 4 b); **paint card**, a card showing a graduated range of paint colours; also *fig.*; **paint frame** (examples); **paint-roller**, a roller covered in an absorbent material which holds paint to be applied to a surface; also *attrib.*; **paint spray**, a device for spraying paint on to a surface; hence *paint-spray* vb. trans., *paint-sprayed* ppl. adj.; **paint-stone**, a stone used as a source of paint.

1725 *New-England Courant* 8–15 Feb. 1/2, I would oblige every Sign-Painter to serve seven Years at College, before he presum'd to handle Pencil or Paint-Box. **1820** Shelley *Posthumous Poems* (1824) 62 Near that a dusty paint box, some old hooks. **1915** Armstrong & Thornber *Field Bk. Western Wild Flowers* 472 Paint Brush. *Castilleja miniata.* Red. Summer. Northwest. This is a very handsome kind, from one to four feet tall. **1968** Mrs. L. B. Johnson *White House Diary* 9 Apr. (1970) 656 Great splashes of wildflowers began to appear along the road and in the pastures..pale pink buttercups, wild verbena, coral paint brush, Indian blanket. **1931** D. Runyon *Guys & Dolls* (1932) xiii. 266 A nervous man ..with a blood pressure away up in the paint cards must live quietly. **1961** [see *grey scale* s.v. *Grey *a.* 8]. **1904** *Windsor Mag.* Jan. 234/1 He handed me a paint-dappled copper stencil-plate, two feet square. **1945** W. de la Mare *Burning-Glass* 54 A paint-daubed woman bound for lonely bed..Stood watching him. **1920** *Blackw. Mag.* Apr. 499 The paint-drums..had been jolted bodily from their lashings. **1901** C. Morris *Life on Stage* v. 31 Run upstairs to the paint-frame (three flights up) and ask the painter to put a little ad-libitum in this bottle for me. **1954** *Archit. Rev.* CXVI. 114 A similar movement may still be seen driving the paint-frame at the Leicester Theatre Royal. **1961** *New Statesman* 20 Jan. 81/1 Backstage, Mr Moro plans..a full-scale paint frame. **1970** *New Yorker* 14 Nov. 59/1 The danger is always that the result will be a paint job, very superficial. **1971** *New Scientist* 22 Apr. 224/3 An imaginative and informed paintjob can achieve things which no camera or CRT could ever do. **1978** G. Vidal *Kalki* xi. 251 The White House (which needs a paint job). **1727** in *Maryland Hist. Mag.* (1923) XVIII. 227 Glass, Paint oile, Druggs and Stationary ware. **1938** N. Marsh *Artists in Crime* i. 4 The painter..found..a handkerchief that had been used as a paint-rag. **1975** N. Freeling *What are Bugles blowing For?* 73 The woman..was wiping dust off her hands with a paint-rag. **1885** *List of Subscribers, Classified* (United Telephone Co.) (ed. 6) 133 Manufacturers..Paint Remover. **1960** *Practical Wireless* XXXVI. 306/1 The enamel can then be softened..by immersing them in a paint remover for perhaps a quarter of an hour. **1962** L. Deighton *Ipcress File* xx. 131 She offered me one of those menthol cigarettes that taste like paint remover. **1973** F. Taubes *Painter's Dict.* 175 Paint remover is sold under various trade names, and each brand is of similar formulation. The fluid is used to remove old paint layers and oil-varnish films. **1951** N. & S. Mager *Amer. Housch. Encycl.* 623 Paint rollers. **1954** *Good Housek. Home Encycl.* 208/1 *Paint Rollers*: Covered in lamb's wool on special felt,..these rollers are simple to use. **1958** *Times* 7 July 6/5 The latest paint roller improvement is a foolproof model which holds and automatically feeds the right amount of paint on to the surface. **1959** *Listener* 23 Apr. 739/1, I have been trying out the self-feeding paint rollers that have recently come on the market. **1971** *Handyman Which?* Nov. 25/2 There are lots of paint rollers and their prices vary..from 28p to about £2. **1866** *Oregon State Jrnl.* 23 June 3/4 Enquire of W. W. Winter, at paint shop under picture gallery. **1899** *Sat. Even. Post* 10 June 795 Eight hundred tons of white lead are ground in the paint shop every twelve months. **1949** J. H. Ousbey *Cellulose Spraying* xv. 61 The bodies, on which the doors have already been hung, come to the paintshop by overhead conveyor and are in bare metal. **1973** *Times* 18 May 26/5 By eliminating the solvents and attendant fumes, Fiat said the danger to the lungs of paintshop employees is sharply reduced. **1922** Joyce *Ulysses* 442 Their paint-speckled hats wag. **1962** *Punch* 11 July 57/1 Twenty thousand men downing paint-sprays at Dagenham. **1967** M. Chandler *Ceramics in Mod. World* iii. 99 Almost all products of this kind..are glazed by spraying, in much the same way as car bodies are paint-sprayed. **1971** *Money Which?* Mar. 69/1 Small tools, such as hand drills, paint sprays, and so on, are strictly speaking capital. **1973** *Scotsman* 13 Feb. 8/6 Dundee's modern shopping precinct has now been further decorated with paint-sprayed gang slogans. **1797** A. Barnard *Let.* 29 Nov. in *S. Afr. a Century Ago* (1901) 121 The 'paint stone' is found in this neighbourhood [*sc.* Paarl] in quantities—namely, an impalpable powder which, mixed with oil, serves the country people with colour to paint their waggons, houses, etc. **1896** *13th Ann. Rep. U.S. Bureau Amer. Ethnol. 1891–92* 115 The articles known as paint-stones scarcely come under the head of implements... Most of them were used merely to furnish paint. **1951** *Engineering* 26 Jan. 100/2 An infra-red paint-stoving plant..is being used for stoving No. 20 gauge aluminium panels for omnibus bodies. **1962** W. D. Hislop in H. W. Chatfield *Sci. Surface Coatings* xviii. 531 All paint stoving ovens can be a hazard as the solvent vapour/air mixture is a potential fire risk. **1971** *Morning Star* 13 Apr. 4/5 While this may only be washing-up liquid,..it could well be something far more dangerous, like bleach or paint-stripper. **1973** J. Rossiter *Manipulators* iv. 37 Paint-stripper fluid poured lavishly over the enamel..had boiled large blisters. **1959** *Sears, Roebuck Catal.* Spring/Summer 1397/6 Paint Thinner. **1960** *Practical Wireless* XXXVI. 313/1 The following will be required:..paint, paint thinners and an artist's small brush. **1977** F. Parrish *Fire in Barley* ix. 96 Two five-gallon drums... One was called paint-thinner, one creosote. **1933** L. A. G. Strong *Sea Wall* i. 3 The mailboat..glided gracefully in, her white paintwork stained a rich orange. **1966** G. N. Leech *Eng. in Advertising* v. 40 Handy Andy shifts grime from paintwork like this. **1973** *Handyman Which?* Aug. 104/1 Curtains..rubbing against a wall could, over a period of time, make marks on the paintwork.

paint, *v.*¹ Add: **9.** *to paint by number*(*s*: to paint in a picture supplied marked out into sections which are numbered according to the colour to be used; hence *paint-by-number*(*s*, *painting-by-number* attrib. phrs.

1970 *Women's Household* July 6/3, I..paint by number and plan to try painting soon. **1971** 'P. Kavanagh' *Triumph of Evil* (1972) vi. 50 He even bought this terrible oil painting... It looked as though it had been painted by numbers. **1976** *Billings* (Montana) *Gaz.* 26 June 14-B/6 But paint-by-number artists have nothing on 11 men at Our Savior's Lutheran Church. Led by Merle Brunsvold, the men put together an electronic pipe organ kit. **1976** *Sunday Post* (Glasgow) 26 Dec. 9/2 Our other products, like playing cards, jigsaws, painting-by-number kits, &c. **1977** J. Hodgins *Invention of World* iii. 42 She..spent the afternoon doing a paint-by-numbers picture, in the living room.

11. paint out. d. *Naut.* = sense 3.

1902 B. Lubbock *Round the Horn* viii. 302 The great day for cleaning and painting out the half-deck has come... The steward also painted out his berth to-day. **1924** 'P. Blundell' *Confessions of Seaman* ii. 28 When was it painted out last, I should like to know? **1963** S. Hayden *Wanderer* (1964) i. 8 You've painted her out—you've even changed her name?

12. a. *trans.* To cause to be displayed or represented on the screen of a cathode-ray tube.

1946 *Jrnl. Inst. Electr. Engin.* XCIII. IIIA. 147/2 When an echo or the background has been painted it rapidly fades away according to a law determined by the properties of the screen. **1949** *Jrnl. R. Aeronaut. Soc.* LIII. 436/1 It will be seen that the area of towering cumulo-nimbus clouds..are clearly painted on the PPI picture. **1960** J. D. Haigh *Radiolocation Techniques* xii. 192 A photographic display of this kind has the advantage that the whole display is of uniform brightness as compared with a normal p.p.i. in which, at any one moment, only that portion of the picture actually being painted by the rotating time-base is at maximum brightness. **1960** *Proc. Inst. Electr. Engin.* CVII. B. Suppl. 19. 54/2 The display console has a double-deflection system consisting of the main deflection circuits..which position a spot on the tube face, and a second deflection coil..which paints a small raster about this position. **1977** *Sci. Amer.* Jan. 60/2 One kind of stimulus we find useful consists of a moving pattern of small, bright dots 'painted' on the screen of a cathode-ray tube with the aid of a computer.

b. *intr.* To show up on the screen of a cathode-ray tube. Also with *up*.

1946 *Jrnl. Inst. Electr. Engin.* XCIII. IIIA. 145/2 To obtain a satisfactory picture it is necessary to arrange that a fixed target 'paints' at approximately the same point on the screen for consecutive rotations of the aerial system. **1949** *Jrnl. R. Aeronaut. Soc.* LIII. 436/2 It was hoped to fly into some of the cloud forms, but the pilot decided that the risk of flying into any area which 'painted up' on the PPI tube was too great.

painted, *ppl. a.* Add: **4. painted beauty**, substitute for def.: a large North American butterfly, *Vanessa virginiana*, with black and white markings on its brownish-yellow wings; (examples); **painted lady**, (*d*) a name used in South Africa for several local species of gladiolus, distinguished by marks of a different colour on some of their petals; **painted terrapin, tortoise, turtle**, substitute for def.: a small American freshwater turtle of the genus *Chrysemys*, distinguished by red and yellow rings on its greenish-brown shell; (examples); **painted top-shell**, a littoral gastropod mollusc, *Calliostoma zizyphinum* (formerly *Trochus* or *Gibbula magus*), which has a vividly coloured conical shell.

1899 *Lippincott's Monthly Mag.* Oct. 631 The Painted Beauty and the Cosmopolitan resemble each other strongly. **1972** Swan & Papp *Common Insects N. Amer.* 237 Painted Beauty: *Vanessa virginiensis*... Also called American painted lady and Virginia lady. **1906** B. Stoneman *Plants S. Afr.* xix. 198 Gladiolus... Painted Ladies and 'Kalkoentjes' belong here. Eighty-one species of this large genus are found in South Africa. **1927** [see *Aandblom]. **1972** *Stand. Encycl. S. Afr.* V. 201/2 The Cape species [of gladiolus] are known as pypies, afrikaners, painted ladies, bells. *Ibid.* 202/1 The painted ladies (e.g. *G. carneus*) have white or pink trumpet-shaped flowers with markings on the lower perianth lobes. **1839** Storer & Peabody *Rep. Fishes, Reptiles & Birds Mass.* 208 *Chrysemys picta*... The painted Tortoise. **1842** J. E. DeKay *Zool. N.Y.* III. 12 The Painted Tortoise..is unquestionably the handsomest of our fresh-water species. **1876** D. S. Jordan *Man. Vertebrates Northern U.S.* 163 *C*[*hrysemys*] *picta* (Herm.) Ag. Painted Turtle. Mud Turtle... One of the most common turtles. **1904** W. T. Hornaday *Amer. Nat. Hist.* xxxvii. 327/1 The Painted Terrapin, hitherto called at random the Painted 'Turtle' and Pond-'Tortoise', is perhaps the most widely distributed species..in the United States. **1949** *Life* 11 Apr. 81 A painted turtle cranes its neck in the spring sun. **1973** M. Crowell *Greener Pastures* 200 We..pick up a painted turtle intent on crossing the road. **1865** J. G. Wood *Common Shells of Sea-shore* xi. 100 The Painted Top-shell ..is rather boldly ridged. *Ibid.* 101 The name of Painted Top is given to it on account of the magnificent hues of the animal. **1901** E. Step *Shell Life* xi. 205 The Painted Top-shell..is a very distinct species, the shape of the three largest of the 8 whorls giving the solid shell a decidedly turreted appearance. **1972** S. P. Dance *Shells* 18 (*caption*) The Painted Top Shell is a gaily coloured species and has long tentacles.

painter¹. Add: **4.** *painter-engraver, -etcher, -graver, -poet*; *painter-engraving, -etching*; **painter-like** *a.*, (*a*) resembling or characteristic of a painter; (*b*) picturesque, artistic; **painter's brush**, (*b*) = *Indian paint-brush* (*Indian A. 4 b); **painter's** (or **painters'**) **mussel**, the fresh-water mussel, *Unio pictorum*.

1879 F. S. Haden *About Etching* I. 18 We shall have to show that the adoption of the tool, (except in the case of the painter-engraver, who..was an original artist,) implies the practice of a secondary art. **1937** *Burlington Mag.* Feb. 77/1 Elevation to the level of painter-engravers such as Hausbuchmeister and Martin Schongauer. **1890** F. S. Haden *Art of Painter-Etcher* 4 This great-master engraving, this original engraving, this painter-engraving. **1880** *Times* 23 Dec. 7/5 Society of Painter Etchers.— ..The society..has been formed to 'promote original etching and the interest of painters practising that branch of art'. **1920** E. H. Hubbard *Etchings* 128 The work of 'painter-etchers' (men who execute original subjects direct on the plate). **1975** *Whitaker's Almanack 1976* 1104/2 Painter-Etchers and Engravers, Royal Society of.., 26 Conduit Street, W.1. **1890** F. S. Haden *Art of Painter-Etcher* 5 The singular restriction of the Fine Arts to three—Oil painting, Sculpture, and Architecture, to the exclusion of Water Colour painting and Painter-Etching. **1879** —— *About Etching* II. 47 Leyden, Lucas Van.. A Flemish painter-graver of great reputation. **1738** J. F. Fritsch tr. *De Lairesse's Art of Painting* I. vi. xv. 331 But since Beauty is attracting, and Deformity offensive, this certainly is true Painter-like, which supposes the best and most agreeable Objects. **1821** Painter-like [in Dict.]. **1845** R. Ford *Hand-bk. for Travellers Spain* II. 595/2 *Villafranca del Vierzo* is truly Swiss-like,.. with painter-like bridges. **1958** *Economist* 8 Nov. (Suppl.) 9/2 His painter-like method of composition 'over the whole surface at once'. **1909** *Daily Chron.* 17 Mar. 3/3 Mr. Graham is what may be termed a painter-poet. **1941** Blunden *Thomas Hardy* 262 The painter-poet insists on being very much in the middle of things time and again in Hardy's verse. **1869** S. Bowles *Our New West* v. 104 The painter's brush, as familiarly called here, is a new flower to me. **1899** S. Hale *Let.* 22 Apr. (1919) 345 Mariposa lilies, painter's-brush, poppies, and dozens of others. **1910** Mrs. H. Ward *Canadian Born* x. 206 Anderson had brought her to a wild garden of incredible beauty... Painter's brush, harebell, speedwell, golden-brown gaillardias. [**1862** J. G. Jeffreys *Brit. Conchol.* I. 34 *U*[*nio*] *pictorum*... Painters'. **1865** L. Reeve *Conchologia Iconica* XXV. s.v. Unio, species 123 The painters' Unio. Shell elongately oblong,..fulvous-olive.] **1896** L. E. Adams *Collector's Man. Brit. Land & Freshwater Shells* (ed. 2) 148 The 'Painters' Mussel' is found in similar localities to *U*[*nio*] *tumidus*. **1901** E. Step *Shell Life* 111 The Painter's Mussel.., so called because the valves were formerly used to hold artists' colours. **1952** J. Clegg *Freshwater Life* xvi. 268 The Painter's Mussel..has a long, thin shell, about two to three inches in length. **1974** M. Saul *Shells* 87 The more solid valves of the Painter's Mussel..were used by artists throughout Europe to hold their colours.

painter³. (Earlier and further examples.)

1803 J. Davis *Trav. U.S.A.* 382 My master..said that I ought to live among *painters* and wolves, and sold me to a *Georgia* man for two hundred dollars. **1834** D. Crockett *Narr. Life* i. 5 This alarmed me, and I screamed out like a young painter. **1940** Arnold & Hale *Hot Irons* 9 You learn, accidentally, that a 'painter' is really a panther. **1972** *Amer. Speech 1968* XLIII. 218 Mountain cuisine consisted almost entirely of meat: rattlesnake,..*painter* (panther), and rabbit.

painterly, *a.* (*adv.*). Restrict *rare* to the advb. and add to def. of the adj.: *spec.* of a style of painting, characterized by qualities of colour, stroke, and texture rather than of contour or line. Also *transf.* (Further examples.)

1932 *Times Lit. Suppl.* 16 June 441/1 A linear style and a 'painterly' (a translation of the German word *malerisch*, which can also mean picturesque). **1942** *Burlington Mag.* Jan. 24/2 A technique which foreshadows the painterly methods of Watteau. **1950** *Eng. Stud.* XXXI. 166 One should not forget the exciting Kiplingesque rhythm..or the painterly qualities of the dramatic, colourful poem of Cummerbund, the monster of India. **1952** H. Read in P. & L. Murray tr. *H. Wölfflin's Classic Art* p. vi, One word, however, calls for comment—the word 'painterly', which has been invented to convey the meaning of the German word *malerisch*... It stands for that depreciation and gradual obliteration of line (outline and tangible surface) and for the merging of these in a 'shifting semblance' of things—it is an attempt to represent the vague and impalpable essence of things. **1958** *Observer* 23 Mar. 16/5 Short lyrics of precision and beauty, painterly poems, touchingly infused with the poet's passion for what is beautiful. **1958** *Times* 24 July 5/2 His recent attempts to exchange a linear for a more painterly manner. **1962** *Punch* 1 Aug. 177/3 Alberta Wheeler, ex-night club dancer and painterly genius. **1963** *Guardian* 11 Mar. 7/2 He is becoming more painterly, his colour more subtle, and his surface dense with matter. **1969** H. E. Bates *Vanished World* xi. 147 It was not only the painterly quality of Crane's prose that attracted me. **1969** R. Mayer *Dict. Art Terms & Techniques* 276/1 *Painterly*, having the quality of expertly brushed workmanship... A term applied to the dominance of tonal masses over line as a means of defining form in painting, sculpture, and architecture. **1973** F. Taubes *Painter's Dict.* 176 Botticelli is a linear painter, whilst Rembrandt's work would be considered painterly. **1973** *Times Lit. Suppl.* 3 Aug. 900/3 The painterly aspect of each drawing is never forgotten: apart from shading with chalk or wash, in order to enliven a line or contour, pastels are employed and often coloured paper used. **1974** A. Dillard *Pilgrim at Tinker Creek* xv. 268 A photograph of earth from space, the planet so startlingly painterly and hung. **1978** P. Porter *Cost of Seriousness* 2 Masts for Woodbridge Crowd three degrees of the horizon, edging A painterly Dutch sky.

Hence **pai·nterliness**, painterly quality.

1955 P. Heron *Changing Forms of Art* xiii. 202 Hilton's *conscious intention* has been to eliminate all charm, all painterliness, even that evidence of mastery of the material which is itself a seductive element in painting. **1958** S. Spender *Engaged in Writing* iii. 43 Marteau's very appearance transformed the atmosphere from the weighted inwardness of Bonvolio-directed programme music, to cracking outdoor painterliness. **1977** *Times Lit. Suppl.* 24 June 761/5 Manet's painterly procedures—as well as his acknowledged painterliness.

paint-in (pēⁱ·nt,in). [f. Paint *v.*¹ + *-in³.] A gathering for the purpose of painting; *spec.* an organized attempt to improve and draw attention to shabby or neglected buildings by cleaning and redecorating them.

1966 *Maclean's Mag.* 14 May 3 What is a Paint-In? Well, it's a protest, like a sit-in, except that people paint and other people gather round. **1969** *Time* 20 June 64 Jane Shay..organized a one-day paint-in by a group of Washington high school art students. **1969** *Parade* (N.Y.) 14 Dec. 12/2 A gigantic clean-up and paint-in began. **1974** *State* (Columbia, S. Carolina) 26 Apr. B-1/3 (*caption*) Gibson Hopper tries his hand at a paint-in during opening day of Fiesta '74.

painting, *vbl. sb.* Add: **6.** *painting-machine*.

1902 *Chambers's Jrnl.* Feb. 125/2 The spray painting-machine is brought into operation where large unbroken surfaces have to be covered. **1966** 'H. MacDiarmid' *Company I've Kept* ii. 59 A painting-machine like Jean Tinguely's to produce unexpected designs.

pair, *sb.*¹ Add: **I. 1. b.** *another* or *a different pair of shoes* (earlier and later examples).

1849 T. Arnold *Let.* 28 Aug. in *N.Z. Lett.* (1966) 135 Nothing is easier than to make a beautiful scheme of education on paper, but to make it work is 'quite another pair of shoes', as they say in New Zealand. **1931** G. B. Shaw *Widowers' Houses* III. 58 in *Works*, Dooty's another pair o' shoes. **1936** W. H. S. Smith *Let.* 9 Aug. in *Young Man's Country* (1977) ii. 22 I've now had a good opportunity to get to know Hill... We haven't got many tastes in common, but I like him. Doha is a very different pair of shoes.

2. b. *Cricket.* = *a pair of spectacles* s.v. Spectacle *sb.*¹ 7 c in Dict. and Suppl.

1862 *Bell's Life in London* 29 June 7/5 Obtained that unenviable score, 'a pair'. **1960** *Times* 22 June 5/3 Willett and Gibson each completed a perfunctory 'pair'. **1974** *Daily Tel.* 12 June 34/1 Engineer, looking to save his pair, would have been run out first ball if Amiss's throw ..had hit the stumps. **1977** *Sunday Times* 27 Feb. 29/7, I wouldn't swop that 'pair' for anything. It taught me much of life and cricket.

3. a. *happy pair*: see *Happy *a.* 3.

4. a. *to be a pair*: of persons, to be two of a kind, to be as bad as one another. *colloq.*

1840 Thackeray in *Fraser's Mag.* Jan. 107/2 'I' faith, I believe you're a pair,' said Mr. Wood. 'Pray, sir, keep your tongue to yourself..' cried Mrs. Catherine, with proper spirit. **1914** G. B. Shaw *Fanny's First Play* I. 178 Dora: We both get a bit giddy when we're lighthearted. Him and me is a pair, I'm afraid. **1931** M. Allingham *Police at Funeral* v. 69 She was a damned bad-tempered old harpy! And so was Andrew—they were a pair. **1967** J. Rosenberg *Double Darkness* I. i. 46 It's only her own respectability she thinks of. Like you. You're a pair. **1976** 'D. Fletcher' *Don't whistle 'Macbeth'* 86 It's a creepy feeling... Aren't we a pair? Come on. Let's go back and cheer ourselves up.

c. Also, such an agreement between opposite sides. (Earlier and later examples.)

1819 C. Arbuthnot *Let.* 14 Mar (1941) 16 It is expected of them all to be there during the whole course of every evening, & that the coming down merely to get a pair will not do. **1965** *New Statesman* 19 Mar. 426/2 One minister.. was flatly refused a pair by his Tory opposite number. **1976** *Southern Even. Echo* (Southampton) 18 Nov. 3/4 Sir Harold had to cancel a flight to Geneva at the last minute because, he claimed, the Tories changed their minds about providing him with a pair in the Commons division.

f. *Mech.* Two mechanical elements that together constitute a kinematic pair (see *Kinematic *a.* b).

1876 A. B. W. Kennedy tr. *Reuleaux's Kinematics of Machinery* i. 43 The kinematic elements of a machine are not employed singly, but always in pairs; or in other words,..the machine cannot so well be said to consist of elements as of pairs of elements. *Ibid.*, If a kinematic pair of elements be given, a definite motion can be obtained by means of them if one of the two be held fast or fixed in position. *Ibid.* xiii. 549 If we put a normal or normally crossed pair in place of one of the parallel pairs we obtain a chain which is constrained, and which contains five cylinder pairs. **1905** Smith & Marx *Machine Design* i. 13 The helical surfaces by which a nut and screw engage with each other are called a twisting pair. **1969** G. D. Redford et al. *Mech. Technol.* ii. 24 Two links which interact directly with, and mutually constrain, each other form a kinematic pair. *Ibid.*, Shafts in plain bearings, pin-jointed links, slide bars and slide blocks, screw and nut assemblies are all forms of lower pairs. **1975** Mabie & Ocvirk *Mechanisms & Dynamics of Machinery* (ed. 3) i. 9 A pair that permits only relative rotation is a revolute or turning pair, and one that allows only sliding is a sliding pair.

g. In basket-making (see quots. 1910 and 1912).

1897 A. Firth *Cane Basket Work* vi. 42 Take No. 3 [spoke]..bringing it down beside No. 1 and behind No. 4.., making one 'pair' of ends turned down. The canes forming these 'pairs' must each in turn be kept side by side..and held perfectly flat under the thumb till the next 'pair' is down. **1904** O. T. Mason *Indian Basketry* I. iii. 94 *Two-rod foundation.*—One rod in this style lies on top of the other; the stitches pass over two rods in progress and under the upper one of the pair below, so that each stitch incloses three stems in a vertical series... The alternate rod, or the upper rod, in each pair will be inclosed in two series of stitches. **1910** *Encycl. Brit.* III. 482/2 The 'pair', two rods worked alternately one over the other, used for filling up bottoms and covers of round and oval baskets. **1912** T. Okey *Introd. Basket-Making* 153 *Pair*, two rods of willow or cane worked alternately over and under each other—the reverse of a fitch. **1953** [see *Fitch *sb.*³].

h. Ellipt. for *a pair of breasts*.

1922 Joyce *Ulysses* 231 Hell's delights! She has a fine pair. **1973** M. Amis *Rachel Papers* 174 'Who was that tart you had round here before?' 'Gloria?' 'Yeah. Tell you what, she's got a right pair on her.'

II. 6. (Further examples of 'a pair of beads'.)

1880 J. H. Shorthouse *John Inglesant* xx. 267 You remind me of some of the rich oratories I have seen..; where everything is beautiful and costly, but where a classic statue of Apollo stands by the side of a crucifix, a Venus with Our Lady, a Cupid near St. Michael, and a pair of beads hanging on Mercury's Caduceus. **1930** *Amer. Speech* V. 427 A necklace is sometimes called a *pair o' beads* in the Ozarks. **1962** A. Jobson *Window in Suffolk* vi. 102 She would refer to a necklace as a pair of beads. **1976** *Publ. Amer. Dial. Soc.* 1973 LX. 16 One informant called it [*sc.* a string of beads] *a pair of beads*.

b. (Further examples.)

1836 Dickens *Let.* c 24 Aug. (1965) I. 170 His notion of the Bedroom is rather more derived,..from his own fourth pair back. **1922** Joyce *Ulysses* 316 And who was he, tell us? A nobody, two pair back and passages, at seven shillings a week.

7*. In roulette (with pronunc. pĕr), an even number, or a number marked 'pair'.

1850, etc. [see *Noir 2 a]. **1902** [see *Impair *sb.*²]. **1953**, **1969** [see *Manque]. **1973** [see *Impair *sb.*²].

III. 8. pair-bond, the relationship formed during the courtship and mating of a pair of animals or two people; so **pair-bonding** *vbl. sb.*, the formation of such a relationship or the patterns of behaviour that help to establish it; **pair-feed** *v. trans.*, to feed two groups of (experimental animals) with a diet identical except for the item whose effects are being tested on one group; so **pair-fed** *ppl. a.*; **pair-formation**, the pairing of animals, esp. birds, in preparation for breeding; **pair-light**, a window of two lights (Light *sb.* 10); **pair-mate** *v. trans.*, to test the sexual compatibility of (experimental animals) by allowing mating within and between each of two groups; also, to control the mating of (experimental animals) so that each male mates with only one female, or vice versa; so **pair-mating** *vbl. sb.*; **pair production** *Nuclear Physics*, the conversion of a gamma-ray photon into an electron and a positron.

[**1939** G. K. Noble in *Auk* LVI. 265 If pairs of birds or fishes are to form a bond between themselves they must develop behavior different from the feeding or locomotion of non-breeding members of the group.] **1940** D. Lack in *Condor* XLII. 282 In some species of birds, the sexes.. form a very temporary pair-bond. **1954** *Behaviour* VI. 279 The intruder male [*sc.* a zebra finch] succeeded in breaking the old pair-bond, won the female over, built her a new nest, and began a fresh cycle with her. **1963** *Listener* 31 Jan. 204/1 The two [*sc.* gannets] perform an elaborate mutual display which is concerned with strengthening the pair bond. *Ibid.* 204/2 The length and intensity of this display is greatest when the pair bond is weakest, that is, when the couple are newly mated. **1969** *Times* 11 Apr. 12/6 The..tree sparrow, a species in which the pair bond is usually strong. **1970** J. Kear in J. H. Crook *Social Behaviour in Birds & Mammals* 358 The pair bond and its stability is obviously of great consequence to parental behaviour. **1974** *Country Life* 7 Mar. 491/1 In few mammals is the pair bond so strong;..a [beaver] couple may remain paired for up to 18 years. **1977** D. Morris *Manwatching* 88 A sign of old friendships or pair-bonds is that two people can sit together in a peaceful silence without feeling the need to keep up a stream of cheerful chatter. **1965** *New Scientist* 17 June 768/1 Pair-bonding..is the ornithologist's in-phrase for procreative conjunction between sexually ardent cocks and hens. **1967** D. Morris *Naked Ape* ii. 62 The pair-bonding mechanism in our species [*sc. Homo sapiens*], although very powerful, is far from perfect. **1978** *Listener* 12 Jan. 35/2 The interplanetary visitor..would quickly conceptualise pair-bonding in what we call marriage. **1972** *Science* 19 May 795/1 Rat litter-mates were pair-fed nutritionally adequate liquid diets. *Ibid.*, The animals fed alcohol had 72 percent more hydroxyproline in hepatic protein than did pair-fed controls. **1974** *Nature* 4 Jan. 48/2 Hamsters were made vitamin A deficient by maintenance on a vitamin A deficient diet starting at day 10–14 after birth; controls were pair-fed with the same diet supplemented with vitamin A... Tracheas were removed from the vitamin A deficient and pair-fed control hamsters. **1940** D. Lack in *Condor* XLII. 269 There is probably more ignorance concerning pair-formation than there is of any other aspect of bird behavior. **1950** *Brit. Birds* XLIII. 392 Pair-formation [of marsh-tits] takes place at all times of the year. **1967** A. Manning *Introd. Animal Behaviour* v. 105 In some birds the female becomes dominant after pair formation. **1970** J. Kear in J. H. Crook *Social Behaviour in Birds & Mammals* 358 Pair-formation itself is influenced by plumage changes, display and maturity. **1971** J. Z. Young *Introd. Study Man* xxxiv. 484 Apparently in baboons and even chimpanzees and gorillas there is no long-lasting pair formation. **1868** G. M. Hopkins *Jrnls. & Papers* (1959) 183 It [*sc.* a tower] is pierced with pair-lights first, higher with a triplet. **1944** *Genetics* XXIX. 526 One hundred and six Azusa wild males [*sc.* fruit-flies] were pair-mated to standard grade 20 Wooster bobbed females. *Ibid.* 529 Two genes were tested by each pair-mating. **1968** R. Rieger et al. *Gloss. Genetics & Cytogenetics* 327 *Pair mating*, a procedure used to determine the degree of sexual isolation between two groups (A and B) of individuals. Separate tests of mating success are made for the four possible mating combinations. **1934** *Physical Rev.* XLV. 137/1 For energies above twenty million volts the predicted pair production is even greater than that computed by Oppenheimer and Plesset. **1958** W. K. Mansfield *Elem. Nucl. Physics* v. 43 The three methods of interaction of γ-rays with matter are Compton scattering, photo-electric absorption and pair production. **1973** L. J. Tassie *Physics Elem. Particles* ii. 9 Electron–positron pair production has a threshold of 1·022 MeV. Pair production cannot occur in free space, because the conversion of a photon into a pair cannot conserve both total energy and momentum... Some other particle must be present.

pair, *v.*¹ Add: **3. b.** Also *absol.*

1841 E. C. Grey *Little Wife* II. vii. 61 If you go on pairing and matching in this manner..you will be the terror of the whole of the male species.

c. In the British Parliament and other legislative bodies: to bring (an opponent) into an agreement to abstain from voting on a given question or for a certain time.

1956 Abraham & Hawtrey *Parl. Dict.* 127 If a member wishes to be absent from the House, he may arrange with a member of the opposite party, who also wishes to be absent, that neither shall attend the House, or at least vote in a division, for an agreed time. They are then said to be 'paired'. **1968** W. Safire *New Lang. Politics* 315/2 When supporters of John F. Kennedy explained that their candidate was seriously ill at the time of the McCarthy censure, liberal Democrats refused to accept the excuse because, they argued, 'the Senator could have been paired against McCarthy'. **1973** *Courier & Advertiser* (Dundee) 21 Feb. 11/3 Mr Teddy Taylor (Cathcart), who did not vote, as he was 'paired' with Mr Ronald King Murray (Leith), said, 'It is a victory for the people.' **1974** *Times* 18 Mar. 2/8 The Conservatives have said that they will only pair sick MPs with sick MPs, and there are no invalids on the Tory side.

4. a. (Earlier and further Parliamentary examples.) Also of other legislative bodies.

1772 *Debates & Proc. Brit. House of Commons 1768–1770* 240 At dinner time many made no scruple, though the cause was not determined, of *pairing off*, as it is called; some pair'd off for every question in the election, others for a day, or a few hours only. **1811** T. Creevey *Let.* 21 Jan. in *Creevey Papers* (1963) iv. 76, I am not to vote to-night... Villiers won't release me from contract of pairing off. **1866** *Harper's Mag.* May 805/2 This vote was given under peculiar circumstances. Mr. Morrill, of Maine, had sometime previously 'paired off' with Mr. Wright of New Jersey, [etc].. **1964** Mrs. L. B. Johnson *White House Diary* 8 Apr. (1970) 102 Those committed as being safe for the bill..those who were absent and might 'pair' if you got to them and could find them an opposite number. **1965** *New Statesman* 19 Mar. 426/2 It is further alleged that one Conservative, at least, saw fit to pair with two Labour members.

c. (Earlier example.)

1803 G. Colman *John Bull* I. 9 Come, Mrs. Brulgruddery, let you and I pair off, my lambkin.

d. In basket-making: to work two rods alternately one over the other.

1901 A. Firth *Cane Basket Work* (ser. 2) iv. 45 Pair round once to divide into twos, still keeping the central side spokes undivided. *Ibid.*, Now turn the basket upside down and pair round once, taking two lots of double spokes together each time, and keeping the row of pairing even with the edge of the weaving.

e. *to pair up*, to form couples, esp. (of birds) to form pairs in preparation for mating; also, to match.

1908 A. W. Myers *Compl. Lawn Tennis Player* 134

The prevalent custom..is for the members to 'pair up' irrespective of style and temperament. **1920** E. O'Neill *Beyond Horizon* II. 37 Don't you think them two'd pair up well? **1937** *Brit. Birds* XXX. 267 As soon as a couple of birds have paired up, they proceed to exclude other Grebes from a certain area. **1951** L. MacNeice tr. *Goethe's Faust* 170 One's bosom finds this paper light to nurse, It pairs up snugly with a billet doux. **1965** D. Lack *Life of Robin* (ed. 4) v. 66 The blackbird apparently pairs up in late autumn.

paired, *ppl. a.* Add: **b.** Special collocations, as **paired associates**, stimulus material presented in pairs to test the strength of associations set up between them at a subsequent presentation of one of the pair; also (freq. in form *paired-associate*) *attrib.*; hence *paired association*; **paired comparison**, a method of testing the discriminations made between different examples of the same type of stimulus by presenting them for comparison in pairs; also *attrib.*

1937 *Jrnl. Exper. Psychol.* XX. 60 (*heading*) The influence of the relative order of presentation of original and interpolated paired associates. **1949** B. J. Underwood *Exper. Psychol.* ix. 287 The number of pairs in a paired-associate list has varied, but from 8 to 15 have commonly been employed. **1971** *Jrnl. Gen. Psychol.* LXXXV. 212 Some are variants of such tasks as those for memory span..and paired-associates learning. **1963** D. T. Campbell in S. Koch *Psychol.* VI. 120 The preexperimental learning of the pro-Communists might be epitomized as a paired association of 'Communist—good'. **1901** E. B. Titchener *Exper. Psychol.* I. I. vii. 154 Smells have enough variety, but are extremely and insistently associative. However, it would be well worth while to apply the method of paired comparisons to them. **1937** G. W. Allport *Personality* (1938) i. 5 The method of paired-comparison recommends itself as an objective and qualitative technique for studying judgments of the affective value of colors. **1951** S. S. Stevens *Handbk. Exper. Psychol.* i. 28/1 Certain criteria of internal consistency, as in the method of paired comparisons. **1970** *Jrnl. Gen. Psychol.* LXXXII. 19 With the use of a paired-comparisons procedure, each group consisting of five animals was exposed to all three pairs.

pair-horse, *a.* Add: (Earlier and later examples.) Hence **pair-horsed** *a.*

1842 *Ainsworth's Mag.* II. 429 The 'Bath pair-horse Invalid' now drew up..for the elderly gentleman. **1902** [see *Differential *a.* 4 b]. **1905** *Daily Chron.* 22 May 6/3 The costly motor-cars of the humble workmen and the pair-horsed carriages of the lordly labourers. **1910** A. Bennett *Clayhanger* I. iii. 19 A couple of pair-horsed trams. **1914** Conrad *Chance* I. vii. 193 Just then the racket was distracting, a pair-horse trolley lightly loaded with loose rods of iron passing slowly very near us.

pairing, *vbl. sb.*¹ Add: (Further examples.) (For quots. 1915, 1969 cf. *Pair *sb.*¹ 4 f.) Also with *off* and *up*.

1792 J. Pearson *Political Dict.* 40 Pairing-off. Two Sneaking Scoundrels, not worth a piece of dog's meat to either party. **1897** A. Firth *Cane Basket Work* ii. 18 To begin by pairing, place each weaver singly behind two consecutive spokes. *Ibid.* vii. 60 This last inch may be all 'pairing'..if preferred, but if woven in the ordinary way, a row or two of pairing must form the edge. **1901** M. White *How to make Baskets* i. 6 Pairing may be used either with an odd or even number of spokes. **1908** *Westm. Gaz.* 19 Nov. 14/3 The question of pairing-up arose, and the other [tennis] players naturally awaited the Prince's choice. **1909** E. Strasburger in A. C. Seward *Darwin & Mod. Sci.* vi. 109 Attention was drawn to the fact that during the reducing division of nuclei which contain chromosomes of unequal size, gemini are constantly produced by the pairing of chromosomes of the same size. This led to the conclusion that the pairing chromosomes are homologous, and that one comes from the father, the other from the mother. **1915** R. F. McKay *Theory of Machines* viii. 89 Line contact is undesirable..in lower pairing. **1926** J. S. Huxley *Ess. Pop. Sci.* 175 In herons and egrets..it is not the male who seeks out territory long before pairing-up, but pairing-up occurs on the communal feeding-grounds. **1939** *Auk* LVI. 265 No bond in the strict sense of the word, that is a pairing off, is formed. **1953** N. Tinbergen *Herring Gull's World* xii. 105 Settling upon a territory will not occur until after pairing-up. **1955** A. W. Boother *Basketry for Beginners* 13 Reverse pairing, used in conjunction with pairing, makes an attractive decoration for a basket... Reverse pairing is also used on cane bottoms and lids of baskets. **1958** B. Hamilton *Too Much of Water* ix. 204 It was, indeed, a day of pairings-off. Annie Maxwell and Fred Upcher seemed to have settled for one another's company. **1964** H. Hodges *Artifacts* x. 146 Pairing was done with two rods woven simultaneously so that they crossed between the stakes to produce an effect similar to twined weave. **1965** D. Lack *Life of Robin* (ed. 4) v. 65 Tradition assigns St. Valentine's Day for the pairing up of wild birds. **1969** G. D. Redford et al. *Mech. Technol.* ii. 24 In higher pairing, contact is usually along a line or at a point, and the motion is that of, or equivalent to, rolling. **1970** Ambrose & Easty *Cell Biol.* x. 326 The result of this process, which is known as synapsis or zygotene pairing, is that there is now a haploid number of chromosome pairs, which are called bivalents. **1971** *Sci. Amer.* Sept. 89/3 Light absorbed by a molecule kicks one of the electrons associated with the molecule into an excited energy state, thereby making the electron available for pairing with an electron from a neighbouring atom or molecule in an electron-pair bond.

b. pairing-call, a call used by birds during the mating season; **pairing-time** (earlier and later examples); also *transf.*

1911 J. A. Thomson *Biol. Seasons* II. 149 The long-drawn-out, modulated pairing-call of many of the waders ..is on the border-line. **1742** W. Ellis *Mod. Husbandman* July xvi. 85 Every Pheasant..that can be proved to be in the Mannor at Pairing-time. **1850** *Punch* 10 Aug. 62/2 Parliamentary Almanack.—Latter end of July, 'pairing' time begins. **1867** O. W. Holmes in *Atlantic Almanac 1868* 2/2 On the 14th of February the windows fill with pictures for the most part odious, and meant for some nondescript class of males and females, their allusions having reference to Saint Valentine's day, the legendary pairing-time of the birds.

pair-oar. Add: Hence **pair-oared** *a.*

1901 *Chambers's Jrnl.* Feb. 129/2 It comes by way of the river, a rotten, old, pair-oared skiff. **1938** C. S. Forester *Ship of Line* i. 7 Hornblower took his seat in a pair oared wherry.

pair-royal. c. (Further example.)

1840 De Quincey in *Blackw. Mag.* XLVIII. 516/2 The year 333 before Christ. Here we have another '*prial*', a prial of threes, for the *locus* of Alexander.

pairwise (pēᵊ·ɹwəiz), *adv.* and *a.* [f. Pair *sb.*¹ + -wise.] In or by pairs; with regard to pairing; forming a pair.

1831 [in Dict. s.v. Pair *sb.*¹ 8]. **1876** [see *Kinematic *a.* b]. **1956** *Nature* 21 Jan. 127/2 In the tetramer, pair-wise engagement of all the CONH groups is again geometrically possible. **1960** H. M. Hoenigswald *Language Change* xiii. 144 Pair-wise reconstructions from three related languages may probe the question of degrees of relationship within the language family. **1965** *Math. in Biol. & Med.* (Med. Res. Council) III. 126 An observer or observers willing to assign a ranking order to the pair-wise resemblances among the objects. **1969** *Word 1967* XXIII. 302 The characters can be anything at all as long as they are pairwise distinguishable. **1971** Powell & Higman *Finite Simple Groups* viii. 260 $K_1, \ldots, K_c$ are pairwise disjoint, non-empty subsets of G. **1972** *Computers & Humanities* VI. 184 Used in roll-call analysis to count pair-wise voting agreement on isolated roll-call votes. **1975** *Language* LI. 378 Even if we allow pairwise (non-linear) extrinsic ordering, the problem is not solved.

paisa (pai·sā). Pl. **paise, -a, paisas.** Now the usu. form of Pice. Now decimalized and equal in value to one-hundredth of a rupee in India (since 1957: see *Naya paisa), Pakistan (since 1966), and Nepal, and to one hundredth of a taka in Bangladesh.

1884 [see *Dam *sb.*³]. **1924** *Regions Beyond* XLV. 44 Flowers are scattered upon the waters, coins are dropped into the depths, perhaps only a paisa (farthing), but the giver is poor and needy. **1956**, etc. [see *Naya paisa.] **1959** [see *Anna]. **1963** *Times* 8 May 21/4 Last year the Pakistan Government made a reduction of 40 per cent—from 25 paise to 15 paise per lb.—in export duty. **1969** *Sunday Tel.* 12 Jan. 7/3 Hashish candy..is available for 50 Nepalese paise. This is about 3d. **1969** *Enact* (Delhi) Dec. 7/3 Oh! You are a munim all over, always counting paisas. **1971** *Femina* (Bombay) 16 Apr. 55/1 Your family won't even know you've been counting every paise [*sic*]. **1975** *Bangladesh Observer* (Dacca) 26 July 2/4 (Advt.), Cement... Taka 7.50 (seven and paisa fifty) only per ton. **1978** L. Heren *Growing up on The Times* v. 166 The more affluent Indians..gave beggars paise, or fractions of farthings.

paisano. Add: **2.** In Spanish-speaking areas: a fellow-countryman; a peasant. Also *attrib.*

1844 G. W. Kendall *Narr. Santa Fé Expedition* II. 230 [He] invariably called me his paisano, or country man. **1890** C. F. Lummis *Land of Poco Tiempo* iv. 88 Every one was out, but they were no longer the friendly *paisanos* we had known. **1935** J. Steinbeck *Tortilla Flat* 11 What is a paisano?.. His ancestors have lived in California for a hundred or two years. **1940** E. Fergusson *Our Southwest* xiv. 247 The Spanish clustered in towns. They fought Indians only when they had to, to assure safety and security. Security was what the *paisano* wanted. **1969** A. Marin *Rise with Wind* x. 118 Carrasco was one of the few *paisanos*..who wore civilian clothes. **1971** *Publishers' Weekly* 18 Oct. 25/2 There are many cookbooks exploring the gourmet or *paisano* delights of foreign countries and places. **1977** H. Fast *Immigrants* 6 No use, paisano. Come back next week, the week after.

Paisley (pēⁱ·zli). The name of a town in Renfrewshire, Scotland, used *attrib.* or *absol.* to designate a garment or material made there or having the curvilinear design characteristic of cloth made there, or the pattern itself.

1834 Paisley shawl [see Shawl *sb.* 2 a]. *c* **1860** [see *Norwich]. **1866** R. S. Charnock *Verba Nominalia* 215 *Paisley*, a shawl made at Paisley, co. Renfrew (Scotland); celebrated also for its manufactures of silk and other shawls, muslin, cotton thread, and ornamental fancy goods. **1898** *Daily News* 5 Mar. 6/4 If black stuffs were chosen, it was only that they might be trimmed with paisleys. **1900** *Ibid.* 28 Apr. 6/6 Paisley velvet is a favourite facing for collars and revers. **1911** *Daily Colonist* (Victoria, B.C.) 23 Apr. 7/1 (Advt.), Paisley silks. **1950** D. Gascoyne *Vagrant* 15 Spread with a soft paisley-patterned cloth. **1951** E. Paul *Springtime in Paris* vi. 125 One of the coffee-coloured men in a flashy paisley robe and wearing gold-bowed pince-nez. **1954** F. Sargeson in C. K. Stead *N.Z. Short Stories* (1966) 5 He hoped to better himself at the fine Paisley work. **1959** H. Hobson *Mission House Murder* xi. 11 My Paisley silk bow-tie. **1964** *McCall's Sewing* iv. 58/2 *Paisley*, any cotton, wool, or rayon which is printed with the traditional scroll design which originated in Paisley, Scotland. **1966** R. Thomas *Spy in Vodka* (1967) vii. 60 He wore a dark-blue flannel sports shirt, a blue and yellow Paisley ascot, a pair of grey flannels that must have cost sixty bucks, and black loafers. **1967** [see *Norwich]. **1975** G. Lyall *Judas Country* xxii. 163 He had his old brownish Paisley-pattern silk scarf folded as a choker. **1976** *Evening Advertiser* (Swindon) 31 Dec. (Advt.), Fur coats, capes and foxes. Victorian nighties, petticoats and camisoles, silk and Paisley shawls, beaded and sequined garments... Buyer calling regularly in the area. **1977** J. Fleming *Every Inch a Lady* xiii. 63 A Paisley-patterned scarf in blue and red silk. **1977** J. Wambaugh *Black Marble* (1978) viii. 105 The long-legged asthmatic..was trying to look dog show respectable in a three-button herringbone coat, gray woolen slacks and a paisley tie.

Paisleyite (pēⁱ·zli,əit), *a.* and *sb.* [f. the name of Ian *Paisley* (b. 1926), Ulster Presbyterian minister and politician + -ite¹.] **A.** *adj.* Of or pertaining to Ian Paisley or his followers. **B.** *sb.* A supporter of Ian Paisley and his advocacy of Protestant interests in Northern Ireland and the independence of Northern Ireland from the Republic of Ireland. So **Pai·sleyism**, the religious and political principles of Paisleyites.

1966 *Guardian* 18 July 4/5 Paisleyism might have been expected to cause alarm and fierce anger in the [Irish] Republic. *Ibid.* 25 July 1/2 A Government order restricting a Paisleyite march. **1966** *New Statesman* 30 Sept. 468/3 The Cormac Square riots, when the Paisleyites marched on the General Assembly of the Irish Presbyterian Church. **1968** *Listener* 19 Dec. 823/2 Through the day, groups of Paisleyites infiltrated into the cordoned-off city. **1969** *Daily Tel.* 26 Feb. 16 Mr. Roy Bradford.. increased his majority in his Belfast constituency.. despite the intervention of a Paisleyite candidate. **1970** *Guardian* 18 Apr. 11/5 The likelihood is that the Paisleyite opposition will increase its numbers in Stormont. **1976** J. Carroll *Madonna Red* (1977) iii. 98 He knew the bitter energy that faith could release against Paisleyites and Jews. **1978** D. Murphy *Place Apart* vii. 148 However people may disagree in their analyses of Paisley the man, everybody recognises the danger of Paisleyism the cult.

Paiute (pəi·ūt), *sb.* and *a.* Also 9 **Pah-Utah, Pah-Utche, Pah-Ute, Pa-Utah Pie-Utaw**; 9– **Piute.** [ad. Sp. *Payuta*, or ad. native name (perhaps *payiutsi* fish people), influenced by *Utah* and *Ute*.] **A.** *sb.* **a.** A Shoshonean Indian people inhabiting parts of Utah, northern Arizona, and southeastern Nevada (more fully *Southern Paiute*); also, a culturally similar Shoshonean people of western Nevada and adjacent parts of California, Oregon, and Idaho (*Northern Paiute*); a member of either of these peoples.

The Southern Paiute and Northern Paiute are not subdivisions of a single people; their languages are distinct, and their territories are not contiguous.

1827 D. T. Potts in D. M. Frost *Notes on Gen. Ashley* (1960) 63 This river [*sc.* the Sevier] is inhabited by a numerous tribe of miserable Indians... They call themselves Pie-Utaws, and I suppose are derived from the same stock [as the Utaws]. **1827** J. Smith in H. D. Carew *Hist. Pasadena* (1930) I. 136 Passing down this river some distance, I fell in with a nation of Indians who call themselves Pah Utches. **1860** Mayne Reid *Odd People* 329 In the western & southern division of the Great Basin, the Digger exists under the name of *Paiute*, or more properly, Pah-Utah—so-called from his supposed relationship with the tribe of the Utahs. **1881** *Encycl. Brit.* XII. 825/2 In California and the south-western States, occupied by the morally debased and physically degraded Pah-Utes. **1910** F. W. Hodge *Handbk. Amer. Indians* II. 186/2 Paiute... In common usage it has been applied at one time or another to most of the Shoshonean tribes of W. Utah, N. Arizona, S. Idaho, [etc.] **1937** R. H. Lowie *Hist. Ethnol. Theory* vi. 55 Had he begun his studies among the Eskimo or Paiute, his general views might have been different. **1947** B. Haile *Prayer Stick Cutting* 43 Neckbands..should be of otter or beaver skin obtained from the..Paiutes. **1952** J. R. Swanton *Indian Tribes N. Amer.* 375 With the Bannock, the Northern Paiute constituted one dialectic group of the Shoshonean Branch of the Uto-Aztecan stock. *Ibid.* 381 The Southern Paiute belonged to the Ute-Chemehuevi group of the Shoshonean branch of the Uto-Aztecan stock. **1959** E. Tunis *Indians* 107/2 The southern Paiute, the Bannock, and the Gosiute were typical 'tribes', though they were not actually organized as tribes. **1973** A. H. Whiteford *N. Amer. Indian Arts* 39 One-rod coiling was done by the Pomo and Paiute. **1974** A. MacLean *Breakheart Pass* i. 12 They say the Paiutes kill every white man on sight. **1977** H. Landar in T. A. Sebeok *Native Lang. Americas* II. III. 327 The term Snake, applied to the Northern Paiute of Oregon, is used of other Shoshonean groups as well.

b. Either of the languages of the Paiute, technically distinguished as *Southern Paiute* and *Northern Paiute*.

1915 *Everybody's Mag.* Oct. 461/2, I talked Piute to him all afternoon and he didn't understand a word of it. **1921** E. Sapir *Language* 67 Paiute, for instance, may compound noun with noun. **1933** L. Bloomfield *Language* iv. 72 The Shoshonean family (in southern California and eastward, including Ute, Paiute, Shoshone, Comanche, and Hopi). **1949** E. A. Nida *Morphol.* (ed. 2) iv. 103 Alternating unvoicing and reduction as in Southern Paiute. **1975** *Language* LI. 124 Consider an alternating stress rule, such as that of Southern Paiute, which stresses every alternate vowel from left to right across a word. **1977** H. Landar in T. A. Sebeok *Native Lang. Americas* II. III. 327 Northern Paiute... 2,000 [speakers] in Nevada, California, Oregon and Idaho.

B. *adj.* Of or pertaining to the Paiute or their languages.

1845 J. C. Frémont *Rep. Exploring Expedition* 260 They rarely carried home horses, on account of the difficulty..of guarding them..from the Pa-utah Indians. **1869** 'Mark Twain' *Innoc. Abr.* xx. 205 Tahoe means grasshoppers. It means grasshopper soup. It is Indian... They say it is Pi-ute—possibly it is Digger. **1938** W. Dyk *Left Handed's Son of Old Man Hat* 11 A Paiute girl came to our place. **1949** *Natural Hist.* June 268/1 Major Powell..named it Tapeats Creek after a Paiute Indian in his employ. **1955** W. Gaddis *Recognitions* II. ii. 388 The Piute Indians followed the sun to that hole where it crawled in at the end of the earth. **1975** *Language* LI. 797 The Southern Paiute suffix *-ʼtü* is restricted to true passives.

paiwari: see Piwarrie in Dict. and Suppl.

pajala (pā·dʒālā). [Malay.] A type of boat used around the Macassar Strait (see quot. 1950).

1937 G. E. P. Collins *Makassar Sailing* 12 When the first stage is completed the ship is a pajala, a low undecked boat of island design. **1950** *Jrnl. Malayan Branch R. Asiatic Soc.* XXIII. 113 The Pajala is a beamy, undecked coasting boat which is normally fitted with a tripod mast setting a single, large rectangular sail. **1964** K. G. Tregonning *Hist. Mod. Malaya* 59 There had been Bugis traders in Malayan waters for centuries. In the sixteenth century Malacca knew well their *pájalas*, their large prahus with a distinctive tripod mast and a deep oblong sail.

Pajarete, var. *Paxarete.

Pajitanian, var. *Patjitanian *a.*

Pak (pæk), colloq. abbrev. of *Pakistan*, *Pakistani *sb.* and *a.*

[**1935** C. Rahmat Ali (*title*) Pakistan, the fatherland of the Pak nation.] **1954** G. S. Rao *Indian Words in Eng.* 134/1 *Pak*, contraction of Pakistan. **1965** P. Robinson *Pakistani Agent* v. 74 It was obvious the Paks were up to some new game. **1967** *Guardian* 24 Aug. 6/6 The official Pakistan news service reported yesterday that 'indecent miscreants' are smuggling Pak grain into India. **1969** *Indian Express* (Bombay) 28 July 11/3 (*heading*) Pak separatist parties merge. **1971** M. Kelly *25th Hour* iii. 214, I don't see all this secrecy and drama. Smuggling us out like a load of Paks. **1971** *Sun* (Ceylon) 17 Sept. 6/4 (*heading*) Pak refugees hit by floods. **1974** *New Society* 13 June 627/3 The chauvinists' scenario runs on about filthy foreigners and Pak shopkeepers (they *do* stay open later). **1975** *Bangladesh Times* (Dacca) 27 July 6/1 (*heading*) Pak flood death toll rises to 49. **1977** *Private Eye* 13 May 7/3 The foreign mission which serves booze in limitless quantities is the Russian Embassy in Islamabad, 200 miles away and many alcoholics are now signing up to join the Pak-Soviet Friendship Societies.

pakapoo, pakapu (pæ·kăpū, pækăpū·). Also **pak-a-peu, puka pu,** etc. [Chinese]. A Chinese gambling game resembling lottery with sheets of paper so marked as to be indecipherable except to an initiate. Phr. *like a pakapoo ticket*, untidy, disordered (*Austral.*).

1911 L. Stone *Jonah* ix. 92 He had come down early to mark a pak-ah-pu ticket at the Chinaman's in Hay Street. **1913** *Chambers's Jrnl.* Feb. 155/1 All kinds of games of chance—'two up', 'pak-a-pu' (the latter a form of lottery imported by the Chinese). **1923** *Daily Mail* 12 Feb. 7 Five Chinese pleaded guilty at Liverpool Assizes to charges of running a gaming house... For the defence it was argued that Pak-a-Peu (or Puck-a-pu) was a game of skill. **1927** *Daily Express* 21 Sept. 7/2 A Japanese ship's captain..appealed against a conviction..for employing two other Japanese to sell chances in an unlawful lottery known as 'Puka pu'. 'It is a favourite game with the Japanese and Chinese and others living in Limehouse,' explained Mr. Horace Fenton. **1932** H. Simpson *Boomerang* x. 275 Brought in evidence two flimsy pieces of printed paper, one a pakapu bet, the other a five-pound-note. **1936** 'R. Hyde' *Passport to Hell* i. 10 Chinese grocery-shops, masonic clubs, and pakapoo saloons. **1951** E. Lambert *Twenty Thousand Thieves* (1952) ix. 89 Henry opened Dooley's pay-book, the pages of which showed liberal sprinklings of the red ink with which fines and convictions were entered. 'What a pay-book!' he sighed. Dooley grinned. 'Like a pak-a-poo ticket,' he agreed. **1959** Baker *Drum* 133 *Marked like a pakapoo ticket*, confusedly or incomprehensibly marked. **1960** *N.Z. Listener* 22 July 9/2 Some of the last of the old Chinese dwellings of the opium-smoking and pakapoo-playing generation are being pulled down in Haining Street in Wellington. **1961** Partridge *Dict. Slang* Suppl. 1212/1 *Look like a pakapu ticket*, to be completely indecipherable: Australian (esp. Sydney) coll.: since ca. 1940. 'Pakapu is a Chinese gambling game, not unlike housie. A pakapu ticket, when filled, is covered with strange markings'. **1964** A. Wykes *Gambling* 330 The only illegal gambling games in New South Wales are *fan-tan*, another Chinese game called *pak-a-p*, and *two-up*.

pakaru, var. *puckeroo.

pak-choi (pakt∫oi·). [Cantonese, lit. 'white vegetable'; cf. *pe-tsai.] A Chinese species of cabbage, *Brassica chinensis*. Also *attrib.*

1847 R. Fortune *Three Years' Wanderings N. Provinces China* xvi. 306 The celebrated 'Pak-tsae', or white cabbage of Shastung and Peking, is a very different plant. **1894** *Bull. Cornell Univ. Agric. Exper. Station* LXVII. 183 The Pak-Choi, commonly called Chinese cabbage and frequently confounded with the Pe-Tsai..is a vegetable which never forms a head. **1900** L. H. Bailey *Cycl. Amer. Hort.* I. 178/1 Pak-Choi Cabbage... This plant is grown by the American Chinese, and is occasionally seen in their gardens. **1931** H. C. Thompson *Vegetable Crops* (ed. 2) xix. 291 The Pak-choi varieties resemble Swiss chard in habit of growth. The leaves are long, dark green and oblong or oval. This type does not form a solid head. **1969** *Oxf. Bk. Food Plants* 134/1 Pak-Choi (*Brassica chinensis*)..is more closely related to rape and swede than to the European cabbages... The plant does not form a heart and in appearance it resembles chard or spinach beet... Pak-choi does best when sown in July or August, to produce an autumn crop. **1974** Y. Lovelock *Vegetable Bk.* 72 The other [Chinese cabbage], Baak-choy (*Brassica chinensis*), is also called Chinese mustard, and is noted for its lack of smell when cooking.

pakeha. Delete ‖ and add earlier and further examples.

1817 J. L. Nicholas *Narr. Voyage to N.Z.* I. v. 139 Many of them had never before..beheld an European, and to see pakaha ... was a novelty. **1838** J. S. Polack *New Zealand* II. iii. 102 He [sc. the chief] said I was a pakeha maori or native white man. **1859** [see *beach-comber]. **1933** *Bulletin* (Sydney) 13 Sept. 8/4 The first pakehas were not at all chivalrous ... **1933** [see *haka-pie n.]. **1938** R. Finlayson *Brown Man's Burden* 1 Rua came from Taupo to the coastal district to work on the farm of a Pakeha. **1959** G. Slatter *Gun in my Hand* xxi. 224 The Maori must smile at the pakeha going all Maori when he's overseas. People on the ship to England wearing tikis and saying good kai this morning. **1960** *Guardian* 13 Sept. 13/1 Race relations in New Zealand..had been based on the absolute equality of Maori and Pakeha (European). **1963** *Evening Post* (Wellington, N.Z.) 25 July, Co-existence between Maoris and Pakehas had seriously affected Maori culture. **1978** *Islands* (N.Z.) Aug. 10 The pakehas' faces floated like white disks in a sea of brown.

pakhal (pā·kāl). [Hind.: see Puckauly.] A vessel for carrying or keeping water, *spec.* a water-skin of leather.

1885 G. C. Whitworth *Anglo-Indian Dict.* s.v. *bhisti*, A double bag called a pakhal, which is carried by a bullock... **1894** W. Wickham *Mules Transport India* xv. 147 The leather pakhals or water bags should ... be dubbined before use. **1930** *Blackw. Mag.* Oct. 464/1 A couple of mules laden with metal pakhals of water. **1925** [see *chagal].

‖ **pakhawaj** (pā·kawādʒ). [Hind.] A double-headed drum used in Indian music, esp. that of the northern part of the country.

1867 E. M. Taylor in *Proc. R. Irish Acad.* IX. 116 Perhaps the pakhawaj is employed more than the other [sc. the tabla] by Hindu professionals. **1921** H. A. Popley *Mus. India* vii. 121 The Pakhāwaj is a drum slightly larger than the mṛidaṅga but similar in shape, which is used in the north of India. **1957** *New Oxf. Hist. Mus.* I. iv. 222 Prominent in our days are the *pakhawaj* and the *tabla*. The former..has a clay body of irregular cylindrical shape, tapering slightly towards the left hand, with a large surface of parchment. **1966** R. Shankar *My Music, My Life* iii. 41/1 The *pakhawaj*, a one-piece drum made of clay with two faces or heads, used to different pitches. **1977** B. C. Deva *Mus. Instruments* 59 The pakhāvaj is the king of drums in Hindustani music, though now it is more a constitutional monarch, respected from a distance.

Pakhto: see Pushtoo *sb.* and *a.* in Dict. and Suppl.

Pakhtun (pă·ktūn), *sb.* and *a.* Also **Pakhtoon, Pukhtun,** etc., and in form **Pashtun, Pushtun.** [Pashto.] **A.** *sb.* A member of a Pashto-speaking tribal people, also called *Pathan, inhabiting parts of south-east Afghanistan and north-west Pakistan; this people collectively. **B.** *adj.* Of or pertaining to this people.

1815 M. Elphinstone *Acct. Kingdom of Caubul* II. i. 151 Their own name for their nation is Pooshtoon, in the plural, Pooshtauneh. The Berdooraunees pronounce this word Pookhtaun. **1867** H. W. Bellew *Dict. Pukkhto* p. vii, To have given place to all the words of those languages used in an unchanged form by the Pukkhtun would have added unnecessarily to the bulk of the work. **1880** — *Races of Afghanistan* vi. 56 The term Pathán is not a native word at all. It is the Hindustani form of the native word Pakhtána, which is the plural of Pukhtún, or Pakhtún...as it is pronounced by the Afridi. And Pukhtún is the proper patronymic of the people inhabiting the country called Pukhtún-khwá, and speaking the language called Pukhtú or Pakhtú. **1885** G. C. Whitworth *Anglo-Indian Dict.* 246/1 *Paṭhān* (Hindustani, from the Pashto *pakhtāna*, the plural of *pakhtūn*, the name of a people inhabiting the country called by Herodotus Pactiya.) **1906** A. Hamilton *Afghanistan* x. 263 After the Afghans the dominant people are the Pukhtun or Pathans, represented by a variety of tribes. **1908** *Encycl. Relig. & Ethics* I. 238/2 The Afghans themselves prefer the designation Pushtūn or Pukhtūn, older form Pashtūn, Pakhtūn (whence their Indian name Pathān). **1940** P. Sykes *Hist. Afghanistan* I. i. 13 The Afghan nomads organized on a tribal system, whose true national name is Pashtun or Pakhtun, generally termed 'Pathan' by Europeans, belong to the Turko-Iranian type. **1955** *Times* 11 May 9/6 Even before 1947 the Pathans (or Pakhtoons, as they are called in their own tongue) had been claiming the right to independence. **1956** *Ann. Reg. 1955* 116 The Afghan Prime Minister..stated that the proposed merger of West Pakistan would never be accepted or recognized either by the 'Pakhtun nation' or by his Government. **1963** *Times* 13 May 9/5 One consequence expected from Sardar Muhammad Daud Khan's resignation was an improvement of relations between Afghanistan and its neighbour Pakistan... Afghanistan is so publicly committed to the cause of the Pakhtuns, however, that no sudden relinquishment can be expected. **1971** *Illustr. Weekly India* 18 Apr. 21/2 West Pakistan in order to consolidate the Baluchis and the Pakhtoons in its north-west, may be forced into a diversionary adventure in Kashmir. **1973** *Times* 27 July 16/5 An attempt was made to raise the Pakhtun flag on the banks of the Indus. **1974** *Encycl. Brit. Micropædia* VII. 783/1 The Pashtuns believe themselves to have originated in Afghanistan and to be descended from a common ancestor.

Paki (pæ·ki), *slang.* Also **Pakki, Pakky.** [Abbrev. of *Pakistani *sb.* and *a.*] A Pakistani, *spec.* an immigrant from Pakistan. Also *attrib.* and in *comb.*, as *Paki-bashing*, wanton physical assault on or other violence directed against Pakistani immigrants (hence *Paki-bash*, *Paki-basher*).

1964 *Guardian* 15 Apr. 8/4 Some big Paki over the water's got her set up for right trouble. **1969** E. Knox *Tallyman* v. 94 All's a Paki—an' you know how it goes. Paki's pretty well look all the same to me. **1970** *Observer* 5 Apr. 3/2 The name of the game is Pakky Bashing... Any Asian careless enough to be walking the streets at night is a target. **1970** *Daily Progress* (Charlottesville, Va.) 13 Apr. 7-C/2 They attack Asian immigrants, and term this 'pakibashing'. **1972** J. Brown *Chancer* xii. 87 Sergeant Burton and me, we broke in the Paki lodging house. **1973** C. Mullard *Black Brit.* iv. 51 ... a practice that has much in common with present-day 'Paki-bashing'. **1973** M. Amis *Rachel Papers* 141 Joe, a young and ambitious cook, was fed up to the teeth with cooking steak and chips for the odd Paki. **1975** J. Symons *Three Pipe Problem* v. 36 He wanted to send the nig nogs and the Pakis back where they belong. **1976** *Daily Tel.* 27 Jan. 3/6 'Paki-bashing' is suddenly a topical phrase in Canada. **1977** *Time* 12 Dec. 19/3 Bands of front backers, swinging fists and banner staves, have sallied into peaceful demonstrations by Indians and Pakistanis in what are cruelly called 'Paki bashes'.

pakihi (pā·kihi). Also **pakahi, paki.** [Maori.] An area of open, swampy, land, esp. characteristic of north-western parts of the South Island of New Zealand; also, the type of waterlogged soil associated with such land. Also *attrib.*

1861 J. von Haast *Rep. Topogr. & Geol. Explor. Nelson Province* iv. 131, I also soon came ... the different pakis, or open tracts of land, and give a short description of them. **1871** C. L. Money *Knocking about in N.Z.* v. 63 We suddenly came out of the bush on to an open pakihi some miles in length. **1896** *N.Z. Alpine Jrnl.* II. 148 The only patch of rata bush on the flat, the rest being partly open 'pakihi' and partly covered with low scrub and timber. **1909** L. J. Wild *Soils & Manures in N.Z.* v. 53 Pakihi Soils of Westland...occur over considerable tracts of sour, swampy, but easily drained terrace lands. **1930** J. Devanny *Bushman Burke* 14 The supplies..had been packed by horse along a track cut out of the bush, and further, up towards the Ridge, the pakahi. **1947** F. Newton *Wayleggo* (1949) x. 110 Little pakihis run up into the bush every here and there. **1959** A. McLintock *Descr. Atlas N.Z.* 41 ... the soils in the scrub-covered terraces are gley podsols (pakihi). **1959** G. Slatter *Gun in my Hand* 76 Green swamp-water, a tangle of blackberry or pakahi beside the twisting railway line. **1970** *N.Z. Listener* 7 Dec. 6/3 The 33 million acres of sour and barren 'pakihi' soil (on the West Coast of N.Z.). *Ibid.* 6/5 So what about all that pakihi? It is red-brown, depressing ... since drainage is blocked by an impervious iron pan. **1972** P. Newton *Sheep Thief* ix. 72 He set off up the creek in search of the horses. He found them grazing in a little pakihi. **1973** *Massey Ferguson Rev.* (N.Z.) Mar.–Apr. 3/1 Two farmers a few miles away from Bald Hill have successfully transformed 170 acres of pakihi by using much the same over-sowing methods as the department.

Pakistani (pakistā·ni, pæk-), *sb.* and *a.* [f. the name *Pakistan* + -i.] **A.** *sb.* A native or inhabitant of Pakistan, an independent state formed in 1947 as a homeland for the Muslims of India from parts of Punjab, Sind, Baluchistan and North-West Province (East Pakistan, formerly East Bengal, achieved independence as Bangladesh in 1971). **B.** *adj.* Of or pertaining to Pakistan, its natives, or its inhabitants.

1941 L. S. Amery *Let.* 25 Jan. in J. Glendevon *Viceroy at Bay* (1971) xvi. 198 Jinnah and his Pakistanis. **1948** *Sunday Times* 2 May 4/5 No Pakistani I have met is yet ready to admit that the achievement was not worth the sacrifice. **1950** *Times* 6 Mar. 5/7 The Pakistani Government soon set about filling the gap, taking care to ensure that the tribal areas and their peoples benefit from the development of West Pakistan as a whole. **1951** W. I. Jennings *Commonwealth in Asia* viii. 117 No Indian—or for that matter Pakistani or Ceylonese—politician wishes to sit at the same table as a representative of the Union of South Africa. **1957** *Times* 19 Dec. 15/2 The agreement signed by the World Bank and Pakistani officials yesterday completes the initial financing of the newly formed Pakistan Industrial Credit and Investment Corporation. **1965** *New Statesman* 30 Apr. 670/1 In neighbouring Sparkbrook, where faded vermilion posters..stare down upon shabbily dressed Pakistanis. **1967** *Listener* 17 Aug. 211/3 Radio comics with their unending imitations of Pakistani bus conductors must find other targets. **1971** *Peace News* 28 Oct. 5/2 We understood that the Pakistani army was burning the villages in the area, in retaliation for the previous day's attack. **1973** M. Amis *Rachel Papers* 186 When I surfaced, dragged along in a tide of fat-legged girls and torpid Pakistanis,..there..was Rachel. **1976** 'W. Trevor' *Children of Dynmouth* iii. 77 He'd seen the Dynmouth Hards beating up the Pakistani from the steam laundry in a bus-shelter.

Pakki, Pakky, varr. *Paki.

pakora (păkō·rā). Also **pakhora.** [a. Hind. *pakorā* a dish of vegetables in grain-flour.] A savoury Indian dish consisting of diced or chopped vegetables coated in batter and deep fried.

1954 J. Masters *Bhowani Junction* xxiii. 192 Our guests will be here in half an hour, and I have forgotten to make pakhoras. **1962** R. P. Jhabvala *Get Ready for Battle* iii. 135 She took a bite from a cheese pakora. **1963** *Guardian* 1 May 6/5 She can make fresh pakoras..dainty morsels of cauliflower, green pepper, onion, or slivers of potato, coated in highly seasoned batter and deep-fried in oil. **1971** *Femina* (Bombay) 30 Apr. 63/1 Even then they get only cold pakoras or oily potato chips. **1972** R. P. Jhabvala *New Dominion* II. 138 The tea..was very nice. They had pakoras and samusas and all sorts of other things. **1978** *Times of India* 18 Mar. 13/6 Delicious smells from the neighbouring *halvai's* shop. He is frying *samosas, pakoras, jalebis* and other mouth-watering delicacies.

‖ **pak pai** (pak pai). [Cantonese, lit. 'white licence'.] In Hong Kong, a car used illegally as a taxi.

1972 *South China Morning Post* (Hong Kong) 4 Dec. 10/6 *Pak Pai*, a car which plies for hire, illegally. **1977** *Ibid.* 13 Apr. 11/7 Triad gangs are involved in the operation of extortion rackets with mini-buses, pak pais and goods vehicles illegally used for passengers in rural areas. **1977** 'J. Le Carré' *Hon. Schoolboy* xix. 464 Collecting gambling debts from the *pak-pai* drivers.

paku, var. Pacu (in Dict. and Suppl.).

pal, *sb.*[1] Add to definition: Now *colloq.* Also applied to a woman. (Further examples.)

1873 Byron *Let.* 30 June in *Works* (1898) I. 130 'Better late than never, Pal,' is a saying..applicable on the present occasion. **1841** S. Bamford *Passages in Life of Radical* (ed. 2) I. xxiv. 151 The thieves and their 'pals', as he termed the repulsive females. **1886** *Lantern* (New Orleans) 27 Oct. 2/3 Reynold Bowers and his pal, Jack Lacoste. **1890** Kipling in *Pioneer Mail* 28 May 698/2, I was great pals with a man called Hicksey. **1924** F. M. Ford *Some do Not* I. ii. 50 Eunice Vanderdecken is a bitterly misjudged woman. She's a real good pal. **1936** M. de la Roche *Whiteoak Harvest* v. 79, I have talked to her..as I couldn't to anyone else... Well, she's been a complete pal—if you know what I mean. **1963** *Listener* 14 Feb. 279/1 The local battalion, the Bradford Pals, was butchered at the Somme. **1972** J. Porter *Meddler & her Murder* x. 128 Be a pal and shove the marge across.

pal, *v.* Add: Also, *to pal around with, up with.*

1915 R. Lardner in *McClure's Mag.* Aug. 21/3, I and Lefty and Mike used to pal round together. **1926** G. Hunting *Vicarion* vi. 103 And I shan't have time to compromise *you* when I can pal around with Charlemagne, or Valentino, or Rameses Second, or Kublai Khan! **1943** I. Wolfert *Tucker's People* viii. 167 All those poor people..were just like the people he palled around with. **1958** B. Hamilton *Too Much of Water* xi. 249, I got tight one night with a chap I'd palled up with. **1975** *High Times* Dec. 24/1 Lenny picked up part of his *schtick* from the characters that he palled around with in New York. **1976** *New Society* 20 May 409/1 Y'know, who to pal up with. **1977** *Time* 28 Mar. 37/1 It has been reported that he occasionally palled around with gangsters on golf courses or in gambling casinos.

pala[1] (pā·lă). *Ent.* Pl. **palæ.** [a. L. *pāla* spade.] (See quot. 1906.) Hence **pa·lar** *a.*, of or pertaining to a pala.

1892 E. Saunders *Hemiptera Heteroptera Brit. Islands* 336 *C*[*orixa*] *Fallenii*... The palæ of the male are truncate at the base. **1906** J. B. Smith *Explanation Terms Entomol.* 95 Pala: the shovel-shaped tarsal joints in many aquatic *Heteroptera*. **1957** Richards & Davies *Imms's Gen. Textbk. Entomol.* (ed. 9) III. 428 The pala is not a stridulatory organ, nor has it been shown conclusively that the peculiar strigil of these insects [sc. Corixidæ] is concerned with sound production. **1959** Southwood & Leston *Land & Water Bugs Brit. Isles* 387 Palar pegs numerous, extending in a row along most of the pala. *Ibid.* 388 Male palae more or less rounded on the top edge.

pala[2]: see Pallah.

palace, *sb.*[1] Add: **1. e.** By metonymy, the monarch or monarchy.

1962 A. Sampson *Anat. Brit.* I. iii. 49 For much of this, it is unfair to blame the palace. Many of the pretensions spring from deeper causes than the monarchy. **1973** *Times* 14 Apr. (Nepal Suppl.) p. i/5 The primacy of the palace in the decision-making process was the principal feature of the constitution that King Mahendra introduced in 1962. **1974** *Listener* 14 Mar. 327/3, I thought the election was going to be a very close thing..actually, the Conservatives have more votes than the Labour Party. But I think the choice made by the Palace was inevitable. **1974** *Times* 6 May 14/7 The Palace..believed it did not *have* to accede to Mr Wilson's request.

4. (Later examples.) Also, *palace of varieties*, a variety theatre.

1899 Beerbohm *More* 125 Oh, for the wasted glories of the old Oxford! Oh, for one hour in the Hoxton Palace of Varieties! **1902** O. Wister *Virginian* xiii. 148, I came upon him one morning in Colonel Cyrus Jones's eating palace. **1933** P. Godfrey *Back-Stage* xiv. 179 Sir Oswald Stoll, by transforming the music-hall into the palace of varieties, achieved the same sort of result that Sir Joseph Lyons reached by converting tea-shops into Corner Houses. **1966** *Economist* 10 Dec. 1144/2 The plush restaurants..have been supplanted by the palaces *à go-go.* **1973** A. MacVicar *Painted Doll Affair* ii. 32 A toilet palace dominates the head of Inveraray pier. **1976** J. M. Brownjohn tr. *Kirst's Time for Payment* 28 There was a big medium-priced restaurant, a porn palace, a hair stylist.

6. b. *palace-bordered* adj.; **c. palace-car** (earlier and later U.S. examples); **palace coup** = *palace revolution*; **palace guard,** (*a*) one who guards a palace; (*b*) one who helps to protect a monarch, president, etc.; **palace-hotel** (earlier U.S. and later examples); **palace revolution** [cf. G. *palastrevolution*], the overthrowal of a sovereign, etc., without civil war, usu. by other members of the ruling group; also *fig.*; **palace style** *Archæol.*, a type of pottery associated with the Minoan palaces, or an imitation of this type.

1893 'Mark Twain' in *Century Mag.* Dec. 234/1 Along the palace-bordered canals of Venice. **1900** J. K. Jerome *Three Men on Bummel* viii. 174 Through Prague's dirty, palace-bordered alleys must have pressed often in hot haste blind Ziska and open-minded Wallenstein. **1868** *Dispatch & Vanguard* (San Francisco) 28 Mar. 1/1, I enjoyed the equivocal luxury of traveling in a 'palace' or 'sleeping car'. **1967** C. O. Skinner *Madame Sarah* viii. 163 They travelled via..three Pullmans..and her own private car, known as a 'Palace Car'. **1970** *Guardian* 13 Jan. 1/2 Some kind of palace coup occurred in Biafra on Friday... The Biafran doves 'invited' their leader to step down. **1970** *Daily Tel.* 16 Feb. 16 This adds another possibility to those of a bid by Reed or a rival—a palace coup which would allow new management to be called in to put through an internal re-organisation. **1887** Palace-guard [in Dict.]. **1948** J. A. Farley *Jim Farley's Story* xxii. 232 Nathan Straus brought me word that the White House 'palace guard' realized the anti-Catholic campaign against me had failed. **1973** *Times* 11 May 1/1 This seemed his [sc. President Nixon's] most direct admission to date that he had allowed himself to be kept too isolated for too long by his departed 'palace guard'. **1870** J. D. Sherwood *Comic Hist. U.S.* 422 By the side of palace hotels, now gleaming along golden bays. **1884** *Century Mag.* Mar. 643/1 It [sc. Washington, D.C.] has no elevated railroads, no palace hotels, no mammoth elevators. **1934** G. B. Shaw *Too True to be Good* 11 Come to our palace hotels. **1969** *Sat. Rev.* (U.S.) 4 Jan. 64/2 Whether the world of the Superjet will allow the survival of the palace hotels is a question facing grand tours and grand tourists. **1904** Palace revolution [listed in Dict., sense 6 a]. **1907** J. London *Iron Heel* xiv. 188 They will be like the guards of the palace in old Rome, and there will be palace revolutions whereby the labour castes will seize the reins of power. **1932** M. Eastman tr. *Trotsky's Hist. Russ. Revolution* I. 83 (*heading*) The Idea of a Palace Revolution. **1935** H. A. L. Fisher *Hist. Europe* I. xii. 143 The [Byzantine] state was shaken by palace revolutions and civil war. **1949** *Mind* LVIII. 500 The ensuing changes must be classed with the 'palace revolutions' of other histories, since they scarcely affected the structure of the state. **1958** *Times* 1 Mar. 7/6 The palace revolution in ski racing technique. **1972** H. Kemelman *Monday the Rabbi took Off* xix. 123 The Persian King feared a palace revolution by Haman and plotted with Esther to bring about his ruin. **1902** A. J. Evans in *Ann. Brit. Sch. Athens 1900–1901* 51 (*heading*) Mycenaean painted pottery of the 'Palace Style'. *Ibid.*, The view that this in fact represents the indigenous 'Palace Style' of Knossos in its highest development is confirmed by the evident parallelism which its motives present to the decorative wall paintings of the building. **1913** R. A. S. Macalister *Philistines* i. 18 In Palestine and elsewhere occasional scraps of the 'palace' styles come to light. **1939** J. D. S. Pendlebury *Archæol. Crete* iv. 180 L.M. II, in fact, was, like M.M. II, a true Palace style, though even more restricted in being confined to Knossos alone. **1974** *Encycl. Brit. Macropædia* XIX. 275/2 Between 1450 and 1375 BC, Mycenaean taste reduced the spontaneity of the early Marine style to a rigid formality, thereby creating the monumental Palace style.

‖ **palacio** (pāla·þyo, pălæ·sīo). [Sp., palace.] A palace, or a country seat, or an official building (in Spain or the Spanish-speaking territories in the Americas); also, the name of a specific building, as the former residence of the Spanish and Mexican governors in Santa Fe, New Mexico, or the former hunting-lodge of the Spanish kings in the Coto Doñana, Sevilla.

1844 J. Gregg *Commerce Prairies* I. 203 These bricks are called *adobes*, and every edifice, from the church to the *palacio*, is constructed of the same stuff. **1885** *Weekly New Mexican Rev.* 18 June 4/1 There he erected a palacio. **1893** Chapman & Buck *Wild Spain* xxxii. 350 The head of our cavalcade sighted the welcome light displayed from the turrets of the ancient shooting lodge of Doñana. Though now in a state of partial ruin, the old Palacio still shows signs of former grandeur, and has been ..a favourite sporting retreat for more than one Spanish king. **1963** *Times* 27 Feb. 11/6 It was frequently visited for hunting parties by the Kings of Spain for whom was created the palacio or hunting-lodge which is still the only substantial habitation many miles from the nearest road. **1968** R. F. Adams *Western Words* (rev. ed.) 218/2 *Palacio*, what the early freighters called the Palace of Governors in Santa Fe. **1969** A. Marin *Rise with Wind* xiv. 170 A four-story white-stucco *palacio*. **1975** N. Luard *Robespierre Serial* xiii. 113 The *palacio* had been..used by the Spanish Kings who'd come to the Coto to shoot. **1977** P. Somerville-Large *Eagles near Carcase* vii. 132, I worked at the *palacio* before I was married.

palæanthro·pic, *a.* Also **palæoanthropic** and with the prefix written **pale-.** [f. Palæo-, paleo- + Anthropic *a.*] Of, pertaining to, or designating extinct prehistoric forms of man.

Quot. 1916 may represent an independent coinage.

1890 *Cent. Dict.* V, Paleo-anthropic. **1916** G. E. Smith in *Amer. Museum Jrnl.* XVI. 325/2 If we refer to the epoch of the modern type of man as the Neoanthropic age,..the Mousterian period and all of man's record that went before it can then be included in a Palæanthropic age. **1935** Huxley & Haddon *We Europeans* ii. 52 Modern types (Neanthropic) of man appear in Europe as the last ice-sheet began to retreat and the earlier types (Palæanthropic) seem to have disappeared. **1954** *Sci. Amer.* Sept. 52/3 Paleoanthropic man is clearly a tool user, a worker in stone and bone. **1962** C. S. Coon *Origin of Races* viii. 334 How many grades..shall we recognize in fossil and living men?.. A compromise nomenclature is Protoanthropic, Paleanthropic, and Neanthropic. **1973** B. J. Williams *Evolution & Human Origins* xi. 175/2 The paleanthropic line has included the finds of: Neandertal, Heidelberg, Peking, Java, Solo, Broken Hill (Rhodesia).

Palæarctic, *a.* Add: Also **Palearctic** and with lower-case initial. (Further examples.) Also *absol.*

The (incorrect) 1857 reference in Dict. is given below in corrected form as quot. 1858.

1858 P. L. Sclater in *Jrnl. Linn. Soc.* (*Zool.*) II. 135, I think we may consider Africa, north of the Atlas, Europe and Northern Asia, to form one primary division of the earth's surface, for which the name Palæarctic or Northern Palæogean Region would be best applicable. **1951** *Antiquity* XXV. 69 If Odysseus began as a bear, as Rhys Carpenter asks us to believe, it is well to know how palæarctic that makes him. **1957** P. J. Darlington *Zoogeogr.* vii. 438 The Palearctic is north-temperate with an arctic fringe. **1974** *Environmental Conservation* I. 7/2 Huge numbers of palaearctic birds overwinter in the savanna zones south of the Sahara.

Palæasiatic, var. *Palæo-Asiatic *sb.* and *a.*

palæencephalon (pæ:-, pēi:li̩ense·fălǫn). *Anat.* Also (chiefly *U.S.*) **paleencephalon.** [a. G. *palaeencephalon* (L. Edinger *Vorlesungen über den Bau der nervösen Zentralorgane des Menschen und der Tiere* (ed. 7, 1908) II. xvi. 241): see *Palæo-, paleo- b and Encephalon.] The phylogenetically older portion of the brain, as contrasted with the neencephalon.

1917 *Jrnl. Compar. Neurol.* XXVIII. 216 It is becoming increasingly evident that the key to this difficult question is to be sought in the subcortical centers of the primitive types, that is, in the 'old brain' (palaeencephalon of Edinger, segmental apparatus of Adolf Meyer). **1972** [see *neencephalon].

palæo-, paleo-. Add: **palæoanthropic,** delete entry and see *palæanthropic *a.*; **palæobathy·metry** *Geol.*, the bathymetric features of an area as they were at some period in the past; so **pa:læobathyme·tric** *a.*; **palæobio·logy,** the biology of fossil plants and animals; hence **pa:læobiolo·gic, -ical** *adjs.*, **palæobio·logist; palæoceanography,** etc., varr. *palæo-oceanography*, etc., below; **Palæocene,** delete entry and see *Palæocene *a.*; **palæochem·istry,** (the study of) the chemical features of something as they were in the geological past; hence **palæoche·mical** *a.*; **pa·læocurrent** *Geol.*, a current, usu. of water, which existed at some period in the past, as inferred from the features of sedimentary rocks; **pa:læoenvi·ronment,** an environment at a period in the past; hence **pa:læoenvironme·ntal** *a.*; **pa·læo-equator** *Geol.*, the equator as it was at some period in the past; hence **pa:læo-equato·rial** *a.*; **pa·læofield** *Geol.*, (the strength of) the earth's magnetic field at a period in the geological past; **palæogeo·graphy** *Geol.*, (the study of) geographical features at periods in the geological past; hence **palæogeo·grapher; pa:læogeogra·phic, -ical** *adjs.*; **pa:læogeogra·phically** *adv.*; **palæogeo·logy,** (the study or reconstruction of) the geological features of an area in past ages; hence **pa:læogeolo·gic, -ical** *adjs.*, **palæogeo·logist; pa:læogeomagne·tic** *a. Geol.*, of or pertaining to the magnetic field of the earth in the geological past; **pa:læogeophy·sics** *sb. pl.*, the study of the physical characteristics of the earth in past ages; hence **pa:læogeophy·sical** *a.*; **palæoge·otherm** *Geol.* [*geotherm* f. Geo- + *-therm*, after Isotherm, etc.], a pattern of temperature variation which existed in the earth's crust at some time in the past; so **pa:læogeothe·rmal** *a.*; **pa·læogra:vity** *Geol.*, the strength of the earth's gravity at some time in the past; **palæohydro·graphy** *Geol.*, (the study of) hydrographic features at periods in the geological past; **pa·læointe:nsity** *Geol.*, the intensity of a palæomagnetic field; **palæola·titude** *Geol.*, the latitude of a place at some period in the past; hence **pa:læolatitu·dinal** *a.*; **palæolimno·logy,** (the study of) the conditions and processes occurring in lakes in the geological past; hence **pa:læolimnolo·gical** *a.*, **palæolimno·logist; pa:læolitholo·gic** *a. Geol.*, applied to a map showing the lithological features of an area at some period in the past; **palæolo·ngitude** *Geol.*, the longitude of a place at some period in the past; **palæomeri·dian** *Geol.*, the meridian of a place at some period in the past; **pa:læometeoro·logy,** the study of atmospheric conditions at periods in the geological past; so **pa:læometeorolo·gical** *a.*, **-meteoro·logist; pa:læooceano·graphy** (also **palæoceanography**), (the study of) the conditions and processes occurring in oceans in the geological past; hence **pa:læ(o-)oceano·grapher; pa:læ(o-)oceanogra·phic, -ical** *adjs.*; **palæopedo·logy** *Geol.*, (the study of) the features of soils in the geological past; hence **pa:læopedolo·gical** *a.*, **palæopedo·logist; pa:læophysio·graphy** *Geol.*, (the study of) the physical and topographical features of the earth's surface in the geological past; hence **pa:læophysiogra·phic, -gra·phical** *adjs.*; **pa:læophysio·grapher; pa·læoplain** *Geomorphol.*, a peneplain which existed at some period in the past and became overlain by other strata, being now buried or re-exposed; **pa·læopole** *Geol.*, a magnetic pole of the earth at a period in the past; **palæopsy·chic** *a. rare*, pertaining to the assumed (prehistoric) origins of behaviour patterns; so (also *rare*) **pa:læopsycho·logy, -ist; pa·læoradius** *Geol.*, the radius of the earth or another planet at some time in the past; **palæosali·nity** *Geol.*, the salinity of the environment in which a sedimentary deposit was laid down; **pa·læoslope** *Geol.*, the former or original slope of a region, or its direction; **pa·læosol** [*-sol], a soil horizon which was formed as a soil in the geological past; hence **palæoso·lic** *a.*; **pa·læospecies** *Palæont.*, a species including a group of fossils from different geological formations that make up a chronological series; **pa·læostructure** *Geol.*, the geological structure of an area at some period in the past; hence **palæostru·ctural** *a.*; **palæotechnic** *a.* (examples); *spec.* (see quot. 1960); **palæotecto·nic** *Geol.*, of or pertaining to tectonic features or events of previous stages in the earth's history; **pa·læotemperature,** the average climatic temperature at a particular place and time in the past; **palæothermo·metry,** the investigation of the temperature of climates and oceans in past ages; **palæotopo·graphy** *Geol.*, the topography of ancient landscapes, esp. as represented today by features that are buried or newly exhumed (cf. *palæogeomorphology); hence **pa:læotopogra·phic, -gra·phical** *adjs.*, **pa:læotopogra·phically** *adv.*; **pa·læowind** *Geol.*, a prevailing wind that existed at some period in the past; freq. *attrib.*; **pa:læozo:ogeo·graphy,** the study of the distribution of fossil animal remains; hence **pa:læozo:ogeogra·phic** *a.*

1945 *Bull. Amer. Assoc. Petroleum Geologists* XXIX. 428 Maps of fossil distribution, with evaluation of the habitat of the organisms, add evidence. Thus from the lithology and fauna, it is possible to construct paleobathymetric maps. **1964** H. W. Menard *Marine Geol. Pacific* vi. 138/1 Since most of the islands existed at about the same time.., a consistent paleobathymetric map can be drawn. **1959** *Jrnl. Paleont.* XXXIII. 944/2 Critical comparative study is also called for of the characteristics of sedimentary rocks in areas where there is possibility of reconstructing a reasonably objective paleobathymetry, as in the..Ventura Basin of California or other areas where the endemic fossils indicate a wide range of depth. **1971** *Nature* 2 Apr. 319/1 The magnitude of the cretaceous–Tertiary hiatus in the deep sea is therefore a function of palaeobathymetry with deeper water sections exhibiting a greater unconformity. **1948** *Bull. Geol. Soc. Amer.* LIX. 1337 (*heading*) Paleobiologic implications of the measurement of paleotemperatures. **1961** Webster, Paleobiological. **1963** D. W. & E. E. Humphries tr. *Termier's Erosion & Sedimentation* ix. 192 Two palaeobiological aspects of this subject should be noted. **1974** *Nature* 15 Feb. 496/1 He [sc. D. M. S. Watson] joined Abel as one of the pioneers of palaeobiological thought. **1900** *Biol. Lect. Marine Biol. Lab. Woods Holl 1899* ix. 132 The method thus elaborated has been and is now in constant use by a number of paleobiologists. **1975** *Nature* 1 May 16/1 Because they have become excited by new biological concepts and wish to apply them to fossils, a number of young researchers would now prefer to call themselves palaeobiologists. **1893** S. S. Buckman in *Q. Jrnl. Geol. Soc.* XLIX. 482 The term 'hemera' will therefore enable us to record our facts correctly; and its chief use will be in what I may call 'palæo-biology'. **1943** *Mind* LII. 127 One can hardly think of a scientific fact better and more impressively documented than the phylogenetic hierarchy, established as it is by the threefold evidence of embryology, comparative anatomy, and paleobiology. **1948** *Jrnl. Paleont.* XXII. 265/1 Paleobiology is..mainly biological in objectives, but many of its techniques are unknown to biology. **1973** P. Tasch (*title*) Paleobiology of the invertebrates: data retrieval from the fossil record. **1854** *Edin. New Philos. Jrnl.* LVI. 9 The palæophysics are hardly studied, and even less the palæochemistry. **1904** *Trans. Canad. Inst.* VII. 535 (*heading*) The palæochemistry of the ocean in relation to animal and vegetable protoplasm. **1926** *Physiol. Rev.* VI. 316 (*heading*) The paleochemistry of the body fluids and tissues. *Ibid.* 331 The high concentration of the salts, 2·852 per cent, in the serum of the lobster would appear to indicate that it is of neochemical rather than of paleochemical origin. **1942** *Proc. R. Irish Acad.* XLVIII. B. 119 An enquiry into oceanic palæochemistry and its bearing on the electrolytes of blood and cells. **1955** *Bull. Geol. Soc. Amer.* LXVI. 1606 (*heading*) Paleocurrents of Lake Superior Precambrian quartzites. **1971** *Nature* 28 May 245/2 A W.N.W. to N. palaeocurrent component predominates in the channel sandstones with pedogenic modification..occurring on most proximal floodplain deposits. **1957** R. J. Braidwood in *Publ. National Research Council* (U.S.) No. 565. 16/2 This field or axis of interrelated disciplines (perhaps 'Pleistocene ecology' or 'paleo-environment' or 'Quaternary geography'—I shall not attempt to name it!) would definitely include Man as an element in, and a factor acting upon, the environmental scene. **1970** *Nature* 29 Aug. 944/2 The palaeoenvironment of the Neolithic occupation site. **1975** *Ibid.* 20 Mar. 187/2 Reconstructions of palaeoenvironments often rely heavily on fossils, and thus engender circular arguments about the habitats occupied by different elements in the fauna. **1961** *Micropaleontology* VII. 366/1 Preliminary results of analyses of these data, in terms of biotic diversity, show interesting parallels with independent evidence concerning paleoenvironmental changes and floral evolution. **1974** *Nature* 29 Mar. p. iv/3 (Advt.), An accompanying text describes each feature and discusses its..preservation and occurrence in sedimentary rocks, and significance for paleoenvironmental reconstructions. **1976** *Ibid.* 20 May 223/1 These palaeoenvironmental similarities may be sufficient to explain the near identity of the two assemblages of microfossils. **1960** *Quaternary Res.* (Tokyo) May 212 (*heading*) The palaeoequator and its relation to the recent distributional area of *Coriaria*. **1962** *Nature* 3 Nov. 427/2 The estimates of palæolatitudes are generally low, ranging from 59° S.–44° N., with 76 per cent of the values lying within 20° of the palæoequator. **1973** *Sci. Amer.* Nov. 111/1 An eastward extrapolation of paleo-equator positions determined from deep-sea drilling, together with a westward extrapolation of crustal age.., enables us to estimate the location and age of a series of points where the East Pacific Ridge and earlier 'paleo-equators' once intersected. **1966** *Nature* 15 Oct. 247 (*heading*) Summary of estimates of palaeoequatorial [magnetic] intensity from igneous rocks in the temperature range 670° C to 500° C. **1968** *New Scientist* 4 Apr. 16/1 Since the newly acquired moment is proportional to the known field, and the natural moment of the virgin rock is proportional to the ancient field, a simple equation allows the 'palaeofield' to be calculated. **1975** *Nature* 27 Feb. 685/2 The obvious ways of obtaining a palaeofield from a rock containing thermoremanent magnetisation require that the rock be heated above its Curie point. **1911** *Bull. Geol. Soc. Amer.* XXII. 262 These results are of the greatest value to the paleogeographer. **1972** *Science* 12 May 665/2 A scholar who, as a paleogeographer, is not narrowly specialized in archeology or geology. **1906** *Bull. Geol. Soc. Amer.* XVII. 248 (*heading*) Paleogeographic charts. **1971** *Nature* 16 July 180/2 The evidence..was taken from two series of palaeogeographic maps showing the distribution of land and sea since the early Cambrian. **1882** E. Hull *Contrib. Physical Hist. Brit. Isles* I. iii. 19 In endeavouring to prepare a series of maps representing the palæo-geographical features of some region,..the requisite number of such maps and their proper order of succession..necessarily corresponds to those of the successive geological formations. *Ibid.* II. i. 55 (*heading*) Palæo-geographical and geological maps of the British Isles. **1956** L. J. Wills *Concealed Coalfields* 6 The palaeogeographical treatment is capable of throwing new light on the problem of where the workable coals may originally have been deposited. **1965** B. E. Freeman tr. *Vandel's Biospeleol.* xvi. 274 A map of the distribution of the cavernicolous sphaeromids can easily be superimposed on the paleogeographical reconstruction of the Miocene epoch. **1934** *Bull. Amer. Assoc. Petroleum Geologists* XVIII. 784 (*heading*) Future research in paleogeographically favorable zones. **1969** Bennison & Wright *Geol. Hist. Brit. Isles* xi. 265 Permian outcrops in Ireland, though not extensive, are palaeogeographically important. **1881** R. Etheridge in *Q. Jrnl. Geol. Soc.* XXXVII. 228 Could we strip off all the Secondary and Tertiary rocks, and reveal or expose the extension of the older or Palæozoic series towards Germany on the east, and France on the south, then the vexed question of the old physical geology and geography (palæogeography) of Britain and the relation and correlation of our area with that of Europe would be revealed. **1946** *Nature* 20 July 89/2 Grabau's widely ranging interests in stratigraphy, palæogeography, palæontology and sedimentation..were synthesized into a whole in his 'Principles of Stratigraphy' (1913). **1969** Bennison & Wright *Geol. Hist. Brit. Isles* i. 3 Changes in the distribution of land and sea, of the physiography of the land, and of climate have taken place and their study is called palaeogeography. **1971** I. G. Gass et al. *Understanding Earth* xx. 292/1 The direction of these currents can be established by studying the sedimentary structures the rocks contain; and this may assist in the determination of the palaeogeography of the times when sedimentation occurred. **1933** *Bull. Amer. Assoc. Petroleum Geologists* XVII. 1113 (*caption*) Paleogeologic map of United States at beginning of Lower Cretaceous or Comanche time, representing areal geology of surface upon which Cretaceous sediments were deposited. **1966** *McGraw-Hill Encycl. Sci. & Technol.* IX. 519/1 Paleogeologic maps, showing the pattern of rocks on the surface at a past time, aid in the interpretation of landforms. **1882** *Sci. Trans. R. Dublin Soc.* I. 257 (*heading*) Palæo-geological and geographical maps of the British Islands and the adjoining parts of the continent of Europe. **1940** *Geogr. Jrnl.* XCV. 208 The island re-emerged during the Tertiary from an ancient, drowned landmass named Palaeonotis, postulated by palaeogeologists..as having extended from Asia as far as New Caledonia. **1933** *Bull. Amer. Assoc. Petroleum Geologists* XVII. 1129 Northeastern Texas..is an excellent example of the application of paleogeology to the problem of the accumulation of oil and gas. **1959** P. C. Badgley *Struct. Methods Exploration Geologist* v. 128 Figure 138 indicates the paleogeology immediately below the pre-Cretaceous erosion surface. **1960** A. I. Levorsen *Paleogeologic Maps* i. 2 Paleogeology ignores all subsequent sedimentation, volcanism, and deformation, and requires that the geologist consider the geology as it was when it was being formed. **1962** *Jrnl. Geophysical Res.* LXVII. 3461/2 Since different parts of formations became magnetized at different times, secular variation of the palaeogeomagnetic field must have produced a certain amount of scatter of the directions of magnetization. **1977** *Sci. Amer.* Dec. 42/1 The maps are computer-generated by rotation around the varying north paleogeomagnetic poles, the amount of rotation controlled by sea-floor magnetic measurements. **1965** *Phil. Trans. R. Soc.* A. CCLVIII. 1 Because no other palaeogeophysical record is comparable in scope, palaeomagnetic surveys in the different continents have led to important conclusions about the Earth's evolution. **1970** *Nature* 18 July 227/1 Palaeogeophysical studies to determine whether present day processes were equally valid in past geological time. **1959** *Geophysical Abstr.* No. 175. 374 'Paleogeophysics' includes all methods which can lead to an understanding of former physical conditions and processes in the earth during its evolution; it is a part of paleogeography. **1970** S. K. Runcorn *Palaeogeophysics* iii. 17 The newest branch of palaeogeophysics is the record presumed to have been found in fossils of the biological rhythms of marine fauna. In the oceans it seems *a priori* reasonable to suppose that these biological clocks set themselves by the exactly periodic variations in their physical environment, which seem only to be the solar day, the synodical month and the tropical year. **1975** *Nature* 3 Apr. 406/2 The palaeogeotherms which existed in the upper mantle immediately before incorporation and transport of the xenoliths by kimberlite eruptions have been derived from data from large numbers of individual xenoliths. **1970** R. G. J. Strens in S. K. Runcorn *Palaeogeophysics* xl. 383 Apart from their usefulness in studying palaeogravity and palaeogeothermal gradients, such geothermometers and geobarometers would be of inestimable value to petrologists. **1978** *Nature* 18 May 221/1 Much debate has centred upon whether some palaeogeothermal gradients are perturbed or not. **1970** R. G. J. Strens in S. K. Runcorn *Palaeogeophysics* xl. 383 Prospects for measuring palaeogravity with an accuracy sufficient to detect major variations (> 10%) over the last 3000 m. yr appear good. **1978** *Nature* 12 Jan. 153/2 Hypotheses involving substantial changes in Earth radius over geological time can be tested by measuring palaeogravity at, or near, the Earth's surface. **1853** *Edin. New Philos. Jrnl.* LV. 298 (*heading*) On the palæohydrography and orography of the Earth's surface. **1933** W. J. Arkell *Jurassic Syst. Gt. Brit.* xviii. 557 To do justice to the palæogeography of the Jurassic period..we should have to proceed systematically from the points of view of..palaeohydrography, palaeoceanography, palaeobiogeography, palaeoclimatology,..and many others. **1971** *Geol. Förening. Stockholm Forhandl.* XCIII. 59 (*heading*) Foraminifera and the paleohydrography of the Arabian Sea. **1965** *Jrnl. Geomagnetism & Geoelectr.* XVII. 417 (*heading*) Preliminary results of investigations made to study the use of Indian pottery to determine the paleointensity of the geomagnetic field for United States 600–1400 A.D. **1974** *Nature* 17 May 227/2 If the Moon were uniformly magnetised, it would need to have had a dipole moment of about 10^{22} gauss cm³ to give an ancient surface field of 2,000 gamma (γ) which is typical of several palaeointensity studies, although a much higher value has been reported. **1959** *Geophysical Jrnl.* II. 307 Thus the shape of the areas of confidence so calculated depends on the palaeolatitude. **1971** I. G. Gass et al. *Understanding Earth* xiii. 174/2 The study of wind systems associated with arid deposits of different ages gives evidence of palaeolatitudes and so has a bearing on continental drift. **1964** *Prof. Papers U.S. Geol. Survey* No. 501-C. 109/1 (*heading*) Paleolatitudinal distribution of ancient phosphorite. **1971** *Nature* 1 Jan. 17/1 The geometrical reconstruction presented here..was obtained using a palaeomagnetic computer program which plotted continental outlines in their palaeolatitudinal position relative to a pole determined palaeomagnetically. **1961** *Mem. Coll. Sci. Kyoto Univ.* B. XXVIII. 68 A paleolimnological study of deep core-samples will indicate not only the developmental history of Lake Biwa-ko but the whole Pleistocene climatic history of the East Asiatic continent. **1973** *Nature* 16 Mar. 184/1 This aspect of the palaeolimnological work is being pursued in greater detail by magnetic, chemical and botanical investigations of the most recent sediments in Lough Neagh. **1960** *Amer. Jrnl. Sci.* CCLVIII. A. 1 (*heading*) Wilmot Hyde Bradley. Geologist, geomorphologist, paleolimnologist, paleontologist, administrator. **1970** *Ann. Acad. Sci. Fennicae* A. 3. CV. 25 The two species *Bosmina coregoni* Baird and *B. longirostris* O. F. Müller..are of particular interest to limnologists and palaeolimnologists. **1942** *Amer. Jrnl. Sci.* CCXL. 237 One particularly significant contribution of paleolimnology to glacial geology may be the derivation of an absolute chronology on the basis of quantitative counts of microfossils. **1948** *Bull. Geol. Soc. Amer.* LIX. 641, I have selected three distinctive features of the Green River formation to interpret in terms of the paleolimnology of the Eocene Green River lakes. These are the carbonate sediments and carbonate minerals, the oil shales, and the

varved sediments. **1973** *Jrnl. Phycol.* IX. 395/2 The algal senescence system may prove valuable in lacustrine paleolimnology as a simple but experimentally accessible analogy to the processes of chlorophyll degradation in the water column. **1945** *Bull. Amer. Assoc. Petroleum Geologists* XXIX. 427 Paleolithologic maps have lines, isoliths, connecting points of similar lithology and separating rocks of differing nature. **1966** *McGraw-Hill Encycl. Sci. & Technol.* IX. 519/1 Paleolithologic maps showing bottom sediment patterns suggest whether rocks were laid in depths of strong wave action or in quieter water of deeps or broad shoals. **1964** K. M. Creer in A. E. M. Nairn *Probl. Palaeoclimatol.* 274 Because of the assumed axial symmetry of the field, palaeomagnetic data cannot yield information about palaeolongitude. **1961** *Nature* 17 June 1097/2 Localities along the same palæo-meridian. **1965** *Phil. Trans. R. Soc.* A. CCLVIII. 27 A palaeomagnetic survey of a suite of rocks representing at least 10⁴y thus provides information from which may be determined: (1) the palaeomagnetic latitude..of the sites, and (2) the palaeomeridian direction. **1901** *Q. Jrnl. Geol. Soc.* LVII. 405 (*heading*) A palaeometeorological explanation of some geological problems. **1962** Rex & Goldberg in M. N. Hill *Sea* I. v. 295 Deep-sea deposits provide excellent source areas for paleometeorological investigations provided eolian materials can be recognised. **1964** K. W. Butzer *Environment & Archeol.* xxii. 334 A great deal of more satisfactory paleoclimatic information must be available before this major barrier to paleometeorological study is removed. **1901** *Q. Jrnl. Geol. Soc.* LVII. 468 The palæometeorologist must work with the best material he can obtain, content with the enunciation of general principles, and with the solution of some of the more simple problems. **1854** *Edin. New Philos. Jrnl.* LVI. 9 We have got very few notions on palæometeorology and palæotemperature or thermics. **1901** *Q. Jrnl. Geol. Soc.* LVII. 469 Among the services which palæometeorology may hereafter render to the geologist..may be that of assisting him to determine the chronological relations of geological zones in different regions where no direct evidence..may be attainable. **1954** *Sci. News* XXXIII. 65 Paleoclimatology, the study of the climates of the past, is considerably more advanced than paleometeorology, the study of past weather. **1957** *Mem. Geol. Soc. Amer.* LXVII. II. xxiv. 684 The development of the method of paleotemperature research..using the O¹⁸:O¹⁶ ratio in carbonates and other solid salts of oxyacids has provided the most powerful single tool yet invented in the equipment of the paleo-oceanographer. **1945** Paleooceanographic [see *palæoclimatologic *a.*]. **1968** *Science* 29 Mar. 1461/1 The distribution of planktonic tests in fossil marine sediments are being used increasingly for paleo-oceanographic reconstructions. **1971** *Nature* 13 Aug. 469/1 Oxygen isotope and palaeontological determination of planktonic foraminifers allow us to make several generalizations on the palaeoceanographic history of the Arctic Ocean back to Mid Pliocene. **1971** *Ibid.* 30 Apr. 563/1 In the interesting palaeoceanographical model proposed by Bandy, a thermal maximum is indicated in marine conditions during the same 9–8 m.y. periods. **1933** Palæoceanography [see *palæohydrography* above]. **1957** *Mem. Geol. Soc. Amer.* LXVII. II. xxiv. 684 (*heading*) Future of the study of paleo-oceanography. **1972** *Nature* 17 Mar. 117/1 The gross geometric relationships of pelagic facies on a ridge system must be interpreted in the light of the spreading process itself before meaningful conclusions can be drawn about palaeoclimatology and palaeo-oceanography. **1927** B. B. Polynov *Contrib. Russ. Scientists to Paleopedology* 1 It is yet scarcely possible to speak of paleopedology, as of a scientific theory that has assumed a wholly definite shape. *Ibid.* 14 As one of the most interesting recent Ukrainian papers on pedology may be especially mentioned the paleopedological map of Ukraina, drawn by Makhov. **1943** *Amer. Jrnl. Sci.* CCXLI. 197 A great deal more must be learned about the soil formation before paleopedologists will understand the true significance of the soil bones. *Ibid.* 199 Because of the nature of the pedogenic fossils it cannot be expected that paleopedology ever will be able to deal with much more than the general types of the ancient soil formation and the changes in the geographical pattern of the zones and regions in which these types prevailed during the different geological periods. *Ibid.* 200 The true paleopedological formations are rather rare. **1971** G. Roeschmann in D. H. Yaalon *Paleopedology* 319 In future, paleopedologists and geologists will only obtain satisfactory results..if they work as a team. **1973** *N.Z. Jrnl. Geol. & Geophysics* XVI. 723 Such paleopedological evidence may be employed to confirm and extend the chronology of the later part of the New Zealand Quaternary. *Ibid.* 735 This identification of these ancient episodes of erosion depends on the identification of tephra formations and paleosols and is a logical application of paleopedology to environmental reconstruction. **1898** J. E. Marr *Princ. Stratigr. Geol.* x. 121 The palæo-physiographer..attempts to restore the physical conditions of greater thicknesses of deposit. **1950** *Bull. Illinois State Geol. Survey* No. 73. iii. 23 (*caption*) Paleophysiographic diagram of the bedrock topography of Illinois. **1954** W. D. Thornbury *Princ. Geomorphol.* ii. 31 (*caption*) A paleophysiographic diagram showing the major features of the preglacial topography of Illinois. **1898** J. E. Marr *Princ. Stratigr. Geol.* x. 122 The utmost that the maker of palæophysiographical maps can expect to indicate, when dealing with considerable thicknesses of strata, is an approximation to the mean position of the shore-lines of the period when these strata were deposited. **1882** E. Hull *Contrib. Physical Hist. Brit. Isles* p. v, I had intended giving the title of 'Palæo-Physiography of the British Islands' to this volume; but certain friends in whose judgment I have the most implicit confidence assured me that no one would have the slightest conception from such a title of the contents of my book. **1915** C. Schuchert *Text-bk. Geol.* II. xx. 450 There is..another record that has so far been almost refused recognition in our time-tables. This is the time evaluation of topographic form at any given stage of development (the physiography of the present, the paleophysiography of the past). **1900** R. T. Hill in *Topogr. Atlas U.S. Geol. Survey* Folio 3. 5/3 Destructional plains are sometimes evolved from constructional plains; the latter, after elevation in long erosion, are reduced in old age to the former. On the other hand, constructional plains are usually established upon areas which were once destructional plains. Ancient buried destructional plains thus veneered by constructional formations might be appropriately termed paleoplains. **1966** *Bull. Amer. Assoc. Petroleum Geologists* L. 2302/2 Farther west along the flank of the Canadian shield..a vast paleoplain..was covered by Lower Cretaceous (locally Upper Jurassic) sediments which include a high percentage of sandstone which could serve as reservoir. **1962** *Geofisica Pura e Appl.* LIII. 52 (*heading*) Rock magnetism and the earth's palaeopoles. **1971** I. G. Gass et al. *Understanding Earth* xv. 212 The evidence favouring drift including geomagnetic data (palaeopole investigations). **1904** G. S. Hall *Adolescence* II. xii. 194 The problem, whether there is any paleopsychic race element, is as inevitable as it is unanswerable. **1916** *N.Y. Med. Jrnl.* CIV. 1077/1 We wish to maintain the idea that there may be other types of fossils to be studied than those derived from plants and animals, namely thought fossils, and that to paleobotany and to paleozoology we may add a science of paleopsychology. The happiest ground of the paleopsychologist is ..in history, in literature, in..possibly diseases. **1960** *Geofisica Pura e Applicata* XLV. 116 The palaeo-radius corresponding to the time *t*. **1978** *Nature* 26 Jan. 316/1 Observations of the surface of Mercury and Mars by spacecraft enable constraints to be placed on the palaeoradius of these extraterrestrial bodies. **1960** *Oil & Gas Jrnl.* 1 Feb. 154/1 A method for determining paleosalinities..would also be of considerable practical value in the search for various types of mineral deposits and petroleum. **1972** *Marine Geol.* XII. 335 (*heading*) Pollen and paleosalinity analyses from a Holocene tidal marsh sequence. *Ibid.* 337 A method for determining paleosalinity was proposed by Nelson (1967) which utilized the relative proportions of iron and calcium phosphate in argillaceous sediments. **1957** *Bull. Geol. Soc. Amer.* LXVIII. 469/1 Mapping of cross-bedding and other primary current structures has proved useful in reconstruction of regional paleoslopes. **1965** *Bull. Amer. Assoc. Petroleum Geologists* XLIX. 341/1 Regional variation in stratigraphic position of the base of the landslide facies establishes the fact that the foot of the paleoslope migrated north-northwestwardly through time. **1975** Read & Watson *Introd. Geol.* II. i. iv. 82 Remarkably constant palaeocurrent directions determined from current-bedding throughout the Athabasca formation indicate a palaeoslope towards the west or north-west. **1950** Hunt & Sokoloff in *Prof. Papers U.S. Geol. Survey* No. 221. 109/1 An ancient soil, hereafter referred to as paleosol, has been dated..as pre-Wisconsin in the Lake Bonneville and Denver basins. **1954** W. D. Thornbury *Princ. Geomorphol.* iv. 82 Paleosols may be found on buried landscapes, on exhumed portions of ancient landscapes, or upon features of the present topography which are relicts of previous geomorphic cycles. **1968** R. W. Fairbridge *Encycl. Geomorphol.* 554/2 A lateritic crust is essentially a paleosol, and reflects a polycyclic regime, usually as a result of repeated alternation of hot, humid conditions developing laterite, with dry, evaporating conditions favoring crust development. **1975** *Nature* 20 Mar. 189/1 Palaeoenvironmental indicators in the sediments include polygonal desiccation cracks, calcretes and other palaeosols. **1956** *Soil Sci.* LXXXII. 441 Erosive stripping of the protective mantle of sediment in many places, however, resulted in the resurrection of the paleosolic profile *in toto*, so that now the paleosol occurs within the continuum of soils on the modern surface. **1965** R. V. Ruhe in Wright & Frey *Quaternary of U.S.* 759/2 These kinds of soil occur adjacent to each other and grade from one to another on the paleosolic surfaces of the older glacial tills. **1954** A. J. Cain *Animal Species* vii. 107 When good series are available, forms that seem to be good species at any one time may become indefinable since they are successive stages in a single evolutionary line... It is convenient to refer to such forms as palaeospecies. **1956** P. C. Sylvester-Bradley *Species Concept in Palaeont.* 2 A third concept is the result of projecting the biospecies into a third dimension, that of time. It is the concept..others have called the 'palaeospecies'. **1964** *Nature* 1 Aug. 451/1 A species in palæontology..is a fundamentally different kind of unit and has been called a chromospecies or palæospecies. **1974** *Ibid.* 1 Nov. 85/2 Physical anthropologists have expanded the concept of variability in palaeospecies making it possible to lump greater ranges of morphological variation within a single species. **1966** *Bull. Amer. Assoc. Petroleum Geologists* L. 2323/1 To demonstrate a logical method of paleostructural mapping by use of carefully selected isopachous intervals. **1972** *Internat. Geol. Rev.* XIV. 1320/1 In any platform region, paleostructural reconstructions should be preceded by a comprehensive study of the conditions of occurrence and distinctive lithologic features of the rocks. **1937** *Jrnl. Geol.* XLV. 51 A paleo-structure map of southeastern Missouri at the end of Lamotte time. **1970** *Israel Jrnl. Earth-Sci.* XIX. 141 Systematic study of the thickness of Senonian strata yields conflicting data regarding palaeostructure. **1904** P. Geddes *City Devel.* xxv. 175/1 Must we not, therefore, call this earlier and crude mechanical civilisation which still predominates among us the 'Palæotechnic' stage, and recognise that the formerly less prominent industrial peoples..are passing more quickly than we into the 'Neotechnic' stage? **1915** —— *Cities in Evolution* iv. 63 We may distinguish the earlier and ruder elements of the Industrial Age as Paleotechnic, the newer and still often incipient elements disengaging themselves from these as Neotechnic. **1934** Paleotechnic [see *eotechnic *a.*]. **1946** *Theology* XLIX. 94 The report..praises the past but its vision for the future is lighted less by a doctrine of the creation than by paleotechnic ideology. **1960** C. Winick *Dict. Anthropol.* 399/1 *Paleotechnic*, in the development of the techniques of western civilization, referring to the period in which mineral resources became prominent and coal power and iron were widely used. It flourished around 1750. The factory system, finance capital, and competition developed. **1973** *New Society* 7 June 543/2 Its automated ticket system..makes London's machines seem paleotechnic by comparison. **1947** *Jrnl. Geol.* LV. 311 (*caption*) Paleotectonic maps of the Cordilleran region in late Paleozoic time. Cross-ruled area is the volcanic archipelago and orogenic belt. Horizontally ruled areas were uplifted and eroded during the period designated. **1957** *Bull. Geol. Soc. Amer.* LXVIII. 151/2 These studies have revealed a paleotectonic framework composed of a geosynclinal basin in central Idaho and a relatively more stable cratonic shelf in southwestern Montana. **1975** *Nature* 10 July 117/1 The positions of Triassic seaways along the southern continental margin of the Tethys 'geosyncline' are defined by the effects of at least two main palaeotectonic events. **1854** Palæotemperature [see *palæometeorology* above]. **1948** *Science* 5 Nov. 492/1 This particular application of the chemical differences in the processes of isotopes occurred to me a year and a half ago, and since that time my colleagues and I have been trying to solve the several difficult problems encountered in making such measurements of paleotemperatures. **1969** *Nature* 4 Oct. 66/2 Recent palaeotemperature measurements obtained from Caribbean cores..suggest that the coldest part of the Weichselian was around 17000 years BP. **1957** *Chem. Abstr.* LI. 12773 (*heading*) Determination of climatic conditions of some regions of the U.S.S.R. in the Upper Cretaceous period by the method of isotopic paleothermometry. **1971** *Nature* 13 Aug. 466/2 Reconstruction of the climatic and hydrologic history of ocean basins has been based on either oxygen isotope palaeothermometry or studies of foraminiferal assemblages preserved in deep sea sediments. **1943** *Jrnl. Sedimentary Petrology* XIII. 111/1 Most paleotopographic maps pass under the name of structural maps but there is a distinct difference. **1960** M. S. Bishop *Subsurface Mapping* ix. 135 Oftentimes the gentle slopes of paleotopographic surfaces are indicative of much steeper dips on the underlying beds, especially if a major unconformity is being mapped. **1960** *Canad. Mining & Metallurg. Bull.* LIII. 535/2 The oil has accumulated beneath the unconformity and the accompanying 'caprock' wherever the subcrop trend of a reservoir bed crosses a paleotopographical ridge. **1966** *Bull. Amer. Assoc. Petroleum Geologists* L. 2296/1 These quantitative concepts are very useful in interpreting an area paleotopographically; *i.e.*, they keep the number of tributaries, lengths of streams, and channel slopes within specific limits according to geomorphological principles. **1943** *Jrnl. Sedimentary Petrology* XIII. 108 The search for new deposits of petroleum is aided by the use of special paleogeographic maps... Other types of maps that may be used show salinity, thickness of sands, paleogeology and paleotopography. **1970** *Jrnl. Geol. Soc. Australia* XVII. 39 The palaeotopography of an east–west valley on the eastern flank of the Great Divide is reconstructed from a consideration of early Tertiary and Palaeozoic rocks. **1957** *Bull. Geol. Soc. Amer.* LXVIII. 1870 (*heading*) Paleo-wind directions in late Paleozoic and early Mesozoic time on the Colorado Plateau as determined by cross-strata. **1964** *Sedimentology* III. 52 The directional features of the dunes suggest that the paleowind pattern was similar to the present-day pattern around Bermuda. **1975** *Nature* 3 Jan. 19/1 We recommend experimental and analytical study of these currents taking into account continental dispersion and palaeowinds. **1967** *Palaeogeogr., Palaeoclimatol., Palaeoecol.* III. 201 Palaeozoogeographic data are thought to be especially useful in establishing a timetable for continental drift. *Ibid.* 210 Many ammonite species were able to disperse widely. While this renders them excellent for purposes of correlation it reduces their importance for palaeozoogeography. **1972** *Nature* 7 Apr. 297/3 Succeeding chapters..examine..the palaeozoogeographic patterns in the fossil record. **1975** *Ibid.* 17 Apr. 556/3 Palaeozoogeography can only make sense as the evolution of organisms in space and time.

b. In anatomical terms designating parts of the brain which are considered to be of relatively ancient development phylogenetically, as *palæencephalon, *palæocerebellum, *palæocortex, *palæopallium, *palæostriatum, *palæothalamus. Cf. *neo- 1 e.

pa:læoanthropo·logy. Also (chiefly *U.S.*) **paleo-.** [ad. F. *paléoanthropologie* (P. Topinard *Élém. d'Anthrop. Gén.* (1885) vii. 177): see Anthropology.] The branch of anthropology concerned with the study of fossil hominids.

1916 *Bull. Amer. Mus. Nat. Hist.* XXXV. 347 Then follows..the conclusion to which all M. Boule's own researches in paleanthropology [*sic*] and palæontology have evidently led him. **1935** *Times Lit. Suppl.* 14 Feb. 84/2 Dr. Leakey also..clearly sums up his results and shows their bearings on general palaeoanthropology and prehistory. **1957** *Antiquity* XXXI. 43 Rapid advances..are being made in paleoanthropology at present. **1965** B. G. Campbell *Nomenclature of Hominidae* 3 New taxa are very rarely presented in palaeoanthropology according to the rules of the more recent zoological congresses. **1974** *Nature* 6 Sept. 10/2 Investigations of the development of locomotion within the hominid lineage provides one of the more active and fascinating areas of palaeoanthropology today. **1978** *Ibid.* 2 Mar. 10/3 Wherever physical anthropology is taught in British universities, palaeoanthropology is recognised as an important component of the courses.

Hence **pa:læoanthropolo·gical** *a.*; **pa:læoanthropo·logist,** an expert or specialist in palæoanthropology.

1934 Webster, Paleoanthropological,..paleoanthropologist. **1935** *Times Lit. Suppl.* 14 Feb. 84/3 The problem of the prehistorical, especially the palaeoanthropological, relations between East Africa and the rest of the continent. **1947** *Sci. News* V. 44 The palæanthropologists [*sic*] went on with their researches..in the hope of one day discovering the hypothetical 'Tertiary man'. **1969** *Times* 17 Jan. 13/6 Dr. Louis Leakey..has for the past 40 years been exploiting the rich palaeoanthropological deposits in the Olduvai Gorge. **1975** *Nature* 10 Apr. 478/3 Although the article admirably reflects recent work in East Africa, it does scant justice to the palaeoanthropological researches in the Transvaal over the past decade. **1978** *Ibid.* 22 June 589/3 In almost all circumstances it now seems important that palaeoanthropologists try to work with a set of alternative hypotheses rather than simply with one favourite interpretation.

Pa:læo-Asia·tic, *sb.* and *a.* Also **Palæoasiatic, Palæasiatic,** and with the prefix spelt **paleo-.** [f. PALÆO-, PALEO- + ASIATIC *a.* and *sb.*] = *PALÆO-SIBERIAN *sb.* and *a.*

1909 A. C. HADDON *Races of Man* 48 The..western, and northern Ural-Altaians form one division, which includes such peoples as the Ugrians (in part), Palæasiatics, some of the Tungus, and the true Mongols. **1925** L. H. D. BUXTON *Peoples of Asia* vii. 191 Some early type may, as one of the components of the highly complex physical form of the Koreans, include as a basis the aboriginal Palæasiatic type. **1929** —— *China* iii. 58 There is a clear division also between the Mongols and the Palaeasiatics, that is the old yellow-skinned tribes of north-eastern Asia, and other tribes on the north. **1932** [see *CHUKCHEE, CHUKCHI *sb.* and *a.*]. **1948** D. DIRINGER *Alphabet* I. ix. 157 The Chukcha..are a..Palæoasiatic, Mongoloid people. **1953** W. K. MATTHEWS *Struct. & Devel. Russian* I. iv. 76 A variety of languages (e.g. Caucasian, Uralian, and Palaeoasiatic). **1962** W. P. LEHMANN *Hist. Linguistics* ii. 44 Far to the east in Siberia the Palaeo-Asiatic or Hyberborean languages are spoken. **1976** *Language* LII. 482 The so-called Palaeo-Asiatic family..is..more a geographical convenience than a genetic grouping.

palæobathymetry, etc.: see *PALÆO-, PALEO-.

pa:læobioche·mistry. Also (chiefly *U.S.*) **paleo-.** [f. PALÆO-, PALEO- + *BIOCHEMISTRY.] The biochemistry of fossils and of organisms of the geological past, esp. as a means of investigating phylogeny; the investigation of the evolutionary development of biochemical processes.

1954 *Carnegie Inst. Yearbk. 1953–4* 97 (*heading*) Paleobiochemistry. **1966** *Science* 6 May 762/2 The Fig Tree organisms are comparable in size, shape, complexity of structure, and isolated habit to many modern bacillar bacteria. Although they may have had a nonphotosynthetic metabolism, there is insufficient information available about their paleobiochemistry to realistically evaluate such a suggestion. **1967** *Compar. Biochem. & Physiol.* XX. 553 (*heading*) Paleobiochemistry of molluscan shell proteins. **1969** *Collier's Encycl. Year Bk.* 406 In one of the first applications of paleobiochemistry it has been found that hydrocarbon compounds in rock 3 billion years old may be composed of fossilized chlorophyll. **1969** G. EGLINTON in Eglinton & Murphy *Org. Geochem.* ii. 21 Two main approaches can be taken to the biochemistry of organisms in past times. The first is to examine fossil specimens in the hope that some of the information is still there in the form of protein structure, secondary metabolites, etc... The second is to examine the biochemistry of present-day living organisms and make phylogenetic comparisons. From these one can attempt to infer evolutionary sequences in the development of biochemical processes..; this approach is known variously as paleobiochemistry, evolutionary biochemistry and chemical paleogenetics. **1969** M. FLORKIN in *Ibid.* xx. 498 Structurally preserved biopolymers kept *in situ* in their anatomical location in fossils..should constitute the most reliable material for studies in paleobiochemistry.

So **pa:læobioche·mical** *a.*

Quot. 1937 is an isolated example, unrelated to the later use of *palæobiochemistry*.

1937 *Bull. Amer. Assoc. Petroleum Geologists* XXI. 770 Theories which appear to be in part applicable to such an accumulation are (1) the existence of paleo-biochemical conditions associated with the shallow-water near-shore deposits which controlled the abundance of petroleum-forming organisms. [Etc.] **1972** *Science* 31 Mar. 1461 (*heading*) Amino acid composition of planktonic foraminifera: a paleobiochemical approach to evolution. **1973** *Nature* 20 July 182/2 The recognition that keratin can survive tens of millions of years..could lead to further discoveries, enabling more paleobiochemical analyses to be undertaken.

pa:læobiogeo·graphy. Also (chiefly *U.S.*) **paleo-.** [f. PALÆO-, PALEO- + *BIOGEOGRAPHY.] The study of the distribution of fossil plants or animals.

1934 in WEBSTER. **1953** *Ecology* XXXIV. 811 (*heading*) A synthesis of paleobiogeography. **1961** P. E. CLOUD in M. Sears *Oceanogr.* ii. 151 The patterns created in the struggle for space..are the concern of paleobiogeography. **1972** *Sci. Amer.* Mar. 12/2 He has written on such subjects as fossil carnivores, dating of early man, ..evolutionary theory and paleobiogeography.

Hence **pa:læobiogeo·grapher,** an expert or specialist in palæobiogeography; **pa:læobi:ogeogra·phic, -ical** *adjs.*

1953 *Ecology* XXXIV. 812/2 Then come the paleobiogeographic maps. **1958** *Bull. Geol. Soc. Amer.* LXIX. 107/1 The amount of detail that can be shown on a paleogeographic or paleobiogeographic map depends on the length of the time interval which it attempts to represent. **1961** P. E. CLOUD in M. Sears *Oceanogr.* ii. 159 For the paleobiogeographer, it is essential not to confuse temperature variations with geodetic terms. *Ibid.* 162 An important paleobiogeographical exception to the direction of size increase is provided in the case of the benthic invertebrates. **1970** *Spec. Papers Geol. Soc. Amer.* No. 124. 23 During the late Paleozoic, unequal distribution and dispersal of biotas created changing paleobiogeographic provinces. **1978** *Nature* 9 Mar. 159/1 Palaeobiogeographic evidence is thus compatible with seafloor spreading and palaeomagnetic data.

palæobiology, etc.: see *PALÆO-, PALEO-.

palæobotany, paleo-. Add: Hence also **pa:læobota·nically** *adv.*, in terms of or as regards palæobotany.

1934 *Bull. Amer. Assoc. Petroleum Geologists* XVIII. 1010 The Jackfork-Stanley is paleobotanically older than the Coal-bearing shale. **1968** R. W. FAIRBRIDGE *Encycl. Geomorphol.* 534/2 The last of the paleobotanically established Holocene stages is..also the historical period.

palæoceanography, etc.: see *palæo-oceanography* s.v. *PALÆO-, PALEO-.

Palæocene, delete entry in Dict. (s.v. PALÆO-, PALEO-) and substitute:

Palæocene (pæ·li,o-, pē·li,osīn), *a. Geol.* Also (chiefly *U.S.*) **Paleocene.** [ad. F. *paléocène* (W. Ph. Schimper *Traité de Paléont. végétale* (1874) III. 680), f. *paléo-* PALÆO-, PALEO + Gr. καινός new, recent (cf. EOCENE, OLIGOCENE, etc.).] Of, pertaining to, or designating the lowest series of the Tertiary system, lying below the Eocene and comprising the Montian and perhaps the Danian stages; (formerly often not recognized as a distinct series but incorporated in the Eocene). Also *absol.*

1877 *Proc. Geol. Soc.* XXXIII. 83 An attempt has been made to establish a zone intermediate between the European Cretaceous and Eocene, under the title of Palæocene. **1877, 1899** [in Dict. s.v. PALÆO-, PALEO-]. **1921** A. W. GRABAU *Textbk. Geol.* II. xlv. 782 The Palæocene is not always recognized by American geologists, who class it as Lower or Basal Eocene. **1949** R. C. MOORE *Introd. Hist. Geol.* xvii. 399 In Europe and in North America, the Tertiary System is recognized as containing main divisions that in upward order are named Paleocene, Eocene, Oligocene, Miocene, and Pliocene. **1957** DUNBAR & RODGERS *Princ. Stratigr.* xvii. 298/1 The Paleocene is commonly not separated from the Eocene in Great Britain and France. **1965** KAY & COLBERT *Stratigr. & Life Hist.* xxiv. 617 There was an evolutionary explosion of mammals..with the advent of the Paleocene Epoch. **1969** [see *MONTIAN *a.*].

pa:læocerebe·llum. *Anat.* Also (chiefly *U.S.*) **paleo-.** [mod.L., f. *PALÆO-, PALEO- b + CEREBELLUM.] A phylogenetically older portion of the cerebellum, comprising mainly the anterior lobe, pyramid, and uvula.

1925, 1954 [see *NEOCEREBELLUM]. **1974** *Encycl. Brit. Macropædia* XII. 990/2 The anterior lobe of the cerebellum represents the paleocerebellum, an area that regulates equilibrium and muscle tone; it constitutes the main mass of the cerebellum in fish, reptiles, and birds.

So **pa:læocerebe·llar** *a.*, of or pertaining to the palæocerebellum.

1936 C. U. A. KAPPERS et al. *Compar. Anat. Nervous Syst. Vertebr.* I. vii. 777 The above mentioned group of authors has separated the cerebellum into a neocerebellar portion and a paleocerebellar portion. **1958** L. HAUSMAN *Clin. Neuroanat.* xxii. 203 (*heading*) Syndrome of paleocerebellar atrophy.

palæochemistry, etc.: see *PALÆO-, PALEO-.

pa·læoclimate. Also (chiefly *U.S.*) **paleo-.** [f. PALÆO, PALEO- + CLIMATE *sb.*] The climate at a period in the geological past.

1924 *Geol. Mag.* LXI. 512 Conclusions regarding palæoclimates. **1954** *Jrnl. Geol.* LXII. 229/1 The organism-induced differentiation in recording varying fractions of the environmental temperature range and the fact that these types of skeletal-temperature records are at present our sole sources of data for paleotemperature investigations introduce uncertainty in evaluation of ambient climatic conditions. **1963** *Times* 3 Jan. 9/7 The geological evidence of palæoclimates. **1974** *Encycl. Brit. Macropædia* XIII. 908/2 Many ingested seeds, hair, feathers, and small animal bones within the coprolites may provide a reliable estimate of the palæoclimate of the time.

So **pa:læoclima·tic** *a.*, of or pertaining to a palæoclimate.

1893 C. A. WHITE in *Rep. U.S. Nat. Museum 1892* 301 This subject relates to what may be designated as paleoclimatic conditions, that is, to formerly existing conditions which in certain parts of the earth were more or less materially different from those which now exist in the same parts. **1946** F. E. ZEUNER *Dating Past* vi. 163 The number of Palaeolithic sites which can be placed on local palaeoclimatic evidence is as yet very small. **1977** A. HALLAM *Planet Earth* 171 The occurrence of ancient calcareous rocks can be used as a palaeoclimatic indicator.

pa:læoclimato·logy. Also (chiefly *U.S.*) **paleo-.** [f. prec. + -OLOGY.] The study or investigation of palæoclimates.

1920 G. W. BERRY in *Ann. Rep. Board of Regents Smithsonian Inst. 1918* 198 In the interpretation of the far distant past paleobotany and..paleozoology contribute fundamental data to geology..; not only as comprising the subject matter of the biology of the past,..but in the elucidation of the climate and other physical conditions of past times, subjects which may be embraced under the terms of Paleoclimatology and Paleoecology. **1946** F. E. ZEUNER *Dating Past* v. 145 Combining the geological evidence with the astronomical time-scale by means of the radiation curves, an absolute chronology is obtained which appears to be sufficiently reliable for the purposes both of palaeoclimatology and of prehistoric archaeology. **1974** *Encycl. Brit. Micropædia* VII. 690/1 Most research in paleoclimatology centres on explaining (1) the warmth of the Northern Hemisphere land masses during at least 90 per cent of the last 570,000,000 years, (2) the semiperiodic occurrence of widespread glaciation.., and (3) the irregular advances and retreats of the ice sheets during the glacial periods.

Hence **pa:læoclimatolo·gic, -lo·gical** *adjs.*, **pa:læoclimatolo·gically** *adv.*; **pa:læoclimato·logist,** one who investigates palæoclimates.

1924 J. G. A. SKERL tr. *Wegener's Orig. Continents & Oceans* vi. 105 Information about deposits of rock-salt, which are so important for palæoclimatological purposes. **1927** *Geogr. Jrnl.* LXX. 167 For the palæoclimatologist.. Wegener's theory is a tempting haven of refuge from his perplexities. **1948** *Bull. Amer. Assoc. Petroleum Geologists* XXIX. 428 (*heading*) Paleoclimatologic and paleo-oceanographic maps. **1957** *Antiquity* XXXI. 78 The field research necessary for a meaningful paleoclimatological, paleobotanical, and paleozoological history of post-glacial south-western Asia is only beginning. **1963** *Amer. Jrnl. Sci.* CCLXI. 282 The hope that palæoclimatologic evidence may support or refute certain of the major rival hypotheses of global paleogeography. **1966** J. A. MABBUTT in G. H. Dury *Ess. Geomorphol.* 83 The zone offers a worthwhile challenge to the climatic geomorphologist,.. partly because of its situation in the palaeoclimatologically little-known Southern Hemisphere. **1968** *Economist* 3 Feb. 42/3 He..leaves one wondering once again whether the ups and downs of ancient history will ever be adequately understood until the palaeoclimatologists have done a proper job on the Mediterranean basin. **1972** *Nature* 17 Mar. 116/1 In either case, significant paleoclimatological or palæo-oceanographic variations might be detectable by analysis of the facies geometry.

palæoco·rtex. *Anat.* Also (chiefly *U.S.*) **paleo-.** [mod.L., f. *PALÆO-, PALEO- b + CORTEX.] A phylogenetically older portion of the cerebral cortex, which is coextensive with the palæopallium.

1909, 1947 [see *NEOCORTEX]. **1964** J. Z. YOUNG *Model of Brain* xix. 300 In the vertebrate brain there are also such tiers of circuits provided, by the hypothalamus, thalamus, palæocortex and neocortex.

Hence **palæoco·rtical** *a.*, of or pertaining to the palæocortex.

1909 *Arch. Neurol. & Psychiatry* IV. 162 The palæo-cortical arrangement..is not nearly as typical as in the ventral grey substance. **1964** J. Z. YOUNG *Model of Brain* vi. 110 The palæocortical centres and basal parts of the forebrain are also concerned largely with 'motivation' as well as being involved in learning.

palæocurrent: see *PALÆO-, PALEO-.

pa:læoeco·logy. Also (chiefly *U.S.*) **paleo-.** [f. PALÆO-, PALEO- + *ECOLOGY.] The ecology of fossil plants and animals.

1898 C. MACMILLAN in *Minnesota Bot. Stud.* I. 950 Paleoecology might be defined as the science of adaptations of fossil organisms. **1916** F. E. CLEMENTS *Plant Succession* xii. 193 Paleo-ecology develops its most fascinating aspect when it reaches the Human period. **1942** *Bull. Amer. Assoc. Petroleum Geologists* XXVI. 1697 Paleoecology, both organic and sedimentary, is a fertile field for geologic research. **1969** *Beaver* Summer 36/2 Too little is known about too few of the insects in Canadian amber to draw conclusions about their palaeo-ecology. **1976** *Nature* 29 Jan. 350/1 Palaeoecology is primarily concerned with the reconstruction of past ecosystems from palaeontological and lithological evidence.

Hence **pa:læoecolo·gic, -ical** *adjs.*; **pa:læoecolo·gically** *adv.*; **pa:læoeco·logist,** an expert or specialist in palæoecology.

1934 WEBSTER, Paleoecologist. **1940** *Bull. Amer. Assoc. Petroleum Geologists* XXIV. 1296 (*heading*) Inferred and possible paleoecological effects of the water medium. **1940** *Jrnl. Paleont.* XIV. 320/2 In a problem of this kind the paleoecologist is greatly handicapped by the fact that every criterion available is susceptible of a variety of interpretations. **1954** *Jrnl. Geol.* LXII. 234/2 Paleoecological factors, such as wide fluctuations in salinity and insolation-induced temperatures, are insufficient for such mean growth temperatures. **1959** *Jrnl. Palæont.* XXXIII. 936/1 More in need of emphasis than paleoecology, the change from water to sediment or between sediment layers may be almost as profound as that from air to water. **1964** [illegible] **1966** POTTER & PETTIJOHN *Palæocurrents & Basin Anal.* i. 1 Knowledge of current direction is also useful in paleoecologic studies. **1975** *Nature* 6 Feb. 400/1 Palaeoecological studies suggest that the species [*sc. Armeria maritima*] was widespread in distribution during the cold and disturbed conditions at the close of the last glaciation. **1977** A. HALLAM *Planet Earth* 218 Because they are far more abundant, the invertebrate fossil faunas of the sea are of more importance to stratigraphers and paleoecologists than are the vertebrates.

palæoenvironment to -geomagnetic: see *PALÆO-, PALEO-.

pa:læogeomorphic (-dʒi,omǭ·ɪfik), *a.* Chiefly *U.S.*, in the form **paleo-.** [f. PALÆO-, PALEO- + GEOMORPHIC *a.*] Of, pertaining to, or formed by buried relief features; palæogeomorphological.

1946 M. KAY in *Bull. Amer. Assoc. Petroleum Geologists* XXIX. 427 (*heading*) Paleogeomorphic maps. **1960** *Prof. Papers U.S. Geol. Survey* No. 400-B. 186/1 Because of the relatively narrow nature of the topographic..formable surface a regional paleogeomorphic pattern has not been developed. [illegible] *Bull. Amer. Assoc. Petroleum Geologists* L. 2279/1 Kay (1945, p. 427), discussing 'paleogeomorphic maps', did not really do justice to Thornbury's later definition of paleogeomorphology, because he omitted all reference to the use of subsurface data in constructing such maps; his main concern was the reconstruction of ancient landscapes. **1972** *Ibid.* LVI. 538/2 Paleogeomorphic traps are formed where

ancient subaqueous or land surfaces and the relief features on them..are buried by younger strata of a different lithology.

pa:læogeomorpho·logy. Also (chiefly *U.S.*) **paleo-.** [f. PALÆO-, PALEO- + *GEOMORPHOLOGY.] The geomorphology of ancient landscapes, esp. as represented today by features that are buried or newly exhumed.

1954 W. D. THORNBURY *Princ. Geomorphol.* ii. 31 It might seem that the recognition of ancient erosion surfaces and the study of ancient topographies does not belong in the field of geomorphology, but the approach of the geomorphologist may be the most logical one. This aspect of geomorphology may well be called paleogeomorphology. **1966** R. MARTIN in *Bull. Amer. Assoc. Petroleum Geologists* L. 2278/1 Enlarging on Thornbury's concept, the writer groups under the term paleogeomorphology the study of all geomorphic phenomena which are recognizable in the subsurface and in outcrops of previously buried landscapes. **1968** R. W. FAIRBRIDGE *Encycl. Geomorphol.* 805/1 Paleogeomorphology is of practical importance insofar as many accumulations of oil, gas, certain ores and fresh water are related to buried relief features. **1972** *Geo Abstr.* A. 727 The paleogeomorphology of the region is dominated by glacial features upon which Holocene processes have been superimposed.

Hence **pa:læogeomorpholo·gical, -logic** *adjs.*; **pa:læogeomorpho·logist,** one who studies palæogeomorphology.

1960 *Canad. Mining & Metallurg. Bull.* LIII. 529/2 Buried relief features are of importance to the petroleum geologist whenever they lead to the trapping of hydrocarbons, either directly or indirectly. In areas where such features abound,..paleogeomorphological features rank in importance with sedimentation and structure as hydrocarbon trapping agents. **1963** *Oil & Gas Jrnl.* 21 Oct. 140/1 It is convenient for the paleogeomorphologist to know that water features at a certain speed can move particles of specific sizes and density. **1966** *Bull. Amer. Assoc. Petroleum Geologists* L. 2279/1 Although these [observations] have received practically no attention from 'modern' geomorphologists, they are of considerable importance to the paleogeomorphologist in the exploration for oil and gas. **1975** *Nature* 14 July 279/2 This work has resulted in the elucidation of the geographic extent, stratigraphic succession, facies change, palaeoenvironments and palaeogeomorphological setting of the basin.

palæogeophysical to -hydrography: see *PALÆO-, PALEO-.

Palæo-I·ndian, *sb.* and *a.* Also **PalæoIndian, Palæoindian, paleo-Indian,** and with the prefix spelt **paleo-.** [f. PALÆO-, PALEO- + INDIAN *a.* and *sb.*] **A.** *sb.* **a.** The culture of the earliest Indian inhabitants of the Americas. **b.** One of these people. **B.** *adj.* Of or pertaining to Palæo-Indians or their culture.

1940 F. H. H. ROBERTS in *Smithsonian Misc. Coll.* C. 51 (*heading*) Developments in the problem of the North American Paleo-Indian. *Ibid.* 77 The morphological significance of the skeletons is considered..in conjunction with the other human remains attributed to the Paleo-Indian inhabitants. *Ibid.* 109 The work of recent years..demonstrates that there actually was a Paleo-Indian. **1941** *Bull. Geol. Soc. Amer.* LII. 2001 Evidences of the Paleo-Indian of the late Pleistocene or Early Recent times are present in Alaska. **1943** *Acta Americana* I. 11. 194 The causes of this rapid and widespread disappearance [of Pleistocene forms of animals] are not yet known..it has been suggested that the Paleo-Indians may have been a contributing factor. **1959** J. J. HONIGMANN *World of Man* xiii. 791 In Union County, New Mexico, a site containing remains of 32 bison, 19 projectile points, and other artifacts indicates a mass kill by Paleoindians. *Ibid.*, The Paleoindian period yields none of the elaborate portable and mural art characteristic of the Upper Paleolithic in Europe. **1968** R. W. FAIRBRIDGE *Encycl. Geomorphol.* 454/2 The lake [*sc.* Lake Agassiz] influenced the movements of early man (paleo-Indian cultures), and artifacts typical of the period 9000–5000 B.C. are found around its margins. **1972** *Science* 16 June 1211/2 Although the literature on North American PaleoIndians is quite extensive, most of this information has been derived from the analysis of kill sites, and only limited data on Paleo-Indian campsites have been available.

palæointensity to -longitude: see *PALÆO-, PALEO-.

palæoma·gnetism. *Geol.* Also (chiefly *U.S.*) **paleo-.** [f. PALÆO-, PALEO- + MAGNETISM.] (The study and interpretation of) the natural magnetism of rocks, which they are believed to have acquired at the time of their formation and is used as evidence for the past relationship of the rocks to each other and to the earth's magnetic field.

1854 *Edin. New Philos. Jrnl.* LVI. 9 The palæomagnetism..will also give rise to most interesting discoveries, and even to magnetical maps in the various geological periods. **1953** T. NAGATA *Rock-Magnetism* vi. 213 Other people might be concerned with rock-magnetism as the medium or the fossil of the geomagnetic field in the past geologic times, namely, as the tool for palaeomagnetism, since the knowledge of the secular variation of geomagnetic field throughout the whole life of the earth is one of the basic subjects in the modern geophysics. **1955** *Nature* 10 Sept. 505/1 (*heading*) Palæomagnetism of sediments from the Colorado Plateau. **1962** *Listener* 17 May 849/1 This new subject of palaeomagnetism enables us to track the movement of the magnetic poles relative to the continents. **1970** *Nature* 6 June 934/1 The motion of these plates with respect to the Earth's spin axis can be determined using the methods of palaeomagnetism. **1971** I. G. GASS et al. *Understanding Earth* xv. 222/2 One may use palaeomagnetism as a tool for verifying continental drift.

Hence **pa:læomagne·tic** *a.*, of or pertaining to palæomagnetism; **pa:læomagne·tically** *adv.*, by means of or as regards palæomagnetism; **palæoma·gnetist,** one who studies palæomagnetism.

1953 T. NAGATA *Rock-Magnetism* vi. 218 It seems that the Königsberger's [*sic*] conclusion quoted above is still substantially right at present and very useful as the criterion of available rock-magnetism for the palæomagnetic purposes. **1957** *Adv. Physics* VI. 166 Drab deltaic sediments of a cool environment appear to be magnetically unsuitable. **1960** *Bull. Geol. Soc. Amer.* LXXI. 763/2 The paleomagnetist would like to know if it is at all possible for a nonaxial dipole field to exist for a long period of time. **1969** BRANSON & WRIGHT *Geol. Hist. Brit. Isles* iii. 51 As well as the marked angular unconformity, there is a difference in the palæomagnetic direction of about 50° which may indicate that this part of the Lower Torridonian is much older than the upper groups. **1976** *New Scientist* 17 Dec. 491/1 A stern criterion which palaeomagnetists have always had to face..is over the long-term stability of the magnetic minerals within their rock samples. **1973** *Nature* 23 Feb. 497/1 It is not difficult to produce paleomagnetically pole positions with polar wandering and continental drift if one or the other acts alone. **1977** A. HALLAM *Planet Earth* 143 Palaeomagnetic studies have been vital in establishing the theory of plate tectonics.

palæomeridian to -meteorology: see *PALÆO-, PALEO-.

palæoniscoid, *sb.* and *a.* [f. mod.L. name of suborder *Palæoniscoidei*, f. generic name *Palæoniscus* (L. Agassiz *Recherches sur les Poissons fossiles* (1833) II. v. 41), f. PALÆO- + Gr. ὀνίσκος a sea-fish resembling cod + -OID.] **A.** *sb.* A fossil fish belonging to the suborder Palæoniscoidei, a group that includes elongate fishes with heterocercal tails and diamond-shaped scales. **B.** *adj.* Of or pertaining to a fish of this kind.

1896 B. DEAN *Fishes Living & Fossil* vii. 166 (*caption*) Palæozoic Palæoniscoid. **1928** *Palæobiologica* I. 87 (*heading*) *Polypterus*, a palæoniscoid? **1963** P. H. GREENWOOD *Norman's Hist. Fishes* (ed. 2) iv. 56 A modified palæoniscoid scale occurs in the superorder Holostei, another group which is best known from fossil forms. **1968** A. S. ROMER *Procession of Life* viii. 161 A typical palæoniscoid was a small, well-streamlined little fish, with thick shiny scales, long jaws with a powerful gape, and..a tail in which the fleshy, scale-covered tip of the body extends upward and backward to the tip of the fin. *Ibid.*, Today there survive only two small groups of ancient palæoniscoid descendants. **1974** D. & M. WEBSTER *Compar. Vertebr. Morphol.* viii. 162 True ganoid scales occurred in the palæoniscoid fishes.

palæontography. Delete *rare*⁻⁰ and add: (Examples.) So **palæonto·grapher,** a person engaged in the description of fossil animals and plants.

1904 P. GEDDES in P. Boardman *Worlds of Patrick Geddes* (1978) x. 357 The study of comparative religion too often readily gives an impression..of but the latest form of palæontography and palæography. **1973** *Nature* 1 June 308/2 The discussions liberally impinge on the cognate sciences of evolutionary palæontography, biogeochemistry and statistics. *Ibid.* 308/3 It..Schopf's *Models in Paleobiology* (1972) should..make the palæontographer who have got stuck in a rut.

palæo-oceanography, etc.: see *PALÆO-, PALEO-.

palæopa·llium. *Anat.* Also (chiefly *U.S.*) **paleo-,** and with hyphen. [mod.L., f. *PALÆO-, PALEO- b + *PALLIUM 3 d.] A phylogenetically older portion of the pallium of the brain, which comprises mainly the pyriform lobe (the hippocampal formation or archipallium is sometimes included).

1909 C. U. A. KAPPERS in *Arch. Neurol. & Psychiatry* IV. 161 This nervous substance, in order to distinguish it from the subventricular grey substance, should be called palæo-pallium. **1933** [see *ARCHIPALLIUM]. **1948** A. BRODAL *Neurol. Anat.* x. 324 (*caption*) In amphibians..a..portion of the hippocampal area (archipallium) and the piriform area (palæo-pallium). **1962** E. C. CROSBY et al. *Correl. Anat. Nerv. Syst.* viii. 411/1 'Paleopallium' is a term applied to the pyriform cortex; that is, to portions at least of the hippocampal gyrus. **1972** T. W. JENKINS *Functional Mammalian Neuroanat.* ii. 35/1 Based on the phylogenetic age, the olfactory cortex at the base of the brain, which is the older cortex is referred to as the archipallium (archicortex), or paleopallium (paleocortex).

Hence **palæopa·llial** *a.*, of or pertaining to the palæopallium.

1936 C. U. A. KAPPERS et al. *Compar. Anat. Nerv. Syst. Vertebr.* II. ix. 1482 In mammals the discharge paths between the archipallial, paleopallial, and archistriatal regions and the diencephalic centers are essentially those which have been described for lower forms. **1962** E. C. CROSBY et al. *Correl. Anat. Nerv. Syst.* viii. 411 The non-olfactory functions of paleopallial and archipallial areas may become dominant in higher mammals.

pa:læopatho·logy. Also (chiefly *U.S.*) **paleo-.** [f. PALÆO-, PALEO- + PATHOLOGY.] (The study of) the pathological conditions found in ancient human and animal remains.

1893 R. W. SHUFELDT in *Pop. Sci. Monthly* XLII. 679 Palæopathology..is a term here proposed under which may be described all diseased or pathological conditions found fossilized in the remains of extinct or fossil animals. **1923** R. L. MOODIE *Antiquity of Disease* iv. 66 Virchow, Mayer, Esper, Schmerling, and the other founders of palæopathology of the caves. **1967** *Science* 11 Aug. 638/3 Since not only skeletal material but disease itself is the subject matter of paleopathology, many disciplines find a meeting ground in these studies. **1975** *Palestine Exploration Q.* CVII. 89 Excavated the cemeteries of four medieval leper hospitals in Denmark... Palaeopathology has been able to contribute to clinical medicine.

Hence **pa:læopatholo·gic, -ical** *adjs.*; **pa:læopatho·logist,** one who investigates or studies palæopathology.

1917 *Johns Hopkins Hosp. Bull.* XXVIII. 261/1 Another American investigator, Dr. Aleš Hrdlička, has developed intensively the paleopathological side in his numerous explorations and studies of the aborigines on this continent. *Ibid.*, A very remarkable exhibition of paleopathologic specimens in San Diego. **1921** *Jrnl. Tuberculosis* XXXIII. 148 Egypt has provided the earliest palæopathological..the archaeologist and the palæopathologist, with unrivalled opportunities and an unprecedented wealth of material for the procession of their respective researches. **1962** O. TEMKIN in S. Jarcho *Human Palaeopath.* 34 Material from Ancient Egypt looms very large in the discussions of palaeopathologists. **1966** W. G. J. PUTSCHAR in *Ibid.* 60 In palaeopathological material we have only the mineralized portion of the affected bone available. **1967** *Amer. Jrnl. Roentgenology* XCIX. 712 The Mochica-Chimú civilizations developed in the coastal deserts of Northern Peru and the dry desiccating sands of these areas have preserved large quantities of their skeletal remains in a remarkably good state for paleopathologic study. **1975** *Palestine Exploration Q.* CVII. 88 To rule out the possibility that they were affected by the disease now known as leprosy..., an opinion supported by the palaeopathological evidence given below.

palæopedological to -salinity: see *PALÆO-, PALEO-.

Palæo-Sibe·rian, *sb.* and *a.* Also **palæo-Siberian, Palæosiberian,** and with the prefix spelt **paleo-.** [f. PALÆO-, PALEO- + SIBERIAN *a.* and *sb.*] **A.** *sb.* **a.** A member of any of several peoples of northern and eastern Siberia who are held to represent the earliest inhabitants of Siberia and whose languages do not belong to any of the major families. **b.** The Palæo-Siberian group of languages.

1914 M. A. CZAPLICKA *Aboriginal Siberia* ii. 15 If we are to provide a name for these unclassified tribes of the extreme north and east of Asia,..we should propose the name 'Palaeo-Siberians'... It implies a comparison and a contrast with the other tribes—Finnic, Mongolic, Turkic, Samoyedic, and Tungusic—who are comparatively recent comers to Siberia. **1965** *Language* XLI. 122 Soviet research in the rather exotic field of Paleosiberian. **1970** *Atlantic Monthly* Feb. 11 The Paleosiberians and a small Mongol enclave in Afghanistan..use these two methods of taking Sona. **1979** *Language* LV. 481 We thus regard the common occurrence of certain Proto-Uralic items to those of other language families (e.g., Indo-European, Indo-Iranian, Turkic, Paleo-Siberian, and Eskimo) is due to an ancient areal affinity, not to linguistic relationship.

B. *adj.* Of, pertaining to, or designating Palæo-Siberians.

1923 R. B. DIXON *Racial Hist. Man* iii. ii. 334 The conclusion cannot be escaped that a relationship of some sort exists between these Palæo-Siberian peoples of Asia and the Indian tribes of the North-west Coast. **1948** D. DIRINGER *Alphabet* 156 The aboriginal or Palaeo-Siberian group..are mainly nomad reindeer breeders and hunters. **1961** L. F. BROSNAHAN *Sounds of Language* viii. 177 The area of the simple stress accent is much larger; it extends..from the west and south of Europe..to include..the..Mongolian, Tungus and Paleosiberian languages of eastern Asia. **1972** *Peoples of Siberia* 56 The characteristics..show an ancient palæo-Siberian race, features of which are observed in other Siberian tribes.

palæoslope to -species: see *PALÆO-, PALEO-.

palæostria·tum (-straiē·tǝm). *Anat.* Also (chiefly *U.S.*) **paleo-.** [mod.L (coined in Ger. by C. U. A. Kappers 1908, in *Anat. Anzeiger* XXXIII. 322), f. *PALÆO-, PALEO- b + *STRIATUM.] The phylogenetically older portion of the corpus striatum, consisting essentially of the globus pallidus. Hence **pa:læostria·tal** *a.*

1913 *Brain* XXXVI. 159 In analogy with the archistriatal commissure which connects the nuclei amygdalæ, and a possible commissure between the palæostriata in Meynert's commissure, such a connexion of the phylogenetically most recent parts of the striatum would not be improbable. **1921** TILNEY & RILEY *Form & Functions*

Cent. Nervous Syst. xliv. 805 In its process of evolution from the lower vertebrates to mammals, the primordial portion of the striate body corresponds to the globus pallidus. This structure may, therefore, be tentatively distinguished as the paleostriatum. *Ibid.* xliv. 819 This is known as the syndrome of the globus pallidus, juvenile paralysis agitans or the paleo-striatal syndrome of Ramsay Hunt. **1936** [see *NEOSTRIATUM]. **1936** C. U. A. KAPPERS et al. *Compar. Anat. Nervous Syst. Vertebr.* II. ix. 1369 The paleostriatal and neostriatal areas of birds and their reptilian equivalents are basal in origin. **1972** [see *NEOSTRIATUM].

palæostructural to -temperature: see *PALÆO-, PALEO-.

palæotha·lamus. *Anat.* Also (chiefly *U.S.*) **paleo-.** [mod.L., f. *PALÆO-, PALEO- b + THALAMUS.] The phylogenetically older portion of the thalamus, usu. taken to include its anterior and medial parts.

1920 S. W. RANSON *Anat. Nervous Syst.* 191/1 (Index), Paleothalamus. See Thalamus, old. **1921** TILNEY & RILEY *Form & Functions Cent. Nervous Syst.* xxxi. 560 The primitive part thalamica is known as the paleothalamus. **1958** L. HAUSMAN *Clin. Neuroanat.* xxii. 242 Phylogenetically, the hypothalamus, the epithalamus and part of the medial division of the dorsal thalamus constitute the old thalamus or paleothalamus. **1973** [see *NEOTHALAMUS].

palæothermometry to -zoogeography: see *PALÆO-, PALEO-.

palæozoology. Add: (Further examples.)

1889 H. A. NICHOLSON in Nicholson & Lydekker *Man. Palæont.* (ed. 3) I. 3 Palæozoology and Palæobotany are inseparably connected with Neozoology and Neobotany. **1933** TWENHOFEL & SHROCK *Invertebr. Paleont.* i. 1 Paleontology..may be divided into paleobotany, treating of fossil plants, and paleozoology, treating of fossil animals. **1953** E. S. BARGHOORN in H. Shapley *Climatic Change* xx. 238 Ecologic interpretation..will always require confirming evidence from the physical geology and paleozoölogy. **1978** D. BLOODWORTH *Crosstalk* iv. 38 A common interest in paleozoology.

palæozoological (earlier and later examples); hence also **pa:læozoo·logist,** a student of extinct or fossil animals.

1866 *Phil. Trans. R. Soc.* CLVI. 672 A complete Monograph of the structure and life-history of that organism [*sc. Pentacrinus*] would be one of the most valuable contributions which Palæo-zoological science could receive. **1909** WEBSTER, Paleozoölogist. **1947** *Jrnl. Paleontol.* XXI. 574/1 This seems to be the favored path of some paleontologists (or should I call them paleozoologists?). **1957** *Antiquity* XXXI. 78 The field research necessary for a meaningful paleoclimatological, paleobotanical, and paleozoological history of post-glacial south-western Asia is only beginning. **1972** D. BLOODWORTH *Any Number can Play* x. 81 Max..was a paleozoologist..doing some work on prehistoric monkeys.

Palaic (pălēi·ik), *sb.* and *a.* [f. *Pala*, appar. a district of Asia Minor + -IC.] **A.** *sb.* The name of an Anatolian language, known from the Hittite archives. **B.** *adj.* Of or pertaining to this language.

1928 C. DAWSON *Age of Gods* xiii. 302 The Hittite archives also refer to three other tongues, Luvian, Palaic, and Harrian or Churrite. **1951** STURTEVANT & HAHN *Compar. Gram. Hittite Lang.* (ed. 2) I. i. 5 In the ritual of the deity Ziparwas certain passages are to be spoken *Pala-um-ni-li* 'in Palaic'. **1966** BIRNBAUM & PUHVEL *Anc. Indo-Europ. Dial.* 237 Watkins has..built a very ingenious hypothesis..on the brittle back of the Palaic hapax *malitanna*. *Ibid.* 243 Among verb stems, Luwian, Hieroglyphic, and Lycian *-s(s)-* contrasts with the Hittite and Palaic 'iterative' *-sk-*. **1972** W. B. LOCKWOOD *Panorama Indo-European Lang.* 263 Palaic was spoken in an area to the north of Hattusa, say in the province later known as Paphlagonia. It occurs solely in interpolations in the Hittite text in connection with the cult of the god Ziparwa. **1973** *Trans. Philol. Soc.* 1971 159 The Palaic particle, unlike the Lycian one, is connective and adversative. **1974** *Encycl. Brit. Macropædia* I. 834/1 An interrogative or relative pronoun *kui-* (compare Latin *quis*) is common to Hittite, Palaic, and Cuneiform Luwian.

palais de danse (pæle dǝ dañs). [Fr.] A public hall for dancing. Also *attrib.*, *fig.*, and *ellipt.* as **palais.**

1919 *Honey Pot* I. 14 The new Palais de Danse, which is to be opened on September 1st, is situated in Brook Green Road, two minutes walk from Hammersmith. **1926** *Punch* 13 Oct. 416/3 The young man you choose [as a dancing partner] out of a pen at sixpence a time at the Palais de Danse. **1928** *Melody Maker* Nov. 1193/3 Its rhythm had all the faults and few of the good points of the heavy 'Palais' style. **1940** HARRISSON & MADGE *War begins at Home* 225 The manager of a large suburban palais. **1946** J. AGATE *Contemporary Theatre 1944–5* 73 Yes, but have they [*sc.* Delibes, Offenbach, and Johann Strauss] the lush, treacly, palais-de-danse Orientalism dear to the British heart? **1958** *New Statesman* 25 Jan. 102/3 The good old-fashioned pit and palais muscians. **1964** W. G. RAFFÉ *Dict. Dance* 368/2 The Palais is an accepted town centre, replacing the mediaeval marriage-market, or the Victorian Assembly Rooms, as a place where eligible young people can meet matrimonial partners. **1966** *Listener* 19 May 711/2 A Tashkent spinning and weaving mill..had its own..palace of culture (depreciated term—shades of the Palais de Danse!) with singing, dancing, and dramatic activities. **1972** P. BLACK *Biggest Aspidistra* I. iii. 30 He [*sc.* Jack Payne] broke up his..group to take a job as pianist with a band at the Birmingham Palais (the huge public-hall fashion of dancing was spreading out from London). **1975** R. BUTLER *Where All Girls are Sweeter* iv. 36 While others studied at night we headed for the loudest Palais de Danse in the piste. **1976** *Times* 7 Aug. 4/5 Mrs Stonehouse, who has been married for 27 years, met her husband at the Hammersmith Palais when he was in the RAF. **1978** *Radio Times* 28 Jan. 69/4 Britain is renowned for its dancers, skaters and has the strongest tradition of *palais de danse* in the world.

‖ **Palais de Justice** (pale dǝ ʒüstis). [Fr., lit. 'palace of justice'.] In France (occas. elsewhere): a law court.

1792 T. BLAIKIE *Diary Scotch Gardener* (1931) 235, I told him..that I must go to the Palais de Justice..but he robbed my house..and the others and I [were] taken to the Palais de justice. **1885** H. JAMES *Little Tour in France* xii. 92 His [*sc.* Jacques Cœur's] house..to-day is used as a Palais de Justice. **1962** N. FREELING *Love in Amsterdam* iii. 133 At the Palais de Justice they made an impressive entrance. **1974** *Dressing of Diamond* 187 Like all public buildings in France, the Palais de Justice is ruled entirely by the concierge. **1978** W. GARNER *Mobius Trip* iii. 65 They drove back to the Palais de Justice, crossed the chill courtyard.

‖ **palais glide** (pæ·le glaid). [f. *PALAIS (DE DANSE) + *GLIDE *sb.* 1 c.] A type of ballroom dance in which large groups dance together.

1938 A. MOORE *Ballroom Dancing* (ed. 2) viii. 251 The Palais Glide can hardly be termed a dance; it is essentially a combination of steps having a great deal more vogue in the Gallop which has been a feature of Palais for many years... It can be danced to any Foxtrot tune..and it is played at a tempo of about 50 bars a minute. **1959** *Britannica Bk. of Year* 196/2 Another instance of the desire to add more tastelessness to British ballrooms has been the occasional introduction, at all kinds of dances, of the old 'Palais Glide'.. This is even more of a romp than the 'Lambeth Walk'. **1969** *Listener* 8 May 649/1 The tune stayed alive in dance-halls where, during the 1930s, it could be used for the 'Palais Glide'. **1970** *Guardian* 24 Dec. 9/3 The girls..engaged in a perpetual Palais Glide, regardless of rhythm. **1974** R. INGHAM *Yoris* xvii. 54 Do you remember the Palais Glide?..Lovely dance.

‖ **Palais Royal** (palę rwayal). The name of a Parisian theatre used *attrib.* to designate a type of indelicate farce said to be typical of this theatre.

1877 *Illustr. Sporting & Dramatic News* 21 Apr. 109/1 In my remarks about *Pink Dominos*..I alluded to the inevitable comparison that must be made between that piece and the other Palais Royal adaptations. *Ibid.* 20 Oct. 109/2, I consider the libretti of three-fourths of the French opéras-bouffes which have been translated..into English, practically and palpably cynical, and they are indecent withal. I may say the same of some Palais Royal pieces, mightily popular amongst us. **1951** N. MITFORD *Blessing* II. v. 195 What a curious thing—intrigues and misunderstandings, just like a Palais Royal farce. **1967** *Oxf. Compan. Theatre* (ed. 3) 715/1 In England the term 'Palais-Royal farce' was applied to the broad suggestiveness of such productions as *The Pink Dominos* (1877) and *The Girl from Maxim's* (1902).

palamino, var. *PALOMINO.

Palamite (pæ·lămǝit), *sb.* and *a.* *Eccl. Hist.* [f. the name of St. Gregory *Palamas*, an intellectual leader of the Hesychasts + -ITE¹.] **A.** *sb.* = HESYCHAST. **B.** *adj.* Of or pertaining to the Palamites or their doctrines; = HESYCHASTIC *a.* 2.

1859 *Encycl. Brit.* XVII. 177/1 At the councils which were severally held in 1341 and 1351 he [*sc.* Gregorius Palamas] pled the cause of his party, and so identified himself with the tenets he advocated that his fellow-sectarians were thenceforth called *Palamites*. **1877** McCLINTOCK & STRONG *Cycl. Bibl. Lit.* VII. 547/2 The peculiar leading tenets of the Palamites were the existence of the mystical light discovered by the more eminent monks and recluses in their long exercises of abstract contemplation and prayer, and the uncreated nature of the light of Mount Tabor seen at the transfiguration of Christ. *Ibid.* 548/1 These alleged heresies were, however, mostly..the inferences deduced by Nicephorus Gregoras and other opponents from the Palamite dogma of uncreated light, and not the acknowledged tenets of the Palamite party. **1900** 'ODYSSEUS' *Turkey in Europe* vi. 252 The quarrel between the Palamites and Barlaamites, after distracting the Eastern Church, was at last settled by a Synod in a sense favourable to the former. **1949** E. L. MASCALL *Existence & Analogy* vi. 151 For the Thomist, supernatural grace means a communication of God himself to the creature... For the Palamite, it means a communication of the uncreated energy of God though not of his uncommunicable essence. **1957** *Oxf. Dict. Chr. Ch.* 633/2 In the second half of the 14th cent. Hesychasm was accepted throughout the Greek Church, its adherents being also generally known as 'Palamites'. **1961** *Times* 24 Nov. 14/4 Moghila did not admit the Palamite doctrine of energies. **1971** *Catholic Dict. Theol.* III. 15/1 This controversy, also known as the Palamite controversy.., was concerned not so much with the spiritual doctrine of the Hesychasts as with its ultimate theological and metaphysical justification.

‖ **palang** (pala·ŋ). *Borneo* and *Philippines.* [Native name.] (See quot. 1974.)

1959 T. HARRISSON *World Within* I. 59 The rapid spread of the *palang*, a cross-piece driven through the male penis, has shown that behind the easy acceptance of many upland women there is a readiness to react in a more elaborate and erotic manner. **1964** *Jrnl. Malaysian Branch R. Asiatic Soc.* XXXVII. 163, I have here chosen the term *palang* as the most widespread of the indigenous Bornean terms (*palang*, *uttang*, etc.) and the one known over parts of the outside world. *Ibid.* 164 It is safest to take the generic term..closely associated with them as the main west Borneo palang users in proto-historical times. **1974** S. E. MORISON *European Discovery of America: Southern Voyages* xvii. The gentle Limasawans, says Pigafetta, went about practically naked... Here, and at Cebu, he describes..'The males, large and small, have their penis pierced from one side to the other near the head, with a gold or tin bolt as large as a goose quill. In each end of the bolt some have what resembles a spur, with points; others are like the head of a cast nail... In the middle of the bolt is a hole, through which they urinate.' *Ibid.* 435 Although certain writers..have either omitted this story or called it one of Pigafetta's tall tales, the *palang* has been found in use in Borneo and parts of the Philippines.

† **palanthropic** (pælænþrǭ·pik), *a. Obs. rare.* [Irreg. f. PALÆO-, PALEO- + ANTHROPIC *a.*] = *PALÆANTHROPIC *a.*

1894 [see *NEANTHROPIC *a.*].

Palantype (pæ·lăntaip). [f. the name *Palan(que* (see quot. 1940) + TYPE *sb.*¹] The proprietary name of a machine for typing in shorthand; also, the system of shorthand used. Hence **pa·lantypist.**

1940 *Trade Marks Jrnl.* 14 Aug. 782/1 Palantype... Typewriters and parts thereof, typewriter inks, typewriter ribbons, typewriting paper, and stands for typewriters..(not being furniture). Clementine Camille Marie Palanque, London,..manufacturer. **1959** *Times* 10 Apr. 3/1 Palantype operator, urgently required by Uganda Government. **1959** *Economist* 23 May 742/1 The institute's courses range from the training of district officers to that of clerks, typists and palantypists. **1964** *Financial Times* 25 Feb. 6/7 The small nucleus of reporting palantypists, the ladies who translate scientific jargon into shorthand hieroglyphs on a small, silent, rubber keyboard—later to convert it back. **1970** *Nature* 12 Sept. 1090/2 Several major hurdles still have to be overcome to transpose the twenty-nine symbols of palantype, the British form of mechanical shorthand, into proper English.

‖ **palari** (pālā·ri). [Malay.] A type of boat (see quot. 1948).

1936 *Times Lit. Suppl.* 7 Nov. 899/3 He wanted a chance of seeing a *palari* working to windward in bad conditions. **1937** G. E. P. COLLINS *Makassar Sailing* 12 During the second stage, which turns a pajala into a palari, the sides are raised and a deck and high stern added. **1948** R. DE KERCHOVE *Internat. Maritime Dict.* 516/1 *Palari*, an East Indian trading boat from southern Celebes and eastern Madura... It is rigged with one or two masts. 2. In Macassar the term 'palari' refers to a fast-sailing pleasure craft with two masts and an elaborately carved stern. **1950** *Jrnl. Malayan Branch R. Asiatic Soc.* XXIII. 112 The Palari is now characterised by the possession of a double-ended hull with a heavy, tripartite bowsprit and a high, overhanging stern. **1964** K. G. TREGONNING *Hist. Mod. Malaya* 59 The Bugis copied their craft from the Portuguese, and the *pajala* was adapted into the famous *palari*, which..is..like a miniature seventeenth century fore and aft rigged galleon.

palarie (pālā·ri), *v.* *Tramps'* and *Circus slang.* Also **palari.** [ad. It. *parlare.*] *trans.* and *intr.* To talk, to speak. Cf. *PARLYAREE.

1846 *Swell's Night Guide* 77 The chanting cadger stalls the chummy's daughter off to a single padded lumber; ranks her of five bob and a bender;..palaries sweat patter, and nobbs on for more blunt. **1893** P. H. EMERSON *Signor Lippo* xx. 91 She looked for all the world like a gippo, and she knew all the cant, and used to palarie thick to the slaveys. **1933** E. SEAGO *Circus Company* iii. 29 A subject not meant for a 'gajo's' ears, to be checked immediately... 'Nante palari before the josher cul.'

palatal, *a.* and *sb.* Add: **A.** *adj.* **2.** (Earlier example.)

1728 CHAMBERS *Cycl.* s.v. T., The *T* is one of the five Consonants which the Abbot de Dangeau calls *Palatal*.

b. Of a sound change: occurring in the environment of a palatal consonant or vowel.

1888 H. SWEET *Hist. Eng. Sounds* 124 This *c*-smoothing is by the Germans called 'palatal-umlaut'. **1894** —— *Anglo-Saxon Reader* (ed. 7) p. xiv, Why then continue..to call the change of *weorc* into *werc* 'palatal mutation', when the change is not a mutation, and is caused not by front, but invariably by back consonants? **1908** J. & E. M. WRIGHT *Old Eng. Gram.* iv. 28 Umlaut is of two kinds: Palatal and Guttural. Palatal umlaut, generally called *i*-umlaut, is the modification (palatalization) of an accented vowel through the influence of an *ī* or *j* which originally stood in the following syllable. **1914** H. C. WYLD *Short Hist. Eng.*, v. 75 'Palatal Mutation'. This term was suggested by Bülbring to denote primarily the loss in Anglian of the second element of the diphthong *ea* (which thus appears merely as *e*) before the consonant-groups *ht*, *hs*, *hþ*, when followed by a front vowel, or when final. **1939** *Trans. Philol. Soc.* 126 By the assumption that 'Breaking', 'Palatal Diphthongization', and 'Back-Mutation' were developments which can be dated within limits, a system of 'sound-changes' has been built up, which in some cases may be purely fictitious, in others only part of a long-drawn-out process. **1959** A. CAMPBELL *Old Eng. Gram.* v. 70 *Sċip* sheep, presumably from **sċēp* with palatal diphthongization of nW-S *sċēp*. **1975** LASS & ANDERSON *Old Eng. Phonol.* iv. 123 'Palatal diphthongization' of *ē*. As we show..there is no very good evidence for such a process in OE.

Hence **pa·latally** *adv.*, towards the palate; by means of the palate.

1934 *Dental Items of Interest* LVI. 206 Any extension of the preparation under the gingiva..palatally is to be avoided as being unnecessary. **1940** J. OSBORNE *Dental Mech.* xiv. 156 The use of black rubber palatally and lingually will give a better appearance to the finished denture. **1963** C. R. COWELL et al. *Inlays, Crowns, & Bridges* iv. 39 The withdrawal path must be inclined palatally. **1970** *Archivum Linguisticum* I. 7 When the following syllable contained an *i* the medial vowel could be palatally coloured.

palatalize, *v.* Add: Also *intr.*, to become palatal, to undergo palatalization.

1943 E. A. NIDA *Handbk. Descriptive Linguistics* II. v. 83 This may be the result of an original *e* which did palatalize but later this *e* changed to *a* and the palatalization remained. **1964** *Language* XL. 28 Cunter does not palatalize before unstressed /a/.., and..Trun does not palatalize before any allophone of /a/. **1973** J. M. ANDERSON *Struct. Aspects Lang. Change* 154 The frontal allophone of /g/ appears to have palatalized first.

palate, *v.* **2.** (Later example.)

1845 [see *BLANC 2].

Palatinate, *sb.* **1. c.** In small-type note, delete 'higher up the Rhine'.

palato-. Add: **palato-alveolar** *a.* (*Phonetics*) (see quots. 1932 and 1962); **palato-velar** (*Phonetics*) *a.*, articulated with the tongue in contact with the palate and velum simultaneously or successively; also either palatal or velar.

1932 D. JONES *Outl. Eng. Phonetics* (ed. 3) ix. 45 *Palato-alveolar*, articulated by the blade of the tongue against the teeth-ridge with raising of the main body of the tongue towards the palate. **1962** B. M. H. STRANG *Mod. Eng. Struct.* iii. 30 *Palato-alveolar* (tongue tip to palatal edge of alveolar ridge). **1964** R. H. ROBINS *Gen. Linguistics* iii. 101 Palato-alveolar fricatives. **1973** J. C. WELLS *Jamaican Pronunc. in London* i. 10 An unexpectedly large number of confusions between post-alveolar and palato-alveolar affricates turned up. **1895** P. GILES *Short Man. Compar. Philol.* II. viii. 113 Osthoff argues that there were originally three series of guttural consonants [in Indo-Germanic], making the velars which are not followed by *u* the third intermediate or 'palato-velar' series. **1902** E. W. SCRIPTURE *Elem. Exper. Phonetics* xxix. 443, 3d Series (middle and back of tongue).. *k*³, *g*³ palato-velar. **1935** W. F. TWADDELL *On defining Phoneme* v. 48 The 'p-phoneme' is therefore the sum of *all* those phonological differentiae which correspond to a bilabial articulation as opposed to alveolar or palato-velar, [etc.]. **1964** R. A. HALL *Introd. Linguistics* I. xvi. 96 In the case of French, we can, on the grounds of complementary distribution, bring bilabial and labio-dental together under 'labial' position; palatal and velar together under 'palato-velar'; and fricative and sibilant together under 'spirant' manner.

palatogram (pæ·lătogræm). *Phonetics.* [f. PALATO- + -GRAM.] A diagram produced by palatography.

1902 E. W. SCRIPTURE *Elem. Exper. Phonetics* xxi. 296 The diagram recording the contact of the tongue with the palate is called a 'palatogram'. *Ibid.* 298 The points at which the tongue touches the palate..in forming sounds can be registered by a mixture of meal and mucilage or by carmine water color or Chinese ink spread over the previously dried tongue. The sound is spoken naturally; the mouth is at once opened and the marks on the palate are observed... The results obtained are called 'palatograms'. **1917** *Nature* 4 Oct. 96/2 Palatograms will be found to corroborate observations of tongue-positions made by other methods. **1932** D. JONES *Outl. Eng. Phonetics* (ed. 3) xxi. 171 Fig. 97 is a palatogram of the sound *s*, as pronounced by me. **1948** J. R. FIRTH in *Bull. Sch. Orient. & Afr. Stud.* XII. IV. 859 Palatograms are not much use for any velar articulation, but throw light on many articulations forward of the soft palate. **1957** [see *linguistic analysis*].

palatography (pælătǫ·græfi). *Phonetics.* [f. PALATO- + -GRAPHY.] A technique of recording the position of the tongue during articulation from its contact with the hard palate. Hence **palatogra·phic** *a.*

1902 E. W. SCRIPTURE *Elem. Exper. Phonetics* xxi. 296 (*heading*) Tongue Contacts; Methods of Palatography; American, Irish and Hungarian Records. **1908** H. SWEET *Sounds of English* 108 There are other methods whose results are obtained only indirectly, such as the palatographic, by which 'palatograms' are made. **1917** *Nature* 4 Oct. 96/1 Palatography..consists in using a special kind of artificial palate, in order to find out what parts of the roof of the mouth are touched by the tongue in the production of different speech sounds. **1935** *Amer. Speech* X. 230/1 Extensive palatographic and kymographic study. **1940** *Ibid.* XV. 102/1 Palatography cannot be used extensively in the correction of foreign brogue and speech defects. **1948** J. R. FIRTH in *Bull. Sch. Orient. & Afr. Stud.* XII. IV. 863 The possibility of a large proportion of palatographic abstractions in any given language. **1957** D. ABERCROMBIE in *Zeitschr. für Phonetik* X. 21, I use the expression '*direct* palatography' to mean the investigation of articulatory movements by means of marks made directly on the roof of the mouth as distinct from the more usual technique which employs an artificial palate. **1960** P. STREVENS in *Lang. & Speech* III. 34 Palatographic studies..show that there are often large variations of the shape of the orifice.., within the speech of a single individual as well as between different speakers. **1970** *Times Lit. Suppl.* 23 July 817/1 The motion of the tongue in speech can be studied by 'dynamic palatography'. **1976** *Language* LII. 508 No mention is made of palatography as an observational technique.

‖ **palatschinken** (pælā·tʃi·ŋkǝn), *pl.* [Austrian Ger. dial., f. Hungarian *palacsinta*, f. Rum. *plăcintă*, f. L. *placenta* a cake.] An Austrian dish of stuffed pancakes.

1929 K. BAEDEKER *Austria* p. xxvii, Farinaceous dishes. *Strudel*,..*Palatschinken*, pancakes [etc.]. **1941** 'R. WEST' *Black Lamb* (1942) I. 144 For the sweet course we were given two apiece of palatschinken, those pancakes stuffed with jam which one eats all over Central Europe. The Balkans inherited the recipe from the Byzantines, who ate them under the name of palacountas. **1952** I. RHODE *Viennese Cookery Bk.* v. 67 The pancakes in this chapter..are strictly Viennese—thin *Palatschinken* and the thick *Schmarren*... Although the name *Palatschinken* is a corruption of the Hungarian *Palacsinta*, these pancakes are more French in origin. **1963** R. CARRIER *Great Dishes of World* xv. 270 Just before serving, dust *palatschinken* with icing sugar. **1971** LYON & BENTON *Eggs, Milk & Cheese* 171 *Palatschinken with Chicken.* This is an Austrian form of pancake which can be used with different kinds of flavouring and stuffing.

Palaung (pālau·ŋ), *sb.* and *a.* Also 9 **Paloung, Poloung.** [Native name.] **A.** *sb.* **a.** A people of the Shan States of Burma; a member of any of the various tribes constituting this people. **b.** The Mon-Khmer language of this people. **B.** *adj.* Of or pertaining to this people or their language. Hence **Palau·ngic** *a.*

1860 F. MASON *Burmah* iii. 69 The Paloungs..are a Shan tribe found north and east of Bamoo. **1885** A. R. COLQUHOUN *Amongst Shans* v. 72 The Paloungs, or Poloungs, are darker and smaller than the Shans. *Ibid.* 75 The slopes of the hills to the west of Thai-nee are cultivated by the Paloungs. **1923** H. T. WHITE *Burma* xiii. 136 Palaungs, timid, peaceable folk, to the number of 144,000, are found principally in the Northern Shan States. *Ibid.* 140 The Austro-Asiatic family is represented by Môn-Hkmer [*sic*] languages, of which the most noticeable are Talaing..and Palaung. **1934** 'G. ORWELL' *Burmese Days* xi. 165 The Palaungs..admire long necks in women. **1942** J. L. CHRISTIAN *Mod. Burma* ii. 15 The Wa and Palaung tribal groups of the Shan hills belong to the Mon Khmer family. **1948** A. KROEBER *Anthropol.* (rev. ed.) x. 424 Asia is particularly rich in tribal societies with 'internally marginal' cultures. Examples are:.. Palaung, [etc.]. **1957** *Encycl. Brit.* XVII. 114/2 The Palaung is darker than the Shan in complexion. **1971** N. BIXLER *Burma* vi. 133 The Palaung, who live less than 3,000 feet above sea level and cannot grow opium, have within this century begun to specialize quite successfully in raising tea. **1974** *Encycl. Brit. Micropædia* VII. 688/2 *Palaungic languages*, branch of the Austroasiatic language family, including..Palaung, [etc.]. **1978** *Language* LIV. 206 The Mon-Khmer languages Palaung, Riang-Lang, and Praok.

palaver, *sb.* Add: Also † **palaber, palava.**

1. Also *fig.*

1951 DYLAN THOMAS in *World Rev.* Oct. 66 In his house on stilts high among beaks And palavers of birds.

2. b. (Earlier example.)

1744 A. HAMILTON *Itinerarium* (1948) 160 The pedlar..sold some dear bargains to Mrs. Williams, and while he smoothed her up with palaber, the Bostoner amused her with religious cant.

† **c.** In West Africa: a dispute or contest. *Obs.*

c **1740** F. MOORE *Trav. Inland Afr.* (ed. 2) 221 They..call a Dispute *a Palaver*. **1744** W. SMITH *New Voy. to Guinea* 32 The English..once had a Settlement upon Charles Island, but having a Pallaver with the Natives they..waded over from the Main; by which Surprize they got Possession, and beat the English off that Island. **1789** *Rep. Lords Comm. Council: Evidence* I. 5 in *Parl. Papers 1731–1800* (Brit. Libr.) LXXXIV, If any Dispute or *Palavers* arose between Two or more Persons, they called a Council of the head Men, where the Persons were tried.

d. Business, concern. *slang* and *African colloq.*

1899 C. J. C. HYNE *Further Adventures Capt. Kettle* 21 It's not your palaver..or mine. **1953** *Eng. Stud.* XXXIV. 286 [West Africa] That's your palaver.

e. Jargon. ?*U.S.* (*rare*).

1909 'MARK TWAIN' *Is Shakes. Dead?* vii. 74, I have been a quartz miner..and know all the palaver of that business.

f. In Africa: trouble.

1953 P. CHRISTOPHERSEN in *Eng. Stud.* XXXIV. 286 *Palaver*..is now part of Standard English in the sense of 'talk' or 'parley'. But in West Africa this meaning is obsolete. In Pidgin and very largely in Coast English the word has come to mean 'business' or 'trouble' (e.g... 'That has caused a lot of palaver' and 'He had some tummy palaver last week'). **1954** [see *D.O.* s.v. *D III. 3]. **1970** *Drum* (E. Afr. ed.) Feb. 39/2, I am a boy of 21 and I am still unemployed. I want a girl friend, someone to pet me. But I feel I should get a job before getting involved with woman palava.

palaver, *v.* Add: **3.** *intr.* const. *to*: to ask (someone) for something, to beg from. *slang.* ?*Obs.*

1858 A. MAYHEW *Paved with Gold* III. i. 255, I thought he was a 'queer gill' (suspicious) at first, and smoked us, from what he palavered to Phil when he gave him his 'deux-wins' (twopence). **1859** HOTTEN *Dict. Slang* 71 'Palaver to the nibs for a shant of bivvy,' ask the master for a quart of beer.

palazzo (pāla·tso). [It.: see PALACE *sb.*¹] **1.** Pl. **palazzi.** A palatial mansion; a large and imposing building.

a **1666** EVELYN *Diary* an. 1644 (1955) II. 228 The Palazzo Barberini..I take to be as..princely an object, as any moderne building in Europ. **1758** M. W. MONTAGU *Let.* May (1967) III. 149 He brags..that you found the Palazzo very clean. **1892** W. JAMES *Let.* 28 Dec. (1920) I. 335 You seem to me something ideal, off there in your inaccessible Cambridge palazzo. **1924** M. ARLEN *Green Hat* v. 143 In drawing-rooms, up and down terraces of palazzos, in clubs and cabarets. **1930** E. POUND *XXX Cantos* xvii. 77 On past the palazzi. **1958** *Oxf. Mag.* 13 Mar. 357/2 What activities it [*sc.* the Institute of Statistics] wishes to pursue in the great *palazzo* which it now demands. **1962** A. SAMPSON *Anat. Brit.* v. 66 The taxis and government Humbers draw up outside the palazzi of Pall Mall, and bowlers and umbrellas disappear through the great stone doorways, acknowledged by reverent porters. **1968** 'E. LATHEN' *Stitch in Time* i. 1 The..Sloan Guaranty Trust, an opulent palazzo of glass, marble and brass that had replaced the stately and venerable edifice on Pearl Street. **1978** W. M. SPACKMAN *Armful of Warm Girl* 12 But now anyhow in her 73rd Street palazzo her phone rang.

2. Usu. as pl. **palazzos.** Loose, wide-legged trousers worn by women.

1972 *New Yorker* 9 Sept. 8/1 (Advt.), Our haltered palazzos come in champagne beige... The no-waistband, back-zip palazzos. **1973** *Harper's Bazaar* Apr. 5 (Advt.), Lollipops loll on a plunging palazzo... Shirring and sashing halt the halter above billowing pyjamas. **1973** *Telegraph* (Brisbane) 6 Sept. 18/2 Female employees..now are being permitted to wear the pants—in the form of trim pants suits, or, we guess, baggies—or palazzos, if they please.

3. *attrib.* and *Comb.*, as (sense 2) *palazzo outfit, pants, pyjamas, shape, suit, trouser*; **palazzo sleeve,** a wide, flowing sleeve.

1973 *Today's Health* Oct. 36/1 Linda stopped at the dry cleaners to pick up the flowered silk pallazzo [*sic*] outfit she planned to wear to the theater on Saturday. **1972** *New Yorker* 7 Oct. 18/1 (Advt.), Velvety fleece palazzo pants. **1973** *Woman's Day* Sept. 168/1 (Advt.), Choose from smashing palazzo pants and clothes with the bare look. **1966** *Listener* 3 Feb. 171/2 The mannequins start with a natural advantage over most other mortals. Who but they could wear palazzo pyjamas..made entirely of ostrich feathers. **1968** *N.Y. Times* 30 Apr. 52 All these varieties continued into the nineteen-sixties, when they were joined by such other forms as palazzo pajamas (wide enough to sweep around a palace in), culottes, pants dresses and pants suits. **1976** *Times* 1 Sept. (Fashion Suppl.) p. viii/5 Emilio Pucci..introduced..the famous palazzo pyjamas. **1973** *New Yorker* 17 Mar. 56 (Advt.), We like the palazzo shape as a slashback culotte. **1972** *Ibid.* 26 Aug. 55 (Advt.), Our fluid navy wool knit with permanent pleated polyester chiffon palazzo sleeves. **1974** *Country Life* 3–10 Jan. 55/1 Drip-dry dresses and palazzo suits. **1972** *N.Y. Times* 3 Nov. 11/4 (Advt.), The palazzo-trouser.

pale, *sb.*¹ Add: **4. e.** From 1791 to 1917, specified provinces and districts within which Russian Jews were required to reside.

The Russ. expression corresponding to 'pale of settlement' is *chertá osédlosti* (lit. 'boundary of settlement').

1890 A. READER *Russia & Jews* viii. 78 The Jews,..as soon as the contract was completed..had to return within the 'pale' of settlement. **1892** I. ZANGWILL *Children of Ghetto* III. 329 The whole history of her strange, unhappy race flashed through her mind... She was overwhelmed by the thought of its sons in every corner of the earth proclaiming to the sombre twilight sky the belief for which its generations had lived and died—the Jews of Russia sobbing it forth in their pale of enclosure. **1897** [see *GOLEM]. **1927** *New Statesman* 6 Nov. 104/1 Bolshevism, whilst destroying the livelihood of the Jewish masses in the so-called 'Pale'—small traders and artisans—has disorganised Russia's economic system. **1969** *Observer* 23 Feb. 23/2 With the Revolution in 1917, the Jews were released from the Pale and allowed to move in great numbers into Russia proper. **1976** *Times* 8 Apr. 11/3 Generally it has been held that Jews had arrived in what was known as the Pale of Settlement, between Russia and Poland, because they were driven there, under expulsion first from England and France, then from Germany. **1977** Y. MENUHIN *Unfinished Journey* i. 4 The Mnuchins..had settled in Gomel, a smallish city..at the very center of the Pale.

8. *pale-fence* (earlier examples), *-gate.*

1834 J. KEMPER in *Wisconsin State Hist. Soc. Coll* (1898) XIV. 423 In walking over the meadow..passed an indian burial place, 2 poles with white flags flying a pale fence partly surrounding the place. **1845** M. M. NOAH *Gleanings* 77 His house is..surrounded with a white pale fence. **1850** H. C. WATSON *Camp-Fires of Revolution* 28 Their ranks looked like a broken pale-fence. **1836** W. DUNLAP *Mem. Water Drinker* (1837) I. 12 It was..a ricketty wooden pale-gate drawn back by a chain and bullet.

pale, *sb.*² (Further example.)

1887 M. E. WILKINS *Humble Romance* 110 'It ain't so much the pale,' said Mrs. Potter, 'but thar's..a kind of a look around..the mouth that I've seen a good many times.'

pale, *a.* Add: **1. d.** *pale ale* (later examples); also *ellipt.*; *pale sherry*, a general term for light-coloured, dry sherries.

1833 C. REDDING *Hist. Mod. Wines* vi. 189 Pale sherry is made from the same grape as the brown, to the wine from which is added a couple of bottles of very pure brandy to each butt. **1846** R. FORD *Gatherings from Spain* xiv. 161 Pure genuine sherry..will stand the importer from 100 to 130 guineas in his cellar... The reader will now appreciate the bargains of those 'pale' and 'golden sherries' advertised in the English newspapers at 36s. the dozen, bottles

PALE

included. 1849 Pale ale (see *brown sherry* s.v. *BROWN *a.* 7). 1853 *Q. Jrnl. Chem. Soc.* V. 173 (*heading*) Alleged adulteration of pale ales by strychnine. 1891 in C. Ray *Compleat Imbiber* (1967) IX. 122 Pale Sherry.. Per. Doz. 20/-. 1965 A. Sichel *Penguin Bk. Wines* III. 231 Intermediate types of sherry are described as brown, light golden, pale, etc., and are for the most part excellent wines, blended to the taste and needs of importers. 1976 'J. Fraser' *Who steals my Name?* ix. 104 Don't guzzle down that Clos de Vougeot as if it was Watney's Pale. That's worth six pounds a bottle. 1977 *Berry Bros. & Rudd Catal.* Apr. 6 South African Sherry.. pale extra dry—per bottle £1.70.

c. Pale Brindled Beauty, a geometrid moth, *Apocheima pilosaria*, usually having light-coloured wings flecked with darker markings.

[1803 A. H. Haworth *Lepidoptera Britannica* 274 (*heading*) The pale Brindle. 1824 G. Samouelle *Entomologist's Useful Compendium* 363 The pale brindle. Trunks of trees.] 1860 H. N. Humphreys *Genera Brit. Moths* 81 (*caption*) The Female of the Pale Brindled Beauty. 1908 R. South *Moths Brit. Isles* (ser. 2) 295 Pale Brindled Beauty... The fore wings of this species are greyish, .. sprinkled with darker grey or brownish. 1955 E. B. Ford *Moths* xiii. 191 A black form of the Pale Brindled Beauty has become well established in some of the industrial areas of the north and round London. 1964 *Sunday Times* (Colour Suppl.) 2 Feb. 33 (*caption*) The male Pale Brindled Beauty moth may have these typical markings, but others have black wings. 1966 *Punch* 30 Mar. 463/1 Some ancient apple trees, .. generous hosts, in season, to the Capsid Bug and.. the Pale Brindled Beauty.

4. *pale-blurred, -breasted, -lipped, -mouthed, -snowed, -starred; pale-gleaming, -glimmering* (example).

1918 D. H. Lawrence *New Poems* 32 Pale-blurred, with two round black drops..my own reflection! 1913 —— *Love Poems* 8 Pale-breasted throstles and a blackbird. 1921 W. de la Mare *Veil* 6 But with set, wild, unearthly eyes Pale-gleaming, fixed as if in fear, She couched in the water. 1954 J. R. R. Tolkien *Two Towers* III. iii. 52 Mist lay there, pale-glimmering in the last rays of the sickle moon. 1920 Blunden *Waggoner* 33 But, alas, she falls in a swoon, Pale-lipped like a withering moon. 1820 Keats *Ode to Psyche* in *Lamia & other Poems* 119 No shrine, no grove, no oracle, no heat Of pale-mouth'd prophet dreaming. *a* 1918 W. Owen *Coll. Poems* (1963) 103 And when the land lay pale for them, pale-snowed. 1929 Blunden *Near & Far* 46 While unamazed I view the siege of pale-starred horror raised By dawn.

b. Special Comb.: **pale crêpe** (or **crepe**) (**rubber**), crêpe rubber of a pale yellowish colour, made by treating the latex with a chemical such as sodium bisulphite to prevent its turning brown.

1913 *India-Rubber Jrnl.* XLVI. 222/1 The preparation of pale crepe..is confined to the plantation rubber industry. *Ibid.*, In the preparation of pale crepe rubber the latex is coagulated in volumes varying from a few gallons to 500 or 600 gallons. 1937 [see *crêpe 2]. 1938 C. F. Flint *Chem. & Technol. Rubber Latex* iv. 126 Pale crêpe rubber may disappear from the market owing to the increasing use of latex for purposes for which pale crêpe was formerly used. 1970 *Encycl. Polymer Sci. & Technol.* XII. 187 Pale crepe is a long-established special rubber used for high-grade shoe soling and for applications needing a very light-colored, pure rubber.

pale, *v.*² Add: **1. b.** In phr. *to pale into insignificance*: to lose importance (freq. in comparison with a greater achievement).

1909 *Daily Graphic* 26 July 10/1 He..made a flight of twenty-five miles across country; but that, of course, pales into insignificance by the side of the Channel flight. 1966 *Listener* 27 Oct. 602/1 This..will..be a standard biography upon a scale which will make..the rest pale into insignificance.

Palearctic, var. Palæarctic *a.* in Dict. and Suppl.

pale-face. Add: **2.** In American Negro use, a contemptuous term for a white man.

1945 Mencken *Amer. Lang.* Suppl. I. 637 The Negroes use various other sportive terms for whites, *e.g.*, *pale-face, chalk* and *milk*. 1964 *N.Y. Times Mag.* 23 Aug. 62/2 *Whitey*, the latest word of contempt for a white person, superseding *ofay* and..*paleface*. 1971 *Publ. Amer. Dial. Soc.* 1969 LI. 33 To Negroes, a white convict is a..*paddy, pale face*. 1971 E. E. Landy *Underground Dict.* 145 *Pale face*, (B) white person.

Palekh (pā·lek). The name of a town in Ivanovo province in northwest U.S.S.R. used *attrib.* to designate the iconography for which the town was renowned in the 18th century, and also a type of miniature painting on such articles as boxes and trays which was developed there during the 19th century.

[1916 R. Newmarch *Russian Arts* iii. 65 Palekh 'the heart-centre of Russian popular ikonography'.] 1960 *Guardian* 1 Dec. 10/6 A Russian and Bulgarian shop opens in London tomorrow... There should be a rush on.. examples of palekh iconography. 1963 M. Chamot *Russ. Painting & Sculpture* i. 7 The famous Palekh work, as it is called from the village where it is produced, is executed with incredible delicacy and carries on to this day something of the inventiveness and charm of early Russian Painting. 1964 *Punch* 16 Dec. p. xii, Gifts at the Russian Shop..include..Palekh painted boxes. 1976 *Times* 19 Mar. 16/6 A sale of Russian and Greek icons totalled £35,609... A private buyer paid £1,400 for an early nineteenth-century Palekh school calendar.

Palermitan (pălɜ·ɹmităn), *sb.* and *a.* [ad. It. *palermitano*.] **A.** *sb.* A native or inhabitant of the Sicilian town or province of Palermo. **B.** *adj.* Of, pertaining to, or characteristic of Palermo.

1673 J. Ray *Observations Journey Low-Countries* 279 There is a great emulation and enmity between the Palermitans and Messanese, which involves the whole Island. 1826 M. Kelly *Reminisc.* I. 60 The Palermitans are all.. 1853 N. P. Willis *Pencillings by Way* I. xxi. 257 The oddity of the Palermitan style of building struck me forcibly. 1847 J. H. Newman *Let.* 25 Aug. (1962) XII. 109 There was a Palermitan father, at dinner.. 1908 *Westm. Gaz.* 6 July 4/2 The editor..is a Palermitan, and his family is the Sicilian branch of the Roman Colonnas. 1936 G. F. H. & J. Berkeley *Italy in Making* II. xiii. 339 Even in Sicily, La Masa and his Palermitans would not rise against Ferdinand with shouts of 'Viva Pio Nono!' 1961 *Times* 14 Jan. 9/1 Palermo's cathedral church..is in its own way more typically 'palermitan' than them all. 1968 D. M. Smith *Medieval Sicily* x. 110 St. Cristina's fair during which Palermitans had had the valuable privilege of not paying customs or excise duty. 1978 *Harpers & Queen* Sept. 42/4 We had..spaghetti con sarda, a Palermitan dish, with sardines, fennel and raisins.

Palestine (pæ·lɛstəin). The name of a territory on the eastern shore of the Mediterranean (see next) used *attrib.* to designate a cream soup made from Jerusalem artichokes.

1854 J. Romilly *Diary* 13 Apr. (1967) 55 He told us that he had given Palestine Soup yesterday: he asked the B. of London the origin of the name: he told him it was because it was made of Jerusalem Artichokes. 1861 Mrs. Beeton *Bk. Househ. Managem.* 912 (*heading*) Dinner for 6 persons... First course. Palestine soup. 1907 *Yesterday's Shopping* (1969) 41/1 Soups..Palestine..0/8½. 1929 ... *Let.* 24 Sept. (1971) 284, I was however agreeably surprised by a Palestine soup which had not the faintest trace of artichokes. 1970 Simon & Howe *Dict. Gastron.* 287/1 *Palestine soup*, the term for a soup made of Jerusalem artichokes.

Palestinian (pælɛsti·niăn), *a.* and *sb.* [f. Palestine (see below) + -ian.]

The name *Palestine* is derived from Gr. Παλαιστίνη (used in early Christian writing, cf. *Philistine*) the name of the Roman province, and designates that territory on the eastern Mediterranean coast which in biblical times comprised the kingdoms of Israel and Judah. There have been many changes in the frontiers in the course of history. It was revived as an official political title for the land west of the Jordan mandated to Britain in 1920. Palestine ceased to exist as a political entity in 1948 when the state of Israel was established, but the name continues to be used to describe a geographical entity, particularly in the context of Arab aims for the resettlement of people who left the area when the state of Israel was established.

A. *adj.* Of, pertaining to, or connected with Palestine.

1875 *Encycl. Brit.* II. 181/1 The books bearing this name are not contained in the Jewish or Palestinian Canon. 1902 D. G. Hogarth *Nearer East* 69 The Palestinian highlands. 1920 *Glasgow Herald* 13 July 8 Mr. Balfour said that for long he had been a convinced Zionist but..he never foresaw..that the great work of Palestinian reconstruction would happen so soon. 1934 [see *Götterdämmerung]. 1936 A. W. Clapham *Romanesque Archit.* v. 113 Certain churches in southern Italy..show evidence of being the product of the Palestinian school. 1956 *Discovery* Aug. 350/1 Palestinian objects of the Bronze Age. 1949 *Radio Times* 15 July 27/2 Musa Alami, a prominent Palestinian Arab. 1969 *New Jewish Encycl.* 476/1 The initial compilation of the Palestinian Talmud is ascribed to Rabbi Johanan ben Nappaha (third century). 1972 *Guardian* 4 Sept. 3/7 Black September operations go to generate a local Palestinian activism. 1973 *Jewish Chron.* 2 Feb. 22/4 Eventually the teachings of the Palestinian Amoraim were gathered together to form the Palestinian Talmud. 1976 *Daily Tel.* 30 June 4/8 He accused the Syrians of collaborating with the Right-wing Christians to suppress the Palestinian commando movement. 1977 M. Mazzawi in *Times* 13 June 14/1 This promise, coupled with 30 years' occupation of the country by Britain, and compounded by a decision of the United Nations in 1947, has resulted in the Palestinian people (who now number 3,500,000) being either in refugee camps, in exile, or under alien military rule.

B. *sb.* A native or inhabitant of Palestine in biblical or later times.

1907 *Daily Chron.* 31 July 5/3 Territorialists..flooded the hall with leaflets declaring that 900 Russian Palestinians were illegally present. 1909 *Ibid.* 9 Sept. 5/6 Those who are for a mass return to the country of their origin..are termed 'Palestinians'. 1920 *Glasgow Herald* 12 July 12 The higher ranks of the Civil Service..would consist mainly of British officials until an increasing number of Palestinians were fully qualified... Other ranks would be open to Palestinians, irrespective of creed. 1968 *Guardian* 12 May 9/3 ...is the Palestinians' actual experience of Israeli occupation. 1973 *Ibid.* 21 Apr. 12/1 The Arabs..[must] put the rights of the Palestinians (to the exclusion of Israel) before anything else. 1974 *Times* 10 Oct. 15/5 King Husain ..is not himself a Palestinian and indeed has a vested interest in preventing the assertion of a distinct Palestinian identity. 1977 *Listener* 19 May 642/1, I do not necessarily equate the Palestinians with the PLO: can you rationally and logically expect me to go along with the covenant that says that the liberation of Palestine means to purge the Zionist presence from Palestine. 1977 M. Mazzawi in *Times* 13 June 14/2 The Palestinians who left their homes were driven out by danger and threats. 1979 *Time* 13 Aug. 13/2 The Administration's first goal then, would be to bring Palestinians, perhaps even some P.L.O. officials, into the talks between the Israelis and the Egyptians on the future of the West Bank and Gaza.

Palestrinian (pælɛstri·niăn), *a.* [f. the name *Palestrin(a* (see below) + -ian.] Of, pertaining to, or in the style of the Italian composer Giovanni Pierluigi da Palestrina (*c* 1525–94).

1954 *Grove's Dict. Mus.* (ed. 5) VI. 511/1 Equally important, in any attempt to re-create Palestrinian style, is it to bear in mind the fact that the order of publication of such of his works as appeared in the composer's lifetime offers no reliable evidence as to date of composition. 1958 *Listener* 2 Oct. 540/1 There always was a Palestrinian tradition..

paletoted (pæ·lĕtōu̇tĕd), *a.* nonce-wd. [f. Paletot + -ed².] Provided with, or wearing, a paletot.

1947 V. Nabokov *Bend Sinister* 34 A mild bore who used to take out his two paletoted dachshunds at nightfall.

palette. Add: **1. c.** Of sounds (*spec.* in music): variety or range of tonal or instrumental colour (see Colour *sb.* 15).

1959 [see *Debussyan *a.* and *sb.*]. 1965 *Listener* 28 Oct. 680/1 If you want to be reminded of his mastery of the orchestra..and the richness of his palette, listen to *Daphnis et Chloé*, and the *Rapsodie espagnole*. 1975 *Broadcast* 3 Nov. 14/3 ...dialogue.. You may find Pinter's subdued palette somewhat baffling. 1977 *Listener* 7 Apr. 452/1 The shock of moving from one tonal palette to another..

2. (Further example.)

1887 D. Maguire *Art of Massage* (ed. 4) 20 The palette, which is also called ferule, or spatula, is an instrument.. ending at one extremity in a handle, and the other in a disc.

5*. In certain card games, a device used by the banker to facilitate the movement of cards and money.

1949 J. Scarne *On Cards* (1955) xxi. 205 The croupier, squatting in the concavity of the kidney table, needs an ebony-handle palette to slide the cash and cards around. 1966 W. Haggard *Power House* xi. 124 The croupier had used on the shoe, the other playing with the palette. 1968 D. Torr *Treason Line* 105 Victoria's mother had invented..by sliding the palette. As *chef de partie* in the high chair he could run a game of baccarat banque or chemin-de-fer as slickly as any professional.

palette-knife. Add to def.: Also used for spreading oil-paint on canvas. (Further examples.) Also *attrib.*

1785 J. Reynolds *Discourse to Students of R. Acad.* 28 Rembrandt, in order to take the advantage of accident, appears often to have used the pallet-knife [*sic*] to lay his colours on the canvas. 1931 L. Richmond *Technique Oil Painting* xi. 86 With palette knife painting..the charm lies to a certain extent in accidental qualities. 1938 D. Sharp *Student's Bk. Oil Painting* vii. 45 A different style of work is palette-knife painting, and this method certainly gives a very fascinating texture to a picture. 1958 *Times* 11 Mar. 3/4 Aix-en-Provence has a 'Rembrandt portrait' with heavy impasto ascribed to Rembrandt (which arrived there in 1863 a year or two before Cézanne painted his palette-knife studies of 'Uncle Dominique'). 1967 W. Gaunt *Compan. Painting* 93 Some equivalent of work with the spatula may be found in earlier pictures of palette knife painting. 1970 *Oxf. Compan. Art* 561/1 Some modern artists, working with thick pigment..

b. A similar instrument used as a culinary tool.

1889 A. B. Marshall *Cookery Bk.* iii. 41 Royal Icing... Put on to the cake with a clean palette knife. 1906 Mrs. Beeton *Bk. Househ. Managem.* xlvii. 1431 Scrape down the sides with a palette-knife, and dip the point of the knife into it all the material scraped down. 1951 E. David *French Country Cooking* 29 A first-class pliable palette knife, ..a selection of wooden spoons. 1975 *Habitat Catal.* 64 Kitchen equipment... Palette knife. Wood handle, steel blade.

Pali, *sb.* and *a.* Add: **3.** *Comb.*, as *Pali-Prakrit, -Pyu.*

1948 D. Diringer *Alphabet* vi. 388 Pali-Prakrit Sinhalese. The language may be dated from the third century B.C. to about the fourth century A.D. *Ibid.* vii. 410 A mixed Pali-Pyu inscription.

paligorskite, obs. var. *Palygorskite.

palilalia (pælilĕⁱ·liă). *Path.* [mod.L., ad. F. *palilalie* (A. Souques 1908, in *Rev. Neurol.* XVI. 340), f. Gr. πάλιν again + λαλιά talk, speech.] A speech disorder characterized by repetition of words, phrases, or sentences.

1908 *Index Medicus* VI. 157/1 (Index), Palilalia. 1927 *Jrnl. Neurol. & Psychopath.* VIII. 26 The palilalia (as well as stammering) disappears during pre-formed speech automatisms, as for instance, when the patient reads aloud, sings or recites. 1934 *Jrnl. Amer. Med. Assoc.* 1 Dec. 1711/2 Those familiar with such symptoms as automatic writing, palilalia, perseveration and verbigeration are inclined to wonder whether or not the literary abnormalities in which she [*sc.* Gertrude Stein] indulges represent correlated distortions of the intellect. 1955 J. M. Nielsen in A. B. Baker *Clin. Neurol.* I. iv. 376 Palilalia is a repetitive disturbance encountered in parkinsonism and in encephalitis (as representatives of organic causes), and in schizophrenia. 1961 W. R. Brain *Speech Disorders* vii. 107, I have once met with palilalia as a temporary phenomenon in a patient who suffered from compression of the medulla.

palimpsest, *sb.* and *a.* Add: **A.** *sb.* **2.** (Later *fig.* examples.)

1918 D. H. Lawrence *New Poems* 31 Darkness comes out of the earth.. Wanes the old palimpsest. 1929 *Oxford Poetry* 17 The world is all a palimpsest That hails the spurious pugilist. 1949 'G. Orwell' *Nineteen Eighty-Four* I. iv. 42 All history was a palimpsest, scraped clean and re-inscribed exactly as often as was necessary. 1962 R. Fuller *Exec. Gardner* x. 194 In Italy every town and house..is a palimpsest of two or three thousand years of building and decay. 1977 *Times* 3 Sept. 4/1 Alan Watts will be principally remembered as the architect of that peculiar theological palimpsest which served as an ideology for the hippie generation: that odd blend of rural fundamentalism and eastern mysticism.

B. *adj.* **2.** *Petrogr.* Of a rock: partially preserving the texture it had prior to metamorphism. Also in *Geol.*, exhibiting features produced at two or more distinct periods.

1912 R. W. Clark tr. *Weinschenk's Petrogr. Methods* x. 198 In the normal case the newly developed subsilicate is combined strictly to the border of the original crystal, but the substance of the old crystal may be recognized excellently, palimpsest [*sic*] structure. 1926 G. W. Tyrrell *Princ. Petrol.* xvi. 271 (*caption*) A palimpsest structure. Garnetiferous biotite hornfels.. Shows alternations of psammitic and pelitic sediments preserved, although the rock is thoroughly hornfelsed with the production of muscovite and biotite. 1951 Turner & Verhoogen *Ign. & Metamorphic Petrol.* xx. 503 It frequently happens..that fabric relics (palimpsest structures), like mineral relics, survive metamorphism and provide valuable indications of the parentage of the metamorphic rock. 1962 A. D. Howard in *Bull. Amer. Assoc. Petroleum Geologists* XLVI. 2255/1 A particularly interesting part of the anomaly is the drainage pattern, an unusual superposition of modern and ancient patterns that is convenient to refer to as palimpsest. In palimpsest drainage, the modern pattern is anomalous with respect to the older; it clearly indicates different topographic and possibly structural conditions at the time of development. 1972 D. J. P. Swift et al. *Shelf Sediment Transport* xxiii. 499 ... 1977 *Nature* ... The central and southern Atlantic shelf is a palimpsest or multiple imprint surface.

palindrome, *sb.* and *a.* Add: **A.** *sb.* Also *transf.*: in *Mus.*, a piece of music of which the second half is the first half in retrograde motion; in *Biol.*, a palindromic sequence of nucleotides.

1947 E. Blom *Everyman's Dict. Mus.* 430/1 *Palindrome*, a word or poem reading the same backwards as forwards. In music a piece constructed in the same way..more or less loosely, as e.g. the prelude and postlude in Hindemith's *Ludus tonalis*. 1962 *Listener* 21 Sept. 445/3 The palindrome is another symmetrical form that he has used several times, most notably and extendedly in the recent *Symphonies* where the strict symmetry of the reversed 'reprise' is relieved only by changes of scoring, etc. 1963 *Ibid.* 28 Mar. 569/3 ...with musical palindromes. 1974 Wilson & Thomas in *Jrnl. Molecular Biol.* LXXXIV. 115 We call these regions in double-chain DNA palindromes, because, given the antiparallel arrangement of the polynucleotide chains, these sequences read the same both backwards and forwards. 1977 *Nature* 3 Nov. 10/2 If these inverted repeats are adjacent (forming a palindrome) renaturation produces a double-stranded hairpin. 1979 *Gramophone* Aug. 327/2 Here is a gorgeous performance, and its orchestral details may be summarized by the moment when Berg moves into a palindrome.

palindromic, *a.* Add: More widely (esp. in *Math.* and *Mus.*), having a structure or composition that reads the same in both directions.

1957 W. J. Reichmann *Fascination of Numbers* ix. 90 The most simple form of palindromic numbers is the one containing a number of identical digits. 1961 *Listener* 23 Nov. 889/1 Schönberg makes his ternary-form prelude in effect a prelude-and-fugue, the middle section being fugal, with a powerfully melodic subject of palindromic construction, and a counter-subject that is also palindromic. 1965 *Ibid.* 17 Jan. 140/3 ...having clear-cut musical ideas worked out logically in a neat palindromic structure. 1964 A. H. Beiler *Recreations in Theory of Numbers* xx. 228 There are an infinite number of palindromic primes. Here are a few: 101, 131, 151, 181, 313. 1972 *Times* 1 Aug. 15/2 The sole point of this letter is the date [*sc.* 27.7.72] at its foot. Apart from the three rather less pleasing palindromes arising on the 27th of the 8th 9th and 11th months of this year, today's is the last palindromic date until 18.1.81. 1974 *Nature* 9 Aug. 467/2 The central assumption of my theory is that the nucleotide sequences at the ends of eukaryote linear DNA molecules are palindromic. 1976 W. Safire *Before the Fall* 181 Prince Norodom Sihanouk, had been deposed by Lon Nol, his palindromic Prime Minister.

2. *Med.* Characterized by frequent, irregular recurrences of short-lived rheumatic attacks.

1941 Hench & Rosenberg in *Proc. Staff Meetings Mayo Clinic* XVI. 814 When an etiologic term cannot be applied, a distinctive descriptive term should be used. Such a designation we found in the term 'palindromic rheumatism', suggested to us by Profs. A. D. Fraser and A. L. Hench, respectively, of the departments of Ancient languages and of English literature of the University of Virginia. 1959 *Ann. Rheumatic Dis.* XVIII. 331/2 When the histories of patients with definite rheumatoid arthritis were reviewed, a palindromic or episodic type of onset was found in 10 to 15 per cent. *Ibid.*, Finger contractures had been a common feature during the palindromic phase. 1972 *Lancet* 5 Aug. 269/2 Palindromic rheumatism is a condition in which an acute arthritis develops over a few hours, lasts for a day or two, and disappears the way it came.

paling, *vbl. sb.*¹ **4.** *paling fence* (later U.S. examples).

1843 *Amer. Pioneer* II. 308 A strong body occupied the yard of Ebenezer Zane..using the paling fence as a cover. 1873 'Mark Twain' & Warner *Gilded Age* v. 60 Hawkins put up the first 'paling' fence that had ever adorned the village. 1901 Merwin & Webster *Calumet 'K'* v. 68 The space was inclosed by a high paling fence which bounded the C. and S.C. right of way. 1925 W. J. Bryan *Mem.* 17 Our yard was enclosed in the old-fashioned paling fence with a baseboard about a foot high.

palingenesis. Add: **3.** *Petrol.* [ad. Sw. *palingenes* (J. J. Sederholm 1907, in *Bull. Comm. Géol. Finlande* XXIII. 89).] The formation of a new magma by the remelting of existing rocks.

1907 J. J. Sederholm in *Bull. Comm. Géol. Finlande* XXIII. 110 The granitic magma, once solidified and, in part, decomposed, undergoes again, when brought into the deeper parts of the earth, a resurrection, or, as the author expresses it, palingenesis. 1965 A. Holmes *Princ. Physical Geol.* (ed. 2) xxx. ... The following list of the radiometric ages of some of the uplifted basement rocks shows how successfully they escaped the widespread palingenesis..that rejuvenated similar old rocks during the Nevadan orogenies farther west. 1974 L. N. Kogarko in H. Sørensen *Alkaline Rocks* vi. 480/1 The significant reductions of the melting temperatures of rocks caused by the volatile components is probably of great importance for palingenesis.

palingenetic, *a.* (Examples corresponding to *Palingenesis 3.)

1942 M. P. Billings *Structural Geol.* xv. 296 Palingenetic granites were actually molten and, in the simplest case, they have the same chemical composition as the rocks from which they were derived. 1974 Borodin & Pavlenko in H. Sørensen *Alkaline Rocks* vi. 523/1 In the Mongol-Tuva province palingenetic and metasomatic formations of alkaline rocks are widespread.

palinspastic (pælinspæ·stik), *a.* *Geol.* [f. Gr. πάλιν again + σπαστ-ός drawing in (see Spastic *a.*).] Of a map or the like: representing layers of rock as returned to their supposed former positions. Hence **palinspa·stically** *adv.*, by means of such maps; **palinspa·stics** *sb. pl.*, the production of such maps.

1937 G. M. Kay in *Bull. Geol. Soc. Amer.* XLVIII. 291 The resulting map displays the several slices in their conceived relative positions... The name *palinspastic*..is proposed for base maps of this character. 1945 *Bull. Amer. Assoc. Petroleum Geologists* XXIX. 426 Palinspastic maps show the locations of rocks in their relative original positions. *Ibid.* 435 Palinspastic maps..have been used to illustrate concepts of continental drift and of intracontinental movements in the development of mountain systems. 1969 J. F. Dewey in M. Kay *N. Atlantic* xxv. 331/2 No attempt was made in Figures 2 and 3 to reconstruct palinspastically the original shape of the paratectonic cover sequence, in terms of thickness or extent. 1971 *Nature* 2 Apr. 310/1 Fig. 2 is an upper Palaeocene palinspastic reconstruction of Fig. 1.

‖ **palio** (pɑ·lio). Also 7 **palio**, 8 **pallio**. [a. It. *palio*, ad. L. *pallium* a covering, cover.] A traditional horse-race held every July and August in the Italian city of Siena; a similar horse-race held in other Italian towns; the cloth or banner of velvet, silk, etc., given as the prize for this race.

1673 J. Ray *Observations Journey Low-Countries* 338 He diverteth and entertaineth the Citizens of Florence..with sports and shows, especially races for prizes (Pallio's they call them). 1791 E. Wynne *Diary* 4 Apr. (1935) I. iv. 60 The King of Naples arrived today [at Padua] and they prepare a Pallio for him. 1863 Geo. Eliot *Romola* I. viii. ... where the richest of Palii, or velvet and brocade banners, ..were..awaiting the winner or winner's owner. *Ibid.* xvi. 291 You are going towards the Piazza della Signoria..we shall perhaps be at the Palio...among these racers. 1873 S. & J. Horner *Walks in Florence* I. xvii. 296 The Corso..was at one time celebrated for horse-races, in which the Florentine youth competed for a piece of cloth of divers colors, called a palio, to give its name to the race. 1881 K. Baedeker *Central Italy* (ed. 7) 21 On July 2nd and Aug. 15th horse-races, called *il Palio*, take place, presenting a very picturesque scene. 1902 E. G. Gardner *Story of Siena* v. 131 The race is still called the Palio, from the rich stuff (now represented as a banner) given as prize. 1904 W. Heywood *Palio & Ponte* 89 In the feast of St John the Baptist, at Florence, another palio is referred to in the *Divina Commedia*. 1925 A. Huxley *Along Road* II. 100 The Palio itself—the painted banner which is given to the *contrada* whose horse wins the race. *Ibid.* 101 The Palio is probably the most dangerous flat-race in the world. 1930 E. Pound *XXX Cantos* xxix. 109 Zoanne da Rimini Has won the palio at Milan with our horse. 1970 C. Fry *Yard of Sun* I. 6 You know it is the first practice-run for the Palio—And you know this parish has got the best horse It has ever had. 1972 *Times* 18 Sept. 19/1 Over the weekend the historic Tuscan town was enlivened with.. the Palio horse race in Renaissance costumes. 1974 *Radio Times* 27 July–2 Aug. 33/2 Old traditions also continue [in Siena], like the famous annual horse-race, the *Palio*, held every summer on the city square.

‖ **paliotto** (pælio̜·to). [It.] The frontal painting on an altar-piece. Also *attrib.*

1937 *Burlington Mag.* July 18/1 The *St. Peter* paliotto of about 1280–85. 1958 *Times Lit. Suppl.* 27 June 356/3 'Judas receiving Payment', the 'Baptism of Christ' from the thirteenth-century St John *paliotto* in the Siena Gallery, the head of Pietro Lorenzetti's St Catherine of Alexandria, [etc.]. 1961 M. Levy *Studio Dict. Art Terms* 82 Paliotto painting.

palisade, *sb.* Add: **3. d.** *Biol.* A region of parallel elongated cells, often at right angles to the surface of the structure of which they form part; *esp.* the palisade parenchyma of a leaf. Freq. *attrib.* (see sense 4).

1914 M. Drummond tr. *Haberlandt's Physiol. Plant Anat.* vi. 263 A remarkable modification, and one which is of great importance for the understanding of the palisade form, is the so-called arm-palisade-cell; in this case the palisade, instead of consisting of entire cells, is made up of groups of cell-branches or -arms. 1956 R. W. Evans *Histol. Appearances of Tumours* vi. 79 In one of Chase's tumours the cells tended to form palisades. 1965 Bell & Coombe tr. *Strasburger's Textbk. Bot.* 349 The outer leaves on the southern sunny side of a tree commonly possess a deeper palisade..than the 'shade leaves' of the northern side. 1971 *Nature* 19 Feb. 561/1 Here the so-called palisades of Vogt are found: fine radial lines about 1 mm long and four per mm of the corneal circumference. 1972 *Arch. Dermatol.* CVI. 865/3 The schwann cells which are closely aggregated often appear to be arranged in palisades.

4. *palisade-trench.*

1935 *Proc. Prehistoric Soc.* I. 124 In this barrow the posts (set in a palisade-trench) were smaller. 1963 W. F. Grimes in Foster & Alcock *Culture & Environment* v. 142 The entrance took the form of a passage between the ends of the bank which was defined by narrow trenches resembling palisade-trenches.

palisade, *v.* Add: **palisaded** *ppl. a.*, (*b*) *Med.*, consisting of, or arranged as in, a palisade (*Palisade *sb.* 3 d); **palisading** *vbl. sb.*, (*b*) *Med.*, arrangement of cells in a palisade.

1951 M. S. McKeehan in *Jrnl. Exper. Zool.* CXVII. 39 During the period of nuclear orientation, the lens ectoderm rapidly changes from a cuboidal to a high columnar epithelium. This phenomenon may be called 'palisading'. 1956 R. W. Evans *Histol. Appearances of Tumours* xvi. 300 When palisading is present it is usually a conspicuous microscopical feature. *Ibid.* xix. 403 This palisaded columnar epithelium lies in contact with the adjacent fibrous tissue. 1966 Wright & Symmers *Systemic Path.* II. xxxviii. 1435/1 The zones of necrosis..are surrounded by radially oriented ('palisaded') connective tissue cells. 1972 *Arch. Dermatol.* CVI. 865/1 The palisaded, encapsulated neuroma is a..benign cutaneous tumor. 1974 *Nature* 18 Jan. 145/1 This 'palisading' gives the lens anlage a regular appearance, but individual cells may deviate considerably from the columnar shape.

palisander (pælisæ·ndəɹ). Also **palissander**. [a. F. *palissandre, palisandre*, prob. of Amer. Ind. origin.] The hard, dark, black-streaked wood of the Brazilian tree, *Dalbergia nigra*, of the family Leguminosæ; also known as Brazilian rosewood. Cf. Rosewood 1. Also *attrib.*

1843 C. Holtzapffel *Turning & Mech. Manipulation* I. 104 The furniture rose-wood..was afterwards imported as Jacaranda, Palisander, and Palaxander-wood, by which names it is still called on the Continent. 1902 G. S. Boulger *Wood* ii. 270 Palisander-wood... A valuable wood, chiefly used in pianofortes. 1930 *Heal & Son Catal.* 14 The timbers that invite the craftsman's hand.. Macassar Ebony, Amaranth and Palisander. 1936 *Burlington Mag.* Aug. 88/2 Furniture in mahogany, palissander, rose- or satinwood. 1941 *Archit. Rev.* LXXXIX. 16 (*caption*) The study, with palisander wood furniture. 1947 J. C. Rich *Materials & Methods Sculpture* x. 294 Brazilian rosewood, also known as Palisander and Jacaranda, is an extremely fine, rare, and expensive variety. 1957 *Times* 4 Nov. 13/1 There is plain Bangkok teak, and palisander wood from South America. 1966 *House & Garden* Dec. 57/1 The Clairtone [hi-fi]..is solid-state encased in rare Brazilian Palisander. 1971 *Country Life* 3 June 1361/1 The top is lined with gilt tooled leather in a surround of palisander wood. 1974 *Ibid.* 3 Oct. 939/1 (*caption*) An 18th-century secretaire bookcase in palissander wood attributed to William Vile. 1977 *Early Music* July 455 (Advt.), The Roessler 'Meister' range [of recorders] is available in rosewood, palisander, olivewood and boxwood.

palissy (pæ·lisi). The name of the French master potter Bernard *Palissy* (*c* 1510–*c* 1590), used *attrib.* to designate pottery made by him, his successors, or his imitators.

1858 Trollope *Three Clerks* I. vii. 137 He..was inclined to ridicule the growing taste of the day for torsos, Palissy ware, and Assyrian monsters. 1867 G. M. Hopkins *Jrnls. & Papers* (1959) 155 Dish of Palissy ware with a pike. 1900 F. Litchfield *Pott. & Porc.* vii. 255 The manufacture of the Palissy ware was continued until the time of Henri IV. 1902 S. Weyman *In Kings' Byways* i. 68 The table gleamed with..Palissy ware and Cellini vases. 1969 S. Sitwell *Gothic Europe* xiv. 162 The rather horrible imitation Palissy pottery made in Portugal with fruit and vegetable and crustacean motifs.

pallasite. For *Min.* read *Geol.* and substitute for def.: Any of a class of stony-iron meteorites which consist largely of iron (usu. with

a small proportion of nickel) and olivine. (Earlier and further examples.) Hence **palla·sitic** *a.*

1868 *Geol. Mag.* V. 78 The arrangement of the meteorites in the museum of Berlin University, by M. G. Rose, is based on their mineral character, and forms two divisions —the metallic and the stony meteorites, the first containing meteoric iron and the Pallasite, the second the Chondrites, Howardites, [etc.]. 1920 *Mineral Mag.* XIX. 59 In most pallasites the iron is poor in nickel,.. and the olivine is correspondingly poor in ferrous oxide. 1956 *Nature* 28 Jan. 156/1 From 60 to 1,600 km. the material of the mantle is identified as dunite and from 1,600 to 3,000 km. it is taken as of the same composition as pallasitic meteorites. 1962 [see *mesosiderite]. 1977 A. Hallam *Planet Earth* 134/2 This led to the speculation that ancient peridots were extraterrestrial, extracted from the meteorites known as pallasites.

pallavi (pāla·vi). Also **pallevi.** [Origin uncertain.] In the music of southern India, the first section of a song.

1891 C. R. Day *Mus. & Mus. Instruments S. India* v. 60 Almost all [melodies] consist of a burden or refrain called Pallevi, a kind of answer to this refrain styled Anupallevi, and stanzas (called Charanam) of which there is usually an uneven number. These parts are in the several compositions arranged in different ways. 1914 A. H. F. Strangways *Mus. Hindostan* iii. 86 A *Pallavi* was then sung round. 1968 *Jrnl. Musical Acad. Madras* XXXIX. 26 A whole line of the Pallavi..had been absent from the piece as current. 1971 *Shankar's Weekly* (Delhi) 11 Apr. 23/1 The pallavi was set to Rupakam with a sub-initial take-off.

pallet, *sb.*³ Add: **3. d.** A portable tray or platform used, esp. in conjunction with a fork-lift truck, for moving or stacking heavy loads in convenient units. Also *attrib.*

1921 R. V. Wright et al. *Material Handling Cycl.* 97 *Pallet*, a flat platform..used to pile material on. 1948 *P.O. Telecomm. Jrnl.* I. iv. 119/2 A pallet is a double-faced wooden or metal platform with a space between the top and bottom faces significantly large to permit the entry of the forks of a forklift truck. 1958 [see *fork *sb.* 16]. 1961 *Times* 10 June 11/6 Tomatoes..packed, 120 at a time, into 'pallets' or metal trays supplied by British Railways. 1963 *Times* 23 July 7/3 The soldiers will sit in what are called 'people pallets' which will be dropped from low-flying Lockheed C130 assault aircraft. The pallets, which hold 12, 24 or 48 men, will be carried on the open cargo ramp at the rear end of the aircraft. 1971 *Power Farming* Mar. 13/2 Five-ton high-lift pallet trailers were used to transport the carrots from the field to the packing station. 1974 *Guardian* 20 Mar. 5/4 Pack your goods onto a standard pallet up to 40" × 48"... We lift the whole pallet and take it to any of..17 different European destinations. 1976 *Farnborough Internat. Exhib.* (Official Programme) 17 Hawker Siddeley Dynamics..has a £6m. contract to build experiment carrying pallets for Spacelab. 1978 S. Brill *Teamsters* vii. 266 Pallets are the wooden trays under which any heavy cargo is loaded.

pallet (pæ·lėt), *v. rare.* [f. Pallet *sb.*²] *intr.* To lie *down* to sleep on or as on a pallet.

1921 G. C. Shedd *Lady of Mystery House* xix. 263 He and I could pallet down on the porch.

palletization (pæ:lėtəizē̍ⁱ·ʃən). [f. *pallet-ize *v.* + -ation.] The action or process of palletizing or of becoming palletized. Also *attrib.*

1946 *Chem. Industries* Aug. 294 (*heading*) Palletization. 1957 *Economist* 19 Oct. 257 They will advise on palletisation, mechanical handling and other modern techniques. 1965 R. B. Oram *Cargo Handling* v. 83 Whisky in cases and cartons, insulation material in bags and fire bricks: all this was cargo that lent itself to palletization. 1967 *Times Rev. Industry* Feb. 71/3 The introduction of more efficient cargo-handling techniques, i.e., palletisation, containerization, packaged timber, roll-on, roll-off, &c.

palletize (pæ·lėtəiz), *v.* [f. Pallet *sb.*³ + -ize.] *trans.* To place on a pallet; to transport on a pallet; to convert a loading system to the use of pallets. (See *pallet *sb.*³ 3 d.) So **pa·lletized** *ppl. a.*; **pa·lletizing** *vbl. sb.*

1954 in Webster Add. 1959 *Times Rev. Industry* Apr. 77/2 Wagons specially designed for palletized loading. 1960 *Farmer & Stockbreeder* 1 Mar. 71/3 Palletizing, widely used in industry, and for fruit and potatoes, has not, it is believed, been used in this. 1964 *Economist* 3 Oct. 61/2 Goods being palletised in advance. 1967 *Jane's Surface Skimmer Systems* 1967–68 21/2 Palletised loads can be 'floated' in on hover-pallets supplied with air from the main compressor. 1970 *Financial Times* 13 Apr. 9/7 A feature of the building is that the structural steel frame is designed as a space frame to perform the function of storing palletised goods. 1977 *R.A.F. News* 22 June–5 July 3 (Advt.), Removals—Home and overseas Storage—Palletised containers.

pallial, *a.* Add: **b.** *Anat.* Of or pertaining to the pallium of the brain.

1901 [see *neopallium]. 1933 *Proc. Nat. Acad. Sci.* XIX. 7 Below the reptiles the entire pallial field is dominated by the olfactory system. 1965 L. B. Arey *Developmental Anat.* (ed. 7) 495 [The commissures] cross partly in the lamina and partly in the fused adjacent portions of the median pallial walls.

pallite (pă·ləit). *Min.* [a. F. *pallite* (Capdecomme & Pulou 1954, in *Compt. Rend.* CCXXXIX. 288), f. *Pall-o*, the name of its locality in Senegal + -ite¹.] A white or greyish ferrian variety of millisite.

1954 *Mineral. Abstr.* XII. 440 Two types of Ca-Al phosphates are distinguished: (1) Pallite..formed by the action of calcium phosphate on montmorillonite; (2) Crandallite..formed by the leaching of (1). 1960 *Amer. Mineralogist* XLV. 257 Uranium is present in amounts up to 140 ppm U in pallite.

pallium. Add: **3. d.** *Anat.* The wall of the cerebral hemispheres (including, in mod. use, the rhinencephalon). Cf. *archi-, *neo-, and *palæopallium.

1890 *Jrnl. Anat. & Physiol.* XXV. 106 When the surface of a cerebral hemisphere is carefully examined, it is seen to be capable of a natural division into two parts: a basal region, or Rhinencephalon, and a superior portion, or Pallium. 1901 *Ibid.* XXXV. 442 And His..freely admits that the rhinal fissure is the line of demarcation between the 'pallium' and the 'rhinencephalon'. 1934 L. B. Arey *Developmental Anat.* (ed. 3) xv. 414 The telencephalon consists of three regional parts. One is the corpus striatum... The second division is the rhinencephalon, or archipallium, while the remainder of the hemisphere makes up the neopallium. The last two portions comprise all of the externally visible hemispheres, and together may be called the pallium. 1948 A. Brodal *Neurol. Anat.* 323 It is only in mammals that the dorsal cortex undergoes a marked development, and increases progressively in the phylogenetic ascent, to reach its peak of development in man, in whom it forms the bulk of the entire pallium. 1972 *Jrnl. Neurochem.* XIX. 2031 During the last third of the gestational period, the cerebral pallium of the rabbit develops from a primitive vesicle with a total weight of about 100 mg to a highly organised structure 10 times as large.

Pallottine (pæ·lŏtəin), *a.* and *sb.* [f. the name of Vincent *Pallotti* (1795–1850), an Italian priest who founded in 1835 the Society of the Catholic Apostolate, a society of R.C. priests, lay brothers, sisters, and associates, known formerly as the Pious Society of Missions, and known popularly as the Pallottine Fathers.] **A.** *adj.* Also **Pallottian, Pallotine.** Belonging or pertaining to the Society founded by Pallotti. **B.** *sb.* A member of the Society.

1890 *Tablet* 19 July 98/2 The English Pallottine Fathers (Society of Missions) have bought the Palazzo Caccia in Rome, and intend converting it into a seminary. 1894 M. E. Herbert *Life Vincent Pallotti* xiv. 153 The Pallottine Sisters have also large schools adjoining the church. 1896 [see *consultor 1 b]. 1903 P. J. Chandlery *Pilgrim-Walks in Rome* xvi. 409 The church is served by the Pallottine Fathers. 1962 J. S. Gaynor *Eng.-Speaking Pallottines* 5 They set about realising the Pallottian ideal in the world. 1962 *Times* 1 Sept. 8/3 Provincial of the Pallottine Fathers. 1971 *Shrewsbury Diocesan Year Bk.* 84 Pallottine Missionary Sisters, Pallotti Hall, Siddington, near Macclesfield. 1976 *Billings* (Montana) *Gaz.* 20 June 12-B/6 Church and state Saturday were ready to crack down on the Pallottine Fathers, an order of Roman Catholic priests dedicated to helping the poor abroad but accused of using millions in charitable funds for land ventures and loans to local politicians.

pally, *a.* See s.v. Pal *sb.*¹ in Dict. Add: (Later examples.) Also in the extended form **pally-wally.**

1915 H. L. Wilson *Ruggles of Red Gap* (1917) i. 9 The Honourable George..had..been almost quite too pally with him. 1922 J. Cannan *Misty Valley* 282 If you cared for me it was not pally to let me go on doing things I didn't know were wrong. 1923 [see *hell-brew* s.v. *hell *sb.* 11 a]. 1929 H. A. Vachell *Virgin* i. 12 She had never been 'pally' with girls. 1936 W. R. Titterton *G. K. Chesterton* i. v. 60 [He] was on pally terms even with small shop-keepers, farmers and country squires. 1951 R. Hoggart *Auden* ii. 38 Auden often wobbles..from the pally to the patronising. 1954 F. Brown in *Astounding Sci. Fiction* Sept. 16/2, I *like* quarrelling. If you're going to go namby-pamby and pally-wally on me, I'll go find someone else. 1974 S. Gulliver *Vulcan Bulletins* 47 Why would Anscudden go along with stealing Javits' shipment? I thought they were supposed to be pally. 1976 *Scottish Rev.* Spring 9 She joined a Whist club and got very pally with another auld maid like herself.

pally (pæ·li). Colloq. abbrev. and spelling of *palais de danse.

1928 D. H. Lawrence *Lady Chatterley* xi. 189 The new girls in their silk stockings, the new collier lads lounging into the Pally or the Welfare. 1947 I. Brown *Say Word* 72 Whether with hurdy-gurdy on the pavement or with boogie-woogie in the 'Pally', still they favour a word with rhyming syllables. 1948 *Amer. Speech* XXIII. 319/2 *Pally*, dance hall (in England abbreviation for Palais de Dance).

palm, *sb.*¹ Add: **7. a.** *palm-frond* (examples). **b.** *palm-flanked, -lined* adjs. **c. palm bottom,** a hollow or valley in which palms grow; **palm-branch** (later examples); **palm-cabbage** (later examples); **palm-heart** = *palm-cabbage* (cf. Heart *sb.* 18); **palm-hut,** a small cabin made from palm-trees; **palm-room,** a room, usu. in a hotel, adorned by potted palms; **palm-soap,** a soap made from palm-oil; **palm-squirrel,** a small, greyish-brown, tree squirrel with three white stripes along its back, belonging to the genus *Funambulus*, esp. *F. palmarum*, which is found in India; **palm-stand,** a stand for supporting a palm grown in a plant-pot; **palm-sugar** (later examples); **palm toddy** (later examples); **palm-wine** (further examples).

1902 D. G. Hogarth *Nearer East* 141 Stony slopes.. only at very rare intervals relieved by palm bottoms. 1914 W. Owen *Let.* 25 Feb. (1967) 235, I had huge success with my Costume: Nothing more elaborate than my Gown on my back a laurel wreath on my head, & a palm branch in the hand. 1928 D. H. Lawrence *Woman who rode Away* 16 Simply marvellous people! And the way they strewed palm-branches under her feet! 1977 H. Kaplan *Damascus Cover* xii. 122 He was tied to the gate of the Al-Frange Synagogue... Then he was beaten with palm branches. 1966 E. J. H. Corner *Nat. Hist. Palms* iv. 93 Not all palm-cabbages are edible. 1972 J. W. Purseglove *Tropical Crops: Monocotyledons* II. 443 The freshly cut terminal bud [of *Cocos nucifera*], known as palm cabbage, is considered a delicacy and may be eaten cooked or raw. 1928 H. Crane *Let.* 31 Jan. (1965) 314 The great palm-flanked arena of Angelus Temple. 1938 M. K. Rawlings *Yearling* i. 5 The palm-frond mill-wheel must just brush the water's surface. 1972 'M. Renault' *Persian Boy* xxvi. 349 He looked up at the waving palm-fronds, and played lazily with my hair. 1973 *Nat. Geographic* Feb. 211/1 On a plain before the fort of Rustaq, I sat on a palm-frond mat with a wiry old sheik. 1974 *Observer* (Colour Suppl.) 13 Oct. 20/3 (*caption*) Swinging from palm fronds like a teenage Tarzan, this Arab boy performs acrobatics for the benefit of admiring tourists at the oasis of Gafsa in Tunisia. [1901 tr. C. G. O. Drude in L. H. Bailey *Cycl. Amer. Hort.* III. 1193/1 From many species are cut out the soft terminal bud (heart), which is eaten as Palm salad.] 1938 M. K. Rawlings *Yearling* xx. 250 He sliced the palm-hearts thinly. 1971 'D. Halliday' *Dolly & Doctor Bird* v. 67 We had palm hearts, a matter of flaccid white tubing, followed by prime rib steak. 1976 *Times* 1 June 6/3 We have found..tinned palm hearts and artichoke bottoms. 1930 R. Macaulay *Staying with Relations* ii. 20 The forest would recede a little, and small clearings and plantations make themselves apparent.., with groups of palm huts dumped among them like beehives. 1936 *Discovery* Dec. 382/1 A riverside palm-hut. 1902 D. G. Hogarth *Nearer East* 74 The abundant waters of its own palm-lined defiles. 1930 R. Macaulay *Staying with Relations* xviii. 265 They rode into the bay, and saw before them the palm-lined harbour front of the Pacific's greatest pearl city. 1903 'C. E. Merriman' *Lett. from Son* 46, I met him in the palm room last night. 1930 E. Pound *XXX Cantos* xxix. 137 'No not in the palm-room.' The lady says it is Too cold in the palm-room. 1931 F. L. Allen *Only Yesterday* i. 11 In the more dimly lighted palm-room there may be a juvenile petting-party or two going on. 1846 *Jewish Manual, or Pract. Information Jewish & Mod. Cookery* Toilette iv. 212 Palm soap, Castille soap..should always be preferred. 1966 J. S. Cox *Illustr. Dict. Hairdressing* 108/1 *Palm-soap*, a soda soap of palm oil. (*Piesse*). 1831 *Proc. Zool. Soc.* I. 103 The Palm Squirrel is very abundant in gardens in Dukhun. 1891 W. T. Blanford *Fauna Brit. India: Mammalia* II. 384 The cry of the palm-squirrel is a shrill chirp, resembling the note of a bird. 1908 *Westm. Gaz.* 15 June 5/3 The workers [*sc.* white ants] are preyed upon by true ants and many other insects;..by rats, mice, and palm-squirrels. 1955 I. T. Sanderson *Living Mammals of World* 115/2 Palm-Squirrels... It is virtually impossible for the non-specialist to identify or even define this tribe. 1926 M. Leinster *Dew on Leaf* v. 55 A large writing-desk and shelf of books,..and a blackwood palm-stand, were some of the surrounding objects. 1937 M. Covarrubias *Island of Bali* (1972) vi. 125 The child is weaned after three birthdays.., when the mother puts a mixture of lime and palm-sugar to her nipples. 1971 *Nat. Geographic* Mar. 344 (*caption*) Train of oxcarts, laden with baskets of palm sugar, rumbles toward a landing on the Irrawaddy River. 1950 'D. Divine' *King of Fassarai* vii. 48 Fishing, canoe-building..and palm toddy. 1974 *Nat. Geographic* Dec. 754 A feast of marinated raw fish..was washed down with palm toddy. 1957 M. Banton *W. Afr. City* iii. 52 Wartime prosperity multiplied the demand for palm-wine in Mende country. 1964 J. P. Clark *Three Plays* 5 A big gourd of palm wine and three heads of kola-nut split before the dead of the land. 1969 J. M. Gullick *Malaysia* i. 31 There may be a shop for the restricted sale of 'toddy' (palm wine). 1976 *Daily Times* (Lagos) 8 July 32/4 But Obadeya told the tribunal that he was in a palm-wine bar with Chidi that night.

palm, *sb.*² Add: **9.** *palm-reader*; **palm-ball** *U.S. Baseball*, a pitch of the ball gripped with the thumb and palm; **palm-print,** the impression left by the palm of the hand.

1948 *Richmond* (Virginia) *Times-Dispatch* 15 Mar. 17/4 The lanky Cincinnati Reds' sideswheeler has added a new pitch to his repertoire—a palm ball. 1950 *Sun* (Baltimore) 28 Aug. 16/2 He carries his own little notebook and can tell you many of the pitches he has made, when he threw the slider, the palm ball and even the gopher ball. 1973 *Times* 15 Aug. 7/3 There are numerous ways of hurling a ball legally to increase its effectiveness, such as the..screw-ball, the fork-ball and the palm-ball. 1976 *Webster's Sports Dict.* 300/2 *Palm ball*,..an off-speed pitch that is gripped between the thumb and the palm instead of with the ends of the fingers and that is thrown so that the fingers do not impart rotation to the ball. 1929 A. C. & C. Edington *Studio Murder Myst.* xv. 202 In wearing gloves the criminal nearly always leaves a very legible palm print. 1946 *Nature* 12 Oct. 526/1 A drawing of a hand emphasizing the features of finger- and palm-print patterns is reproduced from this article. 1954 F. Cherrill *Cherrill of Yard* vii. 69 'Are palm-prints as infallible as finger-prints?' he asked. 1955 *Times* 10 May 4/2 It had now been established that no two human beings had the same palm prints. 1967 N. Lucas

C.I.D. v. 61 In 1930..for the first time, palm-print evidence was accepted in a criminal court. 1977 D. Harsent *Dreams of Dead* 5 Her palm-print shrinks on the mirror as she turns away. 1920 R. Macaulay *Potterism* III. ii. 131 She is the most wonderful palm reader and crystal gazer I have come across.

palm, *v.* Add: **1.** (Further examples.)

1974 *Guardian* 23 Sept. 24/8 Farmer palmed over a header from the impressive Thompson. 1976 *Wymondham & Attleborough Express* 3 Dec. 27/2 From the kick off the ball was put to Chambers on the wing and his hard shot was palmed into the path of Bartrum who put Pollastra 1-0 up in 45 seconds.

4. a. *spec.* in *U.S. Law* = *pass *v.* 64 c.

1880 *Federal Reporter* (U.S.) I. 36 Nor is it necessary, in order to give a right to an injunction, that a specific trade-mark should be infringed, but it is sufficient that the court is satisfied that there was an intent on the part of the respondent to palm off his goods as the goods of the complainant. 1904 *Judicial & Statutory Definitions* VI. 5159/2 'To palm off' means to impose by fraud; to put off by inferior means. The language also imports that plaintiff must have been deceived and cheated by the representations, which he could not have been had he relied upon them. 1939 *Northeastern Reporter* XXI. 837/2 Defendants so conducted their business as to intentionally palm off on the public goods of the defendants as the goods of the plaintiff. 1956, etc. [see *palming *vbl. sb.*¹ 2*]. 1973 *N.Y. Law Jrnl.* 17 Apr. 4/5 A claim that Borden attempted to 'palm off' its dried soup package as that of Lipton's.

b. Delete *rare* and add later examples.

1934 *Punch* 30 May 592/3, I lost seven holes running this morning absolutely and entirely because I had been palmed off with a little swine who sniffed whenever I was about to strike my ball. 1960 B. Kops *Dream of Peter Mann* III. 66 We couldn't have our Superstore just yet and we were palmed off with promises.

Palmach (pa·lmaχ). Also ‖ **Palmakh.** [Heb. shortening of *pᵉluggōt maḥaṣ* striking force.] A commando force of the Jewish *Haganah, active esp. during the war which preceded the formal establishment of the state of Israel in 1948, and incorporated in the Israeli national army in that year.

1945 *Times* 26 Sept. 5/7 The present strength of the *Haganah* itself is variously estimated at 50,000 to 75,000 men, with first-rate equipment,..and a motorized field force (*Palmakh*) capable of throwing in a task force of several thousand men at a few hours' notice at any threatened point of the country. 1949 Koestler *Promise & Fulfilment* III. i. 298 The military cadres of the extreme Left, *Palmach*, became the best and most ferocious shock troops. 1950 H. Levin *Jerusalem Embattled* 288 *Palmach* (Heb.), abbreviation for *Plugot Machatz*. Mobile detachment of the Jewish Defence Organization. 1955 S. D. Goitein *Jews & Arabs* viii. 224 The best boys of the *Palmach*, the 'Commandos' of the *Haganah*. 1960 H. Agar *Saving Remnant* ix. 219 The Jews began the war with five thousand trained men: the *Palmach*, or shock companies of the Haganah. 1966 *New Statesman* 7 Oct. 506/2 He [*sc.* Ben Gurion] risked unpopularity in dissolving the Palmach, the crack commando troops who formed the potential core of a political and military elite. 1968 P. Durst *Badge of Infamy* iii. 27 There are a hundred of my old friends from the Palmach who would jump at the chance. 1971 *Guardian* 21 Oct. 7/5 The Palmach, the underground Jewish commando movement.

palmarosa (pæ:lmărōu̇·să). [It.]. **1.** *palmarosa oil*, an essential oil distilled from the grass *Cymbopogon martinii* var. *motia* and used in soaps, perfumes, and cosmetics.

1897 *Jrnl. Chem. Soc.* LXXII. i. 81 Palmarosa oil... This oil, formerly known as Turkish geranium oil, is prepared in the province of Bombay by distilling with water the leaves of a grass. 1902 C. Salter tr. *Koller's Cosmetics* iv. 64 Palmarosa oil dissolves in 3 parts of 70 per cent. alcohol. 1923 W. A. Poucher *Perfumes & Cosmetics* I. 105 Palmarosa oil is a useful adjunct in preparing almost any perfume of rose type. 1940 H. Trotter *Man. Indian Forest Utilization* xiii. 243 Rosha grass oil..is also known as palmarosa or geranium oil. 1945 E. Sagarin *Sci. & Art of Perfumery* iii. 33 Geraniol..is present in oil of citronella, gingergrass, palmarosa, attar of rose, and a host of others. 1950 E. Guenther *Essent. Oils* IV. 5 In India ..the bulk of commercial palmarosa oil originates. 1972 *Stand. Encycl. S. Afr.* V. 320/2 From others [*sc.* aromatic grasses]..essential oils like palmarosa oil and citronella oil are extracted.

2. In full, *palmarosa grass*. A tropical grass, *Cymbopogon martinii* var. *motia*, cultivated, esp. in India, for the sake of the essential oil it produces; also known as rosha grass.

1950 E. Guenther *Essent. Oils* IV. 9 Distillation of palmarosa grass in direct steam stills increased the yield of oil. *Ibid.* 13 All palmarosa plantings [in Java] were..in excellent condition. 1975 F. Kennett *Hist. Perfume* ix. 183 Geraniol..can be obtained with greater facility and abundance from oils of citronella, geranium, or palmarosa [*sic*] grass.

Palm Beach. The name of a coastal resort in Florida, U.S.A., used to designate a kind of light-weight fabric used for clothing. Also *attrib.*, as *Palm Beach cloth, suit.*

Palm Beach is a proprietary name in the U.S. in this sense.

1915 *Policeman's Monthly* Sept. 21/2 A lady and a daughter thinking to give the husband a pleasant surprise, bought him a fine new silk 'Palm Beach' suit. 1915 *Official Gaz.* (U.S. Patent Office) 26 Oct. 1221/1 Palm Beach... Woolen piece goods, mohair piece goods, and piece goods of combinations of cotton, wool, mohair, alpaca, camel-hair, silk, and artificial silk. 1916 *Daily Colonist* (Victoria, B.C.) 2 July 6/1 (Advt.), Palm Beach Suits in an exceptionally fine quality of mercerized Palm Beach cloth. 1922 H. L. Foster *Adventures Trop. Tramp* i, I had just applied for a job as stoker, but a Palm Beach suit, a Panama hat, and a cane did not seem to be a convincing costume on the figure of an applicant for this position. 1928 —— *If you go to S. America* 17 In..tropical lands white linen or palm beach are desirable. 1928 *Daily Mail* 25 July 12/1 The ideal is a sports shirt with a low open neck, flannel trousers, and a jacket of the material used in Palm Beach suits. 1940 *Chambers's Techn. Dict.* 610/1 *Palm Beach*, a light fabric of plain weave made from cotton warp and lustre worsted weft, or entirely of cotton. 1959 J. Thurber *Years with Ross* vi. 99 Ingersoll had appeared..dressed in a Palm Beach suit. 1973 'I. Drummond' *Jaws of Watchdog* xi. 148 He was elegant in a white Palm Beach suit and a dark pink shirt.

palm court. Also with capital initials. [f. Palm *sb.*¹ + Court *sb.*¹ 1.] A large room or patio, esp. of a hotel, named from the palm-trees used as decoration. Now usu. in *attrib.* use, esp. **palm-court music,** the kind of light music associated with the palm court (also *ellipt.*); **palm-court orchestra,** a small band which plays such music.

These associations are now regarded as old-fashioned.

1908 *Westm. Gaz.* 12 Mar. 10/2 The lounge or palm-court of to-day was merely a revival of the Greek hall. 1910 *Bradshaw's Railway Guide* Apr. 1151 Plymouth, Royal Hotel..Magnificent Palm Court. Orchestra plays daily. 1930 E. Waugh *Vile Bodies* x. 197 The manager of the 'Imperial'..upheld the integrity of British hotel-keeping. Tea, he explained, was served daily in the Palm Court, with orchestra on Thursdays and Sundays, between the hours of four and six. 1945 S. Hughes in C. Madge *Pilot Papers* 89 The 'light' music public thinks of the music it wants in terms..of palm-court orchestras. 1955 *Radio Times* 22 Apr. 13/2 David Galliver sings with the Palm Court Orchestra in Grand Hotel at 9.0 tonight. 1959 *Observer* 21 June 18/4 The sugary tones of the Palm Court orchestra are never far away. 1962 *Ibid.* 28 Oct. 23/7 A retreat into daydream-fantasy that eventually either nauseates or numbs, as palm-court music does. 1966 M. Brewer *Man against Fear* xvi. 171, I recognised a popular Palm Court number from *Traviata*. 1969 *Times* 30 Apr. (Brighton Suppl.) p. i/3 The bingo hall with a palm court bar and lounge. 1970 E. Lee *Music of People* vii. 138 Music which would now be called 'Palm Court' was still widely popular. 1973 J. Ryder *Trevayne* (1974) vi. 50 Trevayne had gone to a corner pay phone to call his wife at the Plaza..but..he was told she wasn't in the Palm Court. 1977 J. Wainwright *Do Nothin'* viii. 124 He sometimes plays pure 'Palm Court'..without that extra lilt which can make a band swing.

palmer, *sb.*¹ **1.** Also *transf.*

1906 *Bungalow* Dec. 8/2 The exodus of these infatuated palmers is ever to the land of Shakespeare.

Palmerstonian (pāmǝɹstōu̇·niăn), *a.* [f. the name of Henry John Temple, Viscount *Palmerston*, English statesman (1784–1865) + -ian.] Of, pertaining to, or characteristic of Lord Palmerston, or the forceful, assertive diplomacy associated with him. Also as *sb.*, a supporter of Lord Palmerston. So **Palmersto·nianism; Pa·lmerstonism.**

1854 *Punch* 17 June 246/2 We also wish he [*sc.* Lord Palmerston] would open a school in Downing Street wherein to furnish instruction in penmanship on the Palmerstonian system. 1858 *Illustr. News of World* 24 Apr. 187/1 Exposed to an attack from Palmerston and the Palmerstonians. *Ibid.* 5 June 283/2 Thunderbolts in the *Times* which made Mr. Disraeli, at Slough, virtuously protest against Palmerstonian corruption of the press. 1866 R. S. Charnock *Verba Nominalia* 217 *Palmerstonism*, ..old-soldierism; soft-soap. 1898 *Westm. Gaz.* 14 Dec. 8/1 The revived Palmerstonianism of Lord Rosebery. 1900 [see *Eldonian *a.*]. 1927 *Observer* 20 Nov. 10/2 Because, in the hard old Palmerstonian phrase, we are not geese. 1928 *Ibid.* 11 Mar. 6/4 In private he was never chary of urging his chief to a more Palmerstonian line of policy. 1946 J. W. Day *Harvest Adventure* xii. 195 One old man, with a Palmerstonian pippin face, in a full-skirted, snuff-coloured greatcoat with a velvet collar, talked of Culford as had done the beater half an hour before. 1954 A. J. P. Taylor *Struggle for Mastery in Europe 1848–1918* 414 Lansdowne did not need much encouragement to abandon the bankrupt Palmerstonian policy. 1966 *Economist* 12 Nov. 683/2 Palmerstonism was already hopelessly bankrupt. 1973 *Listener* 7 June 742/1 You will wonder why such a Palmerstonian view of television cuts any ice with the broadcasting authorities.

palmette. Add: **1.** (Later examples.) Also *transf.* and *attrib.*

1908 *Times Lit. Suppl.* 14 Aug. 260/2 From the tenth to the fourteenth century the palmette-motive disappears. 1931 A. Esdaile *Student's Man. Bibliogr.* vi. 212 Two.. London binders..produced about 1815 some really beautiful bindings decorated with classical palmette borders. 1931 A. U. Dilley *Oriental Rugs & Carpets* iii. 61 A fourth group of superior rugs, distinguished by pattern of palmette and now called Ispahan. 1975 *Ashmolean Mus. Rep. Visitors 1973–4* 17 A fragment of an Attic black-figure amphora decorated with a lotus and palmette pattern, languettes, and part of a battle scene.

palmetto. Add: **c.** *palmetto country, ground* (earlier example), *hat* (later examples), *juice, leaf* (further examples), *swamp, tree* (later example); *palmetto-thatched* adj.; also in sense 'thatched with palmetto leaves', as *palmetto cabin, house* (earlier example), *hut*; **palmetto brush,** a hard brush made from the roots of the palmetto; **palmetto bush,** a young or dwarfed plant of one of the species of palmetto; **palmetto cabbage** = *cabbage palmetto* (see sense b in Dict.); **palmetto flag** (examples); **Palmetto State** (examples).

1913 *Country Life* Nov. 94/3 For the making of palmetto brushes the problem is to remove the pith without destroying the fibres. 1784 T. Hutchins *Hist. Narr. Louisiana & W. Florida* 34 The whole is..covered with thick wood, Palmetto bushes, &c. 1812 *Niles' Reg.* III. 237/1 Many more must have been slain, but were hid from our view by the thick and high Palmetto bushes. 1901 *Scribner's Mag.* Apr. 433/1 Narrow grooves have been worn in the hillsides, divided one from the other by.. pyramids of earth and clay, crested with the stunted stems and roots of palmetto bushes. 1802 J. Drayton *View South-Carolina* 6 Their soil is of very sandy nature; producing..palmetto cabbage, palmetto royal, silk grass. 1870 *Amer. Naturalist* III. 458 With a palmetto cabin, plenty of oysters, game and fish, he lives a free and easy life. 1942 S. Kennedy *Palmetto Country* 24 The Palmetto country rests upon what is geologically known as the Floridian plateau. 1860 in *South Carolina Hist. Mag.* (1964) LXIV. 156 This evening the Palmetto Flag was inaugurated. 1861 *Mitchell's Maritime Reg.* 403/2 The Peter Maxwell sailed off with the Palmetto flag flying at her main. 1744 F. Moore *Voy. Georgia* 124 The Indians were prevailed upon to return to the Palmetto ground. 1877 E. S. Ward *Story of Avis* 410 She looked very young and girlish that day in her palmetto hat and white linen dress. 1889 G. W. Cable in *Century Mag.* Feb. 516/2 Before the end of the month all the women in St. Martinville were wearing palmetto hats. 1974 'B. Mather' *White Dacoit* vi. 64 His hideous palmetto hat..flopped down over his face in a ragged veil. 1741 in *South Carolina Hist. Soc. Coll.* (1887) IV. 42 They came to some Palmetto Houses, where they halted about one hour. [1731 M. Catesby *Nat. Hist. Carolina* I. 69 A man..builds a hut with Palmetto-Leaves, for the shelter of himself and family while they stay.] 1739 W. Stephens *Jrnl.* 29 Dec. in *Colonial Rec. Georgia* (1906) IV. 480, I found them well covered from bad Weather, by a strong Palmeta Hut. 1741 in *South Carolina Hist. Soc. Coll.* (1887) IV. 33 The first Palmetto Hut on the sea beach.. where the Spaniards had once a lookout. 1845 T. J. Green *Jrnl. Texian Expedition* 152 Several were left on the road exhausted for the want of water and here they commenced unfortunately, the use of the palmetto juice as a substitute. 1731 Palmetto-leaf [see *palmetto hut*]. 1763 tr. Du Mont in *Le Page du Pratz' Hist. Louisiana* I. 351 Making thus the form of a house of an oblong square..and cover..with cypress-bark, or palmetto-leaves. 1880 G. W. Cable *Grandissimes* xiv. 89 On it [*sc.* the floor] were here and there in places white mats woven of bleached palmetto leaf. 1891 *Harper's Mag.* Dec. 47/1 Perhaps the colonel would not wave the palmetto leaf too vigorously. 1976 C. Larson *Muir's Blood* xxx. 159 Ferns as wide as palmetto leaves drooped and swayed. 1837 *Globe* (Washington) 14 Jan. 3/3 After exchanging all kinds of civilities..which induced them to believe that the Judge was certain of the Palmetto State. 1843 *Knickerbocker* XXI, 222 The merry days of good old Christmas are still observed in the Palmetto State. 1948 *Sat. Even. Post* 10 July 12/3 Although Palmetto State folks may have hesitated to brag the first year, they're safe now. 1974 *State* (Columbia, S. Carolina) 15 Feb. 1-A/3 Officials of the U.S. Justice Department firmly rejected a disputed reapportionment plan Thursday for the S.C. House of Representatives charging that it could deny equal voting rights to blacks in the Palmetto State. 1853 'P. Paxton' *Stray Yankee in Texas* 56 The '*marais*' or slough,..according to my friend Joe's account, changed into a 'branch'; then after running through a cypress brake or two, ultimately assumed the form of a palmetto swamp. 1888 G. W. Cable *Bonaventure* 86 On a bank of this bayou..[stood] the palmetto-thatched fishing and hunting lodge. 1895 G. King *New Orleans* 34 There is absolutely no seeing of Bienville's group of palmetto-thatched huts by the yellow currents of the Mississippi. 1908 *Daily Chron.* 1 Sept. 7/5 As they strolled together towards the palmetto-thatched, open-face camp fronting on Ruffle Lake. 1971 'D. Halliday' *Dolly & Doctor Bird* xiv. 202 Brady led me..up to the Begum's long palmetto-thatched bar. 1865 'G. Hamilton' *Skirmishes* xiii. 172 If he is concocting..rebellion, can he not go on just as blithely under the Stars and Stripes as under the Palmetto tree?

palmful, *sb.* (Later examples.)

a 1861 T. Winthrop *John Brent* (1883) xxii. 194 They took their water by the throatful, not by the palmful. 1940 C. H. Warren *Corn Country* 3 He pulled out a palmful of unprepared flour.

palmierite (pælmĭ·ĕrəit). *Min.* [ad. F. *palmiérite* (A. Lacroix 1907, in *Compt. Rend.* CXLIV. 1400), f. the name of Luigi *Palmier-i* (1807–1896), Italian meteorologist: see -ite¹.] A sulphate of potassium, sodium, and lead, $(K,Na)_2Pb(SO_4)_2$, found as colourless hexagonal crystals.

1907 *Mineral. Mag.* XIV. 406 Palmierite..found enclosed in aphthitalite amongst the products of the Vesuvian eruption of April, 1906. 1954 *Mineral. Abstr.* XII. 332 Palmierite..was prepared as pearly scales by fusing a mixture of $PbSO_4$ and K_2SO_4.

palmiet (pa·lmit). [Afrikaans, a. Du., f. Sp. and Pg. *palmito*, dim. of *palma* palm.] A South African plant found in swamps and along river-banks, *Prionium serratum* (= *P.*

palmita), of the family Juncaceæ, which has a woody stem, topped with a cluster of long, narrow, serrated leaves two or three feet long, and small, greenish-gold flowers borne in a large panicle.

1785 G. FORSTER tr. *Sparrman's Voy. Cape Good Hope* I. iii. 42 A little river or stream covered with *palmites*, a kind of *acorus* with a thick stem and broad leaves, which grow out from the top, as they do in the palm-tree, a circumstance from which the plant takes its name. 1800 A. BARNARD *Let.* 14 May in *S. Afr. a Century Ago* (1901) xvii. 286, I am living out of town.., removed from all party work, except working parties in our fields, rooting up of palmite roots, and planting of fir trees and potatoes. 1822 W. J. BURCHELL *Trav. S. Afr.* I. iv. 89 The boors believe this brownness [of the water] to be caused by the great quantity of Palmite (Palmiet), which every where grows in these streams. *Ibid.* 91 Most of the rivers which we passed in this excursion, are choked up with the plant called *Palmiet* by the colonists, and from which this one [*sc.* river] derives its name. 1868 J. CHAPMAN *Trav. S. Afr.* I. ix. 193 The flower and root of the bulrush as well as the tsetla root or palmiet..forms the main article of the diet of the Makobas. 1871 H. H. DUGMORE *Reminisc. Albany Settler* 1. 17 The beaver gave way to the home-made palmiet or coffee straw, and the tiger-skin cap. 1944 V. POHL *Adventures Boer Family* xvi. 106 The only hats they possessed were those made by my mother and Sophia from straw, *palmiet* (a water plant) or mealie leaves. 1952 *Cape Times* 20 Sept. 3/2 The boat..was steered to a clump of palmiet. 1973 *Stand. Encycl. S. Afr.* VIII. 439/2 Palmiet... This waterside plant..has a fairly stout, erect or decumbent, woody stem covered with old leaf-bases.

palming, *vbl. sb.*¹ Add: **2*.** *palming off* (U.S. Law) = *PASSING *vbl. sb.* 2 b. Also *attrib.*

1891 *Atlantic Reporter* XXI. 613/2 The language of the court imports an intentional deceit and palming off. 1925 *Federal Reporter* (1926) VII. 604/1 In the case at bar the means are as plainly unlawful as in the usual case of palming off. It is as unlawful to lie about the quality of one's wares as about their maker. 1942 *Ibid.* CXXIV. 706/1 Under Illinois law the 'palming off doctrine' is not treated as merely the designation of a typical class of cases of unfair competition, but as a rule of law itself. 1956 *Dior* v. *Milton* in *N.Y. Suppl.* 2nd Ser. CLV. 452 With the passage of those simple and halcyon days when the chief business malpractice was 'palming off', and with the development of more complex business relationships ..many courts..have extended the doctrine of unfair competition beyond the cases of 'palming off'. 1965 A. BOGSCH in *Ibid.* 329/2 The principle of 'passing off' or 'palming off'.

palm-leaf. Add: **c.** *palm-leaf book, fan, hat* (earlier example), *manuscript, roof*; **palm-leaf pattern,** a device resembling a palm-leaf used in the decoration of oriental carpets.

1937 M. COVARRUBIAS *Island of Bali* (1972) i. 7 The second half of the manuscript [of the Tjatur Yoga] is extremely obscure, full of errors, and appears incomplete, perhaps owing to careless copying of an older palm-leaf book. 1860 J. G. HOLLAND *Miss Gilbert's Career* viii. 132 Then Mrs. Ruggles helped herself to a palm-leaf fan. 1891 *Century Mag.* Mar. 734 Chad substituted a palm-leaf fan from the hall table. 1836 O. W. HOLMES *September Gale* 156 The wind whisked off my palm-leaf hat. 1948 D. DIRINGER *Alphabet* vi. 360 Many palm-leaf manuscripts.. are also written in this [Nagari] script. 1931 A. U. DILLEY *Oriental Rugs & Carpets* Pl. 28 (*caption*) Saraband Rug. Palm Leaf Pattern. 1930 R. MACAULAY *Staying with Relations* ix. 119 They..went to find her, but found nothing except wrecked posts and a smashed palm leaf roof sprawling on the shaken earth.

palm-tree. Add: **c.** Also *Comb.*, as **palm-tree justice,** justice summarily administered, usu. with little regard for legal principle or precedent (with reference to the Islamic cadi (see CADI) administering justice under a palm-tree: see also quot. 1634 for sense a in Dict.).

[1916 A. UNDERHILL in *Shakespeare's England* I. xiii. 383 In Shakespeare's time the Court of Chancery was almost as unfettered by precedent as the typical Cadi under the Palm Tree.] 1959 *Sunday Times* 24 May 5/4 What are the origins and associations of the phrase..'palm-tree justice', which has recently been used several times by Her Majesty's Judges in legal contexts? 1963 *Times* 7 May 9/2 It would be perhaps rather hard luck for Simpson's now to have to part with their North Road site at prices which no longer represented modern prices. It would be no more than palm-tree justice. 1968 *Economist* 3 Feb. 42/3 In this period the [Roman] emperors themselves considered the letters that came up to them and dictated their answers personally: palm-tree justice was still obtainable.

palmy, *a.* Add: **2.** *palmy days* (earlier examples).

1837 DICKENS *Let.* 31 Jan. (1965) I. 232, I hope you will meet with every happiness that you picture to yourself in these palmy days. 1848 TROLLOPE *Kellys & O'Kellys* I. iv. 80 Mrs. Lynch had died before the commencement of Sim's palmy days. They had seen no company in her time.

Hence **pa·lmily** *adv.*; **pa·lminess.**

1886 G. B. SHAW *How to become Mus. Critic* (1960) 112 When old-fashioned people..regret the palmy days of the drama, superstitious ones are apt to take the desirability of palminess for granted... The young London play-goer can hardly judge; for he has no experience of palminess. *Ibid.*, A palmily stall-less pit.

Palmyrene (pæ·lmĭrīn, pælmai·rīn), *sb.* and *a.* Also **Palmyre·nian.** [ad. L. *Palmyrēn-us*, f. Gr. Παλμύρα, L. *Palmyra* Palmyra.] **A.** *sb.* **a.** A native or inhabitant of the ancient city of Palmyra in Syria. **b.** The language and script in use at Palmyra. **B.** *adj.* Of or pertaining to Palmyra, its inhabitants, its language, or its script.

1609 HOLLAND tr. *Ammianus Marcellinus' Roman Hist.* 339 Praysing and extolling of her, as much as the Parthyans do Semiramis, Ægypt Cleopatra..or the Palmyrenes Zenobia. 1609 JONSON *Masque of Queenes* sig. E2ᵛ The ninth, in time, but equall in fame, and (the cause of it) vertue, was the chast Zenobia, Queene of the Palmyrenes. 1695 E. HALLEY in *Phil. Trans. R. Soc.* XIX. 161 But the Palmyrenes being informed of the Design [of M. Antonius to plunder Palmyra], took care to prevent them, and so escaped Plunder. *Ibid.* 172, I have taken care to have the Stone purposely viewed, as also to get from thence the exact Figure of the Syrian or Palmyrene Characters thereon... By the help of these, compared with two others.., I hope we may be able, one day, to make out the Palmyrene Alphabet. 1753 R. WOOD *Ruins of Palmyra* 6 Odenathus, a Palmyrene,..made so proper a use of this situation..as to get the balance of power into his hands. *Ibid.* 25 The ancient inscriptions we found at Palmyra were all Greek, or Palmyrene, except one in Latin. 1776 GIBBON *Decl. & F.* I. xi. 309 The emperor [Aurelian], by his salutory edicts, recalled the fugitives, and granted a general pardon to all who..had been engaged in the service of the Palmyrenian queen. *Ibid.* 310 They..engaged the Palmyrenians in a laborious pursuit. 1875 [see *NABATÆAN *sb.* and *a.*]. 1886 W. P. DICKSON tr. *Mommsen's Provinces Roman Empire* II. ix. 96 Even in votive inscriptions which Palmyrenes set up to their native gods in Rome, and in tombs of Palmyrene soldiers that died in Africa or Britain, the Palmyrene rendering is added. 1900 G. BELL *Let.* 20 May (1927) I. 108 As we drew near Palmyra, the hills were covered with the strangest buildings, great stone towers, four stories high, some more ruined and some less, standing together in groups or bordering the road. They are the famous Palmyrene tower tombs. 1932 D. & T. RICE tr. *Rostovtzeff's Caravan Cities* v. 149 The external aspect of Palmyrene culture strikes the eye with its complexity and peculiarity. 1948 D. DIRINGER *Alphabet* iv. 280 Cantineau considers the Syriac alphabet as related to the cursive Palmyrene, the former having been influenced by the latter thanks to the commercial activities of the Palmyrenians. 1957 *Encycl. Brit.* XVII. 162/2 The Palmyrene princes cherished the idea of an independent empire of their own. *Ibid.* 163/2 The technical terms of municipal government are mostly Greek, transliterated into Palmyrene. 1963 W. F. ALBRIGHT *Biblical Period from Abraham to Ezra* i. 9 Nearly all..parallels come from Nabataean or Palmyrene, as well as later Syrian inscriptions. 1972 P. M. FRASER *Ptolemaic Alexandria* I. i. 12 It was their destruction in the reign of Aurelian, at the time of the Palmyrene invasion, which led to the abandonment of Brucheion. 1978 *Times Lit. Suppl.* 3 Mar. 249/2 There is an interesting section discussing the scattered but far from negligible artistic traces left by Palmyrenes resident *in partibus.*

palo blanco (pa·lo bla·nko). *U.S.* [Amer. Sp., = 'white tree'.] A small tree or shrub of the genus *Celtis*, esp. *C. reticulata*, the western hackberry, belonging to the family Ulmaceæ, native to south-western North America and distinguished by its light-coloured bark, downy leaves, and red berries (see also quot. 1947).

1838 S. MAVERICK *Let.* 30 Dec. in 'R. M. Green' *Samuel Maverick* (1953) v. 83 Griffin & Granville can cut posts or pickets, and haul them to the (palo-blanco) hackberry tree. 1901 J. C. VAN DYKE *Desert* 147 All the common growths like the sage, the mesquite, the palo fierro, and the palo blanco, are blossom bearers. 1926 D. H. LAWRENCE *Plumed Serp.* xix. 326 The car went on, the great lights glaring upon the hedges of cactus and mesquite and palo blanco trees. 1927 —— *Mornings in Mexico* 33 A valley bed, where is..the *palo-blanco*..with big white flowers like pure white, crumpled cambric. 1938 L. N. GOODDING *Notes Native & Exotic Plants* (U.S. Dept. Agric. Soil Conservation Service) 52 Palo Blanco or White Bark Hackberry..furnishes shade along many of the dry washes. 1947 J. C. RICH *Materials & Methods Sculpture* x. 293 *Palo Blanco* means 'white wood' and the name is applied to many woods of different species. 1951 KEARNEY & PEEBLES *Arizona Flora* 220 *Celtis reticulata*... Netleaf hackberry, palo-blanco, sugar-berry. 1958 E. M. REISS *Garden of Chaparral* 140 Palo-blanco; Hackberry (*Celtis laevigata*). A thirty-foot tree with whitish bark which accounts for its local name. 1960 R. VINES *Trees, Shrubs & Woody Vines of Southwest* 203/2 A vernacular name [of *Celtis lindheimeri*] is Palo Blanco.

palo de hierro (pa·lo də hi,ɛ̄·ro). *U.S.* Also **palo fier(r)o.** [Amer. Sp., = 'iron tree'.] The Sonora or desert ironwood, *Olneya tesota*, of the family Leguminosæ, which bears racemes of white flowers; also used as a name for other trees producing particularly hard wood, or the wood itself.

1894 *Amer. Anthropologist* VII. 293 During the rest of the year the Indians devote themselves to..the gathering of the fruit of the cactus, mesquite beans, and the bean of the *palo fiero.* 1912 K. LUMHOLTZ *New Trails in Mexico* 224 All the mules, donkeys, and horses gathered at once around a lone but very large palo fiero tree to eat its dark green juicy leaves. 1931 W. A. DAYTON *Important Western Browse Plants* 86 Tesota (*Olneya tesota*) is variously known as arbol (or palo) de hierro, and desert (Mexican, or Sonora) ironwood. 1949 *Desert Mag.* June 22/2 The great washes along the highway were crowded with palo verde trees and ironwood, or palo fierro. 1951 KEARNEY & PEEBLES *Arizona Flora* 442 *Olneya Tesota*... Known commonly in Arizona as ironwood, or palo-de-hierro. 1963 W. J. SCHALDACH *Path to Enchantment* vi. 74 A tree unique in several ways is the ironwood, or *palo-fierro*. True to its name, the ironwood tree possesses wood so tough that it dulls an ax.

palomino (pælomī·no). orig. *U.S.* Also **palamino.** [Amer. Sp., a. Sp. *palomino* f. L. *palumbīnus* of or resembling a dove.] A light brown or cream-coloured horse with pale mane and tail, believed to have been developed from Arab stock. Also *attrib.*

1914 *Sunset* May 995/1 A Palomino stallion with arching neck and muscle-ridged barrel led the dozen brown and mottled mares of his seraglio up a silent hillside. 1932 H. W. BENTLEY *Dict. Spanish Terms in Eng.* 176 Palomino..A term commonly used in the Southwest and California to describe a horse of a silver yellow color. 1935 J. STEINBECK *Tortilla Flat* x. 180 You will ride a palomino horse. 1936 *New Yorker* 18 Apr. 12/3 He made a fine entrance on his palomino. 1949 *Esquire* Mar. 29/1 They'll reserve a golden palomino horse for you—all yours for the length of your stay. 1955 W. FOSTER-HARRIS *Look of Old West* viii. 239 Palominos are gold-colored horses, light tan or cream, with white manes—a color variant, not a breed. 1958 *Times* 18 Sept. 13/3 Her favourites are not only Arabs, but, also, Thoroughbreds, Percherons..with, as runners-up, Hackneys, Hunters, Palominos. 1959 *Sunday Express* 19 Apr. 5/6 Her Palamino pony-breeding business in Warwickshire. 1967 [see *circus-trick*]. 1973 *Country Life* 8 Mar. 654/3 (*heading*) British Palomino Society annual show. 1976 *Billings* (Montana) *Gaz.* 2 July 11-C/6 (Advt.), 4 year old Palamino mare. Good kid & ranch horse. $300.

2. A pale golden-brown colour.

1951 H. MACINNES *Neither Five nor Three* 1. ii. 27 Her blonde palomino-rinsed head turned towards Mrs. Hershey. 1960 *Times* 31 Oct. 1/3 Mink Coat..and Palamino Stole. 1968 J. IRONSIDE *Fashion Alphabet* vii. 159 The mink breeders.., always searching for different colours...name each variety, the more generally known being:...Palomino..: Honey-gold.

palone (pălō̆u·n). *slang.* [Etym. uncertain; conceivably a phonet. var. of BLOWEN.] A derogatory term for a young woman; also, an effeminate man.

1934 P. ALLINGHAM *Cheapjack* xvi. 202 I'd rather 'andle a man any day than a lot of these silly palones. *Ibid.* 203 Charlie was not a lady's man, and by 'palones' he meant girls. 1947 *Penguin New Writing* XXIX. 99 The authentic flavour of life among the small-time spivs, of caffs and gaffs, grafter and palone. 1969 J. GARDNER *Complete State of Death* v. 66 There was another possibility. Hart could be a palone. Fanny Hart. No.

palooka (pălū·kă). *slang* (chiefly *U.S.*). Also **paluka, palooker.** [Orig. unknown.] An inferior or average prizefighter; any stupid or mediocre person; a lout. Also *attrib.*

1925 H. C. WITWER *Roughly Speaking* (1926) 287 Ben will make at palooka's pan over for you in any style you wish, Reverend Jephtha. 1927 D. HAMMETT in *Black Mask* Feb. 28/1 A paluka who leads with his right. 1932 WODEHOUSE *Hot Water* vi. 112 One of these palookas suddenly pulls out a young carving-knife and sticks me in the wish-bone with it. 1933 *Amer. Speech* Oct. 37/2 He still looks plenty good enough to take the English palooka. 1936 J. TULLEY *Bruiser* v. 48 Don't let these palookers around here laugh you outta seein' me go—all you'll ever get outta these stumble bums is the holes in the doughnuts. 1950 J. DEMPSEY *Championship Fighting* ii. 11 It was only natural that the tide of palooka experts should sweep into the amateur ranks. 1950 A. LOMAX *Mister Jelly Roll* (1952) 139 You won't kick me in the ass, because I can beat this palooka. 1977 *New Yorker* 8 Aug. 12/2 A romantic fable about a Philadelphia palooka who gains his manhood. 1978 *N.Y. Times* 27 Feb. c2/6 Leon Spinks..does not rate highly with at least one former heavyweight title holder. 'He is a palooka,' says Ingemar Johansson.

‖**palourde** (palu·ɹd). [Fr.] A marine bivalve mollusc belonging to the genus *Venerupis*; a Venus clam or carpet-shell. Cf. PULLET 2 in Dict. and Suppl.

[1863 J. G. JEFFREYS *Brit. Conchology* II. 361 According to Collard de Cherres, the Breton designation is 'palourde'.] 1942 E. PAUL *Narrow St.* xxxiv. 303 Baby sea snails to be picked out with a pin; huge *palourdes* and giant snails; the tender coquilles St. Jacques. 1951 —— *Springtime in Paris* xi. 189 Sampling eagerly the oysters, clams, palourdes, cockles, mussels, snails, razor-fish or sea-urchins. 1960 E. DAVID *French Provincial Cooking* 141 We..cannot obtain the various kinds of exquisite little clams, the *praires*, the *palourdes* and the *clovisses* which one gets in France. 1976 *Times* 2 Oct. 10/2 Palourdes on a buttery spinach puree.

palo verde (pa·lo vɛ̄·ɹde). *U.S.* [Amer. Sp., = 'green tree'.] A name used for several small trees or shrubs belonging to the genera *Cercidium* or *Parkinsonia* of the family Leguminosæ, native to south-western North America and distinguished by green bark and racemes of yellow flowers. Also *attrib.*

1854 J. R. BARTLETT *Pers. Narr. Explor. Texas* II. 188 The vegetation consisted of mezquit and palo verde. 1860 *Proc. Calif. Acad. Sci.* II. 129 In the eastern part of the

Papagoria, the country is..covered with a low growth of *mesquite* and *palo verde* brush. 1881 *Amer. Naturalist* XV. 982 The 'Palo[v]erde' of the Mexicans...grows to be some fifteen or twenty feet high. 1891 [see *MESCAL 2]. 1913 *Rep. Brit. Assoc. Adv. Sci.* 1912 533 The valleys [in the Sonora Desert] contain in abundance trees—of which the most common are mesquito (*Prosopis velutina*) and paloverde (*Parkinsonia torreyana*)—shrubs and cacti. 1947 *Southern Sierran* May 4/2 The flaming red of the Ocotillo and the bright lemon yellow of the Palo Verde were outstanding. 1955 *Sci. Amer.* Apr. 72 (*caption*) Desert wash is traced by ironwood (*Olneya*) and paloverde (*Cercidium*) plants. 1969 T. H. EVERETT *Living Trees of World* 200/2 The Jerusalem thorn or palo verde..is not an Old World native but is indigenous from Texas to Argentina. 1972 Y. LOVELOCK *Vegetable Bk.* 1. 177 Another close American relative [of the Judas tree] is the palo verde.., whose Spanish name refers to the distinctive green bark. 1974 P. A. MUNZ *Flora S. California* 437 *Cercidium* Tulasne. Palo Verde. Shrubs or small trees with green bark and ± spinose twigs. *Ibid.* 465 *Parkinsonia* L. Palo Verde. Low trees with green branches and twigs... *P. aculeata* L. Mexican Palo Verde.

palouser (pălū·zəɹ). *U.S. colloq.* [f. the *Palouse*, a region in the north-western U.S.] (See quots. 1918 and 1958.)

1903 *Outing* May 144/2 No, all were not British 'remittance men', Arizona 'palousers', and bank clerks on the trail. 1918 *Dialect Notes* V. 27 *Palouser*, n. 1. A greenhorn; a country fellow. From the fact that the Palouse is a farming country. 2. A lantern made by attaching a bale, horizontally, to an empty can and by inserting a candle through a hole in the side. 3. A gorgeous sunset. From the circumstance that the sunsets in the Palouse are very magnificent. 1958 W. F. MCCULLOCH *Woods Words* 131 *Palouser*, a lantern made by sticking a candle through a hole in a tin can.

palp, *v.* Add: (Further example.)

1967 S. BECKETT *Stories & Texts for Nothing* III. 86 Palp your skull, seat of the understanding.

2. Short for PALPITATE *v.* 1.

1903 'MARJORIBANKS' *Fluff-Hunters* 30 Georgie panted and palped, and the old man gurgled and gasped. *Ibid.* 149 'I am Phyllis Tremayn!' exclaimed the excited bit of fat, palping all over.

palpable, *a.* **1. b.** (Further example.)

1974 *Nature* 22 Mar. 344/2 At 72 and 96 h the lesions were smaller but still palpable.

palpate, *v.* Add: Also *absol.* or *intr.*

1901 W. OSLER *Princ. & Pract. Med.* (ed. 4) 25 There may be early muscle rigidity and increased tension, and spasm on any attempt to palpate. 1963 D. G. W. CLYNE *Textbk. Gynaecol. & Obstetr.* xv. 374 The examiner..then gently palpates with the pulps of the fingers so as to distinguish between the soft, wedge-shaped breech and the hard, round ballottable head.

Hence **palpa·ting** *ppl. a.*

1901 G. R. BUTLER *Diagnostics Internal Med.* xxxiii. 431 The re-enforcement of the palpating hand by the other preserves the perceptive delicacy which the former would otherwise lose. 1974 PASSMORE & ROBSON *Compan. Med. Stud.* III. II. xxxix. 15/1 If the palpating fingers on the appropriate side can reach below the occipital prominence..the head is not engaged.

palping (pæ·lpiŋ), *ppl. a. rare.* [f. PALP *v.* + -ING².] That palps or feels.

1929 R. BRIDGES *Testament of Beauty* iv. 128 It thrusteth out its finely adapted tentacles in their first palping movements to the encounter of life.

palsa (pæ·lsă). *Geomorphol.* [ad. Sw. *palse*, *pals* (pl. *palsar*), introduced as a techn. term (*palse*) by Fries & Bergström 1910, in *Geol. Fören. i Stockholm Förhandl.* XXXII. 195, from Finnish and Lappish *palsa*. The pl. *palsen*, sometimes found in Eng., reflects Ger. usage.] A mound or ridge of peat covered with vegetation and containing a core of frozen peat or mineral soil in which are numerous ice lenses, occurring in subarctic regions (usu. in bogs).

1942 *Geogr. Rev.* XXXII. 420 Peat knobs in swamps and bogs and hillocks in tundra, commonly called *Palsen*, are described from northern Europe and Siberia in an extensive foreign literature. *Ibid.* 421 The solid cores of ice show that the mounds are not due solely to upward movement of fine material, as in the case of most *Palsen*. 1954 W. D. THORNBURY *Princ. Geomorphol.* iv. 89 On surfaces covered by a good growth of tundra vegetation low, rounded mounds composed of fine materials are often found. They are called earth hummocks or palsen. 1973 A. L. WASHBURN *Periglacial Processes* 152 The ice lenses generally distinguish palsas from pingos. 1973 *Nature* 9 Nov. 64/1 In Finland palsas are only found north of the coniferous forest limit.

palship. See s.v. PAL *sb.*¹ in Dict. Add: (Further examples.) Also *attrib.*

1916 [see *NEVER *adv.* 9]. 1936 P. M. CLARK *Autobiogr. Old Drifter* xiv. 200 The wonderful pal-ship of dogs is to me an everlasting delight. 1966 *Listener* 23 June 897/2 The whole feuding quartet had been invited..to join the presidential plane and put on a show of unanimity and palship. 1972 R. D. ABRAHAMS in T. Kochman *Rappin' & Stylin' Out* 236 The protected and licensed confines of palship groupings. 1974 *Publishers Weekly* 11 Feb. 56/2 His long, intimate palship with Marlon 'Bud' Brando.

palsy, *sb.* (*a.*) Add: **1. b. cerebral palsy,** any of various non-progressive forms of paralysis caused by damage to motor areas of the brain before or during birth, manifested in early childhood by weakness and imperfect control of the affected muscles; hence **cerebral-palsied** *a.*, affected with cerebral palsy; also *absol.*

[1888 *Lancet* 14 Apr. 709/1 There are two classes of birth palsies, the 'peripheral' and the 'cerebral'.] 1889 W. OSLER *Cerebral Palsies of Children* i. 2 The cases are usually arranged under the generic terms cerebral palsies—the German *Cerebrale Kinderlähmung*—or spastic palsies, while the specific designation indicates the distribution of the paralysis, whether unilateral, bilateral, or paraplegic. 1940 *Jrnl. Amer. Med. Assoc.* 14 Dec. 2119/1 Treatment depends on..the particular kind of cerebral palsy: spastic, athetoid or ataxic. 1955 *Lancet* 15 Jan. 146/1 Cerebral palsies may occur in mentally normal and in mentally deficient subjects. 1961 *Ibid.* 19 Aug. 433/2 The Pædiatric Research Unit..is contributing to the study of..the special psychological problems of learning in cerebral-palsied children. 1973 *Times* 30 Oct. 2/7 Buildings and land housing St Margaret's School, Croydon, which provides specifically for the cerebral palsied, were leased rent-free to the school by two sisters. 1974 PASSMORE & ROBSON *Compan. Med. Stud.* III. II. xxxvi. 9/1 Many children with mild cerebral palsy require no medical treatment and, if they are of average intelligence, should be considered as normal children, though allowances may have to be made.

palsy (pæ·lzi), *sb.*² (*a.*²) *slang.* Also **palsie** and in extended forms **palsy-walsy, palsie-walsie, palsey-walsey.** [f. PAL *sb.*¹ + -SY.] **A.** *sb.* A friend, a 'pal'; a form of (ostensibly) friendly address.

1930 *Amer. Speech* Dec. 82 Call me Palsy. 1937 J. CURTIS *There ain't no Justice* xxvi. 287 What are you having, palsy-walsy? 1941 H. SMITH *Gang's All Here* 266 There was nothing to do but I must go along with them. I even went into SRO with them. Talk about palsy-walsies! 1945 P. CHEYNEY *I'll say she Does* i. 18 How come, palsie? 1945 E. WILSON *I am gazing into my 8-Ball* 118 Ratoff appealed to him. 'Look, palsy,' he said, 'whawt time I wawz in your house this morning?' 1962 *Coast to Coast 1961–62* 25 'Well, well, if it's not my old palsie-walsie Bert,' one of the detectives said. 1966 'W. HAGGARD' *Power House* x. 111 There was nothing quite so expendable as dear old palsy-walsies who had by now outgrown their usefulness.

B. *adj.* Friendly, 'pally'. So **pa·lsy-wa·lsiness.**

1947 *Philadelphia Bull.* 17 Feb. 8/3 Army planes will drop on them pictures of General MacArthur and Hirohito in palsey-walsey attitudes, to convince them that hostilities have ceased. 1957 'P. QUENTIN' *Suspicious Circ.* vii. 75 What if all that revolting *Daddy Long Legs* palsy-walsiness had been fake? 1959 'J. R. MACDONALD' *Galton Case* (1960) xviii. 149 Him and Pete were palsy-walsy. 1962 BROWN & GILMAN in J. A. Fishman *Readings Sociol. of Lang.* (1968) 268 Very 'palsy' parents may invite their children to call them by first name. 1963 WODEHOUSE *Stiff Upper Lip, Jeeves* xiii. 101 What do you call it when a couple of nations start off by being all palsy walsy and then begin calling each other ticks and bounders? 1969 *Daily Tel.* (Colour Suppl.) 11 Apr. 41/4 The New York police and I are not too palsy right now. 1974 WODEHOUSE *Aunts aren't Gentlemen* vii. 56 Being a Communist, he was probably on palsy-walsy terms with half the big shots at the Kremlin. 1977 J. WAINWRIGHT *Pool of Tears* 218 He's one of those matey types... Very palsy-walsy.

palter, *v.* **1.** (Later example.)

1872 in C. Sumner *Wks.* VI. 34 Some weak-backed quietist, who, afraid to look this thing in the face, would palter weak commonplaces.

paludal, *a.* Add: Esp., of a plant, growing in marshy ground. Also, as *sb.*, a plant requiring a marshy habitat.

1847 H. C. WATSON *Cybele Britannica* I. 65 The proposed series of terms runs thus:—..Paludal. Plants of marshy ground, the roots of which are in water or wet ground most part of the year, or constantly. 1926 *Nat. Hist. Oxford District* 88 Many of these paludals..can grow equally well on either soil. 1932 G. C. DRUCE *Comital Flora Brit. Isles* p. xxv, The Hydrophytes or Water-loving plants..include first the Paludal or Marsh and Bog plants. 1974 *Kew Bull.* XXIX. 542 The paludal species [of giant lobelia]..have little secondary wood.

Paludrine (pæ·liudrīn). *Pharm.* Also **paludrine.** [f. L. *palūs, palūd-em* marsh + *-rine*, after *ATABRINE, *MEPACRINE.] A proprietary name for proguanil hydrochloride, used as an anti-malarial drug.

1944 *Trade Marks Jrnl.* 22 Mar. 127/2 Paludrine... Pharmaceutical substances for the treatment of malaria. Imperial Chemical (Pharmaceuticals) Limited,..Slough, Buckinghamshire; manufacturers and merchants. 1945 *Times* 6 Nov. 6/4 The new drug, which is to be distributed in a form known as Paludrine marks a revolutionary departure in anti-malarial research. 1951 'N. SHUTE' *Round Bend* 288 Connie had taken Paludrine regularly.. and his malaria had not recurred. 1966 D. FORBES *Heart of Malaya* vi. 65, I..pour my tea and take a paludrine tablet with the first gulp. 1974 P. DICKINSON *Poison Oracle* iv. 111 The malaria season was not yet at its height, but he had been giving them both Paludrine. 1977 P. THEROUX *Consul's File* 45 They went off on one of their usual expeditions. No compass, no paludrine, no torch.

palus¹. Restrict *obs. rare* to sense in Dict. and add: **2.** With capital initial and pronunc. (palūs): a wine produced in the Palus region of Bordeaux in France. Also *attrib.*

1833 C. REDDING *Hist. Mod. Wines* v. 144 Bassens and Mondferrand grow the second class of Palus wines. 1861 MRS. BEETON *Bk. Househ. Managem.* 888 The genuine wines of Bordeaux are of great variety..and the principal vineyards are those of Medoc, Palus, Graves, and Blanche. 1953 E. HYAMS *Vineyards in England* 90 Such *palus* wines are good enough of their kind. 1968 *New Statesman* 29 Nov. 744/2 How many little Palus and/or Bourg wines reach the market as Médoc, or even as St Julien?

palygorskite (pæligǭ·ɹskəit). *Min.* Also † **paligorskite.** [ad. G. *paligorskit* (T. v. Ssaftschenkow 1862, in *Verh. d. k. Ges. für d. Ges. Mineral. zu St. Petersburg* 102), f. *Palygorsk*, name of a locality by the Popovka river in the Ukrainian S.S.R.: see -ITE¹.] A silicate of magnesium and aluminium that occurs as soft, light-coloured, fibrous layers and has a structure based on silica tetrahedra arranged in double chains.

1868 J. D. DANA *Syst. Min.* (ed. 5) 406 (*heading*) Paligorskite. 1916 *Chem. Abstr.* X. 581 (*heading*) Studies in the magnesian silicates. The group of zillerite, zermattite and palygorskite. 1921 *Mineral. Abstr.* I. 237 Paligorskite ('mountain-cork') from Billowitz near Brünn [Moravia]. 1968 I. KOSTOV *Mineral.* II. v. 352 Sepiolite can be dense or spongy..whereas palygorskite resembles leather or parchment... Both minerals are secondary products found in altered serpentinitic rocks in association with opal, dolomite, and other minerals. 1973 *Clay Minerals* X. 28 Under a binocular microscope, the palygorskite shows fine interwoven fibres similar to those of coarse-textured paper.

palynology (pælinǫ·lŏdʒi). [f. Gr. παλύνειν to sprinkle (cf. πάλη fine meal = L. *pollen*) + -OLOGY.] The study of the structure and dispersal of pollen grains and other spores, as indicators of plant geography, taxonomic characteristics of plants, fossils used in dating geological formations or archæological remains, or causative agents of allergic reactions. So **palynolo·gical** *a.*, of or pertaining to this study; **palynolo·gically** *adv.*; **palyno·logist,** a student of palynology.

1944 HYDE & WILLIAMS *Let.* 15 July in *Pollen Analysis Circular* 28 Oct. 6 We would therefore suggest palynology ..: the study of pollen and other spores and their dispersal, and applications thereof. We venture to hope that the sequence of consonants p-l-n, (suggesting pollen, but with a difference) and the general euphony of the new word may commend it to our fellow workers in this field. 1944 H. A. HYDE in *Museums Jrnl.* XLIV. 146/1 In view of the admitted inadequacy of the expression pollen analysis it has recently been proposed to substitute for it the new word palynology (Gk. παλύνω (paluno), I scatter; πάλη (pale), meal): the study of pollens and other plant spores and their dispersal, and applications thereof. 1946 *Svensk Bot. Tidskr.* XL. 303 (*title*) Palynological aspects of the pioneer phase in the immigration of the Swedish flora. 1953 *Proc. 7th Pacific Sci. Congr.* V. 172 It is a matter of no small concern to palynologists and plant geographers that our [*sc.* New Zealand's] peat deposits are being greatly modified by draining and burning. 1956 H. GODWIN *Hist. Brit. Flora* iii. 48/1 They [*sc.* pollen-grains] confer much increased sensitiveness and power upon the palynological technique. 1958 *Antiquity* XXXII. 54 Palynology, or the science of pollen analysis, has developed during the past three decades into a major source of knowledge of the past. 1959 *Micropaleont.* V. 27/2 Several samples were investigated palynologically in order to confirm, if possible, the supposed Pleistocene age. 1962 *Courier-Mail* (Brisbane) 10 Sept. 15/9 (Advt.), A major oil company..is setting up a regional stratigraphic laboratory in Perth, Western Australia. A palynologist is required to organise and run the palynological section of this laboratory. 1963 G. ERDTMAN in A. & D. Löve *N. Atlantic Biota* 367 (*title*) Palynology and Pleistocene ecology. 1972 *Courier-Mail* (Brisbane) 15 Apr. 9/7 (Advt.), B.O.C. of Australia Limited..requires a Senior Palynology Technician to supervise its laboratory staff. 1973 *Microscopy* XXXII. 319 This same resistance to attack permits the palynologist to employ selective oxidation techniques to concentrate microfossils from peats and coals. 1973 *Nature* 16 Mar. 187/1 These [*sc.* Pleistocene sediments] were analysed palynologically, and indicated four or five alterations of pluvial and interpluvial conditions. *Ibid.* 8 June 342/1 The date of the appearance of marine grasses is not traceable by palynology because they produce pollen without exine, and so are not fossilized. 1975 *Times* 27 May 14/6 The lake beds [of Hoxne] contain fossil pollen and constitute the palynological type site of the Hoxnian Interglacial.

pamaquin (pæ·măkwin). *Pharm.* Also **-ine.** [f. P(ENTYL + *A(MINO- + *M(ETHOXY- + *-a-* + QUIN(OLINE.] An orange-yellow crystalline salt, $C_{19}H_{29}N_3O.C_{23}H_{16}O_6$, which is a toxic compound formerly used in the treatment of malaria. Cf. *PLASMOCHIN, *PLASMOQUINE.

1941 *Brit. Pharmacopœia 1932* Add. IV. 24 Pamaquin is the 6-methoxy-8-[ω-diethylamino-α-methylbutyl]-aminoquinoline salt of 2:2'-dihydroxy-1:1'-dinaphthylmethane-3:3'-dicarboxylic acid. 1951 A. GROLLMAN *Pharmacol. & Therapeutics* xxxiii. 483 Because of its toxicity and the availability of safer and better drugs, pamaquine is no longer of therapeutic importance. 1961 *New Scientist* 4 May 261/1 The anti-malarial drugs pamaquin and primaquine produce anaemia in certain susceptible persons. 1974 B. G. MAEGRAITH in A. W. Woodruff *Med. in Tropics* ii. 65/1 The 8-aminoquinoline most widely given [for vivax malaria] is primaquine; pamaquine or quinocide are alternatives.

pa·mby, shortening of NAMBY-PAMBY *a.*

1823 BYRON *Let.* 25 Jan. in *Wks.* (1901) VI. 164, I will bet you a flask of Falernum that the most stilted parts of the political *Age of Bronze*, and the most pamby portions of the Toobonai Islands, will be the most agreeable to the enlightened public. 1947 I. BROWN *Say Word* 89 Defeating the accusation that..Ambrose [Philips] composed nothing but 'pamby' stuff.

pampa. 2. pampas deer, substitute '*Blastoceros bezoarticus*' for Latin name in def.; (earlier and later examples); **pampas flicker,** a black, white, and yellow woodpecker, *Colaptes campestris*, found in the eastern part of South America; **pampas fox,** one of several small mammals resembling a fox or a dog, esp. *Dusicyon gymnocercus*, or Azara's fox, found in eastern and southern parts of South America; **pampas woodpecker** = *pampas flicker* above.

1860 MAYNE REID *Odd People* 446 A man on foot can approach much nearer to any game, than if he were mounted upon a horse. This is true..also of the large pampas deer. 1972 G. K. WHITEHEAD *Deer of World* iv. 63 The Pampas deer is the most elegant of all the South American deer. 1912 BRABOURNE & CHUBB *Birds S. Amer.* I. 168 *Colaptes... campestris*... Pampas Flicker. Carpentero. 1926 *Bull. U.S. Nat. Mus.* No. 133. 223 In habit and general appearance the pampas flicker differs little from the familiar *Colaptes aureus* of the eastern United States. 1957 M. H. MITCHELL *Obs. Birds S.E. Brazil* 120 Pampas Flickers, on first sight or hearing, immediately recall to the northerner *Colaptes aureus.* 1923 *Proc. Biol. Soc. Washington* XXXVI. 55 Specimens from the high savanna of Bogotá..indicate that the pampa fox of this area is a depauperate pallid race of the lowland *Cerdocyon thous.* 1956 G. DURRELL *Drunken Forest* v. 105 He was a small, delicately made, grey pampas fox, with slender legs and enormous brush and eager brown eyes. 1972 *Vogue* Jan. 12/2 South American fur rugs..viscacha, guanaco, grey and pampas foxes. 1975 H. J. STAINS in M. W. Fox *Wild Canids* i. 13 The pampas fox is found in Paraguay and southeastern Brazil south through the pampas region of Argentina. 1870 *Proc. Zool. Soc.* 705 (*title*) Notes on the habits of the pampas woodpecker.

pampano, var. POMPANO (in Dict. and Suppl.).

1931 [see *CARANGID *a.* and *sb.*]. 1971 P. CRAMPTON tr. *Heyerdahl's Ra Expeditions* ix. 198 A fat little pampano fish..waggled its tail. *Ibid.* xi. 297 We had pampano among the pilot fish underneath us.

pampas-grass. Add: The Latin name of the plant is now *Cortaderia selloana*. (Further examples.)

1851 LINDLEY & PAXTON *Paxton's Flower Garden* I. 175 This noble plant, now called the Pampas Grass, in consequence of its inhabiting the vast plains of S. America so named, has been introduced within a few years through Mr. Moore, of the Glasnevin Botanic Garden. 1900 L. H. BAILEY *Cycl. Amer. Hort.* II. 703/1 The popular name 'Pampas Grass' is now unchangeable, but the plant does not grow on the pampas or vast grassy plains of South America, but in the mountains. 1934 C. LAMBERT *Music Ho!* II. 76 A Picasso reproduction is not considered 'amusing' unless flanked by pampas grass. 1950 G. BRENAN *Face of Spain* iii. 56 Red branched tamarisks and tufts of pampas grass. 1973 F. A. BODDY *Foliage Plants* vii. 126 The pampas grass..makes such a splendid specimen plant in a lawn, with or without its great feathery plumes in the autumn.

pamphleteer, *v.* Add: **a.** (Earlier and later examples.) Also, to engage in propaganda involving the issue or distribution of pamphlets. Also *fig.*

1698 E. WARD *Trip to Jamaica* 3 The condition of an author is much like that of a strumpet... If the reason be requir'd Why we betake our selves to so scandalous a profession as whoring or pamphleteering, the same excusive answer will serve us both. 1897 *Granta* 24 Apr. 264/2 The result..will be a compromise, which will permanently injure that future of women's education, which has been endangered enough already by the impatient folly of those who try to do a century's work in a week of garrulity and pamphleteering. 1930 T. S. ELIOT tr. *St.-J. Perse's Anabasis* 39 Let me go alone with the airs of the night, among the pamphleteering Princes. 1938 E. WILSON *Triple Thinkers* 284 Long-range literature attempts to sum up wide areas and long periods of human experience... Short-range literature preaches and pamphleteers with the view to an immediate affect. 1943 *Sun* (Baltimore) 22 June 1/4 OWI officials acknowledged that they have 'pamphleteered' several times in the past. 1976 *Time* 27 Dec. 5/1 Since when is pamphleteering per se bad manners or bad art in theatre or film, with such historic examples as Ibsen, Shaw, Eisenstein and Odets?

b. *trans.* To influence or persuade by means of pamphlets. *rare.*

1944 G. B. SHAW *Everybody's Pol. What's What?* iii. 29 The peasants and peasant soldiers, neither of them Communists, but all more or less talked and pamphleteered and journalized into believing that the Bolsheviks were the boys to give them land and peace.

pampsychism (and derivs.): see *PANPSYCHISM.

pan, *sb.*¹ Add: **1. c.** (Further and *fig.* examples of a lavatory.)

1919 R. FRY *Let.* May (1972) II. 451 A real Victorian W.C. with a pull up plug... But..there's no sham Chinese landscape in the pan. 1961 PARTRIDGE *Dict. Slang.* Suppl. 1212/2 *Down the pan*... A Cockney equivalent of *down the drain*, ruined with no chances left. 1972 J. WAINWRIGHT *Requiem for Loser* iv. 75 A race from one shithouse to the next. A lifetime of sitting on the pan. 1974 *Listener* 14 Mar. 347/3 'It's just money down the pan,' said one pensioner.

e. *on the pan* (U.S.), under reprimand or adverse criticism (said of a person).

1923 H. C. WITWER *Fighting Blood* v. 140 Even when the newspapers puts him on the pan..the safe-playing, money-grabbing middleweight king just laughs at us. 1937 C. BOOTHE *Women* I. i. 9 *Edith.* I'll bet you had me on the pan. *Sylvia.* I never say behind my friends' backs what I won't say to their faces. I said you ought to diet. 1939 A. AYLESWORTH in *Better English* May 13/1 Five college professors sitting around a table... A sixth professor who wasn't there because he'd snagged a job at a better institution, was on the pan. MacSnuft leaned across at the rest of us and contributed: 'He's an ignoramus!' 1941 J. SMILEY *Hash House Lingo* 41 *On the pan*, being reprimanded.

f. A metal drum in a West Indian steel band. Also, steel-band music and the way of life associated with it.

1955 *New Commonwealth* 28 Nov. (Suppl.) p. xix/1 To make a 'pan' the end of a metal oil drum is cut off and the bottom of the circular pan so formed is shaped into sections by beating and chiselling. 1958 J. P. HICKERTON *Caribbean Kallaloo* i. 15 The steel bands have reached a high pitch of virtuosity... 'Beating the pan' is a West Indian passion, and we once discovered an old man sitting outside our back gate beating out a rhythm on our upturned rubbish bin. 1960 *Times* 17 Sept. 7/7 In Trinidad a steelband is known as a 'pan-side' and the word 'pan' has two connotations. The first refers to the instrument, the second to a way of musical life. Pan, the instrument, is..a tuned gong, made from the top of a 44-gallon steel barrel. *Ibid.*, The intermarriage of musical cultures—the offspring of which is pan. Pan is the core of national culture and the first expression of a truly West-Indian art-form. 1973 *Trinidad Guardian* 1 Feb. 8/5 The question of having tuners specialise in particular pans.

5. c. = *skid-pan.

1966 T. WISDOM *High-Performance Driving* ii. 30 The first skid pan was introduced at Chiswick [in 1922]. On this 'pan' solid-tyred K and D buses were put through their paces. *Ibid.* 31 'Pans'—circular areas covered with a mixture of grease or oil and water—were by now out of date.

6. c. A face. (Perh. influenced by *to shut one's pan* s.v. sense 4 b. Cf. also *DEAD-PAN *a.*, *sb.*, *adv.*, and *v.*) *slang* (orig. *U.S.*).

1923 [see *HEEBIE-JEEBIE(S)]. 1931 E. LINKLATER *Juan in Amer.* III. viii. 262, I never want to see that pan of yours again! 1943 HUNT & PRINGLE *Service Slang* 50 *Pan*, slang for face. 1944 E. B. WHITE *Let.* 8 Oct. (1976) 260 He starts toward her bearing a bouquet of American Beauty roses, and falls on his pan before he gets there. 1972 *Jazz & Blues* Nov. 11/4 This must have been funny enough when it happened; relayed through the medium of Rich's sourly contemptuous pickled-walnut pan. 1977 *Rolling Stone* 13 Jan. 41/2 As Belushi filled out the registration card, the manager remarked with a deadly pan: 'Write down the name of the person you're staying with.'

10*. Severe or dismissive criticism. orig. *U.S. colloq.*

1936 *Esquire* Sept. 160/3 A pan on a show is a damp blanket. 1958 *Spectator* 24 Oct. 543/1 'Is it a pan?' asked the reporter from *Time* hopefully... 'No, bigod, it's a smashin' crit.' 1960 P. TOMPKINS *To a Young Actress* 60 The notice in *Punch* appeared next to a long pan of *Back to Methuselah.* 1972 *N.Y. Times* 3 Nov. 24/5 This Hunanese restaurant... Appraisals..included the whole possible spectrum of opinion from rave to pan. 1977 *Zigzag* Apr. 34/1 Afterwards they wrote a pan and then had a huge article criticising the pan.

11. a. *pan-load*; (sense *1 f) *pan music.*

1939 J. STEINBECK *Grapes of Wrath* xxii. 414 Ma had taken up a panload of brown pone. 1963 *Which?* 6 Feb. 48/2 Each [dish-washing machine] had to deal..with three separate assortments of dishes, which we call *standard, capacity* and *pan* loads respectively. 1960 *Times* 17 Sept. 7/7 Pan music casts a spell of enchantment on the Trinidadian... The essential feature of pan music is that the melody is carried by one instrument at a time while the others play more or less 'free' variations on the theme. 1977 *Times* 14 May 12/4 'Pan' music, as the drums are called, originated in Trinidad.

b. pan-broiling *vbl. sb.* (hence, as a back-formation, *pan-broil* vb.) (see quot. 1970); **pan-man** (earlier example); (*b*) one who plays the pan (sense *1 f) in a steelband; **pan-scourer, -scrubber,** a scourer, often in the form of a wire pad, for cleaning pans; **pan-side** *West Indian* (see quot.).

1950 L. H. GROSS *Meats, Poultry & Game* 14 Rib steaks —Broil, or pan-broil. 1970 SIMON & HOWE *Dict. Gastron.* 287/2 *Pan-broil.* An American cooking term meaning to cook meat in a hot pan with almost no fat, pouring off fat as it accumulates. 1896 F. M. FARMER *Boston Cooking-School Cook Bk.* ii. 22 When coal is not used, or a fire is not in condition for broiling, a plan for *pan broiling* has been adopted. This is done by placing food to be cooked in a hissing hot frying-pan, turning often as in broiling. 1832 A. DAVIS in P. D. Curtin *Two Jamaicas* (1955) 234 One stocker-man, one pan-man, three boiler-men... These were all slaves. 1959 W. A. SIMMONDS '*Pan*'—*Story of Steelband* 15 In order to understand fully the story of the Pan, one must understand the Panmen, who drew their music from every source that is Trinidadian. 1960 *Times* 17 Sept. 7/7 Some people say that the 'pan-men' take themselves too seriously. 1974 *Trinidad Guardian* 2 Nov. 9/2 Pan men would be trained to become teaching assistants. 1959 'A. GILBERT' *Death takes Wife* ix. 112 'Packet of pan-scourers,' she said. 1960 *Guardian* 19 Sept. 6/4 The very first electric pan scourer. 1971 C. BONINGTON *Annapurna South Face* App. B. 252 Pan-scourers..24. 1926–7 *Army & Navy Stores Catal.* 118/3 *Pan scrubber.* A Metal Sponge for cleaning pots, pans, etc. 1960 *John o' London's* 14 Apr. 440/2 The sound of a wire pan-scrubber chasing grease round a frying pan. 1960 Pan-side [see sense 1 f above].

pan, pán, *sb.*⁶ Add: **2.** *attrib.* and *Comb.*, as *pan-box, -chewing, -garden, -juice*; **panwala, -wallah** [see WALLAH], a person who sells pans.

1922 E. M. FORSTER *Abinger Harvest* (1936) 314 Bidar.. has produced beautiful pan-boxes of a lead alloy inlaid with silver. 1939 R. GODDEN *Black Narcissus* v. 53 A peon..who carried the General's pan-box. 1892 *Chambers's Jrnl.* 14 May 320 After a long course of pawn-chewing, the utterance becomes thick and indistinct, and the teeth black. 1923 *Blackw. Mag.* Dec. 769/1 In one of these pan-gardens, as they are called, a boar had taken up his quarters. 1901 KIPLING *Kim* ii. 47 He spat red *pan*-juice on the floor. 1955 R. P. JHABVALA *To whom she Will* xxvi. 186 He stopped at a pānwala's and..asked for a pān with cardamons and aniseed. *Ibid.* 294 The stalls of pānwalas are found everywhere, and they sell not only pāns but also cigarettes, matches, mineral drinks, biscuits, hard-boiled eggs and anything else suitable for a quick cheap snack. 1969 *Hindusthan Stand.* (Calcutta) 5 Aug. 2/4 As one waits..for the panwallah to stuff a betel leaf with hot or sweet masallah..and roll it into a shapely pan. 1975 O. SELA *Bengali Inheritance* xix. 166 That bloody pan-wallah..lied to you.

‖**pan** (ban), *sb.*⁸ [Chinese *bǎn* slab.] A Chinese percussion instrument (see quots.).

1872 *Catal. Special Exhib. Anc. Mus. Instr. S. Kensington Museum* VIII. 39 Pan. A piece of wood, with a groove cut nearly through its substance, and bamboo sticks for percussion. Used principally by Chinese beggars. 1954 *Grove's Dict. Mus.* (ed. 5) II. 234/1 *P'ai-pan* (or *pan*), percussion clapper. A popular instrument consisting of two slabs of the red wood *huai*, attached by a silk cord, on which a third slab is struck to beat time. 1975 C. P. MACKERRAS *Chinese Theatre in Mod. Times* viii. 131 (*caption*) The *pan* (clapper) consists of three pieces of wood, two of them fastened together (patterned surface visible), the other behind.

pan, *sb.*⁷ *Cinemat.* [Abbrev. PANORAMA or PANORAMIC *a.*] **1.** The action of panning a camera (see *PAN *v.*³); a panoramic sequence.

1922 *Opportunities Motion Pict. Industry* 111 *Pan*.., moving the camera up and down or from side to side to follow the action from one place to another. 1931 J. H. REYNER *Cine-Photogr. for Amateurs* viii. 92 *Rapid pan.* An exception to this slow movement is the case of the rapid panoram which is sometimes necessary in order to keep a moving object within the field of vision. 1937 H. B. ABBOTT *Compl. 9·5-mm. Cinematographer* vi. 88 A satisfactory 'pan' can sometimes be effected by slowly turning the camera on its tripod screw. 1942 *Amer. Cinematographer* May 235/2 The layout-man..figures out all of the camera moves such as..pans (following a character along). 1960 C. MORRIS *Unloved* in D. Wilson *Television Playwright* 447 Camera panning, moves round a stone seat... As the car circles, the pan reveals a boy on the stone seat apparently sketching the front façade. 1962 *Listener* 5 Apr. 596/2 The opening shot of *Exodus* is a huge 200-degree pan across the landscape and coastline of Cyprus. 1970 *Daily Tel.* (Colour Suppl.) 3 July 23 A tripod's a bit of a bind to lug around but it does give a rock steadiness and a touch of professionalism, especially on pans and telephoto shots. 1977 *Spare Rib* June 42/2 A story unfolded in a series of 360° camera pans.

2. *attrib.* and *Comb.*, as *pan shot, -tilt*; **pan-and-tilt** *a.*, used with reference to a tripod or other unit that allows the camera to move in both horizontal and vertical planes; **pan head** (see quot.).

[1937 H. B. ABBOTT *Compl. 9·5-mm. Cinematographer* v. 69 The metal top has both pan and tilt movements with locking device for each.] 1938 G. H. SEWELL *Amateur Film-Making* iii. 34 The ideal tripod..should also incorporate a 'pan-and-tilt' head. 1962 M. BARDWELL *Amateur Cinematogr.* ii. 32 A tripod must not only give firm support—it must also allow controlled horizontal and vertical movement. This is achieved by means of a 'pan-and-tilt' head, which also enables you to lock the camera in position at almost any angle. 1977 *Offshore Engineer* May 97/2 It occupies only a fraction of the space normally dedicated to a camera mounted on a pan and tilt unit. 1940 *Amer. Speech* XV. 359/2 *Pan head*, the mechanism at the top of a tripod which permits the camera to be moved in both horizontal and vertical planes. 1941 STEINBECK & RICKETTS *Sea of Cortez* xxiii. 223 We made jerky little pan shots back and forth. 1956 L. MALLORY *Ciné Camera Secrets* 20 Pan shots must be slow and even. 1975 *New Yorker* 26 May 113/3 An 'atmospheric' tape background such as accompanies pan shots over bleak, windswept headlands. 1970 *Ibid.* 3 Oct. 108/3 Two.. spots..could be directed by means of pan-tilt.

‖**p'an** (pan), *sb. Archæol.* [Chinese *pán*.] A vessel or wash basin.

1958 W. WILLETTS *Chinese Art* I. iii. 154 It is not altogether certain that *p'an* date back as far as Shang-Yin times. 1973 *Genius of China* 57/2 Dish on a high foot *p'an*, of burnished black pottery. 1977 KWANG-CHIH CHANG *Archaeol. Anc. China* (ed. 3) vii. 368 Water utensils: p'an, i, chien, yii, p'en, cheng, chu, wan.

pan, *v.*¹ Add: **1. a.** (Earlier example.)

1839 *Amer. Railroad Jrnl.* VIII. 99 Old machines are invariably burnt up, and the ashes 'panned out' for the fine gold that has lodged in the joints of the wood.

b. (Earlier example.)

1850 N. KINGSLEY *Diary* 27 May (1914) 123 About 200 Indians & squaws came down and began to pan all around us.

4. a. (Earlier examples.)

1849 J. D. DANA *Mineral.* (ed. 2) 317 Gravel or soil..is said to *pan well* or *pan poorly* according to the result. 1866 'MARK TWAIN' *Lett. from Hawaii* (1967) 85 In the mining camps of California..it is etiquette to say, 'Here's hoping your dirt'll pan out gay.'

b. (Earlier and later examples.) Also, to work out, to have a result (not necessarily something favourable).

1868 F. WHYMPER *Trav. Alaska* 282 'It panned out well' means that 'it gave good returns'. 1870 'MARK TWAIN' *Lett. to Publishers* (1967) 31 January and November didn't pan out as well as December. 1884 *Brandon* (Manitoba) *Blade* 24 Jan. 4/3 If the domineering Attorney-General 'pans out' well during the coming session he will probably be the man. 1923 WODEHOUSE *Adv. Sally* xiv. 177 He was hoping all along that this fight would pan out big and that he'd be able to pay you back what you had loaned him. 1925 A. HUXLEY *Let.* 16 Sept. (1969) 253 However, I shall see how things pan out when I get there. 1947 'N. SHUTE' *Chequer Board* vii. 191, I think it may pan out all right. 1956 S. BELLOW *Seize Day* (1957) i. 23 If I don't pan out as an actor I can still go back to school. 1967 *Electronics* 6 Mar. 316/2 Machines that automatically feed and place flatpacks on printed circuit boards haven't panned out, largely because girls with tweezers excel at gently handling the fragile IC packages. 1972 *Daily Tel.* 14 Dec. 5 But Dr Brett cautioned that what sounds exciting from the Moon does not always pan out in the laboratory. 1977 P. DICKINSON *Walking Dead* I. v. 69 They decided to give it a year and see how it all panned out.

c. To speak freely or at length; to expatiate.

1871 [see sense 4 b in Dict.]. 1915 W. J. LOCKE *Jaffery* xxi. 291 I'm panning out about this, because it seems so deuced interesting. 1917 —— *Red Planet* xv. 182, I had.. made up my mind to pan out to you like this. 1928 *Observer* 18 Mar. 9/3 Mr. Lewis..resists even the temptation to 'pan out' about that obviously born temptress.

7. *trans.* To criticize severely; to judge (a performance) to be unsuccessful or inadequate. orig. *U.S. colloq.*

1911 G. ADE in *Chicago Daily News* 16 Dec. 28/2 They would open up on Rufus and Pan him to a Whisper. 1914 *Sat. Even. Post* 7 Mar. 7/3 Kelly got nasty and begun to pan me for quitting and for the way I played. 1926 S. LEWIS *Mantrap* xii. 150 I've never done one single thing to give her any excuse for panning me. 1927 *Vanity Fair* XXIX. 134/3 Will Shakespeare..was panned by the critics because he delved into the argot of his day to put it over. 1935 *Hot News* Apr. 14/3 Don't mistake me. I am not panning 'Hawk' as a player. 1938 G. HEYER *Blunt Instrument* iii. 52 'I have no dealings with actresses.' 'Well, then, stop panning them.' 1939 'N. BLAKE' *Smiler with Knife* v. 78 The lurid headline, 'Famous Woman Explorer Pans Domesticity.' 1947 R. CHANDLER *Let.* 8 Mar. in *R. Chandler Speaking* (1966) 137 MacCarthy panned me, said the toughness was largely bluff. 1960 *Daily Mail* 27 Apr. 8/8 The idea that critics like panning shows is a myth. 1971 *Sunday Times* (Johannesburg) *News Mag.* 28 Mar. 6/1 Dirk de Villiers' latest South African film drama..is being panned by the critics. 1974 *Times* 2 Oct. 14/5 The play was roundly panned by many of our correspondents. 1977 *Time* 26 Dec. 36/2 Colleagues are quick to pan Simon in return: 'The Count Dracula of critics!'

pan (pæn), *v.*³ *Cinemat.* [Abbrev. PANORAMA or PANORAMIC *a.*] **1.** *trans.* **a.** To follow or pass along (a person or object) with a camera.

1913 *Sat. Even. Post* 1 Nov. 64/3 We'll 'pan' you right down the middle of the picture to the raft. 1960 N. KNEALE *Mrs. Wickens in Fall* in D. Wilson *Television Playwright* 167 The Camera pans him away. He calls to the two Englishwomen.

b. To turn (a cine or television camera) in a horizontal plane, esp. in order to keep a moving object in view.

1930 *Electronics* Nov. 373/2 With the advent of sound, the operation of 'panning' the camera to afford a changing point of view became a more complicated process. 1956 *Railway Mag.* Nov. 779/1 Taking up a stance broadside to his target, he 'panned' his camera—that is, swung the camera round with the train—so that the engine remained in the same portion of the viewfinder throughout. 1973 P. L. CAVE *Speed Freaks* v. 43 Gerry panned the camera slowly over the remnants of the once-beautiful TR6. 1977 *Film & Television Technician* Feb. 4/2 The camera is supported by a counterweighted arm, and the arm is attached to a harness worn by the operator. The camera floats in space and the operator can pan, tilt, or crane it with one hand.

2. *intr.* Of a camera: to swing from one scene to another, or along objects or a place; to give a panoramic view in closing up to an object or a place.

1931 R. DYKES *Amateur Cinematographer's Handbk.* iii. 31 The tilting handle..is used to panoram down into valleys... It is also used to 'pan' up cathedral spires. 1932 A. BUCHANAN *Films* vi. 174 The camera 'pans' around the room, bringing to view the shabby furniture. 1936 *Words* Oct. 6/1 One of those inserts in which the camera seems to swing, or 'pan' dizzily from action going on in one place to what is going on in another. 1960 N. KNEALE *Quatermass & Pit* I. 11 The camera pans, to take in all that remains of a little working-class street. 1963 *Listener* 14 Feb. 300/3 The camera pans over photographs of tribesmen. 1971 *Daily Tel.* 21 Aug. 7/2 A camera panned dramatically in on a manorial notice. 1975 *New Yorker* 19 May 81/1 Then the camera moves to a worker with a cart, and pans with him to the end of the assembly line.

pan (pæn), *a.* Abbrev. of *PANCHROMATIC *a.* Also *ellipt.* for *panchromatic film.*

1940 *Amer. Speech* XV. 357 'Going to use pan or N.C.?' 'Neither. Ortho.' 1940 'C. I. JACOBSON' *Developing* 106 On no account must the dark green safelight provided for pan materials be used. 1954 C. WALLACE *Enjoy your Photogr.* iv. 44 Pan films photograph reds well. 1969 J. ELLIOT *Duel* I. iv. 82, I brought ten thousand feet of pan but only three of high speed for interiors.

pan-. Add: **1. pan-Celtic** *a.* (earlier and later examples); hence (as a back-formation) **pan-Celt,** one who believes in the unity of all the Celtic peoples; **pan-Orthodoxy** (earlier example); **pan-Tura·nian** *a.*, embracing all the speakers of Ural-Altaic languages; hence **pan-Tura·nianism, -Tura·nism,** the principle of a union of all speakers of these languages.

1904 *Westm. Gaz.* 3 Sept. 3/2 The Pan-Celts also considered..the question of clothes, and it appears that Ireland is in need of a satisfactory and distinctive national costume. 1955 R. GRAVES *Crowning Privilege* 155, I was introduced to Welsh poetry nearly fifty years ago, when my father became an enthusiastic pan-Celt; and, this noun being new to Merioneth where we lived, he had a famous argument with Mr Postoffice-Griffiths as to whether it would count as one word in a telegram. 1891 YEATS *Let.* Nov. (1954) 181 He [*sc.* Ernest Rhys] has a kind of Pan-Celtic enthusiasm. 1973 *Stornoway Gaz.* 24 Feb. 9/2 (*caption*) Discussing the programme for the Pan-Celtic Week, to be held in Killarney, Ireland, during May 12–20. 1900 'ODYSSEUS' *Turkey in Europe* vi. 286 But Panorthodoxy, if I may use the word, tends to regard Russia as the head, not only of the Slav races, but of all orthodox nations. 1926 *Glasgow Herald* 3 Apr. 5/1 The Pan-Turanian movement..began not more than 40 years ago. 1950 E. H. CARR *Bolshevik Rev.* I. xi. 338 The pan-Turanian aspirations of the 'young Bokhara' movement. 1926 *Glasgow Herald* 3 Apr. 5/1 The alphabet of the West will..give a new aspect to Pan-Turanianism. 1932 *Times Lit. Suppl.* 21 July 525/2 A policy which..has been based in turn upon Panislamism, Ottomanism, Panturanianism. 1974 *Encycl. Brit. Macropædia* XIII. 789/2 Pan-Turanianism developed from a now much-disputed 19th-century theory of the common origin of Turkish, Mongol, Tungus, Finnish, Hungarian, and other languages; in certain very limited circles it looked forward to a great political federation of speakers of these languages. 1977 *Guardian Weekly* 25 Dec. 8/1 Panturanism (the dream of uniting the Turks in Asia with those of Turkey).

2. pananthropism (further examples); **pa:ncytope·nia** (also *erron.* -pœnia) *Path.* [*-PENIA], a condition in which the blood shows a relative deficiency of all three cellular components (erythrocytes, leucocytes, and platelets); **pan-diato·nicism** (see quot. 1937); hence **pan-diato·nic** *a.*; **pa:nencephali·tis** *Path.* [prob. coined in Ger.: cf. G. *panenzephalitisch* adj., *panenzephalomyelitis* (H. Pette 1938, in *Münch. med. Wochenschr.* 29 July 1138/2), *panencephalitis* (Pette & Döring 1939, in *Deutsch. Zeitschr. f. Nervenheilkunde* CXLIX. 32)], a rare form of encephalitis in which both the grey matter and the white matter are affected and there is a gradual but progressive loss of mental and motor functions; **panentheism** (further examples); hence **pane·ntheist** *a.* and *sb.*, **panenthei·stic** *a.*; **panhy:popitu·itarism** *Path.* (see quot. 1941); **pano·ral** *a. Dentistry*, of or pertaining to radiography of the whole mouth in one exposure; **panspo·roblast** *Zool.*, a structure formed by protozoans of the subclass Neosporidia, comprising several sporoblasts and two other cells; **pansysto·lic** *a. Med.*, (of a heart murmur) continuing throughout a systole; **panthelism** (earlier and later examples); **pantropic** (-trǭu·pic, -trǫ·pic) *a. Med.* [*-TROPIC], attacking or affecting many kinds of tissue indiscriminately; **pan-tropical** *a.*, of plants or animals, found in all regions of the tropics; also, including all tropical areas.

1879 J. J. G. WILKINSON *Let.* 20 May in R. B. Perry *Tht. & Char. W. James* (1935) I. 27 Nothing is left to my apprehension but pananthropism, a composite form of pantheism. 1936 *Theology* XXXIII. 265 The poem is an impassioned plea for the truth not of Pantheism but of Pananthropism. 1944 W. DAMESHEK *Leukopenia & Agranulocytosis* ii. 44 The designation of pancytopenia may be applied to conditions in the blood in which there is a well-defined reduction in red cells, white cells and platelets. 1956 M. W. WINTROBE *Clin. Hematol.* (ed. 4) xi. 560 Pancytopenia is not a disease entity but rather a triad which is found under a number of different circumstances. 1974 *Nature* 17 May 263 Three patients, two males and a female, were studied: two suffered from aplastic anaemia and one from a pre-leukaemic state with pancytopoenia. 1937 N. SLONIMSKY *Mus. since 1900* 1938) p. xxii, Pan-diatonicism sanctions the simultaneous use of any or all seven tones of the diatonic scale, with the bass determining the harmony. *Ibid.*, Gebrauchsmusik, proletarian music, and most forms of absolute music make use of pan-diatonic technique. 1963 *Times* 13 May 8/2 The immense profusion of counterpoint was under complete control, and the pandiatonic climaxes were shattering. 1970 *Composer & Conductor* Aug. 6/1, I myself [*sc.* N. Slonimsky] ventured into musical neologism with Pandiatonicism to describe a 20th-century technique in which all seven tones of the diatonic scale are used freely in dissonant combinations. 1950 *Brain* LXXIII. 150 Perhaps the term 'Pan-encephalitis' already adopted by Pette (1942) for forms which attack both grey and white matter could be usefully employed here, i.e. 'Sub-acute sclerosing pan-encephalitis'. 1974 *Sci. Amer.* Feb. 35/1 In 1969..measles virus was isolated from brain cells of patients suffering from the brain inflammation called subacute sclerosing panencephalitis (SSPE). 1940 *Theology* XL. 270 He [*sc.* Eckehart] oscillates between non-duality and modified duality, between pantheism and panentheism. 1959 I. EPSTEIN *Judaism* 244 Cordovero.. safeguards the theistic position by defining his attitude in the formula: 'God is all reality, but not all reality is God'—an attitude which came to be known in modern philosophy as 'panentheism'. 1970 P. BERTOCCI *Person God Is* xi. 207, I shall..argue that a temporalistic form of personalistic theism (not pantheism, not panentheism) can reasonably illuminate what we actually do find in human experience and the world. 1959 W. N. PITTENGER *Word Incarnate* vii. 155 They [*sc.* 'Hartshorne and the other theistic 'process' philosophers']..offer us a *panentheist* view of the world which presents God and his creation, supremely God and man, in continued and intimate relationship. 1974 *Church Times* 22 Nov. 15/1 We regret that, in Canon David Edwards's review of Martin Thornton's *My God*, Fr. Thornton was described as a 'pantheist'. This was a misprint for 'panentheist'. 1918 M. D. PETRE *Modernism* viii. 177, I..began slowly to form an optimistic and panentheistic belief. 1959 W. N. PITTENGER *Word Incarnate* vii. 200 A panentheistic conception, in which God is seen as above and beyond the world yet ceaselessly active in it and intimately related to it. 1970 P. BERTOCCI *Person God Is* xi. 221 But persons have too much autonomy, and God has too much autonomy, in 'essence' and 'content', to fit into a part-whole model, either in the 'organic' sense or the panentheistic. 1941 F. ALBRIGHT et al. in *Trans. Assoc. Amer. Physicians* LVI. 48 By 'panhypopituitarism' is meant a condition in which the anterior pituitary gland as a whole has impaired function. 1954 K. E. PASCHKIS et al. *Clin. Endocrinol.* xiii. 293 Tendency to hypoglycemia..is regularly present in panhypopituitarism. 1972 *Obstetrics & Gynecol.* XXXIX. 397/1 In a woman hypophysectomized at the time a craniopharyngioma was performed and who showed full panhypopituitarism, several regimens of human menopausal gonadotropin..failed to induce ovulation. 1977 *Lancet* 9 Apr. 779/2 The underlying pituitary tumour was not diagnosed until she presented with panhypopituitarism at the age of 76 (serum-prolactin 950 μg/l). 1959 *Dental Practitioner* X. 270 (*heading*) Pan-oral radiology. The most recent advance in dental radiography. 1967 L. M. ENNIS et al. *Dental Roentgenol.* (ed. 6) x. 287 (*caption*) Exposure mechanism and shieldings used ..for the panoral technic. 1893 R. R. GURLEY in *Bull. U.S. Fish Comm.* 1891 408 *Pansporoblast*, the transparent plasma-sphere formed by the condensation of a portion of the plasma around one of the numerous nuclei of the endoplasm of the myxosporidium; in distinction from the sporoblasts which result from the segmentation of the pansporoblast. 1932 BORRADAILE & POTTS *Invertebrata* ii. 93 In the syncytium,..there arise..bodies known as pansporoblasts, each composed of a couple of envelope cells with one or more cells known as sporoblasts. 1973 K. G. GRELL *Protozool.* 464 In most species [of *Myxosporidia*] the sporoblast forms several spores. It is then referred to as a pansporoblast. 1954 *Brit. Heart Jrnl.* XVI. 257 A pan-systolic apical murmur was always associated with some degree of regurgitation at operation. 1966 *Lancet* 24 Dec. 1389/2 There was a harsh pansystolic murmur..radiating into the axilla. 1896 W. CALDWELL *Schopenhauer's Syst.* i. 37 Though Schopenhauer's system has a strong materialistic colouring it is not materialism. It is rather animism or panpsychism (panthelism, in point of fact). 1917 *Encycl. Relig. & Ethics* IX. 612/1 The affirmation that the real is irrational (a blind will, panthelism [πᾶν + ἐθέλω]) in Schopenhauer's pessimism. 1937 *Jrnl. Path. & Bacteriol.* XLIV. 410 The neurotropic virus protects to a certain extent against the pantropic strain. 1967 *Res. Vet. Sci.* VIII. 414 Antibodies will persist..for many years in the sera of cattle and sheep infected with pantropic virus. 1937 *Discovery* Sept. 263/1 Its [*sc.* the fern, *Ceratopteris thalictroides*'] distribution is pan-tropical. 1946 *Nature* 20 July 86/1 The once-popular 'tiger nut' (swellings on the rhizomes of a pan-tropical water sedge). 1953 *New Biol.* XV. 35 The Scale Insect *Aspidotus destructor* is a pan-tropical species which is found wherever the coconut palm grows. 1967 *Palaeogeogr., Palaeoclimatol., Palaeoecol.* III. 203 There are no pan-tropical mammals, as there are pan-tropical plants. 1976 *Nature* 19 Feb. 528/2 A demonstrably Australian flora with many common pan-tropical plants missing.

panache. Add: **2.** *fig.* Display, swagger, verve.

1898 THOMAS & GUILLEMARD tr. *Rostand's Cyrano de Bergerac* v. vi. 294 *Cyrano*... One thing is left, that, void of stain or smutch, I bear away despite you... *Roxane*... 'Tis?... *Cyrano*... My *panache.* 1900 J. T. GREIN *Dramatic Crit.* (1902) III. 65 No one displayed that 'panache' which is the paramount demand of romantic comedy. 1903 G. B. SHAW *Man & Superman* p. xxxi, Shakespear..never conceived how any man who was not a fool could, like Bunyan's hero..with the panache of a millionaire, bequeath 'my sword to him that shall succeed me in my pilgrimage, and my courage and skill to him that can get it'. 1932 R. FRY *Characteristics French Art* III. 53 In real life the fun of soldiering, its bustle, its swagger, its panache, sometimes leads to being mutilated. 1949 F. MACLEAN *Eastern Approaches* III. vi. 370 His must have been, I think, an engaging character, a mixture of southern *panache*, rustic guile, and a childlike desire to please. **1960** D. Walker *Where High Winds Blow* II. viii. 117 Mac wore his flying clothes, the half-laced boots and the old suède jacket; but with his clean blue shirt and a silk handkerchief knotted at his neck, he had a workaday panache. **1972** D. Francis *Smokescreen* v. 58 A certain *panache* about him, but also some of the ruthless cynicism of experienced journalists. **1976** *New Yorker* 22 Mar. 128/3 When he did join the Maquis, late in March, 1944, Malraux exhibited his customary panache. **1978** *Listener* 12 Jan. 49/1 He plays the piano with panache, but cannot read music.

Panadol (pæ·nădǫl). *Pharm.* Also **panadol.** A proprietary name for paracetamol.

1955 *Trade Marks Jrnl.* 14 Dec. 1231/2 Panadol... All goods included in class 5 [*sc.* pharmaceutical, veterinary, and sanitary substances, etc.] for sale in the United Kingdom. Bayer Products Limited,..Kingston-on-Thames, Surrey; merchants and manufacturers. **1959** Wilson & Schild *Appl. Pharmacol.* (ed. 9) xvi. 326 The analgesic activity of N-acetyl p-aminophenol (paracetamol, panadol)..has been shown to be about as great as that of the parent compounds. **1967** M. Culpan *In Deadly Vein* viii. 176 A low table—with..a couple of novels, and a bottle of Panadol tablets. **1971** D. Lambert in C. Bonington *Annapurna South Face* 293 The majority who attended had to be given some form of placebo, and panadol or aspirin were found best for this purpose. **1975** *Sunday Times* 16 Nov. 44/3, I was going crazy trying to find things: the Panadol for my husband's head.

Pan-African, *a.* Add: (Later examples.) Also, of, pertaining to, or comprising all the peoples of Africa generally.

1944 *Ann. Reg. 1943* 132 Sir Godfrey Huggins, Prime Minister of Southern Rhodesia,..foreshadowed a possible Pan-African Council to coordinate problems common to African countries. **1955** [see next]. **1960** *Times* 29 Sept. (Nigeria Suppl.) p. xii/2 Dr. Nkrumah's pan-African way of thinking. **1962** *Listener* 25 Jan. 157/1 The Ghana Government has also tried to promote pan-African schemes of unity. **1967** *Freedomways* VII. 174 It is only by planning along Pan-African lines ourselves can Africa hope to free herself. **1973** *Caribbean Contact* Feb. 16/2 Garvey's views in the 1920's already foreshadow the later Pan African movement. **1975** C. E. Griffith *Afr. Dream* viii. 105 The pan-African advocate was disturbed by contemporary works which assigned Africans last place among the three major races of the world.

Pan-Africanism (pæn,æ·frikăniz'm). [f. Pan-African *a.* + -ism.] A movement which advocates the political union of all the indigenous inhabitants of Africa; the ideals of this movement. Hence **Pan-A·fricanist** *sb.*, an advocate or supporter of Pan-Africanism, also as *adj.*, of or pertaining to Pan-Africanism.

1955 B. Timothy *Kwame Nkrumah* iii. 38 In October, 1945, the fifth International Conference of the Pan-African Congress was held in Manchester... The proceedings of the Conference were conducted under the joint chairmanship of..Dr. P. Milliard, and Dr. W. E. B. Du Bois, who gave birth to Pan-Africanism. **1959** *Cape Times* 7 Apr. 1/7 African leaders from all parts of the Union decided to establish the Pan Africanist Congress. **1960** *Times* 22 Mar. 12/1 The Pan-Africanists' campaign against the pass laws exploded today on the banks of the Vaal river. **1963** *Listener* 17 Jan. 110/1 The fourth political ideal, Pan-Africanism, or continental federation. **1973** S. Henderson *Understanding New Black Poetry* 17 The changing world in which Black Americans of the post-World War II generation found themselves, a world in which articulate men and women rediscovered Africa and Pan-Africanism. **1973** *Black World* Mar. 53/2 An indiscriminate listing of pan-Africanist and Africanist resources. **1975** *Times Lit. Suppl.* 17 Oct. 1238/5 Like many pan-Africanists from the New World, there was often an element of utopianism in Delany's vision of Africa. **1976** *Survey* Summer–Autumn 289 Among black people outside there was often a strong link between Marxism and pan-Africanism. *Ibid.*, Black Americans and West Indians who were pan-Africanists were disproportionately left of centre in their political ideologies.

Panag(h)ia (pænaiyī·ă), now the usual spelling of Panhagia. Also, an image or representation of the Virgin Mary.

1910 *New Schaff-Herzog Encycl. Relig. Knowl.* VIII. 327/1 Panagia ('All Holy'), the usual (though not official) title of the virgin in the Greek Church. **1911** [see *Hodegetria]. **1931** *Times Lit. Suppl.* 19 Mar. 231/2 Devotees who implore the Panagia of Kykko for rain or the Panagia of Tenos for health. **1958** L. Durrell *Balthazar* vi. 135 'I ask you to sleep with him as I would ask the Panaghia to come down and bless him while he sleeps—like in the old ikons.' How..Greek! **1961** D. Attwater *Christian Churches of East* I. 223 *Panagia*.., 'all-holy', used for the Mother of God as we say 'our Lady'. Also another name for the *enkolpion*.

panagraphic, etc., varr. *panographic, etc.

Panama. Add: *Panama disease*, a vascular wilt disease of banana trees, caused by the soil-borne fungus *Fusarium oxysporum* f. sp. *cubense*, and characterized by the yellowing and wilting of the leaves, first described from infected trees in Central America in 1910; *Panama fever* (earlier and later examples); *Panama hat* (further examples); **Panama** *sb.* (earlier and later examples); *Panama hat palm, plant*, the screw-pine, *Carludovica palmata*, which produces leaves used in the manufacture of Panama hats; = *jipijapa a; *Panama red*, a local variety of marijuana grown in Panama.

1910 E. Essed in *Ann. Bot.* XXIV. 488 The Panama Disease.—Preliminary Notice.—This fungoid disease on the *Musa sapientum* var. *Gros Michel* was, it seems, first detected in Central America. **1913** W. Fawcett *Banana* xiii. 87 The true Panama disease also exists in Trinidad. **1934** A. Huxley *Beyond Mexique Bay* 16 That insidious Panama Disease..has ruined so many [banana] plantations throughout the Caribbean. **1949** *Caribbean Q.* I. III. 43 Bananas resistant to Panama disease..are being grown commercially. **1956** H. G. de Lisser *Cup & Lip* x. 119, I instructed him to go to Napleton to see Sampson about the treatment of Panama Disease. **1969** *New Scientist* 16 Jan. 142/2 Panama disease of bananas is not controlled by eliminating the pathogen but by selecting resistant strains of banana. **1972** J. W. Purseglove *Tropical Crops: Monocotyledons* II. 368 Panama disease, also known as banana wilt and vascular wilt,..is one of the world's most catastrophic plant diseases. **1850** J. L. Tyson *Diary of Physician in Calif.* 29 The so-called *Panama fever* rarely occurs, unless previous disease has wasted the powers. **1868** *Overland Monthly* Dec. 561/1 After hearing all about how she felt, his diagnosis was a mild case of fever—Panama fever. **1940** F. Riesenberg *Golden Gate* 109 Complaints charged that the frequent burials at sea resulted from improper care of those who had contracted 'Panama fever' or 'yellow fever'. **1856** C. M. Yonge *Daisy Chain* II. xi. 455 Dr. Spencer was in the hall, with his bamboo, his great Panama hat, and grey loose coat. **1916** 'Taffrail' *Pincher Martin* iii. 34 Vernon Hatherley, the lieutenant-commander (T.), clad in an ancient Panamá hat and a suit of indescribable overalls. **1974** *Country Life* 4 Apr. 816/1 Simple panama hat with gros-grain ribbon. **1931** P. C. Standley in *Publ. Field Mus. Nat. Hist. Bot. Ser.* X. 117 *Carludovica palmata*... Panama hat palm... Common in wet forest; ranging to Guatemala and southward to Peru. **1941** T. H. Goodspeed *Plant Hunters in Andes* v. 146 Along such forest margins small species of bamboo, 'Panama hat' palms, tree ferns, the ginger, and other attractive plants disported themselves. **1954** R. W. Schery *Plants for Man* vii. 176/1 The Panama hat palm..grows wild in most of the American tropics. **1972** J. W. Purseglove *Tropical Crops: Monocotyledons* I. 94 Panama Hat Plant. ..occurs wild in the humid forests of Central America. **1967** *Boston Sunday Herald* 26 Mar. IV. 1/1 Traffic in marijuana—Acapulco Gold and the better quality Panama Red and Yakatanga Purple—out of Mexico has steadily increased in the last three years. **1972** *Last Whole Earth Catalog* (Portola Inst.) 62/3 Acapulco Gold, Panama Red and other strains of grass are reputed to be particularly potent.

1848 *Colburn's United Service Mag.* III. 67 One veteran in a panama and rosette deputed by the body, addressed me in Spanish. **1873** J. Miller *Life amongst Modocs* 44 He could not push his panama any further back. **1975** G. Avery *Childhood's Pattern* ix. 216 School uniform was no badge of servitude... Nobody sat viciously on their Panamas.

Panaman (pæ·nāmăn), *sb.* and *a.* [f. Panama + -an.] = *Panamanian *a.* and *sb.* Also **Pana·mic** *a.*

1901 W. H. Dall in *Proc. U.S. Nat. Museum* XXIII. 285 The northern limit of the Panamic fauna is Point Conception, California. **1904** *Sun* (N.Y.) 25 Feb. 2/6 The constitution settled the question of what the people of that republic are to be called by specifying that they are 'Panamans'. **1906** W. F. Johnson *Four Centuries of Panama Canal* (1907) xx. 360 The Panaman sense of justice is as highly cultivated, and the Panaman sensitiveness to and resentment of injustice are as keen as our own. **1913** E. Peixotto *Pacific Shores from Panama* 26 Verandas..overhang all the thoroughfares, and the indolent Panamans spend much of their time upon them or lounging about the..cafés and hostelries. **1937** *Times Lit. Suppl.* 22 May 397/3 The friction between Yanqui indifference to diplomatic etiquette and Panaman pride are all candidly described here.

Panamanian (pænămēi·niăn), *a.* and *sb.* Also † **Panamenian.** [Irreg., f. Panama + -n- + -ian.] **A.** *adj.* Of or pertaining to Panama. **B.** *sb.* A native or inhabitant of Panama.

The form with medial -*e*- (quots. 1869, 1892) is an adaptation of the Spanish term *Panameño*.

1855 R. Tomes *Panama in 1855* vii. 216, I had no means of judging of the intimate character of the Panamanian dames. **1869** Pim & Seemann *Dottings in Panama* xi. 184 The Panamenians displayed great heroism, but.. the buccaneers could not be repulsed. **1889** W. Nelson *Five Yrs. Panama* 50 The native Panamanians being great stay-at-homes. **1892** J. Bornn in G. S. Minot *Hist. Panama* xv. 74 The Buccaneer..desired..precious metals and stones... But their search for these disclosed to them the fact that the Panamenian had provided against this emergency by placing these aboard a ship, with orders to sail away if the city should fall. **1906** M. A. Chatfield *Let.* 21 Jan. in *Light on Dark Places at Panama* (1908) 45 The best [hotel], the Central, charged $4.00 gold per day, $8.00 Panamanian. **1913** *Chambers's Jrnl.* July 503/2 Travelling without any Spanish and without binoculars puts one wholly at the mercy of the secretive Panamánian or the wily Indian. **1934** [see *Honduran, Hondurean *a.* and *sb.*]. **1959** *Listener* 23 Apr. 718/1 A former Panamanian ambassador in London. **1964** *Daily Tel.* 11 Jan. 16/6 Panamanian claims to sovereignty over the Canal Zone. **1976** *Times* 5 Feb. 20/4 The Lloyd's report shows that 14 of last year's casualties were registered under the Panamanian flag. **1976** *Sci. Amer.* Sept. 140/2 Nicolás Ardito Barletti, a Panamanian, attempted to place a value on the social benefit from research.

Pan-American, *a.* (Further examples.)

1927 *New Republic* 21 Sept. 110/1 The existence of the Pan-American Union, and the calling of an occasional Pan-American Congress, should not deceive anyone as to the predominant position of the United States in this hemisphere. **1934** A. Huxley *Beyond Mexique Bay* 200 Pan-American Airways..are responsible for the long-distance international services. **1966** *Times* 28 Feb. (Canada Suppl.) p. xiv/5 Canada's 1967 Pan-American Games.

Pan-Americanism. Add: (Further examples.) Also, a movement towards better commercial and cultural relations among American nations.

1915 W. Wilson *Public Papers* (1926) III. 409 This is Pan-Americanism. It has none of the spirit of empire in it. **1954** H. C. Allen *Gt. Brit. & United States* xiv. 526 His Pan-Americanism, which aimed at the economic and political consolidation of the Western hemisphere,..led him to leap into action on the Isthmian issue. **1966** *Oxf. Compan. Amer. Hist.* 611/2 Pan-Americanism, a new contribution to U.S. policy during the 1880's,..was formulated by Secretary of State Blaine.

Pan-Arabism (pæn,æ·răbiz'm). [f. Pan- + Arab *sb.* and *a.* + -ism.] The ideal of political union of all the Arab states; a movement advocating such a union. Hence **Pan-A·rab** *a.* and *sb.*, **Pan-A·rabic** *a.*, **Pan-A·rabist.**

1930 *Encycl. Social Sci.* III. 148/2 Pan-Arabism is scarcely more possible when Moslems speaking the Arabic language are ruled in such diverse ways as in French North Africa, Egypt, Syria, Iraq and the divisions of Arabia. **1939** *Asia* Aug. 450 (*heading*) Pan-Arab nationalism. **1958** *Spectator* 7 Feb. 159/1 In a sense pan-Arabism proved Farouk's downfall. *Ibid.* 1 Aug. 155/3 By joining the UAS the ruler [of Kuwait] would preserve his sheikdom and his subjects would have their pan-Arabic aspirations satisfied. **1959** *Times* 10 Mar. 11/2 This uncertainty at the top is bound to encourage others, who know more clearly what they want, to take over, whether they are pan-Arabists or Communists, idealists or self-seekers. **1962** *Listener* 1 Mar. 365/1 To young Arab Nationalists—to young Pan-Arabists everywhere—Egypt under President Nasser seemed destined to unite the Arab world. *Ibid.* 5 Apr. 597/1 The Ba'ath leaders are doctrinaire pan-Arabs of the frontier-smashing variety. **1963** M. Khadduri *Mod. Libya* xi. 330 At the outset, those who advocated Pan-Arabism were limited to the articulate intelligentsia who had received their education in neighboring Arab countries. **1968** *Listener* 15 Aug. 195/2 The commandos themselves are pan-Arab in a new sense... The fedayin have no state. **1974** *Florida FL Reporter* XIII. 52/1 They admitted that each had also learnt (besides Classical Arabic) a pan-Arabic Standard dialect. **1975** N. Luard *Robespierre Serial* iv. 16 Twice Saudi-Arabian delegate to Pan-Arab conferences. **1978** *Times* 10 Aug. 12/7 Messianic Pan-Arabism was rapidly declining, while the Arab national state triumphed.

panarchy. (Later example.)

1948 L. MacNeice *Holes in Sky* 49 He is separate too, who had but now ascended Into the panarchy of created things Wearing his halo cocked.

panary, *a.* (Further examples.)

1875 *Encycl. Brit.* III. 254 The so-called panary fermentation in bread-making is a true alcoholic fermention [*sic*]. **1942** *Proc. Food Group* V. 70/1 During the latter half of the last century the microbiological aspect of panary fermentation attracted considerable attention. **1971** A. R. Daniel *Bakers' Dict.* (ed. 2) 139/2 *Panary fermentation*, the fermentation of bread dough.

panatela (pænăte·lă). Also **panatella, panetela.** [Amer. Sp., a long, thin biscuit, etc.] **a.** A long, slender cigar tapering at the sealed end. Also *attrib.* **b.** A cigarette made of Central or South American marijuana.

1901 'H. McHugh' *Down Line* 32 A young chap..who had been out in the smoking room working faithfully on one of those pajama panatella cigars. **1904** W. Stevens *Let.* May (1967) 74 My idea of life is a fine evening..+ a soft, full Panatela. **1906** L. J. Vance *Terence O'Rourke* II. i. 190 Gravely he inspected the end of the commendable panetela, which he was enjoying by the grace of Chambret; and he puffed upon it furiously, twinkling upon his friend through a pillar of smoke. **1912** G. Frankau *One of Us* v. 41 Apart, unmoved, behind his Panatela, Old Hiram stood, of journalists surrounded. **1928** Ade *Let.* 27 May (1973) 130 For many years after I took up the writing game I smoked whatever was readily obtainable, with a preference for a mild Havana Cigar of the Panatella shape. **1943** R. Chandler *Lady in Lake* (1944) ii. 9 He reached himself a panatela..and lit it. **1946** Mezzrow & Wolfe *Really Blues* xii. 229 Some Spanish guys..rolled it [*sc.* marijuana] in a different sized paper, about half an inch longer than mine and much thinner, and they called their product a 'panatella'. **1956** B. Holiday *Lady sings Blues* (1973) iv. 43 'Girl,' he said, 'come *here*. Jimmy's got the best panatella you ever smoked in your life.' **1969** R. R. Lingeman *Drugs from A to Z* 194 *Panatella*.., bigger, fatter, more potent marijuana cigarette made of Central or South American marijuana. **1970** E. McGirr *Death pays Wages* vii. 156 The Sergeant produced a small box of midget panatellas and a box of matches.

Panatrope (pæ·nătrōᵘp). Also **panotrope** and with small initial. [f. *pana-*, *pano-*, of unknown origin + Gr. τροπή turn, turning.] The proprietary name of a form of (electric) record-player capable of relatively loud reproduction.

1926 *Glasgow Herald* 5 Oct. 5 There was no graduation of musical vibrations that the 'Panatrope' could not reproduce. **1928** E. Waugh *Decline & Fall* II. iii. 168 In a minute the panatrope was playing, David and Martin were dancing, and Peter was making cocktails. **1933** *Punch* 16 Aug. 181/1 Whatever you may lack in the way of plush seats and panotropes you wouldn't see that at an ordinary cinema. *c* **1940** Dylan Thomas & Davenport *Death of King's Canary* (1976) vii. 125 A panatrope sounded over the crack of rifles, the smashing of crockery, the complaining of beasts. **1954** *Trade Marks Jrnl.* 15 Sept. 921/1 Panatrope... Gramophones, radio gramophones, apparatus and instruments for recording and reproducing sound, parts and fittings..for all the aforesaid goods; and gramophone needles and gramophone records. The Decca Record Company Limited,..London,..manufacturers. **1961** *Times* 28 Mar. 12/7 They must now man the ticket office, sell programmes, start the recalcitrant generator, warm up the panotrope. **1968** D. Braithwaite *Fairground Archit.* 165 *Panatrope*, successor to the mechanical organ—gramophone turntables, amplifier and loudspeakers relaying noisy pop records. **1978** C. Humphreys *Both Sides Circle* x. 114, I had a fight with the representative of the firm who had hired us the panatrope, or long-playing record machine.

Panavision (pæ·năviʒən). [f. Pan(orama + *a* + Vision *sb.*] A proprietary name for a type of anamorphic lens; *loosely*, wide-screen cinematography. Also *fig.*

1955 *Jrnl. Soc. Motion Pict. & Television Engin.* LXIV. 233/1 Anamorphic printer lenses used..are the Tushinsky and Panavision. **1963** *Punch* 3 July 30/1 Panavision and colour make the whole thing incongruously cheerful to look at. **1967** *Trade Marks Jrnl.* 24 May 669/2 Panavision... Cinematographic and photographic apparatus..; anamorphic lenses... Panavision Incorporated.., City of Los Angeles. **1973** W. Dancy in S. Henderson *Understanding New Black Poetry* 300 Frail we cringe before Dante's Italic vision Its cineramic focus and panavision scale Swells brain-mind.

pancake, *sb.* Add **1.** Further examples of phr. *as flat as a pancake* (and varr.). Also used with reference to the *fig.* senses of Flat *a.*

1761 Sterne *Tr. Shandy* III. xxvii. 138 He has crush'd his nose..as flat as a pancake to his face. **1830** Marryat *King's Own* I. xvii. 261 Under which it had lain, jammed as flat as a pancake. **1909** *Dialect Notes* III. 411 *Flatter than a pancake*, very flat, of persons and things. **1921** Galsworthy *To Let* I. ix. 79 Fleur was not yet home... Here were her aunt, and her cousins the Cardigans, and this fellow Profond, and everything flat as a pancake for the want of her. **1922** Joyce *Ulysses* 735 The last [stout] they sent from O'Rourkes was as flat as a pancake. **1936** 'G. Orwell' *Keep Aspidistra Flying* i. 15 He was nearly thirty and had accomplished nothing; only his miserable book of poems that had fallen flatter than any pancake. **1959** *Daily Tel.* 14 Mar. 6 His statement to the House of Commons yesterday fell as flat as a pancake.

2. f. A type of flat hat. *U.S.*

1875 E. S. Nadal *Impressions London Social Life* 143 The cap was peculiar, though about the year '56 we had something like it called the 'Pancake'. **1945** *Amer. Speech* XX. 233/1 She had on her duty dress and a French pancake. *Ibid.* 234/1 *French pancake*, flat hat. **1957** M. B. Picken *Fashion Dict.* 241/2 *Pancake beret*, broad flat beret. **1975** G. Howell *In Vogue* 188 (*caption*) Pancake and huge gloves in looped emerald green crochet. **1976** M. & G. Gordon *Ordeal* (1977) xiii. 92 She wore..a pancake Stetson that she could tilt over her face.

g. A vertical descent made by an aircraft in a level position (see quot. 1918¹); the landing of an aircraft in an emergency with the undercarriage retracted (see **pancake landing*).

1912 *Aero* Mar. 66/1 Pride cometh before a pancake. **1913** C. Mellor *Airman* 25 Landings must be 'normal'—not of the 'pancake' order. **1914** Hamel & Turner *Flying* 66 He must be able to learn how to make a fairly safe 'pancake'. **1918** H. Barber *Aeroplane Speaks* (ed. 6) 14 *Pancakes*, pilot's slang for stalling an aeroplane and dropping like a pancake. **1918** Cowley & Levy *Aeronautics* x. 225 Dangerous consequences due to a landing of a pancake type are usually guarded against by a strong under-carriage and by the insertion of shock absorbers. **1974** P. Wright *Lang. Brit. Industry* 5 In the R.A.F. during the last war crash landings were *pancakes*.

h. An opaque facial treatment used as a base for make-up. *Freq. attrib.*, as *pancake make-up*. orig. *U.S.*

1937 *Official Gaz.* (U.S. Patent Office) 13 July 251/1 Max Factor & Co., Los Angeles, Calif... Pan-cake. The word 'cake' is disclaimed apart from the mark. For cosmetic in the nature of a solidified cream used for a make-up base. **1940** *Sears Catal.* Spring/Summer 99 (*caption*) Pan-Cake Makeup. **1946** *Trade Marks Jrnl.* 15 May 244/2 Pancake... Cosmetic preparations for toilet use and for use in theatrical, motion picture, television, and photographic make-up. Max Factor & Co..., Hollywood, United States of America; manufacturers. **1951** H. MacInnes *Neither Five nor Three* I. v. 66 Miss Guttman's face flushed with pleasure even under the pan-cake make-up. **1953** *New Yorker* 13 June 61/1 Like his Cabinet members, he used pancake makeup. **1955** W. Gaddis *Recognitions* III. ii. 737 It's too bad they didn't get some pancake on him before he went up. **1960** L. Cooper *Accomplices* II. ii. 84 A private life that you can put on over your real one..like your pancake make-up. **1962** E. O'Brien *Lonely Girl* ii. 22, I put pancake on Baba's back to hide her spots. **1970** *Sunday Times* 3 May 28/6 Women take hours getting themselves done up to attract men, slapping on pancake, painting their eyes. **1975** J. Crosby *Affair of Strangers* iii. 25 Chantal wore only light pancake, dimming but not obliterating the brown skin. **1975** *Daily Colonist* (Victoria, B.C.) 20 June 4/6 The candidate had ugly mannish hands and, under the heavy pancake make-up, the suspicion of beard stubble. **1978** *Chicago* June 14/3, I didn't used to wear pancake at all—it was a macho thing with me. But now I do.

3. pancake batter, the mixture from which pancakes are made; **pancake coil** *Electr.*, any flat or very short inductance coil (see quots.); **pancake descent, landing** [cf. *pancake *v.* b], the landing of an aircraft in an emergency with the undercarriage retracted; **pancake race,** a race held on Shrove Tuesday, in which the participants are required to toss pancakes as they run; **pancake roll** (see quot. 1967).

1739 E. Smith *Compl. Housewife* (ed. 9) 114 Mix all well together a little thicker than pancake batter. **1747** H. Glasse *Art of Cookery* vii. 69 Make it up into a thick Batter with Flour, like a Pancake Batter. **1965** A. Christie *At Bertram's Hotel* xi. 103 She made herself three pancakes with the pancake batter. **1910** H. M. Hobart *Dict. Electr. Engin.* I. 108 *Pancake coil*, a flat former-wound coil used in the construction of the early smooth-core rotating armatures of alternators. The term is also sometimes applied to the flat separately insulated unit coil used in modern high-pressure transformers. **1921** *Physical Rev.* XVIII. 138 Coursey's curves do not cover the case of coils whose radial dimension exceeds the axial (pancake coils). **1940** *Chambers's Techn. Dict.* 611/1 *Pancake coil*, an inductance coil in which the windings are arranged spirally, in the form of a flat disc. **1960** Cooke & Markus *Electronics & Nucleonics Dict.* 322/2 *Pancake coil*, a coil having a diameter appreciably greater than its length. **1961** *Guardian* 18 Jan. 1/1 The transformer..will be made up of a series of 'pancake' coils of primary and secondary windings. **1914** W. J. Claxton *Mastery of Air* xlviii. 249 It is considered faulty piloting to make a pancake descent where there is ample landing space. **1928** Pancake landing [see *level *v.*¹ 5*]. **1938** *Encycl. Brit. Bk. of Year* 57/2 Nothing better could be expected than a 'pancake' landing which would destroy the undercarriage without seriously injuring the crew. **1960** Wentworth & Flexner *Dict. Amer. Slang* 374/1 *Pancake landing*, Specif., in aviation, the act or instance of landing an airplane on its fusilage rather than on its wheels, done when the landing gear is damaged. **1951** *Sun* (Baltimore) 17 Jan. 3/2 (*caption*) Mrs. Virginia Leete ..takes a spill in the snow during a practice run..in preparation for the annual pancake race scheduled for Shrove Tuesday. **1955** *Ibid.* 23 Feb. 2/3 Pancake races have featured Shrove Tuesday observances in Olney for some 510 years. **1967** D. Brice *Folk-Carol of England* iii. 86 The well-known 'pancake race' that takes place in the Buckinghamshire village [of Olney] every Shrove Tuesday. **1972** *Guardian* 15 Jan. 14/5 Shrovetide brings pancake races like that at Olney in Buckinghamshire, with housewives tossing pancakes as they belt along. **1976** *Times* 3 Mar. 14/6 The annual women's pancake races in Lincoln's Inn Fields. **1967** *Observer* (Colour Suppl.) 30 Apr. 38/4 *Pancake roll*, a pancake with savoury meat and vegetable fillings, deep fried. **1968** R. V. Beste *Repeat Instructions* xi. 121 They had a more adventurous meal than the..vegetable chop suey and pancake rolls he usually ordered. **1969** O. Blakeston *For crying out Shroud* vii. 59 Jim orders fried oysters and crispy pancake rolls. **1976** M. Butterworth *Remains to be Seen* iv. 68 The diligent Chinese..laboured over their crab foo yung and their crispy pancake rolls.

pancake, *v.* **a.** (In Dict. s.v. Pancake *sb.*) Delete *nonce-wd.* and add later examples (chiefly as ppl. adj.). Also *fig.* and in sense *2 h of the sb.

1941 *Time* 6 Oct. 17/1 A..near-hurricane..that killed three people, leveled grain fields, pancaked buildings, blocked highways. *Ibid.* 20 Oct. 2/1 Starting the bill in the House, with a steam roller set to pancake all opposition. **1942** *Capital* (Topeka, Kansas) 15 Mar. (*caption*) Sure! He's pancaked 17 guys in a row! Hits like a train at a grade crossing. **1948** L. MacNeice *Holes in Sky* 13 They tell me report at the first police station. But the station is pancaked—so what can I do? **1953** Dylan Thomas *Let.* 22 June (1966) 409 Sober, airsick, pancaked flat, I saw these intelligent old friends as a warren full of blockish stinkers. **1973** R. L. Simon *Big Fix* vii. 50 His face was pancaked in layers, his hair lacquered. **1974** *Listener* 23 May 678/2 Rows of pancaked Cadillacs and burnt-out Rolls-Royces. **1977** *S. Wales Echo* 18 Jan. 1/4 Police reported 21 confirmed deaths but said it was likely 60 to 70 more bodies remained in a pancaked carriage crushed to a quarter of its bulk by a giant slab of concrete weighing hundreds of tons.

b. *Aeronaut. intr.* Of an aircraft: to descend rapidly in a level position in stalled flight, *spec.* to land in this manner in an emergency with the undercarriage retracted (cf. **pancake landing*). Of the pilot: to cause an aircraft to pancake. Also *transf.* and *fig.* Hence **pa·ncaking** *vbl. sb.*

1911 *Aero* Aug. 136/2 In the meanwhile Conway Jenkins had..'pancaked' badly, and smashed it pretty conclusively. **1912** *Ibid.* Mar. 66/1 He..then shut off his engine, calmly waiting for the machine to return to the ground, which it did with a resultant bump, commonly known to the aviation world as pancaking (falling flatly). **1914** *Aeronaut. Jrnl.* Oct. 316 *Pancake, to*, to descend steeply, with the wings at a very large angle of incidence, like a parachute. **1914** H. M. Buist *Aircraft in German War* 35 The craft pancaking, diving and banking are monoplanes. **1916** C. Winchester *Flying Men* 68 So ..the 'pancaking' of aircraft is not an advisable method of landing. **1920** *19th Cent.* Mar. 570 This pancaking device by which the National Socialists tried at the last moment to save the crash. **1928** C. F. S. Gamble *Story N. Sea Air Station* xv. 263, I took my chance and about 10 feet up 'pancaked'—a horrid crash. **1929** E. W. Springs *Above Bright Blue Sky* 73, I came out of the spin at five hundred feet and pancaked in the reserve lines. **1936** F. Clune *Roaming round Darling* xxv. 271 All at once she [*sc.* a lorry] slithered like hell, and, knifing a corner, pancaked on to a mulga-tree. **1938** *Daily Progress* (Charlottesville, Va.) 30 July 1/8 He suggested the planes might be ordered to comb isolated mountain forests on the Pacific side of Luzon Island on the possibility the 'Clipper' pancaked into the trees. **1943** P. Brennan et al. *Spitfires over Malta* 91, I told my boys to pancake as soon as they had finished engaging. **1950** *Gloss. Aeronaut. Terms* (*B.S.I.*) I. 12 *Pancaking*, the alighting of an aircraft at an abnormally high rate of descent or low forward speed. **1952** M. Tripp *Faith is Windsock* vi. 97 'Beany Able funnels; we are on three engines and must land.' 'Pancake, Able.' 'Able pancaking.' **1962** R. W. Clark *Rise of Boffins* ii. 53 Another great time-saver was the use of a code for passing instructions to the fighters, and such R.A.F. terms as.. 'pancake' (for land), were invented during these experiments. **1977** *Listener* 28 Apr. 559/2 His plane..pancaked into it. The Germans..came out..to take him and the plane.

panchama (pa·ntʃāmā). *India.* [Skr., = fifth.] A member of the fifth division of early Indian society, outside the four main divisions of Brahmin, Kshatrya, Vasiya and Sudra; a pariah, an outcaste. This caste was also called *Pancham Bandam*.

1800 F. Buchanan *Jrnl.* 30 Apr. in *Journey from Madras* (1807) I. i. 19 Their farms they chiefly cultivate by slaves of the inferiour casts, called Súdra, and Panchum Bundum. The Panchum Bundum are by far the most hardy and laborious people of the country. **1874** *Madras Census Rep. 1871* I. xi. 168 We now come to that great division of the people, spoken of by themselves as the 'fifth caste', and described by Buchanan and other writers as the *Pancham Bandam*. **1909** E. Thurston *Castes & Tribes of S. India* VI. 44 The Government ruled that there is no objection to the proposal that Paraiyas and kindred classes should be designated Panchama Bandham or Panchama in future, but it would be simpler to style them the fifth class. *Ibid.*, Panchama students under training as teachers get stipends at rates nearly double of those for ordinary Hindus. **1917** *Rangoon Gaz.* 10 Oct. 12/1 A mass meeting of Panchamas (depressed classes) was held in Madras. **1932** G. S. Ghurye *Caste & Race in India* i. 10 In the Tamil and Malayalam regions..sometimes the village is divided into three parts: that occupied by the dominant caste in the village or by the Brahmins, that allotted to the Sūdras, and the one reserved for the Panchamas or untouchables. **1968** N. W. Ross *Hinduism, Buddhism, Zen* 29 Outside the four main caste divisions,..there has also existed from the earliest times a group familiarly known as the Untouchables. They were called Panchamas, literally 'the fifths'.

panchayat. (Now the normal spelling.) Add: (Later examples.) Also *attrib.* Hence **panchayat samiti** [f. Hindi *samiti* committee].

1881 E. B. Eastwick *Murray's Handbk. Bombay Presidency* (ed. 2) II. 141/1 In order to see the Towers of Silence, permission must be obtained from the Secretary to the Pársí Pancháyat. **1893** Kipling *Many Inventions* 84 Create, further, councils other than the panchayats of headmen, village by village and district by district. **1945** 'P. Woodruff' *Call Next Witness* 14 He was chairman of the village panchayat, the court which could try the smallest local offences. **1955** *Times* 29 Aug. 9/6 Mansingh tried to negotiate peace at a special meeting of the Panchayats, or village councils. **1963** *Times* 11 Mar. 11/7 The emphasis was corrected and laid on agricultural production—but no sooner than the establishment of *panchayat raj* had led villagers to express their needs more outspokenly, and their satisfaction had become the business of the village politicians. **1963** *Economist* 23 Nov. 752/1 His [the King of Nepal's] system of 'panchayat democracy', an elaborate four-tier edifice of indirect elections. **1965** E. Linton *World in Grain of Sand* vi. 73 Were all members of the Panchayat present? No. Then send for them! Panchayats, literally meaning 'councils of five', have existed in villages since ancient times... Numbers need not necessarily be confined to five. **1969** *Listener* 2 Jan. 5/1 The panchayat system is little more popular than Pakistan's basic democracy. **1969** *National Herald* (New Delhi) 29 July 7/4 Mr. Thana Ram, pradhan of panchayat samiti, has criticised the demotion of education extension officers who have completed five years of service. **1971** *Hindustan Times Weekly* (New Delhi) 4 Apr. 8/2 The Agriculture Refinance Corporation will provide Rs 25 lakhs for disbursement as loans among the orange-growers of Halrapatan panchayat samiti in Halawar district. **1971** *Nat. Geographic* Nov. 662/1 Panchayat means 'five elders', a traditional informal council that runs the affairs of Nepalese villages. **1973** *Times* 14 Apr. (Nepal Suppl.) p. i/5 Limited popular representation is permitted through a pyramidal structure of partly elected and partly nominated *panchayats*, or councils, beginning at the village level. **1976** D. Hiro *Inside India Today* 50 What then emerged was a three-tiered system whereby the old district boards..were replaced by zilla parishads (i.e. district councils) with responsibility for co-ordinating development plans to be channelled through panchayat samitis (i.e. council committees) consisting of a number of popularly elected panchayats encompassing one or more villages—all interlinked through indirect elections. This system, popularly known as the panchayat raj, was first introduced in..1959.

panchen pan,tʃĕn). *Tibet.* Also **Banchen, Pantchan,** etc. [Tibetan, abbrev. of *pandita-chen-po* great learned one (cf. Pundit).] A Tibetan Buddhist title of respect, applied *esp.* to the lama of Tashi Lhunpo, who is

held to be the reincarnation of Buddha Amitbha and is next in importance to the Dalai Lama, being styled the *Panchen Lama* or *Panchen Rinpoche* (rinpoche = precious, jewel). Cf. *Rinpoche.

1763 J. Bell *Trav. from St. Petersburg* I. 284 The Kontaysha is of the same profession with the Delay-Lama... I am informed there is a third Lama, called Bogdu-Pantzin, of still greater authority... He lives.. near the frontiers of the Great Mogul. **1784** S. Turner *Let.* 2 Mar. in *Acct. Embassy to Court of Teshoo Lama in Tibet* (1800) III. 373, I was..strongly dissuaded by the Regent Punjun Irtinnee. **1794** A. Dalrymple *Oriental Repertory* II. 273 This Pantchan-lama is the Second Person of Tibeth and of all the Lama-Hierarchy. *Ibid.* 274 The Pan-tchan..asked permission of his Majesty to proceed to the Capital of the Empire. **1800** S. Turner *Acct. Embassy to Court of Teshoo Lama in Tibet* II. viii. 325 *Punjin Rimbochay*, Great Apostolic Master; the mitred professors of religion. **1834** C. Gützlaff *Sk. Chinese Hist.* II. xvii. 64 One of the chiefs of Tibet, on hearing of the death of Banchen Lama at Peking, had gone to Nepaul with an immense treasure. **1851** H. T. Prinsep *Tibet, Tartary & Mongolia* 108 The highest of existing regenerate Boodhs are the Delai Lama of Lassa; the Bandshan Remboochi, of Teshoo Loomboo, the same who was visited by Captain Turner, in the time of Warren Hastings. **1876** C. R. Markham *Narr. Bogle & Manning* p. cxi, The Pundit went..to Teshu Lumbo, to do homage to the Teshu Lama or Panchen Rimboché, a boy eleven years old. **1895** L. A. Waddell *Buddhism of Tibet* x. 235 The Sa-kya Grand Lāmas had been called 'Pan-ch'en', or the 'Great doctor' from the twelfth century. **1925** *Glasgow Herald* 13 Apr. 9 The Panchan Lama is one of the two lama popes, the other being the Dalai Lama, or Ocean Priest, who resides at Lhassa. **1931** C. Bell *Relig. of Tibet* xii. 155 During the reign of the eighth Dalai Lama it is the Panchen Rim-po-che who looms largest in Tibetan history. **1935** *Discovery* Aug. 239/2 On the high altar itself the central position was occupied by an excellent photograph of the Panchen Lama. **1956** K. W. Morgan *Path of Buddha* vi. 256 The present Panchen Lama is the ninth in succession and was selected jointly by the former National Government of China and the followers of the exiled Panchen Lama. **1962** *Listener* 12 July 71/1 People interested in Tibetan institutions will also pay attention to the few pages devoted to the Panchen Lama. **1964** J. P. Mitter *Betrayal of Tibet* 98 The Chinese Amban violated the Trade Regulations of 1908 by forbidding the Pan-chen Lama and his officials to communicate with the British Trade Agent at Gyantse. **1978** *Guardian* 25 Feb. 6/8 The Panchen Lama..remained behind in Tibet when the Dalai Lama and other religious leaders fled to India in 1959.

panchromatic (pæn,krǫmæ·tik), *a.* [f. Pan- + Chromatic *a.*] **1.** *Photogr.* Sensitive (though not equally so) to light of all colours in the visible range. Also *ellipt.*, a panchromatic emulsion or plate.

1903, etc. [see *orthochromatic *a.* 1]. **1906** *Chambers's Jrnl.* May 416/2 This layer..is re-covered with yet another layer of panchromatic, and sensitised. **1921** *Glasgow Herald* 6 Apr. 7 My dark-room lamp has three interchangeable safe-lights,..one a dark green for panchromatics. **1952** *Proc. R. Soc. Edin.* A. LXIII. 206 The usual type of orthochromatic emulsion is a little slow to this radiation, but a panchromatic emulsion might record some red. **1978** *SLR Camera* Aug. 82/1 Panchromatic film—the type almost exclusively used these days for normal photography—..is very much more sensitive to blue and blue-green, than the eye, but less sensitive to green, yellow and orange.

2. = Polychromatic *a.*

1971 J. McClure *Steam Pig* iii. 39 The poser of the panchromatic panties. **1975** M. Kenyon *Mr Big* xix. 180 Two boisterous black girls in patched panchromatic trousers.

Hence **panchro·matize** *v. trans.*, to render panchromatic; **panchro·matizing** *vbl. sb.*

1922 E. J. Wall *Pract. Color Photogr.* ii. 15 Many dyes have been suggested for panchromatizing. **1925** —— *Hist. Three-Color Photogr.* vii. 246 A. Miethe recommended the following mixture for panchromatizing plates. **1960** K. M. Hornsby tr. *P. Glafkidès' Photogr. Chem.* II. xxxv. 729 To make them [*sc.* photographic emulsions] sensitive to the other colours, green, yellow, red and infra red—or to ortho- or panchromatize as we say—it is necessary to incorporate certain special dyes.

panchronic (pæn,krǫ·nik), *a.* *Linguistics.* [tr. F. *panchronique* (F. de Saussure *a* 1913, in *Cours de Linguistique générale* (1916) I. iii. 138), f. Pan- 2 + Chronic *a.*] Pertaining to or designating linguistic study applied to all languages at all stages of their development. Also **panchroni·stic** *a.* Hence **panchro·nically** *adv.*; **pa·nchrony.**

1931 *Amer. Jrnl. Philol.* LII. 79 Scientific grammar must be based on a combination of ideo(syn)chrony and panchrony. **1939** L. H. Gray *Foundations of Lang.* 24 The components of such a panchronic grammar, which may technically be termed *general grammar*, will be few in number. **1949** *Archivum Linguisticum* I. II. 127 On the *panchronistic* plane, there is the usual argument of the complete diversity of words for the same idea in different languages. **1951** S. Ullmann *Princ. Semantics* v. 261 He [*sc.* de Saussure] did admit the possibility of 'panchronistic laws' resembling the universal regularities of natural science, e.g. the ubiquitousness of sound-change. **1952** *Times Lit. Suppl.* 10 Oct. 659/3 A final chapter, devoted to panchronistic or general semantics, is merely a programmatic sketch. **1957** *Archivum Linguisticum* IX. II. 81 Finally, hyper- and hypocharacterization may be used panchronically. **1964** *Ibid.* XVI. I. 23 Clusters so shaped may panchronically tend to undergo just this development. **1966** M. Pei *Gloss. Ling. Terminol.* 192 *Panchronic grammar*, applicable to all languages and at all historical stages of their development. **1969** *Eng. Stud.* L. 417 General phonetics is by definition synchronic, or rather panchronic. *Ibid.* 422 Comparatism was supposed ..to lead to diachrony, not to the establishment of common, general features of language, to panchrony. **1974** R. A. Hall *External Hist. Romance Lang.* 4 The panchronic approach treats those aspects of language for which the passage of time is not relevant. **1978** *Language* LIV. 238/2 In Chapter V, he treats 'Lingua, stile, dialetti'.. from a primarily panchronic point of view.

‖**panchshila** (pʊnʃī·la). Also **panchsheel, panchsila,** and as two words. [Hindi and Skr., f. *panch* five, *shila* foundation.] The five principles of peaceful relations formulated between India and China (and, by extension, other communist countries).

The five principles, stated in the preamble of a treaty signed by India and China in April 1954, are: 1. Mutual respect for each other's territorial integrity and sovereignty. 2. Non-aggression. 3. Non-interference in each other's internal affairs. 4. Equality and mutual benefit. 5. Peaceful co-existence.

1955 *Times* 18 July 7/5 After analysing the popular enthusiasm in Russia over the Nehru visit, the newspaper [*sc.* *Times of India*] says, 'It would be foolish, even dangerous, to work oneself up into a frenzy of apocalyptic fervour and hail those who hailed our Prime Minister as comrades good and true demonstrating in their mammoth enthusiasm the resolve to march in step to the golden melody of Panchshila.' **1958** *Times* 4 July 9/3 India tried to act upon the principles of *panchsila*, and did not wish to interfere in other people's affairs. **1959** *Manch. Guardian* 15 Aug. 5/4 China has slapped India's face, and the Panchshilas (the 'five principles of co-existence') have popped. **1961** *Economist* 2 Dec. 939/2 India was drawing up the *Panch Shila*—the five principles of peaceful co-existence—with the Chinese. **1965** J. Nehru in A. Appadorai *Documents Political Thought* (1976) II. 739 Panchsheel has begun to acquire a specific meaning and significance in world affairs. **1967** L. J. Kavic *India's Quest for Security* iii. 59 On 1 August 1955, a joint communiqué issued in Kathmandu by representatives of the Nepalese and Chinese governments declared that an agreement had been reached which affirmed panch sheel as the basis of Sino-Nepalese relations. **1978** L. Heren *Growing up on The Times* v. 178 Despite the Indian name, the *panchsila* were of Chinese origin, and were written into the preamble of the Sino-Indian Tibetan Treaty on the instance of Peking.

pancreatectomy (pæ:ŋkrĭ,ăte·ktŏmi). *Surg.* [f.Gr. stem παγκρεατ- (Pancreas) + *-ectomy.] Excision of the pancreas.

1900 in Dorland *Med. Dict.* **1903** W. S. Bickham *Text-bk. Operative Surg.* v. 834 Anatomically, complete pancreatectomy is very difficult. **1968** *New Scientist* 27 June 701/2 Pancreas transplantation..might also be useful..where pancreatectomy is needed because of malignancy. **1974** R. M. Kirk et al. *Surgery* vi. 112 Occasionally distal pancreatectomy, the removal of ductal stones, and drainage of the cut end of the pancreas into the jejunum, improves the patient. **1977** *Proc. R. Soc. Med.* LXX. 160/1 One man of 44 died of massive haemorrhage the day after a complicated procedure to relieve intestinal and biliary obstruction, following a pancreatectomy less than three weeks previously.

Hence **pa:ncreate·ctomize** *v. trans.*, to excise the pancreas of; **pa:ncreate·ctomized** *ppl. a.*

1912 *Amer. Jrnl. Physiol.* XXX. 341 The glycolytic action of muscle extracts of both normal and pancreatectomized animals has been tested. **1960** *Recent Progress Hormone Res.* XVI. 503 Rats were fasted and underfed for 8–10 days and then pancreatectomized. **1965** Lee & Knowles *Animal Hormones* vii. 111 If a dog is pancreatectomized and the circulation connected to one, two or three pancreases from normal dogs, the blood glucose is normal in the pancreatectomized animal, irrespective of the number of the pancreases utilized.

pancreatico-. Add: **pancrea:ticoduodene·ctomy** = **pancreatoduodenectomy*.

1941 *Ann. Surg.* CXIV. 612 Until 1935, pancreaticoduodenectomy for cancer involving the pancreas was not attempted. **1973** V. L. Stevenson *Biliary Tract Surg. & Cholangiogr.* xii. 124 Those undergoing pancreaticoduodenectomy generally afforded a longer survival than those undergoing palliation.

pancreato-, comb. form (= Pancreatico-), as in **pancrea:toduodene·ctomy** (see quot. 1928); **pa:ncreato·graphy,** radiological examination of the pancreas.

1928 R. J. E. Scott *Gould's Med. Dict.* (ed. 2) 1044/2 *Pancreatoduodenectomy*, excision of the head of the pancreas with the surrounding loop of duodenum. **1937** *Surg., Gynecol. & Obstetr.* LXV. 681 (*heading*) Resection of head of pancreas and duodenum for carcinoma—pancreatoduodenectomy. **1977** *Proc. R. Soc. Med.* LXX. 153/1 Unfortunately, Whipple's operation or pancreatoduodenectomy for carcinoma of the head of the pancreas was seldom possible. **1971** Rains & Capper *Bailey & Love's Short Pract. Surg.* (ed. 15) xlii. 877 (*heading*) Pancreatography. **1977** *Lancet* 9 July 68/1 A screening test which makes possible the detection of pancreatic disease at an early stage and gives an indication for invasive procedures such as endoscopic pancreatography and selective arteriography is urgently needed.

pancreozymin (pæŋkrĭ,ozəi·min). *Biochem.* [f. Pancre(as + -o + Zymin.] A hormone which stimulates the production of enzymes by the pancreas.

1943 Harper & Raper in *Jrnl. Physiol.* CII. 116 We have obtained preparations which increase the output of enzymes from the cat's pancreas without having any secretin activity... For the active substance producing this effect we suggest the name 'Pancreozymin'. **1956** *Nature* 7 Jan. 22/2 Cholinergic drugs and pancreozymin in the pancreas, and acetylcholine and adrenalin in the salivary glands, all stimulate the secretion of protein. **1965** Lee & Knowles *Animal Hormones* viii. 120 It may be that under normal conditions both secretin and pancreozymin are released together. **1974** R. M. Kirk et al. *Surgery* vi. 110 Pancreozymin..is liberated in response to the presence of protein and fat in the duodenum.

pancuronium (pænkiŭrōᵘ·niəm). *Pharm.* [f. *pan-*, of uncertain etym. + Cur(are + *-onium.] A steroid whose bromide is used as a neuromuscular blocking agent.

1967 *Brit. Jrnl. Anæsthesia* XXXIX. 775/1 Pancuronium bromide (NA97), was first synthesized in 1964 by Hewett and Savage (1966, personal communication). **1976** *Billings* (Montana) *Gaz.* 17 June 1-D/6 Many of the breathing failures at the hospital were caused when unknown assailants injected patients with a potentially lethal muscle-paralyzing drug, pancuronium bromide. **1976** *Lancet* 18 Dec. 1334/1 One group of eight patients received general anæsthesia with thiopenthone, suxamethonium chloride, pancuronium bromide or gallamine, and enflurane ('Ethrane') with nitrous oxide plus oxygen.

panda. Add: **2.** A large, black and white, bear-like mammal, *Ailuropoda melanoleuca*, native to limited, mountainous areas of forest in China, where the first scientific description of it was made by the French missionary, Armand David (1826–1900), in 1869; formerly known as the parti-coloured bear, until its zoological relationship to the red panda was established in 1901.

1901 E. R. Lankester in *Trans. Linn. Soc.* (*Zool.*) VIII. 165 *Æluropus* must be removed from association with the Bears..and is no longer to be spoken of as 'the Parti-coloured Bear', but as 'the Great Panda'. **1928** *Proc. Zool. Soc.* 975 The systematic position of the Giant Panda..is a question about which there has been much disagreement amongst zoologists. **1933** *Discovery* Mar. 91/1 In outward appearance there is considerable difference between these two animals, the giant panda.. being very bear-like, while the little panda is about the size and somewhat the shape of a cat. **1939** *Daily Mail* 12 Apr. 8/4 This sickly sentimental panda plague has infected far more people than can ever hope to eye it in the flesh... Would-be fashionable young women are carrying panda mascots. **1940** N. Mitford *Pigeon Pie* ix. 140 Ming, the panda, would soon eat no food until one of them was played to her. **1943** *Jrnl. Mammalogy* XXIV. 267 The New York Zoological Society has recently acquired a pair of giant pandas... The principal natural diet of the panda is bamboo. **1966** R. & D. Morris *Men & Pandas* vi. 105 There were panda postcards.., panda toys (almost obliterating the teddy bear for a brief period), panda novelties, panda strip-cartoons, panda brooches, and panda hats. **1973** *Times* 2 May 9/8 Children [in Peking] played a multitude of games including 'feed the panda', a variation on 'pin the tail on the donkey'. **1976** *Times Lit. Suppl.* 27 Feb. 231/5 It is rumoured that China has sited her nuclear testing grounds not far away from Panda country.

b. Used *attrib.* to designate a type of pedestrian crossing (see quot. 1962¹). Also *absol.*

1962 *Daily Tel.* 7 Mar. 15/7 'Panda' pedestrian crossings are to be introduced..to supplement zebra crossings. Their warning lights will be operated by push-buttons and they will be given a 12-month trial. *Ibid.*, Differences in appearance between the 'Pandas' and the zebras are that the black-and-white carriageway markings at the 'Pandas' will be altered in shape from rectangles to blunted chevrons. **1962** *Times* 3 Apr. 12/6 Panda crossings, introduced yesterday, held up Croydon's evening traffic. **1963** *Times* 24 May 17/4 The amber lights system used on panda crossings was so complex and ambiguous that the ordinary driver could not understand it. **1965** A. Christie *At Bertram's Hotel* xi. 106 On the whole, the Canon was not what we would call accident prone... Whilst taking no care or thought, they could still survive even a Panda crossing.

c. A police patrol car, so named from the resemblance of a broad white stripe on the car to the markings of the giant panda. Also *attrib. colloq.*

1966 *Guardian* 13 Sept. 8/4 Special one-man patrol cars—painted blue with a broad white stripe and known as 'Pandas'. **1969** J. Wainwright *Take-Over Men* i. 13 What about your Panda Patrols? Your closed-circuit television? **1970** *Times* 17 Mar. 2 Five children, who.. helped catch two thieves, are to be given a ride in a police panda car. **1971** *Daily Tel.* 10 May 2/2 It was felt that panda drivers should be warned that the vehicles were not meant to be pursuit cars. **1974** 'A. Gilbert' *Nice Little Killing* vi. 82 He got out his old second-hand car—the village bobby didn't rate a panda.

pandaite (pæ·ndă,əit). *Min.* Also **pandaïte.** [f. *Panda*, the name of the hill in Mbeya, Tanzania, where it was first found + -ite¹.] A hydrated oxide of barium, strontium, niobium, and titanium, $(Ba,Sr)(Nb,Ti)_2(O,H_2O)_7$, belonging to the pyrochlore group and found as yellow or white octahedral crystals.

1959 E. Jäger et al. in *Mineral. Mag.* XXXII. 24 Some authors prefer to include under pyrochlore all the members of the group. Others give distinct names to the

various members of this group. We prefer the latter and therefore propose to give the mineral described above the name of *pandaite* (after Panda Hill). This name shall be used for those minerals of the pyrochlore group in which Ba predominates over other elements in the *A* positions. **1971** *Mineralium Deposita* VI. 154/2 This pandaite shows large deficiencies in A ions and only 20–25% of the A positions are occupied. *Ibid.* 155/1 The mineral from Bingo [in the Congo] is a hydrated rare-earth variety of pandaïte. **1977** *Amer. Mineralogist* LXII. 407 *Pandaite*.. is a synonym for *bariopyrochlore*. The name should be dropped.

pandal. Add: Now usu. with pronunc. (pændā·l). (Further examples.)

1929 F. T. Jesse *Lacquer Lady* I. xii. 86 Her mother, the Kalawoon's wife, was running the pandal or festival pavilion for Thibaw. **1956** *Times* 13 Jan. 3/3 All the music is amplified, since the temporary *pandal*, which is the equivalent of an Eisteddfod 'tent', seats nearly 1,500 people. **1962** *Housewife* (Ceylon) Feb. 19 It was decided to hold the reception at the 39th lane sports club, where there was ample room for two large pandals to be erected. **1963** *Guardian* 11 Apr. 11/3 The Hindu wedding, celebrated under the flowered palm leaf pandal. **1971** *Weekend* (Ceylon) 8 Aug. 3/3 Permanent pandals will be built to decorate the entrances to sacred cities. **1977** *Oxford Mission Q. Paper* Jan.–Mar. 10 An enormous *pandal* had been erected which covered the whole area on the south side of the church, and the altar was placed on a mound in the middle of it.

pandan² (pæ·ndăn). Also **pandang.** [Malay.] = *Pandanus.

1777 [see Pandanus]. **1783** W. Marsden *Hist. Sumatra* 87 Of the *pandan*, which is a shrub with very long prickly leaves, like those of the pine apple or aloe, there are many varieties. **1935** I. H. Burkill *Dict. Econ. Products Malay Peninsula* II. 1646 The compound pine-apple-like fruit of a *Pandan* is composed of the fused fruits of the individual flowers. **1954** R. E. Holttum *Plant Life Malaya* ii. 23 Pandans have much in common with palms. **1959** 'M. Derby' *Tigress* iii. 125 A clump of *pandan*..edged the near end of the pool. **1972** M. Sheppard *Taman Indera* 158 Pandan leaves are used to make mats of finer quality, for sitting, praying or sleeping on.

pandanus. Substitute for etym. and def.: [mod. L. (G. E. Rumphius *Herbarium Amboinense* (1743) IV. 139/1), f. Malay *pandan*.] A tree or shrub of the genus so called, belonging to the family Pandanaceæ, native to Malaysia, tropical Africa, or Australia, and distinguished by forked trunks with thick aerial roots, long, narrow, prickly leaves arranged in spiral tufts, and large, sometimes edible fruits resembling a pineapple. Also *attrib.* (Add earlier and later examples.)

1830 J. Lindley *Introd. Nat. Syst. Bot.* 285 The seeds of Pandanus are eatable. **1908** E. J. Banfield *Confessions of Beachcomber* I. i. 15 Groups of pandanus palms bearing massive orange-coloured fruits. **1915** *Chambers's Jrnl.* Nov. 698/1 A net..is woven from a strong fibre found in a species of pandanus-tree. **1936** I. L. Idriess *Cattle King* xxxvi. 314 Plenty of water there, rock-bound rivers, pandanus-palm creeks, grass, trees, lily-covered lagoons. **1946** —— *In Crocodile Land* i. 5 They chopped pandanus nuts for breakfast. **1964** R. Perry *World of Tiger* xv. 231 The palms and pandanus wilderness in southern parts of the Island [*sc.* Java]. **1971** *World Archaeol.* III. 140 Unworked river pebbles, used for such purposes as breaking bones and crushing pigments or pandanus nuts. **1977** *Bulletin* (Sydney) 22 Jan. 65/1 The spindly eucalypts and pandanus palms in the Alligator River district were filled with the rasping shriek of millions of cicadas and crickets.

2. The fibre produced from pandanus leaves or the material woven from it. Also *attrib.*

1894 [in Dict.]. **1930** M. Mead *Growing up in New Guinea* ix. 156 A pandanus rain mat is a clumsy thing to carry about. **1963** *House & Garden* Feb. 61/2 Storage unit..teak, with pandanus grasscloth doors. **1971** *Daily Tel.* 23 Dec. 3/7 The children [of Pitcairn Island] will have found their presents in pandanus-leaf baskets suspended by the front porch or above their beds. **1972** M. Sheppard *Taman Indera* 141 The floor is hard and smooth and there is usually a low platform at one end on which spectators can sit, cross-legged, on pandanus mats. **1974** *Nat. Geographic* Dec. 778/1 The scene was one from yesterday—the pandanus-thatched houses under the palms, the circle of grinning, tattooed men.

pandect. Add: **2. b.** A manuscript volume containing all the books of the bible.

1887 F. J. A. Hort in *Academy* 26 Feb. 148/2 There cannot now be a shadow of doubt that the Codex Amiatinus is the 'Pandect' which Ceolfrid sent as a present to Gregory II. **1893** E. G. Browne *Lessons Early Eng. Church Hist.* 68 A pandect means a copy of the whole Bible. **1912** D. S. Boutflower *Life of Ceolfrid* 69 He [*sc.* Ceolfrid] caused three Pandects to be transcribed. **1969** *Jrnl. Brit. Archaeol. Assoc.* XXXII. 1 One of the three pandects, as they were then called (complete bibles in one volume) has survived miraculously intact. This is the Codex Amiatinus.

pandemoniac, *a.* Add: **c.** as *sb.* A pandemoniac person; a denizen of Pandemonium. *rare.*

1923 Galsworthy *Captures* 81 Success, power, wealth —those aims of profiteers and premiers, pedagogues and pandemoniacs.

pandiagonal (pæ:n,dəi,æ·gŏnăl), *a.* *Math.* [f. Pan- + Diagonal *a.*] Used to describe a magic square with the property that, if any number of columns be removed from one side of the diagram and added *en bloc* to the other, another magic square results.

1897 *Amer. Jrnl. Math.* XIX. 99 The square A is magic because each row, column, and diagonal has the same sum, 175; it is pandiagonal because not only the two main diagonals, but also the twelve broken diagonals..have each the same sum. **1919** *Monist* XXIX. 308 Magic squares of order ≡ 2 (mod. 4) made with consecutive numbers cannot be pandiagonal. **1939** H. S. M. Coxeter *Ball's Math. Recreations & Ess.* (ed. 11) vii. 203 A magic pandiagonal square of the fourth order..was inscribed at Khojuraho, India, as long ago as the eleventh or twelfth century. **1976** *Sci. Amer.* Jan. 120/1 And it is pandiagonal (sometimes called Nasik or diabolic), which means that its broken diagonals add up to 65, the constant.

Hence **pa:ndia·gonally** *adv.*

1911 W. W. R. Ball *Math. Recreations & Ess.* (ed. 5) vii. 157 If a pandiagonal square be cut into two pieces along a line between any two rows or any two columns, and the two pieces be interchanged, the new square so formed will be also pandiagonally magic.

pan-drop. *Sc.* [f. Pan *sb.*¹ + Drop *sb.* 10 e.] A hard, peppermint-flavoured sweet, shaped like a flattened sphere.

1877 *Encycl. Brit.* VI. 257/1 A core or centre of some kind is required, and this may consist either of a seed or fruit..; or it may be a small lozenge, as in the case of pan drops. **1904** 'H. Foulis' *Erchie* v. 29, I thoucht it was pan-drops ye cam' oot for, or conversation-losengers. **1927** *Glasgow Herald* 7 Sept. 12/7 An' there's nae mae tears since ye've got him wi' the poke o' pan-draps in his han'. **1956** C. M. Costie *Benjie's Bodle* 106 Wir haean a duff, an' treacle..an twa pan drops. **1964** *Scotsman* 14 Oct. 5 What was described in court as 'a classical line of traditional Scottish sweet—pan drops'. **1966** W. Merrilees *Short Arm of Law* 184 These..were not chocolates at all but hard peppermint sweets..pan-drops as we called them in Scotland.

pandurina (pæ:ndiŭrī·nă). [It., f. *pandura* (see Pandora², pandore) + dim. suffix -*ina*.] A small musical instrument of the mandoline type.

1893 J. S. Shedlock tr. *Riemann's Dict. Mus.* 53/1 *Bandola.* (Span.), Bandolon, Bandora, Bandura, an instrument of the lute family, with a smaller or larger number of steel or catgut strings, which were plucked with the finger like the Pandora, Pandura, Pandurina, [etc.]. **1910** F. W. Galpin *Old Eng. Instruments of Mus.* iii. 40 At this period [*sc.* the sixteenth century] there was another small instrument called by Prætorius *Mandürichen* or *Pandurina*, which could be conveniently carried under the cloak. **1938** *Oxf. Compan. Mus.* 683/1 *Pandurina*, a very small instrument of the lute type, strung with wire—probably the ancestor of the mandoline. **1954** *Grove's Dict. Mus.* (ed. 5) V. 549/1 The pandurina returned to popularity, particularly about 1760–80, under the name Milanese mandoline. **1976** D. Munrow *Instruments Middle Ages & Renaissance* 79/3 Praetorius..also mentions a smaller size [of mandora], the *pandurina*, with four strings tuned to g, d', g', d".

pandy, *sb.*¹ Add: **b.** *attrib.* and *Comb.*, as *pandybat*.

1916 Joyce *Portrait of Artist* (1969) i. 49 Fleming held out his hand. The pandybat came down on it with a loud smacking sound. **1922** —— *Ulysses* 547 Twice loudly a pandybat cracks.

Pandy (pæ·ndi). *Med.* The name of Kalman *Pandy* (b. 1868), Hungarian neurologist, used *attrib.* or in the possessive to denote a reaction or test he devised for globulins in the spinal fluid, in which a sample is treated with a dilute aqueous solution of phenol.

1916 L. F. Barker *Monographic Med.* II. 83 Pandy's test has not received the attention it deserves. **1933** W. R. Brain *Dis. Nervous Syst.* 113 Pandy's reaction is the most sensitive, and may yield a weakly positive result with normal fluids. **1963** *Lancet* 12 Jan. 108/1 Lumbar puncture on the ninth day of the illness yielded clear fluid..; the Pandy test was negative.

pane, *sb.*¹ **II. 4.** Delete † *Obs.* and add later example.

1912 T. D. Atkinson *Eng. & Welsh Cathedrals* 268 The north pane of the cloisters with its sunny aspect.

10. A sheet or page of stamps.

1912 *Chambers's Jrnl.* Nov. 749/1 The print would have represented a 'pane' of one hundred and twenty stamps. **1916** F. J. Melville *Postage Stamps in Making* I. xvi. 173 Where the sheet is in panes, only the pane containing the defective print is discarded. **1971** D. Potter *Brit. Eliz. Stamps* viii. 83 From September 1967 until May 1968 only 6s booklets contained Machin Head stamps, with three panes of 4ds. *Ibid.* xv. 174 In those days British stamps were printed by typography, and the printers' rule placed round the edge of the panes relieved the edges of the plates from the pressure which always falls more heavily on those parts. Marginal arrows..indicate the points of division into counter book panes, less unwieldy than complete sheets.

panel, *sb.*¹ Add: **II. 5. b.** Also, *spec.* a list or group of people called upon to advise, judge, take part in a discussion or contest, etc. (Further examples.)

1934 G. B. Shaw *Too True to be Good* 24 The formation of panels of tested persons eligible for the different grades in the governmental hierarchy. **1947** *Ann. Reg. 1946* 53 The method of forming panels for juvenile courts. **1952** W. J. H. Sprott *Social Psychol.* vi. 103 Another device for assessing the attitudes of special groups of people is to use panels of respondents who are prepared to give their views on expert or general questions. **1958** *New Statesman* 1 Feb. 127/2 Perhaps..he believes the Brainstrusters really are equipped to pronounce themselves upon, virtually, anything... Radio and television have given a great impetus here. 'Do the panel think that there is an after life?' **1958** *Listener* 4 Dec. 916/1 A small panel of experts who were also good broadcasters. **1959** *Times* 28 Feb. 7/4 If one of those contests which require the competitor to list a number of items in order of popularity were to turn its attention to the months of the year, the panel of judges (each one an expert) would surely find February at the bottom of the poll. **1961** *Which?* Sept. 231/2 The assessments were made by a panel of people experienced in listening to tape recorders. **1962** *Listen* 1 Feb. 211/2 It was a panel of architects of many nationalities who sketched out the main design. **1966** *Ibid.* 4 Aug. 168/1, I thought the panel skirted the subject. Why the BBC did not have a child psychologist on it I cannot guess. **1967** C. L. Wrenn *Word & Symbol* 11 The committee of scholars who translated the *New English Bible* New Testament..sought..to weld the whole into agreeable and dignified English with the aid of a 'literary panel'. **1973** *N.Y. Law Jrnl.* 31 Aug. 1/6 In reversing and remanding the case to the Southern District, the Second Circuit panel assigned it to Judge Constance Baker Motley. **1975** *Irish Times* 10 May 3/4 They named a panel of players from which the Ballybofey line-out will be chosen tomorrow. **1976** *Horse & Hound* 3 Dec. 54/3 He introduced a panel of experts for an open forum and considerable discussion ensued. **1977** *Sunday Express* 30 Jan. 31/5 It is customary for the touring side to see the full panel of Test umpires in action in the games outside the Tests.

c. The official list of doctors in a district who accepted patients under the National Health Insurance Act of 1913 (since superseded by the National Health Service Act of 1946). *On the panel*, (*a*) of doctors, registered as accepting patients thus; (*b*) of patients, under the care of a 'panel doctor'; also in extended use.

1913 *Punch* 30 July 101/1 The proposed Laureate was a medical man and not on a panel. **1914** *Times* 12 Feb. 6/5 Of these [doctors] 1500 are already on the panel for the county. **1914** T. Smith *Everybody's Guide Insurance Acts* (ed. 3) 124 Which practitioners are collectively to be known as 'the panel'. **1957** R. Hoggart *Uses of Literacy* i. 21 Almost every worker has been on the 'panel' at the local doctor's. **1964** G. L. Cohen *What's Wrong with Hospitals?* i. 22 Working people still talk about 'going on the Panel' when they're off sick, and don't see why they should use another term. **1974** *Daily Tel.* (Colour Suppl.) 29 Mar. 19/2 The average GP has 2,460 people on his panel. **1975** P. G. Winslow *Death of Angel* v. 117 It's the National Health... If only the government had left things alone, like they always was, with the Panel. **1976** 'J. Bell' *Trouble in Hunter Ward* i. 6 There were thousands of Health Service patients who put themselves upon their doctor's panel because they could no longer, after the war, afford to be private patients.

III. 9. f. A section of a tapestry or other ornamental work, usu. one surrounded by a decorative border. Also, a tapestry regarded as a whole.

1856 O. Jones *Gram. Ornament* xvi, The painter began to usurp the office of the scribe... We have the first stage ..where a geometrical arrangement is obtained with conventional ornament enclosing gold panels, on which are painted groups of flowers. **1911** *Encycl. Brit.* XXVI. 405/1 Other tapestries..are fantastic with schemes of abstract ornament into which are introduced as subsidiary details figure subjects set in panels and medallions. **1918** G. L. Hunter *Decorative Textiles* xii. 243 Tapestry screen panels woven in New York. **1923** F. de Zulueta *Embroideries M. Stuart & E. Talbot* 10 This again is a green velvet curtain, measuring 7½ × 6 feet and mounting twenty-four needle-work panels. *Ibid.* 15 If the centre-piece is not enough, there is the octagonal panel immediately above it. **1946** H. Léjard *French Tapestry* 24 The tapestry panels intended for the decoration of the same room soon came to be composed on related themes. **1953** E. Fisher *Swedish Embroidery* 38 The stimulating colours of this unique hanging panel can be seen in the colour reproductions. **1964** D. DuBon *Tapestries S. H. Kress Coll. at Philad. Mus. of Art: Hist. Constantine* 20 The sarcophagus is framed by an oval wreath of ribbon within an oval panel, bound laurel leaves with a shell form at the top and bottom. *Ibid.* 21 The ornament surrounding the central panels on all of the over-doors is similar. **1965** P. Hentgès tr. *Biryukova's Hermitage, Leningrad: Gothic & Renaiss. Tapestries* Pl. 33 The left-hand panel shows the betrothal of Mary and Joseph. **1974** *Encycl. Brit. Macropædia* XVII. 1055/1 A tapestry set is a group of individual panels related by subject, style, and workmanship and intended to be hung together.

g. One of the shaped sections of a parachute.

1930 O. H. Kneen *Everyman's Bk. Flying* xii. 223 Two men straighten out the twelve 'panels' of silk. **1938** *Flight* 25 Aug. 168c/1 The canopy, which is 24ft. in diameter, is made up of 24 triangular gores cut from high-quality silk. Each gore is composed of four panels, the stitching of which forms a zig-zag pattern round the complete canopy. **1974** *Encycl. Brit. Micropædia* VII. 740/2 The canopy is given extraordinary strength by fabrication from up to 28 separate panels, or gores, each made up of smaller sections.

10. b. Also *transf.*

1898 C. H. Turner in J. Hastings *Dict. Bible* I. 421/1 This picture is cut up, as it were, into six panels, each labelled with a general summary of progress. **1927** A. H.

McNeile *Introd. New Testament* 79 He [*sc.* St. Luke] cuts the history into 'panels'.

IV. 14. a. (Later example in sense of an oil painting on a wooden board.)
1956 Hedström & Taylor tr. *Bergström's Dutch Still-Life Painting* 58 We may now compare Bosschaert's panel with two early works by other artists... The farther edge of the table is considerably more than half way up the panel.

c. A leaf or section of a folding screen or triptych, etc. Also *fig.*
1880 E. Glaister *Needlework* vi. 62 Panel screens..are excellent subjects for fine embroidery. **1896** F. Simmonds tr. *Ricci's Correggio* vii. 122 On the high altar of the oratory..there was once a triptych, the central panel of which represented Christ. **1936** E. G. Troche *Painting in Netherlands* 26/2 Possibly half of a diptych, of which the panel with Our Lady is now lost. **1959** P. & L. Murray *Dict. Art & Artists* 324 Usually the central panel [of a triptych] is twice the width of the wings, so that they can be folded over it to protect it. **1967** N. Amphoux tr. *Troyat's Tolstoy* (1970) ii. viii. 223 In painting the third panel of his triptych he had, as in *Boyhood*, combined the story of his friends, the Islenyevs, with his own. **1970** *Oxf. Compan. Art* 494/1 Panel painting was not developed fully until altars were furnished with painted retables.

15*. A control panel or instrument panel.
1897 E. Wilson *Electr. Traction* x. 219 The panel system of switchboards, whereby the various switches, complete for a given purpose, can be mounted on a panel of slate or marble and placed in line with those already installed. **1923**, etc. [see *control panel* s.v. *control sb. 5]. **1926** *Wireless World* 8 Dec. 760/3 A neat method of mounting a flash lamp bulb so that it may..illuminate the panel and tuning dials at night. **1929** V. W. Pagé *Ford Model 'A' Car* ix. 314 Remove the four screws which hold the instrument panel in place and pull panel back. **1933**, etc. [see *instrument panel* s.v. *instrument sb. 6]. **1940** *Railway Signalling & Communications* xix. 353 Points within 350 yds. of the signal box are mechanically operated by levers and the signals by switches on the panel. **1941** G. E. Irvin *Aircraft Instruments* xvii. 438 Large transport planes carrying two pilots require a dual set of instruments. This necessitates a large panel. **1964** M. Allward *Inside Jet Airliner* v. 39 The main panels contain the indicators and controls for the hydraulic and electrical systems, engine and fuel functioning, anti-icing and air-conditioning. **1969** T. C. Millington *Hillman Imps* x. 117 It is just possible to contrive a panel to mount two 2 in. gauges immediately above the speedometer. **1977** D. Beaty *Excellency* i. 8 He..clambered gingerly inside the fuselage..ran his fingers round the dusty panel.

VI. 20. (sense *5 b) *panel discussion, member*; (sense *5 c) *panel system*; (sense 14 a) *panel painter*; **panel analysis** *Sociol.*, analysis of attitude changes using the panel technique (see below); **panel-back** *a.*, applied to chairs with panelled backs (see quot. 1925); also *absol.* as *sb.*; **panel-beater,** one whose occupation is beating out the metal panels of motor vehicles; hence *panel beating*; **panel board** (see quot. 1954); **panel doctor,** formerly, a doctor registered as accepting patients under the National Insurance Act of 1913; **panel fence** *U.S.*, a fence constructed in panels or sections (see Panel *sb.*¹ 8); **panel fire** = *panel heater; **panel-game** (examples); (*b*) a 'quiz' or similar game played before an audience by a small group of people; hence *panel gamester*; **panel gauge** (see quot. 1966); **panel heater,** an electrically-heated panel mounted on a wall; hence *panel-heated* adj., *panel heating*; **panel-house** (earlier and later examples); **panel patient,** one who received medical treatment from a doctor under the Insurance Act of 1913; **panel pin,** a kind of thin nail, usu. having a tapered head, for securing panels; **panel practitioner** = *panel doctor; **panel-robbery,** the business of a panel-thief; **panel saw,** a fine-toothed saw used for cutting out panels; **panel show** = *panel-game (b); **panel stamp,** a stamp for decorating the panels in the cover of a book; hence *panel-stamped* adj.; **panel study** *Sociol.*, an investigation of attitude changes using a constant set of people and comparing each individual's opinions at different points in time; **panel technique** *Sociol.*, the technique used in panel studies; **panel-thief** (earlier and later examples); **panel truck** *U.S.*, a small lorry or van with a closed body; **panel van** *Austral.* = *panel truck; **panel wall,** (*a*) a division between two panels in a coal mine; (*b*) a wall in a building that does not bear any structural weight; hence *panel-walled* adj.; **panel warming,** warming by means of panel heaters.
1968 *Internat. Encycl. Social Sci.* XI. 371/1 Panel analysis gives rise to the study of an aspect of social change that tends to be neglected in studies of aggregate trends. **1969** J. J. Linz in Dogan & Rokkan *Quantitative Ecol. Anal. Social Sci.* v. 102 The possibility of using ecological units for a kind of panel analysis of aggregate data to explore problems of change over time. **1904** P. Macquoid *Hist. Eng. Furnit.* ix. 223 The late panel-back chair..dated 1691. **1925** Penderel-Brodhurst & Layton *Gloss. Eng. Furnit.* 119 *Panel-back or wainscot chair*, a cumbrous high-seated oak chair with heavy legs, stretchers, and high wainscotted back, in use in Tudor and Jacobean times. **1975** *Oxf. Compan. Decorative Arts* 360/2 The panel-back chair (which was also panelled beneath the arms and seat) was to establish for two centuries the standard pattern of the chair with back of square or rectangular shape. *Ibid.* 361/1 Richly upholstered chairs..were found with more refined types of panel-backs. **1908** *Daily Chron.* 21 Feb. 10/7 (Advt.), Panel beaters, used to hammering landaulette..panels in steel and aluminium. **1973** J. Wainwright *Devil you Don't* 14 The mechanics and panel-beaters working Sunday, double-time. **1978** *Cornish Guardian* 27 Apr. 6/1 (Advt.), Qualified mechanic and/or Panel Beater Sprayer required. **1968** *Gloss. Terms Mechanized & Hand Sheet Metal Work* (B.S.I.) 15 *Panel beating*, a method of roughly forming a hollow body, usually by hammer blows. **1972** K. Bonfiglioli *Don't point that Thing at Me* iii. 21 Moishe Spinoza Barzilai is, as a matter of fact, Basil Wayne & Co., the great coach-builders of whom even you, ignorant readers, must have heard, although not point one per cent of you will ever afford his lovely panel-beating, still less his princely upholstery. **1932** Panel-board [see *corner-block* s.v. *corner sb.¹ 16]. **1954** *Paper Terminol.* (Spalding & Hodge) 43 *Panel boards*, thick, tough, rigid boards made in various ways... Used in the manufacture of cars and in the building trade. **1972** *Gloss. Terms Timber* (B.S.I.) 27 *Panel-board*, fibre building board generally made from wood fibres. **1936** Panel discussion [see *creative a. 1 d]. **1956** W. H. Whyte *Organization Man* (1957) 55 It had started conventionally enough with a panel discussion in which I and two other men spoke. **1971** *Archivum Linguisticum* II. 20 A recent investigation of recorded panel discussions has shown that the average length of a unit of intonation used by the ten Present-English speakers involved was 5·3 (institutional) words. **1913** *Punch* 12 Feb. 127/2 To ask the Secretary of the Treasury if he could state the total population of the island of Canna, and who is the panel doctor. **1932** Kipling *Limits & Renewals* 300 A private party of thirty-two gentlemen and ladies,..all near enough neighbours in Shoreditch to use the same panel-doctor, poured into that man's consulting-room. **1957** R. Hoggart *Uses of Literacy* iii. 63 Working-class people have had years of experience of waiting at labour-exchanges, at the panel doctor's and at hospitals. **1800** W. Tatham *Hist. & Pract. Ess. Tobacco* 10 The *worm* or *pannel* fence,..consists of malled rails. **1858** J. A. Warder *Hedges & Evergreens* 113 A half-acre lot, with a seven foot panel-fence on one side and a hedge on the other. **1949** W. Faulkner *Knight's Gambit* 154 They would ride past mile after mile of white-painted panel fence. **1951** *Southern Folklore Q.* June 130 'Farm Fences'..pictures a panel fence adapted to rocky fields. **1934** *Archit. Rev.* LXXV. 110/1 Panel fires are less than five years old. **1844** G. Wilkes *Mysteries of Tombs* 54/1, I forgot to mention..that Malinda Hoag was convicted..in robbing a countryman of $54 by the panel game. **1857** *Porter's Spirit of Times* 5 Dec. 213/3 Females are employed as decoy-ducks to induce the yokels from the rural districts into places of *un*questionable character, where they are sure to be plundered of their money by the panel-game. **1928** Panel game [see *creep sb. 1 e]. **1953** *Evening News* 2 Jan. 5/3 The first edition of the new TV panel game 'Down You Go' was not an unqualified success. **1957** P. Wildeblood *Main Chance* 55 A singularly witless panel-game in which the contestants, in turn, thought of somebody whom they would like to be and their fellow-panellists had to guess the name. **1971** *Morning Star* 25 June 3/6 The new [radio] shows vary from current affairs, comedy, court dramas, Radio 4's answer to 'World in Action', and panel games. **1976** *Dumfries & Galloway Standard* 25 Dec. 9/3 The weather has continued to play havoc with the football programme and to reduce the 'Pools' to something of a 'panel game'. **1969** *Listener* 6 Sept. 308 Gilbert Harding..brought a compelling viewability to everything he did, whether as panel-gamester..or television cook and general pundit. **1909** *Cent. Dict. Suppl.*, Panel-gage. **1966** A. W. Lewis *Gloss. Woodworking Terms* 34 *Panel gauge*, marking gauge with a long stem and extra-wide stock for gauging the widths of wide boards. **1936** *Archit. Rev.* LXXIX. 109/2 The library is panel-heated, the criss-cross net-work of heating tubes being woven round the slots of the skylights. **1951** *Good Housek. Home Encycl.* 11/1 An electric fire..in the form of a panel heater mounted on the wall. **1928** *Domestic Engin.* XLVIII. 101 (*heading*) The physical and physiological effects of panel heating. **1848** 'N. Buntline' *Mysteries & Miseries N.Y.* iii. 44 This is a panel-house and I have led a bad, bad life for many a year. **1948** [see *lush sb.² 2]. **1967** *Parade* (Austral.) Oct. 61/3 After that Katie Marks and her gang decided to branch out into the panel-house racket—a brothel equipped with sliding panels which allowed thieves to rifle clients' clothing. **1938** *Public Opinion Q.* Oct. 602 A small magazine..which is published expressly for panel members. **1952** *Radio Times* 15 Aug. 37/2 What makes *What's My Line?* so popular? The personalities of the panel members, certainly. **1975** *Listener* 2 Jan. 21/3 What was spent on running the Arts Council? How many artists sat on your panels: who were they: who were the other panel-members? **1911** *Encycl. Brit.* XXVI. 405/2 The earlier painters whether illuminators of MSS. or wall and panel painters. **1937** *Burlington Mag.* Feb. 77/2 The group of Upper Rhenish panel-painters. **1954** M. Rickert *Painting in Brit.: Middle Ages* v. 120 The ability of Matthew Paris as a panel painter. **1913** *Outlook* 23 Aug. 247/1 Green tickets such as are used by ordinary panel patients when temporarily from home. **1924** J. Buchan *Three Hostages* i. 12 He would pay three visits a day to a panel patient, which shows the kind of fellow he was. **1950** T. H. Marshall *Citizenship & Social Class* 57 The early health service added 'panel patient' to our vocabulary of social class. **1964** A. Briggs in S. Nowell-Smith *Edwardian England* ii. 91 Other persons earning less than £160 a year could insure themselves voluntarily and become 'panel patients'. **1977** *Lancet* 8 Oct. 776/1 He took on no panel patients. **1951** *Good Housek. Home Encycl.* 320/1 Secure the glass..with a small sprig or panel pin. **1957** *Practical Wireless* XXXIII. 542/1 Fix the panel to the base with panel pins or small screws. **1960** *Farmer & Stockbreeder* 8 Mar. (Suppl.) 5/2 Take piece P and pin it to the frame with deep-drive panel pins, making sure hole S lines up with the drawer space. **1914** *Aberdeen Univ. Rev.* Nov. 50 The Panel practitioner being obliged to provide only what is termed ordinary medical treatment. **1922** *Encycl. Brit.* XXXI. 384/2 Medical men who act as panel practitioners continued to recommend their panel patients to the hospitals in increasing numbers. **1882** *Harper's Mag.* Feb. 400/1 Stories designed to teach our girls that theft, and arson, and panel-robbery..are the noblest exploits in which they can engage. **1754** *South Carolina Gaz.* 1 Jan. 2/2 Thomas Evance Has just imported..tenent, pannel and compass Saws. **1964** W. L. Goodman *Hist. Woodworking Tools* 151 The Hand, Panel, and Ripping Saws, ranging from 10 in. to 30 in. **1954** G. Marx *Let.* 16 Aug. (1967) 93 The gibbering idiots on panel shows, quiz shows, and other half hours of tripe. **1958** M. Dickens *Man Overboard* vii. 99 That long-lipped ass from the panel show. **1893** *Portfolio* XXIV. 55 John Reynes..often used a large panel stamp, representing the instruments of the Passion treated as a coat-of-arms. **1961** T. Landau *Encycl. Librarianship* (ed. 2) 52/2 Many elaborate panel stamps and roll stamps appear in the 14th and 15th centuries. **1952** J. Carter *ABC for Bk.-Collectors* 130 *Panel-stamped*, a term used by writers on book-binding to describe leather bindings of the 15th and 16th centuries decorated in blind with engraved blocks. **1958** M. Argyle *Relig. Behaviour* iii. 22 Panel studies in which the same subjects are repeatedly studied while their attitudes are changing. **1963** T. & P. Morris *Pentonville* vii. 182 Panel studies by Fiedler and Bass..indicate that inmate attitudes undergo a kind of cyclical change. **1964** M. Argyle *Psychol. & Social Probl.* xiii. 165 Panel studies during election campaigns have shown that there are some individuals who are more likely to change their voting intention than others. **1913** *Act* 3 & 4 *Geo. V* c. 37 § 11 Medical treatment under the panel system. **1926** *Encycl. Brit.* II. 861 At the time of its initiation the panel system met with great opposition from the medical profession. **1938** *Public Opinion Q.* Oct. 596 Instead of taking a new sample for each poll, repeated interviews with the same group of people have been tried. The experiences met with and the problems involved in such a panel technique will be discussed here. **1949** R. K. Merton *Social Theory* I. iii. 107 We may anticipate that the recent introduction of the panel technique—the repeated interviewing of the same group of informants—will in due course more sharply focus the attention of social psychologists upon the theory of attitude formation. **1844** G. Wilkes *Mysteries of Tombs* 48/1 Oh, he's a panel thief. **1947** *True* Nov. 69/1 The two lawyers had in addition the business of every free-lance safecracker, forger,..and panel thief whose business was worth having. **1937** Panel truck [see *bookmobile* s.v. *book sb. 18]. **1966** H. Kemelman *Saturday Rabbi went Hungry* v. 31 A light panel truck bearing the sign Jackson's Liquor Mart drove up. **1973** *Black World* Jan. 58/1 The panel truck followed. **1976** *CB Mag.* June 40/1 The Army Reserve Sergeant and afternoon soap opera buff suddenly switched on his headlights and churned onto the roadway in pursuit of a panel truck. **1969** *Age* (Melbourne) 24 May 60/11 (Advt.), Falcon panel van, 1962 mod[el]. **1977** *Western Morning News* 1 Sept. 6/2 (Advt.), Volkswagen LT 31 Panel Van, white, petrol. **1839** Ure *Dict. Arts* 976 Through the panel walls roads and air-courses are driven. **1962** *Listener* 11 Jan. 64/1 They have shown how sensitively the new and economical materials like concrete frames and panel walls can be handled. **1957** J. Kerouac *On Road* (1958) i. xi. 64 We had our headquarters in the main building, just a wooden contraption with panel-walled offices. **1934** *Times* 19 Feb. 20/5 They have got the latest ideas in panel warming, and the heat required in the type foundry is also supplied by gas.

panel, *v.* Add: **7.** Also *absol.*
1908 *Westm. Gaz.* 14 Mar. 13/2 All the gauzy fabrics will panel well.

panellist (pæ·nĕlist). Also (chiefly *U.S.*) panelist. [f. Panel *sb.*¹ 5 b + -ist.] **a.** A panel doctor. **b.** A member of a discussion panel, committee, group of judges, etc., esp. one taking part in a radio or television programme.
1937 G. Frankau *More of Us* 14 Alas! Came dawn when local panelist Pronounced Jack's Ladye Alice a cadaver. **1952** *N.Y. Times* 14 June 13 Bennett Cerf..television panelist and anthologist, has compiled another book of humor. **1955** *Picture Post* 14 May 10/3 This programme concludes with a gay, exciting parlour variation on musical chairs, in which the panellists exchange faces. **1958** K. Amis *I like it Here* iii. 36 Bowen..had something of the air of a television panellist. **1958** *New Statesman* 19 Apr. 501/1 A member of the audience has a go first; the panellists comment on what he has said; and he has a final brief right of reply. **1959** *Encounter* Dec. 52/1 The panelists at the Socialist meeting. **1971** C. Fick *Danziger Transcript* (1973) 131, I picked up a few bucks being a panelist on one of those Sunday political shows on TV. **1974** H. L. Foster *Ribbin'* vii. 316 Panel presentations were followed by ten small group meetings with panelists, former addicts and parents of addicts leading discussions. **1976** *Nature* 15 Apr. 633/2 Indeed the effect at the lower concentration of bitterness investigated was so marked that four out of the five panellists could detect no bitterness at all after presaturation of their tongues with sucrose.

‖ **panem et circenses** (pæ·nem et sɜːke·nzīz). [L.] = *bread and circuses* (*bread sb. 2 g).
1787 P. H. Maty tr. *Riesbeck's Trav. Germany* I. xxvi. 303 Every thing here cries out *panem et circenses*, and the multitude seem to have no other wishes than to have their paunches well filled, and a theatrical entertainment by way of dessert. **1864** C. M. Yonge *Bk. Golden Deeds* 100 Their [*sc.* the Romans'] cry was that they wanted *panem ac Circenses*. **1903** G. B. Shaw *Man & Superman* p. xxiv, At this moment the Roman decadent phase of *panem et circenses* is being inaugurated under our eyes. **1928** D. H. Lawrence *Lady Chatterley* xiii. 219 The masses are unalterable... Panem et circenses! **1961** D. L. Munby *God & Rich Society* iii. 61 Leaders who..win votes by offering *panem et circenses* to those they despise. **1966** *Listener* 29 Sept. 458/2 Sir Joseph Hutchinson is concerned with the provision of *panem*, but what of *circenses* for a *populus* with vast leisure time created by cybernation?

panetela, var. *panatela.

Paneth (pæ·neþ). *Histology.* The name of Joseph *Paneth* (1857–90), Austrian physiologist, used *attrib.*, in the possessive, and with *of* to designate a secretory cell present at the base of the crypts of Lieberkühn in the small intestine, and also the eosinophilic granules characteristic of their cytoplasm.
1899 *Veterinarian* LXXII. 555 These granular cells of Paneth showed the same reaction, and I was forced to conclude that they were mucus-forming. *Ibid.* 560 The cytoplasm in some instances has been observed to contain granules—Paneth's granules. **1938** *Gray's Anat.* (ed. 27) 1329 The deeper cells, especially in the duodenal glands, contain granules which stain characteristically with phosphotungstic haematoxylin and are termed granules of Paneth. **1968** Passmore & Robson *Compan. Med. Stud.* I. xxx. 21/1 There are only about a dozen Paneth cells in each crypt, but because of the enormous number of crypts the total mass of Paneth cells is large. **1976** *Cell & Tissue Kinetics* IX. 72 The number and localization of Paneth cells was not affected by changes in crypt cell kinetics during recovery after irradiation.

‖ **panettone** (panĕtō·ne). Also **panetone.** Pl. **panetto·ni.** [It.] A rich Italian bread made with eggs, fruit, and butter.
1922 D. H. Lawrence *Aaron's Rod* xvii. 251 He shoved a lump of cake—or rather panetone, good currant loaf—through the window, with a knife to cut it. **1938** Joyce *Let.* 20 Apr. (1966) III. 420, I spent all Easter Day with Lucia,..eating *panettoni di Milano* and fooling generally. **1967** *Economist* 28 Oct. 450/3 Motta's success is inevitably connected to 'panettone', a traditional Milanese Christmas cake. Owing to Motta's successful marketing of this one hand-made product, 'panettone' is now all over Italy, the recognised symbol of Christmas, perhaps even more so than the Christmas tree and the crib. **1969** R. & D. De Sola *Dict. Cooking* 167/1 *Panettone*: (Italian—a kind of cake bread) Often orange flavoured and raisin filled. **1972** L. O'Donnell *Phone Calls* iv. 48 She..pushed forward a plate on which there was a high glossy cake with only one slice cut out of it. 'Please, have some *panetone*. Is very good—.' **1978** *Nieman-Marcus Christmas Bk.* 91 From Italy, the traditional semi-sweet panettone..bread studded with candied fruits and nuts.

Pan-Europe·an, *a.* [f. Pan- + European *a.*] Pertaining to, affecting, or extending over the whole of Europe. Hence **Pan-Europe·anism.**
1901 Conrad & Hueffer *Inheritors* viii. 117 There was an 'All Round the World Cable Company'..., and a 'Pan-European Railway, Exploration, and Civilization Company' that let in light in dark places. **1931** *Ann. Reg. 1930* 172 This same trend in Italian policy was also manifested in Italy's reply to M. Briand's 'Pan-European' proposal in the course of which the Fascist Government declared its disagreement with any scheme of this kind which excluded Turkey and Russia. **1942** L. B. Namier *Conflicts* 2 The two great nations of Central Europe..burdened with Pan-European past..remained in a condition of political disunion and dynastic subdivision. **1961** *Guardian* 9 June 8/6 The later Pan-Europeanism of this loneliest of poets [*sc.* Rilke]. **1966** *New Statesman* 14 Oct. 540/2 This journal repeatedly emphasised that the famous breeze of European competition, or the large accumulations of capital open to pan-European industries, were no substitute for a successful socialist economy. **1979** *Daily Tel.* 20 Jan. 16/6 The new flowers..have been put through their paces under the Fleuroselect scheme which is run by a pan-European organisation for testing them.., with a secretariat in The Hague.

panfan (pæ·nfæn). *Geomorphol.* [f. Pan- + Fan *sb.*¹] = *pediplain.
1915 A. C. Lawson in *Univ. California Publ. Geol.* IX. iii. 33 The surface thus evolved is, in its ideal completion, wholly one of aggradation, a vast alluvial fan surface to which for convenience in discussion I propose to give the name *panfan*. **1931** *Jrnl. Geol.* XXXIX. 138 Pediments are essentially compound graded flood plains excavated by ephemeral streams. As their growth corrodes the mountain mass they eventually coalesce to form smooth graded domes to which Lawson applied the term 'panfan'..—a name that is etymologically unfortunate because these features are not concerned with alluvial fans at all. **1933** *Geol. Mag.* LXX. 345 The rock-floors of the desert, known as pediments and panfans. **1954** W. D. Thornbury *Princ. Geomorphol.* xi. 291 Lawson (1915) proposed the term *panfan* to designate 'an end stage in the process of geomorphic development in an arid region in the same sense that the peneplain is an end stage of the general process of degradation in a humid climate'. He also recognized that both peneplains and panfans represent penultimate rather than ultimate stages of degradation. **1974** [see *pediplain].

panfish. 1. (Earlier and later examples.)
1833 J. F. Watson *Hist. Tales Philad.* 49 Before the house flows a small but deep creek, abounding in pan-fish. **1838** J. F. Cooper *Home as Found* II. v. 71 The Egyptians use them as a pan-fish. **1852** *Trans. Mich. Agric. Soc.* III. 226 These little fish are sometimes used as pan-fish. **1888** G. B. Goode *Amer. Fishes* 36 In season the White Perch is *the* pan-fish, excelled by none. **1969** R. & D. De Sola *Dict. Cooking* 167/1 *Panfish*, any fish suitable for frying, usually hand-caught fish like sunfish or catfish. **1970** *Globe & Mail* (Toronto) 26 Sept. 40/5 In addition to the ancestral stock of Atlantic salmon, Lavalee Lake holds bass, pan fish.

panfry (pæ·nfrəi), *v.* Chiefly *N. Amer.* [f. Pan *sb.*¹ + Fry *v.*¹] *trans.* and *intr.* To fry in a pan with shallow fat. Also *transf.* So **pa·n-fried** *ppl. a.*
1942 M. K. Rawlings *Cross Creek Cookery* Pan-Fried Young Quail or Dove. **1957** J. Kerouac *On Road* (1958) x. 174 That pan-fried chow mein flavored air. **1973** *Sat. Rev. World* (U.S.) 18 Dec. 48/3 It [*sc.* the Kahala's Maile Room] imports filets of trevally from New Zealand..and panfries them with tomatoes. **1974** *Globe & Mail* (Toronto) 20 Mar. 35/1 Ling or freshwater cod can be cooked in a variety of ways—the simplest being to panfry. **1976** *Woman's Day* (N.Y.) Nov. 104 Panfried liver and onions. **1977** *N.Z. Jrnl. Agric.* Jan. 63/1 Grill or pan-fry the bacon.

pang (pæŋ), *sb.*² [Echoic.] Vocal imitation of a short, resonant sound, such as that produced by a drum, a horse's hoof, etc.; a sound of this character.
1925 E. Sitwell *Troy Park* 65 As the hoofèd sound of a drum marched on With a pang like darkness. **1955** E. Pound *Classic Anthol.* iii. 188 Every man eager to pace the stallions, 'Pang, pang' and Rein bells chink. **1958** L. Durrell *Mountolive* vii. 144 There were two excellent hard courts which rang all day to the pang of racquets. **1958** —— *Balthazar* vii. 155 The tattling flutes and the pang of drums.

panga¹ (pā·ŋgă). Also **ponga, pongo.** [a. Amer. Sp. *panga* a boat.] A flat-bottomed boat with rising stem and stern.
1927 G. Bradford *Gloss. Sea Terms* 124/2 *Panga*, a flat-bottomed rowboat of Central America. **1948** R. de Kerchove *Internat. Maritime Dict.* 517/1 *Panga*, a dug-out canoe from Panama, double-ended with rising stem and stern. These craft are about 18 ft. long by 4 ft. 6 in. wide and are cut out of one cedar log. **1970** *National Fisherman* Jan. 18B/4 The chaser-skiffs, also called speed-boats or pongos, are lowered overboard. **1971** *Islander* (Victoria, B.C.) 4 July 15/3 The reds and greens and yellows of the double-ended pongas which the lobster and bonito fishermen took far out to sea. **1978** *Daily Colonist* (Victoria, B.C.) 7 May 17/2 We haul anchor as soon as the ship-to-shore shuttle, called a panga, brings aboard the last of our cruise-mates.

panga² (pæ·ŋgă). [Swahili.] A large knife used in Africa either as an implement or as a weapon. Also *attrib.*
1935 E. Hemingway *Green Hills Afr.* (1936) iv. i. 216 Chopping our way through with the long brush knives that are called pangas. **1952** *Time* 3 Nov. 36 Once pooh-poohed as mere 'press exaggeration', the Mau Mau have already mutilated scores of whites and 'loyal' blacks, with their favorite weapon, the *panga*—a long, machete-like knife. *Ibid.* 10 Nov. 38/3 On the front of the governor's car waved his official flag: two crossed *pangas* (broad-bladed African knives used to chop bananas). The *pangas* seemed symbolic last week, for Kenya Colony, the brightest jewel in Britain's East African Empire, is bleeding badly in a *panga* war. **1953** *Ibid.* 12 Jan. 26/1 There was a noise at the door, a shout, and a gang of Mau Mau thugs, led by the ranch's male cook, burst into the living room, brandishing *panga* knives. **1953** *Newsweek* 6 Apr. 38/3 As the villagers ran from their blazing homes, waiting Mau Maus struck them down with *pangas* (long knives) and hatchets. **1954** D. H. Rawcliffe *Struggle for Kenya* ix. 86 The isolated loyalist groups in the reserves..were still armed with little more than *pangas* and spears. **1955** O. Meeker *Report on Afr.* xii. 199 But until I reached Kenya no-one said a word about the Mau Mau terrorists, secret society oaths, the long knives called *pangas*, or the other Kikuyu specialities. **1964** C. Willock *Enormous Zoo* viii. 138 African butchers sharpened their pangas on stone. **1969** *Daily Tel.* 28 Oct. 16 An African Presbyterian Minister was killed by a panga gang for refusing to take Kikuyu tribal oaths. **1972** J. McClure *Caterpillar Cop* iii. 34 'I hear the murder was sometime around six.'.. 'Usual bit of passion and panga.' **1976** *Maclean's Mag.* 6 Sept. 51/2 Zulu mobs rampaged through Soweto armed with pangas (cane cutters), axes, spears, and knives. **1976** *Survey* Summer-Autumn 303 Forest fighters had..often only the ordinary rural blade of East Africa, the *panga*. **1977** D. Beaty *Excellency* xii. 139 The soldiers had taken out panga knives and were cutting down thorn bush and scrub.

Pangæa (pændʒī·ă). *Geol.* [f. Pan- + Gr. γαῖα land, earth.
The name is freq. stated to have been coined by A. Wegener 1914, in *Die Entstehung der Kontinente und Ozeane*, but it has not been found in the 1st ed. of that book (actually published in 1915); *Pangäa* does occur in ed. 2 (1920), p. 120, but with no indication that Wegener is coining it.]
A vast continental area or supercontinent comprising all the continental crust of the earth which is postulated to have existed in late Palæozoic or Mesozoic times before breaking up into Gondwanaland and Laurasia.
1924 J. G. A. Skerl tr. *Wegener's Orig. Continents & Oceans* xiii. 192 Thus the Pangæa of the Carboniferous era had already an anterior margin (America), which became folded (Precordilleras)..; and a posterior margin (Asia), from which littoral ranges and fragments became detached, and remained fast in the sima of the Pacific as groups of islands. **1928** C. Schuchert in *Theory Continental Drift* (Amer. Assoc. Petroleum Geol.) 106 The rifting of Pangaea and the floating away of Australasia, Antarctica, and the Americas are said to have begun east of Africa in Jurassic time and west of Euro-Africa in early Cretaceous time. **1958** *Continental Drift* (Tasmania Univ. Geol. Dept.) 177 Apart from the dissociation of Pangaea into independent Laurasia and Gondwana, the fundamental picture is much as Wegener saw it. **1971** I. G. Gass et al. *Understanding Earth* xv. 230/2 Before the Pangaea of Permo-Triassic time was formed, one must envisage a different mosaic of continental fragments reassembled in a different way. **1974** *Sci. Amer.* Apr. 81/1 Ancient ocean currents in the vicinity of Pangaea, the single 'supercontinent' that is believed to have existed near the beginning of the Triassic period some 225 million years ago, are indicated here. **1977** *Ibid.* Mar. 92/3 About 600 million or perhaps one billion years ago the ancestors of all the present continents were evidently combined into one immense supercontinent, named Pangaea, which may have come into existence as much as 2.7 billion years ago.

Pangan (pæ·ŋgăn). [Malay, = 'forest country'.] = Negrito.
1839 T. J. Newbold *Pol. & Statistical Acct. Straits of Malacca* II. v. 60 In the interior of Pakaa, an aboriginal race is said to exist, termed Pangan. **1929** A. Chambers tr. *Schebesta's Among Forest Dwarfs Malaya* v. 162 From the Malays in Tadoh I first heard the word *Pangan*. This was the name they gave to the Orang-Utan tribes. **1947** R. O. Winstedt *Malays* 8 The oldest of Malaya's existing races anthropologically is the Negrito, termed Semang in Perak and Pangan in Kelantan. **1965** R. McKie *Company of Animals* i. 35 Pangans had been digging with fire-hardened sticks for tubers, and later we found one of the small camps of these jungle aborigines.

Pangasinan (pāngăsīnā·n). Pl. **Pangasinans, Pangasinanes.** [Native name.] **a.** (A member of) a people inhabiting the central area of the Luzon district of the Republic of the Philippines. **b.** The Austronesian language of this people.
1840 [see *Ilocano]. **1885** [see *Ibanag sb. and a.]. **1948** D. Diringer *Alphabet* vii. 432 The other important vernaculars spoken by the Filipinos are: (1) Pangasinan, [etc.]. **1954** Pei & Gaynor *Dict. Linguistics* 158 *Pangasinan*, a language spoken in the Philippine Islands by almost 400,000 persons; a member of the Indonesian sub-family of the Malayo-Polynesian family of languages. **1964** E. Bach *Introd. Transformational Gram.* v. 89 Some languages that have been worked on from this point of view..are English,..Pangasinan. **1974** *Encycl. Brit. Micropædia* VII. 717/a *Pangasinan*, eighth largest cultural-linguistic group in the Republic of the Philippines. They occupy the central area of the province of Pangasinan in Luzon.

pangeran (pæŋgĕræ·n). Also **pangarang.** [Indonesian.] A Javanese prince or noble.
1817 T. S. Raffles *Hist. Java* I. ii. 79 His [*sc.* the sovereign's] family are called *Pang'érans*. **1821** J. Leyden tr. *Malay Annals* viii. 272 The pangeran of Surabaya..came to Malaca. **1831** *Canton Miscellany* ii. 79 The Princes or Pangarangs who were seated on chairs walked with us. **1900** Conrad *Ld. Jim* xx. 219 He came into the Council-hall where all the rajahs, pangerans, and head-men were assembled, with the queen,..reclining on a high couch under a canopy.

Pan-Ge·rmanist. [f. Pan-German *a.* and *sb.* + -ist.] A supporter of Pan-Germanism. Also *attrib.*
1909 *Daily Chron.* 25 June 4/6 A racing yacht devised, built, and..manned in Germany—an object of idolatry..to the Pan-Germanists. **1914** *Atlantic Monthly* Oct. 448/2 The schemes of the Pan-Germanists indeed reach to the creation of a vast confederation of states. **1939** *Tablet* 3 June 706/1 He defends himself as best he can against the Pan-Germanist propaganda of the turbulent German minority. **1957** Elliott & Summerskill *Dict. Politics* 251 Pan-Germanists have advocated particularly the absorption into Germany of the German-speaking provinces of Austria. **1974** *Encycl. Brit. Micropædia* VII. 717/b Pan-Germanists continued to press for expansion; the most articulate and active force toward that end was Hitler and the Nazi Party.
Hence **Pan-Germani·stic** *a.*, of, pertaining to, or supporting Pan-Germanism.
1915 *World's Work* (N.Y.) Aug. 456/1 To block the Pan-Germanistic plan.

panghulu, var. *penghulu.

panglima (pānglī·mă). Also **penglima.** [Malay *pĕnglima*.] A Malay leader of secondary rank; a Malay chief.
1839 T. J. Newbold *Pol. & Statistical Acct. Straits of Malacca* I. v. 237 For further security, two panglimas sat on each side. **1900** W. W. Skeat *Malay Magic* 169 Two swords were produced and placed crosswise, and a couple of Panglimas selected for the dance. **1939** A. Keith *Land below Wind* xv. 264 The Penglima, or native chief..met our *perahus* with much ceremony.

Pangloss (pæ·nglǫs). Name of the philosopher and tutor in Voltaire's *Candide* (1759) who believes that 'all is for the best in the best of all possible worlds', used allusively for one who is optimistic regardless of the circumstances. Also *attrib.* Hence **Pa·nglossism,** an unrealistically optimistic attitude or saying; **Panglo·ssic** *a.*, characteristic of Pangloss.
1844 Disraeli *Coningsby* I. ii. i. 153 The political Panglosses who..were continually proving that this was the best of all possible governments. **1926** J. S. Huxley *Ess. Pop. Sci.* xii. 147 This conclusion, from its inception with some of the Stoics to its Panglossic latter end in certain evolutionists, he dismisses. **1928** A. Huxley in *Life & Letters* 1 Oct. 345 We are to interpret them, Pangloss fashion, in terms of preconceived philosophy. **1931** *Times Lit. Suppl.* 10 Dec. 994/4 We can best envisage him..still the sworn disciple of Pangloss the optimist. **1959** *Encounter* Nov. 66/2 He includes..some of his pet Panglossisms, such as that learning is happiness. **1964** *Economist* 11 July 128/2 This is cheerfulness raised to the level of Panglossism. **1973** *Listener* 5 Apr. 438/1 The [Egyptian] regime's official Dr. Pangloss, Culture Minister Dr Abdel Khader Hatem.

Panglo·ssian, *a.* and *sb.* [f. prec. + -ian.] **A.** *adj.* Of, pertaining to, or characteristic of the philosophy of Pangloss. **B.** *sb.* One who shares this philosophy.
1831 Disraeli *Young Duke* I. i. iii. 23 He was quite disembarrassed of that Panglossian philosophy, which had hitherto induced him to believe, that the Earl of Fitz-pompey was the best of all possible uncles. **1922** Hardy *Late Lyrics* Apol. p. xi, Hence should anything of this sort in the following adumbrations seem 'queer'—should any of them seem to good Panglossians to embody strange and disrespectful conceptions of this best of all possible worlds, I apologize; but cannot help it. **1936** G. B. Shaw *Simpleton* Pref. 3 If author and journalist are both placid Panglossians, convinced that their civilization is the best of all possible civilizations..there is no trouble. **1937** A. Huxley *Ends & Means* vii. 68 In those who make the identification it induces a kind of busy, Panglossian fatalism. **1957** M. McCarthy *Memories Catholic Girlhood* p. xxiv, Some of my relations philosophize to this effect, in a somewhat Panglossian style. **1967** *Guardian* 17 May 1/3 This somewhat Panglossian interpretation of the significance of General de Gaulle's press conference. **1976** *Nature* 18 Mar. 196/2 The first, and most widely appreciated, is the old Panglossian fallacy that natural selection favours adaptations that are good for the species as a whole, rather than acting at the level of the individual.

Pangola, var. *Pongola.

pangram (pæ·ngræm). [f. Pan- + -gram.] A sentence containing all the letters of the alphabet (see also quot. 1953). So **pangramma·tic** *a.*
1933 M. E. Ohaver *Cryptogram Solving* 31 Pangrammatic, containing all the letters of the alphabet. **1953** W. R. Trask tr. *Curtius's European Lit.* xv. 283 The 'pangrammatic' affection, which consists in having as many successive words as possible begin with the same letters. **1963** *Medium Ævum* XXXII. 149 'Pangrammatic' verses are far older than the thirteenth century. **1964** *Sci. Amer.* Sept. 222 The pangram, an ancient form of word play, is an attempt to get the maximum number of different letters into a sentence of minimum length. **1965** *Time* 17 Sept. 72 Also represented: Sotadic [palindromic] verses, pangrammatic rubaiyat and problems in alphametics (alphabet arithmetic).

panhandle, *sb.* Add: Also applied to territories outside the U.S.A. (Earlier and later examples.)
1856 *Porter's Spirit of Times* 8 Nov. 159/1 He was from old Virginny—from what, he said, they called the *Pan-handle*. **1861** *Vanity Fair* (N.Y.) 25 May 246/2 We will wrap the flag of our fathers around the 'Pan Handle' of Virginia, and upset the entire dish of Old Dominion Secession. **1877** E. E. Hale *G.T.T.* 30 So they..whirled relentlessly across the Pan Handle by which domestic name that funny strip of West Virginia is known which shoots up like an inverted icicle between Pennsylvania and Ohio. **1932** *Atlantic Monthly* Mar. 295/1 Your atlas will show that Idaho has a panhandle jutting up to the Canadian border. **1952** B. Ulanov *Hist. Jazz in Amer.* (1958) iv. 33 Jack Teagarden, a big burly Texan with an infectious Panhandle accent in his singing, has always been associated with the blues. **1960** *20th Cent.* Mar. 248 The 'panhandle' of North Wiltshire near Lechlade. **1967** *Economist* 14 Oct. 141/2 These targets are usually in the 'panhandle', as Americans call the thin southern part of North Vietnam. **1974** *Union* (S. Carolina) *Daily Times* 22 Apr. 1/3 Mt. Hermon straddles the Lebanese-Syrian border and is just north of the Israeli panhandle. **1977** *Time* 7 Mar. 52/1 In a grimy arc, from Nebraska through the plains of Kansas and Colorado, on into the panhandles of Oklahoma and Texas, scenes right out of *The Grapes of Wrath* suddenly materialized in the swirl of dust billowing up to 12,000 feet.
2. The act of begging (cf. next).
1849 J. J. Hooper in *Spirit of Times* 14 Apr. 87/3 The elephant was the great point of attraction. 'I want his hide and frame for a corn crib,' said a fellow from the Pan-Handle Beat. **1894** *Harper's Weekly* 5 May 429/2 'Workin' the pan-handle.' 'Eh?' 'I mean, are you beggin'?' **1900** Ade *More Fables* 142 He usually found some one waiting on the Door-Step to give him the Sign of Distress and work the fraternal Pan-Handle on him. **1942** Berrey & Van den Bark *Amer. Thes. Slang* § 486/2 *A solicitation*,..panhandle.
Hence (sense 1) **pa·nhandled** *a.*
1976 T. Stoppard *Dirty Linen* 65 Enormous women in taffeta dresses stir the air with panhandled fans.

panhandle (pæ·nhænd'l), *v. slang.* (orig. and chiefly *U.S.*). [Back-formation from next.] *trans.* and *intr.* To beg (from); to steal or purloin. Hence **pa·nhandling** *vbl. sb.* and *ppl. a.*
1903 *N.Y. Even. Post* 9 Dec. 1 The prisoners were members of a 'panhandling' corporation which operated extensively throughout the financial district. **1904** G. H. Lorimer *Old Gorgon Graham* 53 A lot of men..who wouldn't think of asking for money, will panhandle both sides of a street for favors. **1907** 'O. Henry' *Strictly Business* (1910) xix. 242 He felt his sleeve grasped and held. Suspecting that he was about to be panhandled, he turned a cold and unprofitable face, and saw that his captor was—Dawe. **1931** 'D. Stiff' *Milk & Honey Route* viii. 84 Domestic panhandling, or hitting back doors,..seems to be on the decline. **1932** L. C. Douglas *Forgive us our Trespasses* (1937) xiv. 276 It's never been in a suppliant attitude, panhandling the brethren and sisters for sixpences to pay the parson's coal-bill. **1951** E. Paul *Springtime in Paris* xi. 198 The street music..was strictly for panhandling—stray accordion players, atrocious old fiddlers, and half-crazed women with cracked voices and maudlin lyrics. **1959** *Times Lit. Suppl.* 6 Nov. p. xxxv/5 A poor art scholar..has his precious manuscript on Giotto destroyed by the panhandling Jewish refugee from Israel. **1973** 'A. Blaisdell' *Crime by Chance* (1974) ix. 173, I ain't had a bite to eat since yesterday. I was in that bar panhandlin'. **1976** *Yellowstone Explorer* July 1/3 The absence of bears along the park's roads is a major concern of visitors, especially those who were here just a few years ago and saw literally dozens of bears panhandling picnic items passing cars. **1977** *It* June 5/2 They tackled the dilemma facing anyone who has been pan-handled—'Why don't I take this poor bastard home with me,'—and they did it on a massive scale.

panhandler (pæ·nˌhæːndləɹ). [f. Panhandle *sb.* in Dict. and Suppl. or *panhandle *v.* + -er¹.] **1.** A beggar. *slang* (orig. *U.S.*).
1897 F. Moss *Amer. Metropolis* II. vii. 393 A party of petty thieves and 'pan-handlers' (able-bodied street beggars), went into the 'Morgue'. **1899** Ade *Doc. Horne* xxiii. 255 The freckled boy then announced that he had 'sized' the hustler for a 'panhandler' from the start. **1909** [see *lone *a.* 3 c]. **1929** *Daily Tel.* 8 Jan. 11/6 Large profits from begging in the rich Fifth Avenue business districts have produced a 'king of the pan-handlers'. **1954** *Manch. Guardian Weekly* 30 Dec. 14/3 The delicatessen stores alone in this city [*sc.* New York] would make the Phoenicians feel like pan-handlers and the merchant princes of Venice like fugitives from a soup kitchen. **1962** H. Hood in R. Weaver *Canad. Short Stories* (1968) 2nd Ser. 202 In a minute they would speak to him. They always did, drunks and panhandlers. **1966** F. Shaw et al. *Lern Yerself Scouse* *A pan'andler*, a scrounger of food. **1973** 'E. McBain' *Hail to Chief* vi. 113 Don't..start screaming if a panhandler taps you on the shoulder. He may only want a quarter for a drink. **1975** *Globe & Mail* (Toronto) 5 July 2/1 Panhandlers..urinate in doorways, sleep in the streets. **1978** G. Vidal *Kalki* vi. 136 First they would approach a well-dressed person of the sort who would normally run a mile to avoid a panhandler for god, any god.
2. (With capital initial.) A native or inhabitant of a Panhandle.
1936 *Sun* (Baltimore) 12 May 12/5 If the gentleman from Texas wants us to revise our lingo so that it meets the approval of the Panhandlers, the least he could do is offer a substitute for the word he wishes eliminated. **1940** E. Fergusson *Our Southwest* xviii. 335, I talked all this over with a native-born Panhandler. **1949** *10 Story Western* May 28/2 He remembered how it was down there in the Panhandle—how the Panhandlers hated these Rio Valley floaters as no Idaho man could.

Panhonlib (pænhǫnli·b). [f. *Pan*ama, *Hon*duras, *Lib*eria.] Designating a merchant ship of Panama, Honduras, and Liberia flying a 'flag of convenience'. Cf. *flag *sb.*⁴ 1 f, *Panlibhonco.
1958 *Times* 1 Apr. 11/5 There has recently been an intensification of the campaign against registration of ships, particularly tankers, under what have come to be known as 'PanHonLib' flags. **1958** [see *Panlibhonco]. **1962** S. G. Sturmey *Brit. Shipping & World Competition* 228 Few British owners have sought Panholib registration.

panic, *sb.*¹ **b. panic-grass** (earlier and later examples).
1597 Gerard *Herball* i. 7 Pannicke grasse is garnished with chaffie and downie tufts. **1870** W. Robinson *Wild Garden* ii. 121 Twiggy Panic Grass..is an elegant plant. **1901** C. T. Mohr *Plant Life Alabama* 355 Wiry Panic-Grass... Exposed places in light soil. **1929** Weaver & Clements *Plant Ecol.* vii. 134 (*caption*) Competition between tall panic grass and evening primrose. **1963** Gleason & Cronquist *Man. Vascular Plants North-eastern U.S.* 101 Panic grass. Spikelets lanceolate or hisiform to ovate.

panic, *a.* and *sb.*² Add: **B.** *sb.*² **2. c.** *fig.* A noteworthy or amusing person, thing, or situation.
1936 R. Ackland *After October* i. 47 Oh, my dear, aren't you *swell*! All grown up and sophisticated. Doesn't she look a panic, Timmy? *Ibid.* 49 My dear, it was a panic! **1946** T. Williams *27 Wagons Cotton* iii. 27 *Flora*: Says—(She goes off into another spasm of laughter.) *Jake*: What ever he said must've been a panic!
3. a. (Earlier U.S. examples.)
1842 *Southern Q. Rev.* I. 88 The sudden and violent contraction of 1833..produced the scenes of what is usually termed the 'panic session'. **1854** T. H. Benton *30 Years' View* I. 369/2 On the second day of December, 1833, commenced the first session of the Twenty-third Congress, commonly called the Panic Session.
3. b. panic bolt, a special bolt for a door designed to unfasten readily in emergencies; **panic button,** a switch or button for operating various devices in emergencies (see quots.); also *fig.* in phr. *hit, press, the panic button*, to become over-excited, take emergency measures (the origin of the expression is discussed in *Amer. Speech* XXXI (1956) no. 3, 240); **panic buying,** the buying in large quantities of goods of which a shortage is threatened or suspected; hence (as a back-formation) **panic-buy** *v.*; **panic party** (see quots. 1929, 1943); **panic stations,** a state of emergency (freq. *fig.*); **panic-stricken** *a.* (earlier and later examples); **panic-striking,** causing, or likely to cause, a panic; **panic-struck** *a.* (earlier examples).
1930 *Aberdeen Press & Jrnl.* 1 May 7/3 When he took the cinema in July, 1928, he put panic bolts on the wooden door..where there were ordinary slip bolts before. **1940** *Chambers's Techn. Dict.* 611/2 *Panic bolt*, a special form of door-bolt which is released by pressure at the middle of the door; commonly used on exit-doors in public buildings. **1964** J. S. Scott *Dict. Building* 223 *Panic bolt*, a door bolt often used at the double exit doors of theatres. It is opened by pressure from inside on to a horizontal bar within the door at waist height. **1972** *Times* 28 Dec. 1/8 London fire brigade checked the new arrangements last night and said that 'panic bolts' on the inside complied with regulations. **1955** *Amer. Speech* XXX. 117 *Hit the panic button*, panic, get excited. **1956** *Ibid.* XXXI. 240 Discussion with several pilots..reveals at least four buttons or switches, each one of which may be referred to as the 'panic button'. **1958** *Ibid.* XXXIII. 183 *Panic button*, the release button on the one-point release harness worn by the smokejumper. When a jumper is hung up in a tree,..he twists a metal knob (the 'panic button') on his harness fifteen degrees to the left and hits it. This releases the webbing straps of the harness, and the entire parachute and harness fall off. 'I hit the panic button and let down about sixty feet.' **1959** *Times* 9 Mar. 13/4 *Hit the panic button*, get over-excited. **1962** *Review & Herald* 25 Oct. 24/3 Dr. Franklin Clark Fry, president of the Lutheran Church in America, warned here that the time is coming for world Christian missions 'to press the panic button', because Christianity is dying out. **1970** *Washington Post* 30 Sept. D4/5, I haven't thrown the panic button yet and I'm not going to. **1970** D. Francis *Rat Race* iv. 45 Someone in the control tower had pressed the panic button. Fire engines screamed up. **1970** *Motoring Which?* July 113/1 As a last resort, pressing the 'panic-button' would 'dump' the craft, but could cause damage on a hard surface. **1972** T. Ardies *This Suitcase* viii. 71 The President reacted reasonably enough. He didn't push the panic button... He..ordered a stand-by alert. **1973** *Listener* 15 Feb. 200/3 The panic-button was pressed: foreign-exchange markets slammed their doors. **1974** *Scottish Daily Express* 14 Oct. 1/6 But one grocery chain manager stressed last night there was no immediate need for customers to panic buy as stocks were high and many stores' supplies were still unaffected by the strike. **1974** *Times* 15 Nov. 18/3, I had already panic-bought five gallons of petrol..saving 42p on next week's prices. **1942** *Washington Post* 23 Nov. 11/6 An anti-hoarding regulation for householders..to curb panic-buying of foods. **1949** *Sun* (Baltimore) 2 Dec. 4/5 There is no question but that panic buying has contributed to the current price rise. **1972** *Guardian* 18 Aug. 13 Concern over the number of mortgages available in the future and fears that selling [houses] could become more difficult have created a flood of panic-selling (which makes a change from panic-buying). **1973** *Times* 7 Dec. 18/8 Panic buying of spirits in the High Street, caused largely by forecasts of the shortage. **1974** *Guardian* 19 Jan. 20/2 Panic-buying would in itself lead to 'artificial shortages'. **1919** *Boy's Own Paper* XLI. ix. 456/2 The 'Farnborough'..disembarked a 'panic party'..to pretend that the officers and crew were abandoning their ship. **1929** *Papers Mich. Acad. Sci., Arts, & Lett.* X. 313 *Panic party*, a feigned demonstration of alarm or panic on board a decoy (mystery) ship in order to lure the commander of a submarine alongside. When a mystery ship was torpedoed, the panic party took to the boats, apparently abandoning the vessel, but always leaving on board another crew to man the guns and finish the submarine if it came near enough. **1929** F. C. Bowen *Sea Slang* 100 *Panic Party*, the men whose job it was to leave a Decoy Ship (Q-boat) in disorder when a German submarine opened fire. **1932** 'N. Shute' *Lonely Road* vii. 144 They shelled the panic party in the boats. **1943** Baker *Dict. Austral. Slang* (ed. 3) 57 *Panic party*, any rush move (Digger slang). **1961** in Partridge *Dict. Slang* Suppl. 1213/1 *Panic stations, be at*, to be prepared for the worst. **1963** 'J. Prescot' *Case for Hearing* iii. 53 Someone has been into Greenwood's again..and got away with another three hundred... The police seem to be at panic stations about it. **1972** A. Draper *Death Penalty* xviii. 113 Let's face it, Caleb, you'll be the first to run to panic stations. **1804** M. Hays *Harry Clinton* xxxii. 199 A ladder was..speedily brought, the panic-stricken family assisted to descend, and charitably conducted to a neighbouring inn. **1904** [see *Dixie² 1 b]. **1977** M. T. Bloom *13th Man* iii. 39 They both turned their faces to me and the car and they looked panic-stricken, stuck in place. **1851** H. Melville *Moby Dick* III. vii. 56 Pip loved life, and all life's peaceable securities; so that the panic-striking business in which he had somehow unaccountably become entrapped, had most sadly blurred his brightness. **1934** R. Campbell *Broken Record* ii. 46 A senseless, dutiful exposure of panic-striking and depressing facts is worse than any amount of mischievous lying. **1776** J. Thacher *Mil. Jrnl.* (1823) 70 [Washington] made every effort to rally them, but without success; they were so panic struck that even the shadow of an enemy seemed to increase their precipitate flight. **1835** J. E. Alexander *Sk. Portugal* vi. 139 The Miguelites at last became panic-struck.

panic, *v.* Delete (*nonce-wd.*) and add later examples.
1917 'Contact' *Airman's Outings* 184 Nothing seems to panic the Boche more than a sudden swoop by a low-flying aeroplane. **1919** H. L. Wilson *Ma Pettengill* iv. 127 He was sure going to annoy Ben from time to time, even if he didn't panic him much. **1932** Kipling *Limits & Renewals* 169 Then I'd come round the corner and hailed him, and that panicked him. **1932** *New Yorker* 4 June 46/1 Gough..had a violence and fervor on the platform which packed them and panicked them everywhere. **1957** *Economist* 30 Nov. 795/1 The markets are healthy and not likely to be panicked into headlong fall. **1966** A. Sachs *Jail Diary* xxi. 189 That will panic the Whites in South Africa even further. **1971** *Daily Tel.* 2 Nov. 2/5 A radio dramatisation of H. G. Wells's 'War of the Worlds'..panicked thousands throughout America in 1938. **1974** 'M. Innes' *Appleby's Other Story* xi. 93 At least it panicked her. She came..to tell me a pack of lies. **1975** *Daily Tel.* 1 May 13/6 When I received a tax demand asking for £600 by Tuesday it panics me completely. **1977** *Time* 23 May 24/3 The Defense Secretary is by no means panicked.

2. *intr.* To get into a panic, to lose one's head.
1910 Kipling *Divers. Creatures* (1917) 310 Jules was, so to speak, panicking in a water-tight flat through his unfortunate lack of language. **1921** 'Sapper' *Man in Ratcatcher* 30 For a few agonizing seconds..she panicked; then..she pulled herself together and tried to stop him. **1924** M. Newman *Consummation* v. xxii. 240 They panicked one night, started rapid fire and killed two of their own men. **1930** J. Cannan *No Walls of Jasper* 196 Martin helped Phyl to unpack. All at once she panicked, rummaging wildly among the paper in the tea-basket and saying, 'Good heavens, Martin! there's no sugar for his tea!' **1946** M. Peake *Titus Groan* 371 Swelter,..thinking the thin man to have panicked, pursued him. **1958** N. Marsh *Singing in Shrouds* (1959) xi. 235 I'd had one or two drinks over the eight and I suppose that's why I panicked. **1971** *Radio Times* 4 Nov. 72/4 Their headmaster..rather panicked at the word 'drug'. **1975** P. Somerville-Large *Couch of Earth* iii. 48, I thought you might have panicked and hared off.
Hence **pa·nicked** *ppl. a.*, stricken with panic; also *fig.*
1916 G. Frankau *Guns* 15 His panicked watchers spy us, a droning threat in the void. **1920** E. Sitwell *Wooden Pegasus* 107 The light falls like a rain of panick'd leaves Through the gold heart of eves. **1978** R. Lewis *Uncertain Sound* iii. 80 With a sense of panicked claustrophobia I felt I had to get out of the office.

panicky, *a.* Add: **b.** quasi-*sb.* That which is panicky.
1924 Galsworthy *White Monkey* i. xii. 96 'That appears to savour of the panicky,' he said.

panicum (pæ·nikǒm). [L. name of a type of millet, adopted by Linnæus (*Species Plantarum* (1753) I. 55) and earlier botanists as the name of a genus.] A grass belonging to the large genus so called, including the European millet, *Panicum miliaceum*, and several other important cereals or fodder grasses; = Panic *sb.*¹
1844 J. D. Hooker *Let.* 10 Feb. in L. Huxley *Life J. D. Hooker* (1918) I. viii. 169 Fancy two new Panicums; I cannot make them agree with any others. **1870** W. Robinson *Wild Garden* ii. 121 Elegant Panicum... America. Annual... Banks, slopes, fringes of shrubberies. **1932** *Discovery* Jan. 24/2 A shade plant grows rankly and below is a mat of *Panicum* to the exclusion of everything else. **1966** C. A. W. Guggisberg *S. O. S. Rhino* iv. 92 *Panicum*, which grows to a fair height, is eaten more especially when short, and the rhino have a tendency to keep it down. **1968** F. W. Gould *Grass Systematics* 214 Bulb Panicum is an important forage grass of the southwestern mountain ranges.

‖ **panier de crabes** (pa·nye də krab). [Fr., lit. basket of crabs.] An internal struggle, a 'rat race'.
1963 *Economist* 30 Mar. 1218/2 The visitor is..not able to move around in this communist *panier de crabes*. **1979** *Dædalus* Summer 137 In what the French would term the *panier de crabes* of sixteenth century Toledan society.

Paninean (pæni·niăn), *a.* Also **Paninian.** [f. Skr. *Pāniniyā* in the same sense (also used), f. the name *Pāṇini*: see below.] Of or pertaining to the Sanskrit grammar of Pāṇini (6th or 5th c. B.C.); adhering to the rules formulated by Pāṇini.
1801 H. T. Colebrooke in *Asiatic Researches* VII. 204 A performance, such as the Pániníya grammar, must inevitably contain many errors... The studied brevity of the Pániníya Sûtras renders them in the highest degree obscure. **1807** W. Jones *Wks.* IV. 107 Asked what he thought of the Pániníya, he answered very expressively, that 'it was a forest'. **1879** W. D. Whitney *Skr. Gram.* ii. 9 The Paninean scheme..classes *a* as guttural. **1900** A. A. Macdonell *Hist. Skr. Lit.* ix. 268 But the most important information we have of pre-Pāninean grammar is that found in Yāska's work. **1915** S. K. Belvalkar *Acct. Syst. Skr. Gram.* 43 For most of those writers who followed Kaiyyata there was very little original work in the Pāṇinīya school that was left to be done. **1929** L. Bloomfield in *Language* V. 273 A brief survey of both the Pāṇinean and the later systems is given by Belvalkar. **1967** S. M. Katre *Pāṇinian Stud.* I. p. viii, With the resurgent interest in linguistic studies..it is appropriate that Pāṇinian studies should form part of the Building Centenary and Silver Jubilee Series. **1977** *Language* LIII. 222 Given the above, it is not impossible that Kātyāyana and Patañjali might have suggested additions or modifications to Pāṇinian rules.

pani-wallah. *India.* [f. Hindi *pānī* water + Wallah.] A water-carrier (see also quot. 1957).
1934 'G. Orwell' *Burmese Days* xxv. 368 Ba Pe is pani-wallah in the same house at sixteen rupees a month. **1936** W. H. S. Smith *Let.* 2 Aug. in *Young Man's Country* (1977) ii. 20 My nice young pani-wallah has had to return to Rajshahi. **1957** D. G. O. Baillie *Sea Affair* 244 *Pani-wallahs*..are not really watermen at all, but oilmen, or greasers. **1960** *Times* 16 Mar. (Canberra Suppl.) p. xiv/6 Greasers are called panniwallahs, where 'panni' means water (once a more popular lubricant than grease). **1971** *Blitz* (Bombay) 6 Mar. 9/2 Householders are cursing him loudly, calling him 'paniwala maharaj'.

Panjabi: see *Punjabi *sb.* and *a.*

panji(e, varr. *punji.

panjrapol, var. *pinjrapol.

Panlibhonco (pænlibhǫ·nko). [f. *Pa*nama, *Lib*eria, *Hon*duras, *Co*sta Rica.] Designating or pertaining to a merchant ship of Panama, Liberia, Honduras, and Costa Rica flying a 'flag of convenience'. Cf. *flag *sb.*⁴ 1 f, *Panhonlib.
1958 *Hansard Lords* 20 Mar. 369 We have to be very careful lest any action taken by Her Majesty's Government may harm British shipping more than 'Panlibhonco' owners. **1958** *Times* 14 Nov. 13/3 This great and growing block of virtually stateless tonnage—the tonnage of what is commonly known as 'PanHonLib' or 'Panlibhonco'. *Ibid.* 15 Nov. 8/2 Many of the big Panlibhonco fleets pay rates well above the average. **1959** P. Padfield *Sea is Magic Carpet* vi. 97 We sunbathed and watched the tankers sweep by on either side—many of them magnificent, streamlined beasts of the PanLibHonCo variety.

pan-loaf. (In Dict. s.v. Pan *sb.*¹ 11 b.) Add: Chiefly *Sc.* (later examples). Also used *fig.* of an affected or cultured accent, or of someone whose behaviour is regarded as pretentious, this usage originating in the fact that a pan loaf, being more expensive, was a sign of affluence. Hence **pan-loafy** *a.*
1906 'H. Foulis' *Vital Spark* i. 5 Four men and a derrick, and a water-butt and a pan loaf in the fo'c'sle. **1907** J. Kirkland *Mod. Baker* I. 112 Tin or Pan Loaves... Scotch pan loaves..are generally baked four in a pan, and to ensure that they separate with a smooth face each loaf is greased on the ends before being placed in the tin. **1922** Joyce *Ulysses* 143 They buy..four slices of panloaf at the north city dining rooms in Marlborough street. **1946** 'D. Twitter' *Tales o' the Toonie* 48, I warned Sarah Amelia no' tae start speakin' pan-loafy fin I wiz wi' her. She thrapit doon my thrapple that if I spak braid Farfar fowk wud tak me for a Turk. **1947** H. W. Pryde *First Bk. McFlannels* vi. 61 An' yer pan-loaf talk! 'Good-marning, Mrs. McTweed,' says you, 'fehn weather we've heving for this tehm of the year!' **1957** *Bulletin* (Glasgow) 25 Feb. 11/2 Pan loaves are coming back into favour again. **1959** M. Pugh *Chancer* 61 'The little kids used to shout, "Grey breeks, grey breeks, let's hear you speaking pan-loafy,"' she said... An all girls' school where they paid fees and wore grey uniforms and didn't talk in the local accent was splendidly ridiculous. **1964** *Weekly Scotsman* 16 June 8 Ah dinna like her, she's oafy pan loaf. **1973** Boyd & Parkes *Dark Number* v. 58 Him and his fancy clothes and his pan-loaf accent and his money! Why couldn't he stay on his own side of the tracks? **1974** *New Society* 28 Feb. 500/3 It was the authentic voice of the people of Glasgow.., and not the 'pan-loaf' (falsely polished) accents of the Scottish BBC. **1976** *Scotsman* 24 Dec. (Weekend Suppl.) 1/7 His pan-loaf accent would make you die. He's gettin' ready, y'see, to be a great sports commentator on the telly after he's made his mark in athletics.

panmixia. Also **panmixis, panmixy.** Add to def.: Now used to mean random mating within a breeding population. Also *transf.* (Further examples.)
1889 [see *Galton]. **1890** W. James *Princ. Psychol.* II. xxviii. 687 The obsolescence of disused organs he [*sc.* Weismann] explains very satisfactorily..by his theory of panmixy. **1896** A. Tille in T. Common tr. *Nietzsche's Case of Wagner* p. x, In a tribe the members of which..assist each other in every kind of danger natural selection must soon come to an end, a kind of panmixy must arise and lead to a rapid decline. **1909** W. M. Urban *Valuation* xi. 341 Social pan-mixia, the breaking down of class barriers, makes impossible that fixity and contrast of ideals. **1943** *Genetics* XXVIII. 114 Mass selection..is all that can occur under panmixia. **1949** Darlington & Mather *Elem. Genetics* 408 Panmixis, unrestricted Random Mating. Properly excluded by any restriction, but sometimes used where random mating is assumed within the operation of a restriction, especially with dioecy in animals. **1955** *New Biol.* XVIII. 35 Panmixia or random mating..does not mean promiscuity. **1968** J. W. Purseglove *Tropical Crops: Dicotyledons* II. 360 Panmixia: by growing a number of c[ulti]v[ar]s together and hand-pollinating between them a highly variable population can be synthesized. **1971** L. N. Morris *Human Populations* (1972) ii. 34 Panmixis does not take place within the total [human] species.
2. In form **panmixis.** A population within which random mating takes place.
1968 J. W. Purseglove *Tropical Crops: Dicotyledons* II. 360 In Uganda a panmixis of 19 [*Gossypium*] *latifolium* c[ulti]v[ar]s from a number of countries was chosen.
Hence **panmi·ctic** *a.*, characterized by panmixia.
1943 *Genetics* XXVIII. 117 The term random breeding or panmictic unit will be used for any local population of the same effective size as the parental group. **1971** *Nature* 12 Feb. 468/1 It is possible that in animal species..well able to migrate, no local population is sufficiently isolated to prevent the entire species or subspecies forming effectively one panmictic population. **1975** S. K. Jain in Frankel & Hawkes *Crop Genetic Resources* ii. 18 All individuals within a colony might be one panmictic group.

panne. Delete ‖ and add earlier and later examples.
1794 A. Young *Trav. France* (ed. 2) I. ii. xix. 550 St. Omers. There is a manufacture of..a kind of stuff called *pannes*. **1919** [see *bud *sb.*¹ 3 d]. **1923** [see *marocain]. **1940** *Chambers's Techn. Dict.* 611/2 *Panne velvet*, a warp pile fabric with silk pile; used for dresses and furnishings. **1972** *New Yorker* 7 Oct. 24/1 (Advt.), Pants are perfect for evening when they're lush panne velvet. **1977** *Vogue* Dec. 124/3 *Big sister*..in chocolate brown panne velvet.

pannier, *sb.*¹ Add: **6. pannier bag,** a bag or similar container (usu. one of a pair) placed above or to the side of the rear wheel of a bicycle or motor cycle; also *ellipt.*; **pannier pocket,** a large pocket attached to the side of a skirt or dress; **pannier tank,** a type of small steam locomotive which has a water tank on each side of the boiler.
1939–40 *Army & Navy Stores Catal.* 783/3 Cycle accessories..pannier bags and carriers. **1959** I. Jefferies *Thirteen Days* (1961) i. 9, I was forced to pull off the road... I had a bottle of cold beer..in the pannier. *Ibid.* xi. 170, I was taking the letters out of the pannier bag when the phone rang. **1975** J. Wood *North Kill* x. 139 The speaker kicked his bike into life... The others were storing their cleaning materials into side pannier bags. **1976** *Times* 27 Aug. 12/8 The world's most expensive bicycle... For nearly £600 you get neither mudguards nor pannier bags, and not even a rear reflector. **1922** Joyce *Ulysses* 501 Those pannier pockets of the skirt..are devised to suggest bunchiness of the hip. **1973** *Times* 11 Dec. 13/3 Bill's new skirt with its slung pannier pockets is pretty. **1949** C. J. Allen *Locomotive Pract. & Performance 20th Cent.* vi. 65 Shunting on all railways is entrusted in large measure to small 0-6-0 tanks (pannier tanks on the Western Region). **1950** H. C. Webster *Railways for All* ix. 83 Frequently such locomotives carry the water in tanks fitted on top of the boiler and are therefore termed 'saddle tanks'. Others carry the water in two tanks secured high up along the boiler sides, and these are known as 'pannier tanks'. **1957** *Railway Mag.* Nov. 751/2 Today 0-6-0 pannier-tanks usually deal with the traffic. **1970** *Ibid.* Oct. 549 (*caption*) Wellington to Much Wenlock afternoon passenger train..headed by 0-6-0 pannier-tank No. 7754. **1973** *Country Life* 8 Mar. 593 A type peculiar to the GWR—the pannier tank..these modest 0-6-0 engines, which carried their water in 'panniers' at either side of the boiler.

panniform, *a.* *Bot.* [f. L. *pannus* cloth + -form.] = Pannose *a.*
1894 J. M. Crombie *Monogr. Lichens Brit.* I. 238 The thalline reactions, in conjunction with the general aspect of the plant, show that it belongs to *P*[*armelia*] *revoluta*, produced no doubt by a panniform condition of this species. **1921** A. L. Smith *Handbk. Brit. Lichens* 141 *Panniform, Pannose*, felted. **1957** Snell & Dick *Gloss. Mycol.* 110/1 Panniform. Having a felted or matted appearance.

pannikin. Add: **a.** (Further examples.)
1908 E. J. Banfield *Confessions of Beachcomber* ii. i. 280 One day a bucket of milk was brought to the camp at dinner-time and served out with pannikins. **1911** C. E. W. Bean *'Dreadnought' of Darling* xxxii. 282 There was a case of pannikins that had come up as freight. **1924** *Truth* (Sydney) 27 Apr. 6 *Pannikin*, small drinking vessel, made out of tin, carried by the sundowner, and tied to his billy. **1936** I. L. Idriess *Cattle King* ii. 14 The boy stared, then snatching his pannikin pushed the reluctant foal away and commenced milking the mare. **1945** Baker *Austral. Lang.* 169 *A friendly pannikin*, a drink with a companion.
b. The head; in slang phr. *off one's pannikin*, off one's head.
1895 C. Crowe *Austral. Slang Dict.* 56 *Off his pannikin*, silly. **1899** 'S. Rudd' in Murdoch & Drake-Brockman *Austral. Short Stories* (1951) 110, I seen 'im just now up in your paddick, an' he's clean off he's pannikin. **1916** C. J. Dennis *Moods of Ginger Mick* 126 Per'aps I'm orf me pannikin wiv' sittin' in the sun. **1934** B. Penton *Landtakers* (1935) iv. vii. 360 He's gone raving off his pannikin in Sydney.
c. = *pannikin-boss*.
c **1926** 'Mixer' *Transport Workers' Song Bk.* 7 My power is such to make or break—I'm a pannikin, get me?
Comb. **pannikin-boss** (later examples).
1930 L. G. D. Acland *Early Canterbury Runs* (ser. 1) v. 105 At 'smoke-oh' he used to take a standing jump over a bale of wool and then Brucksaw, the 'pannikin boss' (head general hand), used to carry it out of the shed single-handed. **1936** A. Russell *Gone Nomad* ix. 70 There in the Silver City a telegram awaited me. It contained the offer of a job, that of 'pannikin-boss' and book-keeper on a sheep run east of Broken Hill, on the Milparinka Track. **1959** D. Hewett *Bobbin Up* viii. 97 All this unshakable distrust of the pannikin boss, the boss's man..the lowest of the low in a world where dog ate dog. **1966** G. W. Turner *Eng. Lang. Austral. & N.Z.* vii. 146 The manager of a sheepstation was called a *pannikin boss*, a term I have heard used..for a foreman on building jobs in New Zealand. **1969** D. Niland *Dead Men Running* iv. 98 Father Vaughan seemed to project himself as no more than the mouthpiece of a pannikin boss of a God who sounded like a brutal and violent pirate.

PANNING

panning, *vbl. sb.* Add: **1.** Also with *out.*
1839 *Amer. Railroad Jrnl.* VIII. 99 This operation is continued until all the sand is removed, and nothing but the gold left. It is called 'panning out'. 1872 'Mark Twain' *Roughing It* lxi. 443 'Panning out' refers to the washing process by which the grains of gold are separated from the dirt.

c. The action of denouncing or criticizing severely. (Cf. *pan *v.*[1] 7.)
1914 *Sat. Even. Post* 15 Aug. 9/1 Speed sure got a pannin' in the clubhouse... Everybody..roasted him, but it didn't do no good. 1946 *Jazz Writings* 21/2 All this sounds like a merciless panning, I'm afraid, and yet it is kindly meant. 1958 *Spectator* 16 May 624/2 The Council of Industrial Design's exhibition..has had a panning from the design critics. Speaking..as a consumer, I agree with the criticism. 1960 *News Chron.* 27 Feb. 4/8 It was about the only thing in the show which did not get a panning from the critics.

c. *attrib.*, as *panning test, trough.*
1850 N. Kingsley *Diary* 17 May (1914) 122 Stoped down to day and made a panning trough to pour quicksilver from the riffler into and fix the pump. 1951 *Oxf. Jun. Encycl.* VII. 21/1 One of the oldest methods of assay ..is the miner's concentration or 'panning' test.

2. *Agric.* The action of Pan *v.*[1] 6: the hardening of a layer of soil.
1939 *Geogr. Jrnl.* XCIV. 468 A heavy black clay of undetermined depth without any evidence of panning, no iron concentrations and no characteristics of laterite. 1960 *Farmer & Stockbreeder* 26 Jan. 78/3, I am completely convinced that a panning effect is negligible compared with that of a plough. 1971 *Power Farming* Mar. 50/2 (Advt.), Even in sticky conditions the Soilmaster acts efficiently without 'bulldozing'—Positively no panning. Before buying a new cultivator get the facts on the Soilmaster range.

pa·nning, *vbl. sb.*[2] [f. *pan *v.*[3] + -ing[1].] The action of *pan *v.*[3] Also *attrib.*, esp. in *panning shot.*
1917 C. N. Bennett *Guide to Kinematogr.* ii. 22 Sweeping round the camera from side to side is called panoraming or 'panning'. 1939 G. Greene *Lawless Roads* vi. 147 If I had moved a camera all round..the little plaza in a panning shot. 1946 *Electronic Engin.* XVIII. 207 The trick of camera panning from one speaker to another. 1953 K. Reisz *Technique Film Editing* ii. 42 A scene.. shot from a large number of set-ups, some of them with a panning or tracking camera. 1967 *Spectator* 6 Oct. 392/2 The producer..was not helped by his camera crews, who, if their panning shots were tremulous, zoomed in on things like homing missiles. 1969 *Amat. Photographer* 23 Apr. 51/4 When you are forced to work more or less broadside-on,..swing the camera as you expose in the same direction as the movement. This technique, known as 'panning', is much used by press photographers. 1977 *Radio Times* 28 May 17/1 A long panning shot across some velvet cushions.

Pannonian, *a.* (Earlier and later examples.)
1605 Bacon *Adv. Learning* ii. f. 70[v] We see a notable example in Tacitus of two Stage-plaiers, Percennius and Vibulenus, who by their facultie of playing, put the Pannonian armies into an extreame tumulte and combustion. 1912 *Q. Rev.* Oct. 335 His happiest days were certainly spent away from Rome in German and Pannonian wars. 1975 R. Browning *Emperor Julian* i. 13 Decius, a Pannonian senator who found himself proclaimed emperor against his will in 249. 1976 *Classical Q.* XXVI. 302 It was established in Dalmatia in or soon after A.D. 9, in the wake of the Pannonian uprising.

pannus. Add: **2.** *Path.* A layer of granulation tissue that forms by a thickening of the synovial membrane and tends to spread over and absorb adjacent cartilage in a joint affected by rheumatoid arthritis.
1904 *Boston Med. & Surg. Jrnl.* 17 Nov. With the swelling which results the membrane at times extends in or over the cartilage, forming a pannus, and wherever this persists for any length of time.., the cartilage under the pannus is more or less absorbed. 1929 R. Pemberton *Arthritis & Rheumatoid Conditions* iii. 53 The proliferation of the synovial membrane causes a layer of granulation tissue to extend in the form of a thin pannus more or less completely over the joint cartilage. 1973 *Nature* 14 Dec. 419/1 The pannus of the rheumatoid joint has the ability to produce an active collagenase and invade and degrade cartilage *in vitro.*

panoche (panǫ·tʃi). *U.S.* Also **panocha, penoche, penuche.** [Amer. Sp., 'brown sugar'.]
1. A type of coarse brown sugar.
1847 *Californian* 10 Apr. 3/2 The cargo consists of 180 bales..of Mexican Sugar; 30 Packages Panoche and one Bale of Zarapies. 1856 L. J. F. Jaeger *Jrnl.* 17 Aug. in *S. Calif. Hist. Soc. Publ.* (1928) XIV. 124 Had to leave 3 cargoes of flour and one fanega of beans & ½ cargo of panocha. 1881 H. T. Williams *Pacific Tourist* 300/2 The ordinary brown sugar (panoche) of the Mexicans is also obtained from this plant. 1887 *Outing* Apr. 10/2 It is ordinarily made of corn, roasted and crushed and slightly sweetened, the most common sweetening being *panoche*, a crude sugar. 1936 J. A. McKenna *Black Range Tales* 52 With ten pounds of jerky, a cone of *penoche* or pressed brown sugar, and a serape, or gray blanket, a Mexican is fitted for a trip of several hundred miles through the mountains.
2. A kind of sweet resembling fudge, made with brown sugar, butter, milk or cream, and nuts. Also *fig.*
1872 *Rep. Indian Affairs 1871* (U.S.) 359, I doubt the good policy of issuing bread, and at times candy (panoche), to the pupils. 1921 M. L. Matthews *Foods & Cookery* 274 Panocha... Mix sugar, milk, and salt. Boil until it reaches the 'soft-ball' stage. 1930 *Sunset* Dec. 30/2 At lower left are seen fondant-dipped and penoche-dipped figs and prunes. 1949 *Arizona Q.* Autumn 255 He made the most luscious panocha and rich syrups, which he sold to the public. 1949 *Chicago Tribune* 28 Sept. ii. 2/2 Our recipe for Butterscotch Cake, and Penoche Icing, is printed on the recipe folder. 1952 J. Steinbeck *East of Eden* xix. 226 Faye twisted around in her chair to reach a piece of brown panocha studded with walnuts. When she spoke it was around a mouth full of candy. 1966 *Cookies & Candies* (Better Homes & Gardens) 79/2 Penuche. Rich and creamy brown-sugar fudge. A favorite of grandmothers (granddads, too!) 1968 *Cook's & Diner's Dict.* (Funk & Wagnalls) 167/2 Pe-nu-che..A candy made with brown sugar..Also called panocha. 1971 *Islander* (Victoria, B.C.) 7 Feb. 14/1 There are recipes for all the classic home candies like fudge, butterscotch, penuche, etc. 1976 *Publishers Weekly* 29 Mar. 59/3 With its story of aggressive hatred.., murder attempts.., a later revealed illegitimate birth, and a love interest as well, it's a well-iced piece of penuche.

panographic (pænogræ·fik), *a. Dentistry.* Also **panagraphic.** [f. Pano(ramic *a.* + Radio)graphic *a.*] Of, pertaining to, or designating radiography of several teeth and the adjacent bones in a single exposure by means of a small X-ray source placed inside the mouth and a film outside it.
1952 *Jrnl. Dental Res.* XXXI. 165 In the evaluation of the panographic technique it is well to remember that it must be judged as an adjunct to conventional radiography and not as a substitute for it. 1963 *Dental Radiogr. & Photogr.* XXXVI. 81/2 (*caption*) Panagraphic equipment and technic for the maxilla and the mandible. 1967 *Amer. Jrnl. Roentgenology* CI. 989/2 (*caption*) Panographic study reveals a large expansile cystic lesion in the right mandible, proven to be a dental cyst.
Hence **pa·nograph,** (*a*) a panographic radiograph; (*b*) an X-ray machine for use in panography; **pano·graphy,** panographic radiography.
1961 *Oral Surg., Oral Med., Oral Path.* XIV. 1178 (*heading*) Panagraphy. *Ibid.* 1179 The x-ray tube used in the Panagraph is the Swiss Panoramix tube. 1963 Blackman & Poyton *Man. Dental & Oral Radiogr.* xvi. 160 (*caption*) Panagraph of lower jaw showing unerupted and supernumerary teeth. 1967 *Amer. Jrnl. Roentgenology* CI. 988/1 Panography is a radiographic technique recently described by Blackman for use predominantly in dentistry. 1969 Wuehrmann & Manson-Hing *Dental Radiol.* (ed. 2) viii. 161 Panagraphy suggests a technic to study structures in a panoramic fashion. The Panagraph or Panoramix is an x-ray machine utilizing an intraoral source of radiation to expose film placed extraorally. *Ibid.* 163 The main advantages of the Panagraph appear to be that it is relatively portable, is easy to operate, and may be used effectively for mass dental radiography.

panoply, *v.* **b.** (Earlier and later examples.)
1873 'Mark Twain' *Gilded Age* xxxii. 290 It would be.. judicious to send her forth well panoplied for her work. —So he had added new and still richer costumes to her wardrobe, and assisted their attractions with costly jewelry. 1880 —— *Tramp Abroad* xxviii. 284, I and my agent panoplied ourselves in walking costume. 1940 'Gun Buster' *Return via Dunkirk* i. i. 11, I was panoplied for War.

panopticon. 1. b. (Earlier example.)
1850 *Deed of Settlement of Royal Panopticon of Science & Art* 1 Royal Panopticon of Science and Art. An institution for Scientific exhibitions and for promoting discoveries in arts manufactures.

panora·m, *v.* [A shortening of Panoramic *a.*] *intr.* = *pan *v.*[3] Hence **panora·ming** *vbl. sb.*
1915 *Wells Fargo Messenger* Oct. 18/3 We are before the Erie cut, and as the camera 'panorams' around, we get a glimpse of our splendid Eleventh Avenue Stable in Jersey City. 1938 G. H. Sewell *Amateur Film-Making* ix. 80 The revolution of a camera on its vertical axis, so that it covers the features of a landscape, is known as 'panoraming' or 'panning'. *Ibid.* 81 To panoram with a hand-held camera, set your feet firmly on the ground,.. brace yourself up and turn from the hips.

panorama. Add: **1. b.** (Earlier examples.)
1813 M. Edgeworth *Let.* 16 May (1971) 56 Pray do not think because I name these fine people..that my poor little head is turned... Be assured that the whole panorama passes before me as a panorama. 1818 'T. Brown' *Brighton* I. p. xi, A novel should be a panorama of life; and we trust that our *views* will be found correct in the present one.
2. (Earlier example.)
1805 M. Wilmot *Russ. Jrnls.* (1934) 151 Up at Cock Crow to drive to Sparrow Mount from which spot Moscow is seen as a Panorama, & a most exquisite view it is indeed.

pan-pie. *U.S.* [f. Pan *sb.*[1] + Pie *sb.*[2]] = Pandowdy.
1723 J. Nott *Cook's & Confectioner's Dict.* sig. Y8[v] This Paste is proper for Pan pies that are set on the Table without a Desert or Banquet of sweet Meats. 1862 'G. Hamilton' *Country Living* 70 No pan-pie with hot brown bread on Sunday morning. 1883 *Rep. Maine Board Agric. 1882* 403 You have all heard of the pan-dowdy, or pan-pie, the pride of our grandmothers.

panplain (pæ·nplēⁱn). *Physical Geogr.* Also **-plane.** [f. Pan- + Plain *sb.*[1]] A plain formed by the coalescence of previously separate flood-plains.
1933 C. H. Crickmay in *Geol. Mag.* LXX. 345 We would expect the growing floodplains to become confluent, making one broad universal plain shared by all the streams of the region. The surface will be flat, and will have a general slope like that which now characterizes the lower floodplains of great rivers. The only relief will be remnants of some of the interfluves which may persist as low monadnocks. This plain, formed of floodplains joined by their own growth, may be called a panplain. 1942 O. D. von Engeln *Geomorphol.* vi. 97 Eventually the level surfaces of adjacent stream floors, concurrently developed by such planation, become confluent and a panplane, a very level plain with a general downward inclination, is produced. 1960 B. W. Sparks *Geomorphol.* xv. 340 There seems little doubt that river planation is an important agency of erosion,..but whether these can extend indefinitely to form a panplain seems more doubtful. 1967 J. Hays in Jennings & Mabbutt *Landform Stud. Austral. & New Guinea* ix. 200 Near Darwin, the panplains of north-flowing and west-flowing streams have coalesced.

panplanation (pæ:nplǣnēⁱ·ʃən). *Physical Geogr.* [f. Pan- + *planation, after *panplain.] The formation of a panplain.
1933 C. H. Crickmay in *Geol. Mag.* LXX. 345 The essential difference between panplanation and peneplanation is that the former starts from the lower floodplains of rivers and grows laterally in all landward directions, whereas the latter is of universal occurrence. 1942 O. D. von Engeln *Geomorphol.* xvi. 356 It is now considered improbable that flood-plain accumulation and panplanation (Crickmay, C. H., 1933) combined..could completely dispose of all relief in a peneplaned region. 1970 R. J. Small *Study of Landforms* v. 168 Many authorities have regarded lateral plantation by running water as a major factor in the formation of desert surfaces.., and a process akin to panplanation has been assumed by some to account for savanna plains.

panpot (pæ·npǫt), *sb.* [f. *pan(oramic) pot(entiometer).*] A kind of potentiometer used to vary the apparent position of a sound source by varying the strengths of the signals to individual speakers without changing the total signal strength.
1941 *Jrnl. Soc. Motion Pict. Engin.* XXXVII. 130 A special 3-circuit differential junction network, nicknamed 'The Panpot', is used to dub one original track onto one, any two, or all three of our Fantasound program tracks. 1964 H. B. Hadden *Pract. Stereophony* iii. 39 To present a monophonic signal, as a point source at a given position.., a control known as a 'panoramic potentiometer' is employed... This control, known as a 'panpot' for short, is a two-gang fader with the two halves arranged back to back, so that as the gain of one half of the particular channel is increased, the other half is decreased. 1975 *Gramophone* Dec. 1122/1 The deck was plugged into a large discothèque type console with a quadrophonic panpot in the middle.
Hence **pa·npot** *v. trans.*, to process by means of a panpot; **pa·npotted** *ppl. a.*, **pa·npotting** *vbl. sb.*
1973 *Jrnl. Audio Engin. Soc.* XXI. 3/2 The recording may..be made by 'pan-potting' sounds into the channels to stimulate [*read* simulate] pickup by directional microphones. 1976 *Gramophone* Jan. 1269/1 This 'panpotting' method also enabled the balance to be adjusted after the recording session. *Ibid.* 1269/2 The medium is largely confined to pan-potted multi-track material.

panpsychism (pænsəi·kiz'm). *Philos.* Also **pampsychism.** [Pan-.] (See quot. 1901.) Hence **panpsy·chic** *a.*, pertaining to or based on panpsychism; **panpsy·chist,** one who believes in panpsychism; also *attrib.* or as *adj.*; **panpsychi·stic** *a.*, connected with or characterized by panpsychism; **panpsychi·stically** *adv.*
1879 G. H. Lewes *Mind as Function of Organism* ii. 34 We must therefore pronounce against the hypothesis of Panpsychism. 1881 W. James in *Unitarian Rev.* Nov. 415 All modern thought converges toward idealistic or, as I should rather call them, pan-psychic, conclusions. 1901 Panpsychism [in Dict. s.v. Pan- 2]. 1903 C. A. Strong *Why Mind has Body* p. vi, Hence I think panpsychists are justified in maintaining that with their principles they are able to explain the connection of mind and body. *Ibid.*, I have chosen my title with the object of putting this panpsychist pretension distinctly on record. 1904 J. McCabe tr. *Haeckel's Wonders of Life* xv. 354 His [*sc.* Fechner's] system is..panpsychistic and at the same time pantheistic. 1911 J. Ward *Realm of Ends* i. 20 To this principle pampsychism appeals. *Ibid.* 21 On this, the pampsychist view, Nature..resolves into a plurality of conative individuals. *Ibid.* iii. 62 The pampsychist.. maintains..that at all events there are no things wholly inert. 1924 B. Edgell *Theories of Memory* 133 The present writer has failed to find any link between M. Bergson's pampsychism and his individualism. 1935 R. B. Perry *Tht. & Char. W. James* II. 443 Whether this 'beyond', this thing-in-itself, shall be interpreted panpsychistically, is questioned and again left unsettled. 1937 A. H. Murray *Philos. of James Ward* v. 98 The pampsychistic solution, at which he arrives, dates back to Leibniz and earlier. 1940 *Mind* XLIX. 43 Berkeley probably saw that pampsychism, for him, spelled pantheism. *Ibid.* 45 With this new background the Berkeleian immaterialism took on a new aspect; it is no longer pampsychist. 1955 H. J. Koren *Introd. Philos. Animate Nature* i. i. 35 The ancient hylozoists..and modern panpsychists..imply or state that all things are alive. 1959 *Chambers's Encycl.* XII. 718/2 Should pampsychism replace the older hylozoism, the truth being that the inner side of everything is its

soul, the outer aspect observable by an outsider being its body? 1968 F. Copleston *Hist. Philos.* VIII. v. xvii. 393 Neither Strong nor Drake meant to imply that stones, for instance, are conscious. Their panpsychism was linked with the idea of emergent evolution. 1973 *Nature* 20 July 183/1 At one time he expressed panpsychic ideas, thinking there could be some consciousness even in a mug, but later was prepared to restrict it to animals with a nervous system. 1975 C. Burt *ESP & Psychol.* vi. 99 Theists and panpsychists..have no qualms about ascribing minds to human beings.

panpygoptosis (pæ:npəigoptōᵘ·sis). *nonce-wd.* [A fanciful formation combining the elements Pan-, Pygo-, Opto-, -osis.] = *duck's disease* s.v. *duck *sb.*[1] 12. Hence **panpy:gopto·tic** *a.*
1938 S. Beckett *Murphy* v. 97 Duck's disease is a distressing pathological condition in which the thighs are suppressed and the buttocks spring directly from behind the knees, aptly described in Steiss's nosonomy as Panpygoptosis. *Ibid.* 104 Miss Dew's control, a panpygoptotic Manichee of the fourth century,..had not..been raised so wholly a spiritual body as yet to sit down with much more comfort than she had in the natural.

Pan-Roman, Panroman (pæ:n,rōᵘ·măn). [Pan- + Roman *sb.*[1] 3.] An artificial language invented for universal use by H. Molenaar; also known as *Universal.
1907 W. J. Clark *Internat. Lang.* ii. vi. 103 The last few years have produced quite a crop of artificial languages..Idiom Neutral; Pan-Roman or Universal, by Dr. Molenaar; Latino sine flexione [etc.]. *Ibid.* 105 The victorious Esperantists..poke fun at these new-fangled schemes. A parody in Esperanto verse..narrates the fickleness of Pan-Roman and how it changed into Universal. 1922 *Nature* CIX. 494/1 A language of the Neo-Latin type, somewhat similar to Neutral Idiom, is the 'Panroman' (or 'Universal') of the German positivist and pacifist, Dr. H. Molenaar. 1927 E. S. Pankhurst *Delphos* vi. 81 Universal (1903), later called Panroman (1906), by Dr. H. Molenaar, is another neo-Latin language.

‖ **pansala** (pɒ·nsālā). [Sinhala, f. *pan* leaf + *sala* dwelling, f. Skr. *parnaśālā*, Pali *paṇṇasālā*.] A Buddhist temple or monastery; orig., a forest hut constructed from leaves.
1850 R. S. Hardy *Eastern Monachism* xiii. 129 The dwelling of the priest is called a pansala, from pan, leaves, and sala, a dwelling, or a place to which any one is accustomed to resort, from a root which signifies to go. 1913 L. Woolf *Village in Jungle* i. 12 He had fallen ill.. and had stayed for a month or two in the priest's pansala. 1927 *Missionary Herald* Sept. 219/2 Mr. Weeraratua went to see him at his pansala and talked to him about Christ. 1956 R. Pieris *Sinhalese Social Organization* ii. ix. 73 There was, in almost every village, temple land, the property of which was vested in..a local *vihāra* or *pansala.*

pan-se·xual, *a.* [f. Pan- 2 + Sexual *a.*] Of or pertaining to pan-sexualism; that is not limited in sexual choice; **pan-se·xualism,** the view that the sex instinct plays a part in all human thought and activity and is the chief or only source of energy. Hence **pan-se·xualist** *a.*, pertaining to the theory of pansexualism; **pan-sexu·ality.**
1917 C. R. Payne tr. *Pfister's Psycho-anal. Method* 60 Which..has brought the reproach of 'pansexualism' against psychoanalysis. 1922 J. Strachey tr. *Freud's Group Psychol.* iv. 39 Psycho-analysis, then gives these love instincts the name of sexual instincts... The majority of 'educated' people have regarded this nomenclature as an insult, and have taken their revenge by retorting upon psycho-analysis with the reproach of 'pan-sexualism'. 1926 W. McDougall *Outl. Abnormal Psychol.* i. 20 It has led Freud.., as Janet has said, to construct 'an enormous system of medical philosophy', the theory of Pansexuality. *Ibid.* vi. 131 Freud, in accordance with his pansexualist tendency, expressed the opinion [etc.]. *Ibid.* xviii. 314 The dogma that the Œdipus complex is present in all men is the principal instrument of the pansexual theory. 1972 *Jrnl. Social Psychol.* LXXXVII. 51 In the beginning the human organism has the potential of pansexuality. 1974 *Observer* 7 Apr. 36/6 Eventually, no doubt, some biographer will tell us how far he [*sc.* H. de Montherlant] was homosexual, heterosexual or—as seems to be suggested by some discreet passages about bestiality and incest—pansexual. 1977 *Guardian Weekly* 7 Aug. 18/2 An exquisitely Victorian taste for extravagant, pansexual erotic fantasy.

pansified (pæ·nzifəid), *ppl. a.* [f. Pansy + -fy + -ed[1].] Excessively stylized or adorned; affected, effeminate.
1941 C. King *Diary* 17 July in *With Malice toward None* (1970) 135 These birds from Portland Place...think anything said with feeling is rather vulgar, so usually give out a very pansified BBC message. 1944 'G. Orwell' *Crit. Ess.* (1951) 145 The rather pansified drawings of youths. 1959 W. D. Pereira *North Flight* iv. 56 You don't know this place. Like a pansified rabbit warren. Architect designed. 1973 *Listener* 20 Dec. 860/3 Hervey's relationship with..the pansified international Algarotti. 1976 *Ibid.* 27 May 679/1 Being not only a Londoner, but a central Londoner,..I can breakfast and dine in my pansified quarters in the heart of the soft South, having spent most of the intervening time in the gritty, bracing, humorous, individualistic North.

panspermy. Add: (The usual form is now **panspermia.**) In recent use applied to the idea that micro-organisms or chemical precursors of life are present in space and able to initiate life on reaching a suitable planet. (Further examples.)
1908 H. Borns tr. *Arrhenius' Worlds in Making* viii. 217 The so-called theory of panspermia really shows a way. According to this theory life-giving seeds are drifting about in space. They encounter the planets, and fill their surfaces with life as soon as the necessary conditions for the existence of organic beings are established. 1938 S. Morgulis tr. *Oparin's Origin of Life* ii. 39 At the beginning of the twentieth century the idea of the transfer of genus from one celestial body to another was again revived in the form of the so-called theory of panspermia, originated by the great Swedish physical chemist S. Arrhenius. 1970 *Daily Tel.* 9 Jan. 19 The theory of 'panspermia', the idea that the seeds of life drift through the Galaxy until they find a planet on which to settle, was eagerly discussed on the..final day of the moon-rock conference in Houston. 1971 I. G. Gass et al. *Understanding Earth* ix. 140/2 The hypothesis of 'panspermia'..is an old one which has never found much favour. 1975 *Times Lit. Suppl.* 25 July 846/3 The possibility of panspermia—the idea, recently revived by Leslie Orgel and Nobel laureate Francis Crick, that life did not originate on earth at all, but arrived here from elsewhere in the universe.

panstick (pæ·nstik). Also **Pan-Stik** (proprietary name). [f. *pan(cake *sb.* 2 h + Stick *sb.*[1]] A kind of matt cosmetic in the form of a stick.
1949 *Trade Marks Jrnl.* 10 Aug. 705/2 Pan-Stik... Non-medicated toilet preparations, cosmetic preparations and perfumes. Max Factor & Company.., Los Angeles, State of California, United States of America; manufacturers. 1962 *New Statesman* 18 May 708/2 We were crushed in the toilets. All round girls smeared on pan-stick. 'I can't go with him, he's too short.' 'All the grey glitter I put on me hair come off on his cheek and I hadn't the heart to tell him.' 1967 *Guardian* 1 Feb. 6/6 The kids I meet..just don't wash enough. And they plaster their faces..with layers of cover creams, and thick tinted foundations and then probably panstick make-up as well. 1973 J. Burrows *Like an Evening Gone* i. 13 Talk..as superficial as panstick.

pansy, *sb.* Add: **1. c.** *pansy-coloured* (further example), *-dark, -purple* (further example).
1909 *Daily Chron.* 18 Mar. 10/3 Lady Kenmare in black and Georgiana Lady Dudley, tall and beautiful in pansy-coloured cloth, were Lady Mayo's assistants. 1862 G. M. Hopkins *Vision of Mermaids* (1929), Their pansy-dark or bronzen locks were strung With coral shells. 1940 C. Day Lewis tr. *Virgil's Georgics* iv. 89 Eridanus, than which through fertile lands no river Rushes with more momentum to the pansy-purple sea.
2. The colour of a pansy; *spec.* a shade of blue or purple.
1914 Joyce *Dubliners* 227 A red-faced young woman, dressed in pansy. 1926 *Eaton's News Weekly* 26 June 13/1 Bathing suit...in blue, cardinal, pansy, black. 1935 *Amer. Speech* X. 193/2 The following terms are applied to blue..: *slate, aster, indigo, pansy*, [etc.].
3. An effeminate man; a male homosexual.
1929 M. Lief *Hangover* 210 'Say, what do you know about this?' he said. 'One of those pansies was trying to date me up!' 1937 M. Allingham *Dancers in Mourning* xiv. 192 You don't want to feel that every other user of the road privately feels that your club is nothing but a pack of pansies on bicycles. 1947 A. P. Gaskell in D. M. Davin *N.Z. Short Stories* (1953) 276 He struck a chesty attitude standing naked on the seat. 'Do I look like a pansy?' 'Not with that thing.' 1956 [see *camp *a.* (and *sb.*[5])]. 1957 [see *boy friend, boy-friend* s.v. *boy *sb.*[1] 7]. 1960 J. Betjeman *Summoned by Bells* ix. 98 There Bignose plays the organ And the pansies all sing flat. 1967 G. Jackson *Let.* 16 May in *Soledad Brother* (1971) 115 They make emotional pansies of the boys with that sanctimonious dogma. 1971 F. Forsyth *Day of Jackal* xx. 335, I don't like you in that stuff. It makes you look like all those nasty pansies back in there. 1975 *Amer. N. & Q.* XIII. 146/2 Evidence that Sir Thopas is a 'pansy' or a 'queer', who would be ridiculed as a homosexual. 1976 Botham & Donnelly *Valentino* vii. 52 A group of degenerate art students, most of whom he considered pansies.
b. *Comb.*, as *pansy-ass, -boy.*
1934 *New Statesman* 15 Sept. 318/2 Reminiscences of the fate of Heines and Röhm were reflected in shouts about 'pansy-boys'. 1976 N. Thornburg *Cutter & Bone* viii. 192 He learned all about the ingratitude and stupidity of the man's..pansy-ass sons.

pa·nsy, *a.* [f. *pansy *sb.* 3.] Effeminate; homosexual; affected. So **pa·nsyish** *a.*
1929 J. Devanny *Riven* xvii. 112 'Thanks. Don't bother.' The voice was warm... A rich telephone voice. To an artist a pansy voice; a purple pansy. 1934 M. Hodge *Wind & Rain* iii. 83 She'll forget all about it, in the arms of Roger Cole! I think he's pansy, anyhow. 1942 E. Paul *Narrow St.* xxxvi. 323 Pierre..started sending affectionate notes to certain homosexual French officers who were rabidly anti-Left... One by one, these pansy reactionaries began to disappear from their units. 1951 'E. Crispin' *Long Divorce* viii. 84 I'd want her to be walking out with a decent lad, not a pansy little foreign gramophone-record. 1953 E. Taylor *Sleeping Beauty* ii. 27 Laurence leant awkwardly against the chimney-piece in a rather pansyish pose. 1971 *Daily Tel.* 21 Aug. 3/4 'Most of these new designs are too pansy, too effeminate,' said Leading Seaman Robert Nelson.

pa·nsy, *v.* [f. *pansy *sb.* 3] **a.** *trans.* To dress or adorn in an affected or effeminate manner. Freq. *refl.* and const. *up.* **b.** *intr.* To act or walk in an effeminate manner.
1946 'Brahms' & 'Simon' *Trottie True* 154 Luke Lovelock had pansied himself into a feature of every fashionable production. Luke had such perfect taste. 1951 N. Marsh *Opening Night* i. 27 The theatre was shut dahn for a long while until they 'ad it all altered and pansied up. 1966 J. Wainwright *Crystallised Carbon Pig* xxxix. 172 Originally, his hair had been mousy brown. He'd tried to pansy himself up—and failed. 1972 M. Kenyon *Shooting of Dan McGrew* i. 10 It's over a month old..and the last word from McGrew before he went pansying off.

pant (pænt), *sb.*[3] [sing. back-formation f. Pants *sb. pl.*] **1.** = Pants *sb. pl. U.S.*
1893 H. A. Shands *Some Peculiarities of Speech in Mississippi* 49 *Pant*.., an abbreviation of *pantaloons*, used by clerks in dry-goods stores. They say: 'I have a pant that I can sell you,' etc. Of course, *pants* is a well-known abbreviation, but I think *pant* is rather a new word. 1936 A. L. Hench (*MS. note dated Feb.*), For six or seven years now, I have been hearing clothes dealers speak of a pair of pants as 'a fine pant'. The reasoning seems to be that if a pile of pairs of pants is *pants*, then one pair is a *pant*. 1962 *L. L. Bean Catal.* Spring 10 A practical and well made pant for general sportswear. 1976 *Billings* (Montana) *Gaz.* 16 June 3-A/4 (Advt.), You just can't beat the value of this 3-piece jacket/dress/pant ensemble at our irresistible price... The pull-on pant makes a good thing better.
2. *attrib.* and *Comb.*, as *pant-leg, look*; **pant-coat,** a women's coat designed for wearing with trousers; **pantdress,** a dress with a divided skirt; **pantskirt,** a divided skirt; **pant suit, pantsuit** = *trouser suit.
1970 *N.Y. Post* 16 Dec. 6 Pantcoat, on duty for wintry weather and for city traffic. 1974 *New Yorker* 25 Feb. 62/1 (Advt.), A great pantcoat: this Weatherbee, with its easy swing and snappy fit. 1964 *Women's Wear Daily* 30 Nov. 44 Julie Isles..likes the pantdress that stops just above the knee. 1967 *Maclean's Mag.* Dec. 35 At far left a Persian pantdress in pure wool (with a matching mini-nightie, about $60). 1968 Pantdress [see *pantskirt* below]. 1956 H. Gold *Man who was not with It* (1965) ix. 73 The pantleg was sticky where it was bruised. 1974 D. Richards *Coming of Winter* i. 6 He moved uphill very quickly, his boots and pantlegs soaking from the water. 1976 *New Yorker* 5 Jan. 22/3 We glanced down the row and saw..the next straightening the pant leg over his right calf in a single motion, and so on. 1970 *Women's Wear Daily* 23 Nov. 31/2, I think another pant look will take over. 1964 *Times* 3 Aug. 11 The pants and pantskirt as shown by Marc Bohan at Dior are for the country and around the house. 1968 *N.Y. Times* 15 July 43 This time, it is a more coordinated trend—pant-skirts, pant-dresses, pant-suits, tops and pants and so on. 1977 *Evening Post* (Nottingham) 27 Jan. 9/5 Stride ahead in this all-season look of a classic shirt paired with a trim pant-skirt. 1966 *Sun* 15 June 3/1 'At least they didn't come in trouser suits.' Two girls did. One wore a pant suit in pink and green linen. 1968 *Telegraph* (Brisbane) 24 Oct. 31/6 Model Kelly shows off the latest pantsuit in white stretch towelling. 1970 *Harrods Christmas Catal.* 14/1 Pant suit hangers. 1975 *N.Y. Times* 8 Sept. 33/4 Another woman.. sits by herself looking aloof, brushing hairs off her well-pressed rust pantsuit.

pant-. Add: **Panta·rchic** *a.*, of or pertaining to a pantarchy.
1883 L. F. Ward *Dynamic Sociol.* I. 466 The cosmopolitan, or pantarchic stage.

pantaleon. (Earlier example.)
1757 A. Linde tr. *Keysler's Trav. Germany* IV. lxxxvi. 125 M. Panthaleon Hebenstreit, never refuses to gratify strangers with a sight of the Panthaleon, a musical instrument called by his own name, who was the inventor of it.

pantalettes, *sb. pl.* Add: Also (in *sing.*) **pantelette.** (Earlier and later examples.)
1834 *Knickerbocker* IV. 117 In the first place, in their blushing girlhood, they assume the pantalettes, or little pantaloons. 1838 *Southern Lit. Messenger* IV. 28/1 Two pretty sisters, in pantalettes, waited on table. 1922 Joyce *Ulysses* 528 The scanty, daringly short skirt, riding up at the knee to show a peep of white pantelette, is a potent weapon. 1975 R. Player *Let's talk of Graves* iii. 78 Little girls in pantalettes bowled their hoops.

pantee, see *panties *sb. pl.*

Panthalassa (pænþălæ·sä). *Geol.* Also **panthalassa.** [f. Pan- + Gr. θάλασσα sea.] A universal sea or single ocean, such as would have surrounded *Pangæa.
1893 E. Suess in *Natural Sci.* II. 186 We might.. rather be induced to infer that in Prepalæozoic times there may have existed a universal hydrosphere or panthalassa covering the whole of the planet. 1924 J. G. A. Skerl tr. *Wegener's Orig. Continents & Oceans* x. 149 It is ..not improbable that in the most ancient 'pre-geological times' the film of sial..could then have been..only about 30 km. thick, and have been covered with a 'Panthalassa' ..which probably left exposed only small proportions, or none at all, of the earth's surface. 1937 A. L. Du Toit *Our Wandering Continents* x. 210 The Drift Hypothesis.. has to view the great geosynclinal seaways of earlier geological history as branches, albeit important ones, of a universal ocean or 'Panthalassa', which was transformed into the present oceans through the horizontal drifting of the crustal blocks. 1970 R. S. Dietz in Johnson & Smith *Megatectonics Continents & Oceans* iii. 36 (*caption*) Accretionary development of two supercontinents (Laurasia and Gondwana)..producing two supercontinents of equal area, surrounded by the equatorial universal ocean of Panthalassa. *Ibid.* iii. 43 The Pacific Ocean may be regarded as the remnant of Panthalassa, with the Atlantic, Arctic, and Indian Oceans being mostly rift oceans.

pantheize (pæ·nþi,əiz), *v. rare.* [f. Panthe(ist + -ize 1.] To imbue with the characteristics and ideas of pantheism; to make compatible with pantheism.
1909 W. James *Pluralistic Universe* iii. 118 Theologians have felt its irrationality acutely, and the 'fall', the predestination, and the election which the situation involves have given them more trouble than anything else in their attempt to pantheize Christianity.

panther. Add: **3. b.** *ellipt.* f. *Black Panther* s.v. *black *a.* 19; also used *attrib.* or as *adj.*
1968 *Guardian* 28 June 20/7 The Panthers wear a uniform consisting of Black Leather jacket and beret. 1968 *Listener* 5 Sept. 290/2 The hotel cops, baited in the beginning not only by rocks and bottles but by language of insane obscenity, and having waded into the advance guard of panthers, militants white and black, perambulating trouble-makers, then went berserk. 1970 G. Jackson *Let.* 4 Apr. in *Soledad Brother* (1971) 220 The young Panther party member, our vanguard, must be embraced, protected, allowed to develop. 1972 J. Mills *Report to Commissioner* 120 The Panthers are worse than the Shylocks. They won't let you alone. 1973 E. Bullins *Theme is Blackness* 141 A Black Panther selling Panther newspapers. *Ibid.*, Say, brother..you Panthers sell papers just like the Muslims..don't cha? 1973 *Black Panther* 3 Mar. 8/1 Yet there are some people who say..that Panther rhetoric invites repression and destruction.
5. panther juice, panther('s) piss, panther sweat, strong liquor, usu. spirits, esp. of local or home manufacture; *ellipt.* **panther,** gin.
1942 Berrey & Van den Bark *Amer. Thes. Slang* § 100/9 *Gin*,..panther. 1960 Wentworth & Flexner *Dict. Amer. Slang* 374/2 *Panther*,..inferior liquor, esp. gin. From 'panther sweat'. 1960 J. Philips *Whisper Town* (1961) i. i. 4 Here's your panther juice, Judge. 1941 Baker *Dict. Austral. Slang* 52 *Panther's p—s*, liquor, esp. spirits. 1946 T. Heggen *Mister Roberts* 134 'That whiskey they make,' said Dowdy, 'is really panther-piss.' 1955 W. Gaddis *Recognitions* ii. i. 308 Yeah? Well did you ever drink panther piss? the liquid fuel out of torpedoes? 1973 'B. Mather' *Snowline* v. 57 Locally distilled stuff of the genus known as panther piss. 1929 *Amer. Speech* IV. 386 Whiskey is sometimes called *donk* or *mule* because of its..'kick'... *Panther-sweat*..and *rat-track whiskey* are less easily classified.

panti-, see next.

panties (pæ·ntiz), *sb. pl.* Also occas. **pantees.** [dim. of Pants *sb. pl.*] **1. a.** Men's trousers or shorts. Usu. in derog. contexts.
1845 *Knickerbocker* XXVI. 433 If your panties weren't sheeted home at the bottom, you'd out-jump a monkey. 1848 W. E. Burton *Waggeries* 19, I hadn't on nothin'.. only a blue cotting shirt and sail-cloth pantys. *Ibid.* 95 I've a Colt's revolver in each pantey's pocket. 1910 Kipling *Divers. Creatures* (1917) 310 The umpires, all in short panties, conferred. 1928 *Weekly Dispatch* 27 May 15/7 Panties for boys and skirts for girls..are being made *very* short. 1930 E. Pound *XXX Cantos* xviii. 82 And he was my gawd scared out of his panties.
b. Short-legged or legless knickers worn by women and girls.
1908 M. Morgan *How to dress Doll* vi. 59 The undergarment is..easily made, for the little waist and panties are cut in one piece. 1932 *New Yorker* 11 June 40/1 There is a lace brassière top on a circular satin slip, and panties.. are built in underneath. 1958 *Times* 25 Apr. 13/3 Brand new ex-Wren officers' silk/wool panties. 1969 M. Pei *Words in Sheep's Clothing* (1970) v. 34 'Panties'..short, abbreviated pants worn by women and children (men, as everyone knows, wear shorts, not panties). 1972 F. Warner *Maquettes* 39 Bra and panties. 1976 J. Crosby *Snake* (1977) xxxv. 216 She..picked up her panties, and slipped them on. After that the sweater, the skirt, stockings, and shoes.
c. Also *transf.* and *fig.*
1909 *Sat. Even. Post* 24 Apr. 15/3 New York..would be inhabited by cow-persons in décolleté leather panties. 1936 L. C. Douglas *White Banners* ii. 42 There won't be any French chops with pink panties and a little bite of meat about the size of a peppermint lozenge. 1949 *Sat. Even. Post* 26 Mar. 35/2 Large bottles in woven straw panties stood on the checked table top.
2. *sing.*, as **pantie, panty.** = sense 1 b above. Also *attrib.* orig. *U.S.*
1932 *New Yorker* 11 June 28 (Advt.), If ever a garment could outwit that old ogre depression, it's the pantie in this picture... It is made of celanese. 1939 *Reader's Digest* May 107/2 There is nothing so mournful as a pantie manufacturer who cannot get space in the pantie section [of the garment-making area of New York City]. 1951 C. W. & P. Cunnington *Hist. Underclothes* 246 For sports, pantie trunks and pantie briefs. 1960 *Harper's Bazaar* Apr. 56/1 Striped pantie; boned perforated foam bra. 1961 *New Yorker* 2 Dec. 48 Rogers places you there in the briefest baby doll and matching panty. 1970 *Focus* June 11/3 No apparent dye loss when washed, but pantie part shrank an inch all round.
3. *Comb.* of **pantee, panti, pantie, panty. a.** In names of women's undergarments combining the function of panties with that of some other undergarment, as *panty-belt, -blouse, -brassière, -girdle, -hose, -stockings, -tights.*
1957 A. Adburgham in *Punch* 27 Mar. 419/1 How pert the little pantie-belt, blue-spotted, with its blue-spotted suspenders to match. 1961 S. Price *Just for Record* iii. 26 The sort of birds I lie me down with..shimmy out of their pantie-belts. 1971 *Guardian* 24 Aug. 9/1 Pantie-blouses are..the newest thing in all-in-one garments. 1925 *Ladies' Home Jrnl.* May 87/3 Panty-brassiere, $12. 1941 Hermer & May *Havana Mañana* 43 We recommend panty-girdles of mesh elastic and net or lace bras. 1946 [see *Lastex]. 1961 *She* Feb. 42 Au Fait's panty girdle.. evenly supports tummy, behind, hips and thighs. 1961 *Housewife* Apr. 103/1 Wear a strong pantie-girdle—not a roll-on. 1968 J. Ironside *Fashion Alphabet* 71 *Panti-girdle*, lightly elasticated briefs, exercising some control over bulges. 1972 E. T. Renbourn *Materials & Clothing in Health & Dis.* xix. 387 The pantie-girdle syndrome has crept into the recent literature. Here, marked constriction of the circulation of the upper thigh..leads to aching legs and..swelling ankles. 1972 J. Wambaugh *Blue Knight* (1973) vii. 98 A micro-mini that showed her red-flowered panty girdle when she sat down. 1963 *N.Y. Times* 22 Dec. 19 Our exclusive panti-hose... She'll enjoy the comfort and freedom of..panty tops and micromesh stockings, all in one. No garters, no seams. 1967 *Vogue* 15 Oct. 16 (Advt.), For 12/11 you can treat yourself to Panti-hose seam tights. 1970 *Focus* June 9/2 Invented ten years ago, tights (or pantie-hose) gained ground with the advent of the mini-skirt in 1968. 1970 *Toronto Daily Star* 24 Sept. 38/8 (Advt.), The first quality panty hose at a sensible price. 1972 G. V. Higgins *Friends E. Coyle* xviii. 108 She don't own no pants... Wears them panty hose. 1975 *Publishers Weekly* 1 Dec. 67/2 The type of wan, lightweight heroine who can't support her own pantyhose let alone a mystery novel. 1976 M. Spark *Takeover* iv. 41 She pulled, through her dress, at the top of her pantyhose, setting herself to rights like a schoolgirl. 1966 P. O'Donnell *Sabre-Tooth* vii. 104 The combined pantie-stockings she always wore. 1968 S. E. Ellacott *Everyday Things in Eng. 1914–1968* iv. 69 For some years previously [to 1966], during cold weather, women had worn 'tights' (pantee stockings), and these were now developed in sheer nylon for everyday wear with mini-skirts. 1968 *Vogue* 15 Apr. 92 Pantie-stockings,..3 gns. 1971 *Sunday Nation* (Nairobi) 11 Apr. 37/1 (Advt.), Stockings, Panty-stockings..Tights. 1971 *Guardian* 24 Aug. 9/1 Stocking tights and pantie stockings. 1970 *Guardian Weekly* 11 Apr. 14 'Sorry,' the man in overalls said. 'We're right out of them. What about a pair of panti-tights?' 1970 G. F. Newman *Sir, You Bastard* viii. 207 Her lunch-break dash to Selfridges for pantie-tights, or whatever.
b. *spec.* **panty leg,** the leg part of a pair of panties; (with hyphen) *attrib.*, having such legs; **panty raid** *U.S.*, a prank involving the raiding of women's rooms for trophies of underwear; also *transf.* or *fig.*; **panty-waist** *U.S.*, (*a*) a sissy, a coward; used *attrib.*, effeminate, weak; (*b*) (*rare*) a garment, usu. for children, consisting of panties attached to a bodice.
1908 M. Morgan *How to dress Doll* vi. 61 Face the armholes, neck and panty legs with a narrow facing. 1963 *New Yorker* 8 June 62 Favorite panty-leg bathing suit. 1966 *Time* 2 Dec. 53 Gold and silver pantyleg stockings.. are selling so fast stores can't keep them in stock. 1952 *Stars & Stripes* (Pacific ed.) 21 May 2/3 A wild wave of panty raids swept a dozen college campuses Monday night... The [Univ. of] Colorado raid was staged by 1500 male students intent on seizing intimate trophies of lingerie. *Ibid.* 2/5 More than 1000 Northwestern university male students Monday went on a gleeful 'panty raid', stealing 'all the underwear in sight' and carrying a bewildered police sergeant on their shoulders. 1953 *Newsweek* 11 May 94/1 Traditionally, American undergraduates seem to regard the coming of spring as the rightful time to raise various kinds of hell. Be it goldfish eating, panty raids, or general disturbance of the town peace, boys will be boys, and, more recently, girls will be girls. 1957 Turner & Killian *Collective Behavior* 208 The fad may also exist as a *permission* to act contrary to the folkways and mores, as in the example of college 'panty raids'. 1968 *Listener* 3 Oct. 428/2 'I think we should organise a political panty raid.' There is little organised opposition to the radicals from other students, although one group did deplore 'holding members of the administration captive and burning irreplaceable files of professors who are unsympathetic.' 1936 *Amer. Speech* XI. 280/1 *Panty-waist*, a sissy. 1937 *Sun* (Baltimore) 28 Apr. 6 (Advt.), Now Mike don't be calling *me* a panty waist. 1939 C. Morley *Kitty Foyle* 15 Some of my pantywaists and nightgowns. 1942 *Short Guide Gt. Brit.* (U.S. War Dept.) 5 The English language didn't spread across the oceans and over the mountains and jungles and swamps of the world because these people were panty-waists. 1951 M. McLuhan *Mech. Bride* (1967) 125/1 No panty-waist humanitarianism here. 1952 N. Spain in C. Asquith *Second Ghost Bk.* 33, I should have said that I was pretty tough... I didn't do too badly in the War. I was in the Marines and they don't encourage panty waists in the Marines. 1971 'A. Burgess' *MF* x. 116 Some goddam British poet with one of these pantiwaist names, like Vere de Vere. 1971 H. A. Smith *View from Chivo* i. 7 He is a pantywaist kind of fellow, very *dainty* about things. 1975 *Daily Colonist* (Victoria, B.C.) 30 July 5/3 When they [*sc.* the police] do lay hands on the culprits and get them to the courtroom they are let down by our pantywaist judges.

pantile. 1. c. Delete † *Obs.* and add later examples.
1831 M. Edgeworth *Let.* 16 Apr. (1971) 525 Yesterday I went to Tunbridge Wells.., saw the Pantiles... The pantiles looked to me wondrous small and narrow and the roof over the row too low. 1907 *Daily Chron.* 29 Jan. 8/5 It is at the east end of the Pantiles that the original spring, ..comes to the surface. 1936 *N. & Q.* 26 Dec. 461/2 The old Chapel of Ease..at entrance to the Pantiles, Tunbridge Wells..was built in 1684. *Ibid.*, When I was at school at Tunbridge Wells..I used to be taken to a church, close to the Pantiles. 1974 *Encycl. Brit. Micropædia* X. 179/1 The Pantiles Parade, with the original chalybeate spring..is preserved.
pantiled *a.* (later examples.)
1951 [see *mansarded* adj. s.v. *mansard]. 1963 *Guardian* 5 Mar. 7/2 A precipitous assembly of orange pantiled houses. 1978 M. Butterworth *X marks Spot* iii. i. 121 They passed over a red pantiled roof of a farmhouse.

panting, *vbl. sb.* Add: **b.** *spec.* In Shipbuilding: the movement of the plates of the ship's hull under stress, esp. occurring at the fore and aft ends of the ship. Also *attrib.*, often in names of structures designed to prevent such movement.
1885 H. Paasch *From Keel to Truck* 24/1 *Panting* (of a ship). 1899 E. L. Attwood *Text-bk. Theoret. Naval Archit.* 212 Panting beams and stringers to be fitted at the after end. 1904 A. C. Holms *Pract. Shipbuilding* I. ix. 105 At the stern, panting stresses are usually unimportant. 1922 C. M. Swainston *Reed's Seamanship* (ed. 23) i. 187 The principal methods to resist panting are:—Closer spacing of frames, double frames, an extra tier of beams called *panting beams*, broadening of the stringers called *panting stringers*, [etc.]. 1927 G. Bradford *Gloss. Sea Terms* 124 *Panting strains*, those produced by..the pressing in and out of a ship's plates due to the pressure of the waves. *Ibid.* 125 The *panting stringers* reinforce the plates against these stresses, along the sides, and *panting beams* and *frames* at the bow. 1961 F. H. Burgess *Dict. Sailing* 156 *Panting*, the vibrations of the forward plates caused by the variable resistance of the water. *Ibid.*, *Panting beam*, a strengthened beam fitted forward to decrease vibration.

pantless (pæ·ntlės), *a.* [f. *pant *sb.*[3] + -less.] Wearing no pants.
1880 S. Lakeman *What I saw in Kaffir-Land* xi. 136 They [*sc.* Cape baboons] shot from branch to branch,.. like flying fish, or as pantless Zazel shoots from the cannon's mouth to her swinging rope. 1948 D. Ballantyne *Cunninghams* 260 Three pantless Maori kids met him. 1969 N. Behn *Shadowboxer* (1970) ix. 79 He stared directly up the plump pantless female thighs. 1971 W. Hanley *Blue Dreams* xxi. 331 Then, bra-ed but pantless, she moved toward Walter.

panto (pæ·nto). [Abbrev. Pantomime *sb.* (*a.*).] = Pantomime *sb.* (*a.*) 3 in Dict. and Suppl.; also *attrib.*
1852 E. L. Blanchard *Jrnl.* 15 Sept. in Scott & Howard *Life E. L. Blanchard* (1891) I. 98 At home till 5 p.m. fixing on titles for pantos... Arrange for panto with Smith for Drury. 1914 C. Mackenzie *Sinister St.* II. iv. ii. 865 'You're on the stage, aren't you?' 'I usually get into panto,' she admitted. 1922 Joyce *Ulysses* 533 Immoral panto boys in flesh tights. 1929 J. B. Priestley *Good Companions* ii. i. 251 That's right. A lot of experience. C.P. work, halls, panto, low comedy in legit., know it all. 1937 G. Frankau *More of Us* xi. 121 Eke tho' of words bard still commands a brainful Which she can still make dance like panto elves. 1962 *Oxford Times* 1 June 23/4 Orsino and Viola..the latter part, although difficult in its mixture of boyishness and charm, should not really be played as a principal boy in panto. 1968 *Listener* 26 Dec. 871/3 Nor has the standard of the witless, panto-type sketches been raised. 1969 *Ibid.* 28 May 663/1, I was glad that Hazel Hughes should play her..like Kenneth Williams as painless panto dame. 1976 *Sunday Post* (Glasgow) 26 Dec. 3/3 She was transporting it from Calderpark Zoo to the Pavilion panto, where it was due to appear in the cartoon spot featuring TV personality Glen Michael. 1977 *R.A.F. News* 11–24 May 3/5 The organizers ran an 'ad lib' version of the panto 'Cinderella'.

pantocain (pæ·ntokēⁱn). *Pharm.* Also **-caine.** [a. G. *pantocain*, f. Gr. παντο- Panto- + *-cain*, after G. *cocain* Cocaine.] The hydrochloride salt, $C_{15}H_{24}N_2O_2{\cdot}HCl$, of a diamino-ester which is used as a local anæsthetic; also called *amethocaine* or *tetracaine hydrochloride.*
1931 *Manuf. Chem.* II. 105/1 The new local anæsthetic of the I.G. Farbenindustrie, Pantocain,..has been systematically evolved from the novocaine series and is the hydrochloride of *p*-butylaminobenzyldimethylaminoethanol. 1937 *Brit. Med. Jrnl.* 20 Nov. 1036/2 Local infiltration with 0·5 per cent. pantocaine caused excessive oedema of the lids and face..and intense itching. 1942 Parsons & Stallard *Dis. Eye* (ed. 10) xxi. 431 Particles of lime must be perseveringly picked out with forceps, after previous application of pantocain. 1970 *Brain Res.* XIX. 102 The neck afferents were eliminated by..the blockage of afferent fibers by Pantocain injection into the intervertebral foramen.

Pantocrator (pæntǫ·krătəɹ). Also **Pantokrator.** [ad. Gr. παντοκράτωρ almighty.] With reference to God or Christ: the Almighty, all-ruler; hence, an artistic representation of the figure of Christ, esp. as a characteristic form in Byzantine art.
1871 Ruskin *Fors Clavigera Letter 12* 11 In the Apocalypse it is 'Lord, All governing'—Pantocrator—which we weakly translate 'Almighty'. 1911 O. M. Dalton *Byzantine Art & Archaeol.* xii. 672 In the Last Judgement, and as the Pantokrator, Christ is bearded, because in his function as Judge he is regarded as merely continuing his earthly mission. 1931 *Antiquity* V. 508 With the exception of the Pantocrator in the dome..the Daphni mosaics belong to the pictorial, representational, Hellenic tradition. 1947 C. Stewart *Byzantine Legacy* v. 111 The condition of the Church of Christ Pantocrator is typical of many. 1950 A. Huxley *Themes & Variations* 174 Young Domenikos received a sound Greek education and studied painting... That indecently human personage—was that supposed to be the Pantocrator? 1962 *New Statesman* 25 May 768/3 It [*sc.* a cathedral tapestry] fails to achieve the commanding presence of a Byzantine pantocrator. 1963 D. T. Rice *Art of Byzantine Era* 88 In Basil's church the bust of Christ Pantocrator dominated the building from the dome. 1970 *Oxf. Compan. Art* 182/1

The Iconoclastic crisis..culminated in the formation of a new religious iconography... It was then that the decorative scheme of the Byzantine church was fixed..: in the dome the Pantocrator (Christ the Ruler) surrounded by archangels. 1974 D. Yarwood *Archit. Europe* iii. 110/2 The mosaics and frescoes..of the dome..illustrate the Christ Pantocrator in all His Glory.
Hence **pantocra·tic** *a. rare.*
1949 Auden *Under Sirius* in *Horizon* Oct. 210 And out of the open sky The pantocratic riddle breaks;—'Who are you and why?'

pantograph, *sb.* Add: **c.** A jointed, self-adjusting framework on the top of an electric locomotive for conveying the current from overhead wires.
1907 F. H. Davies *Elect. Power & Traction* xxiii. 269 The..collecting gear is that known as the pantograph, and the object..is to permit of high speed running and reversal of direction without any corresponding adjustment of the gear. 1920 *Glasgow Herald* 23 Sept. 7 Electric locomotives can..be fitted..with pantograph collectors. 1930 *Engineering* 20 June 793/3 Only one pantograph was used at all speeds, so long as the current did not exceed 1800 amps. 1957 *Railway Mag.* Mar. 159/2 The pantograph is raised by pressing the push-button in the driving trailer. 1970 *Daily Mail* 8 Jan. 1/4 The pantograph.. appears to have jumped on top of the wire instead of running beneath it. 1977 *Modern Railways* Dec. 492/1 The primary reason for the central position was power collection problems at high speed in a two-power-car formation if each power car had its own pantograph.
d. Used of other mechanisms in the form of a movable diamond-shaped trellis or lazy tongs.
1942 *Archit. Rev.* XCII. 46 (*caption*) The early tube carriages of 1906, *seqq.*, had pantograph doors at both ends. 1975 *Observer* 9 Mar. 33/5 The driver's screen-wiper is a pantograph type and sweeps a commendably large area but the passenger's section of the screen has a big unswept corner.

pantograph (pæ·ntǒgrɑf), *v.* [f. the sb.] *trans.* To enlarge by means of a pantograph. Cf. Pantographing *vbl. sb.* in Dict.
1934 in Webster. 1936 J. Agate in *Sunday Times* 12 Apr. 5/1 Can a drop-earring be pantographed to chandelier-size without loss of exquisiteness?

pantographically, *adv.* (Examples.)
1911 E. C. Worden *Nitrocellulose Industry* II. xiv. 697 These enamels may be closely imitated..by taking a given pattern, enlarging it pantographically or otherwise, [etc.]. 1945 F. J. Camm et al. *Newnes Plastics Manual* xii. 135 This machine [*sc.* the 'Keller' die-sinking machine] will produce all kinds of dies and jigs by a process of copying from a pattern pantographically.

pantoic (pæntōᵘ·ik), *a. Chem.* [f. as *panto-yl + -ic.] *pantoic acid*, the unstable parent carboxylic acid $C_6H_{12}O_4$, of the pantoyl radicle.
1945 *Jrnl. Biol. Chem.* CLXI. 513 In connection with studies on the antagonistic action of pantoyltaurine on salicylate inhibition of *Escherichia coli* we have observed that pantoic acid, or, more properly, the pantoate ion, is more active than pantolactone. 1966 *McGraw-Hill Encycl. Sci. & Technol.* XIV. 259/2 α-Ketoisovaleric acid ..is also a precursor of leucine and of the pantoic acid moiety of pantothenic acid.
Hence **pa·ntoate,** the anion, or a salt or ester, of pantoic acid.
1945 [see above]. 1968 *Compar. Biochem. & Physiol.* XXVII. 647 Pantothenate and pantoate inhibit the growth of *Trypanosoma lewisi in vitro.*

pantomime, *sb.* (*a*). **3.** Add to def.: By the 20th century, the traditional form changed, with the loss of the pantaloon and harlequin features. The entertainment, primarily for children, is now based on the dramatization of a fairy tale or nursery story, and includes songs and topical jokes, buffoonery and slapstick, and standard characters such as a pantomime 'dame', played by a man, a leading boy, played by a woman, and a pantomime animal, e.g. horse, cat, goose, played by actors dressed in a comic costume, with some regional variations. (Further examples.)
1896 *Pall Mall Mag.* Oct. 302/1 She was still playing principal boy in the pantomime—a gay, gallant Prince, in plumed cap and tights. 1901 R. J. Broadbent *Hist. Pantomime* xxi. 225 Present-day Pantomime, with the immense sums spent annually on its gorgeous spectacular display and costly dresses..is a subject that is well known to us all... The best parts are, as a general rule, allotted to music-hall 'stars' whose names will draw the most money. 1911 *Times* 7 Nov. 9/4 Some of us would have passed through a childhood of tasteless pantomimes, lacking the most exciting of their three annual thrills, which were always the clown, the 'principal boy', and 'Jimmie' Glover. 1950 *Oxf. Jun. Encycl.* IX. 273/2 Pantomimes of to-day are on a splendidly lavish scale, with hundreds of performers. They all have their stock characters... They provide many of the catch phrases and popular songs of the season during their long runs.., often from Christmas till Lent. 1965 D. Arundell *Story of Sadler's Wells* xii. 149 Phelps began his 1847–8 season on 23rd August..with of course a grand pantomime at Christmas (pantomime now meaning the extravaganza fairy-tale as we understand it today). 1978 *Times Lit. Suppl.* 24 Feb. 247/2 The town is agog over the annual pantomime which opens on Boxing Day.
b. *transf.* and *fig.*
1781 G. Selwyn *Let.* 27 Feb. in *15th Rep. R. Comm. Hist. Manuscripts* App. vi. 464 in *Parl. Papers 1897* (C. 8551) LI. 1, I believe there is no actor upon the stage of either theatre who, repeating what the author has wrote, does not, at the same time, recite his own private sentiments oftener, than our pantomimes in Parliament. 1837 Dickens *Let.* 13 Feb. (1965) I. 236 Oliver and the Pantomime of Life are both finished, and with the Printer. 1941 *Penguin New Writing* III. 109 It's a proper pantomime. The old Tabbies'll have to mind their dignities if they steps out to-day. 1972 *Times* 7 Aug. (Jamaica Suppl.) p. v/4 The uniquely Jamaican 'pantomime'..has evolved into the musical comedy formula incorporating topical allusions..along with folklore characters.
5. Further *attrib.* and *Comb.* examples.
1765 H. Timberlake *Mem.* 80 [The Cherokees] are.. very dexterous at pantomime dances; several of which I have seen performed that were very diverting. 1883 D. Cook *On Stage* I. x. 219 In the pantomime season, or whenever any great pageant or spectacle is to be produced, these plots are of prodigious extent. 1901 R. J. Broadbent *Hist. Pantomime* xix. 196 This clashing.. cannot but..adversely affect the box-office receipts, unless, of course the Pantomime-goer makes a point of 'doing the round'. 1908 G. B. Shaw *Pen Portraits* (1932) 74 A pantomime animal with two men in it is a mistake when the two are not very carefully paired. 1919 —— *Great Catherine* 114 They..produced scene after scene of.. tragic relief in the torture chamber with the monarch as pantomime demon committing real atrocities. 1922 Joyce *Ulysses* 431 In pantomime dame's stringed mobcap, crinoline and bustle, widow Twankey's blouse..and cameo brooch. 1935 T. S. Eliot *Murder in Cath.* i. 41 All things are unreal, Unreal or disappointing: The Catherine wheel, the pantomime cat, The prizes given at the children's party. 1939 —— *Old Possum's Pract. Cats* 37 In the Pantomime season I never fell flat. 1946 M. Dickens *Happy Prisoner* ix. 189 He had been afraid they were going to guy her up like a pantomime dame. 1967 *Listener* 19 Oct. 513/2 Everyone will want to see Alec Guinness do his pantomime-dame act in *Wise Child.* 1973 'A. Hall' *Tango Briefing* viii. 99 Whether you die like a man or the back end of a pantomime horse you're going to stop breathing. 1976 M. Gilbert *Night of Twelfth* xvii. 156 Think of old Dip in a wig and falsies. He'd look like a pantomime dame.

pantomime, *v.* Add: **1.** Also, to behave as though in a pantomime.
1958 *Spectator* 8 Aug. 187/2 The vanity of a Lloyd George or a Ramsay MacDonald, who preferred pantomiming round the world in a continual circus to staying put where they belonged.
Hence **pa·ntomimed** *ppl. a.*
1950 A. Ronell in Manvell & Huntley *Technique Film Music* (1957) iii. 137 His pantomimed thoughts find voice through the inflection of instruments whose colours express Harpo's spirited style.

pa·ntomimish, *a.* [Pantomime *sb.* (*a.*) + -ish[1].] = Pantomimic *a.* 2.
1923 *Glasgow Herald* 3 Feb. 6/6 A few veterans may recall how pantomimish Mr. Gladstone looked when..he borrowed the hat of a colleague.

pantomography (pæntǒmǫ·grăfi). *Med.* [f. Pan(oramic *a.* + *tomography.] A form of tomography for obtaining radiographs of curved layers of an object, *spec.* the teeth and jaws, by rotation of the body and film during exposure; a further modification of the technique is *orthopantomography.
1952 Y. V. Paatero in *Suomen Hammaslääkäriseuran Toimituksia* XLVIII. 7 Pantomography—a tomographic method for roentgenographing curved surfaces. 1954 —— in *Acta Radiologica* XLI. 321 A new method of tomographic roentgenography (which I subsequently named pantomography.) *Ibid.* 326 Since the roentgen beam in pantomography is relatively narrow, it is obvious that secondary radiation emanates from only a narrow portion of the object. 1956 K. C. Clark *Positioning in Radiog.* (ed. 7) xxviii. 536/2 Pantomography..By this method of simultaneous movement of head and film in front of the slit diaphragm of the tube, a progressive panoramic view of the whole jaw is produced. 1969 G. J. van der Plaats *Med. X-Ray Technique* (ed. 3) iv. 104 Pantomography is mainly applied in dental radiography.
Hence **panto·mogram,** a radiograph obtained by pantomography; **panto·mograph** (see quot. 1954[1]); **pa:ntomogra·phic, -gra·phical** *adjs.*
1952 Y. V. Paatero in *Suomen Hammaslääkäriseuran Toimituksia* XLVIII. 7 Divergence of rays in pantomographical roentgenography. *Ibid.*, Taking of lateral roentgenograms of jaw with the pantomographic method. 1954 —— in *Acta Radiologica* XLI. 321 *Pantomograph*, a device used with a normal roentgen apparatus, which produces pantomographic roentgenograms = pantomograms. *Ibid.*, Thus, pantomographic roentgenography must be distinguished from all previous tomographic methods which produce roentgenographic representations of level surfaces. *Ibid.* 326 The pantomograph used rotates around its axis once every 20 seconds.

pantonal (pæn,tōᵘ·nəl), *a.* [Pan- + Tonal *a.*] Used as a synonym for 'atonal', in twelve-tone music; including all tonalities; hence **panto·nalism, pantona·lity.**
1958 R. R. Reti *Tonality, Atonality, Pantonality* iii. iii. 70 The question of how in the realm of pantonality the problem of consonance and dissonance is treated. *Ibid.*, 73 The first example tending towards the pantonal, the second towards the atonal concept. 1961 *Times* 6 Dec. 17/6 It is true enough that a crowd of Muscovites do not sing pantonal fugues. 1963 *Listener* 31 Jan. 220/3 In some sections I have expanded this pantonality into harmonic structures which vertically include up to twenty-three notes. 1966 *Ibid.* 19 May 736/1 Schoenberg disliked the word 'atonal', and preferred the expression 'pantonal', inclusive of all tonalities. 1970 W. Apel *Harvard Dict. Mus.* (ed. 2) 640/1 *Pantonality, pantonal*, the inclusion of all tonalities. The terms are sometimes used instead of atonality and atonal.

Pantopon (pæ·ntopǫn). *Pharm.* Also **pantopon.** [f. Pant- + Op(ium *sb.* + *-on.*] A proprietary name in the U.S. for a mixture of the hydrochlorides of the opium alkaloids. Cf. *omnopon.
1909 *Chem. Abstr.* III. 944 Pantopon..contains according to Sahli..all the opium alkaloids. 1g. = 5g. opium. 1910 *Official Gaz.* (U.S. Patent Office) 1 Feb. 183/1 F. Hoffmann-La Roche & Co., Basel, Switzerland. Filed Sept. 8, 1909. *Pantopon*... A pharmaceutical preparation containing all the alkaloids of opium in an easily-soluble state and in a form suitable for subcutaneous injections. 1940 H. A. McGuigan *Appl. Pharmacol.* 524 It is claimed that pantopon causes less depression of the respiration, less nausea, and has a slower and more prolonged action [than morphine]. 1953 W. Burroughs *Junkie* (1972) xii. 121 Lupita sells her stuff in papers. It is supposed to be heroin. Actually, it is pantopon cut with milk sugar and some other crap. 1972 *Biol. Abstr.* LIV. 1738/2 The changes in histamine may have been affected by the narcotics—amobarbital and Pantopon.

pantothenic (pæntoþe·nik), *a. Biochem.* [f. Gr. πάντοθεν from every side (f. παντο- Panto-) + -ic.] *pantothenic acid*: an oily, optically active carboxylic acid, $HOCH_2C(CH_3)_2$-$CHOH{\cdot}CO{\cdot}NH(CH_2)_2COOH$, which is widely distributed in plant and animal tissues (mainly in combined form as coenzyme A), is essential for the growth of yeast and certain bacteria, and is a member of the vitamin B complex.
1933 R. J. Williams et al. in *Jrnl. Amer. Chem. Soc.* LV. 2925 We can safely regard the activity [in stimulating the growth of yeast] as due to a single acid. Since this acid appears to be of very widespread occurrence..we have tentatively called it 'pantothenic' acid. 1953 J. Ramsbottom *Mushrooms & Toadstools* viii. 82 Some strains of *Saccharomyces cerevisiae* require inositol, biotin, aneurin, pantothenic acid and other factors. 1960 *Times* 2 July 2/6 Certain..vitamins [in royal jelly]..are almost entirely members of the B-complex, and include an unusually high proportion of pantothenic acid. 1974 Passmore & Robson *Compan. Med. Stud.* III. i. xxiv. 30/1 Pantothenic acid deficiency has been considered, on the basis of therapeutic trials, to be the cause of the burning feet syndrome.
Hence **pantothe·nate,** the anion, or an ester or salt, of pantothenic acid.
1941 *Chem. Abstr.* XXXV. 3606 $[\alpha]_D^{25}$ for the pantothenate ion is calcd. to be −26·8°. 1955 *Sci. News Let.* 1 Oct. 211/1 The cancer cells, Dr. Eagle finds, need these seven vitamins: choline, folic acid, nicotinamide, pantothenate, pyridoxal, riboflavin and thiamine. 1974 R. W. Doskotch in W. O. Foye *Princ. Med. Chem.* xxv. 613/2 The commercial synthesis of calcium pantothenate starts with isobutyraldehyde and formaldehyde.

pantothermal (pæntopɔ̄·ɹmăl), *a.* [f. Panto- + Thermal *a.*] Able to stand a wide range of temperature.
1906 *Athenæum* 10 Feb. 175/1 A revision of all captures ..appeared to show one species..as cosmopolitan and pantothermal.

pantoyl (pæ·ntoil, pæ·nto,əil). *Biochem.* [f. *panto(thenic *a.* + -yl.] The optically active radical $HOCH_2C(CH_3)_2CH(OH)CO$—, present in pantothenic acid.
1942 Barnett & Robinson in *Biochem. Jrnl.* XXXVI. 357 Dr. McIlwain..suggested a simplified nomenclature, following which we have adopted a system based on the name 'pantoyl' for the α:γ-dihydroxy-β:β-dimethyl-γ-butyryl radicle. 1971 *Analytical Biochem.* XLII. 8 Pantoyl lactone, the other product of hydrolysis of pantothenates, can be easily and directly evaluated by gas chromatography, providing a simpler method for the quantitation of Vitamin B_5.
Hence **pantoyltau·rine,** a sulphonic acid, $C_8H_{17}O_5NS$, which inhibits the action of pantothenic acid in microorganisms.
1942 *Biochem. Jrnl.* XXXVI. 366 The degree of inhibition by pantoyltaurine depends on the amount of pantothenic acid present in the medium. 1954 E. E. Snell in Sebrell & Harris *Vitamins* II. ix. 680 Addition of pantoyltaurine increased the lag phase and decreased the rate of growth during the first half of the logarithmic phase of *Streptococcus hemolyticus.* 1968 *Compar. Biochem. & Physiol.* XXVII. 649 When pantoyltaurine, an analogue of pantothenic acid, was incorporated into the culture media, inhibition of growth [of *Trypanosoma lewisi*] was observed.

pantry, *sb.* Add: **1. b.** Used in the names of tea rooms and cafés.
1948 J. Betjeman *Sel. Poems* 79 The shops..on the Esplanade—The Circulating Library, the Stores, Jill's Pantry. 1958 J. Cannan *And be a Villain* i. 16 Pam's Pantry started to serve tea at three-thirty. 1967 M. Kenyon *Whole Hog* xvii. 169 They agreed on the Pancake Pantry..Liz said the pancakes were good. 1971 S. Phillips *Death in Sheep's Clothing* iv. 36 A teashop known as *Pam's Pantry*..with fresh chintz curtains, flowers on

each tiny oak table, and a lot of unidentifiable brass objects hanging from the beams.

2. *pantry work*; **pantry-maid**, a maidservant who has duties in the pantry.

1921 *Dict. Occup. Terms* (1927) § 900 *Pantry maid*..washes china and glass and cleans up in pantry. 1928 *Daily Tel.* 5 June 4/7 He had paid her 22s a week as a pantrymaid. 1973 *Courier & Advertiser* (Dundee) 12 Jan. 1/3 (Advt.), Part-Time woman required for general cleaning and pantry work in hotel.

pants, *sb. pl.* Add: **1. a.** No longer 'chiefly *U.S.*'. Now used for trousers, worn by either men or women, and no longer considered vulgar. **b.** Now used for underpants, panties, or shorts worn as an outer garment: cf. *hot pants* (*hot *a.* 12 c). (Earlier and later examples.)

1840 E. A. Poe *Peter Pendulum* in *Burton's Gentleman's Mag.* Feb. 88 Standing on one leg three hours, to show off new-touch strapped pants. 1928 R. Campbell *Wayzgoose* ii. 58 Through pants and vest the God explored. 1930 H. G. Wells *Autocracy of Mr. Parham* ii. i. 95 He grows more and more independent of the idea that his pants are him. 1940 O. Nash *Face is Familiar* 91 Sure, deck your lower limbs in pants. 1951 T. Sterling *House without Door* xiii. 152 She chose her blue underwear... She laid the pants and brassière on her bed. 1956 H. Gold *Man who was not with It* (1965) i. 5 Grack..plucked a tricksie in shorts as she wiggled by. He took the thin pants between his horny fingers. 1964, 1968 [see *pantskirt* s.v. *pant *sb.*[3] 2]. 1971 *New Yorker* 11 Sept. 12/1 (Advt.), The back-zippered tunic is a great topper for skirts and pants. 1973 N. Moss *What's the Difference?* p. ix, I heard an American student at Cambridge University telling some English friends how he climbed over a locked gate to get into his college and tore his pants, and one of them asked in confusion, 'But how could you tear your pants without tearing your trousers?' 1976 *National Observer* (U.S.) 20 Nov. 13/2 The men in flannel shirts and work pants stood in the driveway outside the hillside house and talked about Vivien Kellems.

c. Slang phr. *to be caught with one's pants down*: in a state of embarrassing unpreparedness. orig. *U.S.*

1932 *Amer. Speech* VII. 330 To be caught with one's pants down, to be caught off guard. 1943 N. Balchin *Small Back Room* xiv. 164 We just let him carry on alone. Now, we're caught with our pants down, with nobody knowing anything about the damned thing. 1959 *Times Lit. Suppl.* 10 July 407/2 Here were the 'better elements' caught with their pants down, as Americans coarsely put it. 1963 P. McCutchan *Man from Moscow* iii. 36 There was..four days to go before the arrival of the Foreign Ministers but the West was not going to be caught with its pants down. 1974 'M. Innes' *Appleby's Other Story* xvi. 133 He overheard Maurice Tytherton, in a great fury, say something to his nephew about having caught him with his pants down. 1976 N. Thornburg *Cutter & Bone* v. 129 His record..did not amount to much more than indiscretion, an embarrassing talent for getting caught with his pants down.

d. Slang phr. *to bore* (or *scare, talk*, etc.) *the pants off* (someone): to bring about (the action of the verb) to a state of extremity.

1933 E. O'Neill *Ah, Wilderness!* i. 38, I tell you you scared the pants off him. 1934 E. Waugh *Handful of Dust* iii. 133 She bores my pants off, but she's a good trier. 1937 N. Coward *Present Indicative* iv. 164 Even if I had known then how much time and ink he [sc. a critic] was going to waste in the future roasting the pants off me, I [etc.]. 1939 E. B. White *Quo Vadimus?* ii. 65 And if I did have a butler named Fish, wouldn't I kid the pants off him? 1940 O. Nash *Face is Familiar* 74 And that Mrs Comfitmonger while pounding her beat has dealt with personalities who would scare the pants off Lombroso. 1953 H. Clevely *Public Enemy* i. 2 'Did you win?' 'We took the pants right off them.' 1956 B. Holiday *Lady sings Blues* (1973) i. 6 Once a girl hit me on the nose and it just about finished me. I took my gloves off and beat the pants off her. 1966 R. Ellison in A. Chapman *New Black Voices* (1972) 402 Where does that idea come from?..One place (which almost frightened the pants off me) was in *Commentary*. 1968 M. Woodhouse *Rock Baby* vi. 51 He..told me there was an international athletics meeting at White City..and that Denmark would undoubtedly thrash the pants off us. 1972 G. Bromley *In Absence of Body* ix. 119 Usually they take the pants off us, which is not surprising—they play regularly and it's our only game. 1975 'W. Haggard' *Scorpion's Tale* ii. 23 There's some stupid story the island is haunted...there's something which scares the pants off the local peasantry.

e. In other phrases: *to wear the pants*: to be the dominant member of a household; *to keep one's pants on*: to keep calm, not to panic or get angry; *(by) the seat of one's pants*: in the handling of an aeroplane, car, etc.: (by) human instinct or experience, as opposed to technical aid or scientific knowledge. Also in extended uses and with hyphens (*seat-of-the-pants*) as adj. phr.

1931 *Amer. Mercury* Nov. 331/1 He claimed that Peggy was bossy, that she wore the pants and gave orders to Pal. 1936 J. Steinbeck *In Dubious Battle* vi. 83 'I wish it would start,' Jim said... 'Keep your pants on,' said Mac. 1942 *Harper's Mag.* May 626/1 When you check your instruments you find it is doing a correct job of flying and that the seat of your pants and your eyes would have tricked you had you been allowed to do the 'co-ordinating'. 1947 *Richmond* (Virginia) *Times-Dispatch* 26 Dec. 2/3 Even expert flyers can't tell by the feeling in 'the seat of their pants' when an airplane is about to stall. 1957 L. P. Hartley *Hireling* x. 77 She's older than he is and she wears the pants. 1958 *Listener* 20 Nov. 835/3 That's no help to the man who's driving by the seat of his pants, as we used to say in the R.A.F. police. 1965 R. Sheckley *Game of X* (1966) xxi. 146 Flying was in fact extremely difficult, but..I was just one of those seat-of-the-pants naturals who instinctively do everything right. 1971 *Sunday Nation* (Nairobi) 11 Apr. 29/1 The type of car which I have just put through one of my seat-of-the-pants road tests—the Alfa-Romeo 1300 GT Junior. 1972 *Times* 18 Sept. 20/4 There was a feeling among the workforce that the firm was being run 'by the seat of the pants'. 1973 E. Bullins *Theme is Blackness* 68 [Mother] Now, put it some place where they'll be safe..understand? Safe! [Daddy] Okay..okay..Matilda..just keep your pants on, will ya? 1977 M. Walker *National Front* i. 17 Mussolini had governed by the seat of his pants, guided in part by his early Socialism, in part by his..bombastic nationalism and above all..by his flair for presentation and publicity. 1978 R. Jansson *News Caper* viii. 85 Thackray was not looking at the instruments... Perhaps that was what they meant by flying by the seat of the pants.

3. *attrib.* and *Comb.*, as *pants pocket*; **pants dress**, a dress with a divided skirt; **pants rabbit** *U.S. slang* (chiefly *Mil.*), a body louse; **pants skirt** = *pantskirt* s.v. *pant *sb.*[3] 2; **pants suit** = *pant suit, *trouser suit.

1964 *Women's Wear Daily* 30 Nov. 4 Catherine Deneuve ordered at Heim a pantsdress in multicolored striped chiffon. 1969 *Sears Catal.* Spring/Summer 32 Floral print especially smart in this pants dress because it's done in navy and white. 1931 E. O'Neill *The Hunted* iv, in *Mourning becomes Electra* (1932) 170 He fumbles in his pants pocket. 1951 T. Sterling *House without Door* vi. 78 He walked down the stairs, struggling for change in his pants pocket. 1974 R. B. Parker *Godwulf Manuscript* xii. 99, I took a jackknife out of my pants pocket. 1918 *Nat. Geogr. Mag.* June 499 They call the things 'pants rabbits' and 'seam squirrels'. 1928 W. H. Upson *Me & Henry & Artillery* 11 Some of the wise crackers in the battery used to call them pants-rabbits, which is not real scientific, as they usually roam around your back and shoulders and seldom hit below the belt. 1937 J. Steinbeck *Of Mice & Men* ii. 36 What the hell kind of bed you giving us, anyways? We don't want no pants rabbits. 1964 *Glamour* July 77 A blouse that looks like challis edged in wool lace, with a wine-dark leather pants-skirt. 1969 *Sears Catal.* Spring/Summer 34 Plaid pants skirt. Twill. Front panel hides plaid culotte. 1964 *Glamour* Dec. 112, 1 and 2 [sc. jacket and pants] make a pants-suit that's very current and handsome. 1968 Mrs. L. B. Johnson *White House Diary* 26 June (1970) 691, I changed into a beige pants suit. 1975 *New Yorker* 5 May 81/3 Elizabeth Franz..looked fine in a becoming pants suit.

pantun. Add: (Earlier and later examples.) Also *attrib.*

1783 W. Marsden *Hist. Sumatra* 162 The essentials in the composition of the *pantoon*..are the rythmus and the figure, particularly the latter, which they consider as the life and spirit of the poetry. 1821 J. Leyden tr. *Malay Annals* 83 They sing of it in Pantuns. *Ibid.* 259 There came a Pantun poet, who was famous for his skill in horsemanship. 1964 M. Taib bin Osman in Wang Gung-Wu *Malaysia* iii. xv. 211 The *pantun* can be considered as folk-ditty; it is used on almost all occasions in Malay life. 1970 *New Yorker* 14 Nov. 58/1 She drafted three poems—another rondeau, a pantoum, and a cynghanedd. 1975 'G. Black' *Big Wind* x. 183 She was singing a Malay *pantun*, a verse form tending to the obscene.

panty, see *panties *sb. pl.*

‖ **panung** (pā-nuŋ). [Thai.] A Siamese garment, worn by men and women, consisting of a long piece of cloth draped round the lower part of the body.

1857 J. Bowring *Kingdom & People of Siam* I. 132 Of the garments worn by the Siamese, the *panung*..is worn round the waist and thighs. 1886 *Pall Mall Gaz.* 3 Aug. 4/1 Two men dressed in their ordinary garb—only a dark blue 'panung' about their loins (all Siamese wear the panung). 1928 *Daily Express* 13 Mar. 12/2 The Eton crop in Bangkok is as old as the 'panung'—that peculiar garment worn by men and women alike, which resembles a cross between a 'sarong' and a baby's napkin. 1958 *Listener* 18 Dec. 1039/3 Siamese girls..wear their traditional *panung*, or slim skirt folded in front adapted from their own *pasin*.

panzer (pæ·nzəɪ, ‖ pa·ntsər). [G., 'mail, coat of mail'.] Used *attrib.*, of or pertaining to a German armoured unit; also in G. form *Panzerdivision*. As *sb.*, a panzer unit or a member of such a unit. Also *transf.* and *fig.*

1940 *Economist* 20 Apr. 729/1 It is more than silly..to calculate the speed at which Germany's *panzerdivisionen* could traverse the Hungarian plain. 1940 *Topeka* (Kansas) *Daily Capital* 24 June 1 b Delegations deploying to stop Willkie's 'panzer' wire attacks. 1941 *Hutchinson's Pict. Hist. War* 19 Mar.–13 May 189 An Imperial division concentrated round Bengazi was left in the air when the German Panzer inrush swept aside the British armoured brigade. 1942 *R.A.F. Jrnl.* 3 Oct. 5 You were in danger of breaking your leg by falling over a complete panzer division of dachshunds moving from bar to shove ha'penny table on a broad front. 1943 E. J. Pratt *Coll. Poems* (1958) i. 82 And what specific can unmesh The tangle of civilian flesh From the traction of the panzers? 1945 Koestler *Yogi & Commissar* iii. iv. 250 Geniuses are panzer-spearheads. 1949 —— *Promise & Fulfilment* ii. iii. 234 Just after Glubb's panzers had begun the shelling of the Jewish quarter. 1962 *Listener* 28 June 1117/1 German Panzer troops arriving in Wales last week. *a* 1963 S. Plath *Ariel* (1965) 55, I have always been scared of *you*, with your Luftwaffe..your Aryan eye, bright blue Panzer-man, panzer-man, O You. 1965 *New Statesman* 14 May 759/1 They tried to defend the Book of Genesis against Darwin's panzers. 1969 G. Macbeth *War Quartet* 74 Each panzer was a coiled spring, oiled For instant action. 1973 R. Lewis *Blood Money* iii. 30 He served in North Africa during the Second World War in the Panzer Corps. 1975 *Times* 24 June 4/7 Three new brigades have been added to the West German army... The number of tanks in a Panzer battalion is being reduced. 1976 A. White *Long Silence* ix. 83 A lean panzer major walked from the other half track.

‖ **pao-chia** (bau,dʒyā). Also **pao chia, paochia.** [Chinese *bǎojia*.] In China, a system by which households were organized for the purposes of administration. Also *attrib.*

1937 E. Snow *Red Star over China* ii. i. 50 The *pao-chia* system, an ancient method of controlling the peasantry..is now being widely imposed..by the Kuomintang in China and the Japanese in Manchukuo. *Pao-chia* literally means 'guaranteed armour'. 1959 C. K. Yang *Chinese Village in Early Communist Transition* vii. 103 To increase the effectiveness of governing such a large population..there had developed the system of collective responsibility, or *pao chia*, rooted far back in Chinese history, especially in the 'new policy' of the Sung prime minister Wang An-shih. 1965 J. Ch'ên *Mao & Chinese Revolution* (1967) i. xi. 246 In the 'orderly' areas, they used the age-old system of collective responsibility known as the *pao-chia*—every 100 families were organized into a *chia* and every 1,000 families into a *pao* in which all members were made responsible for each other's actions. 1973 T. R. Tregear *Chinese* iii. 68 Law and order are maintained very largely through a modification of the old *pao chia* system by which every group of ten households was held responsible for the behaviour of its own members.

‖ **pao-tzu** (bau,dzə̄). Also **bao zi.** [Chinese *bāozi*.] A steamed roll with savoury or sweet fillings.

1956 B. Y. Chao *How to cook & eat in Chinese* ii. xx. 228 When *man-t'ou*, or Steamed Bread, is filled with stuffing, then it is *pao-tzŭ*. 1972 K. Lo *Chinese Food* i. 61 The miniature *pao-tzu* (steamed buns with various fillings) are sometimes steamed first and finally fried until they are crisp, and served as crispy savouries. *Ibid.* iii. 233 *Pao-tzu* are meant to be eaten on their own, like hot sandwiches. 1973 T. R. Tregear *Chinese* iv. 93 The staple diet is boiled millet or kaoliang, *pao tzu* (steamed bread), *tu fu* (bean curd) and vegetables. 1978 *Nagel's Encycl.-Guide: China* 376 The same variety of preparation is to be found in the different sorts of *bao zi*, or rolls stuffed with vegetables and meat (*cai rou bao zi*), or meat alone (*tian jin bao zi*).

pap, *sb.*[2] Add: **1.** *fig.* (Further examples.) Also, *spec.* in *U.S.*, a political appointment or grant; patronage, 'graft'.

1825 *Delaware* (Ohio) *Patron* 10 Feb. 3/2 An irresistible desire..to serve the state, and to taste a little of the 'Treasury Pap', impelled us towards the capital. 1841 *Congress. Globe* 26th Congress 2 Sess. App. 300/2 The very new States are nursed from their chrysalis territorial condition into existence upon Federal pap from the Executive spoon. 1862 in J. B. Ranck *Albert Gallatin Brown* (1937) 217 Young, strong men..were feeding on government pap whilst wounded soldiers..were in a state of positive want. 1894 *Voice* (N.Y.) 6 Sept. 1/6 The Prohibition Party is the only party that is not controlled by public pap-seeking politicians. 1965 *Chicago Schools Jrnl.* Feb. 197/2 Most of the new clientele will have little taste for or interest in traditional literary values. This does not mean that we should give them over to pap. 1976 *Sounds* 11 Dec. 34/2 To describe this record as 'maudlin pap' must be extremely hurtful to Bernie Taupin. His lyrics are the result of his own poignant experience of a broken marriage. 1977 *Lebende Sprachen* XXII. 10/2 The spoils of office, the rewards for political activities..are called sweets, fat, spices, pap (baby food), plum, pie, persimmon, melon pork, grease, and gravy. 1977 *Rolling Stone* 21 Apr. 84/3 Spector retreated into an increasingly contrived world of sound, lavishing his skill and money on cutting..such soft-core pap as the Righteous Brothers 'Unchained Melody'.

3. *pap-food*; **pap-warmer**, formerly, a contrivance for keeping food or drink warm, usu. incorporating a night-lamp.

1905 *Daily Chron.* 13 May 4/5 Too prolonged use of artificially digested and 'pap-foods' must be avoided. 1920 W. J. Pountney *Old Bristol Potteries* x. 141 Another most interesting piece..is a caudle cup or pap warmer. 1961 L. G. G. Ramsey *Connoisseur New Guide Antique Eng. Pott., Porc. & Glass* 66 Food warmers, catalogued as pap-warmers, performed the duties of night-light shelter as well as keeping liquids warm enough to drink during the night. 1969 E. H. Pinto *Treen* 122 The majority of wooden night-light holders are combined with pap warmers.

pap (pæp), *sb.*[3] Abbreviation of Papa[1]. Also applied to an older man. Chiefly *U.S.*

1844 *Knickerbocker* XXIII. 15 They said, pap wasn't at home. 1854 M. J. Holmes *Tempest & Sunshine* v. 69 Come here, and shake your old pap's paw. 1886 C. M. Yonge *Chantry House* I. xxi. 207 She never took liberties with him, nor called him Pap or any other ridiculous name. 1899 A. Nicholas *Idyl of Wabash* 53 His pap left him right smart of a lump. 1924 W. M. Raine *Troubled Waters* ii. 24 There can't any of you..run me out the way you did Pap Thomson. 1955 D. W. Maurer in *Publ. Amer. Dial. Soc.* xxiv. 105 A *pappy* (or *pap*) is an elderly man.

Pap (pæp), *sb.*[4] Also **pap.** Abbrev. of *Papanicolaou (used only *attrib.*).

1963 Greisheimer & Troyer *Physiol. & Anat.* (ed. 8) xx. 789 The so-called 'Pap' test is so named after Papanicolaou, who with Stockard called attention to the importance of the changes in the vaginal epithelium first in lower animals in connection with estrous cycles. 1969 *Awake!* 8 Nov. 15/1 A study conducted at the University of Chicago 'reportedly shows a sixfold increase in positive Pap smears..among women who have taken oral contraceptives'. 1973 *Sci. Amer.* July 22/3 Of the estimated 2·6 million women served last year by organized family-planning programs,..eight in 10 had annual breast examinations and Pap tests. 1973 *Nation* (Barbados) 16 Dec. 10/4 I'd like to see all of the women who should be having pap smears coaxed into their doctor's offices. 1977 *Spare Rib* May 20/4 Pap smear and breast cancer tests could have been lost.

pap, *v.*[1] Add: **3.** To make into pap.

1927 *Observer* 6 Feb. 14/4 This does not mean papping food for babes; it means speaking intelligibly to grown-ups.

papa[1]. Add: **2.** *transf.* **a.** A woman's lover or husband. *U.S. slang.*

1904 'No. 1500' *Life in Sing Sing* 261, I blew out and rung in with a couple of penny-weighters. A Tommy and his papa... I left them and went with two expert thieves who make it a practice to rob jewelers, a woman and her lover. 1926 L. Hughes in *New Republic* 14 Apr. 223/2, I met a yellow papa, He took my last thin dime... I give it to him cause I loved him But I'll have more sense next time. 1942 —— *Shakespeare in Harlem* 107 That's one time, pretty papa, You'll sure stay in your place. You was a mighty lover and you Ruled me many years. 1960 Wentworth & Flexner *Dict. Amer. Slang* 375 *Papa*, a male lover.

b. Phr. *tell papa*, confide in the speaker.

1929 E. Wallace *Red Aces* v. 48 'Tell papa,' he said. 1961 A. Christie *Pale Horse* xvi. 166 'Come now, tell Papa', said the odious Bradley.

3. *attrib.* or as *adj.*, paternal.

1900 'S. Grand' *Babs* (1901) lxxv. 350 So long as he does not assume papa airs with me, I don't mind.

papa[3] (pā·pā). [Maori.] A soft bluish clay or mudstone found in the North Island of New Zealand.

1873 J. H. H. St. John *Pakeha Rambles through Maori Lands* xi. 183 We descended a steep slide into..a river with a bed of papa rock. 1892 E. S. Brookes *Frontier Life* xvii. 150 The country [Taranaki]..is principally composed of papa rock. 1905 J. M. Thomson *Bush Boys N.Z.* ix. 62 That Papa Rock is beastly stuff to slip... The Papa Rock, of which many of the cliffs in the bush country in New Zealand are formed, is really a very hard blue clay. 1909 G. H. Scholefield *N.Z. in Evolution* i. 12 Papa country produces the readiest crops of grass. 1911 W. H. Koebel *In Maoriland Bush* xxiii. 239 The track..has dented a passage upon the soft pa-pa rock. 1921 H. Guthrie-Smith *Tutira* ii. 9 The materials of which the station is formed are marl or 'papa', sandstone, sandy marl, limestone, and conglomerate. 1949 F. Sargeson *I saw in my Dream* ii. xiii. 128 Nearly all the soil had slipped away and left only great faces of papa. 1969 —— *Joy of Worm* iii. 105 The river had been willing..to slide in casual ripples over slabs of papa, as though to look more closely at a tiny curve of beach.

‖ **papabile** (papā·bile), *a.* [It.; cf. Papable *a.*] Of a prelate: worthy of becoming Pope; having good prospects of being elected Pope. Also in pl. form *papabili* and in L. form *papabilis*. Also as *sb.*, a prelate regarded as a possible Pope (during a conclave).

1934 G. Seldes *Vatican* v. 94 Certain cardinals always are *papabili*, or in line for the papacy. 1935 V. Pirie *Triple Crown* 9 Not too many virtues were necessary for a candidate to be considered 'papabile', that is, acceptable to a reasonable amount of voters. 1938 *Times Lit. Suppl.* 1 Jan. 6/2 At the next Conclave he was *papabilis*. 1939 *Daily Tel.* 11 Feb. 12 Cardinal Hlond..and Cardinal Copello..are both considered *papabile*. 1958 *Times* 29 Oct. 11/3 A number of the younger Cardinals..will still be *papabili* at the next election. 1958 *New Statesman* 1 Nov. 583/3 This week, there were scarcely half-a-dozen *papabili* and the only suitable 'neutral', Roncalli, was 76. 1963 *Times* 4 June 10/1 Cardinals Montini of Milan, Lecaro of Bologna, and Siri of Genoa remain, however, among the *papabile*. 1963 *Economist* 29 June 1342/1 He, alone among the *papabile* cardinals, made a point of nailing his..colours to the mast. 1973 A. Mann *Tiara* viii. 71 His own integrity, combined with his consummate administrative ability, led many observers to consider him the most *papabile* of the European Cardinals. 1976 *Church Times* 9 July 14/3 The other Italian '*papabile*' figures are numerous. 1978 *Guardian Weekly* 10 Sept. 11/1 The cardinals, who had a wide range of ten papabili to choose from.

papacy. Add: ¶ **2. b.** = Papism.

1914 *Trans. Shropshire Archæol. & Nat. Hist. Soc.* IV. 45 Mr. Jermor seems to have been himself suspected of a leaning towards papacy.

papadam, papadum, varr. Popadam.

Papago (pæ·pāgo, pā·pāgo), *sb.* and *a.* [Sp., ad. native name.] **A.** *sb.* **a.** An Indian people of the south-western U.S. and northern Mexico; a member of this people. **b.** The Uto-Aztecan language of this people and (in some uses) of the closely related Pima Indians; also called Pima. **B.** *adj.* Of or pertaining to the Papago or their language.

1839 A. Forbes *California* 162 On the River Gila..Papaga, 4000. 1864 *Harper's Mag.* Dec. 25/1 The Papago Indians also do good service by..killing the hostile savages. 1875 *Encycl. Brit.* II. 538/2 The inhabitants of Arizona are mostly Indians... 4000 Papagoes, a wandering tribe in the south-eastern part of the territory, have no grounds allotted them. 1878 R. J. Hinton *Hand-bk. Arizona* 227 South and west of these lie the old Papago villages and wells. 1911 H. B. Wright *Winning of Barbara Worth* 101 She pointed to a smoky, copper-colored Papago in a green headcloth and decorated shirt. 1912 K. Lumholtz *New Trails in Mexico* 185 Hydrophobia is called in Papago *nótakik*. 1936 *Univ. Arizona Gen. Bull.* iii. 118 These caterpillars feed on a spreading four-o'clock or on the Papago spinach on the desert. 1946 E. E. Cummings *Let.* 31 Jan. (1969) 170, I sat opposite a Papago Indian. 1964 E. A. Nida *Toward Sci. Transl.* viii. 162 Certain problems of translating magic incantations of the Papago Indians of Southern Arizona. 1968 P. M. Postal *Aspects Phonol. Theory* i. 10 The situation described here is based to a certain limited extent on that existing in Papago, a Uto-Aztecan language of the American Southwest. 1973 A. H. Whiteford *N. Amer. Indian Arts* 13 Modeling and paddling is a techniques..till used by the Papago and Yuma of Arizona. 1977 *Language* LIII. 459 L now recognizes..that Papago *hɨjɨl* and *PaPay*..probably do not belong in this cognate set.

papagoite (pǎpā·gōu,əit). *Min.* [f. prec. + -ite[1].] A basic silicate of calcium, copper, and aluminium, approximately $CaCuAlSi_2O_6(OH)_3$, which is found as blue monoclinic crystals.

1960 Hutton & Vlisidis in *Amer. Mineralogist* XLV. 600 The mineral described herein has been named papagoite after the Indian tribe that inhabited the area in which the active mining center of Ajo is situated. 1965 *Mineral. Abstr.* XVII. 359/2 Papagoite from the type deposit at the Ajo Mine, Arizona.., is monoclinic, with *a* 12·94, *b* 11·52, *c* 4·68Å, β 100°30′; space group *C* 2/*m*.

papaia, var. papaya: see Papaw.

1914 R. Brooke *Let.* Feb. in E. Marsh *Rupert Brooke* (1918) 108 Great squelchy tropical fruits, custard-apples, papaia, pomegranate,..and the rest. 1932 W. S. Maugham *Narrow Corner* xix. 143 Breakfast in the little hotels in the Dutch East Indies is served at a very early hour. It never varies. Papaia, *œufs sur le plat*, cold meat, and Edam cheese.

papain. Add to def.: which is used to assist the digestion of patients suffering from chronic dyspepsia and gastritis, as a meat tenderizer, and in clarifying beverages. **b.** The pure crystalline protease extracted from papaya latex. (Further examples.)

1937 *Science* 22 Oct. 379/1 (*heading*) Crystalline papain. 1943 Sumner & Somers *Chem. & Methods of Enzymes* 158 The impure papain of commerce is a mixture of two distinct enzymes. 1963 *Daily Tel.* 14 June 17/6 British meat traders are concerned about the use of papain, an enzyme preparation, injected into animals before they are slaughtered, to produce artificially tenderised meat. 1969 T. C. Thorstensen *Pract. Leather Technol.* vi. 100 The enzyme papain introduced into this system exhibited very strong unhairing action. 1972 *Materials & Technol.* V. xix. 700 Papain is a relatively small enzyme, with a molecular weight of 21 000 and an isoelectric point of 8·75.

papalist. Add: Also *attrib.*

1903 *Eng. Hist. Rev.* XVIII. 482 What little significance lies in the expression depends on Llywelyn's clerk's serving up to the papalist primate the conventional phraseology of Roman documents. 1964 P. F. Anson *Bishops at Large* ix. 325 The year 1920 saw the first of several Anglo-Catholic Congresses, and with them a rapid increase of baroque and rococo furnishings in Anglican Papalist churches.

Papality. (Later example in pl.)

1826 W. S. Landor *Imag. Conv.* (ed. 2) I. viii. 163 He resisted the authority of the pope, and refuted the doctrine of transubstantiation, with several other papalities.

papaloi (pa·pa,lwa). [ad. Haitian Creole *papalwa*, f. *papa* father + *lwa* *loa[2].] A voodoo priest. Cf. *mamaloi.

1884 S. St. John *Hayti* v. 184 He again consulted the Papaloi or priest. *Ibid.* v. 194 The Haytians have corrupted the compounds Papa Roi and Maman Roi into Papaloi and Mamanloi. 1935 [see *mamaloi]. 1965 J. Von Sternberg *Fun in Chinese Laundry* i. 16 My mind seems full of stealth, it has no goal, no conclusions, and no ambitions, as if a Haitian 'papa loi' had made me into a zombie. 1976 *Billings* (Montana) *Gaz.* 11 July 2-F/5 So he consulted the voodoo 'papaloi', who diagnosed the ailment as a minor spell which had been laid on him by an enemy.

Papanicolaou (pæ:pānikǒlā·ū, pæpāni·kǒlau). *Med.* The name of George Nicholas Papanicolaou (1883–1962), Greek-born U.S. anatomist, used *attrib.* and in the possessive with reference to a technique he devised for examining exfoliated or secreted cells, used chiefly as a means of detecting cancer, esp. of the vagina, cervix, and uterus.

1947 *Surg., Gynecol. & Obstetr.* LXXXV. 275 (*heading*) Modifications of the Papanicolaou technique. *Ibid.* 276/2 A number of flakes..were prepared with Papanicolaou's stain. 1956 *Nature* 18 Feb. 330/1 The cells were smeared on microscopical slides coated with egg-albumin or human serum and fixed in Papanicolaou's fixative. 1958 E. Day in R. W. Raven *Cancer* III. xxii. 450 In taking cervical smears by the Papanicolaou method a cotton-tipped applicator is used. 1966 W. H. Cole in R. W. Cumley et al. *Recent Adv. Diagnosis of Cancer* 183 The Papanicolaou stain (1942) represents one of the greatest advances in diagnosis and treatment of cancer during the past several decades. 1972 Graham & Urbach in J. H. Graham et al. *Dermal Path.* xxx. 675/1 During the past three decades the Papanicolaou technique, or various modifications of this method, has been applied to most organ tissues where secretions, exudates, or exfoliated cells can be obtained.

‖ **paparazzo** (paparā·tso). Pl. **paparazzi.** [It.] A free-lance photographer who pursues celebrities to take their pictures. Also *attrib.*

1968 *Daily Tel.* (Colour Suppl.) 29 Nov. 66/4 The anticipated horde of detested *paparazzi*—those scavenging Italian street photographers whose sole purpose appears to be to make every film celebrity's life a misery. 1972 W. Garner *Ditto, Brother Rat!* xxii. 163 Pik..hoisted his camera and began zip-click-zipping at the delegation like a *paparazzo* who's suddenly found nothing between him and royalty in the nude. 1972 *N.Y. Times* 6 July 1 United States District Court Judge Irving Ben Cooper ruled yesterday that the activities of Ronald E. Galella, the self-styled 'paparazzo' photographer, had 'relentlessly invaded' the right to privacy of Mrs. Aristotle Onassis. 1974 V. Gielgud *In Such a Night* vii. 64 The Roman *paparazzi*..are so frequently the terror of film-actors with thin skins. 1974 *Times Lit. Suppl.* 20 Dec. 1439/4 The London Clinic which she had entered a week before with the defiant exclamation to the vulturine paparazzi at the entrance, 'Don't think I'm coming here to die, I'm not.' 1977 *Maclean's Mag.* 21 Mar. 64/1 If Margaret was troubled by the publicity or the paparazzi that followed her during her New York stay, she certainly didn't show it.

papaumu (papa,u·mu). *N.Z.* Also **papauma.** [Maori.] A small evergreen tree or shrub, *Griselinia littoralis*, belonging to the family Cornaceæ, native to New Zealand, and distinguished by thick, ovate leaves with shiny upper surfaces; = *broad-leaf* s.v. *broad *a.* D. 2.

1882 W. D. Hay *Brighter Britain!* II. vi. 199 The Karamu or Papaumu.., a family of pretty flowering shrubs. 1928 Cockayne & Turner *Trees N.Z.* 57 Papaumu, Broadleaf. A small tree, 30–50 ft. high, or a shrub, with short irregular trunk. 1956 [see *broad-leaf]. 1966 *Encycl. N.Z.* I. 252/1 Broadleaf, papauma (*Griselinia littoralis*)..is a common hardwood tree throughout the mixed and beech forests.

papaw. The spelling **pawpaw** is now more frequent; cf. *papaya.

1. a. (Later examples.)

1902 *Westm. Gaz.* 24 Dec. 1/3 The little mustard and cress seeds out of the paw-paws. 1908 E. J. Banfield *Confessions of Beachcomber* i. i. 43 Until we grew fruit, the papaw, the quickest and amongst the best, vegetables were more necessary. 1918 *Chambers's Jrnl.* Oct. 669/2 The great golden paw-paw..brought in showers to the earth by a shake of the tree. 1936 *Geogr. Jrnl.* LXXXVIII. 330 Such sub-tropical fruits as bananas, pineapples, pawpaws. 1953 G. Durrell *Overloaded Ark* vi. 106 The cook..overbalanced into a basket containing eggs and some very ripe and soft pawpaw. 1972 Y. Lovelock *Veg. Bk.* i. 109 The melon tree.., of Central American origin, bears fruit known as pawpaw or papaya.

b. For 'N.O. Papayaceæ' substitute 'family Caricaceæ'. (Later examples.)

1920 *Nature* 2 Sept. 36/1 A fungus..causes powdery mildew on the leaves of the pawpaw plant. 1948 *Archit. Rev.* CIV. 94 In a forest of bamboo, palm, bread-fruit and paw-paw trees the white temples rear their phenomenal towers. 1958 J. Carew *Black Midas* iv. 64 We walked past banana, paw-paw, and cocoa trees. 1964 D. Varaday *Gara-Yaka* xi. 96, I intended placing thorn fences around the paw-paw trees. 1966 B. Kimenye *Kalasanda Revisited* 23 Anna remained as alien to Kalasanda as an orange in a papaw tree.

2. (Later examples.)

1882 *Cornhill Mag.* May 580 Often we pass by groves of young paw-paws. 1925 C. E. Mulford *Cottonwood Gulch* xi. 148 You let me catch you foolin' 'round this ranch an' I'll turn you into pulp as soft as a paw-paws! 1969 *Northwest* (*Sunday Oregonian Mag.*) 14 Dec. 19/1 The fragrant aroma and rich flavor of the Pawpaw is remindful of many tropical favorites. 1970 B. Miles *Bluebells & Bittersweet* iv. 47/3 *Asimina triloba* (pawpaw)... Large drooping leaves give this a tropical look, and solitary dark-purple, bell-shaped flowers about 2 inches across in May are followed by strange cylindrical fruits 3 to 7 inches long... Pawpaws are difficult to move.

3*. (Usu. with initial capital.) In Jamaica, a slave brought from the region of West Africa so called, in Dahomey, near the town of Ouidah. *Obs.* exc. *Hist.*

1707 H. Sloane *Voy. Jamaica* I. p. liv, The Negros called Papas have most of these scarifications. 1725 *Ibid.* II. 376 Its [sc. belly-ach-weed's] use was first made known in Jamaica, by Papau-Negros, and thence call'd Papau-weed. 1740 C. Leslie *New Hist. Jamaica* xi. 307 They generally believe there are Two Gods..; the first they call Naskew in the Papaw language. 1774 E. Long *Hist. Jamaica* II. iii. iii. 425 In 1769, several new masks appeared; the Ebos, the Papaws, &c. having their respective Connús, male and female, who were dressed in a very laughable style. 1793 B. Edwards *Hist. Brit. Colonies W. Indies* II. iii. 73, I now proceed to the people of Whidah, or Fida. The Negroes of this country are called generally in the West Indies Papaws. 1949 *Caribbean Q.* I. i. 11, 33 were Nagoes and 24 Pawpaws from the Slave Coast.

papaya, now freq. used as an alternative to Papaw 1. (Examples.)

1769 [see Papaw 1]. 1832 W. C. Bryant *Poems* (N.Y.) 82 For thee the wild grape glistens, On sunny knoll and

tree, And stoops the slim papaya With yellow fruit for thee. 1874 E. Lear *Indian Jrnl.* (1953) 57 Bits of palmyra-palm, papaya, and dark clumps of oak-like trees around. 1920 W. Popenoe *Man. Tropical & Subtropical Fruits* vii. 229 The fruit of the papaya, as well as all other parts of the plant, contains a milky juice. 1933 H. Allen *Anthony Adverse* VI. xxxix. 571 The papayas were already prodigious and there were shiploads of bananas. 1937 M. Covarrubias *Island of Bali* (1972) iii. 39 Dark green island of tall palms, breadfruit, mango, papaya, and banana trees. 1962 A. Huxley *Island* xi. 177 Gardens shaded by palms and papayas and bread-fruit trees. 1965 *Austral. Women's Weekly* 20 Jan. 25/1 This same cook..concocted mango and papaya soufflés of a texture I'd never before encountered. 1966 D. Forbes *Heart of Malaya* i. 16 There were..groves of papaya and clumps of coco-nut palms beside them. 1969 *Oxf. Bk. Food Plants* 114/1 Papaya (*Carica papaya*) is also commonly known as 'pawpaw' (sometimes spelt papaw). A native of tropical America, it is now widely planted all over the tropics. 1972 *Kent Life* July 82/2 Fresh juicy pineapple or papayas put out the fires in your mouth. 1974 T. Heyerdahl *Fatu-Hiva* ix. 318 The papaya was another strictly tropical American plant, and two varieties grew in the Marquesas. 1977 *New Yorker* 25 July 20/3 He bought papaya and melon for dinner every day.

Pape (pēip). *Sc.* and *Ulster*. Also **pape.** [f. Pope *sb.*[1] or as shortening of Papist.] An opprobrious term for a Roman Catholic.

1935 L. Kerr *Woman of Glenshiels* iv. 56 Mary...wouldn't click with a 'pape' or a boy who whistled after them. 1939 Joyce *Finnegans Wake* iii. 440 Skim over *Through Hell with the Papes* (mostly boys) by the divine comic Denti Alligator. 1957 *Bulletin* (Glasgow) 11 Oct. 15/3 Lucas had been drinking. When charged he said, 'I set the house on fire to burn these Papes and Niggers.' 1968 J. Braine *Crying Game* v. 118 Adam's a good Catholic... It's smart to be a Pape now. *Ibid.* xii. 217 When I marry..it must be a Catholic... I mean, who else would I marry but a Pape? 1970 G. M. Fraser *General danced at Dawn* 48 Years later, when he led a famous league side out to play Celtic, this same corporal, having said his Hail-Mary and fingered his crucifix, instructed his team, 'Awright, fellas, let's get stuck intae these Papes.' 1972 *Listener* 7 Sept. 304/3 Gerry Fitt has been bawling about the number of Roman Catholics..who have been found shot dead... Gerry deduces that the dead papes have been killed by Protestant guns. 1974 *Socialist Worker* 2 Nov. 11/1 During my childhood it was constantly hammered home to me that I should be a good boy at school, that I shouldn't question what my elders told me, and that I should join the Orange Lodge because the 'Papes' are bastards.

† **papelito** (papělī·to). *Obs.* [a. Sp. *papelito* slip of paper, bit of paper (cf. *papelillo* cigarette).] A cigarette.

1845 R. Ford *Hand-bk. for Travellers Spain* II. 784 So they jogged on, smoking their *papelitos*, to the Escorial. 1861 L. Wraxall tr. *Aimard's Freebooters* xvi. 211 He rolled a husk cigarette..lit his papelito, and was soon surrounded by a dense cloud of bluish and fragrant smoke. 1867 'Ouida' *Under Two Flags* I. i. 15 Something to drink and something to smoke, were it only a glass of brown sherry and a little papelito.

paper, *sb.* Add: **1. d.** (Further examples of *on paper*.) *paper-and-pencil* (*attrib.*): executed in writing, carried out with paper and pencil.

1788 *Amer. Museum* III. 336/2 The form of their constitution, as it is on paper, admits not of coercion. But necessity introduced it in practice. 1948 [see *conniption]. 1965 N. Chomsky *Aspects of Theory of Syntax* i. 10 Let us use the term 'acceptable' to refer to utterances that are..comprehensible without paper-and-pencil analysis. 1971 *Jrnl. Gen. Psychol.* Oct. 308 The five measures described here—three paper-and-pencil inventories and two visual tasks—all appear to be concerned with a central phenomenon, that of the individual's need to maintain an optimal level of stimulus input or variability. 1972 *Jrnl. Social Psychol.* LXXXVII. 156 Test anxiety scales are typically self-report, paper-and-pencil measures. 1973 *Jrnl. Genetic Psychol.* Sept. 35 The measure of approach was a paper and pencil measure. 1973 D. Westheimer *Going Public* i. 19 For every point it goes up you've made yourselves a hundred twenty seven thou five hundred dollars. On paper. I must warn you that paper profits are often illusory. 1976 *Verbatim* Sept. 5/2 Six nifty paper-and-pencil games. 1977 *World of Cricket Monthly* June 82/3 On paper, Hampshire looked a short-odds bet for any of the one-day competitions.

2. Delete †*Obs.* and add later examples.

1966 *Rep. Comm. Inquiry Univ. Oxf.* I. 241 Upon this depends the ability of the Vice-Chancellor to make himself felt, his capacity to think ahead and to give a lead without having continually to submerge himself in paper. *Ibid.* 253 Only thus can it [sc. the General Board] see over the top of the piles of paper and look at those general topics..which ought to be its principal concern. 1977 D. Aitkin *Second Chair* ii. 19 Watch out that you don't get smothered in paper. That foxy little friend of ours wants you to do all his hackwork while he..thinks great thoughts.

3. (Further examples.)

c 1722 Ld. Mar *Legacy to Scotl.* (1896) 201 The paper could not exceed more than a certain quantity..in proportion to the specie in the nation. 1850 Thackeray *Pendennis* II. xxvi. 259 It was whispered among the tradesmen, bill-discounters, and others..that the Captain's 'paper' was henceforth of no value. 1906 *Westm. Gaz.* 20 Oct. 12/2 Without..feeling—as he puts it—that he is 'a pawn with a breech-loader on an open-air chess-board, to be moved at the bidding of a despotic keeper who only takes paper'. 1925 [see *hot *a.* 7 e]. 1930 *Liberty* 5 July 23/2 Next day the news cracks about a twenty-grand payroll robbery at the factory in Sheffield. Bob comes in at night loaded with paper and gives me five hundred bucks of it. 1937 E. Snow *Red Star over China* vi. iv. 234 Only Soviet paper was in use, except in the border counties, where White paper was also accepted. 1949 J. Cary *Fearful Joy* 255 Now they've got some loose paper they're going to make it float. 1969 *Times* 26 Mar. 28/8 A 'proper' level for three-month paper is thought to be about 8⅜ per cent compared with a frequently quoted level of between 9 1/16 and 8⅞ per cent. 1977 *Law Rep.* 5 July 620 The bank..also bought them, in the recognised international market in what was called 'medium-term paper'.

4. a. (Earlier and later examples.)

1785 *Apol. Life G. A. Bellamy* (ed. 2) II. xliii. 114 The piece [sc. Romeo and Juliet at Drury Lane and Covent Garden] was performed so many nights, that the public as well as the performers were tired and disgusted with it. We, [at Drury Lane] however, got the advantage of some nights. But this was not done without a great deal of paper, which was bestowed upon the occasion. 1820 C. Mathews *Let.* in A. Mathews *Mem. Charles Mathews* (1839) III. 165 He had spoken to the cash-taker of the rooms, who said, this is all the money (not much), and there's plenty of paper. 1825 P. Egan *Life of Actor* iv. 144 'Theatrical paper' has been frequently known to *silence* many a harsh tongue; and also to change the looks of an angry creditor. 1927 *Vanity Fair* XXIX. 132/3 'Paper' is a pass. 1951 'J. Tey' *Daughter of Time* ii. 21 Johnny Garson can tell you how much paper there is in the house what time he is sobbing his heart out.

b. (Later examples.)

1929 M. A. Gill *Underworld Slang* 9/2 *Paper*, marked playing cards. 1938 H. Asbury *Sucker's Progress* 37 In the early days of Poker the marked cards used by sharpers were prepared beforehand by the gamblers themselves, and were known as 'paper'; or were marked during the process of the game with the finger nail or a needle point embedded in a ring.

c. *U.S.* A forged cheque or document.

1850 [see *kite-flying *vbl. sb.* 2]. 1925 *Flynn's* 7 Mar. 191/2 *Paper*,..forged notes or checks. 1930 *Liberty* 19 July 27/1, I turn out bills of sale by the dozen. I don't like to do it as it is not playing square with my dealers. They don't know they're getting worthless papers.

d. *U.S.* Posters or similar publicity material. Also individual singular: a poster or placard.

1878 *Harper's Mag.* Mar. 599/2 Struggling families who have 'one room to let', or..a modest paper in a window 'Boarders wanted'. 1896 *N.Y. Dramatic News* 18 July 12/1 Car No. 6..was here July 7–8, posting very attractive paper, which reads 'coming soon'. 1903 W. C. Thompson *On Road with Circus* i. 23 The posters and lithographs sent out in advance are 'paper'. 1942 Berrey & Van den Bark *Amer. Thes. Slang* § 571/6 *Paper*, posters.

6. c. (Earlier examples.)

a 1746 M. Leapor *Poems* (1748) I. 5 Let Isabel unload her aking Head Of twisted Papers, and of binding Lead. 1772 J. Woodforde *Diary* 21 Apr. (1924) I. 114 We..caught my Sister Jane at table with her hair up in papers. 1819 Keats *Let.* 16 Apr. (1958) II. 92 Do you put your hair in papers of a night? 1838 Dickens *Nickleby* (1839) vii. 60 The lady..was dressed in a dimity night jacket with her hair in papers.

d. *U.S.* A playing card. (Cf. sense 4 b in Dict. and Suppl.).

1842 *Southern Lit. Messenger* VIII. 412/1, I found myself..around a table in a corner, and the 'papers' in motion. 1862 O. W. Norton *Army Lett.* (1903) 41 Those whose taste inclines them that way are playing with the 'spotted papers'. 1935 A. J. Pollock *Underworld Speaks* 86/1 *Papers*, playing cards.

7. d. (Later examples.)

1914 E. A. Powell *Fighting in Flanders* ii. 45 One never stirred out of doors in Antwerp without one's papers, which had to be shown before one could gain admission to the post office..or any other public buildings. 1960 O. Manning *Great Fortune* ii. 142 Recalled to his regiment. His papers came yesterday and off he had to go. 1960 *Victorian Studies* June 326 Those, like Captain Gadsby..in the end sent in their papers for the sake of their wife and family. 1963 R. D. Symons *Many Trails* xi. 110 The great steam-engine..was attended by a proper steam engineer, with 'papers'. 1966 *Listener* 24 Feb. 267/2 The Moscovite lives by the rules. 'Without "papers" a man is nothing,' my exasperated interpreter once asserted. 1970 *Globe & Mail* (Toronto) 28 Sept. 29/2 (Advt.), Applicant must possess thorough knowledge of hot water heating systems and preferably possess engineers papers. 1977 H. Kaplan *Damascus Cover* xiv. 140 Some [Nazis]..proceeded to Rome and Geneva where clandestine processing stations..supplied them with the papers necessary for travel to the Middle East.

8. b. = Wall-paper.

1764 in E. Singleton *Social N.Y. under Georges* (1902) 43 The Dining-Room is 14 × 19, hung with genteel Paper; the Entry or Passage from the Door, is hung with the same. 1830 M. Edgeworth *Let.* 17 Nov. (1971) 430 When I went down to the library..I was charmed even with the fuschia-trellis looking paper. 1873 C. M. Yonge *Pillars of House* II. xiv. 53 What our paper may have been in its earlier stages of existence I am not prepared to say; but since I can remember..the wall presented every *nuance* of purplish salmon. 1945 *Catal. Exhib. Eng. & Hist. Wall-papers* (Central Inst. Art & Design) 6 The most beautiful of all..are the hand-painted Chinese papers. 1967 Wodehouse *Company for Henry* iv. 57 His spiritual home would have been some such establishment as Edgar Allen Poe's House of Usher, into which he would have fitted like the paper on the wall. 1975 'A. Hall' *Mandarin Cypher* vii. 104 We had all the paper off the wall at the Hong Kong Cathay [Hotel].

c. *pl.* The publicity afforded by the newspapers; esp. in phr. *to make the papers*: to gain publicity.

1963 'A. Gilbert' *Ring for Noose* viii. 97 At all events he hadn't made the morning papers. 1967 Wodehouse *Company for Henry* iii. 47 One of these days that woman is going to get herself Into The Papers. 1972 *Village Voice* (N.Y.) 1 June 16/2 Through many lonely months McGovern had trouble making the papers.

10. a. (Further examples.)

1594 T. Nash *Unfortunate Traveller* sig. H 4v As if it had beene a candle in a paper lanterne. 1723 J. Nott *Cook's & Confectioner's Dict.* sig. Cc 2v (*heading*) To dry Plums..when they are dry put them into Paper Bags full of small Holes, and hang them up. 1723 *Amer. Weekly Mercury* 7 Nov. 2/2 The natural Situation of these Counties and the Practice of our Neighbours, has laid us [of Penna.] under the necessity of coming into a Paper Currency. *a* 1790 B. Franklin *Autobiogr.* in *Writings* (1905) I. 306 The wealthy inhabitants..being against all paper currency. *Ibid.*, I wrote and printed an anonymous pamphlet..entitled 'The Nature and Necessity of a Paper Currency'. 1796 Jane Austen *Let.* 9 Jan. (1952) 3 We have trimmed up and given away all the old paper hats of Mamma's manufacture. 1808 *Ibid.* 24 Oct. 225 We do not want amusement: bilbocatch,..spillikins, paper ships, riddles, conundrums, and cards..keep us well employed. 1864 *Harper's Mag.* Dec. 58/2 With an umbrella and a shawl, inclosing a box of paper collars. 1867 J. Laing *Theory of Business* vi. 77 With Austria, Russia, America, and other countries using paper currencies..the par is exceedingly uncertain. 1870 J. K. Medbery *Men & Mysteries Wall St.* 23 A broker cannot..innocently fling a 'paper dart' at a neighbour without being amerced ten dollars. 1899 'Mark Twain' in *Forum* (N.Y.) Mar. 31 A billionaire in a paper-collar, a king in a breech-clout. 1905 *Daily Chron.* 15 May 8/5 There is a craze just now for 'paper hats'. It sounds crude, but the smartest and prettiest tricornes, mushrooms, and shady river hats are daintily woven from a soft paper. 1907 *Yesterday's Shopping* (1969) 352 c/1 Paper d'oyleys. Star pattern. 1913 C. Mackenzie *Sinister St.* I. ii. iv. 190 They threw paper darts and paper pellets with unerring aim. 1922 Joyce *Ulysses* 423 A bandy child..with a papershuttlecock. 1934 A. Huxley *Beyond Mexique Bay* 6 He..decreed..that we should be given paper hats, balloons, and cardboard trumpets. 1935 H. Edib *Clown & his Daughter* xlii. 233 Women distributing sweets in coloured paper bags passed in and out of the audience. 1951 'J. Tey' *Daughter of Time* i. 17 She was carrying various paper bags and a small tight bunch of anemones. 1951 R. W. Jones *Thomson's Dict. Banking* (ed. 10) 452/1 *Paper currency*, the paper instruments such as bank notes, cheques, bills, and other forms which take the place of money and act as a currency or circulating medium. 1955 A. Marshall *I can jump Puddles* (1956) xix. 136 'Skeeter couldn't fight his way out of a paper bag,' Joe asserted. 1958 Paper doily [see *doily *sb.* or *a.*]. 1969 'H. Pentecost' *Girl Watcher's Funeral* (1970) ii. i. 94 That first punch..smashed through my guard like a paper doyley. 1972 M. Woodhouse *Mama Doll* xi. 145 Andy Dylan made paper darts. 1973 M. Amis *Rachel Papers* 153 Not, for her, the wet Brillo-pad, nor the paper-bagful of kedgeree. 1973 J. Leasor *Mandarin Gold* ii. 18 Paper lanterns were glittering above shop fronts and over stalls. 1974 *Sunday Times* (Colour Suppl.) 17 Feb. 55/3 Strange how people assume at once the personalities of their paper hats. That man, for instance, in a boiled shirt and three exiguous Red Indian feathers. 1974 M. G. Eberhart *Danger Money* xiii. 134 Shoes, stockings and paper tissues littered the room. 1975 *New Yorker* 3 Mar. 38/3 She never got over her spending a hundred and twenty-five dollars on paper lanterns for the engagement party. 1977 R. Barnard *Death on High C's* ix. 87 She went through men like other women go through paper tissues.

b. (Earlier example of *paper ship* in *fig.* use.)

1854 *2nd Rep. Select Comm. Emigrant Ships* 61 in *Parl. Papers* XIII. 267 These advertisers...have a kind of arrangement with the owners of vessels; they are like recruiters, they collect emigrants; their ships are known in the trade by the name of 'paper' ships.

c. (Further examples.)

1809 'D. Knickerbocker' *Hist. N.Y.* I. iv. iv. 231 This all potent word, which served as his touchstone in politics, at once explains the whole system of proclamations, protests, empty threats, windmills, trumpeters, and paper war. 1826 M. Kelly *Reminisc.* II. 119 Most of the popular songs which he sang in Sacchini's operas were composed by himself, although the credit of them was given to Sacchini; but upon a severe quarrel between them, Rauzzini, in a paper war, actually avowed himself the author of them, and accused Sacchini of the greatest ingratitude. 1963 D. Ogilvy *Confessions Advertising Man* (1964) i. 15, I abhor people who wage paper-warfare. 1970 *Times* 13 Feb. 10/5 In the much larger Arab world, it is a paper war for many people.

d. (Further examples.)

1803 *Deb. Congress U.S.* 23 Feb. (1851) 129 Paper blockades were substituted for actual ones, and the staple commodities of our country lay perishing in our storehouses. 1812 *Boston Gaz.* 20 July 1/5 The paper-blockades, which have justly occasioned so much irritation, are now abandoned. 1932 *Sun* (Baltimore) 15 Aug. 7/5 It was granted there might be a half-dozen 'paper' agreements at least..from the conference. 1941 J. S. Huxley *Uniqueness of Man* xiii. 270 Hedonism, like utilitarianism, is another of those paper schemes, beautifully logical, that just are not true. 1960 *Washington Post* 27 Nov. E 3 At least 70 anti Castro groups have been formed in Miami, but many of them are paper organizations. 1965 A. Nicol *Truly Married Woman* 43 One more week towards the time when Olu Jones and his brethren would take over. He was not worried, however. Perhaps..because I have paper qualifications. 1970 *New York* 16 Nov. 54/3 Abbie is a simple bourgeois basket case, a paper Yippie, and over 30 anyway. 1973 *Time* 25 June 66/2 Lloyd also has a number of paper companies set up in Liechtenstein. 1977 E. Ambler *Send No More Roses* xi. 266 A chain of twenty different corporations..all making paper profits. 1978 *Jrnl. R. Soc. Arts* CXXVI. 607/2 These targets are real, and not just paper objectives.

11. a. *paper-clip* (further examples); so *paper-clip* vb.

1904 Chesterton *Nap. Notting Hill* iii. ii. 151 One of those queer little shops..which must be called toy-shops only because toys..predominate; for the remainder of goods seem to consist of almost everything else in the world—tobacco, exercise-books,..halfpenny paper clips. 1921 V. Sackville-West *Orchard & Vineyard* 97, I watched..And thought how London clerks with paper-clips Had filed the bills of lading of those ships. 1962 A. Nisbett *Technique Sound Studio* ii. 40 Thin paper..individually paper-clipped to thick paper or card. 1963 A. Ross *Australia 63* ii. 49 The two banks of the city paper-clipped together by harp-like steel bridges. 1969 P. Highsmith *Tremor of Forgery* xi. 101 Ingham paper-clipped his notes and put them on a corner of his desk. 1973 *Time* 25 June 44/2 With roads, the North Vietnamese can bring in the stuff of life—the paper clips for a bureaucracy, the beginning of a postal system, school supplies, the works. 1974 H. L. Foster *Ribbin'* vi. 284 He entered, took a tray of paper clips from her desk and sat down on a chair directly in front of her. 1976 *New Yorker* 1 Mar. 64/2 The note was paper-clipped to a collection of President Nixon's Vietnamization speeches. 1978 N. Freeling *Night Lords* xv. 64 A thin pile of letter paper, neatly squared off and paper-clipped by her fair hand.

b. *paper-keeper.*

1927 *Daily Tel.* 3 May 3 In the opinion of the Postmaster-General the paperkeepers were amply remunerated at the present rates. 1970 *Ibid.* 12 Jan. 1/6 There would be a series of strikes by 10,000 messengers and paper-keepers in Whitehall. *Ibid.*, One effect of the strike..will be that documents will not be circulated to departments by the paper-keepers.

c. *paper-bound* (examples) (see also sense 12 below), *-collared* (example), *-covered* (further examples), *-insulated, -soled, -wrapped* adjs.

1882 W. Whitman *Daybks. & Notebks.* (1978) II. 287 Paper-bound L of G with revisions prop. 1901 *Sketch* 24 July 26/2 The unsold paper-bound books in Germany are returned to the publisher, who re-binds them. 1928 A. Huxley *Let.* 12 Dec. (1969) 304 He'd be able to get rid of the paper bound copies at the same time. 1942 E. Waugh *Put out More Flags* i. 32 The Vichy water and the paper bound-volume of Balzac on the table before her..spoke of what..she would have called her 'personality'. 1976 *National Observer* (U.S.) 2 Oct. 14/4, I disappeared into the bathroom..with a jar of bath salts and a paper-bound novel. 1874 A. Bathgate *Colonial Experiences* viii. 86 A section of the community known as the 'paper-collared swells', who are the government officials, medical men, bank employees *et hoc genus omne*. 1867 Geo. Eliot *Let.* 21 Mar. (1956) IV. 354 People write..to tell me of one paper-covered American copy of Felix Holt being brought to Europe. 1923 J. M. Murry *Pencillings* 18 The paper-covered book, in fact, a rough and ready test of literary curiosity. 1952 *Amer. Speech* XXVII. 149 The paper-covered book in the United States is earning a place for itself much above that of the 'paper-back' and 'dime novel'. 1978 *Jrnl. R. Soc. Arts* CXXVI. 702/1 To have a catalogue of all the Institute's *dix-huitième* drawings in the small compass of a paper-covered 8vo volume including a coloured illustration of every one of them, is most valuable. 1900 *Jrnl. Inst. Electr. Engin.* Dec. (Advt., verso rear cover), Diatrine paper-insulated..cables. 1967 *IEEE Trans. Power Apparatus & Systems* LXXXVI. 34 (*heading*) Drying and impregnation of paper-insulated power cables. 1902 Conrad *Typhoon* ii. 19 A mournful,..Chinaman, walking behind in paper-soled silk shoes, and who also carried an umbrella. 1973 M. Amis *Rachel Papers* 32 An old woman passed by surreptitiously dropping paper-wrapped sugar-lumps on to the chair opposite.

d. *paper-hearted, -thin* (freq. *fig.*), *-white* (later examples).

1939 Dylan Thomas *Let.* Mar. (1966) 226 The English poets now are such a..paperhearted crowd you could blow them down with one bellow out of a done lung. 1929 E. Bowen *Last September* vii. 85 The door went paper-thin as they raised their voices. 1939 'N. Blake' *Smiler with Knife* 99 His old, paper-thin voice. 1971 *Guardian* 17 Sept. 4/6, I do want to dedicate what is left of my life..to..this paper-thin thing of law and order. 1977 *Guardian Weekly* 28 Aug. 3/1 Even a paper thin majority in favour of the rule at Congress would ensure TGWU support. 1978 B. Norman *To nick Good Body* vii. 57 There was this thumping in their bedroom. Paper-thin the walls are in those houses. 1976 *Liverpool Echo* 24 Nov. 5/4 The Government's paper-thin majority. 1806 *Curtis's Bot. Mag.* XXIV. 947 (*heading*) Italian or paper-white Narcissus. 1938 [see *grandiflora *a.*]. 1957 L. Durrell *Justine* 66 They [sc. Egyptian women] have become tuns of pleasure, rolling on paper-white blue-veined legs. 1973 J. Burrows *Like an Evening Gone* xi. 128 [He] had turned paper-white when given the news. 1977 P. Moyes *To kill Coconut* v. 66 Paper-white legs proclaimed him a new arrival.

12. paper-board; delete †*obs.* and add later examples; (*b*) *pl.* boards with a paper cover, used in book-binding; **paper boat**, (*a*) a model boat made from folded paper; (*b*) a lightly made vessel; **paperbound** *sb.* (see sense 11 c above) (chiefly *U.S.*) = *paperback; **paper box**, (*a*) a box made of paper; (*b*) a box in which to keep papers; **paper cable**, an electric cable insulated with paper; **paper cap**, (*a*) a cap made of usu. coloured paper and worn at festivals, parties, etc.; (*b*) a cap made of paper worn by carpenters and other workmen; **paper chain**, a chain made of usu. coloured paper as a decoration, esp. at Christmas; **paper-chase**, also *transf.* and *fig.*; **paper-chewing** *slang* (*rare*), official correspondence; **paper chromatogram** *Chem.*, a chromatogram made on a paper support; **paper chromatography** *Chem.*, the separation of substances by chromatography on a paper support; hence *paper chromatographic* adj., *paper-chromatographically* adv.; **paper cover**, (*a*) see quot. 1960; (*b*) = *paperback; **paper cup**, a drinking cup made of thin cardboard; **paper-cutter** (earlier and later examples); **paper doll**, (*a*) a doll-shaped figure cut or folded from a sheet of paper; (*b*) *U.S. slang*: (see quots. 1968–70, 1970); **paper dress**, an inexpensive disposable dress made of paper; **paper dust** *Printing* (see quot. 1964); **paper-feed**, a device for inserting sheets of paper into a typewriter, printing machine, or the like; **paper flower**, (*a*) an imitation flower made from paper; (*b*) *U.S.*, a name used for several plants with flowers of a papery texture, esp. *Psilostrophe cooperi*, a small shrub of the family Compositæ, native to south-western desert areas of the United States and bearing panicles of yellow flowers; **paper-folding**, the making of objects by folding paper, origami; **paper game**, a game played using pencil and paper; **paper gold** = *special drawing rights* (*special *a.* 3d); **paper guide**, an adjustable device on a typewriter for ensuring that the left edge of each sheet of paper is inserted at the same place; **paper handkerchief**, a disposable handkerchief made from soft tissue paper; **paper hankie, hanky** *colloq.*, = **paper handkerchief*; **paper kiosk**, a kiosk at which newspapers are sold; **paper-making wasp** = *paper-wasp*; **paper man**, a musician, esp. a drummer, who plays from written music; **paper-match** = *book match* (*book *sb.* 18); **paper napkin**, a disposable table-napkin made of paper; **paper nylon**, a stiff paper-like form of nylon; **paper pattern**, a pattern cut out of paper; *spec.* a dressmaking pattern printed on paper, now usu. on tissue paper with printed instructions; **paper plate**, (*a*) a disposable plate made of paper or cardboard; (*b*) a specially treated paper used as an offset printing plate in certain office duplicating machines; **paper ribbon**, (*a*) = *paper tape* below; (*b*) = *paper streamer* below; **paper-round**, the job of regularly delivering newspapers; the particular route covered; **paper route** *U.S.*, = *paper-round* above; **paper run** *N.Z.*, = *paper-round* above; **paper sack**, (*a*) *U.S.* a paper bag; (*b*) a large sack-like container made of strong paper; **paper sculpture**, the making of three-dimensional structures from one or more pieces of paper by folding, cutting, etc.; **paper shale** *Geol.*, shale which readily splits into very thin paper-like laminæ; **paper-shell** (earlier and later examples); so **paper-shelled** *a.*, having a very thin shell; **paper shredder**, a machine that tears up esp. secret documents into small unreadable fragments; so *paper-shredding* adj.; **paper streamer**, a long narrow strip of coloured paper used as a decoration, etc.; **paper-taffeta**, a lightweight taffeta with a crisp papery finish; **paper tape**, tape made of paper; *esp.* such on which data is represented by means of holes punched in it; freq. *attrib.*; cf. Tape *sb.*[1] 2 b, *perforated tape* (*perforated *ppl. a.* 1), *punched tape* (*punched *ppl. a.* 2); **paper tiger** [tr. a Chinese expression first used by Chairman Mao], a person, country, etc., that appears outwardly powerful or important but is actually weak or ineffective; **paper towel**, a small disposable towel made of absorbent paper; **paper tower**, the part of a Monotype machine (*monotype *sb.* 3) in which the perforated paper tape is held; **paper town** *N. Amer.*, (*a*) a town that is projected or promoted but not always actually founded; (*b*) a town or city supported by the paper-making industry; **paper ware**, (articles made of) papier-mâché; **paper-work**, (*c*) (earlier example); (*d*) work done on paper, the filling-in of esp. official forms, the keeping of administrative records.

1888 C. T. Jacobi *Printers' Vocab.* 95 *Paper boards*, a term applied to cheap bindings in boards, but with paper instead of cloth sides. 1929 A. J. Vaughan *Mod. Bookbinding* iv. 216 *Paper or Cloth Boards*, a binding consisting of a case made from a paper or cloth cover. 1944 *Sun* (Baltimore) 9 Jan. 3/5 The Robert Gair Company plant, manufacturing paperboard for containers and boxes for war purposes, was shut down today because of a lack of wastepaper. 1959 R. Hostettler et al. *Technical Terms Printing Industry* (ed. 3) 103/1 (Bound) in paper boards. 1959 A. McLintock *Descr. Atlas N.Z.* 59 The value of production of pulp, paper, and paperboard for 1956–7 was £12·4 million. 1961 J. Carter *ABC for Bk.-Collectors* (ed. 3) 143 Notwithstanding that 'original boards'..are in fact covered with paper, the term *paper boards*, if used of any but quite modern books..suggests boards..covered with paper of a plain colour, usually not the original binding. 1965 B. J. Kirkpatrick *Bibliogr. E. M. Forster* 47 Pale green paper boards; printed in black on upper cover. 1971 *New Scientist* 27 Apr. 259/2 Scrap paper..is used to make some forms of paperboard for packaging. ? 1846 Mrs. Gaskell *Let.* (1966) 48 All the children were very kind to Florence, and made her paper boxes, and boats. 1863 Dickens in *All Year Round* 26 Sept. 108/1 My voyages (in paper boats) among savages often yield me matter for reflection at home. 1929 F. C. Bowen *Sea Slang* 101 *Paper boat*, a lightly-built vessel of any sort, but particularly applied to paddle excursion steamers. 1931 R. Campbell *Georgiad* iii. 61 The anecdotes Of Alfred's cakes and Shelley's paper boats. 1961 F. H. Burgess *Dict. Sailing* 156 *Paper boat*, any boat with very thin planking. 1964 M. Clive *Day of Reckoning* viii. 73 Parlour tricks, such as making cocked hats out of newspaper or paper boats from half-sheets of writing paper. 1978 *Listener* 12 Jan. 57/2 A parting gift from his cell-mate..a little paper boat..not more than two or three millimetres long. 1961 *Spectator* 26 May 765 (Advt.), Yale Paperbounds. Yale University Press. 1970 *Scholarly Publishing* I. 419 Yale Fastbacks will be made available as paperbounds, at low cost, and they will often reflect the newest techniques of rapid book production. 1973 *Publishers Weekly* 7 May 10 (Advt.), Now *The Divine Vision* is a Quest paperbound. 1754 *Connoisseur* I. 189 The man of taste takes his Strasburgh veritable tabac from a right Paris paper-box, and the pretty fellow uses a box of polished metal, that by often opening it he may have the opportunity of stealing a glance at his own sweet person reflected in the lid of it. 1757 in S. M. Hamilton *Lett. to Washington* (1899) II. 80 Paper Box of Tarsils..1. 1776 J. Wedgwood *Let.* 14 July (1965) 195 They have not much expectation in those articles (unless in very cheap paper Boxes) from any market where they have the French for their Rivals. 1861 D. G. Rossetti *Let.* 19 Jan. (1965) II. 389 Will you tell her we are very thankful for her paper-box, which is very useful? 1869 *Boyd's Business Directory N.Y. State* 460 H. Lettington, Manufacturer of Paper Boxes. 1913 J. London *Valley of Moon* i. xi. 96 As if I didn't know..how long you worked in..the paper-box factory. 1926–7 *Army & Navy Stores Catal.* 383/2 Paper boxes, japanned tin. 1893 W. J. Hopkins *Telephone Lines* xiv. 209 In the paper cables made by John A. Roebling & Sons Company, two paper strips are laid on lengthwise, as loosely as possible, being held in place by thread wound about them. 1936 *Economist* 22 Feb. 437/1 The board intended to extend the business of the company into new lines, the main business being the manufacture of paper cables. 1973 R. W. Sillars *Electr. Insulating Materials* v. 88 Drying of high voltage oil-filled paper cables..is more critical than transformer drying, for the stress at working voltage is considerably higher and therefore discharge is more difficult to avoid. 1809 'D. Knickerbocker' *Hist. N.Y.* II. vi. viii. 162 Little urchins..followed in droves after the drums, with paper caps on their heads. 1835 Dickens *Sk. Boz.* (1836) 1st Ser. II. 149 An unshaven, dirty, sottish-looking fellow, whose tarnished paper-cap..communicates an additionally repulsive expression to his very uninviting countenance. 1887 Kipling *Plain Tales from Hills* (1888) 231 A big blue paper cap from a cracker. 1967 C. V. Bark *See Living Crocodiles* iii. 44 The guests wore paper caps and pulled crackers. 1974 L. Lamb *Man in Mist* vi. 37, I can well remember when carpenters and masons wore paper caps, as they still do in Italy and in Tenniel's *Alice*. 1943 N. Balchin *Small Back Room* iii. 31 It's a damned shame we haven't got a few paper chains and a bit of misletoe for the old boy. 1971 M. McCarthy *Birds of America* 31 He..preferred the [Christmas] tree..with..the paper chains he cut and pasted. 1973 E. Jong *Half-Lives* 82, I tie you to the bed with paper chains. 1914 Paperchase [see *cycle-car]. 1932 *Times Lit. Suppl.* 9 June 425/4 This is not the usual paper-chase of clues from crime to detection. 1977 *Arab Times* 14 Nov. 2/1 Now, in what local journalists call the great paper chase, three newspapers for blacks are competing at the newsstands. 1934 'G. Orwell' *Burmese Days* ii. 30, I can't stick my bloody office..signing one chit after another. Paper-chewing. *Ibid.* 38 All this paper-chewing and chit-passing. 1944 *Biochem. Jrnl.* XXXVIII. 231/2 The paper chromatograms are by no means united to the separation of aminoacids. 1972 K. Narita in M. Funatsu et al. *Proteins* II. 246 Radioactivity of acetylglycylserine was almost the same as that found in the material which stayed at the origin on the paper chromatogram. 1956 *Nature* 7 Jan. 22/1 He described preliminary studies on the successful paper-chromatographic separation of intact tissue phospholipids. 1971 *Jrnl. Chromatogr.* LX. 381 Almost the entire present knowledge on the composition of human urinary sugar content in health and disease can be ascribed to paper chromatographic studies. 1966 *Bot. Mag.* (Tokyo) LXXIX. 507 Three amino acids are formed that can be identified paper-chromatographically. 1948 *Science* 7 May 483/2 It appears that..paper chromatography will be found to be an increasingly important research tool in analyzing for amino acids. 1971 C. T. Kenner *Analyt. Separations & Determinations* xv. 282 Two-dimensional paper chromatography is used to separate complex mixtures of amino acids produced by hydrolysis of protein. 1856 Geo. Eliot *Let.* 24 Dec. (1954) II. 282 George definitely votes for..boards. He thinks a paper cover for a philosophical book a bad augury. 1903 [in Dict., 10 a]. 1913 T. E. Lawrence *Let.* 5 Apr. (1938) 152 If you have any cheap paper covered copy..(paper-covers are customs free)..I would be exceedingly grateful. 1952 *Amer. Speech* XXVII. 149 It is not clear whether the *Matrix*..uses *limp-cover* in contradistinction to *hard-cover* or *paper-cover*. 1960 G. A. Glaister *Gloss. Bk.* 295/1 *Paper covers*, a style of binding much used for cheap reprints, and, especially on the Continent, for original works in which no boards are used, and the stiff paper cover which encloses the book is adhered to the back. 1907 *Yesterday's Shopping* (1969) 352E/2 Paper drinking cups, complete with holders. 1939 G. Greene *Lawless Roads* xi. 272 The odious child takes all the paper cups from the water-tap by the lavatory. 1971 J. Henderson *Copperhead* (1972) xv. 191 A girl appeared with..the usual paper cups. 1978 R. Hill *Pinch of Snuff* xxiv. 239 She..poured tea from the flask into the paper cups. *a* 1828 D. Wordsworth *Jrnl.* (1941) II. 128, I purchased a ladle and a paper-cutter..made by the peasants of these mountains. 1969 *New Yorker* 14 June 46/3 His overhead is hit with his whole arm—no mere flick of the wrist. The arm comes down like the moving part of a paper cutter. **1849** G. S. Appleton *Mother Goose in Hieroglyphicks* (1963) (Advt., recto rear cover), Chandler's paper dolls of the latest Paris fashions. No. 1—Carrie, with her Dresses and Bonnets. **1903** H. Keller *Story my Life* I. ii. 12 Two little children..were busy cutting out paper dolls. **1968–70** *Current Slang* (Univ. S. Dakota) III–IV. 81 *Make like a paper doll, and cut out*,..to leave. **1970** C. Major *Dict. Afro-Amer. Slang* 89 *Paper doll*,..to play hookey from school; to leave. **1972** T. I. Elliott in *Kawai's Origami* (ed. 4) p. iv, Playing with paper dolls is for girls, and the closest a guy approaches a paper doll is in the famous song where he buys 'a paper doll that other fellows will not steal'. **1976** *New Yorker* 15 Nov. 50/2 Maria undresses her paper doll and deliberately rips a feather off its hat. **1966** *Listener* 15 Dec. 893/1 This ..is the year of the paper dress that can be worn and then tossed in the waste basket. *Ibid.* 893/2 Paper-dress fabrics take colour well, and can be shortened with scissors. **1967** *Observer* 21 May 28/6 The best use so far of so-called 'paper' dresses—as instant summer bargains to be chucked away with no regrets. **1970** *New Society* 5 Mar. 385/2 It's a market that started developing in the fifties, boomed briefly in the mid-sixties with gimmicks like paper dresses, reassessed itself and looks set for a genuine boom in the mid-seventies. **1906** E. Dyson *Fact'ry 'Ands* iii. 27 Over these [sc. side-whiskers] the feathery paper-dust collected till they looked like the wings of an adolescent gosling. **1964** *Gloss. Letterpress Rotary Printing Terms* (*B.S.I.*) 5 *Paper dust or fluff*, fibre (fluff) or loading (dust) which leaves the web as it passes through the printing press or slitters. **1971** *Engineering* Apr. 79 (Advt.), Clean air in every facet of industry... Paper dust. **1920** H. Etheridge *Dict. Typewriting* 177 The paper feed of a typewriter consists of a paper roller (or platen) and two or more small feed rollers, which latter are kept in contact with the platen by means of springs, to allow of different thicknesses of paper being inserted. **1960** *Times* 24 Feb. 9/2 New fast paper-feed with integrated controls to reduce hand-movement. **1961** T. Landau *Encycl. Librarianship* (ed. 2) 120/2 The modern electric duplicator with its automatic paper-feed, counting mechanism and automatic stopping device, is a highly efficient and economical machine. **1967** Karch & Buber *Offset Processes* ix. 369 The delivery paper-feed control..controls the feed of paper from the delivery end. **1854** *Rep. Trans. Pennsylvania State Agric. Soc.* 176 The first premium on paper flowers, is awarded to No. 177. **1892** *Jrnl. Amer. Folk-Lore* V. 99 *Xeranthemum*, *Helichrysum*, paper-flowers. **1915** Armstrong & Thornber *Field Bk. Western Wild Flowers* 542 Paper Flowers. *Psilostrophe Cooperi.* Yellow. Spring. Southwest. **1935** R. Macaulay *Personal Pleasures* 256 Like those Japanese paper flowers which gently unfold and bloom in bowls of water..the mind puts out boughs and sprigs of blossom and ripe fruits, inebriating and enticing the charmed soul. **1972** *Islander* (Victoria, B.C.) 2 Apr. 2/3 Paper-flower and sundrop raced up the hills. **1974** C. Fremlin *By Horror Haunted* 94 Just look at all these mats, and doilies, and paper flowers! **1893** T. S. Row (*title*) Geometric exercises in paper folding. **1905** *Westm. Gaz.* 30 Sept. 13/2 Paper folding has long been a favourite amusement in our Kindergartens. **1908** W. F. White *Scrap-bk. Elem. Math.* 144 (*heading*) Symmetry Illustrated by Paper Folding. **1968** [see *origami]. **1972** T. I. Elliott tr. *Kawai's Origami* (ed. 4) p. i, Origami, paper-folding, can be enjoyed by children and adults alike. **1879** C. M. Yonge *Magnum Bonum* III. xxxv. 759 To listen to an exposition of the microphone, to share in a Shakespeare reading, or worse still, in a paper game, was..such a bore. **1934** E. Waugh *Handful of Dust* i. 21 'No paper games?' 'Oh, no, nothing like that. A certain amount of bridge and backgammon and low poker.' **1961** A. Wilson *Old Men at Zoo* iii. 125, I felt..that it would not be long before we were involved in paper games or even a sing-song round the piano. **1975** R. Player *Let's talk of Graves* iv. 108 We played games—paper games and Ludo. **1966** *Wall St. Jrnl.* 1 Dec. 5/2 World monetary reform negotiators removed some of the major stumbling blocks in their path to creating 'paper gold', although many more still remain. **1971** *Daily Tel.* 10 May 14 'Paper gold'..is a credit, listed in a computer in Washington, for each country which is a member of the International Monetary Fund. The members agree to use it to pay debts among themselves under certain rules, the only one of which need bother us is that no one except governments ever owns 'paper gold'. **1973** *Times* 15 Nov. 25/6 The nations of the world should start doing something to find a more catchy name for the Special Drawing Right, whose popular (well, fairly popular) nickname of 'paper gold' has recently been coming under attack in Britain. **1978** *Guardian Weekly* 19 Nov. 5/1 An agreement..to begin to phase out the world role of the United States dollar and its progressive replacement with so called 'paper gold', the basket of international currencies known as the IMF Special Drawing Rights. **1952** Leslie & Pepe *Methods of Teaching Typing Simplified* i. 9 Paper guide should be set and paper inserted by the teacher. **1962** *Which?* Dec. 354 Only four of the machines..had a paper guide for helping to get the left edge of the paper at the same place each time. **1970** L. Gartside *Teaching Business Subjects* viii. 173 Adjustment of margin stops with special reference to the paper-guide. **1907** *Yesterday's Shopping* (1969) 510/1 Handkerchiefs, Paper (Medicated)—These soft, silky papers are specially prepared for invalids, and are invaluable to sufferers from bronchial affections, catarrh, &c. **1939** R. Stout *Red Threads* (1941) viii. 109 A luxury brand of paper handkerchief, used for wiping creams from the skin. **1954** I. Murdoch *Under Net* iv. 62, I was choking and sneezing and using up a sackful of paper handkerchiefs. **1976** P. Ferris *Detective* i. 2 Rubbing the sweat off his bald head with a paper handkerchief. **1969** *Woman* 19 July 9/2 What man will pick up a dropped paper hanky? **1970** A. Ross *Manchester Thing* 66 Purse, loose change, paper hankies, comb—that kind of stuff. **1974** N. Freeling *Dressing of Diamond* 171 Vera..gave him a paper hanky. Bernard..mopped at his face. **1935** E. Bowen *House in Paris* i. 11 A paper kiosk opened to take its stock in. **1867** *Amer. Naturalist* I. 140 The odor that arises from the Tarantula killer when she uses her sting..resembles the odor of the paper-making wasp (Vespa), only much stronger. **1936** *Metronome* Feb. 21/4 *Paper man*, drummer who plays only what's written. **1936** *Delineator* CXXIX. 10/3 We have heard that no music is used... No, indeed! *Papermen* are not welcome in this esoteric milieu. **1970** C. Major *Dict. Afro-Amer. Slang* 89 *Paper man*, musician..who plays according to written music. **1832** F. Trollope *Dom. Manners Amer.* II. xxvi. 74 Four ink-wipers, three paper-matches, and a paste-board watch-case. **1844** Dickens *Mart. Chuz.* xlv. 519 Little Ruth..had a particular interest in some delicate paper-matches on the chimney-piece: wondering who could have made them. **1895** *Montgomery Ward Catal.* 109/1 Paper Napkins ..for tourists, travelers, lawn parties, lunches, picnics. **1945** G. Brooks in S. Henderson *Understanding New Black Poetry* (1973) 175 Some paper napkins in a water glass. **1972** L. Lamb *Picture Frame* xiv. 122 He put in his order, saw it..served in paper napkins on plates. **1959** J. T. Story *Mix me a Person* iv. 42 The fossilised remains of juke boxes and female frolic skeletons in paper-nylon slips. **1960** C. MacInnes *Mr. Love & Justice* 82 The man in Italian drape..., the woman with..paper nylon petticoat and white stilettos. **1702** Pepys *Let.* 13 Sept. (1926) II. 272 A strict measure cutt in paper of the originall writeing..and..a strict copy taken of the sayd writeing.. one copy thereof and of the paper-patterne, attested by the Doctor, to bee delivered to him. **1833** in A. Adburgham *Shops & Shopping* (1964) iv. 40 A great variety of Morning, Dinner, Evening, Ball & Opera Dresses,..which are made in models and full-size Paper Patterns. **1909** G. Stratton-Porter *Girl of Limberlost* iv. 54 Margaret Sinton was busy with the gingham and the intricate paper pattern. **1969** *Times* 30 Sept. 15/2 Sales of paper patterns —an invention claimed by an American, Mrs. Butterick more than 100 years ago but almost certainly extant in London earlier—are quoted at more than 22 million per year. **1974** M. Cecil *Heroines in Love* iv. 84 Paper patterns for home dressmaking. **1723** J. Nott *Cook's & Confectioner's Dict.* sig. B2v Almond Bisket..bake them on Paper-Plates in a moderate Oven. **1948** R. R. Karch *Graphic Arts Procedures* viii. 242 Three kinds of plates are used on the Multilith Duplicators. On the paper plate provided for one-run jobs, you can type, write, letter, draw, paint, rule, or trace the image desired in special inks. The plate is then placed on the press without further preparation. **1966** 'D. Shannon' *With a Vengeance* (1968) iii. 48 [He] confiscated one of the hot cookies she'd just transferred to the paper plate. **1971** R. K. Smith *Ransom* (1972) iv. 167 Greer took the paper plate of hamburgers in one hand..and walked to the standup counter. **1975** *New Yorker* 16 June 25/2 Some girls went to wash their hands, and others..killed time by wearing paper plates on their heads. **1976** *Times* 1 Apr. 32/9 (Advt.), Multilith paper plates, large surplus stock, variety of sizes. **1876** Paper ribbon [see Perforator 1]. **1888** *Encycl. Brit.* XXIII. 120/2 The paper ribbon R is moved forward by its centre row of holes. **1903** [see *paper tape* below]. **1922** [see *paper tower* below]. **1935** G. Greene *Basement Room* 102 He hadn't blown whistles or thrown paper ribbons. **1965** Paper ribbon [see *monotype *sb.* 3]. **1948** C. Day Lewis *Otterbury Incident* iv. 47, I asked if he'd let me take on part of his paper-round that evening. **1960** C. Dale *Spring of Love* i. 22 If I'm ever hard up for a job I'll come to you for a paper round... So if you ever need a paper boy, remember yours truly. **1971** C. Storr *Thursday* i. 15 The paper-round money had to provide Thursday with the things other children got without having to work. **1868** *Figaro* (San Francisco) 23 July 2/1 A Paper Route—One of the best on the most substantial city daily. **1929** T. Wolfe *Look Homeward, Angel* (1930) xxvi. 349 He found a substitute for his paper route. **1973** *Publishers Weekly* 19 Nov. 60/3 Isabelle gets to use some of that energy taking over her brother's paper route. **1975** *New Yorker* 28 July 28 (*caption*) Thank goodness, Winant has got himself a paper route, so we have a little something coming in. **1948** D. Ballantyne *Cunninghams* i. viii. 48 Gilbert wanted..a bike for Christmas so's he could have a paper run. **1904** *Dialect Notes* II. 420 Put the apples in a paper sack. **1940** W. Faulkner *Hamlet* III. ii. 207 The note was in pencil, on a scrap torn from a paper sack, unsigned. *Ibid.* IV. i. 279 Eck had gone on into the store and emerged with a paper sack, from which he took a segment of cheese. **1944** *Chicago Daily News* 14 July 9/2 She carried a thermos jug of coffee and a couple of sandwiches in a paper sack to eat on the bus. **1955** *Times* 5 July (Reed Suppl.) p. iii/5 In little more than twenty-five years the development of the multi-wall paper sack, in which Medway Paper Sacks Limited has been largely instrumental, has established new standards of efficiency, convenience and cleanliness in the packaging of powdered and granular products. **1971** 'R. Macdonald' *Underground Man* xi. 74, I..had a double hamburger with a paper sack of French fried potatoes. **1974** S. Marcus *Minding Store* (1975) ix. 185 A little old lady in shabby clothes, carrying a paper sack instead of a handbag. **1946** A. Sadler *Paper Sculpture* 17 Paper Sculpture ..is composed of sheets of finished paper, so rolled, bent, scored, cut and folded, that it makes a desired form. **1957** B. Angrave *Sculpture in Paper* 24 A legend has already grown up that modern paper sculpture was born in Poland, and is a development of the folk-art paper-cut tradition in that country. **1973** 'D. Halliday' *Dolly & Starry Bird* viii. 111 A fantasia of pure abstract design; a garden of convoluted plastic as fine as paper sculpture. **1978** *Cornish Guardian* 27 Apr. 33/5 Mrs. Phillips, president, welcomed Mrs. Spargo who gave a talk and demonstration of her paper sculpture. **1877** Paper shale [in Dict., sense 10 b]. **1931** *Prof. Papers U.S. Geol. Survey* No. 168, 7/2 'Paper shale' is the term applied to finely laminated claystones, siltstones, mudstones, and marlstones that show a pronounced tendency to part along the closely spaced bedding planes. **1969** Bennison & Wright *Geol. Hist. Brit. Isles* xiii. 292 An abundant fauna of ammonites and lamellibranchs is present, except in thin paper-shales. **1884** G. B. Goode *Fisheries U.S.: Nat. Hist. Aquatic Animals* 776 The terms 'Soft Crab', 'Paper-shell', and 'Buckler' denote the different stages of consistency of the shell. **1912** *Outing* Oct. 377/1 The only difference between the so-called 'paper-shell' pecan and the fruit from wild trees is that the former has been grown on a budded or grafted tree. **1945** *New England Homestead* 27 Oct. 20/2 Paper-shell pecan in Shell 50¢ per pound. **1948** F. N. Howes *Nuts* 110 Among the many varieties [of almond] cultivated in France are those that range from the thinnest of shells or 'paper shells' to thick hard-shelled forms. **1969** R. & D. De Sola *Dict. Cooking* 167/2 Papershell: Soft-shell pecan. **1974** *Anderson* (S. Carolina) *Independent* 22 Apr. 5A/3 Georgia is the leader in the production of papershell pecans in this country. **1911** Webster, Paper-shelled, *a.* Having a very thin shell, as a paper-shelled almond, or a crab whose shell is beginning to harden. **1946** *Nat. Geogr. Mag.* Apr. 526/2 Immediately after molting the animal [sc. a lobster] is soft-shelled. In a few days it is paper-shelled. **1948** F. N. Howes *Nuts* 110 It is usual to classify almonds according to the thickness of the shell—thin- or paper-shelled, soft-shelled and thick- or hard-shelled. **1962** L. Deighton *Ipcress File* xxx. 195 A small machine like a typewriter carriage. It was a paper shredder. Jay fed the sheet in and pressed a button. It disappeared. **1973** 'D. Jordan' *Nile Green* xiii. 54 The House has an obsession with security so it provides paper shredders on every floor. **1958** S. Hyland *Who goes Hang?* xxxii. 137 John Wintour's bankruptcy..was now officially expunged..by..the paper-shredding machines. **1968** A. Diment *Bang Bang Birds* i. ii. 14, I fed the telex strip into my paper-shredding device. **1930** A. P. Herbert *Water Gipsies* xxv. 373 It was a Gala Night; and the waiters were distributing paper streamers, balloons, dolls, squeakers and fans. **1935** C. Isherwood *Mr. Norris changes Trains* iii. 41 The ruffled plumes of a paper streamer..stirring like seaweed in the draught from an electric fan. **1941** M. Treadgold *We couldn't leave Dinah* xvii. 266 A box of crackers and paper streamers left over from the famous Carnival. **1969** N. Freeling *Tsing-Boum* xvi. 117 Broken paper streamers knee deep. **1957** M. B. Picken *Fashion Dict.* 344/1 *Paper taffeta*, crisp lightweight taffeta with a somewhat papery feel. **1963** *Times* 27 Feb. 12/2 The bride, who was given away by her father, wore a gown of white paper taffeta, the bodice and trained-skirt trimmed with knife-pleated frilling. **1972** *Vogue* 15 Mar. 3/2 Plaid paper taffeta dress. **1890** *Electrician* 4 July 233/2 Each of these styles can be pressed against the paper tape by the armature of a corresponding electromagnet. **1903** C. H. Sewall *Wireless Telegr.* iii. 133 There is a telegraphic apparatus known as the 'Wheatstone', in which a paper ribbon is first perforated and then sent through a machine, recording at the distant end with ink marks upon paper tape. **1924** P. J. Risdon *Wireless* xvi. 135 When the operator depresses the keys,..instead of typing letters and figures, it perforates the paper tape with holes corresponding to the dots and dashes of the Morse code. **1943** *Rep. Progress Appl. Chem.* XXVIII. 131 In the standard test for the adhesive power of gummed paper tape the sample is moistened under standard conditions. **1961** *Times* 3 Oct. (Computer Suppl.) p. vi/4 Each day the paper-tape readers feed into the computer more than nine million characters relating to long-distance calls. **1971** *Ann. Rep. Curators Bodl. Libr. 1969–70* 45 The correction of records..suffered from rapid changes of the staff operating the paper-tape typewriters. **1975** J. B. Harley *O.S. Maps* i. 14 The names are then typed on a 'Monotype' keyboard which reproduces them on punched paper tape to operate the filmsetter. **1952** 'Han Suyin' *Many-Splendoured Thing* v. 313 America is only a paper tiger... That's what the Peking government says. **1958** *Peking Rev.* 11 Nov. 7 In August 1946 Comrade Mao Tse Tung gave an interview to the American journalist Anna Louise Strong and expressed his famous view point that all reactionaries are paper tigers. **1963** *Economist* 12 Jan. 98/1 In the Chinese view, the 'rotten, decadent, paper-tiger nature' of imperialism cannot change. **1973** *Black Panther* 10 Nov. 14/3 Our minority may be powerful now, but in the end even that power will prove to be a paper tiger once the people unite. **1976** J. Snow *Cricket Rebel* 25 We [sc. Sussex] were something of paper tigers when it came to the championship games. **1943** D. Baker *Trio* i. 55 She jerked a paper towel off the roller and did a careful job of drying. **1953** *Times* 31 Oct. 1/11 Many paper manufactures, including..paper towels and handkerchiefs. **1972** *Guardian* 17 May 9/5 Plain coloured paper towels cost 17p for a two-roll pack. **1975** *New Yorker* 8 Dec. 41/2 The stenographers and typists had to make do with paper towels that scratched when new and dissolved when damp. **1916** *Monotype System* (Lanston Monotype Machine Co.) Gloss. p. lxvi, Paper Tower... The mechanism of both the Keyboard and Casting Machine..that carries the paper ribbon and advances it one marginal perforation for each character, or space, struck at the Keyboard or cast at the Casting Machine. *Ibid.* xi. 35 The operator has only to turn the small Valve Handle.. at the left side of the bottom of the Paper Tower. **1922** *Casting Machine Adjustments* (Lanston Monotype Machine Co.) 17 The Paper Tower is the controlling mechanism of the Casting Machine. In it is placed the paper ribbon prepared at the Keyboard by the compositor, and the ribbon compels the Casting Machine to produce the characters required. *Ibid.*, The Paper Tower mechanism provides that the advance of the paper will be absolute. **1951** S. Jennett *Making of Bks.* iv. 68 Above the keybank [of the Monotype] is the paper-tower, in which is a roll of paper perforated along the edges like cine-film. **1819** E. Evans *Pedestrious Tour* 228 On this river too is General Simcoe's paper town called London. **1943** T. Pratt *Barefoot Mailman* xi. 90 How do we know you are not a paper-town shark yourself? How do we know you have not come here to boom our city to false values? **1948** E. N. Dick *Dixie Frontier* 151 Many towns..never got beyond the stage of 'paper towns'. **1957** B. Hutchison *Canada: Tomorrow's Giant* 307 Logging camps, mills, paper towns, the new aluminium town of Kitimat..pour their products into Vancouver. **1959** *Tararua* XIII. 49 *Paper road* and *paper town*..describe those roads and towns which have been surveyed but whose actual existence has got no further than the maps. **1969** H. Horwood *Newfoundland* i. 4 Driving back through the star-studded night towards the paper town of Grand Falls. **1925** G. Dickinson *Eng. Papier-Mâché* i. 3 In 1772 Henry Clay, japanner, of Birmingham, invented a material which had certain heat-resisting properties, that made it suitable for japanning or lacquering processes. The body of the material was made by pasting sheets of paper together, and the articles made from it were called 'paper ware'. **1969** *Canad. Antiques Collector* Jan. 8/1 The craze for making 'Paper-ware' spread from the 17th to the 19th century, when recipes for making and decorating bowls, vases and plates, were

printed in such papers as the '*Gentleman's Magazine*'. **1889** W. Fraser *Words on Wellington* 136, I have heard it said that a vivâ voce examination is not fair upon a young man; and that, what at Oxford we call 'paper work' should be used for military examinations. **1917** R. Kipling *Let.* in C. E. Carrington *Rudyard Kipling* (1955) xvii. 445 Both sides groan together over the enormous amount of unnecessary paper-work. **1958** I. Murdoch *Bell* viii. 120 Time was badly needed to catch up on the paper work of the previous week. **1961** [see *feed-house* s.v. *feed *sb.* 6]. **1969** J. Argenti *Managem. Techniques* v. 24 Too much paperwork. Symptoms: masses of complicated forms. **1971** S. Hill *Strange Meeting* ii. 123 They sat in their dug-out in the evening, reading or doing paperwork, listening to the gramophone. **1973** C. Bonington *Next Horizon* xxi. 290 Back to Base Camp, to start wading through the mass of paper-work which the end of the expedition.. entailed. **1977** *New Yorker* 8 Aug. 68/3 (Advt.), We do the paperwork, you do the legwork.

paper, *v.* Add: **3. e.** Fig. phr. *to paper over*: temporarily to conceal; esp. in phr. *to paper over the cracks* (see *crack *sb.* 7 f).

1955 *Times* 16 Nov. 10/5 This document was treated by the western Ministers as no more than an attempt to paper over the complete divergence in policy. **1957** *Economist* 28 Dec. 1114/2 The party's usual split has been papered over. **1966** J. Deakin *Lobbyists* 79 Not even the impressive legislative accomplishments under Lyndon Johnson can paper over Congress's serious..frailties. **1969** *Sunday Sun* (Baltimore) 16 Mar. K 1/3 The two sides were still able to paper over their differences and agree to a compromise program. **1974** *Times* 6 Apr. 1/2 Mrs Meir has persuaded the party previously to paper over such differences.

4. b. (Earlier and later examples.)

1859 E. Fitzball *35 Yrs. Dram. Author's Life* II. 113 The second night comes, the unfailing 'Lady of Lyons'... House well papered, but badly *gilt*—calls similar. **1959** *Times* 3 Nov. 15/3 'Papering a house' is all very well when a company is playing to empty stalls, but is hardly a good idea when the house is full. **1973** *Courier-Mail* (Brisbane) 26 July 9/5 Surely the theatre could..at least have 'papered' the house with complementary guests for the opening if the box office was looking so poorly. **1978** A. Morice *Murder by Proxy* xx. 155 We should be playing to an audience of approximately thirty-five, about two thirds of it papered.

6. *intr.* and *trans.* To pass forged cheques; to defraud by issuing forged cheques. Also in extended use. *U.S. slang.*

1925 *Flynn's* 7 Mar. 191/2 *Paper*,..to pass worthless or forged checks. **1941** *Amer. Speech* XVI. 248/2 *To paper the burg*, to pass a quantity of forged checks. **1958** *Daily News* (N.Y.) 16 Apr. 60 Helped by phony..credentials and a blonde, a former stable boy..papered Queens and Long Island with $10,000 to $15,000 worth of bum checks. **1976** Scott & Koski *Walk-In* xi. 63, I want to know that this *is* Li we're dealing with and not some ringer they've papered on us.

7. *trans.* To preserve (insects) by storing them in triangular packets made of folded paper.

[**1894** W. F. Kirby *Hand-bk. Lepidoptera* I. p. iii, Collectors abroad generally put their captures into papers folded to resemble a triangular envelope. *Ibid.* p. lv, The cheapest way of buying Butterflies is to buy miscellaneous lots at an auction, especially lots in papers.] **1955** Wagstaffe & Fidler *Preservation Nat. Hist. Specimens* I. 186 Specimens may often have to be stored for considerable periods before they can be set to form part of the permanent collection. The usual method of storing such specimens is to 'paper' them.

‖ **paperasserie** (paparasəri). [Fr.] An accumulation of paper-work; administrative red tape.

1928 *Observer* 22 July 12 The Frenchman loves to make fun of 'paperasserie', the elaborate and meticulous bureaucracy, whose spirit he really admires in his heart. **1955** D. Barton *Glorious Life* 123 Even though I make a great parade of enthusiasm and am far too conscientious about my work.., all this *paperasserie* is not my element. **1965** *New Statesman* 21 May 809/1 The complaints universally made against Indian administrators today—delay and file-passing—seem to be just the complaints that Curzon made against the British administrators of his time. His first battle was against paperasserie.

paperback (pēi·pəɹbæk). Also **paper-back.** [f. Paper *sb.* + Back *sb.*[1]] A book with a paper back or cover. Also *attrib.*, and as *v. trans.*, to publish in a paperback edition.

1899 [in Dict. s.v. Paper *sb.* 12]. **1954** etc. [see *hard-back 2]. **1957** *Times Lit. Suppl.* 12 July i/2 The revolution wrought in the world of books by the emergence of the paperback can best be measured by comparing a classical private library with a contemporary collection of books. *Ibid.* ii/3 In one way or another most English publishers have accepted, and come to terms with, the paperback revolution. **1960** *N.Y. Times Mag.* 5 June 2 Has Emily Post been paper-backed? **1962** A. Huxley *Let.* 11 Feb. (1969) 929 In this context see the list of 112 exercises in awareness..printed at the end of *Zen Flesh, Zen Bones* (now in paperback). **1963** *Sunday Times* (Colour Suppl.) 24 Nov. 23/2 They paperback their *World of Art* series at 18s. **1964** M. McLuhan *Understanding Media* (1967) ii. xxxi. 347 The paperback, especially in its highbrow form, was tried in America in the 1920s and thirties and forties. It was not, however, until 1953 that it suddenly became acceptable. **1968** *Guardian* 27 Apr. 7/1 The worse the rubbish the wider the market: trash is paperbacked in many languages. **1971** *Country Life* 18 Feb. 381/2 His *Highland Year*..which is already in three editions and about to make paperback, is followed by *Highland Deer Forest*. **1971** *New Scientist* 19 Aug. 433/1 It looks as if most people are going to have to wait for the paperback. **1973** 'M. Innes' *Appleby's Answer* xvii. 149 The paperback rights..would pay for far more than Miss Pringle's customary inexpensive holiday. **1974** *Times* 28 Feb. 15/8 Paperback rights bring £27,000 bid... Hardback rights in Britain are held by Alison Press. **1975** *Bookseller* 11 Oct. 2038/2 The book was called *Harris in Wonderland*, and Cape published it. It was adapted for radio, and paperbacked in Hungary, of all places. **1976** H. M. Harrison in Aldiss & Harrison *Decade the 1950s* 12 With the coming of the paperback revolution..SF found new places to go.

So **pa·per-backed** *a.* or *ppl. a.*, having a paper back, published in paper-back form; also (*fig.*), lacking in strength, feeble.

1888 Kipling *Soldiers Three* (1889) 52 'Push, men!' sez Crook; 'Push ye paper-backed beggars!' he sez. **1903** *Work* 14 Mar. 89/3 (*heading*) Handy method of binding paper-backed books. **1909** *Chambers's Jrnl.* Sept. 610/2 The man then buys a paper-backed novel for fourpence-halfpenny. **1932** D. L. Sayers *Have his Carcase* x. 116 One or two paper-backed books. **1957** *Times Lit. Suppl.* 12 July p. ii/3 American bookstalls and bookshops are nowadays crowded with solid, well made, often highbrow but still paperbacked volumes. **1968** *Guardian* 29 July 20/8 Paperbacked D. H. Lawrence floods the bookshops. **1974** C. Fremlin *By Horror Haunted* 58 Years of..reading of lurid paper-backed stories under the desks at school had not provided him with this sort of solid..information.

pa·per-bark. [f. Paper *sb.* + Bark *sb.*[1]] **1.** A name used for several Australian trees distinguished by flaky layers of pale bark, esp. *Melaleuca leucadendron*, the cajeput, and other members of the genus *Melaleuca*, belonging to the family Myrtaceæ. Also *attrib.*

1842 *Western Australia* 81 It [sc. the Melaleuca, or tea-tree] is sometimes known by the name of the paper-bark tree, from the multitudinous layers (some hundreds) of which the bark is composed. These layers are very thin, and are loosely attached to each other, peeling off like the bark of the English birch. **1846, 1866** [see Paper *sb.* 12]. **1908** E. J. Banfield *Confessions of Beachcomber* i. vi. 216 Few of the forest trees are more picturesque than the paper-bark or tea-tree..of free and stately growth, the bark white. **1920** B. Cronin *Timber Wolves* 88 From the paper-bark swamps came the reverberating boom of frogs. **1924** Lawrence & Skinner *Boy in Bush* 100 It was salty paper-bark country. **1946** K. Tennant *Lost Haven* (1947) 2 Wide, soggy moors..from which the great, white paper-barks tower. **1957** P. White *Voss* 59 She walked in the garden, amongst the camellia bushes..and the scurfy native paperbarks. **1961** A. Upfield *Bony & White Savage* ix. 78 A ring of paperbark trees, all grey-white of trunk and branch... He was able to tear off a strip, to find it composed of layer above layer of paper-thin wafers... Under the outer one the wafers were flesh-coloured. **1969** *Northern Territory News* (Darwin) *Focus '69* 85/1, I took up my most pleasant sentry duties amongst the paperbark trees.

2. The durable bark of *Melaleuca leucadendron*.

1857 J. Askew *Voy. Austral. & N.Z.* 433 The dead bodies are burnt or buried, though some in North Australia place the corpse in the paper bark of the tea-tree, and deposit it in a hollow tree. **1908** E. J. Banfield *Confessions of Beachcomber* I. ii. 82 It became necessary..to keep the ridge covering of paper-bark in position. **1941** I. L. Idriess *Great Boomerang* iv. 28 It was she who found him paperbark and fashioned his first sandals.

3. paper-bark maple, *Acer griseum*, a maple native to central China, introduced into Europe in 1901, and characterized by flaky, light brown bark; **paper-bark tea-tree, ti-tree** = sense 1 above.

1927 A. Rehder *Man. Cultivated Trees N. Amer.* 577 Paperbark M[aple]. Tree to 8 m., with cinnamon-brown bark separating in thin papery flakes. **1938** *Amer. Nurseryman* 15 Oct. 3/3 The leaves of the paperbark maple..are finely and thickly hairy underneath. **1969** T. H. Everett *Living Trees of World* 223/1 The paper-bark maple (*A[cer] griseum*) of western China is remarkable for its beautiful rich cinnamon brown bark which peels in broad paper-thin strips. **1974** A. Mitchell *Field Guide Trees Britain* 349 Paper-bark Maple... Now much planted in gardens. **1967** A. Rule *Forests Austral.* xiv. 154 An enterprising firm manufactures cork from the bark of the paperbark tea tree. **1944** *Living off Land* vii. 137 Open-grained Timbers... Paper-bark Ti-tree [etc.].

papered, *ppl. a.* Add: **a.** (Further examples.)

a **1828** D. Wordsworth *Jrnl.* (1941) II. 335 The little room so snug—the carpets—the papered walls. **1846** *Jewish Manual, or Pract. Information Jewish & Mod. Cookery* vi. 158 Bake in a papered tin. **1860** F. Nightingale *Notes on Nursing* ii. 16 Old papered walls of years' standing..are..ready sources of impurity to the air. **1977** M. T. Bloom *13th Man* ii. 13 The papered corridors had the palpable feel of transitoriness, of waiting to be called —up or out.

b. *slang.* Of a theatre, etc.: filled by means of free passes.

1936 *Amer. Speech* XI. 221 The house will be *papered*, which means that free passes to the show will be given away. **1959** *Oxf. Mag.* 12 Feb. 244/1 Last week the B.B.C. was able to fill the Festival Hall decently full (even allowing for 'papered' seats). **1974** *Plain Dealer* (Cleveland, Ohio) 27 Oct. 2-c/3 If there is a large crowd at ringside the suspicion will be a papered house, for the government doesn't want the world to see an empty stadium.

2. Of insects in a collection, stored in triangular packets made of folded paper.

1937 C. Longfield *Dragonflies Brit. Isles* 27 Preserving the brilliant colours after death is extraordinarily difficult, and for that reason a 'pinned' or 'papered' collection of dragonflies is often very disappointing. **1955** Wagstaffe & Fidler *Preservation Nat. Hist. Specimens* I. 187 (*caption*) Box for storing papered insects.

paperer. Add: (Further examples of sense 'a paper-hanger'.)

1837 J. Romilly *Diary* 25 May (1967) 119 Set the Painters & Paperers to work in the new house. **1928** A. Huxley *Let.* 8 Oct. (1969) 302 The house is in a state of chaos—painters and paperers still at work, furniture huddled here and there.

b. *slang.* One who issues or receives free passes to a theatre, etc.

1885 *Referee* 14 June 3/3 Results showed that the 'paperer' understood his business. **1895** *Punch* May 230/1, I took Lil, Dannel's youngest, larst week to the play with some tickets I'd got. Well paperers mustn't be choosers.

paper-hanger. Add: **1.** (Earlier and later examples.)

1796 M. Edgeworth *Parent's Assistant* (ed. 2) 164 A new carpenter and paper-hanger..were appointed. **1969** *New Yorker* 12 Apr. 80/2 'We'll be as busy as one-armed paperhangers,' says Lind. **1975** *Ibid.* 21 Apr. 95/1 According to some twenty separate lists that were obtained from underwriters' manuals..occupations that various insurance companies consider to be grounds for rejection of applications for auto insurance, or acceptance only after careful study or on a restricted basis, included those of.. paperhangers,..sports coaches and assistants, travelling salesmen,..and doctors.

2. *slang.* (orig. *U.S.*). One who passes forged or fraudulent cheques; a forger.

1914 Jackson & Hellyer *Vocab. Criminal Slang* 101 Whoever thoughtlessly leaves his check book in accessible places incurs the jeopardy of..personal loss, seeing that 'paper hangers' are vigilant in the search for these. **1938** *Detective Fiction Weekly* 23 Apr. 73/1 Next to the con man is the 'paperhanger' or a 'writer of sad, short stories', these terms referring to a forger. **1941** J. G. Brandon *Death in Quarry* xiii. 126 'Paper-hanger,' McCarthy echoed. 'That's a new one on me, William.' 'Passin' the snide, sir,' Withers informed him. 'Passing flash paper. Bank of Elegance stuff.' **1945** L. Shelly *Jive Talk Dict.* 31 *Paper hanger*, one who deals in counterfeit money. **1976** *Times Lit. Suppl.* 16 Apr. 457/4 A legendary confidence man and 'paperhanger' (described as 'a criminal who specialises in hanging, or passing on stolen or counterfeit securities on brokerage houses, insurance companies, and individuals').

paper-hanging. Add: **2. b.** The affixing of bills, advertisements, etc., on a bill-board or hoarding.

1851 Dickens in *Househ. Words* 22 Mar. 604/2 [I] Hired a large one [sc. hoarding]..let out places on it, and called it 'The External Paper-Hanging Station'. **1961** *Times* 21 June 13/6 A hundred years ago 'paper hanging' (a term for bill-posting inherited from the eighteenth century) was an irresponsible and pirate trade.

3. *slang.* (orig. *U.S.*). The passing of forged cheques; forging.

1927 *Dialect Notes* V. 458 *Paper hanging*,..passing forged cheques. **1930** *Detective Fiction Weekly* 30 Aug. 697/2 'Paperhanging', or passing fictitious checks..is not always a mere matter of offering a check to a merchant and trusting to luck that he will cash it. Many tradesmen make it a rule never to cash a check for a stranger, unless it be a pay check or a certified check; but a clever paperhanger can get around this easily enough. **1932** *Evening Sun* (Baltimore) 9 Dec. 31/5 *Paper hanking*, forging. **1975** C. Weston *Susannah Screaming* (1976) viii. 42 From paperhanging to murder—that's a pretty big jump. **1976** A. Schroeder *Shaking it Rough* 95 It all involved paperhanging—fake credit cards, rubber checks, [etc.].

papering, *vbl. sb.* Add: **1.** (Further examples.)

1841 Thackeray *Gt. Hoggarty Diamond* (1849) x. 130 The painting, papering, and carpeting of my house. **1855** Mrs. Gaskell *North & South* I. viii. 97 It needed the pretty light papering of the rooms to reconcile them to Milton. **1928** A. Huxley *Let.* 25 Oct. (1969) 302 The main work is done..all painting and papering. **1975** P. Dickinson *Lively Dead* xxi. 125, I do want to get the papering done before my son comes back. He's an absolute devil with glue.

2. (See quot. 1966.)

1798 Jane Austen *Let.* 1 Dec. (1952) 35 My long hair is always plaited up out of sight, and my short hair curls well enough to want no papering. **1815** R. Fenton *Mem. Old Wig* 9 Having undergone the varied ordeal of papering, pinching, crimping, baking, and torture a thousand ways, I was promoted to thatch the cranium of the notorious judge Jefferies. **1966** J. S. Cox *Illustr. Dict. Hairdressing* 108/1 *Papering*, 18th cent. term for placing the paper papillotes around the wound hair preparatory to pinching it with hot pinching irons.

papery, *a.* Add: Also *fig.*

1924 R. Fry *Let.* 27 June (1972) II. 554 Ingres's *Stratonice* is a shock..it's so thin and papery. **1937** D. Canfield *Fables for Parents* (1938) 55 It was..more apparent that it was a papery idea out of a book.

Paphlagonian (pæflăgōᵘ·niăn), *a.* and *sb.* [f. Gr. Παφλαγονία, L. *Paphlagonia*, an ancient region in northern Asia Minor + -an.] **A.** *adj.* Of or pertaining to Paphlagonia or its inhabitants. **B.** *sb.* A native or inhabitant of Paphlagonia.

Also (spelt *Paflagonian*) with reference to the fictional people in Thackeray's *The Rose and the Ring*.

1596 T. Nashe *Have with you to Saffron-Walden* sig. O2v, Procris and Cephalus, and a number of Pamphlagonian things more, that it would rust & yron spot paper, to have but one sillable of their names breathed over it. **1607** Topsell *Foure-f. Beasts* 291 Touching the Paphlagonians about the education of their horsses see more among the Venetians. *Ibid.* 562 The Paphlagonians, which before the Troyan warre were called *Eneti*, and afterwards *Veneti*. **1748** Hume *Philos. Ess. conc. Human Understanding* x. 189 The Impostor..was enabled to proceed, from his ignorant Paphlagonians, to the inlisting of Votaries, even among the Grecian Philosophers. **1855** Thackeray *Rose & Ring* i. 4 Two nations which had been engaged in bloody and expensive wars, as the Paflagonians and the Crimeans had been. *Ibid.* ii. 5 The Paflagonian nobility did not care who was king. **1954** T. Gunn *Fighting Terms* 32 Here in a cave the Paphlagonian King Crouched. **1974** *Encycl. Brit. Micropædia* VII. 734/1 The Paphlagonians were one of the most ancient peoples of Anatolia.

papia, var. *papaya.

1921 *Outward Bound* Feb. 69/1 The natives..kept bringing fresh fruit to our view—mangoes and custard apples and papia.

‖ **Papiamento** (pæpyāme·nto). Also **Papiamentu.** [Sp.] A Spanish-based creole language of Curaçao, Aruba, and Bonaire, in the Caribbean Sea.

1949 *Caribbean Q.* I. iii. 36 The remnants of the African culture, are fast disappearing... There are no signs, however, that their language, the papiamento, which is still in use everywhere, will disappear. **1953** *Ibid.* III. i. 27 Dialects such as Papiamento..were unworthy of the grammarians' labour. **1956** *Publ. Amer. Dial. Soc.* xxvi. 11 The derivative of Spanish spoken in Curaçao (called Papiamentu or Papiamento), of French in Haiti (Creole), of English in Guiana (Ningro-Tongo), in the West Indies, and in South Carolina (Gullah) are some examples of languages which can either be regarded as dialects of the colonial languages or as new languages of their own. **1967** *Language* XLIII. 818 Papiamentu 'may well represent a fusion of two earlier pidgins or creoles'. **1972** J. L. Dillard *Black English* i. 24 Some of them [sc. pidgin and creole languages] are vehicles of education and literature, like Papiamento, whose speakers have profited from an enlightened Dutch attitude toward languages. **1974** R. A. Hall *External Hist. Romance Lang.* vii. 157 In other languages, however, such as Papiamentu, the dominant lexical influence in later centuries has come from a non-Romance language, in this case Dutch. **1976** *San Francisco Examiner & Chron.* 13 June (Travel Suppl.) 8/3 Although Papiamento is spoken on radio and TV, and newspapers and books are printed in it, the official language on the island [sc. Aruba] is Dutch.

‖ **papier** (papye). The French word for 'paper' used in various phrases, as **papier collé** (kǫle) ['gummed paper'], a collage made from paper; also *attrib.*; **papier déchiré** (deʃire), paper torn haphazardly for making collages; a collage made of such paper; **papier poudré** (pūdre), a paper impregnated with face-powder. See also Papier mâché in Dict. and Suppl.

1907 *Yesterday's Shopping* (1969) 539/1 Papier Poudre. In 3 shades, white, rose and rachel... These exquisitely perfumed leaves are designed to supersede the use of 'puff and powder box'. **1919** R. Firbank *Valmouth* xi. 189 'A book is anathema to her.' 'Even a *papier poudré* one; for, when I gave her my little precious volume of *blanc de perle* in order to rub her nose, she started grating her teeth at me.' **1935** D. Gascoyne *Short Survey Surrealism* v. 107 The *papiers collés* of Picasso and Braque (newspaper, wallpaper, cigarette packets). **1959** *Listener* 30 Apr. 766/3 It is a picture as utterly *original* as the first *papiers collés*. *Ibid.* 19 Nov. 868/1 Arp even welcomed the ravages wrought by time on his own works, and started to help nature along by doing the destruction for her, making *collages* of *papiers déchirés*, roughly and haphazardly torn. **1960** E. H. Gombrich *Art & Illusion* x. 356 His [sc. Picasso's] more playful creations, such as his *papiers déchirés*. **1960** D. Holman-Hunt *My Grandmothers & I* iv. 121 Grandmother was scrubbing her nose with a sheet of Papier Poudre. **1962** *Listener* 15 Feb. 305/1 In using the technique of *papier collé* they were substituting fragments of reality itself for the earlier fragmented images of reality. *Ibid.*, Their *papier collé* pictures. **1963** *Times* 8 Feb. 14/3 The reliefs..and the sculpture..are built up of plywood layers treated in the manner of a *papier dechiré* collage. **1965** *New Statesman* 12 Nov. 758/2 The introduction of *trompe-l'œil* and *papier-collé*. **1973** *Country Life* 1 Mar. 556/3 Tiny booklets..containing *papier poudre*, for 12p..very popular years ago for taking the shine off noses. It works.

‖ **papier mâché.** Add: Also *fig.*

1818 'T. Brown' *Brighton* I. i. 13 He has a clay-cold heart, and a mere papier maché mind. **1920** F. M. Ford *Let.* 24 July (1965) 115 The plaster Pillars of the State and the papier maché hearts of men. **1977** *Broadcast* 7 Feb. 7/2 It's unlikely that he would want to be party to a papier maché exercise.

papilio (pāpi·lio). [a. L. *pāpilio* butterfly, adopted as a generic name by Linnæus in *Systema Naturæ* (ed. 10, 1758) I. 458.] A swallow-tail butterfly belonging to the large genus so called, frequently distinguished by tail-like projections on the hind pair of wings; formerly, any butterfly.

1789 *Loiterer* 26 Sept. 9 The Wings of Moths, and Papilios. **1835** J. Duncan *Brit. Butterflies* 92 The word *Papilio* was used by Linnaeus in the comprehensive sense which he was accustomed to attach to such terms, to designate generically all the diurnal lepidoptera. **1932** J. S. Huxley *Probl. Relative Growth* ii. 55 Many of these [sc. holometabolous insects] possess organs which increase in relative size with increase of absolute size of body..[like] the 'tail' on the hind wing of Papilios. **1936** *Discovery* July 213/2 Many large Papilios were flying about. **1965** R. McKie *Company of Animals* xiii. 178 Papilios and pierids and danaids, the Browns, the Blues, the Skippers, and many many more. **1972** L. E. Chadwick tr. *Linsenmaier's Insects of World* 41/1 The males [of *Papilio dardanus*] occur throughout the range in the normal, long-tailed costume of the papilios.

papillœdema (pæpilidī·mă). *Ophthalm.* Also (chiefly *U.S.*) **papilledema.** [f. Papill(a + Œdema, as repr. G. *papillenödem* (A. Elochìng 1895, in *A. von Græfe's Arch. für Opthhalm.* XLI. ii. 276).] Non-inflammatory swelling of the optic disc due to increased intracranial pressure on the optic nerve, usu. as a result of a tumour or abscess of the brain.

1908 J. H. Parsons *Path. of Eye* IV. xxvii. 1349, I suggest the use of the term 'papillœdema' to replace 'choked disc' (Stauungspapille). **1922** R. F. Moore *Med. Ophthalm.* i. 23 In a few cases papillœdema runs its course to complete subsidence in one eye, the other remaining normal throughout. **1950** Berens & Siegel *Encycl. of Eye* 34 Papilledema is a common finding in increased pressure, although it is by no means an early sign. **1961** *Times* 11 May 9/1 The pathogenesis of papilloedema. **1972** *Pediatrics* XLIX. 248/1 Papilledema in a patient with cyanotic congenital heart disease is not necessarily a sign of brain abscess. **1977** *Proc. R. Soc. Med.* LXX. 235/2 Acute blockage of the shunt is usually obvious because of the clinical signs, the most valuable being bradycardia. Papilloedema is not to be relied upon.

papillon (pa·piyoṅ). [a. F. *papillon* butterfly.] A breed of toy dog related to the spaniel, having a white coat with a few darker patches, esp. on the head, and erect ears resembling the shape of a butterfly's wings. Also *attrib.*

1907 R. Leighton *New Bk. Dog* xvii. lxi. 536/1 A very engaging little dog is the Papillon, or Squirrel Spaniel. *Ibid.* 536/2 The name Papillon is obviously given to the dog in reference to its ears, which stand out large and erect like the wings of a butterfly, heavily feathered. **1910** *Encycl. Brit.* VIII. 376/1 At the annual show of the Kennel Club in the autumn of 1905..additional breeds..[were] Chesapeake Bay dogs, Chihuahuas, Papillons and Roseneath terriers. **1924** *Glasgow Herald* 4 Apr. 8 A new toy dog has been brought from Belgium, the Papillon. It is very small, weighing when full grown 3 to 5 lb. **1927** M. B. Cooper in C. C. Sanderson *Pedigree Dogs* 341 The Papillon or Butterfly dog was first introduced into England in any number in 1923. **1929** R. Graves *Poems* 25 Those pugs and papillons and in-betweens. **1960** *Times* 11 Mar. 1/7 (Advt.), Adorable Papillon Puppies. **1971** F. Hamilton *World Encycl. Dogs* 541 The Papillon has bred true to type for some 700 years or more, as can be verified in the art galleries and museums of the world. **1972** *N.Y. Times* 4 June 6/2 Deja..has been best Papillon 125 of the 135 times he has been shown.

papillote. Add: **1.** (Further examples.) Also *fig.*

1848 Thackeray *Van. Fair* xliii. 392 Glorvina, trembling with all the papillotes. **1860** Queen Victoria *Let.* 3 Nov. in R. Fulford *Dearest Child* (1964) 278 We drove over..to Clifden which unfortunately was en papillotte. **1959** S. G. Flitman *Craft of Ladies' Hairdressing* viii. 74 For years the posticheur had used papillottes (small triangles of paper) to hold damp flat curls until they were dried.

attrib. **1960** Cunnington & Beard *Dict. Eng. Costume* 156/1 *Papillotte comb*.., a decorative comb of tortoiseshell. **1966** J. S. Cox *Illustr. Dict. Hairdressing* 108/2 *Papillotte tongs*, pinching irons.

2. A paper wrapper, usu. greased, in which certain types of meat and fish are cooked or served.

1818 T. Moore *Fudge Fam. Paris* v. 38 One's hair and one's cutlets both *en papillote*. **1868** M. Jewry *Warne's Model Cookery* 296/2 Wrap them [sc. mutton cutlets] in buttered papers... The fat of the dressed meat is absorbed in the papillotes. **1959** R. Postgate *Good Food Guide 1959–1960* 109 Tamar salmon en papillotte, ham soufflé, or chicken à la crème, may appear on the menu. **1961** *Guardian* 10 Mar. 10/6 Trout can..be grilled..in paper-bags (*papillotes*)...first spread the inside of the *papillote* generously with maître d'hôtel batter.

Papist. Add: **3.** An imitator or follower of the poet, Alexander Pope. Also *attrib.*

a **1849** H. Coleridge *Ess. & Marginalia* (1851) II. 118 Nor would so many really monotonous jinglers have passed for correct, orthodox Papists. **1902** F. Harrison *John Ruskin* ii. 22 Many a prize poem has had worse couplets in the Papist vein than these on Etna.

papodam, -dum, varr. Popadam (in Dict. and Suppl.).

papolater (pĕipǫ·lătəɹ). [f. L. *pāpa* pope + *-later* (see -latry).] One who practises papolatry.

1913 A. Fortescue *Lesser Eastern Churches* i. 4 We are Creed-tamperers, Papolaters, gross disturbers of the peace by our shameless way of sending missionaries.

papoose. Add: **b.** papoose-root, a perennial plant, *Caulophyllum thalictrioides*, of the family Berberidaceæ, native to eastern North America, and bearing panicles of small yellowish flowers and blue berries; the thick, twisted root was formerly used medicinally. (Earlier and later examples.)

1815 D. Drake *Nat. View Cincinnati* ii. 85 Poppoos root. **1843** J. Torrey *Flora N.Y.* I. 33 Blue Cohosh. Pappoose-root... The root of this plant is in some repute as a diuretic and bitter. **1943** R. Peattie *Great Smokies* 190 The old wives of the mountains today are not averse to..giving their teething children a little papooseroot.

papovavirus (păpōᵘ·vāvəi°rəs). *Microbiol.* Also **papova virus.** [See quot. 1962.] Any of a group of small animal viruses which includes those causing polyoma, papilloma, sarcoma, and warts, the members of which consist of double-stranded DNA in an icosahedral capsid with no envelope.

1962 J. L. Melnick in *Science* 30 Mar. 1128 In the course of work in this laboratory with papilloma (wart) virus of man, polyoma virus of mice, and vacuolating virus of monkeys, I have been made conscious of the similarities in properties between each of them and papilloma virus of rabbits. This has led me to group these viruses together in the pa'po'va virus group, the name being derived from the first two letters of each virus name: *papilloma, polyoma, vacuolating*, in the order in which the viruses became known. *Ibid.* 1129 The papova viruses contain DNA. **1969** S. T. Lyles *Biol. Microorganisms* xxvi. 536 Papovaviruses characteristically produce warts in human infections; the most common type is the *verruca vulgaris*, called the common wart. **1973** R. G. Krueger et al. *Introd. Microbiol.* xxi. 562/1 Members of the papovavirus group can produce tumours readily when injected into tissues of immunologically undeveloped neonatal animals and with difficulty when injected into older animals. **1977** *Proc. R. Soc. Med.* LXX. 393/2 Dulbecco (1976) has recently considered evidence suggesting that cellular mutations may be needed for the full expression of cell transformation by papova viruses.

Pappenheimer (pæ·pĕnhəi·məɹ). *Med.* [The name of Alwin M. *Pappenheimer* (b. 1908), U.S. biochemist, who described such bodies in 1945 (*Q. Jrnl. Med.* XXXVIII. 75).] *Pappenheimer('s) body*: a siderosome that stains with Romanowsky's or Wright's stain.

1947 McFadzean & Davis in *Glasgow Med. Jrnl.* XXVIII. 238 Preparations of the peripheral blood of this case were sent to Dr. Pappenheimer, who..confirmed their apparent identity with those studied by him... We shall in this paper refer to these inclusion bodies as Pappenheimer's bodies. *Ibid.* 255 It would appear that the siderotic granules described by Grüneberg..and by Doniach *et al.*..are probably similar to the Pappenheimer bodies. **1966** J. W. Linman *Princ. Hematol.* iii. 63 Siderocytes are red cells that contain one or more hemosiderin granules; these inclusion bodies..are usually demonstrable only with special stains, such as Prussian blue. Occasionally they are associated with sufficient basophilic material to be evident with Romanowsky stains and are then referred to as Pappenheimer bodies. **1972** W. J. Williams et al. *Hematol.* viii. 80/2 Siderosomes staining with Wright's stain have been called Pappenheimer bodies.

pappy, *sb.*[1] Add: Also **pappie.** Delete 'Now *rare*' and add further examples.

1909 Joyce *Let.* 29 July (1966) II. 230 All are delighted with Georgie, specially Pappie. **1918** —— *Exiles* 20 Do you want to speak to my pappie? **1929** W. Faulkner *Sartoris* iii. iv. 206 Whut you hear, pappy? **1962** L. Deighton *Ipcress File* xx. 130 My pappy used to say, 'Drink Scotch by itself, with rye mix a little water.' **1963** M. Duggan in C. K. Stead *N.Z. Short Stories* (1966) 109 Fanny never chattered much and less than ever in the presence of her pappy. **1978** D. Bagley *Flyaway* xxix. 280 He wanted to find his Pappy's airplane.

paprika (pæ·prikă, pa·prikă, pāprī·kă). [Hungarian, f. Serbo-Croat *pàpar* pepper (see H. H. Bielfeldt 1965, in *Sitzungsber. d. deutsch. Akad. d. Wissensch. zu Berlin: Klasse für Sprachen, Literatur u. Kunst* I. 20).] **1.** A condiment made from the dried, ground fruits of certain varieties of the sweet pepper, *Capsicum annuum*. Also *attrib.* and *fig.*

1896 J. T. Law *Grocer's Manual* 521/1 Paprika, or Hungarian Red Pepper. **1897** *Sears, Roebuck Catal.* 9/3 Pure Ground Spices..Paprika—Hungarian (¼lb boxes), sweet pepper. **1898** C. H. Senn *Culinary Encycl.* 70 *Paprika*, Hungarian red pepper. A kind of sweet capsicum of a brilliant scarlet colour. **1908** *Daily Chron.* 29 Apr. 9/6 Beat together, adding oil every two minutes; paprika to taste. **1918** A. Quiller-Couch *Foe-Farrell* 91 You rubbed a soupsong of garlic into them with three drops of paprika. **1930** *Time & Tide* 14 Feb. 206/2 The Master of the pig-stick had a face and a temper as scarlet as paprika. **1962** H. T. Moore *Coll. Lett. D. H. Lawrence* I. p. x, He [sc. Lawrence] often sprinkled in the paprika of gossip. **1974** *Times* 14 Mar. 11/5 Prawn cocktail... Garnish with a sprinkling of paprika pepper. **1976** *Times* 3 Mar. 13/6 A touch of paprika in the finale was not allowed to lessen appreciation of contrapuntal cunning.

2. One of several European varieties of the sweet pepper, *Capsicum annuum*, bearing mildly flavoured fruits.

1925 J. A. Hammerton *Countries of World* V. 3586/1 Onions, garlic, paprika (Turkish pepper), beans and cabbage are common [in Serbia]. **1941** 'R. West' *Black Lamb* II. 55 Tomatoes and paprikas glowed their different reds. **1960** E. David *French Provincial Cooking* 96 Red or green sweet peppers, also called capsicums. The pimientos of Spain, paprika peppers of Hungary and peperoni of Italy. **1969** *Oxf. Bk. Food Plants* 128/1 In general, the term 'paprika' is applied to European types [of sweet pepper] with large mild fruits; Spanish paprikas are called pimiento. **1978** *Times* 16 Mar. 25/4 The paprikas grow freely [under glass] as though Holland were a tropical country.

3. The orange-red colour of paprika. Also *attrib.*

1934 in Webster. **1938** L. Bemelmans *Life Class* ii. vii. 191 The little rough-haired dachshund, almost paprika red. **1972** *Guardian* 11 Aug. 9/6 The colour combinations are..lemon with navy/lemon plaid, and paprika with wine/yellow plaid. **1976** *National Observer* (U.S.) 22 May 19/3 (Advt.), Any body looks better in our ribknit turtleneck... One size fits all men. Colors: Natural, Yellow, Paprika, Forest Green, Light Blue, Navy.

4. Used *attrib.* to designate various dishes flavoured with either the condiment or the vegetable.

1935 M. Morphy *Recipes of All Nations* 337 Among the most famous of all Hungarian dishes are..their paprikas ..being divided into pörkölt, the paprika dish without sour cream, and the other being paprika dishes containing sour cream. *Ibid.* 345 Pork chops with paprika sauce. *c* **1938** *Fortnum & Mason Price List* 26/2 Paprika Chicken—per tin 2/6. **1963** *Guardian* 25 Jan. 10/6 An excellent Viennese meal which included paprika chicken (strictly speaking a dish of Hungarian antecedents). **1966** *Harrods Food News* Sept. 2/1 Paprika Goulash—2 [Portions] 11/6. **1968** R. V. Beste *Repeat Instructions* xiv. 145 He ordered Bulgarian salata, paprika chicken and a bottle of Riesling. **1969** R. & D. De Sola *Dict. Cooking* 168/1 *Paprika butter*, butter sauce colored and flavored with paprika. **1977** K. Benton *Red Hen Conspiracy* ix. 53 The table was set with dishes..ranging from the delicate flesh tones of Parma ham, the rusty scarlet of paprika sausage, [etc.].

paprikahuhn (pāprī:kăhū·n). Also (pl.) -**hühner.** [G., = paprika chicken.] An Austrian dish, perhaps of Hungarian origin, consisting of poached chicken in a rich cream sauce flavoured with paprika.

1905 K. Baedeker *Austria-Hungary* p. xvii, Some of the Austrian dishes have curious names;..'Gulyás', Hungarian baked meat, peppered; 'Paprikahuhn', fowl prepared in a similar way. **1906** [see *goulash 1]. **1957** S. Strong *Good Food from Vienna* 112 Chicken Paprika (Paprika Hühner)... This is a typical Viennese dish deriving from Hungary and should on no account be missed. **1965** R. Philpot *Viennese Cookery* 95 Paprikahuhn... Get a good spring chicken..two ounces of butter, a medium-sized onion,..very mild paprika,..tomatoes,..chicken stock, double cream. **1969** B. Sias *Chicken Cookbook* 426 Paprika Huhn—Paprika Chicken—Austria's most famous chicken dish, even though Hungary has a claim on it. **1975** *Times* 9 Aug. 8/5 The *Paprikahuhn* proved as bland as baby food.

† **Papua** (pæ·pūă, -iŭă). Also **Papoo, Papu.** *Obs.* [ad. Malay *papuah, pĕpuah* frizzled.] = *Papuan *sb.* 1.

1619 W. Phillip tr. *Schouten's Relation of Voiage from Straights of Magelan* 62 Wee thought those people to be Papoos, for all their haire was short, and they eate Betell and Chalke mingled with it. *Ibid.* 66 A kind of ill favoured people, all Papoos, their haire short, and curled, having rings in their noses and eares, and..hogs teeth hanging about their neckes. **1684** Locke *Jrnl.* 20 Apr. in K. Dewhurst *John Locke* (1963) 242 Amongst the draughts he had of severall Esterne people in their natural colours and habits were the Kakerlacks and Papu's. *Ibid.*, The Papua was an olive coloured man with a taile. **1840** *Penny Cycl.* XVII. 219/2 Along the coast of the western half of the island are the Papuas, who have received that name from the Malays, in which language the word signifies 'frizzled hair'. **1845** *Encycl. Metrop.* XXV. 239/1 The Papúas of New Guinea..seem to approach more nearly to the make and strength of the African. **1885** *Encycl. Brit.* XVIII. 231/1 New Guinea..and other islands peopled by Papuas.

Papuan (pæ·piŭăn), *sb.* and *a.* [f. *Papua* (see prec.), formerly a name for the island of New Guinea and later for a territory consisting of its south-eastern part (now incorporated in the state of Papua New Guinea, independent since 1975): see -an.] **A.** *sb.* **1.** A native or inhabitant of Papua (or Papua New Guinea); also, a member of the racial type found there.

1814 J. Maver tr. *J. Martinez De Zúñiga's Hist. View Philippine Islands* I. p. xii, It is generally allowed that the language spoken by the Papuans, Samangs, and Negritos of the Philippines, and adjacent islands, is totally different from the Malayan. **1869** A. R. Wallace *Malay Archipelago* II. xl. 445 In stature the Papuan decidedly surpasses the Malay. **1876** *Encycl. Brit.* V. 790/2 The rite of circumcision..is still kept up..among the Papuans. **1902** *Chambers's Jrnl.* May 287/2 [With] the Negro..he throws in the Papuans and Malays, who have black or olive skins. **1913** J. G. Frazer *Belief in Immortality* I. ix. 190 The Papuans, a tall, dark-skinned, frizzly-haired race, inhabit apparently the greater part of New Guinea, including the whole of the western and central portions of the island. **1954** M. K. Wilson tr. *Lorenz's Man meets Dog* p. ix, Even highly civilized peoples..were accustomed to treat their prisoners no better than domestic animals... The Papuans eat them even to-day with excellent appetite. **1975** J. van de Wetering *Outsider in Amsterdam* (1976) i. 12 Suddenly de Gier knew what this man had to be. Not a Negro but a Papuan.

2. The Papuan group of languages.

1925 H. Murray *Papua of Today* ii. 33 The Territory shows even a greater variety of languages. These languages are classified as Papuan and Melanesian. **1939** L. H. Gray *Foundations of Lang.* 388 The Dravidian family seems to be isolated within India, all attempts to connect it either with Uralic, Altaic, Elamite, Subaraean, Burushaskī, Andamanese, Australian, or Papuan having proved unsuccessful. **1949** M. Mead *Male & Female* 416 The Arapesh are a Papuan-speaking people. **1960** C. Winick *Dict. Anthropol.* 401/1 *Papuan*, a New Guinea linguistic stock, non-Melanesian or non-Austronesian. **1978** *Language* LIV. 467 This vast area contains two well-established language families (Austronesian and Australian), as well as a bewildering congeries of seemingly diverse speech communities centering in New Guinea which until recent years were known only under the negative collective label 'Papuan' (= non-Austronesian, non-Australian).

B. *adj.* **1.** Of, pertaining to, or characteristic of Papua (or Papua New Guinea) or its inhabitants.

1869 A. R. Wallace *Malay Archipelago* II. xl. 449 These people..are tall and well-made, with Papuan features, and curly hair. **1875** *Encycl. Brit.* III. 739/1 The Papuan Subregion..comprises, besides the large and imperfectly-known island whence its name is derived, three other provinces, which may be named the Timorese, the Celebesian, and the Moluccan. **1930** A. Huxley *Let.* 7 Jan. (1969) 326 The way every trace of beauty, originality, charm, nobility, existing in the various indigenous arts and crafts—from Papuan and Melanesian to Chinese and Indian—had been utterly stamped out. **1957** P. Worsley *Trumpet shall Sound* v. 98 Its leader..declared that he was charged by God to convert his Papuan fellows. **1973** A. Behrend *Samarai Affair* xii. 124 A small roundish lump now shrunken and indeterminate but once the head of a Papuan tribesman.

2. *spec.* Of or pertaining to a group of non-Austronesian languages spoken in Papua (or Papua New Guinea).

1885 *Encycl. Brit.* XVIII. 780/1 Still less known are the Papuan or Negrito languages, belonging to the black race with frizzled hair inhabiting most of New Guinea, and found also in the interior of some of the other islands having been driven from the coasts by superior intruders of the Malay race. **1908** T. G. Tucker *Introd. Nat. Hist. Language* 145 The Papuan languages, in New Guinea and some smaller islands, breaking the geographical continuity of the Malayo-Polynesian family. **1912** J. H. P. Murray *Papua* v. 135 The languages of the islands at the east end of Papua are all classed as Melanesian, with the exception of Rossel Island, the language of which is considered to be Papuan. *Ibid.* v. 137 The language is..unmusical in sound... It seems to be a Papuan language. **1925** —— *Papua of To-Day* ii. 33 The Papuan and Australian languages meet,..in Torres Straits. **1933** L. Bloomfield *Language* iv. 71 The other families of this part of the earth have been little studied; the Papuan family, on New Guinea and adjacent islands, and the Australian language. **1943** *Official Handbk. New Guinea* v. 343 The Papuan languages..are said to differ as much among themselves as do the languages of the Indo-European group. **1957** *Oceania* XXVIII. 159 There is a distant genetic relationship between a group of Papuan (non-Austronesian) languages in the Vogelkop peninsula of Netherlands New Guinea and the equally non-Austronesian languages of North Halmahera. **1971** Laycock & Voorhoeve in *Current Trends in Linguistics* VIII. 509 It is rare for speakers of Papuan languages to have a name for themselves. **1975** *Encycl. Papua & New Guinea* II. 610/2 The Papuan languages are not a single family.

papulate (pæ·piulĕt), *a.* [f. Papula + -ate[2] 2.] = Papulated *a.*

1876 J. S. Bristowe *Treat. Theory & Pract. Med.* ii. ii. 318 Not unfrequently these patches are papulate,.. gyrate, or marginate.

papyro-. Add: **papyro·grapher,** a writer on papyrus; **papyrolo·gical** *a.*, pertaining to or dealing with papyrology; **papyro·logist,** a student of papyrology.

c **1904** *Encycl. Dict.* Suppl., Papyrological. **1906** J. H. Moulton *Gram. N.T. Greek* I. 159 In the less educated papyrographers we find blunders of this kind. **1922** *Glasgow Herald* 14 Apr. 5/2 A most helpful Guide prepared..by the well-known papyrologist Mr. H. I. Bell. **1925** H. S. Jones in *Liddell & Scott's Greek-Eng. Lexicon* (1940) I. Pref. p. viii, Mr. H. Idris Bell..has supplied valuable notes on recent papyrological publications. **1939** A. Toynbee *Study of Hist.* VI. 74 The benefit of half a century of papyrological enterprise and ingenuity. **1968** *Sunday Times* 25 Aug. 3/5 He spent his winters as a young papyrologist excavating the rubbish-mounds of Græco-Roman cities in Egypt for the significant 'waste-paper' of a forgotten civilization. **1977** *Times* 28 Oct. 14/3 A famous lost play of antiquity has been discovered ..among the Oxyrhyncus papyri,..by Professor Eric Turner, the great papyrologist.

par, *sb.*[1] Add: **2. b. mint par**: see *mint *sb.*[1] 6. **c.** (Earlier and later examples.) Also **no par,** having no face value.

1848 W. Armstrong *Stocks* 5 The par value of any stock is that proportion of the capital stock which it represents [etc.]. **1952** *Economist* 27 Dec. 904 No par value shares were not endorsed by the Cohen Committee when it studied the company law nearly ten years ago, and Sir John Barlow's private Bill, which sought to make no par value shares permissible, went virtually unsupported by the City and by the Government. *Ibid.*, The n.p.v. share is distrusted on the Left,..simply because it tells the truth about an equity share in a way that the share with a nominal or unrealistic par value never could. **1960** Nanassy & Selden *Business Dict.* 142 *No par value*, refers to stock issued with no par value printed on the face. Each share represents a fractional part of the total value of the business. **1964** *Financial Times* 25 Feb. 5/2 The shares will have a par value of Kr. 100 each. **1973** D. Westheimer *Going Public* i. 18 There's not a high degree of relationship between par value and what a share of stock will bring on the market.

3. b. (Further examples.)

1776 H. Newdigate *Let.* in A. E. Newdigate-Newdegate *Cheverels* (1898) i. 11 As to my Spirits they are rather above than below par. *c* **1793–4** Jane Austen *Lady Susan* (1925) 111 Sir James is certainly under par. **1880** Geo. Eliot *Let.* 1 Aug. (1956) VII. 308 These conditions found him a little below par from long protracted anxiety and excitement... But..he has been getting strong again. **1934** G. B. Shaw *Too True to be Good* i. 31 There is nothing constitutionally wrong. A little below par: that is all. **1940** Wodehouse *Quick Service* x. 104 Mrs. Chavender's Pekinese..had woken up that morning a little below par, and Sally was driving her and it to the veterinary surgeon in Lewes. **1958** A. Huxley *Brave New World Revisited* (1959) viii. 99 Whenever anyone felt depressed or below par, he would swallow a tablet or two of a chemical compound called Soma.

4. Substitute for def.: The number of strokes which a first-class player should normally require for a hole or course, calculated from the length of the holes with two putts for each green, and in some cases taking account of difficulties and obstacles in the course. Add further examples. Also *fig.*, **par for the course** (see quot. 1961[1]).

1924 J. Braid *Golf Guide* 164 *Par Play*, perfect golf without flukes. Thus, if a green can be reached in two strokes, the hole is a Par four; two putts being allowed on each green. **1947** *Partisan Rev.* XIV. 363 Nancy had married and moved to San Francisco and had had three children immediately. 'Par for the course,' said Seymour to Jasper. **1957** *Encycl. Brit.* X. 507/1 Distance is the chief factor in determining the par for a hole. Following are the divisions: all distances up to 250 yd., par 3; 251 to 445, par 4; 446 to 600, par 5; more than 600, par 6. **1961** Partridge *Dict. Slang* Suppl. 1213/2 *Par for the course, that's (just) about*, that's pretty normal; that's what, after all, you can expect. Canadian c.p.: since ca. 1946. Ex golf. **1961** M. Beadle *These Ruins are Inhabited* (1963) x. 137 While waiting..I caught a fragment of another subscriber's telephone conversation. This is also par for the course in making an Oxford phone call. **1973** A. MacVicar *Painted Doll Affair* viii. 96 Let's see if you can still keep shooting all these pars and birdies. **1974** *Encycl. Brit. Macropædia* VIII. 250/2 Par is essentially a U.S. term that came into use in the early 1900s as a base for computing handicaps under the system devised by L. Calkins of Plainfield, New Jersey. **1975** A. Bergman *Hollywood & Le Vine* (1976) ix. 129 I'm not sure about his sex life... But that's par for the course out here, you don't even give it a second thought. **1976** *Scotsman* 15 Dec. 19/8 A perfect approach gave him another birdie at the thirteenth and with pars at the other inward holes he set a target which was not successfully challenged all day. **1977** *Times* 23 Mar. 16/8 *Mail* readers ..will, it is true, be getting their news a day later than you who take *The Times*, but that is about par for the course.

par, *sb.*[4] (Earlier and later examples.)

1844 E. L. Blanchard *Diary* 27 Sept. in Scott & Howard *Life E. L. Blanchard* (1891) I. 36 Wrote some little pars for Alderton about 'screw Penholder'. **1854** Geo. Eliot *Let.* 17 May (1954) II. 155, I intend to have the first sentence or par[agraph] in every § of the appendix in italics. **1928** D. L. Sayers *Unpleasantness at Bellona Club* i. 9, I am ready to sacrifice my nearest and dearest in order to curry favour with the police and get a par. in the papers. **1969** *Daily Tel.* (Colour Suppl.) 31 Oct. 21/2 A story that the *Daily News* would splash might make only a couple of pars well back in the [*New York*] *Times*. **1973** K. Giles *File on Death* v. 119 There was a par. in the evening papers. **1976** *Listener* 2 Dec. 712/3 My business [as television reviewer] is not to bore the readers, to get 'em in the first par and bounce 'em with a last par they'll remember.

par, *v.*[2] Restrict *rare* to sense in Dict. and add: **2.** *Golf.* To complete (a hole or a course) with a score equal to par.

1961 Webster s.v., *Par.., to make a golf score on (a hole) equal to par. **1974** *Spartanburg* (S. Carolina) *Herald* 19 Apr. B5/1 Heard, who parred the course Wednesday, said he was driving badly 'but I chipped and putted very well.' **1976** *Scotsman* 24 Dec. (Weekend Suppl.), We won every par five we parred. **1977** *Evening Post* (Nottingham) 24 Jan. 16/6 He parred the next nine holes before his second bogey of the day at the 18th.

par, *prep.* Add: **2.** *par exemple* (examples); *par parenthèse* (earlier examples).

1791 A. S. Damer *Let.* 18 Aug. in 'L. Melville' *Berry Papers* (1914) 63 They have seen her, and..admire her talents, and, *par parenthèse*, I do really believe that he means to marry her. **1847** in F. A. Kemble *Rec. Later Life* (1882) III. 264 There are a few expressions I should like to have stricken out of it, *par exemple*, I hate the word *stink*. **1847** C. Brontë *Jane Eyre* I. xii. 204 This, *par parenthèse*, will be thought cool language. **1853** Thackeray *Let.* in H. Ritchie *Lett. A. T. Ritchie* (1924) iv. 49 (This is par parenthèse). **1857** C. Kingsley *Two Yrs. Ago* I. p. ix, 'You shall see enough to-day..*Par exemple*—' And Claude pointed to the clean large fields. **1863** Geo. Eliot *Let.* 4 Dec. (1956) IV. 118 Miss Hennell is staying there and writes me word that Miss Remond is a fellow guest lecturing (par parenthèse) on Slavery. **1867** H. James *Let.* 22 Nov. in R. B. Perry *Tht. & Char. W. James* (1935) I. 251 Tonight, *par exemple*, I am going into town to see the French actors. **1878** —— *Europeans* I. iii. 111

'Ah, par exemple!' cried the young man. 'You deserve that I should never leave you.' **1889** E. Dowson *Let.* 30 Jan. (1967) 30 Is he *the* Coquelin, par exemple or is he another? **1916** E. Pound in *Lett. J. Joyce* (1966) II. 375 And par exemple, the 'practical' Pinker was able to do less than I was.

para¹. (Further examples.)

1907 [see *dinar b]. **1935** H. Edib *Clown & his Daughter* xvii. 90 'Rabia Abla, ten paras' worth of chewing-gum!' shouted a shrill voice from the street. **1960** O. Manning *Great Fortune* I. 12 He took the coins from his pocket... They comprised a few *lire*, *filler* and *para*. **1971** *Daily Tel.* 18 Sept. 7/7 The first stamps issued in 1941 took the form of the Yugoslavian issues overprinted 'NEZAVISNA DRZAVA HRVATSKA' (Independent State, Croatia). Denominations ranged from 50 paras to 5·50 dinars. **1971** *Whitaker's Almanack 1972* 988 Dinar of 100 Old Dinars or 100 Paras.

Pará². Add: Also **Para** and in some collocations **para.** The seaport is now usu. known as Belém, *Pará* being the name of the state in which it is situated.

Pará grass, (*b*) substitute for def.: a forage grass, *Panicum purpurascens*, native to Brazil but widely cultivated in tropical or sub-tropical regions; (later examples); **Pará-nut** (later examples); **Pará rubber** (earlier and later examples).

1916 L. H. Bailey *Stand. Cycl. Hort.* V. 2453/1 Pará-Grass... Intro[duced] from Brazil. *P[anicum] numidianum*, Lam., is a closely related species of the E. Indies, sometimes confused with the true para-grass. **1929** J. W. Bews *World's Grasses* vi. 230 'Para grass' (a perennial, with stout stolons, as much as 15 feet long..)..is cultivated for forage. **1958** J. Carew *Black Midas* iv. 65 Here and there amidst lotus lilies, reeds or paragrass were alligator's eyes. **1968** E. Lovelace *Schoolmaster* xiv. 221 Silence, and the many fingers of para grass at the roadside ..gesturing skyward. **1973** Tothill & Hacker *Grasses S.E. Queensland* I. 17 Para grass..is an introduced pasture grass which is planted in wet places. **1884** *Encycl. Brit.* XVII. 761/1 Para-nut or Brazil-nut oil, yielded by the kernels of *Bertholletia excelsa*, is employed in South America as a food-oil and for soap-making. **1931** B. Miall tr. *Guenther's Naturalist in Brazil* iv. 77 It [*sc.* the sapucaja] yields..edible fruits..whose nuts, known to the trade as Pará-nuts, appear on our Christmas dinner-tables as Brazil-nuts. **1857** T. Hancock *Personal Narr. Caoutchouc* 281 (Index), Para rubber. **1860** *Chem. News* 25 Aug. 125/1 The Para rubber, which is of a superior quality, is generally sent in the shape termed bottle rubber. **1947** J. C. Rich *Materials & Methods of Sculpture* v. 98 Clarke states that the rubber cement can be made by dissolving ½ ounce of caoutchouc (para rubber) in 25 ounces of benzene. **1968** A. S. Craig *Dict. Rubber Technol.* (1969) 112 Para rubber was the best variety of all wild rubber but the advent of plantation rubber steadily reduced its importance until it is now of little significance in world rubber production.

b. Used *absol.* for *Pará rubber*.

1897 [in Dict.]. **1922** [see *overvulcanize v.]. **1954** H. J. Stern *Rubber* i. 17 Apart from some domestic consumption the wild rubber of South America is now of small commercial importance, although the so-called 'fine hard Para' is still favoured in some quarters. **1963** A. S. Craig *Rubber Technol.* iii. 18 As late as 1920, the best quality of Para (pa-rá) rubber (known as 'Fine Hard Para') was the standard by which the newer plantation rubbers were judged.

para³ (pā·rā). [Maori.] A New Zealand name for the large, evergreen fern, *Marattia salicina*, or its swollen rhizome, formerly used as food.

1855 J. D. Hooker *Bot. Antarctic Voy.: Flora Novæ-Zelandiæ* II. 49 *Marattia salicina*... Northern and eastern parts of the Northern Island... Nat[ive] name, 'Para',.. (Cultivated at Kew.). **1890** H. C. Field *Ferns of N.Z.* 153 *Marattia fraxinea*..'Para', 'Para reka', or 'Para tawhiti' of the Maoris. 'Horse-shoe fern' of Europeans. **1906** T. F. Cheeseman *Man. N.Z. Flora* 1026 Para; Parareka... The large starchy rhizome was formerly eaten by the Maoris, and hence the plant was occasionally cultivated near their villages. It is now fast becoming rare. **1921** [see *king fern* s.v. *king sb. 13 c]. **1946** *Jrnl. Polynesian Soc.* LV. 149 If there is no distinguishing suffix para is understood to mean the fern-tuber [of *Marattia fraxinea*].

para⁴ (pæ·rā). Abbrev. of Paragraph *sb.*

1859 J. Blackwood *Let.* 18 Apr. in *Geo. Eliot Lett.* (1954) III. 52 We had better set a paragraph afloat... If you send a para(graph) to me here I will set it afloat among the Edinr. papers. **1885** R. Kipling *Let.* 26 Sept. in C. E. Carrington *Rudyard Kipling* (1955) iv. 70 How am I to tackle your letter... Para. two from the butt end asks me if I know *The City of Dreadful Night*. **1938** 'G. Orwell' in *New English Weekly* 9 June 169/1 Casual half-inch paras in every issue of the newspapers. **1951** Wodehouse *Old Reliable* x. 123 There is a morality clause in my contract..Para Six. **1972** 'G. Black' *Bitter Tea* (1973) viii. 124 After this 'Dealer' para the news of your sunk ship could push them to a decision.

para⁵ (pæ·rā), *a.* (*adv.*) [f. Para-¹.] **1.** *Chem.* (Now usu. italicized.) Characterized by or relating to (substitution at) two opposite carbon atoms in a benzene ring; at a position opposite *to* some (specified) substituent in a benzene ring. Also as *adv.*

1876, etc. [see *ortho a. (adv.) 1]. **1903** A. J. Walker tr. *Holleman's Text-bk. Org. Chem.* II. 446 There remains no possibility, except the *para*-structure, for the third hydroxybenzoic acid melting at 210°. **1938** L. F. Fieser in H. Gilman *Org. Chem.* I. ii. 132 The para coupling of a free phenol is regarded as a 1,4-addition to the conjugated system of the nucleus, followed by loss of water. **1949** [see *orient v. 4 a]. **1968** R. O. C. Norman *Princ. Org. Synthesis* xii. 402 The inductive effect is relayed through one more carbon atom than is the case for *ortho* or *para* substitution. **1972** R. A. Jackson *Mechanism* ii. 12 Explanations based on the resonance effects of the methyl group do not..explain the more pronounced effect of *meta* compared with *para* substitution.

2. *para* (or *Para*) *red*, any of various dyes that consist chiefly of the coupling product of diazotized paranitraniline and β-naphthol and are used in printing inks and paints.

1907 *Jrnl. Soc. Dyers & Colourists* XXIII. 20/2 Para red discharges on indigo have been produced for the last ten years. **1930** A. W. C. Harrison *Manuf. Lakes & Precipitated Pigments* xii. 163 When Para red is present in old water paint on a wall surface, it is again best to remove the old material. **1967** [see *fire-red sb.].

3. See *para-¹ 3.

para⁶ (pæ·rā). *Obstetr.* [the ending of *nullipara, primipara, multipara*.] A woman who has had a specified number of confinements, as indicated by a preceding or following numeral.

1881 *Trans. Edin. Obstetr. Soc.* VI. 70 Of the 48 cases, 26 were primiparæ and 22 multiparæ, as follows: ii. paræ, 11; iii. paræ, 4; [etc.]. **1908** *Practitioner* Aug. 312 Fromme records the case of a vi-para, aged 34, who developed pyæmia after an abortion. **1923** *Jrnl. Obstetr. & Gynæcol.* XXX. 568 In one patient, a iii-para,..the second stage of labour occupied 1½ hours. **1950** *Amer. Jrnl. Obstetr. & Gynecol.* LIX. 737 The second maternal death occurred in a 40-year-old, para ii, gravida iv, whose diabetes was of two years' standing. **1966** *Fertility & Sterility* XVII. 336 A 24-year-old para 2 who had menstrual irregularity prior to insertion of the spiral. **1967** [see *multipara]. **1977** *Lancet* 23 Apr. 910/1 A 36-week gestation 2·2 kg Black male infant was born to a 36-year-old gravida 7, para 5 mother by vaginal delivery.

para⁷ (pæ·rā). Abbrev. of *paratrooper.

In early quots. a. Fr. *para*, abbrev. *parachutiste*.

1958 *Spectator* 20 June 807/2 This has not greatly endeared him to the 'paras'. **1962** A. Buchwald *How Much is that in Dollars?* 15 My son assured me paras could never land in the Parc Monceau. **1966** M. Catto *Bird on Wing* ii. 24 Louis..had been a captain in the *paras*. He had learned certain things in Algeria. **1967** L. Forrester *Girl called Fathom* xii. 148 Commandant Daniel Jules Delavigne, late of the Paras—Indo-China, Algeria. **1972** *Listener* 9 Nov. 625 The First Battalion of the Parachute Regiment pulls out of Northern Ireland at the end of the month... Incidents like Bloody Sunday..have earned the Paras a reputation for toughness. **1973** *Ibid.* 26 Apr. 534/1 A gun battle between the Paras and the Provos. **1977** J. Cartwright *Fighting Men* vii. 95 Right, paras get ready to jump.

para⁸ (pæ·rā). Slang abbrev. of Paraplegic *a.*

1961 Partridge *Dict. Slang* Suppl. 1213/2 *Para*. 2. A *para*plegic (a spinal-cord paralytic): Canadian doctors' and nurses': since ca. 1946. **1969** *Sun* (Melbourne) 18 Apr. 7/3 I'd like to say it's a disgrace that quadras (quadraplegics) and paras (paraplegics) have to wait so long before courts get around to clearing up the mess.

para-¹. Add: **1.** (The less important terms beginning with this prefix, other than those in *Chem.*, are placed below; the description under this sense in Dict. is not applicable to all of them.) **parabro·nchus** *Zool.*, any of the minutest ramifications of the bronchi in the lung of a bird; **parace·llular** *a.*, passing or situated alongside and between cells; **para-ce·rvical** *a.*, pertaining to or designating the region surrounding the cervix; hence **para-ce·rvically** *adv.*; **para-church** (see quot. 1970); **paraconfo·rmity** *Geol.* = *non-sequence; **parafi·scal** *a.*, ancillary to or containing elements not usually regarded as fiscal; **pa:rageosy·ncline** *Geol.*, (*a*) a geosyncline situated at the edge of a continental kratogen (craton) (? *obs.*); (*b*) a geosyncline situated within an older kratogen (craton); [in sense (*b*) ad. G. *parageosynklinale* (H. Stille 1935, in *Sitzungsber. d. preuss. Akad. d. Wissensch.* (*Phys.-mat. Kl.*) 182)]; hence **pa:rageosyncli·nal** *a.*; **paragnath** (pæ·rāgnæþ), **paragnathus** (pāræ·gnāþʊs) *Zool.* (usu. in pl. **-gnaths, -gnatha**) [Gr. γνάθ-ος jaw], (*a*) one of the pair of lobes forming the lower lip in most Crustacea; (*b*) one of the pair of lobes forming the hypopharynx in certain insects; (*c*) one of several paired, tooth-like scales found inside the mouth of certain annelid worms; **pa·ragneiss** *Petrogr.* [a. G. *paragneiss* (H. Rosenbusch *Elem. d. Gesteinlehre* (1898) 467)], gneiss derived from sedimentary rocks; **paragno·sis** [Gnosis], knowledge which is beyond that which can be accounted for by known methods; so **pa·ragnost,** a person possessing or allegedly possessing powers of clairvoyance or foreknowledge; **paragno·stic** *a.*; **paragra·mmatism,** the confused or incomplete use of grammatical structures found in certain forms of speech disturbance; so **paragramma·tic, -gramma·tical** *adjs.*; **parahippoca·mpal** *Anat.* [Hippocampus], a gyrus on the inferior surface of each cerebral hemisphere that posteriorly is continuous via the isthmus with the cingulate gyrus and anteriorly ends in the uncus; **parakerato·tic** *a. Path.*, affected by or symptomatic of parakeratosis; **paralexia** (further example); **paralitu·rgical** *a.*, parallel or ancillary to the liturgy; **parame·nstruum** [Menstruum], the period of eight days consisting of the first four days of each menstruation and the preceding four days; hence **parame·nstrual** *a.*; **parame·trial** *a.*, of or pertaining to the parametrium; **parametritis,** add: [coined in Ger. by R. Virchow 1862, in *Arch. f. path. Anat. u. Physiol.* XXIII. 416: see quot. 1869] (earlier example); **parametrium,** add: [back-formation from prec.] (examples); **paramnesia,** add: [ad. F. *paramnesie* (Lordat *Analyse de la Parole* (1843) 31]; now usu. = *déjà vu a; (examples); hence **paramne·sic** *a.*; **parana·sal** *a. Anat.*, situated beside the nose: the epithet of certain sinuses; **parano·tum** *Ent.* (pl. **-nota**) [Notum], in certain insects, a lateral expansion of the dorsal part of a thoracic segment; so **parano·tal** *a.*; **paraphasia** (further examples; see also quot. 1972); **paraphysical** *a.*, add: of or pertaining to physical phenomena for which no adequate scientific explanation exists; **parapoli·tical** *a.* (see quot. 1965); **parapsy·chic** *a.*, of or pertaining to mental phenomena etc., for which no adequate scientific explanation exists; also **parapsy·chical** *a.*; **parareli·gious** *a.*, parallel to, or outside, the sphere of orthodox religion; **parasagi·ttal** *a. Anat.*, situated adjacent or parallel to the sagittal plane; †**parasyna·psis** *Cytology*, the side-by-side pairing of chromosomes at meiosis; hence **parasyna·ptic** *a.*, **-syna·ptically** *adv.*; †**parasynde·sis** *Cytology* [ad. G. *parasyndese* (V. Häcker 1907, in *Ergebnisse und Fortschritte der Zool.* I. 74), f. Gr. σύνδεσις binding together] = *parasynapsis* above; hence **parasynde·tically** *adv.*; **paratecto·nic** *a. Geol.*, (*a*) accompanying deformation (? *obs.*); (*b*) [ad. G. *paratektonik* sb. (H. Stille *Einführing in den Bau Amerikas* (1940) i. 9)], formed by, or of the nature of, a deformation which is chiefly epeirogenic and produces relatively simple, broad folds such as those in Germany north of the Alps (believed to be characteristic of parageosynclines); cf. *orthotectonic* adj. s.v. *ortho- 1; **parate·rminal** *a. Anat.*, epithet of a strip of cortex in the rhinencephalon that lies immediately in front of the lamina terminalis at the anterior end of the third ventricle and superiorly is continuous with the indusium griseum; chiefly in *paraterminal gyrus* (or † *body*); **parathe·cium** *Bot.* [Thecium], in cup fungi and lichens, the outer, dark-coloured layer of an apothecium; so **parathe·cial** *a.*; **parathyroid,** add: [ad. mod.L. (*glandula*) *parathyreoidea* (coined in Sw. by I. Sandström 1880, in *Upsala Läkareförenings Förhandl.* XV. 466)]; freq. *attrib.* or as *adj.*, esp. in *parathyroid gland, hormone* (= *parathormone); (earlier and later examples); **para·tomous** *a.*, (*a*) (in Dict. as main entry); (*b*) *Zool.*, of or pertaining to paratomy; **para·tomy** *Zool.* [ad. G. *paratomie* (F. von Wagner 1890, in *Zool. Jahrbücher. Abth. für Anat.* IV. 393): see -tomy], in certain annelid worms, asexual reproduction in which new organs are developed before the division of the animal into two or more parts; **paratra·cheal** *a. Bot.*, describing the structure of wood in which the position of the parenchyma depends on that of the vessels; **paraventri·cular** *a. Anat.*, situated next to a ventricle: epithet of (*a*) a nucleus in the hypothalamus situated above the supra-optic nucleus, and (*b*) one of the mid-line nuclei of each thalamus.

1893 A. Newton *Dict. Birds* II. 522 Secondary Bronchi.., besides opening into Air-sacs, send off a number of radially-arranged parabronchia [*sic*], all of which extend to and end blindly near the surface of the Lungs. **1971** *Sci. Amer.* Dec. 75/1 The bird lung is perforated by the finest branches of the bronchial system, which are called parabronchi. **1900** G. Eisen in *Jrnl. Morphol.* XVII. 16, I designate as paracellular bodies numerous non-cellular bodies situated between the regular cells of the testes. **1977** *Lancet* 15 Jan. 139/2 During intestinal secretion considerable ion movement occurs by a paracellular route via lateral intercellular spaces and the so-called tight junctions rather than through the cells. **1922** R. T. Frank *Gynecol. & Obstetr. Path.* xii. 439 Three zones [of pelvic connective tissue spaces] are readily demonstrable—a para-vesical, para-cervical and para-rectal one. **1945** *Amer. Jrnl. Obstetr. & Gynecol.* L. 527 (*heading*) Paracervical anesthesia for the relief of labor pains. *Ibid.* 532 The injection of anesthetic solutions paracervically produces prompt relief from the pain caused by uterine contractions. **1977** *Lancet* 29 Jan. 260/1, I learnt my lesson whilst demonstrating to a colleague how simple is a termination of pregnancy using paracervical block as a local anæsthetic. **1970** *Guardian Weekly* 12 Dec. 14 Groups that don't attract or seek publicity, that meet in upper rooms... This is sometimes called the para-church, the church of the future which is beginning to take shape. **1976** *Church Times* 17 Dec. 6/3 The author shows that the 'underground' churches that sprang up in the late 1960s have rightly given place to a new form—namely the 'para-church', or alternative church—which exists alongside the institutional churches. **1957** Dunbar & Rodgers *Princ. Stratigr.* vi. 119/2 We propose to restrict the term *disconformity* to the third type, in which two units of stratified rocks are parallel but the surface of unconformity is an old erosion surface of appreciable relief, and to introduce a new term *paraconformity* for the fourth type, in which the beds are parallel and the contact is a simple bedding plane. **1975** *Nature* 3 Jan. 15/1 Here we use the term unconformity to refer to a significant gap (demonstrated or inferred) in the stratigraphic record (disconformity or paraconformity). **1968** *Economist* 30 Nov. 66/3 Either it would mean higher prices for French farmers..or else some parafiscal expedient to prevent this which would be a breach in the whole common price principle. **1974** B. Pearce tr. *Amin's Accumulation on World Scale* I. ii. 257 It is not practicable to take a share of their profits away from these enterprises by fiscal or parafiscal measures. **1978** *Guardian Weekly* 26 Mar. 12/1 Sums collected as parafiscal levies..by employer associations. **1956** L. V. de Sitter *Struct. Geol.* xxiv. 346 The blocks or nuclei sometimes became partly nuclear (parageosynclinal) basins, and partly remained continuously above sea level. **1961** *Jrnl. Geol.* LXIX. 650/2 Northern Sakhalin..was characterized during the Tertiary by parageosynclinal conditions. **1923** C. Schuchert in *Bull. Geol. Soc. Amer.* XXXIV. 199 These recording basins can not be grouped into any of the mentioned types of geosynclines, since some of them have oceanic depths, but all are actually a part of the Asiatic continent. They are marginal geosynclines or parageosynclines (geosynclines beside a continent). **1936** tr. H. Stille in *Bull. Amer. Assoc. Petroleum Geologists* XX. 853 Less intense orogenies..may take place in areas prepared by having been 'special basins' (parageosynclines) in regions that had become consolidated earlier. **1941** *Ibid.* XXV. 1403 The influx of orthogeosynclinal clastics into the Artinskian parageosyncline (in the sense of Stille) is comparable with the invasion of Ouachita-derived geosynclinal sediments into the base of the Strawn in the Oklahoman geosyncline. **1968** R. W. Fairbridge *Encycl. Geomorphol.* 446/2 A ring of Paleozoic basins ('parageosynclines') surrounds both the Canadian and the Scandinavian shields. **1888** Rolleston & Jackson *Forms Animal Life* (ed. 2) 170 The sides of the mouth [of the common crayfish] are overhung by the bases of the mandibles, and behind the latter are two small soft lobes united by the posterior margin of the mouth. These lobes are the paragnatha, metastoma, or lower lip. **1921** *Psyche* XXVIII. 86, I would claim that the so-called 'superlinguæ' of insects most emphatically do represent the paragnaths of Crustacea. **1952** R. C. Moore et al. *Invertebr. Fossils* xi. 454/2 (*caption*) Morphologic features of worms... Paragnaths. One or more pairs of minute denticulate distal plates. **1963** R. P. Dales *Annelids* ii. 43 Between this muscular part or 'pharynx' and the mouth [of nereids] is a membraneous buccal tube bearing small immovable teeth or paragnaths. **1902** A. Harker *Petrol.* (ed. 3) xxii. 331 All these have the chemical composition of sedimentary rocks; Rosenbusch styles them 'paragneisses', in contra-distinction to 'orthogneisses', which have the composition of, and are believed to represent, igneous rocks. **1932** Paragneiss [see *orthogneiss* s.v. *ortho- 1]. **1965** *Mem. Geol. Survey Dept. Malawi* No. 3 vii. 102 The dominant paragneisses in the hilly area around the Chaumbwi vent are quartzo-feldspathic granulites which occur as belts up to 1,100 yards wide. **1933** 'W. Carington' *Death of Materialism* viii. 193, I shall..use the words 'paragnosis', 'paragnostic' and the like; the point being that all the phenomena I have in mind..show signs of the possession or acquisition of knowledge (gnosis) which is, *prima facie*, beyond (para) what can be ascribed to the operation of classical law. **1946** G. N. M. Tyrrell *Personality of Man* v. 53 *Paragnosis*, awareness of additional to normal knowledge. **1964** J. H. Pollack *Croiset* (1965) i. 14 His mentor, Professor W. H. C. Tenhaeff, calls him a 'paragnost', a word which he coined in 1932. **1973** *Radio Times* 1 Nov. 67/4 More Things in Heaven and Earth... Gerard Croiset paragnost and healer. **1958** Goodglass & Hunt in Saporta & Bastian *Psycholinguistics* (1961) 449/1 Most authorities have distinguished between an 'agrammatic' form [of aphasia], marked by simplification and loss of grammatical detail, and a 'paragrammatic' form, marked by confused and incomplete, but not necessarily simplified constructions. **1962** Fish & Stanton tr. *Kleist's Sensory Aphasia* v. 71 'Then it is left had'..is paragrammatical. **1924** A. A. Brill tr. *Bleuler's Textbk. Psychiatry* xiii. 397 At times grammar fails them [*sc.* schizophrenics] (paragrammatism). Many words are used incorrectly. **1946** *Brit. Jrnl. Psychol.* XXXVII. 11 Paraphasia and paragrammatism are generally associated with the receptive ('temporal') syndromes. **1961** W. R. Brain *Speech Disorders* iv. 43 Sentence-deafness is characterized by a difficulty in understanding sentences and by 'paragrammatism' in expression, a term intended by Kleist to describe confusion in the use and order of words and grammatical forms. **1962** Fish & Stanton tr. *Kleist's Sensory Aphasia* v. 67 (*heading*) Sentence deafness and its abortive form with paragrammatism. **1958** *Gray's Anat.* (ed. 32) 1031 The parahippocampal gyrus..commences at the isthmus, where it is directly continuous with the gyrus cinguli, and passes forwards bounded on its lateral side by the collateral and rhinal Sulci. **1969** Truex & Carpenter *Human Neuroanat.* (ed. 6) xxi. 522/2 The lateral olfactory stria, the uncus, and the anterior part of the parahippocampal gyrus constitute the..pyriform lobe. **1972** M. L. Barr *Human Nervous Syst.* xiii. 213/1 The parahippocampal gyrus..hooks sharply backward as the uncus. **1943** *Arch. Dermatol. & Syphilol.* XLVII. 9 In an area above a large focal infiltration of the cutis the epidermis was thin and covered by a condensed parakeratotic horny layer. **1973** *Internat. Jrnl. Dermatol.* XII. 153/1 This histologic picture consists of a parakeratotic column that stains lighter than the adjacent stratum corneum on hematoxylin and eosin stains. **1950** *Jrnl. Speech & Hearing Disorders* XV. 291/1 Paralexia is defined as the substitution of any other word or words for the given symbol in reading. **1977** *Times Lit. Suppl.* 25 Feb. 225/1 Much hagiographical material was transmitted through the liturgy. Eventually, miracle plays based on saints' lives made their appearance as a paraliturgical halfway-house between ritual and drama. **1978** G. Wainwright in C. Jones et al. *Study of Liturgy* II. i. i. 38 Paraliturgical activities flourished: the Stations of the Cross, the rosary, the cult of the Sacred Heart. **1968** *Sunday Times* 29 Dec. 3 The para-menstrual failure rate in 'O' level candidates was 17 per cent. for girls whose menstrual loss lasted up to four days. **1966** K. Dalton in *Proc. R. Soc. Med.* LIX. 1015/2 *Paramenstruum* is used in this study for the four days immediately before menstruation and the first four days of menstruation. **1970** *Times* 30 Sept. 14 Recent studies have shown that in women half of all medical and surgical admissions to hospital occur during the paramenstruum. **1976** *Drive* Sept.–Oct. 31/2 The paramenstruum (the four days before menstruation and the first four days of blood-loss). **1903** *St. Louis Med. Rev.* XLVII. 449/2, I advised that the patient submit to examination under anesthesia, when the obstructing mass could be cleared away and the extent of the parametrial involvement approximately determined. **1962** J. W. Huffman *Gynecol. & Obstetr.* vi. 140/2 Cervical tears at labor may extend upward into the..parametrial tissues. **1869** J. M. Duncan *Pract. Treat. Perimetritis & Parametritis* 4 It is..to Virchow that I am indebted for the suggestion of the chief terms I propose to use habitually. Taking example from the heart and other organs, he proposes to use *peri* to imply inflammation of serous membrane, and he uses *para* to imply inflammation of [adjacent] cellular or connective tissue... *Perimetritis*, then, will strictly imply inflammation of the uterine peritoneum. *Parametritis* will imply inflammation of the cellular tissue in connection with the uterus. **1878** tr. *H. von Ziemssen's Cycl. Pract. Med.* VIII. 281 These abnormal conditions in the parametrium are described by various authors under different names..: for example, phlegmon of the pelvis (Erichsen), parametritis phlegmonosa (Virchow), and purulent oedema (Pirogoff). **1908** *Practitioner* Aug. 312 Nine days later rigors commenced, and in the right parametrium there was a diffuse infiltration, though thrombosed veins were not palpable. **1967** G. M. Wyburn et al. *Conc. Anat.* i. 43 The general condensation of tissue around the base of the broad ligament and lower end of the cervix of the uterus is known as 'parametrium'. **1888** *Amer. Jrnl. Psychol.* I. 735 Several philosophers..have noticed that illusions of memory occur in dreams; and, judging from the writer's own experience, such phenomena are not uncommon. Several dreams illustrative of paramnesia have come to my notice. **1941** *Mind* L. 323 However strong the feeling that this has all happened before, it may turn out that one is not remembering, but suffering from paramnesia, a feeling of *déjà vu*. **1961** J. Heller *Catch-22* (1962) xx. 202 The subtle, recurring confusion between illusion and reality that was characteristic of paramnesia fascinated the chaplain. **1897** *Mind* VI. 285, I frequently read a new poem with a vague sense of familiarity, but such an experience never puts on a really paramnesic character, as I quickly realise that it is explainable by the fact that the writer of the poem has fallen under the influence of Heine, or Tennyson, or Rosetti, as the case may be. **1963** *Lancet* 19 Jan. 164/2, I emerged from my paramnesic reverie to see the two attendants pursuing me across the hall. **1909** J. P. Schaeffer in *Univ. Pennsylvania Med. Bull.* XXII. 235/1 While making a study of the sinus maxillaris..my attention was called to some anatomical conditions which to my mind are of great importance in arriving at the etiology, diagnosis, and proper treatment of some affections of this paranasal chamber. **1954** L. B. Arey *Developmental Anat.* (ed. 6) xxvi. 528 Lodged within the adjoining bones, and in communication with the nasal cavity, are several irregular chambers known collectively as the paranasal sinuses. **1973** J. Davies in Paparella & Shumrick *Otolaryngology* I. iii. 166/2 The paranasal air sinuses comprise the maxillary, the ethmoidal, the frontal, and the sphenoidal sinuses. **1916** G. C. Crampton in *Jrnl. N.Y. Entomol. Soc.* XXIV. 8, I would refer to these lateral folds as the 'paranota', regardless of whether they are entirely tergal in origin, or entirely pleural, or a combination of both. The theories dealing with this origin of the wings may therefore be referred to as the paranotal theories. **1935** R. E. Snodgrass *Princ. Insect Morphol.* viii. 158 A third stage was inaugurated with the transformation of the paranotal lobes of the mesothorax and the metathorax into movable organs of true flight. **1964** R. M. & J. W. Fox *Introd. Compar. Entomol.* iv. 112 Although no insect, living or fossil, is known to have paranota that can be clearly interpreted as precursory to wings, paranota are present in certain living insects. **1973** *Nature* 16 Nov. 127/1 He [*sc.* G. C. Crampton] pronounced judgment unequivocally in favour of the paranotal theory. **1946** Paraphasia [see *paragrammatism* above]. **1959** Schuell & Jenkins in H. Schuell *Aphasia* (1974) xi. 212 Jargon and paraphasia were present. **1972** *Sci. Amer.* Apr. 78/2 Verbal paraphasia is the substitution of one word or phrase for another... Literal or phonemic paraphasia is the substitution of incorrect sounds in otherwise correct words. **1933** O. Lodge in T. Besterman tr. *Driesch's Psychical Res.* p. ix, For the experimental establishment of reality the paraphysical stand first, although they are admittedly on a lower plane and have less important consequences than the psychical variety. **1934** *Mind* XLIII. 255 The 'theories' of psychical research in their application to 'paraphysical' and 'parapsychical' phenomena. **1961** *Ann. Reg. 1960* 420 A lady had bought an instrument which was claimed by its maker to be capable of diagnosis and treatment on a para-physical plane. **1965** D. Easton *Framework Polit. Anal.* iv. 52, I shall refer to the internal political groups and organizations as *parapolitical systems* and retain the concept 'political system' for political life in the most inclusive unit being analyzed, namely, in a society. **1968** F. G. Bailey in M. J. Swartz *Local-Level Politics* (1969) xiii. 281 My hope..is a model for all kinds of politics in village India, and beyond that for politics in all parapolitical situations. **1971** P. A. Allum *Politics & Society Post-War Naples* (1973) vi. 166 Party and para-political organisation membership is a *sine qua non* of a successful candidature. **1911** H. Carrington in *Flournoy's Spiritism & Psychol.* i. 39 The most striking case of this character which I have encountered is that of Mme. Guelt, in which parapsychic gifts and tendencies were manifested in four generations of her family. **1930** D. Ibberson tr. *Oesterreich's Possession* II. vii. 207 Accounts of the parapsychic performances of the mediums are not susceptible of subsequent proof. **1918** D. Wright in *Boirac's Psychic Sci.* p. v, Unless we choose to coin a special word for the purpose, such as 'parapsychical', as suggested by Dr. Boirac. **1957** Rhine & Pratt *Parapsychol.* i. 5 The observations and events dealt with in parapsychology—parapsychical phenomena—are associated in some central way with *living* organisms. **1966** *New Statesman* 18 Feb. 229/2 These uncommitted people were unable to take up new religions, but they could adopt a set of parareligious dogma if it was called scientific. **1974** *Daily Colonist* (Victoria, B.C.) 31 May 7/6 What is usually involved in 'black magic' here is not really necromancy or witchcraft, but rather 'anteria', a parareligious form of African origin and related to Macumba in Brazil and voodoo in Haiti. **1909** *Cent. Dict.* Suppl., Parasagittal. **1925** *Jrnl. Compar. Neurol.* XXXIX. 200 (*caption*) Parasagittal section near the median line, showing the tractus olfacto-tuberalis and its connections. **1929** Brain & Strauss *Recent Adv. Neurol.* iv. 57 Parasagittal meningiomas..arise from the wall of the superior longitudinal sinus. **1969** D. Sutton *Textbk. Radiol.* lix. 1051/2 Parasagittal tumours being in the midline, and often bilateral, are better shown by encephalography than angiography. **1975** *Nature* 6 Mar. 48/2 Parasagittal crests relatively far apart and meeting posteriorly almost at posterior border of frontoparietal. **1909** E. B. Wilson in *Jrnl. Exper. Zool.* VI. 84 Pyrrochoris shows a close similarity to Tomopteris... This comparison has convinced me that synapsis occurs at the same period in both—whether by parasynapsis (side to side union) or telosynapsis (end to end union). [*Note*] I have for some years made use of these terms in my lectures on cytology. **1912, 1925** [see *parasyndesis* below]. **1932** *Proc. 6th Internat. Congr. Genetics* II. 319 Parasynapsis may be demonstrated..by observation of actual side-by-side association of homologous chromosomes or chromosome-segments at zygotene. **1956** *Biol. Abstr.* XXX. II. (Index), Parasynapsis. (*See* Chromosomes; Meiosis.) **1910** *Ann. Bot.* XXIV. 727 Grégoire.., while agreeing with the parasynaptic chromosome formation, put a different interpretation on to the 'gamosomes' and 'zygosomes' of Strasburger and Miyake. **1912** *Jrnl. Exper. Zool.* XIII. 394 Bivalent segments, each consisting of two chromosomes in parasynaptic union. **1921** *Ann. Bot.* XXXV. 386 Both the telosynaptic and the parasynaptic methods of synapsis may occur, the latter perhaps more largely in forms with long thready chromosomes and the former with short and stout chromosomes. **1929** *Jrnl. Genetics* XXI. 46 In *Prunus cerasus*..the method of pairing is parasynaptic, judging from the occurrence of diplotene chiasmata and their occasional persistence to metaphase as interstitial exchanges. **1910** *Ann. Bot.* XXIV. 752 Whether these univalent strands join with their homologous pairs telosynaptically or parasynaptically, or by any other intermediate method between these two extremes, resolves itself merely into a question of non-essential detail. **1926** *Genetics* XI. 274 It was thought that one could differentiate in a triploid between splitting and parasynaptic union of chromosome threads, because it seemed possible that the three threads might all unite parasynaptically. **1911** *Jrnl. Morphol.* XXII. 754 The main difference between the views of parasyndesis and metasyndesis lies in the interpretation of the longitudinal cleft of the gemini. **1912** *Jrnl. Exper. Zool.* XIII. 392 Do they [*sc.* the chromatin-elements] conjugate side by side (parasynapsis, parasyndesis), end to end (telosynapsis, metasyndesis) or in both ways? **1925** E. B. Wilson *Cell* (ed. 3) vi. 508 Evidence has steadily accumulated to show that in a large class of cases synapsis involves a side-by-side union of the synaptic mates (parasynapsis or parasyndesis) instead of an end-to-end union ..as was formerly supposed. **1911** *Jrnl. Morphol.* XXII. 750 The chromosomes conjugate parasyndetically. **1929** *Jrnl. Genetics* XIX. 171 She found a continuous spireme composed of parasyndetically paired threads alternating with single ones. **1938** Knopf & Ingerson *Struct. Petrology* viii. 110 Such a structure is a typical paratectonic crystallization under conditions of differential displacement. **1956** L. V. de Sitter *Struct. Geol.* i. 15 Paratectonic Regions contain curved folds, predominantly of concentric type without thickening of the strata in the hinges, accompanied by faulting which is secondary to the folding. **1969** *Mem. Amer. Assoc. Petroleum Geologists* XII. xxiv. 309/1 South of the fundamental Highland Boundary fault zone, inliers of the paratectonic Caledonides consist of Cambrian through Lower Devonian strata typified by simpler upright fold styles and a low degree of metamorphism. **1973** *Nature* 5 Oct. 244/2 Thickening of crust in paratectonic orogeny when two continental plates collide. **1901** G. E. Smith in *Jrnl. Anat. & Physiol.* XXXV. 434 The..'paraterminal body' is a structure of great morphological interest and importance, the essential unity of which has not hitherto been recognised. **1935** *Gray's Anat.* (ed. 26) 964 Immediately in front of the lamina terminalis and almost co-extensive with it, there is a narrow, triangular field of grey matter, which is termed the paraterminal gyrus (paraterminal body). **1951** O. Larsell *Anat. Nervous Syst.* (ed. 2) xvii.

428 The septal or paraterminal area...includes the gray substance of the basal portion of the cerebral hemisphere extending from the region of the anterior commissure to the caudal end of the anterior olfactory nucleus. **1921** A. L. Smith *Handbk. Brit. Lichens* 141/1 Parathecium, layer surrounding the thecium (hymenium). **1973** M. A. Letrouit-Galinou in Ahmadjian & Hale *Lichens* ii. 76 The parathecial apparatus..comprises (1) a parathecium, often cup-shaped, flaring, and composed of filaments which..are elongated and branched [etc.]. **1895** *Jrnl. Physiol.* XVIII. p. xxx, The tissue of the parathyroid gland does not at all resemble that of the thyroid in its adult form. **1925** *Jrnl. Biol. Chem.* LXIII. 395 (*heading*) The extraction of a parathyroid hormone which will prevent or control parathyroid tetany and which regulates the level of blood calcium. **1948** Martin & Hynes *Clin. Endocrinol.* v. 101 Four parathyroid glands are normally situated at the posterior extremities of the lateral lobes of the thyroid. **1960** *Farmer & Stockbreeder* 1 Mar. 136/1 There are two parathyroids in the bird. **1968** *Times* 10 Feb. 5/2 Parathyroid hormone is produced by the parathyroid glands, situated in the neck. **1930** J. Stephenson *Oligochaeta* xiii. 522 It would seem that regeneration after separation is the more primitive form; regeneration before separation a more recent development; the first is called..architomy, the second paratomy. **1963** R. P. Dales *Annelids* viii. 161 The fragmentation of an individual into two or more parts may be referred to as 'scissiparity', either before ('architomy') or after ('paratomy') formation of heads on the parts which break away. *Ibid.* 162 In this genus [sc. *Trypanosyllis*] a series may be traced from *T. coeliaca*, in which simple paratomous stolonization occurs. **1908** Paratracheal [see *metatracheal* s.v. *meta- 4]. **1933** *Tropical Woods* XXXVI. 9 *Paratracheal Parenchyma*, aggregated wood parenchyma in association with the vessels or vascular tracheids. **1965** K. Esau *Plant Anat.* (ed. 2) xi. 258 The phylogenetic sequence among the distributional types of wood parenchyma is from the diffuse arrangement to the other apotracheal and the paratracheal types. **1935** J. C. White *Autonomic Nervous Syst.* iii. 19 There is a good deal of evidence that the paraventricular nuclei [*sic*] preside over the sympathetic system. **1942** F. A. Mettler *Neuroanat.* xiv. 321 More diffuse cells scattered about in the ventricular wall are collectively called the paraventricular nuclei.. and represent what is left of a system which, in lower forms, interrelates the two thalami. **1945** *Jrnl. Compar. Neurol.* LXXXIII. 11 Here [*sc.* laterally] the anterior paraventricular nucleus becomes a thin, vertical strip of cells. *Ibid.*, Ventrally and dorsally the posterior paraventricular nucleus fades into the surrounding periventricular gray. **1972** M. L. Barr *Human Nervous Syst.* xi. 190/2 The precursors of vasopressin and oxytocin appear in the cytoplasm of cells of the supraoptic and paraventricular nuclei as neurosecretory droplets or granules.

2. a. paraca·sein, a phosphoprotein produced as a calcium compound in the form of a curd by the action of rennet on milk; hence **paraca·seinate,** a compound of paracasein with a metal; **paraformaldehyde** (examples); also called **pa·raform; paramy·osin,** a protein which forms the thick filaments of the contractile units of molluscan muscle; **parapro·tein** [a. G. *paraprotein* (K. Apitz 1940, in *Virchows Arch.* CCCVI. 685)], any of various proteins found in the blood only in certain diseases (as myelomatosis); hence **pa:raproteinæ·mia** [ad. G. *paraproteinämie* (K. Apitz, loc. cit.), f. Gr. αἷμα blood], the presence of paraproteins in the blood; **pa:rarosa·niline** [ad. G. *pararosanilin* (E. & O. Fischer 1878, in *Ann. der Chem.* CXCIV. 266)], a colourless, crystalline alcohol $(H_2NC_6H_4)_3COH$, which is used in making triphenylmethane dyes and whose red hydrochloride is used as a biological stain.

1906 Paracasein [see *caseinogen]. **1937** A. L. & K. G. B. Winton *Struct. Foods* III. 184 Paracasein at 5° and 23°C. has 1·5 times as much base-combining power as casein. **1959** Jenness & Patton *Princ. Dairy Chem.* x. 314 In view of the close similarity between the casein and paracasein, it is not surprising that the mechanism of the primary action of rennin has long remained obscure. **1961** Paracasein [see *caseinogen]. **1907** *Chem. Abstr.* I. 1740 Pure neutral sodium caseinate and sodium paracaseinate solutions..are not precipitated by saturation with pure sodium chloride. **1937** A. L. & K. G. B. Winton *Struct. Foods* III. 185 Rennet-casein..consists of calcium paracaseinate, formed from the calcium caseinate associated with dicalcium phosphate carried down mechanically. **1913** *Jrnl. Industr. & Engin. Chem.* June 508/1 We have succeeded..in producing a fusible phenol resin by heating ..paradioxydiphenylmethan with 10 grams of paraform. **1932** I. D. Garard *Introd. Org. Chem.* vi. 79 Paraformaldehyde is made into candles and sold for fumigating purposes under the name of *paraform*. The heat of the burning candle converts the paraform into formaldehyde. **1966** *McGraw-Hill Encycl. Sci. & Technol.* V. 472/1 It [*sc.* formaldehyde] is also sold as the solid hydrated polymer known as paraformaldehyde or paraform. **1894** *Jrnl. Chem. Soc.* LXVI. I. 487 Methylal is readily obtained by this method from paraformaldehyde and methylic alcohol. **1913** J. Walker *Org. Chem. for Students of Med.* 40 Paraformaldehyde is largely used as a source of formaldehyde for the purpose of disinfecting rooms, clothing, etc. **1951** Kirk & Othmer *Encycl. Chem. Technol.* VI. 861 On heating, paraformaldehyde depolymerizes to yield a mixture of monomeric formaldehyde gas and water vapor. **1946** C. E. Hall et al. in *Biol. Bull.* XC. 44 Since this protein can be identified by electron microscope observation and x-ray diffraction it merits a distinguishing name and is therefore designated as *paramyosin*. **1963** *Jrnl. Molecular Biol.* VII. 234 Light scattering, viscosity and sedimentation experiments on solutions of *Venus mercenaria* paramyosin show that the paramyosin molecule is a rod, 1330 Å long and 20 Å in diameter, and has a mass of 220,000 atomic mass units. **1972** *Biochemistry* (Easton, Pa.) XI. 4532/1 The paramyosin molecules align to form a bipolar core of the thick filament which is covered by a surface layer of myosin. **1949** *Chem. Abstr.* XLIII. 9097 The paraproteins have a different structural analysis than plasma protein. **1961** *Lancet* 9 Sept. 603/2 Large errors may arise if the urinary protein consists of a paraprotein, such as Bence Jones'. **1972** *Jrnl. Immunol.* CIX. 511/2 No feline paraproteins have been characterized physically or chemically. **1958** *Arch. Internal Med.* CII. 618/1 This method to this day has been used clinically as a method of evaluating dysproteinemias and paraproteinemias in disease states. **1972** *Clin. & Exper. Immunol.* XI. 488 There were six patients whose disease was associated with an IgM paraproteinaemia..and another six without a paraproteinaemia. **1879** *Jrnl. Chem. Soc.* XXXVI. 385 Pararosaniline is the name applied by the authors to the colouring matter derived from paratoluidine. **1926** J. Read *Text-bk. Org. Chem.* xi. 203 By replacing six hydrogen atoms in the molecule of the magenta dye, pararosaniline, with six methyl groups, a deep violet dye, known as 'crystal violet' is produced. **1971** E. Gurr *Synthetic Dyes* 81 Fuschin basic..is, in fact a hybrid mixture of the chlorides of pararosaniline, rosaniline and new magenta.

b. para-aminosalicy·lic acid, a colourless crystalline compound, $HOOC\text{-}C_6H_3(OH)\text{-}(NH_2)$, which is used (usually with isoniazid) in treating tuberculosis; **pa:ra-dichlor(o)be·nzene,** a colourless crystalline compound, $C_6H_4Cl_2$, that has a low melting point and is used as a moth-proofing agent; **paranitra·niline,** a pale yellow crystalline compound, $H_2NC_6H_4NO_2$, used in making azo dyes; **para-phe:nylenedi·amine,** a colourless, crystalline compound, $C_6H_4(NH_2)_2$, used as a photographic developer, for dyeing hair and fur, and for making safranine and sulphur dyes; **paraxy·lene,** an isomer of xylene that melts to a colourless liquid at room temperature and is now obtained from petroleum naphtha for use esp. as a source of terephthalic acid.

1946 *Lancet* 5 Jan. 15/1 (*heading*) *Para*-aminosalicylic acid in the treatment of tuberculosis. **1954** S. Duke-Elder *Parsons' Dis. Eye* (ed. 12) x. 120 For the common ophthalmological infections, however, the most important compounds are the sulphonamides, the sulphones and para-amino-salicylic acid. **1966** *McGraw-Hill Encycl. Sci. & Technol.* I. 320/1 *para*-Aminosalicylic acid (PAS) was synthesized in 1901. **1876** *Jrnl. Chem. Soc.* XXIX. 81 By heating equal molecules of chlorobenzenesulphonic chloride and phosphorus pentachloride to 200°–220°, paradichlorobenzene, melting at 53°, and boiling at 172°–174°, is formed. **1938** *Forum & Century* (N.Y.) Feb. 96/2 There is probably no better moth protection than paradichlorbenzene crystals. **1965** Zigrosser & Gaehde *Guide to Collecting Orig. Prints* vii. 113 Preventative measures [against microbiological infection of prints], however, can be taken by placing open containers of paradichlorobenzene crystals upon the storage shelves. **1872** Para-nitraniline [in Dict.]. **1918** C. M. Whittaker *Applic. Coal Tar Dyestuffs* vi. 87 The first stable form of diazotized paranitraniline manufactured commercially was nitrosamine red. **1963** A. J. Hall *Textile Sci.* iv. 181 A deep bright red shade developed in the fabric as the result of formation within the fibres of an insoluble pigment by coupling of the beta-naphthol with the diazotized paranitraniline. **1873** *Jrnl. Chem. Soc.* XXVI. 167 Dinitrobenzene was reduced by means of tin and hydrochloric acid; the bromine was eliminated by the nascent hydrogen and paraphenylenediamine was formed. **1906** *Jrnl. Soc. Dyers & Colourists* XXII. 77/1 Paraphenylenediamine is employed by furriers for dyeing pelts, and it has been used also as a hair dye. **1966** L. F. A. Mason *Photogr. Processing Chem.* i. 25 Although these paraphenylenediamine derivatives are stable in acid solution or as salts in the solid state, the free bases..readily oxidise and are not very soluble. **1873** *Jrnl. Chem. Soc.* XXVI. 272 The author, by acting upon isoxylene and paraxylene with benzyl chloride, in presence of zinc, has obtained benzylisoxylene and benzylparaxylene. **1954** R. W. Moncrieff *Artificial Fibres* (ed. 2) xxi. 265 Development of a method for making large quantities of *para*-xylene proved to be difficult and delayed the manufacture of Terylene in the United Kingdom, but in America the Standard Oil Co. have developed a method of making it from petroleum. **1969** Para-xylene [see *feedstock* s.v. *feed sb. 7]. **1975** W. G. Roberts *Quest for Oil* (rev. ed.) ix. 95 Very pure aromatic compounds such as paraxylene, which is used in synthetic fibre manufacture.

c. Names of minerals having a chemical composition the same as or similar to those to the names of which *para-* is prefixed, but a different crystal structure: **parabu·tlerite** [f. *butlerite* (f. the name of Gordon *Butler* (1881–1961), U.S. geologist)], a basic hydrated sulphate of ferric iron, $FeSO_4(OH)\cdot 2H_2O$, found as orange orthorhombic crystals; **parahopeite** (-hōᵘ·pəit), a hydrated zinc phosphate, $Zn_3(PO_4)_2\cdot 4H_2O$, found as colourless, transparent, triclinic crystals; **paralau·rionite,** a hydroxide and chloride of lead, Pb(OH)Cl, found as colourless, transparent, monoclinic crystals; **paramela·conite,** an oxide of copper, CuO, in which some of the bivalent copper (typically about a quarter) is replaced by monovalent copper and which is found as black tetragonal crystals; **parara·mmelsbergite,** an arsenide of nickel, $NiAs_2$, that occurs as white, opaque, tabular crystals that alter to erythrite when exposed; **paratacamite** (pæ:rætăkā·məit), a secondary mineral that is a basic chloride of copper, $Cu_2(OH)_3Cl$, and is found as green to greenish black hexagonal crystals; **parate·llurite,** an oxide of tellurium, TeO_2, found as soft, white or yellow tetragonal crystals with a waxy lustre; **paravau·xite,** a secondary mineral that is a hydrated basic phosphate of ferrous iron and aluminium, $FeAl_2(PO_4)_2(OH)_2.8H_2O$, and occurs as brittle, whitish or colourless, triclinic crystals, usu. in association with vauxite and wavellite; **parawo·llastonite,** the monoclinic form of the calcium silicate, $CaSiO_3$, of which wollastonite is the commoner triclinic form, both occurring as intergrowths with one another.

1938 M. C. Bandy in *Amer. Mineralogist* XXIII. 742 Parabutlerite $Fe(SO_4)(OH).2H_2O$. A basic hydrate of iron of this same composition has already been described as the mineral butlerite and as an artificial compound in the system $Fe_2O_3\text{–}SO_3\text{–}H_2\text{–}H_2O$. **1968** I. Kostov *Mineral.* 499 Metahohmannite, parabutlerite, and fibroferrite..occur as yellow crusts or reniform aggregates with fibrous texture, products of weathering of pyritic deposits. **1907** *Nature* 12 Dec. 143/1 Another new species, named parahopeite, has the same chemical composition as hopeite, $Zn_3P_2O_8.4H_2O$, but is anorthic. **1908** L. J. Spencer in *Mineral. Mag.* XV. 18 Parahopeite. This name I propose to give to a new species of hydrous zinc phosphate, identical with hopeite in chemical composition, but differing from both α-hopeite and β-hopeite in physical and crystallographic characters. **1955** *Mineral. Abstr.* XII. 479 Preliminary note on the occurrence of parahopeite and tarbuttite..in the Kef Semmah mine [in Algeria]. **1974** *Mineral. Mag.* XXXIX. 684 Several uncommon phosphate minerals including tarbuttite, parahopeite, scholzite, and collinsite occur in near-surface gossans in the Reaphook Hill zinc prospect [in South Australia]. **1899** G. F. H. Smith in *Mineral. Mag.* XII. 102 On some of the specimens [of lead slags from Laurium] a new mineral, paralaurionite, was found, which possesses the same chemical composition as laurionite. **1950** *Mineral. Mag.* XXIX. 341 About 1942 paralaurionite was identified in a suite of minerals from the Mammoth mine, Tiger, Arizona. **1891** G. A. Koenig in *Proc. Acad. Nat. Sci. Philadelphia* 289 The tetragonal crystals are so unique in their appearance, that they should be accorded the rank of a very distinct species, and the name *Paramelaconite* is proposed for them. **1941** *Amer. Mineralogist* XXVI. 659 The paramelaconite occurs as stout prismatic crystals up to 3 cm. in length. **1962** *Ibid.* XLVII. 779 On a recent collecting trip to the Algomah mine [in Michigan] a few massive pieces of paramelaconite were found. **1939** M. A. Peacock in *Ibid.* Dec. II. 11 Recently described materials provisionally named rammelsbergite.., from Cobalt, Ontario, and Elk Lake, Ontario, give identical *x*-ray powder photographs unlike those of from Schneeberg and Eisleben... The Canadian mineral is thus a distinct species for which the name *pararammelsbergite* is proposed. **1967** *Canad. Mineralogist* IX. 129 Pararammelsbergite has vacant arsenic lattice positions which are occupied in part by excess metal atoms. **1972** *Amer. Mineralogist* LVII. 1 Pararammelsbergite ($NiAs_2$) is orthorhombic.., space group *Pbca*, $Z = 8$. **1905** *Nature* 13 Apr. 574/2 This new mineral, to which the name paratacamite was given [by Prior and Smith], has the same chemical composition as atacamite. **1950** *Mineral. Mag.* XXIX. 280 It is paratacamite, not atacamite that occurs at the famous Cornish locality, Botallack mine, St. Just. **1960** Switzer & Swanson in *Amer. Mineralogist* XLV. 1272 Orthorhombic TeO_2 is found in nature as tellurite. The tetragonal form, well known as a chemical compound, has been found associated with tellurite and native tellurium at Cananea, Sonora, Mexico. The name paratellurite is proposed for the new mineral. **1973** *Chem. Abstr.* 8 Oct. 489/2 In an acoustooptic light deflector, frequency modulated acoustic shear waves propagate in the [110] direction in paratellurite (TeO_2). **1922** S. G. Gordon in *Science* 14 July 50/1 Preliminary notes on vauxite and paravauxite. Among the mineral specimens collected on the Vaux-Academy Andean expedition of 1921 are two that have proved to be new. **1944** *Proc. Acad. Nat. Sci. Philadelphia* XCVI. 339 Vauxite, paravauxite, metavauxite, and childrenite are secondary phosphates, derived from solution of apatite by supergene waters, and deposited usually upon wavellite. **1968** I. Kostov *Mineral.* 452 Paravauxite and gordonite have perfect {010} cleavage. **1935** M. A. Peacock in *Amer. Jrnl. Sci.* XXX. 525 It thus seems necessary to regard the triclinic modification as the normal one properly entitled to the name wollastonite; the name parawollastonite is, therefore, proposed for the rarer monoclinic modification. **1963** W. A. Deer et al. *Rock-forming Min.* II. 172 The distinction between wollastonite and parawollastonite is based on the extinction angle β:y, which is 3°–5° in wollastonite and 0° in monoclinic parawollastonite. *Ibid.* 173 Parawollastonite, identified by single-crystal X-ray photographs, has been recorded from Monte Somma, Vesuvius, ..from Crestmore, California.., and from Csiklova, Roumania.

3. *Physics* and *Chem.* Of, pertaining to, or designating the form of some homonuclear diatomic molecules in which (as in para-hydrogen) the two nuclei have antiparallel spins (see also quot. 1940² s.v. *ortho- 3); also used similarly with reference to the electronic spins of helium. Also as an independent word.

1927, etc. [see *ortho- 3].

para-³, comb. form of Parachute *sb.* **a.** With sbs., denoting 'dropped by parachute', 'trained or equipped for descending by parachute', as *para-bomb, -cargo, -commando, -girl, -marine, -mine, -nurse, -pa(c)k* (hence *-packed* adj.), *-pooch, -spy*. Also *paradoctor, *paradrop *sb.* and *v.*, *parafoil, *paraglider, Parakite (in Dict. and Suppl.), *paramedic¹, *para-rescue, *parascending *vbl. sb.*, *paratroops *sb. pl.*, *parawing. **b. parabrake,** a parachute which opens behind an aircraft and acts as a brake; **parafrag bomb,** a bomb dropped by parachute which bursts into fragments on hitting its target; **parajump** = *jump *sb.*¹ 1 c; so **parajumping** *vbl. sb.*; **parajute,** a parachute made of jute; **parapants,** women's knickers made from parachute silk; **paraplane,** (see quot. 1942); **parasheet,** (see quot. 1951); **paraski** *attrib.*, (*a*) designating a parachute trooper trained to ski from the point where he lands; (*b*) designating a sport in which skiers ski from a place to which they have dropped by parachute; **paraskier,** a paraski trooper; **paraspotter,** a person who watches for enemy parachute landings.

1943 *Time* 18 Oct. 36/2 Parabombs burst above the ground, spray their fragments with telling effect. **1951** Parabrake [see *drogue 3]. **1954** *Britannica Bk. of Year* 638/1 The verbal shorthand habitual to members of the armed forces produced such contractions as *parabrake*, a parachute used to slow down an aeroplane on landing. **1967** N. E. Borden *Jet-Engine Fund.* 97 Military fighters release a parachute, called a parabrake, from their tail as soon as their wheels contact the runway. **1951** R. Malkin *Boxcars in Sky* 172 As for military paracargo, however, the picture is entirely different. **1965** *Britannica Bk. of Year* (U.S.) 869/2 *Para commando, n.*, a parachute commando, as in the Congolese army. **1967** *Economist* 30 Sept. 1180/1 The Israelis train Mr Mobutu's crack corps, the para-commandos. **1978** *Guardian Weekly* 18 June 16/3 The bigger white community in Likasi..has told the Belgian Government it would leave en masse if the paracommandos go. **1944** *Tuscaloosa* (Alabama) *News* 5 Oct. 1 This..Jap Sally plane went up in smoke a few seconds after this picture was taken—destroyed by parafrag bombs. **1972** *Courier-Mail* (Brisbane) 16 May 1/7 Lucky paragirl Jackie Smith fell from 2400 ft. into the sea when her parachute failed to open during a week-end skydiving show—and lived. **1973** *Jewish Chron.* 9 Feb. 15/2 The nearest your reporter came to hearing fighting words from Israel's paragirls. **1971** *Islander* (Victoria, B.C.) 20 June 3/4 Most parajump clubs in Canada are affiliated with the Canada Sport Parachute Association. **1977** *R.A.F. News* 30 Mar.–12 Apr. 3/3 The £16 fee (lower than that charged by similar, civilian para-jump clubs) covers the cost of membership. *Ibid.* 3/4 A halt had been called to any para-jumping that week-end. **1956** W. Slim *Defeat into Victory* 225 In a month we had a parajute that was eighty-five per cent as efficient and reliable as the most elaborate parachute. It was made entirely of jute. **1973** J. Lucas *Big Umbrella* x. 108 The parajutes were not intended for man-carrying, but about 100,000 of them were used for supply-dropping in the Burma campaign. *Ibid.* 107 The 'parajute'..had no large vent in the apex, but several smaller ones all over, and it was only about one-seventh less efficient than the standard silk parachute. **1944** *Veterans' Weekly* (Lincoln, Nebraska) 15 Dec. 2 Oliver N. Magee, paramarine, son of War Dad and Mrs..Magee [etc.]. **1969** *TV Times* (Austral.) 22 Jan. 27 He is aided by the tough leader of a group of paramarines. **1944** in *Amer. Speech* (1945) XX. 221 Nazi para-mines nearly blocked supply lines. **1942** A. M. Low *Parachutes* p. x, A paranurse is..a nurse dropped by parachute. **1946** *B.B.C. War Report* 234 Suddenly the pilot called our attention to the parapacks coming out from the aircraft in front of us. *Ibid.*, There go the parapacks from the formation ahead of us—yellow, brown, red, drifting down gently, dropping their containers. **1950** in *Amer. Speech* (1956) XXXI. 62 A resupply mission was flown in the afternoon—with a drop of parapaks and other supplies simulated. **1954** *N.Y. Times Mag.* 6 June 79/2 Captain Schweiter slapped the switch controlling the parapacks (belly bundles), bellowed: 'Follow me,' and jumped. **1945** *Birmingham* (Alabama) *News* 27 Mar. 1/5 Allied Airborne troops and parapacked supplies are shown as they plummeted down five miles beyond the Rhine. **1944** *Time* 10 Apr. 12/3 Parapants. In Manhattan, Mrs. Virginia Bell Jack received from her Thunderbolt-pilot husband in England a pair of real silk (German parachute) panties. **1942** A. M. Low *Parachutes* 111 In Russia, some years ago, a young engineer, B. Pavlov, invented what he called a 'paraplane' which consisted essentially of two wings made of heavy linen stretched over a duralumin framework. These were strapped to the back of the parachutist and enabled him to glide before opening his parachute. **1974** *Sport Parachutist* June 18/1 Ten paraplane jumpers who are all..experienced paraplane flyers. **1944** *N.Y. Times* 26 Apr. 5/6 St. Bernard Qualifies as 'Parapooch' for Army. Major [*sc.* a St. Bernard dog] has made seven high altitude jumps with a regular size parachute. **1951** W. D. Brown *Parachutes* 315 *Parasheet*, a parachute constructed from one piece of fabric (or from several pieces with their warps parallel) in the form of a regular polygon, with the rigging lines attached to the apexes of the polygon. **1973** J. Lucas *Big Umbrella* x. 110 Parasheets, which acted like parachutes but were formed of strips of fabric rather than gores, were used for these [*sc.* slow-falling parachute flares]. *Ibid.*, A pair of parasheets were used for each flare to increase stability. **1942** *Christian Science Monitor* 26 Mar. 3 Para-ski troopers of the 503rd Parachute Battalion. **1974** *Sport Parachutist* June 26 (*heading*) German Para-Ski Championships. **1942** *Christian Science Monitor* 26 Mar. 3 'Let 'Er Buck' Is War Cry of Para-Skiers. **1976** *National Observer* (U.S.) 13 Mar. 6/3 Para-skiers make 3,000-foot parachute jumps from a hovering helicopter and attempt to land precisely on a four-inch disk, then they race against the clock down a giant slalom course. **1940** in *Amer. Speech* (1944) XIX. 12/2, Britain has 400,000 paraspotters ready to fight parachutists. **1943** *Daily Express* 10 Sept. 1/5 (*headline*) Para-spies dropped in Germany.

para-aminobenzoic (pæ:rā,ămīnobenzōᵘ·ik), *a.* [f. Para-¹ 2b + *aminobenzoic* s.v. *amino-.] *para-aminobenzoic acid*: the *para*-isomer of aminobenzoic acid, which is sometimes considered a member of the vitamin-B group, is widely distributed in plant and animal tissue, has the ability to neutralize the bacteriostatic effects of the sulphonamides, and has been used in the treatment of rickettsial infections, esp. typhus and Rocky Mountain spotted fever. Abbrev. *PABA.

1906 *Proc. R. Soc.* A. LXXVIII. 82 (*heading*) Methyl derivatives of para-aminobenzoic acid. **1940** *Nature* 28 Dec. 838/2 (*heading*) Para-amino benzoic acid as a bacterial growth factor. **1963** [see *PABA].

para-aminosalicylic acid: see *para-¹ 2 b.

para-aortic (pæ:rā,e,ǭ·ɪtik), *a. Anat.* [f. Para-¹ + Aortic *a.*] Situated beside the aorta; used chiefly (in *para-aortic body*) as an epithet of certain paraganglia.

1927 *Jrnl. Anat.* LXI. 317 The pre- and para-aortic tissue of a 16-week embryo from the region of the suprarenal glands is seen on section to contain the semilunar ganglia, collections of haemo-lymphoid tissue, and other encapsuled masses (the Zuckerkandl anlage). **1952** *Ibid.* LXXXVI. 357 Wrete (1927) observed the continuity of the chromaffin bodies of the para-aortic region with the chromaffin tissue of the adrenal glands. *Ibid.* 358 Because of the lack of uniformity in the interpretation of the 'paraganglia', it has been replaced in the present investigation by the topographical term 'para-aortic bodies'. A para-aortic body is defined as an encapsulated collection of chromaffin cells lying in intimate contact with the sympathetic nervous system and yielding a pressor substance after extraction. **1959** R. D. Lockhart et al. *Anat. Human Body* 488/2 The paraganglia, small brownish masses,..are similar to the suprarenal medulla... The largest masses, the para-aortic bodies, a third of an inch long at birth, lie on each side of the inferior mesenteric artery. **1965** [see *paraganglion]. **1973** *Gray's Anat.* (ed. 35) 144/1 Chromaffin organs are found in connection with certain..of the secondary plexuses of the sympathetic system; the largest members of this series are the para-aortic bodies,..commonly termed paraganglia. *Ibid.* 1381/2 The para-aortic bodies..attain their maximum size in the first three years of post-natal life, when the largest takes the form of two elongated brownish bodies..which lie on each side of the inferior mesenteric artery. *Ibid.*, The chromaffin cells of the paraganglia and para-aortic bodies.

parabasal, *a.* (*sb.*) Add: **2.** *Zool.* [in this sense a. G. *parabasal* (C. Janicki 1911, in *Biol. Centralbl.* XXXI. 321).] Applied to the kinetoplast (sense *a) of protozoa; chiefly in *parabasal body*. Also *ellipt.*

1912 E. A. Minchin *Introd. Study Protozoa* vi. 89 The ring of blepharoplasts in *Lophomonas* is supported on the edge of a membranous structure,..which in its turn is surrounded by a peculiar striated body, the 'collar' of Grassi, or 'parabasal apparatus' of Janicki. **1924** Hegner & Taliaferro *Human Protozool.* v. 127 Most of the members of the genus [*Trypanosoma*] are characterized by the possession of a more or less spindle-shaped body, a central nucleus, and a spherical or rod-shaped parabasal body, with which is closely associated a small blepharoplast and a flagellum. **1925**, etc. [see *kinetoplast]. **1973** K. G. Grell *Protozool.* 17 Trichomonads have only a single, though occasionally branched, parabasal body which is anchored at the base of the flagella and can either wind about an axostyle or follow a straight course to the posterior end.

3. *Med.* Applied to cells from the layers of stratified epithelium just above the deepest (basal) layer of cells.

1948 G. N. Papanicolaou et al. *Epithelia of Woman's Reproductive Organs* ix. 38 The parabasal postpartum cells..show a greater variety in size and form, somewhat larger nuclei, extensive vacuolization, and a tendency to congregate into larger groups. **1960** *Obstetr. & Gynecol.* XVI. 407/2 Deep in the layers of the epithelia are the parabasal cells, which cover the basal germinal layer of cells and are characterized by dense, thick cytoplasm with vesicular nuclei. **1972** *Acta Cytologica* XVI. 382/2 The presence [in amniotic fluid] of parabasal cells in excess of 70 per cent suggests a fetal weight of less than 500 gms.

parabasis. Add: (Further examples.) Also *transf.*, any digression in which the author addresses the audience on personal or topical matters.

1949 *Oxf. Classical Dict.* 216/2 *Parabasis* (addresses to audience separated by brief chants). **1952** F. H. Dudden *Henry Fielding* I. p. v, In the frequent parabases intercalated in his novels..[Fielding] has laid open his innermost self. **1965** *New Statesman* 30 Apr. 694/2 This is Coward's exercise in parabasis form: a lightly fictionalised justification of his way of theatre, acting and life. **1974** *Encycl. Brit. Micropædia* VII. 511/1 The *parabasis*, in which the chorus addresses the audience on the topics of the day and hurls scurrilous criticism at prominent citizens.

Parabellum (pæræbe·lǝm). Also **parabellum.** [f. L. *parā* imp. of *parāre* to prepare + *bellum* war (see quot. 1970).] The proprietary name of a make of automatic pistol or machine-gun. Also *attrib.*

1904 [see *Luger]. **1918** E. S. Farrow *Dict. Mil. Terms* 432 *Parabellum Gun*, a German aeroplane gun of small caliber. It has a belt of cartridges which contains not less than a thousand projectiles. **1924** *Blackw. Mag.* Feb. 157/1 The vessel..carried two old-pattern Maxims and a German Parabellum, mounted for anti-aircraft fire. **1962** E. Ambler *Light of Day* ii. 46 Six Parabellum pistols. **1970** R. A. Steindler *Firearms Dict.* 165 Based on the Latin 'Si vis pacem, para bellum' ('If you want peace, prepare for war.'),..the word Parabellum is the protected tradename of Mauser & D.W.M. **1973** J. Wainwright *Pride of Pigs* 127 A German gun..Official title—the German parabellum P'08, military automatic pistol... the Luger. **1974** S. Gulliver *Vulcan Bulletins* 18 The 9mm jacketed Parabellum rounds used in..Stens and Sterlings.

parabiont (pæræbəi·ǫnt). *Biol.* [f. Para-¹ + Gr. βιουντ-, pr. pple. stem of βιοῦν to live, f. βίος life.] An animal subjected to a parabiotic union.

1935 *Proc. Soc. Exper. Biol. & Med.* XXXIII. 568 Since the serum from each parabiont agglutinated the cells of both Pearlneck and Ring dove, it would seem that cellular differences were present other than those particular to each of the 2 species. **1955** *Anatomical Rec.* CXXII. 225 The lateral one-third of the adjacent kidneys in each parabiont was excised. **1970** *Nature* 19 Dec. 1186/2 The abdominal cavities of the parabionts were connected by a common free canal.

parabiosis (pærăbəi,ōᵘ·sis). *Biol.* [mod.L., ad. F. *parabiose* (A. Forel 1898, in *Bull. de la Société Vaudoise des Sciences Naturelles* XXXIV. 380), f. Para-¹ + Gr. βίωσις way of life (f. βίος life).] The anatomical union of a pair of organisms either natural or produced by surgery; the state of being so joined.

1908 *N.Y. Med. Jrnl.* LXXXVII. 374/1 Sauerbruch and Heyde give the name parabiosis to the new condition produced in animals which are experimented upon when they are organically connected together in an artificial manner. **1930** *Physiol. Rev.* X. 589 Parabiosis is a skin flap union, (to which is usually added peritoneal union), in which two whole organisms, instead of parts of organisms, are joined to each other. **1952** *New Biol.* XII. 46 Referring to the vascular anastomosis normally occurring between dizygotic bovine twins as 'nature's experiment in parabiosis'. **1955** *Anatomical Rec.* CXXII. 226 From parabiotic pairs formed between rats approximately 21 days of age, 12 pairs in successful parabiosis were chosen. **1962** D. J. B. Ashley *Human Intersex* iv. 58 Natural embryonic parabiosis has been observed in three instances in man. **1965** Lee & Knowles *Animal Hormones* iii. 64 If a male and a female salamander are joined by parabiosis (this allows the blood of the two animals to mix), a blood-borne substance from the male will lead to complete involution of the ovaries.

parabiotic (pærăbəi,ǫ·tik), *a. Biol.* [f. Para-¹ + Biotic *a.*] Of, pertaining to, or existing in parabiosis.

1911 *Arch. Internal Med.* VII. 396 The reaction appeared very clearly after a parabiotic union of a tuberculous and healthy guinea-pig. **1930** *Physiol. Rev.* X. 591 Each of the parabiotic individuals continues, therefore, to a certain extent, to live its own life. **1935** *Proc. Soc. Exper. Biol. & Med.* XXXIII. 568 Each member of a pair of parabiotic twins developed antibodies against the red cells of the other. **1955** [see prec.]. **1970** *Nature* 19 Dec. 1186/2 Two months later when the animals became accustomed to the parabiotic condition, they were again anaesthetized with ether and the abdominal cavities opened.

Hence **parabio·tically** *adv.*, so as to produce parabiosis.

1915 *Arch. Internal Med.* XV. 45 Forschbach united dogs parabiotically, bringing peritoneal cavities and homologous layers of the parietes into continuity. **1961** *Lancet* 23 Sept. 707/1 If a cockroach thus rendered 'arhythmic' is joined parabiotically to a normal cockroach, its activity is restored.

parable, *sb.* Add: **e.** *parable-art, -opera, -play.*

1935 Auden in G. Grigson *Arts To-Day* 20 There must always be two kinds of art, escape-art..and parable-art, that art which shall teach man to unlearn hatred and learn love. **1976** *Listener* 8 July 27/1 Auden both devised and named the literary genre—'parable-art'—that the decade [of the 1930s] demanded. **1976** *National Observer* (U.S.) 18 Dec. 18/1 Britten composed in almost all the musical forms available to him—and even invented one, the parable opera, such as *Curlew River* and *The Prodigal Son*, in which a moral lesson was set forth in direct and easily assimilable musical terms. **1941** L. MacNeice *Poetry of Yeats* 187 Thirdly, there are those plays which are near to fable or which might be called parable-plays —*The King's Threshold* (1904).

parablepsia, var. Parablepsis. **parablepsy** (further example).

1913 F. W. Hall *Compan. Classical Texts* 154 Lipography (parablepsia), or simple omission of any kind.

1934 L. F. Powell in G. B. Hill *Boswell's Life of Johnson* II. 370 Powel.. government (*by parabolepsy*).

parabolic, *a.* and *sb.* **A.** *adj.* **2.** (Further examples.)

1955 *Sci. Amer.* Mar. 38/1 A parabolic 'dish', either solid or made of a wire screen, reflects incoming radio waves to a focal point, where a small dipole or rod picks up the energy. 1960 *Practical Wireless* XXXVI. 391/1 The radio telescope, a parabolic mirror of 250ft diameter.. scans the sky. 1962 A. Nisbett *Technique Sound Studio* i. 23 An assembly consisting of a cardioid or omnidirectional microphone fitted at the focus of a parabolic reflector is also strongly directional. 1965 P. Wayne *Wind on Reeds* vi. 74 Separate E.M.I. recording equipment, including.. a microphone which could be used in conjunction with a parabolic reflector. 1969 *Times* 4 Feb. 13/3 He seems to have recorded pulses of energy by means of a large array of parabolic mirrors. 1977 P. Hill *Fanatics* 38 Could we have a parabolic microphone in the control flat?

parabolicalism (pærăbǫ·likăliz'm). *rare.* [-ISM.] Parabolical character; matter which is parabolical.

1854 C. Walton *Notes Biogr. W. Law* 258 The deeply experienced spiritual man.. will be much disappointed.. at finding so much deep experience buried in such a huge mass of parabolicalism and idiocratic deformity.

parabolize, *v.* Add: Hence **para:boliza·tion,** the process of making parabolic or paraboloidal.

1903 *Sci. Amer. Suppl.* 17 Oct. 23232/3 Draper's method of 'parabolization by measure'.

parabutlerite, -casein(ate): see *PARA-[1] 2 c, 2 a.

paracentric (pærăse·ntrik), *a.*[2] *Cytology.* [f. PARA-[1]: cf. *-CENTRIC 2.] Involving only the part of a chromosome at one side of the centromere. Opp. *PERICENTRIC *a.* 2.

1938 H. J. Muller in *Collecting Net* XIII. 187/2 If the breaks were to one side of the centromere, the inversion may be termed 'paracentric', and it will be noted that the proportions of the two arms, and hence the general shape of the chromosome as seen at mitosis, is not changed. But if the breaks included the centromere between them, being 'pericentric', the mitotic chromosome will have the relative sizes of its two arms altered, except in the special case in which the two distal sections are sensibly equal in size. 1957 C. P. Swanson *Cytol. & Cytogenetics* xv. 485 Paracentric inversions are by far the most common type of aberration found in natural populations. 1975 *Nature* 3 July 40/1 Heterozygosity for a paracentric inversion, that is, a structural rearrangement in which a chromosome segment that does not include the centromere is rotated through 180°, results in suppression of recombination in the inversion region.

paracetamol (pærăsī·tămǫl). *Pharm.* [f. *para-acetylaminophenol*, its chemical name.] A white crystalline compound, $C_8H_9NO_2$, with mild analgesic and antipyretic properties; a tablet of this.

1957 *Approved Names* (Brit. Pharmacopœia Comm.), Paracetamol. 1963 *Brit. Pharmaceutical Codex* 564 Paracetamol.. is a suitable alternative for patients sensitive to aspirin. 1971 *Daily Tel.* 18 June 13/4 The active ingredients of pain-killing drugs that can be bought at the chemist are only two, namely paracetamol and aspirin. 1972 J. Gill *Tenant* III. ii. 92 Denis still had his headache when he woke and he went into the bathroom and took Paracetamol. 1976 *Liverpool Echo* 23 Nov. 1/8 Open verdict recorded by Merseyside Coroner at inquest into death of A— C— (32),..who died..of paracetamol poisoning. 1977 *Listener* 28 Apr. 563/3 An obligatory late-night snack for all production staff of toasted cheese and paracetamols..and who knows what new programmes would result.

parachor (pæ·răkǫɹ). *Chem.* [f. PARA-[1] + Gr. χορός (= dance, but taken by the coiner, in mistake for χώρα, as = space).] A numerical quantity (found empirically to be constant over a wide range of temperature) equal to the molecular weight of a liquid multiplied by the fourth root of its surface tension and divided by the difference between its density and that of its vapour.

1924 S. Sugden in *Jrnl. Chem. Soc.* CXXV. I. 1178 The quantity *P* can be regarded as function of chemical composition. For saturated substances, *P* is an additive function... It is proposed to name this quantity the parachor ..to signify comparative volume. 1940 Glasstone *Textbk. Physical Chem.* viii. 517 The mean parachor equivalent of the —NC group, in a number of alkyl and aryl isocyanides, is 66; this corresponds closely to that required for the structure —N⁺≡C⁻, thus N(12·5), C(4·8), triple bond (46·6), making a total of 63·9. The alternative structure —N=C would have a parachor equivalent of only 40·6. 1956 I. L. Finar *Org. Chem.* II. i. 9 A comparison of parachors of different liquids gives a comparison of molecular volumes at temperatures at which liquids have the same surface tension. 1974 *Nature* 22 Nov. 296/2 For a given salt, k_s was proportional to the characteristic volume of the non-electrolyte which in $m^3\ mol^{-1}$ equals the parachor (calculated in the usual way in c.g.s. units) $\times 10^{-6}$.

para-church: see *PARA-[1] 1.

parachute, *sb.* Add: **1.** Now more commonly used for descent from an aircraft. (Further examples.)

1938 *Britannica Bk. of Year* 79/1 Parachuting.. seems to be considered as a kind of popular amusement for everybody in Russia and France if performed with stiff parachutes on ropes from special jumping towers. 1940 *Chambers's Techn. Dict.* 613/1 *Free parachute*, a parachute to be released or opened by the falling person. 1974 *Encycl. Brit. Micropædia* VII. 740/3 Sport parachutes have large holes that permit the air to escape and drive the parachute in the direction opposite the hole, much like a low-power jet engine.

2. (Further examples.)

1930 R. Campbell *Adamastor* 50 The proud White gannet in his parachute of snow. 1947 Auden *Age of Anxiety* v. 112 In pelagic meadows The plankton open their parachutes.

5. Dropped by or attached to a parachute, as *parachute bomb, flare, light, mine, pack, rocket, signal*; designating part of a parachute, as *parachute cord, harness, ring*; using a parachute, as *parachute drop, jump* (so *jumper, jumping* vbl. sb.), *skiing* vbl. sb., *system, troops*; for, involving or consisting of parachute troops, as *parachute aircraft, attack, battalion, brigade, landing, regiment, wing*; resembling or acting as a parachute, as *parachute garment, spinnaker*; used for making parachutes, as *parachute nylon, silk*; **parachute assembly** (see quot. 1951); **parachute course,** a course of instruction in parachuting; **parachute tower,** a tower from which one may make a parachute jump.

1962 G. Chatterton *Wings of Pegasus* 32 There was a very limited number of tug aircraft and parachute aircraft. 1951 *Gloss. Aeronaut. Terms* (B.S.I.) II. 14 *Parachute assembly*, a parachute complete with all equipment for deployment and for harnessing a load. 1978 T. Allbeury *Lantern Network* iv. 36 They clambered into the thick parachute assemblies. 1941 *Hutchinson's Pict. Hist. War* 12 Jan.–18 Mar. 74 We must all be prepared to meet gas attacks, parachute attacks, with constancy, forethought and practised skill. 1942 Parachute battalion [see *PARA *sb.*⁵]. 1912 *Sci. Amer.* 16 Nov. 422/1 A Parachute bomb for Aeronautic Use... The bomb is provided with a small parachute which quickly destroys the horizontal velocity communicated by the airship. 1943 *Hutchinson's Pict. Hist. War* 25 Nov. 1942–16 Feb. 1943 148 Groundstaff of the R.A.F. loading parachute bombs into Hampden aircraft. 1974 *Times* 19 Apr. 15/4 The 1st Parachute Brigade fighting in North Africa. 1941 R. Crompton *William does his Bit* viii. 193 Robert's got a bit of German parachute cord. 1976 A. White *Long Silence* ii. 101 We checked ourselves for climbing... It was very similar to checking ourselves for a parachute jump... I had taken a loop of nylon parachute cord with me. 1946 E. Capell *Simiomata* I. 13 Training, having got round rules excluding men of his age, obtained the privilege of a parachute course. 1977 D. Seaman *Committee* 151 Like every one else in the Department, Walters had done his parachute course. 1928 *Even. News* 9 May 5/1 There will be wing walking and a parachute drop by Miss June. 1974 'H. Carmichael' *Motive* iii. 31 A sky-diver in a delayed parachute drop. 1918 *War Illustr.* 13 July 372/2 We saw flashes far to the south—shrapnel, star-shells, and parachute flares. 1941 A. O. Pollard *Bombers over Reich* 46 So we dropped another parachute flare, which.. showed wreckage lying all over the place. 1974 S. Gulliver *Vulcan Bulletins* 130 Wire-guided missiles, small aerial incendiaries, parachute flares. 1912 C. B. Hayward *Pract. Aeronaut.* 690 (*heading*) Parachute garment as a safeguard. *Ibid.*, A parachute garment has been devised for use should the aviator fall. 1929 F. P. Gibbons *Red Napoleon* xii. 3 I made a last inspection of my parachute harness. 1970 Parachute jump [see *JUMP *sb.*[1] 1 c]. 1977 *Listener* 28 July 104/3, I had hoped to be making my first parachute jump that Saturday. 1912 C. B. Hayward *Pract. Aeronaut.* 691 The parachute jumper insisted on going up at least a thousand feet for the first trial. 1932 Auden *Orators* II. 71 The Mimosa's affair with the parachute jumper. 1954 *Chambers's Jrnl.* May 261/1 Parachute-jumping.. the monopoly belongs to the Soviet Union. 1969 *Listener* 20 Feb. 235/1, I won the Northern Junior Sky-Diving Championship, but have given up parachute-jumping at least for the time being. 1974 *Encycl. Brit. Micropædia* VII. 741/1 The sport of parachute jumping is usually governed by the parachute branch of the national aeronautic club. 1976 A. White *Long Silence* i. 14 Can he climb?.. Parachute jumping? 1940 W. S. Churchill *Into Battle* (1941) 222 If parachute landings were attempted.. these unfortunate people would be far better out of the way. 1942 E. Waugh *Put out More Flags* 247 Parachute landings were looked for hourly. 1876 Voyle & Stevenson *Mil. Dict.* (ed. 3) 285/2 *Parachute light*, a suspended light, invented by Colonel (now General) Boxer R.A., and which is used for the same purpose as *ground light* balls..viz. to light up the enemy's works and working parties. 1918 E. S. Farrow *Dict. Mil. Terms* 432 *Parachute Lights*, rockets or flares fired electrically from the pilot's seat, through a tube. 1940 *Hutchinson's Pict. Hist. War* 20 Dec. 1939–13 Feb. 1940 2 When the 'parachute' and magnetic mines were first used in the war, many people assumed that the Allies were taken by surprise. 1961 B. Fergusson *Watery Maze* i. 44 The Germans dropped some parachute mines into the harbour. 1974 N. Freeling *Dressing of Diamond* 90 It was indeed difficult to see what a human agency could do, short of a few parachute mines. 1972 J. Poyer *Chinese Agenda* (1973) v. 42 Mountain tents of very light-weight, close-woven parachute nylon. 1977 *New Yorker* 10 Sept. 101 Two parachute nylon traveling bags. 1975 T. Allbeury *Special Collection* iv. 20 There was ample room for.. the parachute pack. 1977 P. Way *Super-Celeste* ii. 57 Bosco.. pulled the green apple on the oxygen cylinder attached to his parachute pack. 1972 Parachute regiment [see *PARA[1]]. 1973 *New Statesman* 28 Sept. 410/3 Reaction against the cloistered Hampstead life drove him into the Parachute Regiment. 1977 *R.A.F. News* 11–24 May 20/6 The Dakota.. stands outside the Parachute Regiment's museum at Aldershot. 1930 C. Dixon *Parachuting* 53 He will then pull the parachute ring in the front of his harness which will open the pack to let the parachute fly out. 1935 *Discovery* Feb. 43/2 The multi-tube parachute rocket used for the Harz Mountain experiments. 1976 *Star-Phoenix* (Saskatoon) 23 June 52/4 Since many rescues have to be performed at night or in darkened, stormy conditions, he suggested police and other officials involved in rescues carry illuminating parachute rockets. 1937 *Discovery* June 187/2 The manufacture of marine signals,.. parachute signals,.. railway flares. 1962 M. Duffy *That's how it was* xvi. 132 She was juggling with some pieces of parachute silk she had been given, trying to shape them to a pair of cami-knickers. 1977 J. Cleary *High Road to China* i. 28 A length of old parachute silk was a curtain that hid.. our skimpy wardrobe. 1971 *Bahamanian Rev.* Nov. 15/2 For those who like to be on the water as well as in it, water skiing is available at the larger hotel beaches. The more daring may wish to sample parachute skiing. In this unique sport, the skiers use the wind and motion of the boat to climb on the lift of a parachute and soar perhaps a hundred feet in the air for a thrilling ride. 1932 *Yachting* Oct. 68/1 That the 'parachute' spinnaker—or 'double' spinnaker, if you prefer—has come to stay is pretty evident to those who have given it any kind of a fair trial. 1964 M. Weeks *Compl. Boating Encycl.* 396/1 *Parachute spinnaker*, a large, wide spinnaker introduced in 1927 by the Swedish yachtsman Sven Salén. 1971 *Daily Tel.* 1 July 30/6 After aerodynamically braking in the atmosphere the parachute system was put into action and before landing the soft-landing engines were fired. 1946 A. Lee *German Air Force* 37 This stage [in training, etc.].. parachute tower in the Park of Rest and Culture at Moscow. 1938 *Jrnl. R. Aeronaut. Soc.* XLII. 840 It appears that the landing of parachute troops will require special sighting devices. 1942 *R.A.F. Jrnl.* 18 Apr. 29 Two parachute troops were landed. 1978 *Listener* 20 Nov. 507/1 Malthausen.. held whole armies of capable men and women who hoped that Allied parachute troops.. would drop on, or near.. 1938 P. Kemp *No Colours or Crest* xii. 264 As a further precaution we had to remove our badges of rank and parachute wings. 1973 *Times* 18 Oct. 18 General Amin.. arrived in Amman earlier this week wearing Israeli parachute wings.

Hence **pa·rachutage,** a drop of supplies, etc. by parachute; **pa·rachuter,** a parachutist; **pa·rachutal** *a.*; **pa·rachutic** *a.* (sense 2). (See also next.)

1905 *Spectator* 14 Jan. 47/1 A parachutic arboreal serpent is not an impossible animal. 1930 *Flight* 21 Feb. 340/1 The last part of the lecture was devoted mainly to a discussion of vertical descent and to the parachutal efficiency of the autogiro. 1940 Parachuter [see *CHUTE, *CHUTIST]. 1941 G. Millar ... xxi. 221, I learnt it in case I meet any parachuters. 1945 G. Millar *Maquis* iii. 48 I told myself it was risky to sleep in bed so near the parachutage. 1956 R. Braddon *Nancy Wake* xii. 142 Whenever a parachutage was due, the B.B.C. would insert the group's secret code.

parachute, *v.* (In Dict. s.v. PARACHUTE *sb.*) Add: **a.** *trans.* (Further examples.) (Also **parachuted** *ppl. a.*) **b.** *intr.* (Earlier and later examples.) Also, to use a parachute. **parachuting** *vbl. sb.* (further examples).

1860 W. H. Russell *My Diary in India* II. ix. 174 And thus, with an able-bodied aborigine holding on by my tunic-tail behind, I parachuted down. 1914 G. B. Shaw *Misalliance* 46 *Tarleton*.. Been up much? *Lina*. Not in an aeroplane. I've parachuted; but thats child's play. 1930 E. W. Hendy *Wild Exmoor* 245 Meadow-pipits parachuted down to the brink. 1938 Parachuting [see *PARACHUTE *v.* 1]. 1945 G. Millar *Maquis* xv. 316 One pair of the high brown American parachuting-boots. 1946 *News Chron.* 2 May 3 Brig. Nicholls was parachuted into Albania in October, 1943. 1956 'C. Blackstock' *Dewey Death* ii. 27 He was with the Resistance.. They parachuted him down into France. 1969 N. Freeling *Tsing Boum* xvii. 130 They came out at night to steal parachuted supplies. 1977 *R.A.F. News* 11–24 May 3/6 An instructor in high-altitude parachuting at Abingdon. 1977 *New Yorker* 20 June 90/3 The connecting roads between tanks and parachuted troops are single lanes.

c. *fig.* Const. *in* or *into. trans.* To appoint or elect an outsider in such a way as to disregard the existing hierarchy; *intr.* to obtain a position in such a way. Also **parachuting (in)** *vbl. sb.*

1954 B. & R. North tr. *Duverger's Pol. Parties* II. iii. 357 The 'parachuting' of candidates, so developed in the first proportional elections when some deputies had never set foot in their constituency before being elected, was radically impossible in the arrondissement system. 1968 *Listener* 13 June 759/3 Too many of the existing hierarchy are civil servants 'parachuted' in from outside. 1968 D. Stuebing *Trudeau: Man for Tomorrow* v. 39 Trudeau was accused of parachuting into Mount Royal, the term in this sense implying that the candidate was dropped into the riding under party sponsorship and over the objections

of the riding association. 1977 P. A. Allum *Politics & Society Post-War Naples* (1973) vi. 180 A controversial example of the 'parachuting' of a candidate into the constituency was the transfer and inclusion in the list in 1968 of the Parisian correspondent of *L'Unità*, Maria-Antonietta Macciocchi. 1973 *Globe & Mail* (Toronto) 9 June 6/2 Mr. Roberts prudently concluded that local Liberals would resist if an officer from the Prime Minister's office parachuted into the riding to push aside a respectable candidate, and a woman at that. 1973 *Times* 13 July 5/1 One 'parachuting in' of two young men at a relatively senior level caused some bitter feelings among existing Commission staff, who saw their promotion prospects threatened. 1975 *Globe & Mail* (Toronto) 4 Mar. 6/2 Competent French-Canadians develop a sense of frustration and inferiority. They move more slowly and in smaller numbers up through the middle ranks. To redress the balance, the Government has to parachute French-Canadians from outside the civil service into senior positions. This parachuting produces complaints from English Canada about 'The French' taking over the Government.

parachutist. Add to def.: Now more commonly one who makes a parachute descent from an aircraft, esp. a soldier dropped by parachute. (Further examples.)

1888 *Sci. Amer.* 13 Oct. 231/1 An American Parachutist in England. 1927 *Illustr. London News* 10 Sept. 406 (*caption*) The perilous work of the parachutist. 1936 [see *AMBO-BED]. 1940 *Hutchinson's Pict. Hist. War* 10 Apr.–11 May 114 Another photograph showing large numbers of Red Army parachutists falling from troop-carrying aircraft during Soviet Army manoeuvres. 1940 *B.B.C. War Report* 78 Parachutists were to the job, but in the darkness and bad weather the paratroops were widely scattered and only 150 men reached the rendezvous for the attack. 1972 *Daily Tel.* (Colour Suppl.) 7 Jan. 11/4 Parachutists compete in individual aerial acrobatics or accuracy work. 1974 *Times* 19 Apr. 15/5 A sports parachutist just starting out could expect to spend £500 on his kit. 1976 A. White *Long Silence* i. 14 He'll also need to be a parachutist and rock-climber.

paraclinical (pærăkli·nikăl), *a. Med.* [f. PARA-[1] + CLINICAL *a.*] Of or pertaining to the branches of medicine, esp. the laboratory sciences, that provide a service for patients without direct involvement in their care.

1961 *Lancet* 29 July 235/2 In each case paraclinical laboratories have been included in the main complex of buildings. 1968 *Rep. R. Comm. Med. Educ. 1965–8* 83 in *Parl. Papers* 1967–8 (Cmnd. 3569) XXV. 569 A course of clinical instruction which includes.. such paraclinical subjects as pathology and microbiology. 1971 *Inside Kenya Today* Mar. 13/2 The A.I.D. regional programme has given assistance to the Veterinary Faculty.. including commodities necessary for the construction and equipping of a paraclinical building.

paracone (pæ·răkōᵘn). [f. PARA-[1] + CONE *sb.*] An external cusp on the front, outer corner of a mammalian upper molar tooth.

1888 H. F. Osborn in *Amer. Naturalist* XXII. 1072 (*table*) Proposed terms... Paracone. 1896 *Proc. Zool. Soc.* 363 The first two upper molars [of the hedgehog] are.. provided with two well-developed external cones, the paracone and the metacone. 1922 W. K. Gregory *Origin & Evolution Human Dentition* i. 74 It seems very likely that the high apex of the upper-molar crowns [in the marsupial mole] is really the paracone. 1934 W. E. Le Gros Clark *Early Forerunners of Man* iv. 71 The premolars of a generalized mammal would be of simple form with a single pointed cusp which in the upper teeth is called the paracone and in the lower the protoconid. 1971 W. A. Clemens in A. A. Dahlberg *Dental Morphol. & Evolution* x. 187 The paracone and metacone of Cretaceous marsupials were maintained at approximately equal height.

paraconformity: see *PARA-[1] 1.

paraconid (pærăkōᵘ·nid). [f. *PARACON(E + *-ID[2].] A cusp on a mammalian lower molar tooth corresponding to the paracone on an upper molar.

1888 H. F. Osborn in *Amer. Naturalist* XXII. 1072 As the hypocone develops, the paraconid recedes. 1896 *Proc. Zool. Soc.* 364 The internal pair.. of the lower molars.. 1922 W. K. Gregory *Origin & Evolution Human Dentition* i. 84 It seems probable that the paraconid, metaconid, and entoconid, arose in situ.. 1975 *Nature* 25 July 336/1 The presence of a paraconid in such a position is more characteristic of fossil lemuroids and omomyoids. 1975 [see *METACONID].

paracrine (pæ·răkrīn), *a. Physiol.* [ad. G. *parakrin*, f. Gr. παρα- PARA-[1] + κρίν-ειν to separate (cf. *ENDOCRINE *a.* and *sb.*).] Used of the action of a hormone whose effects are only local, and of the tissues which release and respond to such a hormone.

1972 F. Feyrter in *Endocrinol. 1971* 137 The hypothesis of the peripheral endodermal endocrine (paracrine) glands was confirmed (1) by the evidence of a pressor agent in the enteral carcinoid.., (2) by the discovery of the islet cells of the pancreatic duct with A and B cells, their endocrine function having already been established. 1976 *Nature* 8 July 92/3 The actions of this hormone are paracrine, directed that is to neighbouring cells and tissue, rather than truly endocrine. 1978 *Ibid.* 20 Apr. 730/2 Further investigations related to the possible release of other endogenous substances or peptides, such as vasopressin or those of the paracrine system of the gut, where enkephalin immunoreactivity has been demonstrated, are needed.

paracrystal (pærăkristăl). [f. PARA-[1] + CRYSTAL *sb.* and *a.*] An assemblage of particles that has some degree of order but is not a true crystal.

1933 *Trans. Faraday Soc.* XXIX. 1019 On addition of acetone to a cold saturated solution in acetic ester of phytosterin valerate the latter.. separates out as a paracrystal. 1953 S. E. Luria *Gen. Virol.* v. 94 In the needles, which should more correctly be called 'paracrystals', the individual rods are oriented sideways with great regularity. 1956 *Nature* 10 Mar. 473/1 The great regularity of plant viruses is shown even more strikingly by their ability to form crystals (or paracrystals) which give good X-ray photographs. 1970 *Jrnl. Biochem.* (Tokyo) LXVIII. 885/1 Paracrystals are formed by side-by-side association of F-actin particles in the presence of an excess amount of $MgCl_2$.

Hence **paracry·stalline** *a.*, of the nature of a paracrystal; **paracrystalli·nity.**

1933 *Trans. Faraday Soc.* XXIX. 1027 An especially strong paracrystallinity is to be ascribed to the paracrystalline sperm heads of Sepia officinalis. 1950 W. J. Moore *Physical Chem.* xiv. 408 The compound ethyl-amino-2-paraminocinnamate passes through three distinct paracrystalline phases between 83° and 139°. 1964 G. H. Haggis et al. *Introd. Molecular Biol.* xi. 283 The haemoglobin in cells which had taken up the sickle shape was in a para-crystalline, or tactoid form. 1974 *Amer. Jrnl. Anat.* CXXXIX. 135 Paracrystalline aggregates of microtubules were observed by electron microscopy in some cells of the anterior pituitary glands from ten untreated chinchillas.

parade, *sb.* Add: **I. 2. b.** (Later example.)

1930 E. Raymond *Jesting Army* I. iii. 45 A medical officer.. and the whole of his Sick Parade ran.

3. c. Easter Parade, a crowd of promenaders in new clothes at Eastertime; a parade or pageant held at Eastertime. Hence **Easter-Parading** *vbl. sb.*

1904 O. Henry in *N.Y. World Mag.* 27 Mar. 10/4 Will it tire you to be told again that Aileen was beautiful? Had she.. joined the Easter parade.. you would have hastened to say so yourself. 1933 I. Berlin *Easter Parade* (song), In your Easter bonnet.. you'll be the grandest lady in that Easter parade. 1942 O. Nash *Good Intentions* 5 Life is an Easter Parade. 1968 *New Statesman* 26 Apr. 544/3 In a sense, I suppose, this infernal Easter Parading.. is a middle-class demo. 1973 *Times* 24 Apr. 12/7 Several thousand people turned up at Battersea Park for the Easter Parade.

d. *transf. spec.* of broadcasting, a sequence or recital of forthcoming programmes, events, etc.

1947 *Radio Times* (Scottish ed.) 2 May 8/1 Scottish Programme Parade. 1948 *Broadcasting in Wales* (B.B.C.), Listen to your regional Programme Parade at 8.10 a.m. daily. 1963 *Rep. Comm. Broadcasting 1960* 255 in *Parl. Papers* 1961–2 (Cmnd. 1753) IX. 259 Broadcast of a half-hour 'parade' of new programmes.

5. (Further examples.) Now freq. used of a row of shops in a town, and of the street in which they are situated.

1766 C. Anstey *New Bath Guide* ix. 57 Whether thou art wont to rove By Parade, or Orange Grove,.. In the Circus or the Square. 1791 F. Burney *Jrnl.* 20 Aug. (1972) I. 35 O how I have thought.. of my poor Mrs. Thrale!—I went to look at the House on the North Parade where she dwelt. 1883 *List of Subscribers, Brighton* (South of Eng. Telephone Co.) 5 Viner E.B... 154, Marine-parade. 1968 R. K. Cox *Retail Site Assessment* ii. 15 Most new shopping centres.. have broken away from the old strip parades which usually face each other across heavy inter-town traffic. 1970 *Times* 9 Mar. 15/2 It is convenient for the local shopper to have a compact shopping area.. to provide in modern terms the facilities offered by the small local 'parades' of the 1930s. 1976 *Derbyshire Times* (Peak ed.) 3 Sept. 18/6 (Advt.), A vacant shop having a total area of 710 sq. feet, situated in a parade of shops and in a busy location. 1979 P. Hill *Hunters* viii. 95 The small parade of shops near the village hall.

7. *parade horse*; **parade drum,** a large drum played at a parade.

1953 M. Lawrence in R. Weaver *Canad. Short Stories* (1960) 2nd Ser. 115 From Captain Fossey.. the boy learned how to play the parade drum. 1967 W. Soyinka *Kongi's Harvest* 64 The big parade drum is heard. 1864 *Rep. Vermont Board Agric.* XIV. 123 The descendants of Woodbury Morgan.. possess that proper qualification necessary for the parade horse. 1968 'J. Le Carré' *Small Town in Germany* ii. 21 Lieff, an empty-headed parade horse from Protocol Department, sat on his left. 1974 — *Tinker, Tailor* xvi. 132, I refuse to bequeath my life's work to a parade horse.

parade, *v.* Add: **1. b.** *intr. Mil.* Of troops, to assemble for parade. Also *transf.*

1802 C. James *New Mil. Dict.*, *To parade*, to assemble in a prescribed regular manner, for the purpose of being inspected, exercised, or mustered. 1811 *Gen. Regulations* (Army) 102 All Guards are to parade with shouldered Arms. 1914 T. A. Baggs *Back from Front* xxiv. 122 We paraded for marching at 9 a.m., and set off with two manacled prisoners.. before us. 1916 W. Owen *Let.* 1 Feb. (1967) 377 Of course I 'paraded sick', but having no rash, I just have to crouch in my Hut. 1930 E. Raymond *Jesting Army* III. ii. 292 The working parties parade under the trees at nine o'clock. 1964 M. Banton *Policeman in Community* ii. 15 He 'parades' at the beat box where he meets the constable coming off duty.

parade-ground. orig. *U.S.* [PARADE *sb.* 4] A place where troops parade; now the more usual term than PARADE *sb.* 4.

1724 [see PARADE *sb.* 7]. 1843 N. Boone *Jrnl.* 31 July in L. Pelzer *Marches of Dragoons* (1917) 237 By one o'clock our command was formed on the parade ground of Fort Gibson. 1846 T. L. McKenney *Mem.* I. v. 103 The level of the ground, and its freedom from undergrowth, were such as to give it the appearance of a parade ground. 1891 *Century Mag.* Mar. 715 The rats were so numerous that they were common sights on the parade-ground. 1937 H. H. Symonds *Walking in Lake District* ii. 44 Before reaching the camp you pass.. the 'parade ground', a large square of cleared ground. 1975 J. Cleary *Safe House* i. 18 Camp 93.. long huts, a parade ground with a pock-marked surface. 1977 D. Beaty *Excellency* xx. 221 The Major led the way outside... The parade ground.. was floodlit.

b. *transf.* and *attrib.*, esp. in phr. *parade-ground bearing, manner, voice*, etc.

1863 'G. Hamilton' *Gala-Days* 174 Besides abundance of food and parade ground, these happy fowls have a very agreeable prospect. 1867 A. D. Whitney *Summer in L. Goldthwaite's Life* vi. 129 Clothes-lines like a parade-ground of telegraphs. 1894 *Harper's Mag.* Dec. 137 Both dismounted at the parade-ground gate. 1932 *Times Lit. Suppl.* 29 Sept. 693/4 The preposterous chief constable is of no use at all, and his parade-ground manners make a suitable contrast to Colonel Gethryn's cool and gentlemanly efficiency. 1937 M. Innes' *Hamlet, Revenge!* I. iv. 74 The precocious Bass, favoured on public-school parade-grounds. 1937 J. R. Firth *Tongues of Men* ii. There is a well-known Prussian parade-ground type of voice. 1944 J. D. Carr *Till Death do us Part* i. 8 Major Horace Price.. had made a trumpet of his hands and was addressing them in a parade-ground voice. 1961 M. Beadle *These Ruins are Inhabited* (1963) ii. 28 He expected the boys to have a parade-ground bearing. 1963 S. Plath *Crossing Water* (1971) 23 Pawing like paradeground horses. 1964 L. Deighton *Funeral in Berlin* v. 25 The parade ground of Europe has always been that vast area.. that stretches eastward from the Elbe. 1966 *New Statesman* 9 Apr. 358/3 In a peremptory parade-ground voice General Di Haniel repeatedly bellowed the name of the current occupant of the chair ('Sir Herbert!' 'Sir Samuel!') in an effort to raise points of order. 1973 M. Innes' *Appleby's Answer* xvi. 136 He had a.. parade-ground manner. 1976 E. Condon *Whisper of Axe* I. xxi. 130 A spit-and-polish, parade-ground major. 1978 R. Lewis *Uncertain Sound* vi. 189 He was all tension.. no longer the stiff, military, parade ground bearing.

paradichlorobenzene: see *PARA-[1] 2 b.

paradiddle (pæ·rădid'l). *Mus.* [Echoic.] A basic drum roll, produced by alternate beating with the left-hand and right-hand drumsticks.

1927 *Melody Maker* Aug. 804/3 The ordinary paradiddle is greatly to be recommended. This is a wonderful exercise for getting out of 'one-hand' playing, and there are some variations of this beat which are invaluable to the dance drummer. 1934 *Metronome* Feb. 27 Either the single paradiddle or the flam paradiddle may be used during a marching step. 1941 *Amer. Speech* XVI. 229/2 The ratamacue and the paradiddle are advanced rhythmic drum movements, the words possibly being onomatopoeic. 1956 L. McIntosh *Oxford Folly* 153 Fiona thought hard for a moment of a negro Staff-Sergeant in the American Air Force whom she had once seen throwing the whole of the University Jazz Club into a prolonged ecstasy with an erotic and protracted paradiddle... But the drumsticks in her inexpert hands refused to make a smooth roll. 1960 K. Amis *Take Girl like You* xvii. 210 A great saxophone growl came from the organ and a slow-motion paradiddle from the drums. 1961 A. Baines *Mus. Instruments* 337 At a steady march tempo, with right-hand strokes on the drum in each bar, the sound of the paradiddle is fascinating. 1976 D. Munrow *Instruments of Middle Ages & Renaissance* 33/4 The specialized techniques of the side drum, such as the roll, flam, drag, and paradiddle developed in the first instance to fulfil a practical object: to encourage friend and frighten the foe.

paradigm. Add: **I. b.** *attrib.*, as **paradigm case,** a case or instance to be regarded as representative or typical.

1955 J. L. Austin *How to do Things with Words* (1962) xi. 132 We were content to refer to 'statements' as the typical or paradigm case. 1962 *Listener* 4 Oct. 516/1 Plato's morality is supported and underlined by his theory of Forms, according to which mathematics is the paradigm case of knowledge. 1965 *Mod. Law Rev.* XXVIII. 309 The paradigm case at first instance—the core situation—appears to absorb so much attention that little concern is expended on the appeal process. 1974 *Jrnl. Philos.* LXXI. 337 Nagel employs a fairly standard 'paradigm case argument' in his analysis. 1977 *Canad. Jrnl. Linguistics* XXII. 1. 13 A paradigm case is Anderson's description of Breton vowel lowering.

3. b. *transf.* and *fig.*

1929 C. Day Lewis *Transitional Poem* II. 25, I would be pædagogue—hear poplar, lime And oak recite the seasons' paradigm. 1964 *Listener* 6 Aug. 200/2 If one uses the word 'paradigm' as Wittgenstein himself used it, to denote a logical or conceptual structure serving us as a form of thought within a given area of experience. 1966 A. F. Parker-Rhodes in *Automatic Transl. of Lang.* (NATO Summer School, Venice, 1962) 173 The concept of paradigm thus enables us to approach the problem of mathematizing the process of syntactic description with greatly enhanced resources. 1970 *Eng. Stud.* LI. 18 Although Ohmann determines objective criteria to state the similarity (and at the same time the dissimilarity), there still remains a whole paradigm of related structures out of which the author has to choose the particular alternative(s) to match the marked term with. *Ibid.* 46 But, of

course, for her Edwardian family life is a convenient paradigm of civilisation as a whole. 1973 C. Sagan *Cosmic Connection* (1974) xxiii. 155 There is a generation of men and women for whom.. the Moon was the paradigm of the unattainable. 1973 *Times Lit. Suppl.* 2 Mar. 238/4 The unfolding of terror and duplicity which follows is easily seen as a paradigm of the suppression of Dubček's liberalizing administration. 1973 *Nature* 6 July 59/3 The use of induced epilepsy as an 'interfering technique' in the study of learning and memory paradigms. 1975 *Language* LI. 1009 The publication of Chomsky's *Syntactic structures* provided a new paradigm for linguistics. 1976 T. Eagleton *Crit. & Ideology* i. 19 In the drive for order, proportion and propriety, the demand for socially cohesive categories of Nature and Reason,..history once again selects criticism as both paradigm and instrument of such a project. 1976 *Language* LII. 286 As for the periods in between the quantum leaps, Kuhn contends that each period of normal science in the development of a scientific discipline corresponds to one and only one methodological framework or *paradigm*. In a nutshell, paradigms are 'universally recognized scientific achievements that for a time provide model problems and solutions to a community of practitioners'. 1976 F. Zweig *New Acquisitive Society* II. x. 132 The television set ..is the paradigm of consumer culture, with its disarming passivity prone to desires divorced from action.

paradigmatic, *a.* Add: (Further examples.) **b.** *Linguistics.* Belonging to a set of linguistically associated forms. Cf. PARADIGM 3.

1662 [see *AGOGIC *a.* 1]. 1948 J. R. Firth in E. P. Hamp et al. *Readings in Linguistics II* (1966) 175 Most phoneticians..have continued to elaborate the analysis of words... Such studies I should describe as paradigmatic and monosystemic in principle. 1953 C. E. Bazell *Linguistic Form* 43 A special instance of paradigmatic indiscreteness of phonic character is afforded by English æ. 1964 *Eng. Stud.* XLV. 388 The concept of an active selective function of what has been called a paradigmatic frame does not serve teleological explanation of linguistic history. 1964 R. H. Robins *Gen. Linguistics* 49 Paradigmatic relations are those holding between comparable elements at particular places in structures. 1965 M. I. Finley in *New Statesman* 11 June 926/1 The authors do not distinguish between history as a systematic discipline and Aristotle's or Machiavelli's use of the past as a quarry for data for his social and political theories ('paradigmatic history', that has been called). 1972 W. Labov *Language in Inner City* v. 215 Larry is a paradigmatic speaker of black English vernacular as opposed to standard English. 1973 J. M. Anderson *Struct. Aspects Lang. Change* 124 In some cases syntagmatic influence may be the dominant force..; at other periods and under different conditions, paradigmatic forces may be stronger. 1973 *Black World* Sept. 51 A violence that becomes, in Wright's vision, paradigmatic of the entire spectrum of violence Blacks experience in this country. 1974 *Nature* 16 Aug. 609/1 Most of the philosophers of science..take the Comtean view, of physics as the paradigmatic science. 1975 *Language* LI. 665 Halle 1973 argues that paradigmatic information should be represented in the dictionary. 1976 *Times Lit. Suppl.* 23 Jan. 88/3 To his contemporaries Defoe was insignificant except as the paradigmatic Grub Street hack. 1977 *Church Times* 18 Feb. 6/3 Even if the (New Testament) accounts have been stylised, they have nonetheless a paradigmatic trustworthiness, an incontestable inner truth.

paradigmatically, *adv.* (Further examples.)

1953 C. E. Bazell *Linguistic Form* 7 An allomorph is paradigmatically relevant. 1960 C. Geertz in J. A. Fishman *Readings Sociol. of Lang.* (1968) 283, I offer the accompanying three charts depicting paradigmatically how a single sentence alters within each of the [Javanese] dialects and among them. 1964 R. H. Robins *Gen. Linguistics* 49 The status of a particular case as a grammatical category applicable to certain word endings in a language like Latin is stated..paradigmatically in terms of the number of different cases formally marked in the language. 1967 D. Cooper *Psychiatry & Anti-Psychiatry* p. xii, The third chapter seeks paradigmatically to make intelligible the patient-career of one young diagnosed schizophrenic in terms of the nature of his family world and the key events that have happened in it. 1971 T. F. Mitchell in *Archivum Linguisticum* II. 49 Within such a selection of phrases, *working* is definable as adjectival on the basis of the relationships it accretes paradigmatically.. and syntagmatically. 1976 *Amer. Speech* 1974 XLIX. 79 These relationships are seen..as constituting the basis for lexical sets paradigmatically.

paradise, *sb.* Add: **1. c.** *ellipt.* The plumage of a bird of paradise (cf. BIRD *sb.* 7).

1905 E. Wharton *House of Mirth* II. x. 446 Mrs. Trenor's hat? The one with the green Paradise? 1928 *Daily Express* 24 May 5/3 The same firm was responsible for wonderful curls of shaded paradise,..toning from dark to palest beige tones.

3. c. *Assoc. Football.* (With capital initial.) A name given to Celtic Park, Glasgow, the home ground of the Celtic Football Club.

1946 C. A. Oakley *Second City* III. 168 Celtic Park.. seemed so palatial, in odd comparison with an adjacent graveyard, that it was described as the 'Paradise'. 1958 C. Tully *Passed to You* xxii. 92 One of the best things about being at Paradise is that you're pretty certain to move in good company... You'll go a long way before you meet a better bunch than the Tims of Parkhead.

8. *paradise garden*; **paradise crane,** the blue or Stanley crane, *Anthropoides paradisea*, found in South Africa and distinguished by a cluster of very long, black tail feathers.

1906 *Daily Chron.* 8 May 7/4 His consignment..included.. three paradise cranes, five wolves and seven baboons. 1958 E. T. Gilliard *Living Birds of World* 146/1 Other species [of crane] are named for their ornamental plumage, coloration, wattles or geographical ranges—as, for example..the Paradise or Stanley Crane.. of southern Africa; and the Crowned Crane. 1910 O. Lindemann tr. *Delius's Village Romeo & Juliet* 178, I know another place not very far from here where we'll be quite unknown. 'Tis the Paradise Garden. *Ibid.* 192 Der Paradiesgarten... The Paradisegarden. 1972 *Country Life* 23 Mar. 682/3 It has been suggested that such places as this, in which an attempt is made to bring together plants from all parts of the world, should be known as paradise gardens. 1977 A. Wilson *Strange Ride R. Kipling* iv. 221 The Woolsack [*sc.* their South African house] was a delight to the whole Kipling family. .. For the children..it was clearly a Paradise garden.

paradoctor (pæ·rădǫ:ktəɹ). Chiefly *U.S.* [f. *PARA-[3] + DOCTOR *sb.* 6.] A doctor who is parachuted to patients in remote areas.

1944 *Time* 10 July 92/3 He is one of six paradoctors attached to the Search and Rescue Station of the Second Air Force. 1947 *Birmingham* (Alabama) *News* 30 May 1/7 A paradoctor from Athens..parachuted into the jungle. 1949 *Ibid.* 7 Aug. 1/7 Dr...Little,..former Army paradoctor, gave first aid to the injured.

parador (pæ·rădōəɹ). Pl. **paradores.** [Sp. *parador* inn, hostel.] Formerly an inn or hotel in Spain, now used as the name of a chain of hotels owned and administered by the Spanish government. Also *attrib.*

1845 R. Ford *Hand-bk. for Travellers Spain* II. 569/1 *La de Navarra*, near the Pla. Mayor, is but a mere *parador*. 1927 C. Connolly *Let.* 13 Feb. in *Romantic Friendship* (1975) 257 Roach I feel was..at home in the straw strewn venta or the noisy parador. 1960 *News Chron.* 10 Mar. 8/6 Castles in Spain have become paradores, Government run hotels. 1966 *Vogue* Nov. 182/3 Gran Canaria..here the broad terrace of a government parador looks out across a contorted volcanic landscape. 1973 *Country Life* 22 Nov. 1688/1 The needs of tourists can provide new functions..for historic buildings. Examples are the parador hotels in Spain.

parados. Add: **b.** The rear wall of a trench. Also *fig.*

1917 A. G. Empey *Over Top* 303 *Parados*, the rear wall of a trench which the Germans continually fill with bits of shell and rifle bullets. Tommy doesn't mind how many they put in the parados. 1938 H. G. Wells *Apropos of Dolores* iv. 139 Much of this discourse flowed over me. I did my best to keep my head down beneath the level of parapet and parados. 1957 P. Kemp *Mine were of Trouble* iv. 79, I saw Frejo and Santo Domingo on top of the parados.

paradosis (pāræ·dōsis). *Theol.* [Gr. παράδοσις, a handing down, a tradition.] A historical tradition, *spec.* relating to the teachings of Christ and of his disciples; teaching based on this tradition.

1950 L. S. Thornton *Revelation & Mod. World* ix. 285 The apostolic *paradosis* was embodied, not only in the apostolic writings, but also in the accredited teachers of the Catholic Church who could show their 'didactic successions' from the apostles. 1953 *Scottish Jrnl. Theol.* VI. 117 This is a tradition, a *paradosis* which does not fall under the condemnation which Jesus pronounces with regard to the *paradosis* in general. 1956 *Ibid.* IX. 434 Cullmann's thesis is that Paul was able to accord a higher place to the *paradosis* he had received, despite the fact that Jesus had denounced the high place accorded to *paradosis* in Judaism. 1958 A. Richardson *Introd. Theol. N.T.* xvi. 365 We must not make the mistake of the older NT critics who thought that what was to be interpreted was a number of ancient documents, which could be more objectively judged if they were isolated from the paradosis of the Church. 1973 *Amer. Jrnl. Philol.* XCIV. 320 Modern editors treat χειᾷ as a reasoned Triclinian conjecture. But χειᾷ proves almost certainly to be nothing but a further perversion of an already corrupt paradosis.

paradox, *sb.* (*a.*) Add: **2. a.** *spec.* in *Literary Criticism.*

1939 Brooks & Warren *Understanding Poetry* vi. 637 *Paradox*, a statement which seems on the surface contradictory, but which involves an element of truth. Because of the element of contrast between the form of the statement and its true implications, paradox is closely related to irony. 1942 C. Brooks in A. Tate *Language of Poetry* 37 Few of us are prepared to accept the statement that the language of poetry is the language of paradox... Yet there is a sense in which paradox is the language appropriate and inevitable to poetry. 1947 — *Well-Wrought Urn* 230 Paradox, as a device for contrasting the conventional views of a situation, or the limited and special view of it such as those taken in practical and scientific discourse, with a more inclusive view. 1960 *Commentary* Nov. 369 He had been instructed in *paradox, tension*, and *ambiguity* in a course called 'Introduction to Literature'.

c. *Logic.* A statement or proposition which, from an acceptable premise and despite sound reasoning, leads to a conclusion that is against sense, logically unacceptable, or self-contradictory; freq. distinguished by name, esp. of its propounder or of the type of problem it raises. Cf. *LIAR 1 d, *Russell's paradox* s.v. *RUSSELL.

1903 B. Russell *Princ. Math.* xliii. 358 This paradox, which, as I shall show, is strictly correlative to the Achilles, may be called for convenience the Tristram Shandy. 1921 W. E. Johnson *Logic* I. iii. 45 The paradox of implication assumes many forms, some of which are not easily recognised as involving mere varieties of the same fundamental principle. 1948 H. C. Brodie in Brodie & Coleman *Chwistek's Limits of Sci.* p. xxxiii, Typical of such paradoxes is the contradiction of Burali-Forti. 1950 R. Carnap *Logical Found. Probability* vii. 469 This is an instance of what Hempel calls the paradox of confirmation. 1955 A. N. Prior *Formal Logic* III. i. 224 As with Lewis's paradoxes, these appear less startling when the definitions of the terms used are considered. 1966 W. V. Quine *Ways of Paradox* i. 7 The paradoxes in this class are called antinomies, and it is they that bring on the crises in thought. An antinomy produces a self contradiction by accepted ways of reasoning. 1971 *Brit. Jrnl. Philos. Sci.* XXII. 337 The Nelson-Grelling paradox requires separate notice, because it can be presented with an explicit distinction between predicative phrases and what they are supposed to express. 1972 T. Stoppard *Jumpers* I. 29 Zeno overlooked the fallacy which is exemplified at its most picturesque in his famous paradoxes, which showed in every way but experience that an arrow could never reach its target. 1973 J. L. Mackie *Truth, Probability & Paradox* vi. 237 There is a group of paradoxes..which includes the Epimenides and other forms of the liar, heterologicality, Russell's class paradox..and so on.

paradoxical, *a.* Add: **3. b.** Applied to sleep that is characterized by increased physiological and mental activity (e.g. rapid eye-movements and dreaming in man) and normally alternates with longer periods of orthodox sleep.

[1959 Jouvet & Michel in *Compt. Rend. Soc. de Biol.* CLIII. 422 Cette phase est suivie d'une 'phase paradoxale' dont l'activité est extrêmement caractéristique.] 1962 I. Oswald *Sleeping & Waking* i. 16 (*heading*) 'Paradoxical phase' of sleep in the cat. 1969 *Sci. Jrnl.* Dec. 78/3 The brainwaves of paradoxical sleep are more like those during normal wakefulness. 1970 *New Scientist* 23 Apr. 170/2 After about an hour of this orthodox sleep phase..paradoxical sleep begins and lasts about 10 minutes before orthodox sleep is resumed. The electrical brain waves and many other bodily functions during paradoxical sleep are different—..most muscles become quite flaccid and their reflexes are lost; the penis is erect; the heart, breathing and blood pressure are irregular; [etc.]. 1974 *Times* 4 Sept. 6/7 There was increasing evidence of a 90-minute rhythm in both sleep and waking activity. Paradoxical sleep (when dreaming) occurred every 90 minutes. 1977 D. Morris *Manwatching* 315 So it is safe to assume that the total of 1½ hours of Paradoxical Sleep we have each night really does represent 1½ hours of actual dreaming.

paradoxician (pæ:rădǫksi·ʃăn). *rare.* [f. PARADOX *sb.* + -ICIAN.] One who deals in paradoxes; a paradoxer.

1909 W. J. Locke *Septimus* xvi. 186 Sypher was not convinced by the airy paradoxician.

paradoxling (pæ·rădǫ:ksliŋ). *nonce-wd.* [f. PARADOX *sb.* + -LING[1] 2.] A statement or tenet that is slightly paradoxical.

1863 G. M. Hopkins *Let.* 10 July (1956) 199 Yes. You are a Fool. I can shew it syllogistically, by an Epimediculum or paradoxling.

paradoxographical, *a.* Delete *nonce-wd.* and add later example.

1904 W. H. Stevenson in *Eng. Hist. Rev.* Jan. 139 He assigns the younger periplus to the Alexandrian or post-Alexandrian times on account of its paradoxographical character.

paradrop (pæ·rădrǫp), *sb.* and *v.* [f. *PARA-[3] + *DROP *sb.* 12 g or DROP *v.*] **A.** *sb.* The dropping from aircraft of men or supplies by parachute. **B.** *v. trans.* To drop (men or supplies) in this way. Hence **paradro·pping** *vbl. sb.*

1948 *Shell Aviation News* No. 118. 9/1 Back at the main supply base of Shell-Mera a radio message is received from this headquarters for a paradrop and feverish activity ensues. 1950 *Birmingham* (Alabama) *News* 27 Nov. 28/3 An Iuka man was one of the pilots who para-dropped more than 4,000 men of the 11th Airborne on an arc between the North Korean cities of Sukcon and Sunchow. 1952 *Time* 31 Mar. 71/2 (Advt.), Here, it paradrops vital supplies 'up front'. 1961 *Flight* LXXX. 371/2 The rear ramp allows paradropping. 1971 *Morning Star* 26 June 1 Liberation troops inspecting a U.S. ammunition paradrop captured in the far north of South Vietnam.

parafango (pæ:răfa·ŋgo). [f. PARA(FFIN *sb.* + *FANGO.] A mixture of mud and paraffin wax used for medicinal purposes (see quots.).

1969 *Daily Tel.* 25 June 15/6 Other new treatments recently introduced at Henlow include..the parafango, a mixture of mud and paraffin wax for spot reducing, arthritis and rheumatism. 1970 *Guardian* 4 Apr. 11/7 Parafango is the technical term for warm wax or mud baths.

parafe, *v.* App. var. of PARAPH *v.* 2.

1922 Joyce *Ulysses* 212 Farrell parafes his polysyllables.

paraffin, *sb.* Add: **4.** *paraffin heater, stove, tin*; **paraffin oil** (later examples); **paraffin test** (see quots.).

1939–40 *Army & Navy Stores Catal.* 191/3 Coleman Paraffin Heater..Burns ordinary paraffin oil. 1975 J. McClure *Snake* iv. 59 She had dumped..that very serviceable old paraffin heater, that was only a little rusty, on her new rubbish tip. 1976 *Sunday Mail* (Glasgow) 21 Nov., It is a disgrace that people who have worked hard all their days should be forced to use paraffin heaters because they cannot afford their electricity bills. 1949 Paraffin oil [see *KEROSENE]. 1950 *Sci. News* XV. 99 Serum from the umbilical cord can be guaranteed to increase haemoglobin production in rats only if collected under paraffin oil, that is, when protected from the oxygen of the air. 1966 M. Woodhouse *Tree Frog* xvi. 123 There was a pressure cooker and two large paraffin stoves. 1978 'L. Black' *Foursome* ii. 18 The fug of the small wooden shed heated by a paraffin stove. 1950 *Ellery Queen's Mystery Mag.* Oct. 101/1 'What's a paraffin test?' asked Nicky... 'Every gun..has a certain amount of backfire. Some of the gunpowder flashes back and is embedded in the hand of the man that fires. They coat his hand with hot paraffin and then draw it off like a glove. They then test it for gunpowder..and if it's positive, it means that the man fired the gun.' 1974 R. B. Parker *Godwulf Manuscript* iii. 23 A paraffin test. When you fire a handgun cordite particles impregnate your skin. A lab man puts paraffin over it, lets it dry, peels it off, and tests it. The particles show up in the wax. 1935 H. Edib *Clown & his Daughter* lv. 342 Some of them brought empty paraffin-tins. 1937 K. Blixen *Out of Afr.* I. 12 The Swahili town..was built mostly out of old paraffin tins hammered flat.

paraffiny (pæ·răfini), *a.* [f. PARAFFIN *sb.* + -Y[1].] Of, belonging to, or suggestive of paraffin; covered or smeared with paraffin; smelling of paraffin.

1902 Conrad *Youth* 21 The ascending air was hot, and had a heavy, sooty, paraffiny smell. 1925 B. Beetham in E. F. Norton *Fight for Everest, 1924* III. vi. 368 Paraffiny fingers will taint the whole canteen. 1952 'J. Tey' *Singing Sands* vi. 93 A large wooden tray of tuppenny buns... They were crummy and depressed-looking,..and they smelled very faintly of paraffin...the paraffiny buns and the margarine.

Parafilm (pæ·răfilm). Also **parafilm.** A proprietary name for certain thermoplastic materials (see quots. 1952, 1956).

1934 *Official Gaz.* (U.S. Patent Office) 18 Sept. 520/2 Marathon Paper Mills Co., Rothschild, Wis. Filed July 23, 1934. *Parafilm.* For moistureproof, self-sealing flat wrapper, claims use July 11, 1934. 1952 *Trade Marks Jrnl.* 21 May 472/1 Parafilm... Backing cloth, being piece goods..consisting of textile material coated with a thermoplastic substance containing rubber, the textile material predominating, for use in the manufacture of boots and shoes. Lindsay & Williams Limited,..Manchester..; manufacturers. 1956 *Ibid.* 8 Feb. 131/2 Parafilm... Electrical insulation identification tape. Lindsay & Williams Limited,..Manchester..; manufacturers. *Ibid.* 26 Sept. 947/2 Parafilm... Thermoplastic materials in the form of sheets, ribbons and tapes, none being textiles... Lindsay & Williams Limited,..Manchester..; manufacturers. 1967 K. M. Smith *Insect Virology* xi. 214 Recently..the animal membrane has been replaced by stretched Parafilm. 1974 *Nature* 3 May 85/1 The mouth-parts of a dehydrated tick were inserted through a wax-coated parafilm membrane up to the base of the palps.

parafiscal, *a.*: see *PARA-[1] 1.

parafoil (pæ·răfoil). Also **para-foil.** [f. *PARA-[3] + *AERO)FOIL.] A structure of fabric designed to function as both a parachute and an aerofoil, providing lift that enables the wearer to glide.

1967 *N.Y. Times* 13 Aug. 15 A revolutionary parachute invention..known as the para-foil, would enable pilots bailing out over enemy territory to glide like birds until they reached safety. 1968 *Sunday Times* 28 Sept. 5/1 Air fills the cells of his parafoil and flows over the upper surface, creating 'lift', as with an aeroplane. 1975 *Sci. Amer.* Mar. 122/3 A major breakthrough was the invention of an inflatable multicell airfoil of fabric by D. C. Jalbert of Boca Raton, Fla... It is called the Para-Foil... Its shape in the form of a rigid, low-speed wing, is maintained entirely by air that enters openings at the leading edge to build up internal pressure. On landing the Para-Foil..can be collapsed into a manageable bundle of lines and cloth. One can also jump with it from an aeroplane. 1976 *Listener* 8 July 30/3 We sent up a parafoil instead, an amazing American invention without any rigid structure, a mixture of balloon, parachute, aerofoil and kite, which instantly climbs to the permitted height of 200 feet.

parafolli·cular, *a. Anat.* [f. PARA-[1] + FOLLICULAR *a.*] Situated near to, or around, a follicle: applied to cells found between the follicles of the mammalian thyroid gland, which secrete the hormone thyrocalcitonin.

1932 J. F. Nonidez in *Amer. Jrnl. Anat.* XLIX. 479 In the following pages the large epithelial cells with argyrophile granules will be termed 'parafollicular' cells, since they lie in the interstitial spaces in close proximity to the follicles from the epithelium of which they arose. 1968 H. Rasmussen in R. H. Williams *Textbk. Endocrinol.* (ed. 4) xi. 877/2 Between the follicles [of the thyroid gland] are groups of epithelial cells variously described as interstitial cells, mitochondrial-rich, or parafollicular cells. 1975 Francis & Martin *Introd. Human Anat.* (ed. 7) viii. 267 The ultimo-branchial body, an integral part of the thyroid, in man is represented by the parafollicular or C cells.

paraform(aldehyde: see *PARA-[1] 2 a.

parafovea (pærăfōᵘ·vĭă). *Anat.* Also with hyphen. [f. PARA-[1] + FOVEA, or as back-formation from next.] An annular area of the retina immediately surrounding the fovea centralis. Cf. *PERIFOVEA.

1941 S. L. Polyak *Retina* xvi. 211 The parafoveal region or parafovea is the intermediate belt of the central area [of the retina]. 1944 *Jrnl. Optical Soc. Amer.* XXXIV. 713/1 Rods, though absent from the fovea, appear in the parafovea and increase to a maximum density within the area of the retina to which the bright field image is projected. 1960 R. A. Weale *Eye & its Function* v. 65 The corneo-lenticular system cannot form a sharp and undistorted image outside an area called the para-fovea. This surrounds the macula and would be covered by the image of a circular disc subtending at the eye an angular diameter of some 20°. 1970 J. A. Howard *Aerial Photo-Ecol.* xiii. 141 The fovea is most sensitive to green light at 0·555μ whilst in subdued light the para-fovea responds most to light at 0·515μ.

parafoveal (pærăfōᵘ·vĭăl), *a. Anat.* [f. PARA-[1] + FOVEA + -AL.] Of or pertaining to the parafovea; adjacent to the fovea.

1925 *Brit. Jrnl. Ophthalm.* IX. 53 Frequently [in macular disease] the paracentral or parafoveal elements are mainly involved. 1941 [see prec.]. 1959 S. Duke-Elder *Parsons' Dis. Eye* (ed. 13) xxii. 321 Central serous retinopathy..is presumably caused by exudation from the parafoveal capillaries, probably of toxic or allergic origin. 1971 *Jrnl. Gen. Psychol.* LXXXIV. 48 The considerable amount of stray light..stimulates parafoveal rods.

Hence **parafo·veally** *adv.*, in a parafoveal manner; by means of the parafovea.

1960 *New Scientist* 10 Nov. 1267/2 When an object is seen parafoveally or peripherally, detail is not perceived; the object is seen as a comparatively vague shape, and the eye in these regions is sensitive mainly to motion. 1963 *Jrnl. Psychol.* LV. 394 Brightness enhancement was measurably present for part-spectrum impingements presented parafoveally under the proper conditions of intermittency.

paraganglioma (pæ:răgæŋgliōᵘ·mă). *Path.* Pl. **-omas, -omata.** [ad. F. *paragangliome* (Alezais & Peyron 1908, in *Compt. Rend. des Séances de la Soc. de Biol.* LXV. 746): see next and *-OMA.] A tumour thought to arise from a paraganglion (in its wider sense) or the adrenal medulla; *esp.* one of non-chromaffin tissue. Cf. *PHÆOCHROMOCYTOMA.

1914 *Surg., Gynecol. & Obstetr.* XVIII. 209/1 Whenever a tumor of a paraganglion—a paraganglioma—is suspected, it should be fixed in a solution containing chromic acid or its salts. 1925 *Amer. Jrnl. Anat.* XXXIV. 89 By far the greater number of pathologists believe in a vascular origin for the growths of the carotid body... Some later workers would call these tumors 'paragangliomas'. 1948 Martin & Hynes *Clin. Endocrinol.* vii. 144 Chromaffin tumours. These uncommon tumours include the phæochromocytomata which arise from chromophile cells of the adrenal medulla, and the paragangliomata which originate from chromaffin tissue in the intrathoracic cervical chain, the carotid bodies, or the organs of Zuckerkandl. 1956 H. M. Zimmerman et al. *Atlas Tumors Nervous Syst.* 177/1 The chromaffin tumors are found most commonly in the adrenal medulla... They are called pheochromocytomas (chromaffin tumor, functionally active paraganglioma)... The term 'paraganglioma' is best used for the non-secretory and hence non-chromaffin tumors of the paraganglionic tissues. 1974 Passmore & Robson *Compan. Med. Stud.* III. I. xvii. 37/1 Glomus jugulare tumours. These are another example of non-chromaffin paragangliomata; they..grow in the connective tissue that lies between the bulb of the internal jugular vein and the floor of the middle ear.

Hence **pa:raganglio·matous** *a.*

1965 *Jrnl. Clin. Path.* XVIII. 291/2 Medullary carcinoma may resemble various neural or paragangliomatous tumours.

paraganglion (pærăgæ·ŋgliən). *Anat.* Pl. **-ia.** [a. G. *paraganglion* (A. Kohn 1900, in *Arch. f. mikrosk. Anat.* LVI. 130): see PARA-[1] and GANGLION.] Any of several highly vascular groups of chromaffin cells that are similar to those of the adrenal medulla and in position and development are closely associated with the sympathetic nerve trunks; also applied to some structures now recognized as non-chromaffin (see quots. 1940, 1962).

1907 *Med. Rec.* (N.Y.) 3 Aug. 188/2 These occur in the medullary substance of the adrenal bodies, in the so-called paraganglia of the same organs. 1930 Maksimov & Bloom *Textbk. Histol.* xxxiv. 709 These paraganglia include the carotid gland and widespread, rather small accumulations of cells in the retroperitoneum which are often spoken of as the organs of Zuckerkandl. 1937 *Contrib. Embryol.* XXVI. 17 A study of the development of the carotid body..has left me unconvinced that it can be regarded as a paraganglion in the strict sense of the term. *A priori*..the term paraganglion means a structure of which the essential cells are derived in their entirety from the nervous system. 1940 *Q. Rev. Biol.* XV. 167/2 Recently..a distinction has been made between chromaffin and non-chromaffin paraganglia. 1962 E. C. Crosby et al. *Correl. Anat. Nerv. Syst.* viii. 545/1 Small collections of cells found in relation with certain blood vessels constitute the carotid body, the aortic bodies, the jugular body..and the coccygeal body. .. Although sometimes referred to as paraganglia, as the term is used in its wider sense, these cell masses usually contain no chromaffin cells and are unlike the chromaffin bodies in other ways as well. 1965 Lee & Knowles *Animal Hormones* iv. 85 In mammals additional chromaffin tissue may be found in the lower part of the abdominal aorta (para-aortic glands), or in contact with sympathetic ganglia (paraganglia).

Hence **pa:raganglio·nic** *a.*

1937 *Contrib. Embryol.* XXVI. 27 (*caption*) Reconstruction of the branchial-arch arteries in a 26-mm. human embryo to show positions in which 'paraganglionic' tissue can be found. 1959 W. Andrew *Textbk. Compar. Histol.* xiii. 530 No true carotid body is present in fishes, amphibians, and reptiles. It is only in birds and mammals that the paraganglionic cells in this region form an organ.

paragenesis. Add: Pl. **-geneses** (-dʒe·nĕsīz). **2.** Substitute for def.: The occurrence together of different minerals, esp. as reflecting the conditions of their formation; a set of minerals occurring together or with a given mineral; also, the sequence and periods of formation of the constituent minerals. (Earlier and later examples.)

1853 *Edin. New Philos. Jrnl.* LIV. 324 By the paragenesis of minerals he [*sc.* Breithaupt] understands the more or less definite mode of association, by means of which he endeavours to determine their relative age. 1871 J. H. Collins *Handbk. Mineral. Cornwall & Devon* 71 A more strict paragenesis would deal with those groups of minerals which are immediately associated with each other. 1934 *Q. Jrnl. Geol. Soc.* XC. 338 Sillimanite has been abundantly developed alongside some quartz veins... This paragenesis forms a selvedge between quartz veins and the biotite-muscovite-schist. 1951 *Mineral. Mag.* XXIX. 677 The paragenesis sylvine-halite - magnesite - quartz - anhydrite - (carnallite).. first makes its appearance in the halite zone. 1954 R. L. Parker tr. *Niggli's Rocks & Mineral Deposits* iv. 128 Another method..consists in constructing theoretical (so-called normative) mineral associations..to constitute idealized parageneses under certain physical-chemical conditions. 1966 E. W. Heinrich *Geol. of Carbonatites* vii. 233 (*caption*) Paragenesis (from oldest to youngest): (1) some cerite.., (2a) bastnäsite.., (2b) monazite, (3) parisite, (3b) sahamalite. 1974 *Nature* 22 Nov. 336/1 The mineral parageneses of alkali pegmatites are not considered in detail.

paragenetic, *a.* Add: **c.** *Min.* Involving or pertaining to paragenesis (sense *2). Hence **paragene·tically** *adv.*

1853 *Edin. New Philos. Jrnl.* LIV. 325 The paragenetic phenomena met with in druses..indicates that the deposition of some more recent minerals has taken place more readily upon certain of the pre-existing minerals than upon others. 1963 *Mineral. Abstr.* XVI. 166/1 Paragenetically, pitchblende is associated with calcite, quartz, [etc.]. 1966 E. W. Heinrich *Geol. of Carbonatites* vii. 182 All the other sulfides..are not only uncommon to very rare but, unlike most of the pyrite, are usually paragenetically late. 1974 *Nature* 15 Mar. 261/2 Miyashiro distinguishes, somewhat anomalously, some subfacies on the basis of facies series, even though paragenetic criteria may not be available.

parageosyncline, -al: see *PARA-[1] 1.

paraglider (pæ·rǎgləidəɹ). [f. *PARA-[3] + GLIDER.] A large kite-like structure composed of two flexible triangular sections joined side by side, and designed to glide with a passive load or with a pilot to control its flight.

Whether this is the sense in quot. 1942 is uncertain.

1942 A. M. Low *Parachutes* 223 There have been occasions when Russian pilots with their machines damaged have dived them straight on to their targets. No real importance attaches to these 'human bombs'... Paragliders released from aircraft may prove quite another story. 1960 F. M. Rogallo et al. *Prelim. Investigation of Paraglider* (NASA TN D-443) 4 In evaluating the paraglider concept in a practical application as a reentry vehicle, calculations were made by using equations of motion involving two degrees of freedom. *Ibid.* 8 The results of this study indicate that this flexible-lifting-surface concept may provide a lightweight controllable paraglider for manned space vehicles. 1961 *Flight* LXXX. 651/2 There are two recovery systems that are now being seriously considered for application to the Saturn system —the Rogallo or paraglider wing, and the parachute recovery. 1966 [see *low-altitude* s.v. *LOW *a.* 20]. 1973 *Daily Colonist* (Victoria, B.C.) 3 May 1/7 A 27-year-old father of three was killed Wednesday when his lightweight para-glider plunged 200 feet to the ground.

paragnath(us to -gnostic: see *PARA-[1] 1.

paragogic, *a.* (Further examples.)

1827 [see *ASYLLABIC *a.*]. 1968 W. S. Allen *Vox Graeca* iv. 95 Adding the so-called ν ἐφελκυστικόν (alias 'paragogic ν'). 1972 *Language* XLVIII. 35 The development of such 'paragogic' vowels is known also from Ukrainian and Czech dialects. 1975 *Canad. Jrnl. Linguistics* XX. 61 Portuguese phonotactics generally does not tolerate word-final stops; thus, borrowed words ending in a stop receive a paragogic final *e*: *time* 'team', *clube* 'club', etc.

paragon, *sb.* (*a.*) **6.** Delete † *Obs.* and add later examples. Usu. written **paragone.**

1848 J. D. Dana *Man. Mineral.* vii. 349 The *Nero-antico* marble of the Italians is an ancient deep black marble; the *paragone* is a modern one, of a fine black color, from Bergamo. 1888 G. H. Blagrove *Marble Decoration* 68 In Italy a black marble, sometimes called Paragone, is found mixed with marble of inferior quality at Castle Nuovo, in Piedmont [etc.]. **1894** H. W. Pullen *Handbk. Anc. Roman Marbles* II. 140 The term Paragone has..been loosely applied to several very black columns, such as those at a Tomb in the Winter Choir of St. John Lateran. **1955** M. H. Grant *Marbles & Granites of World* 71 Paragone. [Locality.] Bergamo, Italy. [Characteristics.] A pure, fine black.

paragonimiasis (pæ:răgonimai·ăsis). *Med.* [f. mod.L. *Paragonim-us*, generic name (f. Para-¹ + Gr. γόνιμος productive, fertile, f. root γεν-, γον- to produce) + *-iasis.] Infestation with worms of the genus *Paragonimus*, esp. the lung fluke, which results from eating infected crustacea and is marked at first by abdominal pains and later by a persistent cough and expectoration of blood.

1907 *Allbutt's Syst. Med.* (ed. 2) II. II. 861 The symptoms of paragonimiasis, or endemic hæmoptysis as it is sometimes designated, are a chronic cough—usually worst in the morning, a persistent pneumonic-like sputum in which ova abound, and recurring attacks of more or less profuse hæmoptysis. **1935** *Nature* 26 Oct. 674/1 There is fortunately no reason for anticipating that the crab will introduce into Europe the lung disease, paragonimiasis, of which it is one of the vectors in the Far East. **1961** L. E. Bollo *Introd. Med. & Med. Terminol.* xiv. 167 Paragonimiasis (caused by the lung fluke *Paragonimus westermani*) is common in Korea, Japan, the Philippine Islands, and parts of China. **1973** Kun-Yen Huang in J. R. Quinn *Med. & Public Health China* 258 Although careful roentgenological examinations..may differentiate paragonimiasis from tuberculosis, the diagnosis of paragonimiasis relies primarily on the demonstration of parasite ova in the sputum or an intradermal or CF test with antigen prepared from adult worms.

paragrammatism, etc.: see *para-¹ 1.

paragraph, *sb.* Add: **2. c.** *transf.* A distinct passage or section in a musical composition.

1959 *Listener* 16 July 114/2 The opening paragraph of the Fifth Symphony..takes the old-type dirge..as its model. **1975** *Gramophone* Sept. 466/3 In the slow movements and the cadenzas he shows himself to be capable of shaping long paragraphs with real discrimination. **1977** *Listener* 12 May 628/3 The opening..is one of the most difficult in the symphonic repertory..creating a tension from which the big first paragraph must be felt to spring.

3*. In Ice-skating, used *attrib.* and *absol.* with reference to the manner in which various figures are performed in competitions.

1930 T. D. Richardson *Mod. Figure Skating* xx. 184 Let me give a few suggestions of figures requiring the utmost technique; rockers and counters in eight form; three rocker three, and three counter three in paragraph form, i.e. making an eight formed figure. **1948** — *Compl. Figure Skater* ix. 79 (*caption*) The first of the 'paragraph' figures—one foot eight forward. **1952** E. Jones *Elements Figure Skating* (ed. 2) vi. 127 The complete paragraph consists in order of a half-circle on the right outside edge, a full circle on the right inside edge, then a take-off on to the left foot, a half-circle on the left inside edge and finally a full circle on the left outside edge..this means describing three circles,..all in exact line with one another, all of equal size and symmetrically constructed. **1959** T. D. Richardson *Girls' Bk. Skating* iv. 57 All you have to do..is to apply your knowledge of the components when putting figures into paragraph form. **1967** *Daily Tel.* 1 Mar. 12/6 The powerful East German later narrowed the gap with her more consistent second tracing, the backward paragraph three. **1973** *Times* 7 Feb. 15/8 On the second figure, the paragraph-loop, he was beaten.

4. paragraph mark = Paragraph *sb.* 1.

1855 *N. & Q.* 29 Dec. 521/2 The old paragraph mark, ¶, he [*sc.* Bilderdijk] considers to be the Roman P. **1956** H. Williamson *Methods Bk. Design* ix. 119 If indention is not used, the typographer will have to find some other means of indicating the start of a new paragraph, such as a drop initial or a paragraph mark—¶.

paragraph, *v.* Add: **3.** Also *fig.* Cf. Punctuate *v.* 3b.

1909 H. G. Wells *Ann Veronica* ix. 168 Ramage looked at her, and then fell into deep reflection as the waiter came to paragraph their talk again. **1959** *Vogue* Dec. 91 A soft dress in pure silk is scoop-necked and paragraphed with a lightly tying belt.

pa·ragraphed, *ppl. a.* [f. Paragraph *v.* + -ed¹.] Mentioned or written about in a newspaper paragraph.

1898 G. B. Shaw *Plays Pleasant* Pref. p. ix, The much paragraphed 'brilliancy' of *Arms and the Man.* **1928** *Manch. Guardian Weekly* 17 Aug. 135/2 A new comedy and the first visit to Manchester of a much-paragraphed young actress brought a large and eager audience to the Palace. **1930** *London Mercury* Feb. 319 He realised..that if he ever linked his future with a member of the opposite sex, it would not be with any such perfect and paragraphed ecstasy as Dandylion or Clytemnestra.

paragraphist. Add: (Earlier U.S. examples.)

1790 *Gazette of U.S.* 27 Nov. 655/1 A paragraphist in the General Advertiser of Thursday last. **1792** T. Jefferson *Writings* (1854) III. 467 One of its principal ministers enlists himself as an anonymous writer or paragraphist.

Paraguay. Add: Also with pronunc. (-gwai).

1. (Earlier and later examples of *Paraguay tea.*)

1793 B. Edwards *Hist. Brit. Colonies W. Indies* I. 476 Ilex Cassine. Paraguay Tea. **1825** J. C. Loudon *Encycl. Agric.* I. 200 Paraguay tea..is used [in Brazil] as a substitute for that of China. **1839** W. Parish *Buenos Ayres* xv. 347 Even the yerba-mate, or Paraguay tea..is now introduced from the southern provinces of Brazil. **1856** *Househ. Words* 3 May 377/2 An eligible draught presents itself in the shape of Yerva de Paraguay, or Paraguay tea. **1924** Record & Mell *Timbers Trop. Amer.* II. 92 Holly bushes, the source of the famous Brazilian or Paraguay tea or 'herva mate'. **1937** A. F. Hill *Econ. Bot.* xxi. 510 Paraguay tea..is next to coffee, tea, and cocoa in importance.

Paraguayan (pærăgwē·ăn, -gwai·ăn), *a.* and *sb.* Also 7 **Paragueyan,** 9 **Paraguarian.** [f. Paraguay + -an.] **A.** *adj.* Of or belonging to Paraguay or its inhabitants; produced in or characteristic of Paraguay. **B.** *sb.* A native or inhabitant of Paraguay.

1693 P. Gordon *Geogr. Anatomized* II. iv. 192 The Paraguayans are reported to be a people of very tall and big bodies, yet extraordinarily nimble and much given to running. **1699** *Ibid.* (ed. 2) II. ii. 283 The opposite Place of the Globe to Japan, is that part of the Paragueyan Ocean, lying between 340 and 350 Degrees of Longitude, with 30 and 40 Degrees of Southern Latitude. **1832** J. Bell *Syst. Geogr.* VI. I. 238 The Paraguarians collected an army of 6,000 men. **1856** C. Kingsley *Misc.* (1859) II. 18 Very interesting also..are..scattered hints as to the qualities of the Paraguayans themselves. **1884** R. G. Watson *Spanish & Portuguese S. Amer. Colonial Period* I. xvi. 273 Two Fathers, accompanied by thirty Paraguayan disciples, set out. **1901** *Chambers's Jrnl.* Sept. 623/2 The Paraguayans are ominously polite. **1935** P. de Ronde *Paraguay* iii. 43 From the unions between the Spanish and the natives there evolved in course of time the present Paraguayan race. **1956** G. Durrell *Drunken Forest* iii. 62 Our housekeeper, a dark-skinned, dark-eyed Paraguayan woman. **1957** P. Kemp *Mine were of Trouble* vi. 114 He had smuggled arms to one or other of the belligerents in the Gran Chaco war—I think to the Paraguayans. **1973** G. Greene *Honorary Consul* v. ii. 264 He does not think in terms of Paraguayans, Peruvians, Bolivians, Argentinians. **1977** *Gramophone* Dec. 1106/1 Agustin Barrios, the Paraguayan guitarist and composer.

paragutta (pærăgɒ·tă). [f. Para-¹ + Gutta².] A material derived from rubber and gutta percha used for insulating telephone cables.

1931 A. R. Kemp in *Jrnl. Franklin Inst.* CCXI. 37 An insulation called paragutta has been developed which as the name suggests is derived essentially from rubber and gutta percha. **1959** J. W. Freebody *Telegr.* xvii. 671/2 Paragutta, which has a specific capacitance of 2·6, eventually replaced gutta percha for the insulation of more modern cables.

parahelium (stress variable). Also **parhelium.** [f. *para-¹ 3 + Helium.] The form of helium whose spectrum does not exhibit the fine structure of orthohelium, owing to the spins of the two orbital electrons being antiparallel.

1896 Runge & Paschen in *Astrophysical Jrnl.* III. 18 We come to the conclusion that clèveite gas consists of two elements, one corresponding to the second and third series (single lines), and the other corresponding to the fifth and sixth series (double lines..). If this is true, the name *Helium* should be given only to the second element... The first element Professor Stoney has proposed to call *Parhelium.* **1903** A. M. Clerke *Probl. Astrophysics* v. 58 'Parhelium' may then safely be treated as fictitious. **1922, 1961** [see *orthohelium]. **1974** G. Reece tr. *Hund's Hist. Quantum Theory* iv. 60 Runge and Paschen were able to add series for O, S, Se and He in 1895–7. They discovered the two spectral systems of orthohelium and parahelium in the process.

parahippocampal, -hopeite: see *para-¹ 1, 2 c.

parahormone (pærăhǭ·ɹmǭun). *Physiol.* Also † **parhormone.** [a. F. *parhormone* (E. Gley 1911, in *Rev. Scientifique* 11 Mar. 262): see Para-¹ and *hormone.] A product of metabolism which has a secondary hormonal role.

1918 M. Fishberg tr. *E. Gley's Internal Secretions* 154 At most we might apply the term *parhormones* to the excretory products which play only an accessory rôle as excitants, reserving the name of hormones for the specific glandular products. **1922** R. G. Hoskins in L. F. Barker et al. *Endocrinol. & Metabolism* I. I. 8 For nonspecific regulatory substances such as carbon dioxid, Gley has proposed the designation 'parhormones'. **1935** *Biol. Rev.* X. 440 Water..could presumably be classified as a parahormone for renal function. **1948** C. D. Turner *Gen. Endocrinol.* ii. 65 The synthesis, storage and release of sugar by the liver is not a true endocrine activity, since glucose is not a specific product of this organ. Sugar and such materials are important metabolites and may be designated as parahormones.

parahydrogen (stress variable). Also **para hydrogen** and with hyphen. [f. *para-¹ 3 + Hydrogen, as tr. G. *parawasserstoff* (Bonhoeffer & Harteck 1929, in *Naturwissenschaften* XVII. 182/1), coined on the analogy of *par(a)helium* (see quot. 1935 s.v. *orthohydrogen).] The form of molecular hydrogen in which the two nuclei in the molecule have antiparallel spins, so that the spectrum exhibits no hyperfine structure. Cf. *orthohydrogen.

1929 *Industr. & Engin. Chem.* 10 July 6/1 At room temperatures and atmospheric pressure in glass vessels para-hydrogen is a stable gas. **1935,** etc. [see *orthohydrogen]. **1969** H. T. Evans tr. *G. Hägg's Gen. & Inorg. Chem.* xviii. 452 The melting and boiling points of parahydrogen are approximately 0·1° lower than for ordinary hydrogen.

para-influenza (pæ:ră,influ,e·nză). *Biol.* Also **parainfluenza.** [f. Para-¹ + Influenza.] Any of a group of paramyxoviruses which resemble the influenza viruses and include the one causing croup; (orig. classed as myxoviruses before paramyxoviruses were recognized as a separate group). Usu. *attrib.*, esp. in *para-influenza virus.*

1959 C. H. Andrewes et al. in *Virology* VIII. 129 (*heading*) Para-influenza viruses 1, 2, and 3: suggested names for recently described myxoviruses. *Ibid.* 130 The following names are accordingly proposed: Sendai (including HA2). *Myxovirus para-influenzae 1* (Para-influenza 1). CA virus: *Myxovirus para-influenzae 2* (Para-influenza 2). HA1: *Myxovirus para-influenzae 3* (Para-influenza 3). **1959** *Brit. Med. Bull.* XV. 221/2 It seems..that the Far Eastern strain of para-influenza 1 virus (Sendai) is endemic in laboratory mice and very probably pathogenic for pigs and man also. *Ibid.* 222/1 Para-influenza 2. This virus was called CA (croup-associated) by Chanock (1956). It was independently described by Beale and his colleagues as the virus of acute laryngotracheobronchitis of children. **1965** C. Andrewes *Common Cold* xi. 98 Para-influenza infections are mainly seen in children in whom they cause respiratory infections of all degrees of severity up to fatal pneumonia. **1974** *Nature* 23 Aug. 650/2 As *in vivo* hybridisation seems to have permitted a human tumour to become..unusually lethal in the hamster,..similar mechanisms may be implicated in human cancer, particularly when such fusing agents as parainfluenza viruses are prevalent in man.

parajou·rnalism. orig. *U.S.* [f. Para-¹ past, beyond, contrary to + Journalism.] A type of unconventional journalism not primarily concerned with the reporting of facts. So **parajou·rnalist, parajournali·stic** *a.*

1965 D. Macdonald in *N.Y. Rev. Bks.* 26 Aug. 3/1 A new kind of journalism is being born, or spawned. It might be called 'parajournalism', from the Greek *para*, 'beside' or 'against': something similar in form but different in function. **1966** *New Statesman* 4 Mar. 300/1 The second achievement is 'parajournalism'—Dwight Macdonald's word. **1966** *Atlantic Monthly* June 89 MacDonald..castigates a Wolfe review of Mailer's latest novel and berates the reviewer as again playing parajournalist because his technique was 'to jeer at the author's private life and personality—or rather his *persona.*' **1970** *Americana Ann.* 503/1 A wave of parajournalistic publications, the so-called underground press, was mounting a serious challenge to established dailies. **1972** *Times Lit. Suppl.* 21 Apr. 444/1 (Advt.), 'What is now called parajournalism—writing it the way it happened' (Daily Mirror). **1978** *New York* 3 Apr. 81/2 Charlie, the megalomaniacal parajournalist.

parakeelya (pærăkī·lyă). *Austral.* Also **parake(e)lia, parakilja, parakylia.** [Aboriginal name.] An annual or perennial herb belonging to the genus *Calandrinia*, of the family Portulacaceæ, native to Australia and America and bearing succulent leaves and clusters of reddish flowers. Also *attrib.*

1898 D. W. Carnegie *Spinifex & Sand* vii. 216 Given 'parakeelia' every second night or so a camel would never want to drink at all. **1921** *Trans. R. Soc. S. Austral.* XLV. 11 The common parakeelya at Tarcoola..appears to be this species [*sc. Calandrinia polyandra*]. **1931** I. L. Idriess *Lasseter's Last Ride* xii. 99 The secret was in the Parakelia grass. In appearance something like ice-plant, its thick leaves were stored with water. **1934** [see *claytonia]. **1935** H. H. Finlayson *Red Centre* iii. 33 One can only refer to the sprightly purple and yellow flowered succulents, the parakeelia and munyeroo. **1944** F. Clune *Red Heart* 36 Crushed parakelia weed, broken mulga bough..—it was by signs such as these that the trackers deduced the course of the wayfarer. **1959** *Listener* 7 May 785/1 It [*sc.* the landscape] is ablaze with the scarlet of Sturt's desert pea, with cerise-coloured parakelia. **1965** *Austral. Encycl.* VI. 480/1 Parakeelya (Parakilja), the aboriginal name for several species of fleshy herbs in the genus *Calandrinia* (family Portulacaceae) which can thrive in the most arid parts of inland Australia. **1967** O. Ruhen in *Coast to Coast 1965–66* 189 A few parakeelya leaves can evoke the blessed relief of flowing saliva. **1972** Y. Lovelock *Vegetable Bk.* I. 193 The aborigines bake the succulent leaves of the parakilja..with bark as a food; the plant was once eaten by the early settlers.

parakite. Add: **2.** In modern use applied to a parachute kite which is towed along against the wind by a motorboat, car, or other fast vehicle. So **pa·rakiting,** the sport of soaring when harnessed to a parakite.

1962 *Aeroplane* CII. 233/2 Para kite.—First trials of the Lemoigne parachute kite have taken place in England... Mr. Walter Neumark remained at about 150 ft. for over 5 minutes while moored to the towing car. **1964** T. W. Willans *Parachuting & Skydiving* v. 50 The Lemoigne parakite of the 1960s—a sort of ascending parachute, a

few of which had found their way from France into England. **1970** *Time* 30 Mar. 42/1 In parakiting, the water skier becomes airborne when the trailing parachute pops open. **1978** *Lancashire Life* Apr. 50 (*caption*) Instead of hang gliding's crunch-down, parakiting's finale is a splash-down: on this occasion, in an icy sea. *Ibid.*, Stan Lyons was the guinea-pig—the man setting out to make Morecambe's first parakite flight between the Stone Jetty and West End Pier.

paralanguage (pæ:rălæ·ŋgwedʒ). *Linguistics.* [f. Para-¹ alongside + Language *sb.*¹] The system of non-phonemic but vocal factors in speech, such as tone of voice, tempo of speech, and sighing, by which communication is assisted. Cf. *kinesics.

Some authorities also include in paralanguage such adjuncts of speech as gesture and facial expression.

1958 G. L. Trager in *Studies in Linguistics* XIII. 4 The vocalizations and voice qualities together are being called *paralanguage* (a term suggested by A. A. Hill). **1959** H. L. Smith in *College English* XX. 172/2 Speech does not take place in a vacuum but is surrounded, as it were, by patterned bodily motions—the *kinesic* system—and by systematically analyzable vocalizations, or *paralanguage.* **1964** *Language* XL. 202 Trager's 1958 paper outlined a taxonomic system for the analysis of the phenomena of paralanguage. **1965** [see *morpheme-like* s.v. *morpheme c]. **1967** *Jrnl. Eng. Linguistics* I. 28 *Paralanguage*, the non-linguistic but communicatively significant orchestration of the stream of speech, involving such phenomena as abnormally high or low pitch, abnormally fast or slow tempo, abnormal loudness or softness, drawl, clipping, rasp, openness, and the like. **1972** W. M. Austin in A. L. Davis *Culture, Class, & Lang. Variety* viii. 159 We judge the dynamic and effective speaker by his use of kinesics and paralanguage as well as language. **1978** *Verbatim* May 15/2 That particular methodology called generative-transformational did not include paralanguage, kinesics, or cultural influences.

paralaurionite, -lexia: see *para-¹ 2 c, 1.

paralegal (pæ:rălī·găl), *a.* and *sb.* Chiefly *N. Amer.* [f. Para-¹ + Legal *a.*] **A.** *adj.* Of or pertaining to auxiliary aspects of the law. **B.** *sb.* One trained in subsidiary legal matters, though not fully qualified as a lawyer, etc.; a legal aide.

1972 *N.Y. Law Jrnl.* 22 Aug. 2/7 (Advt.), Para-legals (that is, legal assistants or paraprofessionals) are used by an ever-increasing number of prominent attorneys to reduce their unwanted load of paralegal matters and free up their time to render legal advice more efficiently. **1975** *Daily Colonist* (Victoria, B.C.) 23 Aug. 1/1 Miss Lissell intends to take a two-year para-legal course at one of the Vancouver colleges. **1977** *Times* 12 July 4/6 If legal aid was not available, the staff of the first-tier agency, who would have some paralegal training, would try to help. **1977** *N.Y. Rev. Bks.* 15 Sept. 55/2 (Advt.), Smith College graduate. Economics major. Financial, law and management background. Light skills. Seeks position as paralegal, office manager, social secretary, coordinator, etc.

paralic (păræ·lik), *a. Geol.* [ad. G. *paralisch* (C. F. Naumann *Lehrb. d. Geognosie* (1852) II. II. vii. 452), f. Gr. παράλ-ιος by the sea (see Paralian): see -ic.] Formed or having occurred in shallow water near the sea.

[**1911:** see *limnic *a.*] **1914** H. Ries *Econ. Geol.* (ed. 4) i. 13 A distinction is, however, sometimes made between (1) limnetic coals..; and (2) paralic coals, or those derived from plant remains which collected in marshes near the sea border. **1940,** etc. [see *limnic *a.*]. **1963** D. W. & E. E. Humphries tr. *Termier's Erosion & Sedimentation* i. 30 There was an extension of warm marine faunas, a large distribution of paralic flora and the development of lateritic soils. **1966** *McGraw-Hill Encycl. Sci. & Technol.* XII. 142/2 The molasse is a product of paralic sedimentation. **1977** A. Hallam *Planet Earth* 154 The seams formed from peats in coastal-swamp environments (paralic coals) are characteristically relatively thin after compaction.

paralinguistic (pæ:răliŋgwi·stik), *a.* and *sb.* *Linguistics.* [f. Para-¹ alongside + Linguistic *a.* and *sb.*; cf. *paralanguage.] **A.** *adj.* Of or pertaining to paralanguage; of or pertaining to vocal communication effected non-phonemically by tone of voice, tempo of speech, etc. So **pa:ralingui·stically** *adv.* Cf. *kinesic *a.*

1958 A. A. Hill *Introd. Ling. Struct.* xxi. 409 The paralinguistic area investigated by Birdwhistell has been called by him kinesics. **1959** [see *kinesic *a.*]. **1964** G. L. Trager in D. Abercrombie et al. *Daniel Jones* 267 It has become possible to separate out paralinguistic pitch phenomena from those of language proper. **1964** Crystal & Quirk (*title*) Systems of prosodic and paralinguistic features in English. **1965** *Canad. Jrnl. Linguistics* XI. 36 This is linguistically and paralinguistically irrelevant. **1967** M. Argyle *Psychol. Interpersonal Behaviour* v. 89 People are often quite unaware of the emotive, paralinguistic aspects of their speech—they do not realize how cross they sound, for instance. **1971** M. L. Samuels in A. J. Aitken et al. *Edin. Stud. Eng. & Scots* 5 A paralinguistic feature like an unusual voice-quality is found to accompany an unusual phonetic system. **1976** *Amer. Speech 1974* XLIX. 286 Nonverbal, paralinguistic features related to male/female language involve voice pitch, body language and facial expression, and the place of silence in sex-role communication.

B. *sb.* (in form a *pl.*) [-ics.] The study of paralanguage; non-phonemic factors of vocal communication. Cf. *kinesics.

1958 A. A. Hill *Introd. Ling. Struct.* xxi. 408 A part of communication activity which is outside the area of microlinguistics..is what can be called paralinguistics. **1958** G. L. Trager in *Studies in Linguistics* XIII. 4 The words *paralinguistic* and *paralinguistics* are self-explanatory. **1962** D. H. Hymes in J. A. Fishman *Readings Sociol. of Lang.* (1968) 106 Features of speech, such as tone of voice and hesitation pauses,..phenomena which have recently been systematized in a preliminary way under the heading of 'paralinguistics'. **1972** *Sci. Amer.* Sept. 60/1 The most probable links to investigate exist within human paralinguistics: the extensive array of facial expressions, body postures, hand signals and vocal tones and emphases that we use to supplement verbal speech. **1973** *Archivum Linguisticum* IV. 53 *Linguistic levels*..Grammar, lexis, phonology, semantics, graphology, phonetics, paralinguistics.

parallax. Add: **3.** Special comb.: **parallax error,** error in reading an instrument caused by parallax when the scale and the indicator are not precisely coincident.

1901 C. J. Leaper *Graduated Exercises Elem. Pract. Physics* p. iii/1 (Index), Parallax error. **1906** Bower & Satterly *Pract. Physics* v. 83 The reading, especially of the burette, is very liable to parallax error. **1967** *Electronics* 6 Mar. 117/2 With analog instruments..operator and parallax errors, meter movement wear and aging often reduce their nominal accuracies.

parallel, *a.* and *sb.* Add: **A.** *adj.* **1.** *parallel bars* (earlier and later examples); also *fig.*

1868 Trollope *He knew he was Right* (1869) I. iii. 17 Certain poles and sticks and parallel bars with which feats of activity might be practised. **1878** [see *horizontal *a.* (*sb.*) 2 b]. **1962** H. C. Weston *Sight, Light & Work* (ed. 2) viii. 225 Various objects have been suggested and used, such as the 'parallel-bar' test-object of Luckiesh. **1964** G. C. Kunzle *Parallel Bars* 19 The parallel bars are the most interesting and varied of all the pieces of apparatus. **1973** J. Fleming *You won't let me Finish* vii. 52 He was in the small gym giving exhibitions of walking along the parallel bars on his hands.

b. *parallel text,* one of two or more versions of a literary work, etc., printed in a format which allows direct textual comparison, freq. on facing or consecutive pages of the same volume; a text of different versions of a work set out in such a way; also (with hyphen) *attrib.*; *parallel tracking,* tracking in which the pick-up is kept tangential to the record groove by a rectilinear motion of the arm; freq. *attrib.*; *parallel turn,* a swing in skiing, with the skis kept parallel to each other.

Delete *parallel circuit* and see instead sense A. 1 d below.

1870 F. J. Furnivall *Chaucer Soc. Six-Text Print of Chaucer's Canterbury Tales in Parallel Columns* (verso rear cover of first section), The issue for 1870 is, in the First Series, XIV. The Miller's, Reeve's, and Cook's Tales, with an Appendix of the Spurious Tale of Gamelyn, in 6 parallel Texts. **1871** — (*title*) Chaucer Society one-text print of Chaucer's minor poems, being the best text of each poem in the parallel-text edition. **1877** *Trans. Philol. Soc. 1875–76* 10 The two best features of our editing are minute accuracy and fullness of material. Hence our parallel-text editions. **1889** C. Plummer *Two Saxon Chrons. Parallel* p. viii, This, the first of our two parallel texts, is commonly known as the Parker MS. **1941** *B.B.C. Gloss. Broadcasting Terms* 22 Parallel tracking unit. **1954** Brodermann & McPartlin *Ski-ing for Beginners* ix. 77 The Parallel Turn is a speed turn. **1962** A. Nisbett *Technique Sound Studio* viii. 145 Where parallel tracking arms are used, calibration above the track is a fairly simple matter. (But parallel tracking..has not proved itself in practice.) **1963** *Amer. Speech* XXXVIII. 206 An especially fast kind of short parallel turn. **1971** (*title*) William Wordsworth: The Prelude: A parallel text. **1978** *N. & Q.* Feb. 75/2 Mrs. Bawcutt's own admirable parallel-text edition..will doubtless remain standard.

d. *Electr.* Involving connection in parallel (cf. sense B. 6 in Dict. and Suppl.).

[**1884** *Jrnl. Soc. Telegr. Engin.* XIII. 497 These two machines..may be connected in one of two ways: they may be in parallel circuit with regard to the external conductor..; or they may be coupled in series.] **1886** *Electrician* 19 Feb. 296/2 The three direct systems are the parallel, series, and parallel-series methods of attaching lamps to the main conductors. **1891** J. W. Urquhart *Dynamo Construction* xvi. 232 For many years this difficulty stood in the way of parallel working with series machines. **1940** *Chambers's Techn. Dict.* 614/2 *Parallel feed,* a method of connecting the anode of a thermionic valve to the high-tension supply through a high resistance or inductance, whilst the a.c. circuits are connected through a condenser. The d.c. and a.c. components of the anode current are thereby separated. **1962** [see sense *B. 6].

2. c. (Further examples.)

1953 K. Reisz *Technique Film Editing* i. 20 Porter himself developed this kind of parallel action editing further in his subsequent films. *Ibid.* 281 *Parallel action,* device of narrative construction in which the development of two pieces of action is represented simultaneously by showing first a fragment of one, then a fragment of the other, and so on alternately.

d. *Computers.* Involving the concurrent or simultaneous performance of certain operations; functioning in this way.

1948 *Math. Tables & Other Aids to Computation* III. 149 The use of plugboard facilities and punched cards permits parallel operation (as distinguished from sequence operation), with further gain in efficiency. **1963** W. H. Ware *Digital Computer Technol. & Design* II. xi. 3 Parallel arithmetic tends to be faster than serial arithmetic because it performs operations in all columns at once, rather than in one column at a time. **1974** P. H. Enslow *Multiprocessors & Parallel Processing* i. 1 This book focuses on ..the integration of multiple functional units into a multiprocessing or parallel processing system.

e. S. Afr. *parallel development* = *separate development.

1950 *Ann. Reg. 1949* 140 The conflict between the Southern Rhodesian policy of 'parallel development' with its emphasis on permanent European control, and the United Kingdom policy of partnership leading to self-government. **1971** *Weekend World* (Johannesburg) 9 May 3/4 Chief George agreed with an Opposition claim that parallel development in the Republic meant that the Europeans were in the sky and Africans were in the mud.

f. S. Afr. *parallel-medium,* used *attrib.* to designate schooling or a school in which instruction is given through the medium of more than one language.

1958 *Cape Argus* 10 Dec. 20/5 The classroom instruction given in Afrikaans-medium classes in a parallel-medium school would be as Afrikaans as instruction given in the classes of an exclusively single-medium Afrikaans school. **1971** *Sunday Times* (Johannesburg) (Business Section) 28 Mar. 4/2 (Advt.), Separate English and Afrikaans medium primary schooling, and parallel-medium schooling to matriculation standard is available.

g. Biol. *parallel evolution* = *parallelism 7.

1963 E. Mayr *Animal Species & Evolution* xix. 609 There are numerous cases of..parallel evolution in the animal and plant kingdoms. **1972** *Canad. Jrnl. Earth Sci.* IX. 1032/1 The shell shape of *C[olus] brevis* closely resembles that of *A[tractodon] stonei* and provides a striking example of parallel evolution in response to parallel selective pressures.

3*. *parallel cousin* = *ortho-cousin* (*ortho- 1).

1936 R. Firth *We, the Tikopia* vi. 221 The differentiation between cross-cousin and parallel cousin is certainly not one of the outstanding features of the Tikopia kinship system. **1949** F. Eggan in M. Fortes *Social Struct.* 124 Parallel cousins are treated as siblings, whereas cross-cousins are differentiated. **1970** E. Leach *Lévi-Strauss* 121 A parallel cousin..is a cousin of the type 'mother's sister's child' or 'father's brother's child'. **1972** [see *ortho-cousin* s.v. *ortho- 1].

b. Forming adjectival phrases with sbs., as *parallel-jaw(s), -plate.*

1951 *Good Housek. Home Encycl.* 325/2 The bench..having..a parallel-jaw vice at the other end. **1962** L. S. Sasieni *Princ. & Pract. Optical Dispensing* ix. 214 Parallel jaw pliers. **1971** B. Scharf *Engin. & its Lang.* viii. 60 (*caption*) Parallel jaw vice. **1926** *Encycl. Brit.* II. 331/2 The parallel plate method..for measuring the absolute conductivity of air. **1962** Corson & Lorrain *Introd. Electromagn. Fields* ii. 37 If one plate of a parallel-plate capacitor is charged on one face, an equal and opposite charge must exist on the opposite plate of the capacitor.

B. *sb.* **I. 1.** (Further example.)

1972 *Sci. Amer.* Dec. 102/1 Circles of varying radii that go around the hole or center of the torus on parallel planes are called parallels.

II. 6. *in parallel* (earlier and later examples): in *Electr.* also said of individual circuit components connected by such wires, so that a current is divided between them; also *transf.*

Delete the cross-reference to sense A. 1 b and see instead sense A. 1 d above.

1884 *Jrnl. Soc. Telegr. Engin.* XIII. 529 The two alternate-current generators cannot work in series;..they can work in parallel. **1943** C. L. Boltz *Basic Radio* i. 21 A voltmeter is always connected in parallel with whatever part of a circuit across which we wish to measure the electrical pressure. **1949** E. P. Abraham et al. in H. W. Florey et al. *Antibiotics* II. xv. 644 The basic unit.., of which there were six working in parallel, was a glass tube containing amyl acetate. **1962** D. F. Shaw *Introd. Electronics* iii. 51 The capacitance *C* is connected in parallel with the combination *L* and *R*. **1962** *Newnes Conc. Encycl. Electr. Engin.* 569/2 *Parallel operation.* The operation of generators or transformers in parallel implies equality of terminal voltage. **1971** *Sci. Amer.* June 31/3 This could be accomplished by using a flow rate of 2,300 gallons per minute with two pumps operating in parallel.

7. b. (Further examples.) Also *in parallel* (without *with*), concurrently, simultaneously.

1938 *New Statesman* 7 May 796/2 Nor..does it seem right to attribute to 'the scholastic tradition of the universities' of the time, the failure of social studies to grow..in parallel with the natural sciences. **1957** R. K. Richards *Digital Computer Components* viii. 365 By transmitting all bits of a word to and from storage simultaneously or 'in parallel' a great increase in the speed of operation can be obtained in comparison with the transmission of one bit at a time. **1969** P. B. Jordain *Condensed Computer Encycl.* 373 By searching all (or very many cells) in parallel, the time of the operation is greatly reduced. **1971** *Sci. Amer.* Sept. 45/1 In parallel with the increase in carbon dioxide in the atmosphere there has also been a rise in suspended particulate contamination. **1977** *New Scientist* 21 Apr. 140/1 ILLIAC IV is actually a group of 64 individual computers working in parallel and linked to a one billion bit bulk memory.

parallel, *v.* **6.** Delete ? *Obs.* and add later examples.

1907 *Smart Set* Mar. 52/2 He..recognizes the truth that so easily their paths might have paralleled if events had only favored. **1977** *Zigzag* Mar. 21/1 Then it parallels to R&B in quite a few ways.

9. *trans.* To connect (electrical apparatus) in parallel. Const. *with.*

1902 *Electr. Rev.* 27 June 1056/2 (*heading*) Apparatus for paralleling alternators. **1903** T. Sewell *Elem. Electr. Engin.* (ed. 2) xviii. 379 There is not so much danger in paralleling machines which have iron cored armatures, for their self-induction prevents a dangerous current from flowing. **1921** [see *phase *v.* 5 a]. **1952** G. V. Mueller *Alternating-Current Machines* ix. 339 When a shunt generator is to be paralleled with an operating d-c system, it is driven at its rated speed by a prime mover. **1965** *Wireless World* Sept. 431/1 They [*sc.* thyristors] may be used singly to give a 2A d.c. output or they may be paralleled to give any desired output provided that suitable arrangements are made for simultaneous firing.

parallelepipedal, *a.* (s.v. Parallelepiped). (Earlier and later examples.)

1754 *New & Compl. Dict. Arts & Sci.* II. 1394/1 The capacities of all sorts of vessels.., as cubical, parallelopipedal, cylindrical,..&c. are computed. **1950** L. R. Underwood *Rolling of Metals* I. iv. 83 If a rectangular network of lines on the bar before rolling is still rectangular after rolling, then the total deformation may be regarded as parallelepipedal. **1974** *Chem. Physics* VI. 2/2 The term 'unit cell' will be retained here..to mean the parallelepipedal cell (whether primitive or multiple) used by crystallographers.

parallelism. Add: **2.** (Further examples.)

1962 *Listener* 5 Apr. 606/2 The success of *apartheid* or parallelism or separate development—call it what you will —is dependent on educating the Bantu to take over all their responsibilities themselves. **1968** *Economist* 4 May 18/2 'Parallelism' in the activities of party and state can be eliminated quite simply by emphasising that the party is the boss and the government merely its executive servant [in Romania]. **1972** *Nature* 8 Dec. 339/2 A rough parallelism between the histories of the Iceland and Hawaii plumes is noteworthy.

b. (Further example.)

1955 P. W. Stallmar in *College English* XVII. 25/1 For relationships between works that are not necessarily borrowings of the one from the other, I would use the general label 'parallelism'. The differentia of the parallelism is, I suggest, that a parallelism is not necessarily a conscious borrowing.

3. Also in Anglo-Saxon poetry.

1813 J. J. Conybeare in *Archaeologia* XVII. 269 The parallelism (if I may be so allowed to term it) of the Anglo-Saxon writers... The poems attributed to Cædmon afford innumerable instances of the same figure. **1876** H. Sweet *Anglo-Saxon Reader* p. xcix, There is also a tendency to parallelism, or repetition of the same idea in different words. The last half of one line is often connected with the first half of the next in this way. **1935** A. C. Bartlett *Larger Rhet. Patterns Anglo-Saxon Poetry* iii. 30 Every literary model impelled the Anglo-Saxon toward structural parallelism in pairs. **1938** A. Campbell *Battle of Brunanburh* 38 The sentence structure is essentially that of the older verse, with its free use of parallelism both of expressions and sentences. **1977** J. A. Cuddon *Dict. Lit. Terms* 471 Parallelism is common in poetry of the oral tradition—for instance, in *Beowulf.*

6. *Psychol.* The theory that mental (psychic) and physical processes are concomitant and that any change in the one will be correspondingly reflected in the other.

1860 J. D. Morell tr. *Fichte's Contrib. Mental Philos.* iii. 41 How far into details this parallelism between the mind and the world reaches, it is the province of psychology to show. **1877** *Illustr. London News* 5 May 427/1 As to the relation of mind to matter, he held that there is an exact parallelism of mental and material events..as two aspects of the same thing. **1891** M. E. Lowndes tr. *Höffding's Outl. Psychol.* ii. 64 Both the *parallelism* and the *proportionality* between the activity of consciousness and cerebral activity point to an *identity* at bottom. **1902** *Encycl. Brit.* XXXII. 66/2 The last of these [*sc.* the Monism of Spinoza, which reduced matter and mind to parallel attributes of the One Substance]—severed, however, from Spinoza's metaphysics—is now the prevailing theory, and to it the term *Psychophysical Parallelism* most properly applies. **1925** C. D. Broad *Mind & its Place* iii. 121 Psycho-neural Parallelism has also a positive side. **1956** [see *interactionism]. **1976** *Progress in Sci. Culture* (E. Majorana Centre) Spring 11 If we are to avoid falling into parallelism with its self-stultifying philosophy of determinism, we have to develop a dualist-interactionist philosophy according to which the self-conscious mind has an identity and activity that are not entirely dependent on brain events.

7. *Biol.* The development of similar characteristics by two related groups of animals or plants, in response to similar environmental pressures.

1887 E. D. Cope *Origin of Fittest* ii. 98 Among the higher groups [of Lacertilia Leptoglossa] the parallelisms lie in the arrangement..of the head shields. **1898** A. S. Woodward *Outl. Vertebr. Palæont.* p. xxiii, The case of the horses is often cited as suggesting that such a parallelism in evolution may have occurred. **1907** V. L. Kellogg *Darwinism To-Day* viii. 279 (*heading*) Parallelisms in variation. **1934** W. E. Le Gros Clark *Early Forerunners of Man* i. 6 If this thesis is carried to a logical conclusion, it will necessarily demand a much greater scope for the phenomenon of parallelism or convergence in evolution. **1961** G. G. Simpson *Princ. Animal Taxon.* iii. 78 Parallelism is the development of similar characters separately in two or more lineages of common ancestry and on the basis of, or channeled by, characteristics of that ancestry. *Ibid.* 103 Parallelism may be difficult or practically impossible to distinguish from homology on one hand and convergence on the other. **1967** R. E. Blackwelder *Taxon.* iv. 139 Parallelism..differs from convergence in that the development of the similar features is the result of and is channelled by a common ancestry.

b. *Anthrop.* A similarity between the evolution and achievements of different cultures.

1937 R. H. Lowie *Hist. Ethnol. Theory* xi. 190 Schmidt differs from Morgan mainly in denying universal parallelism, *unilinear* evolution. **1949** G. P. Murdock *Social Struct.* vii. 116 The extraordinary extent of parallelism, both in kinship terminology and in types of kin and local groups. **1958** E. A. Hoebel *Man in Primitive World* (ed. 2) xxxiv. 607 Very little attention has been given to parallelism, or independent invention, in this discussion for the reason that relatively few of the total mass of cultural traits possessed and shared by the peoples of the world have been invented more than once. **1958** F. M. Keesing *Cultural Anthropol.* vi. 148 Anthropologists..have therefore been critical of attempts to read historical connections into what they regard as an instance of parallelism between Old and New World cultural elements.

parallelist. 2. Delete *nonce-use* and add further examples (cf. *parallelism 6, 7). Also *attrib.* or as *adj.*

1903 C. A. Strong *Why Mind has Body* I. i. 23 The parallelist hypothesis. *Ibid.* vii. 126 The two arguments most commonly appealed to by parallelists. **1925** C. D. Broad *Mind & its Place* iii. 124 The orthodox Parallelist ..goes much further. **1937** R. H. Lowie *Hist. Ethnol. Theory* v. 44 McLennan was essentially a parallelist. 'All the races of men have had..a development from savagery of the same general character.' *Ibid.* 48 Apart from the parallelist faith in universal stages, we note the erroneous idea that totemism generally implies worship. **1950** R. Piddington *Introd. Social Anthropol.* I. i. 27 Obviously, claimed the parallelists, there had been no contact between these peoples, and the similarity in custom must be explained by the operation of similar psychological processes in the two widely separated areas. **1976** *Progress in Sci. Culture* (E. Majorana Centre) Spring 11 It is not in question that the happenings in the cerebral cortex are *necessary* for the experience of consciousness... However, it must not be naively assumed that these brain events are *sufficient* for the conscious experiences... This in fact is the parallelist position.

parallelistic *a.* (further examples).

1904 G. S. Fullerton *Syst. Metaphysics* xxi. 341 He quite wrecks his parallelistic formula. **1934** [see *interactionism]. **1946** *Brit. Jrnl. Psychol.* Jan. 52 It was precisely because in the *Manual* Stout endeavoured to exclude philosophical discussion that his parallelistic conclusion remained, as *mere* parallelism must..remain, an exasperating mystery.

parallelity (pærăle·lĭti). [f. Parallel *a.* and *sb.* + -ity.] The state or condition of being parallel (*lit.* and *fig.*); parallelism.

1953 *Cases Court of Session, Scotl.* 290 There is no parallelity or necessary mutual dependence between the two inquiries. **1970** *Physics Bull.* Feb. 71/1 In the spaghetti, adjacent lengths tend to lie somewhere near to parallelity without achieving exact alignment.

parallelization (s.v. Parallelize *v.*). (Further examples.)

1892 J. Nasmith *Students' Cotton Spinning* v. 150 Its result is to effect a much greater parallelisation of the fibres in the carded sliver. **1922** *Bull. Geol. Soc. Amer.* XXXIII. 443 From the meteorological standpoint such a far-reaching parallelization of the fluctuations of the weather from one year to the other, in Sweden and North America, is surprising. **1933** *Trans. Faraday Soc.* XXIX. 211 The process of parallelisation of long thin bodies.

paralog (pæ·rălɒg). [f. Para-¹ + Gr. λόγ-ος word.] (See quot. 1968.)

1951 Trager & Smith *Outl. Eng. Struct.* 85 The instances just cited are examples of the use of different *paralogs*, a paralog being one of the forms constituting an inflectional paradigm. **1968** J. Jung *Verbal Learning* iii. 30 Nonsense syllables and other laboratory learning materials, such as trigrams and paralogs. Trigrams are nonsense syllables, that is, a trigram is any three-letter combination which does not form a word. Paralogs or dissyllables are verbal units containing two syllables and range from meaningless units to actual words. **1970** *Jrnl. Gen. Psychol.* LXXXIII. 55 The nonsense words..were five paralogs with approximately equal, low levels of association value.

paralogia (pærălǭu·dʒiă). *Med.* [f. Para-¹ + *-logia* (see -logy and -ia¹).] (See quots.)

1811 R. Hooper *Lexicon Medicum* 596/1 *Paralogia*, a delirium in which the patient talks wildly. **1857** R. G. Mayne *Expos. Lex. Med. Sci.* (1860) 877/1 *Paralogia*,..term for a slight degree of madness or of delirium. **1900** Gould *Dict. Med.* (ed. 5) 973/1 *Paralogia*, difficulty in thinking logically. **1905** S. Paton *Psychiatry* xiv. 383 Another important symptom [of dementia præcox]..is the grotesque irrelevancy exhibited in replying to questions (Paralogia). **1919** R. M. Barclay tr. *Kraepelin's Dementia Praecox & Paraphrenia* ii. 21 Evasion or paralogia consists in this, that the idea which is next in the chain of thought is suppressed and replaced by another which is related to it. **1923** Stedman *Med. Dict.* (ed. 7) 737/1 *Paralogia*, false reasoning, involving self-deception. **1965** *New Scientist* 25 Nov. 605/1 The disorders of generalization are..subdivided into lowering of the level of generalization and distortion of generalization. The..latter seems to be the same as Kleist's 'paralogia' and Cameron's 'overinclusion'.

paralyse, *v.* Add: **2.** (Later examples.) Also with constr.

1871 L. W. M. Lockhart *Fair to See* II. xxv. 280 He saw all this, quite paralysed out of the power of surprise or wrath. **1890** *Congress. Rec.* 19 May 4933/1 You boast about what you have done for the American farmer... What audacity! It paralyzes me. **1900** *Dialect Notes* II. 47 *Paralyze...* In phrase 'to *paralyze* the professor'; to make a perfect recitation.

paralysedly (pæ·rălaizdli), *adv.* [f. Paralysed *ppl. a.* + -ly².] In a paralysed manner.

1876 R. Broughton *Joan* III. I. xxxiii. 48 As she so paralysedly sits the door opens softly.

pa·ralysingly, *adv.* [f. Paralysing *ppl. adj.* + -ly².] In a paralysing manner.

In quot. used hyperbolically.

1926 *Socialist Rev.* Dec. 21 The paralysingly stupid 70/- a week shipping or insurance clerk.

paralysis. Add: **1. b.** *general paralysis* (earlier and later examples); also called *general paralysis of the insane* [tr. F. *paralysie générale des aliénés* (L. F. Calmeil *De la Paralysie considérée chez les Aliénés* (1826) 9], and now recognized as a form of neurosyphilis. So *general paralytic*, an individual with general paralysis.

1820 *Edin. Med. & Surg. Jrnl.* XVI. 373 Dissection of a case of general paralysis... The disease of the brain seemed to have originated in indolence and chagrin from the sudden loss of fortune. **1847** *Further Rep. Commissioners in Lunacy* 46 in *Parl. Papers* (Brit. Libr.) XLIX. 291 The forms of insanity, which are occasioned by extreme indigence and privation,..are of the worst kind, and..many of them have invariably a fatal termination. [*Note*] This is particularly observable in the very frequent form of paralysis, termed the general paralysis of the insane. **1856** *Jrnl. Mental Sci.* III. 170 General paralytics are not malignant, and although sometimes furious, their passion is gusty and transient. *Ibid.* 172 In general paralysis, the pathological conditions of which involve the whole nervous system, the excito motory sensibility is almost abolished. **1925** *Amer. Speech* I. 24/2 The disease 'general paralysis', also known as paresis, usually is spoken of as softening of the brain. **1930** *Daily Express* 8 Sept. 1/1 Formerly every person developing general paralysis of the insane died after a period of distressing symptoms and agony. **1941** T. Warwick *Handbk. Venereal Dis.* v. 53 The great majority of general paralytics are men. **1964** King & Nicol *Venereal Dis.* v. 62 In all cases of general paralysis, tests of the mental status should be carried out. **1970** W. J. Brown et al. *Syphilis & Other Venereal Dis.* vii. 125 The common types of neurosyphilis are tabes dorsalis, general paralysis of the insane, and meningovascular neurosyphilis.

c. *paralysis agitans* [L. *agitans* shaking], Parkinson's disease, shaking palsy.

1817 J. Parkinson *Ess. Shaking Palsy* in M. Critchley *James Parkinson* (1955) 153 (*heading*) Shaking palsy. (*Paralysis Agitans.*) **1845** *Encycl. Metrop.* VII. 548/2 Paralysis Agitans..consists of a feeble trembling action of the muscles, not amounting to palsy. **1888** W. R. Gowers *Man. Dis. Nervous Syst.* II. 594 The great characteristic of the tremor of paralysis agitans is, as Parkinson pointed out, that it continues during rest. **1909** [see *Parkinson¹]. **1941** *Brit. Jrnl. Psychol.* XXXII. 5 Only a few years after the conclusion of the peace he fell a tragic victim to that incurably fatal disease of the central nervous system, paralysis agitans. **1973** *Neurology* XXIII. 215 (*heading*) Prevalence of neoplasms and causes of death in paralysis agitans.

paralytic, *a.* and *sb.* Add: **A.** *adj.* **1. b.** Of a (form of) disease: characterized by paralysis.

1890 F. Taylor *Man. Pract. Med.* 279 (*heading*) General paralysis of the insane (paralytic dementia). **1948** O. Breland *Animal Facts & Fallacies* i. 45 The bats also transmit diseases... One of the worst is the frequently fatal paralytic rabies which has occasionally been transmitted to human beings. **1951** Whitby & Hynes *Med. Bacteriol.* (ed. 5) xxiv. 408 This preparalytic stage may progress no further... On the other hand paralysis or encephalitic symptoms may appear after a few hours (paralytic poliomyelitis). **1974** H. MacInnes *Climb to Lost World* ii. 33 My first reaction was to telephone our local doctor: 'Dr. Mackenzie, have you any idea where I can get a vaccine for Paralytic Rabies?' **1976** *Yorkshire Even. Press* 9 Dec. 1/5 A seven-month-old baby from Kippax, near Leeds, is in hospital with paralytic polio according to health authorities.

3. d. Intoxicated; incapably drunk. *slang.*

a **1921** E. W. Hornung in *Penguin Bk. Austral. Ballads* (1964) 103 The shanty-keeper *he* was just as steady as a rock, And me as paralytic as a fool. **1927** *Daily Express* 23 Nov. 7 Woman at the Thames Court: I was not drunk. I was suffering from paralysis. Mr. Cairns: I have heard being drunk called being paralytic. **1958** [see *honkers *a.*]. **1966** F. Shaw et al. *Lern Yerself Scouse* 76 Half-dreaming, half-par'latic on me back.

paralytically, *adv.* Add: Also *fig.* (see prec.)

1969 *Sunday Times* (Colour Suppl.) 16 Feb. 38/2, I was paralytically drunk.

paramagnet (pærămæ·gnét). [Back-formation from next, after *magnet* and *magnetic.*] A paramagnetic body or substance.

1909 in *Cent. Dict.* Suppl. **1973** *Nature* 21/28 Dec. 445/1 Only above 100 K does the material become a self-respecting paramagnet, with thermally randomised spins.

paramagnetic, *a.* Add: **1.** (Earlier and additional examples.) This sense is now *Obs.*, having been superseded by the next.

1850 W. Whewell *Let.* (1876) II. 364 [*To Faraday.*] Hence it would appear, that the two classes of magnetic bodies are those which place their length *parallel* or

according to the terrestrial magnetic lines, and those which place their length *transverse* to such lines. Keeping the preposition *dia* for the latter, the preposition *para* or *ana* might be used for the former; perhaps *para* would be best as the word *parallel*..would be a technical memory for it. Thus we should have this distribution—Paramagnetic: Iron, Nickel, Cobalt, &c. Diamagnetic: Bismuth, Phosphor, &c. If you like *anamagnetic* better than *paramagnetic*, as meaning magnetic *according to* our standard, terrestrial magnetism, I see no objection. **1880** [see *ferromagnetic *a.* 1].

2. a. Of a body or substance: very weakly attracted by the poles of a magnet but not retaining any permanent magnetism; having a susceptibility that is positive but small, so that the relative permeability is only slightly greater than one and hysteresis does not occur. **b.** Characteristic of or pertaining to paramagnetism. Cf. *ferromagnetic *a.* 2 and the note there.

1896, etc. [see *ferromagnetic *a.* 2]. **1902** *Encycl. Brit.* XXX. 430/2 The magnetic susceptibility of a vacuum..is 0, that of a diamagnetic substance..has a negative value, while the susceptibility of paramagnetic and ferromagnetic substances..is positive. **1903** *Proc. R. Soc.* LXXI. 239 The law of Curie, that the susceptibility of weak paramagnetic substances is inversely proportional to the absolute temperature. **1931** S. R. Williams *Mag. Phenomena* v. 159 Oxygen, palladium, air, glass, porcelain, and $FeSO_4$ were some of the paramagnetic bodies studied by Curie. **1958** N. Cusack *Electr. & Magn. Prop. Solids* xii. 269 At a critical temperature, called the Curie point.., ferromagnetic matter undergoes a phase transition and becomes paramagnetic. **1966** C. R. Tottle *Sci. Engin. Materials* vi. 133 Curie..deduced the law for paramagnetic susceptibility.

B. *sb.* A paramagnetic body or substance.

1855 *Phil. Mag.* IX. 293 The assertions I have made regarding mutual influence..are confirmed amply by experiment for paramagnetics. **1911** *Encycl. Brit.* XVII. 324/1 Between the ferromagnetics and the paramagnetics there is an enormous gap. **1914, 1962** [see *ferromagnetic *sb.*]. **1966** C. R. Tottle *Sci. Engin. Materials* vi. 134 The group of elements iron, cobalt, nickel demonstrates this. Below a certain critical temperature these elements remain permanently magnetized after removal of the external field... Above it, the materials behave as normal paramagnetics.

paramagnetically, *adv.* (Examples.)

1883 *Encycl. Brit.* XV. 248/2 By virtue of differential action, a body may behave paramagnetically or diamagnetically according as it is placed in a less or in a more permeable medium than itself. **1974** *Nature* 31 May 426/1 The reduction at 25° C of the cytochrome c_3 from *Desulfovibrio vulgaris*..has been studied by following the paramagnetically shifted NMR resonances which lie to very low field.

paramagnetism. Add: (Earlier and later examples.) Now distinguished from *ferromagnetism, but formerly synonymous with it.

1850 W. Whewell *Let.* (1876) II. 365 [*To Faraday.*] Will it not do to talk of iron, nickel, &c., as *paramagnetic*, and glass, phosphorus, &c., as *diamagnetic*? Then this new branch of science, for so, of course, it will soon become, will be *Paramagnetism.* **1930** [see *ferromagnetism]. **1958** N. Cusack *Electr. & Magn. Prop. Solids* xiii. 291 All matter possesses a basic diamagnetism but where there are carriers, e.g. free electrons in metals,..the orientation of the carriers in the field direction gives a paramagnetism which usually masks the diamagnetism. **1964** J. W. Linnett *Electronic Struct. Molecules* iii. 53 The paramagnetism of the ground state of O_2.

paramedian, *a.* (Examples.)

1890 Billings *Med. Dict.* II. 289/1 Paramedian sulcus. **1928** [see *induction 9 b]. **1962** *Lancet* 19 May 1049/2 The abdomen was opened through a long left paramedian incision. **1977** *Ibid.* 28 May 1128/1 Patients in whom a midline or paramedian incision was made were not included.

paramedic¹ (pærăme·dik). orig. *U.S.* Also **para-medic.** [f. *para-³ + Medic *a.* and *sb.*] A person trained to be dropped by parachute to give medical aid.

1951 *Sunday Mirror* (N.Y.) 8 Apr. 3 Para-medics from air-sea rescue squadrons..were in the search planes. **1957** *Time* 2 Sept. 35/2 A C-47 with a paramedic aboard started to track his flight. **1961** *Daily Tel.* 27 Feb. 18/2 The group's nucleus consists of six 'paramedics'. These are practised parachutists with a medical background. **1964** J. Masters *Trial at Monomoy* vii. 225 Why didn't they send a plane down with skis, paramedics, radios, food? **1974** *Spartanburg* (S. Carolina) *Herald* 25 Apr. A2/5 He hopes the state pilot, advanced course for paramedics will be available to local EMS technicians by August or September. **1974** *Courier-Mail* (Brisbane) 17 Sept. 11/4 (*caption*) Dummy United States Air Force para-medic, dressed in full gear, at the jump position of a door of a U.S.A.F. Hercules rescue aircraft.

paramedic² (pærăme·dik). [f. *paramedic(al *a.*¹ and *sb.* (cf. Medic *a.* and *sb.*).] A paramedical worker; = *paramedical *sb.* Also *attrib.* or as *adj.*

1970 *Time* 9 Nov. 38 More than 40 training programs for doctors assistants are under way across the country. The graduates..are tagged with clumsy names—paramedic, clinical associate, health practitioner. *Ibid.*, Paramedic studies are wide-ranging—from community health to bacteriology and psychosomatic medicine, plus techniques such as regulating intravenous infusions and operating respirators. **1974** *Telegraph* (Brisbane) 27 Feb. 32/2 Paramedics tried artificial respiration, but to no avail. **1974** *Aiken* (S. Carolina) *Standard* 22 Apr. 4-A/1 Paramedic Services and the city police will provide first aid and protection during the walk. **1975** *Daily Tel.* 29 Sept. 3/1 The *Lancet* report gives details of 600 such operations, 366 performed by medical auxiliaries or 'paramedics' as they are described. The rest were carried out by qualified doctors. **1976** *Amer. Speech 1973* XLVIII. 195 Also of invaluable help to the nurse are the *paramedic personnel*, frequently shortened to *paramedics.* In their ranks we find those employees trained to read slides in the laboratory or to assist the doctor and nurse with other medical tasks. **1976** *Sci. Amer.* Sept. 72/3 (Advt.), Heart pacemakers, paramedic telecare units and ultrasonic cardioscopes—all are outgrowths of space technology. **1977** *It* May 21/2 They also have a fully-equipped ambulance, manned by State-certified paramedics and emergency medical technicians.

paramedical (pærăme·dikăl), *a.*¹ and *sb.* [f. Para-¹ + Medical *a.* and *sb.*] **A.** *adj.* Supplementary to or supporting the work of medically qualified personnel.

1921 *Lancet* 15 Oct. 814/1 Para-medical research. The report for 1920–21 of the Committee of the Privy Council for Scientific and Industrial Research touches the fringe of several medical problems and describes work linked at several points with that being undertaken under the auspices of the Medical Research Council. **1952** C. P. Blacker *Eugenics* 303 Marriage guidance, premarital examinations, eugenic prognosis, birth control, and treatment of infecundity have recently been described as 'para-medical services'. **1962** *Times* 2 Mar. 5/7 Nurses and other paramedical personnel. **1969** *Nature* 15 Feb. 604/1 If a student failed to complete the medical course, he would at least have a qualification that would enable him to take up a career in some non-medical or paramedical field. **1972** *Sci. Amer.* Feb. 117/1 Intended largely for paramedical students, nurses and technicians, it presents..a review of the general structure of the body. **1977** *Private Eye* 1 Apr. 7/3 New ventures in the paramedical field include Polyglot Schools (UK) limited..who canvass doctors in the Harley Street area with their scheme, 'Arabic for the Medical Professions'.

B. *sb.* A paramedical worker.

1972 *Guardian* 23 Oct. 11/1 It was suggested that abortion be delegated to paramedicals. **1974** *Financial Times* 5 July 1/8 It was announced on Wednesday that the committee, which is looking at nurses' and midwives' salaries, will also make recommendations for the 'paramedicals'. **1975** *Nature* 21 Aug. 610/3 [In China] barefoot doctors..are called 'doctors', not paramedicals or auxiliaries. **1976** *Times Lit. Suppl.* 16 July 872/5 The enlistment of para-medicals to perform the tasks of health care.

paramedical (pærăme·dikăl), *a.*² [f. *para-³ + Medical *a.* and *sb.*] Trained in parachuting and competent to give medical aid.

1962 *Flight Internat.* LXXXI. 227/1 The Near East Air Force 'para-medical' team in Cyprus. **1974** *Telegraph* (Brisbane) 17 Sept. 34/1 An American paramedical airman will parachute into the water off Redland Bay..as part of a simulated search and rescue operation.

paramelaconite: see *para-¹ 2 c.

paramenstruum: see *para-¹ 1.

parameter. Add: **2.** (Examples in *Computing.*)

1954 *Computers & Automation* Dec. 18/1 *Parameter*, in a subroutine, a quantity which may be given different values when the subroutine is used in different parts of one main routine, but which usually remains unchanged throughout any one such use. **1958** *Communications* (Assoc. Computing Machinery) Dec. 16 A procedure..is a closed and self-contained process with a fixed ordered set of input and output parameters. **1965** [see *keyboarded* ppl. adj. s.v. *keyboard *v.*]. **1969** *Computers & Humanities* III. 278 Thus input parameters were included to specify page width and length, and the dictionary was photo-offset. **1973** C. W. Gear *Introd. Computer Sci.* v. 201 A and B are input parameters, so they can be given values by any expression when the subroutine is called. **1973** Murrill & Smith *Introd. Computer Sci.* 585 Through the use of arguments and parameters, subroutines and functions can be used throughout a program to perform identical operations upon many different data items.

b. Delete †*Obs.* (Later examples.)

1963 R. H. Merson in M. Roy *Dynamics of Satellites* 83 The values of the six basic elements at a given time..are determined from a set of observations by a differential correction technique... In addition to the estimates of the orbital parameters, estimates of their variances and covariances are also determined. **1967** *Technology Week* 20 Feb. 13 Orbital parameters were apogee—817·2 mi.; perigee—128 mi.; inclination—48·4 degrees, and period 100·3 minutes. **1975** *Nature* 18 Sept. 184/2 Even though a few comets with hyperbolic orbits have been seen, Whipple's calculations show that for four candidates—the only ones..with well determined parameters—the chance of any one having arrived on such an orbit from interstellar space is less than 1 in 10^4.

d. *Math.* An independent variable in terms of which each co-ordinate of a point is expressed, independently of the other co-ordinates.

1873 G. Salmon *Treat. Higher Plane Curves* (ed. 2) viii. 317 The coordinates x', y', z' are..expressible as quadric functions of a parameter θ. **1907** Grace & Rosenberg *Coordinate Geom.* xvi. 220 If we can find simple expressions for the coordinates of points on a conic in terms of, one variable quantity, a point on the curve may be looked on as determined by a definite value of the variable, the variable being usually called the parameter. **1937** Michell & Belz *Elem. Math. Analysis* I. vii. 401 Taking $x^2/a^2 - y^2/b^2 = 1$, (a, b positive), as the equation of the hyperbola, we can evidently write $x = \pm a \cosh \phi$, $y = b \sinh \phi$, introducing ϕ as parameter. **1969** J. J. Stoker *Differential Geom.* iv. 75 As with curves, the surface parameters are to a considerable degree arbitrary; in fact, parameters u, v can be replaced by new parameters through the equations $\bar{u} = \phi_1(u, v)$, $\bar{v} = \phi_2(u, v)$, provided that the Jacobian $\partial(\bar{u}, \bar{v})/(u, v)$ is different from zero.

e. *Electr.* Any of several numerical quantities that can be used jointly to characterize a network.

1911 *Trans. Amer. Inst. Electr. Engin.* XXX. 885 The impedances required to make a normal type of network of the requisite number of parameters equivalent to the given network under specified conditions of operation. **1930** T. E. Shea *Transmission Networks & Wave Filters* iii. 71 Any network having one pair of input and one pair of output terminals may be completely represented..by a T network (or network having any form providing at least three independent parameters) as far as external current and voltage conditions are concerned. **1962** Simpson & Richards *Physical Princ. Junction Transistors* v. 82 The *h*'s define the following circuit parameters: $h_{11} = h_i$ = input impedance with output short-circuited.., $h_{12} = h_r$ = reverse voltage ratio with input open-circuited. **1966** R. H. Mattson *Electronics* ix. 381 The *y* parameters are used when discussing feedback amplifiers and pentode amplifiers. The *h* parameters are used when discussing transistor circuits.

f. *Statistics.* A numerical characteristic of a population (as distinguished from a 'statistic', which relates to a sample).

1922 R. A. Fisher in *Phil. Trans. R. Soc.* A. CCXXII. 311 The law of distribution of this hypothetical population is specified by relatively few parameters. *Ibid.* 313 These involve the choice of methods of calculating from a sample statistical derivates, or as we shall call them statistics, which are designed to estimate the values of the parameters of the hypothetical population. **1939** A. E. Treloar *Elem. Statistical Reasoning* x. 130 The true sampling error of each sample mean and standard deviation, so far as those statistics form estimates of the corresponding parameters, may be expressed as [etc.]. **1962** E. S. Keeping *Introd. Statistical Inference* v. 95 If.. the population is assumed to be normal, as far as a particular variate is concerned, the density function for this variate will contain two parameters, μ and σ, which are the population mean and standard deviation respectively. **1975** Harnett & Murphy *Introd. Statistical Anal.* i. 5 The numerical characteristics of a sample are used to estimate the parameters of the parent population from which this sample was drawn. A numerical characteristic used for this purpose is referred to as a sample statistic, or usually just a statistic.

3. In extended use: any distinguishing or defining characteristic or feature, esp. one that may be measured or quantified; an element or aspect *of* anything; *loosely*, a boundary or limit.

1927 *Proc. R. Soc.* A. CXIII. 642 In the case of phenacite, the symmetry of the structure imposes no limitations on the position of the seven atoms in the molecule, so that twenty-one parameters are required to define the structure. **1934** *Ibid.* CXLVI. 570 Few of the structures of hydrated salts have as yet been found. This may be due to the large numbers of parameters usually involved, which, in the absence of any general laws concerning water of crystallization, makes the analysis very difficult. **1939** Brevoort & Joyner *Cooling on Front of Air-Cooled Engine Cylinder* (NACA Techn. Rep. No. 674) 1/1 These results are introductory to the study of front cooling and show the general effect of the several test parameters. **1950** J. C. Slater *Microwave Electronics* x. 230 The quantity, *x*..occurs frequently in klystron theory and is called the bunching parameter. **1957** *Times* 11 Sept. 6/2 The principle that it was possible to specify the sounds of speech in terms of six parameters or factors, which might be considered as functions of time. **1961** *Jrnl. Speech & Hearing Res.* IV. 10/1 There is some evidence..that parameters other than the formant frequencies may influence human judgement of vowel qualities. **1962** *Rep. Comm. Broadcasting 1960* 335 in *Parl. Papers 1961–2* (Cmnd. 1753) IX. 259 Many Western European countries..are considering whether there would be advantage in using..the technical parameters they already use for 625-line standards in the 7 Mc/s channels of Bands I and III. **1964** A. Edel in I. L. Horowitz *New Sociol.* xiv. 220 A theory of human society is seen to involve a specific picture of the nature of man. We would then say that a social theory has a human-nature parameter. **1965** *Listener* 9 Dec. 943/2 There remains the bulk of those for whom politics is a parameter of life rather like the weather. **1967** *Economist* 16 Dec. 1157/1 A second, even bigger generator of 1,300 MW was also contracted for... With these orders, practically every existing parameter of power generating practice has been exceeded. **1970** *Time* 3 Aug. 9 The fact that Nixon was willing to make his chastisement public suggests..that the President at least understands 'the parameters of the problem'. **1971** *Jrnl. Gen. Psychol.* LXXXIV. 18 Three phenomena corresponding to the three major parameters of color—brightness, hue, and saturation. **1972** *Nature* 18 Feb. 373/2 In Fig. 2 ozone and radon concentrations together with various meteorological parameters are shown for two consecutive Saturdays. **1973** *N.Y. Times Mag.* 25 Feb. 71/4 It carries, to begin with, the liberal presumption that the mind of man can in fact comprehend the major parameters of the world we inhabit. **1975** D. M. Davin *Closing Times* p. xviii, There are parameters to these recollections which may not be immediately apparent: the world of learning..and the war... My chief parameter, however, is that of art. **1975** *Times* 14 Oct. 15/4 The considerable element of indeterminacy which exists within the parameters of the parole system. **1975** *Publishers Weekly* 27 Oct. 20/1 One disappointment for the publishers is that it [*sc.* the High Court decision] fails to provide any clear guidelines on the larger issue of the parameters of Government secrecy. 1976 *Listener* 30 Sept. 419/3 Carter, who has made the running so far by raising the debate beyond the orthodox economic and financial political parameters. 1976 H. Young *Crossman Affair* i. 19 At this meeting a word was first spoken and a concept first articulated which later came to dominate the Crossman Diaries case. The word was 'parameters'... Sir John Hunt, in giving guidance on the limits within which an edited version of Crossman would have to be prepared, now formalized into a set of rules his interpretation of past practice... These parameters, or limits, excluded four particular areas from detailed report or discussion.

para:meteriza·tion. Also **para:metriza·tion.** [f. as next + -ation.] The action of parameterizing; a parametric representation.
1939 H. Weyl *Classical Groups* ii. 56 (*heading*) Cayley's rational parametrization of the orthogonal group. 1964 L. Wilets *Theories Nuclear Fission* ii. 14 (*heading*) Parameterization of the nuclear surface. 1970 I. E. McCarthy *Nuclear Reactions* I. iii. 69 To facilitate numerical calculations the following parametrization is used. 1972 A. W. F. Edwards *Likelihood* vi. 127 In view of the relatively high conformation of *x̂* and *ŷ* the former parametrization in *p* and *r* is more suitable. 1975 *Physics Bull.* July 323/3 The required degree of accuracy is established at the beginning by a theoretical study using the virial coefficients of the post Newtonian parameterization expansion for the viable gravitation theories. 1976 *European Econ. Rev.* VIII. 287 The TF form, a finite parametrization of the well-known final form, is appropriate for control and forecasting.

parameterize (păræ·mĭtərəiz), *v.* Also **parametrize.** [f. Parameter + -ize.] *trans.* To describe or represent in terms of a parameter.
1940 E. T. Bell *Devel. Math.* xv. 322 The wave surface in optics, parametrized by elliptic functions. 1949 [see *interval sb. 6 b]. 1964 *Ann. Rev. Automatic Programming* IV. 125 A translation algorithm is presented, capable of being conveniently parameterized for various source language-target language pairs. 1970 *New Scientist* 9 Apr. 76/2 The nuclear charge distribution..can be parametrized directly using a suitable mathematical form which does not necessarily have fundamental significance. 1973 *Nature* 14 Sept. 61/1 Cigarette smoking is parametrized by the number smoked daily both before pregnancy and after the fourth month. 1974 *Ibid.* 20/27 Dec. 673/1 The zonal velocity (parameterised by *A*) leads to a secular change in *i* as the value of ω of the satellite orbit changes.
Hence **para·meterized** *ppl. a.*, **para·meterizing** *vbl. sb.*
1962 [see *magnetosonic* adj. s.v. *magneto-]. 1964 *Ann. Rev. Automatic Programming* IV. 125 (*heading*) A parameterized compiler based on mechanical linguistics. 1971 *Physics Bull.* Jan. 24/2 Only the large scale physics of the atmosphere is well represented in our models and the subgrid scale physics..can only be included in some parametrized form.

parametral, *a.* (Further example.)
Except in *Cryst.*, *parametric* is the usual adj.
1973 H. D. Megaw *Crystal Struct.* v. 103 The plane used to define the axial ratios *a*:*b*:*c*, the parametral plane, is (111).

parametrial: see *para-1 1.

parametric, *a.*1 Add: (Further examples.) *parametric curve*, a curve obtained by keeping constant one of the parameters in the parametric equations of a surface; *parametric equation*, one of a set of equations each of which expresses one of the co-ordinates of a curve or surface as a function of one or more parameters (*parameter 2 d).
This, rather than *parametral* or *parametrical*, is the usual adj. (except in *Cryst.*: cf. *parametral a.).
1900 *Trans. Amer. Math. Soc.* I. 461 (*heading*) Parametric representation of the fundamental quadric. 1909 L. P. Eisenhart *Treat. Differential Geom. Curves & Surfaces* i. 1 (*heading*) Parametric equations of a curve. *Ibid.* ii. 55 Upon a surface (2) there lie an infinity of curves whose equations are given by equations (2), when *u* is constant, each constant value of *v* determining a curve... In a similar way, there is an infinite family of curves *v* = *const.* The curves of these two families are called the parametric curves for the given equations of the surface. 1942 C. H. Lehmann *Analytic Geom.* xi. 229 The parametric equations of a specific locus are not unique. *Ibid.* 230 Find the rectangular equation of the curve whose parametric equations are $x = 2 + 3 \tan \theta$, $y = 1 + 4 \sec \theta$. 1969 J. J. Stoker *Differential Geom.* ii. 14 Many important results in differential geometry can often be made direct and easy to achieve once a special parametric representation has been tactfully chosen.
b. *Electronics.* Applied to devices and processes in which amplification or frequency conversion is obtained by applying a signal to a non-linear device that is modulated by a pumping frequency, so that there is a transfer of power from the latter to the output, which in general can include the sum and difference frequencies.
So called because the action of the pumping frequency is to modulate the parameters of the non-linear device.
1957 *RCA Rev.* XVIII. 578 (*heading*) Theory of parametric amplification using nonlinear reactances. *Ibid.* 579 In this paper the parametric amplifier is analyzed in terms of an equivalent circuit using a nonlinear inductance. 1961 *Guardian* 14 Feb. 24/1 The so-called 'parametric amplifiers'..can increase the sensitivity of radio reception over great distances. 1968 Angelakos & Everhart *Microwave Communications* iv. 82 A parametric amplifier converts power at one frequency (from a source generally called the pump) into power at another frequency, the signal frequency... The pump voltage is mixed with the signal voltage by a nonlinear reactance, which in microwave systems is generally a varactor diode. 1971 *Physics Bull.* Aug. 464/3 Parametric conversions of waves to other frequencies are familiar in the field of nonlinear optics. 1972 *Sci. Amer.* Sept. 136/3 In present-day satellite-communication terminals the maser amplifier has given way to the cooled parametric amplifier, which combines low-noise performance with an even wider bandwidth. 1972 Zernike & Midwinter *Appl. Nonlinear Optics* (1973) vii. 153 The parametric up-converter is a special case of sum-frequency generation. Similarly, the parametric amplifier and the parametric oscillator are special cases of difference-frequency generation.

parametrically (pærăme·trikăli), *adv.* [f. Parametric *a.*1: see -ly2.] In terms of a parameter or parameters.
1894 C. A. Scott *Introd. Acct. Plane Analyt. Geom.* v. 89 The possibility of expressing the coordinates of a point on a curve parametrically. 1940 E. T. Bell *Devel. Math.* xv. 322 Kummer's (1864) quartic surface..is the so-called singular surface of the quadratic line complex, and..is represented parametrically. 1962 W. B. Thompson *Introd. Plasma Physics* v. 86 The coordinates (*X*,*Y*) can be given parametrically in terms of the phase velocity *v*(θ) and the angle between the wave vector and the magnetic field. 1968 C. G. Kuper *Introd. Theory Superconductivity* xii. 202 The probability of occupation of a given single-particle state will depend parametrically on the *whole* distribution of quasiparticles, but not on the detailed question of whether some other particular state is occupied.

parametritius, -metrium: see *para-1 1.

parametrize, -metrization: varr. *parameterize *v.*, -meterization.

parametron (pāræ·mitrọn). *Electronics.* [f. Parametr(ic *a.*1 + -on; coined in Jap. by E. Goto 1955, in *Denki Tsûshin Gakkai Zasshi* XXX. 770.] A digital storage element consisting of a parametric oscillator in which the digit is represented by the phase (0° or 180°, corresponding to 1 or 0) of the output signal relative to that of an applied reference signal of the same frequency.
1956 *ETJ of Japan* June 64 A new type of electronic computer component called the 'parametron' was invented by Ei-ichi Goto of the Faculty of Science, University of Tokyo, in spring of 1954. 1957 *Jrnl. Sci. Res. Inst.* (Tokyo) LI. 59 (*caption*) A parametron unit; an exciting current is supplied from 1, causing an oscillation in the L-L′-C circuit. Input and output lines are 2 and 3 respectively. 1960 T. E. Ivall *Electronic Computers* (ed. 2) xiii. 234 The parametron requires no valves or transistors, only passive reactive elements, and being therefore extremely stable, reliable and long-lived, is ideally suited for use in digital computers. The main limitation is that because several cycles of oscillation are required to establish a binary digit,..the digit rate is necessarily low. 1967 R. K. Richards *Electronic Digital Components & Circuits* vi. 337 Parametrons quickly became very popular with Japanese computer manufacturers... However, not one computer employing parametrons is known to have been built or designed in the United States.

paramilitary (pærămi·litări), *a.* [f. Para-1 + Military *a.*] Of or pertaining to an organization, unit, force, etc., whose function or status is ancillary or analogous to that of military forces, but which is not a professional military force. Also as *sb.* Hence **parami·litarism.**
1935 *Ann. Reg. 1934* 96 A difficult problem has been raised in regard to the so-called 'para-military training', *i.e.*, the military training outside the Army of men of military age. 1936 *Punch* 1 Apr. 376/1 Let us at once impale the new and unnecessary mongrel 'paramilitary'—'paramilitary forces (S.A., S.S., Labour Corps and other organisations)'... Why not 'semi-military', 'quasi-military', or even 'sub-military'? 1949 F. Maclean *Eastern Approaches* iii. xvii. 516 The military and paramilitary forces which they [*sc.* the quisling administrations] had raised had either surrendered or else were withdrawing northwards. 1958 *Listener* 14 Aug. 238/2 The Nazi storm-trooper is preceded by a hardly less 'heroic' communist in his paramilitary uniform. 1962 *Times Lit. Suppl.* 16 Feb. 102/1 Lawrence of Arabia was almost a paramilitary symbol as long ago as 1917. 1969 *New Statesman* 11 Apr. 499/2 The Israelis were to receive unofficial guarantees that a UN presence, or paramilitary force as it was termed, would be stationed on the Israeli-Egyptian border. 1970 *New Yorker* 14 Feb. 33 An editorial in the Times..talked about the Panthers' 'Mao-Marxist ideology and Fascist paramilitarism'. 1972 J. McClure *Caterpillar Cop* xii. 199 I'd forgotten you chappies were really a para-military outfit. 1974 J. White tr. *Poulantzas's Fascism & Dictatorship* iii. ii. 102 Party representation was short-circuited by the formation of para-military organizations. 1976 *Times* 9 Mar. 16/8 The para-militaries are no longer prepared to be used..unless they have a say in the running of the country [*sc.* Ulster]. 1976 *Listener* 16 Sept. 325/3 To change the climate within which the paramilitaries operate. 1977 M. Walker *National Front* i. 18 Although Mussolini had specifically warned him against building a para-military force in Britain, Mosley was convinced..that he needed a corps of tough stewards to guarantee order at his meetings.

paramnesia, -ic: see *para-1 1.

paramount, *a.* Add: **1. c.** *paramount chief*, esp. in African countries, a tribal chief of the highest order, whose authority extends over an entire district.
1844 in F. Brownlee *Transkeian Native Territories: Hist. Rec.* (1923) 92 Treaty of Amity entered into..on behalf of Her Britannic Majesty, of the one part, and Faku, Paramount Chief of the Amapondo Nation. *a* 1882 G. W. Stow *Native Races S. Afr.* (1905) x. 183 They acknowledged a Bushman captain..as their great chief, who..was succeeded by 'Khiba, or 'Kheba, who was the paramount chief over the men of the caves. 1885 in F. Brownlee *Transkeian Native Territories: Hist. Rec.* (1923) 20 Some of these clans..depended directly upon the paramount chief, others were grouped under a sub-chief. 1919 G. M. Theal *Ethnogr. & Condit. S. Afr. before 1505* (ed. 2) x. 212 Sometimes the heads of the clans were members of the family of the paramount chief. 1928 G. P. Lestrade in A. M. Duggan-Cronin *Bantu Tribes S. Afr.* I. i. 18 The petty chiefs, each of whom is responsible to the paramount chief for the maintenance of good order in his section of the tribe, share in..privileges at the chief's pleasure. 1948 B. G. M. Sundkler *Bantu Prophets S. Afr.* ii. 38 As the Queen of England was the head of the English Church, so the Paramount Chief of the Tembu should be the *summus episcopus* of the new religious organization. 1954 E. A. Hoebel *Law of Primitive Man* viii. 193 Every village, which belongs to a subclan, has its headman... If an ordinary headman, a lesser chief..his powers extend only to the boundaries of his own village. If he is a full chief..his influence will spread over several villages and their subchiefs. If he is a paramount chief among the full chiefs, it will extend over an entire district. 1957 P. Worsley *Trumpet shall Sound* vi. 119 Three of the Paramount Chiefs failed to report the movement to Government. 1965 A. Nicol *Truly Married Woman* 87 His main preoccupation was to find ways of slighting the neighbouring and more powerful provincial Paramount Chief. 1971 *Rand Daily Mail* 4 Dec. 3/4 No member of your Government should consider his position to be more important and exalted than that of the Paramount Chief. 1974 *Afr. Encycl.* 505/1 The policy of 'separate development'..which led to the establishment of the Transkei has been opposed by the leader of the 'True' Thembu, Paramount Chief Sabata Dalinyebo. 1976 *Times* 26 Oct. 8/4 Dr. Nicolaas Diederichs, the President of South Africa, handed over a copy of the Transkei Act to the country's first Prime Minister, Paramount Chief Kaiser Matanzima.

paramountly, *adv.* (Later examples.)
1962 S. E. Finer *Man on Horseback* v. 65 The so-called 'Free Officers'..regrouped and decided to overthrow the régime, and paramountly the king who headed it. 1971 *Country Life* 19 Aug. 464/1 The tragedy is that man..still constitutes..the paramountly weak link.

paramyosin: see *para-1 2 a.

paramyxovirus (pærămi·ksovəiərŏs). *Biol.* [f. Para-1 + *myxovirus.] Any of a group of related viruses which includes the parainfluenza viruses and those causing mumps, measles, distemper, and rinderpest, and which differ from the myxoviruses in their larger and more variable size and in being hæmolytic.
1962 [see *myxovirus]. 1968 *Times* 27 Nov. 9/3 Mumps, a disease which is caused by one of the paramyxoviruses. 1969 A. Cohen *Textbk. Med. Virol.* vi. 71 Myxoviruses are characterized by particles whose size varies from 800 to 1100 Å in diameter, and whose helical internal component is 90 Å wide and lightly packed inside the envelope. Paramyxoviruses..are characterized by larger sized particles, 1000–5000 Å in diameter, whose internal component is 180 Å wide and more loosely packed within the virus envelope. Morphological differences between myxoviruses and paramyxoviruses are associated with some differences of biological activity. 1974 [see *myxovirus].

Parana (parā·nā). Also **Paraná**. The name of a river and a province in Brazil, used *attrib.*, usu. in **Parana pine**, to designate a large evergreen tree, *Araucaria angustifolia*, of the family Pinaceæ, found in the high plateau region of south-western Brazil, Paraguay, and northern Argentina; also, the light-coloured, softwood timber obtained from this tree.
1923 Dallimore & Jackson *Handbk. Coniferæ* 153 Araucaria brasiliensis, Loudon. Parana Pine; Parana Wood. 1924 Record & Mell *Timbers Trop. Amer.* ii. 92 The Paraná pine is the most extensively exploited timber in South America. 1959 *Archit. Rev.* CXXV. 333/2 Natural wood, parana pine, oak and western red cedar. 1964 *Listener* 17 Sept. 428/3 Paraná pine..—highly polished—forms the glistening verandah floor. 1971 N. E. Hickin *Wood Preservation* 15 *Araucaria angustifolia* from Argentina and South Brazil is Parana Pine. 1977 *Irish Press* 29 Sept. 16/1 (Advt.), Unique solid Parana Pine bunk beds, £59.

paranasal: see *para-1 1.

paranatal (pærănēi·tăl), *a. Med.* [f. Para-1 + Natal *a.*1 and *sb.*] Of or pertaining to the time shortly before and after birth.
Orig. given a broader meaning (quot. 1940).
1940 F. Schreiber in *Jrnl. Pediatrics* XVI. 297 (*heading*) Neurologic sequelae of paranatal asphyxia. [*Note*] The term *paranatal* has been coined to designate the entire period of fetal life, the period of birth, and the neonatal period within which the results of occurrences during the first two periods may become manifest. 1954 *Jrnl. Amer. Med. Assoc.* 19 June 719/1 Proponents of the brain injury hypothesis state that in most cases anoxia during the prenatal and paranatal periods has resulted in brain damage. 1964 *Jrnl. Nerv. & Mental Dis.* CXXXIX. 357 (*heading*) Pre- and paranatal factors in mental disorders of children. 1971 C. B. Courville *Birth & Brain Damage* xi. 197 (*caption*) Gliosis, the most significant tissue reaction in paranatal anoxia. 1976 *Word* 1971 XXVII. 62 The developing human foetus..is 'practicing', in amniotic fluid, those neuromuscular gestures which will lead, in air, to paranatal cry and neonatal cry specifically.

paranemic (pærănī·mik), *a.* [f. Para-1 + Gr. νῆμα thread + -ic.] Of, pertaining to, or designating two or more like helices coiled together side by side in such a way that they may be fully separated without being unwound. Opp. *plectonemic *a.*
1941 [see *plectonemic a.]. 1950 *Biol. Rev.* XXV. 500 Until quite recently almost all observers were agreed that the plectonemic spiral was characteristic of mitosis and the paranemic of meiosis (meaning the first meiotic division). 1953 Watson & Crick in *Cold Spring Harbor Symp. Quant. Biol.* XVIII. 129/2 With paranemic coiling, the specific pairing of bases [in DNA] would not allow the successive residues of each helix to be in equivalent orientation with regard to the helical axis. 1974 F. Crick in *Nature* 26 Apr. 767/2 Looking back, I think we deserve some credit..for our forthright stand against paranemic (as opposed to plectonemic) coiling.

parang. (Earlier and later examples.)
1839 [see *kleywang]. 1966 D. Forbes *Heart of Malaya* v. 60 A full-blooded *amok*, slaughtering every man, woman and child in reach with the snaky kris or the curved parang. 1967 J. Cleary *Long Pursuit* iii. 81 They cut branches from the mangroves with a parang they had found in the boat. 1972 [see *jack sb.4 b]. 1976 'G. Black' *Moon for Killers* ix. 109 The rent in the cloudbank..its edges as rough as a wound made with a *parang*.

parangi (pără·ŋgi). Also 9 **parangy.** [Sinhalese *parangi* (*lede*), lit. '(disease of) foreigners', i.e. the Portuguese (= Skr. *phiranga* 'Frankish', European), f. Pg. *Frangue*, name given by the Moors to Frenchmen, Spaniards, and European Christians generally (cf. Feringhee).] The name given in Sri Lanka to a disease now known to be identical with yaws.
1821 H. Marshall *Notes Med. Topogr. Interior Ceylon* iii. 43 There is a complaint mentioned in the Kandyan medical works, called *parangy lede* (Parangy disease). *Ibid.*, *Parangy lede* seems to have been originally intended to denominate a new disease;..it may perhaps be inferred that the term meant Portuguese disease. There is, however, no tradition among the Kandyans respecting the importation of a disease. *Ibid.* 44 Many of the cutaneous affections which they denominate *parangy*, are evidently herpetic, and cannot be referred to a syphilitic origin. 1882 *Med. Times & Gaz.* 14 Jan. 30/1 The diseases which parangi resembles are syphilis and its various manifestations, lupus leprosy, and frambœsia. 1913 L. Woolf *Village in Jungle* i. 11 There were few in the village without the filthy sores of parangi, their legs eaten out to the bone with the yellow, sweating ulcers.

paranitraniline: see *para-1 2 b.

‖ **paranjah** (parandʒā·). [a. Russ. *parandzhd*, ult. f. Arabic.] A long wide robe with a (horse-)hair veil worn by Muslim women.
1928 *Daily Express* 7 Dec. 9 The women, completely covered from head to foot by the hideous 'paranjas' (veils), watch the proceedings from the far off roof-tops. 1947 *New Times* 20 June 24/2 In the streets of Sarajevo you may meet Moslem women wearing the paranja. But you needn't be surprised if the muffled figure under the black horsehair veil turns out to be a local women's leader. 1954 Koestler *Invis. Writing* ii. xii. 141 The doors of the harems were thrown wide open; the sack-like *paranjahs* and black veils fell.

paranoia, paranœa. Add: The form **paranœa** is now *obs.* The various forms of the disorder are now usu. considered as belonging to the schizophrenic group of mental illness. Also in trivial use. (Earlier and later examples.)
1811 R. Hooper *Lexicon Medicum* 596/2 *Paranœa*, alienation of the mind; defect of judgment. 1848 Dunglison *Dict. Med. Sci.* (ed. 7) 625/2 *Paranœa*, delirium, dementia. 1918 [see *neuropsychosis* s.v. *neuro-]. 1940 Hinsie & Shatzky *Psychiatric Dict.* 395/2 Perhaps no term in psychiatry has undergone wider variations of meaning than the term paranoia. 1954 W. Mayer-Gross et al. *Clin. Psychiatry* iv. 158 Much of the age-old controversy on 'paranoia' has arisen from the difficulty of distinguishing between paranoid reactions and paranoid schizophrenia. *Ibid.* vi. 256 The effort to maintain paranoia as a distinct condition has also failed. 1957 V. Nabokov *Pnin* iv. 96 There is nothing more banal and more bourgeois than paranoia. 1970 Hinsie & Campbell *Psychiatric Dict.* (ed. 4) 540/2 Several types of paranoia have been described on the basis of the type of delusion that predominates: litigious, depressed, persecutory, grandiose, erotic.., and infidelity. *Ibid.* 541/2 The patient with amorous paranoia develops delusions of marital infidelity in relation to his spouse. 1972 *Encycl. Psychol.* II. 366/1 Paranoia...was thought to be distinct from the group of schizophrenias because it did not lead to deterioration. This is not now thought to be the case. 1977 *Time* 16 May 56/3 They constitute what has become a standard trip down paranoia lane.

paranoiac, *a.* and *sb.* Add: (Later examples.)
1902 [see *association 9]. 1914 A. A. Brill tr. *Freud's Psychopathol. Everyday Life* xii. 309 The gap between the paranoiac's displacement and that of superstition is narrower than appears at first sight. 1935 D. Gascoyne *Short Survey Surrealism* v. 102 Dali claims that it is the paranoiac faculty that enables him to discover a head where there was, until he looked at it, only an African village. 1937 *Brit. Jrnl. Psychol.* XXVII. 245 It has frequently been suggested that those who come much in contact with paranoiacs tend themselves in time to exhibit paranoid symptoms. 1952 A. Huxley *Let.* 20 May (1969) 645 Boastful in an altogether childish way, mildly paranoiac, but well-meaning. 1977 A. Sheridan tr. *Lacan's Écrits* i. 3 The social dialectic that structures human knowledge as paranoiac. 1978 J. Blackburn *Dead Man's Handle* i. 20 Paranoiacs and schizophrenics like George Heath and Neville Haigh, and the Boston Strangler.
Hence **paranoi·acally** *adv.*, in a paranoiac manner.
1964 P. F. Anson *Bishops at Large* vi. 213 He continued to build castles in the air,..paranoiacally refusing to face up to reality. 1976 *Listener* 22 Apr. 505/1 The girl next door is already out there, doing her yoga and singing paranoiacally.

paranoic, *a.* Add: (Further examples.)
1952 W. J. H. Sprott *Social Psychol.* 244 The Haida Chief whose relative has died suspects supernatural persecution; with us he would be labelled 'paranoic' and sent for 'treatment'. 1977 *Times Lit. Suppl.* 15 Apr. 462/2 His paranoic dislike not only of Robert Ferguson. *Ibid.* 1 July 799/3 The paranoic fantasies to which Bely was prey in 1916, when he believed that he was a 'human bomb', under surveillance from 'a dark-skinned man in a bowler hat'.
Hence **parano·ically** *adv.*
1976 *Times Lit. Suppl.* 28 May 648/1 The outline of a paranoically hostile Soviet Union.

paranoid (pæ·rănoid), *a.* and *sb.* [Irreg. f. Paranoia + -oid.] **A.** *adj.* Resembling or characterized by paranoia; also used *colloq.*, *transf.*, and *fig.* So **paranoi·dal** *a.*
1904 *Brit. Med. Jrnl.* 15 Oct. 972 The collective grouping of hebephrenia, katanoia, and the paranoid forms makes so vast a congeries that it is impossible to perceive any connecting link between the items of the mass. 1904 tr. *Kraepelin's Lect. Clin. Psychiatry* 151 Paranoidal forms of Dementia Præcox. 1919 R. M. Barclay tr. *Kraepelin's Dementia Praecox & Paraphrenia* 195 Paranoid Weak-mindedness. *Ibid.* 252 Paranoid Forms. *Ibid.* 253 Delusions and hallucinations of quite the same kind, as we see them in paranoid cases, occur also in most of the remaining forms of dementia præcox. 1954 Koestler *Invis. Writing* xxv. 270 The paranoid tendencies inherent in the movement. 1961 L. Mumford *City in Hist.* ii. 39 A paranoid psychal structure was preserved and transmitted by the walled city. 1967 M. Argyle *Psychol. Interpersonal Behaviour* viii. 135 *Paranoid reactions* are closely related to schizophrenia. Where there is personality disintegration but with rather more systematic delusions this is called paranoid schizophrenia. 1970 *Guardian Weekly* 28 Feb. 19/1 The sort of international news letters run by paranoid ex-generals. 1974 *Guardian* 13 June 10/3 Those paranoid placards with their Red Rats and Hot Seats for Traitors and Fry 'em.
B. *sb.* A person afflicted with, or showing symptoms of, paranoia.
1922 *Brit. Jrnl. Psychol.* Oct. 113 In discussing the case of paranoids, Neisser writes, 'the patient is progressively less able to exert an independent control over the course of his presentations'. 1938 S. Beckett *Murphy* ix. 167 Paranoids, feverishly covering sheets of paper with complaints against their treatment. 1950 T. Sugrue in M. Hay *Foot of Pride* p. xix, The rabble-rousing race bigots who peddle hate and fear to simpletons and paranoids. 1958 *Times Lit. Suppl.* 5 Sept. 495/4 The racy blend of anecdote and psychological jargon produce..a basic Hindu paranoid with suppressed homosexual tendencies, who is never precisely related to particular Hindus. 1967 *Listener* 9 Feb. 186/2 Naive Russian advisers might even believe that only the paranoid would see a Russian defence system as threatening the United States. 1967 M. Argyle *Psychol. Interpersonal Behaviour* viii. 136 Paranoids feel that they are being plotted against, spied upon, or otherwise victimized, and that this is the explanation of their failures. 1976 G. McDonald *Confess, Fletch* (1977) xiii. 55 Even paranoids have enemies.

parano·rmal, *a.* [Para-1.] Applied to observed phenomena or powers which are presumed to operate according to natural laws beyond or outside those considered normal or known; also *absol.* Hence **paranorma·lity**, the state or character of being paranormal; **parano·rmally** *adv.* Cf. Supernormal *a.*
1920 Webster Add., Paranormal. 1935 *Discovery* May 138 (*caption*) The paranormal displacement of a handkerchief actuated electrically the flashlight by which this photograph was taken. 1950 A. Huxley *Themes & Variations* 133 The general tendency of ideas to embody themselves at any price and by fair means or foul, normally or paranormally. 1955 —— *Let.* 30 June (1969) 749 Two 'sensitives', one who specializes in paranormal diagnosis, the other a 'healer'. 1958 J. Blish *Case of Conscience* (1959) i. 82 He has no belief in the supernatural—or, as we're calling it in our barbarous jargon these days, the 'paranormal'. 1959 *Listener* 11 June 1032/3 Her judgments will be equally remote from dogmatic incredulity and uncritical acceptance of claims to paranormality. 1961 W. H. Salter *Zoar* ii. 15 'Paranormal' has now in general use taken the place of 'supernormal'. *Ibid.* vi. 69 They were convinced that the moulds were produced paranormally by 'Ideoplasmy'. 1968 *New Scientist* 24 Oct. 209/2 The out-of-the-body or 'ecsomatic' experience has long been familiar to students of the paranormal. 1973 *Daily Colonist* (Victoria, B.C.) 19 Apr. 5/2 Critics who dislike the seemingly paranormal condemn the claims with a vehemence beyond objective scientific candor. 1973 *Daily Tel.* (Colour Suppl.) 30 Nov. 26/1 The individuals who have been at the receiving end of the paranormally-transmitted information. *Ibid.* 31/3 Jung was to claim that if he had his life over again, it would have been devoted to the study of the paranormal. 1975 *Nature* 10 Apr. 470/2 Among the claims made are that plasticity of metal was paranormally produced. 1976 'J. Ross' *I know what it's like to Die* xxv. 157 Rogers' belief..in the paranormality of Pybus's experiment with the two dead bodies. 1977 *N.Y. Rev. Bks.* 13 Oct. 45/2 The enormous qualitative difference between testing psychics for paranormal powers, and experimentation in *all other* branches of science.

paranotal, -notum: see *para-1 1.

Paranthropus (pæræ·nþropŏs, pærænþrō·upŏs). [mod.L. (R. Broom 1938, in *Nature* 27 Aug. 379/1), f. Para-1 + Gr. ἄνθρωπος man.] A fossil hominid first described from remains found at Kromdraai and other sites in southern Africa, formerly included in the genus so called, and now usually included in the species *Australopithecus robustus*.
1941 R. Broom in *Nature* 5 July 13/1 We have a good lower jaw of Paranthropus. Its teeth are almost typically human. 1946 *Transvaal Mus. Mem.* II. iii. 85 In the male Plesianthropus the anterior surface of the maxilla is essentially similar to that in the female, and thus unlike that in the presumably male Paranthropus. *Ibid.* 92 The Paranthropus skull resembles the living anthropoids much more than it does man. 1953 [see *Australopithecus]. 1959 J. D. Clark *Prehist. S. Afr.* iii. 62 Now, however, anatomists are agreed that only one generic form is represented in the Man-ape remains but that two specific forms exist—*Australopithecus* and a later, more specialized form, *Paranthropus*. 1961 *Lancet* 30 Sept. 768/1 The existence of a *Paranthropus*-like form at Olduvai and later at Swartkrans and Kromdraai..seems to suggest that this genus..was moving out peripherally from a central African habitat. 1971 S. A. Barnett *Human Species* (ed. 5) iv. 100 There are..some more recently discovered forms ('Paranthropus') which were probably above the average height of modern man.

paranym (pæ·rănim). [f. Para-1 + *-nym* in Antonym, Paronym, Synonym *sb.*, etc.] **a.** A near synonym. *rare.* **b.** (See quot. 1976.)
1963 L. Hogben *Essential World Eng.* 11 The L.E.S.U. [sc. List of Essential Semantic Units] cites non-essential paranyms for new terms even if definable by reference to terms which appear earlier in the list. 1976 *Listener* 17 June 773/3 A newspaper columnist has recently been collecting what he calls 'paranyms'—words whose meaning is generally the opposite of that intended by the speaker, such as 'provisional' or 'liberation' or 'rationalise'. The writer Brian Aldiss thereupon contributed an example he had found in the New Testament: ' "everlasting life"; in other words "death"'. 1977 *Sunday Times* 2 Jan. 32/4 What a year for paranyms. Never before can so many decent, explicit words have utterly altered their meaning to conceal an evasion or untruth.

parapet. Add: **1.** *spec.* a bank of earth in front of a military trench. Also *fig.*
1916 J. Buchan *Greenmantle* xx. 266 A crump took the parapet of the trench. 1918 'R. West' *Return of Soldier* i. 6 On the war-films I have seen men slip down..from the trench parapet. 1928 Blunden *Undertones of War* 317 It seems, as now I wake and brood, And know my hour's decrepitude, That on some dewy parapet The centuries' spirit gazes yet. 1929 'C. Edmonds' *Subaltern's War* i. 25 A head and shoulders, seen from the sniper's loophole, leaping past a gap in the enemy's parapet. 1954 W. Faulkner *Fable* 25 He saw only a few figures crouching along his own parapet. 1975 *Times* 14 July 13/1 The Leader of the House, to whom the Prime Minister..will yield the honour of going over the parapet first. 1975 *Listener* 14 Aug. 207/1 Some TV performers..have been howitzered into the mud... A few have had the sense to learn..to keep their heads behind the parapet. 1977 H. Fast *Immigrants* iii. 171 The men were leaning high on the parapet, staring out over the ruptured, wire-strewn earth that separated them from the enemy.
4. *parapet mounting.*
1914 *Illustr. London News* 29 Aug. 332/2 A Hotchkiss machine-gun on a parapet mounting.

paraphasia: see *para-1 1; **para phenylene-diamine:** see *para-1 2 b.

paraphilia (pærăfi·liă). [f. Para-1 + Gr. φιλία affection.] Perverted sexual desires. Hence **paraphi·liac, paraphi·lic** *adjs.* and *sbs.*
1925 J. S. van Teslaar tr. *Stekel's Peculiarities of Behav.* II. 341 *Paraphilia*, interest in perversions. 1935 L. Brink tr. *Stekel's Sadism & Masochism* I. iv. 100 If the patient can be brought to normal sexual relations, the paraphiliac impulses recede into the background. 1958 *Times Lit. Suppl.* 31 Oct. 619/2 Stekel..says that pluralistic orgies are tribal in derivation, and that frequently the paraphiliac who engages in them seeks a family combination. 1960 *Arch. Gen. Psychiatry* III. 442/1 He was allegedly heterosexually quite adequate (by using different paraphilic fantasies during coitus). 1962 C. Allen *Textbk. Psychosexual Disorders* i. v. 73 We regard the paraphylics [*sic*] divisible into three degrees. 1971 *Nature* 16 Apr. 433/2 Contact with urine plays little part in human relations, though it may be emphasized in paraphilias. 1977 *Proc. R. Soc. Med.* LXX. 792/2 Some patients have fantasies which involve masochistic or other paraphilic activity. 1977 E. J. Trimmer et al. *Visual Dict. Sex* (1978) i. 12 The common paraphilias that we choose to call sexual perversions today, were defined by the Greeks as being parallel to love.

paraphony, anglicized form of Paraphonia.
1919 H. J. Watt *Found. Music* 157 The term paraphony was used by several later writers, Thrasyllus, Bacchius and Gaudentius. 1924 T. H. Y. Trotter *Music & Mind* 154 The words 'symphony', 'paraphony', and 'diaphony' are used to express more or less complete unity and dissonance.
Hence **paraphonic** *a.* (later example); **parapho·nically** *adv.*
1919 H. J. Watt *Found. Music* 156 For the proper flow of simultaneous melodies intervals must either be themselves actually paraphonic or they must be used paraphonically.

pa:raphrasabi·lity. [f. Paraphrasable *a.* + -ity.] The capacity of being paraphrasable.
1965 *Amer. Philos. Q.* II. 185/2 Paraphrasability is generally regarded as the central issue. 1965 *Times Lit. Suppl.* 25 Nov. 1082/4 He brings the [Wittgensteinian] technique to bear on two aesthetic problems: the paraphrasability of poems and the tonality of atonal music. 1977 *Amer. Speech* 1975 L. 82 Constructions in which the modifier is the author of the headword, such as *Zunser's hymn*, also vary in their paraphrasability with *have*.

paraphrasable, *a.* (Later examples.)
1936 F. R. Leavis *Revaluation* v. 155 It is difficult not to believe, after reading, say, Book II of *The Prelude*, that one has been reading a paraphrasable argument. 1952 *Essays in Crit.* II. 106 If it is self-sufficient (that is, if it yields a paraphrasable meaning answering the normal demands of logic and syntax). 1966 *Listener* 25 Aug. 282/1 Mr Lodge carefully demolishes the view that the novel is reducible to its paraphrasable and translatable content. 1970 *Language* XLVI. 831 The adjectives shown..are normally considered stative..and they are indeed paraphrasable with active verbs.

paraphrase, *sb.* Add: **1.** (Examples of a musical passage.)
1880 Grove *Dict. Mus.* II. 741/2 His [*sc.* Liszt's] transcriptions, paraphrases, and arrangements, comprise not only vocal and orchestral works of German, French, Italian, and Russian composers, but also the national melodies of Europe, Asia, etc. 1900 L. Godowsky *Let.* 24 Dec. in H. C. Schonberg *Great Pianists* (1964) xxiv. 320, I came out to play my seven Chopin paraphrases and Weber's 'Invitation'. 1944 W. Apel *Harvard Dict. Mus.* 554/1 Liszt's paraphrases on Wagnerian operas. 1963 H. C. Schonberg *Great Pianists* (1964) xxiv. 322 Big technicians will occasionally attempt his paraphrase on *Fledermaus*. 1976 *Gramophone* Apr. 1598/2 Liszt's *Eugene Onegin* paraphrase is not one of his most elaborate, nor his finest, but it has the authentic glitter, and a most enjoyable flair.
e. *transf.* In art, the representation of a subject in a realistic or other manner so as to convey its essential qualities.
1951 G. Sutherland in *Listener* 6 Sept. 378/1, I feel that now we can perhaps enlarge the field of painting by setting our emotional paraphrases of reality—they themselves have been conceived more optically—within the ambience of optical reality. 1962 R. G. Haggar *Dict. Art Terms* 246/2 *Paraphrase.* The term is used by Graham Sutherland to explain the nature of his realist art, implying that he seeks to express the character and mood of the landscape or object which inspires him by other and general forms which do not constitute 'views'. 1962 *Listener* 26 July 134/1, I believe that in the case of a portrait, there are two ways of doing it. One..is the real paraphrase such as Picasso does. 1965 *New Statesman* 14 May 775/1 His line overstates and under-praises, as in the paraphrase of a Cranach nude.

paraphrase, *v.* Add: **1.** (Later examples of *fig.* use, *spec.* in art.) Cf. prec.
1961 D. Cooper *Work of Graham Sutherland* ii. 14 He [*sc.* Sutherland] found that when he paraphrased what he saw he had captured more of 'the essence or the *gesture* of reality.' 1962 *Listener* 26 July 134/1, I might have wanted to paraphrase a landscape.

paraphrastical, *a.* Delete 'Now *rare* or *Obs.*' and add later examples.
1960 *Spectator* 30 Sept. 497 No don with simple-minded paraphrastical or reductive tastes [etc.]. 1977 *Archivum Linguisticum* VIII. 10 This paraphrastical relation..does not characterize the respective sentences of [examples] (67)–(72).

paraphrenia (pærăfrī·niă). [ad. F. *paraphrénie* f. Para-1 + Gr. φρήν mind + -ia1.] A form of mental disorder; a term sometimes used to refer to mental disorders of the paranoid and schizophrenic varieties. Hence **paraphre·nic** *a.*
[1846 Dunglison *Dict. Med. Sci.* (ed. 6) 552/1 *Paraphrénie*, insanity.] 1890 in Billings *Med. Dict.* II. 290. 1917 C. R. Payne tr. *Pfister's Psychoanal. Method* 522 Dementia praecox (schizophrenia according to Bleuler, paraphrenia according to Freud). 1919 R. M. Barclay tr. *Kraepelin's Dementia Praecox & Paraphrenia* 2 It seems to me that the term 'paraphrenia', which is now no longer in common use, is in the meantime suitable as the name of the morbid forms thus delimited. 1934 Webster, Paraphrenic. 1952 Purves-Stewart & Worster-Drought *Diagnosis of Nervous Dis.* (ed. 10) xxii. 737 The term paraphrenia is applied to those cases of paranoidal schizophrenia who retain their personality... Paraphrenic symptoms usually develop later in life than those of the ordinary paranoidal type, often as late as the menopause. 1958 *Daily Tel.* 30 June 10/4 The effect was found to be beneficial even with some of the seriously disturbed paraphrenic patients. 1962 Henderson & Gillespie *Text-bk. Psychiatry* (ed. 9) xii. 291 We would..suggest that the term paraphrenia should be discarded, as it does not serve any useful purpose. 1973 T. & R. Millon *Abnormal Behav.* (1974) xvii. 381 The closest approximation to what we have termed paraphrenia may be found in the DSM-II descriptive text of 'schizophrenia, paranoid type'.

paraphysical: see *para-1 1.

paraplegic, *a.* Add: **B.** *sb.* A person with paraplegia.
1890 W. James *Princ. Psychol.* I. ii. 16 Paraplegics draw up their legs when tickled. 1950 *Time* 31 July 39/1 The story of war-wounded paraplegics makes a powerful and moving salute to the human spirit. 1970 C. Hampton *Philanthropist* iii. 26 Like realizing that Socialism is about as much use to this country as—a pogo-stick to a paraplegic. 1975 *Oxford Times* 12 Dec. 4/4 Her son had been a paraplegic since he injured his spine in a fall four years ago.

‖ **parapluie** (pærăplüi). [Fr.] An umbrella.
The word had wide literary currency in the 19th c.
1781 H. Newdigate *Let.* 30 Sept. in A. E. Newdigate-Newdegate *Cheverels* (1898) iii. 35 It has poured all this Day..but with yͤ help of Pattins & Parapluies we got to yͤ Well. 1790 E. Helme tr. *Le Vaillant's Trav. Afr.* I. ii. 32 He that takes the East side of the mountain should carry his *paraplue* [sic], while he that takes the West would have occasion for his *parasol*. 1813 M. Edgeworth *Let.* 31 May (1971) 69 A shower came on and all the dressed groups were forced to take shelter under trees and parapluies. 1827 M. Wilmot *Jrnl.* 21 Aug. in *More Lett.* (1935) 291 Both wet, but not very bad as they had parapluis [sic]. 1847 *Punch* 6 Feb. 62/2 Our umbrellas..bear no announcement of some new pill... The *parapluie* is destined to become a tremendous vehicle for information. 1970 T. S. Crawford *Hist. Umbrella* viii. 158 In the 1880s, the French author, Uzanne, was implying that certain French 'gentlemen' armed with *parapluies* were in the habit of preying on girls..on rainy evenings.

parapolitical: see *para-1 1.

parapraxia (pærăpræ·ksiă). Also **parapra·xis,** *pl.* **-es.** [f. Para-1 + *-praxia.] The faulty performance of an intended act; in psychoanalysis, a minor error said to reveal a subconscious motive.
1912 Stedman *Med. Dict.* (ed. 2) 657/2 *Parapraxia*, a condition..in which there is a defective performance of certain purposive acts. 1937 tr. *Freud's Gen. Sel. Wks.* i. 25 That group of everyday mental phenomena whose study has become a technical help for psycho-analysis. These are the bungling of acts (parapraxes) among normal men as well as among neurotics. 1953 M. Critchley *Parietal Lobes* v. 160 The patient, requested to make a particular movement, may do something quite different (parapraxia or parakinesis). 1959 *Observer* 1 Feb. 19/4 Such forces in scientists may produce quite unpredictable parapraxes in their experimental work. 1969 P. Anderson in Cockburn & Blackburn *Student Power* 261 No appeal to the conventions of drawing-room conversation can controvert the parapraxes of the couch. 1975 *Times Lit. Suppl.* 4 July 713/1 Have we recognized a bit of the Latin Mass?.. An astronomical reference? A Freudian parapraxis? 1975 *New Society* 11 Sept. 600/3 All too many malapropisms and misprints (or are they parapraxes? We get, for instance, 'He apostasises', followed by a quotation from Mill, for 'He apostrophises'.).

paraprofessional (pæ:răprofe·ʃənăl). [f. Para-1 + Professional *a.* (*sb.*).] A person without professional training to whom a particular aspect of a professional task is delegated. Also as *adj.*, of or pertaining to such a person.
1967 *Maclean's Mag.* May 64 There is some talk now of using para-professional help, trained on the job like interns. In some schools, mothers already are supervising lunch hours and study periods. 1968 *Bull. Council for Basic Educ.* (U.S.) June 2 The future student..would discover other quaint educational terms which are currently fashionable... 'Paraprofessional' is one, meaning a mother who helps the kindergarten teacher in putting on children's overshoes. 1971 *Harvard Business Rec.* Nov.–Dec. 39/2 The understandable reluctance of the physician to delegate tasks to paraprofessionals. 1972 *N.Y. Law Jrnl.* 22 Aug. 3/5 Another paraprofessional in our office is a retired police officer, presently responsible for..important trial preparation functions. 1974 *Yale Law Jrnl.* LXXXIII. 806 In many professions greater use of paraprofessional employees could help to solve problems in the delivery of professional services. 1975 *Publishers Weekly* 22 Sept. 89/1 A paraprofessional's view of how to survive breast cancer. 1976 *Maclean's Mag.* 17 May 26/1 Basford also stressed the need for developing native lawyers and para-professional court workers for small settlements. 1976 *Times Lit. Suppl.* 16 July 872/4 The personal service ideologies of medical professionals, paraprofessionals, indigenous non-professionals.

paraprotein(æmia: see *para-1 2 a.

parapsoriasis (pæ:răsorəi·ăsis). *Med.* [mod.L. (coined in Fr. by L. Brocq 1902, in *Ann. de Dermatol. et de Syphiligr.* III. 446), f. Para-1 + Psoriasis.] Any of various rare chronic diseases of the skin which resemble psoriasis and are characterized by red scaly patches, resistance to treatment, and lack of subjective symptoms.
1906 Dorland *Med. Dict.* (ed. 4) 526/1 *Parapsoriasis*, a chronic skin-disease resembling psoriasis and lichen. 1920 J. M. H. Macleod *Dis. of Skin* xxvii. 844 Brocq has divided the parapsoriasis group into three varieties namely: Parapsoriasis en gouttes, P. lichénoide, P. en plaques. 1936 *Practitioners Libr. Med. & Surg.* X. vi. 617 Much of the confusion that has been occasioned by parapsoriasis is attributable to its widely differing clinical types and its multitudinous nomenclature. 1956 D. M. Pillsbury et al. *Dermatol.* xxxii. 744 The original definition of parapsoriasis was a chronic eruption resembling psoriasis and having no acute manifestations. *Ibid.*, Lichenoid parapsoriasis is the parapsoriasis analogue of lichen planus. 1961 R. D. Baker *Essent. Path.* xx. 540 There are other histologically identifiable noninfectious dermatoses in addition to psoriasis and pityriasis rosea. These include parapsoriasis, lichen planus and several other rare conditions. 1968 A. J. Rook et al. *Textbk. Dermatol.* I. 804/2 The terms guttate parapsoriasis and varicelliform parapsoriasis are still frequently used as synonyms for pityriasis lichenoides chronica and pityriasis lichenoides acuta, respectively.

parapsychic(al: see *para-1 1.

parapsycho·logy. [Para-1.] The science or study of phenomena which lie outside the sphere of orthodox psychology. Cf. *metapsychics *sb. pl.* Hence **parapsycholo·gical** *a.*, **parapsycholo·gically** *adv.*; **parapsycho·logist.**
1924 *Times Lit. Suppl.* 10 Jan. 27/2 Its inherent merit..renders the publication a noteworthy and welcome contribution to parapsychological literature. 1927 H. Driesch in C. Murchison *Case Psychical Belief* 164 The philosophic importance of Parapsychology. 1936 *Discovery* Dec. 396/1 The book [sc. *The Modern Dowser*]..should be useful to the would-be dowser and of interest to the parapsychologist. 1954 M. Lowry *Let.* 10 May (1967) 369 You were so sporting as to write us parapsychologically suspecting some Lowry misery-grisery. 1957 Rhine & Pratt *Parapsychol.* i. 12 The term parapsychology was adapted from the German word *Parapsychologie*... It means the same as the older English expression, *psychical research*, and the French, *métapsychique*. 1967 A. Wilson *No Laughing Matter* i. 12 Alas for the limits of our parapsychological knowledge. 1973 *Daily Tel.* (Colour Suppl.) 30 Nov. 33/1 The Russians, who had previously persecuted parapsychologists as bourgeois deviationists, began to encourage research, doubtless with their space communications system in mind. 1975 *Sci. Amer.* Oct. 117/3 Whenever a major experiment, such as the SRI test of Targ's ESP machine, is a conspicuous failure, parapsychologists themselves become strongly motivated to give reasons for the failure. 1976 *Listener* 3 June 698/3 Once you postulate psychic communication..you are really committed to a whole new dimension... You can't sell parapsychology by the yard. 1977 *Church Times* 7 Apr. 6/1 They accept the reality of the 'Easter experience' understood psychologically, perhaps parapsychologically.

paraquat (pæ·răkwǫt, -kwæt). [f. Para-1 2 b + Quat(ernary *a.* and *sb.*
So called from the fact that the bond between the two pyridyl groups is in the *para* position with respect to their quaternary nitrogen atoms.]
Any salt, esp. the dichloride or dimethylsulphate (both crystalline and water-soluble), of the 1,1′-dimethyl-4,4′-bipyridylium ion, $(—C_5H_4\overset{+}{N}\cdot CH_3)_2$, a quick-acting contact herbicide that is rendered inactive by the soil.
1961 *Weed Abstr.* X. 572 *Paraquat di(methyl sulphate), previously PP.910, a new herbicide.* Publ. Plant Protection Ltd., 1961, pp. 6. Paraquat (1,1′-dimethyl-4,4′-bipyridylium) is the name given to the free radicle. Two salts are currently under test, the di(methyl sulphate) (PP.910) and the dichloride (PP.148)... Paraquat salts are very rapidly absorbed into the aerial parts of plants. 1962 *New Scientist* 6 Sept. 490/1 The active ingredient—paraquat—is a member of a family of herbicidal chemicals (quaternary salts of bipyridyls). 1962 *Listener* 15 Nov. 836/1 A new family of weed-killing substances based on 'paraquat' may herald a new agricultural revolution. Paraquat is almost a total weed-killer but is especially useful in killing grasses. 1965 *Economist* 5 June 1193/2 The 'Paraquat' herbicides, used in ploughless cultivation. 1968 *Times* 17 Dec. 10/5 The weedkiller paraquat..is highly poisonous but is swiftly broken down to harmless components by bacteria in the soil. 1972 *Country Life* 20 Jan. 169 The advent of paraquat was one of the most important developments in British farming in the 1960's. 1974 *Daily Tel.* 2 Feb. 1/1 The weedkiller paraquat has been officially declared a poison... Only registered pharmacists or firms will be able to sell paraquat, and only people engaged in agriculture or horticulture will be able to buy it. 1974 *Times* 2 Feb. 4/2 At least twenty deaths in Britain in the past 10 years are known to have been caused by paraquat. In almost all cases death followed the decanting of agricultural paraquat solutions into soft drink bottles. Granular preparations containing less than 5 per cent of paraquat salts are not affected by the new controls. 1977 *Jrnl. R. Soc. Arts* CXXV. 567/2 Paraquat acts..by killing the green top growth of all plants and is inactivated by soil.

pararammelsbergite, -religious: see *para-1 2 c, 1.

para-rescue (pæ·răre:skiu). [f. *para-3 + Rescue *sb.*] A rescue carried out by a parachutist or parachutists. Chiefly *attrib.* and *Comb.*
1950 *N.Y. Times* 30 Jan. 20/2 A para-rescue crew is being held in readiness here and will be sent out at once if the ground party finds any trace of the missing transport. 1961 *Aeroplane* C. 458/1 A ground search party and 12 trained pararescue men are included. 1968 *Washington Post* 4 July A 11/6, I realized there was a guy in green bending over me... He was the chopper's para-rescueman. 1973 *Maclean's Mag.* July 58/3 Gathercole..told Mooddy to direct..a helicopter and a Hercules with a pararescue team, piloted by Captain Neil Toby.

para-rhyme, pararhyme (pæ·rărəim). [f. Para-1 + Rhyme *sb.*] A half-rhyme, with the same consonant pattern but vowel variation.
In English poetry particularly associated with Wilfred Owen (1893–1918).
1931 E. Blunden in W. Owen *Poems* 28 Having discovered and practised this para-rhyme, Owen became aware that it would serve him infinitely in the voicing of emotion and imagination... By means of it he creates remoteness, darkness, emptiness, shock, echo, the last word. 1939 *Eng. Stud.* XXI. 99 He [*sc.* Wilfred Owen] had invented what has been called pararhymes. Choosing words built upon the same framework of consonants but different vowels, he played with this blend of similarity and dissimilarity..placating the ear while disturbing it. 1951 [see *dissonantal a.]. 1961 *Listener* 23 Nov. 863/1 Owen muted the rhythms of the romantics by the use of pararhyme. 1975 *Ibid.* 9 Jan. 69/1 [Wilfred] Owen had taught him pararhyme, but [Keith] Douglas uses both pararhyme and assonance.

pararosaniline, -sagittal: see *para-1 2 a, 1.

parascending (pæ·răsendiŋ), *vbl. sb.* [f. *para-3 + Ascending *vbl. sb.*] A sport in which participants, wearing open parachutes, are towed behind a vehicle to gain height before release for a conventional parachute descent, usu. towards a predetermined target area. Hence **pa·rascender**, one who takes part in this sport.
1970 *Policy, Organisation & Rules of Scout Assoc.* iii. 169 In order to ensure the safe control of the sport of Parascending within the Scout Movement the following safety regulations must be applied. 1973 J. Lucas *Big Umbrella* xiii. 137 The advent of multiple aerofoil parachutes has led to a challenging new kind of fun-jumping—parascending, which is enjoying increasing popularity among children as well as adults. Parascending is practised at sea as well as on land. One girl, towed by a motor boat, travelled more than half-way across the English Channel by parachute. On land, the parachutist takes off with the aid of a towing vehicle, such as a Land-Rover. *Ibid.*, Parascending made its first appearance in Britain in 1962, and has gained a firm foothold in the sporting world. 1973 *Daily Tel.* 28 Feb. 16/4 (*heading*) Parascenders split from parent group. *Ibid.*, Parascending..has become independent of the British Parachute Association. *Ibid.* 16 July 12/4 The parascender..chooses the point of release and tries to hit the centre of the target, five and a half inches across. 1975 *Valiant* 30 Aug. 1 Parascending is a new all-action sport. 1978 *Observer* (Colour Suppl.) 9 Apr. 43/4 Two helpers hold the parachute open, the Land-Rover moves forward and the parascender soars into the air.

parascience (pæ·răsəi,ĕns). [f. Para-1 + Science.] The study of phenomena assumed to be beyond the scope of scientific inquiry or for which no scientific explanation exists. So **pa:rascienti·fic** *a.*, **-sci·entist.**
1953 *Mind* LXII. 360 [The fallacy] that I know about thinking by 'observing' myself thinking, whereby thought is transformed into a paraphysical object studied by a parascience of introspection. 1961 *Times Lit. Suppl.* 1 Sept. 577/3 Time Travel, like hyperspace, is one of the classical Science-Fiction presumptions which, though abstruse mathematical discussion may give it para-scientific justification, is essentially a basis for fantasy. 1974 *Nature* 8 Nov. 129/1 So much of the interpreting, let alone the theorising, of parascientists and parathinkers is based on seemingly natural occurrences which either cannot be fully corroborated according to the rules of evidence, or cannot be properly repeated according to the rules of experiment. 1975 *Ibid.* 6 Nov. 22/1, I lumped them all [*sc.* telepathy, telekinesis, parapsychology, etc.] under the title 'parascience'. 1976 *Word* 1971 XXVII. 77 As we know very little about human language (a fact..little known to most parascientists..), any approach toward its analysis may contribute something to our understanding of the subject.

parasexual (pæræse·ksiuăl), *a. Genetics.* [f. Para-1 + Sexual *a.*] Involving, exhibiting, or being a process by which recombination of genes from different individuals occurs without meiosis.
1954 G. Pontecorvo in *Proc. 9th Internat. Congr. Genetics* 192 In the last ten years, processes other than standard sexual reproduction, and yet resulting in recombination of hereditary properties, have come to light in microorganisms of widely different groups... The purpose of the present paper is to give a summary of the work which has led, in our Laboratory, to the discovery in the filamentous fungi of another of these mechanisms, which could be called 'para-sexual'. This mechanism, based on *mitotic* segregation and recombination, occurs side by side with a standard sexual cycle in one of the three species investigated. 1962 *Ann. Rev. Microbiol.* XVI. 39 Using parasexual processes, 40 markers have been assigned to eight linkage groups in *A*[*spergillus*] *nidulans*. 1962 G. Dalldorf *Fungi & Fungous Dis.* x. 135 Even in asexual fungi, recombinant types may arise through a parasexual process. 1970 J. Webster *Introd. Fungi* 161 The occurrence of parasexual recombination explains how variation can occur in Fungi Imperfecti. 1974 *Nature* 27 Sept. 321/2 Genetic analysis of their development [*sc.* that of cellular slime moulds] has been hampered by the apparent absence of a true sexual cycle, although progress has been made in parasexual genetic analysis. 1977 P. B. & J. S. Medawar *Life Sci.* iii. 39 In some bacteria..the sexual process is restricted to something that amounts to hardly more than the infection of one bacterium by nucleic acid from another—the 'parasexual' process that lies at the root of the enormously important phenomenon of bacterial transformation.
Hence **pa:rasexua·lity**, the parasexual process.
1959 *Mycologia* LI. 109 Somatic recombination or parasexuality..reassorts genetic characters of heterokaryotic components in a vegetative system to yield products that are comparable to the meiotic products of the sexual process. 1974 *Nature* 18 Jan. 142/1 Using an improved version of the methods employing parasexuality.., we have selected axenic resegregants carrying various markers from non-axenic strains.

parashoot (pæ·răʃūt), *sb. rare.* [Alteration of Parachute *sb.* after Shoot *sb.*1, Shoot *v.*] (See quot.)
1940 [see *chutist, 'chutist].

parashot (pæ·răʃǫt). [f. *para-3 + Shot *sb.*1] In the war of 1939–45, a member of the British Home Guard whose task was to shoot down enemy parachutists. Also *attrib.* So **pa·rashoot** *v. trans.*; **pa·rashooter.**
1940 *Star* 14 May 8/5 (*caption*) 'What are you doing with that gun?' 'I'm practising-to-parashoot Germans!' 1940 *Daily Mirror* 17 May 5/2 Over a quarter of a million had applied to join Britain's parashooters by midnight on Wednesday, the War Office stated last night. 1940 Parashot [see *paratroops sb. pl.]. 1940 *Star* 22 May 8/2 Clubs' part in fight against parachute troops... The appeal for parashooters has brought rifle shooting into the news. Clubs are offering free instruction to applicants. 1940 *Economist* 8 June 1005/2 Air-raids will precede invasion; and the transfer of wardens and other civil defence workers to become Ironsides and parashooters..should be stopped immediately. 1940 *New Yorker* 29 June 46/3 A movement to provide parashots with hand grenades. 1941 A. Christie *N or M?* vi. 85 I've got to go to a meeting about this Parashot business, raising a corps of local volunteers. 1942 A. M. Low *Parachutes* p. x, Parashot, the word coined by the newspapers in the summer of 1940 to describe gentlemen with shotguns seeking enemy parachutists. 1944 *Ourselves in Wartime* 129 The village men lined up..to join the local Defence Volunteers. The 'Parashooters' they called themselves in those days, when everyone thought the enemy might drop on us from the sky just in the same way as he had done over the Low Countries.

parasitæmia (pæ:răsəitī·miă). *Path.* Also (chiefly *U.S.*) **-emia.** [f. Parasit(e *sb.* + Gr. αἷμα blood: see -ia1.] The presence of demonstrable parasites in the circulation.
1947 Dorland & Miller *Med. Dict.* (ed. 21) 1066/2 *Parasitemia*, the presence of parasites (especially malarial parasites) in the blood. 1956 *Nature* 21 Jan. 133/1 In December 1952, N'Dama 151 was challenged with a strain of *T*[*rypanosoma*] *vivax*..; a low-grade intermittent parasitæmia persisted until August 1953, after which time no parasites were detected in the blood. 1972 *Ann. Trop. Med. & Parasitol.* LXVII. 390 A direct counting technique for estimating high parasitaemias of *Babesia argentina* is described. 1973 B. J. Williams *Evolution & Human Origins* v. 80/2 He confirmed Beet's observation that the incidence of parasitemia was less in *AS* individuals than it was in *AA* individuals.

parasite, *sb.* Add: **4. parasite drag** *Aeronaut.*, the drag of all parts of an aircraft other than that induced by the lift or due to the lifting surface (quot. 1927 represents a broader use); **parasite (jet) fighter** *Aeronaut.*, an aircraft carried by and operating from another aircraft; **parasite resistance** *Aeronaut.* = *parasite drag.
1927 V. W. Page *Mod. Aircraft* (1928) iv. 134 The parasite drag results from friction of the air on the parts of the airplane, including the wings, tail, fuselage, landing gear, etc., and from the eddies set up by these parts when in motion. 1934 *Jrnl. R. Aeronaut. Soc.* XXXVIII. 459 For aeroplanes of normal design, the ratio of induced drag to parasite drag was very low at high speeds. 1958 *Guided Missiles* (U.S. Dept. Air Force) ii. 24/2 Both parasite and induced drag vary as the square of the velocity. 1965 C. N. Van Deventer *Introd. Gen. Aeronaut.* iv. 59/1 The resistance of parts that do not contribute to lift is called parasite drag. 1948 *Daily Progress* (Charlottesville, Va.) 2 Jan. 10/1 A new jet fighter plane is expected soon to start and end a flight high in the sky for the first time in history. This is the McDonnell XP-85, known as a 'parasite' fighter because it is based on a larger craft. 1948 *Shell Aviation News* No. 121. 6/3 The Air Force has announced the building by McDonnell of a parasite jet fighter..designed to be carried in the front bomb-bay of Consolidated's B-36. It has no landing gear and is launched and picked-up by means of a retractable hook which engages in an eye on the mother plane. 1977 *New Scientist* 25 Aug. 489/4 In the immediate post-war years the US Army Air Force test launched the diminutive McDonnell XF-85 Goblin parasite fighter from the bomb bay of a Boeing B-29 bomber. 1918 Webster Add., Parasite resistance. 1921 *Flight* XIII. 509/2 Great reduction in parasite resistance..is brought about by the cantilever construction. 1929 *Aircraft Engineering* Mar. 10 Giant seaplanes are another story... Aerodynamically they are disappointingly full of parasite resistance.

parasite, *v.* Add: **1.** (Later example.)
1932 [see *gazoomph v.].
2. (Later examples.) Also *fig.*
1882 *Amer. Naturalist* XVI. 150, I had the opportunity of examining a larva..parasited by an allied species. 1963 *Guardian* 14 June 10/2 The cuckoo bees Psithyrus..parasite the Bombex. 1968 *Daily Tel.* (Colour Suppl.) 8 Nov. 12/3 Viruses are incomplete cells. They exist by parasiting 'proper' cells, and getting into them. 1969 K. Giles *Death cracks Bottle* vii. 83 The only worry I had was that he might be parasiting the business stone dry. 1976 *Eastern Daily Press* (Norwich) 19 Nov. 12/2 Beds of pelargoniums ..heavily parasited by this rust.

parasitic, *a.* Add: **3. c.** Applied to trades: † (*a*) see quot. 1909; (*b*) non-productive.
1909 *Q. Rev.* Jan. 83 The so-called parasitic trades—that is, trades in which it is alleged that workers who have incomes or maintenances derived from sources other than their wages underbid those who live entirely on their wages. 1926 *Spectator* 19 June 1032/1 Far too much still goes in what we may call parasitic middlemen's profits.
4. Applied to unwanted subsidiary phenomena and effects in physical apparatus, esp. in electronic devices and electrical machinery.
1889 *Telegraphic Jrnl.* 1 Nov. 497/1 If..the iron core were solid, the E.M.Fs. induced in it would..cause enormous currents to flow... They are currents which could not be utilised... By suitably dividing the iron core..these so-called parasitic currents may be rendered almost negligible. 1921 *Wireless World* 15 Oct. 437/1 The..parasitic noise which abounds in amplifiers wired on the usual principle. 1927 V. W. Page *Mod. Aircraft* (1928) vi. 206 (*heading*) Reduction of parasitic resistance. 1943 F. E. Terman *Radio Engineers' Handbk.* vi. 498 Parasitic oscillations..are very likely to occur when large tubes are employed because of the long leads, large interelectrode capacities, and relatively high values of transconductance involved. 1958 Nayler & Saunders *Handbk. Aircraft Industry* viii. 125 In modern aircraft the 'parasitic' drag due to extraneous items..has been largely eliminated. 1959 *Listener* 25 June 1109/1 They are electric insulators and therefore do not carry parasitic electric currents (so-called eddy currents) when their magnetization changes rapidly. 1967 *Electronics* 6 Mar. 251/1 With discrete components, parasitic capacitance from the diodes' packages increased the over-all capacitance. 1972 *Sci. Amer.* Aug. 19/2 The Wankel operates with less friction than a conventional engine... The lower friction alone means decreased parasitic losses and can be translated directly into an increase in fuel economy.
5. *parasitic bronchitis* = *hoose; *parasitic gastritis, gastro-enteritis*, in cattle, horses, or other domesticated animals, a gastric inflammation caused by parasitic nematode worms.
1925 *Vet. Rec.* 19 Dec. 1137/1 Parasitic bronchitis is also known as husk, hoose, verminous bronchitis, or paper skin. It is a bronchial irritation, arising from the presence in the air passages of nematode parasites. 1942 *Skandinavisk Veterinär-Tidskrift* XXXII. 488 (*title*) On parasitic gastritis in the horse. 1947 *New Biol.* III. 49 The inflammation of the alimentary tract of sheep (and of that of other farm animals also) which is caused by parasitic roundworms is called by the veterinarian parasitic gastro-enteritis. 1951 *Vet. Rec.* 22 Dec. 864/2 The literature presents certain references to the occurrence of parasitic bronchitis in adult cattle. 1965 *Ibid.* 9 Oct. 1196/1 Field investigations into parasitic gastro-enteritis in the west of Scotland showed that *Ostertagia ostertagi* was the predominant parasite. *Ibid.*, In Great Britain, outbreaks of bovine parasitic gastritis have been reported for many years. 1967 W. R. Kelly *Vet. Clin. Diagnosis* vi. 27/2 (*caption*) Moderate degree of dyspnoea in a yearling heifer suffering from parasitic bronchitis.
6. Applied to an aerial that is not electrically connected to a transmitter or receiver.
1936 L. S. Palmer *Wireless Engin.* xii. 469 Antenna B is not directly energised from the power source, but receives its current through the induction and radiation field of the directly energised antenna A. Antenna B..is sometimes called a parasitic antenna. 1947 C. E. Tibbs *Frequency Modulation Engin.* vi. 114 When the parasitic aerial is at resonance.., and the separation between it and the energised aerial is held at 0·1λ, maximum radiation is in the direction of the parasitic aerial, which is then called a 'director'. 1970 J. Earl *Tuners & Amplifiers* v. 113 When a reflector and directors (called parasitic elements because they are not electrically connected to the dipole) are added to the system the centre impedance of the dipole falls well below 75Ω.
B. *sb. pl. Electronics.* Parasitic oscillations.
1943 F. E. Terman *Radio Engineers' Handbk.* vi. 498 Parasitics cause reduction in the power at the desired mode of operation, introduce spurious frequencies, give rise to distortion in linear amplifiers and modulators,..etc. 1967 *Electronics* 6 Mar. 177/2 (Advt.), 'Strays', 'streaks', 'parasitics'..you may forget them all.

parasitism. **1.** (Further examples.)
1974 *Times* 19 June 16/5 One of them [*sc.* Russian Jews], Vitali Rubin, was..told that he had 15 days to get a job, otherwise he would be tried for parasitism, a charge entailing a heavy prison sentence. 1975 *Nature*

17 Apr. 554/3 Methods mentioned include the issue of call-up papers for retraining in the Soviet Army.., prosecution for 'parasitism' (being without employment, although the scientists concerned have been deprived of their jobs as a result of applications for exit visas), [etc.]. **1976** *Ibid.* 3 June 363/1 Some 20 scientists called in to regional police headquarters in Moscow were told that if they did not commence working within two weeks they would be charged under the 'parasitism' law, which covers such offences as prostitution, drug addiction, and alcoholism.

parasitization (pærăsitəizẽⁱ·ʃən, pærăsəitəizẽⁱ·ʃən). [f. Parasitiz(e *v.* + -ation.] The infestation of a plant or animal by a parasite.

1932 J. S. Huxley *Probl. Relative Growth* i. 38 Parasitization of Inachus with Sacculina..actually increases the growth-ratio of the female abdomen. **1946** *Nature* 3 Aug. 174/1 A team of workers is co-operating on different aspects of the work, namely, rate of growth, parasitization-rates and mortality. **1950** *Hilgardia* XIX. 422 There was a slight drop in the population (probably due to *Apanteles* parasitization). **1953** *New Biol.* XIV. 100 Parasitization of the larvae [of wheat blossom midges] does not prevent them from feeding. **1962** Gordon & Lavoipierre *Entomol. for Students of Med.* xxxii. 201 A few days after parasitisation has occurred the site occupied in the skin by the larva assumes the appearance of a boil-like swelling. **1975** *Nature* 29 May 403/1, I examined the effects of parasitisation on tumorigenesis by parasitising *vg tu*^{a4} larvae.

parasitize, *v.* Add: Now usu. with pronunc. (pæ·răsitəiːz). Also *intr.* Also *fig.* (Later examples.)

1915 W. B. Herms *Med. & Vet. Entomol.* ii. 7 That an animal is parasitized does not necessarily involve it in death. **1922** W. M. Wheeler *Social Life Insects* v. 197 Our bodies, our domestic animals and food plants, dwellings, stored foods, clothing and refuse support such numbers of greedy organisms, and we parasitize on one another to such an extent that the biologist marvels how the race can survive. *Ibid.* 198 The thoroughly parasitized host must abandon all hope of being an end in itself. **1934** J. S. Huxley *Sci. Res. & Social Needs* vii. 111 New suburbs, almost all in ribbon development, have been allowed to parasitize the road. **1955** *Sci. Amer.* Apr. 98/3 Viruses.. used to be thought of solely as foreign intruders—strangers to the cells they invade and parasitize. **1956** T. W. M. Cameron *Parasites* 235 In some cases (*Leishmania*) the phagocytic cells themselves are parasitized. **1970** *Nature* 25 July 382/1 Mice were infected with..erythrocytes parasitized with the NK 65 strain of *Plasmodium berghei*. **1974** *Sci. Amer.* Oct. 98/3 The estrildid finches parasitized by these two related widow-bird species are also two species of the same genus, *Uraeginthus*. **1977** *N.Z. Jrnl. Agric.* Jan. 21/3, 90 per cent of the caterpillars were parasitised.

parasitoid (pæ·răsitoid), *sb.* and *a.* [a. mod. L. *Parasitoïdea* (O. M. Reuter *Lebensgewohnheiten und Instinkte der Insekten* (1913) v. 53), f. Parasit(e *sb.* + -oid.] An insect, esp. one belonging to the orders Hymenoptera or Diptera, whose larva lives as an internal parasite which eventually kills its host. Also as *adj.*, of or pertaining to an insect of this kind.

1922 W. M. Wheeler *Social Life Insects* ii. 46 Recent studies of the parasitic, or as I prefer to call them with O. M. Reuter, the 'parasitoid' Hymenoptera. *Ibid.*, The parasitoids exhibit another peculiarity. **1941** J. S. Huxley *Uniqueness of Man* v. 135 There exist fully formed adult insects—a beetle or two, and several parasitoid wasplike creatures—of smaller bulk than the human ovum. **1971** *Nature* 12 Feb. 508/1 Many viruses, bacteria, fungi and insect parasitoids can attack insects. **1972** L. E. Chadwick tr. *Linsenmaier's Insects of World* 300/1 Since most 'parasites' of the Hymenoptera destroy the host, they are not true parasites. Instead they are more accurately called parasitic predators or parasitoids. *Ibid.*, Although it is important to know the distinction between parasites and parasitoids, in actual practice even entomologists tend to use the word parasite for both. **1974** *Nature* 22 Feb. 572/1 A parasitoid fly was suspected as the cause of death.

parasitological, *a.* Add: (Later examples.)

1921 H. G. Ponting *Gt. White South* 125 On the south side..was Dr. Atkinson's parasitological laboratory. **1962** J. D. Smyth *Introd. Animal Parasitol.* ii. 8 The question of the part played by environmental conditions in controlling the development of the parasite is..one scarcely touched on by modern parasitological research. **1971** *Nature* 16 July 209/2 Both these organisms have been widely used in parasitological research.

Hence **parasitolo·gically** *adv.*

1976 *Nature* 15 Apr. 608/2 Our experimental group consisted of 100 male and female white laboratory mice, strain ICR, from 3 weeks to 3 months old, which had been standardised both microbiologically and parasitologically (specific pathogen-free). **1977** *Lancet* 21 May 1095/1 Stool samples can be negative for *Giardia lamblia* even in parasitologically confirmed cases of giardiasis.

parasitopolis (pæ:răsəitǫ·pǫlis). [f. Gr. παράσιτο-ς Parasite *sb.* + πόλις city; see -polis.] A parasite city; a city that is overdeveloped and economically non-productive.

1927 P. Geddes in A. Defries *Interpreter* viii. 199 In all the great cities..you have in progress the history of Rome in its decline and fall. Beginning as Polis—the city, it developed into Metro-polis—the capital; but this into Megalo-polis—or city overgrown... Next, with its ample supply of 'bread and shows' (nowadays called 'budget') it was Parasito-polis, with degeneration accordingly. Thus, all manner of diseases, bodily, mental, moral: hence Patholo-polis, and finally,..Necro-polis—city of the dead. **1945** L. Mumford in *Archit. Rev.* XCVII. 7/1 But one outstanding fact must first be frankly acknowledged: the Parasitopolis of the late nineteenth century has already become the spectral Necropolis of the mid-twentieth century. **1961** —— *City in Hist.* viii. 234 Parasitopolis had become Patholopolis, and even further, Patholopolis had turned into Psycho-pathopolis, with a Nero or a Caligula as absolute ruler.

parasol, *sb.* Add: **2. b.** An aircraft having wings raised above the fuselage. Also *attrib.* and *Comb.*

1914 *Aeroplane* 29 Jan. 110/1 M. Gilbert has been flying another 'parasol', 60-h.p. *Ibid.*, The 'parasol' monoplane ..has been fitted with a new..Gnome [engine]. **1930** *Flight* 8 Aug. 896/1 The machine used by Lombardi was a Fiat 'A.S.1' parasol monoplane. **1940** *Jrnl. R. Aeronaut. Soc.* XLIV. 661 Outboard tanks were employed with the Junkers J.21/22 parasol fighters. **1952** A. Y. Bramble *Air-Plane Flight* xi. 161 The high-wing type was sometimes known as the 'parasol' type, especially when the wing was carried on struts rising above the main structure of the fuselage. **1969** K. Munson *Pioneer Aircraft 1903–14* 131/1 A few R.E.P. parasol-winged monoplanes were in French and R.N.A.S. service during the early part of the war.

para-state (pæ·răstẽit). [f. Para-[1] + State *sb.* IV.] An institution or body which takes on some of the roles of civil government or political authority; an agency through which the state works indirectly. Also *attrib.* Hence **parasta·tal** *a.* and *sb.*

1959 *Listener* 5 Feb. 245/2 He [sc. Lenin] described the proletarian dictatorship as a sort of a para-state.., a state progressively dissolving in society and working towards its own extinction. **1966** *Economist* 24 Sept. 1233/3 Since the [Tunisian] government is only too keen to load off its shares.., there is developing a group of what are called para-state industries, with both government and private capital. **1969** *Nationalist* (Dar es Salaam) 25 Jan. 1/4 The Vice-President told the workers that the aim of opening TANU branches in industries and parastatal organisations was to strengthen the political role of the Party. **1969** *Reporter* (Nairobi) 13 June 20/3, 65 per cent of parastatal investment would also come from within the country. **1971** *Guardian Weekly* 23 Oct. 6 The parastatals were still almost wholly outside central control. **1971** P. A. Allum *Politics & Society Post-War Naples* (1973) i. 38 The public administration groups a plethora of institutions and para-state agencies...prefecture, field offices of ministries, municipalities, state banks (Bank of Naples) ISVEIMER, IRI, ENI, etc. **1974** J. White tr. *Poulantzas's Fascism & Dictatorship* iii. i. 74 A whole series of hidden parallel networks, operating as the channels of real communication of real power and decisions, varying from the emergence of pressure groups and private militia..to the setting up of virtual para-state networks. **1974** *Daily News* (Tanzania) 13 Sept. 5/6 Employees of hotels, bars and parastatal organisations in Shinyanga Region who fail to enrol in adult education courses by the end of this month will be sacked. **1976** *Nigerian Herald* 20 July 5/1 In terms of depth of investigation and the sheer volume of assignment, the Odje Commission of Inquiry, which probed parastatals in the Bendel State, is remarkably unprecedented.

parasymbiosis (pæ:răsimbəiōu·sis, -biōu·sis). *Biol.* [a. G. *parasymbiose* (W. Zopf 1897, in *Ber. Deutsch. Bot. Ges.* XV. 90), f. Para-[1] + Symbiosis.] The relationship between a free-living lichen and an organism (either a fungus or another lichen) which infests it and establishes a symbiotic relationship with the algæ of the lichen. So **pa:rasymbio·tic** *a.*, of or pertaining to such an association; **para·sy·mbiont**, an organism involved in an association of this kind.

1897 *Jrnl. R. Microsc. Soc.* 228 Prof. W. Zopf finds fresh confirmation of his theory that many of the fungi which grow on lichens are not true parasites, but have a kind of symbiotic relationship to the host, which he terms parasymbiosis, the hyphæ of the 'parasite' enveloping the algal constituent of the lichen, without inflicting any injury upon it. **1921** A. L. Smith *Lichens* vi. 264 Zopf found several instances of such parasymbiosis in his study of fungal parasites. *Ibid.*, There occur in lichens, certain parasites classed as fungi which at an early stage are more or less parasymbionts of the host. *Ibid.* 265 Tobler has added two more of these parasymbiotic species on the border line between lichens and fungi. **1952** Parasymbiont [see *lichenized *ppl. a.* b]. **1967** M. E. Hale *Biol. Lichens* x. 155 This [sc. the order Lecanorales] includes..the parasymbiotic lichen fungi. *Ibid.*, Other parasymbionts are classified under appropriate fungal families.

pa:rasympathe·tic, *a.* *Anat.* [Para-[1].] Of, pertaining to, or designating one of the major divisions of the autonomic nervous system, distinguished from the sympathetic system by its place of origin, its use of acetylcholine as a transmitter, and its general association with rest and recuperation rather than alertness. Also *absol.*, the parasympathetic system.

1905 J. N. Langley in *Jrnl. Physiol.* XXXIII. 403, I use the word para-sympathetic for the cranial and sacral autonomic systems. **1920** *Gray's Anat.* (ed. 21) 768 The peripheral part [of the autonomic system] is..arranged in two subsidiary systems—the parasympathetic system and the sympathetic system. **1925** *Jrnl. Physiol.* LX. p. ix, The absence of effect of pilocarpine and atropine on muscle tonus indicates that the parasympathetics do not control the tonus of voluntary muscle. **1940** [see *muscarine]. **1946** *Nature* 10 Aug. 207/1 Most internal organs have a double nerve supply of which one is excitatory (sympathetic) and the other inhibitory (parasympathetic), that is, they work against each other. **1948** A. Brodal *Neurol. Anat.* xi. 340 From the anatomical arrangement of the sympathetic and parasympathetic it is also to be inferred that the former must be involved predominantly as a whole in diffuse reactions, affecting the entire organism, whereas in the latter, the structural organization permits more restricted, localized effects. **1954** S. Duke-Elder *Parsons' Dis. Eye* (ed. 12) v. 68 A second type of drug acts as a parasympathetic stimulant. **1967** G. M. Wyburn et al. *Conc. Anat.* iii. 92/1 Branches from the sympathetic and the vagus (the parasympathetic) form plexuses in relation to the arch of the aorta and the bifurcation of the trachea. **1968** Passmore & Robson *Compan. Med. Stud.* I. xxiv. 90/2 Overall parasympathetic activity corresponds roughly to the picture of an old man asleep after dinner, i.e. slow heart, small pupils, constricted bronchioles and noisy breathing, salivary secretion, vigorous peristalsis. **1972** J. A. Wilson *Princ. Animal Physiol.* xi. 416/2 That part of the autonomic nervous system originating in the thoracolumbar portion of the spinal cord is the sympathetic division; that part originating in the other three segments of the central nervous system is the parasympathetic system. **1977** D. Morris *Manwatching* 166 Whenever we are called upon to perform some intense, violent activity, the sympathetic system takes over and pours adrenalin into the blood, temporarily dominating the keep-calm parasympathetic system.

Hence **pa:rasympathe·tically** *adv.*

1957 *Jrnl. Nerv. & Mental Dis.* CXXV. 463/1 For the most part a situation in which the sympathetic nervous activity is high inhibits parasympathetically innervated consummatory reactions. **1978** A. Wingate in M. U. Barnard et al. *Human Sexuality for Health Professionals* iii. 18 The first phase [of the male sexual response]..is parasympathetically mediated.

pa:rasympatholy·tic, *a.* (*sb.*) *Pharm.* [f. prec. + -o + *-lytic.] Annulling or opposing the physiological action of the parasympathetic nervous system. Also as *sb.*, a substance which does this.

1949 *Jrnl. Pharmacol. & Exper. Therap.* XCV. 53 There is need for a simple *in vivo* assay for the study of parasympatholytic agents. **1960** A. Burger *Medicinal Chem.* (ed. 2) liii. 1147/2 Diphemanil methylsulfate is a parasympatholytic agent which blocks the cholinergic impulse to the eccrine sweat glands. **1969** *Scand. Jrnl. Gastroenterol.* IV. 641/1 The following parasympatholytics were applied: atropine, Isopropamide.., and Gastrixone. **1973** *Jrnl. Pharmacol. & Exper. Therap.* CLXXXVII. 293/2 The mechanism for this parasympatholytic augmentation of dopamine's pressor and inotropic actions is not clear and may involve an interaction between dopamine and acetylcholine within the central nervous system.

parasy:mpathomime·tic, *a.* (*sb.*) *Pharm.* [f. as prec. + -o + Mimetic *a.* (and *sb.*).] Producing physiological effects characteristic of the action of the parasympathetic nervous system by promoting stimulation of parasympathetic nerves. Also as *sb.*, a substance which does this, either by mimicking the action of acetylcholine or by interfering with that of cholinesterase.

[**1940** *Q. Jrnl. Med.* IX. 239 Observations on the effect of parasympathetic-mimetic drugs have been made.] **1946** *Jrnl. Pharmacol. & Exper. Therap.* LXXXVII. 31 A range of doses of eleven parasympathomimetic drugs (acetylcholine, mecholyl, [etc.]..) was administered subcutaneously. **1968** W. C. Bowman et al. *Textbk. Pharmacol.* xxviii. 702 In addition to their important muscarinic actions, many parasympathomimetic drugs exert other actions at different sites. For example, acetylcholine itself..stimulates ganglion cells and striated muscle. **1972** *Exper. Neurol.* XXXIV. 355 (*heading*) The central activity of parasympathomimetics on gastric secretory function. **1974** *Ann. Internal Med.* LXXXI. 49/1 All parasympathomimetic maneuvers and drugs affect arrhythmias by causing local release of acetylcholine in cardiac tissues. **1974** M. C. Gerald *Pharmacol.* v. 103 Analogous terms are used for drugs modifying cholinergic activity, that is, cholinergic or parasympathomimetic agents and cholinergic blocking agents.

parasynapsis to **-tacamite**: see *para-[1].

parataxic (pærătæ·ksik), *a.* *Psychol.* [f. Para- i + Tax(is 6 + -ic.] A term used mainly by H. S. Sullivan to describe the mode in which subconscious attitudes or emotions affect overt interpersonal relationships. See also *prototaxic *a.*, *syntaxic *a.*

1938 H. S. Sullivan in *Psychiatry* I. 125/2 A *parataxic* situation, a much more complicated entity in that two of the *four* or more persons now concerned, while illusory, are real antagonists to any collaboration of A and Mrs. A. Our Mr. A has become multiplex. **1945** P. Mullahy in *Ibid.* VIII. 184/1 With the development of the parataxic mode of symbol activity, the original undifferentiated wholeness, oneness, of experience is broken. **1961** J. A. C. Brown *Freud & Post-Freudians* ix. 168 By a process of what Sullivan describes as 'parataxic distortion' one may attribute to others traits taken from significant people in one's past. **1970** Hinsie & Campbell *Psychiatric Dict.*

(ed. 4) 222/2 One way to learn what is true and what is parataxic in thinking or feelings about another is to compare one's evaluations with those of others. **1973** *Jrnl. Genetic Psychol.* CXXIII. 338 The extent to which a person distorts his perception of the norm for self-acceptance by attributing his own high or low acceptance of self to the average person is termed 'parataxic distortion'.

paratectonic to **-terminal**: see *para-[1].

paratha (parāthā·). [Hindi *parāṭhā*.] In India and among Indian communities outside India: a variety of unleavened bread fried in butter, ghee, etc., on a griddle.

1935 M. Morphy *Recipes of All Nations* 691 *Paratha* (Indian bread)..as Chapati, but the dough..is..fried in butter on the griddle. **1960** R. P. Jhabvala *Householder* i. 46 Her eyes were modestly lowered and she appeared intent on preparing dough for parathas. **1964** S. M. Sadeek *Windswept & Other Stories* (1969) 13 The wife and two daughters of the East Indian headman were busy with the cooking of the curry-mutton..the rice and paratha rotie. **1971** *Hindustan Times Weekly Rev.* (New Delhi) 4 Apr. p. iv/5 Have you ever tried mashing up leftover *gobhi*, *aloo*, or *matar bhujia* and using it as a filling for stuffed *parathas*? **1977** *New Society* 3 Feb. 242/1 While others were still wondering whether chapatis were folded or flat, I was boldly ordering poppadoms and parathas.

parathecial, -ium: see *para-[1] i.

parathion (pæræþəi·ǫn). [f. the elements Para-[1] 2 b and Thio- in its chemical names + -*n*.] An organophosphorus insecticide which is also highly toxic to mammals and is available in commercial preparations similar to those of malathion; diethyl-*p*-nitrophenylthiophosphate, $(C_2H_5O)_2 \cdot PS \cdot O \cdot C_6H_4NO_2$.

1947 *Jrnl. Econ. Entomol.* XL. 915/2 (*heading*) Parathion for control of green peach aphid. *Ibid.*, Parathion may be formulated into emulsions, dusts and wettable powders. **1953** *Sci. Amer.* July 62/2 A slow-acting insecticide—e.g., parathion—may have a devastating effect on a bee colony. **1958** E. Hyams *Taking it Easy* ii. 198, I noticed Red Spider mite on some newly-budded peach trees... I went to my small store of parathion. **1960** *Jrnl. R. Hort. Soc.* LXXXV. iii. 127 A parathion smoke was used and..200 dead beetles were collected from the floor and staging. **1970** *Daily Tel.* 22 Aug. 3/1 On the fruit and vegetable farms of California..it is not uncommon for workers coming in contact with parathion to faint, become nauseated, or go into convulsions. **1977** *Jrnl. R. Soc. Arts* CXXV. 566/2 In 1950 parathion, the first of many powerful organic phosphorus materials, gave positive control of Red Spider Mites.

parathormone (pæræþǭ·ɹmōun). *Physiol.* [Blend of *parath*(*yroid* (s.v. Para-[1] i) and *hormone.] A polypeptide hormone secreted by the parathyroid glands of higher vertebrates, which increases the amount of calcium in the blood by its action on bones, kidneys, and gut.

Registered in the U.S.A. as a proprietary name, although its protection lapsed in 1966.

1925 *Official Gaz.* (U.S. Patent Office) 17 Nov. 591/1 Eli Lilly & Company, Indianapolis, Ind. Filed Oct. 1, 1925. *Para-thor-mone*... Medicine or Pharmaceutical Preparation. Claims use since Sept. 1, 1925. **1926** *Encycl. Brit.* III. 147/2 Collip prepared from the [parathyroid] glands an extract, parathormone, which is capable, when injected subcutaneously into parathyroidectomised dogs, of promptly removing the symptoms of tetany. **1948** Martin & Hynes *Clinical Endocrinol.* v. 101 Parathormone is a protein substance obtainable in solid, but not crystalline, form by acid extraction of fresh parathyroid tissue. **1965** Lee & Knowles *Animal Hormones* vi. 104 It would appear that the rate of secretion of parathormone is controlled by the level of the blood calcium.

pa:rathyroide·ctomy. [f. *parathyroid* s.v. Para-[1] i in Dict. and Suppl. + *-ectomy.] Excision of the parathyroids, or a portion of them.

1903 *Med. News* (Philadelphia) 31 Oct. 823/2 Various detailed studies of the results of complete and partial parathyroidectomy have..been made. **1921** *Times Lit. Suppl.* 12 Nov. 898/2 The primary effect in both parathyroidectomy and guanidine poisoning is an abnormal breakdown of tissue proteins. **1971** *New Scientist* 19 Aug. 437/3, I recently had a parathyroidectomy, which was followed, 48 hours later, by acute hypocalcaemia. The effects of this were according to the text book—tingling limbs and fuzzy-headedness. **1974** F. Ellis in R. M. Kirk et al. *Surgery* viii. 171 Correction of calcium metabolism defects by the use of vitamin D for rickets and osteomalacia, and by parathyroidectomy for hyperparathyroidism.

Hence **pa:rathyroide·ctomize** *v. trans.*, to deprive of the parathyroids; **pa:rathyroide·ctomized** *ppl. a.*

1903 *Med. News* (Philadelphia) 31 Oct. 823/2 The condition into which these parathyroidectomized animals fall after a few days is..as follows. **1922** *Endocrinology* VI. 222 When a litter of eight pure black rats 75 days old..was parathyroidectomized all but two survived the 48 hour period. **1965** Lee & Knowles *Animal Hormones* vi. 107 In the female duck oestrogen administration leads to an elevation of the blood calcium, but it is without effect in the parathyroidectomized animal. **1972** *Jrnl. Endocrinol.* LIV. 107 Bone loss did not occur in parathyroidectomized dogs.

paratomous, *a.* Add: **2.** (See s.v. *para-[1] i.)

paratomy: see *para-[1] i.

paratonal (pærătǭu·năl), *a.* *Linguistics.* [f. Para-[1] + Tonal *a.*] Ancillary to tone; *spec.* of or pertaining to a range of features, excluding pitch, associated with a particular tone.

1966 H. L. Shorto in C. E. Bazell *In Memory of J. R. Firth* 399 The paratonal register distinction is broadly similar to that described for Cambodian by Henderson. Its exponents are distributed throughout the articulatory complex but exclude pitch features.

paratone (pæ·rătǭun). *Linguistics.* [f. Para-[1] + Tone *sb.* i.] A postulated tonal unit of a sequence of simple tone-groups, in which information concerning sentence-type, the speaker's attitude, etc. is conveyed. Also *attrib.* Hence **paratona·lity**; **paratoni·city**.

1973 A. Fox in *Archivum Linguisticum* IV. 21 As far as the paratone-group is concerned, we may identify three kinds of choice which parallel exactly those applicable to the tone-group. By the side of tonality we may have 'paratonality', which is concerned with the number of paratone-groups in the utterance. Since the position of the major tone-group within the paratone-group is variable, we must also have a 'paratonicity' system to locate it..; and parallel to the choice of tone we may have a choice of 'paratone', that is, of the specific tone-sequence to be used.

paratonic, *a.* **2.** Add: In general, applied to plant movements caused by external stimuli (e.g. tropisms and kineses). [First formed in this sense as G. *paratonisch* (J. Sachs *Lehrb. der Bot.* (1868) iii. i. 517).]

1910 J. M. Coulter et al. *Textbk. Bot.* I. ii. 454 The terminal leaflet of *Desmodium gyrans*, like leaves of other members of the bean family, exhibits paratonic movements (i.e. those due to special stimuli, not tonic; opposed to autonomic). **1973** A. Cronquist *Basic Bot.* xxvi. 412/2 Some growth movements are self-controlled (autonomic); others are induced by external stimuli (paratonic)... Paratonic movements are further classified as tropisms and nasties.

paratracheal: see *para-[1] i.

para-transit (pæ·rătræːnzit). Also as one word. [f. Para-[1] + Transit *sb.*] Public transport of a flexible, informal kind (see quots.). Also *attrib.*

1973 *Technology Rev.* July-Aug. 46/2 Modes such as taxis, jitneys, dial-a-ride, car pools, subscription buses, and minicars, the modes we collectively call 'para-transit'. **1974** R. Kirby et al. (*title*) Para-transit: neglected options for urban mobility. **1977** *Chicago Tribune* 2 Oct. XII. 37/7 (Advt.), Transit manager to implement and supervise innovated public paratransit system in growth suburb NW of Chicago. System will consist of subscription mini-bus and dial-a-ride van service. **1978** *Paratransit: Rep. 40th Round Table Transport Econ.* (Europ. Conf. Ministers of Transport) i. 9 'Paratransit'..embraces a series of transport modes, organisational procedures and services falling broadly *midway* between two pre-eminent types of conventional transport, i.e. private cars and public transport services.

paratroops (pæ·rătrūps), *sb. pl.* [f. *para-[2] + Troop *sb.* 2.] A body of soldiers dropped by parachute from aircraft flying over enemy territory. Also, in the singular, = *paratrooper below. Also *attrib.* and *fig.*

1940 *Daily Express* 16 May 3/1 Parashots—the new Local Defence Volunteers now being enrolled to deal with paratroops dropped from planes—will have uniforms of overall material. **1940** *N. & Q.* 29 June 459/1 Parachutists dropped as troops, or to establish themselves in the enemy's country..have now been shortened to 'paratroops'. **1940** *English Digest* Sept. 43 Copford, East Anglia, has a nasty lesson in store for any paratroops who may land there. **1941** *Aeronautics* Apr. 41/1 Paratroops. Better late than never! Parachute troops, originally developed by the Russians..were used by the British early in February. **1941** *Time* 14 July 19/2 Boys and girls..were organized as paratroop detectors. **1942** in *Amer. Speech* (1943) XVIII. 64/2 George Hopkins had made 2,300 parachute jumps..before enlisting as a paratroop. **1942** *Time* 17 Aug. 38 He was transferred to paratroop service. **1943** *Illustr. London News* 17 July 71 (*caption*), A paratroop about to land, with feet together and knees bent ready to take the shock. **1944** *Times* 18 Mar. 4/7 A big-scale exercise with paratroops..was in progress at one point. **1946** A. R. D. Fairburn in Chapman & Bennett *Anthol. N.Z. Verse* (1956) 106 Love's paratroops have hit the ground. **1956** *Nature* 4 Feb. 218/1 During the Second World War he [*sc.* Dr. P. B. Walker] was responsible at the Ministry of Aircraft Production for the supervision of the design of gliders and paratroop aircraft. **1958** *Spectator* 20 June 807/2 The attitude of many French paratroop officers towards General Raoul Salan. **1966** *New Scientist* 22 Sept. 659/3 Emotional sweating.. has been reported in paratroops before jumping. **1973** 'R. MacLeod' *Burial in Portugal* i. 18 She'd been young, she'd married a pair of paratroop wings. **1977** *R.A.F. News* 27 Apr.–10 May 13/1 To give them a detailed refresher course on formation flying and paratroop dropping, the film company enlisted the help of the RAF.

Hence **pa·ratroo:per**, a soldier of this kind; also, an aeroplane transporting paratroops.

1941 *Time* 4 Aug. 20/1 The paratroopers..had never been in an airplane when it was landed. **1943** J. Steinbeck *Once there was a War* (1959) 228 The lieutenant wanted to look back to see whether any of the paratroopers were in sight. **1945** *Times* 12 Jan. 8/3 The trial of..an American paratrooper..has been fixed for Tuesday next. **1946** *R.A.F. Jrnl.* May 180 There were..three distinct types of *Halifax* Glider Tugs and Paratroopers. **1956** *Flight* 8 June 748/2 As a paratrooper the Beverley carries 70 soldiers. **1957** *Economist* 5 Oct. 39/1 The dispatch of paratroopers to Little Rock was a painful blow to Americans. **1962** *Times* 29 Mar. 15/5 For the paratroopers the sheer luck of their landing place meant..the difference between a swift regrouping or death in the air.

paratrophic (pærătrǭu·fik), *a.* *Biol.* [ad. G. *paratroph* (A. Fischer *Vorlesungen über Bakterien* (1897) v. 47), f. Para-[1]: see Trophic *a.* (*sb.*).] Needing live organic matter for nutrition.

1900 A. C. Jones tr. *Fischer's Struct. & Functions Bacteria* v. 49 The paratrophic group includes all those bacteria that can exist only within the living tissues of other organisms. **1902** [see *metatrophic *a.*]. **1919** F. O. Bower *Bot. Living Plant* xxx. 460 On the basis of nutrition Bacteria have been classified into three groups:.. (iii) Paratrophic, those which develop normally only within the living tissues of other organisms, and are true, and obligatory parasites, such as the germs of Tubercle or Diphtheria. **1930** S. Thomas *Bacteriol.* (ed. 2) xiv. 228 The paratrophic organisms..must have living organic matter from which to get all or part of their food requirements.

paratype (pæ·rătəip). *Taxonomy.* [f. Para-[1] + Type *sb.*[1]] A specimen from a group that includes the one designated as the nomenclatural type of a species in its first description, but not the type itself.

1893 O. Thomas in *Proc. Zool. Soc.* 242 Since the other specimens mentioned or enumerated..in the original description are of unquestionably great value in a typical sense, they ought also to have a name and might be called 'para-types' (or side-types). *Ibid.*, A Para-type is a specimen belonging to the original series but not the type, in cases where the author has himself selected a type. It should, however, be one of the specimens mentioned or enumerated in the original description. **1914** *Brit. Mus. Return* 171 in *Parl. Papers* LXXI. 193 Two paratypes of a new species of River-crab from Cochin. **1933** *Jrnl. R. Soc. W. Austral.* XIX. 15 The type specimen is being presented to the Western Australian Museum and a paratype to the British Museum. **1943** [see *isotype 3]. **1951** G. H. M. Lawrence *Taxon. Vascular Plants* ix. 204 A paratype is a specimen cited with the original description, other than the holotype. **1956** *Internat. Code Bot. Nomencl.* ii. 15 A paratype is a specimen cited with the original description other than the holotype or isotype(s). **1962** Gordon & Lavoipierre *Entomol. for Students of Med.* liii. 325 Specimens other than the holotype which the author used when describing the species are known as the 'paratypes'. **1964** *Internat. Code Zool. Nomencl.* xvi. 79 After the holotype has been labelled, each remaining specimen (if any) of the type-series should be conspicuously labelled 'paratype', in order clearly to identify the components of the original type-series. **1971** *Nature* 30 July 311/2 It [*sc.* the skull of a fossil hominid] resembles the paratype from Bed II (Olduvai H. 13) in both cranial and dental characters.

paratyphoid (pærătəi·foid), *sb.* (*a.*) *Path.* [f. Para-[1] + Typhoid *a.* (*sb.*).] A fever similar in symptoms to, but generally milder than, typhoid fever, from which it can be distinguished bacteriologically. Freq. *attrib.* or as *adj.*, esp. in *paratyphoid fever*.

1902 *Amer. Jrnl. Med. Sci.* CXXIII. 977 Paratyphoid fever closely resembles typhoid in its course, temperature curve, and abdominal symptoms. **1904** *Jrnl. R. Microsc. Soc.* June 369 If there is suspicion of para-typhoid (or dysentery) the agglutination of the colonies in question must be investigated with the specific serum of this disease. **1948** Morgan & Cheever in R. J. Dubos *Bacterial & Mycotic Infections of Man* xvii. 384/2 *Salmonella paratyphi* is found only in man where it has been described as one cause of paratyphoid fever. **1965** Hargreaves & Morrison *Pract. Trop. Med.* ii. 91 Paratyphoid infections are caused by *S*[*almonella*] *paratyphosus A.*, *S. paratyphosus B* and *S. paratyphosus C.* In general paratyphoid A occurs in the tropics whilst B infections are met with more frequently in temperate zones. Paratyphoid C occurs in the Indian subcontinent, the Middle East and the Balkans. **1972** Hook & Johnson in P. D. Hoeprich *Infectious Dis.* lv. 588/2 Members of the genus *Salmonella* other than *S. typhi* can produce an illness with all the clinical characteristics of typhoid fever—paratyphoid fever.

parautochthonous (pærǭtǫ·kþǭnəs), *a.* *Geol.* [f. Para-[1] + Autochthonous *a.*; prob. orig. formed in Ger.] Intermediate in character between autochthonous and allochthonous.

1927 L. W. Collet *Struct. Alps* i. vi. 20 The Morcles Nappe and similar parautochthonous folds. **1935** E. B. Bailey *Tectonic Ess.* iii. 34 In parautochthonous folds and thrust-masses (Arnold Heim) the travel of the rock has been considerable or great, according to ordinary standards, but small in the Alpine scale of magnitudes. **1962** Read & Watson *Introd. Geol.* I. x. 580 Parautochthonous granites have travelled only a short distance from their place of origin. **1963** J. G. Ramsay in Johnson & Stewart *Brit. Caledonides* vii. 166 It is not certain where these pieces have their roots. They might be parautochthonous folds, but some evidence seems to indicate that a few might represent the noses of nappes.

paravane (pæ·răvẽⁱn). [f. Para-[2] + Vane.] An apparatus, fitted with vanes to keep it at a constant depth, designed to be towed at the

bows of a vessel in order to clear its path from mines, cut the moorings of submerged mines, or destroy hostile submarines. Also *attrib.* Also *Aeronaut.*, a towing device (see quot. 1959).

1919 *Daily Mail* 6 Jan. 6/2 Like many other great conceptions, the paravane seems a quite simple thing. *Ibid.*, On one occasion last year a flotilla of light cruisers found themselves in a minefield, and by using the paravane they were able to cut their way safely through it. *Ibid.*, The paravane was invented by Lieut. Dennis Burney, R.N. **1920** *Rep. Brit. Assoc. Adv. Sci. 1919* 273 The Paravane has been developed as a weapon to fulfill two purposes:—(1) To attack a submarine. (2) To protect vessels from moored mines. **1920** [see *otter *sb.* 4 c]. **1920** *Glasgow Herald* 8 Oct. 10 Nor was part of the manufacturing profits in respect of paravane systems supplied to the Navy paid by the Government. *Ibid.* 5 Nov. 11 The 'paravane' is a name invented in or about 1916 to denote a particular kind or type of water kite of special shape and material and with special appliances. It was evolved and developed and in 1916 was adopted by the naval authorities as standard. **1922** *Encycl. Brit.* XXXII. 33/1 Paravane, a naval device used in the World War first for attacking submerged submarines and subsequently for protecting vessels against mines and for cutting up hostile minefields. The name of Acting-Comm. C. D. Burney is especially associated with its design and development. **1948** R. de Kerchove *Internat. Maritime Dict.* 519/2 Two paravanes are towed, one on each side of the ship, by means of specially constructed wire towropes. **1959** J. L. Nayler *Dict. Aeronaut. Engin.* 186 *Paravane*, a kite-shaped device by means of which a wire is kept at an angle from a towing vehicle. This device is used at Woomera, towed behind an aircraft, in snatch recovery methods for intercepting guided missiles towards the end of their flights.

paraventricular: see *para-[1] i.

paravertebral (pærăvɜ̄·ɹtɪbrăl), *a.* *Med.* [f. Para-[1] + Vertebral *a.*] Situated or occurring beside the vertebral column or a vertebra.

1893 Dunglison *Dict. Med. Sci.* (ed. 21) 822/1 *Paravertebral*, situate beside or in the vicinity of the vertebral column. **1923** R. E. Farr tr. H. Braun in *Pract. Local Anesthesia* v. 123 Paravertebral anesthesia had been employed quite extensively in abdominal surgery. **1934** *Practitioners Libr. Med. & Surg.* V. xiii. 854 The sympathetic nerves pass from the higher centers in the midbrain down the spinal cord and then out into the paravertebral sympathetic chain. **1963** *Lancet* 12 Jan. 85/2 The patient was given..paravertebral block with 1% procaine. **1967** J. J. Bonica *Princ. & Pract. Obstetr. Analgesia & Anesthesia* I. xxxvii. 643/2 Each thoracic paravertebral space contains loose areolar and fatty tissue.

paravisual (pærăvi·ʒiuăl, -vi·ziuăl), *a.* [f. Para-[1] + Visual *a.*] Conveying information visually but not requiring a person's direct gaze.

1960 *Aeroplane* XCVIII. 764/1 Airborne experience with the Smiths Para-Visual Director system will leave the newcomer with a strong conviction that this new flight-director/instrument-flying concept must inevitably alter present ideas about weather minima and automatic-landing prospects. **1964** M. Allward *Inside Jet Airliner* vii. 53 Three little instruments called Para-Visual Directors mounted in line along the top of the instrument panels in front of each pilot. **1968** G. D. P. Worthington *Airline Instrument Flying* xvi. 256 The Para-visual Director is designed to apply command information in regard to pitch and bank through a rotating 'barber's pole' display which is so located that it is within the pilot's general field of vision but outside the limits of his direct field of vision. The direction of rotation of the pole applies a signal in the pilot's peripheral field of vision which demands fly up or fly down in the case of the pitch bars, or fly left or fly right in the case of the bank presentation.

parawing (pæ·răwiŋ). *Aeronaut.* Also **para wing**. [f. *para-[3] + Wing *sb.*] A parachute device having a flat inflatable wing in place of the more usual umbrella, allowing greater manœuvrability.

1960 R. L. Naeseth *Exploratory Study of Parawing as High-Lift Device for Aircraft* (NASA TN D-629) 1 A wind-tunnel investigation was made of the high-lift capabilities of two supersonic airplane configurations equipped with auxiliary parawings which are lightweight, stowable, fabric wings of parachute-like construction that may be used for take-off and landing. **1964** *Engineering* 7 Feb. 237/1 All of these gliders make use of the fact that the parawing is simple and cheap enough to be expendable. **1966** *Guardian* 12 May 14/5 Inflated triangular 'parawings' are being tested by the US army for possible use in place of parachutes. The parawing, which can be folded, packed and released like a parachute, inflates on opening into a flattened triangular dirigible that glides and maneuvers through the air. **1973** D. Poynter *Hang Gliding* ii. 54 The idea of a limp para wing was presented to NASA as a reentry vehicle and millions of dollars were spent in research and tests... Stiffened Para Wings have performed much better than the limp models.

parawollastonite: see *para-[1] 2 c.

paraxial, *a.* Add: **b.** *Optics.* Situated close to the axis of an optical system, and (if linear) virtually parallel to it; of or pertaining to such a region.

1906 H. D. Taylor *Syst. Appl. Optics* iv. 55 The ultimate or paraxial rays. **1930** L. C. Martin *Introd. Appl. Optics* I. i. 17 Imagine..that the point P is brought so close to the axis that the inclination of the ray to the axis becomes very small; these conditions are characteristic of a 'paraxial' ray. **1934** W. H. A. Fincham *Optics* v. 83 For light in the paraxial region the angles of incidence and refraction will be very small, and may be considered as proportional to their sines in applying the law of refraction. **1974** W. T. Welford *Aberrations Symmetr. Optical Systems* iii. 17 We..define the domain of paraxial or Gaussian optics to be close enough to the axis to ensure that all terms of higher order of magnitude than quadratic in *x* and *y* are to be neglected.

Hence **para·xially** *adv.*, adjacent to or virtually parallel to an axis.

The 1905 reference given in quot. 1909 is incorrect and has not been traced.

1909 *Cent. Dict.* Suppl., *Paraxially*, in *zoöl.*, in such a way as to lie on each side of the long axis (of the body). *Proc. Roy. Soc.* (London), Feb., 1905, p. 318. **1927** V. T. Saunders *Light* iii. 38 If a single ray is incident on the mirror through the principal focus it is seen to be reflected paraxially. **1950** H. H. Hopkins *Wave Theory of Aberrations* 162 The paraxial equations..refer to rays which are limitingly close to the axis. Since, however, the equations are linear in the angles *u*, *α*, *i* rays at finite apertures may be traced paraxially.

paraxylene: see *para-[1] 2 b.

parca, var. *parka.

parcel, *sb.* **4. b.** Delete *local* and add further examples.

1887 R. Hunt *Brit. Mining* (ed. 2) 911/2 A parcel of ore is a pile or heap of copper or lead ore dressed for sale. **1898** *Barrier Weekly Post* 29 Oct. 13 [They] received satisfactory prices for their parcels. **1958** M. D. Berrington *Stones of Fire* 20 They gradually collected a 'parcel' of choice stones. **1965** G. T. Williams *Econ. Geol. N.Z.* viii. 119/1 Increase in the price of gold resulted in renewed activity in 1935 and a certain amount of prospecting and development ensued for over a decade, though apparently only one parcel of 400 tons was treated.

7. d. A large amount of money gained or lost. *slang.*

1903 A. M. Binstead *Pitcher in Paradise* vii. 172 'Aye, it's a pinch for t'pair of 'em, y'r Graace', roared Old Jack, with much warmth; 'an' what's moo-re, if y'r Graace doesn't *pack oop a reglar parcel* over 'em, why—why, A'al *never* speak to y'r Graace on a racecourse agin!' **1922** E. Wallace *Flying Fifty-Five* x. 56 In the argot of his kind he had 'packed up a parcel' over the disqualification of Fifty-Five. **1923** Wodehouse *Inimitable Jeeves* xii. 131 'But if you haven't dropped a parcel over the race,' I said, 'why are you looking so rattled?' *Ibid.* xiv. 162, I think I can put you in the way of winning a parcel on the Mothers' Sack Race.

10. **parcel bomb**, a bomb wrapped up so as to resemble a parcel; **parcel(s) shelf, tray**, a shelf or tray upon which parcels may be placed, esp. in a motor vehicle; **parcel tanker**, a vessel designed to carry various liquids with separate piping and tanks.

1950 *Times* 22 Aug. 3/1 Injured by parcel bomb. A small parcel addressed to Mr. Thomas Rose..blew up when he opened it on Sunday night. **1966** 'A. Hall' *9th Directive* iv. 44 There's a dozen ways—prussic acid..the parcel bomb. **1974** *Guardian* 25 Jan. 24/5 Police scientific experts are examining the remains of a parcel bomb which exploded in an Israeli bank in the City of London yesterday. **1977** *New Society* 27 Jan. 163/3 The weekend parcel-bomb murder of Jason Moyo. **1951** *Motor* 2 May 386/3 There are two useful parcel shelves unobtrusively located beneath the front seat cushions. **1953** *Motor* 9 Dec. 728/3 Large cases..have to be lifted over a somewhat high rear bulkhead, and then slid forward into place below the rear parcels shelf. **1973** *Country Life* 31 May 1547/2 Stowage capacity is also good for a large car... There is..a useful parcel shelf. **1974** 'J. Ross' *Burning of Billy Toober* i. 9 He placed the pipe in the parcels shelf. **1976** *Chrysler World of Motoring* '77 15/4 With the parcel shelf folded away, and the rear seat folded flat, you've suddenly got 49 cu. ft. **1973** *Sea Breezes* May 297/1 The typical modern purpose-built parcel tanker is a complex and expensive investment. **1974** *Times* 31 May (Poseidonia Suppl.) p. iii/4 Parcel tankers have been developed which have the ability to carry incompatible chemicals in separate tanks. **1950** *Motor* 19 Apr. 372/3 There is a parcel tray of generous size below the instrument panel. **1956** *Motor* 24 Oct. 525 (*caption*) Air for de-misting this window is blown through the slots seen here on the parcel tray.

B. *adv.* or quasi-*adv.* or *adj.* **1. c.** Also with vbl. sbs.

1867 [see Parcel-gilding]. **1902** A. H. Hiorns *Metal-Colouring & Bronzing* (ed. 2) iii. 243 (*heading*) Parcel coppering or bronzing as applied to fine zinc castings. **1907** *Handbk. Electro-Plating* (W. Canning & Co.) (ed. 3) 64 (*heading*) Parcel-plating. Plating articles in two or three colours. **1911** S. Field *Princ. Electro-Deposition* xii. 175 Partial deposition..is, in the case of copper, called parcel coppering. **1925** Field & Bonney *Chem. Coloring of Metals* xiii. 166 'Parcel plating'..is applicable to all deposited metals. **1971** T. C. Collocott *Dict. Sci. & Technol.* 855/2 *Parcel plating*, the electrodeposition of a metal over a selected area of an article, the remainder being covered with a nonconductor in order to prevent deposition.

parcelization (pā:ɹsĕləizẽⁱ·ʃən). Also **parcellization**. [f. Parcel *sb.* + -ization.] = Parcellation.

1960 E. R. Goodman in J. A. Fishman *Readings Sociol. of Lang.* (1968) 725 The treatment of the Moslem peoples of the Soviet Union provides the clearest illustration of this policy of parcelization. In an effort to avoid the creation of a large Moslem state in the Volga-Urals area, the Soviet regime created separate Bashkir and Tartar ASSRs. **1963** *Listener* 21 Feb. 322/1 Parcellization gets worse and worse. *Ibid.* 322/2 Even such an obvious reform as a change in the law which permits parcellization seemed to be viewed with suspicion. **1975** Djurfeldt & Lindberg *Behind Poverty* 124 The parcellization of land among many small owners.

pa·rcellate, *v.* *rare.* [f. Parcel *v.* + -ate[3] or as back-formation from next.] *trans.* To divide into separate parcels or portions. So **pa·rcellated** *ppl. a.*

1934 Webster, Parcellate. **1971** *Country Life* 3 June 1374/1 About 50 [vine] growers among the 1,700 whose heavily parcellated strips dominate the district. **1978** *Ibid.* 7 Sept. 642/1 The 100 growers concerned..now have an average of 800 square metres of consolidated vineyard in place of the 200 square metres of parcellated terraces they formerly had to work.

parcellation. (Later example.)

1965 K. H. Connell in Glass & Eversley *Population in Hist.* xvii. 429 Connaught..was..the province where parcellation of the land was most acute.

parcel post. (Earlier examples.)

1837 *9th Rep. Comm. Managem. Post-Office* 28 in *Parl. Papers* XXXIV. 431 Would it not occasion great delay if you made a parcel-post of it to that extent? **1843** *Rep. Sel. Comm. Postage* 41 in *Parl. Papers* VIII. 1 The Banghy post of the East Indies is a parcel post; the maximum of weight appears to be about 15 lb.

parch, *sb.*[1] Restrict *rare* to sense in Dict. and add: **2.** *attrib.*, as **parch mark** *Archæol.*, a localized discoloration of the ground in dry weather over buried remains.

1947 *Antiquity* XXI. 82 The crop's growth had not improved matters, and curiously enough there did not seem to be any 'parch-marks'. **1977** *Times* 19 Sept. 3/2 The latest discovery emerged partly through last summer's drought. Parch marks on the ground disclosed regular lines of Roman trench work.

‖**parch** (pāɹχ), *sb.*[2] *rare.* [W. *parch* reverence, *parchedig* reverend.] In Wales: a clergyman.

1944 Dylan Thomas *Let.* 21 Sept. (1966) 267 Hearing rise slimy from the Welsh lechered Caves the cries of parchs and their flocks. **1953** —— *Under Milk Wood* (1954) 20 A beer-tent black with parchs.

parcheesi: see *pachisi.

parchment, *sb.* Add: **2. b.** A certificate; *spec.* (see quot. 1962).

1888 C. M. Yonge *Our New Mistress* ii. 14 She had been two years from her training college, and had an excellent parchment and report from the place she had left. **1914** 'Bartimeus' *Naval Occasions* xxii. 206 The sailor himself..describes his Certificate as his 'Discharge'. In Accountant circles in which the thing circulates it is known as a 'Parchment'. A Service Certificate..is a double sheet of parchment with printed headings, foolscap size. **1962** Granville *Dict. Sailors' Slang* 86/1 *Parchment*, naval rating's service certificate on which his character and abilities are assessed by the commanding officer of each ship in which he has served.

3*. A colour resembling that of parchment.

1934 Brit. Colour Council *Dict. Colour Standards* 58 Parchment. B.C.C. 165. General representation of samples submitted by Textile and other Colour Using Industries. **1947** J. H. Bustanoby *Princ. Color & Color Mixing* iv. 68 Parchment. A pale tint of brown, resembling the prepared and polished skin of sheep, goats, lambs, young calves, and other animals, used for writing, painting, engraving, etc. **1970** P. West *Words for Deaf Daughter* vi. 117 Old golds, and parchments (the tone, not the stuff). **1974** *Simpson* (*Piccadilly*) *Catal.* Christmas 15 Cable cardigan... Parchment, vicuna colour or light green.

4. a. Also, parchment-coloured.

1939–40 *Army & Navy Stores Catal.* 637/2 Satin, marocain, etc...in Oyster, Pearl and Parchment shades. **1959** *Sears, Roebuck Catal.* Spring/Summer 1224/1 Vitreous china Bathroom Fixtures..Frosty pink, Parchment beige, Sage green. *Ibid.* 618/3 Rayon Fortisan [curtains] ..Frosty pink, Parchment ivory. **1977** *Times* 29 Dec. 17/7 (Advt.), Mercedes-Benz..Magnetic blue/parchment velour.

b. *parchment-coloured* adj.; **parchment window** *Obs. Canad.*, a window-pane made of parchment.

1932 W. Faulkner *Light in August* vi. 115 Always against her eyelids was that..parchmentcolored face watching her. **1936** —— *Absalom, Absalom!* viii. 335 The slight dowdy woman..with parchment-colored skin. **1979** *Country Life* 9 Aug. 431/3, I have included photographs.. of two intarsia sweaters... I particularly like the parchment-coloured one. **1775** Hearne & Turnor *Jrnls.* (1934) 182 The Carpenter Employed nailing on a set of new Parchment windows. **1882** *Royal Readers* (Canada) v. 435 He [*sc.* Robert Campbell] and his companions were forced to use for food the parchment windows of their hut.

parchment-lace. For †*Obs.* read '*Obs.* exc. *Hist.*' and add further examples.

1905 N. H. Moore *Lace Bk.* ii. 62 The English term for this old Guipure was 'Parchment Lace'. **1930** P. G. Trendell *Guide to Collection of Lace* (Victoria & Albert

Mus.) 6 In England there was frequent mention of 'parchment lace' in Queen Mary's reign. **1960** C. W. Cunnington et al. *Dict. Eng. Costume* 268/1 *Parchment lace.* 16th and 17th c's. A lace usually of gold or silver but occasionally of coloured silks.

par-cook (pā·ɹkuk), *v.* *rare.* [After Parboil *v.* 2.] *trans.* To cook partially.

1927 *Daily Express* 17 Nov. 5/2 The chicken was par-cooked and cooled, the stock being set aside for next day.

pardalote. Add to etym. after '*Pardalotus*' (L. J. P. Vieillot *Analyse d'une nouvelle Ornithologie* (1816) 31). (Earlier and later examples); = *diamond-bird* s.v. Diamond *sb.* 12.

[**1826** J. F. Stephens in G. Shaw *Gen. Zool.* XIII. ii. 252 Olive-green Pardalotus, with the back spotted with fulvous.] **1894** A. Newton *Dict. Birds* iii. 684 *Pardalote*, see *Diamond-bird*. **1901** A. J. Campbell *Nests & Eggs Austral. Birds* I. 441 The Red-tipped Pardalote occasionally breeds underground. **1961** *Coast to Coast 1959–60* 63 He pointed to where some pardalotes and sitellas were fluttering about the drooping branch of pepper-tree. **1965** *Austral. Encycl.* VI. 481/1 Pardalotes, a group of very small, insectivorous, tree-haunting birds (genus *Pardalotus*)... Most are brightly coloured and liberally spotted, hence the term 'pardalote' and the alternative name 'diamond-bird'. **1975** I. Rowley *Bird Life* iii. 40 Pardalotes and shrike-tits may turn to bark searching.

pardessus. Restrict *Obs.* to sense in Dict. and add: **1.** (Earlier example.)

1843 *Godey's Lady's Bk.* Nov. 240 *Pardessus*.—At the present moment this most useful appendage to a lady's out-door costume, is more than ever in request.

2. Mus. *pardessus de viole* (də viǫl) [F. *par-dessus* adv.], a small treble viol, played esp. in France during the eighteenth century; also *ellipt.* as *pardessus*.

1889 Grove *Dict. Mus.* IV. 277/2 The five-stringed Treble Viol survived longest in France where it was called 'Quinton' or 'Pardessus de Viole'. **1893** J. S. Shedlock tr. *Riemann's Dict. Mus.* 832/2 The violin quietly took the place of the 'Pardessus de viole' in the band of viols, for the highest part. **1941** N. Bessaraboff *Anc. Europ. Mus. Instruments* iv. 303 The *quinton* is very often confused with the *pardessus de viole*, which also has five strings. These instruments belong to different families: the *quinton* is a true violin, and the *pardessus de viole* (the high treble) is a viol. **1954** *Grove's Dict. Mus.* (ed. 5) VIII. 804/2 A small sopranino viol was evolved, called a *pardessus de viole*, which was in all respects a true viol, but suited to a tuning a fourth above the treble. *Ibid.*, All sizes of viols, even the smallest *pardessus*.., must be held downward resting on or between the knees. **1961** T. Dart in A. Baines *Mus. Instruments* vii. 189 A tiny viol—the *pardessus de viole*—found an admirer or two among the court ladies of France. **1974** *Early Music* Jan. 75/3 Several pardessus de viole, the smallest of the viols, cultivated by court ladies of the time. **1976** *Ibid.* July 361/1 The pardessus is built like a small treble viol, with deep ribs, flat back and frets.

pardner (pā·ɹdnəɹ). *colloq.* (orig. *U.S.*). Also †**pardener**. Var. Partner *sb.* Cf. Pard[2].

1795 B. Dearborn *Columbian Gram.* 137 Improprieties, commonly called Vulgarisms... Pardener for Partner. **1837** A. Sherwood *Gazetteer Georgia* (ed. 3) 71 *Pardner*, for partner. **1847** F. A. Buck *Let.* 23 Feb. in *Yankee Trader in Gold Rush* (1930) 4 Mr. Johnson has taken a pardner, a Mr. Hollister, who appears to be a very smart man. *Ibid.* 5 The young lady who was my pardner gave me a private lesson beforehand so that I walked through it pretty well. **1857** *Observer* 12 Apr. 2/4 The Punch and Judy man..cleared from £2 to £3 a week, he and his 'pardner'. *a* **1861** T. Winthrop *John Brent* (1862) vii. 71, I don't mean sech. I mean jolly dogs, like me and my pardener. *Ibid.* xiii. 140 'Pardners for a kerdrille!' cried Jake. **1864** Dickens *Mut. Fr.* (1865) I. i. i. 4 'Arn't been eating nothing as has disagreed with you, have you, pardner?' 'Why, yes, I have,' said Gaffer. 'I have been swallowing too much of that word, Pardner. I am no pardner of yours.' **1875** J. Miller *First Fam'lies Sierras* (1876) v. 55 That evening Limber Tim..told..what a hero his 'pardner' had become. **1882** D. Pidgeon *Engineer's Holiday* I. xvii. 200 The mine is worked by two 'pardners' who dig and wash by turns. **1907** S. E. White *Arizona Nights* i. xi. 178 It's money I haven't got, and can't get unless I let somebody in as pardner. **1926** *Ladies' Home Jrnl.* Nov. 24 'There', she added as she crouched once more beside her pardner. **1952** M. Allingham *Tiger in Smoke* xiii. 196 'Now, pardner,' he said. They were both great readers of Westerns. **1973** R. Perry *Nowhere Man* iv. 59 'Well, pardner,' he said.. overdoing both the western accent and the sarcasm, 'it seems I'm stuck with you.'

pardon, *sb.*[1] Add: **6. d.** Ellipt. for *I beg your pardon* (see sense 6 in Dict.). *colloq.*

1898 G. B. Shaw *Man of Destiny* 161 *Giuseppe* (coming to the foot of the couch) Pardon. Your excellency is so unlike other great men. **1914** —— *Fanny's First Play* iii. 221 *Knox*... You sit there after carrying on with my daughter, and tell me coolly youre married... *Duvallet.* Pardon. Carrying on? What does that mean? **1930** J. Cannan *No Walls of Jasper* xv. 277 Julian said, 'That's all nonsense. You're drunk.'..'Pardon?' said Eric. **1930** A. P. Herbert *Water Gipsies* xiii. 178 To gain time she said 'Pardon?' and Mr. Baxter had to repeat his question. **1951** [see *grant *v.* 4 e]. **1954** A. S. C. Ross in *Neuphilol. Mitt.* LV. 45 *Pardon!* is used by the non-U in three main ways:—1) if the hearer does not hear the speaker properly; 2) as an apology (e.g. on brushing by someone in a passage; 3) after hiccupping or belching. **1978** I. Murdoch *Sea* 211 'Did you destroy the letter?' 'Pardon?' 'Did you destroy the letter?' 'Yes.'

‖**pare** (pā·re), *sb.*[2] *N.Z.* [Maori.] A lintel in a Maori building.

1897 A. Hamilton *Maori Art* (1901) ii. 156 The small doorway has the usual *pare* or *korupe* over it with a single figure in the centre, and the bird-headed monsters at the ends. **1911** *Dominion Museum Bull.* (N.Z.) No. 3. 106 From the Salem Museum comes also the photograph of a *pare* or door-lintel... This *pare* can be definitely located, from the style of carving, as having been made in the Bay of Plenty District. **1927, 1949** [see *korupe]. **1950** *N.Z. Jrnl. Agric.* May 501 The dark-brown pare or korupe framing the window..is a modern totara carving. **1962** N. Davis in Davis & Wrenn *Eng. & Medieval Stud.* 324 The central group on a lintel (*pare*) from Porangahau, Hawke's Bay.

pared, *ppl. a.* (In Dict. s.v. Pare *v.*[1]) Add: Also with *down*.

1974 *Country Life* 21 Feb. 398/2 The pared-down practicality of mini skirts and boots. **1977** *Rolling Stone* 30 June 98/2 The arrangements are just pared-down versions of the originals.

Paree (pærī·). *colloq.* [Repr. the Fr. pronunc. of *Paris*.] Paris; esp. in phr. *gay Paree*.

1848 F. A. Durivage *Stray Subjects* 116 Walk in, gentlemen, and see the collection.., which beats the Zoological Gardens all holler, and can't be come over by the Gardens des Plantys in Par-*ee!* **1903** Mrs. G. de H. Vaizey *Pixie O'Shaughnessy* xvi. 171 The Major..revived recollections of an old visit to 'Paree'. **1930** E. Waugh *Labels* 21 There are good young men saving up their money for a beano in 'Gay Paree'. **1964** 'P. Loraine' *Day of Arrow* i. 12 The porter, putting on a weary archness reserved for British and American visitors to *Paree*. **1974** *Times* 9 Feb. 10/5 The prize: a second honeymoon in—guess where? You've got it—Gay Paree.

pareiasaur (părəi·ăsǭɹ). Also (erron.) **pariasaur.** [f. mod.L. name of suborder *Pareiasauria*, f. generic name *Pareiasaurus*: see following entry.] A herbivorous fossil reptile of the group once included in the suborder Pareiasauria, known from Permian remains found in southern and eastern Africa, eastern Europe, and Russia, now classified as the superfamily Pareiasauroidea of the suborder Procolophonia. So **pareiasau·rian** *sb.* Also as *adj.*, of or pertaining to an animal of this type.

1905 E. R. Lankester *Extinct Animals* v. 220 These Pariasaurs were about as big as well grown cattle, but not so high on the legs. **1927** Haldane & Huxley *Animal Biol.* xi. 242 (*caption*) Primitive type [of reptile] (Pareiosaurian [*sic*]). **1933** A. S. Romer *Vertebr. Paleont.* vi. 129 A more advanced type..is that of the pareiasaurs of the Middle and Upper Permian of Europe and Africa. **1966** E. Palmer *Plains of Camdeboo* vi. 105 The great 'Cheek-Lizards', the Pareiasaurs,..were found..in the lowest and oldest zone of the Beaufort series. **1971** E. C. Olson *Vertebr. Paleozool.* viii. 315 The main link between procolophons and pareiasaurs is provided by *Rhipaeosaurus*. *Ibid.*, *Rhipaeosaurus* has a dentition that appears somewhat like that to be expected in an ancestral pareiasaurian. *Ibid.*, The skull..has a somewhat pareiasaurian and somewhat procolophonian look. **1975** *Nature* 3 Apr. 415/1 The first..is an exposure of the Madumabisa Mudstones yielding abundant therapsid and some pareiasaur remains. **1977** A. Hallam *Planet Earth* 270 The second group were the pareiasaurs, which were up to 3m..long.

pareiasaurus (pārəi:ăsǭ·rǝs). Also (erron.) **pariasaurus.** [mod.L. (R. Owen *Descr. & Illustr. Catal. Fossil Reptilia S. Afr.* (1876) 7), f. Gr. παρειά cheek + σαῦρος lizard.] = prec.

1876 R. Owen *Descr. & Illustr. Catal. Fossil Reptilia S. Afr.* 10 (*caption*) A vertebra from the hinder part of the dorsal, or from the lumbar, region of a *Pareiasaurus*. **1905** E. R. Lankester *Extinct Animals* v. 218 (*caption*) Photograph of a skeleton of Pariasaurus. **1959** *Times* 19 June 7/5 Those [animals] like the pareiasaurus..have gradually become extinct.

‖**parens patriæ** (pæ:renz pæ·tri,i). *Law.* [mod.L., lit. 'parent of the country'.] The sovereign, or some other authority, regarded as the guardian or protector of citizens who are unable to protect themselves.

1764 T. Cunningham *New & Compl. Law-Dict.* II. s.v. Ideots and Lunaticks § 2, It seems to be agreed at this day, that the King as *parens patriæ* hath the protection of all his subjects, and that in a more particular manner he is to take care of all those who, by reason of their imbecillity and want of understanding, are incapable of taking care of themselves. **1883** *Wharton's Law Lexicon* (ed. 7) 593/2 *Parens patriæ*, the sovereign, as *parens patriæ*, has a kind of guardianship over various classes of persons, who, from their legal disability, stand in need of protection, such as infants, idiots, and lunatics. **1955** *Times* 25 May 15/2 Was it prepared, as regarded a subject of her Majesty and having regard to the jurisdiction which they exercised on her Majesty's behalf as *parens patriae*, to send the child away to become a citizen of another country? **1973** *N.Y. Law Jrnl.* 20 July 1/1 A Court of Appeals has upheld the right of a state to sue to enjoin antitrust violations as parens patriae for its citizens. **1976** *Howard Jrnl.* XV. i. 51 A legal rational was discovered in the ancient doctrine of parens patriae. **1976** *Washington Post* 19 Apr. A22/4 These bills contain a so-called *parens patriae* provision permitting state attorneys general—directly or by hiring outside attorneys—to bring a class action for alleged anti-trust infractions.

parent, *sb.* Add: **1.** *parent-in-law* (further examples).

1932 E. E. Evans-Pritchard in *Sociologus* VIII. 411 There may be difficulty later with his parents-in-law because he has taken no steps to protect their daughter from the perils of child-birth. **1937** R. H. Lowie *Hist. Ethnol. Theory* vii. 78 This typical parallelist cites many instances of African, American, Australian, and Asiatic parent-in-law avoidance. **1972** D. Bloodworth *Any Number can Play* ix. 71 An agent..with parents-in-law in Peking was obviously open to pressure. **1976** *Southern Even. Echo* (Southampton) 10 Nov. 24/7 His wife Josephine and five-year-old son Simon, were at Heathrow Airport yesterday, together with his parents-in-law.

4. b. *Nuclear Sci.* A nuclide that becomes transformed into another nuclide (the 'daughter') by nuclear disintegration.

1905 E. Rutherford in *Phil. Mag.* X. 294 The experiment..was also utilized to prove that radium E was the parent of the *α* ray product radium F. **1950** [see *daughter 6 d]. **1961** G. R. Choppin *Exper. Nucl. Chem.* vi. 82 If the parent is shorter lived than the daughter, the daughter activity will grow to some maximum value, then decay with its own characteristic half life. **1972** Smith & Stokes *Princ. Atomic & Nucl. Physics* xii. 364 The number of atoms of the daughter nuclide decays with either the daughter's or the parent's half-life, whichever is the larger.

5. a. (Further examples.)

1896 [see *division 1 f]. **1905** E. Rutherford in *Phil. Mag.* X. 295 The activity of the successive product, when in equilibrium with the parent substance, can be utilized to determine the period of a substance which itself does not emit rays. **1909** J. Joly *Radioactivity & Geol.* iii. 57 Detrital sediments are 67 per cent. of the total parent igneous rocks. **1914** *Phil. Mag.* XXVIII. 837 It is possible from the disintegration equations of uranium and thorium, and the atomic weights of the parent elements, to calculate the atomic weights of the end products. **1934** M. Bodkin *Archetypal Patterns in Poetry* 329 In the psychological study of poetry it seems to me to have value, partly because it helps us to relate to the facts of poetic experience, those facts which Freud has formulated under the hypothesis of the parent-imago, or super-ego. **1955** [see *daughter atom* s.v. *daughter 7 c]. **1956** *Nature* 11 Feb. 248/1 The fact that the chief expressive movements are the same throughout the world he regards as affording an argument that we are descended from a single parent-stock. **1957** G. E. Hutchinson *Treat. Limnol.* I. xv. 829 One of these isotopes, mesothorium 1, has a half life..long enough for it to undergo some geochemical migration with radium, independently of the parent element thorium. **1967** H. Hellman *Controlled Guidance Syst.* vi. 158 The 'parent' aircraft can carry a larger transmitter and transmitting antenna than the missile can. **1971** I. G. Gass et al. *Understanding Earth* viii. 121/1 The abundance of the common chondrites indicates that most of the parent bodies must have had similar compositions. **1978** *Jrnl. R. Soc. Arts* CXXVI. 686/1 In many cases the mechanical properties of EB welds remain unchanged from those of the parent metal.

c. **parent–child**, used *attrib.* of or pertaining to both a parent and a child, esp. in phr. *parent-child relationship*; **parent company**, a company of which other companies are subsidiaries; **parent figure**, one who is regarded as having some of the characteristics of a parent; **parent language**, a language from which certain other languages are derived; **parents' day**, a day on which parents visit their children's school; **parent ship**, a ship which protects smaller vessels or which acts as a base for ships or aircraft; **parents' meeting**, a meeting of parents with their children's teachers at a school; **parent–teacher**, used *attrib.* of or pertaining to parents and the teachers of their children, chiefly in phr. **parent–teacher association**, a local organization of parents and teachers established to promote closer relations and improve educational facilities; *abbrev.* P.T.A. s.v. *P II.

1928 in *Smith Coll. Stud. Social Work* (1931) I. 411 (*title*) A study of parent-child dependency. **1939** Auden in *I Believe* (1940) 26 The family is based on inequality, the parent-child relationship. **1942** H. Nicolson *Diary* 28 Aug. (1967) 239 Will that child not come to..lose that atmosphere of sacrifice-gratitude which is the best parent-child relationship? **1965** M. Morse *Unattached* i. 70 Strained parent-child relationships were characteristic of the unattached in all areas. **1972** *Guardian* 11 Aug. 9/6 The idea is..for parent-child involvement in getting the maximum from the material. **1869** *Bradshaw's Railway Manual* XXI. 5 They would soon be enabled to declare a dividend equal to that of the parent company. **1943** J. D. Dawson *Tunisian Battle* i. 21 Six Simcas, small cars similar to Fiats but manufactured in France for the parent Italian company. **1970** T. Lupton *Managem. & Social Sci.* (ed. 2) ii. 47 The American parent company..had taken steps to reduce labour costs. **1976** *Times* 1 Mar. 12/4 The British Government did not bind the American parent company [of Chrysler] to continue its operations in Britain for any specific period. **1960** I. Bennett *Delinquent & Neurotic Children* i. 11 The same process that occurs in the rearing of every child, i.e. that of identifying with the parent-figure and incorporating his ideals. **1976** S. Hynes *Auden Generation* ii. 51 In the war, young men ..were faced with a real challenge which was yet like a school game: highly competitive..and earning..the approval of parent-figures. **1905** O. Jespersen *Growth & Struct. Eng. Lang.* ii. 19 The Arian language..was in course of time differentiated into all these languages, or as the same fact is generally expressed in a metaphor of dubious value, was the parent-language from which all these languages have descended. 1933 L. Bloomfield *Language* xviii. 298 In the case of the Romance languages, we have written records of this parent language, namely, Latin. 1965 H. A. Gleason *Linguistics & Eng. Gram.* 33 This reconstructed parent language is now generally called Proto-Indo-European,..abbreviated PIE. 1971 D. Crystal *Linguistics* 154 This parent language (*Ursprache*) was probably more inflected than any of the attested languages. 1973 *Times* 31 Oct. 10/4 This was certainly parents' day with a difference. Small groups were escorted round the buildings and shown classrooms stacked with books on Marxism, on Russian geography and on Cuba. 1973 *Listener* 15 Nov. 675/3 Parents' Day at their son's prep school. 1976 C. Storr *Unnatural Fathers* v. 60 She and Martin..always appeared together at the parents' days at their children's schools. 1933 *Jane's Fighting Ships* 148/3 Beskytteren... Cruising speed is 9 kts... Serves as parent ship for aircraft. 1961 F. H. Burgess *Dict. Sailing* 157 *Parent ship*, a mother ship to several smaller ones. 1972 *Guardian* 17 Oct. 19/8 The parents who never come to a parents' meeting or try..to help or influence their children's school. 1973 'J. Patrick' *Glasgow Gang Observed* viii. 79 He simulated the voice of a form teacher at a parents' meeting. 1915 (*title*) Parent-teacher associations in the rural and village schools of Oregon. 1916 *Ann. Amer. Acad. Pol. & Social Sci.* LXVII. 139 The Congress [of Mothers and Parent-Teacher Associations] assumed the task of organizing Parent-Teacher Associations in every school. 1951 M. McLuhan *Mech. Bride* (1967) 126/2 There is in the parent-teacher relationship a basic violation of the idea of equality. 1957 *Times* 16 Sept. 11/5 The National Federation of Parent-Teacher Associations, formed last year, seeks to promote closer relations between parents and teachers mainly by practical means. 1968 *Daily Tel.* 12 Nov. 21 (*heading*) Parent-teacher link guide for schools. 1973 *Times* 10 Apr. 3/2 In an ideal world all schools would have parent-teacher associations.

parent, *v. trans.* Delete *rare* and add later examples. Hence **pa·rented** *ppl. a.* (cf. Parented *a.* in Dict. and Suppl.).

1904 Belloc *Avril* i. 4 Literary..epochs..are..definitely parented. We know their special stuff and harmony. We can show..the parts meeting and blending. 1954 E. H. W. Meyerstein *Verse Lett. to Five Friends* 2 One so quick, so parented, as you Needed but *time*, to feed her fancies new. 1957 *Times* 7 May 6/5 It is almost heaven-sent that the Suez Canal Company, with its position and its money, should be wanting to parent the idea [of a Channel tunnel]. 1975 *Listener* 13 Mar. 340/3 Over 75 couples..have already been approved as adoptive or foster-parents... Many..are most suitable candidates to parent the child in question.

2. *intr.* To be a parent. Hence **pa·renting** *vbl. sb.*; also *attrib.*

1959 *Britannica Bk. of Year* 547/1 *Parenting*, the supervision by parents of their children. 1970 F. Dodson (*title*) How to parent. 1970 L. B. Ames in *Ibid.* p. xi, New parents have a great deal to learn from those already experienced in what Dr. Fitzhugh Dodson calls the art of parenting. 1972 *Times* 30 Oct. 8/3 The single-minded, unconditional desire..to provide a loving, caring home, which is the hallmark of good parenting. 1973 A. E. Wilkerson *Rights of Children* 305 While making available to parents the range of resources necessary for effective parenting, we need to be more explicit about the social expectations of parents. *Ibid.* 308 The price of waiting indeterminably for parents who cannot or will not develop acceptable parenting requirements is paid by the child. 1975 *N.Y. Times* 16 Sept. 84 Because of all the changes in American society, we are losing our intuitive ability to parent. 1976 *Guardian* 16 Aug. 9/3 Energy gone, parenting is handed over to the parent-substitute, the teak box in the corner.

parentalism (păre·ntăliz'm). [f. Parental *a.* + -ism.] The character or quality of a parent.

1878 W. L. Blackley in *19th Cent.* Nov. 838 What some folk sneer at under the name of 'parentalism'. 1923 *Daily Mail* 4 Oct. 7/2 The parentalism of our laws, with their mixture of foolish prohibitions and foolish laxities.

parentcraft (pē³·rĕntkraft). Also with hyphen. [f. Parent *sb.* + Craft *sb.*] The 'craft' or business of a parent; knowledge of, and skill in, the rearing of children.

1930 *Lancet* 22 Mar. 672/1 (*heading*) The teaching of parentcraft. 1945 *Times* 10 Mar. 8/2 The Ministry were discussing various schemes regarding children under two, and they had lately set up a committee to go into the question of parentcraft. 1958 *Times* 7 July (National Health Service Suppl.) p. xii/1 It [*sc.* maternity care] should also include guidance in parent-craft and in problems associated with infertility and family planning. 1973 *Guardian* 3 Apr. 18/2 A school can teach good parentcraft, both explicitly and implicitly. 1976 *Oxford Times* 10 Dec. 17/1 This state of affairs has certainly been fostered by the voluntary parentcraft classes organised by doctors, midwives and health visitors.

parented, *a.* Add: Also not in comb.: having parents.

1974 *Times Lit. Suppl.* 11 Oct. 1109/1 The orphan..tearfully watching his parented friends go off to camp.

parentela (pærĕntī·la). *Biol.* [a. L. *parentēla* relationship, f. *parent-*, stem of *parens* parent.] The set of all descendants of a particular pair of individuals.

1927 J. F. van Bemmelen in *Proc. Sect. Sci. Kon. Akad. Wetensch. Amsterdam* XXX. 772 The pedigree itself forms a so-called Parentela, which means a survey of all the descendants of a certain pair. 1932 *Proc. 6th Internat. Congr. Genetics* II. 7 The Table of Descendants is better called the parentela. There are two kinds: the definite or closed parentela and the indefinite or open parentela. Only the latter is important for research in heredity. 1946 R. R. Gates *Human Genetics* II. xxix. 1320 The members of this parentela were also small as children.

parentelic (pærĕnte·lik), *a.* [f. as prec. + -ic.] Of or pertaining to relationship based on common ancestry.

1895 Pollock & Maitland *Hist. Eng. Law* II. 294 In a parentelic scheme my great-nephew, since he springs from my father, is nearer to me than my first cousin. 1957 *Jrnl. R. Anthrop. Inst.* Jan. 40 The other entrenched law, åsetasrett, shows the same emphasis but it operates in close conjunction with the law of inheritance which is based on quite different principles, namely the equality of spouses, bilateral affiliation, and equality of siblings. A full discussion of this law and of the parentelic system that it employs is out of place here.

parenteral (păre·ntərăl), *a. Med.* [f. Para-[1] + Gr. ἔντερ-ον intestine + -al.] Involving the introduction of a substance into the body other than by the alimentary tract. Also as *sb.*, a preparation for parenteral administration. Hence **pare·nterally** *adv.*

1910 *Lippincott's New Med. Dict.* 692/1 *Parenteral, -ly*, not by way of the digestive tract. 1912 A. E. Taylor *Digestion & Metabolism* viii. 467 The secretions of the intestine and of the pancreas are toxic on parenteral introduction. 1916 A. J. Smith tr. *O. von Fürth's Probl. Physiol. & Path. Chem. of Metabolism* ii. 48 The toxic effect of the parenterally introduced trypsin. 1949 E. P. Abraham et al. in H. W. Florey et al. *Antibiotics* II. xv. 639 Such a test could scarcely be expected to give an accurate forecast of the effect of a drug administered by mouth or parenterally for the treatment of a generalized infection. 1957 *Obstetr. & Gynecol.* X. 261/2 Esterification of the different steroid preparations resulted in longer-acting, oil-soluble, nonirritating parenterals being made available for clinical use. 1971 *Nature* 12 Nov. 101/2 Excitatory effects of amphetamine administered parenterally in other regions of the brain. 1974 R. M. Kirk et al. *Surgery* iii. 38 When it cannot be given into the gut, it must be administered parenterally. 1974 Passmore & Robson *Compan. Med. Stud.* III. ii. l. 11/1 In experienced hands, complete parenteral nutrition is a safe if time-consuming form of therapy.

parenthesis. Add: **3. c.** *Logic.* Such curved lines (brackets) or other symbols used in the notation of formal logic to punctuate a proposition or to indicate that the expression they contain forms a unit within the whole proposition; also *attrib.* and *comb.* as **parenthesis-free** *a.*, a term referring to notation, esp. that of Łukasiewicz, which eliminates the need for such symbols.

1918 C. I. Lewis *Survey Symbolic Logic* iv. 233 For any function of one variable we here omit any parenthesis around the variable. 1940 W. V. Quine *Math. Logic* i. 40 The parenthesis notation formulated at the beginning of the section is retained. 1954 I. M. Copi *Symbolic Logic* viii. 253 (*heading*) A parenthesis-free notation. 1959 O. Bird tr. *Bocheński's Precis of Math. Logic* v. 82 In this it is better to use the Peano–Russell notation with parentheses, since its similarities to algebra facilitate the 'multiplication'. 1963 O. Wojtasiewicz tr. *Łukasiewicz's Elem. Math. Logic* p. ix, I enumerate here the more important new results whose authorship, I think, I may ascribe to myself. They are as follows: 1. The parenthesis-free notation of expressions in the sentential calculus and in Aristotle's syllogistic. 1965 Dickoff & James *Symbolic Logic* Pref., The parenthesis-free Lukasiewicz notation was especially convenient. 1976 H. Leblanc *Truth-Value Semantics* i. 11 When no ambiguity threatens, we shall omit the outer parentheses of conditionals, conjunctives..and biconditionals.

parenthesize, *v.* Add: **1. a.** So **parenthesized** *ppl. a.*

1940 W. V. Quine *Math. Logic* i. 39 It may be bounded at its other end by the limit of that parenthesized expression. 1971 *Amer. Jrnl. Physics* XXXIX. 501/1 The first parenthesized term in the final member of [equation] (29). 1973 A. H. Sommerstein *Sound Pattern Anc. Greek* i. 6 A schema with parentheses abbreviates a sequence of rules with and without the parenthesized elements, the longest first.

parenthetical, *a.* Add: **B.** as *sb.*

1957 *Publ. Amer. Dial. Soc.* xxviii. 73 We may divide the segmental tentations according to position into three groups: initials, parentheticals, and finals. 1977 *Canad. Jrnl. Linguistics* 1976 XXI. ii. 166 Parentheticals containing 'do'.

parent-in-law: see Parent *sb.* 1 in Dict. and Suppl.

pareo, var. *pareu.

pareœan (pærei·ăn), *a.* and *sb. Anthrop.* [f. Para-[1] + Gr. ἑῷ-ος dawn, eastern + -an.] **A.** *adj.* Designating a Southern Mongol people, esp. those found in and near China, perhaps to distinguish them from the older Chinese stock whose myths spoke of themselves as the people of the dawn. **B.** *sb.* A member of this people.

1904 T. W. Kingsmill in *Jrnl. R. Asiatic Soc.* (North China Branch) XXXV. 95 In the Mantses, and in a less degree in..the Lolos, the yellow pigment, characteristic of the Pareoean [*sic*] races, is either altogether absent or at most slightly developed. 1909 A. C. Haddon *Races of Man* 18 (*heading*) Indo-Chinese, Pareoeans or Southern Mongols. 1929 L. H. D. Buxton *China* iii. 44 The Chinese belong to the southern or 'Pareoean' branch of that great group of humanity which is usually called 'Yellow Man' or 'Mongolian'. *Ibid.* 45 Pareoean (παρηῷος), 'from beside the east', is a term originally proposed by [illegible] Haddon for the group frequently known as 'southern Mongols'. 1931 C. G. Seligman in W. Rose *Outl. Mod. Knowl.* 443 The Southern Mongols, who in recent literature are tending more and more to be called Pareoeans. 1957 *Encycl. Brit.* XVIII. 866/2 In Korea there is a narrow belt which forms a connecting link between the northern and the Pareoean peoples. The latter differ most strikingly from their northern neighbours in having less prominent cheek-bones and a broadish nose.

parera (pa·rĕrā). [Maori.] A New Zealand name for the grey duck, *Anas superciliosa*, found in many islands of the South Pacific region.

1835 W. Yate *Acct. N.Z.* (ed. 2) ii. 57 *Parera*, or Wild Duck—These birds exactly resemble the common English wild-duck. They are of a fine flavour. 1855 R. Taylor *Te Ika a Maui* xxv. 407 Parera, tutuki..the duck; very similar to the wild duck of England. 1864 C. W. Richmond *Let.* 22 Mar. in *Richmond–Atkinson Papers* (1960) II. x. 597 Your Uncle James got a parera (or wild duck). [illegible] B. Oliver *N.Z. Birds* 215 Grey Duck. Parera. 1966 R. A. Falla et al. *Field Guide to Birds N.Z.* 90 Grey Duck *Anas superciliosa*... Maori name: Parera.

parergon. Add: **2.** (Further examples.) Also a work, composition, etc., that is secondary to or a derivative of a larger or greater work; an opusculum.

1928 W. M. Wheeler *Foibles of Insects & Men* p. vii, For some time friends have been urging me to republish in book form some of my more general papers published in various scientific journals. The continued demand for these parerga..suggests that they are perhaps..of interest to readers who do not habitually consult scientific journals. 1957 H. Nicolson *Let.* 21 Oct. (1968) 339, I don't think that you [*sc.* Vita Sackville-West] will really go down to posterity as a writer of gardening articles. You will be remembered as a poet... So your gardening things will be regarded as mere parerga ('a bye-work'), like the flute-playing of Frederick the Great. 1963 *Times* 15 Feb. 6/6 At the end of the programme he and his orchestra played the suite of Symphonic Dances from *West Side Story*, a parergon of the musical show, and a very distinguished one. 1975 A. Mirz tr. N. Kazantzakis *Let.* in *Nikos Kazantzakis* vi. 526, I think that my whole soul..is crystallized in the *Odyssey*. All the other works are parergons. 1975 *Times Lit. Suppl.* 16 May 531/2 Henry Bradley's *The Making of English*..is what it is only because it arose as an inspired parergon to its author's main work as co-editor of the great Oxford Dictionary.

pareschatology (pær:eskăto·lŏdʒi). *Theol.* [f. Gr. παρά- + *eschatology*.] penultimate + -logy.] A term introduced for theories about human life after physical death and before the final resolution.

1976 J. Hick *Death & Eternal Life* i. 22 Whereas eschatology is the doctrine of the *eschata* or last things, and thus of the ultimate state of man, pareschatology is, by analogy, the doctrine of the *pareschata*, or next-to-last things, and thus of the human future between the present life and man's ultimate state. *Ibid.* iv. 34 The rather rare Greek word *pareschatos* means 'penultimate' or 'next-to-last', and pareschatology is..suggested as a name for the study of the next-to-last things, on analogy with 'eschatology', the study of the last things. I am grateful to my colleague Michael Goulder for this useful word. 1977 *Theol. Today* XXXIV. 182 His own constructive proposal is for a 'pareschatology'—for a form of the doctrine of resurrection expanded to include what he calls 'vertical' as opposed to 'horizontal' reincarnation. 1977 *Times Lit. Suppl.* 1 Apr. 390/4 He examines in detail Western and Eastern pareschatologies (*ie*, pictures of what happens between death and an ultimate state).

Paretan (pərē·tən, -ĭ·tən), *a.* and *sb.* Also Paretian (-ʃən). [f. *Pareto (q.v.) + -an, -ian.] **A.** *adj.* Of or pertaining to Pareto or his economic or sociological theories or methods. **B.** *sb.* A follower of Pareto or someone who adheres to his theories or methods. Hence **Pare·tanism.**

1936 Wirth & Shils tr. *Mannheim's Ideology & Utopia* v. 279 From Nietzsche the lines of development lead to the Freudian and Paretian theories of original impulses. 1949 R. K. Merton *Social Theory* viii. 219 Not only ideological analysis, but also..Marxism, semanticism, propaganda analysis, Paretanism and functional analysis have..a similar outlook on the role of ideas. 1965 *McGraw-Hill Dict. Mod. Econ.* 365 A situation is not a Paretian or social optimum if it is possible..to make one person better off without making another person (or persons) worse off. 1969 D. MacRae in Ionescu & Gellner *Populism* 153 For over a century ideologies have been regarded as epiphenomena by sociologists and political scientists. As such Marxists have 'unmasked' them, Paretans treated them as the verbal derivations of non-logical sentiments,..Freudians psycho-analysed them, and so on. 1969 R. Wollheim *Family Romance* 170 A pupil asked me a question about Paretan *optima*.

paretic, *a.* Add: **B.** *sb.* A person affected with paresis.

1882 *Brit. Med. Jrnl.* 12 Mar. 394/1 Local anæsthesia seems..the rule with paretics. 1924 [see *confabulate v.* 2]. 1957 J. Kerouac *On the Road* (1958) ii. vi. 142 An old college schoolmate whose father, a mad paretic, had died [illegible]. *Ibid.*, Paretic [illegible]. 1962 *Brit. Jrnl. Psychiatry* CXXI. 146/1 Quite a few paretics can be rehabilitated following a complete treatment.

Pareto (parē·to, -ĭ·to). The name of the Italian economist and sociologist, Vilfredo *Pareto* (1848–1923), used *attrib.* and in the possessive to indicate his theories or methods and esp. the law, or mathematical formula, in which he claimed that the distribution of income for any society could be expressed.

1920 A. C. Pigou *Econ. of Welfare* v. ii. 699 When these curves are coincident, the general defence of 'Pareto's law' is a law of even limited necessity rapidly crumbles. 1930 E. R. A. Seligman in *Political Sci. Q.* XLV. 341. His *Cours d'économie politique*..contained among other notable contributions the first formulation of the principle which subsequently became known as Pareto's law. This was a generalization which attempted to express the relation between the amount of income and the number of its recipients. 1937 Yule & Kendall *Introd. Theory Statistics* (ed. 11) vi. 100 In economic statistics this form of distribution [*sc.* the extremely asymmetrical] is particularly characteristic of the distribution of wealth in the population at large..and the curve to which it gives rise has been called the 'Pareto line', after Vilfredo Pareto. 1949 [see *heterogeneity]. 1962 *Times Lit. Suppl.* 24 Aug. 634/1 The Pareto coefficients measuring such interspatial and intertemporal differentials of remuneration show 'a very considerable consistency'. 1967 [see *impossibility 4]. 1968 *Internat. Encycl. Social Sci.* XI. 406/1 Thus, Pareto's law is nothing else than the ordinary negative exponential distribution, truncated at the left to log A. 1971 J. L. Gould in *Public Choice* X. 59 The definition of a Pareto-optimal set is a set such that every point outside it is dominated by at least one point inside it, whereas no point inside it is dominated by any other point in it. 1974 *Encycl. Brit. Macropædia* XIX. 680/1 The right-hand tail of the Pareto curve for disposable income distributions is curved downward considerably by taxation. 1974 *Times* 25 Mar. 17/8 We were warned of the dangers of 'Pareto's law'—to spend 80 per cent of the time on 20 per cent of the people.

pareu (pa·re,u). Also pariu, pareo. [Native Polynesian name.] A skirt worn by men and women in Polynesia, made of a single rectangular piece of cloth, usu. of printed cotton. So pareu cloth, the cloth of which this and other Polynesian garments are made.

1860 Mayne Reid *Odd People* 211 There is but one 'garment' to be described, and that is the 'pareu', which will be better understood, perhaps, by calling it a 'petticoat'. 1894 Stevenson & Osbourne *Ebb-Tide* i. i. 13, I saw a man in a *pareu*, and with a mat under his arm, come along the beach from the town. 1914 R. Brooke *Let.* 5 Apr. (1968) 574 I'll wind my pareu [*sic*] tighter round my middle & go & pull out the canoes & we'll all jump in. 1934 *Let.* 24 May (1968) 588 Raymond Buildings must be littered with dropped smocks. May I add a word from *pareo* to the heap on Friday week—a day or two after you get this? 1939 *Century Mag.* Aug. 452/2 The light fell upon..the men,..catching ruddy gleams from red pareus. 1950 W. S. Maugham *Moon & Sixpence* iii. 230 The *pareo* is a long strip of trade cotton... It is worn round the waist and hangs to the knees. 1932 *Amer. Cornel* ix. 46 In a little while a blackfellow, wearing nothing but a pareu, came along. 1949 P. H. Buck *Coming of Maori* (1950) ii. v. 158 The woman's skirt, termed *pareu* in the Cook and Society Islands, was a wider piece of material..to which..was attached by a waistband above the knees for single girls and below the knees for married women. 1962 *Sunday Express* 26 Nov. 11/1 A gorgeous young Polynesian girl..was wearing a native pareu..a sort of wrap-around costume of flowered cloth. 1969 R. T. Wilcox *Dict. Costume* (1970) 259/2 *Pareu*, a Polynesian skirt or loincloth of standard size and colors, a rectangle printed with conventional flower design. 1974 *Islander* (Victoria, B.C.) 16 Apr. 2/1 The Pacific Islanders..brightly colored, boldly designed cloth is used all over the Pacific for virtually all clothing, as well as for bed-covers, drapes, room dividers and tablecloths. 1974 T. Heyerdahl *Fatu-Hiva* i. 41 To walk in the woods and hills of Papenoo was sheer pleasure, for there we wore only an airy *pareu*, and the temperature felt like a pleasant dream.

parfait (pā·ɪfē). [a. F. *parfait* sb., absolute use of *parfait* Perfect *a.*] A rich iced pudding of whipped cream, eggs, etc.; also, ice cream, fruit, syrup, whipped cream, etc., arranged in layers and served in a tall glass. Also *attrib.*

1894 C. Ranhofer *Epicurean* 994 Parfait with Nougat and with Almonds. 1906 [see *mousse]. 1932 N. Smith *Lonely Road* iv. 67 The parfaits came, and proved to be a tinned peach and ice wedged together in a cup. 1938 G. Greene *Gun for Sale* i. ii He stared with distaste at the list of cocktails and iced drinks, of *parfaits* and *sundaes* and *coupes* and splits. 1948 *Good Housek. Cookery Bk.* ii. 490 A Parfait..is usually understood to be a rather rich form of Mousse, light for the reason that it contains whipped whites of eggs. 1953 E. Taylor *Sleeping Beauty* ix. 161 Sundaes, shakes, parfaits, whips, melbas. 1969 R. & D. De Sola *Dict. Cooking* 168/2 *Parfait*, dessert made of alternate layers of ice cream, fruit, syrup, and whipped cream. *Ibid.*, *Parfait glass*, tall narrow glass used for serving parfaits. 1978 *Monitor* (McAllen, Texas) 25 June 1E Mayan Parfait is a smooth, rich mix of vanilla ice cream layered between a banana-nut frozen confection. *Ibid.*, Parfait glasses are perfect for showing off this pretty dessert.

‖ **Parfait Amour** (pa·rfĕt amū·ɹ). [Fr., lit. 'perfect love'.] A sweet, spiced liqueur.

1828 E. Moore *Fudge Fam. Paris* 23 A neat glass of *parfait-amour* which one sips Just as it bottled velvet tipp'd Over one's lips! 1864 C. Schultz *Manual for Manuf. Cordials* &c. 163 *Parfait Amour*... 8 ounces of cedrat rinds, 4 do. lemon peels, 5 do. cloves. Ground; macerate for 24 hours with 5 gallons of alcohol, 95 per cent, and 3½ gallons of water... distil from off the water 5 gallons of flavored alcohol; add 30 lbs. of sugar dissolved in 5½ gallons of water; color deep red, and filter. 1877 [see *orange bitters* s.v. *orange sb.* 7 a]. 1950 O. A. Mendelsohn *Earnest Drinker* xvii. 159 The bright purple colour of such a liqueur as *Parfait Amour* is..an artificial an effect as possible. 1961 *House & Garden* Dec. 90/3 *Parfait Amour*. Sweet and heavy taste and colouring. 1972 N. Freeling *Long Silence* ii. 156 She..was being given a choice between crème de cacao, crème de banane and a mauve concoction known to Holland as Parfait Amour.

parfleche. Add: Also **-flesh**. (Earlier and later examples.) Occas. *fig.* Hence **parfle·ched** *a.*, made or covered with parfleche.

1827 E. Bryan *Enlarges* in *Trans. R. Soc. Canada* 1912 (1913) VI. ii. 110 We embarked with crews and cargoes as follows: viz..1 pack Parfleches. 1845 J. C. Frémont *Rep. Exploring Expedition* 237 Some of us had with the foresight to wear moccasins with parfleche soles, so slippery that we could not keep our feet. 1850 L. H. Garrard *Wah-to-Yah* vii. 106 With a sole of *par-flêche*, hopping over on top of the rock. 1867 *Harper's Mag.* Oct. 584/2 The *teet-sock* or *parfleche* is generally made of a dried buffalo hide, the hair of which has been beaten off with a stone... 1874 Its two ends out in the shape of an envelope... 1928 J. Stuart *40 Yrs. on Frontier* (1925) II. 40 The medicine man] usually had a highly ornamented parfleche, in which he kept one or more fetishes. 1958 P. H. Godsell *Red Hunters of Snows* 54 Their little bucks..stored commodities in painted parfleche bags ready for the forthcoming journey to York Factory. 1940 F. Niven *Mine Inheritance* 61, I saw her bending over another parfleched in front of the tent and taking out two long, gleaming knives. 1952 *Beaver* June 16/1 Side by side stood weathered rafters or were divested of their hair and made into tepee covers, clothing, moccasins, *parfleche* trunks and shields. 1956 V. Fisher *Pemmican* 299 The old man's face was a parfleche of sedans. 1972 D. Kennedy *Recoll. Assiniboine Chief* 92 These parfleches were made from flint hides with the hair scraped off. They were decorated with colours in geometric designs. 1973 A. H. Whiteford *N. Amer. Indian Arts* 78 Parfleches are large envelopes of rawhide used in the Plains to pack dried food and other things. 1974 *Sci. Amer.* Jan. 129/1 These modern ingenuities do not overshadow the photographs of a parfleche, a yurt or an old folding leather-bed-in-a-chest.

parfocal (pāɹfō·u-kăl), *a.* [f. Par *sb.*[1] + Focal *a.*] Having or pertaining to the property that corresponding focal points of different lenses lie in the same plane, so that they may be interchanged without the need to adjust the focus. Const. *with*.

1886 *Microsc. Bull. & Sci. News* Aug. 31/1 Referring to the article in the April issue..on 'changing eye-pieces without altering focus, etc.', we announce that we are prepared to furnish eye-pieces, as there described, made 'parfocal', meaning 'of equal focus'. 1944 C. P. Shillaber *Photomicrography* iii. 234 The objectives are said to be parfocal with each other when one setting of the microscope will serve to focus all four objectives. 1959 *Sci. Amer.* Feb. 124/1 This is called 'parfocal mounting', and it saves a lot of time when, for example, you are searching your collection for a specific type of crystal, surface texture or color to match an unknown mineral. 1965 J. R. Benford in R. Kingslake *Appl. Optics & Optical Engin.* III. iv. 167 The 3·5× objective is a comparatively recent 'telephoto' construction, designed to be parfocal with the high-power objectives. 1970 R. P. Loveland *Photomicrography* I. iii. 111 The rotating objective holder is the most convenient in use when the objectives are parfocal. 1976 *Physics Bull.* Nov. 511/1 To meet the recently increased demand for low magnification objectives, a parfocal distance of 45 mm has been employed. 1979 *Nature* 1 Mar. 22 xii, All optical components are parfocal and permanently centred.

Hence **parfocal·ity**, the property of being parfocal; **parfo·cally** *adv.*

1965 J. R. Benford in R. Kingslake *Appl. Optics & Optical Engin.* III. iv. 165 Tightening one of these and loosening the other provide a precise means of adjusting the objective lens up and down to bring about the desired parfocality setting, so that objectives on a multiple nosepiece can be interchanged without losing focus. 1971 J. H. Richardson *Optical Microscopy for Materials Sci.* i. 41 Parfocality is not an inherent property of an objective, but is rather a convenience often provided on the more modern microscopes. 1974 *Physics Bull.* May 197/2 For optical measurements the instrument has a high pressure xenon arc lamp and six pairs of matched condenser and objective lenses mounted parfocally in turrets.

parfocalize (pāɹfō·u-kăloiz), *v.* [f. prec. + -ize.] *trans.* To make parfocal. So **parfo:calization**, the action or result of parfocalization.

1944 C. P. Shillaber *Photomicrography* iii. 234 As a rule parfocalization can be attained for all microscope objectives except those of great focal length. 1958 G. H. Needham *Pract. Use Microscope* xv. 234/2 On the modern, routine stands these three or four objectives are parfocalized on the revolving nosepiece so that after focusing with the low dry, the high dry may be swung into position without racking the body tube up. 1965 J. R. Benford in R. Kingslake *Appl. Optics & Optical Engin.* III. iv. 165 The second adjustment is to parfocalize the objective.

‖ **parfumerie** (parfümərī). [Fr.] A factory which produces, or a (department of a) shop which sells, perfumes.

1844 *Poe Mystery of Marie Rogêt* in *Ladies' Compan.* (N.Y.) Dec. 97/1 The disappearance of..Marie Rogêt from the parfumerie of Monsieur Le Blanc. 1895 E. Wharton *Helionde* iii. 125 The department of parfumerie would have used up a whole continent of 'sweet gul',..steppes of lavender, fields..of rose. 1951 W. Sansom *Face of Innocence* 138 In Eve's parfumerie shadowy with scent and cool with pale aluminium doors. 1956 'Troy' *Blind Man's Garden* vi. 63 They were sitting in André's office at the *parfumerie*. 1977 *Time* 18 Apr. 19/3 They are fellow clerks in the same Budapest *parfumerie*.

parge, *v.* Delete † *Obs.* and add later examples.

1908 G. P. Bankart *Art of Plasterer* vi. 77 'Lambert's Farm', in the parish of East Grinstead, Essex, has a parged front.

pargeter. Delete † *Obs.*, remove restriction of *pargetter* to 6–9, and add later examples.

1936 S. R. Jones *Eng. Village Homes* vi. 96 Men who dabbed on the clay were 'daubers', and those responsible for working the plaster were 'playsterers' or 'pargetters'. 1951 Lambert & Marx *Eng. Popular Art* iii. 48 In the nineteenth century the pargeter sometimes turned his hand to making plaster casts. 1958 J. Arnold *Shell Bk. Country Crafts* iii. 55 This includes millwrights, masons, thatchers, sawyers, drystone-wallers and pargetters.

pa-rgetry. [f. Parget *sb.* + -ry.] = Parget *sb.* 2.

1908 G. P. Bankart *Art of Plasterer* vi. 64 (*caption*) Pargetry on House in High Street, Maidstone, now pulled down. 1936 S. R. Jones *Eng. Village Homes* vi. 97 Pargetry in design, and capacity to invent suitable tools for working, brought a good deal of variety to pargetry, as may be seen in roughcast and those numerous pricked, combed and scratched arrangements once known as arrowheads, tortoise-shell, zig-zag, herring-bone, basket-work, rustic, interlacing squares and wavy lines.

parge-work (in Dict. s.v. Parge *v.*). Delete † and add later examples. Hence **pa·rge-worker.** Also *parge decoration*, etc.

1908 *Essex Rev.* XV. 162 The unique designs in parge-work on its front. 1908 G. P. Bankart *Art of Plasterer* vi. 58 One form of parge decoration consisted of a simple type of incising, or cutting patterns through the top layer of plaster down to the coating underneath. *Ibid.* 79 The favourite sports of the parge-worker were sheep, grapes, gable ends, and lunettes, [etc.]. 1940 *Chambers's Techn. Dict.* 617/2 *Parge-work* (*Build.*), an ancient form of external plastering with a mixture similar to that used in pargeting; chimneys. 1951 Lambert & Marx *Eng. Popular Art* iii. 48 This 'parge work' ornament was very popular in the sixteenth, seventeenth and eighteenth centuries. At Ipswich there are numerous examples of external parge decorations from about 1557.

parging (pā·ɪdʒiŋ), *vbl. sb.* [f. Parge *v.* + -ing[1].] = Pargeting *vbl. sb.*

1897 F. C. Moore *How to build Home* iii. 34 The parging or plastering of the inside of the flue is permitted. 1903 *Encycl. Dict.* 2/1 (with 3 supplements) [illegible] lining of a chimney. (2) a coating. 1940 *Chambers's Techn. Dict.* 617/2 *Parge-work*... Also called *parging*. 1974 *Sci. Amer.* Dict. *Building* 224 *Parging*.

pargyline (pā·ɹgiliːn). *Pharm.* [Etym. unkn.] A monoamine oxidase inhibitor used in the treatment of benign hypertension, usu. in the form of the hydrochloride, a white crystalline powder; *N*-methyl-*N*-prop-2-ynylbenzylamine, $C_6H_5CH_2N(CH_3)CH_2C:CH$.

1961 *Current Therapeutic Res.* III. 581 Pargyline hydrochloride appears to be a potent antihypertensive agent, the maximal effect of which is manifested slowly. 1963 *Pharm. Jrnl.* 29 Jan. 51/2 [illegible] Pargyline..is the approved name for N-benzyl-N-methyl-2-propynylamine. 1970 Passmore & Robson *Compan. Med. Stud.* II. xiii. 10/2 Pargyline, a monohydrazine compound, possesses modest anti-hypertensive potency and its effect begins several days after the drug is given and continues for 2–3 weeks after treatment is stopped. 1972 [see *methyldopa s.v. *methyl c]. 1977 *Lancet* 6 Aug. 275/1 Plasma-prolactin levels doubled in depressed patients after treatment with the monoamine-oxidase inhibitors, clorgyline and pargyline.

parhelium, obs. var. Parhelion, var. *parahelium.

parhormone, obs. var. *parahormone.

pariah. Add: Also freq. with pronunc. (pəraɪ·ə). **4. b. pariah brig**, a sea-vessel built in India.

1929 F. C. Bowen *Sea Slang* 101 *Pariah brigs*, deep-sea native vessels in India. 1935 M. H. Beattie *On Hooghly* 116 She was what was termed a pariah brig, or native craft, which would find her way to Kedgeree and there pick up a native pilot. 1946 'Shalimar' *Ships & Men* 122 Indian pariah brigs were taken up to replace coasting steamers.

pariahdom (further examples); also **pa·riah-hood, pa·riahism, pa·riahship.**

1887 *Globe* 22 Oct. 1/4 It is astonishing that any person..should regard the national uniform as a badge of pariahism. 1894 *Work & Workers* June 258/2 Ostracism from the class carries with it..hopeless, entire pariahdom. 1897 W. J. Locke *Derelicts* xx. 256 Forgetful of the gaol and his pariahdom. 1906 — *Beloved Vagabond* (1908) vi. 68 They walked on together, and I dropped behind, suddenly realising my pariahdom. 1920 *Edin. Rev.* Jan. 28 The possibility of intermarriage is the crucial test of equality of consideration; its absence sets a stamp of servility and pariahship on the proscribed caste. 1936 W. Faulkner *Absalom, Absalom!* 234 Rode the two horses through that night..in something very like pariah-hood. 1945 R. Hargreaves *Enemy at Gate* 19 This obsessive aggravation of desperadoes and 'poor masterless men', welded into their solidarity of pariahdom which is the outlaws' primary source of strength. 1967 H. Arendt *Orig. Totalitarianism* (new ed.) iii. 68 Disraeli..discovered the secret of how to preserve luck, that natural miracle of pariahdom. 1977 *New Yorker* 15 Aug. 70/2 Moynihan is the strongest force in the attempt to shift New York out of the congressional pariahdom to which it has long been consigned.

Parian, *a.* (*sb.*) Add: **A.** *adj.* **1.** *Parian Chronicle*, add: a further fragment was found in 1897 and is kept on Paros.

3. *Parian cement*, a plaster similar to Keene's cement but prepared with borax in place of alum.

1858 P. L. Simmonds *Dict. Trade Products* 276/1 *Parian-cement*, a fine or coarse cement, according to the price of the article to be used. 1880 *Encycl. Brit.* XI. 351/2 Parian cement is plaster hardened with water containing 10 per cent. of borax. 1950 Kirk & Othmer *Encycl. Chem. Technol.* III. 443 Keene's cement is the best known of the hard-plaster group, others, differing only slightly, being Parian cement and Martin's cement.

B. *sb.* **1.** A native or inhabitant of Paros.

1550 T. Nicolls tr. *Thucydides' Hystory* (Peloponnesian War) iv. xiii. 92 [illegible]. 1840 Hobbes tr. *Thucydides' Hist. Peloponnesian Warre* xv. 170 Thasus (which is an island, and a colonie of the Parians, distant from Amphipolis, about halfe a dayes sayle). 1753 W. Smith tr. *Thucydides' Hist. Peloponnesian War* II. iv. 85 Thasus is an island, a colony of the Parians, and distant about half a day's sail from Amphipolis. 1849 *Encycl. Metrop.* XVII. 487/2 Themistocles made the Parians pay severely for their perfidy to Aristides. 1911 *Encycl. Brit.* XX. 861/1 So high was the reputation of the Parians that they were chosen by the people of Miletus to arbitrate in a party dispute. 1960 R. Carpenter *Greek Sculpture* i. 14 The Parians..had acquired the craft in Egypt. 1978 Carson & Clark *Paros* (rev. ed.) 5 Campers have established a good reputation among Parians.

2. (See sense 2 in Dict.)

3. = *Parian cement.

1880 H. C. Seddon *Builder's Work* vi. 238 Parian is a white cement. 1967 A. G. Geeson *Building Sci. Materials* (ed. 2) II. i. 56 Adhesion on Keene's or Parian is notoriously poor.

paribuntal (pæribə·ntăl). [f. Paracale, Parañaque, *Parang* or *Parasan*, places in the Philippines + *buntal.] A fine straw.

1926 *Vogue* May 105 An attractive summer model is carried out in paribuntal, a new fine straw. 1939 *Times* 17 June 11/3 For a tall girl there is a charming wide brimmed hat in brown paribuntal. 1963 *Harper's Bazaar* May 47 The hat..in cream paribuntal. *Ibid.* 49 The hat..in white paribuntal overlaid with toast parfisisan straw.

parichnos (pari·knɒs). [a. F. *parichnos* (C. E. Bertrand 1891, in *Trav. & Mém. Facultés de Lille* II. vi. 84), f. Para-[1] + Gk. ἴχνος track, trace.] A strand of tissue found beside the leaf traces in fossil plants of the family Lepidodendraceæ. Also *attrib.*

1893 W. C. Williamson in *Phil. Trans. R. Soc.* B. CLXXXIV. 10 Since I agree with M. Bertrand on this point, I shall accept and employ his name of parichnos. 1905 *Ann. Bot.* XX. 269 (*title*) On the presence of a parichnos in recent plants. *Ibid.*, The term Parichnos was used by Bertrand to designate the thin-walled parenchymatous strand of tissue, occurring in *Lepidodendron Harcourtii*, which accompanies the leaf-trace on the posterior side during its outward course. 1917 F. O. Bower *Primitive Land Plants* xii. 234 These lateral pits are connected internally with the parichnos. 1937 R. E. Holttum *Introd. Lower Plants* xi. 235 Two structures entirely unknown in modern lycopods occur on either side of the leaf trace—these are the parichnos scars. *Ibid.*, There may be two other scars beneath the leaf scar which are also parichnos strands branching off those entering the leaf.

parietal, *a.* (*sb.*) Add: **1. a.** *parietal eye*, in the tuatara and many lizards, a structure of unknown function, resembling an eye and situated in the upper part of the skull beneath an opening in the parietal bone.

1886 W. B. Spencer in *Nature* 13 May 35/1 In formation of the paired eyes (of *Hatteria punctata*) invagination to form an optic cup takes place, whilst apparently it does not do so in the case of what may be called the parietal eye. 1911 *Phil. Trans. R. Soc.* B. CCI. 264 The pineal or parietal eye in Sphenodon is..the left-hand member of the original pair of pineal outgrowths. 1937 *Discovery* May 135/1 The so-called third eye of the tuatara..is sometimes known as the parietal or pineal eye. 1969 A. Bellairs *Life of Reptiles* I. vi. 232 In many lizards the parietal eye seems to play some part in regulating the amount of time spent basking.

2. (Later examples.)

1966 'E. Lathen' *Come to Dust* (1969) xiv. 140 Two young women had been discovered at a tree-lined quiet cul-de-sac... 1972 A. Ulam *Fall of Amer. Univ.* iii. 106 In any case in most schools, certainly at Harvard, the formerly idiotically strict parietal rules had been eroded by the sixties to sensibly hypocritical proportions. 1973 E. Taylor *Serfant under It* (1974) xxi. 177 The kinds of things that stir them [*sc.* students] up these days are parietal hours and open admissions or black studies. 1977 *National Observer* (U.S.) 1 Jan. 10/4 Parietal rules were ignored and, later, abandoned.

3. (Further example.)

1918 [see *mural a.* 2].

B. *sb.* **2.** *pl.* (see quot.) *U.S. slang.*

1970 *N.Y. Times* 17 Dec. iv. 9 Yale students..have rejoined the nationwide battle for liberalized 'parietals'—campus terms for women's visiting hours in male dormitories, or vice-versa.

parietin (pərai·ətin). *Chem.* and *Bot.* [f. Parietalis, (*parietinus* ...) of or belonging to walls (f. *paries*, *pariet-* wall), in the form a specific epithet of the lichen *Xanthoria parietina* from which the compound was obtained: see *-in[1]*.] An anthraquinone derivative present as an orange-yellow pigment in some lichens; 1:8-dihydroxy-3-methoxy-6-methylanthraquinone, $C_{16}H_{12}O_5$.

1844 R. D. Thomson in *Proc. Philos. Soc. Glasgow* I. ix. 187, I have succeeded in obtaining the colouring matter, or *Parietin*, as I propose to term it, in the form of needles. 1894 *Jrnl. Chem. Soc.* LXVI. i. 541 The colouring matter..may be extracted by means of benzene without destroying the lichens. It crystallises in small, golden-yellow needles which are soluble in alkalis with blood-red coloration... The authors consider this colouring matter to be a dihydroxyanthraquinone, and propose for it the name *physciol*... The colouring matter was termed parietin by Thomson [etc.]. 1951 A. L. Smith *Handbk. Brit. Lichens* 44 The species [of *Xanthoria*] grow most freely in maritime districts, and are bright-yellow in the open where the acid substance parietin..is freely formed. *Ibid.*, Parietin is produced in more or less abundance in the thallus of most species [of *Placodium*], and in the apothecia of all except *P. rubellum* which is probably an impoverished form. 1966 *New Phytologist* LXV. 211 Unlike the species of *Teloschistes*, those of *Xanthoria* are relatively rich in parietin. 1967 M. E. Hale *Biol. Lichens* vii. 105 The following four pigments are also common to non-lichenized fungi: endocrocin and parietin, both anthraquinones, and polyporic and thelephoric acids, both terphenylquinones.

‖ **parigot** (parigo), *a.* *colloq.* [Fr.] Of an accent, etc.: Parisian. Also as *sb.*, Parisian French.

1974 N. Freeling *Dressing of Diamond* 175 'What sort of accent would you call that, Johnny?' 'Overlaid,' said the technician stolidly. 'Predominance parigot.' 'Parigot hell, that's peasant...' Peasant who went up to Paris and acquired the rhythms.' *Ibid.* 202 A young or youngish woman who speaks with something of a parigot accent. 1977 *Times Lit. Suppl.* 6 May 546/3 French, in this case the most idiomatic 1977 Parigot French.

pari mutuel. Add: Now usu. **pari-mutuel** and sing. (Further examples.) **b.** The booth at which such bets are placed, or the machine which issues tickets recording such bets.

1881 [see *Pan-Eastern a.]. 1891 *Harper's Mag.* Mar. 511/1 For this rough horde of human beings the only interest that the races offered was the betting..by means of the *pool* or *pari mutuel* system. 1913 'F. Somerville' *Spirit of Paris* v. 78 But there is a great keenness on the animals too, and men and women are seen at the pari-mutuel with much earnestness. 1933 E. Hemingway *Three Stories & Ten Poems* 23 The group..the chorus..at the bar and their pari mutuel tickets rattling down. 1931 E. Waugh *Remote People* 71 There had been a horse-race..with a pari-mutuel. 1934 'A. Bridge' *Ginger Griffin* xvi. 220 They strolled towards the *pari-mutuel* windows, which extended in a long row between the paddock and the grand-stand. 1955 S. Ransome' *Deadly Bedfellows* xiii. 123 The legislature..will choose to legalize pari-mutuel horse-racing. 1964 A. Wykes *Gambling* vii. 191 The pari-mutuel system (which was devised in 1872 by a French chemist named Pierre Oller). 1966 C. Drummond *Odds on Death* iii. 43 The big market is not the parimutuel—totaliser, call it what you will—the big market is bookmakers. 1969 L. Koppett *N.Y. Times Guide Spectator Sports* x. 172 Pari-mutuel betting refers to a system in which all the participating bettors are betting against each other. 1973 D. Francis in *Winter's Crimes* 5 115 It meant hanging around the pari-mutuel with your eyes open. 1976 *Maclean's Mag.* 12 Jan. 34/1 The state bicentennial commission raised $845,000 from the taxes on one extra day of betting at pari-mutuel outlets.

paring, *vbl. sb.* Add: **4.** *paring bee*: for U.S. read N. Amer. and add earlier and later examples; *paring-chisel* (later examples), *frolic*, *gouge*, *-knife* (later examples).

1819 J. Pickering *Emigration* 72 A paring 'bee', or 'frolic', [is] an assemblage of neighbours to pare and slice apples and quarters for drying. 1845 *Lowell* (Mass.) *Offering* V. 169 When we were about to have a paring bee one evening, our father invited us to the paring-bee parties that..a few previous. 1850 *Knickerbocker* XXXV. 24 Give me the real paring-bee reels and jigs before all your waltzes and Spanish dances. 1857 *Quinland* I. 191 Went this evening..with the young people to a paring-bee at Squire Carter's. 1933 E. C. Guillet *Early Life Upper Canada* 195 A paring bee produced large numbers of strings of dried apples, and these were suspended from the ceiling of kitchen or attic. 1895 *Montgomery Ward Catal.* Spring & Summer 365/2 Barton's paring chisels. 1940 *Chambers's Techn. Dict.* 617/1 *Paring chisel* (*Carp.*, *Join.*), a long chisel with a thinner blade than a firmer tool, used for finishing off work by hand. 1964 W. L. Goodman *Hist. Woodworking Tools* 33 Plate 1 of the 'Charpente' article shows a workman using the paring-chisel end of this tool. 1968 J. Arnold *Shell Bk. Country Crafts* 287 Each blade produced may be compared very roughly to a carpenter's paring-chisel in having a bevel along its two sides. 1974 P. W. Blandford *Country Craft Tools* v. 76 A longer and thinner chisel for hand pressure only was called a 'paring' or 'heading' chisel. 1931 V. P. Seary *Romance Marit. Provinces* 177 Another sort of frolic was the 'paring frolic', when young men and girls gathered to pare and slice apples, that they might be dried out and kept throughout the winter. 1909 Webster, Paring gouge. 1940 *Chambers's Techn. Dict.* 617/2 *Paring gouge* (*Carp. etc.*), a gouge having the bevel ground upon the inside or concave face of the cutting edge. 1966 A. W. Lewis *Gloss. Woodworking Terms* 39 A paring gouge is a long thin scribing gouge. 1908 *Sears, Roebuck Catal.* 768/1 Kitchen or paring knives... Length of blade, 3 inches. 1925 *Scribner's Mag.* Oct. 430/1 She was rubbing the paring-knife across her fingers to free them from dirt and water. 1968 J. Arnold *Shell Bk. Country Crafts* 139 The box contains hammer, pritchel, buffer, paring-knife, pincers and, of course, a supply of nails.

Paris. Add: Also, in names of articles associated with or designed in Paris, esp. with an implication of being fashionable, exclusive, or expensive, as *Paris cloth, cut, dress, felt, gown, hat, net, shirt*; **Paris binding** (see quot. 1964); **Paris cap** (see quot.); **Paris embroidery** (see quots.); **Paris green** (earlier and later examples); also used as an insecticide.

1918 E. & M. Wallbank *Dress Cutting & Making* x. 65 For binding skirt seams, use ½" lute or Paris binding with one edge folded over rather less than half the width. 1964 C. Penton *ABC of Sewing* 64/1 *Paris binding*, a braid ½" wide used for covering raw edges cut on the straight, and slip-stitched to the fabric beneath. 1966 J. S. Cox *Illustr. Dict. Hairdressing* 109/1 *Paris cap*, a woman's head-dress of the mid-sixteenth century. It fitted the head closely with a jewelled band running over the top of the head and ending in a point on the cheeks. 1960 C. W. Cunnington et al. *Dict. Eng. Costume* 268/1 *Paris Cloth, Toile de Paris*. Med. and 17th c. Originally a fine white linen; later a woollen cloth. 1748 Smollett *R. Random* II. xlv. 80 In the evening, [I] dressed myself in a plain suit of the true Paris cut. 1959 J. Braine *Vodi* xiv. 195 The rather cold-eyed girl with the Paris dress and the real pearls. 1973 H. McCloy *Change of Heart* ii. 17 The black dress was Paris and so at least four years old... Yet old as it was that Paris dress made every other dress in the room look shapeless. 1882 Caulfeild & Saward *Dict. Needlework* 378/1 *Paris Embroidery*.—This is a simple variety of Satin Stitch worked upon Piqué with fine white cord for washing articles, and upon coloured rep silk, or fine cloth with filoselles for other materials. 1957 M. B. Picken *Fashion Dict.* 243/2 *Paris embroidery*, white cord embroidery on piqué. 1969 R. T. Wilcox *Dict. Costume* (1970) 259/2 *Paris embroidery*, a fine white cord embroidery appliquéd in satin stitch on piqué. Used in washable linens and garment accessories. 1853 J. B. Felt *Customs New Eng.* 119 They were at first called *sombreros*,..*slouches*, and *California hats*,..but latterly, by some, *Paris felts*. 1839 F. A. Kemble *Jrnl. Residence Georgian Plantation* (1863) 346 A Paris gown and bonnet might have been in equal danger of shocking his prejudices. 1896 E. Turner *Little Larrikin* xiii. 141 The commanding presence in the Paris gown. 1913 A. Bennett *Regent* ii. viii. 219 Elegant women wearing Paris or almost-Paris gowns. 1965 Roach & Eicher *Dress, Adornment, & Social Order* 396 Writer sees Hollywood fashions beginning to take some of the prestige away from Paris gowns in the 1930's. 1868 *Amer. Agriculturist* XXVII. 321 The following is going the rounds of the press 'Sure death to Potato Bugs: Take 1 lb. Paris green, 2 lbs. pulverized lime. Mix together, and sprinkle the vines.' We consider this unsafe, as..Paris green is a compound of arsenic and copper, and a deadly poison. 1902 *Encycl. Brit.* XXVIII. 530/2 The best fruit farmers spray fruit trees regularly in the early spring..with quassia and soft soap and paraffin emulsions, and a very few with Paris green only. 1966 Lockhart & Wiseman *Introd. Crop Husbandry* vii. 231 Baits such as Paris green. 1845 G. Dodd *Brit. Manuf.* 5th Ser. 172 The plush for the larger number of silk hats is woven in Lancashire; but for 'Paris hats', as they are called, it is woven in France. 1906 E. Dyson *Fact'ry 'Ands* vii. 87 A man in a long frock coat, a high, glittering, Paris hat. 1951 C. Porter *Kiss me Kate* 108 Mr. Harris, plutocrat, Wants to give my cheek a pat, If the Harris pat Means a Paris hat—Bé-bé. 1957 M. B. Picken *Fashion Dict.* 165/2 *Paris hat*, high silk hat worn by men. 1759 *Newport* (Rhode Island) *Mercury* 26 June 4/3 To be sold by Jacob Richardson,..plain Gauze, Parisnets, gimp and floss Garland. 1766 C. Anstey *New Bath Guide* iii. 23 Stomachers and Parisnets. 1882 Caulfeild & Saward *Dict. Needlework* 378/1 *Paris net*, a description of Net employed in Millinery. 1937 N. Coward *Present Indicative* i. 28, I can smell the eau-de-Cologne, see..the stripes on his Paris shirt.

parish, *sb.* Add: **2. e.** *transf.* and *fig.* (also influenced by sense 1).

1940 Auden *Another Time* 52 The ape Is really at home in the parish Of grimacing and licking. 1941 — *New Year Let.* i. 25 However miserable may be Our parish of immediacy. 1951 *Times* 1 Jan. 7/6 Covering the whole of the F.E.A.F. [*sc.* Far East Air Force] area,..two impressions stand out..the vastness of its 'parish' and of its commitments..and the slenderness of the aircraft resources at its disposal. 1958 P. Kemp *No Colours or Crest* viii. 153 My parish includes not only the old frontiers of Albania, but the new regions incorporated into the country by the Axis. 1976 *Shooting Times & Country Mag.* 16–22 Dec. 13/2 Others—in the north—consider they are too far south, while in the south some consider they are too far north! In fact, WAGBI is based almost equally between their interests, bordered by the 'parish' of Northern Ireland; the North of Scotland; Cornwall and Kent. 1977 D. Beaty *Excellency* ii. 25 From British Embassies and High Commissions all over the world came messages reporting reactions of their parishes to the recent events.

7. a. *parish doctor* (examples), *mag*, *magazine* (examples); *parish relief* (further examples). **b. parish communion, Eucharist,** a communion service held as the principal service of the day (usu. Sunday), and at which most of the congregation communicate; **parish mass,** a mass celebrated in a parish church, *spec.* = **parish communion* (quot. 1763 is a fortuitous collocation); **parish pump,** used allusively (often *attrib.* or as *adj.*) to denote political speakers and their speeches, and other matters, that are limited in scope, outlook, or knowledge, or of local importance only; hence **parish pumper,** one who is concerned with parish-pump politics; **parish pumpery,** concern with local matters only, parochialism; **parish-pumpish** *a.*, limited in outlook and interests, parochial; **parish rig** (see quots.); **parish-rigged** *a.* (later examples); **parish school** *Sc.* and *N. Amer.*, = *parochial school* (*parochial *a.* (*sb.*) 1a); so *parish schoolmaster* (the collocation in quot. 1788 may be fortuitous); **parish work,** the work or duty of attending to the poor and sick of a parish; pastoral work in a parish.

1936 *Church Q. Rev.* Oct.–Dec. 103 There will naturally be no later Solemn Eucharist, and only provision earlier for those who cannot possibly attend the Parish Communion. 1937 A. G. Hebert *Parish Communion* i. 3 By 'the Parish Communion' is meant the celebration of the Holy Eucharist, with the communion of the people, in a parish church, as the chief service of the day, or, better, as the assembly of the Christian community for the worship of God. 1953 [see *morcellated *ppl. a.*]. 1968 L. Dewar *Outl. Anglican Moral Theol.* vii. 171 The rise of the 'Parish Communion' in the last thirty or forty years..is one of the most remarkable phenomena in the history of the Church of England. 1972 C. Stephenson *Merrily on High* ii. 36 It was his custom to go to Holy Communion on the first Sunday in the month, but when a new vicar started a parish communion..my father felt it his duty to back him up and became a weekly communicant. 1975 *Church Times* 8 Aug. 6/5 The Sunday Parish Communion at 9.15 a.m. had been pioneered, though not originated, by St. John's Newcastle, in 1927. 1848 *Punch* 12 Feb. 59 (*caption*) Well, young man. So you wish to be engaged as Parish Doctor? *c* 1875 'Brenda' *Froggy's Little Brother* (new ed.) ix. 115 It turned out to be only the parish doctor come to see little Deb Blunt. 1977 R. L. Wolff *Gains & Losses* viii. 425 The parish doctor is a rich man, a scientist. 1936 E. Underhill *Worship* v. 94 The choral Parish Eucharist, the Roman Catholic rite of Benediction,..or any other truly congregational service where the general movement is understood, and hymns, chants, and actions are familiar to all. 1965 C. E. Pocknee *Parson's Handbk.* (ed. 13) p. xii, The whole trend of Sunday morning worship as manifested today in the Parish Eucharist had been foreshadowed by John Wordsworth in *The Ministry of Grace* (1901). *a* 1966 M. Allingham *Cargo of Eagles* (1968) xii. 140 The vicar called to leave 'is compliments and a parish mag. 1888 C. M. Yonge *Our New Mistress* ii. 14 As had been put into the 'Parish Magazine', she had been two years from her training college. 1926 S. T. Warner *Lolly Willowes* i. 67 The parish magazine said: 'The vicar had scarcely left East Bingham when war was declared.' 1955 M. Allingham *Beckoning Lady* ii. 19 He'd ploughed down there through the snow to take her the parish magazine. 1974 *Times* 8 Mar. 16/8 The editor of *The Shoreham Gazette*, a parish magazine, boasts that he is the first to get into print with a prayer for the new government. 1976 *Deeside Advertiser* 9 Dec., In his parish magazine the Vicar of Shotton has warned parishioners to be prepared for disturbances during Midnight Mass on Christmas Eve. 1763 C. Cordell *Divine Office for Use of Laity* I. p. vi, the prayers, publications, and familiar instructions used at the Parish-Mass, on Sundays. 1929 S. Leslie *Anglo-Catholic* xx. 275 The following morning Jasper did not return from saying the Parish Mass in time for his early breakfast. 1958 S. Neill *Anglicanism* xiv. 403 In many parishes the 'Parish Mass' is followed by a parish breakfast. 1965 C. E. Pocknee *Parson's Handbk.* (ed. 13) ii. 22 The parson must think out whether he intends to develop the Parish Mass with a choir as his chief act of Sunday morning worship. 1970 H. Braun *Parish Churches* xvii. 206 While the sermon has always played a part in the parish Mass, no architectural provision for its delivery appears to have been made until late in the Middle Ages. 1915 *Truth* 21 Apr. 620/1 They are the last word in parochialism; but the table is their parish pump and the croupier is the beadle. 1923 *Daily Mail* 12 Mar. 5 Parish pump politicians distort every word they [*sc.* statesmen] utter. 1923 U. L. Silberrad *Lett. J. Armiter* x. 211 The to-dos we make over our own parish pump matters. 1961 A. Wilson *Old Men at Zoo* i. 11 The Treasury job..had called for a good measure of toughness; after it, Regent's Park affairs smelt a little of the parish pump. 1962 *Punch* 14 Feb. 285/1 Parish Pump radio, when it comes, may be the biggest draw yet. 1962 *Radio Times* 22 Nov. 40/2 In 'Talking Point' we introduce a bit of controversy to avoid being too parish-pump. 1963 *Times* 27 Feb. 8/6 Resistance from parish-pump politics is clearly being encountered; hostility to fresh ideas seems inevitable in a country that has some 38,000 communes containing fewer than 500 people. 1973 *Times* 6 June 18/2 Graham Tope, a Liberal, swept to victory..on a platform of community politics, or the politics of the parish pump. 1977 *N.Y. Rev. Bks.* 9 June 32/4 The news they brought did nothing to still those local manifestations around the parish pumps which had been the very essence of the Old Republic in its Golden Age. 1978 G. Greene *Human Factor* i. i. 14, I don't think I've voted since the war. The issues nowadays so often seem—well, a bit parish pump. 1963 *Economist* 29 June 1387 Worthy parish-pumpers would do well on the enclave councils. 1979 G. Pottinger *Secretaries of State for Scotland 1926–76* xv. 163 To combat apathy at local elections it was hoped that the top tier would offer an avenue to aspiring politicians, while parish-pumpers would find their satisfaction in the second one. 1962 *Economist* 5 May 425/1 The rest of the country may simply shrug its shoulders at London's enduring parish-pumpery. 1968 *Listener* 29 Feb. 283/2 It all sounds incredibly parish pumpish, but..the parish pump is extremely important when you live next door to it. 1864 R. A. Arnold *Hist. Cotton Famine* vi. 179 One or two members of the Manchester Committee..evidently considered that all subscriptions should be applied to supplement parish relief. 1949 N. Mitford *Love in Cold Climate* xiii. 134 Uncle Matthew..was quite certain in his own mind that he would end up on parish relief. 1937 Partridge *Dict. Slang* 606/1 *Parish-rig*, a poorly found ship or an ill-clothed man. 1958 J. Bisset *Sail Ho!* 36 These men [*sc.* pierhead jumpers] usually had nothing except the clothes they stood up in—known as a 'parish rig'. 1933 P. A. Eaddy *Hull Down* 135 A couple of the new hands who had been shipped just before we left Portland had come aboard pretty well 'parish rigged', to use an old sailor term for a man going to sea short of clothes. 1962 A. G. Course *Dict. Naut. Terms* 145 When a seaman joins a ship with few clothes and little working gear he is said to be 'parish rigged'... The term came to be used afterwards with reference to the ships themselves, so that if they had a minimum of sails, spares and gear to start a voyage they were said to be 'parish rigged'. 1968 L. Morton *Long Wake* i. 15, I joined *Beeswing* in what in those days was called 'Parish rigged' [*sic*], in other words with little or no kit. 1794 J. Muirhead in J. Sinclair *Statist. Acct. Scotl.* XI. 81 A parish school is now a momentary, or at least a temporary employment, for some necessitous person of ability. 1812 Parish school [in Dict.]. 1875 G. MacDonald *Malcolm* I. vii. 67 A cottage rather larger than the rest, which stood close by the churchyard gate. It was the parish school. 1910 J. Kerr *Scottish Educ.* 196 The name, parish schools, conveys no definite idea of the very varied character of the work done in them. 1964 *Winnipeg Tribune* 27 Feb. 1/5 Parish schools deserve a measure of public support, in keeping always with the overall resources of the community and without detriment to the public school system. 1788 P. M. Freneau *Misc. Wks.* 371 She would have killed the parish schoolmaster with the cluva-stick. 1929 J. B. Philip *Weelum* 11 More than a hundred years ago, the Parish Schoolmaster, who was also a poet, often 'andered here. 1873 Mrs. H. Wood in *Argosy* XVi. 133 Parish work is not to everyone's taste. 1885 C. M. Yonge *Nuttie's Father* i. xiv. 163 She had a practical soul for parish work, and could appreciate..the exertions made for people of the classes she had always supposed too bad or else too well off to come under clerical supervision. 1911 W. Owen *Let.* 18 June (1967) 75 To become the 'assistant' of some hard-worked or studiously inclined parson, helping in parish work, correspondence etc.

parish, *v.* Add: **b.** Of a clergyman: to do parish work. *rare*.

1880 J. Gott *Lett.* (1918) 132 The growth and gymnastics of the mind, the mind with which one prays and parishes.

‖ **parishad** (pāriʃa·d). [Bengali and Hindi, f. Skr. *parishad* an assembly, council, f. *pári* around, about + *sad*, *sídati* to sit down.] In India and Bangladesh: an assembly, group, or council; also *attrib.* Also *zil(l)a parishad*, a district council.

1919 R. Mookerji *Local Govt. Anc. India* i. 29 The communal life of ancient India..sought to express itself through a variety of institutions, civic and municipal, industrial and commercial, political and religious... The following, for example, are the terms we generally come across in our literature, viz. *kula*,..*parişat*, *charana*. 1932 V. R. R. Dikshitar *Mauryan Polity* iii. 97 This passage only corroborates our view that the *parişad* (Council) exercised real executive powers. 1959 R. S. Sharma *Aspects Polit. Ideas & Inst. Anc. India* xiii. 191 The influence of caste is to be also seen in some of the collective institutions such as the *parişad*. 1963 B. A. Saletore *Anc. Indian Polit. Thought & Inst.* v. v. 417 The term *parişad*, which was confined originally in the Vedic days only to a congregation or assembly of learned men, seems to have been used in a wider sense of a council or assembly of ministers by the time of Pāṇini. 1971 *National Herald* (Lucknow) 1 Apr. 3/6 The commission will also go into the wage structure of employees of local bodies, zila parishads, town areas, notified areas, municipalities, [etc.]. 1975 *Bangladesh Times* 23 July 2/3 One Nagoruddin Mondal, member Bilasbari union parishad under Badalgachi Police station has been suspended for selling relief goods in black market and the misuse of test relief fund. 1976 *CRC Jrnl.* July 5/3 There was a very wide representation, from Trade Unions..Vishwa and Parishad groups. 1976 D. Hiro *Inside India Today* 50 What then emerged was a three-tiered system whereby the old district boards..were replaced by zilla parishads (i.e. district councils) with responsibility for co-ordinating development plans to be channelled through panchayat samitis (i.e. council committees). 1977 *Bangladesh Times* 20 Jan. 1 The district-wise break-ups of the Union Parishads where polls took place on Wednesday are Dacca and Sylhet 18 each.

parishionate (pări·ʃənĕit). [f. Parishion(er + -ate[1].] Body of parishioners.

1910 *Tablet* 3 Sept. 363 The archiepiscopal diocese with its parishionate of nearly four million souls.

Parisian, *sb.* and *a.* Add: **A.** *sb.* **b.** The French spoken in or associated with Paris.

1841 M. Edgeworth *Let.* 23 Mar. (1971) 587 Educated at Paris and all proper—'hors les p-s and b-s and c-s' which could not pass surely..for true Parisian. 1846 R. Ford *Gatherings from Spain* xi. 119 Their silly grandees murder the glorious Castilian tongue, by substituting what they fancy is pure Parisian. 1909 W. J. Locke *Septimus* xii. 177 Peculiar vocables which she had learnt at school, and which Hégisippe declared to be the purest Parisian he had ever heard an Englishwoman use. 1932 Kipling *Limits & Renewals* 322 His speech—to suit his hearers—ran From pure Parisian to gross peasant. 1969 [see *Grenadian *a.* and *sb.*]. 1976 'Trevanian' *Main* (1977) iii. 59 Guttmann speaks up in his precise European French, the kind Canadians call 'Parisian', but which is really modelled on the French of Tours.

B. *adj.* Special Combs.: **Parisian cloth** (see quot.); **Parisian French** = *sb.* b. above; **Parisian ivory**, an early type of celluloid; **Parisian matins** (later example); **Parisian pattern** (see quot.); **Parisian stitch** (see quot.).

1960 C. W. Cunnington et al. *Dict. Eng. Costume* 268/1 *Parisian Cloth*, 19th c., an English textile of cotton warp and worsted weft. 1974 E. Ambler *Dr. Frigo* i. 20 He speaks Parisian French. 1976 E. Berckman *Be All & End All* v. 59 Cecil spoke Parisian French and he himself hated the English. 1921 *Daily Colonist* (Victoria, B.C.) 26 Oct. 9/1 (Advt.), A new display of Parisian ivory. 1962 P. O'Brian tr. *Erlanger's St. Bartholomew's Night* iv. 162 At four o'clock in the morning the tocsin in Saint-Germain l'Auxerrois had begun ringing for what history was to call the Parisian matins. 1964 W. L. Goodman *Hist. Woodworking Tools* 34 This tool [*sc.* twybill] is 2ft. 6in. long, has no handle, and is described as the 'Parisian pattern'. 1934 M. Thomas *Dict. Embroidery Stitches* 157 *Parisian stitch*.., a Canvas Stitch, consisting of upright stitches worked alternately over one and three horizontal threads of the canvas.

Parisianize, *v.* Add: Also *refl.* Hence also **Pari·sianized** *ppl. a.*

1897 [see *boulevarded *a.*]. 1913 E. Wharton *Custom of Country* xii. 156 Mrs. Harvey Shallum, a showy Parisianized figure. 1916 W. J. Locke *Wonderful Year* xv. 220 The last thing a solid and virtuous citizen of Central France desires to do in Paris is to Parisianize himself. 1962 *Punch* 28 Nov. 783/1 This fashion should not be a Parisianised version from Bond Street.

parity[1]. Add: **1. b.** A state in which two countries potentially hostile to one another have equal strategic resources, used *spec.* of the capacities in nuclear weapons of the U.S. and the U.S.S.R.

1955 *Bull. Atomic Sci.* Mar. 100/2 To try to achieve parity in conventional weapons would mean such a regimentation of our industry and manpower that we would lose the freedom we seek to preserve. 1965 H. Kahn *On Escalation* 295 The term 'parity' is shorthand for 'nuclear parity' or 'strategic parity'. Parity exists when neither side obtains any important strategic technical advantages..from its central war forces. 1971 *Human World* Nov. 20 In the last five or six years the Russians have achieved nuclear-missile parity with the United States.

c. In phr. *parity of esteem*, the state or condition of being regarded as equal, used *spec.* of the status of administratively comparable educational institutions.

1961 D. Jenkins *Equality & Excellence* vi. 112 The authors of the 1944 [Education] Act..enunciated their principle..that there should be 'parity of esteem' for all forms of secondary education. 1961 *Guardian* 3 Apr. 3/2 The training colleges must aim high... Parity of esteem with the universities must be earned. 1966 D. Jenkins *Educated Society* ii. 63 There has to be 'parity of esteem' as between the various sections of the community. 1966 *Rep. Comm. Inquiry Univ. Oxf.* I. 117 We are convinced that undergraduate and postgraduate education should enjoy parity of esteem. 1974 *Listener* 23 May 661/3 If they [*sc.* polytechnics] could accept the challenge that they can offer a different kind of degree from well-established universities, they might get something like parity of esteem.

4. b. The oddness or evenness of a number, or of the number characteristic of something.

1915 P. A. MacMahon *Combinatory Analysis* I. iii. vi. 139 (*heading*) The parity of the greater index. 1949 G. & R. C. James *Math. Dict.* 258/1 If two integers are both odd or both even they are said to have the same parity; if one is odd and the other even they are said to have different parity. 1960 G. N. Lance *Numerical Methods for High Speed Computers* i. 10 The extended word should always have odd parity unless binary digits have been lost or gained during the transfer of the word from one part of the machine to another. 1966 R. R. Arnold et al. *Introd. Data Processing* xiv. 268/2 Codes that use an odd number of bits are said to have an odd parity. Codes that use an even number of bits are said to have an even parity. 1971 Hunter & Monk *Algebra & Number Syst.* ii. 28 Let us..say that *b* is related to *a*, when the positive integer *b* has the same parity as the positive integer *a*, that is when *a* and *b* are either both odd or both even. 1975 J. Finkel *Computer-Aided Experimentation* xvii. 374 Parity is computed by adding the total number of 'ones' in a word. If the total is an even number and even parity is desired, the parity bit is stored as a 'zero'.

c. *Physics.* The property of having or being a spatial wave function that remains the same (has even parity) or changes sign (has odd parity) when a change of sign is applied to the co-ordinates; also (*charge parity, G-parity*), the same property with respect to certain other symmetry operations (see quots. 1964, 1970); the value of the quantum number (eigenvalue) representing such a property (+1 for even parity, −1 for odd parity).

1939 *Physical Rev.* LVI. 526/2 Only the selection rules for *J* and parity would remain valid ($\Delta J = \pm 1$ or 0, no change in parity allowed). **1953** *Ann. Rev. Nuclear Sci.* II. 240 Means are available..for determining the angular momentum and parity change carried off by the emitted radiation. **1955** R. D. Evans *Atomic Nucleus* iv. 174 The parity of an isolated system is a constant of its motion and cannot be changed by any internal processes. Only if radiation or a particle enters or leaves the system..can the parity change. **1956** *Canad. Jrnl. Physics* XXXIV. 1110 There is..evidence for spin and parity assignment of ½(+) for the C^{15} ground state. **1957** *Times* 11 Nov. 11/5 The launching of an Earth satellite is of less fundamental importance than..the failure of the law of parity conservation. **1964** G. Källén *Elem. Particle Physics* xii. 316 We can describe the *G*-operation as a charge conjugation followed by a rotation [in isospin space] through the angle π around the 1-axis. The eigenvalue of this operator is usually referred to as 'G-parity'. *Ibid.* 317 *G*-parity is a conserved quantum number in all reactions involving strong interactions. **1968** M. S. Livingston *Particle Physics* vii. 134 The parity of a wave function representing a system of particles includes the intrinsic parities of each of the individual particles and also depends on the relative angular momentum of the several particles. **1969** *Observer* 13 Apr. 2/6 Certain particles do not obey..the parity (or P) principle. This P principle says, in effect, that nature does not 'know' the difference between left and right. **1970** Martin & Spearman *Elem. Particle Theory* v. 237 We define an operation *C* as the mapping of a physical system into another physical system in which each particle has been replaced by one with the opposite value of charge, baryon number, (lepton number) and strangeness... The eigenstates of *C* have eigenvalue either +1 or −1, that is they have either even or odd charge parity. **1973** L. J. Tassie *Physics Elem. Particles* iii. 32 When discussing the parity of the π^-, we assumed that the proton and the neutron have the same intrinsic parity. If we assume that the proton and the neutron have opposite intrinsic parities, then we find that the π^- and the π^0 have opposite intrinsic parities... We choose the parities for different charge states to give the simplest scheme.

5. c. An agreed price for agricultural produce, relative to other commodities.

1941 *Time* 2 June (Air Exp. Ed.) 2/3 'Parity' is a political concept which holds that the farmer should receive prices for his products which will give him a purchasing power (in terms of other commodities) equal to that which he had in the period 1909–14. **1977** *Askov* (Minnesota) *American* 31 Mar. 1/2 The milk price support increase to 83 percent of parity will have a positive impact on Minnesota dairy farmers.

d. The value of one currency in terms of another or others, as agreed by the procedures of the International Monetary Fund, which became effective on 1 March 1947.

1945 *Ann. Reg. 1944* I. ii. 41 All transactions between the Fund and members were to be at par, and all transactions in member currencies at rates within an agreed percentage of parity. **1949** *Britannica Bk. of Year* 254/2 Mexico temporarily suspended exchange transactions by the Central bank on July 22, 1948, pending the establishment of a new parity. Although the new parity had not been announced, the Bank of Mexico was authorized to engage in free market transactions. **1971** *Daily Tel.* 10 May 14 The major countries agree..that if..the price of their own currency in relation to any other foreign exchange moves to more than 1 p.c. away from the fixed price (called parity), they will intervene. **1971** *Ibid.* 11 May 1/3 The Swiss franc closed below its new parity of 4·08 to the dollar. **1976** *Shooting Times & Country Mag.* 18–24 Nov. 28/2 Doubtless the parity of the pound sterling against other currencies is leaning..in favour of the visitor.

7. *parity level, price*; **parity bit** *Computers*, a bit that is automatically made 1 or 0 so as to make the parity of the word or set containing it either odd or even, as previously determined; **parity check** *Computers*, a check on the correctness of a set of binary digits that involves ascertaining the parity of a number derived from the set in a predetermined way; so **parity checking** *vbl. sb.*; **parity digit** *Computers* = *parity bit* above.

1957 D. D. McCracken *Digital Computer Programming* xii. 151 A common system is to assign six bits to each character... The six bits and the parity bit are then recorded in a row, which requires a seven-channel tape. **1975** J. Finkel *Computer-Aided Experimentation* xvii. 373 The parity bit is computed by the memory write hardware before storing a word. When a word is fetched from storage the parity bit is recomputed, and if it does not agree with the parity bit appended to the word an interrupt is issued. **1950** R. W. Hamming in *Bell Syst. Techn. Jrnl.* XXIX. 150 The type of check used..will be called a parity check. The above was an even parity check; had we used an odd number of 1's to determine the setting of the check position it would have been an odd parity check. **1958** *IRE Trans. Electronic Computers* VII. 207/1 If the specified number of ONE digits is even the parity check digit is a zero. If the number of ONE digits is odd the parity check digit is a one. **1970** N. R. Scott *Electronic Computer Technol.* v. 220 'Digital' parity checks..depend upon the digits of the number and not upon the significance attached to the digits by virtue of their position... The numerical parity check..is a function of the number represented by the digits but not a function of the digits themselves. **1972** *Computers & Humanities* VI. 149 One channel is reserved for the internal control, the parity check. **1958** *IRE Trans. Electronic Computers* VII. 207/1 Parity checking usually is defined for the binary system in terms of the odd or evenness of the number of ONE digits in a specified block of binary digits. **1966** R. R. Arnold et al. *Introd. Data Processing* xiv. 268/1 Parity checking is a built-in self-checking feature utilized in most magnetic tape as well as paper tape coding methods. **1954** *Computers & Automation* Dec. 18/1 *Parity check*, use of a digit (called the 'parity digit') carried along as a check. **1959** S. H. Hollingdale *High Speed Computing* ii. 27 Some computers are designed to deal with numbers which contain an extra digit, known as the parity digit. **1970** N. R. Scott *Electronic Computer Technol.* v. 217 The correct parity digit is attached to the digit group at the source, and then the augmented group is tested upon reception to determine whether the parity digit and the message digits are still in agreement. **1907** *Daily Chron.* 3 Oct. 2/1 Opening under the parity level prices continued to lose ground every hour up to the close. **1909** *Westm. Gaz.* 20 May 12/4 The parity price of Amalgamated was 85 3–16. **1937** McIntosh & Orr *Pract. Agric. for High Schools* 15 Show the actual and parity prices of the major farm products. **1955** *Times* 24 Aug. 7/2 American farm prices are kept relatively high and often above world prices by the arrangements whereby the Government must buy certain produce for stock when its price falls to certain specified percentages of so-called 'parity prices'.

pariu, var. *PAREU.

park, *sb.* Add: **2. a.** Also, an enclosed piece of ground, of considerable extent, where animals are exhibited to the public (either as the primary function of that 'park' or as a secondary attraction). See also **safari park*, **zoological park*.

1909 Elliot & Thacker tr. *Hagenbeck's Beasts & Men* ii. 40, I wished my new park to be a great and enduring example of the benefits that can be wrought by giving the animals as much freedom and placing them in as natural an environment as possible. **1914** E. Velvin *From Jungle to Zoo* xxiii. 336 There are 139 employees engaged in taking care of the ground and collections of this Park [*sc.* the New York Zoological Park]. **1976** W. Blunt *Ark in Park* i. 15 William the Conqueror established or perhaps took over an already existing animal park at Woodstock, near Oxford. **1977** *Belfast Tel.* 22 Feb., The initiation of compulsory safety programmes at wild life parks and circuses.

b. Add to def.: or for the preservation of wild life. (Further examples.)

1895 Roosevelt & Grinnell *Hunting in Many Lands* 400 The preservation of elk, deer, antelope and the carnivora is assured... Their wide distribution within the Park,..added to the danger attendant on killing them within the Park, is a sufficient protection. **1922** *Baedeker's Dominion of Canada* (ed. 4) vii. 309 About 3M. to the S. of Lamont lies the pretty Elk Island Park.., one of the Canadian National Parks.., 16 sq. M. in area and including elk, deer, moose, and buffalo. **1926** *Encycl. Brit.* II. 1020/1 In Canada since 1910 the following national game preserves, bearing the name of parks, have been established... Nemiskam National Park.., an antelope preserve in southeastern Alberta [etc]. **1957** *Ibid.* XXIII. 603/1 The Serengeti National park.. preserves the finest remaining assembly of the plains game of Africa. **1959** *Chambers's Encycl.* IX. 705/1 In African parks the emphasis is on wild life conservation and public access is strictly controlled. **1974** *Afr. Encycl.* 362/1 *National parks* are areas of land where large numbers of wild animals live in natural surroundings. **1975** *Islander* (Victoria, B.C.) 31 Aug. 4/1, I got to see the first of the wild animals for which Africa is famous. This was in the Game Park at Lake Nakuru. **1978** K. Turner *Serengeti Home* i. 1 This was my first visit to the Serengeti, a vast wildlife park in Tanganyika, now Tanzania.

c. A sports ground or stadium; *spec.* (*a*) in the U.S., a baseball field (cf. *ball park* s.v. *BALL *sb.*[1] 21); (*b*) a football field or stadium; also in the names of football teams.

1867 *Chicago Times* 25 July 5/2 These cars connect with the stock-yards dummy, which runs to within a short distance of the park. **1880** *Times* 8 Nov. 11/3 Notts v. Glasgow Queen's-Park. This match was played at Nottingham on Saturday, and ended in favour of Queen's-park by four goals to three. **1892** J. Higson *Hist. Salford Football Club* 18 Our first ground was the Peel Park cricket ground. *Ibid.* 96 The first game they played with us was against Birkenhead Park. **1902** *'Golden Penny' Football Album 1901–2* 32/3 The record gate at an International match is £4,387 9s. 6d. for Scotland v. England, at Celtic Park, Glasgow, in 1900. **1917** C. Mathewson *Second Base Sloan* 217 Which way is the park from here, please? **1948** S. Matthews *Feet First* xii. 66 The first thought that flashed through my anxious mind was Hampden Park. Would this injury keep me out of the England team to play Scotland? **1974** *Linlithgowshire Jrnl. & Gaz.* 16 Aug. 12/5 Old favourite Paddy Buckley led the Bo'ness team on to the park on Saturday. **1974** *Sunday Tel.* 8 Sept. 33/6 Middlesbrough began as if they were going to sweep Chelsea off the park. **1976** E. Dunphy *Only a Game?* i. 31 He'd been troubled by knee injuries ever since Palace kicked us off the park the year they got promotion. **1976** *Scotsman* 27 Dec. 10/5 The sad sight of fighting on the terracing and terrified youngsters spilling on to the park.

d. *industrial park*: see *INDUSTRIAL *a.* e.

3. c. *U.S.* An enclosure into which animals are driven for slaughter; a corral. *Obs.*

c **1797** in L. F. R. Masson *Les Bourgeois de la Compagnie du Nord-Ouest* (1889) I. 280 The chief of the park thinks that if he were to eat any of this meat thus killed, it would be out of his power to make buffaloes enter his park ever after; so he must have meat killed in the open field for his own use. **1805** M. Lewis in Lewis & Clark *Orig. Jrnls. Lewis & Clark Expedition* (1904) I. 313 There was a park which they had formed of timber and brush, for the purpose of taking the cabrie or antelope. **1820** D. W. Harmon *Jrnl. Voy. & Trav. Interior N. Amer.* 99 The Natives..killed upwards of eighty [buffalo] by driving them into a park, made for that purpose. **1839** Z. Leonard *Adventures* (1904) 224 After travelling a short distance we arrived at a large pen, enclosing about three-fourths of an acre, which they call a park or correll.

4. b. (See quots.)

1950 *Amer. Speech* XXV. 163 The meaning of *park* as 'any grassy piece of level land enclosed by trees, hills, or mountains' has become, or at least is fast becoming, the central meaning of the word [in Colorado]. **1961** *Ibid.* XXXVI. 269 *Park* means either 'mountain meadow' or 'clearing', from the foothills communities westward; especially in the northwest it is likely to mean both.

5. b. An open space, a building, or underground accommodation, in or near a city or town, where cars and other vehicles can be left; = *car-park* (*CAR *sb.*[1] 6). Also *transf.* See also **caravan park*.

1916 A. Bennett *Lion's Share* xxiii. 162 Audrey's motor-car..was waiting in the automobile park outside the principal gates. **1925** *Times* 14 Apr. 8/5 The Automobile Association..has put forward a scheme for the construction of motor parks below ground. **1929**, **1944** [see *air park* s.v. *AIR *sb.*[1] III. 7]. **1970** *Times* 9 Feb. 13/4 Underground there will be a park for up to 2,200 cars. **1972** R. Hill *Fairly Dangerous Thing* I. i. 11 The park was quite full and he would probably never have noticed the two-tone Consul if it hadn't begun to move.

c. In a motor vehicle with automatic transmission: the position of the selector lever in which the gears are locked, preventing the vehicle from moving.

1963 *Which? Car Suppl.* Oct. 116/1 P meant *Park*. This position could only be engaged with the car at rest and it served as a transmission brake which would hold the car on a hill. *Ibid.* 116/2 If the engine was started with the lever at *Park* and it was running quickly, the car would tend to jerk backwards as the lever was passed through *Reverse* on its way to *Drive*. **1965** Priestley & Wisdom *Good Driving* v. 40 In *Park* the gearbox is locked and thus the car is completely immobilized. **1967** *Times* 31 Mar. 3/7 It is obvious that in using a car with automatic transmission it is of the utmost importance that drivers should ensure that the lever is in the 'park' position, while they use the handle. **1972** D. E. Westlake *Cops & Robbers* (1973) i. 15 He shifted into drive, caught up, and shifted back into park. **1977** J. Cleary *Vortex* viii. 208 Stenhouse moved the gear-lever from *Park* to *Drive* and the car started to move forward.

7. park bench, a bench in a park provided for the public; also *attrib.*; **park-ranger, -warden**, an official responsible for the patrolling and maintenance of a national park; **park-way**, add: freq. **parkway** (as one word); (*a*) for *U.S.* read orig. *U.S.*, and add earlier and later examples; (*b*) a name given to a railway station with extensive parking facilities, situated on the outskirts of a city for the use of travellers into the city centre; also *attrib.*

1906 *Daily Chron.* 6 Sept. 3/2 When a Park-bench orator has shouted at you for a quarter of an hour, you cease to be able to attend. **1908** *Busy Man's Mag.* Mar. 63/2 So amiably and with gusto he economized, sleeping in the moonlight upon a park bench and following the marvels of the water front by day. **1946** A. Clarke *Second Kiss* 14 That park bench in the rain, the dredge of leaves, I sat there wrapped in miserable sleeves. **1965** *Listener* 11 Nov. 763/1 Those tendencies of sloth and indolence which..have marked me out as a friend of park-bench philosophers. **1976** 'O. Bleeck' *No Questions Asked* xii. 137 If I weren't here, you could find a priest or a psychiatrist or just somebody on a park bench. **1912** J. B. Bickersteth *Let.* 20 June in *Land of Open Doors* (1914) 150 Next day, after seeing the park ranger about the burial place, the doctor and I went down..to a place where there is a flat stretch of land. **1940** E. Fergusson *Our Southwest* 144 Park rangers now assiduously police almost two million acres of land. **1963** *Weekly News* (Auckland) 15 May 30/2 The maintenance of bush tracks and huts is almost a full-time occupation for the park rangers [at the Fiordland National Park]. **1972** G. Durrell *Catch me a Colobus* ix. 192, I felt that, if we talked to the park rangers, they would be sure to give us information about the whereabouts of the Teporingoes. **1977** *Borneo Bull.* 7 May 4-A/1 Park rangers, who have continued to look for him, found some traces. **1936** D. McCowan *Animals Canad. Rockies* vii. 62 Bill Hartley.. is Park Warden at Glacier in British Columbia. **1964** C. Willock *Enormous Zoo* ix. 159 The park warden is more than an impresario. **1973** *Times* 9 Feb. 9/1 In the party will be..an Ethiopian Government minister, and national park wardens. **1887** *Visit to States* (ser. 1) xxix. 378 This broad parkway has a magnificent drive on either side of a central walk for pedestrians. **1929** *Times* 23 Jan. 20 Parkway system near New York City. **1937** *Times* 13 Apr. (Suppl.) p. x/2 The plan may provide for orbital and radial roads, parkways, viaducts and tunnels, communications to aerodromes, railway stations, and docks. **1938** *Archit. Rev.* LXXXIV. 238/3 The city of Stockholm, which has had the foresight to buy up large tracts of land in its neighbourhood, has been able to plan a system of 'Parkways'. Initially an American development the 'Parkway' represents a form of planning which might with advantage be extensively used in the English countryside. **1939**, etc. Parkway [see *clover-leaf* s.v. *CLOVER *sb.* 4]. **1944** *Ann. Reg. 1943* 58 An easy flow of open space from.. parkway to green wedge, and from green wedge to Green Belt. **1958** *Listener* 23 Oct. 642/1 The development of the

Wythenshawe estate by the City of Manchester..with its beautifully landscaped parkway—the first, I think, in this country. **1972** *Modern Railways* July 271 The Western Region recently commissioned two new stations, one completely new at Bristol Parkway. **1973** *Bulletin* (Sydney) 25 Aug. 11 Canberra's major new freeway-type road which is the subject of a major 'environmental impact investigation' is now styled the Molonglo Parkway. **1976** P. R. White *Planning for Public Transport* viii. 155 The 'parkway' stations opened by British Rail in recent years, notably that at Bristol, offer undoubted evidence of cars being abandoned by their users in favour of a rail journey. **1977** *Evening Gaz.* (Middlesbrough) 11 Jan. 1/4 The route of their return to the Royal Yacht will be made by the Mandale Interchange and the new Parkway.

park, *v.* Add: **2. b.** To place or leave (a vehicle or the like), usu. temporarily, in a park (sense *5 b), at the side of the road, or elsewhere. orig. *U.S.*

1864 J. S. Billings in F. H. Garrison *John Shaw Billings* (1915) 95 The trains are parked along the edge of the river. **1867** A. D. Richardson *Beyond Mississippi* 79 At night the wagons are parked in a circle. **1900** *Congress. Rec.* 2 Feb. 1445/1 No part of said street..shall be used for depot purposes, or railroad yard, or for the purpose of switching, shifting, or parking cars. **1911** *N.Y. Even. Post* 29 Nov. 16 The train was parked near the Union Station and was visited by hundreds of townsfolk and countrymen. **1921** *Daily Colonist* (Victoria, B.C.) 13 Mar. 6/2, I am now in the position of fearing to leave my car anywhere at all in the central part of the city lest I should be parking it where it should not be. **1925** Wodehouse *Carry on, Jeeves!* vi. 156 It was about an hour later that I shoved my way out to where I had parked the car. **1929** J. B. Priestley *Good Companions* I. iv. 139 They were not able to keep the van with them, but had to..park it up a side-street in a line of other cars and carts and caravans. **1938** *Times* 25 Feb. 15/2 The Chinese..state that over 40 Japanese aeroplanes were observed parked on the air-field. **1969** *Highway Code* 22 Make sure you always park your vehicle safely. **1974** *Nature* 15 Nov. 185/3 Down by the river, in the specially built loop for parking punts, the economic crisis seemed a thousand light years away.

c. *transf.* To place or leave (a person or thing) in a suitable or convenient place until required; to put aside for a while. Also *refl.*

1908 *St. George's Rev.* July 282 The children being parked' in their own schoolyards. **1922** *Atlantic Monthly* June 773 High-school girls..'park' their corsets when they go to dances. **1923** Wodehouse *Inimitable Jeeves* iv. 40 At this point the brother, who after shedding a floppy overcoat and parking his hat on a chair had been standing by..gave a little cough. *Ibid.* ix. 94 The policeman, having retrieved a piece of chewing-gum from the underside of a chair, where he had parked it against a rainy day, went off into a corner. **1927** *Evening Standard* 7 Dec. 19/1 Then I suppose I park myself here. **1938** E. Waugh *Scoop* I. v. 81 A voice said in English, 'Anyone mind if I park myself here?' and a stranger stood at the table. **1949** R. Harvey *Curtain Time* 66 So for Hattie the Grand Opera House, Daly's Theatre, Wallack's, Booth's, Tony Pastor's Variety Theatre and Niblo's Garden became familiar enchanted night nurseries where her father would park her, safe and amused, until his meeting was over. **1960** E. W. Hildick *Jim Starling & Colonel* xv. 137 Come on, dad! Park yerself! **1968** [see *BOOSTER 2 c]. **1971** 'A. Burgess' *MF* xiv. 158 His companion had parked his black lenses on his brow. **1972** J. Philips *Vanishing Senator* (1973) III. iii. 139 Peter crawled round to the other side of the bed where his aluminium leg was parked. **1978** G. Greene *Human Factor* III. ii. 117 The girl was parking her gum on the back of the telephone directory while she got down to a long satisfactory conversation.

d. *intr.* To take up a position in or as in a park; to place a vehicle in a park or elsewhere; to occupy a suitable or stationary position; to stay where one is.

1865 O. W. Norton *Army Lett.* (1903) 255 The wagons parked behind the stables to wait orders. **1926** G. Frankau *My Unsentimental Journey* xi. 149 There, Stidger put on his brakes, 'parked', took out the inevitable keys to lock his gear-lever and ignition-switch. **1929** *Strand Mag.* Feb. 183 'I want them' persisted the other and I guess I'm parking right here until I do get 'em'. **1948** *Democrat* 1 Jan. 4/2 Drivers now can park or back into alleys or up to loading platforms with much greater ease. **1959** *Daily Tel.* 24 Mar. 9/3 Besides asking motorists not to park on main roads, he urged them to use alternative routes. **1966** P. Moloney *Plea for Mersey* 51 Not hardened junkies, when deprived of dope, Ere felt such anger..As Scouseville driver seeking space to park.

parka (pā·ɹkă). Also **parkha**; (all †) **parca, parkee, parki**. Pl. **parkas** (northern Canada **parki**). [Aleutian, from Russ. *párka* skin jacket.] An outer garment or long jacket with a hood attached, made of skins and worn by Eskimos; a similar garment, usu. of windproof fabric, worn by mountaineers, skiers, etc. Also *attrib.*

1780 W. Coxe *Acct. Russ. Discoveries* 256 The inhabitants of Alaxa, Umnak, Unalaksha..wear coats (parki) made of bird skins. **1813** G. H. von Langsdorff *Voy. & Trav.* II. ii. 37 They are called *parka*, and are worn sometimes with one side outwards sometimes with the other. **1818** V. M. Golovnin *Narr. Captivity in Japan* I. i. 32 The women wore parkis made of the skin of birds with the feathers outward. **1851** J. Richardson *Arctic Searching Exped.* II. 379 (*heading*) Eskimo vocabulary. English... Parka... Kuskutchewak... atkuk. **1907** R. W. Service *Songs of Sourdough* 56 Talk of your cold! through the parka's fold it stabbed like a driven nail. **1910** —— *Ballads of Cheechako* 25 My eyes were seared, yet thralled I peered through the parka hood nigh blind. **1922** *Chambers's Jrnl.* Feb. 137/1 He had no snowshoes, no parki, and he did not see the dense black blizzard sweeping up from the north-west. **1922** *19th Cent.* Feb. 269 They changed their drill parkees for coats of caribou fur. **1926** *Spectator* 18 Sept. 408/2 The woodsman of the north..wears no fur, unless it be a little trimming round the neck of the 'parca'. **1934** *Sun* (Baltimore) 31 Jan. 3/3 Stocky Indians and Eskimos, fur-trimmed parkas and mukluks, miners with violent red and green plaid shirts,..—these characterized the simpler functions at crossroad taverns. **1948** *Manch. Guardian Weekly* 1 Jan. 9 Your correspondent will now buckle on his parka, latch his skis, and take off. **1955** E. Hillary *High Adventure* 37 We took possession of our own equipment—..double-layered windproof parkas. **1958** *Tararua* XII. 31 *Parka*. This name is the only one used in New Zealand for the hooded garment based on that of the Eskimos. *Anorak* seems to be common in England. **1958** L. Whishaw *As far as You'll take Me* vii. 94, I spent a wonderful couple of hours trying on fabulous parkas (pronounced parkees in the North). **1963** *Times* 25 Feb. (Canada Suppl.) p. xv/1 Eskimo children..Huddled inside their sealskin *parkhas* and warmed by the low flame of a blubber lantern,..may listen spellbound to stories of the powers of Talluliyuk the seal-goddess. **1968** *Globe & Mail* (Toronto) 17 Feb. 1/7 Busloads of parka-clad workers..arrived almost hourly yesterday at Elliot Lake's two hotels. **1973** *Guardian* 10 Apr. 13/2 Fox-trimmed parka with shirred waist. **1973** A. H. Whiteford *N. Amer. Indian Arts* 86 The Kutchin made fine tailored skin skirts, parkas, and one-piece leggings and moccasins. **1976** *Evening Post* (Nottingham) 14 Dec. 1/6 A boy's parka coat worth £4·50 was stolen from the cloakroom of the Chaucer Junior School, Ilkeston. **1978** *Times* 23 Feb. 13/6, I wore a silk Chinese padded jacket under my parkha.

park-and-ride (pā:ɹk,ănd,rai·d), *a.* [f. Park *v.* + And *conj.* + Ride *v.*] Designating or pertaining to a system whereby commuters and other visitors travel by private car to car-parks situated on the outskirts of a city, and continue their journey to the centre of the city by means of public transport. Also as *sb.*

1966 *Leicester Mercury* 30 Dec. 24/3 Park 'n' Ride was the success of the year. **1968** *Economist* 20 Apr. 48 Public transport to those facilities, largely by rail (incorporating park-and-ride stations), must be built up. **1971** *Sunday Times* (Johannesburg) (Business Section) 28 Mar. 7/4 He wants to attract shoppers to the city centre by way of off-street parking and establish 'park-and-ride' schemes outside the city with a speedy bus service to the downtown area. **1973** *Times* 28 Feb. (Victoria Centre, Nottingham Suppl.) p. iii/3 (*heading*) 'Park-and-ride' can beat the traffic problem. **1974** *Drive* Autumn 110/1 Public transport improvements now in operation include a park-and-ride service, catering for the parking of 250 cars..and distribution of the occupants from a site 1½ miles south of the city centre. **1977** *Modern Railways* Dec. 461/3 Perhaps we can hope for improved road/rail interchange or even park-and-ride facilities to woo potential rail commuters.

Park Avenue. Name of a street in New York City, U.S.A., used *attrib.* and as *adj.* to designate the fashionable and luxurious style of life for which it is noted.

1956 R. Macaulay *Towers of Trebizond* vi. 59 Mrs. Van Damm looked very handsome and bland, with blue hair and eyes and Park Avenue clothes. **1971** M. Babson *Cover-up Story* vi. 67 Lou-Ann isn't Park Avenue Hillbilly material... Hillbilly, yes. Park Avenue, no. **1975** J. Hone *Sixth Directorate* iv. 149 She had all the rare finesse of a Park Avenue debutante. **1976** *National Observer* (U.S.) 13 Mar. 7/1 They know we run a Park Avenue operation at Marquette, that we SRO the joint. The kids realize that by playing at Marquette they'll be constantly in the limelight and get good pro contracts.

parked, *ppl. a.* (Further examples.)

1919 C. P. Thompson *Cocktails* 176 The old farm where the V.A.D. drivers were cleaning their parked ambulances. **1932** *Daily Tel.* 23 May 8/6 In view of..the parked cars using the park, it was felt that it was an injustice that the horse-rider should be solely blamed for damage to the turf. **1962** [see *DOUBLE-PARK *v.*]. **1973** *Daily Tel.* 19 June 3/1 A visitor to Gatwick Airport complained yesterday that he had been able to walk unchallenged up to a parked airliner. **1976** C. Bermant *Coming Home* II. i. 113 One can hardly move for tourists and traffic and parked cars.

parker[2] (pā·ɹkəɹ). [f. Park *v.* + -er[1].] **1.** One who parks a vehicle.

1930 M. McClintock *Rep. Parking & Garage Probl. Washington* 78 (*caption*) Street obstruction by a desperate parker. **1959** *Encounter* Aug. 32/2 He's used to pinching parkers on the broad highway. **1959** *Times* 8 Dec. 5/6 Mr. Marples gave an example of the immunity these all-day parkers have enjoyed. **1974** W. Garner *Big enough Wreath* xi. 151 Smith had the luck to find a car leaving the meter. He reversed fast..cutting out another would-be parker.

2. *Austral. colloq.* = **parking light*. Usu. *pl.*

1967 S. H. Courtier *Murder's Burning* xiii. 187 Mr. Proctor switched on his tail lights and parkers. **1971** R. Dentry *Encounter at Kharmel* x. 179 Keep out of the way as soon as you see me switch the parkers on. **1971** *Southerly* XXXI. 71 A prowl car told us to switch our parkers on.

Pa·rker[3]. [Name of the manufacturing company.] The proprietary name of a pen made by the Parker Pen Company.

1906 *Official Gaz.* (U.S. Patent Office) 2 Oct. 1507/2 Parker Pen Company, Janesville, Wis... Used ten years Parker..for Fountain-pens. **1914** *Trade Marks Jrnl.* 5 Aug. 1229 'Parker' Fountain Pens, not of Precious Metal or imitation thereof. The Parker Pen Company..Wisconsin, United States of America. **1923** *Official Gaz.* (U.S. Patent Office) 2 Jan. 18/1 The Parker Pen Co., Janesville, Wis... Parker..Fountain Pens and Mechanical Pencils. Claims use since 1891 on fountain pens; since 1921 on mechanical pencils. **1935** *Trade Marks Jrnl.* 5 June 718/1 Parker... Fountain pens and propelling pencils (none being of precious metal or of imitation precious metal) and desk stands (not of precious metal or imitation precious metal) for pens. The Parker Pen Company..Wisconsin, United States of America. **1959** I. Jefferies *Thirteen Days* (1961) x. 143, I reckon I'll get a Parker 51, a gold one. **1974** 'E. Anthony' *Malaspiga Exit* ii. 44 An assortment of pens from ball-point to gold-nibbed Parkers. **1976** 'Z. Stone' *Modigliani Scandal* I. iv. 38 He filled up the form with the gold Parker in his pocket.

parkerite (pā·ɹkərait). *Min.* [f. the name of R. L. *Parker* (b. 1893) of Zurich + -ite[1].] A sulphide of nickel, bismuth, and lead, $Ni_3(Bi,Pb)_2S_2$, occurring as orthorhombic crystals with a metallic lustre.

1936 D. L. Scholtz in *Publ. Univ. Pretoria* 2nd Ser. I. 186 The new minerals F and G were first recognised towards the end of the year 1932..when the writer benefited greatly by the valued advice and criticism of Professors P. Niggli and R. Parker. The author, therefore, proposes the name 'Niggliite' for the former (F), and 'Parkerite' for the latter (G). **1943** *Amer. Mineralogist* XXVIII. 345 Single grains of parkerite..are bright bronze coloured with brilliant metallic lustre on fresh fractures, becoming darker and dull on tarnished surfaces. **1969** *Canad. Mineralogist* IX. 610 The parkerite in the Great Slave Lake area occurs as small inclusions in niccolite and is associated with native bismuth.

Parkerizing (pā·ɹkəraiziŋ), *vbl. sb.* [f. the name of the *Parker* Rust-Proof Co. of America (incorporated 1915), which introduced the process.] A rust-proofing process in which iron or steel is given a protective coating of phosphate by immersing it for a short time in a hot acidic solution of a metal phosphate (usu. manganese dihydrogen phosphate).

A proprietary name in the U.S.A.

1919 *Chem. & Metall. Engin.* 31 Dec. 787/2 Messrs. Allen and Richards were mainly responsible for changes in the Coslett process, which made its present modification one of the most simple methods of rust prevention. Their patented modification in use today is known as 'Parkerizing'. **1932** *Metal Industry* XL. 369/2 The basis of the Parkerizing process is a solution of a prepared powder, which consists essentially of manganese dihydrogen phosphate. **1949** *Official Gaz.* (U.S. Patent Office) 24 May 984/1 Parker Rust Proof Company, Detroit, Mich... Certification mark. 'Protected by Parkerizing for rust resistance.'.. Claims use since Mar. 15, 1936. **1954** E. Molloy *Electro-Plating & Corrosion Prevention* xiii. 213 'Parkerising' is applied to cameras and instruments, rifles, typewriters, chains, bolts, nuts, tacks, and so on. **1976** J. A. von Fraunhofer *Basic Metal Finishing* xiii. 147 The original Coslett process, based on phosphoric acid containing zinc phosphate, has been superseded by the accelerated Parkerising, Bonderising and similar processes.

So **Pa·rkerize** *v. trans.*, to treat in this way; **Pa·rkerized** *ppl. a.*, a proprietary name applied to articles so treated.

1922 *Raw Material* V. 438/2 Heavy castings of intricate shape, containing irregular cavities, can be completely Parkerized. *Ibid.* 439/1 On structural iron work, heavy Parkerized coatings can be produced. **1932** *Metal Industry* XL. 369/2 The Parkerized surface is..excellent for absorbing and retaining paint, lacquer or enamel. **1939** Burns & Schuh *Protective Coatings for Metals* xvi. 375 It is possible to Parkerize zinc coatings. **1942** *Trade Marks Jrnl.* 20 May 205/2 Parkerized... Hand tools and side arms. The Pyrene Company Limited,..Brentford, Middlesex; manufacturers. *Ibid.* 2 Dec. 505/2 Parkerized... Small domestic utensils and containers, non-electric instruments for cleaning purposes..all being goods of common metal; steel wool. The Pyrene Company Limited,..Brentford, Middlesex; manufacturers and merchants. **1961** *Chem. Abstr.* LV. 8883 (*heading*) The resistance of parkerised iron in polar marine climatic conditions.

† **Pa:rker's ceme·nt.** *Obs.* except *Hist.* [Named after James *Parker*, who patented it in England in 1796.] = *Roman cement* s.v. Roman *a.*[1] 16 c.

1814 *Trans. Geol. Soc.* II. 193 This place is particularly known on account of its furnishing abundance of the septaria, from which that excellent material for building under water and for stucco is made, known by the name of Parker's cement. **1839** *Penny Cycl.* XV. 420/1 The mortars made with them are called hydraulic mortars. Of these, Parker's cement is a well known kind. **1889** W. A. Tilden *Watts' Man. Chem.* I. 337 Parker's or Roman cement is made in this manner from the nodular masses of calcareo-argillaceous ironstone found in the London clay. **1917** E. A. Dancaster in G. Martin *Industr. & Manuf. Chem.: Inorg.* II. 85 The first of the natural cements was prepared by James Parker towards the end of the eighteenth century. It was known at first as Parker's cement, but was afterwards called Roman cement, by which name it is still known.

Parkes (pāɹks). *Metallurgy.* [Name of Alexander *Parkes* (1813–90), English chemist and inventor, who first patented the process in 1850 (*Brit. Pat. 13,118*).] *Parkes* (or † *Parkes'*, ¶ *Parke's*) *process*: a process for removing silver and gold from lead by adding

zinc to the molten lead, so that the precious metals form an alloy with the zinc and collect on the surface.

1857 Phillips & Darlington *Rec. Mining & Metalurgy* 180 (*heading*) Parkes' process for desilverizing lead. **1892** W. Crookes tr. *R. von Wagner's Man. Chem. Technol.* II. 180 A further advantage of the Parkes process is that a minimum proportion of gold, present in the work-lead, can be first extracted by a small addition of zinc..whilst the subsequent main quantity of silver extracted..is free from gold. **1912** J. W. Mellor *Mod. Inorg. Chem.* xxi. 382 Lead can be desilverized by means of Pattinson's or Parkes' process. **1923** U. R. Evans *Metals & Metallic Compounds* IV. 268 Parke's process.. only serves to remove silver and gold, and would leave the objectionable impurities in the lead. **1940** *Chambers's Techn. Dict.* 617/2 Parke's process. **1969** R. F. Lang tr. *Henglein's Chem. Technol.* 531 The refining of crude lead takes place in two steps: 1. refining proper... 2. desilvering by the aid of Zn (Parkes process).

Parkesine (pā·ɹksīn). Now *Hist.* [f. prec. + -ine[5].] A substance more or less identical with celluloid, based on pyroxylin and castor oil or camphor.

1862 *Chem. News* 9 Aug. 75/2 Parkesine.—A number of pretty and useful articles, formed of a material which the inventor, Mr. Parkes, has named after himself, are exhibited in Case 1112, Class IV. The basis of this material is the mixture of collodion and castor oil. **1868** *Chambers's Encycl.* X. 679/1 Parkesine was first shewn in quantity at the International Exhibition of 1862. **1911** E. C. Worden *Nitrocellulose Industry* I. xi. 364 Lewthwaite utilized the patented product of A. Parkes (called Parkesine) by ornamenting, embossing, and printing, and treated it in a manner similar to real leather. **1958** *Times* 25 Mar. (Careers in Industry Suppl.) p. xii/3 'Parkesine' was the forerunner of 'xylonite' and is identical with 'celluloid', the American article which in the 1870's was the first commercially satisfactory plastics product. **1964** V. E. Yarsley et al. *Cellulosic Plastics* xv. 181 The celluloid of the Parkesine process was produced by squeezing out the dough between rolls and subsequently removing the residual volatile solvent.

parkie, parky (pā·ɹki). *colloq.* (chiefly *Sc.* and *north.*). [f. Park *sb.* 2 + -y[6], -ie.] A park-keeper.

[**1939** Joyce *Finnegans Wake* 587 Touching our Phoenix Rangers' nuisance at the meeting of the waitresses,.. and those pest of parkies, twitch, thistle and charlock.] **1953** *Scottish Jrnl. Theol.* VI. 424 [Schopenhauer] one day sat hunched up on a seat in the Tiergarten, sunk in profound reflection. A *Parkaufseher*, or as we should say 'parky', saw him and not unnaturally put him down as at least a suspicious character. **1957** J. Kirkup *Only Child* 124 We lived in mortal terror of being caught by the parky. **1965** *Listener* 4 Nov. 720/1 The Parkie's usually an auld man and he cannie climb the poles. **1971** *Sunday Times* 9 May 34 It's just coats on the ground and the parkies chase us at least once a month for doing something wrong. **1975** *Scottish Field* Apr. 4/3 You had to be on the lookout for the Parky if, as was likely, you had nipped on to the Alexandra Park golf course] without paying.

parkin. (Earlier and later examples.)

1800 D. Wordsworth *Jrnl.* 6 Nov. (1941) I. 71, I was baking bread, dinner, and parkins. **1968** E. R. Buckler *Ox Bells & Fireflies* xix. 268 Two women had brought oatmeal parkins on cake plates that were exactly alike. **1973** *New Society* 20 Dec. 709/3 The re-birth of interest in regional specialities like parkin (a rich, dark gingerbread, eaten with cheese).

parking, *vbl. sb.* **2.** Add to def.: Also, in some regions of the U.S., a strip of grass between the footpath and the curb. Also **parking strip.**

1945 *Amer. Speech* XX. 154/1 *Parking*, the grassed area between curbing and sidewalk. The term is used by some Minnesotans instead of *Boulevard*, by others as an alternate. **1963** R. I. McDavid *Mencken's Amer. Lang.* 667 In the Minneapolis area it [*sc. boulevard*] designates the grass strip between the sidewalk and the curb, elsewhere called a *tree lawn* or a *parking strip*. **1964** *Amer. Speech* XXXIX. 293 That strip of grass and weeds between the sidewalk and the curb..seems to be most commonly called *tree lawn* or *parking strip*... I use the colorless *parking* from the language of relatives in Iowa. **1966** *Inland* (Inland Steel Co., Chicago) Autumn 16/2 For the grass strip between sidewalk and street there is a bewildering array of local terms:..*parking* in Illinois. **1969** *Better Homes & Gardens* (U.S.) Apr. 85 They look best when used in a formal manner, such as in pairs on front parkings or equally spaced along property lines.

3. The placing or leaving of a vehicle or vehicles in a park (*PARK *sb.* 5b), at the side of the road, or elsewhere. Also *transf.*

1926 *Rep. Commissioner Police Metropolis 1925* 17 in *Parl. Papers* (Cmd. 2660) XV. 239 Parking of Cars.—The arrangements tentatively made by Police for parking cars on certain highways have been given statutory effect. **1929** *Minnesota Alumni Weekly* June 619 The new space along with a lot now used for parking will be seeded. **1931** *Times Lit. Suppl.* 15 Oct. 789/2 Let off with a caution at Marlborough-street for improper parking of his car. **1959** *Daily Tel.* 24 Mar. 9/2 A possible ban on parking on main roads was hinted at yesterday. **1961** *Product Engineering* 14 Aug. 34/1 Additional symbols of status are granted, such as reserved parking, distinctive badge passes.., and a difference in the treatment of financial progress through merit. **1970** *R.A.C. Guide & Handbk. 1970–71* 63/2 Where unexpired time is shown on the meter at the time of parking this period may be used without payment. **1977** 'M. Underwood' *Fatal Trip* i. 7 A re-arrangement of her domestic routine including the parking of Simon on a good-natured neighbour.

b. *attrib.* and *Comb.*, as *parking apron, area, attendant, fee, fine, garage, offence, place, space*; **parking bay**, a recess at the side of a road or other space allocated for parking a vehicle; **parking brake**, a brake provided on a motor vehicle or trailer for holding it at rest; **parking deck** [*DECK *sb.*[1] 3 b], a floor of a building used as a parking place for vehicles; also, a multi-storey car park; **parking disc**: see *DISC *sb.* 2 f; **parking lamp, light**, a small (often detachable) light on a motor vehicle for indicating its position when parked at night; = *SIDE-LIGHT; **parking lot** orig. *U.S.*, a plot of ground used for the parking of vehicles; **parking meter** orig. *U.S.*, a coin-operated meter which registers the time a vehicle has been parked; **parking orbit**, an orbit around the earth or some other planet from which a space vehicle can be launched farther into space; **parking strip** (see quots.); **parking tag, ticket** orig. *U.S.*, a ticket attached by an official to a vehicle which has violated parking regulations; **parking warden** = **traffic warden*.

1974 Hawkey & Bingham *Wild Card* i. 20 The mile-long journey to the parking apron. **1961** R. A. Futterman *Future of our Cities* iii. 62 Less ambitious freeway plans may be more successful—especially when the roadways and interchanges are raised, allowing for cross access at many points and providing parking areas below the ramp. **1966** 'A. Hall' *9th Directive* vii. 63 A car had come into the parking area. **1977** E. Leonard *Unknown Man No. 89* xxi. 212 He..crossed the parking area to the front entrance. **1941** B. Schulberg *What makes Sammy Run?* iii. 48 He was on speaking terms with everybody, the parking attendant, the hat-check girl. **1962** J. Brierley *Parking of Motor Vehicles* 298 *Parking attendant*, a person authorized by or on behalf of a council or parking authority to supervise a parking place. **1973** A. Mann *Tiara* ix. 87 He handed the Maserati over to a parking attendant of the Automobile Club. **1962** J. Brierley *Parking of Motor Vehicles* v. 49 A parking bay 18 feet long will be long enough for the majority of modern cars. **1972** *Times* 26 Jan. 6/1 The notice was given to comply with the provisions for rent revision contained in an underlease for a flat on the seventh and eighth floors of Ambassador House and a parking bay. **1976** T. Heald *Let Sleeping Dogs Die* viii. 167 The van ahead berthed in a parking bay. **1944** L. D. Kitchin *Road Transport Law* 41/4 It is an offence to leave a vehicle without stopping the engine..and applying the parking brake. **1959** E. K. Wenlock *Kitchin's Road Transport Law* (ed. 12) 20/1 On all trailers exceeding 2 cwt unladen the braking system must also be capable of acting as a parking brake to prevent at least two wheels from revolving. **1967** *Gloss. Caravan Terms* (*B.S.I.*) 2 *Parking brakes*, brakes for holding the caravan when at rest, usually the overrun brakes or power brakes provided with additional means of manual application. **1974** *New Yorker* 25 Feb. 44/3 He went into a double-parking maneuver that culminated as, with a flourish, he pulled at the parking-brake handle. **1970** Parking deck [see *DECK *sb.*[1] 3b]. **1972** *Graphic* (Tuscaloosa, Alabama) 30 Nov. 12/3 The County Commission Tuesday formally adopted a multi-level parking deck to be constructed adjacent to the County Jail. **1972** *Birmingham* (Alabama) *News* 17 Dec. 4A/1 With two parking deck sites assured..the property surrounding the station could be purchased and the proposed seven-story structure built in an L-shape. **1974** *Tuscaloosa* (Alabama) *News* 17 Feb. 3D/2 The authority was to build a parking deck for the downtown area. **1932** *Autocar* 9 Sept. 469, I have only once, in two years, been asked to pay a parking fee. **1963** P. Hall *London 2000* v. 111 One exception, which parking fees do not meet, is the extra congestion in the very short morning and evening rush periods. **1971** 'S. Smith' *Grave Affair* ix. 134 Horsham Police were surprised to receive payment of the £2 parking fine on the Cortina... It came in a plain manilla envelope,..and inside were two pound notes and the parking ticket. **1972** P. D. James *Unsuitable Job* iv. 117 She daren't risk a parking fine nor the impounding of the car. **1977** D. Williams *Treasure by Degrees* xviii. 166 Miss Stopps was not the type of citizen who would ignore a parking fine demand. **1948** *Sun* (Baltimore) 18 Feb. 10/2 There will undoubtedly be debates as to whether the commission should build and operate parking garages as public enterprises. **1974** M. G. Eberhart *Danger Money* (1975) xi. 118 Greg drove..to a parking garage where he left the car and hailed a taxi. **1976** *Billings* (Montana) *Gaz.* 1 July 3-B/2 Bids to construct Billings first municipally owned parking garage have all been rejected as being at least $300,000 too high. **1926–7** *Army & Navy Stores Catal.* 1134/1 Motor lamps..parking lamp... Burns very small amount of current. **1957** E. K. Wenlock *Kitchin's Road Transport Law* (ed. 11) 60/1 Parking lamps which can be used only on certain vehicles..must show a 1in diameter white light to the front and a 1in diameter red light to the rear. **1972** *Daily Tel.* 27 Apr. 19/6 Motorists will be able to discard their clip-on auxiliary parking lamps after Sunday next, when standardised regulations for parking without lights come into effect throughout the country. **1938** H. A. Tripp *Road Traffic & its Control* III. ix. 160 'Parking lights' of extremely low power (in order to economise current) are fitted to some cars. **1943** R. Chandler *Lady in Lake* (1944) xxiv. 136 A motor purled gently in the car with the parking lights on it. **1973** 'B. Mather' *Snowline* xiv. 171 It was a car running on its parking lights. **1924** H. Croy *R.F.D. No. 3* 172 Some of the people still lingered under the arc light, with its summer collection of bugs still in it, waiting for the two to come from the parking lot. **1930** M. McClintock *Rep. Parking & Garage Probl. Washington* 19 Parking lot rates average substantially lower than those for garages. **1958** *New Statesman* 1 Nov. 590/3 Partial over-building of these parking lots would do little to disguise the inherent visual bleakness of conception. **1972** *Times* 18 Mar. 8/7 People who in Leeds, for instance, will have stepped across the parking lot from rehearsals in the Playhouse to others in the studio. **1976** H. Nielsen *Brink of Murder* ix. 81 Simon found the restaurant..and drove into the parking lot. [**1935** *Pop. Sci. Monthly* Aug. 33/3 (*heading*) Curb-parking meter times autoist's stay.] **1936** *Amer. City* Jan. 95/3 In July..there came to the attention of the officials in Dallas a device known as the parking meter. **1938** *Encycl. Brit. Bk. of Year* 649/2 Many cities have installed parking meters, usually requiring the deposit of five cents per hour for permission to park at the kerb. **1949** *Chicago Daily News* 11 Aug. 3/4 A runaway horse and wagon collided with a parking meter. **1956** *Planning* XXII. 210 There has been much discussion of the merits of coin-in-slot parking meters as a means of enforcing time limits. **1958** *Observer* 13 July 7/8 Six hundred parking meters, the first in Britain, came into operation in Mayfair on Thursday morning. **1966** *New Scientist* 24 Nov. 447/1 The psychological effect of the parking meter is always diluted by the element of gambling involved. **1974** *Guardian* 24 Jan. 28/4 Increases in parking meter charges. **1959** M. Summerton *Small Wilderness* i. 21 Once..he'd let me off a parking offence. **1960** G. Mikes *How to be Inimitable* 60 The most heinous offence known to the Police is officially called 'obstructing the Queen's Highway'. The Queen is brought into it to underline the close connection between a parking offence and high treason. **1960** *Aeroplane* XCIX. 496/1 Such a satellite could be launched for immediate interception, or placed in a 'parking' orbit, always ready to intercept, interrogate or inspect in detail another object orbiting in space. **1961** *Flight* LXXIX. 426/1 Injection into 'parking orbit', which places the vehicle at the proper location in space for departure into the desired lunar trajectory. **1970** *New Scientist* 2 July 21 The spacecraft goes on into a parking orbit and the two boosters coast down to land. **1925** *Act* 15 & 16 *Geo. V* c. 71 §68 (9) In this section the expression 'parking place' means a place where vehicles, or vehicles of any particular class or description, may wait. **1941** Parking-place [see *BINDI-EYE]. **1975** N. Luard *Robespierre Serial* ii. 6 It's the residents coming back from work; they've got permits for the parking-places. **1924** *Collier's* 5 Jan. 17/2 Secretary Mellon has asked permission to move the Washington Monument so as to get more parking space. **1926** Parking space [see *car-park* s.v. *CAR *sb.*[1] 6]. **1941** G. Marx *Let.* 25 July (1967) 29 Parking space is at a premium around this ramshackle building. **1971** A. Price *Alamut Ambush* iii. 35 His mind on a parking space thirty yards ahead. **1961** *New Left Rev.* July–Aug. 57/2 In front of the garages and kitchens are cobbled parking strips which absorb oil stains and from which moving traffic is kept by low bollards. **1966** *Inland* (Inland Steel Co., Chicago) Autumn 16/2 For the grass strip between sidewalk and street there is a bewildering array of local terms:..*parking strip*..in Illinois. **1956** 'E. McBain' *Cop Hater* (1958) iii. 28 Anybody who ever got a parking tag is automatically a cop hater. **1968** *Globe & Mail* (Toronto) 17 Feb. 5/8 More than $6-million of this amount came from the magistrates' courts. About $1,850,000 came from parking tags. **1977** *Time* 22 Aug. 34 (*caption*) The telltale parking tag and Deputy Craig Glassman with hate letters from his neighbor. **1947** *Denver Post* 2 Mar. A1/2 Parking tickets no longer could be fixed. **1951** T. Sterling *House without Door* vii. 80 These old bastards haven't even read a parking ticket for the last twenty years. **1959** W. R. Bird *These are Maritimes* x. 300 He took an old parking ticket out of his pocket. **1971** [see *parking fine* above]. **1973** H. Gilbert *Hotels with Empty Rooms* xvi. 137 The car stood where they had left it. A parking ticket had been inserted under the right-hand windscreen-wiper. **1966** P. Moloney *Plea for Mersey* 51 This sentinel your chariot will keep Till Parking Wardens, roused up from their sleep, Espy the yellow in the monster's face, And mark the car, with symbol of disgrace. **1974** L. Deighton *Spy Story* xiii. 130, I watched two parking wardens clobber a delivery van.

Parkinson[1] (pā·ɹkinsən). The name of James *Parkinson* (1755–1824), English surgeon and palæontologist, used in **Parkinson's disease** [tr. F. *maladie de Parkinson* (J. M. Charcot 1876, in *Progrès médical* 2 Dec. 838/2)], a chronic, slowly progressive disorder of the central nervous system that occurs chiefly in later life as a result of degenerative changes in the brain and produces tremor, rigidity of the limbs, and slowness and imprecision of movement (described by Parkinson, under the names *shaking palsy* and *paralysis agitans*, in *An Essay on the Shaking Palsy* (1817)); **Parkinson syndrome** = *PARKINSONISM.

1877 G. Sigerson tr. D.-M. Bourneville in *Charcot's Lect. Dis. Nervous Syst.* (ser. 1) v. 144 This man, aged 50 years, was attacked by 'Parkinson's disease' in consequence of a strong emotion occasioned by the attempts of the Federalists, during the time of the Commune, to incorporate him in their battalions. **1888** W. R. Gowers *Man. Dis. Nervous Syst.* II. 589 From the fact that it was first fully described by Parkinson in 1817, it has been called 'Parkinson's disease', but the name which he gave to it of 'shaking palsy' is both apt and adequate. **1909** *Practitioner* Feb. 290 In Parkinson's disease (paralysis agitans), the drug produced considerable decrease in all the cases. **1933** [see *HOFFMANN 3]. **1950** A. Huxley *Let.* 19 July (1969) 627 Poor Osbert [Sitwell] has got Parkinson's disease and has started to tremble. **1955** *Sci. News Let.* 20 Aug. 120/2 A drug of the antihistamine class has helped almost half of a group of patients suffering with the Parkinson syndrome, best known to the layman as shaking palsy. **1970** [see *DOPA]. **1973** *Sci. Amer.* July 98/3 Symptoms of Parkinson's disease..include active features such as tremor and muscular rigidity and negative features such as slowness in the initiation of movement and loss of the usual facial expression of emotions.

Parkinson[2] (pā·ɹkinsən). The name of Cyril Northcote *Parkinson* (b. 1909), historian and

journalist, used in the possessive to denote the 'law' propounded by him, that work expands to fill the time available for its completion. Also *transf.*

1955 *Economist* 19 Nov. 635/1 Before the discovery of a new scientific law—herewith presented to the public for the first time, and to be called Parkinson's Law—there has.. been insufficient recognition of the implications of this fact in the field of public administration. **1957** C. N. Parkinson (*title*) Parkinson's law. **1957** *N.Y. Times* 5 May III. 1/7 Parkinson's Law is that British Government employes multiply by about 5 per cent a year even though their total work output does not increase in proportion. **1958** C. N. Parkinson *Parkinson's Law* (U.K. ed.) 4 Parkinson's Law or the Rising Pyramid. Work expands so as to fill the time available for its completion. **1958** N. Mackenzie *Conviction* 15 Our civil servants are bureaucratic slaves to Parkinson's Law. **1958** *Times* 3 July 11/6 An extension of Parkinson's Law to the parliamentary system establishes that the British instinct 'is to form two opposing teams, with referee and linesmen, and let them debate until they exhaust themselves'. **1960** *Guardian* 22 July 10/4 There is some 'Parkinson's law' that cars increase in numbers to fill..any space made available. **1962** *Lancet* 28 Apr. 898/2 The transportation of anything to high camps is subject to Parkinson's law: the greater the weight, the greater the number of porters; the greater the number of porters, the greater the weight of supplies; the greater the weight of supplies, the greater the number of porters. **1964** M. Argyle *Psychol. & Social Probl.* xiv. 172 Studies of factories, hospitals and school districts of different sizes show that larger ones have if anything a lower proportion of administrators, so that Parkinson's law is by no means generally true. **1966** R. H. Robins *Gen. Linguistics within Liberal Educ.* 15 A sort of Parkinson's Law applies in vocabulary: the meaning range of a word expands to fill the available space. **1972** *Daily Tel.* 3 Feb. 14 The engineers will tell you that, once it is built, the traffic will be there; and this, alas, is true, for Parkinson's Law ensures that traffic expands to fill the space available. **1974** L. Deighton *Spy Story* xiii. 127 Before we had the security cards, there had been no delays. I was a victim of some Parkinson's law of proliferating security. **1978** *Cornish Guardian* 27 Apr. 3/1 Having spent the money, they then see the need to take on scores of new planners and technical officers. It was Parkinson's Law gone mad.

Hence **Parkinso·nian** *a.*[2], of or pertaining to Parkinson or his law; also as *sb.*[2], a believer in this law; **Pa·rkinsonism**[2], the principle or doctrine reflected in Parkinson's law; an instance of this.

1957 *Life* 15 Apr. 57/1 Maybe where the squeeze ought to be applied hardest is on 'Parkinsonism', that trend for every bureau to proliferate. **1959** *Times* 3 Mar. 3/7 A port scene by Seurat, compounded of a thousand 'pointillist' dots must seem to Parkinsonians like a troupe of typists at £12 a week and 10 clerks at £20 and a dozen 'high executives' at £5,000 a year all busily employed in the composition and dispatch of a 'Yours of the 9th to hand'. *Ibid.* 5 May 11/5 They seem, perhaps, to be in some danger of falling into the (Parkinsonian?) trap of treating theoretical maxima as practical minima. **1960** *N.Y. Times* 31 Jan. 56/3 Professor Parkinson's book, 'Parkinson's Law', expounding this and other Parkinsonisms, was published in 1957. **1962** *Daily Tel.* 14 Mar. 12/2 The LCC has built up some efficient teams, with high status and pay at the top, for supervising its many services. Though some of these may smack of Parkinsonian internal empire building, many are good. **1964** R. Braddon *Year Angry Rabbit* iv. 33 Committees of otherwise useless Civil Servants (whose numbers had swollen to quite terrifying and parkinsonian proportions since World War II). **1964** *New Society* 29 Oct. 6/2 In spite of the charges of Parkinsonism, Mr. Wilson's administration is fairly orthodox in its size and structure. **1968** A. Diment *Bang Bang Birds* ii. 15 At the outset it was just a little sub-department but by inevitable Parkinsonian growth it had assumed an identity of its own. **1971** C. Russell *Crisis of Parliaments* III. i. 110 The Parkinsonian process by which people administering particular subjects under the Chamber acquired subordinates and a formal organization was accelerated by the increase in government business. **1975** M. Sinclair *Long Time Sleeping* xiv. 166 The best of all Parkinsonian models—when he had little to do he did nothing; when overstretched he always managed to fit extra into the day.

Parkinsonian (pāɹkinsōu·niăn), *sb.*[1] and *a.*[1] *Med.* Also **parkinsonian.** [f. *Parkinson[1] + -ian.] **A.** *sb.* A person affected with Parkinsonism.

1899 Church & Peterson *Nervous & Mental Dis.* vii. iii. 516 The gait of Parkinsonians is strikingly peculiar. **1949** *Lancet* 26 Feb. 364/2 The parkinsonian has difficulty in starting to walk, and in stopping when once started. **1973** M. Riklan *L-Dopa & Parkinsonism* iii. 185 In the usual therapeutic dosages, L-Dopa has a wide variety of behavioral effects, both positive and negative, on the behavior of the parkinsonian.

B. *adj.* Characteristic of or affected with Parkinsonism.

1906 P. Stewart *Diagnosis Nervous Dis.* xvii. 243 The posture and gait of *paralysis agitans* are diagnostic... In a well-marked case the patient stands with the trunk stooping forwards, the face appearing 'starched' and expressionless—the so-called 'Parkinsonian mask', in which there is little or no emotional play of features. **1930** *Arch. Ophthalm.* IV. 364 Neurologic examination revealed a characteristic postencephalitic parkinsonian facies, gait and attitude. **1933** W. R. Brain *Dis. Nervous Syst.* viii. 447 The Parkinsonian gait..is usually slow, shuffling, and composed of small steps. **1970** *Nature* 4 Apr. 23/2 It is likely..that at least 50 per cent of the total Parkinsonian population may obtain benefit from L-dopa.

Parkinsonism (pā·ɹkinsəniz'm). *Med.* Also **parkinsonism.** [f. *Parkinson[1] + -ism.] **a.** The group of symptoms seen esp. in Parkinson's disease, but occurring also in other cerebral disorders. **b.** = **Parkinson's disease.*

1923 *Brain* XLVI. 268 Post-encephalitic Parkinsonism is usually due to degeneration of the substantia nigra. **1949** *Lancet* 12 Feb. 258/1 A patient with advanced parkinsonism..was bedridden owing to extreme rigidity. **1954** H. H. Merritt in L. J. Doshay *Parkinsonism & its Treatment* i. 3 Two striking characteristics of the symptoms of parkinsonism are their relationship to emotional tension and their disappearance in sleep. **1966** Wright & Symmers *Systemic Path.* II. xxxiv. 1292/2 Tremor and rigidity may develop in other diseases, and it is usual to refer to this syndrome as Parkinsonism. Postencephalitic Parkinsonism..is the most familiar example of this disability; it also occurs in chronic manganese poisoning and in patients who have survived severe carbon monoxide or nitrous oxide intoxication. **1970** *Times* 29 Apr. 2/3 There are about 60,000 sufferers from Parkinsonism in England and Wales. **1974** E. Ambler *Dr. Frigo* II. 110 If you can catch Parkinsonism in the early stages treatment nowadays can do a lot to help the patient.

parkland (pā·ɹklænd). [Park *sb.* 7.] **1.** An area of grassland scattered with occasional clumps of trees. Also *attrib.*

1907 H. A. Kennedy *New Canada* 182 Very soon the parklands of the north were all behind us. **1909** E. Warming *Oecol. Plants* lxxxix. 325 We may cite the 'park-lands' of eastern Asia, where grassland has become occupied by trees and shrubs. **1920** H. G. Wells *Outl. Hist.* 84/2 They were forest and parkland peoples without horses. *Ibid.* 267/2 A slow change in climate..was replacing the swamps and forests and parklands of South Russia ..by steppes. **1926** Tansley & Chipp *Study of Vegetation* x. 209 Parkland..aptly describes the great stretches of country which in Africa almost completely surround the Closed Forest... The characteristic feature of the Parkland is grass, interspersed with trees. **1948** A. L. Rand *Mammals E. Rockies* 110 The parklands of central Alberta. **1953** *Canad. Geogr. Jrnl.* XLVI. 240/1 This fact would seem to urge caution in the extension of irrigation northward into the parkland belt. **1960** N. Polunin *Introd. Plant Geogr.* xiv. 441 Savanna-woodlands or 'parklands' are often found where the rainless period is more prolonged and the annual rainfall less heavy than in true closed forest. **1968** R. M. Patterson *Finlay's River* 3 It [*sc.* a relief map] has been carefully painted with the varying greens and yellows that indicate the different kinds of forest, parkland and tundra.

2. Land given over to the cultivation of a park or parks (Park *sb.* 1 b).

1937 *Proc. Prehist. Soc.* III. 73 The edges of the barrow are turfed and merge into undisturbed parkland. **1952** *Antiquity* XXVI. 161 The houses of Stowe or Wimpole or Castle Howard were being absorbed into their parklands. **1954** M. Beresford *Lost Villages* i. 28 The parkland.. was both attractive to the eye and useful to the pocket. **1959** *Geogr. Rev.* XLIX. 28 A thousand acres of parkland was deliberately left unaltered. **1977** *Times* 26 Nov. 28/5 (Advt.), Enjoy a weekend in the country at a beautiful Georgian Rectory in 26 acres of parkland.

3. *Canad.* **a.** A parcel of land required by law to be set aside for public recreation and wildlife conservation.

1957 *Financial Post* (Toronto) 29 June 23/3 Every major Canadian city knows it should be providing 10 acres of parkland for every 1,000 people. **1977** *Little Cataraqui Environment Assoc. Newsletter* Nov. 1 The [conservation] authority will maintain this land as a natural parkland, paying both taxes to the city and maintenance costs.

b. A national or provincial park.

1958 *Maclean's Mag.* 10 May 8/1 We do, of course, sometimes set aside tracts of country for recreation and so forth—and we say these are in perpetuity; but this phrase only seems to mean until the parkland is needed for logging, or mining. **1971** *Weekend Mag.* (Montreal) 21 Aug. 9/2 Quebec..now has an accelerated program attending to some 25,000 square miles of developed and partially developed parkland.

Park Lane (pāɹk lē̆in). The name of a fashionable London street used as a symbol of wealth and breeding and the social attitudes they imply. Hence **Park Lane-ite.**

1880 Trollope *Duke's Children* II. vii. 78 Park Lane was sweeter than the Fifth Avenue. Lord Silverbridge was nicer than the bank clerk. **1906** *Westm. Gaz.* 27 Apr. 1/3 The attitude of this lady of the ballad is a common one among the Park-lanes of every land. **1936** 'J. Tey' *Shilling for Candles* vii. 73 What I'd like to know is if that method goes in Pimlico or if you keep it for Park Lane. **1944** M. Laski *Love on Supertax* ii. 27 Would you yourself be quite happy to see your children playing with scrubbed little brats from Park Lane? **1961** A. Smith *East-Enders* ii. 29 If they want to live like Park Lane-ites they should live in Park Lane.

parklet (pā·ɹklĕt). [dim. of Park *sb.*] A small park in an urban area.

1966 *Listener* 27 Jan. 135/1 The charming little 'parklets' ..have real herbaceous borders. **1968** Mrs. L. B. Johnson *White House Diary* 2 Oct. (1970) 715 These two small parklets on Hobart Street in a low-income area were a gift to the city by David Lloyd Kreeger.

parky, *a.*[1] Delete *rare*, and add earlier and later examples.

1831 M. Edgeworth *Let.* 11 Apr. (1971) 517 We came to a beautiful parky place where Dr. Fitton flourishes for the summer. **1955** R. P. Hobson *Nothing too Good for Cowboy* v. 43 Cow tracks fanned out through the parky poplar country leaving a messy swath many yards wide. **1973** R. D. Symons *Where Wagon Led* III. x. 181 The country around Turtleford was parky, with aspen bluffs and just the odd spruce.

parky, *a.*[2] (Earlier and later examples.)

1895 *Sporting Times* 9 Feb. 1/4 A toff came and ordered a pint of hot, As he said that the weather was parkey. **1916** 'Taffrail' *Pincher Martin* xi. 191 'Strewth!..it's a bit parky, ain't it?' **1969** J. Clarke *Foxon's Hole* xxv. 153 Gawaine ventured to say that it was a fine day... 'Ah. Parky, though. Shouldn't wunner if we 'ad a bit o' snow.' **1970** *Kenya Farmer* Feb. 36/3 All our visitors seem to labour under the delusion that it should be warm and sunny all twenty four hours... They are utterly demoralised by our parky nights on the farm. **1975** T. Heald *Deadline* v. 99 'Cold isn't it?' 'Pretty parky.'

parky: see *PARKIE.

‖ **parlando** (pāɹlæ·ndo), *a.*, *adv.*, and *sb. Mus.* [It.] A direction that a passage is to be played or sung 'as if speaking', in an expressive or declamatory manner.

1876 Stainer & Barrett *Dict. Mus. Terms* 342/2 *Parlando, Parlante* (*It.*), in a declamatory manner, as if speaking. **1880** Grove *Dict. Mus.* II. 650/2 *Parlando, Parlante*, 'speaking'. A direction allowing greater freedom in rendering than *cantando* or *cantabile*, and yet referring to the same kind of expression. **1930** *Time & Tide* 7 June 745/1 Against a dissonant pianoforte accompaniment, the voice disclaims monotonously in the 'parlando' style. **1944** W. Apel *Harvard Dict. Mus.* 554/1 Parlando occurs particularly in rapid tempo when the syllables of the text change with every note... In connection with instrumental music, parlando (*parlante*) calls for an expressive declamation, suggestive of speech or song. **1955** *Times* 27 May 13/4 Miss Nancy Evans sang 'The Water Mill' too slowly and with too much tone—this is a *parlando* song. **1960** *Times* 23 May 16/7 A..lyrical setting of the Bible story, involving..choral parlando. **1970** *Daily Tel.* 28 Sept. 11/3 Where he perhaps carried relaxation a shade too far ..was in a habit, during less important recitative-like passages, of lapsing occasionally into a toneless parlando of too vaguely defined pitch. **1977** *Gramophone* Dec. 1098/3 Brahms to them is a romantic to the hilt, and they go out of their way to extract the maximum *parlando*-type expression from every phrase.

parlary, var. *PARLYAREE.

parlay (pā·ɹlē̆i), *sb. U.S.* [Now the more usual form of Parley *sb.*[4]] A cumulative series of bets in which winnings accruing from one transaction are transferred to the next. Also *attrib.* See also Parley *sb.*[4]

1927 *National Turf Digest* (U.S.) Jan. 19/1 From the standpoint of the bookmaker..no money coming from a winning horse can be 'iffed' to a parlay on that same horse and another. **1928** *Ibid.* Feb. 91/1 A parlay player cannot protect himself against a changed track condition, a change of jockeys, or added starters. **1932** D. Runyon in *Collier's* 26 Mar. 40/2 He has a dispute with Sorrowful about a parlay on the races the day before. **1946** *Sun* (Baltimore) 2 Apr. 26/4 A $2 parlay would have paid $10,906.70. **1957** *New Yorker* 2 Nov. 105/1 Colonel Martingale..has figured out that a parlay on the same horses would have paid $4,045.58. **1963** T. Pynchon *V.* vii. 158 Both, together, were like a parlay of horses, capable of a whole arrived at by some operation more alien than simple addition of parts. **1976** *New Yorker* 23 Feb. 85/2 Well, all the players who made parlays on the Proud horses—Birdie and Delta—clicked.

parlay (pā·ɹlē̆i), *v.* [Now the more usual form of Parley *v.*[2]] **1.** See Parley *v.*[2]

2. *trans.* To increase (capital) by means of gambling; more generally, to exploit (a circumstance) for gain, to transform (an asset, advantage, etc.) *into* something considerably greater or more valuable. Also *absol. U.S. colloq.*

1942 *San Francisco Examiner* 5 May 18/5 As far as the girl who was kicked off the '36 Olympic team and parlayed it into a million dollars or so, is concerned, water these days is strictly for drinking and bathing. **1947** *Life* 20 Jan. 89/2 Only 25 years ago trucking firms consisted of a few men with strong backs and iron nerves who parlayed their savings into creaky trucks and did their bookkeeping on backs of envelopes. **1949** *Sat. Even. Post.* 25 June 32 H. J. Heinz..parlayed a pickle into one of the most valuable family heirlooms in America. **1952** *Sun* (Baltimore) 8 Oct. 11/2 She would like to parlay an original 'People's Union' fund of $100 into lots more and start in motion her mailing plan for errant politicians. **1956** B. Holiday *Lady sings Blues* (1973) vi. 57, I hoped he could shoot enough dice to parlay it into a bill big enough I didn't have to feel ashamed to send home. **1964** 'E. McBain' *Axe* v. 78 Cards... Dice. Anything where he can parlay a small stake into some quick cash. **1972** *Publishers' Weekly* 17 Jan. 56/3 His family had parlayed the sum awarded him by the court into millions. **1973** *Times Lit. Suppl.* 28 Sept. 1131/3 Parlaying his genuine physical tragedy, by unspoken comparison, into a certificate of literary genius. **1975** *High Times* Dec. 106/2 (Advt.), I have parlayed $146 into $90,000. **1976** *Wood Sci.* VIII. 180 These findings may be parlayed into savings in sample size in future experiments on the strength of lumber. **1977** *New Society* 5 May 236/3 Former pupils who had parlayed a third-class degree into £7,000 a year and ten minutes' teaching a week. **1977** *Time* 27 June 15/1 Rainer Werner Fassbinder, 31, the *Wunderkind* of low-budget West German cinema,..is about to parlay his critical acclaim into box office success as an international director.

‖ **parlementaire** (pārlĕmaṅtɛ̄r'). [Fr., f. *parlementer* to discuss terms, to parley.] = Parliamentary *sb.* 3.

1918 E. S. Farrow *Dict. Mil. Terms* 434 *Parlementaire*, in the French services the term meaning a bearer of a flag of truce. **1931** *Economist* 14 Nov. 892/2 The Japanese expeditionary force..began by sending parlementaires to the local Chinese military commander..who agreed to withdraw his troops ten kilometres. **1945** R. Hargreaves *Enemy at Gate* 230 Terms having been refused his *parlementaires*, Osman was left with no option but to submit in unconditional surrender. **1945** C. S. Forester *Commodore* xxii. 253 'French or Russian, sir,' said the parlementaire 'they will die unless they receive speedy aid.' **1950** W. S. Churchill *Second World War* (1951) IV. ii. xxiv. 376 At dawn on the 21st General Klopper sent out a parlementaire with an offer to capitulate, and at 7.45 a.m. German officers came to his headquarters and accepted his surrender.

parley, *v.*² Add: (Earlier and later examples in form *parlee*.)

The usual spelling is now *parlay.

1890 J. P. Quinn *Fools of Fortune* ii. ii. 194 Almost all [faro] bankers will allow a player to 'parlee' as the percentage is largely in favor of the bank. **1903** Ade *People you Know* 110 He wanted to parlee a $2 Silver Certificate and bring home enough to pay the National Debt.

parliament, *sb.*¹ Add: **3. e.** *Act of Parliament clock*, a type of wall clock produced in the 18th century for use in inns and taverns and characterized by a black or green dial with gold numerals over which there was no glass. The name arose from the popular belief that such clocks were acquired by innkeepers in order to attract custom after Parliament imposed a tax on clocks and watches in 1797.

1899 F. J. Britten *Old Clocks & Watches* 337 In country inns and other places Act of Parliament clocks may still occasionally be seen. **1917** A. Hayden *Chats on Old Clocks* iv. 124 It is supposed that these clocks suddenly came into being when private clocks were taxed. Owing to such a deep-seated belief they are always known throughout the country as 'Act of Parliament' clocks. But they were used earlier than the Act of 1797, and were probably ordinary inn clocks in common use about that time. **1952** J. Gloag *Short Dict. Furnit.* 191 *Coaching Inn Clock*..Such clocks have been misnamed Parliament or Act of Parliament clocks, on the assumption that they were introduced and used extensively by innkeepers after Pitt's Act of 1797 (under which clocks and watches were taxed) presumably to save customers..buying a watch or clock and then paying tax on it—not a plausible theory. **1960** *House & Garden* Apr. 100/1 The round-faced Act of Parliament clock..is two feet across. **1962** *Kent & Sussex Courier* 19 Oct. 6/6 The Rose and Crown Hotel boasts the town's oldest public clock. A fine example of an Act of Parliament clock, it was made in 1797 and now stands in the entrance hall.

5. c. (Further examples.)

1905 R. B. Smith *Bird Life* x. 386 Two such 'Parliaments of rooks' I have had the opportunity of watching, from early times. **1939** C. E. Hare *Lang. Field Sports* xxiii. 192 The modern accepted rook term is believed to be *parliament*, but no authority for this can be traced. **1968** J. Lipton *Exaltation of Larks* 57 (*caption*) A parliament of owls.

9. parliament-gingerbread (example); **parliament hinge** (example).

1861 C. M. Yonge *Stokesley Secret* iii. 39 A stall full of parliament gingerbread. **1841** C. Cist *Cincinnati in 1841* 247 The lighter castings kept in hardware stores—butt and parliament hinges, for example—will be made here.

pa:rliamentariza·tion. [f. Parliamentary *a.* + -ization.] The act or process of becoming parliamentary in character or in means of government.

1924 *Contemp. Rev.* Aug. 256 The book deals..with the progress of the ideas of Parliamentarisation and racial self-determination under the stress of war. **1974** J. White tr. *Poulantzas's Fascism & Dictatorship* iii. iii. 132 The peace pact and the parliamentarization of the fascist movement were resented. **1975** *Times Lit. Suppl.* 12 Sept. 1028/2 *Gesellschaft, Parlament und Regierung* is concerned with the painful evolution of German parliamentarism and its weaknesses both before and after the First World War. More specifically, Dr Ritter's own introduction asks why those forces that might, on the strength of their economic resources, have pressed for parliamentarization..failed to mount an effective challenge to the traditional elites.

parliamentary, *a.* Add: **1.** (Further examples.)

Parliamentary Commissioner (for Administration) = *Ombudsman; *Parliamentary Counsel*, barristers employed as established civil servants to draft government bills and amendments; *Parliamentary Private Secretary*, a member of Parliament who acts as assistant to a government minister; the post has no official status and is unpaid.

1833 *Rep. Select Comm. Establ. House of Commons* 163 in *Parl. Papers* XII. 341 You are Parliamentary Counsel to the Treasury?—I am. **1850** *Rep. Sel. Comm. Official Salaries* 8 in *Parl. Papers* XV. 179 Both the Secretaries to the Treasury..are most responsible officers. With regard to the Parliamentary secretaryship, which I once held myself, I do not know so difficult or so disagreeable office in the Government. **1858** H. G. G. Grey *Parliamentary Govt.* vi. 90 Our whole system of Parliamentary Government rests..upon the Ministers of the Crown possessing such authority in Parliament as to enable them generally to direct its proceedings. **1872** Disraeli in *Times* 4 Apr. 5/2, I believe that, without party, Parliamentary government is impossible. **1886** *Whitaker's Almanack 1887* 156/1 Office of Parliamentary Counsel,—Spring Gardens. *Parliamentary Counsel*, Hen. Jenkyns, C.B. **1917** *H.M. Ministers & Heads of Public Departments* (Stationery Office) 1 Parliamentary Private Secretary... Capt. Hon. W. Ormsby-Gore, M.P. **1918** *Act* 8 *Geo. V* c. 3 § 1 (1) A Secretary who shall discharge the functions both of a parliamentary secretary to the Board and a parliamentary under-secretary to the Secretary of State. **1930** W. K. Hancock *Australia* x. 210 The practice of the Australian Labour party makes England's classic philosophy of parliamentary government appear strangely artless and out of date. **1930** A. J. Balfour *Chapters of Autobiogr.* viii. 103 On taking his new post, he [*sc.* Salisbury in 1878 when he became Foreign Secretary] asked me to become his Parliamentary Private Secretary. **1939** W. I. Jennings *Parliament* vii. 229 A department official and the draftsman are seated in the 'box' and communications pass through his parliamentary private secretary or 'fetch-and-carry' man. **1954** H. Morrison *Govt. & Parliament* iv. 66 The life of the Parliamentary Secretary can be interesting and fairly full, or, on the other hand, uninteresting and rather empty. **1966** *Listener* 11 Aug. 194/2 Sir Edmund Compton, Comptroller and Auditor-General, is to be Britain's first Parliamentary Commissioner, or Ombudsman. **1968** T. Stoppard *Real Inspector Hound* 11 I dream of champions chopped down by rabbit-punching sparring partners while eternal bridesmaids turn and rape the bridegrooms over the sausage rolls and parliamentary private secretaries plant bombs in the Minister's Humber. **1969** *Times* 2 May 22/1 The work of the Parliamentary Counsel is not widely known. They draft the Government's Parliamentary Bills. *a* **1974** R. Crossman *Diaries* (1975) I. 42 What was a bit crazy was to put Charles Snow in as his Parliamentary Secretary. *Ibid.* 239 Whereas the parliamentary draftsman who has worked on my Bill is superb, in the Ministry I feel we could have got a far stronger team together if we hadn't relied so entirely on the administrative class. **1976** *Daily Tel.* 20 July 1/2 An issue which has split both Ministers and the different wings of the Parliamentary Labour party. **1976** B. Kemp *Sir Robert Walpole* iv. 79 Walpole and the other ministers cannot be called the 'executive'... Their relation to parliament cannot be called 'parliamentary government', if by this is meant a state of affairs where the 'executive power and the power of legislation are virtually united in the same hands'. **1976** H. Wilson *Governance of Britain* ii. 32 When choosing ministers of state and parliamentary under-secretaries, the prime minister would naturally consult—or at least inform—the Cabinet minister concerned.

2. c. (See quot.)

1886 J. Barrowman *Gloss. Scotch Mining Terms* 49 *Parliamentary pit*, an outlet pit required by statute.

‖ **parloir** (pārlwār). [Fr.: see Parlour.] A room in a monastery or convent used for conversation with people from outside, or among the inmates; = Parlour B. 1. Also, a similar room in a prison.

1728 Chambers *Cycl.* II. 354/2 *Parloir, Parlour*, in Nunneries, a little Room, or Closet, where People talk to the Nuns, thro' a kind of grated Window... Antiently, there were also Parlours in the Convents of Monks, where the Novices used to converse together at the Hours of Recreation. **1924** A. D. Sedgwick *Little French Girl* iii. vi. 276 She might have sat, in her early convent days, giving an account of herself in the *parloir*..to the relative who had come to pay her a weekly visit. **1927** J. Rhys *Left Bank* 43 The old man and the little boy were the last of the queue of people waiting..to be admitted to the *parloir*—a row of little boxes where on certain days prisoners may speak to their friends through a grating. **1955** J. Thomas *No Banners* xxv. 248 His camp bed had been installed in the narrow, steel-barred passage dividing the two sections of the *parloir* or visiting cell.

parlour. Add: **2. e.** In *attrib.* use with the names of outdoor games which have been adapted to a smaller scale for use indoors.

1872 A. Elliot *Within Doors* 45 Numerous Parlour Games have recently been introduced... Such are Parlour Croquet,..Parlour Billiards, [etc.]. **1881** *Cassell's Bk. In-Door Amusem.* 74 The game described in this book as German Balls is sometimes also known as Parlour Bowls. **1887** E. Custer *Tenting on Plains* xv. 501 A game of parlor croquet was proposed. **1895** *Montgomery Ward Catal.* 235/3 Parlor Tennis. This new and fascinating game is arranged for parlor or lawn use and is played with 12 light rubber balls... Parlor Quoits..consists of two turned posts..and four quoits five inches in diameter. **1899** Beerbohm *More* 140 Playing parlour-golf with his only child. **1901** Parlour cricket [see *dart sb. 1 d]. **1926–7** *Army & Navy Stores Catal.* 854/3 The parlour golf hole... For practising 'putting' indoors.

f. In *attrib.* use applied to persons of comfortable or prosperous circumstances who profess support, usu. of a non-participatory nature, for radical, extreme, or revolutionary political movements, as *parlour Bolshevik, communist, socialist*, etc. Hence *parlour Bolshevism, socialism*.

1910 *Ann. Library Index 1909* 273 (*title*) Parlour socialists. **1915** T. Dreiser *Let.* 26 Apr. in *Lett. H. L. Mencken* (1961) 68, I hold no brief for the parlor radical. **1918** [see *Bolshevism]. **1922** R. Nevill *Yesterday & Today* i. 14 What may be called 'Society Socialism' is an entirely modern development, pretty well limited to England and America where the 'Parlour Socialist' has become recognized as a regular type. **1926** G. Frankau *My Unsentimental Journey* iv. 56 The audience..were only 'parlour Bolsheviks'. **1929** F. P. Gibbons *Red Napoleon* 67 Margot was more than a parlour pink; she was an ardent internationalist. **1930** H. G. Wells *Autocracy of Mr. Parham* ii. i. 86 Don't imagine we are that mysterious unseen power, the Money Power, your parlour Bolsheviks talk about. **1938** G. T. Garratt *Shadow of Swastika* 201 Mr. Neville Chamberlain remained..invincible because of his backing amongst the very wealthy and influential parlour fascists outside. **1939** C. Isherwood *Goodbye to Berlin* 105 Wasn't I a bit of a sham..with my arty talk..and my newly-acquired parlour-socialism? **1954** Koestler *Invis. Writing* iii. 40 The most fashionable poet among the snobs and parlour-Communists of the period was Bertold Brecht. **1960** *News Chron.* 22 June 6/5 A wonderfully reactionary view of country life. It makes John Buchan look a 'parlour pink'. **1969** *Times* 24 Mar. 7/7 Cripps..had just come into notoriety before the war as a 'parlour Bolshevist' of a high intellectual order. **1973** K. Giles *File on Death* vi. 156 A parlour pink! Did he have anything to contribute? **1976** S. Hynes *Auden Generation* x. 367 The stock notion of the 'thirties writer as a *New Country* parlour-communist.

4. For *U.S.* read orig. *U.S.* (Examples.)

1884 Ice cream parlor [see *ice-cream *attrib.*]. **1908**, etc. Beauty parlour [see **beauty parlour* s.v. *beauty *sb.* III]. **1912** Manicure parlor [see *manicure *sb.* 2]. **1913**, etc. Massage parlour [see *massage *sb.* b]. **1927** E. Glyn *'It'* xiii. 122 'The Oak Parlour', a new little restaurant. **1928** *Daily Express* 22 Oct. 1/3 The bodies of the boys will be kept in sealed caskets in an 'undertaking parlour' until the mother is well enough to attend the funeral. **1942** H. C. Bailey *Dead Man's Shoes* xxvi. 100 Pat's Parlour, a tea shop for holiday visitors. **1952** S. Selvon *Brighter Sun* iv. 63 He was listless and dull, and attracted very little business, though his was the only well-stocked parlour in that part of Barataria. **1963** H. Garner in R. Weaver *Canad. Short Stories* (1968) 2nd Ser. 41, I tried a couple of beer parlours, but couldn't stand the noise and laughter. **1973** W. McCarthy *Detail* ii. 115 Stuart..went to the adjoining pizza parlour. **1974** *Listener* 8 Aug. 168/3 Model industrial communities, with sun-lamp parlours.

b. *ellipt.* form of *milking-parlour* s.v. *milking *vbl. sb.* 4.

1950 *N.Z. Jrnl. Agric.* June 541/1 Near Davis [California] I visited some dairies using the 'parlour' system of milking. **1973** *Country Life* 28 June 1904/1 Development work on milking machinery..directed at speeding the movement of cows through the parlour.

6. *parlour game* (earlier and later examples), *pastime, politics*; **parlour-boarder** (earlier examples); **parlour-girl** *U.S.*, = Parlour-maid; **parlour-house,** (*a*) a house having a parlour; (*b*) *U.S. slang*, an expensive type of brothel; **parlour-jumper** *slang*, one who robs rooms (see quot. 1938); **parlour man,** a male domestic servant; = *house-parlourman* s.v. *house *sb.*¹ 23; (the sense in quot. 1851 is uncertain); **parlour melodeon** *U.S.*, a kind of parlour organ; **parlour-organ** (examples); **parlour-palm,** the aspidistra; **parlour-piece,** a slight entertainment suitable for performance in a parlour; **parlour social** *U.S.*, = *house-rent party* s.v. *house *sb.*¹ 23; **parlour trick,** (*b*) an amusing 'turn' or trick performed, often by an animal, as entertainment.

1777 P. Thicknesse *Year's Journey* I. ii. 12 The Prieure of this convent..had received, as parlour boarders, some English ladies of very suspicious characters. **1812** *Theatrical Inquisitor* I. 211, I am a parlour boarder at Mrs. Twizzle's school. **1872** Parlour game [see sense 2 e above]. **1923** W. de la Mare *Riddle* 127 She *talks* to you; but it's all make-believe. It's all a 'parlour game'. **1958** Wodehouse *Cocktail Time* xviii. 157 The television set..was now deep in one of those parlour games designed for the feeble-minded trade. **1975** *Listener* 28 Aug. 279/1 At this season of the year, the listener to Radio 4 cannot expect much more than..parlour games. **1863** A. D. Whitney *Faith Gartney's Girlhood* iii. 9 The parlor-girl made her appearance with her mop and tub. **1875** Mrs. Stowe *We & Neighbors* xxxiv. 323 Maggie was parlor-girl and waitress, and a good one too. **1872** E. Crapsey *Nether Side N.Y.* 142 A most deplorable change..greatly decreasing the number of parlor houses, while houses of assignation have multiplied. **1924** in Henderson & Maddock *Housing Acts* (1930) 431 Appropriate normal rents may be fixed for different classes of houses, *e.g.* parlour and non-parlour. **1926** J. Black *You can't Win* iv. 28 Women who kept 'parlor houses' in the Tenderloin district. **1927** St. John Ervine *Wayward Man* i. i. 3 Three shops, four parlour-houses... The front of his house looked like that of any other parlour-house, but if a passer-by had peeped through the window he would have seen, not, as he might have expected, a small d'oyley-covered table, bearing a pot of geraniums or [etc.]. **1927** F. E. Fremantle *Housing of Nation* 40 At Roehampton the cost of a parlour house rose to £1,750. **1975** J. Gores *Hammett* (1976) v. 38 The parlor houses, cribs, brothels and bagnios had disappeared..and a thousand prostitutes had been thrown out of work. **1977** *Belfast Tel.* 19 Jan. 25/3 (Advt.), Parlour house, off Newtonards Rd., good condition. **1898** *Daily Tel.* 4 Aug. 3/2 A constable explained that the prisoner..was known as a 'parlour-jumper'... He went in for robbing rooms. **1938** F. D. Sharpe *Sharpe of Flying Squad* xv. 170 'Parlour Jumpers'..went round knocking on doors until they found a temporarily unoccupied house or flat, when they would force a way in and collect everything of value.., put it in the table cloth ..and walk out. **1851** H. Melville *Moby Dick* II. xiv. 121 Beale's..frontispiece, boats attacking Sperm whales, though no doubt calculated to excite the civil scepticism of some parlour men, is admirably correct..in its general effect. **1922** *Glasgow Herald* 31 Oct. 7/1 The men who have disappointed as 'housemen' and 'parlourmen' are for the most part ex-Service men..prepared to do anything to get a job. **1960** *Times* 5 Jan. 3/3 (Advt.), Cook and Butler/House Parlourman required for Surrey;..excellent kitchen; own redecorated flat. **1909** 'O. Henry' *Roads of Destiny* vii. 107 The natives were panning out enough from the beach sands to buy all the rum, red calico, and parlour melodeons in the world. **1845** in C. Cist *Cincinnati Misc.* I. 179/1 'I was on a visit to Vermont, a few weeks since,' said he, 'and intended to buy a parlor Organ.' **1943** A. G. Powell *I can go Home Again* 96 There was an ordinary parlor organ, but on the days in which Old Lady McCan..attended services the organ in the Baptist

Church could not be used. **1904** *Amateur Gardener's Diary* 145 *Aspidistra* (Parlour Palm), one of the hardiest of in-door plants, as it will survive dust and even the fumes of gas. **1876** G. M. Hopkins *Poems* (1967) 65 And ever, if bound here hardest home, You've parlour-pastime left. **1938** *Amer. Speech* XIII. 255 Most of the music..played on these many pianos was..'light classical', and in anthologies the terms 'salon music' and 'parlor pieces' were used. **1957** T. Hughes *Hawk in Rain* 20 (*title*) Parlour-piece. **1940** H. G. Wells *New World Order* § 1. 18 This is no small affair of parlour politics we have to consider. **1956** Parlour social [see *house-party* s.v. *house *sb.*¹ 23]. **1966** W. T. E. Kirkeby *Ain't Misbehavin'* iv. 39 Another and popular way to meet expenses was the parlour social or, as it later became familiarly known, the 'rent-party'. **1922** R. Leighton *Compl. Bk. Dog* vii. 99 He [*sc.* the Chow Chow]..often has a clever gift for parlour tricks. **1961** *Times* 7 Jan. 7/7 The art of skiing is..gradually suffering conversion into a gigantic outdoor parlour trick.

parlyaree (pāɹliǟ·rī). Also **parlary.** [f. It. *parlare* to speak, talk.] A form of slang used by actors and showmen, particularly in the 18th and 19th centuries, and characterized by Italianate vocabulary. Cf. *palarie *v.*

The word appears to be related to *nanty parnarly* recorded in Barrère & Leland *Dict. Slang* (1890) II. 81/2.

1933 Partridge *Slang To-day & Yesterday* iii. 223 Until about the end of the eighteenth century, actors were so despised that, in self-protection, they had certain words that, properly, should be described as cant and were actually known as Parlyaree. **1933** *Times Lit. Suppl.* 15 June 412/3 Circus slang is a nineteenth-century offshoot from the Parlyaree of the seventeenth and eighteenth centuries. **1950** Partridge *Here, There & Everywhere* 117 It was among showmen and strolling players that parlyaree originated. *Ibid.* 122 Most parlyaree-speakers prefer a *carser* or *carsey*, ie. a house, when they can get it. *Ibid.* 125 Parlyaree..is a glossary, a vocabulary, not a complete language. **1952** Granville *Dict. Theatr. Terms* 132 *Parlyaree* (occasionally *parlary*). Little is known about this language which has neither accidence nor syntax of its own but is built on a base of Italian words and phrases, whereon cant terms and illiteracies are piled. **1960** J. Franklyn *Dict. Rhyming Slang* 106/1 The term [*sc. parlamaree*] is obviously a mispronunciation of Parlyaree, the language of the Circus. **1960** *Spectator* 11 Mar. 355 The canting jargon of the Victorian fairgrounds known as 'parlyaree'.

parm (pāɹm). Colloq. contraction of *pardon* (*me*).

1945 A. Kober *Parm Me* 180 Parm me fa conterdicting. **1957** J. Blish *Fallen Star* vii. 101, I was hun'ry. Parm me. **1967** C. Drummond *Death at Furlong Post* xiv. 171 'Piss off,' said Hart. 'Parm?' **1973** K. Giles *File on Death* v. 144 Parm me, sir, but I've got me good name to consider.

Parma¹ (pā·ɹmă). The name of a city in northern Italy, used *attrib.* or *absol.* to designate products associated with the region, as **Parma ham,** a local type of ham which is eaten uncooked; **Parma violet,** (*a*) a cultivated violet with double, scented flowers, usually light or deep purple, belonging to a group of cultivars of *Viola alba*; a crystallized flower of this kind; (*b*) a perfume manufactured from a flower of this type or imitating its scent; (*c*) a deep or medium shade of purple.

[**1960** E. David *French Provincial Cooking* 134 It is a common misconception that all hams eaten raw, including that of Parma, are smoked.] **1964** A. Launay *Caviare & After* vii. 54 The Bayonne and Parma hams are served raw. **1966** 'J. Melville' *Nell Alone* i. 21 She's doing a lot of eating..Parma ham, tinned stuff, the best coffee. **1971** *Sunday Times* (Colour Suppl.) 27 June 50/3 Parma ham.. and Westphalian smoked ham, all eaten raw, have always been popular with gourmets. **1976** E. Ward *Hanged Man* xx. 118 Eggs and Parma ham and oven-hot bread and coffee. **1856, 1880** Parma violet [see Violet *sb.*¹ 2 a]. **1907** *Yesterday's Shopping* (1969) 521/1 Perfumes... Opoponax. Parma Violet. Peau d'Espagne. **1919** K. Mansfield in *Art & Lett.* II. 155 A stout lady in blue serge, with a bunch of artificial 'parmas' at her bosom. **1922** *Weekly Dispatch* 10 Dec. 15 A leather set in the new shade of violet. Something between purple and parma, this hue will soon be the rage. **1923** *Daily Mail* 7 Feb. 1 (Advt.), Black, Navy,..Parma. *Ibid.* 12 Feb. 1 (Advt.), Shades: Fawn, Grey, Lovat, Parma, Browns. **1932** *New Flora & Silva* IV. 190 The origin of the Parma Violet has never been satisfactorily elucidated... There are many violets growing round Naples, but none..which I have seen suggest themselves as parents of the Parma. **1938** G. L. Zambra *Violets* vii. 62 The family of Parma violets gives the sweetest perfume, the longest flowering season, and the handsomest foliage of all the double violets. **1954** 'M. Cost' *Invitation from Minerva* 96 A frock..parma-violet in colour. **1956** *Punch* 23 May 626/1 If the modest violet is employed it must be..a shameless bunch of Parmas. **1963** R. D. Meikle *Garden Flowers* 105 The delicious Parma Violet, with smaller, rather glossy leaves, and slender-stalked flowers, is probably a form of *Viola alba* Bess., found wild (with innumerable variations) over most of the Mediterranean area and Asia Minor. **1970** *Observer* 1 Feb. 32/7 Dior: Parma violet coat. **1970** I. Origo *Images & Shadows* ii. 60 The ladies in summer gowns..leaving behind them a faint aroma of lavender and Parma violet. **1974** *Country Life* 25 Apr. 1025/2 Using parma violet or electric blue, she sprinkles these vibrating colours through the wardrobe. **1977** *Times* 14 June 18/6 Women in the chocolate factory..place walnuts and Parma violets on to the store's handmade confectionery.

parma² (pā·ɹmă). *Geol.* (See quots.)

1888 *Encycl. Brit.* XXIV. 4/1 The section [of the Urals] between the 64th and 61st parallels has..a wholly distinct character... From the broad plateaus, or *parmas*, which stretch towards the north-west, it might be conjectured.. that the structure is more complicated. **1904** *Amer. Jrnl. Sci.* CLXVIII. 469 This may mean that no true axis or 'parma' was in existence during Richmond time, but it does seem to show that the Wabash parma at least indicates the strike for the then highest land. **1904** tr. *Suess' Face of Earth* I. ii. xii. 601 We see great folded chains merge with gradually flattening undulations into the similar foreland, where they form secondary folds, or 'parmas'—this is the case in the Urals and the Appalachians. **1913** A. W. Grabau *Princ. Stratigr.* xx. 808 Many of the low-dipping domes are perceptible as such only by the erosion which has removed their central portion, often leaving a topographic depression. Such low domes have also been called parmas. **1957** *Gloss. Geol.* (Amer. Geol. Inst.) 212/1 *Parma*, a low dome or quaquaversal.

Parmentier (pāɹma·ṅti,e), *a.* [f. the name of Antoine A. *Parmentier* (1737–1813), French agriculturalist, who popularized potatoes in France.] In cookery, made with or accompanied by potatoes.

1906 *Mrs. Beeton's Bk. Househ. Managem.* 1314 *Parmentier eggs*... Poach the eggs in salted water flavoured with lemon-juice, and place them carefully in the halved potatoes. **1929** A. E. Housman *Let.* 14 Sept. (1971) 284 Food not varied or inventive, especially soup: I do not mind Santé twice in ten days, but Parmentier I do. **1951** *Good Housek. Home Encycl.* 580/2 *Parmentier*, the name of the man who introduced the potato into France... The term is now applied to a number of potato dishes... Parmentier soup. **1963** A. L. Simon *Guide Good Food & Wines* 38/1 In culinary parlance *Parmentier* always means potatoes; it is the homage of French cooks to Parmentier, who introduced potatoes in France. **1965** *House & Garden* Dec. 84/2 *Parmentier*. Named after the man who did so much to improve the cultivation of potatoes in France.. and so, naturally, means potatoes.

Parmesan, *a.* and *sb.* Add: **A.** *adj.* (Later examples.)

1885 [see *Edam]. **1946** G. Millar *Horned Pigeon* iii. 47 Denis Patchett..stole a large lump of Parmesan cheese from one of our guards.

B. *sb.* (Later examples.)

1876 [see *Brie]. **1960** [see *cannelloni *sb. pl.*]. **1971** *Sunday Times* (Colour Suppl.) 28 Mar. 39/2 Parmesan is a cheese with remarkable keeping qualities.

parnas(s (pāɹnā·s). Pl. parnassim (pāɹnasī·m). [Heb.] The lay head of a Jewish synagogue congregation.

1831 [see *Mahamad]. **1892** I. Zangwill *Childr. Ghetto* I. 97 Michael Birnbaum was a great man in the little local synagogue... He had been successively *Gabbai* and *Parnass*, or treasurer and president. **1907** —— *Ghetto Comedies* 76 The Great Synagogue..struck a note of modern English gaiety,..looking towards the box of the *Parnass* and *Gabbai*, she saw it was occupied by officers with gold sashes. *Ibid.* 122 The *Parnass* proffered his presidential hand in pious congratulation. **1932** C. Roth *Hist. Marranos* 247 The community boasted a model organisation. The power of the *Parnasim*, the elected Wardens, was autocratic, as offenders like Benedict Spinoza or Uriel Acosta learned to their cost. **1949** *Spectator* 4 Nov. 595/2 The *Parnas Presidente*, the President of the Wardens of the Synagogue. **1962** B. Abrahams tr. *Life Glückel of Hameln* ii. 18 My father has been *parnass* for many years. [*Note*] In Glückel's time the office of *parnass* (president or warden [of the synagogue]) was a monthly one. **1967** D. T. Kauffman *Dict. Relig. Terms* 346/1 *Parnas steward*, lay president of a synagogue congregation.

parnassia (pāɹnæ·siă). [mod.L. (Linnæus *Hortus Cliffortianus* (1737) 113), f. Parnass(us + -ia¹.] A small perennial herb of the genus so called, belonging to the family Saxifragaceæ, native to temperate or cold regions of the northern hemisphere, and bearing radical ovate or cordate leaves and white or pale yellow flowers; = *grass of Parnassus* s.v. Parnassus c.

1772 R. Weston *Universal Botanist* II. 546 Parnassia, Grass of Parnassus. **1793** J. E. Smith *Eng. Bot.* II. 82 The Parnassia agrees with Saxifraga in the wonderful œconomy of its impregnation. **1883** W. Robinson *Eng. Flower Garden* 207/2 In many of our moist heaths and bogs the Marsh Parnassia..is not unfrequently met with. **1941** W. A. Percy *Lanterns on Levee* ix. 97 In the streams where the iris forgathered there are parnassia, the snow-drop's only kin. **1951** *Dict. Gardening* (R. Hort. Soc.) III. 1485/1 The Parnassias are hardy, and succeed in moist, peat soils.

Parnassian, *a.* and *sb.* Add: **A.** *adj.* **1. c.** Of or pertaining to sense *B. 1 c. **B.** *sb.* **1. c.** *spec.* In the writings of G. M. Hopkins, a second kind of poetry, which can only be written by poets but which is not the language of inspiration.

1864 G. M. Hopkins *Let.* 10 Sept. (1956) 216 *Parnassian* then is that language which genius speaks as fitted to its exaltation, and place among other genius, but does not sing..in its flights. Great men, poets I mean, have each their own dialect as it were of Parnassian, formed generally as they go on writing, and at last,—this is the point to be marked,—they can see things in this Parnassian way and describe them in this Parnassian tongue, without further effort of inspiration. *Ibid.* 217 In Parnassian pieces you feel that if you were the poet you could have gone on as he has done, you see yourself doing it, only with the difference that if you actually try to find you cannot write his Parnassian. *Ibid.* 218, I believe that when a poet palls on us it is because of his Parnassian.

Parnassianism (paɹnæ·siăniz'm). [f. Parnassian *a.* 1 b + -ism.] The Parnassian style in poetry.

1905 *Times* 4 Oct. 6/2 He began to write the sonnets which attracted the attention of the most expert connoisseurs in Parnassianism. **1922** *Freeman* (N.Y.) 26 Apr. 105 Parnassianism means objectivity, impassivity, attention to line and image rather than to colour and music and vague suggestiveness. **1927** *Observer* 11 Sept. 7/3 Parnassianism, Symbolism, and the Ecole Romane have all had their day. **1967** *Guardian* 19 May 7/2 Some of the poems are too deliberate, Parnassian... And very splendid, too, parnassianism at that level.

Parnate (pā·ɹnēⁱt). *Pharm.* A proprietary name for the drug tranylcypromine.

1960 *Official Gaz.* (U.S. Patent Office) 17 May TM112/2 Smith Kline & French Laboratories, Philadelphia, Pa. Filed Oct. 19, 1959. *Parnate* for anti-depressant. First use Oct. 6, 1959. **1960** *Trade Marks Jrnl.* 29 Dec. 1681/1 Parnate... Preparations and substances in capsule or tablet form for use in medicine and pharmacy. Smith Kline & French Laboratories Limited,..Welwyn Garden City, Hertfordshire; manufacturing chemists. **1964** *New Statesman* 6 Mar. 354/1 An American anti-depressant drug—Parnate or tranylcypromine—has been declared unsafe by the US Food and Drug Administration. **1965** J. Pollitt *Depression & its Treatment* iv. 56 The monoamine oxidase inhibitors include several members, among them phenelzine (Nardil), iproniazid (Marsilid),..and tranylcypromine (Parnate). **1970** *Daily Tel.* 7 July 3/3 The result of eating cheese with the drug Parnate was to trigger off fatal side-effects. **1976** *Ibid.* 11 Mar. 19/4 She was taking Parnate, which is a very strong drug. If mixed with alcohol it could have a most terrible effect.

Parnellite. (Earlier and later examples.)

1881 E. W. Hamilton *Diary* 26 Jan. (1972) I. 102 This nettled the Parnellites, and they intimated they would fight to the death. **1882** *Ibid.* 30 Apr. 259 Wednesday was quite an important day, as it was taken up by the discussion of the bill brought in by the Parnellite party, with the name of Parnell himself on the bill. **1892** W. B. Yeats *Lett.* (1954) 222 Parnellite Dublin and the Parnellite young men in the country parts. **1931** *Times Lit. Suppl.* 17 Sept. 708/1 His anecdote of the Parnellite in an Irish theatre. **1936** W. B. Yeats *Poems* (1957) 586 Come gather round me, Parnellites, And praise our chosen man. **1939** Joyce *Finnegans Wake* 307 Are Parnellites Just towards Henry Tudor?

parochial, *a.* (*sb.*) Add: **1. a.** *parochial school* (*Sc.* and *N. Amer.*): a school established and maintained by a religious body. Hence *parochial schoolmaster*. (See also *parish school* s.v. *parish *sb.* 7 b.)

1755 in *Sc. Nat. Dict.* (1968) VII. 37/2 How great importance it would be both to the College and the Parish to have a Parochial School. **1791** A. Murray in J. Sinclair *Statist. Acct. Scotl.* I. 457 The parochial schools are by no means supplied with such enlightened teachers as those that were formerly instrumental in diffusing this knowledge. **1792** D. McDougal in *Ibid.* III. 188 The parochial schoolmaster teaches Latin, English, Gaelic, [etc.]. *a* **1817** [in Dict.]. **1832** *Chambers's Edin. Jrnl.* I. 226/2 Acts of Parliament..have considerably enlarged the salaries of the parochial schoolmasters. **1851** C. Cist *Sk. Cincinnati in 1851* 58 Parochial Schools. The Catholic schools are the only ones which are strictly parochial. **1860** *Nor' Wester* (Red River Settlement) 14 Feb. 2/1 The Parochial Schools of our Protestant population speak for themselves, and I am sorry that they should have found disparagement at the hands of your correspondent. **1876** J. Grant *Hist. Burgh & Parish Schools Scotl.* 100 The teacher of the burgh and parochial school was invariably session clerk and precentor. **1904** F. Crissey *Tattlings Retired Politician* 263 When he [*sc.* the Governor] was renominated the parochial school teachers camped on his trail and made it some hot for him. **1926** J. B. Ritchie *Forres* 74 It was held that the Grammar School of Forres had never been a parochial school. **1955** *Western Star* (Corner Brook, Newfoundland) 10 Mar. 2/3 An amendment to the Municipal Act..would allow municipalities to exempt parochial schools from taxation. **1964** *Calgary Herald* 11 Feb. 15 Premier Roblin Monday proposed a partial solution to the dilemma of Manitoba's parochial schools. **1972** *Lebende Sprachen* XVIII. 35/2 US *parochial school*—BE/US *denominational school*. **1976** *Globe & Mail* (Toronto) 16 Jan. 29/8, I attended parochial school in Winnipeg. That meant I was Roman Catholic, that I ate fishcakes on Friday and smelled of candle wax and incense. **1978** *Times* 30 Jan. 12/5 As soon as you get more than half black in a school then the whites..put their youngsters in to private and parochial schools.

parochially, *adv.* (Later examples.)

1929 S. Leslie *Anglo-Catholic* xiii. 182 He could never tell the truth to his father, since the Canon had parochially decreed that his cryptic daughter-in-law was no fit person to live in the parish. **1971** *Daily Tel.* 19 June 15/1 This [tax] relief has been very limited in its scope and I doubt whether its passing will be more than parochially mourned. **1977** *Ibid.* 19 Apr. 12 For a flower festival to open its golden jubilee celebrations at Addington Palace, Croydon, next month the Royal School of Church Music plans to have desert flowers from Southern California, orchids from Singapore and proteas from South Africa to demonstrate its ecumenism. More parochially, though, displays from this country will be the responsibility of the Surrey Flower Arrangers' Association.

parodic, *a.*² Delete *rare* and add later examples.

1962 D. Lessing *Golden Notebk.* iv. 529 The word *correct* had an echoing parodic twang. **1974** C. A. Patrides *Eng. Poems George Herbert* 9 Herbert looked beyond the Bible for parodic purposes. **1976** *Archivum Linguisticum* VII. 27 H. B. Richardson..follows Meyer-Lübke in suggesting *escuerzo* < *scorteum*, 'made of leather', meaning in 1544c 'toad'; this is possible, given the parodic context (the mock planctus for Urraca).

parodistic, *a.* (Later examples.)

1937 *Scrutiny* VI. 298 His [*sc.* Lord Berners']..measure is by no means summed up in his satirical and parodistic pieces. **1949** Koestler *Insight & Outlook* 427 Freud distinguished between the comic effects of imitation as such, and the parodistic attitude which usually accompanies it. **1975** *Times Lit. Suppl.* 21 Oct. 1305/1, I must record my partial disappointment, both at the parodistic and the confessional level. **1977** *Gramophone* June 106/3 The parodistic 'Danzon' is also, I think, the most entertaining of the three dances from Bernstein's ballet *Fancy Free*.

parœcious, *a.* (Earlier and later examples.)

1882 *Encycl. Brit.* XIV. 718/2 They [*sc.* the antheridia] are usually seated in the axils of modified leaves (perigonial), sometimes appearing..on special branches of the same plant (parœcious). **1912** S. M. Macvicar *Student's Handbk. Brit. Hepatics* p. xix, In some cases they [*sc.* the antheridia] occur below the female inflorescence,..usually making this 'paroicous' [*sic*] form of inflorescence easy to recognize. **1968** E. V. Watson *Brit. Mosses & Liverworts* (ed. 2) 34 Paroecious, bearing antheridia and archegonia close together, but not mixed, the antheridia being in the axils of bracts below those that surround the archegonia.

parole, *sb.* Add: **1. a.** Now generally used for a system of conditional release of selected prisoners before they have completed their sentences. (Further examples.) **b.** (Further example.)

1908 J. M. Sullivan *Criminal Slang* 18 *Parole*, released from prison, not a pardon. **1939** Joyce *Finnegans Wake* 246 So they must have their final since he's on parole. **1966** *Listener* 3 Mar. 301/2 Parole was introduced in the United States—at the Elmira Reformatory in New York—in 1876; today some form of parole is in use in all the fifty States, as well as in the separate Federal prison system. **1972** J. Gores *Dead Skip* (1973) xvi. 116 We go see his parole officer. If he was sent up two years ago and is out now, he's on parole. **1974** *Times* 17 Apr. 14/1 Only a minority of the prisoners are getting a parole. *Ibid.* 14/2 Since the parole system began on April 1, 1968, about 14,000 prisoners have been paroled and about 1,550 are on parole at any one time.

2*. *Linguistics.* With pronunc. (parǫl). [f. *parole* in sense '(spoken) word, utterance'.] The actual linguistic behaviour of individuals, in contrast to the linguistic system (opp. *langue 3).

1935 W. F. Twaddell in *Lang. Monogr.* XVI. 40 The utterance occurs, it is speech, 'parole', the form exists, so to say, it is a part of the language, 'langue'. **1939** L. H. Gray *Foundations of Lang.* ii. 18 The third part of our definition of language..is obviously concerned almost exclusively with *langage* rather than with *langue* or *parole*. **1953**, etc. [see *langue 3]. **1959** W. Baskin tr. *F. de Saussure's Course in Gen. Linguistics* (1960) 13 Execution is always individual, and the individual is always its master: I shall call the executive side *speaking* (*parole*). **1968** J. Lyons *Introd. Theoret. Linguistics* i. 51 Let us follow de Saussure, and say that all those who 'speak English' (or are 'speakers of English') share a particular *langue* and that the set of utterances which they produce when they are 'speaking English' constitute instances of *parole*. **1971** D. Crystal *Linguistics* 162 This leads to the correlative Saussurean concept of *parole*, the actual, concrete act of speaking. **1974** R. Quirk *Linguist & Eng. Lang.* iii. 47 Every individual has a unique *parole*, a unique realization of what is possible in the language of his time and place. But at the same time this is not to deny that the *parole* of some individuals is more interesting than that of others. **1976** *Language* LII. 93 The particular contextual variant of the adjective may vary for each parole application.

3. *parole board, clinic, matron, officer, scheme, sponsor, system*.

[**1908** *Charities & Commons* 26 Sept. 730/2 Clearly the Board of Parole is acting adversely to its own rule.] **1916** *N.Y. Times* 9 Jan. iv. 19/4 There will be weekly meetings ..of the Parole Board. **1938** C. Himes *Black on Black* (1973) 163 He beseeched God to bless the warder and the deputy warder and the chaplain and the guards and the parole board and the outside judges and the governor and the sovereign state itself. **1956** B. Holiday *Lady sings Blues* (1973) xix. 158 If I had been a booster or a petty thief I'd have the parole board helping me to get a job. **1972** 'W. Haggard' *Protectors* v. 63 Jack would be out in around five years, even less if the Parole Board were helpful. **1974** *Times* 17 Apr. 14 Prisoner's cases are considered by the local review committee..and are then referred..to the Parole Board for England and Wales. **1939** *Sun* (Baltimore) 22 Nov. 9/2 The State hospital system will institute an extensive program of parole clinics within the next few weeks to provide a follow-up service for furloughed and discharged patients which authorities believe may tend to curtail readmissions. **1907** *Charities & Commons* 24 Aug. 609/2 Three parole officers and one parole matron have been added to the police department. **1949** *Times-Picayune* (New Orleans) *Mag.* 13 Nov. 23/3 He became chief probation and parole officer for the federal court. **1970** G. Jackson *Let.* Apr. in *Soledad Brother* (1971) 51 Parole officers have sent brothers back to the joint for selling newspapers. **1974** *Guidelines to Volunteer Services* (N.Y. State, Dept. Correctional Services) 36 *Parole officer*, title. A law enforcement officer, peace officer specifically charged with the supervision, and other related duties, of inmates who are released from correctional facilities via parole or some other form of conditional release. They are professional caseworkers in a law enforcement setting. **1973** *Daily Tel.* 20 June 8/6 During the five years of the parole scheme, only 36 of the 11,055 paroled from sentences for crimes of sex or violence had been further convicted of similar offences. **1973** *Philadelphia Inquirer* (Today Suppl.) 7 Oct. 14/3 The Rev. Anthony Velasquez, his parole sponsor, insists that 'Tony has never had a fair trial.' **1900** *Congress. Rec.* 24 Jan. 1130/2 We have in that State what is known as the parole system. Prisoners are put out on their good behavior. **1952** *Manch. Guardian Weekly* 8 May 4 The parole system should be reformed and speeded up. **1973** *Guardian* 31 Jan. 6/1 Lord Hunt..said the parole system had been operating for five years and had shown a low failure rate in terms of recalling of prisoners.

parole, *v.* **2.** (Earlier and later examples.)

1790 D. Fanning *Narrative* (1861) 33, I then parolled the prisoners, except 30, which I sent to Wilmington. **1948** *Chicago Daily News* 27 Feb. 1/6 Another of those parolled..put up $5,000 as a fee. **1973** *Publishers' Weekly* 2 Apr. 61/2 The murders took place in 1934. Mrs. Judd was parolled in 1971. **1974** [see *parole *sb.* 1]. **1975** *Daily Tel.* 3 Apr. 3/4 The two Great Train robbers paroled from jail yesterday had each served more than their legal minimum sentences.

paroled, *ppl. a.* (Earlier and later examples.)

1865 L. N. Boudrye *Hist. Rec. Fifth N.Y. Cavalry* 196 It is remarkable how readily paroled Rebel soldiers affiliate with us. **1908** *Independent* (N.Y.) 16 Jan. 146/2 Of his one thousand and seven or eight paroled men, up to this evening, seventy-seven have fallen. **1925** *Scribner's Mag.* Oct. 410/1 A large proportion of paroled prisoners have been reclaimed from their evil ways by this judicious system. **1966** *Listener* 4 Aug. 155/1 There may be one or more conditions attached to it, and failure to observe them may cause the paroled prisoner to be returned to prison to finish his sentence. **1972** J. Gores *Dead Skip* (1973) xviii. 131 I'm looking for a paroled con named Howard Odum.

parolee (pārǭˢlī·). [f. Parole *sb.* or *v.* + -ee¹.] One who is released on parole.

1916 *Dialect Notes* IV. 327 *Parolee*.., one to whom parole is granted, esp. one sentenced to jail or reformatory or penitentiary who is released 'on parole'. **1934** *Sun* (Baltimore) 3 Dec. 8/3 One of the defects in the system in Baltimore is obvious—lack of supervision of parolees because of inadequate funds. **1937** *Ibid.* 9 Feb. 15/7 (*headline*) Parolee held in murder. **1938** *Yale Rev.* XXVIII. 157 The large majority of parolees are out to-day under such conditions as those which have just been described. **1957** V. Nabokov *Pnin* 46 The Nansen Passport (a kind of parolee's card issued to Russian émigrés). **1964** M. Argyle *Psychol. & Social Probl.* v. 72 Shaw also organized a good deal of local support and sympathy for parolees and ex-convicts. **1973** *Publishers Weekly* 3 Dec. 40/1 In 1971, Long Island University opened its doors to young parolees. The ex-prisoners lived with the other students.. except that they were closely supervised by their parole officer. **1975** *Times Lit. Suppl.* 26 Sept. 1100/2 The process of making parole decisions, and the subsequent supervision of parolees, must be looked at from the viewpoint of all the various participants—prisoners, prison staff, paroling authority and parole supervisors.

parolein (pæ·rǒlīn). *Pharm.* Also **-oleine.** [f. Par(affin *sb.* + L. *ole-um* oil + -in¹, -ine⁵.] = *liquid paraffin* s.v. *liquid *a.* 7.

Formerly a proprietary name in the U.K.

1892 Martindale *Extra Pharmacopœia* (ed. 7) 311 Paroleine, a still more fluid odourless petroleum, or Adepsine Oil..are used..more especially as bases for Laryngeal and Nasal Spray Solutions or Pigments. **1908** *Practitioner* Apr. 441 A parolein spray containing a little oil of eucalyptus. **1964** S. Duke-Elder *Parsons' Dis. Eye* (ed. 14) xv. 184 Local treatment consists in relieving the dryness with methycellulose, parolein or weak alkaline solutions.

paromomycin (pæ:romomai·sin). *Pharm.* [f. Gr. παρόμοιος closely resembling (f. παρ- (see Para-¹) + ὅμοιος resembling): see *-mycin.] A broad-spectrum antibiotic that is a mixture of the sulphates of certain substances (chemically related to neomycin) produced by some strains of the bacterium *Streptomyces rimosus*, and is given orally in the treatment of intestinal infections. Also called *paromomycin sulphate*.

1956 *Brit. Pat.* 797,568 Paromomycin is a stable amorphous white substance which is very soluble in water, moderately soluble in methanol and sparingly soluble in absolute ethanol. **1964** M. Hynes *Med. Bacteriol.* (ed. 8) x. 138 Kanamycin..and paromomycin..are chemically similar to neomycin (and streptomycin) and have much the same properties. **1969** J. H. Thompson in J. A. Bevan *Essent. Pharmacol.* li. 545 Paromomycin sulphate (Humatin) is usually prescribed as 10 to 25 mg per kilogram per day. *Ibid.*, Paromomycin has been used in the treatment of intestinal amebiasis, trichomoniasis, some types of bacterial dysentery, preoperatively to suppress the colonic flora, and in hepatic coma and precoma. **1974** *Indian Jrnl. Med. Res.* LXII. 495 Streptomycin and paromomycin were selected for treatment of cholera carriers at Hongkong.

parotidectomy (pārǫ:tide·ktŏmi). *Surg.* [f. Parotid *sb.* + *-ectomy.] Excision of the parotid gland.

1893 Dunglison *Dict. Med. Sci.* (ed. 21) 825/1 *Parotidectomy*, operation for excision of the parotid gland. **1951** Marshall & Miles in *Surg. Pract. Lahey Clinic* 122 The operation may be altered to extensive subtotal parotidectomy. **1974** *Lancet* 7 Dec. 1353/2 The patient's left parotid gland was explored, found to contain a mixed tumour in the deep lobe and a total parotidectomy was performed. **1974** J. D. Maynard in R. M. Kirk et al. *Surgery* ix. 201 Intraoral duct ligation may lead to parotid atrophy and cessation of symptoms in some cases, otherwise conservative parotidectomy must be considered.

Parousia (pārū·ziă). *Theol.* Also **parousia.** [ad. Gr. παρουσία presence, in N.T. (Matth. xxiv. 27, etc.) used as below.] The Second Coming or Advent of Christ. Also *transf.*

1875 *Expositor* May 385 The feverish expectation of a visible *parousia* was requiring modification. **1895** *Dublin Rev.* Apr. 334 The date of Our Lord's second coming, the Parousia. **1910** W. Montgomery tr. *Schweitzer's Quest Historical Jesus* xix. 360 The Parousia of the Son of Man is to be preceded according to the Messianic dogma by a time of strife and confusion. **1918** J. H. Leckie *World to Come* ii. 66 The Church has held its belief in the Parousia in varying forms throughout the ages. **1927** A. H. McNeile *Introd. New Testament* 112 At the end of his [*sc.* St. Paul's] life, the thought of the Parousia..had practically faded from his mind. **1936** A. M. Ramsey *Gospel & Catholic Ch.* iii. 42 Christ is in us—yet the Parousia is in the future. **1941** Auden *New Year Let.* 39 Thus *Wordsworth*.. Saw in the fall of the Bastille The Parousia of liberty. **1964** *Listener* 27 Feb. 352/1 Your personal Parousia, the Biggest Show on Earth! **1969** A. T. Hanson in A. Richardson *Dict. Christian Theol.* 113/2 The complaint that the *parousia* (coming) had not arrived as expected. **1977** *Illustr. London News* Nov. 50/4 A pale yellow cross is just visible against a background of deeper yellow... Austin Winkley [the architect] calls it a Parousia cross, significant of the mysterious end of a mysterious journey. **1977** G. W. H. Lampe *God as Spirit* iii. 65 The presence of the Spirit is..a substitute for the early *parousia* that had at first been expected.

Hence **Parou:siama·nia,** excitement or frenzy aroused by the thought of the Parousia.

1904 *Amer. Jrnl. Relig. Psychol. & Educ.* May 40 Men chanted, raved, spoke in unknown tongues, prophesied, gazed up into heaven all day, longed for vision, with a real parusiamania.

Parowax (pæ·rowæks). Also **parowax.** [f. *paro-* (f. Para(ffin *sb.*) + Wax *sb.*¹] A proprietary name in the U.S. for paraffin wax.

1909 *Official Gaz.* (U.S. Patent Office) 7 Dec. 301/2 Standard Oil Company... *Parowax*. Particular description of goods.—Paraffin. **1926** *Daily Colonist* (Victoria, B.C.) 4 July 2/1 (Advt.), Parowax, for sealing, lb. 15c. **1945** *Industr. & Engin. Chem.* June 518/2 To extend the series of *n*-paraffins into the range of higher paraffins, Parowax was used. Fractional crystallization..showed the composition to be substantially normal paraffins... The paraffins ranged from C_{20} to C_{34}, with an average number of 23·8 carbon atoms. **1974** J. E. Underhill *Wild Berries Pacific Northwest* 23 Cook until thick, then pour into hot sterilized jars, and seal promptly, with parowax or tight lids.

paroxytonization (parǫ:ksitǒnaizē·ʃən). [f. Paroxytonize *v.* + -ation.] The rule which places stress on the penultimate syllable.

1973 A. H. Sommerstein *Sound Pattern Anc. Greek* v. 177 This rule..changes all properispomenon words ending in *-ks* to paroxytones: (221) Paroxytonization. [+sharp] →[-falling]/—C₀Vks[+WB.].

parp (pāɹp), *sb.* [Echoic.] A honking noise, *spec.* that of a car horn. Hence as *v. intr.*, to make such a noise; also *trans.*

1951 E. Blyton *Big Noddy Bk.* 51 Noddy had a good little hooter on his car. When he pressed it it said 'Pip-Pip', and sometimes 'Poop-Poop', and sometimes 'Parp-Parp'. **1953** —— *Noddy at Seaside* 14 'Parp-parp-parp!' Yes, that was Noddy's little car, longing to be taken out into the sunshine for a drive. *Ibid.* 16 'Parp-parp!' said the little car excitedly. **1958** [see *bleep *sb.*]. **1968** S. Challis *Death on Quiet Beach* v. 69 Outside his hotel, a horn suddenly parped. **1973** M. Amis *Rachel Papers* 124 To break her reverie I parped the horn.

parquet. Add: (Now usu. with pronunc. pā·ɹkēⁱ.) **4.** *parquet floor* (earlier examples); **parquet carpet,** a patterned square of carpet.

1902 Parquet carpet [see *art square* s.v. *art *sb.* V]. **1819** M. Wilmot *More Lett.* (1935) 20 Fruit and flowers dirt cheap—parqué flours—*Carpits* if you chuse to give a daughters dowery for them. **1833** Princess Elizabeth *Let.* 28 Dec. (1898) 213 And the doctor said with those shoes she might now walk about the parquet floors. **1865** M. Eyre *Lady's Walks S. of France* xxxi. 330 The *parquet* floors are undusted, unwaxed, and unswept.

parrot, *sb.* Add: **4.** *parrot-cry* (earlier and later examples), *-learning, -phrase* (examples), *-pie, -shooting, -voice*; *parrot-bright, -plumed, -sharp* adjs.; *parrot-fashion* adj. and adv.; **parrot-bill,** (*c*) applied *attrib.* and *absol.* to a type of cutting-tool the blades of which resemble a beak; **parrot disease, fever** = Psittacosis in Dict. and Suppl.; **parrot snake** (see quot. 1931); **parrot tulip,** substitute for def.: a variety of tulip with fringed and ruffled petals, often of variegated colours; (earlier and later examples).

1971 *Power Farming* Mar. 46/4 Stem cutting...parrot bill shears may be used. **1972** *Country Life* 23 Mar. 690/3 After this major excitement a willow lapses into humdrummery and can be lopped with the parrot-bill. **1920** E. Sitwell *Wooden Pegasus* 24 From her fan, sliding slow, Parrot-bright fire's feathers. **1937** —— *I live under Black Sun* 89 Giving a little girl a forbidden parrot-bright apple. **1837** J. S. Mill in *Westm. Rev.* XXVIII. 3 There would be an end to the parrot cry of 'Do not endanger the Ministry'. **1956** [see *eyebrow 1 c]. **1963** *Times* 20 Feb. 4/7 Such attacks are only worth noticing because they have tended to become a parrot cry. **1977** *Socialist Press* 2 Mar. 7/5 'If you have a case, why don't you go to an industrial tribunal', has become the favourite parrot cry of every barrack room lawyer crossing the picket line. **1908** *Spratt's Parrot Culture* 29 Should a room have become infected with the parrot disease..it will be needful to have it fumigated with sulphur. **1930** *Daily Express* 6 Feb. 11/5 They [*sc.* alarming facts] concerned that dread illness, psittacosis, or parrot disease, of which a number of cases have occurred lately in London and Birmingham. **1955** *Times* 8 June 6/4 A case of psittacosis (parrot disease) has occurred in the aircraft-carrier Centaur, which berthed here to-day on her return from the Mediterranean. **1951** *Mind* LX. 346 People just know it by heart and recite it parrot-fashion. **1956** D. Abercrombie *Probl. & Princ.* 25 Parrot-fashion teaching is apt to result from regarding reasoned explanation as 'unnatural'. **1977** 'F. Clifford' *Ten Minutes on June Morning* 111 Reassurances..were passed on, parrot-fashion, without knowledge or understanding. **1955** *Sci. News Let.* 3 Sept. 148/2 Viral hepatitis, better known to the layman as jaundice; psittacosis or parrot fever; rabies; smallpox; yellow fever; the common cold,..are other of the virus diseases. **1957** O. Breland *Animal Friends & Foes* ii. 63 This malady has also been called parrot fever, because the first known human cases were traced to sick parrots. **1973** 'D. Shannon' *No Holiday for Crime* (1974) i. 9 The stolen goods had been..tropical parrots, and..one of the San Diego detectives had subsequently succumbed to parrot fever. **1901** G. G. Coulton *Public Schools & Public Needs* 312 We cannot prevent..mere parrot-learning, from counting somewhat..against real culture. **1977** *Observer* 20 Mar. 13/3 There seem to be two potent reasons why memorising (or 'parrot-learning', as people ignorant of the mental capacities of parrots sometimes call it) should once more become a staple component of curricula. **1891** O. Wilde *Pict. Dorian Gray* v. 89 'Foolish child! foolish child!' was the parrot-phrase flung in answer. **1958** *People* 4 May 4/2, I can get no comment except the parrot-phrase: 'The Home Secretary is still considering this case.' **1907** P. Fountain *Rambles Austral. Naturalist* ii. 8 Parrot-shooting is a favourite sport in Australia. *Ibid.*, Parrot-pie is as much esteemed in Australia as rook-pie in England. **1923** E. Sitwell *Bucolic Comedies* 25 Who came from the parrot-plumed sea. **1936** —— *Victoria of Eng.* ii. 33 Her dark parrot-sharp face. **1931** R. L. Ditmars *Snakes of World* Pl. 19 (*caption*) Green Tree Snake; Chocoya or Parrot Snake, *Leptophis occidentalis*. Found from Guatemala to northern South America. Uniform leaf-green with two hair-like strips on the back. **1958** J. Carew *Wild Coast* ii. 28 A green parrot-snake slithered down a coconut tree. [**1829** J. C. Loudon *Encycl. Plants* 266 One of the latest London catalogues (Mason's) enumerates six sorts of early blowing tulips; four perroquets or middle blowers; twenty-two double sorts.] **1856** C. M. Yonge *Daisy Chain* ii. xxi. 586 She was nothing better than a parrot-tulip, stuck up in a parterre. **1890** O. Wilde *Pict. Dorian Gray* iii, in *Lippincott's Monthly Mag.* July 22 Some large blue china jars, filled with parrot-tulips, were ranged on the mantel-shelf. **1911** J. Weathers *Bulb Bk.* 441/2 Parrot or Dragon Tulips. These curious-looking and remarkable Tulips are believed to be derived from *T[ulipa] viridiflora*. **1932** A. J. Macself *Bulbs* v. 58 The Cottage, Darwin, and Parrot tulips..require similar general treatment. **1971** R. Genders *Collecting Antique Plants* viii. 191 Early Parrot tulips are depicted in a water colour drawing about 1700... The artist is Herman Henstenburg. **1925** R. Fry *Let.* 1 May (1972) II. 568 A lady with a parrot voice screaming that she wanted a Picasso of the blue period. **1975** C. Fremlin *Long Shadow* v. 37 'Very well, thank you,' she heard her parrot voice saying..to the two or three people who rang up.

parrot, *v.* Add: **1.** (Later example.)

1970 C. Hampton *Philanthropist* iv. 49 Will you please stop parroting on about breakfast?

2. (Later examples.)

1965 *Austral. Women's Weekly* 20 Jan. 48/1 'I'll wait,' he said. 'He'll wait,' she parroted. **1968** *Language* XLIV. 204 School textbooks..had simply parroted a series of rules. **1971** *Nature* 13 Aug. 456/2 Thus a child who produces the correct response when asked the AC question may do so by parroting a verbal label picked up during the initial comparisons. **1976** *Amer. Speech 1973* XLVIII. 259 She quickly muddies the water when she parrots the creed that all languages and dialects are equally fit to express 'concepts such as time or relativity'. **1977** *Spare Rib* Jan. 11/1 The catechism, which we had to be able to parrot, went into the different sorts of sins at great length.

parroting *vbl. sb.* (later examples.)

1934 E. Pound *Let.* 30 Dec. (1971) 263 Mr. Croft seems to me an idiot... His kind of parroting seems to me exactly what does keep people from studying the classics. **1951** G. Humphrey *Thinking* iii. 223 Much 'parroting' is not entirely meaningless. **1971** *Daily Tel.* 1 Nov. 5/7 Rote learning—learning something by constant repetition or 'parroting'—free recall and other memory tests were included.

pa·rroted, *ppl. a.* [f. Parrot *v.* + -ed¹.] That is repeated mechanically in the manner of a parrot.

1927 M. Sadleir *Trollope: a Comm.* 295 Wherever he appears as..waverer from their parrotted idealisms, Sir Thomas Underwood is Trollope himself. **1966** *Listener* 30 June 949/1 This romanticism is probably a reaction from thousands of inane, parroted essays. **1969** *Daily Tel.* 5 Sept. 17 Underpaid teachers whose mission it was to hammer into my brain..the parroted conjugations of Latin verbs.

parrot-house. A building in a zoological garden in which parrots are kept; freq. in *transf.* or *fig.* use *esp.* with reference to loud or raucous noise.

1872 Geo. Eliot *Middlem.* I. ii. xv. 251 Our chat would be thin and eager, as if delivered from a camp-stool in a parrot-house. **1923** H. C. Bailey *Mr Fortune's Practice* vii. 186 It is an old-fashioned orphanage..as noisy as the parrot house. **1929** J. Buchan *Courts of Morning* i. 35 Yanqui youth..is chronically alcoholic and amorous, and its manners are a brilliant copy of the parrot-house. **1945** E. Waugh *Brideshead Revisited* ii. i. 216 The parrot-house fever of my wife's party. **1959** *Times* 2 Mar. 3/3 Without Mias to coordinate them, there was the tell-tale, parrothouse chatter of half a dozen leaders. **1967** 'S. Woods' *And shame Devil* 126 A clamour in many ways reminiscent of the Parrot House at the Zoo. **1977** 'M. Underwood' *Murder with Malice* xxiii. 209 The court-room became a parrot-house of chatter as people pushed towards the exits.

parse, *v.* Add: **a.** In extended use in computational linguistics, to analyse (a string) into syntactic components to test its conformability to a given grammar.

1962 J. J. Robinson *Prelim. Codes & Rules Automatic Parsing of English* (Rand Corp. Memo. RM-339-PR) p. v, This Memorandum presents a set of grammar codes and rules for analyzing, or 'parsing', English sentences automatically on a digital computer. *Ibid.* iii. 34 All the words in the string being parsed have been accounted for. **1963** *Communications Assoc. Computing Machinery* VI. 669/1 It will parse strings describable in essentially Backus Normal Form. **1967** D. G. Hays *Introd. Computational Linguistics* viii. 148 First, we transcribe the grammar, omitting all context restrictions. Obviously, the new grammar parses every string acceptable to the old one and if the restrictions are not vacuous it either assigns extra structures to some acceptable strings or accepts some additional strings, and may do both. **1975** J. S. Rohl *Introd. Compiler Writing* xiii. 226 For syntax analysis it is convenient to have the definitions in an analytic form, so that we can analyse or parse a string of characters to see whether they conform to the grammar.

b. *fig.* (Later example.)

1965 P. Kael *I lost it at Movies* 9 A movie had to tell some kind of story that held together: a plot had to parse.

e. *transf.* To examine or analyse minutely.

1788 F. Grose *Rules for drawing Caricaturas* 14 When a caricaturist wishes to delineate any face..he may commit it to his memory, by parsing it in his mind (as the school-boys term it). **1860** *Leisure Hour* 9 Aug. 507/2 Let him soak and remove the leather covering, parsing his way, as it were, by minute examination. **1931** *Times Lit. Suppl.* 7 May 353/3 Reade's biographer is confronted with the necessity of, as it were, 'parsing' a character which.. does not make sense. **1962** P. Tompkins *Spy in Rome* xxxi. 307 Franco spoke Italian with a slightly foreign (or aristocratic) accent—depending on which way the listener chose to parse it.

Hence **parsed** *ppl. a.*; **parsing** *vbl. sb.* (further examples); also *attrib.* or as *ppl. a.*

1962 J. J. Robinson *Prelim. Codes & Rules Automatic Parsing of English* (Rand Corp. Memo. RM-339-PR) i. 2 The English codes and parsing rules being developed at RAND are essentially a machine grammar. **1963** *Communications Assoc. Computing Machinery* VI. 669/1 The automatic parsing algorithms..simplify compiler construction but contribute little to the production of 'optimized' machine code, for example. **1964** *Ibid.* VII. 131/2 The right side of the algorithm..in effect runs the input string backwards..until the tentatively parsed word is reached. *Ibid.*, An example of a syntactically ambiguous phrase is 'medical schools and hospitals of Boston' which actually has five possible parsings: ((MS)AH)OB; (MS)A(HOB); (M(SAH))OB; M((SAH)OB); M(SA(HOB)). **1967** D. G. Hays *Introd. Computational Linguistics* vi. 107 A systematic answer to questions like this, determining the exact sequence in which reductions will be made to any particular string, is the basis for a parsing strategy. *Ibid.* 114 Only complete parsing of a string reveals that any part of a tentative parsing is correct. **1971** E. Wilson in R. A. Wisbey *Computer in Lit. & Ling. Res.* v. 210 This definition it attempts to transform into an equivalent definition which can be parsed by a one-track algorithm. If it succeeds, SID can be made to generate this parsing algorithm. **1972** J. A. N. Lee *Computer Semantics* v. 273 The form of text which is generated by the analyzer, we shall name the parsed text, which may take the form of a syntactic tree or a phrase marked string.

parse (pāɹz, -s), *sb.* [f. the vb.] The action or result of parsing.

1963 *Communications Assoc. Computing Machinery* VI. 670/1 In the algorithm presented here, all possible parses are carried along as shown below in the progressing parse of *abce* according to the syntax of the earlier example. When the symbol *c* is encountered, Parse 1 cannot be continued and is dropped, leaving Parse 2 as the correct one. **1973** W. M. Waite *Implementing Software for Non-Numeric Appl.* viii. 292 Both pattern-directed and string-directed scans have been used for parsing algorithms. If a pattern-directed scan is used, the procedure is known as a 'top-to-bottom' or 'top-down' parse; a string-directed scan yields a 'bottom-to-top' or 'bottom-up' parse.

parsec (pā·ɹsek). *Astr.* [f. Par(allax + Sec(ond *sb.*¹] A unit of length equal to the distance at which a star would have a heliocentric parallax of one second of arc, viz. $3{\cdot}09 \times 10^{16}$ metres ($19{\cdot}2 \times 10^{12}$ miles, 3·26 light-years), approximately.

1913 F. W. Dyson in *Monthly Notices R. Astron. Soc.* LXXIII. 342 There is need for a name for this unit of distance... Professor Turner suggests *Parsec*, which may be taken as an abbreviated form of 'a distance corresponding to a parallax of one second'. **1921** *Glasgow Herald* 11 July 4 He estimates the distance of the object as 140 parsecs or four times the distance of the Hyades cluster. **1955** *Sci. News Let.* 29 Jan. 71/2 These hydrogen clouds have diameters of several parsecs, one parsec being the distance light travels in 3·26 years. **1962** F. I. Ordway et al. *Basic Astronautics* vi. 290 Within 5 parsecs (16·5 light years) of the Sun, 53 individual stars have been counted, the nearest of which is Proxima Centauri, 25 trillion miles away. **1974** S. V. M. Clube in R. H. Stoy *Everyman's Astron.* viii. 309 The nearest star clusters are the Pleiades and Hyades which are at distances of 130 and 40 parsecs respectively and therefore beyond the reach of trigonometric parallax determinations. **1977** *Time* 30 May 42/2 The four of them are even now setting out to deliver the secret plans to rebel headquarters, light-years and parsecs away.

‖ **parsemé** (pārsĕme), *a.* [Fr., pa. pple. of *parsemer* to sprinkle, strew f. L. *per-* through + *sēmināre* to sow.] Sprinkled or strewn (with); used *esp.* in embroidery with reference to the decoration of fabrics and costumes.

1814 M. Birkbeck *Notes Journey through France* App. 14 The numerous longitudinal ridges..with which this charming country is, 'parsemé', appear to be the venerable remains of the ancient surface. **1832** F. Trollope *Dom. Manners Amer.* I. vi. 74 She was preparing to set to work in a yellow dress parsemé with red roses... I thought it was a pity to spoil so fine a gown. **1883** H. C. Dent *Jrnl.* 9 June in *Year in Brazil* (1886) i. 4 The country on the north bank has a low coastline, *parsemé* with red-roofed white houses. **1890** O. Wilde *Pict. Dorian Gray* ix, in *Lippincott's Monthly Mag.* July 71 A skull-cap *parsemé* with pearls. **1905** *Athenæum* 17 June 760 The rise of a new conception of design, with figures and ornaments parsemé on a velvet ground,..in the fifteenth century.

parser. Add: Also, a computer program for parsing.

1965 *Communications Assoc. Computing Machinery* VIII. 688/1 In testing the program on 300 basic English sentences Knowlton's parser produced 137 correct parsings and 44 incorrect ones. **1967** D. G. Hays *Introd. Computational Linguistics* vi. 116 Whereas the other parsers considered..have generally operated more or less from left to right on the string, or from right to left, this one operates by choosing..the top of the dependency structure, and working downward. **1970** *Computers & Humanities* V. 25 More computational parsers are being written following the generative-transformational model than any other. **1977** E. von Glasersfeld in D. M. Rumbaugh *Lang. Learning by Chimpanzee* v. 121 The parser, however, can handle a lexicon of 250 items.

parsettensite (pāɹse·tĕnzait). *Min.* [ad. G. *parsettensit* (J. Jakob 1923, in *Schweiz. min. und petrogr. Mitt.* III. 227), f. Alp *Parsettens*, name of the locality in Val d'Err, Graubünden, E. Switzerland where it was first found: see -ite¹ 2 b.] A basic silicate of manganese that often contains appreciable potassium and occurs as copper-red masses.

1924 *Mineral. Abstr.* II. 251 Parsettensite..occurs as a filling in rather thick veins. **1962** W. A. Deer et al. *Rock-forming Min.* III. 103 Parsettensite..was shown by Fankuchen..to be a manganese stilpnomelane, and it is proposed to retain this name to describe varieties rich in manganese. **1968** I. Kostov *Mineral.* 362 Transitional towards the vermiculite group are stilpnomelane, ekmanite, and parsettensite, with formulas similar to those of the illites but as a rule without aluminium in tetrahedral coordination.

Parseval (pā·ɹsĕvăl). Also **Parsefal.** [The name of the inventor, the German engineer August von *Parseval* (1861–1942).] A type of non-rigid dirigible airship formerly in use in Germany.

[**1908** W. H. Story tr. *Hildebrandt's Airships* viii. 85 (*caption*) Major Parseval's dirigible balloon.] **1909** *Chambers's Jrnl.* Oct. 660/2 At present the airship fleet consists of three Zeppelins, three Parsefals, and two Gross dirigibles. *Ibid.*, It is of the Parsefal type. **1910** C. C. Turner *Aerial Navig. To-day* xxiv. 295 The 'Parseval' is a non-rigid balloon with a cubic capacity of 190,000 cubic feet... It is rounded at the front and pointed at the rear. **1931** C. St. J. Sprigg *Airship* viii. 106 In its final form the Parseval was not much smaller than the early Zeppelins, and was of a good streamline shape. *Ibid.* 110 Two of the Parseval type were built by Messrs. Vickers under licence. **1957** *Encycl. Brit.* I. 465/1 Maj. August von Parseval in 1906 constructed for the German army the first in a series of some 28 Parseval pressure type airships which were built from that date until 1929. **1959** J. A. Sinclair *Famous Airships of World* iii. 30 In Germany, Major von Parseval..in 1911 constructed the *Parseval No. 3*... The steel tube car carried 12 passengers. **1971** R. Jackson *Airships* v. 112 On the outbreak of war, the Astra-Torres and the Parseval were immediately allocated to coastal patrol duties.

parsimony. Add: **c.** Also, the principle that organisms tend towards economy of action in learning or in fulfilling their needs.

1931 D. K. Adams in *Brit. Jrnl. Psychol.* XXII. 153 This economy upon repetition or, better, the property (we shall call it parsimony) of which it is simply one manifestation, is a fundamental property of a certain class of bodies. *Ibid.*, I think that parsimony is a property of all organisms. **1948** E. R. Hilgard *Theories of Learning* xi. 295 The process of need satiation is regulated by a principle called 'parsimony'. That is a preference for short-cuts, described by others as the principle of least action. **1955** *Sci. Amer.* June 68/1 This is the grand overriding law of the parsimony of nature: every action within a system is executed with the least possible expenditure of energy.

d. (Further examples.) Also *parsimony, principle of parsimony*.

1933 J. C. Flügel *Hundred Years Psychol.* ii. 124 The 'law of parsimony', according to which we must always explain animal behaviour in terms of the simplest mental processes that will account for the facts. **1957** R. K. Merton *Social Theory* (rev. ed.) viii. 259 The theoretical objective of *parsimony*, found whenever several empirical generalizations are derived from a more general formulation. **1970** M. H. Marx *Learning: Theories* I. i. 16/2 The principle of parsimony, often called William of Occam's razor or Lloyd Morgan's canon,..is a rough guideline to the acceptability of hypotheses and principles. *Ibid.* 17/1 The failure to accept the principle of parsimony results in the overloading of relatively untested..ideas. **1972** *Encycl. Psychol.* I. 202/2 The essential similarity between classical and operant conditioning has led students to pose questions about the parsimony or necessity of more than one principle to account for this type of learning.

pars intermedia (pɑːz intəɹmī·diă). *Anat.* [mod.L., = 'middle part'.] A layer of tissue in the hypophysis between the anterior and posterior lobes (sometimes regarded as a part of the anterior lobe).

1908 P. T. Herring in *Q. Jrnl. Exper. Physiol.* I. 132 It is convenient to consider as the anterior lobe only that portion of it which has already been distinguished from the 'pars intermedia'. **1912** H. Cushing *Pituitary Body & its Disorders* I. 4 The posterior lobe comprises the pars nervosa..and its epithelial investment (pars intermedia of Herring). **1926**, etc. [see *neuro-intermediate* adj. s.v. *Neuro-]. **1932** [see *Intermedin]. **1962** E. C. Crosby et al. *Correl. Anat. Nerv. Syst.* vi. 323/2 From Rathke's pouch, an embryonic diverticulum from the roof of the stomadeum, are derived the pars distalis, pars intermedia, and pars tuberalis. **1968** W. C. Bowman et al. *Textbk. Pharmacol.* xiv. 365 The hormone of the pars intermedia (the melanocyte stimulating hormone, MSH) stimulates dispersion of the melanin granules in melanocytes of the skin of fish, amphibia and reptiles, with a resulting darkening of the skin.

parsley. Add: **3.** *parsley-dark* adj.; **parsley butterfly** *U.S.*, the black swallowtail butterfly, *Papilio polyxenes asterius*; **parsley caterpillar** *U.S.*, the larva of the anise swallowtail butterfly, *Papilio zelicaon*, which is a pest of umbelliferous plants in western North America; **parsley frog**, a spadefoot toad, *Pelodytes punctatus*, found in western Europe; **parsley green**, a colouring additive used in cookery; **parsley-leaved elder**, a cultivated variety of the elder, *Sambucus nigra* var. *laciniata*, distinguished by its cut leaves; **parsley sauce**, a white sauce flavoured with parsley; **parsley-worm** *U.S.*, the larva of the parsley butterfly, which is a pest of umbelliferous plants.

1889 S. H. Scudder *Butterflies Eastern U.S.* II. 1353 Papilio Polyxenes.—The black swallow-tail... Parsley butterfly (Emmons). **1926** E. O. Essig *Insects Western N. Amer.* xxvii. 634 The western parsley caterpillar, *Papilio zelicaon*,..is yellow or orange and black. **1962** Metcalf & Flint *Destructive & Useful Insects* (ed. 4) xiv. 598 In the West it [sc. the parsleyworm] is replaced by the western parsley caterpillar. **1920** E. Sitwell *Wooden Pegasus* 31 Face as white as any clock's Cased in parsley-dark curled locks. **1897** *Proc. Zool. Soc.* 577 (*title*) On the structure and development of the hypobranchial skeleton of the Parsley-Frog. **1934** J. Fletcher tr. *Rostand's Toads* vii. 72 In a pond near Paris..the larvae of the Common Frog were in the North and West;..those of the Parsley Frog, the South-east. **1960** R. Mertens *World of Amphibians & Reptiles* ii. 34 The parsley frog (*Pelodytes*)..occurs only in south-western Europe and the Caucasus. **1845** E. Acton *Mod. Cookery* iv. 151 (*heading*) Parsley green, for colouring sauces. Gather a quantity of young parsley,..pound it in a mortar,..set it into a pan of boiling water. **1731** P. Miller *Gardeners Dict.* s.v. Sambucus, The Cut or Parsley-leav'd Elder. **1838** J. C. Loudon *Aboretum* II. 1028 The Parsley-leaved Elder; has the leaflets cut into fine segments. **1904** E. Step *Wayside & Woodland Trees* 125 An Elder with its leaflets deeply cut into very slender lobes...is an escape from cultivation—a garden variety (*laciniata*) known as the Cut-leaved or Parsley-leaved Elder. **1836** E. Copley *Cook's Compl. Guide* II. v. 359 *Parsley Sauce.* Boil a bunch of green parsley in salt and water for five minutes; when done, chop it fine, put in half a pint of bechamel sauce, or good melted butter. **1877** E. S. Dallas *Kettner's Bk. of Table* 329 *Parsley Sauce*..is generally in England given to what in the French kitchen is known as maître d'hôtel sauce. **1965** R. Carrier *Cookbk.* ii. 90 (*heading*) English parsley sauce. **1978** *Listener* 23 Mar. 366/1, I've just dialled for the weather... There was a slight crossed line. The outlook is warm with parsley sauce, Regulo 7. **1842** T. W. Harris *Insects Injurious to Vegetation* 211 In the month of June, there may be found, on the leaves of the parsley and carrot, certain caterpillars, more commonly called parsley-worms. **1972** Swan & Papp *Common Insects N. Amer.* 204 Black Swallowtail (Parsleyworm). *Papilio polyxenes asterius.*

parsnip. Add: **1. c.** In various colloq. or slang expressions: *before you can say parsnips*, very rapidly, 'in the twinkling of an eye'; *to look parsnips*, to look sour or displeased; *I beg your parsnips*, joc. alteration of 'I beg your pardon'.

1803 G. Colman *John Bull* II. ii. 18 'You'll come back again,' says she—'That's what I will, before you can say parsnips, my darling,' says he. **1837** J. Hook *Jack Brag* II. iv. 167 'I'm delighted,' said Jack, looking parsnips. **1886** H. Baumann *Londinismen* 231/2 *Parsnip*,..I beg ~s (statt pardon) bitt' um Entschuldigung. **1922** Joyce *Ulysses* 297 Who said Christ is good?—I beg your parsnips, says Alf.—Is that a good chap?, says Bob Doran, to take away poor little Willy Dignam?

3. *parsnip soup*; **parsnip butterfly, swallowtail** = *parsley butterfly* s.v. *Parsley 3; **parsnip web-worm** *N. Amer.*, the larva of a moth, *Depressaria pastinacella*, a pest of parsnips and related plants.

1867 *Amer. Naturalist* I. 220 The Parsnip butterfly (*Papilio Asterias*) may now be seen flying over the beds of parsnips. **1845** E. Acton *Mod. Cookery* i. 25 (*heading*) Parsnip soup. Dissolve,..four ounces and a half of good butter,..and slice in directly two pounds of sweet tender parsnips. **1943** E. O. Essig *College Entomol.* xxxi. 503 *Papilio ajax* Linn. Parsnip swallowtail. **1888** *Insect Life* I. 94 The Parsnip Web-worm... We found this insect extremely common in the stems of Wild Parsnips. **1928** Metcalf & Flint *Destructive & Useful Insects* xvi. 509 Parsnip Webworm... The flower heads of parsnip and celery are webbed together with silk and devoured by small yellow, greenish or grayish caterpillars covered with small black spots. **1954** Borror & DeLong *Introd. Study Insects* xxvi. 536 The parsnip webworm..attacks parsnips, celery, and related plants.

parson. Add: **6. d. parson's table** *U.S.*, a small, simple, wooden table with a square top supported at each corner by straight legs.

1969 *Sears Catal.* Spring/Summer 1385/1 Parson's Tables. Avocado green. **1971** E. Hayes *Hungarian Game* xlvi. 271 Except for a matched set of Sheraton chairs, a salon mirror and a parson's table, the landing was empty. **1976** *Billings* (Montana) *Gaz.* 24 June 7-F/6 (Advt.), 16" parsons tables.

Parsonian (pɑːsǝʊ·niǎn), *a. Sociol.* [f. the name of Talcott Parsons (1902–), Amer. sociologist + -ian.] Of or relating to the theories of action and change within a society or culture put forward by Parsons, or to his structural-functional method of analysing a social system. Hence **Parso·nianism**, the views or theories of Parsons.

1961 A. Hacker in M. Black *Social Theories of Talcott Parsons* 298 One obstacle to a Parsonian theory of class and power may not be easy to overcome. **1970** Touraine & Pécaut in I. L. Horowitz *Masses in Lat. Amer.* iii. 68 If the Parsonian categories are to be directly applied, it should be possible to define a system of values. **1973** J. Rex *Discovering Sociol.* iii. 117 As Parsonianism developed,.. **1974** R. Jessop *Traditionalism, Conservatism & Brit. Polit. Culture* i. 23 In addition to the well-known Parsonian pattern-variables, there are others developed specifically for political analysis and relevant to the problems in hand. **1977** *Times Lit. Suppl.* 25 Feb. 198/3 The family thus functioned, in the best Parsonian fashion, as the principal agent of 'socialization'. *Ibid.*, Mr. Gutman never does tell us what he means by 'socialization', and his implicit Parsonianism hardly does justice to Talcott Parsons.

parsonical, *a.* (Later examples.)

1902 *Irish Rosary* VI. 771 His [sc. Herrick's] verse..is on the whole more poetical than parsonical. **1972** *Times* 13 Oct. 3/6, I never wear a parsonical grey suit since one asks.. **1976** *Church Times* 5 Nov. 3/6 Since the advent of the tape-recorder has one hopes, eradicated the worst aberrations of the parsonical voice.

parsonify, *v.* (In Dict. s.v. Parson). Add: (Further examples.) Also, to make into a parson.

1737 J. Thomson *Let.* 11 Jan. in W. Goodhugh *Eng. Gentl. Libr. Man.* (1827) 261, I have not yet seen the round man of God to be. He is to be parsonified a few days hence. **1798** tr. *William III's Early Life* iii. 19 He..left all dogmas and creeds severely alone. They were, in his view,..apt to 'parsonify' the grand and simple outline of the Christian Faith.

parsonsite (pɑːsǝnzait). *Min.* [a. F. *parsonsite* (A. Schoep 1923, in *Compt. Rend.* CLXXVI. 173), f. the name of Arthur L. *Parsons* (b. 1873), Canadian mineralogist: see -ite¹ 2 b.] A hydrated phosphate of lead and uranium, $Pb_2(UO_2)(PO_4)_2.2H_2O$, found as crusts and powdery aggregates of transparent or translucent, minute, lath-like crystals.

1923 *Chem. Abstr.* XVII. 1776 (*heading*) Parsonsite, a new radioactive mineral. **1950** *Amer. Mineralogist* XXXV. 247 Parsonsite occurs at the Ruggles pegmatite near Grafton Center, Grafton County, New Hampshire, as crusts of microscopic spicular or lath-like crystals... The parsonsite is associated with autunite and phosphuranylite and all of these have ultimately been derived from the alteration of the primary uraninite. **1964** *Bull. U.S. Geol. Survey* No. 1064. 233 Parsonsite probably is monoclinic, but the possibility of triclinic symmetry cannot be ruled out.

‖ **pars pro toto** (pɑːz prǝʊ tǝʊ·to). [L., = 'a part for the whole'.] A part considered as representative of the whole. Also *attrib.*

1702 [see Synecdochical *a.*]. **1958** W. Stark *Sociol. of Knowl.* 156 The fallacy of *pars pro toto*—taking the part for the whole. **1965** *Eng. Stud.* XLVI. 55 It is a *pars pro toto* figure, just as *rand* is used to refer not only to the metal border or ring round the wooden board but also to the shield as such. **1970** *Jrnl. Gen. Psychol.* LXXXIII. 66 The tendency to combine areas of questionable form into strange or absurd entities or the *pars pro toto* effect.

part, *sb.* (*adv.*) Add: **I. 2. d.** Each of the separate or separable pieces that go to make up a machine or the like. Also *attrib.* in *pl.*

1860 D. Clerk *Gas Engine* 5 Fig. 3 is a sectional elevation of the first engine, showing the principal working parts. *Ibid.* 8 Barnett's second engine..is double-acting, and therefore requires a greater number of parts. **1890** W. Robinson *Gas & Petroleum Engines* ii. 11 The wearing parts can be easily taken out and, when worn, replaced by duplicates. **1897** *Trans. Inst. Naval Archit.* XXXVIII. 217 Set of accessories, spare parts, and securing gear. **1923** *Radio Times* 28 Sept. 6 (Advt.), The user of the 'Geco-phone'..may desire to purchase in his own locality spares and replacement parts. **1939** G. W. Stubbings *Diseases Electr. Machinery* iii. 60 A.C. transformers..are entirely static and have no moving parts. **1939** H. R. Simonds *Industr. Plastics* (1940) ix. 144 The average household refrigerator has more than 20 important plastics parts. **1947** W. W. McCullough *Electr. Motor Maintenance* ii. 13 Worn parts should be replaced promptly. **1968** *Amer. Speech* 1967 XLII. 40 Owner's manuals, parts catalogues, motoring publications from abroad. **1971** *Good Motoring* Sept. 4/1 It was driven by Billy Mackay, a VW Motor parts manager in Scotland. **1974** *Daily Tel.* 4 Dec. 12/3 On average, parts for a Renault 12TL are 77 per cent. more expensive than for the Austin Allegro 1300. **1975** *Sci. Amer.* Feb. 25/2 Industry mass-produces parts in large numbers, each of which carries its own unique part serial number. **1976** *Nature* 1 Apr. 391/3 Two fibres, called Xerxiar and Xerxiar 49 by Dupont,..look likely to be used in tyre belting and in composites for body armour and aircraft parts.

II. 12. (Later *absol.* (*euphem.*) examples.)

1942 Berrey & Van den Bark *Amer. Thes. Slang* §121/37 Genitals,..parts. **1958** S. A. Grau *Hard Blue Sky* 152 The young girls giggled and felt a hot touch in their parts. **1977** J. Le Carré *Hon. Schoolboy* ix. 41 The devil's red-hot wind would burn his parts to a frazzle.

IV. 17. (Earlier and later examples.)

1871 'Mark Twain' in *Galaxy* Aug. 284/1 He..brushed his hair with elaborate care,..accomplishing an accurate 'part' behind. **1933** J. Steinbeck *To God Unknown* (1935) viii. 61 He raked a nervous hand through his hair and destroyed the careful part. **1970** *Globe & Mail* (Toronto) 20 Sept. 10/1 (Advt.), Balding? So you moved your part down over your ear. **1972** D. Ramsay *Little Murder Music* 141 A shoulder-length fall of blue-black hair divided by a snow-white part in the centre. **1976** 'R. Macdonald' *Blue Hammer* xxiv. 126 The part in her hair was white and straight.

V. 19. a. A part of speech: also (with hyphens) *attrib.*

1933 L. Bloomfield *Language* i. 17 Even the fundamental features of Indo-European grammar, such as, especially, the part-of-speech system, are by no means universal in human speech. **1964** *Language* XL. 167 One-syllable words,..have the same part-of-speech..

VI. 29. part-playing, playing in parts (sense 10 in Dict.); **part-score** *Bridge* (see quot. 1936); **part-whole** *a.*, of or pertaining to the relationship of a part or parts to a whole; **part-work**, (*a*) used *attrib.* to designate a system of part-time work; (*b*) a book or the like published in parts; also *attrib.*

1946 R. Blesh *Shining Trumpets* (1949) i. 18 *Tempo*: I. Strict tempo or controlled acceleration. II. Moderate, never too fast. Relaxed, and with room for improvised counter-playing. **1960** *Times* 29 Feb. 15/1 Balanced part-playing and clear harmonic progression appeared to dominate his intentions to the detriment of artistic communication. **1932** *Official Syst. Contract Bridge* 187 (*heading*) Bidding against a side which has a part score. **1936** E. Culbertson *Contract Bridge Complete* 17 *Part-score*, (1) A contract of less than game; (2) the points earned for the making of such a contract. **1973** *Times* 20 Oct. 12/3 You then run the risk of missing a game unless you have a part-score. **1949** M. Mead *Male & Female* iii. 73 The Balinese child..develops a part-whole relationship to the world, in which each part of his body is whole, and yet each is part of the whole. **1953** C. E. Bazell *Linguistic Form* 103 It is likely that, as soon as the part-whole rather than the member-class terminology is used, the ostensible criteria for phonemic analysis should be transferred to the analysis of phonemic parts. **1972** *Language* XLVIII. 452 Semantic analysis based on a strict part-whole conception of meaning. **1966** *Economist* 26 Mar. 1233/1 One of the aims of the part-work system is to increase the sum of technically trained people in the rural areas. **1969** *Times* 13 Mar. 20/7 (Advt.), Part work publishing. *Ibid.*, Part-works now account for more than 10 per cent of all money spent on periodicals. **1971** *Guardian* 22 Oct. 9/2 'World of Wildlife' is the latest in a string of partwork publications... Partworks began in the mid-eighteenth century when the 'Encyclopaedia Britannica' found itself unable to raise sufficient funds to publish all the volumes in one go. **1975** *Nature* 24 Jan. 227/2 Dr Magnus Pyke..currently to be seen on British television, advertising the appearance of a new partwork about science.

B. c. *part-cause* (later examples); **part-exchange**, a transaction in which the owner of an article exchanges it for another (usu. new) article and pays a sum of money to cover the difference between the value of the two articles; hence as *v. trans.*, to exchange (something) in this way; also *fig.*; **part-load** (see quot.

1971); **part-pay**, a part of one's pay; *spec.* that part paid to whalemen before the start of a voyage.

1878 C. Read *On Theory of Logic* 130 It is certain that C is a Cause or Part-Cause of E. **1901** H. H. Joachim *Study Ethics of Spinoza* 169 In our ignorance we attribute these qualities of sensation directly to the external bodies, which are at most their part-causes. **1932** H. H. Price *Perception* viii. 270 One needs..to know that any material thing is a part-cause of the sense-data belonging to it. **1964** M. Argyle *Psychol. & Social Probl.* v. 61 Different part-causes combine together, so that if one is very strong, or if several of them add together, the result is a delinquent. Each part-cause has a known statistical weight over a large sample, so that it is possible to estimate the chances that a given person will be a delinquent. **1926** *Punch* 10 Nov. 505/3 A correspondent writes to know if, when the new wave-lengths come into force, the old ones will be accepted in part-exchange. **1929** *Melody Maker* Jan. 18 (Advt.), Your present instrument in part exchange. **1931** D. L. Sayers *Five Red Herrings* xx. 218 He had something to do with the second-hand motor trade and was taking the bike in part-exchange for something. **1961** *Guardian* 12 June 2/3 The urge to part-exchange my car comes upon me. **1968** *Listener* 30 May 711/1 One way is to resolve to *part-exchange* an idea long before it has been driven down to the rims. **1972** *Daily Tel.* 10 Oct. 3/3 He part-exchanged his old power-boat for the Carnation in 1970. **1973** *Times* 25 Oct. 38/7 (Advt.), Steinway and Sons..are prepared to purchase or take part exchange pianos of their own or other makes. **1976** *Leicester Trader* 24 Nov. 21/2 They do part-exchanges, offer an excellent after-sales service and an insurance repair service. **1969** *Jane's Freight Containers 1968–69* 315/1, 20% will go to wagon-load and 80% to LCL, or part-load. **1971** M. Tak *Truck Talk* 115 *Part load*, a consignment to a destination that is less than a complete trailer load. **1850** H. Melville *White Jacket* II. xxxvii. 234 There were instances of men in the Neversink receiving money in part pay for work done for private individuals. **1851** —— *Moby Dick* III. xlix. 309, I hope my poor mother's drawn my part-pay ere this.

part, *v.* Add: **3.** Also *to part brass-rags*: see *Brass *sb.* 7.

4. c. (*d*) Usu. *to part off*. (Examples.)

1923 C. M. Linley *Lathe Users' Handbk.* vii. 118 In bar work, as each piece is finished, it is parted off. **1945** W. C. Durney *Capstan & Turret Lathes* iii. 78 Mount a parting off tool in one of the remaining stations of the rear square turret and locate a longitudinal stop to part off the bar ⅛ in. from the collet face. **1948** L. H. Sparey *Amateur's Lathe* x. 126/1 When work of large diameter must be parted-off, it is not advisable to make the part in one cut. **1958** C. T. Bower *Aids to Workshop Pract.* viii. 94 The parting-off tool shown..has been designed for use on a 3½-in. centre lathe to enable work gripped in the chuck to be parted off without jamming.

6. d. (Earlier and later examples.) Also const. *up* (*Austral.* and *N.Z. colloq.*).

1864 Hotten *Slang Dict.* 196 *Part*, to pay, restore, give up; 'he's a right un, he is; I know he'd part.'.. The term is in general use in *Sporting* circles. **1889** *Bulletin* (Sydney) 21 Sept. 20/1 An' then they reckoned I'd been usin' 'em [sc. double-headed pennies in a game of two-up] all the time, and they made me part up. **1913** E. Wharton *Custom of Country* II. xiv. 202 People said of him that he 'didn't care to part'. **1933** A. G. Macdonell *England, their England* xv. 267 Might squeeze another hundred [pounds]... Not more. He doesn't part easily. **1943** *Amer. Speech* XVIII. 92 Words and phrases..which deserve to be recorded in any attempt to convey something of the flavour of the colloquial speech of the country [sc. New Zealand]...*to part up*, to pay up. **1946** F. Sargeson *That Summer* 152 Ted got more but of course there was his wife, and he had to part up. **1952** Wodehouse *Barmy in Wonderland* viii. 81 Fanny won't part. She's so tight she could carry an armful of eels up two flights of stairs and not drop one. **1953** K. Tennant *Joyful Condemned* iv. 39, I guess Rene might part up to know who her mum was. **1966** Wodehouse *Plum Pie* i. 17 Uncle Tom..had to foot the bills. He has the stuff in sackfuls, but he hates to part. **1970** J. Aiken *Embroidered Sunset* x. 209 'So where are all the old girl's pictures?' 'Scattered all over the village. Nobody will part; they are thought to be lucky.'

Partaga (pɑːtā·gă). The proprietary name of a brand of Havana cigar.

1862 *Illustr. Catal. Internat. Exhib., Industr. Dept., Brit. Div.* I. No. 869 The exhibitors are manufacturers of the following Cigars:—..Cubas. Partagas. Salvadoras. **1865** *Dublin Univ. Mag.* Apr. 378/1 By-and-by, after smoking two or three Partagas, I may recollect some lunatic who would do such an absurd thing. **1871** M. Legrand *Cambridge Freshman* 145 The O'Higgins was accommodated with a prime Partaga. **1878** *Trade Marks Jrnl.* 17 Apr. 264/1 Design patent. Partagas... Juan Antonio Bances, of and on behalf of the firm of Bances and Co.,..Havana, Cuba; manufacturers of cigars, cigarettes, and picadura. **1908** *Ibid.* 3 June 921 Partagas... Cigars. Cifuentes, Fernandez y Ca.,..Havana, Cuba; cigar manufacturers. **1930** A. Bennett *Imperial Palace* xvi. 100 Evelyn took a cigar out of a box of Partagas. *Ibid.* xxiii. 145 Oldham handed Partaga cigars, and, the table having been cleared of all but finger-bowls, ash-trays and cigars,..disappeared. **1973** J. M. White *Garden Game* 115 Our cigars..were Partagas, and merited careful attention.

partake, *v.* Add: **1. b.** Also *absol.*

1844 D. B. Reid *Illustr. Theory & Pract. Ventilation* 181 Nor was any of the members aware..that they had partaken more heartily than usual. **1870** J. P. Smith *Widow Goldsmith's Daughter* ix. 144 Chris could not touch anything, but the widow partook with the particular relish which a well-spent morning gave her. **1974** J. McClure *Gooseberry Fool* v. 80 He hardly ever drank... He didn't often partake, but then it was also bloody hot.

parthenocarpy (pɑ̄·ɹpĕnokāɹpi). *Bot.* [a. G. *parthenocarpie* (F. Noll 1902, in *Sitzungsber. d. Niederrheinischen Ges. für Natur- und Heilkunde* 160), f. Gr. παρθένο-ς virgin + καρπός fruit + -y³.] The development of a fruit without fertilization having taken place in the plant producing it. Hence **pa:rthenoca·rpic** *a.*, of a fruit, produced without prior fertilization; **pa:rthenoca·rpically** *adv.*

1911 J. M. Coulter et al. *Textbk. Bot.* II. 917 In striking contrast to ordinary fruit production is parthenocarpy, or the development of fruit without the fusion of gametes. *Ibid.*, Plants with parthenocarpic fruits are..propagated vegetatively. **1924** M. Skene *Biol. Flowering Plants* v. 406 An embryo may develop without fertilisation having occurred. Corresponding to this parthenogenesis we have parthenocarpy, where a fruit is produced without any seeds. **1929** *Nature* 13 July 63/2 Triploids occur..with partial parthenocarpy in the cultivated apples. **1949** *Endeavour* VIII. 191/1 Even in plants which normally produce seeded fruits, parthenocarpy may occur spasmodically. *Ibid.* 191/2 Parthenocarpic fruits often differ from seeded fruits of the same variety, not only in size and shape but in chemical composition. **1951** *New Biol.* XI. 72 There is virtually no fertilization in the seedless edible varieties [of banana], the fruits being formed parthenocarpically. **1960** *New Scientist* 11 Feb. 322/3 Such parthenocarpic fruits contain no viable seed. **1965** K. Esau *Plant Anat.* (ed. 2) xix. 586 Formation of a fruit may also occur without seed development and without fertilization, a phenomenon known as parthenocarpy.

parthenogenesis. (Further examples.)

1902 D. H. Campbell *Univ. Text-bk. Bot.* v. 122 In one species of Chara, *C. crinita*, the oöspores are developed without fertilization—one of the few well-authenticated cases of parthenogenesis. **1936** [see *Eutelegenesis]. **1950** *Adv. Genetics* III. 195 In vertebrata normal parthenogenesis is unknown (with the possible exception of certain fish hybrids). **1965** Bell & Coombe tr. *Strasburger's Textbk. Bot.* 203 There are exceptions in which a sexual cell will germinate and undergo development without fertilization. This phenomenon is referred to as parthenogenesis. Habitual parthenogenesis is that occurring when egg cells germinate regularly without fertilization.

parthenogenetic, *a.* **1.** (Further examples.)

1884 *Q. Jrnl. Microsc. Sci.* XXIV. 266 In certain cases, the oospores become normally developed and capable of germination without any male organs being formed at all... Pringsheim himself termed these oospores parthenogenetic. **1936** [see *Amphidiploid *a.* and *sb.*]. **1950** *Adv. Genetics* III. 195 Some big animal groups are wholly (or almost wholly) characterized by parthenogenetic reproduction. **1965** B. E. Freeman tr. *Vandel's Biospeleol.* ix. 110 *E[laphoidella] bidens*, a species with parthenogenetic reproduction, is cosmopolitan.

parthenogenetically, *adv.* (Further examples.)

1884 *Q. Jrnl. Microsc. Sci.* XXIV. 281 *Ascobolus furfuraceus* probably produces its fructification parthenogenetically. **1895** *Ann. Bot.* IX. 638 In *S[aprolegnia] mixta*..we conclude that fertilization frequently takes place, but in default of its occurrence the oosphere may develop parthenogenetically. **1924** *Jrnl. Agric. Res.* XXVII. 513 The stem-mother is always wingless and gives birth to living young parthenogenetically. **1938** [see *Apomict]. **1950** *Adv. Genetics* III. 195 Some vertebrata have been artificially induced to reproduce parthenogenetically. **1972** *Nature* 28 Jan. 196/1 Microtubule proteins are still synthesized in parthenogenetically activated, nucleated half-eggs.

parthenogenone (pɑːɹpĕno,dʒe·nǝʊn). *Zool.* [f. Parthenogen(esis + *-one*.] An organism of parthenogenetic origin, having only one parent.

1957 R. A. Beatty *Parthenogenesis & Polyploidy in Mammalian Devel.* i. 4 Parthenogenesis will be considered here as a term applicable not only to born young but also to embryos and foetuses, the parthenogenetic organism being called a parthenogenone. **1970** *Sci. Jrnl.* June 42/3 The human population, by virtue of its enormous size, could indeed contain a few parthenogenones. **1973** *Nature* 13 Apr. 475/2 [Mouse] eggs were examined under the × 50 magnification of a Wild dissecting microscope to determine the overall frequency and types of parthenogenones induced. **1977** *Ibid.* 6 Jan. 53/2 Diploid and haploid parthenogenones evidently possess the capacity to form teratomas and give rise to differentiated tissues.

Parthenopean (pɑːɹpĕnǒpī·ăn), *a.* [ad. It. *Partenopea*, f. L. *Parthenopēi-us* belonging to Naples (f. *Parthenopē* Naples) + -an.] Of or belonging to Naples; applied *esp.* to the short-lived republic established in Naples by French revolutionary forces in 1799.

1799 Nelson *Let.* 19 Jan. (1845) III. 236 The Parthenopien Republic is forming. **1858** S. Horner tr. *Colletta's Hist. Kingdom of Naples* I. iv. ii. 341 The day had arrived when the Parthenopean Republic was to be abandoned to her own resources. **1879** *Encycl. Brit.* IX. 79/1 The French, entering the city after a furious but undisciplined resistance by the lazzaroni, established with the aid of the citizens and nobles the 'Parthenopean Republic'. **1894** C. E. Clement *Naples* vi. 156 Vesuvius..sent forth a brilliant flame, which the Neapolitans regarded as an omen of future prosperity. Thus was the Parthenopean Republic established. **1900** P. Orsi *Mod. Italy* ii. 45 It was then that the Parthenopæan Republic—so-called from the ancient name of the city—was proclaimed at Naples. **1903** C. Giglioli (*title*) Naples in 1799: an account of the revolution of 1799 and of the rise and fall of the Parthenopean Republic. **1915** C. Mackenzie *Guy & Pauline* 284 They would travel farther south and perhaps come to that Parthenopean shore calling to him still now from the few days he had spent upon its silver heights and beside its azure waters. **1921** *Q. Rev.* Apr. 365 The relationship of the Parthenopean Republic to the Mother-Republic of France, could not be safely dealt with in a public print. **1925** L. V. Bertarelli *S. Italy* 262 Burdensome taxation..aroused the Neapolitans to insurrection under Masaniello in 1647, but his 'Parthenopean Republic' endured only a few months... In 1799..Gen. Championnet, at the head of a Napoleonic army, founded the second Parthenopean Republic. **1956** H. Acton *Bourbons of Naples* xvii. 333 The Parthenopean Republic was installed by conquest, and the French army of occupation was its only solid prop. **1974** *Encycl. Brit. Macropædia* IX. 1156/1 Thus was born the Parthenopean Republic, which..was the most democratic of the Italian states set up between 1796 and 1799.

Parthian, *a.* and *sb.* Add: **B.** *sb.* **2.** The Iranian language of the Parthians; = Pahlavi *a.* and *sb.* Also as *adj.*

1932 W. L. Graff *Lang.* x. 373 For a time Armenian was deemed to be an Iranian dialect on account of the large number of Parthian (Persian) words in its vocabulary. **1933** L. Bloomfield *Language* iv. 63 Discoveries of manuscript fragments in Chinese Turkestan gave us knowledge of other medieval Iranian languages, which have been identified as Parthian, Sogdian, and Sakian. **1939** L. H. Gray *Foundations of Lang.* 320 We have a fair amount of material in some other Middle Iranian dialects, notably *Middle Parthian* north of Persia. **1944** *Trans. Philol. Soc. 1942* 44 Scripts used in Iran in ancient times..derive from the Aramaic script of twenty-two letters (not counting..Chinese for Middle Persian and Parthian). **1954** M. Boyce *Manichaean Hymn-Cycles in Parthian* 1 There are in Parthian three long texts which are divided into sections known as *handāms* or 'limbs'. **1972** W. B. Lockwood *Panorama Indo-European Lang.* 236 The Manicheans borrowed the Estrangelo Syriac script and used it to write other Middle Iranian languages as well, i.e. Parthian and Sogdian.

parti. Add: **2.** *parti pris* (further examples). Also *attrib.* and as *pred. adj.*

1885 [see *Faute de mieux]. **1905** R. Broughton *Waif's Progress* xxiv. 268 It can be no *parti pris* that has dwindled her to half her size. **1923** H. Crane *Let.* 18 Feb. (1965) 125 In his letter his partis pris emotionalism was too evident to convince his readers properly. **1958** L. Durrell *Mountolive* v. 103 Personally I think we both have made a mess of it, and I have no *parti-pris* in the matter. **1959** *Times* 31 Dec. 11/3 Professor Brinton is not *parti pris* as Lecky and Westermarck were. **1973** *Times Lit. Suppl.* 15 June 660/5 With none of the characters in Lawrence's life is he *parti-pris*. **1974** *Broadcast* 9 Dec. 17/2 Most discussion about advertising is conducted on a basis of *parti pris*. **1977** *N.Y. Rev. Bks.* 9 June 21/2 The ideological *parti-pris* of these last two sources does not invalidate their generally rigorous and sound historical documentation.

partial, *a.* (*sb.*) Add: **3. b.** (*a*) (Further examples.) (*c*) *partial fractions*: the simpler fractions as the sum of which a compound fraction can be expressed; (*d*) *partial product*: (i) the product of one term of a multiplicand and one term of its multiplier; (ii) the product of the first *n* terms of a series, where *n* is a finite integer (including 1); (*e*) *partial sum*: (see quot. 1973); (*f*) *partial ordering* or *order*: a transitive antisymmetrical relation among the elements of a set, which is not necessarily informative about each pair of elements; (*g*) *partial pivoting*: see *Pivoting *vbl. sb.* 2.

1889 W. W. Johnson *Treat. Ordinary & Partial Differential Equations* xi. 288 An equation..giving the value of a single partial derivative, or more generally an equation giving a relation between the several partial derivatives of a function of two or more independent variables, is called a partial differential equation. **1975** F. G. Hagin *First Course Differential Equations* i. 32 Another important concept is the partial derivative... Recall that this can be computed simply by treating *y* as a constant and differentiating *F* as a function of *x*.

1816 Peacock & Herschel tr. *Lacroix's Elem. Treat. Differential & Integral Calculus* II. 186 The general method of integrating differentials of the above form, consists in decomposing them into others, whose denominators are more simple, which we designate by the name of partial fractions. **1908** G. H. Hardy *Course Pure Math.* vi. 198 It is very often convenient, in differentiating a rational function, to employ the method of partial fractions. **1975** Flanders & Price *Algebra* vi. 207 In general, $\frac{ax+b}{(x-r)(x-s)} = \frac{A}{x-r} + \frac{B}{x-s}$ for suitable constants *A* and *B*. This expression is called the partial fraction decomposition of the given rational function.

c **1823** *New Pract. Builder* 554 The sum of all the partial products will be the answer. **1959** G. & R. C. James *Math. Dict.* 285/1 *Partial product*, the product of the multiplicand and one digit of the multiplier, when the latter contains more than one digit. **1977** *Sci. Amer.* Sept. 82/2 Since each digit of the multiplier must be either a 0 or a 1, each partial product formed must be equal either to zero or to the multiplicand.

1966 W. Rudin *Real & Complex Analysis* xv. 290 The p_n are the partial products of the infinite product. **1972** A. G. Howson *Handbk. Terms Algebra & Anal.* xxviii. 145 Given a sequence (a_n) of non-zero real or complex numbers we form a second sequence (P_n) whose terms are the partial products $P_n = \prod_{k=0}^{n} a_k = a_0a_1\dots a_n$.

1926 Bromwich & MacRobert *Introd. Theory Infinite Series* (ed. 2) 540 (Index), Partial Sum of Fourier Series. **1928** R. C. Young tr. *Knopp's Theory & Appl. Infinite Series* ii. 99 An infinite series is a new symbol for a definite sequence of numbers deducible from it, namely the sequence of its partial sums. **1973** D. Ball *Introd. Real Anal.* iv. 71 The sum of the first *n* terms of a series $a_1 + a_2 + a_3 + a_4 + a_5 + \dots$ is called the *n*th partial sum.

1941 Birkhoff & MacLane *Survey Mod. Algebra* xi. 326 (*heading*) Partial orderings. **1964** T. O. Moore *Elem. Gen. Topology* vii. 126 Many writers require that a partial ordering be reflexive; some do not. We choose not to do so in this book. **1972** A. G. Howson *Handbk. Terms Algebra & Anal.* iii. 18 A binary relation on a set *X* which is reflexive, transitive and antisymmetric is called a partial order of *X*. A set with a partial order is known as a poset.

f. *partial pressure*: the pressure that would be exerted by a gas in a given mixture if it alone occupied the space.

1857 H. E. Roscoe tr. *Bunsen's Gasometry* 131 The quantity of each constituent gas absorbed, is proportional to the pressure on that constituent part..; and these pressures may be distinguished as 'partial pressures', in contradistinction to the 'total pressures' of the whole mixture. **1899** J. Walker *Introd. Physical Chem.* vii. 55 When a mixture of gases dissolves in a liquid, each component dissolves according to its own partial pressure. **1968** *Brit. Med. Bull.* XXIV. 249/2 The changes in composition of the body when the CO_2 partial pressure..is raised or lowered. **1971** *Physics Bull.* Feb. 83/3 The UKAEA's pulsed electrolytic hygrometer..can measure water in a gas down to a partial pressure of 10^{-7} atm.

g. Dentistry. *partial denture*: a denture that replaces one or more, but not all, of the natural teeth of one set.

1860 J. Richardson *Pract. Treat. Mech. Dentistry* xv. 374 (*heading*) Partial dentures constructed in a base of vulcanizable gums. **1921** D. Gabell *Prosthetic Dentistry* ix. 206 Partial dentures should slide smoothly and tightly into place and rest evenly on their supports. **1975** H. Thomson *Occlusion* xi. 215 With the exception of the canine an abutment tooth for a partial denture should have two roots.

h. Chem. *partial valency* [tr. G. *partialvalenz* (J. Thiele 1899, in *Ann. d. Chem.* CCCVI. 89)]: a partially unsatisfied valency formerly attributed to some atoms in unsaturated compounds to account for the addition reactions of olefins and the stability of the benzene ring.

1899 *Jrnl. Chem. Soc.* LXXVI. i. 554 The author holds the view that, in unsaturated compounds, whilst two affinities of every atom which participates in the double linking are occupied with those of the contiguous atom, the combining energy is not completely absorbed, so that the atoms in question still possess valency (*Partialvalenz*), and it is in this partial valency that the source of additive capacity is to be found. **1937** H. B. Watson *Mod. Theories Org. Chem.* viii. 105 The reactivity of the olefinic linkage was attributed to the incomplete saturation of the affinities of the doubly bound carbon atoms, which were thus regarded as possessing free 'partial valencies'. **1964** N. G. Clark *Mod. Org. Chem.* xix. 394 When applied to a conjugated system, such as occurs in buta-1,3-diene, Thiele postulated the union of the centre pair of partial valencies, thus creating a relatively inert type of double bond between C_2 and C_3, and leaving reactive partial valencies only at C_1 and C_4.

i. Dentistry. *partial veneer*: used *attrib.* to designate a crown consisting of a covering of three or more, but not all, of the surfaces of a tooth (the labial or buccal enamel being left exposed).

1928 *Jrnl. Amer. Dental Assoc.* Oct. 1919/2 The cast restoration is indicated in most other locations where esthetics will permit its use... One of the greatest fields of usefulness is in the construction of abutment pieces of the inlay, partial and full veneer types. **1940** S. D. Tylman *Theory & Pract. Crown & Bridge Prosthesis* xxvii. 332 (*heading*) The preparation of anterior teeth for partial veneer crowns. *Ibid.*, The partial veneer retainer is indicated primarily in bridge prosthesis when two or more missing teeth are restored. **1963** C. R. Cowell et al. *Inlays, Crowns, & Bridges* ix. 98 Although a well-constructed partial veneer crown shows little gold, an alternative preparation should be undertaken if a patient is anxious to avoid showing any gold.

j. *partial title* (see quots.).

1938 L. M. Harrod *Librarians' Gloss.* 113 *Partial title*, one which consists of only a part of the title as given on the title-page. **1967** *Anglo-Amer. Catal. Rules: Brit. Text* 267 *Partial title entry*, an added entry made under a secondary part of the title as given on the title page, e.g. a catchword title, subtitle, or alternative title.

k. *Cryst.* Of a dislocation: such that the displacement involved, as represented by the Burgers vector, is not an integral multiple of the lattice spacing.

1951 F. C. Frank in *Phil. Mag.* XLII. 816 Unlike an ordinary twin-boundary, a translation-twin-boundary need not go right through the crystal. When it does not, its edge in the interior of the crystal is a dislocation. It will be called an imperfect dislocation (alternatively, a partial dislocation) in contrast with perfect dislocations which are surrounded entirely by good crystal. **1960** [see *Glide *sb.* 4*]. **1966** C. R. Tottle *Sci. Engin. Materials* iv. 101 Metals such as copper, silver, and gold have low values of this stacking-fault energy, and so readily form partial dislocations. **1976** M. T. Sprackling *Plastic Deformation Simple Ionic Crystals* iv. 51 A strip of stacking fault extending through a crystal has two opposite sides terminated by partial dislocations.

l. Physics. *partial wave* (see quot. 1971).

1953 R. G. Sachs *Nuclear Theory* iv. 65 The straightforward analysis of a scattering problem involving a short-range potential makes use of the method of partial waves. **1970** I. E. McCarthy *Nuclear Reactions* I. i. 13 Large values of *l* correspond to trajectories which miss the nucleus. Therefore the effect of the nuclear forces is noticeable only in the first few partial waves. **1971** *Physics Bull.* Sept. 516/2 Any wavefunction describing a quantum mechanical system can be expanded in terms of eigenfunctions of angular momentum ('partial waves') characterized by an integer *l* ranging from zero to infinity; usually only a finite number of partial waves is required to specify the system.

m. *partial drought*: see *Drought 2.

B. *sb.* **2.** *Cryst.* A partial dislocation (see sense 3 k above).

1952 Read & Shockley in W. Shockley et al. *Imperfections in nearly Perfect Crystals* ii. 85 It was first pointed out by Heidenreich and Shockley that a dislocation having a $\frac{1}{2}$ [110] slip vector (taking the lattice constant as unit length) in a face-centered cubic crystal could lower its energy by dissociating into two partials (Shockley partials) having slip vectors $\frac{1}{6}$ [211] and $\frac{1}{6}$ [121], respectively, and connected by a stacking fault. **1967** A. H. Cottrell *Introd. Metall.* xvii. 280 The more widely the partials are separated initially, the more the energy required to bring them together to form the constriction and the more rare is the cross-slip. **1969** tr. *Kubo & Nagamiya's Solid State Physics* v. iii. 752 The characteristic of Shockley partials is that they together with their stacking faults can move and slip freely inside the slip planes.

partial (pā·ɹfăl), *v. Statistics.* [f. Partial *a.* (*sb.*)] *partial out* (trans.): to eliminate (a factor or variable) during analysis so as to remove its influence when considering the relationship between other variables.

1932 *Brit. Jrnl. Psychol.* XXIII. 184 This figure [for mean concrete imagery] is reduced..when the influence of visual verbal imagery is 'partialled out'. **1940** G. H. Thomson *Anal. Performance Test Scores* ii. 9 The form of distribution of age is not very important since age was partialled out. **1949** Bruner & Postman in J. S. Bruner *Beyond Information Given* (1974) iv. 73 We cannot partial out the differential effect of serial position of a card, whether first or third or fifth in the series, independently of the kinds of experience the subject had before being presented any given card. **1972** *Visible Language* Winter 57 For eighth graders there is no relation between linear spatial ability and either word or nonsense anagram performance when decentration is partialled out. **1976** *Nature* 24 June 689/1 When the effect of air temperature was partialled out..a rise in adjusted oral temperature was found in the 6 d following.

partially, *adv.* Add: **II. 2. b.** (Further examples.) *partially ordered* (Math.), having a partial ordering (see *Partial *a.* (*sb.*) 3 b (*f*)).

1941 Birkhoff & MacLane *Survey Mod. Algebra* xi. 326 Partially ordered systems with a finite number of elements can be conveniently represented by diagrams. **1949** Partially-sighted [see *dark-ground* s.v. *Dark *a.* 13c]. **1971** *Optometry Today* (Amer. Optometric Assoc.) 15 Partially-sighted and legally-blind persons must first be located. **1974** Hilton & Wu *Course in Mod. Algebra* 2 *Zorn's lemma.* Every inductive partially ordered set has a maximal element.

participant, *a.* and *sb.* Add: **B.** *sb.* **1.** Also *attrib.* and *Comb.* **participant democracy** = *participatory democracy; **participant observer**, a research worker (esp. in the social sciences) who, while apparently belonging to the group under observation, is gathering information about it for the study team; hence **participant observation, observing**, this method of research.

1924 E. C. Lindeman *Social Discovery* II. viii. 191 For experimental purposes the coöperating observers have been called '*participant observers*'. The term implies, not that the observers are participating in the study but that they are participating in the activities of the group being observed. **1933** Hader & Lindeman *Dynamic Social Res.* x. 147 (*heading*) Participant observing as a technique for psycho-social research. *Ibid.* 148 *Participant Observation* is based on the theory that an interpretation of an event can only be approximately correct when it is a composite of two points of view, the *outside* and the *inside*. **1948** *Mind* LVII. 510 Anthropologists, psychologists and administrators will all offer their contributions, making use of social surveys, opinion polls, official statistics, time-budgets, interviews of various types, 'participant-observers', and new techniques of self-observation. **1971** G. K. Roberts *Dict. Polit. Analysis* 144 Ethical and practical problems are raised connected with the concealed 'dual role' of the participant observer. *Ibid.*, Several advantages are served by participant observation rather than external observation or interview techniques. **1971** *Sci. Amer.* Mar. 72 These 'alternative institutions' frequently emphasize values similar to those of a therapeutic community: group cohesion and commitment;..and 'participant democracy', meaning involvement of the entire group in decision-making. **1976** *Word 1971* XXVII. 421 Classical anthropological techniques were used in our research: participant observation, interviews, sociological data surveys, questionnaires, and so on. **1977** *Jrnl. R. Soc. Arts* CXXV. 198/1 Some concern is also paid to academic ethics; and research involving, for example, joining a political or aberrant social group so that one can be a participant observer—the new academic name for *snoop*—or the manipulation of people as subjects for research purposes, may not get very far.

participating, *ppl. a.* (Further examples of the sense 'profit-sharing'.)

1930 *Daily Express* 6 Nov. 14/3 The dividend on the Participating Preferred Ordinary shares is again made up to the maximum of 9 per cent. by the recommendation of a final dividend of 5½ per cent. **1952** *Prentice-Hall Encycl. Dict. Business* 458/1 Participating insurance is a plan of insurance under which the policyholder receives dividends from his insurance company. *Ibid.* 626/1 *Participating* preferred stock gives the stockholder the right to receive dividends beyond the fixed rate. *Ibid.* 626/2 A company with two kinds of preferred stock might have one class participating, and the other non-participating. **1957** Clark & Gottfried *Dict. Business & Finance* 268/2 Practically all mutual insurance company policies, and some stock insurance company policies, are participating policies. *Ibid.* 337/1 Typically the participating stock receives dividends at a fixed rate, after which the common stock receives the excess up to a given amount. **1959** L. E. Davids *Dict. Insurance* 158 *Participating insurance*, insurance on which the policyholder is entitled to share in the surplus earnings of the company through dividends which reflect the difference between the premium charged and actual experience. **1964** *Lebende Sprachen* IX. 100/2 Participating bonds share in the profits of the insuring company in addition to receiving a fixed rate of interest. **1974** *Terminol. Managem. & Financial Accountancy* (Inst. Cost and Managem. Accountants) 61 *Participating preference shares*, shares which usually entitle the holder to a fixed dividend, and to participate in any surplus profits after payment of dividends at a specified rate on the ordinary share.

participation. Add: **2. b.** *spec.* the active involvement of members of a community or organization in decisions which affect their lives and work. Cf. *audience participation* s.v. *Audience 7 d.

1939 *Amer. Speech* XIV. 254 *Teacher placement*,..*pupil participation*,..*language usage* are a few typical examples of the monstrosities I refer to. **1944** H. P. Fairchild *Dict. Sociol.* 213/2 *Participation*, entry into, identification with, as through communication or common activity, some defined social situation. **1948** A. L. Kroeber *Anthropol.* (rev. ed.) ix. 347 The second mechanism making for participation of the young in their culture, and their putting into society, is education. **1964** M. Argyle *Psychol. & Social Probl.* viii. 113 One way of allowing greater participation is through the use of democratic techniques of supervision. **1968** *Guardian* 13 July 9/2 General de Gaulle's social reform programme, summed up by the slogan 'Participation'. *Ibid.* 10 Sept. 8/5 Participation begins when employees at all levels feel that their own supervisors enlist their help to get what has to be done at their own level performed expeditiously. **1970** *Ibid.* 5 May 13/2 Students have shown no more enthusiasm for 'participation' than did French workers when General de Gaulle proposed it to them. **1974** *Listener* 28 Feb. 276/3 Publicity for proposals is easy, but..effective participation is much more difficult..there is a need to involve all sectors of the community in any debate on proposals. **1976** *Which?* May 110/3 So it's important to make sure that the major issues that you think affect your area do get discussed, and one of the purposes of the public participation legislation is that the public should be involved at the formative stage. **1977** *Times* 25 Feb. 1/7 The unions had an essential role in any system for developing participation.

c. A participating bond or share.

1931 *Economist* 5 Dec. 1057/1 In Germany, foreign exchange (by which is meant foreign notes, claims, bills of exchange, cheques and gold in any form, but not foreign bonds or shares or other industrial participations, except bonds acquired after July 12th) must, within three days after acquisition, be sold to the Reichsbank. **1968** *Globe & Mail* (Toronto) 13 Feb. B2/4 In addition to long-term bonds with call provisions, the new mortgage bank would sell participation certificates on the mortgage holdings in its secondary market operations portfolio.

d. participation mystique, imaginative identification with people and objects outside oneself, regarded as an attribute of primitive peoples by the French anthropologist Lucien Lévy-Bruhl (1857–1939); merging of the individual consciousness with that of a group or with the external world.

Discussed by Lévy-Bruhl in *Les fonctions mentales dans les sociétés inférieures* (1910) I. ii. 78, etc.

1927 A. Huxley *Proper Stud.* 77 A subjectivized world, with which the observer lives in a state of what Lévy-Bruhl calls '*participation mystique*', is unamenable to scientific treatment. **1933** Dell & Baynes tr. *Jung's Mod. Man in Search of Soul* viii. 198 The secret of artistic creation and of the effectiveness of art is to be found in a return to the state of *participation mystique*—to that level of experience at which it is man who lives, and not the individual, and at which the weal or woe of the single human being does not count, but only human existence. **1949** D. Macardle *Children of Europe* I. i. 20 Powerful, also, is a deeper source of unrest—a *participation mystique*—the craving, much stronger in some races than in others, which makes individuals long to merge in the herd. **1956** F. Herbert *Dragon in Sea* (1960) 36 *Religious services*, thought Ramsey... *Participation Mystique!* **1957** N. Frye *Anat. Crit.* 295 College yells, sing-songs, and similar forms of *participation mystique*. **1966** J. B. Priestley *Moments* 228 In our early childhood..we exist in a state that a French anthropologist has called *participation mystique*. **1974** *Times* 23 May 10/5 As one who has never visited the Indian sub-continent, nothing I have read has..given me so much the sense of *participation mystique* in a civilisation..the antithesis of our own.

participational (pɑ̄ɹtisipēi·ʃǒnǝl), *a.* [f. Participation + -al.] Involving or requiring participation.

1959 *Times* 18 June (Queen in Canada Suppl.) p. xv/4 Recreational pursuits turn to swimming, fishing and

'participational' outdoor life. **1964** M. McLuhan *Understanding Media* ix. 82 Electric technology seems to favor the inclusive and participational spoken word over the specialist written word. *Ibid.* xvii. 168 This participational and do-it-yourself aspect of the electric technology. **1970** *Americana Ann.* 687 The $30 million Ontario Science Center opened in September in Toronto. The Center features hundreds of 'participational' exhibits that are operated by push buttons. **1971** *Guardian Weekly* 19 June 12 In spite of its premature closing, the participational section of the Morris exhibition was valuable because of the discussion and thought it provoked among artists and public alike.

participative, *a.* Add: (Later examples.)

1975 *Church Times* 22 Aug. 13/4 (Advt.), Many Clergy have benefited from our open consultations on running the local church... Ecumenical and participative methods of learning. *Ibid.* 3 Oct. 6/3 Those who are not so high-minded..might be dedicated to sabotage a new system as it staggered forward to achieve a new, different and participative venture.

b. In business administration: pertaining to or characterized by the sharing of the decision-making process with either (*a*) the lower grades of management, or (*b*) the workers.

1961 B. von H. Gilmer *Industrial Psychol.* v. xiv. 303 Gradually industrial organizations moved towards *participative management*, still leaving the worker in the position of having someone else plan for him... Participative management aimed higher, giving increased status to the people in middle management. **1965** R. C. Sampson *Managing the Managers* xii. 193 In participative coaching, a manager works with the executive, both in problem-solving and learning situations, in such a way that the subordinate retains the decision-making function and learns as the result of the experience. **1966** *New Statesman* 19 Aug. 255/3 There is growing interest in 'participative' and even democratic styles of management. **1972** *Jrnl. Social Psychol.* LXXXVII. 100 They [sc. Blake and Mouton] indicate that..participative leadership..can create conditions where work group standards are high. **1973** *Nature* 6 Apr. 381/1 The view that a boss-and-minion society can be replaced by something more cooperative and participative. **1976** *Birmingham Post* 16 Dec. 6/6 Another important feature of such improvements programmes is participative training to change attitudes, open minds and spread awareness and knowledge of control and improvement techniques. **1977** *Jrnl. R. Soc. Arts* CXXV. 671/2, I think we have now moved out of the participative area, in terms of management; I think Bullock is maybe ten years out of date.

participatory, *a.* Add to def.: *spec.* in government, etc., involving members of the community in decisions; allowing members of the general public to take part, as *participatory art, broadcasting, democracy, radio, television, theatre*.

1968 *N.Y. Times* 6 May 46 Both the Negroes and the antiwar groups have made use of the politics of marches, sit-ins and mass demonstrations. But those who practice this 'participatory democracy' can ultimately achieve their objectives only if they work through electoral processes and win control of Congress and the Presidency. **1968** *N.Y. Rev. Bks.* 11 July 31/2 Those who really are committed to 'participatory democracy' and hence insist on participating directly and fully in all forms of social life that can rightly command their allegiance, are separated by an ideological abyss from those traditional representatives of 'representative democracy'. **1968** *Guardian* 10 Oct. 2/5 The prevailing catechism..requires all 'dialogue' to be 'meaningful' and democracy itself to be 'participatory'. **1970** *Time* 23 Feb. 68 These ventures in dramatic exploration are also intimately related to an attempt to bridge the we–they gap in the actor–audience relationship—what is popularly called 'participatory' theater. **1970** *Sat. Rev.* (U.S.) 30 May 9/2 The most dramatic action of the early New Left was the journey hundreds of young people took in the summer of 1963 to live and work in Mississippi, helping Negroes organize themselves... The main slogan of that movement..was 'participatory democracy'. **1972** *Listener* 17 Aug. 197/3 Participatory broadcasting could be seen as a threat to the impartial provision of facts on the air. **1972** *Guardian* 2 Sept. 10/3 The best of BBC participatory local radio. *Ibid.* 6 Nov. 10/6 The BBC is to start 'participatory' television... A trial run of 50 weekly programmes, each devised by community or pressure groups on their own terms. **1977** *Times* 10 Aug. 14/7 That most trendy of activities, participatory art.

† **participled** (pā·ɹtisip'ld), *a. Obs.* [f. Participle *sb.* 2 + -ed².] Euphemism for 'damned' or 'confounded'.

1887 *Sat. Rev.* 17 Dec. 815 Thucydides.., by the way, was a participled Tory, like Clarendon, Gibbon, Tacitus, and all the greatest historians.

parti-ckler, *sb.* and *a.* (*adv.*) Also **partiklar, partic'lar.** ¶ With distortion of spelling to indicate an uneducated pronunciation of Particular *a.* and *sb.* (*adv.*)

1833 Dickens *Let.* ? Jan. (1965) I. 14, I am so anxious to hear the *partickler*s. **1837** —— *Pickw.* xv. 147 He wants you partickler; and no one else'll do. *Ibid.* xliv. 482 Vich is your partickler wanity? **1871** *Harper's Mag.* Oct. 690 Ef Pat Role, or any other consarned Irishman, kicks up a muss 'bout these yer diggins, he'll kotch *partic'lar* lightnin'. *c* **1875** 'Brenda' *Froggy's Little Brother* (new ed.) ii. 22, I feels quite well, sir..and I wants to go home partiklar. **1901** M. Franklin *My Brilliant Career* xvii. 150 The boss is so dashed partickler too.

particle, *sb.* Add: **1.** (Later example.)

1881 'Mark Twain' *Prince & Pauper* xxxi. 374 Now began a movement of the gorgeous particles of that official group.

2. (Further examples.) In *Physics* now applied esp. to the constituents of atoms and to other sub-atomic entities (some of which are now regarded as likely to be composite in nature). Cf. *alpha particle* s.v. *Alpha 3 e, *elementary particle* s.v. *Elementary *a.* 6.

1898 J. J. Thomson *Discharge Electr. through Gases* 189 The other theory..regards the cathode rays as marking the course of a stream of negatively electrified particles. **1942** H. Dingle *Sub-Atomic Physics* i. 12 We next assume that the atoms of all bodies are constructed from three kinds of particles—one positively electrified (the proton), one negatively electrified (the electron), and one unelectrified (the neutron). **1968** M. S. Livingston *Particle Physics* iv. 77 By this time [sc. 1947] the definition of elementary particles had expanded to include more than the components of atoms. It now included particles created in nuclear decay processes, such as pions and muons..; it also included antiparticles, although the positron was the only example which had been observed... The list of elementary particles jumped to over 30 within the next 8 years. **1969** *Times* 8 Jan. 12/2 This star..would consist entirely of the nuclear particles called neutrons. **1970** P. H. A. Sneath *Planets & Life* i. 24 Modern physics has demonstrated the existence of a large number of 'smaller' subatomic particles, some of them very short-lived, and it has recently been suggested that they are all composed of yet 'smaller' particles, called quarks.

3. (Further examples.)

1924 O. Jespersen *Philos. Gram.* 87, I therefore propose to revert to the old terminology by which these four classes [sc. adverbs, prepositions, conjunctions, and interjections] are treated as one called 'particles'. **1933** O. Jespersen *Essent. Eng. Gram.* vii. 69 Some particles can be used in one capacity only, others may be used now as adverbs, now as prepositions, and now as conjunctions, others again in two of these capacities. **1935** H. Straumann *Newspaper Headlines* 56 Particles, then, are all those words which cannot be looked upon as nominals, verbals, and neutrals. **1964** A. S. C. Ross *Essent. Eng. Gram.* 18 English consists of *words* and *particles*. The main difference between these two things is that the class of words is very numerous and can be added to at will, whereas the class of particles is rather small and cannot be added to at will. **1965** *Eng. Stud.* XLVI. 439 By 'particles' he [sc. A. S. C. Ross] means, not only prepositions, articles, indefinite pronouns, etc., but also prefixes, suffixes and inflectional endings. **1977** M. Cohen *Sensible Words* i. 40 Particles..are the grammatical functions that serve what Arnaud called the reasoning and ordering operations of the mind.

b. The preposition-like word which forms part of a complex (phrasal) verb and which can be optionally separated from the verb in certain constructions. Also *attrib.*

1925 Grattan & Gurrey *Our Living Lang.* xii. 80 When, therefore, such words [as in He has run *up* a bill] differ clearly from the ordinary adverb, it is advisable to give them a more precise label: Verbal Particles. **1957** N. Chomsky *Syntactic Struct.* vii. 76 Further investigation of the verb phrase shows that there is a general verb + complement..construction that behaves very much like the verb + particle construction. **1964** Katz & Postal *Integr. Theory Ling. Descr.* (1965) iii. 41 The particle inversion transformation..inverts the particle of a certain set of complex verbs with their Noun Phrase objects. **1968** Jacobs & Rosenbaum *Eng. Transformational Gram.* xiii. 102 Some sentences contain words known as *verb particles* in their verb phrases. *Ibid.* 103 When question sentences containing verb particles are generated, the particle *must* remain in its original position. *Ibid.* 104 The transformation responsible for introducing a particle segment into the structure is called the *particle segment transformation*, or more simply, the particle transformation. *Ibid.* 106 The particle movement transformation is normally optional. **1968** R. W. Langacker *Lang. & its Struct.* 118 The Particle Shift rule..separates a verb and a particle by placing the particle after the following direct object noun phrase. **1974** McArthur & Atkins *Dict. Eng. Phrasal Verbs* 5 Phrasal verbs..are, usually, combinations of simple, monosyllabic verbs (put, take, get etc.) and members of a set of particles (on, up, out etc.).

4. *attrib.* and *Comb.*, as *particle-size*; *particle-accelerating, -like* adjs.; **particle accelerator** = *Accelerator e; **particle physics**, the branch of physics concerned with the properties, relationships, and interactions of sub-atomic particles; so **particle physicist**. See also sense 3 b above.

1947 *Electronics* Dec. 82/1 An electrostatic particle-accelerating machine called a Van de Graaf generator. **1959** *Sunday Times* 5 Apr. 8/6 The particle-accelerating machines of the nuclear physicist. **1975** *Nature* 2 Oct. 360/2 Work has thrived on the use of pulsed sources, including particle-accelerating machines, for neutron diffraction experiments. **1946** *Physical Rev.* LXX. 91/1 (*heading*) Particle accelerators as mass analyzers. **1968** *Times* 21 Dec. 13/7 Bubble chambers..are used in conjunction with powerful particle accelerators—machines that produce beams of high energy particles—to study how subatomic particles interact. **1977** D. Bagley *Enemy* xvii. 145 Microbiology isn't like atomics; you don't need a particle accelerator costing a hundred million. **1959** G. Troup *Masers* ii. 14 The 'wave-like' aspects of radiation have been stressed rather more than the 'particle-like' aspects. **1973** *Sci. Amer.* Oct. 104/2 Dirac called his hypothetical particl-like holes positrons. **1977** *Ibid.* Apr. 116/1 Allowance must be made for wavelike properties, such as interference, diffraction and polarization, and for particlelike properties, such as the momentum carried by a beam of light. **1969** *New Scientist* 24 Apr. 171/1 What the high-energy particle physicists are witnessing..is evidently a yet further spectrum of some kind. **1971** *Ibid.* 5 Aug. 334 Even particle physicists and molecular biologists would be hard put to point to new discoveries, insights and ideas rivalling those in the field of astronomy..since 1961. **1946** *Proc. Amer. Philos. Soc.* XC. 44/2 Some of the outstanding experimental problems in elementary particle physics—problems concerned with electrons and positrons, nuclear explosions, neutrino physics, and most of all with the meson. *Ibid.* 47/1 Elementary particle physics..will get sources of energetic particles and use these sources to study the transformations of the elementary particles. **1962** Livingston & Blewett *Particle Accelerators* p. v, Phenomena in nuclear physics and high-energy particle physics. **1969** *New Scientist* 24 Apr. 171/1 In 20th century particle physics there has been a continuous progression from the world of the atom,..through the MeV world of nuclear physics..to the present perplexing phenomena of the GeV world created by the big particle accelerators. **1946** *Nature* 21 Dec. 908/2 Particle-size in silts and sands. **1966** D. G. Brandon *Mod. Techniques Metallogr.* v. 249 There is very little to be said about the quantitative analysis of particle-size or grain-size distributions.

particle board. [f. Particle *sb.* 2 + Board *sb.* 1.] (See quot. 1957.)

1957 *Brit. Commonwealth Forest Terminol.* II. 29 *Board, particle*, a board (sheet) constituted from fragments of wood (chips, shavings, sawdust, etc.) and/or other vegetable materials that have been partly or wholly comminuted and then consolidated by pressure, heat, etc., with or without binders and supplementary material. **1959** *Times Rev. Industry* Apr. 91/3 Particle board is a building board made by bonding woodflakes or shavings under heat and pressure with the aid of a resin adhesive. **1966** A. W. Lewis *Gloss. Woodworking Terms* 55 Manufactured boards..include hardboard, chipboard or particle board, plywood, and laminboard. **1973** *Time* 25 June 29/1 (Advt.), We manufacture and distribute building materials, including plywood, particleboard, sidings, prefinished paneling, and adhesives. **1973** *Globe & Mail* (Toronto) 4 Aug. 8/4 In 1970 a lumber company owned by New Brunswick industrialist K. C. Irving obtained cutting rights in the area and was expected to put up a particle-board plant. **1977** *36 Home Handyman Projects* (Austral. Home Jrnl.) 6/1 Here we show how to make some great seating from particle board—it's simple to make, can look elegant or casual and..it's so comfortable.

parti-colour, *a.* (*sb.*) Delete *Obs.* and add: esp. in reference to a dog's coat, marked in patches of two distinct colours. As *sb.*, a dog whose coat is coloured in this way.

1945 C. L. B. Hubbard *Observer's Bk. Dogs* 113 Pekingese... Colour red, tricolour, parti-colour. **1961** C. H. D. Todd *Popular Whippet* vi. 90 Like most judges..I dislike a 'butterfly nose', but it is permissible in a parti-colour. **1971** F. Hamilton *World Encycl. Dogs* 200 Colors [of Tibetan terriers] are white, golden, cream, gray or smoke, black particolor and tricolor. *Ibid.* 262 In particolors [sc. cocker spaniels], the contrasting color must be ten per cent or more.

parti-colour, *v.* Delete †*Obs.* and add later example.

1971 F. Meynell *My Lives* xi. 168 A man..can paint..or decorate a room or parti-colour a motor-car.

parti-coloured, *a.* Add: esp., in reference to dogs, having a coat marked with two or more colours in distinct patches.

1879–81 V. Shaw *Illustr. Bk. Dog* xxvi. 185 Parti-coloured dogs [sc. Pomeranians] are much objected to. **1922** R. Leighton *Compl. Bk. Dog* 285 All colours [of Pekingese] allowable, red, fawn, black, black and tan, sable, brindle, white and parti-coloured.

particular, *a.* and *sb.* Add: **I. 2.** *particular average* (examples).

1773 *Encycl. Brit.* I. 506/1 The simple or particular average..consists in the extraordinary expences incurred for the ship alone, or for the merchandizes alone. **1895** W. Gow *Marine Insurance* xii. 208 The repairs of damage of the nature of particular average are confined to what will put the vessel in the same state of efficiency as she was in before the accident which rendered these repairs necessary. **1960** Dover & Calver *Banker's Guide Marine Insurance of Goods* 287 If incurred as a consequence of a peril insured against, particular average is made good by underwriters subject to the conditions of the policy. **1974** L. E. Davids *Dict. Insurance* (ed. 4) 203 *Particular average*, loss borne by one of a number of carriers in marine insurance, such as partial loss of cargo, hull, or freight, falling entirely on the interest concerned.

7. e. *in a particular condition*, pregnant. (Cf. *Interesting *ppl. a.* 3.)

1922 Joyce *Ulysses* 411 All these little attentions would enable ladies who were in a particular condition to pass the intervening months in a most enjoyable manner.

8. a. (Further example.)

1813 Jane Austen *Let.* 26 Oct. (1952) 359 Your Saturday's Letter..was quite as long & as particular as I could expect.

10. (Later examples.)

1932 *Punch* 2 Nov. 488/1 He was rather particular what he ate and drank. **1933** D. C. Peel *Life's Enchanted Cup* xiv. 176 Mothers who were 'particular' used to see to it that their girls went to dances with other girls well known to them.

10*. *particular integral* (Math.): **a.** A solution of a differential equation obtained by assigning values to the arbitrary constants of the complete primitive of the equation. Also called *particular solution*.

PARTICULARISM

1814 P. Barlow *New Math. & Philos. Dict.*, *Particular Integral*, in the Integral Calculus, is that which arises in the integration of any differential equation, by giving a particular value to the arbitrary quantity or quantities that enter into the general integral. 1885 A. R. Forsyth *Treat. Differential Equations* iii. 49 The primitive then consists of two parts: First, the quantity η, which is called the Particular Integral and is any solution whatever (the simpler the better) of the original equation; Second, the quantity Y, which is called the Complementary Function. 1897 D. A. Murray *Introd. Course Differential Equations* i. 6 The solution which contains a number of arbitrary constants equal to the order of the equation, is called the general solution or the complete integral. Solutions obtained therefrom, by giving particular values to the constants, are called particular solutions. 1958 G. E. H. Reuter *Elem. Differential Equations & Operators* i. 5 General solution = particular solution plus complementary function. 1966 S. Ross *Introd. Ordinary Differential Equations* iv. 95 Consider the differential equation $\frac{d^2y}{dx^2}+y=x$... A particular integral is given by $y_p=x$.

†**b.** A solution of a differential equation which cannot be obtained by assigning values to any or all of the arbitrary constants of the complete primitive; now called *singular solution*. *Obs.*

1820 G. Peacock *Coll. Examples Appl. Differential & Integral Calculus* ii. xi. 477 A particular integral of the original equation, involving only one arbitrary function. 1845 *Encycl. Metrop.* II. 23 This value of *y* satisfies the proposed equation; but as it cannot be derived from the complete integral we have obtained above by assuming a particular value for one of the arbitrary constants, it ought to be considered as a particular integral.

particularism. **1.** (Later example.)

1969 J. E. Dittes in Lindzey & Aronson *Handbk. Social Psychol.* (ed. 2) V. 633 Religious ideology tends to promote a concept of social exclusiveness or 'particularism'..especially with notions of..special election as a member of a divinely chosen group.

2. (Further examples.)

1955 *Bull. Atomic Sci.* Apr. 142/2 The humanitarian theme of the two preceding centuries certainly persisted, but universalism yielded step by step to national particularism which was in many respects naive but which became increasingly noticeable among European scientists and scholars. 1964 *Welsh Hist. Rev.* II. 127 The broader theme of local particularism in general elections..was presented in some quarters. 1973 *Daily Tel.* 4 May 18 He attacks the concept of youth particularism, stating that as a group youth has few common interests or problems. 1975 *New Left Rev.* Nov.–Dec. 1 The intensely uneven and combined development of capitalism and imperialism were bound to intensify..national particularism and antagonism.

3. (Further and *transf.* examples.)

1912 *Q. Rev.* Jan. 212 A recognition of Albanian particularism would create a precedent of which all other nationalities would take advantage. 1965 *Mod. Law Rev.* XXVIII. v. 616 He may well be right in affirming that regional particularism..call[s] for a redrawing along provincial lines. 1976 M. A. Jones *Old World Ties Amer. Ethnic Groups* 13 Particularism..had for centuries kept the German states apart.

4. (Later example.)

1936 *Nature* 25 Apr. 681/1 The advance of science is along two roads: the first is in the direction of greater intensity, particularism and of empiricism, and the second, from intensity, particularism and empiricism, towards extensity, generalisation and synthesis.

5. *Philos.* The fact or quality of being concerned with elements that have a particular (as opposed to universal) application, or to which no general standard is applicable.

1939 J. Dewey in P. A. Schilpp *Philos. J. Dewey* 544 In philosophy there is also the need to find an alternative for that combination of atomistic particularism with respect to empirical material and Platonic *a priori* realism with respect to universals. 1943 *Mind* LII. 140 Almost every philosophical term which connotes a tendency to particularism may be predicated of the Berkeley of this early period. 1963 R. M. Hare *Freedom & Reason* ii. 19 Such a philosopher could indeed embrace..the extremest form of particularism.

b. *Sociol.* and *Econ.* In some analyses of social and economic organization, a name that characterizes the particular or fixed nature of a role or element, as contrasted with the universal, general, or mobile nature of other elements or roles.

1949 T. Parsons *Ess. Sociol. Theory* viii. 197 In all these cases though in different ways and degrees, particularism tends to replace universalism. 1951 Parsons & Shils *Toward Gen. Theory of Action* ii. i. 98 The pattern variables most relevant to the description of the normative patterns governing roles..are achievement-ascription and universalism-particularism. 1959 B. F. Hoselitz *Sociol. Aspects Econ. Growth* (1960) ii. 32 We must not expect the principle of particularism in assigning economic roles to appear in complete purity in all societies on a low level of economic advancement. 1964 T. Parsons *Ess. Sociol. Theory* (rev. ed.) 16 On the level of the research techniques he [*sc.* Weber] used, the broad contrasts, e.g. as between Chinese traditionalistic particularism and Western universalistic 'rationalism', were unmistakable. 1969 I. Deutscher *Marxism in our Time* (1971) 190 The growth of bureaucracy was further stimulated by the breaking down of feudal particularism and the formation of a market on a national scale.

particularist, *sb.* (*a.*) (Further examples.)

1935 *Jrnl. Gen. Psychol.* XII. 55 For the same reason we cannot accept the definition proposed by the extreme particularist. 1939 E. Muir *Present Age* 160 But he remained an inveterate particularist; his philosophy is not an organic whole, but is made up of a number of peculiar ideas. 1963 R. M. Hare *Freedom & Reason* ii. 18 A particularist (if I may use that name for the opposite of a universalist). *Ibid.* 20 It is quite impossible for a naturalist to be, consistently, any sort of particularist. 1968 D. M. Murphy tr. *Gelin's Concept of Man in Bible* iv. 63 In the Bible there is a universalist outlook and a particularist outlook.

particularistic, *a.* (Further examples.)

1937 T. Parsons *Struct. Soc. Action* iii. xv. 551 The whole Chinese social structure accepted and sanctioned by the Confucian ethics was a predominantly 'particularistic' structure of relationships. 1955 M. Gluckman *Custom & Conflict in Afr.* v. 123 She also brings out the particularistic nature of their answer to the problem, what is man? 1956 *Jrnl. Theol. Stud.* VII. 99 Both the universalistic and the particularistic strands in their teaching. 1964 Gould & Kolb *Dict. Social Sci.* 489/1 He may decide to be particularistic, treating them 'in accordance with their standing in some particular relationship to him or his collectivity, independently of the objects' subsumibility under a general norm'. 1970 J. Cotler in I. L. Horowitz *Masses in Lat. Amer.* xii. 426 Through the ..particularistic pattern in which juridical and political authorities are designated, the Pisac mestizos become political figures. 1972 *Science* 9 June 1094/3 The differentiation of scholarship within a field into a variety of highly particularistic specialties reduces the potential for the type of behavior associated with intellectuality. 1974 tr. *Wertheim's Evolution & Revolution* i. 51 Popular emancipation movements do not stress an abstract universalism as their main ideology, but rather strike a note of particularistic loyalties.

So **particulari·stically** *adv.*

1951 Parsons & Shils *Toward Gen. Theory Action* 260 A specific situation vis-à-vis particularistically designated persons. 1963 R. M. Hare *Freedom & Reason* ii. 20 The particularistically inclined non-naturalist.

particularity. Add: **1.** (Further examples.) *spec.* in *Theology*, with ref. to Christ as the incarnation of God as a particular human being at a particular time and place.

1930 E. Hoskyns in Bell & Deissmann *Mysterium Christi* 89 The philosopher should..make sense of it [*sc.* revelation] by some other means than by obscuring the particularity of the Old Testament and by refusing to recognize that in the end the particularity of the Old Testament is only intelligible in the light of its narrowed fulfilment in Jesus, the Messiah, and of its expanded fulfilment in the Church. 1966 G. W. H. Lampe in Lampe & Mackinnon *Resurrection* vii. 92 The Incarnation necessarily involves particularity. If the Word was truly made flesh then he had to be incarnate as a certain individual man in a particular time and place. 1969 T. F. Torrance *Theol. Sci.* iii. 140 God reveals Himself in the contingent particularity and sheer singularity of Jesus Christ. 1975 *Listener* 17 July 92/2 She seems to lack awareness of individuals in the particularity of each. 1979 B. Hebblethwaite in M. Goulder *Incarnation & Myth* iv. 93 The inevitable limitations of that particularity are overcome ..by his spiritual and sacramental presence and activity, by means of which God's personal self-revelation in Jesus is universalized.

5. Delete †*Obs.* and add later examples.

1961 W. Herberg in *Webster* s.v., Fixing exclusively on the particularities of the current situation. 1977 *Times Lit. Suppl.* 11 Feb. 148/3 Sociologists are notorious for their use of generalizing terms that ride roughshod over the particularities of history.

particulate, *a.* Add: (Earlier and later examples.)

1871 *Q. Jrnl. Microsc. Sci.* XI. 325 It may be supposed either that the germinal substance is universally and equally distributed, *i.e.* dissolved in such liquids, or that it is unequally distributed or particulate. 1962 F. I. Ordway et al. *Basic Astronautics* iv. 120 Interstellar and interplanetary particulate matter. 1966 *McGraw-Hill Encycl. Sci. & Technol.* IX. 197/2 Beta rays are particulate radiation consisting of electrons or positrons emitted from a nucleus during β-decay. 1974 *Environmental Conservation* I. 16 The addition of carbon dioxide and particulate matter to the atmosphere by mankind.

b. (Further examples.) *particulate inheritance*, the manifestation in offspring of discrete characters each inherited from one or other of the parents.

1886 F. Galton in *Rep. Brit. Assoc. Adv. Sci. 1885* 1213 To express this aspect of inheritance, where particle proceeds from particle, we may conveniently describe it as 'particulate'. 1889 —— *Nat. Inheritance* ii. 8 The exact meaning of Particulate Inheritance, namely, that each piece of the new structure is derived from a corresponding piece of some older one. 1930 R. A. Fisher *Genetical Theory Nat. Selection* i. 1 The need for an alternative to blending inheritance was certainly felt by Darwin, though probably he never worked out a distinct idea of a particulate theory. *Ibid.* 8 Apart from dominance and linkage, ..all the main characteristics of the Mendelian system flow from assumptions of particulate inheritance of the simplest character, and could have been deduced *a priori*. 1971 J. Z. Young *Introd. Study Man* xxviii. 392 (*heading*) Genes and their mutations. Particulate inheritance.

B. *sb.* A particulate substance; particulate material. Also *attrib.*

1960 *New Scientist* 13 Oct. 1001/3 The future will see the ultra-centrifuge used more and more as a tool to determine the physical structure of specific parts of such macromolecules as the nucleoproteins, coenzymes and cell particulates. 1971 *Nature* 20 Aug. 553/2 Airborne particulate was collected on 0·45 μm 'Millipore' membrane filters. 1973 *Physics Bull.* May 314/3 Methods for the elemental analysis of air particulates. 1974 *Post-Herald* (Birmingham, Alabama) 29 June A7/3 The Jefferson County particulate count has exceeded the 260 level 82 times during the past year. 1976 *Sci. Amer.* July 77 The emission of particulates is reduced by the liberal use of filters.

partier (pā·ɹtiəɹ). *colloq.* [f. Party *sb.* 9 + -er¹.] One who likes to give or attend parties.

1965 J. Hart *File for Death* xii. 94 Jinsie most certainly did not care for the 'partiers'. 1973 *Daily Colonist* (Victoria, B.C.) 16 Sept. 13/3 Thor said he understood the hesitancy of many landlords to rent to male rather than female students. Women are generally neater than men, he concedes, and are not partiers.

pa·rtified, *ppl. a. colloq.* [f. Party *sb.* 9 + -fy + -ed¹.] Dressed up for a party.

1928 *Sunday Express* 13 May 16/4 He couldn't quite get over the queerness of seeing Bobs with feathers on her head and white gloves up to her shoulders—and that strip of train on the ground..the thought of a word used by his childhood's nurse to denote a certain standard of sartorial tribute to 'occasions'. 'You look so frightfully "partified",' he added. 1969 A. Christie *Hallowe'en Party* i. 9 They look good and partified.

parti-generic (pā·ɹti,dʒene·rik), *a. Linguistics.* [f. Parti-² + Generic *a.* (*sb.*)] Referring to an indefinite amount or sub-set of a whole.

1939 P. Christophersen *Articles* ii. 33 Continuate-words have only zero-form and *the*-form. The former is used when the thing meant is viewed as unlimited or having indefinite limits. We can distinguish three different significations of the zero-form: (1) The whole genus everywhere and at all times (*toto-generic sense*)... (2) An indefinite amount of the genus (*parti-generic sense*). *Ibid.* 34 (3) (In negative phrases) nothing of the genus (*nulli-generic sense*). 1962 J. Söderlind in F. Behre *Contrib. Eng. Syntax* 103 They [*sc.* 'uncountables and plural countables' in zero-form] then refer either to the whole genus of their content (toto-generic sense) or to some indefinite amount or number of it (parti-generic sense).

parting, *vbl. sb.* Add: **2. a.** (*b*) *parting off*: the separation of a piece from a longer length (cf. *part *v.* 4 c (*d*)). Freq. *attrib.*

1905 J. Horner *Tools for Engineers & Woodworkers* v. 60 Tools for parting off..have clearance both behind and below. Being generally very thin at the cutting end, this is commonly reduced from a bar of greater width, in order to afford sufficient width and rigidity for clamping in the tool-holder. 1923 C. M. Linley *Lathe Users' Handbk.* v. 88 In capstan work where parting-off tools are in continual use..I have used milling cutters or slitting saws as tools with great success. 1950 C. T. Bower in A. W. Judge *Machine Tools & Operations* II. viii. 186 The parting-off saw shown..has been evolved for cutting off non-ferrous extrusions or bars up to 4 in. by 2½ in. deep. 1977 C. R. Shotbolt *Technician Workshop Processes & Materials* I. vii. 84/1 The draw tube and the back end of the collet are hollow to permit bars to be fed through the spindle for repetition turning and parting-off of workpieces.

7. a. (Further examples of *parting shot*.) Cf. *Parthian shot* s.v. Parthian *a.*

1906 Galsworthy *Man of Property* iii. ix. 372 He could not resist a parting shot. 'H'mm! All flourishing at home? Any little Soameses yet?' 1957 P. Worsley *Trumpet shall Sound* iii. 67 Dasiga..finally left with a parting-shot of 'unintelligible gibberish'. 1962 P. Van Greenaway *Crucified City* xiii. 136 Still Creston permitted himself a parting shot. 'I suppose we're going through the whole farce again?' 1963 *Daily Tel.* 19 Aug. 8/6 It was also something of a parting shot, following a 100 yards victory in 9·9 sec. 1967 T. Stoppard *Rosencrantz & Guildenstern* ii. 57 He smiles briefly at them without mirth, and starts to back out, his parting shot rising again.

parti pris: see Parti 2 in Dict. and Suppl.

‖**Parti Québecois** (parti kebɛkwa). *Canad.* [Fr., f. Parti + *Québecois *a.*] A French-Canadian political party which advocates greater autonomy for Quebec. Also *attrib.*

1968 *Times* 30 Oct. 4/7 After many years as the first official champions of the separatist cause in Quebec, Le Rassemblement pour l'Indépendance Nationale (R.I.N.) has decided to join forces with Mr. Rene Levesque's new Parti Québecois which until recently was Le Mouvement Souveraineté-Association (M.S.A.). 1970 *Globe & Mail* (Toronto) 27 Apr. 4 Rene Levesque, the Parti Quebecois leader. 1972 *Maclean's Mag.* Mar. 8/2 It is interesting to note that the Parti Québecois..has managed to affect the terrorist movement. 1976 *Southern Even. Echo* (Southampton) 16 Nov. 1/2 Parti Quebecois won a clear majority of seats in the Quebec National Assembly. 1978 *Times* 20 Mar. 5/6 On the Parti Québecois social policies, Mr Lévesque said that it was a left-of-centre party.

partisan, *sb.*¹ Add: **2. c.** In the war of 1939–45, a guerrilla, esp. one working in enemy-occupied territory in Eastern Europe and the Balkans, *spec.* in Yugoslavia. Also *attrib.*

1939 C. Gubbins (*title*) Partisan leader's handbook. 1942 *Daily Tel.* 22 May 1/3 Behind the fighting front the Russian 'partisan front' in the German rear forms a skeleton army. 1944 *Hutchinson's Pict. Hist. War* 27 Oct. 1943–11 Apr. 1944 414 In the autumn of 1941 Marshal Tito's partisans began a wild and furious war for existence against the Germans... The partisan movement soon outstripped in numbers the forces of General Mikhailovitch. 1958 P. Kemp *No Colours or Crest* vi. 100 He arrived with thirty Partisans, saying he intended to lay an ambush in exactly the same place as ours. 1965 B. Sweet-Escott *Baker St. Irreg.* vii. 191 Maclean and Velebit were mainly concerned to obtain British training for a Yugoslav tank regiment and a fighter squadron, and to get a fleet of light craft for a partisan navy. *Ibid.* 193 His assignment had been to make contact with the Bulgarian partisans. 1968 *New Left Rev.* Jan.–Feb. 67 During the Second World War I had no doubts about which side I was on in the struggle, let us say, between the Yugoslav Partisans and the Nazi occupation forces. 1974 tr. *Sniečkus's Soviet Lithuania* 47 The Lithuanian people gave every possible aid to the partisans, whom they regarded as true patriots. 1978 A. Price *'44 Vintage* xix. 220 He got back in..in 1939... France in '40, then the Middle East... And finally Yugoslavia as a weapons adviser to a big Partisan outfit.

partisanery (pāɹtizæ·nəri). *rare.* [f. Partisan, partizan *sb.*¹ (*a.*) + -ery.] A partisan feeling or act.

1911 G. B. Shaw *Getting Married* Pref. 119 Such paltry follies and sentimentalities, snobberies and partisaneries, as ignorance can understand and irresponsibility relish.

partisanly, *adv.* (Earlier and later examples.)

1866 H. Sidgwick *Let.* 7 Nov. in A. & E. M. Sidgwick *Henry Sidgwick* (1906) 153 To ensure no..votes be lost, partisanly speaking. 1976 *Church Times* 9 July 12/4, I wish I could understand why so many Christians feel so strongly—or so partisanly—about events in South Africa. 1977 *Times Lit. Suppl.* 27 May 644/3, I may be partisanly over-optimistic about the ability of the Sussex Constabulary to stand up to Mother Ancilla.

partita (pāɹtī·tă). *Mus.* [It.] An air with variations; a suite.

1880 Grove *Dict. Mus.* II. 656/1 He [*sc.* Bach] also wrote three Partitas (in the Suite-form) for the lute. The name has very seldom been used since Bach... But in the modern rage for revivals it may possibly reappear. 1925 *Chambers's Jrnl.* 31 Oct. 755/1 In such a week Mr Samuel has played sixteen preludes and fugues, ten French suites, partitas and English suites. 1948 *Times* 30 July 7/3 Mr. Vaughan Williams's most recent work, a Partita for strings, was given its first concert performance at the Albert Hall last evening. 1967 I. Spink *Hist. Approach Mus. Form* iii. 67 Bach clearly distinguishes the forms in his six sonatas for violin alone by calling those in church style 'sonatas' and those in chamber style 'partitas'.

partition, *sb.* Add: **1. a.** esp. the division of a country into two or more nations; *spec.* (*a*) the division of Ireland into Northern Ireland and the Irish Republic; (*b*) the division of the Indian sub-continent into India and Pakistan in 1947. Hence **parti·tionist**, one who advocates partition; also as *adj.*

1919 *Times* 25 Jan. 9/5 (*headline*) Irish Unionist breach. New League against Partition. *Ibid.* 27 Jan. 9/4 The principles of the [Anti-Partition] League were defined as follows:—To maintain the legislative union of Great Britain and Ireland, to secure Ireland against partition, and to safeguard the liberties and interests of Irish Unionists. 1919 P. S. O'Hegarty *Ulster* 1 In North Down the Independent vote was counted as Partitionist. 1921 *Spectator* 4 June 713/2 Partition has come to be reckoned the unforgivable sin by the Sinn Feiners. The worst thing a man can be called is a partitionist. 1938 *Ann. Reg. 1937* 256 The World Zionist Executive, sitting in Jerusalem, formally resolved that the Zionist Organization would resist any attempt to curtail the rights of the Jews as defined in the Mandate, either by Partition or any other measure. 1941 K. Chandra *Tragedy of Jinnah* xv. 220 Many schemes of partition of India on communal basis [*sic*] were put forward by a few fanatics, off and on. 1945 *Ann. Reg. 1944* 164 This weakening of Mr. Gandhi's hitherto uncompromising opposition to partition was denounced by the Hindu Mahasabha and the Sikhs and caused much misgiving in Congress circles. 1948 *Ann. Reg. 1947* 153 The Muslims having voted for partition of India, the Hindus and Sikhs then voted for the partition of two Provinces, and the frontiers of Pakistan were thus drawn in the midst of the Punjab and Bengal. 1955 R. P. Jhabvala *To whom she Will* vi. 43 Hari Sahni's family.. were Punjabi Indians who in 1947, at the time of Partition, had had to leave their native Lahore. 1959 P. Colum *Arthur Griffith* ii. iv. 121 The Irish people were now shown that this claim would be countered by a move for Partition. 1967 P. M. Hubbard *Custom of Country* (1969) vii. 81, I agreed I was English, and he said, 'Ah, then you will have been a government servant out here before Partition.' 1971 R. Dentry *Encounter at Kharmel* iii. 58 There hasn't been a tribal rising..since the Partition troubles died down. 1972 A. Boyd *Brian Faulkner* i. 12 the issue, as in all previous elections, was partition. 1973 *Archivum Linguisticum* IV. 42 In the Indo-Pakistan subcontinent before partition..Delhi and Lucknow were recognized as the places where 'they spoke the best Urdu'. 1975 *Guardian* 24 Feb. 9/6 Just as in every other country that has fallen for the silly expedient of Partition—Ireland, Korea, Vietnam—nobody profits but the politicians.

d. *Physical Chem.* The distribution of a solute between two immiscible or slightly miscible solvents in contact with one another, in accordance with its differing solubility in each. Cf. *partition coefficient* (sense *10).

1861 *Nat. Philos. for Use of Schools: Chem. & Chemical Analysis* ii. 44 Partition of Elements.—Affinities are sometimes modified by circumstances which might not be supposed likely to produce any effect: thus, the more difficult solubility of one of the compounds present in a solution. 1898 C. L. Speyers *Text-bk. Physical Chem.* v. 119 (*heading*) Partition of substance between two solvents. 1950 F. Haurowitz *Chem. & Biol. Proteins* iii. 21 Synge..has made use of the partition of amino acids between water and..organic solvents which are immiscible with but partly soluble in water. 1970 Sherma & Zweig in G. Zweig et al. *Paper Chromatogr. & Electrophoresis* (1971) II. ii. 11 The paper..acts by a combination of partition, adsorption, and ion exchange.

6. b. *Math.* A collection of non-empty subsets of a given set such that each element of the latter is a member of exactly one of the subsets; a way of dividing a set thus.

1905 J. Pierpont *Lect. Theory Functions Real Variables* I. ii. 82 Let α be any number of ℜ; we can use it to throw all numbers of ℜ into two classes *A*, *B*. In *A* we put all numbers < α; in *B* all numbers > α. The number α we may put in *A* or *B*. This division of the numbers of ℜ into two classes we call a partition. 1937 Michell & Belz *Elem. Math. Analysis* II. xxi. 1051 The notion of partitions of the rationals forms the basis of Dedekind's treatment of real numbers. 1968 E. T. Copson *Metric Spaces* i. 12 If ~ is an equivalence relation on a set *E*, two equivalence classes are either identical or have no common member; the collection of all equivalence classes is a partition of *E*. 1972 A. G. Howson *Handbk. Terms Algebra & Anal.* xxiv. 118 A finite set of points $P=\{a_0,\ldots,a_n\}$ satisfying the above requirements, i.e. $a=a_0<a_1<a_2<\ldots<a_n=b$, is known as a partition of the interval ⩽*a*, *b*⩾. 1959 Perfect & Petersen tr. *Alexandroff's Introd. Theory of Groups* vi. 67 Every group may be partitioned into classes of mutually conjugate elements. 1966 S. Beer *Decision & Control* vi. 106 The set of commercial responsibilities is partitioned, for example, into home sales (heavy), home sales (light) and export sales. 1968 E. T. Copson *Metric Spaces* v. 68 We can partition this set into two separated sets.

10. *partition fence*; **partition chromatography**, chromatography which utilizes the differing solubilities of the components of a mixture in a liquid sorbent (chosen to be immiscible with the carrier if this is a liquid); *spec.* that in which the sorbent is a liquid and the carrier is a less polar liquid; hence **partition chromatogram**, **-chromatographic** *a.*; **partition coefficient** *Physical Chem.* [tr. F. *coefficient de partage* (Berthelot & Jungfleisch 1872, in *Ann. de Chim. et de Physique* XXVI. 398)], the ratio of the concentrations of a solute in each of two immiscible or slightly miscible liquids, or two solids, when it is in equilibrium across the interface between them; **partition function** *Physics*, a sum of the form $\Sigma\, g_i \exp(-E_i/kT)$ (where g_i is the degeneracy of the state with energy E_i, *k* is Boltzmann's constant, and *T* the absolute temperature), or an analogous integral, which enters into the expression for the distribution of the particles of a system among different energy states and other thermodynamic quantities; symbol *Z*; **partition treaty** (earlier example).

1944 *Biochem. Jrnl.* XXXVIII. 286/2 On paper strip partition chromatograms..a number of free peptides travel as reasonably narrow bands whose presence can be revealed by treatment with ninhydrin. *Ibid.* 65/1 We record here some technical aspects of the experience which we have gained in the use of our partition chromatographic method..for the quantitative analysis of amino-acid mixtures. 1968 *Jrnl. Chromatogr.* XXXVII. 97 Partition systems..are much less likely to hinder complete recovery of unchanged, separated components. Hence there is an interest in developing practically useful liquid partition chromatographic systems. 1943 A. H. Gordon et al. in *Biochem. Jrnl.* XXXVII. 79/1 In the present paper we report new applications and developments of partition chromatography* (Martin & Synge, 1941b) in the study of amino-acids and peptides. [*Note*] * We employ the term 'partition chromatography' at the suggestion of Dr. E. Lester Smith, to distinguish it from the classical adsorption chromatography. Our earlier term 'liquid-liquid chromatography' was liable to confusion with the fractional elution procedure sometimes called 'liquid chromatography'. 1966 *McGraw-Hill Encycl. Sci. & Technol.* III. 94/2 Volatile, nonpolar substances such as hydrocarbons may be examined by gas adsorption or gas partition chromatography... Weakly polar substances such as alcohols..are examined by adsorption or partition chromatography. *Ibid.* 95/2 With the phases reversed, that is with the polar phase as the wash liquid and the less polar phase fixed in the support, the method is known as reversed phase partition chromatography. It provides a chromatographic sequence the inverse of that produced by partition chromatography. 1967 M. E. Hale *Biol. Lichens* viii. 130 It is only in the past 15 years that the development of partition chromatography has brought a rapid and sure means of identifying plant products within the reach of taxonomists and physiologists. 1891 *Jrnl. Chem. Soc.* LX. ii. 1148 (*heading*) Relation between affinity and partition coefficients in immiscible solvents. 1925 *Jrnl. Iron & Steel Inst.* CXII. 492 Variations in the concentration of cementite can be studied by means of the position of the Curie point, which allows of ascertaining variations in the partition coefficient of manganese between the carbide and the ferrite according to the annealing temperature. 1964 G. I. Brown *Introd. Physical Chem.* xxi. 235 Suppose that 100 cc. of benzene is available for extracting a solute, X, dissolved in 100 cc. of water, and that the partition coefficient of X between benzene and water is 5. 1639 *Early Rec. Dedham, Mass.* (1892) III. 51 A pticon [*sc.* partition] fence in the same. 1748 *New Hampsh. Probate Rec.* (1916) III. 608 [This land is] to be possess'd and enjoy'd by them..as ye partition fence between them now stands. 1870 *Rep. Comm. Agric. 1869* (U.S. Dept. Agric.) 395 Partition fences must be proof against sheep. 1927 Fisher & Hartree tr. *Born's Mech. Atom* 3 The so-called partition function (Zustandsintegral). 1970 P. W. Atkins *Molecular Quantum Mech.* x. 392 The proportion of molecules in the rotational state *J* is given by $Z^{-1}(2J+1)\times\exp\{-E(J)/kT\}$, where *Z* is the rotational partition function. 1711 Swift *Conduct of Allies* 15 The Violation of the Partition-Treaty, by the French.

partition, *v.* Add: **1. c.** *Math.* To subdivide by means of a partition (sense *6 b).

1943 K. Mather *Statist. Anal. Biol.* xi. 180 A compound χ^2 can be partitioned into simple components each dependent on a single comparison and each taking 1 degree of freedom.

partless, *a.* **3.** (Later examples.)

a 1868 G. M. Hopkins *Poems* (1967) 36 Your parley was not done and there! You went into the partless air. 1904 G. S. Fullerton *Syst. Metaphysics* xxv. 410 An unanalyzable and part-less thing like the percept. 1972 *Sci. Amer.* Aug. 8/2 Democritus..desired that his atoms be partless, a property they were to have in common with geometrical points.

partner, *sb.* Add: **3. e.** *Biol.* Each of a pair or group of symbiotically associated organisms.

1924 J. A. Thomson *Sci. Old & New* xxvii. 147 The fungus partners, which supply the water and salts, sometimes get the upper hand and absorb their partner algae, without which, however, they cannot continue to live. 1925 R. S. Lull *Ways of Life* ii. 39 Both partners [of a lichen] combine, when they come to the formation of sexual reproductive bodies. 1970 *Canad. Jrnl. Zool.* XLVIII. 371 (*heading*) The association of *Calliactis tricolor* with its pagurid, calappid, and majid partners in the Caribbean.

f. partners' (also **partner's, partners**) (**pedestal**) **desk** (see quot. 1952).

1946 *Connoisseur* Dec. p. i (Advt.), Mahogany Partners' Pedestal Desk in style of William Kent. *Circa* 1745. 1950 *Apollo Misc.* p. ix (*caption*) A fine mahogany pedestal partners' desk. 1952 J. Gloag *Short Dict. Furnit.* 351 *Partners' desk*, a large flat-topped kneehole desk, at which two people may sit, facing each other. The term is probably of 19th century origin. 1971 J. Leasor *Love-All* iii. 36 His eighteenth-century partner's desk was covered with polished leather. 1974 *Country Life* 11 Apr. (Suppl.) 70 Handsome Partners Desk... The Desk is fitted with five drawers each side.

partner, *v.* Add: **3.** *intr.* (cf. next). **a.** To associate or work as partners. **b.** To associate or join with as a partner.

1961 A. B. Mayse in *Webster* s.v., Him and me, we partnered once. 1961 R. O. Bowen in *Ibid.*, He still partnered with Tom on the piers. 1968 *Globe & Mail* (Toronto) (Mag.) 17 Feb. 9/2 In 1929 he partnered with a U.S. businessman, Ben Raeburn, to publish a series of 'forbidden' sex books. 1977 *Bangladesh Times* 19 Jan. 5/1 Partnering with Habib against Tipu and Haroon, the foursome played perhaps the most evenly contested matches in Bangladesh tennis to date.

partnering (pā·ɹtnəriŋ), *vbl. sb.* [f. Partner *v.* + -ing¹.] Association as partners; the action or work of a partner.

1897 S. & B. Webb *Industr. Democr.* II. 475 Occasionally the employer has tried to have only one boy-piecer to two spinners. This system, called 'joining' or 'partnering', is always resisted by the union. 1963 *Times* 8 Jan. 4/1 His partnering proved adequate and his dancing showed vitality and breadth. 1975 'M. Fonteyn' *Autobiogr.* i. iv. 41 This impromptu lesson stood me in good stead after I joined the ballet and sometimes danced with boys as new to partnering as I was. 1976 U. Holden *String Horses* vii. 80 Wait until you try my partnering. I'll take a turn with you both.

partnership. Add: **1.** (Further examples.) Now esp. of relationships in industry and politics.

1933 *Planning* 25 Apr. 2 It aims at giving labour effective partnership in industry and at creating a new attitude of mind to replace sterile hostilities. 1941 *Ann. Reg. 1940* 140 The attainment by India of free and equal partnership in the British Commonwealth. 1959 *Times Lit. Suppl.* 31 July 449/2 It has been a cardinal principle in Rhodesia that the way should be open not for *apartheid* but for partnership, and that the new State will avoid the racial impasse of the Union. 1976 *Star* (Sheffield) 30 Nov. 14/7 Dennis Amiss and Brearley, the opening partnership which England hope will serve them throughout the forthcoming Test series with India, made a quiet start to their alliance when they launched the innings. 1977 *World of Cricket Monthly* June 31/2 He and Kallicharran set about repairing the damage with a fourth-wicket partnership of 90 in 85 minutes and 20 overs.

partness (pā·ɹtnės). Also **part-ness**. [f. Part *sb.* + -ness.] The fact or quality of being partial or incomplete.

1925 D. H. Lawrence *Let.* ?26 Jan. (1962) II. 828 Any relation based on the one half—say the delicate spiritual half alone—inevitably brings revulsion and betrayal. It is halfness, or partness, which causes Judas. 1945 Koestler *Yogi & Commissar* 240 Isolation destroys their very character of 'part-ness'. *Ibid.* 253 A 'vertical' approach which brings to the dry concepts of part-ness, love and all-oneness the igniting spark of experienced reality.

parton (pā·ɹtɒn). *Nuclear Physics.* [f. Part(icle *sb.* + *-on¹.] Each of the hypothetical point-like constituents of the nucleon that were invoked by R. P. Feynman to explain the way the nucleon inelastically scatters electrons of very high energy.

The printed coinage was published after the term had been given currency by Feynman in discussions.

1969 *New Scientist* 26 June 679/2 A similar 'current bun' concept of the proton is implied by the so-called 'parton' theory of Feynmann [*sic*]. 1969 R. P. Feynman in C. N. Yang et al. *High Energy Collisions* 241, I call the fundamental bare particles of my underlying field theory 'partons'. 1973 *Sci. Amer.* Aug. 34/3 There is some evidence that partons and quarks are the same, although they have been postulated in different ways. 1974 M. L. Perl *High Energy Hadron Physics* xx. 482 We take the partons to be point particles with fixed mass and fixed internal quantum numbers... The quark model discussed in Ch. 14 is a particular form of the parton model in which the partons have been assigned a particular set of internal quantum numbers.

part-owner. Add: Also stressed pa:rt-ow·ner. Hence (stress variable) **part-own** *v. trans.*; **part-ownership**.

1890 *Act* 53 & 54 *Vict.* c. 39 §2 Joint tenancy, tenancy in common, joint property, common property, or part ownership does not of itself create a partnership as to anything so held or owned. 1969 R. Esser *Hot Potato* 42 He..part-owned a night club. 1974 *Listener* 10 Oct. 460/3 A mainly middle-class constituency was told of his part-ownership of a factory. 1975 D. Bloodworth *Clients of Omega* iv. 31 This mighty organization..covers two thousand factories, banks and businesses... It part-owns a high percentage of them. 1977 M. T. Bloom *Thirteenth Man* (1978) ii. 26 Now he was wealthy enough to have..part-ownership of a private bank.

partridge. Add: **5.** *partridge pie*; **partridge bush** = Partridge-berry b; **partridge plum**, the fruit of the partridge-berry; **partridge** (**berry**) **vine** (examples).

1843 *Amer. Pioneer* II. 125 The vivid green leaves and bright scarlet berries of the 'partridge bush', or 'Checkerberry'. 1723 J. Nott *Cook's & Confectioner's Dict.* sig. Y7 (*heading*) To make a Partridge Pye. 1757 Earl of Buckingham *Let.* Sept. in *Lett. to & from Henrietta, Countess of Suffolk* (1824) II. 239 If the partridge-pie gives you as much pleasure as your letter did to me. 1963 A. L. Simon *Guide Good Food & Wines* 579/2 Partridge pie. 1872 Mrs. Stowe in *Christian Union* 3 Jan. 32/3 Little Love gathered stores of bright checker berries and partridge plums. 1868 H. W. Beecher *Norwood* 91 Here the little queen took on airs, and sent her Ethiop..[for] some partridge-berry vines from the edge of the wood. 1880 *Harper's Mag.* Nov. 864/1 Here are soft beds of rich green moss studded with scarlet berries of wintergreen and partridge vine. 1913 W. P. Eaton *Barn Doors & Byways* 245 We have come upon ferns still flaunting through the snow and partridge berry vines; scratched up into sight by some hungry bird. 1940 *Sun* (Baltimore) 9 Dec. 8/4 In Christmas seasons when holly berries are comparatively scarce, the berries of the smoke bush come as a substitute, and often of the dogwood and of the partridge vines in the woodlands.

part-time (pā:ɹt,təi·m), *a.* [Part *sb.* (*adv.*) B. c.] Employed, occurring, lasting, etc., for part of the time or for less than the customary time. Also as *adv.* Cf. *full time.

1891, 1896 [see Part *sb.* (*adv.*) B. c]. 1931 *Times Lit. Suppl.* 14 May 380/3 The legislation of 1919..has put some 10,000 small farmers and 20,000 men upon part-time and spare-time holdings. 1955 *Times* 1 July 10/7 The chairman need devote only part of his time to the board. Most of the members should be part-time. 1959 *New Statesman* 23 May 710/2 Are married women who worked part-time for years after the war, but are now being squeezed out, to be compensated? 1965 M. Morse *Unattached* i. 33 She began studying part-time for G.C.E. 'O' levels. 1973 *Fisheries Fact Sheet* (Environment Canada Fisheries & Marine Service) No. 1. 4/3 Winter fishing..is carried on by teams of men, many of whom are only part-time fishermen.

Hence **part-ti·mer**, a part-time worker, student, or the like.

1927 *Daily Tel.* 3 May 3 (*heading*) Part-timers employed. 1936 A. G. Street *Gentleman of Party* vii. 134 The dairyman's eldest son, George, went to work as a part-timer at ten years of age. 1939 *Rostrum* (N.Z.) 7 Some grove of Academe where part-timers will have no place. 1952 G. Wilson *Julien Ware* 19 The part-timers were a handful of men engaged by John Cecil for a couple of months at the peak of each year. 1972 *Daily Tel.* 25 Apr. 7/2 Places for at least one million full-time students in 1980 would be planned, with a comparable provision for part-timers. 1976 *Listener* 2 Dec. 712/2 What is often seen as a part-time job is further split among several part-timers: the *Times* has four people who review [TV] programmes.

parturiate, *v.* **a.** (Later example.)

1922 Joyce *Ulysses* 206, I am big with child... Let me parturiate!

parturient, *a.* Add: **B.** *sb.* A parturient woman.

1956 *Amer. Jrnl. Obstetr. & Gynecol.* LXXI. 1251 A clinical program was set up to evaluate the effects of chlorpromazine on the parturient during the first and second stages of labor. 1958 R. Liddell *Morea* ii. viii. 192 No birth or death might take place in the sacred enclosure, and the dying or parturient had to be carried hastily on to the hills. 1974 R. Winston tr. *Wunderlich's Secret of Crete* xxv. 334 Many highborn young ladies brought children into the world in this way; the infants were ascribed to the god in whose sanctuary these women had been temple servants or whose medical men had taken care of the parturient.

part-way (pā·ɹtwēi), *adv.* Also **part way, partway**. [Part *sb.* (*adv.*) B. c.] Part of the way; a certain distance along the way; to some extent; partly.

1859 Mrs. Gaskell *Lett.* (1966) 530 Flossy is gone to school..at Knutsford. I took her part way on Tuesday last. 1875 [see Part *sb.* (*adv.*) B. c]. 1930 Blunden *De Bello Germanico* 15 Half-ruined houses, with sacks stretched partway over some windows. 1954 *Essays in Crit.* July 320 It is unbearable for a man or woman to be faced with anything less than a person—and thus, tragically, even part-way unbearable to be faced only with other human persons, where the personal relationship is inevitably enmeshed in material situations. 1968 C. Helmericks *Down Wild River North* i. ii. 22 Starting off toward Billings, we learned partway there that the road was blocked. 1976 I. Levin *Boys from Brazil* i. 34 He stopped and rewound part-way and replayed a few bits. 1978 *Sci. Amer.* Feb. 68/1 When a peg is partway into a hole, it can wobble back and forth a good deal farther than the clearance itself.

party, *sb.* Add: **II. 6. c.** *spec.* (freq. with capital initial) *the party*: the Communist Party.

1920 *Times* 5 Oct. 14/3 (*heading*) Realities of Russia. Iron Rule of 'The Party'. 1922 E. P. Oppenheim *Great Prince Shan* iv. 38 'Her father at present represents the shipping interests of Russia and England. He is one of the authorised consuls.' 'Is he of the party?' 1928 E. & C. Paul tr. *Stalin's Leninism* I. 168 The central unit of organisation is the Party. 1936 A. Huxley *Eyeless in Gaza* xxii. 316 One joined the Party, one distributed literature. 1943 Koestler *Arrival & Departure* i. 18 He had courage, but he could not adapt himself to changes in the tactics of the movement. That's why he had to leave the Party. 1959 *Times Lit. Suppl.* 11 Sept. 575/4 The recent dramatic dismissal of the Kerala Government by the President of India..is certain to influence the attitude of the Party towards legally constituted authority. 1975 *New Left Rev.* Nov.–Dec. 65 A whole series of import-export co-operatives..pay a tithe of their profits to the party.

9. b. Phrases: *the party is over*: enjoyment must stop; the happy or easy times are at an end; *to keep the party clean*: to act responsibly; to conform to accepted patterns of behaviour.

1937 M. Allingham *Dancers in Mourning* xxvii. 328 'You've made up yer mind to go to-day? It's a lovely day.' His wistfulness was pathetic, and Campion felt sudden sympathy for him. 'I'm afraid so,' he murmured. 'The party's over. Sorry.' 1938 E. Waugh *Scoop* iii. 210 'The party's over,' said Bannister... 'From tomorrow onward, I shall get a daily pile of bumf from the Ministry.' 1959 *Times* 7 Nov. 7/4 Then they [*sc.* young people] should be invited to cooperate in keeping the party clean and so ensuring that the letting off of fireworks is kept within bounds. 1965 D. Francis *For Kicks* vi. 91 'Do your dress up,' I said. 'Why? Are you impotent after all, Danny boy?' 'Do your dress up,' I repeated. 'The party's over.' 1975 *Times* 10 May 1/1 Local government..is coming to realize that, for the time being, the party is over. 1977 *Time* 22 Aug. 5/1 Now in the '70s, the party is over. 1977 Partridge *Dict. Catch Phrases* 131/1 *Keep the party clean!*... Don't talk smut or tell dirty stories; don't act loosely or indelicately. A correspondent..commented thus: 'But the speaker often does not quite mean it. "Give me my hat and knickers," she said. "I thought you were going to keep the party clean."' 1978 D. Murphy *Place Apart* xi. 243 To them [*sc.* Provos] their campaign is..a conventional war, and they want to keep the party clean.

c. An attack, a combat, or fight; an operation, or a unit engaged in an operation. *Armed services' slang.*

1942 'B. J. Ellan' *Spitfire!* v. 23, I just fired when something came into my sights and then turned like hell as something fired at me! What a party! 1942 *R.A.F. Jrnl.* 27 June 7 Confirmation came through that the big party was on. And that the target was Cologne. 1943 P. Brennan et al. *Spitfires over Malta* 18 This party is about 40 miles North, coming South. 1946 J. Irving *Royal Navalese* 131 A good example of naval understatement for, in one sense, a 'party' can mean quite a tough fight while it lasts. 1959 *Times Lit. Suppl.* 7 Aug. p. iii/3 Classic understating metaphors like 'having a party'..had their value in time of war when men had to accept as steadfastly and as mildly as they could the possibility of their own violent and horrible deaths.

III. 14. c. A telephone subscriber; a person using a telephone.

At first used only with reference to party lines (*party line 2), subscriber being the usual term.

1912 J. Poole *Pract. Telephone Handbk.* (ed. 5) xxxii. 531 Party lines up to 4 stations on a line can be worked on the automatic by giving each of the parties on the line a separate number and multipleing the lines on four sets of connectors. 1938 C. W. Wilman *Automatic Telephony* (ed. 2) xix. 185 All incoming calls are received at one of the ordinary telephones, from which they can be transferred to the wanted party. 1955 W. Gaddis *Recognitions* ii. v. 508, I think we got a crossed wire, would the other parties mind hanging up. 1973 *Times* 23 May 8/7, I heard him say: 'I am receiving a report on that right now' to the party on the other end.

V. 19. (sense 6 or *6 c) *party boss, card, caucus, cell, conference, congress, convention, discipline, government* (earlier example), *hack, image, label, leader* (later examples), *list* (later example), *loyalty, machine, manager, meeting, member, membership, -mindedness, passion, platform, point, -political* adj. (examples; also ellipt., a party-political broadcast), *politician, politics* (later example), *press, question, rally, secret, -state, system, worker*; *party-bound* adj.; (sense 8) *party-boat* (N. Amer.) (hence as *v. intr.*); (sense 9) *party call, dress, game, girl, -giving* (earlier and later examples), *-going, mood, piece, record, thrower, trick*; *party-frocked, -like* adjs.; **party manners**, good manners, best behaviour; **party piano**, a boogie-woogie or barrel-house style of piano-playing; **party plan**, a sales strategy by which goods are displayed or demonstrated at a party in a private house; also *attrib.*; **party poop, pooper** *U.S. slang*, one who throws a gloom over social enjoyment; **party spirit**, (*a*) feeling of solidarity with and support for one's political party (in Dict.); (*b*) feeling or atmosphere of festivity.

1937 E. Hemingway *To have & have Not* i. i. 14 You party-boat captains. *Ibid.* 28, I..went party-boating and broke out this sword-fishing in Cuba. 1963 J. T. Rowland *North to Adventure* i. 27 She [*sc.* a sloop] had a long, open cockpit and a small cuddy, since it was his intention to rent her as a party boat. 1909 *Times* 27 Apr. 4/5 There would thus be a good chance of undermining the 'party boss'. 1938 *Ann. Reg. 1937* 320 The men and the movement, the party and the party bosses, the national aims of the Third Reich. 1977 J. Cleary *Vortex* ii. 32 He had never been able to make his mark with any of the state's party bosses. 1908 *Westm. Gaz.* 30 Dec. 4/3 The one honest solution which is constantly shirked by those who are either party-bound or who..hope for a secular solution. 1936 Wirth & Shils tr. *Mannheim's Ideology & Utopia* iii. 136 Those whose standpoints are party-bound are finding it necessary to have a broader perspective. 1910 J. W. Tompkins *Mothers & Fathers* 144 They only came twice, and those were party calls. 1935 N. Mitchison *We have been Warned* ii. 209 He had very few papers..his trade union card, and his Party card. 1970 *Guardian* 10 Mar. 13/1 Union leaders must talk to Ministers, whatever the shade of their party cards. 1977 P. Johnson *Enemies of Society* xiv. 185 Such a brutal division can only be maintained by laws of status and privilege—a nobility by birth, or by party card. 1882 W. M. Thayer *From Log Cabin to White House* 300 Garfield..was nominated by acclamation at the party caucus, and unanimously elected. 1977 *N.Z. Herald* 5 Jan. 1–6/3 The party caucus has long functioned as part and parcel of New Zealand politics. 1949 *Ann. Reg. 1948* 285 There were 25,635 party cells in all sections of the country's structure and economy. 1978 F. Maclean *Take Nine Spies* iv. 132 The Party..[was] using him to organize Party cells and cadres and study-groups. 1935 N. Mitchison *We have been Warned* iv. 381 It was the best Party Conference he'd been at. 1969 *Listener* 12 June 827/1 The Party Conference defeated Gaitskell..on the question of nuclear weapons. 1954 B. & R. North tr. *Duverger's Pol. Parties* ii. i. 273 After 1935 the army played a part of first-rate importance in the party Congresses at Nuremberg. 1977 W. Weaver tr. *Morante's Hist.* (1978) 341 The Party Congress had been held in Rome. 1881 *Nation* (N.Y.) XXXIII. 4 The slipshod method in which the Vice-President is commonly chosen by party conventions. 1976 *National Observer* (U.S.) 1 May 4/5 The rural followers of William Jennings Bryan fought Al Smith's city legions at the party convention in New York City. 1830 *Deb. Congress U.S.* 9 Feb. 149/2 The provisional power of removal from office by a President..[should not] be exercised..in the corrupting spirit of 'party discipline'. 1909 H. Zimmern tr. *Nietzsche's Human, All-too-human* I. 181 The old obsequiousness.. still survives in party-feeling and party-discipline. 1933 *Discovery* Feb. 64/1 The English party system has many advantages, but it also has the disadvantage that party-discipline and party-loyalty do sometimes exercise a prejudicial, cramping or numbing effect on the mind and actions of individual members of the House of Commons. 1961 W. A. Swanberg *Citizen Hearst* iv. iii. 213 He.. came to be regarded..as a professional radical leading a small pack of obedient terriers whose constant snapping was demoralizing to party discipline. 1873 'S. Coolidge' *What Katy did at School* (1874) x. 175 Elsie is much excited over the party dresses which Mrs. Hall is having made for her. 1960 [see *cover-up *attrib.* or quasi-*adj.*]. 1973 G. Greene *Honorary Consul* iv. ii. 195 A pale-faced child of three..dressed in a blue party dress. 1960 *Times* 3 Aug. 5/2 A juvenile delinquent cousin who appears.. party-frocked in the second [act]. 1929 G. Mitchell *Mystery of Butcher's Shop* xvii. 192 Idiotic invitations to play silly party games with pencils and paper. 1972 *Guardian* 14 Mar. 14/6 Party game time. What do the following have in common? 1936 *Sun* (Baltimore) 20 Apr. 3/1 A 'party girl's' flat was hunted by police today as the probable scene of the slaying of Arthur P. Hewitt, rich retired contractor. 1960 A. West *Trend is Up* vii. 298 He had never thought of asking himself what she was, inside of the hard shell of her disguise as the party girl who would go the limit for fun. 1968 *Globe & Mail* (Toronto) 11 July 5/2 The prostitutes and partygirls who hang around the beer parlors are having poor pickings, police say. 1977 *Time* 26 Dec. 36/1 He characterized Sylvia Miles as a 'party girl and gate crasher'. 1864 Trollope *Small House at Allington* II. xxiii. 242 Nor had they any ground on which to stand, except the party-giving ground. 1963 *Times Lit. Suppl.* 10 May 345/2 A sensible guide to party-giving of every sort. 1864 Trollope *Small House at Allington* II. xxvi. 274, I knew it wasn't to be all party-going and that sort of thing. 1973 *Listener* 14 June 805/2, I am probably..cloistered..when it comes to party-going, but I do hear a lot of talk and gossip. 1977 M. Hinxman *One-Way Cemetery* xvi. 119 She didn't feel up to party-going and I sympathized. 1869 R. F. D. Palgrave *House of Commons* 22 This system is called party government. 1899 Beerbohm *More* 89, I should be glad to see..his office held by an artist, not by a party-hack. 1974 E. Ambler *Dr. Frigo* ii. 73 Your father's party needed a Cavour... All they had were party hacks. [1908 G. Wallas *Human Nature in Politics* iii. 84 A party..is primarily a name, which, like other names, calls up when it is heard or seen an 'image' that shades imperceptibly into the voluntary realisation of its meaning.] 1960 Butler & Rose *Brit. Gen. Election 1959* iii. 17 In the two years before the 1959 election the Conservative party engaged in a public relations campaign...the term 'party image' was continually invoked. 1962 *Listener* 19 Apr. 702/2 In the discussion of the Party Image in the same programme, the Liberals appeared to me to come out best. 1931 G. B. Shaw *Fabian Ess. Socialism* Pref. p. xiv, Candidates with a party label, pledged to vote for their party right or wrong. 1882 A. Bain *John Stuart Mill* ii. 60 That his father would have made an able minister or party-leader, we must cheerfully allow. 1974 E. Ambler *Dr. Frigo* i. 56 The loss of our Party leader..was a demoralizing blow. 1976 *Encounter* June 78/2 The party leader believes in himself, and his supporters believe in him because he is *their* leader. 1832 Dickens *Let.* 30 July (1965) I. 7, I give you this early notice not because there is anything formal or party like in the arrangements. 1968 *Listener* 20 June 806/2 Freedom and participation in one's place of work are to replace constituency or 'party list' democratic choice. 1882 A. Bain *John Stuart Mill* iv. 125 He set a good example of perfect party loyalty, combined with the assertion of difference of opinion on particular questions. 1968 W. Safire *New Lang. Politics* 321/1 'Sometimes party loyalty asks too much,' said John F. Kennedy to a Democratic friend, excusing him for supporting Republican Leverett Saltonstall for Senator of Massachusetts. 1891 'O. Thanet' *Otto the Knight* 266 He can't be trusted to run the office as a party machine, and Milton Bedford can! 1918 *Observer* 29 Sept. 6/2 No serious citizen..can doubt..what is imperatively demanded by the interests of the nation, no matter what may be the interests of some caucuses and party-machines. 1939 *War Illustr.* 2 Dec. 375/3 Even in Germany itself there are millions who stand aloof from the seething mass of criminality and corruption constituted by the Nazi party machine. 1972 *Guardian* 8 June 12/2 The wholehearted support of the party machine rather than just a resigned tokenism. 1977 *Cleethorpes News* 6 May 7/4 It is a moot point whether they do much good because, like other party machines, the Conservative local unit functions day-in and day-out pretty successfully. 1881 *Bradstreet's* IV. 305 The voters of Kings county have usually been relied upon by party managers. 1885 W. Harris *Hist. Radical Party* ii. 19 That great body of the people who were too frequently regarded by placemen and party managers as machines to be used and property to be disposed of. 1930 A. Harris tr. *P. Cohen-Portheim's England, the Unknown Isle* iv. 45 When we get into our party clothes we put on our party manners and party conversation with them. 1931 R. R. Marett in W. Rose *Outl. Mod. Knowledge* 422 After much steam has been blown off, all cheerfully assume their party-manners. 1969 'J. Morris' *Fever Grass* ii. 22 A child suddenly remembering its party manners. 1974 G. Jenkins *Bridge of Magpies* xii. 178 Judging from what I've seen of their party manners, it won't be a pretty operation. 1931 H. Nicolson *Diary* 22 Sept. (1966) 91 Party-meeting at 11.30. 1976 S. Hynes *Auden Generation* v. 138 *It's a Battlefield*..is full of Communists, and a long scene is devoted to a Party meeting. 1920 S. Lewis *Main St.* x. 116 He's a socialist..a regular old-line party-member. 1942 [see *fellow-traveller 2]. 1974 'J. Le Carré' *Tinker, Tailor* xvii. 147 His father was a docker..and a Party member. 1972 *Listener* 20 July 69/2 According to the proposed Democratic Party charter, there would be a national dues-paying party membership. 1976 S. Hynes *Auden Generation* ix. 317 For Upward, Party membership seems to have been a mode of salvation in a desperate time. 1957 R. N. C. Hunt *Guide to Communist Jargon* xxxii. 110 *Party-mindedness* ('Partiinost') was given its classical formulation by Lenin in his polemic with Peter Struve of the middle 'nineties, when he laid down that 'Materialism involves party-mindedness, since it compels us in evaluating anything which takes place openly and directly to adopt the standpoint of a specific social group'. 1958 *Ann. Reg. 1957* 209 Leading writers.. were reminded that negative features of Soviet life could be criticized only from the standpoint of 'Party-mindedness' (*partiynost*) i.e. recognition that all defects were being successfully overcome by the Party. 1973 H. McCloy *Change of Heart* ix. 104 I'm hardly in a party mood. My father is still unconscious. 1976 G. Sims *End of Web* xvii. 116 Being quite sober and not in a party mood they shared a feeling of detachment. 1838 J. S. Mill in *Westm. Rev.* XXXII. 246 The 'Standard'..on this occasion has merged the party passions..in the sympathy of talent for talent. 1945 W. S. Churchill *Victory* (1946) 193 If party passions, doctrines, and ambitions were to dominate our life for any lengthy period, [etc.]. 1942 C. E. Smith *Jazz Record Bk.* 81 The 'party piano' style, a growth that owes more to oldtime blues playing than to any other one source, was already a flourishing development in the 1930's. 1946 R. Blesh *Shining Trumpets* xiv. 337 A number of gifted Harlem pianists in the 'party-piano' tradition. 1962 *Listener* 6 Sept. 367/1 A late-night series of party pieces. 1967 'P. Loraine' *W.I.L. One to Curtis* i. 20 It's been my party piece for over a year now: not *family* parties, I hasten to add. 1969 E. H. Pinto *Treen* 86 Although all these spoons could be used,..one feels that they were very much 'party pieces'. 1973 *Times* 25 Apr. 10/1 (Advt.), Scott James of Westminster distribute a superb range of family clothing direct to the consumer, through party plan and catalogue selling techniques. 1976 *Billings* (Montana) *Gaz.* 30 June 5-D/3 (Advt.), Montana owned firm looking for party plan representatives in Billings area. Work your own hours for good commission. 1977 *Evening Gaz.* (Middlesbrough) 11 Jan. 10/2 (Advt.), Party Plan demonstrators, car available for ambitious applicant. 1848 J. R. Lowell *Biglow Papers* 1st Ser. 111 It gives a Party Platform, tu, jest level with the mind Of..honest folks thet mean to go it blind. 1964 Gould & Kolb *Dict. Social Sci.* 484/1 A *party platform* is a general statement of principles, policies, and issues, and a programme of promises which the party pledges to enact into legislation. 1972 *Jrnl. Social Psychol.* LXXXVII. 46 Public opinion should not be a major concern of political leaders in writing party platforms. 1957 *Times* 14 May 11/2 There was an admirable absence of rancour or effort to make party points when the decision to afford British ships the facilities for using the Suez Canal again was announced in the House of Commons yesterday. 1886 G. Whetenall *Echetlus* i. 60 Let us consider shortly some of these results, the common phenomena of our party-political life. 1910 Party-political [see *cartoon *sb.* 2]. 1947 *Radio Times* 14 Mar. 1/1 Party political broadcasts of a controversial character are to be resumed. 1966 *New Statesman* 4 Mar. 289/1, I am.. unable to watch party-politicals at all. 1974 *Guardian* 23 Jan. 1/1 A plan to shorten 'News at Ten' to make way for a party political broadcast. 1977 *Jrnl. R. Soc. Arts* CXXV. 467/1 Such questions are..essentially political, though not necessarily party-political. 1831 J. S. Mill in *Examiner* 9 Jan. 20/2 They who prefer the ravings of a party politician to the musings of a recluse, may consult a late article in Blackwood's Magazine. 1922 *Brit. Jrnl. Psychol.* Oct. 116 The 'party politics complex' in the mind of the party politician. 1942 L. B. Namier *Conflicts* 200 It is of paramount importance that the extent to which our present constitutional system is bound up with party organisations and party politics should be fully understood. 1961 *Middle East Jrnl.* XV. 6 The bulk of the preparation had..proceeded under the supervision of the Ministry of the Interior, whose officials are barred from party activity and probably generally disinterested in party politics. 1969 *New Yorker* 11 Oct. 53/1 They pecked the hostess farewell, apologized in unison for being party poops. 1954 *Amer. Speech* XXIX. 293 Such comic masterpieces as *lounge lizard* and *party pooper* are of American origin. 1956 in Wentworth & Flexner *Dict. Amer. Slang* (1960) 377/1 No one can call Mr. Bulganin and Mr. Khrushchev party poopers... The Russian leaders demonstrated their suavity and cleverness at the party. 1842 Dickens *Let.* 1 Apr. (1974) III. 176 The silly, drivelling, slanderous, wicked, monstrous Party Press. 1950 *Middle East Jrnl.* Apr. 168 With one full-fledged *journal d'information*..a party press..Egyptian journalism can look to an interesting future. 1803 *Deb. Congress U.S.* 6 Jan. (1851) 337 This ought not to be made a party question. 1885 A. Crump *Formation Polit. Opinion* 152 The position was now far too grave to be treated as a party question. 1941 'G. Orwell' *Lion & Unicorn* i. 16 No party rallies, no Youth Movements. 1959 I. Jefferies *Thirteen Days* vi. 75 A single truck stuck out like a top hat at a Party rally. 1964 *Amer. Folk Music Occasional* i. 10 Under-the-Counter 'party' records provide a traditional version of *The Eddystone Light*. 1968 P. Oliver *Screening Blues* vi. 175 The cheapened styles of the party records which have a large illicit sale in the white society. 1855 Tennyson *Maud* xxv. 91 And another, a statesman there, betraying His party-secret, fool, to the press. 1711, 1882 Party-spirit [in Dict.]. 1941 N. Marsh *Death & Dancing Footman* (1942) viii. 151 He's gone into a huddle over the fire and does *not* exactly manifest the party spirit. 1971 S. Jepson *Let. to Dead Girl* xix. 218 I'm not ungregarious but I'm quite incapable of the party spirit in a roomful of people busy smoking themselves into lung cancer. 1974 J. White tr. *Poulantzas's Fascism & Dictatorship* iii. ii. 113 It is unnecessary to go into details of the continuous contradictions between big capital and the Nazi party-state. 1977 P. Johnson *Enemies of Society* vi. 78 Nazi Germany was equally frank in imposing control of labour and eliminating all foci of political and economic power other than the party-state. 1824 J. S. Mill in *Westm. Rev.* I. 530 Here..the evils of the party system are most clearly shown. 1959 B. & R. North tr. *Duverger's Pol. Parties* (ed. 2) ii. 203 With the exception of the single-party states, several parties coexist in each country: the forms and modes of their coexistence define the 'party system' of the particular country being considered. 1961 *How-to-do-it-Encycl.* (Mechanix Illustr.) I. 12 If you are a party thrower, you may need added capacity. 1962 *Listener* 22 Feb. 347/3 One friend of mine, a compulsive party thrower, had four flats in five months. 1929 *Radio Times* 8 Nov. 390/1 Tony Monke..has been persuaded to execute his best party trick of standing on his head. 1892 *Courier-Jrnl.* (Louisville, Kentucky) 2 Oct. 1/7 The 'party workers'..were well represented. 1935 N. Mitchison *We have been Warned* i. 51 The thick sense of urgency and seriousness which was beginning to show among the Party workers. 1974 *Listener* 14 Mar. 330/2 On election day, my wife and I..did the traditional tour..to thank and encourage the party workers.

party, *v.* Restrict †*Obs. rare* to senses in Dict. and add: **3.** *N. Amer.* **a.** *intr.* To give a party; to attend a party; to have a good time. **b.** *trans.* To entertain (someone) at a party; to accompany to a party. Hence **pa·rtying** *vbl. sb.* and *ppl. a.*

1922 E. E. Cummings *Let.* 5 Dec. (1969) 91 Haven't seen Vanity All Is Fair in? but have extensively partyed with Er former Heditor. 1948 *Penguin New Writing* XXXV. 106 Between times, when they were not drinking at the cafés, partying, writing, or making love, they talked a lot and did a certain amount of thinking. 1953 *N.Y. Times Bk. Rev.* 12 July 8/1 The delegates and guests..were partied to a crisp. 1953 W. P. McGivern *Big Heat* iv. 54 A woman..waiting up for a partying teenage daughter. 1957 M. Millar *Soft Talkers* 97 All those times when I was so ill I could scarcely move and he went off partying. 1963 D. B. Hughes *Expendable Man* (1964) ii. 30 You can't imagine the entertainment she's had.. every club on the campus has partied her. 1967 [see *joiner *sb.* 1 b]. 1970 N. Armstrong et al. *First on Moon* ii. 42 There was so much partying that not a few people wondered about the size of the national hangover. 1971 J. Philips *Escape a Killer* (1972) i. iii. 34 You're to drive into New York..stay there a few days... Peter will party you around. 1973 *Islander* (Victoria, B.C.) 25 Feb. 6/3 Everybody crowded in, the home brew arrived and we partied until 4 in the morning. 1976 H. Nielsen *Brink of Murder* xx. 177 A small crowd..who had been partying on another boat. 1977 *Time* 31 Jan. 31/1 Outgoing Democratic National Committee Chairman Robert Strauss partied along with singer Helen Reddy and actor Alan Alda. *Ibid.* 25 Apr. 48/3 Los Angeles rockers do not lack for private places in which to party. 1977 *Daily Tel.* 9 Apr. 15/6 Mrs Trudeau returned to Ottawa after a weekend of partying in Toronto with the Rolling Stones.

partyism. (In Dict. s.v. Party *sb.*) (Earlier example.)

1842 *Amer. Pioneer* I. 278 Partyism or love of party is the vibratory motion.

partyless (pā·ɹtilės), *a.* [f. Party *sb.* + -less.] Not having or not belonging to a political party.

1896 *N.Y. Voice* 9 July 2/4 This means..a reign of straight-out, inexorable, sectless, seminaryless, partyless righteousness in citizenship. 1909 *Westm. Gaz.* 15 Mar. 2/1 In 1901 Francis Ferdinand was practically alone, partyless. 1963 *Times* 2 Feb. 9/2 Mr. Nkomo, the partyless but undoubted African Nationalist leader. 1966 N. Nicolson in H. Nicolson *Diaries & Lett.* 66 Harold Nicolson, being partyless, was one of the first to join him. 1977 *Times* 15 Nov. 15/2 Nepal will stick to its 'partyless panchayat' rather than risk..one man one vote.

party line. [f. Party *sb.* + Line *sb.*²] **1.** A policy adopted by a political party; an area of policy or 'line' separating the policy of one political party from that of another. Also *transf.* and (with hyphen) *attrib.*

1834 T. H. Benton *30 Years' View* (1854) I. 431/2 Look at the vote in the Senate upon the adoption of the resolution, also as clearly defined by a party line as any party question can ever be expected to be. 1904 G. B. Shaw *Common Sense of Municipal Trading* xi. 105 A councillor selected on strict party lines. 1915 Mrs. H. Ward *Eltham Ho.* xiv. 268 The dogged adherence to the traditional party lines and shibboleths on the part of the average British voter. 1937 C. Connolly in L. Russell *Press Gang!* 95 The kind of bullets they keep for reviewers who step across the party line. 1942 Z. N. Hurston in A. Dundes *Mother Wit* (1973) 32/2, I am a mixed-blood,.. but I differ from the party line in that I neither consider it an honor nor a shame. 1948 *Daily Ardmoreite* (Ardmore, Okla.) 2 May 5/4 On party-line votes, this will give the democrats guarantee of a two-thirds majority. 1957 *Times Lit. Suppl.* 27 Dec. 782/3 Mr Dawson's party-line pronouncements on such subjects as contraception or the mediative role of the Church. 1966 *Listener* 10 Mar. 357/2 When any of his whig friends deviated from the party-line, the old pedant laid into them. 1975 J. P. Morgan *House of Lords & Labour Govt.* iv. 134 It can be safely concluded that, for whatever reason, there is such a phenomenon as party-line voting in the Lords. 1976 S. Hynes *Auden Generation* vi. 171 This party-line definition of proletarian art. 1977 A. Ecclestone *Staircase for Silence* v. 88 He was well aware of the faith-destroying outcome of the dogmatism of the party-line, whether in Church or State.

2. [*Party *sb.* 14 c.] A telephone line shared by two or more subscribers. Also *transf.* and (with hyphen) *attrib.*

1893 W. J. Hopkins *Telephone Lines* i. 4 When 'party lines', so called, are used, they should be connected according to the 'bridging-bell' method. 1901 *Ann. Rep. Amer. Telephone & Telegr. Co. 1900* The general adoption of..party lines..has attracted many thousands of subscribers. 1930 *Telegraph & Telephone Jrnl.* XVI. 115/2 The 'party line' is not so rife in this country as in America. *Ibid.* 119/2 The number of rural party lines at Dec. 31, 1929 was 10,322. 1933 *Punch* 10 May 508/1 If you take a real personal interest in your fellow-beings as I do, I should advise all those living in the country to have a party-line. 1959 *Ibid.* 19 Aug. 30/3 The new 'party-line' electronic stethoscope..allows up to four doctors to listen to a patient's heart simultaneously. 1973 M. Russell *Double Hit* vi. 44 He overheard things on our party line.

Hence **pa:rty-li·ner**, one who follows the line adopted by a political party. Also *transf.* So **pa:rty-li·ning** *vbl. sb.*

1940 *Common Sense* (N.Y.) Feb. 21/2 Right now a Communist Party-liner does not have a ghost of a chance in the A.F. of T. 1943 *Sun* (Baltimore) 28 Sept. 12/2 The speech he [*sc.* Earl Browder] has just made at Chicago is fully as brash as anything he said before Marshal Stalin disowned him and other party-liners outside of Russia by liquidating the Communist Internationale in May of this year. 1948 *Daily Mail* 7 Feb. 2/6 When the party-liner squawks an indignant protest he is at once knocked off his perch. 1949 *Virginia Q. Rev.* Winter 44 For the past eleven years he has allied himself with, and never broken on any significant issue from, the 'party-liners'. 1958 *Spectator* 27 June 846/3 Large immigrant colonies of Poles, Lithuanians and others who were not easily taken in by Soviet party-lining. 1966 *Economist* 1 Oct. 39/3 The Chinese here—Nationalists and party-liners alike..are clearly confused. 1969 *Guardian* 31 July 5/3 The..gulf.. between..party-liners and deviationists. 1973 C. Mullard *Black Brit.* ix. 106 Good councillors who find themselves in disagreement with their party's views on race can be removed and replaced with party-liners.

party-man. Add: **2.** (Further examples.)

1933 *Discovery* Feb. 64/1 M. Briand, not at all a 'party' or 'group' man in his later career, also had his own personal following of deputies. 1938 W. S. Churchill *Into Battle* (1941) 15 The Fascists of Italy, the Party-men, are asking themselves whether all this is to the permanent safety..of their native land. 1946 'G. Orwell' *Crit. Ess.* 136 The young G.P.U. man..is the typical 'good party man'.

3. One who frequently attends or gives parties.

1936 *Mademoiselle* Apr. 20/1 The true Party Man possesses a *je ne sais quoi* which lends a rich grace to the most distinguished party. 1963 *Times* 14 Jan. 9/4 The non-party man (speaking in the social rather than the political sense) is saddened by the occasional receipt of invitations... How does one become a successful party man?

partyness (pā·ɹtinės). Also **party-ness**. [f. Party *sb.* + -ness.] **a.** Party-mindedness (see quot. 1976). **b.** The state or condition of being a political party.

1952 *Mind* LXI. 120 It is, of course, a principle of Marxism-Leninism that philosophy should be written in a 'party spirit', with 'partyness'. This principle..originated with Lenin. 1969 P. Worsley in Ionescu & Gellner *Populism* 228 Professor Brough Macpherson has described ..the United Farmers of Alberta,..and its successor, Social Credit... Macpherson..labels this a 'quasi-party' system: not 'non-party', because institutionalized opposition and other..manifestations of 'party-ness' still persisted. **1976** *Russ. Rev.* (N.Y.) XXXV. 69 Lenin advocated the principle of *partiinost*, a concept roughly translatable as 'partyness', the constant referral to party policy as the inspiration and guide to action.

parvenu, *sb.* and *a.* Add: **A.** *sb.* (Further examples.)

1944 PARTRIDGE *Here, There & Everywhere* (1950) 54 A word begins its career as a *parvenu*. **1955** *Times* 20 Aug. 7/7 To such a man the Jews are not only aggressors and common thieves but parvenus too. **1978** P. VAN GREENAWAY *Man called Scavener* vii. 95 Genealogy..attracts a growing body of parvenus.

B. *adj.* (Earlier and later examples.)

1828 J. S. MILL in *Westm. Rev.* IX. 290 No one licked the dust before the *parvenu* emperor with greater *gusto* than the abbé Maury. **1958** *Times* 12 Nov. 3/5 She sounds a *parvenu* proletarian, if such a fancy be comprehended. **1978** *Amer. N. & Q.* XVII. 31/2 Some of the parvenu universities are rapidly approaching maturity.

parvenuess (pāɹvěniu,e·s). *rare.* [-ESS¹.] A female parvenu.

1903 'O. HENRY' in *Everybody's Mag.* July 59/2 As proud and satisfied as a prince that's abjured a two-hundred-dollar crown for a million-dollar parvenuess.

parvovirus (pā·ɹvovəiᵊ·rǝs). *Microbiol.* [f. L. *parv-us* small + -o¹ + VIRUS.] Any of a group of very small animal viruses consisting of single-stranded DNA in an icosahedral capsid without an envelope and occurring in a wide variety of vertebrates.

1965 *Proposals & Recomm. Provisional Committee Nomencl. Viruses* in *Ann. de l'Inst. Pasteur* CIX. 629 (*table*) Parvovirus. **1968** RHODES & VAN ROOYEN *Textbk. Virol.* (ed. 5) III. 352/2 Johnson..has isolated and identified the virus of feline panleukopenia (or feline enteritis) as a parvovirus. **1974** J. A. ROSE in Fraenkel-Conrat & Wagner *Comprehensive Virol.* III. i. 50 The biological significance of most of the parvoviruses is still obscure. Only the feline and related viruses are known to cause disease naturally, but the pathogenicity of rodent viruses under experimental conditions indicates that other disease potentials may exist... Because of their latency and hardiness, parvoviruses must always be considered as potential contaminants of cells, other virus stocks, and certain vaccines. **1976** FENNER & WHITE *Med. Virol.* (ed. 2) xxii. 424 There is current controversy about whether two important human pathogens, hepatitis A virus and the Norwalk agent of gastroenteritis, are parvoviruses or picornaviruses, but the former seems more likely.

parylene (pæ·rilīn). Also **Parylene.** [f. *par(ax)ylene* s.v. PARA-¹ 2 b.] Any of several transparent thermoplastic polymers of para-xylene or its substituted derivatives that are obtained as thin films or particles by condensation of the vapour of the monomer.

1965 *N.Y. Times* 18 Feb. 43 The product, named parylene, has been successfully used as a dielectric..on capacitors. **1965** *Mod. Plastics* Mar. 113/2 The simplest member of the family is parylene N. A chlorinated member, designated parylene C, is also available. **1969** C. A. HARPER *Handbk. Electronic Packaging* vii. 43 The basic member of the series, called Parylene N, is poly-para-xylylene, a completely linear, highly crystalline material. *Ibid.*, The parylenes are produced by vapor-phase deposition in a variety of forms. **1975** J. A. BRYDSON *Plastics Materials* (ed. 3) xxi. 470 With both Parylene materials the polymers have molecular weights of the order of 500,000.

‖**pas.** Add: **2.** *pas d'action*, a thematic dance or mime; *pas de basque*, a dance step derived from Basque national dances; *pas de bourrée*: see *BOURRÉE 2; *pas de chat* (see quot. 1957); *pas de cheval* (see quots.); *pas de deux* (earlier and later examples; also *transf.* and *fig.*); *pas de quatre*, a dance or figure for four persons; *pas de trois*, a dance or figure for three persons; also *transf.* and *fig.*; *pas seul* (examples).

1951 *Ballet Ann.* V. 25 He has a superb plastic imagination which he showed even in that wickedly conceived *pas d'action* to the music of Tristan. **1957** G. B. L. WILSON *Dict. Ballet* 206 *Pas d'action*, lit. action dance. A dance expressive of a theme or telling a story, in opposition to absolute dance—e.g. the Rose Adagio in *Sleeping Beauty*. **1818** T. WILSON *Quadrille & Cotillion Panorama* 100 *Pas de basque*, a Step used in Quadrille Dancing. **1865** MRS. GASKELL *Lett.* (1966) 748 The neat maid servant had performed a sort of 'pas-de-basque', hopping & sliding with more grace than security to the dishes she held. **1892** E. SCOTT *Dancing* ix. 119 Remember, in performing a *gavotte*, that the French *pas de basque* is employed, in which the first two movements are made to one count of the music. **1948** *Ballet Ann.* II. 144 To see a Basque dancer perform the *pas de basque* is to realise the true nature of that step. The soft, cat-like landing gives an impression of immense strength and of power in reserve. **1964** J. F. & T. M. FLETT *Trad. Dancing in Scotl.* viii. 190 In reel tempo the commonest setting step was the pas de Basque. **1968** J. WINEARLS *Mod. Dance* (ed. 2) iii. 88 The pas de basque lends itself to many rhythmic and expressive variations. **1914** T. & M. W. KINNEY *Dance* iv. 90 In this group might appropriately be included *pas de bourrée dessus-dessous* (i.e., in front and behind); *glissades*; *petits battements*; and the devilish-looking little *pas-de-chat*. **1916** *Dancing Times* Sept. 327 [Espinosa's] 'Syllabus of Dancing.'.. Steps.—..pas de basque, pas de chat, pas de cheval. **1922** BEAUMONT & IDZIKOWSKI *Man. Classical Theatr. Dancing* II. v. 197 *Pas de chat*. This *pas* is composed of *three* movements, generally executed *en diagonale*. **1953** E. TAYLOR *Sleeping Beauty* vi. 114 The staggered thumps of children practising a *pas de chat*. *Ibid.* 116 She was still humming the *pas de chat* tune. **1957** G. B. L. WILSON *Dict. Ballet* 207 *Pas de chat*, lit. a cat's step. A springing movement in which one foot jumps over the other, executed on the diagonal. A jumping step in which the feet are one after the other drawn up to the knee of the opposite leg... Also called pas de papillon. **1976** *New Yorker* 15 Nov. 111/1 The pas-de-chat/pirouette ending of the second variation, nowadays done at the New York City Ballet as a line of brisés, gave her no difficulty. **1916** Pas de cheval [see *pas de chat* above]. **1930** CRASKE & BEAUMONT *Theory & Pract. Allegro in Classical Ballet* 41 Pas de cheval. Stand erect in the centre of the room and face 2, with the head upright [etc.]. **1957** G. B. L. WILSON *Dict. Ballet* 207 *Pas de cheval*, a pawing movement of one foot while the dancer hops on the other. **1762** G.-A. GALLINI *Treat. Art Dancing* 282 Venus and Adonis form a *pas-de-deux*, or duet-dance. **1775** H. WALPOLE *Let.* 23 Aug. in *Corr.* (1965) XXXII. 255, I was not so struck with the dancing as I expected, except with a *pas de deux* by the Marquis de Noailles and Madame Noistein. **1802** [see *ENSEMBLE *sb.* 1]. **1806** E. FREMANTLE in *Wynne Diaries* (1940) III. 265 The grand Ballet..has fine effect.. Des Hayes pas de deux with his wife quite beautiful. **1808** T. ASHE *Trav. Amer.* III. xxxv. 118 The Waltz had most votaries; the *Pas de deux* next. **1955** *Times* 5 July 14/1 The choreography is designedly adventurous,..particularly in the long central section which is set as a *pas de deux*. **1960** [see *pas de trois* below]. **1962** *Listener* 13 Sept. 380/2 Dr Adenauer would have to go faster and further in the *pas de deux* to convince himself and his people that 'the Franco-German alliance is what counts'. **1973** *Times Lit. Suppl.* 26 Oct. 1324/1 Between them they perform a ritual examination of conscience, a *pas de deux* in which they are painfully at cross-purposes. **1978** *Times* 27 May 14/7 This was no ordinary *debut* performance in the *corps de ballet*, but was a solo performance in a *pas de deux* specially introduced for her. **1978** *Country Life* 27 July 220/2 She..rode side-saddle in the pas-de-deux dressage with David Hunt. **1882** *Standard* 26 Dec. 5 The famous *pas de quatre* which had the effect of killing the Ballet in England. **1890** [see *GAIETY 4]. **1897** G. BELL *Let.* 28 Jan. (1927) I. iii. 42 There is a rather nice sort of variant of the 'pas de quatre' which they call the 'pas de patineur' which I quickly learnt. **1902** CHESTERTON *Twelve Types* 43 His metre is a bounding 'pas de quatre'. **1978** *Times* 9 Mar. 13/6 Ashton's *pas de quatre* and Neapolitan dance. **1762** G.-A. GALLINI *Treat. Art Dancing* 280 A *pas-de-trois* or trio-dance follows. **1773** H. WALPOLE *Let.* 3 Apr. in *Corr.* (1965) XXXII. 116 Lord Delawar's two eldest daughters and the Ancaster infanta performed a *pas de trois* as well as Mlle Heinel. **1794** [see DIVERTISSEMENT 2]. **1823** T. CREEVEY *Let.* 14 Feb. in *Creevey Papers* (1963) xi. 190 We went to our *own* playhouse, where we saw 1st a *pas de trois* between Wilson, Hobhouse and Canning, and then a *pas de deux* between Broughman and Canning. **1957** G. B. L. WILSON *Dict. Ballet* 208 *Pas de trois* (de quatre, de cinq, de six, etc.), a dance for three (four, five, six, etc.) persons. **1960** A. PODHAJSKY *Spanish Riding School Vienna* 40 The 'Pas de Deux' or 'Pas de Trois' demands perfectly trained horses which allow themselves to be led by the rider both as regards keeping time and performing the steps with great precision. **1976** *New Yorker* 9 Feb. 95/1 There is a gracious and supple, long-limbed pas de trois..in which the boy holds an invisible lute or mandolin. **1977** D. WILLIAMS *Treasure by Degrees* vi. 55 The Dean's violent about-turn led him straight into the arms of Amelia Hatch... She, in turn, recoiled on to Witaker so that all three had momentarily become engaged in a curious *pas de trois* for ill-matched performers. **1813** BYRON *Waltz* 14 Not decent David, when, before the ark, His grand pas-seul excited some remark. **1830** R. BARTON tr. *Blasis's Code of Terpsichore* v. 354 From time to time these bacchanalia are interrupted by *pas seuls* and *pas de deux*. **1832** F. TROLLOPE *Dom. Manners Amer.* I. xiii. 188 Had Mercury stepped down, and danced a *pas seul* upon earth, his godship could not have produced a more violent sensation. **1839** DICKENS *Nickleby* xlviii. 473 A castanet pas seul by the Infant Phenomenon. **1849** THACKERAY *Pendennis* I. xxvii. 260 Now the young ladies went over to Pen's side, and Cornet Perch performed a *pas seul* in his turn. **1863** MRS. GASKELL *Sylvia's Lovers* I. xiii. 291 Philip hardly knew what he said in reply, the mention of that *pas-seul* lifted such a weight off his heart. **1939** JOYCE *Let.* 28 Jan. (1966) III. 436, I am afraid the traditional *pas seul* with high kicking effects..will be beyond my powers this year of grace. **1955** *Times* 29 Aug. 10/5 If Mr. Ashton has not succeeded in making the love-duet the climax of his work he has almost compensated for it in Juliet's pas seul after the marriage in Friar Laurence's cell. **1978** S. ROSENFELD *Temples of Thespis* iv. 71 Little Chatterley, as the black slave, was admired for his *pas seul*.

pasa doble: see *PASO DOBLE.

‖**pasar** (pæ·sāɹ). [Indonesian *pasar*, perh. f. Pers. *bāzār* market.] A market in Indonesia and Malaysia. So **pasar gelap** [Indonesian *gelap* dark], black market; **pasar malam** [Indonesian *malam* night], a fair.

1958 H. FORSTER *Flowering Lotus* ii. 26 The women ..walked miles from their villages..to the big *pasar*, the covered market. **1959** 'M. DERBY' *Tigress* iv. 154 The *pasar gelap* or black market trade with Sumatra was risky but profitable. **1961** CONYN & MARTEN *Bali Ballet Murder* xvi. 175, I shall ask the cook to go to the *pasar*. **1972** *Malay Mail* (Kuala Lumpur) 25 May 4/5 The pasar malam in Gurney Drive will end on Tuesday.

Pascal (pæska·l). [The name of Blaise *Pascal* (1623–62), French scholar and scientist.] **1.** Math. *Pascal's triangle*: a triangular array of numbers in which those at the ends of the rows are 1 and each of the others is the sum of the nearest two numbers in the row above, the apex, 1, being at the top. [Described in Pascal's *Traité du Triangle Arithmetique* (1665).]

1886 G. CHRYSTAL *Algebra* I. iv. 65 (*heading*) Pascal's triangle generalised. **1977** R. E. MEGILL *Introd. Risk Analysis* vii. 67 You can construct a triangle, called Pascal's triangle, from the binomial expansions just reviewed.

2. Written **pascal** (pæ·skăl). The unit of pressure in the M.K.S. system (now incorporated into the International System of Units), equal to one newton per square metre (approximately 0·000,145 p.s.i., $9{\cdot}9 \times 10^{-6}$ atmosphere, or 0·102 kilogramme-force per sq. m.). Abbrev. Pa.

1956 KAYE & LABY *Tables Physical & Chem. Constants* (ed. 11) 7 (*table*) Pascal. **1964** H. S. HVISTENDAHL *Engin. Units* iii. 29 In the French decree of May 3rd, 1961, the name pascal (Pa), is adopted for the N/m²,..although that name has not yet received international recognition. **1972** *Physics Bull.* Jan. 40/1 The 14th General conference of Weights and Measures (CGPM) met in Paris on 4–7 October 1971... Amongst the main decisions taken by the conference were the final adoption..of the 'pascal' (Pa) as the SI unit of pressure (N m⁻²) and the 'siemens' (S) as the SI unit of conductance (Ω^{-1}). **1974** *B.S.I. News* Sept. 12/2 The pascal is advised by the aircraft industry..as the best unit to use in their fields. **1975** *Sci. Amer.* Feb. 110/2 The manual uses the MKS pressure unit, the pascal, almost throughout. **1976** *Nature* 29 Apr. 745/3 A Pascal of pressure on top of your head Is the same push that butter exerts on sliced bread.

Pascalian (pæskā·liăn), *sb.* and *a.* [f. prec. + -IAN.] **A.** *sb.* An admirer or adherent of Pascal (see prec. (etym.)); an interpreter of his works. **B.** *adj.* Of or pertaining to Pascal or to his ideas and philosophy.

1929 A. HUXLEY *Do What you Will* 235, I now propose to write of Pascal. As a positivist first of all... More sympathetically next, in the guise of a Pascalian. **1930** tr. *Chevalier's Pascal* i. 26 Charron's work contained, as it were, the germ of Pascalian apologetics; at any rate we find in it a great deal of material that Pascal was to utilize after changing its order. **1937** A. HUXLEY *Ends & Means* xv. 304 There is, however, an element of truth in the Pascalian doctrine. **1937** *Sunday Times* 25 Apr. 8/3 The truth is Pascalian Christianity is only fit for heroes and saints; it is not a possible life for ordinary men. **1948** *Scottish Jrnl. Theol.* I. 98 A full length study of Pascal by an established Pascalian scholar. **1971** G. STEINER *In Bluebeard's Castle* iii. 64 We are forced now to return to an earlier, Pascalian pessimism, to a model of history whose logic derives from a postulate of original sin. **1973** *Listener* 6 Sept. 298/1 The man who finds himself suddenly independent in this way will then fall prey to Pascalian despair, existential anguish, dereliction. **1977** *N.Y. Rev. Bks.* 14 Apr. 26/4 Sennett is thus an atheist about social science, but a Pascalian about society.

Pascarete, var. *PAXARETE.

Pa·schaltide. [f. PASCHAL *a.* + TIDE *sb.*] The period following Easter Sunday, esp. Easter Week.

1894 G. F. X. GRIFFITH tr. *Fouard's St. Paul* xv. 349 The octave of Paschal-tide, a rite of Jewish origin. **1974** *Oxf. Dict. Chr. Ch.* (ed. 2) 1038/1 *Paschaltide*, the period in the ecclesiastical year immediately after Easter.

Paschen¹ (pæ·ʃǝn). *Physics.* The name of L. C. H. Friedrich *Paschen* (1865–1947), German physicist, used *attrib.* and in the possessive to designate certain phenomena he discovered, as **Paschen('s) curve,** a graph of the breakdown voltage in a gas against the product of pressure and interelectrode distance; **Paschen's law,** the law that at a constant temperature the breakdown voltage in any given gas depends only on the product of the pressure and the interelectrode distance (formulated by Paschen in *Ann. d. Physik und Chem.* (1889) XXXVII. 90); **Paschen series,** a series of lines in the infra-red part of the spectrum of atomic hydrogen, with wave numbers represented by $R(1/3^2 - 1/m^2)$ (where R is the Rydberg constant and $m = 4, 5, \ldots$), of which the first line has a wavelength of 1857 nanometres and the series limit is at 821 nanometres.

1903 *Proc. R. Soc.* LXXI. 375 A general application of Paschen's law demands that the minimum spark potential must be a physical constant for each gas. **1922** Paschen series [see *LYMAN]. **1957** A. F. MONYPENNY tr. *Penning's Electr. Discharges in Gases* vi. 32 (*caption*) Paschen curves for breakdown between flat iron plates in various gases and gas mixtures. **1967** *IEEE Trans. Electr. Insulation* II. 82/1 In accordance with Paschen's curve, equally large discharge voltages could also result for very small voids. **1967** Paschen series [see *LYMAN]. **1968** G. F. WESTON *Cold Cathode Glow Discharge Tubes* ii. 59 For the case of a non-uniform field Paschen's law relating breakdown to gap width and pressure no longer holds.

Paschen² (pæ·ʃǝn). *Med.* The name of Enrique *Paschen* (1860–1936), Mexican-born bacteriologist, used *attrib.* in **Paschen body,** each of the particles making up one of the cytoplasmic inclusions found in epithelial cells in cases of smallpox (described by Paschen in *Münch. med. Wochenschr.* (1906) 4 Dec. 2391).

1931 *Jrnl. Path. & Bacteriol.* XXXIV. 122 A preparation is shown which demonstrates the presence of elementary bodies precisely similar to the Paschen and Borrel bodies. **1974** J. D. ACTON et al. *Fund. Med. Virol.* xi. 167 Although the poxviruses are DNA viruses, they replicate in the cytoplasm and produce characteristic cytoplasmic inclusions called Guarnieri bodies; the inclusions are composed of a dense aggregation of many virus particles. Virus particles are also referred to as 'elementary bodies' or Paschen bodies.

Paschen–Back effect (pæ:ʃǝn bæ·k). *Physics.* [f. *PASCHEN¹ + the name of Ernst E. A. *Back* (1881–1959), German physicist, who jointly published a description of the effect in 1913 (*Ann. der Physik* XXXIX. 897).] An effect observed when a source of spectral lines is in a magnetic field so strong that the resultant splitting of each line is comparable in magnitude to the separation of the lines in a multiplet, the spacing of the lines corresponding to the normal Zeeman effect rather than the anomalous Zeeman effect generally observed at lower field strengths.

1923 H. L. BROSE tr. *Sommerfeld's Atomic Struct. & Spectral Lines* vi. 389 The Paschen–Back effect links together only such lines as belong together in a series as multiplicities. **1964** G. W. KING *Spectrosc. & Molecular Struct.* iii. 109 A magnetic field that would be strong enough to produce a resolvable splitting of nuclear Zeeman lines is usually more than sufficient to uncouple **I*** from **J*** and cause them to precess independently around the *z* axis, giving a nuclear Paschen–Back effect instead. **1970** G. K. WOODGATE *Elem. Atomic Struct.* viii. 150 This strong-field limit of the Zeeman effect is called the Paschen–Back effect.

pascual, *a.* Add: esp. describing plants growing in pasture or grassland. Also as *sb.*, a pascual plant.

1847 H. C. WATSON *Cybele Britannica* I. 65 The proposed series of terms runs thus:—1. Pratal... 2. Pascual. —Plants of pastures and grassy commons, where the herbage is usually less luxuriant than in the meadowlands, [etc.]. *Ibid.* 67 The pratal plants are occasionally pascual plants, as Phleum pratense; the pascuals are in turn ericetals, as Prunella vulgaris. **1926** G. C. DRUCE in J. J. Walker *Nat. Hist. Oxf. Distr.* 96 *Carex tomentosa* L., usually classed as a paludal, is a pascual in its only known Oxford locality near Burford.

Pascuan (pæ·skiuăn), *a.* and *sb.* [a. F. *Pascuan* (H. Lavachery 1935).] **A.** *adj.* Of or pertaining to Easter Island in the South Pacific, its inhabitants, or its script. **B.** *sb.* **a.** A native or inhabitant of Easter Island. **b.** The script used by the inhabitants of Easter Island. So **pascuense** *a.* and *sb.*

[**1935** H. LAVACHERY *Ile de Pâques* i. 16 Ici la naiveté des Espagnols apparaît non moins admirable que l'ingéniosité des Pascuans.] **1940** B. BONNERJEA in *N. & Q.* 17 Feb. 110/2, I have used the term 'Pascuans'—following Lavachery—in place of the more cumbersome English expression 'Easter Islanders'. *Ibid.* 111/1 It is an undeniable fact that the Pascuans are Polynesians. *Ibid.*, Since the..discovery of the Pascuan script, numerous articles have been written on the subject. **1948** D. DIRINGER *Alphabet* 137 According to local traditions, Hotumatua, an ancestor of the Pascuans..came to the island.. and brought with him 67 inscribed wooden tablets. **1960** J. FUENTES *Dict. & Gram. Easter Island Lang.* 591 The spoken language keeps some of the old Pascuense characteristics, that is to say, the Pascuense before assimilating modern foreign expressions.

‖**pas devant les enfants** (pā dǝvaṅ lezaṅfaṅ). [F.] Not in front of the children, used of an expression or action that requires discretion. Also *transf.* and *ellipt.*, *pas devant*, and as *adj.*

1951 N. MITFORD *Blessing* I. xv. 153 'Adultery is for when you're older, darling.' 'Oh I see. A sort of pas devant thing?' **1953** D. PARRY *Going Up—Going Down* i. 21 Mrs. Tyndale was holding forth in that extraordinary *pas-devant* French which the upper-class English used to speak so confidently. **1955** L. P. HARTLEY *Perfect Woman* vii. 65 'There's something I want to say to you. *Mais pas devant les enfants*,' he went on heavily. **1958** M. SPARK *Robinson* v. 51 'Pas devant,' said Tom Wells, casting his eyes towards the child. **1962** *Listener* 7 June 975/2 There is a sickening folklore of *pas devant*—'not in front of the children'—which excludes children from some of the very conversations that would show their parents as people with strong interests. **1962** *Times* 6 Aug. 9/3 It will be a mistake for the War Office to adopt a *pas devant les enfants* attitude. **1968** J. FLEMING *Kill or Cure* vii. 87 *Pas devant les enfants*; such bad form having rows before one's—friends. **1971** O. NORTON *Corpse-Bird Cries* v. 85 '*Pas devant les opérateurs*,' I said. I knew what manual [telephone] exchanges can be. **1974** J. JOHNSTON *How Many Miles to Babylon?* 7 'Dr. Desmond is an ass.' 'Frederick, pas devant —.' **1978** D. BLOODWORTH *Crosstalk* xxxix. 311 You mean *pas devant les enfants terribles*.. Not in front of the CIA?

pasear (paseā·r), *v.* and *sb. slang* and *U.S. dial.* [See *PASEO.] **A.** *vb. intr.* To take a paseo or walk. **B.** *sb.* = *PASEO. Also *attrib.*

1840 R. H. DANA *Two Yrs. before Mast* xxviii. 313 He was going to paseár with our captain a little. **1847** *Calif. Star* (San Francisco) 24 July 2/3, I am told this *pasear* over the mountains, will cost the Commodore [Stockton] five thousand dollars. **1878** B. HARTE *Man on Beach* 112, I was reck'nin' on taking a little *pasear* with you. **1892** STEVENSON & OSBOURNE *Wrecker* xii. 192, I tell you, Mr. Dodd, it was a queer thing to see me and the old lady taking a *pasear* in the garden, and the old man scowling at us over the pickets. **1903** CONRAD & HUEFFER *Romance* III. iii. 141, I just come from taking a pasear that way. **1914** *Sunset* July 64/1 It was the pasear madness that made despairing feet give way to auto tires—it is the undiminishing nature of pasear joys that is stretching the eighteen-million-dollar highway through the state. **1948** *Popular Western* June 16/2 Yuh're takin' a little *pasear* to the penitentiary in Walla Walla.

pasel, var. *PASSEL.

‖**paseo** (pāse·o). [Sp. *paseo* walk, *pasear* to walk.] In Spain and southwestern parts of the United States, a walk taken at a leisurely pace for exercise, amusement, or the like; any trip or outing of a similar nature; (concretely) a street or promenade; a parade, a procession, *spec.* at a bull-fight.

1832 W. IRVING *Leg. Alhambra* 111 An alameda, or public walk..not so fashionable as the more modern and splendid paseo of the Xenil. **1840** R. H. DANA *Two Yrs. before Mast* xxii. 219 The theme of..conversation..in our afternoon's *paséo* upon the beach, was the ship. **1897** 'H. S. MERRIMAN' *In Kedar's Tents* xviii. 205 He..proposed to Julia that they should take a 'paseo' in the garden. **1902** *Out West* Dec. 683 Such bosoming of motherly hills, knee-deep with winter wild flowers, as you may have unrolled to you in an afternoon's *paseo* from the metropolis. **1920** *Glasgow Herald* 22 Sept. 8/5 The greater companies..have built model towns, intersected with finely laid-out paseos and plazas, for their employees. **1927** E. HEMINGWAY *Men without Women* (1928) 29 They formed up for the paseo as soon as the bull had gone through. **1950** G. BRENAN *Face of Spain* ix. 187 It was the hour of the evening *paseo*: the girls were in their best frocks; the young men had oiled and smoothed their hair. **1967** MCCORMICK & MASCAREÑAS *Compl. Aficionado* iii. 72 You will now show me how to walk in the *paseo*. **1974** *Times* 2 May 17/2 Less than a decade ago, the *paseos* and boulevards were crowded on sunny days with strollers... Madrid has sacrificed these shady strolling places. **1976** E. P. BENSON *Bulls of Ronda* iv. 25 The stewards..led the *paseo*; the matadors were next, followed by the *bandilleros*... The bull..came charging into the arena. **1977** P. SOMERVILLE-LARGE *Eagles near Carcase* v. 91 Evening was being heralded..by the emergence of the *paseo*, if the three youths..and the four fat girls in black could be so described.

pash (pæʃ), *sb.* and *a. colloq.* **A.** *sb.* Abbreviation of PASSION; esp. in phr. *to have a pash for*, to be infatuated with, to have a 'crush' on; *transf.* a person who is the object of an infatuation. **B.** *adj.* Abbreviation of PASSIONATE *a.*

1914 *N.Y. World Mag.* 1 Nov. 5/6 There wasn't much 'pash' about it. **1920** F. SCOTT FITZGERALD *This Side of Paradise* (1921) 52 That isn't as pash as some of them. **1922** C. E. M. JOAD *Highbrows* iii. 102, I have met such a duck of a man. You'll never believe! I've quite a pash for him. **1924** P. MARKS *Plastic Age* 24 Let's go the movies... Gloria Nielsen is there and she's a pash baby. **1930** A. HUXLEY *Brief Candles* 217 Miss Figgis, the classical mistress, had been her pash for more than a year. **1930** WODEHOUSE *Very Good, Jeeves!* x. 257 The last bloke in the world..who you would think would ever fall a victim to the divine pash. **1934** G. GREENE *It's a Battlefield* 168 When you've got a pash for someone like I have, anybody's better than nothing. **1937** AUDEN & MACNEICE *Lett. from Iceland* 17 Sometimes containing frank demands for cash, Sometimes sly hints at a platonic pash. **1952** [see *CRUSH *sb.* 2 d]. **1955** AUDEN *Shield of Achilles* i. 30 As when past Iseult's tower you floated The willow pash-notes of wanted Tristram. **1972** J. MCCLURE *Caterpillar Cop* xiii. 209 'You know her then?' 'Oh sure. Had a pash for her big sister once.' **1975** J. HITCHMAN *Such a Strange Lady* iii. 36 In her efforts to get over her 'pash' on Dr. Allen she became extremely bossy.

Pashto. Now the usual spelling of PUSHTOO, -TU *sb.* and *a.* in Dict. and Suppl.

Pashtun, var. *PAKHTUN.

pasmo (pæ·zmo). [Amer. Sp., f. Sp. *pasmo* spasm.] A disease of flax, first reported from the Argentine in 1911, caused by the fungus *Mycosphaerella linorum*, and distinguished by circular brown or yellowish lesions on the leaves and stems of the plants affected. Also *attrib.*

1926 W. E. BRENTZEL in *Jrnl. Agric. Res.* XXXII. 25 (*title*) The pasmo disease of flax. *Ibid.*, The disease of flax called 'pasmo' in South America..caused great injury to flax in Argentina. **1942** *N.Z. Jrnl. Sci. & Technol.* A. XXIV. 102 Pasmo disease of flax is of almost world-wide occurrence. **1946** *Nature* 16 Nov. 723/1 The dreaded Pasmo disease was first described from the Argentine in 1911, reached Europe in 1936, and spread to five countries by 1942. **1958** *New Biol.* XXVII. 21 A number of fungi responsible for causing diseases of flax are seed-borne. The principal parasites in this group and the diseases they cause are..*Botrytis cinerea* Fr. (grey mould),..and *Sphaerella linorum* Wollenw. (pasmo). **1974** E. C. STACEY *Peace Country Heritage* ii. 121 They [sc. flax varieties] had been bred for resistance to pasmo and other diseases.

paso (pa·so). [Sp.] An image, or group of images, representing Passion scenes, carried in procession as part of Holy Week observances in Spain.

1923 *Chambers's Jrnl.* Apr. 213/1 Away goes the paso across the square. **1939** SPENDER & GILI tr. *Lorca's Poems* 19 (*title*) Paso. *Ibid.* 141 [Note] *Paso*, image, or group of images, representing a scene from the Passion of Christ, carried in procession during Holy Week, particularly in Andalusia. **1950** G. BRENAN *Face of Spain* viii. 178 Making our way into another church, we found the *pasos*, or floats holding images, pulled out into the nave in preparation for the Easter processions. *Ibid.*, One particularly large *paso*, supported on cart wheels, showed the scene of Christ being whipped by Roman soldiers. **1970** 'D. HALLIDAY' *Dolly & Cookie Bird* x. 155 The procession..got moving again...the paso of the poor hunting bishop [*sc.* St. Hubert] came rollicking past.

paso doble (pa:so do·ble). Also (erron.) **pasa doble,** and as one word. [f. Sp. *paso* step + *doble* double.] A quick Spanish dance-step; the music for such a dance.

1927 V. SILVESTER *Mod. Ballroom Dancing* 121 The *Paso Doble* is danced very little in this country, but it is popular in certain parts of the Continent. It is often referred to as the Spanish one-step. The walk is short and springy, not unlike a very modified Quickstep. **1934** C. LAMBERT *Music Ho!* II. 94 The Russian folk dance gives way to the pasodoblé of the street band. **1939** E. AMBLER *Mask of Dimitrios* vi. 101 An accordion band was playing a *paso-doble*. **1948** F. BORROWS *Theory & Technique Lat.-Amer. Dancing* iv. 156 Paso Doble music is in march time. **1955** W. GADDIS *Recognitions* III. iii. 771 An old man with a battered guitar..had two tunes, one a vaguely recognizable pasa doble. **1959** *Listener* 18 June 1059/2 The band plays a tinny *paso doble*. **1971** *New Scientist* 22 Apr. 219/1 The only sensuality indulged.. would be a quick routine or two from the pas a doble [*sic*] or the slow foxtrot. **1973** *Times* 23 Apr. 6/2 The baby-faced couples who dance the samba..and Pasodoble..are sometimes a bit short of Latin-American fire. **1974** *Times* 7 Jan. 5/8 There is Italian smooch-song and Spanish light music, pasadoble and tango. **1975** 'M. FONTEYN' *Autobiogr.* I. iii. 35 At every tea-dance in the..Hotel we waltzed and fox-trotted and danced the *paso-doble*, he so tall and me a little shrimp of eleven years.

paspalum (pæ·spălǝm). [mod.L. (Linnæus *Systema Naturæ* (ed. 10, 1759) II. 855), f. Gk. πάσπαλος a kind of millet.] An annual or perennial grass of the genus so called, native to warm regions, especially South America, and cultivated elsewhere for fodder. Also *attrib.*

1772 R. WESTON *Universal Botanist* III. 547 Round-flowered American Paspalum. **1857** *Ann. Rep. Mass. Board Agric.* IV. 84 Hairy Slender Paspalum..has an erect or decumbent, slender culm, from one to two feet high. **1884** G. VASEY *Agric. Grasses U.S.* 31 *Paspalum laeve.* (Smooth paspalum.) This species grows from 2 to 4 feet high. ..*Paspalum dilitatum.* (Hairy-flowered paspalum.) **1906** *Chambers's Jrnl.* 24 Feb. 207/2 Paspalum Grass..has been known in the United States since 1880, where it is called hairy and flowered paspalum. **1926** *Brit. Weekly* 12 Aug. 392/3 There's a corner of paspalum down there on the creek that would do your two eyes good to see. **1929** *Contrib. U.S. Nat. Herbarium* XXVIII. 1. 4 Paspalum-grass..was introduced into the southern United States from Uruguay or Argentina about the middle of the last century. *Ibid.* 5 The seeds of paspalum are eaten by a large number of birds. **1940** A. UPFIELD *Bushranger of Skies* x. 111 Bony stepped off the veranda.. and..trod the yielding paspalum grass lawn to arrive at the bottom fence. **1962** *Coast to Coast 1961–62* 94 In no time weeds and paspalum had hidden the grave. **1970** A. T. SEMPLE *Grassland Improvement* 381/2 Brownseed paspalum. **1977** *N.Z. Jrnl. Agric.* Jan. 52/3 Rhizomatous grasses—such as paspalum..—are not controlled with paraquat.

pasquinade, *sb.* (Later examples.)

1934 W. GERHARDI *Resurrection* xv. 43 A man famous for his evil tongue came up and..delivered himself of a long pasquinade at the expense of my friend. **1946** W. S. MAUGHAM *Then & Now* xxxv. 204 A cold shiver ran down his spine at the thought of the pasquinades, the epigrams that his misadventure would suggest. **1977** *Times Lit. Suppl.* 6 May 572/1 The famous pasquinade: 'quod non fecerunt barbari fecerunt Barberini'.

pass, *sb.*¹ Add: **3. b.** *to sell the pass*: see SELL *v.* 7 g.

pass, *sb.*² Add: **I. 1. a.** (Further examples.)

1966 *B.B.C. Handbk.* 53 Previously we had to snatch in a hit or miss fashion at 'passes' as *Telstar* or one of its cousins streaked from horizon to horizon. **1967** *Technology Week* 23 Jan. 28/2 Primary function of the subsystem is to correct the flight trajectory to assure a close pass by Mars. **1968** M. WOODHOUSE *Rock Baby* xvi. 153, I don't say it was impossible to spot, but..., short of an overhead pass by a very low-flying helicopter, I thought it would pass muster. **1974** *Nature* 4 Jan. 24/1 The orbit is near-polar with the north-going passes on the nightside at about 2230 LT, and the south-going passes on the day-side at about 1030 LT. **1977** *Offshore Engineer* Apr. 28/1 It is followed up by an external welder or 'bug', which completes a 3 mm weld thickness in one pass over wall thicknesses above 9·6 mm.

c. *Bridge.* The act of declining to make a bid.

1923 M. C. WORK *Auction Bridge of 1924* 497 Business Pass, a pass which indicates to the partner, who has made an Information Double, that the existing declaration will be remunerative. **1927** [see *AUCTION *sb.* 2 c]. **1958** *Listener* 25 Dec. 1094/3, I agree with West's opening pass and with his next bid of Two No Trumps. **1959** *Ibid.* 15 Jan. 146/1 After two passes, South opened Three Spades.

1959 REESE & DORMER *Bridge Player's Dict.* 165 West's double is for a take-out and East's pass is a penalty pass. **1977** *Times* 14 May 12/7 Hands which were so freakish that an unorthodox pass was the only road to safety.

d. *Computers.* A passage of data through a computer for processing; a single cycle of reading, processing, and writing; the performance of a particular kind of operation on each in turn of a set of data.

1954 *Jrnl. Assoc. Computing Machinery* I. 131/1 [illegible] pass of the data through the 702 may be enough to carry out..a statistical analysis. **1961** *Ibid.* VIII. 26 Passes are continued until no item is exchanged during a given pass... This signals that all items have been sorted. **1968** *Amer. Documentation* Jan. 78/1 During one pass all elements which have already been classed in a particular category are retrieved. A second retrieval pass is then made to retrieve all remaining elements [illegible] of the previous pass. **1968** G. EMERY *Electronic Data Processing* ii. 10 Figure 1.5 shows a single job, made up of three separate runs... Runs are sometimes subdivided further into separate passes—sorting runs are an example. **1976** H. LORIN *Sorting* ii. 6 A pass of Linear Selection involves selecting the element with the lowest key on the list to be sorted and placing it on a growing output list. Additional passes are made until the output list is complete.

II. 7. a. Also phr. *(to come to) a pretty pass*, (to reach) a regrettable state of affairs.

1824 [see PRETTY *a.* 3 c]. **1843** DICKENS *Mart. Chuz.* (1844) xiv. 176 [illegible] a pretty pass [illegible]. **1876** H. MELVILLE *Clarel* II. iv. xvi. 501 [illegible] The mysteries dishonored by an ass!' **1900** *Dialect Notes* II. 39 *Pretty pass*, [illegible] an abnormal condition: used in derogatory sense. **1965** *Times* 3 Oct. 9/4 Things, one felt, must have reached a pretty pass if the big banks, or any one of the big banks, had run into the same squalid staffing problems as the National Coal Board or the chain grocers. **1970** *Brewer's Dict. Phr. & Fable* (rev. ed.) 808/1 *A pretty pass*, a difficult or deplorable state of affairs.

III. 8. b. (Further examples.)

1939 [see *LAP *sb.*ⁿ]. **1939** JOYCE *Finnegans Wake* ii. 507 Such my bellet. Buy a barrack pass. **1955** 'N. SHUTE' *Requiem for a Wren* iii. 67 Before she had been a year at Ford Janet came to look forward to her next pass with something close to apprehension. **1955** [see *BIND *v.* 23].

c. (Examples.)

1838 *Actors by Daylight* I. 141 Give the Mounseer a pass to the pit. **1944** J. W. KRUTCH *Samuel Johnson* (1945) viii. 210 Johnson..complained that Garrick had just refused a pass to the theater for Miss Williams because [illegible]. **1961** BOWMAN & BALL *Theatre Lang.* 250 *Pass*, [illegible] a theatre without a ticket. British: complimentary ticket. **1977** *Times* 22 Apr. 15/1 Forging free passes to movie theatres.

f. In South Africa and (formerly in) Rhodesia, a document issued to non-white residents (now usu. one which it is obligatory for them to carry), authorizing and regulating their movement and residence in particular areas.

1828 J. PHILIP *Res. S. Afr.* I. 167 Among the many hardships to which the Hottentot is subject by this proclamation, one not adverted to is the Law of Passes, contained in the 16th article. [illegible] **1948** *Rep. Native Laws Comm. 1946–48* (Dept. Native Affairs, S. Afr.) 26/1 It is not always clear whether a particular document can rightly be described as a pass or not. Neither from European nor from Native witnesses did we receive a satisfactory definition, but we think it would be correct to say that in the mind of the Natives a document is a pass, to which they object, if it is a document—(*a*) which is not carried by all races, but only by people of a particular race; and which either (*b*) is connected with restriction of the freedom of movement of the person concerned; or (*c*) must at all times be carried by the person concerned on his body, since the law lays the obligation on him of producing it on demand by the police and certain other officials and the mere failure to produce it is by itself a punishable offence. **1949** *Handbk. Race Relations S. Afr.* 271 The tendency was to equate a pass with a document controlling movement. Clearly however, the pass has acquired a wider connotation to-day. **1952** *Statutes Union S. Afr.* 1013 To repeal the laws relating to the carrying of passes by natives; to provide for the issue of reference books to natives. **1956** D. JACOBSON *Dance in the Sun* [illegible] the police and told them to chase him [illegible] at his passes and said his passes [illegible] *Ann. Reg. 1963* 327 All adults were [illegible] vided that they had registered and [illegible] been recorded on their identity cards or [illegible] **1968** R. GRIFFITHS *Man of River* 137 But they [illegible] they must carry their passes! **1971** T. SHARPE *Riotous Assembly* ii. 10 Verkramp confessed him[illegible] experiment and the milk delivery [illegible] being out without a Pass. **1972** P. DRISCOLL *Wilby Conspiracy* (1973) iii. 40 The black man [illegible] automatically for his pass, and [illegible] through the green booklet. **1977** [illegible] tribal homelands are henceforth to [illegible] to their nationals, a more digni[illegible] But these, even if upgraded to pass[illegible] ependence, will serve the police and [illegible] Affairs as well as the passes. Pos[illegible] needed to apply for a job in white [illegible]

IV. 11. b. An amorous advance, esp. in phr. *to make a pass at* (someone). *colloq.*

1928 J. P. MCEVOY *Show Girl* 27 Almost all parties [illegible] off—a few high balls, a few dunces, [illegible] merry and making preliminary pas[illegible] *Not so Deep as a Well* 70 Men seldom make passes At girls who wear glasses. **1938** G. GREENE [illegible] He made no immediate pass at Ida [illegible] R. COOPER *Designs in Scarlet* ii. 14 [illegible] her drink, swallowing with it her [illegible] be forced to stick with a deado [illegible] making a pass at a 'gentleman'. **1942** [illegible] *More Flags* 173 What are you doing [illegible] a pass at Susie? **1944** AUDEN *For* [illegible] progress from outrage to outrage [illegible] the gross climax of His making [illegible] a pass at her virgin self. **1945** J. [illegible] xix. 190, I been over there. He [illegible] **1952** W. PLOMER *Museum* [illegible] visit to the painting room he had [illegible] hidden from me the feelings which [illegible] to make a pass at me. **1957** J. [illegible] vii. 70 A little gentle flirtation, even [illegible] would have changed her attitude [illegible] BUCKINGHAM *Wind Tunnel* xv. 127 [illegible] new to Janet. **1972** J. GORES [illegible] 89 As for making a pass at her, he'd [illegible] at his five-year-old daughter. Sex [illegible] D. SHANNON *No Holiday for Crime* [illegible] rted to get fresh, threw a pass, she [illegible] **1978** D. DEVINE *Sunk without Trace* [illegible] ulterior motives. Was she perhaps [illegible]

12. a. (U.S. examples.)

[illegible] *Calculus at Heaven* in E. Seaver *Cross-Section* [illegible] Did I ever tellya how I got to [illegible] Bank game one year?' **1961** J. S. [illegible] *Sports* 319 *Pass* (baseball, softball), [illegible] *Washington Post* 30 Sept. D4/3 [illegible] accurate this year, completing 37 [illegible] rds and a 47 per cent average. **1972** [illegible] He has been throwing passes in [illegible]

b. In real tennis, a service which drops in the pass-court; in lawn tennis, a shot which [illegible] the racket of an opponent.

1888 *Encycl. Brit.* XXIII. 179/2 A pass counts for nothing but annuls a previous fault. **1900** in A. E. T. Watson *Young Sportsman* 614 A 'pass' shall not neutral[illegible] **1911** *Encycl. Brit.* XXVI. 627/2 [illegible] ball drops beyond the pass[illegible] case does not count, but a 'pass' [illegible] fault, as was once the case. **1961** [illegible] *Gloss.* 259/2 *Pass*...Passing shot. [illegible] A mixture of lobs and angled [illegible] *Sports & Games* 825/1 The re[illegible] area is the 'pass court' and, if the [illegible] is called and a let played. **1978** [illegible] combe actually reached set point [illegible] cross-court pass off second [illegible]

[illegible] (Further examples.)

[illegible] e]. **1967** A. H. COTTRELL *Introd. Metall.* [illegible] the first few passes the draught (i.e. [illegible] on) is light... Reductions of 10–50 [illegible]

13*. In full, *food-pass*. Among certain birds of prey, the habit of passing food from one bird to another while in flight.

1931 D. NETHERSOLE-THOMPSON in *Brit. Birds* XXV. [illegible] stages of incubation the food 'pass' [illegible] occasionally witnessed. **1940** H. F. WITHERBY et al. *Handbk. Brit. Birds* III. 19 During this [illegible] male feeds hen both by food pass [illegible] to adjoining perch. **1948** B. H. RYVES *Bird Life in Cornwall* vi. 93 The aerial 'pass' of a [illegible] **1956** D. A. BANNERMAN *Birds Brit.* [illegible] views of the food-pass of the cock [illegible] also witnessed. *Ibid.* 195 [illegible] business of the 'pass' will be [illegible] **1970** E. HOSKING *Eye for Bird* xi. 157 They [*sc.* a [illegible]] flew over, swung up their feet and [illegible] the other. It is the speed and [illegible] which makes the hobby's food-pass [illegible]

V. 15. (Further examples.)

1930 *Engineering* 12 Dec. 759/2 The drawing of the strip through the pass of the rolls. **1939** [see *LAP *sb.*² 2 e]. **1960** D. J. O. BRANDT *Manuf. Iron & Steel* (ed. 2) xxxiii. 249 The most commonly used passes in bar and [illegible] are the diamond, the diagonal, the oval, and to [illegible]

VI. 17. a. (*a*) (sense 4) *pass class, coach, degree* (further examples), *examination* (earlier example), *mark, moderations* (colloq. *mods*), *rate*; (*b*) (sense *8 f) *pass-bearing* adj.; *pass office, regulation, system.*

1943 E. H. BROOKES *Bantu in S. Afr. Life* v. 11 The provisions of the Act under which Industrial Councils are formed do not apply to 'pass-bearing Natives'. **1953** K. AMIS *Lucky Jim* viii. 82 Of course, their problems down there are very different... The Pass classes in particular. **1908** E. M. SNEYD-KYNNERSLEY *H.M.I.* xi. 110 He had taken refuge in a Hall. There he exhausted the pass-coaches of Oxford. **1915** W. S. MAUGHAM *Of Human Bondage* xxvi. 109 When he only got a pass degree his friends were astonished. **1919** MENCKEN *Amer. Lang.* 105 His [*sc.* an English university man's] daily speech is full of terms unintelligible to an American student, for example, *wrangler, tripos, head, pass-degree* and *don*. **1948** M. LASKI *Tory Heaven* i. 5 After reading for a Pass Degree at Oxford, James had..been sent to..an uncle's rubber-plantation in Malaya. **1861** MILL *Repr. Govt.* xiv. 259 A mere pass examination never, in the long run, does more than exclude absolute dunces. **1911** W. OWEN *Let.* Sept. (1967) 79 Have I passed? I do think so! even in Arith. & Geometry. At least I have done enough right to score pass marks. **1957** *Economist* 12 Oct. 104/1 It is often assumed that there is a common pass mark for the whole of each county, but there can hardly be that until the school bus is replaced by each child's own jet propelled transport. **1977** P. STEVENS *New Orientations Teaching Eng.* ii. 34 In many countries..the effective criterion for success in an English-language course has been to achieve the pass mark in an examination set and marked in Britain. **1969** G. SMITH in *Lett. Aldous Huxley* II, 1913... October: H. enters Balliol and prepares for Pass Moderations. **1912** *Rep. Brit. Assoc. Adv. Sci. 1911* 219 The opinion has been expressed that Pass Mods. is not a bad thing. **1928** Pass office [see sense 8 f above]. **1948** *Rep. Native Laws Comm. 1946–48* (Dept. Native Affairs, S. Afr.) 66/2 Any Native *entering* the Province must proceed to the nearest Pass Office (or to the nearest Pass Office to his destination if entering by rail) and take out an *Inward Pass*. **1953** P. ABRAHAMS *Return to Goli* IV. ii. 129 He was soon picked up outside the Pass Office by one of the touts or 'runners' who look out for unemployed Africans. **1973** *Morning Star* 18 May 4/2 In the schools in the countryside the average pass rate was more than 95 per cent. **1976** *Daily Mail* (Hull) 16 Dec., Their pass rate of 79 per cent from 77 entries was better than most he had seen. **1948** *Rep. Native Laws Comm. 1946–48* (Dept. Native Affairs, S. Afr.) 33/1 If in course of time, as matters develop, the pass regulations fall into entire disuse, the laws authorising their application may well disappear from the statute book. **1859** *Queenstown Free Press* 2 Mar. (Pettman), Upon more occasions than one I have endeavoured to bring to the notice of the public the evils of the Pass system. **1936** *New Statesman* 11 July 46/1, I badly wanted to know whether God approved of the colour bar and the Pass system for natives. **1939** J. S. MARAIS *Cape Coloured People* p. viii, The pass system as applied to the Coloured north of the Orange, was abolished by the Natives (Urban Areas) Act of 1923. **1972** P. DRISCOLL *Wilby Conspiracy* (1973) vi. 74 A national campaign among Africans against the pass system.

b. passband, a frequency band within which signals are transmitted by a filter without attenuation; **pass-burner** *Southern Africa*, an African who burns his pass in protest against the pass laws; hence **pass-burning** *vbl. sb.*; **pass check** (examples); **pass court** *Real Tennis* (see quots.); **pass door** (further examples); **pass duty,** a duty levied on goods entering a territory; **pass lamp,** a lamp on a motor vehicle for use in fog; **pass law** *Southern Africa*, a law regulating the carrying of passes (see sense 8 f above); **pass light** = **pass lamp*; so **pass-lighting**; **pass line** *Real Tennis*, the line between the pass-court and the service court; **pass pawn,** a passed pawn; **pass play** *Amer. Football*, a sequence of passes between members of the same team; **pass raid,** in South Africa, a raid on African premises by the white authorities to check that passes are in order.

1922 G. A. CAMPBELL in *Bell Syst. Techn. Jrnl.* Nov. 15 The pass band and stop band characteristics of wave-filters are concretely illustrated..by the curves of Figs. 8–13. *Ibid.* 26 Each state holds for one or more continuous bands of frequencies; these bands have been distinguished as stop bands and pass bands. **1965** *Wireless World* Sept. 459/1 The low-frequency passband of the filter. **1970** J. EARL *Tuners & Amplifiers* v. 102 A BBC station at ±75kHz deviation with a modulation frequency of 15kHz...would call for a passband of about 180kHz for the best results. **1970** *Nature* 12 Dec. 1069/2 A 30 Å passband Fabry–Perot type filter was used to isolate the line. **1953** P. ABRAHAMS *Return to Goli* v. i. 190 Strikers and pass-burners were jailed and shot down in large numbers. **1961** *Economist* 4 Nov. 415/1, 2,072 [Northern Rhodesians] were convicted on charges of stoning, pass-burning, arson, road-blocking, and the like. **1844** J. COWELL *30 Yrs. among Players* I. xi. 27 We agreed to pay the extra three and sixpence and go into the boxes; but as to obtaining a pass check, it was impossible. **1858** G. A. SALA *Twice round Clock* in *Welcome Guest* 22 May 60/1 There is a theatrical pass-check, and the thumb of a white kid glove, very dirty, lying..at the back [of the hansom]. **1961** BOWMAN & BALL *Theatre Lang.* 250 *Pass check*, a re-admission pass for a spectator who leaves the theatre temporarily. British: *pass-out check*. **1900** G. E. A. Ross in A. E. T. Watson *Young Sportsman* 608 The pass-court is the area enclosed by the pass-line, the service-line, the end-wall and the main-wall. **1961** J. S. SALAK *Dict. Amer. Sports* 320 *Pass court* (court tennis)—the part of the floor on the hazard side that lies between the main wall, the grille wall, the pass line (but not including the pass line), and the service line (including the service line). **1975** Pass court [see sense 12 b above]. **1937** N. COWARD *Present Indicative* VIII. 313 At the final curtain there was booing... I dashed through the pass door and on to the stage as quickly as I could. **1939** JOYCE *Finnegans Wake* I. 146 The little passdoor, I go you before, so, and you're at my apron stage. **1950** 'J. TEY' *To love & be Wise* xviii. 233 Marta..came down to the edge of the stage... 'Come through the pass door, will you?' **1970** A. MORICE *Death in Grand Manor* x. 103 Be an angel, Annie, and show Miss Crichton to the pass door. **1909** *Daily Chron.* 20 Feb. 4/4 Opium..is grown in the Native States, the Government levying a heavy pass duty on its entrance to British territory. **1948** *Times* 14 Jan. 2/5 It has been found that the low-mounted passlamp is especially liable to cause dazzle. **1897** A. MILNER *Let.* 12 Aug. in C. Headlam *Milner Papers* (1931) I. vi. 194 If they do exempt Cape Boys from the degrading provisions of the Pass Law. **1901** Pass law [see sense 8 f above]. **1921** *Outward Bound* May 46/2 Soon after that, all my own boys cleared one night... I would not report them, that was not my way, let the pass-law say what it pleased. I don't hold with pass-law slavery. **1948** *Rep. Native Laws Comm. 1946–48* (Dept. Native Affairs, S. Afr.) 26/1 They regard even the certificates of exemption, which are issued to them as evidence of exemption from the pass laws, as passes; for only Natives require them, and they must be produced at any time on demand by an authorised official. **1956** T. HUDDLESTON *Naught for your Comfort* ii. 30 Another consequence of the pass laws—a consequence known to every intelligent South African at all interested in penal reform—is that it leads to an absolute contempt for the law. **1960** *Observer* 27 Mar. 16/4 Throughout South Africa's history, the pass laws have caused more bitter resentment than any other grievance. **1971** *Rand Daily Mail* 4 Dec. 1/1 It is where family life—the very basis of stability in traditional African societies—is being devastated by the pass laws and other apartheid regulations. **1974** A. WILLIAMS *Gentleman Traitor* xiii. 213 Two African youths had been arrested for breaking the Pass Laws. **1977** *Times* 16 Feb. 8/3 Legislation tabled in Parliament [in South Africa] today proposes the doubling of fines for violations of the country's pass laws. **1977** *Time* 2 May 22/3 The 'pass laws'—which control the movement of blacks into white areas—cost South Africa no less than $130 million a year to administer. **1938** *Times* 20 July 12/5 A rubber footrest for the clutch foot in which is the switch to cut out the head-lamps and put on a pass-light. **1948** *Times* 14 Jan. 2/5 Three-quarters of the cases of dazzle are caused by dipped headlights and pass lights. **1959** *Motor Manual* (ed. 36) vi. 181 The so-called spotlight ..provides a concentrated beam, but of more general use is the pass-light or fog-lamp, which is designed to throw a broad, flat beam with a fairly sharply defined top edge. **1965** PRIESTLEY & WISDOM *Good Driving* ii. 21 Many cars are fitted with 'fog' or 'pass' lights. **1938** *Times* 14 Oct. 11/1 Radio sets, permanent jacking systems, and an arrangement of pass-lighting which conduces to safety are outstanding features in the standardization this year. **1888** *Encycl. Brit.* XXIII. 179/2 A 'pass' [is called] if the ball has gone beyond the pass line. **1911** *Ibid.* XXVI. 627/1 The pass-line is drawn 7 ft. 8 in. from the main wall. **1935** *Encycl. Sports* 619/2 On the 'hazard side' of the court..is traced the 'service line'..and at right angles to this again is the 'pass line'. **1961** J. S. SALAK *Dict. Amer. Sports* 321 *Pass line* (court tennis)—the line on the floor nearest the grille and extending from the service line to the grille wall. **1908** *Daily Chron.* 29 Sept. 1/6 Lasker has a pass pawn, but..a draw appears probable. **1968** *Globe & Mail* (Toronto) 11 July 32/3 Turek scored on a 71-yard pass play with Joe Zuger. **1969** *Official Playing Rules Canad. Football* 29 For interference on a forward pass play see Rule 6, Section 4. **1970** *Toronto Daily Star* 24 Sept. 17/7 Eckman added Ticats' clinching touchdown a few minutes later when he shook off a blitzing B.C. lineman and ran seven yards on a broken pass play. **1958** *New Statesman* 8 Nov. 619/3 'Pass raids' are so commonplace in Johannesburg that even few liberal whites experience any real shock when they see a group of 10 or 20 Africans under police guard on a street corner, waiting to become part of the more than a thousand of their kind who, every day of the year, spend at least a day in custody because their papers are not in order. **1971** *Rand Daily Mail* 27 Mar. 1/1 To this is added perpetual insecurity, the harassment of pass raids and the miseries of life below the poverty datum line.

pass, *v.* Add: **A.** *intr.* **II. 5. a.** (Later examples.)

1933 M. GRANT *Conquest of Continent* 269 This enables some of these light Negroes to 'pass' as Whites. **1948** *Time* 16 Feb. 25/1 Possibly as many as 5,000,000 people with 'a determinable part' of Negro blood are now 'passing' as whites.

c. (Later example.)

1895 *Century Mag.* Sept. 676/2 It was a poor thing for the Bruce boys to do, to try to pass upon him like this.

d. To be held or accepted as a member of a religious or ethnic group other than one's own. Used esp. of a person of Negro ancestry who is held to be or regards himself as a white person.

1935 H. W. HORWILL *Dict. Mod. Amer. Usage* 224/2 In Am. there are many persons with a strain of Negro blood in whom this heritage of colour is so inconspicuous that they might easily be supposed to be of pure white lineage. If such persons leave their Negro associations and succeed in becoming accepted as Whites, they are said to *pass*. **1938** I. GOLDBERG *Wonder of Words* vi. 118 There are other Jews who resembling, psychologically, the troubled Negro—try to 'pass', which is the Negro term for being taken as a White. **1952** M. STEEN *Phoenix Rising* vii. 170 Coloured people who hide their origin and live as whites are said to 'pass'. **1953** E. H. BROOKES *S. Afr. in Changing World* vii. 147 Because of the permutations of nature, a coloured man white enough to 'pass' can have children or grandchildren who look 'Coloured'. **1955** D. VIKLUND tr. *Tingsten's Probl. S. Afr.* xiii. 148 There are two coloured brothers. One of them manages to 'pass' and becomes a barrister, a respectable member of white Cape society. **1960** D. JACOBSON *Evidence of Love* 92 He could 'pass' in the Transvaal, very easily, she thought. **1963** M. MCCARTHY *Group* xiv. 319 'Freddy's parents were trying to pass,' she went on sombrely. 'Like so many rich German Jews.' **1966** K. L. MORGAN in A. Dundes *Mother Wit* (1973) 606/2 The children used to love to come to Philadelphia with Caddy because there they could pass and have fun. **1971** *Encycl. Judaica* X. 60 At the height of 19th-century liberalism in Europe and America, it was possible for some Jews to 'pass' without doing anything more than simply ceasing to function in any Jewish association. **1976** *Wilson Q.* Autumn 87 Acceptable pigmentation plus wealth and influence were necessary for persons who hoped to obtain the *limpeza-de-sangue* document. But how dark one could be and still 'pass' varied from region to region.

VI. 15. (Further examples.)

1907 G. B. SHAW *Major Barbara* II. 222 Youd pass still. Why didnt you dye your hair. **1962** J. BRAINE *Life at Top* xv. 191 She said: 'He's a lovely little boy.' 'He'll pass,' I said absently.

IX. 21. c. (Later *transf.* example.)

1917 ADE *Let.* 29 July (1973) 68 If you want me to pass upon the sub-titles, I shall be glad to do so and if I change them it will be to make them more compact and not more intricate.

XI. 26. b. (Further examples.) Also, in Bridge and other card games, to decline to make a bid. Also *fig.*

1869 'MARK TWAIN' *Innoc. Abr.* xxxiv. 375 Jack said, 'I pass'—he plays euchre sometimes—and we all passed in turn. **1908** R. F. FOSTER *Auction Bridge* 29 The player on his left must either pass, or make a better declaration, or 'double'. **1918** [see *GO *v.* 78 k]. **1929** M. C. WORK *Compl. Contract Bridge* 42 South..bids one No Trump: North..passes. **1959** REESE & DORMER *Bridge Player's Dict.* 163 North, not being obliged to keep the bidding open, might pass. **1964** *Official Encycl. Bridge* 414/2 *Pass*, a call by which a player indicates that, at that turn, he does not choose to contract for a number of odd tricks at any denomination, nor does he choose, at that turn, to double a contract of the opponents or redouble a contract by his side that opponents have already doubled. **1976** L. SANDERS *Hamlet Warning* (1977) xxv. 216 'If you want to run up and take a look..' 'I'll pass. If you've seen one cannonball, you've seen them all.' **1976** *Washington Post* 15 Oct. A 23/3 The Washington Post passed. It made no mention whatsoever of what the committee had to say about the news media.

d. In dominoes, to miss a turn when one does not have a suitable number.

1960 R. C. BELL *Board & Table Games* I. 169 When one player has drawn all but the last two dominoes from the pool, and can neither make a ten, nor play a matador, he says, 'pass'. *Ibid.* 170 When a player is unable to make a match he calls 'pass', and the next player tries.

B. *trans.* **I. 29. c.** (Earlier and later examples.)

1870 J. K. MEDBERY *Men & Mysteries Wall St.* 137 To 'pass' a dividend... A dividend is said to be passed when the directors vote against declaring it. **1903** *Forum* (N.Y.) Oct. 209 Concerns which not only passed dividends..but went bankrupt. **1965** PERRY & RYDER *Thomson's Dict. Banking* (ed. 11) 422/2 *Passing a dividend*, a term used when a company decides not to pay a dividend.

C. Causative uses. **III. 46. a.** *to pass the buck*: see *BUCK *sb.*⁸; *to pass the word* (colloq., orig. *Naut.*), to convey information orally; to issue an oral order or instruction.

1843 S. LEECH *30 Yrs. from Home* ix. 186 Her officer.. so exasperated our captain that he passed the word to fire into her. *Ibid.* x. 218, I heard the order from an officer, of 'Pass the word for the boy Leech'. **1884** *Naval Encycl.* 636/2 *To pass the word* for a man is to summon him by name, the cry being repeated by the boatswain's mates on all decks. **1901** G. B. SHAW *Caesar & Cleopatra* II. 133 Pass the word to the guard; and fetch my armor. **1910** H. Y. MOFFAT *From Ship's-Boy to Skipper* iii. 39 He called to the sentry: 'Pass the word for the boy Moffat.' **1924** J. BUCHAN *Three Hostages* vi. 96, I will visit it as a man..to see about the meter... Macgillivray will pass the word for me. **1946** E. O'NEILL *Iceman Cometh* III. 184 They know I was framed. And once they've passed the word, it's as good as done, law or no law. **1961** *Sat. Even. Post* 3 June 60/2 Hundreds of men are required to pass the word to the button pushers and to push the buttons.

IV. 52. a. *to pass a* (or *the*) *remark*: to make an observation or comment, esp. one that is gratuitously sarcastic or depreciatory; freq. *pl. colloq.*

1899 R. WHITEING *No. 5 John St.* xxi. 218, I didn't sye I 'ad nothin' to sye to 'im. I only passed the remark. **1924** KIPLING *Debits & Credits* (1926) 157 'E could pass remarks, too!' Humberstall recited..a fragment..ending with, 'You lazy-minded, lousy-headed, long-trousered, perfumed perookier.' *Ibid.*, Macklin had a wonderful way o' passing remarks on a man's civil life; an' he put it about that our B.S.M. had run a dope an' dolly-shop with a Chinese woman. **1929** J. B. PRIESTLEY *Good Companions* II. v. 370 'Dewn't you be so personal', said Mr. Jerningham... 'You're always passing remawks.' **1933** C. MACKENZIE *Water on Brain* x. 142, I only passed the remark to Mr Wigmore yesterday, 'Mr Wigmore,' I said, 'people are getting pre-war again.' **1934** R. FERGUSON *Celebrated Sequels* 134 Were they or were they not laughing and passing remarks? **1939** JOYCE *Finnegans Wake* 463 He is looking aged with his pebbled eyes, and johnnythin too, from livicking onpidgins' ifs with puffins' ands, he's been slanderising himself, but I pass no remark. **1941** E. BOWEN *Look at Roses* 103, I merely passed the remark. There's no harm in my passing a remark occasionally. **1958** J. CANNAN *And be a Villain* iii. 70 I'll tell him you've broken off your engagement and not to pass any remarks. **1962** —— *All is Discovered* ii. 43, I passed a remark about her being indoors. I don't like her. **1975** C. STORR *Chinese Egg* xxvi. 168 He called her Fatty..and passed remarks about her figure.

c. *to pass the time of day* (further examples).

1851 [see TIME *sb.* 28 b]. **1936** 'F. GERALD' *Millionaire in Memories* xii. 316, I was riding home after doing some work on my racecourse and I pulled up at Swanson's swagger camp to pass the time of day. **1960** G. DURRELL *Zoo in Luggage* viii. 192, I was passing the time of day with the garage man. **1965** N. GULBENKIAN *Pantaraxia* xi. 230 Joe Boyle, one of the directors of the Royal Dutch Shell..went to see my father, as though merely to pass the time of day. **1965** *Listener* 23 Sept. 453/2 The English chaps would pretend..to be very friendly and jovial and pass the time of day and that sort of thing. **1978** M. GIROUARD *Life in Eng. Country House* v. 147 The King.. was then conveyed to the prince's bedchamber, where *he* passed the time of day for a few minutes.

V. 54. (Earlier example.)

1859 L. WRAXALL tr. *Mem. J. E. Robert-Houdin* I. viii. 157 He also said, in allusion to the ace of hearts, which he had 'passed' on one of the most beautiful women in the room: 'Will you be kind enough, madam, to lay your hand on your heart?'

D. With prepositions and adverbs.

I. With prepositions. **57. pass over—. a.** (Further examples.)

1749 *Apol. Life B.-M. Carew* 125 Passing over this Ferry they came into Rhode Island. **1874** J. W. DRAPER *Hist. Conflict Relig. & Sci.* vi. 161 The distance passed over in a voyage from Italy to the Gulf of Guinea.

58. pass through—. e. To follow a ritual of going through a narrow opening in a natural object such as a cleft tree or rock, etc., in the belief that the object will prove a barrier to evil or will absorb illness. Also *trans.*, in causal senses. Hence **passing-through** *sb.*

1804 *Gentl. Mag.* Oct. 909/1 Rowe's son was passed through the present tree in 1792, at the age of one or two. **1846** *Athenæum* 5 Sept. 909/1 A second point arises upon the 'passing through'... The passing through a cleft or aperture in a rock is a medical superstition..found in many countries. **1900** C. HOSE in *Geogr. Jrnl.* XVI. 45 The funeral procession climbed the mound on which the ceremony was situated, passing through the V of the cleft stick in single file. **1913** J. G. FRAZER *Golden Bough: Balder the Beautiful* (ed. 3) II. xi. 176 The words uttered by the mourners in passing through the cloven stick shew clearly that they believe the stick to act as a barrier or fence, on the further side of which they leave behind the ghost. *Ibid.* 184 In Scotland children who suffered from hectic fever and consumptive patients used to be healed by passing thrice through a circular wreath of woodbine. **1961** C. HOLE *Encycl. Superstitions* 21 A young ash-sapling was split, and the child was passed through it, three or nine times. **1968** *Proc. Amer. Philos. Soc.* CXII. 388/1 Passing-through rituals in the earth usually involved grassy terrain, where sod could be cut, piled, and arched to make a passageway. *Ibid.* 391/1 Prevention of disease by passing through seems more pronounced in passing through stones than in other kinds of rituals involving passing through.

II. With adverbs. **63. pass in.** *to pass in one's cheques*: see also *CHECK *sb.*¹ 15; *to pass in one's chips*: see *CHIP *sb.*¹ 2 e; also *absol.*; *to pass in one's marble*: see *MARBLE *sb.* 4 b.

1904 *N.Y. Even. Jrnl.* 3 May 2 'I may die' he told friends, 'and I want to breathe American air again before I pass in.'

b. *trans.* To withdraw from an auction sale because of failure to reach the reserve price. *Austral.* and *N.Z.*

1973 *Sun-Herald* (Sydney) 26 Aug. 23/2 The house was passed in at $37,000—the reserve was $42,000. **1977** *N.Z. Herald* 5 Jan. 1-11/10 What seemed to be unrealistic prices asked for by vendors in general resulted in 60 of the 99 lots on offer at a thoroughbred sale at Alexandra Park yesterday being passed in.

64. pass off. b. (Earlier examples.)

1787 J. WOODFORDE *Diary* 4 Dec. (1926) II. 356 Everything passed of [*sic*] as agreeable as one might expect from such a Meeting. **1788** E. SHERIDAN *Jrnl.* (1960) 107 All pass'd off very well—My Father a little stately at first but soon thoroughly cordial with his Son. *Ibid.* 141 Our day pass'd off very well—no awkwardness or unpleasantness of any kind.

c. *spec.* in *Law*, to represent (goods, etc.) as those of another.

1884 [in Dict.]. **1900**, etc. [see *PASSING *vbl. sb.* 2 b]. **1902** *Encycl. Brit.* XXXIII. 388/1 No trader is entitled to 'pass-off' his goods as those of another... Even if the 'passing-off' is done innocently it will be restrained. **1972** WHITE & JACOB *Kerly's Law of Trade Marks* (ed. 10) xvi. 362 The question whether the use of particular words or badges is calculated to pass off the defendant's goods as those of the plaintiff is often one of difficulty.

65. pass on. c. *intr.* To proceed from one existence or activity to another; *spec.* to die.

1804–1820 W. BLAKE *Jerusalem* III, in *Compl. Writings* (1972) 714 So Men pass on: but States remain permanent for ever. **1884** *150th Anniv. Settlement of Boscawen & Webster, New Hampsh.* 44 They have all passed on to become soldiers of the unseen army. **1920** R. MACAULAY *Potterism* III. iii. 131 If I have to pass on before Percy, he will be left bereaved indeed. **1923** *Amer. Mag.* June 15/1 The murderer took poison and so the two passed on. **1925** *Nat. Geogr. Mag.* XLVII. 489/2 His mother now dwells in comfort..while his father has passed on. **1928** L. NORTH *Parasites* 77 When my dad passed on there was just enough insurance to have let me finish up. **1930** 'E. QUEEN' *French Powder Myst.* xxxv. 274 Bernice [should] come in for a good share of Cyrus's estate when Cyrus should pass on. **1945** A. HUXLEY *Time must have Stop* vi. 66 'They don't die,' said Mrs. Gamble. 'They pass on.' **1946** E. O'NEILL *Iceman Cometh* I. 89 Too late! The old Doc has passed on to his Maker. **1960** S. KAUFFMAN *If it be Love* i. 12 'Margaret passed on.' William winced.. when someone used a euphemism for 'died'. **1973** A. BROINOWSKI *Take One Ambassador* i. 14 I'm Mrs Bert Norrice, well, Mrs Fern Norrice these days, since Mr

Norrice passed on last year. 1977 B. Pym *Quartet in Autumn* i. 10 Old Snowy had long since died, 'passed on' or 'been taken', however one liked to put it.

66. pass out. c. *intr.* (*a*) to die; (*b*) to faint, become unconscious; also, *to pass out cold*. *colloq.*

1899 *Westm. Gaz.* 6 Mar. 2/1 Another [spirit] who passed out with consumption is heard coughing. **1924** P. Marks *Plastic Age* i. 12 He left us a whole lot of jack when he passed out. **1927** *Hutchinson's Myst. Story Mag.* Feb. 80 First, the name of the dead relative of friend, then where they passed out. **1939** Joyce *Finnegans Wake* 627 I am passing out. O bitter ending! **1968** *Publ. Amer. Dial. Soc.* 1966 XLVI. 28 *Pass out*, die—'When he passed out, he was buried with his dogs.'

1915 C. Mathewson *Catcher Craig* i. 9 They sure do work you hard over there. I worked in the stock-room one summer and nearly passed out! **1918** J. M. Grider *War Birds* (1927) 97 We..carried him home after he passed out. **1924** P. Marks *Plastic Age* xxi. 254 A man 'passed out cold' and had to be carried from the gymnasium. **1936** J. T. Farrell *Guillotine Party* 191 Young Johnny Herbert had gotten drunk for the first time in his life and he'd passed out like a light. **1936** Wodehouse *Laughing Gas* ii. 26, I broke off here, because she had fainted... She slid sideways along the seat and quietly passed out. **1939** J. B. Priestley *Let People Sing* vii. 167 He's a big success in the cocktail bar... Does a few little tricks now and then, an' they nearly pass out. **1942** D. Powell *Time to be Born* (1943) x. 245 'What'll we do if he passes out in your place?' Corinne asked. 'I'll look after him,' said Vicky. **1947** 'N. Shute' *Chequer Board* i. 3, I suddenly passed out cold... **a 1953** E. O'Neill *Hughie* (1962) 22, I got scared he'd pass out with excitement. **1974** D. Gray *Dead Give Away* ix. 95 'How has she taken it?' 'Badly... She passed out when I told her.'

d. *trans.* To knock unconscious. *Austral. slang.*

1906 E. Dyson *Fact'ry 'Ands* xii. 151 He promised to show Feathers a 'boshter knack for passing out gazobs'.

e. *intr.* To complete a course of instruction, etc., successfully; to graduate. Also *trans.*, to undergo (a course of instruction); to allow or enable (someone) to pass out.

1916 *Daily Colonist* (Victoria, B.C.) 28 July 4/7 Midshipman Robert W. Wood...passed out first in 1914 both in the college and in ships. **1920** *Discovery* Mar. 77/2 Airship pilots..are required to pass-out a course in free ballooning, which includes a night flight and a solo flight. **a 1968** S. Fassler in R. Weaver *Canad. Short Stories* (1968) 2nd Ser. 335 My plan, after being passed out of Grade VIII, was to go..to Harbord Collegiate. **1968** J. Sangster *Touchfeather* ii. 8, I romped through the training, passing out with the highest marks anyone could remember.

f. To hand out or distribute.

1926 *Publishers' Weekly* 10 July 116 Librarians...are eager and willing to pass out catalogs that have won their confidence. **1927** *Ibid.* 12 Feb. 599 On that day she had passed out 130 books. **1964** Mrs. L. B. Johnson *White House Diary* 5 June (1970) 154, I was impressed that Charlotte Ford said, 'I'll do anything you want me to do. Type, pass out cards, anything.' **1973** P. Hill *Ruling Passion* III. iii. 179 He passed out some photostatted sheets. **1976** *Detroit Free Press* 1 Apr. 8B/1 Allow the ATF to lift licenses from dealers at any time in the future for the most minor of errors in complicated bookkeeping procedures, as well as pass out $10,000 fines.

g. *Bridge.* Of the players: to make three consecutive passes following (a bid), the auction thus ending and this bid becoming the contract; to make no positive bid at all (in a hand). Usu. in *passive*.

1959 *Listener* 19 Mar. 530/3 The fourth hand bid Three Spades which was passed out. **1960** T. Reese *Play Bridge with Reese* 91 Partner now surprises us by going five hearts. That is passed out. **1964** A. Truscott *Gr. Bridge Scandal* 307 A two notrump opening bid that has been passed out.

67. pass over. b. (Later examples.)

1794 W. Blake *Urizen* iv, in *Compl. Writings* (1972) 228 And a second Age passed over, And a state of dismal woe. **1795–1804** —— *Vala* v, in *Compl. Writings* (1972) 307 Till many a morn & many a night pass'd over in dire woe. **1840** R. H. Dana *Two Yrs. before Mast* xxv. 250 The night passed over without any trouble. **1878** R. L. Stevenson in *London* 6 July 10/1 The journey passed over without much incident.

h. To convey across; to transport.

1832 N. Wyeth *Jrnl.* 18 June in *Corr. & Jrnls.* (1899) 157 Reached the place for fording the plains, 1889. Passed over my goods during a severe wind without accident.

i. To die.

Associated esp. with spiritualism.

1909 *Review of Reviews* Feb. 123/2 His automatic writing..came..through the same friend through whom he has constantly communicated ever since he passed over. **1928** *Daily Mail* 7 Aug. 16/5 This doctor was now with my mother, who had recently passed over. **1930** D. L. Sayers *Strong Poison* xvi. 209, I have had the most marvellous talks with the dear ones who have passed over. **1938** Auden & Isherwood *On Frontier* III. i. 89 Those who have passed over are all very happy. He said the Other Side was difficult to describe. **1958** C. Watson *Coffin scarcely Used* ii. 18 If the living's good and bad mixed, then those who've passed over are two sorts as well. **1974** J. Stubbs *Painted Face* vii. 108, I wonder if her poor husband ever read those diaries, after she passed over?

69. pass up. To give up or abandon (a course of action, etc.); to decline or refuse to have (further) dealings with (someone or something); to reject (something) on the terms offered; to renounce or forgo (an opportunity, prospect, etc.). orig. *U.S.*

1896 Ade *Artie* i. 5 If he gets on a street-car where I am, I get off and walk. That ain't no lie. I pass him up. *Ibid.* xii. 112 Well, I guess I'll pass up the whole thing. **1906** H. D. Pittman *Belle of Bluegrass Country* vii. 108 'I know,' he continued, 'when I've got enough. I pass this little town up.' **1923** R. D. Paine *Comrades of Rolling Ocean* iii. 33 My duty is to stand by the family... That is why I passed up college. **1926** *N.Y. Times Mag.* 13 Aug. 1 He was sent to the U.S. Senate, but on his own terms, having previously passed up the Senatorship rather than take it on the terms of others. **1932** W. G. McAdoo *Crowded Yrs.* xxiii. 503 This would have been congenial work..but I had to pass it up. **1932** Wodehouse *Hot Water* xi. 188 And those jewels! You're really going to pass them up? **1939** —— *Uncle Fred in Springtime* i. 19 Your name...will be mud if you pass up an excellent bet like old Horace Davenport. **1943** [illegible] *Let.* Feb. (1972) 241, I am lucky to be getting lectures at all here and there a college and feel I mustn't pass any of them up. **1948** [illegible] v. 98, I don't like to see a good man waste himself. Pass up his opportunities. **1951** E. Paul *Springtime in Paris* xvi. 310 'You mean that miserable bookworm turned her down, when she offered herself? He's a cad.' 'Don't say that... He didn't pass her up.' **1956** 'A. Gilbert' *And Death came Too* xiv. 146 Eventually he agreed to take the case [illegible] being an inability to pass up a chance of wiping the official eye). **1962** *Listener* 26 Apr. 722/2 They passed up opportunities of wealth and property. **1964** Mrs. L. B. Johnson *White House Diary* 5 July (1970) 175 When Lyndon insisted that I go over to the Scharnhorst Ranch [illegible] I was not about to pass it up. **1969** Auden *City without Walls* 48 If you pass up a dame, you've yourself to blame, For shame is neurotic, so snatch! **1972** J. Potts *Trouble-Maker* (1973) x. 72 All false leads so far, but they can't afford to pass any of them up. **1975** *Times Lit. Suppl.* 11 Apr. 401/1 When the book does put reproductions of two states of the same etching side by side...Mr Passeron passes up the opportunity to discuss the meaning of the changes made. **1976** *Daily Tel.* 26 Oct. 17/2 He had passed up a job offer with a large accounting firm to cruise the South Seas.

passable, *a.* Add: **8.** as *sb.* A person or thing that is passable.

1908 *Westm. Gaz.* 26 June 2/1 Among such offspring there would be a small class of 'desirables', a large class of 'passables', and a small class of 'undesirables'.

passade. Add: **4.** (With pronunc. [pasād].) A transitory love affair; a passing romance.

1931 *Times Lit. Suppl.* 19 Feb. 134/2 His [sc. Mérimée's] singular *passade* with George Sand. **1934** H. G. Wells *Exper. Autobiogr.* II. vii. 465 The French..distinguish between the *passade*, a stroke of sexual impulse that may happen to any couple, and a real love affair. In theory, I was now to have *passades*. *Ibid.* 466 For women even more than for men, the frequent *passade* seems unattractive. **1937** —— *Brynhild* ix. 176 This lady had experienced two *grand passions* and a vast number of minor *passades*... The term *passade* was new to Brynhild. She tried to imagine the technique of a *passade*. **1938** —— *Apropos of Dolores* vi. 158 She has a wonderful French word '*passades*'. She may have *passades*. Possibly with rather scared youngish men. **1973** Wodehouse *Bachelors Anonymous* viii. 88 'Sure I did, the first moment I got here,' said Mr Llewellyn, feeling it unnecessary to complicate things by mentioning his *passade* with Miss Vera Dalrymple.

passage, *sb.* Add: **I. 1. e.** (*b*) *bird of passage* (further examples); esp. = *passage-migrant* (sense 16b below); *fig.* (earlier and later examples). Also (*slang*), a tramp.

1748 Chambers *Cycl. Suppl.* 1787 Birds of Passage, are such as only come at certain Seasons, and then disappear again. **1763** J. Bell *Trav. from St. Petersburg* I. 188 As for water-fowl, they are also birds of passage. **1771** Smollett *Humph. Cl.* I. 150 The..entertainments of Bath are over for this season; and all our gay birds of passage have taken their flight to Bristol-well, Tunbridge..&c. **1785** E. Sheridan *Jrnl.* (1960) 52 Our young Man is I find only a Bird of passage as the lady will be our only dinner companion. **1789** G. White *Selborne* I. xv. 29 Mr. Stillingfleet makes a question whether the blackcap (*motacilla atricapilla*) be a bird of passage or not. **1853** Dickens *Bleak Ho.* xl. 398 My Lady has been but poorly, and is a bird of passage as it were... **1897** Mrs. C. Cook *Comic Hist. N.S.W.* 47 It was a speculation by a bird of passage, one (poor little) Sparrow. **1893** A. Newton *Dict. Birds* ii. 560/1 Others again—and these are strictly speaking the 'Birds of Passage'—which show themselves but twice a year, passing through the country without staying longer. **1926** [illegible] 4 June (1967) 365 People arrive daily; most of them, however, birds of passage. **1943** Baker *Austral. Lang.* 103 *Traveller* and *commercial traveller*, together with *food inspector*, *bird of passage*, *wallaby tracker*, *tourist*, *footman* and *professional pedestrian*, are often applied to itinerants. **1960** *Guardian* 9 Oct. 12/3 As a 'bird of passage' the graduate tended to be supernumerary in the various departments. **1963** *Times* 20 Apr. 3/4 There is a need to differentiate between the birds of passage and those who give a lifetime to the profession.

2. c. *rites of passage*: see *Rite 1 d.

4. b. *to work one's passage*: see Work *v.* 12h in Dict. and Suppl.

7. (Further examples.)

1931 J. T. Adams *Epic of Amer.* ix. 240 The passage of a more stringent fugitive-slave law. **1968** *Globe & Mail* (Toronto) 3 Feb. 3/5 She rejected a request from Edward Schreyer (NDP, Springfield) that passage of the clause be postponed until members had had a chance to think about it. **1974** *Encycl. Brit. Macropædia* VII. 875/2 He could not prevent the passage of the bill.

10. b. *Med.* and *Biol.* A stage in the maintenance of a strain of micro-organisms or cells, from inoculation into a host organism or culture medium, through a period of multiplication, to extraction; the process of passing micro-organisms or cells through a series of hosts or cultures in this way, so as to maintain them or modify their virulence. Freq. with pronunc. (pæsā·ʒ).

1892 [illegible] **1906** [in Dict., sense 10]. **1926** G. H. Smith tr. *F. d'Herelle's Bacteriophage* iv. 160 The virulence of a bacteriophage may be exalted by successive passages in suspensions of a susceptible bacterium. **1944** [illegible] A chorio-allantoic suspension of the 250th chick embryo passage. **1947** [illegible] LXXXV. 24 The virus was maintained by occasional lung passage in albino Swiss mice. **1973** *Nature* 18 May 163/1 These tumours are transplantable and have been transplanted for up to six consecutive passages.

III. 13. b. (Further example.)

1920 *Brit. Mus. Return* 115 in *Parl. Papers* XXXVI. 673 The threatened extermination of the Elephants in the Addo Bush...has led to the passage of a number of letters between the Museum and various writers.

14. e. Add to def.: *spec.* an area of a picture where one tone merges into another; the technique of achieving this effect.

1961 M. Levy *Studio Dict. Art Terms* 84 *Passage*... Also used to describe the transition from one tone to another, by means of a half-tone. **1962** *Listener* 15 Feb. 304/1 They [sc. the Cubists] exaggerated his [sc. Cézanne's] use of his device known as *passage* by which the near end of a plane is clearly defined while the far end dissolves into space. **1962** R. G. Haggar *Dict. Art Terms* 247/2 *Passage*, a term used to describe a certain area of paint on a picture where one colour or tone merges into another, where some special technique has been used, or where there may be over-painting by another hand. **1967** J. N. Barron *Lang. of Painting* 143 These passages are used to relate volumes or three-dimensional forms to the two-dimensional picture frame.

V. 16. b. passage-free *adv.*, free of charge for passage or conveyance; **passage grave** *Archæol.*, an underground burial chamber connected with the surface by a passage; also *attrib.*; **passage-migrant**, a bird that stays for a short time in an area on the route of its migration to summer or winter quarters; also *fig.*; **passage-work** *Mus.*, a passage of a composition which calls for virtuosic display; the execution of such a passage.

1918 *Daily Express* 28 Aug. 3/7 She..brings to Canada almost passage-free any man of good health and physique who is an experienced agriculturist. **1888** F. H. Woods tr. *Montelius's Civilisation of Sweden* i. 30 The graves of this period are commonly described as 'dolmens' (*stendösar*), 'passage-graves' (*gånggrifter*), and 'stone cists' (*hällkistor*). **1939** [see *boat-axe* s.v. *Boat *sb.* 3]. [illegible] marked by great stone tombs, first dolmens then passage-graves, and lastly long stone cists. **1943** J. & C. Hawkes *Prehist. Brit.* ii. 45 The men who introduced passage-grave architecture and art seem to have reached Ireland from Portugal and southern Spain. **1958** F. E. Zeuner *Dating Past* (ed. 4) 81 This leaves only something like 200 years for that part of the Neolithic in Denmark which is clearly anterior to the arrival of the Bronze Age Beaker folk in Britain, namely all that preceding the middle of the Passage Grave period. **1963** E. S. Wood *Collins Field Guide Archæol.* I. iv. 57 Passage Graves have a [illegible] connected by a narrow passage to the outside of the [illegible] mound or cairn which covers it. [illegible] **1934** Landsborough & Salmon *Birds in Brit. To-day* xiii. 115 Records of small parties of Scandinavian lesser black-backed gulls... occurring as passage-migrants, especially in spring, should be considered extremely doubtful. **1940** H. F. Witherby et al. *Handbk. Brit. Birds* III. 12 [illegible] British Isles.—Resident and passage-migrant. [illegible] **1977** *Times* 18 Aug. 14/6 Formby Point is a haunt of passage-migrants. [illegible] *Musical Times* [illegible] passage-work [illegible] **1923** *Gramophone* Sept. 42/1 All three of these concertos are well constructed and melodious, sometimes with brilliant passage-work for the soloist.

passage, *v.*[2] Add: **3.** *Med.* and *Biol.* To subject a strain of (micro-organisms or cells) to a passage (sense *10 b). In *Med.* usu. with pronunc. (pæsā·ʒ).

1927 *Brit. Jrnl. Dermatol.* XXXIX. 7 Although the herpetic strain has been submitted to intracerebral passages for 4 years, it is still far from being as virulent or 'neurotropic' as the lethargica strain which has been passaged for 18 months only. **1952** *Jrnl. Exper. Med.* XCV. 260 All [influenza strains] were prepared from allantoic fluid passaged in 10 to 11 day old embryos. **1973** R. G. Krueger et al. *Introd. Microbiol.* xix. 514/1 The

higher the donor species lies on the evolutionary scale, the more times one can passage that species' tissue *in vitro*. **1973** *Nature* 18 May 163/2 Non-infected, non-transformed hamster cells passaged *in vitro* fifteen times failed to develop tumours in 1-day-old hamsters.

passage-money. (Further examples.)

1833 *Chambers's Edin. Jrnl.* 25 May 140/3 Hereabout the captain collects the passage-money. **1889** F. D. Lugard *Diary* 31 Dec. (1959) I. i. 60 Rs. 300 is I think very moderate. He asks no passage money &c. and speaks the language. **1966** *Times* (Australia Suppl.) 28 Mar. p. xv/1, I am not talking about cultural encouragement, only passage money.

passage-way. (Earlier N. Amer. examples.)

1649 *Rec. Early Hist. Boston* (1877) II. 98 Wm. Franklin is fined 20s. for disabling the passage way. **1725** in *Cambridge* (Mass.) *Reg. Bk.* (1896) 276 It is Neither Needfull nor convenient for to have a passage Way thro' Said Dickson's lott. **1846** T. L. McKenney *Mem.* I. ix. 191 On reaching the War Department I was met in the passage-way by the doorkeeper, James Barbour.

passalid (pæ·sălid), *a.* (*sb.*) *Ent.* [f. mod.L. family name *Passalidæ*, f. the generic name *Passalus* (J. C. Fabricius *Entomologia Systematica* (1792) I. ii. 240), f. Gr. πάσσαλος peg, in allusion to the shape of the insect: see -id[3].] Belonging to the family Passalidæ, which includes black or dark brown beetles with slightly flattened bodies, found in decaying wood in warm forest regions. Also as *sb.*, a beetle belonging to this family.

1905 G. C. Arrow in *Trans. Entomol. Soc. London* 748 The Passalid beetle stridulates by the opposition of certain stout spines upon the wings to other spines upon the antero-lateral parts of the dorsal segment. **1916** *Records Indian Museum* XII. 138 Madame Merian's larva can no longer be interpreted as a Passalid. **1927** *Chambers's Jrnl.* 20 Aug. 601/1 The whole Passalid family lives in rotten tree-stumps. [illegible] ance to the comparative study of social behaviour of insects that the reader may wish to know more about them. **1972** L. E. Chadwick tr. *Linsenmaier's Insects of World* 155/2 Famous for their family life are the passalid beetles.

Passamaquoddy (pæsămăkwɒ·di). Also 8 Passamaquoda, Pesmaquady. [Micmac, = 'place where pollack are plentiful', with reference to Passamaquoddy Bay.] (A member of) a tribe of North American Indians nearly identical in language and culture to the Maliseet and inhabiting parts of south-east Maine and (formerly) south-west New Brunswick; the Algonquian language of this tribe. Also *attrib.* or as *adj.*

1726 J. Gyles in *Maine Hist. Soc. Coll.* (1853) III. 357 Memorandum of ye No of Indians in each tribe from 16 years of age... Pesmaquady Indians 30. **1759** T. Pownall in *Ibid.* (1857) V. 371 Zacharie was asked what Tribe he was of, answ Passamaquoda. **1842** *Wasp* (Nauvoo, Illinois) 3 Sept. 3/1 Our Passamaquoddy Indians are divided into two political parties, between which a good deal of acrimonious feeling exists. **1857** *Porter's Spirit of Times* 11 July 291/2 Much interest was felt in the birch canoe race, between some Indians of the Passamaquoddy, Penobscot and Micmac tribes. **1910** F. W. Hodge *Handbk. Amer. Indians* II. 207/1 Passamaquoddy... A small tribe belonging to the Abnaki confederacy... They formerly occupied all the region about Passamaquoddy bay and on St. Croix and Schoodic lake, on the boundary between Maine and New Brunswick. [illegible] **1974** *Encycl. Brit. Micropædia* VII. 786/1 In 1969 there were two Passamaquoddy reservations in Maine. **1977** C. F. & F. M. Voegelin *Classification & Index World's Lang.* 17 Passamaquoddy... D[ialect] also Malecite... 9,000–10,000 [speakers]. New Brunswick, Maine, and adjacent Quebec. **1978** *Times* 11 Jan. 7/3 In Maine, the Passamaquoddy and Penobscot Indians are claiming..nearly two thirds of the entire state.

passback (pɑ·sbæk). [f. Pass *v.* + Back *adv.*] **a.** *Amer. Football.* = *snap-back. **b.** *Association Football*, etc. A defensive pass directed backwards to a team-mate, usu. the goalkeeper.

1934 in Webster. **1947** *Richmond* (Virginia) *Times-Dispatch* 16 Nov. 8B/7 For the first time in the game Filer called play No. 43. The passback came directly to the big fullback. **1976** *Scottish Daily Express* 23 Dec. 12/5 Goalkeeper Ted McPheat dropped a passback at the feet of Mike Lafnach. **1976** *Sunday Post* (Glasgow) 26 Dec. 33/3 If there's a better fetch-and-carry midfield man in the country than Lex Richardson (as long as he doesn't attempt passbacks) I'd like to see him.

pass-book. Add: **1.** (Further examples.)

1902 G. Meredith *Let.* 20 Apr. (1970) III. 1441, I do not see in my pass-book the Swedish £15. **1916** *Banking Publicity* Oct. 1/1 It is not more necessary for the banker to advertise who would have men and women leave their money with nothing more than a passbook to carry home in place of it? **1922** Joyce *Ulysses* 707 A bank passbook issued by the Ulster Bank, College Green branch showing statement of a/c for halfyear ending 31 December 1903. *Ibid.* 710 The endowment policy, the bank passbook, the certificate of the possession of scrip. **1949** G. B. Shaw *Buoyant Billions* iii. 30 What money I need appears to my credit in my bank passbook as cash or dividends on the few investments my stockbroker has advised. **1954** R. Gittings *John Keats: Living Year* v. 49 A small red leather wallet with a tuck-in flap, rather like a bank passbook. **1977** *Time* 28 Feb. 38/1 If the salesman had $2,000 or more in a savings account, he could borrow on his passbook and pay 1% more in interest on his loan than he received in interest on his savings.

2. (Earlier and later examples.)

1833 *Chambers's Edin. Jrnl.* II. 321/1 [Are there] no unheard-of overcharges in those pass-books I see flying about like evil spirits? **1861** Mrs. Beeton *Bk. Househ. Management* 1108 The 'pass-books' employed backwards and forwards between bakers, butchers, and the like [illegible] their customers. **1972** *Lebende Sprachen* XVII. 34/1 US passbook—BE credit sales book, [illegible]

3. *S. Afr.* = *Pass *sb.*[2] 8 f.

Usu. written as two words without hyphen.

1961 T. Matshikiza *Chocolates for my Wife* 88 The sergeant..thumbed querulously through each ninety-six paged pass book. **1963** K. Mackenzie *Dragon to Kill* xvi. 181 The police came to know about the evidence he could give in this case when they arrested him again, on a charge involving forged pass books. **1971** *Post* (Golden City, S. Afr.) 24 Mar. 7 Bishop Alpheus Zulu, who was arrested a week ago for not carrying his pass book. **1972** P. Driscoll *Wilby Conspiracy* (1973) vi. 74 People were advised..to destroy the pass books which the law required them to carry. **1973** *Times* 21 Sept. 15/5 Sizwe appropriates the dead man's name and pass book.

pass-by. Add: **3. a.** (Further examples.)

1911 *Act 1 & 2 Geo. V* c. 50 §43(3) Where, in the case of a pass-by, the tubs or trains of tubs are coupled or uncoupled at the face, or at the pass-bye next the face. **1967** *Gloss. Mining Terms* (B.S.I.) x. 11 *Pass-by*, 1. A siding in an underground haulage track having a turnout at both ends to the main line... 2. A loop of track around the shift at the face.

b. A place on a plate-rail (Plate *sb.* 8) where vehicles can pass. Also *attrib.*

1797 J. Curr *Pract. Coal Viewer* 26 Pass bye plates. Useful for 2 horses going contrary ways and passing each other with a draught of corves. *Ibid.* 27 Supposing a branch of road is required to be made to a new pit, one end of the above described pass bye..will accommodate such purpose.

passé, *a.* Add: **B.** as *sb.* In Ballet (see quot. 1948). Also *attrib.*

1948 A. Chujoy tr. *Vaganova's Basic Princ. Classical Ballet* vi. 65 Passé corresponds to its French meaning 'passed'. It is an auxiliary movement which transfers [passes] the leg from one position into another. If you are standing in développé effacé forward and you wish to bring the leg back into arabesque without doing a grand rond de jambe, you bend the leg at the knee, leaving it at a height of 90°, brush the toe of that foot past the standing leg and then proceed to arabesque. The passing the leg through this path is called passé. **1957** T. Mara *Second Steps in Ballet* vi. *Passé*... This means to pass the foot to a position at the knee in preparation for opening it in the *développé*. The right foot is strongly pointed at the little hollow of the supporting knee. **1959** B. & P. Fletcher *How to improve your Ballet Dancing* ix. 100 Relevé to passé position on left toe with right toe touching inside of supporting knee. **1967** G. Grant *Tech. Man. & Dict. Classical Ballet* (ed. 2) 82 *Passé*,..is an auxiliary movement in which the foot of the working leg passes the knee of the supporting leg from one position to another (as, for example, in développé passé en avant) or one leg passes the other in the air (as in jeté passé en avant) or one foot passes the supporting foot in back or in front of the supporting leg (as in chassé passé).

‖**passe** (pas). [Fr., f. *passer* to pass: see quot. 1903.] In roulette, the section of the cloth covering the numbers 19 to 36; a bet placed on this section.

1850, etc. [see *noir 2 a]. **1902** *Encycl. Brit.* XXXII. 304/1 *Pair* indicates even numbers, *impair* odd numbers; *manque* includes the numbers from 1 to 18; *passe*, from 19 to 36. **1903** 'L. Hoffmann' *Card & Table Games* (ed. 3) 649 So called because in this event the ball 'fails' (Fr. *manque*) to fall into a higher number than 18. *Passe* is so called because the ball 'passes' that number. **1923**, etc. [see *manque]. **1937** G. Frankau *More of Us* vii. 76 By plying rake on impair, manque, passe, parity, or any number you might put your fish on. **1971** P. O'Neil-Dunne *Roulette for Millions* iv. 38 The roulette table is divided into six areas which, starting from the top nearest the wheel, are labelled: *manque* for low; *passe* for high; *impair* for odds; *pair* for evens.

passe-. Add: **passe colmar** (earlier and later examples); usu. with capital initials.

1837 *Amer. Q. Rev.* XXI. 377 We could speak of others just coming into notice, such as the passe colmar. **1860** R. Hogg *Fruit Manual* 204 Passe Colmar... An excellent pear. Ripe during November and December. **1928** H. B. Tukey *Pear* x. 108 Passe Colmar..and other winter sorts sometimes shrivel and rot before reaching an edible condition.

passed, *ppl. a.* Add: **3.** *passed-out*: unconscious, *spec.* through alcoholic drink. *colloq.*

1927 *Amer. Speech* II. 277 *Passed out*, intoxicated. **1939** G. Greene *Lawless Roads* iii. 95 The blue soda-water bottles and the passed-out Mexican.

‖**passeggiata** (pasėdʒiā·tā). [It., = walk, promenade.] A leisurely walk; a regular stroll.

1967 *Listener* 30 Mar. 434/2 They step out together like a mother and children on a walk, or an Italian family making a *passeggiata*. **1969** Hurd & Osmond *Smile on Face of Tiger* v. 141 In the canyons of the Chinese town the deafening *passegiata* [*sic*] of the poor, dressed in their holiday best. **1971** N. Fisher *Rise at Dawn* viii. 137 We drove into Viareggio one evening. It was the hour of the *passeggiata*... The pavements thronged with strolling families. **1973** *Sat. Rev. Society* (U.S.) Feb. 19/3 The *passeggiata* after working hours of citizens along the Via dei Caizaiuoli. **1978** 'A. Stuart' *Vicious Circles* 4, I found myself peacefully watching the flow of people..taking their passegiata, or shopping.

‖**passéisme** (paseiz'm). [Fr.] Adherence to and regard for the traditions and values of the past, esp. in the arts. So **passéist(e)**, a traditionalist, one who is backward-looking (also *attrib.* or as *adj.*).

1914 Wyndham Lewis *Let.* 14 June (1963) 62 There are certain artists in England who do not belong to the Royal Academy nor to any of the passéist groups, and who do not on that account agree with the futurism of Sig. Marinetti. **1927** R. Fry *Let.* 4 Aug. (1972) II. 603 I've never been a Passéist—I was a Futurist but I have gradually trained myself to be a Presentist. **1943** Wyndham Lewis *Let.* 15 Oct. (1963) 368, I remember that clown Marinetti (the 'father of fascism') and his bellowings about 'passéisme' and his proposal to destroy all the pictures and buildings reminding people of the Past in Italy. **1953** —— *Let.* 23 Nov. (1963) 552 In editing *Blast* I regarded the contributions of Ezra as compromisingly passéiste. **1962** *Times* 12 Apr. 18/5 Delacroix was far more of a *passéiste* than the author suggests. **1968** *Listener* 29 Feb. 259/1 The old conflict between *passéiste* faculty and *antipasséiste* students is over.

passel (pæ·səl). Also **pasel, passle, pazil.** *dial.* and *U.S. colloq.* var. Parcel *sb.* 6.

1835 A. B. Longstreet *Georgia Scenes* 195 'How did you come on raisin' chickens this year, Mis' Shad?' 'La Messy, honey! I have had mighty bad luck. I had the prettiest pasel you most ever seed till the varment took to killin 'em.' **1865** W. B. Forfar *Kynance Cove* vii. 43 She ax'd about 'n a fine passle more than she wud ef he'd ben a stranger. **1871** E. Eggleston *Hoosier Schoolmaster* 169 A passel of thieves. **1881** *Atlantic Monthly* June 740/1 A passel o' folks. **1889** T. E. Brown *Manx Witch* 16 She knocked two dishes And a pazil of plates there off the dresser. **1890** S. S. Buckman *John Darke's Sojourn in Cotteswolds* ii. 6 Lor, thur wur quite a passel o' volk altogither. **1893** H. A. Shands *Some Peculiarities of Speech in Mississippi* 49 *Passle* (pæsel). Used to some extent by all classes, but principally by the uneducated, to mean a *parcel*, not in the sense of a small bundle or a small quantity, but in that of a considerable number; as, 'There was a whole *passle* of hogs in the yard'; i.e. there were a good many. The word has, perhaps, a somewhat larger meaning than *a good many*, but denotes less than a multitude. This word is used in Kentucky, but is becoming rare there. **1903** K. D. Wiggin *Rebecca* xix. 202 Then you can explain, if you can, who gave you any authority to invite a passel of strangers to stay here overnight. **1906** Kipling *Puck of Pook's Hill* 264 'Twas a passel o' nosense talk. **1935** Z. N. Hurston *Mules & Men* I. vii. 151 A man had a wife and a whole passle of young 'uns. **1936** [see *no-'count *a.*]. **1948** *Sun* (Baltimore) 3 Dec. 14/2 Who wants to gamble that a passel of bureaucrats in a planned economy could have shown similar bounce in the face of adversity? **1957** W. C. Handy *Father of Blues* vi. 80 We had to absorb a 'passel' of oratory of the brand served by some Southern politicians just this side of the turn of the century. **1972** M. J. Bosse *Incident at Naha* iii. 134 'He'll forget,' I declared, thinking of the wives and the passel of kids. **1973** *Science* 12 Jan. 162/1 But the AAAS did succeed in having a passel of young activists evicted from the meeting's main registration area. **1973** D. Westheimer *Going Public* iv. 54 How'd you like to make yourself a passel of money without hardly havin' to do any work? **1977** *Time* 20 June 47/2 She plays a small rancher who pools her resources with neighboring Landowner James Caan to fight off greedy Cattle Baron Jason Robards and a passel of oil companies lusting after their range land.

passementerie. Add: Also **pasmentier.** (Earlier and later examples.)

1794 A. Young *Trav. France* (ed. 2) I. xix. 550 They assert their *pasmentiers* of silk and cotton mixed, to be cheaper than any similar fabric in England. **1903** [see *dangly *a.*]. **1933** M. de la Roche *Master of Jalna* viii. 100 'Put a frill on it,' he suggested. 'A frill! A frill of what?' 'Would *passementerie* do?' he asked. **1936** M. Mitchell *Gone with Wind* xii. 232 Her stout bosom heaved violently beneath its glittering passementerie trimmings.

passenger. Add: **3. b.** Ellipt. for *passenger train*. *colloq.*

1886 H. Baumann *Londinismen* 132/1 *Passenger*,.. passenger-train. **1920** 'O. Douglas' *Penny Plain* xxiii. 259 He could spend ecstatic days watching every 'passenger' and every 'goods' that rushed..along the permanent way. **1962** 'D. Shannon' *Extra Kill* ix. 145 I'd just taken a couple to the Union Station, I guess to make the Owl for San Francisco—only passenger I know of leaving about then.

6. For *slang* read *colloq.* and add to def.: Also, in extended uses, one who takes a passive role in a group activity, enterprise, etc., making no personal contribution and requiring the continuous support of the rest. (Earlier and later examples.)

1852 J. F. Bateman *Aquatic Notes* iii. 23 Some University scratch Four-oared Races were rowed... Here would be seen three good oars endeavouring to row along a 'passenger', of some eleven stone weight. **1908** *Animal Managem.* 297 A sick or lame ox should be removed from the span at once, as he..is in fact 'a passenger', and has to be dragged along by the others. **1914** *Daily Mail* 6 Apr. 9/4 There was not a passenger in any division of the winning team. **1932** Auden *Orators* i. 19 We simply can't afford any passengers or skrimshankers. **1932** *Times Educ. Suppl.* 6 Aug. 301/2 It is a lucky school that has no passenger on its staff. **1946** D. Hamson *We fell among Greeks* iv. 47 We nearly had to shoot both of them later on at different times, because they were lazy and untrustworthy and we had no room for 'passengers'. **1948** 'N. Shute' *No Highway* xi. 288 As a scientist from Farnborough I was expected to be a passenger, useless in the woods. **1949** 'J. Tey' *Brat Farrar* xx. 185 You cannot expect to carry two adults as passengers in the estate... They are both capable of earning their own living. **1951** J. B. Priestley *Festival at Farbridge* II. ii. 319 Theodore plodded on, like a man walking on a hot afternoon to some place he hated. 'I'm beginning to feel I'm almost a passenger.' **1957** *Listener* 28 Nov. 882/2 More women have had to learn to become self-reliant in the mountains, or at least not to be mere passengers. **1961** A. Wilson *Old Men at Zoo* ii. 79 If you haven't any appreciation at all for serious research work, then the sooner you get out..the better. We're carrying enough passengers already. **1964** J. Masters *Trial at Monomoy* vii. 219 I'd want to do my share of work. I don't want to be a passenger. **1971** L. P. Davies *Shadow Before* v. 49 The scheme was really yours in the first place. Jack Latham wasn't much more than a passenger. **1971** R. J. White *Second-Hand Tomb* vii. 72 It was..an affection born of patronage. Pamela was so obviously a passenger.

7. a. *passenger cabin, car* (earlier and later examples), *carriage* (examples), *depot* (examples), *door, elevator, lift, line, list, lounge* (spec. in an airport), *manifest, plane, seat, side, terminal, trade* (examples), *traffic* (earlier and later examples), *train* (earlier and later examples), *way, window*; *passenger-carrying* adj.; *passengerless* adj. and adv.

1946 R. A. McFarland *Human Factors Air Transport Design* xi. 492 The general arrangement of the passenger cabin will naturally vary with the type of service for which the plane will be used. **1952** *Shell Aviation News* No. 166. 24 (*caption*) Cutaway drawing showing interior arrangement of the 44-passenger Convair-Liner 340. Forward of the passenger cabin is a large compartment for baggage and cargo. **1832** *Amer. Railroad Jrnl.* I. 305/3 Arrived, 9 passenger cars with 71 passengers. **1847** *Hunt's Merchants' Mag.* XVI. 211 Attached to this station, are also..two wood and water stations..a brass foundry, passenger car house, passenger rooms, offices &c. **1968** *Globe & Mail* (Toronto) 3 Feb. B 3/3 Extensive absenteeism by members of local 444 of the United Auto Workers forced a shutdown of the passenger car lines yesterday at the Windsor plant of Chrysler Canada Ltd. **1975** A. Bergman *Hollywood & LeVine* xiv. 211 Some Indian children waved at the train. They..stopped to stare at the silver blur of passenger cars. **1977** *Reinforced Plastics* XXI. 22/3 Engineering plastics materials have been used for many years for the production of cooling fans for passenger car engines. **1838** *Mechanics' Mag.* 13 Oct. 32/1 The *passenger carriages* of the American railways are extremely large and commodious. They are seated for 60 passengers, and are made so high in the roof, that the tallest person may stand upright in them without inconvenience. **1879** *Harper's Mag.* July 165 A bustling little locomotive with one passenger-carriage comes whistling down the valley. **1978** N. Marsh *Grave Mistake* vii. 210 A slow train with a passenger carriage. **1909** *Daily Chron.* 8 Sept. 1/6 He was placed third in the speed contest, fourth in the Gordon-Bennett Cup contest, and second in the passenger-carrying competition. **1928** *Manch. Guardian Weekly* 21 Sept. 224/4 It is with this passenger-carrying airship that the Germans hope to fly across the Atlantic. **1937** *Discovery* Sept. 270/2 The ultimate aim of all such experiments is the production of a passenger-carrying rocket plane. **1967** *Jane's Surface Skimmer Systems 1967–68* 87/2 The PT 20, a 27-ton boat for 75 passengers, is considered by Supramar to be the smallest size hydrofoil suitable for passenger-carrying coastal services. **1849** *Hunt's Merchants' Mag.* XX. 342 A spacious freight and passenger depot..has been completed in the lower part of Detroit. **1958** *Amer. Speech* XXXIII. 145 Though both units are likely to be in the same building, separate reference is sometimes made to the *freight depot* and the *passenger depot*. **1952** *Shell Aviation News* No. 165. 24 It is authorized that 87 occupants may be carried in aircraft with six exits, and one passenger door in the passenger area. **1968** K. Bird *Smash Glass Image* xii. 149 The dark-green Seat swung into the kerb... The passenger door opened. **1974** 'J. Le Carré' *Tinker, Tailor* xxxv. 311 In one of these blaring side-streets..Ricki Tarr would unlock the passenger door and hold him up at gun-point. **1886** J. A. Porter *New Stand. Guide Washington* 205 The building..is furnished with passenger-elevator, steam-heating, deposit-vaults, speaking tubes. **1919** C. Morley *Haunted Bookshop* iv. 87 Maybe he had no right to be riding in the passenger elevator. **1926** *Scribner's Mag.* Aug. 196/1 We ain't no trunks. Take us up in a passenger-elevator. **1906** *Westm. Gaz.* 11 Dec. 6/3 One would almost imagine that the running of a passengerless train from station to station,..would 'grow' on the conductors. **1952** R. Finlayson *Schooner came to Atia* 110 He was soon back, passengerless. **1907** *Shipping World* 16 Jan. 111 (Advt.), S.S. 'Lusitania' is being fitted with Passenger Lifts. **1931** *Times* 16 Mar. 21/7 An unique and beautifully-fitted flat, ideally situated and equipped with constant hot water, central heating, passenger and service lifts. **1938** M. Allingham *Fashion in Shrouds* xv. 248 They came up in a small handworked passenger lift to a front door. **1975** D. Bloodworth *Clients of Omega* xxii. 207 Dr. Moondance..will take the fast number one passenger lift to the fiftieth floor. **1851** *Min. Proc. Inst. Civil Engin.* X. 255 Owing to the different value now put upon the resistances to railway trains at high velocities,..good gradients had become relatively of less importance on passenger lines. **1843** Dickens *Mart. Chuz.* (1844) xvi. 196 I've just now sent a boy up to your office with the passenger-list. **1869** 'Mark Twain' *Innoc. Abr.* xx. 199 We had the whole passenger list for company. **1928** Kipling *Bk. of Words* 268 H.M.S. Great Britain carries a passenger list..of forty-five millions. **1931** —— *Limits & Renewals* (1932) 189 It's on the back of the passenger-list. **1958** B. Hamilton *Too Much of Water* ii. 23 A copy of the Passenger List which he had found in his cabin. **1973** A. Price *October Men* ii. 31 He was on the passenger list... It was an ordinary scheduled flight. **1969** D. Barron *Man who was There* i. 14, I..was considering looking for Gregory beyond the confines of the passenger lounge when they began to broadcast the announcement of our impending departure. **1970** M. Pereira *Pigeon's Blood* viii. 95 The Hall..for obvious reasons, was the only way of leaving the Passenger Lounge. **1971** Passenger manifest [see *manifest *sb.* 3]. **1976** B. Jackson *Flameout* (1977) ix. 149 I've..got the list of cargo... And the passenger manifest. **1931** W. L. Smith *Air Transport Operation* i. 5 The cost per passenger mile is about 7 cents for the largest passenger planes. *Ibid.* 4 In about the same period [12 yrs.], the capacity of passenger planes has increased from 2 to 30 passengers. **1937** *Discovery* May 164/1 The Pan-American Grace Company converted one of its passenger planes into a cargo plane. **1970** 'D. Holliday' *Dolly & Cookie Bird* viii. 124 Every now and then a big passenger plane would come droning in. **1975** S. Johnson *Urbane Guerilla* III. 114 The US..has banned shipment of nuclear material by passenger plane. **1937** M. Allingham *Dancers in Mourning* xxiv. 299 The body lay doubled up on the floor with..its head jammed against the front of the passenger seat. **1962** D. Francis *Dead Cert* xvii. 187 Pete himself poked his big bald head out of the passenger seat and called to me. **1972** *Guardian* 5 Dec. 7/2 Travel organisers..[must] keep within the terms and conditions of their licences... The licence will cost £250 a year plus 2p for each one-way passenger seat authorized by it. **1977** B. Pym *Quartet in Autumn* x. 91 Ken stuck in the passenger seat of a car on test. **1969** J. Gardner *Compl. State of Death* x. 216 He removed the bar, gently placed it on the passenger-side floor. **1972** *Country Life* 15 June 1577/2 A large glove box on the passenger side and a smaller one for the driver. **1976** *New Yorker* 16 Feb. 76/2 More personnel are needed at passenger terminals than at freight terminals relative to the value of their respective services. **1780** A. Young *Tour in Ireland* II. vi. 30 At the ports of Belfast, Derry, &c. the *passenger trade* as they called it, had long been a regular branch of commerce, which employed several ships, and consisted in carrying people to America. **1866** 'Mark Twain' *Lett. from Hawaii* (1967) 21 The sailing vessels..[are] too slow and uncertain to build up the passenger trade. **1972** 'G. Black' *Bitter Tea* (1973) vii. 110 The public is held back by a long counter..only we don't have much public really, not being in the passenger trade. **1836** *Mechanics' Mag.* 15 Oct. 30/1 No credit whatever was taken in the Eastern Counties Railway estimates for any of the passenger-traffic from *transmarine* sources, as that traffic was, at best, of a contingent character. **1883** W. H. Maw *Recent Pract. Marine Engin.* I. 117/2 In the case of a larger launch 40 ft. long fitted with engines of twenty indicated horse-power, and employed for passenger traffic on the River Dart, the average [fuel] consumption per week is but 10 cwt. **1933** S. L. Miller *Inland Transportation* xxi. 361 Efforts to increase the volume of railway passenger traffic have been in the past decade increasingly active. **1963** N. Wymer *Behind Scenes at London Airport* i. 17 The 'Short-haul' Building, opened in 1955, is for passenger traffic to and from Europe. **1976** Passenger traffic [see *private motoring* s.v. *private *a.* (*sb.*) 4c]. **1836** *Mechanics' Mag.* 5 Nov. 83/1 What is the usual weight you carry in one of your trains; your passenger train?—Forty or fifty tons; no, not more than thirty to forty tons, carriages and passengers together. **1937** *Discovery* Mar. 88/2 Each [van]..is vacuum fitted and fully equipped for working on fast passenger trains. **1976** *Daily Tel.* 20 July 4/1 Last Thursday he left Nairobi in a passenger train for Kampala. **1908** *Westm. Gaz.* 14 Mar. 2/1 Using the parallel of the street..he claimed that the river should be regarded as a passenger way. **1971** R. Petrie *Thorne in Flesh* xiv. 175 There was a large, dark saloon parked at the kerb... The nearside passenger window slid down and Tina called to him. **1977** P. Hill *Fanatics* 145 Dice..wound down the passenger window.

b. Special Comb., as **passenger-mile**, a unit of measurement representing one passenger travelling a distance of one mile; hence *passenger-mileage*; also *passenger-kilometre* (abbrev. *km.*).

1900 *Geogr. Jrnl.* XVI. 221 The number of passengers carried in 1898 was 126 millions, the number of passenger-kilometres amounting to 4439 millions. **1903** E. Johnson *Amer. Railway Transportation* x. 140 An equal mileage of road accommodates a much greater traffic in Europe than in the United States. This is shown by dividing the total number of miles traveled by all passengers (the 'passenger miles') by the miles of railroad. **1930** *Flight* 28 Nov. 1381/2 Taking paying passengers only, the figure becomes 41 passenger-miles. **1936** *Jrnl. R. Aeronaut. Soc.* XL. 850 The criterion then being the number of passenger miles flown per period. **1943** *Ibid.* XLVII. 249 An expectation that the accident rate can..be reduced to a long-term average of no more than 1·0 fatality per 100,000,000 passenger-miles. **1962** R. B. Fuller *Epic Poem on Industrialization* 219 Passenger miles per capita per annum By all modes of transport Represents a smoothly ascending Curve. **1971** *Guardian* 13 July 2/2 TWA on its own rates as the world's second largest airline—in terms of passenger-miles—behind United, the American domestic operator. **1974** *State* (Columbia, S. Carolina) 1 Apr. 2-A/2 (Advt.), If your carpool cars average, say, 13 miles to the gallon and you share costs with three others, you are paying for only one gallon in four, and you're getting 52 passenger miles-per-gallon. **1976** P. R. White *Planning for Public Transport* viii. 175 Of the 29000 million or more passenger-km generated on the BR system each year about 45 per cent are classified as intercity. **1978** *Jrnl. R. Soc. Arts* CXXVI. 427/1 In London Transport, we have adopted as our corporate aim the maximization of passenger-mileage within the financial resources available to us from fares and grants.

pa·ssengered, *a. rare.* [f. Passenger + -ed[2].] Of a vehicle, ship, etc.: carrying or occupied by passengers.

1929 R. Graves *Poems* 27 That was the hospital-boat of twelve years back, Passengered as before with doubt and dying. **1955** W. Gaddis *Recognitions* III. i. 723 Traffic often consists only in the gay orange garbage carts, passengered by black vultures.

passenger-pigeon. Add to def.: The bird became extinct in 1914. (Later examples.)

1907 W. B. Mershon *Passenger Pigeon* p. ix, As recently as 1880 the Passenger Pigeon was thronging in countless millions through large areas of the Middle West. **1955** A. W. Schorger *Passenger Pigeon* ix. 229 The passenger pigeon became extinct through such constant persecution that it was unable to raise sufficient young to perpetuate the race. **1967** D. Goodwin *Pigeons & Doves of World* 203 The Passenger Pigeon shares with the Dodo the doubtful fame of being one of the two best known species that have been exterminated by man. **1972** G. Durrell *Catch me a Colobus* x. 213 It was impossible, everyone thought, that the Passenger pigeon (so delicious to eat and so plentiful) could ever be exterminated... The last Passenger pigeon in the world died in the Cincinnati zoo in 1914. **1975** *New Yorker* 7 Apr. 45/1, I saw the last passenger-pigeon flocks in Iowa.

passe-partout, *sb.* Add: **2. c.** A kind of adhesive tape or paper used for framing photographs and for other purposes. Also *attrib.* and as *vb. trans.*

1909 *Cent. Dict.* Suppl. 952/1 *Passepartout*,..to place in a passe-partout frame. **1910** V. Tree *Let.* 13 Nov. in *Castles in Air* (1926) 54, I have found a manufacturer of *passe-partouts* for my flower and French costume prints. **1910–11** *T. Eaton & Co. Catal.* Fall & Winter 144/1 Passepartout binding, black, green, brown, grey, red and white. **1928** *Daily Express* 17 May 9/4 A favourite occupation is evidently to 'passe-partout' their pictures. **1954** *Paper Terminol.* (Spalding & Hodge) 44 *Passe-partout*, a strong embossed paper, gummed on one side and sold in coils about 1 in. wide. It is made in many colours and is used for picture mounting and the binding of lantern slides. **1969** R. Blythe *Akenfield* viii. 136 Passe-partouted photographs of their sons and daughters..hang on long strings from the picture rail. **1978** J. Goodman *Last Sentence* iii. 112 Haphazardly hung photographs, all framed amateurishly with passe-partout.

passer. Add: **3. b.** In ball games, a player who passes the ball to another player (cf. Pass *v.* 46 b).

1905 *Westm. Gaz.* 12 Dec. 9/2 From a clever pass—the 'passer' could not be distinguished in the fog—Parker feinted and swerved cleverly, scored behind the posts, and an easy goal resulted from Nesbitt's kick. **1970** *Washington Post* 30 Sept. D 4/3 If he played the whole season, he would rank among the top passers in the country. **1972** J. Mosedale *Football* ii. 18 Clark was his team's leading passer. **1973** *Times* 25 Jan. 12/8 Ian's a much better passer of the ball on the ground.

c. In various trades, a person who examines materials or manufactures to ensure that they are of the required quality, workmanship, etc. (see Pass *v.* 45 b).

1921 *Glasgow Herald* 21 June 9/7 The proposed reduction is 3d. per hour in respect of measure cutters,..fitters-up, tailors' pressers, machinists, passers, etc.

d. One who receives and passes on counterfeit money. *slang.* (Cf. Pass *v.* 46 c.)

1929 *Detective Fiction Weekly* 25 May 683/1 The dealer calls the carrier..they meet..and..a few bills are handed out as a sample. The carrier calls the 'shovers', sometimes known as 'passers' or 'pushers' who begin to operate. To this class belong the men who actually place the bogus money in circulation. **1955** W. Gaddis *Recognitions* II. v. 490 I'm going out to meet a passer, to hand this stuff over to him. It's all arranged and paid for. *Ibid.*, It's always trouble with the middleman and the passers that get you pulled in.

4. One who 'passes' as a member of an ethnic group other than his own. (Cf. *Pass *v.* 5 d.)

1953 P. Abrahams *Return to Goli* II. iv. 64 If the passing is successful even the parents of the passer cease to know him or her. **1956** L. Kuper *Passive Resistance in S. Afr.* I. ii. 66 The psychology implicit in apartheid legislation is the psychology of the coloured man who has successfully 'passed' into the white group. Over the generations, there would have been inculcated into the successful 'passer' the dominant value that 'whiteness is all'.

5. One who sells drugs illicitly.

1955 [see *meet *sb.* 1b]. **1956** *Sun* (Baltimore) 26 July 14/1 This is an Act of the utmost severity, even providing the death penalty for 'passers' under certain circumstances.

passe-temps. ? Delete †*Obs.* and add later examples.

1840 Geo. Eliot *Let.* 20 July (1954) I. 60 It [sc. *Don Quixote*] is a charming passe-temps (a word by the bye that I ought to have outlawed long ago). **1848** T. Arnold *N.Z. Lett.* (1966) 27 One cannot be always reading and this sail-making forms a very agreeable passe-temps. **1891** W. Fraser *Disraeli & his Day* 344 After many dreary weeks of solitude at the 'Three Cups' hotel, the only *passe-temps* being to watch the corpses, removed from the sunk-fort, of soldiers who died daily of small-pox, the Election took place.

pass–fail (equal stress), *a.* [f. Pass *v.* 17 + Fail *v.* 14.] Of or pertaining to passing and failing (in an examination or the like); applied *spec.* to methods of assessing examination performance in terms simply of success or failure, without further reference to the individual standards attained.

1959 *Psychol. Rev.* LXVI. 62/2 Using pass-fail scoring on all eighteen tests, the test-subject matrix was 91.8 per cent reproducible. **1963** *Times* 2 Dec. 11/7, I would like to suggest a Use of English paper on a five-point scale with no pass-fail line drawn. **1966** *Crimson-White* (Univ. of Alabama) 1 Dec. 4/3 On the last SRI poll, students were asked to give their opinion on a system known as 'pass-fail electives'. Under this system a student could take 2 or 3 electives while he was here at the University and not have to worry about his grade. **1970** *Times* 20 Apr. 2, I believe we have gained a great deal in suggesting grading instead of somewhat more rigid pass-fail standards. **1972** *Sat. Rev.* (U.S.) 4 Mar. 49/2 All courses are given on a pass/fail basis. **1973** *Harvard Law School Bull.* Apr. 18/2 A strictly pass-fail grading system.

‖**passglas** (pa·sglas). Also **pass glas,** and in anglicized form **pass glass** (pɑs glɑs). Pl. ‖**passgläser, pass glasses.** [Ger.] A tall, cylindrical drinking glass decorated with parallel rings or a spiral down its length.

1897 A. Hartshorne *Old Eng. Glasses* 68 Touching the Passglas, its usual type was a tall cylindrical vessel..spaced, by fine wheeled stringings, into divisions for measuring and controlling the drinking. *Ibid.*, Rembrandt,..in his portrait by himself at Dresden, holds up exactly such a Rhine-land Passglas. *Ibid.* 81 Certain rare Pass glasses are ornamented with figures of knaves from playing cards, painted in enamel on white grounds. **1907** E. Dillon *Glass* xvi. 273 The tall *pass*-glass..is dated 1662. **1926** W. Buckley *European Glass* 54 The Pass-glas is a typical German form which was made chiefly during the 17th century. It is always cylindrical, and is decorated by a stringing which sometimes encircles the glass spirally and at other times forms a number of rings more or less equidistant from one another. **1942** *Burlington Mag.* Dec. 299/1 The vessels depicted in Dutch still-life paintings are in fact as often of German as of Italian form—the *Roemer*, the *Passglas*,..are as commonly seen as the Italianate 'winged' goblets. **1960** H. Hayward *Antique Coll.* 211/1 *Passglas*,..a German glass beaker encircled by a notched spiral thread dividing it into equal parts so that the glass could be passed from guest to guest allowing each his allotted share. **1965** A. von Saldern *German Enameled Glass* 145 Cylindrical Stangen with horizontal rings in multicolored enamel—so-called Pass glasses..are derived from similar glasses with applied rigaree bands or spirals; each horizontal ring is called a Pass... In drinking, the trick was to bring the wine or beer down exactly to the level of one of the rings.

‖**pas si bête** (pā si bę̄t). [Fr.] Not so foolish; 'not that stupid' (said of the speaker or of someone other than the speaker). Also (in quot. 1924), not so bad.

1840 Thackeray in *Fraser's Mag.* June 727/2, I am not holding up the whole affair as a masterpiece—*pas si bête*. **1849** —— *Pendennis* I. xxix. 283 'Emily was always as stupid as an owl,' said Miss Blenkinsop. 'Eh! pas si bête,' the old Peer said. **1862** W. Collins *No Name* II. 82, I am not fool enough to open it. *Pas si bête*, as we used to say in the English circle at Zurich... *Pas si bête!* **1923** Galsworthy *Captures* 247 Why suppose one's family superior to other people's? *Pas si bête!* **1924** J. Buchan *Three Hostages* xii. 179 'What about the weather?' I asked anxiously. '*Pas si bête*,' he said, sniffing. 'The wind is pretty sure to go down.' **1939** N. Marsh *Overture to Death* xvi. 183 'Try us,' suggested the young man. 'Pas si bête,' said Alleyn, 'I want my lunch.' **1965** N. Freeling *Criminal Conversation* II. i. 102 You see how the parallel fails?.. *Pas si bête*.

passimeter (pǎsi·mɪtəɹ). Also (erron.) **passimetre.** [f. Pass *v.* or Pass(enger + -meter.] An automatic machine for supplying public transport passengers with tickets and recording the number of people who pass through.

1921 *Railway Gaz.* 2 Dec. 860/2 There are three separate registering cyclometers... The first registers the dating and cancellation of each ticket, setting the passenger free to proceed through the 'passimeter', whereby another cyclometer automatically records his passage. **1923** *Westm. Gaz.* 11 Aug. 6/4 Fifteen passimeters will be installed, and..it is hoped..to eliminate booking queues. **1927** *Observer* 16 Jan. 18/2 The work of substituting passimeters for the booking offices is now in progress. **1928** *Daily Express* 10 May 9 What a contrast with the modern 'passimetre', the glitter of green and white tile, the silent motion of escalators! **1928** J. P. Thomas *Handling London's Undergr. Traffic* xii. 156 The first passimeter booking-office was introduced at Kilburn Park in 1921. **1964** *Guardian* 26 June 5/2 Central London is to have six experimental 'standee' buses, on which a single coin fare will be placed in a 'passimeter'. *Ibid.* 1 Dec. 4/2 On the six fast 'standee' buses which London Transport hopes to introduce next year..the fare..will be placed in a 'passimeter'.

passing, *vbl. sb.* Add: **1. a.** (Examples in sense *5 d of the vb.)

1926 C. Van Vechten *Nigger Heaven* 286 *Passing*, passing for white. **1952** L. Marquard *Peoples & Policies S. Afr.* iii. 77 Though..it is impossible to say how many coloured have succeeded in 'passing'—that is, in being fully accepted as European—'passing' has certainly occurred. **1952** M. Steen *Phoenix Rising* vii. 176 Those who succeed in 'passing' live their lives in mortal terror of being found out. **1953** [see *passer 4]. **1958** S. E. Hyman in A. Dundes *Mother Wit* (1973) 46/2 The account of lynching, passing, discrimination, or varieties of resistance. **1961** *Guardian* 4 May 10/4 'Passing', the word used to describe Negroes merging indistinguishably into a white community in America. **1973** G. D. Berreman *Caste in Mod. World* 10/2 The response of *Burakumin* to their birth-ascribed status is that common to all low castes: accommodation on the most part, and occasional 'passing'... As in all societies passing is difficult and stressful, for the fear of discovery and the necessary loss or attenuation of family contacts are traumatic.

b. *spec.* in ball games, the action of transferring (the ball) to another player (see sense 46 b of the vb.). Also *absol.*

1882 *Blackburn Times* 1 Apr. 6/3 While the Rovers worked their way towards their opponents' goal by passing, the Etonians did so by rushes. **1889, 1892** [in Dict.]. **1906** Gallaher & Stead *Compl. Rugby Footballer* 85 A player's object in passing the ball is to give possession of it to a colleague who is in a better position for making further headway with it than he is himself. **1952** J. B. Pick *Phoenix Dict. Games* I. 63 This..ensures that the passing will be precise. **1960** E. S. & W. J. Higham *High Speed Rugby* 5 A method of passing the ball in which the outside elbow is bent upwards.

2. b. With *off*: see sense 64 c of the vb. in Dict. and Suppl.

1900 *Rep. Patent Cases* (Patent Office) 15 Aug. 482 In the *Yorkshire Relish* case the inference was drawn that the mere use of a name implied passing off. **1902** [see *pass *v.* 64 c]. **1908** *Westm. Gaz.* 30 Apr. 9/2 Damages were claimed by the plaintiff for alleged libel, and the passing off of a story written by her in 1895 and 1896. Defendants denied that there was any libel or passing off. **1959** *Times* 6 Feb. 17/1 His action against them for alleged libel and passing off. **1960** *Times* 20 Sept. (Pure Food Suppl.) p. i/5 There still remains..the possibility of verbal passing-off at the time of sale. **1962** *Listener* 15 Mar. 457/1 We might ..be called upon to change our law in respect of 'passing-off'. Broadly speaking, this is selling one's own goods or services in a way which deceives the public into thinking that they are those of some other trader. **1970** White & Jacob *Patents* III. viii. 65 In an action for passing-off the plaintiff must in practice prove that he has extensive enough goodwill for his goods to be recognised by members of the public.

3. (Earlier and later examples.)

1848 E. C. P. in C. H. Hartshorne *Eng. Medieval Embroidery* 113 A rich gold thread, called passing, or tambour. **1880** L. Higgin *Handbk. Embroidery* i. 8 *Gold and silver passing*, a very fine kind of thread. **1957** M. B. Picken *Fashion Dict.* 349/2 *Passing*, smooth, flattened thread made by twisting strands of gold or silver around a strand of silk. **1960** B. Snook *Eng. Hist. Embroidery* 48 A very fine flexible metal thread called 'passing' was known to the Elizabethans.

4. a. passing door *Mining*, an arrangement of doors in a gallery that enables people to pass through while preventing the free passage of air currents; **passing novel** (see quot.); **passing place** (*c*) on a narrow or single track road.

1839 Ure *Dict. Arts* 989 Passing doors..may be substituted in any place for a passage where there is a stopping. **1964** J. H. Clarke *Harlem* 345 James Weldon Johnson..wrote 'passing' novels, i.e. novels about Negroes who were able to pass as whites. **1951** N. M. Gunn *Well at World's End* xxiv. 210 The lorry swung into a 'passing place' with squealing brakes. **1963** P. MacTyre *Fish on Hook* v. 73 He pulled ponderously into the next passing-place, to allow the vehicle behind to overtake. **1972** *Times* 16 Oct. 13/6 It's a single track road, with passing places not very close together—and of course heavy lorries can't back. **1973** G. Moffat *Lady with Cool Eye* vii. 74 A narrow tarred track..only wide enough for one vehicle. At intervals there were passing-places.

b. With advbs., as *passing-off action*; *passing-out examination, inspection, parade*; *passing-through ritual*.

1925 F. I. Schechter *Hist. Found. Law Trade-Marks* i. 10 Lord Chancellor Halsbury, in analyzing the plaintiff's pleadings in a 'passing-off' action, stated that such an action 'has been a well recognized cause of action, certainly for the last two hundred and fifty years'. **1946** *Nature* 2 Nov. 604/1 Traders were obliged to rely on cumbersome and expensive passing-off actions to protect their name and goods. **1970** White & Jacob *Patents* III. viii. 66 Many businesses do not keep their trade marks fully and validly registered, so that resort to a passing-off action may always be necessary to cover flaws in the trade-mark position. **1976** *Century of Trade Marks* (Patent Office) i. 2/2 In France ..there is at least one recorded instance in the 16th Century of what would be called today a passing-off action, in which the defendant was subjected to a perpetual injunction and made to pay heavy costs. **1916** *Daily Colonist* (Victoria, B.C.) 28 July 4/7 He came out first in the recent passing-out examinations. **1973** *Soviet Weekly* 5 May 2/3 Moscow School of Choreography..has begun its passing-out examinations. **1930** *Times Educ. Suppl.* 2 Aug. 340/3 The passing-out inspection of aircraft apprentices..was held on Tuesday. **1955** *Times* 11 Aug. 6/5 The Sudan defence force to-day held the first passing-out parade of officer cadets since the Sudanization of the force. **1971** B. W. Aldiss *Soldier Erect* 156 The C.O. of Kanchapur spoke to us on passing-out parade. **1973** 'S. Harvester' *Corner of Playground* I. v. 38 Officer cadets due to receive their commissions at the passing-out parade. **1969** *Times* 19 Mar. 4/4 The wedding ring and the custom of the groom carrying his bride over the threshold may be relics of the 'passing through' rituals.

passing, *ppl. a.* Add: **2. c.** *passing show*, the spectacle of contemporary life; an entertainment using as material current events and interests, a revue.

1908 *Sears, Roebuck Catal.* 1047/3 A whole passing show... Your friends grotesquely photographed... By getting a focus on passing pedestrians, horses, cars, etc., the most ludicrous pictures are witnessed. **1915** (*title of journal*) The passing show. **1915** A. Wimperis (*title of revue*) Passing show. **1956** *B.B.C. Handbk. 1957* 31 To keep the listener fully in touch with the 'passing show' of contemporary life. **1968** *N.Y. City* (Michelin Tire Corp.) 81 In 1864, Tony Pastor's Opera House made vaudeville fashionable...the 'revue' or 'passing show' caught on in the 90's. **1976** *Scottish Rev.* Summer 9 Reason tends to be hostile to the passing show, tends to view reality as an eternity of fixed types amenable to the eye of reason.

passing-by. Restrict †*Obs.* to sense in Dict. and add: **b.** The action of ignoring or neglecting.

1909 *Westm. Gaz.* 6 May 8/3 The ousting of home-bred meat, and the passing-by of the market by the great importers.

passing-note. (Earlier example.)

1730 *Short Treat. Harmony* v. 28 We make use of the Second, the Seventh, and of the Fourth as Discords or Passing Notes.

passion, *sb.* Add: **1. c.** Also, a dramatic setting of the Passion of Christ; = *passion-play*.

1823 W. Hone *Anc. Mysteries Described* 169 In 1298, the passion was played at Friuli. **1903** E. K. Chambers *Mediæval Stage* II. xxii. 129 There were performances of Passions in Reading in 1508, in Dublin in 1528, [etc.]. **1962** R. Southern *Seven Ages of Theatre* 107 The *Passion* of Mons may well have run to ninety-eight separate representations of 'scenes'.

2. b. A narrative account of the passion of a martyr.

1904 T. Shearman *Veneration of S. Agnes* 90 Helen of Rossow, or Roswitha, a Benedictine nun of the Convent of Gandersheim, Saxony, wrote poems in the 10th century, 'to replace,' as she says in her preface, 'the pagan passions which dishonour the profane drama, by the triumphs of the Christian heroines, the chaste spouses who are admitted to the Nuptials of the Lamb.' **1913** E. R. Barker *Rome of Pilgrims* xiii. 183 In an eighth-century manuscript there is a note that Passions are to be read at Office in the Church of S. Peter. *Ibid.* xiv. 192 It is always the conventional version of a Passion which is reproduced in numerous manuscripts. *Ibid.* 195 For this saint ..there exists not only the contemporary Passion, but also a series of records. **1927** F. J. E. Raby *Hist. Christian-Latin Poetry* ii. 56 His poem was used as a basis for later prose passions of Cassian.

11. a. *passion-clouded, -dimmed, -frantic, -pale, -pastured, -plunged, -wearied* adjs.

1925 W. B. Yeats *Vision* III. 183 Aristophanes' passion-clouded eye. **1899** —— *Wind among Reeds* 26 Because your crying brings to my mind Passion-dimmed eyes and long heavy hair. **1916** A. Huxley *Burning Wheel* 29 So, troubled, passion-frantic, The poet's mind boils gold and amethyst. **1889** O. Wilde in *19th Cent.* Jan. 47 The passion-pale face of Andromeda. **c 1865** G. M. Hopkins *Voice from World* in *Poems* (1967) 125 How turn my passion-pastured thought To gentle manna and simple bread? **1876** —— *Wreck of Deutschland* xxxiii, in *Poems* (1967) 62 Our passion-plungèd giant risen, The Christ of the Father compassionate, fetched in the storm of his strides. **1881** O. Wilde *Poems* 4 With passion-wearied face.

b. passion-fruit, add to def.: esp. *Passiflora edulis*, the granadilla, which produces egg-shaped fruit with reddish-purple, slightly wrinkled skin and sweet yellow pulp surrounding small black seeds; (further examples); **passion killers** *slang* (see quots.); **passion play** (earlier and later examples); also *transf.*; **passion vine** (earlier and later examples); **passion wagon** *slang* (see quot. 1948).

1867 R. Henning *Let.* 18 Feb. (1966) 234, I have also been making some passionfruit jelly. **1908** E. J. Banfield *Confessions of Beachcomber* I. vi. 192 There may be some who do not know that the humble papaw..belongs to the passion-fruit family. **1934** T. Wood *Cobbers* xvii. 217 Passion fruit, squeezed into a wineglass, mixed with cream and sugar and a spoonful of sherry, has a rich smoothness. **1961** L. van der Post *Heart of Hunter* I. ii. 27 His old lady, dark and wrinkled with age like a passion fruit about to fall. **1969** *Oxf. Bk. Food Plants* 98/1 Passion Fruit or Purple Granadilla (*Passiflora edulis*). A perennial climbing plant, originally native to Brazil but now widely planted in the tropics, it is also sufficiently hardy to be grown in some Mediterranean countries. **1974** *Herald* (Melbourne) 5 Apr. 23/1 Pavlova,..a crusty meringue-like sweet-cake made from egg whites and sugar and topped with whipped cream and, usually, passionfruit. 1976 *Observer* 17 Oct. 36/3 (Advt.), Easy to grow delicious passion fruits. Our own specially cultivated pot-grown species of Granadilla for fruiting in Britain. 1977 'E. Crispin' *Glimpses of Moon* xii. 235 The infant Grand Duchess..lisps a request for a glass of..passion-fruit juice. 1943 C. H. Ward-Jackson *Piece of Cake* 47 *Passion killers*, service knickers issued to air-women. 1946 J. Irving *Royal Navalese* 136 An elastic-bound bifurcated undergarment said to be worn in the women's Services and known..as 'passion-killers'. 1974 *Times* 17 Dec. 12/5 Stout beecy lined drawers, which would have been called by this generation 'passion-killers'. 1870 in J. Brown *Lett.* (1912) 328 I was very much touched by the Passion-play, and wrote some very bad verses at Ammergau. 1966 B. Sweet-Escott *Baker St. Irreg.* iii. 90 It turned out to be..the ritual passion play on the 10th of the month of Muharram which commemorates the death of Hassan. 1975 *Listener* 10 Apr. 472/3 Going to Oberammergau to the Passion Play. 1853 'P. Paxton' *Stray Yankee in Texas* 57 The passion vine with its singular flower and luscious fruit. 1862 R. Henning *Let.* 23 Sept. (1966) 100 A veranda covered with passion-vine and a garden full of petunias in most brilliant flower. 1946 *Coast to Coast 1945* 64 Let his girls dig in the orchard or chip around the passion-vines. 1977 M. West *Kundu* ii. 19 A passion vine trailing over a bamboo summer-house. 1969 *West Australians* 5 July 41/7 (Advt.), Nellie Kelly the amazing grafted passion vine. 1948 Partridge *Dict. Forces' Slang* 137 *Passion wagon*, truck taking men out for a day's, or part of a day's, leave, into a town or place of entertainment. 1961 *New Left Rev.* Jan.–Feb. 24/2 He knows every girl who comes out the base on Saturday on the passion-wagon.

passionful, *a.* Add: **2. b.** Subject to or susceptible to passion.

1902 *Amer. Anthropologist* Jan.–Mar. 33 The savage man conceived the diverse bodies collectively constituting his environment..to be living, thinking, willing, passionful beings.

Hence **passionfulness,** the state or quality of being passionful.

1922 *Glasgow Herald* 16 Dec. 10/6 Several members..by their passionfulness of heart and uncontrollable spirit had broken the order and decorum of the House of Commons.

Passionist, *sb.* (*a.*) Add: **1. a.** (Earlier and later examples.)

1832 G. Spencer *Let.* 7 June in C. R. Leetham *Luigi Gentili* (1965) iii. 40 The General of the Passionists. 1839 Ld. Shrewsbury *Let.* 16 Apr. in E. S. Purcell *Life & Lett. A. P. de Lisle* (1900) I. vi. 105, I have seen Lord Clifford, Father Glover and the Passionists. 1957 *Oxf. Dict. Chr. Ch.* 1022/2 In 1841 the Passionists came to England, where they were the first religious after the Reformation to lead a strict community life and wear their habit in public. 1967 D. T. Kauffman *Dict. Relig. Terms* 347/1 Passionists wear black garments and heart-shaped badges symbolizing the Passion. 1975 R. Player *Let's talk of Graves* iii. 89 Father Dominic the Passionist who received our dear Newman into Christ's Church.

b. (Earlier and later examples.)

1844 A. P. de Lisle *Jrnl.* 8 July in E. S. Purcell *Life & Lett. A. P. de Lisle* (1900) I. vii. 118 The Superior of the Passionist Monks called upon us. 1911 *Encycl. Brit.* XX. 887/1 The order of Passionist Fathers..was founded by St Paul of the Cross..in 1720. 1975 *Irish Times* 10 May 16/1 The fact that in Ireland strong religious piety co-exists with fierce sectarian hatreds is lamentable and points to a need for a questioning of the reality of religious beliefs, President O Dalaigh states in an interview in the current issue of the Passionist magazine, the Cross.

Passion Sunday. Add: (In the R.C. Ch. suppressed as a separate observance in 1969.)

passion-tide. (Earlier examples.)

1847 *Dublin Rev.* Mar. 25 The physical cause of our Lord's death is a subject..adapted to the season of Lent and Passion-tide. 1849 J. H. Newman *Discourses to Mixed Congreg.* xv. 323 Though at this season [*Note* Passion-tide] many words would be out of place.

passival, *a.* Restrict *rare* to sense in Dict. and add: **b.** Semantically passive, *spec. passival verb*, an intransitive verb with a quasi-passive meaning.

1892 H. Sweet *New Eng. Gram.* I. 90 *The book sells well, meat will not keep in hot weather*.. We call *sells* and *keep* in such constructions passival verbs. 1926 H. Poutsma *Gram. Late Mod. Eng.* II. II. xlvi. 64 Sweet..calls the verbs thus used passival verbs. 1950 *Eng. Stud.* XXXI. 163 It is generally said that in a sentence like *His books don't sell* the verb is active in form, but passive in meaning, what is 'really' meant being *His books are not sold*. In accordance with this theory verbs used in this way are sometimes called active-passive or passival. 1961 Y. Olsson *On Syntax Eng. Verb* vii. 180 That article, which..rightly rejects the analysis of such collocations as 'active-passive' or 'passival'. 1969 F. T. Visser *Hist. Syntax Eng. Lang.* II. 152 (*heading*) Intransitive verbs used to represent the action as quasi-automatic, or self-originated. (Sweet's 'passival verbs'; Jespersen's 'activo-passive' use.)

passivate (pæ·sivēit), *v.* [f. Passiv(e *a.* and *sb.* + -ate[3].] **a.** *trans.* To render (metal) passive (Passive *a.* 7 b in Dict. and Suppl.), e.g. to prevent corrosion.

1913 *Chem. News* 28 Nov. 259/1 We assume the passive state of metals (which can be passivated) to be the normal state. 1916 *Jrnl. Chem. Soc.* CIX. 1365 Hittorf has passivated chromium by anodic treatment in hydriodic acid.

1961 *Flight* LXXIX. 747/1 All surfaces that are to be in contact with HTP [*sc.* High Test Peroxide] must be 'passivated' to prevent surface decomposition. 1967 *Times Rev. Industry* Apr. 80/1 Unpainted non-stainless steel parts are cadmium plated and passivated for use at temperatures below 200 degrees centigrade. 1973 *Sci. Amer.* June 119/2 The brass or iron piping and fittings you use must be degreased, nitrogen-purged and passivated in a dilute stream of the fluorine, allowed to enter slowly enough to form a metal-fluoride film on the surface.

b. *trans.* To give (semiconducting material) a protective coating of some relatively inert material in order to protect it from contamination.

1964 *IBM Jrnl. Res. & Devel.* VIII. 368 (*heading*) Space-charge model for surface potential shifts in silicon passivated with thin insulating layers. 1969 Gray & Searle *Electronic Princ.* v. 176 The oxide layer is said to passivate the semiconductor surface. 1972 Planer & Phillips *Thick Film Circuits* xi. 134 Active devices..are passivated with a layer of glass or silicon dioxide.

Hence **pa·ssivated** *ppl. a.*, **pa·ssivating** *vbl. sb.* and *ppl. a.*

1914 *Trans. Faraday Soc.* IX. 261 The amount of hydrogen..is too small to counteract the passivating action of the potassium hydroxide. 1919 H. P. Tiemann *Iron & Steel* (ed. 2) 364 Iron in this condition has been termed passive, passivated, passivified, inactive, altered, or prepared. 1941 *Nature* 28 June 803/1 An electro-formed wedge..of passivated nickel..made and a coat of nickel..applied to the outer passivated nickel face. 1950 C. F. Coyte in Coyte *in Electroplating* iii. 57 Nickel passivating dips. 1963 *Engineering* 27 Dec. 814/1 Passivated and sheathed stainless steel is used for the housings. 1967 *Electronics* 6 Mar. 281/2 The silicon planar passivated construction employed provides long-term parameter stability. 1974 J. A. von Fraunhofer *Concise Corrosion Sci.* viii. 79 Nitrates are poor passivating agents since they..are only sluggishly reduced. 1978 *Grumophone* Jan. 1334/3 The main chassis is constructed from passivated steel.

passivation (pæsivēi·ʃən). [f. as prec. + -ation.] The process of passivating.

1913 *Trans. Faraday Soc.* VIII. 234 (*heading*) Direct experiments on the passivation of metals. 1925 *Jrnl. Iron & Steel Inst.* CXI. 60 The anodic passivation of iron in sodium sulphate solution is demonstrated experimentally. 1956 *Nature* 17 Mar. 508/1 The mechanism of metallic corrosion, oxidation and passivation processes. 1964 *IBM Jrnl. Res. & Devel.* VIII. 385/1 Films..used for surface passivation of semiconductor devices. 1969 *Jane's Freight Containers 1968–69* 484/2 Approximate cost £1,450, depending on optional equipment fitted..e.g. ..separate compartments, ..passivation, fork-lift pockets, etc. 1972 Planer & Phillips *Thick Film Circuits* xi. 133 Passivation with silicon nitride yields a considerably increased protection against ionic contamination.

passivator (pæ·sivēitə). [f. as prec. + -or.] A passivating agent.

1935 R. N. Speller *Corrosion* (ed. 2) ix. 378 Passivators may be used effectively to retard corrosion in refrigerating systems or anti-freezing mixtures. 1951 *Engineering* 28 Dec. 819/3 The oxidation, which causes sludging and the formation of acids, may be prevented by using an inhibited oil. The substances employed for this purpose include catalyst passivators and de-activators. 1974 J. A. von Fraunhofer *Concise Corrosion Sci.* viii. 79 Passivators or Type III inhibitors... Generally these substances are oxidising agents with redox potentials that are more noble than that of the metal and they are readily reduced.

passive, *a.* and *sb.* Add: **A.** *adj.* **7. b.** Also, applied to substances that are normally reactive. (Earlier and later examples.)

1836 *Phil. Mag.* IX. 54 The third wire can make no difference if passive a fourth one, and so on. *Ibid.*, Direct contact between the two wires..is not an indispensably necessary condition for communicating chemical activity from the active wire to the passive one; for any metal..renders the same service. 1946 *Nature* 19 Oct. 506/1 The addition of a sufficient amount of chromium to iron confers upon the iron the property of producing spontaneously upon its surface a passive and resistant film as a result of contact with the atmosphere or with certain aqueous environments. Such passive films, if mechanically damaged, are self-repairing. 1969 D. Ashbury *Inorg. Chem.* x. 447 Nitric acid renders some metals completely passive, e.g. iron and nickel are rendered passive by the concentrated acid.

c. Of a trade balance: unfavourable; with debits greater than credits; *passive rate* (see quot. 1972).

1930 *Economist* 15 Mar. 587/2 The passive balance of Kr. 25 million was quite normal for the time of the year, as there is always an excess of imports during the winter months. 1936 K. A. H. Egerton *Bower's Dict. Econ. Terms* (ed. 10) 119 *Passive trade balance*, a balance of imports in excess of exports—another term for 'unfavourable trade balance'. 1969 G. Crowther *Outl. Money* viii. 258 (Following the usual terminology we can say that the balance of trade between the two countries was unfavourable (or 'passive') to Switzerland. 1972 *Times* 19 Dec. 15/1 It has also been accepted that the annual rate of monetary growth should be curbed gradually rather than suddenly until it reaches a 'passive' rate—the planned rate of economic growth plus the going rate of inflationary expectations.

d. *Immunol.* [tr. G. *passiv* (given this sense by P. Ehrlich 1892, in *Zeitschr. f. Hygiene und Infectionskrankheiten* XII. 189).] Produced by or involving the introduction into the body of antibodies of external origin.

1895 *Science Progress* III. 204 'Passive immunity', a term first employed by Ehrlich. 1898 R. T. Hewlett *Man. Bacteriol.* v. 123 'Passive immunity' is soon lost, but..is transmitted to the fœtus. 1935 F. P. Gay et al. *Agents of Dis.* xxii. 450 Passive transfer occurs naturally from immune mothers to their offspring. 1970 W. H. Parker *Health & Dis. in Farm Animals* ix. 114 Natural passive immunity is acquired from the first meals of colostrum. 1974 R. M. Kirk et al. *Surgery* ii. 33 Passive immunity is often transmitted to humans by using the serum from a highly immunised horse.

e. *Chess.* Designating a sacrifice (*q.v.*) in which a piece or pawn attacked by an opponent's move is left to be captured; (*b*) that an opponent need not accept.

1907 S. S. Blackburne *Terms & Themes Chess Probl.* 87 *Sacrificing*, offering a White man to be captured. If a man or men already *en prise* be left so, this may be called a 'passive sacrifice'. 1924 A. Emery *Chess Sacrifices & Traps* ii. 40 In general, 'passive' sacrifices like that in No. 1—where, the Queen being attacked, Alekhine calmly allows it to be taken—are more pleasing than the 'active' variety. 1945 J. du Mont tr. *Spielmann's Art of Sacrifice in Chess* i. 2 Under the heading 'form', there are two types, namely active and passive. In making a distinction between these two types, the deciding factor, from a scientific point of view, would be whether the sacrifice arises from a move made for the purpose, or from a raid by the enemy... For reasons of practicability, however, it has seemed to me better to make the distinction a different one, namely, whether acceptance of the proffered sacrifice is compulsory or not. Those which must be accepted I call active, the others passive. 1968 P. H. Clarke tr. *Vukovic's Chess Sacrifice* i. 12 Passive sacrifices have the drawback that they can be declined; they can, as it were, be ignored.

f. *Psychol.* Of, relating to or characteristic of the female or the inactive role in a sexual relationship, freq. associated with masochism in psychoanalytic theory; that fails or refuses to respond with, or shows an abnormal lack of, activity.

1916 A. A. Brill tr. *Freud's Leonardo da Vinci* ii. 39 Strangely enough this phantasy is altogether of a passive character; it resembles certain dreams and phantasies of women and of passive homosexuals who play the feminine part in sexual relations. 1921 *Internat. Jrnl. Psycho-Anal.* II. 439 The author comes to the conclusion that masochism has to be considered as the result and expression of the primacy of passive partial impulses. 1930 *Ibid.* XVI. 337 The sexual aims of the little boy's incestuous wishes are clearly passive. 1940 Henderson & Gillespie *Text-bk. Psychiatry* (ed. 5) 112 Predominantly inadequate or passive, this agent is aggressive or hostile. 1969 R. L. Kelley in Solomon & Patch *Handbk. Psychiatry* xii. 521 Severe characterologic problems such as sexual perversion, alcoholism..and passive dependent personality. 1973 L. C. Kolb *Mod. Clin. Psychiatry* (ed. 8) vi. 95/1 In this type the personality contains a considerable element of aggression.. expressed by passive measures, such as sullenness, stubbornness, procrastination (etc.).

g. *Electronics.* Containing no source of e.m.f.

1924 K. S. Johnson *Transmission Circuits Teleph. Communication* xi. 121 The transmission properties of passive networks may often be best determined by considering them as equivalent to lines having smoothly distributed constants. 1930 T. E. Shea *Transmission Networks & Wave Filters* ii. 43 A network composed only of inductances, capacitances, and resistances is a passive network. 1966 *Wireless World* July 332/1 In the so-called hybrid circuit, the active elements are formed in the silicon slice by the normal planar process, but..the passive elements (resistors, capacitors and conductors) are deposited as thin films on to the thermally grown silicon dioxide protective coating. 1970 J. Earl *Tuners & Amplifiers* ii. 29 A junction diode is equivalent to the thermionic valve diode, a transistor to a thermionic valve and an IC to a multiplicity of active and passive devices and components.

h. *Linguistics.* Of vocabulary, etc.: that is recognized and understood but through inability, lack of assurance, or for some other reason, is not used by the auditor or reader himself.

1935 G. K. Zipf *Psycho-Biol. of Lang.* (1936) v. 220 The auditor's passive vocabulary (i.e. the words which the auditor can understand). 1966 E. Derrick *Teaching Eng. to Immigrants* ii. 90 Most stories will contain far more material than the pupils are expected to reproduce themselves (i.e. relying on and helping to build up their 'passive' or recognition vocabulary). 1976 *Word* 1971 XXVII. 85 Grammont reported and demonstrated that 'passive language' precedes 'active language'.

i. Of radar, homing systems, etc.: relying on radiation generated by the target. Of a satellite, space relay station, etc.: not generating any signal.

1954 K. W. Gatland *Devel. Guided Missile* (ed. 2) iii. 67 A missile can be homed on to its target...by 'passive' homing (whereby the missile homes on to a source of energy radiated by the target). 1960 *N.Y. Times* 9 Oct. E 9/6 Echo, a 'passive' satellite, reflects or bounces radio signals sent from one station back to another station. 1962 J. Clemow *Missile Guidance* ii. 45 It is possible with a passive radar system where a receiver carried in the missile detects the direction of the source of radar signals from the target. 1966 *McGraw-Hill Encycl. Sci. & Technol.* XII. 504/2 Modern submarines carry passive sonar apparatus which is believed to be capable of detecting ships as far away as 100 miles. 1967 H. Hellman *Controlled Guidance Syst.* vi. 161 Additional advantages of passive systems are that they emit no telltale

signals to aid the enemy, and the guidance equipment is kept to a bare minimum. 1969 *Proc. IEEE* LVII. 427/1 Passive remote sensing at microwave frequencies has applications which range from meteorology to oceanography and geology.

B. *sb.* **1. b.** *pl.* In pillow-lace making, the bobbins holding the threads which correspond to the warp threads in weaving.

1907 Mincoff & Marriage *Pillow Lace* vii. 89 The other pairs which these [*sc.* the wefts] must cross are called the 'passives'. 1953 M. Powys *Lace & Lace-Making* iv. 20 This makes the connection and the worker bobbins pass back again across the passives. *Ibid.* xi. 166 The bobbins hanging straight down are called the 'passives'.

passive resistance. (In Dict. s.v. Resistance 1 b.) Add: (Further examples.) Now mostly used of refusal to comply with demands or legal requirements imposed by a government or other authority.

1883 *Encycl. Brit.* XX. 147/2 The student of English constitutional history will observe the success with which they [*sc.* the Quakers] have, by the mere force of passive resistance, obtained from the legislature and courts indulgence for all their scruples and a recognition of the legal validity of their customs. 1909 J. J. Doke *Indian Patriot in S. Afr.* ii. 5 The Passive Resistance movement had come into prominence [in 1907]. 1927 C. Bergmann *Hist. Reparations* xxiii. 173 (*heading*) Force and passive resistance. *Ibid.* 180 The military invasion of the Ruhr District was met by the German people and the German Government with an attitude which has become generally known as 'passive resistance'. 1928 V. G. Desai tr. *Gandhi's Satyagraha in S. Afr.* xii. 172 None of us knew what name to give to our movement. I then [*sc.* in 1906] used the term 'passive resistance' in describing it. 1930 *Economist* 22 Mar. 631/1 The Educational Bill, which was the cause of the bitter religious controversy that gave rise to the passive resistance movement. 1936 [see *conscientious a.*]. 1940 E. Paul *Narrow St.* xi. 46 Unable to collect reparations, on account of German poverty and 'passive resistance' which became spasmodically impassive, Poincaré declared an embargo on iron and steel into Germany. 1950 G. Brenan *Face of Spain* vii. 167 The land is difficult to legislate for, and those who own it are past masters in passive resistance. 1974 J. White tr. *Poulantzas's Fascism & Dictatorship* iv. ii. 170 In July 1923, with..the failure of passive resistance in the Ruhr. ..the party was in an objective crisis.

Hence **passive-resi·stant, -resister,** one who practises passive resistance; **passive-resi·stful** *a.*, expressive of passive resistance. Also **passive-resist** *v. intr.* (*rare*), to practise passive resistance.

1904 G. B. Shaw in *Daily Mail* 27 Feb. 4/4 They look at me with volumes of reproach in their earnest, passive-resistful eyes. 1906 Passive resister [see *lippy a.* 2]. 1907 G. B. Shaw *John Bull's Other Island* Pref. p. xxx, The passive resister. The Passive Resisters is Voltaire's watery, 'Ecrasez l'Infâme'. 1936 M. Plowman *Faith called Pacifism* 86 It was as a passive resister that the greatest revolutionary in the world became the greatest social force in the world. 1949 Koestler *Promise & Fulfilment* i. xii. 131 The soldiers were confronted with the grotesque task of dragging the passive resistants..into barbed-wire cages for internment. 1952 B. Wolfe *Limbo* (1953) xvii. 235 Couldn't you just lie down and passive-resist? 1968 *Punch* 2 Oct. 474/1 He incurred much unpopularity as a Passive Resister, which is what people called those who refused to pay their rates towards the upkeep of Church Schools.

† passivication (pæsivikēi·ʃən). *Obs.* [Irreg. f. Passive a. and *sb.* + *-ication*, after nouns of action like *application*, *publication*.] = *passivation.

1922 *Trans. Faraday Soc.* XVIII. 4 (*heading*) Anodic passivication. 1942 R. T. Rolfe *Steels for User* (ed. 2) x. 300 Owing to passivication of the surface by nitriding, no corrosion was found to occur.

† passivification (pæsi:vifikēi·ʃən). *Obs.* [f. as prec.: see -fication.] = *passivation.

1907 *Jrnl. Soc. Chem. Industry* 31 Aug. 903/1 The passivification of iron is brought about..by raising the liquid in question less oxidizing. 1937 *Jrnl. Iron & Steel Inst.* CXXXV. 315 Ammonia or aqueous solutions activation of chromium-iron alloys.

† passivifier (pæsi·vifaiə). *Obs.* [f. next + -er[1].] = *passivator.

1911 J. N. Friend *Corrosion of Iron & Steel* 298 Passivifiers are oxidizers. 1921 *Jrnl. Chem. Soc.* CXIX. 646 The passivifiers are not oxidizers, hydronitric acid being a case in point.

† passivify (pæsi·vifai), *v.* *Obs.* [f. Passiv(e *a.* and *sb.* + -ify.] *trans.* = *passivate *v.* a.

1907 *Jrnl. Soc. Chem. Industry* 31 Aug. 902/1 All the solutions enumerated above that passivify iron contain an oxy-anion. 1934 *Jrnl. Iron & Steel Inst.* CXXIX. 619 The addition of easily passivified metals, such as chromium and nickel, to iron yields alloys which are more easily passivified than iron itself.

Hence **passi·vified** *ppl. a.*, **passi·vifying** *vbl. sb.* and *ppl. a.*

1907 *Jrnl. Soc. Chem. Industry* 31 Aug. 903/2 The passivifying of an iron anode. 1911 J. N. Friend *Corrosion of Iron & Steel* xii. 189 (*heading*) The characteristic properties of passivified iron. 1915 *Jrnl. Physical Chem.* XIX. 644 It is..possible that both nitrous acid and nitrogen peroxide may be passivifying agents. 1919 [see *passivated *ppl. a.*]. 1934 [see above]. 1934 *Jrnl. R. Aeronaut. Soc.* XXXVIII. 425 An alkaline pigment, such as zinc oxide or a passivifying pigment, like zinc chromate. 1938 *Jrnl. Iron & Steel Inst.* CXXXVIII. 494A, Strong nitric acid acted primarily as a cleaner, but had some value as a passivifying agent.

passivism (pæ·siviz'm). [f. Passiv(e *a.* and *sb.* + -ism.] An abnormal state of passivity, esp. that of a male who accepts or desires the passive role in a sexual relationship.

1903 H. Ellis *Stud. Psychol. Sex* III. 93 Stefanowsky, who also discussed this condition [*sc.* masochism]..termed it passivism. 1940 Hinsie & Shatzky *Psychiatric Dict.* 403/2 *Passivism*, a form of sexual perversion in which the subject, usually male, is submissive to the will of the partner in the unnatural sexual practices. 1961 *Brit. Med. Dict.* 1060/1 *Passivism*, a type of sexual perversion characterized by submission of one partner (usually the male) to unnatural sexual practices desired by the other partner, who may be of either sex.

passivist (pæ·sivist). [f. Passiv(e *a.* and *sb.* + -ist.] One who or that which is characterized by passivity, *spec.* (*a*) a (male) person who abnormally accepts or desires a passive sexual role (see *passivism); (*b*) an opponent of active participation in war. Also *attrib.* or as *adj.*

1895 [see *masochist]. 1942 Partridge *Usage & Abusage* (1947) 222/1 There is, in cultured usage, a tendency to make *pacifist* an active, *passivist* an inactive, indeed a negatively passive, opponent of war. 1945 K. R. Popper *Open Society* II. xviii. 201 It continues on the lines of Kant's criticism of what we have called the 'passivist' theory of knowledge. 1955 *Bull. Atomic Sci.* Sept. 265/3 But if armaments are not acceptable to the pacifist, does this mean that he will submit to the aggressor and meekly resign himself to what he considers evil? The answer is emphatically no. This is to confuse the pacifist with the passivist. The pacifist is definitely not a passivist. 1957 P. Worsley *Trumpet shall Sound* 236 The basic division..is not between millenarian and non-millenarian movements but between activist and passivist movements. 1961 *Brit. Med. Dict.* 1060/1 *Passivist*, anyone who is affected by passivism.

passivity. Add: **2. a.** (Later examples.)

1968 T. D. Weldon *Kant's Critique of Pure Reason* (ed. 2) iv. 132 Thus there really is a marked distinction between the activity of the understanding and the passivity of sensibility. 1967 G. H. von Wright in N. Rescher *Logic of Decision* 130 He [*sc.* the agent] can do nothing to produce a change..or to prevent a change which..independently of him happens. We may call this case [of impotence] (forced) passivity.

4. b. Substitute for def.: The state of chemical inactivity that is produced in some normally reactive metals following a slight initial attack. (Further examples.)

1881 *Jrnl. Chem. Soc.* XL. 344 Passivity may be induced in a rod of iron by the immersion of a part only in concentrated nitric acid. 1940 Glasstone *Text-bk. Physical Chem.* xii. 1010 The resemblances between a metal rendered passive by chemical and electrochemical methods is very marked, and there is little doubt that the fundamental cause of passivity is the same in each case. *Ibid.* 1011 In certain cases the deposition of an oxide is prevented by a visible film, e.g. lead dioxide on a lead anode in dilute sulphuric acid: this effect has been called 'mechanical passivity', but it is probably not fundamentally different from electrochemical passivity. 1966 *McGraw-Hill Encycl. Sci. & Technol.* III. 488/1 Passivity is usually due to surface films which act as barriers between the metal and its environment... Aluminum is a reactive metal, but it is widely used for corrosion applications because of protection by a stable aluminum oxide film.

5. *Psychol.* The state or condition of being abnormally inactive or lacking in normal responsiveness (see *passive *a.* 7 f); also *attrib.*

1927 Henderson & Gillespie *Text-bk. Psychiatry* 88 In other cases the patient believes that someone reads his thoughts... These latter conditions are examples of 'passivity'. 1952 W. Wolff *Threshold of Abnormal* xviii. 436 Passivity may appear as apathy, as lack of interest... 1955 *Psychiatric Q.* XXIX. 604 Nowhere is it more difficult to decide whether 'passivity' is an ego defense mechanism or an instinctual gratification than in the study of masochism. 1958 M. E. Spiro *Children of Kibbutz* iv. 146 Play among the two younger groups is marked by an aimlessness and passivity that are a far cry from their earlier creative activity. 1973 *Times* 10 Oct. 15/7 Where the dreams are better than real life, and tension passivity increase withdrawal and passivity where they have no power to control their own dreams. 1969 W. Mayer-Gross et al. *Clin. Psychiatry* (ed. 3) vi. 271 The passivity phenomena in which this loss of [self] is best seen are indeed very characteristic of schizophrenia.

passivizable (pæsivai·zăb'l), *a.* *Gram.* [f. as next + -able.] Able to undergo passivization; meeting the structural analysis for the passive transformation. So **passivizabi·lity.**

1970 N. Chomsky in Jacobs & Rosenbaum *Readings Eng. Transformational Gram.* 203 Passivizability is a property of verbs. 1972 A. Makkai *Idiom Struct. Eng.* 152 The tourists *kick the bucket* are not passivizable. 1978 *Language* LIV. 172 Phrases are apt to be passivizable if 'they perform the same function as single verbs'.

passivization (pæ:sivaizēi·ʃən). *Gram.* [f. next + -ation.] Conversion into the passive form.

1965 N. Chomsky *Aspects of Theory of Syntax* ii. 104 It accounts automatically for the restriction of passivization

to Verbs that take Manner Adverbials freely. *Ibid.* 226 The sentences 'I regard John as pompous', 'it struck me blind'..are freely subject to passivization. 1968 *Language* XLIV. 230 The syntactic motivation for this treatment comes from the fact that the verbs that can undergo passivization are restricted to those that take manner adverbials freely. 1970 *Ibid.* XLVI. 466 Whatever the machinery may be to handle these cases, it can clearly also handle constraints on the passivization of verbs like *condescend*. 1971 J. M. Anderson *Gram. of Case* ix. 132 Both 'primary' and 'secondary' passivization..as well as the two kinds of object ('direct' and 'indirect') are allowed for in this way. 1976 S. Greer *Lexical Struct. Syntax & Semantics* I. iv. 96 The deletion of *for* in certain complement constructions apparently applies after passivization.

passivize (pæ·sivaiz), *v.* [f. Passiv(e *a.* and *sb.* + -ize.] **† 1.** = *passivate *v.* a. *Obs.*

1910 *Jrnl. Physical Chem.* XIV. 731 When the current is interrupted after the anode has been passivized, the potential of the anode rises slowly. 1912 *Trans. Faraday Soc.* VIII. 279 One of the slabs was 'passivized' and then allowed to regain its normal state.

2. *Gram.* **a.** *trans.* To convert into the passive form. **b.** *intr.* To be subject to conversion into the passive.

1965 N. Chomsky *Aspects of Theory of Syntax* 229 Such sentences as 'John strikes me as pompous', 'his remarks impressed me as unintelligible' do not passivize. 1972 R. A. Palmatier *Gloss. Eng. Transformational Gram.* 120 The application of the passive transformation, by which the sentence is said to have been passivized. 1973 *Amer. Speech* 1969 XLIV. 212 There are certain transitive phrasal verbs which are phrasal verbs although they do not passivize. 1977 *Verbatim* Feb. 10/1 Thus, transitive verbs in idiomatic expressions frequently will not passivize (*the cowboy kicked the bucket, but not the bucket was kicked by the cowboy*). 1978 *Language* LIV. 49 Causation is a transitive process which can be passivized.

Hence **pa·ssivized** *ppl. a.*, **pa·ssivizing** *vbl. sb.*

1975 *Language* LI. 792 Under Hasegawa's analysis, the passivized clause originates as an object complement. 1977 *Trans. Philol. Soc.* 1975 103 It is clear that the passivized/intransitivized cover section for none. *Ibid.*, Assuming that both reflexive and passivizing/intransitivizing functions passed into the Sudkx verbs..we have a source for the four functional types having middle-voice forms which are found in Lowland Eastern Cushitic now. 1978 *Amer. Speech* LIII. 25 Between the passivized versions of the following sentence, only the one in which the deep-structure subject has been deleted is passivizing.

passle, var. *passel.

passman. Add: **b.** [see Pass *sb.*[1] 8.] A prisoner who is allowed to leave his cell. *slang.*

1965 H. Knox *Taste of Proof* iii. 49 The passmen, the privileged, trusted prisoners. 1973 J. Patrick *Glasgow Gang Observed* ix. 82 He had envied the 'passmen', boys who in the afternoon were permitted to leave their cells.

pass-out (pɑ·s,aut). [Pass-.] **a.** (A document given permission to leave and re-enter a theatre, etc.; also *attrib.* in *pass-out check*, *ticket*.

1894 *Harper's* 1896 in Dict. s.v. Pass-]. 1907 *Daily Chron.* 29 Aug. 3/6 He begged for a pass-out ticket to see how his old friends were doing down below. 1922 Joyce *Ulysses* 106 No fear of anyone getting out, no passout checks. 1952 Granville *Dict. Theatr. Terms* 133 *Pass-out checks*, a ticket given by the usher to permit a patron to re-enter the theatre if he, or she, leaves during the interval. 1959 K. Waterhouse *Billy Liar* vii. 159 The Roxy was unguarded. They dodged in, giggling... 'Where's yer pass-outs, you two?' yelled Stump. 'Hey, mister, they're getting in too!' 1964 J. Wain *Strike Father Dead* 60, I stuck to Lucille till the dance ended. At one point she suggested that we should get pass-outs and go out for a drink. 1968 C. Drummond *Death & Leaping Ladies* iii. 61 No need for pass-outs... I'm the Manager. The pub's okay.

b. A fit of unconsciousness. Also, one who has become unconscious (cf. *pass *v.* 66 c).

1946 E. O'Neill *Iceman Cometh* ii. 115 He puts his head back on his arms and closes his eyes, but this time his habitual pass-out has a quality of hiding. 1949 *Esquire's Handbk. for Hosts* 272 There is small pleasure to be derived from finding yourself with an eighteen-year-old passout on your hands. 1967 E. Fenwick *Passenger* iv. 31 She slept like a pass-out, a total, never making a sign of any kind of consciousness.

passover. Add: **4. b.** A path or pass over hills.

1839 Z. Leonard *Adventures* (1904) 230 We..continued all day without any interruption, and in the evening encamped at the foot of the passover.

passo-verish, *a.* [f. Passover + -ish[1].] Suggestive of the passover.

1921 A. Huxley *Crome Yellow* xxviii. 299 The disorganized, passoverish meal that took the place of dinner on this festal day. 1930 H. G. Wells *Autocracy of Mr. Parham* ii. 106 After an exceptionally passoverish dinner at Marmion House.

passport, *sb.* Add: **6.** *passport holder, number, office, officer, official*; **passport control,** (*a*) regulation of the issuing and inspecting of passports; (*b*) the department or office at a port, airport, etc., which checks passports; **passport photo(graph),** (*a*) the identification

photograph in a passport; (*b*) a photograph of the size required for passports.

1947 Auden *Age of Anxiety* i. 17 An ordered world Of planned pleasures and passport-control. 1948 M. Laski *Tory Heaven* i. 6 A lifetime of devoted service in Passport Control. 1960 'R. East' *Kingston Black* vi. 60 Passport control would report when she left the country and returned. 1966 C. Mackenzie *My Life & Times* V. 43 We were lucky to have a Minister like Sir Francis Elliot..; he agreed to this experiment in passport control. 1973 W. McCarthy *Detail* iii. 150 He walked across the airport towards passport control. 1976 *CRC Jrnl.* July 3/2 The UK passport holders from East Africa have a right to settle in Britain. 1971 M. Kelly *25th Hour* i. 15 He opened the other page..closed the passport and held it out... He took a pen..and wrote something down. I said, 'What's that?' 'Your passport number.' 1976 'M. Barak' *Secret List H. Roehm* xiv. 139 He checked names, addresses, and passport numbers. 1849 I. Spencer *Let.* 14 Aug. in U. Young *Life Fr. Ignatius Spencer* (1933) iii. ii. 165, I write from the Belgian passport office. 1975 *Times* 14 June 6/3 Long queues form at passport offices [in Angola] and 40,000 passports are already on order from Lisbon. 1950 P. Bottome *Under Skin* i. 16 The passport officers are in the dining saloon. 1958 L. van der Post *Lost World of Kalahari* viii. 193 There was a group of vigilant painted animals assembled on a ledge rather like passport officers at a frontier. 1975 D. Bloodworth *Clients of Omega* xxiv. 235 The Passport Officer said..'May I see your passport, please?' 1922 M. Arlen *Piracy* iii. vi. 89 The passport officials at the ports. 1961 J. Barlow *Term of Trial* i. vi. 127, I got my passport photo to show you. 1975 'A. Hall' *Mandarin Cypher* x. 150 Passport photos are only ever good for a giggle. 1935 J. Buchan *House of Four Winds* iii. 84 The passport photograph isn't unlike him. 1939 G. Greene *Confid. Agent* i. ii. 50 Life seemed determined to make him look less and less like his passport photograph. 1965 G. Lyall *Midnight plus One* ii. 17 There is one passport photograph only. 1966 N. Freeling *King of Rainy Country* 85 Switzerland confirmed that nobody would bother checking such well known passport photographs.

passportless, *a.* (Further examples.)

1919 J. Buchan *Mr. Standfast* v. 96 It seemed to me that, in spite of being passportless, I might be able somehow to make my way. 1968 *Punch* 16 Oct. 525/2 The Common Market..envisages a common nationality and a passportless society for most of Western Europe. 1970 *Guardian* 24 Mar. 11/3 Europe's airport lounges are still littered with passportless Kenyan Asians. 1973 *Daily Tel.* 5 Apr. 8/3 Passportless, on the run, he escapes into the Connecticut woods.

pass-through (pɑ·sþrū), *a.* and *sb.* [f. vbl. phr. *to pass through*: s.v. Pass *v.* 58.] **A.** *adj.* **1.** Through which something may be passed.

1955 *Sun* (Baltimore) 24 Aug. 12/4 An arm length's [*sic*] away via the 'pass-through' window is the kitchen range. 1976 C. Larson *Muir's Blood* (1978) xxix. 153 The phone rang..while Blixen was lifting it from his crowded breakfast table to the pass-through bar.

2. Of costs, etc.: that are passed on to the buyer; that are chargeable to the customer.

1972 *Time* 17 Apr. 44/1 The commission may order an end to 'pass-through' profits. At present, businessmen are allowed to pass along to customers not only their increases in costs, but also to tack on their standard profit margins. 1976 *National Observer* (U.S.) 14 Feb. 17/3 'Political' ads are also excluded from pass-through expenses in most states, such as those that promote offshore oil and gas drilling.

B. *sb.* **1.** A passage; a means of passing through; *spec.* a hatch through which food, etc., is passed.

1958 *Washington Post* 16 Aug. B3/2 (Advt.), Over-size dining-family room served by louvered pass-through from kitchen. 1959 'S. Ransome' *I'll die for You* iii. 37 Anne..began piling the breakfast dishes in the pass-through to the kitchen. 1971 *Daily Colonist* (Victoria, B.C.) 21 Oct. 15/1 Como sawed through two vertical bars in his cell into the passageway, and then 'enlarged a food pass-through' to get into an outside corridor. 1976 C. Weston *Rouse Demon* (1977) xx. 94 Through the bar-type pass-through into the kitchen they could see a rusty stove. 1977 *Austral. House & Garden* Jan. 58/1 Above the stove the pass-through overlooks the informal dining room.

b. *fig.* A route for money, profits, or investments. *N. Amer.*

1968 *Economist* 18 May 77/2 Last March Ottawa promised to make sure that its banks and other financial intermediaries here would not permit American investors to use Canada as a pass-through for funds destined for the Euro-dollar market.

2. An act of passing through or passing on.

1975 *Sci. Amer.* June 55/1 The beams collide at two regions on the perimeter of the ring... The probability of even a single e^-e^+ annihilation in any one pass-through of the bunches is quite low. 1976 *Billings* (Montana) *Gaz.* 16 June 5-A/3 O'Leary took the position that a Supreme Court decision of last Dec. 30, which permitted a temporary pass-through of Montana Power's higher Canadian-gas costs, has the effect of also authorizing reduced rates on an interim basis.

passway (pɑ·swēi). [Pass *sb.*[1]] **a.** A means of passing; a passage or gangway. **b.** = Pass *sb.*[1] 3.

1825 *Mechanics' Mag.* IV. 203/1, I hope we shall have a better passway than the present, otherwise we shall have the bridge down again. 1835 A. B. Longstreet *Georgia Scenes* 99 These were the only passways to the interior. 1874 J. W. Long *Amer. Wild-Fowl Shooting* 161 There is a good passway for flight shooting. *Ibid.* 245 Good sport may then be had on the passways. 1889 'C. E. Craddock' *Despot of Broomsedge Cove* xiv. 267 Through the broad passways he could see the white frost gleam responsive upon the expanse of the fields. 1889 *Harper's Mag.* Aug. 390/2 Our family carriage..is left out in the streets along with many others to block up the passway. 1920 *Blackw. Mag.* June 817/1 There is only one pass-way through the wild hills at the back..—a narrow defile.

past, *ppl. a.* and *sb.* Add: **A.** *ppl. a.* **4.** *past tense*: also *attrib.* and in extended uses; *past participle* (examples); *past-participial* adj.

1798 J. H. Tooke *Diversions of Purley* (ed. 2) I. viii. 263 The adjective *Less* and the comparative *Less* are the imperative..; and the superlative *Least* is the past participle. 1892 W. W. Skeat *Primer Eng. Etym.* ix. 104 The suffix so common in Lat. past participles, as in *amā-tus*.., loved. The corresponding past participial suffix in E. is *-d*, as in *dea-d*.., orig. a past participle. 1937 *Jrnl. Eng. & Gmc. Philol.* XXXVI. 474 The strong vowel is divided into classes..according to the vowel of the past and ignoring the past participle vowel. 1961 R. B. Long *Sentence & its Parts* vii. 167 Where main predicators are past-tense forms, common-mode predicators in subordinate clauses..are likely to be past-tense too if a choice between past forms and presents is possible. *Ibid.* xviii. 406 Latin past-participial stems are commonly marked..by the use of either the letter *t* or the letter *s*. 1961 B. Malamud *New Life* (1962) 290, I did—ah—see one but that's all past tense. 1976 H. MacInnes *Agent in Place* xxviii. 286 No need to think anything. It's all past tense now. 1976 *National Observer* (U.S.) 27 Mar. 19/3 Vidal's spicy past-tense peep show adds seasoning to this year's version. 1976 G. L. Brook *Lang. Shakespeare* iii. 110 The present participle is normally active and the past participle is normally passive. 1977 *Word* 1972 XXVIII. 98, 89·5 percent of all past-tense environments preserve lenition.

B. *sb.* **1.** Phr. *a thing of the past.*

1863 *Athenæum* 15 Aug. 200/3 Even in America the woman-doctor is an eccentricity, and most probably will in a few years be a thing of the past. 1903 G. B. Shaw *Man & Superman* III. 92 Do not ask me how old I was—as if I were a thing of the past. I am 77. 1952 E. O'Neill *Moon for Misbegotten* I. 43 Who told you I fall for the dainty dolls? That's all a thing of the past. 1961 *New Eng. Bible Luke* xvi. 9 When money is a thing of the past. 1977 C. Allen *Raj* i. 21/2 By the 1880s the discomfort of travelling by *palkee* (palanquin)..was already becoming a thing of the past.

C. Also from B, as *past-coloured, -dissecting, -done* adjs.; **past-future** *a.* (Gram.), of a tense: expressing an action or a state viewed as future in relation to a given past time; **past-president,** one who has been a president.

1939 S. Spender *Still Centre* 24 In the past-coloured pigment of the mind's eye They feed and fly and dwell. 1939 L. MacNeice *Autumn Jrnl.* 18 The final cure is not in his past-dissecting fingers. 1762 Sterne *Tr. Shandy* VI. xxi. 89 Chatting..upon past-done deeds. 1925 Grattan & Gurrey *Our living Lang.* 230 The Past-Future Tense (a)..(a) He said he should (would) write, etc. *Ibid.*, The Past-Future-Continuous Tense (b)..(b) He said he should (would) be writing, etc. *Ibid.*, The Past-Future-Perfect Tense (c)..(c) He said he should (would) have written, etc. *Ibid.*, The Past-Future-Perfect-Continuous Tense (d)..(d) He said he should (would) have been writing, etc. 1961 R. B. Long *Sentence & its Parts* v. 127 Progressive-aspect forms..sometimes emphasize..in past-perfect and past-future tenses closeness to a past time that is central in the attention at the moment. 1903 *Nature* 12 Feb. 348/2 James Glaisher..was also a past-president of the Royal Meteorological Society. 1961 *Newark* (New Jersey) *Even. News* 22 Mar. 25 He regretted Hughes had made a personal attack on a past president.

past, *prep.* and *adv.* Add: **A.** *prep.* **1. c.** (later example.)

1967 M. Forster *Trav. Maudie Tipstaff* I. v. 98 In Maudie's opinion, no woman could get past forty and still have those needs.

2. b. (Further examples.) spec. *to get past*, to pass, to overtake.

1857 T. Hughes *Tom Brown's School Days* I. v. 109 They're the bounds. As soon as the ball gets past them, it's in touch, and out of play. 1906 Gallaher & Stead *Compl. Rugby Footballer* xix. 262 We don't think Scoular was to blame for allowing Smith to get past him and score. 1971 *Croquet Gaz.* July 14/2 The..Singles..were won by Lady Ursula Abbey who just got past Mrs. Temple in a close and protracted final. 1977 *Daily Express* 29 Mar. 32/4 England put nine goals past Luxemburg 16 years ago in the away leg of a World Cup qualifier.

3. a. *past praying for* (later examples); also loosely, beyond hope (of cure, recovery, etc.).

1881 G. M. Hopkins *Lett. to R. Bridges* (1955) 126, I have become very musical of late... I could make great progress—not in execution: that is past praying for—but in composition and understanding. 1897 *Cornh. Mag.* June 830 The man who can deliberately set aside his own personal knowledge and the gift of reason and common-sense with which God has endowed him,..is indeed 'past praying for'. 1902 H. James *Wings of Dove* I. i. 5 The precious name..in spite of the harm her wretched father had done it..was not yet past praying for. 1909 Kipling *Rewards & Fairies* (1910) 46 I've seen her walk to her own mirror by bye-ends, and the woman that cannot walk straight there is past praying for. 1939 C. Day Lewis *Child of Misfortune* III. i. 264 Everything's past praying for. 1962 J. Ludwig in R. Weaver *Canad. Short Stories* (1968) 2nd Ser. 255 Sidney was past praying for, she herself couldn't have kids.

b. Esp. in phr. *not to put it* (or *anything*) *past* (*someone*), to think (a person) quite capable of performing a specific action, or behaving in a specified way.

1870 G. M. Hopkins *Jrnls. & Papers* (1959) 198 Br. Yates gave me the following Irish expressions—*I wouldn't put it past you* or *I wouldn't doubt you* = It is just what I should expect of you. 1894 Somerville & 'Ross' *Real Charlotte* I. v. 63, I wouldn't put it past Charlotte to be trying to ketch Mr. Dysart. 1912 J. N. McIlwraith *Diana of Quebec* xvii. 259, I did not put it past her to have a desire to meet the scoundrel once more, since I had assured her he was really a Green Mountain Boy. 1916 H. L. Wilson *Somewhere in Red Gap* vi. 272, I wouldn't put it past him that he had old Jerry kicked on purpose to-day! 1922 Joyce *Ulysses* 733, I wouldn't put it past him. 1929 W. Faulkner *Sound & Fury* 251 I'm not surprised though... I wouldn't put anything past you. 1930 J. B. Priestley *Angel Pavement* v. 214, I believe he waits until he has the tickets, then rings you up that morning and makes it up... I wouldn't put it past him. 1946 K. Tennant *Lost Haven* (1947) xv. 240 'Bracewell is a boy any mother could trust.' 'Don't you be too sure.. I wouldn't put anything past that little devil.' 1953 S. Beckett *Watt* 16 Poor woman, God forgive her, said Telty. Faith I wouldn't put it past him, said Mr Hackett. 1961 'E. Lathen' *Banking on Death* (1962) i. 5, I wouldn't put it past him to do this deliberately. 1976 M. Birmingham *Heat of Sun* ix. 159 'Do you think she could possibly consider killing justified for the sake of her deprived flock?' 'I wouldn't put it past her'.

2. *past it* (slang), incompetent through senility, etc., no longer competent, ineffective after long use; (quot. 1864) dead. Also (with hyphen) *attrib.*

1864 C. M. Yonge *Trial* II. xi. 197 'He is almost past it,' said Tom, 'but..he may be roused by my voice.' 1928 E. Wallace *Flying Squad* xv. 130 He was a handy old chap—but he was getting rather past it. 1950 'J. Guthrie' *Is this what I Wanted?* ii. 37 One never dreamed of going to them for advice. The fact was they were past it; they had lived their lives. 1959 *Listener* 22 Jan. 154/1 They never knew much about it anyway... Ramsay was past it then. 1972 K. Bonfiglioli *Don't point that Thing at Me* xiv. 118 The faded allure of portly, past-it Mortdecai. 1974 M. Babson *Stalking Lamb* xix. 138 'You're getting past it, Ma.' Aaron seemed obscurely satisfied by her display of weakness. 1978 *Times Lit. Suppl.* 1 Dec. 1388/2 Not for him the slumped envy of the past-it fantasizer.

C. a. *past-prayer*; **past-pointing** *Med.* [tr. G. *vorbeizeigen* (R. Bárány 1910, in *Wien. med. Wochenschr.* LX. 2036)], pointing to one side of an object that a person intends to point at, e.g. after being spun round, as a diagnostic test.

1916 *N.Y. Med. Jrnl.* 15 July 100/2 Movement of the endolymph in the semicircular canals in a given direction, stimulates the sensitive hair cells in these canals, and produces definite phenomena. These phenomena are: 1, A twitching of the eyes or nystagmus of a certain type; 2, vertigo; 3, so-called 'past pointing'; 4, falling reactions. 1934 R. R. Grinker *Neurology* xiii. 372 In cerebellar disturbances if a past pointing does occur it is outward, no matter where the lesion. 1977 J. Macleod *Davidson's Princ. & Pract. Med.* (ed. 12) xiv. 662 If a movement is attempted with the eyes closed the finger overshoots towards the side of the cerebellar lesion ('past-pointing'). 1876 G. M. Hopkins *Wreck of Deutschland* xxxiii, in *Poems* (1967) 62 A vein for the visiting of the past-prayer, pent in prison, The-last-breath penitent spirits.

pasta (pæ·stă). [It., = Paste *sb.*] A generic name for various forms of Italian dough mixtures or 'pastes', as macaroni, spaghetti, vermicelli, etc. Also *attrib.* and *Comb.*

[1673 J. Ray *Observations Journey Low-Countries* 405 *Paste* made into strings like pack-thread or thongs of whitleather (which if greater they call *Macaroni*, if lesser *Vermicelli*) they cut in pieces and put in their pots as we do oat-meal to make their *menestra* or broth of.] 1874 R. H. Busk *Folk-Lore of Rome* 118 There was a dish of 'pasta' heaped up like a mountain. 1927 F. Stark *Lett. from Syria* (1942) i. 13 They were preparing us a gigantic *pasta asciutta*. 1934 *Discovery* Aug. 215/1 What these country folk do not know about the preparation of *pasta* in its various forms of macaroni, spaghetti, tagliatelli and so on,..is not worth knowing. 1946 C. P. Stewart *Her Husband's House* xx. 134 She slapped a sheet of pasta down upon the..table. 1958 *Woman* 8 Feb. 23/1 British manufacturers are now successfully making pasta products. 1959 *Sunday Times* 22 Mar. 5/4 Pasta-loving Italian girls carry too much weight around the hips. 1973 *Times* 1 Dec. 6/5 The pasta-makers have stopped delivering spaghetti and other products to the shops. 1974 *Guardian* 31 Jan. 12/4 Trials have been made here to grow the durum or 'pasta' wheat, from which macaroni is made. 1976 *New Yorker* 15 Nov. 51/3 'George would be pleased to hear that,' Lenore says, lifting a small piece of pasta to her lips.

pastance. (Later *arch.* example.)

1906 *Outlook* 7 Apr. 471/2 During the dog-days..the grouse, the pheasant, and the fox are..withdrawn by the needs of Nature from their altruistic task of providing pastance for the gentlemen of England.

paste, *sb.* Add: **8.** *paste-brush* (also in sense 1); *paste-pot* (earlier examples); **paste-bodied** *a.*, of porcelain (see quots.); **paste-cutter** (earlier example); **paste grain,** also, occas. used of morocco; (further examples); **paste-wash** *sb.* (examples); hence **paste-wash** *v.*, to apply a mixture of paste and water to leather bindings, in order to improve the surface before lettering or decoration is applied; hence *paste-washing* vbl. sb.; **paste-water** (further examples).

1915 R. L. Hobson *Chinese Pott. & Porc.* II. ix. 141 Steatitic porcelain,..with the body..composed of *hua shih*..is light to handle, and opaque; and the body has a dry, earthy appearance, though it is of fine grain and unctuous to touch. It is variously named by the Chinese *sha-t'ai* (sand bodied) and *chiang-t'ai* (paste bodied). 1936 *Burlington Mag.* Jan. 10/1 Many fine specimens of a 'soft paste' white porcelain (paste-bodied the Chinese call it). 1964 M. Medley *Handbk. Chinese Art* 63/1 *Chiang-t'ai*, 'paste bodied' wares made from a fine-grained white firing clay, often miscalled 'soft paste'. These wares occur mainly from the 18th century onward. 1846 *Jewish Manual, or Pract. Information Jewish & Mod. Cookery* vi. 106 Beat the yolk of an egg, dip a paste-brush into it, and lay it on the crust before baking. 1845 E. Acton *Mod. Cookery* i. 5 Divide the bread into dice, or cut it with a deep paste-cutter into any other form. 1923 H. A. Maddox *Dict. Stationery* 59 Pastegrain.—Also abbreviated to P.G., but more correctly specified as pastegrain roan. Comprises the thin grain side of a split sheepskin, mechanically grained with a cracked or fissured pattern and stiffened slightly by pasting on the back... In the fancy trade P.G. roan is elaborately but erroneously described as French morocco. 1963 B. C. Middleton *Hist. Eng. Craft Bookbinding Technique* xi. 122 In the 30s and 40s of the nineteenth century hard- and paste-grain morocco replaced straight-grain morocco and russia for use on fine bindings. 1845 *Ainsworth's Mag.* VII. 27 Elliston now demanded the paste-pot. 1857 N. W. T. Root *School Amusements* 207 A large scrap-book is prepared, a committee of selection is chosen, a paste-pot made ready, and contributions are invited. 1880 J. W. Zaehnsdorf *Art of Bookbinding* 116 The porous varieties [of leather] must be paste-washed carefully. *Ibid.* 174 Paste-wash.—Paste diluted with water. 1946 E. Diehl *Bookbinding* II. xxii. 327 When leathers such as calf, sheep, or russia are used, they should be paste-washed before tooling, in order to fill up the many fine pores. *Ibid.*, Calf and other very porous leathers require a heavier paste wash than morocco or levant. 1963 B. C. Middleton *Hist. Eng. Craft Bookbinding Technique* xiv. 211 Paste-wash, glaire, glue-size, yolk of egg and boot and furniture polishes have been used on leather, but mainly to improve the surface. 1921 *Librarian* Nov. 74 Pastewashing and varnishing tends to crack and destroy the leather, and does not soften or preserve it in the slightest degree. 1946 E. Diehl *Bookbinding* II. xxii. 327, I do not approve of paste-washing or sizing leather if it can be avoided. 1880 J. W. Zaehnsdorf *Art of Bookbinding* 116 The non-porous leathers need only be washed with thin paste water or vinegar. 1901 D. Cockerell *Bookbinding* xiv. 198 Paste-water is paste and water well beaten up to form a milky liquid.

paste, *v.* Add: **1.** *to paste down*, to line the cover of a book by attaching half the end-paper to it.

1835 'J. A. Arnett' *Bibliopegia* i. 67 When dry, the end papers are pasted down, and the work finished. 1880 J. W. Zaehnsdorf *Art of Bookbinding* viii. 35 When the book is to be pasted down, the ends [*sc.* end-papers] are lifted from the book. 1901 D. Cockerell *Bookbinding* xvii. 254 To paste down end papers, the book is placed on the block with the board open. 1963 B. C. Middleton *Hist. Eng. Craft Bookbinding Technique* xiii. 203 Until, roughly, the end of the first quarter of the nineteenth century the endpapers of all grades of bindings were pasted down without being trimmed.

4. (Earlier and later examples.) Also, to inflict heavy damage on by bombing or shelling; in *Cricket*, to hit (a ball, the bowling) hard.

1846 *Swell's Night Guide* 58 They pasted his nibs, and scarpered rumbo. 1911 *Daily Colonist* (Victoria, B.C.) 30 Apr. (Mag.) 10/1 As the ducks kept coming round the point the shooters in the canoes had a great opportunity of pasting them. 1924 A. C. MacLaren *Cricket Old & New* xiii. 128 Many and many a short ball bowled by Gregory in the Test Matches of 1921, would have been pasted to the square leg boundary in the days 'when Plancus was consul'. 1930 *Strand Mag.* May 348/1 The Major was gazing at Trout's forehead, where the stick had pasted him. 1934 D. L. Sayers *Nine Tailors* iii. 275 Deacon..pastes the fellow one. 1942 *Hutchinson's Pict. Hist. War* 18 Mar.–9 June 23 (*caption*) The nose of a Westland Whirlwind single-seater fighter... The Whirlwind has been used with much success for 'pasting' enemy aerodromes. 1955 M. Allingham *Beckoning Lady* ix. 127 He came to this part of the country in 1942 when London was being pasted. 1973 A. Mann *Tiara* ii. 17 [He] used to play for Yorkshire. Let's go down and watch him paste the bowling. 1977 *New Yorker* 25 July 70/3 She guessed correctly each time just where Mrs. King, in charge of the forecourt, would be pasting her volley.

pasteboard, *sb.* (*a.*). **3. c.** (Earlier examples.)

1856 'Ocksidе' & 'Doesticks' *Hist. & Rec. Elephant Club* 29 Putting his physiognomy before the seven by nine aperture through which the money goes in and the pasteboard comes out. 1873 J. H. Beadle *Undevel. West* xxxvi. 771 The call of 'Tickets, gents', showed one man without the pasteboard.

paste-down (pēi·st,daun). [f. Paste *v.* 5 + Down *adv.* 1.] In modern bindings, that part of the end-paper which is pasted to the inside of the cover of a book; in earlier bindings, a piece of paper or parchment used as a lining inside a cover.

Replaces def. s.v. Paste *v.* 5 in Dict.

1888 C. T. Jacobi *Printers' Vocab.* 97 *Paste-downs*, the blank flyleaves, sometimes coloured, at either end of a book which are pasted down on the covers. 1901 D. Cockerell *Bookbinding* xvii. 254 One of the paste-down papers is then stretched over the board. *Ibid.* 256 All rubbing down must be done through paper, or the 'paste-down' will be soiled. 1934 *Yorkshire Archæol. Jrnl.* XXXI. 338 The paste-down and recto of the fly-leaf are blank. 1954 N. R. Ker (*title*) Fragments of medieval manuscripts used as pastedowns in Oxford bindings. 1963 *Times* 11 June 5/7 A collection of leaves and fragments from manuscripts..which had been used as pastedowns in later bindings went to Quaritch. 1972 P. Gaskell *New Introd. Bibliogr.* 148 Then the endpapers were sewn on... Their purpose was, as paste-downs, to reinforce the joints of the covers.

paste-in (pēi·st,in), *a.* and *sb.* [f. Paste *v.* 5 + In *adv.* 1.] **A.** *adj.* Pasted in, inserted by pasting. Also, of a scrapbook, etc., containing blank paper on which cuttings or pictures may be pasted. **B.** *sb.* A correction or illustration printed separately from the main text of a book, and attached to the relevant page by its inner edge being pasted to the margin.

1902 [see Paste *v.* 5]. 1907 *Yesterday's Shopping* (1969) 436A (*heading*) Paste-in Albums for Photographs, &c. 1939 *Library* XIX. 486 When, more than a month after the publication of the *Lyrical Ballads*, 1800, Wordsworth received copies of Volume II, he was greatly irritated to find that the lower half of page 210..was blank... There exist at least two specimens of a paste-in supplying the fifteen missing lines... Each of the two specimens is pasted to the inner margin of page 211 to open out onto the blank on page 210. 1960 *Times* 13 Oct. 15/7 Where paste-ins have become puckered, wavy lines must be expected. 1971 L. M. Harrod *Librarians' Gloss.* (ed. 3) 488 Paste-in. 1. A correction or addition to the text supplied after the sheets have been printed, and tipped into the book opposite the place to which it refers... 2. A separately printed illustration or map, cut to the size of the book, and the inner edge pasted into the text before gathering.

pastel[2]. Add: **3.** Also applied to soft or subdued shades used generally in textiles, interior decoration, etc.

1926–7 *Army & Navy Stores Catal.* 678/1 Gloves.. washable suède, light grey, pastelle, or white. 1934 *Archit. Rev.* LXXV. 14/2 There is, for instance, a stone known as Quartzite, which..when worked to a rough surface provides an everlasting wall lining in a variety of cool pastel shades. 1951 *Good Housek. Home Encycl.* 19/1 Upholstered fibre furniture..is available in various pastel shades. 1970 H. Ratcliffe *Home Decorating* xiii. 84/2 White plus a colour is a tint, white plus colour plus grey or black is a pastel shade. 1974 N. Freeling *Dressing of Diamond* 96 A glimpse of pastel green tiles. 1976 *Star* (Sheffield) 26 Nov. 24 (Advt.), 1976 (P) Cavalier 1600 L 4-door. In pastel beige. Fitted radio.

4. *pastel-coloured, -tinted* adjs.

1932 *Daily Tel.* 25 Apr. 4/4 Pearls on Pastel Gowns. Pearls are to be worn even more than usual this season. The reason is the vogue for pastel coloured gowns. 1978 B. Freemantle *Clap Hands* iv. 26 A fine-featured.. man who affected pastel coloured shirts with matching socks. 1922 H. Crane *Let.* 2 Mar. (1965) 80 Delicate pastel tinted flowers. 1952 *Granta* 29 Nov. 8/1 We find it agreeable to place a Victorian chair or print among the blond woods and foam rubber, or to hang pastel-tinted antlers on the wall near a mobile.

paster. Add: **2.** For *U.S.* read 'orig. *U.S.*' (Earlier example.) Also, a piece of adhesive paper used for various other purposes.

1870 *Congress. Globe* 13 Apr. 2659/3 There were ten tickets..which were scratched and had pasters with the name of Caleb N. Taylor. 1882 *Nation* (N.Y.) 6 July 7/2 The Erie and Central Railroads have made the attempt to rid themselves of all liabilities..by putting a 'paster' on their bills of lading. 1906 *World's Work* (N.Y.) May 7511/1 You may read upon its label that it has been 'U.S. Government Inspected'. The paster on the box from which it came assures us again of that fact.

paste-up (pēi·st,ʌp). [f. Paste *v.* 5 + Up *adv.*[1] 21.] **1.** A plan of a page or group of pages, with the position of text, illustrations, etc., indicated. Also *attrib.*

1930 Freshwater & Bastien *Pitman's Dict. Advertising & Printing* i. 186/2 A proof with blocks, blue prints, or odd pieces of type matter pasted up in position is termed a paste-up. 1948 H. Missingham *Student's Guide Commercial Art* ii. 76 Paste-up. The various components of the complete layout pasted together on a card to form the complete advertisement. 1949 Melcher & Larrick *Printing & Promotion Handbk.* 202/1 Extreme neatness is not necessary on a paste-up dummy. *Ibid.*, There are many tricks to paste-up technique. 1967 E. Chambers *Photolitho-Offset* iii. 35 This key drawing or paste-up, as it is often called, should at least include all type pulls. 1975 J. Butcher *Copy-Editing* v. 61 In the paste-up the illustrations are represented by photocopies, rough sketches or just empty rectangles. 1977 *Sci. Amer.* Sept. 95/3 (Advt.), In May, 1977 we announced KODAK MX-929 Resist, for use in an arrangement for translating a paste-up of a newspaper page to metal by scanning with a He-Ne laser and modulating an argon ion laser to expose the resist.

2. A piece of paper or cardboard with a newspaper or book clipping, etc., pasted to it. Also *transf.*

1944 *Daily Progress* (Charlottesville, Va.) 11 Mar. 8/6 Simple paper paste-ups a child could do. 1953 *Amer. Speech* XXVIII. 81 The D[*ictionary of*] A[*mericanisms*].. had the benefit of but a mere 300,000 quotation slips... Nor did this number represent fresh materials only, for it included thousands of paste-ups from its predecessors. 1972 D. H. Laurence *Coll. Letters G. B. Shaw* II. p. xviii, Making paste-ups of previously published correspondence. *a* 1974 R. Crossman *Diaries* (1975) I. 401 The following weekend my red box contained the paste-up of all the extracts and they included one or two absolutely unequivocal statements. 1977 *Times Lit. Suppl.* 3 June 682/1 His novel is a carefully constructed, defiantly random paste-up of *faits divers*.

Pasteur (pɑstö·ɹ). The name of *Pasteur* (see Pasteur *v.*) used *attrib.* and † in the possessive to designate apparatus he devised and effects he discovered, as **Pasteur effect,** the effect of oxygen of inhibiting fermentation in favour of respiration in certain organisms and tissues; formerly called the **Pasteur reaction*; **Pasteur('s) flask,** a glass flask with an elongated neck bent downwards so that micro-organisms in the air cannot contaminate its contents; **Pasteur('s) pipette,** a sterile pipette which at one end has a plug of cotton-wool or the like and a rubber bulb and at the other terminates as a capillary tube whose end is sealed at the time of drawing and not broken until the pipette is to be used; also, in mod. use, a pipette bought for similar purposes but not sterilized, plugged, or sealed; **Pasteur reaction** [after G. *Pasteursche reaktion* (O. Warburg 1926, in *Biochem. Zeitschr.* CLXXII. 435)] = **Pasteur effect*.

1935 *Nature* 15 June 995/2 (*heading*) Mechanism of the Pasteur effect. 1942 O. Meyerhof et al. *Symposium Respiratory Enzymes* 48 Most doubly equipped organisms possess in the Pasteur effect a regulatory device that enables them to use, as occasion demands, either their aerobic or their anaerobic systems. 1953 Fruton & Simmonds *Gen. Biochem.* xx. 483 Under anaerobic conditions the rate of consumption of carbohydrate by muscle tissue is approximately six to eight times that observed under aerobic conditions... This inhibition, by oxygen, of the rate of carbohydrate breakdown is frequently termed the Pasteur effect. 1971 I. G. Gass et al. *Understanding Earth* x. 146/2 In many primitive organisms, the change-over from fermentation to respiration occurs when oxygen reaches about 1 per cent of its present concentration in the atmosphere. Pasteur pointed out this effect..during his study of the spoilage of wines, so it is known as the 'Pasteur effect'. [1876 tr. *Schützenberger's On Fermentation* ii. iii. 327 (*heading*) M. Pasteur's flask to deprive the air of its germs.] 1882 W. W. Cheyne *Antiseptic Surg.* i. 17 Mr. Lister has found that in Pasteur's flasks with the long open necks, no floating dust is present after what was originally there has settled. 1913 G. Martin *Industr. & Manuf. Chem.: Organic* v. i. 231 A pure culture in sterilised wort is obtained by inoculating from each separate colony nutrient sterilised wort contained in a Pasteur's flask. *Ibid.*, A small amount of pure yeast culture from a Pasteur flask. 1969 R. K. Das *Industr. Chem.* II. xvi. 221 The sterile wort is inoculated with pure culture (from the Pasteur flask). 1902 J. W. H. Eyre *Elem. Bacteriol. Technique* i. 21 (*heading*) Capillary pipettes or Pasteur's pipettes. 1903 Swithinbank & Newman *Bacteriol. of Milk* 547 (*heading*) Method of making a Pasteur pipette. 1972 *Jrnl. Endocrinol.* LIV. 108 Fluid from control and test dishes was removed with a sterile Pasteur pipette. 1930 *Biochem. Jrnl.* XXIV. 1302 Ethyl *iso*cyanide is therefore said to be a specific inhibitor of the Pasteur reaction, *i.e.* of the reaction between respiration and fermentation, using the latter term to include glycolysis. 1966 Irvine & James tr. *Lundegårdh's Plant Physiol.* iv. 193 The exact position of the Pasteur reaction in the glycolysis chain..cannot be very easily determined.

pasteurella (pɑstöre·lă). Pl. **-ellæ, -ellas.** [mod.L. (V. Trevisan 1887, in *Rendiconti Reale Istituto Lombardo di Scienze e Lettere* XX. 94), f. prec. + *-ella*, diminutive ending.] A bacterium of the genus so called, which includes those causing plague and other acute infectious diseases in man, other mammals, and birds.

1913 H. J. Hutchens tr. *Besson's Pract. Bacteriol.* xxviii. 446 Lignières..came to the conclusion that varieties of the Pasteurella could be distinguished by means of their pathogenic properties. *Ibid.* 447 Organisms of the pasteurella group are..very pleomorphic. 1929 G. F. Petrie in *Syst. Bacteriol.* (Med. Res. Council) III. 139 The plague bacillus..is generally considered to be a member of the Pasteurella group. 1929 H. Schütze in *Ibid.* IV. 447 (*heading*) Pasteurellas in man. 1949 *Vet. Rec.* 5 Feb. 64/1 (*title*) Pneumonia in sheep associated with infection by a pasteurella-like organism. 1961 *Lancet* 7 Oct. 812/2 Organisms of the pasteurella group are non-motile aerobic gram-negative rods. 1963 Jubb & Kennedy *Path. Domestic Animals* I. iii. 172/1 A complete separation of the different strains of the pathogenic pasteurellae has still to be determined. 1974 Q. N. Myrvik et al. *Fund. Med. Bacteriol. & Mycol.* xxiv. 309 The pasteurellae..are parasites and pathogens for an unusually wide range of animal species.

pasteurellosis (pɑ:störelōu·sis). *Med.* and *Vet. Sci.* Pl. **-oses.** [ad. F. *pasteurellose* (J. Lignières 1901, in *Ann. Inst. Pasteur* XV. 734), f. prec. + -osis.] An infection produced by a bacterium of the genus *Pasteurella*.

1902 *Nature* 29 May 120/1 The name pasteurelloses is applied to a group of diseases of the same type, including typhoid fever and pneumonia of the horse, chicken cholera and hæmorrhagic septicæmia of the sheep, ox, and pig. 1913 H. J. Hutchens tr. *Besson's Pract. Bacteriol.* xxviii. 447 (*heading*) The pasteurelloses and plague in animals. 1925 *Jrnl. Hygiene* XXIV. 66 Deaths due to Pasteurella and deaths in which no evidence of pasteurellosis was found..were grouped together. 1929 [see *barbone]. 1963 Jubb & Kennedy *Path. Domestic Animals* I. iii.

272/1 In the various hosts, pasteurellosis may take a variety of forms. **1966** *Daily Tel.* 4 Nov. 13/4 He has pasteurellosis too, or 'pseudo-tuberculosis', also found in voles, sparrows, magpies, wrens, swallows, and, especially wood pigeons. More important, children can be affected—it causes a swelling of the glands that may be mistaken for appendicitis.

pasteurize, *v.* **1.** Delete the first clause of def. and add: To kill most but not all of the micro-organisms present in (food) so as to render its consumption safe and to improve its keeping quality (as by heat treatment or irradiation). (Add further examples.)

Distinguished from *sterilize*, which implies the killing of all the micro-organisms.

1945 *A B C of Cookery* (Ministry of Food) vi. 19 Unless milk has come from a herd of tubercolin tested cows or has been pasteurized it should be scalded or boiled before drinking. **1959** *Internat. Jrnl. Appl. Radiation & Isotopes* VI. 205/2 Packaging does not present a problem for foods 'pasteurized' by radiation. **1966** *Proc. Internat. Symp. Food Irradiation* (Internat. Atomic Energy Agency) 842 The cost of pasteurizing food (300 000 to 500 000 rads) should be less than one-tenth the cost of radio-sterilization. **1970** *Preservation of Fish by Irradiation* (United Nations F.A.O.) 3 A test was..developed for evaluating whether or not there exists a botulinum Type E hazard in fish fillets pasteurized with gamma radiation.

pasteurization (further examples); **pasteurized** *ppl. a.*, also *fig.*; also **pa·steurizing** *ppl. a.*

1957 *Jrnl. Appl. Bacteriol.* XX. 286 For meat processing, two different types of treatment have been envisaged. One is complete sterilization... The other process, which is becoming known as 'pasteurization', consists in decreasing the bacterial numbers sufficiently to give a substantial increase in storage life, usually under refrigeration. Doses from 5 × 10⁴ to 5 × 10⁵ rads have been suggested as giving such results. **1959** *Internat. Jrnl. Appl. Radiation & Isotopes* VI. 143/1 Some species of fish stand up better to irradiation than others and..these when subjected to only pasteurizing levels suffer none of the undesirable changes produced by sterilization. **1963** *Listener* 13 Sept. 406/2 After Mr Kokoschka's verve and spontaneity our average spoken word sounds pasteurized and insipid both in substance and expression. **1966** *Proc. Internat. Symp. Food Irradiation* (Internat. Atomic Energy Agency) 527 With the exception of thiamine, no significant changes occurred in the water-soluble vitamins of haddock fillets and clam meats subjected to pasteurizing doses of radiation. *Ibid.* 549 (*heading*) Investigations on pasteurization of cold marinades by ⁶⁰Co gamma rays. **1973** Stewart & Amerine *Introd. Food Sci.* v. 203 Radiation dosages for pasteurization are only about a tenth of that [*sic*] required for sterilization. **1976** *National Observer* (U.S.) 19 June 20/5 The pasteurized, homogenized, tanned, toothy, breasty, false-lashed, shade-goggled, honey cascade-haired beings thrust upon us as representatives of young womanhood. **1976** N. Freeling *Lake Isle* vi. 98 Castang, though a youngish cop with a university degree, didn't think much of this more pasteurized colleague.

pasticcio. Add: Also in the orig. It. sense, a pie containing numerous ingredients, of which macaroni and some form of meat are the chief constituents. (Further examples of senses in Dict.)

1907 Joyce *Let.* 18 Feb. (1966) II. 216, I have made a lovely pasticcio, it seems. **1923** *Mrs. Beeton's Bk. Househ. Managem.* 1323 *Pasticcio Maccaroni.* (*Macaroni Pie*). *Ingredients.*—½ of a lb. of cold beef or mutton, ¼ an onion, 3 or 4 tomatoes, ½ of a lb. of macaroni, breadcrumbs, grated cheese, stock, salt, pepper, nutmeg. **1937** M. Morphy *Good Food from Italy* 63 Pasticcio di maccheroni (Ferrara). **1967** *Listener* 19 Jan. 107/1 The pasticcio, the medley of movements contributed by different composers. **1973** *Guardian* 14 Apr. 5/2 Anyone at sea among unfamiliar Italian dishes would be well rewarded by going straight for such specialities as Pasticcio. **1973** J. Wainwright *Touch of Malice* 65 Fernlea Autumn Fair..had ended up as a crummy, screaming pasticcio of trash lights and belching noise. **1974** *Listener* 28 Feb. 282/3 Next year Handel supplied..*Deborah*, largely a *pasticcio* over which he took little trouble. **1976** *Survey* Summer-Autumn 322 The contents turn out to be a pasticcio of themes rather tightly connected.

pastiche, *sb.* Add: Now in more general use than *pasticcio*. (Further and *attrib.* examples.)

1934 C. Lambert *Music Ho!* i. 20 The illogicality of some of the present-day pastiches may give you 'a rare turn'. **1955** *Times* 11 May 7/5 At the Players' we are still basking in the glory of having started *The Boy Friend* on its historic career, and to attempt pastiche again would have been an affront to the theatrical gods. **1961** *Listener* 7 Sept. 335/1 The subject that lies at the back of popular writing on contemporary themes is sociology: it is the Americans who have sold best—Whyte, Riesman, Galbraith, the pastiche merchants Wright Mills and Vance Packard. *Ibid.* 5 Oct. 508/1 The new examination is not conceived as a pastiche of the G.C.E. **1964** *Ibid.* 16 Apr. 649/1 Henze is sometimes called an eclectic composer.. an inference that he is the worse for being a pasticheur... Henze's interest in pastiche has long since died a natural death. **1978** *Listener* 2 Feb. 158/2 Such genre pastiches as the recent film, *Chinatown*.

pasti·che, *v.* [f. the sb.] **a.** *intr.* To create pastiches. Also const. *about.* **b.** *trans.* To copy or imitate the style of. Hence **pasti·ching** *vbl. sb.*

1957 *Listener* 26 Dec. 1082/2, I pastiche of course and exaggerate, but that is the kind of thing. **1962** D. Lessing *Golden Noteb.* III. 375 Well, Anna, and how do you describe all this pastiching about? **1965** R. G. Cohn *Toward Poems of Mallarmé* 6 Rimbaud..'pastiched' Mallarmé's *Les Fleurs*. **1970** *Times* 17 Oct. 20 The unfortunate Victorian habit of 'reviving', that is, pastiching, the Renaissance, the Baroque and just about every other style of the past. **1973** *Nature* 11 May 49/2 A device for selecting as university students only those who have already demonstrated that they can survive a university course as pastiched at school.

‖**pasticheur** (pastiʃör). [Fr., f. Pastiche.] An artist who imitates the style of another artist.

1912 R. Fry *Let.* 3 Feb. (1972) I. 353 One doesn't like to be called a *pasticheur*... I've always been searching for a style to express my *petite sensation* in. **1913** *Nation* 2 Aug. 676 A skill which might lead him [*sc.* C. Brancuzi], in default of any overpowering imaginative purpose, to become a brilliant *pasticheur*. **1934** C. Lambert *Music Ho!* v. 329 The dog-Latin classicism of the post-war pasticheurs. **1944** *Burlington Mag.* Mar. 71/2 Steer, on the other hand, was a *pasticheur* and to say of him that he 'was beyond doubt the greatest of our landscape painters since Turner and Constable'..is as meaningless as it is ridiculous. **1958** *Observer* 7 Dec. 16/4 The obvious accusation against Betjeman is that he is a pasticheur—and it is certainly true that he has scarcely ever written a poem which was not, quite consciously, in the manner of someone else. **1964** [see *Pastiche *sb.*]. **1973** *Daily Tel.* 9 Apr. 5 For many he [*sc.* Picasso] was one of the great creators of his age. To others he was at best an accomplished pasticheur and at worst a mountebank. **1976** *Gramophone* Apr. 1643/3 Karg-Elert (1877–1933) was a pasticheur, too, and even a consciously humorous parodist in some of his miniatures.

pastie (pē·sti). [f. Paste *v.* + -y⁶, -ie.] (Chiefly *pl.*) A covering for the nipple of a woman's breast.

1961 *Washington Post* 17 May A3/6 Miss Mason was lying on the floor with nothing on except the scantiest of brassieres, known in the trade as 'pasties'. **1964** *Punch* 29 July 164/3 The young ladies [of the *Folies Bergère*] have been bidden [in New York]..to wear, in the cause of innocence, two 'pasties'. **1969** *Sunday Truth* (Brisbane) 2 Nov. 10/4 Stripper Sharon was promoting a Valley night club, wearing nothing on top but a couple of pasties to keep her modest. **1973** 'E. McBain' *Let's hear It* viii. 12 Topless dancing, in this city, was something more than topless—the something more being pasties or filmy brassieres. **1975** *New Society* 23 Oct. 198/3 Go-go dancers in New York are not allowed to dance completely nude; they wear G-strings and mini bras and nipple-covering 'pasties'... For the dancers to get a tip, the 'pasties' usually have to come off.

‖**pastiglia** (pastī·lyā). *Art.* Also pastiglio. [It., = paste.] A kind of stucco (see quots.). Also *attrib.*

1927 Eberlein & Ramsdell *Pract. Bk. Ital., Span., & Portuguese Furnit.* 36 The process of *pastiglia* decoration ..consisted in covering the *cassone*, cabinet, or whatever the piece of furniture might be with coarse linen or fine canvas, strained or glued over both the moulded and flat surfaces. Over this were painted successive coats of *pastiglia*, a thick creamy mixture of *gesso*, which dried and hardened rapidly. **1938** L. E. Cotchett *Evol. Furnit.* ii. 21 Canvas was..painted with several coats of *pastiglia* or gesso. **1959** L. A. Boger *Compl. Guide Furnit. Styles* iii. 27/2 Pastiglia work as a form of decoration is chiefly identified with the 15th century. **1969** R. Mayer *Dict. Art Terms & Techniques* 284/2 (*caption*) Italian casetta (casket) of the latter half of the 15th century, ornamented in pastiglia relief. **1977** *Technical Bull.* (National Gallery) Sept. 3/2 The top corners of the eight panels..were decorated in pastiglio.

pastil, pastille, *sb.* Add: **2. b.** *Med.* A small disc of barium platinocyanide whose gradual change of colour when exposed to X-rays was formerly used as an indication of the dose delivered.

1906 *Sci. Abstr.* A. IX. 49 (*heading*) Use of platinocyanide pastilles in radiotherapy. **1922** G. W. C. Kaye *Pract. Appl. X-Rays* iv. 71 Of all the various intensity measurers the pastille finds the most favour with medical men in this country. *Ibid.*, The pastille is placed at a specified distance from the anticathode of the bulb, and the colour is matched against one of a number of standard tints. **1956** C. W. Wilson *Radium Therapy* (ed. 2) iv. 74 Sabouraud.. made use of the fact that a pastille of barium platinocyanide turns from apple green to reddish brown when irradiated by X-rays.

5. pastille-burner = Cassolette 2; **pastille dose** *Med.*, an obsolete unit of radiation dose corresponding to a change from one standard colour to another of a pastille (sense *2 b).

1904 A. Hayden *Chats on Eng. China* i. 11 Crown Derby Pastille-Burner. **1957** Mankowitz & Haggar *Conc. Encycl. Eng. Pott. & Porc.* 32/2 There is..a rare script mark 'P. Bradley 1828' on a pastille burner probably of Coalport manufacture in the National Museum of Wales. **1971** *Country Life* 7 Oct. 922/2 The larger sizes [of pottery model houses] have a hole in their back, being intended for use as nightlights or pastille burners. **1909** *Lancet* 15 May 1380/1 A Sabouraud pastille dose with the anticathode at 6¼ inches from the nearest point of the scalp is given to the vertex, occiput, lower occiput, right side and left side in succession. **1935** *Nature* 14 Dec. 960/2 The ionisation unit of X-ray quantity, the röntgen, has practically replaced the older arbitrary standards such as the pastille dose. **1950** Walter & Miller *Short Textbk. Radiotherapy* x. 243 The 'pastille dose'..has now been completely superseded and there is no justification for its continued use.

pastillage (pæ·stilėdʒ, ‖pastiyāʒ). [Fr., lit. compression of paste into blocks.] **1.** *Pottery.* (See quot. 1940.)

1901 W. P. Rix tr. *Bourry's Treat. Ceramic Industries* II. ix. 417 This method of decoration is known as 'pastillage'. **1940** *Chambers's Techn. Dict.* 619/2 *Pastillage*, dot and dribble designs made with coloured slip, which is dribbled from a container with a flexible base and a spout. **1947** *Horizon* June 369 There must be hundreds of years difference between the coarse idols obtained by a process of *pastillage* and the delicately coloured and modelled figures of the field reproduced here. **1962** E. Kissell tr. *Lehmann's Pre-Columbian Ceramics* i. ii. 19 The *pastillage* technique used in the decoration. *Ibid.* ii. 40 The eyes, almost always slanted, are indicated by incisions or pastillage. **1974** Savage & Newman *Illustr. Dict. Ceramics* 214 *Pastillage*,..decoration applied by trailing through a quill.

2. A type of icing.

[**1951** *Good Housek. Home Encycl.* 499/1 The special stiff royal icing..(called pastillage in French).] **1963** *Times* 20 Apr. 10/3 The cake is decorated with 4½ lb. of marzipan and 23 lb. of royal icing and pastillage.

pastilled (pæ·stild), *a. rare.* [f. Pastil, pastille *sb.* + -ed².] Subjected to the effect of a medicated pastille.

1936 R. M. Fortner *Hunger Harvest* (1936) 102 The clearing of pastilled throats.

pasting, *vbl. sb.* Add: **1.** (Earlier example of *pasting-lace.*)

1846 G. Dodd *Brit. Manuf.* 6th Ser. vi. 132 Coach-trimmings... 'Pasting-lace', about half an inch broad..is employed to cover and hide rows of tacks.

3. (Further examples.) Also *fig.*

1922 G. B. Shaw *Let.* in *To a Young Actress* (1960) 24 There is nothing for your soul in Shakespear.. but there is plenty of pasting for your ear; and you need an exquisite ear for tragedy. **1950** *Collier's* 1 Feb. 44/3 Instead of the usual pasting, the Giants beat Lillian and giving her a pasting for such goings on, Wilbur is somewhat pleased. **1952** *R. A. F. Jrnl.* 33 Oct. 33 We were getting a pasting. Everything going up in smoke and everything coming down with a bang. **1950** J. D. Macdonald *Brass Cupcake* (1955) x. 102 Fictional heroes..can take brutal pasting from a pasting that should have put them in hospital beds. **1953** in *Webster's Third New Internat. Dict. Amer. Slang* (1960) 377/2 [The] Philadelphia A's took a 13–0 pasting from the Washington Senators. **1974** *Times* 14 Jan. 3/7 Some managers have taken a real pasting when things have gone wrong. **1976** *Listener* 1 Apr. 405/2 'More embarrassing interviews, and more feeble comedy sketches'... Was it fair to give a show such a pasting in advance?

pastis (pæ·stis). [Fr.] An aperitif made from aniseed.

1926 C. Barry *Detective's Holiday* vi. 51 'Have your tea if you like, but I will drink a "pastis".' 'A matter of taste, *mon ami*,' Gilmartin replied. 'To me your terrible aniseed smells like a pharmacy.' **1943** W. Sansom in *Penguin New Writing* XIV. 72 Jean was saying, 'Damn that waiter! Why must he bring me quinquina instead of pastis?' **1955** *Times* 8 Aug. 8/7 No longer the friendly gossip at the pump, no longer the fraternal glass of pastis on the terrace of the Auberge de la Papesse. **1958** X. Fielding *Corsair Country* viii. 165 Even obvious Europeans like the pastis-drinkers in the bar. **1959** *Encounter* Oct. 9/2 Spain for the bull-fights, Provence for the pastis. **1967** A. Lichine *Encycl. Wines* 365/2 Pastis, the aperitif of Marseilles. Pastis is made with an alcohol base with herb flavourings, notably liquorice. **1973** P. O'Donnell *Silver Mistress* iii. 24 He had called at the auberge to drink a pastis.

pastism (pɑ·stiz'm). [f. Past *sb.* + -ism.] Memory of past things; nostalgia for the past.

1921 W. de Morgan *Old Man's Youth* xxvii. 258 The late fifties were still under the spell of Pastism. **1962** *Listener* 30 Aug. 316/3 The legend-mongers was sterile 'past-ism' by contrast with contemporary futurism.

pastless (pɑ·stlės), *a.* [f. Past *sb.* + -less.] Without a past; having no history.

1954 W. Faulkner *Fable* 12 The perfect soldier: pastless, unhampered, and complete. **1965** P. Rogers in *Grand Summer* 12/1 It's all very well being pastless in a moment of existential dilemma. **1971** S. Cavel *World Viewed* 97 His people are without fantasy (hence pastless and futureless, hence presentless).

past-mistress. Add: (Later examples.) Also written without hyphen.

1940 [see *cock-and-bull 3]. **1939** A. Thirkell *Before Lunch* iii. 65 Mrs. Middleton was a past mistress in the art of keeping a staff. **1974** G. Butler *Coffin for Canary* vi. 72, I would never reproachful gazes from Sarah. She was a past mistress in the art of giving me such a look.

pastness. Delete *rare* and add later examples.

1890 W. James *Princ. Psychol.* I. xv. 605 They are associated with other things which for us signify pastness. But how do these things get *their* pastness? What is the *original* of our experience of pastness, from whence we get the meaning of the term? **1919** T. S. Eliot *Sel. Essays* (1951) 14 The historical sense involves a perception, not only of the pastness of the past, but of its presence. **1925** C. D. Broad *Mind & its Place* v. 262 The very same characteristic..we took to be 'pastness' in other situations. **1941** F. Matthiessen *Amer. Renaissance* XIV. iv. 654 He did not so often think of the presentness of the past as of the pastness of the present, of its illimitable shadowy extensions backward to the roots of history. **1954** K. Tillotson *Novels of Eighteen-Forties* i. 111 The Brontës..turn from the present, emphasizing their choice by a specified..'pastness'. **1973** A. H. Sommerstein *Sound Pattern Anc. Greek* ii. 63 The verb whose pastness brings the rule into operation may be in the infinitive or participial form. **1977** *Theology* LXXX. 201 The faith of the communion of saints..confers on the death of Jesus at least one kind of capacity to survive 'pastness' and to become in a valid sense objectively present for us.

pastoral. Add: **A.** *adj.* **2. b.** *pastoral lease*, in Australia and New Zealand, a lease of land for sheep or cattle farming.

1850 *Papers Rel. Crown Lands in Austral. Colonies* 95 in *Parl. Papers* XXXVII. 287 You are empowered to grant pastoral leases for eight years... No leases.. whether pastoral or tillage leases, are to convey a perpetual right of renewal. **1894** W. Epps *Land Syst. Australasia* 154 In the event of a renewal of any pastoral lease being determined upon, it must be offered at auction 12 months before the expiry of the term. **1924** S. H. Roberts *Hist. Austral. Land Settlement* 293 Thus, after ten years of waste, the resumed areas of 1884, both in the centre and the west, were again placed under pastoral lease. **1948** V. Palmer *Golconda* iii. 18 Nominally the country around was held on pastoral lease by an old cattleman named Gourlay. **1950** *N.Z. Jrnl. Agric.* Nov. 413/2 At this period of Canterbury's development it would not have been possible for many of the so-called squatters to make a profit out of grazing sheep if they had been required to freehold the land. Though the pastoral lease was a partial answer, it did not give any security of tenure. **1953** A. Upfield *Murder must Wait* xix. 169 Prospecting pastoral leases in the far north of South Australia. **1959** A. McLintock *Descr. Atlas N.Z.* 38 With the growth in sheep numbers, regulations covering the granting of pastoral leases were laid down.

4. *pastoral letter* = Pastoral *sb.* 5 b.

1885 *Encycl. Brit.* XVIII. 351/2 *Pastoral letter*, a letter addressed, in his pastoral capacity, by a bishop to his clergy, or the laity of his diocese, or both. **1957** *Oxf. Dict. Chr. Ch.* 1023/2 *Pastoral letters*,..official letters addressed by a bishop to all members of his diocese.

B. *sb.* **3. a.** (Later examples.)

1949 *Poetry* (Chicago) LXXXIII 245 Pastoral. In Empson, a frequent literary device 'of putting the complex into the simple' by a process of inversion; e.g., the last shall be first, a little child shall lead them, etc. 'The essential trick of the old pastoral, which was felt to imply a beautiful relation between rich and poor, was to make simple people express strong feelings (felt as the most universal subject, something fundamentally true about everybody) in learned and fashionable language (so that you wrote about the best subject in the best way).' **1957** N. Frye *Anat. Crit.* 43 The pastoral of popular modern literature is the Western Story.

pastoralia (pastōrā·liă). [L., neut. pl. of *pastōrālis*, Pastoral *a.*] Things having relation to spiritual care or guidance; the duties of a pastor.

1959 *Listener* 26 Feb. 379/3 Mr. McCulloch seems,..to be reaching towards not a blank rejection but..towards an existential theology and pastoralia. **1962** G. Lawton *John Wesley's English* 261 It is..a little classic of pastoralia. **1973** *Church Times* 20 July 2/2 The Rev. John Elford..specialises in ethics, philosophy and pastoralia. **1975** M. Sullivan *Watch how you Go* iii. 57 The new Warden lectured in Church History and Doctrine and gave us a full course in Pastoralia. **1977** *Church Times* 7 Apr. 11/5 Pastoralia was the soft edge of the traditional three-year course [for ordinands].

pastoralism. (Further examples.)

1959 *Economist* 7 Feb. 489/2 The economy of West Germany today is something very different from the 'pastoralism' of the Morgenthau Plan. **1973** *Nature* 28 Sept. 194/2 The human response to the low and highly variable rainfall..is nomadic pastoralism, which is a very functionally adapted form of life. **1974** F. Emery *Oxfordshire Landscape* i. 36 Neolithic settlers..liked to graze their cattle, pigs and sheep in the woods and clearings; together with this strong pastoralism they retained a high degree of skill in hunting and snaring wild animals. **1976** *Sci. Amer.* Sept. 170/2 In traditional agriculture and pastoralism ruminants also represent a means of storing and transporting food supplies.

pastoralist. 2. (Further examples.)

1911 J. Collier *Pastoral Age in Australasia* iii. 16 The pastoralist, however, needs something more than level and spacious lands; he needs water and grass. **1926** *Daily Colonist* (Victoria, B.C.) 10 Jan. 1/5 A sheep calendar is to be issued, illustrating the various acts to be engaged in by the pastoralist in care of his flock. **1941** I. L. Idriess *Great Boomerang* v. 35 Coastal pastoralists [in Australia] generally speak of a beast or beasts to the acre. **1973** *Sci. Amer.* July 74/3, I obtained more responses from the herders of Kapsirika than from the farmers of Sasur. This was partly because polygamous marriages are twice as common among the pastoralists. **1975** *Nature* 15 May 180/2 A highland ridge, supporting an agricultural population of some 2.2 million, cuts into a semi-desert lowland ('Issa and Ogaden) which is seven times greater in area, with a population in normal times of up to 0.5 million pastoralists. **1977** *Caravan World* (Austral.) Jan. 107/1 The road..proved a lifeline to the pastoralists of the west.

pastoraliza·tion. [f. Pastoralize *v.* + -ation.] The fact or process of pastoralizing land, *spec.* the restoration of an industrial area to agriculture.

1949 I. Deutscher *Stalin* 501 Even as late as September 1944 both Roosevelt and Churchill were still to favour plans for the 'pastoralization' of Germany which were to deprive Germany of her heavy industry. **1964** F. Crouzet in *Jrnl. Econ. Hist.* Dec. 573 Because of the permanent injury inflicted on many Continental industries by interruption of overseas trade, the war brought about a lasting de-industrialization or pastoralization of large areas. **1975** T. Balogh in *Times Lit. Suppl.* 10 Oct. 1212/2 Reparations based on dismantling factories—pastoralization—a policy on which the Russians and Americans at first agreed, could of course not really be pursued.

‖**pastorela, -ella** (pastŏre·lă). [Prov., Pg.: see Pastorel, Pastourelle.] = *Pastourelle b.

1878 F. Hueffer *Troubadours* i. viii. 77 Guiraut Riquier is the Provençal representative of the 'Pastorela', or 'Pastoreta', the shepherd's song. **1885** *Encycl. Brit.* XIX. 555/2 The effects of Diniz's influence pervade the whole of Portuguese poetry, for not only was he in his pastorellas the forerunner of the great pastoral school, but..he..perpetuated..lyric forms of great beauty. **1899** J. H. Smith *Troubadours at Home* II. 451 The *pastorela*, *alba*, and *balada* were popular forms, and for that reason pretty much ignored by the troubadours. **1910** E. Pound *Spirit of Romance* iii. 58 The Pastorella has a peculiar interest in so far as it is one of the roots of modern drama. **1922** A. F. G. Bell *Portuguese Lit.* ii. 53 Airas Nunez.. wrote a *pastorela* in the manner of the trouvères. **1931** [see next]. **1960** Beckson & Ganz *Reader's Guide to Lit. Terms* (1961) 159 *Pastourelle* (*Pastorella*), a type of medieval lyric in dialogue form in which a knight or a man of equivalent social rank attempts to court a shepherdess. **1976** M. Bogin *Women Troubadours* 40 The refrain.., common in some of the more popular song forms such as..the *pastorela*.

‖**pastoreta** (pastŏre·tă). [Prov.] = prec.

1867 *Chambers's Encycl.* IX. 560/2 They [*sc.* the troubadours] even descended to depict the life of the peasantry, and sang their adventures with shepherdesses, &c. in *pastoretas* and *vaqueyras*. **1878** [see prec.]. **1931** W. P. Jones *Pastourelle* i. 6 The pastourelle takes its name.. from its principal actor, the shepherdess herself... The name was perhaps applied first to the Provençal specimens, the *pastorela* or *pas*[*t*]*oreta*.

‖**pastorie** (pastori·). *S. Afr.* [Afrikaans, = Du. *pastorie*, ad. med.L. *pastoria*.] The dwelling of a pastor of the Dutch Reformed Church or one of its sister churches: a parsonage.

[**1856** E. M. Murray *Let.* 20 Sept. in J. Murray *Young Mrs. Murray* (1954) 11 Bloemfontein Parsonage (or Pastorie as the Dutch call it).] **1934** E. A. Walker *Gt. Trek* vii. 217 A house was provided for public worship at the capital, and granted an erf for the metropolitan church and another alongside it for the *pastorie*. **1944** L. Smit *Sudden South-Easter* 8 He had received a telephone call asking him to supper at the Pastorie. **1946** *Cape Times* 26 Jan. 11/7 Smoke was..seen coming from the thatched roof of the Dutch Reformed Pastorie. **1958** *Ibid.* 19 Nov. 5/6 On the site..a second pastorie will be built.

pastourelle. Add: **b.** A medieval lyric whose theme is love for a shepherdess. Also *attrib.*

1882 A. B. Grosart *Spenser's Compl. Wks.* III. p. li, Equally vital to an adequate apprehension and comprehension of Spenser as a pastoral poet, is a like vision and grasp of the mediæval *pastourelle*, or courting of a shepherdess by a man of rank. **1908** *Mod. Philol.* VI. 33 A *pastourelle* is a simple poem set in a rustic scene, graceful and trifling in tone, describing the meeting of a man of culture and an *ingenue*, generally a shepherdess. **1929** C. R. Baskervill *Elizabethan Jig* i. i. 26 The dominance of the pastourelle figure Marian in the Robin Hood cycle would argue for song drama. *Ibid.* i. viii. 253 The rôle of the Gentleman, recalling that of the gentleman or knight of the pastourelles who pursues the sweetheart of the shepherd or clown. **1933** R. Tuve *Seasons & Months* iv. 97 The lady is the one figure in the spring landscape that we may come to expect..in *aube*, *pastourelle* or *chanson de danse*. **1965** P. Dronke *Med. Latin & Rise Europ. Love-Lyric* I. iv. 214 The motif of the young girl scolded or beaten by her parents..on account of her lover is most frequent in the *pastourelle*.

pastrami (pæstrā·mi). [Yiddish, f. Rum. *pastramă*, f. *păstra* to preserve.] A smoked beef, usu. prepared from a shoulder cut, highly seasoned and eaten hot or cold. Also *attrib.*

1940 G. Marx *Groucho Lett.* (1967) 45 The catering was delegated to Levitoff, the demon pastrami prince. **1941** L. G. Blochman *See you at Morgue* (1946) xxiv. 164 The hot pastrami sandwich on rye. **1945** A. Kober *Parm Me* 110 Mr. Freidkin was..eating a plate of pastrami and eggs. **1953** W. P. McGivern *Big Heat* ix. 111 He liked Jewish food..sour red cabbage, pastrami, cheese cake. **1961** E. Hunter *Mothers & Daughters* ii. 160 She..unwrapped the pastrami sandwich. **1962** J. Ludwig in R. Weaver *Canad. Short Stories* (1968) 2nd Ser. 239 The air inside was like a home-made mist—garlic pickles, pastrami, salami. **1967** M. Waldo *Internat. Encycl. Cooking* II. 477/2 *Pastrami* (Jewish), a cut of beef, usually the shoulder, highly seasoned (especially with garlic) and smoked. Cut into thin slices, it is served either hot or cold. **1973** *New Yorker* 24 Feb. 114/2 The sandwich makers at the counter always maintain rigid queue discipline while hand-slicing a high-quality pastrami on rye. **1976** *Time* 27 Dec. 31/1 One child conquered her stage fright by downing a quick pizza and a hot pastrami.

pastry. Add: **1. b.** = Paste *sb.* 1.

1845 E. Acton *Mod. Cookery* xvi. 402 This glazing answers also very well,..if used before the pastry is baked. **1892** A. B. Marshall *Larger Cookery Bk.* xiv. 528 Prepare a pastry fleur case with paste as below and cook it as for Fleur of Apples. **1906** *Mrs. Beeton's Bk. Househ. Managem.* xxxi. 883 The pastry may be used at once, but it will be lighter if allowed to stand for 1 hour in a cool place before being used. **1948** *Good Housek. Cookery Bk.* ii. 408 Fold the pastry in three, folding the first lap away from you and the second one towards you. *Ibid.* 409 Roll the pastry about ½ inch thick. **1952** F. Dale *Ambitious Cook* 51 Roll out the pastry thinly into a strip 4 in. wide. **1970** Simon & Howe *Dict. Gastron.* 290/1 *Pastry*, a paste usually a fat-and-flour mixture with water or other liquid added to make a dough for shaping and baking. Also the product of the baking.

c. A small confection made wholly or partly of pastry.

1906 *Mrs. Beeton's Bk. Househ. Managem.* xxxi. 879 Since the dinner *à la Russe* banished almost everything of an edible nature from the table, any talent in this direction has been chiefly expended on small pastries. **1928** E. J. Kollist *French Pastry* ii. 15 This cream is used for various pastries (Tartelettes), gâteaux, petits-fours, etc. **1935** D. L. Sayers *Gaudy Night* xix. 395 A plateful of synthetic pastries in Ye Olde Worlde Tudor Tea-Shoppe. **1938** H. Robertson *Cassell's New Cookery Bk.* 420 Pastries are baked in a hot oven. **1957** S. Strong *Good Food from Vienna* 181 (*heading*) How to enter earthly heaven by way of Viennese pastries, cakes, bakery and torten. **1960** E. David *French Provincial Cooking* 435 The ingredients of good pastries..are very expensive. **1972** 'M. Yorke' *Silent Witness* vi. 125 They carried a box of cream pastries..and said they had come to tea. **1973** 'J. Ashford' *Double Run* viii. 60 In a café..he ate a stale pastry and half drank a cup of very peculiar-tasting coffee.

4. *pastry-brush, chef, -scraper, shop* (later examples).

[**1907** *Yesterday's Shopping* (1969) 115/1 Brushes, &c., Pastry, tin bound or boxwood.] **1948** P. Harben *Way to Cook* (rev. ed.) x. 203 Moisten right round the edge of your sheet of paste (a pastry brush is the best tool for the job). **1976** M. Patten *Barbecue* 25/2 Glaze (brush) with beaten egg; use a pastry brush for this. **1961** *N.Y. Times* 3 July 13/5 A dedication to decoration worthy of a pastry chef creating a wedding cake for [etc.]. **1971** R. Thomas *Backup Men* ix. 82 Check out the new pastry chef. **1977** *John O'Groat Jrnl.* (Wick) 6 Sept. 8/1 *Pastry chef*, Post House Hotel, Aviemore, requires a Pastry Chef. **1963** *Listener* 22 Nov. 887/1 Pastry scraper, mixing spoon, and pastry brush. **1782** J. Woodforde *Diary* 25 May (1924) II. 26, I dined at a pastry shop on 3 Cheese Cakes. **1977** 'S. Leys' *Chinese Shadows* (1978) ii. 68 The local pastry shop..sells..strudels and little pink marzipan pigs.

pasture, *sb.* Add: **4. b.** Also phr. *pastures new.*

[**1638** Milton *Lycidas* in *Iusta Eduardo King* ii. 25 To morrow to fresh woods and pastures new.] **1906** G. Ade (*title*) In pastures new. **1922** Joyce *Ulysses* 645 Wrapped in the arms of Murphy, as the adage has it, dreaming of fresh fields and pastures new. **1975** L. Gillen *Return to Deepwater* i. 6 He had always resolutely refused to leave his native heath..and..had never felt the need to seek pastures new. **1977** *It* June 5/2 Some of this third group came to London to lead the winos and the rejects out to pastures new and they used 66 as a clearing house.

6. *pasture-field* (later examples), *ground* (U.S. examples).

1874 G. M. Hopkins *Jrnls. & Papers* (1959) 251, I sat down in the lap or fold of a steep slanting pasturefield. **1922** Joyce *Ulysses* 407 A region where grey twilight ever descends, never falls on wide sage-green pasturefields. **1949** K. M. Wells *Owl Pen Reader* (1969) ii. 196 Even the mill-pond is vanished. The old pond bottom is now part of a tree-spangled pasture field. **1668** in *Connecticut Hist. Soc. Coll.* (1912) XIV. 21 All that percell of pasture Ground lyeing on the east Side. **1775** J. Nourse in *Jrnl. Amer. Hist.* (1925) XIX. 351 Water enough for Cattle may be kept all the year here for pasture grounds. **1789** J. Morse *Amer. Geogr.* 381 On the north end it subsides gradually into extensive pasture-grounds. **1841** H. S. Foote *Texas & Texans* I. 14 The tide of indiscriminate havoc..[marked] its dreadful course with..the spoliation of her fair plantations and pasture-grounds.

pasture, *v.* Add: **3.** Also, to use (land) as pasture; to feed cattle on (land).

1901 J. Muir *Our National Parks* 5 The great Central Valley of California..is ploughed and pastured out of existence, gone forever. **1976** L. Sanders *Hamlet Warning* (1977) iv. 39 The ranch seems in good shape. He pastures about four hundred head of cattle on it.

pasture land. Also pasture-land, pastureland. Grass land used or suitable for the grazing of cattle or sheep; pasturage. Also *fig.*

1591 [in Dict. s.v. Pasture *sb.* 6]. **1669** J. Worlidge *Systema Agriculturae* 15 Meadow and Pasture Lands are of..considerable use and advantage to the Husband-man. **1718** *New Hampsh. Probate Rec.* (1914) II. 4, I give and bequeath unto my son..two thirds of my pasture land. **1786** G. Washington *Diaries* (1925) III. 108, I..directed the best plowman at it to break up about 10 Acres of Pasture land. **1867** 'T. Lackland' *Homespun* i. 65, I believe in my heart that this same huckleberry field..is real pasture-land for the spirit of the boy. **1885** *Outing* VII. 58/1 Fortunately I live within a mile of real pasture land and forest. **1920** W. D. Howells *Vacation of Kelwyns* 146 They all struck through the woods into a piece of pasture-land beyond. **1922** Joyce *Ulysses* 124 Of bosky grove and undulating plain and luscious pastureland of vernal green. *Ibid.* 290 Sheep and pigs and heavyhooved kine from pasturelands of Lush and Rush and Carrickmines. **1966** *New Statesman* 23 Dec. 928/2 The pasturelands of higher output and industrial efficiency and competiveness.

pasty, *sb.* Add: (Further examples.) Now usu. a small pastry turnover containing meat and vegetables (see **Cornish pasty*), or fruit. Also *transf.*

1877 *N. & Q.* 14 Apr. 297 *The Divisions of an Orange*... The word 'pasty' is used in Cornwall, from the likeness to the shape of the Cornish pasty baked without a dish. **1880** M. A. Courtney in Courtney & Couch *Gloss. Words Cornwall* 41/2 *Pasty*, a meat and potatoe or fruit turnover. **1906** J. H. Harris *Cornish Saints & Sinners* (ed. 2) xx. 194 When small, a pasty is a snack; when large, it's a meal... The home of the pasty is Cornwall. **1966** *Times* 28 Dec. 9/7 Cornwall is as protective about its pasties as Devon is about its cream. **1972** K. Stewart '*Times*' *Cookery Bk.* xvi. 213 (*heading*) Chicken liver and bacon pasties. **1978** R. Busby *Garvey's Code* x. 119 Cooper collected a couple of pasties in a paper bag.

pasty, *a.* Add *fig.* examples.

1909 *Westm. Gaz.* 28 Jan. 4/1 The pasty feeling of exhaustion usually experienced at the end of a long [railway] journey. **1926** E. O'Neill *Great God Brown* 100 A little dab of pasty resignation here and there—and even broken hearts may be repaired to do yeoman service!

pat, *sb.*¹ Add: **I. 2. b.** (Further fig. examples.)

1933 E. O'Neill *Ah, Wilderness!* (1934) i. 41 He gives him an approving pat on the back. **1952** M. Allingham *Tiger in Smoke* iv. 81 He had liked the boy... The present assignment had been in the nature of a personal pat on the head for him. **1969** *Listener* 15 May 698/2 A pat on the back for the regional bulletin *South-East*, which gave Eastbourne's maligned medical officer a chance to explain ..what he really meant about the artificial prolongation of life into living death. **1973** L. Meynell *Thirteen Trumpeters* vi. 100 If I could dive like that—well, I'd give myself a big pat on the back.

II. 3. c. = **cow-pad, -pat.* Also of other animals.

1940 F. Sargeson *Man & Wife* (1944) 22 On cold mornings we'd watch out, and whenever a cow dropped a nice big pat we'd race for it, and the one who got there first wouldn't let the others put their feet in. **1957** V. Nabokov *Pnin* iii. 63 The bright pat of dog dirt somebody had already slipped upon. **1959** C. T. M. Herriot tr. *Voisin's Grass Productivity* ii. 115 The scattering of dung pats over the whole area of a pasture could make the cattle anything but willing to graze. **1966** 'J. Hackston' *Father clears Out* 181 A cow,..will,..in a most flagrant manner, deposit a pat fair bang in the bail. **1973** G. Mitchell *Murder of Busy Lizzie* iv. 51 The rough road.. was muddy with the tramplings of cattle and plentifully endowed with large pats of cow-dung.

Pat, *sb.*² (Earlier example.)

1806 *Port Folio* (Philad.) 11 Oct. 221/2 A company of honest Pats in the purlieus of St. Giles's.

pat, on one's: see *Pat Malone.

pat, *v.*¹ Add: **5.** *to pat on the back* (earlier and later examples).

1821 M. Edgeworth *Let.* 18 Nov. (1971) 270 Sydney Smith who wrote the review..first patted his friend Holford on the back and then cut him up. **1866** *Galaxy* I. 750 It seems as if John Bull would never have done patting us on the back for our performance of a very plain and simple duty. **1946** E. O'Neill *Iceman Cometh* (1947) iii. 148 Dey all pat me on de back and say, 'Joe, you sure is white.' **1973** *Washington Post* 13 Jan. B 8/6 'We broke our arm, trying to pat ourselves on the back,' is the way Wood now describes the ad.

pat, *adv.* and *a.* **3. b.** (Earlier example.)

c **1868** *How Gamblers Win* 51 When quick work is to be made with a victim, 'pat hands', in other words, hands which fall complete,..are given out.

pata (pa·tā). *Indian Art.* [Skr. *pata.*] Cloth, canvas; a picture painted on a scroll of canvas.

1948 T. Bhattacharyya *Study on Vāstuvidyā* 261 A mere representation of a temple, building, idol and the like, either in sculpture..or a painting on paṭa. **1963** *Canons of Indian Art* xxx. 373 The Kātyāyana Saṃhitā prescribes worship of image (Pratimā) or drawings on Pata (paint on canvas). *Ibid.* 382 Paintings on canvas having the form of a scroll were known as paṭas. *Ibid.* 383 Such paṭas were commonly shown along with explanatory songs in East Bengal fairs. **1970** *Centennial Acquisitions Mus. Fine Arts, Boston* 43 (*caption*) Mandala of Eight Bodhisattvas..probably the earliest known example of Nepali painting of the type known as pata. **1974** *Times* 30 Apr. 18/6 The top price was paid for..a Nepalese pata depicting the mandala of a Tantric deity. **1976** *Sunday Standard* (Bombay) 5 Sept. 9/7 The patas, or painted scrolls, are specimens of folk-art in Bengal.

pataca (pătā·kă). [Sp. and Pg.] **1.** = Patacoon. *Obs.* except *Hist.*

1875 C. Schreiber *Jrnl.* (1911) I. 415 We found him to be a stupid old man,..selling bars of iron to people who paid him in heavy patacas. **1948** C. R. Boxer *Fidalgos in Far East 1550–1770* vi. 102 The 1625 Voyage..netted Dom Francisco Mascarenhas a personal profit of 26,000 *patacas* (silver dollars).

2. The monetary unit of Macao, and, formerly, of Timor; also, a coin of this value; also *attrib.*

1928 *Whitaker's Almanack 1929* 804/2 *Macao*, in China, on the Canton River..total trade 1926–27, *Patacas* 25,057,898. *Portuguese Timor*..(the budget includes a loan to be realized of *Patacas* 1,120,074); total trade 1926–27, *Patacas* 3,178,439. **1967** 'A. Cordell' *Bright Cantonese* x. 109 One..came over..with his girl..a pataca woman... For her the big black Mozambiques would pay about two or three patacas a night. **1968** *Economist* 16 Nov. 39/2 In exchange for the grant of.. franchises to import gold, the government now charges the syndicate a tax of 7.60 patacas an ounce. **1977** *Times* 16 Apr. 11/3 The Macanese *pataca* is at par with the Hongkong dollar... But get rid of any *patacas* before returning to Hongkong, where they are not generally accepted.

Patagonian, *a.* and *sb.* Add: **A.** *adj.* **2.** *Patagonian cavy, hare*, a large rodent, *Dolichotis patagona*, belonging to the cavy family and found in southern parts of South America; = Mara¹.

1833 [see Mara]. **1910** *Encycl. Brit.* V. 586/2 A very different animal is the Patagonian cavy, or mara.., the typical representative of a genus characterized by long limbs, comparatively large ears, and a short tail. **1961** G. Durrell *Whispering Land* i. 43 Small, desert-like areas seemed to be favoured by that curious animal, the Patagonian hare... They had blunt, rather hare-like faces, small, neat, rabbit-shaped ears, neat forequarters with slender forelegs. But the hindquarters were large and muscular in comparison, with powerful hind-legs. **1965** D. Morris *Mammals* 226 The Mara, or Patagonian Cavy, is the most hare-like of all the rodents. **1971** L. H. Matthews *Life of Mammals* II. vii. 205 When they run.. they hop with a gait much resembling that of a rabbit—this gait, and the long ears and legs, have earned them the misnomer of Patagonian hare.

pataka (pā·tākā). [Maori.] In New Zealand, a Maori storehouse for food, raised off the ground on poles.

1842 W. R. Wade *Journey N. Island N.Z.* vi. 151 In the pa, a large elaborately carved pataka, or kumara store, supported on four strong wooden pillars, attracted my attention. **1843** E. Dieffenbach *Trav. N.Z.* II. v. 70 A third sort of structure are the provision-houses (pataka), which are built on poles to prevent rats from entering them. **1949** P. H. Buck *Coming of Maori* (1950) ii. i. 106 In order to store the preserved food, special storehouses (*pataka*) had to be built. *Ibid.* ii. 132 The best *pataka* were much more extensively carved than the meeting houses, to outward appearance. **1950** *N.Z. Jrnl. Agric.* May 502/3 The carved pile of a pataka, a storehouse, [is a] choice example of age-old Native arts. **1967** A. & D. Reid *Paddle Wheels on Wanganui* 33 A few other scattered cottages, whares, and patakas belonging to the Maori residents and a few odd ancillary sheds and outbuildings completed the layout.

pataphysics (pætăfi·ziks), *sb. pl.* (const. as sing.) Also 'pataphysics. [ad. Gr. τὰ ἐπὶ τὰ μεταφυσικά 'the (works) imposed on the Metaphysics' (see Metaphysics *sb. pl.*).] The study of a realm additional to metaphysics, a concept introduced by Alfred Jarry (1873–1907), French writer and dramatist of the absurd. Hence **pataphy·sical** *a.*, **pataphysi·cian.**

[**1911** A. Jarry (*title*) Gestes et opinions du docteur Faustroll, pataphysicien.] **1945** C. Connolly tr. *Jarry's Ubu Cocu* in *Horizon* Dec. 375 What's all this? Monsieur Ubu, sometime King of Poland and Aragon, Professor of Pataphysics? That makes no sense at all. *Ibid.* 376 Remember that you are conversing with a famous pataphysician... Pataphysics is a branch of science which we have invented and for which a crying need is generally experienced. **1960** *Evergreen Rev.* May–June 131 'Pataphysics..is the science of that which is superinduced upon metaphysics... 'Pataphysics is the science of imaginary solutions. **1971** Lennon & McCartney in A. Aldridge *Beatles' Illustr. Lyrics* II. 106 Joan was quizzical studied pataphysical Science in the home Late night all alone with a test-tube. **1973** *Times Lit. Suppl.* 13 Apr. 415/4 That dogged pedantry which is perversely common among 'pataphysicians like himself. **1975** *Physics Bull.* Feb. 61/1 Perhaps it should also be mentioned here the existence of another 'physics'—'pataphysics', an invention of Alfred Jarry, probably the father of the Dada movement.

Patarin, -ene, *a.* (Examples.)

1926 A. L. Maycock *Inquisition* iv. 89 In February 1231 a number of Patarin heretics were arrested in Rome. **1934** R. Macaulay *Going Abroad* ii. 33 It was odd how those dear people inclined to the Patarine heresy. **1968** *Trans. R. Hist. Soc.* XVIII. 25 After 1056, its [*sc.* Milan's] order and independence were rudely challenged when the Patarene movement gave rise to nineteen years of civil strife.

Patau (pæ·tau). *Path.* [Name of K. *Patau* of Germany, who with others described the condition in 1960 (*Lancet* 9 Apr. 790).] *Patau*('s) *syndrome*: a syndrome due to trisomy and marked by mental retardation, seizures, cleft lip and palate, and other congenital anomalies, and usu. resulting in death soon after birth.

1964 Gorlin & Pindborg *Syndromes of Head & Neck* xvii. 80 (*heading*) Trisomy 13–15 syndrome (Patau's syndrome, trisomy D_1). **1971** Levitan & Montagu *Textbk. Human Genetics* iii. 76 (*caption*) The shortness of the neck and low-set malformed ears are also found frequently in the Patau (D_1) syndrome. **1976** W. L. L. Rees *Short Textbk. Psychiatry* (ed. 2) xxviii. 255 Patau's syndrome is due to an extra chromosome of the D group. **1978** D. A. P. Evans in R. B. Scott *Price's Textbk. Pract. Med.* (ed. 12) iv. 354/2 Mongolism (Down's syndrome) is caused by trisomy 21 and the much rarer Edwards and Patau syndromes..by trisomy 18 and 13 respectively.

Patavinian (pætăvi·niăn). [f. L. *Patavin-us* (see Patavinity) + -ian.] A native or inhabitant of Padua (formerly Patavinium); *spec.* Livy. Also †**Pa·tavin**(e *obs.*

1611 Coryat *Crudities* 129 The Romanes..priviledged them..that the Patauines should give their suffrages in the election of the Romane Magistrates. *Ibid.* 131 Three [statues] are of that famous Historian Titus Livius, who was borne and brought vp in Padua: the other three of other worthy Patauins. **1771** C. Burney *Present State of Mus. France & Italy* 124 He died universally regretted by the Patavinians. **1924** *Glasgow Herald* 6 Sept. 4/2 Livy.. does not make the actual route clear, for among the Patavinians excellent geographical accuracy is not included. **1932** E. Weekley *Words & Names* 149 Virgil the Mantuan, Livy the Patavinian.

pat-ball. Substitute for def.: A game in which a ball is hit back and forth between two players. Also used as a contemptuous name for lawn tennis, especially when not played vigorously; also, tactical slow and gentle play in lawn tennis. (Further examples.) Also *fig.*

1890 S. W. Gore in *Tennis, Lawn Tennis, Rackets, Fives* (Badminton Libr.) 282 This derisive name of 'pat-ball' was applied to lawn tennis by tennis and racket players, who maintained that, from the absence of back- or side-walls, it was impossible to hit hard without sending the ball out of court. *Ibid.* 285 The final blow to the 'pat-ball' game was given by the brothers Renshaw when they discovered that they could stand back at the service-line, and..volley. **1900** *Captain* IV. 26/1 In the summer, a great form of 'eccer' is 'patters', a corruption of 'patball', *i.e.* tennis. **1904** J. P. Paret *Lawn Tennis* ii. 13 The next three years have been aptly described by an English historian as 'the era of pat-ball'. **1923** *Daily Mail* 28 Apr. 11 At this stage Mishu played 'pat ball'..and Norton wisely did likewise, for in slowing the pace he affected Mishu's game. **1927** *Daily Tel.* 21 June 12/3 No one would have guessed..that the mild game derisively nicknamed 'pat-ball'..would develop so rapidly into this highly exacting and arduous exercise. **1944** C. Dilke in *Wine & Food* No. 41. 16 He might have been back at the Eureka tennis club..watching the nymphs who played patball on summer evenings. **1955** *Times* 20 June 9/4 Lawn tennis was still known as pat-ball when I was an undergraduate. **1959** *Economist* 18 July 150/3 An astonishing amount of rather owlish erudition and a no less astonishingly patball standard of controversy. **1963** *Times* 23 Apr. 13/4 No contestant..at Wimbledon would care to have his sport spoken of as pat-ball. **1973** P. Geddes *Ottawa Allegation* xii. 160 When they spoke during the journey, it was like pat-ball. **1977** J. Wainwright *Do Nothin'* vi. 102 A kiosk means a telephone call. Where to go next. Maybe another kiosk, for another call. A crazy, pat-ball game that could stretch forever.

patch, *sb.*¹ Add: **1. d.** (Further examples.)

1912 'Aurora' *Jock Scott* i. 4 He passed out of the *Britannia* a midshipman and was wearing his patches the day he left... Naval cadets wear a little bit of white twist on their coat collars, while a full-blown midshipman has a patch of white cloth about two inches square instead. **1970** N. Armstrong et al. *First on Moon* v. 107 John Young's wife..designed a handsome red Roman numeral X for the patch he and Mike Collins wore on Gemini 10. **1974** R. B. Parker *Godwulf Manuscript* i. 8 A fatigue jacket with a staff sergeant's stripes, a Seventh Division patch.

3. e. The area which is assigned to a policeman as his responsibility; a policeman's 'beat'. Also *transf.*

1963 *T.V. Times* (Austral.) 18 Apr. 10/2 *Patch*, a police area: as in 'It's on my patch'. **1965** 'W. Haggard' *Hard Sell* xii. 126 I'm a foreign official on another man's patch. I'm quite without standing. **1965** J. Wainwright *Death in Sleeping City* ii. 123 My patch in the city. I'm not like you—a county officer. **1969** D. Devine *Death is my Bridegroom* xi. 113 Smith was from the south and had never before turned up in Christie's patch. **1975** J. Symons *Three Pipe Problem* xvi. 157 Either he gets off your patch or he finds his reputation as an art dealer ruined.

5. b. *transf.* With qualifying adj.: a period of time with a particular characteristic.

1926 Wodehouse *Heart of Goof* i. 32 How like life it all was!.. We strike a good patch and are beginning to think pretty well of ourselves, and along comes a George Parsloe. **1928** *Daily Express* 6 July 9/3 He dreaded to think what would happen to Kent if those players had a month's bad patch. **1958** *Daily Sketch* 2 June 12/4 A friend helps you over a sticky patch in the afternoon. **1974** I. Murdoch *Sacred & Profane Love Machine* 213 If there were bad patches I've simply forgotten them. **1976** J. M. Brownjohn tr. *Kirst's Time for Payment* v. 114 All new businesses go through a sticky patch, but it's only temporary.

6. f. A temporary electrical connection. Usu. *attrib.* (see *patch bay*, etc., in sense *8).

1923 *Bell Syst. Techn. Jrnl.* II. 123 A temporary [telephone] connection made in this manner..is called a 'patch'. **1937** *Printers' Ink Monthly* May 40/1 *Patch*, a temporary and removable connection on studio equipment. **1977** R. L. Duncan *Temple Dogs* (1978) ii. 259 He just called computer, requested a patch on the Metro interface. He wants to know what the Tokyo police have picked up.

7. patch-up. Also as *adj.*

1904 *Westm. Gaz.* 30 Sept. 2/1 The kind of patch-up policy which he would accept for the next election. **1971** 'H. Calvin' *Poison Chasers* v. 74 There may be some value in a patch-up operation. **1974** P. Flower *Odd Job* ii. 16 His patch-up jobs on furniture.

8. patch bay, an area in an analogue computer that receives one or more removable patch-boards; **patch-board,** a plug-board, esp. one in an analogue computer or similar device; **patch cord,** an insulated lead with a plug at each end, used for making connections between the sockets of a patch-board or different pieces of electronic apparatus; **patch fox,** a North American red fox, *Vulpes Fulva*, in its yellowish colour phase, or the skin of such an animal;

= Cross-fox; **patch lead** = *patch cord* above; **patch panel** *sb.*² = *patch-board* above; (cf. Patch-panel *sb.* and *a.* in Dict.); **patch-plug** = *patch cord* above; **patch pocket,** a pocket consisting of a piece of cloth sewn on like a patch.

1948 Patch bay [see *patch cord* below]. **1962** Huskey & Korn *Computer Handbk.* iv. 26 Most multipurpose electronic analog computers are programmed by means of a patchboard system which comprises (1) a patchbay with spring-contact terminations for the computing elements and (2) interchangeable removable problem boards which carry the actual interconnecting patchcords. **1949** *Math. Tables & Other Aids to Computation* III. 512 Receptacle patch-board for 500 element connections and 25 main busses. **1961** G. Millerson *Technique Television Production* i. 17 Tape recorders, racks of audio amplifiers, patchboards and an electronic reverberation unit, complete the general set-up. **1971** J. H. Smith *Digital Logic* ii. 28 The inputs and outputs of the logic elements, and the connections to the switched inputs, external terminals and indicator lamps are brought out to a programming device consisting of a patchboard and detachable panel. **1938** G. E. Sterling *Radio Manual* (ed. 3) vii. 442 The input may be connected to a terminal board in rear or to normal-through standard double patch-cord jacks at front of panel. **1948** *Electronics* July 119/3 To set up the computer the elements are connected..by means of patch cords joining the proper inputs and outputs through the patch bay. **1967** [see *patch panel* below]. **1971** J. H. Smith *Digital Logic* ii. 28 Switching problems are set up by using patch cords on the front panel. **1835** Patch fox [see *Lord *sb.* 16]. **1930** *Economist* 4 Jan. 10/2 The highest priced fur in 1927–8 was silver fox..; cross or patch fox was second. **1942** M. Bosanquet *Saddlebags for Suitcases* 48 The most common of these [variations] is the 'cross' or 'patch' fox, which is yellow with a dark cross or patch across the shoulders. **1964** C. P. Gilbert *Design & Use Electronic Analogue Computers* vi. 364 The dotted line in Fig. 6.7(*d*) encloses all the sockets within reach of amplifier 3 using short patch leads. **1971** J. H. Smith *Digital Logic* ii. 18 The reader is advised to have at least 50 patch leads to connect the circuits together. **1952** G. A. & T. M. Korn *Electronic Analog Computers* viii. 339 It should..be possible to provide so-called removable patch panels of the type used in the IBM punched-card machines. **1967** *N.Y. Times* 9 Jan. 140 Basically, an analog computer consists of an assembly of individual electronic computing elements that can be interconnected by means of a 'patch panel' outside the machine. This panel is a terminal board with holes, each hole facing an internal contact. The computer operator uses 'patch cords' (wire connectors) to interconnect specific holes for the kind of operation he wants the machine to perform. **1973** *Physics Bull.* Aug. 500/2 The new system, PB100, is a large, sophisticated unit..for rapidly simulating complex digital, analogue or hybrid systems. It features removable patch panels each of which will accommodate up to 44 dual-in-line integrated circuits. **1962** *Gloss. Terms Automatic Data Processing* (*B.S.I.*) 82 *Patchcord*, [deprecated synonym] *patchplug*, in a.d.p. a connector used to interconnect the sockets of a plugboard. **1964** G. A. & T. M. Korn *Electronic Analog & Hybrid Computers* xi. 443 With a separate copper-bar signal ground behind the patchbay.., we can use the patchbay shielding as a relay ground and return each relay-coil connection through a single patchplug grounded to its shield. **1895** *Montgomery Ward Catal.* 556/1 Coat, three-button sack, four patch pockets. **1908** *Times* (Weekly ed.) 14 Aug. p. iii/3 There are two deep patch pockets..for carrying fly-book and sandwich-case. **1928** *Daily Express* 22 May 5/2 A plain, collarless coat.. with two large patch pockets. **1973** 'D. Rutherford' *Kick Start* iv. 86 He wore an expensive tropical suit with patch pockets and a waist belt. **1976** *Horse & Hound* 3 Dec. 24/1 (Advt.), Two large patch pockets,..and adjustable rear belt make this a smart comfortable coat.

patch, *v.* Add: **1.** Also *absol.*

1870 J. P. Smith *Widow Goldsmith's Daughter* xviii. 287, I could patch and darn for you.

8. *Electronics.* **a.** *trans.* To connect temporarily; also with *in, into*; similarly to *patch out* (see quot. 1940²). **b.** *intr.* To be temporarily connected. **c.** *trans.* To represent or simulate by means of temporary connections.

[**1923**: implied in *Patching *vbl. sb.*¹ 1.] **1937** *Printers' Ink Monthly* May 40/1 *Patch it in*, to tie together various pieces of apparatus to form a circuit. **1940** *Chambers's Techn. Dict.* 619/2 *Patch*, to join together units of apparatus..by flexible cords terminated on plugs, which are inserted into break-jacks bridged across the terminations of each unit. *Ibid.*, *Patch in* and *patch out*, the temporary connexion (*patching in*) of spare apparatus in a circuit with patch cords, defective apparatus being thereby *patched out*. **1948** *Electronics* July 120/2 Two inverting or summing amplifiers in the computer unit..are patched to the servo as illustrated. **1962** Huskey & Korn *Computer Handbk.* iv. 36 The small extra cost of duplicate resistors for a few plug-in patchboards is negligible compared with the almost incredible nuisance of patching, say, a summing integrator with patchcord connections alone. **1964** C. P. Gilbert *Design & Use Electronic Analogue Computers* vi. 363 In Fig. 6.7(*d*) unit 2 can patch directly into units 1 and 3, unit 3 can patch directly into units 2 and 4, and so on. **1966** *Times* 21 Sept. (Ascension Island Suppl.) p. iv/3, I recently picked up this telephone and asked the communications centre..to patch me in on the network. **1971** J. H. Smith *Digital Logic* v. 79 Electronic control systems can be quickly developed by patching the required design on a simulator. **1975** J. Grady *Shadow of Condor* (1976) xii. 190 Kevin used the powerful radio in his car to call CIA headquarters in Langley, Virginia. The technicians there patched his radio call into the old man's office phone.

patched, *ppl. a.* Add: *patched-up* (later examples). So *patched-together.*

1905 *Daily Chron.* 13 Feb. 5/1 Japan will tolerate no patched-up peace. **1916** H. G. Wells *Mr. Britling* iii. i. 385 It was the queer halting telling of a patched-together tale. **1935** T. S. Eliot *Murder in Cath.* i. 16 Peace, but not the kiss of peace. A patched up affair, if you ask my opinion. **1954** Koestler *Invis. Writing* xxii. 249 To emerge in the end with a more or less successfully patched-up personality. **1972** *Listener* 21 Dec. 872/1 A young man who deals in patched-up cars.

patchily, *adv.* (In Dict. s.v. Patchy *a.*) (Further examples.)

1972 *Daily Tel.* 24 Apr. 14/8 Anyone who bites or licks her lips to concentrate will soon look patchily fungoid rather than trendy. **1974** I. Murdoch *Sacred & Profane Love Machine* 144 His face patchily pink. **1975** G. Lyall *Judas Country* xxx. 220 I stepped cautiously out into the patchily-lit alley.

patchiness. (Earlier example.)

a **1828** D. Wordsworth *Jrnl.* (1941) II. 93 The character of the Lake..is of free chearfulness...it is not *disturbed* by buildings. *I* complained not of *patchiness* or *spottiness*.

patching, *vbl. sb.*¹ **1.** (Further examples: cf. *Patch *v.* 8.)

1923 *Bell Syst. Techn. Jrnl.* II. 123 In open-wire installations it has been the practice to equip each line circuit.. with a full complement of jacks suited to provide the maximum degree of flexibility in 'patching'. *Ibid.* 132 With this arrangement of apparatus, any 'ringer'..may be connected temporarily to the system by means of 'patching' cords. **1948** J. Atkinson *Herbert & Procter's Telephony* (new ed.) I. xvii. 346 (*caption*) Patching jack at incoming end of order wire. *Ibid.* xxi. 473 The junctions are routed through break jacks on the test jack frame and a small cord shelf accommodates a number of patching-out cords to facilitate temporary changes. **1964** C. P. Gilbert *Design & Use Electronic Analogue Computers* vi. 361 To interconnect the amplifiers, leads with a plug at each end would be inserted into the appropriate sockets giving what is known as simple direct 'patching'. *Ibid.*, The patching is carried out on the plug board before it is inserted into the 'patch' panel. **1964** G. A. & T. M. Korn *Electronic Analog & Hybrid Computers* xi. 443 Iterative differential analyzers..require frequent patching of integrator and memory control circuits.

3. (Earlier example.)

1835 A. B. Longstreet *Georgia Scenes* 286 He..drew out his patching, found the most even part of it, [etc.].

patchless (pæ·tʃlės), *a.* [f. Patch *sb.*¹ + -less.] Not having or exhibiting patches. (In quot. fig.)

1927 *Observer* 1 May 6 Hundreds of different..things are mentioned; but the mentionings are all woven into a seamless, patchless, and nowhere ragged history of the subject.

patchouli'd (pătʃū·lid), *a. poet.* Also patchoulied. [f. Patchouli + -ed².] Perfumed with patchouli.

1925 H. Acton in *Oxf. Poetry* 7 The pulsing cafés and patchouli'd vamps. **1929** W. de la Mare in H. Granville-Barker *Eighteen-Seventies* iii. 51 The patchoulied nosegay of artifice. **1943** D. Gascoyne *Poems 1937–42* 42 Tenebrous lanes Down which at times patchouli'd ghosts flit by.

patch panel²: see s.v. *Patch *sb.*¹ 6.

pa·tch test. *Med.* [f. Patch *sb.*¹ + Test *sb.*¹] A test for determining a patient's sensitivity to a substance, by applying to his skin a patch made of or containing it, and noting whether erythema is produced. Hence (with hyphen) as *vb. trans.*, to subject to such a test; also **pa·tch-testing** *vbl. sb.*

1933 R. W. & R. M. B. MacKenna *Dis. Skin* (ed. 3) xv. 320 Recently the 'patch' or 'contact eczema' test has been much used to determine whether or not a patient is susceptible to an external irritant. **1956** *Brit. Med. Jrnl.* 21 Jan. 148/1 She was patch-tested to nylon stockings of three shades and gave positive reactions to all of them. **1963** *Lancet* 5 Jan. 61/2 The patient was patch-tested with the sample that he had provided, and whereas the control patch remained unaffected, the test patch became reddened in three days, and redder and scaly in a week. **1963** V. J. Fontana in F. Speer *Allergic Child* xv. 223 Patch testing is a helpful procedure in eliciting the specific cause in many types of contact dermatitis. **1967** *Oceanogr. & Marine Biol.* V. 366 Patch testing with pieces of the polyzoon..have confirmed that the *Alcyonidium* is the causative agent. **1968** A. Rook et al. *Textbk. Dermatol.* I. xiv. 301/1 The patch-test was first devised by Jadassohn 1896 and later brought into general use by Bloch. *Ibid.* 310/1 Any patch-test involves a risk of sensitization.

patchwork. Add: **4. a.** and **b.** Also *fig.*

1905 *Daily Chron.* 30 Aug. 5/6 A rupture is preferable to a patch-work peace. **1933** Chesterton *All I Survey* ix. 45, I am afraid of the Patchwork Peril, which is all colours and none. **1951** M. McLuhan *Mech. Bride* (1967) 144/2 Everyone is intellectually and emotionally a patchwork quilt of occupied and unoccupied territory. **1956** H. J. Paton in H. D. Lewis *Contemp. Brit. Philos.* (ser. 3) 348, I never had any use for the patchwork theory popular at that time, which supposed his [*sc.* Kant's] work to be a mass of contradictions. **1977** P. Scupham *Hinterland* 3 Now, the celebrated Ride stiffly through a patchwork multitude.

Hence **pa·tchworker,** a maker of patchwork; also *fig.*; **pa·tchworky** *a.* [-y¹], resembling or suggestive of patchwork.

1844 *Ainsworth's Mag.* VI. 112 A patchworker of the piquant anecdotes of the newest French memoirs. **1884** E. W. Hamilton *Diary* 29 Mar. (1972) II. 585 The exterior [of the house, Coombe Warren] is picturesque, but almost *too* unsymmetrical and 'patch-work-y'. **1888** *Advance* (Chicago) 3 May 275/1 It would quicken the zeal of the little patchworkers also, if they could see how pretty their gay quilts look upon the beds. **1906** *Speaker* 20 Oct. 71/2 To a foreign student London presents..a patchworky spectacle. **1972** *New Yorker* 26 Aug. 50/1 (*caption*) Tie on this patchworky, wrap-around skirt. **1977** R. Richardson *Discovering Patchwork* (rear cover), This book..should..have much to offer the experienced patchworker.

patchy, *a.*¹ Add: (Further examples.) Also, occurring only in patches, or at separate points; irregular, spasmodic.

1905 *Sat. Westm. Gaz.* 16 Dec. 15/1 That evening I noticed a peculiarity in the pit's applause. It was 'patchy'. **1921** *Ampleforth Jrnl.* Jan. 139 Scent was patchy and a good deal of lifting was necessary to maintain the line. **1957** [see *Contagious *a.* 8]. **1967** W. Carr *Daily Tel. Beauty Bk.* 23 Those who freckle have only groups of pigment-producing cells, providing patchy, inefficient protection. **1971** *Nature* 30 July 352/2 The survey is patchy and by no means as up to date as one expects. **1976** *Shooting Times & Country Mag.* 16–22 Dec. 38/1 Many countries report varied and patchy scent.

pâté. Add: **1.** *pâté de foie gras*, now used chiefly of the goose-liver filling only: cf. next (further examples).

1863 G. Meredith *Let.* 19 July (1970) I. 217 Did you have salmon at Strasbourg—and pâté de foie gras—it's not a good place for the latter. **1876** Geo. Eliot *Dan. Der.* III. vi. xlviii. 343, I can't eat *pâté de foie gras*. **1892** A. B. Marshall *Larger Cookery Bk.* vi. 253 Cut up the contents of a small tin of *pâté de foie gras*. **1960** E. David *French Provincial Cooking* 35 His *pâté de foie gras* and *mousse de foies de volaille*, smooth, pink, and marbled with green pistachio nuts and black truffles,..were as good and delicate as they looked.

b. A paste made from liver or other meats. So *pâté de campagne*, a coarse pork and liver pâté; *pâté en croûte*, a pâté baked in a pastry surround; *pâté maison*, a pâté cooked according to the recipe of a particular restaurant.

1901 C. H. Senn *New Cent. Cookery Bk.* xxii. 399 Fleurettes de Foie Gras (Foie Gras with Mayonnaise).—1 medium-sized tin or terrine of foie gras pâté, 1 truffle [etc.]. **1931** X. M. Boulestin *What shall we have To-day?* 84 Let the *pâté* get cold in the larder with a weight over it. Serve in the *terrine* in which it has cooked. **1948** *Good Housek. Cookery Bk.* ii. 74 Liver and other savoury pâtés ..may be served as an hors d'oeuvre. **1931** J. Berjane *French Dishes for Eng. Tables* x. 152 Pâté de Campagne (*foie de porc*) Gascon... Chop up the liver and the fat [etc.]. **1966** P. V. Price *France: Food & Wine Guide* 101 A *pâté de campagne* will be a rather rough-cut *pâté*. **1967** V. Canning *Python Project* ii. 31 He was not English. The French accent was as thick and meaty as *pâté de campagne*. *c* **1938** *Fortnum & Mason Price List* 29/2 Pâté en croutes (October to March). **1961** S. Beck et al. *Mastering Art of French Cooking* ix. 569 The recipe we have chosen to illustrate *pâté en croûte* is boned duck stuffed, reformed, surrounded with decorated pastry, and baked. **1967** A. Christie *Endless Night* xiii. 113 We had brought *pâté en croûte* with us and French bread. **1947** M. McCarthy *On Contrary* (1962) 6 *Open City*,..Oscar Wilde, a reprint of Henry James were *pâté de maison* to this lady who wanted the definitive flapjack. **1956** M. Laski in *Observer* 14 Oct. 17/4 It is..misleading to say that..*pâté de foie gras* is known under the names *pâté de foie*, *pâté de campagne*, *pâté maison*. **1963** R. Carrier *Great Dishes of World* 40 Every French restaurant boasts its *pâté maison*. **1967** *Listener* 20 Apr. 533/3 The standardization and mass-production of food extends beyond the same tinned '*pâté maison*'..wherever one goes.

‖**pâte** (pāt). [Fr., = Paste *sb.*] **a. pâte** (erron. pâté) brisée (pāt brīze), short pastry.

1845 E. Acton *Mod. Cookery* xvi. 405 Pate brisee, or French crust for hot or cold meat-pies. **1960** E. David *French Provincial Cooking* 205 This is one version of the *pâte brisée*..used for most open tarts in French cookery. **1978** *N.Y. Times* 29 Mar. c. 6/6 A pâté brisée is one of the best sweet pastry doughs.

b. The clay from which porcelain is made. So **pâte dure** (pāt diūʳ), hard clay; **pâte-sur-pâte** (pāt siūʳ pāt), clay applied in layers to form relief decoration; **pâte tendre** (pāt taṅdr), soft clay.

1863 W. Chaffers *Marks Pott. & Porc.* 163 Porcelain of the *pâte tendre* has the appearance of an unctuous white enamel, like cream;..the *pâte tendre* is also soft in another sense, being unable to bear so great a degree of furnace heat as the hard porcelain. *Ibid.*, The *pâte dure*, or true porcelain, is of the whiteness of milk, and feels to the touch of a hard and cold nature. **1870** C. Schreiber *Jrnl.* (1911) I. 71 We found an exquisite pâte tendre St. Cloud group. **1881** C. C. Harrison *Woman's Handiwork* ii. 104 To Minton's unmatched artist, Solon, the world is indebted for an exquisite style of ceramic ornamentation, low relief carving in clay, known as pâte-sur-pâte. **1890** [see *Kutani]. **1899** R. Glazier *Man. Hist. Ornament* 83 Porcelain is technically known under the terms 'hard paste' ('pâte dure') and 'soft' ('pâte tendre'). **1904** H. James *Golden Bowl* I. viii. 149 He had handled nothing so precious as the Principino..whom he could manipulate.. as he couldn't a correspondingly rare morsel of an earlier *pâte tendre*. **1932** R. Fry *Characteristics of French Art* iii. 66 He loved to feel the *pâte tendre* of a piece of fine pottery. **1947** J. C. Rich *Materials & Methods of Sculpture* ii. 44 *Pâte sur pâte* is the phrase applied to a method of modeling very low reliefs with slip. **1959** *Times* 3 Oct. 9/5 Her sister Florence, who specialized in birds..and also painted in *pâte sur pâte*. **1964** H. Hodges *Artifacts* i. 33 In a process known as pâte-sur-pâte, clay made to the consistency of a thin paste may be painted on, layer by layer, to produce low relief modelling. **1972** *Country Life* 20 Jan. 152/2 They are early Minton pâte-sur-pâte, their maker the technically gifted Marc Louis Solon. **1974** Savage & Newman *Illustr. Dict. Ceramics* 215 *Pâte dure*, the French term for hard-paste porcelain. **1976** *Times Lit. Suppl.* 3 Sept. 1074/4 A display of Minton *pâte-sur-pâte* wares.

c. pâte de verre (pāt də vɛr), powdered glass that has been refired.

1907 E. Dillon *Glass* xxii. 359 Of quite another nature is the *pâte de verre*, a substance somewhat of the nature of a glass frit, which has been made use of by the French sculptor, M. Henri Cros, in the modelling of polychrome reliefs and friezes. **1961** E. M. Elville *Collector's Dict. Glass* 178/2 [James] Tassie's medium was a finely powdered potash-lead glass, or pâte-de-verre. **1978** *Guardian Weekly* 9 Apr. 14/4 Sofas are covered in bright prints, lamps are pink or blue *pate de verre*.

patellectomy (pætĕle·ktŏmi). *Surg.* [f. Patell(a + *-ectomy.] Surgical removal of the patella.

1940 *Q. Cumulative Index Medicus* XXVIII. 706/1 Patellectomy in therapy of recidivating dislocations. **1944** *Surg., Gynecol., & Obstetr.* LXXIX. 536/1 These observations..suggest that routine total patellectomy for fracture of the patella in the human may be folly. **1977** *Proc. R. Soc. Med.* LXX. 258/2 All had been considered for surgical treatment, including patellectomy.

patellofemoral (pătēːlofe·mŏrăl), *a.* *Anat.* [f. Patell(a + -o + Femoral *a.* and *sb.*] Of or pertaining to the patella and the femur.

1934 in Webster. **1964** J. McM. Mennell *Joint Pain* xii. 126 (*heading*) The patellofemoral joint. **1978** *Sci. Amer.* Jan. 44/3 The [knee] joint includes two articulations, or movable parts, that transmit force and relative motion. One is the patellofemoral bearing surface between the patella (the kneecap) and the femur (the thighbone).

patent, *a.* Add: **2.** (Further examples.) *patent house, theatre*, a theatre established by Royal Patent; *spec.* in London, the theatres of Covent Garden and Drury Lane, whose Patents were granted by Charles II in 1662.

1835 Dickens *Sk. Boz* (1837) 2nd Ser. 166 Why were they not engaged at one of the patent theatres? *c* **1844** C. Gore in M. R. Booth *Eng. Plays of 19th Cent.* (1973) III. 16 The disproportion and caricature *established into the custom of the stage* by the exigencies of our colossal patent theatres. **1891** [in Dict.]. **1932** *Times Lit. Suppl.* 24 Nov. 888/3 In 1832, however, the fashion [of stalls] spread at last to the patent houses. **1973** M. R. Booth *Eng. Plays of 19th Cent.* III. 1 Social and political change, the enlarging of the patent theatres, the broadening and inevitable coarsening of audience tastes—all this brought with it a demand for a new kind of comedy. **1973** *Times Lit. Suppl.* 19 Oct. 1272/2 During the period 1740–80 there were only two patent theatres in London: Drury Lane and Covent Garden. Unlicensed theatres were confined to musical performances, and dubiously legal public 'rehearsals' of plays.

3. *patent food*, a proprietary food preparation; *patent insides, outsides* (see quot. 1970); *patent leather* (see Leather *sb.* 1); also *ellipt.*, a patent leather boot or shoe; also *fig.*; *patent log*, a mechanical device for measuring the speed of a ship; *patent medicine*, a proprietary medicine manufactured under patent; *patent sail*, an automatically controlled windmill sail (see quots.); *patent still* [patented by Aeneas Coffey in 1830], a type of still for the continuous production of alcohol of greater strength and purity than is obtainable in a pot still, steam being used to heat the wash directly and carry off the alcoholic vapour; freq. *attrib.*

1871 *London Jrnl.* Apr. (Advt.), Dr. Ridge's patent food. **1903** 'A. McNeill' *Egregious English* 56 Mammas.. who suckle their children out of patent-food tins. **1925** *Scribner's Mag.* Sept. 274/2 Even the 'quick lunch' takes time, so a widely advertised patent food is put up in tablet form, to be eaten at the business desk itself. **1929** Galsworthy *Exiled* ii. 66, I have recently had to make a series of pictures for a patent food called Vital. **1964** M. Laski in S. Nowell-Smith *Edwardian England* iv. 205 The richer parents..were more likely to feed the infant on one of the many patent foods. **1882** I. M. Rittenhouse *Jrnl.* 8 June in *Maud* (1939) 103 The funny places in all the old patent-insides of newspapers talk about the sweet girl graduate. **1931** *Sat. Even. Post* 28 Feb. 129/2 Some publishers bought patent insides, which were the interior pages of the newspaper ready printed for use. **1968** E. Russenholt *Heart of Continent* iv. xiii. 243 The 'St. James Leader'..is an 8-page weekly (including 'patent insides'). **1970** R. K. Kent *Lang. Journalism* 98 *Patent insides* (or *outsides*), features or other syndicated material that come to a newspaper already printed on inside (or first and last) pages; readyprint pages. **1829** *Poulson's Amer. Daily Advertiser* (Philad.) 25 Apr. 3/5 Just received, an extensive assortment of Japanned Patent Leather, of superior quality. **1846** A. J. H. Duganne *Daguerreotype Miniature* 7 A pair of patent-leather boots and a Polka hat were the extremes of his apparel. **1849** G. G. Foster *New York in Slices* 64 Our young gentlemen..thus preserve their patent-leathers. **1882** *Encycl. Brit.* XIV. 387/2 *Patent or Enamelled Leather.*—Leather finished with a brilliant, smooth, and glossy surface, used for dress boots and shoes,..is known under a variety of names, as lacquered, varnished, japanned, and enamelled leather, &c. **1890** O. Wilde *Pict. Dorian Gray* i, in *Lippincott's Monthly Mag.* July 8 Lord Henry..tapped the toe of his patent-leather boot. **1905** H. A. Vachell *Hill* ii. 29 He had to varnish Grieve's patent-leathers for Sunday. **1910** J. W. Tompkins *Mothers & Fathers* 356 [He was] humbly removing the overshoes that covered Mr. Hammond's patent leathers. **1955** [see *lair *sb.*⁴]. **1976** Botham & Donnelly *Valentino* iv. 35 Rodolpho reached up a hand to pat his patent leather hair into place. **1876** Patent log [see Log *sb.*¹ 5]. **1940** *Chambers's Techn. Dict.* 509/1 The modern *patent* (or *taffrail*) *log* mechanically indicates the rate of travel by means of a submerged fly or rotator, whose revolutions are conveyed to a register on the rail of the vessel by a braided hemp line secured to the rotator. **1961** F. H. Burgess *Dict. Sailing* 157 *Patent log*, a mechanical device with which a rotator is used, to work a dial indicating the distance run through the water. **1770** *Essex Gaz.* (Salem, Mass.) 17 Apr. 4/4 To be sold by Benjamin Eaton..in Marblehead..a collection of genuine patent medicines. **1799** [in Dict.]. **1830** Scott *Lett. Demonology* v. 144 The proprietor of a patent medicine, who should in those days have attested his having wrought such miracles as we see sometimes advertised. **1866** 'Mark Twain' *Speeches* (1923) 7 It is said by some..that Kanakas won't lie, but I know they *will* lie..lie like patent-medicine advertisements. **1887** [in Dict.]. **1888** G. B. Shaw *Let.* 20 Sept. (1965) I. 198 Nearly all the citizens..buy immense quantities of charms called patent medicines. **1901** *Chambers's Jrnl.* Jan. 63/1 Soaps, patent medicines, chocolates..are the things most advertised. **1914** *Rep. Sel. Comm. Patent Medicines* p. xii, in *Parl. Papers* IX, Patent and proprietary medicines differ very widely in character. At one end of the scale is the valuable scientific preparation; at the other end is the mere vulgar swindle. **1961** *Today's Health* Feb. 30/2 The medical device pirate of today, of course, is a far more sophisticated operator than his predecessor of yesteryear—the gallus-snapping hawker of snake oil and other patent medicines. **1978** P. Bailey *Leisure & Class in Victorian Eng.* i. 15 Street preachers, stump orators and patent medicine salesmen. **1871** *Lancaster Intelligencer* 3 Apr., The editor who surrenders control of one-half of his paper to some manufacturer of patent-outsides, may make a slight reduction in his current expenses, but in the end he will lose both money and influence. **1890** *Boston Jrnl.* 7 Mar. 4/5 He was running his patent outsides for country newspapers. **1945** *Archit. Rev.* XCVIII. 72/2 In 1807 Sir William Cubitt invented what has even since been called 'patent sail'. **1968** J. Arnold *Shell Bk. Country Crafts* 170 This was the patent sail and enabled adjustment to the wind to be made without interrupting the milling. **1973** J. Vince *Discovering Windmills* (ed. 3) 21 The most significant improvement in sail design came about in 1807 when William Cubbit [*sic*] invented his patent sail. This retained Meikle's shutters, but they were controlled automatically by a weight suspended outside the mill. **1887** A. Barnard *Whisky Distilleries of U.K.* 12 Blenders without number can be found who will strenuously affirm that to give the public a moderate priced article with sufficient age, there is no way but to use good old Patent Still Grain Spirit as a basis. *Ibid.* (Advt.), Flemming, Bennet & McLaren... Makers of Fire Stills, Steam Stills, Coffey's Patent Stills,..&c. **1906** *Daily Chron.* 27 Feb. 1/7 The North London magistrate held that patent-still spirit was not whisky. **1934** J. I. Davis *Beginner's Guide Wines & Spirits* viii. 85 Irish Whisky is always 'pot-stilled'... Some Scotch Whisky is so made, but most of it is manufactured in a patent still, which completes the distillation in one operation. **1937** *Thorpe's Dict. Appl. Chem.* (ed. 4) I. 178/1 The spirit produced in these 'patent stills'..has been used only to a limited extent for the production of spirit for whisky blending... The great bulk goes into industry for use as a raw material or as a technical solvent. **1968** I. C. Taylor *Highland Whisky* (An Comunn Gaidhealach) 3 The practice of blending pot-still and patent-still whiskies began about 1860.

patent, *sb.* Add: **6.** *patent agent.*

1860 Bartlett *Dict. Amer.* (ed. 3) 310 *Patent Agent*, one who procures patents for inventors. **1884** *List of Subscribers* (London & Globe Telephone Co.), Haseltine, Lake & Co... Patent Agents. **1957** *Encycl. Brit.* XVII. 371/1 Patents are usually, although not necessarily, obtained through the intervention of 'patent agents'. **1959** *Digest of Patent & Other Cases* (Patent Office) II. 778 A person who lodges and signs a complete specification as Agent for Applicant does not thereby describe himself as a Patent Agent.

patent, *v.* Add: **1.** (Earlier examples.)

1789 J. Morse *Amer. Geogr.* 261 They..patented away to their particular favorites, a very great proportion of the whole province. **1831** J. M. Peck *Guide for Emigrants* iii. 319 The Military Bounty Tract..was set apart by Congress and patented for soldiers who served in the last war.

3. (Earlier examples.)

1675 *Calendar Virginia State Papers* (1875) I. 8 Major Lawrence Smith..did patent foure thousand six hundred acres of land. **1815** D. Drake *Nat. View Cincinnati* i. 51 The following is the course pursued in locating and patenting these lands.

5. *Metallurgy.* To subject to the process of 'patenting' (see *patenting *vbl. sb.* b).

1922 J. W. Urquhart *Steel Thermal Treatment* xi. 271 The process known as 'patenting' wire is of recent origin. **1932** Barr & Honeyman *Steel* xix. 102 The rods require to be patented once only.

patented *ppl. a.* (examples in sense *5 of the vb.).

1891 J. B. Smith *Wire* i. 60 'Patented or improved steel wire' implies that which has been treated by a patented or special 'improving process' of annealing, hardening, and tempering. **1916** D. K. Bullens *Steel & its Heat Treatment* xvii. 402 The high strength and toughness of patented wire are due to its carbon condition and its peculiar structure. **1956** A. K. Osborne *Encycl. Iron & Steel Industry* 307/2 Best patented steel wire is wire intended for the production of steel cables which has been drawn to a maximum stress of between 80 to 90 tons per sq. in.

patentability. (In Dict. s.v. Patentable *a.*) Further examples.

1946 *Nature* 23 Nov. 726/1 In these circumstances the requirements for patentability as a selection from the general disclosure or claim are discovery of a previously unrecognized advantage.., and limitation to a manufacture based on that discovery. **1972** *N.Y. Law Jrnl.* 10 Oct. 5/2 The general level of innovation necessary to sustain patentability remains the same. **1972** *Sci. Amer.* Oct. 45/3 Novelty, though essential for patentability, does not guarantee utility. **1978** *Nature* 22 June 584/1 Patent law has not yet got to grips with the problem of genetically-engineered bacteria, both in terms of their patentability and of the particular difficulties there might be in storing bacteria whose claim to originality lies in the properties possessed by a plasmid.

patentable, *a.* (Earlier and later examples.)

1817 *Niles' Reg.* XII. 283/2 The improvement relied on by Witness was not useful, and consequently not a patentable improvement under the patent law. **1977** *New Scientist* 7 Apr. 24/2 Although computer programs are not as such patentable, computer hardware can be protected.

patentably (pēi·tĕntăbli), *adv.* [f. Patentable *a.* + -ly².] In a way that satisfies the conditions for patenting anything.

1903 *Sci. Amer.* 28 Feb. 159/1 Patents have been granted in Great Britain..without any enquiry to learn whether the inventions were patentably new.

patenter. For *Obs.* *rare*⁻¹ read *rare* and add further example.

1883 E. W. Hamilton *Diary* 25 Mar. (1972) II. 413 The only other person at dinner was Lane Fox, the inventor and patenter of an electric lighting system.

patenting, *vbl. sb.* Add: **b.** *Metallurgy.* In the manufacture of wire, a process for improving ductility similar to normalizing, but involving cooling in either air or molten lead or salt.

The subject of *Brit. Patent 1104* (granted to J. Horsfall in 1854).

1891 J. B. Smith *Wire* i. 60 'The patenting or improving' of steel wire is..an occult process..for each manufacturer has his own..special methods. **1916** D. K. Bullens *Steel & its Heat Treatment* xvii. 405 The structure obtained by patenting permits much further cold drawing than does the structure obtained by annealing. **1921** L. Aitchison *Engin. Steels* viii. 275 The patenting furnace is a long, open-ended chamber. **1945** *Engineers' Digest* VI. 169 (*heading*) Direct resistance heating of salt baths for the patenting of steel wire. **1961** *Engineering* 14 Apr. 516/2 A batch of 24 wires is fed through the 60 ft patenting furnace at a time, being heated to 950°C.

pater. Add: **3. b.** spec. *Anthropol.* The legal father.

1949, etc. [see *genitor¹ b]. **1951** E. E. Evans-Pritchard *Kinship & Marriage among Nuer* 113 The children clearly regard themselves as members of a legal family to which the brother of their pater does not belong, although he is their foster-father, and may also be their genitor. **1955** M. Gluckman *Custom & Conflict in Afr.* iii. 71 In these institutions physiological paternity is distinguished from social fatherhood—as anthropologists put it, the *pater* need not be the *genitor*.

pateras, var. *pattress.

Pateresque (pēiːtəre·sk), *a.* [f. the name of Walter Horatio *Pater* (1839–94), English writer and critic.] Resembling the style of Pater's writing or his method of criticism. So **Paterian** (pĕtiə·riăn), *a.*

1903 Beerbohm *Around Theatres* (1953) 274 One reason why this book is so fresh and welcome is that we see for the first time the Pateresque manner and method of criticism applied to current dramatic art. 'Pateresque' is no slight. **1905** *Westm. Gaz.* 24 July 14/1 The definition of the priest's 'triumph' as that of 'achieving as much faith as possible in an age of negation' recalls a passage from Bishop Blougram, but is truly Pateresque in its expression. **1938** L. P. Smith *Unforgotten Years* viii. 207 The record of an attempt to see things elegantly and nobly in the Paterian spirit. **1939** *Scrutiny* VII. iv. 442 For years he cultivated a Paterian manner in his own writing. **1944** W. Stevens *Let.* 27 Mar. (1967) 463 He considers my poems to be expressions of Paterian hedonism. **1965** K. Graham *Eng. Crit. of Novel* iii. 93 An interesting Pateresque attempt to explain the flaws..in terms of form alone. **1977** R. L. Wolff *Gains & Losses* ii. 188 This Paterian Christianity is astonishingly like Paterian paganism.

paternal, *a.* Add: **1. b.** paternal roof, the home of one's father.

1828 Carlyle in *Edin. Rev.* Dec. 293 But now, at this early age, he quits the paternal roof. **1845** J. Porter *Thaddeus of Warsaw* (new ed.) xliii. 475 Longing earnestly for a temporary sanctuary under his friend's paternal roof. **1861** J. W. Carlyle *Let.* 22 Sept. (1883) III. 91 If you are returned to 'the paternal roof', no need almost of this letter.

paternalist (pătə̄·ɪnălist), *a.* [see -ist.] = Paternalistic *a.*

1928 *Britain's Industr. Future* (Liberal Industr. Inquiry) iii. xviii. 237 'Welfare work' has an unpleasantly paternalist and patronising sound.

paternalistic, *a.* Add: (Later examples.) Hence **paternali·stically** *adv.*

1918 *Nation* (N.Y.) 7 Feb. p. xii/2 It is perhaps as well that a paternalistic Government, with unlimited power of taxation to make good the deficit, is behind it. **1918** E. H. Griggs *Soul of Democracy* xviii. 125 The breakdown of paternalistically achieved efficiency has been evident in Germany's utter failure to understand, [etc.]. **1961** *Daily Tel.* 24 Apr. 12 They are humble people, Europeans, half-castes and Africans, whose only crime is to have pursued their livelihoods unprotestingly under a paternalistic Government which is now challenged by the spread of black nationalism. **1966** *Listener* 16 June 889/1 I felt the face of Britain was reasonably presented in this paternalistic atmosphere. **1976** *Daily Mirror* 16 July 5/7 That puts him on the side of the Labour unions, who are uneasy about Carter, a paternalistic employer.

paternalized (pătə̄·ɪnăləiz'd), *ppl. a.* [f. as Paternalism + -ize + -ed¹.] Characterized by or subjected to paternalism.

1903 *Electr. World & Engin.* 11 Apr. 597 The socialistic spirit..that would have every man on the pay roll of the State or the municipality in a vast series of paternalised institutions.

paternity. Add: **5.** *attrib.* and *Comb.*, as *paternity case, leave, suit*; **paternity test**, a blood test used to assess or discount the likelihood of paternity in a particular case; hence **paternity-testing** *vbl. sb.*

1940 D. MacCarthy *Drama* 104 In the opening scene we have seen him confronted, as a soldier in authority, with a paternity case he cannot solve. **1973** *Guardian* 4 June 9/1 Imagine the day when firms allow their male employees paternity leave to be with their wives during the last part of their pregnancies and the first days in the life of their children. **1975** *Times* 12 Mar. 3/1 The Greater London Council..agreed tonight that fathers on their staff should normally be allowed up to five days paternity leave. **1977** *Spare Rib* June 18/3 We're also now pressing for paternity leave, realising it's a joint thing. **1975** J. McClure *Snake* viii. 108 See a doctor, get some pills... Paternity suits I don't contest, madam. **1977** D. Anthony *Stud Game* iii. 21, I can imagine the damage a paternity suit would do. But I'm broke, and the baby's due in less than five months. **1926** R. J. E. Scott *Gould's Med. Dict.* 773/2 Mayoral and Jiminez paternity test. **1943** D. Harley *Medico-Legal Blood Group Determination* vi. 44 (*heading*) Paternity tests. **1960** I. A. Stanton *Dict. for Med. Secretaries* 113/1 *Paternity test*, blood comparison tests to rule out the possibility of paternity. **1968** F. E. Camps *Grandwohl's Legal Med.* (ed. 2) xiii. 171/1 Anti-S [antibody] has been found sufficiently often for it to be used economically for paternity testing. **1972** B. Knight *Legal Aspects Med. Practice* xxiii. 196 Certain haemoglobin-binding proteins in human serum, called haptoglobins, can be detected by electrophoresis... Their inheritance is..determined on fixed genetic principles, thus making them of use in paternity testing.

paternoster, *sb.* Add: **4. c.** A lift consisting of a succession of doorless compartments on an endless chain in continuous slow motion that allows entry at any time. Also *paternoster lift.*

1912 *Engineering Index* XXXI. 72/2 The 'Paternoster' continuous elevator (Ascensori a movimento continuo, o ascensori 'Paternoster'). U. Quintavalle. An endless chain elevator with several compartments... *L'Industria*—Mar. 10 1912. **1937** *Discovery* Dec. 388/2 Lifts and conveyors of the 'Paternoster' type have been installed. **1971** R. Lewis *Error of Judgment* vi. 205 He and the victim of his attack were on a small square landing at the end of the corridor and directly in front of the paternosters. **1973** *Times* 24 Dec. 4/3 Instead of orthodox lifts, the architects..use 'paternosters', continually moving, doorless two-person lifts which reduce waiting. **1977** *Times* 19 Feb. 2/3 Passenger-carrying 'paternoster' lifts consisting of continuous chains of small cars.

6. paternoster lake *Physical Geogr.*, each of a line of lakes in a glaciated valley.

1942 C. A. Cotton *Climatic Accidents in Landscape-Making* xix. 256 If the step-tread basins are occupied by lakes..these may follow one another like beads upon a string—'paternoster lakes'. **1968** R. W. Fairbridge *Encycl. Geomorphol.* 468/2 Since the floor of a glaciated valley is often undulating..or with a step-like rock floor.., it is sometimes marked by a succession of lakes.. rather fancifully named paternoster lakes.

Paterson (pæ·təɹsən). *Austral.* The surname of an Australian family, used in *Paterson's curse*, a blue-flowered weed, *Echium plantagineum*, of the family Boraginaceæ which is said to have spread from their garden near Albury, New South Wales, in the 1890s.

1905 *Agric. Gaz. New South Wales* XVI. 268 That 'Paterson's Curse' produces some feed is undoubted, but it is a smothering, rough, coarse plant, whose room is far better than its company. **1922** G. M. Thomson *Naturalisation of Animals & Plants in N.Z.* iii. 445 It [*sc. Echium plantagineum*] is a serious pest in some parts of Australia, where it is known as 'Paterson's Curse'. **1926** F. W. Hilgendorf *Weeds N.Z.* vi. 144 Paterson's Curse..is the local name of this weed where it is commonest in New Zealand, viz. in the Thames and neighbouring districts. **1930** A. J. Ewart *Flora of Victoria* 971 Paterson's Curse... A weed proclaimed for the whole state. **1970** R. M. Moore *Austral. Grasslands* xii. 189/2 Paterson's curse..has a wide distribution in South Australia, Victoria, and southern New South Wales.

patesi (pătēⁱ·si). [erron. transliteration of Sumerian *ensi*.] The ruler of a Sumerian city-state; a petty sovereign or priest-king. Hence **pate·siship**, the office or position of a patesi.

1894 A. H. Sayce *Primer of Assyriol.* iii. 45 At a later date Tello lost its independence, and its rulers became merely patesis or high-priests. **1910** L. W. King *Hist. Sumer & Akkad* iv. 101 The human kings and patesis were nothing more than ministers, or agents, appointed to carry out their will. **1927** Peake & Fleure *Priests & Kings* 178 The Sumerian patesi was a magistrate who performed sacred or priestly functions. **1928** C. L. Woolley *Sumerians* v. 138 Only in Nippur did the patesiship continue to descend from father to son.

path, *sb.* Add: **1. d.** *Physiol.* = *pathway 2 a.

1881 A. L. Ranney *Appl. Anat. Nervous Syst.* iii. 299 It is as certainly proved that the motor impulses travel along the anterior half of the spinal cord, while the path of sensory impressions is intimately associated with the posterior half. **1902** H. E. Santee *Anat. Brain & Spinal Cord* (ed. 2) x. 190 The sensory paths conduct two varieties of impulse. *Ibid.* 195 There are two auditory paths, the Cochlear and the Vestibular. **1942** F. A. Mettler *Neuroanat.* x. 201 There are thus two paths for delicate tactile sensations: a long, crossed path and an uncrossed one. **1950** *Physiol. Rev.* XXX. 461 An extrapyramidal path..appears to diverge from the cortico-spinal tract in the pons.

e. *Biochem.* A metabolic pathway (*pathway 2 b).

1909 A. E. Garrod *Inborn Errors Metabolism* i. 7 This conception of the permanency of the metabolic paths is no new one. **1927** M. Bodansky *Introd. Physiol. Chem.* xi. 270 The possible paths of metabolism of this amino acid are the following. **1935** C. F. & G. T. Cori in Harrow & Sherwin *Textbk. Biochem.* xx. 553 With the enzyme systems existing in muscle it seems more probable..that cleavage products of glycogen are led into the lactic acid path.

f. A schedule which is allotted to or is available to an individual train over a given route.

1961 *Guardian* 28 Apr. 30/4 The code..will indicate the class of train..and the number of its 'path' on operational timing. **1971** D. J. Smith *Discovering Railwayana* x. 60 *Train path*, vacant line in the timetable which may be used by an extra train. **1977** *Modern Railways* Dec. 463/3 Through freight traffic can be a problem too, some of these trains have paths at the busiest times.

3. b. *Math.* A continuous mapping of a real interval into a space.

1939 M. H. A. Newman *Elem. Topology of Plane Sets of Points* vi. 143 Paths and loops are not merely sets of points. **1956** E. M. Patterson *Topology* iv. 76 A path in a topological space T is a continuous mapping $\alpha : C \rightarrow T$ of the space C, that is the set of real numbers u satisfying $0 \leqslant u \leqslant 1$ with the usual topology, into T. **1974** L. Loomis *Calculus* xi. 437 A parametric representation of a curve is often called a path. Sometimes we call the curve itself a path.

5. path-breaker [tr. G. *bahnbrecher*], one who or something which breaks open a path; a pioneer; so *path-breaking* adj. [G. *bahnbrechend*]; **path difference** *Physics*, difference in path length; **path-finder**, (*a*) (further examples); also *fig.*; (*b*) an aircraft or its pilot sent ahead of bombing aircraft to locate and mark out the target for attack; **path-finding**, the state of being a path-finder; also *attrib.*; **path length** *Physics*, the length of the path followed by a light ray, sound wave, or the like (in the case of light usu. after allowing for the retarding effect of the medium: cf. *optical path* s.v. *optical *a.* 6); **path-master** *N. Amer.* (see quot. 1869).

1905 *Daily Chron.* 15 Nov. 4/3 A gentle path-breaker in her chosen..field of the delineation of child life and child millinery. **1913** *Q. Rev.* Oct. 407 The late Frederic Seebohm's 'English Village Community' was literally a path-breaker. **1914** R. M. Jones *Spiritual Reformers 16th & 17th Cent.* iv. 46 A man of heroic spirit and a path-breaking genius. **1973** *Times Lit. Suppl.* 27 Apr. 466/3 Not an epoch-making or path-breaking book. **1978** *Sci. Amer.* Feb. 131/3 His words apply equally well to other path-breaking discoveries in physics. **1929** J. K. Robertson *Introd. Physical Optics* ix. 186 If the thickness of the film gradually increases, the path difference between..pairs of interfering rays will do so also and consequently the face of the film will be alternately light and dark. **1962** A. Nisbett *Technique Sound Studio* i. 28 (*caption*) The effective path difference (i.e. the extra distance travelled by the [sound] wave to reach the back of the ribbon 1) is equivalent to the distance D round the magnet pole-piece 2. **1866** *Harper's Mag.* June 28/1 The great Path-finder, unfortunately for himself, took the wrong path. **1898** W. James *Coll. Ess. & Rev.* (1920) 408 Philosophers are after all like poets. They are path-finders. **1932** C. Fuller *Louis Trigardt's Trek* 18 These were the 'path-finders' of the greater movement that followed in their wake. **1943** *Times* 25 Nov. 4/4 Red tracer bullets were continually fired from the ground at the pathfinders' flares. **1944** [see *marker 3 c]. **1946** *R.A.F. Jrnl.* May 168 For the crews of Bomber Command's Pathfinder Force it was all a question of time. **1959** R. Collier *City that wouldn't Die* i. 22 As pathfinders their function was to spotlight the target..with thousands of chandelier flares and incendiary bombs. **1973** *Nature* 2 Mar. 67/1 He was an inventor and an innovator, a path-finder and prognosticator. **1977** 10 June 16/3 A former wartime Pathfinder with DFC and Bar. **1888** Path-finding [in Dict.]. **1943** *Jane's All World's Aircraft 1942* p. iv/2 The success.. was contributed to by the introduction of the system of path-finding. **1944** *Living off Land* iv. 85 Accurate path-finding in the bush..tests your common sense. **1948** H. Brighouse in J. Marriott *Best One-Act Plays 1946–47* 145 Well, pathfinding for Bomber Command..so a bit of marine navigation wasn't beyond me. **1961** *Listener* 30 Nov. 908/1 The development of radio aids and path-finding techniques was proceeding. **1978** R. V. Jones *Most Secret War* xviii. 148 KGr 100's pathfinding role. **1949** S. Silver *Microwave Antenna* iv. 123 Fermat's principle: The optical ray or rays from a source at a point P_1 to a point of observation P_2 is the curve along which the optical path length is stationary with respect to infinitesimal variations of path. **1956** *Nature* 10 Mar. 469/1 The summary of section A contains a useful review ..of resonance and pulse velocity tests... An important omission..is mention of the practical difficulty of measuring path-lengths sufficiently accurately in testing in-situ concrete structures by the pulse velocity method. **1971** *New Scientist* 3 June 566/1 Long path-length spectroscopy is made plausible by the combination of high intensity and directionality. **1799** *Upper Canada Gaz.* (York, Ontario) 29 June 3/2 The public are much indebted to Mr. John McDougal, who was appointed one of the path-masters at the last town meeting, for his great assiduity and care in getting the streets cleared of the many and dangerous (especially at night) constructions therein. **1869** *Rep. Comm. Agric. 1868* (U.S. Dept. Agric.) 348 The immediate supervision of construction and repairs is generally under the direction of local 'road supervisors', or 'path masters', as they are termed in some districts. **1959** *Maclean's Mag.* 20 June 83/2 The name was changed about 40 years ago because Gottlieb Watts, town pathmaster, could not spell it. **1968** E. S. Russenholt *Heart of Continent* iv. xi. 206 The Department of Agriculture orders that pathmasters cut weedy grain, and have the work charged to landowners.

path. (pæþ). Also written without full stop. Abbrev. of Pathology. Freq. *attrib.*

1937 'J. Bell' *Murder in Hospital* xii. 233 He does not appear to have had any path. tests done. **1944** H. Ashton *Yeoman's Hospital* xi. 239, I always hoped he'd land that job he wanted..in the Path. Lab. there. **1965** K. Giles *Some Beasts no More* v. 137 Sir Shelly diagnosed leukemia and it was confirmed by path. Prognosis was twelve months. **1968** C. Watson *Charity ends at Home* vii. 90 He's like a kid in the bath when he gets into that path lab. **1972** *Guardian* 20 June 16/4 People rather miss out on their psychi, because it happens to coincide with path finals. **1978** N. Freeling *Night Lords* xiv. 61 You find a suicide which..is a phony. We wait for the path. report.

Pathan (pătā·n). Also 7 **Pattan, Puttan,** 8 **Patan,** 9 **Puthan.** [Hind., f. Pashto *Pakhtun.*] The name of a Pashto-speaking people inhabiting parts of south-east Afghanistan and north-west Pakistan, also called *Pakhtun. Also *attrib.* or as *adj.*

1638 T. Herbert *Trav.* (ed. 2) 64 Most of her Pattans were slaine. **1787** C. Hamilton *Hist. Relation Rohilla Afgans N. Provinces Hindostan* 53 The tribe of Afgans denominated Rohees, or Rohillas, (so termed from Roh, which in the Pâtan dialect signifies a mountainous country.) **1792** *Asiatick Researches* III. 6 The principal inhabitants of the mountains, called Pársici,..seem to have been destroyed or expelled by the numerous tribes of Afghâns or Patans. **1851** H. B. Edwardes *Year on Punjab Frontier* I. 78, The people whom we geographically call Afghans, style themselves nationally Puthâns. **1864** *Athenæum* 5 Nov. 598/3 Pathans and Sikhs inspired them with feelings very little removed from panic. **1901** Kipling *Kim* vii. 173 Except for Mahbub Ali, and he is a Pathan, I have no friend save thee, Holy One. **1903** *Strand Mag.* May 530/1 A swarthy Pathan face grinned wickedly over a rubble heap. *Ibid.* 530/2 To compete with the..Pathan in his own hills. **1915** R. W. Campbell *Private Spud Tamson* iii. 23 Once on the Frontier of India he had slaughtered ten bloodthirsty Pathans in the space of an hour. **1930** *N.Y. Times* 6 Feb. 22/6 His [*sc.* Kipling's] commissioners, syces, Pathans, and budmashes continue to be as well known as the fascinating denizens of the animal kingdom celebrated in 'The Jungle Books'. **1958** A. Toynbee *East to West* 128 That British century had already brought the modern world to the threshold of the Pathan highlander's hovel. **1968** *Economist* 16 Nov. 31/1 Extremists..argue that the Pathans, like the Nagas, were never part of pre-partition India and should therefore be allowed to determine their own future. **1971** R. Dentry *Encounter at Kharmel* iii. 58 Your Pathan tribesman is a realist. He enjoys cutting throats. **1973** *Times* 27 July 16/3 The Pathans are estimated to account for half the present population of Afghanistan. **1976** *Spare Rib* Nov. 12/1 It was the case of a Pathan family. The Pathans are often very orthodox Muslims—sometimes religious fanatics.

pathetic, *a.* (*sb.*) Add: **A. 4.** So in phr. *pathetic fallacy*, first used by John Ruskin (see quot. 1856 in Dict.) to describe the attribution of human response or emotion to inanimate nature.

1856 [in Dict.]. **1856** Geo. Eliot in *Westm. Rev.* Apr. 631 Mr. Ruskin..enters on his special subject, namely, landscape painting. With that intense interest in landscape which is a peculiar characteristic of modern times, is associated the 'Pathetic Fallacy'—the transference to external objects of the spectator's own emotions. **1895** C. H. Herford *Spenser's Shepheards Calender* p. xlviii, Pastoral nature is founded upon the 'pathetic fallacy'. **1906** W. W. Greg *Pastoral Poetry & Pastoral Drama* 93 We have here a specific inversion of the 'pathetic fallacy'. **1930** H. S. V. Jones *Spenser Handbk.* iii. 64 Each elegy opens with an apostrophe to the poet's verse, and each illustrates what Ruskin called the 'pathetic fallacy'. **1930** L. Powys (*title*) The pathetic fallacy: a study of Christianity. **1959** *Listener* 6 Aug. 223/2 Many awaited death.. while the pathetic fallacy laboured away with ill winds and rain. **1968** *Ibid.* 18 Jan. 68/3 To believe that a television station should be part of its audience could seem like the Pathetic Fallacy (Communications Division), a piece of electronic anthropomorphism. **1975** M. C. Davis *Near Woods* v. 83 The next morning I fell under the spell of literature's pathetic fallacy at nearly every step. **1977** D Watkin *Morality & Archit.* ii. ii. 38 Le Corbusier's argument..combines succinctly the pathetic fallacies we are investigating: that particular types of architectural form are morally regenerative and physically health-giving.

6. Miserably inadequate; so poor as to be ridiculous. *colloq.*

1937 Partridge *Dict. Slang* 609/2 *Pathetic*, ludicrous. **1969** *Listener* 10 July 41/1 The military government clearly thinks it is established for good. The alleged plots against it are either mythical or, when genuine, pathetic. **1974** *Liverpool Echo* (Football ed.) 26 Oct. 3/2 The standard of refereeing in English soccer is pathetic. There is no consistency.

Pathet Lao (păte·t lau). [Laotian Thai, lit. 'land of the Lao'.] A communist guerrilla movement and political party (after December 1975 the ruling party) in Laos. Also *attrib.*

1954 *Times* 10 May 6/3 Earlier in the day statements by the two leaders—Mr. Son Ngoc Minh for the Khmer and Mr. Souvannouvong for the Pathet Lao—had been distributed to the Press... The Pathet Lao claim was to control half the territory. **1966** 'A. Hall' *9th Directive* xxvi. 236 The territory..is in the hands of the Pathet Lao, which takes orders from Peking. **1968** H. Tove *Laos* v. 136 It was not long before they reached an area where the local Pathet Lao organization could protect them. **1970** Langer & Zasloff *N. Vietnam & Pathet Lao* vi. 100 Prior to the Geneva Conference of 1962, the Soviet advisory role to the Pathet Lao had been very brief. **1970** *Peace News* 8 May 3/2 A Cambodian Communist Party has existed since the end of the second world war, when it fought shoulder-to-shoulder with the Pathet Lao, and the Vietminh. **1977** *Bangladesh Times* 20 Jan. 7/4 The Laotian authorities have set up a committee to censor every thing written, sung, recited or danced, the Pathet Lao news agency Khao San reported today.

pathic, *sb.* and *a.* Add: **A.** *sb.* **1.** (Later example.)

1972 Stedman *Med. Dict.* (ed. 22) 930/1 *Pathic*, a person who assumes the passive role in any abnormal sexual act.

B. *adj.* **1.** (Later example.)

1959 *N. & Q.* Dec. 435/2 The Pardoner's pathic role in the perverted relationship thus suggested is clearly indicated in A 691.

2. Delete *rare*⁻⁰ and add later examples. Also, that suffers from disease or disorder.

1902 W. A. Hammond *Aristotle's Psychol.* p. lxviii, Desire, as Aristotle employs it, is not a purely pathic or affective element. **1940** Hinsie & Shatzky *Psychiatric Dict.* 404/1 *Pathic*,..pertaining to or affected by disease or disorder. **1951** S. F. Nadel *Found. Social Anthropol.* xi. 291 Vague states of feeling ('pathic states', as they have been called). **1973** M. Amis *Rachel Papers* 148 Rachel received this idiot outpouring with a pathic nod.

Hence **pa·thically** *adv.*, in a passive manner.

1934 *Mind* XLIII. 300 The root notion..seems to be that the given must be passively, or (should I say?) pathically received.

patho-. Add: **pathobio·logy**, the study of the biological processes associated with diseased or injured tissue.

1971 LaVia & Hill *Princ. Pathobiol.* i. 4 This is the subject of pathobiology—the alterations in normal biological mechanisms that occur in response to injury. **1972** *Lancet* 20 May 1104/1 Pathobiology (a trendy name for general pathology) seems to be a fashionable subject in the United States. **1975** *Amer. Jrnl. Path.* LXXIX. 183 Pathobiology of an endocrine disease.

pathogen. Substitute for def.: Any agent that causes disease, esp. a micro-organism. Add further examples.

1942 E. C. Stakman in *Aerobiology* (Amer. Assoc. Adv. Sci. Publ. No. 17) 2/1 The possible aerial dissemination of pathogens of domestic animals by air currents is a relatively unexplored field. **1970** M. Treshow *Environment & Plant Response* ii. 17 Contrary to common belief, a pathogen technically does not have to be an organism or virus. A pathogen can be any component of the physical environment, including adverse climate, soil, or air relations. **1972** *Daily Colonist* (Victoria, B.C.) 8 Feb. 6/2 Diener said the new agent, which he described as a 'novel type of pathogen' or disease-causing agent, is about 80 times smaller than the smallest known virus.

pathogenesis. Add: Hence **pathogene·tically, pathoge·nically** *advs.*, as regards pathogenesis.

1904 *Brit. Med. Jrnl.* 10 Sept. 559 The cells pathogenically affected by a toxin may not be the cells of origin of antitoxin. **1928** *Amer. Jrnl. Path.* IV. 632 Primary and secondary contracted kidneys in this respect are pathogenetically identical. **1972** Aronson & Elliott *Ocular Inflammation* xi. 258/2 Pathogenetically, the diffuse fundus lesion presents as a disseminated metastatic choroiditis,..relatively early in life.

pathography. (In Dict. s.v. Patho-.) Add: **b.** The, or a, study of the life and character of an individual or community as influenced by a disease.

1917 C. R. Payne tr. *Pfister's Psychoanal. Method* xxvii. 573 The history of the Catholic sainthood affords the analytic pathography an inexhaustible material. **1959** *New Statesman* 28 Nov. 760/2 The founder of Protestantism with his fits of melancholia, his anxiety attacks, night sweats, and anguish about concupiscence, is an obvious subject for historical pathography. **1972** Q. Bell *Virginia Woolf* II. i. 20 The Japanese psychiatrist Mme Miyeko Kamiya is, I believe, preparing a pathography of Virginia Woolf.

Hence **patho·grapher**, one who writes a pathography.

1974 *Times Lit. Suppl.* 15 Mar. 256/4 The pathographer must allow for the language and limitations of the theory.

pathological, *a.* Add: **1. b.** (Later examples.) Also in more general use.

1921 R. S. Woodworth *Psychol.* (1922) i. 16 The pathological method..traces the decay or demoralization of mental life instead of its growth. **1933** E. Glover in S. Lorand *Psycho-Anal. Today* 192 (*title*) Pathological character formation. **1949** *Sat. Rev. Lit.* (U.S.) 25 June 11/1 For Ford, you see, was what those who were not charmed by him insisted upon describing as a pathological liar. **1951** V. Nabokov *Speak, Memory* ii. 32 Age had developed in her a pathological stinginess. **1951** M. Lowry *Let.* 25 Aug. (1967) 256 A himself is an almost pathological liar. **1971** *Jrnl. Gen. Psychol.* LXXXV. 64 In fact.. pathological anxiety might..represent a heightened susceptibility to over-arousal.

2. (Earlier examples.)

1796 F. A. Nitsch *Gen. View Kant's Princ. concerning Man* 195 The pathological interest aims at the agreeable and pleasing consequences of an action. **1798** A. F. M. Willich *Elem. Critical Philos.* 101 A rational observance of pathological laws.

3. *Math.* Grossly abnormal in properties or behaviour, as compared with the well-behaved functions normally encountered in classical applications (see quots. 1946, 1960).

1939 I. S. Sokolnikoff *Advanced Calculus* iv. 105 Such pathological behavior of continuous functions led to a careful inquiry into the meaning of such geometrical concepts as the area under a curve. **1946** H. & B. S. Jeffreys *Methods Math. Physics* i. 17 We can speak of a function of x that is equal to 1 if x is rational but to 0 if x is irrational. Such a function would be fairly regarded by a physicist as pathological, and he is interested in a much narrower class of functions, roughly speaking such as can be represented by graphs. **1960** *New Scientist* 25 Aug. 537/2 The term 'pathological functions' does not name any specific functions... It is applied rather informally to functions whose behaviour appears delinquent to those who expect mathematical functions to correspond only with physical events or, conversely, to possess only properties that can be represented by a graph or in a computer. Often the 'pathological' features concern continuity and smoothness. **1968** Fox & Mayers *Computing Methods for Scientists & Engineers* ix. 179 Examples could be constructed of pathological repeated inaccurate consistency. **1971** D. W. Sciama *Mod. Cosmol.* viii. 113 In this case the geometry of space is said to be hyperbolic... The volume of the space is infinite except in pathological cases that need not concern us.

pathologically, *adv.* Add: **1. b.** Corresponding to sense *1 b of the adj.: morbidly; abnormally.

1933 *Mind* XLII. 138 In the expression of his feelings and appreciations he was almost pathologically reticent. **1978** P. G. Winslow *Coppergold* 75 She was pathologically jealous.

pathologist. Add: (See quots.)

1971 *Lancet* 29 May 1124/1 A pathologist (O.E.D., probably about 1905) is 'One versed in pathology: a student of or writer upon diseases' and pathology 'treats of the causes and nature of diseases'. Even at that time, when departments of pathology were beginning to be usual in medical schools and several specialised journals included the word in their titles, that was a little out of date... The word was already used chiefly for practitioners of laboratory medicine. By the thirties it had differentiated further: non-specialised laboratory doctors called themselves 'clinical pathologists', but used alone the word was increasingly limited to practitioners of the oldest branch of the trade, morbid-anatomy-histopathology. The N.H.S. has partly reversed this trend, advertising posts, for instance, as 'pathologist with a special interest in bacteriology'. The Royal College of Pathologists goes even further, being anxious to include the non-medical biochemists, though there is..a backwoods faction which challenges the propriety of calling anyone non-medical a pathologist. **1977** *Role & Membership of College* (R. College of Pathologists) 5 The pathologist can be defined as a medical (or dental) graduate who has ultimate responsibility for the diagnostic tests performed in the pathology laboratory and for the advice given to clinicians on the selection and interpretation of tests.

pathology. 1. c. (Examples in further extended senses.)

1972 W. Labov *Language in Inner City* iv. 134 There is evidence from linguistic pathology that a deep-seated knowledge of this fact may be present in native speakers. **1972** T. Kochman *Rappin' & Stylin' Out* p. xi, This book is an attempt to get beyond what Albert Murray has called the 'fakelore of black pathology' and its corollary, the 'folklore of white supremacy'. **1977** D. M. Smith *Human Geogr.* x. 278 The major metropolitan states with their high general *average* affluence levels clearly experience high levels of social pathology.

pathopolis (pæþǫ·pŏlis). [f. Gr. παθολογικός, f. πάθος suffering, disease + πόλις city; see Pathologic *a.*, -polis.] A diseased or morally degenerate city.

1927, 1961 [see *parasitopolis].

pathophysiology (pæːþofiziǫ·lŏdʒi). *Med.* [f. Patho- + Physiology.] The physiological processes associated with disease or injury; the study of such processes.

1952 *Adv. Internal Med.* V. 428 Abnormalities in protein metabolism form the central problem in the pathophysiology of multiple myeloma. **1962** *Lancet* 2 June 1172/2 An analysis of the pathophysiology of these disturbances may further our understanding of cerebral function. **1972** *Ibid.* 20 May 1104/1 It is impossible to keep up with all trends in pathophysiology. **1974** M. C. Gerald *Pharmacol.* xvi. 296 We may anticipate that the etiology of these disorders will be better understood with future advances in the pathophysiology and biochemistry of mental disease. **1975** *Nature* 3 Jan. 77/2 A symposium on the pathophysiology and clinical aspects of pain.

Hence **pa:thophysiolo·gic, -physiolo·gical** *adjs.*, of or involving abnormal physiological processes; pertaining to pathophysiology, **pa:thophysiolo·gically** *adv.*, as regards pathophysiology.

1952 *Adv. Internal Med.* V. 398 From the pathophysiologic point of view, the most interesting part of the myeloma problem is the disturbance of protein metabolism. **1960** (*title*) Works of the Institute of Higher Nervous Activity. Pathophysiological series. (Academy of Sciences of the U.S.S.R.) **1962** H. G. Keitel *Pathophysiol. & Treatment of Body Fluid Disturbances* i. 12 In many illnesses the anatomical, biochemical, and pathophysiological characteristics of the associated fluid derangement and their successful treatment have been fairly clearly defined. **1973** Wasserman & Brown in Holland & Frei *Cancer Med.* xix. 1367/1 The pathophysiologic mechanisms responsible for these symptoms. **1973** *Sci. Amer.* Sept. 121/3 Selective errors in the metabolism of the monamines in specific areas of the brain may be the pathophysiological basis for psychotic disorders. **1976** *Nature* 29 July 397/2 This..results in a chronic murine hepatosplenic disease which closely resembles, pathophysiologically, human chronic schistosomiasis.

pathosticate (păþǫ·stikēⁱt), *v.* *nonce-wd.* [joc. f. Pathos + Pathe)ticate *v.*] *intr.* To induce pathos by melodramatic writing.

1901 G. B. Shaw *Cashel Byron's Profession* (rev. ed.) Pref. p. xii, In novel-writing there are two trustworthy dodges for capturing the public. One is to slaughter a child and pathosticate over its deathbed for a whole chapter.

pathway. Add: **2. a.** *Physiol.* A chain of nerve cells forming a continuous route along which impulses of a particular kind habitually travel.

1924 R. M. Ogden tr. *Koffka's Growth of Mind* iv. 235 Von Kries pointed out that the arousal of associations can not be explained on the basis of a mere 'pathway'-hypothesis which assumes that nervous excitations travel along fixed paths. **1934** J. H. Globus *Neuroanat.* (ed. 6) 63 There will now be no difficulty in linking up the chain of nuclei and tracts, which form the optic pathways. **1952** W. F. T. Tatlow et al. *Synopsis Neurol.* ix. 162 Hyperalgesia.—Occurs when visceral and somatic (deep or superficial) afferents share pain pathway, so that on simultaneous stimulation, summation of subliminal stimuli can occur. **1971** *Sci. Amer.* July 48/2 The pathways extending from the brain to the lower spinal cord are some two feet long. **1972** *Science* 5 May 536/2 The descending cortical and brainstem pathways to the spinal cord represent the main instrument by which the brain controls movements.

b. The sequence of reactions undergone by a compound or class of compounds in a natural environment, esp. a living organism.

1927 M. Bodansky *Introd. Physiol. Chem.* xi. 271 Another pathway of metabolism has been suggested, namely one involving the conversion of arginine into guanidine-butyric acid,..which by β-oxidation would yield guanidine acetic acid. **1947** *Growth* XI. 232 The pathway of galactose fermentation [in yeast]. **1961** *Proc. Nat. Acad. Sci.* XLVII. 378 (*heading*) Transformation studies on the linkage of markers in the tryptophen pathway in *Bacillus subtilis*. **1967** M. E. Hale *Biol. Lichens* viii. 106 The biosynthetic pathways by which they appear to be synthesized. **1971** *Sci. Amer.* Sept. 45/2 The carbon dioxide pathways in our biosphere. **1971** *Nature* 17 Sept. 163/2 His latest experiments are an attempt to identify the pathway of incorporation of ³H-thymine into DNA. **1973** *Ibid.* 13 Apr. 453/1 This information will be required..to elucidate the pathways and interactions of mercury in the estuarine and marine environment.

-pathy. Add: **2.** Forming the names of bodily disorders of a specified part (as Myopathy) or kind (Idiopathy).

patience, *sb.* Add: **6.** *patience board, card* (further examples).

1907 *Yesterday's Shopping* (1969) 379/2 Patience and 'Pigmy' Cards. Rounded corners, printed backs, two packs in a box. *Ibid.* 380/1 Patience Board With Cards. **1916** J. Buchan *Greenmantle* ii. 24 From his pocket he had taken a pack of Patience cards and had begun to play the game called the Double Napoleon. **1926–7** *Army & Navy Stores Catal.* 392 Patience Board. For playing the game of Patience off the table.

patience-dock. 1. (Earlier and later examples.)

[**1640** J. Parkinson *Theatrum Botanicum* ii. iii. 154 Garden Patience is a Docke bearing the name of Rhubarbe, for some small purging quality therein.] **1820** T. Green *Universal Herbal* II. 498 *Rumex Patientia*; Patience Dock, or Rhubarb. **1822** J. C. Loudon *Encycl. Gardening* 715 Herb Patience, or Patience Dock..is a hardy perennial plant, a native of Italy, introduced in 1573. **1859** *Trans. Illinois Agric. Soc.* III. 513 Patience dock comes early, and makes good greens. **1972** Y. Lovelock *Vegetable Bk.* i. 218 The spinach dock..is used in the early spring..and goes by such names as patience and patient dock. **1973** C. A. Wilson *Food & Drink in Britain* vi. 204 Many other large-leaved green plants were employed in this pottage, including orache, clary, mallows, patience dock, borage and bugloss.

patient, *a.* and *sb.* Add: **A. 4.** *spec.* in *Grammar* (see quot.).

1939 L. H. Gray *Foundations of Lang.* xii. 374 A distinction is drawn between the ergative case as the logical subject of a transitive verb and the patient case as the subject of an intransitive verb.

5. patient Lucy orig. *U.S.*, = *busy Lizzie* s.v. *busy *a.* 11.

1946 M. Free *All about House Plants* xvii. 161 The names Patience Plant and Patient Lucy are interesting examples of how the original meaning of a plant name can be reversed. The vernacular names are derived from the botanical name *Impatiens*,..referring to the sudden bursting of the seed pods. **1956** [see *busy *a.* 11]. **1977** K. & G. Beckett *Illustr. Encycl. Indoor Plants* 110/2 Busy Lizzie; Patient Lucy; Sultana. A familiar house plant, native to tropical Africa.

B. *sb.* **4.** *spec.* in *Grammar*.

1968 J. Lyons *Introd. Theoret. Linguistics* viii. 342 But this conflicts with the notion of the subject as the 'actor', rather than the 'goal' (or 'patient'). **1971** J. Anderson *Grammar of Case* ix. 140 Accounts of transitivity conducted in terms of 'actor-action-patient'. [*Ibid.* iv. 52 The labels 'ergative' and 'nominative' are usually used with respect to..an inflectional..system; alternative terms are 'actif'/'nominatif' (Lafitte, 1962) 'agens'/'patiens' (Troubetzkoy, 1929).] **1975** *Language* LI. 806 It is significant that the suffix has come to designate the agent or instrument of the verbal activity in certain daughters—but not the patient, product, or location of the verbal activity.

pa·tienthood. [f. Patient *sb.* 2 + -hood.] The state or condition of being a patient.

1970 *New Yorker* 14 Nov. 108 Millions of boys..live, as Captain Ahab says, with half of their heart and with only one of their lungs, and the world is the worse for it. Now and again, however, an individual is called upon..to lift his individual patienthood to the level of a universal one and to try to solve for all what he could not solve for himself alone. **1971** *Harper's Mag.* May 111/2 Hers [*sc.* R. Mackenzie's] is the best account of the psychology of patienthood in a modern hospital I've ever read.

‖**patiki** (pā·tiki). *N.Z.* [Maori.] A local name for a flat-fish belonging to one of several species of *Rhombosolea* found in New Zealand waters; cf. Flounder *sb.*¹ 1.

[**1820** *Gram. & Vocab. Lang. N.Z.* (Church Missionary Soc.) 190 Patiki, *s.* a fish so called.] **1838** J. S. Polack *New Zealand* I. ix. 322 The *pátiki*, between the large flounder and the sole, is equally excellent with the European fish. **1843** *N.Z. Jrnl.* No. 92 177/2 The natives.. brought in an abundant supply of fine fish; mullet,.. patiki, (the flounder of Europe) and other delicious species. **1855** R. Taylor *Te Ika a Maui* xxv. 412 *Patiki*, common name for the sole and flat-fish. **1879** *Trans. N.Z. Inst.* XII. 316 Large *Patiki*, flat fish, are occasionally speared up the river. **1949** P. H. Buck *Coming of Maori* (1950) ii. xiii. 321 One [rafter-painting] design, in which the field was roughly divided into lozenge-shaped areas, was named *patiki* (flounder) from its outline.

patina. Add: **2.** Also *fig.*

1933 H. Nicolson *Diary* 24 Feb. (1966) 140 He says what the Americans lack is *patina*. **1955** Koestler *Trail of Dinosaur* 79 This tendency prompts people to have their wall-brackets and picture-frames artificially dirtied to lend them the patina of age; so let us call it the 'patina-snobbery'. **1957** R. Hoggart *Uses of Literacy* ix. 227 He develops a strong patina of resistance, a thick and solid skin for not taking notice. **1967** N. Mailer *Cannibals & Christians* i. 13 It gives them all a high instant patina, their skin responding to the call of the wild. **1977** *Time* 18 July 28/2 Defending the rights of homosexuals.., she mingled with and took on some of the patina of the loony left. **1978** J. Thomson *Question of Identity* x. 97 The disorder was not a mere evening's untidiness. It had taken time to build up that rich patina of squalor.

Hence **pa·tinaed** *a.*, covered with a patina, patinated; also *fig.*

1948 W. Faulkner *Intruder in Dust* (1949) xi. 244 Took from the inside coat pocket a leather snap-purse patina-ed like old silver and almost as big as Miss Habersham's handbag. **1968** D. Murphy tr. *Gelin's Concept of Man in Bible* vii. 117 This is what I call a 'patinaed' reading.

patinate (pæ·tinēⁱt), *v.* [f. Patina 2 + -ate³.] *trans.* To cause to develop a patina; to cover with a patina. Also *fig.* So **pa·tinated** *ppl. a.*, **pa·tinating** *vbl. sb.*

1880 Patinated [in Dict. s.v. Patina]. **1920** Webster, Patinate. **1934** J. & C. Gordon *Portuguese Somersault* x. 237 Roofs of rich red-tile, patinated with lichens of varied tints. **1957** *Encycl. Brit.* IX. 382/2 When patinated flint occurs in gravels containing iron salts a yellow staining, producing the well-known ochreous patina, results. *Ibid.* XX. 231/2 The slower the patinating the more artistic the results is a good general rule to follow. **1957** *New Yorker* 6 Apr. 31 The riches Ian had inherited were so blindingly brand-new that it had taken time and pains to patinate them with good manners and good taste. **1969** L. R. Rogers *Sculpture* vi. 210 Most sculptors today prefer a patina to a clean bronze surface and use acids..to patinate their work. **1970** G. Savage *Dict. Antiques* 312/2 Handsomely patinated bronzes are much sought. **1974** K. Clark *Another Part of Wood* iii. 103, I learnt a great deal from some lectures on the fascinating subject of Aristotle's *Poetics* by a richly patinated character named Farquharson.

patination. (Further examples.)

1936 *Bull. Raffles Museum, Singapore* Ser. B. No. 1. 53 The neolithic implements of Puming and Pajitan..have practically no patination at all. **1947** J. C. Rich *Materials & Methods Sculpture* vii. 199 The patination of metals is a highly specialized art. **1960** H. Hayward *Antique Coll.* 212/2 Patination and colour pose problems to a faker. **1970** *Cabinet Maker & Retail Furnisher* 30 Oct. 301/2 The optional antique patination is hand-dyed. **1973** *Country Life* 27 Sept. (Suppl.) 91 Jacobean oak court cupboard of excellent colour and patination. **1975** *Nature* 7 Aug. 469/1 The bifaces from that site do not, on the whole, have a dark-brown polish but have a grey to white, calcareous patination or no patination at all.

patine. (Further examples.)

1910 G. B. Shaw *Let* 18 Sept. (1972) II. 942 Back in the dark [in a rowing-boat], without compass,..49 strokes to the minute striking patines of white fire from the Atlantic. **1940** *Harper's Mag.* Nov. 566 Her [*sc.* a pilgrim hawk's] back was an indefinable hue of iron; only a slight patine of the ruddiness of youth still shone on it. **1966** M. M. Pegler *Dict. Interior Design* (1967) 326 *Patina or Patine*, a greenish coating on the surface of old bronze.

patine (pæ̆tī·n), *v.* [f. Patina.] *trans.* To coat or cover with a patina (sense 2). Also *fig.* Hence **pati·ning** *vbl. sb.*

1896 A. H. Keane *Ethnol.* v. 84 Many [flints] have been deeply patined and rusted sometimes even right through. **1936** L. P. Smith *Reperusals & Re-Collections* i. 2 Time and history adds to their significance; it patines and mellows them. **1947** J. C. Rich *Materials & Methods Sculpture* vii. 200 Brass can be patined in several colors. *Ibid.* 209 Iron and steel can be patined a blue-black by the application of a hot solution composed of 10 grains of sodium thiosulphate to each ounce of water. *Ibid.* 403/2 (Index) Chemical methods of patining bronze.

patinize (pæ·tĭnəiz), *v.* [f. as prec. + -ize.] *trans.* To coat with or as with a patina. Hence **pa·tinizing** *vbl. sb.*

1904 *Sci. Amer. Suppl.* 27 Feb. 23548/1 The patinizing of zinc articles has become a very important question in the art industry.

patio. Add: **1.** Also in extended uses not in Spanish or Spanish-American contexts, a courtyard or enclosure; a paved area belonging to a house and usu. adjoining it. (Further examples.)

1891 Kipling *City of Dreadful Night* vi. 36 The central square, the *patio* or whatever it must be called, reeks. **1941** B. Schulberg *What makes Sammy Run?* xi. 271 He was..dancing with her out on the patio to the rhumba orchestra. **1947** *Chicago Maroon* 25 July 2/1 Tables will be set up in the patio with free cokes, and a faculty member or new student from each state will be on hand to welcome new students. **1949** *Here & Now* (N.Z.) Oct. 22/2 The patio is really an extra room without a roof. It is protected on three sides by the house itself and on the north by the bank. **1955** J. Cannan *Long Shadows* i. 16 The addition of two bathrooms built over a sun-room, which reduced the sooty little triangular garden to a size convenient to the construction of a *patio*. **1957** C. Brooke-Rose *Languages of Love* 23 Bernard sat now in an imitation patio under an imitation palm-tree. **1959** *Observer* 22 Mar. 1/1 But behind is a small golf course and a patio where a naval commander could be seen arranging the cushions. Filipino mess boys have polished up the interior. **1959** N. Mailer *Advts. for Myself* (1961) 378 Eitel is smoking a cigarette on the patio. **1968** *Globe & Mail* (Toronto) 17 Feb. 45 (Advt.), Detached 4-bedroom home on large fenced lot, beautiful patio. **1969** *Daily Tel.* 11 Nov. 17 Shops, dotted about among tree-strewn patios, fountains and open-air cafés [in Paris]. **1972** [see *limbo³]. **1973** *Irish Times* 2 Mar. 22/8 (Advt.), Bungalow: Dundrum—Scandinavian style, central heating, living room.. looking onto secluded patio. **1976** E. Scarrow *N.Z. Vegetable Gardening Guide* 5 A number of vegetables can be grown..in tubs or pots on a patio. **1977** *Austral. House & Garden* Jan. 39/2 This sunny family room has one wall of glass looking out to the plant-hung pergola and patio and the lush small garden.

2. (Earlier and later examples.) So *patio process.*

1845 *Chem. Gaz.* 1 Sept. 373 The preparation of magistral, an indispensable agent in the amalgamation by *patio*. **1856** *Hutching's Mag.* Sept. 104/1 The ore deposited on the *patio*, another set of laborers engage in separating the large lumps and reducing them to the size of common paving stones. **1863** *Proc. Calif. Acad. Nat. Sci.* II. 133 No experiments have been made in working this ore by the *patio* or Spanish-American amalgamation process. **1867** J. A. Philips *Mining & Metall. Gold & Silver* xvi. 334 At the hacienda of Regla..the patio, comprising an area an acre and a half in extent, is carefully covered with a wooden floor. **1896** C. H. Shinn *Story of Mine* 82 He treated them [*sc.* silver sulphurets] with the chemicals of the *patio* process. **1913** *Trans. Inst. Mining & Metallurgy* XXII. 661 The ore is received in the 'patio' or yard adjoining the mill, where it is weighed. **1974** *Encycl. Brit. Micropædia* VII. 664/1 The patio, or Mexican, process of separating silver from the ore by amalgamation with quicksilver was perfected in Pachuca by Bartolomé de Medina in the 16th century.

3. *attrib.* and *Comb.*, as *patio block, furniture, garden, model, stand*; **patio door**, a large, glass, sliding door leading to a garden or balcony.

1973 *Houston* (Texas) *Chron. Mag. People, Places, Pleasures* 14 Oct. 3 (Advt.), 7½″ × 16″ concrete patio blocks. **1979** *Sci. Amer.* Apr. 110/2 The concrete slabs were 'patio blocks' that weigh 6·5 kilograms and were 40 centimeters long, 19 centimeters wide and four centimeters thick. **1973** *Daily Tel.* 10 July 15/1 Usually the patio door is a replacement for the rather clumsy French doors which were put in by the thousand between the wars... The new sliding patio doors are vastly superior. A complete wall of glass, they can be five feet wide or even extend, with several panels, up to 24ft wide. Usually they are around 7ft high. *Ibid.* 21 July 3 (Advt.), We are Patio Door specialists. **1975** *Oxford Jrnl.* 31 Jan. 9 (Advt.), Add a new dimension to your living. With patio doors that slide open at a touch on warm summer days and close snugly to seal out the rain and draughts of winter. Their slim aluminium frames won't warp or need painting. **1976** *Homes & Gardens* July 116/2 Although patio doors are becoming an increasingly popular feature in houses, as yet there is no safety legislation governing their manufacture. **1969** *Sears Catal.* Spring/Summer 9 Moderately priced all-wood patio furniture. **1973** C. Williams *Man on Leash* (1974) i. 5 The standard small swimming pool and sundeck with patio furniture and umbrellas. **1912** J. London *Let.* 18 Jan. (1966) 360, I should like to see a novel of mine come out from the presses back of the patio garden. **1976** *Outdoor Living* (N.Z.) I. ii. 44 For the patio garden in the ideal location on the sunny side of the house the list of plants is unending. **1969** *Sears Catal.* Spring/Summer 13 Patio model [gas-fired grill] with its own 18-inch diameter mounting base. *Ibid.* Barbecue on patio stand.

patisserie. Add: Also **pâtisserie.** (Further examples.)

1802 [see *confiture]. **1912** R. K. Wood *Tourist's Russia* ii. 34 All the larger cities have good patisserie shops where afternoon tea is served à la Russe. **1960** E. David *French Provincial Cooking* 434 Elaborate *pâtisserie* and confectionery require practical experience and knowledge of an art quite distinct from that of normal household cookery. **1974** *Times* 25 Feb. 8/7 She gives a Viennese tea concert, at which the audience..claps with fingers sticky from the patisserie. **1976** *Times* 1 June 6/2 For patisseries we can recommend Toutabeurre.

b. A shop which sells *pâtisserie*; a pastry-cook's.

[**1884** F. E. A. Gasc *Dict. French & Eng. Lang.* (ed. 3) 410/2 *Pâtisserie*,..pastry; pastry-work *or* making *or* business; biscuit-baking *or* bakery.] **1903** F. B. Smith *How Paris amuses Itself* ii. 48 One of the most doleful sights I have seen in Paris was a sad-looking gentleman in black sitting at a cold marble-topped table of an expensive *patisserie* lunching on a weak cup of tea and a plate of cream-puffs. **1927** E. Bowen *Hotel* xv. 181 Will you come to tea with me and Victor at the Pâtisserie? **1930** *Chambers's Jrnl.* 15 Nov. 800/1 'Let's go and get tea at a café.' The idea found instant acceptance; three or four hundred yards from the hotel we found a good *pâtisserie* and we were soon..drinking excellent tea. **1949** A. Wilson *Wrong Set* 161 It was so cool inside the patisserie that Jeremy would gladly have stayed on there for ever. **1966** A. Christie *Third Girl* i. 5 A brioche..from the fourth shop he had tried. It was a Danish *patisserie* but infinitely superior to the so-called French one near by. **1975** N. Luard *Robespierre Serial* ii. 6 The patisserie at the corner of the Rue Vaumar.

‖**pâtissier, patissier** (pati·sie). [Fr.] One who makes *pâtisserie*; a pastry-cook.

1924 A. D. Sedgwick *Little French Girl* iv. vi. 346 She had to buy a *baba-au-rhum*..and asked André to drive them across to the *pâtissier's*. **1960** E. David *French Provincial Cooking* 434 A French housewife..is able to order what she requires..from that local *pâtissier* whom she knows to be most skilful and to use only the finest ingredients. **1961** A. Wesker *Kitchen* 59 *Paul.* But then I think: I should stop making pastries?... *Kevin.* Hush pâtissier! Hush! **1972** *Times* 29 July 9/3 Try the specialities of Catalan pâtissiers. **1975** *Sat. Rev.* (U.S.) 15 Nov. 50/1 Michel, the *chef de cuisine*, and André, the *pâtissier*.

Patjitanian (pædʒitā·niăn), *a.* *Archæol.* Also **Pajitanian.** [f. *Pajitan*, a town on the south coast of central Java + -ian.] Of or pertaining to Pajitan or the Early Palaeolithic chopper culture discovered near there in 1935. Also *absol.*, the Patjitanian chopper industry.

[**1936** G. H. R. von Koenigswald in *Bull. Raffles Museum, Singapore* Ser. B No. 1. 52 In October, 1935, the author was travelling with Mr. M. W. F. Tweedie, Curator of the Raffles Museum, in Central Java. During this trip we discovered on the 4th of October, a new site with big stone implements of various types, including, for the first time in Java, hand-axes... The site is near Pajitan, a town on the south coast of Central Java.] **1943** H. De Terra in *Trans. Amer. Philos. Soc.* XXXII. 458/2 It was from the basal stratum of the 10-m. terrace that we extracted a few rolled implements of 'Patjitanian' type. *Ibid.* 459/1 My impression is that the question of the absolute age of the Patjitanian industry can be determined only when the terrace geology of the region is known. **1944** H. L. Movius *Early Man & Pleistocene Stratigr. S. & E. Asia* 90/2 The Patjitanian should be placed probably in the second half of the Second Interglacial. *Ibid.* 91/1 The Patjitanian implements include many large, massive forms which have been crudely worked. **1949** K. P. Oakley *Man the Tool-Maker* viii. 70 No implements were found with the remains of the Java Man, *Pithecanthropus erectus*, but beds of slightly later age in Java have yielded the Patjitanian industry.., which recalls some of the artifacts of the related Pekin Man. **1964** M. W. Thompson tr. *Semenov's Prehist. Technol.* ii. ii. 36/2 The rough hand-axes of Java (Pajitanian)..are very inexpensive stone objects. **1969** Coles & Higgs *Archaeol. Early Man* xx. 390 A small number of Patjitanian artefacts might well be described as handaxes in African contexts. *Ibid.* 391 The stratigraphical position of the Patjitanian is not well-established, because much of this material is in a derived state in recent gravels of the

Basoka River. **1974** *Encycl. Brit. Micropædia* II. 887/1 These traditions [of stone tools] include..the Patjitanian industry, Java (associated with Java man at Sangiran and Trinil).

Pat Malone (pæt mălōᵘ·n). *Rhyming slang* (orig. and chiefly *Austral.*). = OWN *a.* 3. Also *ellipt.* **pat.**

1908 MRS. A. GUNN *We of Never-Never* xii. 146 He travels day after day and month after month, practically alone—'on me Pat Malone', he calls it. **1908** *Austral. Mag.* 1 Nov. 1251 'On my own' (by myself) became 'on my Pat Malone' and subsequently 'on my Pat' a very general expression nowadays. **1916** C. J. DENNIS *Moods of Ginger Mick* 110 But, torkin' straight, the Janes 'as done their bit. I'd like to 'ug the lot, orl on me pat! **1930** *Bulletin* (Sydney) 12 Mar. 47/1 On your pat now, aren't you? When did the old man go away? **1937** E. HILL *Great Austral. Loneliness* i. 22 If I was out there on me Pat Malone for too long—. **1943** N. MARSH *Colour Scheme* ix. 156 We're dopey if we let that bloke go off on his pat. **1948** V. PALMER *Golconda* xv. 120 Perhaps if I start off [singing this duet] on my pat there's some of you will take pity and not see me left. **1952** J. CLEARY *Sundowners* 276 First the missus died, then a coupla months later he went, and I was left on me pat malone. **1959** C. MACINNES *Absolute Beginners* 58 Standing there all on his Pat Malone. **1966** 'L. LANE' *ABZ of Scouse* 110 *On me tod*, by myself. An alternative phrase is *on me pat*. **1966** G. W. TURNER *Eng. Lang. Austral. & N.Z.* vi. 119 *On one's pat* 'alone'..has reached New Zealand. **1971** *Private Eye* 2 July 16 Pat malone again! Cripes I am cheesed.

Patna (pæ·tnă). The name of a district in north central Bihar, India, used *attrib.* in **Patna rice,** a small-grained rice, used principally in curries and other savoury dishes. Also *ellipt.*

1845 [see *CAROLINA]. **1861** MRS. BEETON *Bk. Househ. Managem.* 473 Well wash 1 lb. of the best Patna rice. **1868** *M. Jewry Warne's Model Cookery* 75/2 One and a half pound of whole Carolina or Patna rice. **1888** KIPLING *Story of Gadsbys* 48 A spattering gale of Best Patna... Throws half-a-pound of rice at G. **1902** 'KETAB' *Indian Dishes for Eng. Tables* 3 Patna rice is the best for boiling and should increase in boiling to about three times its bulk when raw. Good Patna rice has fine, rather long grains, and should be of a pale straw colour when rubbed to remove the dust. **1948** *Good Housek. Cookery Bk.* 400 For plain boiled rice, curries, risotto, and such dishes, in which the aim is to keep the grains separate, the Indian varieties of rice, such as Patna and Burma, are best. **1952** F. WHITE *Good Eng. Food* II. vi. 145 Wash ½ lb. of Patna rice. **1960** E. DAVID *French Provincial Cooking* 97 The long-grained rice which we call Patna is usually known in France as *riz caroline*. **1965** *Guardian* 6 Aug. 6/1 Long thin Patna for its dry separate grains to accompany curries. **1970** SIMON & HOWE *Dict. Gastron.* 327/1 What we call Patna rice is Patna seed type. Very little of so-called Patna rice comes from Bihar, but the name is reserved for good quality long grain rice.

patois. Add: **b*.** The folk or Creole speech of the English-speaking Caribbean (esp. Jamaica).

1934 J. RHYS *Voy. in Dark* I. vi. 83 She said something in patois and went on washing up. **1953** *Caribbean Q.* III. I. 24 The hybrid dialects of French origin which in philology come under the heading *Creole*. In Trinidad the word used to denote these dialects is *Patois*. **1970** *Caribbean Stud.* July 108 *Patois*, used by many Jamaicans in reference to Jamaican Creole. **1971** *Caribbean Q.* XVII. II. 13 Same name, different referent..patois.

patootie (pătū·ti). *U.S. slang.* [Perh. a corruption of *potato* (see quot. 1921).] A sweetheart, girl-friend; a pretty girl.

1921 *Dialect Notes* V. 110 *Patootie*,..Sweetheart. Reported from four different localities [in California]. Etymology unknown. Suggested, by sweetheart and sweet potato. **1923** G. MCKNIGHT *Eng. Words* iv. 61 In the vocabulary of modern youth, chivalry is dead... A girl.. if she is popular..is a *darb*, a *peach*,..a *sweet patootie*. **1935** *Nation* (N.Y.) 15 May 562/2 He calls the object of his affection a 'hot patootie'. **1948** LAIT & MORTIMER *New York: Confidential!* I. vi. 61 New Yorkers..tell their patooties how pretty they are. **1950** *Times-Herald* (Washington) 27 Jan. I. 14/3 A batch of pretty-panned patooties. **1958** P. DE VRIES *Mackerel Plaza* 149 You like to shake a leg with a hot patootie now and then, do you? **1977** *New Yorker* 26 Sept. 32/1 She was, successively,..the wife and/or sweet patootie of the quartet.

patrass, var. *PATTRESS.

patri- (pæ·tri, pēi·tri), used as the combining form of L. *pater* (*patr-is*) father, in words used in connection with the prominence of males and the importance of relationship on the male side in social organization. Some examples are given below as main words. Cf. also PATRIARCH *sb.*, PATRIARCHAL *a.*, etc., in Dict.

‖ **patria** (pæ·triă, pēi·triă). [ad. L. *patria*, fatherland.] Native country; homeland; also, heaven, as the region from which the soul is exiled while on earth and to which it longs to return.

a **1914** JOYCE *Stephen Hero* (1944) xix. 64 He refused therefore to set out for any task if he had first to prejudice his success by oaths to his patria. **1919** G. B. SHAW *What I really wrote about War* (1931) 352 As all the delegations have a different patria, and every patria has moral pretensions intolerable to and incompatible with the moral pretensions of all the other patrias, patriotism has to be dropped before any discussion is possible. **1936** H. G. WELLS *Anat. Frustration* iv. 46 The causes and devotions, the churches and organizations, the patrias and gangs, the family honour and the caste duty, to which the imagination of man..has clung. **1957** G. V. SMITHERS in *Medium Ævum* XXVI. 151 The use elsewhere of *ælþeodig(-nes)* to render *peregrinus*, *peregrinatio*..shows that *elþeodigra eard* here means the heavenly home (*patria*) of good Christians (*peregrini*). **1959** C. S. LEWIS *Let.* 5 Nov. (1966) 289 It is just when there seems to be most of Heaven already here that I come nearest to longing for a patria. **1965** J. C. POPE in *Franciplegius* 182 The word *elþeodig* here is used with reference to the idea that good Christians are exiles and aliens on earth, destined to travel as *peregrini* toward their *patria* in heaven. **1977** *N.Y. Rev. Bks.* 26 May 30/3 The attachment of the creoles to what they had come to regard as their *patria*—a land of eternal spring as eulogized by local poets in baroque extravaganzas—required spiritual patrons which they could genuinely call their own.

patrial, *a.* (*sb.*) Restrict *rare* to senses in Dict. and add: **3.** *spec.* Having the right of abode in the U.K. Also as *sb.*, one who has this right.

1971 *Sunday Times* 24 Jan. 11/1 The draft Bill..lays down that a patrial basically is: 1. A person born in the UK or one of whose parents or grandparents were; 2. A naturalised citizen; 3. A former citizen of the Commonwealth already resident in Britain. **1971** *Times* 25 Feb. 4/4 Anyone who is patrial will be exempt from deportation. *Ibid.* 15/1 A new distinction is being drawn between patrials, who will be free of immigration control, and non-patrials who will require work permits. **1971** *Sunday Times* 28 Feb. 12/1 Conferring full patrial status on grandchildren of people born here has some strange implications. **1973** C. MULLARD *Black Brit.* v. 62 [The 1971 Immigration Bill] created a new 'right of abode' for a certain category of Commonwealth immigrant, 'patrials'. **1973** LD. BOYLE in N. Fisher *Iain Macleod* 20 Secondly, Macleod's compassion, and his strongly felt concern for the fair treatment of individuals, extended beyond those fellow-citizens who have come to be defined as 'patrials'. **1976** *Equals* Dec. 7/5 Foreign nations and people from the old Commonwealth (excluding 'patrials' and those on working holidays) made up the remainder.

So **pa·trialism, pa·trialization.**

1971 *Sunday Times* 24 Jan. 11/2 At the end of the probationary period, the work permit holder will be able to apply for 'patrialisation'. *Ibid.*, 'Patrialism' is intended to clear it all up once and for all. **1971** *Times* 25 Feb. 4/4 Patrialism is a thin disguise for whites, albeit from the Commonwealth.

patriality (pēitriæ·lĭti). [f. prec. + -ITY.] Eligibility for or right to patrial status (*PATRIAL *a.* (*sb.*) 3).

1971 *Guardian Weekly* 6 Mar. 8 In some cases, where patriality depends on ancestral connection, a certificate issued through the British High Commissioner will be needed as proof of that right. **1971** *Guardian* 9 Mar. 12/4 Mr Maudling denied that the 'patriality' concept was racialist. **1971** *Times* 19 Mar. 3 The jurists were also concerned that people arriving without work permits or certificates of patriality would be sent back to their countries of origin. **1973** *Daily Tel.* 22 Sept. 18 A woman cannot transmit United Kingdom citizenship or patriality to a man by marriage. **1974** *Ibid.* 31 Oct. 7/3 Although a passport was not essential legally to leave or re-enter the country, it had strong evidentiary value of 'patriality'. **1978** R. MAUDLING *Mem.* xi. 159 The solution we found was the introduction of the new concept of patriality.

patria·rchalist. [f. PATRIARCHAL *a.* + -IST.] One who advocates or approves of a patriarchal system of society or government.

1923 *Contemp. Rev.* Oct. 450 The mutual contempt of the patriarchalist and the feminist is identical in its sources..with the mutual contempt of the 'tough' and the 'tender' races.

Patriarchist (pēi·triāɹkist). [f. PATRIARCH *sb.* + -IST.] A supporter of the Patriarch of Constantinople against the Exarch of Bulgaria during the schism of 1872–1945. Also *attrib.*

1903, etc. [see *EXARCHIST]. **1903** *Daily Chron.* 23 Sept. 3/5 They [*sc.* the Vlachs] are attached to the Greek or Patriarchist party. **1907** A. FORTESCUE *Orthod. Eastern Ch.* IV. x. 320 The Patriarchists..stand by the Patriarch of Constantinople. **1921** *Contemp. Rev.* May 587 Bulgarian Patriarchists—*i.e.* Bulgarians who affect the Greek religion..are numbered with the Greek inhabitants. **1972** D. DAKIN *Unification of Greece* ix. 129 Vatikiotis.. reported that Greek intervention would be welcomed by the Slav-speaking patriarchists. *Ibid.* xii. 160 It was precisely in those regions where the exarchist and patriarchist school populations were evenly matched that the struggle between the two Churches was fiercest.

patriate (pēi·triēit), *v. Canad.* [f. RE)PATRIATE *v.*] *trans.* To bring (legislation) under the constitutional authority of an autonomous country, used with reference to laws passed on behalf of that country by its former mother-country.

1966 *Deb. Commons Canada* 28 Jan. 373/2 *Mr. T. C. Douglas* (*Burnaby-Coquitlam*):..would the Prime Minister care to indicate to the house what action the government now proposes to take with a view to having a constitution in Canada amendable by Canadians? *Right Hon. L. B. Pearson* (*Prime Minister*):..we intend to do everything we can to have the constitution of Canada repatriated, or patriated. **1976** *Daily Gleaner* (Fredericton, New Brunswick) 12 Apr. 3 (*heading*) Trudeau wants serious bid to patriate constitution. *Ibid.*, He set out three possible ways of patriating the constitution, the British North America Act. **1978** *Globe & Mail* (Toronto) 2 Jan. 10/3 These things 10 years ago were..almost academic exercises and when Victoria failed nobody saw it as a.. tragedy—so Trudeau didn't succeed in patriating the constitution.

Hence **patria·tion,** the act or process of patriating; also *attrib.*

1976 *Globe & Mail* (Toronto) 20 Apr. 6/2 Haven't there been hundreds of spontaneous demonstrations across the country in support of unilateral patriation of the constitution? *Ibid.* 16 Aug. 5/6 The talks will be the most extensive on patriation of the BNA Act since the inconclusive Victoria conference in 1971. **1976** *Daily Colonist* (Victoria, B.C.) 3 Oct. 1/7 (*heading*) Patriation formula. **1978** *Globe & Mail* (Toronto) 13 Feb. 7/5 Mr. Ryan urged Premier Robert Bourassa to take a firm stand, and to refuse any patriation of Canada's constitution.

Patrician, *a.*[2] Delete *rare* and add earlier and later examples.

1872 A. T. DE VERE *Legends St. Patrick* p. x, In the legends of the Patrician Cycle the chief-loving old Bard is ever mournful. **1932** *Universe* 26 Feb. 3/2 The Patrician sites in Co. Down. **1933** *Clergy Rev.* May 382 A normal development of the Patrician system of organization. **1950** *Month* May 379 Five chapters..with an introduction on Patrician scholarship past and present. **1963** *Times Lit. Suppl.* 19 Apr. 264/2 The Patrician controversy.

patricentric (pætrise·ntrik), *a.* [f. *PATRI- + CENTRIC *a.*] Emphasizing, organized around, or dominated by the father or male line.

1949 R. K. MERTON *Social Theory* viii. 244 Freud himself was in his patricentric character, a typical representative of a society which demands obedience. **1957** V. W. TURNER *Schism & Continuity in Afr. Society* vii. 222 The patricentric family..is attached to a locality.

patri-clan (pæ·triklæn). [f. *PATRI- + CLAN *sb.*] A patrilineal kin group or clan; also, *occas.*, the clan of the father.

1937 W. E. LAWRENCE in G. P. Murdock *Stud. Sci. of Society* 319 'Patri-clan' and 'matri-clan', denoting small exogamous kin-groups with patrilineal and matrilineal descent respectively, are suggested to eliminate monotonous repetition of adjectives. **1957** *Contrib. Indian Sociol.* I. 52 Incidentally, 'patri-clan' is not taken here in the sense of 'patrilineal clan', but of the father's clan. **1959** G. D. MITCHELL *Sociol.* iv. 69 The Tallensi seem to make a distinction between offences committed by a man with a member of the same patri-clan, such as with a paternal aunt, daughter, or sister,..and offences committed by a man with the wife of a member of the same patri-clan, such as the wife of a father, brother, or son. **1975** G. A. COLLIER *Fields in Tzotzil* iii. 61 The second, not so widely accepted by scholars, is that Mayah social structure was and is characterized by patrilocal extended families, patrilineages, and, in some cases, fully developed patriclans. **1978** *Language* LIV. 214 It seems likely that vocabulary replacement and differentiation relate to the development of the lexical encoding or indexing of other sociolinguistic variables, e.g. patri-clan affiliation.

patrilateral (pætrilæ·tĕrăl), *a.* [f. *PATRI- + LATERAL *a.*] (See quot. 1964.)

1949 N. FORTES *Social Struct.* 70 Ashanti say..that a person can claim house-room in the house of a patrilateral kinsman. **1957** V. W. TURNER *Schism & Continuity in Afr. Society* vii. 222 Kafumbu's..followers were a group of uterine siblings and their children, related to him by ties of marriage and patrilateral cross-cousinship. **1964** GOULD & KOLB *Dict. Social Sci.* 486/2 *Patrilateral* is sometimes used as a synonym for *patrilineal*... It is more usual nowadays to use the term for relationships traced through the father in a matrilineal system..or for relatives on the father's side in a non-unilineal kinship system. **1969** M. FORTES *Kinship & Social Order* (1970) II. xi. 201 Patrilateral connections are, as a rule, recognized among the offspring of men whose own fathers were the sons of one man, that is, among the children of parallel cousins, but rarely beyond that stage. **1973** *Times Lit. Suppl.* 6 July 787/2 That ideal marriage partner, the father's brother's daughter... It looks well, in Kabyl society, for a man to marry his patrilateral parallel cousin.

patriline (pæ·trilain). [f. *PATRI- + LINE *sb.*[2] 24.] A patrilineal line of descent.

1957, etc. [see *MATRILINE]. **1972** P. LASLETT *Household & Family in Past Time* i. 18 A patriline, that is a succession of male heads of household directly descended from each other.

patrilineage (pætrili·niĕdʒ). [f. *PATRI- + *LINEAGE 2c.] Patrilineal lineage.

1949 J. F. HOLLEMAN *Pattern of Hera Kinship* i. 1 The relationships and their respective terminology have to be seen in the framework of patrilineages. **1955** [see *LINEARITY]. **1957** V. W. TURNER *Schism & Continuity in Afr. Society* vii. 222 A system of patrilineages..requires virilocal marriage. **1969** *Tanzania Notes & Rec.* July 10 All minor local government posts, with the exception of the necessarily localised village headmen, had been consistently awarded not only to men resident in the chief's settlement, but also only to men from his own immediate patrilineage.

patrilineal (pætrili·niăl), *a.* [f. *PATRI- + LINEAL *a.*] Of, pertaining to, or recognizing kinship with and descent through the father or the male line.

1904, 1906 [see *MATRILINEAL *a.*]. **1923** A. L. KROEBER *Anthropol.* xiii. 356 Within each area or type of culture the matrilineal tribes manifest superiority over the

patrilineal tribes in a preponderance of cultural aspects. **1936** R. FIRTH *We, the Tikopia* vi. 226 In this, such aggregations differ from the ordinary *paito*..which are patrilineal kinship groups of exclusive membership. **1951** [see *MATRILINEAL *a.*]. **1957** V. W. TURNER *Schism & Continuity in Afr. Society* viii. 236 This virtual equality between family and lineage as principles of local organization is at least partially responsible for the merging of patrilineal and matrilineal kin as joint members of a single genealogical generation. **1966** *Punch* 17 Aug. 262/2 There is much good sense in a matrilineal society. One suddenly finds oneself asking if it isn't perhaps the patrilineal one which is out of date? **1974** *Encycl. Brit. Micropædia* III. 484/3 Patrilineal (or agnatic) systems, in which the relationships through the father are emphasized.

Hence **patrili·neally** *adv.*

1934 in WEBSTER. **1954** J. LAYARD in E. E. Evans-Pritchard et al. *Inst. Primitive Society* v. 55 Even in England, which is predominantly patrilineal, we now have a Queen, and though she succeeds to the throne patrilineally, her son will succeed matrilineally, through her. **1955** HOMANS & SCHNEIDER *Marriage, Authority & Final Causes* 13 The only difference is that the men of B lineage, defined either patrilineally or matrilineally [etc.].

patrilinear (pætrili·niăɹ). [f. *PATRI- + LINEAR *a.*] = *PATRILINEAL *a.*

1913 [see *MATRILINEAR *a.*]. **1926** *Contemp. Rev.* Apr. 528 Among the Bakitara, a patrilinear people, Canon Roscoe shows that on a man's death the sister of the heir entered [etc.]. **1939** JOYCE *Finnegans Wake* 279 Pot price pon patrilinear plop. **1943** *Nature* 18 Sept. 317/2 The historic change from matrilinear to patrilinear inheritance has also led to important conflicts. **1950** [see *MATRILINEAR *a.*].

patriliny (pæ·trilini). [f. *PATRI- + LIN(E *sb.*[2] + -Y[3].] The observance of patrilineal descent and kinship.

1906 N. W. THOMAS *Kinship Organisations & Group Marriage Austral.* ii. 21 It is unnecessary to go into the complicated question of the relation of brother-inheritance to matriliny and patriliny. **1949** M. FORTES *Social Struct.* 76 Patriliny is not legally recognized. **1957** *Jrnl. R. Anthrop. Inst.* Jan. 40 If it were the only rule determining the rights of descendants we should be inclined to describe Norwegian kinship as based strongly on patriliny and primogeniture. **1965** [see *MATRILINY]. **1974** *Encycl. Brit. Micropædia* III. 484/3 Under a patriliny, a man (or woman) is linked to his (or her) sons and daughters, his sons' children, his sons' sons' children, and so on.

patrilocal (pætrilōᵘ·kăl). [f. *PATRI- + LOCAL *a.*] Applied to the custom in certain social groups for a married couple to settle in the husband's home or community.

1906 N. W. THOMAS *Kinship Organisations & Group Marriage Austral.* iii. 30 When the husband removes and lives in his wife's group the marriage is *matrilocal*; if the wife removes it is *patrilocal*. **1920** *Q. Rev.* July 168 How could it be otherwise wherever..patrilocal marriage occurs in conjunction with matrilineal descent? **1949** M. FORTES *Social Struct.* 70 Households with male heads can be..'patrilocal', that is, made up of a man and his dependants by marriage and paternity. **1965** L. STONE *Crisis of Aristocracy* XI. iii. 634 To use the terminology of the sociologists, temporary patrilocal residence was the norm. **1971** *Sci. Amer.* Dec. 93/3 Marriage is patrilocal among the peoples of the Upper Congo.

Hence **patriloca·lity,** the custom of patrilocal residence; **patrilo·cally** *adv.*, in a patrilocal manner.

1949 M. GLUCKMAN in M. Fortes *Social Struct.* 146 Members of these tribes, who..usually marry patrilocally, have been migrating into central Barotseland. **1951** E. E. EVANS-PRITCHARD *Social Anthropol.* ii. 34 He [*sc.* McLENNAN] suggested how patriliny might have developed out of matriliny through a combination of the customs of polyandry and patrilocality. **1952** M. N. SRINIVAS *Relig. & Society among Coorgs of S. India* v. 125 Men, thanks to patrilocality, are assured of continuous residence. **1969** *Language* XLV. 468 In this large household, inheritance by males followed their right to reside patrilocally.

patrimonial, *a.* Add: **1. b.** *patrimonial seas, waters*, etc., an area extending beyond territorial waters, the natural resources of which are the property of the coastal nation though ships and aircraft of other countries have freedom of passage through it.

1973 *Caribbean Contact* Jan. 12/3 One such [new] concept is that of 'patrimonial' waters which Foreign Minister Calvani [of Venezuela] defined thus: 'A coastal nation exercises rights of sovereignty over its natural resources, both renewable and non-renewable, that are found in the waters, the ocean floor and the subfloor of a zone adjacent to territorial waters'. **1973** *Nature* 14 Sept. 63/1 In essence next year's conference is intended to sort out the whole of the law of the sea, a brief that not only includes fishing and mineral and hydrocarbon rights in territorial waters, the so-called patrimonial waters, on the high seas and on the deep ocean floor, but also [etc.]. *Ibid.* 63/2 A number of developing nations took the stance..that patrimonial seas should be established that extend for 200 miles or to the outer edge of the continental shelf, whichever is the greater. **1973** *Internat. & Compar. Law Q.* XXII. IV. 668 The patrimonial sea can be briefly defined as an economic zone not more than 200 miles in breadth from the base line of the territorial sea (the limit of which shall not exceed 12 miles), where there will be freedom of navigation and overflight for the ships and aircraft of all nations, but in that zone the coastal state will have an exclusive right to all resources.

2. *Sociol.* A term used by Max Weber to designate a traditional type of social structure in which the chief or ruler maintains authority through his officials, army, etc., who are retained by him and whose loyalty is to him personally. Hence **patrimo·nialism,** a system of patrimonial authority.

1946 GERTH & MILLS tr. M. Weber in *From Max Weber* (1947) xi. 297 As a rule, this meant that princely prerogatives became *patrimonial* in nature. Patrimonialism can also develop from pure patriarchism through the disintegration of the patriarchical master's strict authority. **1947** HENDERSON & PARSONS tr. *Weber's Theory Social & Econ. Organization* iii. 318 The primary external support of patrimonial authority is a staff of slaves, coloni, or conscripted subjects, or, in order to enlist its members' self-interest.., of mercenary bodyguards and armies. **1968** E. FISCHOFF et al. tr. *Weber's Econ. & Society* I. iii. 232 Where domination is primarily traditional, even though it is exercised by virtue of the ruler's personal autonomy, it will be called *patrimonial authority*. *Ibid.*, Patrimonial authority under which the administrative staff appropriates particular powers and the corresponding economic assets. *Ibid.* III. xvi. 1366 The patrimonial structure of the Roman ruling stratum. **1968** M. ILFORD tr. *Freund's Sociol. of Max Weber* iv. 236 The old bureaucracies were essentially patrimonial in character. **1968** *World Politics* XX. 195 Patrimonial rulers..endeavour to maximize their personal control. *Ibid.*, Lately, some attempts, primarily in the field of African studies, have been made to remember the meaning of patrimonialism. **1970** H. BIENEN in Huntington & Moore *Authoritarian Politics in Mod. Society* 119 His [*sc.* Zolberg's] party-state emerges as a system where bureaucratic and patrimonial features coexist. **1974** tr. *Wertheim's Evolution & Revolution* i. 27 Weber's 'patrimonial bureaucracy', as a sub-type of a feudal political structure, comes much nearer to historical reality.

patriot, *sb.* Add: **2. d.** In the war of 1939–45, *spec.* a loyal inhabitant of a country overrun by the enemy, esp. a member of a resistance movement. Also *attrib.*

1945 *News Chron.* 7 May 1/5 The formal liberation of Denmark had begun. Actually the patriots had started it much earlier... When we landed the Danish patriots.. had the situation under complete control. In patriot cars,..we drove down streets lined with cheering Danes. **1959** *Listener* 23 Apr. 727/2 Wingate's leadership of the ill-found 'Patriot' forces [in Ethiopia] was audacious.

3. *U.S.* **Patriots' Day,** anniversary of the Battle of Lexington and Concord in the American Revolution, 19 April 1775, observed since 1894 as a legal holiday in Maine and Massachusetts.

1897 *Boston Even. Transcript* 18 Apr. 8/4 Lowell mill agents, having heard the indignant protest against the running of machinery in the mills Patriots' Day, have decided to reconsider their action. **1909** *Springfield* (Mass.) *Weekly Republ.* 22 Apr. 11 The celebration of Patriots' day, the 134th anniversary of the battles at Lexington and Concord. The day is a legal holiday in Massachusetts and Maine. **1925** *Boston Even. Transcript* 21 Apr. 10/1 Sesquicentennial of Patriots' Day passes into history. **1948** *Daily Ardmoreite* (Ardmore, Okla.) 18 Apr. 14/7 They are down to play a second game in the afternoon, since it's Patriot's day in Boston.

patrioteer (pēitriǫtīᵊ·ɹ). [f. PATRIOT *sb.* + -EER.] One who makes a public display of patriotism; one whose patriotism is spurious and insincere. Also *attrib.*

1928 *Amer. Speech* III. 262 The second camp is made up of nationalists, or, if you will, of patrioteers. **1935** *Evening Sun* (Baltimore) 11 Apr. 27/2 Some British patrioteer has grown terribly excited over the possibility that Soviet arguments broadcast from Moscow may be so strong..as to sweep the masses of Englishmen from their political moorings. **1939** *Time* 27 Feb. 9/1 By patrioteer *Time* means to describe the professional patriot, the kind of refuge-seeking scoundrel who waves a red-white-&-blue handkerchief when he should be wiping his own nose. **1941** 'FANFARLO' in *Penguin New Writing* X. 112 A patrioteer is very simply a man who will cut his country's throat to an old tune. **1954** *Manch. Guardian Weekly* 18 Feb. 9 Talking in 'patrioteer' terms. **1954** *Birmingham* (Alabama) *News* 14 Apr. 10 They are quick to detect the phony and they can distinguish a patriot from a patrioteer. **1954** D. RIESMAN *Individualism Reconsidered* III. vii. 137 Mr. MacLeish has suffered greatly from patrioteers. **1956** C. W. MILLS *Power Elite* xii. 271 The American elite have not remained as patrioteer essayists have described them to us.

patripotestal (pætripǭte·stăl), *a.* [f. *PATRI- + POTESTAL *a.*] Characterized by the exercise of authority by the father or his relatives in a family or household.

1906 N. W. THOMAS *Kinship Organisations & Group Marriage Austral.* i. 8 Three main types of family may be distinguished: (1) patripotestal, (2) matripotestal, (*a*) direct, and (*b*) indirect, in which the authority is wielded by the father, mother, and mother's relatives, in particular her brothers, respectively. **1944** H. P. FAIRCHILD *Dict. Sociol.* 215/2 *Patripotestal*, characterized by the exercise of authority, especially in the family or household, by the father or paternal grandfather. **1952** A. R. RADCLIFFE-BROWN *Struct. & Function Primitive Society* i. 22 A society may be called patriarchal when descent is patrilineal.., marriage is patrilocal.., and the family is patripotestal (i.e. the authority is in the hands of the father or his relatives). **1964** GOULD & KOLB *Dict. Social Sci.* 486/2 *Patripotestal* modes of social organization. **1972** D. DAVIES *Dict. Anthropol.* 144/1 *Patripotestal*, the holding of authority over a family by the father or his relative

patrist (pæ·trist). *Psychol.* [f. *PATR(I- + -IST.] A term applied to someone whose behaviour or attitude is influenced or dominated by the father. Also *attrib.*

1953 G. R. TAYLOR *Sex in Hist.* iv. 77 Though I am no great lover of jargon, it would be tedious to refer continually to persons who have modelled themselves on their fathers. I shall therefore speak of them as patrists. **1958** —— *Angel-Makers* xiii. 272 The patrist accepts learning in the sense of mere erudition..but deeply distrusts all original inquiry. **1958** *Sunday Times* 27 Apr. 6/3 The 'Patrists', in the shape of the commercial middle class. **1977** E. J. TRIMMER et al. *Visual Dict. Sex* (1978) xxii. 248 Male-dominated or patrist societies.. tend to a certain stern prudishness.

patroclinous (pætroklai·nəs), *a. Biol.* [f. Gr. πατήρ, πατρ- father + κλίν-ειν to lean + -OUS.] Resembling the male rather than the female parent in inherited characters; involving or possessing a tendency to inherit a character or characters from the male parent only. So **pa·trocliny,** patroclinous inheritance.

1913 *Jrnl. Exper. Zool.* XV. 590 For comparison I give the F_1 from normal males by sisters such as those to which I mated the patroclinous males. **1917** *Genetics* II. 247 The hybrids in these reciprocal crosses resemble the pollen parent strongly, i.e., they are strongly patroclinous. *Ibid.*, So far as the vegetative characters are concerned patrocliny can not be ascribed to either of the hybrids. **1936** *Ibid.* XXI. 592 It has also been shown that X chromosome inversion heterozygotes give rise to patroclinous males among their progeny. **1973** *Nature* 26 Oct. 439/2 Stimulated by Richard Goldschmidt's erroneous explanation of patroclinous inheritance in *Oenothera.*

patrol, *sb.* Add: **1. c.** A reconnaissance flight by military aircraft.

1917 *Flying* 19 Dec. 347/3 A low patrol over the Fleet was carried out by three Flight-Lieutenants in Sopwith machines, during which they encountered and attacked a number of hostile craft. **1957** *Economist* 7 Dec. 836/2 To guard against surprise attack, bombers flying on patrol from Britain carry hydrogen bombs.

3. b. A unit of scouts or guides consisting of from six to eight members.

1908 R. S. S. BADEN-POWELL *Scouting for Boys* 22 A troop consists of not less than three patrols... A patrol consists of six scouts. **1908** *Scout* 18 Apr. 20/2 Several patrols together can form a 'Troop' under an officer called a Scout-master. **1917** R. E. PHILIPPS *Patrol System* ii. 13 Here is the Patrol, consisting of six, seven, or eight boys. **1946** C. CHRISTIAN *Seventh Magpie* xix. 216 Her bulging haversack bumping against her, to the extreme peril of the patrol milk supply she was carrying. **1974** *Policy, Organisation & Rules of Scout Assoc.* (ed. 3) 124 The Troop is composed of Patrols, each consisting of six to eight Scouts, including the Patrol Leader and Assistant Patrol Leader. **1977** *Guider* July 315/2 Estimated publication date is 1st September 1977, so you need to tell your Patrols about it *now* and encourage them to have a money-raising effort during the school holidays.

4. *patrol-craft, system, watch*; **patrol car,** a motor car employed by the police on patrol; **patrol leader,** (*a*) the boy scout in charge of a patrol (sense *3 b); (*b*) the leader of a military patrol; **patrol officer,** a representative of the Australian government in Papua New Guinea; **patrol-wagon:** for *U.S.* read *N. Amer.*; (*a*) (earlier and later examples).

1931 *Chicago Police Problems* v. 88 Each district normally has two small patrol cars. **1951** A. MARTIENSSEN *Crime & Police* iv. 49 In the Aberdeen system, the patrol cars and the beat constables have been formed into teams. **1967** N. LUCAS *C.I.D.* vi. 70 The Information Room put out an all car call for any disengaged patrols to join in the pursuit of the stolen Rover. Five patrol cars responded to the call. **1977** 'E. MCBAIN' *Long Time no See* i. 9 A radio motor patrol car was angle-parked into the curb. **1930** *Times Lit. Suppl.* 9 May 379/2 Officers who served in the French mine-sweepers and patrol-craft during the War. **1908** *Scout* 18 Apr. 20/2 One boy is then chosen as Patrol Leader to command the patrol. **1918** E. S. FARROW *Dict. Mil. Terms* 438 Patrol leaders. **1929** E. K. WADE *Twenty-One Years of Scouting* vi. 193 Patrol leaders have taken closer command of their Patrols. **1973** *Guardian* 11 Apr. 11/4 One of my best friends used to be a patrol leader. **1977** M. JANCATH *Seatag* II. i. 63 He rifled through the files... 'Dishonourable discharge..Patrol leader Aden.' **1924** 'R. DALY' *Outpost* i. 11 'In my district,' put in Jessel amiably, 'there's a regulation that no patrol-officer shall be a married man.' **1935** *Discovery* Nov. 346/1 Local conditions fully justify the title of knights errant to the patrol officers and other members of the administration of Papua. **1964** *Mod. Encycl. Austral. & N.Z.* 790/1 Districts are divided into sub-districts in charge of an Assistant District Officer (A.D.O.), who has assistance from Patrol Officers (P.O.). **1880** A. W. TOURGÉE *Fool's Errand* II. xii. 507 The old 'patrol' system of the antebellum days..was also one of the active causes of the rapid spread of the Klan. **1966** *Listener* 20 Oct. 571/1 Baden-Powell..dreamed up the patrol system: autonomous groups of boys, whose leaders plan the troop programme together with the scoutmaster. **1887** *Courier-Jrnl.* (Louisville, Kentucky) 22 Jan. 3/5 The patrol wagon, filled with officers, was driven to the place at a breakneck speed. **1921** *Daily Colonist* (Victoria, B.C.) 5 Apr. 7/1 Dr. Tomalin was called and attended to the unfortunate man and Sergeant Blackstock and Constable Walton attended with the patrol wagon and removed him to the Jubilee Hospital. **1974** *Amer. Speech 1971* XLVI. 78 Large police van: paddy wagon, van, Black Maria, patrol wagon, patrol. **1810** *Boston* (Registry Dept.) *Records* (1904) 33rd Rep. 426 Return of the patrole watch read. **1821** *Ibid.* (1909) 39th Rep. 227 Granting permission for private patrole watches.

patrol, *v.* Add: **1.** Also, to act as patrol in an aircraft.

1940 [see *dive-bomb* vb. s.v. *DIVE *v.*].

patrolette (pătrōᵘle·t). [f. PATROL(MAN + -ETTE.] A woman or girl on patrolling duty.

1960 *Oxford Times* 1 Jan. 1/7 The Oxford office of the Royal Automobile Club invites applications from Young Ladies for the duties of patrolette. **1973** *Daily Tel.* 3 Jan. 10/3 Patrolettes and radio rescue mechanics also have new gear.

patroller. Delete *rare* and add earlier and later examples.

1744 *Bristol* (Va.) *Vestry Bk.* (1898) 118 To Burwell Green for his Levy, Being a patroler, 50. **1901** B. T. WASHINGTON *Up from Slavery* 77 The 'patrollers' were bands of white men..organized largely for the purposes of regulating the conduct of the slaves at night. **1938** W. FAULKNER *Unvanquished* 18 Into this room they would be fetched to face the Patroller. **1968** [see *KETCH *v.*[2]]. **1972** R. ADAMS *Watership Down* xlix. 401 Without Campion, probably not one rabbit would have got back to Efrafa. As it was, all his skill as a patroller could not bring home half of those who had come to Watership.

patrolman. For 'Chiefly *U.S.*' read 'orig. *U.S.*' and add: **a.** (Further examples.)

1902 *Chambers's Jrnl.* Oct. 673/2 Nor is this all. He is well off, even if he never rises beyond the grade of patrolman. **1955** W. GADDIS *Recognitions* II. vii. 644 The patrolman turned his attention to his charge. **1970** P. LAURIE *Scotland Yard* iii. 70 Very occasionally a patrolman meets crime in progress. **1974** *Anderson* (S. Carolina) *Independent* 20 Apr. 1B/4 Participating in the investigation and arrests were..Patrolmen Jerry Gambrell and Jimmy McKinney of the Anderson police. **1977** 'J. FRASER' *Hearts Ease* ii. 12 The two police patrolmen were..directing traffic.

c. In general senses.

1867 J. M. CRAWFORD *Mosby* 330 [They] captured five patrolmen, from whom..they succeeded in obtaining the countersign. **1878** *Harper's Mag.* Feb. 331/2 Each patrolman will carry a beach lantern. **1900** J. LONDON *Let.* 31 Jan. (1966) 86, I was..a fish patrolman, a longshoreman, and a general sort of bay-faring adventurer. **1945** *Seafarers' Log* 22 June 6/1 The SS Prospector of the Alcoa SS Company, paid off here in an Army Base, and two Patrolmen managed to get aboard her. **1959** *Times* 14 Aug. 13/2 A beach patrolman said, [etc.]. **1965** M. SPARK *Mandelbaum Gate* iii. 69 The hill road to the Potter's Field bordered on disputed territory, and wanderers in the area were likely to be shot at by the patrolmen of either country. **1966** R. & D. MORRIS *Men & Pandas* v. 90 An exhausted zoo patrolman. **1974** *Country Life* 5 Dec. 1772/3 All AA patrolmen carry a copy.

patron, *sb.* Add: **III. 6.** (Further examples.) Also, a similar person in N. American waters.

1777 P. THICKNESSE *Year's Journey* I. viii. 59 The *Patron* of the barge..affected to shew how much skill was necessary to guide it through the main arch. **1817** J. BRADBURY *Trav. Interior Amer.* 192 Her crew consisted of five French Creoles, four of whom were oarsmen, and the fifth steered the boat, he is called the *patron*. **1849** T. T. JOHNSON *Sights in Gold Region* 12 The Creoles..were generally the patrones or captains, and the owners of the boats. **1906** [see *BOSMAN]. **1968** R. F. ADAMS *Western Words* (rev. ed.) 222/1 *Patron*, a trader's name for the head of a barge engaged in the Missouri River fur trade. In river boating, a rudder man on a mackinaw.

8. (Later examples.) Also with reference to countries other than Spain.

1973 *Times* 25 Aug. 12/4 To them a good restaurant without the patron's presence is a paradox. **1978** T. ALLBEURY *Lantern Network* iii. 46 They..sat..in the warmth of a small restaurant... The *patron* moved among his customers.

8*. *U.S.* and *Canad.* (with capital initial). A member of a political association, in full *Patrons of Husbandry* (or *Industry*), founded in the U.S.A. in 1867, or of a similar association founded in Canada in 1891, for the promotion of farming interests. Now chiefly *Hist.*

1873 *N.Y. Times* 3 July 1/2 The organization known as the Patrons of Husbandry originated in Washington in 1867, and the National Grange was organized in December of that year in this city. **1880** [see GRANGE *sb.* 5]. **1894** *Weekly Globe* (Toronto) 23 May 1 Mr John A. Leitch, the Conservative candidate for West Middlesex.. said:—'The Patron Order was originated in the Western States, and was imported into Canada by dissatisfied politicians.' **1903** J. S. WILLISON *Sir Wilfrid Laurier & Liberal Party* II. xxvi. 281 The position of the Liberals was also measurably affected by their practical alliance for the campaign of 1896 with Mr. D'Alton McCarthy and the Patrons of Industry... The Patrons were an off-shoot from the farmers' organizations of the United States, and their demands embraced simplification of the laws and machinery of government, limitation of public subsidies, protection against industrial combinations, and a tariff for revenue. **1914** W. S. WALLACE in Shortt & Doughty *Canada & its Provinces* XVII. 173 Patrons, as they were called *tout court*, were representatives of the farming class. **1932** A. BRADY *Canada* iii. 104 The new organization, known as the Patrons of Industry, succeeded in 1894 in electing sixteen of its nominees to the legislature of Ontario. **1962** *New Democrat* Oct. 3/2 In 1894 the Patrons of Industry elected 14 members to the Ontario legislature. **1963** A. S. MORTON *Kingdom of Canada* 382 The Patrons did everything the Grangers did, but they added a special emphasis on co-operation. **1977** *Canad. Hist. Rev.* LVIII. 401 McCarthy gravitated further from the Conservatives, emerging as an Independent in 1893 and an ally in all but name of the Liberals and Patrons in 1896.

patronage, *sb.* Add: **3. f.** *Rom. Antiq.* The rights and duties or position of a patron (sense 2 b).

1697 [see CLIENTSHIP]. **1885** *Encycl. Brit.* XVIII. 413/1 The patronage and the clientage were alike hereditary.

6. (sense 4) *patronage curse, -monger, polity, system*; **Patronage Secretary** (earlier examples); so *Patronage Secretaryship.*

1907 *Daily Chron.* 18 July 3/6 The patronage curse.. has received the benediction of a Liberal Government. **1968** *Economist* 28 Dec. 21/2 It seemed unquestionably right to establish the teaching profession as a separate civil service beyond the reach of politicians and patronage-mongers. **1971** P. A. ALLUM *Politics & Society Post-War Naples* (1973) iv. 98 The patronage polity has been absorbed within the parliamentary system despite the contradictions between them. **1852** DISRAELI *Lord George Bentinck* xvii. 314 Sir Robert appointed the man of the world financial secretary of the treasury..and entrusted to the student, under the usual title of patronage secretary of the treasury, the management of the house of commons. **1873** TROLLOPE *Phineas Redux* (1874) I. xvi. 127 Roby.. was at this moment Mr. Daubeny's head whip and patronage secretary. **1909** *Westm. Gaz.* 16 Sept. 9/2 When he laid down the Patronage Secretaryship he assumed the offices of Lord Privy Seal and Chancellor of the Duchy of Lancaster. **1802** *Deb. Congress U.S.* 18 Feb. (1851) 580 A variety of circumstances..gave the patronage system the preponderancy, during the first three Presidential terms of election. **1976** *National Observer* (U.S.) 24 Apr. 16/4 The patronage system in the nation's fourth-largest city remains intact, and it is expected that the power it wields will be utilized with considerable impact.

‖ **patronat** (patronā). [Fr.] An organization of industrial employers in France; French employers collectively.

1958 *Economist* 15 Nov. 616/1 There are certainly two distinct attitudes in France. One group, which includes the *patronat* or employers' federation, seems opposed to any sort of free trade area on terms possible for Britain. **1963** *Times* 13 Feb. 9/1 Against this it is argued that in 1958 the Common Market seemed to the French *patronat* a dangerous experiment, whereas now it is an established international entity. **1971** *Guardian* 15 July 16/4 We still have to decide whether the enlarged EEC will..take the place of the UK in EFTA..or whether it will insist on rules leading to tighter harmonisation... The Patronat has declared itself firmly in favour of the second solution. **1972** LD. GLADWYN *Mem.* xvii. 306, I dwelt at length on the very considerable opposition in France to the entry of the UK into the Common Market... For their part, the Patronat, having accepted the Common Market, still maintained an uncompromising opposition..to any reduction of the Common External Tariff.

patronite (pæ̆trōᵘnait). *Min.* [f. the name *Patron* (see quot. 1906) + -ITE[1].] A black, lustrous, fine-grained mixture of vanadium sulphides that occurs in Peru and is exploited as a source of vanadium.

1906 F. HEWETT in *Engin. & Mining Jrnl.* LXXXII. 385 There was discovered, on Nov. 20, 1905, in the neighborhood of Cerro de Pasco, Peru, a new material containing vanadium... The material..was taken to Señor Antenor Riza Patron, metallurgist of the Huaraucacar smelter, nine miles from Cerro de Pasco... It is suggested, in appreciation of the work of the discoverer of the material, Señor Antenor Riza Patron, that it be given the name 'patronite'. **1922** *Amer. Jrnl. Sci.* CCIII. 200 The real patronite is a black mineral of fine grain and metallic luster. In polished section it is shown to consist of a very fine-grained mixture of three minerals. **1946** J. R. PARTINGTON *Gen. & Inorg. Chem.* xxi. 644 Vanadium is widely distributed, the principal ores being carnotite.., vanadinite.., and especially the impure sulphide patronite found at 17,000 ft. in the Peruvian Andes and in North Rhodesia. **1968** I. KOSTOV *Mineral.* 146 Patronite is found as graphite-like masses associated with bravoite, native sulphur, bituminous substances, quartz and other minerals in Minas Ragra in Peru.

‖ **patronne** (patrǫn). [Fr.] A woman who is the owner or the wife of the owner of a business, esp. a café, hotel, or restaurant (in quot. 1777, a barge).

1777 P. THICKNESSE *Year's Journey* I. vii. 51 Your female *Patronne*..for they are all conducted by females. **1898** W. J. LOCKE *Idols* xxiii. 324 The little inn came in sight... The buxom *patronne*..was grinding coffee. **1921** *Spectator* 9 Apr. 465/1 The *patronne* came in, and gave me a liqueur glass of rum. **1942** E. PAUL *Narrow St.* v. 40 He had found that Madame Sara, as he called his *patronne*, was one of the gentlest and most patient women alive. **1955** *Times* 8 Aug. 8/7 Madame Ribard, the patronne, had surpassed herself; an omelette aux champignons that melted in the mouth, [etc.]. **1973** P. O'DONNELL *Silver Mistress* i. 15 'What about your reputation with the patron?' 'Patronne. What will really shock Mme. Martine is giving us separate rooms.'

patronomate (pætrǫ·nǒmēit). [f. Gr. πατρονόμ(ος (f. πατήρ father + νέμειν to rule), the title of certain magistrates at Sparta + -ATE[1].] The office of a *patronomos* in Hellenistic Sparta.

1910 *Year's Work Class. Stud.* 68 The election of 'Divine Lycurgus' to the eponymous patronomate at Sparta for a series of years.

patsy (pæ·tsi), *sb. slang* (orig. *U.S.*). [Origin unknown.] A person who is ridiculed, deceived, blamed, or victimized.

1903 'H. MCHUGH' *Back to Woods* 68 I'm the Patsy, oh, maybe! **1920** ADE *Hand-Made Fables* 76 Sometimes they ask him to come back and be the Village Patsy once more. **1927** [see *BUILD-UP a]. **1953** WODEHOUSE *Performing Flea* 205 That gentle pity which the kind-hearted always feel when they regard the fellow whom Fate has called upon to be the Patsy, the Squidge or, putting it another way, the man who has been left holding the baby. **1954** J. STEINBECK *Sweet Thursday* vii. 45 She's making a patsy of you. **1960** *Analog Science Fact/Fiction* Oct. 151/1 We had to have a patsy—some one to put the blame on. **1967** *Punch* 8 Feb. 211/3 Blamey blunders about, the perfect patsy, while we watch the real sex-maniac at work. **1970** J. H. GRAY *Boy from Winnipeg* 57, I had grown somewhat more able to take care of myself and hence had ceased to be the school patsy. **1971** *Times Lit. Suppl.* 26 Nov. 1467/3 The police are what they would call 'the patsies', the focus for popular discontent. **1974** *Daily Tel.* 17 Apr. 1/8 [He] said yesterday he was not going to be turned into a scapegoat. 'Whatever happens I am not going to be the patsy in this business.' **1977** *Rolling Stone* 13 Jan. 32/2 He felt Silkwood had possibly been a pawn or a patsy. **1977** *Time* 9 May 24/1 Or would the politically inexperienced Frost prove a patsy and let Nixon filibuster with those same skillful diversions that always seemed to be the answers but never were?

patsy (pæ·tsi), *a. U.S. slang.* [Origin unknown.] Satisfactory, all right.

1930 *Amer. Mercury* Dec. 457/1 *Patsy*, all right. 'The mutt offices patsy and we walk into a collar.' **1935** A. J. POLLOCK *Underworld Speaks* 86/1 *Patsy*, satisfactory; O.K.; when high pressure salesmen guarantee stock purchaser not to lose and get back money invested with profit in (90) days. **1941** J. SMILEY *Hash House Lingo* 42 *Patsy*, all right. **1950** H. E. GOLDIN *Dict. Amer. Underworld Lingo* 153/1 *Patsy*, a. (Rare) All right; okay; trustworthy.

pattawalla (pa·tāwǫlā). Also **pattawala, puttiwala,** etc. [ad. Hind. *paṭṭā-wālā* one wearing a belt: see PUTTEE, WALLAH.] In India, a messenger or servant.

1878 MONIER WILLIAMS *Mod. India* 34 Here and there a belted Government servant (called a Patti-wālā, or Patta-wālā, because distinguished by a belt)—all within call—all ready to answer..to the Sahib's summons, and eager to execute his behests. **1881** E. M. GUTHRIE *Life in Western India* I. vi. 12 Behind M— stood the tall and handsome Jew, G—'s writer, and the putthawaller, with his badge of office. **1949** R. LAWRENCE *Indian Embers* 23 Pattawallas in stiff white, with scarlet belts and turbans, were salaaming deeply. **1971** A. D. GORWALA *Queen of Beauty* 79 Give me a cup of coffee too and tell your pattawalla not to come in and not to let anyone else in.

patte. Add: **3.** *patte de velours* (də vəlūr), the velvet paw (of a cat; i.e. a paw with the claws held in): used *fig.* indicative of resolution or inflexibility combined with apparent softness or gentleness. Cf. *IRON *a.* 3c.

1853 C. BRONTË *Villette* III. xxxiii. 84 She played before me the amiable; offered me patte de velours; caressed, flattered, fawned on me. *a* **1855** —— *Professor* (1857) I. xi. 185 The soft touch of a patte de velours. **1859** LYTTON *What will he do with It?* III. VII. xviii. 150, I always felt that she had the claws of a tigress under her *patte de velours*! **1881** *Atlantic Monthly* Jan. 137/2 An innocent-looking creature, with feline manners, *pattes de velours*, and such claws! **1904** P. PENNINGTON *Diary* 1 Jan. in *Woman Rice Planter* (1913) ii. 59 One is so apt to forget that the 'patte de velours' which every one uses in polite society is even more of a help in dealing with the most ignorant.

patten, *sb.* **6. patten-shoe:** delete † and add later examples.

1957 R. LISTER *Decorative Wrought Ironwork* 231 *Patten shoe*, in farriery, a shoe used for a hip-shot horse. Its underside is forged into a hollow hemisphere. **1963** *Times* 25 Feb. 1/7 Sometimes it was necessary to rest a leg that was strained, so a Patten shoe was obtained which had a raised heel to relax the back tendon of the leg while the horse was resting after muscular injury.

patter, *sb.*[1] Add: **2.** Now *spec.*, the speech of a comedian or a stage magician.

1949 *Amer. Speech* XXIV. 40 Anything he says while performing is *patter*, and he almost never says silk handkerchief, but simply calls it a *silk*. **1952** GRANVILLE *Dict. Theatr. Terms* 134 *Patter*, quick speeches uttered between their songs by music-hall comedians. **1965** G. MELLY *Owning-Up* vi. 59, I can still remember some of the abysmal patter which he delivered. **1976** *Liverpool Echo* 6 Dec. 10/5 Songs and patter formed the mainstay of his senior citizens' act which has already won awards and will doubtless claim more.

3. *patter-act*; **patter-song** (earlier and later examples).

1972 *Times* 24 June 11/3 A patter act is usually written so that the lines sometimes don't finish. **1823** C. MATHEWS *Let.* 23 Feb. in A. Mathews *Mem. Charles Mathews* (1839) III. xvii. 385 The only striking subject for a patter-song is the inordinate love of title. **1839** J. ADOLPHUS *Mem. John Bannister* I. xii. 234 It formed the precedent for what are now, in the technical slang, called 'patter songs'. **1965** *Listener* 23 Dec. 1050/2 The Barber himself a buffo bass,.. has a famous Gilbertian patter song, with compound rhyming. **1975** *Ibid.* 13 Mar. 328/3 His characters, his funny voices, his patter song.

patter, *sb.*[2] Add: freq. in phr. *the patter of little* (or *tiny*) *feet*: used to suggest the presence of young children or the expectation of the birth of a child. See PATTERING *vbl. sb.*[2] and *ppl. a.*[2] in Dict. and Suppl.

1863 LONGFELLOW *Tales of Wayside Inn* 209, I hear in the chamber above me The patter of little feet, The

sound of a door that is opened, And voices soft and sweet. **1883** [in Dict.]. **1924** N. COWARD *Rat Trap* III. 68 And we're to expect little clinging fingers and the patter of tiny feet. **1945** E. BOWEN *Demon Lover* 66 They knew there was going to be the patter of little feet. I wasn't actually *born*..till 1918. **1966** *Guardian* 29 Dec. 14/1 At any time now, the patter of little feet is expected to indicate that Helga's dream has at last come true... Helga is a polar bear. **1972** *Daily Tel.* 8 May 12 Amen Court.. resounds with the patter of tiny feet... The Rev. Patrick Tuft, the Succentor, has become the father of triplet boys. **1977** *Times* 29 Oct. 20/1 Expectant motherhood these days is marked less by the patter of tiny feet than the tinkling of cash registers.

patter, *v.*[1] Add: **4.** Also *transf.*

1905 B. TARKINGTON *In Arena* 259 Between the acts the orchestra pattered ragtime and inanities from the new comic operas.

pattering, *vbl. sb.*[2] and *ppl. a.*[2] Add: Freq. with allusion to *PATTER *sb.*[2]

1849 LONGFELLOW *Kavanagh* xi. 40 With these daydreams mingled confusedly the pattering of little feet. **1884** [in Dict.]. **1903** BEERBOHM *Around Theatres* (1953) 257 Napoleon may..drill a squad of small children. But ..his motive..was not a delight in pattering feet and chubby cheeks. **1955** M. EWER *No Abiding Place* vii. 112 No pattering feet on the way? **1962** A. LEJEUNE *Duel in Shadows* vi. 81 All was sweetness and light,..and the pattering of tiny feet round the Christmas tree.

pattern, *sb.* Add: **2. b.** A model or design in dressmaking, *spec.* a paper pattern from which material for a garment can be cut out and sewn together.

1792 JANE AUSTEN *Catharine* in *Wks.* (1954) VI. 207, I expect a new Cap from Town... Every Body will be longing for the pattern. **1811** —— *Sense & Sens.* I. xxi. 281 Taking patterns of some elegant new dress. **1890–1** *T. Eaton & Co. Catal.* Fall & Winter 52/1 Ours is the only store in Toronto where you can get Butterick's dress patterns. *Ibid.* 52/2 By means of a system invented..by the Buttericks, each pattern is graded to suit every size in which it is furnished. **1911** O. ONIONS *Widdershins* 183 A mass of tissue-paper patterns and buckram linings. **1964** *McCall's Sewing* ii. 15 Without patterns, home-sewing would probably be a lost art... Not every pattern style will look equally well on everyone. **1974** D. KYLE *Raft of Swords* vi. 60 The women sew mini-dresses from McCall patterns.

8. c. *fig.* An arrangement or order of things or activity in abstract senses; order or form discernible in things, actions, ideas, situations, etc. Freq. with *of*, as *pattern of behaviour* = *behaviour pattern* (see *BEHAVIOUR 6), and as second element with defining word.

1901 G. B. SHAW *Admirable Bashville* II. i. 309 Fates That weave my thread of life in ruder patterns Than these. **1906** C. S. SHERRINGTON *Integrative Action Nervous Syst.* v. 176 (*caption*) The cutaneous fields of the 'scratch-reflex', the 'flexion-reflex', the 'extensor-thrust', are areas which in nowise fit in with the pattern of the cutaneous fields of the afferent spinal roots. **1915** V. WOOLF *Voyage Out* xxii. 366 According to him, too, there was an order, a pattern which made life reasonable, or, if that word was foolish, made it of deep interest anyhow, for sometimes it seemed possible to understand why things happened as they did. *Ibid.* xxiv. 385 Perhaps, then, everyone really knew as she knew now where they were going; and things formed themselves into a pattern not only for her, but for them, and in that pattern lay satisfaction and meaning. **1922** JOYCE *Ulysses* 562 Arabesquing wearily, they weave a pattern on the floor. **1927** E. O'NEILL *Marco Millions* III. 152 The young boys and girls take up their censers and dance their pattern out backward, preceded by the musicians. **1933** H. G. WELLS *Shape of Things to Come* III. §6. 301 Old habits of thought, old values, old patterns of conduct. **1933** T. S. ELIOT *Use of Poetry* 88 There is something integral about such greatness, and something significant in his place in the pattern of history. **1936** *Nature* 18 Jan. 87/2 In these sections..there are interesting analyses..of the technological and economic patterns observable in material culture. **1936** A. HUXLEY *Olive Tree* 290 Our habits are not those of the Romans, Greeks and Hebrews... Patterns of behaviour change. **1937** —— *Ends & Means* iii. 22 Every culture is full of arbitrary and fortuitous associations of behaviour-patterns, thought-patterns, feeling-patterns. *Ibid.* 23 Thought-patterns, feeling-patterns and action-patterns..have seemed in their time inevitable and natural. **1945** T. S. ELIOT *What is a Classic?* 32 Each literature has its greatness, not in isolation, but because of its place in a larger pattern, a pattern set in Rome. **1951** J. M. FRASER *Psychol.* III. xx. 236 Different patterns of relationships will develop according to what kind of task the group happens to be engaged in. **1956** A. C. GUYTON *Textbk. Med. Physiol.* xlvii. 591 Sensory impulses from the eyes, the ears, the proprioceptors, etc.,..assess whether or not the nail is being hammered and, if not, change the pattern of movement so that it will be hammered. **1956** *B.B.C. Handbk.* 1957 115 Reference has already been made..to the pattern of viewing by those having sets equipped for both BBC and ITA programmes. **1958** *Spectator* 30 May 692/1 A study..of the patterns of marriage. **1958** *Listener* 12 June 964/1 How the railways ought to behave in fixing the pattern of their charges. **1958** *Spectator* 8 Aug. 204/1 The pattern of supply is constantly changing. **1959** H. GARDNER *Business of Crit.* II. iii. 148 Both a study of the patterns of images, and their part in the structure of a poem, and the knowledge of ideas, theories, and beliefs current in a period are of great value as tools in an interpreter's hands. **1968** P. B. WEIZ *Elem. Zool.* viii. 123/2 A given external stimulus usually leads to the completion of several or many simultaneous reflex responses, all occurring as a single, integrated pattern of activity. **1976** *Sci. Amer.* Jan. 96/2 The limited evidence on other mammals suggests a different pattern of evolution.

d. *Linguistics.* A discernible order or arrangement in some branch of language, esp. in phonology.

1921 E. SAPIR *Language* iii. 56 Every language, then, is characterized as much by its ideal system of sounds and by the underlying phonetic pattern (system, one might term it, of symbolic atoms) as by a definite grammatical structure. **1926** *Germanic Rev.* I. I. 49 The Indo-European consonant pattern differed radically from that of Sanskrit. **1933** L. BLOOMFIELD *Language* 136 The structural pattern leads us to recognize also compound phonemes. **1935** G. K. ZIPF *Psycho-Biol. of Lang.* v. 195 The only difference between a *pattern* and a *configuration* is that the former is the more generic and collective term. One infers the nature of speech-patterns from the exemplifications of the patterns, i.e. the configurations of speech-elements. **1951** *Language* XXVII. 295 This explicit talk about the fact of patterning makes possible the distinction between the grammar (specific pattern) and grammaticalness (degree of patterning) of language. **1960** *Language Learning* X. I. 59 No two languages have the same set of patterns of pronunciation, words, and syntax. **1963** C. FRIES *Linguistics & Reading* ii. 67 The habits of pronunciation that the child develops in learning his native language are not habits of producing and hearing the separate sounds as isolatable items in individual words but rather habits of patterns of functioning contrasts in the unique structured system of a particular language. **1968** CHOMSKY & HALLE (*title*) The sound pattern of English. **1972** M. L. SAMUELS *Linguistic Evol.* xiii. 160 Noticeable in the Middle and Early Modern periods is the start of a new pattern of quantitative gradation in the verb.

e. *Physiol.* A particular sequence or arrangement of nerve impulses, in time and space, that is correlated with a particular sensation.

1947 W. E. LE GROS CLARK *Anat. Pattern* 7 The multiple nerve fibres approach the spot from different directions through the cutaneous nerve plexus, so that stimulation of a sensory spot gives rise to nerve impulses which reach the central nervous system by different routes, and thus lead to some specific pattern of excitation there. **1955** *Brain* LXXVIII. 586 There has been a revulsion from the ..idea of the nervous system as a telephone exchange, and this has found one expression in the suggestion that a specific cutaneous sensation results when the brain receives from the skin impulses which make up a characteristic pattern. **1961** T. L. PEELE *Neuroanat. Basis Clin. Neurol.* (ed. 2) xix. 448 A 'touch' pattern requires more large fibers than a 'pain' pattern. **1969** MELZACK & WALL in K. H. Pribram *Brain & Behav.* II. 145 The pattern theory proposed by Weddell and Sinclair, then, fails as a satisfactory theory of pain. **1975** —— in M. Weisenberg *Pain* i. 12/1 There can no longer be any doubt that temporal and spatial patterns of nerve impulses provide the basis of our sensory perceptions.

11. (Earlier and later examples.)

1859 'STONEHENGE' *Shot-Gun and Sporting-Rifle* I. ii. 14 So much depends on the pattern made at thirty and forty yards by the gun intended to be used. *Ibid.* III. i. 175 A gun can only be made to combine a certain amount of strength with regularity of pattern. *Ibid.* 176 They shall give such a pattern on the target as will prevent the escape of a partridge or grouse. **1961** C. WILLOCK *Death in Covert* ii. 34 Pattern..is the spread of the shot. **1972** *Shooting Times & Country Mag.* 1 July 14/2 The more the manufacture of shotgun ammunition is influenced by the needs of competitive clay pigeon shooting, the more it will repay game and pigeon shooters to check that they are not getting unduly close patterns. **1976** *Shooting Mag.* Dec. 52/2 A new game cartridge,..features the exclusive Monowad, claimed by the manufacturers to deliver up to 10 per cent more pellets inside the pattern.

13. b. *pattern discrimination, -quality, recognition*; **pattern baldness,** baldness in which there is a gradual loss of hair in accordance with a characteristic pattern, as in the receding hair-line that commonly occurs in men as they grow older; **pattern body** *rare*, a dress pattern taken from an existing dress; **pattern-bomb** *v.*, to bomb a target from aircraft according to a prescribed pattern in order to obtain maximum effect; so **pattern-bombing** *vbl. sb.*; **pattern book** (*a*) (earlier and later examples); also *transf.* and *fig.*; **pattern card** (earlier and later examples); also *fig.*; **pattern congruity** *Linguistics*, conformity to the structure of a language, esp. the phonological structure; **pattern darning,** a type of embroidery in which darning stitches are used to form a design, freq. as a geometric background; also *pattern darn*; **pattern-maker,** (*b*) (earlier example); **pattern-making** (further examples); **pattern-paper,** the paper from which a pattern (sense *2 b) is made; **pattern practice,** in learning a foreign language, intensive repetition of its distinctive constructions and patterns; **pattern setter** (*b*), anyone or anything that establishes a pattern or precedent; so *pattern-setting* adj.; **pattern shop** (examples); **pattern variable** *Sociol.*, a term used by Talcott Parsons in his attempt to define social action as the choice between five main dichotomous patterns of behaviour; also *attrib.*; **pattern-welding,** a technique used by the Anglo-Saxons for forging sword blades; also, a piece of pattern-welded metal; so **pattern-welded** *a.*

1916 *Jrnl. Heredity* VII. 349/2 Congenital baldness must not be confused with pattern baldness. **1956** C. AUERBACH *Genetics in Atomic Age* 16 The so-called pattern-baldness of men is due to a mutated gene which acts most effectively on the background of a male constitution. **1974** *Jrnl. Clin. Endocrinol. & Metabolism* XXXIX. 1012/1 Androgens may paradoxically cause male pattern baldness in individuals with a genetic predisposition. **1819** M. EDGEWORTH *Let.* 28 Jan. (1971) 165 The gown..is made by the very best dressmaker in Paris by a pattern body which I got my dear Fan to take from a gown of yours. **1943** *Jane's All World's Aircraft* 23a/1 The air targets could be pattern-bombed so that a bomb fell in every area of 50 square yards. **1944** *Britannica Bk. of Year* 770/1 *Pattern-bomb*,..to bomb, from a number of aircraft, in such a way that the relative position of the craft determines the 'pattern' of the bombs when they strike, so as to cover the target in a desired manner. **1947** *Sun* (Baltimore) 29 Mar. 2/2 Sending a fleet of jetpropelled planes from the Kurile islands (north of Japan) to pattern-bomb all of America. **1940** *War Illustr.* 5 Jan. 555 (*caption*) Wellingtons preparing for 'pattern bombing'. **1941** E. C. SHEPHERD *Mil. Aeroplane* 4 Anti-aircraft fire can..break up the formations so that mass bombing or pattern bombing becomes impossible. **1948** E. WAUGH *Loved One* 78 You couldn't really get away from the war even there. The ladies didn't seem to have a mind for anything higher than pattern-bombing. **1973** *Times* 6 June 19/5 It was subjected to air bombardment..by a process of deliberate and sustained 'pattern bombing' which wiped out the little town. **1774** N. CRESSWELL *Jrnl.* 7 Apr. (1925) 9 Spent the evening with Mr. Longsdon, who gave me a pattern Book and desires me to do some business for him. **1821** P. EGAN *Real Life in London* I. vi. 91 And was followed by a servant with pattern-books, the other apparatus of his trade. **1846** [see *EAST-END]. **1950** E. H. GOMBRICH *Story of Art* x. 141 It was in the thirteenth century that artists did occasionally abandon their pattern book altogether. **1959** *Times* 5 Nov. 15/2 Mr. Busch knows his job..but his people are pattern-book and never suprise. **1978** CADOGAN & CRAIG *Women & Children First* viii. 165 Lorna moves in a world of Women's Institute whist-drives..and Weldon's pattern books. **1773** J. WEDGWOOD *Let.* 21 Nov. (1965) 155 Voyer's Seals are sad trash, but Boden & Smiths were mounting half a Groce of them..to be sent..as Pattern Cards. **1821** P. EGAN *Life in London* II. i. 136 Mr. Primefit, of Regent-street, was..ordered to attend upon Mr. Hawthorn, with his pattern-card, to take orders. **1881** [see *BEVELLER]. **1970** G. HEYER *Charity Girl* i. 17, I shall attend Hetta's wedding... I daresay Hetta will be better off with her pattern-card. **1934** *Language* X. 124 The criterion of pattern congruity. Particular formulations must be congruous with the general phonemic pattern of the given language. **1941** *Ibid.* XVII. 229 That /č, ǰ, š, ž/ are unit phonemes appears partly from their distribution.., partly from their behavior in clusters... There may be some dialects in which they can be analyzed as /tj, dj, sj, zj/ respectively, but considerations of pattern congruity make this unlikely. **1964** E. BACH *Introd. Transformational Gram.* viii. 178 It seems natural to identify simplicity with the number of symbol tokens.. in the grammar. We would exclude from our count symbols of metatheory..and count as single symbols..the primes of the various parts of the grammar. This consideration seems to underlie many statements about 'pattern congruity' and the like. **1906** MRS. A. H. CHRISTIE *Embroidery & Tapestry Weaving* ix. 197 The second kind is called pattern darning; in it the stitches are picked up in some regular order, so that they form various geometrical patterns over the surface. *Ibid.* 199 Samplers ..may be seen entirely filled with these pattern darns. **1915** M. SYMONDS *Elem. Embroidery* xii. 138 Pattern darning is also used for backgrounds, in which cases the linen..should be strong but not woven too closely. **1932** D. C. MINTER *Mod. Needlecraft* 14/1 Pattern darning.. consists of the regular picking up of threads in such a way as to cover a background of a design with a pattern. **1967** E. SHORT *Embroidery & Fabric Collage* i. 34 Bead embroidery must give a raised texture, shadow work or pattern darning a relatively smooth surface. **1951** S. S. STEVENS *Handbk. Exper. Psychol.* xx. 764/2 So far we have dealt only with pattern discriminations: the capacity to tell the difference between a triangle and a circle or between an upright triangle and an inverted triangle, and so on. **1851** C. CIST *Sk. Cincinnati in 1851* xv. 297 He engaged in the foundry..as pattern-maker. **1934** C. LAMBERT *Music Ho!* III. 143 We must not think, however, that the modified internationalism of the eighteenth century is any more a permanent and integral part of musical tradition than the objective pattern-making of the period. **1937** H. READ *Art & Society* i. 23 But from the normal point of view we have to explain..the almost complete atrophy of the artistic impulse in man—at least the disappearance of the individual work of art in an undifferentiated mass of pattern-making as monotonous as the standardized products of our own machine age. **1926** J. MASEFIELD *Odtaa* iii. 59 She picked up some pattern-paper..snipped it with scissors..and then shook it out as a sort of cape or shawl of lace. **1944** C. FRIES *Intensive Course in Eng. for Latin-Amer. Students* VI. 1 A..class hour is given to the 'pattern practice' and drill. *Ibid.* 2 (*heading*) Pattern practice in conversation. **1948** *Language Learning* I. I. 27 This type of completely oral pattern practice approximates the language activity involved in free conversation while..it provides the concentrated practice of simple imitation. **1960** N. BROOKS *Language & Lang. Learning* iv. 49 Pattern practice, which opens the door to analogy, may be called the antithesis of paraphrasing. **1932** H. H. PRICE *Perception* viii. 243 AB.. is a *sensible* complex... It means also that the complex AB has a certain characteristic which we may call *sensible pattern-quality*. **1938** R. G. COLLINGWOOD *Princ. Art* x. 233 A new pattern-quality emerging from a particular way of combining psychical experiences. **1976** *Shooting Times & Country Mag.* 16–22 Dec. 14/2 It is not a high-velocity cartridge: a type which, in certain guns, so often gives hostages to fortune in the shape of pattern quality, so vital for satisfactory full-range work. **1959** *Proc. Eastern Joint Computer Conf.* 225/1 These approaches prove..to center upon analysis of the specific characteristics of patterns into parts, followed by a synthesis of the whole from the parts. In these studies, pattern recognition of the whole, that is, Gestalt recognition, was chosen as a more fruitful avenue of approach. **1964** J. Z. Young *Model of Brain* xix. 312 There is every reason to think that similar arrangements are an essential part of the pattern-recognition systems of the brain. **1970** O. Dopping *Computers & Data Processing* xi. 173 Automatic recognition of characters is a special case of a more general problem, called pattern recognition. **1974** W. R. Adey et al. *Brain Mechanisms & Control of Behav.* xi. 474 There has been a very earnest search for computer methods of pattern recognition. **1976** *Gloss. Documentation Terms (B.S.I.)* 47 *Pattern recognition*, machine-sensing or identification of visible patterns. **1899** W. James *Talks to Teachers* 213 We, here in America, through following a succession of pattern-setters whom it is now impossible to trace,..have at last settled down collectively into what, for better or worse, is our own characteristic national type. **1902** —— *Var. Relig. Exper.* i. 6 It would profit us little to study this second-hand religious life. We must make search rather for the original experiences which were the pattern-setters to all this mass of suggested feeling and imitated conduct. **1973** *Tucson* (Arizona) *Daily Citizen* 22 Aug. 2 The UAW has made good use of the 'strike target' strategy it dreamed up in 1955 to put added pressure on one auto company to agree to a pattern-setting contract. **1878** *Harper's Mag.* Apr. 648/1 Here is the great hall of the pattern shop fragrant with new wood. **1916** 'Taffrail' *Pincher Martin* xiv. 256 Before joining the destroyer he had been at the College at Dartmouth, teaching the naval cadets their business in the pattern-shop. **1964** S. Crawford *Basic Engin. Processes* (1969) x. 234 A finished component drawing is sent to the pattern shop providing all essential information. **1951** Parsons & Shils *Toward Gen. Theory Action* II. 48 The pattern-variable scheme defines a set of five dichotomies. Any course by any actor involves (according to theory) a pattern of choices with respect to these five sets of alternatives. **1959** D. Martindale in L. Gross *Symposium Sociol. Theory* II. ii. 76 In explaining this surprise, Parsons is led to assign importance to all sorts of factors not even mentioned in his set of pattern variables. **1964** I. L. Horowitz *New Sociol.* 15 The long trek from an action context to a paradigm for describing all types of action in a four-part pattern variable is no better..than Hegel..ending with..the perfect equation of Reason equating itself. **1948** H. Maryon in *Proc. Cambr. Antiquarian Soc.* XLI. 76 The welding of these swords represents an excessively difficult operation. I do not know of finer smith's work... I have named the technique 'pattern welding'... Examples of pattern-welding range in date from the third century to the Viking Age. **1956** *Nature* 29 Dec. 1432/2 (*caption*) Modern pattern-welded sword: experiment No. 7. *Ibid.*, Welding also came to be used successfully..in delicate work, involving the 'piling' of many sheets into a composite laminate, and developing into pattern-welding. *Ibid.* 1433/1 Most pattern-welded swords are so corroded that..[metallographic study] is not possible. **1962** H. R. E. Davidson *Sword in Anglo-Saxon England* i. 25 The ninth-century sword from the Palace of Westminster..had a pattern-welded blade. *Ibid.* 29 A means of re-using old strips of pattern-welding from worn swords to make a new blade. *Ibid.* 30 Short swords made by the pattern-welded technique. *Ibid.* 32 By the ninth century the art of pattern-welding was on the decline. **1964** H. Hodges *Artifacts* v. 88 In this process, known as pattern welding, case-hardened bars of iron were piled or faggotted white hot and forged. **1975** *Anglo-Saxon England* IV. 179 The technique of pattern-welding died out during the tenth and eleventh centuries.

pattern, *v.* Add: **6.** Also const. *after*, to take (someone or something) as a model or example (*absol.* use of sense 2), *U.S.* Now *rare*.

1878 J. H. Beadle *Western Wilds* xxii. 356 That was a nice family for us Americans to pattern after, wasn't it? **1884** 'C. E. Craddock' *In Tennessee Mts.* I. 4 They dunno what he patterned arter.

7. b. To order or arrange (a number of things) into a pattern; to design or organize (something) for a specific purpose. Also *intr.*, to form or cast a pattern (*rare*).

1931 W. Stevens *Harmonium* 133 A pale silver patterned on the deck And made one think of porcelain chocolate And pied umbrellas. **1967** *Boston Sunday Herald* 14 May II. 13/2 (Advt.), The s.s. Rotterdam.. patterned for epicures,..art connoisseurs, and other bon vivants. **1967** *Times Rev. Industry* July 89/1 Organisations tend to be patterned for a variety of reasons. **1971** J. Z. Young *Introd. Study Man* iii. 47 The essence of the operation that we call coding is that events patterned in one medium are made to correspond to events patterned in another. **1972** *Where* Feb. 40/1 The borders of the new jigsaw are becoming clear, even if we haven't found and patterned all the pieces yet. **1977** *Sci. Amer.* Sept. 124/3 The uppermost layers of integrated circuits are formed by depositing and patterning thin films.

c. *intr. Linguistics.* To make, fall into, or form part of a pattern (*pattern *sb.* 8 d).

1942 *Amer. Speech* XVII. 147 They pattern congruently with the similarly distributed varieties of /p/ and /k/. **1951** Trager & Smith *Outl. Eng. Struct.* ii. 53 The distributional gaps are often found to pattern as if they were themselves partials with phonemic content. **1963** *Amer. Speech* XXXVIII. 53 Most natural languages, including English, do not pattern on the finite-state model, which cannot handle certain regular processes of sentence formation. **1965** *Word Study* Feb. 7/2 We should distinguish between prepositions (which always have an object, occasionally elliptical) and verbal particles (which pattern with transitive and intransitive verbs alike). **1971** D. Crystal *Linguistics* 89 The way words pattern in sequences to form sentences.

patternation (pætəɹnēi·ʃən). [f. Pattern *sb.* + -ation.] The fact or action of forming, or conforming to, a pattern; *spec.* non-uniformity in the distribution of spray from a jet.

1946 M. Peake *Titus Groan* 399 Her hips..swayed when she talked.., they did all but chime as her sharp, unpleasant voice..dictated their figure-of-eight (bird's-eye view, cross-section) patternations. **1947** *Shell Aviation News* No. 110 21/3 Atomizers may, in some cases, be subjected to radical distribution tests, a specialized species of patternation with the main object of measuring the variation of spray density at different radial distances from the spray axis. **1949** *Jrnl. R. Aeronaut. Soc.* LIII. 161/1 The use of specialised atomising jets with their accompanying troubles of penetration and patternation is avoided. **1955** *Jrnl. Brit. Interplanetary Soc.* XIV. 218 By this time patternation tests were carried out [on a combustion chamber injector] as well as calibrations.

patterned, *ppl. a.* Add: **a.** (Further examples.) Also, conforming with, or forming, an arrangement or pattern.

1930 E. Pound *XXX Cantos* v. 19 Ecbatan, City of patterned streets. **1961** *Lancet* 29 July 259/2 The release from patterned behaviour forced the choice between good and evil. **1964** Gould & Kolb *Dict. Social Sci.* 480/2 These..represent potentialities for the most varied outcomes, yet outcomes extremely patterned culturally. **1967** E. Short *Embroidery & Fabric Collage* iii. 64 When wall papers, patterned fabrics and carpets are used any embroidery must be more carefully considered to avoid a 'messy' effect. **1970** G. A. & A. G. Theodorson *Mod. Dict. Sociol.* 293 *Patterned evasion*, a regularised way of deviating from an established social norm. **1973** *Technical Translation Bull.* XIX. 103 *Patterned glass* is the name used where the patterns are distinctive and fancy, e.g. Arctic, Hammered, Moroccan (traditional patterns still going strong) or modern types such as 'Mersey' or 'Manhattan'. **1973** J. M. White *Garden Game* 64 A bright flower-patterned cretonne. **1977** *Jersey Even. Post* 26 July 10/1 Her bridesmaid..wore a long, tiered empire-line voile dress, made of a yellow and red floral patterned material.

b. *patterned ground* (Physical Geogr.): ground showing a definite pattern of stones, fissures, vegetation, etc. (commonly polygons, rings, or stripes), such as is typical of periglacial regions.

1950 A. L. Washburn in *Revue Canad. de Géogr.* IV. 8 The terms *Rutmark, Strukturboden..stone polygons, mud circles, soil circles, mud polygons, soil polygons, fissure polygons, tundra polygons, stone stripes, soil stripes, solifluction stripes* and others have all been used to describe features here collectively named *patterned ground* for want of a satisfactory collective term in English... The writer would restrict the use of *patterned ground* to more or less symmetrical features. **1956** *Bull. Geol. Soc. Amer.* LXVII. 846/1 Frost wedging in bedrock is capable of developing several varieties of patterned ground, all intimately associated with bedrock structure. **1973** *Nature Physical Sci.* 4 June 85/2 This mechanism creates the *gilgai* of clay-rich and commonly alluvial soils in many hot sub-humid to semi-arid regions of the world. These structures are forms of patterned ground, having a surface expression as roughly polygonal to rectilinear-parallel systems of low ridges between hollows.

patterning, *vbl. sb.* Add: Also, the fact or process of forming (part of) an abstract pattern, as of behaviour, speech, etc. (Further examples.)

1921 E. Sapir *Language* iv. 61 It also has a definite feeling for patterning on the level of grammatical formation. **1937** B. L. Whorf in *Language* (1945) XXI. 1 This view loses sight of various word-classes that are marked not by morphemic tags but by types of patterning. **1939** J. Dollard et al. *Frustration & Aggression* vii. 152 Traditional patterning identifies another group of circumstances in which the aggression may be expressed. **1952** *Internat. Jrnl. Psycho-Anal.* XXXIII. 411/2 Certain fundamental relationships, resulting in characteristic pregenital patterning..are offered here. **1961** R. B. Long *Sentence & its Parts* i. 22 In *he's sort of nice* the relationship is similarly upside down: *he's rather nice* shows the syntactically more ordinary patterning. **1963** *Lancet* 12 Jan. 67/1 The inseparability of genetic and environmental influences should not deter us from the study of isolated aspects of growth and patterning. **1964** *Amer. Speech* XXXIX. 140 Nineteenth- and twentieth-century works that discover and elucidate structural patterning in one or more languages. *Ibid.*, It is..in the gradually developed concept of patterning that the structural teachings of our own day are rooted. **1964** M. Critchley *Developmental Dyslexia* xiii. 79 The evolution of behaviour can be 'conceptualised' as the process of development of intersensory patterning. **1973** *Word 1970* XXVI. 122 A deep sense of the presence of patterning in the phenomena of man. *Ibid.*, To extend the scope of linguistic inquiry..to include..verbal art, cultural symbolism and patterning, and so on.

patternism (pæ·tǝɹniz'm). [f. Pattern *sb.* + -ism.] A name given (chiefly by its critics) to a way of describing religions (esp. those of the ancient Near East) not on the basis of historical development but on the basis of common and recurrent patterns; also, a mode of literary appreciation based on recurrent patterns. Hence **pa·tternist**, a proponent of this theory (also *attrib.* or as *adj.*).

1951 H. Frankfort *Problem of Similarity in Anc. Near Eastern Religions* 10 One may admit the close relationship between the myths and rites of a religion..without falling into the error of those 'functionalist' and 'patternist' authors who declare that a myth is merely the spoken accompaniment of ritual. **1956** *Jrnl. Theol. Stud.* VII. 276 The absurdity of labelling scholars holding such divergent views on the subject of 'patternism'..as the 'Scandinavian' school of patternists. *Ibid.* 277 It must be admitted that some of the Scandinavian patternists have carried their interpretation of the Old Testament to lengths which invite criticism. **1957** *Scottish Jrnl. Theol.* X. 95 This volume is all the more welcome because since Professor Hooke edited *Myth and Ritual*, he has often been unjustly held responsible for all the vagaries of the wildest of the patternists. *Ibid.*, Not all will agree with Dr Hooke's patternist explanation of Elijah on Carmel and away to Horeb. **1961** *Times Lit. Suppl.* 17 Feb. p. xiii/3 Another interesting feature of Dr. Carrington's 'patternism' is his suggestion that Mark has divided his narrative into five divisions.

patternization (pæ:tǝɹnaizēi·ʃən). [f. Patternize *v.* + -ation.] Arrangement in a pattern. Cf. *patternation.

1938 *Mind* XLVII. 379 When he discusses the patternisation of a square of dots he makes a series of remarks which are so significant that we must quote them at length. **1960** R. Carpenter *Greek Sculpture* viii. 217 the patternization of the..underlying muscular structure.

patternless, *a.* Add: **c.** Formless; conforming to, or possessing, no discernible arrangement or pattern.

1960 R. Carpenter *Greek Sculpture* ix. 243 The patternless tangle of shadow in hair and beard. **1963** *Times* 20 May 4/7 Tambling was the man for the quick break and almost stole two more goals near the end to make nonsense of all Sunderland's patternless hammering. **1975** *Sci. Amer.* May 47/2 Clearly a more sensible definition of randomness is required, one that does not contradict the intuitive concept of a 'patternless' number.

patteroller (pæ·tǝrōuləɹ). Also **pateroller, patter(-)roller.** Southern U.S. varr. Patroller; *spec.* a person who watched and restricted the movements of Blacks by night. *Obs.* except *hist.*

c **1862** J. C. Harris *Uncle Remus & Friends* (1892) 196 He sing en he play—oh, gals, go 'way! Whar de patterroller never kin see. **1893** *Nation* (N.Y.) 7 Sept. 173/1 Ability to write meant ability to counterfeit passes which would outwit the ignorant midnight 'patterrollers'. **1899** B. W. Green *Word-bk. Virginia Folk-speech* 268 *Patteroller*, a patroller. **1917** *Dialect Notes* IV. 385 *Pateroller*, Night-guard over negro slaves on an ante-bellum plantation. Ky. **1928** S. V. Benét *John Brown's Body* 40 He's friends with de ha'nts and steel won't touch him But the paterollers is sure to cotch him. **1936** M. Mitchell *Gone with Wind* xvii. 307 What are you boys doing so far from Tara? You've run away, I'll be bound. Don't you know the patterollers will get you sure? **1964** R. Hayden in *Negro Digest* June 47 Moon so bright and no place to hide, The cry up and the patterollers riding.

patters (pæ·təɹz). *University slang.* [f. first syllable of Pat-ball (in Dict. and Suppl.) + *-er⁶.] A university students' name for tennis.

1900 *Captain* IV. 26/1 In the Summer, a great form of 'eccer' is 'patters', a corruption of 'patball', *i.e.* tennis. **1912** A. F. Wilding *On Court & Off* v. 103 A tremendous amount of 'patters', as tennis is popularly called, is played at the University.

Pattinson (pæ·tinsən). *Metallurgy.* [Name of Hugh Lee *Pattinson* (1796–1858), English metallurgical chemist, who patented the process in 1833 (*Brit. Pat. 6497*).] *Pattinson('s) process*: a process formerly used for desilverizing and purifying lead (see quot. 1881).

1856 W. A. Miller *Elem. Chem.* II. xvi. 994 (*heading*) Concentration of silver in lead by Pattinson's process. **1881** [see Pattinsonize *v.*]. **1912** [see *Parkes]. **1946** *Thorpe's Dict. Appl. Chem.* (ed. 4) VII. 224/1 The value of the Pattinson process as a means of purifying lead is shown by the fact that lead to be used for white-lead making was frequently Pattinsonized, although the amount of silver present was too small to repay the cost of its recovery. **1964** H. Hodges *Artifacts* vi. 93 This method of enriching a lead alloy is now known as Pattinson's process.

pattoon, obs. var. Patten *sb.*

1715 *Boston News-Let.* 17 Oct. 2/2 All Persons may have Boots, Shoes, Pattoons, or anything belonging to that Trade mended. **1743** *Ibid.* 3 Feb., To be Sold.. Women's & Children's Shoes & Pattoons.

pattress (pæ·trės). Also **pat(t)rass, pateras.** [(Corruption of) *pateras*, pl. of Patera.] A wooden or plastic block attached to a surface to carry a gas bracket, electric light switch, ceiling rose, or the like; the base of a wall socket. Also *pattress block, box.*

1886 J. Black *Gas Fitting* v. 35 Screw on the pattress blocks, pendants and brackets. **1900** P. N. Hasluck *Pract. Gas-Fitting* iv. 80 The wooden block or pattress is now placed over the tube-bit, the screwed end being passed through the hole in the centre. **1905** C. C. Metcalfe *Prac. Electr. Wiring* ii. 33 Casing brought from skirting board to pateras..will hardly be noticeable. **1928** Maycock & Kemp *Electr. Wiring* (ed. 6) iii. 292 Single- and two-circuit fixture blocks,..are virtually ceiling-rose blocks embedded in hard-wood pattrasses, and without covers. **1934** *Pract. Electr. Engineer* II. 310/2 The pattrasses are marked off and drilled, for the fixing and cable holes on their faces, and the conduit entry holes at their sides. **1969** A. J. Coker *Electr. Wiring* (ed. 7) v. 81 Pattress boxes are also available to convert flush-type to surface mounting. *Ibid.* 83 For fixing and wiring a surface

socket-outlet, the circuit cables are first passed through a suitable knockout in the pattress block. **1973** G. A. T. Burdett *Householder's Electr. Guide* xi. 85 These switches are mounted on moulded plastic surface pattress boxes or metal boxes which are sunk into the wall flush with the plaster. **1976** *Pract. Householder* Nov. 66/3 The ceiling roses will, or should be, mounted on pattress blocks but there is unlikely to be an earth.

patty-cake. 2. Delete '¶ Error for Pat-a-cake' and substitute 'U.S. var. Pat-a-cake'. (Later examples.) Also *fig.*

1950 O. Nash *Family Reunion* (1951) 11 All of Granny's muscles ache From half an hour of patty-cake. **1972** *Bankers' Mag.* (Boston, Mass.) Winter 93/2 The crushing price increases that are putting the older fixed incomers on welfare don't emanate from the big firms that are assumed to be playing patty cake with the unions. **1976** *Word 1971* XXVII. 34 Laura loves to play patty-cake. She laughs and flaps her hands together. She will respond to the words alone, but she does not respond to the motions alone. **1976** M. Machlin *Pipeline* xxxviii. 412 Ever since we started this damn pipeline, Golconda has been playing patty-cake with Friends of the Earth and all the rest.

‖patu (pā·tu). *N.Z.* Also 8–9 **pat(t)oo.** Also in redupl. form. [Maori.] A short club-like weapon with sharpened edges made of stone, whalebone, or nephrite, used for striking rather than thrusting.

1769 J. Cook *Jrnl.* 12 Nov. (1955) I. 200 They have short Truncheons about a foot long, which they call Pattoo Pattoos, some made of wood some of bone and others of stone. **1770** J. Banks *Endeavour Jrnl.* (1962) II. 26 *Patoo patoos* as they calld them, a kind of small hand bludgeon of stone, bone or hard wood most admirably calculated for the cracking of skulls. **1817** J. L. Nicholas *Narr. Voyage to N.Z.* I. iii. 89 The men in the canoe.. exhibited..mats, spears, hooks, fishing-lines, thread, *pattoo pattoos* (war implements). **1882** T. H. Potts *Out in Open* 82 It [*sc.* fern-root] was soaked, roasted, and repeatedly beaten with a small club (patu) on a large smooth stone, till it was supple. **1921** H. Guthrie-Smith *Tutira* x. 77 Tua Kiaki pulled out a *patu* concealed beneath his mat, and with it there and then slew Te Mautaranui. **1949** P. H. Buck *Coming of Maori* (1950) II. xi. 277 The short clubs (*patu poto*)..are made in three types: the *mere, kotiate*, and *wahaika*..the whalebone clubs carrying the descriptive name of *patu paraoa* (*patu*, club; *paraoa* whalebone). **1974** *Nat. Geographic* Dec. 760 (*caption*) This Tahitian war club is older than the similar patu of the New Zealand Maori.

patulin (pæ·tiŭlin). *Biochem.* [f. L. *patul-um*, specific epithet of the mould, neut. of *patulus* (see Patulous *a.*): see -in¹.] A colourless crystalline antibiotic compound, $C_7H_6O_4$, that was obtained from the mould *Penicillium patulum* and afterwards found to be identical with *clavacin and *claviformin.

1943 H. Raistrick in *Lancet* 20 Nov. 625/1 Some time ago a metabolic product of *Penicillium patulum* Bainier which had not been previously encountered here was isolated and shown to have antibacterial properties; it has now been identified as anhydro-3-hydroxymethylene-tetrahydro-γ-pyrone-2-carboxylic acid, and has been given the shorter name of 'patulin'. **1947** *Sci. News* V. 98 In 1943 there was a report that a substance..called patulin had proved powerful in treating colds. Further tests on more people showed no significant improvement, however. **1953** [see *clavatin]. **1965** *New Scientist* 28 Oct. 253/2 He also worked out the structure of the remarkable mould product patulin, which, because of its unusual properties, was at that period a substance of considerable interest.

Patum Peperium (pā·tǝm pepi^ǝ·rium). [Invented name based loosely on L. *piper* pepper: *patum* is fanciful.] A commercial name for a savoury paste; = *Gentleman's Relish* s.v. *gentleman 7 c.

1884 M. L. Allen *Breakfast Dishes* 82 That made by C. Osborne of London, 'Patum Piperium, or Anchovy Paste,' recommended by author as the very best made. **1907** [see *Gentleman's Relish*]. **1935** *Discovery* Dec. 364/1 There used to be in Victorian times—perhaps it still exists—a breakfast compound known as 'Patum Peperium, the Gentleman's Relish'—salted anchovy with a dash of pepper. **1976** *Times* 26 Mar. 14/4 Patum Peperium, the anchovy paste known as gentleman's relish.

paturon (pæ·tiurǫn). [a. Fr. (P. Lyonet *Recherches sur l'anatomie et les métamorphoses de différentes espèces d'Insectes* (*a* 1789, published 1832) 76, f. Gk. πατ-εῖν to tread + οὐρ-ά tail + -on.] = *falx.

1926 T. H. Savory *Brit. Spiders* v. 44 The poison fangs or chelicerae..are two-jointed and unchelate. Various names have been given to these joints; we prefer Lyonnet's terms, unguis and paturon. **1951** Locket & Millidge *Brit. Spiders* I. iii. 25 The chelicerae consist of two segments, the basal, called the paturon, and an apical one called the fang. **1970** K. R. Snow *Arachnids* iv. 38 The appendages of the first segment, the chelicerae, are composed of two podomeres: a basal podomere or paturon and a distal unguis or fang.

patwari (pætwā·ri), var. Putwary.

1819 F. Hamilton *Acct. Kingdom Nepal* II. i. 155 A Patwari or clerk..has one-half ana on the rupee of rent. **1873** E. Balfour *Cycl. India* (ed. 2) IV. 457/2 *Patwari*.., a village accountant, responsible for keeping the accounts of the village, noticing changes in the list of proprietors, [etc.]. **1913** J. H. Morrison *On Trail of Pioneers* xiii. 60 The headman of the village is the patél, and, assisting him, is the patwari or village clerk, usually a Brahmin, who keeps a record of the village lands. **1931** E. A. H. Blunt *Caste System N. India* xi. 222 The numerous class of *patwaris* (keepers of the village revenue records) consists almost entirely of Kayasthas: and as the *patwari* has a bad name for chicanery, the better class Kayasthas affect to despise this occupation. **1948** 'P. Woodruff' *Whatever Dies* 180 The patwari, who is the local representative of the government and looks after some sixty or eighty villages. **1958** O. Lewis *Village Life N. India* iii. 95 The plots were never..officially registered, because the *patwari* [sic] wanted 30 rupees for the registration, which the villagers refused to pay.

patzer (pæ·tsəɹ). *slang.* [Origin uncertain: cf. G. *patzen* to bungle.] In chess: a poor player, a 'rabbit'.

1959 *S. Afr. Chess-Player* VII. 11 That patzer Grivainis got Evans with an opening trap, but now the difference in strength begins to show. **1960** Wentworth & Flexner *Dict. Amer. Slang* 378/1 *Patzer*, an inferior chess player. Although said to be from the Yiddish, there is no Yiddish, German, or Hebrew word or word combination to suggest it. Prob. from 'patsy' with the familiar '-er' ending added. **1965** tr. *A. D. de Groot's Thought & Choice in Chess* p. v, Why do masters find the good moves that patzers overlook? **1972** *Daily Tel.* 28 July 15/4 So Fischer after beating off a ferocious attack..'played like a patzer', said one American Grandmaster, 'went to sleep on the job', said another. **1978** *New Statesman* 27 Oct. 556/2 He appears (or perhaps pretends) to be as tempted as the average patzer, by any old poisoned pawn, and has to have his folly explained to him.

paua (pā·wā). *N.Z.* Also **pawa.** [Maori.] A large gastropod mollusc of the genus *Haliotis*, esp. *H. iris*, which attaches itself to rocks by suction and is sometimes collected and used for food. Also *attrib.* Cf. *abalone, Ormer.

[**1820** *Gram. & Vocab. Lang. N.Z.* (Church Missionary Soc.) 191 Paua, *s.* a shell-fish so called.] **1846** C. Heaphy *Jrnl.* 6 Apr. in N. M. Taylor *Early Travellers N.Z.* (1959) 211 The mutton fish, or *pawa*, although resembling india rubber in toughness and colour, is very excellent and substantial food for explorers. *Ibid.* July 244 At Tunupoho we obtained twenty *paua* or 'mutton fish'. **1949** P. H. Buck *Coming of Maori* (1950) I. ii. 13 The shores [of the Chatham Islands] yielded quantities of shell fish which included the *paua*. **1959** M. Shadbolt *New Zealanders* 120 He would feel underwater, knife in hand, for the pauas. *Ibid.* 125 He heard them talking quietly in the kitchen as they prepared the meal, hammering soft the paua-steaks. **1963** *Times* 12 Jan. 1/5 This exotic New Zealand delicacy prepared from Paua (Par-War) Clams delicately flavoured with Asparagus is now obtainable from leading delicatessen and high-class food stores throughout the United Kingdom. **1966** J. K. Baxter *Pig Island Lett.* 13 A corrugated shack With fried pauas in the pan.

2. In full, *paua shell.* The oval shell of this mollusc, which may be as much as six inches long and two deep, distinguished by the row of holes along the back and the blue, green, and pink nacreous lining, which is used to make jewellery or other ornaments.

1873 J. H. H. St. John *Pakeha Rambles through Maori Lands* vii. 131 The eyes [of a carving] are formed of the inner coating of the 'pawa' shell, a kind of blueish mother of pearl. **1882** T. H. Potts *Out in Open* 162 Immense heaps of paua shells..show how largely these substantial mollusks were consumed. **1920** 'K. Mansfield' *Bliss* 43 Chocolate custard which she had decided to serve in the pawa shell. **1931** *Times Educ. Suppl.* 11 July p. ii, (*caption*) Elaborate carving on a gateway of a Maori village... The white spots are pieces of pawa shell let into the wood. **1936** 'R. Hyde' *Check to your King* xvii. 203 The women cut their flesh with thin paua shell. **1936** N. Marsh *Death in Ecstasy* ii. 22 A figure carved in wood with protruding tongue and eyes made of pawa shell. **1949** E. de Mauny *Huntsman in Career* ii. 104 On top of the bookcase were pawa shell ashtrays. **1958** S. Ashton-Warner *Spinster* 162 The released mind revolves, flashing different colours like a paua. **1959** M. Shadbolt *New Zealanders* 157 The large clean paua-shell ashtray gleaming copper and purple colours.

paucal (pǭ·kăl), *a. Gram.* [f. L. *paucus* few: see -al.] Applied to a 'number' or inflected form denoting more than two but fewer than the number denoted by the plural. Hence **pauca·lity.**

1964 R. H. Robins *Gen. Linguistics* vi. 247 A few [languages] have four [numbers], singular, dual, trial or 'paucal'..and plural. **1966** J. E. Buse in C. E. Bazell *In Memory of J. R. Firth* 52 These systems relate ultimately ..to the activity of counting and the resultant concepts of singularity, duality, plurality, paucality and multiplicity. *Ibid.* 58 Filling Place 1 in the structure is a commutation of *t-* and absence, the latter marking the possessed as paucal. **1973** *Archivum Linguisticum* IV. 39 The..'little plural' or 'paucal' form xoxaat 'a few peaches'. **1977** *Canad. Jrnl. Linguistics 1976* XXI. II. 217 Faced with a system like than of Fijian, where there is a further distinction between trial (or paucal) and multiple within plural (Milner 1956), the reapplication of non-minimal membership would have to operate within two branches of the resulting tree diagram.

pauci-. Add: Also in *Min.*, as **paucili·thionite** [Lithionite], a hypothetical end-member of the lepidolite system (see quot. 1942).

1942 A. N. Winchell in *Amer. Mineralogist* XXVII. 117 The second end-member [of the lepidolite system] $(K_2Li_3Al_3Si_6O_{20}F_4)$ has no name and no varietal name in the literature seems to be appropriate. In these circumstances the writer would suggest that it be called paucilithionite. **1963** *Mineral. Abstr.* XVI. 189/2, 142 Chemical analyses of lithium micas from the literature have been transformed to the molecules polylithionite, paucilithionite, muscovite, and siderophyllite.

Pauillac (po,i·yak). [Fr., f. the name of a commune in the department of Gironde, France.] Claret produced in Pauillac. Also *attrib.* passing into *adj.*

1858 [see Margaux]. **1897** A. Beardsley *Let.* c 27 Apr. (1971) 310 The Pauillac at Lapérouse is *excellent*! only 2 fr. a bottle. **1920** G. Saintsbury *Notes Cellar-Bk.* iv. 66 Margaux..and Pauillac..have very much fewer tricks played on them. **1966** H. Yoxall *Fashion of Life* xxv. 246, I..could separate, among the Médocs, the red Graves, Pauillacs, Margaux, St Juliens and St Estèphes. **1968** *Guardian* 29 Mar. 9/4 Ch. Ponter Canet, the best so far, with a fine Pauillac bouquet and taste. **1977** B. Roueché *Fago* I. ii. 19, I slipped the wine out of its paper bag. It was a good Bordeaux, a Pauillac of the Haute-Médoc.

Paul. Add: **3. Paul Pry**: substitute for def.: name of a very inquisitive character in a U.S. song of 1820; often used allusively (also *attrib.*). (Later examples.)

1928 E. Wallace *Double* xiii. 208 There are lots of quiet little nooks and places where a fellow can sit without a lot of Paul Prys seeing him. **1934** *Sun* (Baltimore) 27 Apr. 12/2 The Senate's theory that the way to enforce the tax laws is to give the Paul Prys of every community access to the private details of every man's gross and net income. **1956** H. G. de Lisser *Cup & Lip* ix. 109 It would be ruinous to a doctor to be known as a paul pry. **1978** H. C. Rae *Sullivan* I. ii. 24 Twenty-five thousand dollars?..It's the going rate for a quiet investigation, a straight Paul Pry?

b. Hence **Paul-Prying** *vbl. sb.*; **Paul Pryism,** the conduct of a Paul Pry.

1927 *Daily Express* 6 Oct. 8/2 These restrictions were imposed during the war... Their maintenance to-day is simply part of that fussy Paul Pryism which covers the State with ridicule. **1960** *Times* 4 Mar. 13/7 The straitest champion of marital fidelity would, surely, not defend such monstrous Paul Prying.

4*. Paul Jones [the name of John *Paul Jones* (1747–92), Scottish-born naval officer noted for his victories for the Americans during the War of Independence]: a ballroom dance during which the dancers change partners after circling in concentric rings of men and women. Also *attrib.* and *fig.*

1920 *Atlantic Monthly* July 89/1 The whole sprightly, smiling, hand-clapping population seems engaged in one vast 'Paul Jones'..with no one..refusing to join the dance. **1934** *Punch* 14 Feb. 174/1 There was nothing doing in the matter of Paul Joneses, from which even the most emphatic protestations could not give us release. **1938** *Times* 10 Jan. 10/4 The 'party' began with 'I've been to Harlem', a change-partner dance of the 'Paul Jones' type sung to the pure English harvest-home tune, 'I've been to France and I've been to Dover'. **1942** M. Dickens *One Pair of Feet* vii. 147 A blond A.C.2 whom I had picked up in the Paul Jones. **1958** L. Durrell *Balthazar* xiii. 233 But now the band had begun to play a Paul Jones (perhaps the very dance in which Arnauti first met Justine?). **1967** *Times Rev. Industry* May 58/3 Driving a private car is often a death-defying Paul Jones with an endless succession of lorries. **1972** V. Canning *Rainbird Pattern* iv. 66 Harriet..was seduced in the back of the officer's car while Grace was dancing a Paul Jones.

Paul–Bunnell (pǭl bʊne·l). *Path.* The names of J. R. *Paul* (1893–1936) and W. W. *Bunnell* (1902–1965), U.S. physicians, used *attrib.* and *absol.* with reference to a test first described by them in 1932, in which the presence of an antibody reaction to the red blood cells of sheep confirms a diagnosis of infectious mononucleosis (glandular fever).

1938 *Amer. Jrnl. Med. Sci.* CXCVI. 79 The diagnosis was established by the clinical course, confirmed by a characteristic blood smear and a positive Paul–Bunnell test. **1952** *Brit. Med. Jrnl.* 22 Mar. 637/1 The Paul-Bunnell reaction was positive in a dilution of 1 in 28. **1958** *Woman* 27 Sept. 70/3 A special blood test, a Paul–Bunnell test, proved that it could only be gladular fever. **1970** Passmore & Robson *Compan. Med. Stud.* II. xviii. 116/2 In some cases [of glandular fever], a heterophile antibody capable of agglutinating sheep erythrocytes appears (Paul–Bunnell antibody). **1976** *Lancet* 11 Dec. 1297/1 Paul Bunnell, blood cultures, electrocardiogram, and chest X-ray were all normal.

Pauli (pau·li). *Physics.* The name of Wolfgang *Pauli* (1900–58), Austrian-born physicist, used *attrib.* and in the possessive to designate the **exclusion principle*, which he enunciated in 1925 (*Zeitschr. f. Physik* XXXI. 765–83).

1926 *Jrnl. Optical Soc. Amer.* XIII. 10 By means of schemes (7) and (8) we easily can write down the $\bar{m}_r$ and $\bar{m}_k$ values observing Pauli's principle. *Ibid.*, the period is closed with the *N*th electron where the Pauli principle gives only a $^1S_0$ term. **1926** *Physical Rev.* XXVIII. 339 Whether the impossibility of obtaining coordination for

equivalent electrons is directly connected with Pauli's exclusion principle is difficult to say. **1928, 1930** [see **exclusion principle*]. **1946** J. R. Partington *Gen. & Inorg. Chem.* x. 257 For an atom containing more than one electron, the maximum number of electrons in each shell is fixed by Pauli's exclusion principle. **1968** C. G. Kuper *Introd. Theory Superconductivity* ix. 157 The wave function (9.51) is antisymmetric against interchange of particles, and therefore automatically satisfies the Pauli exclusion principle. **1974** G. Reece tr. *Hund's Hist. Quantum Theory* xiii. 181 The spin of the electron had hitherto been taken into account only in so far as it had no consequence beyond the Pauli principle.

Paulicianism. (In Dict. s.v. Paulician *sb.* and *a.*) (Examples.)

1839 *Penny Cycl.* XIV. 385/1 The Manichæan doctrines..continued to have supporters, under their new name of Paulicianism, till a very late period. **1874** J. H. Blunt *Dict. Sects* 414/2 From the close of the eleventh century Paulicianism as such ceases to be significant. **1941** [see *Bogomil, -mile]. **1967** N. G. Garsoïan (*title*) The Paulician heresy. A study of the origin and development of Paulicianism in Armenia and the eastern provinces of the Byzantine empire.

Paulinism. (In Dict. s.v. Pauline *a.* and *sb.*) Add: (Later examples.)

1910 J. Moffatt *Paul & Paulinism* II. 33 Paulinism.. was the outcome of the apostle's attempt to hunt out for himself the relations of the Lord Jesus Christ to God, the Law, the universe, and the church. **1916** G. B. Shaw *Androcles & Lion* Pref. p. xci, The Christianity of Jesus failed completely to establish itself politically and socially, ..whilst Paulinism overran the whole western civilized world.

b. An expression or feature characteristic of St. Paul.

1927 A. H. McNeile *Introd. New Testament* iv. 65 The presence in *Mark* of 'Paulinisms' or other features thought to be secondary on subjective grounds.

Paulist (pau·list), anglicized f. *Paulista. Hence **Pauli·stic** *a.*

1900 Paulist [see Mameluco]. **1942** A. St. James tr. *Zweig's Brazil* 214 Anyone still desirous of seeing something of the Paulistic type of the nineteenth century habitation had better hurry.

Paulista (pauli·stă). Also with lower-case initial. [Pg., f. *São Paulo* (see below) + *-ista* -ist.] **a.** A person of mixed Portuguese and Brazilian Indian descent; *spec.* one of the explorers and settlers of the hinterlands of southern Brazil. (*Obs.* exc. *hist.*). **b.** A native or inhabitant of the city of São Paulo in southern Brazil.

1817 Southey *Hist. Brazil* II. xxiii. 300 The Paulistas have acted so memorable a part in Brazil and Paraguay that it becomes of importance to trace their history distinctly, and clear it from fables and misrepresentations. **1884** R. G. Watson *Span. & Portug. S. Amer.* II. xi. 169 The search for the precious metals had long shared with slave-hunting the efforts of the *Paulistas* and others. *Ibid.* 174 *Minas Geraes* soon acquired the unenviable notoriety..of being the most turbulent settlement in *Brazil*. Its people were divided into two classes, called.. *Paulistas* and *Florasteiros* or strangers. **1896** A. H. Keane *Ethnol.* vii. 152 In Brazil the famous '*Paulistas*' (so called from the province of São Paulo), a cross between the first Portuguese immigrants and the aborigines, have always been the most vigorous and enterprising section of the community. **1910** *Encycl. Brit.* VII. 677/2 Cuyabá was founded in 1719 by Paulista gold hunters. **1932** W. S. Robertson *Hist. Lat.-Amer. Nations* (ed. 2) vi. 123 'Paulistas', as the half-breed adventurers from São Paulo were called, gradually penetrated farther and farther into the interior. **1942** A. St. James tr. *Zweig's Brazil* 211 During the seventeenth and eighteenth centuries on the banks of the little river Ticté there lies a small unimportant town, more headquarters and camp than a permanent settlement of those roaming gangs, the Paulistas, who roved through the whole country..in search of loot. *Ibid.* 218 The Paulistas have a most competitive attitude toward Rio de Janeiro, and the desire not to appear inferior or less artistic. **1944** S. Putnam tr. *E. de Cunha's Rebellion in Backlands* ii. 63 The Paulista—and this name, in its historic signification, takes in the sons of Rio de Janeiro, Minas, São Paulo, and regions south—now arose as an autonomous type, adventurous, rebellious, freedom-loving. **1966** *Economist* 3 Sept. 902 Pleasure-loving *cariocas* (those who live in Rio), economically eager *paulistas* (those who live in São Paulo). **1976** R. Perry *One Good Death* v. 75 The daily nightmare that many Paulistas called travelling home from work.

paulownia. Also (erron.) **pawlonia.** Substitute for etym. and def.: [mod.L. (P. F. von Siebold & J. G. Zuccarini *Flora Japonica* (1835) I. 25), f. the patronymic of Anna *Paulowna* (1795–1865), daughter of Tsar Paul I and wife of William II of the Netherlands.] A deciduous tree of the genus so called, esp. *P. tomentosa*, belonging to the family Scrophulariaceæ, native to China or Japan, and bearing panicles of bell-shaped blue or lilac flowers. Also *attrib.* (Earlier and later examples.)

1843 *Paxton's Mag. Bot.* X. 7 *Paulownia imperialis* (Imperial Paulownia)... A considerable quantity of this noble tree has lately been introduced to Britain. **1901** L. H. Bailey *Cycl. Amer. Hort.* III. 1224/1 The Paulownia is one of the most conspicuous flowering trees in spring. **1971** *Country Life* 1 Apr. 741/1 A potentially large shrub or tree,..like the..tree of heaven (*Ailanthus*) or pawlonia. **1973** C. Lloyd *Foliage Plants* x. 189 A group of paulownia seedlings, grown and pruned for their enormous, furry, heart-shaped leaves. **1976** P. Quennell *Marble Foot* v. 182 The local carpenter made me a large knee-desk of silvery pawlonia wood.

paunch, *v.*¹ Restrict 'Now *rare* or *dial.*' to senses 1, 3 and 4 and add: **2.** (Further examples.)

1906 *Chambers's Jrnl.* Sept. 681/2 The animals [*sc.* rabbits] have to be killed, bled, and paunched. **1952** F. White *Good Eng. Food* II. 111 One of the things I had to do before I was twenty was to paunch and skin a hare.

pauperization. (Earlier and later examples.)

1838 W. Howitt *Rural Life Eng.* II. III. ii. 149 The working classes..what distress, what pauperization.. they have gone through. **1955** *Times* 6 July 10/3 The recent C.G.T. congress..insisted that revolutionary action was the only remedy to the 'progressive pauperization' of the working class under capitalism. **1971** J. J. Shapiro tr. *Habermas's Toward Rational Society* vi. 110 Underprivileged groups are not social classes... Their *disfranchisement* and pauperization no longer coincide with *exploitation*. **1974** J. White tr. *Poulantzas's Fascism & Dictatorship* v. iii. 259 In this process of pauperization, artisans and traders lost almost half their income.

‖ paupiettes (popiet), *sb. pl.* Also in *sing.* [Fr.] The current form of Poupiets. (See quots.)

1889 A. Filippini *Table* 241 Panpiette [*sic*] of Veal à la Faubonne. **1892** A. B. Marshall *Larger Cookery Bk.* iv. 86 Eel paupiettes à la Française. **1906** *Mrs. Beeton's Bk. Househ. Managem.* lxii. 1666 *Paupiettes* (Fr.), slices of meat or fish rolled with forcemeat. **1936** Lucas & Hume *Au Petit Cordon Bleu* 174 *Paupiette*, a thin slice of meat or..fish, sometimes spread with a farce, then rolled up and tied with white cotton. **1948** *Good Housek. Cookery Bk.* II. 141 Glazed paupiettes of sole. **1970** Simon & Howe *Dict. Gastron.* 291/2 *Paupiettes*, thinly sliced pieces of meat used as wrappers for various meat or force-meat fillings. Also, in Escoffier's usage, fillets of sole and cabbage leaves used in the same way. The meat slices are spread with a well-spiced forcemeat and rolled up. Wrapped up in thin slices of fat bacon, they are tied up with cotton thread to be braised, baked or casseroled. Bacon and thread are removed before serving. **1975** J. Symons *Three Pipe Problem* xvi. 159 The soup was followed by what Sue called paupiettes de porc, pancakes with some sort of minced pork filling. **1976** *Evening Advertiser* (Swindon) 31 Dec. 13/1 (Advt.), Scottish restaurant menu at £7.00 inclusive with..Paupiette of Plaice 'Auchenshuggle'.

pause, *sb.* Add: **1. d.** (See quot. 1966.) Also *fig.*

1962 *Listener* 29 Mar. 549/2 The point is clearly made by General Norstad, who says..that what he needs in order to impose what he calls the pause, in order to identify a threat, in order to relieve himself of the intolerable choice between retreating and using nuclear weapons, is thirty divisions. *Ibid.* 19 Apr. 674/1 The time has come to declare a 'pause' on the culture-front. **1966** Schwarz & Hadik *Strategic Terminol.* 85 *Pause*, in the defense of Western Europe, a moment of reflection imposed on any aggressor before the defense resorts to nuclear weapons.

2. (Later examples.) In *Linguistics*, spec. the break marking juncture, sometimes regarded as having phonemic status.

1933 L. Bloomfield *Language* xii. 185 Since the constituents of phrases are free forms, the speaker may separate them by means of *pauses*. Pauses are mostly non-distinctive; they occur chiefly when the constituents are long phrases; in English they are usually preceded by a pause-pitch. **1948** *Language* XXIV. 19 Some utterances contain a perceptible time-interval during which none of the vocal organs perceptibly articulates... Such a time-interval is an *internal pause*... The absence of speech before or after an utterance is an *external pause*. **1950** *Ibid.* XXVI. 97 Pause may be regarded as a kind of zero phone, characterized by a complete lack of qualities. **1951** *Ibid.* XXVII. 520 An interrupted sequence is considered to include the phoneme of *pause*. **1952** W. P. Lehmann *Proto-Indo-European Phonol.* ii. 10 Choice of positional variant in PIE was determined by preceding and following phoneme, group of phonemes, or pause. **1957** B. Deutsch *Poetry Handbk.* (1958) 35 The pause in the last foot of the second line is made more emphatic because the words conclude the line and the poem as well. **1968** J. Lyons *Introd. Theoret. Linguistics* v. 200 The native speaker is able to actualize the 'potential pauses' in his utterances when he wishes to, even though he does not do this normally. **1972** M. L. Samuels *Linguistic Evol.* ii. 13 Unvoicing of final consonants (i.e. voicing is not maintained till the end of the word, especially before a pause).

5. *pause-filler, -linking, -marker, -pattern, rhythm, -substitute; pause-giving* ppl. adj.; **pause-pitch,** the pitch pattern which characteristically precedes a pause in utterance.

1967 A. Laski *Seven Other Years* i. 13 It sounded like the kind of remark which is made as a pause-filler. **1887** A. Seth *Hegelianism* ii. 74 [T. H.] Green..constantly assumes a stream of sensations as the material upon which the pause-giving and rationally constitutive activity of thought is exercised. **1963** *Economist* 9 Nov. 577/1 Papers with..pause-giving titles. **1970** *Canad. Jrnl. Linguistics* XV. 112 Thus, if the pause-linking rule was solely phonological we would have no way of determining whether or not it applies to these phrases. **1956** *Kenyon Rev.* XVIII. 433 The lesser pause-markers: comma, colon and semi-colon. **1965** *Times Lit. Suppl.* 25 Nov. 1070/3 New sentence-shapes, new pause-patterns. **1933** L. Bloomfield *Language* vii. 115 We must recognize *pause-pitch* or *suspension-pitch* [,], which consists of a rise of pitch before a pause within a sentence. It is used.. to show that the sentence is not ending at a point where otherwise the phrasal form would make the end of a sentence possible: *I was waiting there* [,] *when in came the man.* **1902** E. W. Scripture *Elem. Exper. Phonetics* xxxvi. 517 A tone may be sounded for a definite time at definite intervals. The result is a 'rhythm of sound and pause,' or..a 'pause rhythm'. **1964** J. L. M. Trim in D. Abercrombie et al. *Daniel Jones* 375 Major tone-groups.. are followed by a pause, or pause-substitute.

pause, *v.*¹ Add: **3.** *trans.* To cause to stop temporarily.

a **1542** Wyatt *Coll. Poems* (1969) 104 Sorowfull david.. yt..pausid his plaint and layd adown his harp. **1908** A. S. M. Hutchinson *Once aboard Lugger* II. ii. 101 The strain on his invention paused him.

pauseful, *a.* Add: **b.** That causes a pause. (Cf. Pausefully *adv.*)

1958 *Times* 5 May 12/5 A pauseful finger was being laid upon his life.

pauw (pāu, ‖pǭu), var. Paauw. Cf. *pouw.

1800 G. Yonge in G. M. Theal *Rec. Cape Colony* (1898) III. 197 Pauws or Wild Peacocks are become extremely scarce. **1838** [see *gompaauw, gompauw]. **1886** G. A. Farini *Through Kalahari Desert* v. 62, I saw a large *pauw* out in the open. **1939** S. Cloete *Watch for Dawn* iii. 39 To make this feast there had been a great killing: of oxen,..of wildfowl, guineas, pauws, and pheasants.

pav¹. (pæv). Abbreviation of Pavilion *sb.* (sense 6): *spec.* (*a*) the London Pavilion (a music-hall and theatre, later a cinema); (*b*) a cricket pavilion.

1864 Hotten *Slang Dict.* 197 *Pav.*, the Pavilion Theatre,—sometimes called the P.V., *i.e.*, pe-ve. **1892** *Idler* Mar. 127 One Saturday night, I wended my way to the 'Pav.'. **1901** *To-Day* 26 Sept. 266/1 The retiring victim [*sc.* a stumped batsman] came back to the Pav. **1903** [see *cart *v.* 1 d]. **1934** *Observer* 1 Apr. 18/1 That long-established landmark, the London Pavilion (better known to its habitues as 'the Pav.') is following the example already set it by the Empire and the Alhambra and 'going over to the pictures'. **1978** *Guardian* 20 Mar. 9/1 St. B's always had them, especially behind the cricket pavs. **1978** L. Meynell *Papersnake* i. 10, I had gone over to the sports pavilion, 'the Pav'..to make sure the gear was all right for the match.

pav². Abbreviation of *pavlova. *Austral.* and *N.Z.*

1966 G. W. Turner *Eng. Lang. Austral. & N.Z.* viii. 173 *Pavlova cake*, a meringue sweet, sometimes shortened to *pav*. **1974** *Herald* (Melbourne) 5 Apr. 23/1 The pavlova (now fondly abbreviated to the 'pav').

pave, *v.* Add: **1. c.** To form a pavement for; to be a pavement under.

1821 Shelley *Epipsych.* 15 The air-like waves Of wonder-level dream, whose tremulous floor Paved her light steps.

2. c. To write interlinear or marginal translations in (a Latin or Greek text-book). *School slang.*

1888 [implied in *paving *vbl. sb.*]. **1897** A. Sidgwick in P. A. Barnett *Teaching & Organisation* 308 Cases of dishonesty are pretty certain to turn up..to 'pave' the text, *i.e.* write the English down at the side. **1940** M. Marples *Public School Slang* 52 A common word of special meaning is pave, which denotes the practice of writing the English meaning above words in a Greek or Latin text.

pave, *sb.* (Earlier examples.)

1835 *Southern Lit. Messenger* I. 357, I met a friend on the *pave* last week. **1859** [see Nymph 2 b]. **1880** J. Bowick *Montrose Characters* 138 Gaun pauchlin' alang the pave.

pavement, *sb.* Add: **1.** (Further examples.) Also (without *a*), paving.

a **1817** [see *knowing *ppl. a.* 3]. **1843** A. B. Blackie *Wood Pavement* 13 The efficient labour of a horse on Wooden Pavement, compared with that of the same horse on a perfectly consolidated Macadamized road, being as 42 to 17. **1900** T. Aitken *Road Making* ix. 300 Streets of many English towns are still paved with cobbles, but these are being gradually replaced by better descriptions of pavement. **1947** *Engineering News-Record* 16 Oct. 534/3 The surface of the runways, which it is hoped will be adequate for the pavement base, will be finally solidified by a 'super-compactor' unit. **1952** *Jrnl. R. Aeronaut. Soc.* LVI. 879/1 The pavement must be capable of carrying with safety the heaviest aircraft. **1977** *Bitumen* (Shell Internat. Petroleum Co.) 7 Shell companies' main interest in bitumen technology has been the engineering properties of bitumen and the structural design of roads and airfield pavements.

b. For '*obs.*' read '*obs.* exc. *techn.* and *U.S.*' Hence (*techn.* and *U.S.*), the roadway as distinct from the adjacent footway. (Further examples.)

1877 D. K. Clark in Law & Clark *Construction of Roads* 12 The surface of the pavement soon became very uneven, and not unfrequently sunk so much as to form hollows, which rendered it..dangerous to horses and carriages. **1918** E. Poole *Dark People* i. 5 You could see the sidewalks on either hand, but the dark wooden pavement of the street was almost lost in shadows. **1935** H. W. Horwill *Dict. Mod. Amer. Usage* 226/1 When preparations were being made in Washington for a procession.., the newspapers of the city complained that in Pennsylvania Avenue the grandstands filled the sidewalk and compelled pedestrians to walk on the pavement. **1939**

Liverpool Daily Post 9 Nov. 3/2 When any road works are about to be undertaken, a notice headed 'Reconstruction of Pavement' is exhibited near the scene of operations, stating that the 'pavement of this thoroughfare' is shortly to be reconstructed. It is invariably the roadway, and not the sidewalk, which in these cases receives attention. **1958** *Engineering* 4 Apr. 441/3 It will also be a double carriageway,..comprising 7 in of granular fill on which will be laid 11 in of reinforced concrete, placed in a single pass for the full 24 ft width of each pavement. **1966** R. Ashworth *Highway Engin.* x. 171 The modern road pavement is usually composed of several layers of material of differing quality; the strongest material being placed uppermost and forming the actual running surface. **1971** D. Hamilton *Poisoners* x. 78 She was barreling out on.. the road that'll take you clear down Baja California to La Paz, if you and your vehicle are tough enough to make it... The pavement ends about ninety miles south of Ensenada at present. After that, things get pretty rough. **1976** *Billings* (Montana) *Gaz.* 17 June 1-A/4 A southern Indiana woman died when her car skidded out of control on wet pavement.

f. (See quot. 1965.)

1899 P. Dearmer *Parson's Handbk.* v. 128 The thurifer and boat-bearer..go to the right of the priest, as he stands on the pavement. **1908** *Ritual Notes* (ed. 5) II. 59/3 When the Deacon descends to the pavement, the Thurifer will stand at his right. **1922** C. Mackenzie *Altar Steps* xix. 212 The baldacchino was given by one rich old lady, the pavement of the church by another. **1936** *Server's Manual* 6 Go and stand in some convenient place on the 'pavement' of the sanctuary. **1965** C. E. Pocknee *Parson's Handbk.* (ed. 13) ii. 23 The pavement, i.e. the level of the sanctuary between the lowest step before the altar and the communicants' rail, should extend to six feet at the very least. *Ibid.* ix. 98 He remains standing on the pavement swinging the censer until the hymn or psalm is finished. **1978** *Church Times* 20 Jan. 3/4 An application was made for a faculty to remove the sanctuary pavement and transfer from a columbarium underneath 177 caskets containing..cremated remains.

2. (Further examples.)

1931 J. R. Norman *Hist. Fishes* vii. 126 They [*sc.* the teeth of Nurse Sharks and Hounds] are arranged in pavement fashion, and all or most of the rows are in use at the same time. **1971** P. J. P. Whitehead tr. *Budker's Life of Sharks* iii. 41 The only sharks having teeth in a 'pavement' are those belonging to the genera *Mustelus, Hexanchus* and *Heterodontus*.

b. *Geol.* A horizontal or gently sloping expanse of bare rock.

1827 G. P. Scrope *Mem. Geol. Central France* vii. 154 The lower portion of this bed is very beautifully columnar, the upper obscurely so; this latter has been in parts destroyed, and a pavement or causeway left, formed by an assemblage of upright and almost geometrically regular columns fitted together with the utmost symmetry. **1932** C. R. Longwell et al. *Physical Geol.* vii. 157 Many gentle slopes above the levels of the playas are floored with 'desert pavements' consisting of pebbles fitted so closely together and with their top surfaces so even that the general effect suggests a mosaic. **1937** Wooldridge & Morgan *Physical Basis Geogr.* xix. 288 Bare limestone surfaces commonly show a widening of joints by solution, or in extreme cases, a complex fretting or fluting of the surface... Limestone pavements tending to this type are called 'clints' or 'grykes' in the North of England. **1954** J. F. Kirkaldy *Gen. Princ. Geol.* vi. 69 The direction of ice movement can also be proved if glaciated pavements can be found. These are surfaces of rock, hard enough to be smoothed and polished by the ice and showing striations caused by the harder rocks dragged across them by the ice. **1965** *Proc. Geologists' Assoc.* LXXVI. 421 Carboniferous limestone in Great Britain and Ireland is frequently exposed in broad, curiously sculptured plateaux termed pavements. **1977** *Oxf. Diocesan Mag.* Nov. 10/1 We have botanized over a limestone 'pavement' in Westmoreland, several bogs in Scotland, [etc.].

4. *pavement café*; **pavement-pounder** *slang*, a policeman; **pavement princess** *Citizens Band Radio slang* (see quots.); **pavement-toothed** *a.*, having teeth arranged in a pavement (see sense 2).

1953 *Observer* 10 May 5/4 It isn't worth..doing much about pavement cafés or open-air dancing. **1972** J. Aitken *Butterfly Picnic* i. 12 Several pavement cafés were thronged with elderly men. **1942** Berrey & Van den Bark *Amer. Thes. Slang* § 460/17 *Policeman*..ossifer, pavement pounder, P.D. **1947** K. Jaediker *Tall, Dark & Dead* vii. 102 Neal had put out a teletype for my car, and some Brooklyn pavement-pounder had spotted it. **1959** I. & P. Opie *Lore & Lang. Schoolch.* xvii. 369 There are, in the London area, at least thirty nicknames... Nobby, Pavement Pounder, Peeler, Robert, [etc.]. **1976** L. Dills *CB Slanguage Dict.* (rev. ed.) 53 *Pavement princess*, roadside or truckstop prostitute. **1976** *Time* 10 May 79/2 Prostitutes ('pavement princesses') who plug their charms on CB have become so common that there is even a song about them. **1976** *Daily Tel.* (Colour Suppl.) 16 July 10/2 Car-borne prostitutes..describe themselves as 'a dream for sale' in the West..a 'pavement princess' and a 'snuff-dipper' in the East. **1904** *Nature* 5 May 13/1 He discusses the affinities of the pavement-toothed genus Endothiodon.

pavement, *v.* Add: (Later example.) Also *transf.* Hence **pa·vementing** *vbl. sb.*

1930 R. Clements *Grey Seas* 126 The blown, empty sky, pavemented by the tossing sea. **1977** *Lancet* 20 Aug. 402/2 At 30 and 60 min, these showed an inflammatory reaction to both solutions with pavementing of polymorphs and a perivascular infiltrate.

paver. Add: **1. b.** A machine for depositing and spreading material for a road, etc., as it travels.

1947 *Engineering News-Record* 16 Oct. 535/3 A paver was used to mix the soil-cement and to deposit it along the pipe line as required. **1955** *Concrete Roads* (Road Research Lab.) xv. 279 For large works, pavers which consist of a non-tilting mixer mounted on crawler tracks are sometimes convenient. **1972** *Travelling* Autumn 33/2 The concrete slab will be laid with a very high accuracy using a purpose built slip form paver. **1977** *Bitumen* (Shell Internat. Petroleum Co.) 5 The mixes are laid by various types of mechanical paver.

Pavian (pā·viăn), *sb.* and *a.* [f. *Pavia* name of a city of northern Italy (L. *Ticinum*) + -an.] **A.** *adj.* Of, pertaining to, or characteristic of Pavia or its people. **B.** *sb.* A native or inhabitant of Pavia.

1856 O. Jones *Gram. Ornament* xvii. 5 Our woodcuts, selected from the Piscina of the High Altar, furnish some idea of the general style of the Pavian arabesques. **1888** W. Benham tr. *Platina's Lives of Popes* I. 260 John the Fourteenth, a Roman, or, as some will have it, a Pavian. **1936** A. W. Clapham *Romanesque Archit.* iii. 62 In some of the Pavian churches an acanthus decoration is to be met with, consisting of a regular diaper or brocade pattern of palmette leaves. **1940** G. F.-H. & J. Berkeley *Italy in Making* III. viii. 202 The Pavian student volunteers had occupied Colà.

pavilion, *sb.* Add: **6.** (Earlier examples of use in *Cricket*.)

1799 *Times* 1 June 3/4 The colours..were presented.. to the corps in Lord's cricket ground... After the military ceremony was over, the Earl and Countess..partook of a cold collation provided for them in the pavillion. **1853** F. Gale *Public School Matches* 10 All of a sudden the bell from the Pavilion strikes up, and the ground is gradually cleared.

7. b. (Later examples.)

1938 *Amer. Speech* XIII. 228/1 A *ward* is a unit or division in the hospital, often called a *floor, pavilion*, or by the number or letter under which it is listed. **1973** *Lancet* 7 July 33/1 Princess Alexandra Eye Pavilion, Royal Infirmary of Edinburgh.

pavillon. Add: **2.** *pavillon chinois* (ʃīnwa) = *Chinese pavilion, *jingling Johnny* (*a*).

1876 Stainer & Barrett *Dict. Mus. Terms* 347/1 *Pavillon chinois*, an instrument consisting of little bells attached to a staff. **1920, 1970** [see *jingling *ppl. a.*].

paving, *vbl. sb.* Add: **a.** (Examples in sense *2 c of the vb.)

1888 H. Logeman *Rule of S. Benet* p. xxxvi, Dr. Thompson..said that the Rugby boys' slang term for this process was *paving*—paving smooth (I suppose) the rough road of learning Latin. **1914** 'I. Hay' *Lighter Side School Life* v. 138 He is greatly addicted to a more venial crime known as 'paving'. The paver prepares his translation in the orthodox manner, but whenever he has occasion to look up a word in a lexicon he scribbles its meaning in the margin of the text, or, more frequently, just over the word itself, to guard against loss of memory on the morrow. **1958** L. Foster in *Aspects of Translation* 10 The 'paving' of books by schoolboys and the old-fashioned classical 'literal crib' are rather different cases of translations intended to facilitate comprehension of the original text, not to supplant it.

b. *paving-block, -sand* (later example).

1911 *Daily Colonist* (Victoria, B.C.) 1 Apr. 7/3 The city will purchase from the Michigan Puget Sound Lumber Company a quantity of wood paving blocks sufficient to complete the pavement on View Street. **1968** J. Arnold *Shell Bk. Country Crafts* xxxi. 332 The Scots pine is planted cultivated and felled..for the primary purpose of providing pit-props and telephone poles, railway sleepers and at one time for paving-blocks. **1934** *Ledger-Dispatch* (Norfolk, Va.) 11 June 7/8 There is glass sand, moulding sand, building, paving, grinding and polishing sand.

pavisand (pæ·visænd), *v.* [f. Pavisade.] *intr.* To display a formidable array of clothing and ornament; to flaunt one's appearance. Hence **pa·visander.**

1910 Kipling *Rewards & Fairies* 297 Forth she come pavisanding like a peacock—stuff, ruff, stomacher and all. **1950** I. Brown *Having Last Word* 90, I can picture Queen Elizabeth pavisanding at times. *Ibid.*, Among great pavisanders also was Milton's Delilah.

Pavlov (pa·vlǫf, pæ·vlǫv). Also **Pavloff,** etc. The name of the Russian physiologist Ivan Petrovich *Pavlov* (1849–1936), used *attrib.* or in the possessive to designate aspects of his work, esp. those connected with conditioning the salivary reflexes of a dog to the mental stimulus of the sound of a bell.

1911 Stedman *Med. Dict.* 644/2 *Pavloff method*,..a quantitative study of the modifications of the salivary reflexes caused by psychic reactions. **1922** K. Dunlap *Elements Sci. Psychol.* xiv. 303 The development of the auditory-salivary reaction in Pavloff's dog. **1933** J. C. Flügel *Hundred Years Psychol.* xi. 208 He became much concerned with Pavlov's conditioned reflex and the psychology of food. **1949** Koestler *Insight & Outlook* xxviii. 379 The Pavlov-trained dog, when faced with an ambiguous stimulus,..becomes deranged in all his reflexes. **1967** *Listener* 3 Aug. 138/2, I had a kind of Pavlov-dog reaction, shaking with nerves, because I'd been very nervous when I'd done that first film nine years before. **1974** *Sunday Times* (Colour Suppl.) 27 Oct. 30/4 Such de- or re-conditioning is quick, cheap, and, like Pavlov's bells, it gets results. **1976** A. White *Long Silence* ii. 19 You're not a Pavlov dog, trained to bark when I ring a bell. You have a mind of your own.

pavlova (pævlōu·vă). *Austral.* and *N.Z.* [f. the name of Anna *Pavlova* (1885–1931), Russian ballerina.] A dessert or cake, now usually one made with meringue, whipped cream, and fruit. Also *attrib.*

1927 *Davis Dainty Dishes* (ed. 6) (Davis Gelatine, N.Z., Ltd.) 11 Pavlova... Dissolve all but a teaspoonful of Gelatine in the hot water, and all the sugar except a dessertspoonful [etc.]. **1929** K. McKay *Pract. Home Cookery* 155/1 *Pavlova cakes*... Cook like meringues... They are delightful and simple to make besides being a novelty. **1952** *Weekly News* (Auckland) 30 July 14/4 (*heading*) Soft-centred Pavlova Cake. *Ibid.*, R.J.S. (Nelson) writes:—Could you give me a recipe for making a pavlova with a soft centre? *Ibid.*, Most good pavlova recipes are soft inside. **1952** B. Nilson *Penguin Cookery Bk.* xviii. 396 Pavlova Cake (to use as a cake or cold sweet). **1957** *Daily Mail* 7 Oct. 11/4 Pavlova. *Ingredients*: 3 egg whites 6 oz. castor sugar 1 teaspoonful vanilla 1 teaspoonful vinegar 1 teaspoonful cornflour ½ pint double cream (whipped and flavoured) 16 oz can Australian pineapple or apricots cherries angelica. **1958** *N.Z. News* 2 Dec. 10/2 Supper included some renowned New Zealand dishes such as pavlova, whitebait, and oysters. **1960** I. Cross *Backward Sex* 85 I'll give you some of Mum's pavlova cake for supper. **1964** *Guardian* 18 Apr. 5/4 A Pavlova..is a meringue basket so called because it spreads out like the skirts of a ballerina. **1968** *N.Z. News* 11 Dec. 11/5 Pavlova cake—the New Zealand and Australian sweet—is believed to have been created as a compliment to the famous dancer when she visited those countries. **1972** V. C. Clinton-Baddeley *To study Long Silence* v. 191 A *Pavlova*—a New Zealand speciality of choice fresh fruit and whipped cream wrapped in meringue-like base. **1975** *Times* 16 Dec. 12/4 A Pavlova, an Australian dessert..a meringue with cream, passion fruit, ice cream and strawberries.

Pavlovian (pævlōu·viăn), *a.* [f. *Pavlov + -ian.] Of, pertaining to, or connected with Pavlov, his theories, experiments, or methods. Also in extended and weakened senses.

1931 A. Huxley *Let.* 24 Aug. (1969) 351 The effects of such sociological reforms as Pavlovian conditioning of all children. **1951** H. Humphrey *Thinking* 6 In the Pavlovian system,..we have an attempt to account objectively for all psychological facts in terms of the primary interaction of organism and environment. **1952** V. Nabokov *Nabokov's Dozen* (1959) 206 A mad Pavlovian world where..variations in simple visual values influence and gradually replace flavour. **1963** A. Heron *Towards Quaker View of Sex* 67 Some recoveries with 'deconditioning' treatment along Pavlovian lines. **1974** *Daily Tel.* 8 Feb. 8/2 The report does not hesitate to name names, a procedure that will inevitably touch off a Pavlovian response from Leftist circles to deride it as a 'Reds under the Beds' scare. **1976** *Survey* Winter 24 The Soviet Union.. managed to exploit the South African issue in order to create the well-known Pavlovian reflex all through Black Africa. **1977** *Meanjin* XXXVI. I. 108 The adroit orchestration of comedy..that is, the patrons were not treated as a collection of complete Pavlovian half-wits.

‖ pavor. Restrict *Obs. rare*—⁰ to sense in Dict. and add: **b. pavor nocturnus** [L. *nocturnus* nocturnal], a sudden and inexplicable terror which may afflict a sleeping person, esp. a child, in the night; = *night-terror* s.v. Night *sb.* 14; similarly **pavor diurnus** [L. *diurnus* belonging to the day] (see quot. 1940).

[**1848** Dunglison *Dict. Med. Sci.* (ed. 7) 634/1 *Pavores nocturni seu dormientium*, fear during sleep.] **1889** *Albany Med. Ann.* June 200 The victim of night terrors, or *pavor nocturnus*, experiences an awful, unpleasant, terror stricken disturbance of the mind. **1889** J. Thomson tr. *Henoch's Lect. Children's Dis.* I. 241 One of the rare cases of pavor diurnus which I have seen affected the son of an actor (7 years old), a nervous, anaemic, delicate child. **1900** *Lancet* 3 Feb. 292/1 Two cases of pavor diurnus..have recently been under my care. **1915** J. N. Hall *Borderline Dis.* I. iii. 96 Nightmare in adults and pavor nocturnus in young children are vivid dreams with sensations of oppression in the chest, of horror, and inability to escape some impending catastrophe. **1927** W. P. Lucas *Mod. Pract. Pediatr.* II. xii. 698 Pavor diurnus is more rarely seen than night terrors but is more significant of a disorder of the nervous system. **1940** Hinsie & Shatzky *Psychiatric Dict.* 406/1 *Pavor diurnus*, fear reactions which occur in the young child during the afternoon nap, similar to night terrors but not so frequently taking place as the latter. **1950** J. Strachey tr. *Freud's Totem & Taboo* iv. 128 Phobias of this type..are, in my opinion, at least as common in childhood as *pavor nocturnus*. **1966** *Sci. & Psychoanal.* IX. 176 The purposiveness of the rumination is seen in the fact that..the event does ultimately become commonplace... The suggestion is strong that the *pavor nocturnus* of the child may have a similar accustomizing function.

pavvy (pæ·vi). Also **pavy.** Abbreviation of Pavilion *sb.* Cf. *pav¹.

1899 Kipling *Stalky & Co.* 159 Forty shillin's or a month for hackin' the chucker-out of the Pavvy on the shins. **1900** Farmer *Public School Word-Bk.* 146 *Pavvy, The* (Harrow).—The pavilion on the cricket-ground. **1905** H. A. Vachell *Hill* v. 117, I say, there's going to be a ruction in front of the Pavvy. Come on! **1961** Partridge *Dict. Slang* Suppl. 1215/1 *Pavy, the*,..The pavilion: Harrovian:..Hence, at certain other Public Schools': c. 20.

paw, *sb.*¹ Add: **4.** *paw-mark, print* (also *fig.*); also *pawful*.

1925 F. M. Ford *No More Parades* ii. 73 She resembled a white Persian cat luxuriating, sticking out a tentative pawful of expanding claws. **1964** D. Varaday *Gara-Yaka*

PAW

xix. 173 The invaders replied very effectively to the warnings, and the Prodigal and his family [of lions] had a pawful of trouble. **1894** 'Mark Twain' in *Century Mag.* June 234/2 Are you going to ornament the royal palaces with nigger paw-marks? **1929** D. H. Lawrence *Lett.* (1932) 833 Such dark paw-marks of the wind on the sea! **1975** *Sunday Times* 16 Nov. 44/4 Every pawmark shows up on those virgin white surfaces. **1925** *Scribner's Mag.* July 33/1, I saw..the curious paw print of a porcupine, with its little pebbled markings. **1938** M. K. Rawlings *Yearling* iv. 35 All about were the paw-prints of the small things. **1963** *Times* 25 Jan. 12/7 The badger, whose paw-prints are square-fronted and easily recognizable, is better off. **1968** C. Nicole *Self Lovers* vii. 88 The whole thing has his pawprints all over it. His speciality is taking his victims swimming at dawn. **1977** D. Harsent *Dreams of Dead* 55 Dark ooze by the apple tree stippled with massive paw-prints.

paw (pǭ), *sb.*[4] *U.S.* = Pa.
1903 *Dialect Notes* II. 324 *Paw, maw*, nouns. Father; mother. (In the North *pa*; *ma*.) **1919** E. O'Neill *Rope* in *Moon of Caribees* 183 Come on back to the house, Paw. It's gittin' near supper time. **1929** W. Faulkner *Sound & Fury* 46 Your paw told you to stay out that tree. **1933** J. V. Allen *Cowboy Lore* iv. 67 He said he had to leave his home, his paw had married twice. **1935** Z. N. Hurston *Mules & Men* i. vii. 163 His paw said, 'Son, Ah don't see how you gointer do dat.' **1939** in *Jrnl. Amer. Folklore* LII. 108 I am Peetie Wheet Straw, the devil's son-in-law, The woman I married, old Satan was her paw. **1942** Ade *Let.* 1 Feb. (1973) 228 The little red school-house is a thing of the past but don't forget that it turned out some of our best people, including possibly your paw and maw and, certainly, your grandparents. **1968** E. J. Gaines *Bloodline* 247 He follow his mom and paw out the house **1975** J. Gores *Hammett* (1976) xiv. 102 'What's the brother's name?' 'Don't rightly know. May be my paw—.'

paw, *v.* **3. a.** Now esp. to fondle (usu. a woman) lasciviously. Also const. *about, over.*
1902 Ade *Girl Proposition* 58 He told himself that he was a Chump for continuing to worship one who could be pawed over and man-handled by anything that wore a Derby Hat. **1918** H. G. Wells *Joan & Peter* xi. 387 A fellow had to..watch..Joan being ordered about and.. *pawed* about. **1928** A. Huxley *Point Counter Point* xi. 176 Other men were liable to pounce on you and try to paw you about and kiss you. **1934** E. O'Neill *Days without End* ii. 70 Walter was drunk, pawing over his latest female. **1942** A. Christie *Body in Library* ii. 24, I hate to see a girl..who..lets a disgusting Central European paw her about. **1955** G. Freeman *Liberty Man* i. iii. 51 Maureen had been mad for him to go on pawing her for hours last night. **1959** 'C. Carnac' *Death of Lady Killer* xii. 136 A real dirty tyke he was, always trying to paw any woman within reach. **1975** J. I. M. Stewart *Young Pattullo* vii. 153 Fish, who had decent feelings, would have preferred to be pawed in privacy. **1978** D. Bloodworth *Crosstalk* xxxi. 240 The outraged shopper said..that when he saw her looking at him he had winked and pawed her.

pawa, var. *paua.

‖ **pawang** (pā·waŋ). Also (*rare*) **puwang.** [Malay.] A Malay sorcerer or medicine-man; a wizard or witch; a wise man, expert.
1821 J. Leyden tr. *Malay Annals* 51 He immediately ordered an artificer to be sent for, named Pawang Bentan. **1839** T. J. Newbold *Pol. & Statistical Acct. Straits of Malacca* II. viii. 98 They have 'wise men', or Puwangs, who pretend to be able to ascertain the most favourable spots for sinking a mine. **1893** F. A. Swettenham *About Perak* 33 A Malay Páwang (medicine-man) has the same sort of nose for tin that a truffle dog has for truffles. **1906** *Macm. Mag.* Aug. 778 An old *pawang*, or sorcerer, stepped forward with a bunch of twigs of a tree for which a tiger is thought to have a peculiar dread. **1907** F. A. Swettenham *Brit. Malaya* vii. 156 When a patient becomes dangerously ill..it is common..to call in a *páwang*, a kind of wizard or witch, who tries by incantations and other forms of the black art to lure the evil spirit from his prey. **1933** L. Ainsworth *Confessions Planter in Malaya* 128 A 'Pawang' or witch doctor was called, and for the sum of twenty dollars he agreed to perform the complete ceremony. **1965** C. Shuttleworth *Malayan Safari* ii. 34 Contact with the spirit world is only made through the medium of the tribal *pawangs* or medicine-men. **1972** M. Sheppard *Taman Indera* 94 *Pawang*, an expert in any art believed to involve the use of magic.

pawing, *vbl. sb.* and *ppl. a.* (Further examples, esp. of sense *3a).
1749 J. Cleland *Mem. Woman Pleasure* II. 134 The tiresome pawing and toying. **1906** E. Nesbit *Railway Children* xii. 256 Like most boys..[he] hated..kissing and holding of hands. He called all such things 'pawing'. **1931** W. Faulkner *Sanctuary* xix. 205 Impassable, swinging hands with their escorts, objects of casual and puppyish pawings, they dawdled up the hill toward the college. **1935** *Scrutiny* IV. 128 Revolutionary feeling is for her bound up with an incessant kissing and pawing between and among the sexes. **1951** J. C. Fennessy *Sonnet in Bottle* iii. v. 82 Their pawing hands stretched out..to feel and finger their prisoners. **1977** H. Osborne *White Poppy* xxix. 194 A certain amount of kissing and pawing, but absolutely no more. **1978** *Times* 7 Sept. 13/3 Ladies never touch their gentlemen in public... Pawing and clinging, with its nasty carnal implications, is reserved for..foreign adventuresses.

pawkily, *adv.* Add later non-dial. examples.
1963 *Times* 21 May 4/5 The Essex batsmen, only partially inhibited by the cold, pushed the score pawkily along, sending up the 50 in just over the hour. **1971** *Guardian* 8 Sept. 8/3 His pawkily British sense of humour.

pawky, *a.* Add later non-dial. examples.
1935 K. A. Porter *Flowering Judas* 112 He went on in his pawky way trying to make clear to her his mystical faith in these men who went ragged and hungry. **1966** *Listener* 3 Feb. 171/2 With these advantages, and his convivial, pawky wit, which was enjoyed as much over the port wine as it was over tea in the withdrawing room, Radcliffe's progress was remarkable. **1970** *Daily Tel.* 10 Nov. 12/3 Beethoven's Trio in B flat, Opus 11, on the other hand, was given a rather pawky performance, and the first movement, especially, was robbed of weight by the over detached articulation. **1973** *Daily Record* (Glasgow) 6 Aug. 6/3 Her other pawky comments include: T is for *Training*: This is entered into with particular vigour by the Reserve Team, so that they may escape relegation to the First Eleven. **1976** W. Gérin *E. Gaskell* xiv. 146 Mr. Brontë could..be witty and pawky as his later letters to Mrs. Gaskell show.

pawl, *v.* Add: Also **pall. 2. a.** Also, to detect.
1859 Hotten *Dict. Slang* 71 *Pall* to detect. **1975** H. R. F. Keating *Remarkable Case* viii. 92 She's been palled once..a-trying ter get a look at that door.

pawlonia, var. *paulownia.

pawn, *sb.*[1] Add: **c.** *pawn end-game*; **pawn chain**, an unbroken diagonal line of pawns extending across several adjacent files; **pawn skeleton**, the structure of the pawns at the end of a chess opening; **pawn storm**, an attacking advance of pawns against a castled king.
[**1818** W. S. Kenny *Pract. Chess Exercises* 45 This is better than breaking his chain of pawns.] **1937** M. Euwe *Strategy & Tactics in Chess* iii. 50 An immobile sequence of Pawns is called a pawn-chain. **1957** L. Barden *Guide to Chess Openings* v. 105/2 White is hoping to safeguard the base of his pawn-chain. **1973** *Sci. Amer.* June 104/2 The program now is sophisticated enough to execute very simple positional ideas such as creating pawn chains, striving for control of the center and so forth. **1939** A. Alekhine *My Best Games of Chess* 86/2 K-K4, P-QKt4!.. with a won Pawn end-game. **1914** J. du Mont tr. *Lasker's Chess Strategy* (1915) iv. 26 Each opening is characterised by a well-defined pawn formation... Naturally the formation of a pawn skeleton is not an independent factor. **1927** *Brit. Chess Mag.* XLVII. 170 First you must learn the normal Pawn skeleton of the opening very thoroughly. **1950** R. N. Coles *Chess-Player's Week-End Bk.* 11 If the pawn skeleton remained sound, the game could be continued from one phase to another. **1926** *Brit. Chess Mag.* XLVI. 134 Herr Bachmann makes it plain that such 'modern' tactics as..the Pawn-storm against the opponent's Castled position..are not 'modern' at all. **1957** L. Barden *Guide to Chess Openings* iv. 69/2 If White can carry through a pawn storm and advance his KBP to KB6 then Black will be completely throttled.

pawn, *sb.*[2] **4.** Add: †**pawn party**, app. a game resembling blindman's buff. *Obs.*
1831 H. Smith *Festivals Games, & Amusem.* (N.Y.) 330 The village and country lasses enjoy their *spinning* and *quilting* bevies, singing-schools, and *pawn* parties, with at least an equal zest. **1952** *Amer. Speech* XXVII. 47 A pawn party must have been something like blindman's buff.

pawn, *v.* Add: **d.** *Stock Exchange*. To deposit (stock) with a bank as security.
1902 *Encycl. Brit.* XXXII. 866/1 So much stock is 'pawned' with banks that the conclusions arrived at by the jobbers from examining only what they are carrying over themselves are liable to be falsified.

Pawnee (pǭnī·), *sb.*[2] Also 8–9 **Pane.** [ad. Canad. F. *Pani*, f. Ioway-Oto *panyi*.] † **a.** The Wichita Indians, *spec.* the Wichita and Tawehash bands of the Red River of Oklahoma and Texas. *Obs.*
1770 P. Pittman *Present State Europ. Settlem. on Mississippi* 40 The Arcansas or Quapas Indians..bring in very frequently young prisoners and horses from the Cadodaquias, Paneise, Podoquias, &c. of which they dispose to the best advantage. **1803** in C. E. Carter *Territorial Papers U.S.* (1940) IX. 74 The length of Red River is not known, it is Six or seven hundred Miles to the Pawnie or Towiash Indians. **1830** in *Ibid.* (1954) XXI. 215 The settlers are in continual alarm from the Pawnee Indians.

b. A confederacy of Caddoan Indians, formerly inhabiting the Loup, Platte, and Republican River valleys in Nebraska; a member of this group of tribes. Also *attrib.* or as *adj.*
1778 J. Carver *Trav. N.-Amer.* 118 This is the road they [*sc.* Indians] take when their war parties make their excursions upon the Pawnees. **1794** [see *Mandan *a.* and *sb.*]. **1806** Z. M. Pike *Jrnl.* 22 Sept. in *Acct. Expeditions Sources Mississippi* (1810) ii. 140 [I] met a Pawnee hunter. **1810** [see *Kiowa *sb.*]. **1827** J. F. Cooper *Prairie* II. xi. 177 He will never see a Pawnee become a Sioux. **1841** J. Williams *Narr. Tour Indiana to Oregon* (1921) 31 The Caws..told me that the Pawnees were a bad nation, and that they had a battle with them. **1868** *N.Y. Herald* 31 July 5/3 A large band of the Sioux and Cheyennes had attacked a small party of the Pawnee scouts. **1890** J. G. Frazer *Golden Bough* I. iii. 381 The Pawnees annually sacrificed a human victim in spring when they sowed their fields. **1901** 'Mark Twain' in *N. Amer. Rev.* Feb. 163 The oldest Americans are the Pawnees. **1925** Z. A. Tilghman *Dugout* 13 The Pawnees were late going south that year. **1946** G. Foreman *Last Trek of Indians* 242 The Sioux made two more raids, killing a Pawnee each time. **1959** E. Tunis *Indians* 84/1 Many tribes had broken off from them in that time and some had moved northward up the river valleys—the Pawnee to Nebraska, the Arikara to North Dakota, for instance. **1969** *Observer* (Colour Suppl.) 18 May 32/2 Mrs Hines was brought up in a Pawnee Indian school where bull-hide boots were compulsory. **1972** *N.Y. Times* 3 Nov. 78/4 Earlier, Mrs. Martha Gras, a 71-year-old Pawnee Indian, appeared to sum up the sentiment of the Indian gathering.

c. The Caddoan language of these people.
1806 Z. M. Pike *Let.* 2 Oct. in *Acct. Expeditions Sources Mississippi* (1810) App. ii. 48, I asked..[for] a Tetau prisoner who spoke Pawnee, to serve as an interpreter. **1821** J. Fowler *Jrnl.* (1898) 55 Mr. Roy—He Spoke Some Pane and (in) that language our Councils Ware Held. **1877** L. H. Morgan *Anc. Society* iii. iii. 440 In Mandan my brother's wife is my wife, and in Pawnee.. the same. **1965** *Canad. Jrnl. Linguistics* X. 105 The structure of Pawnee as compared with Oneida. **1968** R. W. Langacker *Language & its Struct.* viii. 231 Other families found in this area [*sc.* the Great Plains] are Caddoan (Wichita, Caddo, Arikara, Pawnee), [etc.].

d. Special combs.: (in sense a) **Pawnee pic, pique** (see also quot. 1916); (in sense b) **Pawnee lands**, reservations of Pawnee Indians; **Pawnee Loups, Mohas, Republics, Republicans**, sub-tribes of the larger Pawnee confederacy.
1931 *Amer. Speech* VII. 4 The 'Indian reserves' were frequently referred to as the 'Pawnee lands', 'Otoe Lands', or 'Indian territory'. **1806** in *Deb. Congress U.S.* (1852) 9th Congress 2 Sess., App. 1046 Pānias Loups (or Wolves)... These are also a branch of the Panias proper. **1823** E. James *Acct. Expedition Rocky Mts.* II. 165 The camp had been occupied by a war party of Skeeree or Pawnee Loup Indians. **1847** D. Coyner *Lost Trappers* (1859) 62 As Doranto proved to be a son of a grand chief of the Pawnee Loups, he was greatly prized as a captive. **1843** N. Boone *Let.* 11 Aug. in L. Pelzer *Marches of Dragoons in Mississippi Valley* (1917) 187 Whether this was caused by a fear that we'd frighten off the buffalo, or not, they kept up a continual alarm of Pawnee Mohas. **1806** in *Deb. Congress U.S.* (1852) 9th Congress 2 Sess., App. 1075 Pania Piqûe... [Also called] Paunee Piqûe. **1856** J. P. Beckwourth *Life* ii. 26 The Pawnee Pics or Tattoed Pawnees. **1916** Thoburn *Hist. Oklahoma* I. 124 The confusion of the two tribes was doubtless due to the French traders and trappers, who called the Wichitas 'Pawnee Piques', i.e. 'Tattooed Pawnees', hence the corrupted American term, Pawnee Pict. **1836** L. Ford in *Army & Navy Chron.* 19 May 312/1 The Pawnee Loups, Pawnee Republics, [etc.].. lie upon the Loup fork of the Platte, twenty or thirty miles distant from the Grand Pawnee village. **1917** Will & Hyde *Corn among Indians of Upper Missouri* 145 The Pawnee Republics had only enough corn to thicken their soup. **1806** in *Deb. Congress U.S.* (1852) 9th Congress 2 Sess., App. 1045 Pānias proper and Pānias Republican live in the same village. **1810** Z. M. Pike *Acct. Expeditions Sources Mississippi* App. ii. 14 On the La Platte, reside the grand Pawnee village, and the Pawnee loups on one of its branches, with whom the Pawnee Republicans are at war. **1823** E. James *Acct. Expedition Rocky Mts.* I. 159 They arrived about noon, seventy in number, consisting of individuals of each of the three tribes called *Grand Pawnees, Pawnee Republicans*, and *Pawnee Loups*.

pawn-haus, var. *ponhaus.

pax[1]. Add: **1. a.** (Examples of *pax Romana*.)
Cf. Pliny, *Nat. Hist.* xxvii. 3 immensa Romanae pacis maiestate.
1884 W. J. Clarke tr. *Duruy & Mahaffy's Hist. Rome* II. i. xxxiv. 201 Later, another form expressed the advantage, which was the compensation for this imperious sway, *pax romana*, that 'Roman peace' destined to draw the nations together and blend all languages,.. whose boundless majesty, *immensa romanæ pacis majestas*, the nations will honour with sincere homage. **1888** *Encycl. Brit.* XXIII. 591/1 The *pax Romana* died with the empire. **1934** G. B. Shaw *On Rocks* Pref. 183 We Romans have purchased the *pax Romana* with our blood. **1956** A. Toynbee *Historian's Approach to Relig.* ii. xvi. 212 The final collapse of the *Pax Romana* in the third century. **1974** *Encycl. Brit. Macropædia* II. 371/2 This Pax Romana..ensured the survival and eventual transmission of the classical heritage.

a*. Also combined freely with Latinized adjs. to form phrases on the model of *pax Romana*, as *pax aeronautica, Americana, atomica, Britannica, Communistica, Egyptiana, hispanica, Sovietica.*
1933 M. Arlen *Man's Mortality* ix. 180 These privately-owned stations..were of immense service to inter-communication throughout the world... They were the visible symbols of the pax aeronautica. **1967** 'R. Raine' *Wreath for Amer.* xxv. 182 The whole Western world..is living under..a Pax Americana, just as the world once lived in peace under a Pax Britannica. **1971** *Newsweek* 18 Oct. 17 In the opinion of many European experts, the Kremlin has a more ambitious objective: to alter the status quo by exchanging the *Pax Americana* for a *Pax Sovietica*. **1976** *Times Lit. Suppl.* 26 Nov. 1488/5 The 'Pax Americana', a phrase which several contributors toss about carelessly, is and was a myth, if meant to take in the world as a whole. There has not been so much peace since the Second World War for the right term to be either Pax or Americana. **1966** R. Ardrey *Territorial Imperative* (1967) viii. 283 Pessimism, under the rule of a *Pax Britannica*, was a dirty little luxury which any could afford; under a *pax atomica* it carries small selective value. **1969** W. Garner *Us or Them War* xxv. 203 I'll give you a toast, girl. The *pax atomica*! Long may it last! **1886** *Jrnl. R. United Service Inst.* May 865 We should see Pax Britannica far transcend what Pliny called the 'immensa Romanæ Pacis Majestas'. **1899** [in Dict.] **1939** E. H. Carr *Twenty Years' Crisis* v. 104 In the past, Roman and British imperialism were commended to the world in the guise

of the *pax Romana* and the *pax Britannica*. **1960** *Times* 29 Sept. (Nigeria Suppl.) p. x/3 As a result of the *Pax Britannica*..the population has increased steadily. **1967** J. Cleary *Long Pursuit* v. 113 We brought peace to this part of the world, Pax Britannica is more than just a phrase. **1969** J. Mander *Static Society* ii. 83 It brought greater stability..than did the *pax britannica*. **1973** *Listener* 1 Feb. 146/2 In the 19th century, the British Navy dictated the Pax Britannica and more or less guaranteed a century of peace. **1946** H. Nicolson *Diary* 22 Aug. (1968) 75 They [*sc.* the Russians] believe that.. people will say in the end, 'Anything for peace', and will accept the Pax Scythica or Communistica. **1961** L. Mumford *City in Hist.* iii. 88 There may well have been [in ancient Egypt] a long period, a Pax Egyptiana, that relaxed both the internal tensions and the need for external protection. **1969** J. Mander *Static Society* ii. 83 The centuries of peace under the *pax hispanica* that shielded Latin Americans. **1945** A. Huxley *Let.* 7 May (1969) 522 One can only hope that the Pax Sovietica may last as long as the Pax Romana. **1969** *Sunday Times* 19 Oct. 49 There won't be a Pax Sovietica or anything like it, because the world is too big to be governed by anybody. **1977** *Time* 29 Aug. 19/3 Bell-bottom denims, miniskirts and platform shoes have turned Ulan Bator's girls into the prettiest within the Pax Sovietica.

d. *Pax Romana*, also the name of an international organization of Roman Catholic students.
1957 M. P. Fogarty *Christian Democracy in W. Europe* xvii. 263 Catholic graduates of all faculties are linked up through *Pax Romana* which also includes students. **1967** D. T. Kauffman *Dict. Relig. Terms* 349/1 *Pax Romana*, world organization of Catholic students. In Europe it is called the International Movement of Catholic Students.

Paxarete (paχāre·tə). Also **Pajarete, Pascarete, Paxarette.** [f. *Paxarete*, a small town in the Jerez district of Spain.] A mixture of fortified wine and boiled-down grape juice, formerly drunk as a sherry, now used primarily for colouring or sweetening sherry or whisky.
1827 Scott *Chron. Canongate* 1st Ser. II. vi. 151 When they were comfortably seated over a bottle of Paxarete. **1846** R. Ford *Gatherings from Spain* xiv. 152 It is of this grape that the rich and luscious sweet wine called *Pajarete* is made. **1891** in C. Ray *Compleat Imbiber* (1967) IX. 122 Sherry... Paxerette. **1920** G. Saintsbury *Notes on Cellar-Bk.* ii. 18 Light Spanish wines..for instance, the lighter Paxarettes..which most literary people to-day associate only with Sir Telegraph in 'Melincourt'. **1958** A. L. Simon *Dict. Wines* 122/1 The *Pajarete* or *Paxarete* liqueur wine..is mostly used for sweetening high-class sweet sherries. **1965** O. A. Mendelsohn *Dict. Drink* 254 *Paxarete*, Spanish compound wine, also known as Pedro Ximénèz, made by fortifying *arrope* (concentrated grape juice)... Also known as *pascarete* and *pajarete*.

paxillar, *a.* (Example.)
1900 *Proc. Zool. Soc.* 292 Paxillar crowns are very large and oval.

paxillose, *a.* Add: **c.** Provided with paxillæ; paxillate.
1900 *Proc. Zool. Soc.* 290 The abactinal surface is paxillose, each paxillus having a circular crown of about eight papillæ, the centre being usually smooth.

Paxolin (pæ·ksǒlin). Also **paxolin.** A proprietary name for a type of laminated plastic widely used as an electrical insulating material.
1918 *Trade Marks Jrnl.* 23 Jan. 86 Paxolin... Electrical insulating preparations... The Micanite & Insulators Company, Limited,..London,..manufacturers. **1924** *Harmsworth's Wireless Encycl.* III. 1529/1 Paxolin is an insulating material made of paper impregnated with a varnish consisting of phenol and formaldehyde. **1944** *Electronic Engin.* XVI. 385/3 The objective lens was mounted inside a piece of paxolin tubing. **1958** *Official Gaz.* (U.S. Patent Office) 28 Oct. TM 135/2 The Micanite & Insulators Company Limited, London, England, Filed Feb. 14, 1958. Paxolin... For electrical insulating materials. **1960** *Practical Wireless* XXXVI. 413/1 To make such a screen, a piece of thin insulating material such as paxolin is wound with 26 s.w.g. wire. **1962** N. H. Codling in G. A. T. Burdett *Automatic Control Handbk.* viii. 16 Characteristics of grade S5 silicone glass Paxolin are given in Table 9. **1965** *Wireless World* Aug. 377/1 Each unit has been assembled on a paxolin board using connecting pins.

pay, *sb.* Add: **4.** † **b.** In early N. Amer. colonial use, any article used as a medium of payment. *Obs.*
1663 *Early Rec. Portsmouth, Rhode Island* (1901) 118 To sell the tounes cow..for wompom or other pay. **1681** *Town Rec. Topsfield, Mass.* (1917) I. 34/2 Twente pownd of it in siluer forti five pownd in other pay as namli in Corne porke and beefe. **1704** S. Knight *Jrnl.* (1825) 42 Pay is Grain, Pork, Beef, &c. at the prices sett by the General Court that year. **1767** in *Essex Inst. Hist. Coll.* (1912) XLVIII. 75 And if you should purchase light pay, then proceed for Turks island.

5. (Later examples.)
1873 *Trans. Illinois Dept. Agric.* X. 249 Many farmers were very slow pay. **1926** J. Black *You can't Win* iv. 28 They were good pay, but he could not get away from his work at the right hour to find them. **1973** R. Thomas *If you can't be Good* (1974) xii. 103 Everybody in town is falling over each other to give him credit—even though he's a slow pay.

6. (Earlier and later examples.) Also, one of oil or natural gas. Also *concr.*, the bed itself.
1857 J. D. Borthwick *Three Yrs. in Calif.* vii. 140 After prospecting a little, we soon found a spot on the bank of the stream which we judged would yield us pretty fair pay for our labor. **1868** *Rep. J. Ross Browne Mineral Resources West of Rocky Mts.* (U.S. Treasury Dept.) 101 In 1866 they struck into pay and erected a 10-stamp mill, which is driven by a hurdy-gurdy wheel. **1933** E. Caldwell *God's Little Acre* i. 2 I've been digging in this land close on fifteen years now, and..I figure we're going to strike pay pretty soon. **1975** [see *pay zone* s.v. *pay- 2]. **1977** R. E. Megill *Introd. Risk Analysis* ix. 107 Every oil or gas field has a certain areal size, a thickness of the producing formation and a recovery per foot of pay.

7. Short for Paymaster. *slang.*
1878 F. Davenport *On Man-of-War* 114 While the boy went forward after the steward, Pay regarded the omelet gloomily. **1914** 'Bartimeus' *Naval Occasions* xxiii. 216 Give it a shake, Pay, and put it on like a man! **1916** 'Taffrail' *Pincher Martin* x. 173 Cashley, the fleet paymaster, was vainly endeavouring to get up a four at auction bridge... 'Going to take a hand?'..'Bridge,..not to-night, Pay; thanks, all the same.' **1942** [see *boffin 1]. **1944** K. D. McCracken *Baby Flat-Top* 57 The Head of the Supply Department is known universally as 'Pay'.

pay, *v.*[1] Add: **2. d.** *not if you paid me, him,* etc.: under no circumstances; not at all.
1896 Kipling *Seven Seas* 153 He couldn't lie if you paid him, and he'd starve before he stole! **1910** E. M. Forster *Howards End* xxiv. 201, I couldn't live near her if you paid me. **1952** M. Allingham *Tiger in Smoke* iv. 67, I shan't sleep, you know, I couldn't if you paid me. **1959** M. Summerton *Small Wilderness* i. 8 Six months ago I was all set to read law, now I wouldn't go back to swotting if you paid me. **1979** R. Jeffries *Murder begets Murder* iii. 19, I wouldn't stay on here if you paid me to. The moment I've sorted everything out I'm off.

3. b. *pay out* (later examples).
1914 D. H. Lawrence *Widowing of Mrs. Holroyd* iii. 66 He'd say to himself he'd pay me out. That's what he always does say, 'I'll pay thee out for that bit—I'll ma'e thee regret it.' **1940** G. D. H. & M. Cole *Counterpoint Murder* v. 51 He told Best to do it just in order to annoy him, to pay him out. **1951** M. Kennedy *Lucy Carmichael* i. v. 36 I'm paid out for saying I wouldn't have a wreath. I thought a sort of vague cloud round my head would look nice. **1978** J. Thomson *Question of Identity* xiii. 140 It was his way of paying out Maguire for giving him the push.

5. b. *to pay in* (later examples.)
1951 R. W. Jones *Thomson's Dict. Banking* (ed. 10) 468/2 Paying-in slips may also contain a notice requesting customers to cross all cheques before paying them in. **1961** *New Eng. Bible Matt.* xvii. 27 Take the first fish that comes to the hook, open its mouth, and you will find a silver coin; take that and pay it in; it will meet the tax for us both. **1978** N. J. Crisp *London Deal* v. 93, I didn't pay that money in, and no one else paid it in..and handed me the slip.

c. *to pay with the fore-topsail*: to leave port without paying one's debts or creditors. *Naut. slang.*
1843 J. F. Cooper *Ned Myers* 149 We sailed next morning, and I paid for the poor 'nigger' with the foretopsail. **1850** H. Melville *White Jacket* I. ii. 5 The *middies* were busy raising loans to liquidate the demands of their laundress, or else—in the navy phrase—preparing to pay their creditors *with a flying fore-topsail*. **1910** D. W. Bone *Brassbounder* 262 Paid 'ee wi' tawps'l sheets, didn't 'e? **1929** F. C. Bowen *Sea Slang* 101 *Pay debts with the fore topsail*, to slip away to sea in debt.

6. c. *pay up* (examples in *absol.* use).
1941 J. D. Carr *Case of Constant Suicides* xiv. 190 The insurance companies would have been compelled to pay up. **1972** *Daily Tel.* 13 Jan. 13/3 More than 100,000 TV licence bilkers have paid up since an anti-evasion campaign was launched last October. **1978** N. Marsh *Grave Mistake* ii. 43 I'm always having to yank him out of trouble... I'll go on paying up, I suppose.

9. b. *pay in*: to make (regular) contributions *to* a fund.
1911 *Rep. Labour & Social Conditions in Germany* (Tariff Reform League) III. 71 Men must pay in to the trade society to which they transfer their labour. *Ibid.* 82 Employees..commence to pay into State fund when 16 years old.

10. Also *to pay out.*
1909 'O. Henry' *Roads of Destiny* xii. 193 Nobody in the bank knows those notes as I do. Some of 'em are a little wobbly on their legs, and some are mavericks without extra many brands on their backs, but they'll most all pay out at the round-up.

c. *pay off*: to succeed; to be profitable or advantageous; to show results.
1951 R. Malkin *Boxcars in Sky* 118 It was the Big Briefing which, eventually, would pay off in plenty of free space outside the advertising columns. **1953** J. Wain *Hurry on Down* iv. 73 Like a good many insane actions, it paid off. **1957** *Listener* 7 Nov. 757/1 Still, the cool piece of blackmail and bluff paid off. **1959** *Ibid.* 4 June 969/2 There are signs, already, that this policy of patience is paying off. **1962** A. Shepard in *Into Orbit* 104, I could feel that all the training we had gone through with the blockhouse crew and booster crew was really paying off down there. **1967** *Technology Week* 23 Jan. 61/1 (Advt.), Our aim is to make current space hardware and experience pay off for the national space program. **1971** R. Dentry *Encounter at Kharmel* ix. 150 Your hunch had better pay off. The Tal's the last possibility. **1978** T. Allbeury *Lantern Network* viii. 93 By mid-May the training had begun to pay off.

13. Also *transf.*
1962 H. Hood in R. Weaver *Canad. Short Stories* (1968) 2nd Ser. 211 He paid out a little string and began to run across the parking lot towards the main building. **1976** D. Clark *Dread & Water* ii. 52 'Roped together?' 'Just like the book says... He went first and I payed out... I hadn't enough hands to cling tightly and pay out the rope'.

pay-. Add: **1. b.** *pay-check, -cheque, code, -envelope* (examples), *packet, -scale.* **c.** *pay-box* (esp. = *box office 1), *-desk, -night* (earlier and later examples), *-table* (later examples), *-window.* **d.** *pay-bed, -school, -toilet, ward.*
1895 *Brit. Med. Jrnl.* 2 Mar. 501/2 In opening..certain pay beds the Committee of Management had simply felt that they were carrying out what they were bound to do under the constitution of the hospital. **1928** *Daily Express* 19 July 9/4 The special committee appointed..to consider the needs of the professional and middle classes recommend that additional 'pay-beds' should be provided for them by the hospitals. **1934** *Lancet* 23 June 1371/2 To establish pay-beds for patients of moderate means. **1969** 'W. Haggard' *Doubtful Disciple* v. 52 In his pay-bed in the hospital Jacky D was depressed. **1974** *Times* 6 Nov. 2/8 The Owen working party..is discussing the new consultant contract and the pay-bed issue. **1976** H. Wilson *Governance of Britain* iv. 86 Fresh concern arose about government legislation and action in connection with pay-beds and private practice, in October and December 1975. **1851** Dickens in *Househ. Words* 30 Aug. 531/2 He darts upon my luggage..pays certain francs for it, to a certain functionary behind a Pigeon Hole, like a pay-box at a Theatre. **1889** J. K. Jerome *Three Men in Boat* xix. 313 We attracted a good deal of attention at the Alhambra. On our presenting ourselves at the pay-box we..were informed that we were half-an-hour behind our time. **1952** V. Gollancz *My Dear Timothy* 20 The practice was to open this door, and let people up the long staircase to a point a little short of the pay-box. **1975** N. Luard *Travelling Horseman* iii. 71 The butler who'd..contact[ed] them from a street pay-box. **1904** 'O. Henry' in *Everybody's Mag.* Aug. 240/1 Joe Wheeler signs the voucher for his pay-check. **1964** Mrs. L. B. Johnson *White House Diary* 24 Apr. (1970) 122 How sensible these girls are to be starting on a skill that they can exchange for a paycheck almost anywhere. **1977** *Time* 10 Oct. 56/2 Chicagoans tell of a local executive who has supposedly never spent a paycheck in 30 years but lives entirely off the expense account. **1930** J. Collier *His Monkey Wife* xvii. 243 You take your pay cheque at the end of the week without making a lot of fuss about the size of it. **1973** J. Wainwright *Touch of Malice* 71 The inconvenience went with the rank and the pay-cheque. **1976** *Sunday Mail* (Glasgow) 21 Nov., And that wouldn't come into line with the Government's pay code which..limited everyone to £6 a week. **1898** A. Bennett *Man from North* vii. 50 Jenkins eagerly drew Richard's attention to the girl at the pay-desk. **1919** W. Deeping *Second Youth* vii. 64 Nearly always she sat at the same table near the pay-desk. **1932** D. L. Sayers *Have his Carcase* xxx. 395 Bunter was four behind him in the queue at the pay-desk. **1909** *Sat. Even. Post* 5 June 17/1 (*heading*), The pay envelopes of the stars. **1911** E. Ferber *Dawn O'Hara* iv. 46 My bank account has always been an all too small pay envelope at the end of each week. **1973** E. Taylor *Serpent under It* (1974) xi. 172 Mr. Ramsay finally rrremembered that the rrrest of us like to eat, and left the pay envelopes. **1820** C. Mathews *Let.* 25 July in A. Mathews *Mem. Charles Mathews* (1839) III. vii. 148 The common outcry was against Saturday for a second performance, as it is pay-night, and the worst night in the week. **1970** G. Greer *Female Eunuch* 288 They have taken the pay packet out of the old man's pocket when he has finally arrived home on pay-night. **1941** 'N. Blake' *Case of Abominable Snowman* ix. 101 A stoppage of work and less money in the pay-packet on Fridays. **1973** *Times* 1 Jan. 3/2 Miners' institute libraries..were financed by pennies from miners' pay packets. **1976** *Norwich Mercury* 19 Nov. 1/4 Many firms find that a hamper is much appreciated by the workforce rather than '£10 in the pay packet', particularly as payment is subject to tax but a hamper is not. **1961** *Lancet* 9 Sept. 595/1 Our results make it possible to establish a pay-scale, based on production. **1976** F. Muir *Frank Muir Bk.* 95 Teaching..became a respectable profession..with its own union to keep an eye on pay scales. **1856** X. D. MacLeod *Biogr. F. Wood* 191 The cost to us in taxation is not one fifth the usual expense for an ordinary pay-school education. **1936** M. Mitchell *Gone with Wind* v. lii. 905 There were no free schools. Few had money to send their children to pay schools. **1915** Kipling *New Army* 29 The men..saluted emphatically at the pay-table, and fell back with their emoluments. **1965** 'Lauchmonen' *Old Thom's Harvest* xvi. 181 Every Saturday..the men were at the paytable in the ball-field. **1977** *Times Lit. Suppl.* 25 Mar. 334/5 His now receptive mind drifts back into the past where he is, once again, leader and paymaster of a racially mixed crew... The ghosts of the men again come forward to the paytable. **1946** E. Hodgins *Mr. Blandings builds his Dream House* ix. 132, I wouldn't come back here if you had a pay toilet. **1947** Auden *Age of Anxiety* (1948) iii. 72 The scene has all the signs of a facetious culture, Publishing houses, pawnshops, and pay-toilets. **1968** *Listener* 7 Nov. 610/1 There's a song about not having a dime for a pay toilet, and another one about the sexual relevance of Kleenex. **1976** *New Yorker* 15 Nov. 66/2 There was a character named Exotica A La Carte, who lived on the proceeds of a string of pay toilets. **1895** *Brit. Med. Jrnl.* 2 Mar. 502/1 It is desirable that the system of pay wards now in operation be so modified that the patients in those wards may..be attended in the hospital by outside practitioners of their own selection. **1934** *Archit. Rev.* LXXV. 92/1 The first doorway has a glass window which serves as a pay-window. **1977** A. C. H. Smith *Jericho Gun* i. 16 It's a winning ticket... If you present it at the Tote pay window, even you might get some money back.

2. *pay-dirt* (further examples); also *fig.*; *-ground, -rock* (examples), *-streak* (examples), *zone* (later examples).
1866 *Dublin Rev.* Jan. 10 Even officers of men-of-war were seized by the gold mania, and 'ran' to soil their white hands in the precious 'pay-dirt'. **1930** E. Rice *Voy. to*

Purilia xvii. 249 As one elderly prospector expressed it: 'That's pay dirt in them thar hills.' **1935** *Motion Picture* Nov. 46/2 Frances Langford, the cutie whose 'pay dirt' reaches $1,000 per week. **1948** A. Huxley *Ape & Essence* (1949) 132 We see Loola standing in a three-foot hole wearily digging... 'When you hit the pay dirt..come and report to us.' **1953** *Economist* 24 Oct. 248 Senator McCarthy may have struck pay-dirt at last. **1967** *Ibid.* 26 Aug. 721/3 Senator Dirksen..applauds the President for permitting the bombers to attack targets within ten miles of the Chinese frontier: 'We seem to be getting close to paydirt.' **1972** *Lebende Sprachen* XVII. 34/1 US pay dirt—BE/US workable ore-bearing soil. **1973** C. Callow *Power from Sea* ii. 60 The thickness of the 'pay dirt'—that is the width of the sand in which the gas was contained. **1977** *Rolling Stone* 21 Apr. 88/3 Bowie hits celestial pay dirt on one of the pieces. **1927** *Daily Tel.* 25 Oct. 2/7 The results, taken in conjunction with the pay ground passed through the haulage, gives promise of..an important shoot. **1862** 'Mark Twain' *Let.* 8 Feb. (1920) 51 We'll have a mill-site, water power, and pay-rock, all handy. **1947** W. A. Chalfant *Gold, Guns, & Ghost Towns* 141 Thompson and Ramsay prospected over the same ground ..and struck some pay rock that was almost the pure stuff. **1856** *Daily Even. Bulletin* (San Francisco) 11 Oct. 1/1 These lucky miners worked one, two, and even three years, to reach the pay streak. **1910** R. W. Service *Ballads of Cheechako* 68 Late in the year he struck it rich, the real pay-streak at last. **1965** G. J. Williams *Econ. Geol. N.Z.* vii. 80/1 The gold is distributed over the schist bottom in more or less orientated paystreaks trending north-westerly. **1977** *New Yorker* 20 June 85/1 A pay streak appeared to be there, and what was needed now was a means of moving gravel in a major way. **1973** C. Callow *Power from Sea* ii. 61 The yield was 3.6 million cubic feet a day and the 'pay zone' was 64 feet thick. **1975** *Offshore* Sept. 75/2 It encountered three pay zones between 8,468 ft and 9,595 ft, with about 130 ft of net oil pay and a smaller zone of about eight ft of gas pay.

3. pay-back, the fact or action of paying back; reward or return, *spec.* the net return in profits from an investment project equal to the initial capital outlay; also *attrib.*; also as *adj.*, retaliatory.
1959 *Wall St. Jrnl.* (Eastern ed.) 31 Mar. 13/1 The payback order affects only the Star-Bulletin and not the union. **1965** H. I. Ansoff *Corporate Strategy* ii. 14 Three common methods for evaluation are the payback period, the internal rate of discount, and the net present worth. **1970** M. Kelly *Spinifex* iii. 63 'I can remember him taking part in a big pay-back raid a few years later.' 'Payback?' 'Pidgen for vendetta——.' **1971** C. R. W. Wysock Wright in B. de Ferranti *Living with Computer* iii. 22 This necessary investment in good education is essential although the payback may only be in the long term. **1972** *Accountant* 28 Sept. 391/2 You simply cannot say that.. discounted cash flow is superior to payback. **1973** *Sunday Times* (Colour Suppl.) 10 June 46/2 Such 'payback', or revenge killings, are common in the Highlands. **1977** *Irish Times* 8 June 10/7 It would, he said, depress cash flow and would lengthen the payback period for the mine's capital cost.

b. pay-as-you-earn, applied to a system of collecting income tax where the tax payable is deducted by employers from current earnings; **pay-as-you-enter**, applied to a system of collecting fares in public transport where passengers pay the driver on entering the vehicle; **pay-as-you-go**, applied to a system of paying or discharging obligations as they are incurred; **pay-as-you-see**, applied to a system of supplying an extra television channel to viewers paying either by subscription or by inserting coins into a box attached to the television receiver; **pay-as-you-use**, applied to a system of paying for an article while it is being used; **pay-as-you-view** = *pay-as-you-see.*
1943 *Daily Tel.* 22 Apr. 3 As to deduction of income-tax from wages he warned the House of the experiences of other countries which had attempted the 'pay-as-you-earn' idea. **1972** *Accountant* 6 Apr. 445/2 Pay-as-you-earn rates for 1972 are so much higher than last year that they could raise the tax take by the equivalent of some £1,500 million. **1908** *Sci. Amer.* 1 Feb. 76/2 (*heading*) A new type of fare register to be used on the New York pay-as-you-enter cars. *Ibid.*, The pay-as-you-enter cars which it is proposed to install on the New York city lines will be equipped with a device for collecting and automatically registering fares. **1913** *Chambers's Jrnl.* Aug. 623/1 This [system of tram-car operation] is the 'pay as you enter', or as it is termed briefly the 'paye' system. **1966** *Daily Tel.* 12 Sept. 20/3 A bus driver in Brighton collecting a fare from a passenger yesterday when the country's first pay-as-you-enter, one-man operated double deck bus service began for a experimental period. **1840** *Farmers' Cabinet* 15 May 319/1 Pay as you go..is the truest economy. **1888** J. Bryce *Amer. Commonwealth* II. lxvi. 507 'Pay as You Go' Convention! **1936** M. W. Childs *Sweden* xii. 214 The central government has consistently followed a pay-as-you-go tax policy. **1969** *Times* 30 Apr. 26/2 The rate of contribution needed from employees and employers jointly to finance the proposed State pensions on a 'pay-as-you-go' basis would increase from 7.7 per cent of earnings in 1972–73 to 11.4 per cent in 2002–03. **a 1974** R. Crossman *Diaries* (1977) III. 176 We must therefore really accept pay-as-you-go. **1978** *Dumfries & Galloway Standard* 21 Oct. 5/7 (Advt.), It's pay-as-you-go; no big quarterly shocks. **1955** *Times* 6 Apr. 6/6 Sir Alexander Korda intends to seek permission of the United Kingdom authorities to operate 'pay as you see' television. **1962** *Variety* 22 Aug. 3 It's perhaps significant that all, or just about all, film distributors are leasing pictures to the Paramount-owned International Telemeter pay-as-you-see video in Etobicoke, Toronto suburb. **1929** *Radio Times* 8 Nov. 453/1 Place your order.. immediately, 'Pay as you use' terms can be arranged. **1961** *Engineering* 1 Sept. 257/1 Pay-as-you-use techniques covering the leasing of vehicles, machinery and equipment of all kinds are to be introduced into Britain. **1958** *Spectator* 11 July 61/3 The vital issue of whether a Pay-As-You-View television service is allowed to operate in this country. **1963** *Ann. Reg. 1962* 443 Other recommendations of the [Pilkington] Committee were that pay-as-you-view television..should be rejected.

c. *pay-on-* —: denoting a system of paying for a service when the action expressed has been performed or fulfilled.
1899 J. London *Let.* 29 July (1966) 46 And with these pay-on-acceptance fellows, did you ever get your check at the same time you were notified of acceptance? **1960** *Guardian* 11 Apr. 8/3 London's new..bus which..has the pay-on-entry system. **1961** *Daily Tel.* 23 May 17/6 (*heading*) 'Pay-on-reply' telephoning. *Ibid.*, Post Office engineers yesterday demonstrated the new 'pay-on-answer' coin boxes in seven telephone kiosks in the centre of Dartford, Kent.

4. Special combs. of Pay *sb.*, as **paybob** *slang*, = Paymaster; **pay freeze** = *pay pause*; **pay pause**, a period during which no wage or salary increases should occur, esp. as proposed in 1961 in the U.K.; **pay phone** orig. *U.S.*, a telephone that operates or connects the caller when coins are inserted; **pay-rise**, a wage or salary increase; **pay station** *U.S.*, (*a*) a public call-office; (*b*) a public telephone-booth; **pay-telephone** = *pay phone*; **pay television** = *pay-TV*; **pay-tone** (see quot. 1962); **pay-train** (see quot. 1969); **pay-TV**, a system of television based on the pay-as-you-see principle (see sense *3b).
1916 'Taffrail' *Pincher Martin* viii. 125 Carn't yer get a hadvance o' money from th' paybob? **1962** Granville *Dict. Sailors' Slang* 87/1 *Paybob*, senior accountant officer. **1978** *Navy News* Dec. 6/1 The paybob and his chum never batted an eyelid as I signed my chit and I often wonder if they paid the difference. **1961** *New Scientist* 7 Sept. 568 The Chancellor of the Exchequer's pay-freeze. **1972** *Listener* 23 Nov. 690/3 The pay and wage freeze is a shoddy compilation. Over two million workers demanding pay rises are trapped in the freeze. **1975** *Guardian* 21 Jan. 26/3 The Chancellor..said there was no question of a pay freeze in the foreseeable future. **1961** *Daily Tel.* 19 Sept. 1/8 The increases for primary and secondary schools will cost £42 million a year, £5½ million less than the Burnham recommendation rejected by the Government on account of the 'pay pause'. **1969** C. Booker *Neophiliacs* vi. 158 On 25 July [1961], the Government mustered all the paraphernalia to meet a major crisis, from a 7 per cent Bank Rate to Selwyn Lloyd's celebrated 'Pay Pause', the first recent peacetime attempt by a Government to impose an overall regulation of wage increases. **1977** *Evening Post* (Nottingham) 27 Jan. 5/1 When the pay pause was relaxed for doctors earning over £8,000, it was decided not to pay the increment for another year. **1936** L. Duncan *Over Wall* ii. 31 He would then go to one of the pay phones. **1952** W. R. Burnett *Vanity Row* (1953) xiii. 90 'Where you calling from?' 'A pay-phone.' **1973** W. McCarthy *Detail* iii. 215 He gave the agent the number of a pay-phone in Burbank. **1975** *New Yorker* 27 Oct. 32/1 There are six pay phones downstairs and only five of them work. **1976** *Daily Express* 1 July 14/2 In April 1976 I asked the Post Office to change my business phone to a pay phone. **1936** J. Steinbeck *In Dubious Battle* iii. 26 Hell, we don't want only temporary pay-rises. **1957** P. Worsley *Trumpet shall Sound* x. 197 The soldiers of the Pacific Islands Regiment had achieved pay-rises. **1923** M. Watts *L. Nichols* 209 [He] rushed off to the nearest telephone pay station to call up the Grace house. **1948** *Time* 21 June 2 When you drop a nickel in a pay station and dial a call..as many as 1000 telephone relays go into action. **1973** W. McCarthy *Detail* iii. 215 He would like you to call this number...it is a pay station. **1974** *Spartanburg* (S. Carolina) *Herald* 24 Apr. A3/4 Police said Norton had answered a call from two men using a pay station, asking that a cab pick them up at a motel. **1963** A. Orlov *Handbk. Intelligence* xiii. 153 The agent avails himself of the forthcoming 'pay telephone' conversation with his superior. **1964** *Punch* 13 May 702/2 A man known to have installed a pay-telephone. **1971** A. Hunter *Gently at Gallop* viii. 89 There was a pay-telephone for customers. He fed in a coin and dialled the police-station. **1976** *Billings* (Montana) *Gaz.* 28 June 6-A/3 The chairman of the American Telephone and Telegraph Co. said Sunday a pay telephone call probably will cost a quarter in most places. **1957** *Economist* 28 Sept. 1028/2 The Federal Communications Commission has tentatively opened the airwaves to 'pay television'. **1962** *Rep. Comm. Broadcasting 1960* 262 in *Parl. Papers 1961–2* (Cmnd. 1753) IX. 259 'Subscription television' (or 'toll' or 'pay' television, as it is variously called) involves a third method of paying. **1973** *Listener* 31 May 706/1 'The big news in cable now is Pay.' He meant pay television. **1962** *Sunday Express* 4 Feb. 12 The pips are the pay-tone signal, which tells the caller that you have picked up the receiver and it is time for him to put in the coins. **1972** J. Wainwright *Night is Time to Die* (1974) 167 On S.T.D., a call-box is a dead give-away by the pay tone! **1968** D. I. Gordon in *Regional Hist. Railway Gt. Brit.* V. xii. 232 The Eastern Region is making a gallant fight. Conductor guards and the whole concept of the 'basic' railway with its Pay Trains, fast and well-integrated services, [etc.]..are all features of the current scene. **1969** *Railway Mag.* CXV. 169/2 Conductor/Guard workings ('pay trains') began on January 19 between Newcastle and Carlisle. *Ibid.* 473/2 From June 15, 'paytrains'—on which tickets are issued by the guard—were introduced by the Eastern Region. **1972** *Times* 15 June 3/1 Public transport, even in an age of one-man buses, 'pay-trains', and automatic fare collection, is much more labour intensive than other industries. **1976** P. R. White *Planning for Public Transport* ii. 53 Little detail is recorded of trip patterns on 'paytrain' services. **1956** *Britannica Bk. of Year* 492/2 Also introduced from the United States—though not yet fully accepted into British English—were such expressions as Pay TV, [etc.]. **1960** *News Chron.* 29 June 4/6 'Pay-TV', a system whereby you put some coins in a slot at the top of the set and are rewarded by several hours uninterrupted viewing. **1972** *Listener* 8 June 773/2 Pay-TV was expensive and offered very little that was new. **1976** J. Lund *Ultimate* vi. 57 He looked at the pay TV programme—'Jaws' showing on odd days it said.

payability. Delete *rare* and add later examples.
1933 *Flight* 30 Mar. 296/1 We have on occasion discussed with Mr. C. C. Walker the subject of 'payability' of aircraft. **1955** *Times* 3 May 17/5 The 170 feet sampled assayed 182.32 dwts. over 5.12 inches or 933 inch-dwts. at a payability of 100 per cent. **1971** *Daily Tel.* 11 Oct. 17 The payability is low as advance development is now confined largely to the flanks of the mine where values are inclined to be more scattered.

payables (pē̆i·ăb'lz), *sb. pl. Comm.* [f. the adj.] Debts that one owes; liabilities. Cf. *receivable *sb.* 3.
1972 *Accountant* 12 Oct. 445/2 Money and receivables and payables. **1977** *Sci. Amer.* Apr. 95/1 (Advt.), You also get a complete package of programs for general ledger, payables, receivables, payroll, and inventory. **1977** *Detroit Free Press* 11 Dec. 18-D/7 (Advt.), Bookkeeper part-time. Full charge, receivable, payables, invoicing, and payroll.

pay-day. Add: (Further examples.)
1897 [see *account day]. **1895** [see *name-day 3]. **1916** 'Taffrail' *Pincher Martin* viii. 147 'Our ship's company made a bit of a pay-day over it.' 'Pay-day!' 'Ow d' yer mean?'..'Bettin'!' **1930** R. Campbell *Adamastor* 33 We attend the Great Inspection, The Roll-call of the Resurrection, The pay-day of Eternity. **1944** Dylan Thomas *Lett.* (1966) 262 I'll be up in London..always on payday Fridays. **1972** *Daily Tel.* 4 July 32/6 She laughed and replied 'They will have to do. It's the only pair I've got till pay day.'

b. Wages; the amount paid to a person on pay-day. *Naut.*
1915 D. W. Bone *Broken Stowage* 239 We had fondly hoped to be strutting on Liverpool streets with our women folk, a twelve-months' 'pay day' in our pockets. **1922** E. O'Neill *Anna Christie* (1923) ii. 60 'Tis no more drinking and roving about I'd be doing then, but giving my pay-day into her hand and staying at home with her as meek as a lamb. **1932** J. W. Harris *Days of Endeavour* 20 The Bos'n, his fat pay-day having dwindled away..had sailed.

paying, *vbl. sb.*[1] Add: **2. b.** *paying-in slip*: in *Banking*, a form listing cash, cheques, etc. which are paid in to the credit of an account; *paying-off pennant* (or *pendant*): a pennant flown by a homeward-bound naval ship or by a merchant vessel that will shortly discharge all hands.
1898 H. T. Easton *Work of Bank* iv. 26 The 'paying-in' slip can be utilised for pasting in the cash-book and ledger. **1968** 'C. Aird' *Henrietta Who?* x. 91 According to the paying-in slips, she always handed it over herself. **1974** 'S. Woods' *Done to Death* 50 When she makes up her paying-in slip for the Bank whatever cash there is has been taken over the counter. **1977** *Grimsby Even. Tel.* 14 May 5/8 Mrs. —— said she had signed other names 'on bank paying-in slips and things like that'. **1914** 'Bartimeus' *Naval Occasions* xxiii. 212 The paying-off pendant looks as if it were impatient. **1927** B. M. Chambers *Salt Junk* 50 At last the great day came: the hoisting of the paying-off pendant, a yard for every day of the commission. It was so long that the gilded bladder on the end fouled the houses which bordered the creek. **1954** Bradford & Quill *Gloss. Sea Terms* 142/2 *Paying-off pennant*, flown by a naval vessel when homeward bound from a foreign station. **1961** F. H. Burgess *Dict. Sailing* 158 *Paying off pennant*, an extremely long pennant hoisted as a sign that a ship is shortly 'paying off'. **1977** *Navy News* Aug. 40 H.M.S. Matapan enters Portsmouth Harbour for the last time, paying off pennant suspended from a large balloon.

paying, *ppl. a.* Add: (Earlier and later examples.) *paying guest*: a lodger.
1853 E. Clacy *Lady's Visit Gold Diggings Austral.* 111 The two holes were 'bottomed' before noon with no paying result. **1871** *Trans. Illinois Agric. Soc.* VIII. 238 We need not expect to get a paying crop from stiff clays. **1872** Geo. Eliot *Middlem.* IV. viii. 363 His skill was relied on by many paying patients. **1879** Trollope *John Caldigate* I. x. 135 'It's a paying concern, I suppose,' said Caldigate. 'It has paid;.. Whether it's played out or not, I'm not so sure...' **1895** G. Gissing *Paying Guest* i. 7 It's a very common arrangement nowadays, you know; they are called 'paying guests'. **1900** Paying guest [in Dict.]. **1929** E. F. Benson *Paying Guests* i. 9 Never had Wentworth and Balmoral and..Belvoir entertained so continuous a complement of paying guests. **1957** J. Braine *Room at Top* i. 16 Her voice paused perceptibly at the word *lodger* as if considering..the euphemisms—paying guest..and so on. **1958** J. Cannan *And be a Villain* vii. 148 Next pew—paying patients..then the doctors, the colleagues 'showing up'.

pay-load (pē̆i·lōud). [f. Pay- + Load *sb.*] **a.** The part of an aircraft's load from which revenue is derived; the passengers, mail, or cargo carried by an aircraft.

1930 *Times* 12 Nov. 11/4 Her [*sc.* a flying-boat's] range is determined by the amount of 'pay' load she has to carry. **1933** *Discovery* Dec. 367/2 There is no reason why the pay load should not be decreased to give an increased range but it could not carry a paying load for 2,000 miles. **1936** *Economist* 25 Apr. 190/1 From the operational aspect, the key factor of seasonal pay-loads is still present, and this must necessarily limit the prospects of large profits. **1946** R. A. McFarland *Human Factors Air Transport Design* x. 406 It is often contended that extra crew members replace passengers and payload. **1951** 'N. Shute' *Round Bend* 145 They pulled out a revised crew accommodation ..and added a hundred and ten pounds to the payload. **1955** *Sci. Amer.* Jan. 38/2 Helicopter designers are now busily engaged in harnessing jet engines to helicopters to raise their speed, range and payload. **1970** H. A. Taylor *Airspeed Aircraft since 1931* 170 It is practicable to use short, lightweight loading ramps which can be carried in the aircraft with a minimum loss of space and payload. **1978** *Jrnl. R. Soc. Arts* CXXVI. 685/1 The benefits of such scaling will be greatest on aircraft having the lowest percentages by weight of payload.

b. The bombs, warhead, etc., carried by an aircraft or rocket; the instruments, equipment, manned capsule, etc., carried by a rocket.
1936 *Sky* Nov. 7/3 Let the littlest and middlesize rocket serve as the 'pay load' of the biggest rocket. **1941** D. Garnet *War in Air* ii. 12 The British Bomber..could carry less petrol and more bombs, that is to say a higher 'pay-load' of bombs than the German Bomber. **1946** J. P. Baxter *Scientists Against Time* ii. 35 Allied scientists and intelligence officers racked their brains to determine what would be the pay load in these long-range missiles. **1955** *Times* 6 Aug. 6/1 The payload would be instruments capable of recording conditions outside the atmosphere and a telemetering apparatus to transmit findings to base. **1962** F. I. Ordway et al. *Basic Astronautics* iv. 121 Sounding rockets..with 50-lb payloads designed to analyze the results of these atomic explosions. **1970** *New Scientist* 8 Oct. 77/3 The proposed payload of the US space shuttle has come down some 50 per cent in the last three months. **1974** *Globe & Mail* (Toronto) 22 Oct. 7/1 In 1972, former President Richard Nixon.. promised that NASA would launch all foreign payloads intended for peaceful purposes under basically the same conditions that apply in the United States.

c. Goods, etc., carried on a road vehicle. Also *transf.* and *fig.*
1938 *Nation* (N.Y.) 13 Aug. 144/1 The trucks have to carry all their own gas, oil, and water..which does not leave much room for pay load. **1960** E. L. Cornwell *Commercial Road Vehicles* ix. 250 The articulated vehicle, weight for weight, generally has a rather lower payload capacity than a rigid vehicle. **1967** G. F. Fiennes *I tried to run a Railway* vii. 84 We have the pay load of some 500 tons transferred between road and rail by some six men. **1968** D. Braithwaite *Fairground Archit.* 88 But the electric 'Scenic'..constituted the heaviest pay load ever and often required the attendance of two traction engines. **1971** M. Tak *Truck Talk* 115 *Pay load*, the cargo or freight that a trucker hauls. **1971** C. Bonington *Annapurna South Face* vii. 85 Another factor one had to reckon with was that people moving back up the mountain, as opposed to ferrying, could carry only a very limited payload of supplies, if any, since they would be carrying their own personal equipment. **1976** P. R. White *Planning for Public Transport* x. 215 Although offering similar weight: payload characteristics as does an internal combustion engine and its fuel tank, such a 'fuel' would require large amounts of energy for its creation. **1978** J. B. Hilton *Some run Crooked* xiii. 132 Once the water-table overflows, there are gullies adding their pay-load from both sides.

So **Payloader, pay-**, the proprietary name of a type of heavy mobile machine used for lifting and loading.
1953 *Sun* (Baltimore) 12 Oct. 14/5 The Coast Guard informed Rukert Terminals Corporation that pay loaders could not be used in the bottom of the holds to remove the fertilizer because of possible fire hazard. The loaders are scoops operated by gasoline. **1957** *Muck Shifter* June 340/1 (*heading*) 'Four-in-one' attachment. The entire line of four-wheel-drive Hough 'Payloader' tractor-shovels will now offer Drott '4-in-1' buckets as optional equipment. **1959** *Trade Marks Jrnl.* 27 May 548/1 Payloader... Lifting, loading and mechanical handling machines and mechanisms and parts of all these goods.. The Frank G. Hough Co. Libertyville, State of Illinois, United States of America; manufacturers. **1963** T. Pynchon *V.* xii. 347 Someday there would be cranes, dump trucks, payloaders, bulldozers to come and level the neighbourhood. **1965** *Courier-Mail* (Brisbane) 10 Nov. 1/6 A man was killed at Mica Creek power station yesterday when a payloader he was driving overturned. **1969** *Jane's Freight Containers 1968–69* 267/1 The ancillary equipment is composed of..two 1.8 cu m payloaders, one 25 ton side-loader and several heavy-duty fork-lifts. **1977** G. A. Browne *Slide* xxii. 180 A heavy-duty ditch digger..called a payloader because of its combination digging-conveying system.

Payne's grey (pē̆inz grē̆i). [f. the name of William *Payne* (fl. 1800), English artist, + Grey *sb.*] A composite pigment composed of blue, red, black, and white permanent pigments, used esp. for watercolours.
1835 [see *neutral *a.* 3 c]. **1886** H. C. Standage *Artists' Man. Pigments* vi. 68 Payne's grey resembles neutral tint in being a compound colour. **1888** [see Grey *sb.* 4 e]. **1924** [see *neutral *a.* 3 c]. **1934** H. Hiler *Notes Technique Painting* ii. 133 As to the prepared greys,.. Payne's Grey, etc., they are usually mixed by the manufacturers. **1959** *Listener* 13 Aug. 254/3 The embellishment of the same green, sometimes with a touch of Payne's grey, fields. **1979** *Dryad Crafts* 49/4 Oil Colours... Payne's Grey.

pay-off. Also **payoff.** [f. vbl. phr. *to pay off* (in various senses): see Pay *v.*[1]] **1. a.** Winnings from gambling or the paying of these. Also *attrib.*
1905 F. Hutchison *Johnnie the Gent* 63 An' then there's the know-it-all-bloke that has just beat a couple of races, wit' about an ounce each way or maybe a deuce to peek. Oh, he's the wisest guy that ever give the pay-off gazebo the lofty leer when he reached for his dough. **1938** G. Greene *Brighton Rock* iv. i. 149 'I've won, Pinkie. A tenner.'.. A young man with oiled hair stood on a wooden step paying out money... Spicer called out to him..: 'Well, Sammy, now the pay-off.' **1943** *Sun* (Baltimore) 22 Apr. 18/1 This crowd backed New Moon confidently, the final payoff being $3.90 for $2. **1964** A. Wykes *Gambling* vi. 142 If he throws a natural or crap,.. the payoff odds will be considerably smaller. **1967** *Listener* 10 Aug. 168/2 A zero-sum game is one in which the pay-offs cancel out to zero—I've won sixpence, you're sixpence down. **1970** *Globe & Mail* (Toronto) 28 Sept. 20/2 How about the $800 daily double payoff the track made one day on a bet that never was made. Is that not bookmaking?

b. *Criminals' slang.* A confidence trick in which the victim is encouraged to venture a large sum of money by the success of a bet, investment, etc., involving a small sum or one furnished by the confidence tricksters. Also, one who employs this confidence trick (= *pay-off man* (*a*) (sense 5 below)).
1915 G. Bronson-Howard *God's Man* iii. iii. 197 Specialists in check-raising, wireless wire-tapping, 'the match', 'the pay-off', and cards. **1928** [see *creep *sb.* 1 d]. **1935** *Evening News* 29 June 3/2 The sucker is induced to put a small sum into one venture. His winnings are promptly paid and he has visions. This is the '*pay-off*'. **1938** P. J. Smith *Con Man* ix. 192 It is to his genius that the successful swindle known as the 'Pay Off' was attributed. **1943** *Police Jrnl.* Mar. 69 *Pay off*, a confidence trick—Stock Exchange fraud.

c. The return on an investment; profit. Also, the point at which an investment begins to yield profit.
1955 *Times* 5 Aug. 9/7 Countries which entered on the first stage would be relying on the second stage for their 'pay off'. **1969** *Daily Tel.* 11 Mar. 6/1 Profits in the past two years have been held back by the Woolco development and the pay-off here still could be a long way away. **1974** *Nature* 1 Feb. 248/1 The spending proposals entail vast concentration of resources in areas which are likely to have a quick payoff. **1974** *Information Handbk. 1974–5* (Shell Internat. Petroleum Co.) 64 A major oilfield can represent an investment as high as £400 million and it may take several years to reach pay-off. **1977** *Jrnl. R. Soc. Arts* CXXV. 58/2 The second problem..was to ensure that a fairly early pay off was secured from the new investment—and this in an industry where long lead times seemed inevitable. **1978** *N.Y. Times* 30 Mar. D. 5/3 The payoff in high-technology fields is extraordinarily great.

2. The payment of bribes; graft. Also, a sum of money given as a bribe. Also *attrib.*
1930 (*film title*) The Big Payoff. **1935** D. Lamson *We who are about to Die* xi. 193 Witnesses, juries, pay-off, fixin's—don't matter what it is... There ain't nothin' he won't do, long as you got the potatoes. **1938** R. Chandler in *Dime Detective Mag.* Jan. 62/1 He took my gun and his payoff money. **1950** *Sat. Even. Post* 27 May 20/1, I saw that in the Navy there's a lot of pay-off [graft]. **1958** S. Ellin *Eighth Circle* (1959) ii. iii. 44, I never took a penny of pay-off money since I got into the Department. **1971** R. Dentry *Encounter at Kharmel* xii. 219 *Money* for everything you've been through—a piddling wee payoff to close our mouths? **1976** *National Observer* (U.S.) 4 Dec. 2/2 Tanaka is one of several Japanese officials accused of receiving $12 million in pay-offs from Lockheed for promotion of the company's sales in Japan. **1977** *New Yorker* 22 Aug. 70/1 The scandal involving alleged political payoffs by agents of the South Korean government..first broke last year.

3. *Criminals' slang.* The division of the proceeds after a robbery.
1931 G. Irwin *Amer. Tramp & Underworld Slang* 141 *Pay off*, the division of spoils after a robbery. **1935** N. Ersine *Underworld & Prison Slang* 56 *Payoff*,..the end of a job and the splitting of the loot. **1935** A. J. Pollock *Underworld Speaks* 86/2 *Pay off joint*, place where the plunder (loot) is divided.

4. *transf.* and *fig.* Result, outcome, conclusion; return, recompense; punishment; the settling of accounts (in criminal contexts, esp. by murder). Also, a decisive or crucial factor; 'the ultimate'; 'the limit'. Also *attrib.*
It proved unrealistic to attempt to separate the examples that follow into clearly distinct sections. Many of them stand contextually at the border of at least two senses or embrace more than one sense.
1927 *Vanity Fair* Nov. 67/2 Conway's 'That's the pay off!' is swiftly making the rounds. It is employed when one enthusiastically describes anything that is first-rate: the acme, the last word! **1928** J. P. McEvoy *Show Girl* xiii. 195, I thought show business was all laughter and applause... It's a headache. It's a pain anywhere you sit. And then imagine falling in love on top of it. That's the pay-off. **1930** P. Annixter in *Flynn's* 11 Oct. 690 (*title*) The pay-off. **1932** *News* (San Francisco) 6 Aug. 12/3, I wanted to take one of those pictures which show how the greatest pitchers of the game hold their pay-off deliveries. **1937** N. Coward *Present Indicative* vii. 295, I had..lived far too strenuously. This [*sc.* a nervous breakdown] was the pay-off. **1937** *Sun* (Baltimore) 4 Aug. 14/1 The white-hulled defender isn't as impressive out of the water as the polished blue challenger but that isn't the payoff in this million-dollar sport. **1940** J. G. Brandon *Gang War!* ix. 93 It's a gang 'knock-off', or 'pay-off', whichever you like. **1944** *Sun* (Baltimore) 1 Nov. 1/5 The payoff is that there is no such thing as an effective Japanese fleet today. **1952** C. Day Lewis tr. *Virgil's Aeneid* ix. 189 The Fates and Venus have had their pay-off, in that the Trojans Have reached our fertile land of Ausonia. **1953** R. Lehmann *Echoing Grove* 42 The final pay-off, the practical one that always has to be gone through when there has been a death. **1957** *Times* 1 Oct. 11 It may be true, as Mr. Wilson said, that this economic crisis is 'the pay-off for the Government's policy on the past six years'. **1958** K. Amis *I like it Here* 200 He'd carried on in the same sort of way before, explaining he was part of the history of the English novel and all the rest of it, but this was really the pay-off. **1958** *Times* 8 May 11/6 To these crews, their terrible bomb loads are the devastating pay-off element of a daring formula: at war—to prevent war. **1970** G. Greer *Female Eunuch* 156 All that they have offered in the name of generosity and altruism has been part of an assumed transaction, in which they were entitled to a certain payoff. **1970** G. F. Newman *Sir, You Bastard* v. 149 All the inconvenience and suffering, and this was the pay-off. **1971** R. Dentry *Encounter at Kharmel* xii. 199 There was nothing to be gained from beating the hell out of this foul-mouthed creep... The pay off could wait. **1976** *Survey* Summer-Autumn 45 Among the less apparent payoffs, one might well ponder the changes in regard to superficially non-political facets of Soviet life. **1977** *Time* 28 Feb. 8/2 Danish Premier Anker Jørgensen wagered his political future in January and last week collected the payoff.

b. The climax or dénouement of a story, play, etc.; the point or crux of a story, etc. Cf. *pay-off line* (sense 5 below).
1947 Wodehouse *Full Moon* vii. 141 A raconteur of established reputation expects something better than silence when he comes to the pay-off of one of his best stories. **1962** W. Nowottny *Lang. Poets Use* iv. 96 Marvell's poem has its 'pay-off' in the ambiguity of the *da capo* with which the poem comes to a conclusion. **1969** *Listener* 15 May 698/1 Some failed even to detect the snook being cocked at them in Mary's climactic confession that she'd 'always worry about Jim'—a pay-off one could take nostalgically or ironically.

5. Special combs.: **pay-off line**, the point of a story; the 'punch-line' of a story, limerick, etc.; **pay-off man** *Criminals' slang*, (*a*) a confidence trickster; (*b*) the cashier of a gang of criminals; **pay-off matrix, table** *Game Theory*, an array specifying the utilities to the players of all the possible outcomes of a game.
1934 J. O'Hara *Appointment in Samarra* (1935) i. 16 And they always knew when to laugh, even when it was a Catholic joke, because Reilly signalled the pay-off line by slapping his leg just before it was delivered. **1944** [see *bar-fly* s.v. *bar *sb.*[1] 30]. **1965** N. Coghill in J. Gibb *Light on C. S. Lewis* 61 We all laughed at this pay-off line. **1927** *Fresno* (Calif.) *Bee* 9 June 1/3 The complaint asserts that Justice of the Peace Murphy was introduced to Frank E. Howell, former deputy sheriff, who was alleged to have been the 'pay-off' man of the contractors. **1928** M. C. Sharpe *Chicago May* 286 *Pay-off men*,..confidence men (or women). **1932** *Sun* (Baltimore) 27 Apr. 15/7 There were frequent references to checks made payable by Plummer to an unidentified 'pay-off' man. **1934** [see *number *sb.* 3 e]. **1935** A. Squire *Sing Sing Doctor* v. 59 They surrounded themselves with bodyguards, flunkies, killers, fixers, pay-off men. **1938** D. Castle *Do Your Own Time* 287 *Pay-off man*, cashier of a mob. **1952** J. C. C. McKinsey *Introd. Theory of Games* i. 7 We shall henceforth describe such a game as this by giving merely the payoff matrix. **1971** D. C. Hague *Managerial Econ.* vii. 138 The columns show the result of each act... The rows show the events leading to these possible results. Table 11 is therefore known as a payoff matrix (or table). **1974** Anton & Kolman *Appl. Finite Math.* viii. 361 Games of the type described in this example are called matrix games and the matrix is called the payoff matrix. **1976** *Nature* 8 Apr. 481/1 He..assigned payoffs for winning, losing, getting injured in an escalated fight and so on, and used these values to construct a payoff matrix for each strategy against all others.

payola (pē̆i,ǒu·lă). *colloq.* (orig. *U.S.*). [f. Pay *sb.* or *v.*[1] + *-ola.] A secret or indirect payment or bribe to a person to use his position, influence, etc., to promote a commercial product, service, etc.; *spec.* such a payment to a disc-jockey for 'plugging' a record or song.
1938 *Variety* 19 Oct. 41 (*heading*) Plug payolas perplexed. *Ibid.*, The payola element had made their deals with bandleaders on the expectation that they continue to get 19c, thereby making it profitable to do business with the plug at a rate of around 10c a point. **1953** *Time* 23 Feb. 56/3 A world where *cut-ins* (giving a performer a share of a song's profits), *hot stoves* (open bribes) and other forms of *payola* were standing operating procedure. **1958** S. Ellin *Eighth Circle* (1959) ii. vii. 84 The unerring way he gravitated toward the graft, the payola, the swindle. **1966** T. Pynchon *Crying of Lot 49* iii. 61 They got the contracts. All drawn up in most kosher fashion, Manfred. If there was payola in there, I doubt it got written down. **1969** N. Cohn *AWopBopaLooBop* (1970) v. 51 There was a huge scandal about in 1959, the payola fuss, and a lot of people came crashing down. **1969** [see *dropsy *sb.* 4]. **1972** P. Black *Biggest Aspidistra* i. iii. 30 Gramophone records began to take over from sheet music and presented fresh temptations... Plugging re-emerged under the name of payola. **1974** *Guardian* 5 Nov. 7/2 Miss Squires and Mr Dabbs appeared in the dock together accused of 'payola' corruption. **1974** *Daily Tel.* 18 Nov. 6/1 There is growing talk of abuse of power by leading figures within the country's political establishment..at the expense of both Western businessmen and of the Kenyan people, involving payola for 'services rendered'. **1976** *Observer* 7 Nov. 1/6 (*heading*) Lockheed payola. A Spanish air force general and a colonel secretly managed the Lockheed Aircraft Corporation's sales in Spain and

earned commissions worth £170,000 each. **1977** *Time* 31 Jan. 13/1 In the process of popularizing their products, the competing firms created a payola monster. They began slipping money as well as footwear to the stars of their choice.

pay-out. [f. vbl. phr. *to pay out* (see PAY *v.*¹ 5 b, 13).] The fact or action of paying out; the amount paid out (see also quot. 1904). Also, a place (in a shop, etc.) where payment is made.

1904 *Dialect Notes* II. 386 *Pay out*, n. Said of a well which more than returns the expense of drilling and pumping, or the capital invested in it. **1959** W. S. EVANS *Petroleum in E. Hemisphere* 19 As a result of the relatively low 'going' price for crude, his payout may be considerably longer than will be satisfactory, particularly if the price weakens further. **1960** *Guardian* 18 Dec. 5/1 The payout of stage artists by manufacturers for an advertising plug slipped into the artist's act. **1968** *Nature* 23 Nov. 752/1 Ross identified the bottom by a decrease in the rate of pay out of line from the free-spooling reel... An abrupt decrease in the rate of pay-out was indicative of bottom contact. **1970** *Globe & Mail* (Toronto) 25 Sept. B10/6 Profit..will not increase above last year's $1.05 a share because of payouts on long-range marketing programs. **1977** J. WAINWRIGHT *Day of Peppercorn Kill* 70 At the payout, she wrote a cheque. **1977** A. MORICE *Scared to Death* iv. 25 She had set off towards the pay-out windows to collect her winnings.

pay--roll, payroll. [PAY- 1.] The total amount to be paid to employees in a specified period; also, (a list of) employees receiving regular pay. Freq. in phr. *on the payroll*: employed by a particular company or person. Also *fig.*

1740 C. CIBBER *Apol. for Life* xiii. 257 The Rate of their respective Sallaries were only enter'd in our daily Pay-Roll. **1775** *Rec. New Hampsh. Comm. Safety* in *New Hampsh. Hist. Soc. Coll.* (1863) VII. 26 Examined and allowed Capt. Crafford's pay Roll of his men engaged for fourteen days. **1780** *Calendar Virginia State Papers* (1875) I. 387 Enclose pay roll & account for purchase of kettles and dutch-ovens. **1840, 1898** [in Dict. s.v. PAY-1 b]. **1921** C. E. MULFORD *Bar-20 Three* viii. 96 Looks like some Greaser had a grudge agin' him—somebody he's mebby fired off his payroll, or suspected of cattle-liftin'. *Ibid.* ix. 113, I ain't on Kane's payroll—yet. **1958** *Punch* 23 July 98/1 Skilled British agents are showing cause why they should be retained on the pay-roll. **1964** MRS. L. B. JOHNSON *White House Diary* 11 Jan. (1970) 39 On to Goldsmith Mill where a payroll of about a hundred had been saved by the joint work of the Community and the ARA. **1968** *Time* 17 May 66 Despite all the fuss at Columbia over IDA, none of its professors are actually on the IDA payroll. **1973** 'H. HOWARD' *Highway to Murder* viii. 98 Nobody on my payroll had anything to do with Vince Portelli's killing. **1977** *New Yorker* 29 Aug. 87/1 The Italian anarchists who were convicted of murdering two men while stealing a payroll in South Braintree, Massachusetts, in 1920 were executed on August 23, 1927.

2. *attrib.*, as *pay-roll index*; **pay-roll tax,** a tax levied on businesses according to the number of persons employed or on the wages-bill of the business.

1934 *Planning* II. xxxi. 7 The United Kingdom possesses nothing comparable to the payroll index, which is one of the major instruments of economic measurement in the United States. **1935** *N.Y. Times* 16 Jan. 2/6 The plans sent to the President by his Cabinet Committee included: unemployment insurance, financed, in part, at least by a payroll tax. **1937** M. NEWCOMER in *Stud. in Current Tax Probl.* 39 The new payroll taxes for unemployment insurance and old-age benefits will eventually become an important factor in the tax burden. **1971** *Sunday Australian* 8 Aug. 5/4 Australia's State schools paid about $15 million in payroll tax in the last financial year. **1976** in R. Crossman *Diaries* II. 58 Selective Employment Tax. A payroll tax paid by employers with some rebate for industrial enterprises.

‖**Pays du Tendre** (pei dü tãndr'). Also with lower-case initials. [ad. Fr. *pays de Tendre*, with ref. to *Tendre*, an imaginary country whose topography symbolized aspects of love, devised by Madeleine de Scudéry (1607–1701) in her novel *Clélie* (1654–60).] Matters concerning love; the 'region' of the affections.

1910 W. J. LOCKE *Simon* viii. 113 A crock..with one foot in the grave has no business to put the other into the *Pays du Tendre*. **1913** —— *Stella Maris* xv. 200 Herold.. adventured with her into the Land of Tenderness—the *Pays du Tendre* of the old French romanticists. **1938** *Times Lit. Suppl.* 15 Jan. 43/3 The heroine's first excursion into the *pays du tendre* ends abruptly with the young man's betrothal to another girl. **1939** *Ibid.* 1 July 387/2 Mastery of this familiar type in the *pays du tendre*.. is a handsome equipment for story-telling. *a* **1976** A. CHRISTIE *Autobiogr.* (1977) IV. i. 168 The art of flirtation.. was an approximation, I think, to what the old troubadours called '*le pays du tendre*'.

Pazand (pā·zænd). Also **Pāzend.** [f. Pers. *pā-zand* interpretation of the Zend: see ZEND-AVESTA.] A transcription of, or the method of transcribing Persian sacred texts from Pahlavi (see PAHLAVI *a.* and *sb.* in Dict. and Suppl.) into the script of the Avesta. Freq. *attrib.*, designating this mode of transcription.

[**1700** T. HYDE *Hist. Relig. Vet. Pers.* xxvi. 338 Literæ.. quæ..apud incolas vulgò audiunt..Character Zundicus, vel si Anglicè loquimur, the Zund Character; à quo aliquantulum differt Character Pazendicus.] **1772** J. SWINTON in *Phil. Trans. R. Soc.* LXI. 354 It not a little resembles that endued with the power of the short A, deduced from the Zend and Pazend, by Dr. Hyde. **1871** E. W. WEST (*title*) The book of the Mainyo-i-Khard. The Pazand and Sanskrit text... With an English translation, a glossary of the Pazand text.., a sketch of Pazand grammar, and an introduction. **1928** E. G. BROWNE *Lit. Hist. Persia* I. ii. 77 Hence the so-called Pázend and Pársí books, which are merely transcriptions of Pahlawí books into the unambiguous Avestic and Arabic characters respectively, all the Huzvárish, or Aramaic, words being replaced by their Persian equivalents, or supposed equivalents. *Ibid.* 81 Just as *Zend* is the 'explanation' of an Avestic text in Pahlawí, so is *Pázend* (= *paiti-zainti*) a re-explanation' of a Pahlawí text by transcribing it into a character less ambiguous than the Pahlawí script, and substituting the proper Persian words for their respective Huzvárish equivalents. When the Avesta character is used for this transcription, the result is called 'Pázend'; when the Persian (*i.e.*, the Arabic) character is adopted, the term 'Pársí' is often substituted. In either case the product is simply an archaic or archaistic..form of 'modern' (*i.e.* post-Muhammadan) Persian, from which the whole Aramaic element has disappeared. **1934** *Trans. Philol. Soc.* 1933 56 There are reasons for believing that these Pāzand writers were active in the Central region from Kāšān to Yazd and Kirmān. **1939** L. H. GRAY *Foundations of Lang.* 319 This 'Book-Pahlavi'..falls into two types: *Huzvarišn*, in which Semitic (Syriac) words are written, but with Iranian inflexions..; and *Pāzand*, which uses only Iranian. **1948** D. DIRINGER *Alphabet* 308 The most famous of the Persian indigenous scripts is the Pazand or Avesta alphabet, the script of the sacred Persian literature. **1968** M. BOYCE in *Handbuch der Orientalistik* IV. i. 47 It survives only in a mediaeval Sanskrit version, and in Pazand i.e. Middle Persian transcribed in mediaeval times out of Pahlavi into Avestan script. **1968** P. VAN POPTA-HOPE tr. *Rypka's Hist. Iranian Lit.* 34 In the Arab period Iranian Zoroastrian writers turned to new alphabets and attempted to use them for writing down Middle Persian texts phonetically. The writings in the Avestan alphabet are called *Pāzand*, those in the Arabic consonantal writing are called Pārsī. **1972** W. B. LOCKWOOD *Panorama Indo-European Lang.* xii. 235 Sometimes Pahlavi texts are found transcribed into vocalised Avestan script.., such texts being called Pazend.

pazazz, pazzazz. varr. *PIZZAZZ.

pazil, var. *PASSEL.

pea¹. Add: **I. 1. d.** In the West Indies and southern U.S.A., a name for the seeds of various other legumes, including the red pea, *Vigna unguiculata*, and the Gungo pea, *Cajanus cajan*; esp. in phrase *pea(s) and rice*, the name of a local dish.

1928 FREEMAN & WILLIAMS *Useful & Ornamental Plants Trinidad & Tobago* (ed. 2) 166 Vigna sesquipedalis... Yard Bean. Bodi... The young pods are edible as French beans and also the ripe dry beans. V[igna] sinensis. Cow Pea. Black Eye Pea. **1930** B. S. RHETT *200 Yrs. Charleston Cooking* 59 (*heading*) Peas and rice pilau. **1969** *Daily Tel.* 11 Jan. 14/1 [In Jamaica] 'peas' are not peas at all but kidney beans. **1970** M. SLATER *Caribbean Cooking* 32 'Peas and Rice'..is cooked on every island. **1971** *Bahamas* XXIII. III. 33/1 Being a true native son [of the Bahamas], Sidney Poitier sometimes has irresistible urges to devour such disastrous delicacies as pea 'n' rice or pea soup. **1972** C. D. ADAMS *Flowering Plants of Jamaica* 364 Vigna..cultivated in many varieties; native of tropical Asia; Black Eye Pea, Cow Pea, etc. **1973** *Advocate-News* (Barbados) 22 Jan. 13/3 (Advt.), Today's menu: Fried and boiled chicken,..dry peas or split peas and rice.

e. [In allusion to the pea used by a thimble-rigger.] A favourite; a horse likely to win. *Sporting slang* (*obs.* exc. *Austral.*). Also *transf.* someone in a favoured or favourable position, a person with authority (*Austral. colloq.*).

1888 *Sporting Life* 11 Dec. 4/4 Sweeny..forced the fighting, and was still the pea when 'Time!' was called. **1891** *Licensed Victuallers' Gaz.* 20 Mar. 187/3 Well, Albert, now what is the pea? we asked, hurrying towards the paddock. **1900** E. WELLS *Chestnuts* xxiii. 227 Informed me that the right pea for the race was 'L'Abbesse de Jouarre'. **1911** E. DYSON *Benno* xvi. 206 Mr. Dickson ..ran his eye down the card and chanced it. 'Dandy's the P,' he said. 'Put yer whole week's wash on Dandy, 'n hold me responsible if the goods ain't delivered.' **1953** BAKER *Australia Speaks* v. 118 Other expressions used by racing fans include: *pea*, a horse that is being ridden to win, especially when there is a doubt about the genuineness of other runners. **1958** F. HARDY *Four-Legged Lottery* xxv. 190 I've got the tip about it. Old Dapper Dan earwigged at the track. Swordsman is the pea. **1969** M. CALTHORPE *Defectors* iii. 17 'For the time being, I'm satisfied.' 'You're the pea,' Mick said. **1973** A. BUZO *Rooted* III. iii. 92 He's had his eye on her for some time, you know, but I'm the pea, she said. **1974** *Sun-Herald* (Sydney) 1 Sept. 15 The usual assumption has been that the Social Security Minister, Mr Hayden, 41, would move into the Treasury... Recent events have cast some doubts on that. The Deputy Prime Minister and Overseas Trade Minister, Dr Cairns, now seems the 'pea' for any change at the Treasury.

II. 5. b. = *pea-coal* (see sense 7 in Dict.); *pl.*, coals of a very small size.

1880 [see *EGG *sb.* 3 b]. **1886** J. BARROWMAN *Gloss. Scotch Mining Terms* 50 *Peas*, coal a grade smaller than nuts. **1905** A. S. CUNNINGHAM *Rambles in Scoonie & Wemyss* 260 Most of the trebles, nuts, beans and peas produced at Wemyss colliery are treated by the washer. **1930** *Engineering* 5 Dec. 708/1 The employment of anthracite duffs in place of washed grains and peas. **1949** *Black Diamond* 26 Feb. 54/3 Prices range as follows:..nut and pea, $3.50–$4.50.

III. 7. *pea-crop* (earlier example), *-field*, *patch*, *-seed*; **pea coffee** *U.S. obs.*, a beverage made by boiling roasted peas; **pea-combed** *a.*, of poultry, having a pea-comb; **pea-crab** (later examples); **pea-dodger** *Austral.* = BOWLER³; **pea-flour** (later examples); **pea gravel,** gravel consisting of particles similar in size to peas; **pea-lamp,** a very small, round electric lamp such as is often used as an indicator light; **pea-moth,** substitute for def.: the larva of the moth *Cydia nigricana*, which feeds on peas; (examples); **pea-shooter** (*a*) (earlier examples); **pea-stick** (earlier and later examples); **pea-viner,** a machine for picking, washing, and grading peas.

1805 T. E. WHITE *Jrnl.* 14 July (1904) 24, I drank three or four cups of pea coffee and then went to bed. **1818** 'A. BURTON' *Adventures J. Newcome* II. 112 Pea-coffee, Hurry-hush, and Chowder. **1851** H. MELVILLE *Moby Dick* I. ii. 14 The very spot for cheap lodgings, and the best of pea coffee. **1868** DARWIN *Variation of Animals & Plants under Domestication* I. vii. 254 In some breeds the comb is double..; it is triple in the pea-combed Brahmas. **1922** R. C. PUNNETT *Mendelism* (ed. 6) 32 The pea-combed bird contains the factor for pea but not that for rose. **1901** M. NEWBIGIN *Life by Seashore* x. 202 The tribe Catometopa..includes the curious pea-crab, *Pinnotheres pisum*, found inside the bivalves. **1960** C. M. YONGE *Oysters* vii. 118 The hundred or so species of Pea Crabs, most of which inhabit the mantle-cavity of various species of bivalves. **1978** G. DURRELL *Garden of Gods* iv. 88 It was he..who had got me the biggest clam shell in my collection and, moreover, with the two tiny parasitic pea-crabs still inside. **1732** W. ELLIS *Pract. Farmer* 39 This [weed] I cannot say will utterly destroy the Pea-Crop, but will so cripple it, as not to be a quarter Value. **1933** *Bulletin* (Sydney) 5 Apr. 12/3 'Elizabeth Owen':.. the different terms applied to 'bowler' hats—I have also heard them called 'egg-boiler' and 'pea-dodgers'. **1959** BAKER *Drum* (1960) II. 133 *Peadodger*, a bowler hat. [**1677** W. HUBBARD *Narr. Troubles with Indians in New-England* I. 24 If there were Indians in the Neck they should send them about a Peas-field not far off.] **1972** D. HASTON *In High Places* iii. 44 The trip was planned in the pubs of the pea-fields of Kent, where we had been supplementing our meagre incomes. **1881** Pea-flour [see *FLADBRÖD, -BRÖD]. **1915** D. H. LAWRENCE *Rainbow* ii. 46 The vicar put pea-flower [*sic*] into the crocuses, for his bees to roll in. **1928** E. E. BAUER *Highway Materials* xii. 122 After the bituminous material has been applied, the entire surface must be covered with an application of pea gravel. **1962** R. PAGE *Educ. Gardener* x. 277, I devised a very simple arrangement of areas of fine pea gravel and panels of grass. **1963** *Times* 16 Feb. 11/3 A bed of well weathered clinker ash, or small well washed pea gravel say 8 in. deep is placed on the floor of the greenhouse. **1973** C. WILLIAMS *Man on Leash* (1974) ix. 140 Bare planks..and then another two steps down onto the grating crunch of pea gravel. **1938** G. H. SEWELL *Amateur Film-Making* ii. 20 The glowing filament of a pea lamp. **1950** *Electronic Engin.* XXII. 413/1 The output of this amplifier was used to light a pea-lamp mounted alongside the microphone. [**1859** S. J. WILKINSON *Brit. Tortrices* 230 The larva [of *Endopisa nigricana*] feeds on the growing and unripe seeds of the Pea.] **1881** E. A. ORMEROD *Man. Injurious Insects* 131 Pea Moth... The caterpillars of this Moth cause the 'worm-eaten' or 'maggoty' Peas often found in old pods. **1931** G. S. CHAPPELL *Gardener's Friend* 153 The lupines..have inherited from their lowly ancestry an appeal to..wireworms and pea moths. **1964** F. G. W. & M. G. JONES *Pests of Field Crops* v. 93 The pea moth has become increasingly important with the intensive cultivation of peas in certain areas of Britain. **1972** *Arable Farmer* Feb. 55/2 Folithion (insecticide for pea moth control from Bayer Agrochem). **1834** *Knickerbocker* III. 35 Didn't I turn that pied heifer of yourn into my pea patch. **1863** 'G. HAMILTON' *Gala-Days* 34 No premonition floated over from that adjoining pea-patch. **1941** J. STUART *Men of Mountains* 120 Tear off the damn Dingus silk shirts..for to make skeery-cows out'n for the pea patch! **1744** W. ELLIS *Mod. Husbandman* Feb. v. 29 When Horse-bean and Pea-seed are to be sown together.. the stated Allowance..is..two-third Parts Beans, and one-third Part Pease. **1946** *Nature* 31 Aug. 293/1 The fungus was grown in various modifications of Czapek-Dox medium with addition of manganese sulphate, in some cases with pea-seed extract. **1960** *Farmer & Stockbreeder* 16 Feb. 97/1 It is in practice difficult to get the fertilizer below the peas without forcing the pea-seed coulters into a too-shallow position. **1803** J. RUSSELL *Jrnl.* 23 Sept. in S. Walpole *Life Ld. John Russell* (1889) I. i. 6 The boys play at hoops, peg-tops, and pea-shooters. **1833** *Boy's Week-Day Bk.* 210 When you shoot a pea through your pea-shooter, it is quite as well to know that the natives of Macoushie shoot arrows in the same manner. **1745** J. MACSPARRAN *Letter Bk.* (1899) 27 Harry is come home..& has bro't home Pea Sticks. **1971** *Country Life* 2 Sept. 580/2 Most of the cut wood has its uses with the seasonal demand for pea sticks and bean poles. **1973** *Daily Tel.* 7 July 7/8 What can we stake our herbaceous plants with..now that you can no longer get pea sticks? **1943** C. CROW *Great Amer. Customer* 179 The pea viner is perhaps the most marvelous of them all. Vines fresh from the field are fed into the robot which hulls the peas, grades them as to size, and sends them on their way to the cooker. **1952** J. W. DAY *New Yeomen of Eng.* ii. 32 This giant pea-viner..deals with a ton of peas an hour. It picks them from the vines, washes, and grades them.

Peabody (pī·bǫdi). *U.S.* The name Peabody used *absol.* or *attrib.* in **Peabody bird** to designate the white-throated sparrow, *Zonotrichia*

albicollis, whose call is said to resemble 'Sam Peabody, Peabody, Peabody'.

1865 E. A. SAMUELS in *Rep. Comm. Agric. 1864* (U.S. Dept. Agric.) 422 White-throated Sparrow—Peabody Bird—Wheat Bird... This beautiful sparrow arrives in Massachusetts by the first week in April. **1903** *N.Y. Even. Post* (Sat. Suppl.) 24 Oct. 1/1 That lovely bird, the white-throated sparrow, which under the name of 'Peabody bird' is well known in the North. **1917** W. P. EATON *Green Trails* 16 All day long in this pasture the Peabodies, or White-throated sparrows, sing their flutelike call. **1939** R. T. PETERSON *Field Guide Birds East of Rockies* 159 Song, several clear, pensive whistled notes, easily imitated. The White-throat is often called the 'Peabody bird' by New Englanders who fancy these whistled notes sound like Old Sam Peabody, Peabody, Peabody. **1940** E. T. SETON *Trail of Artist-Naturalist* 224 The night singer of the Assiniboine, was neither more nor less than the white-throated sparrow, the Peabody bird of New England, the nightingale of the farther north. **1964** A. WETMORE et al. *Song & Garden Birds N. Amer.* 369/2 The Peabody bird is in best voice in his summer home.

peace, *sb.* Add: **I. 4. c.** (Later examples.) Also, an action symbolizing the kiss of peace. Now usu. a light embrace, a hand-shake, or a bow.

1935 D. H. HISLOP *Our Heritage in Public Worship* xi. 243 Here, either before or after the Peace, in many rites is placed the Creed. **1957** *Oxf. Dict. Chr. Ch.* 771/1 Originally an actual kiss, the form of the Peace has been modified in all rites. **1974** R. J. HALLIBURTON in R. C. D. Jasper *Eucharist Today* vii. 90 The rubric in Series 3 gives no directions as to how the Peace is to be given. **1975** C. F. BAZLEY in C. O. Buchanan *Further Anglican Liturgies* x. 183 The Peace (in Chile called the Holy Kiss or the Love Embrace) is not included as a formal part of the service itself. **1976** *Church Times* 9 July 13/1 In recent years we have been bombarded with all kinds of fanciful and eccentric changes in the liturgy, of which 'the peace', in all its constantly changing forms, is surely the most ludicrous. *Ibid.* 8 Oct. 9/2 Staid churchwardens embraced others in the congregation during the giving of the Peace with a warmth and friendliness that they would have found difficult to express in a parish church. **1977** *Theology* LXXX. 175 The representation of 'the giving of the peace' by, rather oddly, shaking hands with one's neighbour.

7. b. In alliterative association with *plenty*.

1393, etc. [see PLENTY *sb.* 3]. **1596** SPENSER *Prothalamion* vi, Let endlesse Peace your steadfast hearts accord, And blessed Plentie wait vpon you[r] bord. **1703** *New Hist. Trojan Wars* III. 17 Now the Wars are done, and Peace and Plenty are pouring in upon us. **1713** POPE *Windsor-Forest* 2 And Peace and Plenty tell, a Stuart reigns. **1823** BYRON *Age of Bronze* xv. 32 How rich is Britain! not indeed in mines, Or peace, or plenty, corn or oil, or wines. **1949** W. S. MAUGHAM *Writer's Notebk.* 306 The world has always been a place of turmoil. There have been short periods of peace and plenty, but they are exceptional.

7*. With initial capital. A vigorous hybrid tea rose bearing large yellow flowers shaded with pink, belonging to a variety developed by Francis Meilland, French nurseryman, in 1939, and introduced into cultivation in 1942. Also *attrib.*

1944 R. PYLE *Let.* in A. Ridge *For Love of Rose* (1965) xii. 210 We are persuaded that this greatest new rose of our time should be named for the world's greatest desire: Peace. **1945** *Los Angeles Times* 30 Apr. I. 11/1 Today's main event [at the Pacific Rose Society show] was the official christening of the newly developed Peace rose. **1945** *Horticulture* 15 Sept. 409/2 The new rose Peace, which seems to be causing a furore, is another of many good roses coming to us from France. **1950** A. S. THOMAS *Better Roses* ix. 93 Mme A. Meilland was listed and sold in France in 1942 under this, its original and therefore correct name, but in Germany it has always been called Gloria Dei, and in Italy Gioia, while it was introduced into America in 1945 and sold under the title of Peace, its fourth name. **1963** W. BLUNT *Of Flowers & Village* 89 *R[osa] Andersonii.* Like a phlox-pink wild rose. Peace. For Mrs. Benham, who has promised to plant it where it can't be seen from the house. **1965** L. MEYNELL *Double Fault* II. v. 174 A heaped profusion of Peace roses looked superb in a large silver bowl. **1976** E. B. LE GRICE *Rose Growing Compl.* (ed. 2) xii. 192 Abroad this [*sc.* 1946] was the year of 'Peace' (Meilland) which was to have a profound effect on all rose breeding everywhere, giving a new standard in growth, health and size of bush and flower. **1978** M. DUFFY *Housespy* i. 9 The dusk was still full of the scent of Peace roses.

II. 8. a. (Earlier and later examples of attrib. use.)

1882 E. W. HAMILTON *Diary* 3 Sept. (1972) I. 332 Mr. G. wrote the other day a capital letter to Mr. Richard putting the war in Egypt in a light which would be likely to convince the strongest 'peace-at-any-price' person. **1910** BEERBOHM *Around Theatres* (1953) 579, I myself am not such a peace-at-any-price man as to be frightened away.

b. (Later example.)

1973 *Washington Post* 13 Jan. A23/3 President Nixon stated in Kentucky that we now had in our grasp a peace with honor instead of a peace with surrender.

14*. *no peace for the wicked* [Isaiah xlviii. 22, lvii. 21]: no rest or tranquillity for (the speaker); incessant anxiety, responsibility, or work.

1944 A. THIRKELL *Headmistress* iv. 86 'It's for Dr. Perry..' 'No peace for the wicked,' said Dr. Perry. **1953** D. PARRY *Going up—Going Down* iv. 128 'No peace for the wicked,' she said. 'But I expect it will be quiet from now on.' **1967** 'F. CLIFFORD' *All Men are Lonely Now* I. v. 89 The painters are descending on us tomorrow... Couldn't happen at a worse time, but there's no peace for the wicked is there?

III. 15. a. Freq. in senses: founded, held, organized, propounded, etc., to promote peace or end a specified war; advocated by pacifists; as *peace activist, advocate, aim, area, bloc, campaigner, conference, convention, crank, demonstration, -feeler, -fighter, formula, -front, march, marcher, -mediator, meeting, -mentality, mission, move, movement, negotiation, offensive, offer, petition, plan, propaganda, rally, society, symbol, terms, walk.* **b.** *peace-commanding, -conferring, -inspiring* (earlier example), *-loving* (later examples), *-promoting* adjs. **c.** *peace-calm, -complacent, -inspired, -minded* adjs. **d. Peace Corps,** orig. *U.S.* (see quot. 1962); **peace economy,** an economy, characteristic of peace-time, in which a large part of the labour force produces goods for export (as opposed to being engaged in arms production, etc.); **peace establishment** (earlier example); **peace-game** [after *war-game* s.v. WAR *sb.*¹ 11 in Dict. and Suppl.], an exercise in the maintenance of international peace; hence **peace-gaming** *vbl. sb.*; **peace-guild** (example); **peace line,** a line of demarcation drawn to avert conflict; **Peace People** (see quots.); **peace pledge,** (*b*) an undertaking to abstain from fighting, or to seek peace (in industrial relations); **peace prize,** an award (as a *Nobel prize*) presented for a contribution to the prevention of war; **peace sign,** a sign of peace made by holding up the hand with palm out-turned and the first two fingers in a V-shape; **peace talk,** conversation or discussion about peace or the ending of hostilities; *spec.* in *pl.*, a conference or series of discussions aimed at achieving peace in particular circumstances; hence **peace-talker, peace-talking** *vbl. sb.*

1968 *Listener* 31 Oct. 566/1 Those peace activists who at last year's Pentagon demonstration announced that they were going to make the Defence Department levitate 300 feet in the air. **1906** *Westm. Gaz.* 24 Mar. 6/3 Master of counsel sage and fluent pen, Peace-advocate, averse from warlike act. **1910** W. JAMES *Mem. & Stud.* (1911) xi. 284 Our socialistic peace-advocates all believe absolutely in this world's values. **1940** *Economist* 6 Jan. 9/2 The French have not fed much fuel to the idealist fire which has raged in this country over the discussion of peace aims. **1957** *Ibid.* 28 Sept. 1000/1 Replace alliances by security pacts—create 'peace areas' (ranging from southern Asia to the Baltic). **1939** W. S. CHURCHILL *Into Battle* (1941) 94 The first [step]..is the full inclusion of Soviet Russia in our defensive peace bloc. **1939** R. CAMPBELL *Flowering Rifle* I. 15 And loosing these in turn to drink and graze The peace-calm waters and the flowery ways. **1954** B. & R. NORTH tr. *Duverger's Pol. Parties* I. ii. 109 A large number of Europeans, very far removed from Communism, are..vulnerable to the attacks of the Peace Campaigners. **1807** J. BARLOW *Columbiad* x. 359 Enlighten'd interest, moral sense at length Combine their aids to elevate her strength, Lead o'er the world her peace-commanding sway. **1928** S. SASSOON *Heart's Journey* 31 Paid, with a pile of peace-complacent stone. **1889** *Cent. Dict.* Peace convention or *conference*, same as *Peace congress*. **1899** *Hazell's Ann.* 1900 462/1 A Peace Conference was held at the Hague in May, June, and July '99. **1919** G. B. SHAW *Peace Conference Hints* i. 7 Before the Peace Conference can be discussed with any profit, it must be approached in the light of the facts. **1933** *Radio Times* 14 Apr. 75/1 The war was newly over... There was a tremendous fuss of coming and going across the Channel..politicians..busy with the Peace Conference. **1978** *Times* 26 July 13/3 A strong performance of *The Green Table*..Kurt Jooss's allegory of the sometimes fatal effects of peace conferences. **1909** W. JAMES *Pluralistic Universe* iii. 128 It [*sc.* the absolute] might, with all its defects, be, on account of its peace-conferring power and its formal grandeur, more rational than anything else in the field. **1885** W. P. & F. J. GARRISON *Life W. L. Garrison* II. 230 The Peace Convention held in Boston, September.., 1838. **1960** H. HUMPHREY in *Congress. Rec.* 15 June 12634/3 Mr. President, I introduce..a bill to establish a Peace Corps of American young men to assist the peoples of the underdeveloped areas of the world to learn the basic skills necessary to combat poverty, disease, illiteracy and hunger. **1962** *Ann. Reg. 1961* 175 The Peace Corps was officially set up on 1 March as an organization to train and send American volunteers for service in foreign countries to help meet the need for skilled manpower. **1967** *Economist* 2 Sept. 785/2 Japan's version of the peace corps is getting into its stride. **1970** *Times* 19 Mar. (Liberia Suppl.) p. i/4 Sturdy American girls in Bermuda shorts earnestly go about the business of the Peace Corps. **1974** 'G. BLACK' *Golden Cockatrice* iii. 48 The Peace Corps phase in Manila..followed by free-lancing in refugee camps outside Kowloon. **1916** D. H. LAWRENCE *Let.* 11 Dec. (1962) I. 491 Fusty, fuzzy peace-cranks and lovers of humanity are the devil. **1935** A. HUXLEY *Let.* 5 June (1969) 395 His hobby is congresses and he has already organized one peace demonstration—at Amsterdam in 32, I think. **1910** W. JAMES *Mem. & Stud.* (1911) xi. 283 The military party..says..that mankind cannot *afford* to adopt a peace-economy. **1948** G. CROWTHER *Outl. Money* (rev. ed.) viii. 267 A peace economy..is chiefly interested in selling to foreign countries, a war economy in buying from them. **1766** in *Rep. on MSS. Mrs. Stopford-Sackville* III in *Parl. Papers* 1904 (Cd. 1892) XLVII. I, I see no end of the load imposed upon us, as our peace establishment is so far beyond the ordinary supplies. **1942** H. NICOLSON *Diaries & Lett.* (1967) 225 There are all sorts of rumours... Germany is putting out peace-feelers. **1972** J. WILLIAMS *Home Fronts* v. 94 The Allies' curt rejection of peace feelers put out by the Chancellor. **1958** *New Statesman* 23 Aug. 230/1 Until last year Paul Robeson and Howard Fast ranked with the Dean of Canterbury as the foremost 'peace-fighters' of the English-speaking world. **1974** *Times* 16 Feb. 1/4 The chances of a peace formula emerging from the Pay Board's enquiry into the relative position of miners' pay. **1977** *Times* 17 Aug. 2/7 Postal workers..agreed on a peace formula after a five-day dispute. **1939** H. NICOLSON *Diary* 22 Aug. (1966) 411 This smashes our peace-front. **1968** *Guardian* 2 Dec. 9/8, I.. believe..that..practical pacifism [is] a reasonable goal. I am currently engaged in talking to computers about this in an elaborate peace game. **1968** *Economist* 23 Nov. 72/2 More attention should be paid to using these techniques for what he calls 'peace gaming'. **1913** *Encycl. Relig. & Ethics* VI. 215/1 The *frith gild*, or *peace gild*, so called, refers to an occasional feature of town life in Northern Europe from the 6th century. **1909** *Westm. Gaz.* 1 Sept. 9/1 The reactionary Holy Alliance was also a peace-inspired measure. **1828** DISRAELI *Voy. Capt. Popanilla* x. 119 The calm and peace-inspiring crosier. **1957** *Economist* 19 Oct. 209/2 When Japanese fishermen are apprehended in Peace Line violations, they are brought to Korea and given fair and open trials. **1969** *Times* 10 Sept. 1/1 (*heading*) Army 'peace line' to replace barricades across Belfast. *Ibid.* 1/2 General Sir Ian Freeland..said that troops would start building 'peace-line' barricades today. **1972** *Belfast Tel.* 12 Oct. 6/6 No doubt when the other volumes make the shops, Ulster will get its dishonourable mentions—no-go area, provo, peaceline. **1930** R. A. KNOX *Caliban in Grub St.* xiii. 212 Telling us to be honest and sincere and peace-loving. **1944** *Ann. Reg. 1943* 83 A general international organisation, based on the principle of the sovereign equality of all peace-loving States. **1955** *Ann. Reg. 1954* 191 Broadcasts to foreign audiences continued to extol the full religious freedom allegedly existing in the U.S.S.R., and to court the 'peace-loving' Catholics. **1970** V. CANNING *Great Affair* xii. 207 Basically I am a simple, peace-loving, even a dull person. **1977** T. ALLBEURY *Man with President's Mind* xiii. 131 The peace-loving American people. **1977** *Ann. Reg. 1976* 485 There are still certain forces who are bent on a return to cold war politics, which led to the division of the continent into opposing blocs. Communist Parties and other democratic and peace-loving forces have fought against and continue to fight against these policies. **1961** A. WESKER *Kitchen* 58 Did you go on that peace march yesterday? **1975** P. HARCOURT *Fair Exchange* 46, I proposed to wait until the Peace March had gone by. **1961** *Listener* 12 Oct. 548/2 The Western peace marchers approaching Moscow. **1972** J. WAMBAUGH *Blue Knight* (1973) vi. 77 Lieutenant Hilliard..was a cool old head and wouldn't get into a flap over fifteen peace marchers. **1884** Peace-mediator [see *HONEST *a.* 3 e]. **1939** C. DAY LEWIS *Child of Misfortune* III. i. 260 Sitting on the platform at peace meetings. **1935** A. P. HERBERT *What a Word!* 243 Get your angles straight on air-mindedness, class-consciousness, peace-mentality, stomach-awareness. **1939** *War Illustr.* 14 Oct. p. ii/3 To-day an inoffensive peace-minded majority of the German people are being unwillingly dragged by a bloody-minded minority of their race along the road of doom. **1967** *Freedomways* VII. 118 Miss Barbara Deming..has been very active in the Peace Movement (she has taken part in a 'peace walk'..and a peace mission to Saigon in order to protest the war). **1976** R. MOORE *Dubai* i. 10 We [Americans] have a peace mission actively working to avoid war in the Middle East. **1940** H. NICOLSON *Diary* 16 July (1967) 102 We may be faced at any moment by a German peace move, and..a purely negative attitude is not sufficient. **1977** *Listener* 18 Aug. 195/1 If..the peace moves collapse..what will happen to the insecure régimes in Egypt and Syria? **1953** M. MCCARTHY in *Reporter* (N.Y.) 22 Dec. 33/3 My fiancé.. had known the organizer for years, perhaps from the peace movement. **1975** *Guardian* 2 Dec. 6/6 Defence Counsel..accused the prosecution of conducting a smear campaign against the British peace movement. **1912** Peace negotiation [see WILL *v.*¹ 46]. **1918**, etc. Peace offensive [see *OFFENSIVE *sb.* 2]. **1977** *Time* 3 Jan. 46/1 While Assad and Sadat have captured world headlines with talk of a peace offensive, Rabin has been attacked at home for being timid and indecisive. **1918** C. P. SCOTT in D. Ayerst *Guardian* (1971) xxvii. 410 The peace offers of the Emperor Karl, so insultingly and stupidly turned down by Clemenceau. **1966** N. NICOLSON in H. Nicolson *Diaries & Lett. 1930–39* 412 Hitler opened his attack on Poland as planned, claiming that the Poles had rejected his peace-offer. **1976** *Times* 19 Oct. 12/1 The Peace People, this misleadingly trendy title now adopted by the [Ulster] women's peace movement. **1977** *Time* 22 Aug. 6/2 A year ago last week, Catholics Betty Williams and Mairead Corrigan founded the so-called Peace People's Movement, which has attracted mass support from both Catholics and Protestants. **1977** *Arab Times* 14 Dec. 2/6 Mrs. Williams added: 'I am angry, the Peace People are angry that war at home dribbles on, and around the world we see the same stupidity gathering momentum for far worse wars than the little one which the little population of Northern Ireland has had to endure.' **1940** H. NICOLSON *Diary* 19 Sept. (1967) 116 Already the Communists are getting people in the shelters to sign a peace-petition to Churchill. **1955** *Treatm. Brit. P.O.W.'s in Korea* (H.M.S.O.) 11 From lectures on 'peace'..it was but a short step..to the production of 'peace' propaganda and 'peace' petitions by prisoners. **1968** *Listener* 3 Oct. 429/1 The process has apparently forced the Israelis to clarify their position with a 'peace plan'. **1977** A. WILSON *Strange Ride R. Kipling* vii. 306 The Pope put forward his peace plan in 1917. **1935** H. R. L. SHEPPARD *We say No* p. x, To give Pacifist opinion a chance to crystallize, I launched my Peace Pledge. **1956** A. H. COMPTON *Atomic Quest* 296 The 'Stockholm Peace Pledge.' In this 'peace pledge' a prime point was again made of outlawing atomic weapons. **1974** W. FOLEY *Child in Forest* II. 247 This group of young left-wing idealists..belonged..to the Peace Pledge Union. **1976** *Daily Mirror* 12 Nov. 2 Jack Jones yesterday declared: 'There will be *no* dock strike.' The Transport Union leader's peace pledge came amid dockland fury at the crippling of the Dock Work Regulation Bill in the Commons. **1902** *Idler* Nov. 244/2 The com-

mission appointed to decide the peace prize. **1974** *Encycl. Brit. Micropædia* IV. 875/2 [Dag Hammarskjöld] was posthumously awarded the Nobel Peace Prize for 1961. **1906** *Westm. Gaz.* 12 Mar. 2/2 We certainly hope and believe that the Government will act in a peace-promoting spirit. **1929** D. H. LAWRENCE *Pansies* 85 Loud peace propaganda makes war seem imminent. **1955** Peace propaganda [see *peace petition*]. **1951** 'A. GARVE' *Murder in Moscow* iii. 43, I attended a great 'peace' rally at the Bolshoi Theatre. **1975** P. HARCOURT *Fair Exchange* 41 It's not a political rally. It's a peace rally. **1969** *New Yorker* 30 Aug. 21/2 Just as it finished, an Army helicopter flew over. The whole crowd..looked up and waved their forefingers in the peace sign. **1973** D. WESTHEIMER *Going Public* iv. 67 She raised her hand in a peace sign... He realized it was not the peace sign at all. To those of the old woman's generation it was V for Victory. **1976** *Scotsman* 20 Nov., (*caption*) Brian Robertson (right) gives the 'peace sign' as he leaves the High Court handcuffed to Jeremy Salmon. **1816** N. WORCESTER *Friend of Peace* I. VII. 30 (*heading*) First annual Report of the Massachusetts Peace Society. **1970** *Time* 2 Nov. 6 *American Opinion*..compared the familiar peace symbol to an anti-Christian 'broken cross'. .. The peace design was devised in Britain for the first Ban-the-Bomb Aldermaston march in 1958. **1972** *Times* 4 Aug. 1/2 He..is..said to have cut the peace symbol (the nuclear disarmament insignia) into the ice with his skates. **1789** J. STEELE *Papers* (1924) I. 51, I only mean to hold a peace talk. **1800** B. HAWKINS in *Georgia Hist. Soc. Coll.* (1848) III. i. 72 Peace talks are always addressed to the cabin of the Mic-co. **1852** J. REYNOLDS *Pioneer Hist. Illinois* 165 All the 'peace talks' ever presented to the red men, could not have kept them in peace, under these circumstances. **1918** W. OWEN *Let.* 10 Oct. (1967) 583 Tonight I must stand before them [*sc.* my company] & promulgate this General Order: 'Peace talk in any form is to cease in Fourth Army. All ranks are warned against the disturbing influence of dangerous peace talk.' **1930** J. CANNAN *No Walls of Jasper* 63 Look at the newspapers! Nothing but peace talk. **1958** *Times Lit. Suppl.* 10 Jan. 21/4 The campaign ended, un-Napoleonically, in the feeble peace-talks of Villafranca. **1973** *Times* 9 Nov. 1/1 Israel and Egypt have accepted a ceasefire agreement that could lead to Middle East peace talks by the end of the year. **1968** *Punch* 15 May 693/1 This is no surprise to the Korean peace talkers. **1917** D. H. LAWRENCE *Look! We have come Through!* 126 Everything was tainted with myself..nations, armies, war, peace-talking. **1935** MRS. BELLOC LOWNDES *Diary* 19 Dec. (1971) 136 Wickham Steed..talked with great excitement of what has happened over the offer of peace terms to Italy and Abyssinia. **1976** *Classical Q.* XXVI. 272 Andocides does not deal explicitly with the question of the status of the Greeks of Asia in the proposed peace-terms. **1967** Peace walk [see *peace mission*].

peaceful, *a.* Add: **4. a.** Not violating or infringing peace; used esp. of methods for effecting purposes for which force, violence, or war, is an alternative or more obvious means.

1876 G. HOWELL *Handy-Bk. of Labour Laws* (ed. 2) ii. 29 Peaceful picketing is no longer prohibited, for, although the Government refused..to legalise 'peaceful persuasion', yet it was distinctly declared that it was legal under the Act. **1902** *Engineering* 28 Mar. 413/3 The methods of trade unions run within the line of the law, with the single exception of the form of intimidation known as 'peaceful picketing'. **1903** E. GREY in *Hansard Commons* 18 Feb. 245 Russia seems undoubtedly..to be carrying on a process of absorption in Persia, and it is being done by what, I think, a French writer has called peaceful penetration. **1906** *Act 6 Edw. VII* c. 47 § 2 *marg.*, Peaceful picketing. **1909** *Westm. Gaz.* 7 Aug. 7/2 English proposals for the peaceful penetration of China from Burma have varied considerably since 1831. **1916** *Q. Rev.* Apr. 571 A regular system of 'peaceful picketing' was set up; and wounded heroes in mufti found white feathers thrust upon them by well-meaning females. **1930** *Economist* 18 Jan. 122/2 If Canada has been subjected to peaceful penetration by the economic forces of the United States it is because [etc.]. **1935** *Discovery* Nov. 320/1 It [*sc.* the spread of grasshoppers from Angarà] is rather a matter of 'peaceful penetration', of infiltration, proceeding slowly and imperceptibly through the ages. **1950** *Chambers's Encycl.* XIII. 726/1 The Conspiracy and Protection of Property Act..in effect conceded the right to strike and reinstated peaceful picketing in a modified form. **1961** *U.S. Peace Corps Fact Bk.* 22 If it is decided to make a small shift which may be required from military aid or special assistance funds, in order to carry out the purposes of the Mutual Security Act through this new peaceful program, this will be a hopeful sign to the world. **1962** N. S. FALCONE *Labor Law* xi. 343 Peaceful picketing conducted at the premises of neutral employers was for the purpose of informing the public of a primary dispute and was not intended to bring about any work stoppage among the neutral employees.

b. *peaceful coexistence*: in the foreign policy of Soviet Russia, a concept of varying emphasis referring to relations with the capitalist West. Also *transf.* and *fig.*

In recent years *peaceful coexistence* has implied avoidance of nuclear confrontation. (Quot. 1920, cited from Carew Hunt *A Guide to Communist Jargon* (1957) 27, could not be verified as no copy of the newspaper appears to have survived.)

1920 *N.Y. Even. Jrnl.* 18 Feb., [Interview with Lenin] Our plans in Asia? The same as in Europe: peaceful coexistence with the peoples, with the workers and peasants of all nations. **1952** *Times* 2 Apr. 3/3 (*heading*) The invitations are..careful to emphasize the peaceful coexistence theme... The conference is based on the assumption..that mutually advantageous economic co-operation between Socialist and capitalist countries is quite possible. **1954** *Time* 4 Oct. 32/1 Attlee's men replied with one of those ingenious compromises that make peaceful coexistence possible between the two wings of the Labor Party: a policy favoring arms and sovereignty for the Germans but also offering to 'consider' some more Big Three talks with the Russians. **1956** R. MACAULAY *Towers of Trebizond* vi. 62 Turks do not believe in peaceful co-existence with Russia, they never have, and Father Chantry-Pigg agrees with them. **1959** *Listener* 26 Mar. 539/2 Both the Communist Party and the Church [in Poland] benefit immensely from the unprecedented type of peaceful coexistence they have worked out together. **1961** *Times* 2 Jan. 8/3 Mr. Khrushchev,..at a New Year banquet..in the Kremlin,..raised his glass and bade the whole company drink to peaceful coexistence. **1973** *Time* 25 June 31/3 Articles have appeared..applauding Brezhnev's peaceful-coexistence policy. **1977** *Lancet* 29 Oct. 903/2 Many microorganisms have been accused; so far, none have fulfilled Koch's postulates and some are known to live in peaceful coexistence with the healthy gut.

peacefully, *adv.* (Examples in sense *4 a of PEACEFUL *a.*)

1904 *Minutes of Evid. R. Comm. Trade Disputes* 262/1 in *Parl. Papers* 1906 (Cd. 2826) LVI. 137, I have never seen picketing conducted peacefully yet. **1920** *Act 10 & 11 Geo. V* c. 55 § 2 (1) No such regulation shall make it an offence for any person or persons..peacefully to persuade any other person or persons to take part in a strike. **1935** E. WEEKLEY *Something about Words* i. 25 Unfortunately, 'refined' American is also penetrating peacefully, and I confidently expect that the English undertaker will soon describe himself as a *mortician*. **1961** *Nation* (N.Y.) 11 Nov. 371/3 Khrushchev..would like, above all, the Berlin and German problems to be settled peacefully.

peace-keeper. Add: Also, an organization that keeps or maintains the peace; one regularly employed in the maintenance of peace between nations or communities; a soldier in a force so employed. So **pea·ce-keeping** *vbl. sb.* (freq. *attrib.*).

1961 *Times* 2 Oct. 13/3 Budgetary procedures..including the cost of peace-keeping operations. **1961** *Guardian* 24 Oct. 8/3 The problems of a disarmed world, peace-keeping machinery, etc. **1963** *Times* 22 Apr. 11/3 The development of the United Nations as a peacekeeping organization will be seriously hampered before the summer is out if its present financial difficulties are not solved. **1964** *Daily Tel.* 17 Jan. 12/2 The sending of some international peacekeeping force has become a matter of extreme urgency. **1964** *Manch. Guardian Weekly* 5 Mar. 1 If an international truce force went to the island [*sc.* Cyprus], Turkey would forgo its right to intervene for three months... This is..a rare declaration of faith in the United Nations as a peace-keeper. **1965** *Spectator* 15 Jan. 76/2 Peace-keeping is the basic function of the United Nations. **1965** *Maclean's Mag.* 1 Dec. 16 Lieutenant-General Burns..was the first of a new breed of international trouble-shooters who now try to halt the escalators of war by policing cease-fires—the Canadian peacekeepers. **1973** *Times* 17 Sept. 16/8 Peacekeeping is..a purely temporary role; a permanently-active peacekeeper must in the end become an irritant. **1977** R. HOLLAND *Self & Social Context* ix. 274 The forces of capital and labour face each other unmediated by the normal peace-keeping functions of police intermediaries. **1978** *Globe & Mail* (Toronto) 20 Oct. 10/6 Although Mr. Jamieson refused to commit Canadian forces to a peacekeeping force in the territory, he indicated he was moving in that direction.

peacemonger. Add: (Further examples.) So **pea·cemongering** *vbl. sb.*

1928 *Funk's Stand. Dict.* II. 1815/2 Peacemongering, *n.* **1949** *N.Y. Times* 26 June IV. 1/2 Peacemongers, apparently in the hope that East-West tension would be lessened and the West's position softened. **1967** *Economist* 1 Apr. 26/3 He need not seek the political kudos that holders of other offices may hope to gain by public peacemongering. **1969** *Guardian* 18 Sept. 3/1 Peace-mongers who prolong the Vietnam war.

peacenik (pī·snik). [f. PEACE *sb.* + *-NIK.] A member of a pacifist movement, esp. when regarded as a 'hippy'; freq. *spec.* an opponent of the military intervention of the United States in Vietnam. Also *attrib.*

1965 *Time* 23 Apr. 13/2 (*heading*) War & Peaceniks. **1965** *San Francisco Examiner* 6 Sept. 14/2 Dean Plapowski..described himself as a 'peacenik'. This, he explained, 'is probably a beatnik who's got himself hung up in pacifist and non-violent activity'. **1966** *Guardian* 17 Aug. 9/2 The 60-year-old Judge is not suspected of 'peacenik' sympathies. **1971** *Ibid.* 14 July 11/5 The air base appears to treat its peaceniks with an easy tolerance. **1974** K. MILLETT *Flying* (1975) I. 117 A tenement crammed with peaceniks who have painted the word *love* across their brickfront. **1975** P. HARCOURT *Fair Exchange* I. 56 Who is this peacenik anyway?

peace-officer. (Later examples.)

1959 JOWITT *Dict. Eng. Law* II. 1320/1 *Peace officer*, a constable, coroner, justice or sheriff. **1965** *Economist* 7 Aug. p. xi/3 The officers in charge of these [S. African local labour] bureaus are now being designated 'peace officers', and have powers of arrest and search of houses. **1973** *Black Panther* 16 June 12/3 The scores of uniformed 'peace officers'. **1976** *Columbus* (Montana) *News* 10 June 1/3 They had hoped more residents would attend to express their wishes about having a full-time peace officer.

peace-time, peacetime. Also peace time. [PEACE *sb.* 15.] The time when a country is not at war. Also *U.S.* (with pl.), a period during which no declaration of war is in force. Also *attrib.*

1551 R. ROBINSON tr. *More's Utopia* I. sig. C v recto, The hole realme is fylled and besieged wyth hierede soldiours in peace tyme, yf that be peace. **1917** A. G. EMPEY *Over Top* 311 *Territorial*, a peace-time soldier with the same status as the American militiaman. **1923** *Man. Seamanship* (Admiralty) II. 5 In peace time at sea. **1938** *Sun* (Baltimore) 16 Apr. 8/3 His [*sc.* Roosevelt's] savage attacks in 1932 upon Mr. Hoover's Administration as the greatest spender of all peacetimes. **1940** *Economist* 5 Oct. 426/2 Already its price is more than four times the peace-time level. **1942** *Short Guide Gt. Brit.* (U.S. War Dept.) 23 Most British food is imported even in peacetimes. **1952** R. KNOX *Hidden Stream* xii. 106 I should want to take a good look round, as we did in the days of old-fashioned, peace-time shopping. **1955** *Times* 5 July 10/1 After transfer the Royal Navy will continue to enjoy the facilities of the Simonstown base in peace-time. **1968** S. HYNES *Edwardian Turn of Mind* viii. 273 That curious last-minute plunge toward the twentieth century that marked so strikingly that last peace-time summer. **1971** S. HILL *Strange Meeting* iii. 188, I hope all the rest will not be spoilt for leave, or peace-time. **1971** W. H. MCNEILL in A. Bullock *20th Cent.* 47/1 War-born techniques of administrative mobilization were carried over into peacetime.

peach, *sb.*¹ Add: **1. b.** *slang.* Someone or something of exceptional worth or quality; someone or something particularly suitable or desirable, esp. an attractive young woman.

1754 E. TURNER *Let.* 16 Aug. in Dickins & Stanton *18th-Cent. Corresp.* (1910) 238, I had almost forgot that orange Peach, your Niece. **1863** B. HARTE in *Daily Even. Bull.* (San Francisco) 9 Dec. 5/3 Phrases such as camps may teach..Such as 'Bully!' 'Them's the Peach!' **1888** *Puck* (U.S.) XXII. 415/2 An' two young darters—one eighteen. A reg'ler peach. **1904** W. H. SMITH *Promoters* vii. 134 You're a brick! You're a peach! **1907** *Punch* 2 Jan. 13/2 Prof. Br—ce: H'm! Nice pleasant expression! One who was not a purist in language might almost describe him as a '*peach*'. **1917** WODEHOUSE *Man with Two Left Feet* 62 Opinions differ about girls. One man's peach..is another man's poison. **1919** H. L. WILSON *Ma Pettengill* iv. 111, I..landed a hard right on the side of his jaw and dropped him just like that. It was one peach I handed him and he slumped down like a sack of mush. **1924** 'J. SUTHERLAND' *Circle of Stars* xii. 126 It's a peach of a storm, and it's getting worse every moment. **1925** F. SCOTT FITZGERALD *Let.* June (1964) 484 He's a peach of a fellow and absolutely first-rate. **1930** 'R. CROMPTON' *William—the Bad* i. 19 Now would you think that a peach like her would fall for a fat-headed chump like that? **1943** E. B. WHITE *Let.* 1 Jan. (1976) 236 You were a peach to give me such a good present. **1949** *Sunday World-Herald Mag.* (Omaha, Nebraska) 1 May 2/1 The new recipe for making a peach cordial: Buy her a drink. **1974** *Times* 1 Apr. 12/4 She had, of course, a peach of a subject. **1976** *Derbyshire Times* (Peak ed.) 3 Sept. 22/6 (Advt.), 1972 Peugeot 504, white, 34,000, a real peach, £1,395. **1976** P. DICKINSON *King & Joker* iv. 46 Louise had a history essay, a real peach for which she'd only needed to look up a few dates. **1977** D. FRANCIS *Risk* xiv. 179 Dad's brought the detestable Lida... Actually I would have liked it..if he'd fallen for a peach.

c. *peaches and cream*: used *attrib.* and *absol.* to designate a fair complexion characterized by creamy skin and pink cheeks.

1901 ADE *Forty Mod. Fables* 188 Give me some perfumed Dope that will restore a Peaches and Cream Complexion. **1967** 'D. SHANNON' *Rain with Violence* (1969) i. 10 Carole had very blonde hair..and a peaches-and-cream skin. **1969** 'J. ASHFORD' *Prisoner at Bar* vii. 62 She had the perfect peaches-and-cream beauty that was often called classical English. **1975** *New Yorker* 9 June 46/3 She had a real peaches-and-cream complexion and a trim figure. **1978** J. W. WAINWRIGHT *Jury People* lxii. 211 His complexion..was pure 'peaches and cream'.

4. (Earlier examples.)

1809 M. L. WEEMS *Life Gen. F. Marion* viii. 74 Suppose you take a glass of Peach. **1845** J. J. HOOPER *Some Adventures Simon Suggs* v. 53 Thar's koniac, and old peach, and rectified.

6. *peach-orchard* (examples); also with names of colours; designating that shade of the colour which is shown by the peach, as *peach-beige, -green, -pink, -red*; **peach aphid, aphis,** one of several aphides infesting peach trees, esp. the peach-potato aphis, *Myzus persicae*; **peach-black** (examples); **peach-brandy** (earlier and later examples); **peach cobbler** *U.S.*, a cobbler (sense 4) made with peaches; **peach fly** = *peach aphid*; **peach leaf-curl** = *leaf-curl* (*b*) s.v. *LEAF *sb.*¹ 17; **peach Melba:** see *MELBA; **peach oak** (examples); **peach-potato aphid, aphis,** an aphis, *Myzus persicæ*, which causes leaf-curl in peach trees and other plants, and also transmits many plant virus diseases; **Peach State,** a sobriquet of the State of Georgia in the U.S.; **peach-worm** (examples); **peach yellows,** substitute for first part of def.: a virus disease affecting cultivated peach-trees, esp. in the United States; (examples).

1909 F. V. THEOBALD *Insect & Other Allied Pests* 324 Peach Aphides... At least four species of aphis attack the peach in this country. **1937** A. M. MASSEE *Pests of Fruit & Hops* vii. 163 The Peach Aphis has been recorded on a very large number of host plants, including fruit trees. **1942** *Phytopathology* XXXII. 93 (*title*) A virosis-like injury of snapdragons caused by feeding of peach aphid. **1963** *Jrnl. Insect Physiol.* IX. 875 (*title*) Some amino acid requirements of the green peach aphid, *Myzus persicae*. **1927** Peach-beige [see *GRÈGE *a.* and *sb.*]. [**1835** G. FIELD *Chromatogr.* 265 (Index), Black..Peach-stone 180. *Ibid.* xxi. 180 Similar blacks are prepared of vine twigs and tendrils,..also from peach-stones, &c. whence Almond black.] **1869** T. W. SALTER *Field's Chromatogr.* (new ed.) xxi. 407 *Peach Black*, or Almond Black, made by burning the stones of fruits, the shell of the cocoa-nut, &c., is a violet-black, once much used by Parisian artists. **1948** F. A. STAPLES *Watercolour Painting* (1951) i. 3 You will want to add the following to the palette: Raw Sienna,.. New Blue and Peach Black. **1963** *Times* 6 May 16/3 Many students will remember how much 'Corfi' enjoyed laying a wash of thickly sedimented peach black and raw sienna on a drawing that had taken weeks to prepare. **1711** W. BYRD *Secret Diary* 9 Sept. (1941) 403 After drinking two drams of peach brandy we returned to Mrs. Randolph's. *c* **1780** [see *apple-brandy*]. **1965** AMERINE & SINGLETON *Wine* xvii. 268 Wines were made also from peaches and distilled into peach brandy. **1976** J. MCCLURE *Rogue Eagle* vii. 129 As peach brandy goes, this is among the best *sluks* I've ever tasted. **1859** BARTLETT *Dict. Amer.* (ed. 2) 90 Cobbler... According to the fruit, it is an apple or a peach cobbler. **1880** [see *COBBLER 4]. **1947** *Reader's Digest* Apr. 130/2 You could smell a peach cobbler all through dinner. **1976** *National Observer* (U.S.) 28 Aug. 14/3 Peach cobbler (*recipe follows*). **1865** *Our Young Folks* I. 715 The peach-fly was thus kept from laying its eggs in the soft bark at the surface of the ground. **1905** *Chambers's Jrnl.* May 368/1 The peach..is not now obtainable, through the inroads of the peach-fly. **1971** J. DRUMMOND *Farewell Party* 8 A great sunset..a wash of peach-green that ran across the sky. [**1887** *Bot. Gaz.* XII. 216 (*title*) The 'Curl' of Peach Leaves. *Ibid.* Pl. XIII. (*caption*) Knowles on Peach Curl. **1888** *Amer. Naturalist* XXII. 738 *T[aphrina] deformans* Tul., causing the 'peach curl' of the leaves of the peach tree.] **1899** *Bull. Cornell Univ. Agric. Exper. Station* CLXIV. 371 Peach leaf-curl is a disease which has long been known to the orchardist as well as to the botanist; and since the seasons of 1897 and 1898 there are probably very few peach growers..who are unfamiliar with the disease. **1904** *Westm. Gaz.* 6 Oct. 10/2 A fungus disease called peach leaf-curl..does injury to the extent of £600,000 annually in the United States. **1920** P. J. FRYER *Insect Pests & Fungus Dis.* xxxv. 557 Peach Leaf Curl... Plants Attacked. Peaches, nectarines and almonds. **1955** H. WORMALD *Dis. Fruit & Hops* (ed. 3) vii. 172 Peach Leaf Curl..is found not only on peaches but also on nectarines and almonds. **1976** *Country Life* 18 Mar. 685/3 Garlic is said to protect peaches from peach-leaf curl. **1835** J. MARTIN *New Gazetteer Virginia* 209 Peach oak (so called from the resemblance of its leaves to that of the peach tree). **1897** G. B. SUDWORTH *Nomencl. Arborescent Flora U.S.* 177 *Quercus phellos* Linn. Willow Oak... Common Names... Peach Oak (N.J., Del., Ohio). **1676** T. GLOVER in *Phil. Trans. R. Soc.* XI. 628 Here are likewise great Peach-Orchards, which bear..an infinite quantity of Peaches. **1758** *Calendar Virginia State Papers* (1875) I. 257 We..overtook them at a peach orchard. *c* **1805** D. MCCLURE *Diary* (1899) 68 Between the house & the bank of the River was a..peach orchard. *a* **1936** KIPLING *Something of Myself* (1937) vi. 170 A bull-kudu.. would jump the seven-foot fence round our little peach orchard. **1955** W. MOORE *Bring Jubilee* xix. 185, I made my way towards a farm on which there was a wheat-field and a peach orchard. **1974** *Sat. Rev. World* (U.S.) 2 Nov. 32/3 A long valley, green and golden with peach orchards —Canada's peach heartland. **1926** M. LEINSTER *Dew on Leaf* I. vii. 97 Ah Dai fingered and thumbed a fragment of the peach-pink silk he had unfurled for her inspection. **1934** A. HUXLEY *Beyond Mexique Bay* 2 The last word in cocktail bars and peach-pink sanitary fittings. **1956** R. MACAULAY *Towers of Trebizond* xiv. 171 Through the windows I saw the circle of the Circassian mountains, indigo and brown and peach-pink in the sunset. [**1931** K. M. SMITH *Textbk. Agric. Entomol.* vi. 50 (*heading*) Potato and Peach Aphis.] **1951** *New Biol.* XI. 51 The peach-potato aphid..is the main carrier of the known plant virus diseases throughout the world. **1959** *Times* 27 July 9/5 The aphids responsible for spreading the viruses—mainly the peach-potato aphid *Myzus persicae* —are able to multiply on the [potato] crop during the summer. **1975** D. S. HILL *Agric. Insect Pests of Tropics* v. 163/1 Peach-Potato Aphid... Peach (primary host)... Potato (secondary host), and polyphagous on many other crop plants and weeds. **1926** M. LEINSTER *Dew on Leaf* 114 My unborn son waits to clutch my heart-strings with peach-red fingers, with the call of flesh to flesh. **1935** W. DE LA MARE *Poems, 1919–1934* 377 Peach-red carnelian, apple-green chrysoprase, Amber and coral and orient pearl! **1941** G. E. SHANKLE *State Names* (rev. ed.) ii. 110 Georgia was nicknamed *The Peach State* in 1939 because 'peaches have been an important product of Georgia since the middle of the sixteenth century'. **1954** *Nat. Geogr. Mag.* Mar. 318/1 Georgia's automobile license plates carry the legend, 'Peach State'. **1970** G. PAYTON *Webster's Dict. Proper Names* 515/1 *Peach State*, a nickname for Georgia, where peaches are now a valuable crop in the center and south. **1976** *S. Wales Echo* 26 Nov. 2/5 With out-of-state tourists flocking to Jimmy Carter's home town in Plains, Georgia, officials are looking for ways to lure the visitors to the peach state's other attractions. **1814** *Cramer's Pittsburgh Mag. Almanac 1815* 55 (*heading*) Remedy for the Peach Worm. *a* **1817** T. DWIGHT *Trav. New-Eng.* (1821) I. 76 The Peach Worm has been known here for about fifty years; and is now become very common. **1856** *Rep. Comm. Patents: Agric. 1855* (U.S. Dept. Agric.) 299 The ravages of the peach-worm have proved more extensive than usual. [**1808** R. PETERS in *Mem. Philad. Soc. for Promoting Agric.* I. 23 Mr. H. begins to suffer by the disease, I call the 'yellows'. *Ibid.* 24 The 'yellows' are seen making destructive ravages in Mr. Heston's peach plantation.] **1888** *Bull. U.S. Dept. Agric. Bot. Div.* IX. 9 Peach yellows appears to be confined exclusively to the Eastern United States. **1928** F. T. BROOKS *Plant Dis.* iii. 23 The only means of checking the spread of Peach Yellows is to destroy affected trees as soon as seen. **1956** H. W. ANDERSON *Dis. Fruit Crops* vii. 265 Peach yellows is undoubtedly of American origin. **1974** K. M. SMITH *Plant Viruses* (ed. 5) i. 2 Two years later [*sc.* 1888], Erwin F. Smith proved that the disease known as 'peach yellows' was also communicable and could be transmitted by budding.

peach, *v.* Add: **2.** Also const. *on*.

1881 *Punch* 26 Nov. 241/2 Eve flirted with Jerrem; Adam, enraged, 'peached' on Jerrem. **1927** KIPLING *Limits & Renewals* (1932) 170 Will and I wouldn't have peached on him. **1966** *New Statesman* 1 July 9/2 The other members of the gang..would not hesitate to peach on him if it would serve their purpose. **1976** *National Observer* (U.S.) 17 July 17/3 Middle-level bureaucrats cravenly peach on their bosses everytime one of them does something the tiniest bit illegal, like violate the Constitution. **1978** P. LOVESEY *Waxwork* 123, I shan't ask you to peach on one of your neighbours... What I want from you is the name and address of the supplier.

peach-bloom. Add: **b.** (Example.)

1923 D. H. LAWRENCE *Birds, Beasts & Flowers* 13 Why, from silvery peach-bloom—..This rolling, dropping, heavy globule?

c. A pink colour characteristic of the monochrome glazes on some Chinese porcelain; the glaze itself; = PEACH-BLOW a, b. Also *attrib.*

1886 [see PEACH-BLOW]. **1898** W. G. GULLAND *Chinese Porc.* I. 139 The following are the names by which some of the colours met with are generally indicated:—..Lavender Clair de lune Peach bloom [etc.]. **1900** F. LITCHFIELD *Pott. & Porc.* vii. 109 Some charming results were obtained in many of those beautiful self colours that collectors delight in; amongst others,..coral, lilac, peach bloom, crushed strawberry [etc.]. **1902** W. G. GULLAND *Chinese Porc.* II. 360 Here we have..peach bloom employed along with other coloured glazes in the decoration of white porcelain. **1906** [see *CRUSHED *ppl. a.* 3]. **1937** *Burlington Mag.* Oct. 195/1 Thus, such apparent mysteries as double crackle,..the achievement of peach bloom, the splashes in Chün glazes..are all comfortably disposed of. **1970** *Oxf. Compan. Art* 235/2 Many people consider the most admirable wares of the K'ang Hsi period to be those with monochrome glazes... The brilliant red *sang-de-bœuf* and soft pink 'peachbloom' celadon.

peach-blow. Add: **a.** (Earlier and later examples.)

1829 T. FLINT *George Mason* 32 The Red Bud in a thousand places was one compact tuft of peach-blow flowers. **1837** J. L. WILLIAMS *Terr. Florida* 75 This bird is of a peach-blow colour. **1922** H. SHOEMAKER *Allegheny Episodes* xxv. 344 Her skin was transparently white, and the delicate peach-blow color in her cheeks was too hectic to betoken good health.

d. A type of glass of a similar colour.

[**1886** *Official Gaz.* (U.S. Patent Office) 20 July 236/2 13,523 Glass Table-Ware and Fancy Glass Articles.. Frederick S. Shirley, New Bedford, Mass... Used since July 1, 1885 'The word "peach".'] **1886** *Pottery & Glassware Reporter* 4 Nov. 21/2 In recent years four novelties have been brought out by the New England Glass Works, under Mr. Libbey's administration. We refer to the Amberina, Pomona, Peachblow, and Agata grades of goods. **1930** L. W. WATKINS *Cambridge Glass 1818–1888* 156 The peach blow glass took its name from the Chinese peach blow porcelain... The New England [Glass Co.] peach blow is one of the loveliest and most unusual things in American glass. In color it shades by imperceptible degrees from white to a deep rose. **1933** *Antiques* (U.S.) Aug. 48 (*heading*) Peachblow Glass. *Ibid.*, Among those glass collectors who are more susceptible to color than to antiquity, 'peachblow items' are highly valued. **1944** R. W. LEE *Victorian Glass* xxii. 561 None of the glass factories at which Peachblow was produced adhered either to the shapes or to the purposes of the Peachblow porcelains. **1972** K. M. WILSON *New England Glass* vii. 352 Both the Mount Washington Peachblow and the New England Wild Rose are to be found in their natural state and also with a 'plush', or satin finish. **1975** *Daily Colonist* (Victoria, B.C.) 26 Oct. 30/6 Peachblow originated with Hobbs Brockunier around 1883. It is lined with white and shades from yellow to peach.

peacherino (pītʃərī·no). *slang* (chiefly *U.S.*). [Fanciful f. *PEACH *sb.*¹ 1 b.] = *PEACH *sb.*¹ 1 b. Also **peacheri·ne, peacheroo·.**

1900 *Dialect Notes* II. 48 *Peacherine*,..synonym for peach. [**1900** *Polk's Kansas State Gazetteer* IX. 695 (Advt.), As a hero you'll forever Take the 'peacherino' yam.] **1905** A. M. BINSTEAD *Mop Fair* iii. 47 Archie, who had undertaken not only to give the Dragoon a merry evening, but to ring in miracles on him in the peacherino line,..repented the loss of all his addresses. **1908** G. H. LORIMER *Jack Spurlock* iv. 71, I went up in the air like an old wife happening by the office and discovering her husband dictating to a new blonde peacherino instead of old reliable. **1910** S. E. WHITE *Rules of Game* III. xv. 273 Plant has a drag with Chairman Gay; don't know what it is, but it's a good one, a peacherino. **1928** *Chambers's Jrnl.* Feb. 98/2 Though Captain Reginald saw little of her except at meals, he realised that here indeed was a 'peacherino'. **1966** M. WOODHOUSE *Tree Frog* xxiii. 171 'She [*sc.* an aeroplane]'s a peach,' he said. 'A real peacheroo.' **1967** C. ROUGVIE *When Johnny Died* iv. 88 When I was his age, they were hauling them out from under me... And all young peacherinos, too. **1970** R. GOLDMAN *Sob Story* in *This Side of Parodies* (*National Lampoon*, U.S.) (1974) 78 'What do you think?' I asked. 'Isn't she a peacheroo?'

peachy, *a.* Add: (Later examples not referring to the complexion.)

1973 *Country Life* 15 Feb. 425/2 Their frosted bronzed peachy-pink lipstick. **1976** *Vogue* 15 Mar. 19/1 (Advt.), One (Barley Sugar) is a peachy sort of shade.

2. *slang.* Attractive, outstanding, marvellous, etc.

1926 E. GLYN *Love's Blindness* ii. 25 He..whispered to the man behind him——'Peachy bit in the eighth row—— Look at the pearls.' **1929** S. ANDERSON in *Mercury Story Bk.* 228 It was a peachy time for me. **1932** J. T. FARRELL *Young Lonigan* iv. 193 He told himself that Airedales were peachy dogs, they were fighters, they could swim and liked the water, and they were smart. **1942** O. NASH *Good Intentions* 27 Do you know a picture program that Mr Oglethrip would find simply peachy? **1973** D. WESTHEIMER *Going Public* xi. 166 How about it, fellows?.. Isn't it a peachy idea? **1976** 'W. TREVOR' *Children of Dynmouth* v. 115 Your mum has a touch of style, Kate. I heard that remarked in a vegetable shop. I'd call her an eyeful, Kate. Peachy. **1977** *Time* 10 Jan. 28/3 Carter vowed that his Administration would not let New York City go into bankruptcy... Asked for a reaction to the meeting, Carey beamed: 'Peachy'.

b. *peachy-keen* adj. *U.S. slang* (see quot. 1960).

1960 WENTWORTH & FLEXNER *Dict. Amer. Slang* 379/2 *Peachy-keen adj.* 1 Excellent; fine... 2 All right; fair; not good enough to warrant enthusiasm but adequate. **1969** N. COHN *A WopBopaLooBop* (1970) v. 50 We dig America. We think it's really peachy-keen. **1975** J. GRADY *Shadow of Condor* (1976) iii. 52 Everything in Montana is peachy keen.

peacify (pī·sifəi), *v. rare.* [f. PEACE *sb.* + -IFY, influenced by PACIFY *v.*] To make calm, to pacify.

1845 [see *nursery language* s.v. *NURSERY 8 a]. **1922** JOYCE *Ulysses* 335 Joe and little Alf round him..trying to peacify him. **1942** BERREY & VAN DEN BARK *Amer. Thes. Slang* § 269/1 *Peacify*, to pacify.

pea-coat. (Earlier and later examples.)

1790 *Pennsylvania Packet* 4 Jan. 2/2 There are now lodged in the said Office..1 pea coat;..1 coatee [etc.]. **1842** DICKENS *Amer. Notes* II. viii. 244 The hoarse pilot, wrapped and muffled in pea-coats and shawls. **1974** *New Yorker* 25 Feb. 80/2 Neatly dressed in a sort of modified peacoat of generally Edwardian cut. **1976** *National Observer* (U.S.) 17 July 16/6 His youth was not exactly the Andy Hardy story. He got into trouble with teen-age drinking. At one point, he lived in a car; at another he fenced hot peacoats.

peacock, *sb.* Add: **4*.** Short for *peacock-blue*. Also *attrib.* or as *adj.*

1873 L. TROUBRIDGE *Life amongst Troubridges* (1966) viii. 60 A peacock grosgrain and white lace bonnet. **1881** C. C. HARRISON *Woman's Handiwork* III. 165 Peacock, turquoise, celestine, drake's neck, Damascus blue and robin's-egg blue. **1897** W. B. YEATS *Tables of Law & Adoration of Magi* 35 When the peacock curtains had closed behind us. **1922** *Daily Mail* 11 Dec. 14 (Advt.), Frock... In Brown, Lemon, Peacock, Rose, Mauve. **1924** C. MACKENZIE *Heavenly Ladder* i. 11 He..sat for awhile on the sweet short grass of Pendhu cliffs, contemplating the peacock sea below. **1963** *New Yorker* 29 June 44 Sizes 8–18. Cranberry, peacock, olive. **1971** 'D. HALLIDAY' *Dolly & Doctor Bird* xii. 166 The sea lay clear as shellac underneath us, jade and turquoise, cerulean and peacock.

5. a. *peacock colour* (earlier example), *-grey*.

1598 R. HAYDOCKE tr. *Lomazzo's Tracte containing Artes of Curious Paintinge, Caruinge, Buildinge* III. x. 110 The shaddowes of the simple and immixt colours of the thirde degree, suppose the aggate colour, are burnt oker, darke blew, peacocke colour [etc.]. **1935** DYLAN THOMAS *Sel. Lett.* (1966) 153 You write better when you've got someone..sneering when you go purple & using a cruel pencil over your choicest peacock-greys.

b. Peacock Alley *U.S.*, the name given to the main corridor of the original Waldorf-Astoria Hotel in New York, where fashionable people promenaded; hence the main corridor of other hotels; also *attrib.*; **peacock-blue** (earlier and later examples); **peacock butterfly,** for '*Vanessa Io*' substitute '*Inachis io*'; (earlier and later examples); **peacock copper,** substitute for def.: iridescent copper ore ('peacock ore'), esp. chalcopyrite or bornite; (examples); **peacock mottle** (see quots.); **peacock ore,** add: = *peacock copper* (in Dict. and Suppl.); (further examples); **peacock pheasant,** a small, south-east Asian pheasant belonging to the genus *Polyplectron*, whose markings resemble those of a peacock.

1906 *N.Y. Times* 2 Dec. III. 7/2 The Waldorf-Astoria is the New York headquarters of Kalamazoo, Michigan, and Brassband, Wisconsin. Its main corridor is known as Peacock Alley. To sit there five minutes makes a man a representative American. **1925** E. HUNGERFORD *Story of Waldorf-Astoria* vii. 139 The outstanding feature of this ground floor was a huge corridor—in after years to be known, somewhat irreverently, as 'Peacock Alley'— which was to run practically the entire length of the building, parallel to Thirty-fourth Street. **1930** J. DOS PASSOS *42nd Parallel* II. 156 Seedy-respectable or Peacock Alley clothes. **1932** L. C. DOUGLAS *Forgive us our Trespasses* (1937) x. 212 The peacock-alley of the hotel, stuffily scented, was daily, gravely, tallied off for Joan. **1974** *Washington Post* 20 Dec. D8/5 The Carr project included: Total restoration of the Willard's..public rooms and the famous Peacock Alley. **1881** C. C. HARRISON *Woman's Handiwork* I. 65 The curtains made of peacock blue, are bordered with..bands of turquoise blue serge. **1882** H. P. GRATTAN in *Theatre* June 348 Fashion..was carried to the verge of caricature. Crimson and peacock blue stocks, three layers of different coloured underwaistcoats, [etc.]. **1968** R. H. R. SMITHIES *Shoplifter* (1969) vii. 151 An improbable peacock blue evening jacket. *c* **1760** B. WILKES *Eng. Moths & Butterflies* III. i. 55 The peacock-butterfly. You must look for the Caterpillar that produces this Fly in the great Stinging-Nettle. **1906** R. SOUTH *Butterflies Brit. Isles* 73 Usually the Peacock butterfly assumes the perfect state but once in the year. **1965** P. WAYRE *Wind in Reeds* ix. 114 Tortoiseshell and peacock butterflies feed on the nectar. **1976** *Cumberland News* 3 Dec. 8/7 Mr Jack Thirlwell showed his prize winning films on the life style of the swallow tail and peacock butterflies. **1858** GREG & LETTSOM *Man. Mineral. Gt. Brit. & Ireland* 340 At Great Crinnis, St. Austell, in the neighbourhood of which town the mines produce the finest iridescent massive variety [of chalcopyrite], known as peacock copper. **1897** *Slocan* (Brit. Columbia) *Pioneer*

4 Sept. 1/6 The Michigan claim on Toad mountain is showing up well, some very fine grey copper and peacock copper having been encountered. **1937** A. F. Rogers *Introd. Study Minerals* (ed. 3) 300 Chalcopyrite... Color brass yellow, often with an iridescent tarnish, hence the name 'peacock copper'. **1924** G. O. Wheeler *Old English Furnit.* (ed. 3) xii. 278 Another variety [of mottle in mahogany] was once termed peacock mottle from its supposed resemblance to the tail of that bird. **1968** *Canad. Antiques Collector* Aug. 24/2 *Honeycomb or peacock mottle.* This is a variety of figure remarkable for its fine appearance; it is associated almost entirely with the mahoganies. **1911** *State* (Cape Town) Nov. 487 The ore is principally bornite—peacock-ore as it is often called on account of its beautiful iridescent colouring. **1964** H. Hodges *Artifacts* iv. 65 This group includes..the iron sulphide minerals chalcopyrites (copper pyrites, $Cu_2Fe_2S_4$) and bornite (peacock ore, Cu_5FeS_4). **1977** A. Hallam *Planet Earth* 124 (*caption*) Bornite, an important ore of copper, is often called 'peacock ore' for its iridescent tarnish. **1864** *Proc. Zool. Soc.* 373 From Calcutta... 1 Peacock Pheasant. **1906** *Macm. Mag.* Aug. 779 A peacock-pheasant..ceased its clamour. **1922** C. W. Beebe *Monogr. Pheasants* IV. 55 Peacock pheasants..are birds of the lowland forests. **1964** A. L. Thomson *New Dict. Birds* 627/2 The peacock pheasants *Polyplectron* spp. form a very distinct genus of small pheasants with long tails and a grey or brown plumage marked with metallic green and purple ocellae on the mantle, wings, and tail.

peacock, *v.* Add: **3.** *Austral.* (See quot. 1898.) *Obs.* exc. *hist.*

1898 Morris *Austral. Eng.* 344/2 To *peacock* a piece of country means to pick out the *eyes* of the land by selecting or buying up the choice pieces and water-frontages, so that the adjoining territory is practically useless to any one else. **1928** 'Brent of Bin Bin' *Up Country* xxi. 347 They had been able to 'peacock' their runs and safeguard their holdings. **1959** Baker *Drum* (1960) i. 14 Droughts and the activities of small selectors who 'peacocked' or 'picked the eyes out of' the country..capped the pioneers' woes.

Peacockian (pīkǫ·kiăn), *a.* and *sb.* [f. the name of Thomas Love *Peacock* (1785–1866), English novelist and poet + -ian.] **A.** *adj.* Pertaining to or characteristic of Thomas Love Peacock or his works.

1886 *Macm. Mag.* Apr. 424/2 It is not necessary..to be a believer in education, or in telegraphs, or in majorities, in order to feel the repulsion which some people evidently feel for the Peacockian treatment. **1904** A. B. Young *Life & Novels T. L. Peacock* 31 The squire's chief interest in the novel..seems to be mainly 'pushing' the bottle round, while that of Dr. Gaster, the first of many clergymen who figure in the Peacockian novels, is that of emptying it. **1927** J. B. Priestley *T. L. Peacock* vi. 160 His manner and talk are all his own and have the Peacockian sparkle and salty tang. **1930** *Times Lit. Suppl.* 13 Mar. 210/2 The Peacockian Mr. Dottery easily wins our affections. **1954** R. Macaulay *Last Lett. to Friend* (1962) 145 It is a kind of Peacockian set of imaginary discussions between a group of people. **1963** *Listener* 21 Mar. 531/1 The visitors are all rather Peacockian. **1972** *Guardian* 20 Jan. 11/4 The Open House by Michael Innes... Sir John Appleby..intrudes on Peacockian mansion, and, overnight, identifies killer and unravels inheritance tangle. **1978** P. Van Greenaway *Man called Scavener* iii. 34 A Peacockian world where conversation grows more polished than the superb sideboard.

B. *sb.* An admirer or devotee of Peacock and his writings.

1886 *Macm. Mag.* Apr. 420/2 One piece of verse..the 'War-song of Dinas Vawr'..has had some vogue, but the rest is only known to Peacockians. **1911** C. Van Doren *Life T. L. Peacock* xi. 277 Peacockians are wont to plume themselves upon a taste denied to the vulgar. **1973** *Times* 22 Mar. 12/3 An actor and Peacockian of some wit and taste.

peacocking, *vbl. sb.* Add: **b.** *Austral.* (See quots.) *Obs.* exc. *hist.*

1894 W. Epps *Land Syst. Australasia* iii. 28 When the immediate advent of selectors to a run became probable, the lessees endeavoured to circumvent them by dummying all the positions which offered the best means of blocking the selectors from getting to water. This system, commonly known as 'peacocking', was assisted by the use of Volunteer Land Orders. **1965** *Austral. Encycl.* V. 234/1 Many of the counter-tactics employed by pastoralists were equally indefensible, notably the purchase of sites to prevent selectors from getting to water ('peacocking').

peacockly, *a.* and *adv.* For *Obs.* read '*Obs.* exc. *arch.*' and add: **b.** (Later example.)

1941 E. R. Eddison *Fish Dinner* (1968) viii. 132 His coming hither but yesterday, most peacockly strained to the height of your philosophy and at an undue hour of eleven o'clock in the night.., was with purpose.

peacockry. [f. Peacock *sb.* + -ry.] = Peacockery.

1909 *Daily Chron.* 14 Aug. 2/3 At Siena you leave Pinturicchio's 'peacockry' in the Cathedral library only to meet a bevy of youths with striped legs, tight doublets, and feathered caps tossing banners down the street of St. Catherine. **1932** *Times Educ. Suppl.* 6 Aug. 305/4 A fault among boys was what was called 'peacockry', when their vanity caused them to pinch anything to adorn their person. **1967** D. Beyfus in L. Deighton *London Dossier* 229 Traditional British lairs of masculine peacockry.

pea-flower. a. (Later examples.)

1946 D. C. Peattie *Road of Naturalist* i. 20 In the innocent phase of spring there had bloomed an astragalus, very like a lupine, but straggling, crazy, clouded, its pea flowers sickly pink. **1977** M. Allan *Darwin & his Flowers* iv. 78 He found six adesmias-shrubs with pea-flowers.

pea-jacket. (Earlier and later examples.)

1721 *Amer. Weekly Mercury* 23 Mar. 2/2 Clothed with a double-breasted Pee-Jacket. **1848** A. Brontë *Tenant of Wildfell Hall* III. xv. 301 A dubious, sidelong glance at my splashed, grey trousers and rough P jacket. **1898** F. H. Smith *C. West* iii. 36 He had left his pea-jacket in the cabin. **1922** W. S. Maugham *On Chinese Screen* xlvii. 186 A pea-jacket such as you see in Leech's pictures of the sea-faring man. **1968** *Wall St. Jrnl.* 19 Feb. 1/1 A long-haired Berkeley student wearing a *Kill for Peace* button in the lapel of his black pea jacket. **1976** *Time* 20 Dec. 17/1 In his Navy pea jacket and worn brown boots,.. Jordan loped down the Senate halls, looking like the country boy he tries hard to remain.

Peak, *sb.*[1] **3.** (Later examples applied to millstones.)

1933 *Times Lit. Suppl.* 14 Dec. 891/1 We are given..a notice of the mill's character..down to the quality of millstones—Peak or French burr. **1936** *Ibid.* 16 May 416/4 A very individual vocabulary whose words, such as..peak or French burr-stones..will soon be a dead if not a forgotten language.

peak, *sb.*[2] Add: **I. 1. f.** Delete *Obs.* and add later examples.

1938 A. Morris *Step-by-Step Method Water Waving* 39 (*caption*) Hair line showing peak and receding part over eyes. **1951** A. Seton *Foxfire* ii. 42 She had soft hair.., and it curled all around her heart-shaped face. Really heart-shaped, because a widow's peak cleft the white forehead. **1971** W. Cooper *Hair* vii. 206 In English folklore, if a woman's hair grows to a point low on the brow, it is said to indicate that she will live to be a widow, and so it is often called a 'widow's peak'. **1978** 'M. M. Kaye' *Far Pavilions* xxiii. 339 A small muslin turban..covered her hair and showed only the deep widow's peak in the angle where its folds crossed.

II. 5. c. *spec.* one on a graph. Cf. sense *5 e.

1922 *Encycl. Brit.* XXXII. 1024/1 The potential difference of the arc electrodes is an irregular curve with sharp peaks. **1926** W. R. Inge *Lay Thoughts* II. i. 89 If we look at a chart of the births and deaths in Germany for the two generations before the Great War we shall see that each war is marked by a peak in the line showing the death rate and a ravine in the line showing the birth rate. **1968** *Brit. Med. Bull.* XXIV. 212/1 The corresponding histogram of conjugated bilirubin is markedly bimodal, and also shows an artificial peak at 0·5 mg. (100 ml.).

e. A highest point in a period of any varying quantity, as electric power, traffic flow, prices, etc.; the time when this occurs; a culminating point or climax. Cf. sense *5 c.

1902 *Encycl. Brit.* XXV. 35/1 Accumulators will take the peaks of the load, relieving the machinery from sudden jerks. **1923** *Daily Mail* 28 May 4 We have long since passed the peak in this unpleasant business. **1923** *Westm. Gaz.* 11 Aug. 6/4 During the morning, evening and theatre peaks, two escalators in each group can be run in either an upward or downward direction. **1943** *Sun* (Baltimore) 2 July 17/1 Steels enjoyed a last minute upswing and assorted favorites emerged..a number at three year peaks. **1962** A. Nisbett *Technique Sound Studio* iii. 63 The closer one gets to an open piano, the more the transients associated with the strike tone will be apparent; at their strongest..they may be difficult to control without ..risking momentary distortion on the peaks. **1967** *Listener* 23 Mar. 386/2 The nuclear disarmament campaign was already past its peak. **1968** *Brit. Med. Bull.* XXIV. 219/2 This is a continuous-flow analyser with its output arranged so that individual results are produced as voltage peaks following one another at intervals. **1971** *Hi-Fi Sound* Feb. 68/2 It is a basic hi-fi requirement that peaks should be accommodated without serious distortion. **1976** *Daily Tel.* 20 July 1/5 At the peak of the wages rush last year, the annual rate of increase of earnings reached 30 per cent.

f. *Phonetics.* The most prominent sound in a syllable with regard to sonority.

1935 J. S. Kenyon *Amer. Pronunc.* (ed. 6) 69 The phonetic center, or 'peak' of a syllable is its point of greatest sonority. **1942** Bloch & Trager *Outl. Linguistic Analysis* 22 The sounds which constitute the peaks of sonority are called syllabic. **1960** E. Sivertsen *Cockney Phonol.* ii. 23 Stressed simple syllable peaks do not occur before juncture, and there are other limitations in the distribution of unstressed peaks in this position. **1964** E. Palmer tr. *Martinet's Elem. Gen. Linguistics* ii. 52 A consonant like [l] when placed between consonants of lesser perceptibility or audibility such as [p] and [k], may function as syllable peaks. **1965** W. S. Allen *Vox Latina* 2 Sounds which may function either as peaks or as valleys of prominence, whilst classified as vowels in their peak (or 'nuclear') function, are generally termed semi-vowels..in their valley (or 'marginal') function.

g. *Surfing.* The highest point of a wave.

1963 *Surfing Yearbk.* 42/2 *Peak*, the highest point of the wave. **1965** Farrelly & McGregor *This Surfing Life* iv. 44/2 On most occasions in this sort of surf you take off straight down the peak. **1965** J. M. Kelly *Surf & Sea* ii. 26 The wind blows gently into the faces of the white-crested waves. It holds up their peaks giving you time to speed away on the clear green slopes before they break. **1968** *Surfer Mag.* Jan. 48/1 The way the peak was breaking didn't offer many rights.

III. 6. (sense *5 e) *peak-clipping, -limiting*; (sense *5 f) *peak nucleus, satellite*; *peak-bearded, -capped* (examples), *-crested* adjs.; **peak experience**, the momentary awareness of joy or fulfilment akin to ecstasy, of a higher and different quality from ordinary life, experienced by some people; **peak factor** *Electr.*, the ratio of the maximum value (or the difference between the maximum and minimum values) of a wave to the r.m.s. value; **peak listening** (see sense *7 a); **peak programme meter** (see quot. 1941); **peak shaving**, storage of part of the gas produced when demand is low so that it can be used to increase the supply at times of peak demand; **peak-to-peak** *a.* and *adv.*, (measured or expressed as the difference) between extreme values of a periodically varying quantity; also called **peak-to-valley** (*rare*); similarly **peak-to-mean** *a.*; **peak viewing** (see sense *7 a); **peak voltmeter** *Electr.*, a voltmeter that measures the peak value of an alternating voltage.

1905 *Daily Chron.* 12 Aug. 5/2 At one carriage a little baby-girl was held up by its mother to kiss farewell to a peak-bearded gloire bluejacket. **1972** *Drive* Spring 147/1 A peak-capped driver at the helm of a Rolls-Royce Corniche. **1976** *Field* 30 Dec. 1275/3 A peak-capped figure on one knee beside a folded stretcher. **1961** *Which?* July 156/1 There are two accepted methods for achieving loudness compression. One is called A.V.C. and the other peak clipping. **1975** G. J. King *Audio Handbk.* iii. 60 For example, if the 1 kHz input sensitivity is 2 mV and peak clipping..occurs at 20 mV, the overload margin is said to be 10:1 or 20 dB. **1879, 1881** Peak crested [see *blondinette]. **1962** A. H. Maslow *Toward Psychol. of Being* III. vi. 69 An attempt to generalize in a single description some of these basic cognitive happenings... These and other moments of highest happiness and fulfilment I shall call the peak-experiences. **1969** H. Geiger in Sutich & Vich *Readings Humanistic Psychol.* xvii. 307 If..there should come a kind of 'social' peak experience..a new rhythm of humane historical relationships could be established in the world. **1975** *Sat. Rev.* (U.S.) 22 Feb. 20/2, I underwent a religious-like peak experience in which the presence of divinity became almost palpable. **1976** N. Postman *Crazy Talk* 85 Bombing the Vietnamese back to the Stone Age was quite possibly a 'peak experience' for millions of Americans. **1914** Peak factor [see *crest *sb.*[1] 7 e]. **1963** Williams & Prigmore *Electr. Engin.* vii. 185 When deciding whether a particular voltage can be safely applied to an insulator, the r.m.s. value must be multiplied by the peak factor. **1970** *IEEE Trans. Information Theory* XVI. 86/1 The perceptual quality of synthetic speech signals depends to some extent on the 'peak-factor' (defined here as the difference between the maximum and minimum amplitudes of a signal divided by its root-mean-square value). *Ibid.*, FM signals have low peak factors. **1959** *B.S.I. News* Dec. 14 Recommendations regarding automatic gain control or peak limiting have also been excluded. **1960** E. Sivertsen *Cockney Phonol.* ii. 13 A simple peak consists of one of the six vowels. A complex peak consists of one of the six vowels as peak nucleus plus one of the peak satellites [h j w]. **1941** *B.B.C. Gloss. Broadcasting Terms* 23 *Peak programme meter*, instrument used (especially for the purpose of facilitating control) to measure the volume of programme peaks, averaged over a period of less than one-hundredth of a second. **1962** A. Nisbett *Technique Sound Studio* v. 94 There are several types of meter that can be used to line up equipment or check for overmodulation; but a 'peak programme meter' (PPM) seems to be the most satisfactory instrument. **1960** Peak satellite [see *peak nucleus* above]. **1960** *Wall St. Jrnl.* 5 Oct. 10/2 Pilot plants were being planned in this country a quarter century ago..looking for economic means of 'peak showing.' This is the practice in which standby sources of supply are used to meet demand at peak periods. **1973** *Times* 30 July 11/3 Both are designed to absorbe the stresses imposed on the gas supply system by a very cold day throughout the country in the depth of winter. The remedy is termed peak shaving. **1965** *Wireless World* July 329/1 A recording level indicator should essentially be a peak registering type because music has a large peak-to-mean ratio. **1962** Simpson & Richards *Physical Princ. Junction Transistors* ix. 219 The shift due to the rise in ambient temperature is thus relatively small and can be tolerated for peak-to-peak output-current swings of about 7 mA. **1967** *Electronics* 6 Mar. 80/2 (Advt.), Model 900 Nanovolt Galvanometer. Noise: Less than 2 nV or (2 pA) peak-to-peak for all source resistances. **1973** *Nature* 9 Nov. 72/2 During the eclipse itself the stability of the aircraft was excellent: pitch ≲0·1°, roll ≲0·5° (peak-to-peak values). **1974** Harvey & Bohlman *Stereo F.M. Radio Handbk.* ii. 21 The peak-to-peak amplitude of the second harmonic between the collector and the tap on the coil is limited to approximately twice the line supply voltage. **1957** *Physical Rev.* CV. 1416/2 Nuclear emulsion as a target was found to have a significantly weaker asymmetry (peak-to-valley ratio of 1·40 ± 0·07). **1924** *Jrnl. Sci. Instrum.* I. 281 A compact peak voltmeter, using a thermionic rectifier for measuring positive and negative peak voltages up to 600 volts, is described. **1967** *IEEE Trans. Electr. Insulation* II. 80/2 The peak voltmeter may find wide application in corona routine measurements.

7. Passing into *adj.* **a.** Characterized by or pertaining to a greatest value or largest number; *peak-listening, -viewing*, listening to the radio, or viewing of television, by the largest audience of the day; freq. *attrib.* (from a false analysis of phrases like *peak listening-period* as *peak-listening period*).

1903 *Electr. World & Engin.* 9 May 789/1 The direct-current ends of these rotary converters are often worked in multiple with an old generating station..during the peak-hours. **1924** *Westm. Gaz.* 8 Aug. 3/4 A drop of nearly £40,000,000 in pensions expenditure since the 'peak' year of 1920–21 is mentioned. **1937** *Archit. Rev.* LXXXII. (Suppl.) 1 Traffic congestion at the 'peak hours' is deplorable. **1946** *Vogue* June 2/2 The Sunday evening peak-listening series, 'The Challenge of our Time'. **1948** E. Waugh *Loved One* 57 It was as though..his speech came from some distant and august studio; everything he said

might have been for a peak-hour listening period. **1949** *Radio Times* 15 July 31/2 Leonard Hooper, a..dining-car attendant, tells you something about his work..especially during the holiday peak periods. **1960** M. O'Conor et al. *Children & Television Programmes* iii. 8 Pressures of different kinds and degrees exist to compel the television organizations to seek very large audiences for at least some of the programmes placed within the peak viewing period. **1962** L. Deighton *Ipcress File* xxv. 158 One long fluorescent day punctuated by interrogations like TV commercials in a peak hour play. **1966** *B.B.C. Handbk.* 14 A serious endeavour to improve the range of peak-time programmes. **1966** *Listener* 5 May 643/2 The peak age [for juvenile crime] is during the last year at school. **1969** G. Rees *St Michael* xv. 184 It was not until 1948 that the figures of turnover exceeded those of the previous peak year of 1941. **1974** *Times* 19 Dec. 2/7 The fare rises..will not arrest traffic growth. Peak services to holiday areas.. are being increased. **1976** *Jrnl. R. Soc. Arts* June 360/2 It would have been stupid and arrogant to think that 'Nobody will want all that news at the peak hour' or 'They ought to have half an hour's news, it is good for them.' **1977** *Herald* (Melbourne) 18 Jan. 1/1 A packed peak-hour express train tore down an overhead bridge.

b. Greatest; that is a maximum.

1903 [see *load *sb.* 3 f]. **1930** *Daily Express* 6 Sept. 10/1 Ordinary shares..reached a peak price of 26s. 10½d. during the 'boom'. **1946** *R.A.F. Jrnl.* May 180 At peak production, the Halifax group turned out one complete aircraft every working hour. **1949** R.-M. S. Heffner *Gen. Phonetics* II. v. 79 Two resonators with peak resonances below 1,200 cycles per second. **1950** *Engineering* 10 Feb. 168/2 Each cylinder of a multi-cylinder engine may be fitted with one of these gauges, and the peak pressures attained..are then read at a glance. **1958** *Times Rev. Industry* May 24/3 Important for the overall economy of gas supply are the processes used to produce for peak loads. **1959** *Ann. Reg. 1958* 431 Mr. John Davis..anticipated attendance at the cinemas would have dropped during 1958 to..just over half the peak audiences achieved in the years immediately after the war. **1973** S. Fisher *Female Orgasm* vii. 202 Ideation and fantasy do not..play a large or consistent role during the peak arousal phase.

peak, *v.*[2] Add: **1.** Also const. *up.* (Further examples.)

1929 R. Bridges *Testament of Beauty* i. 23 Untill the pyramid in geometrical enormity peak'd true. **1962** T. Masters *Surfing made Easy* 65 *Peak up*, when a swell begins to break. **1965** J. M. Kelly *Surf & Sea* iii. 39 This is where the wave peaks up and first starts breaking. **1968** W. Warwick *Surfriding in N.Z.* 10/3 If you find the wave you have caught is peaking up further along the beach from you, paddle towards the peak. **1976** *Woman's Day* (N.Y.) Nov. 100/2 Don't overpluck, overpencil or change the place where your brows peak.

b. *fig.* To reach the highest point; to attain maximum intensity, activity, etc.

1958 *Bird Migration* I. 1. 2 Common and Black-headed Gulls were usually present, the former peaking at 32 on 14th, the latter at 27 on 30th. **1961** T. H. White *Making of President 1960* xii. 299 There were now eighteen days left to the campaign, and Mr. Nixon was free to take the gloves off and 'peak' in his own manner. **1966** *Punch* 24 Aug. 283/1 Athletes are an awkward squad... Why does a young man fail to reach his potential on the day?.. His anxiety level is so high that he peaks too early. **1968** *Guardian* 21 Sept. 1/1 My campaign, according to the polls and surveys, has not peaked too soon. **1971** *Nature* 29 Jan. 304/2 Instead, the spectrum peaks around 4 keV and falls rapidly to both higher and lower photon energies. **1973** *Times Lit. Suppl.* 6 Apr. 366/5 That wild, speculative spirit peaked in 1929. **1974** *Sci. Amer.* Sept. 143/2 Only at relatively high occupational levels do the average earnings of men peak at the same time that the needs of their families are also peaking, that is, when the children are adolescent and of college age. **1975** *Listener* 8 May 615/1 Like so many of his ink, Man Ray peaked early, and turned dilettante. **1976** *Publishers Weekly* 26 Apr. 52/3 Growth, the very life-blood of corporate capitalism, has peaked in our time and now begun a decline towards what Jones calls 'a permanent recession'. **1977** *Horse & Hound* 25 Mar. 55/1 The eight [riders] named have been asked to programme their potential horses to 'peak' at this time but without too many competitions in advance.

c. To level *out* after reaching a peak.

1958 *Washington Post* 2 June A12/5 The Commissioner of the Bureau of Labor Statistics says the cost of living index is 'peaking out'. He follows this with the even more remarkable statement that the index may creep up further this summer after peaking out now. **1967** *Technology Week* 23 Jan. 55/3 When we learn just a trifle more about the hormonal control of brain development, these phenomena will peak out in human interest. **1971** *Daily Tel.* 2 Mar. 18/6 Since margins peaked out in the latter half of 1969 returns have not been so impressive and the rate of profit growth between the two halves has slackened from 7 p.c. to 5 p.c. **1977** *Time* 29 Aug. 46/3 General Motors shares peaked out at almost 114 in 1965 and are now down to around 65.

d. To have a peak experience (see *peak *sb.*[2] 6).

1970 J. Howard *Please Touch* 20 People who 'peak' can transcend the mundane and feel ecstatically fulfilled. **1972** *Village Voice* (N.Y.) 1 June 78/4 The hill with the tall fir cross, only 30 yards from where Michael and Ellen and I had peaked on the acid.

2. Add to first part of def.: to bring to a peak or maximum. Also const. *up.* (Examples.)

1957 *Practical Wireless* XXXIII. 718/2 When a station is found, the trimmers of range 5 are adjusted to peak it up. **1960** *Ibid.* XXXVI. 375/2 Trimmers can be peaked for minimum meter reading. **1961** T. H. White *Making of President 1960* xii. 296 He might move his campaign into its third, or final phase, 'peaking' it for impact on the week end before election. **1962** A. Nisbett *Technique Sound Studio* ix. 158 As the scene comes to a close the speech is faded down and the effects are lifted to swamp the line. Then after the effects have been peaked for a few seconds they too can be slowly faded out. *Ibid.* 263 *Peak up*, lift the volume either of an individual component of a mix, or of the entire programme.

peak, *v.*[3] **b.** (Earlier examples.)

1836 N. Isaacs *Trav. E. Afr.* II. 347 They immediately hauled down their sail, peaked their oars. **1849** J. F. Cooper *Sea Lions* I. xi. 136 The men now 'peaked' their oars, as it is termed; or they placed the handles in cleets made to receive them, leaving the blades elevated in the air, so as to be quite clear of the water. **1851** H. Melville *Moby Dick* II. vi. 42 The boat's five oars were seen simultaneously peaked.

peaked, *a.* **1.** (Further examples.) Cf. *peak *sb.*[2] 5 e.

1952 [see *kurtosis]. **1965** *Wireless World* July 329/1 The frequency of the peaked response is accordingly altered by switching each arm capacitance of the parallel-T network. **1968** A. J. Merrett *Executive Remuneration in U.K.* p. xiii, The indivisible nature of many executive roles..necessitates great personal involvement,..and the strain of an erratic and highly peaked work load.

3. *peaked-faced, -looking* adjs.

1891 'L. Malet' *Wages of Sin* I. I. iii. 56 It 'ud aggravate a saint, that it would, to hear you so taken up with a little peaked-faced bit of a maid. **1889** C. King *Queen of Bedlam* xiv. 188 Randall M'Lean, very white and 'peaked' looking, was sitting propped up in bed.

pea·kiness. [f. Peaky *a.*[1] + -ness.] Peaked or pointed character.

1921 L. B. Turner *Wireless Telegr.* iii. 22 The 'peakiness' of this curve measures the 'sharpness of tuning'. **1924** W. Deeping *Three Rooms* ii. 12 That slight peakiness about the chin, the ugly lines in the throat.

Peakland (pī·klænd). [See Peak *sb.*[1]] The Peak District of Derbyshire; = Peak *sb.*[1]

924, 1610 [see Peak *sb.*[1] 1]. **1891** J. Leyland *Peak of Derbyshire* i. 1 The 'Peak of Derbyshire' is a term which, to many, does not carry with it a very definite signification, for although most of the favourite resorts of tourists are known to lie within Peakland, few have inquired as to the boundaries of that district. **1909** *Westm. Gaz.* 5 Apr. 8/1 The death occurred..on Sunday night of..one of the best known figures among Peakland agriculturalists. *a* **1917** R. M. Gilchrist *Peakland Faggot* (1926) 97 The moon..foresaw a tragi-comedy in Peakland. **1926** E. Phillpotts in *Ibid.* p. viii, His [sc. Gilchrist's] incomparable pictures of Peakland were only won from long and self-denying service..in the courts and sanctities of the place and people. **1931** H. Walker (*title*) Peakland poems. **1974** *Country Life* 12 Dec. 1867/1 The strange disappearance of a Peakland river.

peaky, *a.*[1] Add: **3.** Special collocations, as **peaky blinder**, formerly, a hooligan active in the Birmingham area and distinguished by his hat, worn pulled into a peak over the eyes.

1896 *Birmingham Daily Argus* 17 Nov. 2/3 Is there.. any Volunteer officer who will come down and captain a company of budding 'peaky blinders'? **1898** *Daily News* 22 Oct. 2/5 The woman..saw the two 'peaky blinders' leaving on Thursday morning. **1901** *N. & Q.* Feb. 94/2 *Peakyblinder*.., the 'larrikins', 'rufflers', or 'hoodlums' of the Midlands are thus known from a custom they adopted of wearing the peak of their cap drawn down over their eyes when at their nefarious practices. **1971** B. Sleigh *Smell of Privet* x. 82 A mission in..Birmingham which took in some of the notorious Peaky Blinders.

peaky, peeky, *a.*[2] Add: (Earlier example.)

1821 M. Edgeworth *Let.* 23 Oct. (1971) 240 Poor young Worthington himself is rather peeky-weakee. He has a sore throat.

b. *Comb.*, as *peaky-faced* adj.

1906 *Westm. Gaz.* 12 May 11/2 A peaky-faced boy of about nine. **1910** *Chambers's Jrnl.* Jan. 53/1 He looked at the peaky-faced boy with the scared black eyes.

pealer[2] (pī·ləɹ). *U.S.* [var. *peeler[1] 3.] A person who displays exceptional aptitude or enthusiasm *for* an activity.

1834 S. Smith *Sel. Lett. J. Downing* 142 Pennsylvany chaps are real pealers for electing folks when they take hold. **1869** [see Staver *sb.*[1]].

peanut. For 'a native of the West Indies and West Africa' substitute 'native to South America'; = Ground-nut 2. Add earlier and later examples.

1807 *Salmagundi* 27 June 240 Young seniors go down to the flag-staff to buy peanuts, and beer. **1937** A. F. Hill *Econ. Bot.* xvi. 358 The peanut is a native of Brazil but was early carried to the Old World tropics by the Portuguese explorers. It was brought to Virginia from Africa by the slaves and is now one of the most important crops of the south. **1967** N. Freeling *Strike Out* 96 His wife was.. reading Proust and eating peanuts. **1968** J. W. Purseglove *Tropical Crops: Dicotyledons* I. 225 More than half the edible peanut stocks of the United States are used for peanut butter.

b. *peanut farmer* (so *-farming*), *oil, -seller, -shell, vendor*; *peanut-brained* adj.; **peanut boy** *U.S.*, a boy who sells peanuts and other wares; **peanut brittle**, a brittle toffee with roasted peanuts in it; **peanut butter**, a paste made with ground roasted peanuts; also *attrib.*; **peanut candy** *U.S.*, candy with roasted peanuts in it; **peanut gallery** *U.S. slang*, the top gallery in a theatre; **peanut-parcher** *U.S.*, = *peanut-roaster (b)*, also *attrib.*; **peanut politician** *U.S.*, one who deals in peanut politics; **peanut roaster** *U.S.*, (*a*) a machine in which peanuts are roasted; (*b*) *fig.* a piece of machinery that puffs or hisses; **peanut stand** *U.S.*, a booth, stall, or stand where peanuts, etc., are sold; **peanut valve** *Electronics*, a type of small thermionic valve (see quots.); also *ellipt.*

1857 *Porter's Spirit of Times* 5 Sept. 12/1 At length the mare reached the quarter pole, where a little pea-nut boy had stationed himself. **1873** 'Mark Twain' *Gilded Age* xxxvi. 333 In the cars,..the peanut-boy..always hands you out a book of murders if you are fond of theology. **1922** Joyce *Ulysses* 421 Come on, you doggone, bullnecked, beetlebrowed, hogjowled, peanutbrained, weaseleyed fourflushers, false alarms and excess baggage! **1903** *N.Y. Even. Post* 2 Oct. 7 To prescribe that all records [of great eating] henceforth shall be measured in peanut brittle. **1947** J. Bertram *Shadow of War* 336 The same jasmine tea, the same peanut brittle. **1965** Mrs. L. B. Johnson *White House Diary* 14 Feb. (1970) 244 Luci came bounding in this morning, to give her daddy a box of peanut brittle for his Valentine. **1976** *Monitor* (McAllen, Texas) 30 Sept. 8C/7 Crunchy peanut brittle, chock-full of tasty, nutritious peanuts, isn't difficult to make if you remember to spread the candy thinly on the cookie sheet. **1903** *Harper's Mag.* Oct. 981 Four sandwiches... Two of wholewheat bread with peanut butter. **1926–7** *Army & Navy Stores Catal.* 2/2 Peanut Butter jars—each 1/-. **1974** 'R. B. Dominic' *Epitaph for Lobbyist* xiii. 113 A carnival with peanut butter fudge made by the Soroptimists. **1977** *Time* 14 Mar. 42/2, I grew up on peanut butter sandwiches. **1856** Mrs. Stowe *Dred* I. iv. 51 Dancing, flirting, writing love-letters, and all other enormities down to eating pea-nut candy. **1901** B. Matthews *Notes on Speech-Making* 53 Some postprandial addresses..resemble the peanut candy where you cannot see the candy for the peanuts. **1976** *Time* 27 Dec. 23/3 He met Carter at a commission conference in 1973 and was one of the few who early took the peanut farmer's presidential aspirations seriously. **1977** *Time* 3 Jan. 13/2 As President Harry Truman was saved from haberdashing by failure, Jimmy Carter was saved from peanut farming by success. **1888** *Lippincott's Monthly Mag.* XLII. 734 Go to the lowest theatre in any of our large cities, or..mark what is called the 'Family Circle' by theatre proprietors and to the general world is more felicitously known as the 'Peanut Gallery'. **1945** *New Yorker* 5 May 15/1 We were sitting in the peanut gallery of the Opera House. **1975** *Audubon* Nov. 26/3, I can hear the laughter down in the pit and up in the peanut gallery. **1976** *National Observer* (U.S.) 21 Aug. 12/1 Sitting in the peanut gallery with her two young daughters, she admitted: 'I wouldn't miss it for the world. I grew up watching it.' **1882** H. H. Kane *Opium-Smoking* 35 A small glass lamp with a glass cover, perforated just above the flame, and in which sweet or peanut oil is burned. **1912** A. H. Lewis *Apaches N.Y.* 132 He trimmed the peanut-oil lamp. **1973** R. Thomas *If you can't be Good* (1974) ii. 13 The reports of mine that did surface were major scandals... The Peanut Oil King Affair comes to mind. **1929** W. Faulkner *Sartoris* (1932) III. 261 'Narcissa'll take you..in her car.'..'In that little peanut-parcher?' **1942** —— *Go down, Moses* 228 The diminutive locomotive and its shrill peanut-parcher whistle. **1954** —— *Fable* 393 The shrill peanut-parcher whistle which did not presage the lurch [of the train]. **1931** W. G. McAdoo *Crowded Yrs.* xii. 191 Any Democratic Cabinet, if not actually deficient mentally, consists of adolescents and small peanut politicians. **1977** *Lebende Sprachen* XXII. 10/2 A politician who is merely interested in small advantages is a *peanut politician*. **1902** *Sears, Roebuck Catal.* 589/3 The Boss Peanut and Coffee Roaster is the only successful roaster on the market. **1904** 'O. Henry' in *N.Y. World Mag.* 22 May 4/2 The whistle of a peanut-roaster puffed a hot scream into his ear. **1939** *These are our Lives* (Federal Writers' Project, U.S.) 283 [He] drew out a gallon at a time as needed for his peanut roaster. **1942** Berrey & Van den Bark *Amer. Thes. Slang* § 769/2 *Peanut roaster*, an intake manifold with a leak. **1960** Wentworth & Flexner *Dict. Amer. Slang* 379/2 *Peanut-roaster*,..a small locomotive..an old or ramshackle automobile. **1971** M. Tak *Truck Talk* 116 *Peanut roaster*, an intake manifold with a leak. **1971** 'D. Halliday' *Dolly & Doctor Bird* iii. 33 Peanut-sellers and newsvendors have free access to the front door. **1856** Mrs. Stowe *Dred* I. i. 9 'There isn't one of the train that I would give *that* for!', said she, flirting a shower of peanut-shells into the air. **1862** 'Mark Twain' *Let.* 8 Feb. (1917) I. iii. 67 He hasn't business talent enough to carry on a peanut stand. **1864** T. Pastor '*444' Combination Songster* 66 A blackguard by the name of McCarty..was book-keeper to a peanut-stand, And sold apples by the dozen. **1919** [see *frankfurter]. **1947** *Time* 27 Jan. 58/2 He was always dabbling shrewdly in dry cleaning stores and peanut stands. **1923** G. Parr *Princ. & Pract. Wireless Transmission* viii. 119 A valve which has proved very popular in America both on account of its size and low current consumption is that known as the Polar 'Pea-nut' valve. The bulb is tubular, its dimensions being 2 in. long by ½ in. diameter, and is fitted with a bayonet cap similar to the ordinary incandescent bulb. The valve requires only ·25 ampere at ·08 to 1 volt, and has a life double that of the ordinary tungsten valve. **1924** [see *detector 3]. **1930** *B.B.C. Year-bk. 1931* 448/2 *Peanut Valve*, a type of three-electrode receiving valve requiring low filament current and anode voltage. The dimensions of the valve are very small and it is therefore of use where space and small battery consumption are a consideration. **1910** G. B. McCutcheon *Rose in Ring* 56 The lowliest peanut-vender was laughing in his sleeve at the sleuth. **1978** S. Naipaul *North of South* II. v. 209 The peanut vendor was guilty of waging capitalist war against socialist society.

2. a. *pl.* Something small, trivial, or unimportant; *spec.* a small sum of money, esp. when regarded as inadequate payment. orig. *U.S. slang.*

1934 H. N. Rose *Thesaurus of Slang* iii. 35/2 *Small robbery*,..peanuts; ex: The job was peanuts. **1936** *Metronome* Feb. 21/4 *Peanuts*, any pay from a nickel a night and down. **1941** B. Schulberg *What makes Sammy Run?* viii. 176 They got you working for peanuts. **1946** J. B. Priestley *Bright Day* x. 285 'How was the poker game?' 'Peanuts. All I got was about twenty-five dollars and a headache.' **1959** J. Osborne *World of Paul Slickey* II. x. 87 There's a thousand pounds a week from record sales..it ain't peanuts. **1968** *Globe & Mail* (Toronto) 13 Jan. 29/2 All this is peanuts compared to the steady core of 3,600 ministers who make up the United Church clergy. **1973** *Scotsman* 13 Feb. 8/5 A salary of £3000 a year is peanuts for a man at the top of his profession. **1975** *New Yorker* 26 May 23 (Advt.), Being in a region may be your only chance to sample the local delicacies. And the cost? Peanuts. **1977** *Time Out* 17 June 15/1 The *New Review*'s share of the budget is a much criticised 10% of the total, yet it's peanuts.

b. A small or unimportant person. Also in more specialized contexts (see quots.).

1942 Berrey & Van den Bark *Amer. Thes. Slang* § 389/1 *Insignificant or petty person*,..palooka, peanut, person of straw, picayune, [etc.]. *Ibid.*, Peanuts, rubbish, small fry, small potatoes *or* punkins, trash, truck, *persons of no consequence*. *Ibid.* § 429/4 *Small person*,..peanut, peewee, picayune. **1945** Baker *Austral. Lang.* viii. 157 *Peanut*, a simple-minded soldier. **1963** *Australasian Post* 14 Mar. 51/1 'And what,' I asked cheerfully, 'was this peanut's particular whinge?' **1968** *Daily Mail* 16 Mar. 6/4 Mods are the traditional enemies of Rockers, but there were no Mods that night... 'They're scared of us... We call them peanuts.' **1969** *Daily Mirror* 3 Sept. 12/4 The youths were peanuts, or skin-heads. **1970** *Observer* (Colour Suppl.) 12 Apr. 46/3 Once me and my mates used to go around robbing peanuts—mod girls.

3. *attrib.* passing into *adj.* Trivial, worthless.

1836 W. Dunlap *Thirty Years Ago* II. iii. 25 They were your pea-nut fellows, I suppose. **1854** *Congress. Globe* 19 May 1230/3, I know them—a set of peanut agitators and Peter Funk philanthropists. **1892** *Congress. Rec.* 18 June 5394/2 This country is not a peanut institution; it is a great country. **1910** G. B. McCutcheon *Rose in Ring* 203, I suppose that peanut aristocrat friend of yours has told you it ain't swell or proper to wear tights.

pear, *sb.* Add: **5. pear-apple,** (*b*) the fruit of a prickly pear, a cactus belonging to the genus *Opuntia*; **pear-blight** (*a*), add: = *fire-blight* s.v. *fire *sb.* B. 5 (earlier and later examples); **pear-drop,** (*c*) used *attrib.* of parts of furniture, etc., shaped like pears; **pear midge**, a small gall midge, *Contarinia pyrivora*, whose larvae damage the fruit of pear trees; **pear-wood,** (*b*) the wood of one of several West African trees, esp. *Guarea cedrata*.

1898 H. S. Canfield *Maid of Frontier* 205 He knew.. which of the 'pear apples' were good to eat. **1854** E. Emmons *Agric. N.Y.* V. 165 Atmospheric Blight..proves itself to be independent of the cause that sometimes produces the pear blight. **1924** *Phytopathology* XIV. 478 (*title*) Experiments in the control of cankers of pear blight. **1961** A. Schoenfeld tr. *Stapp's Bacterial Plant Pathogens* II. 134 This disease, variously called 'fire blight', 'blossom blight', 'fruit blight', 'twig blight', 'apple blight', or 'pear blight', according to the place affected, is one of the most dangerous and dreaded tree diseases of North America. **1978** A. Huxley *Illustr. Hist. Gardening* v. 176 Pear blight was specifically described as 'a vegetable apoplexy'. **1914** Eberlein & McClure *Pract. Bk. Period Furnit.* i. 27 It is necessary for us to know whether a chest or cupboard ought to have knobs, pear drop or bail handles. **1925** Penderel-Brodhurst & Layton *Gloss. Eng. Furnit.* 121 *Pear-drop-handle*, a small pendent brass handle in pear-shape form, which came into use in England in the Restoration period. *Ibid.* 122 *Pear-drop ornament*, an ornament usually decorating the upper portion of a plain frieze,..consisting of a series of Gothic arches in relief with drops at the lower points suggesting capitals. **1960** H. Hayward *Antique Coll.* 213/1 *Pear-drop moulding*, a moulding, found below a plain cornice on late 18th cent. bookcases, carved in a repetitive design of inverted pear-shaped forms. **1884** E. A. Ormerod *Rep. Observations Injurious Insects 1883* 53 The Pear Midge is a very small two-winged gnat fly. **1920** P. J. Fryer *Insect Pests & Fungus Dis.* xvi. 234 The pear midge..is a destructive pest on fruit. **1956** Peairs & Davidson *Insect Pests* (ed. 5) xvii. 442 The pear midge is an introduced insect, present in the northeastern states for over 50 years. **1973** H. Martin *Sci. Princ. Crop Protection* (ed. 6) xv. 358 He [*sc.* F. V. Theobald] observed the extermination of the pear midge..by fowls penned under the attacked trees. **1915** J. H. Holland *Useful Plants Nigeria* III. 419 One of the finest timber trees in W. Africa [is *Mimusops Djave*], sold in Europe as 'African Pear Wood'. **1922** W. Schlich *Man. Forestry* (ed. 4) I. II. 320 The most prominent species [in Sierra Leone] are given as:—..red ironwood, very common. A species of Mimusops known as pearwood [etc.]. **1937** J. M. Dalziel *Useful Plants W. Trop. Afr.* 357 M[*anilkara*] *lacera*...African pearwood. *Ibid.*, M[*imusops*] *djave*... African pearwood. **1950** C. W. Bond *Colonial Timbers* 77 Guarea (Nigerian pearwood). **1961** F. R. Irvine *Woody Plants of Ghana* 519 *Guarea cedrata*... The names 'Nigerian Pearwood', 'Nigerian Cedar', and 'Cedar Mahogany', formerly given for this tree, are dropped as likely to cause confusion. **1971** F. H. Titmuss *Commerc. Timbers of World* (ed. 4) 134 Two species of *Guarea*..may be included in consignments of this timber, which has occasionally been sold as Nigerian Pearwood.

pearceite (pīə·ɹsəit). *Min.* [f. the name of Richard *Pearce* (1837–1927), British-born metallurgist and chemist + -ite[1].] A black, lustrous, brittle sulphide of silver and arsenic, $Ag_{16}As_2S_{11}$, that occurs as tabular crystals and usu. contains copper in place of some of the silver.

1896 S. L. Penfield in *Amer. Jrnl. Sci.* CLII. 18 The author proposes that hereafter the name polybasite shall be restricted to the antimony compound..and to make of the corresponding arsenic compound..a distinct species. For the arsenical mineral he takes pleasure in proposing the name *pearceite* as a compliment to his friend, Dr. Richard Pearce, of Denver. **1942** *Econ. Geol.* XXXVII. 491 Pearceite, recognizable only in polished section, is almost invariably associated with pyrargyrite. **1963** *Amer. Mineralogist* XLVIII. 567 The traditional polybasite and pearceite actually are members of two separate solid solution series in which Sb and As substitute mutually. **1968** I. Kostov *Mineral.* 173 The crystals of pearceite, polybasite, and xanthoconite are pseudohexagonal platy {001}.

pearl, *sb.*[1] Add: **I. 1. d.** (Earlier example of *Roman pearl.*) *essence of pearl*: now usu. called *pearl essence* (see sense *18); also prepared from the scales of other fish (as the herring).

1805 C. Wilmot *Let.* 26 Aug. in *Russ. Jrnls.* (1934) I. 170 A dress of white crape & roman pearls & white cameo ornaments.

II. 12. (Earlier example.)

1872 *Young Englishwoman* Oct. 543/1 *Ether pearls*, small round capsules about the size of a pea, are of marvellous efficacy in instantly calming attacks of Asthma.

16. b. Applied to an electric light bulb that is frosted on the inside so as to diffuse the light.

1930 *Engineering* 21 Mar. 393/1 The panels are illuminated from behind by a standard 40-watt 'pearl' lamp backed by aluminium reflectors. **1938** *Times* 3 Feb. 17/1 The conclusion reached is that lamps of the pearl type give a 7½ per cent better light than the opal type at present in use. **1972** 'R. Crawford' *Whip Hand* I. iii. 12 A single pearl bulb lit the store.

III. 17. a. *pearl necklace, -rope, string, stud.* **b.** *pearl-making, -sliding* adjs. **c.** *pearl-enamelled, -flushed, -hung, -paved* adjs. **d.** *pearl-tinted* adj. **e.** *pearl-bright, -grey* (later examples); also as sb., *-pale* adjs.

1914 G. Frankau *Tid'apa* (1915) v. 29 Pearl-bright under purple eyelids the unshed dew of a tear. **1943** D. Gascoyne *Poems 1937–42* 60 Like Some priceless pearl-enamelled toy. **1952** R. Campbell tr. *Baudelaire's Poems* 193 While Phoebe sheers Through pearl-flushed hours, To rain down tears In glittering showers. **1875** J. Blackwood *Let.* 30 Nov. in *Geo. Eliot Lett.* (1956) VI. 195, I rather lean to the pearl grey which he has told me is the name of the colour. **1931** V. Woolf *Waves* 110 Among the lustrous green, pink, pearl-grey women stand upright the bodies of men. **1969** in Halpert & Story *Christmas Mumming in Newfoundland* 182 Costumes used for disguise are listed:..fancy vests, beaver hats, pearl-gray spats. **1976** *Eastern Even. News* (Norwich) 9 Dec. 18/2 (Advt.), 1973 Capri 1600XL, pearl grey, 44,934, £1280. **1965** F. Sargeson *Mem. Peon* vi. 181 A short distance from my Leonora's pearl-hung ear. **1924** J. A. Thomson *Science Old & New* xx. 110 It seems highly probable that the walls of the pearl-making sac are in a state of inflammation. **1708** J. Lovett *Let.* 1 May in M. M. Verney *Verney Lett.* (1930) I. xii. 202 Tell Deare Bess her Pearle Necklas is come. **1819** M. Wilmot *Let.* 8 Dec. (1935) 33, I can wear my pearl necklace clasped with the amathyst behind. **1972** J. Wilson *Hide & Seek* ii. 35 He reached for her imitation pearl necklace. **1895** W. B. Yeats *Poems* 6 A pearl-pale, high-born lady. **1908** *Westm. Gaz.* 30 Dec. 2/3 There, in 'the Garden' roofed with glass, He flutes on the pearl-paved ridge of dawn. **1925** E. Sitwell *Troy Park* 75 And there the pearlropes fall like shawls. **1948** C. Day Lewis *Poems 1943–47* 15 What unseen clue Threads my pearl-sliding hours. **1939** L. M. Montgomery *Anne of Ingleside* xxi. 142 'Aren't those pearl strings pretty?'..'You'd almost think they were real.' **1927** E. Glyn '*It*' xix. 172 His shirt with its incomparable two pearl studs. **1975** S. Lauder *Killing Time on Corvo* i. 8 Her plain grey suit and hat, pearl studs, gloves. **1908** *Daily Chron.* 29 Sept. 7/5 He..entered the deserted garden where pearl-tinted spikes of iris perfumed the air.

18. pearl-berry, also *fig.*; **pearl essence,** **pearlessence,** essence of pearl (see Pearl *sb.*[1] 1 d in Dict. and Suppl.), or a synthetic imitation of this; **pearl-fish,** (*c*) = *fierasfer; **pearl-white** *a.*, add to def.: orig. used of *pearlware; also *ellipt.* as *sb.* = *pearlware (earlier and later examples).

1924 E. Sitwell *Sleeping Beauty* xvii. 67 And the pearl-berries of the snow upon dark bushes freeze. **1921** *Sci. Amer.* 12 Mar. 213/3 'Essence d'Orient' was easily manufactured, which was readily given the name of 'pearl essence' by the Bureau of Fisheries. **1946** Simonds & Bregman *Finishing Metal Products* (ed. 2) xxviii. 279 Pearl essence is used to substitute for metal powder finishes on everything from women's compacts and cosmetic containers to heavy industrial equipment and machinery. **1961** *Soap, Perfumery & Cosmetics* XXXIV. 60/3 By crushing or pulverising, natural and synthetic pearl essences lose part of their brilliance and become dull or greyish. **1972** P. G. I. Lauffer in Balsam & Sagarin *Cosmetics* (ed. 2) I. xii. 370 Natural pearlessence, consisting of a castor oil suspension of guanine crystals prepared from fish scales, produces a beautiful pearly luster when added to lipsticks. **1905** D. S. Jordan *Guide to Study of Fishes* II. xxix. 522 In the little group of pearl-fishes, called Fierasferidæ or Carapidæ, the body is eel-shaped with a rather large head. **1972** J. Binyon *Physiol. of Echinoderms* vi. 72 The famous pearl fish *Carapus* (= *Fierasfer*) *bermudiensis*..seems to live indefinitely in the cloaca and respiratory trees of this holothurian without ill-effect. **1779** J. Wedgwood *Let.* 19 June (1965) 236, I thank her majesty for the honor she has done to the *Pearl White*, and hope it will have due influence upon all her loyal subjects. **1866** E. Meteyard *Life J. Wedgwood* II. x. 482 An Italian order furnishes the accompanying tureen and plate in pearl-white ware. **1884** [see *creation 5c]. **1914** G. Frankau *Tid'apa* (1915) vi. 34 Pearl-white 'gainst the darkling lustre at the black-pearl plinth of the capes. **1937** V. Woolf *Years* 295 The road stretched pearl-white in front of them. **1956** G. Durrell *My Family* xvi. 218 The curve of pearl-white sand was backed by the great lily-covered dune behind.

pearl, *v.*[1] Add: **5. b.** To refine (potassium carbonate) in the preparation of pearl-ash.

1850 *Rep. Comm. Patents 1849* (U.S.) I. 176 The process of first roasting or heating the ashes..and then pearling in the pearling oven.

8. b. *Surfing.* (See quots.)

1962 [see *pearling *vbl. sb.* 4]. **1967** J. Severson *Great Surfing* Gloss., *Pearl* or *pearling*, while riding, the nose of the surfboard goes beneath the surface and continues downward, usually throwing the rider off (originally taken from pearl diving). **1970** *Studies in English* (Univ. Cape Town) I. 31 A milder form of wipe-out occurs when the surfboard *pearls*, in other words, the nose of the surfboard knifes under the water surface, usually throwing the surfer off. Variations on this expression include to *pearl-dive*, to *nose*, and to *plough*. **1971** *Ibid.* II. 27 If he is too far forward on the board the nose may dig in and the board *pearl* (shortened form of *pearl-dive*).

pearl-diver. Add: **b.** A person who washes crockery in a café or restaurant. So **pearl-diving** *vbl. sb.* (in this sense). *slang* (orig *U.S.*).

1913 E. A. Brown *Broke* iii. 29, I am in line for a pearl-diver's (dishwasher's) job tomorrow. **1930** J. Dos Passos *42nd Parallel* v. 402 He got a job pearldiving in a lunchroom. **1956** *Amer. Speech* XXXI. 151 *Pearl diver*, a dishwasher in a logging cookhouse. **1970** *Daily Progress* (Charlottesville, Va.) 15 Jan. 3B/5 Euphemism has upgraded other jobs. In parts of the country, dishwashers have been called utensil maintenance men, though back at the sink they still may be 'pearl divers'. **1970** C. Kersh *Aggravations M. Ashe* viii. 100 In the catering pearl diving is slang for dish washing.

pearler (pɔ̄·ɹləɹ), *sb.*[2] *N.Z.* and *Austral. slang.* [Var. Purler.] Something excellent or outstanding; a 'beauty'.

1941 Baker *N.Z. Slang* vi. 51 Expressions..in constant use by our youngsters..swinjer pearler stunner snorter. **1965** *Telegraph* (Brisbane) 5 July 8 *Ripsnorter* (or pearler, bosker, boshter). **1980** *Weekend Australian* 16 Aug. 13 Flo's 35-minute speech was a pearler.

pearlescent (pɔ̄ɹle·sĕnt), *a.* and *sb.* [f. Pearl *sb.*[1] + -escent.] **A.** *adj.* Having or producing an appearance of mother-of-pearl.

1949 *Industr. Finishing* (Indianapolis) Oct. 115 (Advt.), Chem-Scale pearlescent finishes are very effective for artificial pearls, costume jewelry,..etc. **1966** *Vogue* Dec. 84/3 New pearlescent lipsticks. **1972** A. Maruszewski in Balsam & Sagarin *Cosmetics* (ed. 2) I. xi. 357 The pearl material is not to be roller milled or colloid milled, since this tends to destroy the pearlescent property of the material.

B. *sb.* A pearlescent material or finish.

1960 *Times* 6 Jan. 15/4 The Carinex range includes basic polymers... A wide range of colours..include such special effects as pearlescents, tinsels and fluorescents. **1972** A. Maruszewski in Balsam & Sagarin *Cosmetics* (ed. 2) I. xi. 357 There is another pearlescent available under the tradename of Bilite. Chemically it is bismuth oxychloride deposited epitaxially on thin platelets of white mica. **1975** P. Browne *Bodywork Maintenance* ix. 102/1 Acrylics loaded with pearlescents also give an iridescent effect.

So **pearle·scence**, a pearlescent effect or material.

1953 *Organic Finishing* July 12/1 There is now more attention being directed to the production of 'Pearlescence' than to the duplication of the natural product known..as Pearl Essence. We may tentatively define Pearlescence as that action on light which results in a pearly lustrous appearance. **1969** *New Scientist* 2 Oct. 26/2 One of M. Pariat's most appealing products is known as 'pearlescence'. Essentially it consists of including crystals of polycarbonate of bismuth or lead during the final calendering stage of cardboard production. **1973** J. B. Wilkinson et al. *Harry's Cosmeticology* xvi. 205 Attempts to counteract this tendency..tend to decrease the pearlescence of the enamels.

Pearl Harbour. [The name (*Pearl Harbor*) of a U.S. naval base on Oahu, one of the Hawaiian Islands: tr. Hawaiian *Wai Momi*, lit. 'pearl waters'.] Used with direct allusion to the military attack by Japanese aircraft on Pearl Harbour on 7 December 1941, which, delivered without a declaration of war, severely damaged the surprised U.S. Pacific fleet and began the Pacific phase of the war of 1939–45. Also *transf.* and *fig.* Hence as *v. trans.*, to attack suddenly and effectively.

1942 *Progressive* 31 Jan. 272/2 Compare these recently abandoned myths in Britain with the pre-Pearl Harbor folklore about the Japanese which prevailed in the United States. **1942** *Capital* (Topeka, Kansas) 20 Mar. 15/3 Delay along this line is the delay that spells Pearl Harbor to the vital industrial nerve centers of our economy. **1945** Koestler *Twilight Bar* II. 44 Maybe they are doing a Pearl Harbour on us. **1955** *Times* 22 Aug. 7/2 This dangerous local situation could be the result of military aggression or of political subversion. In fact, the real danger is not now so much of an 'atomic Pearl Harbour' as of a new Sarajevo murder, building up into a major atomic war. **1959** *Economist* 10 Jan. 99/2 It will put a premium on

'Pearl Harbour' tactics to knock out opposing missiles before they leave the ground. **1963** *Guardian* 8 Jan. 8/4 No aggressor would dare to Pearl-Harbour any member nation of this club. **1974** *Ibid.* 25 Mar. 15/4 Hornby was working for the Japanese Ministry of Education when Pearl Harbour came. **1975** *Listener* 14 Aug. 211/1 In 1970, Aston Villa were relegated to the third division. Eric Woodward, the commercial manager, describes that as 'our Pearl Harbour'. **1978** *Times* 20 May 14/2 Mrs Thatcher was caught with the Sunday morning Pearl Harbour attack by Mr Peregrine Worsthorne..in last week's *Sunday Telegraph*.

pearling, *vbl. sb.* Add: **1. b.** *pearling lugger* (see quot. 1948).

1924 G. H. P. Muhlhauser *Cruise of Amaryllis* v. 233 A fair number of Australian 'black fellows' are to be found among the crews of the pearling 'luggers', as they are called, though really ketches. **1930** *Mariner's Mirror* XVI. 192 The pearling 'luggers' of the Torres Straits are usually ketch-rigged. **1948** R. de Kerchove *Internat. Maritime Dict.* 525/1 *Pearl lugger, pearling lugger*, a local name given in northwest Australia to small ketch-rigged boats employed in pearl fisheries. **1978** O. White *Silent Reach* vi. 71 It takes a different kind of knowhow to cut oyster rafts adrift or..sink pearling luggers..from the knowhow needed to dynamite windmills.

2. c. Decoration of furniture or architecture with pearl-shaped carving.

1899 R. Glazier *Man. Hist. Ornament* 43 Later Norman work is very rich,..the lozenge and the beading or pearling. **1914** Eberlein & McClure *Pract. Bk. Period Furnit.* 222 Besides these we find reeding, fluting, beading, pearling, spandrel fans, rosettes, and ribbons. **1925** Penderel-Brodhurst & Layton *Gloss. Eng. Furnit.* 122 *Pearling*, a series of rounded forms of the same size or graded, in more or less relief, used as a decoration on furniture. **1966** M. M. Pegler *Dict. Interior Design* (1967) 328 *Pearling*, a series of rounded forms of the same size, or graduated like a string of beads. The pearling was used as a furniture embellishment, either in straight lines, arced, or swagged.

4. *Surfing.* (See *pearl *v.*[1] 8 b.)

1962 T. Masters *Surfing made Easy* 65 *Pearling*, when the nose of the surfboard goes under the water during a ride. **1965** J. Pollard *Surfrider* ii. 18 The tendency of the nose of the board to dip into the water is 'pearling'. **1968** *Surfer Mag.* Jan. 56/1 The nose was curved to avoid pearling.

pearlite. Add: **2.** (Earlier and later examples.)

1888 H. M. Howe in *Engin. & Mining Jrnl.* 18 Aug. 132 Minerals which compose iron. Name suggested here... Pearlyte. A mixture of about ⅔ ferrite and ⅓ cementite. **1966** C. R. Tottle *Sci. Engin. Materials* x. 228 A two-phase alloy will tend to corrode more readily because an electrochemical cell will be set up between the different areas. Mild steel, containing ferrite (pure iron) and pearlite (a eutectoid, itself two-phase), shows this. **1977** *Sci. Amer.* Oct. 127/1 The presence of pearlite in an iron artifact is a clear indication that the artifact has been carburized.

Hence **pearli·tic** *a.*

1904 *Proc. Inst. Mech. Engin.* Nov. 1149 'Sorbitic' steels..are produced by a much more rapid cooling than is necessary to produce the 'pearlitic' steels. **1947** *Nature* 11 Jan. 50/1 Ordinary cast iron has a mixed ferritic-pearlitic matrix, and the engineering irons are usually fully pearlitic. **1962** *B.S.I. News* June 12/2 Draft proposals considered by the committee related to chemical composition, mechanical properties and methods of test for blackheart, whiteheart and pearlitic malleable iron.

pearlized (pɔ̄·ɹləizd), *ppl. a.* [f. Pearl *sb.*[1] + -ize + -ed[1].] Treated so as to resemble mother-of-pearl or to convey a suggestion of mother-of-pearl.

1955 *Britannica Bk. of Year* (U.S.) 681/2 Pearlized and luster leathers were extremely important in women's shoes and colours ranged all the way from pastels to black. **1955** *Wall St. Jrnl.* 21 Nov. 5/2 Retailers note that round, perforated pearlized beads are taxable, and oval ones aren't. **1957** *New Yorker* 1 June 70/1 For town as well as the seashore, there are hats the size of beach umbrellas—sailors of white pearlized straw with straight brims, and sailors of white starched straw curled up at the edges. **1958** *Observer* 17 Aug. 7/4 A coat of pearlised suede lined in cotton. **1969** *Guardian* 12 Aug. 7/2 Transparent make-up..available in three pearlised tan shades. **1972** 'J. Melville' *Ironwood* vi. 89 The names slid easily between her pearlised lips. **1976** *Morecambe Guardian* 7 Dec. 5/1 For a children's party, she made a table arrangement using a soft toy toad, on an oasis shaped like a toadstool, decorated with apples, seed pods and pearlised foliage.

pea·rlware. Also **pearl ware.** White-coloured pottery ware, orig. manufactured by Josiah Wedgwood.

1922 W. Burton *J. Wedgwood & his Pottery* iv. 37 This 'pearl' ware, as Wedgwood made it, differed somewhat in composition from his cream ware, for it contained a larger proportion of ground flint and china-stone. **1948** W. B. Honey *Wedgwood Ware* ii. 10 The 'pearl-ware' was chiefly used for tea-services made in rivalry with porcelain. **1953** W. Mankowitz *Wedgwood* ii. 44 The peak production period for pearlware is the first quarter of the nineteenth century. **1961** *Times* 8 Apr. 11/6 Cream-coloured earthenware and the harder, whiter pearlware. **1969** *Canad. Antiques Collector* Feb. 14/2 Round about the year 1800 this model must have been turned out by the dozen, sometimes, as in this instance, in pearlware, [etc.]. **1975** *Daily Tel.* 5 Feb. 12/5 A Leeds pearlware figure of a horse, 17 ins high, was bought anonymously for £1,350 at Sotheby's yesterday.

pearly, *a.* (*adv. sb.*) Add: **3. b.** *pearly gates*: the gates of heaven as described in Rev. xxi. 21, used allusively.

1853 [in Dict.]. **1927** H. Crane *Let.* 29 May (1965) 300 If I can avoid the pearly gates long enough I may do better. **1953** [see *dodger *a.*]. **1969** J. Wainwright *Big Tickle* 72 Dago said: 'A shiv kick—and pearly gates.' **1973** J. Porter *It's Murder with Dover* xvi. 162, I heard somebody'd pushed that Marsh cat through dem pearly gates. **1977** *Gay News* 24 Mar. 21/4 Perfection's death to me, the pearly gates, and I do think pearl's awfully vulgar for gates.

c. *Pearly King* (or *Queen*): a leading London costermonger, dressed in festive costume covered with pearl-buttons.

1933 *Times* 26 Aug. 9/5, I wrote..to 'Snowy Tabram, Pearly King, Islington,' asking him when and where the annual meeting of the Pearlies would take place. **1934** *Times* 1 June 13/3 The Rev. A. D. Belden..unveiled yesterday..a statue of Henry Croft, the 'pearly king', who died four years ago. The statue represents Mr. Croft in his 'pearly king' clothes with top hat. **1935** F. W. Tickner *London through Ages* xiv. 286 (*caption*) 'Pearly King and Queen' of the Costermongers. **1942** Wyndham Lewis *Let.* 27 Jan. (1963) 315 They go about talking to themselves—in the purest idiom of the Pearly King. **1963** *Times* 16 May 15/5 Reminding me of one of the 'Pearly Kings' or button-covered costermongers that I had seen in London when I was a boy. **1967** E. Short *Embroidery & Fabric Collage* ii. 41 Most pearly kings and queens design and make their own costumes, the decoration being made up entirely of pearl buttons sewn on the fabric of the suit or dress. **1975** *Evening News* 26 Apr. 4/1 Pearly King Bill Davison raised thousands of pounds for charity. **1977** *Times* 28 Jan. 16/6 The silver-painted London double decker, decorated with Cockney slogans, pictures of buskers and an advertisement for an insurance company, is the contribution of the Pearly kings and queens of London to the Queen's silver jubilee, and will be used to raise money for a recently-formed Pearlies' charity appeal.

4. (Earlier and later examples of *pearly grey.*)

1845 *Punch* VIII. 247 The following terms..may be used pretty much at random: 'Chiaroscuro', 'texture', 'pearly greys', 'foxy browns'. **1978** I. Murdoch *Sea* 437 A thick clammy pearly-grey mist surrounded the house.

B. as *adv.* (Further examples.)

1891 Kipling *Light that Failed* xiii. 249 Maisie lifted up her face, and it was pearly white. 'No! No! Not blind! I won't have him blind!' **1952** A. G. L. Hellyer *Sanders' Encycl. Gardening* (ed. 22) 241 H[*ymenanthera*] *crassifolia*, yellow, pansy-like flowers, succeeded by pearly-white berries in autumn.

C. *sb.* Add to def.: Also, the pearl-buttons themselves. (Further examples.)

1898 J. D. Brayshaw *Slum Silhouettes* 142 He'd..a blue coat and weskit with the artfullest little pearlies you ever see. **1914** 'Bartimeus' *Naval Occasions* xviii. 156 What time the citizen ashore donned 'pearlies' or broadcloth and shut up shop. **1935** P. Cohen-Portheim *Spirit of London* viii. 99 There are still costermongers in their 'pearlies' (one of the prettiest dresses to be found anywhere). **1949** R. Graves *Seven Days in New Crete* 123 'But you should see my gala suit! It is in rain-grey linen, covered with little pearl-buttons.'.. He went off to buy his pearlies. **1957** *Encycl. Brit.* VI. 511/1 (*caption*) London costermonger family of the Victorian age. These costumes, with their rich embroidery of pearl buttons or 'pearlies', are now used rarely, and only on festive occasions.

b. A costermonger or a pearly king (or queen).

1928 *Daily Express* 27 June 13/3 It is given to few men to be popular alike with princes and 'pearlies'. **1959** *Manch. Guardian* 4 Aug. 4/7 Hampstead Bank Holidays.. felt the loss of..the beery men, the pearlies, and the 'Stout Parties'. **1974** *Times* 20 Aug. 12/2 There will be 'three Pearlies including the Pearly King of London'. **1977** [see sense 3 c above].

Pearson (pīə·ɹsən). *Statistics.* The name of Karl *Pearson* (1857–1936), English mathematician, used *attrib.* and in the possessive to designate: **a.** The members of a family of curves described by him in 1895, which include many probability distribution functions.

1908 *Biometrika* VI. 4 Consequently a curve of Prof. Pearson's Type III may be expected to fit the distribution of s^2. **1927** H. L. Rietz *Math. Statistics* iii. 58 The method of moments plays an essential rôle in the Pearson system of frequency curves. **1936** *Statistical Res. Mem.* I. 41 A better approximation to $p(L_1')$ might be obtained if a Pearson Type I curve were fitted with the correct first four moment coefficients. **1974** P. Lumb in I. K. Lee *Soil Mech.* iii. 51 Some of the most useful standard forms are members of the Pearson family of distributions defined by

$$\frac{d}{dx}\{\log g(x)\} = \frac{x - c_0}{c_1 + c_2x + c_3x^2}.$$

b. A measure of the skewness of statistical distributions, proposed by him in 1895.

1911 G. U. Yule *Introd. Theory Statistics* viii. 150 There is, however, only one generally recognised measure of skewness, and that is Pearson's measure..—

$$\text{skewness} = \frac{\text{mean} - \text{mode}}{\text{standard deviation}} \ldots$$

The numerator of the above fraction may..be replaced approximately by 3 (mean – median). **1925** F. C. Mills *Statistical Methods* v. 168 Pearson's formula for measuring skewness. **1962** *Lebende Sprachen* VII. 114/1 Pearson measure of skewness.

c. The product–moment correlation coefficient (see *product *sb.*[1] 6).

1912 *Jrnl. R. Statistical Soc.* LXXV. 609 Professor Pearson's coefficient..was described and its use illustrated in two memoirs published in 1900. **1957** Kendall & Buckland *Dict. Statistical Terms* 215 The product-moment coefficient of correlation is sometimes referred to as the Pearson coefficient of correlation because of K. Pearson's part in introducing it into general use. **1973** L. D. Phillips *Bayesian Statistics* x. 207 The correlation coefficient (or Pearson product-moment correlation coefficient, as it is sometimes called) is usually designated by r and is defined as $r = \frac{\Sigma Z_x Z_y}{N - 1}$. *Ibid.* xii. 294 Transform the ranks into normal scores, compute the Pearson-r between the pairs of scores and use the inference method just discussed.

d. The chi-square test.

1912 *Jrnl. Exper. Zool.* XIII. 203 Pearson's test depends upon a variable $\chi_2[sic] = S\{(m_r - m_r')^2/m_r\}$ where m_r is the theoretical frequency and m_r' the observed. **1969** H. O. Lancaster *Chi-Squared Distribution* ix. 175 With several degrees of freedom, for class frequencies of 5 or more, the distributions of the Pearson χ^2 approximate satisfactorily to the asymptotic or theoretical χ^2 distributions.

e. A set of formulæ described by him in 1899, used for estimating human stature from the length of limb bones.

1925 S. Smith *Forensic Med.* v. 53 (*heading*) Pearson's formulæ. **1947** *Sci. News* IV. 40 Pearson..compiled tables relating the length of certain of the arm and leg bones to the total height of the body. These Pearson formulas are quite remarkable in their accuracy. **1966** R. Jackson *Crime Doctors* 25 Simpson measured the only intact bone he had—the upper part of the left arm—and, using Pearson's formula, concluded that the bones were those of a woman who had been slightly over five feet tall.

Pearsonian (pīəɹsōu·niăn), *a. Statistics.* [f. prec. + -ian.] Of or originated by Karl Pearson (see prec.); used esp. with ref. to senses a and b s.v. *Pearson.

1907 F. B. Wyatt in W. P. Elderton *Frequency-Curves* p. v, In January, 1903, Mr. W. Palin Elderton read before the Institute of Actuaries an interesting paper dealing with the application of the Pearsonian frequency-curves to the graduation of a mortality experience. **1927** H. L. Rietz *Math. Statistics* iv. 82 The degree of correlation is often measured by the Pearsonian coefficient of correlation represented by the letter r. **1969** *Computers & Humanities* III. 145 The correlation coefficient used here is the Pearsonian product-moment correlation. **1972** R. B. Cain *Elem. Statistical Concepts* xix. 163 The Pearsonian coefficient of skewness brings out the relation between the mean and the median in a skewed distribution which has exactly one mode.

pearten (pīə·ɹtən), *v. dial.* [f. Peart *a.* + -en[5].] *trans.* and *intr.* To cheer up; to become more lively or sprightly; to hasten. Freq. const. *up.*

1879–81 G. F. Jackson *Shropshire Word-bk.* 317 *Pearten, peartle*,..to revive; to enliven; to cheer. (1) 'Oh! yo'n soon *pearten* up, yo' beginnen to look better a'ready'. (2) ''Er quoite *pyurtled* 'im ŏop w'en 'er come wŏam'. **1895** [see *druther]. **1896** G. F. Northall *Warwickshire Word-bk.* 169. **1909** *Dialect Notes* III. 356 *Pearten*, *v.i.* and *tr.* To hasten, go faster: often with *up*. 'We will have to *pearten up* if we expect to get there on time.' **1949** H. Hornsby *Lonesome Valley* v. 70 Crit seemed to pearten up at that. **1951** H. Giles *Harbin's Ridge* xvi. 139 But when he peartened up, we all took fresh heart.

peasant, *sb.* Add: **1.** Substitute for def.: One who lives in the country and works on land which he has the right to use, relying for his subsistence mainly on the produce of his own labour and that of his household, and forming part of a larger culture and society in which he is subject to the political control of outside groups; also, loosely, a rural labourer. (Further examples.)

Circumstances have changed considerably since the publication of the Dictionary and although modern sociologists agree that a 'peasant' works the land, the more wealthy peasants may also be land-owners, rentiers, hirers of labour, etc., and in these capacities share interests with completely different social groups. Hence in the analysis of many rural societies divisions within the class frequently have to be made.

1850 H. Macfarlane tr. *Marx & Engels's Manifesto of German Communist Party* in *Red Republican* 16 Nov. 171/2 The small manufacturers, shopkeepers, proprietors, peasants, &c., all fight against the Bourgeoisie, in order to defend their position as small Capitalists. They are, therefore, not revolutionary but conservative. **1926** E. & C. Paul tr. *Marx's Eighteenth Brumaire of L. Bonaparte* vii. 137 The interests of the peasants no longer coincide, as during the reign of the first Napoleon, with the interests of the bourgeoisie, with the interests of capital. There is now a conflict of interests. The peasants, therefore, find their natural allies and leaders in the urban proletariat, whose mission it is to subvert the bourgeois order of society. **1927** M. J. Olgin tr. *Engels's Peasant War in Germany* 18 The small peasants (bigger peasants belong to the bourgeoisie) are not homogeneous. They are either in serfdom bound to their lords and masters,..or they are tenants, whose situation is almost equal to that of the Irish. **1933** E. & C. Paul tr. *Stalin's Leninism* II. 205 In the conditions prevailing in our country, the peasantry consists of various social groups,..the poor peasants, the middle peasants, and the kulaks. It is obvious that our attitude to these various groups cannot be identical. The poor peasant is the *support* of the working class, the middle peasant is the *ally*, the kulak is the *class enemy*—such is our attitude to these respective social groups. **1934** *Encycl. Social Sci.* XII. 48/2 The term peasantry has undergone many changes in meaning in the past and is still subject to various interpretations. Common to all the shifting meanings, however, is a view of the peasant as a tiller of the soil to whom the land which he and his family work offers both a home and a living. **1938** tr. *Lenin's Sel. Wks.* X. 223 The big peasants (*Grossbauern*) are the capitalist *entrepreneurs* in agriculture who..employ several wage workers and are connected with the 'peasantry' only by their low cultural level, habits of life and the manual labour they themselves perform on their farms. These constitute the largest of the bourgeois strata, and they are the..enemies of the revolutionary proletariat. **1948** A. L. Kroeber *Anthropol.* (rev. ed.) vii. 284 Peasants are definitely rural—yet live in relation to market towns; they form a class segment of a larger population which usually contains also urban centers, sometimes metropolitan capitals. **1951** Bonner & Burns tr. *Marx's Theories of Surplus Value* B. v. 192 In the capitalist mode of production the independent peasant or handicraftsman is sundered into two persons. As owner of the means of production he is capitalist, as worker he is his own wage worker. **1954** tr. *Mao Tse-tung's Sel. Wks.* I. 139 The rich peasant as a rule possesses land... The exploitation the rich peasant practises is chiefly that of hired labour. *Ibid.*, In many cases the middle peasant possesses land. The middle peasant relies wholly or mainly on his own labour as the source of his income. *Ibid.* 140 As a rule the poor peasant has to rent land for cultivation and, exploited by others, has to pay land rent and interest on loans and hire out a small part of his labour. **1956** R. Redfield *Peasant Society & Culture* i. 27 Those peoples are to be included in the cluster I shall call peasants who have..this in common: their agriculture is a livelihood and a way of life, not a business for profit. **1966** E. Wolf *Peasants* 3 Peasants..are rural cultivators whose surpluses are transferred to a dominant group of rulers that uses the surpluses both to underwrite its own standard of living and to distribute the remainder to groups in society that do not farm but must be fed for their specific goods and services in turn. **1969** *Internat. Social Sci. Jrnl.* XXI. 286 A peasant may be at one and the same time owner, renter, share-cropper, labourer for his neighbours and seasonal hand on a near-by plantation. Each different involvement aligns him differently with his fellows and with the outside world. **1971** tr. E. Feder in T. Shanin *Peasants & Peasant Societies* vii. 90 But while the landed elite [in Latin America] has no interest in the peasants' aspirations and keeps aloof from their world, it is still keenly aware of its obligations to keep the peasants in check and subservient. **1975** J. A. Hellman tr. *Stavenhagen's Social Classes in Agrarian Societies* v. 65 A useful distinction is sometimes made between tribal peoples, peasants, and modern farmers... In contrast to tribal or primitive peoples, peasant societies do form part of wider economic, social, and political units.

β. Later examples of *paysan* (now a gallicism). Also *attrib.* or as *adj.*

1801 C. Wilmot *Let.* 5 Dec. in *Irish Peer* (1920) 9 We met a young Paysan Savant. **1872** *Young Englishwoman* Oct. 542/1 The *paysan* blouse, so much in fashion now, has brought into favour the plain broad stitched hem. **1891** E. Dowson *Let.* 1 July (1967) 206 This is the queerest little auberge imaginable, quite *paysan*, but full of excellent folk. **1949** W. Stevens *Let.* 13 Oct. (1967) 650, I shall feel sorry about paysans and tepid by the time this reaches you.

c. Delete †*Obs.* and add later examples. Now *slang*, usu. implying ignorance, stupidity, or boorishness. Cf. *farmer[2] 7 c.

1943 Baker *Dict. Austral. Slang* (ed. 3) 58 *Peasant*, an ordinary rank. (R.A.A.F. slang.) **1947** S. Bellow *Victim* i. 10 She showed such a dread of hospitals that at last he exclaimed, 'Don't be such a peasant, Elena!' **1957** *Sunday Mail* (Glasgow) 10 Feb. 11 *Peasant*—an older person who does not..understand the goings-on of teenagers. **1958** *Listener* 14 Aug. 247/3 People who dress conventionally are called [by teddy-boys] 'peasants'. **1961** G. Smith *Business of Loving* v. 161 Laura took me out riding... I'm a complete peasant in this, but she's an expert. **1964** G. Lyall *Most Dangerous Game* xix. 146 Alone? Of course I'm not alone, you—you peasant. D' you think I drive myself? **1966** H. Kemelman *Saturday the Rabbi went Hungry* (1967) iii. 27 Well, let the big boys worry, I'm just one of the peasants.

2. a. *peasant-boy* (earlier example), *-farmer* (earlier and later examples), *girl*, *-novelist*, *owner*, *-poet*, *-proprietor* (examples), *-soldier*, *tenant*, *woman*, *-worker*; *peasant farming*; *peasant proprietorship* (further examples).

1852 Dickens *Bleak Ho.* xxxi. 307 Mr. Skimpole..sang one [*sc.* a ballad] about a Peasant boy. **1848** Mill *Pol. Econ.* I. I. iv. 72 When a peasant farmer or proprietor lives on the produce of his land. **1906** *Daily Chron.* 21 Mar. 6/6 Of peasant-farmer stock, the elder Bunsen..became tutor in an English family. **1978** R. Mitchison *Life in Scotland* iii. 47 If a man had failed as a peasant farmer there might well be no work for him as a farm labourer. **1936** tr. *Lenin's Sel. Wks.* III. 183 Landlord farming evolves in a capitalist way... Peasant farming also evolves in a capitalist way and gives rise to a rural bourgeoisie and a rural proletariat. **1952** *Oxf. Jun. Encycl.* VI. 135/2 Peasant farming is the general rule in France, Germany, the countries of northern Europe, and all through central Europe, as well as in the greater part of Asia. **1883** C. M. Yonge *Stray Pearls* I. ix. 99 She kept the cows and knitted like a peasant girl. **1915** D. H. Lawrence *Rainbow* x. 250 Peasant-girls with wreaths of blue flowers in their hair. **1976** Scott & Koski *Walk-In* (1977) xix. 124 The second [woman] giggled like some empty-headed peasant girl. **1907** *Westm. Gaz.* 21 Nov. 12/2 A new book by Peter Rosegger, the Styrian peasant-novelist,..will be published on November 25. **1951** D. Mitrany *Marx against Peasant* xii. 201 In both countries [*sc.* France and Italy] the Communists have made great efforts to win influence in the countryside, especially among landless labourers and peasant tenants, who themselves hope to become peasant owners. **1974** J. White tr. *Poulantzas's Fascism & Dictatorship* vi. i. 275 The small peasant owners are the 'rural petty bourgeoisie' *par excellence*. **1857** Bagehot *Coll. Wks.* (1965) II. 24 The eager peasant-poet was at fault in the..refinements of the..drawing-room. **1794** A. Young *Trav. France* (ed. 2) I. xix. 542 *Caussade*.—This country is full of peasant proprietors of land. **1899** G. B. Shaw *Let.* c 23–24 Dec. (1972) II. 121 You have to deal with a war [*sc.* the Boer war] declared by a peasant-proprietor State. **1974** *Encycl. Brit. Macropædia* VI. 1064/2 Indispensable though his [*sc.* the local craftsman's] services may be, they do not give him equality with his peasant-proprietor neighbours. **1960** R. K. Webb *Harriet Martineau* xi. 338 The abuses of peasant proprietorship in France. **1974** *Encycl. Brit. Macropædia* VI. 256/1 In agriculture, peasant proprietorship or large private estates—particularly for export products—remained the general rule [in developing non-Communist countries]. **1966** *Sociol. Rev.* XIV. 21 By cultural intercourse and intermixture, if not by indoctrination, the peasant-soldier is taught to think in wide national, and not village-limited terms. **1978** D. Bloodworth *Crosstalk* xvi. 130 Wang, a peasant soldier..cunning in detail..but devoid of broad perception. **1951** Peasant tenant [see *peasant owner* above]. **1856** Dickens *Dorrit* (1857) II. i. 323 The child carried in a sling by the laden peasant woman..was quieted with picked-up grapes. **1891** Hardy *Tess* II. v. xxv. 209 You are an unapprehending peasant woman. **1977** G. Butler *Brides of Friedberg* vi. 155 In the woods..a group of peasant women were gathering wood. **1962** E. Snow *Other Side of River: Red China Today* (1963) xviii. 134 Another peasant-worker dictatorship was proclaimed as a local soviet on November 18 in an area far removed from Tsalin. **1972** H. C. Stevens tr. *Galeski's Basic Concepts Rural Sociol.* vii. 174 The most numerous social categories in the Polish village today are (1) peasant owners of family farms, and (2) the highly diversified category of 'peasant-workers', i.e. families living in the village and..combining work on a farm with regular employment outside the village.

b. *peasant art, class, community, family, group, league, mind, revolution, society, style*; **peasant economy**, an economy in which the family is the basic unit of production.

1934 A. Huxley *Beyond Mexique Bay* 197 It is a typical semi-sophisticated peasant art. **1961** L. G. G. Ramsey *Connoisseur New Guide Antique Eng. Pott., Porc. & Glass* 72 The simple vigour and ingenuousness of a 'peasant' art. **1866** *Chambers's Encycl.* VIII. 379/1 Communal government is the fundamental principle of all the rights of the peasant class. **1954** B. & R. North tr. *Duverger's Pol. Parties* II. i. 265 It [*sc.* the Communist party in the Soviet Union] liquidated the 'Kulaks' and the land-owning middle class, and for a long time it gave the working class of the towns preponderance over a peasant class that was actually more numerous. **1974** *Encycl. Brit. Macropædia* VIII. 1164/2 The [European feudal] aristocrats considered both their own and the peasant class to be permanent, God-given arrangements of hereditary status. **1951** R. Firth *Elem. Social Organiz.* v. 166 In various peasant communities in parts of Africa..new art sanctions have been provided in modern workshops. **1974** *Encycl. Brit. Macropædia* VIII. 1164/1 In Africa, scattered peasant communities occur in the upland areas of the Mediterranean shore. **1951** R. Firth *Elem. Social Organiz.* iii. 87 The term peasant has primarily an economic referent. By a peasant economy one means a system of small-scale producers, with a simple technology and equipment, often relying primarily for their subsistence on what they themselves produce. **1966** D. Thorner et al. *A. V. Chayanov on Theory of Peasant Econ.* p. v, The most sophisticated and best documented studies of the theory and problems of peasant economy in the half-century from 1880 to 1930 were written by Russians. **1975** J. A. Hellman tr. *Stavenhagen's Social Classes in Agrarian Societies* v. 65 Peasant economy tends towards self-sufficiency and the household is the main unit for production and consumption, based on the intensive use of family labor. **1926** E. & C. Paul tr. *Marx's Eighteenth Brumaire of L. Bonaparte* vii. 133 In so far as millions of families live in economic circumstances which distinguish their mode of life, their interests, and their culture, from those of other classes, and make them hostile to other classes, these peasant families form a class. But in so far as the tie between the peasants is merely one of propinquity, and..the identity of their interests has failed to find expression in a community..or in a political organization, these peasant families do not form a class. **1974** tr. *Sniečkus's Soviet Lithuania* 75 Inefficient, semi-natural farms, unbearable tax burdens, and constant anxiety over the future—such was the lot of many, many peasant families. **1934** *Encycl. Social Sci.* XII. 52/2 Even within a single country such as Germany local differences—physiographic, economic, historical—make for a very considerable variation among peasant groups. **1974** *Encycl. Brit. Macropædia* III. 151/1 Through illegal peasant leagues, founded in the late 1950s, and legitimate rural unions, which were authorized in 1962, many [Brazilian] peasants were able for the first time to make their needs known to the political leaders. **1975** *New Left Rev.* Nov.–Dec. 65 Their peasant leagues performed well in the clerical dominated countryside. **1911** J. London *Let.* 8 Jan. (1966) 330 Charmian has no peasant-mind. **1941** Wyndham Lewis *Let.* 22 Nov. (1963) 310 The stuffy conservatism of the land-locked peasant-mind. **1974** M. B. Brown *Econ. of Imperialism* xiii. 327 We could still see bourgeois and peasant revolutions which fall far short of socialism. **1949** E. Coxhead *Wind in West* vii. 176 Hermia..would frequently deplore the Fascist trend latent in peasant societies. **1964** Gould & Kolb *Dict. Social Sci.* 490/2 The term *peasant society* has long referred to Europe, changing in meaning as the industrial revolution has brought about changes in European life. Usage of the term..has recently been extended to native populations of much of the world, especially as former primitive societies have come to resemble the old European peasantry. **1974** *Encycl. Brit. Micropædia* VII. 823/3 In peasant society ultimate control of the means of production is usually not in the hands of the primary producers. **1952** G. Bemrose *19th Cent. Eng. Pott. & Porc.* vi. 31 A good deal of our unlettered art, especially the so-called peasant style, appears to lie dormant in our racial consciousness. **1973** *Times* 24 Aug. 2/2 She was wearing a cream-coloured peasant-style blouse with blue smocking at the neck and cuffs.

c. In crafts, fashion, etc.: in the style of articles produced by peasants or of clothes worn by them, as *peasant blouse, dress, skirt, sleeve, tapestry, weave.*

1953 'T. Sturgeon' *More than Human* III. 167 Janie in a peasant blouse, with a straight spear of morning sunlight bent and moulded to her bare shoulder. **1963** 'E. McBain' *Ten plus One* (1964) xii. 137 She wore one of these very low-cut peasant blouses. **1970** *New Yorker* 15 Sept. 1 (Advt.), The cultivated peasant dress is news. **1960** C. W. Cunnington et al. *Dict. Eng. Costume* 158/1 *Peasant skirt.* 1885. A full round tennis skirt made with 2 or 3 wide tucks and a fall of lace. **1965** 'M. Neville' *Ladies in Dark* viii. 74 She was wearing a peasant skirt and blouse. **1911** *Daily Colonist* (Victoria, B.C.) 8 Apr. 20/2 (Advt.), The new peasant sleeves are featured in the waist part of these garments. **1900** *Archit. Rev.* June p. xxii/2 A twin bedstead..is covered with peasant tapestry designed by Mr. Godfrey Blunt. **1962** L. Deighton *Ipcress File* vii. 45 The rugs..of simple dark-toned peasant weaves. **1975** I. S. Black *Man on Bridge* v. 66 A peasant-weave curtain covered the window.

3. *peasant-minded* adj.

1961 J. Barlow *Term of Trial* I. iii. 53 The louts are essentially peasant-minded about sex.

peasantism (pe·zăntiz'm). [f. Peasant *sb.* + -ism.] **a.** The doctrine that political power should be in the hands of the peasants; also = *Narodnikism. Hence **pea·santist** *sb.* and *a.*

1894 P. Milyoukov in *Athenæum* 7 July 23/3 'Peasantism' [in Russia] puts its faith exclusively in the character and 'spirit' of the people. *Ibid.*, The programme of the 'peasantists' (*Narodnuichestvo*)..was subjected to analysis and discussion. *Ibid.*, Another section of our Radical party..contributed much to the criticism of the 'peasantist' programme. **1963** *Times* 3 May 12/6 Colonel Turkesh listed nine principles which..constituted the doctrine of his group. Many of these appeared to be vague abstractions, including such curiosities as 'peasantism', 'freedomism', and 'developmentism'. **1969** G. Ionescu in Ionescu & Gellner *Populism* 99 Peasantism, born in Eastern Europe in the twentieth century,..takes the individual peasant explicitly as its social prototype..blends its social-economic doctrines with a strong nationalistic concern..and it claims that the peasantry is entitled as a class to the leadership of the political society. **1969** A. Walicki in *Ibid.* 104 Denmark was and remained the model..for most East European peasantists. *Ibid.* 106 A dictatorial peasantry-class of some of the peasantist régimes in power after the First World War (and especially that of Stambolisky in Bulgaria).

b. = Peasanthood.

1901 M. Franklin *My Brilliant Career* v. 25 My parents ..had dropped from swelldom to peasantism.

c. A proposal or movement for the diffusion of art among the peasants.

1903 L. F. Ward *Pure Sociol.* 454 There is probably something in the doctrine of 'peasantism', which seeks to rescue art from the exclusive control of the leisure class.

peasantize (pe·zăntaiz), *v. rare.* [-ize.] *refl.* To make (oneself) into a peasant.

1904 G. S. Hall *Adolescence* II. 513 They go West, to the colonies, the slums; devise new enterprises, sometimes almost want to peasantize themselves and fall in love with wheel-grease and the smell of the barnyard.

peasantry. Add: **1.** (Further examples.)

See *peasant *sb.* 1 and note there.

1933 E. & C. Paul tr. *Stalin's Leninism* II. 206 Why is the peasantry described [by Lenin] as the last capitalist class? Because of the two main classes of which our society is composed, the peasantry is a class whose economy is based on private property and small commodity production. Because the peasantry, as long as it remains a peasantry, living by small commodity production, will throw up capitalists from its ranks. **1934** *Encycl. Social Sci.* XII. 51/2 In the strongholds of European peasantry—the states formed in the western provinces of the former Russian Empire, the Danube basin and the Balkans—the peasants won a leading position after the World War when the great landed proprietors were compelled to surrender both their land and their dominant political status. **1936** tr. *Lenin's Sel. Wks.* III. 183 The better the condition of the 'commune', the greater the prosperity of the peasantry in general, the more rapid is the process of differentiation among the peasantry into antagonistic classes of capitalist agriculture. **1951** R. Firth *Elem. Social Organiz.* iii. 87 It is this close economic and social..attachment to the soil that is historically one of the main distinguishing features of a European peasantry. **1964** A. R. Holmberg in H. F. Dobyns et al. *Peasants, Power, & Appl. Social Change* (1971) ii. 33 More than 50 percent of the world's population is peasantry, the large majority of which lives in the so-called underdeveloped countries or newly emerging nations, under natural conditions and social structures that have denied peasants effective participation in the modernization process. **1966** *Sociol. Rev.* XIV. 6 The peasantry consist of small producers on land who, with the help of simple equipment and the labour of their families, produce mainly for their own consumption, and for the fulfilment of their duties to the holders of political and economic power. **1966** E. Wolf *Peasants* 11 It is only when a cultivator is integrated into a new society with a state—that is, when the cultivator becomes subject to the demands and sanctions of power-holders outside his social stratum—that we can appropriately speak of peasantry. **1971** I. Deutscher *Marxism in our Time* i. 28 We see all over the West the disappearance of those

middle classes that were supposed to constitute the conservative foundation of capitalism; of the small-owning, small-holding peasantry. **1975** J. A. Hellman tr. *Stavenhagen's Social Classes in Agrarian Societies* v. 66 For some students, peasantries are merely holdovers from precapitalist times, which tend to disappear as capitalism develops. While this may be the case in Western Europe, the situation in the underdeveloped countries is quite different. Here we still have hundreds of millions of peasants ..who are well integrated into the colonial or underdeveloped capitalist system and who, if present trends continue, will maintain peasant characteristics for many generations.

2. (Later examples.)

1937 *Observer* 17 Oct. 19/2, I find it hard to believe that anybody, having read or seen 'The Power of Darkness', would not instantly swear..to remove from himself any traces of peasantry that he might possess.

pea·santy, *a.* [-y[1].] = Peasantly *a.*

1933 *Times Lit. Suppl.* 8 June 400/2 Dark peasanty men. **1970** *Daily Tel.* 4 May 13 The peasanty voile top we've photographed over calico trousers is in fact sold as a mini dress.

pease, *sb.* Add: **5. pease-brose**: see Brose b.

1811 W. Aiton *Gen. View Agric. Ayr* 271 A few [late peas] are thrown in among the beans when sown broadcast. They are..made into meal for a species of pottage called 'pease-brose'. **1861** R. Leighton *Rhymes & Poems* (ed. 2) 12 'Pease Brose to dinner! brose alone! With neither boil nor stew! But say, what did you breakfast on?' They answer '*Pease Brose too.*' **1922** R. Thomas *Sandie McWhustler's Waddin'* v. 52 He was sittin' in his sark sleeves an' suppin' his pease-brose. *c* **1965** *Rebels Ceilidh Song Bk.* No. 2. 15 It'll be pease brose again.

pea-soup. Add: **a.** *pea-soup fog* (earlier and later examples); also *absol.*

1849 H. Melville *Jrnl. Visit to London & Continent* (1948) 45 Upon sallying out this morning encountered the oldfashioned pea soup London fog. **1887** *S. Austral. Advertiser* 8 Jan. 4/7 A month or two ago London experienced a succession of 'pea-soup fog' days. **1965** Mrs. L. B. Johnson *White House Diary* 7 Apr. (1970) 256 We flew in pea soup and uncertainty. **1976** J. Lee *Ninth Man* 10 He couldn't see more than fifteen or twenty feet through this pea soup. **1978** 'M. M. Kaye' *Far Pavilions* xix. 283 Londoners groping through a pea-soup fog.

b. A French Canadian; the French spoken in Canada. *N. Amer. slang.*

1896 G. Parker *Pomp of Lavilettes* 60 Yes, an' dey call us Johnny Pea-soups. **1912** B. Heeney *Pickanock* 22 Pea-soup! I never drink with the likes of you, Pauquett! **1931** 'D. Stiff' *Milk & Honey Route* iii. 38 A Canadian Frenchman is a 'Canuck' or sometimes a 'pea soup'. **1937** Partridge *Dict. Slang* 612/1 *Talk pea-soup*, to talk French-Canadian. **1945** H. MacLennan *Two Solitudes* 49 Listen, you goddam peasoup, you're too fast with your mouth. **1959** J. W. Godsell *I was no Lady* x. 170 I'm going to the Halfway right now to fix that damned Pea-soup. **1965** *Globe & Mail* (Toronto) 13 Oct. 6/3 Our childhood forays in Ottawa between pea-soup and English-speaking gangs.

pea-sou·per. *colloq.* [f. Pea-soup + -er[1].] **1.** A pea-soupy or thick yellow fog. Also *pea-souper fog.*

1890 J. Payn *Notes from 'News'* 8 The fogs we have had this year have been made too much of... You could see something in them if you looked long enough, which is not the case of a genuine Peasouper. **1907** *Daily Chron.* 30 Nov. 4/4 A country cousin who wishes to see and breathe and mingle in a metropolitan pea-souper. **1926** *Chambers's Jrnl.* Mar. 192/1 The fog..became dense—a real pea-souper. **1954** M. Sharp *Gipsy in Parlour* v. 62 The coal-burning London of my childhood was undoubtedly foggier than the London of to-day: the legend of the pea-souper, like all legends, has roots in fact. **1973** H. Carvic *Miss Seeton Sings* (1974) 196 Fog had clamped down on Paris... The scene reminded Miss Seeton of the London pea-soupers of her childhood. **1975** G. Howell *In Vogue* 4/2 There were high prices and strikes, peasouper fogs, and precious little coal. **1978** *Jrnl. R. Soc. Arts* CXXVI. 490/2 Pea-souper fogs are no longer experienced in London or in other cities.

2. = *pea-soup b.

1942 Berrey & Van den Bark *Amer. Thes. Slang* §385/5 Pea-souper, *a French-Canadian.* **1962** *Maclean's Mag.* 2 June 51/2 And *then* we can highstick those pea-soupers. **1966** [see *Joe *sb.*[2] 7]. **1968** P. C. Newman *Distemper of our Times* xix. 257 You're selling Canada to the pea-soupers!

peat[1]. Add: **2*.** A dark brown resembling the colour of peat.

1971 *Homes & Gardens* Sept. 84 Quite a lot of dark browns (anything from donkey to peat.) **1975** *Times* 7 Oct. 11/4 Long-sleeve sweater..in colours loganberry, peat, brown, and mid-blue. **1978** *Country Life* 16 Nov. 1685/1 There is nothing brash about Biba colours. Moss, peat, [etc.].

3. a. *peat-ditch, -land, -pulp, -smoke* (later examples). **b.** *peat-cutter* (later examples), *-digger* (examples). **c.** *peat-stained* adj. Also similative, as *peat-black, -brown* adjs. and sbs.

1961 R. S. Thomas *Tares* 35 Nerves strengthened with tea, Peat-black. **1898** B. Kirkby *Lakeland Words* 160 *Peat-broon*, t' colour of a dried peat, er bit of undyed woo'. **1906** *Westm. Gaz.* 9 Aug. 10/1 A rush-grown pool of peat-brown water. **1962** D. Francis *Dead Cert* xv. 173 My suit was a filthy peat-brown. **1963** *Times* 10 June 14/6, I followed a track that was once used by peat cutters. **1969** E. H. Pinto *Treen* 94/2 A peat cutter is a square-ended spade of normal length, with a right-angled, forward projecting blade at one side, like the breast plough. **1774** T. Pennant *Tour in Scotl. & Voy. Hebrides 1772* I. 66 By the imprudence of the peat-diggers, who were continually working on that side [of the moss]. **1894** A. Gordon *Northward Ho!* 202 The peat-digger was the most notorious carouser in Carglen. **1978** *Maledicta* II. 167 *Peat-digger*, any rural Irish person. **1903** G. W. Hartley *Wild Sport* i. 11 Jumping in and out of crumbling peat-ditches. **1907** *Daily Chron.* 1 Oct. 8/1 They wandered hand in hand across the peatland and the marshes. **1973** Moore & Bellamy (*title*) Peatlands. **1977** *Undercurrents* June–July 40 They are tackling the problems of what to do with bull calves, a large acreage of peatland and the need to produce extremely high quality winter fodder in a high rainfall area. **1908** *Chambers's Jrnl.* Jan. 122/2 The latest development in the production of peat-pulp is being made in Sweden. **1922** Joyce *Ulysses* 184 The peatsmoke is going to his head. **1971** *Country Life* 4 Nov. 1226/1, I may be prompted by something to recall the scent of peat-smoke. **1903** G. W. Hartley *Wild Sports* ix. 193 Its contents were for the thick-set, peat-stained beast standing a little to the right.

d. peat-bog, also *attrib.*

1965 G. M. Brown in *New Statesman* 9 July 52/2 Being under age And wringing peatbog whisky from a clout Into a secret kettle.

pea-time. *U.S. colloq.* [f. Pea[1] 7.] The season in which peas ripen. So *fig.* in phr. *the last of pea-time*: the last stage of anything, the end of one's life; *pea-time is past*: a thing is finished.

1834 [see *last *a.* 1 d]. **1850** 'M. Tensas' *Odd Leaves Life Louisiana Swamp Doctor* 174 It war the last of pea-time with me, sure, if I didn't rise 'fore bar did. **1862** J. R. Lowell in *Atlantic Monthly* Jan. 128 Ther' 's ollers chaps a-hangin' roun thet can't see pea-time's past. **1867** —— *Biglow Papers* 2nd Ser. I. p. lviii, *Last of pea-time*, to be hard up. **1889** 'C. E. Craddock' *Despot of Broomsedge Cove* x. 174 Ye oughter git some air an' light, Marcelly; ye look like the las' o' pea-time. **1893** M. A. Owen *Voodoo Tales* 199 'Deed my gyarden am a-lookin' mighty bad. Hit look mo' lak de las' o' pea-time den de fust o' truck-time. **1904** [see *last *a.* 1 d]. **1911** R. D. Saunders *Col. Todhunter* 108 'What on earth's the matter, Bill?' he asked. 'You look like the last of pea-times.' **1923** *Dialect Notes* V. 238 Utterly worn out. 'He looks like the last o' pea-time.'

‖**peau de chagrin** (pō: də ʃagræ·n). [Fr., lit. 'skin of grained leather'.] The title of a novel by Balzac (1831), in which a piece of shagreen diminishes in size as wishes are granted through its magic power, used *fig.* or allusively to indicate the progressive diminution of the human life-span.

1861 Geo. Eliot *Let.* 31 Dec. (1954) III. 475 The years of retrieval keep shrinking—the terrible *peau de chagrin* whose outline narrows..with our ebbing life. **1910** W. J. Locke *Simon* v. 56, I see my little allotted span of life shrinking visibly, like the *peau de chagrin*. **1946** L. P. Hartley *Sixth Heaven* v. 111 Money..[is] really like the Peau de Chagrin, and dwindles with every wish.

peau-de-soie. (Further examples.)

1912 C. Mackenzie *Carnival* iv. 38 But when his granddaughter,..all chiffon and ostrich plumes, took her upon a *peau de soie* lap..Jenny thought she had never experienced any sensation half so delicious. **1930** R. Cuthill tr. *Schober's Silk* iv. 272 *Peau de Soie.*—A close, firm silk fabric. Weave: eight- or ten-shaft twill. **1958** *Times* 16 Oct. 14/3 The bride..wore a gown of pearl coloured peau-de-soie with a train. **1974** *Times-Picayune* (New Orleans) 15 Aug. v. 6 The bride..wore a peau de soie gown.

‖**Peau d'Espagne** (pō:despa·ny). Also with lower-case initials. [Fr., lit. 'skin of Spain'.] **a.** A perfumed leather. Cf. *Spanish leather* s.v. Spanish *a.* 2 b. **b.** A scent supposedly suggestive of the aroma of this leather.

1855 G. W. S. Piesse *Art of Perfumery* vii. 143 Peau d'Espagne, or Spanish skin, is nothing more than highly perfumed leather. *Ibid.*, Finally, each double skin, now called peau d'Espagne, is to be enveloped in some pretty silk or satin, and finished off to the taste of the vender. **1898** *Junior Army & Navy Stores Catal.* 190/1 Atkinson's Perfumes..Peau d'Espagne, in Fancy Ribbon Sachets.. each 1/6. **1906** 'O. Henry' *Four Million* (1916) 112 A 200-pound woman breathing a flavour of Camembert cheese and Peau d'Espagne. **1907** *Yesterday's Shopping* (1969) 37/2 *Soaps*..Peau d'Espagne—box of 3 tablets 4/9. **1912** *Maclean's Mag.* Jan. 262/2 When his mother swept into the dining-room at meal-times, her hair faultlessly arranged, and wafting *peau d'Espagne* as she moved, Eugene followed in her wake like a small dog. **1919** [see *otto[1]]. **1922** Joyce *Ulysses* 747 That cheap peau despagne that faded and left a stink on you. **1942** E. Bowen *Seven Winters* 31 She used *Peau d'Espagne* scent. **1972** T. McLaughlin *Gilded Lily* vi. 70 The combination of perfume and leather could be obtained also in the form of *peau d'Espagne*, small perfumed pieces of leather that were used like lavender sachets or pot-pourri bags to hang in cupboards or tuck into pockets of clothes.

‖**peau d'orange** (pō dorā·ŋʒ). *Med.* [Fr., = 'orange-peel'.] A characteristic pitted appearance of the skin of the breast in some cases of breast cancer.

[**1875** *Nouveau Dict. de Méd. et de Chirurgie Pratiques* XXI. 569 La peau, au lieu d'être d'un blanc rosé, est.. comparable, comme l'a si bien fait remarquer Nélaton, à la peau d'orange. **1882** A. Nélaton *Élements de Path. Chirurg.* (ed. 2) V. xxi. 148 Quelquefois la peau de la mamelle, suivant la comparaison du Dr Mauduyt, ressemble à une peau d'orange.] **1898** A. M. Sheild *Clin. Treat. Dis. Breast* ix. 346 The skin..has that coarse pitted aspect to which the terms 'peau d'orange' and 'pigskin saddle' appearance have been well applied. **1948** C. English *Dis. Breast* xi. 75 In the early stages the growth is entirely unattached to the overlying skin, but later dimpling of the skin occurs, and a condition known as 'peau d'orange' follows. **1974** R. M. Kirk et al. *Surgery* iv. 62 Invaded lymphatics are blocked causing overlying lymphoedema of the skin known as peau d'orange.

pea-vine. For 'Pea Vine of California' substitute 'a valuable fodder plant'. (Earlier and later examples.)

1675 J. Fenwick in *Pennsylvania Mag. Hist. & Biogr.* (1882) VI. 89 You have Grass as high as a Man's Knees.. interlac'd with Pea-Vines, and other Weeds that Cattel much delight in. **1880** *Harper's Mag.* June 23/1 A search was instituted—under the bed, in the bed,..behind the wood-pile and in the pea-vines. **1910** *Chambers's Jrnl.* June 364/1 A little beetle has climbed up the pea-vine and laid its eggs in the pod. **1973** R. D. Symons *Where Wagon Led* vi. xvii. 271 Both horses and cattle thrived on the good grass and plentiful pea vine.

2. pea-vine hay, the dried stalks and foliage of a pea-vine.

1846 V. Aimé *Plantation Diary* (1878) 111 Peavine hay is excellent this year, the vine being still green and juicy. **1860** *Southern Cultivator* XVIII. 211 A little corn with Pea Vine Hay will keep them fat. **1932** W. Kelley *Inchin' Along* 55 Everybody knows there is nothing a mule likes better than pea-vine hay with the peas left on.

peb (peb), abbrev. *pebble *sb.* 1 c. *Austral. slang.*

1903 R. Bedford *True Eyes* 129 The session broke up—pebs and donahs wandered off in couples. **1916** C. J. Dennis *Moods of Ginger Mick* 114 They wus pebs, they wus norks, they wus reel naughty boys. **1916** —— *Songs Sentimental Bloke* 127 *Peb*, a flash fellow, a 'larrikin'. *a* **1943** L. Esson in *Penguin Bk. Austral. Ballads* (1964) 233 'E's the bloke to fite, 'E's the peb, gorblime. **1959** Baker *Drum* (1960) 133 *Peb*, a larrikin.

pebble, *sb.* **1. c.** For (*Australian slang*) read *slang* (chiefly *Austral.*) and add earlier and later examples. Also, a term of affection (applied to a person or animal). Phr. (*as*) *game as a pebble* (see quot. 1959) (*obs.*).

1829 P. Egan *Boxiana* 2nd Ser. II. 20 Hudson, as *game* as a pebble, stuck to his man like glue. **1851** 'W. T. Moncrieff' *Scamps of London* III. i. 56, in *Sel. Dram. Wks.* I, Now, my pebbles, I'll give you a toast. *c* **1863** T. Taylor *Ticket-of-Leave Man* I. 11 Doctor? Nay; I'm as game as a pebble and as stell as a tree! **1874** M. Clarke *His Natural Life* iv. vii. 415 'You're not such a pebble as folks seem to think,' grinned Frere. **1888** 'R. Boldrewood' *Robbery under Arms* III. ix. 123 Then the Turon favourite—a real game pebble of a little horse—began to show up. **1893** K. Mackay *Out Back* 188 Cabbage Tree Ned is as game as a pebble, and may try to dash through in spite of us. **1901** E. Dyson in *Bulletin Story Bk* 134 The Imp.., game as a pebble, despite his age and infirmities. **1918** C. Featherstonhaugh *After Many Days* 227 Traveller was game as a pebble, and he just passed Quadrant on the post and no more. **1945** Baker *Austral. Lang.* iv. 88 *Game as a pebble*... (a *pebble* is a person, especially a larrikin, or an animal hard to control.). **1959** —— *Drum* (1960) 133 *Pebble*, a person (occasionally a horse) hard to control. Whence, *game as a pebble*, extremely courageous. **1974** D. Stuart *Prince of My Country* 166 He was as hard as nails and as game as a pebble.

d. Colloq. phr. *the only pebble on the beach* and varr., the only person or thing to be considered in a particular situation; used *spec.* (usu. in negative contexts) with reference to an eligible man or woman.

1896 H. Braisted (*song-title*) You're not the only pebble on the beach. *Ibid.*, If you want to win her hand, Let the maiden understand That she's not the only pebble on the beach! **1906** E. Dyson *Fact'ry 'Ands* x. 128 S' elp me shicker, Twenty, you was the on'y pebble. **1924** H. De Sélincourt *Cricket Match* v. 156 There were other pebbles on the beach beside him. **1927** S. Spaeth *Weep some more, my Lady* x. 259 Another phrase that became a recognized part of the English language was 'You're not the only pebble on the beach', made into a song by Harry Braisted and Stanley Carter. **1930** M. Kennedy *Fool of Family* vii. 66, I won't look at your damned things... You think you're the only pebble on the beach. **1933** E. O'Neill *Ah, Wilderness!* I. 19 What do I care for him? He's not the only pebble on the beach. **1933** D. L. Sayers *Hangman's Holiday* 193 The Birmingham–London express reached Rugby at 10.24, departing again at 10.28. But, swift and impressive as it was, it was not the only, or the most important, pebble on the station beach, for over against it upon the down line was the Irish Mail. **1936** G. B. Shaw *Simpleton* II. 73 The British Empire was not the only pebble on the beach. **1945** 'P. Wentworth' *Clock strikes Twelve* ii. 15 A heart-to-heart talk with someone who makes me feel I'm the only pebble on the beach. **1977** *World of Cricket Monthly* June 87/1 He was not the only pebble on the Middlesbrough beach.

5. a. *pebble brooch* (earlier example); *pebble (eye)glasses, lens, spectacles.*

1818 M. Edgeworth *Let.* 21 Dec. (1971) 150 This little pebble brooch will reach you I hope just in time for you to give it..to Harriette for a Christmas box. **1933** C. St. J. Sprigg *Fatality in Fleet Street* i. 14 His beady eyes regarded the Chief coldly behind his pebble eyeglasses. **1938** E. Ambler *Cause for Alarm* xvi. 206 The pair of thick pebble glasses..rendered me practically blind. **1958** P. Mortimer *Daddy's gone a-Hunting* ii. 11 A stern little girl in thick pebble glasses. **1972** M. Woodhouse *Mama Doll* ii. 4 A short fat man with mad bushy hair and pebble glasses. **1977** J. Aiken *Last Movement*

i. 14 His eyes myopic behind thick pebble glasses. **1955** H. Spring *These Lovers fled Away* 437 The spectacles now had pebble lenses. **1958** *Spectator* 30 May 687/3 The sadistic young slug of a lodger with dirty fingernails and pebble-lens spectacles. **1964** R. Church *Voyage Home* viii. 166 Thick, pebble lenses, flashed at me like headlamps. **1955** H. Spring *These Lovers fled Away* 142 He was a dark, robust-looking youth, with flaring nostrils and pebble spectacles. **1976** 'W. Trevor' *Children of Dynmouth* v. 115 He wore thick pebble spectacles, behind which his eyes were unnaturally magnified.

b. *pebble-lensed, -like* adjs.; **pebble-beached** *a. slang*, (*a*) penniless, destitute; (*b*) dazed, absent-minded; **pebble-bed,** (*a*) *Geol.*, a conglomerate that contains pebbles, esp. one from which they readily work loose with weathering; (*b*) *Nuclear Sci.*, used *attrib.* to designate a nuclear reactor in which the fuel elements are in the form of pellets having an outer layer of moderator; **pebble chopper** *Archæol.*, a primitive chopping tool made from a pebble (see *pebble tool* below); **pebble culture,** name given to a culture which uses pebbles as materials for tools, identified orig. in Africa but now known to have existed also in America, Asia, and Europe; **pebble dash** (examples); **pebble dashing** (examples); **pebble grain,** a patterned grain produced by pebbling (see Pebble *v.* 3); **pebble-grained** *adj.*, (*a*) having a pattern produced by pebbling; (*b*) of lenses etc., having the appearance of pebble (see sense 2 b in Dict.); **pebble tool** *Archæol.*, a primitive tool made by chipping and shaping pebbles, thought to be the earliest use of stone tools by man; **pebble weave,** a weave producing a rough surface.

1890 in Barrère & Leland *Dict. Slang* II. 120/2 He had arrived at a crisis of impecuniosity compared to which the small circumstance of being pebble-beached and stony-broke might be described as comparative affluence. **1897** *Sporting Times* 3 July 1/4 She was absolutely stony, pebble-beached to all the world. **1934** Wodehouse *Right Ho, Jeeves* xvii. 216 Gussie..switched on that pebble-beached smile again and tacked down to the edge of the platform. **1851** *Q. Jrnl. Geol. Soc.* VII. 77 The valley is.. so encumbered with detritus of the overlying and reaggregated pebble beds, that except at the town of Vichy it is almost impossible to make good observations. **1914** J. Park *Text-bk. Geol.* xxviii. 377 The three main divisions of the Trias recognised in the British Isles are:—3. Rhætic... 2. Keuper... 1. Bunter... The Bunter consists of red and variously hued sandstones and conglomerates or pebble beds of fluviatile or fluvio-lacustrine origin. **1961** *New Scientist* 16 Mar. 695/1 W. Germany is to have what is probably the first operational 'pebble-bed' reactor... The fuel elements are balls of graphite with centres of uranium carbide. **1969** *Financial Times* 9 Jan. 7/6 The experimental 'pebble bed' reactor at Juelich more closely approaches the idea of a nuclear system into which fuel is simply shovelled. **1969** Bennison & Wright *Geol. Hist. Brit. Isles* xii. 274 The Bunter Pebble Beds characteristically occur in the Midlands... The name Pebble Beds is a descriptive term and the formation comprises sandstone with pebbles. **1959** J. D. Clark *Prehist. S. Afr.* v. 115 Some of these flakes were struck from cores and were not simply the waste flakes removed in preparing the pebble-chopper. **1964** Jennings & Norbeck *Prehist. Man in New World* 164 At the basal level Butler places the Congdon I complex—peripherally flaked pebble choppers. **1974** *Encycl. Brit. Micropædia* VII. 825/3 *Pebble chopper*, or pebble tool, primordial cutting tool, examples of which have been dated at over 2,500 years BC. **1931** J. D. Solomon in L. S. B. Leakey *Stone Age Cultures Kenya Colony* 264 Mr. C. van Riet Lowe..has studied the terraces of the Vaal River, and finds... (i) High terrace, with coarse gravel containing what may be a pebble-culture. **1955** *Sci. Amer.* Aug. 50/2 He has found two teeth of the man-ape in a deposit together with crudely chipped stone tools of the 'pebble culture'. **1970** J. D. Clark *Prehist. Afr.* ii. 68 It is a common misconception that the Oldowan tools were made from pebbles and the term 'Pebble Culture' has been used as synonymous with Oldowan. **1902** *Ann. Rep. Board of Regents Smithsonian Inst. 1901* 106 A cheap frame construction was used, the sides of which were treated with pebble-dash and the roof made of asphalted felt covered with crushed slag. **1911** *Encycl. Brit.* XXI. 785/1 *Rough-cast* or *Pebble-dash* plastering is a rough form of external plastering in much use for country houses. **1960** [see *God-wottery]. **1973** G. Moffat *Deviant Death* iv. 54 A dull collection of pebble dash houses and wood trim in cardinal colours. **1940** *Chambers's Techn. Dict.* 621/1 *Pebble-dashing* (*Plast.*), a rough finish given to a wall by coating it with plaster, on to which, while it is still soft, small stones and liquid lime are thrown. **1978** *Lancashire Life* Sept. 3/2 (Advt.), A detached house.., built approximately 1938 of brick with pebbledashing to most elevations. **1897** *Sears, Roebuck Catal.* 194/3 Child's pebble grain school shoe. **1931** W. Faulkner *Sanctuary* xv. 133 Disembodied voices blaring from imitation wood cabinets or pebble-grain horn-mouths above the rapt faces. **1969** R. T. Wilcox *Dict. Costume* (1970) 263/2 *Pebble grain*,..a leather, imitation leather or fabric given a grained surface by running it between rollers under pressure. **1971** 'O. Bleeck' *Thief who painted Sunlight* (1972) xiv. 123 His black, pebble-grained loafers were burnished and gleaming. **1973** W. M. Duncan *Big Timer* i. 10 Pale blue eyes which stared myopically from behind thick pebble-grained glasses. **1968** F. Mullally *Munich Involvement* iii. 21 The woman..hesitated, blinking at Sullivan through pebble-lensed glasses. **1975** D. Pitts *This City is Ours* xix. 65 [He] peered..through heavy, pebble-lensed glasses. **1977** 'E. Crispin' *Glimpses of Moon* xiii. 284 Long-haired, pebble-lensed cissies in white coats. **1924** W. de la Mare *Ding Dong Bell* 56 The pebble-like tattling of a robin. **1960** C. Day Lewis *Buried Day* 220 His stony eyes behind pebble-like glasses. **1931** J. D. Solomon in L. S. B. Leakey *Stone Age Cultures Kenya Colony* 263 We thus seem to have the following succession of events in East Africa during the Pleistocene: (i) Wet period..which coincided with the epoch of the man who made pebble-tools (Kafuan). **1959** J. D. Clark *Prehist. S. Afr.* iv. 74 It is an open question whether *Australopithecus* could have made pebble-tools, the earliest examples of intentionally made implements known. **1961** *Times* 5 Sept. 13/5 In fact..*Australopithecinae* had been found at four or five sites in association with pebble-tools. **1973** B. J. Williams *Evolution & Human Origins* ix. 148/2 These are often referred to as 'pebble-tools', as they usually represent river pebbles that have been fractured in such a way that a cutting edge is produced. **1941** *Archit. Rev.* LXXXIX. 40 The floor is close-carpeted with a grey 'pebble-weave' material. **1958** *Woman's Jrnl.* Mar. 3 (Advt.), Classic casual in pebbleweave. **1969** R. T. Wilcox *Dict. Costume* (1970) 263/2 *Pebble weave*, a rough-surfaced fabric produced by weaving together shrunken, twisted yarns.

pebbled, *a.* Add: **1. b.** Of spectacles, etc.: made with or having the appearance of pebble lenses. Also *fig.*

1939 Joyce *Finnegans Wake* 463 He is looking aged with his pebbled eyes. **1959** *Listener* 5 Mar. 406/2 Eccentric bookworms with pebbled glasses and bulging foreheads. **1969** R. Petrie *Despatch of Dove* I. 10 He watched minutely from behind thick-pebbled lenses. **1978** M. Butterworth *X marks Spot* 18 Shrewd eyes glittered behind pebbled lenses.

pebbly, *a.* Add: **2.** *fig.* Resembling pebbles; uneven; *esp.* of textiles: having a rough surface.

1793 H. Fuseli in *Analytical Rev.* Jan. 2 The hoarseness of Northern language bound in pebbly monosyllables. **1923** [see *armure]. **1936** Chesterton *Autobiogr.* 148 They had shiny pebbly eyes. **1955** *New Yorker* 19 Mar. 100 Done in pebbly navy wool, with deep raglan sleeves. **1964** *McCall's Sewing* i. 8/1 A pebbly crêpe will not noticeably add pounds, but a very rough, nubbly wool tweed may have a decided effect.

Pebidian (pĕbi·diăn), *a. Geol.* [See quot. 1877 and -an.] Epithet of a thick sequence of volcanic rocks of Pre-Cambrian age exposed in Preseli, SW Wales. Also *absol.*

1877 H. Hicks in *Q. Jrnl. Geol. Soc.* XXIII. 230, I propose now to divide the Pre-Cambrian rocks [in the vicinity of St. David's] into two distinct series under the local names of Dimetian..for the lower, and Pebidian (*Pebidiauc* being the name of the division or hundred in which the rocks are chiefly exposed) for the upper series. **1880** *Ibid.* XXXVI. 538, I have long been disposed to consider the felspathic series in Shropshire, or at least the Lilleshall group, as representing the Pebidian. **1937** Pringle & George *S. Wales* ii. 15 In the neighbourhood of St. David's, the Pebidian tuffs occupy a broad anticlinal area which extends inland for several miles in an east-north-east direction. **1969** Bennison & Wright *Geol. Hist. Brit. Isles* iv. 80 The Caerfai Series commences with a reddish basal conglomerate, some of the pebbles being derived from the adjacent Precambrian volcanics, the Pebidian.

pébrine. Add to etymology: (A. de Quatrefages 1858, in *Comptes Rendus* XLVII. 530). Substitute for def.: A disease of silkworms, caused by the microsporidian parasite *Nosema bombycis* (cf. *nosema), and characterized by the appearance of dark spots on the insect, whose growth is stunted. (Further examples.)

1873 A. S. Packard *Our Common Insects* iii. 47 The still more formidable disease called pebrine is thought to be of vegetable origin. **1911** *Proc. Zool. Soc.* 625 The parasite *Nosema apis* was closely allied to that of pébrine, the silkworm disease due to *Nosema bombycis*. **1912** E. A. Minchin *Introd. Study Protozoa* xvi. 412 The Microsporidia first attained an unenviable notoriety through the ravages caused by *Nosema bombycis*, the cause of 'pébrine' or silkworm-disease. **1964** T. C. Cheng *Biol. Animal Parasites* v. 141/2 *Nosema bombycis* is parasitic in silkworms and causes the fatal pébrine disease. **1972** Swan & Papp *Common Insects N. Amer.* 291 The microsporidian disease is called *pebrine* in France for the telltale spots on the infected silkworm. **1973** M. A. Sleigh *Biol. Protozoa* x. 245 *Nosema bombycis*, is responsible for the pebrine disease of silkworms, in which tissue cells of any type may be infected at any stage of growth.

pec (pek). *N. Amer. slang.* Also **peck.** [abbrev. Pectoral *sb.* and *a.*] = Pectoral *sb.* 3. Usu. *pl.*

1966 L. Cohen *Beautiful Losers* (1970) II. 159, I saw you with massive lower pecs and horseshoe triceps, with bulk and definition simultaneously. **1972** B. Rodgers *Queens' Vernacular* 147 *Pecks*, pectoral muscles. **1977** *Gay News* 24 Mar. 23/1 He sought only the most virile men—the brutes with watermelon-sized pecs and pea-sized heads.

pecan. Add: Also pecon. For '*Carya olivæformis*' substitute '*Carya illinoensis*'. (Further examples.)

1797 A. Ellicott *Let.* 1 Apr. in C. V. Mathews *Andrew Ellicott* (1908) 152, I have a large Keg of Pecon Nuts put up for you. **1818** G. Flagg *Let.* 12 Sept. in *Trans. Illinois State Hist. Soc. 1910* (1912) 158, I have seen some [hogs] as fat upon Hickorynuts, Acorns, Pecons & Walnuts as ever I did those that were fatted upon Corn. **1969** *Oxf. Bk. Food Plants* 28/2 Nowadays, pecans can be bought in many British shops. *Ibid.*, The Pecan is a large tree up to 170 feet high, with grey, furrowed bark. **1975** *New Yorker* 3 Feb. 25/2 The cake..was made of thin layers of yellow sponge cake and filled with a whipped cream laced with brandy, pecans, and a special coffee extract.

c. *pecan pie.*

1936 F. M. Farmer *Boston Cooking-School Cook Bk.* (new ed.) 633 Pecan Pie... 3 eggs... 1 cup light corn syrup... 1 cup finely chopped pecans. **1976** *Express-News* (San Antonio, Texas) 2 Oct. 3-E/4 The specialty-of-the-house dessert—which is made every day—is pecan pie. **1976** *Listener* 22 July 86/1 Ordinary Southern food..fried chicken and pecan pie.

peccaminous, *a.* Delete †*Obs.* and add later examples.

Still *rare* It is the kind of word that Joyce may have picked up from the *O.E.D.*

1922 Joyce *Ulysses* 720 A volume of peccaminous pornographical tendency entitled *Sweets of Sin*. **1939** —— *Finnegans Wake* 288 To put off the barcelonas from their peccaminous corpulums.

pêche à la Melba, pêche Melba: see *Melba.

pecia (pī·siă). Pl. peciæ, pecie, pecias. [a. med.L. *pecia* Piece *sb.*] A gathering of a manuscript, usu. a gathering of four leaves. **pecia system,** a system of copying pecia by pecia.

1908 H. Hall *Stud. in Eng. Official Hist. Documents* III. 380 A rarer term [in Palæography] occurs in the use of *peciae* to denote the component parts of a file of loose documents. **1912** E. M. Thompson *Introd. Gr. & Lat. Palaeogr.* 68 A survival of the ancient method of calculating [*sc.* the remuneration of scribes] has been found in the practice..in the middle ages, of paying by the *pecia* of sixteen columns, each of sixty-two lines with thirty-two letters to the line. **1930** R. Steele in *Library* XI. 230 The pecia is a unit, devised mainly for estimating the rate of payment to the copyist, which seems to have arisen in the University of Bologna early in the thirteenth century. **1934** Little & Pelster *Oxford Theol. 1282–1302* I. v. 56 (*heading*) The pecia and some characteristics of Oxford scholastic manuscripts. *Ibid.* 57 The *pecia* is originally, according to the researches of Savigny and Destrez, a piece of parchment which when folded contains two sheets, i.e. 4 leaves on 8 pages. As copyists were frequently paid according to the number of *pecie* which they had written, the beginning of a new *pecia* in the original examples was not infrequently marked in the copies by a *p* or *p*[a] with the corresponding numeral. **1935** *Times Lit. Suppl.* 14 Dec. 858/3 The *pecia* was the loose quire, generally of four folios or eight pages, lent out by an approved 'stationer' to be copied. **1958** C. H. Talbot in Wormald & Wright *Eng. Library before 1700* iv. 67 The unit of their [*sc.* the university scribes'] work was the *pecia*, a technical term (borrowed no doubt from the tanners and parchment makers) designating a sheep-skin which could be treated for writing on. *Ibid.* 68 The size of the *pecia* appears to have differed somewhat in the manuscripts we know, but the general conclusion seems to be that it was a sextern. **1963** R. A. B. Mynors *Catal. MSS. Balliol Coll.* 43 The care with which it is written and its *peciae* are marked might suggest that it was intended to be kept..as a standard copy. **1964** G. Pollard in *Beiträge zum Berufsbeurusstsein des Mittelalterlichen Menschen* (Miscellanea Mediaevalia III) 338 (*heading*) The pecia system. **1972** E. J. Dobson *Eng. Text of Ancrene Riwle* p. x, It must have been one of two copies made simultaneously from a single exemplar by a form of the *pecia* system. **1976** A. G. Judy *Kilwardby's De Ortu Scientiarum* p. xxi, There would have been exactly eighteen pecias in this copy.

peck, *sb.*[3] Add: **4. peck-right,** in a group of birds, the way in which those of higher rank are able to attack those lower in the hierarchy, without provoking an attack in return; cf. *peck-order, *pecking-order.

1931 W. C. Allee *Animal Aggregations* xix. 344 Hens with this power [of pecking inferiors] are said to have the 'peck-right' over those submitting to the pecking. **1954** Fisher & Lockley *Sea-Birds* vii. 170 There is [among gulls], as in domestic hens, a definite order of precedence, or peck-right. **1962** J. C. Welty *Life of Birds* x. 184/2 The dominant bird is said to possess a peck right over the subordinate bird.

peck (pek), abbrev. *peckerwood b. *U.S. Black slang.*

1932 *Evening Sun* (Baltimore) 9 Dec. 31/5 *Peck*, a white person. **1964** O. Harrington in J. H. Clarke *Harlem* 96 Every member loved Old Snakes and every school-child knew that he'd once been caught in a Texas mob which was joyously barbecuing another Negro. And when old Snakes began laughing the pecks stared in amazement and let him walk right through. **1969** C. Brown in A. Chapman *New Black Voices* (1972) 183 A poor white peck will cuss. A poor white peck will cuss worse'n a nigger. I am talking about white men who ain't poor like them pecks. **1970** C. Major *Dict. Afro-Amer. Slang* 90 *Peck*, white person.

peck, *v.* Add: **1. c.** To kiss perfunctorily; to give a peck (Peck *sb.*[3] 1) to.

1969 *New Yorker* 11 Oct. 53/1 They pecked the hostess farewell. **1977** C. McCullough *Thorn Birds* xv. 343 Meggie leaned over to peck her brothers on their cheeks self-consciously.

peckawood: see *peckerwood b.

pecked, *ppl. a.* Add: *spec.* in *Archæol.*: consisting of or characterized by pecked strokes or marks (see Peck *v.*[1] 5); *pecked curve, line* (further examples).

1959 J. D. Clark *Prehist. S. Afr.* Pl. 12 (*caption*) Fine example of an eland in a pecked engraving style, Transvaal. **1963** *Field Archaeol.* (Ordnance Survey) (ed. 4) 43 The earth-fast boulder decorated with pecked marks and designs of uncertain age and significance is a common feature in many parts of Britain north of the Trent. *Ibid.*, Other pecked figures of axes are known elsewhere. **1971** *Nature* 21 May 160/1 This ungual must be rotated through an angle of 180° around its longitudinal axis (pecked line in Fig. 2*F*). **1975** J. B. Harley *O.S. Maps* iii. 33 Overhead features, distinguished by pecked lines, are shown when they are of a size and character to be useful features. **1976** *Nature* 29 Apr. 772/1 In Fig. 2 the pecked curve represents the 3-block running mean of February maximum temperatures for six stations surrounding the tree location.

pecker. Add: **1. c.** Abbrev. of *peckerwood b. *slang.*

1966 R. G. Toepfer *Witness* xvi. 127 Not a chance. These peckers know that as well as me.

3. a. (Further examples.)

In view of sense *3 b a use commonly avoided by British travellers in the U.S.

1901 R. C. Lehmann *Anni Fugaces* 56 Weighed down by debt they yet keep up their pecker. **1918** Galsworthy *Five Tales* iv. 162 Keep your pecker up, and get off abroad. **1922** Joyce *Ulysses* 320 Keep your pecker up, says Joe. She'd have won the money only for the other dog. **1928** A. Merritt *Seven Footprints to Satan* xiii. 180, I was talkin' loud to keep my pecker up. **1934** F. W. Crofts *12.30 from Croydon* xix. 263 Charles could not eat, in spite of the rough kindness of one of the warders, who adjured him to keep his pecker up. **1973** N. Moss *What's the Difference?* 44/1 The Englishman in North America should beware of using the phrase 'keep your pecker up'.

b. The penis. *slang* (chiefly *U.S.*).

1902 Farmer & Henley *Slang* V. 289/2 The *penis*.. pecker. **1936** H. Miller *Black Spring* 142 Ought to stand on Times Square with my pecker in my hand and piss in the gutter. **1949** —— *Sexus* I. iii. 114 Walking towards the house I open my fly and let my pecker out. **1958** N. Levine *Canada made Me* ii. 57 Ground sunflower seeds..This will make your pecker stand up to no end of punishment. **1960** E. L. Wallant *Human Season* ix. 97 You know how it is when the ol' pecker ain't what it used to be, hah, hah! **1964** *Amer. Folk Music Occasional* I. 12 There is a house down in New Orleans, They call it The Rising Sun, When you want to get your pecker spoilt, That's where you get it done. **1967** I. A. Baraka in W. King *Black Short Story Anthol.* (1972) 125 Oh, Lucasta, find me here on the bed, with hard pecker and dirty feet. **1969** K. Vonnegut *Slaughterhouse-Five* v. 107 Billy took his pecker out, there in the prison night, and peed and peed on the ground. **1971** F. Norman *Dodgem-Greaser* iii. 45, I unzipped my fly and exposed my pecker to the night air with one deft flick of my finger. **1975** R. H. Rimmer *Premar Experiments* (1976) i. 27 Tony's pecker would no sooner get inside me than he'd pop off.

4. pecker-head *U.S. slang*, an aggressive objectionable person.

1955 *Time* 14 Nov. 116/2 When the girl's husband.., a got-rich peckerhead, finds out about that hotel visit, he ravishes his wife, just to even the score. **1977** E. Leonard *Unknown Man No. 89* xxiii. 239 Them peckerheads'd never make it.

peckerwood (pe·kəɹwud). *U.S.* [f. Woodpecker with reversal of the two elements.] **a.** A woodpecker. *dial.*

1859 Bartlett *Dict. Amer.* (ed. 2) 314 *Peckerwood*, Western for Woodpecker. **1893** H. A. Shands *Some Peculiarities of Speech in Mississippi* 49 *Peckerwood*, woodpecker. Bartlett says that this word is Western. It is also heard very frequently in Mississippi, as in Tennessee. **1909** F. B. Calhoun *Miss Minerva* 140 A big, red-headed peckerwood. **1926** E. M. Roberts *Time of Man* 153 Not had the sense of a pecker-wood. **1935** Z. N. Hurston *Mules & Men* I. vi. 137 Ah wasn't gointer kill no ole tough peckerwood for you to eat, baby. **1940** R. Wood *Amer. Mother Goose* 55 Peckerwood a-settin' on a swingin' limb, Blue-bird a-buildin' in the garden. **1947** J. H. Brown *Outdoors Unlimited* 76 Sitting there watching the jays and peckerwoods and listening to the distant shrillings of a hawk, I thought Lucius looked tougher than a pine-knot. **1968** *Harper's Mag.* Sept. 61/1, I even liked what the kids at Tuskegee Institute called 'those crazy li'l ol' peckerwoods'.

b. *slang.* Also *peckawood.* A white person, esp. a poor one. Also *transf.*

1929 T. Gordon *Born to Be* 236 *Peckerwood*, a poor white man. **1929** C. McKay *Banjo* 217 For example, we have words like ofay..peckawood..and so on. **1941** W. C. Handy *Father of Blues* vi. 86 Mound Bayou had no such words addressed to 'peckerwoods' or 'rednecks'. **1942** W. Faulkner *Go Down, Moses* 255 Even a Delta peckerwood would look after even a draggle-tail better than that. **1947** *Sat. Rev. Lit.* (U.S.) 29 Mar. 13/2 There is a difference between white plantation people and the white 'peckerwoods' of the hills. **1966** R. G. Toepfer *Witness* ii. 11 They was three peckerwoods had the boy. **1967** C. Major in A. Chapman *New Black Voices* (1972) 299 How come so many Of us niggers Are dying over there In that white Man's war They say more of us Are dying Than them peckerwoods & it just Don't make sense. **1967** *Trans-Action* Apr. 27/2 There is a social distinction here between the white elite and the 'redneck' or 'peckerwood' families. **1967** *N.Y. Times* 7 Sept. 42/4 When I tried to get into the black caucus, they said, 'No peckerwoods allowed in here, Sonny.' **1974** *New Yorker* 29 Apr. 94/2 A horse can be well one day and a sick peckerwood the next. **1974** E. Brawley *Rap* (1975) II. xxiv. 374 And we done, peckerwood, we finished.

Peckham (pe·kăm). [Name of a suburb of London.] **a.** Used in various joc. phrases, esp. with play on peck = food, to eat (Peck *sb.*[3] 3, Peck *v.*[1] 4). (See quots.)

Neither this sense nor the next seems to have much currency outside dicts.

1788 Grose *Dict. Vulgar T.* (ed. 2) s.v. *All Holiday*. It is all holiday at Peckham, or it is all holiday with him; a saying signifying that it is all over with the business or person spoken of or alluded to. **1823** 'J. Bee' *Slang* 134 *Peckham* (going to), dinner. 'All holiday at Peckham'—no appetite. 'No Peckham for Ben, he's been to Clapham,' i.e. is indisposed, in a certain way. Peckish—hungry. **1864** Hotten *Slang Dict.* 198 *Peckham*, a facetious meaning of the name of this district, implying a dinner; 'all holiday at Peckham', *i.e. nothing to eat.* **1902** Farmer & Henley *Slang* V. 157/1 *Peckham.* To have (or spend) a holiday at Peckham, *verb phr.* (old)—To have nothing to eat. Going to Peckham = going to dinner. **1922** A. M. Hyamson *Dict. Eng. Phr.* 267/2 *Peckham, To go to*, to go to dinner. **1970** *Brewer's Dict. Phr. & Fable* (rev. ed.) 814/2 *All holiday at Peckham*,..no appetite, not peckish.

b. *Peckham rye*, tie. *Rhyming slang.*

1925 Fraser & Gibbons *Soldier & Sailor Words* 221 *Peckham rye*, tie. (Rhyming slang). **1960** J. Franklyn *Dict. Rhyming Slang* 106/2 *Peckham Rye*, tie (necktie). 19 C., and by far the most usual term. **1973** B. Aylwin *Load of Cockney Cobblers* 84 Tie. Peckham Rye.

peck horn (pe·k hǫɹn). *Jazz slang.* Also with hyphen and as one word. [Origin uncertain.] A mellophone, saxophone, or similar instrument.

1936 *Amer. Mercury* May p. x/2 *Peck horn*, mellophone. **1942** Berrey & Van den Bark *Amer. Thes. Slang* § 377/8 (*Alto*) *peck horn*, an alto horn or mellophone. **1948** Mencken *Amer. Lang.* Suppl. II. 705 It [*sc.* jive] arose in the honky-tonks and tingle-tangles of the pre-jazz era, and many of its current names for musical instruments go back to that era or even beyond, *e.g.*,..*pretzel* or *peckhorn* for a French horn. **1961** *Down Beat* 13 Apr. 37 This set contains a good sampling of his abilities on trumpet, alto, baritone, and the oddball peckhorn. **1966** *New Yorker* 25 June 44/2 The band generally had two trombones, three trumpets, a bass horn and a baritone horn, a peck horn, a clarinet. *Ibid.* 46/3 From the age of eight I played the upright alto—the peck horn—in my father's band. **1970** C. Major *Dict. Afro-Amer. Slang* 90 *Peck horn*,..mellophone or saxophone. **1975** *Sunday Times* (Colour Suppl.) 12 July 47/1 Straight band singers were unknown in the Twenties—everyone, even Bing Crosby, had an instrument to hold. 'I had a peckhorn, like a flugelhorn.'

pecking order (pe·kiŋ ǫːɹdəɹ). [f. Pecking *vbl. sb.*[1] + Order *sb.*, tr. G. *hackliste* (T. J. Schjelderup-Ebbe 1922, in *Zeitschr. für Psychologie* LXXXVIII. 227).] **1.** A pattern of behaviour first observed in hens and later recognized in other groups of social animals, in which those of high rank within the group are able to attack those of lower rank without provoking an attack in return.

1928 A. Huxley *Point Counter Point* xxvi. 438 Observing the habitual and almost sacred 'pecking order' which prevails among the hens in his poultry yard..the politician will meditate on the Catholic hierarchy and Fascism. **1939** *Auk* LVI. 263 A position in the pecking order is not..always determined at the first meeting. **1952** M. K. Wilson tr. *Lorenz's King Solomon's Ring* xi. 147 This can be convincingly demonstrated by the existence of an order of rank, known to animal psychologists as the 'pecking order'. **1965** *Listener* 10 June 861/1 Normal monkeys, like many other kinds of animals, form a sort of pecking order. **1972** *Country Life* 9 Mar. 541/1 The fat little bantam hen, who proved first in the pecking order..quickly asserted herself. **1974** *Times Lit. Suppl.* 4 Oct. 1088/4 The term 'pecking order' has become part of common parlance... After a brief trial of strength, every animal in the flock or the herd learns to know its place.

2. *transf.* A hierarchy based on rank or status.

1955 H. Nicolson *Good Behaviour* i. 7 In a perfect classless society..similar pecking orders must exist. **1957** L. Durrell *Justine* I. 64 There is a Pecking Order among diplomats as there is among poultry. **1959** G. Endore *Detour through Devon* 11 The pecking order obtains among cars on the road as well as among chickens in the barnyard. **1959** T. H. White *Godstone & Blackymor* 172 The ghosts..may have taken precedence in a kind of pecking-order, by virtue of the number of pipes which they could claim. **1961** *Times Lit. Suppl.* 29 Dec. 925/2 Man has animal instincts, instincts for desiring and holding territory and property, for creating a pecking order of dominance and hence status. **1962** *Observer* 25 Nov. 3/2 One of the most disastrous weaknesses of our whole educational system is its insistence on preserving formal hierarchies, a sort of academic pecking order. **1967** J. Potter *Foul Play* xvi. 184 The inspector had a pretty low rating in the CID's pecking order... His office overlooked an unbroken expanse of sooty wall. **1967** M. Argyle *Psychol. Interpersonal Behaviour* iv. 70 Groups develop definite 'pecking orders' in terms of amount of speech and influence permitted. **1973** *Nature* 30 Nov. 318/1 One of the interpenetrating pecking orders that bedevil life in universities is the hard science/soft science hierarchy. **1976** H. Wilson *Governance of Britain* ii. 23 My own procedure was to rely on the order of precedence within the Cabinet, the so-called 'pecking-order', making clear that the second in the list chairs Cabinet.

peck-order (pe·k,ǫɹdəɹ). [f. Peck *sb.*[3] + Order *sb.*, tr. G. *hackliste* (T. J. Schjelderup-Ebbe 1922, in *Zeitschr. für Psychologie* LXXXVIII. 227).] = prec., sense 1.

1931 W. C. Allee *Animal Aggregations* xix. 344 The 'peck-order' decides which birds may peck others without being pecked in return. **1939** G. K. Noble in *Auk* LVI. 264 A peck order..does not appear unless the birds [*sc.* night herons] are crowded together in a strange area. **1955** *Brit. Jrnl. Animal Behaviour* III. 94/2 It is now recognized that the peck-order forms the basis of all group behaviour in adult chickens. **1966** *New Scientist* 26 May 536/1 Wolves live in groups, in which certain individuals are dominant: that is, they have prior access to food, females and other amenities. This sort of arrangement may be called a peck order.

2. *transf.* = prec., sense 2.

1953 A. Upfield *Murder must Wait* xvi. 138 Amid the lower Australian peck order..wines are imbibed from the bottle. **1962** A. Sampson *Anat. Brit.* I. x. 150 The Inns [of Court] have their own elaborate snobberies and peck-order. **1965** *Punch* 17 Mar. 389/1 Dons have always had an instinctive feeling for prestige, not to say 'peck-order'. **1971** W. J. Burley *Guilt Edged* ix. 149 The human peck-order is far more subtle than that of the hen-house.

Peck's bad boy, *phr. U.S. slang.* The name of a fictional character created by George Wilbur Peck (1840–1916) used allusively of an unruly or mischievous child. Also *attrib.* or as *adj.*

1883 *Peck's Sun* 14 July 1/6 The [cuff] buttons are.. gold..and on the back they are engraved 'Geo. W. Peck, Milwaukee, from his newsboy friends of Chicago', and on top of the Russia leather case are the words, 'From Peck's Bad Boys'. **1933** E. O'Neill *Ah, Wilderness!* I. 19 Sid Davis, his brother-in-law, is forty-five, short and fat, bald-headed, with the Puckish face of a Peck's Bad Boy who has never grown up. **1946** *True* Apr. 34/1 Insulting the advertisers has become Big Business with the Peck's Bad Boy of Radio. **1967** *Atlantic Monthly* Feb. 4 [Governor George] Wallace's motives—ego, a Peck's-bad-boy desire to make trouble, a yen to see just what would happen if a presidential election were thrown into the House of Representatives, or a combination of all these—do not actually matter. **1970** *Time* 22 June 78/2 The book is an earnest effort by Della Femina to buttress his reputation as the Peck's Bad Boy of Advertising. **1977** *Redbook* Mar. 226/3 'I'm going to be forty years old.' He laughs with disbelief. 'It blows my mind.' Yet he sounds curiously unchanged, still the exuberant Peck's Bad Boy of whatever revolution comes to hand, still taunting the establishment. **1977** *Time* 19 Sept. 56/1 Now, in an attempt to break the longest winning streak in modern sports history, a new challenger from Down Under named *Australia* is squaring off with the 1974 U.S. defender, *Courageous*, skippered by Turner—the Peck's Bad Boy of yachting—in the waters of Newport, R.I.

Pecksniff. Add: Also as *v. intr.* Hence **Pecksniffery** (later example); **Pecksniffian** *a.* (earlier and further examples); **Pecksni·ffianly, Pecksni·ffingly** *advbs.*, in a manner resembling Pecksniff. Also *Comb.*, as *pecksniff-shark* (nonce use).

1851 J. Chapman *Diary* 1 July in G. S. Haight *Geo. Eliot & J. Chapman* (1940) 188 M[r] B. put on such an exquisitely saintlike Pecksniffian aspect as to seem nearly in heaven. **1894** Lloyd George *Family Lett.* (1973) 73 That pecksniffian young gent will have to climb down a peg or two. **1903** G. B. Shaw *Let.* 26 Dec. (1972) II. 386 And that you are to come Pecksniffing at me in this fashion. **1913** Wyndham Lewis *Lett.* (1963) 50 A new form of fish in the troubled waters of Art..the Pecksniff-shark, a timid but voracious journalistic monster. **1914** Kipling *Lett. of Travel* (1920) 232 A government which is not a government but the disconnected satrapy of a half-dead empire, controlled pecksniffingly by a Power which is not a Power but an Agency. **1915** G. B. Shaw *Pen Portraits* (1931) 116 To lie—lie impudently, snobbishly, spitefully, Pecksniffianly, Tartuffily. **1926** A. Huxley *Let.* 25 Dec. (1969) 278, I would have made him survive into the Pecksniffian epoch. **1945** R. Hargreaves *Enemy at Gate* 212 The Tsar himself, temporarily assuming the mantle of Mr. Gladstone's best Pecksniffian manner, continued to give whining expression to pious hopes for some sort of accommodation. **1947** K. S. Sorabji *Mi contra Fa* x. 87 He is apparently unaware..that he is..submitting himself to the judgment of his own upper classes... This is surely non-conformist Pecksniffery gone mad. **1959** J. Braine *Vodi* ix. 131 At least it was recognisably the work of a human being, even if he were a greedy Pecksniffian old get who used to sneak around the factory in carpet slippers. **1963** *Times* 22 Apr. 11/1 Disgust at the Pecksniffian pose that the ruling party has shown in denying the blatantly obvious. **1971** M. S. Howard *Jonathan Cape* vii. 111 Edward Garnett poured scorn on the Pecksniffery of these British officials.

Peclet (pe·kle). Also **Péclet.** [Name of J. C. E. *Peclet* (1793–1857), French physicist; adopted in this context by H. Gröber *Die Grundgesetze der Wärmeleitung und des Wärmeüberganges* (1921) ii. 168.] *Peclet number*: a dimensionless parameter used in calculations of heat transfer between a moving fluid and a solid body, equivalent to the product of the Reynolds number and the Prandtl number, viz. Dvc_p/k, where D is a characteristic length of the body, v is the speed of the fluid past it, c_p is the heat capacity of unit volume of the fluid at constant pressure, and k is its thermal conductivity.

1933 W. H. McAdams *Heat Transmission* vii. 173 Equations of this form, expressing hD/k in terms of the

Peclet number DGc_p/k are used by many German writers such as Nusselt and Grober. **1956** GLASSTONE *Princ. Nucl. Reactor Engin.* xi. 679 For the correlation of heat-transfer data it is now appropriate to use another dimensionless modulus, namely, the Peclet number, Pe. **1958** [see *NUSSELT]. **1974** F. M. WHITE *Viscous Fluid Flow* ii. 89 In low-speed flows (V_e less than 30 percent of the speed of sound), we can usually neglect both the pressure term and the dissipation term, leaving only the Peclet number (*RePr*) as the single important parameter.

pecon: see *PECAN.

pecoraite (pĭkǭ·rā,əit). *Min.* [f. the name of William T. *Pecora* (b. 1913), U.S. geologist + -ITE[1].] A nickel silicate, $Ni_3Si_2O_5(OH)_4$, found as green monoclinic crystals in the Wolf Creek meteorite in Australia.

1969 G. T. FAUST et al. in *Science* 4 July 59/3 A green mineral of the serpentine group, rich in nickel, was described..from the Wolf Creek meteorite in the desert of Western Australia. This mineral is the nickel analog of the well-known magnesium silicate, clinochrysotile. The new mineral is named pecoraite after Dr. William T. Pecora, Director of the U.S. Geological Survey, in recognition of his contributions..to the mineralogy and geology of nickel silicate deposits in North America and South America. **1973** *Mineral. Mag.* XXXIX. 113 Pecoraite.. was found as fracture fillings in the Wolf Creek meteorite of Western Australia.

‖**pecorino** (pekŏrī·no). [It., f. *pecora* sheep.] An Italian cheese made from ewes' milk.

1931 C. L. T. BEECHING *Law's Grocer's Man.* (ed. 3) 242/1 '*Pecorino Romano*' type, '*Moliterno*' type... They must be kept in a dry and fresh room, turned every week and oiled with linseed oil when they begin to get mouldy. **1954** E. DAVID *Italian Food* 301, I always buy *pecorino* when I see it in London, for its taste evokes memories of good country food..and coarse red wine. **1958** J. LODWICK *Bid Soldiers Shoot* vi. 166, I was given a slice of.. Italian *pecorino* cheese, some maps, and five gold sovereigns, two of them Victorias. **1967** T. A. LAYTON *Wine & Food Soc. Guide Cheese & Cheese Cookery* 84 When cured for only 5–8 months Pecorino is eaten as a table cheese. **1970** I. ORIGO *Images & Shadows* ix. 205 The delicious sheep's-milk cheese, *pecorino*, which is a speciality of this region. **1973** *Country Life* 8 Mar. 570/3 Women used to come..to milk the ewes, and to make..the Roquefort..or the *pecorino*.

pectase. Add: (Further examples.) Now usu. called *PECTINESTERASE.

1945 [see *PECTINESTERASE]. **1972** [see *PECTINASE].

pectic, *a.* Add: Also, of or pertaining to pectin. (Further examples.)

1930 [see *GALACTURONIC *a.*]. **1964** D. D. DAVIES et al. *Plant Biochem.* iii. 147 Pectic acid is the simplest pectic substance and is the basis of the others. Pure pectic acid appears to be an unbranched chain of α-1,4 linked D-galacturonic acid units which are present in the C-1 chair form..of the pyranose ring. Any compound of this structure which is large enough to possess colloidal properties is classed as a pectic acid. Most pectic acids appear to contain about 100 units with a minimum of approximately 5 units. **1972** [see *PECTINASE].

b. Applied to a class of substances that includes the pectins and other colloidal polymers of galacturonic acid.

1889 *Chem. News* 1 Nov. 221/2 The author [*sc.* L. Mangin] shows the presence of pectic compounds in vegetable membranes. **1913** HAAS & HILL *Introd. Chem. Plant Products* II. 126 Comparatively little is known about the chemistry of pectin or the pectic bodies, as there appear to be several of these substances. **1957** E. V. MILLER *Chem. Plants* i. 14 Pectin is perhaps the pectic compound with which the layman is most familiar. **1966** R. M. DEVLIN *Plant Physiol.* vii. 137 During the ripening of the fruit, protopectin is converted into the more soluble pectic substances—pectin and pectic acid. **1973** J. A. GOSS *Physiol. Plants* xvi. 344 In contrast to the other pectic substances, pectin is water-soluble, and is located in the protoplasm of the cell.

pectin. Add: Now recognized as having a variable composition but being principally a high-molecular-weight polymer of partially methylated galacturonic acid in which galactose and arabinose residues are also present. (Further examples.)

1913, etc. [see *PECTIC *a.* b]. **1917, 1930** [see *GALACTURONIC *a.*]. **1964** B. NILSON *Pears Family Cookbk.* 223/2 Over-ripe fruit will not set so well because it will contain less pectin. **1972** *Materials & Technol.* V. xix. 703 Pectins, which occur only in certain plant tissues, are polymers of galacturonic acid linked by α-1,4-glycosidic units, with about two thirds of the carboxylic acid groups esterified with methanol. **1975** D. GREEN *Food & Drink from your Garden* i. vii. 65 Flowers yield very little sugar, but their musts will release no pectins (the glue-like substances holding plant tissues together) and so will clear easily.

b. Any of the pectic substances.

1896 *Jrnl. Chem. Soc.* LXX. I. 7 Pectin Substances... Many attempts have been made to establish a relationship between pectins and vegetable gums (carbohydrates). **1929** R. A. GORTNER *Outl. Biochem.* xxvii. 584 The pectin producing these jellies is the only water-soluble member of a group of related compounds known as the 'pectic substances', or sometimes called the 'pectins'. **1931** E. C. MILLER *Plant Physiol.* i. 12 It is now fairly well established that these substances are not chemically combined with cellulose, although they are closely associated with it, so that they are classified under the general name of the 'pectic compounds' or 'pectins'. **1966** NOWAKOWSKI & CLARKE tr. *Kretovich's Princ. Plant Biochem.* ii. 86 The ripening of fruits is characterized by the conversion of insoluble into soluble pectin.

† **2.** *attrib.* in the sense of *PECTIC *a.* b. *Obs.*

1877 *Jrnl. Chem. Soc.* XXXII. 502 The formula of the group of pectin bodies. **1895** F. E. WEISS tr. *Sorauer's Pop. Treat. Physiol. Plants* vi. 123 Together with gums and acids we often find substances consisting also of carbon, hydrogen, and oxygen, which form gelatinous masses on boiling, and are termed pectin substances. **1900** A. J. EWART tr. *Pfeffer's Physiol. Plants* I. viii. 477 Many gums..seem to be in part allied to pectin substances.

pectinase (pe·ktinēiz, -s). *Biochem.* [a. F. *pectinase* (E. Bourquelot 1899, in *Jrnl. de Pharm. et de Chimie* IX. 567): see PECTIN and *-ASE.] An enzyme found in plants and in certain bacteria and fungi which hydrolyses pectin to its constituent monosaccharides.

1899 *Jrnl. Chem. Soc.* LXXVI. I. 652 In all probability it is a new ferment and the name pectinase is suggested for it. **1929** R. A. GORTNER *Outl. Biochem.* xxvii. 593 Pectinase hydrolyzes pectin (and possibly pectic acid) to its simple components, sugars and galacturonic acid. **1953** F. T. BROOKS *Plant Dis.* (ed. 2) x. 156 Brown..has indicated that the enzyme (pectinase) or enzyme complex of *B. cinerea* which softens the cell walls probably also kills the host protoplasm. **1972** *Materials & Technol.* V. 703 They [*sc.* the pectic enzymes] can be broadly classified,.. into two sub-groups. These are the polygalacturonases, which cleave glycosidic linkages between adjoining galacturonic acid units, and the pectic methyl esterases... The two groups were formerly known, respectively, as pectinases and pectases.

pectinesterase (pektine·stərēiz, -s). *Biochem.* Also **pectin esterase.** [f. PECTIN + *ESTERASE.] An enzyme found in plants and in certain bacteria and fungi, which hydrolyses pectin to pectic acid and methanol; = PECTASE, *PECTINMETHYLESTERASE.

1945 LINEWEAVER & BALLOU in *Arch. Biochem.* VI. 373 The enzyme that acts on pectin and causes gelation in the presence of calcium has been known as pectase... Since the enzyme hydrolyzes the ester bonds in pectin we propose the name pectinesterase, which indicates the esterase character of the enzyme. **1956** *New Biol.* XX. 96 It was concluded that the enzyme pectinesterase was responsible, at least in part, for the development of disease symptoms [in the tomato shoots]. **1966** *McGraw-Hill Encycl. Sci. & Technol.* IX. 609/1 With the enzyme, pectin esterase, obtained from such sources as roots, leaves, and fruits of all higher plants and also from a number of microorganisms, the ester groups are quickly removed. **1975** *Biochim. & Biophys. Acta* CCCLXXVII. 408 The ability of *B. cinerea* polygalacturonase to degrade pectin N.F. was connected with the presence of pectinesterase in the preparation.

pe:ctinme:thyle·sterase. *Biochem.* Also **pectin methylesterase,** and with hyphen. [f. PECTIN + METHYL + *ESTERASE.] = *PECTINESTERASE.

1945 MCCOLLOCH & KERTESZ in *Jrnl. Biol. Chem.* CLX. 149 Two enzymes which act on soluble pectic substances are recognized at the present time; namely, pectin-polygalacturonase (PG),..and pectin-methylesterase (PM), which catalyses the hydrolysis of the methyl ester groups. **1954** *Biol. Abstr.* XXVIII. 27349 The purpose of the study was to determine the extent to which several plant pathogens produced pectinmethylesterase..and polygalacturonase. **1966** [see *PECTINESTERASE]. **1975** *Phytochemistry* XIV. 109/2 No attempt was made to determine if pectin was being degraded..by the two step reaction catalyzed by pectinmethylesterase and endo-polygalacturonase.

pectoral, *sb.* **1. c.** For †*Obs.* read '*Obs.* exc. *Hist.*' and add: also, a piece of armour to protect the breast of a horse: cf. sense 1 b.

1656 [see PEITREL, PEYTREL, PETREL *sb.*]. **1786** F. GROSE *Treat. Anc. Armour & Weapons* 30 The Poitrinal, Pectoral, or Breast Plate was formed of plates of metal rivetted together. **1821** in G. F. Laking *Catal. European Armour & Arms in Wallace Collection* (1900) 315 Armour of the Elector Joseph of Bavaria on horseback. This superb suit of black and gold, with the pectoral, chanfron, and other trappings for the horse, of the finest workmanship of the time of Henry VIII. **1824** S. R. MEYRICK *Crit. Inquiry Antient Armour* III. Gloss. s.v. Pectorale, Sometimes the ends of the pectoral were raised so high as to protect the abdomen of the knight.

peculant (pe·kiŭlănt), *a. rare.* [ad. L. *peculant-em*, pres. pple. of *peculāri* to embezzle.] = next.

1920 *Glasgow Herald* 16 Aug. 8/3 Conveying large sums of money into their own pockets without having to resort to the clumsy methods practised by peculant contractors ..in the Napoleonic wars.

peculative (pe·kiŭlătiv), *a.* [f. PECULATE *v.* + -IVE.] That practises embezzlement or peculation.

1909 GALSWORTHY *Fraternity* xii. 102 Unlike other clubs, it..had special arrangements for the safety of umbrellas and such books as had not yet vanished from the library; not, of course, owing to any peculative tendency among its members, but because, after interchanging their ideas, those members would depart,..each grasping some material object in his hand. **1921** *Times Lit. Suppl.* 10 Feb. 84/3 The taxes so rapaciously collected by a host of peculative Turkish officials.

peculiar, *a.* and *sb.* Add: **A.** *adj.* **1. c.** *Astr.* Applied to the motion or velocity of a celestial object relative to a group of objects of the same kind; *spec.* that of a star in the frame of reference in which the average velocity of the stars in the neighbourhood of the sun is zero.

[**1890** A. M. CLERKE *Syst. Stars* xxii. 340 The motus peculiaris itself is only a projection upon the sphere of a line of travel which may make any angle with the line of sight.] **1927** H. N. RUSSELL et al. *Astron.* II. xix. 668 The peculiar motions of the fainter stars are more rapid than those of the brighter. **1936** E. HUBBLE *Realm of Nebulæ* v. 106 It was expected that, when the solar motion was removed, the residual, peculiar motions of the nebulæ would be much smaller than the observed velocities and.. that they would be distributed at random. **1967** R. KURTH *Introd. Stellar Statistics* ii. 14 The parallactic and peculiar motions of the stars close to the Sun..can be estimated if the hypothesis..is adopted that the average of the peculiar motions vanishes. **1975** G. O. ABELL *Exploration of Universe* (ed. 3) xx. 396/1 The space velocity of a star, its motion with respect to the sun, is made up of both the star's peculiar velocity and a component due to solar motion. **1978** PASACHOFF & KUTNER *University Astron.* iii. 62 Most peculiar velocities are a few tens of km/sec; very few are above 100 km/sec. *Ibid.* 64 If we observe a large number of stars, their peculiar motions, since they are random, will average to about zero.

4. b. *Astr.* Of a galaxy: not belonging to any of the types, elliptical, spiral, and irregular, which include almost all galaxies.

1936 E. HUBBLE *Realm of Nebulæ* ii. 47 The remaining irregulars might be arbitrarily placed in the regular sequence as highly peculiar objects. **1959** *Listener* 31 Dec. 1152/1 There are a few galaxies that do not fit conveniently into this classification of spirals, ellipticals, and irregulars. These are the 'peculiar' galaxies. **1972** JASTROW & THOMPSON *Astron.* 213 In all the cases that have been examined thus far, the unusual event that altered the appearance of the peculiar galaxy seems to have been either a *collision* with another galaxy, or a gigantic *explosion* within the galaxy. **1973** *Sci. Amer.* Dec. 39/1 One or 2 percent, however, do not conform. Because of their bizarre appearance or unusual spectra they are known to astronomers as 'peculiar' galaxies.

ped[2]. Add: (Further examples.) Now chiefly *U.S.*

1897 *National Police Gaz.* (U.S.) 26 May 10/4 The consensus of opinion is that the Irish-Scots ped. came to the mark in the pink of condition. **1962** *N.Y. Times* 26 Jan. 1/3 'Peds' is short for pedestrians in traffic engineers' jargon. **1973** D. BARNES *See the Woman* I. 134 A ped about three-quarters of a block away.

ped[3] (ped). *Soil Science.* [f. Gr. πέδ-ον ground, earth.] (See quot. 1958.)

1951 *Soil Survey Manual* (U.S. Dept. Agric. Handbk. No. 18) 225 An individual natural soil aggregate is called a ped, in contrast to (1) a clod, caused by disturbance.., (2) a fragment caused by rupture.., or (3) a concretion caused by local concentrations of compounds that irreversibly cement the soil grains together. **1958** *U.S. Dept. Agric. Yearbk.* 1957 764/1 *Ped*, an individual natural soil aggregate such as a crumb, prism, or block, in contrast to a clod, which is a mass of soil brought about by digging or other disturbance. **1971** R. L. DONAHUE et al. *Soils* (ed. 3) ii. 41 There are four principal types of soil structure: 1. Platy. Peds exhibit a matted, flattened, or compressed appearance... 4. Spheroidal. Peds are imperfect spheres like marbles, but are usually smaller. **1972** J. G. CRUICKSHANK *Soil Geogr.* iii. 79 Clayey soils also tend to develop large columnar peds which on shrinking produce non-capillary pores.

ped-, form of *PEDO- used before a vowel (as in *PEDALFER).

pedagoguery, (*b*). (Earlier example.)

1872 F. HALL *Recent Exempl. False Philol.* 31 It is not because of any poverty of matter for remark in the heading sciolism of the one and in the piddling pedagoguery of the other.

pedagoguette (pedăgǫge·t). *nonce-wd.* [f. PEDAGOGUE + -ETTE.] A school-mistress.

1960 V. NABOKOV *Invitation to Beheading* viii. 87 The strident commands of the red-haired 'pedagoguette'.

pedal, *sb.* Add: **1. b.** *soft pedal*: see also as main entry in Suppl.

2. b. *spec.* Such a lever forming one of the controls in a motor vehicle; often *attrib.*

1902 W. W. BEAUMONT in A. C. Harmsworth et al. *Motors* x. 219 The friction of the band on the drum..pulls on the band in the same direction as the pedal. *Ibid.* 220 A good form of brake is that..in which the pull on the rod F from pedal E pulls the arm D. **1909** *Westm. Gaz.* 4 Feb. 4/2 It seems to me that to dismantle the universal joint and pedal-gear means taking them out. **1926** *Scribner's Mag.* Aug. 153/1 The girl was satiated with speed. She relaxed the pressure of her foot on the pedal, and leaned back in her seat. 'Some car!' she said. **1929** *Times* 5 Nov. 7/5 The clutch pedal-shaft bearings have grease valves. **1962** *Which? Car Suppl.* Oct. 128/2 Most drivers liked the pedal controls in the VW 1500. *Ibid.* 129/1 The pedal control layouts..were generally liked. **1974** *Country Life* 21 Mar. 659/2 Pedals are nicely positioned for heel and toe driving. **1976** 'E. MCBAIN' *Guns* (1977) vi. 139 Great day for a robbery... Your driver hit the gas pedal and off you went.

4. Also *transf.*

1892 G. B. SHAW *Let.* 21 Apr. (1965) I. 337 Her voice has become much more powerful—quite Hyde Parkian in its pedal notes. **1905** *Daily Chron.* 25 Sept. 6/4 He did all that the pedal-notes of his magnificent voice could do toward realising..a character not wholly suited to his temperament.

b. On some brass and wind instruments, = FUNDAMENTAL *sb.* 2; usu. *attrib.*, as *pedal note, tone.*

1856 M. C. CLARKE tr. *Berlioz's Treat. Mod. Instrumentation* 153/2 All trombones..possess..three notes; which are..called pedals... Supposing that the bass trombone possesses the first only of these pedal notes..it would still be of great value for certain effects. *Ibid.* 154/1 The vibrations of the pedal notes are slow, and require much wind. **1938** *Oxf. Compan. Mus.* 953/2 The actual fundamental notes are not so easy to produce [on a trombone] as the harmonics above them; they are spoken of as *Pedal Notes*—possibly because they are considered to be useful for the holding of a 'pedal', in the sense in which the word is used in the terminology of harmony.., or else, as Berlioz suggests, from their resemblance to the low pedal notes of an organ. **1944** W. APEL *Harvard Dict. Mus.* 340/1 Owing to the narrow bore..the lowest tone of this series (pedal tone) is practically unobtainable. *Ibid.* 817/1 Another term for the fundamental tone is pedal tone... A distinction is made between whole-tube instruments.. producing the pedal tone, and half-tube instruments. **1954** *Grove's Dict. Mus.* (ed. 5) VI. 608/1 These prime notes, the lowest proper tones of the instrument,..are known as pedal notes... A pedal note always stands for the first note or No. 1 in the harmonic series. On trumpets and some other instruments the pedal notes are practically impossible. **1959** *Collins Mus. Encycl.* 491/2 *Pedal*,..the fundamental (or first note of the harmonic series) on a brass instrument. A few of these notes can be produced with a slack lip on the trombone, the tuba and the B♭ section of the double horn. **1961** C. W. MONK in A. Baines *Mus. Instruments* xi. 283 This gives *E* as the bottom note of the trombone's continuous scale, leaving a gap down to the less-used fundamental or 'pedal' notes from the first-position *B'*♭ downwards.

6. (Earlier example.)

1849 H. MELVILLE *Mardi* II. xliv. 204 To cool his heated pedals, he established..stopping-places.

7. *pedal-bike, cycle, cyclist, work; pedal-operated* adj.; **pedal bin,** a rubbish bin with a lid which is opened by means of a pedal; **pedal boat,** a boat, usu. a pleasure boat, propelled by means of pedals; also **pedal-driven boat; pedal car,** a car, usu. a child's toy, propelled by means of pedals; **pedal clarinet** (or **clarionet**), a clarinet sounding an octave below the bass clarinet; **pedal-craft,** pedal boats; **pedal entry,** in organ music, a point where a theme or figure is introduced on the pedal stops; also *transf.* and *fig.*; **pedal point** (earlier and later examples); also fig.; **pedal pusher,** a cyclist; hence in *pl.* (orig. *U.S.*), a type of girls' or women's trousers, reaching only just below the knee, suitable for wearing when cycling; also (with hyphen) *attrib.*; **pedal steel guitar,** an electric guitar fixed on a stand and connected to pedals by which the tension of the strings can be altered to produce glissando effects; also **pedal guitar, pedal steel; pedal wireless,** a small radio transceiver, with a generator powered by a foot-pedal, providing a means of communication in the Australian out-back; also **pedal radio, pedal set.**

1974 'A. GILBERT' *Nice Little Killing* vi. 83 I've been travelling around on a pedal-bike. **1951** *Catal. of Exhibits, South Bank Exhib., Festival of Britain* 52/1 'Binette' pedal bin. *Ibid.* 124/1 Pedal bin for soiled napkins. **1958** *Engineering* 7 Mar. 320/2 Polythene is used widely: for pedal bins, trug baskets,..and watering cans. **1966** N. FREELING *Dresden Green* I. 14 He..picked up his pedal bin (emptied that and every morning into one of the big dustbins). **1977** J. HEALD *Just Desserts* i. 14 An empty magnum of Château Waitrose by the pedal bin in the kitchen. **1951** *Go* Apr.–May There are little safe pedal-boats for venturing on to the lake. **1958** *New Yorker* 13 Sept. 130/3 (Advt.), The Yacht club with pedal boats, water skiing, skin diving. **1959** *Encounter* Oct. 17/2 The animated statue who looks after the pedal-boats. **1951** *Catal. of Exhibits, South Bank Exhib., Festival of Britain* 126/2 Pedalcar, 'Austin 40'. **1968** *Radio Times* 28 Nov. 6 A pedal car, a doll's pram, a trike or a bike. **1973** *Guardian* 11 June 6 The event for Formula One pedal cars was part of the RAC L-Driver of the Year finals. **1892** *Orchestral Times* Jan. 6/2 Messrs. F. Besson & Co.'s inventive faculty seems inexhaustible... That eminent firm has recently given us..the 'Pedal' Clarionet (contra-bass clarionet). *Ibid.* 7/2 The 'Pedal' Clarionet is the deepest-voiced instrument ever constructed for orchestral use... The fingering of the pedal clarionet is similar to the ordinary clarionet. **1911** *Encycl. Brit.* XXI. 36/2 *Pedal clarinet*, a contrabass instrument invented in 1891 by M. F. Besson to complete the quartet of clarinets..; it is constructed on practically the same principles as the clarinet, and consists of a tube 10 ft. long..doubled up twice upon itself. **1959** *Collins Mus. Encycl.* 491/2 *Pedal Clarinet*, another name for the Contrabass Clarinet. **1961** J. A. MACGILLIVRAY in A. Baines *Mus. Instruments* x. 260 Of those [*sc.* contrabass clarinets] in B♭, an octave below the bass clarinet, Fontaine-Besson's 'Pedal clarinet' was the first to arouse much interest. **1966** *Listener* 27 Oct. 632/2 He includes such rare instruments as..pedal-clarinet. **1957** G. BELLAIRS *Death in High Provence* ix. 105 Families.. sporting on the beaches, little pedal-craft skimming across the water. **1937** R. F. BROAD *Motor Driving made Easy* (ed. 6) vii. 94 A moment ago we used the word 'cyclist'. By this term is generally understood the pedal-cycle user. **1963** *Times* 5 Mar. 6/3 Motor cycle casualties fell by 7,632..and pedal-cycle casualties by 3,974. **1973** J. WAINWRIGHT *Pride of Pigs* 69 Sykes arrived in a Cortina. The postman arrived *with* a pedal cycle—pushing it. **1974** *Times* 8 Feb. 15 The local authority provides a pedal cycle speedway track. **1931** *Highway Code* 12 Certain of the rules for drivers of motor vehicles..also apply ..to pedal cyclists. **1974** L. LAMB *Man in Mist* vii. 43 The early evening shoal of pedal cyclists. **1927** *Sunday Express* 14 Aug. 1 Mr. E. P. Tierney..collapsed yesterday when attempting to carry out a test in his 12-foot pedal-driven boat Carrie, with which he had planned to cross the Atlantic. **1928** AUDEN in *Oxf. Poetry* 2 We sang Our descant until love one day, That pedal-entry in the fugue Roared in, swept soul and knees away. **1932** —— *Orators* I. 30 The bowel tremors at the pedal-entry. **1970** A. BELLOW *Illustr. Hist. Guitar* vii. 154 An even more unusual guitar, known as a 'pedal guitar', was constructed by Eduard Bayer. **1908** *Westm. Gaz.* 20 Oct. 4/3 A similar powered car..with patent pedal-operated plate clutch. *Ibid.* 10 Dec. 4/2 (*heading*) Pedal-operated car. **1936** *Discovery* July 224/2 A pedal-operated volume-control [on an electronic piano]. **1960** *Farmer & Stockbreeder* 5 Jan. 95/2 This pedal-operated device counteracts wheel-spin. **1964** W. L. GOODMAN *Hist. Woodworking Tools* 153 Holding his work in a vertical, pedal-operated vice or 'horse'. **1852** J. HULLAH *Gram. Musical Harmony* xxvii. 71 From a very obvious and effective mode of using these (foot) keys has arisen the term *pedal point*, by which is understood a note maintained during several successive changes of chords, or passages of melody. **1974** *Publishers Weekly* 16 Dec. 49/1 These tales are generally of high standard... But the insistent pedal point theme, the overwhelming atmosphere of misery.., makes this one to be taken in small doses. **1977** *New Yorker* 16 May 139/1 Hindemith's Requiem opens with a slow instrumental prelude, a four-note tolling ostinato over a pedal point. **1934** M. H. WESEEN *Dict. Amer. Slang* xvii. 262 *Pedal pusher*,—a bicycle rider, especially in a race. **1942** [see *PEDALLER]. **1944** *Life* 28 Aug. 65/2 When college girls took to riding bicycles in slacks, they first rolled up one trouser leg, then rolled up both. This..has now produced a trim variety of long shorts, called 'pedal pushers'. **1945** *Liberty* 1 Sept. 68 Miss McCardell..borrowed little-boy pants for 'pedal-pushers', the knee length shorts. **1957** H. ROOSENBURG *Walls came tumbling Down* vii. 189 His trousers had shrunk to the size of pedal-pushers. **1959** M. STEEN *Tower* I. vii. 92 A girl in..blue pedal-pusher jeans. **1960** S. PLEYDELL *Festival for Gilbert* i. 10 Black pedal-pusher trousers. **1971** Pedal-pusher [see *HAPPI-COAT]. **1959** *Manch. Guardian* 5 Aug. 5/4 All the aeroplane, the pedal-radio..and, latterly, the road-train have done to break down the isolation of life in inland central Australia. **1944** *Living off Land* iv. 77 Cars, trucks, aircraft, and radio pedal-sets, have put an end to the old era of isolation. **1949** H. M. MADELEY *Australia* xxxiv. 138 These pedal sets are in mining camps, in lonely houses, in police stations, in nursing homes. **1976** *Gramophone* Aug. 351/3 Paul Cotton..and Rusty Young (pedal steel, dobro, mandolin etc.) excel instrumentally. **1977** *Zigzag* Mar. 30/1 This coupled with Mike Utley's delicate piano and Al Perkins' unobtrusive pedal steel makes the track one of the standout cuts on the first side. **1969** *Listener* 20 Mar. 398/3 It features another splendid electronic invention, the pedal steel guitar which..specialises in vertiginous slides and swoops. **1969** *Rolling Stone* 28 June 17/1 Today's added use of drums, pedal steel guitar,..even harpsichord have [*sic*] made some arrangements far more complete. **1940** BAYNE & LAZARUS *Austral. Community* II. 76 Messages are sent out and received by means of the pedal wireless or transceiver (invented by the Australian Inland Mission and now much improved). **1944** F. CLUNE *Red Heart* 7 The combination of pedal-wireless sets, to call the doctor, and of aeroplanes, to bring him to the patient, is a triumph of modern times. **1963** A. LUBBOCK *Austral. Roundabout* 25 The hundreds of transceiver or pedal-wireless sets. **1944** R. LEHMANN *Ballad & Source* 96, I was already practising my technique for Bicyclists' Dashing Hill—a piece of frantic momentum-gathering pedal work.

pedal, *a.* Add: **1. b.** *pedal bone*, the lowest phalangeal bone in a horse's foot; = *coffin-bone* (COFFIN *sb.* 13).

1881 *Encycl. Brit.* XII. 178/1 A powerful tendon..passes down over the..phalanges, to be inserted mainly into the upper edge of the anterior surface of the last phalanx or pedal bone. **1920** F. T. BARTON *Horse* xiii. 107 Two tendons pass down the back of the foot,..the former being attached to the lower surface of the pedal-bone. **1973** EDWARDS & GEDDES *Compl. Bk. Horse* III. iii. 113/1 The end bone of the leg is the pedal bone..which corresponds to the last bone in the second finger of man. *Ibid.*, The pedal bone is roughly the shape of the foot and is of pumice-like consistency.

pedal (pe·dăl), *a.*[2] [ad. It. *pedale*: see PEDAL *sb.*] Applied to the lower and thicker part of a kind of straw grown in Italy for plaiting; *ellipt.*, a special plait made from this straw, usually having five or seven strands. (Hence *five-* or *seven-ends pedal.*)

1887 *Encycl. Brit.* XXII. 593/2 The straw of Tuscany, specially grown for plaiting, is distinguished into three qualities,—..from the third quality, *Santa Fioro*, only 'Tuscan pedals' and braids are plaited. **1907** *Yesterday's Shopping* (1969) 894 Fine pedal 'Homburg'... Fine pedal straw. **1923** *Daily Mail* 7 Feb. 1 (Advt.), Italian pedal straw hats... These hats are very soft and pliable, made of seven ends pedal. **1930** *Times* 13 Mar. 11/6 Fine pedal straw has returned, and is seen in a brimmed hat.., with the new sweeping line. **1962** A. SOUTHERN *Millinery* ii. 30 The fancy band straws and pedal come into this [*sc.* real Italian straws] group, the latter being dyed into fashion shades.

pedal, *v.* Add: **a.** Also of a pianoforte or similar instrument. Also *trans.*, to use the pedals in playing (a passage of music, etc.); also, to use the pedals of.

1889 *Cent. Dict.* s.v., *Pedal..v.i.*,..to work a pedal; use the pedals, as of a piano, organ, bicycle, etc. **1909**, etc. [see *PEDALLING *vbl. sb.*]. **1922** S. GREW *Art of Player-Piano* iv. 23 We correct the first condition..by ceasing to pedal for a moment or two. *Ibid.* xviii. 112 An attempt..to 'pedal' the piece in the way it is 'fingered' for the pianist. **1922** JOYCE *Ulysses* 259 Upholding the lid he..gazed..at the oblique triple (piano!) wires. He pressed.., soft pedalling a triple of keys. **1938** F. C. RAUSER tr. *Leimer & Gieseking's Rhythmics* vii. 48 All of Beethoven's Sonatas can be properly pedalled by means of the time-tread. **1954** T. A. JOHNSON *Princ. Pianoforte Pedalling* 10 It is sometimes permissible to pedal certain staccato passages. **1961** C. CLUTTON in A. Baines *Mus. Instruments* v. 93 John Field..was the first person to develop the use of the sustaining pedal as part of his technique, and, like Chopin, he pedalled after the note. **1975** H. FERGUSON *Keyboard Interpretation* ix. 66 The passage must be pedaled thus. **1978** *Gramophone* July 231/1 The ending of 'Ende vom Lied'..is deeply impressive—I only wish he had not pedalled.

b. Also in *refl.*, *pass.*, and *fig.* uses.

1896 *Queen* 25 Jan. 169/2 If young ladies are to be allowed to pedal themselves about in..London, then it will certainly be necessary to provide..some proper escort. **1924** GALSWORTHY *White Monkey* II. xi. 208 'Well,' said Michael, '*I* think we shall pedal through yet.' **1955** G. GREENE *Quiet American* IV. ii. 243, I found a trishaw and was pedalled home. **1973** D. MAY *Laughter in Djakarta* viii. 132 A betjak came along the street, pedalled by a very young smooth-faced boy. **1977** *Daily Express* 29 Jan. 35/2 Derek Underwood pedalling slowly backwards before clutching it in his hands to send Gavaskar miserably away to an accompaniment of boos and jeers from the 40,000 crowd.

pedalfer (pėdæ·lfəɹ). *Soil Sci.* [f. *PED- + AL(UMINIUM + L. *fer-rum* iron.] A soil in which there is no layer of accumulated calcium carbonate but in which oxides of iron and aluminium have tended to accumulate (generally acidic and characteristic of humid climates). Cf. *PEDOCAL.

1928 [see *PEDOCAL]. **1930** L. A. WOLFANGER *Major Soil Divisions of U.S.* iv. 100 The agriculture of the pedalfers is characterized by its wealth in crop variety and its extended development of crop centers. **1965** B. T. BUNTING *Geogr. Soil* viii. 95 Accumulation of Fe and Al oxides at depth is characteristic..of pedalfers, especially of podzols. **1972** C. B. HUNT *Geol. Soils* viii. 167 The significant feature of pedalfers is that moisture is sufficient to wet the soil to its capacity..and to allow excess water to.. remove the soluble constituents. **1977** A. HALLAM *Planet Earth* 179 These ideas were accepted in the USA, and the concepts of pedalfers and pedocals added: pedalfers are leached soils in humid areas where aluminium and iron accumulate in the B horizon.

Hence **pedalfe·ric** *a.*

1928 [see *PEDOCALIC *a.*]. **1930** L. A. WOLFANGER *Major Soil Divisions of U.S.* iv. 101 Although several of these crops are also grown in noteworthy quantity in the pedalferic zone, the chief production areas..are included within the pedocalic division.

Pedaline (pe·dăləin, -īn). Also with lower-case initial. [f. *PEDAL *a.*[2] + -INE[4].] A synthetic straw (see quot. 1957). Also *Pedaline straw.*

1927 *Daily Express* 16 Feb. 4/2 (Advt.), Women's becoming hats of Pedaline Straw, trimmed and bound with Ribbon Velvet. *Ibid.* 14 Mar. 5/3 (*heading*) Fashion's new vocabulary... The following are a few additions made to her vocabulary this spring by Dame Fashion:—..*pedaline*, a new type of plait for millinery, having a polished effect. **1942** J. G. DONY *Hist. Straw Hat Industry* v. 96 (*caption*) Pedaline, a Japanese imitation of a Swiss braid... Coburg pedaline, a Japanese braid. **1957** M. B. PICKEN *Fashion Dict.* 337 *Pedaline*,..synthetic straw made of hemp fiber covered with Cellophane and woven between cotton threads. Made chiefly in Japan.

pedaller, pedaler. Add: (Further examples in var. senses.) Also in *pl.*, a name for *pedal pushers* s.v. *PEDAL *sb.* 7.

The spelling *pedaler* is predominantly *N. Amer.*

1922 S. GREW *Art of Player-Piano* 17 Effects in the music which you cannot hope to create until you are an experienced pedaller. **1942** BERREY & VAN DEN BARK *Amer. Thes. Slang* § 426/10 *Bicyclist*,..pedaler, pedal pusher. **1945** *Consumers' Guide* (U.S.) July 6 Another part of the Clothes Magic show illustrated what had been done in a 'make over and make do' program... Pedalers (bicycle shorts) from G.I. pants, and a stunning beach ensemble..from a discarded evening dress were a few of the other things shown. **1953** BERREY & VAN DEN BARK *Amer. Thes. Slang* (1954) (ed. 2) § 87/15 *Pedalers, pedal pushers*, bicycle shorts that end just below the knee. **1972** *Village Voice* (N.Y.) 1 June 26/2 In an era of precious pedalers, back to nature sentimentalists, and ecology moralizers, Jack La Lanne is an honest, straightforward salesman. **1974** *Times* 12 Nov. 14/7 It seems against the interests of the environment to chase away harmless pedallers with a polluting petrol-engined monster. **1976** A. HILL *Summer's End* ii. 25 I'd got a pedaller that would run alright. No brakes, no mudguards. But I could pedal it. **1977** *Times* 14 June 16/7 Readers..keep complaining about the 'tedious' routine of having to apply to the Cyclists Touring Club for tickets to take their bicycles free on trains. More than 10,000 pedallers have done it without making a fuss.

pedalling, *vbl. sb.* Add: (Further examples.) Also *fig.* Also with reference to propelling a sled. Also as *ppl. a.*

1909 J. HOFMANN *Piano Playing* 42 Harmonic clarity in pedalling is the basis, but it is only the basis; it is not all that constitutes an artistic treatment of the pedal. **1911** H. BROWER *Art of Pianist* xiv. 170 Imagine what a performance of the compositions of Chopin or Liszt would sound like, deprived of proper pedaling. **1916** JOYCE *Portrait of Artist* (1969) v. 190 'It may be uphill pedalling at first. Take Mr. Moonan. He was a long time before he got to the top.'.. 'I may not have his talent.' **1925** E. H. YOUNG *William* iv. 38 A ringing of bicycle bells was heard, and then two pedalling figures.. appeared. **1936** Y. BOWEN (*title*) Pedalling the modern pianoforte. **1938** F. C. RAUSER tr. *Leimer & Gieseking's Rhythmics* vii. 57 It is possible to play a *glissando* with uninterrupted pedalling. **1948** A. FOLDES *Keys to Keyboard* (1950) viii. 50 Pedalling is one of the most complicated processes in piano playing. **1954** T. A. JOHNSON *Princ. Pianoforte Pedalling* 9 (*heading*), Pedalling of Arpeggios. **1954** *Grove's Dict. Mus.* (ed. 5) VI. 345/2 We must now consider the subject of pedalling... To look at the pedal-board is often impossible and rarely helpful. **1972** *Evening Telegram* (St. John's, Newfoundland) 24 June 14/3 Pedalling is an art and must be done smoothly, in order not to jerk the sled. **1973** *Guardian* 17 Feb. 15/2 In Britain..pedalling for pleasure is undergoing a sort of mini-renaissance. **1973** K. A. LEIBOVITCH tr. *Neuhaus's Art of Piano Playing* 165 How sad that such a great pianist as Schnabel..showed his lack of understanding in his inappropriate pedalling.

pedalo (pe·dălo). [f. PEDAL *sb.* + *-O[2].] A small pedal-operated pleasure boat used on lakes and at the sea-side.

The spelling in quot. 1962 is unusual.

1959 *News Chron.* 4 Dec. 3/7 One rescue team..manned a four-seat 'Pedalo' found on an untouched part of the beach. **1962** 'O. MILLS' *Headlines make Murder* xviii. 199 A shilling-an-hour peddallo. **1969** *Daily Tel.* 21 Apr. 11/5 So many beaches today are crowded with water-skiers, canoeists, *pedaloes* and sailing boats that accidents are on the increase. **1973** W. FAIRCHILD *Swiss Arrangement* v. 61 The water was glass calm and children in pedalos glided in all directions. **1976** *Daily Tel.* 20 July 15/5 A man armed with a bottle defied an RAF helicopter and two lifeboats which tried to rescue him from a drifting pedalo four miles off Margate last night. **1977** *Navy News* Feb. 18/1 Glancing through a scuttle after reading the papers lately, a member of the Andrew might have expected to see a pedalo forging ahead of his fuel-saving warship.

‖**pedanda** (pedæ·ndă). Also **padanda, padenda,** and with capital initial. [Balinese, = 'bearer of the staff'.] In Bali, a Brahman priest. Also *attrib.*

1817 T. S. RAFFLES *Hist. Java* II. p. ccxxxix, The duties and ceremonies of religion, the conducting of which is in the hands of..learned Brahmins called *Padénda*. **1877** *Jrnl. R. Asiatic Soc.* IX. 113 The *Padandas* are Brahmans who have received a complete education from another *Padanda* (their *Guru*). They must be thoroughly acquainted with religion and with literature. In order to become a Padanda, they undergo all kinds of tests... The mark of the dignity is a staff, *danda*, which they receive from the Guru... After this staff they are called *Padanda*, that is, 'bearing a staff.' **1924** T. DE KLEEN *Mudrās* 23 Although the Balinese in general are a polite and hospitable people, the *Pedandas* showed themselves very reserved, sometimes obviously hostile. *Ibid.* 27 My *Pedanda* friends in Sidemen. **1936** R. GORIS in *Netherlands Indies* 16 Feb. 71/1 There are in Bali both Padandas 'Shiva' and Padandas 'Buda'. *Ibid.* 77/1 Every bench of magistrates in Bali must have at least one Padanda sitting on it as a judge. **1937** M. COVARRUBIAS *Island of Bali* (1972) ix. 293 The *pedandas* still exert a powerful influence on Balinese life despite the fact that their relations with the people were never intimate; they represent the law, and the judges of the high native courts..are still *pedandas* in the majority. **1953** J. BELO *Bali: Temple Festival* 4 The temple priests who are the guardians of these temples have not studied as have the Brahmana high priests, the *pedanda*, who are often scholars and learned men. **1961** P. KEMP *Alms for Oblivion* vii. 116 Priests wear white, and a high priest or *pedanda* goes bareheaded and carries a staff surmounted by a crystal ball.

‖**pedang** (pĕdæ·ŋ). [Malay.] A type of sword (see quots.).

1817 T. S. RAFFLES *Hist. Java* vi. 296 The *pedáng*, *bandól*, [etc.]..are varieties of sword. **1911** *Encycl. Brit.* XVII. 477/1 The Malays use..long swords of ordinary pattern called *pedang*. **1936** G. B. GARDNER *Keris & Other Malay Weapons* iii. 69 Theoretically this is any type of sword of foreign origin... The *pĕdang* may be straight or curved.

peddler (pe·dləɹ). [var. PEDLAR *sb.*] In this spelling, used to denote anyone who peddles goods in some way illicitly, as stolen goods, forged notes, illegal drugs, etc.

The spelling was mainly confined to the U.S. until the 1960s, and the word is still commonly written as *pedlar* in the U.K. even in the sense defined above.

1872 G. P. BURNHAM *Mem. U.S. Secret Service* p. vii, *Peddler*, an itinerant counterfeit money-seller. **1929** M. A. GILL *Underworld Slang*, *Peddlers*, drug bootleggers. **1930** [see *drug-peddler* s.v. *DRUG *sb.*[1] 1 b]. **1935** *Jrnl. Abnormal Psychol.* XXX. 363 *Peddler*, an inmate who steals and sells state property. **1935** N. ERSINE *Underworld & Prison Slang* 57 *Pint peddler*, a petty bootlegger who carries a number of pints of liquor about his person. He usually hangs around poolrooms. *a* **1953** E. O'NEILL *Hughie* (1959) 27 Take my tip, pal, and don't never try to buy from a dope peddler. **1953** W. BURROUGHS *Junkie* (1972) ii. 29 In fact, a peddler should not come right out and say he is a peddler... Everyone knows that he himself is the connection, but it is bad form to say so. *Ibid.* iv. 41 A peddler..was pushing Mexican H on 103rd and Broadway. **1978** T. WILLIAMSON *Technicians of Death* vi. 44 They're ready to deal in junk [*sc.* drugs]... Fringe groups..will start feeding the peddlers. **1978** *Guardian* 25 Aug. 11/6 The officials even wanted me to identify the street peddler from whom I bought a copy of the English book.

pedestal, *sb.* Add: **1. b.** Used *fig.* in phr. *to put* (or *place*, etc.) *on a pedestal*, to regard as highly admirable or important; to accord an important place to; to exalt or magnify.

1859 [see SEAT *v.* 1 a *fig.*]. **1882** R. L. STEVENSON *Familiar Stud. Men & Bks.* 158 This is to put friendship on a pedestal indeed. **1916** G. B. SHAW *Pygmalion* 205 She wishes she could get him alone..and just drag him off his pedestal and see him making love like any common man. **1922** JOYCE *Ulysses* 638 They discovered..that their idol had feet of clay, after placing him upon a pedestal. **1930** A. ROOSEVELT in H. Powell *Last Paradise* p. xiii, In the United States we are so used to work that we can't conceive of life without it. We have placed work on a pedestal. It is our God. **1957** D. ROBINS *Noble One* xvi. 152 He had unconsciously put her on a pedestal. **1968** *Listener* 3 Oct. 440/3 When somebody becomes prime minister they're immediately put on a pedestal... Well, Ramsay fell off his pedestal. **1975** *Ibid.* 18 Dec. 815/1 The doctor is on a pedestal? Yes, unapproachable, because his expertise sets him apart.

3. c. (Later examples.)

1896 *Heal & Son Catal.: Bedsteads, etc.* 170 Washstand.., Chamber Pedestal, Towel Horse. **1959** W. S. SHARPS *Dict. Cinematogr.* 116 *Pedestal*, a simple vertical support for a film projector. **1960** O. SKILBECK *ABC of Film & TV* 95 *Pedestal*, a simple, high Dolly, used in T.V. **1961** G. MILLERSON *Technique Television Production* iii. 23 A one-man camera mounting, the pedestal has high manoeuvrability. *Ibid.* 24 The pedestal column is telescopic. **1971** *Nature* 9 July 98/1 The fines were mounted on the electron microscope pedestal with a drop of absolute methanol. **1975** J. RATHBONE *Kill Cure* I. iv. 32 The cracked W.C. pedestal did not bear looking into.

d. *Television.* The level of the video signal voltage during line blanking; also, this part of the signal.

1937 *Electronics* June 15/2 From the shading panel the signal goes to a control amplifier where the pedestal level is set. This pedestal is a voltage level corresponding to black, or slightly 'blacker than black', on which the synchronizing impulses are placed and which exists throughout the return trace of the cathode beam. **1951** R. B. DOME *Television Princ.* ix. 214 The width of the horizontal synchronizing pulse is approximately half the width of its pedestal. **1956** M. SLURZBERG et al. *Essent. Television* ii. 24 The blanking signal is often referred to as a pedestal on which the synchronizing pulses are mounted. **1972** F. H. BELT *How to interpret T.V. Waveforms* 101 Video looks okay at first glance, although actually it's compressed a bit at the black end (up near where the sync pedestals should be). Also, blanking (the sync pedestal) appears widened.

4. *pedestal base, cupboard, dancing, desk*; **pedestal (wash) basin,** a wash basin with a single columnar support; **pedestal mat,** a mat which fits around the base of a lavatory pedestal; **pedestal table** (examples); also used of other types of table; also *pedestal writing table*; **pedestal vase,** a vase with a pedestal base.

1948 A. LANE *Greek Pott.* ii. 9 Wide and shallow with pedestal bases. **1978** *N.Y. Times* 30 Mar. B21/9 (Advt.), Pedestal base dining tables. **1967** J. MORRISON in *Coast to Coast 1965–66* 140 There was no pedestal basin and no tap. **1896** *Heal & Son Catal.: Bedsteads, etc.* 194 Pedestal Cupboard. **1959** G. SAVAGE *Antique Collector's Handbk.* 122 Pedestal cupboards, surmounted by urns, appear quite frequently. **1906** *Daily Chron.* 12 Mar. 7/1 Mrs. Lannon said pedestal dancing was not a speciality, but they had introduced it. **1952** J. GLOAG *Short Dict. Furnit.* 353 Small pedestal desks were introduced early in the 18th century. **1957** V. NABOKOV *Pnin* iii. 69 It [*sc.* a room] had come with two ignoble chairs..and a humble pedestal desk of indeterminable wood. **1970** *Country Life* 1 Oct. (Suppl.) 48B/1 A very fine quality Hepplewhite mahogany Pedestal Desk with bow front and dummy drawers. **1962** *Guardian* 5 Dec. 6/3 Matching bathroom sets..bath mat, pedestal mat, lavatory seat cover. **1972** *Ibid.* 23 Feb. 18/5 Matching bathmats..pedestal mats, and loo seat covers. **1939** *Army & Navy Stores Catal.* 1032 Bedside pedestal table..With half door..or full door. *Ibid.*, Kidney Shape Pedestal Table, fitted with six drawers. **1952** J. GLOAG *Short Dict. Furnit.* 355 *Pedestal table*, a term sometimes used for round, oval, square, or rectangular tables that are supported by a single pillar or column which rests upon a stabilising base. **1966** M. M. PEGLER *Dict. Interior Design* (1967) 329 The pedestal table is also a popular modern design with the table surface resting on a thin support which flares outward as it reaches the floor. **1975** *Habitat Catal.* 35 Pedestal table. Spun aluminium base..melamine laminate top. **1960** K. M. KENYON *Archaeol. in Holy Land* vii. 171 Neither pedestal vases..nor flaring carinated bowls..are found. **1967** *Gloss. Sanitation Terms* (*B.S.I.*) 62 *Pedestal wash basin*, a wash basin supported from the floor by a column-shaped base. **1883** *Heal & Son Catal.: Bedsteads, etc.* 193 Pedestal Writing Table,..Leather Top.

pedestrian, *a.* and *sb.* Add: **B.** *sb.* **b.** *rare.* One who is dull, prosaic, or uninspired.

1969 D. F. HORROBIN *Sci. is God* iii. 24 These..purveyors of ideas..irritate intensely those pedestrian men who feel that they never have enough information for the proposal of a hypothesis. Unfortunately, for the pedestrians, it is the dreamers who tend to steal most of the scientific glory.

c. *attrib.* and *Comb.* *pedestrian-operated* adj.; **pedestrian crossing,** a marked section of the roadway where pedestrians crossing the road are given precedence over vehicular traffic; **pedestrian deck,** a series of pavements or walk-ways, usu. built above ground level and often roofed, reserved for pedestrians; **pedestrian precinct,** an area reserved for pedestrians only, usu. in a town centre or shopping centre.

1935 *Punch* 3 Apr. 374 (*caption*) Sorry, old man, but my wife has just signalled that she has thought of a name for our new dog, and I'm dashing along to the next pedestrian crossing so that I can go over and hear it. **1936** [see *BELISHA BEACON]. **1950** *Ann. Reg. 1949* 473 The plaintiff..was crossing a street..by means of a controlled pedestrian crossing. **1973** D. MILLER *Chinese Jade Affair* xvii. 164 The light at the intersection caught us..and my glamorous chauffeur brought the little Fiat to a halt just a yard or so the wrong side of the pedestrian crossing. **1976** *Evening Post* (Nottingham) 15 Dec. 2/5 We have suggested to the City Planners that a pedestrian crossing.. would be desirable on safety grounds. **1962** *Listener* 24 May 903/2 The pedestrian deck planned for the now abandoned project for a new town in Hook, Hampshire. **1963** *House & Garden* Mar. 35/2 Andover will also have its pedestrian deck, 15 feet above a ground level complex of footways and underpasses. **1937** *Daily Herald* 21 Jan. 3/7 No pedestrian operated signals for crossing places. **1938** H. A. TRIPP *Road Traffic* xii. 269 (*heading*) Pedestrian-operated signals. Signals fitted with buttons to be pressed by pedestrians desiring to cross the road are employed. **1953** F. GIBBERD *Town Design* v. 121 A system of pedestrian precincts as short cuts between shopping streets can be developed in a large centre. **1960** *New Left Rev.* July–Aug. 23/1 He proposed to ring the centre with an elaborate motor road, and to turn the entire central area..into a pedestrian precinct. **1971** P. GRESSWELL *Environment* 183 Pedestrian precincts include roads and other areas closed to traffic. **1977** *Belfast Tel.* 24 Jan. 5/1 The police officers chased the gunmen through a pedestrian precinct into Water Street.

pedestrianize, *v.* Add: **b.** *intr.* To produce something commonplace or unremarkable; *trans.* to make (something) commonplace or prosaic.

1838 W. HOWITT *Rural Life Eng.* II. I. iii. 57 We want a designer for wood-cuts..who would pedestrianize in simple style. **1945** W. DE LA MARE in *Trans. R. Soc. Lit.* XXII. 99 Genius originates what talent pedestrianizes.

c. To make accessible only to pedestrians; to make into a pedestrian precinct.

1963 *Observer* 15 Dec. 6/6 He [*sc.* Prof. Buchanan] even suggested that some of these central streets..should be closed to traffic and, in the jargon, pedestrianised. **1969** *Daily Tel.* 15 Nov. 7 (*caption*) Chancellors Way..which will become pedestrianised when the new building..is completed. **1971** *Hansard Lords: Official Rep. Comm. Highways Bill, 1st Sitting 28 Apr.* 11/1 The need for this provision is particularly clear where the street pattern in historic towns is being redesigned, and where it is designed to 'pedestrianise' a street and to construct a new road to give access to shop premises. **1973** *Guardian* 12 Oct. 15/2 The great effort to pedestrianise Carnaby Street. **1977** *Listener* 17 Feb. 207/3 All the streets around here—for the first time in Paris—have been pedestrianized.

Hence **pede·strianized** *ppl. a.*; **pedes:trianiza·tion.**

1964 *Listener* 3 Sept. 341/1 We have a maze of medieval alleys..—ideal for pedestrianization. **1966** *Economist* 2 Apr. p. xv/1 After a look at what is for sale in those urbanely pedestrianised shops, one is tempted to scream it. **1967** *Times Rev. Industry* Oct. 52/3 The medieval road pattern in the city centre..is 'natural for pedestrians' and he has an experimental scheme for pedestrianization of the upper High Street. **1971** P. GRESSWELL *Environment* 182 Pedestrianisation is a clumsy word to describe a simple process: closing shopping streets to traffic. **1972** *Times* 30 Dec. 16/4 The £45,000 scheme will provide 370 yards of 'pedestrianized' roadway. **1973** *Country Life* 15 Nov. 1575/2 It is to be hoped that extensive pedestrianisation and a one-way baffle system may be considered. **1974** *Oxford Times* 18 Oct. 4 Members agreed to stick to their policy of banning all but emergency ambulances from pedestrianised streets and bus lanes. **1977** *Lancashire Life* Aug. 34/1 The proof of Blackburn's pedestrianisation is in its perambulation.

pedgill (pe·dʒ'l), *v. dial.* [Cf. PEGGLE *v.*]. To work hard and painstakingly *at*; to plod or persevere.

1913 D. H. LAWRENCE *Let.* 12 Jan. (1962) I. 175 The thought of you pedgilling away at the novel frets me. **1915** —— *Rainbow* i. 8 But at drawing, his hand swung naturally in big, bold lines, rather lax, so that it was cruel for him to pedgill away at the lace designing, working from the tiny squares of his paper, counting and plotting and niggling.

pediatric, etc.: see *PÆDIATRIC *a.*, etc.

pedicab (pe·dikæb). Also **peddicab.** [f. PEDI- + CAB *sb.*[3]] A small pedal-operated vehicle, usu. a tricycle, serving as a taxi in countries of the Far East. Also *attrib.*

1948 *Time* 22 Nov. 15/2 The usual crush of pedicabs surged down the street. **1951** *N.Y. Herald-Tribune* 26 Mar. 1/7 The rickishaw and peddicab boys. **1953** *Here & Now* (N.Z.) July 17/1 Trishaw, or pedicab: 'a tricycle affair for one or two passengers, pedalled by a driver. In the larger cities [in China], the trishaw has almost completely replaced the rickshaw.' **1966** D. FORBES *Heart of Malaya* x. 119 He hired his pedicab for one dollar a day. **1971** *Nat. Geographic* Jan. 21 Java's tricycle taxis—rear-driven pedicabs called *betjaks*—also haul freight through city streets. **1974** *Times* 30 Apr. 8/3 In Saigon a pedicab driver set fire to himself outside the President's palace. **1977** *Time* 14 Feb. 41/3 A pedicab parked in the lobby.

pedicure, *sb.* Add: **1.** Also *fig.*

1918 A. QUILLER-COUCH *Stud. in Lit.* 1st Ser. 271 Against this positive deed of friendship and thirty years of devotion little is set by sneering at Watts as 'a pedicure of the Muses'.

2. (Further examples.)

1900 [see *face-massage* s.v. *FACE *sb.* 26]. **1907** *Yesterday's Shopping* (1969) 538/2 Pedicure Scissors—2/6. Pedicure Cases—each 7/6. **1953** J. GORDON *Beauty Bk.* xvi. 167 The instruments you use for a pedicure should not be used for a manicure. *Ibid.*, *Pedicure routine*.. Remove all varnish with a pad of cotton-wool. **1974** *Times* 27 Aug. 9/2 Services..include waxing, manicure, pedicure.

pedigree, *sb.* Add: **2. d.** *colloq.* The 'life history' of a person or thing. Also, a person's criminal record. **pedigree-man** (see quot. 1923).

1903 *Dialect Notes* II. 324 *Pedigree*,..history. 'If he doesn't go straight I'll tell his pedigree.' Not applying to family descent, but to personal history. *a* **1911** D. G. PHILLIPS *Susan Lenox* (1917) II. xvii. 397 'I run her in myself.' 'Oh, she's got a record... Why the hell didn't you say so?' 'I thought you remembered. You took her pedigree.' **1923** J. MANCHON *Le Slang* 220 *Pedigree-man*,..un récidiviste, un cheval de retour. **1942** BERREY & VAN DEN BARK *Amer. Thes. Slang* § 477/5 *Pedigree*, a prisoner's police record. **1964** 'D. SHANNON' *Root of All Evil* (1966) ix. 123 Dorothy had a little pedigree for shoplifting. **1969** C. IRVING *Fake!* (1970) xii. 147 Another element in the 'pedigree' of the painting was a certificate from a prior owner, usually one of several well-known collectors. **1975** *Times* 22 Aug. 3/3 It has been decided to establish a national register to check the pedigrees of vehicles, particularly their milages.

5. pedigree-stick, a stick carved to record the genealogy or history of a tribe.

1893 *Jrnl. Anthrop. Inst.* XXII. 319 Had the Polynesians any means of recording degrees of descent?.. *Aufau fetii* is 'the genealogy of a family', and must have been a staff bound in some especial manner to serve the purpose of a pedigree-stick. *Ibid.* 320 An undoubted pedigree-stick..is figured..in Roth's translation of Crozet's 'Voyage to Tasmania', where it is described as 'a staff recording the history of the Ngati-Rangi Tribe' of New Zealand. **1895** A. C. HADDON *Evol. Art* II. iv. 273 These carved shafts of sacred paddles and adzes were pedigree-sticks. **1908** *Encycl. Relig. & Ethics* I. 826/2 Colley March first suggested that the carved shafts of the sacred paddles and adzes were pedigree-sticks, the patterns being 'the multitudinous human links between the divine ancestor and the chief of the living tribe'.

pedigreed, *a.* Add: (Later examples.)

1916 *Mem. N.Y. Bot. Garden* VI. 353 The possibility of cross-sterility between sister plants of a seed progeny was proven and the interrelations of sterility studied in a pedigreed seed progeny. **1971** *Farmer & Stockbreeder* 23 Feb. 53/3 (Advt.), Ayr Bull Sale..Pedigreed, Milk Recorded and Brucellosis Accredited Ayrshire Bull Stirks. **1976** J. VAN DE WETERING *Tumbleweed* vii. 65 He is a pedigreed cat. **1977** *New Yorker* 11 July 70/1, I saw a sled being pulled by an Irish setter, another by a mongrel collie, a third by a pedigreed scottie.

b. Having a criminal record. Cf. *PEDIGREE *sb.* 2 d.

1935 *Amer. Speech* X. 13/1 To have one's record in the possession of the police;..*pedigreed*. **1942** BERREY & VAN DEN BARK *Amer. Thes. Slang* § 461/3 *Pedigreed crook*, one with a police record. **1971** 'A. BLAISDELL' *Practice to Deceive* viii. 112 Rodriguez and D'Arcy were out again hunting the pedigreed sex fiends.

pediment[1]. Add: **1. b.** *Geomorphol.* A broad, gently sloping, eroded rock surface that extends outwards from the abrupt foot of a mountain in arid and semi-arid regions and is usu. slightly concave and partly or wholly covered with a thin layer of alluvium.

This is not the sense in quots. 1882, where the word denotes steep rock slopes roughly triangular in shape, more like architectural pediments.

[**1882** C. E. DUTTON *Tertiary Hist. Grand Cañon District* v. 85 Just opposite to us the pediments seem half buried, or rather half risen out of the valley alluvium. *Ibid.*, Between the alcoves the projecting pediments present gable-ends towards the valley-plain.] **1897** W. J. MCGEE in *Bull. Geol. Soc. Amer.* VIII. 92 The tide-carved coast cuts a typical granitic butte..rising sharply from the inclined foot-slope of Sierra Seri, yet the rugged-faced knob is seen to surmount a granite pediment nearly half a mile across in the line of section. **1922** *Bull. U.S. Geol. Survey* No. 730. 52 The mountains of the Papago country rise from plains which are similar in form to the alluvial plains that commonly front mountains of an arid region, but large parts of the plains are without alluvial cover and are composed of solid rock. These plains are called 'mountain pediments', a term suggested by McGee's usage. *Ibid.* 58 A mountain pediment buried in alluvium may be called a concealed pediment. **1933** [see *PANFAN]. **1935** [see *PEDIPLAIN]. **1960** B. W. SPARKS *Geomorphol.* xi. 257 The sharp break of slope between the pediment and the mountain front seems to point to a change of operative process, but there is no agreement as to the nature of the processes involved. **1974** [see *PEDIPLAIN]. **1977** A. HALLAM *Planet Earth* 85 The low-angle (generally less than about 8°) concave surfaces which coalesce to form the pediplains are called pediments. *Ibid.*, More recently it has been argued that pediments develop through surface and subsurface weathering.

3. pediment pass *Geomorphol.* (see quot. 1930).

1930 C. SAUER in *Univ. Calif. Publ. Geogr.* III. 370 Under less advanced conditions of pediment development we find narrow, flat, rock-floored tongues extending back from the general pediment, but still penetrating along the mountain sufficiently to meet another pediment slope extending into the mountain front from the other side... To distinguish this less advanced feature from the broad saddle plains..we may call it a pediment pass. **1974** C. H. CRICKMAY *Work of River* viii. 206 In a few places, the upper end of the pediment is the smoothly rounded summit of a pediment pass, or rock floored gap through a low mountain ridge.

pedimentation (pedimentēi·ʃən). *Geomorphol.* [f. PEDIMENT[1] + -ATION.] The formation of a pediment.

1948 *Bull. Geol. Soc. Amer.* LIX. 372 Down the slope from the knickpoint pedimentation rapidly reaches stability. **1962** L. C. KING *Morphol. Earth* v. 146 Together, scarp retreat (by gully-head erosion and mass slipping of material) and pedimentation (by bedrock levelling and sheet-waste removal) are the most potent agents modifying epigene landscapes. **1973** *Nature* 9 Nov. 75/2 Recession of valley nickpoints inland from the coastal margin was followed by valley widening, scarp retreat and pedimentation, leading ultimately to the formation of gently undulating pediplains or erosion surfaces.

pedimented, *a.* Add: **2.** *Geomorphol.* Characterized by the presence of pediments (*PEDIMENT[1] 1 b).

1949 *Geol. Mag.* LXXXVI. 249 Pedimented landscapes consist essentially of hillslopes and of pediments. **1960** B. W. SPARKS *Geomorphol.* xi. 258 Landscapes resembling the pedimented areas of the arid south-western parts of the United States are found on an altogether larger scale in Africa.

pediocratic (pediokræ·tik), *a. Geol.* [f. Gr. πεδίον a plain + κράτ-ος strength + -IC.] Characterized by an overall lessening of relief as a result of erosion predominating over crustal upheaval.

1924 W. RAMSAY in *Geol. Mag.* LXI. 155 During anorogenic periods, again, the continents become more or less peneplained. I call such a condition pediocratic. **1929** L. J. WILLS *Physiogr. Evol. Britain* ii. 7 Even in the pediocratic periods, the non-spectacular phenomena of erosion and deposition, and the ordered changes in the life of the world, given the requisite length of time, produce equally profound modifications in its physiography. **1961** [see *OROCRATIC *a.*].

pediplain (pe·diplēin). *Geomorphol.* [f. PEDI(MENT[1] + PLAIN *sb.*[1]] An extensive plain formed in a desert by the coalescence of neighbouring pediments (believed to represent a late stage in the cycle of erosion in arid and semi-arid climates). Cf. *PEDIPLANE.

1935 MAXSON & ANDERSON in *Jrnl. Geol.* XLIII. 94 Widely extending rock-cut and alluviated surfaces of this type formed by the coalescence of a number of pediments and occasional desert domes may be called 'pediplains'. **1968** C. R. TWIDALE *Geomorphol.* xi. 308 Pediplains,.. like peneplains, are surfaces of low relief, but generally differ from the latter in a number of respects. They are smoother, and less dissected.., they are mostly dominated by concave profiles, and meet the adjacent uplands in a piedmont angle. **1973** [see *PEDIMENTATION]. **1974** C. H. CRICKMAY *Work of River* viii. 211 Where pediments are sufficiently extensive to be conjoint and broadly continuous, the whole is termed a pediplain or a panfan. **1977** A. HALLAM *Planet Earth* 190 The continents as a whole were, however, almost completely worn down to pediplains.

Hence **pe·diplained** *a.*

1970 R. J. SMALL *Study of Landforms* ii. 28 This is one of the explanations of the development, on the pediplained surfaces of South America, Africa and Australia, of very hard weathering crusts.

pediplanation (pediplănēi·ʃən). *Geomorphol.* [f. *PEDI(PLANE, *PEDI(PLAIN + *PLANATION.] Erosion to, or the formation of, a pediplain (or pediplane).

1942 A. D. HOWARD in *Jrnl. Geomorphol.* V. 11 *Pediplanation* may be applied as a general term to the process of formation of pediplanes. **1948** *Proc. Geol. Soc. S. Afr.* L. p. xxvii, Every one of these features is ubiquitous on the terrains of Southern and Central Africa, clear proof that the landscape of those regions has evolved through pediplanation. **1963** D. W. & E. E. HUMPHRIES tr. *Termier's Erosion & Sedimentation* ii. 40 It is on plateaus, sometimes uplifted and peneplained (often those which have undergone initial pediplanation) that desert dunes (Sahara) and ice caps (Scandinavia) have developed. **1975** *Nature* 31 July 442/1 Part 2 discusses the character and development of the tropical terrain and includes chapters ..on pediments and pediplanation.

pediplane (pe·diplēin). *Geomorphol.* [f. PEDI(MENT[1] + PLANE *sb.*[3]] A piedmont slope in arid and semi-arid regions comprising a pediment and a peripediment (or just one of these, if the other is not present). Cf. *PEDIPLAIN.

1942 A. D. HOWARD in *Jrnl. Geomorphol.* V. 11 The writer proposes the term *pediplane* as a general term for all degradational piedmont surfaces produced in arid climates which are either exposed or covered by a veneer of contemporary alluvium no thicker than that which can be moved during floods. *Ibid.* 13 In some localities the upland rocks may be extremely resistant or the basin rocks may be extremely weak... In such regions the pediplane will truncate basin sediments almost exclusively and will consist largely of peripediment. **1948** *Bull. Geol. Soc. Amer.* LIX. 370 In regions of rising base level the pediplane will be made up for the most part of a peripediment, and the pediment as a near-mountain zone cut on solid rock is sometimes overlooked. **1955** *Proc. Geologists' Assoc.* LXVI. 173 The combined pediment and peripediment may then be described as a pediplane (Howard, 1942).

pediscope, var. *PEDOSCOPE.

pediunker (pedĭ,ʊ·ŋkəɹ). [Etym. unknown]. A name for the grey petrel, *Procellaria cinerea*, originally used in the island of Tristan da Cunha, where this bird is common.

1910 K. M. BARROW *Three Yrs. in Tristan da Cunha* 275 The 'Pediunker' lays in May and June; it is like a petrel. **1923** A. H. MACKLIN in F. Wild *Shackleton's Last Voy.* xii. 253 Gordon Glass had with him his dog, which occasionally discovered a 'pediunker', a species of seabird which frequents the island [*sc.* Tristan da Cunha]. **1952** J. FISHER *Fulmar* xvi. 381 Barnacles on birds are rare, though they have been found in the tail-feathers of the sub-antarctic *Adamastor cinereus*, the pediunker or grey petrel. **1967** B. ROBERTS *Wilson's Birds of Antarctic* 139/2 Grey Petrel or Pediunker, *Procellaria cinerea*. Drawings of a recently killed adult ♂.

pedo- (pe·do, pedǫ-, pīdǫ-), or before a vowel **ped-**, repr. Gr. πέδον ground, earth, is used in the sense 'soil' in *PEDOLOGY and other technical terms.

pedocal (pe·dokæl). *Soil Sci.* [f. *PEDO- + CAL(CIUM.] A soil that contains a layer of accumulated calcium carbonate (generally characteristic of dry climates). Cf. *PEDALFER.

1928 C. F. MARBUT in *Proc. & Papers 1st Internat. Congr. Soil Sci.* IV. 20 A formal definition of the two groups is as follows: I. Includes all soils in whose maturely developed profiles no higher percentage of lime carbonate is found than in the parent material beneath them and in which either a shifting or an accumulation of sesquioxide and in many cases both, has taken place. II. Includes all soils with fully developed profiles in which lime carbonate is found on some horizon in the solum in higher percentage than in the parent geological formation beneath... The name *Pedalfers* is suggested as a designation for the soils of group I, and *Pedocals* for those of group II. **1938** *Nature* 21 May 925/2 It has been generally accepted that the climate of Great Britain is too wet for pedocals to develop. **1949** F. J. PETTIJOHN *Sedimentary Rocks* xii. 384 The pedalfers and pedocals are subdivided into many soil types. **1968** [see *halomorphic* adj. s.v. *HALO-]. **1972** J. G. CRUICKSHANK *Soil Geogr.* v. 158 Marbut..established a classification that became the most widely known of the schemes based on soil genesis, in which he made the two-fold primary sub-division between the leached soils (pedalfers) and the non-leached (pedocals) freely drained soils. **1977** A. HALLAM *Planet Earth* 179 These ideas were accepted in the USA, and the concepts of pedalfers and pedocals added: pedalfers are leached soils in humid areas where aluminium and iron accumulate in the B horizon, and pedocals occur in more arid regions.

Hence **pedoca·lic** *a.*

1928 *Proc. & Papers 1st Internat. Congr. Soil Sci.* IV. 23 Most of the groups of Pedalferic soils in category V were defined by Russian pedologists, but they made no attempt to differentiate the Pedocalic soils on the same basis since the Eurasian conditions do not present an opportunity for doing it. **1938** *Nature* 21 May 925/1 (*heading*) Pedocalic tendencies in soils of southern England. **1941** H. JENNY *Factors Soil Formation* vi. 191 Not all soils that are embraced by the area labeled as pedocals contain lime horizons. Only those soils which have the required normal topography and which have attained sufficient maturity deserve the attribute pedocalic.

pedogenesis (pedodʒe·nėsis). [f. *PEDO- + -GENESIS.] Soil formation.

1936 J. S. JOFFE *Pedology* vi. 134 From the point of view of pedogenesis, the classification of the soil-forming processes, in their broad aspects, should hinge on the elements of climate. **1943** *Proc. Soil Sci. Soc. Amer.* VII. 187/1 Most investigations on mechanical separates of the soil have dealt with the physical aspects of the subject, with no relation to pedogenesis and agropedologic implications. **1963** D. W. & E. E. HUMPHRIES tr. *Termier's Erosion & Sedimentation* vi. 135 During pedogenesis (soil formation), the nature of the alteration of the titanium minerals depends more upon chemical diagenetic environmental conditions than on the nature of the parent rock. **1972** J. G. CRUICKSHANK *Soil Geogr.* ii. 42 As part of the role of climate in pedogenesis, its moisture and temperature components also affect the rate of decay and incorporation into the soil of surface organic litter.

Hence **pedogene·tic, -gene·tical** *adjs.* = *PEDOGENIC *a.*; **pedogene·tically** *adv.*

1946 *Soil Sci.* LXI. 389 Each soil series..is thus pedogenetically connected with ten other series. **1950** *Ann. Assoc. Amer. Geographers* XL. 218 Within these pedogenetic regions no one of the weathering processes occurs exclusively. **1963** D. W. & E. E. HUMPHRIES tr. *Termier's Erosion & Sedimentation* vi. 153 Butterlin (1958) believes that these soils are formed by the pedogenetic alteration of the limestones. **1978** *Nature* 30 Mar. 477/3 Work on pedogenetical formations and Quaternary deposits in central and northern Dobrudja. *Ibid.* 28 Sept. 285/2 Deeper, pedogenetically differentiated profiles in relict areas immune from rapid denudation.

pedogenic (pedodʒe·nik), *a.* [f. *PEDO- + *-GENIC.] Soil-forming.

1924 *Geol. Mag.* LXI. 448 (*heading*) The pedogenic processes. [*Note*] I am indebted to my friend Dr. Edward Greenly, F.G.S., for this convenient term, which aptly describes the processes of soil formation and metamorphism. **1936** *Nature* 24 Oct. 729/2 The effect of human interference as a pedogenic factor was raised in the discussion. **1943** [see *palæopedology* s.v. *PALÆO-, PALEO-]. **1975**

Nature 20 Feb. 617/2 The carbonate nodules which occur in the Pleistocene but not in the Holocene portions of the core are regarded as pedogenic and as symptomatic of a soil climate drier than the present one.
Hence **pedoge·nically** *adv.*
1972 J. G. CRUICKSHANK *Soil Geogr.* iii. 74 The upper boundary is obvious, but the lower one may be either the maximum plant rooting depth or the base of the pedogenically altered material.

pedology (pe-, pīdǫ·lǒdʒi). [f. *PEDO- + -OLOGY. Cf. G. *pedologie* (e.g. F. A. Fallou *Pedologie* (1862) I. 9), Russ. *pedológiya* (e.g. *Entsikl. Slovar'* (1898) XXIVA s.v. *pochvovedenie*; *Pochvovedenie* (1900) II. 140, (1902) IV. 1; the Fr. title of this periodical was *La Pédologie* from its inception in 1899).
The usual Russ. word for the subject has always been *pochvovédenie*, lit. 'soil science' (cf. G. *bodenkunde*, given by Fallou as a synonym of *pedologie*). The Eng. word *pedology* occurs in the galley proofs of an unpublished dict. of *c* 1900–10, according to L. D. Stamp *Gloss. Geogr. Terms* (1961) 358, but prob. only in reference to foreign equivalents.]
The scientific study of soil, esp. its formation, nature, and classification; soil science.
[1923 M. M. MCCOOL et al. in *Soil Sci.* XVI. 106 It places soil study on a natural basis and in fact lays the foundation of a new science which we might name, podology.] 1924 G. W. ROBINSON in *Geol. Mag.* LXI. 444 (*heading*) Pedology as a branch of geology. [*Note*] The writer ventures to hope that this convenient term (Gr. πέδον = soil or earth) will be more generally used to describe the scientific study of soils. 1926 TANSLEY & CHIPP *Study of Vegetation* vii. 114 The science of the soil (sometimes called pedology) has made very great strides within the past quarter of a century. 1938 A. B. YOLLAND tr. *A. A. J. de Sigmond's Princ. Soil Sci.* 1 The first to try to liberate soil science from this position was the German geologist Frederick Augustus Fallou, who, however, by basing his soil classification upon geological-petrographic principles unconsciously subordinated pedology to geology. 1958 I. W. CORNWALL *Soils for Archaeologist* 13 For information about the earth the archaeologist turns first to the sciences of geology, geography and pedology (soil-science). 1973 *Nature* 27 July p. ii/1 (Advt.), Scientists interested in sediments and in allied fields such as pedology, geomorphology, soils engineering and cement technology will find in this book a valuable research tool.
Hence **pedolo·gic, -lo·gical** *adjs.*, of or pertaining to pedology or soil; **pedolo·gically** *adv.*, in pedological terms; as regards pedology; **pedo·logist**, one who studies pedology.
1924 *Geol. Mag.* LXI. 450 Among mature soils, i.e. among soils which have reached a state of pedological equilibrium. *Ibid.* 454 The remaining class in Glinka's scheme is only of limited interest for western European pedologists. 1927 C. F. MARBUT tr. *Glinka's Great Soil Groups* 5 The distribution of such soil units also would be the same as that of the rocks from which derived and would be petrographic rather than pedologic. 1932 *Proc. & Papers 2nd Internat. Congr. Soil Sci.* V. 1 If it be pedologically justifiable to grant the type status to podzols. 1945 *Antiquity* XIX. 172 The pedological characters of Anglesey by which vegetation would be affected and to a large extent controlled. 1963 D. W. & E. E. HUMPHRIES tr. *Termier's Erosion & Sedimentation* ix. 192 Sedimentation in these basins..depends to a large extent on the pedologic evolution of the continents. 1964 R. FEYS in A. E. M. Nairn *Probl. Palaeoclimatol.* iii. 68 The sandstone is derived from sand which has been pedologically altered and mechanically sorted. 1972 J. G. CRUICKSHANK *Soil Geogr.* v. 159 The Marbut classification also influenced pedologists on an international level because it was discussed at international soil congresses in 1927, 1932, and 1935. 1974 *Nature* 4 Jan. 74/1 There was no known method by which termites or pedological processes could bring about the observed accumulation of calcium carbonate in termite mounds.

pedon (pe·dǫn). *Soil Sci.* [a. Gr. πέδον ground, earth.] A notional column of soil that extends vertically from the surface to the parent material and is of sufficient area (usu. at least a square metre) to be representative of the surrounding soil and its horizons.
1960 SIMONSON & GARDINER in *Trans. 7th Internat. Congr. Soil Sci.* IV. 128 Nature of the Pedon. It has recently been proposed that the term, pedon.., be used as a generic term for small basic soil entities. [*Note*] The term was proposed by Guy D. Smith. *Ibid.* 129 A pedon consists of a small volume of soil which includes the full solum and the upper part of the unconsolidated parent material.., is usually less than 2 meters in depth, and has a lateral cross section that is roughly circular or hexagonal in shape and between 1 and 10 square meters in size. 1972 J. G. CRUICKSHANK *Soil Geogr.* iii. 76 Pedons have such minimal horizontal area that often they fail to reflect important features of the larger soil body. 1973 THOMPSON & TROEH *Soils & Soil Fertility* (ed. 3) xvi. 369 Soils with large variations in their properties within short horizontal distances are allowed to have pedons as large as 10 sq m.. so that their properties may be fully represented.

pedophilia, etc., varr. *PÆDOPHILIA, etc.

pedoscope (pe·dǒskōup). Also **pediscope.** [f. PEDI- (erron. *pedo-*) + -SCOPE.] An X-ray machine for showing the fitting and movement of the feet inside shoes (formerly common in shoe shops).
1923 *Chambers's Jrnl.* Aug. 560/1 The machine used is called the pedoscope. 1959 *Times* 30 Apr. 6/5 The use of pedoscopes, although frowned on by The Medical Research Council, was not diminishing. 1967 *Punch* 22 Nov. 792/3 In one of the earliest scenes Harry, given a vacuum flask to smuggle to Helsinki, finds out what is inside by putting it under the X-ray of a shoe-shop's pediscope.

pedosphere (pe·dosfīəɹ). *Geol.* [f. *PEDO- + SPHERE *sb.*] The earth's soil layer.
1938 S. MATTSON in *Lantbrukshögskolans Ann.* V. 261 Soils might be said to represent the sum of the mechanical and the products of the chemical interaction of the four spheres: the lithosphere, the hydrosphere, the atmosphere and the biosphere. They constitute a dispersed system in which the material from these four spheres alternate as dispersed phase and dispersion medium. To this sphere of spheres the name of pedosphere has been given. 1946 *Nature* 6 July 31/1 G. A. Maximovich..has calculated the average porosities of different types of rocks and has also calculated the average porosities of the geospheres. These, in percentages, are as follows: pedosphere (soils), 55; [etc.]. 1968 R. W. FAIRBRIDGE *Encycl. Geomorphol.* 416/1 On top of this [sc. the lithosphere] the continents have an additional soil layer or pedosphere with a depth generally of only 1–2 meters.

Pedrail (pe·drēil). *Obs.* exc. *Hist.* [f. PED(I- + RAIL *sb.*²] A type of traction engine, patented in 1899 and 1902 by B. J. Diplock; it moves forward by means of articulated feet, instead of wheels, which support rollers upon which run rails fixed to the vehicle itself.
1902 *Daily Chron.* 29 Dec. 6/6 A recent invention of Mr. Diplock—the 'Pedrail'—was exhibited. 1903 *Ibid.* 23 Nov. 6/4 The Pedrail was primarily designed as a traction engine for the transport of heavy goods on common roads. 1908 *Chambers's Jrnl.* Mar. 270/1 The ped-rail traction-engine—an engine equipped with a number of feet placed around the periphery of its driving-wheels, thereby imparting a walking action. 1916 *Ibid.* Feb. 83/1 This device is called the pedrail, and consists of a flat chain round the wheels.., the chain being armed with discs offering a flat surface to the ground. 1959 W. J. HUGHES *Cent. of Traction Engines* xvi. 212 At the beginning of 1902 some interesting experiments were being carried out with Diplock's 'Pedrail' engine... The particular engine under notice..besides having the Pedrail wheels,..had four-wheel drive. 1960 R. H. CLARK *Devel. Engl. Traction Engine* xiv. 291 In Diplock's Pedrail the rail (the bottom concave run) runs over the feet by virtue of the rollers interposed between them.

pedro (pe·drǒ). Also with capital initial. [f. Sancho *Pedro*, the name of a U.S. card game.] **a.** A variant of the card game Sancho Pedro in which the sancho, or nine of trumps, does not count. **b.** The name for the five of trumps in such games. Cf. *CINCH *sb.* 3, *SANCHO².
1874 *Reno* (Nevada) *Crescent* 8 May 2/3 The five of trumps is 'pedro'. 1876 *Avalanche* (Silver City, Idaho) 15 Mar. 2/3 Whist, pedro, pool.., etc., seem to be flourishing pastimes. 1880 W. B. DICK *Amer. Hoyle* (ed. 13) 210 Pedro may be taken with any trump higher than the Five. 1929 R. S. & H. M. LYND *Middletown* 281 The growing rigidity of the social system today is centering parties more and more upon cards, pedro among the workers and bridge among the others. 1944 A. H. MOREHEAD *Pocket Bk. Games* 199 Auction Pitch has many variants. Most of them—Pedro, Pedro Sancho, Dom Pedro, Snoozer—have been swallowed up by their own ultimate creation, Cinch. 1947 *Sat. Even. Post* 17 May 102/3 He operates a pedro game.

pee (pī), *v.*² *colloq.* [f. initial letter of PISS *v.*] **1.** *intr.* To urinate. Hence **pee·ing** *vbl. sb.*
1879–80 *Pearl* (1970) 216 Your private parts, or cunny, Should not be let for money, They're only meant to pee with. 1886 in F. T. ELWORTHY *West Somerset Word-Bk.* 1929 C. CONNOLLY *Let.* Nov. in *Romantic Friendship* (1975) 329 It [sc. a kinkajou] seemed just a machine for shitting and peeing. 1932 AUDEN *Orators* II. 78 The boys, out of control, imbibe Vimto through india-rubber tubing, openly pee into the ink-pots. 1948 M. MCCARTHY in *Partisan Rev.* Mar. 227 'My God', you yell..'can't a man pee in his own house?' 1960 C. DAY LEWIS *Buried Day* 74 How to leave the room during lessons—I solved..by peeing in my trousers. 1963 M. MCCARTHY *Group* iii. 56 Anybody..who wandered in to pee during a cocktail party. 1965 J. R. HETHERINGTON *Selina's Aunt* 50, I could laugh till I peed. 1966 A. LA BERN *Goodbye Piccadilly* xi. 109, I kept wanting to pee every few minutes. 1969 P. ROTH *Portnoy's Complaint* 132 Here is how I learnt to pee into the bowl like a big man. 1971 N. SAUNDERS *Alternative London* (ed. 2) 236 As above for ladies, not so easily cured though peeing just after sex is meant to help. 1973 M. AMIS *Rachel Papers* 74 She looked fretful, importunate, almost bouncing up and down, like a little girl wanting to pee. 1975 *Sunday Times* (Colour Suppl.) 23 Feb. 26/2 The guys were forever peeing over the side so there was piss everywhere.
2. a. *trans.* To wet by urinating.
1788 E. PICKEN *Epitaph Favourite Cat* in *Poems & Epistles* 47 He never stealt, though he was poor, Nor ever pee'd his master's floor. 1948 D. BALLANTYNE *Cunninghams* 219 She nearly pees her pants every time he kids to her.
b. *refl.* To urinate into one's clothes; esp. (hyperbolically) in phr. *to pee oneself laughing.*
1946 G. KERSH *Clean, Bright & Slightly Oiled* i. 4 Even the Sarn-Major peed 'imself laughing. 1962 K. ORVIS *Damned & Destroyed* xv. 104 You did wrong to hit..so much... I peed myself. 1978 R. BUSBY *Garvey's Code* xii. 168 He must've realized what was going to happen.. because he peed himself right there.

pee (pī), *sb.*⁵ [f. prec.] **1.** An act of urination.
1902 R. MACLAGAN *Evil Eye* 51 The milk has gone along with the pee. 1946 G. MILLAR *Horned Pigeon* xii. 156 The urinals alone would have a monument the size of London's Cenotaph... When they opened the sewer trap in the monumental pee-house great clouds of mosquitoes..rose from the opening. 1951 S. SPENDER *World within World* 273 In Russia it's so cold that when you do a pee, you can break it off in sticks. 1958 L. DURRELL *Mountolive* xiii. 244, I must just do a pee. 1966 J. CHAMIER *Cannonball* xiii. 119 Best go and have a pee, lad. 1973 *Daily Tel.* (Colour Suppl.) 23 Feb. 54/1 If people came in just to use the lavatory, he would ask them for their address 'in case I need a pee when I'm passing your house'.
2. Urine.
1961 H. RUDD *Shores of Schizophrenia* in *My Escape from CIA* (1966) 10 That gloomy, hideous building, with its smell of cedar sawdust and stale, infantile pee. 1968 R. P. WARREN *Incarnations* 43 Jesus, Wouldn't just *being* be enough without Having to have the pee..knocked out of You by a 1957 yellow Cadillac. 1971 J. OSBORNE *West of Suez* I. 42 Do you know what the cure for chilblains was then? Soaking your feet and hands in your own pee. 1976 P. CAVE *High Flying Birds* ii. 16 Sarcasm runs off on them like pee on a plastic bedsheet.

pee (pī), *sb.*⁶ *colloq.* [f. initial letter of PENNY.] Representing the pronunciation of the initial letter of 'penny', i.e. a new penny, a unit of decimal currency introduced in Britain on 15 Feb. 1971. See also *p* s.v. *P II, *PENNY I. 1.
The pronunciation of *p.* as (pī) is common in everyday speech but is avoided by many people, who prefer (pe·ni) (as singular) and (pens) (as pl.).
1971 *Observer* 14 Feb. 9/5 Everyone at the Decimal Currency Board has taken to calling new pence 'pee'. 1972 *Daily Tel.* 9 Aug. 16 If Mr Broca had contacted us we would have sold him one not for 'seventy-eight pees', but for 69p. 1974 *Punch* 6 Mar. 362/2 The Scandinavian revenue men..intimating that it's either an immediate fifty pee in the £ or chuck the belongings back into the red-spotted hankie and ring up a mini cab. 1974 R. RENDELL *Face of Trespass* ii. 23 May I trouble you for forty-two pee? 1974 *Observer* (Colour Suppl.) 15 Dec. 13/4 Few could be bothered to say 'new pence' for the decimal stuff, so we used 'pee', and that is what we are lumbered with today. 1976 *Times Lit. Suppl.* 2 Apr. 388/1 A small boy on the loose in London with a million pound picture in a laundry bag and 'two pee' in his pocket. 1977 *Transatlantic Rev.* LX. 187 He was accosted by a group of rotting cider bums: 'Mister..we just need ten pee to get ourselves another bottle.'

peek, *sb.*¹ (Earlier and later examples.)
1844 'J. SLICK' *High Life N.Y.* II. 41, I jest give a peak in for a minit, and streaked it upstairs. 1869 L. M. ALCOTT *Little Women* II. xx. 300 'Ain't it a sight to see her settin' there,'..muttered old Hannah, who could not resist frequent 'peeks' through the slide. 1938 E. AMBLER *Cause for Alarm* vii. 116 Supposing you take an occasional peek at these other guys' hands, tell me what you see. 1969 *Daily Mail* 16 Jan. 5/1 After insertion into Earth orbit I had a lot of tests to perform on the spacecraft systems but, like all the rookies before me, I must confess to a sneak peek out of the window. 1973 P. EVANS *Bodyguard Man* 10, I poked my head out and took a peek. 1974 *Black World* Sept. 40 When you get inside cock your eyes ace-deuce and snatch a peek around. 1976 *National Observer* (U.S.) 24 Jan. 19/2 He was to proceed to Cairo, take a peek at Mecca, and head west across the bulge of Africa until he either sighted the Niger or fell into it.
b. *Comb.*, as *peek-hole*, a peep-hole.
1910 'MARK TWAIN' *Speeches* 222, I peeked through the little peek-holes they have in theatre curtains. 1927 G. ADE et al. *Let.* 4 Mar. (1973) 117 Gibraltar..is honeycombed with tunnels, which the visitor is not permitted to see, and from those tunnels there are peek-holes out of which guns can be pointed in any direction. 1927 *Sat. Even. Post* 24 Dec. 12/2 That's Fred's peek hole, where he sees out of. 1930 E. POUND *XXX Cantos* xvii. 79 One eye for the sea, for that peek-hole.

peek, *v.*¹ Add: (Further examples.) Also, to glance *at*. Hence **pee·king** *vbl. sb.*
1928 *Publishers' Weekly* 22 Sept. 1120/1 One cannot escape the temptation to peek at prices however and I found one marked six shillings and took it. 1947 AUDEN *Age of Anxiety* (1948) ii. 38 Pushing through brambles, I peeked out at Her fascination. 1953 *Manch. Guardian Weekly* 29 Jan. 4 A coloured butler peeked out of a cream-coloured door. 1962 K. ORVIS *Damned & Destroyed* vii. 49 When I went in [to jail] they didn't even ask for peekings. 1968 *Globe & Mail* (Toronto) 15 Jan. 21/7 Jerry Tighe, of Whitworth College in Spokane, came close to a fall in the men's two miles when he peeked over his shoulder to check the field. 1973 'S. HARVESTER' *Corner of Playground* III. i. 165 Are you so bored with me you must peek at other men? 1973 *Black World* June 10/2 The vain Southern belle who undresses in front of the stable boy and then tells her 'Daddy' that he peeked. 1973 *Radio Times* 20 Dec. 100/4 What's he getting for Christmas?..'Absolutely no peeking before Christmas morning.' 1976 D. HEFFRON *Crusty Crossed* ix. 71 'It's time you learned the difference,' she called out after me, 'between high-class spying and low-down peeking.'

peek-bo, peek-a-boo. Delete 'now chiefly U.S.' and add further examples. Also *attrib.* and as *vb.*
The spelling *peek-a-boo* is now usual.
1903 [see *high sign* s.v. *HIGH *a.* 21]. 1922 JOYCE *Ulysses* 46 Peekaboo. I see you. 1940 S. O'CASEY *Let.* 19 May (1975) I. 862 I'm..kicking football with Niall; & peek a boo with Shivaun. 1950 G. BARKER *True Confession* ii. 10 But Memory flirts with seven veils Peekabooing the accidental. 1960 *Times* 22 June 5/5 Covering up behind his high 'peek-a-boo' guard, he [sc. Floyd Patterson] rode out the brief crisis. 1967 S. BECKETT *Stories & Texts for Nothing* 124 But peekaboo here I come again, just when most needed.

b. Used *attrib.* and *absol.* of various garments decorated with holes (see quots.).
1895 *Montgomery Ward Catal.* 306/3 Child's Sun Bonnet, made of printed lawn, peek-a-boo front. 1906 *Daily Chron.* 16 Aug. 8/1 The dreamer is embowered in soft cushions, and being punted by a River Girl, in a peek-a-boo blouse—all grace and lissomness and tan and bare arms. 1906 *Springfield* (Mass.) *Weekly Republ.* 10 May 13 In San Francisco there is no winter suit and summer suit. The same medium-weight garment is worn the year round and the peek-a-boo waist is unknown. 1907 *Westm. Gaz.* 21 Sept. 3/2 'The Peek-a-boo blouse'..is as popular here as in America. 1910 'MARK TWAIN' *Speeches* 87 Why not adopt some of the women's styles?.. Take the peek-a-boo waist, for instance. 1913 *Maclean's Mag.* Aug. 103/1 The wars upon cigarettes, bridge whist and peekaboo waists are passing madnesses. 1930 M. SULLIVAN *Our Times* III. 499 A shirt-waist, supposed to be an extreme of daring, in which embroidered perforations permitted sight of female epidermis upon the arms and as much as two inches below the nape of the neck, was called the 'peek-a-boo'. 1957 M. B. PICKEN *Fashion Dict.* 246/2 *Peek-a-boo waist*, shirt-waist of eyelet or sheer fabric. 1959 J. BRAINE *Vodi* xiv. 195 The peekaboo blouse..now seemed vulgar, a barmaid's uniform. *Ibid.* 196 A naughty little girl who wore peekaboo blouses. 1969 R. T. WILCOX *Dict. Costume* (1970) 263/2 *Peek-a-boo blouse*, a shirtwaist of sheer lawn, voile or eyelet embroidery, a popular fashion of the 1890's and the first decade of the twentieth century. It was worn over a lacy, frilled corset cover. 1972 'E. MCBAIN' *Sadie when she Died* x. 107 Carella..wondered if he should tell Teddy about the brunette in the peekaboo blouse.
c. Used *attrib.* and *absol.* of a woman's hairstyle (see quots.).
1948 M. M. MILLER *Winchester's Screen Encycl.* 117/2 Her 'peek-a-boo' style secured major part for her, under new name Veronica Lake. 1966 J. S. COX *Illustr. Dict. Hairdressing* 109/2 *Peekaboo locks*, hanging hair that partly conceals the face and eyes. 1968 J. IRONSIDE *Fashion Alphabet* 195 *Peek-a-boo*, hair style that originated..in 1941. The hair was cut so that it hung about 8 inches below the shoulders; in front it covered one eye and was softly waved to the end. 1973 *Daily Tel.* 9 July 17/3 Veronica Lake, the filmstar who was famous in the 1940s for her 'peek-a-boo' hairstyle, died in Vermont, America, on Saturday of acute hepatitis. *Ibid.*, The peek-a-boo hairstyle—long blonde tresses hanging over her right eye—disappeared in 1943 after the United States Government made an official request to Paramount to cut her hair.

peel, *sb.*¹ Add: **4.** (Further examples in spelling *pele.*)
1888 W. W. TOMLINSON *Comprehensive Guide Northumberland* 156 Jutting crags..and lofty precipices, constitute the natural defences of the pele on three sides. 1921 P. S. ALLEN *Let.* 21 Sept. (1939) 180 It is built out of the Prior's Lodgings, and includes a regular 'pele tower' such as line the Border: low towers with immensely thick walls—this one 7 ft. thick at the bottom, but with mural chambers and staircases. 1964 *Dumfries & Galloway Standard* 8 July 7, The lost Dalswinton, Conggleton and Wigtown, the pele at Lochmaben are a story in themselves. 1973 FEDDEN & JOEKES *National Trust Guide* 656 *Pele, pele tower*, a small or moderate-sized tower or keep which is easily defended. Peculiar to houses or castles on both sides of the Scottish border. Generally built between the 13th and 15th centuries.
6. peel-tower (further examples of spelling *pele-tower*).
1935 *Hist. Northumberland* XIV. 79 This [sc. Henry VIII to Charles I] forms a continuously evolving period, at first characterised by the erection of pele towers and bastle houses. 1965 N. RIDLEY *Portrait of Northumberland* 21 Scattered over wide areas of Northumberland are the Pele towers... These Pele towers were fortified dwellings, where the cattle could be driven into the ground floor rooms, to protect them from the Reivers and Moss Troopers. 1973 FEDDEN & JOEKES *National Trust Guide* 325 Consequently a wide area is studded with massive pele towers, commonly with a vaulted chamber on the ground floor into which cattle might be driven at the first warning of raiders, and upper storeys arranged for human occupation. 1980 *Historic Houses, Castles & Gardens* 48/3 Hutton-in-the-Forest..14th cent. Pele Tower with later additions.

peel, *sb.*³ Add: **2.** *Rugby Football.* The action of peeling from a set formation (see *PEEL *v.*¹ 5 g).
1973 *Scotsman* 21 Feb. 18/6 Thus, when it comes to deflecting the ball for a peel, Strachan prefers to operate from No. 8 where there is a little more scope for manoeuvre. 1978 *Sunday Express* 19 Mar. 31/4 From yet another winning line-out by Martin, Graham Price slipped him a pass on the peel and Edwards lofted over a towering drop goal.

peel, *sb.*⁴ (Later examples.)
1890 J. KERR *Hist. Curling* 240 He absolutely refused to play his stone on one occasion when the game stood peels. 1918 *Kelso Chron.* 4 Oct. 3 The players were 'peels' at 12. 1956 in *Sc. Nat. Dict.* (1968) VII. 70/2 The sides finished peels.

peel (pīl), *sb.*⁶ [see *PEEL *v.*³] In Croquet, the action of peeling.
1907 C. D. LOCOCK *Mod.-Croquet Tactics* xi. 145 The croquet term 'Peel' is derived from the eminent player of that name. *Ibid.* 146 This peeling generally..takes place in the course of a single break; but sometimes, and especially in Doubles, it is worth while to go for a peel on the captain of the other side. 1914 LD. TOLLEMACHE *Croquet* xviii. 111 In Handicap games it is frequently impossible to win without a Peel of some sort. 1932 *Times Lit. Suppl.* 18 Aug. 580/2 Like golfers its [sc. croquet's] devotees discuss it in mystical terms of 'tices', 'peels' and 'three-ball breaks'. 1953 *Times* 17 Feb. 3/7 Cotter used the turn to its fullest advantage and finished the match with a triple peel. 1974 *Observer* 23 June 40/3 Triple peels (three in a row) were commonplace at Cheltenham.

peel, *v.*¹ Add: **3. b.** (Further examples.) Also, to remove or separate (a label, bank-note, etc.) by peeling. Also *fig.*
1868 BROWNING *Ring & Bk.* I. 17 Also he peeled off that last scandal-rag Of Nepotism. 1896 'MARK TWAIN' in *Harper's Mag.* Aug. 358/1 He peeled off one of his bulliest old-time blessings, with as many layers to it as an onion. 1946 E. O'NEILL *Iceman Cometh* (1947) I. 75 He pulls a big roll from his pocket and peels off a ten-dollar bill. 1971 M. MCCARTHY *Birds of America* 181 With his thumbnail, unobtrusively, he peeled off the price-tag. 1977 *Monitor* (McAllen, Texas) 31 May 1B/1 She got out of the car and peeled off her gloves. 1977 H. FAST *Immigrants* I. 35 He..took out a wad of bills, peeling off two fives and two singles. 1978 D. MURPHY *Place Apart* xii. 259 We *must* peel off the terrorist labels and look at the individuals underneath.
c. *to peel one's eyes*: to keep one's eyes peeled. See PEELED 4 b in Dict. and Suppl.
1875 J. G. HOLLAND *Sevenoaks* xii. 161 An' peel yer eyes, Mike, for I'm goin' to show ye some thin' that'll s'prise ye. 1947 A. MILLER *All my Sons* I. 12 Now go out, and keep both eyes peeled... A policeman don't ask questions. Now peel them eyes! 1976 T. HEALD *Let Sleeping Dogs Die* v. 99 I've been peeling my eyes... There are some funny goings-on going on.
5. a. (Further examples.) Also const. *off.*
1879 *Boy's Own Paper* 18 Jan. 2/1 'Look sharp and peel!' cried our captain. So we hurried to the tent and promptly divested ourselves of our outer garments. 1922 [see *DISHYBILLY]. 1922 JOYCE *Ulysses* 490 Come and I'll peel off. 1950 *Variety* 13 Dec. 1/5 The gals are peelin' in 23 clubs through Los Angeles County.
e. *to peel off, away* (Aeronaut.): to veer from a straight course, esp. one alongside another aircraft; to break away from an airborne formation. (Said of the aircraft or of the pilot.) Also *transf.*
1941 *Christian Science Monitor* 6 Mar. 4/8 Other fanciful R.A.F. Terms include..'peeling off', for veering away from another aircraft. 1941 *N.Y. Times* 27 July 21/2 To 'peel off' is to curve away from another aircraft—the movement as one machine comes up close to another and then slants away is supposed to resemble the act of peeling off the skin of a banana. 1941 *Reader's Digest* Dec. 59/2 Our fighters seemed to be doing a good job on the Huns because only one peeled off to attack us. 1943 HUNT & PRINGLE *Service Slang* 51 *Peel off*,..to break away from a formation in order to meet an attack, or to leave a squadron to initiate an attack. 1945 *News Chron.* 7 June 3/1 The Spitfire came out of the clouds high above us, peeled off in our direction, circled around for a bit and then, apparently satisfied, made off. 1953 'N. SHUTE' *In Wet* vii. 212 He dismissed the escort [of fighter aircraft].., and they peeled away up into the clear blue sky. 1976 A. WHITE *Long Silence* vii. 53 We had picked up our fighter escort... Every so often, one of them would peel off and sweep an observation circuit. 1976 *Shooting Times & Country Mag.* 16–22 Dec. 29/2, I shot at another goose which peeled off, flew fluttering for a hundred yards, and then windmilled down stone dead. 1977 M. BABSON *Lord Mayor of Death* xviii. 191 The police car..peeled away from the procession and sped for open ground.
f. To leave, depart; to move *off* in another direction. Also *transf.* and *fig.*
1952 S. SELVON *Brighter Sun* iii. 52 'Well, one for de road, Ah peeling off now,' and after the drink he waved his hand and departed. 1958 *Times Lit. Suppl.* 31 Oct. 621/3 The drug-peddling charlatan artist peels off with the German homosexual who has more money than he knows what to do with. 1960 *Tamarack Rev.* xiv. 25 The way he have it figure out, if he stay in the work he have now, he going to be able to peel off and spend the summer on the Continent. 1968 *Surfer Mag.* Jan. 47/3 Ten-foot waves that peel off in good right and left slides. 1968 W. WARWICK *Surfriding in N.Z.* 21/2 When waves break over a sandbar they tend to either dump or peel off along the beach depending on the level of the tide. 1970 N. ARMSTRONG et al. *First on Moon* ii. 38 Just short of the point where a main highway peels off westward.., motels and cocktail lounges..nest alongside gleaming new buildings. 1974 D. GRAY *Dead Give Away* v. 54 'What do we do between tea and dinner?' asked Tony. 'I peel off to my room and read,' said Bob. 1976 *Sounds* 11 Dec. 29/4 The rhythm picks up and Ponty and Stuermer run up and down a soaring, emotive riff together before peeling off to indulge in their own fantasies in turn.
g. *spec.* In Rugby football, to move *off* in various directions from a set formation. Also const. without *off.*
1960 *Times* 24 Oct. 14/1 It was a joy to watch the smooth way in which they peeled off from a tight scrummage. 1960 *Rugby World* Nov. 6/2 From the line-out.. they aim to burst through frontally, either by the now familiar 'peeling off' method introduced by their late captain, Lucien Mias, or by extending the line and creating gaps through which a break can be made. 1963 *Times* 28 Jan. 4/6 Abetted by little Lacroix at their heels, they peeled off in all directions in concerted changes of the focal point of attack. 1970 *Times* 13 Apr. 6 Wiltshire peeled from the lineout for a try after five minutes of incessant pressure. 1977 *Western Mail* (Cardiff) 5 Mar. 18/1 They were trailing only 6-7 at the interval after No. 8 Roger Lane had peeled from a close range scrum to put Adrian Jones diving at full stretch for a try.

peel, *v.*² Add: Also in wider use in *Curling.* (Later examples).
1921 *Glasgow Herald* 25 Aug. 4/7 The Scottish Tourists ..played a two-rink game [of bowls] at Balham yesterday, 'peeling' at 19 on one and losing the other by 12. 1950 *Scotsman* 9 Aug. 7/3 The players peeled at several stages in the game and were 17–17 at the seventeenth end. 1962 *Even. Dispatch* (Edinburgh) 29 Jan. 7 The last named fought back in the closing stages to peel the game [sc. curling] at 11 at the 13th. 1969 R. WELSH *Beginner's Guide Curling* iv. 33 *Peels*, to be equal in shots. 1971 *Rand Daily Mail* 27 Mar. 23/3 Thomson..came back into the game and peeled 15–15 on the 20th end.

peel (pīl), *v.*³ [f. the name of Walter H. *Peel*, founder of the All England Croquet Assoc. and a leading exponent of the practice.] In Croquet, to send a ball other than one's own through a hoop. So **pee·ling** *vbl. sb.*², **peeled** *ppl. a.*²
1899 L. B. WILLIAMS *Croquet* iv. 123 Closely allied to this idea that the partners must be kept together at all hazards and, in a measure, dependent upon it, is the notion that it is a player's duty to put his partner through a hoop when the balls are both for the same point. This manœuvre is occasionally necessary, especially at the last hoop. It is called 'peeling', after its greatest exponent, the late Mr. Walter Peel... Mr. Peel practised it with great accuracy and success. 1914 LD. TOLLEMACHE *Croquet* xviii. 110 The attempt is sometimes made in the second break to 'Peel' your first ball through its remaining Hoops during the course of your second break. 1960 E. P. C. COTTER *Tackle Croquet this Way* 77 To peel is to cause a ball other than the striker's ball to run its hoop in order. *Ibid.* 78 You can peel firmly and confidently, and there is less chance of Black sticking in the hoop. 1961 *Croquet* ('Know the Game' Series) 36/1 *Peel*, to send a ball other than your own through its own hoop. 1966 MILLER & THORP *Croquet* ix. 86 The better players are capable of peeling a ball through two or three hoops during a break... A ball peeled through its hoop in order scores that hoop. 1976 *Denbighshire Free Press* 8 Dec. 12/5 (Advt.), Do you find Croquet, the Croquet of Breaks and Bisques, baffling? Or do you Peel with consummate ease?

peelable (pī·lăb'l), *a.* [f. PEEL *v.*¹ + -ABLE.] Suitable for peeling, capable of being peeled. Hence **peelabi·lity.**
1958 [see *CELLULOSE *sb.* c]. 1966 G. N. LEECH *Eng. in Advertising* xx. 178 *Peelability*..is not the sort of quality one would normally have occasion to mention, in talking about oranges or anything else. 1971 D. POTTER *Brit. Eliz. Stamps* xv. 163 A good-quality hinge is almost completely peelable. 1976 *Wireless World* Nov. 11 The reverse side..is covered by a peelable, tough vinyl insulation.

peeled, *ppl. a.*¹ Add: **4. b.** Now also *U.K.* (Further examples.)
1853 *Daily Morning Herald* (St. Louis, Missouri) 6 Jan. 2/1 Young man! Keep your eye peeled when you are after the women. 1872 E. EGGLESTON *End of World* xxvii. 186 [It would] teach the fellow to let monte alone, and keep his eyes peeled when he traveled. 1886 H. STEVENS *Recoll. J. Lenox* 45 In reading catalogues and reports from all parts of the world, one eye at least was always kept peeled for his desiderata. 1906 *Springfield* (Mass.) *Weekly Republ.* 20 Sept. 16 The carpenters..are keeping their 'eyes peeled' for the many coins which have..slid between the planks. 1918 E. O'NEILL in *Smart Set* June 96 We'll have to keep an eye peeled from now on. I know 'em. 1923 L. J. VANCE *Baroque* xvi. 95 He sent a request to the door-porter to keep an eye peeled and let him know if the cab.. seemed disposed to tarry in the offing. 1945 E. NEWHOUSE in *55 Short Stories from New Yorker* (1949) 238 Keep your eyes peeled and let me know if you come across Aladdin's lamp. 1956 B. HOLIDAY *Lady sings Blues* (1973) xxi. 167 If you're doing something wrong, you *know* it and you've got at least one eye peeled looking for trouble. 1974 'R. TATE' *Birds of Bloodied Feather* viii. 163 Keep your eyes peeled for a break in the mist.

peeler¹. **2. b.** Delete '*Sc.* and *north. Irel.*' and add: Also *attrib.* (Later examples.)
1911 C. O. MINCHIN *Sea-Fishing* xvii. 245 When the cold outer shell has cracked and is ready to fall, and the new soft shell beneath is growing to replace it, the crab is called a 'peeler', and is then in the best condition for bait. 1934 *Sun* (Baltimore) 27 July 9/8 Hook-and-line fishermen can scarcely buy the popular peeler crab bait for fishing. 1956 M. KENNEDY *Salt-Water Angling* x. 321 The crab is perhaps most useful as a bait shortly before it moults—when it has developed the new shell beneath the old one,..the angler anticipating nature by stripping the old shell off it. In this stage it is known variously as a peeler, peel, pill or shedder crab. 1971 *Angling Times* 10 June 3 Van driver Colin drifted his peeler crab bait around the rocks on float tackle. 1976 *Scottish Daily Express* 24 Dec. 12/6 For the smaller fish, rag and lugworm, soft peeler crab and strips of herring or mackerel are usually acceptable.
3. *U.S.* An exceptional or noteworthy example of anything; *spec.* a violent storm.
1823 J. F. COOPER *Pioneers* I. xv. 212 It's a peeler without, I can tell you, good woman. 1834 C. A. DAVIS *Lett. J. Downing* 331 The Captain was all the while boastin of [his boat] the 'Two Pollies'; and well he might, for she was a peeler. 1861 *Entertaining Things* I. 197 The gale..was a steady hard blow, what sailors call a peeler.
b. A person of exceptional or unusual qualities; a lively or energetic person.
1833 S. SMITH *Life & Writings J. Downing* 218 Them are Pennsylvany chaps are real peelers for electing folks when they take hold. 1834 C. A. DAVIS *Lett. J. Downing* 88 If he does turn broker, you'll hear more on him; for he's a peeler I tell you. 1844 'J. SLICK' *High Life N.Y.* I. 82, I was talking with a rare peeler of a gal. 1869 MRS. STOWE *Oldtown Folks* 117 She was spoken of with applause under such titles as 'a staver', 'a pealer', 'a roarer to work'. 1881 W. M. THAYER *From Log Cabin to White House* xiii. 207 He's a peeler for work, too; ain't afraid to dirty hisself.

4. One who removes his clothing: *spec.* a pugilist ready to strip for a fight.
1852 *As Good as Comedy* iv. 56 'I know you hain't got the teeth to raise the skin of that varmint.' 'Hain't I, then? Just you try it, then,..and see if I ain't a peeler.' 'Will you peel?' 'Won't I, then?' 'Jake, my boy, I've come here to-day to strip the skin off you altogether.'
b. A strip-tease artist; a stripper.
1951 *Variety* 20 June 53/2 (*heading*) Old Peelers Fade Away in St. Louis Crackdown. *Ibid.*, A steady exodus from the city of strippers in..St. Louis has been reported... It results from the crackdown of the law, with the peelers being replaced by jugglers, singers and hula dancers. 1951 GREEN & LAURIE *Show Biz* 570/2 *Peeler*, stripteaser. 1955 *Variety* 9 Mar. 62/4 Chicago, for years a stripper's haven, is now seriously beset by a dearth of peelers. 1961 A. BERKMAN *Singers' Gloss. Show Business* 66 *Peeler*,..strip-teaser.
5. *U.S.* A cowboy.
1894 O. WISTER *Out West: Jrnls. & Lett.* (1958) 198 *Peeler*, cowpuncher. 1902 *Out West* June 623 The 7 TX peelers is all in on this play,..and the 10 EC outfit will want a hand too. 1903 [see *ACT *v.* 9 f]. 1914 B. M. BOWER *Flying U Ranch* 7 This is Mr. Mig-u-ell Rapponi, boys—a peeler straight from the Golden Gate. 1937 *Dialect Notes* VI. 618 Used in its strictest sense, the *peeler* refers to the cowpuncher who rides into the herd and 'cuts' out the horse desired. 1943 L. V. HAMNER *Short Grass* 163 The driver, or 'peeler', rode the wheel horse and guides the whole team with one line.
6. In full, *peeler log.* The trunk of a tree, esp. a softwood one, suitable for the manufacture of veneer by the use of a rotary lathe, which peels thin sheets of wood from the log.
1935 *Timberman* Dec. 3/2 Some studies should be undertaken at once to determine the economic limits of the Douglas fir peelers... The log scarcity has brought forth the Veneer specialist, who makes a business of producing and preparing peeler logs for the plywood trade. 1942 WOOD & LINN *Plywoods* vi. 179 From the [North American Pacific Coast] forests the 'peeler' logs, in lengths up to 40 feet, are conveyed by water, rail or road to the plywood mills. 1948 *Q. Jrnl. Forestry* XLII. 33 Poplar plywood could be manufactured if more peeler logs were grown. 1958 W. F. MCCULLOCH *Woods Words* 133 Peeler... A log suitable for plywood. 1966 A. W. LEWIS *Gloss. Woodworking Terms* 65 *Peeler log*, log from which veneer is cut by the rotary process of peeling. 1973 *Nature West Coast* 62 Wood [of grand fir] is used for pulp, lumber and peelers (core stock for plywood).

peeling, *vbl. sb.* Add: **1. d.** (Further example.)
1938 [see *OUTSTRIP *v.*²].

peel-off (pī·lǫf), *sb.* and *a.* [f. phr. *to peel off*: see PEEL *v.*¹ 3, 5 in Dict. and Suppl.] **A.** *sb.* The action of peeling off. **B.** *adj.* Designating or pertaining to this action. Also *fig.*
1939 *Flight* 18 May 504a (*caption*) The peel-off. Westland Lysander monoplanes..start a mock dive-bombing attack. 1962 D. WARNER *Death of Bogey* iv. vi. 168 The Lanes did the peel-off. Ran it by lorry. 1972 D. E. WESTLAKE *Cops & Robbers* (1973) xi. 149 They'd busily switched the license plates on the squad car and put the new peel-off numbers on its sides.

peely-wally (pī·liwǫli), *a.* *Sc.* Also **peelie-wallie** and as one word. ['Orig. prob. imit. of a whining, feeble sound' (*Sc. Nat. Dict.*). Cf. Eng. dial. *pee-wee* whining, small (E.D.D.), and WALLYDRAG.] Pale, feeble, sickly, ill-looking.
1832 A. HENDERSON *Proverbs* 208 *Peelie*, thin; meagre. —*Peelie-wallie*, thin; sickly. 1833 J. KENNEDY *Geordie Chalmers* 81 But may I ride on the win' wi' auld Nanee Logan, the witch o' Glenteerie, when I gang to siccan a peely-wally concern again! 1895 W. STEWART *Lilts* 104 The sun sen's forth its flickerin' rays, Fu' peely-wally wan. 1904 'H. FOULIS' *Erchie* xii. 73, I was a kind o' eccentric peely-wally soul, because I sometimes dried the dishes. 1932 'L. G. GIBBON' *Sunset Song* 278 And damn it, if before a twelvemonth was up she didn't have a bairn, a peely-wally girl. 1945 B. FERGUSSON *Lowland Soldier* 25 Ye'd say he was thin, Peelywally, bow-leggit and shilpit. 1962 A. MACLEOD *Eighth Seal* vi. 71 The wee Englishman is too peelywally to start any scrapping. 1966 K. WHITE *Lett. from Gourgounel* 96 A snail..her long peelie-wallie neck with two horns slowly prodding the air.

peen, *sb.* Add: Now usu. written **pein** (pīn, pēin). Add to def.: Hence, the other end of a hammer-head from the face, whether sharp-edged or rounded. (Further examples.)
1885 *Spons' Mechanics' Own Bk.* 84 This process is termed 'paning' or 'pening', from the pane or pene of the hammer being generally used to perform it. 1904 W. H. VAN DERVOORT *Mod. Machine Shop Tools* (ed. 4) i. 17 The machinist hammer..is made of high-grade steel, carefully tempered on head and pene. 1939 *Specification for Hand Hammers* (*B.S.I.*) 7 On completion, the forgings shall be hardened on the striking faces and peins only. *Ibid.* 12 (*caption*) Engineer's ball pein hammer. 1943 D. J. SWARTZ et al. *Fund. Shopwork* (1945) i. 27 The peen hammer is widely used. The head is so shaped that one end, called the peen, can be used to produce dents or depressions or to set up tension in metal. 1947 J. C. RICH *Materials & Methods Sculpture* vi. 182 In working the heavier thicknesses of $\frac{3}{16}$ inch and the $\frac{1}{4}$ inch, larger ball pein hammers are required. 1961 *B.S.I. News* Oct. 24/1 Most of the trouble was due to..heads too hard on either the striking face or the pein. 1964 S. CRAWFORD *Basic Engin. Processes* i. 16 The three types of hammer most generally used by the fitter are (a) Ball Pein, (b) Cross Pein, and (c) Straight Pein... All three are of a standard shape at the striking face end but can be readily identified by the shape of the opposite end known as the pein. 1971 B. SCHARF *Engin. & its Lang.* ix. 63 Ball-pein hammers have one ball-shaped end (pein, peen) for hand-riveting or burring over.

peen, pene, *v.* Delete '*Obs.* exc. *dial.*' and add examples of the sense: To strike with the pein of a hammer. (Usu. written **peen.**)
1888 J. ROSE *Mod. Machine-Shop Pract.* II. xxv. 162/2 The side faces of the arms would require to be pened. 1902 *Internat. Libr. Technol.* IV. § 44.22 Iron may be stretched by peening it; that is, by striking it with the peen of a hammer. This method is often used to loosen a collar or a nut. 1941 A. C. DAVIES *Sci. & Pract. of Welding* ii. 118 In the arc welding of cast iron the risk of fracture is definitely reduced if the short beads of weld metal are lightly peened immediately after they have been laid. 1950 *Engineering* 31 Mar. 345/3 Each run of weld is peened before the next run is applied. 1964 S. CRAWFORD *Basic Engin. Processes* i. 16 The bench fitter frequently uses different types of hand hammers for fitting operations requiring force, e.g...peening over the end of rivets. 1977 *Good Motoring* Mar. 32/2 A leak between two cylinders caused through a plug in the underside of the head which had been 'very poorly peened in place'.
Hence **pee·ning** *vbl. sb.* (see also *shot peening* s.v. *SHOT *sb.*¹ 30).
1885 [see *PEEN *sb.*]. 1905 W. S. LEONARD *Machine-Shop Tools & Methods* (ed. 3) ii. 33 The peening affects only the outer surface of the shaft, and the turning removes this outer surface, thereby partially neutralizing the effect of the peening. 1907 *English Mechanic* 23 Aug. 49/1 After the bottom is slipped on the paning down can be done with a paning hammer.., or..with a sheet-metal worker's common hammer. 1960 *Times Rev. Industry* Jan. 36/2 The peening of springs for increasing the fatigue life, particularly for the motor industry, can be done in air operated plant as well as the airless type. 1964 J. G. TWEEDDALE *Mech. Properties of Metals* vii. 125 Residual stresses may be induced in several ways, such as surface rolling and peening. Peening can take the form of hammer peening or shot peening.

peent (pīnt), *sb.* *N. Amer.* Also **peeent.** [Echoic.] A representation of the high whistling sound emitted by a woodcock. Also *attrib.*
1895 F. M. CHAPMAN *Handbk. Birds of Eastern N. Amer.* 153 [The Woodcock] begins on the ground with a formal, periodic *peent, peent*, an incongruous preparation for the wild flight that follows. 1925 E. H. FORBUSH *Birds Massachusetts* I. 386 Voice [of woodcock].—A nasal 'peent' something like common note of Nighthawk. 1931 *Sun* (Baltimore) 11 Apr. 4/7, I heard a woodcock, 'peent' sing, And I saw a hum-bird rise. 1944 L. A. HAUSMAN *Illustr. Encycl. Amer. Birds* 438/2 This song [of the woodcock] is introduced, and concluded, by a series of nasal peeent notes, much like the flying calls of the Nighthawk. 1953 MURPHY & AMADON *Land Birds Amer.* iii. 83/1 Both before and after these flights, they [sc. Woodcocks] utter a nasal call, usually written as peent. 1965 A. WETMORE et al. *Water, Prey, & Game Birds N. Amer.* 330/1 The woodcock zigzags rapidly down to earth with chirping notes. He struts across a patch of open ground, sounds his peent notes again, then soars to repeat the aerial act.

peent (pīnt), *v.* *N. Amer.* [f. the sb.] **a.** *intr.* To emit a high whistling sound characteristic of a woodcock. **b.** *trans.* To utter by making such a sound. So **pee·nting** *vbl. sb.*
1897 F. M. CHAPMAN *Bird-Life* vii. 103 Unless disturbed, he [sc. the woodcock] will probably return to near the spot from which he started and at once resume his *peenting*. 1956 PETERSON & FISHER *Wild Amer.* v. 56 Over downtown roof-gardens the 'peenting' of cruising nighthawks can often be heard. 1973 M. CROWELL *Greener Pastures* 197, I hear a woodcock 'peenting' over on apple tree hill. 1976 *Kingston* (Ontario) *Whig-Standard* 22 Oct. 14/1 Often, shortly after sunset, the male [woodcock] peents a nasal sound. *Ibid.*, Once landed, the peenting resumes.

peeny-wally (pī·niwǫli). Also **peeni(e)wally, peeny(-wauly).** [Etym. uncertain; perh. *PEELY-WALLY infl. by Jamaican dial. *peeny* small (see also E.D.D.).] A Jamaican name for a firefly, esp. the largest of those found on the island, *Pyrophorus plagiophthalmus.*
1907 W. JEKYLL *Jamaican Song & Story* II. 184 'Peeny' is the Candlefly, which shines like my donkey's coat. 1961 F. G. CASSIDY *Jamaica Talk* xiii. 292 The big one [sc. firefly] is also *peeny* or *peeny-wauly*, an unexplained word, though *wauly* may refer to its shining 'eyes'. 1962 S. WYNTER *Hills of Hebron* i. 23 When the rains stopped, the 'peeniwallies', hundreds of them, fluttered transparent wings against the lamp. 1971 *Islander* (Victoria, B.C.) 17 Jan. 2/3 These fireflies are affectionately called 'peeniewallies' by the islanders [sc. Jamaicans].

peep, *sb.*¹ Add: **2. b.** A slight sound or utterance; a single item or piece of information, chiefly in neg. contexts. Cf. PIP *sb.*² 1 b.
1903 in *Eng. Dial. Dict.* 1908 R. W. CHAMBERS *Firing Line* xxiv. 411 Nobody's heard a peep from you. What on earth do you mean by this? 1928 S. LEWIS *Man who knew Coolidge* I. 13 I'd never made a peep about how maybe it'd be a good stunt for him to go out and maybe earn a little money on the side. 1954 *Picture Post* 2 Jan. 34/3 'One more peep out of you, Mister, and I'll get the boys to push you and your b— stall in the oggin', which was a nearby canal. *a* 1974 R. CROSSMAN *Diaries* (1975) I. 275 Not a peep came out of any of them. 1974 T. P. WHITNEY tr. *Solzhenitsyn's Gulag Archipel.* I. i. 10 They take you from a military hospital with a temperature of 102, as they did with Ans Bernshtein, and the doctor will not raise a peep about your arrest. 1977 P. COSGRAVE *Cheyney's Law* v. 47, I know coppers... They're not supposed to be able to take it... But there's not been a peep out of him... Maybe..he's tougher than you are.
4. a. Delete *Obs.* and substitute for def.: A young chicken. (Later *U.S.* examples.) Cf. *PEE-PEE¹.
1931 *Amer. Speech* VII. 20 Peeps. Little chickens. 1935 *Ibid.* X. 171/1 Peep. A chick just hatched, or a small chicken. 1943 *Sun* (Baltimore) 10 Aug. 10/3 The Boonsboro *Times* reports the birth there of a 'peep' with no eyes and no sign of an eye.

peep, *sb.*² Add: **1. d.** *dial.* and *U.S.* After a negative, a short interval (of sleep), a wink.
1905 R. BEACH *Pardners* (1912) ii. 49 Most people called him crazy, 'cause he had fits of goin' for days without a peep.
4. peep-joint, a place where striptease is performed; **peep-machine,** a machine through which a peep-show is seen; **peep-toe** *attrib.*, designating a kind of shoe whose tip is cut away allowing the toes to be seen; also *absol.*; also **peep-toed** *a.*
1960 *News Chron.* 23 Sept. 10/1 Jayne is..head stripper in the Pink Flamingo, a gilded peep-joint. 1938 G. GREENE *Brighton Rock* III. ii. 117 The motor-track, the shooting booths and peep machines. *Ibid.* III. iii. 130 Framed snapshots of King Edward VII (Prince of Wales) in a yachting cap and a background of peep machines. 1939 *Vogue* 3 May 5 (Advt.), Blue calf thonging and heel sets off this peep-toe sandal in blue calf, 45/9. 1940 *Ibid.* June 14 (Advt.), Only Joyce..would think of using duckskin and calf in this peep-toe creation for your every leisure hour. 1957 *Observer* 25 Aug. 11/7 A young French lady named Tracy, in peeptoe shoes and ice-blue, skin-tight gown with a good deal of cleavage. 1960 *News Chron.* 6 June 5/2 A blade in a tweed coat and peep-toe sandals. 1968 J. IRONSIDE *Fashion Alphabet* 133 *Peep-toe*, a rather coy name for any shoe, usually court or sling-back, from which the toe-cap is cut away to expose part of the toes. 1969 E. WILSON *Hist. Shoe Fashions* xx. 247 The war years of 1939–45 saw..a heavy restriction in shoe-making materials... Peep-toes and big chunky trimmings decorated the very popular court shoe. 1976 *Vogue* 15 Mar. 77 Pale blue and navy peep-toe sandals. 1953 K. TENNANT *Joyful Condemned* xix. 173 A really pretty pair of peep-toed, high-heeled sandals.

peep (pīp), *sb.*⁴ *U.S.* [see quot. 1943.] = *JEEP *sb.*
Sources are divided as to whether peep should designate a vehicle larger or smaller than a jeep, but are agreed in referring both terms to the same type of vehicle.
1941 *N.Y. Herald Tribune* 28 June 14/3 Peep (son of a jeep) means a bantam car. 1941 *N.Y. Times* 26 Oct. xx. 3/2 The one-half ton 'jeep' command reconnaissance car, its name taken from the model designation 'G.P.', and one-quarter ton 'peep' reconnaissance cars are combat vehicles. 1943 *Amer. N. & Q.* III. 137/1, I laid down an editorial ukase that the $\frac{1}{2}$-ton truck was thereafter to be the 'jeep' and the $\frac{1}{4}$-ton the 'peep'. 1946 *Amer. Speech* XXI. 245 In the Armored Force..the $\frac{1}{4}$-ton 4 by 4 is always a *peep*, the term 'jeep' being applied to the command and reconnaissance car. 1953 M. BURY *Rolling Wheels* 242 G.I.'s called a command car a *jeep*, called what you call a jeep, a *peep*—and bicycles were called *creeps*! 1962 *Amer. Speech* XXXVII. 78 The peep proved as versatile as the jeep, and had the advantages of a lower silhouette, less gasoline consumption, cheaper production, and the occupation of less cargo space when shipped.

peep, *v.*¹ Add: **2. b.** To betray a confidence; to inform. Also *trans.*
1911 J. LONDON *Let.* 6 Mar. (1966) 340 The convicts are few and far between who come out and dare to peep a word of what they know. *Ibid.* I have known ex-cons who became dead for peeping. 1950 H. E. GOLDIN *Dict. Amer. Underworld Lingo* 154/2 *Peep*, to betray associates; to give information to the police.

peep-bo. Add: (Further examples.) Hence **pee·p-boing** *vbl. sb.*
1818 'A. BURTON' *Adventures J. Newcome* II. 114 His Toes played peep-boh through his Shoes. 1828 A. ROYALL *Black Bk.* II. 137, I was not disposed to play at peep-bo with him. 1853 MRS. GASKELL *Ruth* I. vi. 145 After some 'peep-boing', she was about to snatch a kiss, when Harry ..hit Ruth a great blow on the face. 1974 P. DICKINSON *Poison Oracle* vi. 153 Dorah and Peggy were playing peep-bo round a hut.

pee-pee¹ (pī·pī). *U.S.* [Perh. onomatopœic, f. the sound made by the birds.] A young chicken, or, esp. in Jamaica, a young turkey.
1890 *Dialect Notes* I. 74 Pee-pee..a very small chicken. Eastern Pennsylvania. 1927 ANDERSON & CUNDALL *Jamaica Negro Proverbs & Sayings* 73 Cuss John Crow 'peel-head', turkey pee-pee bex [i.e. little turkey angry]. 1950 *Publ. Amer. Dial. Soc.* XIV. 51 Pee-pee: *excl.* A call to turkeys. Imitative of the cry of the poults. 1953 *Amer. Speech* XXVIII. 252 Pee-pee..A newborn or very young chick. 1961 F. G. CASSIDY *Jamaica Talk* v. 104 Young fowl, usually turkeys, are called pee-pees.

pee-pee² (pī·pī). [redupl. *PEE *sb.*⁵ Cf. Fr. *pipi.*] = *PEE *sb.*⁵
1923 J. MANCHON *Le Slang* 220 To do peepee. 1941 E. P. O'DONNELL *Great Big Doorstep* x. 143 Commado said, 'When them twins get in the show like lass time, one's gotta make pee-pee, the udda one gotta climb on the seat [etc.].' 1962 B. J. FRIEDMAN *Stern* I. 58 Do you still make peepee in your pants? 1972 H. C. RAE *Shooting Gallery* iv. 263 Time stopped. The pee-pee was cold and swollen in her tummy.

peeper¹. **2. b.** Substitute for def.: A small tree-frog of the genus *Hyla*, esp. *H. crucifer*, found in eastern North America. (Earlier and later examples.)
1857 S. H. HAMMOND *Wild Northern Scenes* 30 All is still now, save the piping notes of the little peeper along the shore. 1906 M. C. DICKERSON *Frog Bk.* 139 There are few people in the eastern United States who do not know the voices of the Spring Peepers... The Peepers have spring in their hearts. 1938 J. W. LIPPINCOTT *Animal Neighbours* (1940) xx. 192 The ridiculously small spring frog or peeper, an inch and a quarter long, comes first, sometime in March. 1961 D. M. COCHRAN *Living Amphibians of World* 113/1 The spring peeper, *Hyla crucifer*, is one of the earliest of the frogs to appear in spring. 1976 *Yankee* Apr. 78/2 Nearly every Easterner..has heard the spring peeper's song.

peeper². Add: **1. b.** A private detective or investigator; occas., a policeman.
1940 R. CHANDLER *Farewell, my Lovely* xxiii. 177 Peeper, huh, pally? From the big bad burg, huh? 1943 —— *Lady in Lake* (1944) xxxv. 185 Don't bother to call your house-peeper... I'm allergic to house-peepers. 1968 R. CLAPPERTON *No News on Monday* iii. 25 Police protection for a peeper. What kind of advertisement is that? 1968 E. MCGIRR *Lead-Lined Coffin* ii. 64 He..flipped the wallet open. 'A peeper,' he said. 1970 G. F. NEWMAN *Sir, You Bastard* ii. 54 'I have an arrangement with the peeper across the road. He's had a lean time lately.' His voice was level. Moving towards him, Angie said: 'Policemen are different.' 1975 R. L. SIMON *Wild Turkey* xxii. 160 What kind of bullshit are you throwing around, peeper? You want me to slap you in the can?
2. (Further examples.)
1858 A. MAYHEW *Paved with Gold* II. xii. 191 They went to work slogging, Jack delivering a 'head-acher'.. and Ned one not to be worked at on the 'peepers'. 1901 'H. MCHUGH' *John Henry* 93, I hate to have a girl plant her pleading peepers on me. 1908 A. N. LYONS *Arthur's* I. ii. 10, I jerked me peeper round, an' there was a girl squattin' on a tar barrel. 1928 E. WALLACE *Double* xiii. 204 Unless your poor old peepers are going wrong you would have seen them. 1937 G. FRANKAU *More of Us* iv. 42 She winked one peeper; But Innocent's eyes, downcast, still withheld her man Glances he craved. 1968 A. DIMENT *Bang Bang Birds* vii. 108 There was a leer in his rheumy peepers as he gave Marianne a mental strip. 1978 J. SYMONS *Blackheath Poisonings* III. 153 He had lost an eye long ago in the Ashanti campaign... 'I'm lucky to be alive, but I lost one of my peepers.'

peep-hole. Add: (Further examples.) Also *transf.* and *attrib.*
1952 E. F. DAVIES *Illyrian Venture* xi. 209 Every sentry looked in frequently through the peep-hole in the door. 1962 *Gloss. Terms Glass Industry* (*B.S.I.*) 17 *Peephole*, a small opening in a furnace wall through which observations are made. 1963 *New Yorker* 15 June 98/3 (Advt.), These fully washable shorties..sport dainty floral sprays of peephole petals. 1978 W. H. JOBSON *To die a Little* ix. 148 Every now and again someone lifted the flap of the round peephole in the [cell] door.

peepie-creepie (pī·pikrī·pi). [f. PEEP *v.*² + -IE and CREEP *v.* + IE.] A portable television camera used for close shots on location.
1952 *Time* 14 July 22/3 Most startling TV innovation was a portable camera known as the walkie-lookie, or peepie-creepie, with which the enterprising TV reporter could sneak up to Mr. Delegate and catch him yelling his head off or scratching his nose. 1952 *Newsweek* 21 July 53 NBC's walkie-lookie, or 'peepie-creepie', was not a singular success in its debut. 1953 A. K. C. OTTAWAY *Educ. & Society* 82 Here it was that the 'peepie-creepie' also called the 'walkie-lookie' began to achieve prominence. 1954 F. G. CASSIDY *Robertson's Devel. Mod. Eng.* (ed. 2) viii. 188 *Walky-talky* (a portable radio..), and its recent offspring *peepie-creepie* (a portable television camera).

peeping, *ppl. a.*² Add: (Later allusive examples of *Peeping Tom.*) Now freq. written *peeping Tom* and usu. applied to a prying person, esp. with connotations of prurience; one who obtains gratification from furtively observing women not fully clothed or the sexual activity of others, = *VOYEUR. Also *transf.*, *fig.*, and *attrib.* Hence *peeping Tommery*, the activity of a peeping Tom; also *peeping Tom-ism.*
1915 R. FROST *Let.* 11 Nov. (1964) 17 She executes a frightfulness. Somewhere else she brings in the Peeping-Tom idea. 1926 G. HUNTING *Vicarion* ii. 38 What sort was a man who did not instinctively respect the privacy of others?—not mere physical privacy, on which any Peeping Tom might intrude, but the infinitely more intimate thing, to spy upon which was violation. 1933 *Week-End Rev.* 8 July 34/2, I can assure you that neither reporters nor sub-editors find satisfaction in playing the rôle of Paul Pry or Nosey Parker or Peeping Tom. 1955 *Sun* (Baltimore) 7 Nov. 30/7 The House Judiciary Committee is sending its general counsel here from Washington today to attend the City Council hearing on a bill to outlaw electronic 'peeping toms'. 1958 *Times* 12 Nov. 3/4 The curiously prosaic *bizarreries* of Magritte, and the peeping-Tom eroticism of such basically crude pictures as those of Labisse and Delvaux, are of the 'photographic' kind. 1960 *Spectator* 21 Oct. 602/3 Those semi-nude gambols [in a cinema film]..break any feeling of reality second by second by giving one a sense of peeping-tommery. 1963 A. HERON *Towards Quaker View of Sex* 67 *Exhibitionism* and the associated disorder *Voyeurism* (peeping-tom) involve sexual pleasure obtained respectively from displaying the naked body..or from observing the sexual acts or organs of other people. 1966 AUDEN *About House* 42 Peeping Toms Are never praised, like novelists or bird watchers, For their keenness of observation. 1966 *Guardian* 26 Mar. 1/6 They've had Peeping Tom cameras and now they've got eavesdropping microphones. 1972 W. P. MCGIVERN *Caprifoil* i. 9 Surveillance was a constantly expanding and proliferating industry... He did have a professional view of this constantly expanding peeping Tom-ism, and it was a sour one. 1974 M. KELLY *That Girl in Alley* vii. 120 To avoid bringing myself into suspicion of peeping-tommery I moved a few steps down the yard, out of sight of the drilling class. 1977 *Transatlantic Rev.* LX. 38 At various times during the preceding year, there had been complaints of Peeping Toms.

peep of day. Add: **2.** *peep-of-day boys* (earlier and later examples). Also *transf.*
1780 A. YOUNG *Tour in Ireland* II. vi. 30 In England we have heard much of whiteboys, steelboys, oakboys, peep-of-day-boys, &c... All but the whiteboys were among the manufacturing protestants in the north. 1922 JOYCE *Ulysses* 44 Raw facebones under his peep of day boy's hat.

peep-show. Add: (Earlier and later examples.) Also *attrib.*
1851 *Househ. Words* II. 290/1 There were tambourines, books, work-boxes..peep-show boxes, all kinds of boxes. 1894 [see *donkey-ride* s.v. *DONKEY 3 a]. 1914 G. B. SHAW *Misalliance* 67, I did a cheap trip to Folkestone. I spent sevenpence on dropping pennies into silly automatic machines and peepshows of rowdy girls having a jolly time. 1937 L. MACNEICE *Poems* 110 It's no go the merry-goround, it's no go the rickshaw, All we want is a limousine and a ticket for the peepshow. 1941 J. MASEFIELD *In Mill* 110, I had seen him box in one of the little primitive peep-show cinemas. 1951 M. MCLUHAN *Mech. Bride* (1967) 48/1 *Sexual Behavior in the Human Male* is a penny arcade peep show given the chrome treatment of scientific charts and figures. 1973 *Times* 22 Mar. 8/7 The bookshops, the peep shows, the pornographic cinemas..are all nasty. 1976 *National Observer* (U.S.) 18 Dec. 1/3 It sits in leased quarters on the eighth floor of a commercial building in a neighborhood of numerous adult book-stores and peep-show parlors on 9th street in Washington.

peepy, *a.* (Earlier and later examples.)
1833 *Chambers's Edin. Jrnl.* 5 Jan. 385/1 He is then a poor, peepy wretch, with blear eyes. *Ibid.* 27 Apr. 97/3 The contrast..between the meaning of the words, and the poor, peepy..voice in which they are given. 1939 N. MARSH *Overture to Death* xii. 117 It was the pot-boy, very tousled and peepy, and accompanied by a gust of stale beer. Alleyn thought that he looked like all pot-boys at dawn throughout time and space.

peer, *sb.* Add: **2. b.** *Anthropol.* and *Sociol.* An equal; a contemporary; a member of the same age-group or social set. Also *attrib.* (see also sense 6a below).
1944 C. M. TRYON in *Nat. Soc. Study Educ. Yearbk.* (U.S.) I. xii. 233 Although in meeting some adult standards girls must undergo less change than boys, in their relation to their own peer culture they must often be more adaptive in relation to changing requirements. 1953 A. K. C. OTTAWAY *Educ. & Society* 109 What is called 'peer culture', in other words the community of the same age, has a very great influence on the individual. 1958 W. J. H. SPROTT *Human Groups* 71 Wherever there is an adolescent 'peer-culture' it will have an influence which competes with that of the home. 1964 MINTURN & LAMBERT *Mothers of Six Cultures* 288 Nuclear family cultures are least punitive of peer-to-peer aggression. 1966 BEREITER & ENGELMANN *Teaching Disadvantaged Children* i. 18 In reading down the list of contrasts between the home and the nursery school environment, one finds..the same contrast could be drawn between the upper-middle-class and the lower-class child's environment: adult versus peer contacts, [etc.]. 1966 *Listener* 14 Apr. 535/2 With their peers—that is, people of the same age, sex, and status—adolescents really experience for the first time relationships embodying equality and democracy. 1971 *New Society* 18 Nov. 975/2, I have discovered that wild-born monkeys spend more time watching a live peer, than themselves, in mirrors. 1972 *Where* Mar. 95/3 You need to consider the quality of the relationship which exists between your son and the teachers, your son and his peers, and between you and the teachers. 1972 *Jrnl. Social Psychol.* LXXXVI. 111 A subject receiving positive evaluations from a group of his peers was more active in the group. 1973 *Jrnl. Genetic Psychol.* CXXII. 263 Studies of human peer-social behavior in the preschool period. *Ibid.* 275 Retrospection and Peer-Perception scores did not differ from each other. *Ibid.* CXXIII. 124 Firstborn children have greater verbal facility, and there is evidence that they have more successful relationships with their teachers than do the later born although later born are more often successful in their peer relationships. 1975 *Nature* 4 Dec. 382/1 At the heart of the inquiry is the so-called peer-review system, which is used in some shape or form by virtually every government agency which supports academic research. 1977 *Listener* 7 Apr. 427/3 In the relatively public competition between rival research groups seeking financial support, there is no practical alternative to 'peer review' by committees of experts in the relevant fields. 1977 *Sci. Amer.* Sept. 177/3 Alternatively a distributed-processing system can be organized into a peer structure. All the computers..can communicate with one another on an equal footing.
4. Now also a man elevated to the peerage on a non-hereditary basis; = *life peer* s.v. LIFE *sb.* 17 in Dict. and Suppl. Also = *Peeress in her own right.*
1869 [see *life-peer* s.v. LIFE *sb.* 17]. 1958 *Oxford Mail* 21 July 1/7 The announcement will enable the lifepeers to take their seats in the Lords in time for the opening of the next session of Parliament in November. It will be the first time women peers have been allowed to sit in the Lords. 1958 *Times* 24 July 8/7 The main point of this interim constitutional reform [sc. introduction of life peerages] was to enlarge Labour representation in the Upper House with working peers. 1974 *Observer* (Colour Suppl.) 24 Mar. 30/1 'We are very passionate that we are not peeresses; peeresses are the wives of peers...' Now lavatories are discreetly marked 'Peers' and 'Women Peers'.
6. a. (sense *2 b) **peer group,** a group of people, freq. a group of adolescents, of the same age or social status.
1943 BRECKENRIDGE & VINCENT *Child Devel.* xiii. 463 Although the adolescent declares his independence of adult standards and controls, he is actually very dependent upon conformity with his peer group. 1948 J. H. S. BOSSARD *Sociol. of Child Devel.* vi. xxi. 494 This element of antagonism between peer groups and adults has been fostered through the years in a great many ways. 1957 P. LAFITTE *Person in Psychol.* viii. 104 Their standing in their peer group at school. 1959 *Listener* 19 Feb. 324/2 Each man's situation is slightly different from those of his peer-group. 1961 [see *INNER *a.* (*sb.*²) 1 n]. 1964 GOULD & KOLB *Dict. Social Sci.* 297/1 One of these concepts is that of *peer group* which commonly refers to a group of homogeneous age composition. There is no reason, however, why this term cannot be applied to a group whose members are equal in some respect other than age. 1964 M. ARGYLE *Psychol. & Social Probl.* xv. 187 Young people between puberty and marriage are consciously aware of belonging to this peer-group, wear distinctive clothes, meet in groups at coffee-bars, [etc.]. 1966 *Listener* 14 Apr. 535/2 The gang is a relatively authoritarian form of peer group. 1972 J. L. DILLARD *Black English* i. 34 In the ghetto culture, peer group relations govern the social activity, including language, of the child to a degree far beyond its importance among middle-class whites. 1976 *Broadcast* 29 Mar. 19/1 A desire..to be seen to be involved with peer-group heroes.

peerie (piə·ri), *a.* (*sb.*) *Sc.* (now only *Orkney* and *Shetland*). [Etym. uncertain: see S.N.D.] **A.** *adj.* Small, diminutive. Also in special collocations (see quots. and S.N.D.).
1808 JAMIESON, *Peerie, adj.*, little, small. *A peerie foal*, a small bannock or cake, Orkn. Shetl. 1868 D. GORRIE *Summers & Winters in Orkneys* 12 An *Oyce* or inlet, locally termed the 'Peerie Sea'. 1871 R. COWIE *Shetland* II. iv. 149 The *peerie* steamer—as the natives call her, in contradistinction to the larger one trading to the south. 1884 *Rep. H.M. Comm. Inquiry Crofters Highlands Scotl.* 1404 in *Parl. Papers* (C. 3980) XXXIV. The possession of even four ures of land constitutes the proprietor a small, or, in the vernacular, 'peerie' laird. 1891 J. J. H. BURGESS *Rasmie's Buddie* 80 An sün her peerie winkie haands O' cockaloories bricht wis fu. 1916 G. W. HOGGAN in B. Thynne *Shetland Sheep-Dog* 14 The 'peerie' Shetland Collie... These 'peerie things' are seldom seen outside—probably for fear of being trampled on. 1921 *Glasgow Herald* 1 Jan. 6/7 The mill stood..right on top of the sluice through which the waters of the bay rushed into the 'Peerie (little) Sea'. 1922 J. FIRTH *Reminisc. Orkney Parish* (ed. 2) 75 Sometimes as many as half-a-dozen women were called to the house..to keep away the 'peerie-folk'—those unearthly visitants who were particularly busy when a new arrival came. 1929 E. LINKLATER *White-Maa's Saga* 72 The world's nae comin' to an end for a peerie while yet. 1948 C. L. B. HUBBARD *Dogs in Brit.* xviii. 205 In its home islands it [sc. the Shetland sheepdog] is known as the Tounie Dog or Peerie Dog. 1956 C. M. COSTIE *Benjie's Bodle* 58 Afore he kent whar he was the Peerie Laird wad be pitten tae the horn. 1959 I. & P. OPIE *Lore & Lang. Schoolch.* x. 182 There she goes, there she goes, Peerie heels and pointed toes. 1978 *New Shetlander* Summer 22/2 Whaur is du gaun, du peerie boy, Wi dee face laek a smoored fire? *Ibid.* 28/2 Da Maister sat in a peerie hoose Wi Jordan's water rinnin near.
B. as *sb.* A Shetland sheep-dog.
1949 K. M. WELLS *Owl Pen Reader* (1969) II. 211 Shetland Sheep Dogs do possess the seeing eye,.. Shetland Sheep Dogs were once called 'Peeries' or 'Fairy Dogs'. 1971 F. HAMILTON *World Encycl. Dogs* 96 Shetland Sheepdog, Shetland Collie, Miniature Collie, Shetland, Sheltie, Toonie or Peerie—at some time or other the Sheltie has answered to all these names.

peeta, var. *PITA².

peeve (pīv), *v.* orig. *U.S.* [Back-formation f. PEEVISH *a.*] *trans.* and *intr.* To irritate, exasperate; to grumble, complain petulantly. So **peeved** (pīvd) *ppl. a.*, irritated, annoyed.
1908 G. V. HOBART *Go to It* 31 It may be interesting to some people, but it gets me peeved. 1912 ADE *Knocking Neighbors* 10 The Waiter peeved at being slipped a paltry $1.60. 1919 WODEHOUSE *Damsel in Distress* xxi. 242 About a million other people who'll be most frightfully peeved at my doing the Wedding Glide without consulting them. 1923 H. G. WELLS *Men like Gods* I. i. 5 Liberalism would never do anything more for ever than sit..grumbling and peeving. 1926 *Bulletin* (Glasgow) 17 June 5, I have a letter to-day from a peeved smoker..on the huge profits disclosed by the big tobacco combines. 1934 R. MACAULAY *Going Abroad* xvii. 139, I suppose he'd peeved me in some way. 1937 G. FRANKAU *More of Us* v. 62 'Just look at that,' peeved she, one slim hand plucking High from her ankles' grace, the limbs' frail sheath. 1945 *Sun* (Baltimore) 7 Dec. 8/1 Mighty peeved at that, Lorena outs with a derringer and marches the bachelor to the nearest J.P. 1966 J. JEFFERIES *House-Surgeon* v. 101 'They were peeved, like little school-kids.'..'Who peeved them, Harry?' 1975 *Daily Mail* 13 June 7/4 The agency won't talk about the work; its executives are rather peeved that the news has got out.

peeve (pīv), *sb.* orig. *U.S.* [f. the vb.] A grumble, a (cause of) complaint or irritation; a peevish humour; *pet peeve*, a special or recurring source of irritation.

It is not certain that the sb. is intended in quot. 1911.

1911 *Dialect Notes* III. 549 Some common cases of 'back-formations', or 'back-shortenings', are..*peeve*. **1919** C. H. Darling *Jargon Bk.* 25 *Pet peeve*, the thing that provokes you the most. **1920** S. Lewis *Main St.* 293 You're simply hot and tired, and trying to work off your peeve on me. **1936** R. E. Sherwood *Idiot's Delight* iii. 141 Every time you get yourself into a peeve, you take it out on us. **1945** *Southwest Rev.* Summer 330/2 Such misrepresentation of this lovable American character is my pet peeve. **1963** B. S. Johnson *Travelling People* i. 21 Henry thought he could afford to indulge in a Peeve... This was very quickly achieved by concentrating his mind on as many of the disasters of his past as were conveniently on short call from his memory. **1973** 'Trevanian' *Loo Sanction* 12 His peeve lasted only a second. **1975** *Bibliognost* Aug. 27 My pet peeve? Scouts and dealers who don't identify themselves and expect you to pull out the best books in the store. **1976** *National Observer* (U.S.) 7 Feb. 8/4 Poorly designed parking garages have riled me for a long time, but they've become a full-fledged pet peeve in recent years.

peever (pī·vəɹ). *Sc.* Also **peaver, peevor, peiver.** [Origin unknown.] The stone, piece of pottery, or the like, used in hopscotch. Also (freq. pl.) the game itself.

a**1850** in *Sc. Nat. Dict.* (1968) VII. 78/1 s.v. *peever*. **1856** J. Strang *Glasgow & its Clubs* 218 The young misses indulged in scoring the flagstones with their peevers, for the purpose of playing at pal-lall. **1887** D. Donaldson *Jamieson, Suppl.*, Add. 314/2 *Peever*, the pitcher or flat stone with which the children's game of *beds* or *pallall* is played; the game is therefore sometimes called '*peever*' or '*the peever*'. West of S. **1901** R. C. Maclagan *Games Argyleshire* 134 Pieces of broken pottery are by Lowlanders called *Lalies*, and the broken bottom of a bowl, a laly, is also called a *peaver*. **1921** *Edinburgh Even. News* 13 June 4 Chalking and disfiguring the street playing 'peevers'. **1931** A. J. Cronin *Hatter's Castle* iii. ix. 616 We'll be lookin' on at a game o' peever next if we're not careful, like a band o' silly lassies. **1962** *3rd Stat. Acct. Scotl.* XII. xii. 147 Peevors is still played on traditional stances. **1966** *Daily Tel.* 5 Nov. 13/8 Mr. Alex Lynch says that in Scotland hopscotch is known as 'peevers'.

pee-wee. Add: **2.** For '*Grallina picata*' substitute '*Grallina cyanoleuca*'. (Examples.)

1898 [see *mudlark *sb.* 3c]. **1911** Lucas & Le Souëf *Birds Austral.* 340 The familiar shrill and rather whistling cry gives it [sc. the magpie lark] the common name of the Pee-wee or Pee-wit. **1947** *Coast to Coast 1946* 27 He can see a black-and-white bird, a pee-wee, that runs along the path between the rushes. **1956** A. Bell *Some Common Austral. Birds* 96 The Pee-wee is vigorous and unusually widespread. **1968** K. Weatherly *Roo Shooter* 119 There was a peewee that visited him often through the day.

3. (Further examples.) Also, usu. *attrib.* or as *adj.*, small, tiny, composed of children, and *transf.* of animals or things, and *fig.* (*attrib.*). *N. Amer.* and *dial.*

1877 F. Ross et al. *Gloss. Words Holderness* 106/1 *Pee-wee*, adj. small; diminutive. **1930** *Amer. Speech* VI. 133 *Peewee* soft song man, petty confidence worker. **1935** *Ibid.* X. 271/2 *Peewees*, small, stunted pigs or lambs. **1936** *Time* 6 Apr. 45/1 Merciless, ignorant, lecherous, full of peewee patriotics, he wages senseless war against the Hill People. **1946** Mezzrow & Wolfe *Really Blues* (1957) vi. 69 A peewee jockey whose onliest riding crop was a stick of marihuana. **1952** *Daily News* (N.Y.) 13 Aug. C 16/3 The Bantams, an organization of men who are what they themselves call 'peewees and proud of it' with 5 [foot] 4 [inch]..Lenny Herman as spokesman. **1958** *Cut Knife* (Saskatchewan) *Grinder* 6 Feb. 1/5 The Cut Knife PeeWees were hosts to the Unity team and defeated them. **1965** *Western Wonderland* (Seattle) Apr. 22/1 The nine sorting categories [of logs]..are: hemlock; sawlogs..peewees.. and lastly, boomsticks. **1968** *Globe & Mail* (Toronto) 17 Feb. 44 Suburban Duberger defeated nearby Vanier 4-2.. in the ninth annual international pee-wee hockey tournament. **1970** C. Major *Dict. Afro-Amer. Slang* 90 *Pee wee*, small; a very narrowly rolled marijuana cigarette. **1973** *New York* 26 Mar. 52/2 Why should the Rangers concern themselves with Cub Scouts and peewee players. *Ibid.* 54/3 With the sportswriters giving better notices to the peewees. **1975** *New Yorker* 12 May 113/1 All he comes up with is a cosmetic trick—a peewee Watergate analogy—by using the blunder as evidence that the producer whom Tod works for is corrupt.

b. *spec.* A small marble.

1848 Bartlett *Dict. Amer.* 246 *Pee-wee*, the name given by boys to a little marble. **1862** C. C. Robinson *Dial. Leeds* 383 *Pey-wey*, a very small marble. **1902** Farmer & Henley *Slang* V. 160/2 *Pee-wee*... 3. (school).—A small marble. **1941** Baker *Dict. Austral. Slang* 53 *Peewee*, a small yellow marble. **1951** J. Frame *Lagoon* 122, I had pee-wees and bully taws and changers that weren't made of glass.

peeweep, peweep, piewipe. Add: **b.** *transf.* (See quot. 1966.) *rare.*

a**1902** S. Butler *Way of all Flesh* (1903) xl. 179 Looking for a watch and purse on Battersby piewipes was very like looking for a needle in a bundle of hay. **1966** S. H. Burton in *Ibid.* (new ed.) 391 'Piewipes' is a dialect word for lapwings—birds that are noted for building nests so well camouflaged that they are very difficult to see. The sense seems to be that the Battersby 'piewipes' are moorlands; barren featureless land affording natural camouflage.

peg, *sb.*¹ Add: **1. e.** *off the peg* adv. phr. and (with hyphens) adj. phr.: said of (the purchase of) ready-made clothes from the peg on which they hang in a shop; available for immediate purchase or use. Also *transf.* and *fig.*

[**1879–81** G. F. Jackson *Shropshire Word-bk.* 44 *Bought off the pegs*, said contemptuously of second-hand or 'slop-made' clothing.] a**1916** R. Asquith in Spender & Asquith *Life H. H. Asquith* (1932) II. xlviii. 234 Love ready-made or glamour off the peg. **1922** *19th Cent.* June 1026 Before the war the average undergraduate was..garbed in a 'sports' coat obviously 'off the peg'. **1926** M. A. von Arnim *Introd. to Sally* vi. 74 The poor Pinners would have to buy clothes off the peg. **1954** J. Betjeman *Few Late Chrysanthemums* 84 An officer's lady—and more so Than those who buy off the peg. **1957** *Economist* 28 Dec. 1120/2 There simply is no such thing as suitable spectacles 'off the peg'. **1959** [see *bar-tacking* s.v. *bar *sb.*¹ 30]. **1959** *Daily Tel.* 4 May 13 (*heading*) Off-the-Peg thrills of the Law. **1963** *Times* 19 Feb. 18/2 Shipowners may before long buy vessels 'off the peg' instead of each vessel's being an individual tailor-made job. **1971** *Daily Tel.* 8 Nov. 12/3 Her clothes are mostly bought 'off the peg' in Zurich boutiques. **1974** *Country Life* 28 Mar. 732/1 Georgian sash windows..have been replaced by off-the-peg glass rectangles. **1975** *Times* 11 Mar. (Italian Industry Suppl.) p. vii/4 It was the equivalent of the passage from bespoke to off-the-peg tailoring. **1977** *Navy News* June 22 (Advt.), Every Officer and Rating may be fitted immediately 'off the peg'.

f. *West Indies.* A segment of a citrus fruit. (Perh. a different word: cf. Pig *sb.*¹ 8c.)

1909 *Cent. Dict. Suppl.* 960/2 *Peg*... One of the cells or natural divisions into which an orange may be separated after removing the skin. [West Indies.] **1956** J. Hearne *Stranger at Gate* i. 9 He moved his hand and closed it on.. the..tangerine..stripping the soft skin from the fruit and cramming the pegs into his mouth. **1971** *Caribbean Q.* XVII. ii. 14 Different name, same referent..feg/fig/peg/plug/sprig (of orange). **1973** *N.Y. Times* 3 June L 19/1 A section of an orange..is..called..'peg'.

g. *Railways.* A semaphore signal.

1911 C. E. W. Bean *'Dreadnought' of Darling* xxxiii. 288 Recollec' that cove with a red beard we come on camped by the railway peg near Nine Mile Tank..? **1971** D. J. Smith *Discovering Railwayana* x. 58 *Peg*, signal.

2. f. *Cricket.* A stump.

1865 *Bell's Life in London* 1 July 9/2 Griffith then bowling Davis's centre peg. **1901** H. Bleackley *Tales of Stumps* i. 31 Little Tommy tossed a slow long-hop on the leg peg,..and Nat managed to fumble it away for two. **1909** *Westm. Gaz.* 15 July 12/1 He was beaten by another fine ball from Smith, which, after pitching well outside the off peg, broke across the wicket and hit the top of the leg stump. **1972** R. Robinson *Wildest Tests* xi. 120 Cunis swung one so late and so far that it hit Gandotra's leg peg.

g. *Mountaineering.* = *piton 2. Also in *Comb.*

1920 G. W. Young *Mountain Craft* iv. 201 My party has taken pegs three times in all, as a precaution, and used one once (on a new descent). **1946** J. E. Q. Barford *Climbing in Brit.* ii. 25 Pegs or large nails with rings in one end, which are driven into rocks to provide an anchor where no natural one exists. **1957** *Times Lit. Suppl.* 1 Nov. 655/2 The 'peg-bashers' have concentrated their main attention on short ferocious overhangs. **1971** [see *ironmongery 1 d]. **1973** C. Bonington *Next Horizon* viii. 120 We fully expected our route to be high-standard peg-climbing the whole way.

3. b. *Mil. slang.* A charge. Usu. in phr. *on the peg*, on a charge, under arrest.

1890 Barrère & Leland *Dict. Slang* II. 122/2 *On the peg* (military), to be under arrest... The expression is also used when a soldier is put under stoppages. **1919** [see *for *prep.* 13 e]. **1923** J. Manchon *Le Slang* 220 *To whip..on the peg*, mettre..aux arrêts. **1941** *Amer. Speech* XVI. 186/2 *On the peg*, under charge for misdemeanor.

5. (Further examples.) Also *absol.* and const. *for.*

1852 Geo. Eliot *Let.* 24–25 July (1954) II. 50 The publishing world seems utterly stagnant—nothing coming out which would do as a peg for an article. **1909** J. R. Ware *Passing Eng.* 194/1 *Peg* (*Theatrical*, 1884). Sensation point or effect of a piece. Something upon which the actors ..can build up a scene. **1929** D. H. Lawrence *Phoenix II* (1968) 590 And as for his relation to his women..they are pegs to hang clothes on, and there's an end of them. **1930** E. Pound *XXX Cantos* vii. 26 The Elysée carries a name on And the bus behind me gives me a date for peg. **1953** A. Huxley *Let.* 19 July (1969) 678, I have been asked..to do an article on recent developments in parapsychology, using your forthcoming book as the peg on which to hang my remarks. **1966** *Listener* 10 Feb. 221/2 The event itself seemed too slight a peg on which to hang a sixty-minute programme. **1972** *Guardian* 11 Sept. 8/6 Newspapermen..with their contempt for yesterday's story, their embarrassment when confronted with masterpieces lacking any evident 'peg'. **1976** *Times Lit. Suppl.* 13 Feb. 174/3 Dr Hacking uses this theme as a series of pegs on which to hang his discussions of particular authors' philosophies of language.

11. peg-bag, a bag used as a container for clothes-pegs; **peg-basket,** a repository for clothes-pegs; **peg-board,** (*a*) (further examples); (*b*) a board having regularly spaced holes for holding hooks on which objects can be hung; hence **peg-boarding; peg-box, pegbox,** a structure at the head of instruments of the lute or violin type, where the strings are attached to the tuning-pegs; **peg doll,** a doll made from a clothes peg or similar piece of wood; **peg-house** *slang*, (*a*) a public-house (see sense 6 in Dict.); (*b*) a brothel or meeting-place for male homosexuals (U.S.); **peg leg** (earlier and later examples); also *transf.* and as *v. intr.*, to move with a limp or a stiff gait (see also quot. 1932); **peg-legged** *a.*, having a peg leg (also *transf.*); **peg-legger,** one having a peg leg; *slang*, a beggar; **peg-man** (*b*), a workman who lasts pegged boots or shoes; **peg-pot** = *peg-tankard*; **peg-rent,** cloak-room charges; **peg rhizoid,** in certain liverworts of the order Marchantiales, a rhizoid distinguished by peg-like processes on the inner surface; **peg-tankard** (later example).

1951 J. Fleming *Man who looked Back* xl. 145 She gathered up her arty peg-bag. **1972** P. Flower *Cobweb* iii. 92 Already a faithful clientele. What d'you think about tea cosies and table-mats, peg bags, teapot stands and so on? **1914** D. H. Lawrence *Widowing of Mrs. Holroyd* i. i. 12 Jack, can you go and take the stockings in for me?..Minnie, you take the peg-basket. **1951** *Catal. of Exhibits, South Bank Exhib., Festival of Britain* 152/1 Pegboard; Educational Supply Association Ltd. **1960** *Guardian* 24 Feb. 13/5 A peg board lining to the door..for extensive hanging purposes. **1962** E. Godfrey *Retail Selling & Organization* ii. 19 It [sc. the counter] will be replaced by shelving, pegboard or tiered stands. **1967** R. Whitehead in Wills & Yearsley *Handbk. Managem. Technol.* 69 Those wonderful pegboard systems and elaborate charts designed along foolproof lines to check the flow of work are notorious for the way they fail to work. **1971** *Femina* (Bombay) 30 Apr. 27/3 Picture puzzles, pegboard, block building..sharpen motor skills. **1975** P. G. Winslow *Death of Angel* iv. 96 The shed was deep... Piles of what looked like sheets of metal lay along one side. Tools hung from a pegboard in the back. **1960** *Woman's Realm* 2 Apr. 10/3 Sheets of pegboarding..make a..useful space for hanging utensils. **1962** *Friend* 3 Aug. 946/2 The most expensive item was peg-boarding for the ceiling. **1883** Grove *Dict. Mus.* III. 81/2 It [sc. the rebec] was shaped like the half of a pear, and was everywhere solid except at the two extremities, the upper of which was formed into a peg-box identical with that still in use, and surmounted by a carved human head. **1938** *Oxf. Compan. Mus.* 524/1 The head [of the lute], containing the peg-box, is generally bent back at an angle from the neck. **1961** M. W. Prynne in A. Baines *Mus. Instruments* vii. 158 For balance, the bent-back pegbox has to be as light as possible with very slender pegs or 'lute-pins'. **1976** D. Munrow *Instruments Middle Ages & Renaissance* 25/4 From the twelfth to the fourteenth century, the instrument is regularly referred to in French literature as *mandoire* or sometimes *mandola* and illustrations often reveal the instrument's most recognizable feature; its sickle-shaped pegbox. **1950** *Dryad Catal.* 96 Peg doll. Height 11"—Each 4/6. **1951** *Notes Coll. Dolls & Figurines* (Wenham, Mass., Hist. Assoc.) 50 The popular wooden play doll commonly called 'Peg-doll', 'Penny-wood', was born in the Thuringian forest. **1969** [see *Flanders baby]. **1972** *Times* 30 June 18/2 Nanny..is one of a series of pretty peg dolls all dressed in the fashions of 1870 servants... They are..made from genuine dolly pegs, hand painted and dressed by Somerset craftsmen. **1922** C. Aiken *Jig of Forslin* 40 And once I murdered, by the waterfront: A drunken sailor, in a peg-house brawl. **1931** 'D. Stiff' *Milk & Honey Route* 211 *Peg house*, a place where, if the hobo wishes, he may meet Angelina. **1942** Berrey & Van den Bark *Amer. Thes. Slang* § 507/4 *Peg house*, a male homosexual brothel or gathering place. **1972** R. A. Wilson *Playboy's Bk. Forbidden Words* 222 A 'peg-boy' is a young male who prostitutes himself to homosexuals; 'peg-house', a homosexual brothel. There is an unsubstantiated story that boys in East Indian peg-houses were required to sit on pegs between customers, giving them permanently dilated anuses. **1769** J. Wedgwood *Let.* 6 Dec. (1965) 88 The peg leg is much wanted. **1932** *Amer. Speech* VII. 269 When cable-tools alternately strike bottom and miss in drilling, they are said to *peg-leg*. **1969** D. Francis *Enquiry* xiv. 190, I peg-legged up the back drive. **1971** M. Tak *Truck Talk* 116 *Peg leg*, (1) the first rear axle of a tandem-axle tractor when it has only single tires... (2) a three-legged trailer... (3) a trailer with one broken dolly. **1974** *Daily Tel.* 3 Sept. 14/6 During the 1914–18 war more than 20,000 men with legs amputated were fitted with clumsy wooden 'peg-legs'. **1967** B. Patten *Little Johnny's Confession* 18 It should scatter woodworm into the bedrooms of all peglegged men. **1974** H. L. Foster *Ribbin'* vi. 245 The current trend is..ankle-length, peg-legged jeans and platform shoes in the city. **1937** Partridge *Dict. Slang* 615/2 *Peg-legger*, a beggar... Either rhyming s[lang] or ex the preceding [sc. peg-leg]. **1943** *Bournemouth Daily Echo* 28 Oct. 2/3 Another 'peg-legger'—as the one-legged men call themselves—was Marine Edgar Saunders. **1897** S. & B. Webb *Industr. Democracy* I. 418 'Lasters'..(in hand-sewn work these are known as 'makers', in 'pegged work'..they are called 'pegmen' or 'rivetters'). **1903** *Athenæum* 24 Jan. 122/1 In 1873 a peg-pot similarly engraved..was offered to the city, but declined. **1911** *Chambers's Jrnl.* Feb. 115/1 The man who likes to eat a meal without worry lest somebody should exchange hats with him..must pay peg-rent. **1911** J. M. Coulter et al. *Textbk. Bot.* II. i. 516 In the Marchantiaceae, rhizoids are of two kinds, plane rhizoids.. and peg rhizoids, in which the cell wall grows out internally into peg-like or antler-like projections. **1958** *New Biol.* XXVII. 90 One of the peculiarities of *Marchantia* is that these rhizoids are of two kinds, smooth rhizoids ..and 'peg rhizoids' which have numerous peg-like thickenings on the inner surface of the wall jutting into the cell cavity. **1969** F. E. Round *Introd. Lower Plants* viii. 102 (*caption*) Section through a pore region of *Conocephalum*, note photosynthetic filaments, peg and smooth rhizoids and amphigastrium. **1967** *Times* 7 Mar. 21/6 (Advt.), A York peg tankard, by John Plummer, circa 1695.

peg, *v.* Add: **1. c.** Also, in extended uses, to fix (a price, wage, etc.) at a certain level; to set a value on (a currency) in relation to gold or another currency; to set a numerical or quantitative limit on (something). (Further examples.)

1933 *Ann. Reg. 1932* 26 Shortly afterwards the Bank rate was considerably reduced, and sterling was effectively 'pegged' at a gold value of about 15s. **1933** *Sun* (Baltimore) 13 July 10/2 The British went off gold after a strenuous effort to maintain gold parity in 1931 and it can hardly be proved that England's trade balance has improved sufficiently to permit pegging the pound at any such figure. *Ibid.* 21 July 2/4 The market was 'pegged' once before this year... The fluctuations in wheat at that time were limited to 5 cents a bushel. **1940** *New Statesman* 13 Apr. 485 The dual system..of rationing by amount and of pegging prices is anomalous. **1944** Auden *Sea & Mirror* in *For Time Being* i. 5 When he learns the price is pegged to his valuation. **1949** Koestler *Promise & Fulfilment* 292 It is a kind of aseptic social operation, and as wages are pegged to the cost-of-living index, there is no occasion even for local strikes. **1954** *Birmingham* (Alabama) *News* 17 Feb. 1/6 The old method pegged unemployment at 2,259,000 the first week in January. **1959** *Punch* 19 Aug. 30/1 Inflation can be licked by other means, by pegging wages and profits, increasing productivity and output,.. and so on. *Ibid.* 30/2, I have already encouraged my teen-agers to peg their consumption of sweets, soft drinks, records, cosmetics and cycle accessories in the hope of bringing manufacturers to heel. **1965** *Listener* 20 May 728/1 There is room for scepticism about the machinery which is supposed to peg these rents at a 'fair' level. **1970** *Times* 18 Dec. 2/5 The orchestras have had their grants pegged for three years and are in a serious position. **1971** *Morning Star* 10 May 1/5 The mark will be allowed to float, meaning its parity will no longer be pegged against the dollar. **1977** *Evening Post* (Nottingham) 27 Jan. 5/4 Planners hope to peg the cost of the tour at £60 or less per tourist.

d. *fig.* To categorize; to form an opinion of (someone, occas. something). Freq. in phr. *to have* (*someone* or *something*) *pegged*: to have a fixed opinion of (that person or thing). *N. Amer. colloq.*

1920 *Collier's* 31 July 29/2, I had him pegged from the go in for what he is—one of them tea-room boys which will stop at nothin' but work! **1926** J. Black *You can't Win* xxi. 320 The expression, 'I have him pegged', which has crept into common usage, is thieves' slang..and has nothing to do with the game of cribbage. **1940** J. O'Hara *Pal Joey* 175, I tho't I could peg a joint like that from 2 mi. away. **1949** *New Yorker* 12 Nov. 28/3 An elderly lady..who pegged him looked back, and then winked. **1959** J. Ludwig in R. Weaver *Canad. Short Stories* (1968) 2nd Ser. 232 Naturally Mitchell has her pegged: doesn't he know this shopping trip is a fake..? **1967** M. Reynolds *After Some Tomorrow* 12, I peg him as a holier than thou cloddy. **1968** H. Waugh *Con Game* ix. 91, I always knew she was a slut. An ignorant, stupid child. I pegged her from the start. **1972** D. Lees *Zodiac* 40, I had her pegged as a bit of a nut. **1976** *Publishers Weekly* 18 Oct. 52/1 Combine the pace of a Disney film with the reassurance of characters clearly good or evil and you will quickly peg what Nicolaysen's chase adventure has going for it.

2. d. To fasten the soles on to (boots or shoes) with wooden pegs.

1850 *Rep. Comm. Patents 1849* (U.S.) 295 Improvement in Machines for Pegging Boots and Shoes. **1858** [see Pegged *ppl. a.*]. **1895** *Montgomery Ward Catal.* 511/1 Ladies' Heavy Pegged Shoes. **1940** *Chambers's Techn. Dict.* 622/2 *Pegged-sole shoes*, a type of footwear in which the outer sole is attached to the inner sole and the upper by two or more rows of pegs; used for sea boots.

e. To insert small wooden pegs into the stalks of (tobacco).

1850 *Rep. Comm. Patents: Agric. 1849* (U.S. Dept. Agric.) 321 'Pegging' tobacco..is done by driving little pegs, about six inches long and half an inch or less square, into the stalk about four inches from the big end of the stalk. **1968** *Publ. Amer. Dial. Soc. 1966* xlv. 19 It's hard work to peg tobacco.

f. *Cricket.* To drive pegs into (the face of a bat) (see quot. 1934).

1853 F. Gale *Public School Matches* 17 The captain is going in. An old bat, well pegged but very clean, looks like business. **1906** A. E. Knight *Compl. Cricketer* ii. 48 Pegging down the bat is simple, but destructive and ineffective. **1934** W. J. Lewis *Lang. Cricket* 186 *To peg a bat*, to drive a few small pegs into the face of a bat as a remedial measure when the grain of the wood has risen with use.

4. b. *peg back*, (*a*) (Racing): of a horse, to pull past, overtake (another horse); also, to gain on another horse by (a specified distance); (*b*) in a game: to pick up (a point or advantage) or to change (the score) so as to reduce or eliminate an opponent's lead.

1928 *Sunday Express* 24 June 22/3 He came..in the last furlong to peg back the flying French colt. **1932** *New Yorker* 14 May 52/2 Burgoo King pegged him back three furlongs from home. **1971** *Sunday Nation* (Nairobi) 11 Apr. 44/4 Owen de Souza pegged one back for Blues in the dying minutes when he converted a penalty-push. **1977** D. Francis *Risk* ii. 18 Open ditch next; Tapestry met it just right and we pegged back a length in mid-air. **1978** *Rugby World* Apr. 4/1 The Irish side had given all Wales a fright by pegging the score back to 13–13 after the visitors had gone 13–3 in front.

10. *peg along* (further U.S. example); *peg away* (earlier and later examples).

1818 Keats *Let.* 24 Jan. (1958) I. 216 The musicians began pegging & fagging away at an overture. **1882** 'Mark Twain' *Lett. to Publishers* (1967) 158, I still lack about 30,000 words... I shall peg along, day by day. **1956** *People* 13 May 1/6, I have just kept pegging away year after year. **1978** L. Deighton *SS-GB* xiv. 115 How I envied you doing Greats, while I pegged away at my Civil Law.

13. g. To hang (washing) with pegs from a clothes-line.

1922 D. H. Lawrence *England, my England* 102 Helped his wife to peg out the washing on the clothes line in the meadow. **1974** *Country Life* 21 Mar. 633/1 The foreground girls selfconsciously pegging out the washing.

1978 J. Thomson *Question of Identity* xii. 115 Betty Lovell was pegging out sheets on a washing-line.

Pegamoid (pe·gămoid). Also **pegamoid.** A trade name for a kind of waterproof cloth or imitation leather used in upholstery, bookbinding, etc. Also *attrib.* and *fig.*

1895 *Current Hist.* V. 731 It is claimed for 'pegamoid', a product recently placed on the markets of Europe, that it will render materials of any kind absolutely impervious to water. **1896** *Westm. Gaz.* 10 June 5/2 By imposing it on cotton cloth or impregnating it 'Pegamoid' can be turned to many and varied purposes. **1909** *Pract. Upholstery* 12 Pegamoid Cloth. This is one of the better class imitation leathers, and is obtainable in a large variety of 'grains', colours, and qualities. **1922** *Daily Mail* 10 Nov. 15 (Advt.), Seats in red, brown, green or blue rexine, pegamoid or velvet. **1957** L. Durrell *Justine* i. 21 He is a pegamoid sloth of a man, a vast slow fellow. **1959** B. Ruck *Romantic Afterthought* xxxiv. 180 The pegamoid-covered chair.

Pegasean, *a.* (Later example.)

1923 'R. Crompton' *William Again* iv. 59 'I feel'—his Pegasean imagination soared aloft on daring wings—'I feel 's if I might die if I went to church this mornin' feelin' 's ill as I do now.'

pegasse (pĕgæ·s). Also **pegass.** [Etym. unknown.] A kind of peaty soil found in the Caribbean and northern S. America.

1924 *Chambers's Jrnl.* July 475/1 Suckers should be planted..15 feet apart on pegass or peat soil. **1949** *Caribbean Q.* I. iii. 42 Experiments in Surinam seem to show that the pegasse lands..can be used to grow bananas. **1959** *Listener* 17 Sept. 435/1 The north-west of Guiana is a place of countless ridges of hills islanded in a sea of pegasse swamp. **1974** *Encycl. Brit. Macropædia* VIII. 507/1 Pegasse soil, a type of tropical peat, occurs behind the coastal clays and along the river estuaries.

pegged, *ppl. a.* (Later examples.)

1960 H. Hayward *Antique Coll.* 214/1 *Pegged construction*, furniture made by joiners constructed with mortice and tenon joints, the tenon being held in the mortice by a square peg, riven or split from green wood. **1966** A. W. Lewis *Gloss. Woodworking Terms* 100 *Pegged or pinned tenon*, tenoned joint which is fixed by having a hole drilled through it near the shoulder and a wood pin inserted. **1967** *Wall St. Jrnl.* 9 Feb. 1/4 Mr. Glick..wears pegged pants, boots and an Army jacket. **1968** *Globe & Mail* (Toronto) 17 Feb. 46 (Advt.), Recreation room with bar and pegged floor. **1973** *N.Y. Law Jrnl.* 19 July 16/8 (Advt.), Parquet and pegged oak floors. **1974** *Times* 7 Mar. 2/6 Growers in Holland and Germany received government help through pegged oil prices.

pegging, *vbl. sb.* Add: **1. a.** (Later examples.)

1926 T. E. Lawrence *Seven Pillars* lxvii. 352 They told me something of Davenport's work: of his continual pegging away in Abdulla's sector. **1931** *Observer* 11 Jan. 19/1 Mr. Scultin's views about the pegging of wages are unknown. **1948** G. Crowther *Outl. Money* (rev. ed.) viii. 248 'Pegging' has usually meant 'pegging up', but we can invent the term 'pegging down' for the practice that grew up among some governments in the nineteen-thirties of maintaining their currencies at a fixed undervaluation. **1977** *Times* 24 Dec. 9/3 Continental motoring will be given added impetus..by the pegging of most cross-Channel ferry prices.

b. *level-pegging*: see as main entry in Suppl.

peggotty (pe·gŏti). Also **peggitty, peggoty, pegity,** and with capital initial. [Fanciful extension of Peg *sb.*¹] A children's board game in which four players in turn place pegs in holes, the object being to complete a row of five pegs.

Pegity is registered in the U.S. as a proprietary name.

1925 *Official Gaz.* (U.S. Patent Office) 31 Mar. 996/2 Parker Bros... Pegity. Game of the Type of 'Go-Bang'. Played with Pegs, for Adults and Children. **1929** *Games & Toys* July 66/1 Pegity..is one of the most interesting board games ever devised. **1950** *Oxf. Jun. Encycl.* IX. 73/1 Peggotty, a very simple modern game on the Noughts and Crosses principle, is played by four players who take it in turn to stick coloured wooden pegs into holes in the board, the winner being the first to get five pegs in a row. **1958** M. Stewart *Nine Coaches Waiting* iv. 45, I brought you a game..called Peggitty. **1974** P. Dickinson *Poison Oracle* ii. 60 Tribesmen..played their stone-age version of peggotty in noisy groups.

peggy, *sb.* Add: **5.** *Naut. slang.* A ship's mess-steward or menial.

1902 A. B. Lubbock *Round the Horn* iii. 105 An institution on board a sailing-ship is 'peggy'. Each of us take it in turn, and peggy has to fetch the grub from the galley. **1929** F. C. Bowen *Sea Slang* 102 *Peggy*, the man who looks after the seamen's and the firemen's messes in a modern liner. In the old sailing days it was applied to the ship's boy or to a hand who was called upon to do all the odd jobs in a watch. **1930** 'Greenhorn' *Tinker, Tailor* viii. 191 The sailors' Peggy is a kind of fo'c'sle steward, and.. he has to wash up the pots and pans in the forrard galley, clean out the fo'c'sle, make the bunks, and generally keep the sailors' quarters swept and garnished. **1946** J. Irving *Royal Navalese* 133 *Peggy*, an Ordinary Seaman detailed to act as Petty Officers' mess-man. **1967** S. Waters *Indentures Indorsed* i. 11, I was initiated into the mysteries of acting as 'Peggy'. As the name implies this menial does all the domestic chores. **1972** *Courier-Mail* (Brisbane) 19 July 24/3 Some waterfront employers think Peggy should do other work..between tea breaks and lunch.

6. (See quots.) [Perh. a different word.]

1940 *Chambers's Techn. Dict.* 622/2 *Peggies*, slates 10–14 in. long. **1959** *Archit. Rev.* CXXV. 292 *Peggies*, small sizes of slates of random sizes which are sold by weight.

7. peggy bag, a style of women's hand-bag orig. having side handles and outside pockets; **peggy-work** *Naut. slang*, the work of a peggy (sense *5).

1920 *Scot at Hame an' Abroad* 1 July 5/3 Mirren had a wheen peppermints in her peggy bag. **1922** *Daily Mail* 11 Dec. 13 (Advt.), Peggy bags with side handle and two outside pockets. **1939–40** *Army & Navy Stores Catal.* 874/4 Peggy Bag, in plain Calf..with divided inner compartment and mirror. **1974** *Trafford Catal.* Spring–Summer 182/2 Classic 'peggy' bag with twin easy-to-get-at compartments..twin handle length straps. **1959** Scott-Shawe & Wykes *Mariner's Tale* i. iii. 29, I was escaping my peggy-work in the *Ashmore's* galley.

pego (pī·go). *slang.* [Origin unknown.] The penis.

1680 in Rochester *Poems* 36 As oft as Finger, Dildoe, Pego, Rape, The Virgin Hymen, she repaires the Gap. **1691** E. Ward *Poet's Ramble* 10 Pego, like an upstart Hector,..Would fain have Rul'd as Lord Protector: Inflam'd by one so like a Goddess, I scarce could keep him in my Codpiece. c**1763** J. Wilkes *Ess. on Woman* 19 Then shall man's pride and pego comprehend His actions and erections, use and end. **1788** Grose *Dict. Vulgar T.* (ed. 2), *Pego*, the penis of man or beast. **1879–80** *Pearl* (1970) 250 This made Neddy's Pego, accustomed to sprout, Shrink into his belly, and turn up his snout. **1974** H. R. F. Keating *Underside* ix. 90 There's some as likes..her dirty old fingers round their pego.

peg-top, pegtop. Add: **1. a.** (Earlier U.S. and later examples.)

1788 *Massachusetts Spy* 3 Apr. 4/3 Children's Books... Memoirs of a Pegtop. By the Author of, The Adventures of a Pincushion. **1923** D. H. Lawrence *Birds, Beasts & Flowers* 171 Chieftains, three of them abreast, on foot Strut like peg-tops. **1978** *Country Life* 14 Dec. 2103/2 Tops: humming tops, peg tops, optical tops, gyroscopes.

b. (Earlier and later examples.)

1803 [see *pea-shooter* s.v. *pea¹ 7]. **1931** P. Guedalla *Duke* i. iv. 22 Nor do the martial virtues thrive upon a simple diet of peg-top and battledore.

2. (Earlier example.)

1858 *Punch* 15 May 202/2 The fashion of trowsers improves, if anything, in ridicule. Henceforth pegtops are split.

3. peg-top pants, = *peg-top trousers*; **peg-top skirt,** a skirt that is wide at the top and narrow at the bottom.

1917 *Times* 24 Jan. 9/5, I noticed yesterday in the leading papers the leading couturiers announced the 'new peg-top skirt, cut on most becoming lines'. **1923** E. B. White *Let.* Feb. (1976) 63 A stunning suit—real western cut with peg top pants and everything. **1956** *Punch* 15 Aug. 190/3 This Autumn Collection..emphasizing a belted natural waist-line, short peg-top skirts, and a wrapped-up look. **1970** *Daily Tel.* 23 July 15 Pegtop pants were Patou's big contribution to yesterday's fashion scene. **1972** J. Minifie *Homesteader* xvii. 146 The baggy end of a drummer's blue peg-top pants. **1975** G. Howell *In Vogue* 222 Schiaparelli's slim silhouette with..peg-top skirt.

pegtopped, *ppl.a.* Add: **pegtopped skirt** = *peg-top skirt.

1959 *News Chron.* 22 July 3/3 Gently fitted jackets over stiffened pegtopped skirts. **1973** *Country Life* 8 Mar. 632/3 If this peg-topped, tapered, pencil skirt does catch on there will be resounding cheers from the girdle manufacturers.

Peguan (pe·giuăn), *sb.* and *a.* Also **Pegue, Peguer.** [f. the place-name *Pegu* (see below) + -an.] **A.** *sb.* **a.** A native or inhabitant of the city or district of Pegu in southern Burma, the ancient capital of the Mon (*Mon *sb.*² and *a.*) people. **b.** The language of the people of Pegu. **B.** *adj.* (Also **Pegu,** the place-name used *attrib.*) Of or pertaining to these people.

c**1591** R. Fitch in Hakluyt *Princ. Navigations* (1599) II. i. 262 The Pegues if they have a sute in the law which is so doubtfull that they cannot well determine it, put two long canes into the water..and there sit men to iudge, and they both do diue vnder the water. **1727** A. Hamilton *New Acct. E. Indies* II. xxxvi. 36 The Peguer finding that he could not recover his Lands without foreign Aid and Assistance, invited the Portuguese. *Ibid.* 37 Neither the Siamers not the Peguers..understood the use of Fire Arms. c**1759** [see *Karen *sb.* 1]. **1800** M. Symes *Acct. Embassy to Ava* 12 The power of the Peguers now seemed hastening to its wane. **1828** [see *Mon *sb.*² and *a.*]. **1834** *Chinese Repository* II. 504 The most important is the Peguan. **1851** *Ibid.* XX. 345 Frequent wars with the northern Siamese and the Peguans, or Mons. **1858** C. T. Winter *Six Months Brit. Burmah* ii. 15 The inhabitants consist principally of Burmans and Peguans. **1874** J. M. Haswell *Gram. Notes & Vocab. Peguan Lang.* p. vi, I sent a list of over sixty Peguan words to missionaries among the Kohls. **1910** L. Milne *Shans at Home* i. 26 A vassal of the Pegu King. *Ibid.* 27 The conquering army of the Pegu king. **1920** G. A. Grierson *Linguistic Survey India: Index of Lang.-Names* 164 Peguan..According to the Burma Linguistic Survey, a form of Mōn (3) spoken in Amherst District. **1932** J. G. Scott *Burma & Beyond* i. 16 The Khmēr went, or were pushed, farther east,..but the Mōn remained behind, and came to be known to the early merchant adventurers as Peguans. **1942** J. L. Christian *Mod. Burma* ii. 12 The Mons, known also as Talaings or Peguans, are an ancient race. **1955** E. Mannin *Land of Crested Lion* xv. 190 The gift of a wealthy Pegu donor. **1956** *Introducing Burma* 38 The Mon Khmer Branch consists of the Mon Group comprising—Talaing (Mon, Peguan, Martabanese) which are spoken in the Pegu and Tenasserim Divisions. **1973** J. M. & E. G. Maring *Hist. & Cultural Dict. Burma* 147 *Mon*, A lowland Mon-Khmer ethnolinguistic group of Mongoloid racial stock... Synonyms: *Moñ, Mun, Peguan, Talaing.*

peh, var. *p'o.

Peierls (pəi·əɹlz). *Physics.* The name of Sir Rudolf *Peierls* (b. 1907), German-born physicist, used *attrib.* with reference to a spatially periodic distortion of a linear chain of atoms or molecules in certain solids, adduced to explain an observed change from a conducting to a semiconducting or insulating state at low temperatures.

Described by Peierls in *Quantum Theory of Solids* (1955) 108.

1973 *Solid State Communications* XII. 1130/2 The important question is whether or not one can stabilize such a system just above the Peierls transition.., so that on lowering the temperature the superconducting state becomes truly stable. **1973** *Physical Rev.* VIII. B. 574/2 The experimental superperiod of 8 × 2·88 Å due to the platinum atoms is..in exact agreement with the expected superperiod as due to a 'Peierls distortion'. **1979** *Sci. Amer.* Oct. 54/3 The Peierls transition..comes about through a periodic distortion of the lattice.

pein, var. Peen *sb.* (in Dict. and Suppl.).

‖ **peineta** (pĕine·tă). [Sp.] An ornamental comb worn by Spanish women with a mantilla.

1935 *Discovery* Oct. 307/2 The *peineta*, or ornamental high shell-comb which together with the *mantilla* constitutes the national head-dress of Spanish women. **1939** Spender & Gili tr. *Lorca's Poems* 23 *Calañés* and tall *peinetas*.

Peirce (pɜ̄ɹs). The name of the American philosopher and logician Charles Sanders *Peirce* (1839–1914) used in the possessive to indicate his theories or methods, esp. as **Peirce's Law**, a logical formula relating to implication (see quot. 1967).

1918 C. I. Lewis *Survey Symbolic Logic* i. 100 By a further important modification of Peirce's method, a theoretically adequate logic of mathematics may be obtained. **1934** W. V. Quine *Syst. Logistic* vii. 64 $p \supset q. \supset p: \supset p$ This is Peirce's law... The theorem has been so named by Łukasiewicz. **1949** *Jrnl. Philos.* XLVI. 513 This paper assumes Peirce's semiotic as the basis for a discussion..of the logical paradoxes. **1954** I. M. Copi *Symbolic Logic* iii. 64 Assuming Peirce's Law false leads to a contradiction. **1955** A. N. Prior *Formal Logic* i. iii. 50 And so *p* is true (by 'Peirce's law'). **1967** S. C. Kleene *Math. Logic* § 3. 13 $((P \supset Q) \supset P) \supset P$. (Peirce's law, 1885). **1972** Lambert & van Fraassen *Derivation* 49 The following principle is known as Peirce's Law.

Peircian (pɜ̄·ɹsiăn), *a.* [f. prec. + -ian.] Of or relating to the American philosopher, Charles Sanders Peirce, his theories or methods.

1905 *Mind* XIV. 237 The Peircian pragmatist. **1949** *Mind* LVIII. 130 Mr. Feibleman is a well-known Peircian scholar. **1952** *Mind* LXI. 205 This surely manifests a thoroughly Peircean attitude. **1965** *Amer. Philos. Q.* II. 115/2 We mobilize the Peircean idea of an inductive community. **1978** N. Jardine in Hookway & Pettit *Action & Interpretation* 123 Such a picture of scientific progress may encourage, though it does not entail, a Peircean vision of an ultimate science, an omega of human knowledge in which the essences of all things are revealed.

Peisistratid, var. *Pisistratid *sb.* and *a.*

peitrel, *sb.* For †*Obs.* read '†*Obs.* except *Hist.*' and add: The spelling **peytral** is now preferred by the best authorities. (Later examples.)

1894 H. A. L. Dillon in W. J. Loftie *Authorised Guide to Tower of London* (ed. 2) 147 Note also the horse armour consisting of the chanfrein for the head, the crinet for the neck, and the bard protecting the body, that is, the peytral on the chest, the flanches on the sides, and the crupper covering the hinder portion of the horse. **1926** *Fine Armour* (Christie's Sale Catal. 29 June) 4 A Peytrel for the Horse, of three plates of plain bright steel. *Ibid.* 15 A Horse Chanfron and Peytrel Plates. **1929** *Archaeologia* LXXIX. 242 The two sales of the armoury of Prince Radziwill..included no less than 22 chanfrons and 15 peytrals of Maximilian date. **1953** *Proc. Prehistoric Soc.* XIX. 175 The object was claimed to be a 'peytrel' or chest ornament for a small horse. **1962** J. Mann *Wallace Coll. Catal.: European Arms & Armour* I. 12 *Peytral* of five large plates. **1967** *Punch* 29 Nov. 809/3 If some hulking great brute..were to straddle an already overburdened war-horse, the whole caboodle would collapse in a sorry ruin of pauldron and salade, breastplate and roundel, cantle, crupper, greave, flanchard, peytral, chanfron and crinet.

pejoration. Add: **b.** *Linguistics.* The development of a less favourable meaning or of unpleasant connotations of a word.

1889 in *Cent. Dict.* **1939** L. H. Gray *Foundations of Lang.* ix. 259 Their [sc. words'] degeneration (technically termed *pejoration*;..) is often due to a selection and specialisation of some ethically lower connotation which may be implied in them. **1956** *Archivum Linguisticum* VIII. 74 And is not pejoration a general feature of semantic development? **1966** *Word Study* Dec. 7/1 Perhaps Walt Disney would be interested in the pejoration and 'spread' of the name for his major cartoon character to a word now so loosely defined that it might some day take three dictionary columns to list. **1975** *Amer. Speech 1972* XLVII. 295 In the case of two English words, borrowing [in German] has resulted in pejoration.

pejorism. Delete *nonce-wd.* and add later example.

1910 R. Brooke *Let.* 7 Nov. (1968) 261 Pejorism is the Art of thinking things Worse than they Are in order that you may thereby be the more powerfully impelled to Better them.

pejorist (pī·dʒŏrist, pe·dʒŏrist). [f. as Pejorism + -ist, after *pessimist.*] One who believes that the world is becoming worse.

1919 W. de Morgan *Old Madhouse* ii. 3 Are we the better or the worse off by the change? The Optimist says better, the Pessimist says worse. I think the present writer must be sitting on a fence—a pejorist, suppose we say, since jargon is in vogue nowadays. **1932** *Times Lit. Suppl.* 14 July 508/4 Between optimists and pessimists there are those who style themselves..bonists, malists and pejorists. **1933** A. E. Housman *Let.* 5 Feb. (1971) 329, I am not a pessimist but a pejorist (as George Eliot said she was not an optimist but a meliorist). **1958** *Listener* 31 July 171/3 Why he [sc. A. E. Housman] was impelled to rape the major pejorists of five literatures for phrase and form to enable him to make poetry of his own predicament.

pekan. For '*Mustela pennanti*' substitute '*Martes pennanti*'. (Earlier and later examples.)

1760 T. Jefferys *Nat. & Civil Hist. French Dominions* i. 37 The fur of this animal [sc. a kind of pole-cat], as also that of the Pekan, another creature of the wild-cat kind,.. are what is called the *Menuë Peleterie*, or lesser furs. **1771** T. Pennant *Synopsis Quadrupeds* 224 The *Pekan* and *Vison* of M. de Buffon resemble each other so nearly, that I do not separate them. **1870** J. Yeats *Nat. Hist. Commerce* ii. 270 The pekan inhabits North America, and is also called Hudson's Bay Sable. **1910** E. T. Seton *Life-Hist. Northern Animals* II. 926 The name 'Pekan', first recorded by Charlevoix (1744) and popularized by Buffon 1765, is the Abenaki name, adopted without change. **1963** R. D. Symons *Many Trails* xvii. 177 It was one of the loveliest and rarest of fur-bearers—a pekan or fisher. **1973** R. Fiennes *Headless Valley* v. 93 Pekans are better known as Fishers which is strange since they never fish.

Peke (pīk). Also **Pek, Pekie.** Abbreviation of *Pekin(g)ese *sb.* c.

1915 *Vanity Fair* (N.Y.) Jan. 52/1 England has known the Peke for some time. **1920** *Chambers's Jrnl.* 21 Feb. 177/2 Adjoining were the kennels where the Pekies live. **1922** W. J. Locke *Tale of Triona* ix. 105 Instead of pulling your weight you think it's your right to sit on a cushion, a passenger—or a Pekie dog—and let other people pull you. **1924** Galsworthy *White Monkey* ii. iv. 157 I'll see what I can do, if you'll lend me your Peke for an hour or so. **1926** *Spectator* 22 May 859/2 A young lady of fashion happens to be travelling to-day..with a couple of wardrobe trunks and a fortune in sables, satchels, vanity cases, also a 'Pek'. **1936** [see *ickle *a.*]. **1956** E. Berckman *Beckoning Dream* xiii. 95 The vacant eyes, round like a Peke's. **1973** *Listener* 9 Aug. 181/3 He gave me my first dog..a tiny Pekinese puppy... He came and broke the news to me about the death of that particular peke.

Pekin. Add: Also (now more usually) **Peking.**

The distribution of the spellings *Peking* and *Pekin* is uneven: for convenience, in sense 3 below, the spelling *Peking* is given but in some of the *attrib.* and *Comb.* uses *Pekin* occurs with equal or greater frequency.

1. a. (Further example.)

1954 'M. Cost' *Invitation from Minerva* 105 The mahogany bed was draped with Pekin.

b. A type of Chinese rug or carpet.

1904 M. B. Langton *How to know Oriental Rugs* viii. 223 The Peking, Tientsin, and Samarkand are the only varieties known in this country. **1913** G. G. Lewis *Pract. Bk. Oriental Rugs* (rev. ed.) 303 Those [rugs] which reach our own shores are generally divided into three classes according to the districts from which they came, namely, Pekin, Tientsin and Thibet. **1962** C. W. Jacobsen *Oriental Rugs* 270 The Peking is a medium priced rug, costing much less than the Sarouk and half that of a good Kirman.

c. = **Peking duck* (a).

1885 *Encycl. Brit.* XIX. 647/1 The Pekin, a white breed with pale yellowish tint in the plumage, and a very bright orange bill. **1902** *Ibid.* XXXI. 882/2 Some duck-farmers in England have..also adopted the Peking. **1972** *Guardian* 24 May 9/1 You've got your Peking, your Aylesbury, your Muscovy.

3. *attrib.* & *Comb.*, as *Peking carpet, crepe, rug*; **Peking duck,** (*a*) a large white duck with yellow bill and legs, belonging to a breed imported from China to Britain and the U.S.A. in 1873; (*b*) a speciality of Chinese cuisine; **Peking Labrador** (see quot.); **Peking man,** a fossil hominid, *Homo erectus pekinensis*, first described in 1926 from remains found in caves near the village of Choukoutien; **Peking opera,** a stylized form of opera which evolved in China during the nineteenth century; *Peking point* (see quot.); **Peking spaniel** = *Pekin(g)ese *sb.* c; **Peking stripe** (see quots.).

1969 G. Sims *Sand Dollar* i. 14 He slumped down further in the chair, letting his feet sink deep into the Pekin washed-silk carpet. **1971** *Nat. Geographic* Oct. 548/2 The Westerner who makes his way past the propaganda finds Shantung silk, Swatow lace, Peking carpets. **1972** P. L. Phillips tr. *Formenton's Oriental Rugs & Carpets* 237 (*caption*) Typical Pekin carpet woven in the second half of the nineteenth century. **1895** *Montgomery Ward Catal.* 7/2 Pekin Crepe or Momie Cloth, 29 inches wide, used for fancy party dresses, draperies, etc. **1880** L. Wright *Illustr. Bk. Poultry* (rev. ed.) xxxv. 541 The Pekin Duck is one of the most valuable of recent introductions... The Pekin Duck differs from all others in the shape and carriage of its body, which is a peculiar boat or barge shape. **1902** *Encycl. Brit.* XXXI. 882/2 In America the Peking duck is universally used, and has been made by selection both larger and a better layer. **1955** F. Oliver *Chinese Cooking* ii. 53 Peking Duck..is one of the great Chinese dishes and was always on the banquet menu in Peking. **1973** L. Hellman *Pentimento* (1974) 185, I was driving back to the farm trying not to listen to the noise that came from two crates of Pekin ducks. **1964** Lee Su Jan *Fine Art of Chinese Cooking* xii. 94 The recipe for the authentic Peking Duck appears in this chapter. The incomparable flavour of the duck is a result of the vapours of the wine, soy sauce, and other condiments penetrating the meat. **1973** *Country Life* 15 Mar. 699/1 The infinite variety of Chinese food, with classic dishes such as Peking duck and shark's fin soup. **1974** *Encycl. Brit. Macropædia* VII. 943/2 The greatest of all delicacies of this region is of course the Peking duck. This elaborate, world-renowned dish requires lengthy preparation and is served in three separate courses. **1975** *Nature* 6 Mar. 12/3 The Peking duck originated in China and was first noted in the U.S.A. around 1870. **1960** C. W. Cunnington et al. *Dict. Eng. Costume* 268/2 *Pekin Labrador*,..a Pekin silk flowered in wreaths. **1926** *Peking Leader* 24 Oct. in *Bull. Peking Soc. Nat. Hist.* (1928) II. iv. p. xvii (*heading*) Notable archaeologists deliver addresses at 'Peking Man' meeting. *Ibid.* p. xix, His [sc. J. G. Andersson's] work..culminated in the discovery of the two teeth of the 'Peking Man' (Dr. Andersson did not call the ancient inhabitant by that title, but probably this will be his popular designation). **1929** *Times* 30 Dec. 9/4 The Peking man is considered to antedate Neanderthal man, and is held to be nearer the genus *Homo* than the Piltdown and Java types. **1937** *Discovery* Jan. 27/1 'Peking Man' Skulls. A fifth skull, which may prove to be the most important of all, has been found at Choukoutien. **1946** J. S. Huxley *Unesco* i. 10 They [sc. man's innate mental powers] certainly were improved..in the earliest stages of his career, from Pekin man through the Neanderthalers to our own species. **1955** *Proc. Prehist. Soc.* XXI. 39 The oldest undoubted hearths are those recorded in the Choukoutien caves, occupied by Peking Man at the beginning of the second interglacial period. **1973** *Listener* 10 May 605/2 The classical find of *Homo erectus* was..made in China. He is Peking man, about four hundred thousand years old. **1976** *Times Lit. Suppl.* 7 May 544/5 There is an interesting, if speculative, account of the nature, habits and affinities of Peking Man. **1954** *Folk Arts of New China* 41 Peking Opera is traditionally played by men even to the female characters. Peking Opera has no sets or scenery. **1965** S. Knight *Let.* 1 Oct. in *Window on Shanghai* (1967) xv. 65 Extracts from Peking opera—three modern and one ancient, the story of how the Monkey King ate the peaches of Immortality. **1968** *Guardian* 4 Nov. 7/5 Peking opera.. is extremely formal, a mixture of song, dance, music, and acting. Each movement and gesture and every word sung or recited is synchronised with the music. **1970** W. Apel *Harvard Dict. Music* (ed. 2) 153/1 The Peking opera has a short history (c. 100 years). Although the highly stylized singing and acting demands a cultivated taste, the Peking opera remains the most popular musical art form. **1971** *Guardian* 28 Dec. 12/3 The essence..is communicated to the children through slides, films..revolutionary Peking opera, books, posters. **1973** *Listener* 19 July 84/3 After 1949, a great attempt was made to play down what had become the Court opera—Peking opera—and to encourage the local traditions. **1974** *Encycl. Brit. Macropædia* XII. 675/2 Credit for the beginning of Peking opera is given to actors from Anhwei appearing in Peking in the 1790s. But Peking opera really combines elements from many different earlier forms and, like Western grand opera, can be considered to be a 19th-century product. **1960** C. W. Cunnington et al. *Dict. Eng. Costume* 268/2 *Pekin point*,..a very rich white silk painted with flowers or bouquets with foliage, with a light mixture of gold in the pattern. **1962** C. W. Jacobsen *Oriental Rugs* 270 A Peking rug is a Chinese design rug, hand woven in Japan. **1913** A. Conan Doyle *Poison Belt* (ed. 2) v. 163 There were three gaily dressed women, all young and beautiful, one of them with a Peking spaniel upon her lap. **1908** *Dry Goods Economist* 13 June 81/3 Pékiné, or pekin stripes.—A design in stripes of alternating colors, the stripes usually being of equal widths. **1960** C. W. Cunnington et al. *Dict. Eng. Costume* 268/2 *Pekin*,..a silk textile of the nature of taffeta, having fine stripes running through it; hence 'Pekin stripes'.

Pekin(g)ese (pīkinī·z, pīkiŋī·z), *sb.* and *a.* [f. *Pekin(g)* (see Pekin) + -ese.] **A.** *sb.* **a.** A native or inhabitant of Peking. **b.** The form of Chinese used in Peking. **c.** A Pekingese dog. **B.** *adj.* Of or pertaining to Peking; applied esp. to a breed of dwarf pug-dogs with long, silky hair, obtained originally from the Imperial Palace at Peking; *Pekinese stitch*, a stitch in embroidery.

1849 *Ann. Propag. Faith* Mar. 94, I have been informed that you speak..Pekinese. **1866** *Leisure Hour* XV. 45/2 (*heading*) Peking and the Pekingese. **1879** *Good Words* 745/1 It is old and stiff like Pekingese Buddhism. **1888**

Peel City Guardian 14 Apr. 281/5 A singular Pekingese New Year custom is mentioned in the Shen Pao. **1898** *Kennel Gaz.* June 266/2 Foreign Dogs..Pekin Primula.. (Pekinese Spaniel). *Ibid.* July 299/3 Foreign Dogs. Jock (Pekinese Pug). **1902** *Ibid.* Jan. 19/1 In the hurried departure of the Court from Pekin during the late war.. not a single Pekinese Spaniel was left to..be annexed as loot. *Ibid.* Apr. 127/3 In future the breeds will be known as 'Japanese' and 'Pekinese' only, and neither the term spaniel nor pug added. A separate Classification on the Register of Breeds has also been given to 'Pekinese'. **1904** H. Compton *20th Cent. Dog* I. 261 It is now more than forty years since the first specimens of the Pekinese dog, or lion-dog of China, were imported into England. *Ibid.* 262 Thirty years had to elapse before the Pekinese dog became—I cannot call it popularised..but rather say, in evidence. **1906** *Field* 20 Oct. 663/2 Pekingese were forward in strong numbers, the best dog weighing [etc.]. **1910** *Encycl. Brit.* VI. 216/2 Cantonese is supposed to approximate most nearly to the primitive language of antiquity, whereas Pekingese perhaps has receded farthest from it. **1914** R. B. Lee *Hist. & Descr. Mod. Dogs Gt. Brit. & Ireland* (*Non-Sporting Division*) (ed. 3) Add. p. i, Many additions to the classification..have been made... The most remarkable instance..is that of the Pekingese. *Ibid.* p. v, An ideal Pekingese must have a long coat with a thick undercoat, straight and soft, neither curled nor wavy, the feather on thighs and legs long and profuse. The tail should be carried high on the loins in a loose curl. **1919** G. B. Shaw *Heartbreak House* ii. 81 She has the Ancient Mariner on a string like a Pekinese dog. **1920** *19th Cent.* Sept. 384 The Grand Duchess Tatiana carried in her arms her little Pekingese dog. **1931** A. C. Dixey *Lion Dog of Peking* iii. i. 150 The Pekingese is above all fearless and sporting. **1932** D. C. Minter *Mod. Needlecraft* 46/1 Pekinese stitch..is..most effective... A study of Chinese work will show this stitch worked..in fine silks to give beautiful graded effects. **1934** M. Thomas *Dict. Embroidery Stitches* 159 Pekinese Stitch may be used as a line stitch or as a solid filling where shading is necessary, and is very pretty and braid-like in effect. **1935** [see *crisp *sb.* 7]. **1950** D. Jones *Phoneme* vi. 17 In Pekingese, too, we find that the difference between the first (high-level) and third (low-rising) tones is accompanied by a difference of vowel quality and of length. **1958** *Proc. 8th Internat. Congr. Linguists* 764 Language reform in China covers the reform of the Chinese script and the standardization of Pekingese as the national language. **1960** B. Snook *Eng. Hist. Embroidery* 50 The stitches used [sc. in Black work] include chain, back, sword, Pekinese,..and speckling. **1969** *Queen* 17–30 Sept. 14/2 The hot sour soup ..and Peking duck are authentic examples of the subtle art of Pekingese cuisine. **1970** *Language* XLVI. 675 Even though Pekingese is hardly a rugged testing-ground for methods of describing phonotactic distributions..it might offer a very real test for certain kinds of prosodic analysis. **1976** A. Powell *Infants of Spring* iii. 51 A conspicuous member of the household was Pekoe, my mother's pekinese, so huge in size that he was once mistaken for a St Bernard puppy.

Pekin(g)ology (pīkinǫ·lŏdʒi, pīkiŋ-). [f. *Pekin(g)* (see Pekin) + -ology.] The study of Chinese politics and current affairs.

1962 *Economist* 21 Apr. 228/1 Pekinology is the murkiest of arts, and it takes something like extra-sensory perception to glean much about Chinese policy from the session. **1966** *N.Y. Times* 3 Apr. iv. 6 Practitioners of the recondite art of Pekinology say that if Chairman Mao died tomorrow the party leadership probably would fall. **1976** *Pacific Affairs* XLIX. 126 Inevitably, a study of China's past zone of expansion leads geographically up to and into the zone of Russia's more recent expansion, and here power-politics, geo-politics, Kremlinology and Pekinology raise their hydra heads.

So **Pekin(g)o·logist,** an expert on or student of Chinese politics and current affairs.

1962 *Economist* 21 Apr. 228/2 He [sc. Chou En-Lai] mentioned light and heavy industry in that order (which is the kind of clue that Pekinologists have to fall back on). **1966** *Guardian* 27 July 8/6 If Mao wasn't actually ailing through all these months of secluded speculation, what was he doing? Our tame Pekinologist suggests: training for his epic swim. **1968** W. Safire *New Lang. Politics* 76/1 A new synonym for China watcher is 'Pekingologist', coined on the analogy of Kremlinologist. **1969** *Punch* 2 Apr. 478 Fleet Street is said to be desperate for reliable Pekinologists who can churn out a thousand weekly words on the Chinese enigma. **1972** D. Bloodworth *Any Number can Play* xvi. 150 They've had the pekinologists declaring Mao dead, blind, gaga.

peko, var. *pikau *sb.*

pelagic, *a.* Add: Also applied *spec.* to the environment in any part of the sea away from the littoral and benthic regions and to marine life at any depth that is independent of these regions. (Further examples.)

1882 *Nature* 5 Oct. 559/1, I have spoken of pelagic life as belonging to the surface waters of the ocean..; but,.. it is impossible as yet to limit definitely the range of pelagic forms in depth, and we shall even have to refer to some connections of the fauna of the deep ocean bottom with that of the surface. Pelagic life then includes the inhabitants of the whole ocean waters, excluding those belonging to the bottom and shores. **1891** Murray & Renard in *Rep. Sci. Results Voy. H.M.S. Challenger: Deep-Sea Deposits* iv. 251 We would suggest that the term oceanic Plankton be subdivided into pelagic Plankton for the animals living in the waters from the surface to 100 fathoms, zonary Plankton for those living in the intermediate zones between 100 fathoms from the surface and 100 fathoms from the bottom,..and abyssal Plankton for those living within 100 fathoms from the bottom over pelagic deposits. **1912** Murray & Hjort *Depths of Ocean* ix. 562 The conception of a 'pelagic' mode of life, originally associated with the animal-life of the ocean-surface, thus gradually proved to hold true for life in mid-water also... The main characteristic of pelagic life is its independence of the bottom. **1954** N. B. Marshall *Aspects Deep Sea Biol.* v. 89 Swimming and floating between the surface and the deep-sea floor are the pelagic animals. **1957** *Mem. Geol. Soc. Amer.* No. 67. I. xxii. 643 The distinction between pelagic and benthic species in these depths is difficult in many cases and depends on how far from the bottom a species must live in order to be considered pelagic. *Ibid.* 645 As yet nothing is known about pelagic animals from depths greater than 6000 meters—*i.e.*, from the trenches. *Ibid.* II. vi. 97 The subdivision of the water above the bottom, the great pelagic region, is more satisfactorily established. **1962** K. F. Lagler et al. *Ichthyol.* xiv. 468 Marine fishes may be classed into three main categories: (*a*) the shore or shelf fauna..; (*b*) the pelagic or open sea fishes, generally living near the surface of tropical and warm-temperature seas..; (*c*) deepsea or abyssal forms, inhabiting depths greater than 100 fathoms. **1969** [see *nekton]. **1970** D. A. Ross *Introd. Oceanogr.* vi. 140 The pelagic realm can be subdivided into the neritic environment (the water that overlies the continental shelf) and the oceanic environment (the water of the deep sea). **1974** [see *neritic *a.*].

b. Similarly applied to whaling and whalers.

1941 J. S. Huxley *Uniqueness of Man* viii. 184 With the advent of pelagic whaling it seemed certain that, unless international regulation of the industry were achieved, whales would certainly become exceedingly scarce, and some species might be wiped out. **1958** *Times* 12 Nov. 11/6 The total number of whales caught last year by pelagic or factory ship expeditions in the Antarctic in 69 catching days was 35,977. **1974** *Nature* 4 Oct. 367/2 If the 1974–75 quotas for pelagic whaling in the Southern hemisphere are reached..the catch of all species next season may reach 380 thousand tons.

c. Of sea-bed material: formed within the sea itself, rather than transported from the land.

1884 Murray & Renard in *Proc. R. Soc. Edin.* XII. 515 The following table shows the nomenclature we have adopted:—Terrigenous deposits... Pelagic deposits. **1891** —— *Rep. Deep-Sea Deposits* iii. 185 From the point of view of their composition, as well as of their geographical and bathymetrical position, Marine Deposits may be separated into two great divisions, viz. (I.) Pelagic Deposits—those formed towards the centres of the great oceans, and made up chiefly of the remains of pelagic organisms along with the ultimate products arising from the decomposition of rocks and minerals; and (II.) Terrigenous Deposits. **1959** A. Hardy *Fish & Fisheries* v. 103 Pelagic deposits..cover the floor of the great oceans in the depths beyond the edge of the shelf. **1970** D. A. Ross *Introd. Oceanogr.* viii. 297 Deep-sea sediments can be divided into two major groups: pelagic sediments and terrigenous sediments.

d. = *limnetic *a.*

1899 G. C. Whipple *Microsc. Drinking-Water* viii. 105 The plants and animals that inhabit lakes and ponds may be classified according to their habitat... The limnetic or pelagic organisms are those that make their home in the open water. **1955** C. C. Davis *Marine & Fresh-Water Plankton* i. 11 Lakes may be subdivided into horizontal and vertical portions. Horizontally, the relatively shallow area close to shore..is called the littoral region, while the region of open water is known as the limnetic (or pelagic) region. **1975** G. A. Cole *Textbk. Limnol.* ii. 9/1 Beyond the pond weeds..is the open water. This is the limnetic or pelagic zone, a region in the lake where shore and bottom have lessened influence.

e. Of birds: inhabiting regions of open sea beyond the edges of a continental shelf, feeding on plankton and other marine organisms, and returning to shore only in the breeding season.

1935 V. C. Wynne-Edwards in *Proc. Boston Soc. Nat. Hist.* XL. 240 The typical species of these three communities [of birds] in the temperate North Atlantic might be separated as follows: 1. Inshore... 2. Offshore... 3. Pelagic. *Ibid.* 241 Pelagic birds..feed chiefly on plankton. **1936** R. C. Murphy *Oceanic Birds S. Amer.* I. ii. 323 South America..is the longest mass of land lying in the relatively open oceans of the southern hemisphere... It is natural..that nearly all the genera, and a large proportion of the species of southern-hemisphere pelagic birds should occur within the limits of the field. **1954** Fisher & Lockley *Sea-Birds* vii. 170 Little is known of the construction of pelagic flocks. **1974** A. Dillard *Pilgrim at Tinker Creek* x. 165 The oceanic breeding grounds of pelagic birds are as teeming and cluttered as any human Calcutta.

pelagically (pĭlæ·dʒikăli), *adv.* [f. prec.: see -ly².] In pelagic regions.

1966 R. M. Lockley *Grey Seal, Common Seal* x. 129 Seals which breed pelagically on floating icefields are threatened by fishers who hunt them today by boat, helicopter and light aircraft. **1975** *Nature* 8 May 100/3 Introductions, the American smelt and the alewife, dominated the main trench and were also common pelagically.

pelandok (pəlændǫ·k). Also **plandok.** [Malay.] The lesser mouse-deer, *Tragulus javanicus*, native to parts of south-east Asia.

1821 J. Leyden tr. *Malay Annals* 89 One of his dogs roused a white pelandok. **1836** J. Low *Diss. Soil & Agric. Penang* iii. 188 The *plandok*, or cheurotin [*sic*] of Buffon, or hornless deer, about the size of a hare, is plentiful. **1839** T. J. Newbold *Pol. & Statistical Acct. Straits of Malacca* I. vii. 436 The Plandok is a favourite animal among the Malays, and frequently alluded to both in their prose compositions and poems. **1900** W. Maxwell in W. W. Skeat *Malay Magic* v. 113 There lived a man whose wife.. was seized with a violent longing for the meat of the *pelandok* (mouse-deer). **1905** *Outing* Jan. 469/2 It may be anything from a *plandok* (mouse deer) to a tiger. **1958** [see *kidang]. **1962** *Listener* 22 Nov. 878/2 The spider being here [sc. in Africa] what the *plandok* is in Malaya, a jungle Till Eulenspiegel. **1965** C. Shuttleworth *Malayan Safari* ii. 26 The flesh of the pelandok is extremely edible, and he is hunted constantly. **1969** Ld. Medway *Wild Mammals Malaya* 106/1 Lesser Mouse-deer. Pelandok, Kanchil... Largely nocturnal; generally solitary.

pelargonidin (pelăɹgōu·nĭdin). *Chem.* [a. G. *pelargonidin* (R. Willstätter 1914, in *Sitzungsber. d. k. preuss. Akad. d. Wissensch.* 405), f. as next: see *-idin.] An anthocyanidin (usu. isolated as the chloride, $C_{15}H_{11}O_5Cl$) that is the aglycone of pelargonin and many other red plant pigments.

1914 *Chem. Abstr.* VIII. 3421 The pelargonin of the scarlet pelargona flower on hydrolysis gave 2 mols. dextrose and 1 mol. of pelargonidin, $C_{15}H_{11}O_6Cl$ [*sic*]. **1934** *Jrnl. Chem. Soc.* 1612 The orange-scarlet nasturtium contains a pelargonidin 3-bioside, the red *gloxinia* flowers are coloured by a pelargonidin rhamnoglycoside, and pelargonidin 3-glycosides occur in scarlet carnations, strawberries,..and in other flowers. **1937** [see *delphinidin]. **1965** [see *peonidin]. **1966** [see *cyanidin]. **1974** *Phytochemistry* XIII. 2002 The major anthocyanins were dentified as the 3-glucoside, 3-galactoside, 3-rutinoside and 3-robinobioside of pelargonidin.

pelargonin (pelăɹgōu·nin). *Chem.* [a. G. *pelargonin* (R. Willstätter 1914, in *Sitzungsber. d. k. preuss. Akad. d. Wissensch.* 405), f. G. *pelargonie* or mod.L. Pelargonium: see -in¹.] An anthocyanin (usu. isolated as the red chloride, $C_{27}H_{31}O_{15}Cl$) that is the colouring matter of zonal pelargoniums and on hydrolysis gives pelargonidin and glucose.

1914 [see *pelargonidin]. **1934** C. C. Steele *Introd. Plant Biochem.* xix. 221 The different species of *Primula* contain different delphinidin derivatives, while cyanin and pelargonin predominate in different coloured varieties of *Pelargonium zonale*. **1956** I. L. Finar *Org. Chem.* II. xv. 585 Pelargonidin chloride, $C_{15}H_{11}O_5Cl$. This is formed, together with two molecules of glucose, when pelargonin chloride is hydrolysed with hydrochloric acid.

Pelasgian, *a.* and *sb.* Add: **b.** (Earlier and later examples.) Also, the Indo-European language attributed to the pre-Hellenic population of Greece and the Aegean. Also *attrib.*

a**1490** J. Skelton tr. *Diodorus Siculus' Bibliotheca Historica* (1956) I. iv. 321 Summe call theym Pelasgians by encheson that they first proceded from theym of Pelasgye which colaterallyth vnto the Grecians. **1875** *Calcutta Rev.* LXI. 4 The Celts..were followed by the so-called Pelasgians, who then separated into Greeks and Latins. **1939** L. H. Gray *Foundations of Lang.* 376 N. Marr..held that Caucasian was..cognate with Basque, Etruscan, Pelasgian, and other 'Mediterranean' tongues. **1954** Pei & Gaynor *Dict. Linguistics* 163 *Pelasgian*, an extinct language of southern Europe, variously described as Mediterranean or Japhetic, and said to have been linked with Caucasian, Basque and Etruscan. **1966** *Lingua* XVI. 277 The discussion of Pelasgian phonology offers nothing new. **1972** W. B. Lockwood *Panorama Indo-European Lang.* 12 Pelasgian was apparently a living language locally in the Aegean until the fifth century B.C.

Hence **Pela·sgianist,** a student of the Pelasgian language.

1965 *Lingua* XIII. 349, I propose to examine..in section E those 'Pelasgian' etymologies which have been agreed by three or more Pelasgianists with no dissentient. **1967** *Ibid.* XVIII. 148 Errors made with 'Pelasgian' are characterized by the effort *to cling to the customary conceptions*..both on the side of the adversaries of 'Pelasgian' and on the side of the 'Pelasgianists' themselves. **1970** *Trans. Philol. Soc. 1969* 84 This [sc. reconstruction of a language from its loan-words in another language] is the desperate hazard attempted by the school of 'Pelasgianists'.

Pelasgic, *a.* Add: **b.** *sb.* = *Pelasgian b above.

1966 E. P. Hamp in Birnbaum & Puhvel *Anc. Indo-European Dial.* 114 It is convenient here to reproduce Georgiev's subgrouping of Indo-European..: Central: Greek, Daco-Mysian.., Indo-Iranian, Phrygian-Armenian, Thracian, Pelasgic.

pele, var. *peel *sb.*¹ 4.

Peléan (pĕlē¹·ăn), *a.* *Geol.* Also **Peléean, Pelean,** and with small initial letter. [f. the name of Mount *Pelée*, a volcano on the island of Martinique, W. Indies, which erupted in this way in 1902: see -an.] Of, pertaining to, or designating a type of volcanic eruption characterized by the lateral emission of *nuées ardentes* from a point of weakness in the flank and the vertical extrusion of very viscous lava in the centre which tends to become consolidated as a solid plug.

1903 Anderson & Flett in *Phil. Trans. R. Soc.* A. CC. 499 We propose to adopt the term 'A Peléan Eruption' to designate this group of phenomena. **1934** C. R. Longwell et al. *Outl. Physical Geol.* x. 189 The Peléean phase is the most violently explosive of all. *Ibid.* 191 The extraordinary features of a Peléean cloud are that it is emitted as a horizontal blast from beneath the lava plug in the summit of the volcano; that it carries with it an enormous amount of rock fragments, [etc.]. **1935** *Publ. Carnegie Inst. Washington* No. 458. 97 The general morphological features of the Peléan summit are those of an ancient crater wall, remaining in place on all sides but one. **1938** [see *bread-crust bomb* s.v. *BREAD *sb.* 10]. **1944** C. A. COTTON *Volcanoes* xii. 183 The chief product of both Vulcanian and Pelean eruptions is ash. **1965** A. HOLMES *Princ. Physical Geol.* (ed. 2) xii. 310 There are *nuée ardente* eruptions that differ in certain fundamental ways from those described as Peléan. **1972** G. A. MACDONALD *Volcanoes* x. 226 Two of the greatest Peléean eruptions of recent years are those of Bezymianny, Kamchatka, in 1956, and Mt. Lamington, New Guinea, in 1951. **1976** P. FRANCIS *Volcanoes* iii. 116 The presence of such a dome is sometimes considered to be a characteristic feature of Peléean eruptions.

pelerine. Add: **b. pelerine stitch** (see quots.).
1926 J. CHAMBERLAIN *Hosiery, Yarns & Fabrics* vi. 128 The Pelerine Stitch has a limited application as a shawl stitch, but modern variations of it are used for producing (1) cellular fabric; (2) in-turned welt... This stitch or combination of stitches is based on the transferring of a formed sinker loop to one or two needles. **1950** '*Mercury*' *Dict. Textile Terms* 392/1 *Pelerine stitches*, these stitches are produced by meshing the upper and lower components of a single loop together at the next course. Modifications of these stitches produce raised designs and places and eyelet holes. By meshing the lower components of the first row of loops to a subsequent row a hem or welt is produced on a plain fabric.

pelican. Add: **5.*** With capital initial: the proprietary name of a series of non-fiction books published by Penguin Books; a book in this series.
1937 *Bookseller* 3 Feb. 147 (Advt.), A glance at the magnificent first list below will reveal the general nature of *Pelican Books*, a new series of popular books on science, astronomy, archæology, politics, economics, history, etc. **1942** *Scrutiny* X. iv. 385 Enough passages like the above could be found..to yield Professor Stebbing cannon-fodder for at least a chapter, if not for a whole Pelican. **1948** G. V. GALWEY *Lift & Drop* vi. 158 He..began to study the book-shelves... There was..a fair showing of Pelicans. **1953** E. SIMON *Past Masters* I. i. 15 He published a Pelican on his work under Chrichton in the Islands, which I thought you might have seen. **1957** R. HOGGART *Uses of Literacy* viii. 210 There are the dailies and weeklies,..the Penguins and the Pelicans. *Ibid.* xi. 261 They are in the habit today of buying copies of Pelicans. **1957** *Trade Marks Jrnl.* 22 May 532/1 Pelican... Printed books, being literary, dramatic, musical or artistic works, but not including books relating to pelicans. Penguin Books Limited,..Harmondsworth, County of Middlesex; manufacturers and publishers. **1966** 'L. BLACK' *Bait* viii. 128 He was reading a paper-back... It appeared to be a Pelican. **1968** *Guardian* 26 Sept. 11/1 By the time of the first Pelicans Penguins (the mother birds) were, of course, well established... Pelicans were born in 1937, the inheritors of a half century and more of self-education.
5.** In full *pelican crossing*: a pedestrian-crossing controlled by push-buttons (see quot. 1966).
1966 *Evening Standard* 26 May 11/6 We hope the Ministry will install 'pelicans'..in the town. Pelicans would be safer than zebras and easily understood by the public—the pedestrian just pushes a button which operates red, amber and green lights telling motorists when to stop. **1969** *Daily Tel.* 9 July 17/7 Another type of push-button pedestrian-controlled crossings—to be known as Pelicans—will come into operation on Monday. *Ibid.*, Instead of the white 'X', the Pelican (*pedestrian light controlled*) crossing will show a green signal to a driver until a pedestrian presses the button to start the signal sequence. **1974** *Country Life* 30 May 1332/2 A Minister stating that 'pelicans will be interchangeable with zebras' ..was referring to the replacement of the sturdy and familiar orange beacons at the kerbside by the flashing lights and little coloured men... A circular from the DoE elucidates some of the mysteries of the pelican crossing. **1975** *Times* 22 Sept. 12/5 Question time at council meetings is peppered with inquiries and entreaties about zebra and pelican crossings and such like. **1976** *New Scientist* 24 June 702/1 The GLC survey..studied 40 pelicans which had been converted from zebras. **1976** *Flintshire Leader* 10 Dec. 1 Residents of Oakenholt are ready to block the main road through the village in their campaign for a pelican crossing.
6. Pelican flag *U.S.*, the flag of the State of Louisiana; **Pelican State** *U.S.*, a sobriquet of the State of Louisiana.
1860 *Charleston* (S. Carolina) *Mercury* 25 Dec. 4/5 The Pelican flag of Louisiana was unfurled in the streets, amid tremendous cheering... The Pelican flag consists of a red star upon a white field, with the ancient Louisiana emblem of a Pelican feeding her young. **1865** A. D. RICHARDSON *Secret Service* 40 There were Pelican flags, and Lone Star flags, and devices, unlike anything in the heavens above. **1859** *Harper's Mag.* May 853/2 A well-known writer in the Pelican State writes us a good thing from one of his little folks. **1934** G. E. SHANKLE *State Names* 119 The name, the *Pelican State* was given to Louisiana from the fact that this bird is so frequently seen along the streams..which fact caused it to be chosen as the emblem in the state coat of arms. **1949** *Times-Picayune* (New Orleans) *Mag.* 27 Nov. 53/2 Now Mississippi has nosed out the Pelican State. **1959** E. A. DAVIS (*title*) Louisiana: the Pelican State. **1974** *Encycl. Brit. Micropædia* X. 272 (*table*) State nickname(s) or slogan... Louisiana..Pelican state.

pe·lican, *v. rare*—[1]. [f. the sb.] *trans.* To swallow or eat like a pelican.
1953 DYLAN THOMAS *Under Milk Wood* (1954) 65 And she bursts into tears, and, in the middle of her salty howling, nimbly spears a small flatfish and pelicans it whole.

pelike (pe·liki, peli·ki). *Gr. Antiq.* Also **pelice, pelice.** Pl. **pelikai.** [ad. Gr. πελίκη a wooden bowl, pitcher.] A type of amphora with an ovoid body, wide mouth, and broad base used for holding wine or water.
1873 S. & J. HORNER *Walks in Florence* II. 433 On a Pelice, a vase rarely found in Etruria, and belonging to the most perfect style of art,..is the figure of a man binding up the arm of a youth. **1891** NETTLESHIP & SANDYS *Seyffert's Dict. Classical Antiquities* 685 (*caption*) Various shapes of Greek vases... pelikē. **1902** J. H. HUDDILSTON *Lessons from Greek Pott.* i. 75 The largest group is found on the Lotos pelike, which although [illegible] Homer,..are two women, introduced..to remind us of Briseis. **1948** J. D. BEAZLEY *Greek Vases in Poland* iii. 64 His pictures, which are chiefly on small hydriai, small pelikai, and marriage-vases, are usually taken from the life of women. **1956** *Burlington Mag.* May 253/1 A pelike with snarling lion and Amazons. **1967** R. S. FOLSOM *Handbk. Greek Pottery* 159 The name pelike was applied by early archaeologists to the one-piece amphora with a sagging belly and broad neck. In fact the name appears to have no justification, but is retained for convenience. The form first appeared about 520 B.C. and lasted until the 4th century B.C. **1971** *Ashmolean Mus. Rep. Visitors 1970* 9 A notable addition to the collection of Athenian red-figure vases is a pelike, of the 4th century B.C.

Pelion (pī·liən). The name of a mountain in Thessaly, used in phr. *to pile* (or *heap*) *Pelion upon Ossa* (or *Ossa upon Pelion*) [tr. Virgil *Georgics* I. 281 *imponere Pelio Ossam*]: to add to what is already great; to add difficulty to difficulty. Also in similar phrases.
1589 A. FLEMING tr. *Virgil's Georgiks* i. 10 Thrise did they trie and indevor to heape upon mount Pelius, To lay the mountaine Ossa. **1594** NASHE *Unfortunate Traveler* sig. C4v Whosoeuer seekes by headlong meanes to enter into Heauen..shall with the Gyants..be ouer-whelmed with Mount Ossa and Pelion. **1609** DEKKER *Guls Horne-Booke* 30 By talking and laughing..you heape Pelion vpon Ossa, glory vpon glory. **1628** T. MAY tr. *Virgil's Georgicks* i. 34 Thrice they indeavour'd with strong hand to place The mountaine Ossa on high Pelion. *a* **1734** R. NORTH *Examen* (1740) ii. v. 336 It is Pelion upon Ossa to set Philips above Power. **1919** G. B. SHAW in *Shaw on Shakespeare* (1962) 257, I might pile Pelion on Ossa with illustrations from both camps that might very well be cut out of Shakespeare's plays. **1927** C. A. & M. R. BEARD *Rise Amer. Civilization* I. xii. 565 In piling Ossa on Pelion, Webster did not overlook mundane considerations—the economic and political substance of the pending issue, the sale of those annoying western lands. **1957** PARTRIDGE *English gone Wrong* i. 22 To *categorise* and *categorization*, and those Pelions-upon-Ossa horribles, *recategorize* and *recategorization*. **1967** *Word Study* Mar. 3/1 We discover Mount Pelion piled on Ossa, however, when we note that our teachers are even more conservative than their conservative textbooks. **1976** *Times* 16 Apr. 14/6 Piling Pelion upon Ossa (a nasty habit that foreigners are more given to) they..calculate the rate of inflation in Britain.

peliosis (peliō·sis). *Path.* [mod.L. (F. Swediaur *Novum Nosologiæ Methodicæ Systema* (1812) I. p. xxiii, II. 173), f. Gr. πελίωσις extravasation of blood, f. πελιός livid.] = PURPURA 1; *peliosis rheumatica* [Gr. ῥευματικός subject to a discharge], an uncommon disease characterized by localized eruptions of the skin and mucous membranes, and associated with arthritis; (now usu. regarded as the same as *Henoch–Schönlein purpura* s.v. *HENOCH).
1839 DUNGLISON *Dict. Med. Sci.* (ed. 2) 771/2 Peliosis, or Purpura haemorrhagica. **1850** A. T. THOMSON *Pract. Treat. Dis. affecting Skin* v. 338 Under the term Peliosis, or Purpura [illegible] which has been termed by others Roseola rheumatica. In this disease local extravasations occur in the skin, and erythematous patches about the joints, which are swollen and painful. **1860** *New Sydenham Soc. Yr.-bk. 1860* 403 (*heading*) On the peliosis rheumatica of children. **1868** C. H. FAGGE et al. tr. *Hebra's Dis. of Skin* II. xxvii. 422 Albert ..makes unnecessary distinctions between peliosis (purpura) and petechiæ. **1911** *Brit. Med. Jrnl.* 11 Feb. 331/1 That pelosis rheumatica is more of rheumatic origin is hardly capable of proof. **1968** CHAMPION & WILKINSON in A. J. ROOK et al. *Textbk. Dermatol.* I. xvii. 428/1 The separation of Schönlein's 'peliosis rheumatica' from the Henoch-Schönlein syndrome] is no longer justified.

pelisse. 2. b. (Earlier and later examples.)
1805 T. FREMANTLE in *Wynne Diaries* (1940) III. 154 Mistress Tittler with a black Velour pelisse, tell her I desire she will not spoil it until I come home. **1807** JANE AUSTEN *Let.* 7 Jan. (1952) 173 Caroline's new pelisse depended upon her mother's being able or not to come so far in the chair. **1828** M. O'BRIEN *Jrnl.* 20 Oct. (1968) I. 19 Mary set to work on a pelisse for the baby. **1922** JOYCE *Ulysses* 20 In babylinen and pelisse, bighcaded, with a caul of dark hair.

Pell (pel). *Math.* [See PELLIAN *a.*] *Pell('s) equation*: any Diophantine equation of the form $ax^2 - y^2 = 1$ (*a*, *x*, and *y* being integers). Also *absol.*
1910 *Encycl. Brit.* I. 617/2 Although Pell had nothing to do with the solution, posterity has termed the equation Pell's Equation. **1912** E. E. WHITFORD (*thesis title*) The Pell equation. **1966** OGILVY & ANDERSON *Excursions in Number Theory* x. 129 It turns out that the equation $y^2 - Nx^2 = 1$ known as Pell's equation, has solutions in integers whenever N is not a perfect square. **1974** *Sci. Amer.* July 116/3 Whenever the coefficient is not a square the Pell equation has an infinity of solutions. *Ibid.* 120/2 The general Pell equation, a key to so much of this kind of number analysis.

Pellegrini–Stieda (pelĕgrī·ni,ʃtī·da). *Med.* The names of Augusto *Pellegrini* (c 1880–c 1940), Italian physician, and Alfred *Stieda* (1869–1945), German physician, used *attrib.* and in the possessive to designate a condition described by them in 1905 and 1908 respectively, in which ossification of the tibial collateral ligament of the knee occurs as a result of injury. [Named by R. Pettrignani 1930, in *Contributions à l'Étude de la Maladie de Pellegrini-Stieda* (Thèse, Paris) 14.]
1932 *Amer. Jrnl. Roentgenology* XXVIII. 97/2 Pellegrini-Stieda's disease is of importance from its medicolegal aspect. **1945** *Surg., Gynecol. & Obstetr.* LXXXI. 212/1 The imposing name of 'Pellegrini-Stieda disease' is misleading. It implies that the one finding of the ossification over the adductor tubercle is a disease complex. **1971** A. A. MICHELE *You don't have to Ache* v. 128 A fairly common condition in men between the ages of 24 and 40 years, is hardening and calcification of the ligament at the inner side of the knee joint (Pellegrini Stieda [*sic*] disorder).

pellet, *sb.*[1] Add: **2. c.** = CAST *sb.* 19.
1802 G. MONTAGU *Ornith. Dict.* I. s.v. *Owl, White*, Their food is chiefly mice, which they swallow whole, and..eject the bones and fur in large pellets, which are termed castings. **1834** R. MUDIE *Feathered Tribes Brit. Islands* I. 141 Mice are preferred to birds, the feathers being more untractable than the fur, both in swallowing, and in casting pellets or quids. **1895** *U.S. Dept. Agric. Yearbk. 1894* 217 These masses, known as 'pellets' are regurgitated before fresh food is taken. **1909** *Daily News* 5 Jan. 4/3 The brown owl's pellet very rarely contains the remains of shrews. **1948** *Brit. Birds* XLI. 290, I found a secluded coomb..which, judging by the quantity of droppings and pellets below the ledges, had been used as a roosting place by a number of Ravens. **1964** A. L. THOMSON *New Dict. Birds* 608/2 Pellets are best known in respect to birds-of-prey, but..birds of very widely differing species regularly eject pellets. **1971** S. HILL *Strange Meeting* i. 31 The small, bleached bones from the owls' pellets.
d. The droppings of various small animals, esp. the rabbit, which resemble pellets in shape.
1916 'W. N. P. BARBELLION' *Diary* (1920) 129 Those sand dunes! Their characteristic feature was rabbits' skulls..and the little round dry pellets of rabbits, more numberless than the sand-shells. **1958** *Nature* 10 June 982/2 Numerous coccidial oocysts were also found in the stomach pellets of the control rabbit..which appeared to prove the faecal origin of the pellets. **1966** THOMPSON & WORDEN *Rabbit* iii. 27 A blinded rabbit took fresh pellets from its anus. **1964** R. M. LOCKLEY *Private Life Rabbit* ii. 32 At night their pellets were scattered along the boundary fence. **1972** R. ADAMS *Watership Down* i. 405 Under snow they [*sc.* rabbits] may stay underground for days at a time, feeding only by chewing pellets.
3. (Later example.)
1953 *Country Life* 27 May 1303/2 A purely secular [illegible], at..the time 1664, its so-far unidentified maker's mark 'P.D.' with three pellets above, a cinquefoil below.
5. pellet bomb, a type of anti-personnel bomb; **pellet mill,** an apparatus for pelleting powders.
1970 *Guardian* 10 Mar. 12 Anti-personnel bombs..the same pellet bombs which figured so prominently in North Vietnamese protests. They are dropped in canisters..the bombs, each a bit bigger than a man's fist, might go on exploding for two days. **1970** *Peace News* 3 Apr. 5/3 An unexpected pellet bomb she had touched while digging in the fields. **1950** J. H. PERRY *Chem. Engineers' Handbk.* (ed. 3) 1189/1 Pellet mills are designed to agglomerate permeable free-flowing materials into pellet form. **1964** *Power Farming* Mar. 22/1 Demand for cubed and pelleted feed has led to the development of the Cubamix single unit mobile feed plant... The equipment comprises the grinding and mixing section as already used on the company's Grindamix model and a separately driven Simon Barron pellet mill.

pellet, *v.* Add: Pples. **pelleted, -eting** (occas. incorrectly **-etting**). **a.** Restrict †(*obs.*) to the second def. and add later examples of the first sense; to form or shape into pellets; *esp.* to coat (plant seed) with soluble nutritive and protective substances to facilitate handling and promote growth.
1936 L. M. T. BELL *Making & Moulding of Plastics* xi. 185 For convenience and ease in the handling of the powder..moulding powders are frequently pelleted cold prior to moulding. **1944** *Sugar Beet Jrnl.* Jan. 41 A process of 'pelleting' sugar beet seed segments has been developed. The new process coats the rough segments with a water-soluble layer of beneficial and inert material, making the seed pieces smooth, spherical, and about the size of small seed peas. **1949** *Chem. Abstr.* XLIII. 2154 [The apparatus] is particularly adapted to pelleting Pb alloys into uniform fine shot. **1950** J. H. PERRY *Chem. Engineers' Handbk.* (ed. 3) 1189/1 Although originally designed for pelleting animal feeds, it is now being adapted to the handling of many other products. **1958** *Times* 24 Nov. 15/3 These low seed rates can be achieved..in the case of timothy, by using seed pelleted with basic slag. **1970** *Country Life* 17–24 Dec. 1187/1 Pelleting hay has been tried without more than local acceptance. **1973** *Daily Tel.* 30 June 8/5 (Advt.), Seed pelleter with sufficient Seedex Plant Food Compound to pellet hundreds of seeds. **1974** *Nature* 22 Nov. 327/1 After incubation for 30 min at 37°, the microsomes were pelleted by centrifugation and the DNA reisolated and purified.
Hence **pe·lleting** *vbl. sb.* and *ppl. a.*
1936 L. M. T. BELL *Making & Moulding of Plastics* xi. 185 Figure 56 shows a crank pelleting machine..with the powder hopper removed. **1944** *Business Week* 26 Aug. 52/2 Western companies still regard the process as experimental, one question being whether the pelleting material, which easily melts from around the seed in damp weather, may not have more restraint in the dry western soils. **1945** H. T. BIRKBY *Phenolic Plastics* vii. 78 In the Plastics industry the term tabletting means the compression of moulding powder into a variety of small, easily-handled blocks, each of a pre-determined weight. Two other terms often used for this process are 'preforming' and 'pelleting'. **1950** J. H. PERRY *Chem. Engineers' Handbk.* (ed. 3) 1189/1 Pelleting of dusts, fumes,..and the like can be accomplished in a rotating-drum device known as the Dwight-Lloyd Segregating Pelletizer. **1963** P. FINN-KELCEY in A. N. Duckham *Farming* III. xi. 404 Pelleting machines introduced in recent years are already proving popular with farmers who compound their own stockfoods.

pelletable (pe·lĕtăb'l), *a.* [f. PELLET *v.* + -ABLE.] Capable of being formed into pellets. Hence **pelletabi·lity.**
1970 *Nématologica* XXIV. 296 (*heading*) A simplified protocol for electron microscopy processing of 'pelletable' material. **1973** *Biol. Abstr.* LV. 2932/2 Bull sperm of 296 animals was tested for pelletability. **1976** *Nature* 29 Apr. 324/2 Unknown amounts may have been associated with organelles or plastid fragments of a lower pelletability than intact etioplasts. *Ibid.* 325/1 The pelletable phytochrome investigated by many other workers seems to bind to membranous material only in the Pfr form.

pelleted, *ppl. a.* Add: **2.** Formed into or supplied as pellets.
1960 *Los Angeles Times* 25 Nov. 13/5 Tests being conducted here with newly developed 'pelleted' seed show that..complete seed control can be obtained and expensive thinning..eliminated by the use of mechanical thinners or thinning eliminated altogether. **1951** *Sci. News Let.* 28 Apr. 269/ One of the most valuable aids to the blind gardener..is the recent development of pelleted seeds. **1958** *Listener* 5 Dec. 90/3 Slugs are controlled with the aid of pelleted metaldehyde. **1970** *Nature* 10 Oct. 178/2 [illegible] pelleted diet of chopped lucerne hay and oats. **1973** *Guardian* 17 Mar. 8/5 Every effort should be made to sow the seed thinly. Pelleted seeds, of course, make this much easier. **1975** *N.Z. Jrnl. Agric.* Sept. 69/1 They were all fed to three weeks of age on pelleted rations.

pelleter[3]. [f. PELLET *v.* + -ER[1].] An apparatus for pelleting.
1960 *Times Rev. Industry* July 51/1 BIPEL now sells its [illegible] outside the plastics industry. This includes pelleters for explosives, punch presses for metal work, etc. **1963** P. FINN-KELCEY in A. N. Duckham *Farming* III. xi. 405 Whether an installation is to consist of a mill only, a mill and a mixer or the full layout of mill, mixer and pelleter,..careful thought must be given to the positioning of each. **1973** [see *PELLET *v.*].

pelletization (pelĕtaizē·ʃən). [f. next + -ATION.] The action or process of forming into pellets.
1959 *Jrnl. Iron & Steel Inst.* CXCIII. 75/3 The article deals with..determining the optimum conditions for making use of 'cold' agglomeration, i.e. pelletisation in which the fine material is transformed into small spheres by mechanical rolling. **1966** KIRK & OTHMER *Encycl. Chem. Technol.* (ed. 2) IX. 435 One of the more interesting developments which [illegible] fluidized calcination is the technique of pelletization. **1971** *Physics Bull.* Mar. 136/2 Major cooperative projects under consideration included an iron mine and an iron ore pelletization plant in the USSR. **1977** *Jrnl. R. Soc. Arts* CXXV. 388/1 Taconite ores..are more amenable to pelletization than higher grade Mesabi-type, deposits.

pelletize (pe·lĕtaiz), *v.* [f. PELLET *sb.*[1] + -IZE.] *trans.* To form or shape into pellets. Cf. PELLET *v.* a in Dict. and Suppl.
1953 KIRK & OTHMER *Encycl. Chem. Technol.* IX. 900 The peat is excavated, macerated, pelletized in a long drum, and dried. **1963** *Engineering* 1 Feb. 180 The wood [illegible] to the Australian [illegible]. **1964** *Economist* 23 May 855/1 In America ore [illegible] of 'pelletising' the burden. **1973** D. R. PEHLKE *Unit Processes Extractive Metall.* ii. 19 The final temperatures during induration reach 2300–2400°F, depending upon the concentrate being pelletized.
So **pe·lletized** *ppl. a.* = *PELLETED *ppl. a.* 2; **pe·lletizing** *vbl. sb.* and *ppl. a.*
1951 *Industr. & Engin. Chem.* June 1390/1 It is not [illegible] pelletized [carbon] black directly for [illegible]. **1951** *Engineering* 22 June 761/2 During a recent visit to the United States, [illegible] that was being done in [illegible] of iron ores. **1964** [illegible] stainless steel tubes. **1968** *Sci. Jrnl.* Oct. 23/3 Bentonite is used..as an agent for the pelletizing of iron ore. **1969** *Times* 2 May (Mining Suppl.) p. iv/7 Savage River produces magnetite from a relatively low grade deposit, grinds and concentrates this to a high grade pellet feed and pipes it 53 miles to the pelletizing plant at the port. **1970** *Times* 24 Aug. 15/6 Kawasaki is working on a system using oxygen and crude oil or liquefied natural gas combined with pre-treated, pelletized iron ore. **1977** *Cork Examiner* 4 June 17/4 The pellets will get harder with time... At no time was it considered that metal ground.. would be included in the pelletised mix. **1977** *Jrnl. R. Soc. Arts* CXXV. 394/1 Attempts to encourage the industry to construct at least some processing plants near to its raw materials have been most successful in the case of iron ore pelletizing plants.

pelletizer (pe·lĕtaizəɹ). [f. prec. + -ER[1].] = *PELLETER[3].
1942 *Ann. Rep. Progr. Rubber Technol.* 1941 V. 120 After discharge from the internal mixer the stock passes through a 'Hale Pelletizer' and the pellets are conveyed to storage bins. **1950** [see *PELLETING *vbl. sb.*]. **1961** *Engineering* 9 June 788/2 Pan pelletizers are used for granulating materials in the chemical and fertiliser industries. **1970** *Materials & Technol.* III. iii. 148 Rotary kiln pelletizer.

Pellian, *a.* Add: (Earlier and later examples.) Also *absol.*
[**1767** EULER in *Novi Commentarii Acad. Sci. Imp. Petropolitanæ* XI. 28 (*heading*) De usu novi algorithmi in problemate Pelliano solvendo. *Ibid.* 31, *p* = *aq* + 1... Atque hoc est illud problema olim quidem a Pellio Anglo propositum ... a solutionis ingeniosissimae auctore Pellianum vocatum.] **1862** *Rep. Brit. Assoc. Adv. Sci. 1861* 314 There does not seem to be any ground for attributing either the problem or its solution to Pell... Nevertheless the equation $T^2 - DU^2 = 1$ is often called the Pellian equation after him, probably upon Euler's authority. **1911** *Encycl. Brit.* XIX. 853/1 This is usually called the Pellian equation, though it should properly be associated with Fermat, who first perceived its importance. **1974** *Sci. Amer.* July 118/3 The Pellian for square root of 3, $3x^2 + 1 = y^2$, which is solved by finding the convergents of the continued fraction for the square root of 3.

pellotine (pe·lŏtīn). *Chem.* [ad. G. *pellotin* (A. Heffter 1894, in *Ber. d. Deut. Chem. Ges.* XXVII. 2977), f. Mexican *pellote* *PEYOTE: see -INE[5].] An alkaloid, $C_{13}H_{19}NO_3$, obtained from peyote and formerly used as a hypnotic; 1,2,3,4-tetrahydro-8-hydroxy-6,7-dimethoxy-1,2-dimethylisoquinoline.
1895 *Jrnl. Chem. Soc.* LXVIII. I. 120 Pellotine, ..so called from the Mexican name 'pellote', of *Anhalonium Williamsi*, crystallises in colourless, transparent, anhydrous plates, and melts at 110°. **1900** A. R. CUSHNY *Text-bk. Pharmacol. & Therapeutics* II. iii. 220 Pellotine has been introduced as a hypnotic, and has been favorably commented on. **1972** *Science* 9 June 1132/3 Tetrazotized benzidine spray..aided in the identification of the major components [from the Mexican cactus *Pelecyphora aselliformis* Ehrenberg] as hordenine.., pellotine.., anhalidine.., and an unknown alkaloid.

Pelman (pe·lmăn). The name of Christopher Louis *Pelman*, founder (in 1899) of the Pelman Institute for the Scientific Development of Mind, Memory and Personality in London, used *attrib.* to designate the system of memory training taught by this Institute.
1900 *Phonetic Jrnl.* 15 Dec. 790/2, I may say that I speak of the Pelman system from intimate personal knowledge. **1939** R. CAMPBELL *Flowering Rifle* iv. 113 All his Pelman course and monkey's glands With Charlie intellect expands. **1972** *Listener* 17 Aug. 210/3 I did a Pelman course, which hasn't done very much good because I've got the worst memory of anybody.
So **Pe·lmanism,** the system taught by the Pelman Institute; also *transf.*; a card game in which the cards lie face down and pairs must be selected from memory as successive cards are turned up; **Pe·lmanist, Pe·lmanite,** a student or advocate of Pelmanism; **Pe·lmanize** *v. intr.* and *trans.*, to practise Pelmanism; to teach (someone) Pelmanism.
1919 *Honey Pot* I. 4 (Advt.), Lots of our fellows are Pelmanising out here. **1920** *Pelman Pie* 25 (Advt.), A very large proportion of its readers are 'Pelmanists'. **1920** *Blackw. Mag.* Nov. 561/2, I fear I must be suffering from what the Pelmanites call 'mind-wandering'. **1921** *Chambers's Jrnl.* Mar. 176/1 In some Oriental way he had Pelmanised his memory. **1922** E. WALLACE *Flying Fifty-Five* xxvi. 154 Bill had begun, as he described it, to Pelmanize his horses. He was super-imposing upon the memory of one evil whistle another which held the most kindly association. **1923** *Hoyle's Games Modernized* 230 Pelmanism is..a splendid exercise for the memory besides a source of amusement to the players, of whom there may be any number. **1931** R. CAMPBELL *Georgiad* ii. 17 Broadcast your love and Pelmanise your Passion. **1934** 'E. M. DELAFIELD' *Provincial Lady in America* 142 Brief exercise in Pelmanism enables me to connect woman in red hair with first name, which is Marcella. **1934** *Brit. Weekly* 11 Jan. 319/3 (Advt.), Pelmanise for promotion, for progress, for success. *Ibid.*, There are in the world to-day more than half a million men and women who have become Pelmanists. **1950** *Oxf. Jun. Encycl.* IX. 114/1 In Pelmanism, a memory game, the cards are spread face downwards on the table. The players turn up two at a time, trying to find pairs. Cards which do not make pairs are turned down again, and the players must remember where they are. **1958** R. GODDEN *Greengage Summer* vii. 80 We stayed on at our table, playing cards with old packs from the bar, Racing Demon or Pelmanism or Snap. **1963** R. M. HARE *Freedom & Reason* iv. 60 Ought I not rather to take a course in Pelmanism to correct my absent-mindedness? **1972** *Guardian* 15 Nov. 10/6 There is a kind of common ground here between pulp fiction, science fiction pop mysticism, and a new sort of Pelmanism that goes beyond mere self-improvement and proposes the ways and means of self-transcendence. **1975** *Daily Tel.* 11 July 11/3 Between games of ping-pong or pelmanism they spend the night reminiscing about old times over a continuous supply of drinks.

‖ **pelmeny** (pe·lmeni), *sb. pl.* Also **-ni.** [a. Russ. *pel'méni.*] In Russian cookery, small pastry cases stuffed with meat, etc.
1943 E. M. ALMEDINGEN *Frossia* x. 376 A whole mound of *pelmeny*, little lumps of specially made pastry stuffed with meat and onions and boiled in water. **1958** N. & G. J. FROUD *Home Bk. Russian Cookery* 23 Left-over pel'meni are delicious fried in a little butter. **1962** K. PETROVSKAYA *Secrets of Russian Cooking* 183 The best, real way of preparing Siberian pelmeny is to use two kinds of meat for stuffing—beef and pork or veal and pork—and plenty of seasoning. **1972** *Times* 12 Apr. 9/2 Such Russian specialities as Pelmeny (a kind of ravioli).

pelmet (pe·lmĕt). [Prob. ad. F. *palmette* (see PALMETTE), formerly a conventional ornament on window cornices.] A valance or horizontal strip of curtain, wood or other material, fitted across the top of a door or window, usu. to conceal curtain fittings. Hence **pe·lmeted** *a.*, fitted with a pelmet.
1904 P. N. HASLUCK *Upholstery* 149 This pelmet is fixed to a wood lath screwed underneath the door casing. **1908** *Ladies' Field* 24 Oct. 318/2 (*caption*) Window Treatment: Pelmet and Curtains in Shadow Damask. **1922** *Daily Mail* 21 Nov. 14 If a pelmet is used it should be straight-edged. **1925** PENNEFEATHER-BRABOURNE & LAYTON *Gloss. Eng. Furnit.* 123 *Pelmet*, a word used by upholsterers and sometimes by art dealers, who prefer the word 'palmette', [illegible] rod, rings and headings of the hanging curtain decorating a door, window, bed, etc. **1932** G. B. STERN *Little Red Horses* iv. xxxix. 340 [illegible] pelmet. **1960** C. G. TOMKELEY *Furnishing your Home* xvi. 109 Pelmets [illegible] are an improvement to most windows. **1964** *McCall's Sewing* xvi. 286 The depth of the pelmet depends on the height of the window, usually one-sixth to one-ninth the length of the curtain. **1970** *Kay & Co. (Worcester) Catal.* Autumn/Winter 544 Plywood pelmet. Made in white. **1974** M. CECIL *Heroines in Love* viii. 189 These magazines ..gave them [*sc.* readers] the latest way with pelmets and casseroles for winter evenings. **1974** D. FRANCIS *Knock Down* v. 59 Faded and pelmeted curtains and silk lamp-shades..spoke of enough money.

pelo-. Add: **pelobatid** (example); **pe·lophile** *Ecol.* [a. F. *pélophile* (J. Thurmann *Essai de Phytostatique* (1849) I. xiii. 268): see -PHIL], a plant growing on mud or clay; **pelophilous** *a.* (later example).
1956 *Nature* 18 Feb. 342/2 I have examined tadpoles of the pelobatid *Megophrys major*. **1909** B. D. JACKSON *Gloss. Bot. Terms* (ed. 2) 345/1 *Pelophile*..., a clay-loving plant. **1966** N. POLUNIN *Introd. Plant Geogr.* xvi. 528 It is also possible..to categorize marine Algae according to..the nature of their substratum (such as epiliths attached to rocks, or pelophiles growing on mud). **1909** E. WARMING *Oecol. Plants* xviii. 230 (*heading*) Pelophilous halophytes.

peloid (pe·loid). [f. Gr. πηλ-ός clay, mud + -OID.] **1.** (See quots.)
1933 S. J. LEWIS in *Arch. Med. Hydrol.* XI. 181 (*heading*) Semi-solid bath media or 'peloids'. *Ibid.* 165/1 The word Peloid (Greek, πηλός) should be used as a generic name for all forms of muds, moors, etc... Artificial preparations might be described as artificial peloids. **1953** W. KOSMATH *Brit. Pat.* 693,916 This invention relates to a method for the manufacture of therapeutically effective agents for peloids, which term is here intended to mean bog peat, organic slimes, muds, clays formed by weathering, marl and the like.
2. *Geol.* A particle of microcrystalline or cryptocrystalline carbonate.
1963 E. MCKEE *Prof. Papers U.S. Geol. Survey* No. 475-C. 215 Grains within the limestone range in size from fine to very coarse and consist of both bioclasts and peloids (intraformational clastic particles), commonly in an aphanitic calcite matrix. **1967** *Mem. Geol. Soc. Amer.* CXIV. 24 Peloids are defined as ovoid particles of microcrystalline or cryptocrystalline material. *Ibid.* 555 Beds of peloids that seem to be the result of lime-mud deposits having been disturbed and broken up into particles are uncommon. **1972** *Nature* 25 Feb. p. ix/1 (Advt.), Petrography of carbonate grains: ooids, peloids, pisoliths, pellets and other biotic fabrics. **1976** R. C. SELLEY *Introd. Sedimentology* v. 117 Studies of modern carbonate sediments show that peloids form by a variety of processes. *Ibid.*, In..lagoons and sheltered embayments, peloids are sufficiently abundant to be a dominant rock builder.
Hence **peloi·dal** *a.*
1933 S. J. LEWIS in *Arch. Med. Hydrol.* XI. 182/1 Peloid..is applicable in all languages in the same way as colloid or alkaloid, giving plurals and adjectives (e.g. 'peloidal') by the same models. **1969** MCKEE & GUTSCHICK in *Mem. Geol. Soc. Amer.* CXIV. 555 Peloidal limestone is defined as rock composed dominantly of particles of cryptocrystalline or microcrystalline material, commonly ovoid in shape. **1976** *Nature* 29 May 221/2 A thin oolitic iron formation is interbedded with peloidal and oolitic ferruginous chert.

pelong (pī·lǫŋ). [Derivation uncertain: perh. ad. Malay *pĕlang* striped.] A kind of material used for gowns worn in southern India.
1687 *Charter to Fort St. George* (E. India Co.) f. 5v The said mayor and aldermen may always upon such solemn occasions wear scarlet serge gowns all made after one form or fashion..and that all the burgesses may upon such solemn occasions wear white pelong or other white silk gowns after one form or fashion to be agreed. **1798** *Acct.*

Interview Teshoo Lama & Capt. Samuel Turner 6, I advanced, and, as is the custom, presented a white pelong handkerchief. **1922** C. ILBERT *Govt. of India* 23 The burgesses are, on these occasions, to wear white 'pelong', or other silk gowns. **1924** E. SITWELL *Sleeping Beauty* i. 11 Pelongs, bulchauls, pallampores. **1928** —— *Five Poems* 15 Panōpe..Sees Asia, Parthenope, Eunomia, Euphrosyne, Urania, Ausonia In feathered head-dresses as bright as sleep,..In pelongs, chelloes, and great palampores.

peloothered (pelū·pəɹd). *rare*—[1]. [Fanciful formation.] Drunk.
1914 JOYCE *Dubliners* 197 It happened that you were peloothered, Tom.

Peloponnesian (pelŏpǫnī·ziăn), *sb.* and *a.* Also 6–8 **Peloponesian.** [f. Gr. Πελοπόννησ-ος, L. *Peloponnes-us* Peloponnesus + -IAN.] **A.** *sb.* A native or inhabitant of the Peloponnesus (or Peloponnese), a peninsula forming the southernmost part of the Greek mainland. **B.** *adj.* Of or pertaining to the Peloponnesus or its inhabitants. *Peloponnesian war*, a war fought between Athens and Sparta from 431 to 404 B.C., in which Sparta and its Peloponnesian allies were victorious.
a **1490** J. SKELTON tr. *Diodorus Siculus' Bibliotheca Historica* (1956) I. iv. 275 Oonly the Boecians and the Peloponnesians, and non other, be permytted and licenced to take away of this golde with theym. **1550** T. NICOLLS tr. *Thucydides* (*title*) The hystory..of the warre, whiche was betwene the Peloponesians and the Athenyans. **1579** NORTH tr. *Plutarch's Lives* 184 Pericles..was thought the only original cause & author of the Peloponnesian warres. **1629** HOBBES tr. *Thucydides' Eight Bks. Pelop. War* i. 14 This Warre, which began from the time that the Athenians and Peloponnesians brake the League. **1709** I. LITTLEBURY tr. *Herodotus' Hist.* II. ix. 370 When the Lacedemonians were advanc'd to the Isthmus, and encamp'd with their Army; the other Peloponesians..thought they could not stay behind without Disgrace. **1752** HUME *Polit. Discourses* x. 226 When they were all chas'd into town, by the invasion of their territory during the Peloponnesian war, the city was not able to contain them. **1808** W. MITFORD *Hist. Greece* II. xv. 95 The Athenians.. a little before the beginning of the Peloponnesian war, sent Phormion with thirty triremes to their assistance. **1827** J. R. MAJOR *Questions Mitford's Hist. Greece* 277 The answer of the Pythoness was understood to import that the Peloponnesians would be victorious. **1890** C. W. C. OMAN *Hist. Greece* xxvii. 293 Great battles on shore were very rare during the Peloponnesian war. **1911** *Encycl. Brit.* XXI. 73/2 In 429 the Peloponnesians were deterred by the plague from invading Attica. **1959** N. G. L. HAMMOND *Hist. Greece* 167 Modern scholars have called it [*sc.* the Spartan Alliance], rather misleadingly, 'the Peloponnesian League'. **1960** A. R. BURN *Lyric Age Greece* ix. 176 The Peloponnesian Argives defeated Sparta on the field of Hysiai. **1969** C. M. WOODHOUSE *Philhellenes* iv. 115 Stanhope..adopted the idea of organising a grand conference of national unity... He..visited the seat of the provisional government at Nauplia, to persuade the Peloponnesian leaders to take part... But all was in vain. The Peloponnesians would not come. **1972** D. DAKIN *Unification of Greece* x. 143 Deliyannis..had become the eloquent and popular leader of the disgruntled Peloponnesian peasants. **1972** R. MEIGGS *Athenian Empire* x. 181 Peace..was to be made with the Peloponnesians. **1976** *Classical Q.* XXVI. 233 It is doubtful whether the Peloponnesian detachment was dispatched during the actual celebration of the Olympic games. *Ibid.* 247 The Peloponnesians began construction of the Isthmian wall immediately upon hearing of the outcome of Thermopylae.

pelorus (pĕlō^ə·rŏs). Also **Pelorus.** [Name of the supposed pilot of Hannibal when he left Italy.] A compass rose equipped with one or two sighting arms, used for taking the relative bearings of sighted objects.
1854 FRIEND & BROWNING *Brit. Pat.* 2652 This invention has for its object the construction and use of an instrument or apparatus which we denominate a pelorus, for determining the amount of magnetic aberration occasioned by local attraction in ships or vessels of every description,..and by the use of which..the true course.. may be from time to time accurately ascertained. **1881** S. T. S. LECKY '*Wrinkles*' *in Pract. Navig.* x. 67 Having a Pelorus, the first thing to do is to provide suitable stands for it in various parts of the ship, so that it may be moved from one to the other as may be found convenient. **1904** WILSON-BARKER & ALLINGHAM *Navigation* (ed. 2) iii. 17 For the purpose of obtaining bearings when the object is not visible from the standard compass..a Pelorus is very useful. **1934** J. IRVING *Navig. Small Yachts* 284 As the ship's head is steadied on a point by the steering compass the bearing of the distant object is noted by the 'pelorus', and this bearing angle (from right ahead) is applied to the compass direction of ship's head. **1943** REDPATH & COBURN *Air Transport Navig.* iv. 63 A pelorus (a dummy compass rose equipped with sighting vanes) is usually mounted on top of the hatch above the cockpit in the airplane. **1973** *Country Life* 27 Nov. (Suppl.) 66/2 Mid 18th century French brass Pelorus... With original travelling case.

pelosity, *rare* var. PILOSITY.
1922 JOYCE *Ulysses* 687 To Bloom: the problems of irritability, tumescence, rigidity, reactivity, dimension, sanitariness, pelosity.

pelota. (Earlier and later examples.)
1844 *Colburn's United Service Mag.* Apr. 492 To see a Spaniard at a festival, a bull-fight, or the rustic game of *pelota*, no one would conceive him given to melancholy. **1929** C. CONNOLLY *Let.* in *Romantic Friendship* (1975) 332 She wants to build a modern Corbusier house with a flat roof which can be used as a pelota court. **1934** R. MACAULAY *Going Abroad* xxix. 251 He was showing Hero how to hold a pelota racquet. **1974** *Spain* (Michelin Tyre Co.) 108 The overriding Basque passion..is *pelota*... There is even a Pelota University.

pelotherapy (pīlo̞þe·răpi). *Med.* [f. PELO- + THERAPY.] The application of mud to the body as a therapeutic measure.
1933 S. J. LEWIS in *Arch. Med. Hydrol.* XI. 178/1 The Greek word *Pelos*..would combine with terms of Greek origin forming derivations such as Pelotherapy. *Ibid.* 182/1 'Pelotherapy' would be the medical treatment by Peloids. **1970** *Telegraph* (Brisbane) 12 June 5/1 The mud treatment—or pelotherapy—is claimed to heal body ailments.

‖ **pelouse** (pelūz). [Fr.] An area of grass, a lawn; *spec.* a public enclosure at a French racecourse. Also *transf.*
1923 E. HEMINGWAY *Three Stories and Ten Poems* 36 The procession of them went around on the other side past the pelouse. **1928** *Observer* 11 Mar. 13/3 The Longchamp tote..will unify the betting in all three parts of the course—paddock, pavilion and pélouse. **1953** E. E. CLARK *Indian Legends* 222 A large area of land in south-eastern Washington [State] thought to have been called *pelouse*, 'the grass lands' by French-Canadian voyageurs. **1957** *N. & Q.* June 270/1 On the *pelouse* of open race-courses in and near Paris. **1967** *Punch* 27 Sept. 464/3 A troutstream flows east to west towards the chateau through beautiful Dornford Yatesian pelouses.

pelt, *v.*[1] Add: **3. b.** Also of missiles.
1916 'BOYD CABLE' *Action Front* 210 Maxim and rifle bullets were still pelting from somewhere in half enfilade at long range.
5. (Further example.)
1916 'BOYD CABLE' *Action Front* 116 A heavy rifle and machine-gun fire which was pelted across from the opposite parts of the British line.

pelt, *v.*[2] **a.** Delete '*Obs.* exc. *dial.*' and add later examples.
1919 W. T. GRENFELL *Labrador Doctor* ix. 176 Then having killed, 'sculped', and 'pelted' the seal, the exciting return to the vessel. **1936** D. MCCOWAN *Animals Canad. Rockies* xxiv. 211 In Canada alone in the first decade of this century, ten millions of the animals [*sc.* mink] were pelted for the sake of the satiny fur. **1948** A. L. RAND *Mammals E. Rockies* 116 It [*sc.* a wolf] is a large, awkward animal to pelt on the trap-line. **1950** *N.Z. Jrnl. Agric.* Sept. 256 (*caption*) Pelting skins on the killing chain. **1972** 'M. YORKE' *Silent Witness* iii. 51 The Derringtons had a mink farm. 'We pelt them in November.'

pelta. Add: **3.** *Archit.* An ornamental shield motif (see quots.).
1828 A. W. CLAPHAM in *Archaeologia* XXVII. 227 The *Pelta Ornament*..consists essentially of pairs of half-circles set back to back in such a way that the combination forms a series of figures like a double-headed axe or an Amazon-shield. **1936** —— *Romanesque Archit.* iii. 61 Of much less frequent occurrence is the Carolingian pelta-ornament, used as a diaper on roll-mouldings at Piacenza, Modena, Cremona, and Ferrara Cathedrals. **1942** *Oxoniensia* VII. 70 The eye is at once attracted to the small field of decoration on the head allotted to an engraved and enamelled pelta-scroll flung haphazard on the surface. **1954** M. RICKERT *Painting in Brit.: Middle Ages* 230 *Pelta*, a classical ornamental motif formed by two converging curves joined by another line. The ancestor of the trumpet pattern. **1967** *Antiquaries Jrnl.* XLVII. 167 On the extreme right is a border of black and white composed of small alternating semi-circular forms producing the effect of a pelta pattern. **1965** R. KRAUTHEIMER *Early Christian & Byzantine Archit.* 362 *Pelta*, a shield-shaped design composed of a convex curve joined to two concave curves.

pelter, *sb.*[1] For †*Obs.* read *rare* and add later example.
1922 JOYCE *Ulysses* 130 That old pelters [*sic*], the editor said.

pelter, *sb.*[2] Add: **1. c.** (Earlier and later examples.)
1791 J. BYNG *Torrington Diaries* (1935) II. 360 Tho' it rain'd all the way, so as to hurry me, yet it was not a pelter. **1816** JANE AUSTEN *Let.* 9 July (1952) 459 We were obliged to turn back..but not soon enough to avoid a Pelter all the way home. **1966** T. H. RADDALL *Hangman's Beach* I. vi. 83 Boats' crews and carpenters..came out in the cold pelter to McNab's island.
d. Phr. *in a pelter*: in a hurry; at speed.
c **1889** 'F. LESLIE' *Let.* in W. T. Vincent *Recoll. Fred Leslie* (1894) II. xxiii. 97 Dear, dear! I have wasted time and ought to have been at work on our burlesque. Now I am determined to go in a pelter.
e. An old or slow horse. *U.S. colloq.*
This may belong to PELTER *sb.*[1]: see *Dict. Amer.*
1856 *Knickerbocker* XLVIII. 314 When his earthly tenement yields his soul no shelter, May it animate the corpse of an ancient pelter. **1896** ADE *Artie* i. 4 It's like hitchin' up a four-time winner 'longside of a pelter. **1902** H. F. DAY *Pine Tree Ballads* 147 He'd..take a wheezy old pelter with a hopity gait and he'd make you believe.. there were all kinds of pedigrees tied up in him. **1931** D. RUNYON in *Hearst's International* Sept. 84/1 Mahogany.. is..not such a bad old pelter.
3. Something exceptionally large. *dial.*
c **1780** M. LONSDALE in S. Gilpin *Popular Poetry of Cumberland* (1875) 61 An' dall! but it's a pelter. **1892** E. J. MILLIKEN *'Arry Ballads* 70/1 Their ain't nothink the nobs is fair nuts on but wot these 'ere bellerers ban. Wy, they're down upon Sport, now, a pelter. Perposterous, ain't it, old man?

pelter, *v.* Add: **1.** (Later example of *intr.* use.)
1939 DYLAN THOMAS *Map of Love* 12 Chimes cheat the prison spire, pelter In time like outlaw rains on that priest, water, Time for the swimmers' hands, music for silver lock And mouth.
2. *intr.* To run with rapid steps; = PELT *v.*[1] 7.
1906 W. S. MAUGHAM *Bishop's Apron* xix. 297 The strange spectacle of a comely young woman and an ecclesiastical dignitary..peltering towards the Achilles Statue as fast as they could go. **1923** *Chambers's Jrnl.* Apr. 240/2 Rawlins..peltered up on deck to recover his composure.
Hence (in sense 1) **pe·ltering** *ppl. a.*
1858 [in Dict.]. **1927** *Glasgow Herald* 27 Aug. 8 The peltering rains (which were certainly general) made the grass so wet that the..cow ate far too much juice.

Peltier (pe·ltie). *Physics.* The name of J. C. A. *Peltier* (1785–1845), French amateur scientist, used *attrib.* with reference to an effect he discovered in 1834, whereby heat is given out, or absorbed, when an electric current passes across a junction between two materials; *Peltier coefficient*, (*a*) the quantity of heat liberated or absorbed at a junction between two conductors when unit charge passes between them; (*b*) of a material, the quantity of heat liberated or absorbed when unit electric charge traverses a junction between that material and a reference conductor.
1856 *Proc. R. Soc.* VII. 54, I hope also to be able to make determinations in absolute measure of the amount of the Peltier effect for a given strength of current between various pairs of metals. **1896** FOSTER & ATKINSON *Elem. Treat. Electr. & Magn.* xiv. 135 This coefficient, known as the Peltier coefficient,..represents for a junction of any two metals the amount of energy converted into or produced from heat per second when a current of unit strength traverses the junction. **1904** J. S. AMES *Text-bk. Gen. Physics* xliv. 681 These forces at the surface of contact of two substances are called 'Peltier electro-motive forces', having been first discovered by him. **1916** F. B. PIDDUCK *Treat. Electr.* vi. 210 In addition to the Peltier heat there is the usual heat developed by the mere passage of the current through the wires. **1958** CONDON & ODISHAW *Handbk. Physics* IV. vi. 84/2 The Peltier coefficient for a junction AB is representable as the difference of two intrinsic Peltier coefficients characteristic of each material separately. **1966** *McGraw-Hill Encycl. Sci. & Technol.* XIII. 581/2 If 1 coulomb of positive charge moves slowly around the circuit, there will be Peltier cooling in its passage from Bi to Cu at T and Peltier heating at T_0. **1972** R. W. URE in Willardson & Beer *Semiconductors & Semimetals* VIII. ii. 68 The Peltier effect at the junction between the liquid and solid phases has been used to remove the latent heat of crystallization in crystal growth.

Pelton (pe·ltən). Also **pelton.** The name of L. A. *Pelton* (1829–1908), U.S. engineer, used *attrib.* to designate an undershot water-wheel he invented that has divided buckets fixed to the rim, which deflect a jet of water directed at them.
1885 *Engineering* 30 Oct. 433 The Pelton wheel is said to have used 162·98 cubic feet of water a minute, and to have given 107·49 horse-power, showing an efficiency of 90·2 per cent. **1916** J. PARK *Text-bk. Pract. Hydraulics* xiv. 242 A Pelton water-wheel 3 feet in diameter, formed of a solid steel disc with phosphor-bronze buckets riveted to the rim, is working in California under a head of 2100 feet. **1962** *Times* 21 May (Commonwealth Chambers of Commerce Suppl.) p. xii/5 (Advt.), Cast steel runners and pelton wheels for hydro-electric plants. **1972** J. M. K. DAKE *Essent. Engin. Hydraulics* vi. 163 It is possible to have more than one jet operating on a Pelton wheel and two jets are quite common.

pelure[2]. Add: **2.** *pelure d'oignon* (dǫnyǫ̃) [lit. 'onion-skin'], a tawny colour in wines; a wine of this colour; *spec.* the name of a wine produced in the Jura region of France; also *attrib.*
1935 SCHOONMAKER & MARVEL *Compl. Wine Bk.* i. 46 The red and *pelure d'oignon* wines of Arbois were among the many life-long favourites of that..lover of charming ladies and good wines—Henri IV. *Ibid.* viii. 188 A good red wine is red, a deep crimson when young,..a rather more delicate red when old, with that faint but unmistakable brownish tinge which the French call '*pelure d'oignon*'. **1951** H. W. ALLEN *Natural Red Wines* vii. 263 The depth and brightness of its divine reds, purples and browns, the perfect example of the *pelure d'oignon* hue. **1952** A. LICHINE *Wines of France* xiii. 175 When old, the red Hermitages are enormously rich in aroma and aftertaste... Instead of having the complete red spectrum, the wines have a brownish tinge, which the French call *pelure d'oignon*, or onion skin. **1962** *Harper's Bazaar* Oct. 159/2 You will also find some unexpected Rhône wines..and Pelure d'Oignon (onion skin colour). **1965** A. SICHEL *Penguin Bk. Wines* III. 160 The best red wine of the Jura..is light in colour and is often referred to as '*pelure d'oignon*'—onion skin. **1965** O. A. MENDELSOHN *Dict. Drink* 255 *Pelure d'oignon*,..used to denote the characteristic shiny brown hue that aged red wines may acquire. **1967** P. PURSER *Twentymen* xxvii. 187 The day's third bottle of Pelure d'Oignon, beaded with condensation from the refrigerator. **1971** *Guardian* 12 Nov. 9/3 The Montrichard..is a pale pink—nearer

the rosy yellow of pelure d'oignon than red. **1976** 'J. FRASER' *Who steals my Name?* xvi. 188 Aveyard examined his crystal glass appreciatively, then replenished it with the delicious Pelure d'Oignon Rosé wine.

pelurious (pelʲū·riəs), *a. rare*—[1]. [f. PELURE[1] + -IOUS.] Furred, hairy.
1922 JOYCE *Ulysses* 289 Sieves of gooseberries, pulpy and pelurious.

pelvic, *a.* Add: **1. b.** *pelvic thrust*: the repeated thrusting movement of the pelvis during sexual intercourse.
1953 A. C. KINSEY et al. *Sexual Behav. Human Female* xv. 618 The..rhythmic pelvic thrusts during sexual activity are among the distinctive characteristics of the class Mammalia. **1966** MASTERS & JOHNSON *Human Sexual Response* xviii. 297 Male pelvic thrust and female pelvic accommodation initially are voluntary muscular attempts at sex tension increment. **1969** W. B. POMEROY *Girls & Sex* vii. 105 Intercourse continues with a series of pelvic thrusts. **1969** D. R. REUBEN *Everything you always wanted to know about Sex* (1970) vi. 90 A reasonable yardstick for male potency is the ability to continue intercourse for five to ten minutes. During that time a normally potent male will deliver from fifty to one hundred pelvic thrusts.

† **pelvigraph** (pe·lvigrɑf). *Obs.* [f. PELVI(S + -GRAPH.] An instrument for recording measurements of the pelvis. So † **pelvi·graphy,** the use of the pelvigraph.
1890 BILLINGS *Med. Dict.* II. 306/1 *Pelvigraphy*, obtaining a contour of the wall of the pelvis. **1892** F. P. FOSTER *Med. Dict.* IV. 2540/1 *Pelvigraph*, a device adopted by Pinard for recording automatically the measurements of dried pelves. **1903** J. C. EDGAR *Pract. Obstetr.* II. 184 The principle involved in the construction and application of the pelvigraph is that of the parallel rulers, one number representing a palpator for the localization of points within the pelvis, while the other is provided with a water-level and a dial index. **1904** *Lancet* 18 June 1728/1 A description is given of pelvigraphy or the method of taking a series of measurements of certain pelvic diameters and thus plotting out the size of the various pelvic planes. **1913** *Jrnl. Obstetr. & Gynæcol.* XXIV. 257 Martin's pelvigraph, introduced in 1827, was..intended for use in the dry pelvis.

pelycosaur (pe·likosǭɹ). Also **pelycosaurian.** [f. mod.L. name of order *Pelycosauria* (E. D. Cope 1878, in *Proc. Amer. Philos. Soc.* XVII. 511), f. Gk. πέλυξ, πέλυκ- bowl, cup + σαῦρος lizard.] A fossil reptile of the order Pelycosauria, known from Permian remains, and sometimes distinguished by bony spines developed from some of the vertebrae. Also *attrib.* or as *adj.*
1880 [in Dict. s.v. PELYCO-]. **1930** H. H. SWINNERTON *Outl. Palæont.* (ed. 2) xii. 334 *Dimetrodon* was a typical Pelycosaurian. **1933** A. S. ROMER *Vertebr. Paleont.* xi. 222 The neural arches of pelycosaurs were (as in most reptiles) narrower than those of the cotylosaurs. *Ibid.*, The pelycosaur palate was constructed on a primitive pattern. **1954** W. E. SWINTON *Fossil Amphibians & Reptiles* vi. 29 A group of very important reptiles, the Therapsida, are descended from the *Dimetrodon*-like Pelycosaurs. **1968** A. S. ROMER *Procession of Life* xiv. 235 The central types among the therapsids were carnivores, continuing the flesh-eating tradition of the main pelycosaur stock. **1973** *Nature* 16 Mar. 203/2 *Dimetrodon grandis* was the end form of an evolutionary series of pelycosaurs that had tended to develop increasingly large sails. **1977** A. HALLAM *Planet Earth* 212 A bizarre characteristic of a number of the pelycosaurs was the development of a huge sail-like structure adorning the back.

Pembroke. Pembroke table (later examples).
1853 [see **conversation card*]. **1925** *Scribner's Mag.* July 94/1 A Pembroke table with unusual stretchers was in the wood-shed. **1973** [see *HAREWOOD]. **1975** J. SYMONS *Three Pipe Problem* xviii. 181 She sold..a damaged Pembroke table skilfully repaired by Fritz to a woman who was actually looking for a brass fender.

pemoline (pe·molīn). *Pharm.* [Perh. f. P(H)E(NYL + *I)M(INO(-) + *OXAZ)OL(ID)INE, elements in its chemical name.] 2-Imino-4-oxo-5-phenyloxazolidine, $C_9H_8N_2O_2$, a white, crystalline, tasteless powder that is a stimulant of the central nervous system.
1961 *MIMS Monthly Index* Jan. 44 Kethamed Medo-Chemicals Pemoline: 20 MG. tablets. Fatigue, convalescence, etc. **1968** *Sunday Times* 27 Oct. 3 Pemoline has been known sometime to be a mild brain stimulant, mid-way between Amphetamine and coffee. **1973** *Lancet* 10 Nov. 1091/1 Six amphetamine users who have been taking large doses of pemoline..have become aggressive, excited and disorganised.

pemphis (pe·mfis). [mod.L. (J. R. & J. G. A. Forster *Characteres Generum Plantarum* (1776) 67), f. Gr. πεμφίς cloud.] A small tree of the genus so called, esp. *Pemphis acidula*, belonging to the family Lythraceæ, and found in coastal, tropical areas of Africa and Southern Asia.
1911 *Trans. Linn. Soc.* (*Zool.*) XIV. 403 Throughout the northern section [of Aldabra] the *Pemphis* jungle continues. *Ibid.* 404 Near the sea the trees were all much stunted by the prevalent winds, even the hardy *Pemphis* trees being dwarfed into a low thick scrub. **1919** *Nature* 9 Oct. 118/2 The vegetation [of Aldabra] consists of four types:— (1) Mangrove swamp... (2) Pemphis bush, a dense growth of the hard-wooded *Pemphis acidula* (Lythraceae), a widely distributed sea-coast plant. **1958** *Times* 4 Nov. 12/6 The island [*sc.* Aldabra] rises only some 15 ft. above the sea and is for the most part clothed with a terrible tangle of pemphis bush. **1971** *Phil. Trans. R. Soc.* B. CCLX. 359 Isolated *Scaevola* and *Pemphis* trees have browse lines. *Ibid.* 476 One of them [*sc.* a Malagasy bulbul] was in a *Pemphis* bush.

pempidine (pe·mpidīn). *Pharm.* [f. PE(NTA- + M(ETHYL + PI(PERI)DINE.] An alkaline liquid, $C_{10}H_{21}N$, which has been used (usu. in the form of its hydrogen tartrate, a white, crystalline powder), as a ganglion-blocking agent in the treatment of severe hypertension; 1,2,2,6,6-pentamethylpiperidine.
1958 *Nature* 21 June 1717/2 In the simple..tertiary amine, 1:2:2:6:6-pentamethylpiperidine.., for which the common name pempidine has been suggested, we have a potent ganglion-blocking drug. **1960** *Lancet* 21 Jan. 143/2 In 1958 a new ganglion-blocking agent, pempidine tartrate, was made available under the trade names of 'Perolysen' and 'Tenormal'. **1961** *Ibid.* 12 Aug. 334/2 A Jamaican..was admitted to hospital and stabilised on pempidine and chlorothiazide. **1974** *Prescribers Jrnl.* XIV 47 Ganglion blocking drugs, such as pentolinium..and pempidine.., have been omitted from this short review because in the drug treatment of hypertension they have now been largely superseded.

pen, *sb.*[1] Add: **1. c.** *spec.* A division in a sheep-shearing shed. Also, the work associated with a sheep-shearing pen. *Austral.* and *N.Z.*
1891 R. WALLACE *Rural Econ. Austral. & N.Z.* xxix. 381 On the outside of the smaller pens, and near to the outer side-walls the shearers are placed. **1900** H. LAWSON *Verses, Pop. & Humorous* 168 The shearers squint along the pens, they squint along the 'shoots'. **1905** —— *When I was King* 38 The shed was cooled by electric fans that was over every shoot; The pens was of polished ma-ho-gany. **1933** L. G. D. ACLAND in *Press* (Christchurch, N.Z.) 11 Nov. 15/7 *Pen.* (1) A small yard; a division in the sheep-holding part of a wool-shed. (2) Shearers catch out of a p[en], and when they apply for work they ask for a p[en]. 'Will you keep me a p[en] for next year, boss?' is often their farewell. **1945** BAKER *Austral. Lang.* iii. 65 A shearer gets a *cut* (it is also called a *stand* or *pen*) when he is employed. **1965** *Austral. Encycl.* VIII. 86/2 Shearing became the work of a nomadic band of men who travelled..from shearing shed to shearing shed... The fortunate ones 'got their pen' at the commencement of a shearing whilst the others moved on in the hope of getting work at some other station. **1966** G. W. TURNER *Eng. Lang. Austral. & N.Z.* iii. 48 An Australian instance of this [*sc.* metonymy] is the shearer's request for a *pen*, i.e. work, some sheep to shear.
2. c. A prison; a cell in a prison. orig. *U.S.* (cf. *BULL-PEN 1 a)
Often indistinguishable from *PEN *sb.*[5].
1845 W. G. SIMMS *Wigwam & Cabin* (1846) 2nd Ser. 93 Laughter..ceased on my part, as I got in sight of the 'pen' in which I was to be kept secure. **1853** F. W. THOMAS *John Randolph* 286 If I had not caught him in Baltimore ..and put him in the pen there for debt, I never should have got the money. **1867** W. L. GOSS *Soldier's Story* 144 Every batch of prisoners sent into the 'pen' were accompanied by a spy in U.S. blue. **1904** *N.Y. Even. Jrnl.* 10 May 2 A panic was caused among the prisoners in the pen of the Ewen Street Police Court jail. **1948** PARTRIDGE *Dict. Forces' Slang* 139 *Pen*, a prisoner-of-war cage. (Mostly Army.)
e. A covered dock forming a berth for a naval vessel, esp. for a submarine.
1917 W. S. CHURCHILL *Second World War* (1949) II. I. xii. 216 A..harbour..with regular pens for the destroyers and submarines. **1932** A. R. BRADSHAW *Eng.-French Naval Terms* v. 76 *Destroyer pens*, les appontements. **1942** *R.A.F. Jrnl.* 16 May 19 (*caption*) These..photographs illustrate the constructional development of a possible new Naval Base and set of Submarine Pens. **1944** *Hutchinson's Pict. Hist. War* 12 Apr.–26 Sept. 229 Aircraft of Bomber Command launched a heavy attack on the 8-ft. thick E-boat pens at Le Havre. **1946** *War Report* (B.B.C.) 111 When the famous submarine pens of Cherbourg were inspected they revealed a twenty-foot thickness of reinforced concrete, designed to be absolutely impregnable to air bombardment. **1959** *Economist* 30 May 816/1 Russia does possess a foothold (complete with submarine pens) on the Mediterranean shore. **1961** F. H. BURGESS *Dict. Sailing* 158 *Pen, penns*, the spaces between a series of piers, so built that vessels may berth four to six deep between them. **1974** *Sci. Amer.* Mar. 117/1 The waste and the sorrow glow too hot to be concealed yet, even under the 5.5-meter reinforced-concrete slab over the submarine pens at Doenitz' headquarters port of Brest. **1975** *Ibid.* Oct. 6/1 There are a number of hard military targets other than missile silos, such as buried command posts, nuclear-weapons storage facilities and submarine pens that nuclear weapons may not be effective against unless they are accurate.
4. pen-mate *Austral.* and *N.Z. slang*, a shearer who catches sheep out of the same pen (as another shearer).
1933 L. G. D. ACLAND in *Press* (Christchurch, N.Z.) 11 Nov. 15/7 Two shearers usually catch out of one p[en] and are called p[en]-mates. **1965** J. S. GUNN *Terminol. Shearing Industry* II. 3 'Pen mates' catch their sheep out of the same pen.

pen, *sb.*[2] Add: **4. c.** (Later example.)
1922 JOYCE *Ulysses* 134 Gallaher, that was pressman for you. That was a pen.

7. a. *pen-line, -painting, set, -spray, -stalk, -stand, -stroke* (further examples). **c.** *pen-nibber.* **d.** *pen-painted* adj.
1895 E. M. THOMPSON *Eng. Illum. MSS.* ii. 38 The features of the human face are indicated by very light pen-lines alone without any attempt at modelling. **1978** *Times* 2 June 13/4 White [wall]papers with a thin pen line stripe of bright green or pink or blue. **1823** *Trans. Soc. Arts* XL. 252 This operation..may be performed still more accurately by the Pen-nibber here represented. **1902** *Chambers's Jrnl.* Nov. 692/2 He had in everyday use: (1) wash-hand tray..(13) pen-nibber, (14) ruler. **1929** E. BOWEN *Last Sept.* II. ix. 104 Cushions with pen-painted sprays. **1862** TROLLOPE *Let.* 28 June (1951) 115 The character of Romola..is the perfection of pen painting. **1934** M. ALLINGHAM *Death of Ghost* xi. 130 The shop..which sold her pen paintings 'phoned her..and she spent a busy hour..getting off a consignment of table centres. **1963** L. DEIGHTON *Horse under Water* xvii. 69 On da Cunha's simple mahogany desk was a porcelain-and-gold pen set. **1977** R. LUDLUM *Chancellor Manuscript* ix. 105 Quinn..sat down behind his desk...His eyes fell on..his pen set. **1905** J. W. BRADLEY *Illum. MSS.* II. iii. 142 The greater part of this volume is in the.. 'Berry' style, *i.e.* the fine pen-sprays of ivy leaf of burnished gold. But the first grand border is..transitional, consisting of the pen-sprays of golden ivy leaf alternating with sprays of natural flowers. **1907** K. D. WIGGIN *New Chron. Rebecca* iii. 78 Last night I dreamed that the river was ink and I kept dipping into it and writing with a penstalk made of a young pine tree. **1933** D. GASCOYNE *Opening Day* i. 12 On the flap of the desk were bottles of inks, a pen-stand in red and white mottled marble..and a profuse litter of papers. **1969** E. H. PINTO *Treen* 260/2 Horizontal tiered pen stands, as opposed to vertical pen holders, only became practical when the straight steel-nibbed pen replaced the curved feathered quill. **1928** O. E. SAUNDERS *Eng. Illumination* I. 66 The shading on the draperies is executed by penstrokes, one thick line being regularly bordered by two thin ones. **1977** *Times Lit. Suppl.* 25 Mar. 365/1 In the animation process, where the energy-charge of single penstrokes is naturally sacrificed to the blur of movement.
8. pen-form, the shape of hand-written letters, esp. those influenced by the writing instrument or the way it is used; so **pen-formed** *a.*; **pen-picture** (later example); **pen recorder,** an instrument for producing a continuous graphical record of a variable measured quantity by means of a pen; so **pen-recorded** *a.*, **pen recording; pen-trial,** something written by a scribe on a manuscript to test his pen.
1906 E. JOHNSTON *Writing & Illuminating* iii. 63 For the practical study of pen-forms use a cane or a reed pen. *Ibid.* xiv. 238 The Pen-formed letters are more easily practised. **1955** J. R. R. TOLKIEN *Return of King* 397 They [*sc.* dwarfs] adhered to the Cirth, and developed written pen-forms from them. **1973** *Times* 15 Dec. 4/2 Later the pupils were asked to rate the instructor on such things as intelligence, likableness, popularity and honesty, and give a short pen picture of him. **1973** *Nature* 27 Apr. 601/1 The pen-recorded chart appears in Fig. 1. **1947** *Canad. Jrnl. Res.* B. XXV. 397 In conjunction with a linear direct current amplifier and a Leeds and Northrup Speedomax pen recorder, this unit allows rapid and accurate recording of a given mass spectrum. **1964** *Listener* 27 Feb. 344/2 The radiations from Jupiter.. may also be recorded as a trace, using a pen-recorder, though great care has to be taken with identification. **1972** *Physics Bull.* Jan. 24/3 A permanent record of any defects can be obtained on a chart, the motion of which is synchronized with the position of the transducer.., if the output signal is fed to a pen recorder. **1942** *Rev. Sci. Instruments* XIII. 218 (*heading*) Direct pen recording of galvanometer deflections. **1945** *Ibid.* XVI. 70/2 It was felt to be desirable to employ a pen recording system. **1953** K. SISAM *Stud. Hist. Old Eng. Lit.* vii. 109 At the extreme top of folios 119*a*, 121*a*, 123*a*, 126*a*..is a prayer at the beginning of the sitting..; but it is also, I think, an inconspicuous pen-trial to make sure that pen and ink will go smoothly. **1978** *N. & Q.* Oct. 405/1 Additions to..the manuscript include the clarification of texts, pen-trials, [etc.].

pen (pen), *sb.*[5] *U.S.* Abbrev. of PENITENTIARY *sb.* 7, with allusion to *PEN *sb.*[1] 2 c (from which some early uses are indistinguishable).
1884 'C. E. CRADDOCK' *In Tennessee Mts.* 68 He b'lieved the Pen could claim it ez convict labor. **1889** *Provo* (Utah) *Amer.* 28 Mar. 1/4 What John got was eighteen months in the pen. **1908** [see *COLD *a.* 1 e]. **1910** 'O. HENRY' *Whirligigs* xvii. 202 One year after I got to the pen, my daughter died. **1924** W. M. RAINE *Troubled Waters* xxvii. 273 He escaped from the pen four days ago. **1939** J. STEINBECK *Grapes of Wrath* x. 123 I'm a-gonna tell you somepin about bein' in the pen. **1940** R. CHANDLER *Farewell, my Lovely* vi. 43 We got a wire from Oregon State Pen on him. **1956** B. HOLIDAY *Lady sings Blues* (1973) xix. 159 This was the first time I met anybody from the federal pen. **1960** *Times Lit. Suppl.* 16 Sept. 589/4 A junkie..who was..busted by the nailers and after a stretch in the pen cold turkeyed. **1967** *Punch* 22 Nov. 796/2 Semple makes a pass at Carole, is rebuffed, strangles her in a demented fit, and does eighteen years in the pen for it. **1972** 'H. HOWARD' *Nice Day for Funeral* iii. 40 He was her meal-ticket. Why should she want him sent to the pen? **1973** M. CAMPBELL *Halfbreed* xviii. 123, I had only been in Vancouver a few days when I met a guy just out of the Pen. **1975** *High Times* Dec. 9/1 Right now I'm in east Tennessee facing a five-to-15 year term in the state pen for something I haven't done—mainly for selling a schedule-one drug to a narc. **1977** *Time* 20 June 33/1 Another escape try from Missouri state pen, on March 10, 1966.

pen, *v.*[3] Add: **c.** *intr.* To use a pen; to write.
1904 T. Hardy *Dynasts* I. II. ii. 64 He pens in fits, with pallid restlessness. 1939 Joyce *Finnegans Wake* 301 He would pen for her, would pen for her.

penal, *a.* Add: **1. c.** *ellipt.* as *sb.* (*a*) (a sentence or period of) imprisonment; (*b*) a school punishment.
1867 [see *capper sb.*[1] 1c]. 1892 *Daily News* 17 Nov. 6/6, I was speaking to a youth who had undergone two penals.. for picking pockets. 1906 D. Coke *Bending of Twig* xii. 200 He can write lines, or 'penals' instead of going to detentions. 1908 *Outlook* 26 Dec. 901/2 The very interesting article on the Milton tercentenary..reminds me of a public school where some years ago boys had 'penals' selected from 'Paradise Lost'. 1927 W. E. Collinson *Contemp. Eng.* 77 A convict doing penal or doing time i.e. sentenced to penal servitude, is sometimes called a lag. 1937 J. M. West *Shrewsbury* v. 71 It was at this time that 'penals' came to be written from Milton's *Paradise Lost*. 1938 F. D. Sharpe *Sharpe of Flying Squad* xvii. 191 Nick-name got four years' penal and his companion eighteen months' hard.
c. (Further examples.)
1908 A. de Horsey in *Times* 15 Aug. 14/3 It is bad enough that the Isle of Wight at Parkhurst should have been selected as a penal station for the convicts of other parts of England. 1963 T. & P. Morris *Pentonville* xi. 225 Their reference group remains very firmly their 'own' prison, which some of the older men still refer to as their 'penal station'.
2. Delete †*Obs.* and add later examples. Now usu. of taxation and other financial burdens.
1953 'M. Innes' *Christmas at Candleshoe* i. 14 His father.. declares that penal taxation is ephemeral, and that the reality big English properties the ownership has not changed. 1957 *Sunday Times* 14 July 13/6 The motor and aeroplane, cinema, radio and television, penal taxation of large incomes and compulsory social services have transformed life. 1958 *Spectator* 15 Aug. 234/1 We have had to struggle with..a penal Bank rate.

penalty. Add: **2. b.** (Further examples.)
1972 *Lebende Sprachen* XVII. 134/2 Increases in strength or stiffness have always brought a weight penalty. *Ibid.* 135/1 The aircraft will accept bulky cargo with little penalty in payload.
c. (Further examples.) *spec.* in *Football*, (the award of) a free kick at goal (see also sense 5 in Dict.).
1897 *Encycl. Sport* I. 434/2 Within the twelve yards line, a referee must enforce law 13, and has no power to mitigate the penalty. 1898 A. E. T. Watson *Turf* 249 When any race is in dispute, both the horse that came in first and any horse claiming the race shall be liable to all the penalties attaching to the winner of that race till the matter be decided. 1899 A. Budd *Football (Rugby)* 33 Free-kicks by way of penalties shall be awarded on claims by the opposite side. 1951 E. Rickman *Come racing with Me* ii. 12 The weights are varied in individual cases by 'penalties' (extra weight) for previous wins. 1969 B. Lakes *England v Scotland* x. 229 Mr Skrankowski awarded no fewer than 55 free kicks and three penalties. 1974 *Rules of Game* 4. 1975 *Great Moments in Sport: Soccer* v. 60 Dorsett shot the penalty home like a thunderbolt. 1974 *Rules of Game* 260/4 Condition (or allowance) races, in which a basic weight allowance for age and sex of horse is varied by added weight penalties for past successes (measured in prize money terms). 1977 *Horse & Hound* 14 Jan. 2/4 Early Spring..was making an 8 lb penalty look very ordinary indeed.
e. *Bridge.* A number of points added to the opponents' score when the declarer fails to make his contract, or to the declarer's score when his call is doubled and he makes his contract.
1908 R. F. Foster *Auction Bridge* 37, 50 points penalty for each of the two tricks by which the bidder failed. 1935 *Encycl. Sports* 174/1 Penalties reasonably incurred may be cheaper than allowing your opponents to make game. 1964 *Official Encycl. Bridge* 417/2 Penalty. (1) An obligation or restriction imposed upon a side for violation of the Laws of Bridge.. (2) An increased score above the line by the declarer's opponents when the declarer fails to make a contract. 1976 *Field* 30 Dec. 1293/2 He again doubled, but, of course, this time for a penalty.
5. (from sense 2c) *penalty bully, corner, flick, goal, kick* (further examples), *point, stroke, trick, try*; **penalty area,** the area in front of the goal on football and other pitches within which offences can incur the award of a penalty; **penalty bench,** in ice-hockey, seating for match officials and penalized players; also **penalty box,** (*a*) the area taken up by a penalty bench; (*b*) = *penalty area*; **penalty card** *Bridge* (see quot. 1964); **penalty carrier** *Golf*, a player who has a number of strokes added to his total as a handicap; **penalty clause,** a clause in a contract stipulating a penalty for failure to fulfil any of its obligations; **penalty double** *Bridge*, a double made to increase a score if an opponent's contract is defeated; **penalty envelope** *U.S.*, an official envelope which may only be used for its designated purpose, under penalty of a fine stated on it; **penalty killer,** in ice-hockey, a player responsible for preventing the opposing side from scoring while his own team's strength is reduced through penalties; hence **penalty killing** *ppl. a.* and *vbl. sb.*; **penalty line,** a line marking a penalty area on a football pitch; **penalty pass** *Bridge* (see quot. 1964); **penalty rate** *Austral.*, an increased rate of pay for overtime; **penalty spot,** the spot from which penalty shots or kicks are taken.
1905 P. Walker *How to Play Assoc. Football* 12 Lines shall be marked 18 yards from each goal-post at right angles to the goal-lines for a distance of 18 yards, and these shall be connected with each other by a line parallel to the goal-lines; the space within these lines shall be the penalty area. 1910 *Encycl. Brit.* X. 611/2 If such infringement take place within the penalty area on the part of a player on the side then defending the goal, a 'penalty kick' is awarded to the attacking side. 1929 *Daily Express* 7 Nov. 19/1 The full-backs were often guilty of dribbling the ball in their own penalty area. 1970 A. Wade *Coach Yourself Assoc. Football* 14 When he reaches the penalty area attacker A can only shoot or pass to B. 1972 G. Green *Great Moments in Sport: Soccer* i. 25 Again Morten-sen reached a spot right of the penalty area. 1934 Webster, Penalty bench. 1962 *Amer. Speech* XXXVII. 126 Stammlach goes to the penalty bench for two minutes on account of a check. 1974 *Rules of Game* 188/1 Penalty bench with space for eight players and extra seating for timekeepers, scorer, and announcer. 1931 *Vancouver Province* 17 Jan. 7/1 Two Vancouver players [were] in the penalty box. 1964 F. C. Avis *Soccer Reference Dict.* 90 *Penalty Box*: see Penalty Area. 1963 *Calgary Herald* 11 Nov. 9/2 Alex Faulkner was in the penalty box serving a major penalty for high-sticking Montreal's Ralph Backstrom and drawing blood. 1974 'E. Lathen' *Murder without Icing* vi. 61 Paul Imrie fought for the shortest part of it. He was a constant occupant of the penalty box, he was always being thrown out of games. 1976 R. Dunphy *Only a Game?* ii. 30 Both goals were breakaways, starting from their own penalty box. 1897 *Encycl. Sport* I. 516/2 A penalty bully is given to the attacking side.. 1909 *Westm. Gaz.* 12 Oct. 12/2 It is rarely that a penalty bully is given in first-class hockey. 1935 *Encycl. Sports* 347/2 If necessary, time of play shall be extended to admit of a penalty bully being played, or completed. 1974 *Rules of Game* 183/5 Extra time is allowed to take a penalty bully if half or full time is already completed. 1958 *Listener* 25 Dec. 1094/2 The card would have been a penalty card as well. 1959 *Ibid.* 30 Apr. 767/1 Declarer can, in fact, treat the remaining cards of either defender as penalty cards. 1963 *Ibid.* 24 Jan. 180/1 Declarer could have..treated the card led as a penalty card. 1964 *Official Encycl. Bridge* 418/1 *Penalty card*, a card that has been prematurely exposed by a defender, and must be left face up on the table until legally played or permitted to be picked up. 1908 *Westm. Gaz.* 12 June 9/3 Mr. Hunter is the only surviving 'penalty-carrier'. His handicap is plus 2, while Mr. Scrut-ton..receives an allowance of fourteen strokes. 1935 Wode-house *Luck of Bodkins* xv. 170 The first thing she would do, if she was a sensible kid, would be to go to her lawyer and have a contract drawn up and signed, with penalty clauses. 1967 J. Woodward in Kirk & Yearsley *Handbk. Managem. Technol.* 74 Some costs increase as the project duration increases, e.g. overheads, penalty clauses, lost revenue, etc. 1969 K. Giles *Death cracks Bottle* i. 13 The new part was accomplished in one cyclonic burst—five days ahead of schedule, ahead of the penalty clauses. 1976 E. Ward *Hanged Man* xxviii. 179 The main contractors had a big penalty clause on me and hired Dieter to frighten off skilled labour so they could collect. 1935 *Encycl. Sports* 147/1 Rule 18 shall also apply to a penalty corner. 1967 J. Potter *Foul Play* xii. 152 The defence conceded one goal, after the opposition had been awarded a much deserved penalty corner. 1974 *Rules of Game* 183/5 Penalty corners are awarded against defenders for deliberately playing the ball over the goal line [etc.]. 1977 *Guardian* 10 Mar. 27/3 Manchester will consider themselves unlucky to have been denied a penalty stroke when the score stood at 1–1. 1959 *Listener* 30 July 190/1 The theory of the responsive double is that the hand on which one would want to make a penalty double..is of much lower frequency than the hand..when one might prefer to give a picture of general values. 1964 *Official Encycl. Bridge* 135/2 The two main categories [of doubles] are penalty doubles and take-out doubles. 1879 *Postal Laws* U.S. §147 Requisitions for postage-stamps, stamped-envelopes..and official penalty-envelopes are required to be made upon printed forms. 1903 N.Y. *Times* 29 Aug. 3/1 The officials of the District Government were not entitled to the use of the mails like other Federal officials who use penalty envelopes. 1977 J. A. Moss *Officers' Manual* (ed. 6) xxv. 272 Official letters are mailed in penalty envelopes. 1967 J. Potter *Foul Play* ii. 25 One of the umpires.. the one, in fact, who had awarded a short corner instead of a penalty flick. 1977 *Cleethorpes News* 6 May 33/2 Skegness..lost..on penalty flicks to Scunthorpe last year. 1951 *Sport* 27 Apr.–3 May 4/1 The Oakwell Reds would have welcomed the compensation of a penalty-goal in their 6–0 setback at Maine Road. 1979 *Times* 12 Dec. 9/6 Penalty goals rather than tries continue to decide most matches. 1897 *Encycl. Sport* I. 434/1 The referee shall award the opposing side a penalty kick. 1925 [see *equalize v.* 4b]. 1960 E. S. & W. J. Higham *High Speed Rugby* 180 There may come a stage in a game when taking a penalty kick at goal is waste of time. 1971 *Referees' Chart* (Football Assoc.) 30 When a penalty-kick is being taken the Referee must not give the signal for the restart until the players have taken up the position ordered by the Law. 1974 *Rules of Game* 159/4 The penalty kick is taken from or behind where the offense occurred. 1962 *Kingston* (Ontario) *Whig-Standard* 14 Dec. 10/1 Not only was Westfall one of the best defencemen in the league, a good point man on the power play and a penalty killer of the first order, but he took over a left wing position just before his departure. 1966 *Hockey News* 1 Jan. 13/2 He is an accomplished man on the power play and is among the top penalty killers in the league. 1968 *Globe & Mail* (Toronto) 15 Jan. 21/1 Three seasons I spent in the National Hockey League exclusively as a penalty killer. 1963 *Kingston* (Ontario) *Whig-Standard* 6 May 11/4 Winger Bill Glashan stepped up from his penalty-killing role to score twice for Flyers. 1963 *Hockey Illustr.* Dec. 38/2 If I took him off penalty killing and put him on the power play he'd score 40 goals a year. 1970 *Globe & Mail* (Toronto) 26 Sept. 35/1 He is the centre on a checking line, a headliner on the penalty-killing unit. 1975 *Cleveland* (Ohio) *Plain Dealer* 6 Apr. 10–C/1 Their power play, keyed by Gordie, is one of the best in the league and the penalty killing, once again led by Howe, is one of the best in the league. 1929 *Evening News* 18 Nov. 13/3 The penalty lines and the touch lines were not visible. 1959 *Listener* 2 Apr. 613/3 We would not consider a penalty pass as his trick-taking capacity was far too slender. 1964 *Official Encycl. Bridge* 420/1 *Penalty pass*, a pass by a player after a take-out double from his partner and a pass by right-hand opponent. 1974 *Rules of Game* 49/2 The competition is won by the finalist with the least penalty points. 1977 *New Yorker* 10 Oct. 150/2 The umpire could have assessed him a penalty point for sportsmanlike conduct. 1956 S. Hope *Diggers' Paradise* 98 All workers when they 'work back'—do overtime—come on to 'penalty rates' as they are called. 1973 *Bulletin* (Sydney) 25 Aug. 3/3 We will expect to be dealt with on exactly the same basis as any other Commonwealth public servant, i.e. a 36-hour week, penalty rates for overtime, annual leave, sick leave, [etc.]. 1937 F. N. S. Creek *Assoc. Football* vi. 126 Penalty spot. 1948 B. Steel *How to play Football* xvi. 144 As the shooter is to take up position for the kick, the wind blew the ball from the penalty spot. 1960 G. Green in Fabian & Green *Assoc. Football* III. viii. 55 Goalkeepers now have to remain stationary on their line until the ball is actually struck from the penalty spot. 1974 *Rules of Game* 183/5 A penalty stroke is taken 7 yards from the goal (the penalty spot by an attacker). 1975 *Evening News* (Edinburgh) (Sports Final ed.) 15 Mar. 1/2 Fairley netted from the penalty spot. 1977 *Daily Mirror* 12 Apr. 26/3 Masson had scored from the penalty spot after Coventry's Brian Roberts and Alan Dugdale combined to push Peter Eastoe to the ground. 1895 W. T. Linskill *Golf* (ed. 3) 45 A penalty stroke shall not be counted the stroke of a player, and shall not affect the rotation of play. 1970 H. Taylor *Golf Dict.* 153 *Penalty stroke*, an additional stroke debited to a player (e.g. for unauthorized touching of the ball). 1974 *Rules of Game* 183/4 If the ball halts inside the circle or passes out of it, the penalty stroke is ended. 1977 *Sunday Times* 9 Jan. 30/6 England..gave away two silly goals, and were then denied an obvious penalty stroke. 1909 *Westm. Gaz.* 20 Mar. 14/2 Penalty-trick scores incurred during the play of a rubber are irretrievably gone. 1923 P. Trevor *Rugby Union Football* xi. 149 The awarding of a penalty try is an occasional happening. 1936 H. B. T. Wakelam *Rugby Football* i. 4 The only other means by which the score can be increased is the very rare penalty try, awarded only under exceptional circumstances by the referee, upon his deciding that a proper try would have been put on had it not been for some exceptionally flagrant breach of the laws by the opposition. 1959 in V. Jenkins *Lions Down Under* (1960) xiv. 206 A penalty-try,..the first (as far as is known) ever awarded to a touring team visiting New Zealand. 1960 Penalty-try [see *in-goal]. 1969 W. Reyburn *All about Rugby Football* vii. 109 *Penalty try*, awarded when, in the opinion of the referee, a player would have scored had he not been obstructed by a defender; most commonly when both are chasing a ball kicked over the goal line. Conversion of a penalty try is taken from in front of the posts, no matter where it is awarded.

pen and ink, pen-and-ink, *phr.* Add: **A.** as *sb.* **2.** (Earlier and later examples.)
1860 D. G. Rossetti *Let.* 28 Sept. (1965) I. 375 Now one of his commissions is for a £50 pen-and-ink. 1976 *National Observer* (U.S.) 10 Apr. 16/2 Mrs. Fisher had once done us a pen and ink of the Poplar Branch library.
3. A stink. *Rhyming slang.*
1859 Hotten *Dict. Slang* 145 *Pen and ink*, a stink. 1935 A. J. Pollock *Underworld Speaks* 86/2 *Pen and ink*, a stink. 1972 G. F. Newman *You Nice Bastard* 347 *Pen (and ink)*, stink.
B. as *adj.* **1.** Delete 'Now *rare* or *Obs.*' and add later examples.
1967 A. L. Lloyd *Folk Song in England* 14 When [Cecil] Sharp was..in search of song..a wide gulf separated the pen-and-ink man from the man with bowyangs of binder twine. 1972 [see *lensman* s.v. *lens sb.* 3].
C. as *vb.* **2.** To stink. *Rhyming slang.* Also *ellipt.*, as *pen*.
1892 *Sporting Times* 29 Oct. 1/2 The air began With language to pen and ink. 1972 G. F. Newman *You Nice Bastard* iv. 137 'I don't mind, provided he takes a bath.' 'Yeah, he does pen a bit.'

pen and inkery, *phr. nonce-use.* [f. prec. + -ery.] The use of pen and ink; an author's business.
c 1909 W. De Morgan in A. M. W. Stirling *W. De Morgan & his Wife* (1922) xiv. 328, I do wish I had paid more attention to them [*sc.* birds, flowers, and trees] in my time—they would come in so useful in these later days of pen-and-inkery.

‖**penanggalan** (pĕnæ·ŋgālăn). Also **penangalan.** [Malay.] A female vampire (see quots.). Cf. *langsuir.
1839 T. J. Newbold *Pol. & Statistical Acct. Straits of Malacca* II. xii. 191 The Penangalan takes up its abode in the forms of females, and afflicts them with an unnatural craving for human blood. 1900 W. W. Skeat *Malay Magic* vi. 320 The Pĕnanggalan..is believed to resemble a trunkless human head with the sac of the stomach attached to it, and which flies about seeking for an opportunity of sucking the blood of infants. 1972 *Daily Tel.* (Colour Suppl.) 12 May 58/3 The Malayan vampire family includes..the flamboyant Penanggalan, a monstrous vampire which also sucks at children's blood.

penatin (pe·nātin). *Biochem.* [f. the L. generic name Pen(icillium + the L. specific epithet *not-at-um* (see *notatin) + -in[1].] = *notatin (*penicillin 2).
1942 W. Kocholaty in *Jrnl. Bacteriol.* XLIV. 143/1 'Penatin' (a newly discovered antibacterial substance produced by *Penicillium notatum*). *Ibid.*, While penicillin..has only very weak antibacterial properties against Gram-negative organisms, and practically none against *Escherichia coli*, penatin is highly bactericidal and bacteriostatic against those two organisms. 1949, 1963 [see *notatin].

Penbritin (penbri·tin, pe·nbritin). *Pharm.* A proprietary name for ampicillin, $C_{16}H_{19}O_4S$, a semi-synthetic penicillin that resembles penicillin G (benzylpenicillin) in its action against Gram-positive organisms but is more effective against Gram-negative ones, and is used esp. in treating infections of the urinary and the respiratory tracts.
1959 *Trade Marks Jrnl.* 18 Nov. 1197/1 Penbritin.. Antibiotic preparations and substances.. Beecham Research Laboratories Limited,..Brentford, Middlesex; manufacturers. 1960 *Official Gaz.* (U.S. Patent Office) 26 July 1113/1 Beecham Research Laboratories Limited, Brentford, England. Filed Mar. 15, 1960. Penbritin. For antibiotic preparations and substances. 1961 *Daily Tel.* 21 July 24/5 A new synthetic penicillin, Penbritin, promises to be very effective against certain infections so far immune to previous penicillins and to be of increased potency against others. 1968 J. H. Burn *Lect. Notes Pharmacol.* (ed. 9) 108 Broad spectrum penicillins; one of these is ampicillin (Penbritin). 1968 *Economist* 6 July 66 The product responsible more than any other for more than doubling profits over the last four years is Beecham's penicillin mutant, Penbritin, which took a decade to develop. 1975 D. Fishlock *Business of Sci.* viii. 111 The science..set Beecham firmly on the path to its big semi-synthetic penicillins of the 1960s: Celbenin..., Orbenin and Penbritin.

pence. b. Applied *colloq.* as *sing.*, orig. to a 'new penny' of the decimal currency introduced in 1971 (see *penny 1), and hence *gen.*
1971 *Record* (Oxf. Univ. Press) Dec. 10/2 The computer was found to be rounding up to the nearest pence the Bank Code Numbers on the Wages Slips. 1973 *Daily Tel.* 24 Oct. 16 In our village shop a customer asked for some small change but the shopkeeper was unable to oblige as she was very short of 'two pences and one pences'. 1974 *Ibid.* 19 Dec. 12 In shops and elsewhere I often hear the non-grammatical term 'one pence'. I presume this is the cause the occurrence of a single penny is becoming a thing of the past. 1975 M. Bradbury *History Man* i. 3 She leaks her daily deputation to the manager with comparatively..up-to-the-minute lists showing how Fine Fare, on lard, is one pence up on Sainsbury's, or vice versa. 1977 *Times Lit. Suppl.* 29 Apr. 528/3 The new and the supplemented lexical terms equally reflect the times, with 'p' (but not the singular use of *pence*) for *new penny*. 1979 *Daily Tel.* 11 Apr. 21/3 Fares..is to go up..You will have to pay an extra charge of one pence on his fare..was killed by the driver, police said in Manila.

pencil, *sb.* Add: **I. 2. d.** *to have the pencil put on one* (Criminals' slang), to be reported to the prison authorities.
1929 *Sat. Even. Post* 13 Apr. 50/3 A prisoner who is reported for some violation is written up and 'get the pencil put on him'. 1934 H. N. Rose *Thes. Slang* 54/2 Reported for violation of rules. *to be written up*; *have the pencil put on*; *have the number taken*.
II. 4. (Examples referring to radiation other than light.)
1913 *Phil. Mag.* XXV. 604 A narrow pencil of a particles fell on a zinc sulphide screen. 1938 R. W. Lawson tr. *Hevesy & Paneth's Man. Radioactivity* (ed. 2) 192 With this apparatus pencils of many micro-amperes of deuterons..have been obtained. 1967 *Listener* 30 Mar. 419/3 The picture is positioned, and a tiny pencil of X-rays is emitted at the spot where we want to analyse a pigment.
6. d. The penis. *slang.*
1937 Partridge *Dict. Slang* 616/2 *Pencil*, the male member. 1942 Berrey & Van den Bark *Amer. Thes. Slang* §121/39 Male pudendum..pencil, pencil and tassel. 1976 P. Haining *Penis Book* v. 58 That Purple Emperor strain is as soft as an old man's pencil.
d. Phr. *lead in one's pencil*: see *lead sb.*[1] 3.
III. 7. a. *pencil-mark* (earlier and later examples as *fig.*). Also with sense 7b below; *pencil-slim, -thin* adjs. Also with sense 'resembling a pencil in shape', as *pencil flash(light), microphone, pants, skirt, stripe, torch*.
1929 E. Queen *Roman Hat Mystery* iv. xxii. 307 Barry examined them..by the same pencil flashlight—a tiny streak of illumination. 1935 R. Chandler in *Black Mask* Jan. 13/1, I tried throwing the beam of my pencil flash along the floor. 1824 J. S. Mill in *Westm. Rev.* II. 399 Hume was not..without authority, for Mr. Brodie saw his pencil marks opposite to this story, in the copy of Perinchief belonging to the Advocate's library. 1855 D. G. Rossetti *Let.* 25 June (1965) I. 257 Ruskin has been reading those translations since you, and says he could wish no better than to ink your pencil-marks as his criticisms. 1962 A. Nisbett *Technique Sound Studio* ii. 36 A 'pencil' microphone just peeping out of a well in the centre of the table. 1973 P. Evans *Bodyguard Man* xix. 120 Her right hand was..holding a pencil-microphone. 1960 *News Chron.* 26 Sept. 9 This one is equally good with pencil pants. 1952 C. W. Cunnington *Eng. Women's Clothing* ii. 245 Coatees... Worn with plain pencil skirt. 1960 [see *kick-pleat* s.v. *kick sb.*[1] 8]. 1974 P. Haines *Tea at Gunter's* ix. 96 She was wearing a pencil skirt so tight she could barely walk. 1949 *Women's Wear Daily* 24 Jan. 6/5 Jacqueline Vienne features..long double-breasted jacket over pencil slim skirts. 1976 'J. Ross' *I know what it's like to Die* xiii. 144 A pencil-slim ochre-coloured cigar. 1897 *Sears, Roebuck Catal.* 181/2 Boys' two-piece wash goods suit, made of heavy navy blue with pencil stripe. 1960 *House & Garden* June 21/1 (Advt.), Alternating bands of closely spaced pencil stripes. 1970 *New Yorker* 17 Oct. 170 (Advt.), Pencil stripes, yes. Bankers pencil stripes. 1972 K. Owen *Damned & Destroyed* i. 12 A..pencil-thin Frenchman. 1963 *Times* 25 Feb. (Canada Suppl.) p. vii/4 The huge, pencil-thin Great Lakes ships. 1978 T. Gifford *Glendower Legacy* (1979) 298 He came down the stairs, Dapper and pencil-thin. 1937 M. Allingham *Dancers in Mourning* iv. 59 He had a pencil-torch in his pocket. 1948 M. Gilbert *They never looked Inside* i. 7 A glow-worm came and went ten feet away, and Rod guessed that 'Gumm' was using his pencil-torch with discretion. 1964 'E. Peters' *Flight of Witch* x. 166 One of those thin pencil-torches that clip in a breast pocket. 1976 'Z. Stone' *Modigliani Scandal* iv. iv. 182 He..shone a pencil torch inside.
b. pencil-arm, the arm of a pair of compasses that carries the pencil; **pencil beam,** a narrow, nearly parallel beam; *spec.* in *Radar*, one which in addition has an approximately circular cross-section; **pencil beard** (see quot.); **pencil box,** a box for holding pencils; **pencil cedar,** add: esp. *Juniperus virginiana*; also, any of several Australian trees resembling these kinds of juniper or yielding wood suitable for making pencils; (further examples); **pencil flower** (earlier and later examples); **pencil knife,** a knife for sharpening pencils; **pencil-line,** a line drawn with a pencil or resembling one so drawn; also *attrib.*, esp. in **pencil-line moustache** (see quot. 1966); **pencil mark** = *pencilling vbl. sb.* 1; **pencil moustache** = *pencil-line moustache* above; **pencil pusher** *U.S.*, a derogatory term for one whose occupation involves much writing with a pencil; **pencil tablet,** a notebook of rough paper suitable for writing in pencil but not in pen.
1899 E. Rowe *Hints on Chip-Carving* i. 9 Do not use the left hand to move the pencil-arm of the compasses. 1946 *Jrnl. Inst. Electr. Engin.* XCIII. IIIA. 25/2 A Yagi aerial.. producing a narrow in two perpendicular planes, from an end-fire array of comparatively small dimensions. 1955 *Sci. Amer.* Mar. 38/1 This antenna has a pencil beam which at a wavelength of 65 centimeters is five degrees wide. 1965 A. Nicol *Truly Married Woman* 98 Bradshaw..turned quickly down the road, his headlights two pencil beams in the darkness. 1966 J. S. Cox *Illustr. Dict. Hairdressing* 110/1 *Pencil beard*, a narrow strip of beard from the lower lip to the chin. 1907 *Yesterday's Shopping* (1969) 132/2 Pencil Box, 9 by 2½ in., 1/0. 1912 'C. F. Ashley' *Fairs & Fetes* 118 School-supplies also may find a place here—slates and pencils..pencil-boxes and rulers. 1946 'E. Crispin' *Holy Disorders* xi. 235 Ties an' jet black, with rich mahogany tan markings. 1966 E. H. Pinto *Treen* 422 The Bavarian manufacturers of children's pencil boxes certainly knew how to bring joy to the hearts of late-Victorian and Edwardian school boys and girls. 1977 D. Clark *Gimmel Flask* vii. 134 Remember the old pencil-boxes with sliding lids and sections inside? 1884 A. Nilson *Timber Trees New South Wales* 53 D[ysoxylon] (*Dysoxylum Muelleri*)..is also sometimes known as 'Pencil Cedar'. 1889 *Encycl. Brit.* XXVI..*D. rufum*.—Bastard Pencil Cedar. *Ibid.* 110 *Flindersia*..—Colonial Deal; White Pine; Pencil Cedar. 1908 E. J. Banfield *Confessions of Beachcomber* i. v. 184 A huge log of pencil cedar had been cast among the boulders. 1932 R. H. Anderson *Trees New South Wales* 89 Bermuda Pencil Cedar..often makes a fine tree. *Ibid.* 143 Red Bean (*Dysoxylum Muelleri*)..is also sometimes known as 'Pencil Cedar'. 1961 *Australian Encycl.* I. 222/1 *Tieghemopanax elegans* (syn. *Panax elegans*), the black pencil cedar, is an ornamental tree with large, divided leaves, known from Queensland. *Ibid.* VII. 106/1 *Glochidion Ferdinandii* (variously known as white beech, pencil cedar, rain tree and 'towwar')..is a medium-sized, rain-forest tree. 1968 W. E. Willis *Timber* i. 6 Virginian Pencil Cedar..is a species of Juniper and a favourite wood for the manufacture of pencils. 1847 A. Eaton *Man. Bot.* 85 *Stylosanthes*..Pencil flower. 1901 C. T. Mohr *Plant Life Alabama* 570 Stylosanthes... Pencil flower. 1958 R. O'Neill *Great God Brown* III. iii. 81 Waving his pencil knife with grotesque flourishes. 1905 Pencil-line [listed in Dict. sense 7a]. 1957 R. Hoggart *Uses of Literacy* ix. 235 A neat moustache in dark thin pencil line. 1966 J. S. Cox *Illustr. Dict. Hairdressing* 110/1 *Pencil Line Moustache*.. (2) A moustache consisting of a very thin line of hairs... (2) A thin line moustache drawn on the upper lip with a coloured cosmetic. 1968 R. C. Galway *Assignment Goalbreak* ix. 88 He had..a vicious-looking pencil-line moustache. 1971 B. Malamud *Tenants* 98 Jacob..had uneasy eyes and a pencil-line mustache. 1976 *New Yorker* 15 Nov. 53/2 Greta Garbo, with pencil-line eyebrows above sex-drugged lids, plays a bored, sensual, wicked woman. 1880 H. Dalziel *Brit. Dogs* ii. 332 Feet tanned, but the knuckles with a clear black line, called the 'pencil mark', up the ridge. 1931 A. C. Smith *About our Dogs* xvi. 245 The standard of the Black-and-Tan Terrier Club:..the forelegs tanned up to the knees, with black lines (pencil marks) up each toe. 1961 W. Brown *Bedeviled* 77 He was a tall, thin youth with slicked-down, black hair, a pencil moustache and shaggy sideburns. 1965 P. Robinson *Pakistani Agent* ii. 8 Parulekar was..in his late twenties, with a pencil moustache. 1973 H. Gilbert *Hotels with Empty Rooms* xv. 128 A cinematographical convention, like crooks with pencil-moustaches. 1881 *Harvard Lampoon* 20 Apr. 42/2 After various chilling repulses, our pencil-pusher discovered a man smaller than himself. 1917 F. D. O'Sullivan *Enemies of Underworld* xcvi. 666/1 *Pencil pusher*, clerk. 1926 Maines & Grant *Wise-Crack Dict.* 12/2 *Pencil-pusher*, office employe. 1952 in Wentworth & Flexner *Dict. Amer. Slang* (1960) 382/1 The number of pencil pushers and typists has increased in the past 25 years out of proportion to the increase in factory workers. 1959 *Amer. Speech* XXXIV. 79 *Pencil pusher*, a camp book-keeper. 1961 B. James *Night of Kill* ix. 120 He'd be damned if he was going to be for a mid-watch pencil-pusher just to please his ulcerated pro-tem captain. 1895 *Montgomery Ward Catal.* Spring & Summer (Index), Pencil Tablets. 1944 T. D. Clark *Pills, Petticoats & Plows* 45 Across the way there was an assortment of school-books, pencil tablets,..and epsom salts.

pencil, *v.* Add: **2. c.** *to pencil in* (fig.), to note, register, or arrange provisionally or tentatively.
1942 Berrey & Van den Bark *Amer. Thes. Slang* §628/5 Be pencilled in, to be tentatively, but not finally hired. 1959 *Times* 22 June 4/2 He is a name pencilled in for the future. 1967 *Punch* 31 May 783/1 We have pencilled in an arrangement to throw open our houses to one another, I'm not sure. 1971 Wodehouse *Much Obliged, Jeeves* xii.128 You and I, regarding Florence coolly, pencil her in as too bossy for human consumption. 1977 *Daily Mirror* 10 May 31/1 The MCC team to play Australia at Lord's from May 25–27, due to be announced later this week and traditionally a Test trial, might still be very different to the one originally pencilled in.

pencil and paper, pencil-and-paper, *phr.* Usu. *attrib.* or as *adj.* Requiring (only) pencil and paper.
1930 J. B. Priestley *Angel Pavement* viii. 385 You had to pretend you were having a marvellous time because you were wearing hats from crackers and playing pencil and paper games ('Let me see, a river beginning with "V"?'). 1952 J. B. Pick *Phoenix Dict. Games* 295 (heading) Pencil and paper games. 1957 L. Fox *Numerical Solution Two-Point Boundary Probl.* p. v, I have included most of the worthwhile methods known to me which are suitable for pencil-and-paper and desk-machine computation. 1973 *Nature* 27 Apr. 594/2 Table 4.1 only lists some values of a simple function..which is therefore amenable to pencil-and-paper arithmetic.

pencil-case. Add: **b.** *Bookbinding.* (See quot.)
1885 W. J. E. Crane *Bookbinding* xvi. 132 This [pressing in at the joint] is very necessary, or the boards may not properly 'go home' and adhere hence an unsightly protuberance of loose paper at the joint will be the result, which is generally termed a 'pencil case', and is a clear mark of bad bookbinding.

pencilling, *vbl. sb.* Add: **1.** (Further *transf.* examples.) Also, natural marking on animals.
1909 W. Bateson *Mendel's Princ. Heredity* ii. 42 Pencilling [in fowls] is a dominant to its absence. 1936 C. L. Morgan *Sparkenbroke* vi. vii. 551 He saw a pencilling of light shine under her door. 1948 C. L. B. Hubbard *Dogs in Brit.* xx. 309 Colour [of the Manchester terrier] is jet black, with rich mahogany tan markings..('pencillings' on the toes to be black). 1971 F. Hamilton *World Encycl. Dogs* 659/1 Pencilling: Dark lines divided by strips of tan on feet of Manchester Terrier.
2. (Earlier and further examples.)
1803 *Lett. Miss Riversdale* I. 325 The friction of his pocket had so completely defaced the pencillings. 1835 N. P. Willis (*title*) Pencillings by the way. 1846 Geo. Eliot *Lett.* (1954) I. 187 Thank you for the pencillings. I see that many are obvious and important emendations. 1923 J. M. Murry (*title*) Pencillings: little essays on literature.

pendeloque. Add to def.: *spec.* a gemstone esp. a diamond cut in the shape of a drop and used as a pendant. Also *attrib.* (Further examples.)
1945, 1949 [see *navette]. 1955 M. Gilbert *Sky High* ii. 25 A pair of pendeloque-cut diamond earrings set in platinum filigree. 1959 *Times* 24 Feb. 18/7 A large fancy golden pendeloque diamond. 1960 H. Hayward *Antique Coll.* 214/2 *Pendeloque*, a gemstone of drop-shaped form and used as a pendant. 1973 *Times* 25 Aug. 17/3 There are nine well-recognized styles of lapidary work upon the diamond: brilliant-cut, step-cut, emerald-cut, rose-cut, marquise, briolette, baguette, cabochon and pendeloque. 1974 *Encycl. Brit. Macropædia* VII. 979/2 Pendeloque among the gemstones are an 86-carat pendeloque diamond, as well as large emeralds and red spinels.

pendency. 1. (Later examples.)
1850 E. S. Seymour *Sk. Minnesota* xii. 208 Out of this arose one or two unsuccessful arbitrations. During their pendency the mills came into the possession of the Boston Company. 1972 *N.Y. Law Jrnl.* 22 Aug. 3/2 Section 1830.12 Petitions; content; pendency of prior proceeding. 1979 'A. Hailey' *Overload* ii. iii. 119 During the pendency of the case, surely there were backstage discussions with commission staff?

pendente lite. (Earlier and later examples.)
1726 J. Ayliffe *Parergon Juris Canonici Anglicani* 235 Because he came in *Pendente lite*. 1964 R. W. Hansen in A. E. Wilkerson *Rights of Children* (1973) 240 This conflict may be revealed at the pre-trial hearing before the Family Court commissioner on *pendente lite* support orders, or at the time of trial before the court. 1972 *N.Y. Law Jrnl.* 10 Oct. 1/7 Judge Motley also refused to enjoin the defendant..pendente lite from distributing its similarly designed loudspeakers. 1973

Ibid. 31 Aug. 18/3 The cross motion for an order for renewal of plaintiff's motion for alimony pendente lite is denied, no sufficiently persuasive ground in support of same having been demonstrated.

pending, *ppl. a.* and *prep.* Add: **A.** *ppl. a.* **1.** (Later examples); so *pending basket, tray*: a basket or tray for correspondence or other papers awaiting attention or decision.
1946 R. Graves *Poems 1938–45* 30 World patents pending; tested in the shops. 1955 E. Waugh *Officers & Gentlemen* i. viii. 102 Guy turned over the papers in the 'pending' tray. 1957 M. Summerton *Sunset Hour* i. 18, I threw the folder into my pending basket. 1961 'T. Hinde' *For Good of Company* i. 18 Out-tray to his right, in- and pending-trays to his left. 1973 J. Wainwright *Pride of Pigs* 50 The day clerks..booked it in—this avalanche of paper—and..tossed it into a series of 'pending' trays. The night clerks *dealt* with it. 1975 *Times* 5 Dec. 2/1 An extradition warrant..was in a tray marked 'pending'.

Pendleton (pe·nd'ltən). Orig. and chiefly *U.S.* The name of the *Pendleton* Woolen Mills (named after Pendleton, a town in Oregon) used to denote garments made by them, esp. a brightly coloured checked sports shirt. Also *attrib.*
1940 *Esquire* Dec. 204/1 (Advt.), Wear a Pendleton—then you'll know why Pendleton shirts have been a religion with sportsmen. 1947 E. S. Gardner in *Amer. Mag.* Aug. 152/2 Our party is a man thirty-eight years old, bronzed, wears cowboy boots, a five-gallon hat, leather jacket, Pendleton trousers, rather chunky, and has a wide firm mouth. 1948 *Official Gaz.* (U.S. Patent Office) 28 Dec. 966/2 Pendleton [Trade Mark]. For outer shirts, lounge robes, trousers and slacks, jackets and blazers, and men's coats. No. 535,993. 1960 *Sports Illustr.* 5 Sept. 22 (Advt.), On campus, a man worth watching will have at least one Pendleton. 1961 'A. A. Fair' *Shills can't cash Chips* iii. 71 The guy looked like a tall Texan. He was wearing Pendletons and cowboy boots. 1963 *Pix* 28 Sept. 62/3 *Pendletons*, bright plaid wool shirts one size too big. One point. 1977 *Guardian Weekly* 9 Oct. 16/2 She had a long, red Pendleton, a wool lumberjack shirt still smoke-musty from camping trips in the North Woods.

‖ **pendopo** (pendō͝u·po). Also **mendopo, pendapa.** [ad. Javanese *pendâpâ*.] In Java: a large, covered porch or veranda in front of a house.
1927 H. S. Banner *Romantic Java* iv. 56 Its [*sc.* the residence's] most arresting feature is the *Mendopo*, a huge, open-sided verandah in the front. 1958 H. Forster *Flowering Lotus* iii. 41 We entered the *pendopo*, skirted the piled desks and reached the 'common room'. 1961 H. Geertz *Javanese Family* i. 7 The greater portion of this building is the *pendapa*, a kind of oversized summer-house about a hundred feet square and always cool, dark, and dignified looking. 1971 *Nat. Geographic* Jan. 27/1 Within its gates are large, low, and lovely buildings, perfectly adapted to the hot and humid climate. The best of these are the *pendopos*, huge open-sided structures covered by four-sided roofs, supported above gleaming marble floors by columns of carved teak.

Pendred (pe·ndred). *Path.* [The name of Vaughan *Pendred* (1869–1946), English physician who described the condition in 1896 (*Lancet* 22 Aug. 532).] *Pendred('s) syndrome*: a recessively inherited condition in which an enzyme deficiency leads to goitre and usu. to deafness.
1960 G. R. Fraser *Deafness with Goitre* (*Syndrome of Pendred*) (Ph.D. thesis, Univ. of London) 3 The name Pendred is suggested as a suitable eponym for this syndrome... A..study of sixty-two cases of Pendred's syndrome in forty-one sibships is described. 1966 J. B. Stanbury et al. *Metabolic Basis of Inherited Dis.* (ed. 2) x. 234/2 The patients with the Pendred syndrome do not have large goiters. 1970 J. F. Sotos in R. M. Goodman *Genetic Disorders Man* xvi. 666/1 In a few reports thyroid disease without deafness has occurred in relatives of patients with the Pendred syndrome. 1964 J. D. Maynard in R. M. Kirk et al. *Surgery* xii. 250 Affected children are goitrous, hypothyroid, and sometimes deaf (Pendred's syndrome after the general practitioner who reported the first family in County Durham).

† **pendulation** (pendiŭlēi·ʃən). *Obs.* [f. L. *pendul-us* Pendulous *a.* + -ation.] An oscillatory motion formerly ascribed to the poles of the earth.
1909 *Westm. Gaz.* 1 May 12/3 According to Professor Simroth, pendulation, which is the periodical oscillation of the earth's axes,..explains nearly every observed fact in the development and distribution of all animals. 1924 J. G. A. Skerl tr. *Wegener's Orig. Continents & Oceans* vi. 95 Unfortunately, Reibisch clothed his ideas..in the singular straight-jacket of a strict 'pendulation' of the poles in an 'orbit of swings', which is probably false.

pendule, *sb.* Add: **2. b.** *Mountaineering.* = *pendulum sb.* 1 d.
1957 E. A. Wrangham *Sel. Climbs Range Mont Blanc* 170 Rappel down from the summit on the Verte side, go over a gendarme by a small *pendule* on the Charpoua side. 1967 Collomb & Crew *Sel. Climbs Mont Blanc Range* I. 180 Traverse L across a very steep ice couloir to a thin and high rock island and continue the traverse by a pendule from a peg high on this island. 1973 C. Bonington *Next Horizon* xvi. 232 We had a spectacular pendule by Tom Patey, as he swung out from the base of the overhanging rock. 1973 J. Bunting *Climbing* 90 *Pendule*, a horizontal *abseil* effected by a sideways swing.

pendule (pe·ndiŭl), *v. Mountaineering.* [f. the sb.] *intr.* To swing to and fro like a pendulum. Also *refl.*
1883 G. MacDonald *Princess & Curdie* xvi. 128 He dropped himself a little below its level, gave the rope a swing by pushing his feet against the side of the cleft, and so penduled himself into it. 1973 C. Bonington *Next Horizon* x. 148 He was now sixty feet up. He paused,..let out some rope and penduled back and forth across the face, trying to work out the best line.

pendulize (pe·ndiŭləiz), *v.* [f. as Pendulous *a.* + -ize.] *intr.* To poise oneself or hover in the air; to be pendant.
1869 E. Newman *Illustr. Nat. Hist. Brit. Moths* 12 He who has not seen this fairy creature pendulizing over a purple patch of the common bugle..has a delight yet to come.

pendulum, *sb.* Add: **1. c.** Used of similar bodies that oscillate but are not similarly suspended: **horizontal pendulum,** an approximately horizontal rod having a heavy weight at one end and pivoted at the other so that it can swing freely in an approximately horizontal plane, supported by a thread or wire passing from the weighted end to a fixed point almost vertically above the pivot; **inverted pendulum,** a vertical rod having a heavy weight at its upper end and resting on a bearing at the other, and held in position by springs which allow it to oscillate in a vertical plane.
1844 *Trans. R. Soc. Edin.* XV. 219 The elegant inverted Pendulum or Noddy contrived by the late Mr Hardy. 1872 F. Zöllner in *Phil. Mag.* XLIII. 491, I explained the principles of such a method and its practicability in an apparatus for which I proposed the name of 'Horizontal Pendulum', in order to distinguish it from other pendulum-like instruments, also suspended by two threads. 1908 C. G. Knott *Physics Earthquake Phenomena* iv. 61 The nearer the point of attachment to a truly vertical position above the pivot the more delicate and the less stable will the horizontal pendulum be, and the better fitted for recording small motions. 1937 D. Kennedy tr. *Imamura's Theoret. & Appl. Seismol.* xi. 254 The form most extensively used is one that weighs a ton. It is shown in Fig. 119. The heavy bob is an inverted pendulum. Its lower point, the end of the supporting rod, rests in a socket, but as it is unstable in this condition, an arm..extends laterally from the upper end of the pendulum, and connects with a steel spring. 1972 R. B. Gordon *Physics of Earth* vi. 124 An instrument suitable for recording horizontal ground motion is the horizontal pendulum.
d. *Mountaineering.* A swinging movement like that of a pendulum, often used as a deliberate move by a climber using his momentum to swing to a new position. Also *attrib.*
1945 G. W. Young *Mountain Craft* (ed. 4) v. 179 The second [artificial aid] is the *pendulum*. Like most modern technical devices, it is a perfecting of an old alpine fashion... By attaching a rope to the highest point convenient upon our first line, we..can swing across upon it, and reach the new set of holds. 1949 A. Roch *Climbs of my Youth* xiv. 115 The pendulum was definitely unpleasant, and a few stones fell loose. 1965 A. Blackshaw *Mountaineering* ix. 268 If there is no suitable transverse crack it will be necessary to do a pendulum or a horizontal rappel. The former involves abseiling..from a piton and swinging over to the desired new position. 1971 C. Bonington *Annapurna South Face* x. 120 He therefore had to climb without any protection from pitons, though if he had fallen off he would have had a punishing pendulum back into the gully. 1972 D. Haston *In High Places* x. 110 There are some devious pendulums to reach the start of the main crack system. A pendulum is an exciting move, very common to Yosemite climbing. When one line of cracks runs out and there is blank wall before another can be reached, often this gap can be filled in by fixing a piton as high as possible, going down for some way hanging on the rope, then running back and forth to get up enough momentum to make a swing into the next crack system.
4. b. pendulum position *Billiards*, a position of the two object balls beside the cushions on either side of a corner pocket which makes a large number of cannons possible; **pendulum saw** (see quot. 1958); **pendulum swing,** a swing or swinging movement like that of a pendulum; also *fig.*
1927 *Daily Express* 26 Apr. 9/4 Reece..made a record break of 1,151, including 568 cannons by what is known as the 'pendulum position'. 1957 Pendulum saw [see *goose saw* s.v. *goose sb.* 8]. 1958 *N.Z. Timber Jrnl.* Jan. 46/1 *Pendulum saw*, a machine cross-cut saw that is drawn across the stationary wood in the process of cutting by swinging from the point of suspension like a pendulum. 1926 *Amer. Speech* I. 632/2 *Pendulum swing*, applied to a type of putting stroke. 1947 C. Day Lewis *Colloq. Element Eng. Poetry* 9 The verse of the Romance poets, of the early Elizabethan lyricists,..and of the Pre-Raphaelite poets represents a series of pendulum-swings towards the formal, esoteric ideal of poetic diction. 1968 J. Winearls *Mod. Dance* (ed. 2) ii. 57 The principle of the outside fall and pick up of a Pendulum Swing can be used in isolated leg and trunk movements.

pe·ndulum, *v.* [f. the sb.] *intr.* To hang or swing like a pendulum. Also *fig.*
1885 W. F. Crafts *Sabbath for Man* vi. 458 The Sabbath of our fathers..was far better than the extreme laxity to which we have pendulumed. 1949 A. Roch *Climbs of my Youth* xiv. 115 We had to drive a piton into a slab and then pendulum across over the ice of the couloir. 1969 J. Elliot *Duel* i. iv. 79 A stop-watch on a long black string pendulummed from her neck. 1971 C. Bonington *Annapurna South Face* xii. 145 My progress dropped to a single push of my jumar clamps at a time; my feet slipped on the snow steps, and I pendulumed clumsily across the arête into the gully. 1973 — *Next Horizon* x. 146 But what if..you miss the other side and go penduluming back against the sheer ice wall, to be left hanging in the void? 1974 H. MacInnes *Climb to Lost World* xii. 212 'I'll belay you from here,' said Don. 'Then if the stone comes away, at least you'll pendulum over to this side and shouldn't hurt yourself too much.'

pene-. Delete penecontemporaneous and peneplain, qq.v. as main entries below. Add: **penesei·smic** *a.* [ad. F. *pénéséismique* (De Montessus de Ballore *La Geogr. Séismologique* (ed. 2, 1906) 11)] (see quot.).
1921 C. Davison *Man. Seismol.* x. 161 Peneseismic countries, in which earthquakes are severe, but fall short of destructive power.

pe:necontempora·neous, *a. Geol.* [f. Pene- + Contemporaneous *a.*] (See quot. 1972.)
1901 [in Dict. s.v. Pene-]. 1939 W. H. Twenhofel *Princ. Sedimentation* x. 375 The present trend of opinion considers that much chert and flint were deposited penecontemporaneously with the enclosing rocks. The views of origin may be placed in three classes of contemporaneous (syngenetic), penecontemporaneous, and subsequent (epigenetic). 1963 *Jrnl. Sedimentary Petrology* XXXIII. 64 Geologic data allow an assignment of possible penecontemporaneous (early diagenetic) origin to the dolomites of this group. 1969 *Nature* 22 Nov. 821/2 There is evidence of post-depositional weathering penecontemporaneous with deposition. 1972 *Gloss. Geol.* (Amer. Geol. Inst.) 525/2 *Penecontemporaneous*, said of a geologic process, or resultant structure or mineral, occurring immediately after deposition but before consolidation of the enclosing rock.
Hence **pe:necontemporane·ity,** the fact of being penecontemporaneous; **pe:necontempora·neously** *adv.*
1933 P. G. H. Boswell *On Mineral. Sedimentary Rocks* i. 1 The mode of formation of matrices and minerals which have developed penecontemporaneously or subsequently in the rocks. 1958 *Q. Jrnl. Geol. Soc.* CXIV. 4 The speaker had been unable to detect any sign of overprinting between these two fold systems..; this was interpreted as proof of penecontemporaneity of the folding. 1964 *Sedimentology* III. 136 The observations of uniform directions in certain penecontemporaneously slumped beds. 1979 *Nature* 7 June 485/1 The Entrance and the Exit Quarries, where sedimentation occurred penecontemporaneously in two separate depositories.

Penelope. Add: **3.** *attrib.* **Penelope canvas,** a double-thread canvas used for needle tapestry work. Also *absol.*
1882 Caulfeild & Saward *Dict. Needlework* 387/2 *Penelope canvas*, a description of cotton canvas made for Berlin woolwork, in which the strands run in couples vertically and horizontally, thus forming squares containing four threads each. 1895 *Montgomery Ward Catal.* 124/3 Java Canvas... Penelope Canvas. 1926–7 *Army & Navy Stores Catal.* 663/3 Canvas for wool work and embroidery... White 'Penelope'. 1975 *Islander* (Victoria, B.C.) 9 Mar. 7/2 The other type of needlepoint canvas is known as Penelope which is made up of two pairs of vertical and horizontal threads.

penelopize, *v.* Delete (*nonce-wd.*) and add earlier and later examples.
1841 *Congress. Globe* 27th Congress 1 Sess. App. 43/2 Diplomacy was still drawing out its lengthened thread, and still weaving its long and dilatory web—still Penelopizing. 1956 'H. MacDiarmid' *Stony Limits* 39 Nor to their new-tongued can we penelopise.

peneplain (pī·niplēin), *sb. Geomorphol.* Also **-plane.** [f. Pene- + Plain *sb.*[1]] A low, nearly featureless tract of land of undulating relief, esp. one held to be the product of long-continued subaerial erosion of land undisturbed by crustal movement and to represent the penultimate stage in the cycle of erosion in a humid climate; also, a former surface of this kind as it exists today (e.g. after uplift and dissection, or buried as an unconformity).
1889 W. M. Davis in *Amer. Jrnl. Sci.* XXXVII. 430 Given time enough, and the faulted ridges of Connecticut must be reduced to a low base-level plain. I believe that time enough has already been allowed, and that the strong Jurassic topography was really worn out somewhere in Cretaceous time, when all this part of the country was reduced to a nearly featureless plain, a 'peneplain', as I would call it, at a low level. 1893 *Bull. Dept. Geol. Univ. Calif.* I. iv. 158 The tilting of the Sierra Nevada peneplane was also a post-Pliocene event. 1933 *Geogr. Jrnl.* LXXXI. 331 These 'residual mountains' rise from a peneplain, that is, a plain of erosion produced during a stable period interrupting an upward movement. 1934 C. R. Longwell et al. *Outl. Physical Geol.* xv. 309 The highest ridges of the present Appalachians are remnants of a former peneplane. 1946 L. D. Stamp *Britain's Struct.* xiii. 150 The magnificent modern sea-cliffs are where the

peneplanes of the past meet the seas of to-day. 1954 W. D. Thornbury *Princ. Geomorphol.* viii. 178 Although the peneplain still remains an important concept with most geomorphologists, it is now recognised that many topographic surfaces have been erroneously called peneplains. 1960 B. W. Sparks *Geomorphol.* xv. 335 Good examples of peneplains are extremely rare and some would say that they do not exist. 1968 J. Arnold *Shell Bk. Country Crafts* xxi. 255 The greater part of the many miles of walls [in Pembrokeshire] is composed of small boulders from the peneplain, which is mainly Cambrian. 1970 R. J. Small *Study of Landforms* v. 164 Most peneplains in the British Isles take the form of 'hill-top surfaces', and are so fragmented that they are by no means easy to identify. 1977 A. Hallam *Planet Earth* 301 He assumed a standard life-cycle for a river valley, marked by youth (steep-sided valleys), maturity (flood-plain floors), and old age as the river valley was worn lower and lower into a 'peneplain'. *fig.* 1964 *Listener* 27 Feb. 353/2 In the eyes of a stray bitch Ribbed with hunger, heavy with young, I saw the peneplain of all imagined Misery.

peneplain (pī·niplēin), *v. Geomorphol.* Also **-plane.** [f. prec. (after Plane *v.*[1]), or a back-formation from next.] *trans.* To erode to a peneplain.
1923 [see *kratogen]. 1931 N. M. Fenneman *Physiogr. Western U.S.* iv. 172 The original folds of all these mountains were approximately peneplaned. 1969 Bennison & Wright *Geol. Hist. Brit. Isles* viii. 181 The Old Red Sandstone of south-western Ireland rests on an irregular erosion surface which had not been peneplaned so that the beds are of variable thickness. 1970 R. J. Small *Study of Landforms* v. 164 The English Chalk country..may well have been effectively peneplaned during the Pliocene period.

peneplained (pī·niplēind), *a. Geomorphol.* Also **-planed.** [f. *peneplain *sb.* + -ed[2].] Made into a peneplain.
1904 Chamberlin & Salisbury *Geol.* (1905) I. iii. 85 (*caption*) A peneplaned surface where the elevations are small but steep-sided. 1922 *Bull. U.S. Geol. Survey* No. 730. 2 The older peneplaned surface was elevated, tilted, and dissected. 1963 D. W. & E. E. Humphries tr. *Termier's Erosion & Sedimentation* ii. 34 (*caption*) Both figures show the Natal monocline, where most of the peneplaned surfaces are dated by marine beds which they cut across near the coast. 1965 G. J. Williams *Econ. Geol. N.Z.* i. 2/2 The peneplained basement was deeply weathered under podozolising influences. 1978 *Nature* 13 July 131/2 Middle Jurassic sandstones were here laid down directly on the peneplaned Caledonian basement.

peneplanation (pī:niplănēi·ʃən). *Geomorphol.* [f. Pene- + *planation.] Erosion to a peneplain.
1899 W. M. Davis in *Amer. Geologist* XXIII. 210 The unevenness of the uplands of to-day is a natural result of imperfect peneplanation followed by submature dissection. 1912 *Proc. Amer. Philos. Soc.* LI. 513 Peneplanation of the bituminous region had become far advanced. 1936 *Geogr. Jrnl.* LXXXVII. 22 A complicated history of peneplanation, uparching, faulting, and erosion. 1970 R. J. Small *Study of Landforms* v. 163 In an area such as the British Isles, perfectly preserved surfaces of peneplanation do not exist at or near present sea-level.

penetralium (pen/trēi·liəm). [erron. back-formation from Penetralia *sb. pl.*] The interior of a building. Also *fig.*
1817 Keats *Lett.* (1958) I. 194 Coleridge..would let go by a fine isolated verisimilitude caught from the Penetralium of mystery, from being incapable of remaining content with half knowledge. 1847 E. Brontë *Wuthering Heights* I. i. 5, I had no desire to aggravate his impatience, previous to inspecting the penetralium.

penetrameter (penĭtræ·mɪtəɹ). *Radiography.* Formerly also **penetrometer.** [f. Penetra(tion + -meter.] An instrument for determining the wavelength, intensity, or total received dose of X-rays by measuring photographically their transmission through layers of metal of known thickness.
1907 M. K. Kassabian *Röntgen Rays & Electro-Therapeutics* III. iv. 431 (*heading*) Skiameters and penetrometers. 1912 J. M. Martin *Pract. Electro-Therapeutics* xiv. 242 Holzknecht's chromoradiometer, Walter's or Benoist's skiameter or penetrameter will be of service in determining the amount of radiance given off from the tube. 1923 Glazebrook *Dict. Appl. Physics* IV. 607/2 Among medical men Benoist's radiochromometer or penetrometer enjoys extensive use as a measurer of hardness. 1950 *Engineering* 7 Apr. 373/2 A penetrameter in steps of 0·010 in., with holes in each step, to calibrate the X-ray film. 1966 *McGraw-Hill Encycl. Sci. & Technol.* XI. 304b/2 When the smallest hole in this penetrameter can be seen in the radiograph, the penetrameter sensitivity is 2%. 1975 *Physics Bull.* Feb. 80/2 (Advt.), This issue also reports on ..theoretical aspects of measuring kilovoltage by penetrameter.

penetrance. Restrict †*Obs. rare*[−1] to sense in Dict. and add: **2.** *Genetics.* [ad. G. *penetranz* (O. Vogt 1926, in *Zeitschr. f. ges. Neurol. und Psychiatrie* CI. 809).] The extent to which a particular gene or set of genes is represented in the phenotypes of individuals possessing it, measured by the proportion of carriers having the phenotype characteristic of the gene.
1934 [see *expressivity b]. 1946 R. R. Gates *Human Genetics* I. ii. 15 Lack of penetrance..has become a very important principle in human genetics... The gene is present in the germplasm, as shown by its transmission to the next generation, but for some reason it has completely failed to express itself in the soma. 1965 *Punch* 10 Nov. 689/2 The second complication is known as penetrance. When a gene affects every individual who carries it usually it is said to have 100 per cent penetrance. 1973 *Nature* 2 Mar. 64/2 An hereditary disease controlled by an autosomal Mendelian recessive gene with full penetrance. 1973 B. J. Williams *Evolution & Human Origins* ii. 30/1 Environmental differences may cause one carrier of a gene to pass this threshold and show the trait whereas another carrier of the same gene does not. This is referred to as incomplete penetrance.

penetrant, *a.* (*sb.*) Add: **A.** *adj.* **3.** *Genetics.* Producing in the phenotype the characteristic effect of the gene or combination of genes.
1955 R. B. Goldschmidt *Theoret. Genetics* III. v. 379 An incompletely penetrant effect is often based upon a system of multiple factors. 1973 B. J. Williams *Evolution & Human Origins* ii. 30/1 Genes are spoken of as fully penetrant or incompletely penetrant, but this is, in fact, as much a property of the environment as of the gene. 1978 *Nature* 27 Apr. 755/1 Gajdusek originally suggested that transmission was due to an autosomal dominant gene, fully penetrant in heterozygous females, but rarely penetrant in males unless they were homozygous.
B. *sb.* **2.** A penetrating coloured or fluorescent liquid used in a technique for detecting surface defects, in which the liquid is applied to the surface, excess removed, and developer applied to bring out the liquid left in the cracks and pores. Freq. *attrib.*, as *penetrant inspection, testing*.
1951 *Materials & Methods* Feb. 92/2 Fluorescent penetrant inspection makes use of a water-washable penetrant of high fluorescence and unusual wetting or penetrating properties. 1958 H. Etherington *Nucl. Engin. Handbk.* x. 152 Penetrant-inspection methods are applicable to all metals as well as glazed ceramics, plastics, and other non-porous materials. *Ibid.*, The penetrant is drawn into surface discontinuities by capillary action. 1959 J. F. Hinsley *Non-Destructive Testing* xiv. 323 (*caption*) Crack in casting flange revealed by fluorescent penetrant. 1973 A. Parrish *Mech. Engineer's Ref. Bk.* viii. 13 Penetrant testing is the modern version of the old oil and whiting technique and is applicable to all metals and many non-metallic materials. 1975 Bram & Downs *Manuf. Technol.* ii. 67 Inspection with penetrant is probably the oldest of the major non-destructive testing methods in use today. *Ibid.*, The surface of the component is examined in order to locate the penetrant indications which have been formed in the developer coating. 1977 *Hot Car* Oct. 58/2 One wheel per batch is pressure tested and a percentage of the batch is checked by a penetrant dye for porosity.

penetrate, *v.* Add: **1. c.** To insert the penis into the vagina of (a woman). Also *absol.*
1953 H. M. Parshley tr. *S. de Beauvoir's Second Sex* iv. iii. 377 Woman, once penetrated, has no such sense of danger; but in return she feels trespassed upon in her flesh. 1960 J. Rodney *Handbk. Sex Knowledge* iv. 55 Deep penetration must not be aimed at for a number of sessions, and when the woman is ready for it she will move her body in such a way that the penis will penetrate deeply. 1963 M. McCarthy *Group* i. 24 Kay had had an awful time with Harald; five times, she insisted, before she was penetrated. *Ibid.* vi. 123 My mother..told me that a gentleman never penetrated his bride on the first night. 1975 *Times Lit. Suppl.* 21 Mar. 293/1 A character..endeavours in the course of one week to penetrate the female offspring of the entire Cabinet.
d. To infiltrate (an organization, esp. an enemy espionage network) as a spy.
1962 L. Deighton *Ipcress File* 221 He organized a train-wrecking group until it was penetrated and the survivors fled. 1967 L. James *Chameleon File* (1968) xvi. 201 We have penetrated the Cuban g-2, Mr. Wilson. We shall know if you do not do as we have asked. 1972 H. MacInnes *Message from Malaga* xix. 261 Are you forgetting my job? It is to penetrate the Lucas set up. 1977 C. McCarry *Secret Lovers* xix. 270 Did you arrange for him to double her, give him the illusion that he was penetrating your network?
4. b. Also, to be understood or fully realized.
1955 E. Coxhead *Figure in Mist* iv. 134 'You're the most frightful woman I've ever met.' 'All right, it's penetrated.' 1973 G. Moffat *Deviant Death* vii. 109 We didn't notice that the gates were open..it wasn't till we came back hours later that it really penetrated.

penetrating, *ppl. a.* Add: **1. b.** Passing readily through matter.
1902 *Nature* 31 July 318/1 Villard..first drew attention to the existence of some very penetrating rays from radium non-deviable by a magnetic field. 1928 [see *cosmic a. 3c]. 1938 R. W. Lawson tr. *Hevesy & Paneth's Man. Radioactivity* (ed. 2) xxv. 280 A penetrating particle had traversed both counters and the Wilson chamber. 1947 *Radiology* XLIX. 358/2 The malignant response elicited by penetrating radiations, irrespective of type, consisted of hemopoietic tissue tumors. 1968 M. S. Livingston *Particle Physics* iv. 74 Muons are the 'penetrating' component of the ionizing particles in cosmic radiation observed beneath great layers of earth in salt mines.

penetration. Add: **1. c.** The insertion of the penis into the vagina in copulation.
1613 J. Chamberlain *Let.* 9 Sept. (1939) I. 475 What would you say yf you shold heare a churchman in open audience demaund of him..whether he had affection, erection, application, penetration, ejaculation. 1729 G. Jacob *New Law-Dict.* s.v. *Rape of women*, There must be Penetration and Emission, to make this Crime. 1848 Wharton *Law Lexicon* 569/2 In Scotland..the following facts are necessary to be proved on a charge of rape: 1, penetration..; 2, actual force in the consummation. 1957 L. Durrell *Justine* III. 185 The whole portentous scrimmage of sex itself, the act of penetration. 1960 [see *penetrate v. 1c] 1963 A. Heron *Towards Quaker View of Sex* 65 'Frigidity'..implies more than failure to have orgasm: it is the inability to enjoy love-making and penetration.
d. The infiltration of a country, organization, etc., by political, financial, etc., means in order to gain influence, power, or information. Also used as a marketing term.
1931 F. L. Allen *Only Yesterday* vii. 176 In general the country extended its empire not by military conquest or political dictation, but by financial penetration. 1951 E. E. Cummings *Let.* 3 Aug. (1969) 214 You should glimpse a huge map..to show the Soviet penetration of Greece—all but a little strip in the middle. 1964 L. Deighton *Funeral in Berlin* 257 The detection and penetration of intelligence. 1971 I. Deutscher *Marxism in our Time* 155 The fall in the ideological level of most of the militants..had facilitated to a certain extent the penetration of police agents into the Party. 1976 *National Observer* (U.S.) 21 Feb. 1/2 The relative decline in circulation of daily newspapers and their absolute decline in 'household penetration', has been accompanied by a growth of other printed media that duplicate one of the local daily's functions, but often better.
4. penetration agent, a spy sent to penetrate an enemy organization; **penetration aid,** an object released from a missile as a decoy to draw off any attacking missiles; **penetration twin** *Cryst.*, a twin crystal that presents the appearance of two interpenetrating crystals; so *penetration twinning*.
1966 'A. Hall' *9th Directive* ii. 22 This is a police job. I'm a penetration agent..they've made a mistake. 1976 P. Henissart *Winter Quarry* III. xxvi. 270 As soon as I learned of McGuire's existence, I mistrusted him—a shiftless contract-agent for the CIA... I know your taste for penetration agents. 1966 *N.Y. Times Mag.* 20 Mar. 59 The decoys—'penetration aids' in the missilemen's jargon—would range from dummy warheads to metalized-plastic warhead shapes. 1967 *Electronics* 6 Mar. 50/2 It is researching the key problem: how to distinguish the lethal reentry vehicle from the penetration aids. 1868 J. D. Dana *Syst. Min.* (ed. 5) 191 Penetration-twins, the forms not corresponding to a regular revolution, but to an irregular interpenetration of unlike parts of the crystal. 1971 C. S. Hurlbut *Dana's Man. Min.* (ed. 18) ii. 97 Twin crystals are usually designated as either contact twins or penetration twins... Penetration twins are made up of interpenetrating individuals having an irregular composition surface, and the twin law is usually defined by a twin axis. 1953 F. H. Pough *Field Guide Rocks & Minerals* I. iv. 53 We also find attractive penetration twinning in phenakite.

penetrometer (penĭtrǫ·mĭtəɹ). [f. Penetr(ation + -ometer.] **1.** An instrument for determining the consistency or hardness of a substance (as asphalt, soil or snow) by measuring the depth or rate of penetration of a rod or needle driven by a known force.
1905 C. Richardson *Mod. Asphalt Pavement* ix. xxvi. 533 (*heading*) New York testing laboratory penetrometer. 1913 Blanchard & Drowne *Text-bk. Highway Engin.* x. 293 Penetrometer to be used in accordance with standard method on materials solid at above temperatures. 1930 *Engineering* 18 July 61/3 The penetrometer is simply a ½-in. square bar fitted loosely in guides and loaded with weights giving steps of about ¼ ton per square foot. 1966 C. Fraser *Avalanche Enigma* iv. 71 The first measurement taken is a penetrometer profile. 1971 B. Buck tr. *Ludewig's Polyester Fibres* iv. 124 In order to determine the melting point, a grain of the polyester is placed into the container of the penetrometer and the needle is centred on the surface of [*printed* on] the substance under test. 1975 D. Bagley *Snow Tiger* iv. 49 This is a penetrometer... It measures the resistance of the snow.
2. Obs. var. *penetrameter.

penfieldite (pe·nfīldəit). *Min.* [f. the name of Samuel L. *Penfield* (1856–1906), U.S. mineralogist + -ite[1].] A basic lead chloride, Pb_2Cl_3OH, occurring as very small, usu. prismatic crystals that are colourless and transparent when pure.
1892 F. A. Genth in *Amer. Jrnl. Sci.* CXLIV. 260 While examining a lot of minerals, formed by the action of sea water on ancient slags which Mr. Geo. L. English collected at Laurion, Greece, I noticed a *very few* hexagonal crystals which proved to be a new species, for which I propose the name: Penfieldite, in honor to Prof. Sam'l L. Penfield the indefatigable worker in mineralogy and crystallography. 1954 *Mineral Abstr.* XII. 453 The stocks..of two Roman anchors recovered from the Mediterranean are covered with a crust of small crystals of anglesite, phosgenite, penfieldite, with less cerussite, hydrocerussite, and specks of metallic copper and copper stains. 1969 *Ibid.* XX. 144/2 Penfieldite,..from Laurium, Greece, and from Sierra Gorda, Chile, has *a* 11·28, *c* 48·65 Å.

pen-friend (pe·nfrend). [f. Pen *sb.*[2] + Friend *sb.*] A friend or contact with whom a regular correspondence is conducted. (See also *pen-pal.) Hence **pen-friendship,** the relationship existing between pen-friends.
1933 *Boy's Mag.* XLVII. 106/1 Any reader at home or abroad who would like to correspond with a 'pen-friend'

in another land is invited to send his name, address, and age to the Editor. **1943** F. THOMPSON *Candleford Green* iv. 65 The correspondence languished, then ceased, in the usual manner of such pen-friendships. **1945** 'O. MALET' *My Bird Sings* I. ii. 18 She had returned to England..but none the less they had kept up one of those rambling pen-friendships. **1957** R. MASON *World of Suzie Wong* I. ii. 15 It had caused my predecessor, a Pole, to propose through the post to a pen-friend in Glasgow whom he had never met. **1960** *Woman's Own* 19 Mar. 77/4, I cannot help you to find pen-friends, but..I will send you the name of a pen-friendship organisation. **1974** *Times* 2 Jan. 24/4 (Advt.), Elderly person desiring a pen friend please reply to Box 2901 B, The Times. **1975** *Bangladesh Times* 27 July 7/7, I am a boy of 18 and studying in the 2nd year Commerce. I am very interested in establishing Pen-friendship with boys and girls from all over the world, including Bangladesh. **1977** *Navy News* June 26/4 Readers seeking penfriends in the Royal Navy are listed here.

‖ **penghulu** (pĕŋghū·lu). Also **panghulu, pengulu.** [Malay.] In Malaysia, a head-man or chief.

1821 J. LEYDEN tr. *Malay Annals* 49 Under the bandahara immediately was the panghulu bandahari. **1894** N. B. DENNYS *Descr. Dict. Brit. Malaya* 285 The Penghulus are elected by the neighbours and confirmed by the Government. **1900** W. W. SKEAT *Malay Magic* v. 232 After my arrival with the *Pĕnghulu* the ceremony began. **1906** —— & BLAGDEN *Pagan Races Malay Peninsula* I. 500 When a village migrates, the Penghulu conducts the migration. **1927** H. M. TOMLINSON *Gallions Reach* xxvii. 219 Norrie addressed himself to the penghulu; the chief answered him with gentle explicitness. **1928** L. R. WHEELER *Mod. Malay* v. i. 194 The peasant class has various grades, from the 'penghulu' type, accustomed to exercise authority on a limited scale in his neighbourhood, to the backward natives. **1971** *Lady* 15 July 88/3 The Pengulu, the Headman, shook us by the hand and welcomed us to his domain. **1972** A. AMIN tr. *Ahmad's No Harvest but Thorn* viii. 85 Jeha remembered how Lahuma had asked her to go to the *Penghulu*'s house. **1977** P. THEROUX *Consul's File* 131 The *penghulu*, the headman, pointed out that..the rice was planted.

penglima, var. *PANGLIMA.

pengo (pe·ŋgō). Pl. **pengo, -oes.** [Hungarian *pengő*, lit. 'ringing'.] The basic monetary unit of Hungary from 1927 to 1946.

1926 *Glasgow Herald* 4 Jan. 12/2 To-day the new Hungarian currency, the pengo, is quoted for the first time. **1927** *Times* 28 Feb. 11/7 The Hungarian government has assigned a sum of 322,820 pengoes..for..a new Hungarian Legation building in London. **1930** *Observer* 23 Feb. 11/2 He earns twenty pengo a week. **1932** *Daily Express* 2 July 12/1 The amounts due to bondholders will be deposited in pengoes in the Hungarian National Bank. **1947** *Whitaker's Almanack* 917/1 The *Pengo* (of 100 *Filler*) was superseded in August, 1946, by a new currency, the *gulder*.

penguin. Add: **2. b.** A machine like an aeroplane but incapable of flight, used in the early stages of an airman's training. Also, a non-flying member of an air force. *Air Force slang.*

1915 G. BACON *All about Flying* vi. 104 A 'penguin'—a machine with engine not powerful enough to raise it from the ground. **1917** J. R. MCCONNELL *Flying for France* 143 The student is put on..a low-powered machine with very small wings... It could not leave the ground. The apparatus is jokingly and universally known as a Penguin. **1918** *Everybody's Mag.* Jan. 113/2 An officer of flying status, but who for some reason does not fly, is called a 'penguin'. **1918** *Sphere* 4 May 76/2 The three official corps familiarly known as the Waacs, the Wrens, and the Penguins. **1919** *Athenæum* 11 July 582/2 Members of the W.R.A.F. were called 'Penguins' because they were 'flappers' who did not fly. **1925** FRASER & GIBBONS *Soldier & Sailor Words* 221 Penguin was also a name for a type of low-powered aeroplane with small planes or wings, used for instructional purposes. **1942** *Gen* 1 Sept. 14/2 No flier spares his contempt for the 'penguins', the nonflying administrative officers in the RAF. **1944** G. GIBSON *Enemy Coast Ahead* (1946) vi. 96 In the average Bomber Officers' Mess,..while penguins sing loudly in the mornings as they get up to shave, it was rather hard for the boys who had been up all night to get a good day's rest. **1950** PARTRIDGE *Here, There & Everywhere* 53 When an airman refers to himself as a *penguin*, he is resorting to the specialized slang of the Air Force; all he means is that he is a member of the ground staff and therefore does not fly.

c. (With capital initial.) The proprietary name of Longman Penguin Limited, formerly Penguin Books Limited (1936–1966) and The Penguin Publishing Company Limited (1966–1972), used *attrib.* and *absol.* to designate paper-backed books or series of books published by this company. Also (*rare*) as *v. trans.*, to publish as a Penguin book.

1935 *Times Lit. Suppl.* 1 Aug. 491/1 We shall look forward to more Penguin Books, and we wish the experiment—a bold one—all success. **1938** 'G. ORWELL' *Homage to Catalonia* x. 177, I..spent hours reading a succession of Penguin Library books. **1939** *Trade Marks Jrnl.* 13 Dec. 1630/1 Penguin... Printed publications, stationery and bookbinding, but not including publications on birds, or shaped sheets of paper for display purposes. Penguin Books Limited,..West Drayton, Middlesex; manufacturers and publishers. **1940** GRAVES & HODGE *Long Week-End* xxv. 426 Penguins were first published in 1936. **1948** J. BETJEMAN *Sel. Poems* 84 Whether we like to sit with Penguin books In sheltered alcoves farther up the cliff. **1950** D. E. STEVENSON *Music in Hills* vi. 50 Miss Douglas..appeared from behind a large bookcase with two Penguins in her hand. **1950** W. STEVENS *Let.* 15 Aug. (1967) 687 At the moment I am reading a Penguin Classic. **1951** R. MACAULAY *Lett. to Friend* (1961) 209 *Mary Lavelle* is Penguin'd. **1956** J. SYMONS *Paper Chase* x. 69 Hedda Pont was at a table reading a Penguin thriller. **1959** J. BRAINE *Vodi* i. 11 You can have those Penguins in the bottom of the clothes locker. **1962** I. MURDOCH *Unofficial Rose* v. 51 There were a few Penguin novels, but they looked dull English tea-party stuff. **1974** *Times Lit. Suppl.* 11 Dec. 22/4 With the publication of *Barnaby Rudge* all but two of Dickens's novels..are now Penguined. **1976** *New Yorker* 15 Nov. 187/1 Michael Hamburger, introducing the Penguin anthology of Enzensberger in translation, writes of the poet's 'moral purpose at variance with his personal needs and perceptions'.

d. A man wearing evening dress (cf. *penguin suit* (*a*) in sense 3 below). *rare.*

1967 *Melody Maker* 1 Apr. 9 Good Music had the sort of melody and clipping beat that even Victor Sylvester didn't have to alter so that the Brylcreemed penguins and their sequined partners could jig about in the ballrooms. **1976** B. BOVA *Multiple Man* (1977) v. 56 These stuffed penguins and their bejeweled ladies.

3. penguin suit, (*a*) evening dress; (*b*) a type of suit worn by astronauts.

1967 PARTRIDGE *Dict. Slang* Suppl. 1289/1 *Penguin suit*, a dinner jacket. **1968** R. JEFFRIES *Traitor's Crime* iv. 46 Some smooth bastard in a penguin suit. **1971** *N.Y. Times* 10 June 18 The astronauts donned the tight-fitting overalls, known as a penguin suit, in which tension is produced by several layers of rubberized material. **1971** *Daily Tel.* 1 July 30/5 During the Soyuz 9 and 11 flights, Russian cosmonauts..wore special suits, called 'penguin' suits. **1979** K. M. PEYTON *Marion's Angels* vi. 101 Geoff'd better go home for his penguin suit. I'll go up and get my tails.

penguinery. Add: Also **penguinry.** (Later examples.)

1921 H. G. PONTING *Gt. White South* 55 The Adélie penguinry was but a mile or two away. **1979** *Nature* 11 Jan. 88/2 The Adelie and chinstrap penguins feed mainly on krill, but they take different sized prey, either by active selection or by the Adelie feeding further from the rookery (or penguinery).

pengulu, var. *PENGHULU.

pen-gun. Add: **1.** (Earlier example.)

1807 T. CAMPBELL *Let.* 3 Oct. in W. Beattie *Life & Lett. T. Campbell* (1849) II. v. 121 We crack'd, as the Scotch say, like pen-guns.

2. A small cylindrical gas bomb (see quot. 1965[2]).

Registered in the U.S. as a proprietary name.

1962 *N.Y. Times* 26 May 27/8 Miss Dunsmore told him that the pen-gun had dropped from her pocket-book and a pellet had ruptured, releasing the tear gas. **1965** *Official Gaz.* (U.S. Patent Office) 4 May TM42/1 Penguin Associates, Inc., Malvern, Pa... *Pengun*. For Tear Gas Projectors and Distress and Highway Flare Projectors. First use January 1961. **1965** *Consumer Bull.* Sept. 18/2 The *Pengun* itself is a stainless steel cylinder encased in a gold-anodized aluminum tube about 4 inches long with an interior diameter of ⅜ inch. The 'cartridge' of gas fits into the tube and is released by a firing pin or plunger. **1973** J. DI MONA *Last Man at Arlington* viii. 73 Before he could say another word, she got him with a pen gun filled with mace. 'Ah Christ,' he yelled, bending over, rubbing his eyes. **1978** P. NIESEWAND *Underground Connection* 24 A pen gun which fired bursts of paralysing gas.

penholder. Add: **2.** Used attrib. to denote a grip in table-tennis in which the bat is held between thumb and forefinger.

1935 M. A. SYMONS *Table Tennis* ii. 14 There are still a large number of players who favour the 'Pen-holder' style of grip. *Ibid.*, Even top-class 'Pen-holder' grip players foozle such a shot. **1959** *Sunday Times* 5 Apr. 36/8 An aggressive opponent who smashed strongly with a pen-holder grip. **1973** *Advocate-News* (Barbados) 11 Dec. 13/3 The Chinese main armoury proved to be not their unorthodox 'pen-holder' style..but their service.

-penia (pi·niă), repr. Gr. πενία poverty, need, is used in *Med.* to denote a deficiency, esp. of a constituent of the blood, as in *GRANULOCYTOPENIA, *pancytopenia* s.v. *PAN- 2. Also (*erron.*) **-pœnia.**

1971 *Lancet* 10 July 108/1 May I add *-penia* to the list of alternatives to *hypo-* suggested by Dr. Ell? In Greek it meant 'poverty' and it is currently used in such words as granulocytopenia and thrombocytopenia.

penicillamine (penisi·lămīn). *Chem.* and *Pharm.* [f. *PENICILL(IN + AMINE.] An amino-acid, $(CH_3)_2C(SH)CH(NH_2)COOH$, produced by the hydrolysis of penicillins and used pharmacologically as a chelating agent; 2-amino-3-methyl-3-mercaptobutanoic acid.

1943 E. P. ABRAHAM et al. in *Nature* 23 Jan. 107/1 The properties of this substance, which we propose to term penicillamine, show that it represents a novel type of naturally occurring base. *Ibid.*, Penicillamine is obtained by hydrolysing barium penicillin..at 100°C. for one hour by means of *N*/10 sulphuric acid. **1947** *Sci. News* IV. 70 The synthesis of penicillin G starts with a benzyl oxazolone and with penicillamine, and attempts to recombine them. **1969** [see *hypogeusia* s.v. *HYPO- II]. **1977** *Davidson's Princ. & Pract. Med.* (ed. 12) 733 This disorder [*sc.* Wilson's disease]..can now be arrested by giving copper-binding drugs (chelating agents)... The most valuable of these is penicillamine.

penicillanic (penisilæ·nik), *a. Chem.* [f. as prec. + *-an-* + -IC.] *penicillanic acid*: an acid, $C_8H_{11}NO_3S$, whose molecular structure is the nucleus of the various penicillins and consists of a β-lactam ring fused to a molecule of 5,5-dimethylthiazolidine-4-carboxylic acid.

1953 J. C. SHEEHAN et al. in *Jrnl. Amer. Chem. Soc.* LXXV. 3293 By a cyclization procedure we have synthesized a β-lactamthiazolidine (VI)... We have chosen to call this compound methyl phthalimidopenicillanate. [*Note*] As a convenience in naming VI and similar analogs of the penicillins we suggest the terms 'penam' and 'penicillanic acid' for the following ring system and substituted ring system. **1959** *Times* 27 Oct. 6/7 The new penicillin, which is the potassium salt of 6-(-α-phenoxypropionamido)penicillanic acid, has the same range of antibacterial activity as previous penicillins. **1969** *Adv. Appl. Microbiol.* XI. 26 (*heading*) Antistaphylococcal activity of 6-substituted penicillanic acid derivatives.

penicillin (penisi·lin). *Pharm.* [f. PENICIL-L(IUM + -IN[1].] **1.** Orig., the antibiotic agent obtained from cultures of the mould *Penicillium notatum*; hence, any of a group of antibiotics that are all derivatives of 6-amino-penicillanic acid in which a radical replaces one of the amino hydrogen atoms, some being acids produced naturally by the growth of various moulds of the genera *Penicillium* and *Aspergillus*, whilst others are acids, salts, or esters prepared synthetically from these; they are active against many kinds of bacteria but virtually harmless to persons not allergic to them.

1929 A. FLEMING in *Brit. Jrnl. Exper. Path.* X. 227 In the rest of this article allusion will constantly be made to experiments with filtrates of a broth culture of this mould, so for convenience and to avoid the repetition of the rather cumbersome phrase 'Mould broth filtrate', the name 'penicillin' will be used. This will denote the filtrate of a broth culture of the particular penicillium with which we are concerned. **1941** H. W. FLOREY et al. in *Lancet* 16 Apr. 188/2 Enough evidence has now been assembled to show that penicillin is a new and effective type of chemotherapeutic agent, and possesses some properties unknown in any antibacterial substance hitherto described. **1947** *Sci. News* IV. 69 There are a number of naturally occurring penicillins, with somewhat different medical effectiveness. **1951** A. GROLLMAN *Pharmacol. & Therapeutics* xxii. 445 The amorphous preparations of penicillin are yellow powders with a characteristic odor and bitter taste. The crystalline pure preparations are white, odorless and practically tasteless. **1953** J. RAMSBOTTOM *Mushrooms & Toadstools* xxiii. 287 Soon after the chemical study was begun it was unexpectedly found that there was more than one kind of penicillin. The first two to be recognised were called Penicillin I and Penicillin II in this country, Penicillin F and Penicillin G in America... Later Penicillin III (X) and Penicillin IV (K) were recognised as being produced in greater or lesser amounts depending on cultural conditions. *Ibid.*, There are now five known 'natural' penicillins. All are dipeptides with the formula $C_9H_{11}O_4N_2S\,R$—the difference between them being in the constitution of the side chain..represented by *R*. **1958** *Listener* 9 Oct. 552/1 For our modern accomplishments are genuine, from the early work we achieved in the realm of atomic science, to the discovery of penicillin and modern progress on jet-propelled aircraft. **1970** PASSMORE & ROBSON *Compan. Med. Stud.* II. xx. 12/2 A method has been devised to use these natural penicillins as a source of a new range of compounds which can be made by chemical synthesis.., the new or semi-synthetic penicillins. *Ibid.* 13/1 These semi-synthetic penicillins have a range of novel properties which make them a major addition to the armoury of antibacterial agents. **1973** M. AMIS *Rachel Papers* 93 Just two jabs of penicillin up the bum and much humiliation at the local clinic. **1974** M. C. GERALD *Pharmacol.* xxvii. 465 The penicillins were the first antibiotics discovered and remain today the second most widely used class of drugs for the treatment of bacterial infections. *Ibid.*, The most important natural penicillin is benzylpenicillin, which is more commonly designated penicillin G. Although penicillin G is the most potent of all penicillin derivatives, it suffers from several major disadvantages.

2. *penicillin A* or *B*: disused names for *NOTATIN (*PENATIN), which is chemically unrelated to the penicillins proper.

1941 C. E. COULTHARD et al. *Brit. Pat.* 552,619 A substance having bacteriostatic activity, which is hereinafter termed 'Penicillin A',..is produced by selecting a strain of Penicillium notatum which possesses the property..when grown on a suitable culture medium,..of maintaining the pH of the culture medium on the acid side for a considerable period. **1942** [see *NOTATIN]. **1943** E. C. ROBERTS et al. in *Jrnl. Biol. Chem.* CXLVII. 47 We wish to report an antibacterial substance, penicillin B, produced by *Penicillium notatum*, which is insoluble in lipid solvents but readily separated from the culture medium by adsorption on benzoic acid. *Ibid.*, We shall refer to our product as penicillin B to differentiate it from the product obtained by Abraham *et al.*..which we shall call penicillin A. Although penicillin B may be similar to 'penatin'.., the absence of a discussion of its chemical properties prevents a comparison of the two products. **1949, 1963** [see *NOTATIN].

3. *attrib.* and *Comb.*, as *penicillin-insensitive, -like, -resistant, -sensitive* adjs.; **penicillin unit,** a unit of penicillin which since 1944 has been the amount having the same antibiotic

activity as a certain quantity (very nearly, and orig. exactly, 0.6 microgramme) of a standard preparation of the sodium salt of benzylpenicillin (penicillin G), and which is approximately equivalent to the Oxford unit that it superseded.

1929 *Brit. Jrnl. Exper. Path.* X. 234 In some [cultures] there were a few diphtheroid bacilli which were always penicillin sensitive, and in others there were Gram-negative bacilli which were penicillin insensitive. **1953** J. RAMSBOTTOM *Mushrooms & Toadstools* xxiii. 284 *Penicillium notatum* grows best in nearly neutral media, which are equally favourable for penicillin-insensitive organisms..which produce the enzyme penicillinase. **1946** *Nature* 28 Sept. 446/1 Penicillin-like antibiotics are produced by a number of moulds besides *Penicillium notatum*. **1942** *Proc. Soc. Exper. Biol. & Med.* LI. 387 (*heading*) Development of penicillin resistant strains of *Staphylococcus aureus in vitro*. **1971** *Nature* 4 June 284/1 Penicillin-resistant strains of staphylococci emerged in the 1950s. **1929** Penicillin-sensitive [see *penicillin-insensitive* above]. **1959** *Times* 6 Mar. 13/7 The proportion of penicillin-sensitive individuals is small. **1943** *Jrnl. Bacteriol.* XLVI. 189 To eliminate the day-to-day deviation, the Oxford group introduced the concept of the penicillin unit and the use of a standard penicillin preparation. **1947** *Jrnl. Biol. Chem.* CLXVII. 554 In the Casamino acid medium, maximum yields were obtained upon addition of 200 penicillin units per ml. of medium initially. **1953** HEILBRON & BUNBURY *Dict. Org. Compounds* (rev. ed.) IV. 51/2 Penicillin-F... Antibiotic activity: 1,490 Penicillin units per mg.

penicillinase (penisi·linēiz, -s). *Biochem.* [f. prec. + *-ASE.] Any of the enzymes (produced by certain bacteria) which cause the breaking up of the carbon–nitrogen bond in the lactam ring of some penicillins (so rendering them ineffective as antibiotics). Cf. *LACTAMASE.

1940 ABRAHAM & CHAIN in *Nature* 28 Dec. 837/1 The activity of the enzyme, which we term penicillinase, is slight at pH5, but increases considerably towards the alkaline range of pH. **1958** *Ann. Rev. Biochem.* XXVII. 176 Cephalosporin-C possesses the interesting property of being resistant to the action of some penicillinases including that produced by penicillin resistant staphylococci. **1971** [see *LACTAMASE]. **1974** M. C. GERALD *Pharmacol.* iv. 72 Recent research has made available new penicillins that are resistant to acid hydrolysis and penicillinase attack. **1975** *Nature* 12 June 526/1 Penicillinases, or as they are now known, β-lactamases, are enzymes which specifically deactivate penicillin by hydrolysis of the β-lactam ring.

penicilloic (penisilōu·ik), *a. Biochem.* [f. *PENICILL(IN + *OIC.] *penicilloic acid*: any of the acids produced when a penicillin is hydrolysed (as by a penicillinase) and the C–N bond of the lactam ring broken.

1945 *Science* 21 Dec. 628/2 The dicarboxylic acid obtained by hydrolysis of penicillin at the site of the potential carboxyl is termed penicilloic acid. This acid is produced in the form of salts by treatment of penicillin with alkalies and is presumably the product of the action of the enzyme penicillinase on penicillin. **1949** MOZINGO & FOLKERS in H. T. Clarke et al. *Chem. of Penicillin* xviii. 542/1 Many of the penicilloic acids..were made in order to study their reactions with various reagents in the interest of their dehydration to the penicillins. **1959** *Observer* 11 Jan. 14/7 It [*sc.* penicillinase] acts by rapidly breaking down penicillin to penicilloic acid, which has no antibiotic activity. **1970** *New Eng. Jrnl. Med.* 16 July 119/1 Penicilloic acid appears to have an important role in some of the allergic reactions caused by penicillins.

Hence **penicillo·ate** [-ATE[1]], a salt or ester of a penicilloic acid.

1946 *Industr. & Engin. Chem.* (*Analytical Ed.*) Oct. 619/1, 8·97 equivalents of iodine per mole..is in reasonably good agreement with the range given..for penicilloates (8·5 to 8·9). **1969** MANHAS & BOSE *Synthesis of Penicillin, Cephalosporin C & Analogs* iii. 38 When this route..met with only limited success the cyclization of penicilloates that could not form oxazolones was investigated.

peninsula·tion. [f. PENINSULATE *v.*] The process of making into a peninsula; the condition of being peninsulated; peninsularity. (In quots. *fig.*)

1923 G. BARKER in Drinkwater & Orpen *Outl. Lit. & Art* (1924) I. x. 194/1 From this peninsulation of the stage several things follow. **1959** *Times* 5 Sept. 9/6 It is the prospect..the isolation or peninsulation.

penis. Delete ‖ and add: Also with pl. **peni** (*erron.*), **penises.** (Earlier and later examples.)

1676 T. BROWNE *Let.* 14 June in *Wks.* (1964) IV. 61 You may observe..the flattish heart, the Lungs,..the penis, the multiple stomack &c. **1926** F. Z. SNOOP *Reproduction & Sexual Evolution* 83 Havelock Ellis quotes other cases, even butterflies (if insects may be here included) who possess excrescences on their penes, which of necessity must cause pain, or something very like pain, during coition. **1929** D. H. LAWRENCE *Paintings*, What do you paint with, Maitre?—With my penis, and be damned! **1933** R. L. DICKINSON *Human Sex Anatomy* vi. 74/2 The long penises are in general narrow and the short broad. **1943** *Amer. Jrnl. Dis. Children* LXV. 541 The penis is primarily an organ of copulation, so that its true physiologic size would be gaged by measurement of the erect phallus. *Ibid.*, The length of the fully stretched penis is practically identical with the length of the erect phallus. **1962** E. L. COCKRUM *Introd. Mammalogy* iii. 76 According to Asdell (1946), Slijper has studied the structure of the penis in its relation to the duration of coitus and developed the following classification of morphological types of peni. **1965** E. W. JOHNSON *Love & Sex in Plain Lang.* ii. 8 A man's most obvious sexual organ is his penis. **1970** H. W. & L. R. LEVI tr. *Kaestner's Invertebr. Zool.* III. v. 130 The paired penes are introduced deep into the posterior opening between the valves, reach the copulatory openings, and mating is accomplished in several minutes. **1972** *Times Lit. Suppl.* 29 Sept. 1156/4 Sexologists now make a clear distinction between 'penis' and 'phallus'; the former being used to define the flaccid organ and the latter the erect organ. **1973** C. PINCHER *Sex in our Time* viii. 92 Just as there are men with large or small penises so there are women with large or small vaginas. **1975** M. SEYMOUR-SMITH *Sex & Society* x. 282 The phallus is the *image* of the *erect* (never flaccid) penis. **1975** FRETTER & GRAHAM *Functional Anat. Invertebr.* (1976) xi. 352 In these sessile forms [*sc.* barnacles] the sperm are filiform and motile and transferred from one individual to the next by a long flexible penis.

b. *penis-cover, -extender, holder, -sheath*; **penis-envy** *Psychoanalysis*, an envy of the male's possession of a penis, postulated by Freud to occur in girls and possibly resulting in a castration-complex (see *CASTRATION 1) or in the adoption of masculine behaviour; **penis gourd** (see quot.).

1966 *New Statesman* 22 Apr. 589/2 Even more remarkable are the penis-covers made of narrow gourds a foot or more in length. [**1920** B. LOW tr. *Freud's Psychogenesis of Case of Female Homosexuality* in *Internat. Jrnl. Psycho-Anal.* I. 146 She had..developed a pronounced envy of the penis.] **1924** *Internat. Jrnl. Psycho-Anal.* V. 58 Its relation to the 'penis-envy' complex is twofold. **1946** *Mind* LV. 354 A girl's penis-envy might also be very much hidden. **1964** A. BOROWIK *How Many Miles to Babylon?* xv. 96 When I hear a guy talking to me about penis envy..I feel an overpowering urge to laugh. **1972** F. WARNER *Lying Figures* III. 20 You'd find penis-envy in a peach. **1973** K. VONNEGUT *Breakfast of Champions* xv. 147 One time Dwayne Hoover got an advertisement through the mail for a penis-extender, made out of rubber. **1971** *World Archaeol.* III. 136 The traditional male dress is the penis gourd: a piece of the stem end of the gourd, varying in length from about three to more than twelve inches, into which the penis is inserted, leaving the scrotum exposed. **1976** *Times* 31 May 7/8 Went native in 1681 living with the Cuna Indians, coated in vegetable dies [*sic*] and clothed only in a conical silver penis holder suspended from the waist by a thong. **1925** C. K. MEEK *Northern Tribes Nigeria* II. 116 Among those people who wear the penis-sheath, the sheath is usually removed before the body is laid in the grave. **1978** *Sunday Times* (Colour Suppl.) 18 June 40/1 Their chief Manuel, denied access to Government chambers in Brasilia for not wearing a suit, has forbidden any Brazilian deputy to set foot on Xavante land unless wearing a penis sheath, hair feathers and body paint.

penitent, *a.* and *sb.* Add: **B.** *sb.* **4*.** *Geogr.* [See quot. 1954[1].] A spike or pinnacle of compact snow or ice which results from differential ablation of a snow or ice field exposed to the sun, occurring esp. in high mountain ranges and freq. in large groups containing specimens of similar size and orientation. Freq. *attrib.* or as *adj.*

[**1910** *Geogr. Jrnl.* XXXV. 125 Among the variety of views that have been advanced, observers have practically agreed that one factor essential to the production of penitentes,..is the unequal melting of *névé* under the application of heat in some form, principally that of the sun.] **1922** WRIGHT & PRIESTLEY *Glaciology* viii. 288 Plate CXCV shows an example of penitent-ice from the Ferrar glacier. **1936** G. SELIGMAN *Snow Struct.* vi. 131 It has been postulated..that the ablative effect in penitent snow has been intensified by the presence of solid matter to absorb the sun's heat. **1941** *Amer. Jrnl. Sci.* CCXXXIX. 382 'Penitent' ice-forms, modelled in some degree by evaporation processes and associated with the same structures, have been described from Antarctica. **1954** *Jrnl. Glaciology* II. 331 We venture to translate the words used by Chileans and Argentinians into English: *penitentes* (noun), *campo de penitentes* (field of penitents). *Nieve penitente* is not used, and *nieve de los penitentes* means 'snow from the penitents'. Both expressions have been introduced into international literature by..glaciologists who did not know Spanish very thoroughly. *Ibid.* 336 When the snow field lies directly upon the ground, the channels between the penitents often succeed in reaching the ground, and the penitents, detaching themselves from one another, assume the vague appearance of an Easter procession of white-cowled Spanish penitents. **1954** W. NOYCE *South Col* v. 83 The ice..had ribbed and wrinkled into bigger honeycomb, more like the ice pinnacles called 'penitents'. **1959** R. E. HUSCHKE *Gloss. Meteorol.* 416 Penitent ice is most developed on low-latitude mountains, especially the Chilean Andes, but has been found even in polar regions. **1972** *Cambridge Mountaineering* 38 An additional reason for travelling to Afghanistan had been to study certain snow formations, called penitents. *Ibid.* 39 Our 'penitents'..were spread all over the place both on the snowfields and sometimes also on the rock surfaces... Their only use turned out to be on steep snow slopes where they provided useful handholds—provided one didn't put too much trust in them.

5. penitent-form (earlier examples).

1865 *Wesleyan-Methodist Mag.* Nov. 484 She was the first to come to the penitent form. **1881** *Doctrines & Discipline Salvation Army* §28 Bring them out to the penitent form before the people, and so test them further, and pledge them publicly.

penk (peŋk), *v. rare.* [? var. PANK *v.*] *intr.* To palpitate; to throb or heave violently or rapidly.

1890 KIPLING *Barrack-Room Ballads* (1892) 27 Wot makes the soldier's 'eart to penk, wot makes 'im to perspire? **1898** —— *Stalky* in *Land & Sea Tales* (1923) 137 They bullocks drove like that—all heavin' an' penkin' an' hotted!

pe·nlight, pen-light. [f. PEN *sb.*[2] + LIGHT *sb.*] An electric torch shaped like a fountain-pen.

1958 *Practical Wireless* XXXIV. 58/2 (Advt.), Uses only one small 8d. penlight battery. **1962** E. AMBLER *Light of Day* x. 210 Miller had a pen-light in his hand and was looking at his watch. **1969** *New Yorker* 12 Apr. 78/2 There are pockets in unlikely places, such as at the ankles and shoulders, for things like scissors, penlights, and checklists. **1971** D. BAGLEY *Freedom Trap* ix. 206, I took out a pen-light and risked a flash. **1972** J. HURTT et al. *Compr. Rev. Orthoptics & Ocular Motility* xviii. 182 The patient is instructed to fixate a wall light at 20 feet while a penlight is held directly in front of his nose. **1977** 'E. MCBAIN' *Long Time no See* ii. 18 Carella took a small penlight from his coat pocket and flashed it over the mailboxes.

penman. Add: **1. c.** *Criminals' slang.* One who commits forgery.

1865 *Sessions Papers* 11 Apr. 519 For being concerned with, *Jemmy the Penman*, and others, now in custody, [etc.]. **1887** J. HAWTHORNE (*title*) An American penman: from the diary of Inspector Byrnes. **1938** F. D. SHARPE *Sharpe of Flying Squad* xxix. 297 As soon as they get some cheques or 'kites', as they call them, these are rushed off to the 'penman' or 'scribe', whose task is that of taking out crosses on them, enlarging the figures and preparing suitable letters to the bank asking them to cash the cheques. **1974** H. MCLEAVE *Only Gentlemen can Play* (1975) II. 97 You'll need a passport... I've got a penman who can doctor it.

pennant[1]. Add: **2. c.** *N. Amer. sport.* A flag symbolizing a league championship; hence, the championship itself. Also *attrib.*

1880 N. BROOKS *Fairport Nine* 188 Billy Hetherington ..was entrusted with the championship pennant. **1886** *Outing* Aug. 572/2 Questions by the dozen come in..in regard to the probable issue of the pennant races in the professional arena. **1915** *Lit. Digest* 21 Aug. 360/3 The Cincinnati Reds..have never yet won a pennant. *Ibid.*, The New York Giants..are not often far from the pennant class. **1924** [see *CINCH *v.* 2 b]. **1947** *Partisan Rev.* XIV. 258 The funeral was the most serious event of Samuel's life,..but this did not prevent him from getting the evening paper when the family returned from the cemetery and studying the final scores in the major league pennant races. **1967** W. S. AVIS et al. *Dict. Canad. Eng., Senior Dict.* 821/2 *Pennant*... **2** any flag taken as emblem of superiority or success, especially in an athletic contest. **1971** L. KOPPETT *N.Y. Times Guide Spectator Sports* i. 35 The pennant winner in each league turned out to be the team with the best regular-season record. **1975** *New Yorker* 22 Sept. 98/2 The Los Angeles Dodgers have won five pennants and three World Series since their relocation.

pennantite (pe·năntəit). *Min.* [f. the name of Thomas *Pennant* (1726–98), Welsh zoologist and mineralogist + -ITE[1].] A basic aluminosilicate of manganese, approximately $Mn_9Al_6Si_5O_{20}(OH)_{16}$, most specimens of which are pleochroic and orange in thin section.

1946 W. C. SMITH et al. in *Mineral. Mag.* XXVII. 217 (*heading*) Pennantite, a new manganese-rich chlorite from Benallt mine, Rhiw, Carnarvonshire. *Ibid.* 220 The names manganchlorite (Hamberg) and manganese-chlorite (Eckermann) have been applied to chlorites containing only a low percentage (1·02–2·28) MnO, so it is desirable to avoid the use of either of these names for a chlorite so rich in manganese as the mineral here described, and we therefore propose for it a new name, *pennantite*. **1954** *Ibid.* XXX. 280 The manganese-bearing chlorites include the remarkable species pennantite, chemically a klementite with the magnesium almost wholly replaced by manganese. **1970** *Mineral Abstr.* XXI. 249/2 (*heading*) The ferruginous and magnesium varieties of pennantite from the Atasui deposits in Central Kazakhstan. *Ibid.*, Fe-pennantite occurs in the Ushatan I deposit in veinlets with pyrosmalite and calcite. It is dark green. *Ibid.*, Magnesium pennantite in the Zhumast deposit occurs in marbles and in carbonate layers in braunite ore... Pleochroism α orange pink, γ light orange.

penned, *a.* **2. a.** (Later example.)

1968 B. HINES *Kestrel for Knave* 79, I started training Kes after I'd had her about a fortnight, when she was hard penned, that means her tail feathers and wing feathers had gone hard at their bases.

penner[4] (pe·nəɹ). [f. PEN *v.*[1] + -ER[1].] One who pens cattle; also (*Austral.* and *N.Z.*) **penner-up,** one who pens sheep ready for the shearers in a shearing shed.

1897 D. McK. WRIGHT *Station Ballads* 101 The penner-up is cursing at the back, The boss is looking savage at a long Australian card. **1904** *Daily News* 2 Dec. 5/1 There are 42 different men in this gang—'penners', 'shacklers', 'hoisters', 'gutters', and so on. **1911** J. COLLIER *Pastoral Age in Australasia* xxix. 216 Besides the shearers, there are penners-up, wool-rollers, pickers-up. **1940** E. C. STUDHOLME *Te Waimate* (1954) xv. 130 The 'sheep-oh' (penner-up)..in addition to filling up the catching-pens.. weighed the bales and recorded them..in the wool book. **1952** [see *DAGGER *sb.*[1] b]. **1955** *People* (Austral.) 30 Nov. 20/1 The penner-up bustles more sheep up the race. **1965** J. S. GUNN *Terminol. Shearing Industry* II. 8 *Penner-up*. This shedhand keeps the sheep moving into the shed and is ready to fill the catching pen when a shearer calls 'sheep-oh'. One man would be full-time 'penner-up' in sheds of six or more shearing stands. **1972** E. HARGREAVES *Fair Green Weed* ii. 24, I had to go down to the cow-pen... One of the penners has had an accident... His mule shied.

penni (pe·ni). Pl. **penni, penniä** (-iă). [Finn.] **a.** A Finnish monetary unit, equal to $\frac{1}{100}$ markka. **b.** The name of the coin equal to this amount.

1893 W. C. HAZLITT *Coinage of European Continent* 222 *Penni*, pl. *pennia*, a Russian copper coin struck for Finland = a French centime. **1903** [see *MARKKA]. **1957** *Whitaker's Almanack 1958* 964 Finland... *Markka* of 100 *Penni*. **1970** R. A. G. CARSON *Coins* (ed. 2) 397 In the later nineteenth century a distinctive coinage began to be issued by the czars of Russia as grand-dukes of Finland. The monetary unit was the silver mark, divided into 100 pennia... A monetary reform in 1963 introduced a new markka equivalent to 100 old markkaa. The types of the former 1 and 5 markkaa coins have been retained for the new 1 and 2 pennia pieces in bronze. **1974** *Encycl. Brit. Macropædia* VII. 306/2 The markkaa (made up of 100 pennia) was devaluated in 1949, 1957, and 1967.

pennif (pe·nif). *slang* (*rare*). [Back-slang f. FINNIP.] = FINNIP; hence, any bank-note.

1862 [see *JUG *sb.*[1] 2 b]. **1891** 'F. W. CAREW' *No. 747* xxxv. 416, I gets clean off with the scawfer and 'bout 'er thirty quid in single-pennifs and silver.

pennill. Add: The correct pl. is **penillion** (quot. 1898, sense b). (Further examples.)

1894 *Wales* Aug. 170/2 As he was so famous a poet, I thought he must be either a writer of hymns or of *penillion* to be sung with the harp. **1938** *Oxf. Compan. Mus.* 1009/2 A certain limited group of harp tunes are habitually used for Penillion performances. **1962** *Times* 31 Jan. (Wales Suppl.) p. xi/6 Penillion Singing (lit., 'singing of verses') to harp accompaniment. **1962** *Listener* 26 Apr. 740/3 The *englynion*, the *penillion telyn*.

Pennsylvania (pensilvēi·niă). One of the middle Atlantic states of the United States, named after Admiral Sir William Penn (1621–70), in 1681. Used *attrib.* to denote articles, inhabitants, products, or varieties of plants characteristic of, or growing in, Pennsylvania, as *Pennsylvania anemone, ash, cap, corn, division, dwarf mountain maple, mountain laurel, salve, wagon, wind flower*; **Pennsylvania German** *sb.* and *a.* = *PENNSYLVANIA DUTCH *sb.* and *a.*

1900 B. B. SMYTH *Plants & Flowers Kansas* ii. 54 On the low prairies may be found plenty of Pennsylvania anemone, a plant with..numerous branches, each terminated by a flower with five broad white sepals. **1810** P. WAKEFIELD *Excursions N. Amer.* (ed. 2) xxviii. 191 The black fir, the Weymouth pine, the red cedar, the common fir, the red maple, the Pennsylvania ash..are also common. **1971** M. TAK *Truck Talk* 116 *Pennsylvania caps*, recapped tires with an unbroken tread line. **1739** in *Colonial Rec. Georgia* (1905) III. 429 We all were disappointed by..planting the yellow Pensilvania Corn. **1929** *Papers Mich. Acad. Sci., Arts & Lett.* X. 314/1 *Pennsylvania Division*, the Twenty-eighth Division. **1785** H. MARSHALL *Arbustrum Amer.* 2 *Acer pennsylvanicum*, ..Pennsylvania Dwarf Mountain Maple. **1869** *Nation* (N.Y.) 30 Dec. 583/2 The Pennsylvania German is a South German dialect. *Ibid.* 584/1 Divine service among the Pennsylvania Germans is held in High German. **1875** A. R. HORNE (*title*) Pennsylvania German manual. **1956** *Publ. Amer. Dial. Soc.* XXVI. 30 The Norwegian spoken among the immigrants showed the same combination of archaism and levelling as other American languages, though the relatively brief period of its life has not permitted the kind of consolidation found in Pennsylvania German or the colonial languages proper. **1970** *Globe & Mail* (Toronto) 25 Sept. T3/3 Mennonite and Pennsylvania-German societies will be in attendance [at a festival]. **1972** H. KURATH *Stud. Area Linguistics* 105 This divergence between Pennsylvania German (essentially a Rhine Frankish folk dialect of west central Germany) and Standard German..would tend to keep the two apart. **1785** H. MARSHALL *Arbustrum Amer.* 127 *Rhododendrum maximum*, Pennsylvania Mountain Laurel. **1899** 'J. FLYNT' *Tramping with Tramps* IV. 396 *Pennsylvania salve*, apple-butter. **1810** M. DWIGHT *Journey to Ohio* (1912) 39 This line is the shape of a Pennsylvania waggon. **1869** J. G. FULLER *Uncle John's Flower-Gatherers* 28 [The anemone] blooms later, in May and June, and is called the Pennsylvania Wind Flower.

Pennsylvania Dutch, *sb.* and *a.* **A.** *sb.* **a.** *pl.* [DUTCH *sb.* 3 a.] The descendants of the original German settlers in Pennsylvania. **b.** [see DUTCH *sb.* 1.] A Pennsylvanian dialect derived from the High German of a great number of the early settlers, with a considerable admixture of English elements. **B.** *adj.* [DUTCH *a.* 1.] Of or pertaining to the Pennsylvania Dutch or their dialect. So **Pennsylvania Dutchman.**

a1824 J. GUILD in *Proc. Vermont Hist. Soc.* (1937) V. 293, I came across a Pennsylvania Dutch man, and I made a bargain with him. **1831** *Canad. Freeman* (Toronto) 19 May 2/3 Let Mackenzie stick to the Central Committee, the Saddlebags, & the Pennsylvania Dutch of the Home District. **1856** *Spirit of Times* 4 Oct. 71/1 But *revenons a mouton*, which, in plain Pennsylvania Dutch means how Fogie Antique caught..catfish. **1868** H. W. BEECHER *Norwood* 468 Them Pennsylvania Dutch think more of their horses than they do of themselves. **1882** P. H. GIBBONS *Pennsylvania Dutch* (ed. 3) 401 A 'Pennsylvania Dutch' remedy for whooping-cough. **1943** *Amer. Speech* XVIII. 112 The diphthong *oi* [ɔi] in the Pennsylvania Dutch dialects is interesting both phonemically and historically because of its infrequent occurrence. **1948** W. STEVENS *Let.* 1 Dec. (1967) 624 A true Pennsylvania Dutchman. **1974** E. MCGIRR *Murderous Journey* 98 He lapsed into Pennsylvania Dutch. The quaint idioms and all. **1976** *New Yorker* 16 Feb. 58/2 The sort of packaging that makes pancake-mix pancakes served in a Pennsylvania hotel 'Pennsylvania Dutch pancakes' instead of just pancakes. **1976** R. CONDON *Whisper of Axe* I. xvii. 103 He was served *schnitz un knepp*: apples, dumplings, and ham from the Pennsylvania Dutch.

Pennsylva·nian, *sb.* and *a.* [f. *PENNSYLVANIA + -AN.] **A.** *sb.* **1.** A native or inhabitant of Pennsylvania.

1685 [see QUAKERISTICAL *a.*]. **1747** G. WHITEFIELD *Let.* 6 May in *Wks.* (1771) II. 94 The Pensylvanians I am sure will soon regret the loss of you. **1755** in S. M. Hamilton *Lett. to Washington* (1898) I. 99 The Road upon which the Pennsylvaneans were Employ'd. **1782** J. H. ST. JOHN DE CRÈVECŒUR *Lett. from Amer. Farmer* 58 Europeans.. become..either Pennsylvanians, Virginians, or provincials, under some other name. **1838** *Southern Lit. Messenger* IV. 165/1 Mr. Ingersoll, being a Pennsylvanian, stands impartial between the two extremes of the Union. **1862** *Evening Post* (N.Y.) 21 May 1/2 Words of warm congratulation were sent to the dashing Pennsylvanian by the Commanding General. **1910** *Harper's Mag.* Aug. 473/1 Georg Shock..is a Pennsylvanian. **1939** [see *INSOMNIAC]. **1953** W. MOORE *Bring Jubilee* (1955) xx. 194 It was the Fourth of July, and a day of victory and rejoicing for all Pennsylvanians. **1967** *National Observer* (U.S.) 3 July 12/2 True to tradition, the Pennsylvanians are the best shots in the brigade. **1974** *Encycl. Brit. Macropædia* XIV. 28/2 Pennsylvanians still tend toward the Republican Party in state elections. **1976** *Billings* (Montana) *Gaz.* 1 July 3–B/2 Schapp urged all Pennsylvanians to listen to their message with an open mind.

2. *Geol.* The Pennsylvanian period or system.

1906 CHAMBERLIN & SALISBURY *Geol.* II. x. 556 In the arctic regions of America, the Mississippian and Pennsylvanian are not differentiated. **1960** J. M. WELLER *Stratigr. Princ. & Pract.* xii. 441 Fossils are rare and of little service in separating the Mississippian from the Pennsylvanian. **1969** [see *MISSISSIPPIAN *sb.* and *a.* A. 2]. **1977** A. HALLAM *Planet Earth* 207 The Pennsylvanian started with a new transgression of the sea over the low land of the central North American continent.

B. *adj.* **1.** Of, pertaining to, or characteristic of Pennsylvania or its people.

1698 G. THOMAS *Hist. & Geogr. Acct. Pensilvania & W. New-Jersey* 2 They (as the Pensilvanian Indians) observe the New Moons with great Devotion. **1785** H. MARSHALL *Arbustrum Amer.* 51 Pennsylvanian Sharp-keyed Ash (*Fraxinus pennsylvanica*). **1853** A. BUNN *Old Eng. & New Eng.* I. viii. 167 Mr. Nicholas Biddle..issued the notorious Pennsylvanian bonds. **1959** *Chambers's Encycl.* X. 536/2 Pennsylvanian oil production is relatively much less important although quality is high.

2. *Geol.* Of, pertaining to, or designating a period and system of the Palæozoic Era in North America that succeeded the Mississippian and preceded the Permian, and corresponds more or less to the Upper Carboniferous in Europe.

1891 *Bull. U.S. Geol. Survey* No. 80. 5 The Coal Measures or Pennsylvanian series. **1906** CHAMBERLIN & SALISBURY *Geology: Earth History* II. x. 539 (*heading*) The Pennsylvanian (coal measures, Carboniferous proper) period. *Ibid.*, The need of a name to distinguish this system of rocks from those which have been described under the name Mississippian has long been felt, and the name Pennsylvanian, which has recently come into wide use in this country, was adopted because the system is well developed and well known in Pennsylvania. **1933** [see *MISSISSIPPIAN *sb.* and *a.* B. 2]. **1945** *Bull. Amer. Assoc. Petroleum Geologists* XXIX. 128 The coal, oil, gas, ceramic clays, and other minerals found in rocks of Pennsylvanian age greatly exceed the value of the mineral resources found in any other system. **1960** J. M. WELLER *Stratigr. Princ. & Pract.* vi. 176 Pennsylvanian coal up to nearly 100 feet thick has been mined from several ancient Missouri sink holes of moderate or small size.

penny. Add: **I. 1.** Since 15 Feb. 1971 of the value of $\frac{1}{100}$ of a pound, and for a while known as the *new penny* (see *NEW *a.* 4). Denoted (after a numeral) by *p* (see *P II, *PEE *sb.*[6]). (Further example.)

1971 *Daily Tel.* 8 Mar. 12 He partly pre-empted his Budget last autumn..by promising six old pennies off the income tax.

2. b. (Earlier and later examples of U.S. usage.) Also Canad. *colloq.*, a Canadian cent.

1831 *Constellation* 12 Mar. 133/4 He meant cents, but they call em pennies in New York. **1902** 'R. CONNOR' *Glengarry School Days* 166 'Six pennies and two dimes', was Hughie's disconsolate reply. **1925** E. GLASGOW *Barren Ground* xi. 303 The price had seemed extravagant, for selling directly to her customer she had asked thirty cents a pound, while butter in Pedlar's store was never higher than ninepence in summer and a shilling in winter, measured in the old English terms which were still commonly used in Queen Elizabeth County. **1966** *New Statesman* 16 Dec. 896/3 Florin..is only used, like the American 'penny', to describe the actual lump of metal. **1971** *Daily Colonist* (Victoria, B.C.) 28 Dec. 7/4 Persons posting first class letters will be nicked another penny starting Jan. 1. **1974** H. MCCLOY *Minotaur Country* (1975) xvi. 186 You recall the penny that was found..? And the dimes found ..after the fire?

IV. 9. a. So *penny for them* ('*em*); also *ellipt.* as *penny*. **f.** (Further examples.) **l.** *pennies from heaven*: money acquired without effort or risk; also *sing.*, a windfall, a godsend. **m.** *to spend a penny*: to visit a lavatory, to urinate (from the former price of admission to public lavatories). **n.** *the penny has dropped*: a situation or statement has belatedly been comprehended; one has reacted belatedly. (With allusion to the mechanism of a penny-in-the-slot machine). **o.** *two* (also *ten*) *a penny*: commonplace, easily obtainable, occurring frequently.

a. 1900 H. G. WELLS *Love & Mr. Lewisham* xxv. 242 'Penny,' she said after an interval. Lewisham started and looked up. '*Eh?*' **1914** C. MACKENZIE *Sinister St.* II. iv. iii. 895 'You're very silent, kiddie,' she said. 'I'll give you a penny for them.' **1921** N. KENT *Quest M. Harland* II. iv. 169 'Penny for 'em, old man,' said Dickie presently, after Michael had eaten in silence for nearly five minutes. 'My thoughts?' Michael started and laughed. **1959** J. BRAINE *Vodi* xiv. 190 Harry's voice broke into her thoughts. 'Penny for 'em, old girl.' **1965** L. MEYNELL *Double Fault* II. v. 175 'Penny,' Lucian said. She laughed... 'Far too rich and rare for a penny to buy them.' **1973** J. THOMSON *Death Cap* x. 142 Finch was sitting looking thoughtfully at the report... 'Penny for them?' suggested the Sergeant.

f. *c*1882 W. S. GILBERT *Iolanthe* II. 33 In for a penny, in for a pound—It's love that makes the world go round! **1906** L. STRACHEY in *Lit. Ess.* (1948) 142 The emendator is on an inclined plane which leads him inevitably from readjustments of punctuation to corrections of grammar, and from corrections of grammar to alterations of rhythm; if he is in for a penny, he is in for a pound. **1976** 'J. FRASER' *Who steals my Name?* xii. 149 He seemed to be having some kind of inner conflict which resolved itself. 'All right,' he said, 'in for a penny in for a pound!' **1977** *Transatlantic Rev.* LX. 189 The cabbie steamed up to Notting Hill Gate with an In for a penny, In for a pound expression on his face.

l. 1936 J. BURKE (*song-title*) Pennies from heaven. **1965** J. D. MACDONALD *Bright Orange for Shroud* xvi. 191 'Sweetie,' I said, 'you are a penny from heaven.' **1971** P. DICKINSON *Sleep & his Brother* v. 117 Hard money is what your hospital pays you... Soft money is pennies from heaven, some dirty big company deciding to earn a bit of tax relief by financing medical research. **1972** 'W. HAGGARD' *Protectors* xiii. 154 He hadn't planned it that way... But when the pennies from heaven fell down he'd seize them.

m. 1945 H. LEWIS *Strange Story* iv. 27 'Us girls,' she said, 'are going to spend a penny!' **1955** J. CANNAN *Long Shadows* iii. 59, I wasn't sure that Trudy [a dog] had spent her penny. *Ibid.* vii. 127 We'll go indoors and pay for tea and spend a penny. **1960** M. CECIL *Something in Common* xxii. 239 It's tricky about the bathroom, but it's amazing how one can train oneself to spend a minimum of pennies. **1973** *People's Jrnl.* (Inverness & Northern Counties ed.) 28 July 10/1 Anyone on the Islands..after that time who wants to 'spend a penny' must make a 10-minute walk.. to the public toilets.

n. [**1942** N. BALCHIN *Darkness falls from Air* IV. 70 The penny seems to have stuck in the machine that time. The proper answer to that is, 'I'm flattered.'] **1951** —— *Way through Wood* xv. 214, I sat and thought for a moment and then the penny dropped. **1959** *Sunday Express* 13 Dec. 1/4, I had seen Vivienne before, but the penny didn't drop until I met her that night. **1961** S. CHAPLIN *Day of Sardine* viii. 174 It took a second or two for the penny to drop. I gave myself a shake and made over to the maybe Old Man. **1973** *Times* 1 Dec. 14 The penny had begun to drop even before the present fuel crisis.

o. 1960 *Times* 11 Jan. 17/1 Penalties were two a penny at Upper Park on Saturday. **1961** NEW ENG. BIBLE *Matt.* x. 29 Are not sparrows two a penny? **1966** *Listener* 27 Oct. 612/3 He found in India that subalterns were two a penny and invited nowhere. **1973** *Nature* 20 Apr. 492/2 Recommendations on the so-called energy crisis are by now two a penny. **1973** A. MANN *Tiara* iv. 34 Hunches are two a penny in this business. **1973** C. L. BARNHART et al. *Dict. New Eng.* 355/2 *Ten a penny* is also used in England.

V. 11. a. *penny arcade, awful, bazaar, bun* (earlier example), *club* (later example), *dreadful* (further examples; also, with hyphen, *attrib.*), *ice* (also *comb.*), *magazine, nap* (examples), *novel, novelette, press* (earlier and later examples), *reading* (further examples), *steamboat, steamer* (earlier and later examples), *toy, whistle* (further examples).

1908 C. E. GRIFFIN *Four Years in Europe* ix. 87 The numerous penny arcades and moving picture shows.. were another new wrinkle in American showmanship. **1961** GETLEIN & GARDINER *Movies, Morals, & Art* I. iv. 48 The penny arcade..can still be found in such urban areas as Times Square. **1889** E. DOWSON *Let.* 15 Mar. (1967) 49 It is very bad, very long, & distinctly 'penny awful' not 'shilling shocking'. **1897** H. JAMES *Spoils of Poynton* xiii. 154 An assortment of pen-wipers and ash-trays, a harvest he had gathered in from penny bazaars. **1966** *Guardian* 29 Aug. 4/4 The Shields tram.. full of early homecomers. They got off at the Penny Bazaar. **1976** *Times* 8 Nov. 14/5 If the Church wants..the responsibility for the appointment of its chief pastors it must do much better than this. We are in danger of reducing a great institution to the level of a penny bazaar. **1824** E. WEETON *Jrnl.* 21 July (1969) II. 309 Having had no dinner..but some cakes and one or two penny buns. **1844** C. M. YONGE *Abbeychurch* xiii. 278 Elizabeth..went to the school to receive the penny-club money. **1884** *World* 20 Aug. 9/2 The wicked noblemen of the transpontine melodrama or of penny dreadfuls. **1906** M. CORELLI *Treasure of Heaven* 55 The proper way for him to behave at this juncture..would be that he should take her ten-

derly in his arms and murmur, after the penny-dreadful style of elderly hero, 'My darling'. **1925** T. DREISER *Amer. Trag.* II. III. xix. 222 By him sold to a penny-dreadful publisher of Binghamton. **1941** V. NABOKOV *Real Life S. Knight* x. 91 He did not mind in the least 'penny dreadfuls' because he wasn't concerned with ordinary morals. **1963** *Times* 18 Feb. 5/3 He was perfectly happy with a 'penny dreadful', a warm fire, a friendly dog, and a good meal inside him. **1872** B. JERROLD *London* xv. 127 We have found the penny ice-man doing a brisk trade. *Ibid.*, The penny ice has proved too strong for the ancient ginger-beer bottle. **1896** G. B. SHAW *Our Theatres in Nineties* (1932) II. 133 Some appalling tenor from I know not what limbo of street-piano padrones, penny-icemen, and broken choristers. **1914** —— *Fanny's First Play* 169 You should be eating penny ices and enjoying yourself. **1835** DICKENS *Sk. Boz* (1837) 2nd Ser. 145 When penny magazines shall have superseded penny yards of song. **1852** Penny magazine [see *penny lecture* s.v. PENNY 11]. **1889** J. K. JEROME *Three Men in Boat* xix. 306 We played penny nap after supper. **1950** *Hoyle's Games Modernized* (ed. 20) I. 138 If a man calls three at 'penny Nap', he receives 3d. **1861** *Punch* 5 Jan. 3/1 A weakness for..reading penny novels. **1896** G. B. SHAW *Our Theatres in Nineties* (1932) II. 213 You would never dream of asking why Morris did not read penny novelettes, or hang his rooms with Christmas-number chromolithographs. **1840** *Picayune* (New Orleans) 15 Sept. 2/2 The six-penny journals have latterly grown wise enough to drop the naughty habit in which they used to indulge of swearing at the penny press. **1843** *North Amer. Rev.* Jan. 227 They [*sc.* Dickens's *American Notes*] have been scattered all over the country by the penny press. **1932** T. S. ELIOT *Sel. Essays* VI. 341 Those sections about which readers of the penny press are most ready to excite themselves. **1871** F. KILVERT *Diary* 3 Feb. (1938) I. 301 This evening we had our 4th Penny Reading. The room was fuller than ever. **1907** A. HUXLEY *Let.* 17 Nov. (1969) 26 Mr. Taylor wants to know for his *penny reading*. **1969** *Telegraph* (Brisbane) 25 Mar. 8/4 The provisional school was used for monthly 'penny readings'..at that period. **1976** *Trans. Yorks. Dial. Soc.* LXXVI. 33 Traditions which must go back to 'Sir Gawain' and even 'Beowulf' when the 'oral literature' of 600 A.D. and Victorian 'Penny Readings' may seem to be not all that far apart. **1859** G. A. SALA *Twice round Clock* 11 The river glideth in peace, undisturbed by penny steamboats. **1862** A. J. MUNBY *Diary* 7 May in D. Hudson *Munby* (1972) 121 His enjoyment of the Thames from the deck of a penny steamer. **1933** *Radio Times* 14 Apr. 72/3 The 'penny-steamers', those cheerful launches that trail along behind the [Boat Race] umpires and the B.B.C. **1905** *Daily Chron.* 18 Dec. 4/5 The first gutter penny-toy merchant. **1955** *Times* 11 May 12/4 Penny toys (how many of them could be bought for that coin today?). **1931** N. DOUGLAS *London Street Games* (ed. 2) 29, I went down the lane to buy a penny whistle, A copper came by and pinch my penny whistle. **1967** W. SOYINKA *Kongi's Harvest* 64 Penny whistles blow to the tune of the Carpenter's Song. **1978** W. HJORTSBERG *Falling Angel* (1979) vi. 71 Someone played a pennywhistle. Shrill, piping notes.

12. a. *penny-cautious, -conscious, -grubbing, -picking* adjs. **c. penny ante** *U.S.*, the game of poker when the ante is fixed at one penny or a similarly insignificant stake; also *attrib.* or as *adj.*, contemptible, trivial; **penny black,** (a specimen of) the first one-penny postage stamp issued in the United Kingdom, on 6 May 1840; also *fig.*; **penny-boy** *slang* (see quot. 1902); (in quot. 1914, a term of mild contempt); **penny-daisy,** prob. = *ox-eye daisy* (OX-EYE 3 b); **penny-in-the-slot** *a.* (earlier and later examples); also *ellipt.* as *sb.*; **penny loafer, pennyloafer** *N. Amer.*, a type of casual shoe with a slot in which coins can be placed; **penny number,** (*a*) a cheap periodical; (*b*) *pl.* insignificant quantities (*colloq.*); **penny packet,** (*a*) = *penny steamer* s.v. PENNY 11; (*b*) a small number of persons or things; (*c*) (with hyphen) *attrib.*, contemptible, insignificant; **penny pawn** (see quot.); **penny-peeler,** an avaricious or niggardly person; **penny-piece** (further examples); **penny-pincher,** a niggardly person; **penny-pinching** *a.* (*colloq.*), niggardly, parsimonious; also as *vbl. sb.*; hence (as a back-formation) *penny-pinch* vb. trans. and intr.; *penny-pinched* adj.; **penny plain** *a.*, plain and unpretentious; hence **penny-plainness; penny stock** *U.S.*, a common stock of value less than one dollar, and therefore highly speculative.

1855 'Q. K. P. DOESTICKS' *Doesticks, what he Says* 259 Napoleon spends most of his time playing penny 'ante' with the three Graces. **1935** A. J. POLLOCK *Underworld Speaks* 87/1 *Penny ante league*, small town racketeers. **1936** L. HELLMAN *Days to Come* III. 88 We always used to play penny-ante there. **1946** *Negro Digest* Aug. 48/1 Compared to the man Bilbo, 63-year-old John Rankin is strictly penny ante and colorless. **1976** M. MACHLIN *Pipeline* v. 63 Prices were offered that made Royal American's earlier bids seem like penny ante. **1976** M. MAGUIRE *Scratchproof* ix. 140 I'm not a penny-ante hood. [**1920** E. D. BACON *Line-Engraved Postage Stamps Gt. Brit.* I. 167 In 1864 an application was made to the Board of Inland Revenue for specimens of the One Penny black.] **1922** A. B. CREEKE in C. Nissen *Plating of Penny Black Postage Stamp* p. i, The cult of the 'Penny Black'—and no true British philatelist is so pedantic as to call the first postage stamp the 'Black Penny'—has..spread amongst..many collectors. **1936** [see *BLACK *sb.* 7 e]. **1972** *Daily Tel.* Colour Suppl.) 12 May 62/1 These are so rare that they are referred to as the Penny Blacks of the cigarette card world. **1973** R. HILL *Ruling Passion* II. ii. 99 'What about the stamps?'..'No penny blacks, I'm afraid.' **1977** *Western Mail* (Cardiff) 5 Mar. (Rugby Suppl.) 7/1 Still, the days when English rugby victories were as rare as the Penny Black seem over. **1902** FARMER & HENLEY *Slang* V. 168/1 *Penny-boy* (old), a boy who haunted the cattle markets on the chance of driving beasts to the slaughter-house. **1914** JOYCE *Dubliners* 273 He saw himself as a ludicrous figure, acting as a pennyboy for his aunts. **1939** DYLAN THOMAS *Let.* July (1966) 233 People forced..to be so penny-cautious. **1964** *Economist* 27 June 1481/2 The penny-conscious President. **1920** D. H. LAWRENCE *Lost Girl* I. 24 Big penny-daisies grew in tufts on the brink of the yellow clay. **1942** *New Statesman* 11 July 25/1 The Jews of Poland, on the whole, I have found penny-grubbing, cunning, and given to circumlocution. **1891** KIPLING *Light that Failed* xiv. 281 They've got one of them penny-in-the-slot cash-machines. **1922** D. H. LAWRENCE *England, my England* 255 I'll just put it aside o' the penny-in-the-slot. **1948** *Training of Doctor* (B.M.A.) xv. 74 The fault lies as frequently with the harried and harassed physician seeking a 'penny in the slot' diagnosis. **1954** L. FAIRFIELD *Epilepsy* i. 23 In spite of the brilliant contributions made by encephalographers to the study of epilepsy, it would be wholly wrong to give the impression that they can provide a 'penny in the slot' diagnosis. **1970** *Daily Tel.* 21 May 6/5 Hoisted on his brother's shoulder, he watches a striptease in a penny-in-the-slot machine at a fair. **1970** *Globe Mag.* (Toronto) 26 Sept. 5/3 Chicks..who aren't really hippie, wear really good jeans... Some have penny loafers. **1973** *Maclean's Mag.* (Toronto) Feb. 28/1 They're classic: he's wearing checkered pants and brogues, she's got a skirt and pennyloafers. **1976** T. GIFFORD *Cavanaugh Quest* (1977) i. 15 Two highly polished penny loafers with virgin tan soles. **1901** G. B. SHAW *Capt. Brassbound's Conversion* III. 297 He got his romantic nonsense out of his penny numbers. **1958** *Listener* 5 June 937/1 Pupils arrive from the depot in penny numbers. **1972** *Shooting Times & Country Mag.* 27 May 12/1 The beats of Spey near Grantown were either returning penny numbers of fish or blanks for hard fishing. *c*1846 J. R. PLANCHÉ *Invisible Prince* I. ii. 14 Fierce whiskered gents, as ever in pea jackets, Smoked bad cigars, on board the penny packets. **1943** HUNT & PRINGLE *Service Slang* 51 *Penny packets*, small parties of soldiers, less than a platoon, as seen from the air. **1961** *Daily Tel.* 21 Apr. 17 Lord Bridges, chairman of the Royal Fine Art Commission, yesterday condemned piecemeal, penny-packet planning of new towns and replanning of older ones. **1979** G. C. PEDEN *Brit. Rearmament* 119 The Treasury principle of locating the Air Force centrally instead of in penny packets around the Empire was observed. **1907** *Westm. Gaz.* 16 Dec. 10/1 What are known as 'penny pawns' abound in the district. A broker who keeps one of these can purchase an article of any value from a penny upwards. He is compelled to keep it for seven full days. **1925** J. GREGORY *Bab of Backwoods* xxi. 269 Willoughby, skinflint, penny-peeler and nickel grabber that he was, smelled a deal and asked them five thousand dollars for ten acres! **1920** D. H. LAWRENCE *Lost Girl* vi. 99 This grubby penny-picking England. **1938** M. K. RAWLINGS *Yearling* ii. 17 Why, you leetle ol' penny piece, you. You're good money, a'right, but hit jest don't come no smaller. Leetle ol' Penny Baxter. **1963** *Times* 25 Apr. 17/1 Coyle, he maintained, did not receive 'a penny piece' in the transactions and his only motive in disposing of the bicycles was to please his immediate superiors. **1975** E. PAGE *Element of Chance* vi. 66 In another three years..you can divorce her without her consent... I very much doubt that..you'd have to pay her a penny piece. **1961** J. YAFFE in *Webster* s.v. *penny-pinch*, A sinister but fascinating kind of joy in..penny-pinching his own family. **1961** S. N. BEHRMAN in *Ibid.*, Penny-pinched himself out of..millions of dollars. **1977** P. G. WINSLOW *Witch Hill Murder* II. v. 92 He penny-pinched on the candles..so the servants had to spend their few hours off in practically pitch darkness. **1979** *Jrnl. R. Soc. Arts* CXXVII. 126/1 The dismally penny-pinched equipment compared with foreign institutions. **1934** WEBSTER, *Penny pincher*, a niggardly or parsimonious person. **1956** F. CASTLE *Violent Hours* (1966) xviii. 174 If Forhaan had really been a penny-pincher, he would never have offered so quickly to pay for Hazel's care. **1967** *Guardian* 22 Feb. 4/7 A sparkling dialectic arose between the risk takers on the boards and the penny pinchers. **1973** 'D. SHANNON' *Spring of Violence* (1974) ix. 150 Typical of the penny-pincher miser. **1905** Penny-pinching [listed in Dict.]. **1920** S. LEWIS *Main St.* xi. 144 The penny-pinching old land-thief. **1951** *Ann. Reg. 1950* 192 Mr Johnson had come under heavy fire as the man whose 'penny-pinching' with military expenditures had paved the way for the reverses in Korea. **1953** C. S. FORESTER *Hornblower & Atropos* xiii. 190 The penny-pinching clerks of a penurious government at home would scrutinize those expenditures in time. **1960** *Guardian* 20 July 6/6 Penny-pinching on new roads is inexcusable. **1971** *Ibid.* 7 Aug. 10/1 Mr Rowley's decision is both penny-pinching and short-sighted. **1973** E. BERCKMAN *Victorian Album* 34 The better-off the woman was, the more apt she'd be to play those penny-pinching tricks. **1977** *Time* 30 May 24/2 Under increasing criticism—from liberals, who regard him as too much of a pennypinching conservative,..Carter took to the hustings again by making a whirlwind, campaign-style tour of California. **1859** G. A. SALA *Twice round Clock* 253 The Scala..with its rabbit-hutch-like private boxes, whose doors are scrawled over with the penny plain and twopence coloured-like coats of arms of the..Lombardian nobility. **1884** R. L. STEVENSON in *Mag. of Art* Apr. 227 (*title*) A penny plain and twopence coloured. **1920** 'O. DOUGLAS' *Penny Plain* vi. 60 Having been all her life so very 'twopence coloured' she wants the 'penny plain' for a change. **1974** 'S. WOODS' *Done to Death* 71 A track led to a penny-plain stone cottage. **1920** 'O. DOUGLAS' *Penny Plain* vi. 60 There is no mistake about our 'penny-plainness'—it jumps to the eye! **1932** C. M. ALSAGER *Dict. Business Terms* 261 *Penny stocks*, a term applied to stocks that sell below one dollar per share and that are usually quoted in cents, not in fractions of a dollar, on an exchange or over the counter. **1935** *Sun* (Baltimore) 25 Oct. 22/1 More often than not..an increased turnover in what Wall Street calls the 'penny stocks' has been a sign that the market is heading toward a reaction. **1942** *Richmond* (Virginia) *Times-Dispatch* 29 Dec. 18/1 Turnover of 1,201,522 shares, propped by belated tax offerings in sizable blocks of 'penny' stocks, was the second largest of the year to date. **1967** *Economist* 28 Oct. 419/1 This is a peculiarly American worry...nowhere else are there so many thousands of 'penny' stocks ripe for speculation.

penny-a-line, *a.* (Later example.)

1930 *Argosy* Apr. 15/2 It's my last bit of freedom before I sink into eternal penny-a-line slavedom.

penny-a-line, *v.* Add: Also *intr.*

1851 MRS. GASKELL *Lett.* (1966) 172, I swore I would penny-a-line and have nothing to do with publishers never no more.

So **pe·nny-a-li·ning** *vbl. sb.* and *ppl. a.*

1850 *Punch* 28 Sept. 140/1 (*heading*) Penny-a-lining under difficulties. **1874** GEO. ELIOT *Let.* 2 Nov. (1956) VI. 87 A penny-a-lining literary affair. *a***1941** V. WOOLF *Captain's Death Bed* (1950) 10 Penny-a-lining came into fashion. **1946** 'G. ORWELL' *Crit. Ess.* 169 In France, all kinds of petty rats—police officials, penny-a-lining journalists, women who have slept with German soldiers —are hunted down.

pe·nny-a-li·neism. [f. PENNY-A-LINE *a.* + -ISM.] The practice of writing in the inflated style of a penny-a-liner; an instance of such writing.

1854 *Punch* 25 Nov. 208/2 That renowned traveller and Protestant champion [*sc.* the Editor of the *Morning Advertiser*] has accepted the appointment of Regius Professor of Penny-a-lineism. **1890** W. JAMES *Princ. Psychol.* I. ix. 263 The whole genus of penny-a-line-isms and newspaper-reporter's flourishes give illustrations of this.

penny-a-liner. (Further examples.)

1871 [see *CENTURY 3 b]. **1889** E. SAMPSON *Tales of Fancy* 22 These effusions were usually written by 'Old Willey', printer, penny-a-liner, and pugilistic scribe. **1930** E. WEEKLEY *Saxo Grammaticus* 12 That type of journalist who used to be rudely called a penny-a-liner. **1947** W. S. MAUGHAM *Creatures of Circumstance* 9 He wouldn't have liked it if some damned penny-a-liner had made fun of Evie's effort in one of the papers.

pe·nny-a-wee·k, *a.* [The phrase (*a*) *penny a week* used attrib.] That collects, receives, or subscribes a penny a week.

1895 G. B. SHAW *Let.* 15 Dec. (1965) I. 576 Penny-a-week men enrolled on the spur of the moment. **1914** JOYCE *Dubliners* 206 Many a good man went to the penny-a-week school. **1943** *Times* (Weekly ed.) 19 May 3 The Red Cross and St. John Fund has exceeded £20,000,000... More than one fourth has been contributed through the penny-a-week committee. **1972** *Evening Telegram* (St. John's, Newfoundland) 5 Aug. 6/5 A school in the upstairs flat, the famed 'Penny-a-week' school.

penny farthing, *a.* and *sb.* **A.** *adj.* Ineffective; insignificant.

1887 KIPLING *Plain Tales from Hills* (1888) 78 It was pleasant to watch her unhappiness, and the penny-farthing attempts she made to hide it. **1967** S. BECKETT *Eh Joe* 17 You know that penny farthing hell you call your mind. *a***1974** R. CROSSMAN *Diaries* (1976) II. 60 Harold Wilson's penny-farthing report on the scandals of Party organization had come out long before the 1959 election. **1977** J. WAINWRIGHT *Day of Peppercorn Kill* 68, I.. kicked hell out of every penny-farthing crook I could lay my hands on.

B. *sb.* An early form of bicycle having a large front wheel and a small rear one. Also *attrib.*

This kind of bicycle was introduced in the 1860s and was known by various names, including *ordinary* (see ORDINARY *sb.* 17 b), *bone-shaker* (see BONE *sb.* 17) for the wooden-wheeled machine, and BICYCLE for the wire-spoke machine. The name *penny farthing* does not seem to have been used until the 1920s, by which time this type of bicycle was obsolete.

1927 LD. BIRKENHEAD in *Sunday Express* 6 Nov. 11/6, I once rode more or less continuously on a high bicycle (called a 'penny farthing') from my native town of Birkenhead to Edinburgh. **1932** G. M. BOUMPHREY *Story of Wheel* xxiv. 89 These improvements had turned the bone-shaker into what was sometimes called the 'penny-farthing'. **1963** *Times* 20 May 11/1 To do without it would be like giving up our helicopters and going back to penny-farthings. **1976** *Country Life* 27 May 1385/1 Nostalgia for a world..of Norfolk jackets, muttonchop whiskers, penny-farthing bicycles.

pe·nnyweighter. *U.S. criminal slang.* Also **penny weighter, penny-weighter.** [f. PENNYWEIGHT + -ER[1].] One who steals jewellery or precious stones or metals.

1899 'J. FLYNT' *Tramping with Tramps* IV. 396 *Pennyweighters*, jewelry thieves. **1905** *Daily News* 26 July 9 In the American description of her she was said to be a 'penny weighter'... That is, one who goes into a jeweller's shop, inspects jewellery, and by means of some sticky substance on the fingers, manages to palm an article, and deposits it beneath the counter for a confederate to pick up. **1916** [see *HEEL *sb.*[1]]. **1935** *Amer. Speech* X. 19/1 *Pennyweighter*, one who steals gold or silver plate. Still survives in mining camps to designate one who steals very small quantities of gold, as opposed to a high-grader who appropriates any big nuggets which he sees in the sluice boxes. Present usage restricts it to a jewel thief, or a jeweler who substitutes paste gems for genuine ones. **1950** H. E. GOLDIN *Dict. Amer. Underworld Lingo* 155/1 *Penny-weighter*, a thief who specializes in stealing uncut and unset diamonds.

So **pe·nny-weighting** *vbl. sb.*

1903 H. Hapgood *Autobiog. Thief* iii. 56 Penny-weighting is very 'slick' graft... A man..enters a jewelry store and looks at some diamond rings... Then he goes to a fancy shop [imitation jewelry] and buys a few diamonds which match the real ones he has noted. Then he and his pal, usually a woman, enter the jewelry store and..one of them..substitutes the bogus diamonds for the good ones. **1924** G. S. Dougherty *Criminal as Human Being* iii. 89 Such a performance is sometimes staged in 'penny-weighting', but..only a single quick one can be taken by substitution.

pennywinkle, dial. var. Periwinkle².

Penobscot (penǫ·bskǫt), *sb.* and *a.* Also **Penobscote**. [Native name: see F. W. Hodge *Handbk. Amer. Indians North of Mexico* II. 226.] **A.** *sb.* An Algonquian Indian people of the valley of the Penobscot River in Maine, U.S.A.; also, a member of this people.

[**1616** J. Smith *Descr. New Eng.* 8 The principall habitation Northward we were at, was Penobscot..though most [Indian peoples] be Lords of themselves, yet they hold the Bashabes of Penobscot, the chiefe and greatest amongst them.] **1624** —— *Gen. Hist. Virginia* vi. 240 The Masachusets call their great God Kiehtan, and their Kings there abouts Sachems: The Penobscotes their greatest power Tantum, and their Kings Sagomos. **1813** in *Maine Hist. Soc. Coll.* (1859) VI. 253 In Witness whereof, We,..by name, Kireborah, Iteansis, and Jackott, for Penobscot,..have hereunto set our hands & seals. **1910** F. W. Hodge *Handbk. Amer. Indians* II. 226/2 The Penobscot took an active part in all the wars on the New England frontier up to 1749, when they made a treaty of peace, and have remained quiet ever since. **1942** *Fading Trails* (U.S. Dept. Interior Nat. Park Service) vii. 59 Many an aged Penobscot..still wistfully awaits the day when the..peculiar 'click-click' of caribou's feet will return once more in the frosty air of the muskeg. **1959** E. Tunis *Indians* 25/2 The Penobscots kept certain of their young men in strict training for running down deer. **1965** *Canad. Jrnl. Linguistics* Spring 135 There are some Algonquian Indians still living along the New England coast,..but except for a few older Penobscot, all speak English. **1974** *Encycl. Brit. Micropædia* VII. 855/1 The Penobscot assisted the French against the English in all the wars on the New England frontier until 1749, when they made peace with the English.

b. The language spoken by the Penobscot people.

1891 J. C. Pilling *Bibliogr. Algonquian Lang.* 110/2 A short catechism in Penobscot begins on page 47. *Ibid.* 199/1 Comparative vocabulary..in the..Penobscot. **1902** [see *animate *ppl. a.* 4b]. **1933** L. Bloomfield *Language* 72 The Algonquian family..includes the languages..of New England (Penobscot...). **1972** *Regional Lang. Stud.—Newfoundland* iv. 2 Penobscot, a dialect closely related to Abenaki.

B. *adj.* Of or pertaining to the Penobscot people or their language.

1727 J. Heath *Let.* 7 July in *Maine Hist. Soc. Coll.* (1859) III. 409 If Capt. Gyles..can steer the Penobscot Indians to Falmouth, it seems as though these may follow. **1779** S. Lovell *Orig. Jrnl. Penobscot Exped.* (1881) 97 We are visited by some Penobscot Indians who are determined to proceed with us. **1819** *N. Amer. Rev.* IX. 185 A specimen of the Penobscot dialect (which we obtained from a friend in the District of Maine). **1831** *Boston Even. Transcript* 7 Oct. 1/2 Last week, a wretched and destitute Penobscot Indian was seen traversing our streets. **1854** Thoreau *Walden* 35, I have seen Penobscot Indians, in this town, living in tents of thin cotton cloth. **1893** *Chicago Tribune* 5 July 1/5 The Penobscot tribe of Indians is represented now by ten members from Old Town, Maine. **1973** A. H. Whiteford *N. Amer. Indian Arts* 91 Northeastern beadwork is lacy, as shown by the Penobscot collar. **1976** *Time* 27 Dec. 25/2 In Maine, the Passamaquoddy and Penobscot tribes have laid claim to 2.5 million acres of land—two-thirds of the state.

pen-pal (pe·npæl). orig. *U.S.* [f. Pen *sb.*² + Pal *sb.*¹] = *pen-friend. Hence **penpalmanship**, **penpalship**, the relationship existing between pen-pals.

1938 *Educational Method* Nov. 83/1 Mary Jones of Portland, Maine, has just requested the Student Letter Exchange of Waseca, Minnesota, to supply her with the name of a 'pen pal' in Mexico City. *Ibid.*, Since its inception in July, 1936, the Student Letter Exchange has provided hundreds of thousands of 'pen pals' around the world. **1941** J. C. Furnas *How Amer. Lives* 272 Pen-pals are not the only piece of news for the writer in the Kriebel boys' mail. **1960** *20th Cent.* June 517 There are some rather elaborate plans for forming relationships with individual Alphas, on the lines of the children's pen-palships in the twentieth-century newspapers. *Ibid.*, You need to be young and to do your own pen pal for this type of correspondence. **1962** J. Kirkup tr. *Brunner's Trouble in Bravada* iii. 26 The girls..had written..to a school class in Zurich with which they regularly corresponded, and..their pen-pals would help them to bear the cost. **1966** *Guardian* 9 June 8/1 The affair was carried on entirely by correspondence... Their penpalmanship had already begun. **1967** *Boston Herald* 1 Apr. 22/6 The children at the Dedham Country Day School are pen pals of children in Dedham, England. They've exchanged Christmas presents and a few have even visited families over there. **1972** M. Mead *Blackberry Winter* vii. 81, I also had two pen pals whose names I got through *St. Nicholas Magazine*. **1972** C. Short *Naked Skier* xxii. 123 Maybe we could be pen-pals... I'll send you a postcard from New York. **1974** 'P. B. Yuill' *Hazell plays Solomon* ii. 25 'Why pay top money for information and a photograph of a child whose surname they don't know?' 'Perhaps they want a penpal?'

pen-pusher (pe·npu:ʃəɹ). [f. Pen *sb.*² + Pusher.] One who is engaged in writing or desk work; a clerk; a writer (freq. derogatory).

1911 *Busy Man's Mag.* Jan. 65/2 That fellow was a pen-pusher in a dough joint—I mean a bank clerk. **1930** *Time & Tide* 24 May 665/2 Clerking! My God, any tuppenny ha'penny pen-pusher can be a clerk. **1940** *Manch. Guardian Weekly* 22 Mar. 228 From his point of view it is much better to be a pen-pusher in..obscurity than a corpse. **1948** N. Shute *No Highway* vii. 173 Who the hell are you, anyway? Just a bloody pen-pusher. **1952** A. Grimble *Pattern of Islands* 177, I do not suppose that George was particularly interested in Stevenson as a writing man—he never had much time for pen-pushers, as he called them. **1954** Wodehouse *Jeeves & Feudal Spirit* ix. 80 Florence tells me that La Morehead is one of the more costly of our female pen-pushers and has to have purses of gold flung to her in great profusion before she will sign on the dotted line. **1957** J. Braine *Room at Top* iii. 31, I saw myself, compared with him, as the Town Hall Clerk, the subordinate pen-pusher, halfway to being a zombie, and I tasted the sourness of envy. **1972** *Guardian* 23 Oct. 9/5 The more assiduous pen-pushers among London Transport's 6,000 administrative, technical, clerical and control staff.

So **pe·n-pushing** *vbl. sb.*, writing by hand.

1936 G. Orwell *Keep Aspidistra Flying* iii. 61 He dreaded..going to work... Pen-pushing in some filthy office—God! **1952** A. Grimble *Pattern of Islands* 179, I knew that Cas Workman would have made a better job of the pen-pushing than I did. **1972** W. A. Pantin *Oxf. Life* iv. 53 These volumes represent a great mass of praiseworthy industry.., whether pen-pushing or type-writing.

penroseite (pe·nrǝʊzait). *Min.* [f. the name of Richard A. F. *Penrose*, Jr. (1863–1931), U.S. mining geologist + -ite¹.] A selenide of nickel, ideally $NiSe_2$, which usu. also contains copper, lead, or silver, probably as impurities, and occurs in grey reniform masses.

1926 S. G. Gordon in *Proc. Acad. Nat. Sci. Philadelphia* LXXVIII. 317 Penroseite was obtained in a small collection of minerals which the writer purchased from a local merchant at Colquechaca, Bolivia. *Ibid.* 321 It is a pleasure to name this new mineral in honor of Dr. Richard A. F. Penrose, Jr. **1957** *Amer. Mineralogist* XLII. 32 The minerals most likely to be intergrown with penroseite and blockite are cubic in symmetry, such as clausthalite,..naumannite..and argentiferous galena. *Ibid.*, Penroseite and blockite may therefore be regarded as identical minerals and the latter name may be discarded. **1968** I. Kostov *Mineral.* 124 Penroseite usually contains some copper and cobalt.

‖ **pensiero** (pensiɛ̄·ro). Pl. **pensieri**. [It.] A thought, an idea; an anxiety. *spec.* in Art, a sketch.

1909 Joyce *Let.* 7 Sept. (1966) II. 231, I am dreadfully nervous from all the worry and *pensieri* I have had, very very nervous indeed. **1959** *Times* 2 Oct. 5/6 Most of the drawings are *pensieri*. **1961** M. Levy *Studio Dict. Art Terms* 80 *Pensiero*, an alternative term for a sketch. **1975** *Times Lit. Suppl.* 4 Apr. 378/3 Each artist was supplied with a design, or *pensiero*, which laid down the pose, proportions and drapery pattern of the statue.

pension, *sb.* Add: **9.** (sense 4) *pension act*, *benefit*, *book* (later examples), *fund*, *law* (examples), *money* (later examples), *plan*, *right*, *roll* (further examples), *scheme*.

1839 *Southern Lit. Messenger* V. 314/1 A few tardy pension acts..are all the tributes their worth has received. **1968** G. D. Gilling-Smith *Compl. Guide Pensions* (ed. 2) i. 16 The 1908 Old Age Pensions Act provided a small amount for those who were already old and unable to provide for themselves. **1945** *Release & Resettlement* (H.M. Govt.) 12 Any..pension benefits... **1966** *Listener* 17 Mar. 391/2 Did you ever see your mother's pension book? **1974** *Times* 6 Mar. 4/3 A child's printing outfit was used in a scheme to cash stolen pension books with a face value of more than £750,000. **1869** *Bradshaw's Railway Manual* XXI. 561 Donation to the pension fund, 18,573*l*. **1907** G. B. Shaw *Major Barbara* iii. 265 Have you gone into the insurance fund, the pension fund, the building society, the various applications of co-operation? **1965** H. I. Ansoff *Corporate Strategy* (1968) vi. 104 At the other extreme..is a company which primarily buys and sells. This may be an investment trust, a pension fund. **1973** A. Behrend *Samarai Affair* iv. 51 'Will the Company look after him?' 'I suppose so. Their pension fund's all right.' **1976** *New Yorker* 15 Nov. 176/1 In addition to making contract concessions, the union members, through their pension funds..have become the city's bankers, lending it nearly two billion dollars in the past year. **1838** *Southern Lit. Messenger* IV. 766 When the revolutionary pension-law was enacted, a majority of the war-worn veterans had travelled..beyond the reach of human reward. **1749** *New Hamps. Probate Rec.* (1916) III. 733, I give..all my Waidges Prize money Pention money [etc.]. **1854** B. P. Shillaber *Life & Sayings Mrs. Partington* 190 The old lady had presented a check for a quarter's pension-money. **1953** *Stroud's Judicial Dict.* (ed. 3) III. 2143 Pension money reduced into possession by the pensioner or his agent loses its character of pension. **1957** Clark & Gottfried *Dict. Business & Finance* 262/1 *Pension plan*, in business, a plan established and maintained by an employer to provide in a systematic manner for the payment of regular pension amounts to retired or disabled employees. **1961** *Factory* Nov. 101/1 Choose carefully between contributory or non-contributory pension plans. **1973** *N.Y. Law Jrnl.* 31 Aug. 4/4 (Advt.), Metropolitan Life can help you set up a pension plan that will entitle you to the same Federal Income Tax deductions available to other corporations. **1956** G. A. Hosking *Pension Schemes* xii. 183 The existence of a pension scheme acts as a deterrent when withdrawal means loss of pension rights. **1974** O. Manning *Rain Forest* i. vi. 87 Temporary and without pension rights, the appointment carried a special salary. **1828** A. Sherburne *Mem.* vi. 239 This gentleman..forwarded to me certificates of the continuation of my name on the pension roll. **1907** *Westm. Gaz.* 5 Apr. 10/1 Miss Robb..was the posthumous child of Captain Robb,..and was put on the State pension-roll at birth. **1892** C. S. Loch *Old Age Pensions & Pauperism* 3 The returns of pauperism in England and Wales are frequently quoted as absolute evidence in favour of some kind of National Pension Scheme. **1928** Robertson & Samuels *Pension & Superannuation Funds* viii. 79 Pension schemes are growing up not only in industry proper; many institutions..have their own schemes of superannuating those in their service. **1962** P. Parker *Twisting Lane* 16, I was in a good pension scheme too. **1973** L. Holcombe *Victorian Ladies at Work* iii. 39 The government in 1875 revived its original teacher's pension scheme, established in 1846 but abolished in 1861. **1977** W. McIlvanney *Laidlaw* xxx. 140 He's worried about his family... They don't have a great pension-scheme for house-breakers.

pension, *v.* Add: **2.** *to pension off* (earlier and later examples). Also *fig.* So **pensioned-off** *ppl. a.*

1842 Geo. Eliot *Let.* 8 Mar. (1954) I. 254 Certainly our decayed monarchs should be pensioned off. **1880** G. Meredith *Let.* 27 Apr. (1970) II. 595 Owing to the attack I suffered under last year, I have been pensioned off all work of any worth for late. **1896** G. B. Shaw *Let.* 1 Sept. (1972) II. 60 My mother and I had to pension off and get rid of a relative of hers—a woman who was an incorrigible drunkard. **1916** —— *Androcles & Lion* Pref. p. lxvi, Poor people are cancers in the commonwealth, costing far more than if they were handsomely pensioned off as incurables. **1953** E. O'Neill *More Stately Mansions* (1965) iii. ii. 154 We'll pension off your mother, and give her the house to live alone in. **1965** D. Francis *Odds Against* xi. 134, I..asked him to find her his wretched hack, a pensioned-off old steeplechaser. **1968** *Listener* 5 Sept. 290 The conversion system..is an old and cunning barridan, as irrelevant as Mayor Daley, and ought to be pensioned off. **1973** C. Bonington *Next Horizon* ix. 131 Gone was the crusty old pensioned-off guide, who lived in a little cubby-hole at the end of the..living room. **1973** *Times* 14 July 8/8 What the BBC publication seems to me to need is the attention of a first-class design consultant who might suggest for example that the ruler which, like a rectangular noose, strangles each page, should be pensioned off. **1974** J. Cleary *Peter's Pence* v. 144 Cork had had no gangsters, just some pensioned-off IRA boys. **1977** *R.A.F. News* 11–24 May 18/5 This is my last match report before being pensioned off.

pensionable, *a.* Add: **a.** (Further examples.)

1906 *Westm. Gaz.* 25 Jan. 4/2 There is nothing radically pensionable about old age; grey hairs are not in themselves a claim to society. **1920** *Act* 10 & 11 *Geo. V* c. 67 Sched. viii. 4 This provision shall apply to the pensionable assistants of the petty sessions clerks at Cork and Belfast.

b. (Further examples.)

1908 *Daily Chron.* 3 Jan. 6/7 They will have the ultimate prospect..of appointment to permanent and pensionable establishments... **1945** *Release & Resettlement* (H.M. Govt.) 45 ..you require..treatment on account of the pensionable disablement, treatment will be provided. **1955** *Times* 13 Aug. 6/6 The Civil Service is to offer pensionable jobs to men and women aged between 40 and 60. **1976** *Ulverston (Cumbria) News* 5 Dec. 9/4 (Advt.), The post is pensionable with promotion prospects to Executive Officer grades and above.

c. Related to, connected with, or affecting a person's pension.

1909 *Westm. Gaz.* 8 Feb. 8/3 Saturday's deputation asked for the recognition of colour service in the Forces for pensionable purposes..reaching the age of sixty. **1920** *Act* 10 & 11 *Geo. V* c. 67 Sched. ix, The allowance awarded..shall in no case exceed two-thirds of his actual pensionable salary. **1970** *Money Which?* Mar. 55/1 If, as well as being self-employed, you have an additional job in which you are employed (i.e. one providing you with some pensionable earnings) the £750 limit will be reduced.

Hence **pensionabi·lity**, entitlement to a pension.

1930 *Times Educ. Suppl.* 24 May 235/4 Sick leave privileges, and pensionability. *a*1966 'M. na Gopaleen' *Best of Myles* (1968) 300 Pensionability of entire local populations in respect of military service, notwithstanding international convention as to non-combatancy of juveniles, children and women.

‖ **pensione** (pensiǭ·ne). [It.] In Italy, a small hotel or boarding house.

1938 E. Ambler *Cause for Alarm* v. 85, I must make a real effort to find a *pensione*. **1957** R. Macaulay *Last Lett. to Friend* (1962) 252 We stayed in a small and pleasant *pensione* on the Giudecca Canal. **1967** R. Sawkins *Snow in Paradise* ii. 23 He lived for a while at a *pensione* round the corner. **1969** R. Airth *Snatch!* iv. 30 We got to this *pensione* off Via Margutta. **1975** C. Mott-Radclyffe *Foreign Body* ii. 27 They lived in a *pensione* nearby.

pensioneer (pe:nʃəniə·ɹ), *v.* [f. Pension *sb.* + -eer, after Electioneer *v.*] To bid for votes in an election by promising higher pensions. Hence **pe:nsionee·ring** *vbl. sb.*

1959 *Daily Tel.* 9 June 17/4 The word 'pensioneering'..has recently been added to the political vocabulary... 'Pensioner, you can't trust the Labour Party. They are simply pensioneering, trying to buy your votes on a promise of 10s.' **1960** *Times* 24 Oct. (Financial Rev.) p. xiii/3 That the paradox of comparative insecurity in the midst of affluence is a disturbing thing..has become apparent in the political field..in the shape of competitive 'pensioneering'. **1963** *Times* 23 Jan. 11/7 As the number of pensioners who could not recover financially from the great depression of the 1930's falls away, what *The Times* called 'pensioneering' becomes politically less productive. *a* **1974** R. Crossman *Diaries* (1975) I. 276 Under cross-examination they got into a horrible mess and confirmed my fear that the whole strategy of our pensioneering, worked out for years before the election, had been jettisoned almost without noticing it by the Minister under the *diktat* of Douglas Houghton.

pensionless, *a.* (Later examples.)

1969 *Guardian* 22 Oct. 10/6 If your husband dies before you are 50, and you have no children, you are thrown pensionless into a hard world. **1974** 'R. Tate' *Birds of Bloodied Feather* iii. 74 Pensionless, I looked round for a job.

‖ **pensionnat** (pãnsyona). [Fr.] **a.** In France and other European countries, a boarding-school. **b.** = Pension *sb.* 6b.

1840 J. R. Hope-Scott in R. Ornsby *Mem.* (1884) I. xiii. 247 Their pupils [in Germany] might come to their classes, but at night they go home, and in their pensionnats..it was impossible to prevent the parents coming every fifteen days or so. **1853** C. Brontë *Villette* I. vii. 121 As I spoke English, she concluded I was a foreign teacher come on business connected with the Pensionnat. **1867** J. A. Symonds *Let.* 26 Sept. (1967) I. 761 We do not see very much of other Pensionnats. They are formed chiefly of Germans, Russians & Americans. **1896** C. Shorter *C. Brontë & her Circle* iv. 100 The girls were day boarders at the Pensionnat. **1933** *Times Lit. Suppl.* 28 Sept. 653/1 A mildly amusing story in mildly bad taste about a *pensionnat* in Switzerland. **1963** *Listener* 10 Jan. 73/1 There are increasing numbers of hotels..called *pensionnats*. These differ from regular hotels in providing only the bare facilities of rooms and meals. **1967** R. Petrie *Foreign Bodies* ii. 27 'Then this house is a *pension*? A boarding establishment?.. And not, definitely not, a *pensionnat*?'.. 'Ah, Monsieur..we are a boarding-house, not a boarding-school.' **1972** A. Christie *Elephants can Remember* v. 78 They had children... A boy at school in England and a girl at a *pensionnat* in Switzerland.

penstock¹. Add: **2.** For *U.S.* read 'Orig. and chiefly *U.S.*' and add earlier and later examples.

1799 *Trans. Amer. Philos. Soc.* IV. 349 Let ABCD Fig. 1 represent a large cistern or penstock, and MKLN an orifice made in one of its sides. **1933** *Discovery* Apr. 110/2 The station is several miles below Niagara Falls, the water being led to the 'pen-stocks' (the tubes which guide the water to the turbines) by means of a concrete canal from a point above the falls. **1955** *Times* 14 May 13/1 With the expected completion in 1955 of a modern wood-handling system and major repairs to its older penstocks, the company will have reached the end of the rehabilitation programme. **1963** *Weekly News* (Auckland) 21 Aug. 27 Six parallel penstocks made from prestressed concrete will carry the water from Lake Benmore. **1965** E. L. Myles *Emperor of Peace River* i. xiii. 135 They whip-sawed lumber for the roof and the penstock flue [of a water mill].

penta-. Add: **pe:ntachlor(o)e·thane** *Chem.*, a colourless liquid, C_2HCl_5, that is an intermediate in the industrial production of certain chlorinated hydrocarbons and is used as a solvent; **pe:ntachlor(o)phe·nate** *Chem.*, a salt of pentachlorophenol, esp. sodium pentachlorophenate, C_6Cl_5ONa, a white crystalline solid; **pe:ntachlor(o)phe·nol** *Chem.*, a colourless, crystalline solid with acidic properties, C_6Cl_5OH, which is widely used (often as its sodium salt) in insecticides, fungicides, weed-killers, wood preservatives, etc.; **pentacyclic** *a.*, (*b*) *Chem.*, containing five rings in the molecule; **pentaga·strin** *Pharm.*, a synthetic pentapeptide having the same action as the hormone gastrin; **pentahy·drate**, a hydrate that contains five molecules of water in each molecule; so **pe:ntahydra·ted** *a.*; **pentahy·dric** *a. Chem.*, containing five hydroxyl groups in a molecule; **pentahy·drite** *Min.*, native magnesium sulphate pentahydrate, $MgSO_4 5H_2O$; **pe:ntahydrobo·rite** *Min.* [ad. Russ. *pentagidroborit* (S. V. Malinko 1961, in *Zapiski Vsesoyuz. Min. Obshchesvta* XC. 673)], a hydrated calcium borate, $CaB_2O_4.5H_2O$, occurring as small, colourless triclinic crystals; **pe:ntahydroca·lcite** *Min.* [ad. Russ. *pentagidrokal'tsit*" (P. N. Chirvinskiĭ 1906, in *Ezhegodnik*" *po Geol. i Mineral. Rossii* VIII. 241)], a pentahydrate of calcium carbonate, $CaCO_3.5H_2O$, the natural occurrence of which is uncertain; **pe·ntamer** *Chem.* [*-mer], a polymeric unit or molecule made up of five monomers; hence **pentame·ric** *a.*; † **penta·meride** *Chem.* [after Isomeride] = *pentamer; **pentame·thylene** *Chem.*, (*a*) a cyclic hydrocarbon, C_5H_{10}, usu. called cyclopentane, which is a colourless volatile liquid found in petroleum; (*b*) the bivalent straight chain radical $—(CH_2)_5—$; **pentame:thylenediamine** (-dəiˌæ·min) *Chem.*, a syrupy, fuming liquid, $H_2N(CH_2)_5NH_2$, now usu. called cadaverine, which is a product of the putrefaction of animal proteins; **penta-nu·cleotide** *Biochem.*, an oligonucleotide in which the number of nucleotides is five; **pentape·ptide** *Biochem.*, an oligopeptide in which there are five amino-acid residues in the molecule.

1872 *Jrnl. Chem. Soc.* XXV 232 (*heading*) Action of bromine on pentachlorethane. **1930** T. H. Durrans *Solvents* vii. 119 The following azeotropic mixture is known: Pentachlorethane 85%, glycol 15%, B.P. 154·5°. **1975** *Internat. Jrnl. Chem. Kinetics* VII. 331 The rate of the inhibited pyrolysis of pentachloroethane was studied over the temperature range of 820 to 865°K using the toluene-carrier technique in a stirred-flow reactor. **1938** *Jrnl. Rubber Res. Inst. Malaya* VIII. 325 The material used in the trials..consists of sodium pentachlorphenate and is known by the trade names of 'Santophen 20 S' or 'Santobrite'. **1959** *Times* 24 Sept. 7/2 (Advt.), The most effective chemical for preventing sapstain is Santobrite, Monsanto's sodium pentachlorophenate. **1971** F. C. Ford-Robertson *Terminol. Forest Sci.* 190/1 Its sodium salt (Na pentachlor(o)phenate) is water soluble and is used for preventing fungal stain and surface mould in unseasoned timber and in eradicating dry rot from buildings. **1879** *Jrnl. Chem. Soc.* XXXVI. 463 When heated at 230° with alcohol it [*sc.* perchlorophenol chloride, $C_6Cl_5\cdot OH$] yields pentachlorophenol, C_6Cl_5OH (m.p. 183–184°). **1960** E. L. Delmar-Morgan *Cruising Yacht Equipment & Navigation* xxiv. 231 Rotproofing... Lauryl pentachlor phenol..is colourless, odourless, and very effective, and is available in an emulsion or solution form. **1972** *Timber Trades Jrnl.* 3 June 44/1 The formulation is based upon the independently established fungicides tributyl tin oxide and pentachlorophenol. **1977** *Time* 4 Apr. 56/3 They discovered that cattle in his herd, and those on at least seven other farms in the state, have been ingesting a wood preservative called pentachlorophenol (PCP)—probably when the animals licked the sides of their feed bins. **1899** *Jrnl. Chem. Soc.* LXXVI. i. 742 It is one of the first cases observed of the conversion of a pentacyclic into a hexacyclic carbon compound. **1972** *Science* 16 June 1230/1 *Tetrahymena*..contains a pentacyclic triterpenoid which has not been found in other animals. **1967** *Lancet* 11 Feb. 291/1 (*heading*) Pentagastrin as a stimulant of maximal gastric acid response in man. **1970** Passmore & Robson *Compan. Med. Stud.* II. x. 3/1 For gastric function studies, pentagastrin has the important advantage over histamine in having no circulatory effects. **1974** *Nature* 15 Mar. 238/2 Because changes in fundic mucosal cGMP could result from release of the hormone gastrin from the antrum in response to vagal stimulation, the effect of pentagastrin in concentration sufficient to produce acid secretion..was tested. **1916** *Amer. Jrnl. Sci.* CXCI. 493 It is claimed that crystals of hydrated carbonate have been found in wells and pumps. [*Note*] F. Pfeiffer..who considered them to be pentahydrate. **1975** *Nature* 28 Aug. 718/2 Hydrated offretite..contains a K ion in each cancrinite cage, a pentahydrate of Mg in each gmelinite cage, and hydrated Ca ions in the main channels. **1851** H. Watts tr. *Gmelin's Hand-bk. Chem.* V. 430 Penta-hydrated.—The ordinary form of cupric sulphate. **1951** C. Palache et al. *Dana's Syst. Min.* (ed. 7) II. 487 Chalcanthite and the not well-established minerals pentahydrite and siderotil are isostructural with a number of artificial salts variously including the pentahydrated sulfates and selenates of Mn, Co, Cu, Zn. **1892** *Jrnl. Chem. Soc.* LXII. 29 Xylitol is..an open-chain pentahydric alcohol, of which xylose is the aldehyde. **1952** J. K. N. Jones in E. H. Rodd *Chem. Carbon Compounds* Ib. xix. 1197 Two pentahydric alcohols, adonitol (ribitol) and D-arabitol occur in nature. **1968** J. A. Monick *Alcohols* v. 426 Ribitol..is a crystalline, 5-carbon pentahydric alcohol. **1951** C. Palache et al. *Dana's Syst. Min.* (ed. 7) II. 492 Pentahydrite *Frondel* (priv. comm., 1948). **1972** *Acta Crystallogr.* XXVIII. B. 1448/2 Magnesium sulfate pentahydrate has even been reported..to occur as a mineral (pentahydrite). **1962** *Amer. Mineralogist* XLVII. 1482 (*heading*) New boron minerals—uralborite and pentahydroborite. **1971** [see *nifontovite]. **1973** *Soviet Physics: Doklady* XVIII. 102/1 Pentahydroborite is the final member in the series of natural water-containing metaborates of calcium, all members of which are characterized by a constant ratio of CaO:B_2O_3 = 1:1 with the water content increasing from korzhinskite to pentahydroborite. **1910** *Mineral. Mag.* XV. 427 Pentahydrocalcite... Hydrated calcium carbonate, $CaCO_3.5H_2O$, occurring as a mould-like encrustation on chalk-marl near Nova-Alexandria, govt. Lublin, Russian Poland. **1928** *Ann. Rep. Progr. Chem.* XXIV. 308 The minerals hydroconite, hydrocalcite (trihydrocalcite, pentahydrocalcite), and lublinite periodically come to be regarded as doubtful minerals, because when re-examined on museum material they are found to be merely calcite. **1957** G. E. Hutchinson *Treat. Limnol.* I. x. 660 The deposition of hydrates, supposedly $CaCO_3.3H_2O$, trihydrocalcite, and $CaCO_3.5H_2O$, pentahydrocalcite, in nature has been recorded. **1968** I. Kostov *Mineral.* 531 Trihydrocalcite..and pentahydrocalcite..are unstable and easily change into calcite. **1929** *Chem. Abstr.* XXIII. 3213 From the MeOH ppt. were obtained another 2 g. of the pentamer and 6, 4, and 6 g. of the hexa-, hepta- and octamers resp. **1955** *Jrnl. Polymer Sci.* XVI. 455 The ACA monomer, a considerable part of the linear oligomers up to approximately the pentamer..and a small portion of the sparingly soluble cyclic oligomers dissolve. **1971** *Nature* 30 July 297/3 Thus electron microscopy has shown macroglobulin (immunoglobulin M) to consist of a cyclic pentamer of γG-like (7S) subunits. **1940** Pentameric [see *hexameric* adj. s.v. *hexa-]. **1971** *Nature* 11 June 361/1 Only the monomer 'IgMs' could be detected inside the cell and only pentameric IgM outside. **1940** *Jrnl. Chem. Soc.* 1171 When the proportion of sulphuric acid in the Bertram-Walbaum reagent was 1–1·8% the yield of polymerides was: dimeride, 29·0; trimeride, 19·5; tetrameride, 18·4; pentameride, 15; higher polymerides, 17%. **1887** *Ibid.* LI. 241 This acid is the orthodicarboxylic acid of pentamethylene, corresponding with phthalic acid of the benzene series. **1909** C. A. Keane *Mod. Org. Chem* v. 63 The simplest cyclic compounds containing four and five carbon atoms are the hydrocarbons tetramethylene, C_4C_8, and pentamethylene, C_5H_{10}. **1929** I. W. D. Hackh *Chem. Dict.* (1930) 538/1 *Pentamethylene*, the bivalent radical $—CH_2(CH_2)_3CH_2—$. **1946** E. G. Rochow *Introd. Chem. Silicones* iii. 48 Those pentamethylene groups which are joined to two different silicon atoms form organosilicon polymers resembling those obtained with phenylene groups. **1951** I. L. Finar *Org. Chem.* I. xi. 203 Pentamethylene glycol (pentane-1:5-diol), $CH_2OH\cdot(CH_2)_3\cdot CH_2OH$,..can be obtained from pentamethylene bromide. **1958** *Nomencl. Org. Chem.* (I.U.P.A.C.) A. 16 Pentamethylene $—CH_2—CH_2—CH_2—CH_2—CH_2—$. **1883** *Jrnl. Chem. Soc.* XLIV. 910 Pentamethylenediamine, $C_5H_{10}(NH_2)_2$ is produced by the action of zinc and hydrochloric acid on an ethereal solution of trimethylene dicyanide. **1964** N. G. Clark *Mod. Org. Chem.* xii. 247 Some [aliphatic diamines] occur in nature as a result of bacterial decomposition of proteins; for example,..cadaverine (pentamethylenediamine). **1931** Levene & Bass *Nucleic Acids* x. 303 On warming a solution of the supposed pentanucleotide in 2 per cent solution of sodium hydroxide.., Feulgen split the substance into the two component parts. **1975** *Nature* 6 Mar. 83/2 Statistical considerations indicate that coincidence among oligonucleotides of length six or more (pentanucleotides are marginal) provides strong evidence for primary structural homology in a sequence of 1,600 nucleotides. **1907** *Jrnl. Chem. Soc.* XCII. i. 901 Characteristic of this pentapeptide and of the preceding tripeptide is the property of being precipitated from aqueous solution by ammonium sulphate. **1946** *Biochem. Jrnl.* XL. p. xliv, Their results are only compatible with the presence in the crystals studied either of a simple pentapeptide molecule or of a decapeptide which has crystallographic two-fold symmetry. **1960** *Ibid.* LXXVI. 16P/2 Appreciable hydrolysis..took place yielding, as one of the products, a pentapeptide containing arginine, proline, glycine, and phenylalanine. **1975** *Nature* 2 Oct. 415/1 It probably does not permeate lysosomal membranes, as might be expected for a pentapeptide.

pentad. Add: **2. a.** (Further example.)

1978 *Nature* 26 Jan. 322/2 South of lat 45° S, however, they conclude that average annual temperatures increased between the 1960–64 and 1970–74 pentads.

b. *Meteorol.* A period of five days.

1906 W. Marriott *Hints to Meteorol. Observers* (ed. 6) 67/2 *Pentad*, a period of five days. **1935** *Nature* 12 Oct. 614/1 There has been great discussion of the relative advantages of the 5-day period, or pentad, and the week, with the result that both units have received international approval. **1959** R. E. Huschke *Gloss. Meteorol.* 416 *Pentad*, a group of five. In climatology, it is applied to a period of five consecutive days. It often is preferred to the week for climatological purposes since it is an exact factor of the 365-day year.

pentaerythritol (pe:ntāˌeri·þritǫl). *Chem.* Formerly also **penta-erythritol, penterythritol.** [ad. G. *penta-erythrit* (Tollens & Wigand 1892, in *Ann. d. Chem.* CCLXV. 316): see Penta- and *erythritol.] A white, crystalline, tetrahydric alcohol, $C(CH_2OH)_4$, that is prepared by the condensation of acetaldehyde and formaldehyde and is widely used in the manufacture of paints and varnishes.

1892 *Jrnl. Chem. Soc.* LXII. i. 127 (*heading*) Penterythritol: a tetrahydric alcohol obtained from formaldehyde and acetaldehyde. **1912** *Ibid.* CI. 2091 The condensation between pentaerythritol and aldehydes in general takes place readily in the presence of sulphuric acid of from 30 to 50 per cent. concentration. **1947** Winding & Hasche *Plastics* iii. 78 The ester of penta-erythritol has become an important resin within the last few years. **1958** *Times Rev. Industry* Feb. 19/3 A large plant has been erected at Dumfries for the production of..pentaerythritol, a material for which there is an increasing demand in the field of paints. **1972** *Materials & Technol.* IV. x. 364 Acrolein is also used for preparing resins such as the glass-clear polymer obtained when acrolein is condensed with pentaerythritol.

b. pentaerythritol tetranitrate, a white crystalline solid, $C(CH_2NO_3)_4$, used as an explosive and also as a vasodilator in the treatment of coronary ailments.

1923 *Jrnl. Chem. Soc.* CXXIII. 75 Pentaerythritol tetranitrate. **1958** A. Grollman *Pharmacol. & Therapeutics* (ed. 3) xx. 456 Pentaerythritol tetranitrate, N.N.R.., although of no value for the immediate relief of anginal attacks, may by its more prolonged action reduce the number or severity of the attacks. **1972** *Materials & Technol.* IV. viii. 301 The most important [reaction]..is the esterification with nitric acid, which gives PETN or pentaerythritol tetranitrate, a very powerful detonating agent, exploding when shocked or exposed to heat.

pentagon, *a.* and *sb.* Add: **B.** *sb.* **2.** (With capital initial.) The name given to a pentagonal building in Washington, D.C., the headquarters of the U.S. Department of Defense. Hence used allusively for the U.S. military leadership.

1945 *Amer. N. & Q.* July 54/1 (*heading*) Pentagon pip: an affliction common among Army officers and enlisted men stationed in Washington; brought on by an Army order proposing overseas duty for men whose war work has kept them heretofore in the United States. **1951** *Business Week* 29 Dec. 40 To help answer this recurring question the Pentagon set up its own watchdog—the Industrial Relations Division of Assistant Secretary Anna Rosenberg's office. **1952** *Brewer's Dict. Phr. & Fable* (rev. ed.) 697/2 *Pentagon*.., a vast five-sided building erected in Washington, D.C., to house government officials. It is said to be so great that newcomers who leave their offices never find them again. **1952** *Observer* 30 Nov. 5/4 The Pentagon, that immense monument to modern man's subservience to the desk. **1957** *Listener* 24 Oct. 664/3 The Pentagon hoped that Seato would produce a maximum morale-building effect with minimum demands on the American armed forces. **1959** *Ibid.* 4 June 972/2 The converting of Iran into a United States rocket ordnance depot is an integral part of the Pentagon plan. **1964** M. McLuhan *Understanding Media* v. 51 Life at the Pentagon has been greatly complicated by jet travel. **1972** A. Price *Col. Butler's Wolf* xiii. 142 The students didn't approve of the Kremlin any more than the Pentagon. **1974** *Jrnl. Politics* XXXVI. 82 In cases where appropriations had been provided to cover broad categories, the Defense Department should keep faith with the committee and with Congress by adhering to the detailed justifications presented in support of the Pentagon's budget. **1976** H. MacInnes *Agent in Place* ix. 85 The Pentagon might start investigating its own security.

Pentagonese (pe:ntăgǫnī·z). [f. prec. + -ese.] (See quot. 1961.)

1951 *Collier's* 24 Nov. 33/3 The great virtue of Pentagonese is the facility it provides in conveying meanings briefly. **1961** *Guardian* 20 Mar. 1/4 Pentagonese—the Defence Department's penchant for turning nouns into verbs by the addition of a suffix. **1977** P. Howard *New Words for Old* 58 *Low profile* is Pentagonese, or American defence jargon.

pentagram. Add: **2.** A series of five letters or characters.

1972 *Computer Jrnl.* XV. 260/2 The peak frequencies are steadily reduced, from one occurrence of the space symbol in seven characters in the case of single characters, to a maximum frequency of approximately 600 in 1,000 documents in the case of the most frequent pentagram, TIONV. **1974** *Sci. Amer.* Jan. 108/3 Rows 1, 2, 3, 4 and 5, in 32 parts, give the 32 pentagrams.

pentagrid (pe·ntăgrid). *Electronics.* [f. Penta- + Grid.] A thermionic valve having five grids; a heptode. Freq. *attrib.* or as *adj.*

1933 *Wireless World* 12 May 347/1 The Pentagrid Converter..has been recently developed in America. *Ibid.* 347/3 With the Pentagrid..the necessary coupling occurs within the valve and, by virtue of the screening, is entirely electronic in nature. **1950** P. Parker *Electronics* xiv. 549 Occasionally a frequency-changing valve..has a suppressor grid; it is then called a heptode or pentagrid. **1953** A. H. W. Beck *Thermionic Valves* x. 308 In the pentagrid mixer, the local oscillation is generated by a separate valve. **1966** H. J. Reich et al. *Theory & Applications Active Devices* vii. 165 Another useful tube structure is the five-grid or pentagrid tube, in which the five grids of helical construction are mounted between the cathode and the plate. This tube affords considerable flexibility in design of certain electronic circuits. Perhaps the most frequent application..has been in the mixer-oscillator section of radio receivers.

pen-tail (pe·ntēil). [f. Pen *sb.*² + Tail *sb.*¹] In full, **pen-tail(ed) tree-shrew.** A species of tree-shrew, *Ptilocercus lowii*, found in Malaysia, Sumatra, and Borneo, and distinguished by rows of long, stiff hairs fringing the end part of its tail.

[**1848** J. E. Gray in *Proc. Zool. Soc.* XVI. 23 Mr. Low brought with him from Borneo some mammalia and reptiles in spirits; amongst them..was 'a rat-like animal with a pennated tail'.] **1883** *Encycl. Brit.* XV. 402/1 Pen-tail (Ptilocercus lowii). **1910** *Ibid.* XIV. 639/2 In the pen-tailed tree-shrew..the fringes of long hair are confined to the terminal third of the tail. **1926** *Proc. Zool. Soc.* 1179 The Pen-tailed Tree-shrew, *Ptilocercus*, has not been hitherto described except for its general appearance. **1927** *Glasgow Herald* 14 May 4/2 The probability is that the pen-tail and tupaia represent two successive phases in the evolution of a Lemurid. **1967** *Jrnl. Zool.* CLII. 375 The Pentail tree-shrews..are more insectivorous although meat is occasionally eaten. **1968** E. P. Walker et al. *Mammals of World* (ed. 2) I. 400/1 Pen-tailed tree shrews probably feed mainly on insects and fruit. **1969** Ld. Medway *Wild Mammals Malaya* 47/2 (*heading*) Pentail Treeshrew. *Ibid.*, Unlike other treeshrews, the Pentail is nocturnal; it is also largely arboreal.

pentamery (pentæ·mĕri). *Biol.* [f. Penta- + Gr. μέρος part + -y³.] A condition in which structures are present in groups of five.

1902 *Encycl. Brit.* XXV. 433/2 In the pentamery and dimery of Dicotyledons there is usually a posterior sepal with a pair of lateral prophylls. **1962** D. Nichols *Echinoderms* i. 14 The adult members [*sc.* echinoderms] show a body pattern having structures present in fives (pentamery).

pentamidine (pentæ·midīn). *Pharm.* [f. Pent(ane + *amidine* (f. Amid(e + -ine⁵).] A diamidine that is used, usu. in the form of its isethionate (a white, hygroscopic, crystalline solid), for the prevention and treatment of certain tropical diseases, esp. sleeping sickness; 1,5-di(4-amidinophenoxy)pentane, $(H_2N)(HN)C\cdot C_6H_4\cdot O\cdot(CH_2)_5\cdot O\cdot C_6H_4\cdot C(NH)\cdot(NH_2)$.

1941 Fulton & Yorke in *Ann. Trop. Med. & Parasitol.* XXXV. 229 The name 'stilbamidine' has been given to 4:4'-diamidino stilbene, 'propamidine' to 4:4'-diamidino diphenoxy propane, 'pentamidine' to 4:4'-diamidino diphenoxy pentane, and 'phenamidine' to 4:4'-diamidino diphenyl ether. **1951** A. Grollman *Pharmacol. & Therapeutics* xx. 417 This led to the trial of other diamidine derivatives of which stilbamidine, propamidine and pentamidine are most widely used. **1974** *Jrnl. Protozool.* XXI. 324/2 Gutteridge..found that *C. fasciculata* mitochondrial respiration was inhibited 71% by 1 mM pentamidine. **1977** *Lancet* 3 Sept. 510/2 A man of 66 with non-Hodgkin's lymphoma was suspected of having *Pneumocystis carinii* infection, and pentamidine was given for 5 days.

pentane. Add: Also **pentano·ic** *a.*, in *pentanoic acid*, valeric acid; **pe·ntanol**, amyl (pentyl) alcohol.

[**1899** *Jrnl. Chem. Soc.* LXXVI. i. 742, 2-Chlorocyclopentanol, $C_5H_8Cl\cdot OH$. *Ibid.* ii. 1135/1 (Index), *cyclo*-Pentanol.] **1927** Pentanoic acid [see *hexanoic *a.*]. **1937** F. C. Whitmore *Org. Chem.* i. 129 Pentanol-2, made by either of these methods, always contained pentanol-3. *Ibid.* 298 *n*-Valeric acid, *n*-valerianic acid, pentanoic acid, *n*-propylacetic acid, $CH_3CH_2CH_2CH_2CO_2H$, is made by the oxidation of *n*-amyl alcohol. **1951** I. L. Finar *Org. Chem.* I. vi. 105 Three amyl alcohols, *viz.*, *n*-pentanol, *iso*pentanol and 'active' amyl alcohol, have been isolated from fusel oil. **1964** N. G. Clark *Mod. Org. Chem.* viii. 139 A mixture of higher-boiling alcohols ('fusel oil') is obtained as a by-product and forms a useful source of pentanols (amyl alcohols), $C_5H_{11}\cdot OH$. **1965** *Nomencl. Org. Chem.* (I.U.P.A.C.) C. 112 Saturated aliphatic monocarboxylic acids... Pentanoic. [*Note*] The trivial name is normally preferred.

pentaploid (pe·ntăploid), *a.* and *sb. Biol.* [f. Penta- + *-ploid.] (Made up of somatic cells) containing five sets of chromosomes. Also as *sb.*, a pentaploid organism.

1921 *Ann. Bot.* XXXV. 185 Among the roses examined were diploid, tetraploid, pentaploid, and hexaploid forms. **1921** [see *hexaploid *a.* and *sb.*]. **1946** *Nature* 19 Oct. 536/1 The cytology of species of *Magnolia* has proved the existence of diploids.., tetraploids.., and pentaploids. **1968** *Canad. Jrnl. Genetics & Cytol.* X. 910 Of the progeny obtained by backcrossing the pentaploid hybrid (AABBD) from the synthetic wheat to its tetraploid *T[riticum] durum* parent as male 49% had 28 chromosomes and 28% had 29 or 30.

pentaprism (pe·ntăpriz'm). Also **penta prism, penta-prism.** [f. Penta(gonal *a.* (*sb.*)+ Prism.] A prism whose cross-section is a pentagon with one right angle and three angles of 112½°, so that with silvered reflecting surfaces any ray entering it through one of the faces forming the right angle is deflected through 90°.

1937 *Magneto-theodolite* (Ordnance Survey) 5 The telescope is eccentrically mounted and the change over from theodolite generation to magnetometer operation..takes place in one second by throwing into the line of sight inside the telescope tube a penta-prism which reflects through 270° rays from the illuminated diaphragm of the eyepiece. **1943** D. H. Jacobs *Fund. Optical Engin.* xi. 160 The penta prism..makes possible the modern high-precision rangefinder. This prism has the unique property of deviating a beam exactly 90° in the plane of the rays shown..even if the beam should not strike the end faces exactly normal. **1954** *Amat. Photographer* 19 May 628/2 The Rectaflex is..a..miniature of the eye-level reflex type, the camera having a manually raised 45-deg. mirror..and brilliant-type screen viewed via a pentaprism. **1977** J. Hedgecoe *Photographer's Handbk.* 11 During the past twenty years..the dominant design trend has been the single lens reflex. This includes both the universally popular 35 mm pentaprism types and roll-film versions.

pentastome. Substitute for first part of etym.: [ad. mod.L. generic name *Pentastoma* (C. A. Rudolphi *Entozoorum Synopsis Mantissa* (1819) 123) or *Pentastomum* (F. C. H. Creplin *Novæ Observationes de Entozois* (1829) 76).] Delete 'an aberrant group of Arachnida'. (Later examples.)

1937 Craig & Faust *Clin. Parasitol.* iii. xxxi. 511 (*heading*) 'Tongue worms', Pentastomes (Linguatulida). **1956** T. W. M. Cameron *Parasites* 149 Pentastomes or 'tongue-worms' were until recently regarded as aberrant arachnids.

pentastomid (pentăstǝʊ·mid), *sb.* and *a.* [f. mod.L. name of class *Pentastomida*, f. generic name *Pentastoma*: see prec. in Dict. and Suppl. and -id³.] A parasitic worm-like arthropod of the class Pentastomida; = *tongue-worm* (*b*) s.v. Tongue *sb.* 16. Also as *adj.*, of or pertaining to a parasite of this kind. Cf. *linguatulid.

1909 A. E. Shipley in *Cambr. Nat. Hist.* IV. xx. 488 Pentastomids are unpleasant-looking, fluke-like or worm-like animals. **1943** *Trans. Amer. Microsc. Soc.* LXII. 194 (*title*) Observations on the pentastomid *Kiricephalus coarctatus*. **1957** *Jrnl. Parasitol.* XLIII. 195 Several pentastomids were removed from a series of mammals and reptiles. **1964** T. C. Cheng *Biol. Animal Parasites* xx. 549/1 Adult pentastomids are usually parasitic in the respiratory tract and lungs of vertebrates. *Ibid.* 552/2 When pentastomid eggs are fed to albino rats,..seven immature stages can be identified within the intermediate host. **1978** *Nature* 2 Mar. 93/1 The remaining 23 chapters deal with metazoan parasites (ten on the platyhelminthes,..and a chapter each on the..pentastomids and parasitic crustacea).

pentathlete. Add: **b.** A competitor in a modern penthathlon (see next). Hence **pentathle·tical** *a.*

1968 *Guardian* 19 Mar. 16/6 Athletes, pentathletes and weightlifters had taken advantage of grants..for altitude training. **1973** *Country Life* 10 May 1296/3 The heartening recruitment of pentathletes from the pony club. **1975** *Times Lit. Suppl.* 12 Dec. 1487/1 These figures [of Burne-Jones] are almost always pentathletical types: imposingly tall, long in the thigh, slim and gradual of hip, and very firmly fleshed. **1976** *Field* 26 Aug. 426/3 In saluting the modern pentathletes on 29 July for their Olympic victory, I remarked the difficulties of a team dividing a single solid prize.

pentathlon. Add: **b.** In modern times, a series of five athletic or sporting events imitative of the ancient pentathlon, *spec.* (*a*) (in full *modern pentathlon*) a competition consisting of fencing, shooting, swimming, riding, and cross-country running; (*b*) a competition for women, consisting of sprinting, hurdling, long jump, high jump, and putting the shot. Also *attrib.*

1905 *Olympic Games Athens* 3 *Pentathlum*, consisting of the five following events: 1) Flat race one Olympic Stade..2) Broad jump..3) Throwing the discus..4) Hurling the javelin..5) Wrestling. **1912** *Olympic Games Stockholm* 10/2 *Modern pentathlon*, duel-pistol shooting..swimming..fencing..riding..cross-country race. **1929** F. A. M. Webster *Athletics of To-day* vii. 107 Eighty events were decided..including a pentathlon contest and several of the field events. **1932** *New Yorker* 23 July 8/3 The Hungarian swimming and pentathlon teams arrived. **1948** E. A. Bland *Olympic Story* xxii. 140 In 1912, when the Swedish Olympic Committee was deliberating upon the programme of events for the Vth Olympiad, they sought a test which would produce the best all-round sportsman in the world... The contest known as the Modern Pentathlon was the result. **1961** C. Willock *Death in Covert* x. 189 An Olympic Pentathlon..[is] a kind of marathon in which the competitors run, ride, shoot and so on without a break. **1964** M. Watman *Encycl. Athletics* 130/1 The pentathlon is a five-event test of all-round ability... The penthathlon has long been a most popular women's event, and was introduced into the Olympic schedule in 1964. **1970** B. Tulloh *World Athletics Handbk.* vi. 99 With the pentathlon being omitted..the 1928 Games..reached much the same form for the men's events as it has today. **1976** *Gazette* (Montreal) 19 July O-8/4 Bromont—Thirteen riders of 47 scored a perfect 1,100 points on the modern pentathlon equestrian course. **1979** *Daily Tel.* 2 Oct. 19/4 Pentathlon is a comparatively new sport for women.

Pentathol, var. *Pentothal.

pentatonic, *a.* Add: Also as *sb.*, a scale with five different notes to the octave.

1909 F. R. Burton *Amer. Primitive Mus.* ii. 41 This scale is not what is generally known as the pentatonic, although it consists of the same tones. **1921** H. A. Popley *Mus. India* iii. 28 The pentatonic was the more primitive scale among all peoples. **1928** *Grove's Dict. Mus.* (ed. 3) IV. 100/2 If we continue..to call them gaps, we may notice that they always occur at the interval of a fourth, (or fifth); and that is the distinguishing mark of the true pentatonic. **1936** E. Blom et al. tr. *Einstein's Short Hist. Mus.* 5 In China the development from the non-semitonal to the seven-note scale is certainly traceable, even though the old pentatonic always remained the foundation of its music. **1962** [see *heterophony].

Hence **pentato·nically** *adv.*, according to a pentatonic scale.

1965 *New Statesman* 17 Dec. 980/3 The almost complete disintegration of traditional controls in those pentatonically whirling winds, those parallel-third woodwinds that are sighing trees. **1967** H. Porter in *Coast to Coast 1965–66* 174 'Good af-ter-noon, Mis-ter Pel-lot,' melodiously and pentatonically in duet chanted the Misses Wee.

pentazocine (pentæ·zosin). *Pharm.* [f. Pent(ane + Az(o- + *-ocine* (f. Oc(ta- + -ine⁵).] A tricyclic heterocyclic compound that is a non-addictive analgesic, given as the hydrochloride in tablet form or as the lactate by injection; 1,2,3,4,5,6-hexahydro-8-hydroxy-6,11-dimethyl-3-(3-methyl-2-butenyl)-2,6-methano-3-benzazocine, $C_{19}H_{27}NO$.

1964 *Jrnl. Pharmacol. & Exper. Therap.* CXLIII. 142/1 (*caption*) Compound II has been assigned the generic name 'pentazocine' and Compound IV, 'cyclazocine'. **1967** *Observer* 5 Mar. 13/1 Pentazocine, which in pain-killing potency falls between morphine and pethidine, becomes available in an injectable form (under the name Fortral) on 5 April. **1974** M. C. Gerald *Pharmacol.* xiii. 248 Pentazocine is an effective analgesic agent that also possesses modest narcotic antagonistic properties. **1977** *Sci. Amer.* Mar. 47/2 The prototype is pentazocine (Talwin), which is widely used in the U.S. and is the only powerful opiate analgesic that is not subject to stringent 'dangerous drug' regulations.

Pentecost. 2. Delete '*arch.* or *Hist.*' and add later examples.

1889 H. M. Luckock *Divine Liturgy* xlix. 394 He mentions Epiphany as one of the three days, and omits Pentecost. **1953** A. A. McArthur *Evolution of Christian Year* iv. 165 What does the following of the year actually mean in practice in the periods after Epiphany and Pentecost? **1957** *Oxf. Dict. Chr. Ch.* 1483/1 In the RC Church the Sundays from Pentecost to Advent are usually numbered 'after Pentecost' but in the Anglican Church they are reckoned..'after Trinity', i.e. the Sunday after Pentecost. **1969** *Calendar & Lessons* (Church of England Liturgical Comm.) 5 Trinity Sunday is historically a late feast, exceptional as proclaiming a doctrine rather than commemorating an event. It is only in northern Europe that the Sundays of the unorganized second half of the Church's year have been dated from it, the octave of Pentecost, instead of from Pentecost itself. The calendar proposes that they should be called Sundays after Pentecost. **1976** B. Barker *When Queen was Crowned* 4 Many of the acts, words and anthems in Westminster Abbey in 1953 were seen and heard on the feast of Pentecost in St Peter's church at Bath in 973 when King Edgar was crowned. **1978** P. G. Cobb in C. Jones et al. *Study of Liturgy* vi. i. 418 Pentecost is given a new prominence by becoming a third focus of the Christian Year.

4. The particular day that the Christian feast of Pentecost commemorates, when the Holy Spirit descended upon the apostles (Acts ii).

1882 G. Smeaton *Doctrine of Holy Spirit* ii. 48 The Pentecost was the great day of the Holy Ghost, the opening of the river of the water of life. **1913** W. H. G. Thomas *Holy Spirit of God* i. v. 42 To the disciples the gift of the Holy Spirit at Pentecost may be said to be analogous to the descent of the Holy Spirit on Christ at his baptism. **1925** Chesterton *Everlasting Man* ii. iv. 250 This learned scholar says that Pentecost was the occasion for the first founding of an ecclesiastical, dogmatic and despotic Church utterly alien to the simple ideals of Jesus of Nazareth. **1938** *Doctrine in C. of E.* 161 They would lay stress upon the idea of the Spirit as guiding the Church into all truth, and as therein revealing more fully after Pentecost the significance of what our Lord did at the Last Supper. **1977** *Christian* IV. 33 Jesus' appointment of the twelve, and their special consecration at Pentecost, are not functional devices, otherwise irrelevant to the preaching of the Kingdom.

Pentecostal, *sb.* and *a.* Add: **A.** *sb.* **b.** = *Pentecostalist *sb.*

1904 in C. R. Paige *Alma White's Evangelism* (1939) I. 81 The Pentecostals refrain even from circulating an insinuating hat. **1946** in S. H. Frodsham *With Signs Following* xxiv. 273 The Pentecostals today are receiving the same kind of treatment that the Brethren received in those early days... I have heard as clear a presentation of the gospel in Pentecostal meetings as I ever heard among the Brethren. **1958** E. P. Paulk *Your Pentecostal Neighbor* 7 Pentecostals have spread..until today they compose a major portion of the Christian body. **1969** K. & D. Ranaghan *Catholic Pentecostals* vii. 247 There is no discord between the charismatic movement and the liturgy... There has been no tendency on the part of Catholic pentecostals to substitute prayer meetings or any of the gifts for the sacramental life of the Church. **1971** D. Gelpi *Pentecostalism* v. 132 Non-Pentecostals find it [*sc.* glossolalia] strange and exotic. **1972** S. Durasoff *Bright Wind of Spirit* (1973) xi. 190 In the past Catholics have been looked on with favor only as a fertile field of evangelism by many Pentecostals. **1974** *Encycl. Brit. Micropædia* VII. 858/1 Pentecostals thus hold that a Spirit-baptized believer may receive one or more of the supernatural gifts that were known in the early Church. **1975** *Christian Order* XVI. 419 For the proverbial mess of pottage, the Pentecostals claim, the Church sold her soul to Caesar.

B. *adj.* **b.** Resembling the mixture of nationalities in Jerusalem at Pentecost (Acts ii. 9–11); heterogeneous.

1896 Kipling *Five Nations* (1903) 90, I have watched them in their tantrums, all that pentecostal crew, French, Italian, Arab, Spaniard, Dutch and Greek, and Russ and Jew.

c. Pertaining to or designating Christian sects, movements, and individuals who emphasize the gifts of the Holy Spirit as recorded in Acts ii, seek to express their religious feelings uninhibitedly (e.g. clapping, shouting, and speaking with tongues) and often are fundamentalist in outlook and maintain that a 'baptism in the Spirit' manifested by speaking with tongues is to be sought by all Christians; freq. (with capital initial) forming part of the name of a sect.

1904 *Daily Graphic* 8 Dec. 4/1 Camberwell has received this new form of worship with mingled feelings of tolerance, indifference and hostility, but the Pentecostal Dancers..have maintained their steadfast resolution of dancing themselves into the hearts of their audiences—with but little success so far. **1906** *Apostolic Faith* Oct. 1/1 The waves of Pentecostal salvation are still rolling in at Azusa Street Mission. **1910** *Latter Rain Evangel* Oct. 12/1 A Pentecostal convention will be held..God willing, November 18 to Dec. 4 1910. **1924** *Ibid.* Oct. 10/1 As usual in a real Pentecostal camp meeting, it was harder to stop than to start. **1928** *Amer. Mercury* Oct. 184/2 The Pentecostal Nazarenes have largely duplicated the Methodist system of government by district conferences. **1932** M. Mead *Changing Culture of Indian Tribe* 108 Most of the poor whites go to the Pentecostal Church. **1946** S. H. Frodsham *With Signs Following* (rev. ed.) v. 41 This Pentecostal visitation became so universal in 'The Holiness Church' that it had to be renamed. It is now known as 'The Pentecostal Holiness Church'. **1958** [see *holiness 4b]. **1966** *Listener* 26 May 754/1 Let us consider the West Indian Pentecostal sects. **1970** P. Oliver *Savannah Syncopators* 56 The hand-clapping..remained a familiar characteristic of the services of the 'Sanctified' and 'Pentecostal' churches. **1974** *Amer. Speech 1971* XLVI. 70 It is too soon to know..whether or not the term [*sc.* jackleg preacher] is used in other urban areas where the storefront pentecostal-and-holiness churches abide. **1974** *Observer* (Colour Suppl.) 17 Nov. 34/1 There are the various 'pentecostal' movements which go in for fervent, excited forms of worship.

Pentecostalism (pentīkǫ·stăliz'm). [f. prec. + -ism.] The beliefs and practices of the Pentecostal movement or Pentecostal sects.

1932 F. C. Martin *Holy Ghost versus Mod. Tongues* xix. 159 God forbid that modern Pentecostalism with the tongues deception be allowed to continue. **1936** H. J. Stolee *Pentecostalism* ii. 11 Modern pentecostalism also has its type in the problems of the early Church. **1946** C. B. Nervig *Christian Truth & Relig. Delusions* v. 75 Pentecostalism does teach the great essentials of the Christian religion. **1958** *Eternity* Apr. 9/3 In some ways, Pentecostalism..has moderated its extremes in the past fifty years. **1961** B. R. Wilson *Sects & Society* 8 Some expressions of early Pentecostalism. **1966** *Listener* 26 May 754/2, I am thinking of a movement something like the Black Muslims, which is logically possible as an alternative, or in addition to, Pentecostalism. **1971** D. Gelpi *Pentecostalism* i. 5 Catholic Pentecostalism has spread through wide segments of the American Church.

Pentecostalist (pentīkǫ·stălist), *sb.* and *a.* [f. as prec. + -ist.] **A.** *sb.* A member of any Pentecostal sect; an adherent of the Pentecostal movement. **B.** *adj.* Of or pertaining to Pentecostalism.

1925 *Forum* (N.Y.) Feb. 152 'Pentecostalists' brought to light our partial neglect of the Holy Ghost. **1928** *Amer. Mercury* Oct. 190/1 The epic case of Holy Roller healing is that of the Rev. David Wesley Myland of the Latter Rain Pentecostalists. **1956** *Gordon Rev.* Dec. 131 The Pentecostalists assert that deliverance from physical sickness is provided for in the Atonement. **1958** M. Argyle *Relig. Behaviour* iv. 34 The Baptists and other Evangelical groups were rather similar in 1850 to the Pentecostalists of today. *Ibid.* ix. 110 Boisen gives a case-study of a Pentecostalist leader who was an epileptic. **1961** B. R. Wilson *Sects & Society* i. i. 18 Pentecostalist stress on the Pauline declarations. **1965** *Guardian* 8 Sept. 4/3 The West Indian Pentecostalist sect drew most of its members from the lowest group of Jamaican working class. **1966** *Listener* 26 May 754/2 If Pentecostalists improve their position and become house owners or landlords, will they leave to join a middle-class denomination, or will they restructure their Pentecostalist beliefs, so that a sect becomes a denomination? **1979** R. Blythe *View in Winter* ix. 300 I'm waiting [on God] not listening. Listening is Pentecostalist, and I find it very hard to do.

Pentel (pe·ntēl). Also **pentel**. The proprietary name of a type of felt-tip pen.

1964 *Observer* 8 Nov. 40/3 What is the particular fascination of a Pentel? It's the new kind of thick-flowing pen—something between a felt marker and a ball-point. **1965** *Guardian* 4 Jan. 5/8 The Japanese 'Pentel' pen, has gained a foothold in British markets... The secret of the 'Pentel' is its tip... The 'Pentel'..will write on anything including glass. **1966** *Trade Marks Jrnl.* 30 Mar. 418/2 Pentel..writing instruments, drawing instruments, marking instruments, chalks,..crayons and artists' brushes. Dainihon-Bungu Kabushiki Kaisha.., Tokyo, Japan; manufacturers and merchants. **1969** 'V. Packer' *Don't rely on Gemini* (1970) ix. 76 Gamble was..making notes with a Pentel. **1976** L. Deighton *Twinkle, Twinkle Little Spy* vi. 51 Several of the hi-fi magazines were marked with a red pentel.

penthouse, *sb.* Add: **B. 1. d.** A separate flat, apartment, etc., situated on the roof of a tall building.

1921 *Country Life* Apr. 65/1 Two of the elevators were designed to run to the roof, where a pent-house..was being built. **1937** *Sunday Dispatch* 28 Feb. 2/7 You all know from American lyric writers that a pent-house is a thing stuck on a roof. It may comprise one or two floors. **1945** E. Waugh *Brideshead Revisited* i. viii. 194 They're going to build a block of flats, and..Rex wanted to take what he called a 'penthouse' at the top. **1948** *National Home Monthly* Feb. 21/2 Back in London in 1932..they built London's first penthouse. **1955** A. Huxley *Let.* 18 Mar. (1969) 738 After that expect to be in NYC until mid-June, when I am to be lent a pent-house on Park Avenue. **1956** 'N. Shute' *Beyond Black Stump* ii. 52 They live in lovely sort of flats called penthouses on the top of skyscrapers. **1958** *Times Lit. Suppl.* 4 July 371/3 After years of travel they built in 1936 what the author calls a 'penthouse' in Park Lane. It was eighty feet above street level. **1978** *Country Life* 3 Aug. (Suppl.) 29/1 A Penthouse with magnificent Thames views..to be sold on a 995 years lease at the rental of one Red Rose on Midsummer Day.

2. b. (Further examples.)

1911 *Encycl. Brit.* XXVI. 626/2 [Royal tennis] is now played in a walled and roofed court, 110 ft. by 38 ft. 8 in., the floor, however, measuring but 96 ft. by 31 ft. 8 in., the difference being the width of a roofed corridor, the 'penthouse', which runs along the two end walls and one of the side walls. **1935** *Encycl. Sports* 619/1 This sloping roof is called the 'penthouse', and is, perhaps, the most characteristic feature of a tennis court. **1963** *Times* 25 May 4/5 Those first four games had cost Aberdare and Warburg the first set, but they were now going well, with Aberdare the best of the four on the floor and Warburg, although he put too much on the penthouse, setting up a powerful attack.

3*. (Properly with capital initial.) The name of a theatre at the University of Washington, used *attrib.* to denote a style of theatre production in which the audience sits in a circular formation around a central acting area. (Cf. *arena 5.)

1940 *Nation's Schools* Nov. 19/1 The penthouse style was first employed by us in the autumn of 1932. *Ibid.* 20/3 Modern comedy and farce, preferably with a single interior setting, are the most successful plays for a penthouse theatre. *Ibid.* 21/3 Without the aid of scenic background and conventional stage atmosphere..the ordinary amateur is not an effective instrument and in penthouse productions he must be effective. **1942** G. Hughes *Penthouse Theatre* iii. 17 Plays which could not very well be done Penthouse style. *Ibid.* viii. 53 We designed a Penthouse Theatre because we had created a tradition of arena production. **1959** W. C. Lounsbury *Backstage from A to Z* 5 *Arena stage*... Also known as arena staging, circular staging, theatre in the round, central staging, Penthouse staging, etc.

4. *penthouse apartment, flat, suite*; fig. *penthouse eyebrow, nab*.

1935 A. Squire *Sing Sing Doctor* v. 59 He developed a taste for lavish penthouse apartments. **1948** *Sun* (N.Y.) 30 Dec. 8 (*caption*) Construction view of the eighteen-story and penthouse apartment building being erected by Samuel Rudin on the former Temple Beth-El site at the south corner of Fifth avenue and 76th street. **1947** Auden *Age of Anxiety* iii. 95 Peasants with penthouse eyebrows. **1972** K. Bonfiglioli *Don't point that Thing at Me* vii. 61 Fifth-floor penthouse flats in Upper Brook Street. **1977** *Wandsworth Borough News* 16 Sept. 15/2 Planning Proposals... Star and Garter Mansions, Lower Richmond-road—erection of penthouse flat on the roof of the building situated between two existing false gables. **1699** B. E. *New Dict. Canting Crew*, *Pentice Nab*, a very broad-brmd Hat. **1785** Grose *Dict. Vulgar T.*, *Penthouse nab*, a broad-brimmed hat. **1948** *Time* 8 Nov. 6/1 The eleventh-floor penthouse suite. **1973** *Times* 5 Dec. 18/1 Sir Lew Grade..is occupying the penthouse suite of the plush Century Plaza hotel.

‖ **pentimento** (pentime·nto). Also **pe·ntiment.** Pl. **pentime·nti.** [It. *pentimento*, repentance.] In a painting (particularly in oils), a trace of an earlier composition or of alterations that has become visible with the passage of time. Also *transf.*

1903 R. Fry *Let.* 6 Mar. (1972) I. 204 What looks like a retouch above the man's left shoulder turns out on closer inspection to be an original *pentimento*. **1933** *Burlington Mag.* May 212/1 Holbein's portraits of the 1530's; a pentimento indicates that originally the *décolleté* was narrower. **1935** *Ibid.* June 259/2 There are signs of *pentimenti* on the heads and hands. **1939** *Ibid.* Sept. 96/1 The pentimenti confirm that we are dealing with an original work of the master. **1945** *Ibid.* Apr. 82/2 A striking *pentimento* is visible, in the form of a segment of a circle, over the red dress under the hands of the Virgin. **1951** R. Mayer *Artist's Handbk.* ii. 100 In galleries one may frequently find a picture in which, by reason of changes wrought by time, oxidation, etc., the refractive index of the oil film has changed and a thin coat of paint..has become sufficiently transparent to allow underpainting or drawing to show through. The effect is called pentimento. **1961** L. G. G. Ramsey *Connoisseur New Guide Antique Eng. Pott., Porc. & Glass* 26 On this somewhat rough base the painter has to carry out his designs without benefit of rubbings-out or *pentimenti*. **1962** *Sunday Times* (Colour Suppl.) 8 Apr. 20 A pentiment in the muzzle of the greyhound who looks back over his shoulder shows that the design of the Cleveland picture was followed at first in Mr. Getty's by an assistant, and then changed by Rubens. **1966** *Listener* 9 June 845/1 X-rays reveal *pentimenti*—those first thoughts—in both edges of the ruff, between the right forearm and the chair and, above all, in her right hand, which was originally both higher up and further to the right. **1971** *Guardian* 23 Aug. 5 One of the things that distinguish an original from a copy is the existence of *pentimenti* alterations..done while the work was being painted. **1973** L. Hellman *Pentimento* (1974) 3 Old paint on canvas, as it ages, sometimes becomes transparent. When that happens it is possible, in some pictures, to see the original lines... This is called pentimento because the painter 'repented', changed his mind. **1975** *Times Lit. Suppl.* 31 Oct. 1278/5 He [*sc.* an Italian writer] has recently announced that, after completing all six volumes, he intends to begin all over again with a revised edition to include any *pentimenti* where he has changed his mind.

pentitol (pe·ntitǫl). *Chem.* [f. Pent(ose: see *-itol.] A pentahydric alcohol.

1907 J. B. Cohen *Org. Chem. Adv. Students* I. viii. 310 (*table*) Pentitol. **1954** S. S. Cohen in D. M. Greenberg *Chem. Pathways of Metabolism* I. v. 211 Two naturally occurring pentitols, D-arabitol and ribitol (adonitol), are known. Both appear to be quite rare but are encountered in plant tissues. **1966** *New Phytologist* LXV. 219 Carbon fixed by *Xanthoria aureola* in photosynthesis appears at first mainly in the pentitol. **1971** J. F. Stoddart *Stereochem. Carbohydrates* ii. 20 There are four pentitols.

Pentland (pe·ntlănd). The name of the Pentland Hills, in Midlothian, Scotland, used *attrib.* in **Pentland Crown, Dell,** etc., to designate varieties of potato developed at the Scottish Plant Breeding Station, which is located there.

1959 *Rep. Scottish Plant Breeding Station* 11 (*heading*) Potatoes.—The Registration Trials conducted by the Department of Agriculture for Scotland in 1958 contained thirteen of the Station's seedlings... One..was registered as a new variety suitable for commercial cultivation. It has been named Pentland Crown. **1961** *Ibid.* 16 As a result of its performance in the Merit Trials conducted by the Department of Agriculture and Fisheries for Scotland the [potato] Seedling 2319(a)3 has received 'Commendation'. It has been named 'Pentland Dell', a name that will be registered in compliance with the provisions of the International Code of Nomenclature. *Ibid.* 18 A red variant has been selected from the cultivar 'Pentland Beauty' and as a preliminary to giving it the name 'Red Pentland Beauty' this variant has been entered for the Wart Immunity and Identity Tests. **1962** *Ibid.* 17 An early maturing [potato] seedling, Reference No. 2299 (10), has successfully completed the Merit Trials conducted by the Department of Agriculture and Fisheries for Scotland and..has been named 'Pentland Envoy' a name that will be registered by the National Registration Authority. **1969** *Dict. Gardening* (R. Hort. Soc.) *Suppl.* 150/1 Pentland Beauty. Oval with pale lemon flesh. *Ibid.* 150/2 Pentland Crown. White oval shape. White flesh. Pentland Dell. Shapely white kidney—shows considerable resistance to Blight. **1971** *Arable Farmer* Feb. 3/1 Our variety was Pentland Crown planted in 30" rows. **1973** *Times* 16 Oct. 6/6 Britain's plant breeders.. are striving to keep their brand names in Europe. To our grain growers Maris means another winner from the Plant Breeding Institute at Cambridge, Pentland a good Scottish potato, and Malling an outstanding new fruit. **1976** *Cumberland News* 3 Dec. 34/4 (Advt.), Pentland Crown seed potatoes for sale, once grown from Scotch seed, blight free, excellent quality.

pentobarbital (pentobæ·ɹbitæl). *Pharm.* The equivalent in the U.S. Pharmacopeia of *PENTOBARBITONE. Also *pentobarbital sodium, sodium pentobarbital.*

1931 *Jrnl. Amer. Med. Assoc.* 30 May 1871/1 Pentobarbital-sodium is a nonproprietary name given to the mono-sodium salt of ethyl-(1-methylbutyl)barbituric acid. **1935** *Proc. Mayo Clinic* X. 536 The next step forward was the use of the isomer..of amytal, which was first called 'embutal'. The sodium salt was called 'nembutal' and finally, 'pentobarbital sodium'. **1955** *Sci. News Let.* 2 Apr. 219/2 The Boston doctors tested the effects of morphine, heroin, amphetamine, the sleeping medicine pentobarbital and as a control, sodium chloride, or salt. **1962** *Times* 20 Dec. 9/7 Pentobarbital anaesthesia in lions has been studied with special reference to pre-anaesthetic medication. **1965** *Pharmacopeia U.S.* (ed. 17) 633 Sodium pentobarbital. [*In ed.* 16 *as* pentobarbital sodium.] **1974** M. C. GERALD *Pharmacol.* xi. 204 Deep sedation is induced by the intravenous injection of amobarbital (Amytal) or pentobarbital (Nembutal).

pentobarbitone (pentobæ·ɹbitəʊn). *Pharm.* [f. PENT(ANE + -O + *BARBITONE.] The synthetic compound 5-ethyl-5-(1-methylbutyl)-barbituric acid, which is widely used as a sedative-hypnotic and anticonvulsant drug, usu. in the form of its sodium salt, $C_{11}H_{17}N_2O_3Na$, a white crystalline powder often known by the proprietary name *NEMBUTAL; = *PENTOBARBITAL. So *pentobarbitone sodium, sodium pentobarbitone.*

1938 S. ALSTEAD *Poulsson's Text-bk. Pharmacol. & Therapeutics* (ed. 2) i. 41 Nembutal, pentobarbitone, sodium ethyl-methyl-butyl barbiturate. **1950** *Brit. Med. Jrnl.* 25 Mar. 706/1 The child is premedicated with atropine.. and pentobarbitone. **1960** *Times* 6 Jan. 9/5 Once the rhino is dazed, two more rangers will move in with syringes to pump 600 c.c. of pentobarbitone sodium into the hide. **1962** *Lancet* 29 Dec. 1379/1 Progressive fall in temperature down to cardiac arrest, seen in two rabbits subjected to surface cooling in a deep-freeze at 0°C while under pentobarbitone anæsthesia. **1971** *Nature* 21 May 182/1 The dog was anaesthetized with pentobarbitone and heparin was used as an anticoagulant. **1975** *Ibid.* 4 Sept. 62/1 Each rat was killed with an overdose of sodium pentobarbitone.

pentode (pe·ntəʊd). *Electronics.* [f. PENT(A- + *-ODE.] A thermionic valve having five electrodes. Also *pentode valve.*

1919 W. H. ECCLES in *Electrician* 18 Apr. 475/2 To be systematic I suggest 'tetrode' and 'pentode' for vacuum tubes with four or five electrodes. **1928** *Daily Express* 14 Dec. 5 One pentode valve will do the work of two ordinary valves, but it will be a heavy drain on dry batteries. **1932** E. V. APPLETON *Thermionic Vacuum Tubes* ix. 111 The five electrode tube or pentode possesses a screen grid as does the tetrode, but also has an additional grid situated between the anode and the screen grid. **1965** *IEEE Trans. Electron Devices* XII. 350/2 Conventional beam pentodes can be designed on the computer to have uniform cathode current density. **1973** WHITTLE & YARWOOD *Exper. Physics* xi. 296 The values recorded..are for a pentode valve used for audio-frequency operation, generally as the output stage of an audio-frequency amplifier.

pentolinium (pentoli·niəm). *Pharm.* [f. *pent(amethylene* (s.v. *PENTA-) + *PYRR)OL-(ID)IN(E + *-IUM b.] A white, crystalline powder which has been used as a ganglion-blocking agent in the treatment of severe hypertension; pentamethylenebis(1-methylpyrrolidinium hydrogen tartrate), $C_{23}H_{42}N_2O_{12}$; also called *pentolinium tartrate.*

1954 *Lancet* 27 Nov. 1097/2 The preparation used was 'Ansolysen' (M. & B. 2050A),..termed pentolinium tartrate by the British Pharmacopœia Commission in 1954. **1958** J. H. BURN *Lect. Notes Pharmacol.* (ed. 5) 20 Pentolinium (Ansolysen) is similar in action to hexamethonium, but it is more powerful. **1974** [see *PEMPIDINE].

Pentomic, pentomic (pentǫ·mik), *a. Mil.* [f. PENTA- + *ATOMIC *a.*] Divided into five battle groups armed with nuclear weapons.

1956 *Washington Post* 28 Dec. 1/8 The Army announced last night that it will begin reducing the manpower of its divisions early next year and revamp them into new-style 'Pentomic' units geared for atomic warfare. **1958** *Times* 4 Nov. 9/3 The plans are based on the American concept of 'pentomic' divisions. **1959** *New Statesman* 29 Aug. 238/1 Two and possibly more of the US divisions are now Pentomic (i.e. nuclear-armed), but the process of training the first British regiment of nuclear artillery has only just begun. **1961** *Observer* 4 June 10/2 Orders have been given to break up the so-called 'Pentomic' divisions in the United States Army in Europe, in which nuclear weapons have been integrated with conventional arms. **1963** *Times* 10 Jan. 11/7 The Australian Army had adapted for its own use the American 'pentomic' formation. **1972** [see *low-yield* adj. s.v. *LOW *a.* 23]. **1972** *Sat. Rev.* (U.S.) 6 May 30/3 It was considered necessary to shrink the [US] army combat division from a force of three regiments containing three battalions of five companies each to a 'Pentomic' force of five battle groups of four companies.

pentomino (pentǫ·mino). [f. PENT(A- + D)OMINO by deliberately false analogy: see quot. 1961.] Any of the twelve distinct planar shapes that can be formed by joining five identical squares by their edges.

In the U.S. the pl. *Pentominoes* is registered as the proprietary name of a board game involving these shapes.

1954 S. W. GOLOMB in *Amer. Math. Monthly* LXI. 681 There are twelve distinct pentominoes. **1961** *New Scientist* 2 Nov. 316/2 A domino is formed from two adjacent squares: he [*sc.* Golomb] argues that practical needs and false etymology will justify our calling a square a monomino—whence we have a series monomino, domino, tromino, tetromino,..., and Maestro pieces then become pentominoes. **1964** *Listener* 11 June 975/3 The jigsaw pattern is made from a set of twelve pentominoes. **1965** S. W. GOLOMB *Polyominoes* 13, I learned of the true antiquity of pentominoes, one kind of polyomino. Although the name was coined in my lecture of 1953, the first pentomino *problem* was published in 1907. **1975** *Official Gaz.* (U.S. Patent Office) 21 Jan. TM279/2 Solomon W. Golomb, La Canada, Calif. Filed Sept. 13, 1972. *Pentominoes.* For equipment, consisting of the twelve distinct five-celled square figures and a playing board, for use in various combinatorial puzzles and in competitive board games... First used November 1953. **1975** A. C. CLARKE *Imperial Earth* xxiv. 173 He replaced the titanite cross in its setting between the F, N, U and V pentominoes.

penton (pe·ntǫn). *Biol.* [f. PENT(A- + *-ON[1] 2.] A capsomere which occupies any of the twelve vertices of the icosahedral capsid of an adenovirus.

1966 H. S. GINSBERG et al. in *Virology* XXVIII. 782/2 The unit at the twelve corners of the icosahedron should then be termed a penton because each has 5 neighbouring units (hexons). The penton corresponds to the B or cell-detaching antigen; thus the soluble antigen should be called the penton antigen. **1970** *Nature* 17 Jan. 226/2 Adenovirus has a more complex capsid containing 240 hexons and twelve pentons each with a molecular weight of about 400,000. **1972** *Ibid.* 14 Apr. 348/1 The DNA-specific endonuclease associated with the penton bases of the virion may affect the host genome in an apparently random manner.

pentosan (pe·ntosæn). *Biochem.* Also †**-ane.** [a. G. *pentosan* (Schulze & Tollens 1892, in *Ann. d. Chem.* CCLXIX. 55), after *glucosan, hexosan*, etc.: see *PENTOSE.] Any of the class of polysaccharides, occurring widely in plants, of which the constituent monosaccharides are pentoses.

1892 *Jrnl. Chem. Soc.* LXII. 1420 (*heading*) The pentosans of woody vegetable fibre. **1913** HAAS & HILL *Introd. Chem. Plant Products* II. 51 Gums: (*a*) Natural gums and pentosanes $(C_5H_8O_4)_n$. (*b*) Mucilages and pectic bodies. **1931** E. C. MILLER *Plant Physiol.* viii. 420 The pentosans have the general formula $(C_5H_8O_4)_n$ and occur, for the most part, in the cell walls of various plant parts. **1938** [see *ARABAN]. **1963** R. R. A. HIGHAM *Handbk. Papermaking* ii. 34 The presence of pentosans in wood pulps helps to produce increased fibre bonding in paper. **1973** R. W. BAILEY in Butler & Bailey *Chem. & Biochem. Herbage* I. iv. 165 Hemicelluloses may be divided into two broad classes; pentosans, based largely on pentoses and non-cellulose hexosans which are pentose-free.

pentose. Add: Any of the monosaccharides with the formula $C_5H_{10}O_5$, among which are ribose and several other naturally occurring sugars. (Earlier and later examples.) [First formed as G. *pentose* (E. Fischer 1890, in *Ber. d. Deut. Chem. Ges.* XXIII. 934).]

1890 [see *heptose* s.v. *HEPTA-]. **1916** A. P. MATHEWS *Physiol. Chem.* ii. 30 The pentoses generally occur in nature in gums and..polysaccharides. **1927** *Jrnl. Biol. Chem.* LXXIII. 18 Any method of pentose estimation.. must prove to be efficacious, not only for xylose and arabinose, but especially for *d*-ribose, for it is as ribose compounds that one finds pentose in the body. **1953** FRUTON & SIMMONDS *Gen. Biochem.* xvii. 381 A number of pentoses have been found in nature; perhaps the most important of these is D-ribose,..a constituent of nucleic acids and of several nucleotides (ATP, DPN, etc.). **1973** R. G. KRUEGER et al. *Introd. Microbiol.* viii. 263/2 Pentoses (five-carbon sugars) are carbon and energy sources for many microorganisms.

2. Special combs.: **pentose nucleic acid,** a nucleic acid in which the sugar is a pentose; (effectively synonymous with *ribonucleic acid*, RNA); **pentose phosphate cycle, pathway,** or **shunt,** a cyclic pathway in the body and in higher plants by which glucose phosphate is converted to a pentose phosphate with the reduction of NADP, the pentose phosphate being afterwards converted into phosphates of a hexose and a triose or else incorporated into nucleotides.

1934 *Biochem. Jrnl.* XXVIII. 2108 The pentose nucleic acid of the pancreas gland. **1947** *Thorpe's Dict. Appl. Chem.* (ed. 4) VIII. 622/1 The idea has arisen that deoxypentose nucleic acid is present in the nucleus, pentose nucleic acid in the cytoplasm. **1953** FRUTON & SIMMONDS *Gen. Biochem.* vii. 184 Although it was once thought that the pentose nucleic acids were characteristic of plant tissues whereas the deoxypentose nucleic acids were confined to animal cells, this separation has been shown to be incorrect. **1968** I. L. FINAR *Org. Chem.* (ed. 4) II. xvi. 724 The nucleic acids are classified according to the nature of the sugar present: the pentose nucleic acids or ribonucleic acids (R.N.A.), and the deoxypentose nucleic acids or deoxyribonucleic acids (D.N.A.). **1960** *McGraw-Hill Encycl. Sci. & Technol.* II. 40/1 The pentose phosphate pathway of glucose decomposition involves hexose monophosphates and pentose monophosphates. **1963** C. H. DOERING tr. *Karlson's Introd. Mod. Biochem.* xv. 269 (*heading*) Glucose oxidation through the pentose phosphate cycle. **1964** W. G. SMITH *Allergy & Tissue Metabolism* viii. 85 Supplies of $NADPH_2$ can be made available by shunting some of the available glucose-6-phosphate into reactions which form ribulose-5-phosphate. This in turn can be converted back to glucose-6-phosphate... The whole process is sometimes referred to as the pentose phosphate shunt. **1970** R. W. MCGILVERY *Biochem.* xxiii. 568 The adrenal cortex has an active pentose phosphate pathway that can provide NADPH in the cytosol by oxidizing glucose-6-phosphate.

pentoside (pe·ntosəid). *Chem.* [f. PENTOS(E, after GLUCOSIDE.] A glycoside which yields a pentose on hydrolysis.

The pentosides of most interest are now referred to as nucleosides.

[**1909**: see *GUANOSINE]. **1910** [see *GLUCOSIDE]. **1916** A. P. MATHEWS *Physiol. Chem.* iv. 171 Guanosine is, therefore, a pentoside. *Ibid.* 172 From yeast another pentoside was isolated, an adenine pentoside called adenosine.

pentosuria (pentosiū³·riă). *Med.* [mod.L., ad. G. *pentosurie* (E. Salkowski 1895, in *Berl. klin. Wochenschr.* 29 Apr. 364): see PENTOSE and -URIA.] The presence of an excess of pentoses in the urine.

1902 *Med. Rev.* V. 94/2 (*heading*) The clinical importance of chronic pentosuria. **1936** *Nature* 7 Nov. 805/2 There are known to be at least two types of the rare chronic pentosuria, differing in the nature of the pentose sugar found, one being optically inactive *dl*-arabinose and the other *l*-xyloketose. **1970** PASSMORE & ROBSON *Compan. Med. Stud.* II. xxxi. 9/1 [Garrod] suggested that in alkaptonuria, albinism, cystinuria and pentosuria there was a block in a metabolic process due to an inherited deficiency of a specific enzyme.

Hence **pentosu·ric** *a.*, of, pertaining to, or having pentosuria; *sb.*, an individual with pentosuria.

1906 J. L. SALINGER tr. F. Blumenthal in R. C. Cabot *Mod. Clin. Med.* 267 The researches..have..made it appear unlikely that pentoses are formed in the pentosuric patient by an imperfect nuclein decomposition. *Ibid.* 271 Have pentosurics..a special liability to become diabetics? **1933** CAMERON & GILMOUR *Biochem. of Med.* vi. 91 Diabetic treatment is unsuitable to the pentosuric. **1964** D. Y. HSIA in G. A. Duncan *Dis. Metabolism* (ed. 5) v. 349 In the 1950's, Lasker reported that she had identified L-xylulose in all 72 pentosuric urines that she had tested. **1968** MARKS & SAMOLS in F. Dickens et al. *Carbohydrate Metabolism* II. xiii. 351 The existence of this pathway of glucose metabolism was first recognised as a result of studies on the biosynthesis of L-ascorbic acid in animals and of L-xylulose in pentosurics.

Pentothal (pe·ntoþæl). *Pharm.* Also (*erron.*) **Pentathol,** and with small initial. [Refash. of *THIOPENTAL.] A proprietary name for thiopentone sodium; also called *Pentothal sodium, sodium Pentothal.*

1935 *Proc. Mayo Clinic* X. 744 Sodium ethyl 1-methyl butyl thiobarbituric acid (thionembutal or pentothal sodium) and sodium allyl secondary butyl thiobarbituric acid (thiosebutal) induce a satisfactorily deep but relatively transitory anesthesia. **1936** *Official Gaz.* (U.S. Patent Office) 11 Feb. 238/2 Abbott Laboratories, North Chicago, Ill. Filed Dec. 6, 1935... Pentothal... For pharmaceutical product having hypnotic, sedative and anesthetic properties. Claims use since Nov. 15, 1935. **1936** *Trade Marks Jrnl.* 28 Oct. 1326/1 Pentothal... Chemical substances prepared for use in medicine and pharmacy. Abbott Laboratories.., City of North Chicago,..State of Illinois, United States of America; manufacturers of pharmaceutical products. **1940** *N. & Q.* 22 June 448/2, I observe in the press mention of two drugs to induce sleep with new names, Evipan and Pentothal. **1946** M. DICKENS *Happy Prisoner* viii. 153 Elizabeth..suggested Pentathol therapy. **1967** *Punch* 1 Mar. 321/3 One could only ask—..after the mystics, the ESP experts and the computing machines, after the sodium pentathol examinations and the lie detector tests..—after all that was he to turn out to be this nondescript house-painter and handyman who killed at random? **1969** *Daily Tel.* 16 Dec. 15/1 Interviewed under the truth drug pentathol, he told doctors he did not realise he had done the baby any harm. **1973** G. GREENE *Honorary Consul* v. iii. 273 Dying is a wonderfully effective truth drug, better than pentathol. **1974** M. C. GERALD *Pharmacol.* xi. 195 Nonvolatile anesthetics include the intravenously administered barbiturate thiopental (Pentothal).

pentryl (pe·ntril). [f. PENTA-, after *TETRYL 2 (prob. from there being five rather than four nitro groups in the molecule).] A crystalline compound similar to tetryl in chemical structure, explosive power, and sensitivity to detonation; $C_6H_2(NO_2)_3{\cdot}N(NO_2)CH_2CH_2ONO_2$.

1933 L. V. CLARK in *Industr. & Engin. Chem.* Dec. 1385/1 In this paper the writer presents the results of an investigation of *sym*-trinitrophenylnitraminoethyl nitrate, hereafter termed 'pentryl'. *Ibid.* 1386/1 Pentryl detonates when struck a sharp blow. **1967** M. JURECKI tr. *Urbański's Chem. & Technol. Explosives* III. iii. 71 Pentryl is remarkable for its high explosive power which, according to various authors, is equal to or slightly higher than that of tetryl.

penuche, var. *PANOCHE.

penumbral, *a.* (Further *fig.* examples.)

a **1922** T. S. ELIOT *Waste Land Drafts* (1971) 37 Within this penumbral consciousness. **1965** *Mod. Law Rev.* XXVIII. 510 The American concern with the judicial process and the creative element in the common law is particularly appropriate in analysing the role of appeal courts, for appeal courts..are primarily concerned with penumbral issues.

penumbrous, *a.* For *rare*[−1] read *rare* and add further examples.

1959 *Times* 16 May 8/6 To proceed to some stygian cellar,..and from the penumbrous interior gathering them [*sc.* mushrooms] by the hundredweight, is not, of course, sport. **1964** *Listener* 19 Mar. 484/2 He's more readable than Grimm, and much more concerned to give a solid idea of those contradictory, penumbrous figures.

Penutian (pĕnū·ʃən, -tiăn), *sb.* and *a.* Also **Penuti.** [f. Maiduan *pen* two + Miwokan and Costanoan *uti* two + -AN.] **A.** *sb.* **a.** A North American Indian language stock comprising the Miwokan, Costanoan, Wintuan, Maiduan, and Yokutsan families of California. Also, a proposed language phylum comprising the Penutian stock (distinguished as *California Penutian*), the Oregon Penutian and Plateau Penutian stocks, the Chinookan and Tsimshian families, and (in the usage of some scholars) certain other language groupings of North and Central America. **B.** *adj.* Of or pertaining to Penutian.

1912 DIXON & KROEBER in *Amer. Anthropologist* XIV. 692 The new larger families and their components are: *Penutian*, comprising the groups formerly known as Maidu, Wintun, Miwok, Costanoan, and Yokuts. [Etc.] **1914** *Ibid.* XV. 649 There is available enough information on the structure of the five Penutian languages to prove their genetic affinity beyond a doubt. **1932** W. L. GRAFF *Lang.* 431 The most important North American branches: Eskimo..Penutian. **1959** *Chambers's Encycl.* VIII. 360/1 The 22 language-families of North American are: Algonquin,..Penuti, [etc.]. **1965** *Canad. Jrnl. Linguistics* Spring 78 Suggestions of affiliation with ..Penutian. **1968** R. W. LANGACKER *Lang. & its Struct.* viii. 232 Continuing north, we find Takelman..Chinookan, and Plateau Penutian (including Klamath, Modoc, Nez Percé, Cayuse, and Yakima). **1972** *Language* XLVIII. 378 Chinook Jargon, once the means of communication among the (as yet) unrelated Indo-European, Athapaskan, Salishan, Penutian, and Wakashan-speaking peoples of the Pacific North-west. **1973** H. LANDAR in *Current Trends in Linguistics* X. 1294 California Penutian. Rubruz for Yokuts, Maidu, Miwok-Costanoan, and Wintun. **1974** *Encycl. Brit. Micropædia* VII. 859/2 The Penutian languages are sometimes grouped into a yet larger stock, called either Penutian or Macro-Penutian, that includes several Meso-American Indian languages. **1977** C. F. & F. M. VOEGELIN *Classification & Index World's Lang.* 287 Their [*sc.* Dixon and Kroeber's] grouping did not extend Penutian beyond the so-called California Penutian. *Ibid.*, A dozen separate language families beside nine different language isolates..are said to be remotely related to each other within the Penutian phylum.

Hence **Penu·tianist,** a student of Penutian.

1965 *Language* XII. 173 Sapir's influence extends also to Penutianists.

penwiper. Add: **a.** (Earlier examples.)

1826 E. BROWN *Let.* in C. Oman *Ayot Rectory* (1965) iv. 90 Brown holland pockets containing housewives,..pencils, penwipers, knives. **1838** DICKENS *Diary* 2 Jan. in *Lett.* (1965) I. 629, I wrote to Mrs. Hogarth yesterday,..sending as a New Years' token a pen-wiper of poor Mary's. **1840** —— *Sk. Young Couples* 88 All manner of presents, such as pocket-books, pencil-cases, pen-wipers. **1841** C. RIDLEY *Let.* Nov. in *Cecilia* (1958) vi. 74, I want to know whether you like penwipers with white linen inside? They are all so here and the housekeeper renews them now and then.

b. A handkerchief. *slang.*

1902 FARMER & HENLEY *Slang* V. 170/1 *Penwiper*..a handkerchief. **1942** BERREY & VAN DEN BARK *Amer. Thes. Slang* §88/4 Handkerchief..penwiper.

pe·nwork. Also with hyphen and as two words. [PEN *sb.*[2]] **a.** Work done with a pen; writing. **b.** The decoration and ornamental lettering of illuminated books and manuscripts done with a pen; also, decoration drawn with a pen on the surface of furniture. Also *attrib.*

1644, etc. [in Dict. s.v. PEN *sb.*[2] 7]. **1844** H. N. HUMPHREYS *Illum. Bks. Middle Ages* 6/1 The letters..terminating in the margin in long tails,..only, instead of being solid, they are formed of light lines slightly enriched with simple and delicate penwork. **1901** J. W. BRADLEY *Illum. Lett. & Borders* iv. 49 The execution [of Celtic illumination] chiefly consists in pen work in black or coloured inks. **1906** E. JOHNSTON *Writing & Illuminating* xiii. 218 Many of the most beautiful MSS. were made in pen-work throughout. And it is well that the penman should stick to his pen as much as is possible. **1928** O. E. SAUNDERS *Eng. Illumination* I. 110 What distinguishes these borders especially from earlier work are the feathery sprays, indicated by pen-work lightly touched with green and ending in flourishes like the tendrils of a vine. **1969** N. R. KER *Medieval MSS. in Brit. Libraries* I. p. xiii, The terms 'decoration'..and 'ornament'..have been used in describing the surrounds of initials to make the important distinction between work with a brush and penwork flourishings. **1973** *Times* 13 Apr. 18/6 The curiosity of the sale was a monumental Empire ebonized secretaire-cabinet, the top shaped as a temple supported on lightly draped female figures and the whole lavishly applied with classical ormolu motifs and penwork. **1976** *Codicologica* I. 78 Practically no research has been done on the history of decoration of the medieval manuscript and of the numerous elements included in that word: initials,.. penwork flourishes on initials or paragraph signs, [etc.]. **1978** *Bodl. Libr. Rec.* IX. 326 Their simple penwork interlace is frequently enlivened by the inclusion of a bird or monster head, a flower or a peacock.

Hence **pe·nworker,** one who works with a pen; **pe·nworked** *a.* (in sense b).

1901 Penworker [in Dict. s.v. PEN *sb.*[2] 7]. **1965** *Harper's Bazaar* June 76/3 A Regency working-table with an exquisite black pen-worked top..£70.

Penzance (penzæ·ns). The title of James Plaisted Wilde, Lord *Penzance* (1816–99), English lawyer and amateur horticulturist, used *attrib.* in **Penzance briar, rose,** to designate a rose belonging to a group of hybrids developed by him from the sweet briar, *Rosa rubiginosa*, and distinguished by scented foliage and single flowers.

[**1891** *Jrnl. R. Hort. Soc.* XIII. p. cxviii, One of the most interesting features in the exhibition was a stand of seedling Sweet Briar hybrids raised by the Right Hon. Lord Penzance... In the case of Lady Penzance the perfume of the leaves was retained in conjunction with small single flowers of a rosy-salmon hue.] **1902** A. FOSTER-MELLIAR *Bk. of Rose* (ed. 2) ii. 20 It is a great mistake to plant these 'Penzance' or any other Sweet-briars in Rose-beds; for they are very strong growers. **1907** *Gardeners' Chron.* 15 June 382/3 Rambler, Wichuraiana, and Penzance Roses will all submit to gentle forcing. **1912** E. WILLMOTT *Genus Rosa* II. 455 With the introduction of the Penzance Briars a new race of roses came into being. **1935** N. MITCHISON *We have been Warned* II. 196 There were Penzance briars and species roses growing unpruned. **1956** B. PARK *Collins' Guide to Roses* xi. 195 Lord Penzance produced by hybridizing many distinct seedling varieties at the end of the XIXth century which he called Penzance Briars; they represent a distinct series. **1969** C. LLOYD *Gardening on Chalk & Lime* vi. 60, I should avoid the Penzance briars... They are victims of black spot.

peola (pī,ōu·lă). *U.S. Black slang.* [Etym. unknown.] A light-complexioned Black person, esp. a girl.

1942 Z. N. HURSTON in A. Dundes *Mother Wit* (1973) 224/2 Dat broad I seen you with wasn't no pe-ola. **1944** C. CALLOWAY *Hepsters Dict.*, *Peola*, a light person, almost white. **1970** C. MAJOR *Dict. Afro-Amer. Slang* 90 *Peola*, a light-skinned Afro-American girl.

peon. Add: **1.** Also in Bangladesh, Sri Lanka (Ceylon), and Malaysia. **c.** Also, a person who does minor work in an office.

1913 L. WOOLF *Village in Jungle* vii. 211 The peon and the interpreter told Babu to hold his tongue. **1927** R. J. H. SIDNEY *In Brit. Malaya To-Day* 136 Postmen, Government *peons* (messengers), prisoners themselves, all wear materials either made or made-up in Singapore prison. **1931** *Times* 17 Feb. 13/5 Five postal *peons* [servants] were injured while on duty, and the City Post Office was closed for three days... Instead of deliveries by *peons* people are asked to inquire at the post office for letters. **1969** [see *DURZEE]. **1969** *Pioneer* (Lucknow) 13 Aug. 6/2 A peon in the office of the regional transport authority in Bombay attends office in his own car daily. **1971** *Ceylon Observer* (Mag. ed.) 19 Sept. 7/5 A peon has to come to the rescue of the official. **1972** *Straits Times* 26 Apr. 7/6 F and N workers have been picketing the factory..in support of a claim for free shoes for three peons. **1973** *Archivum Linguisticum* IV. 91 Peons, low-caste strangers, and their wives. **1975** *Bangladesh Times* 23 July 1/1 Six persons including a professor, a B.Sc examinee and a college peon, were arrested by the Lalbagh police on Tuesday on charge of adopting unfair means in the examination and abetment of the crime.

2. Also in extended use. (Further examples and earlier *attrib.* example).

1826 W. B. DEWEES *Lett. from Early Settler Texas* (1852) 56 The Peons, or lower class—are a sort of slaves, who are employed by the aristocracy. **1826** [see *ARRIERO]. **1847** W. S. HENRY *Campaign Sk. War with Mexico* xii. 134 This 'peone' system is fully equal to our slavery. **1945** J. L. MARSHALL *Santa Fe* 9 In 1850 New Mexico came into the Union as a free soil territory—and went on buying and selling slaves as of old, calling them *peons* and *peonas.* **1962** N. MAXWELL *Witch-Doctor's Apprentice* vii. 81 They weren't Cotos, only a white man named Rodríguez, his partner, Juan Gómez, and five peons. **1970** S. L. BARRACLOUGH in I. L. Horowitz *Masses in Lat. Amer.* iv. 125 On some traditional Andean haciendas, peons still kneel to kiss the corner of the 'patron's' poncho to show respect. **1972** P. DANIEL *Shadow of Slavery* ix. 177 In New Orleans a couple..took the peons in, gave them directions on how to proceed to..Chicago. Perhaps thirty persons escaped on this modern version of the underground railroad. **1977** *Time* 6 June 42/2 He [*sc.* Elvis Presley] periodically tossed a sweat-stained scarf to the peons below.

‖**3.** *pl.* **peones.** = BANDERILLERO.

1932 [see *BANDERILLERO]. **1957** A. MACNAB *Bulls of Iberia* v. 50 Each matador's team consists of five assistants: two picadors, mounted, and three peones or capemen on foot—also called *banderilleros* because they plant the sticks called *banderillas.* **1967** MCCORMICK & MASCAREÑAS *Compl. Aficionado* ii. 52 The senior matador who knows and likes this breed, has told his peones not to show a cape until he gives the signal. **1976** E. P. BENSON *Bulls of Ronda* iv. 26 Navarro's peon played the bull for a few minutes before El Zorro gave the bull to Navarro. The boy began a series of passes with his cape.

peonage. Add: **1.** (Further examples). Also, in parts of southern U.S., an arrangement whereby convicts are leased to contractors. Also *transf.* and *attrib.*

1850 *Ex. Doc. 31st U.S. Congress 1 Sess. Senate* No. 64. 49 From this cause, and the miserable system of 'peonage' that prevails, the products of agriculture are barely sufficient to support the inhabitants [north of El Paso]. *a* **1889** J. J. WEBB *Adventures Santa Fé Trade* (1931) 101 The system of peonage, or voluntary servitude, was a fixed institution. **1903** *Nation* (N.Y.) 3 Dec. 436/3 More peonage revelations in various portions of the South must be opening the eyes of those editors who criticised us last spring for believing that the Alabama cases were other than sporadic and unparalleled happenings. **1934** A. WOOLLCOTT *While Rome Burns* 291 The Guild must often wonder why..it should ask..a company to enter into so benumbing a peonage. **1969** A. G. FRANK *Latin Amer.* xix. 302 After the French intervention and under the thirty-year reign of Porfirio Díaz, peonage returned in full force and concentration of landownership became worse than ever. **1972** P. DANIEL *Shadow of Slavery* ii. 33 The root of the peonage problem was anchored in the long-practiced abuse of black laborers. **1972** 'E. LATHEN' *Murder without Icing* (1973) xi. 107 The player..contended that, if he could be forced to..work for another team at the whim of his employer, he had been reduced to an unconstitutional state of peonage. **1972** *National Observer* (U.S.) 27 May 7/4 Federal Judge J. Robert Martin, Jr., had told the jury that Federal law required that proof of 'peonage' must be based on evidence of indebtedness. **1973** L. G. FORER in A. E. Wilkerson *Rights of Children* 31 Legality of the commitment would then be the issue—not the procedures by which he was committed or the quantum of care that he received or the existence of institutional peonage.

peonidin (pīǫ·nidin). *Chem.* [ad. G. *päonidin* (Willstätter & Nolan 1915, in *Ann. d. Chem.* CDVIII. 136), f. *päonin* *PEONIN: see *-IDIN.] An anthocyanidin (usu. isolated as the reddish brown chloride, $C_{16}H_{13}O_6Cl$) that is the aglycone of peonin.

1915 *Chem. Abstr.* IX. 1307 Hydrolyzed by boiling 2·5 min. with 20% HCl, peonin gives 2 mols. glucose and peonidin chloride. **1956** I. L. FINAR *Org. Chem.* II. xv. 587 Peonidin is the monomethyl ether of cyanidin. **1965** J. B. HARBONE in Bonner & Varner *Plant Biochem.* xxiv. 624 While over a hundred different flavonoid aglycons have been isolated from plants, only eleven of these occur at all commonly... Six are anthocyanidins: the scarlet pelargonidin, the crimson cyanidin, the mauve delphinidin, and the three simply derived methyl ethers, peonidin, petunidin, and malvidin.

peonin (pī·ōnin). *Chem.* [ad. G. *päonin* (Willstätter & Nolan, 1915, in *Ann. d. Chem.* CDVIII. 136), f. *päonie* PEONY: see -IN[1].] An anthocyanin (usu. isolated as the reddish violet chloride, $C_{28}H_{33}O_{16}Cl$) that is the colouring matter of red peonies and on hydrolysis gives peonidin and glucose.

1915 *Chem. Abstr.* IX. 1307 The deep violet-red peony meal contains 3–3·5% peonin. **1946** *Nature* 7 Sept. 342/1 An acid aqueous extract of the yellow 'Hofmann's Glory' added to synthetic peonin or malvin makes the red solutions much bluer, and they then simulate the appropriate flower extracts. **1956** I. L. FINAR *Org. Chem.* II. xv. 687 Peonidin chloride..is produced..when peonin chloride is hydrolysed with hydrochloric acid.

peonism. (Examples.)

1851 D. WEBSTER *Wks.* V. 351, I suppose there is no slavery of that description in California now. I understand that *peonism*, a sort of penal servitude, exists there, or rather a sort of voluntary sale of a man and his offspring for debt. **1857** W. W. H. DAVIS *El Gringo* ix. 231 Another peculiar feature of New Mexico is the system of domestic servitude called peonism, that has existed, and still exists, in all the Spanish American colonies.

peony. Add: **2. b.** *peony-flowered* (later examples), *-pink, -red* adjs.

1813 M. EDGEWORTH *Let.* 19 Apr. (1971) 21 Her color is less of the peony red than it used to be. **1905** *Daily Chron.* 15 May 3/3 A pretty hat.., made in soft Manilla straw, in the natural colouring, and trimmed with one large pink peony, in soft satin and chiffon, and folded draperies of Louisine silk ribbon, in peony-pink shot with white. **1906** *Ibid.* 4 Oct. 6/5 A new variety of peony-flowered, or art dahlia, in crimson and pink, is shown. **1907** *Ibid.* 18 Sept. 3/5 Some very fine examples of the new peony-flowered variety of dahlia. **1927** *Eaton's News Weekly* 12 Mar. 20 This pretty..frock..comes also in gooseberry green and palmetto green, or in peony red tones. **1957** T. R. H. LEBAR *Dahlias for Everyone* ii. 19 The peony flowered varieties..had comparatively flat petals. **1976** W. E. SHEWELL-COOPER *Basic Bk. Dahlia Growing* ii. 16 Paeony-flowered Dahlias have blooms with two or more rings of generally flattened ray florets, the centre forming a disc.

people, *sb.* Add: **1. c.** *Peoples of the Sea*: name given in Egyptian records of the 19th and 20th Dynasties to various sea-borne migrant peoples who invaded and settled parts of Egypt, Syria, and Palestine. See also *Sea Peoples* s.v. *SEA *sb.* 23.

1906 J. H. BREASTED *Hist. Egypt* VI. xxiii. 477 The restless and turbulent peoples of the northern Mediterranean, whom the Egyptians designated the 'peoples of the sea', were showing themselves in ever increasing numbers in the south. **1950** H. L. LORIMER *Homer & Monuments* v. 150 On the monuments of Ramses III the most conspicuous of the Peoples of the Sea, the Shardana and Pulesati, are uniformly represented with round shields with single hand-grips. **1973** K. A. KITCHEN in D. J. Wiseman *Peoples Old Testament Times* iii. 57 The Lukka..appear as raiders in the Amarna letters *c.* 1370 B.C., as Hittite allies against Ramesses II at Kadesh *c.* 1286 B.C., and then in Libya with Libyans and others in the first attack by 'Peoples of the Sea' on Egypt, repulsed by Merenptah *c.* 1220 B.C.

2. c. *People of the Book*: a body or community whose religion entails adherence to a book of divine revelation, *spec.* [tr. Arab. *Ahl al-Kitāb*] Jews and Christians as regarded in Muslim thought.

1834 A. BURNES *Bokhara* I. x. 313 The Vizier took a cup, and said, 'You must drink with us; for you are people of the book, better than the Russians.' **1861** J. M. RODWELL tr. *Koran* 635 O people of the Book! now hath our Apostle come to you to clear up to you The cessation of Apostles. **1885** T. P. HUGHES *Dict. Islam* 280/2 *Kitabi*, a term used for one of the *Ahlu 'l-Kitāb*, 'the people of the Book', or those in possession of the inspired word of God, as Jews or Christians. **1900** 'ODYSSEUS' *Turkey in Europe* v. 178 According to strict [Muslim] theology, Jews and Christians are called 'People of the Book' (Ehlu-'l-kitab), and enjoy a position superior to that of heathen polytheists. *a* **1936** KIPLING *Something of Myself* (1937) viii. 224 It is true the Children of Israel are 'people of the Book', and in the second Surah of the Koran Allah is made to say: 'High above mankind have I raised you.' **1959** [see *KITAB]. **1967** *Guardian* 19 June 8/3 It..pains me, as it pains most Jews, when 'the people of the book' are compelled to wield the sword. **1976** Y. MENUHIN in D. Villiers *Next Year in Jerusalem* 334 A love of improvisation..has never been lost to the people of the Book. **1977** B. GASCOIGNE *Christians* v. 106 Jews, Christians, Muslims.. are all, in the powerful phrase of the Koran, 'people of the book'.

3. c. (Further examples.)

1822 C. LAMB *Let.* 23 Dec. (1935) II. 356, I rather grudge that S[outhe]y has taken up the History of your People. **1886** *Hist. Sk. Foreign Missions Seventh-Day Adventists* 20/1 Eld. Lindermann after a time became estranged not only from our people but also from a large share of those whom he had been instrumental in leading to the observance of the Sabbath. **1900** F. VON HÜGEL *Let.* 7 July (1931) 86 But, as to the Preface, he says he would, on the one hand, even selfishly like to do so, to prevent the book seeming to appear without any support or knowledge of any of our people. **1916** A. HUXLEY *Let.* 7 Aug. (1969) 109 I've arranged to be with my people in the country during August. **1971** 'M. INNES' *Awkward Lie* viii. 133 You know about my wife's people. **1977** *Belfast Tel.* 28 Feb. 3/7 She used to collect it [*sc.* silver paper] for the Multiple Sclerosis people, but they don't take it now.

4. a. (Later examples.) Cf. *man of the people* s.v. *MAN *sb.*[1] 18.

1953 E. SIMON *Past Masters* IV. ii. 229 Which of them is the scion of the upper classes and which the son of the people? **1969** A. G. FRANK *Latin Amer.* xx. 328 The bourgeoisie develops at the cost of exploiting the people. **1973** *Freedomways* XIII. 11 In China, India, the Soviet Union, even in the pre-Nazi Germany of 1932 Robeson traveled, acted and sang and everywhere he met with the people. **1973** *Black Panther* 17 Nov. 9/3 If the people (and when I say 'the people' I mean the oppressed people) control Malcolm X University, if they control it without reservation or without having to answer for what is done there or who speaks there, then Malcolm X University is progressive. **1976** M. J. LASKY *Utopia & Revolution* (1977) xiv. 496 One of the essential preconditions of the establishment..of a revolutionary tradition and its associated components of utopian hope and militant temper is the creation of 'The People' as a political factor. **1977** *Private Eye* 13 May 14/3 It..won't encourage the people to work any harder.

c. (Usu. with capital initial.) The prosecution in a law case as designated in certain States of the U.S.A., the equivalent of the Crown in a British law case.

1801 *Cases of Pract. Supreme Court New-York, 1791–1800* 34 Ludlow *ads.* The People. **1810** *Rep. Cases Supreme Court New-York* II. 301 The People *against* Olcott. **1849** *New York Superior Court Rep.* III. 193 J. McGay for the defendant, cited *The People v. Koeber.* **1898** *Misc. Rep. Courts of Record New York* XXV. 599 The People of the State of New York, Respondent *v.* Irving Mulkins, Appellant. **1926** *Michigan Rep.* CCXXX. 485 People *v.* Lorde. The people's testimony tends to show that..the defendant..went to the store of one John Kay. **1936** E. S. GARDNER *Case of Stuttering Bishop* xiv. 210 You may proceed..with the testimony in the preliminary hearing in the case of People versus Julia Branner. **1960** *California Reporter 1959* I. 245/1 People of the State of California, Plaintiff and Appellant, *v.* One 1952 Mercury 2-door Sedan..Defendant, Gregorio H. Nunez, sole owner of the above described vehicle, Respondent. **1973** *N.Y. Law Jrnl.* 4 Sept. 4/7 The prosecutor mentioned that he had provided defense counsel with pre-trial statements made by the People's witnesses.

5. Also used in the possessive (*spec.* in the terminology of Communism and Socialism) to designate institutions and concepts which are regarded as belonging to, derived from, or benefiting the people considered as the source of power or the basis of society. See also sense 9 below.

1811 *Weekly Reg.* 7 Sept. 9/2, I will attach myself, as an editor, to no party but the People's Party, whose wish is '*peace, liberty and safety*'. **1834** J. J. STRANG *Diary* 3 Mar. in M. M. Quaife *Kingdom of St. James* (1930) 218, I find myself nominated on what is called the people's ticket for constable. **1854** C. FOX *Let.* 21 Nov. in *Jrnls.* (1972) 217 F. Maurice was much cheered by the good beginning of his People's College. **1896** *Rep. on Labor Movement U.S.A. to Internat. Socialist & T.U. Congr.* 2/2 The Socialist Labor party is steadily advancing, the so-called 'People's party'..is not less steadily passing out of sight. **1900** [in Dict., sense 4.]. **1927** H. DOBBS in *Lett. Gertrude Bell* II. 558 On the part of the Opposition, now definitely constituted under the name of the People's Party, with Yasin Pasha as leader, doubts were expressed as to the advantage to Iraq of the extension of the 1922 Treaty for 25 years. **1942** *Ann. Reg. 1941* 16 A number of pacifists, including leading Communists, announced that they were organising a 'People's Convention' to demand 'a People's Government' which should bring the war to an end. **1953** *Encounter* Nov. 69/1 Looking over into East Berlin, one could see only a group of six People's Police in their new grey uniforms. **1958** *Listener* 30 Oct. 682/2 The policy of adventure and provocation of People's China. **1958** *Ibid.* 25 Sept. 452/1 Lloyd George was forty-six when he introduced his 'People's' Budget. **1958** [see *CO-OPERATIVE B. 2 bj. **1959** *Exchange* (N.Y. Stock Exchange) Aug. 2/1 Today we have 12,490,000 [shareholders] plus an estimated 1.4 million owners of private corporations. We have the most broadly owned, most dynamic people's capitalism ever seen on the face of the earth. **1961** *Sunday Bull.* (Philadelphia) 15 Jan. 1. 5/1 Pianist Svyatoslav Richter..has been given the Soviet Union's top artistic award: 'People's Artist'. **1966** R. E. PICKERING *Himself Again* xxii. 162 This must be Bratislava. I had escaped from Hungary into Czechoslovakia... Already I imagined my stubborn silence in a windowless room with the People's Police. **1966** 'H. MACDIARMID' *Company I've Kept* v. 148 Gaeldom, but for the English, gave good promise many centuries ago of evolving an ideal 'people's state'. **1969** C. DAVIDSON in Cockburn & Blackburn *Student Power* 331 The classless society of America's 'people's capitalism'. **1972** *Buenos Aires Herald* 4 Feb. 9/4 In Buenos Aires, police continued the hunt for the 'People's Revolutionary Army' (ERP) extremists who staged the record robbery of over 400 million old pesos at the National Development Bank. **1973** *Black Panther* 21 July 7/1 Elaine Brown,..the first, genuine People's Artist America has produced. **1974** L. DEIGHTON *Spy Story* xviii. 197 It took a long time before the Russians would let the D.D.R. have submarines. But the People's Navy are all ten-year men. **1974** tr. *Sniečkus's Soviet Lithuania* 108 In recent years, an important form of ideological education—the people's universities—has become widespread. **1975** *New Yorker* 28 Apr. 99/1 When I asked one economist what models the Chinese revolution might provide for Vietnam, the man stared at me for a few moments.., and then said, 'Well, what would *you* suggest? The Cultural Revolution? People's Communes?' **1977** K. BENTON *Red Hen Conspiracy* xviii. 143 He'd have to set up..what we call a people's prison to hold the Sheikh safely.

6. In phr. *of all people*, an expression suggesting that no one more surprising could be involved.

1851 S. SPENCER *Let.* 1 May (1912) 410 The *Times* yesterday contained some fine tho' rather enthusiastically loyal verses about the opening of the Exhibition by Thackeray of all people. **1922** CHESTERTON *Man who knew too Much* Why should you, of all people, be so passionate about it? **1965** *Radio Times* 2 May 15/2 Stan and Ollie..cause some hilarious surprises when Stan becomes Lord Paddington with Ollie of all people as his manservant.

e. An individual, a person. *U.S. colloq.*

1926 J. BLACK *You can't Win* ix. 105 He's good people and I want to get him fixed up for a cell with the right folks. **1934** *Detective Fiction Weekly* 28 Apr. 113/1 'Stick yer four-bits in yer shoe', he snorted, 'I'm people.' **1949** 'N. R. NASH' *Young & Fair* I. ii. 14, I guess she's people of good heart. **1956** B. HOLIDAY *Lady sings Blues* (1973) x. 98 A lot of creeps have been dogging Orson Welles ever since but they can't touch him. He's a fine cat... And a talented cat. But more than that, he's fine people.

8. *people-centred, -oriented* adjs.; **people mover,** any of several means of conveying people from one place to another; **people sniffer,** a device that can detect the presence of a person by chemical analysis of the air around him (see also quot. 1977).

1968 *Guardian* 5 Aug. 6/6 The interests of Africa will be best served..by a people-centred system, and African socialism is that system. **1970** *Ibid.* 23 Dec. 7/4 There is such a gulf between books that are system-centred and those that are people-centred. **1971** J. P. ROMUALDI in *Science Year 1972* 375 A 'people mover', a vehicle smaller than a streetcar..will provide continuous service between the old campus in town and the new campus in the suburbs. **1972** *N.Y. Times* 1 June 30/1 Henry Ford 2d announced yesterday that the Ford Motor Company would build rapid-transit systems based on its driverless, rubber-tired people mover system being demonstrated at the Transpo '72 exhibition here. **1974** *Times* 22 Mar. (Buses Suppl.) p. i/2 Magnetic levitation, vacuum tubes, vertical take-off aircraft, and small-scale automatic and semi-automatic 'people-movers' of all kinds for urban situations. **1970** *New Society* 5 Mar. 392/2 They exhibited the familiar people-oriented value-pattern detected many times among prospective teachers. **1975** *Nature* 27 Nov. 286/1 What more logical project for a rebuilt people-oriented science programme than to attempt to predict earthquakes. **1965** *Daily Tel.* 5 Oct. 22/8 A person being examined is placed in a 'people sniffer', a glass cylinder, and an analysis of the outgoing air discloses the chemical make-up of the subject. **1968** *N.Y. Times* 18 Aug. 1. 3 United States troops refer to the gadget as the 'people sniffer'. It leads American officers here in the Mekong delta to enemy hide-outs by 'sniffing out' the kind of ammonia odors given off by the human body. **1973** *Times* 24 Jan. 8/6 There has been use of the Manpack Personnel Detector, or 'people-sniffer'—picking up the enemy by the smell of his sweat. **1977** *Time* 2 May 44/1 Their principal piece of equipment is a 'people sniffer', an electronic sensing device developed to catch the prowling Viet Cong. Despite its name, the instrument actually detects the minute seismic vibrations caused by a person walking.

9. Special combs. with *people's* (chiefly in sense *5): **people's army,** (*a*) an army organized on egalitarian or communist principles; (*b*) an army composed of the common people; **people's car,** an inexpensive motor car designed for popular sale; **people's choice,** a popular favourite; **People's Court,** (*a*) a court set up by the Nazi regime in Germany to deal with political offences; (*b*) a court in the Soviet and similar legal systems; also *transf.*; **people's democracy,** a political system in which power is regarded as being invested in the people, *spec.* a Communist state, esp. in Eastern Europe; **People's front** = *POPULAR FRONT; **People's Palace,** a centre for the recreation and entertainment of the people, *spec.* a former East London institution with library, theatre, educational classes, etc., opened in 1887; also *fig.*; **people's park,** a park intended to be used by all members of a community; **people's republic,** name assumed by a number of left-wing or Communist states, as *People's Republic of China*; also in gen. allusive use; **people's theatre,** a theatre run on socialist lines for the use of the community; **people's war,** (*a*) a war in which the people are regarded as fighting against the ruling classes or foreign aggression; (*b*) a war in which all members of the community are involved, a total war.

1937 E. SNOW *Red Star over China* VI. i. 211 The Kuominchun, the 'People's Army' of General Feng Yu'-hsiang. **1941** 'G. ORWELL' *Coll. Ess.* (1968) II. 116 The Home Guard is..a sort of People's Army officered by Blimps. **1969** A. G. FRANK *Latin Amer.* xxiv 366 Nowhere does Debray suggest how the guerrilla band is later to develop into the people's army. **1970** A. SINCLAIR *Guevara* iii. 37 The regular army must be disbanded and a people's army created..of peasants and workers and soldiers. **1938** *Sun* (Baltimore) 7 Sept. 1/1 Award winners are Prof. Ferdinand Porsche, designer of the 'Volkswagen', Germany's new 'people's car'. **1939** *War Illustr.* 4 Nov. p. iii/1 A scheme by which German artisans paid in advance by weekly instalments for their long-promised 'people's car' would appear to have fallen through, for the great works at Fallersleben, the supposed factory for these cars, are now stated to be turning out munitions. **1958** [see *BUBBLE *sb.* 2c]. **1972** *Buenos Aires Herald* 2 Feb. 7/6 The rise of nationalism has brought demands for inexpensive 'people's cars' in Chile, Peru and Venezuela. **1953** WODEHOUSE *Performing Flea* 205 In Dormitory 309 the People's Choice was good old George Travers. **1961** —— *Service with Smile* (1962) v. 80 'Why is he the people's choice?' 'Because she's got the goods on him.' **1934** H. GRIFFITH *People's Court* 9 Comrade Chernov only turned up just in time yesterday not to make the People's Court look ridiculous in the eyes of its clients. **1935** *Ann. Reg. 1934* I. 191 A law of May 3 constituted a new and extraordinary Court of Justice, the so-called People's Court, for all political offences. This tribunal as well as the old regular courts in numerous cases passed excessively severe sentences on opponents of the Government. **1938** *Ann. Reg. 1937* 181 In April the People's Court sentenced several Catholic priests to long periods of detention. **1946** *Ann. Reg. 1945* 205 Twenty-four People's Courts were established by Decree in Bohemia and Moravia. **1970** H. TREVELYAN *Middle East in Revolution* 145 The notorious Colonel Medhawi, a cousin of Qasim, presided over the People's Court [in Iraq]. **1972** *N.Y. Law Jrnl.* 10 Oct. 1/5 The three-level federal system which emerged was composed of People's Courts with jurisdiction in rural areas; Regional Courts which are courts of first and second instance with appellate jurisdiction; and the Supreme Courts which are divided on a territorial basis into Supreme Courts of the autonomous republics, Union Republics and the U.S.S.R. **1977** *Listener* 15 Dec. 779/1 The Provisionals are now attempting to develop 'People's Courts'. **1947** *New Times* 3 Dec. 3 (*heading*) The people's democracies—a fresh breach in the imperialist system. *Ibid.* 4/2 In the people's democracies, power has passed from the hands of the exploiting classes—the landlords and bourgeoisie—into the hands of the people. **1958**

F. W. NEAL *Titoism in Action* i. 1 The Communist leadership which came to power in Yugoslavia in 1945 organized the country along the lines prescribed by the Soviet Union for an Eastern European 'people's democracy'. **1974** *Times* 22 Jan. 14/2 Mr Jack Jones.. made an unfortunate remark about his desire to see Britain become a People's Democracy, which is the title given to those Eastern European states which have nothing to do with the people and are not democratic; but it turned out that he meant, not a People's Democracy but a people's democracy, which is not at all the same thing and better hadn't be. **1977** P. JOHNSON *Enemies of Society* viii. 109 Social democrat means a Right-wing anti-Communist politician of the centre; a People's Democracy or a People's Democratic Republic is a form of Communist totalitarian state. **1937** 'G. ORWELL' *Coll. Ess.* (1968) I. 271 The worker and the bourgeois.. are fighting side by side. This uneasy alliance is known as the Popular Front (or, in the Communist press, to give it a spuriously democratic appeal, People's Front). **1937** E. SNOW *Red Star over China* III. iii. 99 If the Chinese People's Front is powerfully homogeneous..the war will be short. **1958** *Spectator* 13 June 777/3 Mr. Grant's badly camouflaged appeal for a new People's Front. **1854** *Punch* 24 June 266/1 The People's Palace will become a misnomer if the people are so confined in workshops [etc.]..that none but the comparatively idle can visit what is expressly designed for the appreciation of the industrious. **1889** G. B. SHAW *London Music 1888–89* (1937) 210 Covent Garden is a people's palace compared with Bayreuth. **1890** HARDY *Let.* 13 Mar. (1978) I. 210 We cannot do better than what you propose—purchase a library of fiction for the People's Palace. **1892** [in Dict., sense 4]. **1964** D. OWEN *Eng. Philanthropy* (1965) III. x. 293 In the early 1930's the original People's Palace was destroyed by fire. It was rebuilt and since 1954 has formed a part of Queen Mary College of London University. **1967** *Guardian* 29 July 12/4 Sir John Wardlaw-Milne..has left £100,000 to the State of Jersey..to build a 'people's palace' where tourists with children can shelter on wet days. **1970** E. J. HIRST *Seeing Glasgow* 41 The Old Glasgow Museum, People's Palace, Glasgow Green, provides a visual record of the development of Glasgow. The City's history is shown in pictures and prints. **1971** P. J. KEATING *Working Classes in Victorian Fiction* iv. 96 The People's Palace was built in Whitechapel... In May 1887 the Queen travelled to the Mile End Road to declare the Palace officially open. **1975** *Scottish Field* Mar. 44/1 (*caption*) A bowl traditionally attributed to the old Delftfield pottery, now in the People's Palace Museum, Glasgow. *Ibid.* 45/1 It could mean more interest being taken in the People's Palace exhibit. **1863** DICKENS *Uncomm. Trav.* (1866) xxiii. 163/1 The People's Park near Birmingham..was crowded with people from the Black Country. **1873** C. M. YONGE *Pillars of House* III. xxiv. 8 One of the brothers took her out in the street, or to the 'People's Park'. **1963** *Guardian* 19 Nov. 9/7 Rumford..suggested..in 1789 the laying-out of a great 'people's park' along the Isar. **1970** *Time* 23 Nov. 81 Last year he [*sc.* the chancellor of Berkeley] was unfairly blamed for the way police handled student demands that one of the university's empty lots be turned into a 'people's park'. **1976** *Times* 4 Sept. 10/5 In Bremen's 'people's park' the food is tasty. **1949** *Times* 8 Apr. 5/3 Mr. Rákosi..said..that 'the Hungarian Republic must be developed into a People's Republic'. **1972** J. POYER *Chinese Agenda* (1973) vi. 63 The khaki uniforms and green collar tabs of the Army of the People's Republic of China. **1974** *Encycl. Brit. Micropædia* III. 80/2 Congo (Brazzaville) *Official name*: République Populaire du Congo (People's Republic of the Congo). **1975** *Bangladesh Times* 18 July 2/6 The Magura Journalists Association in a meeting held recently at Magura hailed the Government of the People's Republic of Bangladesh for creating the new district of Magura. **1978** LD. HAILSHAM *Dilemma of Democracy* xxii. 148 Probably our own monarch would not survive the institution in Britain of..a people's republic. **1920** D. H. LAWRENCE *Touch & Go* 5 A nice phrase: 'A People's Theatre'. But what about it? There's no such thing in existence as a People's Theatre. **1929** S. W. CHENEY *Theatre* xxii. 506 In Germany a real 'democratic' theatre was developed before the war, and today functions as a 'people's theatre' of a unique sort. It is, perhaps, historically important as a step toward a theatre for a new Socialistic time. The *Volksbühne* in Berlin is owned by its audiences. **1904** L. HALE (*title*) The 'People's War' in France 1870–1871. **1942** E. WAUGH *Put out More Flags* i. 86 This is all that anyone talks about, thought Ambrose; jobs and the kind of war it is going to be..people's war, total war, indivisible war, war infinite, war incomprehensible. **1947** J. BERTRAM *Shadow of War* 9 'How about China's United Front?' I tried to tell them of what I had seen of this people's war. **1968** K. MARTIN *Editor* xv. 301 It was not until after Dunkirk that the idea of a People's War began to be understood. **1972** P. BLACK *Biggest Aspidistra* II. v. 133 The broadcasts did not cheat; the voices and sentiments were those of the people's war. **1976** M. GREEN *Children of Sun* viii. 311 Official propaganda presented this as a 'people's war', and emphasised the proletarian ordinariness of its heroes.

peo·plehood. [f. PEOPLE *sb.* + -HOOD.] The condition or state of being a people; the consciousness or awareness of being a people.

1909 in WEBSTER. **1969** V. FERDINAND in A. Chapman *New Black Voices* (1972) 379 We the young birds of blk poetics throw our songs brilliant Against the beautiful black sky of an emerging peoplehood & with the love of us..in our hearts. **1971** *Jrnl. Ecumenical Stud.* VIII. 1. 53 This need not..lead to a kind of reductionism which limits the significance of attempts to find historical embodiment for the biblical notion of peoplehood. **1973** *Black World* Mar. 71 For Gwen Brooks, peoplehood or race is not limited to continental boundaries. *Ibid.* Apr. 87 A sense of peoplehood is growing. **1977** *Guardian Weekly* 31 July 7/2 The mini-State, as the Palestinians conceive it, is not just a matter of territory. It is a recognition..of their peoplehood.

pep (pep), *sb. colloq.* (orig. *U.S.*). [Abbrev. PEPPER *sb.*] Vigour, energy, spirit, forcefulness.

1912 *Collier's* 13 Apr. 19/1 'Sure, the good old pep,' interposed Callahan. **1916** *Daily Colonist* (Victoria, B.C.) 19 July 9/4 (Advt.), This newest Overland Four has more power, pep, punch, and speed than any other low priced..car. **1923** WODEHOUSE *Inimitable Jeeves* xv. 187 That seems to be all the poor fish is able to do, dash it. He can chafe all right, but there he stops. He's lost his pep. He's got no dash. **1930** R. MACAULAY *Staying with Relations* iv. 56 What a family!.. They don't have any *pep.* Only little Meg has pep and *she* won't be let grow up with it. **1930** E. POUND *XXX Cantos* xi. 51 And came back with no pep in him. **1931** E. F. BENSON *Mapp & Lucia* ix. 234 A new hanging committee..full of pep and pop and vim. **1972** *Jrnl. Social Psychol.* LXXXVIII. 279 Lively, active, full-of-pep, energetic, peppy, vigorous, activated.

2. *attrib.* and *comb.*, as **pep rally** *U.S.*, a meeting to inspire enthusiasm, esp. before a sporting event; **pep speech** = *PEP TALK. See also *PEP PILL, *PEP TALK.

1945 *Boulder* (Colorado) *Daily Camera* 24 Nov. 4/2 Speaking at a pep rally..Dr. Allen told his audience.. other schools are spending money for players. **1974** *State* (Columbia, S. Carolina) 15 Feb. 3-B/1 University of South Carolina cheerleaders will conduct a pre-game pep rally in Carolina Coliseum Saturday beginning 12:15 as a prelude to the USC-Notre Dame battle. **1977** *New Yorker* 27 June 35/1 Some activity at the college—a pep rally, a football game, a dance. **1946** D. HAMSON *We fell among Greeks* vi. 73, I thought of all the 'pep' speeches we would have to deliver.

pep, *v. colloq.* [f. the sb.] (Const. *up.*) *trans.* To fill or inspire with energy or vigour, to enliven, invigorate, excite, cheer up. Also (*rare*) *intr.*, to improve, to find new life. So **pe·pped-up** *ppl. a.*; **pepper-u·p, pepper-u·pper,** someone who or something which enlivens or stimulates; **pepping-u·p** *vbl. sb.*

1925 H. L. FOSTER *Trop. Tramp with Tourists* 56 'Just leave them to me,' said the Social Manager. 'I'll get them started, and all pepped up, and the rest will be easy.' **1928** *Daily Express* 30 Nov. 19 University athletics are undergoing a strenuous process of 'pepping up', on strictly scientific lines. **1929** J. P. MCEVOY *Hollywood Girl* ii. 28, I was all pepped up about going out there [*sc.* to Hollywood] until I met you the other night. **1931** W. HOLTBY *Poor Caroline* vi. 202 Keep it vivid. Pep it up with a bit o' farce. **1936** 'P. QUENTIN' *Puzzle for Fools* iv. 24 We all had our daily treatments and mine consisted of a thorough pepping up. **1936** *Scrutiny* V. III. 270 Our ears are periodically offended by the ghastly and blasphemous harmonic monstrosities perpetrated by the popular peppers-up of Ye Olde Englysshe Folke Musicke. **1937** *N.Y. Post* 29 Jan. 24/3 (Advt.), Pepperupper. **1938** in Mencken *Amer. Lang. Suppl.* (1945) I. 259 A *cocktail* was therefore what I suppose today would be called a *pepper-upper.* **1945** [see *HOT ROD]. **1959** *Encounter* Sept. 13/1 A kind of forced jollity of a pepped-up Government White Paper. **1965** LEITNER & LANEN *Dict. French & Eng. Slang* 98/1 *Pepper-upper*, stimulant. **1967** N. FREELING *Strike Out* 71 You're looking a bit peaked, old chap... Have a drink to pep you up. **1971** *Daily Tel.* 19 Oct. 15/2 These 'insert' tiles are the newest idea to pep up a block of solid colour. **1972** W. P. MCGIVERN *Caprifoil* (1973) iii. 62 A chemical intoxication that stemmed from pepped-up endocrines. **1976** *Time* 20 Dec. 40/1 To pep up sluggish sales, American Motors Corp. is offering a $253 rebate to customers who purchase its 1977 compact Pacer.

peperite (pe·pərəit). *Geol.* [ad. F. *pépérite* (P. L. A. Cordier *Mémoire sur les Substances minérales dites 'en Masse'* (1815) ix. 66), prob. f. *péper(ine* PEPERINO + *-ite* -ITE[1].] A brecciated volcanic material consisting of fragments of lava and sedimentary rock, regarded by some as formed by the intrusion of lava into wet sediment, freq. under water, and by others as a product of the mixing of volcanic ejectamenta with such sediments. Cf. PEPERINO.

1839 *Civil Engin. & Arch. Jrnl.* II. 195/1 The geological structure of Mr. Hamilton's line of route is simple, being composed of only schistose rocks,..tertiary sandstones and limestones, granite, peperite, trachyte, basalt, and other igneous rocks. **1903** A. GEIKIE *Text-bk. Geol.* (ed. 4) II. 1254 These intercalations of tuff from the 'Peperites' of Auvergne, regarding which so much difference of opinion has been expressed. **1953** *Prof. Papers U.S. Geol. Survey* No. 454-B. 4/2 On Unalaska, pillows are separated from each other by peperite, fragmental igneous rock, argillite, or crustified chalcedony and quartz... Peperite is a mottled light-gray to pale-green breccia of mixed sedimentary and primary igneous debris. **1969** *Proc. Geologists' Assoc.* LXXX. 177 It is concluded that the peperites [of the Auvergne] are the product of explosive projection of basaltic ejectamenta into steadily accumulating lime-mud.

peperoni, var. *PEPPERONI.

pepful (pe·pful), *a. colloq.* [f. *PEP *sb.* + -FUL.] Full of life or verve, vigorous.

1923 *Weekly Dispatch* 22 Apr. 5 The film is..an American conception of the novel, with Richard Dix as a 'pep-ful' parson hero. **1933** *Punch* 29 Mar. 359/2 Mr. Joe Hayman's all-too-brief appearance as the hustling pepful American manager was, as usual, delightful.

pepino (pepī·no). *Physical Geogr.* [Sp., = 'cucumber'.] A small, conical hill characteristic of tropical and subtropical karstic regions, esp. one of those in Puerto Rico. Also *pepino hill.* Cf. *HUM *sb.*[4], *MOGOTE.

1899 *Nat. Geogr. Mag.* Mar. 100 Along their inner border these [hills] are of remarkable pointed character,.. and are appropriately termed by the natives 'Pepinos' or cucumbers. **1915** *Ann. N.Y. Acad. Sci.* XXVI. 16 This type of topography is represented by the small level tract surrounded or dotted over with small hills, called 'pepino hills' locally, standing like haystacks above the plain. **1934** *Jrnl. Geol.* XLII. 545 The development of pepino topography is characteristic of limestones in humid tropical countries. **1954** [see *MOGOTE]. **1968** R. W. FAIRBRIDGE *Encycl. Geomorphol.* 161/2 Also known as 'haystack hills', or 'tit hills', the Pepino hills are a characteristic feature of a mature karst landscape in tropical to subtropical latitudes.

pepless (pe·plės), *a. colloq.* [f. *PEP *sb.* + -LESS.] Without energy or spirit.

1926 WOOD & GODDARD *Dict. Amer. Slang* 15 *Dud*,..a failure; a pepless person. *Ibid.* 17 *Fish, poor*, pepless creature. **1942** BERREY & VAN DEN BARK *Amer. Thes. Slang* §99/12 Nonalcoholic,..pepless. *Ibid.* 249/2 Lazy, indolent,..pepless. *Ibid.* 276/8 Dull, spiritless,..pepless. **1967** *Observer* 30 Apr. 11/3 The 'vice' in Saigon..has a pepless *Playboy* flavour.

peplum. Add: **2.** (Earlier and later examples.) Also, that part of a long jacket or tunic which hangs below the waist, covering the skirt; hence, a jacket or tunic of this design; also *attrib.* Hence **peplumed** (pe·plŏmd) *a.*

1866 *Queen* 28 Apr. 330/1 No dressy afternoon toilette is now considered complete without a peplum. **1906** *Daily Chron.* 5 Apr. 8/1 A pointed tunic of peplum shape in soft white chiffon hemmed with..mirror velvet. **1930** *Morning Post* 8 Aug. 6 Some dresses have peplums or pleated basques. **1970** *Daily Tel.* 23 Feb. 13 The midi suit which doesn't have either a short cropped waist-length jacket or peplum is likely to be long..and belted. **1972** *Daily Tel.* 29 June 7/7 Or you can play it in strictly neat little Ginger Rogers suits, with flaring skirts, peplumed jackets. **1973** *Times* 15 May 20/5 Big-sleeved black seersucker blouse, the one that belts into a tight waist and then flares into a peplum over a full, long-tiered skirt. **1977** *New Yorker* 11 July 84/1 We see nothing of the longer, fuller skirt, the wasp waist, the peplum. **1977** *Time* 8 Aug. 37/1 Over these dresses or satin pants are worn gargantuan wraps—a peplum coat sewn with real gold thread cloth.

pepper, *sb.* Add: **2. b.** esp. the sweet pepper, *Capsicum annuum*, originally native to tropical America but now widely cultivated elsewhere, and its red, green, or yellow bell-shaped fruits; cf. CAPSICUM.

[**1728** R. BRADLEY *Dictionarium Botanicum* I. s.v. *Capsicum siliqua olivaria propendens*, This Pepper hath small and long round Cods.] **1884** tr. *A. de Candolle's Orig. Cultivated Plants* II. iv. 289 It [*sc. Capsicum annuum*] was one of the peppers that Piso and Marcgraf saw grown in Brazil. **1923** H. C. THOMPSON *Vegetable Crops* xxv. 399 Peppers are grown in very much the same way as egg-plants and tomatoes. *Ibid.* 400 Of the large-fruited, sweet peppers, Ruby King, Bell or Bullnose,..and Golden Queen are well-known varieties. **1949** *Nat. Geogr. Mag.* Aug. 166/2 When we say 'peppers' without any qualifying word, we usually mean sweet or nonpungent kinds that are eaten as a vegetable, either cooked or raw in salads. **1963** L. PHILLIPS *Recipes from Guardian* 124 Crisp red and yellow pepper rings dressed with lemon juice and olive oil and seasoned well make a good salad. *Ibid.*, Scrambled eggs take on a new look if small pieces of different coloured peppers are added. **1972** Y. LOVELOCK *Vegetable Bk.* III. 320 Capsicums, or peppers, are the various-sized seed-pods of a solanaceous plant, having nothing in common with the true peppers (Piperaceae) except the hot quality of some, but by no means all, of them. **1975** I. & A. MANCINELLI tr. *Bianchini & Corbetta's Fruits of Earth* 94 Peppers originated in South America, probably Brazil.

4. (Further examples.)

1922 JOYCE *Ulysses* 171 And here's himself and pepper on him, Nosey Flynn said. **1960** *Encounter* Feb. 34/1 'Pepper', that characteristically American streak of free-wheeling, urban critical scepticism. **1966** C. ACHEBE *Man of People* vii. 81 If you insult me again I will show you pepper. **1973** J. WAINWRIGHT *Touch of Malice* 136 'Are you calling me a liar, constable?' Ripley sprinkled pepper on his voice.

4*. Used *ellipt.* for PEPPER-POT.

1897 *Sears, Roebuck Catal.* 439/1 Silver plated Pepper or Salt. **1908** *Ibid.* 360/3 Handsome Salt sand Peppers, hand decorated. **1966** J. DOUGLAS *How to Collect* ii. 9 A plain Queen Anne or early Georgian silver pepper can cost you a great deal. **1976** 'D. HALLIDAY' *Dolly & Nanny Bird* ii. 30 He turfed out my things on to the table. The sugar... A miniature pepper and salt. A pack of fruit gums.

4.** In skipping, the speed at which the rope is turned as quickly as possible.

1901 R. C. MACLAGAN *Games Argyleshire* 229 The skipping may be done rapidly or slowly... In Kintyre slow skipping is called 'salt'; quick skipping, 'pepper'. **1948** *Jrnl. Amer. Folk-Lore* LXI. 65 In Pennsylvania the term 'pepper' means a very rapid, strenuous rhythm. In most other places it is called 'hot pepper' or 'hot peas'. **1972** F. B. MAYNARD *Raisins & Almonds* 56 Sometimes they skipped to the tune..with the spinning rope. (Hazel did pepper faster than anybody I knew).

5. *pepper-coloured* adj.; **pepper-dulse,** correct the date of quot. 1778 to 1777, and add further examples; **pepper gas**, an anti-personnel 'gas' that produces irritation of the throat and nasal passages; also as *vb. trans.*, to attack with pepper gas; **pepper-mill** (examples); **pepper shaker** *N. Amer.* = PEPPER-CASTOR, -CASTER 1; **pepper soup**, a West African soup made with red pepper and other hot spices; **pepper steak** (see quot. 1970).

1962 I. Murdoch *Unofficial Rose* x. 99 Mildred..patted her fluffy pepper-coloured hair into place. **1978** H. Jobson *To die a Little* v. 89, I picked up a pepper-coloured jacket. **1849** D. Landsborough *Pop. Hist. Brit. Seaweeds* 254 It is called pepper-dulse, and it certainly has, especially when young, a very pungent smell and peppery taste... It was formerly eaten in Scotland. **1931** L. Newton *Handbk. Brit. Seaweeds* 340 This species [sc. *Laurencia pinnatifida*] often has a hot biting taste, and was formerly eaten in Scotland under the name of Pepper Dulse. **1972** Y. Lovelock *Vegetable Bk.* i. 209 Pepper dulse (*Laurencia pinnatifolia*) was once eaten in Scotland but never gained great popularity. The name refers to the fact that it has often (though not always) a hot biting taste. **1970** *Times* 9 July 5 About 225 State and city policemen, armed with pepper gas, submachine guns, rifles and shotguns, repelled the mob. **1973** R. Hayes *Hungarian Game* xlvi. 281 When Michael didn't open the door, they began filling the room with pepper gas. **1973** *Black Panther* 17 Nov. 5/4 The 38 have been beaten, peppergassed, maced, isolated, harrassed and now intimidated. **1976** *New Yorker* 26 Jan. 74/2 The police dispersed the demonstrators with tear gas and pepper gas. **1884** G. Meredith *Let.* 13 May (1970) II. 735 My table is the richer for a pepper-mill. **1907** *Yesterday's Shopping* (1969) 210/2 Pepper Mills for Table. Walnut..1/3... Ivory—12/11. **1972** J. Burmeister *Running Scared* iv. 63, I feel like smoked salmon with four hefty winds on the pepper mill. **1895** *Montgomery Ward Catal.* Spring & Summer 543/3 Salt and pepper shakers, made of crystal blown glass, extra large capacity, and well adapted for kitchen as well as table use. Specify salt or pepper when ordering. **1911** *Daily Colonist* (Victoria, B.C.) 22 Apr. 2/1 (Advt.), Table Necessities..Pepper Shakers of Cut Glass, sterling silver tops. **1977** *Transatlantic Rev.* LX. 89 We divide up nine individual packets of sugar, six of ketchup, three rippled pepper shakers. **1964** J. P. Clark *Masquerade* in *Three Plays* 76 Why, Only this morning I opened my fishbasket To have stock for our pepper soup. **1966** C. Achebe *Man of People* xii. 148 My father was..eating pounded yams and pepper soup. **1951** E. David *French Country Cooking* 114 *Pepper steaks*... Score the steaks..rub them with garlic and then with a thin coating of pounded peppercorns. **1965** K. Giles *Some Beasts no More* v. 125, I had something to eat, they still do a good pepper steak. **1970** Simon & Howe *Dict. Gastron.* 297/1 *Pepper steak*, beef steak rubbed liberally with freshly-ground black pepper before cooking. **1976** K. Thackeray *Crownbird* v. 91 The barbecue area where an African was cooking pepper steaks and kebabs.

pepper, *v.* Add: **3.** Also *intr.* in same sense.

1857 [see *GEEWHILLIKINS *int.*]. **1945** C. Mann in Murdoch & Drake-Brockman *Austral. Short Stories* (1951) 263 They would be routed..by the driven sand and salt peppering into their eyes. **1947** W. de la Mare *Coll. Stories for Children* 14 A few hollow cockled-up bean seeds peppered down from out of their dry shucks.

4. b. Const. *away* (earlier example).

1884 'Mark Twain' *Huck. Finn* xviii. 167 The Grangerfords..peppered away at him.

pepper-and-salt. Add: **1.** (Further examples.) Hence, someone wearing pepper-and-salt clothes.

1844 *Ainsworth's Mag.* VII. 370 [The boy] is..installed in the usual pepper-and-salt, with the black velveteens for Sundays. **1900** E. Glyn *Visits of Elizabeth* 236 At dinner I sat between Charlie and one of the pepper-and-salts... They are going to shoot partridges tomorrow. **1907** *Westm. Gaz.* 11 Oct. 3/2 Husband small, fussy,..violent-tempered, pepper-and-salt-check-trousered, and something..in the City. **1915** W. S. Maugham *Of Human Bondage* xxvi. 107 Very neat in his black coat and pepper-and-salt trousers. **1934** G. B. Shaw *Village Wooing* 119 All this it sits writing. He wears pepper-and-salt trousers of country cut, with an apron. **1978** N. Marsh *Grave Mistake* ii. 36 The trousers were unmistakable: pepper-and-salt, shapeless, earthy.

b. *transf.* Used of greying hair, a moustache, or of objects resembling pepper-and-salt cloth in colour. Also, *pepper-salt*.

1853 Mrs. Gaskell *Ruth* II. vi. 124 My hair is nearly white. The last time I looked it was only pepper-and-salt. **1909** *Dialect Notes* III. 414 *Pepper and salt*,..painted with white spatters over a darker background. **1930** [see *LAW *sb.*[1] 16a]. **1934** E. Bowen *Cat Jumps* 83 Matthew had fluffy pepper-and-salt hair. **1951** T. Capote *Grass Harp* (1952) i. 3 A whip-thin, handsome woman with shingled pepper-salt hair. **1955** J. Thomas *No Banners* xx. 174 A scowling Militiaman with an enormous pepper-and-salt moustache walked up. **1957** E. Hyams *Into Dream* 118 Clear grey eyes, pepper-and-salt hair worn in the kind of bob which was fashionable about 1925. **1976** A. Price *War Game* ii. 219 Nayler's fashion had..put into an acceptable scholarly stoop to which his thick pepper-and-salt thatch added distinction.

3. *fig.*

1887 *Lantern* (New Orleans) 9 July 2/2 But let me commence my assault on the offending ones and give them pepper and salt. **1958** *Times* 24 Oct. 15/4 In Prokofiev's work Mr. Kroll emphasised the lyrical element rather more than the composer's characteristic pepper and salt.

4. pepper-and-salt fundus, a symptom of congenital syphilis, the fundus of the eye having a speckled appearance.

1940 S. Duke-Elder *Text-bk. Ophthalm.* III. iii. 2273 The finely pigmented or pepper-and-salt fundus, a picture characteristic of hereditary lues, wherein the entire fundus is dusted with innumerable small bluish pigmented spots between which lie round depigmented areas of a yellowish-red colour. **1975** Martin-Doyle & Kemp *Synopsis Ophthalm.* (ed. 5) vi. 76 Syphilitic Retinitis... In congenital cases a form of peripheral pigmentation known as 'pepper and salt' fundus is common.

5. Forming various nonce-wds.: **pepper-and-salted** *a.*, wearing pepper-and-salt clothing; **pepper-and-saltiness**, pepper-and-salt colour; **pepper-and-salty** *a.*, pepper-and-salt coloured.

1846 R. Ford *Gatherings from Spain* xxiii. 331 This pepper-and-salted Amphion. **1880** R. Broughton *Second Thoughts* I. i. vi. 83 Snow..speckled with blacks into an ugly pepper-and-saltiness. **1952** 'W. Cooper' *Struggles of Albert Woods* i. iii. 34 Dibdin's hair was the colour that is first called sandy and later pepper-and-salty.

pepper-box. Add: **2. b.** The name given to an early type of revolver in which five or six barrels revolve round a central axis; freq. *attrib.*

1861 *Richmond* (Va.) *Examiner* 7 Dec. 3/2 The pistol is one of the old-fashioned pepper-box sort—self-cocking, and..is regarded as dangerous at either end. **1872** 'Mark Twain' *Roughing It* ii. 4 An old original 'Allen' revolver, such as irreverent people called a 'pepper-box'. **1887** [see *coffee-mill* s.v. *COFFEE *sb.* 5 b]. **1901** W. Churchill *Crisis* ii. xviii. 280 Out of his pocket hung the curved butt of a big pepper-box revolver. **1915** W. B. Yeats *Reveries* (1916) 48 An old pepper-box revolver. **1920** C. W. Sawyer *Our Rifles* 65 The rifle was made about 1855, when pepper-box pistols were in everyday use. **1969** *Canad. Antiques Collector* Feb. 29/1 An early (1851) English 'Improved Revolver' which is part revolver and part pepperbox.

5. (Earlier and later examples.)

1771 T. Pennant *Tour in Scotl. 1769* 203 A slender square tower with a pepper-box top. **1948** J. R. Firth in E. P. Hamp et al. *Readings in Linguistics II* (1966) 178 The Romans and the English managed to dispense with those written signs called 'accents' and avoided pepperbox spelling.

peppercorn. 2. b. Substitute for def.: used *attrib.* or as *adj.* to designate the tufted style in which Hottentots and Bushmen wear their hair; also *transf.*

1868 J. Chapman *Trav. S. Afr.* I. i. 16 Bushmen with peppercorn heads. **1893** [in Dict.]. **1935** L. G. Green *Great Afr. Mysteries* x. 121 Reconstructions of these bones suggested that the Strandloopers were never more than five feet in height. Some authorities declare they had peppercorn hair. **1948** H. V. Morton *In Search of S. Afr.* viii. 251, I could see the road running ahead, disappearing for a while and emerging again upon the face of the greyish-brown plain, which was dotted with small peppercorn bushes like a Hottentot's hair. **1958** L. van der Post *Lost World of Kalahari* i. 12 His [sc. the Bushman's] hair was black and grew in thick round clusters which my countrymen called, with that aptitude for scornful metaphor they unfailingly exercised on his behalf, 'pepper-corn hair'. **1959** J. D. Clark *Prehist. S. Afr.* i. 17 The Hottentot closely resembles the Bushman except in stature. He..has black hair which grows in spirals and is known as 'pepper-corn hair'.

3. peppercorn shrub, tree = PEPPER-TREE in Dict. and Suppl.

1830 *Hobart Town Almanack* 65 A thick grove of the pepper shrub..grows in a close thicket to the height of from six to ten feet. **1901** M. Franklin *My Brilliant Career* i. 2 The stringybark roof of the salt shed..peeped out picturesquely from the musk and peppercorn shrubs. **1954** *Coast to Coast 1953–54* 76 Who do you think we see sittin' under a pepper-corn tree but this old sundowner. **1973** *Bulletin* (Sydney) 25 Aug. 43/3 Dejected peppercorn trees by the tennis courts. **1978** G. White *Silent Reach* ix. 104 A line of scrawny peppercorn trees.

peppered, *ppl. a.* **b. peppered moth**, substitute for def.: the popular name of the geometrid moth, *Biston betularia*, which is usually light-coloured with darker flecks. (Further examples.)

1903 F. E. Hulme *Butterflies & Moths* vi. 225 The insect we now represent..is..called the Peppered Moth, or, in the words of some old entomologists, the Pepper and Salt. **1970** [see *MELANIC *a.* 2]. **1972** *Countryman* LXXVII. ii. 135 Bernard Kettlewell's classic study of the peppered moth as an example of natural selection in action is finding its way into the text books. **1975** *Sci. Amer.* Jan. 90/1 Of more than 100 species of larger moths found throughout the British Isles, the peppered moth (*Biston betularia*) is surely the best-known to students of evolution.

c. peppered steak = *pepper steak.*

1966 J. Dixon *Classic French Cuisine* v. 130 (*heading*) Steak au Poivre (Peppered Steak). **1973** D. MacKenzie *Postscript to a Dead Lt.* 90 A Czech people who serve the best peppered steak in town. **1978** *Times* 4 Mar. 11/5 Stars in Soho..does a good peppered steak (£3·25).

pepperet (pepəre·t). Also **pepperette.** [f. PEPPER *sb.* + -ET.] *rare.* A pepper-pot.

1927 W. Deeping *Kitty* i. 11 Regency salt-cellars, mustard-pots and pepperets. **1975** *Country Life* 20 Feb. 427/1 A condiment set..mustard pot with liner, a salt-cellar and two pepperettes. **1977** *Times* 26 July 9/7 (Advt.), This pug-dog pepperette..London, 1881..realised £105.

pepperidge. Add: **2.** esp. *Nyssa sylvatica* (cf. *NYSSA). Also *attrib.* (Earlier and later examples.)

1689 *Huntington* (N.Y.) *Town Rec.* (1888) II. 56 A piperage tree marked faceing eastward and southward. **1743** J. Hempstead *Diary* 22 Feb. (1901) 406 Wee Sawed of a pr Peperage wheels for my Stone Cart. [**1810** F. A. Michaux *Hist. Arbres Forestiers de l'Amérique Septentrionale* I. 30 *Peperidge* fréquemment usitée par les Hollandois du New Jersey.] **1821** J. F. Cooper *Spy* I. ix. 133 A lieutenant of cavalry..whose captain is as tough as a peperage log. **1876** *Field & Forest* I. 66 This parasitic shrub [sc. American mistletoe] has been found growing on several Pepperidge or Sour-gum trees. **1900** J. de F. Shelton *Salt-Box House* ix. 67 A certain tract of land,..beginning at the highway near my present house..to a pepperidge tree. **1969** T. H. Everett *Living Trees of World* 257/2 The pepperidge, black gum, or tupelo (*N. sylvatica*) of eastern North America ranges from Maine to Michigan, Florida and Texas. Up to 100 feet in height, it forms a flat-topped columnar or pyramidal head of usually somewhat pendulous branches and has blunt, obovate or elliptic, lustrous leaves that turn brilliant cellar and two pepperettes. **1977** *Times* 26 July 9/2 (Advt.), This pug-dog pepperette..London, 1881..realised £105.

pepperina (pe:pərī·nă). *Austral.* [f. PEPPER(-TREE + *-ina*, as in CASUARINA.] = PEPPER-TREE in Dict. and Suppl. Also *attrib.*

1930 V. Palmer *Men are Human* xviii. 166 Nothing grew save the drooping pepperina that trailed its sheeny leaves over the kitchen roof. **1941** *Coast to Coast* 145 There was a pepperina-tree in the corner of the adjacent yard. **1967** *Southerly* XXVII. 204 Guinea-hens..roosting at night in the pepperina tree beside the back door.

peppermint. Add: **2. b.** (Earlier examples). Also, any peppermint-flavoured sweetmeat.

1829 G. Griffin *Collegians* I. iii. 56 'He gave me an O'Dell-cake when he was last here,' said one. 'And me a stick of peppermint.' **1835** J. Todd *Student's Man.* (ed. 2) 281 A handful of hot peppermints. **1883** *Harper's Mag.* Sept. 534/2 The windows were decorated..with..glass jars in which were sticks of striped candy, the half-moist peppermint, and the brown sugary squares.

c. Used as a flavouring in drinks: hence, the name of a cordial, and of a liqueur (= *crème de menthe* s.v. *CRÈME 1b).

c **1770** in de Vries & Fryer *Venus Unmasked* (1967) 31 She allows gin and peppermint in the room. **1825** P. Egan *Life of Actor* ii. 54 Now to avail me of the Friar's hint, He bade me take it in some peppermint. *c* **1863** T. Taylor *Ticket-of-Leave Man* i. 8 Four penn'orth of brandy,..and a little peppermint..(*stirring and sipping his brandy and peppermint*). **1865** E. H. Greenhow in E. R. Pike *Human Documents Victorian Golden Age* (1967) 203 Wine, gin, peppermint, and other stimulants are..often given.., their actual effect being..to stupefy the child. **1870** D. J. Kirwan *Palace & Hovel* xxxv. 516 Glass o' nice peppermint! this cold morning—ha'penny a glass! **1900** E. Glyn *Visits of Elizabeth* 77, I could bear most of it, if it wasn't for the peppermint glasses at the end, which the men have. **1963** A. L. Simon *Guide Good Food & Wines* 742 The highly scented leaves of *Mint* are used in the making of liqueurs usually sold under the name of *Crème de Menthe* or *Peppermint*. **1972** C. Drummond *Death at Bar* i. 32 A little drop of peppermint with a dash of lemonade. **1972** M. Gilbert *Body of Girl* xii. 107 Make it a gin and peppermint and I might join you.

d. Used as the name of various colours (esp. green) associated with peppermint-flavoured drinks or sweets.

1868 D. G. Rossetti *Let.* 17 Nov. (1965) II. 676, I wonder has Scotus's peppermint-and-mud tint been applied yet to the wall [illegible]. **1934** A. Huxley *Beyond Mexique Bay* 23 The buttery glossiness of acetate silk [illegible], yellow, or peppermint green. **1962** A. Christie *Clocks* xiii. 123 The door was opened by an elderly woman with..a black skirt and a rather unexpected peppermint-striped jumper. [illegible] **1974** (*caption*) Peppermint washable satin nightdress..£24. **1976** *Milton Keynes Express* 28 May 48/2 (Advt.), [illegible] peppermint green. £475. **1976** N. Thornburg *Cutter & Bone* i. 4 Bone slipped into his peppermint-stripe shirt.

3. (Later examples.) Also, *peppermint gum.*

1911 E. M. Clowes *On Wallaby* iv. 149 In the Wombat Forest, [illegible] peppermint, and swamp-gum. **1936** F. Clune *Roaming round Darling* vi. 51 He found the poor thief by the sight of his boots poking out between two stones that made him a house under a peppermint tree. **1963** W. Blunt *Of Flowers & Village* 94 The Australian peppermint tree (*Eucalyptus amygdalina*) is the tallest of all trees. **1966** *Southerly* XXVI. 107 The stringybarks and peppermint gums at the edge of the encircling scrub. **1967** T. Keneally *Bring Larks* xxv. 189 A half-uprooted peppermint hung over the road.

4. peppermint cake = *mint cake* (b) s.v. MINT *sb.*[3]; **peppermint cordial**, a cordial flavoured with peppermint; **peppermint cream**, a cream sweet flavoured with peppermint and often covered with chocolate; **peppermint-drop** (earlier examples); **peppermint geranium**, a variety of *Pelargonium tomentosum*, with downy, scented leaves and white flowers; **peppermint gum**: see sense 3 above; **peppermint lump**, a type of sweet flavoured with peppermint; **peppermint oil** (examples);

peppermint-scented *a.*, with a scent of peppermint; *spec.* **peppermint-scented geranium** = *peppermint geranium*; **peppermint-water** (further example).

1863 Mrs. Gaskell *Sylvia's Lovers* II. xii. 220 Here's a bit o' peppermint cake; he's main and fond on it. **1917** *Harrods Gen. Catal.* 1255 Schweppes cordials... Peppermint cordial 1/6. **1924** *Guardian* 24 Jan. 6/8 Dr McGill later celebrated his acquittal with peppermint cordial. **1907** *Yesterday's Shopping* (1969) 413 Peppermint cream. [illegible] 'Thank you, dear,' said Mrs. Brown, selecting a peppermint cream. [illegible] **1799** F. Boscawen *Let.* 20 Jan. in C. Aspinall-Oglander *Admiral's Widow* (1942) 177, I shall be glad..if you could send me 3 ounces of pepper mint drops such as I am used to have. **1818** [see *DROP *sb.* 10e]. **1849** Dickens *Dav. Copp.* (1850) vii. 68 He was so kind as to..dissolve a peppermint drop in it. **1922** A. L. Jekyll *Kitchen Ess.* iv. 56 Make a quart of good lemon jelly... Whilst warm add a handful of those large green peppermint geranium leaves, thick as a fairy's blanket, soft as a vicuna robe, and to be found in most old-fashioned gardens. **1931** E. S. Rohde *Scented Garden* vii. 177 The leaves of this kind were large and soft... I know it now for *Pelargonium tomentosum*, usually described as the peppermint geranium. **1966** G. B. Foster *Herbs for every Garden* iv. 123 Peppermint geranium, with its velvety, grape-like leaves and small white blossoms is *P[elargonium] tomentosum*. **1926** *Army & Navy Stores Catal.* 55/2 Sweets... Peppermint lumps. **1932** H. H. Price *Perception* viii. 230 The taste of a peppermint lump may linger on when the lump itself is no longer tactually present in our mouth. **1892** *Analyst* XVII. 14 Commercial samples of peppermint oil differ in quality as well as composition. **1966** *McGraw-Hill Encycl. Sci. & Technol.* VIII. 227/2 Menthol, found as the main constituent in peppermint oil. [illegible] **1907** *Westm. Gaz.* [illegible] **1946** *Nature* [illegible] **1967** *Boston Sunday Globe* [illegible] **1969** [illegible] peppermint water. [illegible] extra dollop of peppermint water.

pepperoni (pepərō·ni). Also **peperoni**. [f. It. *peperone* chilli.] Beef and pork sausage seasoned with pepper.

1934 in Webster. **1960** A. E. Bender *Dict. Nutrition* 113/1 Pepperoni (It.)..are slowly dried to a hard condition. **1967** *Boston Sunday Globe* 23 Apr. (Advt. section) 7/2 Pepperoni is another ready-to-eat of pork and beef with ground red pepper and the usual dry sausage seasonings. It is a little under 2 inches in diameter. [illegible] Cookbook 440/2 Dot the surface of the pizza with one sliced pepperoni sausage. **1974** *Time Out* 27 Sept. 33 [illegible] **1976** *Times* 26 June 12/5 Pizza..with generous topping of good tomato sauce, Mozzarella cheese, pepperoni and fresh mushrooms. **1977** *Custom Car* Nov. 97/3 The only action required is a wink, and wham! out comes his pepperoni!

pepper-pot. Add: **2. a.** (Further examples.) Also *attrib.* in *pepperpot soup*.

1904 [illegible] *Trip to Jamaica* 15 They make a rare Soop they call Pepper-Pot. **1890** [see *FOO-FOO]. **1949** *Caribbean Q.* I. i. 22 The use of cassava juice..was devised by the aboriginal South American Indians to conserve the food they have got by hunting or fishing. **1958** R. Howe *Cooking from Commonwealth* 367 (*heading*) Jamaica. Pepper-pot soup. **1961** F. G. Cassidy *Jamaica Talk* xvi. 336 Cassareep..imparts a peculiar flavour and is the basis of the traditional *pepperpot soup*—is not however favoured. [illegible] **1970** M. Slater *Caribbean Cooking* [illegible] Coalpot..used in the country, and [illegible] **1973** *Advocate-News* (Barbados) 17 Sept. [illegible] **1974** *Jrnl. R. Soc. Arts* [illegible] choicest Continental cuisine to the spicy Barbadian—from the generous cut of Roast Sirloin to the traditional Barbadian pepperpot soup—the food is the best.

b. (Examples.)

1794 *Massachusetts Spy* 13 Mar. 1 A wag in my neighbourhood, a lover of pepper pots. **1800** C. Macpherson *Mem.* 205 'And what have you got for dinner..?'—'Me have got peppa pot, Massa.' **1825** J. K. Paulding *John Bull in Amer.* [illegible] **1930** J. Williamson *Amer. Hotel* 217 A..list of American culinary creations..would include such concoctions as..Philadelphia pepper-pot. **1946** S. Hibben *Amer. Regional Cookery* 25 Lay in Dumplings for Pepperpot and cook as directed.

pepper-tree. b. Substitute for def.: Either of two Australasian evergreen trees, *Drimys aromatica* or *Pseudowintera axillaris* (*HOROPITO), both belonging to the family Magnoliaceæ, and bearing small dark fruits once used as a substitute for pepper. (Further examples of both senses.)

1857 B. L. Hayes *Pioneer Notes* (1929) 183 When I was at San Bernardino last, I obtained two small fir trees and two pepper trees. **1882** W. D. Hay *Brighter Britain!* II. vi. 195 The Horopito, or 'Pepper-tree' (*Drimys axillaris*) [illegible]. **1911** C. E. W. Bean *'Dreadnought' of Darling* xv. 141 The thick shady wilga—like a pepper tree in the distance. **1939** W. Faulkner *Wild Palms* 192 The pepper trees had been [illegible]. **1949** W. Hertrich *Huntington Bot. Gardens* [illegible] of Huntington Drive was necessary, it was a great misfortune to have to eliminate all of these pepper trees. **1950** G. Brenan *Face of Spain* iv. 92 The smooth-trunked rubber trees spread out their [illegible] trail their feathery tendrils. **1971** *Southerly* XXXI. 5 The impressions that will remain, transfigured, in the memory: the pepper tree [illegible] Duffield's yard. **1977** *Austral. House & Garden* Jan. 17/1 Check these and spray them [illegible] Chinese tallow tree, Pepper tree (schinus), [etc.]. **1978** 'M. M. Kaye' *Far Pavilions* ii. 16 Hey..kissed her behind a kindly screen of pepper trees.

pepper-up(per): see *PEPU.

pepperwood. (Examples.)

1856 *U.S. Naut. Mag. & Naval Jrnl.* V. 228 The timber used..is pepperwood, and was cut from the land close by the prison [sc. San Quentin, California]. **1858** C. E. De Long *Jrnl.* 5 Mar. in *Calif. Hist. Soc. Q.* (1930) IX. 253 [illegible] pepper [*sic*] wood tree. **1882** *Humboldt Times* (Eureka, Calif.) 7 Jan. 1/4 The schooner Alaska, lying in the bay ready for sea, has on board 50,000 feet of pepperwood, commonly called California laurel. **1894** R. H. Hodgson *Amer. Woods* V. 30 *Xanthoxylum Clava-Herculis*, L. Prickly Ash, Sea Ash, Toothache Tree, Pepperwood. **1949** *Natural Hist.* Mar. 130/2 Under a pepperwood tree by the name of Pepperwood..in a pile of glass flakings.

peppery, *a.* Add: **2. c.** In extended uses: unpleasant, objectionable; strong, powerful.

1897 P. Egan *Boxiana* 2nd Ser. II. 189 This was a short round, but *peppery*. **1901** M. Franklin *My Brilliant Career* ix. 77 Gertie, the boys, and myself performed our ablutions in a leaky tin dish on a stool outside the kitchen door, which on cold frosty mornings was a pretty peppery performance. **1928** P. Garmond *Decca* [illegible] 'We even have to hire the tractor of the Hebrews at two pounds and a half per diem', [illegible] 'a peppery price, by Mohammed.' **1949** E. Condon [illegible] *Jazz* iv. 55 What was then known as 'The Original Dixieland Jass Band—Untuneful Harmonists Playing Peppery Melodies'. **1958** *Which?* I. [illegible] The overhead-valve engine..is small in displacement, but peppery for its size.

d. Pepper-coloured.

1962 I. Murdoch *Unofficial Rose* xxxvi. 343 Her soft peppery hair..was cut short in a neat yet raffish style which somehow became her face.

pep-pill (pe·p pil). *colloq.* (orig. *U.S.*). [f. *PEP *sb.* + PILL *sb.*[2]] A stimulant drug dispensed in the form of a pill.

1937 *Time* 10 May 45/1 (*heading*) Pep-pill poisoning. The use by college students of a new, powerful but poisonous brain stimulant called Benzedrine last week threw University of Minnesota officials in dithers of worry. *Ibid.*, Students who, while cramming for final examinations, collapsed..are under suspicion of using the substance. They call it 'pepper-up', 'pep pills'. **1955** [see *AMPHETAMINE]. **1955** *Sci. News Let.* 2 Apr. 219/2 Amphetamine, or Benzedrine, known as 'pep pills',..is most likely to produce pleasant sensations in normal persons. **1959** *Times* 29 May 14/6 The performance of athletes may be improved through 'pep pills' by as much as 4 per cent. **1959** *Guardian* 18 Nov. 1/4 They had..tomato sandwiches, with coffee and tea, but no 'pep' pills. **1960** *Times Lit. Suppl.* 20 May 323/2 The managerial class gets its kicks from pep pills instead of art. **1960** *Spectator* 22 July 120 The campaign then being waged by some newspapers against the pep-pill Preludin. **1965** *New Scientist* 22 July 26/3 Amphetamines and barbiturates, variously called stimulants and depressants, pep pills and 'goof balls'. **1967** *Spectator* 30 June 758/1 The Rolling Stone Mick Jagger, was found guilty of possessing pep pills. **1974** E. Ambler *Dr. Frigo* ii. 98 As for that movie star, how do you know he isn't on pep pills?

peppy (pe·pi), *a.* orig. *U.S.* [f. *PEP *sb.* + -Y[1].] Full of pep or vigour; spirited, energetic, lively, forceful.

1915 S. Lewis *Babbitt* vi. 86 Wouldn't it be a good idea if I could go off to Chicago or some peppy place, and study engineering or something by mail? **1922** E. E. Cummings *Let.* 3 May (1969) 84 Thru being more or less true to the peppy theories..I feel more pep, pleasure in living. **1924** Wodehouse *Bill the Conqueror* xx. 303, I said it was the peppiest scheme of the age, a lallapaloosa. **1926** *Picture-Play Mag.* July 32/2 (Advt.), How I used to envy..Billy jazzing up a party with his peppy banjo! **1927** *Melody Maker* Sept. 883/1 They have a red 'hot' peppy dance rhythm and are thoroughly bright and interesting. **1930** E. Waugh *Vile Bodies* xii. 129, I like your page. It's peppy. **1932** T. S. Eliot *Sel. Essays* iii. 197 There is nothing in the play to which could be applied the term appropriately used in the advertisements of some films; the 'peppy situation'. **1934** J. O'Hara *Appointment in Samarra* (1935) 15 Everyone from out of town thought it was the peppiest place in the country at Christmas. **1939** *Melody Maker* 10 June 5/3 This peppy, blonde, stage-and-concert-party artist. **1956** M. Stearns *Story of Jazz* (1957) xvi. 180 Most of these bands recorded a novelty now and then that might be called 'peppy'. **1969** *New Yorker* 10 May 33/2 He did a peppy foxtrot. **1974** J. Heller *Something Happened* 14 She was peppy and direct, always laughing and teasing. **1977** *Time* 14 Feb. 39/3 A.M.C...will give the Pacer a peppier engine. **1977** *Rolling Stone* 7 Apr. 69/3 It's an interesting combination and forces all the songs on *Blondie* to work on at least two levels: as peppy but rough pop, and as distanced, artless avant-rock.

Pepsi-Cola (pe·psi,kou·lă). orig. *U.S.* Also in shortened form **Pepsi.** The proprietary names of a popular soft drink, and of the syrup preparations from which it is made.

1903 *New Bern* (N. Carolina) *Jrnl.* 25 Feb. (Advt.), Pepsi-Cola. At Soda Fountains... Aids Digestion. **1903** *Official Gaz.* (U.S. Patent Office) 16 June 1891/2 Flavoring-syrup for soda-water. Caleb D. Bradham, Newbern, N.C. Filed Sept. 23, 1902 Pepsi-Cola. **1906** *Ibid.* 12 June 2342/2 Tonic Beverage. Caleb D. Bradham, New Bern, N.C. Pepsi-Cola. Filed Apr. 15, 1905. **1916** *Ibid.* 9 May 676/1 The Pepsi-Cola Co., Newbern, N.C. Filed Mar. 19, 1915. Pepsi..A Flavoring-syrup for soda water. *Claims use* since Nov. 21, 1911. **1940** *Life* 7 Oct. 79/1 (*caption*) They [sc. A. B. Kent and A. H. C. C. Johnson]..are best known for these immortal lines which have even been translated into Yiddish. They are basically a swing version of *John Peel*: Pepsi-Cola hits the spot! Twelve full ounces, that's a lot, Twice as much for a nickel too—Pepsi-Cola is the drink for you! **1949** N. Streatfeild *Painted Garden* xiv. 156 He gave her a pepsi-cola. **1953** *Trade Marks Jrnl.* 13 May 417/2 Pepsi-Cola... Non-alcoholic drinks and preparations for making such drinks, all containing cola extract..Pepsi-Cola Limited, Pepsi-Cola Factory,..Brentford, Middlesex; manufacturers and bottlers. **1957** C. MacInnes *City of Spades* i. 107 'Some orange juice or Coke?' 'Ta, Guv, I'll have a Pepsi.' **1959** *Encounter* Sept. 50/2 Mugs bearing Shakespeare's image and sold over the counter with root beer and pepsi cola. **1960** *Trade Marks Jrnl.* 17 Aug. 998/1 Pepsi... Non-alcoholic drinks and preparations for making such drinks, all containing cola extract..Pepsi-Cola Limited,..Feltham, Middlesex; manufacturers and bottlers. **1961** *Western Folklore* XX. 182 Pepsi Cola hits the spot Turn the rope and give her hot. **1966** P. Willmott *Adolescent Boys E. London* 207 We went back to the club and had a drink of Pepsi and a game of darts. **1973** *Sat. Rev. World* (U.S.) 25 Sept. 22/1 She smokes a cigarette, sips a Pepsi. **1976** *National Observer* (U.S.) 21 Feb. 5/5 All that *detente* brings the United States, Reagan says, is 'the right to sell Pepsi-Cola in Siberia'.

pep talk (pe·p tǫk). [f. *PEP *sb.* + TALK *sb.*] A speech or address intended to revive morale or promote energy or enthusiasm in its hearers. So **pep talker**, one who delivers a pep talk.

1926 B. Reynolds *Cocktail Continentale* xiv. 107 Where the great Cicero propounded his philosophy to the ages and from which our 'Pep' talks of to-day are copied. **1931** K. K. Rockne *Coaching* (rev. ed.) xxii. 256 One coach I know has his football team, before a game, gather in a room which is painted red, for their so-called 'pep-talk'. **1934** J. O'Hara *Appointment in Samarra* (1935) iv. 92 'Let's get through these holidays without any more mess... I don't want to give you a pep talk—' 'I know you don't. I don't blame you.' **1935** Wodehouse *Luck of Bodkins* vi. 66 I've seen that expression on her face a hundred times when she was giving us a pep talk before a match. **1940** R. Chandler *Farewell, my Lovely* xxxvi. 169 'I'm afraid of dying, of being nothing, of not finding a man named Brunette.' He chuckled... 'You sure give yourself a pep talk'. **1943** J. B. Priestley *Daylight on Saturday* xxx. 243 I'm wondering if we couldn't invite him to lunch and ask him to give the workers a pep talk in the canteen. **1945** [see *FLANNEL *sb.* 1 f]. **1951** in M. McLuhan *Mech. Bride* (1967) 36/1 Dr. Starch does not merely point out..essential qualities of executive leadership. He does not preach or deliver 'pep' talks! **1957** J. Braine *Room at Top* v. 47 Worse still, there were what he called Pep Talks which were made specially gruesome by the fact that, since he seemed to be able to speak and scarcely open his lips, his..voice seemed to come from nowhere. **1969** J. Wainwright *Take-Over Men* vii. 121 Forget the fancy trappings the politicians and pep-talkers hide it behind. **1971** S. Hill *Strange Meeting* iii. 167 Every now and again Coulter gives us his pep talk, about how we are going to 'go out there and show 'em'. **1973** J. Wainwright *Pride of Pigs* 120 Which was what he was supposed to be doing..taking chances. The reason..for Harris's pep talk. **1976** *New Yorker* 15 Nov. 137/2 The students were given pep talks to build up their confidence and strengthen their 'will'.

peptic, *a.* and *sb.* Add: **A.** *adj.* **1.** *peptic ulcer*, an ulcer that is situated in a part of the alimentary tract bathed by the gastric juice, or that is attributed to its digestive action; so *peptic ulceration*.

1900 Dorland *Med. Dict.* 726/1 Peptic u[lcer]. **1903** tr. *Riegel's Dis. Stomach* ii. 543 Round ulcer of the stomach. Syn.—Ulcus ventriculi simplex,..peptic ulcer, perforating gastric ulcer. **1929** Hurst & Stewart *Gastric & Duodenal Ulcer* x. 496 Other forms of œsophageal ulcer..must be excluded before a diagnosis of peptic ulcer can be made. *Ibid.*, Peptic ulceration of the œsophagus. **1955** *Sci. News Let.* 30 July 3 A child's peptic ulcer may be confused with abdominal migraine, food allergy or other intestinal conditions. **1974** H. J. Dworken *Alimentary Tract* vi. 88 Certain other elements determine the ability to withstand formation of peptic ulcers, although the mechanism for their action is not known. Susceptibility to gastric ulcer increases with age; duodenal ulcers are more common in men than in women. *Ibid.*, People who live in the north of England..or at high altitudes such as in the Peruvian Andes show increased frequencies of peptic ulceration.

peptidase (pe·ptidēiz, -s). *Biochem.* [f. *PEPTID(E + *-ASE.] Orig., an enzyme which

hydrolyses peptides; now usu. restricted to enzymes (exopeptidases) which hydrolyse the terminal peptide bonds of peptides, liberating amino-acids.

1918 *Jrnl. Infectious Dis.* XXII. 148 Normal human serum contains peptidase and maintains a relatively uniform titer. **1923** *Jrnl. Chem. Soc.* CXXIV. i. 496 The proteases can be provisionally classified as follows. A. True proteases which break down protein to the peptide stage... B. Peptidases or ereptases which only split peptides or peptones. **1936** [see *endopeptidase* s.v. *ENDO-]. **1958** Dixon & Webb *Enzymes* v. 228 By no means all peptide links are hydrolysed by all peptidases. **1970** R. W. McGilvery *Biochem.* xv. 307 They [sc. lysosomes] are particulate structures in the cell..loaded with a battery of hydrolytic enzymes: peptidases to attack proteins, esterases to attack lipids, [etc.]. *Ibid.* xxvii. 658 Exopeptidases, sometimes simply called peptidases without a clarifying prefix,..attack terminal peptide bonds.

peptide (pe·ptəid). *Biochem.* [ad. G. *peptid*, back-formation from *di-*, *tripeptid*, etc. (E. Fischer 1902, in *Chemiker-Zeitung* XXVI. 940/2), *polypeptid* (E. Fischer 1903): see *POLYPEPTIDE.] **1.** Any compound in which two or more amino-acids are linked together by peptide bonds (see sense 2 below); according to the number of amino-acid residues such compounds are dipeptides, tripeptides, etc., oligopeptides, or polypeptides. Also *attrib.* or as *adj.*

1906 *Jrnl. Chem. Soc.* XC. ii. 293 (*heading*) The fate of certain amino-acids and peptides in the organism of the dog. **1927** Haldane & Huxley *Animal Biol.* iv. 107 We can make..a sugar or peptide (part of a protein molecule) which only differs from the natural variety in that its molecules are related to the natural molecules as a left hand to a right. **1949** Abraham & Florey in H. W. Florey et al. *Antibiotics* I. vii. 331 The tomato-wilting agent produced by *Fusarium lycopersici* Sacc. was isolated by Plattner and Clauson-Kaas.., who later..showed that it was a peptide and named it lyco-marasmine. **1953** Fruton & Simmonds *Gen. Biochem.* v. 135 Various strains of microorganisms elaborate substances of peptide nature which have antibacterial activity for other microorganisms. **1972** F. M. Menger et al. *Org. Chem.* xiv. 359 The distinction between proteins and peptides is arbitrary. Compounds which have molecular weights greater than 10,000 are generally referred to as proteins. **1976** *Sci. Amer.* Feb. 32/1 There are two large classes of hormones, the peptides and the steroids. *Ibid.* 32/2 A typical peptide hormone is insulin.., human insulin consists of 51 amino acid units. **1977** *Time* 21 Nov. 40/3 Drs. Frank Ervin of U.C.L.A.'s Neuropsychiatric Institute and Roberta Palmour of the University of California at Berkeley described the substance as a variant of a peptide—a short chain of amino acids—that belongs to a family of newly discovered opiate-like brain hormones called endorphins.

2. Special comb.: **peptide bond**, a carbon–nitrogen bond of the type —CO·NH— in an organic molecule; *spec.* one between the carboxyl group of one amino-acid residue and the amino group of another; **peptide chain**, a linear sequence of amino-acid residues joined by peptide bonds; **peptide linkage** = **peptide bond*.

1935 *Jrnl. Biol. Chem.* CXI. 249 Aminopeptidase, carboxypeptidase, and dipeptidase need in addition to a peptide bond a free amino group or a free carboxyl group. **1960** *New Biol.* XXXI. 12 There are twenty main different kinds of amino acid involved in protein make-up, though they are basically alike enough to be all connected to each other by the same kind of chemical link (the peptide bond) to form the chain. **1964** N. G. Clark *Mod. Org. Chem.* xiii. 252 Many important natural products contain the 'amide-linkage' or peptide bond, —CO·NH—; thus, the peptides and proteins..are long-chain polyamides. **1931** *Nature* 2 May 664/2 There is a strong probability that..many proteins will be based on a roughly constant weight of peptide chain. **1935** *Jrnl. Biol. Chem.* CXI. 245 Lysine is coupled in the long peptide chains of proteins with its α-amino and carboxyl groups. **1970** R. W. McGilvery *Biochem.* ii. 9 A protein may be only a single, long, peptide chain, but most proteins are made of several peptide chains associated together. **1925** *Proc. R. Soc.* B. XCVIII. 59 This procedure was adopted in order to bring about the scission of the peptide linkages as rapidly as possible. **1964** N. G. Clark *Mod. Org. Chem.* xvii. 349 Proteins and polypeptides consist of chains of amino-acids linked via amide formation between the carbonyl group of one acid and the α-amino group of the next (the typical linkage, —CO·NH—, is often called a peptide-linkage or -bond).

peptidic (pepti·dik), *a.* *Biochem.* [f. *PEPTID(E + -IC.] Of, pertaining to, or being a peptide. Hence **pepti·dically** *adv.*, by means of a peptide bond.

1949 F. Lipmann in *Federation Proc.* VIII. 597/1 To simplify the following discussion, the term 'peptidic link' is introduced as generic name for a —NH·CO-link between any amino or carboxyl group. The term 'peptide link' is then reserved for the 1-carboxyl, 2-amino-link between two alpha amino acids as it occurs in protein. **1964** *Adv. Enzymol.* XXVI. 212 Muropeptides, depending on the murein type, are known to differ by containing either Lys or one or more of the DAP stereoisomers to provide the free NH_2 group for peptidic linking. *Ibid.*, One enzyme..found..in *E. coli* cells..disconnects two C6 units tied together peptidically into a C3 molecule. **1973** *Nature* 12 Oct. 288/1 G. Sterba..suggested that peptidic neurohormones might be involved in the generation of emotions by effects in the limbic system.

peptidoglycan (peptəi:doglǝi·kæn). *Biochem.* [f. *PEPTID(E + -O + *GLYC(O- + -AN.] = *MUREIN; also, the mucopolysaccharide which forms the strands of this.

1966 *Biochem.* V. 82/1 The biosynthesis of peptidoglycan (mucopeptide) in cell-free extracts of Staphylococcus aureus..has been described by Chatterjee and Park. *Ibid.* 3091 The mechanical strength of the peptidoglycan polymer which forms the rigid network of all bacterial cell walls depends on a high degree of cross-linking between peptide and polysaccharide chains. **1968** A. White et al. *Princ. Biochem.* (ed. 4) xli. 910 Murein synthesis..may be regarded as proceeding in three stages: synthesis of the precursor units, synthesis of the linear peptidoglycan strands, and cross-linking. **1969** *New Scientist* 10 July 64/1 A major component of the cortex layer [of a bacterial spore] is the polymer murein (or peptidoglycan). **1975** *Nature* 10 Apr. 482/2 The walls of all penicillin-sensitive organisms contain a structural component called peptidoglycan, which consists of glycan chains of alternating residues of N-acetylglucosamine and its 3-*O*-D-lactyl ether..and D-amino acids.

peptidolysis (pe:ptəidǫ·lisis). *Biochem.* [f. *PEPTID(E + -O + *-LYSIS.] The degradation of a polypeptide into smaller peptides or amino-acids. Cf. *PEPTOLYSIS. So **pe:ptidoly·tic** *a.*

1970 *Nature* 25 July 337/2 Studies on the mechanism of action of peptidolytic enzymes..have made much use of their ability to hydrolyse esters as well as amides. **1971** *Ibid.* 25 June 495/2 Both sets of workers compared these data with kinetic results from peptidolytic reactions. **1972** *Biochim. & Biophys. Acta* CCLXV. 70 Cleavage, for example, of a 50 000-dalton peptide into fragments of 44 000 and 6000 daltons, respectively, can clearly be detected... Such peptidolysis can be observed only after dissolution of the protein by the detergent, since separated peptide fragments otherwise tend to remain associated through hydrophobic interactions.

peptization (peptəizēi·ʃən). *Chem.* [f. PEPT(ONE + -IZATION.] The transformation of a solid or semi-solid colloid into a fluid form by chemical means.

1864 T. Graham in *Proc. R. Soc.* XIII. 340 Liquid silicic acid may be represented as the 'peptone' of gelatinous silicic acid; and the liquefaction of the latter by a trace of alkali may be spoken of as the peptization of the jelly. **1916** E. F. Burton *Physical Prop. Colloidal Solutions* ii. 24 The large number of colloidal solutions prepared by the method known as peptization are examples of the resolution of a moist coagulum. **1934** *Industr. & Engin. Chem.* Nov. 1190/2 (*heading*) Peptization of lightly vulcanised rubber. **1948** A. Lane *Greek Pott.* i. 5 The coagulation can..be broken down by a process known as 'peptization', by the disintegrating action on the clay of certain chemicals. **1960** R. G. Haggar *Conc. Encycl. Cont. Pott. & Porc.* 210/2 Modern scholars have..demonstrated the understanding which they [sc. the Greeks] had of clay processes, of oxydising and reducing firing techniques, and what to-day are called 'protective colloids' and 'peptization'. **1963** D. W. & E. E. Humphries tr. *Termier's Erosion & Sedimentation* vi. 135 The peptization of colloidal ferric hydroxide occurs at pH 6·6, under the influence of humus and colloidal silica which is present in the soil. **1972** Moeller & O'Connor *Ions in Aqueous Syst.* v. 113 Conversion of precipitate into a colloidal suspension is called 'peptization'.

peptize (pe·ptəiz), *v.* *Chem.* [f. PEPT(ONE + -IZE.] *trans.* To convert into a sol; to cause to undergo peptization.

1864 T. Graham in *Proc. R. Soc.* XIII. 340 The pure jellies of alumina, peroxide of iron, and titanic acid, prepared by dialysis, are assimilated more closely to albumen, being peptized by minute quantities of hydrochloric acid. **1934** H. N. Holmes *Introd. Colloid Chem.* iii. 26 Glue, gelatin, soap, gum arabic, and dextrin are said to be soluble in water. In reality they are merely peptized by water—they are subdivided into particles far larger than molecules. **1939** [see *DISPERSE *v.* 9]. **1955** R. K. Iler *Colloid Chem. Silica & Silicates* v. 92 Various processes have been described which involve making silica gel from acid and silicate, washing out the salts, and peptizing the wet gel by heating it under pressure in the presence of a small amount of alkali. **1972** Moeller & O'Connor *Ions in Aqueous Syst.* v. 116 Freshly prepared and washed precipitates can be peptized in many cases by adding water and a little of the original precipitating reagent.

Hence **pe·ptized, pe·ptizing** *ppl. adjs.* Also **pe:ptizabi·lity, pepti·zable** *a.*

1921 W. D. Bancroft *Appl. Colloid Chem.* v. 167 We may have peptization by a liquid, by a non-electrolyte, by an adsorbed ion, by a salt, or by a peptized colloid. *Ibid.* 170 Water-peptizable colloids like gelatine, gum arabic, dextrine, [etc.]..will peptize many precipitates. **1925** tr. J. M. van Bemmelen in E. Hatschek *Foundations Colloid Chem.* 129 The peptizing agent may be removed from the solution and the colloid remains dissolved. **1934** *Industr. & Engin. Chem.* Nov. 1190 Rubber which has been lightly vulcanized with sulfur can be dissolved in benzene with the help of peptizing agents... The action of soluble zinc compounds on peptized rubber sulfur causes a gelling action. **1938** *Proc. Rubber Technol. Conf.* 289 Rubber 'softeners'..are commonly used to increase the plasticity of rubber, and it is often assumed that they act through some lubrication, swelling, or 'peptising' action on the rubber. **1963** *Ceylon Vet. Jrnl.* XI. 42 The solubility of the protein of groundnut meals has been determined by the salt-peptizability of its nitrogen.

peptizer (pe·ptəizəɹ). *Chem.* [f. *PEPTIZ(E *v.* + -ER[1].] A substance which causes peptization, or which serves to prevent the coagulation of a colloid suspension; *spec.* a catalyst which facilitates the process of mastication or vulcanization of rubber, by preventing the recombination of broken polymer chains.

1931 E. S. Hedges *Colloids* iii. 16 If the freshly precipitated ferric hydroxide is treated with..ferric chloride solution it disperses immediately to form a dark reddish-brown colloidal solution. We call..ferric chloride the peptizer or peptizing agent. **1946** *Shell Aviation News* No. 100. 15/3 These [sc. additives in oil]..may act as 'peptizers' which help to keep in suspension in the oil any insolubles which are formed. **1961** *New Scientist* 2 Mar. 549/1 Other substances, so called 'peptisers', can be added [to rubber] to ensure this simple form of stabilizing the reactive ends of the broken chains. **1963** H. van Olphen *Introd. Clay Colloid Chem.* viii. 109 When a small amount of peptizer is added to a pure clay gel, the yield stress decreases drastically. **1972** *Materials & Technol.* V. xiv. 474 Effective peptizers [of rubber] are hydrazine derivatives and organic sulphur compounds like thio-β-naphthol, pentachlor thiophenol, dibenzoyl disulphide.

peptolysis (peptǫ·lisis). *Biochem.* [f. PEPTO(NE + *-LYSIS.] The degradation of a peptone or polypeptide into smaller peptides or amino-acids. Cf. *PEPTIDOLYSIS.

1904 S. H. Vines in *Ann. Bot.* XVIII. 290 Accepting this connotation of 'proteolysis', the successive stages of the process may, I would suggest, be conveniently distinguished as—(*a*) peptonization, the conversion of the higher proteins into albumoses and peptones; and (*b*) peptolysis, the decomposition of peptones into nitrogenous but non-proteid substances. **1949** H. Tauber *Chem. & Technol. Enzymes* vi. 130 The end products of peptolysis are mostly proteoses and peptones and small quantities of amino acids.

Hence **peptoly·tic** *a.*, **peptoly·tically** *adv.*

1904 *Ann. Bot.* XVIII. 290 Trypsin..forms tryptophane as one of the products of its peptolytic activity. *Ibid.* 299 It was ascertained that a filtered watery extract of yeast was always peptolytically active, however short the period of extraction might be. **1915** *Chem. Abstr.* IX. 2910 The formaldehyde titration for free amino groups in polypeptides and their hydrolytic products is a method practically adapted to the study of the action of peptolytic ferments on polypeptides. **1949** H. W. Florey et al. *Antibiotics* I. i. 47 Gram positive proteolytic bacteria such as..staphylococci when grown with gram positive peptolytic bacteria..in a nitrogen-free medium secreted a bacteriolytic substance which dissolved the latter, thus making their nitrogen available.

Pepysian (pī·psiăn), *a.* [f. the name of Samuel *Pepys*, diarist (1633–1703) + -IAN.] Of, pertaining to, or characteristic of Pepys, his writings, his library, or the age in which he lived. So **Pepysia·na** *sb. pl.*

1786 in *Wks. James I of Scotl.* 98 Dr Percy informs us, that this poem is preserved in the Pepysian Library, at Magdalen College, Cambridge. **1847** E. F. Rimbault *Nursery Rhymes* p. viii, The popular rhyme 'Three Children sliding on the Ice'..is part of a ballad preserved in the Pepysian collection. **1899** H. B. Wheatley *Diary of Samuel Pepys: Suppl. Vol.* (*title*) Pepysiana. **1920** *Glasgow Herald* 21 Sept. 8 Amid the distractions of business he cultivated his early literary bent and kept a diary wherein, with Pepysian frankness,..he chronicled his doings. **1927** W. Deeping *Kitty* xxx. 385 He was a Pepysian soul and kept a diary. **1927** J. S. Huxley *Relig. without Revelation* iv. 119 The Pepysian interest in complete record of all facts centring on self. **1927** W. H. Whitear (*title*) More Pepysiana: being notes on the Diary of Samuel Pepys and on the genealogy of the family, with corrected pedigrees. **1934** *Punch* 19 Dec. 699/3 Builders' models of the Pepysian and later periods, rigged models of Dutch Indiamen,..reconstructed models by present-day experts, all find a place in this fascinating gallery.

Pequot (pī·kwǫt), *sb.* and *a.* Also **Pecoate, Pequod, Pequoitt.** [prob. f. native word *paquatanog* destroyers.] **A.** *sb.* **a.** An Indian of an Algonquian tribe of southern New England. **b.** The language spoken by the Pequots. **B.** *adj.* Of or pertaining to the Pequot Indians, or the language spoken by them.

1631 in *New Hampsh. Hist. Soc. Coll.* (1834) IV. 226 Wee heare their numbers exceed any but the Pecoates and Nawagansets. **1637** *Public Rec. Colony of Connecticut* (1850) I. 10 To parle w[i]th the bay aboute o[u]r settinge downe in the Pequoitt Countrey. **1654** E. Johnson *Hist. New-England* ii. vi. 109 The English sought by all means to keepe these [fighting men] at least from confederating with the Pequods. **1714** S. Sewall *Diary* (1882) III. 12 Commissioners met to give Govr. Sattonstall an Opportunity to vindicate himself relating to the Pequot and Mohegan Indians. **1849** O. W. Holmes *Poems* 256 He heard the Pequot's ringing whoop. **1851** H. Melville *Moby Dick* I. xvi. 110 Pequod..was the name of a celebrated tribe of Massachusetts Indians, now extinct as the ancient Medes. **1871** C. M. Yonge *Pioneers & Founders* i. 6 The Pequot Indians, a tall, well-proportioned, and active tribe, belonging to the great Iroquois nation. *Ibid.* 8 The Pequots were..at war with the Dutch. **1903** Prince & Speck in *Amer. Anthropologist* V. 195 Their language, of course, remained Pequot, a dialect which shows a..striking kinship..with the present speech of the Canadian Abenakis. **1945** C. M. Webster *Town Meeting Country* 11 The Pequots were probably the bravest and most ferocious of all the New England tribes. **1979** *Arizona Daily Star* 1 Apr. A 2/2 Among their guests is to be John Hamilton, Chief Rolling Cloud of the Mohegan and Pequot American Indian tribes.

per, *prep.* Add: **I. 4. per contra. a.** (Further examples.)

1903 R. Fry *Let.* 16 Mar. (1972) I. 206 The article won't do the magazine much good, but *per contra* it won't do anyone..any harm. **1919** J. Stephenson *Princ. & Pract. Commerc. Corr.* iii. iii. 181 Your cash remittance of £1,000 of the 1st inst. came duly to hand and per contra we have purchased Frs. 25,000 French Gold Rente, which we send you enclosed. **1924** A. Huxley *Let.* 9 Aug. (1969) 231 The people who have deliberately set out to put great thoughts into verse have generally been the worst poets on record. But, per contra, the best poets have generally implied or directly expressed great thoughts. **1940** G. F.-H. & J. Berkeley *Italy in Making* III. iii. 52 The Sicilians had only one great enthusiasm—to free their island from the hated Neapolitans... *Per contra*, the Neapolitans' first aim was to maintain their hold on Sicily. **1976** *Listener* 12 Aug. 172/1 Writers from this centre of England..suggest a centrality of English experience... *Per contra*, some explorers..have found a too-muchness in this green country.

b. Also as *adj.*

1972 *Times* 11 May (Spain Suppl.) p. iv/3 A similar list prepared by the *The Banker*, but based on deposits less *per contra* accounts, of world bank groups included two Spanish banks only.

5. per diem. a. (Later examples.)

1906 *Arch. Roentgen Ray* XI. 18/2 When the method of treatment by fractional doses is carried out—for instance, ½X to 1X per diem—the quantimetric method alone is impossible. **1920** A. Huxley *Let.* 23 Dec. (1969) 193, I have to go to at least two and sometimes three theatres per diem and write about them afterwards. **1979** C. Dexter *Service of All Dead* xxxiii. 194 We're all ageing at the standard rate of twenty-four hours *per diem*.

b. (Earlier and later examples.) Also as *adj.*

1809 *Deb. Congress U.S.* 13 Feb. (1853) 350 Officers of the United States..have received..the per diem allowance fixed by law. **1812** *Weekly Reg.* 18 Jan. 361/2 The *per diem* of the members has been raised to *four* dollars. **1839** *Congress. Globe* 25th Congress 3 Sess. App. 66/1 In that case, had he asked for his mileage and per diem, all would have considered it an insult. **1846** T. L. McKenney *Mem.* I. ix. 192, I referred to him the making up of my account for my per diem allowance. **1946** E. Hodgins *Mr. Blandings builds his Dream House* xiii. 195 He would..happily replace the tubs at a per diem rate. **1973** R. Hayes *Hungarian Game* xxxviii. 228 If I made the flight I could crib another twenty dollars on my per diem. **1974** R. Thomas *Porkchoppers* v. 39 He got another $10,000 a year from the union in per diem and expenses. **1977** D. James *Spy at Evening* iv. 21, I liked the job. I accepted. 'A *per diem* of fifty pounds—travel expenses in addition.'

6. per mensem. (Later examples.)

1886 Kipling *Departmental Ditties* (ed. 2) 3 A nice retaining fee Supplied, of course, *per mensem*, by the Indian Treasury. **1916** 'Taffrail' *Pincher Martin* vi. 88 The sum of ten shillings *per mensem* was supposed..to suffice for the midshipmen's needs in the way of extras. **1965** *New Statesman* 24 Sept. 466/3 (Advt.), The amounts mentioned below refer to Malayan dollars (per mensem) and their approximate sterling equivalent. **1974** *Nature* 30 Aug. p. xxvii/1 (Advt.), This scholarship carries a stipend of M$700 per mensem for the first 12 months followed by M$800 per mensem beyond that period.

10. per aliud, by or in another entity; extrinsically; with reference to anything else; **per anum**, by the anus, applied esp. to anal sexual intercourse; **per capita**, (*b*) = **per caput*; **per caput**, per person or head (of population); also as *adj. phr.*; **per curiam** (*Law*), 'by action of the court', applied to a judgement, of concise and peremptory character, formulated by the whole bench; freq. *attrib.*; **per impossible** (*Logic*), 'as is impossible', a qualification governing a proposition which can never be true; **per incuriam** (*Law*), 'by carelessness', applied to a judicial decision evidently contrary to the law or facts; also *transf.*; **per mil, per mille,** in every thousand; **per primam** *Med.*, in full **per primam intentionem**, 'by first intention' (see INTENTION 10b).

1890 W. James *Princ. Psychol.* II. xvii. 42 To say that we feel a sensation's seat to be 'in the brain' or 'against the eye' or 'under the skin' is to say as much *about* it and to deal with it in as non-primitive a way as to say that it is a mile off. These are all secondary perceptions, ways of defining the sensation's seat *per aliud*. **1948** *Mind* LVII. 127 St. Thomas's proof [is] that the existence of finite beings, since it is *per aliud*, must be derived from something that exists *per se*. **1838** *Guy's Hosp. Rep.* III. 340 The constant symptom..was the passing of blood per anum. **1972** P. Green *Shadow of Parthenon* 160 Their liking for intercourse *per anum*, perhaps to preserve their virginity for the marriage market. **1972** *Mod. Law Rev.* XXXV. 107 Sexual intercourse *per anum*. **1926** Fowler *Mod. Eng. Usage* 428/2 *The entire production of opium in India is two grammes per capita yearly.* This use is a modern blunder, encouraged in some recent dictionaries. **1941** Wyndham Lewis *Let.* 9 Nov. (1963) 306 It [sc. Canada] reads less per capita than any other known civilised population. **1942** J. S. Huxley in *Harper's Mag.* Sept. 340/2 The U.S.S.R., in spite of its low per capita wealth, [etc.]. **1952** [see *GROWTH[1] 1 c]. **1955** *Times* 6 July 8/4 Saving *per capita* varies considerably from territory to territory, and in each territory from year to year, according to the study. **1965** *New Statesman* 30 Apr. 672/3 For the bulk of humanity per capita consumption remains the same. **1974** *Times* 25 Apr. 17/7 Let us have..an electoral system that secures proper proportionate representation both *per capita* and by party. **1975** 'D. Jordan' *Black Account* 246 Geneva Airport, I always feel, is the richest airport in the world..in terms of *per capita* elegance, tailoring, luggage, comfort. **1975** *Sci. Amer.* Nov. 56/1 Per capita incomes have declined in recent years. **1919** W. T. Grenfell *Labrador Doctor* (1920) iii. 60 By special arrangement with the railway and other friends, and by very simple living, the per caput charges were so much reduced that many of the boys not only paid their own expenses, but even helped their friends. **1962** *Times* 21 May (Commonw. Chambers of Commerce Suppl.) p. v/5 India plans to..raise the present *per caput* income in the country. **1970** K. J. Parker in G. G. Birch et al. *Glucose Syrups* v. 77 The current total *per caput* consumption of refined sugar is higher in Britain (112 lb) than in the USA (99·7 lb). **1976** *Lancet* 13 Nov. 1050/2 It may be argued that the per-caput cigarette consumption is not a good measure of the cigarette consumption in young women. **1978** *Jrnl. R. Soc. Arts* CXXVI. 651/2 The Harbin Transistor Plant was achieving very high *per caput* sales. **1890** *Cent. Dict.*, *Per curiam*, in *law*, by the court: a phrase prefixed to judicial opinions indicating the sanction of the court to the statements therein, as distinguished from the individual opinions of a particular judge. **1955** *Bull. Atomic Sci.* Oct. 309/2 It was a brief per curiam decision in a case which involved no contested issue. **1959** Jowitt *Dict. Eng. Law* II. 1327/2 *Per curiam, per cur.*, by the court. **1972** *N.Y. Law Jrnl.* 24 Oct. 2/1 Per curiam: Order reversed, with..costs and defendant's motion..granted. **1973** *Ibid.* 20 Feb. 4/4 If a court writes a per curiam opinion like this, what justification is there for demanding better writing from attorneys? **1847** A. De Morgan *Formal Logic* vii. 132 The moods *Baroko* and *Bokardo* do not admit of reduction to the first figure, by any fair use of the phrase: but the logicians were determined that they should do so, and they accordingly hit upon the following plan, which they called reduction *per impossibile*. **1883** F. H. Bradley *Princ. Logic* 217 If, I say, *per impossibile* this phantom could be real—..the above chance of irregularity would vanish. **1896** L. T. Hobhouse *Theory of Knowl.* 199 It was a onesided error to suggest that the immediate object of vision is colour or rows of coloured points from which, *per impossibile*, extension was conceived as removed. **1912** A. Lang *Shakespeare, Bacon & Gt. Unknown* xii. 267 If he knew that the author was Bacon, and knew it under pledge of secrecy, and was asked (*per impossibile*) 'Who wrote these plays?' he had only to say, 'Look at the title-page.' **1923** H. W. B. Joseph *Labour Theory of Value in Karl Marx* vi. 145 And supposing the equilibrium in an equal exchange were *per impossibile* between satisfaction on one side and sacrifice on the other, such equilibrium could only be said to exist in single exchanges independently. **1935** *Mind* XLIV. 237 Butler's 'let it be allowed' that virtue could not be justified if (*per impossibile*) it were contrary to self-interest. **1963** J. Lyons *Structural Semantics* iii. 42 Suppose, *per impossibile*, that we were transported as investigating linguists to Athens of the fifth century B.C. **1972** *Times Lit. Suppl.* 22 Dec. 1550/4 Have you for the moment forgotten *per impossibile* in which opera the chorus 'Upon our sea-girt land' occurs. **1867** Wharton *Law Lexicon* (ed. 4) 709/1 *Per incuriam*, through want of care. **1925** F. Newbolt *Out of Court* iii. 162 To attain this object you should first succeed in life, and have your portrait, or *per incuriam* that of a confusing namesake, inserted in a good picture paper. **1963** *Times* 24 Apr. 5/2 Mr. Puntan now appealed against that order; and his Lordship was afraid that the Divisional Court had acted *per incuriam* in making the order they did. **1970** *Internat. & Compar. Law Q.* XIX. 340 Parts of the judgement must be regarded as given *per incuriam*, and cannot be relied upon. **1976** *Phipson's Law of Evidence* (ed. 12) xxix. 677 A document omitted *per incuriam* was allowed to be put in by the prosecution during the reply. **1902** *Encycl. Brit.* XXXI. 404/2 The bottom waters have almost uniformly a salinity of 34·8 per mille, corresponding closely with the bottom waters of the south Atlantic. **1957** A. Grimble *Return to Islands* 102 Perhaps five or six per mil of his parishioners at most. **1957** L. F. R. Williams *State of Israel* 189 Infant mortality among the Arab community has fallen steadily to about 60 per mille of live births. **1972** *Nature* 25 Feb. 417/1 The seasonal amplitude is reduced to around 2 per mille. **1972** *Science* 22 Sept. 1099/3 The sulfur isotopic values on the west side averaged around 5·3 per mil. **1907** *Practitioner* Sept. 335 The wound healed *per primam*, except at the drain opening, and this was quite closed on the eighteenth day. *Ibid.* 336 Union *per primam* without any trouble. **1957** H. N. Harkins in J. G. Allen et al. *Surgery* ii. 9/1 Healing of wounds can be divided into 3 types: (1) Healing by first intention (per primam intensionum [*sic*]: primary union).

III. 1. (Later examples.) *per margin*: (earlier example); (*as*) *per usual* (later examples); also with ellipsis of *usual*. Also, other humorous and extended uses.

1782 *Town & Country Mag.* Dec. 669/1, I stood with the squadron, as per margin, to the southward, all that night. **1922** Joyce *Ulysses* 343 As per usual somebody's nose was out of joint. **1923** 'K. Mansfield' *Bad Idea* in *Doves' Nest* 146 So I took her up a cup of tea..as per usual on her headache days. **1938** J. Phelan *Lifer* xxi. 212 That's right,..no grounds, as per. **1959** N. Marsh *False Scent* (1960) i. 12 He'll be bringing his present later on, as per usual. **1960** S. Barstow *Kind of Loving* ii. vii. 263, I reckon after tonight we can't carry on as per. **1960** 'B. Mather' *Pass beyond Kashmir* xviii. 240 It'll have to be per boot again—and across country at that. **1966** 'J. Hackston' *Father clears Out* 115 The Butler boys returned home (as per precedent) from breaking-in up north. **1966** *Rev. Mod. Physics* XXXVIII. 221/2 Look at one of the product tableaux obtained as per previous instructions with the markings 1, or 2,..or *N* in the various squares inserted. **1972** 'A. Armstrong' *One Jump Ahead* i. 13, I came back as per usual about five o'clock. **1972** *Mod. Law Rev.* XXXV. 58 It cannot make a bare declaration (see *per* Lord Hailsham L.C., H.L.Deb., Vol. 318, col. 936). **1977** J. Bingham *Marriage Bureau Murders* i. 9 I'll stay in a pub... As per usual.

2. Also with ellipsis of *cent, head, hour, week*, etc.

1899 G. W. Peck *Uncle Ike* iii. 31 Listened to a heavenly choir that is paid a hundred dollars per. **1901** *Hide & Leather* 24 Aug. 30/2 He now sits in the Usher's box near the entrance to the jobbing house and draws $10 per. **1903** 'J. Flynt' *Rise of R. Clowd* iii. 111 The percentage that Ruderick was to receive excited the liveliest interest... 'I wouldn't give any kid more'n twenty-five per.' *a* **1911** D. G. Phillips *Susan Lenox* (1917) II. iv. 86 We'll get married as soon as he has a raise to twelve per. **1911** J. London *Let.* 18 Oct. (1966) 353 To exploit the mediocre for the consumption of mediocrity at so much exultantly per. **1935** J. N. Chance *Wheels in Forest* viii. 153 The road is clear at eleven ten, and Lombard is here at eleven nineteen. That's nine minutes; an average of eighty miles per. **1946** [see *BEEF *sb.* 6]. **1973** E. McGirr *Bardel's Murder* ii. 52 '[He] thinks his two daughters have been got at by the chauffeur.'..'I phoned up old Sir Omicron Pie, who kicks us back fifty per on the surgery.' **1976** *New Yorker* 23 Feb. 28/2 Many of the chain hotels are run by managing directors who have wives who think they are interior decorators and get on the payroll at twenty-five thousand per.

peracarid (perăkæ·rid), *sb.* (and *a.*) [f. mod.L. name of division *Peracarida* (W. T. Calman 1904, in *Ann. & Mag. Nat. Hist.* 7th Ser. XIII. 150), f. Gr. πήρα pouch + καρίς shrimp, prawn: see -ID[3].] A crustacean belonging to the division of the subclass Malacostraca so called, including sand-hoppers and woodlice possessing brood pouches. Also as *adj.* So **peraca·ridan, peracari·dean** *a.*

1931 W. Schmitt *Crustaceans* in *Shelled Invertebr. of Past & Present* ii. iii. 156 The two final orders comprising the peracarids. **1961** H. Schöne in T. H. Waterman *Physiol. Crustacea* II. xiii. 486 Female peracaridans carry eggs and young in a marsupium between their legs. **1965** B. E. Freeman tr. *Vandel's Biospeleol.* ix. 117 *Spelaeogriphus* is most certainly the last relict of a primitive peracarid type. **1965** W. Schmitt *Crustaceans* iii. 88 This [sc. Spelaeogriphacea] is one of the most recently established of the six peracaridean orders. *Ibid.* 89 The Spelaeogriphacea differ from all other peracaridans in having vesicular, oval gills. **1967** *Oceanogr. & Marine Biol.* V. 518 The first group includes the peracarids *Gnatha oxyurea*, *Eriopisia elongata*..and also many (better known) decapods.

peracetate. *Chem.* Restrict †*Obs.* to sense in Dict. and add: **2.** An ester or related derivative of peracetic acid.

1901 *Jrnl. Chem. Soc.* LXXX. i. 308 Acetic anhydride furnishes an oily product, probably ethyl peracetate. **1949** *Chem. Rev.* XLV. 8 Ethyl peracetate has been reported... This compound cannot be classed as a derivative of peracetic acid and ethyl alcohol, but rather as a derivative of acetic acid and ethyl hydroperoxide. **1967** L. F. & M. A. Fieser *Reagents for Org. Synthesis* I. 790. The reaction may proceed through the hemiacetal peracetate.

peracetic (pərăsi·tik), *a.* *Chem.* [f. PER-[1] 5 b + ACETIC *a.*] *peracetic acid*: $CH_3CO·O·OH$, a colourless, corrosive, pungent liquid that is explosive when hot and is widely used, usu. dissolved in acetic acid, as an oxidizing agent in synthesis, as a bleach and as a sterilizing agent, etc.

1903 *Jrnl. Chem. Soc.* LXXXIV. i. 398 The aqueous solution gradually suffers hydrolysis with formation of molecular proportions of acetic and peracetic acids. **1938** [see *EPOXIDE]. **1965** *Economist* 20 Feb. 780/1 The new process for the oxidation of cyclohexanone to caprolactone with peracetic acid. **1970** *New Scientist* 15 Jan. 102/1 Peracetic acid is a powerful oxidizing agent that readily kills bacterial spores. **1972** Norman & Waddington *Mod. Org. Chem.* xxi. 321 It [sc. nylon] can be bleached with a dilute solution of peracetic acid.

peracid (pəræ·sid). *Chem.* Also **per-acid.** [f. PER-[1] + ACID *sb.*, as tr. G. *persäure* (von Baeyer & Villiger 1900, in *Ber. d. Deut. Chem. Ges.* XXXIII. 2480).] An acid which contains a peroxide group, esp. (in *Org. Chem.*) the group —CO·O·OH.

1900 *Jrnl. Chem. Soc.* LXXVIII. i. 626 The authors suggest that hydrogen peroxide, its acyl derivatives, and their peroxides should be called 'hydroperoxide', 'per-acids' and 'peroxides' respectively. **1922** J. W. Mellor *Comprehensive Treat. Inorg. & Theoret. Chem.* I. xiv. 959 The true peracids are either formed by the action of hydrogen peroxide on ordinary acids or their derivatives, or else they furnish hydrogen peroxide when hydrolyzed with dilute sulphuric acid. **1950** N. V. Sidgwick *Chem. Elements* II. 871 Other compounds with the O—O link are numerous per-acids (or peroxy-acids), such as persulphuric, perboric, percarbonic, pertitanic, perchromic, etc. **1972** *Materials & Technol.* IV. xiii. 477 All preparations of, and reactions with, organic peroxide and peracids should be conducted behind safety shields, because a reaction occasionally proceeds with uncontrollable violence.

peracute *a.* Delete 'Now *rare*' and substitute 'Now chiefly *Vet. Med.*'. Add later examples.

1963 *Daily Tel.* 3 Dec. 23/1 A highland steer which went down with peracute pneumonia early yesterday recovered sufficiently to win a first prize in its class later in the morning. **1970** W. H. Parker *Health & Dis. in*

Farm Animals ix. 102 The word subacute is used to describe a condition between acute and chronic while a disease which kills very quickly..is called peracute.

perahera (perāhe·rā). Also **perahar, perahāra.** [Sinhalese *perahera* protection, safety.] In Sri Lanka: a procession, orig. of a religious (Hindu, later also Buddhist) character, of praise or thanksgiving, or of intercession.

1681 R. Knox *Hist. Relation Ceylon* III. iv. 78 That they may..honour these Gods, and procure their aid and assistance, they do yearly in the Month of June or July, at a New Moon, observe a solemn feast and general Meeting, called Perahar. **1817** in R. Pieris *Sinhalese Social Organization* (1956) III. 135 *Perhāra*..is a very ancient ceremony in commemoration of the birth of the god Vishńu. *Ibid.* 136 Five days having expired, another ceremony, an important and essential part of the *perahāra*, commences..which lasts five days more. *Ibid.* 137 The ceremony of *perahāra* is continued up to the day of the full moon... On the night of the full moon..the shrine is carried in the procession. **1913** L. Woolf *Village in Jungle* v. 113 Last night we took him in the perahera, and called upon the god to hear us. **1923** D. H. Lawrence *Birds, Beasts & Flowers* 170 But the best is the Pera-hera, at midnight, under the tropical stars..the Pera-hera procession, flambeaux aloft in the tropical night. **1971** *Ceylon Daily News* 17 Sept. 1/3 He will be taken in a perahera to the avasa where a felicitation meeting is to be held. **1974** *Oxf. Jun. Encycl.* (rev. ed.) I. 102/2 The principal occasion is the Perahera, a great annual pageant in Kandy, when a relic, reputed to be a tooth of Gautama Buddha, is carried about the town in grand procession on the back of a gorgeously caparisoned elephant.

‖**perahu,** var. Proa.

1939 A. Keith *Land below Wind* III. xi. 186 The river travel would be accomplished in small native canoes known as *perahus*. **1958** J. Slimming *Temiar Jungle* ii. 19 The kit was..stowed away in the boat—a thirty-five-foot *perahu* with a thirty-horse-power engine. **1965** R. McKie *Company of Animals* i. 1 The Malay perahu..was thirty feet long with a four-foot beam, a thin slice with a stern flattened just enough to hold an outboard motor, and a bow so sharp that it parted the river like a comb. **1966** *Festival Malaysia 1966: Calendar of Events* 6/2 The intricate carvings that decorate the racing perahu can be seen.

peralkalic (pəræ·lkălik), *a. Petrol.* [f. Per-[1] 4 + Alkalic *a.*] = next.

1902 W. Cross et al. in *Jrnl. Geol.* X. 592 The divisions in classes I, II, and III are fivefold: Rang 1: $\frac{K_2O' + Na_2O'}{CaO'} > \frac{7}{1}$, peralkalic. **1976** *Nature* 10 June 482/1 The Saint Francois Mountains form a distinctive unmetamorphosed igneous complex comprising chiefly alkalic to peralkalic rhyolite and granite.

peralkaline (pəræ·lkălain), *a. Petrol.* [f. Per-[1] 4 + Alkaline *a.*] Of a rock: containing a high proportion of soda and potash; now *spec.* (see quot. 1931).

1913 A. N. Winchell in *Jrnl. Geol.* XXI. 210 Along this co-ordinate igneous rocks are classified as normal or alkali-calcic.., alkaline, and peralkaline. *Ibid.* 211 Peralkaline rocks are characterized mineralogically by the presence of feldspathoids (or lenads)... Chemically they are distinguished by insufficient silica to combine with the abundant alkalies to form feldspars after saturation of other available bases as orthosilicates. **1927** S. J. Shand *Eruptive Rocks* vii. 128 The following groups of rocks stand out as chemically distinct:— (a) A peraluminous group, characterised by primary muscovite, biotite, corundum, tourmaline, topaz, almandine, or spessartite. (b) A peralkaline group, characterised by soda-pyroxenes or soda-amphiboles,..and by the virtual absence of anorthite... (c) A group characterised by common pyroxenes, amphiboles, olivine, [etc]. **1931** —— *Study of Rocks* iv. 52 Peraluminous rocks... The molecular proportion of alumina exceeds the molecular proportions of soda, potash and lime combined... Peralkaline rocks... The molecular proportion of alumina is less than that of soda and potash combined. **1950** *Rep. 18th Internat. Geol. Congr. 1948* II. 129 Peralkaline rocks..only occur in stable parts of the earth's crust, outside active orogenic zones. **1974** *Nature* 24 May 315/1 Alkaline and peralkaline igneous rocks were intruded and extruded in distinct nodes. **1974** [see *peraluminous *a.*].

Hence **pe:ralkali·nity,** the state of being peralkaline.

1969 *Amer. Jrnl. Sci.* CCLXVII. 242 A quadrilateral diagram in which molecular alumina is plotted against soda/potash ratio can then be employed to show the variation in alkali ratio in whole rocks..with changes in silica content and peralkalinity. **1974** Bowden & Turner in H. Sørensen *Alkaline Rocks* IV. viii. 334/2 The variation in the coloured mineral content is dependent on the peralkalinity of the granites.

peraluminous (pərāliu·minəs), *a. Petrol.* [f. Per-[1] 4 + Aluminous *a.*] Of a rock: (see quots. 1974[1] (and 1931), 1972).

1927, 1931 [see *peralkaline *a.*]. **1964** *Mineral. Abstr.* XVI. 488/2 A peraluminous granite stock and related pegmatites..have been emplaced in pelitic and quartzo-feldspathic schists. **1972** *Gloss. Geol.* (Amer. Geol. Inst.) 527/1 *Peraluminous*, said of an igneous rock in which the molecular proportion of aluminum oxide is greater than that of sodium and potassium oxide combined. **1974** I. S. E. Carmichael et al. *Igneous Petrology* ii. 31 This leads to four more classes of rocks, each independent of silica saturation: 1. Peraluminous rocks, in which the molecular proportion of Al_2O_3 exceeds (CaO + Na_2O + K_2O)... 4. Peralkaline rocks, in which Al_2O_3 < (Na_2O + K_2O). **1974** W. C. Luth in H. Sørensen *Alkaline Rocks* VI. vi. 506/1 Several key factors provide limiting conditions. These include..the relatively large amounts of the peraluminous hydrates required.

perambulate, *v.* Add: **3. a.** *intr.* Of a (light) vehicle: to be in motion, to move about. *rare.* **b.** *trans.* To wheel, convey, or conduct (*about*) in or as in a perambulator (sense 3); to travel on or traverse in a perambulator.

1856 *Chambers's Jrnl.* 23 Aug. 116/2 The young brother..can hardly reach to the bar, but nevertheless the light crrriage perambulates obediently under his guidance. **1865** P. H. Gosse *Year at Shore* iv. 87 The open gate of a villa reveals a little girl 'perambulating' a baby. **1902** *To-Day* 30 Apr. 8/1 Babies..are not allowed to 'perambulate' the pavement two or three abreast. **1909** M. B. Saunders *Litany Lane* xxii. 295 The Princess Max, having opened the affair, was being perambulated about as usual. **1922** J. A. Dunn *Man Trap* i. 9 Jovial of mouth and eyes despite the handicap that reduced him to being perambulated. **1929** P. Gibbs *Hidden City* xi. 50 Four acres of garden in which some neat nursemaids were perambulating the pink-cheeked babies of the well-to-do.

perambulating, *ppl. a.* (Further examples.)

1926 W. J. Locke *Stories Near & Far* 280 Then he walked round his perambulating property [*sc.* a caravan]. A big-boned brown horse ceased his munching as he approached. **1938** P. W. Sergeant *Championship Chess* i. 26 There is little to be said for perambulating chess matches —except that they bring in more money. **1949** E. Coxhead *Wind in West* vi. 165 We're..all products of what Rory calls the book-learning. We're his perambulating text-books.

perambulator. 1. a. For Now *rare* or *Obs.* read *rare* and add further examples. Also *fig.*

1870 Hazlewood & Williams *Leave it to Me* 3 Joe's a perambulator;..a perambulating greengrocer, called by vulgar people a costermonger. **1925** J. Bone (*title*) The London perambulator. **1930** R. Campbell *Adamastor* 64 Speed, motion, flight!..Perambulator of the Bored And ambulance of broken hearts! **1971** *Daily Tel.* 18 Oct. 10 (Advt.), Dickens was a determined perambulator of London, either in search of material..or simply wandering the streets.

2. Delete †*Obs.* and add later examples.

1855 J. Butler *Trav. & Adventures Assam* I. v. 56 Some idea may be formed of the impassable nature of the country we travelled over this day, when I state that we only came eight miles one furlong, by the perambulator, in eight hours. *a* **1877** [see *delineator 3]. **1913** Close & Cox *Text Bk. Topogr. & Geogr. Surveying* (ed. 2) iv. 76 A perambulator should not be used over very rough ground, and both it and the cyclometer should be checked over known measured distances. **1964** D. Greenhood *Mapping* ix. 262/2 For doing the same kind of measuring on the ground itself, there are various makes of distance meters, sometimes called 'perambulators', which are wheels that you push by a handle, like a roller toy. They register distances up to 10,000 ft. **1969** Tooley & Bricker *Hist. Cartogr.* 42 (*caption*) Colles's map..was probably made with compass, plane table, and perambulator.

3. (Earlier and later examples.) Also *attrib.*, *Comb.*, and *fig.*

1856 *Chambers's Jrnl.* 23 Aug. 116/2 The *Perambulator*..has given us children, looking on with their grave smooth faces at the business of life,..as they lean back philosophically in their carriages. **1856** *Punch* 22 Mar. 118/2 (*caption*) I shan't play no more with that Matilda Jenkins. —'Er doll ain't got no Perambylatur. **1861** *Temple Bar* I. 539 These creatures [*sc.* kangaroos]..have the power of carrying their delicate, prematurely born young about with them wherever they go. They have this condition, viz. a soft, warm, well-lined portable nursery-pocket, or 'perambulator'. **1866** *Leisure Hour* XV. 347/2 Certain ill-tempered bachelors did indeed protest against them, complaining that perambulator-drivers did occasionally drive their new-fangled machines against their shins. **1936** P. M. Clark *Autobiogr. Old Drifter* iv. 47 Some time after this I was on my way to Rondebosch to meet a married cousin whom I had not seen since my perambulator days. **1972** *Daily Tel.* 3 June 32/4 They came with shopping bags, picnic baskets, babies in perambulators and pushchairs, babies in arms.

perambulatory, *sb.* **a.** (Later example.)

1843 *Knickerbocker* XXII. 85 Let..the temperance-halls and root-beer perambulatories make answer.

perbenzoic (pəɹbenzōu·ik), *a. Chem.* [f. Per-[1] 5b + Benzoic *a.*] *perbenzoic acid*: a colourless crystalline solid, $C_6H_5CO\cdot O\cdot OH$, which is a widely used oxidizing agent, esp. for epoxidation reactions.

1903 *Jrnl. Chem. Soc.* LXXXIV. I. 397 Benzoic acetic peroxide undergoes hydrolysis in aqueous solution with production of perbenzoic and acetic acids, together with benzoic peroxide. **1938** *Thorpe's Dict. Appl. Chem.* (ed. 4) II. 373/1 Benzoyl peroxide is comparatively inert, but treatment with alcoholic sodium ethoxide yields the reactive perbenzoic acid,..which is used very widely for oxidising ethylenic compounds. **1967** L. F. & M. A. Fieser *Reagents for Org. Synthesis* I. 791 The reaction of perbenzoic acid with an olefin usually proceeds smoothly at a low temperature (0–25°) and affords an epoxide in high yield.

perborate (pəɹbō·ə·rēit). *Chem.* [a. F. *perborate* (A. Étard 1880, in *Compt. Rend.* XCI. 932): see Per-[1] 5 and Borate.] Any of a number of strongly oxidizing derivatives of boric acid which contain peroxo-anions, and are usu. prepared by the action of hydrogen peroxide on borates; *esp.* the sodium salt, a white crystalline solid of empirical formula $NaBO_3\cdot 4H_2O$, which is widely used as a bleach and is a constituent of washing powders.

1881 *Chem. News* 14 Jan. 25/2 He [*sc.* A. Étard] has obtained barium perborate,—$Bo_2O_5BaH_4$ + H_2O, a white amorphous insoluble salt. **1898** *Jrnl. Chem. Soc.* LXXIV. II. 427 The heat of decomposition by sulphuric acid was determined in the case of sodium perborate and ammonium perborate. **1916** *Chem. Abstr.* X. 2803 Heretofore bleaching with persalts, such as perborate, has been more costly than bleaching with chloride of lime. **1959** *Observer* 6 Sept. 18/4 Perborate bleach works only at high temperatures. **1959** *Guardian* 28 Sept. 3/6 The woollen cardigan..had..changed colour..due to the perborate (a stain-removing bleach) added..to a soap powder. **1967** E. L. Muetterties *Chem. Boron & its Compounds* iii. 192 The X-ray crystal structure determination of 'sodium perborate'..reveals a dimeric tetrahedral configuration with dihedral angle equal to 64°..and the anionic formula $B_2(O_2)_2(OH)_4{}^{2-}\cdot 2Na^+\cdot 6H_2O$. **1974** *Sci. Amer.* Jan. 125/2 Sodium perborate is the cheapest and safest of all peroxy salts and is much used in detergents, particularly for very-hot-water washing.

perboric (pəɹbō·ə·rik), *a. Chem.* [ad F. *perborique* (A. Étard 1880, in *Compt. Rend.* XCI. 931), f. *per-* Per-[1] 5b + *borique* Boric *a.*] *perboric acid*: the supposed parent acid of perborates, which was formerly thought to have the formula HBO_3 and is only known in acidic solutions containing perborate anions.

1881 *Chem. News* 14 Jan. 25/2 Whilst an equimolar mixture of magnesium sulphate, ammonium chloride, and ammonia is not rendered turbid either by oxygenated water or by boric acid, a mixture of the two precipitate it abundantly, acting as perboric acid. **1924** J. W. Mellor *Comprehensive Treat. Inorg. & Theoret. Chem.* V. xxxii. 116 Perboric acid itself has not been made... In ethereal soln. and an excess of hydrogen peroxide, the partition coeff. of boric acid increases a little corresponding with the formation of free perboric acid in the etheral soln. **1973** N. N. Greenwood *Chem. Boron* (1975) vi. 887 Reaction of orthoboric acid with hydrogen peroxide gives perboric (peroxoboric) acid solutions which probably contain the monoperborate anion [$HOOB(OH)_3^-$].

perbromic (pəɹbrōu·mik), *a. Chem.* [f. Per-[1] 5b + Bromic *a.*] *perbromic acid*: $HBrO_4$, a strong acid with oxidizing properties that was first prepared in 1968. Hence **perbro·mate,** a salt of this acid.

Claims for the preparation of the acid and its salts in the 19th c. proved to be mistaken.

1864 *Chem. News* 30 Apr. 205/2 Perbromic acid has been fruitlessly investigated by many chemists, but M. Kaemmerer has obtained it in the most simple manner by treating perchloric acid with bromine. *Ibid.* Perbromate of potash is more soluble than the perchlorate, and less so than the bromate. **1866** [in Dict. s.v. Per-[1] 5b]. **1912** *Chem. News* 2 Aug. 50/1 It seems that it must be finally concluded that perbromic acid and its salts are incapable of existence. **1968** *Jrnl. Amer. Chem. Soc.* XC. 1900/2 These results indicated the formation of a relatively unreactive perbromate ion and suggested that a determined effort might lead to the preparation of macro amounts of perbromates. *Ibid.* 1901/2 As expected, the volatility of perbromic acid is less than that of perchloric acid. **1973** Downs & Adams in J. C. Bailar et al. *Comprehensive Inorg. Chem.* II. xxvi. 1451 First obtained in studies of the β-decay of $^{83}SeO_4{}^{2-}$, perbromates are also formed in the γ-radiolysis of crystalline bromates. *Ibid.* 1452 On very rapid evaporation, crystallization of perbromic acid solutions occurs (possibly to give $HBrO_4\cdot 2H_2O$) just before decomposition sets in.

Perbunan (pəɹbiu·nǎn). Also **perbunan.** [a. G. *Perbunan*, f. *per-* + *buna* *Buna + *N*, chem. symbol for nitrogen.] A proprietary name for a nitrile rubber first made in Germany and originally called Buna-N.

1938 *Trade Marks Jrnl.* 4 May 551/2 Perbunan... Compositions consisting mainly of reaction products obtained by the polymerisation of butadiene hydrocarbons, sold in the form of sheets, blocks, tubes, [etc.]... I. G. Farbenindustrie Aktiengesellschaft.., Frankfort-on-Main, Germany; manufacturers. **1938** *Chem. Abstr.* XXXII. 3663 The mech. and elec. properties of Buna-S and Perbunan (formerly Buna-N,..)..are described... The elec. properties of both are excellent, and Buna-S is less permeable to water than are Perbunan and natural rubber. **1940** *Jrnl. R. Aeronaut. Soc.* XLIV. 159 A cable consisting of rubber covered with a thin perbunan sheath and an oil varnished ozone-resisting braid would probably give the best results. **1959** *Times* 27 Apr. (Rubber Industry Suppl.) p. ii/4 These basic synthetic rubbers, buna S, neoprene and perbunan, developed in the 1930s, were the forerunners of the main synthetic rubbers we use to-day. **1959** *Official Gaz.* (U.S. Patent Office) 18 Aug. TM 82/1 Farbenfabriken Bayer Aktiengesellschaft, Leverkusen-Bayerwerk, Germany. Filed Feb. 12, 1959. Perbunan... For rubber and rubber substitute materials. **1973** *Nature* 14 Sept. 93/1, I have carried out some measurements on an alkaline perbunan latex dispersion.

perc, var. *perk *sb.*[3], *perk *v.*[3]

perceived, *ppl. a.* **2.** (Later examples, esp. in sense of *perception 9.)

1943 M. Farber *Found. Phenomenology* xii. 335 If the eruption is therefore to be critically judged as a deception,..illusion, etc., then the perceived, seen color of the object also does not exist. **1971** *Nature* 19 Feb. 518/1 Mr Stein's apparently innocent bill to limit noise at New York airports to 108 perceived noise decibels by July.. could be..a serious threat to the viability of Concorde. **1973** *Jrnl. Genetic Psychol.* CXXII. 269 The study of perceived (subjective) age changes. **1976** *Times Lit. Suppl.* 3 Sept. 1080/2 In all these instances the KKE was reacting to actual or perceived efforts to destroy it.

perceivedness. (Later example.)

1967 S. Beckett *Film* 32 Anguish of perceivedness.

perceiver, *sb.* **1.** (Later examples.)

1947 G. Murphy *Personality* xiv. 331 (*heading*) The perceiver. **1971** *Jrnl. Gen. Psychol.* LXXXIV. 158 The history of the gestalt psychology is filled with many other instances in which missing parts are ignored by the perceiver. **1972** *Sci. Amer.* Nov. 85/1 The students who were three-dimensional perceivers spent more time looking at the ambiguous trident. **1973** *Nature* 14 Dec. 434/2 He stresses that painters and sculptors, like poets, expect the perceiver to do at least half the work.

per cent, *phr.* (*sb.*) Add: Now freq. without full-stop, and as one word. **A.** *Phrase.* **a.** Also with preceding numeral as an approximate estimate of extent in unquantifiable contexts. *a* or *one hundred per cent*: see *hundred *sb.* and *a.* 2c.

1939 F. Pratt *Secret & Urgent* 258 It will be noted that more than fifty per cent of all English words end in E, S, D or T. **1961** *Information & Control* IV. 65 The study of this particular data collection showed that over 95 percent of the vocabulary could be represented by 13 characters. **1973** *Times* 21 Dec. 14/6 This trouble's cut my social life by about 35 per cent. **1973** [see *percentage b]. **1975** *Sci. Amer.* Feb. 15/3 The two reasons most often given for emigrating are 'Better training in the U.S.' (69 percent) and 'Political factors' (8 percent). **1976** *Ibid.* May 108/3 Io is particularly red, and its colour, together with its high reflectivity (about 62 percent, roughly equivalent to white sand), make it a unique object in the solar system. **1976** *Daily Tel.* 20 July 1/5 Average earnings in May were 19·4 per cent. up on the previous year, compared with a 15·4 per cent. rise in retail prices.

c. As quasi-*sb.* Percentage; one per cent.

1905 G. W. Rolfe *Polariscope* 96 The per cent of sucrose in the sample. **1934** *Jrnl. Sedimentary Petrology* IV. 68/2 In setting up a histogram, we are in effect setting up a series of separate 'bins', each of which contains a certain per cent of the grains. **1960** *Anatomical Rec.* CXXXVIII. 395/2 The statistical significance..was calculated by determining the median for the combined population and comparing the percent of the animals that had cataracts in each group. **1966** T. Pynchon *Crying of Lot 49* v. 122 Always just that little percent on the wrong side of breaking even... Why don't I quit? **1970** *Nature* 7 Nov. 546/2 If the relative abundances of ^{15}N and ^{17}O in the sample are found to be a few tenths of a percent higher than normal, [etc.]. **1971** *Daily Tel.* 27 Oct. 1/3 The retail price index has risen by only three-quarters of a per cent. in the three months since June. **1972** *Science* 12 May 595/2 The magnitude of the effect is again a few tenths of a percent. **1977** *Daily Tel.* 23 Dec. 2/7 Those provincial journalists whose immediate settlement hopes have foundered over ·56 of a per cent.

percentably (pəɹse·ntăbli), *adv. rare.* [f. Per cent *phr.* (*sb.*) + -ably.] By an appreciable percentage.

1928 *Sunday Dispatch* 2 Sept. 10 Men..who are bent on reducing the moufflon population percentably.

percentage. Add: **a.** (Later *attrib.* examples.) Freq. equivalent to 'per cent' qualifying the sb., as *percentage error* (= error per cent), *point* (= point per cent), etc.

1906 *Westm. Gaz.* 25 Jan. 8/1 Both first- and third-class passengers showed a percentage increase. **1920** H. Crane *Let.* 30 July (1965) 41 A drawing account at the bank..in addition to a good percentage commission on everything I sell. **1928** *Britain's Industr. Future* (Liberal Industr. Inquiry) v. xxxi. 444 The choice in particular cases between block grants and percentage grants. **1941** J. S. Huxley *Uniqueness of Man* v. 144 It warns us not to be too hasty in drawing conclusions as to intelligence from *percentage* brain-weight, or as to the efficiency of circulation from *percentage* heart-weight. **1948** Mencken *Amer. Lang.* Suppl. II. 766 Percentage man. A news photographer who makes a large number of exposures, hoping that chance will give him a few good pictures. **1961** C. C. T. Baker *Dict. Math.* 233 Percentage error = true error/actual error × 100. **1964** *Economist* 18 Jan. 182/2 The common American practice of 'percentage rents'—gearing rents and ground rents over the initial period to a percentage of turnover. **1969** 'R. Crawford' *Cockleburr* I. vii. 71 I'm a percentage man... He's paying me ten per cent to get him clear with the money. **1971** *Gloss. Electrotechnical, Power Terms* (*B.S.I.*) I. iv. 7 *Percentage error*, the relative error multiplied by 100. **1971** *Times* 3 Sept. 13/1 Yesterday's action by the Bank of England in cutting Bank rate by a full percentage point. **1972** *Fremdsprachen* XVI. 61/2 Percentage point, full—*ein ganzes Prozent*. **1973** *Computers & Humanities* VII. 134 Q measures the percentage improvement, in terms of the function F. **1974** 'A. Garve' *File on Lester* xxxvi. 129 The poll figures..were shattering—a big percentage lead for the Government. **1974** *News & Press* (Darlington, S. Carolina) 25 Apr. 16/2 Primary metal industries showed the greatest percentage increase (267 percent) over 1967.

b. *fig.* Advantage, gain; probability of successful outcome (*in* a situation, course of action, etc.). *colloq.* (orig. *U.S.*).

1862 B. Harte *Notes 'by Flood & Field'* in *Golden Era* 14 Sept. 5/3 What's the per centage—workin' on shares, eh? **1911** *Chicago Daily News* 2 Mar. 6/6 Johnny Coulon is unable to see the percentage in taking on Frankie Conley for another pummelling in the adjacent future. **1925** *College Humor* Aug. 117/2 No percentage in staying on in this house. Darn thing's too big. **1938** D. Runyon *Furthermore* xiii. 255 There is no percentage in hanging around brokers [*sc.* people who are broke]. **1940** *Woman* (U.S.) Sept. 69/1 Marge was courageous and a straight shooter but there was no more percentage in taking her out than one of the other guys. **1950** R. Moore *Candlemas Bay* VI. 302, I don't see how you figure that what peas you can shell with one thumb makes any percentage to me. **1952** B. Malamud *Natural* 90 He decided there might be some percentage to all these comparisons. **1952** H. Waugh *Last seen Wearing* (1953) 71 Well, hell, she turned down my dates and..there's no percentage in that. **1960** C. Hatton in *Pick of Today's Short Stories* XI. 151 Roxy..could be relied on to stir up trouble anywhere even if there was no percentage in it for Roxy. **1966** J. Porter *Sour Cream* xii. 162 There was no percentage in hanging around the airport terminus. I had to get away. **1973** E. McGirr *Bardel's Murder* i. 27 He plays better bridge than ninety-eight per cent of the population... He doesn't cheat: he doesn't have to because he's got the percentage. **1976** A. Price *War Game* I. 123 There was no percentage in rushing him.

c. Slang phr. *to play the percentages*: to play safely or methodically with regard to the odds in favour of success.

1964 A. Wykes *Gambling* i. 22 A considerable number of women gamblers take a strictly 'professional' approach. Ignoring 'intuition', they attack with expertise; if their game is horse racing, they are vastly knowledgeable about horses' and jockeys' past records. Others may 'play the percentages' in casinos. **1973** *Daily Pennsylvanian* 9 Oct. 6 Houston knows the game and its angles thoroughly and plays the percentages to perfection. **1977** *Tennis World* Sept. 17/3 To 'out-steady' someone is to play a superior defensive game, and 'playing the percentages' is the art of going for shots that are cheap and clean, rather than costly and glorious. Statistically it pays off better than 'going for broke'—risking everything on breath-taking winners.

perce·ntage-wise, *adv.* Also **percentagewise.** [Wise *sb.*[1] II.] (Expressed) as a percentage; also more generally, (regarded) relatively.

1912 F. Soddy *Matter & Energy* ii. 53 Quite a large number of commonly occurring compounds had been analysed by chemists and the composition expressed percentage-wise as above. **1944** *Sun* (Baltimore) 5 Apr. 10/2 In that State the number of Negroes is, percentage-wise, even smaller than the number of Negroes in Maryland. Here Negroes vote regularly. **1945** Nelson & Wright *Tomorrow's House* vi. 72/1 Very few families, percentagewise, have ever been able to afford hired help. **1955** J. A. Wheeler in W. Pauli *Niels Bohr* 176 Z^2/A does not have to change much percentage-wise to carry fission half lives from values too long to observe to values too short for reasonable stability. **1960** *Guardian* 8 Mar. 1/2 The number of deaths percentagewise is far less than other deaths..by other heating means. **1967** E. S. Gardner *Case of Queenly Contestant* (1973) xiii. 151 She told me..I could inherit a very substantial sum of money and asked me what it would be worth to me percentage-wise if [etc.]. **1972** *Amer. Speech 1968* XLIII. 211 On January 15, 1967, during the..telecast of the Super Bowl Game.., the announcer..said, 'Bart Starr has had a terrific year, percentagewise, as well as touchdownwise.'

percental (pəɹse·ntăl), *a.* Also **per cental.** [f. Per cent *phr.* (*sb.*) + -al I.] Reckoned by the hundred; calculated as a percentage.

1895 *Daily News* 18 Dec. 9/5 In wheat a fair extent of business was put through at ½d per cental decline. **1897** *Geogr. Jrnl.* IX. 319 A map showing, by means of six colours distinguishing different percental proportions, the distribution of German-speaking people in the lands of the Hungarian crown.

percenter. Also **per center, per-center.** [f. Per cent *phr.* (*sb.*) + -er[1].] Following a number (occas. with hyphen): that on (or from) which a percentage (specified by the number) is reckoned; a chance, situation, etc., the value of which is reckoned as a specified percentage. Also, one who lends or deals in money involving a certain rate of interest or commission.

c **1863** T. Taylor *Ticket-of-Leave Man* III. 52 *Moss* (*at the counter, getting out his bills*)... For two hundred at two months—drawn by Captain Crabbs—accepted the Honourable Augustus Greenway: that's a thirty per center. **1897** [in Dict. s.v. Per cent *phr.* (*sb.*).] **1949** [see **influence pedlar*]. **1950** E. Hemingway *Across River* xxxiii. 208 The brown-nosers, the five and ten and twenty percenters and all the jerks. **1959** *Punch* 10 June 781/1 We are tempted to deduce that her age bracket..is a sure hundred percenter. **1960** *Times* 5 Feb. 16/5 One explanation was the rush by the public to buy the new five percenter Treasury bonds. **1976** 'Z. Stone' *Modigliani Scandal* IV. i. 154 Louis was 99 per cent sure the caller was a nutcase: but it was by following up the one-percenters that great exclusives were found.

percentile, *a.* and *sb.* (Later examples.)

1956 W. H. Whyte *Organization Man* (1957) 407 Don't strive to get yourself in the 70th or 80th percentile for extroversion. **1957** G. E. Hutchinson *Treat. Limnol.* I. vi. 384 (*caption*) Mean percentile absorption at different wave lengths. **1970** *Jrnl. Gen. Psychol.* LXXXIII. 154 The high dependency groups consisted of those 13 Ss whose autonomy minus deference scale percentile ranks score was 20 or above.

percentual (pəɹse·ntiuăl), *a.* [irreg. f. Per cent *phr.* (*sb.*) + *-ual*, after *accentual, eventual*, etc.]

1937 H. Tingsten *Political Behavior* iii. 147 The voting frequency of a group increases with the percentual strength of the group in question. **1949** Koestler *Promise & Fulfilment* I. iii. 31 The striking percentual increases in the statistics quoted.

Hence **perce·ntually** *adv.*

1942 L. B. Namier *Conflicts* 73 'Janus' in *The Spectator* of March 7, 1941, shows that percentually the losses of the last war do not justify the talk about a 'missing generation'.

percept. Add: Also *Psychol.* **1.** (Further examples.)

1964 M. Critchley *Developmental Dyslexia* xiii. 78 Not only is it a matter of defective perception, but it is also one of inadequate association of lexical percepts. *Ibid.*, This process of linking one percept with another is where the principal fault may lie. **1973** *Nature* 6 July 54/2 The Necker cube has been viewed as an ambiguous figure whose configuration and accompanying instructions usually limit the number of percepts to two. **1974** *Sci. Amer.* Jan. 126/3 No one else smelled it [*sc.* poison gas], she was assured. Her enemy was so ingenious, she retorted, that his gas was odorless! Her experience was thus no percept at all, but a projection from internal ideas. **1976** Smythies & Corbett *Psychiatry* v. 55 This man had a normal percept but attached a special, personal meaning to it which was quite false.

2. (Later examples.)

1949 *Mind* LVIII. 450 William James sometimes used the word 'percept' to refer to the content of consciousness during perception; it is this fact which has made the name 'Percept Theory' seem to me appropriate for the particular theory of perceptual consciousness which he himself supported. **1970** *Jrnl. Gen. Psychol.* LXXXIII. 66 Sex responses are numerous... The number of human percepts is low. **1972** *Sci. Amer.* Sept. 47/2 It is along this pathway that the visual image formed on the retina by light rays entering the eye is transformed into a visual percept, on the basis of which appropriate commands to the muscles are issued. **1976** *Word 1971* XXVII. 226 Each physical stimulus, after interpretation by the mental processes, will result in a percept.

perception. Add: **6.** (Further examples.)

1943 M. Farber *Found. Phenomenology* xiii. 396 The perception realizes the possibility of the development of the *intending-this* with its definite relation to the object. **1962** Macquarrie & Robinson tr. *Heidegger's Being & Time* I. iii. 130 When the experience of hardness is Interpreted this way, the kind of Being which belongs to sensory perception is obliterated, and so is any possibility that the entities encountered in such perception should be grasped in their Being. **1965** *New Statesman* 3 Sept. 327/4 He [*sc.* Merleau-Ponty] held..that the higher forms of human behaviour—art, science, political life—could only be understood in their genesis from original perception.

9. *Psychol.* The neurophysiological processes, including memory, by which an organism becomes aware of and interprets external stimuli or sensations (closely related to Perception 4 and 6). So *attrib.* and *comb.*, as **perception psychology,** that branch of psychology which is concerned with the study of perception.

1875 A. J. Ellis tr. *Helmholtz's On Sensations of Tone* I. iv. 99 There are several much more complicated cases in which many sensations must concur to furnish the foundation of a very simple perception. **1913** A. A. Brill tr. *Freud's Interpr. of Dreams* vii. 426 We assume that a first system of apparatus takes up the stimuli of perception, but retains nothing from them—that is, it has no memory. **1949** D. O. Hebb *Organization of Behavior* i. 16 According to these ideas, perception does depend on exciting specific parts of the receptor surface. **1951** Licklider & Miller in S. S. Stevens *Handbk. Exper. Psychol.* 1040/1 Although the perception of speech is a psychological problem, it remained for telephone engineers..to develop procedures for the quantitative investigation of speech perception. **1956** J. R. Smythies *Anal. of Perception* p. ix, In order to construct a comprehensive theory of perception..it would be necessary to have at least a good working knowledge of epistemology and the philosophy of sense perception, neurology, neuroanatomy and neurophysiology, psychiatry and psychopathology with particular reference to the effects produced by the hallucinogenic drugs, anthropology, physics and experimental psychology. **1958** M. E. Spiro *Children of Kibbutz* VI. xvi. 435 Assumptions derived from an adaptation of psychoanalytic, learning, and perception theories. **1964** M. A. K. Halliday et al. *Linguistic Sci.* iii. 60 Psychological phonetics..has arisen out of modern developments in perception psychology. **1968** R. N. Haber *Contemp. Theory & Res. Visual Perception* (1970) p. vi, While it will be clear throughout this book that memory and what is traditionally known as perception cannot be distinguished by any but the most arbitrary of rules. **1974** *Drive* Autumn 3/1 At least one perception expert considers they could, indeed, be doing far more in teaching distance assessment in a following situation at speeds over 30mph. *Ibid.* 29/2 The university's department of perception-psychology aims to produce a lighting system that gives clearer warning of a car's presence to pedestrians and on-coming traffic.

perce·ptionalist, *a. rare.* [f. Perception + -al I + -ist.] Of or pertaining to philosophical theories of knowledge that are based on perception.

1847 J. D. Morell *Hist. View Philos.* (ed. 2) I. i. 130 On this perceptionalist controversy, consult Sir W. Hamilton's admirable article.

perceptron (pəɹse·ptrǫn). [f. Percept *sb.* + -tron.] A model or machine devised to represent or simulate the ability of the brain to recognize and discriminate, orig. based on statistical concepts.

1958 F. Rosenblatt in *Psychol. Rev.* LXV. 386 (*heading*) The perceptron: a probabilistic model for information storage and organization in the brain. *Ibid.* 387/2 The theory has been developed for a hypothetical nervous system, or machine, called a perceptron. The perceptron is designed to illustrate some of the fundamental properties of intelligent systems in general, without becoming too enmeshed in the special..conditions which hold for particular biological organisms. **1962** W. S. Holmes in G. L. Fischer et al. *Optical Character Recognition* 213 A program of pattern recognition research based on early concepts of Dr. Frank Rosenblatt..has turned toward the application of perceptrons to useful tasks such as the recognition of printed characters. **1966** Y. Bar-Hillel in *Automatic Transl. of Lang.* (NATO Summer School, Venice 1962) 22 Though certain electronic devices (such as perceptrons) have been built which can be 'trained' to perform certain tasks..and though computers have been programmed to do certain things..it would be disastrous to extrapolate from these primitive exhibitions of artificial intelligence to something like translation. **1971** *New Scientist* 2 Sept. 525/2 There are some 50 approaches to neuron modelling and machines were constructed based on some of these models. Perhaps the most famous..was the Perceptron device built at the Cornell Aeronautical Laboratory. **1978** A. Bundy et al. *Artificial Intelligence* v. 184 The important point about a perceptron is that it makes a global decision about a figure by weighing only local evidence.

perceptual, *a.* Add: (Earlier and later examples.) *perceptual defence*, a raising of the threshold of perception when the stimulus is emotionally charged in an unfavourable way; *perceptual-motor* adj., involving motor behaviour as guided by or dependent on perception.

1878 S. H. Hodgson *Philos. of Reflection* I. 315 The perceptual order being the obverse aspect of the perceptual. **1948** L. Postman in *Jrnl. Abnormal Psychol.* XLIII. 152/1 Value orientation may..raise thresholds for unacceptable stimulus objects. We shall refer to this mechanism as perceptual defense. **1950** *Jrnl. Personality* XIX. 85 These data were interpreted as indicative of a perceptual defense to emotional stimuli originating at a level which precedes full conscious awareness. **1951** *U.S. Human Resources Research Center Research Bull.* No. 51–7 (*title*) The influence of types of instructions on the performance of a perceptual-motor task. **1955** [see *meclozine]. **1962** J. G. Taylor *Behavioral Basis of Perception* vi. 130 It is no more possible for one person to make comparisons with another person's perceptual field than it is to describe the dimensions of personality in terms of centimeters, grams, and seconds. **1968** J. B. Oxendine *Psychol. of Motor Learning* i. 14 A forehand stroke in tennis would be classified as a perceptual-motor skill, and swimming would not. **1969** Freeman & Giovannoni in Lindzey & Aronson *Handbk. Social Psychol.* (ed. 2) V. 688 Studies of schizophrenics indicate that perceptual thresholds and estimation of size are dependent on the emotional concomitants of the visual stimuli. **1970** M. J. Meldman *Dis. Attention & Perception* v. 95 The phenomenon of perceptual defense and the relationship of emotion to perception. **1972** *Village Voice* (N.Y.) 1 June 36/5 Dr. Krippner told me just a few of the non-drugging approaches that could and should be used with children whose classroom behavior is 'divergent'— vitamin therapy, perceptual-motor therapy, [etc.]. **1976** *Classical Q.* XXVI. 39, I shall be illustrating the perceptual case, so let me simply note here a few examples from the case of pleasure and pain. **1977** *Times Lit. Suppl.* 29 Apr. 529/2 We do not observe the seventeenth-century butcher directly vivid though the picture is; we see him through the perceptual schemata of the artist and his patron.

Hence **perce·ptualize** *v.*, to express in perceptual terms; **perce:ptualiza·tion.**

1896 W. Caldwell *Schopenhauer's Syst.* iii. 167 A highly interesting feature..is his pronounced tendency to *perceptualise intellection*, to assimilate all real knowledge to the type of perception and immediate apprehension. **1968** P. McKellar *Experience & Behaviour* iv. 121 This hallucination represents an instance of a perceptualization of the man's realization of what had happened to him.

perceptually (pəɹse·ptiuăli), *adv.* [f. Perceptual *a.* + -ly[2].] In a perceptual manner.

1878 S. H. Hodgson *Philos. of Reflection* I. 394 We might pick out those differents which are most strongly contrasted perceptually. **1909** W. M. Urban *Valuation* xiv. 394 When the object is neither perceptually verifiable nor continuous with other truth judgments, a readjustment of reality-meanings takes place. **1922** A. G. Hogg *Redemption from this World* vi. 197 Thus miracles.. render perceptually obvious both the personality and the attitude of the Divine will. **1932** H. H. Price *Perception* vi. 164 We attribute a characteristic to some material thing which is now being perceptually accepted. **1972** *Science* 9 June 1149/2 We reported this ability to perceptually synthesize missing phonemes.

perceptum (pəɹse·ptəm). Pl. **percepta.** [L. *perceptum* (a thing) perceived, neut. of pa. pple. of *percipĕre* to Perceive.] = Percept 1.

1887 S. H. Hodgson *Let.* 10 Dec. in R. B. Perry *Tht. & Char. of W. James* (1935) II. 82 An immediate perception of spatial extension as part..of the visual perceptum. **1913** *Mind* XXII. 14 To instinctive action there corresponded a *perceptum* or percept which was its object. **1920** S. Alexander *Space, Time & Deity* II. 92 The perceived object or thing, the perceptum, is a contemplated synthesis. **1929** A. N. Whitehead *Process & Reality* 22 When science deals with emotions, the emotions in question are percepta. **1936** W. F. R. Hardie *Study in Plato* ii. 12 Plainly its plausibility will depend on what characteristics of *percepta* are held to preclude their being known.

perceptuo- (pəɹse·ptiŭo), combining form of L. *perceptu-s*, pa. pple. of *percipĕre* to perceive, as *perceptuo-motor* adj., = *perceptual-motor* adj. s.v. *perceptual *a.*

1973 K. Wedell (*title*) Learning and perceptuo-motor disabilities in children. **1974** *Nature* 8 Nov. 121/1 A series of experiments in perceptuo-motor adaptation that provide fairly direct evidence favouring the existence of such a mechanism. *Ibid.* 122/2 Perceptuo-motor effects were found for the voiceless stop consonants but not for their voiced counterparts.

perch, *sb.*[1] Add: **3. perch-hole,** a hole in which perch are found.

1906 *Macm. Mag.* June 574 Agatha by the side of the perch-hole, very erect, with a still more erect fishing-rod, surprised by the..angler.

perch, *sb.*[2] Add: **2. c.** *Theatr.* A platform from which lights are directed on to the front of the stage; *pl.*, the lights placed on this platform.

1933 P. Godfrey *Back-Stage* i. 18 The stops controlling the amber circuits in No. 1 batten, floats, and P. and O.P. perches slide up to full. *Ibid.* vii. 90 'What's in your perches?' 'Ambers, sir.' **1934** A. P. Herbert *Holy Deadlock* 215 From time to time he gave a quiet order to an invisible person called Joe about Batten Number One, about a border or a perch, a flat or the floats. **1957** *Oxf. Compan. Theatre* (ed. 2) 472/2 The mobility of the bridge is allied in control with the 'perches' or ladder-type boomerangs, which can be moved on or off stage according to the width of the scene. *Ibid.* 474/2 The spot batten and perches will, when a cyclorama is used, illuminate adequately the acting area to within a certain distance of the cyclorama. **1959** Rae & Southern *Internat. Vocab. Techn. Theatre Terms* 58 *Perch*, platform for tormentor spot. **1967** *Punch* 16 Aug. 242/3 For *Figaro* and Verdi's *Macbeth*.., John Christie had to bring in a lighting bridge and sixty floods and perches from Glyndebourne.

3. e. *to knock off one's perch*: also in weakened senses: to disconcert, humiliate, snub; *to come* (or *get*) *off one's perch*: to climb down, to adopt a less arrogant or condescending manner.

1896 *Dialect Notes* I. 421 'Come off your perch,' stop being fresh. **1900** 'Flynt' & 'Walton' *Powers that Prey* 238 'It's up to you to do the talking... All I've got to do is just to sit quiet.'.. 'Sure! I'll say it fast enough. But you can come off your perch just the same.' **1915** Mrs. Belloc Lowndes *Diary* 24 Mar. (1971) 60 The American said: 'What you've first got to do is to come off your perch—and listen to what we want...' The great man gave in and got off his perch. *a* **1916** 'Saki' in *Coll. Short Stories* (1930) 316 Mrs. Quabarl, to use a colloquial expression, was knocked off her perch. **1923** G. McKnight *Eng. Words* iv. 65 *Stop crowing* becomes *come off the perch*. **1931** M. Allingham *Look to Lady* xv. 156 For Gawd's sake come off yer perch and listen to this seriously. **1936** S. Sassoon *Sherston's Progress* III. 151 Tells Hooper to come off his perch and put the kettle on, which isn't well received by the golden-haired one. **1976** D. Clark *Dread & Water* v. 104, I reckon..that he's been knocked off his perch by our form of investigation.

4. Also, a horizontal bar used in softening leather.

1898 *Hide & Leather* 24 Sept. 21/3 After drying they [*sc.* skins] are softened, dry, over a perch with a moon-knife. **1902** *Mod. Amer. Tanning* I. 201 When the pelt is about half dry, it must be worked over what is called a perch. **1903** H. R. Procter *Princ. Leather Manuf.* 188 'Perching'..[consists] in fixing the skins on a horizontal pole (the 'perch'), and working them with..a tool formed somewhat like a small shovel with a semicircular blade. **1909** H. G. Bennett *Manuf. Leather* 359 In perching the mechanical treatment is less violent, the goods being fixed on a 'perch'—a horizontal pole about 5 feet above the ground—and scraped by means of the 'moon-knife'. **1940** [see *perching *vbl. sb.*[1]].

perched, *ppl. a.*[1] **1.** Add: (Further examples.) In *Geol.* also applied to blocks left in such a position by other causes; more generally, having an elevation that is exceptionally high in relation to the immediate locality; applied esp. to ground water separated from an underlying saturated zone by an intervening unsaturated zone.

1900 H. James *Little Tour in France* (ed. 2) vi. 62 In the matter of position Amboise is certainly supreme in the list of perched places. **1901** *Bull. Mus. Compar. Zool.* XXXVIII. 134 (*heading*) Perched boulders. **1906** A. C. Veatch et al. in *Prof. Papers U.S. Geol. Survey* No. 44. 57 There are..a number of more or less limited areas of saturated beds above the main one. These perched ground-water tables are for the most part confined to the moraine. **1923** *Water-Supply Papers U.S. Geol. Survey*, No. 494. 42 If water poured into the well all drains out the well evidently ends in an unsaturated bed and the overlying ground water is perched. *Ibid.* 57 The term *perched* may be applied to streams in the same way as it is applied to ground water. **1956** W. Edwards in D. L. Linton *Sheffield* i. 8 All the igneous rocks are basaltic... Their clayey tops support 'perched' water-tables in parts of the limestone uplands. **1968** R. W. Fairbridge *Encycl. Geomorphol.* 740/1 'Cirques'..are formed at the heads of glaciers both large and small—including a great many perched, or hanging, glaciers of small dimensions. *Ibid.* 823/2 (*caption*) Perched block of Bluff Sandstone.., due to slumping, followed by slope retreat. **1972** J. G. Cruickshank *Soil Geogr.* iii. 84 Any cemented or compacted horizon..can function in the same way and support a perched water table with accumulating soil water.

percher[1]. Add: **8.** *Cricket.* A ball that 'perches' or hangs in the air; *spec.* = *bouncer 6.

1913 *Cricket* 14 June 305/2 Every bowler pitches short sometimes, and..the resultant 'rib-roaster', 'percher', 'flier', 'bouncer',..is no more than an ordinary risk. **1961** *Times* 23 June 4/1 As big a percher as can ever have been missed in a match between England and Australia.

perching, *vbl. sb.*[1] Add: **2.** A process for softening skins in leather-making; cf. *perch *sb.*[2] 4. Freq. *attrib.*

1897 C. T. Davis *Manuf. Leather* (ed. 2) 361 There are.. above the perching room.., two large logwood tanks. *Ibid.* 362, 12 Slocomb perching machines. These perching machines take the place of hand work. **1903, 1909** [see *perch *sb.*[2] 4]. **1940** *Chambers's Techn. Dict.* 625/2 *Perching* (*leather*), a process for stretching and softening a skin by working over it with a crutch stake, on the flesh side, while it is fixed to a horizontal pole (*perch*).

perchist (pɜ̄·ɹtʃist). *poet. nonce-wd.* [f. Perch *sb.*[2] + -ist.] A trapeze artist.

1938 L. MacNeice *Earth Compels* 18 *Perchists*... They rise into the tent's Top like deep-sea divers... Hang by their teeth Beneath the cone of canvas.

perchloroethylene (pəɹklō·ə:rōue·pilin). *Chem.* Also **perchlorethylene.** [ad. G. *perchloräthylen*, f. *per-* Per-[1] 5 + *chlor-* Chloro- + *äthylen* Ethylene.] A colourless, nonflammable, toxic liquid, C_2Cl_4, which is widely used as a solvent, esp. in dry cleaning fluid, and medicinally for the treatment of worm infestations; = *tetrachloroethylene.

1873 *Jrnl. Chem. Soc.* XXVI. 866 The action of sodium ethylate on perchloroethylene has been studied by Fischer and Geuther. **1875** *Ibid.* XXVIII. 746 The distillate which comes over slowly consists of perchlorethylene holding in solution aniline and carbon sesquichloride. **1954** A. K. Doolittle *Technol. Solvents & Plasticizers* xii. 720 Perchloroethylene finds use as a dry-cleaning, metal-degreasing, and rubber solvent, and as an anthelmintic. **1963** *Economist* 12 Jan. 115/2 The customer can watch her clothes tumbling through a solvent (perchlorethylene). **1971** *Daily Tel.* 10 June 1/4 Perchloroethylene can cause drowsiness, coma and death if taken in sufficient quantities. **1976** *Sci. Amer.* May 52/2 The apparatus ..basically consists of a tank of 100,000 gallons of the common dry-cleaning fluid perchloroethylene.

percipient. B. *sb.* **b.** (Later examples.)

1955 *Sci. Amer.* Oct. 116/3 As a result of extensive tests, mainly involving two extraordinarily gifted 'percipients' in the business of card guessing, Soal and his co-worker Bateman were converted from doubt to ardent belief. **1966** K. R. Rao *Exper. Parapsychol.* i. 6 The hypothesis that the *percipient*, not the agent, is the likely initiator of the psi experience. **1974** *Listener* 3 Jan. 22/2 Some of the target images were lantern-slides... The guessers or 'percipients' were asked to try to draw the image.

percolate, *v.* Add: **1. b.** (Later examples.)

1970 P. Laurie *Scotland Yard* iv. 92 Churchill's funeral was ten years' planning, and it probably contained, percolated through a succession of intermediate heroes, elements of Nelson's. **1978** *Time* 6 Nov. 28/2 Connecticut's Ella Grasso, the first woman to win a governorship in her own right, says these victories will percolate women into office in a few years.

c. To prepare (coffee) in a percolator.

1966 *New Statesman* 3 June 819/1 First found percolating stale morning coffee in his office. **1974** 'J. Ross' *Burning of Billy Toober* xiv. 127 Rogers made the mortuary in twenty minutes, not stopping to shave or percolate coffee. **1978** N. J. Crisp *London Deal* v. 72 The manservant was percolating coffee.

2. b. (Later examples.)

1934 C. Lambert *Music Ho!* III. 185 Oriental influences..have percolated naturally through these racial frontiers. **1935** B. Malinowski *Coral Gardens* II. VI. 244 The magic percolates,..so that practically everybody in the village knows it. **1977** P. D. James *Death of Expert Witness* III. 128 News percolated through a village community by a process of verbal osmosis.

3. (Later *fig.* example.)

1965 *New Statesman* 7 May 737/2 One reason why this has so slowly percolated British consciousness..is British reporting.

4. *intr.* To walk, to stroll. *U.S. slang.*

1942 Z. N. Hurston in A. Dundes *Mother Wit* (1973) 223/1 Then he would..percolate on down the Avenue. **1945** L. Shelly *Jive Talk Dict.* 15/2 *Percolate*, to meander.

percolater, var. Percolator.

1861 Mrs. Beeton *Bk. Househ. Managem.* 875 Let the coffee be freshly ground..; put it into a percolater, or filter,..and pour *slowly* over it..boiling water. **1958** *Times* 6 Nov. 12/6 If we wanted a pint of essence of ginger, we just put 10 oz. of powdered Jamaica ginger into a percolater, and allowed sufficient rectified spirit to pass through to collect a pint.

percolating, *vbl. sb.* Add: *percolating filter*, a type of filter used in the treatment of sewage, usu. after the removal of suspended solids, consisting of a bed of inert, porous material such as crushed rock through which the sewage is allowed to percolate, so that noxious organic matter is removed by aerobic micro-organisms.

1901 S. Barwise *Bacterial Purification of Sewage* iv. 37 The Commissioners in their Report speak of two artificial filtration processes—Contact Beds and Continuous Filtration. In this book I have adopted the phrase 'Percolating Filters', instead of that of 'Continuous Filters', because some of the continuous filters are worked intermittently, and intermittent continuous filtration is a verbal contradiction. **1936** [see *bacterium 2]. **1972** *Water Research* VI. 781 In the United Kingdom conventional sewage treatment by sedimentation plus secondary treatment by percolating filters or activated sludge plants is not normally adequate to provide an effluent acceptable for re-use.

percolator. Add: **b.** (*a*) (Earlier and later examples.) Also *attrib.*

1845 E. Acton *Mod. Cookery* xxvii. 647 It will be stronger if slowly filtered in what is called a percolator, or coffee-biggin, than if it be boiled. **1869** *Pattern Bk. Copper, etc., Ware* (H. Loveridge & Co.) 103 Coffee Percolators. **1963** *B.S.I. News* May 35 Electric coffee percolators. **1978** L. Deighton *SS-GB* xv. 126 She tipped the coffee into the percolator top, closed the lid and set it on the heat.

c. *fig.* (*a*) A carburettor; (*b*) a house-rent party (*house *sb.*[1] 23); loosely, any party. *U.S. slang.*

1942 Berrey & Van den Bark *Amer. Thes. Slang* §82/4 *Carburetor*, carb, jug, juicer, juice pot, mixer, percolator, pot, sifter. **1946** R. Blesh *Shining Trumpets* xiii. 303 The great South Side institution of 'rent party' (locally known as 'skiffle', 'shake', or 'percolator'). **1956** S. Longstreet *Real Jazz* xvi. 120 You could always..get together..and charge a few coins and have..a percolator. **1967** Partridge *Dict. Slang* Suppl. 1259/2 '*To have a shake, rare or percolator*..to have a party' (Anderson): beatniks': since ca. 1959. **1971** M. Tak *Truck Talk* 117 *Percolator*, the carburetor. **1974** H. L. Foster *Ribbin'* iv. 141 In Chicago, these parties were called a 'parlour social', 'gouge', 'struggle', 'percolator', 'too terrible party', or the 'skiffle'.

percussion, *sb.* Add: **2. a.** Also *concr.*, a percussion gun.

1821 P. Egan *Real Life in London* I. i. 8 My new patent double-barrelled percussion.

c. (Earlier example of *collect.* use.)

1889 G. B. Shaw *How to become Mus. Critic* (1960) 164 Brass and percussion [are] behind the wood wind and under the stage.

e. *bulb of percussion*: see *bulb *sb.* 4b.

5. percussion lock (earlier U.S. example).

1829 in *Reg. Deb. Congress U.S.* (1831) 21st Congress 2 Sess. App. p. xcii/2 I have used the percussion locks but little, but believe them admirably well constructed for general use.

percussionist (in Dict. s.v. Percussion *sb.*). Restrict † (*obs.*) to sense in Dict. and add: **b.** A player of a percussion instrument.

1950 Webster Add. p. cxix, *Percussionist*, one skilled in the playing of percussion instruments. **1955** L. Feather *Encycl. Jazz* 256/1 His use of the top cymbal..was imitated by countless other percussionists. **1962** *Times* 18 Apr. 7/6 In the course of the piece everybody is a percussionist. **1969** *Daily Tel.* 15 Feb. 15/1 A truly professional percussionist rather than just another polished, expert drummer. **1977** *Listener* 18 Aug. 216/1 In the second programme they had Tristram Fry, the percussionist.

percussiveness (in Dict. s.v. Percussive *a.*). (Later examples in *Mus.*)

1958 *Times* 9 Oct. 7/1 The tense percussiveness of Bartok's concerto. **1970** *Daily Tel.* 16 June 14/2 He lets unfold..a veritable rhapsody of percussiveness (by no means just Brahms's in E flat) and reveals the soul in staccato. **1976** *Gramophone* Sept. 424/1 Queffélec, however, never plays with the almost brutal percussiveness of Bernstein.

Percy (pɔ·ɹsi). The masculine Christian name used, freq. with connotations of weakness or effeminacy, as a representative name for **a.** a conscientious objector; **b.** (see quot. 1932); **c.** in the armed services, an officer or an educated man.

1916 G. B. Shaw in *Nation* 27 May 258/2 Mobbed and pilloried and photographed in the 'Daily Sketch' as 'Percy' (all Percies are now—shade of Hotspur!—supposed to be cowards). **1932** E. Weekley *Words & Names* 91 *Percy*.. is still used in the United States of the typical young Englishman. **1961** Partridge *Dict. Slang* Suppl. 1217/2 *Percy*, in the Royal Navy, has, since ca. 1925, meant an effeminate man; but since ca. 1940, also and esp. a studious, quiet, educated man as opposed to an uncouth 'tough'. **1974** 'B. Mather' *White Dacoit* xxi. 216 Most of the young Percies in our mob had to take their boots off to count up to twenty. *Ibid.* xxiii. 237 There's a young Percy in charge... The Percy says to one of them, 'What's your unit, my man?'

percylite. (Further example.)

1974 *Mineral. Rec.* V. 286 We believe that type percylite has been shown to be a mixture [of boleite and pseudoboleite] and that the same is true of percylites from other major localities.

perdure, *v.* Delete 'Now *rare*' and add later examples.

1963 J. Wiesenfarth *Henry James* iv. 91 The romp of Aggie and Petherton perdures through the conversation. **1973** Boilès & Horcasitas tr. *M. León-Portilla's Time & Reality in Thought of Maya* ii. 33 For longer than a millennium and a half, not a little of Maya symbolism has perdured. **1979** *Nature* 22 Mar. 348/1 Thus enough maternal gene products (mRNAs or proteins) may perdure in embryonic cells to allow normal segmentation and cuticular differentiation.

perduring *ppl. a.* (Further examples).

1951 C. Kluckhohn et al. in Parsons & Shils *Toward Gen. Theory Action* iv. 399 A value or values restrain or canalize impulses in terms of wider and more perduring goals. **1977** *Dædalus* Summer 63 The assignment and reassignment of meaning must be investigated as processes in the domain of resilience possessed by each population recognizing itself to be culturally perduring.

‖**père** (pɛ̄r). Also **pere.** [Fr., = father.] **1.** Applied as a prefix to the name of a French priest; = Father *sb.* 6c.

1619 J. Chamberlain *Let.* 30 Oct. (1939) II. 270 The Jesuites hold a chapter of theyre order at Rome whether Pere Coton is sent out of France. **1699** M. Lister *Journey to Paris* 96, I bought the works of Pere Pezaron, a Bernadin, now Abbot de Charmoyse near Rheims. **1777** P. Thicknesse *Year's Journey* I. xx. 174 Nor did the whole community afford but a single member (pere tender, a Fleming) who could speak French. **1879** R. L. Stevenson *Trav. with Donkey* 119 There was Père Apollinaire hauling his barrow. **1978** *Times* 3 Nov. 7/1 (Advt.), The Sayings of Père Patriarche... 'To buy wine with uncertainty lacks any amusement.'

2. The father, senior: appended to a name to distinguish between a father and son of the same name. Cf. *fils.

1802 M. Edgeworth *Let.* 8 Dec. (1979) 58 M. Delessert père at a card table with another gentleman. **1858** O. W. Holmes *Autocrat of Breakfast-Table* 28, I have not taken the trouble to date them, as Raspail, *père*, used to date every proof he sent to the printer. **1893** E. Dowson *Let. c* 28 Nov. (1967) 299, I am dining with Horne & Horne Père at the Constitutional tonight. **1907** 'Elizabeth' *Fraulein Schmidt & Mr. Anstruther* xxxiii. 120 Collins *père* is a person who makes nails in Manchester. **1948** E. S. Turner *Boys will be Boys* v. 80 It will be seen that there was a Harkaway *père* and a Harkaway *fils*. **1964** *Guardian* 18 June 9/3 Everything about Tabarly père gives a kind of ageless impression of the sea... All this is highly relevant in considering Tabarly fils. **1972** J. Wain in Cox & Dyson *20th-Cent. Mind* I. xi. 364 Yeats *père* moved his household many times, and William was educated in London or Dublin by fits and starts. **1977** *N.Y. Rev. Bks.* 10 Nov. 10/2 Zola père was a Venetian of distinguished family.

3. In phrases *père de famille* (də famiy) [see *famille], father of a family, family man; *père et fils* (e fīs) [see *fils], father and son.

1820 M. Edgeworth *Let.* 14 May (1979) 124, I could not see in either his figure or face any hint of the young père de famille. **1853** *Thackeray Let.* in *Lett. A. T. Ritchie* (1924) iv. 52 The quantity of acquaintances and ½ acquaintances that as père de famille I did not care to make whole acquaintances. **1862** —— *Philip* II. ii. 33, I am secretly of the disposition of the time-honoured *père de famille* in the comedies. **1871** D. G. Rossetti *Let.* 7 Aug. (1967) III. 963 Constant calls on Urizen père et fils..in the nature of supplications. **1962** *Observer* 1 July 19/4 Through this gate came Agamemnon the victim and Orestes the avenger, prototypes of the two Hamlets, *père et fils*. **1964** *Ibid.* 27 Sept. 24/6 A rather stuffy English *père de famille*. **1964** *House & Garden* Oct. 111/1 In this dual promotion *père et fils* certainly succeeded. **1973** I. Butler *Eldest Brother* viii. 78 The rôle of '*père de famille*' and respectable bourgeois. **1978** N. Marsh *Grave Mistake* iv. 113 You should hear the Rattisbons, *père et fils*, on the subject.

Père David (pɛ̄r davī·d). The name of Armand *David* (1826–1900), French missionary naturalist, used *attrib.*, in the possessive, or *absol.*, in **Père David**('s) **deer** to designate *Elaphurus davidianus*, a large, long-tailed deer discovered by him in China in 1865, named after him by A. Milne-Edwards in 1866, and now extinct in the wild in its native land, although it survives in zoological gardens and parks, esp. in a large herd at Woburn Abbey, Bedfordshire, established by the 11th Duke of Bedford soon after 1900.

[**1871** P. L. Sclater in *Trans. Zool. Soc.* VII. 333 This fine animal is one of the many zoological discoveries which are due to the researches of M. le Père Armand David, Missionary of the Congregation of Lazarists at Pekin.] **1898** R. Lydekker *Deer of all Lands* 237 The general appearance of Père David's deer, when roaming in the park at Woburn Abbey, being quite unlike that of any other member of the group. **1927** G. Jennison *Nat. Hist.: Animals* 290 It is known to Europeans as the Père David Deer from the Naturalist Missionary who discovered it in 1865. **1955** *Times* 11 May 12/7 Chief among them [*sc.* the Woburn deer] is the herd of 300 Père David's deer—a species now extinct in its native China and existing only at Woburn and in a few small offshoots (for example, at Whipsnade) from the Woburn herd. **1973** G. Durrell *Beasts in my Belfry* v. 96 Undoubtedly the rarest animals in our care, a pair of young Père David deer. *Ibid.* 100 An outbreak of foot-and-mouth disease.. could have exterminated the Père David very successfully. **1975** *New Yorker* 24 Mar. 34/2 The Père David deer, six hundred and forty extant, twenty-seven in the Bronx.

peregrinatory, *a.* (Later example.)

1906 *Chambers's Jrnl.* Feb. 150/1 One sees in the streets ..peregrinatory makers of sugar puppets.

peregrine, *sb.* Add: **4.** Usu. with initial capital. A red-skinned variety of peach with white flesh, developed and introduced by the Rivers nursery in 1903.

1903 *Jrnl. R. Hort. Soc.* XXVIII. p. cxcii, Award of Merit... To Peach 'Peregrine'..from Messrs. Rivers, Sawbridgeworth. **1907** *Daily Chron.* 13 June 6/4 A specimen peregrine peach tree grown in quite a small flower-pot is seven feet high, and bears much fruit. **1929** E. A. Bunyard *Anat. Dessert* 89 In mid-August we have Peregrine, that finest of recent peaches, combining flavour, appearance, and good crop in a manner rarely found. **1958** *Listener* 20 Nov. 853/3 The most outstanding variety of peach is Peregrine. **1971** G. E. Whitehead *Grow Fruit in your Greenhouse* viii. 94 The peach 'Peregrine' and the nectarine, 'Pine Apple', make an ideal pair.

pereiopod. (Further examples using this spelling, which is now the usual one.)

1932 J. S. Huxley *Probl. Relative Growth* iii. 87 Those pereiopods which are used as walking legs..have a definite but slight growth-gradient. **1964** *Oceanogr. & Marine Biol.* II. 467 In these displays, serving a threat function, the crab stands higher on its pereiopods. **1970** *Nature* 16 May 661/2 The pereiopods of *M*[*erguia*] *rhizophorae* are more robust than in most marine shrimps.

perennate, *v.* Restrict *rare* to sense a and add further examples in *Bot.* of sense b. Hence **pere·nnating** *ppl. a.*; **perennation** (further example).

1905 I. B. Balfour tr. *C. E. von Goebel's Organogr. Plants* II. 689/1 Perennating, geophilous shoot. **1913** [see *chamæphyte* s.v. *chamæ-]. **1926** Tansley & Chipp *Study of Vegetation* ii. 21 The perennating buds..continue the growth of the plant from season to season. **1927** *Forestry* I. 108 Besides overwintering in the buds, the fungus perennates on dead oak-leaves. **1969** F. E. Round *Introd. Lower Plants* xiii. 153 Vegetative propagation of the gametophyte from gemma-like outgrowths is common in some genera [of ferns] and in a few these even become thickened and act as perennating organs. **1971** *Country Life* 16 Dec. 1736/3, I nearly always get a casual crop of mushrooms, though where the spawn comes from is a mystery. Perhaps it perennates in the ground below the [compost] heap. **1977** J. L. Harper *Population Biol. of Plants* iii. 62 For perennial plants the seed is an alternative means of perennation. *Ibid.* xxi. 651 In most plants the seed is a perennating organ.

perennial, *a.* Add: **2. a.** *perennial philosophy* = *philosophia perennis.

1933 W. R. Inge *God & Astronomers* i. 13 The perennial philosophy..is the only system which will be found ultimately satisfying. **1945** A. Huxley (*title*) The perennial philosophy. *Ibid.* p. vii, Rudiments of the Perennial Philosophy may be found among the traditionary lore of primitive peoples in every region of the world, and in its fully developed forms it has a place in every one of the higher religions. **1962** E. Wynne-Tyson *Philos. of Compassion* 3 The most fundamental difference between the teachings of the western religions and those of the perennial philosophy and of the original Creed of Christ. **1974** R. C. Zaehner *Our Savage God* 12 'All is One, and One is All..', seems to have been..what Aldous Huxley considered to be the kernel of..the 'perennial philosophy'.

perenniality. (Further example.)

1977 J. L. Harper *Population Biol. of Plants* xviii. 543 It is often possible to extend the perenniality of a biennial by continually preventing seed formation.

perentie (pĕre·nti). *Austral.* Also **parenti, perenty.** [Aboriginal name.] A large, burrowing, monitor lizard, *Varanus giganteus*, which may be as much as eight feet long and is found in desert areas of central and northern Australia. Also *attrib.*

a **1928** E. R. Waite *Reptiles or Amphibians S. Austral.* (1929) v. 125 Perentie or Sjonba... The discovery of the gigantic goana on the Komodo Islands illustrates the inadvisability of naming specimens by comparative measurements. The Perentie is, however, the largest Australian species. **1942** C. Barrett *On Wallaby* iii. 43 Perenties three yards long were said to exist in the ranges. **1944** *Living off Land* ii. 24 The large Parenti lizard provides about the best food supply of the reptiles. **1963** E. Worrell *Reptiles Austral.* 83 Perenty... Length to 8 feet. This is Australia's largest monitor. **1968** V. Serventy *Wildlife Austral.* v. 126 The goannas range from the short-tailed goanna, a seven-inch long sprite, to the eight-foot giant of the inland, the perentie. **1973** *Panorama* (Austral.) Oct. 5/3 (*caption*) Dick Lang with a perentie he chased and caught.

perequation. Delete *rare*—[0] and add later examples.

1920 *Times* 4 Aug. 4/3 This price we pay for English coal has obliged us to force up the interior price for our own coal, by a 'perequation', so as not to handicap those.. who are obliged by geographical reasons to burn English coal only. **1954** *Economist* 19 June 983 A perequation levy is collected from coal sold by countries where the average costs of production are lower than the weighted average. **1973** H. Trevelyan *Diplomatic Channels* ix. 132 It was egalitarianism run mad, the doctrine of the perequation of misery.

perequitate, *v.* For *rare*⁻¹ read *rare* and add later example.
1937 P. M. Kendall *Warwick Kingmaker* iii. vi. 158 A poor clerk of Evreux, leaving the royal cavalcade to take a message to his Chapter and immediately setting forth with their reply, had to ride for sixty-six days before he caught up with his perequitating sovereign [*sc.* Louis XI].

perester (pəre·stəɹ). *Chem.* [f. Per-¹ 5 + *ester.] An ester of a peracid.
1933 *Jrnl. Amer. Chem. Soc.* LV. 351 The perester on standing at room temperature..hydrolyzes to yield methyl hydroperoxide and the original monomethyl ester of camphoric acid. **1946** *Ibid.* LXVIII. 642/1 Organic peresters cannot be classified as derivatives of organic peracids and alcohols, but rather as derivatives of organic acids and hydroperoxides. **1975** *Org. Reaction Mechanisms 1973* iii. 80 Induced decompositions of peresters have been brought about by trialkyltin radicals.

perfect, *a.* Add: **B. 4. d.** *a perfect day* (*colloq.*), a day of which every part has been enjoyable; *esp.* in phr. *the end of a perfect day.*
1910 C. Jacobs-Bond *Perfect Day* (song) 6 For mem'ry has painted this perfect day With colors that never fade, And we find at the end of a perfect day The soul of a friend we've made. **1923** *Liverpool Echo* 13 Sept. 6/3 (*heading*) The boy and the balloon. The sad end of a perfect day. **1976** S. Kaufman *Master & Other Stories* (1977) 193 Nothing crossed her mind as she floated, a thing made of air, and dreamily listened to the carrying voices from shore, except the one thought: What a perfect day.

5. c. (Further examples.) Also in phrases *perfect gentleman, lady.*
1807 Wordsworth *Poems* I. 15 A perfect woman, nobly plann'd, To warn, to comfort, and command. **1818** Byron *Beppo* xxxii. 17 In short, he was a perfect cavaliero, And to his very valet seem'd a hero. **1834** G. Crabbe *Jrnl.* in *Poet. Wks. G. Crabbe* I. vi. 147 Miss Waldron..could sing a jovial song like a fox-hunter,..and yet there was such an air of high ton, and such intellect mingled with these manners, that the perfect lady was not veiled for a moment. **1856** C. M. Yonge *Daisy Chain* i. xxiii. 245 Her instinct showed her that she was talking to a man of high ability. A perfect gentleman she saw him to be. **1872** Geo. Eliot *Middlem.* II. iv. xvi. 299 Rosamond..was active..in being from morning till night her own standard of a perfect lady. **1903** G. B. Shaw *Revolutionist's Handbk.* i, in *Man & Superman* 182 This..is a great advance on the popular demand for a perfect gentleman and a perfect lady. **1949** E. Coxhead *Wind in West* i. 25 You'll like the Fletchers—Hermia is a wonder, the perfect wife. **1967** A. Wilson *No Laughing Matter* iii. 320 All perfect ladies..eat messily, don't they? **1972** J. Porter *Meddler & her Murder* iv. 49, I never knew old Adam was a womaniser! Must say, he's always behaved like a perfect gentleman with me. **1978** H. MacInnes *Prelude to Terror* xv. 34 The perfect hostess..a woman putting a guest at ease with food and drink.

f. Complete, utter (referring to a person in neutral or favourable contexts).
1903 G. B. Shaw *Man & Superman* i. 40 You seem to understand all the things I dont understand; but you are a perfect baby in the things I do understand. **1919** T. K. Holmes *Man from Tall Timber* xxiv. 292 That Anabelle Whitman is a perfect scream. **1927** C. Connolly *Let.* 27 Jan. in *Romantic Friendship* (1975) 231 Thou art heavenly he a 'perfect scream'. **1931** T. E. Lawrence *Let.* (1938) 713 The Coroner was a perfect pet. **1959** *Listener* 4 June 998/2 He [*sc.* Harry Secombe] is indeed a Perfect Scream in both senses of that noun. **1962** Parks & Leighton *My Thirty Years Backstairs at White House* xiii. 190 Rob Roy was a perfect angel with the First Family.

8. (Later examples.)
1901 *Ann. Math.* II. 103 By a perfect number is meant a number which is equal to the sum of those of its divisors which are less than the number. Thus 6 = 1 + 2 + 3 is a perfect number. *Ibid.* 104 His [*sc.* Sylvester's] proof of the non-existence of odd perfect numbers. **1958** R. V. Andree *Sel. Mod. Abstr. Algebra* i. 30 The first four perfect numbers were discovered by the end of the first century. By 1870, only four more had been found. Between 1870 and 1950, four additional even perfect numbers were discovered.. Since then, five more perfect numbers have been found. **1971** *Sci. Amer.* June 56/2 Whether or not there are any odd perfect numbers is still undecided.

10. e. *perfect pitch*, the ability to judge pitch absolutely, and hence recognize the pitch of any individual note. (Cf. *absolute pitch* (b) s.v. *absolute *a.* 16.)
1949 F. Towers *Tea with Mr. Rochester* 19 She is very musical, and has perfect pitch. **1958** *Gramophone* Oct. 17 Listeners with perfect pitch are warned that, this being the Schnitger organ at Cappel, all the works here sound a semitone higher than usual. **1971** *Nature* 2 Apr. 337/1 Few people possess 'perfect pitch' and it is not known whether it is learned or inherited. **1975** *Sunday Times* (Colour Suppl.) 13 July 47/4 Although he does not sight read he has perfect pitch. **1976** *Gramophone* Aug. 318/3 Listeners with perfect pitch should be warned that the present issue sounds a semitone lower than normal.

15. Also applied to sheets that have been printed on both sides.
1841 W. Savage *Dict. Art of Printing* 701 The reader, in revising the second form, then sees the sheet perfect, which is necessary to ascertain that the matter follows. **1960** G. A. Glaister *Gloss. Bk.* 303/2 In edition binding the printed sheets are said to be perfect as soon as some or all of the sheets (and plates) have been printed.

16. *Bot.* Applied to the stage in the life cycle of a fungus at which sexual spores are produced, and to a fungus in that state.
1891 G. Massee *Brit. Fungi* 32 The incomplete form is considered as belonging to the same genus as the 'perfect' form. **1909** *Mycologia* I. 115 A single boil [of cotton]..was examined in the laboratory and found to be covered with the perfect stage of the *Colletotrichum.* **1945** G. R. Bisby *Introd. Taxon. & Nomencl. Fungi* xvi. 87 The perfect state is that which ends in the ascus stage in the *Ascomycetes*, in the basidium in the *Basidiomycetes*, in the teleutospore or its equivalent in the *Uredinales*, and in the spore in the *Ustilaginales.* **1950** E. A. Bessey *Morphol. & Taxon. Fungi* i. 18 In fungi with several stages of development to which different names have been given, the species epithet that is to be retained is the one applied to the 'perfect' stage of the fungus. **1967** M. E. Hale *Biol. Lichens* iii. 45 The parasymbiont *Abrothallus smithii* is the perfect stage of the imperfect fungus *Phoma.* **1971** [see *imperfect *a.* 8 b].

17. Applied to a form of bookbinding in which the single leaves of a book are attached individually to the spine by an adhesive, instead of the printed sheets being folded and sewn.
1893 *Amer. Bookbinder* July 86 (*heading*) Perfect library binding. *Ibid.*, Mr. Crawford is the inventor of what is known as the 'perfect library binding'. **1926** G. A. Stephen *Commerc. Bookbinding* 9/2 A revolution in the method of binding monthly magazines was inaugurated by the invention of the Sheridan 'Perfect Binder'. **1926** *Amazing Stories* July 359/1 We..took immediate ways and means to do away with the old-fashioned binding, and you now hold in your hand a magazine bound with the so-called 'perfect' binding. **1956** H. Williamson *Methods Bk. Design* xiv. 332 Sewing, rounding, backing, and lining can all be dispensed with in the unsewn or 'perfect' methods of binding. **1960** *Times Lit. Suppl.* 1 Jan. p. xvi/2 The so-called 'perfect' binding, in which the backs of the quires having been guillotined away, the resultant single leaves are held together by adhesive. **1977** *Ibid.* 28 Jan. 104/2 The pages are now smaller, the paper thinner,..and the binding is perfect (i.e., imperfect). **1977** *Special Libraries* Feb. 64/2 Perfect bound ('newspeak' for 'unsewn') bindings on books have caused librarians grief and libraries money (for rebinding) since they fall apart so readily... I urge publishers not to utilize this type of binding until they have really perfected the process.

18. *Econ.* Designating (notional or actual) ideal market conditions in which adverse factors are removed; *perfect competition*, competition in which all elements of monopoly are absent and the market price of a commodity is beyond the control of individual buyers and sellers.
1897 *Q. Jrnl. Econ.* XI. 175 In passing from the study of perfect monopoly to that of perfect competition, Cournot considers also the intermediate case of a few, say two, competitors. **1906** *Ibid.* XX. 211 Perfect competition is the fundamental hypothesis of economics in the sense that perfect competition is postulated in nearly every argument as to economic equilibrium. **1922** H. A. Silverman *Substance Econ.* vi. 72 Perfect competition..is not usually realised in practice. **1939** Lyons & Emmet *Introd. Econ.* iv. 38 The characteristics of a perfect market are, firstly, full information... Secondly, complete accessibility... Thirdly, full freedom of choice. **1944** A. Cairncross *Introd. Econ.* xix. 180 The first requirement of perfect competition..is that the market..should be perfect; a perfect market being one in which buyers have no preferences as between the different units of the commodity offered for sale, sellers are quite indifferent to whom they sell, and both buyers and sellers have full knowledge of prices in other parts of the market. **1969** D. C. Hague *Managerial Econ.* iv. 59 There is an established market where all buyers and sellers can keep in close touch with each other, and have become used to doing so. Economists sometimes say that there is then a *perfect market.* **1971** I. Deutscher *Marxism in our Time* (1972) xii. 259 When, on the other hand, [Marx] assumed perfect competition, he did it only in order to prove that that competition was necessarily self-destructive. **1974** M. B. Brown *Econ. of Imperialism* ii. 30 The classical vision of an economy where there is perfect competition reaches its apogee in the Theory of Free Trade. **1976** *Economist* 16 Oct. 21/2 Foreign exchange markets do not quite match up to the ideal of perfect competition described by theory.

19. *Math.* **a.** Of a set of points: closed, and such that every neighbourhood of each point of the set contains at least one other point of the set.
1906 *Q. Jrnl. Math.* XXXVII. 23 *P* is a rim point, so that the rim is closed, and, being dense in itself, is perfect. **1926** J. E. Littlewood *Elem. of Theory of Real Functions* (ed. 2) iv. 50 A perfect set is an existent set which is closed and dense-in-itself. **1957** J. R. Aumann et al. tr. *Hausdorff's Set Theory* vi. 133 The null set is everything: isolated, dense-in-itself, closed, perfect. **1970** C. A. Rogers *Hausdorff Measures* ii. 61 A set is perfect if it is closed, non-empty and dense in itself.

b. Of a group: such that the subgroup generated by the set of commutators of the group is the group itself.
1898 *Amer. Jrnl. Math.* XX. 277 Since a perfect group is identical with its derivatives, it cannot be isomorphic to any Abelian group whose order exceeds unity. **1908** H. Hilton *Introd. Theory Groups of Finite Order* x. 134 Every simple group is perfect. **1940** D. E. Littlewood *Theory of Group Characters* x. 174 A group is perfect if it is identical with its commutator subgroup. Hence the condition that a group is perfect is that it possesses no character satisfying $\chi_1 = 1$ save that character which is unity for every operation. **1959** J. S. Lomont *Applications of Finite Groups* ii. 2 Let us call a group (of order > 1) perfect if it is identical with its commutator subgroup. Every perfect group is then insolvable.

D. 4. [tr. med.L. *perfectus* in the same sense (also used).] Among the Catharist heresy of the Albigenses in the 12th and 13th centuries, one who had received the *consolamentum or spiritual baptism, thereby accepting all the precepts of Albigensian doctrine.
1742 L. Brown tr. *Bossuet's Hist. Variations Protestant Churches* II. xi. cxl. 156 In regard of those four thousand Cathari,..none were understood by that name but the perfect of the Sect... When the Sect was weaken'd, tho' there were but four thousand perfect *Cathars*, yet..the multitude..of simple Believers, was then infinite. **1826** in J. C. L. S. de Sismondi *Hist. Crusades against Albigenses* 3 They were divided into two classes, the perfect and the believers. **1832** S. R. Maitland *Albigenses & Waldenses* x. 271 Those men who assert that they alone are good christians whom the most Holy Roman Church persecutes, and condemns, and calls *perfecti*, or *consolati* (more properly *desolati*) heretics. **1888** H. C. Lea *Hist. Inquisition Middle Ages* I. ii. 84 The Perfects would die rather than violate the precept. *Ibid.* iii. 103 If the Perfect is exhorted by the God in whom he believes to tell all about his life, he will faithfully detail it without falsehood. **1926** A. L. Maycock *Inquisition* ii. 40 The 'Perfect' were forbidden to eat meat, eggs, cheese or anything that was the product of sexual generation. **1957** N. Cohn *Pursuit of Millennium* (1970) viii. 150 The Catharist *perfecti* dominated the religious life of a large part of southern France for half a century or more. **1962** [see *endura]. **1970** [see *consolamentum]. **1978** *Times* 24 Feb. 15/5 Fon had regarded himself a Perfecti, with no sense of sin... Penn represents another type of Cathar.

perfect, *v.* Add: Now usu. with pronunc. (pəɹfe·kt). **1. b.** (Later examples.)
1899 J. Southward *Mod. Printing* III. xii. 117 Rotary web machines also perfect the paper..before it is delivered. **1927** R. B. McKerrow *Introd. Bibliogr.* i. ii. 21 It is often evident..that the printer printed the whole number of impressions on one side before starting to perfect. *Ibid.* iii. i. 261 It is possible that a sheet may be perfected from a wrong forme. **1964** F. Bowers *Bibliogr. & Textual Criticism* iii. i. 71 The specific example of *Match Me in London* and its sheet that was perfected out of phase. **1972** P. Gaskell *New Introd. Bibliogr.* 132 The sixteenth-century account of Le Roy suggests that the heap was normally printed as white paper in the morning, turned at the midday break, and perfected in the afternoon.

perfecta (pəɹfe·ktă). orig. *U.S.* [abbrev. Amer. Sp. *quiniela perfecta* perfect quinella.] A method of betting in which the bettor must pick the first and second finishers of a race in the correct order. Also *attrib.*
1971 *New Yorker* 20 Feb. 107 For horseplayers who hopefully bet on exactas, perfectas, quinellas, doubles, and such, I can report that..an exacta paid $25,257. **1971** L. Koppett *N.Y. Times Guide Spectator Sports* x. 184 In the 'perfecta'..you must pick the first two [finishers] in the correct order. **1972** *Telegraph* (Brisbane) 2 June 13 His target was to hit races which were the subject of 'perfecta betting'. **1974** *Cleveland* (Ohio) *Plain Dealer* 26 Oct. 6-D/4 He also had the $13.50 perfecta in the seventh. **1975** *Ibid.* 6 Apr. 13-C/1 A new ruling allows Northfield four (instead of three) perfecta races nightly, which will be raced as the third, fifth, seventh and tenth races. To pick a winning perfecta combination, you must pick the first two horses that finish a race, in order.

perfectability, var. Perfectibility.
1874 [see Perfectibility 2]. **1970** *Sci. Jrnl.* Apr. 4/1 The U.S places more reliance on technological perfectability and has more faith in human perfectability than any other nation today. **1975** *Christian* II. 229 The potential for perfectability.

perfecting, *vbl. sb.* Add: **a.** (Later examples, in sense 1 of the vb.)
1951 S. Jennett *Making of Bks.* vii. 101 The printing of the second side of the sheet is known as perfecting. **1972** P. Gaskell *New Introd. Bibliogr.* 133 It appears to coincide with the practice of consecutive perfecting at different presses.

b. perfecting cylinder, machine (later examples), **press** (later examples).
1967 V. Strauss *Printing Industry* vi. 298/2 A third cylinder..is denoted as the 'perfecting cylinder' because of its paramount importance to the perfecting operation. **1880** F. J. F. Wilson *Typogr. Printing Machines* xi. 61 Perfecting machines are of three kinds—1. The Web. 2. The Drop-bar. 3. The Gripper. The above terms signify the manner by which white paper is conveyed into the tapes. **1973** J. Moran *Printing Presses* ix. 133 Between 1836–63 Napier made at least eighty-seven perfecting machines. **1902** R. Hoe *Short Hist. Printing Press* 17 In 1814 Koenig patented a continuously revolving Cylinder Press... This press, termed a 'perfecting press', was afterwards improved by Applegath and Cowper. **1967** V. Strauss *Printing Industry* vi. 279/1 Perfecting presses, or perfectors..print both sides of the sheet in one color in one pass through the press.

perfectionism. Add: **2.** Refusal to accept any standard short of perfection.
1937 *Nation* (N.Y.) 30 Oct. 465/2 Labor..cannot afford perfectionism. **1945** F. D. Roosevelt *Public Papers & Addresses 1944–5* (1950) 498 Perfectionism, no less than isolationism or imperialism or power politics, may obstruct the paths to international peace. **1957** *Listener* 24 Oct. 642/1 Sir Lewis Namier once gave up writing a book he had planned because some manuscripts in private hands were not made available to him. This is a fine example of scientific perfectionism. **1968** P. B. Austin *On Being Swedish* iii. 21 Perfectionism always implies, at a deeper level, a fear of failure.

perfectionist. Add: **2. a.** One who is only satisfied with the highest standards.
1934 in Webster. **1951** 'J. Tey' *Daughter of Time* ii. 28 A worrier: perhaps a perfectionist. A man..anxious over details. **1969** 'R. Gordon' *Facts of Life* ii. 14, I hated perfecting anything badly... Like so many women doctors, I was a perfectionist. **1978** *Vogue* 1 Mar. 114/1 Bette Davis's misfortune is to be a perfectionist in an industry run by opportunists.

b. *adj.* Demanding perfection or perfectionism (sense *2).
1958 *Times Rev. Industry* Aug. 50/3 The extreme sensitivity of the Tabyazi and its companion instrument aroused some resistance among engineers at first. They claimed that the methods were too perfectionist for everyday practical purposes. **1977** *Listener* 17 Mar. 333/1 Let us not be elitist, perfectionist... Singapore does provide its people with a decent..existence. **1978** P. Boardman *Worlds of Patrick Geddes* vii. 226 P. G. certainly could not be called a perfectionist parent. He urged them from early years to take notes.

perfectionistic, *a.* Add: **b.** Tending towards perfectionism (sense *2).
1961 P. B. Austin *On Being Swedish* xvii. 124 In its heaviness of spirit, a bleakness of insight so intense that all its perfectionistic arrangements can obviously only be oil on the stormy waters. **1977** W. J. Bate *Samuel Johnson* x. viii. 117 Johnson's fears of insanity..[were] a fateful delusion resulting from a perfectionistic notion of 'sanity'.

perfectivation (pəɹfektivē·ʃən). [f. Perfective *a.* + -ation.] The action of rendering a verb perfective.
1926 G. W. S. Friedrichsen *Gothic Version of Gospels* vii. 100 This is an instance of the colourlessness of verbal prefixes when used as an instrument of perfectivation. **1954** Pei & Gaynor *Dict. Linguistics* 163 *Perfectivation*, the transformation of an imperfective verb into a perfective one by a morphological change. **1962** R. W. Zandvoort in F. Behre *Contrib. Eng. Syntax* 15 Mossé..believes in perfectivation in O.E. by means of the prefix *ge-*.

perfective, *a.* (*sb.*) For 'Now *rare*' read 'Now *rare* except in *Gram.*' and add: **3.** (Further examples.)
1889 [see *dative *a.*]. **1924** [see *aspect *sb.* 9b]. **1968** J. Lyons *Introd. Theoret. Linguistics* viii. 396 The English 'perfect with *have*'..was at first restricted to transitive verbs, and thus preserved its relationship with the perfective passive without 'have' (still current in such sentences as *The work is done, The house is built*). **1975** *Language* LI. 444 The tenses were divided into an imperfective and a perfective set with three tenses in each.
B. *sb.* **2.** (Further examples.)
1949 *Archivum Linguisticum* I. 176 Perfectives are not always easy to recognize formally. **1968** J. Lyons *Introd. Theoret. Linguistics* viii. 314 Many perfectives [in Russian] are derived by prefixation of the corresponding imperfectives. **1969** B. M. H. Strang *Hist. English* ii. 100 The tendency to regularise verb-forms also appears in the use of *has*, etc. to form perfectives.

perfectiveness. (Later example.)
1964 *Philol. Rev.* LXXIII. 20 Rye..was perhaps led to this opinion partly by the perfectiveness of 'to see'.

perfe·ctivizing, *ppl. a.* Rendering perfective.
1908 *Expositor* July 91 The function of the perfectivising preposition is to supply a present answering to the past δεῖπνον. **1949** *Archivum Linguisticum* I. 7 Let us suppose that we have a set of distinct prepositions each identical with a perfectivizing prefix. **1961** *Brno Studies in English* III. 99 The function the Czech perfectivizing verbal prefixes may play.

perfecto (pəɹfe·kto). orig. *U.S.* Also **Perfecto, Perfectos.** [Sp. *perfecto* perfect.] A type of cigar, thick in the centre and tapering at each end.
Also used in the names of various brands of cigar and cigarette.
1894 *Harper's Weekly* 5 May 429/2 A minute later two of the raggedest-looking tramps you ever saw were trudging westward through the rain, each smoking a Carolina Perfecto costing thirty-five cents apiece. **1897** *Sears, Roebuck Catal.* 24/2 La Flor de Portuondo, Perfectos. A 5 inch cigar. **1898** H. E. Hamblen *Gen. Manager's Story* 3 The old gentleman..blowing the smoke from his 'perfecto' out into the cool starlight. **1904** 'O. Henry' *Four Million* (1906) 74 He always..handed out real perfectos to the delighted boys. **1931** E. S. Gardner in *Detective Fiction Weekly* 14 Mar. 302/2 He took a box of perfectos..and selected one. **1944** S. Bellow *Dangling Man* (1946) 149, I sucked tranquilly at my Perfecto Queen and said to myself, 'It's in the bag.' **1953** *Trade Marks Jrnl.* 5 Aug. 707/2 Perfectos Finos... Cigarettes for export. British-American Tobacco Company, Limited, Westminster House, 7, Millbank, London, S.W.1; Tobacco Manufacturers. **1960** R. K. Heimann *Tobacco & Americans* 248/2 Where fat perfectos and 'banker' sizes had once ruled the glassed counters, slim panetelas and palmas moved to the front row. **1961** C. Willock *Death in Covert* iv. 78 Crumbe-Howard took a Perfectos Finos from a gold case. **1968** D. MacKenzie *Three minus Two* 85 Smith chose a Perfecto and lit it. **1973** 'D. Jordan' *Nile Green* xiii. 54, I took a Perfectos from the silver box and tried to light a match on my thumbnail.

perfector. Add: **2.** *Printing.* Also **perfecter.** = *perfecting machine, press* s.v. Perfecting *vbl. sb.* b in Dict. and Suppl.
1899 J. Southward *Mod. Printing* III. xii. 124 (*caption*) Dawsons' Patent Fast Gripper Perfecter. **1940** *Chambers's Techn. Dict.* 626/1 *Perfector*, a type of machine which prints both sides of the paper before delivery. **1951** S. Jennett *Making of Bks.* vii. 101 These machines are in effect two presses combined in one, printing one side of the sheet first and the other side immediately afterwards. They are called perfectors. **1975** *Bookseller* 6 Dec. 2590/3 Paper of 100 per cent recycled content can now be used for printing on offset litho perfectors.

perfervidly (pəɹfɜ·ɹvidli), *adv.* [f. Perfervid *a.* + -ly².] In a perfervid manner.
1906 *Macm. Mag.* Oct. 884 The General was gripping de Pélotin's hand perfervidly. **1922** Joyce *Ulysses* 709 Symposium of incoordinately abstract, perfervidly concrete mercantile coexreligionist excompatriots.

‖ **perfide** (pɛrfid), *a.* [Fr., = treacherous.] In phr. *perfide Albion*, with reference to the Fr. phr. *la perfide Albion*, 'perfidious Albion' (see *Albion).
1864 [see *Albion]. **1899** Beerbohm *Around Theatres* (1924) I. 52 Had Mr. Kipling been born a Frenchman.. he..would be known to us only as a..fulminator against '*perfide Albion*'. **1926** Galsworthy *Silver Spoon* II. iv. 141 Perfide Albion! Well! We always wait till the last moment to declare our policy... Gives the impression that we serve the cause. **1947** H. Nicolson *Diary* 2 Jan. (1968) 88 We always go through three stages of being enemies... It is this that has earned us the reputation of *perfide Albion.* **1973** R. Mayne *Europeans* v. 119 The phrase '*la perfide Angleterre*', later to become '*perfide Albion*', may have been coined by Jacques Bérigne Bossuet in a sermon on New Year's Day, 1654.

perfidiousness. (Later examples.)
1935 E. R. Eddison *Mistress of Mistresses* iii. 54 Perfidiousness is a common water in most princes' courts. **1963** *Times* 30 Jan. 9/4 Mr. Macmillan's journey to Moscow..resurrected in Dr. Adenauer's mind all remembered fears of the perfidiousness of Albion.

perfidity. Delete †*Obs.* and add later example.
1903 J. Kelman *Honour towards God* iii. 22 Instances are only too common in which Pagan and Mohammedan honour has shamed the perfidity of so-called Christians.

perfluorinated (pəɹfluːərine̊ited), *ppl. a.* *Chem.* [f. Per-¹ 5 + *fluorinated *ppl. a.*] Applied to a (usu. organic) compound, radical, etc., in which fluorine has replaced hydrogen to the maximum extent short of altering the characteristic functional groups of the species.
1947 *Chem. Rev.* XL. 52 This method cannot in general be used successfully..to prepare fluorocarbons or other perfluorinated compounds which are of much significance. **1972** *Materials & Technol.* IV. vii. 237 A number of perfluorinated aromatic and heteroaromatic compounds have been synthesized in recent years. **1974** *Physics Bull.* June 226/1 Electrolytic generation from water has made progress with the use of solid electrolytes such as perfluorinated sulphonic acid, which could allow efficiencies of up to 95% for the conversion from electrical energy to energy stored in the form of hydrogen.

perfluoro- (pəɹfluː·əro), *pref.* *Chem.* [f. Per-¹ 5 + Fluor- + -o.] Used to designate organic compounds, radicals, etc., in which hydrogen has been replaced by fluorine to the maximum extent short of altering the characteristic functional groups of the species.
1947 *Industr. & Engin. Chem.* Mar. 236/2 Perchlorobenzene is a high melting solid, whereas perfluorobenzene is a liquid boiling at about the same temperature as benzene. Also, the boiling points of the perfluoroalkanes having more than five carbon atoms are actually lower than those of the hydrocarbon. *Ibid.* 241/1 The members of the panel..made the recommendation that the prefix 'perfluoro' be used to denote complete substitution by fluorine of all positions attached to the carbon skeleton. **1951** I. L. Finar *Org. Chem.* I. v. 89 When catalysts other than copper..are used,..perfluorocompounds are obtained, e.g. *n*-heptane gives perfluoroheptane. **1962** P. J. & B. Durrant *Introd. Adv. Inorg. Chem.* xxiii. 913 Perfluoroethylene polymerises under pressure in contact with an aqueous solution of a persulphate to yield the polymer teflon. **1971** *Nomencl. Org. Chem.* (I.U.P.A.C.) (ed. 2) C. 145 Halogen-containing compounds or radicals in which all hydrogen atoms, except those whose replacement would affect the nature of characteristic groups..present, have been replaced by halogen atoms of the same kind may be named by adding the prefixes 'perfluoro-', 'perchloro-',..[etc.] to the name of the corresponding non-halogenated compound or radical. **1972** *Materials & Technol.* IV. vii. 252 The excellent electrical properties of perfluoropolymers, which have been known for many years, are shared by the simple fluorocarbons. **1976** *Nature* 4 Mar. 8/1 Perfluoropropane, C_3F_8, appears to be an ideal refrigerant.

Hence **perflu:oroca·rbon,** any binary compound of carbon and fluorine, analogous to a hydrocarbon.
1947 *Industr. & Engin. Chem.* Mar. 292/1 Until October 1941 no practical general method for the synthesis of perfluorocarbons had been reported in the literature. **1961** G. H. Beaven et al. *Molecular Spectrosc.* I. iii. 101 Organic solvents for use down to 200 mμ are limited, in practice, to the saturated hydrocarbons and the aliphatic alcohols and ethers, although perfluorocarbons may be used more, as they become more readily available. **1972** *Materials & Technol.* IV. vii. 231 The perfluorocarbons are synthesized either by direct fluorination using elemental fluorine or by vapour phase fluorination techniques using hydrogen fluoride and a suitable catalyst. *Ibid.* 251 One of the most successful ways to coat a surface is to incorporate the fluorochemical in a polymer in which the perfluorocarbon chains constitute a series of side chains.

perforable, *a.* (Further example.)
1926 J. M. Robertson *Mr. Shaw & 'The Maid'* v. 46 The real question is simply whether his shield is perforable.

perforated, *ppl. a.* Add: **1.** *perforated tape*, tape in which data are recorded by means of the pattern of holes punched in it; cf. Tape *sb.*¹ 2b, *paper tape* s.v. *paper *sb.* 12, *punched tape* s.v. *punched *ppl. a.* 2.
1890 *Electrician* 4 July 235/1 The key-board no longer gives direct electric contacts, but produces mechanically a perforated tape on which the signs are represented by holes at their respective places. **1964** N. N. Biswas *Princ. Telegr.* iv. 93 Transmission on the line may be made either from the keyboard or by a perforated tape on the tape transmitter. **1973** Goacher & Denny *Teleprinter Handbk.* ii. 35/2 The perforated tape forms a permanent record of all the code signals involved in a particular message in a form suitable for storage.

perforation. Add: **4. perforation plate** *Bot.* (see quot. 1933).
1933 *Tropical Woods* XXXVI. 7 *Perforation plate*, a term of convenience for the area of the wall (originally imperforate) involved in the coalescence of two members of a vessel. **1953** K. Esau *Plant Anat.* xi. 223 The perforations of vessel members commonly occur on the end walls, but they may be present on the lateral walls too. The wall bearing the perforation is called the perforation plate. *Ibid.* 233 A pitted wall part became a scalariform perforation plate, which changed into a simple perforation plate. **1975** S. Carlquist *Ecol. Strategies of Xylem Evol.* i. 13 There are even dicotyledons with simple perforation plates and scalariform lateral-wall pitting.

perform, *v.* Add: **7. e.** *intr.* To display extreme anger or bad temper; to swear loudly; to make a great fuss. *Austral. slang.*
1901 M. Franklin *My Brilliant Career* xix. 163 Bad-tempered is a tame name for it. You should have seen the dust he raised the other day with old Benson. He just did pefrorm. **1911** L. Stone *Jonah* v. 45 Ow'l Chook perform, if 'e ain't at Ada's? **1959** Baker *Drum* (1960) ix. 68 We say that a man *performs* when he is indulging in a frenzy of anger or vituperation.

performability (pəɹfǭːɹmăbi·lĭti). [f. Performable *a.* + -ility.] The capability of being performed.
1947 A. Einstein *Mus. Romantic Era* xv. 222 He [*sc.* Alkan] lost his sense of proportion, with regard to both the performability and the dimensions of his piano works. **1962** H. C. Weston *Sight, Light & Work* (ed. 2) vi. 185 Levels of illumination which are supposed to give equal visibility or 'performability' for widely differing visual tasks.

performance. Add: **2. a.** (Further examples.) *spec.* the capabilities of a machine or device, now esp. those of a motor vehicle or aircraft measured under test and expressed in a specification. Also used *attrib.* to designate a motor vehicle with very good performance.
Cf. quots. 1766, 1825 in Dict.
1832 *Mechanics' Mag.* 30 June 224/2 Extraordinary railway performances... On two occasions, a load amounting to 100 tons was drawn by one engine..a distance of above 30 miles, in an hour and a half. **1840** *Min. Proc. Inst. Civil Engin.* 34 The paper is accompanied by..tabular statements of their performances during the year 1839, showing the number of miles traversed by each engine, the weight conveyed, [etc.]. **1883** W. H. Maw *Recent Pract. Marine Engin.* I. 216/1 On the first regular journey from Queensborough to Flushing, the Prinses Marie made the journey of 101½ knots in 372 minutes, giving a mean speed of 17·12 knots..and the coal consumption..being 1·92 lb. per indicated horse-power per hour, an excellent performance considering that the cylinders are not steam jacketted. **1907** *Proc. Incorporated Inst. Automobile Engineers* I. 235 A formula of this kind..is most useful in comparing the performance of small engines with large. **1931** R. N. Liptrot in *Handbk. Aeronautics* (R. Aeronaut. Soc.) ii. 98 The top speeds and best rates of climb are now plotted against altitude, giving the required performance of the aircraft. **1952** W. W. Balding in *Bk. of Motor Car* 98/2 If one carefully studies the reports of technical performance.., one quickly recognises the importance attached to vehicle-acceleration. **1961** *Times* 7 Nov. 19/1 In Britain a thriving business has grown up in tuning and modifying the engines of existing models to give more performance. **1968** Miller & Sawers *Technical Devel. Mod. Aviation* vi. 173 The superior performance offered by the first bombers with swept-back wings won them the orders. **1969** *Daily Tel.* 1 Oct. 16/8 Performance: Acceleration: 0–60 mph 12·3 seconds; 50–70 mph in top gear 13 seconds. Speed in gears, 1st 30 mph, 2nd 48 mph, 3rd 75 mph... Fuel consumption: 26·3 mpg (overall); 30·6 mpg (touring). **1976** *Ibid.* 18 Feb. 10 (Advt.), Today we give you details of a high-performance FM tuner. **1976** *Daily Mail* (Hull) 30 Sept. 11/3 A performance car with a top speed in excess of 100 mph, it runs at over 30 mpg on two-star petrol. **1977** P. Harcourt *At High Risk* i. 90 If you're interested in performance cars, as I am, you're also mildly interested in the people who drive them.

d. *Psychol.* The observable or measurable behaviour of a person or animal in a particular, usu. experimental, situation.
1898 E. L. Thorndike in *Psychol. Rev. Monogr. Suppl.* II. iv. 39 The best interpretation of even the most extraordinary performances of animals has been that they were the result of accident and association or imitation. **1901** *Psychol. Rev. Monogr. Suppl.* III. vi. 2 All human performances, when objectified in units of space, time, etc., seem to follow certain laws of variability. **1912** *Psychol. Rev.* XIX. 73 (*title*) The influence of caffein on the speed and quality of performance in typewriting. **1938** R. S. Woodworth *Exper. Psychol.* vi. 138 Several varieties of performance preferential have been discovered. **1949** Stouffer & DeVinney in S. A. Stouffer et al. *Amer. Soldier* I. iii. 85 Morale is presumed to be an important element in performance. **1951** *Mind* LX. 2 The performance or non-performance of a certain act.. we shall call performance-values. **1959** E. Ginzberg et al. *Patterns of Performance* xiii. 272 We did not choose the concept of performance potential because of its predictive value. **1964** Cofer & Appley *Motivation* xi. 520 This distinction between learning and performance and the effects of drive on performance constitute the second problem.

e. *Linguistics.* (See quots.) Opp. *competence.*
1963 N. Chomsky in R. D. Luce et al. *Handbk. Math. Psychol.* II. 326 A generative grammar..can be regarded as a partial theory of what the mature speaker of the language knows. It in no sense purports to be a description of his actual performance. **1964** —— in *Proc. 9th Internat. Congr. Linguists* 915 Clearly the description of linguistic competence provided by the grammar is not to be confused with an account of actual performance. **1964** *Harvard Educ. Rev.* XXXIV. 263 The distinction between competence and performance or language and speech is quite crucial for understanding at least three goals related to linguistic descriptions proper. **1966, 1969** [see *competence 4f]. **1971** B. L. Liles *Introd. Transformational Gram.* i. 8 Another way of stating this is to say that he is interested in the speaker's competence, or knowledge of the language, rather than in his performance, or actual use of it. **1976** *Word 1971* XXVII. 144 He simply recalls our linguistic habits, our performances, as Chomsky would say.

3. d. A display of temperament, anger, or exaggerated behaviour; a fuss or 'scene'; a difficult or annoying action or procedure. *colloq.*
1936 G. B. Shaw *Six of Calais* 94 They tear a piece of linen from the back of his shirt, and bind his mouth with it. He barks to the last moment. John of Gaunt laughs ecstatically at this performance, and sets off some of the soldiers. *Ibid.* 100 *Peter* (growling in his face like a dog) Grrrr!!! *The King* (returning the growl chin to chin) Grrrr!!!!!! They repeat this performance, to the great scandal of the Queen, until it develops into a startling imitation of a dog fight. **1962** A. Nisbett *Technique Sound Studio* x. 176 Dialling is rather more of a performance than being on the receiving end of a phone call. **1964** J. Symons *End of Solomon Grundy* iii. iii. 180 For Christ's sake don't let's make a performance out of it. **1971** A. Price *Alamut Ambush* i. 16 He had to come here and tell you all about it and make a great performance of it.

5. *attrib.* and *Comb.*, as **performance art,** a form of visual art in which the activity of the artist forms a central feature, combining static elements with dramatic performance; so **performance artist; performance bond,** a bond issued by a bank or other financial concern, guaranteeing the fulfilment of a particular contract; **performance test,** (*a*) *Psychol.* (in sense *2 d), a non-verbal test of capability or intelligence based on the performance of certain manual tasks; (*b*) the measurement of weight gain, food conversion, and other heritable characteristics of farm animals, as a guide to selective breeding; also **performance testing;** so **performance-tested** *a.*, having had heritable qualities evaluated.
1971 *Rolling Stone* 24 June 37 *Performance art* is basically an extension of art into the theater, often involving more or less set programs at specified times and places... A..work of performance art was staged by..cars whose drivers all sounded their horns according to a prearranged score, and the noise was broadcast by a local radio station. **1976** *National Observer* (U.S.) 7 Feb. 20/2 Not quite the same as theater or dance, though it combines elements of both, performance art grew out of *avant-garde* movements, particularly in poetry and painting, that swept Europe early this century. **1978** *Times* 23 Nov. 18/1 Three fine arts graduates were discovered walking around East Anglia with a pole on their heads, supported by a council grant for a work of performance art. **1976** *Observer* 11 Apr. 2/3 Despite his outlandish name, P-Orridge..is one of Britain's leading young 'performance' artists, a type of art in which the artist includes himself in his work. **1976** *Loughborough Monitor* 26 Nov. 7/4 Mr. Richards, who describes himself as a performance artist and lives in Nottingham, had applied for a grant for his Christmas show. **1979** *Listener* 22 Feb. 293/2 There were performances by three performance artists... Kevin Atherton stripped off all his clothes on the gallery steps. **1938** *Sun* (Baltimore) 3 June 4/3 It is.. a general practice that the solicitor writing the 'bid bonds' also writes the 'performance bond' of the same contractor if his bid is accepted. **1965** Perry & Ryder *Thomson's Dict. Banking* (ed. 11) 436/1 Perhaps the most common type of performance bond is that given on behalf of a customer who is entering upon a housing contract for a local authority. **1970** *Globe & Mail* (Toronto) 26 Sept. B1/7 A performance bond with the amount of $10-million issued by the Commercial Union Insurance Co. in favor of Prince Albert [Pulp] is in existence. **1977** *Offshore Engineer* Apr. 18/1 Venezuelan conditions included a 25% performance bond and no progress payments. **1917** Pintner & Paterson (*title*) A scale of performance tests. *Ibid.* p. v, The work grew directly out of the psychological examination of deaf children... This work was begun in 1914 with the standardization of a few performance tests. **1921** R. S. Woodworth *Psychol.* (1922) xii. 275 Language plays little part in a performance test. **1932** J. L. Lush in *Proc. Soc. Animal Production* xxiv. 52 The swine Record of Performance test is regarded as a progeny test of the sire and dam rather than as a performance test of the individual pigs which are fattened and slaughtered. **1954** *Jrnl. Animal Sci.* XIII. 215 Few sheep breeders have the necessary assistance, time and facilities to conduct a progeny and performance testing programme. **1959** *Ibid.* XVIII. 1464 Seven years' data..were analyzed to determine the effectiveness of using final weight as a criterion in the selection of performance tested bulls. *Ibid.* 1465 Fifteen bulls and twenty-three heifers were placed on experiment to compare three systems of performance testing. **1970** Preston & Willis *Intensive Beef Production* iii. 132 Performance testing involves the measurement of traits in the live animal... The major advantage of performance testing is that it permits evaluation of the animal at a much earlier age than is possible with progeny testing. *Ibid.* 137 Bulls on performance test should not be given hormone treatment. *Ibid.*, The object of any system of performance testing must be to evaluate.. genetic differences between animals in terms of their phenotypic expression. **1971** *Farmer & Stockbreeder* 23 Feb. 13/3 Co-operative breeding schemes, designed to get the most impressive, performance-tested bulls widely used. **1972** *Encycl. Psychol.* II. 379/2 Performance tests are especially useful for subjects with some speech disability. **1977** *N.Z. Herald* 8 Jan. 2-8/7 (Advt.), Modern methods of management are being used to produce performance-tested rams and bulls for the department's other blocks throughout the province.

performative (pəɹfǭ·ɹmătiv), *a.* and *sb.* [f. Perform *v.* + -ative, as in *imperative.*] **A.** *adj.* Of or pertaining to performance; *spec.* designating or pertaining to an utterance that effects an action by being spoken or written or by means of which the speaker performs a particular act. **B.** *sb.* Such an utterance. Hence **perfo·rmatively** *adv.*, **perfo·rmativeness.**
1955 J. L. Austin *How to do Things with Words* (1962) i. 6 What are we to call a sentence or an utterance of this type? I propose to call it a performative sentence or a performative utterance, or, for short, 'a performative'. **1955** A. J. Ayer in B. I. Evans *Stud. in Communication* 27 There are very many uses of language, prescriptive, ritualistic, playful, or performative, which are not fact-stating. **1956** J. Hollander in *Jrnl. Aesthetics* XV. 239 A *performative* system of scansion..would present a series of rules governing a locutionary reading of a particular poem, before a real or implied audience. It would end up by *describing* not the poem itself, but the unstated canons of taste behind the rules. Performative systems of scansion, disguised as descriptive ones, have composed all but a few of the metrical studies of the past. **1960** *Proc. Aristotelian Soc.* LX. p. v, The most famous of his [*sc.* J. L. Austin's] discoveries in this field was of the element of performativeness that enters into many kinds of utterance ordinarily classified as statements. **1963** J. Lyons *Structural Semantics* ii. 33 The philosophers have accustomed us to the wide variety of uses which the verb *know* can have... Among them they distinguish what has been called a 'performative' use. **1964** R. H. Robins *Gen. Linguistics* 23 It is best to regard knowledge of the meaning or meanings of a word as a performative knowledge (like knowing how to ride a bicycle). **1970** *Language* XLVI. 35 It is probable that the notion of the declarative sentence can be defined in terms of performatives. *Ibid.* 100 *For* clauses..provide important evidence for the performative analysis. **1973** *Archivum Linguisticum* IV. 82 Being an imperative construction, it would presumably require the postulation of an underlying performative. **1973** *Times Lit. Suppl.* 5 Oct. 1161/5 Illocutionary acts of *x*-ing could be made explicit..by use of the 'performative' first-person present-indicative form 'I *x*'—e.g., acts of asking by prefacing the question by the phrase 'I ask you:..'. **1973** G. W. Turner *Stylistics* vii. 208 A performative utterance has validity if the speaker's social position entitles him to make it; it is therefore often part of an occupational language. **1976** *Archivum Linguisticum* VII. 69 The same applies to the everyday use (as opposed to the analysis) of speech act labels, whether these are used performatively or not. **1976** P. Donovan *Relig. Lang.* vii. 80 Performative words are used, for instance, when we vote for a motion by saying 'Aye', bid in an auction by shouting 'Fifty pounds', or adjourn a meeting with the words 'The meeting is adjourned'. **1978** *Listener* 30 Mar. 396/3 When I say, 'I promise', or 'I bet' or 'I apologise' or 'Thanks'..these he called 'performatives'.

performatory (pəɹfǭ·ɹmătəri), *a.* and *sb.* [f. as prec.: see -ory².] = prec. Hence **perfo·rmatorily** *adv.*
1949 *Mind* LVIII. 359 To make a promise is to perform an *act* in which language is involved as an integral part... Mr. J. L. Austin distinguishes this as the 'performatory' use of language. I am indebted to him for this point. **1949** *Philosophy* XXIV. 90 There are..performatory sentences like 'I name this ship Shamrock'. **1951** *Aristotelian Soc. Suppl.* Vol. XXV. 207 Group I [*sc.* the verbs 'advise', 'order', 'command', 'tell'] can be used performatorily, but Group II [*sc.* the verbs 'persuade', 'induce', 'cause', 'get'] cannot; thus I can say 'I advise you to make yourself scarce before he comes', but not 'I persuade you to make yourself scarce before he comes'. **1955** J. L. Austin *How to do Things with Words* (1962) i. 6 Formerly I used 'performatory': but 'performative' is to be preferred. *Ibid.* ii. 12 A few simple utterances of the kind known as performatories or performatives. **1966** L. J. Cohen *Diversity of Meaning* (ed. 2) i. 1 Their arguments have often relied on such distinctions as those between customary and indirect meaning, logical words and object words,..or performatory and non-performatory verb-uses. **1967** *Listener* 2 Feb. 162/1 The listener.. may have expected a more substantial work than was due to be played... If he perseveres with it, what sounded prefatory can come to sound performatory.

performer. Add: **2. b.** One who 'performs' (see *perform *v.* 7e); one who causes trouble or disturbance. *slang* (chiefly *Naut.*).
1937 Partridge *Dict. Slang* 619/1 *Performer*,..one who is apt to make a great fuss or noise. **1946** *Seafarers' Log* 6 Dec. 10/5 You get a performer aboard a ship who makes it bad not only for himself but for the crew and the Union. **1958** E. S. Land *Winning War with Ships* xiv. 193 In the early months of World War II unions were battling against a threat of the Navy taking over merchant marine personnel because of reported incidents of drunkenness, insubordination and trouble making by individuals whom the unions called 'performers'.

performing *vbl. sb.* Add: **4. performing art,** an art (such as the dance, drama, etc.) involving public performance (chiefly *pl.*); **performing right,** add: usu. *pl.* (further examples).
1929 J. B. Priestley *Good Companions* iii. iii. 534 Performing rights, sheet music, gramophone records... There's bags of money in it. **1946** *N.Y. Times* 7 June 21/5 The High School of the Performing Arts will differ from other vocational or academic schools in several important details. **1961** V. Krepela in *Webster* s.v. performing adj., Project an image of the U.S. through displays, films, publications, fine arts, and the performing arts. **1967** Chujoy & Manchester *Dance Encycl.* 719/2 The function of Performing Arts School is not to create artists, but rather to prepare competent craftsmen. **1968** *Listener* 8 Aug. 177/3 When I did my first film..someone said: 'It may not be so much money they are paying you down, but you get an income from it for performing rights.' **1971** *Times* 22 Feb. (Canada Suppl.) p. vii/3 It is much more comfortable politically to subsidize the performing arts (where results can be seen and enjoyed by large numbers of voters) than to support the creative individuals. **1974** *Times* 9 Oct. 18/1 All that is best in our performing arts depends for its survival on money. **1976** *National Observer* (U.S.) 23 Oct. 15/2 You'll find some kind of alliance developing among performing-arts centers in cities like Washington, Denver, Los Angeles. **1977** *Rolling Stone* 7 Apr. 35/2 The performing-rights organization negotiates license fees with *commercial* users (nonprofit organizations, such as churches and schools, are exempt from paying for use of copyrighted material).

performing, *ppl. a.* Add: **2.** (Earlier and later examples.) Esp. designating a flea trained to perform tricks; also *fig.*
1854 Dickens *Hard T.* i. iii. 14 Signor Jupe was that afternoon to 'elucidate the diverting accomplishments of his highly trained performing dog Merrylegs'. **1889** G. B. Shaw *London Music 1888–89* (1937) 205 The only artist who never gets accustomed to his part is the performing flea who fires a cannon. **1922** E. Wallace *Flying Fifty-Five* vii. 45 Your last stable was a stable of performing fleas, for I swear you know nothing about horses. **1953** Wodehouse (*title*) Performing flea: a self-portrait in letters. **1966** M. Woodhouse *Tree Frog* xviii. 134 Driver came me a look as though I were a performing seal. **1973** 'E. Peters' *City of Gold & Shadows* ii. 29 The secret of success with performing fleas..is to synchronise your orders with their hops. **1978** D. Bloodworth *Crosstalk* xxi. 167 I'm a cop. I run a team of performing dogs—the best in the business.

perfume, *sb.* Add: **3.** *perfume atomizer, spray; perfume-sprayed* adj.
1897 *Sears, Roebuck Catal.* 329/1 A fine imitation cut glass perfume atomizer. **1942** E. Paul *Narrow St.* xxviii. 254 He began to throw Jeanne's toilet articles at her, perfume atomizer, box of powder, lipstick, mirror. **1973** M. & G. Gordon *Informant* xxxvi. 142 Out spilled a wallet, lipstick,..a perfume atomizer, nail file. **1898** *Illustr. London News* 22 Jan. 126 (Advt.), These concentrated perfume sprays give a delightful refreshing coolness. **1926–7** *Army & Navy Stores Catal.* 498/2 Perfume sprays. **1975** D. Beaty *Electric Train* 42 Daphne was sitting at the dressing-table, her perfume spray in her hand. **1922** Joyce *Ulysses* 523 As they are now, so will you be, wigged, singed, perfumesprayed, ricepowdered, with smoothshaven armpits.

perfusate (pəɹfiū·zēit). *Med.* [f. Perfus(e *v.* + *-ate*, after *filtrate, precipitate*, etc.] Any fluid used for perfusion.
1915 *Amer. Jrnl. Physiol.* XXXVIII. 201 Winterstein perfused new-born rabbits with saline solution..and concluded that the hydrogen-ion concentration of the perfusate governed the function of the respiratory center. **1938** *Nature* 29 Oct. 800/2 The frog's liver perfused with a saline solution can secrete natural and artificial pigments in concentrations many hundred times those in the perfusate. **1961** *Jrnl. Clin. Invest.* XL. 1079/2 Solutions pumped into the oral end of the tube entered the intestine at a known site; 15 cm distally the perfusate was aspirated and collected from the anal end of the tube. **1968** *Sci. Jrnl.* Nov. 63/3 The use of balanced salt solution perfusates without red cells (solutions having a composition similar to extra-cellular fluid). *Ibid.* 64/2 Blood and a variety of other perfusates were studied.

perfuse, *v.* Add: **3.** *Med.* To supply (an organ, tissue, or body) with a fluid artificially by circulating it through blood vessels or other natural channels; to pass a fluid through (a hollow organ).
1903 *Jrnl. Physiol.* XXIX. 266 The method of examining the physiological action of an organ or tissue by perfusing it by blood after its removal from the body has already proved of great value. **1920** *Amer. Jrnl. Physiol.* LII. 101 Using a mixture of red cells and whole serum or diluted serum or modified Locke's solution, the investigator should be able to perfuse satisfactorily the various organs or combinations of organs and tissues. **1962** *Lancet* 6 Jan. 13/2 Radiography will then reveal whether anastomoses exist, by showing the radio-opaque material in arteries which were not perfused. **1965** *Gut* VI. 387/2 After perfusing the intestine for 30 minutes a steady state is reached, and intraluminal contents are then aspirated through a hole. **1966** *Maclean's Mag.* 2 Apr. 18/3 The current belief is that before freezing the body should be perfused—that is, the blood should be replaced with a chemical that would prevent, or at least minimize, cellular damage during freezing. **1971** *Jrnl. Physiol.* CCXVI. 735 The lumen was perfused with a raffinose-electrolyte solution having a low sodium concentration.
Hence **perfu·sed** *ppl. a.*, (of an organ, etc.) kept supplied with a flow of fluid; (of a fluid) supplied as a substitute for blood; **perfu·sing** *vbl. sb.*
1903 *Jrnl. Physiol.* XXIX. 271 Difficulties in the way of obtaining a sufficiently rapid rate of flow through the perfused organ. **1906** *Ibid.* XXXV. 54 The amount of oxygen in the perfused liquid and the rate of perfusion were as far as possible controlled. **1916** *Amer. Jrnl. Physiol.* XL. 516 With the perfusing fluid running..a cannula is passed into the popliteal below the nutrient artery. **1962** *Lancet* 6 Jan. 13/2 Another unreported source of error is the perfusing of hearts still affected by rigor mortis.

perfusion. Add: **d.** *Med.* The process of passing through an organ or tissue a fluid, esp. treated blood or a substitute for blood; the treatment of a patient by a continuous transfusion of prepared blood. Freq. *attrib.*
1903 *Jrnl. Physiol.* XXIX. 266 Special arrangements were provided for quantitatively determining the changes of the blood gases in the perfusion. **1910** *Ibid.* XL. 297 Perfusion experiments with tortoise hearts were made by Gaskell in 1883, his perfusion liquid being a mixture of 1 part of sheep's blood and 2 of saline. **1940** C. S. Sherrington *Man on his Nature* iii. 87 The perfusion-fluid itself is chemical nutriment for all the cells of the body, supporting their energy-needs. **1963** *Gastroenterology* XLIV. 134/2 Absorption was studied by transintestinal intubation with polyvinyl tubing..with a perfusion technique. **1969** *Jrnl. Physiol.* CCIV. 22P In order to show that it was the ionic composition of the perfusion fluid which produced the changes..rather than the experimental procedure, perfusion of both vasa deferentia with Krebs-Ringer preceded perfusion of one vas deferens with the test solution. **1972** *Brit. Med. Jrnl.* 1 Jan. 23/1 Two patients in deep hepatic coma due to fulminant viral hepatitis were treated by extracorporeal baboon liver perfusion. *Ibid.* 23/2 The liver was aseptically removed from healthy baboons..; it was immediately cooled by perfusion with a chilled electrolyte solution..and then taken to the perfusion apparatus in the patient's room. Perfusion was started after 30–40 minutes of cold ischaemia and was maintained for a period of 13–16 hours.

perfusionist (pəɹfiū·ʒənist). *Med.* [f. prec. + -ist.] The member of a surgical team responsible for the perfusion of a patient while his circulation is interrupted.
1964 *Sunday Mail Mag.* (Brisbane) 25 Oct. 2/3 The doctor who assembles the machine is a perfusionist. He will operate the machine during the operation. **1975** *Islander* (Victoria, B.C.) 21 Sept. 3/4 Although there are five anaesthetists, Mr. Kemna is the only perfusionist and assists at every open heart operation.

Pergamene (pɜ̄·ɹgămīn), *sb.* and *a.* Also **Pergame·nian** (6 **Pargamenian,** 9 **-onian**), **Pergame·nic.** [f. L. *Pergamum*, Gr. Πέργαμος, -ον city and capital of an ancient kingdom in Asia Minor.] **A.** *sb.* An inhabitant of Pergamum. **B.** *adj.* Of or pertaining to Pergamum, the school of sculpture that flourished there in the third and second centuries B.C., or the Church founded there in the first century after Christ.
1579 North tr. *Plutarch's Lives* 1056 He wrote vnto the Pargamenians in this sorte. [**1608** Topsell *Serpents* 122 Wee doe read that the Pergameni did buy..certaine peeces of a Cockatrice.] **1774** T. Reid in Ld. Kames *Sk. Hist. Man* II. iii. 168 He [*sc.* Aristotle] was the first we know, says Strabo, who composed a library. And in this the Egyptian and Pergamenian kings, copied his example. *a* **1823** R. Culbertson *Lect. Revelation* (1826) I. xvi. 214 Pergamos..together with all that territory over which the Pergamonian princes bore rule, was bequeathed by Attalus..to the Romans. **1865** J. B. Lightfoot *Galatians* 5 The Pergamene prince Attalus the first effectually curbed their power. **1867** C. M. Yonge *Pupils of St. John* vi. 87 The sharp piercing two-edged sword of the Word of God..was held up threateningly to warn the Pergamenes. **1896** *Pall Mall Gaz.* 19 May 3/2 A victorious king..is to be met by a troop of priests, and conducted to the Pergamenic altar. **1903** *Westm. Gaz.* 23 Sept. 3/2 Visitors..will be delighted to recognise the Pergamenian type of the so-called 'Dying Alexander'. **1904** W. M. Ramsay in *Expositor* June 407 The honourable history and the steadfast loyalty of the Pergamenian Church..had been tarnished by the error of a small part of the congregation. *Ibid.* 409 We shall find that both in the Thyatiran and in the Pergamenian letter St. John exalts the dignity, authority and power that shall fall to the lot of the victorious Christian. **1926** *Chambers's Jrnl.* 11 Sept. 649/1 It is this temple..which is referred to as 'Satan's throne' in St John's letter to the Pergamene Church. **1947** E. V. Hansen *Attalids of Pergamon* vi. 157 Five envoys were sent by the Pergamenes to both cities to obtain..the evidence of the contending parties. *Ibid.* viii. 277 Another sculptor who had a long career at the court of the Attalids was a native Pergamene, Epigonus the son of Charius. **1960** R. Carpenter *Greek Sculpture* vii. 191 The Pergamenian rulers conceived the ambition to bring that golden period to life. **1968** G. E. Bean *Turkey's Southern Shore* vi. 81 The figure of Nike, goddess of victory, which appears on..coins of Side may commemorate a victory won by the city against the Pergamenes. **1972** P. M. Fraser *Ptolemaic Alexandria* I. iii. 98 A Pergamene inscription tells us that the supreme magistrates of the city..were appointed by the King at least on one occasion. **1972** 'M. Innes' *Open House* xiv. 138 This revolting masterpiece of the Pergamene school.

pergelisol (pɜ̄·ɹdʒe·lisǫl). *Geomorphol.* [f. Per-¹ + L. *gel-āre* to freeze + -i- + *-sol.] = *permafrost.
1946 K. Bryan in *Amer. Jrnl. Sci.* CCXLIV. 635 It is impossible to make a verb or a verbal noun from 'permafrost' as 'permafrosting' and 'permafrosted' imply that a permanent surface or coating has been applied... Further, the term cannot be easily converted into other European languages. These various objections can be met by a new term... Such a word is 'pergelisol'. **1963** *Geomorphol. Abstr.* 38 Canada, in the opinion of Dr. Hamelin, can be divided into eleven periglacial 'provinces'. The first four provinces: Elizabeth, Victoria, Keewatin and Innuit are closely associated with continuous pergelisol. **1968** R. W. Fairbridge *Encycl. Geomorphol.* 1185/1 The south- to west-facing slopes thaw earlier and more deeply than the opposite ones, where the pergelisol remains near the surface. **1972** Sparks & West *Ice Age in Brit.* iv. 99 With this wholly admirable intensification of study [of periglacial phenomena] has come a wholly regrettable spate of jargon, not all of which fortunately has found favour. Thus, most people still speak of frost shattering rather than congelifraction, active layer rather than mollisol, permafrost rather than pergelisol.

perhapser (pəɹhæ·psəɹ). *slang.* [f. Perhaps *adv.* (*sb.*) + -er¹.] A risky stroke in cricket.
1954 J. Fingleton *Ashes crown Year* xxiii. 247 Morris somewhat luckily got Bedser fine for 4... It was what cricketers know as a 'perhapser'. **1957** D. Stivens *Scholarly Mouse* (1958) 86 Did you ever see such a p'rapser—he pushed a yorker away for four!

perhexiline (pəɹhe·ksilīn). *Pharm.* [Arbitrary blend of Pi)peridine and Hexyl.] A white crystalline solid which is a vasodilator, tablets of the maleate of which are given for the relief of angina pectoris; 2-(2,2-dicyclohexylethyl)piperidine, $C_{19}H_{35}N$.
1969 *Current Therapeutic Res.* XI. 99 A clinical study was undertaken to explore the effectiveness of perhexiline in relief of angina pectoris. **1970** *Chest* LVIII. 579/2 There was slowing of the heart rate as a result of the infusion of perhexiline. **1977** *Lancet* 30 July 260/1 The Committee on Safety of Medicines has circulated an adverse-reactions warning about perhexiline maleate ('Pexid'). *Ibid.* 12 Nov. 1028/1 During treatment with perhexiline maleate hypoglycæmia, hyperlipidæmia, and liver dysfunction have been reported.

peri-, *prefix.* Add: **1. a. perianal** (examples); **peria·pical,** situated or occurring around the apex of the root of a tooth; **pe:riaquedu·ctal,** situated around the aqueduct of the midbrain; **pericapi·llary,** around a capillary blood vessel; **periceme·ntal,** of or pertaining to the pericementum; **perico·lic,** situated or occurring around the colon; **perige·nital,** situated in the area around the genitals; **perigona·dial,** situated around a gonad; **perihæ·mal** *Zool.* [ad. G. *perihämal* (H. Ludwig 1877, in *Zeitschr. f. Wissensch. Zool.* XXX. 123)], used to designate certain vessels and cavities in echinoderms and other invertebrates (see quots.); **periodontal:** hence **periodo·ntally** *adv.*; **pe:riurete·ric,** around one or both ureters.
1890 Billings *Med. Dict.* II. 311/1 Perianal. **1897** *Q. Jrnl. Microsc. Sci.* XL. 291 [In the larval form of *Phoronis*] there are three prominent ciliated bands... Of the three the perianal band is the most prominent. **1971** G. H. Bourne *Ape People* xi. 246 Goodall suggests that the function of the swollen perianal region of the sexually receptive chimpanzee..is to signal to males..that the female is in fact in estrus. **1977** *Lancet* 20 Aug. 403/2 The perioral and perianal zones are bright and red. **1920** Endelman & Wagner *Gen. & Dental Path.* xxxi. 422 The periapical tissues may be invaded by bacteria which from the start give rise to chronic symptoms. **1974** H. P. Hitchcock *Orthodontics for Undergraduates* xxviii. 480 Periapical lesion around a replanted tooth. **1950** *Physiol. Rev.* XXX. 460 Excitable foci may be followed caudalward..through the periaqueductal grey. **1973** *Nature* 26 Oct. 447/2 Binding in the periaqueductal area of the midbrain was about the same as that of the posterior amygdala. **1928** *Anatomical Rec.* XXXIX. 45 The impression seems to be general..that there is but one type of pericapillary cell. **1953** *Jrnl. Appl. Physics* XXIV. 1424/1 They [*sc.* the endothelial cells] possess..a large number of vesicles concentrated immediately under the cell membranes facing both the capillary lumen and the pericapillary spaces. **1977** *Lancet* 25 June 1364/2 The basal ganglia, thalamus, corpus callosum, and cerebral white-matter are peppered with innumerable pericapillary and periarteriolar microinfarcts, which sometimes coalesce into larger areas of softening. **1886** Pericemental [see *implantation 5 b]. **1940** H. K. Box *Twelve Periodontal Studies* v. 84 The formation, in the pericementum, of a new tissue..changes the normal arrangement of the pericemental structures, and..replaces the bone substance of the alveolar process. **1907** *Allbutt's Syst. Med.* (ed. 2) III. 1015 Pericolic inflammation may be excited by external violence. **1939** *Times* 20 Feb. 12/7 Lieutenant-Colonel Anderson I.M.S., successfully operated on Lord Brabourne, the Governor of Bengal, at Government House yesterday for pericolic inflammation. **1965** *Arch. Surg.* XCI. 407/2 The final diagnosis was pericolic abscess surrounding a solitary diverticulum o. the ascending colon. **1962** *Science Survey* III. 261 The perianal and perigenital glands of the *Mustelidae* like the stoat, skunk, civet, and others..play an important part in the sexual and social life of their carriers. **1971** *Nature* 7 May 50/1 These results clearly established a sex difference in the development of perigenital adipose tissue. **1888** *Ibid.* 22 Mar. 498/2 The perigonadial spaces (so-called generative glands) and the pericardial space..are, then, the cœlom of the mollusca... In Cephalopods..the pericardial and perigonadial cœlomic remnants are continuous, and form one cavity. **1942** Grove & Newell *Animal Biol.* xiv. 220 Each gonad then acquires a cavity—the primary gonadial cavity—and becomes almost completely surrounded by a secondary or perigonadial cavity (gonocoel). **1881** *Q. Jrnl. Microsc. Sci.* XXI. 171 The space..between the water-vessel above and the ambulacral epithelium below, which is traversed by the perforated longitudinal septum, was named by Ludwig the 'perihæmal canal'. It had been previously called the nerve-vessel or nerve-canal, and was supposed to form an integral part of the blood-vascular system. Now, however, it is regarded by Ludwig merely as a derivate of the body-cavity. **1897** *Ibid.* XL. 321 The front dorsal part of the trunk cœlom is produced into a pair of perihæmal spaces, embracing the dorsal blood-vessel. **1962** D. Nichols *Echinoderms* ii. 27 Perihaemal system. As its name suggests, this system normally surrounds the haemal complex, though some recent authors prefer to call it the hyponeural sinus system, referring to its relation to one part of the nervous system. **1955** J. Osborne *Dental Mech.* (ed. 4) ix. 148 Periodontally diseased teeth. **1975** H. Thomson *Occlusion* xi. 215 Periodontally disturbed abutment teeth..may have to be crowned and splinted to sound adjacent teeth. **1900** Dorland *Med. Dict.* 496/1 *Periureteric*, about the ureter. **1962** *Lancet* 6 Jan. 31/2 The causes of hydronephrosis included prostatic hypertrophy, bladderneck obstruction, periureteric fibrosis, and aberrant renal vessels.

b. pericementum (earlier and later examples); **periodontium,** add: in mod. use, all the tissues surrounding a tooth, including the alveolar process, the cementum, and the gingiva, as well as the periodontal membrane; (examples); (*periodontum* is now *Obs.*).
1879 C. F. W. Bödecker in *Dental Cosmos* XXI. 593 The pericementum (root membrane, or alveolo-dental periosteum, etc., as it has been termed by former writers) is a formation of connective tissue, identical with the periosteum which covers all bones. **1940** Pericementum [see *pericemental* above]. **1881** T. E. Satterthwaite *Man. Histol.* viii. 108 The development of the cement takes place precisely as bone is produced, viz., from the periosteum, or..from the fibrous tissue of the tooth-sac, the periodontium. **1922** K. H. Thoma *Oral Roentgenol.* (ed. 2) iv. 196 *Periodontium*, the pericementum and all investing structures of the teeth. **1927** O. E. Inglis *Burchard's Text-bk. Dental Path. & Therapeutics* (ed. 7) xvi. 536 All those tissues which invest the teeth including the pericementum.., the alveolar process and the gingivae, particularly the marginal and cemental gingivae are now generally understood as included in the term periodontium. **1940** H. K. Box *Twelve Periodontal Studies* ii. 29 In 1920, the writer first used the term 'periodontium' to designate the supporting tissues of the teeth, and to embrace as the three essential tissue-components, the gingivae, the periodontal membrane, and alveolar process. As the word 'periodontium' was occasionally used to denote the periodontal membrane, it was felt that the usage of the term in this new sense would make for exactness in terminology. **1969** *Gloss. Terms Dentistry* (*B.S.I.*) 67 *Periodontium*, the collective term for the tissues immediately surrounding the teeth.

c. periarteritis, add: [coined in Ger. as *periarteritis nodosa* (Kussmaul & Maier 1866, in *Deutsch. Arch. f. klin. Med.* I. 484)]; = *polyarteritis* s.v. *poly- 1; *periarteritis nodosa* (nodōu·zā) [L. *nōdōsus* knotty, f. *nōdus* a knot], an often fatal form of periarteritis characterized by the formation of aneurysms; (earlier and later examples); **pe:ricementi·tis,** inflammation of the pericementum of a tooth.
1876 Dunglison *Dict. Med. Sci.* (rev. ed.) 773/2 *Periarteritis*, inflammation of the sheath of an artery. **1892** F. P. Foster *Med. Dict.* IV. 2547/1 *P*[*eriarteritis*] *nodosa*, a thickening of the intima and infiltration of the adventitia of an artery, producing a nodular prominence. **1933** *Practitioners Libr. Med. & Surg.* III. cxii. 1136 Periarteritis nodosa is a rare disease affecting medium-sized arteries in any portion of the body, most commonly in males between the ages of twenty and forty. **1961** R. D. Baker *Essent. Path.* viii. 136 The interaction of antigen and antibody may produce lesions in the heart,

liver and kidneys resembling those of periarteritis nodosa. **1974** R. M. KIRK et al. *Surgery* ii. 9 Generalised diseases such as infections, uraemia, diabetes, scurvy, periarteritis..nodosa..and food allergies may be associated with frailty of the capillary walls. **1882** *Dental Rec.* II. 441 A little sensitiveness becomes noticeable..indicating a beginning of pericementitis. **1969** LUEBKE & MULLANEY in Morris & Bohannan *Dental Specialities in Gen. Practice* viii. 357/2 Pericementitis..sometimes follows overinstrumentation or overmedication of a noninfected canal.

periapsis (peri,æ·psis). *Astr.* [f. PERI- + APSIS, after PERIGEE, PERIHELION, etc.] That point in the path of a natural or artificial satellite at which it is closest to a primary.

1964 J. L. NAYLER *Dict. Astronautics* 194 *Periastron* (*periapsis*), the nearest point on the orbit of a stellar satellite to the star, or in a binary star the point at which the companion is nearest to the primary. **1971** *Nature* 26 Nov. 168/3 The manoeuvre, which changed the orbit of the spacecraft so that periapsis (closest approach to Mars) is now 868 miles. **1972** *Daily Colonist* (Victoria, B.C.) 25 Feb. 5/2 Pioneer will be close to Jupiter for about four days (100 hours) and will see the planet in full sunlight 50 hours before periapsis, or the closest point of the fly-by. **1976** *Sci. Amer.* June 59/1 Those commands will..place the spacecraft in an elliptical orbit around Mars that will vary in altitude from 33,000 kilometers at apoapsis down to 1,500 kilometers at periapsis.

periauger. Read: Var. PIRAGUA 2.

1898 *Rudder* Dec. 407/2 Let go our hook just ahead of a large periauger-rigged sharpie, called the Pirate. **1899** *Ibid.* Feb. 53/1 Her rig is that of a perianger [*sic*]—or, as some call it, a double cat—having two masts with a boom and gaff sail on each.

pericardium. Add: Hence also **pe:ricard(i)e·ctomy** [*-ECTOMY], surgical removal of all or part of the pericardium; **perica:rdiocente·sis** [Gr. κέντησις pricking], surgical puncturing of the pericardium.

[**1900** DORLAND *Med. Dict.* 491/2 Pericardicentesis.] **1913** STEDMAN *Med. Dict.* (ed. 2) 672/2 Pericardectomy... Pericardiectomy. **1938** M. THOREK *Mod. Surg. Technic* II. xxxi. 1206 Technic of pericardiocentesis. *Ibid.* 1212 (*heading*) Pericardiectomy in the treatment of the Pick syndrome. **1956** W. P. CLELAND in Bailey & Love *Short Pract. Surg.* (ed. 10) xliii. 892 At operation (pericardectomy) it is essential to remove the thickened pericardium from the ventricles. **1967** S. TAYLOR et al. *Short Textbk. Surg.* xvi. 223 Constrictive pericarditis... Pericardiectomy is the treatment of choice, and is best done through a vertical incision splitting the sternum, which affords excellent exposure. **1977** *Lancet* 6 Aug. 301/2, 15 underwent pericardiocenteses (with recurrent of effusion afterwards) before definitive therapy by pericardial drainage and local steroid instillation. *Ibid.* 15 Oct. 817/1, 3 of these 17 had tamponade which was successfully treated by pericardectomy.

pericentric, *a.* Add: **2.** *Cytology.* [cf. *-CENTRIC 2.] Involving parts of a chromosome at both sides of the centromere. Opp. *PARACENTRIC *a.*

1938 [see *PARACENTRIC *a.*]. **1962** *Lancet* 6 Jan. 21/2 Chromosome analyses in a girl with mongolism and in her parents have disclosed abnormalities which we attribute to pericentric inversion of a maternal 21st chromosome. **1973** *Nature* 3 Aug. 260/2 The very fact that some X chromosomes, such as that of the mouse are acrocentrics reveals that pericentric inversions did occasionally occur. **1977** *Ibid.* 6 Jan. 65/2 The aberrant chromosome was interpreted as an aneusomic recombinant from the father who was heterozygous for a pericentric inversion of chromosome 4.

periclinal, *a.* (*sb.*) Add: **2.** (Further examples.) More widely, occupying or occurring in a layer parallel to the surface of an organ.

1965 BELL & COOMBE tr. *Strasburger's Textbk. Bot.* 69 (*caption*) Each segment becomes divided by a periclinal wall..into an inner and an outer (cortical) cell. **1965** K. ESAU *Plant Anat.* (ed. 2) iv. 76 The lateral meristems are particularly distinguished by divisions parallel with the nearest surface of the organ (periclinal divisions). **1975** M. E. MCCULLY in Torrey & Clarkson *Devel. & Function of Roots* vi. 111 The daughter cells of these periclinal divisions lack the intense basophilia of the parent epidermal cells.

b. Applied to a type of chimæra (see quot. 1968). Also as *sb.*, a periclinal chimæra. [ad. G. *periklinalchimäre* (E. Baur 1909, in *Zeitschr. f. induktive Abstammungs- u. Vererbungslehre* I. 344).]

1916 *Jrnl. Genetics* VI. 78 Proof that a plant is a periclinal chimaera may..be obtained from adventitious buds arising in internodes of the stem, as well as from those formed on roots. **1925** *Ibid.* XVI. 44 Simple white-over-green periclinals would presumably..give only white seedlings. **1959** *New Biol.* XXX. 39 In periclinal chimeras there is often great variation in the pigment distribution pattern. **1963** *Heredity* XVIII. 270 The striped varieties of *Commelina*, *Tradescantia* and *Zebrina* have also been shown to be periclinal chimeras with a rather specialised development. **1968** R. RIEGER et al. *Gloss. Genetics & Cytogenetics* 59 Chimeras..in plants may be classified..according to their structure into a) sectorial (different tissues grow side by side and occupy distinct sectors of varying size), b) periclinal (different tissues are disposed one with the other..), c) mericlinal (actually an interrupted periclinal..).

periclinally *adv.*, (*b*) *Bot.*, in the manner of a periclinal division or chimæra.

1916 *Jrnl. Genetics* VI. 79 Plants in which the variegation affected the skin periclinally. **1963** *Heredity* XVIII. 281, L I which does not divide periclinally in the apex may do so frequently during leaf development. **1975** M. E. MCCULLY in Torrey & Clarkson *Devel. & Function of Roots* vi. 111 When the young primordium is about one-third of the way across the cortex a few epidermal cells at its tip divide periclinally and thus produce the root cap initials.

periculous, *a.* Delete †*Obs.* and add later examples.

1932 W. E. D. ALLEN *Hist. Georgian People* iii. 27 The land is poor and snowbound, craggy and periculous. **1959** A. A. MACGREGOR *Phantom Footsteps* iii. 70 The mountain's precipitous upper slopes at the time were treacherously encased in ice, its ravines deceptively filled with periculous snow.

pericyclic (perisəi·klik), *a. Chem.* [f. PERI- + CYCLIC *a.*] Of a reaction: involving a concerted rearrangement of bonding in which all the bonds broken or formed in the reaction lie on a closed ring, whether or not a cyclic molecule is involved.

1969 WOODWARD & HOFFMANN in *Angewandte Chemie* (Internat. Ed.) VIII. 848/1 In our development of the theme of orbital symmetry control of concerted chemical changes, we have laid the basis for a general consideration of all pericyclic reactions—that is, reactions in which all first-order changes in bonding relationships take place in concert on a closed curve. **1974** GILL & WILLIS *Pericyclic Reactions* iv. 99 The basic tenet of the Woodward-Hoffmann theory is that orbital symmetry is conserved in concerted pericyclic reactions. *Ibid.* vi. 162 The six-electron cyclo-addition and reversion processes are by far the most common of all pericyclic changes... Within this category falls the well-known Diels-Alder reaction. **1975** W. R. DOLBIER in Buncel & Lee *Isotopes in Org. Chem.* I. ii. 27 The class of reactions which are presently known as pericyclic or multi-centered reactions..include cyclo-additions, electrocyclic reactions and sigmatropic processes.

pericynthion (perisi·nþiǫn). *Astr.* [f. PERI- + Gr. Κύνθιον, neut. of Κύνθιος, adj. designating Mt. Kynthos on Delos, the birthplace of Artemis, goddess freq. associated with the moon.] That point at which a spacecraft in lunar orbit is closest to the moon's centre: applied esp. if the spacecraft was not launched from the moon.

1959 SPITZ & GAYNOR *Dict. Astron. & Astronautics* 295 *Pericynthion*, that point in the orbit of a moon rocket which is closest to the moon. **1969** *Times* 17 May 8/6 Then the descent engines of the L.M. will be fired..to propel it into an orbit in which it will swing around the moon in an elliptical orbit. At its nearest (pericynthion) it will be eight nautical miles from the surface.

pericyte (pe·risəit). *Histology.* [ad. G. *pericyt* (K. W. Zimmerman 1923, in *Zeitschr. f. Anat. und Entwicklungsgeschichte* LXVIII. 67): see PERI- and -CYTE.] One of many flattened branching cells found around capillary blood vessels.

1925 *Amer. Jrnl. Anat.* XXXV. 257 The 'pericytes' of mammals which Zimmerman has described. **1928** *Anatomical Rec.* XXXIX. 45 The designations Rouget cells, adventitial cells, and pericytes are used indiscriminately on the assumption that they connote the same structures. **1965** HAM & LEESON *Histol.* (ed. 5) xxii. 592 Opinion soon changed, and it was generally conceded that such cells as could be seen just outside the endothelium of the capillaries and the venules of most of the bodies of mammals were not contractile. Moreover, the noncontractile cells in this location came to be known by the terms of perivascular cells or pericytes. **1976** W. J. CLIFF *Blood Vessels* iv. 68 Pericytes are non-contractile cells which have well-developed phagocytic powers. *Ibid.* 71 Pericytes have primarily a mechanical supporting function within the walls of minute blood vessels.

peridinian (peridi·niăn). *Biol.* Also **peridinean, peridiniean.** [f. mod.L. generic name *Peridinium* (C. G. Ehrenberg *Organisation, Systematik und geographisches Verhältniss der Infusionsthierchen* (1832) II. 74/2), f. Gr. περιδινής whirled round: see -AN.] A dinoflagellate, usually from a marine habitat, belonging to the order Peridinales. Also *attrib.* or as *adj.*

1912 MURRAY & HJORT *Depths of Ocean* x. 674 Among plants the peridineans..are noted for their power of emitting light. **1928** RUSSELL & YONGE *Seas* i. 25 The Peridinians..can be considered either plants or animals as they have certain characteristics of both. **1935** F. E. FRITSCH *Struct. & Reprod. Algae* I. 664 The Peridiniean flora of warmer zones..differs very markedly from that of colder seas. **1963** D. W. & E. E. HUMPHRIES tr. *Termier's Erosion & Sedimentation* xi. 236 Most of the flagellates giving rise to 'waterbloom' are Peridineans. **1976** P. BOUGIS *Marine Plankton Ecol.* i. 11 Generally, a peridinian has chlorophyll and is autotrophic. *Ibid.* 12 *Noctiluca* is now placed in the Peridiniales... Its spores are of the peridinian type.

peridotic, *a.* (Examples.)

1880 *Mineral. Mag.* III. p. ix (*heading*) On some peridotic rocks from the island of St. Paul's. **1891** MURRAY & RENARD in *Rep. Sci. Results Voy. H.M.S. Challenger: Deep-Sea Deposits* vi. 374 Manganese is rarer in these rocks, but is found as a constituent of pyroxenic, amphibolic, and peridotic minerals.

peridotite. For *Min.* read *Petrogr.* and substitute for def.: Any of the group of plutonic rocks containing little or no feldspar but substantial olivine, usu. with pyroxene, amphibole, or other mafic minerals. [In this sense ad. G. *peridotit* (H. Rosenbusch *Mikrosk. Physiogr.* (1877) II. 522; previously F. *péridotite* had been used for an olivine-basalt (Cordier & d'Orbigny *Descr. des Roches* (1868) II. 118).]

1878 J. D. DANA *Man. Mineral. & Lithol.* (ed. 3) 451 Doleryte... There are two series: A. Ordinary, B. Chrysolitic, and for the latter the name Peridotyte has been used. **1882** A. GEIKIE *Text-bk. Geol.* 151 A series of crystalline rocks composed essentially of olivine, with usually one or two other magnesium silicates..has been classed by Rosenbusch under the general name of Peridotites. **1931** S. J. SHAND *Study of Rocks* vii. 114 Kimberlite.., the matrix of all African diamonds.., is a highly serpentinised peridotite. **1950** E. E. WAHLSTROM *Introd. Theoret. Igneous Petrol.* xiii. 291 Dunite, a variety of peridotite, is almost 100 per cent olivine. **1963** [see *melanocratic* adj. s.v. *MELANO-]. **1972** *Mineral. Mag.* XXXVIII. 437 The peridotite consists dominantly of coarse-grained (2 mm) olivine with minor interstitial clinopyroxene and amphibole of similar grain size.

Hence **peridoti·tic** *a.*, containing, consisting of, or resembling peridotite.

1886 J. J. H. TEALL *Brit. Petrogr.* iv. 68 Peridotitic, hornblendic, and augitic rocks have been serpentinised. **1935** BRANSON & TARR *Introd. Geol.* iii. 13 Magmas occurring to depths of 10 miles are of one type, those that are 10 to 40 miles (the maximum thickness of the crust) down are of another type, and those in the layer (approximately 900 miles thick) next below the crust are of still another. These three regions of the earth's body are known as the granitic, gabbroid or basaltic, and peridotitic zones. **1971** *Nature* 22 Oct. 522/2 The general hypothesis of sea-floor spreading envisages the upwelling of peridotitic mantle material beneath mid-oceanic ridges. **1974** *Mineral. Mag.* XXXIX. 798 Peridotitic rocks in the Massif du Sud..have suffered about 30% serpentinization.

perifoveal (perifōu·viăl), *a. Anat.* [f. PERI- + FOVEA.] Applied to the part of the retina regarded as the periphery of its central region, surrounding the parafovea.

1926 *Brit. Jrnl. Ophthalm.* X. 229 (*heading*) The perifoveal circulation. **1941** S. L. POLYAK *Retina* xv. 196 From the axial center to the anterior retinal boundary there are..seven regions: the central area, composed of (*I*) central fovea, (*II*) parafoveal region, and (*III*) perifoveal region or areal periphery; and the extra-areal periphery, composed of (*IV*) near periphery, (*V*) middle periphery, (*VI*) far periphery, and (*VII*) extreme periphery. *Ibid.* xvi. 213 The perifoveal region or the periphery of the central area is its most outward belt... Inwardly it begins at points where the ganglion cell layer still contains four rows of cells. Outwardly it ends where the ganglion layer becomes reduced to a single row of closely packed cells. **1966** *Jrnl. Physiol.* CLXXXVII. 456 There is considerable antagonistic interaction between rod and cone signals at the ganglion cell layer in the perifoveal retina of dark-adapted monkeys.

Hence (as a back-formation) **perifo·vea,** the perifoveal region.

1963 H. DAVSON *Physiol. of Eye* (ed. 2) II. i. 86 (*heading*) Parafovea and perifovea. **1964** J. Z. YOUNG *Model of Brain* ix. 160 The larger parasol bipolars and garland and giant ganglion cells make contact with hundreds or thousands of bipolars and thus cones (and rods) of the fovea and perifovea. **1975** *Nature* 6 Feb. 406/2 When fovea and perifovea are stimulated together (as in everyday life) foveal responses are relatively depressed suggesting that different retinal regions interact abnormally.

perifusate (perifiū·zēˡt). *Med.* [f. next + *-ate*, after *PERFUSATE.] The liquid that results from perifusion.

1969 *Lancet* 25 Oct. 883/1 In the perifusion system, 4–6% of the total pancreatic I.R.I. leaked into the perifusate during the initial 'washing' of the freshly cut tissue. **1978** *Nature* 19 Jan. 272/2, 6 mM L-glucose was added to a perifusate containing 3 mM D-glucose without significantly changing the rate of $^{42}K^+$ efflux as compared with controls maintained in 3 mM D-glucose alone.

perifuse (pe·rifiūz), *v. Med.* [f. PERI- after *PERFUSE *v.* 3.] *trans.* To subject to an enveloping flow of liquid. So **pe·rifused** *ppl. a.*

1969 *Lancet* 25 Oct. 882/1 (*heading*) Dynamic aspects of proinsulin release from perifused rat pancreas. **1972** *Diabetes* XXI. 989/2 The islets were perifused with glucose (1·0 mg./ml.). **1974** [see *PERIFUSION]. **1975** *Nature* 1 May 71/1 Cycloheximide (5 μM) was added to the medium used to perifuse one group of halved pituitaries whereas the other group received only the control medium. **1978** *Ibid.* 19 Jan. 272/1 (*caption*) Effect of D-glucose on potassium efflux.., and insulin release..from perifused rat islets.

perifusion (perifiū·ʒən). *Med.* [f. PERI- after *PERFUSION d.] The action or process of perifusing.

1969 *Lancet* 25 Oct. 882/1 We used a perifusion system in which small pieces of pancreas are washed by a continuous flow of buffer. Although, unlike perfusion systems, this method does not utilise an intact vascular bed it has the advantage of permitting simultaneous control and test experiments to be done on the same pancreas. **1974** *Jrnl. Endocrinol.* LXIII. 23P The glands were perifused with KRBGA only for a period of 1 to 2 h and then challenged by 5 min perifusion with hypothalamic extract in KRBGA. **1978** *Nature* 19 Jan. 272/2 A perifusion system was used to monitor continuously the dynamics of $^{42}K^+$ efflux from preloaded isolated rat islets.

perigee. Add: **1.** Also used with reference to artificial satellites. (Further examples.)

1962 J. GLENN in *Into Orbit* 142 We planned for an apogee, or high point, of about 145 miles and a perigee or low point, of about 85 miles. **1966** *Electronics* 3 Oct. 179 It will orbit out to an apogee of 138,000 miles for the interplanetary readings and then dip back to a perigee of 120 miles.

periglacial (periglēˡ·siăl, -ʃăl, -ʃiăl), *a. Geomorphol.* [ad. G. *periglazial* (W. Łoziński 1909, in *Bull. internat. de l'Acad. des Sci. de Cracovie: Classe des Sci. math. et nat.* I. 16): see PERI-[1] and GLACIAL *a.*] Characteristic of or being a region where the influence of an adjacent ice sheet or glacier, or of frost action, is important in forming or modifying the landscape.

1928 *Amer. Jrnl. Sci.* XVI. 163 The complete realization that continental glaciation implies that a zone of periglacial climate borders the ice should lead to interesting and fruitful studies of the zone immediately south of our extensive glacial boundary. **1936** *Proc. Prehist. Soc.* II. 61 Although the Thames Valley was not actually invaded by ice..it was subject to peri-glacial phenomena. **1954** *Sci. News* XXXIII. 67 The processes of earth sculpture operating in periglacial conditions, perhaps a considerable distance from the ice-front, produce deposits and landforms of a special type. **1967** *Jrnl. Glaciol.* VI. 551 The limited earlier work in Southern Africa suggested that the Pleistocene climate was too dry for glaciation to have occurred, but the existence of oversteepened slopes, solifluction slumps and cirques is indicative of a periglacial environment. **1973** *Boreas* II. 9 Bonafide 'periglacial' forms and deposits of late Pleistocene age are present in the Drakensberg and adjacent parts of the northeastern Cape Province.

Hence **perigla·cially** *adv.*, in or by a periglacial environment.

1941 *Trans. R. Soc. Edin.* LX. 406 The Thames lay periglacially to the ice-sheets of East Anglia. **1962** *Proc. Yorkshire Geol. Soc.* XXXIII. 336 Dartmoor..preserves the finest set of periglacially formed tors in Britain. **1972** *Trans. Norfolk & Norwich Naturalists' Soc.* XXII. 229 (*heading*) Periglacially modified chalk and chalk ridge-diapirs from Norwich, Norfolk.

pe:riglacia·tion. *Geomorphol.* [f. PERI- + GLACIATION, after prec.] The state of being subject to a periglacial climate; periglacial processes collectively.

1957 M. T. TE PUNGA in *Tijdschr. Koninklijk Nederlandsch. Aardijkskundig Genootschap* LXXIV. 408 Wind-worked pebbles and boulders..are well known as relics of periglaciation in many parts of the world. **1963** *Geomorphol. Abstr.* 38 (*heading*) Periglaciation of Canada. **1968** *Geogr. Abstr.* A. 30 The scarp..was formed during the Quaternary when the region underwent severe periglaciation with immense frost riving and modification of plateau valleys. **1970** R. J. SMALL *Study of Landforms* i. 2 Our knowledge of the vital role of periglaciation in shaping the landscapes of present-day temperate areas is based almost wholly on analysis by geomorphologists of the distribution and character of solifluxion gravels, 'head', 'coombe rock' and the like.

Périgordian (perigǭᵊ·ɹdiăn), *a. Archæol.* Also **Perigordian.** [tr. F. *Périgordien* (D. Peyrony 1933, in *Bull. Soc. Préhist. Française* XXX. 558), f. the place-name *Périgord* (see PÉRIGORD) + -IAN.] A name given to an Upper Palaeolithic culture represented by flint tools of the kind found at Laugerie-Haute in the Périgord region of Dordogne (see quot. 1938). Also *absol.* as *sb.*

1938 *Proc. Prehist. Soc.* IV. 4 Peyrony..has found at Laugerie Haute an industry of blunted-back blades... He concludes that in the Chatelperron–Laugerie Haute–La Gravette succession we are dealing with a culture totally different from the so-called Middle Aurignacian,..and he proposes to group all those industries characterised by the blunted-back blade under the title of Perigordian. **1941** *Ann. Reg. 1940* 342 A remarkable painted cave of Perigordian art was found at Lascaux in the Dordogne. **1949** M. C. BURKITT *Old Stone Age* (ed. 2) iv. 64 The *Audi* and *Châtelperron* knife blades form an evolving series characteristic of the Périgordian element in the early Aurignacian. **1956** H. READ *Art of Sculpture* ii. 34 There can be little doubt that the so-called *Venus* of Laussel attributed by the Abbé Breuil to the Périgordian period, is such a cult object. **1970** BRAY & TRUMP *Dict. Archaeol.* 174/1 *Périgordian*, French terminology for a series of Upper Palaeolithic flint industries which are thought to represent a continuing technological tradition. *Ibid.* 174/2 No known site has a complete and unbroken 'Périgordian' sequence. **1977** D. K. BHATTACHARYA *Palaeolithic Europe* iv. 168 In 1933, Dr. Peyrony declared that what was earlier thought to be lower and upper Aurignacian, is in reality a tradition different from Aurignacian and he named this Perigordian.

Périgordine. Add: **B.** *adj.* Of or pertaining to Périgord, esp. to the gastronomic specialities of the region.

1931 A. DE CROZE *What to eat & drink in France* 230 Sanguète périgourdine..blood of chicken browned in goose dripping with garlic, shallots, and verjuice. **1951** E. DAVID *French Country Cooking* 192 A kind of soufflé eaten cold, a Périgordine speciality. **1959** *Times* 7 Feb. 9/1 The food..tends to be lighter, though quite as delicious as the famed Périgourdine fare farther west. **1973** *Times* 22 Mar. 15/8 (Advt.), Lovely Perigourdine house XVIIIth century.

perigraphe. Restrict *Obs.* to sense a and add: **b.** (Later example in *poet.* use.)

1956 'H. MACDIARMID' *Stony Limits* 38 We turn in vain This way and that and but changing perigraphs gain.

Périgueux (pe·rigǒ). *Cookery.* The name of a city in the Périgord region of south-west France used *attrib.* to designate a type of sauce made with truffles.

1846 A. SOYER *Gastronomic Regenerator* 25 Sauce Perigeux. Put four middling-sized truffles..into a stewpan with a glass of sherry. **1877** E. S. DALLAS *Kettner's Bk. of Table* 340 Perigueux sauce..is the best brown sauce, with a glass of sherry or Marsala added to it, and a quantity of chopped truffles. **1937** X. M. BOULESTIN *Finer Cooking* II. 84 For the Sauce Périgueux.—Chop the trimmings of the truffles finely and cook them in a small saucepan with a glass of Madeira. **1956** C. BROWN et al. *Four-in-One Bk. Continental Cookery* 337 Dish up, and serve with Périgueux Sauce. **1972** G. A. BROWNE *11 Harrowhouse St.* (1973) v. 51 Next came Tournedos Rossini. Filets of beef..covered with Périgueux sauce.

perikaryon (perikæ·riǭn). *Anat.* Pl. **-karya.** [f. PERI- + Gr. κάρυον nut, kernel.] The cell-body of a neurone; that part of a nerve cell which contains the nucleus.

1897 M. FOSTER *Text Bk. Physiol.* (ed. 7) III. i. 928 It will be convenient to distinguish by a separate name between the processes whether axon or dendrite, and the part from which these processes start, namely the body of the cell surrounding the nucleus; the latter might be called the perikaryon. **1961** *New Scientist* 12 Oct. 117/3 In the invertebrates the [nerve] fibre is not so long and does not usually end on a blood vessel; the perikaryon instead provides more or less all the available storage space. **1975** *Nature* 25 Dec. 746/2 Brain tissue..after this treatment exhibits only weak nonspecific fluorescence with some bright yellow autofluorescent granules in large neuronal perikarya. **1977** P. B. & J. S. MEDAWAR *Life Sci.* xv. 122 The most obviously cell-like part of the neurone is called the cell-body or perikaryon, and this houses the nucleus.

peril, *sb.* Add: **5. peril point** *U.S. Econ.* (see quot. 1965).

1948 *Congress. Rec.* 26 May 6503/2 No foreign trade agreement could be entered into until the Tariff Commission reports to the President its findings as to the so-called peril-point below which tariffs may not be cut. **1949** *Sun* (Baltimore) 11 July 10/2 The main innovation in the Republican program is the so-called 'peril-point' report which must be made to the President by the Tariff Commission. **1949** *Economist* 17 Sept., *Peril Points.* This year's battle over American tariff policy opened just as the Administration was assuring Sir Stafford Cripps and Mr Bevin that the United States would pursue policies appropriate to a great creditor nation. **1961** *Ibid.* 9 Dec. 1025/2 The President's authority to lower tariffs being renewed grudgingly but limited by 'peril points' and 'escape clauses'. **1965** *McGraw-Hill Dict. Mod. Econ.* 376 *Peril point*, the maximum cut in a U.S. import duty which could be made for a given commodity without causing serious injury to domestic producers or to a similar commodity.

perilune (perilū·n). *Astr.* [f. PERI- + L. *lūna* moon; cf. PERIGEE, PERIHELION, etc.] That point at which a spacecraft in lunar orbit is closest to the moon's centre: applied esp. if the spacecraft was launched from the moon.

1960 *Aeroplane* XCIX. 638/2 Lunar gravity should draw the probe into an orbit which has a period of about 10 hr., an apolune (farthest distance from the Moon) of approximately 3,000 miles and a perilune (closest distance) of 1,500 miles. **1968** *Sci. Jrnl.* Oct. 5/2 The orbital velocity at perilune—altitude 100 km, 2°N latitude—was about 7200 km/hour relative to the Moon. **1969** *Nature* 12 July 129/2 Choosing the major axis of the orbit to pass through the mascon..is equally effective in nullifying the orbital instability represented by cumulatively increasing displacements of perilune and apolune.

perimenopausal (pe:rimenǒpǭ·zăl), *a. Med.* [f. PERI- + *MENOPAUSAL *a.*] Occurring at around the time of the menopause.

1961 J. K. FROST in Novak & Jones *Textbk. Gynecol.* (ed. 6) xxxvi. 774 Except postpartum, the parabasal type cell does not exfoliate normally until the later years of the reproductive period, or the perimenopausal period. **1961** *Obstetr. & Gynecol.* XVII. 331/1 Many of our colleagues use ovarian hormones, especially estrogens, for the control of the perimenopausal syndrome. **1970** TE LINDE & MATTINGLY *Oper. Gynecol.* (ed. 4) xxxvi. 795/2 With the current liberal use of estrogen and progesterones in the perimenopausal age group for pregnancy protection and menopausal symptoms, the physician must be particularly alert to the camouflaged symptoms of endometrial carcinoma in patients using steroid hormones. **1977** *Lancet* 8 Oct. 762/2 There was a 6% (P < 0·001) excess of left-sided tumours, primarily at perimenopausal and postmenopausal ages.

So **perime·nopause,** the perimenopausal period.

1962 J. K. FROST in Novak & Woodruff *Gynecol. & Obstetr. Path.* (ed. 5) xxxv. 608/2 Following the initial period of endocrine adjustment in the establishment of menses (perimenarche) and extending into the disruption of endocrine interplay (perimenopause), the hormonal pattern varies widely within each lunar cycle. **1969** *Obstetr. & Gynecol.* XXXIII. 581/2 More than 120 women ..were seen initially for, and found to have problems relating to, the perimenopause or menopause.

perimeter. Add: **1. c.** *Mil.* A defended boundary of a troop position. Also *transf.*, the boundary of an airfield or civil airport.

1943 J. DUFFY *Australians in Malaya* 41 Possibly this was true of an attack from the sea, but on the northern perimeter were no adequate defences. **1958** *Spectator* 11 July 52/3 A can tell you what really happened in the fighting round the perimeter. **1967** *Boston Herald* 1 Apr. 14/6 The enemy opened up with mortar and howitzer fire, hitting American troops who had pulled back into defensive perimeters. **1974** *Guardian* 1 Aug. 1/7 Soldiers.. moved on to the airport perimeter. **1977** *Times* 9 Sept. 15/7 After six hours without food in a plane on the perimeter at Heathrow the flight was cancelled.

3. *attrib.* and *comb.*, as (sense *1c) *perimeter fence, road*; **perimeter track,** a runway round an airfield. See also sense 2 in Dict.

1974 A. PRICE *Other Paths to Glory* III. i. 249 Soldiers.. patrolling the perimeter fence. **1976** *Scottish Daily Express* 27 Dec. 1/2 Two wolves were shot dead only yards from the perimeter fence by zoo director Roger Wheater. **1974** *Times* 7 Jan. 1/4 There were five road blocks of troops and police on the airport perimeter road where it passes close to runways. **1946** *Happy Landings* (Air Ministry) July 10/1 The Spitfire was taxying out for take-off, along the perimeter track. **1959** J. L. NAYLER *Dict. Aeronaut. Engin.* 187 *Perimeter track*, a taxi track round the perimeter of an airfield for the use of aircraft and motor vehicles.

perimetral, *a.* (Later example.)

1971 J. NEEDHAM *Sci. & Civilisation in China* IV. III. xxviii. 94 (*caption*) Diagrams to elucidate Chinese and Western building construction... Comparison of the fundamental Chinese building design with that of Greek and Gothic building... Normally the Greek gable covered the perimetral colonnade.

perinatal (perinēˡ·tăl), *a. Med.* [f. PERI- + NATAL *a.*[1] and *sb.*] Of or pertaining to the period comprising the latter part of fœtal life and the early postnatal period (commonly taken as ending either one week or four weeks after birth: see quots.).

1952 *Amer. Jrnl. Publ. Health* XLII. 505/2 Such conditions as birth injury, congenital malformation, cerebral palsy, and epilepsy..may possibly be traceable to perinatal causes. **1958** *Economist* 22 Feb. 654/2 The phrase perinatal mortality has been coined to describe the combined death rate of babies born dead and those who die in the first week of life. **1966** *Ann. Rev. Med.* XVII. 213 The perinatal period extends from the completion of embryonic differentiation into recognizable organ structures to the end of the first month of postnatal extrauterine existence. **1973** *Where* Apr. 102/3 If they smoke after the third month of their pregnancy there is an excess of peri-natal deaths of about 30 per cent. **1976** *Lancet* 30 Oct. 941/1 The French Government aimed to reduce the burden of handicap caused by conditions arising in the perinatal period (defined as the period from 28th week of gestation to the end of the first week of life). In 1968 40 000 French children who had survived the first year of life were judged to have a handicap of perinatal origin. **1977** *Ibid.* 25 June 1357/2 The W.H.O. recommendation reads: 'It is recommended that national perinatal statistics should include all fetuses and infants delivered weighing at least 500 g (or, when birthweight is unavailable, the corresponding gestational age (22 weeks) or body length (25 cm crown-heel)), whether alive or dead.

period, *sb.* Add: **I. 3. b.** Also *sing.* and *attrib.*

1891 I. ELLIS *Essentials of Conception* 28 (Advt.), Ladies' 'period' towels. **1922** JOYCE *Ulysses* 368 Some women for instance warn you off when they have their period. **1939** M. SPRING RICE *Working-Class Wives* vi. 145 She suffers from 'period pains'. **1953** H. MILLER *Plexus* (1963) iii. 112 Between times I wondered what was eating her. Maybe her period coming on. **1956** R. M. LESTER *Towards Hereafter* x. 123 In January 1952 I missed a period and with high hopes I went to see my G.P. and told him the glad news. **1970** G. GREER *Female Eunuch* 52 The genteel..'I've got my period'. **1976** W. H. CANAWAY *Willow-Pattern War* xv. 156 She'd cried off at the last minute with a period pain.

4. c. *spec.* (*a*) a portion of an artist's life characterized by a particular style; (*b*) freq. with poss. adj.: the particular historical or cultural portion of time with which one is concerned.

1891 O. WILDE *Pict. Dorian Gray* xix. 319 What has become of that wonderful portrait he did of you?.. It belonged to Basil's best period. **1921** W. S. MAUGHAM *Circle* I. 13, I want you to look at this chair I've just got. .. About 1750, I should say... It's exactly my period. **1925** R. FRY *Let.* 1 May (1972) II. 568 Lady Cunard.. wanted a Picasso of the blue period. **1952** 'W. COOPER' *Struggles of Albert Woods* II. iv. 100, I hardly know Picasso's rose period from his blue. **1958** M. KELLY *Christmas Egg* I. 29 He had been at a loss for the date of the battle of Agincourt, and had excused himself with the plea that it was outside his period. *a* **1966** M. ALLINGHAM *Cargo of Eagles* (1968) i. 16 I'm certain there's an early

fortress..just waiting to be uncovered... That's not my period and so is not my province. **1973** M. MACKINTOSH *King & Two Queens* ii. 27 The Picasso Museum..has very good examples of his pink and blue periods? **1978** J. HANSEN *Man Everybody was Afraid Of* xii. 93 She handed Dave one of the mugs. 'From my potting period.'

d. (Earlier and later examples.) Also *out of period*: anachronistic.

1859 A. J. MUNBY *Diary* 19 July in D. Hudson *Munby* (1972) 39 An Englishman of the period, smoking a cigar; his dress is 'civilized'—he wears gloves. **1902** G. B. SHAW *Mrs. Warren's Profession* Pref. p. xv, Both plots conform to the strictest rules of the period. **1933** E. O'NEILL *Ah, Wilderness!* I. 15 Scene—Sitting room of the Miller home in a large small-town in Connecticut—about 7:30 in the morning of July 4th, 1906. The room is..furnished with the scrupulous medium-priced tastelessness of the period. **1961** C. WILLOCK *Death in Covert* iii. 66 A serving-wench.. asked him: 'Sack, mulled claret, or Madeira?' Mr Goss felt that two at least of these were out of period. **1969** Y. CARTER *Mr. Campion's Farthing* ii. 15, I hate getting out of period but sometimes one has to be modern. **1976** *Listener* 20 May 647/4 Wesley..had the imaginative and technical powers to transcend the humdrum idiom of so much church music of the period without, at the same time, discarding its essential nature.

e. *Educ.* A portion of time set aside for a lesson or other activity; cf. *free period* s.v. *FREE *a.* D. 2.

1876 C. M. YONGE *Womankind* xiii. 92 Most people's breakfast hour coincides with this only period permitted [in National Schools] for religious teaching. **1930** *Times Educ. Suppl.* 18 Jan. 21/2 The pupil has five periods a week..for French. **1948** 'N. SHUTE' *No Highway* vi. 167 'What about the school?' 'I've only got one period tomorrow.' **1955** E. BLISHEN *Roaring Boys* I. 30, I had an odd period of history with one of the first-year classes. **1966** J. PARTRIDGE *Middle School* i. 24 Nearly every class has a 'library period'. **1974** H. L. FOSTER *Ribbin'* I. 8 Hey, Teach, we work a period, read comics a period, and then take off the last period—OK?

f. One of the intervals into which the playing time of a sporting fixture is divided.

1898 *Encycl. Sport* II. 128/2 The duration of play in a match shall be one hour, divided into three periods of twenty minutes. **1935** *Encycl. Sports* 359/2 Two time-keepers..inform the referee..that the end of each period or rest has arrived. **1968** *Globe & Mail* (Toronto) 5 Feb. 17/2 Pit Martin also scored for a three-goal first period for Chicago and Wayne Maki tallied early in the second. **1974** *State* (Columbia, S. Carolina) 3 Mar. 1-D/1 The Paladins opened 11-point leads on three occasions in the final period.

III. 11. b. Also added to a statement to emphasize a place where there is or should be a full stop, freq. (*colloq.*) with the implication 'and that is all there is to say about it', 'and it is as simple as that'.

1934 J. O'HARA *Appointment in Samarra* (1935) viii. 248 'An unscrupulous woman can make a man—' 'Period.' **1946** *Sun* (Baltimore) 2 Oct. 8 (Advt.), A cigarette is supposed to give you *pleasure*. Period. **1947** *Mind* LVI. 65 The empirical evidence suggests the generalisation and supports it. If it does, it does. Period. **1948** H. LAWRENCE *Death of Doll* i. 21 'Lucky Monny to have her own pocket.' 'Stop that. Lucky Monny, period.' **1951** C. ARMSTRONG *Black-Eyed Stranger* (1952) xvii. 150, I don't want to think you are a romantic young thing, period. **1956** J. L. AUSTIN in *Proc. Brit. Acad.* XLII. 113 It does *not* follow either that 'I panted whether or not I ran' or that 'I panted' period. **1958** RICE & 'MCBAIN' *April Robin Murders* (1959) xxii. 245 But Browne doesn't care... He wants the money, period. **1960** 'M. CRONIN' *Begin with Gun* xii. 141 'You know how nosy I am about unsolved crimes.' 'Nosy. Period.' **1964** V. NABOKOV *Defence* x. 156, I can't abandon him. And I won't. Period. **1972** *Science* 12 May 638/1 Don't know, but we have exceeded it: 7. Don't know, period: 10. **1974** H. L. FOSTER *Ribbin'* vi. 285 It is wrong for any teacher to have an affair with a student, period. **1976** *Shooting Times & Country Mag.* 16–22 Dec. 18/1 So far as the Spey was concerned this year, however, the fish did not arrive—period! **1977** *Language* LIII. 409 If this is the view R got from 'On generative semantics', he is illiterate, period.

15. *Chem.* A horizontal row in the periodic table of the elements; the set of elements occupying such a row, usu. comprising an alkali metal and those elements of greater atomic number up to and including the next noble gas. Cf. *GROUP *sb.* 3c (ii).

1879 *Chem. News* 5 Dec. 268/1 In the first [table] the elements are placed in large periods, with their atomic weights. In the second they are arranged in groups and series, that is to say, in small periods, in such a manner that the differences between the odd and even series become very apparent. *Ibid.* 19 Dec. 291/1 We see..that the members at the beginning of the large periods (as well as the small periods commencing with Na and Li) are metals of a very strongly pronounced alkaline nature. **1946** [see *LANTHANIDE 1]. **1965** PHILLIPS & WILLIAMS *Inorg. Chem.* I. ii. 40 (*caption*) The series of elements from Li to Ne will be referred to as the first row or the first short period, and similarly the series Rb to Xe as the fourth row or the second long period. **1974** D. M. ADAMS *Inorg. Solids* ii. 37 Across each period of the Periodic Table, ionic radii decrease with increasing charge and atomic number.

IV. 16. *attrib.* or as *adj.* in sense 'belonging to, characteristic, imitative, or representative of, a particular (past) period' esp. in style or design in architecture, dress, furniture, literature, etc.

Freq. in inexact or euphemistic use.

1905 (*title*) Borgia: a period play. **1906** G. KOBBÉ *How to appreciate Music* 47 A pianoforte has no business in a 'period' room. If the person is rich enough to afford 'period' rooms, he also can afford a music room. **1908** *Westm. Gaz.* 17 Dec. 4/1 Some of them..may be said to be striving to create a 'period' type of carriage for themselves. **1914** EBERLEIN & MCCLURE (*title*) The practical book of period furniture. **1920** W. R. LETHABY in *Form in Civilization* (1922) 12 That which now professes to be designed in a style, or, as the still more disgusting slang runs, to be 'period work', has not the essence of life. **1925** in F. Madan *Oxford outside Guide-Bks.* (ed. 2) 202 (Advt.), Write for new illustrated Catalogue containing fullest particulars of Minty Bookcases, including various 'period' styles at moderate prices. **1927** *Times* 28 Oct. 17/3 The bride..wore a period gown of cream chiffon velvet, trimmed with seed pearls. **1928** *Daily Sketch* 2 Aug. 15 (*caption*) 'Period' pages... The little pages wore replicas of the old-time uniform of the Queen's Bays. **1931** *Times Lit. Suppl.* 26 Mar. 249/2 The contrast between these two volumes makes an interesting study. Neither is 'period-printing'; but the Chapman is properly 'spacious' for an Elizabethan work; and the Pope [etc.]. **1935** N. MITCHISON *We have been Warned* II. 154 'You'll be saying you like the "Idylls of the King" next.' 'Oh, but I do. They're so deliciously period.' **1937** D. L. SAYERS *Busman's Honeymoon* xviii. 346, I was going to.. have the spit turned by electricity. And an electric cooker for the days when we didn't feel so period. **1940** L. MACNEICE *Last Ditch* 18 Cranks, hacks, poverty-stricken scholars, In pince-nez, period hats or romantic beards. **1958** B. NICHOLS *Sweet & Twenties* 79 The word 'treasure' is charmingly period. In the upper classes it implied the perfect Jeeves or the ideal nanny. **1960** R. A. KNOX *Occasional Sermons* xl. 331 It is all quite convincing, and beautifully period. Why is it so period? **1965** *Listener* 25 Nov. 869/3 A series of Edith Wharton's novels are being reprinted..; great period interest, and well worth re-examination. **1967** E. SHORT *Embroidery & Fabric Collage* iii. 74 The heavy curtains associated with the period four-poster. **1974** *Listener* 10 Jan. 59/3 *Whose body?*..one of Dorothy Sayers's Lord Peter Wimsey stories..is period stuff, thick with now discarded or at least unfashionable snobberies. **1976** *Liverpool Echo* 24 Nov. 5/4 The undeniable pleasure of watching sport, natural history, travel, ballet, period dramas and even the news is taken for granted. **1977** *N.Y. Rev. Bks.* 29 Sept. 12/4 Baum was a handsome young man with gray eyes, straight nose, dark brown hair, and a period mustache that looked to be glued on. **1977** *Radio Times* 12 Nov. 4/3 It was great fun but it was 'period' and we wanted to get back to the present.

b. Special Combs., as **period–luminosity** *a. Astr.*, relating the period of a variable star, esp. a Cepheid, to its luminosity; **period-piece,** a work of art, furniture, literature, etc., considered from the aspect of its associations with or evocativeness of a past period of time; contemptuously, such a work possessing interest only from such associations or evocativeness.

1918 H. SHAPLEY in *Contrib. Mt. Wilson Solar Observatory* No. 153. 2 For parallaxes obtained with the period-luminosity curve the accuracy appears to surpass that of direct measures on any object for which the parallax is less than 0″·01, and is essentially independent of distance. **1950** *Sci. News* XV. 46 We can..get from the [light] curve the mean apparent magnitude and the period. From the period-luminosity relation we then find the absolute magnitude, and so, knowing both *m* and M we can find the distance. **1964** *Listener* 21 May 831/2 It was by studying the short-period variables in the Small Cloud, fifty years ago, that Miss Henrietta Leavitt, at Harvard..made the discoveries that led on to the 'period-luminosity law' of Cepheid stars. **1927** S. ERTZ *Now East, Now West* iii. 32 She saw she would have the pleasure of buying certain things—period pieces—that she could either sell at the end of their stay or take back to America. **1931** *Times Lit. Suppl.* 24 Dec. 1033/2 Spenser supported the remote splendour of his mythical heroes with deliberately archaic language, turning his poem into a period piece. **1934** C. LAMBERT *Music Ho!* III. 172 The English folk song..is nothing more than a very pretty period piece. **1943** H. PEARSON *Conan Doyle* v. 81 Nowadays we can see that the facile saga of Sherlock Holmes is far more valuable even as a 'period piece' than the diligent epic of Edward the Third. **1957** *Essays & Stud.* X. 54 But I am here concerned with Robert Elsmere not primarily as a work of art but as a symbol or symptom of a certain phase of liberal religious thought in the later nineteenth century—as a 'period piece' if you will, only that I prefer not to use a phrase which might suggest (wrongly, as I think) that the book is a mere antique. **1961** L. MUMFORD *City in Hist.* xiii. 408 Washington..might have been a miracle of the solo town planner's art: a final period-piece to close the epoch. **1972** S. HYNES *Edwardian Occasions* xv. 188 The essays..are belles-lettres of the most inoffensive kind..period pieces even when [Maurice] Hewlett wrote them. **1975** M. DRABBLE *Realms of Gold* IV. 269 Mays Cottage was a period piece, completely unrestored, which in these days seemed to be an asset.

periodate, per-iodate. Add: *periodic acid–Schiff* (Biol.), phr. used *attrib.* and *absol.* to designate a procedure for the detection of carbohydrates by first oxidizing them to polyaldehydes with periodic acid and then staining with Schiff's reagent.

1947 *Jrnl. Laboratory & Clin. Med.* XXXII. 911 McManus reported the use of a periodic acid-Schiff technique for the demonstration of mucin. **1956** *Nature* 3 Mar. 432/2 Secretory inclusions which are positive to the periodic-acid-Schiff test are conspicuous in the non-ciliated iodine-binding cells of the endostyle of *Ciona*. **1960** E. GURR *Encycl. Microsc. Stains* I. 274 The author describes..results obtained with silver staining and periodic acid-Schiff. **1974** H. C. COOK *Man. Histol. Demonstr. Techniques* i. 16 They [*sc.* basement membranes] may be demonstrated..by the periodic acid-Schiff (PAS) technique and a variant of this, the Allochrome method.

periodic, *a.*[1] Add: **2.** (Earlier example in *Math.*) *periodic classification* or *system* (Chem.): an arrangement or classification of the chemical elements according to the periodic law; *periodic law* (Chem.): add to def.: now recognized to be a function of atomic number rather than of atomic weight, thus removing certain discrepancies in the original scheme; (examples); *periodic table* (Chem.): a table of the elements arranged according to the periodic law; *spec.* one in which they are arranged in order of atomic number, usu. in rows, such that groups of elements possessing analogous electronic structures, and hence exhibiting similar properties, form vertical columns ('groups') of the table.

1850 J. F. W. HERSCHEL in *Phil. Trans. R. Soc.* CXL. 400 If A_x, B_x, &c. be simply constant, the function may be termed a periodic one, since it assumes in periodic and constantly recurring succession the values A, B, C..N, A, B, &c. *ad infinitum*. **1872** *Phil. Mag.* XLIII. 251 The regular progression in physical and chemical properties observable in members of the same family..are either consequences of, or closely related to, these 'periodic laws'. **1875** *Chem. News* 24 Dec. 293/1 The periodic law indicates the gaps which still exist in the system of the known elements, and enables us to predict the properties of the unknown elements. *Ibid.* 294/1 These characters of [gallium]..have been obtained..by considering its place in the periodic system of the elements. **1881** *Ibid.* 14 Jan. 16/1 The criticism of Prof. Wurtz upon the periodic classification. **1895** THOMSON & BLOXAM *Bloxam's Chem.* (ed. 8) 278 The Periodic Table has found a twofold application. **1913** Periodic system [see *atomic number* s.v. *ATOMIC *a.* 1]. **1919** *Jrnl. Chem. Soc.* CXV. 11 The occupant of a separate place in the periodic table of elements. **1930** *Engineering* 21 Mar. 372/2 The sequence of the elements in the old periodic table with its eight groups and seven periods. **1957** Periodic table [see *LANTHANIDE 1]. **1969** J. W. VAN SPRONSEN (*title*) The periodic system of chemical elements. **1974** *Encycl. Brit. Macropædia* IV. 116/2 The periodic classification places elements with similar electron arrangements in vertical columns. **1974** GOLDBERG & DILLARD *College Chem.* ii. 38 The periodic law..states that the properties of the elements are periodic functions of their atomic numbers.

4. (Further example.)

1930 H. G. WELLS *Autocracy of Mr. Parham* II. iv. 128 The need of a stronger and clearer guidance in our periodic literature.

6. as *sb. pl.* = *PERIODICAL *sb.* 3.

1920 C. E. MULFORD *Johnny Nelson* xvii. 222 That's th' worst of them periodics! You can't never tell when they'll start.

periodical, *sb.* Add: **1.** Also *attrib.*

1878 *Harper's Mag.* Jan. 192 He used to look into the windows of the periodical stores. **1910** A. E. BOSTWICK *Amer. Publ. Library* 282 In some New York branches periodical reading rooms may be used as assembly rooms. **1938** L. M. HARROD *Librarians' Gloss.* 99 *Magazine case*, a cover for periodicals, usually having some contrivance for holding the magazine-cord, rod, etc. Also called 'Periodical case' and 'Reading case'. **1961** T. LANDAU *Encycl. Librarianship* (ed. 2) 275/2 *Periodical stack*, a stack constructed to display periodicals, with space on lower shelves for storage of back numbers.

3. *pl. U.S. slang.* Recurring drinking bouts or sprees.

1890 in BARRÈRE & LELAND *Dict. Slang* II. 124/1 Are you in the book business?.. Ma and pa were talking last night about your having your little periodicals. **1902** H. L. WILSON *Spenders* x. 107 They telegraphed the Butte National to wire his description, and the answer was 'tall and drunk'. Well, son, his periodicals wa'n't all.

periodicity. Add: **1. b.** *Chem.* The complex periodic variation of the properties of the chemical elements with increasing atomic number.

1879 ROSCOE & SCHORLEMMER *Treat. Chem.* II. II. 506 The law of periodicity was afterwards further developed by Meyer and Mendelejeff. **1907** *Westm. Gaz.* 4 Feb. 6/3 Gallium, scandium, and germanium, all subsequently discovered, did fit into the scale of 'periodicity'. **1969** A. J. IHDE in J. W. Van Spronsen *Periodic Syst. Chem. Elem.* p. x, The periodic table reached its final forms before atomic structure revealed the basis for periodicity.

c. The frequency of a periodic phenomenon, esp. an alternating current.

1900 *Jrnl. Soc. Arts* XLVIII. 848/2 The other carrying about 20 amperes with a periodicity of 60 cycles per second. **1913** *Chambers's Jrnl.* Jan. 102/1 The electricity used for wireless telegraphy is..used under different conditions as to pressure and periodicity. **1938** *Times* 13 Oct. 8/3 The periodicities of the front and back springs are arranged at a predetermined variance so that coincidence should not occur. **1943** *Gloss. Terms Electr. Engin.* (*B.S.I.*) 11 *Frequency* (*periodicity*, deprecated), the number of cycles per second. The reciprocal of the period.

periodiza·tion. [f. PERIODIZE *v.* + -ATION.] Division into periods of time; *spec.* the grouping of historical and cultural events in chronological periods (see PERIOD *sb.* 4c in Dict. and Suppl.) for the purposes of discussion and evaluation.

1938 [see *CONCEPT *sb.* 2 b]. **1952** K. R. POPPER *Open Society* (ed. 2) II. xii. 59 No doubt, his [*sc.* Hegel's] vast historicist generalizations, periodizations, and interpretations fascinated some historians. **1957** K. A. WITTFOGEL *Oriental Despotism* ix. 395 This periodization appeared again in an article in 1916. **1963** R. M. GRANT *Hist. Introd. New Testament* i. 14 The question of periodization arose in the second century... Generally speaking, historians have differentiated three periods. **1967** L. DEUEL *Conquistadors without Swords* xxi. 285 Uaxactún yielded a continuous series of pottery which made it possible to establish a complete overall Maya stratigraphy and periodization. **1970** B. BREWSTER tr. *Althusser & Balibar's Reading Capital* (1975) II. iv. 94 On this level, then, the whole problem of the science of history would consist of the division of this continuum according to a *periodization* corresponding to the succession of one dialectal totality after another. **1973** *Times Lit. Suppl.* 16 Nov. 1386/1 Accepting the traditional periodization of world history according to the days of creation. **1974** J. WHITE tr. *Poulantzas's Fascism & Dictatorship* IV. 233 The general line which was progressively dominant in the USSR and in the Comintern can allow us to make a relatively clear periodization of the Comintern, a periodization which can also be very useful for the history of the USSR. **1976** *Brit. Jrnl. Sociol.* XXVII. 301 In as far as sociology defines its object of investigation as 'society', 'the social system' or some such general unspecified synonym, it denies itself any rigorous principle of historical periodization.

periodize, *v.* Restrict †*Obs.* to sense in Dict. and add: **2.** To divide (a portion of time) into periods; to assign (historical and cultural events) to specified periods. Cf. prec. So (*rare*) **pe·riodizer,** one who periodizes in this way; **pe·riodizing** *vbl. sb.*

a **1943** R. G. COLLINGWOOD *Idea of Hist.* (1946) II. 54, I will take a single example of medieval periodizing. In the twelfth century Joachim of Floris divided history into three periods. **1959** *Listener* 20 Aug. 291/2 The fifteenth century has been a favourite hunting ground for the periodizers of history. **1965** K. CHARLTON *Educ. Renaissance Eng.* ii. 40 The dangers of periodizing history and of ignoring the carry-over of medieval ideas alongside and within humanistic thought have already been mentioned. **1970** B. BREWSTER tr. *Althusser & Balibar's Reading Capital* (1975) II. iv. 103 This is antipodal to the empirically visible history in which the time of all histories is the simple time of continuity and in which the 'content' is the vacuity of events that occur in it which one later tries to determine with dividing procedures in order to 'periodize' that continuity. **1972** *Language* XLVIII. 423 Lyons' manner of 'periodizing' Chomsky's intellectual history has its justification, but a somewhat different distribution of emphasis might be fairer to the histories of both linguistics and rationalism. **1973** *Sci. Amer.* Sept. 194/2 The bulk of the text summarizes and criticizes the theories for Bode's law in the context of an overall view of the origin of the solar system, an event that Nieto broadly periodizes along the lines of Sir Fred Hoyle's nebular-plus-magnetic theory.

periodogram (piᵊriǭ·dǒgræm). [f. PERIOD *sb.* + -O + -GRAM.] A diagram or method of graphical representation which is designed to detect or display any periodicity (usu. with time) in a set of measurements of a quantity; *spec.* one in which the results of harmonic analysis of the data, performed on the assumption in turn of different periods of variation, are plotted as a function of the period. Freq. *attrib.*, as **periodogram analysis,** the analysis of data by means of a periodogram.

1898 A. SCHUSTER in *Terrestr. Magn.* III. 24 It is convenient to have a word for some representation of a variable quantity which shall correspond to the 'spectrum' of a luminous radiation. I propose the word *periodogram*. *Ibid.* 25 The periodogram of the sound emitted by an organ pipe or a violin string consists of a series of equidistant 'lines'. **1906** *Proc. R. Soc.* A. LXXVII. 141 The periodogram..is the diagram representing the intensity of periodic variations as determined from the sum of the squares of the two Fourier coefficients belonging to each assumed period. **1919** *Nature* 26 June 338/1 A periodogram analysis of the Greenwich temperature records. **1939** J. A. SCHUMPETER *Business Cycles* I. iv. 165 This is the most successful application so far made of the periodogram analysis to economic data. **1957** G. E. HUTCHINSON *Treat. Limnol.* I. v. 334 Olson (1950) examined the matter by means of periodogram analysis and found no evidence of a period near 12 hours. **1974** *Nature* 8 Feb. 339/3 The apparently erratic optical fluctuations have been analysed by various investigators using power spectra and the so-called periodogram techniques.

periodograph (piᵊriǭ·dǒgrɑf). [f. PERIOD *sb.* + -O + -GRAPH.] **1.** A periodogram; orig. *spec.* a curve drawn in a periodogram.

1899 A. SCHUSTER in *Trans. Cambr. Philos. Soc.* XVIII. 108 With T as abscissa and S^2 as ordinate, draw a curve, which may be called the 'Periodograph'... It will be seen that the 'Periodograph' corresponds exactly to the curve which represents the distribution of energy in the spectrum. **1936** *Rev. Econ. Statistics* XVIII. 63/2 Like the Fourier periodogram, this periodograph will show distinct peaks in the vicinity of real periodicities. **1950** CONRAD & POLLAK *Methods in Climatol.* (ed. 2) xi. 370 The name 'periodograph' is no longer used and Schuster's 'periodograph' is now called a 'periodogram'.

2. An instrument for automatically making a periodogram analysis or Fourier analysis of a curve by optical means.

1930 G. A. R. FOSTER in *Jrnl. Textile Inst.* XXI. T18 The grating periodograph now described is an instrument which has been..designed for the examination of the irregularities in cotton spinning products... The periodograph is simply a method of carrying out automatically the periodogram analysis..of series of observations for hidden periodicities. **1946** *Suppl. Jrnl. R. Statistical Soc.* VIII. 44 In the correlation periodograph, due to Martindale, the grating of the grating periodograph is replaced by a replica of the curve on a reduced scale. **1973** *Physics Bull.* Mar. 154/2 The 'Periodograph' was never applied to the analysis of electron micrographs until Warren (1972) had the ingenious idea of replacing the grating by a series of point sources such as flashlamp bulbs and interposing a micrograph between the bulbs and the final screen.

periodontia (peri,ǒdǫ·ntiă). *Dentistry.* orig. *U.S.* [f. *periodont(ium* s.v. PERI- 1b + -IA[1].] = *PERIODONTICS.

1914 *Items of Interest* July 429 We..deem it for the best interests of the public and the profession that a society should be formed, to the end that those especially interested may meet and work together..for the scientific investigation of periodontoclasia and caries, and that the practice of oral prophylaxis and periodontia as an exclusive speciality may be encouraged. **1924** *Glasgow Herald* 5 Mar. 9/4 Harold Box..holds the position of Professor of Dental Pathology and Periodontia at the Royal College of Dental Surgeons at Toronto, Canada. **1960** S. SORRIN *Pract. of Periodontia* p. vii/1 The treatment for periodontal disease is of interest not only to the specialist in periodontia but also to the general practitioner of dentistry. **1960** *Times* 27 July 3/2 (Advt.), Applications are invited for the appointment of Head of Department of Periodontia.

Hence **periodo·ntic** *a.* = *periodontal* adj. s.v. PERI- 1a.

1926 R. J. E. SCOTT *Gould's Med. Dict.* 984/1 *Periodontic*, same as periodontal. **1978** *N.Y. Times* 30 Mar. B18/2 (Advt.), Dentist. Experienced for general practice. .. Crown and bridge, periodontic, endodontic, pedodontic.

periodontics (peri,ǒdǫ·ntiks), *sb. pl.* (const. as sing.). *Dentistry.* [f. as prec.: see -IC 2.] The branch of dentistry concerned with periodontal tissue, disorders, etc.

1948 L. I. GROSSMAN *Handbk. Dental Pract.* v. 83/1 Periodontics is that branch of dentistry which deals with the science and treatment of periodontal disease. **1960** S. SORRIN *Pract. of Periodontia* i. 1 The practice of periodontics is based on an understanding and recognition of the healthy periodontium. **1969** BOHANNAN & SAXE in Morris & Bohannan *Dental Specialities* vii. 260/1 Time devoted to periodontics in the traditional dental educational program has been woefully inadequate. **1975** J. E. CHASTEEN *Essent. Clin. Dental Assisting* xi. 200/1 Periodontics is the branch of dentistry that deals with the diagnosis and treatment of diseases which destroy the supporting tissues of the teeth.

periodontist (peri,ǒdǫ·ntist). *Dentistry.* [f. *PERIODONT(IA + -IST.] A specialist or expert in periodontics.

1920 STEDMAN *Med. Dict.* (ed. 6) 755/2 *Periodontist*, a dentist who specializes in periodontia. **1954** *Sydney Morning Herald* 26 Oct. 1/3 The judges were an orthodontist (for straightness), a periodontist (for condition of the gums), and a children's dental specialist. **1963** E. B. WHITE *Let.* 6 May (1976) 500 A periodontist.. proposes to remove a small section of my gum. **1965** J. C. MUHLER *Fifty-Two Pearls* ix. 93 In most cases the family dentist will..refer the patient to a periodontist (a dental specialist who treats diseased gingiva or bone). **1969** BOHANNAN & SAXE in Morris & Bohannan *Dental Specialities* vii. 259/2 There are..not enough practicing periodontists to treat all periodontal disease. **1978** *Detroit Free Press* 16 Apr. 14c/4 (Advt.), A local Don Juan periodontist is murdered.

periodontoclasia (pe:ri,ǒdǫntoklæ·siă). *Dentistry.* [f. as *PERIODONTIA + Gr. κλάσ-ις breaking (f. κλάειν to break) + -IA[1].] Destruction or degeneration of periodontal tissue.

1914 [see *PERIODONTIA]. **1960** S. SORRIN *Pract. of Periodontia* iv. 128/2 Pulpal disease may invade the periodontal space, producing inflammation and periodontoclasia.

periodontology (pe:ri,ǒdǫntǫ·lǒdʒi). *Dentistry.* orig. *U.S.* [f. as *PERIODONTIA + -OLOGY.] = *PERIODONTICS.

1914 *Items of Interest* July 529 (*heading*) A statement to the profession. From the American Academy of Oral Prophylaxis and Periodontology. **1920** *Jrnl. National Dental Assoc.* Feb. 159 (*heading*) Fundamentals of periodontology. **1960** S. SORRIN *Pract. of Periodontia* p. vii, Constant ferment of research and practical application is vital to progress in the art and science of periodontology. **1962** R. BRADLAW in Blake & Trott *Periodontology* p. v, Periodontology has not had the emphasis in dental education that it should have had. **1972** *Science* 2 June 1033/1 Department of Periodontology, Harvard School of Dental Medicine, Boston, Massachusetts.

periodontosis (pe:ri,odǫntōu·sis). *Dentistry.* [f. as *PERIODONTIA + -OSIS.] A periodontal disorder; *spec.* one in which there is a loss of alveolar bone, leading to displacement or loosening of teeth, without inflammation.

1936 DUNNING & DAVENPORT *Dict. Dental Sci.* 422/2 *Periodontosis*, a diseased condition of the periodontal membrane. **1942** ORBAN & WEINMANN in *Jrnl. Periodontology* XIII. 31 (*heading*) Diffuse atrophy of the alveolar bone (periodontosis). **1962** BLAKE & TROTT *Periodontology* iii. 28 Incidence of periodontal associated with alveolar bone destruction and pocket formation. This type of disease has been divided into chronic periodontitis.., which is inflammatory, and periodontosis .., which is principally degenerative. Periodontosis is often complicated by inflammation and the condition is then called periodontitis complex. **1975** J. D. MANSON *Periodontics* (ed. 3) xviii. 222 (*heading*) Juvenile periodontitis ('periodontosis'). *Ibid.*, Orban and Weinmann (1942) coined the term 'periodontosis' believing that the condition represented a degeneration of the periodontal ligament. However, there is no evidence for that belief.

periost (pe·ri,ǫst). *Anat.* [f. mod.L. PERIOST(EUM.] = PERIOSTEUM.

1900 in DORLAND *Med. Dict.* **1902** *Proc. Zool. Soc.* I. 212 The perisclerium is continuous with the periost of the pedicle portion [of the horn]. **1927** HALDANE & HUXLEY *Animal Biol.* ix. 185 The basal joint..was removed and a piece of healthy bone with its bone-forming membrane (periost) grafted in from another situation. **1959** *Jrnl. Exper. Zool.* CXLII. 631 Both the periost and hypertrophied areas show a strong black deposit of cobalt sulphide. **1973** *Biol. Abstr.* LVI. 719/1 In the periost, described were fibroblasts, precursor cells, [etc.].

peripediment (pe·ripediměnt, peripe·diměnt). *Geomorphol.* [f. PERI- + PEDIMENT[1].] A broad, gently sloping surface that is the top of a thickness of detrital alluvium and either extends outwards from a mountain-foot in an arid or semi-arid region or else smoothly continues the line of an intervening pediment.

1942 A. D. HOWARD in *Jrnl. Geomorphol.* V. 11 The term peripediment is suggested for a pediplane which levels an earlier basin fill. If both elements of the pediplane are present, the peripediment is always beyond and peripheral to the pediment. **1970** R. J. SMALL *Study of Landforms* ix. 308 The smaller and steeper bajadas tend to mask only the upper part of the rock pediment, burying the piedmont angle, but the larger and more gently sloping fans may be so extensive as to grade into the alluvial deposits of the peripediment. **1975** *Nature* 7 Aug. 468/1 This gravel..seems to have been deposited as a series of large confluent fans derived from the escarpment, which form an extensive..peripediment.

peripeteia, peripety. Add: (Further examples.) Also, according to the theory of Jung, the third stage, culmination, or turning point of a dream.

1911 BEERBOHM *Zuleika D.* ix. 151 For him to fall in love was a violent peripety, bound to produce a violent upheaval. **1942** K. W. BASH tr. *Jacobi's Psychol. C. G. Jung* iii. 79 *Peripetie*, which forms the 'backbone' of every dream, the weaving of the plot. The intensification of events to a crisis or to a transformation, which may also consist in a catastrophe. **1950** *Brit. Jrnl. Psychol.* XL. 236 Jung goes on to discuss the structure of dreams... He distinguishes..the culmination or peripety, and..the final lysis or solution. **1960** R. F. C. HULL tr. *Jung's On Nature of Dreams* in *Coll. Wks.* VIII. 295 The third phase brings the culmination or *peripeteia*. Here something decisive happens or something changes completely. **1964** M. MCLUHAN *Understanding Media* x. 103 So sudden an upsurge of academic training into the marketplace has in it the quality of classical peripety or reversal. **1976** S. HYNES *Auden Generation* vii. 193 In that pattern, 1936 is the peripeteia, the point where the action turned.

peripheral, *a.* Add: (Further examples.) In *Anat.* also with especial reference to the circulation, as *peripheral resistance*. Also *fig.*, marginal, superficial, on the fringe.

1877 M. FOSTER *Text Bk. Physiol.* I. iv. 92 It is this peripheral resistance (in the minute arteries and capillaries)..which gives the circulation of the blood its peculiar features. **1909** *Jrnl. Physiol.* XXXVIII. 237 Peripheral reference is the earliest phenomenon of recovery. **1915** W. M. BAYLISS *Princ. Gen. Physiol.* viii. 242 The peripheral resistance of the arterial system, resulting from the division into small arterioles, is due entirely to the internal friction of the blood, not to friction against the walls of the vessels. **1949** E. A. NIDA *Morphol.* (ed. 2) 84 A peripheral morpheme never consists of a root and is always structurally 'outside' of the nuclear constituent. *Ibid.*, Note, however, that infixes and some replacives are 'peripheral' even though they are formally included within the nuclear constituent. **1956** D. L. ABRAMSON *Diagnosis & Treatm. Peripheral Vascular Disorders* p. xiii, The physiology of the peripheral circulation. **1962** *Lancet* 27 Jan. 193/2 Chromosome analysis of peripheral-blood leucocytes grown in tissue culture. **1962** *Listener* 29 Mar. 542/2 We must distinguish between the kinds of conventional forces useful in Europe and those that might be used in the peripheral areas. *Ibid.* 10 May 825/2 Dr O'Leary mentions it in his conclusion, but it is peripheral to his main interest. **1965** M. MORSE *Unattached* i. 44 He said little and was essentially peripheral, mildly amused, and always following. **1972** R. HARTENSTEIN *Princ. Physiol.* ix. 421 Peripheral resistance and the force and rate of cardiac contractions are the major overall factors that determine blood pressure, velocity of flow, and the distribution of blood. **1973** *Word 1970* XXVI. 101 Discussion of the ultimate phonetic output of /S/ (or /z/) is peripheral to the core of this paper. **1976** *Vancouver Province* 18 June 21/2 Canadian cultural expression will remain peripheral to Canadians—unless we tackle the economics.

2. *Computers.* Applied to equipment that is used in conjunction with a computer without being an integral or necessary part of it, and to operations involving such equipment.

1956 *Proc. Eastern Joint Computer Conf. 1955* 67/2 An important current trend is toward peripheral equipment with corresponding flexibility. **1962** *Gloss. Terms Auto-*

matic Data Processing (B.S.I.) 25 Peripheral transfer, the process of transferring a block of data between peripheral equipment and a store or between two units of peripheral equipment. 1963 A. M. Hilton Logic, Computing Machines, & Automation vi. 156 Electric typewriters and machines to perforate paper tape are among the most widely used items of peripheral equipment, particularly for small computing machine systems. 1967 Times 6 May 17/1 Many vital pieces of equipment that go into a computer installation are not made in Europe. These units, which come under the heading of peripheral equipment, are becoming more and more important. 1970 Sci. Amer. Oct. 102 East Germany will probably supply peripheral equipment; Hungary, magnetic memories and software (programs).

B. *sb. Computers.* A peripheral device. Usu. *pl.*

1966 Economist 10 Sept. 1048/2 It just has not got the sort of money needed to develop and market a complete line of data processing equipment and the associated 'peripherals'. 1970 Physics Bull. July 306/2 External storage is made up of a variety of bulk or file storage units of very large capacity, which are operated as peripherals to the central processor. 1977 [illegible] To prevent the CPU from slowing to the speed of a peripheral, a buffer may be used so that the peripheral transfers information to the buffer at its own speed while the CPU does other work. 1973 [illegible] Computer peripherals...the bits and pieces you hang on and plug into computers. 1977 [illegible] A small computer with a variety of input and output peripherals including an X-Y plotter.

peri:pheriza·tion. *rare.* [f. Peripher(y + -ization.] Obscurity or indirectness of expression.

1926 E. Pound Let. 15 Nov. (1971) 202 Ms. arrived this a.m... I will have another go at it, but up to present I make nothing of it whatever. Nothing – short of divine vision or a new cure for the clap can possibly be worth all the circumambient peripherization.

periphonic (perifǫ·nik), *a.* [f. Peri- + Phonic *a.*] Such as to reproduce the vertical as well as the horizontal distribution of sound that has been recorded, by means of one or more loudspeakers above the level of the listener in addition to ones around him at his own level.

Whether this is the sense in quot. 1970¹ is uncertain.

1970 Times 25 June 7/2 The French pieces were almost as uneventful, even with benefit of this excellent multi-channel, so-called periphonic, sound. 1970 M. Gerzon in Studio Sound Aug. 338/1 The second part of this article is devoted to the use of these considerations in obtaining a system of Periphonic (Greek: peri-, around) sound reproduction, i.e. the reproduction of sound in all spatial directions. 1974 Nature 13 Dec. 537/1 A minimum of four loudspeakers are geometrically necessary to surround the listener in three dimensions and give periphonic reproduction. 1976 Gramophone Feb. 1398/1 The real potential of four channels lies in periphonic reproduction using at least six loudspeakers. Ibid., Although four channels are in a sense ideal for periphonic reproduction, it can be realised using only three channels.

Hence **peri·phony**, periphonic reproduction.

1970 M. Gerzon in Studio Sound Sept. 380/1 While Granville Cooper has recently described a system of periphony called 'tetrahedral ambiophony', this is only one of many possible periphonic techniques. 1974 Nature 13 Dec. 537/1 To satisfy the psycho-acoustic criteria sufficiently well, however, the practical minimum is..six [loudspeakers] for periphony.

periplasm. Add: **2.** *Microbiology.* The region of a bacterial or other cell immediately within the cell wall, outside the plasma membrane. Hence **periplasmic** *a.*

1961 P. Mitchell in Goodwin & Lindberg Biol. Struct. & Function II. 590 Observations forced us to the conclusion that the glucose-6-phosphatase of intact Escherichia coli is enclosed in a region between the cell wall and the surface of the osmotic barrier component which we might appropriately call the 'periplasm'. 1967 Science 16 June 1453/3 Some time ago Mitchell proposed that glucose-6-phosphatase activity is located in such a 'periplasmic space'. 1974 Jrnl. Bacteriol. CXIX. 243/2 To determine whether the H(II)-reducing activity is present in the cytoplasm, the periplasm, or both. Ibid. 244/1 Alkaline phosphatase is one of the periplasmic enzymes. 1978 Sci. Amer. Oct. 74/2 The penicillin would then 'hitch-hike' with the bacterial penicillinase into the periplasmic space, from which it could be extracted and then assayed with an antibody technique.

periplum (pe·riplɒm). [L., neut. f. Periplus.] In the poetry of E. Pound (see quot. 1940).

1940 E. Pound Cantos lix. 83 Periplum, not as land looks on a map But as sea bord seen by men sailing. 1948 — Pisan Cantos (1949) lxxiv. 7 The great periplum brings in the stars to our shore. Ibid. 13 Under the grey cliff in periplum. Ibid. lxxxii. 118 Three solemn half notes Their white downy chests black-rimmed On the middle wire Periplum.

Perique (perī·k). Also **perique.** [Louisiana F. (see quot. 1931).] In full *Perique tobacco.* A strong, dark, Louisiana tobacco.

1882 Congress. Rec. 6 Apr. 2642/2 Perique tobacco may be sold by the manufacturer or producer..in the form of carottes..without the payment of tax. 1885 E. Custer Boots & Saddles 84 The officers gave this chief tobacco—Perique I think it is called. 1931 W. A. Read Louisiana-French 57 Perique is said to have been the popular pseudonym of Pierre Chenet, an Acadian who first produced this variety of tobacco. 1941 E. P. O'Donnell Great Big Doorstep iv. 63 [illegible] Louisiana products..'Rice, cotton, perique tobacco, and fur.' 1949 Tobacco 7 Apr. 15/1 Perique is the only tobacco steeped in its own juice, and has a mildly fermented smell, like wine. 1976 National Observer (U.S.) 12 June 17-A/5 (Advt.), It's Perique, a zesty, dark and aromatic tobacco.

periscope. Add: **II. 3.** Also, a similar kind of tube-and-mirror or -prism apparatus used on land, as in trench warfare. See *trench-periscope.*

1917 A. G. Empey Over Top 303 Periscope, a thing in the trenches which you look through. 1951 'M. Innes' Operation Pax v. iii. 251 Remnant was fiddling with a long forceps and a couple of mirror-like stainless-steel plates. 'First-rate periscope,' he said. 1976 Sci. Amer. Dec. 32/3 The crane operator, protected by heavy shielding and observing his task through a periscope, could remove and install any of the equipment by using impact wrenches to manipulate the connectors at the ends of the jumpers.

4. *attrib.* and *Comb.*, as **periscope-wise** adv.; **periscope depth** (see quot. 1928); **periscope level** = *periscope depth.*

1928 C. F. S. Gamble Story N. Sea Air Station xviii. 309 German submarines, when travelling awash, could reach 'periscope depth' (that is, the depth at which the fully extended periscope just reaches to the surface—normally 45 feet) in 14 minutes. 1974 L. Deighton Spy Story xviii. 184 'Periscope depth,' said Ferdy... The Captain..took us up to periscope level. 1923 J. S. Huxley Ess. Biologist iii. 116 It [sc. the grebe] lifts its head and neck above the water, periscope-wise, to assure itself of its direction.

pe·riscope, *v. poet.* [f. the sb.] *intr.* To look as if through a periscope.

1934 Dylan Thomas 18 Poems 12 Where fishes' food is fed the shades Who periscope through flowers to the sky.

perish, *v.* Add: **1.** Jocose phr. *or perish in the attempt.*

1861 T. Hughes Tom Brown at Oxf. I. ii 23 However, he addressed himself manfully to his task; savage indeed, and longing to drive a hole in the bottom of the old tub, but resolved as ever to get to Sandford and back before half time, or perish in the attempt. 1870 L. M. Alcott Old-Fashioned Girl xvii. 337 He..seemed resolved to be an honor to his family, or perish in the attempt. 1978 Country Life 30 Nov. 1915/4 Too often publishers are determined to illustrate or perish in the attempt.

c. Esp. of rubber.

1884 Queen 29 Mar. (Advt.), A flat elastic section (which, unlike rubber elastic, will not heat the person or perish in wear). 1926 Bradshaw's Railway Guide Apr. facing p. xv (Advt.), Self-filling fountain pen...No rubber to perish. 1957 [illegible] Familiar to all is the liability of rubber to 'perish', to harden and crack or soften to a sticky residue.

c. Now only in phr. *perish the thought.*

[1700 C. Cibber King Richard III v. 52 Perish that thought!] [illegible]

2. Also in *pres. pple.* with the auxiliary *be* (now the more usual form with reference to exposure to (cold) weather).

1884 A. Edwardes Girton Girl III. i. 11 [illegible] 1930 W. S. Maugham Cakes & Ale [illegible]

3. (Further example.)

1898 W. W. Rowe [illegible] Em'ly xx. 228 Chrise, I'll perish you, if you ain't careful.

perish, *sb.* Restrict *rare*⁻¹ to sense in Dict. and add: **2.** *Austral.* A state of near starvation, great thirst, or any kind of deprivation or destitution; esp. in phr. *to do a perish*: to come to such a state. Also trivially (see quot. 1941).

1894 Argus (Melbourne) 28 Mar. 5/4 When a man or party has nearly died through want of water he is said to have 'done a perish'. 1903 R. Bedford True Eyes 312 If Xavier Quinn hadn't found this show three months ago ye'd have done a perish. 1924 Truth (Sydney) 27 Apr. 6 Perish, doing a, to shiver; to be cold. 1929 K. S. Prichard Coonardoo v. 60 But we hear done a perish. You. 1935 H. H. Finlayson Red Centre iii. 28 The constant struggle out of one 'perish' into another. 1941 Baker Dict. Austral. Slang 53 Do a perish, suffer greatly from thirst, hunger or destitution. 'In a city, to sleep out in parks, to be homeless.' 1942 'M. Innes' Daffodil Affair ii. v. 64 'He didn't die,' said Hudspith. 'He perished.' 'He did a perish,' said Appleby corroboratively and idiomatically. 1944 F. Clune Red Heart 19, I did a thousand miles in eleven weeks on camel-back to find Lasseter's track, and follow it into the country where he did a perish. 1953 D. Stivens Gambling Ghost 3 You'll do a perish, mate, and no mistake. You're two hundred miles as the crow flies from anywhere. 1959 H. P. Tritton Time means Tucker 15/1 The train-crew who had made us do a perish on the Galathera Plain. 1969 'A. Garve' Boomerang ii. 71 His intention was to enjoy this trip..not to 'do a perish' in the Never Never.

perishable, *a.* (*sb.*) **1.** (Further examples.)

1810 E. Weeton Jrnl. Apr. (1969) I. 257 [illegible] perishable household articles [illegible]. 1930 F. C. Bowen Sea Slang 102 Perishable Cargo. In the 18th century, slaves or fruit. 1958 M. Roberts Gustavus Adolphus II. [illegible] Gustav Vasa..[illegible] Lübeck in goods, and sometimes in goods of a dangerously perishable nature: at least one instalment was paid in butter.

perished, *ppl. a.* (Further examples.)

1747 H. Glasse Art of Cookery xxx. 161 If any soft or perished Place appear on the Outside [of cheese], try how deep it goes. a 1922 T. S. Eliot Waste Land Drafts (1971) 75 Over perished plains, stumbling in cracked earth. 1922 H. P. Stevens in S. Morgan Prep. Plantation Rubber xxi. 306 After a time, vulcanised rubber tends to harden, cracks appear on the surface when the article is bent or stretched, and eventually the rubber becomes rotten and 'perished'. 1923 B. D. W. Luff Chem. of Rubber vii. 86 [illegible] 'perished', to use the term now applied to such a change. 1950 J. Cannan Murder Included vii. 158 A perished washer might account for the dripping. 1964 [illegible] The half-perished suspenders and wrinkled stockings. 1966 D. Francis Odds Against xiii. 178 Weather forecasts are as reliable as a perished hot-water bottle. 1967 Leyland & Watts in J. A. Brydson Devel. Rubber Technol. v. 67 [illegible] a perished condition in a relatively short time. 1978 Country Life 30 Nov. 1913/3 [illegible] perished rubber releasing corroded wire.

perisher. Add: **a.** Also, more generally used of persons, with an overtone of pity. Also (*little*) *perishers* pl., children.

1908 A. N. Lyons Arthur's ii. i. 106 Poor ole perisher's 'fraid to come in, 'e is. [illegible] 1935 Wodehouse Luck of Bodkins xv. 181 If you ask me, they don't learn the little perishers nothing. 1944 — Money in Bank (1946) xx. 183 Most modern young men are squirts and perishers. [illegible] Not one of the little perishers knew it.

b. *Austral.* = *Perish sb.* 2.

1903 R. Bedford True Eyes 292 Of course that country we went to on the Peak was a shicer. Just well you didn't come—we near did a perisher there. 1903 'T. Collins' Such is Life i. 7 You will understand that the bullock drivers' choice of accommodation lay between the selection, the run-paddock, and a perisher on the plain... [illegible] It looked as if we were in for what the Inlander calls a 'perisher'.

c. *Naval slang.* A periscope. Hence in extended uses (see quot. 1962).

1925 Fraser & Gibbons Soldier & Sailor Words 221 Perisher, periscope. 1946 Partridge Dict. Forces' Slang 139 The Perisher, the C.O.'s course for submarine commanders. 1952 E. Young One of our Submarines viii. 115 At one time the course on which we were embarked had been called the Periscope School; hence the grimly humorous construction 'Perisher'. Now it was officially the C.O.Q.C., or Commanding Officers' Qualifying Course. [illegible] 1962 Granville Dict. Sailors' Slang 87/2 Perisher, the periscope course for officers selected to command submarines. A 'perishing' difficult examination. [illegible] 1973 D. Reeman Go in & Sink! ii. 16 We did our Perisher together, and even when I got Tristram he was given Tryphon.

perishing, *vbl. sb.* Add: **c.** *spec.* of rubber.

1913 B. D. Porritt Chem. of Rubber i. 12 The oxidation of rubber..is technically known as 'perishing'. [illegible] 1954 H. J. Stern Rubber v. 152 The 'perishing' of rubber, particularly of manufactured rubber articles, has been a source of trouble to all concerned from the time of Hancock down to the present day. 1961 D. W. Huke Introd. Natural & Synthetic Rubbers v. 80 One of the most important problems a rubber compounder has to face is the perishing—usually referred to as the ageing—of rubber.

perishing, *ppl. a.* Add: **2.** Also *Austral.*, with reference to *Perish *sb.* 2.

1941 I. L. Idriess Great Boomerang xvii. 124 Sixty miles to water, along a perishing track on a perishing day.

3. *colloq.* Applied disparagingly to anything: insignificant; troublesome; also as an intensive and as a filler.

1847 E. Brontë Wuthering Heights II. xiii. 266 Do you imagine..that healthy, hearty girl, will tie herself to a little perishing monkey like you? 1903 Kipling Five Nations 196 We were sugared about by the old men (Panicky, perishin' old men). 1926 'Taffrail' Pincher Martin iii. 32 A long coaling in the winter is the 'perishin' limit', as some one put it. 1928 [see *Blighty, blighty sb. c]. 1930 [see flat spin s.v. *flat a. 15]. 1952 M. Allingham Tiger in Smoke iv. 75 These perishing crooks, who do they think they are all of a sudden?

4. as *adv.* Excessively, perishingly.

1888 E. Marshall Bristol Diamonds ix. 106 It is perishing cold to-day. 1906 Westm. Gaz. 26 Feb. 4/2 I'm perishing hungry. [illegible] 1962 G. Millar Maquis i. 21 They all say it's perishing cold in the aircraft.

b. Used as a mere intensive.

1959 M. Gilbert Blood & Judgement v. 54 He..turns right at the top, because it's the only way he perishing well can turn.

perishless (pe·rifles), *a.* [f. Perish *v.* + -less.] That cannot perish; imperishable.

1609 Sylvester tr. Du Bartas's Weekes & Workes 648 We must, to make vs blest, propose our selues that perfect, perish-les, That true vnfained good, that good all [illegible]. [illegible] Wallace of perishless renown. 1918 Times 31 Mar. 9/3 [illegible]

peristeronic, *a.* Add: Also, suggestive of pigeons.

1931 J. Cannan High Table 21 A discourse..which Anne and Cecilia punctuated with polite little peristeronic sounds.

peritectic (perite·ktik), *a.* and *sb.* [ad. G. *peritektisch* (W. Guertler *Metallographie* (1912) I. vi. 276), f. Gr. περί around, about + τηκτικός able to dissolve (f. τήκειν to melt): cf. Eutectic *a.* and *sb.* in Dict. and Suppl.] **A.** *adj.* Of, pertaining to, or designating a reaction that occurs between the solid phase and the liquid phase during the cooling of a mixture, with the formation of a new solid phase; *peritectic point*, the state at which all three phases coexist in equilibrium, the composition being such that a fall in temperature results in the disappearance of the two phases that exist above that temperature; also, the point representing this state in a constitutional diagram.

1924 Jeffries & Archer Sci. of Metals ix. 323 There are different solid phases in equilibrium with the melt above and below the peritectic temperature. During the peritectic reaction there are three phases (two solid and one liquid) in equilibrium, so that the temperature must remain constant until at least one of the phases disappears. 1936 Nature 18 Apr. 657/2 [illegible] 1966 Phillips & Williams Inorg. Chem. I. viii. 302 [illegible] 1967 A. H. Cottrell Introd. Metall. xv. 230 The composition at r..is the peritectic composition and the isothermal line pq marks the peritectic temperature. Just above this temperature there is no β-phase in any alloy; just below, all alloys from p to q contain some β. The silver-platinum system has a simple peritectic diagram. 1968 [illegible] 1973 J. G. Tweeddale Materials Technol. I. vi. 163 The reverse type of change to a eutectic reaction called a peritectic reaction can occur in many materials—this cannot be used in their heat-treatment but may be a means for producing a stable solid solution structure.

B. *sb.* A peritectic point or temperature.

1929 Jrnl. Iron & Steel Inst. CXIX. 337 In carbon steels, between the temperature of freezing and that of the peritectic, iron containing a maximum of 0·07 per cent of carbon is the only solid to separate. 1970 Nature 20 Nov. 220/2 Any model of the igneous processes which produced the ocean floor must account for this preferential generation of liquids at peritectic A (Fig. 1).

Hence **perite·ctically** *adv.*, by a peritectic reaction.

1935 E. G. Doan Princ. Physical Metall. iv. 158 At 1500° the melt and δ crystals react peritectically to form γ-iron solid-solution crystals. 1967 A. H. Cottrell Introd. Metall. xv. 233 Fig. 15.12 shows an example in which the γ primary solution also forms peritectically.

peritectoid (perite·ktoid), *a.* [f. prec. + -oid, after *eutectoid *a.* and *sb.*] Of, pertaining to, or designating a reaction analogous to a peritectic reaction but involving three solid phases.

1936 H. L. Alling Interpretative Petrol. Igneous Rocks viii. 97 The diagram shows that hedenbergite, $CaFeSi_2O_6$, on heating dissociates before reaching the solidus, forming a solid solution. I suggest this may be called a 'peritectoid' reaction. 1967 A. H. Cottrell Introd. Metall. xv. 235 Eutectoid and peritectoid changes occur in the solid state which are the exact analogues of the eutectic and peritectic forms of the liquid-solid change. 1973 J. G. Tweeddale Materials Technol. I. vi. 163 There is also a form of reaction in the solid state, called the peritectoid reaction which is similar to the peritectic reaction.

peritoneoscopy (peritǫnī,ǫ·skǫpi, pe:ritǫni,ǫ·skǫpi). *Med.* [f. Peritoneo- + *-scopy.] Visual examination of the peritoneal contents by means of a narrow instrument passed through a small incision in the peritoneum.

1936 in Stedman Med. Dict. (ed. 13). 1959 Bailey & Love Short Pract. Surg. (ed. 11) [illegible] For many years peritoneoscopy has been advocated for inspecting intraperitoneal organs instead of recourse to laparotomy, but comparatively few surgeons employ it. 1974 R. M. Kirk et al. Surgery vi. 108 The diagnosis can be confirmed by needle biopsy, surgical biopsy, or at peritoneoscopy using an endoscope inserted through a small abdominal incision.

peri-urban (pe·ri,ɜ̄·ɹbǎn), *a.* [f. Peri- + Urban *a.* and *sb.*] Esp. in Africa: immediately adjoining a city or conurbation.

1948 Rep. Native Laws Comm. 1946–48 (Dept. Native Affairs, S. Afr.) 5/1 Johannesburg is still grappling desperately with its problem of peri-urban squatters, of whom over 90,000 have already been collected in the controlled squatters' camps of Moroka and Jabavu. 1954 L. Marquard Peoples & Policies S. Afr. 51 In 1947 the Department of Native Affairs estimated that 154,000 urban and peri-urban houses were required for Africans. 1961 Times 25 July 8/3 The fashionable peri-urban district some 11 miles from Johannesburg. [illegible] 1974 Times 17 Sept. 15/5 Unified Family Courts have been set up in peri-urban areas of Vancouver. [illegible]

periwinkle¹. Add: **3. b.** A blue colour like that of the periwinkle flower. Also *attrib.* and as *adj.*

1902 [in Dict., sense 4]. 1922 Daily Mail 20 Dec. 1 (Advt.), Lingerie crepe... In a full range of charming colours, including Pink, Sky, Jade, Flesh, Shrimp, Saxe, Fawn, Periwinkle and Cham. [illegible] 1977 C. Storr Tales Psychiatrist's Couch 83 'Plenty of time to think of getting married later, that's what I say,' Liz said, fixing me with those bulging periwinkle eyes.

perk, *sb.*² For *slang* read *colloq.* and add earlier and later examples.

1824 J. MacTaggart Scottish Gallovidian Encycl. 383 Perks, any kind of perquisite. 1869 J. Greenwood Seven Curses of London xx. 169 [illegible] 'perquisites' [illegible] It applies to such unconsidered trifles as wax candle ends, and may be stretched so as to cover the larcenous abstraction by our man-servant of forgotten coats and vests. [illegible] 1976 Daily Mirror 16 July [illegible] British Rail are cutting down on travel perks for their top management [illegible]. 1977 Times 8 Feb. 17/1 [illegible] the arts merely as a perk provided by all for the esoteric pursuits of the few.

perk, *sb.*³ *colloq.* Also **perc.** [Abbrev. of Percolator.] **a.** A coffee percolator. **b.** Coffee made in a percolator.

1934 F. E. Baily Fleet St. Girl iii. 59 Amazing coffee Charles makes in that electric perc. 1941 [illegible] Hash House Lingo 48 Perk, coffee. 1945 Baker Austral. Lang. [illegible] 1965 H. Gold Man who was not with It [illegible] But don't try to use the perc, you're too stupid. 1960 Wentworth & Flexner Dict. Amer. Slang 384/1 Perk, perc n., percolated coffee, as opposed to that boiled in a pan. Orig. cowboy use; later hobo use.

perk, *v.*¹ Add: **I. 1.** (Further example.)

1957 M. Spark Comforters viii. 197 Her whole body seemed to perk with delight.

f. For *dial.* read *colloq.* (Further examples.)

1936 Wodehouse Laughing Gas iii. 31 As the days went along, I found myself perking up a bit. I seemed to be making progress. 1936 M. Mitchell Gone with Wind vii. 132 Dr. Fontaine admitted that he was puzzled, after his tonic of sulphur, molasses and herbs failed to perk her up. 1957 M. Spark Comforters i. 18 Then she perked up. 1957 E. Eager Magic by Lake iv. 89 It seemed to do him good. For he perked up noticeably, and the flush of health began to appear on his wan cheek. 1962 Listener 20 Dec. 1041/1 The rate of growth dropped to 3 per cent. between 1958 and 1961. It is only in the past year or so that it has perked up. 1977 Time 10 Jan. 44/2 The Christmas results constituted fresh evidence that consumers are starting to spend again and the sluggish economy is perking up. 1977 Lancet 6 Aug. 291/1 Within days of stopping propranolol the patient perked up and became once more his old self.

II. 2. b. To enliven; elevate; stimulate interest in. Usu. with *up*.

1965 Amer. Speech XL. 287 The plentiful examples...are often entertaining enough..to perk up the laziest student. 1968 Globe & Mail (Toronto) 17 Feb. 9/3 [illegible] 1973 Publishers Weekly 1 Jan. [illegible] 1976 Scotsman 15 Dec. [illegible] 1976 Time 20 Dec. [illegible]

c. To say or comment in a brisk, lively, or self-assertive manner. *rare.*

1940 W. Empson Gathering Storm 15 [illegible] 1973 [illegible] Black on Black 135 [illegible] she perked hopefully.

perk, *v.*³ *colloq.* (orig. *U.S.*). Also **perc.** [Abbrev. of Percolate *v.*] **a.** *trans.* To make (coffee) in a percolator; to boil (coffee) up. Also *absol.* **b.** *intr.* Of coffee: to percolate, bubble, or boil (also *out of* the vessel). Also *transf.* and *fig.* Hence **perked** *ppl. a.*, **perking** *vbl. sb.*

In fig. uses not easily distinguished from *perk *v.*¹ 1f.

1934 in Webster. 1936 Mencken Amer. Lang. (ed. 4) v. 192 To perc (to make coffee in a percolator). [illegible] 1972 Newsweek 10 Jan. 19 By summer, so the calculations go, the economy will be perking quite nicely, and the President will go to the people as the agent of newfound prosperity. 1973 Daily Colonist (Victoria, B.C.) 25 June 7/4 [illegible] 1973 K. Giles File on Death iii. 23 [illegible] 1977 [illegible] At High Risk i. 37 While I waited for the coffee to perc, I made my bed.

Perkin² (pɜ̄·ɹkin). *Chem.* [The name of Sir William *Perkin* (1838–1907), English chemist.] **a.** *Perkin's mauve, purple,* or *violet*, a dye that was first prepared by Perkin in 1856 by oxidizing crude aniline with potassium dichromate, and was the first synthetic dye to be used commercially; = Mauve *sb.*

1860 Chem. News 21 Jan. 74/2 Notices of some of the patents taken out for the preparation and use of the new purple dyes generally known as the 'mauve' or 'Perkins' [sic] Purple, [etc.]. 1886 Roscoe & Schorlemmer Treat. Chem. III. iii. 319 [illegible] Perkin's violet, [etc.]. 1908 Jrnl. Chem. Soc. XCIII. 2147 [illegible] 1964 N. G. Clark Mod. Org. Chem. xxi. 454 Perkin's Mauve, as the substance was called, became the first synthetic dye. 1971 E. Gurr Synthetic Dyes 123 Mauveine (mauve; Perkin's violet) is a basic dye of the azine group... Mauveine has been obsolete for several decades past.

b. *Perkin('s) reaction* or *synthesis*, any of a number of types of reaction discovered by Perkin, esp. that in which, typically, an arylacrylic acid is formed by the condensation of an aromatic aldehyde with the anhydride of an aliphatic acid, in the presence of the sodium salt of the latter.

1882 Jrnl. Chem. Soc. XLII. 190 (heading) Interpretation of syntheses by Perkin's reaction. 1908 Ibid. XCIII. 2226 About 1867 he [sc. Perkin] must have commenced these researches..which..culminated in that beautiful method of synthesising unsaturated acids now known as the 'Perkin synthesis'. 1960 Gowan & Wheeler Name Index Org. Reactions (ed. 2) 189 Perkin synthesis of alicyclic compounds. Compounds containing active methylene groups react with polymethylene dihalides in the presence of a base..to yield alicyclic compounds. 1972 Materials & Technol. IV. xiii. 462 Acetic anhydride is the most applicable reagent in the Perkin reaction. 1973 B. J. Hazzard tr. Organicum 479 In the Perkin

synthesis, aldehydes or ketones are treated with anhydrides of aliphatic carboxylic acids, giving rise to α, β-unsaturated carboxylic acids.

perk test (pɜ̄·ɹk test). *U.S.* [f. Perc(olation + Test *sb.*¹] The act of percolating water through earth as a test of suitability for a septic tank. Hence **pe·rk-tested** *ppl. a.*

1974 State (Columbia, S. Carolina) 1 Apr. 9-B/6 (Advt.), Near Hilton, 158' on water, 208' in depth, 161' on rear, water system & perk test approved (restricted). 1976 Washington Post 19 Apr. C20/3 (Advt.), Surveyed, staked, recorded & 'perk' tested. 1976 Billings (Montana) Gaz. 30 June 7-D/2 (Advt.), Beautiful view, live stream, good road, perk test, soil samples comp. Protective restrictions. Ready to build on.

perlative (pɜ̄·ɹlātiv). *Linguistics.* [f. L. *perlatus*, pa. pple. of *perferre*, to carry through, convey + -ive.] A grammatical case signifying movement alongside or means of transportation.

1953 Trans. Philol. Soc. 1952 72 The use of the perlative case..is of some interest. Ibid., The perlative would seem to indicate going 'alongside' so as to be 'at'. 1966 G. S. Lane in Birnbaum & Puhvel Anc. Indo-European Dial. 217 A distinguishes formally between an instrumental in -yo and a so-called perlative in -ā.

perleau (pɜ̄·ɹlōᵘ). *U.S. dial.* Also **perlo.** = Pilau.

1933 in Amer. Speech VIII. iii. 40/2 A month later I went down to Uncle Rich's hunting cabin for a perleau supper. 1935 Z. N. Hurston Mules & Men i. i. 31 There was plenty of chicken perleau and baked chicken. 1955 This Week 25 Sept. 32/1 Perlo in local parlance [north Florida] is a chicken pilau.

perlemoen (pɛ̆·ɹləmu·n). *S. Afr.* Also **paarlmoer.** [Afrikaans, f. Du. *paarelmoer* mother of pearl.] = *klipkous.

1853 L. Pappe Synopsis Edible Fishes Cape Good Hope 12 Amongst the mollusca, none are more eagerly caught, and none have such a deserved reputation as Haliotis Midae Lin. (Klipkous; Sea-ear), and a species of Stomatia (Paarlmoer). 1911 J. D. F. Gilchrist S. Afr. Zool. ix. 192 The transposition of the organs [of unequal-sided molluscs]..may be so complete that the visceral loop is pulled into the form of the figure 8. Forms in which this occur..are the ordinary Limpet.., the Perlemoen or Klipkoes of South African Seas.., and the common periwinkle. 1947 [see *klipkous]. 1950 Cape Times 5 July 14/3, I have already mentioned having enjoyed a perlemoen dish out at Blaauwberg. 1958 L. G. Green S. Afr. Beachcomber xi. 127 South Africans are compensated for the scarcity of scallops by an abundance of perlemoen. This is the abalone of America and the ormer of the Channel Islands. They grow up to nine inches, and one of that size makes a rich meal indeed. 1966 [see *klipkous]. 1970 G. Croudace Scarlet Bikini ii. 14 Perlemoen was said to have the effect of oysters upon the human libido.

perlite. Add: *spec.* Fine or coarse grains of this mineral used, with appropriate nutrient solutions, as a medium for the growth of plants.

1956 T. M. Morrison in N.Z. Jrnl. Agric. XCIII. 503/1 The coarse grade of expanded perlite now on the market in New Zealand is ideal for plant growth. Ibid. 503/3 Perlite is cheap enough to be thrown out after several crops have been raised in it. 1971 'D. Halliday' Dolly & Doctor Bird viii. 106 We saw the trucks going by from the big netted nursery... Trucks full of potted plants, and bags of horticultural perlite, and Canadian sphagnum peat moss. 1976 A. C. Bunt Mod. Potting Composts ii. 38 Perlite is an alumino-silicate of volcanic origin and is widely used in the USA and New Zealand, both countries having large natural deposits of this mineral.

perlo, var. *perleau.

perlocution (pɜ̄ɹlokiū·ʃən). Restrict † *Obs. rare* to sense in Dict. and add: **b.** *Philos.* A speech act, such as persuading or convincing, that may or may not be successfully achieved by an illocutionary act such as entreating or arguing. Also *attrib.* So **perlocu·tionary** *a.*

1955 J. L. Austin How to do Things with Words (1962) viii. 101 Act (C. a) or Perlocution. He persuaded me to shoot her. Ibid. 102 We can similarly distinguish the locutionary act 'he said that..' from the illocutionary act 'he argued that..' and the perlocutionary act 'he convinced me that..'. 1969 J. Kaminsky Lang. & Ontology viii. 111 It could be said to have perlocutionary act potential in that it is clearly meant to have an effect on Jones. 1973 Times Lit. Suppl. 5 Oct. 1161/5 Austin called a speaker's act perlocutionary so far as, by saying something, the speaker produced some intended or unintended effect on the feelings, thoughts or actions of someone. 1976 Archivum Linguisticum VII. 67 Perlocutionary effects have to do with whether or not one actually succeeds in apologizing, persuading, or whatever, over and above communicating the fact that one is attempting to do this. 1977 Language LIII. 197 Both getting the officer to evict them and getting them to leave are perlocutionary sequels to the illocutionary act of asking the question.

Perlon (pɜ̄·ɹlǫn). Also **perlon.** A proprietary name (first used in Germany) for nylon 6, a type of nylon produced by the polymerization of caprolactam.

1941 Chem. Abstr. XXXV. 7200 In Germany the polymer is known as Igamide, the fiber as Perlon. 1954 Brit. Rayon & Synthetic Fibres Man. i. 45 Perlon (nylon 6) is a polyamide chemically resembling nylon 66 very closely. 1958 Official Gaz. (U.S. Patent Office) 28 Jan. TM 110/1 Perlon-Warenzeichenverband E.V., Frankfurt (Main), Germany. Filed Nov. 21, 1955... Perlon... For nylon fibers, filaments, strands, [etc.]. 1958 Trade Marks Jrnl. 12 Mar. 265/2 Perlon... synthetic textile fibres. Perlon-Warenzeichenverband Eingetragener Verein.., Frankfurt/Main, Germany; merchants. 1960 Vogue Pattern Bk. Autumn 45 Girdle in lightweight Perlon elastic with a Perlon taffeta front panel. 1973 C. Bonington Next Horizon xii. 168 Economising on weight, we had used 7 mm perlon,..with a breaking strain of 2,000 lb. 1976 M. & G. Gordon Ordeal (1977) xvii. 181 They carried pitons, nuts, stoppers, braided perlon cord..and other mountain-climbing equipment.

perlustration. Delete *Obs.* exc. in techn. use and add later examples.

1946 L. P. Hartley Sixth Heaven v. 98 The interest of seeing whether he was before or behind his schedule.. helped..the process of perlustration. 1967 Times 15 Mar. 6/5 Mr. Hugh Fraser..asked the Prime Minister whether cables and radio telegrams sent by M.P.s were privileged from perlustration by the security services. Ibid., Perlustration was in common use in the secret police of the Tsarist regime. 1972 Oxf. Univ. Gaz. CII. Suppl. No. 8. 47 The Curators conducted a perlustration of the Library on 29 May—the first ever at Rhodes House.

perm (pɜ̄ɹm), *sb.*¹ *colloq.* Abbrev. of *permanent wave* s.v. *permanent *a.* (*sb.*) 1 d.

1927 Home Chat 22 Oct. 200 How long does a 'perm' last? 1929 N. Royde-Smith Summer Holiday 113 The old girl's had a perm. Look at the waves. 1932 Modern Weekly 30 Apr. 136 You can have a 'perm.', or you may find a tong wave best. 1937 G. Frankau More of Us xv. 162 All perfume, perm and pearls and prominent teeth. 1943 J. B. Priestley Daylight on Saturday vii. 43 There was a girl called Elsie, who was a fake blonde, had a terrific perm. 1954 Wodehouse Jeeves & Feudal Spirit xii. 109 This aunt who sat before me clutching feverishly at her perm. 1960 M. Spark Ballad of Peckham Rye iii. 45 She moved her hand across her perm, nipping each brown wave in turn between her third and index fingers. 1976 National Observer (U.S.) 31 Jan. 8/2 After carrying around a headful of perfectly straight hair, I now sport a crown of exotic, wavy, fluffy, jaunty, tousled, perky, spirally, bouncy curls. I have a perm.

perm, *sb.*² *colloq.* Abbrev. of *permutation 4*.

1956 News Chron. 1 Nov. 10/8 Perms and plain paper bets accepted. 1958 L. Gibbs Gowns & Satyr's Legs xix. 128 He studied it [sc. a printed paper] intently, being in the process of constructing what..he called a 'perm'. 1958 Punch 27 Aug. 265/3 No pools investor of quality would seek advice from hacks who write:..Middlesbrough rate high in homes perm. 1971 E. Afr. Standard (Nairobi) 10 Apr. 2/7 (Advt.), See our collector for coupons, plans and perms. 1973 Weekly News (Glasgow) (Football Suppl.) 11 Aug. 3/1 And perm fans especially welcome the chance to cover more matches that the lower stakes give. 1974 Guardian 26 Nov. 27/2 Littlewoods have introduced a new perm—the £½ plan—using 12 selections which gives tight cover for only 50p. One is guaranteed 24 points if any nine matches result as score draws, and there are several other combinations which would provide a first dividend with only eight.

perm, *v.*¹ *colloq.* [f. *perm *sb.*¹] *trans.* To give a permanent wave to (the hair). So **permed** *ppl. a.*, having a permanent wave.

1928 Daily Express 17 Mar. 9/5 These girls took their chairs at 7.30 p.m... Three hours later they rose 'permed', as one says in the profession. 1931 S. Holme in Repertory 6 June 10/2 Thea spent nearly all day at the hairdresser's having her hair 'permed'. 1936 A. Christie ABC Murders xi. 83 Her hair had evidently recently been permed, it stood out from her head in a mass of rather frizzy curls. 1952 News Chron. 8 July 4/1 Children's hair differs from adults in that it is finer in texture and less elastic. Meaning that what perms Mother won't necessarily perm daughter. 1956 I. Murdoch Flight from Enchanter 192 Her elaborately permed hair. 1959 News Chron. 11 Aug. 6/2 There is a growing tendency for British mothers to perm children's hair. 1960 F. Raphael Limits of Love iii. v. 328 Newly permed brown hair. 1970 'D. Halliday' Dolly & Cookie Bird iv. 52 His golden, permed sideburns glistened. 1973 J. Thomson Death Cap v. 75 She had her brother's..gingerish, greying hair, although hers was permed into a lot of little waves.

perm, *v.*² *colloq.* [Abbrev. Permute *v.* or f. *perm *sb.*²] To make a selection of (so many) from a larger number; to make a permutation of (*permutation 3 b).

1959 Times 9 Mar. 11/4 One trusting citizen, having duly filled in a complex formula 'perming' a number of matches, looked upon his labour as a guarantee that dividends would automatically come pouring in. 1968 Guardian 11 July 9/7 Perm any six of these reasons to find out why an aircraft..crashed last week. 1972 Ibid. 3 Apr. 9/3 He went about it..composing 50 horoscopes in a single night... Then daily he permed any 12 from 50 so that 'conflict within the home' would befall Leos one week and Scorpios another. 1973 Milestones Summer 21/1 There are 12 all told, if you perm the alternatives of 2- or 4-door body shells and four engine sizes (of which two are also available in hotted up 'Sport Special' form). 1976 J. Snow Cricket Rebel 51 At the start he would 'perm' any two from three with the new ball, each bowler taking a turn at bowling with the wind at his back. 1976 Shooting Mag. Dec. 18/1 Gilstone Manor—I dare not use its real name—sits on the edge of a National Park (perm any one from nine) and is a hybrid i.e. a cross between Cold Comfort Farm and 'Upstairs Downstairs'. 1976 Sunday Mail (Glasgow) 26 Dec. 35/3 Pools Guide... Perm any 3 from 6 in first and second columns with any 2 from 3 in third column—20 × 20 × 3—1200 lines at 1/8p per line = £1.50 staked.

permafrost (pɜ̄·ɹmăfrǫst). [f. Perma(nent *a.* (*sb.*) + Frost *sb.*] Subsoil or other underground material that is at a temperature of less than 0°C throughout the year, as in Arctic regions; permanently frozen ground.

1943 S. W. Muller Permafrost or Permanently Frozen Ground 3 The expression 'permanently frozen ground'..is too long and cumbersome and for this reason a shorter term 'permafrost' is proposed as an alternative. 1952 Sci. News Let. 2 Aug. 70/3 Currently it is studying permafrost, permanently frozen ground that creates many problems in construction work. 1955 Peterson & Fisher Wild Amer. (1956) xxxiii. 357 Here on the tundra the permafrost forbids any digging and the Eskimos bury their dead above the ground. 1958 New Biol. XXVI. 90 In the sub-arctic the peat has a permafrost layer, that is a layer, usually a foot or two below the surface, which remains frozen for the whole year round, and acts as an impermeable layer. 1968 R. W. Fairbridge Encycl. Geomorphol. 838/1 Many other engineering problems result from changes in the mechanical properties of permafrost caused by thawing or freezing of its moisture beneath such structures as roadways or heated buildings. 1971 New Scientist 8 Apr. 70/1 The heated pipeline would be laid for several hundred miles over or through a peaty soil where the permafrost may reach as much as 500 metres depth. 1974 T. P. Whitney tr. Solzhenitsyn's Gulag Archipel. I. i. ii. 24 This wave poured forth, sank down into the permafrost, and even our most active minds recall hardly a thing about it. 1975 Nature 1 May 27/1 A possible reason is that the increasing permafrost in the soil forced the late Vikings to change their burial customs. 1977 New Yorker 9 May 95/2 There is ice under the tundra, mixed with soil as permafrost, in some places two thousand feet deep.

permalloy (pɜ̄·ɹmăloi). [f. Perm(eability + Alloy *sb.*] Any of a series of alloys consisting chiefly of nickel and iron, which have very high magnetic permeability and are widely used in electrical equipment, esp. in telecommunications.

1923 Arnold & Elmen in Jrnl. Franklin Inst. CXCV. 621 For convenience we call these peculiarly magnetic alloys by the general name 'permalloy', which serves at the same time to recall their characteristic capability of attaining high initial permeability. 1925 Chambers's Jrnl. Apr. 220/1 The conductor of this distortionless cable is composed of a special alloy, known as permalloy. 1926 Glasgow Herald 9 Sept. 8/4 The Western Union Telegraph Company announce that the laying of the new permalloy cable between England and New York was successfully completed on Sunday last. 1932 [see *Mumetal]. 1946 [see nickel-iron s.v. *nickel sb. 3 b]. 1965 Wireless World Sept. 431/1 The auto-transformer core should be of magnetically soft material such as permalloy, mumetal, H.C.R. alloy or some similar grade. 1969 E. N. Simon Dict. Alloys 127 Permalloy 45 contains basically 45 nickel, 55 iron, per cent. Ibid., Permalloy 4-79 contains basically 79 nickel, 4 molybdenum, the balance iron, per cent...and is used for audio coils, transformers and magnetic shields.

permanent, *a.* (*sb.*) Add: **1. d.** *permanent blue* (examples); *permanent dye*, a long-lasting dye used in hairdressing; *permanent hardness*, hardness of water that is not removable by boiling; *permanent magnet* (earlier example); so *permanent magnetism*; *permanent pasture*, land left unploughed for a long period, used for growing grass; *permanent press*, a process designed to produce lasting creases in materials for clothing; the fabric treated in this way; *permanent revolution*, a concept orig. attributed to L. D. Trotsky (1879–1940) which envisaged the dependence of Russia's bourgeois and proletarian revolution on a continuing process of European revolutions; also *transf.*; *permanent secretary*, a senior grade of the civil service, now normally denoting a permanent administrative officer of the highest grade in a Department of State; *permanent set* (Mech.), a deformation that remains after the removal of the stress that produced it; *permanent tint* = *permanent dye* above; *permanent under-secretary*, (*a*) a senior permanent adviser to a minister who is a Secretary of State; (*b*) a senior grade of the civil service below that of a permanent secretary and normally applied to the head of a division within a Department of State; *permanent wave*, a special process designed to produce a lasting wave in the hair (also as *vb.*); the wave so produced; so *permanent waver, waving*; *permanent way* (earlier and later examples); also *attrib.*; *permanent white* (later examples).

1886 H. C. Standage Artists' Man. Pigments iii. 32 Permanent Blue is a pale ultramarine with a cobalt hue. 1895 Montgomery Ward Catal. Spring & Summer 252/3 Artists Tube Oil Colors... Permanent Blue. 1939–40 Army & Navy Stores Catal. 372/2 Moist water colours... Permanent mauve..Permanent blue. 1966 J. S. Cox Illustr. Dict. Hairdressing 110/2 Permanent dye, a dye in which the susceptibility to lose colour under normal conditions has been reduced to a minimum. 1888 Encycl. Brit. XXIV. 409/1 The remaining or permanent hardness consists of sulphate of lime and other soluble salts. 1969 H. T. Evans tr. Hägg's Gen. & Inorg. Chem. xxvi. 666 Carbonate precipitation on boiling causes the water to lose its carbonate hardness or temporary hardness while a permanent hardness remains. 1828 F. Watkins Pop. Sk. Electro-Magnetism 12 If a steel needle be inserted in a coil and removed again immediately, it will become a permanent magnet. 1827 J. Cumming Man. Electro Dynamics 259 If it be possible to give permanent magnetism to steel by this species of electricity. 1965 A. H. Morrish Physical Princ. Magn. i. 1 Permanent and induced magnetism represent two of man's earliest scientific discoveries. 1861 M. H. Sutton in Jrnl. R. Agric. Soc. XXII. 416 (title) Laying down land to permanent pasture. Ibid. 421 We offer the foregoing hints, on laying down permanent pastures, &c., founded on our own experience and observation during full thirty years. 1897 W. S. Everitt Pract. Notes Grass & Grass Growing E. Anglia iii. 44 First select the land which is suitable for a permanent pasture. 1924 W. J. Malden Grassland Farming viii. 101 If permanent pasture seeds are sown in the spring without a corn crop, they may or may not.. make growth which will be valuable to feed during the first summer. 1968 F. W. Gould Grass Systematics i. 5 Geese feed extensively on rice..and also take significant amounts of forage from temporary and permanent pastures of the Gulf Coastal Prairie. 1964 Mod. Textiles Mag. Dec. 55 (heading) Garments with permanent press finishes. Ibid. 55/2 The concept of permanent press finish. ..is being widely..exploited in the men's and boys' casual and work slacks market. Ibid. 59/1 There are a number of competing processes now being offered..with claims of permanent creasing, permanent pressing, permanent shape retention, etc. 1965 Observer 10 Apr. 3/4 This new material..is generally known as 'permanent-press', a description which applies to a new process for treating cotton-and-Dacron goods so that they will never lose their shape or pleating. 1967 Family Herald 6 July 35/1 Permanent press involves treating the fabric chemically with a resin compound and then setting or 'curing' it by high heat in an oven. 1969 Time 18 Apr. 61 Foreign competition is most severe in man-made-fiber textiles, the most rapidly growing segment of the industry since advancing technology gave the world wash-'n'-wear shirts and permanent-press pants. 1978 N.Y. Times 30 Mar. A6/1 (Advt.), Made just for us by this top maker, a handsome selection of short sleeve shirts in solids, stripes, plaids and patterns. All in easy care permanent press polyester and cotton. 1928 Permanent revolution [see *Menshevism]. 1942 S. Neumann (title) Permanent revolution. 1964 I. Deutscher in Trotsky's Age of Permanent Revolution 18 Trotsky is deeply committed to one element in classical Marxism, its quintessential element: permanent revolution. 1972 D. Bloodworth Any Number can Play xx. 206 My fidelity to the Maoist theory of permanent revolution is absolute. 1975 Guardian 22 Jan. 10/1 Trotsky was the hero..and the notion of 'permanent revolution' was the great hope. 1867 E. E. Bridges Treasury (1964) 233 My Lords are of opinion that the office should now be given a more substantive character than that of Assistant Secretary and they are pleased to direct that its title shall be that of 'Permanent Secretary to the Treasury'. a 1974 R. Crossman Diaries (1975) I. 25 She continued the war, capturing Fred Willey and putting him in a room by himself in our Ministry while she got hold of his new Permanent Secretary, Mr Bishop, and lectured him. 1976 H. Wilson Governance of Britain iv. 99 The Treasury took the lead in setting up official inter-departmental committees, some at permanent-secretary level. 1822 T. Tredgold Pract. Ess. Strength of Cast Iron ii. 24 The second table..is intended to show the greatest weight a beam of cast iron will bear in the middle of its length, when it is loaded with as much as it will bear, so as to recover its natural form when the load is removed. If a beam be loaded beyond that point, the equilibrium of its parts is destroyed, and it takes a permanent set. 1888 Permanent set [see Set *sb.*¹ 16 j]. 1935 Jrnl. R. Aeronaut. Soc. XXXIX. 554 The highest stress in service should be below the true elastic limit of the material in order to avoid 'permanent set'. 1972 E. N. Simons Testing of Metals i. 21 The proof stress..is the load or stress that, applied for a minimum period of 15 min, gives a plastic extension or permanent set of 0·1% (or 0·5% with certain alloys). 1966 J. S. Cox Illustr. Dict. Hairdressing 110/2 Permanent tint, a euphemism for Permanent Dye. 1968 J. Ironside Fashion Alphabet 191 One very rarely hears the word 'dyeing' nowadays; it is always 'tinting'. There are permanent tints (which last through the life of the hair); semi-permanent tints [etc.]. 1904 Rep. War Office (Reconstruction) Comm. ii. 9 in Parl. Papers (Cd. 1932) VIII. 101 The Council should consist of seven members—four military and three civil—with the Permanent Under-Secretary as Secretary. 1917 G. Bell Let. 20 July (1927) II. 420, I wish you would go and see Sir A. Hirtzel, the Permanent Under-Secretary. 1959 P. Fleming Siege at Peking vii. 106 The fullest account of what they did do is contained in a confidential letter written on 4 September by Sir Claude MacDonald to Mr Bertie, the Permanent Under-Secretary of State at the Foreign Office. 1974 P. Gore-Booth With Great Truth & Respect 324 The Permanent Under-Secretary of any department of Government is the senior official or civil servant in that department. The word 'permanent' is accurate in the sense that, if there is a change of government or of Secretary of State, you do not have to move out or even go through a formal process of resignation and reappointment. 1978 Illustr. London News Nov. 134/3 He certainly was an extraordinary figure—the last of the permanent under secretaries (PUS in professional jargon) to spend the whole of his effective career in Whitehall. 1909 Hairdresser June 6/1 Children who have undergone the X-ray treatment for ringworm are growing curly hair... Would it be possible for a lady who desires to secure a permanent wave to undergo the treatment? 1919 Honey Pot I. ii. 1 (Advt.), Permanent waves... T. Vasco Ltd. 1922 U. Sinclair They call me Carpenter ix. 27 Would you like to see how we make eet—the permanent wave? 1925 Scribner's Mag. July 20 (Advt.), You will enjoy your permanent wave at Nestle's. 1928 Daily Express 23 Aug. 3/6 The curling irons and the tentacles of the permanent waver will be busier than ever in the autumn season. 1928 R. Macaulay Keeping up Appearances viii. 70, I want a permanent wave at twelve o'clock to-day. 1932 Woman's Pictorial 12 Mar. 14/2 What is a permanent wave? Answer (so far as my experience goes): a wave that is anything but permanent. 1946 Mod. Beauty Shop Dec. 128/2 Some hair which I permanent waved did not take the permanent. 1964 N. G. Clark Mod. Org. Chem. xvi. 352 If these linkages are broken (by reducing agents), the fibres bent to an artificial shape, and new disulphide linkages formed (by oxidizing agents), the fibres retain their new shape. This is the basis of 'permanent-waving' hair. 1967 O. Wynd Walk Softly v. 70 The Katsugis, celebrated for having dyed their long black hair bright orange, then permanent-waving it. 1968 J. Ironside Fashion Alphabet 196 At first permanent waving tended to give tight waves and curls, but over the past twenty years or so the tendency has been for a softer more natural looking permanent. 1975 Times 26 Aug. 12/8 Beauty salons..in the hectic flush of the permanent wave boom. 1838 Mechanics' Mag. 13 Oct. 32/2 The permanent way between Deptford High-street and the Greenwich terminus, is in a very forward state. This portion of the line is laid on longitudinal wooden sleepers, with three-feet bearing. 1869 Bradshaw's Railway Manual XXI. 392 Upwards of 4,000,000 L. was expended in England for permanent way materials, locomotives, stores, &c., sent out from that country. 1879 E. J. Simmons Mem. Station Master (1974) iii. 37 The station and line were occasionally enveloped in a very thick fog. On these occasions, we were allowed to employ one of the permanent-way men to stand all night on the line beyond the signals with his hand lamp. 1906 Daily Chron. 31 Jan. 4/5 Instructions had been issued to the permanent-way staff to adhere..to the regulations. 1926 T. E. Lawrence Seven Pillars vi. lxxviii. 409 It was impossible to leave them joined up to the exploder in the proper way, since the spot was evident to the permanent-way patrols as they made their rounds. 1957 J. Bledlow Cotswolds in Colour 47 Work was being done on the permanent way. 1967 G. F. Fiennes I tried to run a Railway iv. 40 They took Ticket Inspector Whipp at Liverpool Street and threw him off the platform on to the permanent way. 1973 Railway Mag. Mar. 129/2 Five of these engines were kept for permanent-way and shunting duties at Neasden. 1854 Permanent white [see *constant a. 4 f]. 1860 C. M. Yonge Hopes & Fears I. i. iv. 108 The front was all over scaffolds and cement, in all stages of colour, from rich brown to permanent white. 1934 H. Hiler Notes Technique Painting ii. 89 Permanent white (a special preparation of oxide of zinc prepared for oil painting).

3. d. = *permanent wave* above.

1926 Hairdressing 10 Sept. 241/1 This can only be done by superior work; namely, excellent setting of the finished permanent. 1932 New Yorker 4 June 64/3 (Advt.), A deep-wave marcel permanent styled for you alone in the modern manner. 1939 A. Huxley After Many a Summer ii. i. 6 Facials, Permanents, Manicures... Next door to the beauty shoppe was a Western Union office. 1941 J. C. Furnas How Amer. Lives 324 Twice a year she goes to a beauty shop to have a new 'permanent'. 1948 E. Waugh Loved One 78 Permanents, facials, wax—everything you get in a beauty parlour. 1951 E. Paul Springtime in Paris v. 117 Is there anything wrong with Yvette's beauty parlour..? I'm going there this afternoon—for a permanent. I hope it lasts a week. 1974 New Yorker 3 June 90/2 It is not uncommon for a woman to be receiving a permanent in one barber chair while a man is shaved in the other. 1976 N. Thornburg Cutter & Bone xii. 283 Their women were..constantly fussing with yesterday's permanents.

permanentize (pə·ɹmăněntəiz), *v.* [f. Permanent *a.* (*sb.*) + -ize.] To make permanent. So **pe·rmanentizing** *vbl. sb.*

A word of little value and rarely found in serious writing.—Ed.

1961 Guardian 20 Mar. 1/4 The latest word to be added to Pentagonese..is 'permanentise'. 1963 Punch 6 Nov. 665/1 The Ferrets have set about permanentising their idyll. 1966 Ibid. 21 Dec. 911/3 Meanwhile jeopardising our chance of permanentising the Sino-Soviet *détente*, and welding the Communist bloc more dangerously together. 1975 M. Bradbury History Man ii. 20 Society's technique for permanentizing the inherent contingency of relationships..that is to say, they got married.

permanently, *adv.* (Further examples.)

1885 Geol. Mag. Decade III. II. 516 On the depth of the permanently frozen stratum of soil in British North America. 1903 G. B. Shaw Man & Superman iii. 103 With such a majority as mine I cannot be kept permanently out of office. 1921 Daily Colonist (Victoria) 18 Mar. 9/4 (Advt.), Ladies come and have your hair permanently waved... We have the latest machines and are expert in the art. 1929 A. Noyes Return of Scare-Crow i. 11 There were pictures of permanently waved young women, with carefully arranged flowers in their hair. 1930 J. Cannan No Walls of Jasper 192 She stood..with..the sun in her eyes, and the wind in her permanently-waved hair. 1959 Housewife June 46 Shirt and permanently-pleated skirt. 1962 Guardian 15 Jan. 4/4 Drip-dry, uncrushable, and permanently pleatable cottons. 1969 Jane's Freight Containers 1968–69 105 A new permanently-coupled articulated flat car developed by Union Pacific. 1975 G. Howell In Vogue 208/2 Nylon transformed the fifties wardrobe..the fake furs and the permanently pleated nightdresses.

permansive (pəɹmæ·nsiv), *a. Gram.* [f. L. *permans-um* supine of *permanēre* to remain (see Permanent *a.* (*sb.*)) + -ive.] Applied to a tense in certain languages which is used to denote a more or less permanent state.

1866 E. Hincks in Jrnl. R. Asiatic Soc. Dec. 485 The verbal forms belonging to each conjugation may be divided into two great classes, which I call permansive and mutative. The former denotes continuance in the state which the verb signifies in that conjugation; the latter denotes change into that state. 1872 A. H. Sayce Assyrian Gram. 52 The Assyrian verb is rich in tenses. It possesses a Permansive, or Perfect as it is generally called in Semitic grammars, of comparatively rare occurrence in the historic inscriptions, but sufficiently common in the tablets; besides four more other tenses. 1924 O. Jespersen Philos. Gram. xx. 269 It [sc. the perfect tense] is a present, but a permansive present: it represents the present state as the outcome of past events, and may therefore be called a retrospective variety of the present. 1939 L. H. Gray Foundations of Lang. 359 In the historic period, Semitic possesses..two aspects (telic and atelic, commonly called perfect and imperfect..), to which Akkadian adds a permansive. 1972 Hartmann & Stork Dict. Lang. & Linguistics 21/1 Permansive aspect expressing a permanent state as a result of a completed action etc.

permeability. Add: Also, the degree to which a solid allows the passage of fluid through it, measured by the *coefficient of permeability* (or *permeability coefficient*), the volume of fluid flowing through unit cross-section in unit time under a unit (pressure or concentration gradient).

1902 Sci. Abstr. V. 856 (heading) Permeability of animal membranes. Ibid. 857 Inactive membranes..offer but little resistance to the passage of the different ions, and the permeability does not alter by contact with the solution. 1917 Rep. & Mem. Advisory Comm. Aeronaut. No. 317. 4 This permeameter is readily adapted to the determination of temperature coefficients of permeability. 1920 Ibid. No. 360. 4 The average permeability of rubbered airship fabrics..is not usually much less than 10 litres per sq. metre per day. 1931 Jrnl. Gen. Physiol. XIV. 408 This definition of permeability is seen to possess a definite physical meaning viz., the number of cubic micra of water entering the cell per minute per unit area of membrane, per atmosphere of difference in osmotic pressure between interior and external medium. 1960 Ibid. XLIII. 523 (heading) Experimental study of the independence of diffusion and hydrodynamic permeability coefficients in collodion membranes. 1962 R. C. S. Walters Dam Geol. xiv. 67 The permeability [of the Oxford clay] ranged from 12×10^{-8} to $4 \cdot 1 \times 10^{-8}$ cm per sec. 1966 McGraw-Hill Encycl. Sci. & Technol. XII. 453/2 The permeability coefficient varies from 100 cm/sec for clean gravel to 10^{-9} cm/sec for heavy clay.

b. (*magnetic*) *permeability*, also, one of the physical parameters of a medium, equal to the ratio of the magnetic induction B to the magnetic field strength H at any point in it; also (more fully *relative permeability*), the ratio of the permeability of a medium to the permeability of free space; *permeability of free space*, a constant μ_0 which in the C.G.S. electromagnetic system of units is unity and in the International System of Units is defined as a base quantity with the value 4π (12·57) × 10^{-7} henry per metre. (Further examples.)

Of the different permeabilities mentioned in quot. 1872 in Dict., the magnetic kind is the only one that gained currency.

1892 J. A. Ewing Magn. Induction ii. 56 Prof. Knott has proposed to call this quantity the 'differential susceptibility'; similarly dB/dH may be called the differential permeability. 1939 L. F. Bates Mod. Magn. ii. 67 Pure iron has an initial permeability of about 250. 1942 Phil. Mag. XXXIII. 488 The universal constant $1/ac^2$ occurring in (12.2) has the value..$1/10^7$ ohm sec./m... and is called the permeability of empty space. Consistency practically compels us to call the constant μ/ac^2 in (12.3) the permeability of the medium. 1944 [see *permittivity]. 1962 Corson & Lorrain Introd. Electromagn. Fields v. 177 The constant μ_0 is called the permeability of free space and is arbitrarily taken to be exactly $4\pi \times 10^{-7}$ newton/ampere² in rationalized m.k.s. units. Ibid. vii. 284 Figure 7-17 shows magnetization data for several different kinds of iron, with lines of constant relative permeability indicated. The detailed shape of the magnetization curve and the maximum permeability achieved with a given sample of iron depend on the purity, the method of annealing, and on the thickness of the sheets. 1971 Nature 16 July 208/2 The indiscriminate use of the symbol μ..both for the permeability of free space and that of ferromagnetic materials may temporarily confuse the unwary.

2. Special Comb.: **permeability tuning** *Electronics*, tuning in which the resonant frequency of a circuit is changed by moving a magnetic core into or out of a coil forming part of it, so as to change its inductance.

1933 W. J. Polydoroff in Proc. IRE XXI. 694 The apparent inductance is increased to tune to lower frequencies by introducing a magnetic core into the field of the coil. As the core is inserted into the coil, more lines of the magnetic field are intercepted by the core, and, in effect, the average apparent permeability of the medium surrounding the coil increases from 1 (for air) to a certain maximum... For this reason, and for other reasons.., it is appropriate to describe this method as 'Permeability Tuning'. 1968 Radio Communication Handbk. (ed. 4) iv. 35/2 Although permeability tuning has been most successfully used for many years by one major American company, the mechanical and electrical complications involved in band-switched receivers have resulted in little progress in this field by amateur constructors.

permeabilize (pə̄·ɹmiăbiləiz), *v.* *Biol.* [f. L. *permeābil-is* Permeable *a.* + -ize.] *trans.* To make permeable. So **pe·rmeabilized** *ppl. a.*
1971 *Nature* 3 Dec. p. ix/2 DNA synthesis—Observation in permeabilized yeast mutants. 1973 *Developmental Biol.* XXXV. 382/2 A technique..to permeabilize eggs for cytochemical and metabolic studies. *Ibid.* 384/1 The permeabilized eggs were..very sensitive to the composition of the incubation medium. 1974 *Jrnl. Bacteriol.* CXVIII. 1186/1 Several techniques have been used to permeabilize bacterial cells.
Hence **pe:rmeabiliza·tion,** the action of permeabilizing.
1973 *Nature New Biol.* 2 May 18/1 The synthesis observed is due to the growth of newly initiated chains, not merely the extension of chains initiated during normal growth before permeabilization. 1973 *Developmental Biol.* XXXV. 386/2 The permeabilization of the vitelline membrane of *Drosophila* eggs with octane offers the opportunity to study the influence of different substances on development and facilitates cytochemical investigations of the egg. 1974 *Jrnl. Bacteriol.* CXVIII. 1186 A cell permeabilization procedure is described that reduces viability less than 10%.

permeameter (pə̄ɹmi,æ·mītəɹ). [f. Perme-a(bility + -meter.] **1.** An instrument for measuring the magnetic permeability of a substance or object.
1890 S. P. Thompson in *Jrnl. Soc. Arts* 12 Sept. 885/2 *Permeameter Method*, this is a method which I have myself devised for the purpose of testing specimens of iron... For carrying it out a simple instrument is needed, which I venture to denominate as a permeameter. 1931 *Bureau of Standards Jrnl. Res.* (U.S.) VI. 355 The permeameters in general use for commercial magnetic testing have an upper limit of about 300 gilberts per centimeter. 1936 C. E. Webb in Vigoureux & Webb *Princ. Electr. & Magn. Measurements* II. xii. 309 The average magnetic properties of a non-uniform specimen may often be found more accurately in the Fahy permeameter than in permeameters dependent on the production of equality of flux or of magnetic potential at a limited number of points in the magnetic circuit. 1969 *IEEE Trans. Magnetics* V. 662/1 Straight magnetic samples are usually tested on hand compensated permeameters.
2. An instrument for measuring the permeability of a substance, esp. soil, to fluids.
1917 G. A. Shakespear in *Rep. & Mem. Advisory Comm. Aeronaut.* No. 317. 3 The following is a brief account of a permeability tester which was designed for the rapid testing of balloon and airship fabrics... The apparatus (called a 'permeameter' for short) consists of a shallow cylindrical vessel of cast iron, about 13 cm. in diameter [etc.]. 1941 D. P. Krynine *Soil Mech.* iii. 54 The general idea of permeameters or apparatus to determine the coefficient of permeability *k* in the laboratory is (*a*) to create a flow of water through a sample of a thickness *L* and having a definite cross section *A* [etc.]. 1971 B. F. Curtis in R. E. Carver *Procedures Sedimentary Petrol.* xiv. 353 Although liquid permeameters are used when thought necessary, it is far more common to measure permeabilities using gas as the moving fluid. 1974 A. Kézdi *Handbk. Soil Mech.* I. vi. 129/1 Constant head permeameters are particularly suited to the testing of highly pervious coarse grained soils. For soils with medium to low permeability, the falling head permeameters are used.

permease (pə̄·ɹmi,eiz, -s). *Biochem.* [ad. F. *perméase* (H. V. Rickenberg et al. 1956, in *Ann. de l'Inst. Pasteur* XCI. 843), f. *permé(able* Permeable *a.*: see *-ase.] Any enzyme which assists the passage of a substrate into a cell through the cell wall.
1957 *Bacteriol. Rev.* XXI. 169/1 During the past few years, however, definite proof of the existence, in bacteria, of stereospecific permeation systems, functionally specialized and distinct from metabolic enzymes, has been obtained... The generic name 'permeases' has been suggested for these systems. 1970 *New Scientist* 1 Jan. 8/1 This so-called 'permease' has since been found to be a protein, located in the cell membrane, which somehow transfers lactose from the outside of the cell to the interior. 1975 D. V. Parke *Enzyme Induction* i. 7 Maltase and maltose permease are induced together by maltose in yeast.

permeator (pə̄·ɹmi,ēitəɹ). [a. L. *permeātor*, f. *permeāre* (see Permeate *v.*).] **1.** One who or that which permeates; in quot. 1944, ? an infiltrator.
1944 G. B. Shaw *Everybody's Pol. What's What?* xxxi. 271, I, a Fabian permeator, knew the questions and had doctrinaire answers ready for some of them. 1969 *Ann. Rev. Plant Physiol.* XX. 602 Wartiovaara..compared the permeability constants of homologue series of permeators (e.g., alcohols..) and concluded that two parallel and alternative pathways for permeation of molecules through the protoplasm did not exist. *Ibid.*, The breaking of hydrogen bonds between permeator molecule and water during the passage of the permeator through the membrane is considered to be the important feature.
2. A vessel divided into two by a semi-permeable membrane, used in the large-scale removal of solutes from a liquid by reverse osmosis.
1975 M. J. Hammer *Water & Waste-Water Technol.* vii. 266 In addition to the permeators, a basic reverse osmosis system consists of pretreatment pumps.., tanks and appurtenances for cleaning and flushing, and a disposal system for waste brine. *Ibid.* (*caption*) A module of 24 reverse-osmosis permeator units used to desalt a municipal ground-water supply. 1977 *Design Engin.* July 6/3 A new high-capacity permeator..can convert thousands of gallons of seawater into potable water every day.

Permian, *a.* (*sb.*) Add: **2.** The name of the people of Perm and the language spoken by them.
1886 [see *Cis-Uralian* adj. s.v. *cis- 1]. 1908 T. G. Tucker *Introd. Nat. Hist. Lang.* 132 Permian, embracing Permian proper, Siryenian and Votiak. These are spoken by sparse populations near the Urals in the E.N.E. of European Russia. 1932 W. L. Graff *Lang.* 405 Permian, with its two varieties Zyrian and Votyak. 1933 L. Bloomfield *Lang.* 68 Four further branches of the Finno-Ugrian stock.. Permian, consisting of Votyak and Zyrian. 1961 L. F. Brosnahan *Sounds of Lang.* viii. 177 The languages of this family.., Permian, Votyak.

Permic (pə̄·ɹmik), *a.* [f. as prec. + -ic.] = Permian *a.* (*sb.*) in Dict. and Suppl.
Permian is the more usual term.
1921 A. W. Grabau *Textbk. Geol.* II. xxxviii. 506 (*heading*) The Permian or Permic system. 1964 *Language* XL. 96 Komi..and Udmurt..comprise the Permic group of the Uralic family. 1967 *Ural-Altaische Jahrbücher* XXXIX. 163 (*title*) Split, shift and merger in the Permic vowels.

permillage. Delete *rare* and add further examples.
1900 *Fortn. Rev.* Jan. 62 It should, perhaps, be remarked that I have reduced Dr. Galton's results to permillages. 1960 *Amer. Mineralogist* XLV. 6 (*caption*) The structure of perrierite projected on (010). The numbers give the y-parameters as permillage of the b-length. 1967 *Oceanogr. & Marine Biol.* V. 152 See Table VII in which enrichment of ^{18}O is expressed in terms of $\delta^{18}O$ the permillage enrichment relative to Standard Mean Ocean Water.

permineralization (pəɹmi:nĕrăləizēi·ʃən). *Geol.* [f. Per-¹ + Mineralization.] The action or result of fossilization by the deposition of minerals from solution in the interstices of hard tissue. Hence **permi·neralize** *v. trans.*, **-mi·neralized** *ppl. a.*
1893 C. A. White in *Rep. U.S. Nat. Museum 1892* 264 There are seven different natural conditions in which fossil remains are recognizable, three of which relate to substance, three to form, and one to both. To those relating to substance I have applied the terms permineralization, histometabasis, and carbonization... The term permineralization applies to that condition of fossil remains of animals which differ least from their original condition as parts of living animals. 1915 C. Schuchert *Text-bk. Geol.* II. xvii. 436 The great majority of fossil specimens preserve more or less of the original hard or mineral substance of the individual plant or animal, and to this may have been added in the organic interstices more or less of other mineral substances during the process of fossilization, forming the permineralized fossils. 1952 R. C. Moore et al. *Invertebr. Fossils* i. 4/1 Shells and bones, which are somewhat porous, may be made more dense by deposition of mineral substances by ground water. Hard parts altered in this way are permineralized; the process of alteration is termed permineralization or petrifaction. 1958 C. L. & M. A. Fenton *Fossil Bk.* i. 5 This process [*sc.* petrification]..takes place in two related ways. The simpler, termed permineralization, takes place when fat and other organic substances decay while water containing dissolved mineral matter soaks into every cavity and pore of hard—especially limy—structures. *Ibid.* 7 (*caption*) Permineralized corals that preserve both shape and structure. 1979 *Nature* 19 Oct. 640/2 Microfossils detected in the upper Brioverian dolomitic limestones..are permineralised in carbonate rather than silica.

perminvar (pə̄·ɹminvāɹ). Also **Perminvar.** [f. Perm(eability + Invar(iable *a.* (*sb.*).] Any of a series of alloys containing nickel, iron, and cobalt which have an approximately constant magnetic permeability over a range of field strengths.
The name is proprietary in the U.S.
1928 *Official Gaz.* (U.S. Patent Office) 12 June 277/1 Western Electric Company..N.Y. Filed Nov. 8, 1927. Perminvar..For ferromagnetic alloys comprising nickel, iron, and cobalt. Claims use since Aug. 23, 1927. 1928 G. W. Elmen in *Jrnl. Franklin Inst.* CCVI. 318 We have chosen 'perminvar' as the name for alloys in the iron-cobalt-nickel series, which are characterized, when properly heat treated, by constancy of permeability for a considerable range of the lower part of the magnetization curve. 1939 *Chem. Abstr.* XXXIII. 6775 Rapidly cooled Perminvar (45% Ni, 30% Fe, 25% Co) exhibits a clear increase in specific heat over a rather broad temp. range. 1962 N. H. Codling in G. A. T. Burdett *Automatic Control Handbk.* viii. 21 Perminvar, comprising nickel, cobalt and iron, subjected to heat treatment, has nearly constant permeability for inductions below 1000 G. 1968 S. J. Rosenberg *Nickel & its Alloys* 131/2 Useful alloys in the Perminvar class include the nickel-iron-alloys Conpernik (50% Ni) and Isoperm (40% Ni).

‖**permis de séjour** (pẽrmi də sezūr). [Fr.] Permission to stay in a country; a permit allowing this, a residence permit. Also *fig.*
1884 A. Forbes *Chinese Gordon* iv. 130 All persons residing in Darjour must have a *permis de séjour*. 1885 W. James *Lit. Remains H. James* 14 The ordinary empirical ethics of evolutionary naturalism can find a perfect *permis de séjour* under the system's wings. 1923 *Michelin Guide Gt. Brit.* (ed. 7) 861 Passengers not of French nationality must present themselves to the local authorities to obtain a 'Permis de séjour' if staying in France over 14 days. 1951 E. Ambler *Judgment on Deltchev* xviii. 219, I propose to have your visa and *permis de séjour* cancelled. 1960 O. Manning *Great Fortune* xxiv. 256 Should his *permis de séjour* be cancelled, Foxy or Fitzsimon would see that it was renewed. 1972 J. Aiken *Butterfly Picnic* iv. 65, I will see that your *permis de séjour* is extended.

permissible, *a.* Add: *permissible dose* (see quot. 1954).
1954 *Brit. Jrnl. Radiol.* XXVII. 245/2 Permissible dose is defined as that dose of ionizing radiation that, in the light of present knowledge, is not expected to cause appreciable bodily injury to a person at any time during his lifetime. 1960 *Lebende Sprachen* V. 163/2 The pilots who fly the modern airplane are subjected to a rather heavy dose (very close to the permissible dose) from the self-illuminating control-instruments.

permissive, *a.* **1.** Add to def.: Freq. in modern use in the sense 'tolerant, liberal, allowing freedom, *spec.* in sexual matters'; freq. in phr. *permissive society*. Hence as *sb.*, a permissive person. (Add later examples.)
1934 *Sun* (Baltimore) 25 May 6/2 All he asked was that the 'permissive' features of the pay-off provisions be made 'directive'. Later he explained that this would mean simply that the FDIC be 'authorized and directed' instead of 'authorized and empowered' to appraise the assets of the closed banks. 1946 *Amer. Psychologist* I. 416/2 If the counselor creates a warm and permissive atmosphere in which the individual is free to bring out any attitudes and feelings which he may have [etc.]. 1956 C. A. Tonsor in *Clearing House* XXX. v. 289, I realize that in the face of the permissive tendencies of the age, there is not much respect for rules. 1967 *Punch* 15 Mar. 372/2 If, in the nineteenth century, sadists in mortar-boards, and often in dog-collars also, belaboured little Johnny's bottom in order to knock Latin grammar into his head, some of today's jeans-and-gimmicks permissives may go too far the other way. 1968 *Listener* 4 Jan. 18/3 This dreadful dilemma of the puritan in a permissive society. 1970 *Times* 5 Feb. 9 It [*sc.* a proposal for a world-wide cricket tour] also irritates the extreme cricket-establishment people, some of whom seem to relish the thought of the tour, barbed wire and truncheons and all, to show that they are not going to be dictated to by the long-haired permissives. 1970 G. Greer *Female Eunuch* 45 The permissive society has done much to neutralize sexual drives by containing them. 1971 *Publ. Amer. Dial. Soc.* 1969 LII. 14 Americans tend to be permissive in matters of pronunciation and vocabulary. 1971 J. Wainwright *Last Buccaneer* II. 106 He was nineteen years old and there are few nineteen-year-old male virgins in the permissive age. 1972 *Guardian* 6 July 14/1 The charge against the permissive society is that the controls have slipped: things are being permitted that ought not to be permitted. 1976 U. Holden *String Horses* ix. 106 My kiddies all have their Daddies. My Herb belongs to Jim. I'm not permissive. 1976 F. Zweig *New Acquisitive Society* I. v. 47 The prevailing ethic of today is the ethos of the permissive society. 1977 *Time* 10 Oct. 24/3 Neither a cajoling arm twister like Lyndon Johnson nor a permissive parent like mild-mannered Mike Mansfield, Byrd is distinguished by his ability to gauge correctly what a majority of the Senate wants.
2. *permissive waste* (later example).
1971 *Mod. Law Rev.* XXXI. 698 A student who uses a conventional law textbook may be excused if he gains the impression that much of his professional career will be spent in advising about perpetuities, permissive waste and the principal mansion house.
3. Expressing permission or exhortation: applied to the verbal mood which expresses permission or an exhortation. Also as quasi-*sb.*
1845 [see *adhortative *a.*]. 1892 H. Sweet *New Eng. Gram.* I. 108 The combination of *may* and its preterite *might* with the infinitive (*may see, might see*) is called the permissive mood, as in *may you be happy!* 1898 *Ibid.* II. 116 Thus the present permissive is used in independent sentences to express wish: *may you succeed!* 1924 O. Jespersen *Philos. Gram.* xxiii. 320 As a tentative scheme of the purely notional ideas expressed more or less vaguely by the verbal moods and auxiliaries of various languages we might perhaps give the following list... 1. Containing an element of will... Permissive: you may go if you like. 1955 J. L. Austin *How to do Things with Words* (1962) xii. 158 In the special case of permissives we might ask whether they should be classified as exercitives or as commissives. 1976 J. S. Gruber *Lexical Struct. Syntax & Semantics* I. vi. 167 *Let* can be used as a fairly general Permissive Agent of Motion.

permissiveness. Add: (Later examples.) Cf. prec.
1946 *Amer. Psychologist* I. 420/2 When genuine acceptance and permissiveness are your tools it requires nothing less than the whole complete personality. 1958 B. Spock *Baby & Child Care* (new ed.) 56 Strictness or permissiveness?.. This looms as a big question for many new parents. 1966 *Listener* 6 Oct. 492/2 Permissiveness can rarely have gone further than it does today, and it may be..that we shall soon be due for a reaction, and a return to stricter standards. 1969 J. Mander *Static Society* ix. 303 The *permissiveness* of Brazilian society is an important element in her reputation for tolerance in racial and social matters. 1971 *Daily Tel.* 21 July 14 Perhaps it is time..for Parliament to have another look at the whole subject of abortion, family planning and perhaps permissiveness in general. 1973 M. Amis *Rachel Papers* 130 The so-called new philosophy, 'permissiveness' if you like, seen from the right perspective, is only a new puritanism, whereby you're accused of being repressed or unenlightened if you happen to object to infidelity, promiscuity, and so on. 1974 F. Warner *Meeting Ends* II. i. 35 With the new laws I suppose we've priced ourselves out of the marriage

market. With 'permissiveness' a man can have it all, so why should he take on a bloodsucking estate as well?

permissivism (pəɹmi·siviz'm). [f. Permissiv(e *a.* + -ism.] Attitudes or beliefs that are regarded as excessively tolerant or permissive.
1968 *Manch. Guardian Weekly* 17 Oct. 3 But the most impressive tributes to 'the high priest of permissivism', as he [*sc.* Benjamin Spock] once described himself, were casual. Not a single baby cried during the 90 minutes of protest. 1972 *Sat. Rev.* (U.S.) 20 May 30/2 Permissivism—i.e., the notion that all art is good art, etc.

permissivist (pəɹmi·sivist). [f. Permissiv(e *a.* + -ist.] A person considered excessively indulgent toward generally unacceptable or unconventional behaviour or attitudes, *spec.* in sexual matters.
1966 *Times* 16 Feb. 13 Theatrical permissivists should ask themselves whether, if there must be censorship, it is not better from their point of view that it should remain with a rationally indefensible institution, which is in no position to enforce for long unpopular or unjustifiable standards. 1970 [see *pablum]. 1972 *Daily Tel.* 13 Apr. 8/6 Since sex is a mere function away from all feeling it may be discussed, say the permissivists, in whatever detail, without anyone suffering any harm. 1973 *Church Times* 16 Nov. 9 The permissivists meanwhile demand..evidence, confident that it is..unobtainable.

permit, *v.* Add: **3.** *spec.* in phr. *weather permitting*, if the weather permits or allows, and in similar phrases.
1840 C. Brown in H. E. Rollins *Lett. J. Keats* (1958) I. i. 422 'Weather permitting', unless of the bad and excessive kind, was not of much force in our agreement. 1922 Joyce *Ulysses* 611 It was not so dear, purse permitting, a few guineas at the outside, considering the fare to Mullingar where he figured on going was five and six there and back. 1957 R. W. Zandvoort *Handbk. Eng. Gram.* I. ii. 36 Compare also the phrase *weather permitting*, where the meaning implied is one of condition. [*Note*] Also *funds permitting*, and similar combinations. 1978 T. Allbeury *Lantern Network* iii. 32 Arms..will be dropped to your instructions, weather permitting.

permit, *sb.* Add: **1.** (Further *attrib.* examples.) Also *Comb.*
1774 in *14th Rep. R. Comm. Hist. Manuscripts* App. x. 393 in *Parl. Papers* 1895 (C. 7883) LIX. 1 Have coaxed the people to part with their money and give paper in return to keep up armies of placemen, permit men, custom house officers, pensioners and soldiers. 1921 *Daily Colonist* (Victoria, B.C.) 11 Mar. 4/1 It will be a method of checking bootlegging, and preventing those who are not permit-holders from coming into possession of liquor. 1926 T. E. Lawrence *Seven Pillars* v. lvii. 300 A mixed body of Egyptian and British military police came round the train... It was proper to make war on permit-men, so I replied crisply. 1933 *Brit. Birds* XXVII. 138 Last year and again this year, this colony was wiped out,.. and so permit-holders visited the colonies on the gravel bed instead. 1945 *Seafarers' Log* 13 Apr. 7/1 To members in full standing, who bring in their friends for permit cards, study the last weeks LOG on the Agents Conference pertaining to permit men. 1977 *Evening Gaz.* (Middlesbrough) 11 Jan. 14/5 But it will not affect last year's winner, Tsuru, one of two representatives for Somerset permit-holder and wholesale butcher Tony Cobden who has also declared Rio.

permittance. (Later example.)
1912 *Housemaster's Lett.* 124 The wilful misuse of them or the callous permittance of them to go blunt and to rust.
†**2.** *Physics.* = *capacitance. *Obs.*
1887 O. Heaviside in *Electrician* 3 June 79/2 A telegraph circuit, when reduced to its simplest elements,.. still has no less than four electrical constants, which may be most conveniently reckoned per unit length of circuit—viz., its resistance, inductance, permittance, or electrostatic capacity, and leakage conductance. 1890 [see *elastance]. 1908 *Jrnl. Inst. Electr. Engin.* XL. 58 Dr. Heaviside's permittivity..is measured in terms of the permittance of unit volume.

permitted, *ppl. a.* Add: **b.** *permitted hours*: the hours during which the sale of intoxicating liquor is legal.
1923 W. B. Capper *Licensed Houses* III. lxxii. 211 In Wales and Monmouthshire, however, there are no permitted hours for licensed premises on Sundays. 1946 H. Dugdale *Managem. of Public-House* x. 67 Occasionally, the permitted hours applicable to a public-house do not sufficiently meet the requirements of the neighbourhood wherein the 'house' is situated. 1964 V. Heaton *Pub of your Own* xvi. 115 Though licensed premises may not sell intoxicating liquor after the permitted hours, nor may it be consumed later than ten minutes after permitted hours, there is no compulsion either to close the premises or to hide any display of liquor.

permittivity (pə̄ɹmiti·vīti). *Physics.* [f. Permit *v.* + -ivity.] One of the physical parameters of a medium, equal to the ratio of the electric displacement *D* to the electric field strength *E* at any point in it; also (more fully *relative permittivity*), the ratio of the permittivity of a medium to the permittivity of free space: = *dielectric constant; *permittivity of free space*, a constant ϵ_0 which in the C.G.S. electrostatic system of units is unity and in the International System of Units is $1/\mu_0 c^2$ (=8·854 × 10^{-12}) farad per metre, where μ_0 is the permeability of free space and *c* is the speed of light.
1887 O. Heaviside in *Electrician* 17 June 124/1 Nomenclature Scheme... Permittance. Permittivity. 1890 [see *elastance]. 1938 G. P. Harnwell *Princ. Electr. & Electromagn.* i. 12 If the charges are measured in coulombs and if *r* is in meters and **F** in newtons, the constant κ_e has the value κ_e = ..8·85 × 10^{-12} farads/meter... This constant..is known as the permittivity of free space. 1944 *Phil. Mag.* XXXV. 83 Guggenheim has suggested the term *permittivity* for the fundamental quantity and *specific inductive capacity* or *dielectric constant* for the derived quantity. In conformity with the other suggestions in this paper it would be more uniform to adopt *permittivity* and *relative permittivity*... Similarly in the magnetic case we have *permeability* and *relative permeability*. 1946 *Nature* 6 July 33/2 The paper contains a description of apparatus used for measurements at frequencies of about 3,000 Mc./s...and a statement of typical experimental results obtained with a specimen of polythene, which at the above frequency had a permittivity of 2·27. 1962 Corson & Lorrain *Introd. Electromagn. Fields* ix. 317 The permittivity of free space ϵ_0 can..be determined directly from measurements involving electrostatic phenomena. 1971 *Engineering* Apr. 44/1 The capacitance C (microfarads) of a parallel plate capacitor..is..proportional to the relative permittivity ϵ_r of the dielectric medium. 1974 Harvey & Bohlman *Stereo F.M. Radio Handbk.* vii. 148 Because of the presence of water vapour in particular, the permittivity of the air in the troposphere is greater than unity.

Permocarboniferous, *a.* Add: Also, pertaining to or including the Permian and the Carboniferous systems or periods. Also *absol.* Usu. written **Permo-Carboniferous.** (Earlier and later examples.)
1874 *Q. Jrnl. Geol. Soc.* XXX. 217 Taking into consideration the great thickness of the Carboniferous in Nova Scotia and the large development of this Upper Permo-Carboniferous member, it would not be surprising that in this last we may have a chronological equivalent of part at least of the European Permian. 1897 *Trans. Manchester Geol. Soc.* XXV. 207 (*heading*) On the Permo-Carboniferous boundary. 1928 C. Dawson *Age of Gods* i. 5 The vast glaciation of Permo-Carboniferous times..marks the end of the Primary Palæozoic world. 1955 G. G. Woodford tr. *Gignoux's Stratigr. Geol.* v. 156 The great variety of facies of the Permo-Carboniferous in Europe and the harmonious way in which they can be grouped make of this system, seen as a whole, a magnificent illustration of stratigraphic synthesis. 1971 A. G. Smith in I. G. Gass et al. *Understanding Earth* xv. 224/1 The distribution of Permo-Carboniferous tillites.

Pe:rmo-Pennsylva·nian, *a.* *Geol.* [f. as next + *Pennsylvanian *sb.* and *a.*] Belonging either to the lowest Permian or the uppermost Pennsylvanian. Also *absol.*
The equivalent in the U.S. of prec.
1937 *Bull. Amer. Assoc. Petroleum Geologists* XXI. 1252 Late Pennsylvanian and Permo-Pennsylvanian. Figure 2 represents a composite of late Pennsylvanian and early Permian time. 1965 *Ibid.* XLIX. 1572/2 An early sequence of Permo-Pennsylvanian sediments accumulated in this bay.

Permo-Triassic (pə̄:ɹmotrai,æ·sik), *a.* *Geol.* [f. *Permo-*, used as comb. form of Permian *a.* (*sb.*) + Triassic *a.*] Of or pertaining to the Permian and the Triassic systems or periods. Also *absol.*
1876 [see Silurian *a.* 2 a]. 1903 J. Le Conte *Elements Geol.* (ed. 5) III. v. 619 Some indications of glaciation have been reported from other horizons than the Permo-Triassic. 1926 *Proc. Geologists' Assoc.* XXXVII. 11 By the close of Permo-Triassic times the climate would have altered considerably from that at the commencement of the period. 1956 W. Edwards in D. L. Linton *Sheffield* i. 3 They are being exploited, for oil as well as for coal, below the 'cover' of Permo-Triassic rocks. 1969 Bennison & Wright *Geol. Hist. Brit. Isles* xii. 272 The drawing of the Permo-Triassic boundary is fraught with difficulties. 1977 *Offshore Engineer* May 51/1 Effective basement in the Gulf of Valencia area appears to be hard Carboniferous sandstones and clays, overlain in some areas by Permo-Triassic sands and evaporites.
So **Permo-Tri·as,** the Permo-Triassic system or period.
1926 *Proc. Geologists' Assoc.* XXXVII. 1, I am convinced that the divisions of the Permo-Trias, in Britain, are not time-divisions, but represent, to a considerable extent, merely different conditions of deposition. *Ibid.*, In Britain the Permo-Trias formations are always unconformable to and rest upon an uneven surface of the rocks below. 1969 Bennison & Wright *Geol. Hist. Brit. Isles* xi. 255 Strata of continental facies were laid down over wide areas in the succeeding Permian and Triassic Periods... Frequently considered together as the Permo-Trias, there are however significant palaeontological differences.

perm. s(ec)., permsec, abbrevs. of 'permanent secretary' (esp. in an African country).
1942 Partridge *Dict. Abbrev.* 78/2 *P.S.*..Permanent Secretary. Also *Perm.S.* 1975 J. Wyllie *Butterfly Flood* xxii. 95 The Perm. Sec's. house was..much pillared and porticoed. 1976 *Daily Times* (Lagos) 8 July 5/3 (*heading*) Permsec files counter-claim. *Ibid.* 16 Oct. 19/1 (*heading*) Why retired permsec's daughter stole in London.

permselective (pə̄ɹmsīle·ktiv), *a.* *Chem.* [f. Perm(eable *a.* + Selective *a.*] Of an ion-exchange membrane: permeable to anions but not to cations, or *vice versa*; more generally, selectively permeable to certain molecules or ions. Hence **pe:rmselecti·vity.**
1953 A. G. Winger et al. in *Jrnl. Electrochem. Soc.* C. 178/2 For the purpose of this paper, an ideal permselective membrane will be defined to be a membrane which, when subjected to a potential gradient, permits passage of cations to the exclusion of anions, or vice versa. *Ibid.* 179/1 In any electrolytic process ion concentrations are appreciable and perfect permselectivity is an idealized situation of little practical interest. 1962 S. B. Tuwiner *Diffusion & Membrane Technol.* ix. 177 The most important factor which affects permselectivity for a given membrane is the concentration of the electrolyte. 1968 *Encycl. Polymer Sci. & Technol.* VIII. 636 The ability of an ion-exchange membrane to discriminate between ions of different sign is called its permselectivity. 1969 W. R. R. Park *Plastics Film Technol.* viii. 197 A process which may come to have the universal utility of distillation is the use of permselective membranes for the separation of mixtures of liquids or gases. 1971 A. F. Stancell in Tobolsky & Mark *Polymer Sci. & Materials* xii. 263 The preferred permeation of one molecule through a polymer with respect to other diffusing molecules is termed membrane permselectivity.

permutate, *v.* Delete *rare* and add: **2.** (Later examples.) Also *absol.*
Probably regarded by those who use it as a back-formation from *permutation*.
1969 *Daily Tel.* 5 May 14/3 One gets the impression that..three basic dress designs were drawn up and three basic jacket patterns, and ever since designers have been ..endlessly permutating these like resigned pools punters. 1971 *Ibid.* 18 June 22 If your baldness cure is a mixture of known ingredients, however you permutate their quantities, it won't get patent protection. 1975 *Times Lit. Suppl.* 18 Apr. 416/3 The plot, involving revelations about hitherto unsuspected relationships and passions permutating with unforeseen power. 1978 *Gramophone* Jan. 1321/2 Coleman really did reject the old package of chords; but what Coltrane did was to find a new way of permutating them, running through the scales they suggested.

permutated (pə̄·ɹmiutēited), *ppl. a.* [f. Permutat(ion + -ed¹.] In *Football Pools*: subjected to permutation.
1947 [see *banker² 5].

permutation. Add: **2. c.** *Logic.* A form of immediate inference from a proposition by negating it and substituting a contradictory predicate; obversion.
1851 W. H. Karslake *Aids Study Logic* I. 64 The third form of Immediate Inference which we have to speak of is, what may be called *Permutation*. 1906 H. W. B. Joseph *Introd. Logic* 214 In Permutation, or (as it has been also called) Obversion, there is no transposition of terms, but the quality of the proposition is changed. 1931 R. M. Eaton *Gen. Logic* 206 Obversion, also known as permutation differs from conversion in that the subject and predicate do not change places.
4. b. In the semantic theories of Nils Gustaf Stern: see quot. 1931.
1931 G. Stern *Meaning & Change of Meaning* xiii. 361 Permutations are unintentional sense-changes in which the subjective apprehension of a detail—denoted by a separate word—in a larger total changes, and the changed apprehension (the changed notion) is substituted for the previous meaning of the word. 1933 *Mod. Lang. Notes* XLVIII. 386 The linguistically conditioned changes are.. analyzed into shifts of..'permutation' and 'adequation'. 1965 *Eng. Stud.* XLVI. 405 The type of semantic change involved is that called by Gustaf Stern 'permutation'.
4*. *Football Pools.* A system whereby any combination of a specified number of entries drawn from a larger, selected, number of chances may be considered for a dividend.
1952 *Times* 16 May 7/5 The whole business of forecasting, study of form, permutations, and all the rest of it is, in fact, pure nonsense. 1954 M. Croft *Spare the Rod* III. v. 198 Football pools—that's the safest bet... Once you've worked up an interest in permutations you can sit down and leave them alone for a whole day. 1959 *Listener* 19 Feb. 347/3 The complications of this revenge put as much strain on one's attention as does the filling of a 'Pools' coupon with recommended permutations. 1960 *Comp* 20 Feb. 7/2 Find the straightforward permutation of allowing for any 3 from 8.

permute, *v.* Add: **2. a.** (Later example.)
1846 *Proc. Philos. Soc.* III. 1 In certain cases a letter may have been permuted, that is, changed to some kindred letter.
b. *Logic.* To submit to the process of permutation or obversion.
1906 H. W. B. Joseph *Introd. Logic* 215 The process of permuting and then converting is called Conversion by Negation.
3. Also in *Linguistics.*
1967 D. G. Hays *Introd. Computational Linguistics* 153 We may be required to permute those elements. 1968 *Language* XLIV. 31 To account for the fairly flexible ordering of major constituents in a German sentence.., we must have a number of optional rules that permute the subject, object, and adverbials. 1975 N. Chomsky *Logical Struct. Linguistic Theory* x. 422 In actually formulating $\Phi_1{}^P$ as a grammatical transformation we must be careful to indicate that the element *K*..and the following verb are not permuted when *K* belongs to the preceding noun phrase.

permuted (pəɹmiū·tĕd), *ppl. a.* [f. Permute *v.* + -ed¹.] Subjected to permutation; transposed.
1846 *Proc. Philol. Soc.* III. 2 In Irish orthography, the permuted letter instead of being displaced by its substitute is merely preceded, or as the Irish grammarians express it, eclipsed by it. 1970 O. Dopping *Computers & Data Processing* xxiii. 373 Provided that the titles of the material one wishes to register are sufficiently descriptive, the permuted title systems have many advantages. 1971 *Computers & Humanities* V. 309 Scope of present study: to produce permuted indices. 1971 D. I. Slobin in W. O. Dingwall *Survey Linguistic Sci.* 352 Structures requiring permutation of elements will first appear in non-permuted form. 1976 *Gloss. Documentation Terms* (*B.S.I.*) 47 *Permuted title*, a title in which the significant words have been rearranged, so that they can be used in indexing.

permutite (pə̄·ɹmiūtəit, pəɹmiū·təit). [ad. G. *permutit*, f. L. *permūt-āre* to exchange + *-it* -ite¹.] **a.** Any of a class of artificial zeolites which are widely employed as ion-exchangers, esp. for the softening of water. **b.** Written **Permutit** or **permutit** (-it). A proprietary name for ion-exchange materials and equipment utilizing such zeolites. Also *attrib.*, as *permutit(e) process*, the softening of water by treatment with any of these substances.
1907 *Chem. Abstr.* I. 2755 He [*sc.* A. Feldoff] reports the use of Permutite (after Gans) in the refinery at Glogau where much better yields of sugar were obtained by its use than without it. 1910 *Trade Marks Jrnl.* 12 Jan. 44 Permutit... Chemical substances..being artificially prepared compounds for the purification of water, molasses and saccharine juices. *Ibid.* 53 Permutit... Filters and engineering and building contrivances for supplying and distributing water,.. J. D. Riedel Aktiengesellschaft,.. Berlin, Germany; manufacturers. 1911 *Chambers's Jrnl.* 29 Apr. 352/1 For the removal of iron and manganese and for the destruction of germ-life, manganese permutit is used in place of the sodium permutit. 1913 *Official Gaz.* (U.S. Patent Office) 4 Nov. 266/2 The Permutit Company, New York, N.Y. Filed Aug. 19, 1913. Permutit. ..water purifying and treating materials. 1917 A. Smith *Introd. Inorg. Chem.* (ed. 3) xxxv. 724 In the permutite process the water is simply filtered through sodium silico-aluminate. 1925 *Glasgow Herald* 13 Feb. 6/2 Under the head of artificial silicates mention should be made of glass, Portland cement, water glass, and the permutites. 1943 *Thorpe's Dict. Appl. Chem.* (ed. 4) VI. 276/2 After partition between 70% alcohol and hexane the alcoholic layer is filtered through 'permutite' to remove adrenalin. 1963 R. R. A. Higham *Handbk. Papermaking* x. 245 This process [*sc.* base exchange] relies on the exchange of sodium ions for those of calcium and magnesium and is referred to as the Zeolite or Permutit method of water softening. 1974 D. M. Adams *Inorg. Solids* vii. 255 This process [*sc.* cation exchange] forms the basis of the Permutit water-softening process in which Ca^{2+} is removed from 'hard' water by exchange for sodium. 1974 D. W. Breck *Zeolite Molecular Sieves* i. 11 The accepted term for synthetic aluminosilicates which are crystallographically amorphous and are prepared for the ion exchange properties is 'permutite'... The chemical composition of most permutites is represented by an empirical formula in terms of the oxides: $Na_2O{\cdot}Al_2O_3{\cdot}xSiO_2{\cdot}yH_2O$ in which *x* often has a value of 5–6.

pern, perne, varr. Pirn *sb.*²

pern (pə̄ɹn), *v.*³ Also **perne.** [f. *pern* var. Pirn *sb.*²] *intr.* In the poetry of W. B. Yeats: to move with a winding motion.
It has been suggested (1961 *N. & Q.* Jan. 9/1) that Pern *v.*³ is the same word.
1920 W. B. Yeats *Michael Robartes* (1921) 17 Though I had long perned in the gyre, Between my hatred and desire, I saw my freedom won And all laugh in the sun. 1920 [see *gyre *v.* 3]. 1928 [see *gyre *sb.* 1]. 1938 W. B. Yeats *New Poems* 34 Those new dead That come into my soul and escape Confusion of the bed, Or those begotten or unbegotten Perning in a band.

Pernambuco (pə̄ɹnæmbu·kō). Also **Fernambuk.** The name of a state in Brazil, used *attrib.* and *absol.* to designate the hard, reddish timber of the tree *Cæsalpinia echinata*, of the family Leguminosæ, which is used for dyeing as well as decorative woodwork. Cf. Brazil¹ 1.
1794 H. Barham *Hortus Americanus* 23 The true Brasil is called Pernambuca, being the place from whence they come in Brasil. 1829 C. Sealsfield *Tokeah* II. 124 The carriage whirled along the levee, through oyster and orange shops; mountains of Pernambuco logs and bricks. 1870 J. Yeats *Nat. Hist. Commerce* II. 219 Brazil wood is imported principally from Pernambuco, and is also known by the name of Fernambuk wood in allusion to the place of importation. 1902 G. S. Boulger *Wood* II. 271 Peachwood (*Cæsalpinia echinata* Lam: Order *Leguminosæ*). Central and South America. Known also as 'Lima, Nicaragua or Pernambuco-wood'. 1920 A. L. Howard *Man. Timbers of World* 40 Used as a dye-wood..there is nothing that will yield the same result as the Pernambuco or brazil-wood. 1943 Record & Hess *Timbers of New World* 275/2 The principal foreign demand for the timber at present is..for the manufacture of violin bows and in this trade it is known as Pernambuco, taking its name from the Brazilian state where the best grades originate. 1976 *Early Music* Oct. 521/1 (Advt.), Suppliers of wood for stringed and wind instruments: Poplar, Lime, Maple, Fruitwood, Ebony, Box, Rosewood, Walnut, Beech, Holly, Oak, Ash, Mahogany... African Blackwood and Pernambuco.

pernettya (pəɹne·tiă). Also **pernettia.** [mod.L. (C. Gaudichaud-Beaupré 1825, in *Ann. Sci. Nat.* V. 102), f. the name of A. J. *Pernetty* (1716–1801), French explorer.] A small evergreen shrub of the genus so called, belonging to the family Ericaceæ, native to South America, New Zealand, or Tasmania, and bearing leathery leaves and white or pink flowers, followed by white, red, or purplish berries.
1835 *Edwards's Bot. Reg.* XX. 1675 (*heading*) Pointed-leaved Pernettia. 1840 *Ibid.* XXVI. 63 (*heading*) Narrow-leaved Pernettya. 1888 *Garden* 4 Feb. 106/2 Where cut flowers are needed sprays of the Pernettyas are very useful. 1916 L. H. Bailey *Stand. Cycl. Hort.* V. 2555/1 The pernettyas are low much-branched shrubs with dense and small evergreen leaves and small nodding flowers, followed by very decorative berries varying in color from white to purplish black. 1962 *Listener* 6 Sept. 370/3 If you..want one of the heaviest berried shrubs I know, get some pernettya seedlings. 1973 C. D. Brickell in A. Gemmell *Sunday Gardener* iii. 85 *Pernettya* has already been mentioned for winter use, but the berries are well coloured by October.

pernicious, *a.*¹ Add: *pernicious anæmia*: this is now susceptible to treatment; (earlier and later examples): [tr. G. *perniciöse* (now *perniziöse*) *anämie* (A. Biermer 1868: see *Correspondenzblatt für schweiz. Aerzte* (1872) 1 Jan. 15).]; *pernicious contrary*, in paper-making, a substance difficult to detect in the raw material, which inhibits the pulping process. Cf. *contrary *sb.* 3 d.
1874 *Med. Times & Gaz.* 21 Nov. 581/2 (*heading*) Pernicious anæmia: a new disease. *Ibid.* 582/1 Under the name of 'Progressive Pernicious Anæmia', Dr. Biermer, of Zürich, has described an affection which differs from ordinary simple anæmia in a marked manner, and which ..appears to be a disease *sui generis*. 1936 *Discovery* Apr. 123/1 Pernicious anaemia, a disease which was reckoned incurable until 1926, and which, since then, it has been found possible to keep under control by large and frequent doses of fresh liver. 1961 [see *contrary *sb.* 3d]. 1963 R. R. A. Higham *Handbk. Papermaking* ii. 50 (*heading*) Pernicious contraries in waste paper. 1968 Passmore & Robson *Compan. Med. Stud.* I. v. 15/2 The patient with pernicious anaemia formerly died, because he was unable to transport one-millionth of a gram of the vitamin [*sc.* B_{12}] daily across one or two millimetres of the gut wall. Nowadays he receives an injection of the vitamin at fortnightly or monthly intervals. With this, he is in every respect a normal healthy person. 1972 *Listener* 21 Sept. 383/2 Some 5 per cent or so of waste paper is tainted by what the industry calls 'pernicious contraries' —substances like the bitumen used to waterproof paper or board.

pernio (pə̄·ɹnio). *Med.* Pl. **perniones** (pə̄ɹni,ōu·niz). [a. L. *pernio*, *perniōn-* chilblain (f. *perna* haunch or ham).] A chilblain.
1676 R. Wiseman *Several Chirurgicall Treat.* I. xiii. 62 Pernio is a peculiar Inflammation, and belongeth to Bloud; it raiseth a thick red Swelling with itching pain in the Hands and Feet. Those affecting the Hands are generally called Chilblanes... When they affect the Feet they are called Kibes. 1822 J. M. Good *Study of Med.* II. 313 That the pernio or chilblain belongs to the genus erythema is perfectly obvious. 1885 in tr. *H. von Ziemssen's Handbk. Dis. Skin* 169 Such is the mode in which chilblains (perniones) make their appearance. 1941 J. Masefield *Gautama* 19 Cures for all from pernio to spasms. 1959 H. L. DuVries *Surg. of Foot* v. 103 Repeated mild frostbite produces vasomotor instability resulting in chilblains, or perniones. 1974 *Arch. Dermatol.* CIX. 57/1 Our cases best fit into the classification of acute pernio or chilblains.

perniosis (pə̄ɹni,ōu·sis). *Med.* [mod.L., f. prec. + -osis.] A chilblained condition of the skin.
1896 N. Walker tr. *Unna's Histopath. of Dis. of Skin* I. 20 In rare cases the point of the nose..is the seat of a perniosis. 1952 M. K. Polano *Skin Therapeutics* iii. 97 A good cream in perniosis. 1968 R. J. Cairns et al. in A. J. Rook et al. *Textbk. Dermatol.* I. xv. 337/1 Papular perniosis may closely mimic erythema multiforme. 1974 Passmore & Robson *Compan. Med. Stud.* III. I. xvii. 27/1 (*heading*) Perniosis (chilblains).

pernoctate, *v.* Delete †*Obs. rare*—⁰ and add examples.
1923 *Blackw. Mag.* Aug. 250/1 Families of Oriental pilgrims, pernoctating within the Church, will squat down in front of the Tomb of Christ. 1975 *Wadham Coll. Gaz.* I. 1. 6 The second event that demands to be recorded is the arrival of the first women undergraduates. The first to pernoctate as a member of the college came from Kingston, Ontario, to read law. 1978 'M. Innes' *Ampersand Papers* iii. 28 A bedroom..and sitting-room were in permanent readiness for him should he be minded to pernoctate at Treskinnick.

Pernod (pẽrno, pə̄ɹno). Also **pernod.** The proprietary name of a drink manufactured by the firm of *Pernod* Fils and used as an aperitif; a glass of Pernod. Also *attrib.*
Before 1918 the name referred to a brand of absinthe.
[1876 *Trade Marks Jrnl.* 11 Oct. 574 Pernod Fils... Louis Pernod, on behalf of self and partner, Fritz Pernod, trading as Pernod, Fils, Pontarlier, Doubs, France; manufacturers of absinthe.] 1908 C. E. Griffin *Four Years in Europe* vi. 58 Men and women sit day and night, sipping their wine or *pernod* (absinthe). 1914 *Blast* June 13 Oh blast France... Blast aperitifs (Pernots, Amers picon) Bad change Naively seductive Houri salon-picture Cocottes. 1919 *Century Mag.* Aug. 446/2 He.. poured a glass of Pernod. 1928 J. Rhys *Postures* xvi. 163 Then she would drink a couple of Pernods..to deaden the hurt. 1931 *Daily Express* 23 Sept. 9/3 There are small cafés, with men and women drinking the yellow-greenish pernod, the near-absinthe, now that the manufacture of absinthe is forbidden, but as like absinthe as one pea is like another. 1936 Bentley & Allen *Trent's Own Case* xi. 130 Trent called for a Pernod. 1942 'N. Shute' *Pied Piper* i. 19 He would take a little glass of Pernod with Madame? 1958 E. Dundy *Dud Avocado* I. i. 9 Glancing down at my Pernod, I discovered..that I'd already finished it. 1964 *Sun-Herald* (Sydney) 21 June 11/2 Five men unconscious on the floor of the hold and five empty pernod bottles. 1967 S. Beckett *Stories & Texts for Nothing* vii. 110 But what is this I see..at pernod time. 1975 *Times* 29 July 5/8 Fendi shapes Pernod coloured mink..into trench coats. 1976 J. van de Wetering *Corpse on Dike* vi. 72 The first French I read here was on the label of the Pernod bottle.

peroba (pĕrōu·bă). [Pg., f. Tupi.] Any of several Brazilian hardwood trees, esp. *Aspidosperma peroba* and other members of this genus, belonging to the family Apocynaceæ, or *Paratecoma peroba*, of the family Bignoniaceæ; also, the wood of these trees.
1875 T. Laslett *Timber & Timber Trees* xxv. 184 Peroba is stronger than teak, but not so heavy. 1920 A. L. Howard *Man. Timbers of World* 209 Peroba Branca... The wood is light greyish-yellow in colour, close and fine in the grain. *Ibid.* 210 Peroba Rosa... This wood is of a pale rose colour with some darker streaks. 1936 *Nature* 9 May 790/2 First Class restaurant [of the Queen Mary]: Peroba, with feature panels in selected maple burr. 1956 *Handbk. Hardwoods* (Forest Prod. Res. Lab.) 185 Red peroba (Great Britain). The name peroba rosa is applied to a group of species of *Aspidosperma*, of which the principal one is *A. peroba*. 1971 F. H. Titmuss *Commerc. Timbers of World* (ed. 4) 236 White Peroba is essentially a timber for high-quality furniture, panelling and veneers. 1978 *Sunday Times* (Colour Suppl.) 18 June 47/1 Once lush rain forest, ablaze with jacaranda and peroba trees, settlement is rapidly turning into a wasteland.

perofskite. Add to def.: and usu. containing lanthanides or alkali metals in place of much of the calcium and often niobium in place of some of the titanium; also, a particular variety or specimen of this mineral, or (more widely) any mineral having the same crystal structure. Freq. *attrib.*, with reference to its structure. (The usual form is now *perovskite*.) (Earlier and further examples.)
1840 *Edin. New Philos. Jrnl.* XXIX. 418 (*heading*) Perowskite, a new mineral species. *Ibid.* It is named Perowskite, in honour of M. von Perowski, an intelligent Russian mineralogist. 1906 J. P. Iddings *Rock Minerals* 463 Perovskite occurs in basic igneous rocks, peridotite, melilite-basalt, alnöite, and in nepheline- and leucite-bearing rocks, that is, in titaniferous rocks low or comparatively low in silica. 1939 R. C. Evans *Introd. Crystal Chem.* viii. 204 The structure of perovskite corresponds to the composition ABO_3 and has the cubic or pseudo-cubic unit cell shown in Fig. 46. *A* ions are situated at the corners of the cell and a *B* ion at its centre, while the faces are centred by oxygen. *Ibid.* 205 The appearance of KIO_3 and $RbIO_3$ among the compounds with the perovskite structure is of particular interest. 1955 *Physical Rev.* XCVIII. 1201/2 $KNbO_3$ and $KTaO_3$ are ferroelectrics of the perovskite type. 1956 *Electronic Engin.* XXVIII. 132 The remarkable and technically important properties of $BaTiO_3$ and other perovskite-type ceramics. 1962 W. A. Deer et al. *Rock-Forming Min.* V. 50 Hevesy *et al.* (1929) reported that for a number of perovskites the ratio Ti : Nb was about 3,000 : 1 and Ti : Ta 7,000 : 1. 1970 *Mineral. Abstr.* XXI. 155/1 The composition of synthetic complex perovskites can be expressed in terms of simpler formula units... E.g. $Pb(W_{1/3}Fe_{1/4}Li_{1/4})O_3 = \frac{3}{4}[Pb(W_{1/3}Fe_{2/3})O_3] + \frac{1}{4}[Pb(W_{3/5}Li_{2/5})O_3]$.

Peronism (pe·rǫniz'm). [ad. Sp. *Peronismo* (also used), or f. the name *Perón* (see below) + -ism.] The political ideology of Juan Domingo Perón (1895–1974), president of Argentina from 1946 to 1955 and from 1973 to 1974, advocating nationalism and the organization of labour in the interests of social progress; the political movement supporting Perón or his policies.
1946 *Times* 4 June 5/6 Although the Labour Party is the left wing and the Peronista Radicals are the right wing of Peronismo, the schism is more a struggle for the fruits of office than a conflict of ideology. 1953 G. I. Blanksten *Perón's Argentina* III. xii. 281 Peronism.. was not naziism. It was not fascism. It was not communism... It was *justicialismo*, or the 'Third Position'. 1957 *Economist* 21 Dec. 1063/2 Peronism is near its end as an organized movement. General Perón has suffered two or three severe defeats since he was overthrown. 1962 *Listener* 28 June 1102/2 The late Frondizi government encouraged a revival of Peronism. 1964 G. Germani in I. L. Horowitz *New Sociol.* 408 A comparative analysis of fascism and peronism..may be found in Gino Germani, *La integración política de las masas y el totalitarismo*. 1964 J. R. Scobie *Argentina* ix. 233 'Peronismo', a doctrine and a political movement which emphasized the importance of the working man. 1972 *Buenos Aires Herald* 1 Feb. 7/2 Members of the youth movement charged..that death of militant member Enrique Castro

was due to groups in Peronism that followed the moderate line of former party chief Jorge Daniel Paladino. 1975 *New Yorker* 28 Apr. 18/1 (Advt.), A didactic, explosive, semi-documentary on Peronism and the necessity to create national consciousness in Argentina and other Latin-American countries. 1977 *Time* 11 Apr. 37/2 We are not ready for elections until Peronism is dismantled and forgotten.

Peronist (pe·rǫnist), *a.* [ad. Sp. *Peronista* (also used), or f. as prec. + -ist.] Of, pertaining to, or advocating Peronism. Also as *sb.*, a supporter of Perón or of Peronism.
1946 [see prec.]. 1955 *Times* 7 May 7/4 A group of 10 Peronista deputies belonging to the General Confederation of Labour submitted a Bill yesterday in the Argentine Congress. 1958 *New Statesman* 8 Mar. 291/2 Though Frondizi's victory was won with the help of the Communist and Peronist parties, he would have won even without their support. 1960 *Economist* 15 Oct. 243/2 The xenophobia of the Peronist years. 1963 *Times* 26 Feb. 11/6 The Peronists, despite their internal bickerings, retained their fundamental unity. 1969 J. Mander *Static Society* viii. 245 Borges never concealed his contempt for the *peronista* régime. 1970 G. Germani in I. L. Horowitz *Masses in Lat. Amer.* xvi. 595 The basis of the new political movement was provided by..Peronist penetration of the older unions. 1971 *Guardian* 17 Apr. 10/3, I was only 12 when Peron fell in 1955, but I have always been a Peronist. 1974 *Times* 15 Oct. 9/3 Three car loads of men..shot up the headquarters of the Peronist youth organization.

peroperative (pərǫ·pĕrĕtiv), *a.* *Surg.* [f. Per-¹ 1 + Operative *a.* and *sb.*] Given, performed, or occurring during the course of an operation. Hence **pero·peratively** *adv.*
1976 *Lancet* 14 Aug. 325/2 (*heading*) Single-dose peroperative antibiotic prophylaxis in gastrointestinal surgery. 1977 *Ibid.* 2 July 6/1 None of the four patients with preoperative or peroperative peritoneal soiling had post-operative infection. *Ibid.* 6 Aug. 304/2 There are several papers demonstrating the efficacy of a combination of lincomycin and gentamicin (or tobramycin) given in one or two doses peroperatively.

peroral (pərōə·răl), *a.* [f. Per-¹ + Oral *a.* (*sb.*)] Occurring or carried out by way of the mouth.
1908 *Lancet* 18 Apr. 1183/2 Sauerbruch's low-pressure apparatus, Brauer's high-pressure apparatus, and Kuhn's method of producing high pressure by per-oral tubage had now made the way for the surgeon an easy one. 1974 *Jrnl. Infectious Dis.* CXXX. 225 The susceptibility of the newborn mouse to peroral infection with a group B coxsackievirus (B_3) was compared with susceptibility to parenteral infection with this agent.
So **pero·rally** *adv.*
1934 in Webster. 1969 *Canad. Jrnl. Physiol. & Pharmacol.* XLVII. 841/2 The rats were administered water perorally. 1976 *Lancet* 18 Dec. 1320/1 Jejunal-biopsy specimens were taken perorally with the Crosby capsule.

perorative (pe·rorĕitiv), *a.* [f. Perorat(ion + -ive.] Appropriate to or suggestive of a peroration.
1921 *Glasgow Herald* 29 Oct. 4 Messrs. Hart (said Cleland in a perorative phrase..) are of that class in society who have found their way to philosophy without the aid of regular tuition.

peroratorical (pe:rǫrătǫ·rikăl), *a.* [f. Perorator after Oratorical *a.*] Characteristic of a peroration; perorational.
1927 C. E. Montague *Right off Map* vi. 56 His voice was taking on the peroratorical note.

perosis (pĕrōu·sis). *Vet. Sci.* [ad. Gr. πήρωσις maiming, f. πηροῦν to maim.] A disease of poultry (see quot. 1937).
1931 *Science* 4 Sept. 249/2 Several of the stations reported that when the 'uniform diet' was fed to chicks kept in confinement a high percentage of them became afflicted with perosis (deforming leg weakness). 1937 *Jrnl. Nutrition* XIV. 155 Perosis..is an anatomical deformity of the leg bones of young chickens, turkeys, pheasants, grouse and quail... The symptoms generally found are gross enlargement of the tibial-metatarsal joint, twisting or bending of the distal end of the tibia and of the proximal end of the metatarsus, and slipping of the gastrocnemius tendon from its condyles. 1950 *New Biol.* VIII. 104 There is some evidence that the disease in hens known as perosis may be due to manganese deficiency. 1970 *Poultry Sci.* XLIX. 1753/1 Perosis is an extremely complex condition that can be produced experimentally by an inadequate intake of several different nutrients.

perosmate (pəɹǫ·zmĕit). *Chem.* [f. Per-¹ 5 + Osmate.] A salt of osmium containing the anion $[OsO_4(OH)_2]^{2-}$, in which osmium has an oxidation state of 8. Formerly represented as $M_2O.OsO_4$ (where M is an alkali metal) and called osmiates (see Osmiate in Dict. and Suppl.).
The passage in ed. 7 (1890) of *Bloxham's Chem.* corresponding to quot. 1895 occurs s.v. *osmiate.
1895 Thomson & Bloxam *Bloxam's Chem.* (ed. 8) 470 By dissolving perosmic anhydride in potash, potassium perosmate is supposed to be formed, but this has not been isolated. 1949 *Thorpe's Dict. Appl. Chem.* (ed. 4) IX. 134/1 Addition of alcohol to the clear solution and washings precipitates the ruthenium..and reduces the perosmate to violet osmate. 1973 S. E. Livingstone in J. C. Bailar et al. *Comprehensive Inorg. Chem.* III. xliii. 1232 Potassium osmate $K_2[OsO_2(OH)_4]$..is best prepared by reduction with alcohol of potassium perosmate $K_2[OsO_4(OH)_2]$, which can be prepared by treating OsO_4 with cold KOH. *Ibid.*, The octavalent state..occurs in the tetroxide OsO_4, the perosmate $[OsO_4(OH)_2]^{2-}$, [etc.].

perovskia (pərǫ·vskiă). Also **perowskia, perowskya.** [mod.L. (G. Karelin 1841, in *Bull. Soc. Imp. des Naturalistes de Moscou* 15), f. the name of V. A. *Perovski* (1794–1857), once governor of the Russian province of Orenburg.] A herb or sub-shrub of the genus so called, belonging to the family Labiatæ, native to temperate regions of west and central Asia, and bearing panicles of deep blue flowers.
1907 *Gardeners' Chron.* 21 Dec. 426/1 In the case of Perovskia there is added charm in the pretty inflorescences late in autumn. 1961 *Amat. Gardening* 25 Nov. 8/3 It is best not to cut down the long stems of perowskia until the spring. 1973 C. D. Brickell in A. Gemmell *Sunday Gardener* iii. 85 *Perovskia*, with tall upright growths with grey-green aromatic leaves, will reach four feet, with long sprays of lavender-blue in September.

perovskite, var. Perofskite in Dict. and Suppl.

peroxidase (pərǫ·ksidĕiz, -s). *Biochem.* Formerly also **peroxydase.** [a. F. *peroxydase* (G. Linossier 1898, in *Compt. Rend. des Séances de la Soc. de Biol.* L. 373), f. *peroxyde* Peroxide: see *-ase.] Any of a large class of iron-containing enzymes found esp. in plants which catalyse the oxidation of a substrate by peroxides, usu. hydrogen peroxide.
1903 *Jrnl. Chem. Soc.* LXXXIV. I. 378 The substance is a very powerful peroxydase, and renders hydrogen peroxide..very active towards pyrogallol, gallic acid, aniline, [etc.]. 1907 J. B. Cohen *Org. Chem. Adv. Students* I. ix. 357 Many other vegetable oxidases have been described, including the so-called 'peroxidases' which activate hydrogen peroxide and possibly other peroxides. 1925 *Jrnl. Chem. Soc.* CXXVIII. I. 615 Peroxydases are found to be present in abundance in most common dried seeds in the resting condition. 1956 *New Biol.* XXI. 46 The same haem is shared by all haemoglobins and myoglobins, as well as..other compounds of biological importance, notably the enzymes catalase and peroxidase. 1971 *New Scientist* 25 Feb. 411/2 The presence of ethylene in a young pea cell causes an increase in the level of the enzyme peroxidase which is capable of tacking the OH group onto proline.
Hence **peroxida·tic** *a.*, characteristic of a peroxidase.
1945 *Biochem. Jrnl.* XXXIX. 300/2 It is conceivable that in addition to alcohols catalase may promote a peroxidatic oxidation of other biologically important substances. 1954 A. White et al. *Princ. Biochem.* xvi. 368 The catalatic splitting of hydrogen peroxide to water becomes merely a special case of a peroxidatic reaction, where hydrogen peroxide serves both as substrate and as acceptor. *Ibid.*, Catalase also exhibits peroxidatic activity. 1972 I. Yamazaki et al. in Åkeson & Ehrenberg *Struct. & Function Oxidation-Reduction Enzymes* 326 Peroxidatic reactions are catalyzed in the presence of various non-specific catalysts, such as transition metal ions and their coordination complexes.

peroxide. Add to def.: Now usu. restricted to those oxides which have at least one pair of oxygen atoms bonded to each other in the molecule, or which contain the anion $O_2{}^{2-}$. **b.** Any organic compound containing two linked oxygen atoms in its molecule. (Further examples.)
1858 B. C. Brodie in *Proc. R. Soc.* IX. 362, I have to add a new term to this series, of which hitherto no analogue has existed. This term is the peroxide of the organic radical,—the body which in the series of acetyl corresponds to the peroxide of hydrogen or barium in the series of the metal. 1881 *Chem. News* 20 May 233/1 The peroxide of ethyl remains as a dense syrupy liquid, miscible with water. 1922 H. G. Denham *Inorg. Chem.* xiii. 198 True peroxides are held to possess the linking present in hydrogen peroxide, —O—O—. 1950 N. V. Sidgwick *Chem. Elements* II. 871 From hydrogen peroxide are derived a large number of compounds containing the O—O link; not only organic derivatives such as the alkyl and acyl peroxides, the percarboxylic acids.. but also many inorganic derivatives, in which one or more oxygen atoms of a basic or inorganic oxide, or an oxy-acid, are replaced by O—O groups. The binary inorganic compounds are commonly known as peroxides: this name should be confined to O—O compounds, but is often extended to include any metallic oxides with an unusually large amount of oxygen, such as PbO_2 and MnO_2. 1971 *Nomencl. Org. Chem.* (I.U.P.A.C.) (ed. 2) C. 161 Ethyl phenyl peroxide $C_6H_5O{-}OC_2H_5$. 1973 E. A. V. Ebsworth et al. in J. C. Bailar et al. *Comprehensive Inorg. Chem.* II. xxii. 763 Organic peroxides decompose readily to give free radicals. *Ibid.* 784 Many transition metal peroxides are dangerously explosive.
2. Short for *peroxide blonde. colloq.
1918 G. Frankau *Poetical Wks.* (1923) II. 108 Thy merchant-princes, whose week-end peroxides Hastened, safe-screened in many a Triplex-glass car. 1919 E. Jordan *Girl in Mirror* (1925) ii. 45 'She's probably a peroxide,' he said. 'Even if she isn't, she can't hold a candle to your sister.'
3. Special Combs., as **peroxide blonde** *colloq.*, a woman with peroxided hair; **peroxide bond,** a single bond between two oxygen atoms in a molecule; **peroxide group,** the divalent group —O—O—; **peroxide hair,** hair bleached with hydrogen peroxide; **peroxide shampoo** (see quot.).
1920 S. Lewis *Main St.* 314 Have you heard about this awful woman that's supposed to have come here to do dressmaking—a Mrs. Swiftwaite—awful peroxide blonde? 1927 A. Christie *Big Four* xiv. 187 (*title*) The peroxide blonde. 1947 N. Marsh *Final Curtain* vi. 90 An old man ..doting on a peroxide blonde. 1974 *Times* 26 Oct. 8/8 The corny peroxide blondes with their plucked eyebrows. 1949 *Chem. Rev.* XLV. 399 Whether the decomposition of a given hydroperoxide involves radical or cationic intermediates probably depends on the amount of polarization of the peroxide bond. 1961 A. G. Davies *Org. Peroxides* x. 143 Familiarity with the homolysis of the peroxide bond delayed the recognition that the above reaction is often heterolytic. 1899 *Jrnl. Chem. Soc.* LXXVI. II. 659 The authors have obtained two new series of compounds which contain the peroxide group. 1939 F. A. Philbrick *Inorg. Chem.* xvii. 274 In structure the true peroxides are distinguished by containing in the molecule one or more peroxide groups, —O—O—, such as are present in hydrogen peroxide. 1968 *Jrnl. Chem. Soc.* A. 397/1 The great majority of transition-metal peroxide complexes involve a co-ordinated bidentate peroxide group, but a few are known in which the peroxide functions as a bridging ligand. 1937 W. S. Maugham *Theatre* i. 2 Notwithstanding her cropped peroxide hair and her heavily-painted lips she had the neutral look that marks the perfect secretary. 1966 J. S. Cox *Illustr. Dict. Hairdressing* 111/1 *Peroxide shampoo*, a soft soap shampoo incorporating a small quantity of 20 vol. peroxide of hydrogen and two or three drops of ·880 ammonium hydroxide.
Hence **peroxi·dic** *a.*, having the properties of a peroxide; containing or forming part of a peroxide group.
1945 *Jrnl. Org. Chem.* X. 416 This liberation of iodine indicated that peroxidic compounds had been formed in these solvents during ozonization. 1949 *Chem. Rev.* XLV. 385 In the decomposition of organic peroxides, rupture of the bond between the two peroxidic oxygens is often accompanied by cleavage of an adjacent carbon-to-carbon bond. 1956 *Nature* 21 Jan. 129/2 The first product is probably the hydroperoxide.., and indeed very labile, strongly peroxidic, solids have been detected, but isolation has not been possible. 1961 A. G. Davies *Org. Peroxides* iii. 56 Peroxylauric acid..loses about 15 per cent of its peroxidic oxygen after 1 week at 25°.

pero·xided, *ppl. a.* [f. Peroxide + -ed¹.] **a.** Treated with (hydrogen) peroxide. **b.** Having bleached hair. Also *absol.*
1906 B. von Hutten *What became of Pam* x. 71 Miss Vesey had highly peroxided hair and a manner of suspicious dignity. 1910 Baroness Orczy *Lady Molly* ix. 234 An over-dressed, much behatted, peroxided young woman. 1930 *Observer* 18 May 15/3 Simon, and Sylvia whom he loved, and Sylvia's peroxided mamma were agreeable company. 1946 Wodehouse *Joy in Morning* iv. 29 What was it? Blackmail? Does he want you to pinch damaging correspondence from the peroxided? Has some quick-thinking adventuress got him in her toils? 1947 D. M. Davin *Gorse blooms Pale* 205 The hair, peroxided it was. 1976 *Times* 5 Mar. 9/5 Two tanned and peroxided matrons in their fifties.

peroxisome (pərǫ·ksisōum). *Cytology.* [f. Peroxi(de + *-some⁴.] An organelle present in the cytoplasm of many kinds of cell which contains the reducing enzyme catalase and usu. some oxidases that produce hydrogen peroxide.
1965 C. de Duve in *Jrnl. Cell Biol.* XXVII. 25A The name *peroxisome* is proposed for the microbodies of rat liver and for the particles of similar biochemical nature existing in kidney and in *Tetrahymena pyriformis*. The peroxisomes contain large amounts of catalase and several hydrogen peroxide-producing oxidases. 1970 *Sci. Amer.* Sept. 113/3 Recently the cells of higher organisms have been found to contain organelles called peroxisomes, whose major function is thought to be the protection of cells from oxygen. 1975 *Science* 21 Nov. 787/1 Peroxisomes (microbodies), cytoplasmic constituents characterized morphologically by a single limiting membrane and a finely granular or homogeneous matrix, have recently been recognized as ubiquitous structures in animal and plant cells.
Hence **peroxiso·mal** *a.*
1969 *Jrnl. Protozool.* XVI. 430/2 The 6 acid hydrolases studied have almost identical distribution patterns, differing clearly from those of the mitochondrial malate dehydrogenase, peroxisomal catalase and urate oxidase. 1972 *McGraw-Hill Yearbk. Sci. & Technol.* 321/1 Although peroxisomal respiration will be considered primarily, the entire process of photorespiration involves the mitochondrion. 1975 *Science* 21 Nov. 789/1 Recent studies.. failed to demonstrate the presence of peroxisomal oxidases in these catalase-containing particles isolated from Harder's gland.

peroxo(-) (pərǫ·kso), *prefix* and quasi-*adj.* *Chem.* [f. Per-¹ 5 + *oxo(-): cf. Peroxide.] **A.** *prefix.* Used in inorganic chemistry in the names of compounds, complexes, etc., that contain a peroxide group. **B.** Hence as quasi-*adj.*: containing or being a peroxide group. Cf. *peroxy(-).
*peroxy(-) is the usual form in organic chemistry.

1910 *Jrnl. Chem. Soc.* XCVIII. II. 858 Tetraethylene-diamine-μ-ammoniumperoxocobalticobalte salts. *Ibid.* 869 On heating with concentrated sulphuric acid, it is decomposed with evolution of oxygen and nitrogen, the volumes of these gases obtained showing that the compound contains a peroxo-group and tervalent and quadrivalent cobalt. **1930** *Chem. Abstr.* XXIV. 2672 In the presence of air and light, peroxo compds. possessing unusual oxidizing powers are produced in addn. to the usual Fe^{+++} and Fe^{++} complex ions. **1943** THORNE & ROBERTS tr. *Ephraim's Inorg. Chem.* (ed. 4) xi. 321 The simplest..of the multinucleate cobaltammines are the peroxo-salts $[(NH_3)_5Co—O—O—Co(NH_3)_5]X_4$. **1959** R. S. CAHN *Introd. Chem. Nomencl.* 19 Peroxoacids, in which —O— is replaced by —O·O—, are similarly distinguished by the prefix peroxo- (peroxy- or simply per- have frequently been used in the past), as, for example, in HNO_4 peroxonitric acid. **1960** HESLOP & ROBINSON *Inorg. Chem.* xxxvii. 527 There are two peroxo-acids of sulphur, $H_2S_2O_8$, peroxodisulphuric acid, and H_2SO_5, peroxomonosulphuric acid. **1966** COTTON & WILKINSON *Adv. Inorg. Chem.* (ed. 2) xiii. 376 The peroxo acids are useful oxidants and sources of free radicals. **1971** *Nomencl. Inorg. Chem.* (I.U.P.A.C.) (ed. 2) 33 The prefix peroxo-, when used in conjunction with the trivial names of acids, indicates substitution of —O— by —O—O—. **1973** [see *PERBORIC *a.*].

peroxy(-) (pərǫ·ksi), *prefix* and quasi-*adj.* *Chem.* [f. PER-¹ 5 + OXY-: cf. PEROXIDE.] **A.** *prefix.* Orig. used in the names of compounds containing a larger proportion of oxygen than the parent compound, now in the names of compounds, radicals, etc., that contain a peroxide group. **B.** Hence as quasi-*adj.*: Containing or being a peroxide group. Cf. *PEROXO(-).

*PEROXO(-) is now the usual form in inorganic chemistry.
1878 *Jrnl. Chem. Soc.* XXXIV. 237 The relation between hæmatin and peroxyhæmoglobin must be sought by investigating the bearing of the albumin. **1900** *Chem. News* 16 Feb. 83/2 We thus obtain a microcrystalline compound, the peroxysulphate of silver, which is decomposed by warm water. **1912** *Jrnl. Chem. Soc.* CII. II. 156 The reaction may be considered a general one for differentiating between three peroxy-salts and hydrogen peroxide additive products. **1956** *Nature* 28 Jan. 182/1 Hydrogen in and on the metal now becomes anodic, and by reaction with oxygen doubtless goes into transient peroxy compounds. **1961** A. G. DAVIES *Org. Peroxides* iii. 55 There are three important types of acyl peroxides—the peroxyacids.., the peroxyesters.., and the diacyl peroxides. **1961** E. G. E. HAWKINS *Org. Peroxides* p. xii, Peroxy (RO_2·) radicals,..give substitution products with many reactive molecules. **1964** *Economist* 25 July 356/2 A virulent family of pollutants known as peroxy-acylnitrates (PANs). **1971** *Nomencl. Inorg. Chem.* (I.U.P.A.C.) (ed. 2) 41 In conformity with the practice of organic nomenclature, the forms peroxy [etc.]..are also used but are not recommended. **1973** E. A. V. EBSWORTH et al. in J. C. Bailar et al. *Comprehensive Inorg. Chem.* II. xxii. 780 The peroxy-group may replace an oxygen atom as a bridge between two other groups, giving compounds such as $Me_3SnOOSnMe_3$. **1975** J. O. SCHRECK *Org. Chem.* iii. 60 Acyl radicals are precursors of peroxyacyl nitrates, which are lacrimators often associated with smog.

Perp, perp. (pɜ̄rp), abbrev. PERPENDICULAR *a.* 3. Now freq. in colloq. allusive use. Also *ellipt.*
1867 *Murray's Handbk. Trav. Yorkshire* 116/1 The great features of the exterior..are the North Porch, and the West Front... Both of these are Perp... The West Front..is as fine an example of a Perp. composition. **1894** K. BAEDEKER *Gt. Brit.* (ed. 3) 407 The most striking features of the exterior [of York Minster] are the noble *W. Façade* (Dec.; towers, 201 ft. high, Perp.) [etc.]. **1933** J. E. MORRIS *Northumberland* (ed. 3) 247 Much of the other work is modern and misleading,..the 'Perp.' W. window of the S. aisle, *c.* 1832. **1937** A. CHRISTIE *Dumb Witness* vii. 68 Though an attractive specimen of what the guidebook calls Early Perp. it [*sc.* a church] had been so conscientiously restored in Victorian vandal days that little of interest remained. **1945** J. BETJEMAN *New Bats in Old Belfries* 46 Grey-blue of granite in the small arcade (Late Perp). **1951** N. PEVSNER *Middlesex* 143 The tomb in the aisle chapel is purely Perp. **1957** 'J. WYNDHAM' *Midwich Cuckoos* I. 10 The church is mostly perp. and dec., but with a Norman west doorway and font. **1967** 'M. HUNTER' *Cambridgeshire Disaster* ix. 60 The usual slab of village with a nice old Perp. church. **1974** SHERWOOD & PEVSNER *Oxfordshire* 118 The wall-shafts of the nave [of Christ Church Cathedral] have Perp. shafts with concave-sided capitals. **1974** *Times* 10 Jan. 16/7 His love of Gothic architecture made him befriend cathedrals, abbeys and churches, particularly the Perp and Dec so abundant in East Anglia.

perpend, *v.* **1. a.** (Later examples.)
1930 J. BUCHAN *Castle Gay* xi. 172 He retired to the inn ..to write out his notes and perpend the situation. **1939** JOYCE *Finnegans Wake* I. 187 Perpending that Putterick O'Purcell pulls the coald stoane out of Winterwater's and Silder Seas sing for Harreng our Keng. **1966** 'H. MACDIARMID' *Company I've Kept* vii. 172 The chapters in Professor Hofrichter's book..should be carefully perpended by those who fail even yet to see the full significance of Ezra Pound's great slogan: 'Make it new!'

perpension. Delete †*Obs.* and add later example.
1890 R. L. STEVENSON *Let.* 13 July (1911) III. 165 Upon these points, perpend, and give me the results of your perpensions.

perpetual, *a.* (*adv.* and *sb.*) Add: **1.** Phr. *perpetual student* (also *perpetual scholar*): used of one who stays on as a student at a university or similar institution far beyond the normal period.
1924 G. CALDERON tr. *Tchekhof's Cherry Orchard* I. 132 Yes, I expect I shall be a perpetual student. *Ibid.*, The 'Perpetual Student' has become a common type in Russia during the last fifteen or twenty years. **1960** G. BUTLER *Death lives Next Door* i. 10 The perpetual scholar; the man who is always proceeding to the next and then the next degree. **1967** E. GRIERSON *Crime of one's Own* xix. 154 He looked a little on the young side..to spend his life in lending libraries. The perpetual student? **1975** 'J. BELL' *Victim* vi. 71 The real university-type fanatics and loud-mouthed perpetual students.

c. *perpetual calendar*, substitute for def.: (*a*) a calendar which can be adjusted to show any combination of day, date, and month; (*b*) a set of tables from which the day of the week can be reckoned for any date; (*c*) (in full *perpetual calendar clock*) a clock which indicates the date and automatically takes account of the length of each month.
1895 *Montgomery Ward Catal.* 254/3 Perpetual Calendar, made of celluloid and..imitation silver..silk ribbons with dates, etc. **1960** H. HAYWARD *Antique Coll.* 215/1 *Perpetual calendar clock*, clock including a calendar, which corrects itself for the short months... The mechanism consists of a slotted wheel revolving once a year..with slots of varying length which control the movement of a lever, allowing it to pass one or more teeth of the calendar wheel at a time. **1962** E. BRUTON *Dict. Clocks & Watches* 130 *Perpetual calendar*, calendar worked by a clock which corrects for months of different lengths (and sometimes for leap years also). **1971** L. P. DAVIES *Shadow Before* iii. 30 A perpetual calendar confirmed his earlier estimate. **1973** W. J. BURLEY *Death in Salubrious Place* v. 105 The perpetual calendar said Wednesday August 25th. Clarissa had changed it that morning.

f. Of an investment: irredeemable. Also *ellipt.* or as *sb.* Cf. ANNUITY 3.
1869 *Bradshaw's Railway Manual* XXI. 30 Perpetual 4 per cent. stock... Present terminable 4½ per cent. stocks. **1882** R. BITHELL *Counting-House Dict.* 224 Consols—commonly called Bank Annuities—are perpetual. **1948** G. CROWTHER *Outl. Money* (rev. ed.) ii. 75 The mediums, the longs and the perpetuals are liquid at a price. **1965** PERRY & RYDER *Thomson's Dict. Banking* (ed. 11) 436/2 *Perpetual annuity*... The purchaser cannot obtain the principal back, but he can sell his right to the annual payment to someone else.

2. c. *perpetual check*: in Chess and related games, a situation in which one player cannot prevent the other from making an unlimited sequence of checking moves (and thus obtaining a draw). Also ellipt. as *perpetual*.
1820 J. S. BINGHAM tr. *E. Dal Rio's Incomparable Game of Chess* 44 At the end of the game..the Rook sometimes draws the game against the Queen, and is admirable in making a drawn game by perpetual check. **1856** C. TOMLINSON *Chess Player's Ann.* 120 'You might have drawn the game,' said he, 'by perpetual check.' **1960** R. C. BELL *Board & Table Games* I. ii. 60 [Shatranj.] *Perpetual check* was considered a drawn game. **1973** *Correspondence Chess* Spring 385/2 Black either has to take a perpetual or allow White to queen a P with a dangerous check.

d. *perpetual-flowering carnation*, a variety of carnation with a long flowering season, usually grown in a cool greenhouse, as it is not hardy out of doors.
1885 T. BAINES *Greenhouse & Stove Plants* 93/2 Carnation (Perpetual Flowering)... With a sufficient number of plants they may be had in bloom all the year round. **1900** J. DOUGLAS in W. D. Drury *Bk. Gardening* iii. 55 Perpetual-flowering Carnations are generally propagated by slips. **1926** M. C. ALLWOOD *Carnations* viii. 55 The Perpetual-flowering Carnation can be planted in May. **1960** *Times* 29 June 17/3 Perpetual-flowering carnations and pinks form a very attractive exhibit. **1971** S. BAILEY *Perpetual-Flowering Carnations* (rev. ed.) 18 The perpetual-flowering carnation owes its origin to at least two *Dianthus* species, namely *D. caryophyllus* and *D. sinensis.*

perpetualism (pəɹpe·tiu,ăliz'm). [f. PERPETUAL *a.* + -ISM.] Lasting, perpetual, or universal quality, *spec.* as a doctrine in political science or religion.
1885 *Encycl. Brit.* XIX. 391/1 Cosmopolitanism..and what has been called perpetualism, or the assumption of a system applicable to every social stage, were alike discredited. **1905** *Westm. Gaz.* 31 July 6/7 Perpetualism abolishes both hell and heaven. **1931** *Observer* 8 Nov. 24 The hybridist who can develop even a slight measure of perpetualism in this plant will sweep the horticultural board of its best cups and medals.

perpetuana. For † *Obs.* read *Hist.* and add later example.
1972 A. PLUMMER *London Weavers' Company 1600–1970* xiv. 292 Indian wrought silks and painted and dyed calicoes became extremely fashionable, taking the place of English silks, half-silks, slight silks, worsted stuffs, says and perpetuanas.

perpetuant. Add: **2.** = PERPETUANA.
1753 E. BOWEN *Map of Devon*, The chief Manufactures [of Devon] are Kerseys, Serges, Druggets, Perpetuants, fine and coarse Cloths, and Lace, in all which many Families are employd.

perpetuate, *v.* Add: Also *absol.*
1894 E. FAWCETT *New Nero* ii. 26 That soulless and mysterious will-to-live, which for ever creates, protects, and perpetuates, though blindly and dumbly, unconscious that she does either.

perpetuative (pəɹpe·tiu,ătiv), *a.* [f. PERPETUATE *v.* + -IVE.] Having a tendency or inclination to perpetuate.
1957 P. WORSLEY *Trumpet shall Sound* 273 Such movements can be either revivalistic—stressing the readoption of customs fallen into desuetude—or perpetuative—seeking to maintain the existing order. **1977** *Dædalus* Summer 61 My personal view is that anthropology is shifting from a stress on concepts such as structure, equilibrium, function, system to process, indeterminacy, reflexivity..but with a tender perpetuative regard for the marvellous findings of those who, teachers of the present generation, committed themselves to the discoveries of 'systems' of social relations and cultural 'items' and 'complexes'.

‖ **perpetuum mobile** (pɜ̄ɹpe·tuŏm mōᵘ·bili, -um mōᵘ·bile). [f. L. *perpetu-us* continuous + *mobil-is* movable, after PRIMUM MOBILE.] **1.** = *perpetual motion* s.v. PERPETUAL *a.* 1 b. Freq. in allusive use.
a **1688** CUDWORTH *Treatise Freewill* (1838) 28 This is an ever bubbling fountain in the centre of the soul, an elater or spring of motion, both a *primum* and a *perpetuum mobile* in us, the first wheel that sets all the other wheels in motion, and an everlasting and incessant mover. **1904** B. RUSSELL in *Mind* XIII. 337 'There is no *perpetuum mobile*' does not mean 'whatever exists differs from the *perpetuum mobile*'. **1933** J. N. FINDLAY *Meinong's Theory of Objects* iii. 89 That China is a Republic, that there is no *perpetuum mobile*, that dirigible airships exist, all these may indifferently be called circumstances or states of affairs. **1953** R. F. C. HULL tr. *Jung's Psychol. & Alchemy* in *Coll. Wks.* XII. II. iii. 172 This theme was already hinted at in dream g, with its pendulum clock, a *perpetuum mobile*. **1964** V. NABOKOV *Defence* x. 162 The passionate chess player is just as ridiculous as the madman inventing a *perpetuum mobile*.
2. *Mus.* = *moto perpetuo* s.v. *MOTO.
1893 J. S. SHEDLOCK tr. *Riemann's Dict. Music* 588/2 *Perpetuum mobile*..the name given to pieces written from beginning to end in notes of equal, and short value. **1938** *Oxf. Compan. Mus.* 708/2 *Perpetuum mobile*,..a rapid instrumental composition that proceeds throughout in notes of the same value. **1969** W. S. NEWMAN *Sonata since Beethoven* xii. 508 The rondo is a virtual *perpetuum mobile*. **1972** H. TISCHLER tr. *Apel's Hist. Keyboard Music to 1700* xix. 664 One of the four pieces..is a kind of *perpetuum mobile* in 12/8. **1978** *Times* 6 Apr. 16/8 Ravel asked Enesco if he would mind running through the whole sonata with him. They played it through, including the punishing perpetuum mobile.

perphenazine (pəɹfe·năzīn). *Pharm.* [f. PI)PER(IDINE + PHEN(YL + *AZINE, constituent parts of the systematic name.] A whitish powder that is a derivative of phenothiazine and has actions and uses similar to, but stronger than, those of chlorpromazine, being used as a sedative and anti-emetic and in the treatment of alcoholism; 2-chloro-10-{3-[4-(2-hydroxyethyl)piperazin-1-yl]propyl}-phenothiazine, $C_{21}H_{26}N_3OClS$.
1957 *Jrnl. Pharmacol. & Exper. Therap.* CXX. 376 Sch3940 is now known by the generic name of perphenazine and is marketed under the trade name, Trilafon. **1971** P. K. BRIDGES *Psychiatric Emergencies* vi. 158 Chlorpromazine.., trifluoperazine..and perphenazine (Fentazin, Trilafon) are commonly given to patients with neurotic illnesses. **1978** *Nature* 23 Mar. 331/2 Chlorpromazine and perphenazine were found..to be as effective as conventional tricyclic antidepressant drugs in the treatment of certain forms of depression.

perrhenic (pəɹ,rī·nik), *a.* *Chem.* [f. PER-¹ 5 b + *RHEN(IUM + -IC.] *perrhenic acid*, a strong acid, $HReO_4$, which is known only as a colourless aqueous solution and is an oxidizing agent.
1929 *Chem. Abstr.* XXIII. 1833 Yellow oxide, Re_2O_7,.. is water sol[uble], hygroscopic, forming a strong acid, perrhenic acid. **1962** P. J. & B. DURRANT *Introd. Adv. Inorg. Chem.* xxiv. 1019 Rhenium heptoxide dissolves freely in water to form perrhenic acid. *Ibid.*, Many salts of perrhenic acid are known. **1973** R. D. PEACOCK in J. C. Bailar et al. *Comprehensive Inorg. Chem.* III. xxxix. 946 Perrhenic acid is a strong acid; this is shown by the pH of its aqueous solution, its attack upon metals, [etc.].
Hence **perrhe·nate** [-ATE⁴], a salt of this acid; the anion ReO_4^-.
1929 *Chem. Abstr.* XXIII. 4632 Perrhenates of non-volatile bases can be ignited in O_2 without decompn. **1950** N. V. SIDGWICK *Chem. Elements* II. 1295 Perrhenates are formed with great ease by the action of oxidizing agents..on metallic rhenium or its lower oxides and their derivatives. **1962** COTTON & WILKINSON *Adv. Inorg. Chem.* xxx. 801 After concentration, the perrhenate is precipitated by addition of potassium chloride as the sparingly soluble salt, $KReO_4$. **1972** *Inorg. Syntheses* XIII. 219 This difficulty has been overcome for ReH_9^{2-} by a synthesis of the disodium salt in which an ethanol solution of sodium perrhenate is reduced with sodium metal to give the hydride in *ca.* 35% yield.

Perrier² (pe·rie). The proprietary name of an effervescent natural mineral water from the South of France.

1907 *Trade Marks Jrnl.* 30 Oct. 1928 Perrier... Natural mineral water obtained from the spring known as 'Source Perrier' situated at Vergeze in France. St. John Harmsworth, trading as Perrier,..London,..merchant. **1907** *Yesterday's Shopping* (1969) 27 Mineral waters (natural). ..Perrier..Table water. **1928** A. HUXLEY *Point Counter Point* xiii. 244 If only Grace could be bottled like Perrier water. **1957** [see *ITALIAN *sb.* 6]. **1960** I. FLEMING *For your Eyes Only* 14 He always stipulated Perrier, for in his opinion expensive soda water was the cheapest way to improve a poor drink. **1973** D. MACKENZIE *Postscript to Dead Let.* 173 Tully was drinking beer. There was a bottle of Perrier in front of Misty. **1975** *New Yorker* 30 June 33/3 She measures his orange juice and Perrier water as carefully as if it were a Martini. **1977** *Rolling Stone* 30 June 73/4 At the Caffe Tartufo, while I lunch, she orders a Perrier water and tosses a carefully rolled stick of Doublemint into her mouth.

perrierite (pe·riərait). *Min.* [a. It. *perrierite* (Bonatti & Gottardi 1950, in *Atti d. Accad. naz. d. Lincei: Rendiconti. Classe di Sci. fis.*, etc. IX. 361), f. the name of Carlo *Perrier* (1886–1948), Italian mineralogist: see -ITE¹.] A silicate of lanthanides, titanium, iron, and other elements, occurring as black or reddish brown, prismatic, monoclinic crystals.
1951 *Chem. Abstr.* XLV. 7923 (*heading*) Perrierite, a new mineral. **1962** *Mineral. Mag.* XXXIII. 45 These experiments..suggest that perrierite is probably an oxidized form of chevkinite. **1966** *Amer. Mineralogist* LI. 1394 Recent *x*-ray studies of metamict minerals which occur in Virginia have shown that some of the materials heretofore reported as allanite or chevkinite are actually perrierite. **1971** *Ibid.* LVI. 308 Chevkinite and perrierite exhibit thermal polymorphism only within certain compositional ranges.

Perrier-Jouët (pe·rie,ʒue). The proprietary name of the champagne produced by the firm of Perrier-Jouët of Epernay.
1891 in C. Ray *Compleat Imbiber* (1967) IX. 122 All Brands of Champagne in stock..Pommery Greno, Perrier Jouet, Heidsieck, Giesler. **1899** O. WILDE *Importance of being Earnest* III. 136 He drank..an entire pint bottle of my Perrier-Jouet, Brut, '89. **1914** C. MACKENZIE *Sinister St.* II. III. xii. 745 Forty Good Eggs drank forty-eight bottles of Perrier Jouet '93. **1920** G. SAINTSBURY *Notes on Cellar-Bk.* v. 71 The very best [champagne] I ever had was a Perrier-Jouet. **1971** J. DOXAT *Drinks & Drinking* 165 *Perrier Jouet*, this champagne has a fine 1961 extra dry *cuvée reserve*. **1975** N. FREELING *What are Bugles blowing For?* xvi. 93 The waiter had brought a 'Belle Epoque' bottle of Perrier-Jouet.

perryite (pe·ri,əit). *Min.* [f. the name of Stuart H. *Perry* (1874–1957), U.S. mineralogist + -ITE¹.] A nickel-rich mineral also containing silicon and phosphorus that is reported to have been found in several meteorites.
1965 FREDRIKSSON & HENDERSON in *Trans. Amer. Geophysical Union* XLVI. 121/2 The Horse Creek, Baca County, Colorado, iron meteorite, found in 1937, is unique for four reasons:..(4) this iron contains about 3% of Perryite, nickel silicide, a new mineral... The approximate chemical composition of Perryite is Ni 81%, Si 12%, Fe 3% and P 5%. **1968** *Mineral. Mag.* XXXVI. 852 The South Oman enstatite chondrite also contains perryite, but its abundance is much less than in Kota-Kota... In this meteorite, perryite occurs as small laths, usually associated with kamacite. **1970** *Geochim. et Cosmochim. Acta* XXXIV. 169 Electron-microprobe analysis of Si, P and Ni in the metal grains and associated schreibersite and perryite of eight enstatite achondrites shows that the compositions of these phases are relatively constant within a given meteorite, but show substantial variations between meteorites.

persecution. Add: **1. d.** *Psychol.* The irrational sense of being victimized by malign forces which features in many forms of mental disorder and is now commonly considered paranoid.
1881 *Jrnl. Nerv. & Mental Dis.* VIII. 28 We may have delusions of persecution which are systematized and such which are unsystematized. **1883** T. S. CLOUSTON *Clin. Lect. Mental Dis.* vi. 255 The third great class of delusional cases are those of suspicion and persecution. **1926** W. MCDOUGALL *Outl. Abnormal Psychol.* xx. 337 The two forms of delusion mentioned.., delusions of persecution and of grandeur, are the fundamental and most frequent. **1970** HINSIE & CAMPBELL *Psychiatric Dict.* (ed. 4) 542/1 A form of paranoia characterized by more or less incessant quarrelsomeness due to alleged persecution.
4. (esp. in sense *1 d) *persecution complex*, *fantasy*; **persecution mania** (earlier and later examples); also *transf.* and in extended uses; hence **persecution maniac**, a person of unbalanced mind suffering from delusions of persecution.
1961 J. HELLER *Catch-22* (1962) xxvii. 294 You've got a bad persecution complex. You think people are trying to harm you. **1966** 'H. MACDIARMID' *Company I've Kept* v. 139 The delusions of a man unbalanced by a persecution-complex. **1971** *Rand Daily Mail* 3 Apr. 5/8 We shall develop a persecution complex and go round moaning that we are misunderstood. **1950** T. WIESENGRUND-ADORNO et al. *Authoritarian Personality* IV. xvi. 615 The persecution fantasy of what the Jews *might* do to her, is used..as a justification of the genocide committed by the Nazis. **1892** D. H. TUKE *Dict. Psychol. Med.* II. 934/2 From a medico-legal point of view, cases of persecution-mania afford matter for consideration of the greatest importance. **1903** R. FRY *Let.* 21 Jan. (1972) I. 200 Langton Douglas..has the persecution mania..and I need hardly say B.B. looms large. **1934** *Punch* 7 Mar. 278/1 The present [Russian] régime's..persecution-mania, misrepresentation of the outside world and progressive debasement of the arts are..set-backs to the advantage of what Miss Hamilton..calls 'the Insect State'. **1942** E. WAUGH *Put out More Flags* iii. 203 He had left his persecution mania downstairs with his hat and umbrella. **1968** C. RYCROFT *Crit. Dict. Psychoanal.* 115 Persecution mania is an obsolete term for paranoia. **1978** I. MURDOCH *Sea* 69, I have never gone in for persecution mania and do not propose to start now. **1943** J. S. HUXLEY *Evolutionary Ethics* iii. 19 Unable to bear the condemnation of his super-ego, the persecution-maniac projects this into society, thus..being able to accuse the world of cruelty or oppression. **1955** KOESTLER *Trail of Dinosaur* I. 19 The persecution-maniacs in the West who still lived on the Red Scare had by that time become a dwindling, reactionary minority.

persecutory, *a.* **1. a.** (Later examples.) **b.** (Later examples.) Now used esp. with reference to *PERSECUTION 1 d.
1881 *Jrnl. Nerv. & Mental Dis.* VIII. 33 We have such [*sc.* delusions] of a depressive erotic character, usually persecutory. **1936** W. S. SADLER *Theory & Pract. Psychiatry* xix. 334 The child can very early acquire a 'persecutory complex' built up out of his memory feelings of being many times unjustly treated. **1952** *Brit. Jrnl. Psychol.* XLIII. 81 Religion is a form of psychotherapy which promotes a belief in the existence of idealized good objects as a defence against persecutory and depressive guilt. **1962** *Lancet* 19 May 1065/1 He soon developed persecutory ideas and was admitted to hospital. **1963** *Observer* 2 June 18/6 Cameras planted all over the place like persecutory eyes.

perseverate (pəɹse·vərĕit), *v.* *Psychol.* [Back-formation from PERSEVERAT(ION; cf. L. *persevērāre* to persevere.] *intr.* To repeat a response after the cessation of the original stimulus, in various senses of *PERSEVERATION 2.
1915 *Brit. Jrnl. Psychol.* VII. 388 The varying degree in which their ideas after disappearing from consciousness continue to 'perseverate' unconsciously. **1927** C. SPEARMAN *Abilities of Man* xvii. 293 They seem not even to have cared to inquire whether..he who perseverates in one kind of operation may be expected to do so..in others. *Ibid.* 306 A tendency for mental processes to have a certain lag or inertia and in this meaning to 'perseverate'. **1968** P. MCKELLAR *Experience & Behaviour* xii. 324 One subject started to draw an analogy with a gramophone record, and perseverated. **1976** SMYTHIES & CORBETT *Psychiatry* vii. 105 He may be talkative but difficult to understand, and perseverates—that is replies to questions with responses appropriate to previous questions. **1977** *Word 1972* XXVIII. 214 Often, the archiunit is one which was used appropriately some time ago in the conversation. It then pops up again and again—both in the patient's own speech and in his misunderstanding of the speech which he hears. In the latter case, the patient is said to *perseverate* on the archiunit.
Hence **perse·verating** *ppl. a.*; **perse·verative** *a.*; **perse·verator**, one who perseverates.
1915 *Brit. Jrnl. Psychol.* VII. 389 This inhibition of a succeeding by the effect of a preceding experience must also be expected to reveal itself in the relative slowness of the perseverator. **1923** E. JONES *Papers on Psycho-Anal.* (ed. 3) xxxiii. 445 The perseverating influence of this last reaction is also to be noticed in the next succeeding one. **1924** R. M. OGDEN tr. *Koffka's Growth of Mind* iv. § 5. 178 This 'perseverative tendency' of certain methods deserves special consideration. **1927** C. SPEARMAN *Abilities of Man* xvii. 292 The perseverator has been assumed to be stable in his emotions and steadfast in his purposes. **1943** H. READ *Educ. through Art* v. 150 On the other hand we have slight ability to direct attention, inclination to concentrate on colour to the neglect of form, perseverative talent—all symptoms of the tetanoid type. **1964** M. CRITCHLEY *Developmental Dyslexia* vi. 23 Words may be repeated in a perseverating fashion, e.g. *The cat the cat.* **1972** *Science* 16 June 1227/1 Attempts to reproduce such effects [*sc.* of medial temporal lesions] in animals have been largely fruitless; perseverative or disinhibitory effects have been the most frequent outcome. **1977** *Lancet* 10 Dec. 1227/2 She had a generalised memory disorder, momentary confusion, and mental and motor inertia and perseverative responses.

perseveration. **1.** Delete † *Obs.* and add later examples.
1915 *Brit. Jrnl. Psychol.* VII. 388 The 'Perseveration'-qualities of character, *i.e.* perseverance or persistency of will. **1971** *Where* Nov. 333/2 The *Pinky and Perky* annual shows this nauseating little pair in stories, pictures, strip cartoons, puzzles, crosswords and as objects to paint. Faced by this merciless perseveration, an adult reader can be excused for thinking longingly of an efficient bacon-slicer.
2. *Psychol.* **a.** The tendency for an activity to be persevered with or repeated after the cessation of the stimulus to which it originally responded, studied as an aspect of behaviour.
[**1901** *Brain* XXIV. 620 G. E. Müller and Pilzecker have shown that an image..that has occupied consciousness tends to rise again to consciousness spontaneously,.. a tendency to which they give the name 'Perseverationtendenz'.] **1915** *Brit. Jrnl. Psychol.* VII. 388 Is Perseveration a general factor,—comparable with General Ability, influencing the entire range of mental activity? **1916** A. A. BRILL tr. *Freud's Leonardo da Vinci* v. 96 We call such a repetition a perseveration. It is an excellent means to indicate the affective accentuation. **1927** C. SPEARMAN *Abilities of Man* iv. 42 He [*sc.* Müller] writes: 'Consistency of thought and action that extends beyond the immediately given is based to an essential degree upon perseveration.' *Ibid.* 52 As most fundamental of all concepts involved, we may pick out that of 'perseveration'. **1951** J. C. FLUGEL *Hundred Yrs. Psychol.* (ed. 2) xi. 322 One of the most remarkable [general characteristics] concerns a factor known as *p* ('perseveration'), manifesting itself as a general inertia, which..makes it difficult for the subject to pass rapidly from one kind of mental operation to another. **1963** T. ALCOCK *Rorschach in Pract.* xi. 172 He is also showing some perseveration of theme on winged objects, whether creatures or emblems. **1973** P. E. VERNON in J. R. Royce *Multivariate Anal. & Psychol. Theory* 127 Tests of different kinds of perseveration, inertia or rigidity, often showed little or no correlation with one another.
b. The mechanical and involuntary repetition of a motor or verbal response, despite a change of stimulus, as a result of brain damage or organic malfunction; usu. distinguished from the stereotypy associated with schizophrenia.
1910 *Lippincott's New Med. Dict.* 715/1 *Perseveration*, the senseless repetition of a word just pronounced or an act accomplished: either a functional or an exhaustion psychosis. **1937** *Jrnl. Mental Sci.* LXXXIII. 144 Perseveration, an extremely common symptom of organic disease..can be present in various degrees, and may extend to words, phrases, actions or to the total behaviour. **1961** W. R. BRAIN *Speech Disorders* v. 62 Perseveration may lead to difficulties, the whole or part of a previous word being repeated and thus persisting to contaminate the new word which should be evoked. **1966** I. B. WEINER *Psychodiagnosis in Schizophrenia* iv. 36 In 'fixed concept' perseveration a response that accurately corresponds to a blot when first given is repeated on subsequent blots without regard for actual blot qualities. **1976** M. HAMILTON *Fish's Schizophrenia* (ed. 2) iii. 58 Perseveration consists in continuing to carry out a goal-directed activity after the need for this activity has ceased. It..is different from a stereotypy because it has been initiated by a goal-directed activity.

Persian, *a.* and *sb.* Add: **A.** *adj.* **1.** Also, of or pertaining to a Persian cat.
1889 H. WEIR *Our Cats* 28 Tabby is not a Persian colour. **1972** ING & POND *Champion Cats of World* II. 73 When the two varieties [*sc.* the Angora and the Persian] were mated together by the early fanciers it was found that the Persian characteristics were dominant, so gradually all the Longhairs were classified as 'Persian'.
2. *Persian carpet* (earlier and later examples), (*flower*) *pattern*; **Persian cat**, a long-haired cat, formerly one belonging to a variety introduced from Persia, distinguished by broad, round heads, small ears, stocky bodies, and thick, rather woolly fur; (earlier and later examples); **Persian cord** (earlier example); **Persian lamb**, the skin or pelt of the karakul lamb, which has a silky, tightly-curled appearance; a coat made from this fur; **Persian lilac**, a shrub, *Syringa persica*, of the family Oleaceæ, bearing panicles of fragrant, mauve or white flowers; **Persian sheep**, a southern African breed of sheep, kept for the meat it produces; also called the blackhead Persian; **Persian ware** (examples); **Persian wheel** (later example); **Persian Yellow** (rose), a variety of the Austrian briar, *Rosa foetida* var. *persiana*, which bears fragrant, double, yellow flowers and was brought back to England from Persia by Sir Henry Willock in 1837.
1616 T. ROE *Jrnl.* 11 Mar. in *Embassy to Court of Gt. Mogul* (1899) I. 143 Vnder foote it is layd with good Persian Carpetts of great lardgnes. **1621** J. CHAMBERLAIN *Let.* 10 Nov. in *Mem. Amer. Philos. Soc.* (1939) XII. II. 406 The Moscovie ambassador..brought divers presents of ermins, sables, blacke foxe, Persian carpetts wrought with gold..besides a faire Persian tent. **1844** A. W. KINGLAKE *Eothen* xxvii. 394 A few Persian carpets (which ought to be called Persian rugs, for that is the word which indicates their shape and dimension,) are sometimes thrown about near the divan. **1969** *Times* 12 Dec. (Kenya Suppl.) p. viii/3 A friend..drove all the way to Rhodesia, with her savings invested in a dozen Persian carpets supporting three dogs in the back of a station wagon. **1978** *Times* 22 Nov. 2/1 Public auction sale of genuinely rare and valuable Persian Carpets on behalf of foreign creditors. **1821** BYRON *Don Juan* III. xviii. 157 Two parrots, with a Persian cat and kittens, He chose from several animals he saw. **1824** M. R. MITFORD *Our Village* I. 289 The white kitten..has succeeded to his lamented grandfather, our beautiful Persian cat. **1889** H. WEIR *Our Cats* 24 The Persian cat..differs somewhat from the Angora. **1903** F. SIMPSON *Bk. of Cat* vii. 98/1 In classing all long-haired cats as Persians I may be wrong, but the distinctions between Angoras and Persians are of so fine a nature that I must be pardoned if I ignore the cat commonly called Angora, which seems gradually to have disappeared from our midst... It is my intention to confine my division of cats to long-haired or Persian cats, and short-haired or English and foreign cats. **1935** E. B. SIMMONS *Cats* xxvii. 133 In the beginning there were Persian cats..and Angora cats... Interbreeding has made the two one. **1972** ING & POND *Champion Cats of World* II. 72 Eventually all Angora and Persian cats were referred to as Longhairs. **1873** *Young Englishwoman* June 312/1 Two good merinos, one Russell or Persian cord. **1889** Persian lamb [see LAMB *sb.* 5 b]. **1892, 1899** [see *BROADTAIL]. **1937** J. LAVER *Taste & Fashion* xv. 215 All kinds

of combinations of furs were tried: squirrel collared with fox, Persian lamb trimmed with mink kolinsky. **1959** J. LUDWIG in *Tamarack Rev.* Summer 6 She..walks.. not in her Persian lamb..but in that worn cloth coat. **1972** C. DRUMMOND *Death at Bar* v. 123 Mrs. Gaukroger, wearing a purple pyjama suit under her Persian lamb coat. **1640** J. PARKINSON *Theatrum Botanicum* 1468 This Persian Iasmine (or Persian Lilac, whether you will) is a shrub, or shrubby plant. **1712** [see LILAC 1 b]. **1800** *Curtis's Bot. Mag.* XIV. 486 The Persian Lilac is a shrub of much humbler growth than the common sort. **1847** F. A. KEMBLE *Let.* 12 Dec. in *Rec. Later Life* (1882) III. 301 Her maid has been with me this morning, with..a bunch of delicious Persian lilac. **1861** [see LILAC 1 b]. **1959** A. MOOREHEAD *No Room in Ark* ii. 62 The incredibly sweet scent of Persian lilac hung in the air. **1975** V. NABOKOV *Tyrants Destroyed* 47 A bright breeze ruffled the Persian lilacs. **1870** D. ROCK *Textile Fabrics* p. lxvi, Though there be seen the 'homa', the 'cheetah', and other elements of Persian patterns, still the discordant two-handled vase..betrays the textile to be not Persian, but Syrian. **1895** *Montgomery Ward Catal.* Spring & Summer 37/2 Ladies' Shirt Waists... New fancy Persian patterns. **1971** *Guardian* 30 Mar. 9/2 Printed Crimplene sweaters in a Persian flower pattern. **1912** R. LYDEKKER *Sheep & its Cousins* x. 209 The Persian fat-rumped sheep..is a well-known breed, which has been carried to Cape Colony and Rhodesia, where it is now bred to a considerable extent. **1932** S. ZUCKERMAN *Social Life Monkeys* xii. 206, I have seen Chacma baboons playing about and foraging in the midst of a flock of Persian sheep. **1966** E. PALMER *Plains of Camdeboo* xvi. 256 'I guess it is a black-head Persian from the Karoo.' 'You're right,' replied Sir Abe [Bailey] with delight. 'It is Karoo Persian.' When he travelled he took with him live Persian sheep from the Karoo to be slaughtered when he needed them. **1903** M. L. SOLON *Hist. Old French Faïence* iv. 52 This..bears a distant likeness to the Persian ware. **1971** L. A. BOGER *Dict. World Pott. & Porc.* 165/1 The earliest extant examples of Persian underglaze blue and white wares date from the 15th century. **1972** H. HODGES *Pottery* 53 Throughout the twelfth century the larger part of Byzantine polychrome pottery is best seen as a poor copy of Persian wares. **1864** J. A. GRANT *Walk across Afr.* xvii. 403 The Persian wheel, with its hanging earthen jars, overhangs the river [Nile] and..raises the water to the heights of the fields and gardens. **1843** *Gardeners' Chron.* 25 Feb. 121/1 The new Persian Double Yellow Rose..is an entirely different variety from *Rosa Harrisonii*; it is very like the old double yellow. **1848** W. PAUL *Rose Garden* 97 Persian Yellow; flowers of the deepest yellow, large and full; form, globular... Introduced from Persia by Sir H. Willock in 1837. **1898** M. A. VON ARNIM *Elizabeth & her German Garden* 17 The Persian Yellows look as though they intended to be big bushes. **1911** E. WILLMOTT *Genus Rosa* II. 271 It is easy to understand the popularity of the beautiful Persian Yellow Rose. **1971** 'L. BLACK' *Death has Green Fingers* iv. 37 Two other strains of roses were added to the family tree, one species called the Persian Yellow.
b. In names of colours associated with Persia or its products, e.g. *Persian blue, green, red.*
1869 *Bradshaw's Railway Manual* XXI. 460/2 (Advt.), Reds..Persian Red..Venetian Red. **1873** *Young Englishwoman* Aug. 390/1 Bright but delicate shades, such as sky blue, Persian green, and blush rose. **1886** H. C. STANDAGE *Artists' Man. Pigments* v. 51 Besides the Persian red obtained from the ochres, there is another which is a chromate of lead. **1903** M. L. SOLON *Hist. Old French Faïence* ix. 139 Some of the earliest pieces are painted upon Persian blue ground. **1934** H. HILER *Notes Technique Painting* ii. 116 *Persian green*, synonym for emerald green. **1951** R. MAYER *Artist's Handbk.* ii. 57 *Persian orange*, lake made of aniline colour on a barytes or blanc fixe base. Not permanent. **1963** tr. *Kornerup & Wanscher's Handbk. Colour* 177/1 *Persian blue*, the colour of Persian porcelain... Before about 1912, this name referred to the colour of certain Persian fabrics dyed with indigo blue. *Ibid. Persian orange*, this name probably corresponds to Persian yellow, a pigment derived from a compound of arsenic and sulphur.
B. *sb.* **4.** Insert 'usu.' before 'used' and add further examples.
1818 M. EDGEWORTH *Let.* 8 Sept. (1971) 85 Little Lady Louisa flying about with her green persian sash floating. **1853** MRS. GASKELL *Ruth* I. ii. 39 Miss Hilton! where have you put the blue Persian?
6. = *Persian cat* in sense A. 2 in Dict. and Suppl.
1871 *Graphic* 22 July 75/3 A Persian, direct from Persia,..'a very amiable beast'. **1879** M. E. BRADDON *Vixen* III. i. 52, I have a Persian who has been my attached companion for the last ten years. **1902** A. BENNETT *Anna of Five Towns* vii. 152 The Persian with one ear met them in the lobby, his tail flying. **1921** C. VAN VECHTEN *Tiger in House* i. 15 Jessie Pickens had a very remarkable brown tabby Persian. **1934** M. V. HUGHES *London Child of Seventies* x. 113 The parlour cats were Persians, sat on laps and best chairs. **1956** G. DURRELL *My Family* ix. 116 The consul was a great cat-lover, and he possessed three large and well-fed Persians to prove it. **1972** ING & POND *Champion Cats of World* II. 72 The fur texture also differed, that of the Angoras being soft, fine and silky, while the Persians were more woolly.
7. Other misc. ellipt. or substantival uses, esp. = *Persian carpet* in sense *A. 2.
1897 [see *FLORAL *sb.* 2]. **1903** W. D. ELLWANGER *Oriental Rug* (1904) ii. 15 The Persians came first, and perhaps in the following order of excellence: Kirman, Sehna, [etc.]. **1905** *Daily Chron.* 9 Sept. 3/2 'Persians' and other East Indian skins are stripped of their original tannage, and then finished as morocco. **1915** J. WEBSTER *Dear Enemy* 43 New rugs on the floor (my own prized Persians). **1957** M. MCCARTHY *Memories Catholic Girlhood* viii. 197 In the winter, she would have on her mink or her Persian or her squirrel or her broadtail. **1960** *News Chron.* 12 Sept. 6/4 Persians can be geometric or floral. **1964** *House & Garden* Dec. 42/3 The rugs are Persians and Kelims **1970** *Times* 12 May 11/8 The design of the print is a combination of paisley and persian.

Persianization (pɜ̄ːɹʃănəizēⁱ·ʃən). [f. PERSIAN *a.* + -IZATION.] The process of making Persian in appearance, structure, or other attributes.
1910 *Encycl. Brit.* XIII. 479/2 This extreme Persianization of Urdu was due rather to Hindu than to Persian influence. **1916** G. A. GRIERSON *Linguistic Survey India* IX. I. 45 This extreme Persianisation of Hindōstānī is not ..the work of conquerors ignorant of the tongue of the people. *Ibid.*, Like Urdū it [*sc.* Dakhinī] is written in the Persian character, but is much more free from Persianisation. **1924** P. BROWN *Indian Painting under Mughals* 18 The earlier Mughal emperors..introduced the art of miniature painting into Hindustan from Persia. Their action..represents..only one comparatively small item in a fairly wide movement. From the nature of this movement it may be termed the 'Persianization' of Northern India. **1937** *Scrutiny* V. 446 Mr. Chib..attributes the Persianization of Urdu and the Sanskritization of Hindi to communal jealousy. **1948** D. DIRINGER *Alphabet* II. vi. 362 *Hindi* is the modern development of Hindustani which is free from Persianization.

persicary. Delete *Obs.* and add later examples.
1938 M. HADFIELD *Everyman's Wild Flowers & Trees* 131 Spotted Persicary, *Polygonum Persicaria*... A branching annual up to 2 ft. high... Pale persicary..is a very similar plant. **1952** *Common Farm Weeds Illustr.* (Plant Protection Ltd.) I. 60 Willow-weed (*Polygonum persicaria*). Other names: Redshank, Persicary. **1972** R. ADAMS *Watership Down* xlv. 375 The weeds of harvest —knot-grass and pimpernel..heartsease and persicary.

persimmon. Add: **3.** (Further U.S. examples.) Phr. *to be a huckleberry to* (or *over*) *someone's persimmon*: see *HUCKLEBERRY 4.
1841 *Spirit of Times* 18 Dec. 499 They had not forgotten that the game little mare had put Sarah up to 7:45–7:40, in March last, and it seemed as if it was now their turn to 'shake down the persimmons'. **1844** in Sperber & Trittschuh *Amer. Pol. Terms* (1962) 313/2 David Tod should go there and repeat that original remark of his about the longest pole knocking down the persimmons. **1845** *Knickerbocker* XXV. 425 Wall now, that are's a jump above my tallest persimmons. **1857** *Call* (San Francisco) 3 Apr. 4/2 He will deal himself four aces and his opponent four queens, so that your honor will perceive he must 'rake the persimmons'. **1861** in W. H. Russell *My Diary North & South* (1863) II. iii. 62 Let both parties meet where there will be no interruption at the scalping business, and the longest pole will knock the persimmon. **1900** F. P. DUNNE *Mr. Dooley's Philos.* 68 'I'll jus' move me music back a mile,' he says, 'an' peg away, an' th' longest gun takes th' persimmons,' he says. **1946** *California Folklore Q.* July 240 That's the ripe persimmon. That is just right, or taken at the best moment.
3*. *U.S.* **a.** The colour of persimmon fruit, yellow to red-orange. **b.** The colour of persimmon wood, reddish brown. Also *attrib.* and *Comb.*
1928 S. V. BENÉT *John Brown's Body* 150 Grievin' yaller gals always does all right. Next time I'se gwine to git me a coal-black gal. I'se tired of persimmon-skins. **1975** *Vogue* Dec. 103 Persimmon lipstick. **1977** *Time* 27 June 50/1 The thickly painted figures with features eroded by light, the sharp eupeptic color—emerald, persimmon, rust, ultramarine. **1977** *New Yorker* 10 Oct. 132/2 They looked forward eagerly to sporting their persimmon outfit, say, in the first round of the club championship.
4. *persimmon-beer* (earlier and later examples), *-bush*.
1737 J. BRICKELL *Nat. Hist. N. Carolina* 38 The following are made in the Country, *viz.* Cyder, Persimmon-Beer, made of the Fruit of that Tree, [etc.]. *a* **1941** P. B. BARRINGER *Natural Bent* (1949) xxvi. 189 In the early seventies alcohol was everywhere in the South, and cut glass decanters stood on every sideboard... Beer was just coming, unless we except 'persimmon beer' and 'locust beer' made on every plantation and in many village homes. **1950** *Publ. Amer. Dial. Soc.* XIV. 51 *Persimmon beer*, a beverage made from ripe persimmons. **1786** G. WASHINGTON *Diary* 8 Aug. (1925) III. 102 A parcel of small Persimon bushes. **1944** G. WILSON *Passing Institutions* 177 Many an upland field not good for cultivation formerly had its flock of sheep, browsing among the sassafras and persimmon bushes.

persistence. Add: **2. b.** *persistence of an impression* (also simply *persistence*): used chiefly with reference to vision (so *persistence of vision*). (Further examples.)
1902 *Encycl. Brit.* XXVII. 95/1 Cinematograph... This apparatus shows in rapid sequence a series of views representing closely successive phases of a moving object, and persistence of vision creates the illusion that the object is in motion. **1924** J. P. C. SOUTHALL tr. *H. von Helmholtz's Treat. Physiol. Optics* II. 228 The best way to realize this persistence of the impression is to turn the eye to a perfectly dark field after it has been gazing at bright objects. **1944** R. W. MONCRIEFF *Chem. Senses* iii. 56 Although it has not been measured the persistence of odour must be short like that of taste. **1952** PIRENNE & ABBOTT tr. *Piéron's Sensations* IV. iii. 283 It is generally stated that visual sensations have a persistence of a tenth of a second. **1966** C. W. WILMAN *Seeing & Perceiving* vi. 52 Persistence of vision is one reason..why a series of pictures presented at short intervals produces a satisfactory effect in the cinema.
c. (The duration of) the emission of light by a luminescent substance after the cause of the luminescence has ceased; *persistence characteristic* (see quot. 1950¹).
1917 *Physical Rev.* IX. 297 It appears that the persistence of luminescence is due to the consistency of the substance and disappears as the fluidity increases. **1935** *Proc. IRE* XXIII. 1325 Slight changes in either the screen material or the method of manufacture may result in large changes..in the persistence with which the light continues to emanate from the excited portion of the screen after the electron beam has been removed. *Ibid.*, Screens which have a short persistence are desired for recording on moving film. *Ibid.* 1341 The persistence characteristic is taken by the use of some form of stroboscopic apparatus by means of which one is able to measure the relative brilliance of the screen at definite times after excitation. **1950** RIDER & USLAN *Encycl. Cathode-Ray Oscilloscopes* vii. 157/1 A plot of the brightness.. versus the decay time of the phosphorescence is termed the persistence characteristic of the screen. **1950** H. W. LEVERENZ *Introd. Luminescence of Solids* v. 150 For (conventional) luminescence emissions, all persistences longer than about 10^{-8} sec are called phosphorescence to indicate an abnormal delay. **1967** *Electronics* 6 Mar. 127/1 In the 710B, the top scope is coated with P7 phosphor for good persistence.

persister. Restrict *rare* to sense in Dict. and add: **b.** *Biol.* A bacterium which continues to live in the presence of enough antibiotic to kill almost all members of its species.
1944 J. W. BIGGER in *Lancet* 14 Oct. 498/1 These abnormal cocci have been termed 'persisters', to denote their power of surviving in the presence of sufficient penicillin to be lethal for the normal forms. **1949** E. CHAIN et al. in H. W. Florey et al. *Antibiotics* II. xxxv. 1143 The proportion of persisters in cultures of staphylococci is not constant, and the factors governing their numbers are not yet fully understood. **1970** *Jrnl. Med. Microbiol.* III. 669 Another type of persister is a cell in a non-replicating phase on which the antibiotic cannot act because of the absence of cell-wall synthesis. **1977** *Lancet* 15 Oct. 822/2 Electron microscopy was the only technique to demonstrate successfully the presence of small numbers of bacterial persisters in the vegetations of this treated case of streptococcal endocarditis.

persnickety (pəɹsni·kĕti), *a.* (*adv.*) *U.S. colloq.* Also **persni·kity.** varr. PERNICKETY *a.* Hence **persni·cketiness.**
1905 *Dialect Notes* III. 63 *Persnickety*,..disagreeable, or snippy. 'They acted mighty *persnickety*.' **1915** *Ibid.* IV. 215 Fern is more persnikity about her clothes than either of the other girls. **1922** W. STEVENS *Let.* 5 May (1967) 227, I have no desire to be persnickety about the arrangement of the group. **1940** *Sat. Even. Post* 10 Feb. 25 Dad was too old to begin being what he called 'persnickety'. **1960** WENTWORTH & FLEXNER *Dict. Amer. Slang* 384/1 *Persnickety, pernickety* adj., fussy; fastidious; punctilious; snobbish. Since *c.* 1890. 'Pernickety' is now obs.; 'persnickety', once dial., is now colloq. **1967** R. STEIN *Great Cars* 222/1 That archetype of persnickety Yankee toolmakers. **1977** *Time* 17 Oct. 48/2 Billy compensates for his brother's sweet-eyed psalm-singing and persnicketiness.

person, *sb.* Add: **I. 1.** Also, *persons of the drama* [tr. DRAMATIS PERSONÆ]. *lit.* or *fig.*
1895 G. B. SHAW *Our Theatres in Nineties* (1932) I. 39 The persons of the drama belong rather to the world of imagination than of reality. **1948** M. SHARP *Flowering Thorn* II. iii. 72 Thus admitted, so to speak, among the persons of the drama, the young American rose to his feet.
II. 2. d. Also, of a woman.
1935 *Punch* 18 Dec. 678 This Pearl person has neglected to say whether [etc.]. **1939** *Punch* 23 Aug. 198/1 She was a sort of secretary person down at the works.
f. Used (*a*) as a substitute for MAN *sb.*¹ (esp. sense 4 p: also for BOY *sb.*¹, etc.) as second element in numerous *Combs.* relating to offices which may be held by a member of either sex, as *chairperson, salesperson*; (*b*) with preceding defining word, as *marketing person*, and in other fanciful formations of this type, as *henchperson.*
In practice usually employed to avoid alleged sexual discrimination and widely regarded as having amusing connotations.
1971 *Sci. News* 11 Sept. 166 A group of women psychologists thanked the board for using the word 'chairperson' rather than 'chairman'. **1971** *Sci. Amer.* Dec. 37/1 (Advt.), If there is any doubt at the counter, let him show the salesperson this ad. **1972** *Listener* 24 Aug. 232/1 Two young black women will almost certainly join Representative Shirley Chisholm in Congress..putting up the number of black 'Congresspersons' to at least 14. *Ibid.*, Yvonne Brathwaite Burke..the stunning and extremely saucy 'Vice-Chairperson'. **1973** *Ibid.* 1 Mar. 286/3 Chairperson Mitchell and her henchpersons looked at the way education brainwashes girls. **1974** *Black Panther* 23 Feb. 9/2 Brother Malcolm Kelley, chairperson for the Committee for Justice for Tyrone Guyton. **1976** 'L. BLACK' *Healthy Way to Die* ii. 19 You're a newspaperwoman—or, as we have to say in these days of female emancipation, a newspaper person. **1976** *Publishers Weekly* 16 Feb. 81/1 The author [*sc.* Jeanne Wilson] is a solid craftsperson who tells her old-fashioned story in a winning manner. **1976** *Oxford Times* 23 July 32 (Advt.), Builders' merchants require yardperson. **1976** *Jrnl. R. Soc. Arts* CXXIV. 510/1 The exercise known amongst marketing men, or should I say marketing persons, as market segmentation. **1977** *Times Lit. Suppl.* 29 Apr. 506/5 A pair of homosexual network anchorpersons. **1977** C. SAGAN *Dragons of Eden* ii. 41 A group of literary Englishpersons, immobilized in the Alps by inclement weather. **1978** *Amat. Photographer* 29 Nov. 119/3 We saw nothing in cine to rival the spectacular application of high-technology design to still cameras for everyman (sorry, everyperson).

IV. 6. b. Euphemistically, the genitals.
1824 *Act 5 Geo. IV* c. 83 § 4 Every Person wilfully, openly, lewdly and obscenely exposing his Person in any Street, Road or public Highway, or in View thereof, or in any place of public Resort, with intent to insult any Female.., shall be deemed a Rogue and Vagabond within the true Intent and Meaning of this Act. **1853** MR. JUSTICE MAULE in *Law Jrnl. Rep.* XXXI. III. 123/1 What do you mean in law by exposing his *person*? The indictment should have been for exposing his *private parts.* **1911** *Straits Times* (Singapore) 13 June 7/3 He let go my arms, held me round the waist with his right arm and used his left hand. He stooped to do it. He put his hand on my person. **1973** R. E. MEGARRY *Second Miscellany-at-Law* ii. 165 Few readers of the newspapers can be unaware of the curious convention whereby for many years past the word 'person' was used anatomically in prosecutions for indecency.
8. (Later examples.)
c **1850** [see APOCOPATE *ppl. a.*]. **1905** [see THOU *pers. pron.* 2 b]. **1951** V. NABOKOV *Speak, Memory* i. 17 In addressing me, a small boy, he used the plural of the second person. **1962** J. G. BENNETT *Witness* xxv. 333 She never said 'I', but always referred to herself in the third person as 'Madame' or even 'she'. **1966** J. DERRICK *Teaching Eng. to Immigrants* ii. 89 The teacher must proceed slowly and patiently, and again has good reason for confining questions to 1st and 2nd person forms only at first.
VIII. 14. *Comb.* **person-object** (*a*) *Gram.*, a personal object of a verb; (*b*) in psychoanalytic theory, the choice of a person as the object of one's libidinal energy; also *attrib.*; **person-oriented** *a.*, of that in which interest or concern is centred on the person as contrasted with (by implication) a theory or thing; **person-perception,** perception which leads to or constitutes awareness and understanding of another person or persons.
1928 H. POUTSMA *Gram. Late Mod. Eng.* (ed. 2) I. I. iii. 176 It will, therefore, often be useful to distinguish person-objects and thing-objects. Even when both objects, considered apart from the context, are the names of things, one of them more or less distinctly suggests, through its connexions, thoughts of personal qualities. **1949** M. MEAD *Male & Female* vii. 154 The distinction between mother's body and the own body..in person-object terms, is an important one. **1954** *Essays in Crit.* IV. 316 The pre-Hellenic nature cults are accused..of failing to own that the person-to-person drive must push on past the person-object situation to find a response which plays back. **1964** E. BECKER in I. L. Horowitz *New Sociol.* 123 The schizophrenic is..someone who has been accustomed to relating to symbol-objects rather than to person-objects. **1972** *Encycl. Psychol.* II. 336/1 An object of experience, especially a 'person-object'. **1967** M. L. KING JR. in *Freedomways* VII. 114 We must rapidly begin the shift from a 'thing-oriented' society to a 'person-oriented' society. **1972** *Jrnl. Social Psychol.* LXXXVI. 135 The concept of dependency has been used..to describe person-oriented behavior. **1973** J. LYONS *Experience* IV. viii. 241 Its major practitioners [*sc.* of encountering], who are likely to be ex-theologians, teachers, counselors, actors..—whatever field can furnish perceptive, person-oriented leaders—rather than formally trained psychotherapists. **1958** TAGIURI & PETRULLO (*title*) Person perception and interpersonal behaviour. *Ibid.* p. x, We propose using the term *person perception* whenever the perceiver regards the object as having the potential of representation and intentionality. **1964** M. ARGYLE *Psychol. & Social Probl.* iii. 36 Person perception has been made the object of considerable research. **1972** *Jrnl. Social Psychol.* LXXXVI. 23 Accurate person perception is repeatedly identified as an essential component of effectiveness in the research literature concerned with interpersonal functioning.
b. With a period of time, as *person-day*, *-month*, units equivalent to one day, month, of one person's work or life. Cf. *MAN *sb.*¹ 20 b.
1970 *Sci. Amer.* Feb. 91 In that year people in California spent some 235 million person-days in specified outdoor recreational activities, primarily swimming, picnicking, fishing and boating. **1975** *Nature* 30 Oct. 733/2 Under an agreement signed recently in Stockholm, scientists and experts from the two countries will exchange visits of 10 person-months a year.

persona. Add: **1. persona grata** (examples). Also **persona non grata** (pl. **personæ non gratæ**), an unacceptable or unwelcome person.
1882 *Standard* 20 Dec. 5 At a supper of criminals in full work in their profession he might be welcomed as a *persona grata*. **1884** E. W. HAMILTON *Diary* 2 Sept. (1972) II. 679 Malet did well at Versailles and is a *persona grata* to Bismarck, who was intimate with Malet's mother. **1904** CONRAD *Nostromo* I. vi. 86 See that, Mr. Gould? *Persona non grata*. That's the reason our government is never properly informed. **1928** 'BRENT OF BIN BIN' *Up Country* iv. 52 He was admirably suited to his calling at that time, and his education and personality made him *persona grata* to all his superiors, from the Surveyor-General downwards. **1928** D. L. SAYERS *Lord Peter views Body* vi. 151 Oh, I'll keep out of it... I shan't be exactly persona grata, don't you know. **1935** H. EDIB *Clown & his Daughter* xlii. 238 The fact that he is his father's son makes him *persona grata* in the Sultan's eyes. **1958** *Oxford Mail* 15 Feb. 1/2 The BMC management should have known that the introduction of two or three people who are persona non grata with the other 350 men in the shop would create difficulty. **1964** M. GOWING *Britain & Atomic Energy, 1939–1945* v. 172 He made it clear that Mr Akers was *persona non grata* to the Americans on account of his industrial connections with I.C.I. **1965** C. D. EBY *Siege of Alcázar* (1966) i. 30 At such times Army officers were *personae non gratae*. **1968** *Listener* 7 Nov. 603/3 Gandhi was..always *persona grata* with the high-ups. **1972** *Daily Tel.* 2 June 3/3 Your recent book has.. caused a lot of annoyance here, and you would not be *persona grata* at Eton on the Fourth of June. **1973** *Times* 15 Feb. 7/8 In view of Pakistan's 'violation', the spokesman added, the Iraq Government had decided to retaliate by declaring the Pakistan Ambassador and a Second Secretary as *personae non gratae* and warning them against returning to Iraq. **1974** *Times* 23 Jan. 15/1 Moscow intercepts a Peking-bound Chinese diplomat.. claims he is carrying espionage material, and declares him persona non grata. **1976** A. GREY *Bulgarian Exclusive* xvi. 109 With half a dozen telephone calls..you will be *persona non grata*..throughout Eastern Europe.
3. *persona designata* [Law L.], a specified person; one who is individually denominated, as opp. one who is included in a legal category or class consisting of several persons.
1875 *Law Rep. Chancery Appeal Cases* X. 359 The legatees in this case, although described as a class, are in fact *personæ designatæ* as much as if they were mentioned by name. **1876** H. S. THEOBALD *Conc. Treat. Construction of Wills* xiii. 114 My nephew Joseph is clearly *persona designata*, and the question then is whom did the testator mean to point out? **1955** *Times* 4 May 4/3 It was against him as the person designated to carry out certain functions prescribed by Act of Parliament of laying before each House of Parliament a draft scheme, and that was done as a *persona designata*. **1973** *Deb. Senate Canada* 19 June 4889/1 There have been examples of an appeal court challenging the actions of an appellate court judge when he was acting *persona designata*.
4. Pl. **personæ, personas.** **a.** A character deliberately assumed by an author in his writing; also *transf.*
1909 E. POUND (*title*) Personae. **1958** *Times Lit. Suppl.* 20 June 345/1 To this extent, Lewis Eliot is, as it were, a convenient and comfortable persona for his author. **1962** W. NOWOTTNY *Lang. Poets Use* ii. 42 So far as a particular kind of *persona* is necessary to the poem, the poet's diction must create it. **1963** AUDEN *Dyer's Hand* 401 The more closely his [*sc.* Byron's] poetic *persona* comes to resemble the epistolary *persona* of his letters to his male friends..the more authentic his poetry seems. **1976** *Gramophone* Dec. 965/3 George Logan and Patric Fyffe in the *personae* of Dr Evadne Hinge and Dame Hilda Bracket are on EMI One-Up OU2125 (7/76).
b. In Jungian psychology, the set of attitudes adopted by an individual to fit himself for the social role which he sees as his; the personality an individual presents to the world; also *loosely*. Opp. *ANIMA.
1917 C. E. LONG tr. *Jung's Coll. Papers Analytical Psychol.* (ed. 2) xv. 466 The persona is always identical with a *typical* attitude, in which *one* psychological function dominates, *e.g.* feeling, or thought, or intuition. **1923** [see *ANIMA]. **1931** H. G. WELLS *Work, Wealth & Happiness of Mankind* (1932) viii. 298 A man's guiding and satisfying idea of himself is what Jung calls his '*persona*'. *Ibid.* xii. 617 There was nothing in their personas to prevent it. **1935** *Trans. Philol. Soc.* 66 We are born individuals. But to satisfy our needs we have to become social persons, and every social person is a bundle of rôles or *personæ*. **1936** 'M. INNES' *Death at President's Lodging* iii. 51 In the Dean's *persona* the episcopal idea had of late been rapidly developing. **1940** H. G. WELLS *Babes in Darkling Wood* IV. ii. 333 Some austerer element in his make-up was putting his *persona* on trial. **1966** COX & ROLFE tr. *Herzog's Psyche & Death* xv. 193 The dreamer has to answer for himself in his own right—he cannot claim the protection of the persona of his office. **1972** *Observer* 30 Jan. 9/6 He can be a pompous, contentious man, yet his private persona sometimes contrasts sharply with his more abrasive public image... He has also kept his dignity, consistently refusing to exploit or trivialise his public persona in the lucrative entertainment field.

personal, *a.* (*sb.*) Add: **1. a.** Also, *personal friend, hygiene.*
1782 LD. AUCKLAND *Let.* 22 Aug. (1861) I. 29 Lord North, too, could on *very easy* terms answer for thirty or forty, quite as personal friends and followers. *a* **1794** GIBBON *Mem.* (1796) I. 71 Mr. Allamand, Minister at Bex, was my personal friend. **1853** C. BRONTË *Villette* I. xv. 304 Had that audience numbered as many personal friends and acquaintance for me, as for him, I know not how it might have been. **1915** T. F. A. SMITH *Soul of Germany* 54 Opponents are often personal friends, but that makes no difference. **1938** AUDEN & ISHERWOOD *On Frontier* I. i. 24 The Valerian School..will educate your dear little kiddies in Patriotism and Personal Hygiene. **1970** *Guardian* 17 Aug. 7/5 Much propaganda about 'personal hygiene' is on the wrong track... Why the obsession with stopping *fresh* perspiration? **1979** D. ATTENBOROUGH *Life on Earth* xi. 248 The sloth..pays such little attention to its personal hygiene that green algae grow on its coarse hair.
c. Designating an official or employee attached to one's person in a subordinate capacity, as *personal assistant, maid*, etc.
1928 *Radio Times* 2 Nov. 301/3 My personal maid.. [was] sent to service at 11 years. **1941** in G. Howell *In Vogue* (1975) 162 A personal maid..is absent, on munitions. **1956** *Times* 21 Jan. 7/5 An adequate type of skilled secretarial assistant could be ensured by certain responsible bodies..holding examinations for executive secretaries/personal assistants (as opposed to company secretaries). **1958** P. SCOTT *Mark of Warrior* I. 21 'These chaps are your personal servants,' the receiving officer explained. **1964** M. LASKI in S. Nowell-Smith *Edwardian Eng.* iv. 144 Personal servants at least must wait up until their masters and mistresses chose to go to bed. **1972** T. P. MCMAHON *Issue of Bishop's Blood* iv. 44 Frank Velandi..three arrests for assault... Said at one time to be Streppelli's personal bodyguard. **1977** A. SCHOLEFIELD *Venom* v. 198 M. Michel Blanchet, the hotel millionaire wanted a personal maid for his new wife. **1978** W. GARNER *Möbius Trip* i. 21 Hand over the mouthpiece she called 'Prime Minister's personal assistant'.
4. e. Of newspaper advertisements: small, on private matters (see also quot. 1902); esp. in *personal column*.
1888 W. WHITMAN *Daybks. & Notebks.* (1978) II. 448 Letter from J. G. Bennett, N.Y. Herald, ask'g me to write for 'Personal' col Herald. **1902** *Encycl. Brit.* XXXI. 173/2 'Personal journalism', *i.e.*, paragraphs about the private life or personal appearance of individuals..of note or notoriety in society or public affairs, has become far more marked. **1936** *Discovery* Dec. 386/1 The 18th century 'personal' advertisement, dealing with such wants as wives, lost umbrellas, or menservants 'of black complexion and sound principles'. **1948** *Chicago Daily News* 30 Aug. 19/3 German newspapers have 'personal' columns filled with advertisements for mates. **1966** *Listener* 6 Jan. 14/2 Most of its members were gathered by putting an advertisement in the personal column of a daily newspaper. **1978** J. WAINWRIGHT *Ripple of Murders* 11 A small ad. in the Personal Column ..will read 'J.D. Message received'.
f. Of a letter or other communication: intended for the attention of a particular recipient.
1934 G. B. SHAW *Too True to be Good* II. 50 Is this a personal letter to be sent on to him, or is it a dispatch? **1940** R. S. LAMBERT *Ariel & all his Quality* ix. 244 A letter was delivered..addressed 'H. Brown, Esq., Broadcasting House'. It was not marked 'Personal' or 'Private'. **1973** 'D. JORDAN' *Nile Green* i. 9 A stack of letters was open on my desk including..two which must have been marked 'Personal' on the envelope. **1978** J. SYMONS *Blackheath Poisonings* v. 241 The rectangular package was addressed..to Mr George Collard, and marked *Personal.*
g. Of a (transistor) radio or television: small (see quot. 1962).
1962 *Which?* Feb. 36/1 The two main categories are: personal radios, which will slip into a pocket—or are just too large to do so—and portable radios, which have to be carried. **1973** *Philadelphia Inquirer* 7 Oct. 16 (Advt.), Personal size 16" diagonal screen.
6. c. *personal representative* (earlier and later examples).
1796 A. ANSTRUTHER *Reports* I. 131 The personal representative is *in general* considered as trustee of the property devised from the testator, undisposed of, *as belonging to him.* **1832** RUSSELL & MYLNE *Rep. Cases Chancery* 1829–30 I. 589 The ordinary sense of the words 'personal representatives' is, executors and administrators. **1967** E. RUDINGER *Wills & Probate* 40 The people who deal with what you own when you die are called your personal representatives.
10. Special collocations, as *personal appearance*, (*a*) the appearance or presence of an individual (esp. a celebrity) in person; (*b*) the visual aspect or looks of a person, considered in terms of dress, grooming, and expression; *personal bill*, a private bill (BILL *sb.*³ 3) usu. introduced in the House of Lords, relating to the estate, status, or other personal concern of an individual (see also quot. 1844); *personal call*, a telephone call in which the caller specifies to the operator the person to whom he wishes to speak, and only the time so spent is charged for (in addition to a fixed service charge); *personal caller*, a prospective client who establishes personal contact with a business; *personal explanation*, a statement made by a Member of the House of Commons in explanation or mitigation of recent conduct; *personal god*, a god possessing personal attributes; *personal government*, autocratic rule in which effective power is vested in the person of a monarch (commonly associated with the reign of George III); *personal idealism*, philosophical idealism which emphasizes the essential role of the conscious person in relation to perception of external reality and, usu., also in relation to God as supreme Person; akin to *PERSONALISM b; hence *personal idealist*; *personal loan*, a loan made to an individual for his private requirements by a finance company or bank; *personal name*, the name by which an individual (occas. a thing) is distinguished or identified; *personal shopper*, one who shops in person, as opposed to by mail-order; *personal touch*, a personal element introduced into something otherwise institutional or impersonal.
1610 S. RID *Martin Mark-all* sig. A 3, They presently send to the Beadle of the Hall to make his personal appearance at the Swan with five necks in Kings Streete. **1736** M. W. MONTAGU *Let.* Dec. (1966) II. 111 Halfe those aspirations to the B.V. would deserve her personal appearance to encourage so sincere a Votary. **1842** DICKENS *Amer. Notes* I. viii. 277 Comparing notes on my personal appearance with as much indifference as if I were a stuffed figure. **1883** [see APPEARANCE 2]. **1914** G. B. SHAW *Fanny's First Play* 152 Mr Trotter.. assisted the make-up by which Mr Claude King so successfully simulated his personal appearance. **1951** S. J. PERELMAN in *New Yorker* 2 June 26/3, I caught

your personal appearance at the Mastbaum in Philly. 1962 L. Deighton *Ipcress File* xvi. 92 A cinema where a nineteen-year-old rock-an'-roll singer was making a personal appearance for £600. 1972 *N.Y. Law Jrnl.* 22 Aug. 8/1 *Personal Appearances*..Ping Keun Chu v. Jade Fountain of Paramus, Inc. [etc.]. 1976 T. Heald *Let Sleeping Dogs Die* i. 7 She had little time for clothes and cared nothing for her personal appearance. [1683 Personal bill: see Personal *a.* 1 a.] 1844 T. E. May *Treat. Parliament* xxviii. 457 All private bills, during their progress in the commons, are known by the general denomination of private bills; but in the lords the term 'private' is applied *technically* to estate bills only, all other bills being distinguished as 'local' or 'personal', although in the standing orders no such distinction is expressed. 1929 G. F. M. Campion *Introd. Proc. House of Commons* ix. 274 Personal Bills, *i.e.* Estate, Divorce, Naturalisation, Restitution and Name Bills..always originate in the Lords. 1973 *Jrnl. House of Lords* (1974) 14 Nov. CCVI. 29/2 Personal Bills: Select Comee appointed. 1930 *Telegr. & Telephone Jrnl.* XVI. 70/2 Another important event of the past year was the introduction of the 'personal' call system in the Inland and Anglo-Continental trunk services. 1947 Auden *Age of Anxiety* (1948) ii. 43 To be held waiting in A packed lounge for a Personal Call From Long Distance. 1960 [see *person-to-person *adj.* (and *adv.*) *phr.* a]. 1967 E. Lemarchand *Death of Old Girl* ii. 21 There's a phone call for you... A personal call from London. 1978 F. Durbridge *Tim Frazer gets Message* iv. 61 Can Miss Thackery take a personal call? 1966 *Listener* 25 Aug. 291 (Advt.), Send now for a free copy... For personal callers—235 Grand Buildings, Trafalgar Sq., W.C.2. Tel.: WHItehall 8377. 1976 *Norwich Mercury* 19 Nov. 12/5 (Advt.), Personal callers will be welcomed by a receptionist at Prospect House. 1844 T. E. May *Treat. Parliament* xi. 195 (*heading*) Personal explanation. 1857 *Sat. Rev.* 14 Feb. 152/2 That green oasis in the desert of legislation—that dainty morsel in the sessional banquet—a personal explanation, which, in Mr. Disraeli's hands, was pretty sure to include also a personal attack. 1886 J. Bailey *Let.* 11 Apr. (1935) 24, I might say a good deal on this subject but as the House of Commons has lately made personal explanations vulgar, I don't think I will. 1974 *House of Commons Man. Procedure Publ. Business* (ed. 11) viii. 116 By the indulgence of the House, a Member may make a personal explanation, although there is no question before the House, but in this case no debatable matter may be brought forward, and no debate can arise. 1860 Personal God [in Dict., sense 5 b]. *a* 1902 S. Butler *Way of All Flesh* (1903) xlix. 225 There is not one of you here who doubts the existence of a Personal God. 1921 G. B. Shaw *Back to Methuselah* p. xlii, We had been so oppressed by the notion that everything that happened in the world was the arbitrary personal act of an arbitrary personal god of dangerously jealous and cruel personal character, so that even the relief of the pains of childbed and the operating table by chloroform was objected to as interference with his arrangements which he would probably resent, that we just jumped at Darwin. 1963 J. A. T. Robinson *Honest to God* iii. 48 The difference..can perhaps best be expressed by asking what is meant by speaking of a *personal* God. Theism..understands by this a supreme Person, a self-existent subject.., who enters into a relationship with us comparable with that of one human personality with another. 1976 P. Hill *Hunters* vi. 73 An allegedly all-powerful and personal God. 1909 W. Toynbee *Glimpses of Twenties* i. 1 George the Third..ascended the throne with a fixed determination to re-establish 'personal government', which quickly aroused the misgivings of even his best-affected subjects. 1954 *Proc. Brit. Acad.* XXXVIII. 225 We no longer believe that George III's system of personal government came to an end in 1784, and that power was then transferred from the King to the Prime Minister. 1901 G. H. Howison (*title*) The limits of evolution, and other essays illustrating the metaphysical theory of personal idealism. 1921 *Encycl. Relig. & Ethics* XII. 229/2 Where personal idealism means spiritual pluralism of a theistic type, the concept of purpose applied to the interpretation of the universe yields a conclusion that satisfies. 1966 F. Copleston *Hist. Philos.* VIII. iii. xiii. 296 It is so often religiously minded people who are attracted in the first instance to personal idealism. 1902 H. Sturt *Personal Idealism* p. vi, Naturalism and Absolutism, then, are the adversaries against whom the personal idealist has to strive. 1966 F. Copleston *Hist. Philos.* VIII. iii. xiii. 296 Unless the personal idealist equates ultimate reality with the system of finite selves, as McTaggart did, he must be a theist. 1914 *Laws State of N.Y.* II. 1435 When authorized by the superintendent of banks..three or more persons.. may form a corporation to be known as a personal loan company. 1957 D. Karp *Leave me Alone* v. 80 Individual enterprisers offered..dry cleaning, baked goods..personal loans. 1958 *Times* 29 Aug. 8/7 Barclays are after all to be the first of the banks to bring a personal loans scheme into operation. 1961 *Which?* May 103/1 The banks have started 'Personal Loan' schemes, whereby you can borrow money from a bank without providing security. 1979 *Guardian* 5 Feb. 15/4 Basically the clearing banks offer three kinds of packages: overdraft facilities; loan accounts, and personal loans. 1748 B. Martin *Institutions of Lang.* 27 Of nouns or names there are three sorts, common, proper or personal, and relative... Proper or personal names are such as denote the individuals of each species; as Cæsar,..London, Paris, &c. 1871 E. B. Tylor *Primitive Culture* I. viii. 276 Up from this savage level the same childlike habit of giving personal names to lifeless objects may be traced, as we read of Thor's hammer, Miölnir. 1911 J. G. Frazer *Golden Bough: Taboo* (ed. 3) vi. 318 (*heading*) Personal names tabooed. 1925 O. Jespersen *Mankind, Nation & Individual* ix. 172 The Araukans carefully conceal their personal-name from strangers. 1950 *Funk's Stand. Dict. Folklore* II. 782/2 It is believed that if one is sick one's name is possibly not agreeing with one, hence the name is 'washed off', and a new personal name given. 1972 *Harrods Christmas Catal.* 58/1 *Fresh pâté de foie gras*... For personal shoppers only. 1976 *Field* 18 Nov. 1023/2 (Advt.), Send SAE for descriptive leaflet of our full range including a How-to-find-us map for personal shoppers. 1887 S. A. Barnett *Let.* 12 Oct. in H. Barnett *Canon Barnett* (1918) II. xxxiii. 63 We talked of how workmen could be made at home in Toynbee. I am sure that attractions won't bring them, but only the personal touch. 1936 A. Christie *ABC Murders* i. 11, I had various affairs to see to in England that I felt could only be successful if a personal touch was introduced. 1967 A. Hunter *Gently Continental* vi. 75 We try to make people feel they belong here. The personal touch, you know. 1976 'O. Jacks' *Assassination Day* iii. 48 Butcher hadn't overplayed his hand. He'd been crafty, relying on the personal touch.

B. *sb.* **2. c.** For def. read: An item in a newspaper about a person or group of persons; a classified advertisement addressed to an individual person. orig. *U.S.*

1861 in A. Sterling *Belle of Fifties* (1904) 238, I inclose you a 'personal' from Brother Clement, published in yesterday's *Enquirer.* 1873 F. Hudson *Journalism in U.S.* 472 Take the 'personals' of the *Herald* any day, and they will set one to thinking. 1875 J. G. Holland *Sevenoaks* viii. 103 Returning..to look over the papers, his eye was attracted, among the 'personals', to an item [etc.]. 1888 [in Dict.]. 1901 *Daily Colonist* (Victoria, B.C.) 16 Oct. 7/1 (*heading*) Personals. 1913 *Collier's* 1 Feb. 17/3 He inserts a 'personal' in a New York newspaper under her initials. 1968 L. Durrell *Tunc* iii. 166 He had invented what he called the mnemon which he insisted was a literary form... *Times* Personals of a slightly surrealist tinge. 1977 R. E. Harrington *Quintain* viii. 76 The classified section was his favorite part of the paper. He enjoyed reading the disguised, plaintive little cries for help in the Personals.

5. A personal friend. *colloq.*

1961 *Listener* 21 Dec. 814/3 Reynard La Spoon, the choreographer—he's a close personal, ent he, Jule?

6. *Basketball.* A foul involving bodily contact with an opponent.

1961 in Webster. 1969 *Eugene* (Oregon) *Register-Guard* 3 Dec. 1D/3 Love played only the first eight or nine minutes of the first half before collecting his third personal. 1974 *State* (Columbia, S. Carolina) 3 Mar. 2-D/1 He fouled out in the game's final minute after playing almost 13 minutes with four personals.

personalia (pǝ̄rsonēi·liă), *sb. pl.* [ad. L. *personālia*, neut. pl. of *personālis* personal.] Personal matters; personal allusions; personal mementoes.

1903 'Sigma' (*title*) Personalia: political, social, and various. 1909 H. G. Wells *Tono-Bungay* iii. ii. 314 My aunt received these personalia cheerfully. 1920 *Glasgow Herald* 27 Mar. 8 His speech on Wednesday contained some quite superfluous personalia. 1928 J. Bailey *Let.* 26 Oct. (1935) 292 The best letters of you, as of other masters of the art epistolary, are the ones full of personalia—*from you to me, etc.* 1969 *Daily Tel.* 25 Oct. 8/1 On the theme of Ruskin and Venice this exhibition includes watercolours and drawings by him, various editions of his works and personalia. 1974 M. Z. Lewin *Enemies Within* xxxi. 140 That lovely desk, which I would have filled to the brim with personalia. 1974 *Nature* 3 May 6/1 With pieces of original equipment..and personalia including Marconi's swordstick.

personalism. Add: **b.** A philosophical view, usually theistic and positing God as supreme Person, that reality has meaning only through the conscious minds of persons; a view of social organization that places primary emphasis on the person and his involvement in it rather than on the material means necessary for achieving such organization.

1908 B. P. Bowne (*title*) Personalism. *Ibid.* iii. 111 We have now to consider the phenomenality of the physical world. This is the next step in the establishment of personalism. 1917 *Encycl. Relig. & Ethics* IX. 771/2 Aristotle laid the foundation for personalism by affirming self-consciousness as the highest being. 1938 tr. *Mounier's Personalist Manifesto* i. 1 Personalism is for us at present a sort of general pass-word. We are using it as an inclusive term for various doctrines that in our present historical situation can be made to agree upon the elementary physical and metaphysical conditions of a new civilization. 1947 *Partisan Rev.* XIV. 396 Berdyaev developed into one of the outstanding religious writers of our time, preaching a synthesis of socialism, personalism, and corporate Christianity. 1957 M. P. Fogarty *Christian Democracy* iii. 29 Personalism, as distinct from individualism, is held by Christian Democrats to imply a certain 'solidarist' conception of the individual's responsibility to and for the society around him. 1959 *Pacific Affairs* XXXII. 77 Since 1956 the Ngo brothers and officials high and low in the government have referred to Personalism (and humanism) as the philosophical basis of the national revolution. 1966 F. Copleston *Hist. Philos.* VIII. iii. xiii. 296 The basic tenet of personalism has been stated as the principle that reality has no meaning except in relation to persons; that the real is only in, of or for persons. 1971 A. R. Caponigri *Hist. Western Philos.* V. vi. v. 349 He [*sc.* the person] finds the plenitude of his self-affirmation and of the affirmation of his vertical relation to God in the recognition of, and cooperation with, other persons. This has been called the social dimension of Stefanini's personalism.

c. Allegiance to a person, esp. a political leader, rather than to a party or ideology.

1937 *Times* 4 Sept. 11/6 Personalism is a characteristic of Argentine politics. A party is the personal following of a man. 1964 M. C. Needler *Polit. Syst. Lat. Amer.* xxi. 518 Personalism stands in inverse relation to permanence of party organization, the extreme case of personalism being found in the party organized solely to support the candidacy of one individual. 1970 N. A. Victoria in I. L. Horowitz *Masses in Lat. Amer.* xv. 557 One of the conditions which may favor the emergence of such leadership is an element of 'personalism', or the extraordinary extension of the personal and emotional sphere. 1976 *Encounter* June 79 General Franco instinctively sensed something of the communal dangers inherent in Spanish personalism ('individualism' is the conventional term, but it fails to convey the whole meaning).

‖ **personalismo** (pęrsonalizmo). [Sp. and Pg.] = *personalism c.

1962 G. Blanksten in M. A. Kaplan *Revolution in World Politics* iii. v. 120 *Personalismo* may be regarded as the custom of following or opposing a political leader on the basis of his personality rather than on ideological grounds. 1964 L. B. Lott in M. C. Needler *Polit. Syst. Lat. Amer.* xii. 244 A strong tendency on their [*sc.* the people's] part to associate themselves with powerful personalities rather than with programs and issues. This tendency we may label *personalismo.* 1969 *Language* XLV. 464 Joshua Fishman's comparison of those two categories in terms of cross-cultural indices reveals a most interesting association between linguistically heterogeneous polities and 'personalismo' or 'charisma' in follower-leader relationships. 1969 J. Mander *Static Society* i. 66 Mexico has cast off the incubus of *personalismo.*

personalist. Add: **b.** Further examples, esp. with reference to *personalism b and c. Also *attrib.*

1917 *Encycl. Relig. & Ethics* IX. 771/2 In this sense Eucken, Howison, Bergson, James.., and others of the modern school may be called personalists. 1937 *Times* 4 Sept. 11/6 The Radicals who remained faithful to the 'Chief' were known as Personalist or Irigoyenist Radicals. 1938 tr. *Mounier's Personalist Manifesto* i. 1 We shall apply the term *personalist* to any doctrine or any civilization that affirms the primacy of the human person over material necessities and over the whole complex of implements man needs for the development of his person. 1938 G. Reavey tr. *Berdyaev's Solitude & Society* 33 Personalist philosophy, as I understand it, has nothing in common with the subjectivist, individualist, empirical or nominalist currents of to-day. 1962 S. E. Finer *Man on Horseback* ix. 131 Paraguay..the parties—Colorados and Azules—are personalist cliques. 1970 B. Brewster tr. *Althusser & Balibar's Reading Capital* (1975) ii. viii. 172 This is a stumbling-block for all the interpretations of Marxism as a 'philosophy of labour', whether ethical, personalist or existentialist. 1970 D. Goldrich et al. in I. L. Horowitz *Masses in Lat. Amer.* v. 190 APRA..the personalist party of former dictator General Odría. 1976 H. A. Williams *Tensions* vi. 99 Personalist pastors, perhaps more than public campaigners, sometimes try (unsuccessfully) to impose a totalitarian tyranny upon scholars and thinkers.

Hence **personali·stic** *a.*, of or pertaining to a person considered as different and separate from other people, esp. of the psychological study of the individual in relation to his personal experience (see quots.). Also occas. **personali·stics** *sb. pl.* (treated as *sing.*).

1929 H. Klüver in G. Murphy *Hist. Introd. Mod. Psychol.* xxv. 424 But returning..to personalistic theory, some further implications of the 'psychophysical neutrality' of 'person' should be considered. 1936 *Mind* XLV. 247 The 'personalistic' Psychologists go so far as to insist on the uniqueness of every combination of dispositions. 1938 H. D. Spoerl tr. *Stern's Gen. Psychol. from Personalistic Standpoint* p. vii, In spite of this basic concern with the whole fabric of psychological specialties, our book will maintain a thoroughly distinctive and novel point of view... This is the *personalistic* point of view, which here finds its first occasion to demonstrate its fitness to formulate and interpret a particular empirical science. *Ibid.* xxvii. 494 The author..points to this personalistic basis of transfer without knowing anything about personalistics. 1939 *Brit. Jrnl. Psychol.* XXIX. 411 Stephenson.. proposed to substitute a modified system of his own, which, he maintains, should lead to 'an entirely new branch of psychometry, for which the term Personalistics may be coined'. *Ibid.* XXX. 65 The roots of 'personalistics' clearly go back to the classical researches embodied in the *Differentielle Psychologie.* 1956 *Jrnl. Theol. Stud.* VII. 165 He [*sc.* Luther] maintains a thoroughly personalistic idea of grace, which is received by faith alone. 1969 C. Davidson in Cockburn & Blackburn *Student Power* 359 The hippy movement has served to make many of our people withdraw into a personalistic, passive cult of consumption.

personality. Add: **2. a.** (Further *fig.* examples.) Also in phr. *to have personality*, to have qualities or traits of character to an unusual or noteworthy degree.

1902 W. D. Howells *Lit. & Life* 249 How many houses now have character—personality? 1934 C. Fox in *Proc. 1st Internat. Congr. Prehist. & Protohist. Sci.* 27 Position, outline, and structure..; the climate resulting from position, and the soil derived from structure, determine the vegetable life which she nourishes and the animal life which she harbours. The whole represents Man's environment, and Britain's 'Personality'. 1940 R. S. Lambert *Ariel & all his Quality* iv. 116 To attract a solid core of permanent readers to any literary paper presupposes that that paper shall have a 'personality' to distinguish it from others of its kind. *a* 1960 E. M. Forster *Maurice* (1971) xxxiv. 155 So he was in a way, but evidently he had personality. 1973 *Times* 31 Jan. (Mediterranean Suppl.) p. i/3 The Mediterranean islands could easily lose their personality in a short time.

c. *Psychol.* and *Sociol.* The unique combination of psychophysical qualities or traits, inherent and acquired, that make up each person as observable in his reactions to the environment or to the social group; also, psychological study concerned with such aspects of the person, and with the similarities and differences that exist between persons.

1879 H. Maudsley *Path. of Mind* (ed. 3) i. 12 It is this physiological unity of organic functions, which is something deeper than consciousness and constitutes our fundamental personality. 1906 M. Prince (*title*) The dissociation of a personality. *Ibid.* i. 3 A more correct term is *disintegrated* personality, for each secondary personality is a part only of a normal whole self. 1921 Park & Burgess *Introd. Sci. of Sociol.* ii. 144 In sociology, personality is studied, not only from the subjective standpoint of its organization, but even more in its objective aspects and with reference to the rôle of the person in the group. 1930 *Psychol. Bull.* XXVII. 677 The methods and problems of contemporary research in personality. 1947 G. Murphy *Personality* p. x, The approach to personality is made chiefly in terms of origins and modes of development on the one hand, interrelations or structural problems on the other. 1949 Kluckhohn & Murray *Personality* i. 6 In trying to remedy these failures, there emerged the first comprehensive dynamic theory of personality—psychoanalysis. 1950 E. Frenkel-Brunswick in T. Wiesengrund-Adorno et al. *Authoritarian Personality* ix. 291 (*heading*) An approach to the prejudiced personality. 1964 J. Strachey et al. tr. *Freud's New Introd. Lect.* xxxi, in *Compl. Psychol. Wks.* XXII. 57 (*title*) The dissection of the psychical personality. 1969 J. W. Getzels in Lindzey & Aronson *Handbk. Social Psychol.* (ed. 2) V. xlii. 463 A social psychology of education is concerned with the interaction of role and personality in the school or classroom. 1975 J. Plamenatz *K. Marx's Philos. of Man* xiv. 401 There are ideas of love and freedom common to liberal 'bourgeois' society and to its Utopian critics which differ from these older ties and forms. There are ideas of personality, of personal relationships, and of their social conditions, peculiar to this society and to its critics.

3. b. A person who stands out from others either by virtue of strong or unusual character or because his position makes him a focus for some form of public interest.

1889 G. B. Shaw in *Church Reformer* Mar. 68/1 Individuality is concentrated, fixed, gripped in one exceptionally gifted man, who is consequently what we call a personality, a man pre-eminently himself, impossible to disguise. 1919 V. Woolf *Night & Day* iv. 46 I've only seen her once or twice, but she seems to me to be what one calls a 'personality'. 1933 *Radio Times* 14 Apr. 82/3, I apply what may seem a whimsical test to broadcasting personalities. I ask myself if I would care to meet and talk with them in the flesh. 1947 *Sat. Rev. Lit.* (U.S.) 26 Apr. 4/1 In Elizabeth Ann McMurray, John McGinnir, Jimmie Albright, and Lon Tinkle, it harbors four of the outstanding personalities in the American book world. 1959 *Language Learning* IX. iv. 79 It is used with the greatest aplomb and ease by radio and T.V. personalities. 1962 *Listener* 22 Mar. 503/1 He is a local councillor in a small town, and one of its prominent personalities. 1973 *Birmingham* (Alabama) *News* 10 June E-7/4 More recently she has tried, fairly unsuccessfully, for a career as television personality. 1973 D. Miller *Chinese Jade Affair* xviii. 176 The woes of being a secret policeman during the visits of V.I.P. personalities. 1976 *National Observer* (U.S.) 13 Nov. 1/1 'In movies they buy personalities,' says Billy Dee.

7. *attrib.* and *Comb.*, as *personality assessment, clash, defect, disorder, problem, test, theory*; **personality cult**, devotion to a leader that is deliberately fostered by the emphasis placed on certain aspects of his personality; **personality dynamics**, a term used for the active, though not necessarily conscious, adaptation effected by a person of his personality to his environment; **personality factor**, a trait considered as sufficiently distinct and general in the study of personality to be measurable by factor analysis; **personality integration** (see quot. 1970); **personality inventory**, a questionnaire designed to assess personality traits; **personality pattern**, the pattern of personality that is formed by the inherited and acquired traits of an individual; **personality structure**, the combination of traits that make up a personality; **personality system**, a sociological term for individual personality in its dynamic social context; **personality trait**, a particular feature or characteristic that can be considered as relatively stable in an individual personality; **personality type**, a classification of personality according to the preponderant features or traits found either in a person or in a society; **personality variable** = *personality factor.

1956 G. G. Stern et al. (*title*) Methods in personality assessment. 1964 *Eng. Stud.* XLV. 50 The intellectual and personality assessment of boys. 1969 *Playboy* July 102/1 These range from personality clashes..to on-the-job incompetence. 1956 *Canadian Forum* May 25/1 The spread of the 'personality cult' diminished the role of collective leadership within the party and sometimes led to serious defects in our work. 1957 *Economist* 21 Sept. 912/2 The election campaign was marked by a personality cult, but its inspiration and techniques owed infinitely more to Madison Avenue than to Dr Goebbels. 1959 *Encounter* July 80/2 The emphatic condemnation of the 'personality cult' at the 20th Congress of the CPSU by Mr. Krushchev. 1960 *20th Cent.* Apr. 342 A big factor in the sale of the more popular 'name' records is the personality cult. 1971 *Black Scholar* Apr.–May 7/2 We must be careful to avoid the tendency of building personality cults around specific individuals. 1971 I. Deutscher *Marxism in our Time* (1972) xv. 290 The crudities and cruelties of the 'personality cult' must have made him shudder more than once. 1973 *Times* 4 Dec. 16/7 Signs of new trouble in the Kremlin are evident in the rapid growth of the Brezhnev personality cult. 1927 *New Republic* 21 Sept. 129/1 The understanding thus gained spreads to all the slighter estrangements, the problems of discipline,..the normalization of those with personality deficits or defects. 1936 W. S. Sadler *Theory & Pract. Psychiatry* xxiv. 393 About one-quarter of all school children carry definite personality defects. 1970 'T. Coe' *Wax Apple* (1973) vii. 56 A naturally offensive man who had found a way..to turn a personality defect to advantage. 1938 L. P. Thorpe *Psychol. Found. Personality* viii. 338 Investigators..have attempted to ascertain the degree of relationship obtaining between glandular disturbances and personality disorders. 1969 *Guardian* 22 July 11/7 The 11-year old girl.. sentenced to life detention..for strangling two small boys ..suffers from..a severe personality disorder, for which there is no organic cause. 1976 Smythies & Corbett *Psychiatry* xvii. 290 Some patients who ask his help.. suffer from a different kind of illness—a personality disorder. 1954 B. R. Sappenfield (*title*) Personality dynamics: an integrative psychology of adjustment. 1958 M. Argyle *Relig. Behaviour* v. 48 M. B. Smith and others (1956) distinguish those people whose attitudes are primarily an adjustment to group standards from those whose attitudes are based more on internal personality dynamics. 1960 J. C. Coleman (*title*) Personality dynamics and effective behavior. 1932 P. M. Symonds *Diagnosing Personality & Conduct* xi. 438 The creatinine concentration of the urine was also found by Rich to be associated with personality factors. 1957 R. B. Cattell *Personality & Motivation* ix. 335 A personality factor will have a series of predictive validities against specific cultural performances. 1971 Lanyon & Goodstein *Personality Assessment* iv. 89 A system of five relatively orthogonal (independent) and easily interpreted personality factors. 1938 L. P. Thorpe *Psychol. Found. Personality* ix. 434 (*heading*) Definition and nature of personality integration. 1970 G. A. & A. G. Theodorson *Mod. Dict. Sociol.* 297 *Personality integration*, the harmonious coordination of the various aspects of the personality with each other, and of the personality as a whole with its environment... Perfect personality integration is.. certainly not normal for persons in social interaction. 1932 P. M. Symonds *Diagnosing Personality & Conduct* v. 208 An integration of these various questionnaires designed to measure adjustment has been effected in a 'Personality Inventory' constructed by Bernreuter. 1933 *Jrnl. Social Psychol.* IV. 389 The test which was constructed has been entitled the Personality Inventory and is referred to herein as the P-I test. 1950 E. A. Suchman in S. A. Stouffer *Measurement & Prediction* v. 162 Evidence of the quasi-scale pattern in the case of personality inventories, information tests, and measures of intensity of feeling. 1968 Blum & Naylor *Industr. Psychol.* iv. 113 Numerous reasons have been suggested to account for the general lack of success in industrial situations of personality inventories. 1971 'D. Halliday' *Dolly & Doctor Bird* xiii. 193 I've done an Eysenck personality inventory on you both... You wouldn't suit. 1949 C. E. Thompson *Thematic Apperception Test: Manual* 3 When cultural prejudices or antagonisms are part of the personality pattern of the Negro they are likely to reduce the subject's identification with the white figures of the TAT. 1960 R. F. Peck et al. *Psychol. of Character Devel.* iv. 89 There are distinct personality patterns which characterize each type group, and which differentiate one group from another. 1973 E. B. Hurlock *Personality Devel.* (1974) ii. 19 The personality pattern is not the product of learning exclusively or of heredity exclusively..it comes from an interaction of the two. 1963 A. Heron *Towards Quaker View of Sex* 51 Symptoms of deeper personality problems. 1939 R. Linton in A. Kardiner *Individual & his Society* p. vi, Basic personality structure, as the term is used here, represents the constellation of personality characteristics which would appear to be congenial with the total range of institutions comprised within a given culture. It..is, therefore, an abstraction of the same order as culture itself. 1947 G. Murphy *Personality* xxviii. 664 Problems of generality of the conditioned response being interwoven with problems of personality structure in the true sense. 1957 R. B. Cattell *Personality & Motivation* viii. 281 A comparatively new world of personality structure..has become visible to psychologists... On this foundation of measurable functions, the psychology of the second half of the twentieth century may proceed to build its laws and theories of personality. 1972 *Jrnl. Social Psychol.* LXXXVI. 151 Individuals will be encouraged to hold attitudes of intolerance largely irrespective of their basic personality structure. 1951 Parsons & Shils *Toward Gen. Theory Action* ii. i. 55 A personality system is a system of action... Social systems, personality systems, and cultural systems are critical subject matter for the theory of action. 1961 L. Thompson *Toward Sci. of Mankind* i. iv. 68 The personality system and the core values function as covert connecting links between the other four systems. 1971 F. Hollis in Roberts & Nee *Theories Social Casework* 59 This means attention to both the interpersonal system—parent-child, husband-wife, family—and to the personality systems of the individuals who compose the interpersonal system. 1927 *Psychol. Bull.* XXIV. 419 A battery containing a mixture of intelligence and personality tests was used by Gallup. 1964 M. Argyle *Psychol. & Social Probl.* xi. 139 The contribution of psychology.. has been to devise various objective tests and measures which are better able to select those who are good at the job. These include intelligence tests, personality tests, [etc.]. 1957 Hall & Lindzey *Theories of Personality* xiv. 538 It seems appropriate..to pause and attempt to identify general trends which exist in spite of the tremendous differences among personality theories. 1977 R. Holland *Self & Social Context* v. 131 The emergence of new personality theories is accompanied by ambivalence towards predecessors as the new theorists filter out what they need from the past and construct around it a new position. 1921 F. H. & G. W. Allport in *Jrnl. Abnormal Psychol.* XVI. 6 (*title*) Personality traits: their classification and measurement. 1931 T. H. Pear *Voice & Personality* ii. 18 In the speech of some persons, there are sounds which really symbolise personality-traits. 1948 *Mind* LVII. 511 The gesture in question is a personality trait of a given individual if it is performed by him, say, six out of every ten times when he might have performed it. 1972 *Jrnl. Social Psychol.* LXXXVI. 30 We deduced that the personality traits of leaders would be more varied in the nonconformity than in the conformity situation. 1919 *Psychol. Rev.* XXVI. 374 Personality-type A..is an individual rated as especially intelligent, prompt, persistent,..sensitive, not at all loquacious. 1936 W. S. Sadler *Theory & Pract. Psychiatry* liv. 845 The most important etiologic factor..is to be found in the personality type of these patients. 1949 MacIver & Page *Society* iii. 58 These studies have rather convincingly demonstrated..that each culture tends to create and is supported by a 'basic personality type'. 1966 *Philosophy* XLI. 299 Different social norms result in different modal personality-types. 1968 A. Etzioni *Active Society* xxi. 627 Most studies of efforts to affect 'deep' personality variables—especially psychoanalysis, 'brainwashing', and psychological experiments—show these efforts to have little effect. 1972 *Encycl. Psychol.* II. 385/1 Such personality variables as aggression, anxiety or authoritarianism. 1972 *Jrnl. Social Psychol.* LXXXVI. 121 Choosing groups dichotomized along some single personality variable, such as aggression, would allow far too many sources of variance.

personalization. (Further examples.)

1911 W. W. Fowler *Relig. Experience Roman People* vii. 149 It is not..*a priori* probable that the process of personalisation (if I may coin the word) should have proceeded..so far as to ascribe to these named deities..the characteristics of human beings. 1957 R. Hoggart *Uses of Literacy* vi. 163 The quite unusual degree of 'personalisation' in the newspapers designed particularly for working-class people. 1968 S. Brittan *Left or Right* viii. 172 The personalisation of politics round the leaders. 1973 *Times* 24 Jan. (Security Printing Suppl.) p. i/9 A developing sphere of operation for the security printer is in the realm of cheque printing and personalization. The latter is the process whereby the account holder's name and initials are printed on to the cheque.

personalize, *v.* Add to def.: to make (some impersonal object or thing) more obviously related to, or identifiable as belonging to, a particular individual; also *fig.*

1935 *Advt. for Mohawk Sheets* (Miller & Rhoads, Richmond, Va.), Now personalized with smart needlecrest initials. 1947 *Amer. Speech* XXII. 71/1 Personalize your luggage, personalized stationery. 1961 *Daily Tel.* 28 Feb. 24/6 He [*sc.* Mr. Gaitskell] probably appreciates that any further move towards 'personalising' the defence issue would merely play into Mr Crossman's hands. 1966 *Electronics* 31 Oct. 42/3 In the CP and EP, the memory is a plug-in unit that can be replaced in a few minutes, so that the design of either model can be quickly personalized for a special application. 1967 *Autocar* 27 Dec. 9/2 All the multitude of accessories offered (and bought) for the embellishment of ordinary cars is enough to show that many buyers want, in the American phrase, to 'personalize' their transport. 1970 *Daily Tel.* (Colour Suppl.) 13 Nov. 21 (*caption*) Dealer delegate is chatted up by an oil executive... Moments like these help to 'personalise' a giant company with the employees. 1972 *Daily Tel.* 24 Nov. 6/7 Outstandingly clever children often pass through school unnoticed by their teachers... 'One major reason is the failure in schools to personalise the learning environments which they provide.' 1977 *New Yorker* 9 May 143/2 Crossman blames the mass media for personalizing politics.

personalized, *ppl. a.* (Further examples.)

1947 *Forum* (Johannesburg) 26 Apr. 46/1 (Advt.), Fare is inclusive of personalised steward service throughout journey. 1959 *Economist* 4 Apr. 68/1 A 'personalised' cheque, one that carries an identifying account number and so can be used only by the owner of the account. 1974 *Times* 14 Dec. 24/6 Personalized crystal glasses, initials, names, dates, crests, etc., hand engraved to your requirements. 1976 *National Observer* (U.S.) 13 Mar. 11/2 (Advt.), Personalized attention, experienced tutors, relaxed atmosphere. 1977 *Theology* LXXX. 191 The man Jesus Christ as the personalized instrument..of the self-expressive activity of God. 1978 *Bull. Amer. Acad. Arts & Sci.* Jan. 10 Scholars with no prior knowledge of computers will be able to obtain a personalized index to the material most relevant to their particular researches.

personalizing, *vbl. sb.* and *ppl. a.* (Further examples.)

1957 R. Hoggart *Uses of Literacy* vi. 165 As the 'personalising' technique becomes yearly more machine-tooled, so a good instinct is pulled out of shape. 1961 *Daily Tel.* 28 Feb. 24/6 (*heading*) No witch hunt. Against 'Personalising'. 1977 *Theology* LXXX. 191 Such a union would be marked by *eudokia*, or divine good-pleasure, and by *synapheia* or true and personalizing relationship. *a* 1977 *Harrison Mayer Ltd. Catal.* 39/3 The Cerama-pen is a fibre tip pen filled with precious metal ink. Ideal for thin line decorating and personalising work.

personhood. [f. Person *sb.* + -hood.] The quality or condition of being an individual person.

1959 *Times Lit. Suppl.* 3 Apr. 197/3 From there he proceeds to the machine-like properties of animals and so on up to responsible human personhood. 1971 *Time* 13 Dec. 36 The United Church of Christ has in hand a statement written by six Christian education executives which maintains that sex is moral if the partners are committed to the 'fulfilling of each other's personhood'—pointedly omitting marriage as a prerequisite. 1973 *Austral.*

Humanist XXVI. 10/1 The locus of power needs to move from institutionalization towards man in his personhood, thus freeing man to respond to life with his own body, his own thinking, feeling and acting, expressing his abilities by working creatively with his human, natural and material environments. 1973 *Black World* Mar. 71 'Collage' defines her Black womanhood first in relation to her personhood, selfhood, or humanness and then in relation to her femininity. 1974 K. Millett *Flying* (1975) i. 34 There is something healing in her talk: its religious care for personhood. 1976 *Church Times* 12 Nov. 9/1 My unreflective believers have been denied the home atmosphere which encourages them to find their own personhood and value it.

personless, *a.* [f. Person *sb.* + -less.] **a.** Unrecognized as a person; denied individuality. **b.** Making no distinction of persons.

1909 E. Hill in Hill & Shafer *Gt. Suffragists* 11 The slaves of ancient empires, like women of to-day, were not recognised as 'persons', but they built the hanging gardens of Babylon and her mammoth buildings, and the material glory of Athens, just as the 'personless persons' of to-day weave the great moral fabric of the universe and, departing, bequeath their ideals to the willing souls ready to receive them. 1932 H. S. Walpole *Fortress* iv. ii. 607 He was by temperament intensely cautious and by training suspicious, and, mingled with these two strains, there was an odd element of personless, rather noble philanthropy.

personnel. Delete ‖ and add: Now almost invariably with pronunc. (pǝ̄ɹsŏne·l). (Earlier and later examples.)

1837 J. S. Mill in *Westm. Rev.* XXVIII. 25 In moments of general enthusiasm it is enough that a party carries the favourite banner; but in the intervals between those moments, its importance depends upon the confidence inspired by its *personnel.* 1899 H. James *Awkward Age* x. xxxvii. 397 The *personnel*, as the newspapers say, of the saloon will shift and change. 1943 *R.A.F. Jrnl.* Aug. 3 The new basis of distribution is one copy to every twenty-five personnel. 1944 *Return to Attack* (Army Board, N.Z.) 15/1 British forces then rushed the landing ground..capturing nineteen aircraft together with a number of somewhat bewildered air-force personnel. 1969 [see *jawan]. 1972 J. Mosedale *Football* v. 56 Sammy Baugh of the Washington Redskins, playing with inferior personnel overall, lasted 16 years.

b. *spec.* The members of an orchestra, band, etc.

1956 *Gramophone* June 25/1 The sleeve [of a record] gives personnels. 1962 *Oxford Mail* 19 Feb. 6/5 The personnel of the nine bands on this important release includes many of the most famous names in jazz. 1967 *Melody Maker* 16 Dec. 10/5 It is certainly a very good band with above average solo strength and a personnel which includes a mass of arranging talent. 1968 *Blues Unlimited* Nov. 10 It is owing to him that personnels and dates for so many blues/gospel recordings by this company are now wellknown. 1973 *Melody Maker* 4 Aug. 50/6 When the Humphrey Lyttelton Band makes an appearance at the Lancastrian Hall, Swinton, Manchester, ..the personnel will include Kathy Stobart (tenor), [etc.].

c. (Usu. with capital initial.) = *personnel department.

1960 M. Spark *Ballad of Peckham Rye* v. 86 I'm just mentioning a factor that Personnel keep stressing. 1970 'D. Craig' *Young Men may Die* viii. 127, I rang Personnel at the office and asked for Stephen's full name. 1973 *Clarendonian* XXVII. i. 21 It was not therefore surprising that when a vacancy occurred, J.L.A. asked for her in Personnel.

2. Personal appearance. *rare.*

1861 T. McWeeney in D. Crow *Theresa* (1966) xvi. 182 Mrs Yelverton is still in the possession of an exceedingly agreeable personnel, and, without being positively handsome, she is most prepossessing and ladylike. 1909 W. De Morgan *It never can happen Again* I. x. 139 Contrast it with the dowdy personnel and awkward manners of the political gentleman's wife.

3. *attrib.* and *Comb.*, as *personnel audit, car, carrier, department, management, manager, officer, policy, procurement, secretary, transfer capsule*; *personnel-designating* adj.; *personnel-wise* advb.

1967 Coulthard & Smith in Wills & Yearsley *Handbk. Managem. Technol.* 211 *Personnel audit*—in the same way that a firm's accounts are audited, or the stockrooms checked periodically, it is desirable to maintain an accurate analysis of the company's personnel needs—in managerial, clerical, sales, and manufacturing areas—together with an assessment of its strengths, as made up by the individuals concerned. 1914 *Illustr. London News* 17 Oct. 540/2 The personnel-car of the Schneider gun-train. 1945 *Finito! Po Valley Campaign* (15th Army Group) 51 A German convoy of two 170 mm cannon pulled by prime movers and followed by personnel carriers swung out of a side road. 1975 *N.Y. Times* 8 Sept. 2/2 A modern armored personnel carrier is one of Israel's obvious military needs. 1976 *Globe & Mail* (Toronto) 24 Sept. 9/1 The Post Office says there are now 908 'mailmobiles' or 'personnel carriers' operating across Canada. 1977 *Time* 25 July 37/1 A wave of Soviet tanks and armored personnel carriers rolls across the northern German plain. 1943 J. B. Priestley *Daylight on Saturday* vi. 32 Mr. Cheviot..was very keen on the personnel department and welfare generally. 1962 S. E. Godfrey *Retail Selling & Organization* vii. 55 The personnel department is also responsible for keeping detailed records of each employee and for authorizing salary payments. 1973 E. Pace *Any War will Do* (1974) i. 15 All the famous benefits the personnel department was always talking about. 1963 F. G. Lounsbury in S. Koch *Psychol.* VI. 570 Lexical units in kinship terminologies can differ, as between different societies, in two aspects of their meanings: the 'personnel-designating' aspect and the 'role-symbolizing' aspect. 1957 Clark & Gottfried *University Dict. Business & Finance* 265/1 *Personnel management*, the branch of business management concerned with the administration and direction of all of the relations between a company and its employees. 1959 *Cambr. Rev.* 6 June 575/1, I have heard personnel management described as a career that the liberal-minded, not unintellectual arts graduate of this university can take up without being accused of selling himself to mammon. 1973 'M. Underwood' *Reward for Defector* v. 38 He would need a bread and butter job... He mentioned personnel management. 1926 M. S. Leuck *Fields of Work for Women* vii. 74 (*heading*) Personnel manager. 1951 J. B. Priestley *Festival at Farbridge* i. i. 21, I have a friend..who's now the personnel manager at Whatmore's. 1973 Personnel manager [see *labour relations* s.v. *labour *sb.* 8]. 1957 C. Smith *Case of Torches* xi. 132 The women's personnel officer was tall, smart and superior. 1958 A. Wilson *Middle Age of Mrs Eliot* iii. 430 Old Shuffler suggested that I got a job as secretary to a personnel officer at some big works in the London area. 1976 M. Hinxman *End of Good Woman* x. 139 'Work your way up, lad,' the personnel officer told him. 1949 M. Mead *Male & Female* xvii. 349 The personnel policy is protecting the firm, not the married men and women. 1950 *N.Y. Times* 20 Apr. 1/3 Donald Dawson, an administrative assistant responsible for personnel procurement, who has the never-ending task of seeking and investigating candidates for important vacancies. 1923 H. Crane *Let.* 10 June (1965) 136, I don't know many at the office yet..but I've already been invited out to tea by the personnel secretary. 1965 *Tuscaloosa* (Alabama) *News* 11 Sept. 8/1 Doctors say a swift rise to the surface would probably be fatal. So the aquanauts will enter a special personnel transfer capsule that will take them to the surface where they will enter a decompression chamber on the mother ship's deck. 1975 *Offshore* Aug. 54/3 The personnel transfer capsule (PTC) was launched from the YDT to carry out open sea dives to 1,000 ft. 1963 R. I. McDavid *Mencken's Amer. Lang.* 250 The practice [of forming adverbs from nouns by adding *-wise*] quickly spread from the admen to the bureaucrats and the educationists and soon yielded *curriculumwise,.. personnelwise* and *weatherwise.*

personology (pǝ̄ɹsŏnǫ·lŏdʒi). *Psychol.* [f. Person *sb.* + -ology.] A term sometimes used for the study of personality. So **persono·logical** *a.*; **persono·logist**, one who studies personality.

1926 J. C. Smuts *Holism & Evolution* x. 262 Personality is, in fact, largely an unexplored subject and requires a discipline to itself as a real factor in the universe. 'Characterology' has been suggested as a name for the new discipline, but there are objections to it, and Personology is suggested as a better name. The 'Person' is a concept of the Roman law, not of Greek philosophy, and the hybrid is therefore justified. 1938 H. A. Murray *Explorations in Personality* i. 4 The branch of psychology which principally concerns itself with the study of human lives and the factors that influence their course, which investigates individual differences and types of personality, may be termed 'personology' instead of 'the psychology of personality' a clumsy and tautological expression. *Ibid.* 8 Since the latter [*sc.* psychic impulses] are intangible, personologists must imagine them. 1951 J. S. Bruner in Blake & Ramsey *Perception* v. 121 The perception-centered approach takes as its primary focus of interest the variables of perception and studies the way these are affected by various learnings, motivational states, personological structures, etc. 1957 Hall & Lindzey *Theories of Personality* v. 157 The focus of this theory is upon the individual in all his complexity and this point of view is highlighted by the term 'personology'. 1967 R. R. Holt in Lazarus & Opton *Personality* 48 Personologists have increasingly begun to recognize that all the error-terms of standard psychological equations are their own happy hunting grounds. *Ibid.* (*heading*), The logic of the Romantic point of view in personology. 1980 *Underground Grammarian* Mar., Since personology must be too subtle a science for the likes of us, we cannot explain how 'personological' variables might be different from differences in persons.

person-to-pe·rson, *adj.* (and *adv.*) *phr.* **a.** Designating a personal telephone call: see *personal call* s.v. *personal *a.* 10. orig. and chiefly *U.S.*

1919 *N.Y. City Telephone Directory* 20/2 A Person-to-Person Toll Call is one made by name for a particular person reached through a telephone which is located outside the local service area and at a point to which there is a person-to-person toll rate. 1933 *Sat. Even. Post* 11 Feb. 57 There are two classes of Long Distance calls: station-to-station and person-to-person... When the caller asks the operator for a specific individual, the person-to-person rate applies. 1960 C. Fitz Gibbon *When Kissing had to Stop* viii. 128 I've put through a personal call, what they call a person-to-person call over there. 1973 C. Egleton *Seven Days to Killing* xx. 222 They stopped by the call-box..and Tarrant got the operator to put through a person-to-person call. 1973 *Sat. Rev. Society* (U.S.) Mar. 70/2 Person-to-person calls are, of course, more expensive than station-to-station calls. 1977 *Transatlantic Rev.* lx. 50 A representative of the Columbus Police Department placed a person-to-person call to Dr. Edward Hudson in Padananim.

b. Taking place directly between individuals; interpersonal.

1951 R. Firth *Elem. Social Organiz.* i. 30 Some anthropologists have argued that a social structure is the network of all person-to-person relations in a society. 1954 [see *high-potential* s.v. *high *a.* 22 a]. 1960 *Times* 12 Feb. 4/6 Person-to-person communication between specialists is much less satisfactory in applied science. 1965 H. Kahn *On Escalation* xiii. 248 The person-to-person meeting of Khrushchev and Kennedy. 1970 *Hospital Tribune* 26 Jan., Person-to-person spread probably accounted for the introduction of new cases into communities. 1977 *Proc. R. Soc. Med.* LXX. 553/1 Leukaemia and Hodgkin's disease are unusual among neoplastic disorders in man in often having aroused suspicion of person-to-person transmission or an infective aetiology.

As *adv. phr.* (usu. without hyphens).

1971 G. Cuttle in B. de Ferranti *Living with Computer* i. 4 This does not mean that in future all contact will be person-to-person over the telephone and viewing screen. 1977 B. Lucas tr. *C. De Foucauld's Lett. from Desert* vii. 128 Apostolic work..as I envisage it consists in talking person to person with infidels.

persorption (pǝɹsǭ·ɹpʃən). *Chem.* [f. Per(meation + *sorption.] Sorption in which molecules of a gas enter pores in a solid that are only a little larger than themselves.

1930 McBain & Britton in *Jrnl. Amer. Chem. Soc.* LII. 2220 The sorbed molecules of nitrogen must be so intimately surrounded by the atoms of carbon with which they are in contact that the designation 'adsorption', which implies a surface, becomes a misnomer. There is very little difference in such a case between adsorption and absorption; and if a precise word to describe sorption by charcoal were required, the new term 'persorption' might be coined. 1948 Glasstone *Textbk. Physical Chem.* (ed. 2) xiv. 1204 It is probable that persorption is operative in the highly active charcoals, and perhaps to some extent in silica gel; the occlusion of hydrogen by palladium..may well be an extreme case of combined activated adsorption and persorption. 1960 A. W. Adamson *Physical Chem. of Surfaces* xi. 523 What might be considered to be a limiting or extreme case of persorption is the formation of clathrate compounds.

persp. (pǝ̄ɹsp). Also **persp** (without full point). Colloq. abbrev. of Perspiration 4.

1923 Wodehouse *Inimitable Jeeves* ii. 24 The good old persp. was bedewing my forehead by this time in a pretty lavish manner. 1966 —— *Plum Pie* i. 19 It was with quite a few beads of persp bedewing the brow that I went back to the dining room. 1974 —— *Aunts aren't Gentlemen* ii. 12 He said 'Phew' and removed a bead of persp. from the brow.

perspectival (pǝɹspe·ktivăl), *a.* [f. Perspectiv(e *sb.* + -al.] Of or pertaining to perspective, esp. in a *fig.* sense of a mental perspective. Hence **perspe·ctivally** *adv.*

a 1866 J. Grote *Exploratio Philosophica* (1900) II. xii. 121 The '*perspectival*' fact (so to call it) of foreshortening..is opposed to his argument in one respect. 1932 H. H. Price *Perception* vii. 196 The group would be a *series* of shapes having a certain limit... The series will consist primarily of perspectival distortions. 1957 *Scottish Jrnl. Theol.* X. 126 It is now clear that all historiography is perspectival, originating from some given point in the economic, political, cultural and moral structures of society and in the course of history itself. 1967 *Philos. Rev.* LXXVI. 193 Seeming to behold our universe(s) *from a place*, perspectivally. 1970 Todes & Dreyfus in J. M. Edie et al. *Patterns of Life-World* xviii. 363 Corresponding to each of these ideas of truth is a certain degree of awareness that knowledge is 'perspectival', i.e., that what we know reflects our ways and means of knowing it. 1977 *Dædalus* Summer 75 The perspectival view from the infrastructure may be at least as false as any superstructural cosmology.

perspective, *sb.* Add: **3. d.** Hence the point of view itself; a way of regarding (something).

1907 H. Adams *Educ. Henry Adams* ii. 20 Time and experience, which alter all perspectives. 1934 M. Bodkin *Archetypal Patterns in Poetry* 307 Writing from the psychological standpoint, I intend this statement less as criticism than as recognition of the limitations of the vital perspective present in these essays. 1949 R. K. Merton *Social Theory* iii. ix. 262 Mannheim's inconsistency..stems from an indefinite distinction between incorrectness (invalidity) and perspective ('onesidedness'). 1963 J. T. Waterman (*title*) Perspectives in linguistics. 1964 Gould & Kolb *Dict. Social Sci.* 262/1 There has been much discussion from many perspectives as to the origins and 'causes' of fascism.

e. An apparent spatial distribution or extent in perceived sound. Freq. preceded by a qualifying word, as *auditory, sound*, etc.

1934 Steinberg & Snow in *Electrical Engineering* (N.Y.) Jan. 12/1 An audience..senses the spatial relations of the instruments of the orchestra. This spatial character of the sounds gives to the music a sense of depth and of extensiveness, and for perfect reproduction should be preserved. In other words, the sounds should be reproduced in true auditory perspective. 1949 Frayne & Wolfe *Elem. Sound Recording* xxxii. 674 Adjustment of the gain of the individual channels also helps in preserving the acoustic perspective. 1961 G. Millerson *Technique Television Production* i. 17 He can warn boom operators against dipping into shot..while assisting them in achieving sound perspective to suit the transmitted picture. 1963 *Times* 12 Jan. 11/3 The sound is all too full and forward, and badly lacking in aural perspective.

5. b. (Later example.)

1965 *Economist* 6 Mar. 989/2 Only a general negotiation can offer the perspective of a return to peace and a real independence.

8. *perspective control*; *perspective-free, -suggesting* adjs.; **perspectiveless** *a.* (*lit.* example).

1971 C. Bonington *Annapurna South Face* App. F. 285, I used 2 Nikon bodies, with 24-mm. and 35-mm. perspective-control lenses, [etc.]. 1966 R. L. Gregory *Eye & Brain* ix. 163 It would be interesting to bring animals up in a perspective-free environment. 1970 *Nature* 4 July 93/1 From the perspectiveless drawing of the Ancient Egyptians to the deliberate mixing of contradictory signs of depth by Hogarth and later artists. 1880 W. James *Let.* 12 Dec. in R. B. Perry *Tht. & Char. W. James* (1935) I. 727 Metaphors and epigrams which, witty and striking and perspective-suggesting as they often are,..may be in danger of having the changes rung on them too long.

Hence **perspe·ctive** *v.*, to set in perspective. Also *intr.*, to draw a plan of the perspective of a drawing, etc. Hence **perspe·ctived** *ppl. a.*

1812 B. R. Haydon *Jrnl.* 4 Apr. in *Autobiogr.* (1853) I. x. 171 Began my picture—perspectived the greater part of the day—felt a sort of check in imagination at the difficulties I saw coming. 1908 *Westm. Gaz.* 8 Aug. 4/1 A certain aimlessness, casualness almost, has suddenly been perspectived into purpose and plan. 1949 Brooks & Warren *Mod. Rhetoric* xiii. 442 It is the prose of a mind which is arranging its world, by delicate adjustments and careful discriminations, into a perspectived pattern. 1978 *Gramophone* July 181/1 These arrangements even make more comfortable and perspectived listening than the harpsichord originals.

perspective, *a.* Add: **2.** (Later examples.)

1852 Thackeray *Esmond* II. x. 174 We have but to change the point of view, and the greatest action looks mean; as we turn the perspective-glass, and a giant appears a pigmy. 1859 Dickens *T. Two Cities* v. 59 If a girl..swoons within a yard or two of a man's nose, he can see it without a perspective-glass. 1867 *Atlantic Almanac 1868* 9 So thought Lonson Nash,..who saw it [*sc.* a sea-serpent] through a perspective-glass in the year 1817.

3. (Further examples.)

1777 P. Thicknesse *Year's Journey* I. p. xiii, I might raise..money sufficient to pay for engraving a perspective view of Montserrat. 1911 *Encycl. Brit.* XX. 470/2 A fresco of 'The Flood' at Florence is even more naive in its parade of the painter's [*sc.* Uccello's] newly won skill in perspective science. *Ibid.* XXI. 257/2 A horizontal plane on which we suppose the objects to rest of which a perspective drawing is to be made. 1935 *Burlington Mag.* Apr. 200/1 This artist's work, including his portraits, perspective pieces and genre subjects. 1942 D. D. Runes *Dict. Philos.* 230/2 In epistemology: the perspective predicament, the limited though real viewpoint of the individual; the plight of being confined to the experience of only part of actuality. 1959 W. C. Lounsbury *Backstage from A to Z* 34 Perspective drawings of sets executed to dimension can be of great value, but avoid misleading, haphazard drawings which merely confuse. 1959 P. & L. Murray *Dict. Art & Artists* 236 The basic assumption of all perspective systems is that parallel lines never meet, but that they appear to do so. 1961 *Architect & Building News* 21 June 822/2 The perspective drawing realistically illustrates this interesting feature of the design. 1961 G. Millerson *Technique Television Production* iii. 31 (*caption*) Viewing from too close or too far makes its perspective look unnatural. This is termed perspective distortion. 1970 *Oxf. Compan. Art* 843/1 Shortly after Brunelleschi made his perspective demonstrations his fellow architect Alberti devised a perspective construction for the special use of painters, which he described in detail in his famous treatise *On Painting* (1436)... This is the first known written account of a fully scientific perspective construction. 1972 *Jrnl. Social Psychol.* LXXXVII. 143 The average number of illusion supporting responses was scored for each of five geometric illusions... Muller-Lyer..; Sander parallelogram..; perspective drawing.

perspectivic (pǝɹspe·ktivik), *a. rare.* [f. Perspectiv(e *sb.* + -ic.] = Perspective *a.* 3.

1949 Koestler *Insight & Outlook* xxii. 315 The laws of perspectivic geometry make the painter *see* differently.

perspectivism (pǝɹspe·ktiviz'm). [f. Perspectiv(e *sb.* + -ism.] **1.** *Philos.* The theory that knowledge of a subject is inevitably partial and limited by the individual perspective from which it is viewed; also, the partiality and limitation inherent in knowledge on this view.

1910 T. Common tr. *Nietzsche's Joyful Wisdom* 299 Fundamentally our actions are in an incomparable manner altogether personal, unique and absolutely individual—..but as soon as we translate them into consciousness, they do not appear so any longer... This is the proper phenomenalism and perspectivism as I understand it. 1949 R. K. Merton *Social Theory* ix. 261 Mannheim's conception of 'perspectivism' is substantially the same as the Rickert-Weber conception of *Wertbeziehung* (which holds that values are relevant to formulation of the scientific problem and choice of materials but are not relevant to the validity of the results). 1954 A. Huxley *Let.* 17 Jan. (1969) 693 Von Bertalanffy..calls it 'Perspectivism', and points out that the unity of science is to be sought..in the isomorphy of explanatory laws in the different fields and disciplines of science. 1965 A. Danto *Nietzsche* iii. 80 Does Perspectivism entail that Perspectivism itself is but a perspective, so that the truth of this doctrine entails that it is false? 1969 C. O. Schrag *Experience & Being* viii. 276 Is it possible for a philosophy of experience..to proceed beyond the perspectivism of point of view philosophizing? 1973 J. P. Stern *On Realism* v. 64 This perspectivism to which all experience is subject.

2. The practice of regarding and analysing a situation, work of art, etc., from different points of view and on different levels.

1948 L. Spitzer *Linguistics & Lit. Hist.* ii. 50 We may assume that the linguistic perspectivism of Cervantes is reflected in his invention of plot and characters... Cervantes' perspectivism, linguistic and otherwise, would allow him qua artist to stand above, and sometimes aloof from, the misconceptions of his characters. 1949 Wellek & Warren *Theory of Lit.* xii. 158 The unsound

thesis of absolutism and the equally unsound antithesis of relativism must be superseded and harmonized in a new synthesis which makes the scale of values itself dynamic, but does not surrender it as such. 'Perspectivism', as we have termed such a conception, does not mean an anarchy of values, a glorification of individual caprice, but a process of getting to know the object from different points of view which may be defined and criticized in their turn. 1962 M. McLuhan *Gutenberg Galaxy* 125 The habit of a fixed..'point of view'..gave popular extension to the avant-garde perspectivism of the fifteenth century.

perspectivist (pǝɹspe·ktivist). [f. Perspectiv(e *sb.* + -ist.] An artist who specializes in perspective effects; one who studies the principles of perspective.

1942 *Burlington Mag.* Jan. 24/2 In the case of Fouquet the traditional methods of the master of the Bedford Hours..gives way to the influence of Fra Angelico, of Jacopo Bellini and of the Florentine perspectivists. 1955 H. Read *Icon & Idea* 149 The 'Commentaries' of Ghiberti.., which are a kind of commonplace book in which we find all the ancient texts and recorded observations that occupied the minds of these fifteenth-century perspectivists, depend a good deal on Arabic sources. 1958 R. Myerscough-Walker *Perspectivist* i. 7 A perspectivist never, under any condition, begins with an image behind the transparency. *Ibid.* 11 There are very few professional perspectivists in England at this moment. 1970 M. H. Pirenne *Optics, Painting & Photogr.* vii. 85 On 'perspectivists', that is artists who made such illusionistic paintings [*sc.* painted ceilings], see Maffei... This interesting article..does not refer to the works of perspectivists extant in Great Britain.

perspectivistic (pǝɹspe:ktivi·stik), *a.* [f. Perspectiv(e *sb.* + -istic.] Of or pertaining to perspectivism.

1948 L. Spitzer *Linguistics & Lit. Hist.* 67 We are offered basically the same perspectivistic pattern that we have noted in the case of the *baciyelmo.* 1950 J. Meserve tr. E. Auerbach in *Partisan Rev.* May–June 416 And so the excursus does not begin until two lines later, when Eurycleia has discovered the scar—the possibility for a perspectivistic connection no longer exists, and the story of the wound becomes an independent and exclusive present.

perspectivity (pǝɹspekti·viti). [f. Perspectiv(e *sb.* + -ity.] The quality or condition of being limited by or confined to a particular perspective or point of view.

1910 A. M. Ludovici tr. *Nietzsche's Will to Power* II. 20 (*heading*) Biology of the instinct of knowledge. Perspectivity. 1930 A. O. Lovejoy *Revolt against Dualism* iii. 92 The necessary diversity of the characters experienced by percipients having different standpoints, we may call 'perspectivity'. 1933 *Jrnl. Philos.* XXX. 63 Such perspectivity is what the organism escapes to the extent that it becomes aware of..the logical dimension of nature. 1969 C. O. Schrag *Experience & Being* viii. 276 In becoming clear about the meaning of perspectivity, a rather firmly entrenched prejudice needs to be suspended.

Perspex (pǝ·ɹspeks). Also **perspex.** [Irreg. f. L. *perspect-*, ppl. stem of *perspicere* to look through (1st pers. perfect *perspexi*), f. Per *prep.* + *specere* to look (at).] A proprietary name for polymerized methyl methacrylate, a tough transparent thermoplastic that is much lighter than glass and does not splinter. Freq. *attrib.* and in *Comb.*

In the U.S. sold under the names of *Plexiglas and *Lucite.

1935 *Trade Marks Jrnl.* 9 Jan. 48/2 Perspex... Synthetic resins sold in the form of sheets, rods, tubes and shaped pieces, and as moulding powders. I.C.I. (Fertilizer & Synthetic Products) Limited, London,..manufacturers. 1937 *Nature* 20 Feb. 336/1 The lenses..were made of a particular form of the plastic material known by the trade name of 'Perspex'. 1943 L. Cheshire *Bomber Pilot* i. 9, I asked Percy to dim his light; it reflected on the perspex in front and I couldn't see out. 1946 *Electronic Engin.* XVIII. 224 The needle holder is in the form of a small perspex block. 1951 *Official Gaz.* (U.S. Patent Office) 13 Nov. 336/1 Imperial Chemical Industries Limited, London... Perspex... Claims use since 1934; and since November 1949 in commerce between Great Britain and the United States. 1957 *Times* 21 Dec. 9/7 The sculpture is dominated by..designs for a fountain and a triptych in Perspex. 1959 G. Freeman *Jack would be Gent.* i. 13 There were two large, ornate chandeliers made of wrought iron and bits of Perspex. 1960 *Practical Wireless* XXXVI. 397/1 Behind a small sheet of perspex screwed to the front above the speaker aperture. 1961 *Lancet* 23 Sept. 680/2 After infection, groups of mice were kept in air-tight 'Perspex' boxes. 1963 *Times* 11 May 4/7 The perspex windscreen is faired down almost to the driver's seat. 1975 N. Luard *Travelling Horseman* iii. 81 A damn great Perspex-faced wall map of London.

persuade, *v.* Add: **2.** Also const. *away from* (a belief, etc.), *down to* (a place, etc.), *off* (an intention, place, etc.).

1777 C. Reeve *Champion of Virtue* 30 Let it be kept from my two cousins.., if they offer to be of the party I will persuade them off it. 1895 'G. Mortimer' *Like Stars that Fall* ii. 17, I wish you'd just try to persuade Lou off a silly idea she's just got hold of. 1941 A. L. Rowse *Tudor Cornwall* xiv. 348 The object of which was to persuade him away from his stand for catholicism. 1959 M. Shadbolt *New Zealanders* 12 He..persuaded his brother off the waterfront to look after the herd for six months. *Ibid.* 218 I've been trying to persuade Izzy down to the city for the last five years.

persuader. Add: **b.** (Further examples.)

1862 *N.Y. Tribune* 3 June 3/5 In the South heavy guns are called persuaders. 1884 W. L. Rede *Sixteen String Jack* ii. iv. 14/2 It's no use resisting, 'cause ve has the persuaders. *Ibid.*, Dick, out vith your persuader! (Draggle puts a pistol to his head.) 1900 Ade *More Fables* 54 The Colonel arose and pulled his Persuader, expecting to make it a Case of Justifiable Homicide. 1925 Fraser & Gibbons *Soldier & Sailor Words* 221 *Persuader*, a nickname for the club, or knob-kerry, carried by trench raiders. Also, bayonet. 1930 *Detective Fiction Weekly* 10 May 48/1 Papers..held down by what crooks call a 'sap' or a 'persuader'. 1935 A. J. Pollock *Underworld Speaks* 87/1 *Persuader*, a pistol. 1964 L. Deighton *Funeral in Berlin* xlii. 258 'Do you have a pistol or a knife or a persuader?' 'I have a persuader... Two hundred dollars in singles.' 1974 P. Cave *Mama* (new ed.) xiv. 113 'How the hell are you gonna persuade the guy to pull off the road?' asked Ethel... 'I've thought of that,' answered Mama coldly. 'And that's one of the little changes... I'm gonna have to take along a little persuader.'

c. *Printing.* A tool used as a lever by a compositor when type matter and furniture is being fitted into a chase.

1898 J. Southward *Mod. Printing* I. xli. 257 Fit the quoins, using the 'persuader' to squeeze in the pages, and tap up all round. The 'persuader' is usually a tool made by the compositor himself... The tool is a lever whereby the space between the type matter and the chase can be expanded.

d. *Television.* An electrode in the image orthicon camera tube which deflects the returning beam of scanning electrons into the electron multiplier. Freq. *attrib.*

1946 *RCA Rev.* VII. 361 The persuader electrode is tied electrically to the first stage. *Ibid.* 362 There is no need to adjust the persuader voltage for controlling uniformity of gain. 1953 Amos & Birkinshaw *Television Engin.* I. 108 The persuader potential can be set to any value up to +300 volts. 1967 H. A. Cole *Basic Television* i. 56 When they get near the electron gun, the returning electrons are diverted on to the first dynode of the electron multiplier by an electrode aptly named the persuader grid.

pert, *a.* Add: **2. b.** (Later example.)

1952 W. G. Hardy *Unfulfilled* iii. v. 246 Actually, as he looked her over, she was rather cute—slim young body and pert young breasts.

4. b. (Further examples.)

1898 G. B. Shaw *Candida* i. 82 In a black merino skirt and a blouse, rather pert and quick of speech, and not very civil in her manner. 1924 —— *St. Joan* vi. 87 We are not so foolish as you think us. Try to resist the temptation to make pert replies to us.

6. (Further examples.)

a. 1565, etc. [see Pearmonger]. 1859 W. Warner *Albion's Eng.* ii. xxxi. 135 As peart as bird. 1774 P. V. Fithian *Jrnl.* (1900) 241 Ben seems a little more pert today. 1832 W. Irving *Jrnl.* 10 Nov. (1919) III. 171 My horse goes quite peart. 1873 A. J. Munby *Diary* 11 July in D. Hudson *Munby* (1972) 335 Sir Arthur Helps with his bristling white hair and 'peart' white face, so like Mazzini. 1891 L. T. Meade *Sweet Girl Graduate* xxii. 187 I'll be glad to lie down... I'll be as pert as a cricket in the morning. 1943 *Amer. Speech* XVIII. 67/2 *Peart* (lively, in good health, especially of older people).

b. 1772 in *Maryland Hist. Mag.* (1919) XIV. 272 The Corn looks pert & green.

C. *adv.* (Later examples.)

1902 W. N. Harben *Abner Daniel* 230 Well, I'm glad I won't have to go furder'n Darley... By ridin' peert I can let you out before sundown. 1972 *News & Observer* (Raleigh, N. Carolina) 30 Dec. 4/3 We aren't journey proud, and few of us get the big head or act pert.

Perthes(') disease (pǝ̄·ɹtəz). *Med.* [Named after Georg Clemens *Perthes* (1869–1927), German surgeon, who described the condition in 1910.] A disease of the hip occurring in children, probably owing to an interrupted blood supply, in which necrosis of part of the head of the femur leads to progressive deformity of the joint.

1915 *Amer. Jrnl. Orthopedic Surg.* XII. 557, I shall try to reproduce these features which give a character of its own to Perthes disease, and to describe it as a new and typical disease of the hip. *Ibid.* 564 One must conclude that Perthes disease has a tendency to spontaneous reparation and that all treatment is useless. 1958 *Jrnl. Bone & Joint Surg.* XL-B. 173 Stigmata of pre-existing Perthes' disease: a large [femoral] head, decreased height of the epiphysis, slight subluxation, shortening or broadening of the femoral neck. 1974 *Israel Jrnl. Med. Sci.* X. 230 The results may indicate that the deformity of the acetabulum in Perthes disease, generally recognized as secondary to femoral head deformity, may in fact develop in the initial stages of the disease.

perthitic, *a.* Add: (Examples.) Hence **perthi·tically** *adv.*, in the manner of perthite.

1906 J. P. Iddings *Rock Minerals* ii. 236 Perthitic intergrowth. 1930 Peach & Horne *Geol. Scotl.* 112 The alkali-felspar includes both orthoclase and albite, which may be present separately, perthitically intergrown. 1949 *Econ. Geol.* XLIV. 174 In the diorite facies the microcline is generally only slightly perthitic..whereas in the syenitic rock it is ordinarily highly perthitic. *Ibid.* 179 The sodic solutions soaked into the microcline and replaced it perthitically by albite and then by grains along margins and fractures. 1965 G. J. Williams *Econ. Geol. N.Z.* xiii. 208/2 A radioactive biotite-granite with quartz, perthitic orthoclase, microcline and 20 per cent of dark minerals was also found in the Big River area.

PERTURB

perturb, *v.* Add: **2.** (Later example of absol. use.)
1902 *Daily Chron.* 23 Apr. 3/3 It is the unexpected that perturbs.
3. *Physics* and *Math.* To subject (a physical system, or a set of equations, or its solution) to a perturbation (sense *4).
1901, etc. [implied in *PERTURBED *ppl. a.* 2]. 1931 *Physical Rev.* XXXVIII. 875 The $^{1}P_{1}$ sequence is perturbed by $X^{1}P_{1}$. 1973 *Nature* 17 Aug. 416/1 If the initial potential is that of a hard sphere, this can be 'perturbed' into a realistic form by adding an attractive term and softening the repulsion.

perturbation. Add: **2. b.** (Further example.)
1946 H. & B. S. Jeffreys *Methods Math. Physics* xvi. 464 Without the disturbance due to other planets, the motion of any planet would be an ellipse, specified by six constants... To allow for perturbations these constants are taken as variables.
4. *Physics* and *Math.* A slight alteration of a physical system, esp. of the conditions which a solution of Schrödinger's equation must satisfy, or of a set of equations, from a relatively simple form to one which is to be studied by comparison with the simpler form. Freq. *attrib.*, as *perturbation calculation, expansion, method, series*; **perturbation theory,** the method of investigating solutions of equations of state by relating them to solutions of similar but simpler equations which can be solved directly.
[1868 *Phil. Mag.* XXXVI. 135 The motions [of molecules] are, however, not altogether free from perturbation.] 1899 *Q. Jrnl. Math.* XXX. 47 In this paper it is proposed to follow the theory of perturbations in the problems of mechanics in the order of its historical development from Lagrange to Lie. 1926 *Proc. R. Soc.* A. CXI. 301 If the [magnetic] field is weak we may use perturbation theory, according to which the change of energy of the stationary states is given, to the first order, by the constant term in the Fourier expansion of the energy of the perturbation in terms of the uniformising variables for the undisturbed system. 1937 E. C. Kemble *Fund. Princ. Quantum Mech.* xi. 380 In quantum mechanics, as in the Bohr theory, perturbation methods are of fundamental importance due to the fact that so few problems can be rigorously solved by direct methods. *Ibid.* xiv. 526 The usual method of approach to the problem of the many-electron atom is through a perturbation calculation in which the unperturbed problem..is of the central-field type. 1956 R. H. Atkin *Math. & Wave Mech.* xi. 241 The method of perturbations is a practical technique for approximating to such solution. 1957 *Technology* Apr. 73/2 Perturbation theory and electron spin are studied, and used to explain the periodic table of elements and chemical bonds. 1961 Powell & Crasemann *Quantum Mech.* xi. 381 We shall begin by deriving..the result of Section 5–8, giving the effect of a small perturbation on the energy levels of a system with discrete stationary states. *Ibid.* 403 A system subject to a weak time-dependent perturbation can be described approximately by a Hamiltonian of the form $H = H^{0} + V(t)$. *Ibid.*, Since the perturbation is time-dependent, the system does not, in general, have stationary states. 1968 Fox & Mayers *Computing Methods for Scientists & Engineers* ii. 17 We accept the 'solution'..and try to find a 'neighbouring' problem, a 'perturbation' of the given problem. 1972 G. E. Brown *Many-Body Probl.* ii. 25 The two types of perturbation theory most commonly used are the Brillouin-Wigner perturbation expansion and the Rayleigh-Schrödinger one. 1973 Alonso & Valk *Quantum Mech.* v. 202 Expressions (5.8–19) through (5.8–23) constitute the Rayleigh–Schroedinger perturbation series.
Hence **perturbation-theoretic, -theoretical** *adjs.*, of, pertaining to, or involving perturbation theory.
1964 *Physical Rev.* CXXXIII. A1070 (*heading*) Perturbation theoretic calculation of polaron mobility. 1968 C. G. Kuper *Introd. Theory Superconductivity* i. 2 Early attempts to construct a perturbation-theoretical model based on Fröhlich's interaction encountered severe mathematical difficulties.

perturbative, *a.* (Examples in *Physics.*)
1971 *Ann. Physics* LXIV. 383 The change in the initial set of site amplitudes is sufficiently small so that a perturbative solution of the set of equation[s] remains valid. 1973 *Physics Bull.* Dec. 734 A perturbative approach to the valence charge density in tetrahedrally bonded semiconductors.
Hence **pe·rturbatively** (or *perturba·tively*) *adv.*
1977 *Nature* 21 July 205/2 Since α is about 1/127, things may be calculated perturbatively.

perturbed, *ppl. a.* Add: **2.** *Physics* and *Math.* Subjected to a perturbation (sense *4).
1901 *Phil. Mag.* II. 268 By a kinematical analysis he (Stoney) shows that such perturbed elliptic motion may be regarded as resultant of two or more circular motions of different amplitudes and frequencies. 1927 *Proc. R. Soc.* A. CXIII. 639 The wave equation of the perturbed system. 1937 E. C. Kemble *Fund. Princ. Quantum Mech.* xi. 380 We designate the problems based on the two operators H_0 and H as the unperturbed and the perturbed problems, respectively. 1949 *Q. Jrnl. Math.* XX. 155 (*heading*) Perturbed functional equations. 1968 Fox & Mayers *Computing Methods for Scientists & Engineers* ii. 17 We can often say that we have obtained an *exact* solution of the perturbed problem.

perturber. Add: **2.** *Physics.* A particle which interacts with a radiating atom or ion, affecting the wavelength of the emitted radiation.
1932 *Physical Rev.* XL. 401 R_1 means, crudely speaking, the distance of closest approach between the excited Hg-atom and its perturber. 1962 M. Baranger in D. R. Bates *Atomic & Molecular Processes* xiii. 505 It often happens that the interaction of the perturbers with the atom in the lower state of a given line is much weaker than their interaction with the upper state. 1974 H. R. Griem *Spectral Line Broadening by Plasmas* ii. 72 The principal difference between the broadening of neutral-atom and positive-ion lines lies in the presence of the long range Coulomb interactions between radiators and charged perturbers.

Perugian (pĕrū·dʒiăn), *sb.* and *a.* [f. *Perugia* the name of a city and province in central Italy + -AN.] **A.** *sb.* A native or inhabitant of Perugia. **B.** *adj.* Of or pertaining to Perugia; *spec.* of or relating to a division of the Umbrian school of painting having Perugia as its centre.
1759 A. Butler *Lives Saints* IV. 62 He with several others was carried away prisoner by the Perugians. *c* 1863 Mrs. Gaskell *Lett.* (1966) 934 The first thing..[is] to tell you how capitally our Perugian journey answered. 1864 Crowe & Cavalcaselle *New Hist. Painting Italy* II. vii. 187 The fragment of a recovered fresco..explains the rise and progress of the Perugian school out of that of Gubbio. 1885 *Encycl. Brit.* XVIII. 680/1 In the centre rises the great marble fountain constructed about 1277 by Bevignate, Frate Alberto (both Perugians), and Boninsegna (a Venetian). 1887 A. H. Layard *Kugler's Handbk. Painting: Italian Schools* (ed. 5) I. vii. 212 That branch of the Umbrian school which we may term the 'Perugian', was developed at a later period..it culminated in *Raphael.* 1914 Brown & Rankin *Short Hist. Italian Painting* ii. 155 Pleasing as is this early phase of Perugian painting, it is chiefly valuable as a factor in the education of less local men. 1934 E. Bowen *Cat Jumps* 193 Over the bed hung a panel of leafy Perugian damask. 1936 G. F.-H. & J. Berkeley *Italy in Making* II. vii. 108 Dr. Luigi Masi..was a young Perugian. 1970 A. P. Oppé *Raphael* ii. 29 The irregularity of Raphael's advance is shown in the two pictures of the Madonna which alone give certain evidence of the characteristics of his Perugian period.

Peruginesque (pĕrūdʒine·sk), *a.* [f. *Perugino* (see below) + -ESQUE.] Resembling the style of the Italian painter Pietro Vannucci (*c* 1450–1523), known as Pietro *Perugino* after the town Perugia.
1842 tr. *Kugler's Handbk. Hist. Painting* I. iv. iv. 172 The figures are beautiful and dignified, but without constraint or Peruginesque mannerism. 1863 G. M. Hopkins *Let.* 10 July (1956) 202 There are the most deliciously graceful Giottesque ashes..here—I do not mean Giottesque though, Peruginesque, Fra-Angelical(!), in Raphael's earlier manner. 1874 E. Eastlake tr. *Kugler's Handbk. Painting: Italian Schools* (ed. 4) I. iv. ii. 297 Various pictures, more or less weak, of a Peruginesque class, bear his name. *Ibid.*, His [*sc.* Lo Spagna's] style is a mixture of the Peruginesque and Raphaelesque. 1936 *Burlington Mag.* Sept. 130/1 The two saints..have indeed something Umbrian, even late Peruginesque, in them. 1956 K. Clark *Nude* vi. 232 Drawings in a Peruginesque style, which show the holy women weeping over the stretched-out body of the dead Christ.

peruke-maker. (Later examples.)
1905 T. Auden *Shrewsbury* viii. 201 Brought up at Manchester as a barber and peruke-maker, he adopted the Jacobite principles. 1966 J. S. Cox *Illustr. Dict. Hairdressing* 111/2 s.v. Peruke 'Riot', As the distressed peruke-makers marched through the streets it was seen that most of them were without wigs themselves.

peruse, *v.* Add: **5.** Also *absol.* or *intr.*
1886 Hardy *Mayor Casterbr.* II. xviii. 254, I have tried to peruse and learn all my life; but the more I try to know the more ignorant I seem. 1909 H. G. Wells *Ann Veronica* i. 25 Her father..appeared not to observe her entry. 'Sit down,' he said, and perused..for some moments.

Peruvian, *a.* (*sb.*) Add: **a.** *Peruvian lily* = *ALSTRŒMERIA.
1883 W. Robinson *Eng. Flower Garden* 10/1 Alstrœmeria (Peruvian Lily)... One or two kinds..are hardy and charming as any flowers on warm soil. 1931 M. E. Stebbing *Hardy Flower Gardening* v. 100 *Alstrœmerias*, called 'Peruvian Lilies', do curiously well in Scotland, considering they come from such a warm climate. 1970 *Sunday Tel.* 3 May 19/2 Among many plants which can be grown out of doors for cut flowers, excitement has been aroused by the new Peruvian lilies, or alstroemerias, bred in Holland.
c. *Peruvian Jew* = sense B. 2 below.
1899 in C. Pettman *Africanderisms* (1913) 370 Peddling Peruvian Jews were mulcted in sums from £10 downwards ..and compelled to contribute to the Pretorian war-chest.
B. *sb.* **2.** [Prob. f. acronym *P.R.U.* Polish and Russian Union.] In South Africa, a contemptuous name for a Jew, esp. from Central or Eastern Europe.
1898 L. Searelle *Tales of Transvaal* 4 A 'Peruvian' standing by, whose name was Schadrach Levi. 1900 *Rand Daily Mail* (Pettman), Behold one of the most striking types of Johannesburg life—the Peruvian. 1936 'Idler' *Rolling Home* 385 He called me one day to a little Jew of the worst type which comes from Eastern Europe—the type of 'Peruvian' in South Africa. 1956 H. M. Bate *S. Afr. without Prejudice* iii. 59 Kruger and other equally stubborn of his advisers saw in this a deliberate move by Rhodes to dominate the polls with mine employees and, as they so ungallantly added, 'Peruvians' (a term of contempt which is applied to Jews of low class). 1972 E. Rosenthal *Let.* 23 May in *Voorloper* (1976) 638 According to Max Sonnenburg, the expression originated in the early days of Kimberley, where a body was set up, called 'The Polish and Russian Union', the initials of which 'P.R.U.', gave rise to the word 'Peruvian'.

perv (pɜːv), *v.* *Austral. slang.* Also **perve.** [f. Pervert *sb.*] *intr.* To act as or like a sexual pervert; to indulge in eroticism. Phr. *to perv at, on*: to look at with sexual or erotic interest. Hence **pe·rving** *vbl. sb.*
1941 Baker *Dict. Austral. Slang* 53 *Perve, to*, to act as a sexual pervert. 1944 L. Glassop *We were Rats* xxxiii. 183 'Doing a bit of perving again?' I asked, looking at the gallery of nudes he had gathered from all sorts of magazines. 1959 Baker *Drum* 134 *To perve at* (a girl), to extract pleasure from looking at her, esp. if she is scantily dressed as on a beach. 1964 B. Hesling *Dinkumization* vi. 116 What they get you for, perving? 1966 Baker *Austral. Lang.* (ed. 2) vii. 154 *Perve on*, to contemplate with erotic interest. 1969 *Truth* (Melbourne) 18 Oct. 2/3 They caught me perving on the nurses at the Austin Hospital. 1972 I. Hamilton *Thrill Machine* iii. 17 She's a cheap thrill machine for the boys to stare at and perve on. 1973 A. Broinowski *Take One Ambassador* iii. 30 'Paper'd fold without me.' 'Yeah, I'll bet. Nothing for old Hastings to perve at.' 1974 K. Cook *Bloodhouse* 65 The little poofter perving on her hand ought to have his balls kicked in.

perv (pɜːv), *sb.* *Austral. slang.* Also **perve.** [Shortened from Pervert *sb.*] A sexual pervert. Also *attrib.* or as *adj.*
1944 L. Glassop *We were Rats* xxxi. 177 Bluey brought a perv book back from Cairo with him. 1949 R. Park *Poor Man's Orange* (1950) v. 51 That dirty old cow, always making up to kids... Merv, Merv, the rotten old perv. 1959 E. Lambert *Glory Thrown In* 18 He was a perv. Special attention given to small boys. 1964 B. Hesling *Dinkumization* vi. 116 Two cops, according to the inquiry, booked nearly two hundred 'pervs' a year. 1967 H. Storey in *Coast to Coast 1965–66* 203 It's that bloody old perve from next door. 1968 D. Ireland *Chantic Bird* x. 101 He might have been a perve or a copper's nark. 1973 A. Broinowski *Take One Ambassador* x. 163 My god, the number of pervs there must be in this country.
2. Someone given to 'perving'; the act of 'perving'.
1963 J. Cantwell *No Stranger to Flame* 15 'Never even saw him. Might have been a spook.' She did up the top button on the green blouse. 'Even spooks like a bit of a perv.' 1974 Stackpole & Trengrove *Not just for Openers* 38 After the next bowl had been bowled, the blokes' heads would turn around unobtrusively so they could have a 'perv' at a bird in a mini-skirt walking down the aisle.

pervaginal (pɜːvădʒəi·năl), *a.* [f. Per-¹ + Vaginal *a.* and *sb.*] Done or performed along the vagina.
1922 Joyce *Ulysses* 483, I have made a pervaginal examination and..I declare him to be *virgo intacta.*

pervaporation (pɜːvæpŏrēi·ʃən). *Chem.* [f. Per(meation + E)vaporation.] The evaporation of a liquid through a semi-permeable membrane with which it is in contact.
1917 P. A. Kober in *Jrnl. Amer. Chem. Soc.* XXXIX. 944 My assistant..called my attention to the fact that a liquid in a collodion bag, which was suspended in the air, evaporated, although the bag was tightly closed... Further experiments..soon forced us to the conclusion that the aqueous vapor is given off through the membrane, as though the water were suspended as a solid without any membrane present. This phenomenon we have named pervaporation. 1934 H. N. Holmes *Introd. Colloid Chem.* ii. 19 Dilute hydrochloric acid solutions were concentrated by pervaporation to the acid of constant boiling point. 1956 *Science* 13 July 77/2 We have used pervaporation to dehydrate mashed potatoes. 1964 *New Scientist* 26 Nov. 591/1 For the separation of mixtures of non-ionic liquids considerable interest is being shown in what is called the method of 'pervaporation' using thin polymer films. 1968 *Encycl. Polymer Sci. & Technol.* VIII. 629 In pervaporation a mixture of liquids, often heated, on the upstream side of a membrane is driven through the film by a vacuum applied to the downstream side.
Hence **perva·porate** *v. trans.* and *intr.*, to evaporate in this way; **perva·porated, perva·porating** *ppl. adjs.*
1917 *Jrnl. Amer. Chem. Soc.* XXXIX. 944 After fanning these containers with an ordinary office fan for 24 hours the aqueous layer had pervaporated to dryness. *Ibid.* 945 The pervaporating surface decreased with the sinking of the water level. *Ibid.* 947 A similar container was filled with..sodium chloride solution and pervaporated. 1934 H. N. Holmes *Introd. Colloid Chem.* ii. 18 A protein digestion residue containing strong hydrochloric acid, histidine, and enough humin to make it black was pervaporated. 1956 *Science* 13 July 77/3 The filtrate from a mixture of mashed potato, ethanol, and water was pervaporated... In pervaporating solutions containing class-I solutes, the percentage of water in the pervaporated vapor varied somewhat.

perveance (pɜːviăns). *Electronics.* [Perh. f. the sound of Pervious *a.* + -ance (after

resistance, conductance, permittance, etc.).] A valve parameter which in the case of a diode is equal to the anode current divided by the three-halves power of the anode voltage.
1928 Y. Kusunose in *Res. Electrotechn. Lab.* (Japan) No. 237. 3 The constant *G* which is called 'perveance' may be determined from the electrode configurations. 1951 D. V. Geppert *Basic Electron Tubes* iv. 155 The perveance of a parallel-plane tetrode is equal to the perveance of the triode portion of the tube considering the screen grid as the plate. 1962 C. Susskind *Encycl. Electronics* 229/2 It is convenient to classify electron guns in terms of their perveance, a parameter that remains invariant when the gun is geometrically scaled. 1973 Ferrari & Jonscher *Probl. Physical Electronics* i. 20 It is interesting to note that the ratio $\Delta V/V_0$..depends upon the beam perveance $P = I_0V_0^{-3/2}$ but not upon the beam radices, suggesting that the quantity *P* is the deciding factor as to whether a beam will be perturbed by its own space charge, independently of any particular beam geometry.

pervenche (pɛrvã·nʃ). [Fr., = Periwinkle¹.] **1.** A shade of light blue, resembling the colour of the flowers of the periwinkle. Also *attrib.* or as *adj.*
1899 *Westm. Gaz.* 30 Mar. 3/1 Pervenche and navy are the opposite points of the cold tone of blue. 1909 *Daily Chron.* 6 July 4/5 She wore a beautiful dress of blue embroidered net in a shade of pervenche blue. 1923 *Daily Mail* 26 Apr. 9 The Queen wore a gown of pervenche blue.
2. = Periwinkle¹ 1.
1948 E. Pound *Pisan Cantos* (1949) lxxvi. 44 And in spite of hoi barbaroi Pervenche and a sort of dwarf morning-glory.

perversion. Add: **1. b.** *Psychol.* A disorder of sexual behaviour in which satisfaction is sought through channels other than those of normal heterosexual intercourse.
1892 D. H. Tuke *Dict. Psychol. Med.* II. 1156/2 *Sexual perversion*, an innate perversion or 'inversion' of the sexual feelings with consciousness of its morbid nature... A passion for the sex to which the sufferer belongs, instead of the normal inclination to the opposite sex. 1894 H. Ellis *Man & Woman* xvi. 365 Sexual perversions, again, are more common in men than in women. 1937 J. S. Plant *Personality* ii. viii. 223 A perversion in sexual expression has nothing at all to do with the form of the act but only with its purpose. 1948 A. C. Kinsey et al. *Sexual Behavior Human Male* viii. 264 Perversions are defined as unnatural acts, acts contrary to nature, bestial, abominable, and detestable. Such laws are interpretable only in accordance with the ancient tradition of the English common law which..is committed to the doctrine that no sexual activity is justifiable unless its objective is procreation. 1949 J. Strachey tr. *Freud's Three Ess. Theory of Sexuality* i. 28 Even in the most normal sexual process we may detect rudiments which, if they had developed, would have led to the deviations described as 'perversions'. *Ibid.* 37 Extreme cases of masochistic perversion. 1967 Brussel & Cantzlaar *Chambers's Dict. Psychiatry* 88 Exhibitionism, a form of sexual perversion in which erotic gratification is obtained from the exposure of parts of the body that have sexual significance. 1968 C. Rycroft *Crit. Dict. Psychoanal.* 88 Masochism, sexual perversion in which the subject claims to get erotic pleasure from having pain inflicted upon himself. 1973 I. Singer *Goals of Human Sexuality* 156 It would be erroneous to assume that the so-called perversions are merely alternative attitudes, as desirable as any other sexual possibility.

pervert, *sb.* Add: **2.** *Psychol.* One who suffers from a perversion of the sexual instinct.
1897 H. Ellis *Stud. Psychol. Sex* I. i. 11 A pervert whom I can trust told me that he had made advances to upwards of one hundred men. 1906 *Jrnl. Abnormal Psychol.* Apr. 28 Subconscious feelings which represent, in embryo, the grosser manifestations of the most abandoned sexual perverts. 1924 D. Bryan tr. *Freud's Hysterical Phantasies* in *Coll. Papers* II. v. 51 The strange conditions under which certain perverts carry out their sexual gratifications—either in imagination or in reality. 1972 *Encycl. Psychol.* II. 388/1 In psychoanalytic theory it is postulated that the child shows perversions or is a 'polymorphous pervert'. 1977 *Gay News* 24 Mar. 27/1 The word 'pervert' hardly seems apt to describe Douglas, in the light of such facts.

pervicacious, *a.* Delete 'Now *rare*' and add further examples.
1973 *Daily Tel.* 16 Apr. 13/4 Are audiences the only thing wrong with the theatre?.. At once funky and firm, a pervicacious horde of floating voters, they rush confidently to support the worst candidate on offer. 1973 *N.Y. Law Jrnl.* 7 Aug. 4 The language of the bureaucrats and administrators must needs be recognized as an outgrowth of legal parlance. There is no other way to explain its pervading, pervicacious and pernicious meanderings. 1978 W. M. Spackman *Armful of Warm Girl* 37 Must she like a pervicacious angel think that because he loved her with every beat of his heart, [etc.].

pervy (pɜ̄·ɹvi), *a.* *slang.* [f. Perv(erted *ppl. a.* + -y¹.] Sexually perverted; erotic.
1944 L. Glassop *We were Rats* xxxi. 178 Listen to this..'He buried his head in the warm fragrance of her bosom.' So-and-so, so-and-so. It gets pervy again here. 'His hungry kisses were returned with passionate abandon.' 1970 G. F. Newman *Sir, You Bastard* viii. 223 Twenty maximum security, the lights never out, pervy screws watching every movement.

perylene (pe·rilīn). *Chem.* [ad. G. *perylen* (R. Scholl et al. 1910, in *Ber. d. Deut. Chem. Ges.* XLIII. 2202), f. *per(i-di-naphth)ylen*, f. *peri-* Peri- (used in a spec. chemical sense) + *di-* Di-² + *naphthylen* Naphthalene.] A yellow, crystalline hydrocarbon, $C_{20}H_{12}$, consisting of five fused aromatic rings, which occurs in coal tar and from which certain organic pigments are derived.
1910 *Jrnl. Chem. Soc.* XCVIII. i. 616 To establish the constitution as a *peri*-derivative, 1:8-naphthalenediamine was converted into the azimide, this into 8-iodo-α-naphthylamine, and further into 1:8-di-iodonaphthalene, which last when heated with copper powder yielded..peri-dinaphthalene, which it is proposed to term perylene. 1946 *Nature* 10 Aug. 209/2 It has also been found that the position and number of fluorescence bands of anthracene, perylene, phenanthrene and naphthacene in benzene are independent of the wave-length of the exciting radiation. 1966 *New Scientist* 29 Dec. 735/3 An electron donor such as perylene (which consists of a series of fused benzene rings) is combined with an electron acceptor such as iodine. 1974 *Environmental Sci.* IX. 352/2 A one-year study of the amount of..benzo(*a*)pyrene, benzo(*k*)fluoranthrene, and perylene, was carried out for the particulate matter collected at the York [Ontario] sampling station.

‖ **Pesach** (pe·saχ). [Heb.] = Passover 1.
1613 Purchas *Pilgrimage* ii. xviii. 173 From the second night of their Pesach they number to their Pentecost fifty daies inclusively. 1887 [see *haggadah 2]. 1893 I. Zangwill *Ghetto Tragedies* 94 'Passover is over,' she said... 'Is *Pesach* over?' he said mournfully. 1905 [see *maror]. 1928 *Daily Express* 9 Apr. 10 Passover, or 'Pesach', is the most interesting ceremony in the Jewish calendar. 1950 M. Hay *Foot of Pride* iii. 77 The date of Easter corresponded approximately to the period of the Jewish *Pesach* when Jews were obliged by their law to eat unleavened bread. 1960 *Jewish Chron.* 8 Apr. 35/1 The Rev. S. Black..told the boys and girls of Pesach and its meaning. 1970 I. Sieff *Mem.* ii. 20 On the eve of this great day, the most cherished of Jewish holidays, called the *Pesach*, a great family *Seder* is held. 1972 [see *maror]. 1973 *Jewish Chron.* 19 Jan. 43/3 (Advt.), Book now for Pesach & Summer season.

Pesaro (pesā·ro). The name of a city in northern Italy, used *attrib.* to designate majolica made there in the fifteenth and sixteenth centuries, and the potters who made it.
1856 O. Jones *Gram. Ornament* xvii. 12 *Renaissance ornament.* As early as 1486 the Pesaro ware was considered so superior to all other Italian ware, that a protection was granted to it by the lord of Pesaro. 1885 [see *Gubbio]. 1960 R. G. Haggar *Conc. Encycl. Cont. Pott. & Porc.* 345/1 A Pesaro potter, Jacomo de Pesaro, was working in Venice in 1542.

pesewa (pesī·wā). [Fante *pesewa* penny.] A monetary unit of Ghana, equivalent to one hundredth of a cedi.
1965, 1970 [see *cedi]. 1976 M. Birmingham *Heat of Sun* iv. 52, I paid him the few pesewas he asked.

peskily, *adv.* (In Dict. s.v. Pesky *a.*). (Further examples.)
1834 C. A. Davis *Lett. J. Downing* 139 The Post Office accounts was the next bother; and that puzzled all on us peskily. 1877 *Atlantic Monthly* July 77/2 It does rile him peskily.

pesky, *a.* Add: (Further examples.) Also *U.K. colloq.*
1775 S. Deane in *Connecticut Hist. Soc. Coll.* (1870) II. 224 What reply, think ye, these heroes of five companies of the invincible Royal Irish, gave to this pesky Yankey? 1830 *Massachusetts Spy* 13 Oct. 4/1 I'm plagu'd most to death with these ere pesky sore eyes. 1860 Hotten *Dict. Slang* (ed. 2) 189 *Pesky*, an intensitive expression, implying annoyance. 1901 M. Franklin *My Brilliant Career* xiii. 113 He had always considered Harold as too sensible to neglect his business to stand grinning at a pesky youngster in short skirts and a pigtail. 1909 L. M. Montgomery *Anne of Avonlea* xxvii. 318 Them pesky hens are in my pansy bed again. 1942 E. Paul *Narrow St.* xxiii. 193 Sometimes I wish he would take over this pesky garden and let me manage the restaurant. 1956 D. Karp *All Honorable Men* 252 Just stay away from reporters. And if you can't—you have no comment. If they get real pesky, tell them to talk to me. 1959 I. & P. Opie *Lore & Lang. Schoolch.* ix. 161 Juvenile repugnance continues to be expressed by the old standbys:..pesky [etc.]. 1974 *Times* 23 Jan. 1/8 Dr Benjamin Spock..says that an inability to be firm with their children is the commonest problem of parents in America today, and that it can lead to a child's personality becoming 'balkier and peskier' as the months and years go by. 1974 *Sunday Express* 30 June 6/4 The pesky thing didn't come anywhere near to working. 1977 *Time* 8 Aug. 39/2 But a pesky psychological climate is overhanging the securities markets.
b. (Later examples.)
1901 *Harper's Mag.* Dec. 228 Pesky few Democrats ever I see. 1939 L. M. Montgomery *Anne of Ingleside* xxii. 153 O' course I don't believe in fairies... I've heard they were pesky mischievous.

pessary. Add: **3.** A contraceptive device which is placed in the vagina, now *esp.* a suppository.
1886 H. A. Allbutt *Wife's Handbk.* (ed. 2) vii. 48 Dr. Mensinga, of Flensburg, has invented a preventive pessary, to be worn by the woman, which..will,.. properly adjusted, be a real preventive of conception. 1922 *Brit. Med. Jrnl.* 19 Aug. 327/2 In cases unable to maintain themselves or their children the woman should be temporarily sterilized by compulsion..—for example, by the insertion of the spring wish-bone pessary. 1935 H. B. Whitehouse *Eden & Lockyer's Gynæcol.* (ed. 4) 211 Of mechanical devices to prevent conception three are in common use to-day, the male sheath or condom.., the female vaginal occlusive pessary.., and the cervical cap pessary. 1957 T. N. A. Jeffcoate *Princ. Gynaecol.* xxxviii. 583 Some pessaries dissolve to form a 'foam' which creates a mechanical barrier between the spermatozoa and the cervix. 1973 B. Law *Family Planning in Nursing* ii. 44 Pessaries, vagitories and foaming tablets are other forms in which spermicides are presented.

pessimum (pe·simŏm). [neut. sing. of L. *pessimus* worst.] The most unfavourable condition in the habitat of an animal or plant. Also *attrib.* and *transf.*
1931 R. N. Chapman *Animal Ecol.* viii. 189 It is possible..to conceive of survival potential as representing the actual position on the temperature scale where a species would experience its optimum and pessimum conditions. 1937 *Nature* 16 Oct. 663/2 The first part [of the Russian book under review] contains..a clear presentation of basic ecological principles, namely, factors of existence, ecological valency, optimum and pessimum, habitat concept, biological types (life-forms) and biocœnoses. 1947 N. Balchin *Lord, I was Afraid* 180 Assume that the radius of effectiveness..is 100 feet for a two-ton bomb—which is in many ways the pessimum hypothesis. 1976 *Nature* 19 Feb. 537/2 Above ~10 Ci s (4 rad to 1 kg) there is a '*pessimum*' number of particles.

pest. Add: **3.** *pest control; pest-free* adj.; **pest officer,** one who is responsible for the control or extermination of animal pests.
1931 J. S. Huxley *What dare I Think?* i. 30 Dr Tillyard, now in charge of pest control. 1947 *Nature* 4 Jan. 32/1 The use of a highly refined petroleum oil for application to orchard trees..is firmly established as a valuable pest-control treatment with citrus. 1978 A. Huxley *Illustr. Hist. Gardening* v. 179 Pest control by chemicals is often called warfare. 1944 J. S. Huxley *On Living in Revolution* x. 111 Once more the goal is in sight.., the goal of pest-free stores and stored materials. 1950 *Mind* LIX. 161 An apple..of pleasing taste, high vitamin content and pest-free. 1963 *Times* 9 Mar. 12/4 The hunting of deer was likely to be a more economic and efficient way of controlling them if the hounds were followed only by a small number of qualified foresters or pest officers. 1976 A. Price *War Game* i. vi. 115 He was..a cross between a high class refuse collector and the municipal pest officer.

pesta (pe·stā). [Malay, f. Pg. *festa* feast.] In Malaysia, a festive gathering, a festival.
1964 K. G. Tregonning *Hist. Mod. Malaya* xiv. 298 The dominant Malay element..was gathered together in a highly successful three day *Pesta* or Festival. 1972 S. Backhouse *Singapore* iv. 95 In 1966 the first *Pesta Sukan*, the festival of sport took place, and now accompanies the National Day celebrations held in August. *Ibid.* 96 The *Pesta Sukan* holds two chess tournaments. 1972 *Malay Mail* (Kuala Lumpur) 27 May 2/5 The pesta will have 50 games stalls run by the various societies in the university. 1972 M. Sheppard *Taman Indera* 93 The Cultural Festival, or 'Pesta', organized in Kuala Lumpur in 1956.., gave a fresh stimulus to older dance forms.

Pestalozzi (pestălǫ·tsi). The name of Jean Henri *Pestalozzi* (see Pestalozzian *a.*) used *attrib.* in *Pestalozzi (children's) village* to designate any of several communities of refugee and homeless children established on Pestalozzian principles in Switzerland and elsewhere in Europe after the war of 1939–45.
1947 *Internat. Child Welfare Rev.* I. 253 Pro Juventute.. on two occasions made collections in favour of the Pestalozzi Village. 1949 D. Macardle *Children of Europe* xv. 219 The most lasting of Swiss projects for child victims of the war was initiated in 1946 and opened in October in Trogen... The Pestalozzi Children's Village is a great experiment in education for peace. 1950 T. Brosse *Homeless Children* i. 5 In Switzerland,.. in the Canton of Appenzell, a Children's Village was built in 1946, on the initiative of Walter Robert Corti. *Ibid.* ii. 16 The Pestalozzi School-Town was founded in January 1945 [in Florence]. 1969 *Guardian* 8 Mar. 5/3 The 20 Tibetan children from the Pestalozzi village in Sussex came to dance in traditional style. 1974 *Radio Times* 14 Mar. 25/4 *Down Your Way.* Brian Johnston recently visited Pestalozzi Children's Village, Sedlescombe, Sussex.

peste, *v.* Restrict *rare* to sense in Dict. and add: also used as a curse or exclamation of annoyance (*arch.*).
1768 Sterne *Sentimental Journey* I. 143 La Fleur.. began to search for the letter..*Diable!*—then sought every pocket..not forgetting his fob—*Peste!*—then La Fleur emptied them upon the floor. 1858 Thackeray *Virginians* I. ii. 15 Peste! I don't know why my father gave up such a property. 1896 G. A. Henty *Through Russian Snows* x. 193 *Peste!* these Russians are obstinate brigands. 1898 S. Weyman *Shrewsbury* xlv. 393 '*Peste!*' he said, taking snuff with a droll expression of chagrin. 'Will anyone else ask a question.' 1932 G. Heyer *Devil's Cub* xv. 240 'But I do not know!' cried madame... 'Oh, *peste!*' said Léonie impatiently.

pester (pe·stəɹ), *v.*² [ad. Romany *pessa* to pay.] To pay. So **pe·stering** *vbl. sb.*
1936 J. Curtis *Gilt Kid* v. 53 She had to pester up herself out of the pound you give her. *Ibid.* viii. 88 Tell him to go out and get me a new shirt... Tell him to pester about seven and six for it. *Ibid.* xi. 116 'It's his flat. He pays the rent.' 'Sure. I know he does the pestering.'

pestersome (pe·stəɹsŏm), *a.* [f. Pester *v.* & *sb.* + -some¹.] Annoying, troublesome.
1843 *Amer. Pioneer* II. 439 All innocent enquiries, by infants and children..should be indulged and encouraged, how pestersome soever they may seem. 1906 *Dialect Notes* III. 150 *Pestersome*, bothersome, annoying.

pesticide (pe·stisəid). [f. Pest + -i- + -cide 1.] A substance for destroying pests, esp. insects.
1939 *70th Ann. Rep. Entomol. Soc. Ontario* 16 A special committee..known as the Pesticide Supply Committee is being set up. 1943 *Farm Jrnl. & Farmer's Wife* Sept. 71/1 A new word, 'pesticide', has crept into garden literature this year. 1947 *Times* 9 May 10/1 The demand for 'Gammexane', a pesticide which we discovered, has grown rapidly. 1955 *Sci. News Let.* 24 Sept. 197/3 A pesticide that kills injurious plant mites, but leaves beneficial honeybees and other insects alive has been developed. 1958 *Manch. Guardian* 13 Sept. 2/5 Chemical weed-killers should generally be regarded only as elements in a management programme, not as specific pesticides to to be used when weeds became a nuisance. 1964 *Daily Tel.* 3 Jan. 19/2 Pesticide residues in the fat of birds, animals and human beings were absorbed into the body and gradually built up in the body fat. 1969 *Times* 8 May 12/7 The chief threat to birds of prey is the use of pesticides. 1971 *Power Farming* Mar. 5/1 Shell, like most other companies in this field, recognize that the indiscriminate use of pesticides is highly undesirable. 1978 *Dædalus* Spring 43 The development of a specific sweetener, pesticide, or weapon could be prevented with little generalized effect.
Hence **pestici·dal** *a.*
1950 in Webster Add. 1956 *Nature* 25 Feb. 350/1 The United States Department of Agriculture has a list of thirty thousand pesticidal formulations. 1971 *Ibid.* 3 Sept. 72/1 The properties, functions, utility and contributions of pesticidal chemicals to human welfare.

pesto (pe·stǫ). [a. It. *pesto*, contracted form of *pestato*, pa. pple. of *pestare* to pound, to crush.] A pasta sauce of crushed herbs, garlic, and olive oil.
1937 M. Morphy *Good Food from Italy* 166 When used with pastes, such as macaroni,..the *pesto* is diluted with 3 or 4 tablespoons of boiling water. 1953 R. Howe *Italian Cooking* 61 Prepare a vegetable soup..just before it is ready stir into it a garlic paste or *pesto*. There are several recipes for making *pesto*. 1954 E. David *Italian Food* 288 (*heading*) Pesto. *Ibid.*, 1 large bunch of fresh basil, garlic, 1 handful of pine nuts, 1 handful of grated Sardo or Parmesan cheese, 1½–2 oz. of olive oil... When the *pesto* is a thick purée start adding the olive oil. 1962 M. Soper *Encycl. European Cooking* 398 Pesto is a sauce of Genoese origin... Any left-over pesto may be placed in a small jar, covered with olive oil and kept for some days. 1976 R. Condon *Whisper of Axe* i. x. 59 Enid grew basil for making *pesto.* 1976 *Times* 6 Mar. 12/3 Home-made pesto followed by aïoli garni at the Carved Angel in Dartmouth. 1976 *Publishers Weekly* 20 Sept. 83/1 Sauces such as pesto, aioli, mayonnaise.

pestology (pestǫ·lŏdʒi). [f. L. *pest-is* Pest + -ology.] The scientific study of pests and methods of dealing with them. Hence **pestolo·gical** *a.*, of or pertaining to pestology; **pesto·logist,** an expert or specialist in pestology.
1921 *Glasgow Herald* 26 Nov. 6 Lieut.-Colonel Nathan Raw, M.P., has become President of the newly formed Institute of Applied Pestology. 1927 *Daily Express* 23 Sept. 3/3 The pestological exhibition and conferences.. opened yesterday... There were insect powders, sprays, pastes and—this will show you how far a pestologist goes —automatic fire-arms. 1927 *Times* 27 Sept. 12/5 An exhibition organized by the College of Pestology..was opened on Thursday. 1971 *New Scientist* 3 June 554/2 Our war-weary pestologists may perhaps become disenchanted with the scientific approach.

pesty (pe·sti), *a.* *U.S. colloq.* [f. Pest + -y¹.] Obnoxious, troublesome, annoying.
1962 E. Lacy *Freeloaders* iv. 65 My last pesty question —can you spare a hundred bucks? 1974 *Spartanburg* (S. Carolina) *Herald* 24 Apr. (Sky City Advt. Suppl.) 5 Twin flaming torches $\$3^{00}$ pr... Kills pesty bugs and mosquitoes. For outdoor lighting, everywhere. 1976 B. Bova *Multiple Man* (1977) xiii. 139 Our pesty friend here..has found out.

pet, *sb.*¹ Add: **2. b.** Also, a sweet or obliging person. † Also as a name for a favourite boxer (*obs.*). Phr. *teacher's pet*: a derogatory term for a teacher's favourite pupil; also *transf.*
1841 Dickens *Let.* 9 Feb. (1969) II. 208 'The Pet of the Fancy', or 'the Slashing Sailor Boy', or 'Young Sawdust'. 1848 Thackeray *Van. Fair* xxxiv. 303 James Crawley had met the Tutbury Pet, who was coming to Brighton to make a match with the Rottingdean Fibber; and enchanted by the Pet's conversation, had passed the evening in company with that..man. 1859 J. Blackwood *Let.* 8 July in *Geo. Eliot Lett.* (1954) III. 113 A dive kept.. by Dick Curtis the pet of the Fancy. 1914 B. Tarkington *Penrod* xii. 89 'Teacher's pet!' whispered Penrod hoarsely. He had nothing but contempt for Georgie Bassett. 1922 Wodehouse *Girl on Boat* iv. 82 Do be a pet and go and talk to Jane Hubbard. I'm sure she must be feeling lonely. 1930 J. Dos Passos *42nd Parallel* iii. 237 The other employees in the department hated her and nicknamed her Teacher's Pet. 1952 S. Kauffmann *Philanderer* (1953) xii. 199 He was not only the 'teacher's pet', he was the 'rich kid'. He was doubly isolated. 1957 J. Kirkup *Only Child* xi. 139 So immediately after lessons were over, I would not linger in the classroom 'sucking up to teacher' as the 'teacher's pets' did. 1968 *Guardian* 16 Mar. 11/5 The anxious child was usually the conventional teacher's pet, always well dressed and obedient. 1976 T. Heald *Let Sleeping Dogs Die* ix. 184 Be a pet and fetch me a Tom Collins. 1976 H. Wilson *Governance of Britain* i. 12 It [*sc.* the phrase 'prime minister'] was used to denote 'court favourite', with connotations similar to 'teacher's pet'.
d. Used as a term of endearment or familiar vocative.
1849 J. Ruskin *Let.* 24 Apr. in M. Lutyens *Ruskins & Grays* (1972) xxi. 185 Do you know, pet, it seems almost a dream to me that we have been married. 1939 L. M. Montgomery *Anne of Ingleside* xxxvii. 293 There is a parcel I want to send up to Thomasine Fair... Will you run up with it this afternoon, pet? 1972 G. Sereny *Case M. Bell* i. ii. 33 Mary smiled and asked to see Martin. I said, 'No, pet, Martin is dead.' 1975 J. Wainwright *Square Dance* 186 He..spoke to the policewoman on duty... 'Now then, pet—can you help me?' 1977 *Daily Mirror* 22 Mar. 24 Sounds like just the job for you, pet, eh?
3. b. (Earlier and later examples.) *pet aversion* (earlier example); also, *pet hate*; *pet peeve*: see *peeve *sb.*
1826 *Blackw. Mag.* XX. 53/1 Men of the most different habits and characters in other respects, resemble each other in the practice of nursing in secret some pet superstition. 1880 'Mark Twain' *Tramp Abroad* xxvi. 262 For years my pet aversion had been the cuckoo clock. 1920 [see *assassinate *v.* 3]. 1939 *Sun* (Baltimore) 21 Apr. 28/2 Hill-passers, he said, were one of his 'pet hates'. 1949 *Proc. Inst. Electr. Engin.* XCVI. ii. 629/1 Engineers will always have their pet ideas and want their special sizes of cables. 1969 *Morning Star* 19 Nov. 4 Many of you will have your own pet dishes... With your help we could give the cookery column a real international flavour. 1974 'R. Tate' *Birds of Bloodied Feather* vi. 118 No doubt you have one of your pet theories. 1977 *National Observer* (U.S.) 22 Jan. 12/2 Another pet hate is the 'News Flash' that breaks into a program with total disregard for its distracting impact on the show.
c. Also **pet-form,** an adaptation of a name used as a pet-name.
1932 E. Weekley *Words & Names* X. 138 *Christopher* may have implied stupidity, as its German pet-form *Stoffel* is synonymous with blockhead. 1956 *Archivum Linguisticum* VIII. 70 *Ned* and *Nanny* are..mere pet-forms like *Ted.* 1960 P. H. Reaney *Orig. Eng. Place-Names* i. 8 We must believe..that *Brihtling* was a pet-form for *Brihtric.*
d. *pet-vendor*; **pet cemetery** = *pets' cemetery*; **pet-day** (further example); **pet-food,** food for pet animals; **pets' cemetery,** a burial-ground for domestic pets; **pets' corner,** a part of a display, zoo, etc., reserved for the display of animals normally kept as pets or suitable for keeping as pets; **pet-shop,** a shop selling animals to be kept as pets.
1967 A. Lewin *Unaltered Cat* ii. viii. 180 He telephoned the pet-cemetery... Mr Carpenter agreed to pick up the cat-corpse. 1973 *Post-Herald* (Birmingham, Alabama) c1/1 The Los Angeles Pet Cemetery has a small 'slumber room' where owners may view their pet lying in state on a blue satin covered stand. 1939 L. M. Montgomery *Anne of Ingleside* ii. 14 Such a lovely day... I'm afraid it's a pet day though—there'll be rain to-morrow. 1961 A. Wilson *Old Men at Zoo* i. 48 If I'd been made Director, Beard would be getting a thumping great subsidy from some of those big pet food people. 1968 *Observer* (Colour Suppl.) 25 Feb. 35/1 Pet foods come sixth in the consumer top ten. Baby food lags way behind. 1973 R. Hill *Ruling Passion* ii. vi. 132 A man was unloading trays of meat and made-up pet food from a blue van. 1908 Hardy *Let.* 23 Dec. in *One Rare Fair Woman* (1972) 138 Our very old cat 'Comfy' died two days ago... He is buried in our pets' cemetery. 1948 E. Waugh *Loved One* 27 He took a job at the pets' cemetery. 1940 Pets' corner [see *dumb *a.* 1 b]. 1961 *Guardian* 5 May 15/4 In the children's corner [of the park] there is..a pets' corner. 1968 J. Rathbone *Hand Out* xiv. 113 Gee, Elmer, it's better'n pets' corner back home. 1976 *Star* (Sheffield) 29 Oct. 14/6 Now he is hoping to open a pet's corner and leisure centre there, with a pride of lions as the star attraction. 1928 Kipling *Limits & Renewals* (1932) 47 Mr. Wilham's fashionable West End pet-shop. 1942 D. Powell *Time to be Born* (1943) xii. 295 In front of the pet-shop window a man stood watching half a dozen infant Siamese kittens. 1976 W. Greatorex *Crossover* 35 He called at the pet shop... There were whining puppies and mewing kittens. 1924 *Glasgow Herald* 21 Nov. 10/7 A London pet-vendor has had about 2,500 snakes through his hands within the last few months.

pet, *v.*¹ Add: **b.** *intr.* To have erotic physical contact with another person by kissing, caressing, and sexual stimulation. orig. *U.S.*
1924 P. Marks *Plastic Age* vi. 53 I'm a bad egg. I drink and gamble and pet. I haven't gone the limit yet.. —but I will. 1953 A. C. Kinsey et al. *Sexual Behav. Human Female* ix. 389 The most responsive females may be the ones who most often pet to orgasm before marriage. 1959 N. Mailer *Advts. for Myself* (1961) 230 The game she cherished was to play the bobby-soxer who petted with a date in the living room and was finally seduced. 1969 E. M. Brecher *Sex Researchers* (1970) v. 113 Some lower-level boys also occasionally pet. 1977 C. Storr *Tales Psychiatrist's Couch* 84 Haven't you ever reached a climax when you've been out with a boy? When you'd be petting?

peta- (pe·tă), *prefix.* [Said to be f. Pe(n)ta-, the mode of formation having been suggested by *tera-/Tetra-.] Prefixed to the names of units to form the names of units 10^{15} times larger (symbol P).
1975 *Physics Bull.* Mar. 105/1 The Committee [*sc.* the International Committee of Weights and Measures (CIPM)] also agreed to recommend that the 15th and 18th powers of 10 be assigned the names 'peta' (symbol P) and 'exa' (symbol E) respectively. 1975 *B.S.I. News* Dec. 13/3 The Conference [*sc.* the General Conference of Weights and Measures (CGPM)]..adopted the name peta, symbol P, for 10^{15} and exa, symbol E, for 10^{18}.

petal, *sb.* Add: **3.** *petal-shower, -tambourine; petal-soft* adj.; **petal collar,** a collar on a woman's garment cut in the shape of petals; **petal ware,** a type of pottery (see quot. 1960).
1957 M. B. Picken *Fashion Dict.* 74/1 *Petal collar*, collar made of overlapping petals of fabric. 1969 *Times* 24 Mar. 13/8 (Advt.), A button-through coat dress for Spring into Summer. Short sleeves, petal collar. *a* 1918 W. Owen *Coll. Poems* (1963) 117 Stirs Of leaflets in the gloom; soft petal-showers. 1945 P. Larkin *North Ship* 4 Whose every hall The light as music fills, and on your face Shines petal-soft. 1947 *Sun* (Baltimore) 6 Aug. 9 (Advt.), Generously cut to the new hem lengths..and pretty as you please, Colony Club's pettiskirt... In petal-soft Bur-Mil rayon crepe, white, pink or black. 1922 Blunden *Shepherd* 26 And petal-tambourines shall earn A largess this May morn. 1930 D. T. Rice *Byzantine Glazed Pott.* 9 In dealing with the Constantinople finds,..six main groups were distinguished; plain glazed ware; stamped ware; white inscribed or sgraffito ware; 'petal' ware [etc.]. 1952 —— *Eng. Art 871–1100* viii. 248 The glazes, however, are of a type not so far known from Germany, and suggest, rather Byzantine prototypes; one group, indeed, where the glaze is thick and dark, and where a decoration of blobs has been added, is extremely close to a Byzantine group, usually classed as 'petal' ware. 1960 R. G. Haggar *Conc. Encycl. Cont. Pott. & Porc.* 345/2 '*Petal*' *ware*, ware found at Constantinople, characterized by the addition of horizontal bands of lumps of clay, pressed firmly to the body at one end but left free at the other, which resemble crude petals, hence the name.

petal (pe·tăl), *v.* *poet.* [f. the sb.] *trans.* To provide or scatter with petals. Also *fig.*
1907 *Westm. Gaz.* 3 June 2/3 Sigh, little wind.., Winnow the lilacs pink and white, Petal the shining grass. 1930 E. Pound *XXX Cantos* iv. 17 Saffron sandal so petals the narrow foot. 1955 —— *Section: Rock-Drill* (1957) xci. 76 The water-bug's mittens petal the rock beneath.

petalled, petaled, *a.* Add: Also, formed like or resembling a petal or petals.
1862 G. M. Hopkins *Vision of Mermaids* (1929), Betwixt ten thousand petall'd lips. 1929 G. C. Allen *Oxf. Poetry* 1 Let's pick the petals of all joy apart, And launch them uncontrolled on the wind-stream With gleam of petalled gold. 1937 Blunden *Elegy* 74 The petalled cloud and the blue brook aflow. 1975 G. Howell *In Vogue* 30 Delysia looking curious in a petalled evening dress.

petalodic (petălǫ·dik), *a.* *Bot.* [f. Gr. πεταλώδ-ης leaf-like (cf. Petalody) + -ic.] Exhibiting petalody.
1909 W. Bateson *Mendel's Princ. Heredity* xi. 198 In the hose-in-hose *Campanula*, which has the sepals petaloid, the well-formed anthers contain plenty of pollen (some may be petalodic).

‖ **pétanque** (petaṅk). [Fr.] A game, resembling bowls, played orig. and chiefly in southern France. Also *attrib.* Cf. *boule⁴ 2.
1955 *Times* 8 Aug. 8/7 No longer the fraternal glass of pastis..no longer..the earnest game of *pétanque*, that travesty of bowls. 1956 N. Ryan tr. *Simenon's My Friend Maigret* vii. 112 Two old men were playing bowls, *pétanque* style, that is without sending the jack more than a few yards from their feet. 1963 N. Freeling *Because of Cats* vii. 105 Cunning little French butane stoves and the chi-chi of le camping, huge rubber water-wings, pétanque sets and elaborate German beach toys. 1967 G. Bellairs *Single Ticket to Death* i. 7 Littlejohn and Dorange were playing *pétanque*... The players, with a twist of the wrist, hurled their steel balls at the small wooden jack. 1972 A.H. Haynes *Story of Bowls* i. 7 In..Petanque..a small marker ball is thrown a distance of up to ten yards; occasionally the distance is doubled and in this event, remarkable skill is demonstrated by the players in putting top and side spin on the heavy steel balls bowled underarm towards the marker ball. 1977 *Times* 25 Apr. 16/7 The finals of the London pétanque championship.

petara(h, varr. Pitarah.

petasma (petæ·zmă). *Zool.* [a. Gr. πέτασμα something spread out.] In prawns of the family *Penæidea*, a membranous appendage attached to the first pair of pleopods.
1888 C. S. Bate in *Rep. Sci. Results Voy. H.M.S. Challenger: Zool.* XXIV. 230 The pleopods [of *Penæus*] are large and powerful, terminating in two foliaceous branches in every pair except the first, which in the male carries attached to the base a large membranous appendage that I call 'petasma'. 1909 A. E. Shipley in A. Sedgwick *Student's Text-bk. Zool.* III. v. 527 In *Penæus, Lucifer* and *Sergestes* a lobe projects inwards from the base of the first pleopods in the male, and may unite with its fellow to form the petasma (or curtain). 1966 *McGraw-Hill Encycl. Sci. & Technol.* V. 105/1 The male [genital]

system [of Eumalocostraca] consists of the testis, vas deferens, ductus ejaculatorius, and sometimes a penis, modified thoracic limbs (pleopods), or petasma.

Pete, pete (pīt). [Dimin. of the name *Peter* (see Peter *sb.*).] **1.** *slang.* (With small initial.) **a.** A safe. Cf. *Peter *sb.*¹ 6 b.
1911 G. Bronson-Howard *Enemy to Society* iv. 73 All the time a man equipped with burglar's tools would be kneeling behind the safe and drilling it open; those 'petes'—as cracksmen call them—in people's houses are generally very easy to open. 1932 Wodehouse *Hot Water* i. 32 Show me the pete I can't open with my eye-teeth and a pin, and I'll eat it. 1938 D. Runyon *Furthermore* viii. 153 This is a very soft pete. It is old-fashioned, and you can open it with a toothpick. 1951 Wodehouse *Old Reliable* x. 129 You think I'm scared to bust that pete?
b. Nitroglycerine, as used for safe-breaking.
1931 D. W. Maurer in *Writer's Digest* Oct. 29/2 *Soup*, nitroglycerine or 'pete'. 1948 Mencken *Amer. Lang.* Suppl. II. 668 Among the cant terms of the *jug-heavies* are..*soup* or *pete*, nitroglycerine.
c. *attrib.* and *Comb.*, as **pete box** = sense *1 a; **pete-man** = *peterman 3 c.
1931 D. Runyon *Guys & Dolls* (1932) iv. 77 Nobody opens pete boxes for a living any more. They make the boxes too good, and they are all wired up with alarms. 1911 G. Bronson-Howard *Enemy to Society* v. 105 You've already caught four 'pete-men' who attempted to drill the safe. 1931 *Everyman* 21 May 522/1 All my safe-blower pals used..'pete-men'.
2. (With capital initial.) Used in various mild exclamations and phrases expressive of exasperation or annoyance; esp. in phr. *for Pete's sake.*
1924 *Dialect Notes* V. 274 For the love of Pete, for Pete's sake. 1942 N. Balchin *Darkness falls from Air* ix. 170 Why in the name of Pete didn't you say so? 1949 N. Marsh *Swing, Brother, Swing* iv. 59 Carlisle heard Mr. Bellairs whisper under his breath: 'For the love of Pete!' 1959 W. Golding *Free Fall* vi. 129 Marry me, Taffy, for Pete's sake marry me. 1973 'B. Mather' *Snowline* xviii. 223 For Pete's sake don't ask bloody fool questions. 1975 *Listener* 24 July 115/1 For Pete's sake when will so-called 'experts'..get it into their noddles that rising wages and prices *are* inflation.

Peter, *sb.*¹ Add: **1. b.** [Imitative.] The cry of various tits.
1874 C. M. Yonge *Lady Hester* ii. 28 The tomtits were calling 'peter' in the trees. 1892 —— *Old Woman's Outlook* ii. 37 Sunshine, setting the thrushes and robins to sing, and the ox-eyes to cry Peter.
2. (Later examples.)
1926 *Times* 7 Jan. 9/6 Martin and Martin had been in low water for a long time and had recourse to the method of robbing Peter to pay Paul. 1961 D. Woodward tr. *Simenon's Premier* iii. 84 After the disastrous experiments made by previous governments, which had lived from day to day, robbing Peter to pay Paul, the only solution was a large-scale devaluation. 1976 *Star* (Sheffield) 29 Oct. 13/7 A Sheffield man who tried to set up a travel agency business was accused of 'robbing Peter to pay Paul', at Sheffield Bankruptcy Court.
6. a. Now also *Taxi-drivers' slang.* (Further examples.)
1930 'A. Armstrong' *Taxi* xii. 164 'Peters' are pieces of luggage,—a threepenny extra for the driver. 1939 H. Hodge *Cab, Sir?* iii. xv. 221 The driver calls each package a 'Peter'.
b. *Criminals' slang.* A safe or cash-box; a cash register, a till.
1859 G. W. Matsell *Vocabulum* 66 *Peter*, a portmanteau; a travelling-bag; a trunk; an iron chest; a cash-box. 1862 [see *peter-cutter*, sense 8 a]. 1868 [see *peter-screwing*, sense *8 a]. 1869 *Macm. Mag.* Oct. 506/1 After we left the course, we..got a peter (cashbox) with very near a century of quids in it. 1889 Clarkson & Richardson *Police!* xxv. 351 In order to 'ready' these places, they watch the shops at closing time, to learn if the swag is placed in the 'Peter', or safe. 1935 A. J. Pollock *Underworld Speaks* 87/1 *Peter*, a cash drawer; a money box; cash register. 1936 J. Curtis *Gilt Kid* xvii. 171 There was no safe. 'There ain't no peter.' 'No.' 1943 *Penguin New Writing* XVII. 66 He pushed me over one more double gin which he only pretended to ring up on the peter, but there wasn't a chance to talk. 1945 Baker *Austral. Lang.* viii. 140 *To tickle the peter*, to rob a till. 1958 F. Norman *Bang to Rights* iii. 121 If some poor unsuspecting manageing [*sic*] director left his peter open well, I ask you? 1965 M. Shadbolt *Among Cinders* xvi. 143 'Did he tickle the peter?'.. 'To the tune of two thousand quid.' 1967 K. Giles *Death & Mr Prettyman* v. 98 But level, old boy, was Prettyman tickling the peter? 1970 G. F. Newman *Sir, You Bastard* viii. 211 There was s'posed to be some dough in the Peter. 1978 *Daily Mail* 25 Jan. 12/2 While most of the country's police call the safe a 'peter', in London it's a 'gas'.
c. *slang.* A cell in a prison, a police station, a court of law, etc.; a lock-up. orig. *Austral.*
1890 Barrère & Leland *Dict. Slang* II. 125/1 *Peter*.. (Australian prison), punishment cell. 1953 K. Tennant *Joyful Condemned* vi. 55 The doors of the peters just crash open at the name of McGarty. 1955 D. Niland *Shiralee* 69 They could throw you in the peter stone-cold sober. 1958 F. Norman *Bang to Rights* i. 9, I turned and walked down the stairs to the peter under the court. 1960 C. Hatton in *Pick of Today's Short Stories* XI. 147 He snapped up any likely bloke just as soon as he..had come out of the peter. 1965 *Guardian* 10 Oct. 21/6 'Hurry up and slop out'—'Get back in your f— Peter'. 1973 'J. Patrick' *Glasgow Gang Observed* ix. 81 He..had spent the first night in a police cell (or 'peter' as they call it).
d. *U.S. slang.* A stupefying drug.
1899 'J. Flynt' *Tramping with Tramps* iv. 396 'Knock-out drops' are also 'peter'. 1933 *Amer. Speech* VIII. ii. 27/1 Among the addicts dope in general is known as *gow, junk*, or *peter* (any kind of *knockout drops*). 1971 E. E. Landy *Underground Dict.* 148 *Peter*..Chloral hydrate.
e. *slang.* The penis.
1902 Farmer & Henley *Slang* V. 177/1 *Peter*... 4. (venery).—The *penis:* also *St. Peter.* 1928 *Dialect Notes* VI. 61 The proper name *Peter*..is so universally used by children and facetious adults as a name for the penis that it never quite loses this significance. Very few natives of the Ozarks will consider naming a boy *Peter.* 1940 C. McCullers *Heart is Lonely Hunter* i. ii. 18 There was one fellow who had had his peter and his left leg blown off in a boiler explosion. 1970 R. D. Abrahams *Positively Black* ii. 41, I fuck your mammy on a red hot heater; I miss her pussy and burn my peter. 1975 M. Kenyon *Mr Big* v. 48 In New Jersey peter was childhood slang for penis, equivalent to willie and john thomas. 1977 J. Wambaugh *Black Marble* (1978) vi. 75 If you look very closely you can see a gerbil's dick, but not a parakeet's peter.
7. Also (*Naut.*) simply *Peter.* In *Bridge*, = *echo *sb.* 8.
1891 Kipling *Barrack-Room Ballads* (1892) 205 See the shaking funnels roar, with the Peter at the fore. 1939 [see *echo *sb.* 8]. 1945 Phillips & Reese *How to play Bridge* iii. xiii. 115 East's play of the 8 is the commencement of what is known as a 'peter'. 1959 *Listener* 27 Aug. 334/2 There are those who advocate the peter to indicate length at virtually all times in defence. 1966 *Sunday Tel.* 14 Aug. 9/6 Every bridge player knows the principle of high-low defence as a signal of encouragement. Generally known as the 'peter' in Britain and the 'echo' or 'come-on' in America, it is most frequently used against an opponent's trump contract to indicate strength or a doubleton. 1974 *Country Life* 28 Mar. 750/3 South played a Diamond to the King, East starting a peter.
8. a. Peter Grievous (also **Peter Grievance**), one who complains; a whining child; freq. *attrib.* or as *adj.*, complaining, fretful, miserable (*dial.* and *slang*); (sense 6) *peter-ringer*; (sense *6 b) *peter-popping, -screwing* (see quots.).
[1724 (*title of play*) Valentine and Orson, with the comical whining humours of Peter Pitiful.] 1774 F. Hopkinson (*title*) A pretty story written in the year of Our Lord 2774. By Peter Grievous, Esq. 1777 H. L. Thrale in *Thraliana* (1942) I. 155 Cradocke had written a Tragedy a very *deep* one they said..We'll call it *Peter* said he—the scene was in Russia, at least said I let it be *Peter Grievous.* 1875 W. D. Parish *Dict. Sussex Dial.* 86 *Peter-grievous*, fretful; whining. 'What a peter-grievous child you are! Whatever is the matter?' 1894 *Southward Ho!* I. 338 (E.D.D.), A peter-grievous wot shrapes an makes a rookery, an a ranky chimley be pettigues fer to send a feller in de crazy-house. 1896 G. F. Northall *Warwickshire Word-bk.* 172 *Peter Grievous, sb.* A grumbler: a grievance-monger. Oxf. (*Peter Grievance*, a cross, fretful child). The word is used as an adjective also, as 'He's a regular peter-grievous fellow.' Glouc., Worc. 1932 E. Weekley *Words & Names* vii. 90 *Peter grievous*, a lachrymose individual, belongs to the class of *simple Simon.* 1960 *Observer* 24 Jan. 5/2 The expertise of peter-popping consists in knowing just how much gelignite to use on the safe in question. 1863 *Once a Week* 7 Nov. 555/2 Well, my friend, a Peter-ringer is one who tries to get his living by stealing carpet-bags. 1868 *Temple Bar* XXIV. 537 'What do you mean by 'lob-sneaking', and 'Peter-screwing'?' 'Why, 'lob' means the till, and 'Peter' means a safe. Stealing the till and opening the safe is what we call 'lob-sneaking' and 'Peter-screwing'.'

Peter (pī·təɹ), *sb.*² The name of Dr. Laurence Johnston *Peter* (b. 1919), Canadian-born U.S. educationalist and author, used *attrib.* in *Peter principle* (see quot. 1968). Also *Peter's principle.*
1968 L. J. Peter *Peter Principle* (1969) i. 25 My analysis of hundreds of cases of occupational incompetence led me on to formulate *The Peter Principle*: In a Hierarchy Every Employee Tends to Rise to His Level of Incompetence. 1969 *Sunday Times* (Colour Suppl.) 21 Dec. 30/2 What is now called the Peter Principle: the system by which a man is always promoted until he reaches a job which he cannot do. 1970 *Guardian Weekly* 25 July 6 Much blame must attach to the administrative system,.. which religiously follows the Peter Principle of promoting mediocrities. 1976 *Publishers Weekly* 15 Mar. 58/3 Most flibbertigibbet corporate president (maybe the Peter Principle applies here). 1978 *Jrnl. R. Soc. Arts* CXXVI. 273/2 In in-company situations, assessment centres can hopefully help avoid Peter's principle, whereby everyone is promoted to his level of incompetence. 1980 *Observer* 27 Jan. 4/3 There is a school of Westminster and Belfast thought which sees Mr Atkins as a classic example of the Peter Principle—promoted above his level of competence.

peter, *v.*² Add: **2.** Also simply *peter.* (Earlier and later examples.) Hence **pe·tering-out** *vbl. sb.*
1846 *Quincy* (Illinois) *Whig* 6 Jan. 1/4 When my mineral petered why they all Petered *me.* If so be I gets a lead, why I'm Mr. Tiff again. 1854 H. H. Riley *Puddleford* vi. 84 He 'hoped this 'spectable meeting war n't going to Peter-out'. 1923 R. Macaulay *Told by Idiot* iii. 221 The year and the government petered towards their end. 1926 E. F. Spanner *Naviators* 100 Lucky your engine petered out, Sterne. 1944 F. Clune *Red Heart* 6 The fabulous silver-lead wealth..has enticed a city of 15,000 inhabitants to arise in the desert wastes—and there they will continue to dwell until the lode peters. 1949 'J. Tey' *Brat Farrar* xix. 170 The..petering-out of the poorer suburbs. 1955 *Times* 28 June 3/3 With the end of this partnership, however, the innings virtually petered out. 1976 *Quoddy Tides* (Eastport, Maine) 13 Aug. 4/4 Hurricane 'Belle'..petered out before reaching the Quoddy area.
b. *trans.* To exhaust; to cause or allow to peter out; const. *away*, to squander. Freq. as ppl. adj.
1869 *Overland Monthly* III. 127 After a long desert journey the oxen become much 'petered'. 1878 C. Hallock *Amer. Club List & Sportsman's Gloss.* p. viii/1 *Peter-out*, to fail; to exhaust; to collapse. 1943 *Amer. Speech* XVIII. 67/2 *Petered out* (exhausted), S.C., N.C., Tenn., La., Tex. 1956 B. Holiday *Lady sings Blues* (1973) vi. 58 He wouldn't give us nothing but a lecture on how he saved his money and how we petered ours away. 1971 'D. Halliday' *Dolly & Doctor Bird* xvi. 228 It was another petered-out trail.
3. *intr.* *Whist* and *Bridge.* To play a high card followed by a low one. Cf. Peter *sb.*¹ 7 in Dict. and Suppl.
1887 *N. & Q.* 29 Oct. 356/1 The Blue Peter..is..used when a ship is about to start... Calling for trumps at whist or 'petering' is derived from this source. 1901 C. J. Melrose *Bridge Whist* 38 Another whist convention, which may occasionally be employed with advantage in Bridge..is known variously as 'Petering', the 'trump signal' or the 'call for trumps'. 1939 N. de V. Hart *Bridge Players' Bedside Bk.* 141 At one time..a player petered to show two cards only of a suit. 1961 *Times* 7 June 8/3 My partner had not petered in hearts. 1976 *Country Life* 22 Jan. 202/1 West started by cashing two top diamonds, on which East petered.
4. *slang.* **a.** *intr.* (See quot. 1925.) **b.** *trans.* To blow open (a safe).
1925 *Flynn's* 7 Mar. 192/1 *Peter, v.*, to use knock-out drops; to use nitroglycerine. 1962 B. Knox *Little Drops of Blood* iii. 65 The Dolman boys are going to peter a pawnshop safe tonight.

Peterborough¹ (pī·təɹ,bʌrŏ). Also **Peterboro.** *N. Amer.* The name of a town in Ontario used *attrib.* and *ellipt.* to designate a type of canoe (orig. built there), made entirely of wood.
1882 *Forest & Stream* 2 Nov. 277/1 There is the open Peterboro' canoe, so familiar to the Canadian eye. 1895 *Rudder* Sept. 215 Eleven paddling canoes—eight Peterboro's and three Rushton—and the Cruiser, complete the fleet. 1897 J. W. Tyrrell *Across Sub-Arctics of Canada* ii. 20 We launched our handsome 'Peterboroughs' in the great stream. 1901 *Daily Colonist* (Victoria, B.C.) 30 Oct. 6/2 We portaged around it on the way up by packing our Peterboro up a hill one hundred feet high, taking us a whole day. 1966 *Canad. Geogr. Jrnl.* Sept. 78/3 John Stephenson..later came up with the excellent cedar rib craft known the world over as the Peterborough Canoe. 1973 D. Andersen *Ways Harsh & Wild* iii. 86, I set a net in an eddy and later paddled Ted's Peterborough canoe down to check.

Peterborough² (pī·təɹ,bʌrŏ). The name of a town in eastern England, site of a phase of the Neolithic Age: used to denote the type of civilization of that period, and the materials or people associated with its culture.
1910 *Archaeologia* LXII. 346 The characteristic decoration of the drinking-cup or beaker is well known, and full justice has been done to the Peterborough series by our Fellow, Mr. Praetorius. 1922 *Antiquaries Jrnl.* II. 231 The Peterborough pottery seems to bear all the signs of its makers advancing in ceramic skill by gradual stages. 1935 Huxley & Haddon *We Europeans* vii. 237 The 'Peterborough' ware was brought from the Baltic by long-headed people who buried their dead in long barrows. 1939 V. G. Childe *Dawn European Civilization* (ed. 3) xviii. 306 In Norfolk and even Wiltshire Peterborough folk..were associated with its exploitation. *Ibid.* 317 Skara Brae and Peterborough traditions survived. 1943 J. & C. Hawkes *Prehist. Brit.* ii. 47 These are the Peterborough people, largely descendants of the old Mesolithic inhabitants of Britain, who, while adopting certain Neolithic accomplishments such as potting, herding and simple husbandry, continued to follow the old mode of life as hunters and fishers. 1947 V. G. Childe *Dawn European Civilization* (ed. 4) xviii. 323 In the standard Peterborough ware the rims are thickened and the shallow bowls are richly decorated. 1951 *Proc. Prehist. Soc.* XVII. 119 The associations and chronology of these Group VI factories and their products are firmly fixed in the Neolithic with emphasis in favour of the Peterborough phase. 1954 S. Piggott *Neolithic Cultures* xi. 303 We can now see that Peterborough pottery is distinctive of only one variant within a group of Secondary Neolithic cultures. *Ibid.* 312 (*heading*) Relationships of the Peterborough culture. 1967 *Antiquaries Jrnl.* XLVII. 201 Its sharply formed tripartite upper half, and the decoration of this so lavishly by whipped-cord impression, support those traits in strongly recalling Neolithic antecedents, among the Mortlake bowls..of British 'Peterborough' ware.

peterman. Add: **3. b.** (See quots.) *slang.*
1897 Eldridge & Watts *Our Rival* xi. 289 The rogue's name for the professional users of 'knockout drops' is Peter-men or Peter-players. 1904 'No. 1500' *Life in Sing Sing* 256/2 *Peter man*,..person who administers a drug for the purpose of robbery. 1908 *Sun* (N.Y.) 2 Mar. 2/2 A peterman is one who uses knockout drops as an aid to robbery.
c. A safe-blower. *slang.*
1900 'Flynt' & 'Walton' *Powers that Prey* 176 Sliger's record, both as 'peter-man' and convict, was produced. 1936 J. G. Brandon *Pawnshop Murder* i. 5 Your flash 'peterman' is as gentle-natured as the average curate. 1950 R. Chandler *Let.* 18 May in *R. Chandler Speaking* (1966) 80 Opening a good safe (without a time lock) requires expensive and heavy tools, the finest drills either to drill out the lock or to get in the nitro if he is a peterman. 1960 *Observer* 24 Jan. 5/1 Gelignite in plastic sticks..was the British peterman's basic material. 1968 P. N. Walker *Carnaby & Gaolbreakers* i. 13 Well, you'd never

be a peter man, or a breaker in a month of Sundays. **1973** 'B. Graeme' *Two & Two make Five* vii. 68 The wall safe..would [not] have presented much difficulty to an expert peterman. **1977** J. Wainwright *Nest of Rats* I. vii. 46 The genuine peterman—the safe-breaker who takes a personal pride in pitting his wits against those of the safe-makers.

Peter Pan. [The name of the boy hero of J. M. Barrie's play *Peter Pan, the boy who wouldn't grow up* (1904).] **1.** Used *attrib.* to designate various styles of clothing, esp. **Peter Pan collar** (also with lower-case initials), a flat collar with rounded ends, often white or light-coloured.

1908 S. Ford *Side-Stepping with Shorty* iv. 60 She was sportin' a Peter Pan peekaboo that would have made Comstock gasp. **1909** *Westm. Gaz.* 10 July 15/1 The collars and cuffs are what we call 'Peter Pan'.., edged with Valenciennes. **1923** *Daily Mail* 12 June 1 (Advt.), Can also be had with Robespierre or Peter Pan collar. **1933** M. Lutyens *Forthcoming Marriages* 204 She was wearing a white Peter Pan collar over a little blue cape. **1948** 'J. Tey' *Franchise Affair* xiv. 148 Miss Tuff had worn peter-pan collars over her dark frock for twenty years. **1958** *Vogue* Sept. 133 Clothes for children with puff sleeves, peter pan collars. **1960** *Times* 15 Jan. 14/3 A white silk shirt with a frilled Peter Pan collar. **1975** *Listener* 10 July 46/1 The maids, they used to wear what we call their print dresses in the mornings. In the afternoon they changed into blue alpaca or black, with white Peter Pan collar and little cuffs.

2. Used allusively for an immature adult (usu. a man); one who is emotionally (occas. physically) retarded. Also *attrib.*

1914 G. B. Shaw *What I really wrote about War* (1930) ii. 109 It is frightful to think of the powers which Europe ..left in the hands of this Peter Pan [*sc.* the Kaiser]. **1927** A. Huxley *Proper Stud.* 163 An electorate composed in a great part of mental Peter Pans. **1931** J. S. Huxley *What dare I Think?* ii. 62 The Peter Pan type of semi-dwarf who, though perfectly proportioned, never grows up fully. **1956** I. Bromige *Enchanted Garden* ii. iii. 99 'Fiona..has the wide-eyed simplicity of eternal youth.' 'A Peter Pan.' **1958** *Sunday Express* 15 June 15/2 Bogarde has still got his hooks into the Peter Pan racket... He is still prowling the screen, demanding mother love from his millions of female fans. **1971** *Guardian* 21 Oct. 6/4 Professor Desmond Pond... Professor of psychiatry at the London Hospital,..told of highly-educated students who became 'Peter Pans' and never managed to leave adolescence. **1976** J. Wainwright *Bastard* v. 76 He is something of a museum-piece..Peter Pan, in person. The fink who never grew up..never acknowledged the responsibilities of adulthood. **1978** *Time* 3 July 48/2 Warren's conquests of women are not totally successful... But the Peter Pan quality in Warren is very attractive to some. He teaches them to fly, and they have extraordinary experiences with him. Then they grow up and go on, and he keeps flying. Like Peter Pan, he always comes back to another little girl who's ready to fly off with him to never-never land.

Hence **Peter Pa·nic** (*nonce*) [joc. f. Panic *sb.*[2]], confused, childish behaviour; also as *adj.* [-ic], characteristic of a Peter Pan; **Peter Pa·n(n)ish** *a.* [-ish[1]] = **Peter Panic* adj.; hence *Peter Pan(n)ishness*; **Peter Pa·nnery, peter-pannery,** immaturity; childish quality or behaviour; **Peter Pa·nning** (see quot. 1974).

1914 'I. Hay' *Knight on Wheels* (ed. 2) xiv. 143 Mr. Mablethorpe remained as incorrigibly Peter Pannish as ever. Although his hair was whitening..he declined to grow up. **1928** F. E. Baily *Golden Vanity* xvi. 252 They were all in the early twenties except Joe, and he had an eternal Peter Pan-ish-ness which made the passage of time as marked by the calendar quite immaterial in his case. **1934** R. Campbell *Broken Record* 160 Though not predisposed to this Peter Panic, I had considerable time to see it at work at first hand. **1937** *Times Lit. Suppl.* 10 July 502/3 If we are seeking to know what gives a thinness to much of his fiction..it was not 'Peter Pannishness' in the sense of shrinking from adult reality. **1958** *Listener* 12 June 987/3 The Peter-Panish English is not entirely the fault of the translator. **1960** *Spectator* 2 Sept. 345 An occasional embarrassing lapse into peter-pannery. **1962** A. Huxley *Island* ix. 152 A year in jail won't cure a Peter Pan of his endocrine disbalance... For Peter Panic delinquency, what you need is early diagnosis. **1962** *John o'London's* 2 Aug. 115/3 General air of Peter Pannery. **1974** *Daily Tel.* 12 Mar. 3/2 So many people have invented fictitious children to evade income tax that Inland Revenue officials have introduced their own catch phrase for it 'Peter Panning'. **1975** *Listener* 9 Oct. 464/3 To use one's children's slang is Peter Pannish. **1978** *Radio Times* 28 Jan. 13/4 You could say that Lewis was a latter-day Lewis Carroll.. There is a Peter Panishness to his heroes and echoes of Ratty and Badger in his anthropomorphics.

Petersen grab (pī·təɹsən). *Marine Biol.* [Named after its inventor, Carl Georg Johan *Petersen* (1860–1928), Danish marine biologist, who first described it in 1911 (*Rep. Danish Biol. Station* XX. 47).] A kind of grab for obtaining a sample of the bed of a body of water, consisting of two jaws semicircular in vertical cross-section and hinged along the top, arranged so as to close automatically on contact with the bottom.

1923 *Sci. Investigations, 1922* (Fishery Board Scotland) iii. 1 (*title*) Preliminary survey of the Scottish waters of the North Sea by the Petersen grab. **1950** *Biol. Rev.* XXV. 309 The Petersen grab will only produce reliable results on the softer bottoms. **1970** tr. *J. Schwoerbel's Methods Hydrobiol.* iv. 118 In larger rivers the Petersen grab can be used... The grab weighs 40 kg.

Petersen graph (pī·təɹsən). *Math.* [Named after Julius *Petersen* (1839–1910), Danish mathematician, who first devised it (*L'Intermédiaire des Mathématiciens* (1898) V. 227).] A graph having ten vertices and fifteen lines, which may be drawn as a pentagram disposed symmetrically within a pentagon, each vertex of the latter being joined by a line to the nearest vertex of the former.

1947 *Proc. Camb. Philos. Soc.* XLIII. 460 The 5-cage is the 'Petersen graph'. **1962** O. Ore *Theory of Graphs* xv. 240 The so-called Petersen graph has the form indicated in Figure 15.1.3. It is regular of degree 3 and order 10. It was first introduced by Petersen as an illustration of a graph with $\rho = 3$ which is not the sum of 3 subgraphs of first degree. **1976** *Sci. Amer.* Apr. 127/1 The other class of uncolorable trivalent maps Tait missed are all nonplanar (impossible to draw on the plane without at least one intersecting edge). The simplest example, known as the Petersen graph, is shown in the upper illustration at the right.

Petersham. Add: **b.** (Examples.)

1873 *Young Englishwoman* Mar. 147/2 Waistband *Petersham*, and all the odds and ends of needlework. **1930** V. Sackville-West *Edwardians* ii. 62 Miss Wace..affected a dress of heliotrope serge with a stiff petersham belt. **1957** *Terms & Definitions* (Textile Inst.) (ed. 3) 82 *Petersham ribbon* (*millinery*), a ribbon usually with silk or rayon warp and having single picks of relatively coarse weft. *Ibid.*, *Petersham ribbon* (*skirt*), a narrow-fabric having a pronounced rib weft-way composed of one or more picks per rib and having lateral stiffness. **1972** *Daily Tel.* 18 May 9/2 The jacket was edged in navy petersham and fastened with navy and gold buttons.

c. A style of hat.

1825 H. Wilson *Mem.* III. 65 His little Petersham hat seemed to have been remit de nouveau, for the third time, at least.

Petertide (pī·təɹtaid). [f. Peter *sb.*[1] + Tide *sb.*] The 29th of June (the feast of St. Peter in the Church of England and of St. Peter and St. Paul in the Roman Catholic Church), or the period round about it.

1912 C. Mackenzie *Carnival* xxxiv. 349 Well, being in the city, I suppose we must follow city manners, but darn 'ee, I never thought to go gazing at dancing like maidens at Petertide. **1974** *Daily Tel.* 3 July 16 Southwark..has worked out a fair system. This accounts for the large number of ordination candidates over Petertide. **1975** *Church Times* 7 Mar. 12/5, I look forward to my ordination at Petertide conscious of the numerous people who have made it possible.

pethidine (pe·þidīn). *Pharm.* [f. P(iper)idine with insertion of *eth* from Ethyl, Methyl.] A narcotic analgesic (usu. given, orally or intramuscularly, as the hydrochloride, a colourless crystalline compound) which has actions similar to those of morphine but of shorter duration and is less addictive; ethyl 1-methyl-4-phenylpiperidine-4-carboxylate, $C_{15}H_{21}NO_2$.

1942 *Lancet* 22 Aug. 234/2 Pethidine hydrochloride, a synthetic compound closely allied to morphine in composition, and combining the analgesic effect of morphine with the spasmolytic effect of the atropine group and papaverine, is issued by Roche Products. **1942** *Ibid.* 24 Oct. 487 Fairly recent..is the carboxylic acid ethyl ester of 1-methyl-4-phenylpiperidine, officially named pethidine by the GMC a year ago and known on the continent as 'Dolantin' and in the United States as 'Demerol'. **1947** *Daily Tel.* 1 Mar. 3/4 Several bottles found in the man's room were marked 'pethidine'. **1955** *Times* 2 July 4/3 The annual Government report to the United Nations.. estimates that there are 317 known drug addicts... Heroin and pethidine were used by 17 per cent. and 16 per cent. respectively. **1962** [see *meperidine]. **1972** *Nature* 15 Dec. 411/2 Hysterotomy patients received pethidine 100 mg postoperatively for analgesia. **1974** M. C. Gerald *Pharmacol.* xiii. 247 Meperidine or pethidine, as this drug is more commonly called outside the United States, was fortuitously discovered in 1939 as the result of a search for atropine-like agents possessing antispasmodic activity. This compound was the first totally synthetic narcotic analgesic agent.

‖ **pétillant** (petiyaṅ), *a.* [Fr.] Crackling, sparkling, lively; *spec.* of semi-sparkling wine (see quot. 1965).

1881 C. C. Harrison *Woman's Handiwork* iii. 215 Permit your wood-fire to sink into the *pétillant* stage upon the tiled hearth. **1902** G. Meredith *Let.* 19 Apr. (1970) III. 1440, I wish I were about in the world to give you communications more *pétillantes*. **1938** *Times Lit. Suppl.* 15 Jan. 43/1 Her spirit is..more fiery, its flame richer, more consuming and far less *pétillant* than before. The sparkle, which made her such an entertaining, quick-witted.. character on her first appearance has now almost vanished. **1955** A. Lichine *Wines of France* (ed. 2) 214 Mousseux and pétillant wines are rarely imported into the United States, largely because these lesser wines carry the same high duty as fine Champagnes, and the resulting prices make them poor value. **1959** *Times* 14 Nov. 9/7 The light pétillant white wines of Neuchâtel are famous. **1960** *House & Garden* Aug. 71/3 It [*sc.* a wine] has a slight prickle—it is what the French call *pétillant*. **1965** O. A. Mendelsohn *Dict. Drink* 257 *Pétillant*, French term, meaning crackling or sparkling, applied to slightly sparkling wines, much less heavily charged than mousseux. **1971** *Homes & Gardens* Aug. 102/1 These pétillant wines are now all the rage in France. **1974** *Times* 16 Feb. 15/4 The only successful *pétillant* (semi-sparkling) wine so far produced in the Eastern Mediterranean is Bellapais.

So **pétilla·nce** *sb.*

1951 R. Postgate *Plain Man's Guide to Wine* iv. 87 *Pétillance* occurs when the wine in bottle is still working, and produces some natural gas. **1974** *Times* 18 May (Summer Drinking Suppl.) p. i/2 A good muscadet..with its faint *pétillance*, more felt than seen. **1975** *Times* 28 June 13/3 A deliciously fragrant rosé... There is a slight liveliness, verging on *pétillance*.

petit, *a.* (*sb.*) Add: **5. petit battement** (sur le cou-de-pied, sur le talon), (see quots. 1957); **petit beurre,** a sweet butter biscuit; **petit bleu,** a telegram in France, esp. one sent by the pneumatic post in Paris; **petit déjeuner,** breakfast in France, and by extension, elsewhere; **petit four,** a small fancy cake, biscuit, or sweet, usu. served with the dessert course of a meal; also *fig.*; **petit nègre** (see quots.); **petit nom,** a pet-name; **petit pain,** a small bread roll; **petits paquets,** name of a game of some kind; **petit poussin,** a young chicken; **petits** (erron. **petit**) **pois,** young sweet green peas; **petits soins,** small attentions or services [*lit.* small cares]; **petit suisse,** a cream cheese (see quot. 1966); **petit tranchet** (*Archæol.*), a small stone artefact whose blade is produced by transverse-flaking; **petit verre** (earlier and later examples).

1914 Petit battement [see *pas de chat* s.v. *pas 2]. **1930** Craske & Beaumont *Theory & Pract. Allegro in Classical Ballet* 66 Bring the *right* foot *sur le cou de pied devant* and execute four *petits battements devants* (that is, beaten without passing). **1957** G. B. L. Wilson *Dict. Ballet* 48 *Petit Battement sur le Talon*, lit. small beating on the heel, a movement in which the danseuse, on the point and supported by her partner, lightly beats the heel of the supporting foot with the sole of the working foot. *Ibid.*, *Petit Battement sur le Cou-de-pied*, an exercise at the bar in which the dancer beats lightly with the working foot, which may be fully extended or partially relaxed, to the front and back of the cou-de-pied of the supporting leg. **1906** *Mrs. Beeton's Bk. Househ. Managem.* facing p. 1432 Biscuits... Cream Toast... Wine... Petit Beurre. **1913** C. Mackenzie *Sinister St.* I. i. ix. 130 They all sat down at midnight,..not at all too much tired to sip grenadine sucrée and to crunch Petit Beurre biscuits. **1937** J. Betjeman *Continual Dew* 22 He gives his Ovaltine a stir And nibbles at a 'petit beurre'. **1958** I. Murdoch *Bell* ii. 39 She indicated a large biscuit tin... 'No more Petit Beurre,' Peter Topglass was saying meditatively to himself. **1967** V. Nabokov *Speak, Memory* (rev. ed.) ii. 46 A couple of broken Petit-Beurre biscuits she had found on a plate. **1975** *Times* 13 Mar. 14/7 An expensive-looking parcel..revealed a pound of *petit beurre* biscuits. **1908** W. S. Maugham *Magician* x. 171 The note..was a *petit bleu* sent off from the Gare du Nord. **1920** D. H. Lawrence *Touch & Go* 7 It may be that coal-owners are like the *petit bleu* arrangement, a system of vacuum tubes for whooshing Bradburys about from one to the other. **1924** W. J. Locke *Coming of Amos* xvi. 204 She had just finished dressing when a *petit-bleu*—a letter sent by pneumatic post, was delivered. **1933** 'G. Orwell' *Down & Out* vi. 44 An agency..sent me a *petit bleu*. **1978** R. Grayson *Murders at Impasse Louvain* xvi. 113 He had sent her..a 'petit bleu' or message by the pneumatic telegraph. **1895** E. Dowson *Let.* 9 Dec. (1967) 326 After a petit déjeuner at the Crémerie I get hungry again. **1909** W. J. Locke *Septimus* iii. 37, I skip afternoon tea and dinner and supper, and *petit déjeuner*. **1914** R. Brooke *Lett.* (1968) 561 Up at 6 and bathe. *Petit déjeuner* of coffee and fruit 6.45. **1926** F. W. Crofts *Inspector French & Cheyne Mystery* xv. 213 The 4.50 fr. for *petit dejeuner* suggested a fairly good hotel. **1936** J. Buchan *Island of Sheep* xii. 228 The fashion of the household was for a skimpy *petit déjeuner* and then an elaborate midday meal. **1963** N. Marsh *Dead Water* (1964) v. 125, I took my *petit dejeuner* in my room. **1972** L. Bachmann *Ultimate Act* v. 39 Your *petit déjeuner* is waiting. **1884** L. Troubridge *Life amongst Troubridges* (1966) 172 We all went to Charbonnel..for iced coffees and *petits fours*. **1898** H. A. de Salis *Housewife's Referee* 266 *Petits Fours*, small fancy biscuits. **1904** H. James *Golden Bowl* II. xlii. 367 Amerigo..selected for presentation to the other visitor a plate of *petits fours*. **1908** [see *crystallized *ppl. a.* 3]. **1948** *Good Housek. Cookery Bk.* 59 *Petits fours*, very small fancy cakes, iced cakes and biscuits served at the end of the formal dinner. Crystallised fruits, and caramellised and fondant-coated fruits are often included in the petits fours. **1961** [see *kugelhupf]. **1970** *New Yorker* 19 Sept. 31/1 The French served champagne and *petits fours*. **1974** *Publishers Weekly* 25 Mar. 55/2 Reichler's little prose petit fours accompanying the book's 100-plus photos..are tooth-some and on-the-nose. **1964** E. Palmer tr. *Martinet's Elem. Gen. Linguistics* v. 156 '*Petit nègre*' is the nearest equivalent in French to pidgin. **1972** J. L. Dillard *Black English* i. 22 Pidgin versions of French (still represented in West Africa by *Petit Nègre*) and English began developing in the factories. *Ibid.* iii. 78 Pidgin French (called *Petit Nègre*)..came to the New World at about the same time. **1974** *Times* 7 Jan. 16/3 Many [immigrants] get by on a pidgin language which the French call 'petit negre'. **1867** O. Logan in *Galaxy* Aug. 442 'Well, you see,' replied he, referring to her familiarly by her *petit nom*, 'Leo hates the leg business as much as anybody but, bless you, nothing else pays now-a-days; so what can she do?' **1939** W. Fortescue *There's Rosemary* lxxi. 368 After dinner she said gaily: 'And now for your important business with B.D. (her *petit nom* for General Long).' **1841** Thackeray in *Fraser's Mag.* June 718/2 I..swallow..the greater portion of my petit pain, too, before my second dish

arrives. **1924** A. D. Sedgwick *Little French Girl* i. i. 3 The long buttered *petits pains*. **1963** J. Creasey *Depths* vi. 47 Sitting on the terrace, eating croissants or *petits pains*. **1977** E. Ambler *Send no more Roses* vii. 141, I had one of the petits-pains just delivered by the village bakery. **1821** M. Edgeworth *Let.* 19 Dec. (1971) 298 Went to the *hall of the marble table* and there played at *petits paquets* (not time to describe) a great deal of running and laughing among pretty men and pretty maids. **1874** L. Troubridge *Life amongst Troubridges* (1966) 88 After tea we..played at Petit Pacquet in the field outside. **1902** Petit poussin [see *milk chicken* s.v. *milk *sb.* 10]. **1926–7** *Army & Navy Stores Catal.* 69/1 Poultry..Petit Poussin (English)—each 3/-. **1927** *Daily Express* 1 Apr. 5, English poultry is dear, but there are spring chicken and petits poussins, ducklings and plovers' eggs. **1961** *Harrods Food News* 16/2 Poultry..English Petit Poussin. **1820** M. Edgeworth *Let.* 4 June (1979) 144, I give you one dinner..2d service—œufs au jus—petits pois (stewed)—lettuce (ditto). **1855** F. Duberly *Jrnl.* 21 Mar. in E. E. P. Tisdall *Mrs. Duberly's Campaigns* (1963) v. 134 We made purchases of chickens, carrots, *petits pois verts* and various other necessaries. **1916** A. Bennett *Lion's Share* xi. 87, I shall like very much to hear the details of this story of *petits pois*. **1951** *Good Housek. Home Encycl.* 591/1 Peas vary in size from the small 'petit pois' to large ones. **1952** A. Grimble *Pattern of Islands* 81 An exquisite little shoulder of frozen lamb, and some onions, and potatoes, AND a tin of real French petits pois. **1961** J. Creasey *Follow Toff* xx. 171 He served the duck..the *petit pois*, tiny new potatoes. **1820** A. Opie *Tales of Heart* IV. 292 Melville dined at home that day, and paid her voluntarily all those *petits soins* which she had demanded of Arthur. **1825** H. Wilson *Mem.* III. 50, I never..heard of one.. who was so eternally au [*sic*] petits soins, and paid a woman the unremitting attention which I received from Worcester. **1847** F. A. Kemble *Let.* 17 Dec. in *Rec. Later Life* (1882) III. 318 The 'small attentions', *les petits soins* of affection. **1857** C. Kingsley *Two Yrs. Ago* II. iii. 59 Elsley showed Lucia no *petits soins*. **1959** *Times* 31 Oct. 7/4 All the barometer ever gets in the way of *petits soins* is this allegedly deleterious tapping. *c* **1906** A. John *Let.* in *Listener* (1972) 6 July 10/1 He is always asking for petits-suisses which are unheard of in this country. **1951** [see *Gervais]. **1962** J. Braine *Life at Top* iii. 53 He selected a piece of Gorgonzola, then..pointed out a *petit Suisse*. **1966** P. V. Price *France: Food & Wine Guide* 284 *Petit Suisse*, a small, round, fresh cream cheese, evolved by a farmer's wife..near Gournay, about the middle of the nineteenth century. **1979** H. McCloy *Smoking Mirror* ii. 12 A delicious cream cheese called *petit suisse*. **1939** W. B. Wright *Tools & Man* ix. 76 The microliths are predominantly trapezes or *petits tranchets* for use as arrow heads. **1949** *Proc. Prehist. Soc.* XV. 127 A flint assemblage including petit-tranchet derivative arrowheads and one leaf-arrowhead. **1954** S. Piggott *Neolithic Cultures* ii. 44 An arrowhead of the 'derived *petit-tranchet*' type. **1963** H. N. Savory in Foster & Alcock *Culture & Environment* iii. 33 In South Wales, the main overlap of this new element was with the Western Neolithic rather than the Beaker Culture, as is shown by the regular appearance of *petit-tranchet* derivatives in surface collections where leaf arrowheads predominate. *a* **1855** C. Brontë *Professor* (1857) I. viii. 141 They proposed a 'petit verre', I declined. **1860** *Once a Week* 23 June 606/2 He must be an unfortunate Frenchman indeed who cannot contrive to get a *bouillon* and a *petit verre* at the railway station. **1939** Auden & Isherwood *Journey to War* 292 Self-confident among the laughter and the *petits verres*.

‖ **petit bourgeois** (pəti būɹʒwa). Also fem. **petite bourgeoise**; pl. **petits bourgeois**; fem. **petites bourgeoises.** [Fr., lit. 'little citizen': see Bourgeois *sb.*[1] and *a.*] A member of the middle or commercial classes in a society; freq. in derogatory use, one judged to have conventional or conservative political and social attitudes. Also *fig.* and (freq. with hyphen) *attrib.* or as *adj.* See also *Petty Bourgeois.

1853 C. Brontë *Villette* III. xliii. 338 Should you object to beginning with three petite bourgeoises, the Demoiselles Miret? **1859** M. Arnold *England & Italian Question* iii. 16 How indignant he was with the townspeople, the *petits bourgeois*. **1887** F. Engels in Moore & Aveling tr. *Marx's Capital* I. i. 39 It is, of course, highly desirable in the eyes of the petit bourgeois, for whom the production of commodities is the ne plus ultra of human freedom and individual independence, that the inconveniences resulting from this character of commodities not being directly exchangeable, should be removed. *Ibid.* ii. 59 We may form an estimate of the shrewdness of the petit-bourgeois socialism, which, while perpetuating the production of commodities, aims at abolishing the 'antagonism' between money and commodities, and consequently,.. at abolishing money itself: We might just as well try to retain Catholicism without the Pope. **1897** G. B. Shaw *Our Theatres in Nineties* (1932) III. 79 She suddenly drops from an Egyptian warrior queen into a naughty English *petite bourgeoise*. **1931** *Times Lit. Suppl.* 1 Oct. 755/1 'Elise Hermann'..takes the reader into a *petit bourgeois* world. **1939** S. Spender tr. *Toller's Pastor Hall* ii. 82 We called freedom a petit bourgeois phrase, so little did we know what slavery is. **1942** E. Waugh *Put out More Flags* iii. 216 The provincial *petit-bourgeois* youth floundering and groping in the gloom of Teutonic adolescence. **1943** C. Gray *Contingencies* (1947) i. 11 Arthur Rimbaud..sought refuge in the existence of a typical French *petit bourgeois* from the terrible and terrifying realities which confronted him. **1958** *Times Lit. Suppl.* 19 Sept. 525/3 Comrade Nemeth is an ambitious bureaucrat, anxious to allay the guilt of his *petit bourgeois* birth by utter conformity to proletarian party standards. **1959** N. Mailer *Advts. for Myself* (1961) 163 Sam and Eleanor do not think of themselves as really belonging to a class, and they feel that the Sperbers and Rossmans are petit-bourgeois. **1973** *Sat. Rev.* (U.S.) 25 Sept. 22/3 The children of an ideology dominated by *petit bourgeois* sexual repression. *a* **1974** R. Crossman *Diaries* (1976) II. 160 His [*sc.* Harold Wilson's] natural modesty has remained unchanged. So have his modest tastes, his simple liking of high tea, his completely unaffected petit-bourgeois habits. **1975** A. Beevor *Violent Brink* iii. 56 She was a fraud; a *petit* [sic] *bourgeoise* in revolt..for emotional reasons. **1976** J. M. Brownjohn tr. *Kirst's Time for Payment* iii. 58 You're putting a petit bourgeois cat among the pigeons.

Hence **petite** (erron. **petit**) **bourgeoisie** (pətīt būɹʒwazī), the middle classes collectively (see Bourgeoisie, *petty bourgeoisie).

1916 A. E. Gallatin *Certain Contemporaries* 53 In all his studies of bohemians, vagabonds and the *petite bourgeoisie*, there exists only tenderness and sympathy. **1928** F. Stark *Let.* 26 Oct. (1974) I. 179 They are all good—Scotch of the very petite bourgeoisie—city folk. **1930** *N. & Q.* 10 May 325/1 The intellectual man is.. frequently drawn from the peasantry or the *petite bourgeoisie*. **1949** N. Marsh *Swing, Brother, Swing* xi. 250 He..and our Mr. Eton-and-Oxford Detective-Sergeant Salis got into a discussion about the *petit bourgeoisie*. **1950** A. L. Rowse *England of Elizabeth* v. 174 The petite-bourgeoisie were a strong element in Cromwell's support. **1970** *Times* 11 Mar. 11/2 There is today a strong Conservative government in power, to which the petite bourgeoisie naturally looks. **1976** *Brit. Jrnl. Sociol.* XXVII. 50 It is a good idea to bear in mind Rosenberg's taunt: 'Amateur sociologists have generally concluded that the petit-bourgeoisie was that mysterious class with the aid of which Hitler and Mussolini won their victory.'

petite, *a.* Add: **2.** (Further examples.) Also used *absol.*

1901 [see *Japanesey, Japanesy *a.*] **1935** H. Edib *Clown & his Daughter* vii. 33 Durnev Hanim did come in, a *petite* person with large innocent brown eyes and very black eyebrows. **1958** *Times Lit. Suppl.* 10 Oct. 573/3 Being American, of course, fifteen-year-old Franzie, petite and bouncy, has begun her sentimental education long before she meets the 'surf-bums' on Malibu beach. **1960** *News Chron.* 12 Sept. 6/1 The dress is from a newish range for the petite. **1972** M. Kaye *Lively Game of Death* ii. 9 Some men would probably dismiss her as 'small'... 'Petite' is *le mot juste*.

2. b. Used of small sizes in women's clothing. Also used *absol.*

1929 *Radio Times* 8 Nov. 439/2 This Stylish Coat.. From Petite to Matrons' Sizes. **1960** *Harper's Bazaar* Oct. 5 Afternoon dress..in 'petite' sizes for the 5'2" and under. **1974** *Times* 26 Apr. 7/7 The tights..are in three sizes—Petite, Medium or Large. **1978** *N.Y. Times* 29 Mar. A6 (Advt.), You'll find whatever you're looking for in misses (6 to 18), petites (6 to 16), juniors (5 to 13) and women's (16½ to 24½) in the brightest spring colors.

3. petite amie (see quot. 1966); **petite culture,** small farming; **petite marmite,** soup of meat and vegetables served in a marmite; **petite noblesse,** the lesser nobility in France; **petite vitesse,** slow train.

1848 Mill *Pol. Econ.* I. i. ix. 179 The working of the *petite culture* cannot be fairly judged where the small cultivator is merely a tenant. **1883** C. M. Yonge *Stray Pearls* I. iv. 32 He had only known of two ladies who had followed their husbands to the wars, and both of them only belonged to the *petite noblesse*. **1887** F. Engels in Moore & Aveling tr. *Marx's Capital* II. xxvi. 739 The labourers of the towns [in Northern Italy] were driven *en masse* into the country, and gave an impulse..to the *petite culture*, carried on in the form of gardening. *Ibid.* xxvii. 741 Japan, with its purely feudal organisation of landed property and its developed *petite culture*, gives a much truer picture of the European middle class than all our history books. **1896** E. Dowson *Let.* *c* 24 Apr. (1967) 355 Here, I have no petites Amies. **1905** *Spectator* 7 Jan. 13/2 France is notoriously a country of *petite culture*. **1906** A. Filippini *Internat. Cook Bk.* 250 Petite Marmite. .. It is very important that during the two and a half hours it should simmer exceedingly slowly. **1913** T. E. Lawrence *Home Lett.* (1954) 277 On a railway with one train a day in four hours should take three days and *petite vitesse* ten days is a mystery. **1921** Beerbohm *Lett. to R. Turner* (1964) 258, I..told him that you had ordered the book for me, and that I expected it had been sent by *petite vitesse*. **1923** A. Huxley *Let.* 2 Sept. (1969) 219 The bulk of the luggage..is still on its way from England, coming by *petite vitesse* which appears to be extremely petite. **1924** Petite noblesse [see *haute bourgeoisie]. **1945** A. L. Simon *Conc. Encycl. Gastron.* vii. 93 *Petite marmite*, the name given in restaurants to a *consommé*.. served in the earthenware pot in which it was made. **1966** A. J. Bliss *Dict. Foreign Words & Phrases* 279 *Petite amie*.., the female friend of a middle-aged man. 20c. Always with the implication that the friendship is not wholly Platonic. **1972** R. Mayne *Europeans* v. 102 Gay Paree..home of the *grisette*, the *petite amie*, the bedroom farce. **1978** W. Garner *Möbius Trip* v. 111 His *petite amie*..had raised the subject of marriage. **1978** G. Vidal *Kalki* viii. 191, I narrowly avoided a lapful of petite marmite as one bemused agent's ladle missed the soup plate.

4. *Biol.* Used, freq. as *sb.*, to designate certain variant strains of yeast that are characterized by the cytoplasmically heritable lack of mitochondrial constituents and tend to form small colonies. [The sense is due to B. Ephrussi et al., who used F. *petite colonie* (*Annales de l'Inst. Pasteur* (1949) LXXVII. 64).]

1951 *Genetics* XXXVI. 572 In many strains of yeast apparently non-genetic, cytochrome-deficient variants exhibiting a single strong alpha absorption band at 550 mu occur frequently ('petites'). **1968** *Jrnl. Molecular Biol.* XXXVII. 493 In cytoplasmic petite mutants no changes were observed in the major band. **1971** D. J. Cove *Genetics* viii. 115 It is possible to get mutant strains of yeast which are incapable of metabolising sugars oxidatively; such strains grow almost as well as the wild type on glucose, which they ferment, but on carbon sources such as acetate, which can only be metabolised oxidatively by the tricarboxylic acid cycle, they are unable to grow. These strains, called petite, can often be shown to be abnormal in their mitochondrial constituents. Some have certain mitochondrial enzymes and cytochromes absent, and others have the relative proportions of these components altered. **1978** *Nature* 23 Feb. 750/2 Cytoplasmically inherited respiratory deficient mutants termed *petites*, were first described in baker's yeast over 20 years ago.

petitionable (pĭti·ʃənăb'l), *a.* [f. Petition *sb.* + -able.] That allows, justifies, or involves, the making of a petition.

1898 *Westm. Gaz.* 14 Mar. 2/1 A few suggestions for amending the Bankruptcy Act... 1. Reduce the petitionable amount from fifty pounds to five.

petitionary, *a.* Add: **2.** (Earlier example.)

1604 Marston *Malcontent* i. v. sig. B4 Petitionarie vassailes licking the pauement with their slauish knees.

petitive (pe·titiv), *a.* [f. L. *pĕtītus* pa. pple. of *pĕtĕre* to ask: see -ive.] Of, relating to, or expressing a prayer or request; = Precative *a.*

1923 J. S. Huxley *Ess. Biologist* vii. 297 Although the value of prayer persists in so far as it is meditative and a self-purification of the mind, yet its commonly accepted petitive value must fall to the ground. **1964** E. A. Nida *Toward Sci. Transl.* ix. 201 Mode (or mood), which defines the psychological background of the action, involves principally such categories as possibility..necessity.. and desire (desiderative, optative, and petitive). **1977** *Maledicta* Summer 33 The more active attitude of seeking or desiring we call *petitive*.

petit-maître. Add: (Further examples.) Also as *adj.*

1883 C. M. Yonge *Stray Pearls* I. x. 115 He would be ashamed to count kindred with that effeminate *petite maître*! *c* **1905** E. Newman in H. Van Thal *Fanfare for E. Newman* (1955) x. 145 It is the *petit-maître* Mozart, tripping along with his manneristic little elegancies of walk and gesture but scarcely conscious of the bigness of the world around him or of the real nature of the humanity that strives in it. **1939** 'A. Bridge' *Four-Part Setting* vi. 63 His manners were slightly *petit-maître*. **1948** D. Cecil *Two Quiet Lives* 88 At moments..one is tempted to dismiss him [*sc.* Horace Walpole] as an affected petit-maître who happened to be gifted with a talent for letter writing.

2. A 'minor master' with reference to musicians, writers, etc. (usu. derog.).

1856 J. B. Waring in O. Jones *Gram. Ornament* (1865) xviii. 132 As regards another main feature in Elizabethan ornament, viz. the complicated and fanciful interlaced bands, we must seek its origin in the..designs of the class of engravers known as the 'petits maîtres' of Germany and the Netherlands. **1934** C. Lambert *Music Ho!* iii. 168 Liadoff, a real petit-maître, produced at rare intervals a few miniatures of extraordinary felicity. **1960** *Times* 8 Mar. 4/7 Michel is accounted a *petit-maître* and in French comparison does not reach the same height as Théodore Rousseau. **1963** *Listener* 3 Jan. 45/3 He [*sc.* Puccini] was a musician of great artistic integrity who clearly recognized his limitations... But such awareness does not make him a *petit maître* any more than it did Chopin, Bizet, Ravel. **1975** *Daily Tel.* 10 June 11/4 Pierre Prins is one of those French petits maîtres whom, if we only saw their best work, we would rate rightly.

petit mal. Add: There is only momentary confusion or unconsciousness without general convulsions or other major manifestations. (Earlier and later examples.)

[**1842**: see *grand mal.] **1874** J. Cunningham *New Theory of Knowing* 173 Some persons afflicted with the *petit mal* continue their work..after they have sunk into unconsciousness. **1879**, etc. [see *grand mal]. **1927** *Daily Express* 6 Aug. 7/6 She was suffering from a disease known as petit mal, which occasionally makes persons unconscious of their surroundings, and they then perform normal actions. **1957** L. Durrell *Justine* iii. 160 An occasional headache only proved him to be a victim of *petit mal*—or some other such customary disease of the rich and idle. **1974** *Daily Tel.* (Colour Suppl.) 18 Jan. 19/1 In a 'Petit Mal' a child looking out of the window may see the nose of a passing bus come into his line of vision and then, straight away, the disappearing tail of the same bus. The only ill effect is that he probably gets ticked off for daydreaming.

‖ **petits chevaux** (pəti ʃəvŏ). [Fr., lit. little horses.] A gambling game in which bets are placed on the performance of mechanically operated horses made to spin round a flag placed at the centre of a specially prepared circular table.

1891 *Clown* 9 June 13/2 Some of the most naughty of us will go and tempt fortune with the 'petit[s] chevaux'—at Dieppe. **1905** W. J. Locke *Morals M. Ordeyne* xii. 147 She has a consuming passion for *petits chevaux*. I speak sagely of the evils of gambling. **1911** [see *boule[4] 1]. **1912** 'Saki' *Chron. Clovis* 181 It was just before *petits chevaux* had been supplanted by *boule*. **1929** R. Aldington *Death of Hero* iii. vi. 330 He disapproved of baccarat, roulette, and *petits chevaux*. **1964** C. Mackenzie *My Life & Times* III. i. 22, I do not know what the rules are to-day for *petits chevaux* but in 1900 horse number 5 of the 9 horses meant that the bank won. **1964** A. Wykes *Gambling* ix. 211 Other variations of roulette..*Petits chevaux*—which still flourishes in Northern France and Ireland.

pe·t-name, *v.* [f. *pet name* s.v. Pet *sb.*[1] 3 c.] *trans.* To give (a person) a pet-name; to call by a pet-name.

1915 E. Corri *30 Yrs. Boxing Ref.* 183 Men of the most human type are usually pet-named by the public in some way. **1942** C. Morley *Thorofare* (1943) xix. 66 He had pet-named the two funnels for his twin daughters, Alma and Sophie. **1973** *Times Lit. Suppl.* 1 June 608/3 His young bride Nadia petnamed him Klop (which is Russian for 'bug').

‖ **peto** (pe·to). [Sp.] A padded or stuffed protective covering for a picador's horse.

1957 A. MacNab *Bulls of Iberia* v. 52 The number of pics is not rigid..for with the *peto* the contact may last some time, and not be broken at once as in the pre-*peto* days. **1967** McCormick & Mascareñas *Compl. Aficionado* ii. 35 Before 1928, when the peto was introduced, the work of the picador was more prominent than it is in the modern corrida. **1968** *Economist* 17 Feb. 32/1 It was she [*sc.* Queen Victoria Eugenia] who insisted on the use of the *peto*, the ugly mattress-like covering that protects picadors' horses, though not infallibly, from disembowelment.

petrean, *a.* Restrict *rare.* ? *Obs.* to sense in Dict. and add: Also **petræan. 1.** (Later example.)

1849 J. Forbes *Physician's Holiday* xx. 294 The same petræan desert continues beyond Swarenbach.

2. (With capital initial.) Of or pertaining to Petra, a city in ancient Arabia Petræa. Also as *sb.*, an inhabitant of Petra.

1852 E. A. Andrews *Copious Latin-Eng. Lex.* 1133/2 *Pĕtra*,.. a city in Arabia Petraea, now the ruins of Wadi Musa... Hence..*Pĕtreus*, a, um, adj., Petrean. **1923** A. Forder *Petra: Perea: Phœnicia* 10 One would gather from Bible records that the Petreans were always a proud and turbulent people. *Ibid.* 20 A feasible explanation..was given to me by a modern Petrean. **1925** A. B. W. Kennedy *Petra* iii. 34 Very little is known about the religion of the Nabataeans as a nation, and even less, perhaps, about the religion of those of the nationality whom we may call particularly Petraeans, and who are responsible for the monuments of Petra. *Ibid.* 35 We know from the dated tombs at Madaïn Salih that the most complex and highly developed of the Petraean designs were already in existence before the commencement of our era. **1957** *Encycl. Brit.* XVII. 652/2 The chief god of Petra was Dhūsharā..he was worshipped under the form of a black rectangular stone, a sort of Petraean Ka'ba. **1965** N. Glueck *Deities & Dolphins* (1966) x. 331 The Petraean feminine figure on a sea animal could.. be identified with an Atargatis or Aphrodite mounted on a fish-tailed Capricornus.

petrefact (pe·trĭfækt). Also **petrifact.** [f. L. *petra* rock, stone after *artefact*.] An object made of stone; also *fig.*, something that has become hardened or fixed.

1911 Beerbohm *And Even Now* (1920) 39 He..does strive, by day and by night, poor petrefact, to rip off these fell and clownish integuments... He forgets that after all he is only a statue. **1932** *Brit. Jrnl. Psychol.* Apr. 313 Those in whom space perception is still of the early, plastic form frequent among children, in whom the later 'petrefacts' have not yet been crystallized out.. should be primary colour perceptors. **1975** *Encounter* Sept. 86 In Germany, in any case, he destroyed numerous outdated structures, did away with old classes, smashed to pieces revered petrifacts.

Petri (pe·tri, pī·tri). Also **petri.** [Name of R. J. *Petri* (1852–1922), German bacteriologist, who first proposed the use of such a dish (*Centralbl. f. Bacteriol. und Parasitenkunde* (1887) I. 279).] *Petri* (or † *Petri's*) *dish*: a shallow, circular, flat-bottomed glass (or plastic) dish with vertical sides and with a cover of the same shape but slightly larger, which is used particularly for growing cultures of bacteria or the like. Also *fig.*

1892 A. C. Abbott *Princ. Bacteriol.* viii. 87 Petri's dishes are flat, double dishes of glass. *Ibid.* xx. 212 Place the tissue in a sterilized Petri dish. **1897** Muir & Ritchie *Man. Bacteriol.* 57 The latter are known as Petri's dishes. **1903** *Univ. Nebraska Stud.* Oct. 2 The hydroids were cut.. into the desired lengths and placed in watch glasses, petri-dishes, finger bowls, etc. **1946** F. Schneider *Qualitative Organic Microanalysis* vii. 187 Stopper the flask with a small cork, place the side arm in the petri dish, and fill the latter with alkaline permanganate solution. **1973** J. Goodfield *Courier to Peking* xv. 211 I've got a test I want to do. Get me a Petri dish smeared with gelatine and the following serums. **1976** *Nature* 24 June 701/1 Both cell types were plated on plastic Petri dishes (5 cm diameter).

petrichor (pe·trikǫɹ). [f. Petr(o- + Ichor.] A pleasant, distinctive smell that frequently accompanies the first rain after a long period of warm, dry weather in certain regions; in quot. 1975, applied to an oily substance obtained from the ground in which this smell was concentrated.

1964 Bear & Thomas in *Nature* 7 Mar. 993/2 The diverse nature of the host materials has led us to propose the name 'petrichor' for this apparently unique odour which can be regarded as an 'ichor' or 'tenuous essence' derived from rock or stone. This name, unlike the general term 'argillaceous odour', avoids the unwarranted implication that the phenomenon is restricted to clays or argillaceous materials; it does not imply that petrichor is necessarily a fixed chemical entity but rather it denotes an integral odour. **1971** *Listener* 4 Nov. 612/3 No matter what kind of rock or earth was used, the oily essence always possessed the aroma of petrichor—the smell of rain falling on dry ground. **1975** *Sunday Mail* (Brisbane) 2 Nov. 32/2 CSIROs Melbourne mineral chemistry division discovered that it was not fungi or dead vegetation which produced the smell, but small yellowish-gold oily globules. The globules, nicknamed 'petrichor' or 'essence of rock' by the researchers, contained at least 50 different compounds, not unlike a perfume and were absorbed into the ground from the air.

petrified, *ppl. a.* Add: Esp. in phr. *petrified forest.* (Further examples.)

1830 *Illinois Monthly Mag.* Oct. 31 The earth's surface is literally covered with stumps, roots and limbs of petrified trees; presenting the appearance of a 'Petrified Forest'. **1873** S. W. Cozzens *Marvellous Country* 76 We came upon the remains of a petrified forest,..converted by some chemical process into specimens of variegated jasper. **1937** M. Huxley *Let.* 13 Oct. in *Lett. A. Huxley* (1969) 425 We drove through all the view places and the petrified forests and the grand-canyons. **1969** E. H. Pinto *Treen* 196 Petrified wood. The most interesting and highly prized results of wood transformed into rock, are examples from the several petrified forests which exist in various parts of the world. *Ibid.* 197 Two of the most famous of the petrified forests are situated in the U.S.A.—one at the Yellowstone National Park.., the other in Chalcedony Park, Arizona.

2. Add to def.: terrified, extremely frightened; also const. *of.* (Further examples.)

1869 'Mark Twain' *Innoc. Abr.* xlvi. 481 A great herd of curious-looking Syrian goats and sheep were gratefully eating gravel. I do not state this as a petrified fact. I only *suppose* they were eating gravel, because there did not appear to be any thing else for them to eat. **1875** —— *Sk. New & Old* 219 Here, here, you petrified fool. **1963** R. Wolff *I, Keturah* (1964) ii. xvii. 229, I was petrified of all the gadgets. **1968** *Observer* 7 Jan. 28/6 [They] *think* they want the foreign flavour; but actually couldn't be more petrified of it. **1974** W. J. Burley *Death in Stanley St.* xi. 186 He was trying to get at the gun... I was petrified!

Petrine, *a.* Add: **1.** *Petrine claims* (example).

1930 tr. *F. von Hügel's Some Notes on Petrine Claims* vii. 30 Four contentions which make up the Catholic Petrine claims.

2. Of, pertaining to, or characteristic of Peter I the Great (1672–1725), tsar and emperor of Russia.

1908 R. N. Bain in *Cambr. Mod. Hist.* V. xvii. 549 At least half the Senate (though that was a purely Petrine institution). **1927** E. M. Almedingen in *Magnificat* Mar. 235/2 No really great men of letters lived in Peter's time: Theophanus..and Prince Kantemir..were the two men who left any mark at all in the annals of Petrine literature. **1956** E. Munzer *Solovyev* iii. 41 It was he [*sc.* the Patriarch Nikon] who paved the way to the Petrine perversion, or pseudomorphosis, of the Russian soul. **1964** *Listener* 5 Mar. 392/1 [Alexis] Tolstoy's Peter the Great modifies the Petrine myth in accordance with an up-to-date and almost professional conception of history. **1974** *Times Lit. Suppl.* 13 Sept. 984/3 No extracts on eastern Europe;.. none on Russia, where the Petrine reforms were nothing if not military.

Petrinist (pī·trinist). [f. Petrine *a.* + -ist.] A follower of St. Peter; a student of Petrine theology.

1922 Jackson & Lake *Beginn. Christianity* II. i. v. 123 Gfrörer thought that the compiler of Acts..used a collection of unhistorical legends arranged by a zealous Petrinist.

‖ **pétrissage** (petrisāʒ). [Fr., f. *pétrir* to knead.] A kneading process used in massage.

1886 W. Murrell *Massage* iii. 11 Next [after *effleurage*] comes *pétrissage* which is more important and is by no means easy to acquire. **1906** *Practitioner* Dec. 769 *Pétrissage* is performed by grasping the tissues to be operated on, lifting the mass thus seized, and alternately loosening and tightening the grasp. **1957** *Encycl. Brit.* XV. 32/1 Pétrissage..consists of a seizure by the hands of muscles or other tissues which are then subjected to a process of kneading. **1961** [see *malaxation c].

petro-[2] (pe·tro), combining form of Petroleum, in recent formations as *petro-politics, -power, -resources, -wealth*; freq. with ref. to revenue, esp. foreign exchange, deriving from petroleum exports, as *petro-billion, -naira, -pound*. Also *petrodollar. Cf. *petrochemistry 2.

1973 *Time* 3 Dec. 44/1 The energy crisis..may have been artificially imposed, but its implications stretch far beyond petropolitics. **1974** *Newsweek* 7 Oct. 52/3 Top Arab leaders have now decided not to put their petro-billions into U.S. Treasury bonds..but to invest in American industry instead. **1975** *Publishers Weekly* 19 May 99/1, I understand they're rolling in petropounds since oil was discovered in the North Sea. **1976** *Ibid.* 5 Jan. 60/1 An Arab emirate saturated with petro-resources. **1976** *Daily Times* (Lagos) 5 May 7/1 As the tenth largest oil producer in the league, Nigeria has every reason to tout her petro-wealth. **1976** *Daily Colonist* (Victoria, B.C.) 20 June 35/1 'There are two kinds of power,' he said, 'petropower and agripower.' **1976** *Daily Times* (Lagos) 20 July 7/1 Nigeria's foreign exchange reserve zoomed to N2,047 million through oil boom, thus projecting an over-sized petro-naira bubble which beclouded the vision of some former military rulers. **1977** *Time* 11 July 36/3 The Saudis can be expected to wield their petropower prudently.

petrochemical (petroke·mikăl), *a.* and *sb.* Also **petro-chemical.** [f. next, after *chemistry, chemical*.] **A.** *adj.* **a.** Of or pertaining to petrochemistry (sense *1).

1913 *Chem. Abstr.* VII. 3949 (*heading*) Petrochemical studies. **1952** *Geochim. et Cosmochim. Acta* II. 283 (*heading*) A petrochemical study of the Tertiary lavas of north-east Ireland. **1960** *Internat. Geol. Rev.* II. 273 (*heading*) Petrochemical study of the Cenozoic basaltic rocks in eastern China. **1975** *Nature* 27 Feb. 691/1 Another study of the intrusion was initiated in the hope that new geophysical and petrochemical techniques might provide the added insight needed to resolve some of these questions.

b. Of or pertaining to petrochemistry (sense *2) or petrochemicals.

1942 *Oil & Gas Jrnl.* 30 Apr. 47 The long experience of the Badger Company..enables it..to meet the requirements of the rapidly expanding Petro-Chemical Industry. **1947** *Chem. & Engin. News* 8 Dec. 3634/1 Increasing demands for organic chemicals..have led to rapid development of the petrochemical industry. **1959** *Times Rev. Industry* Sept. 99/3 The Monsanto Chemical Company..will build a petro-chemical factory..to produce styrene monomer from which polystyrene plastics are made. **1961** G. R. Choppin *Exper. Nucl. Chem.* xi. 185 Further reduction is effected by using solvents of petro-chemical origin. **1969** *Times* 30 Apr. 23/2 (Advt.), Dutch-American firm, specialising in design and fabrication of unusual petrochemical heat transfer equipment,..seeks experienced man. *a* **1974** R. Crossman *Diaries* (1976) II. 290 There was a likelihood of petro-chemical and associated development in the Invergordon area.

B. *sb.* Any chemical compound or element obtained or derived industrially from petroleum or natural gas. Freq. *attrib.* in *pl.*

1942 *Oil & Gas Jrnl.* 25 June 180 (Advt.), Typical 'petro-chemicals' already in production: alcohols, butadiene, benzol, [etc.]. **1950** *N.Y. Times* 19 Nov. 11/3 Petrochemicals which are derived from the vast supplies of petroleum and natural gas available in the South, are the bellwether of the country's economy. **1959** *Engineering* 2 Jan. 24/1 The new petrochemicals plant at the Fawley refinery of Esso. **1967** *New Scientist* 16 Feb. 385/3 Chemical engineering has now advanced to the point whereby the idea of using this gas [*sc.* carbon monoxide] as the key building block for petrochemicals—such as plastics and pesticides..—is no longer a pipe-dream. **1972** Harker & Allen *Fuel Sci.* iii. 35 Crude oil also provides the source of raw material for the rapidly growing petrochemicals industry. **1975** *Petroleum Rev.* XXIX. 381/2 (*caption*) 65,000 tons of petrochemicals..are stored under an inert blanket.

Hence **petroche·mically** *adv.*

1952 *Trans. Edin. Geol. Soc.* XV. 82 Ophitic texture is the result of delayed crystallisation of augite from magma characterised petrochemically by marked undersaturation in the pyroxene-components. **1963** *Economist* 28 Sept. 1140, 60 per cent of world nitrogen is petrochemically derived. **1973** *Nature* 27 Apr. 566/1 'Petrographically' the dredged lavas are typically submarine fine grained, vesicular and porphyritic basalts.. 'Petrochemically' the basalts are classified as tholeiites according to the Yoder and Tilley normative scheme.

petrochemistry (petroke·mistri). [f. Petro- (in sense 2 repr. *petroleum*: cf. *petro-[2]) + Chemistry.] **1.** *Geol.* The chemistry of the composition and formation of rocks (as distinct from minerals, ore deposits, etc.), esp. igneous and metamorphic ones.

1937 *Bull. Volcanologique* I. 59 (*heading*) Petrochemistry of the Scottish Carboniferous-Permian igneous rocks. **1958** *Q. Jrnl. Geol. Soc.* CXIII. 393 (*heading*) The petrochemistry of the Ardara aureole. **1971** *Internat. Geol. Rev.* XIII. 561/2 In petrochemistry we are often required to compare associations of rocks of different regions and assign them to definite petrographic groups. **1974** *Nature* 18 Oct. 581/2 There is, as yet, insufficient documentation of the relationship between precisely measured Sr isotope ratios and the petrochemistry of Icelandic rocks.

2. The chemistry of petroleum and natural gas, and of their refining and processing.

1942 *Oil & Gas Jrnl.* 18 June 96/1 The section in next week's Journal devoted to petro-chemistry is something you'll not want to miss. **1954** *Chem. & Engin. News* 20 Sept. 3719/3 Entrance of one company into field of petro-chemistry in 1925 was followed by dozens of others. **1959** *Ibid.* 22 June 41 (Advt.), Exciting new products through petro-chemistry.

petrodollar (pe·trŏ,dǫlaɹ). [f. *petro-[2] + Dollar.] A notional unit of currency available in a petroleum-exporting country. Freq. *attrib.* of the surplus of petroleum exports over imports of all other goods, and in *pl.*

1974 *Globe & Mail* (Toronto) 27 Aug. 7/1 What emerged in Washington..was a growing concern over the swiftly accelerating petrodollar holdings of the Arabs. (Petrodollars are defined as the excess foreign-exchange assets of the oil-producing countries)... There simply isn't sufficient data on the flow of petrodollars. *Ibid.* 7/6 Still, the question of petrodollars and their potential recycling into U.S. industries..is something officials and analysts are studying. **1974** Ld. Gladwyn in *Hansard Lords* 5 Nov. 310 The loan to Italy, the sugar deal, which they approved, the beginning of petro-dollar recycling are all instances in point. **1974** *Financial Times* 21 Nov. 23/3 The size of the 'petrodollar' surplus—defined as the current earnings of the OPEC countries over and above what they can spend on imports—has been estimated at about $60 bn. [*sc.* thousand million] this year. **1975** *Times* 8 Jan. 1/6 Two suggested schemes for recycling surplus petrodollars to deficit countries. **1976** *Time* 27 Dec. 21/1 Problems of inflation, devaluation and petrodollars intimately bind our economy and that of other nations into a common system. **1977** *Time* 18 July 39/1 In Aspen, Colo. the Empress..danced away the Fourth of July at a local night-spot and dropped some petrodollars in Aspen shops.

petrofabrics (petrofæ·briks), *sb. pl.* *Geol.* [f. Petro- + Fabric *sb.*: see -ic 2.] The texture and microscopic structure of a rock or rocks, or the study of these, esp. as guides to the movements to which they have been subjected. Cf. *petrotectonics *sb. pl.*

1934 B. Sander in *Amer. Jrnl. Sci.* XXVIII. 37 (*heading*) Petrofabrics (Gefügekunde der Gesteine) and orogenesis. *Ibid.*, Petrofabrics denotes the study of the internal space relations of a rock. **1949** *Q. Jrnl. Geol. Soc.* CV. 111 The greatest contribution of the Sander–Schmidt school of petrofabrics has been its emphasis on the movements influencing fabric rather than on the forces responsible for the movements. **1960** *Liverpool & Manch. Geol. Jrnl.* II. 503 (*heading*) New work on petrofabrics. **1974** *Nature* 12 Apr. 621/3 Papers on..the petrofabrics of the peridotite suite xenoliths.

So **petrofa·bric** *a.*

1934 B. Sander in *Amer. Jrnl. Sci.* XXVIII. 38 Petrofabric analysis is..important for purely petrographic investigations that seek only descriptive characteristics of rocks. **1942** M. P. Billings *Structural Geol.* xviii. 341 Many petrofabric diagrams show the attitude of the cleavage rather than the orientation of the optic axes. **1956** E. W. Heinrich *Microsc. Petrogr.* vii. 208 Petrofabric studies reveal that the optic axes of the quartz grains usually are well oriented in one of several patterns. **1968** B. Bayly *Introd. Petrol.* xxv. 303 There is..a large field known as petrofabric analysis, in which the orientation of many individual grains is studied.

Petro-Forge (pe·trofǭɹdʒ). Also **-forge.** [f. Petro- (taken as repr. Petrol) + Forge *sb.*] A forging machine powered by a petrol engine.

1964 *Engineering* 2 Oct. 438/2 A session at the fifth MTDRC was devoted entirely to high energy rate forming... The paper read last: 'Petro-Forge Mk I-DRD',.. gave evidence of industry's keen interest in the development of a method of high energy operation which uses the internal combustion process for its power source. **1965** *Economist* 18 Dec. 1354/3 The Petro-Forge has been sufficiently developed (it started under a DSIR grant in 1962) to show that even when working at slow speeds it will be cheaper and better than existing forging machinery. **1970** Davies & Austin *Devel. High Speed Metal Forming* ii. i. 54 Multi-blow operation at one second intervals is.. possible and enables the Petro-Forge to be used for large volume production of components. **1973** E. C. Rollason *Metall. for Engineers* (ed. 4) vii. 129 In the Petro-forge the energy is derived from the combustion of a fuel/air mixture in the cylinder of an internal combustion engine.

petrogenesis (petrŏdʒe·nĕsis). *Petrol.* [f. Petro- + -genesis. Cf. G. *petrogenese* (R. Th. Simler *Ueber die Petrogenese* (1862)); *petrogenesis* was used by F. Zirkel (*Lehrb. der Petrogr.* (1866) I. 159).] (The study of) the formation of rocks, esp. igneous and metamorphic rocks.

1901 *Q. Jrnl. Geol. Soc.* LVII. p. lxv, Hutton was in advance of his time on matters relating to petrogenesis. **1911** *Geol. Mag.* Decade V. VIII. 248 (*heading*) The fundamental problems of petrogenesis, or the origin of the igneous rocks. **1956** *Nature* 14 Jan. 68/1 He has achieved high distinction for his researches on petrogenesis, notably on rock transformations in the granitic suite. **1960** *Times* 13 Feb. 8/4 Lyell Medal to Dr. Doris Reynolds for her work on the rocks of Northern Ireland and her contributions to petrogenesis. **1971** *Nature* 24 Sept. 260/1 Romey..suggests that theories of lunar petrogenesis are developing too fast. **1977** A. Hallam *Planet Earth* 313 *Petrogenesis*, an envelope term embracing all aspects and features of the formation of rocks.

Hence **pe:trogene·tic** *a.*, **-gene·tically** *adv.*

1911 *Geol. Mag.* Decade V. VIII. 251 Such a conception exists, but we can oppose to it several geological data and petrogenetic considerations. **1950** A. K. Wells in F. H. Hatch et al. *Petrol. Igneous Rocks* (ed. 10) 5 The modern tendency to explain many petrogenetic problems in terms of emanations, ionic migration and 'granitization'. **1954** *Jrnl. Geol.* LXII. 172 (*heading*) The trace elements of the plutonic complex of Loch Doon..and their petrogenetic significance. **1970** *Mineral. Abstr.* XXI. 77/1 Petrogenetically it is suggested that a basalt differentiated to produce an olivine-augite peridotite which subsequently assimilated a potassic granite.

petrogeny (petrǫ·dʒĕni). *Petrol.* [f. Petro- + -geny.] = *Petrogenesis. Chiefly in *petrogeny's residua system.*

1937 N. L. Bowen in *Amer. Jrnl. Sci.* CCXXXIII. 1 It is concluded..that in mixtures containing all of these (SiO_2, Al_2O_3, iron oxide, CaO, MgO, Na_2O and K_2O) the residual liquid from fractional crystallization will show this same character of great enrichment in alkali-alumina silicate. Since the oxides listed above make up some 97 per cent of the composition of the average igneous rock the alkali-alumina silicate system ($NaAlSiO_4$–$KAlSiO_4$–SiO_2) is here referred to as petrogeny's 'residua system'. **1948** [see *magmatism a]. **1950** E. E. Wahlstrom *Introd. Theoret. Igneous Petrol.* i. 3 Petrogeny places emphasis on the manner of origin or genesis of rocks. **1966** *Jrnl. Petrol.* VII. 115 It is necessary to know something of the residual liquids that might be produced in the course of crystallization of silicate melts, and a considerable part of the efforts of experimental petrology has been directed to this end. The classic system in this regard is 'petrogeny's residua system', $NaAlSiO_4$–$KAlSiO_4$–SiO_2..but.. even this is a residua system only for liquids in which the alkali/alumina ratio [*printed* ration] is unity.

So **petroge·nic** *a.* = *petrogenetic *a.*

1908 *Amer. Jrnl. Sci.* CLXXVI. 45 In a fully represented petrogenic cycle at a batholithic area..the oldest intrusion should be a rock of gabbroid (basaltic) composition. **1947** A. Knopf *Pirsson's Rocks & Rock Minerals* (ed. 3) vi. 144 A period in which a genetically related series of igneous rocks was formed is called a petrogenic epoch, and as this concept is more definite than 'petrographic province', it is supplanting the older idea. **1971** I. G. Gass et al. *Understanding Earth* 12 The rise of understanding of rocks, and particularly of petrogenic processes, was initially highly dependent on the way in which microscopy assisted in the study of the constituent minerals.

petroglyph. (Further examples.)

1952 V. G. Childe *New Light Most Anc. East* ii. 23 Petroglyphs, like microliths, attest hunting over many areas now uninhabitable. **1955** *Sci. News Let.* 9 July 31/2 The crescent is not a common figure among petroglyphs and pictographs of northern Arizona. **1958** E. A. Armstrong *Folklore of Birds* i. 10 In the Californian deserts I have seen petroglyphs made by Red Indian initiands. **1972** *Sci. Amer.* June 91/1 The use of a line to depict a contour may well have been one of the earliest developments in art, as exemplified by the 'line drawings' in the pictographs and petroglyphs of prehistoric artists. **1974** *Environmental Conservation* I. 8/2 The desert is dotted with..fossil remains..of domestic animals whose present range lies well outside the areas of the petroglyphs.

petrographic, *a.* Add: *petrographic province*: see *province 6 b.

petrographical, *a.* Add: *petrographical province*: see *province 6 b.

petroil (pe·troil). [f. Petr(ol + Oil *sb.*[1]] = *mixture 3 e. Freq. *attrib.*

1927 A. W. Judge *Mod. Motor Engineer* III. vi. 164 In the case of two-stroke engines using the 'petroil' (i.e. petrol and oil mixture) system of lubrication, it is important to keep the float chamber and filter clean. **1959** *Motor* 21 Oct. 344/1 The petroil system of lubrication has been so refined that only one pint of oil need be added to each 5 gallons of petrol. **1967** P. E. Irving *Two-Stroke Power Units* vii. 129 With crankcase compression, it is manifestly impossible to have large quantities of oil splashing around inside the engine... The most generally accepted way out of this situation is to dissolve a small amount of oil in the fuel, the resulting mixture being termed 'petroil' (in England, anyway). **1972** J. Stevens *Scooter* iv. 98 Another cut in running costs comes from the scooters which run on a petroil mixture containing only 2% of oil. **1977** *Good Motoring* Oct. 8/1 Mixed a gallon of petroil to fill the tank of a two-stroke motor cycle.

petrol. Add: Now only with pronunc. (pe·trǫl). **3.** (Earlier example.)

For an attribution of the Eng. word to Frederick Simms, 1893, see E. Liveing *Pioneers of Petrol* (1959) 42. However *pétrole* was current in Fr. by 1892 (cf. G. Richard *Les nouveaux Moteurs à Gaz et à Pétrole* (1892)).

1895 F. R. Simms in *Sat. Rev.* 3 Aug. 144/1 The Daimler carriage can easily take a gradient of 1 in 10... The fuel used is rectified petroleum or petrol or benzoline, of a specific gravity of 0·680 to 0·705, which has many advantages over common petroleum, and is obtainable almost anywhere.

3. b. *petrol adjustment, bowser* [*bowser[2]], *bus, can,* † *car, consumption, costs, engine* (further examples), *feed, -filler, fumes, lamp, motor, ration, rationing, tank, tanker, tap, tin, vapour*; *petrol-driven, -engined, -propelled* adjs.; **petrol attendant,** one who attends to the petrol pumps at a petrol station; **petrol blue,** a shade of blue similar to the colour of petrol; **petrol coupon,** a petrol rationing coupon; **petrol-electric** *a.*, (*a*) driven by electric motors powered by current from a generator, which in turn is driven by a petrol engine (cf. **diesel-electric* adj.); (*b*) applied to a petrol-driven electricity generator; **petrol gauge,** a meter indicating the quantity of petrol in a tank; **petrol injection,** fuel injection, where the fuel is petrol; **petrol lighter,** a cigarette lighter employing petrol; **petrol pump,** (*a*) a pump at a petrol station for supplying motor vehicles with petrol; (*b*) a pump which delivers petrol from the petrol tank of a motor vehicle or aircraft to the engine; **petrol sniffer,** one who inhales petrol fumes for a narcotic effect; hence **petrol-sniffing,** the activity of a petrol-sniffer; **petrol station** = *filling station* s.v. *filling *vbl. sb.* 4.

1920 Petrol adjustment [see *idle *v.* 4 a]. **1963** M. Levinson *Taxi!* iv. 47 He can very easily get a part-time job..as a petrol attendant. **1975** *Evening Herald* (Dublin) 8 May 12/3 (Advt.), Petrol attendants required. **1949** *Dict. Colours Interior Decoration* (Brit. Colour Council) III. 21/1 *Petrol blue*, a colour name introduced into seasonal ranges by B.C.C. in 1943. **1971** *Sunday Mirror* 21 Mar. 15/1 You can't sell them bright or pastel colours. In summer they just want petrol blue, bottle green or navy. **1973** *Guardian* 10 Apr. 13/3 (*caption*) Chunky petrol blue mohair cardigan. **1943** T. R. St. George *C/o Postmaster* 58 We learned, presently, that it wasn't a gas station at all, for that matter, but a 'petrol bowser'. **1943** P. Brennan et al. *Spitfires over Malta* 24 A petrol bowser on the perimeter track caught fire..and blazed fiercely. **1911** *Chambers's Jrnl.* Jan. 77/1 The electrically propelled vehicle..is smoother in running, more silent, and more convenient to handle than the petrol-bus. **1966** M. Woodhouse *Tree Frog* xxii. 159 The jeep bounced... Yancy cursed the petrol cans fluently. **1975** M. Russell *Murder by Mile* xiv. 151 The petrol cans that she's just been emptying. **1900** Petrol car [in Dict., sense 3 a]. **1902** A. C. Harmsworth et al. *Motors* x. 183 (*heading*) The petrol car. **1908** *Westm. Gaz.* 16 June 4/2 Petrol-consumption is a factor which is taken into serious consideration by purchasers of cars nowadays. **1936** 'N. Blake' *Thou Shell of Death* ii. 37 Your defence planes of the future..must, above all, have a very low petrol consumption. **1974** A. Ross *Bradford Business* 15 We covered the hundred and seventy miles..at an average speed of 91·5 miles an hour... The petrol consumption was..sixteen imperial gallons. **1955** *Times* 10 May 6/3 Their petrol costs, according to a branch official, are being met by some of the voluntary miners' funds. **1939** *Times* 14 Oct. 5/3 (*heading*) Supplementary petrol coupons. **1963** D. Hughes in Sissons & French *Age of Austerity* iv. 86 Forged petrol coupons changed hands at half a crown a time. **1971** A. Diment *Think Inc.* iv. 59 This man concerned himself with rough copies of ration books and petrol coupons. **1973** T. Allbeury *Choice of Enemies* viii. 35 Bring..your car—we'll see you get enough petrol coupons. **1905** *Westm. Gaz.* 2 Jan. 2/1 Purely petrol-driven auto-cars on exactly the same principle as a road automobile. **1937** Blunden *Elegy* 50 The televisionary world to come, The petrol-driven world already made. **1960** *Farmer & Stockbreeder* 22 Mar. (Suppl.) 8/1 You can pay as little as 33gs for a 14in petrol-driven mower. **1964** J. J. Walsh *Understanding Paraplegia* xvi. 113 Popularity of the fast and more powerful petrol-driven tricycle. **1905** *Westm. Gaz.* 2 Jan. 2/1 The North-Eastern, Great Northern, and Brighton Railways. The first-named has some petrol-electric coaches. **1909** *Installation News* III. 117/2 The petrol electric set..is increasing in popularity every day; at the moment however the danger does not present itself..as a source of loss to the Central Station. **1930** *Times* 25 Mar. 25/1 With regard to the replacement of our motor omnibuses,.. the petrol-electric machine.. has given us very good service. **1940** *Chambers's Techn. Dict.* 631/1 *Petrol-electric generating set* (Elec. Eng.), a small generating plant using a petrol engine as the prime-mover. **1902** R. J. Mecredy in A. C. Harmsworth *Motors* vii. 105 'Petrol Engine' is a slang term for an engine driven by a series of explosions of a mixture of the vapour of a light spirit of petroleum with air. **1916** A. Garrard (*title*) Gas, oil, and petrol engines. **1933** Petrol engine [see *injection 1 b]. **1974** *Country Life* 3 Oct. 948/1 Daimler's petrol engine was used in vehicles constructed between 1885 and 1888. **1906** *Daily Chron.* 27 June 3/5 The invention of a petrol-engined torpedo-boat. **1936** *Discovery* Sept. 289/1 The new L.M.S. 'Coventry' petrol-engined railcar. **1971** *Daily Tel.* 28 Jan. 3/1 A regulation requiring new petrol-engined vehicles to be fitted with means to prevent fumes escaping from the crank-case is to be laid before Parliament. **1913** *Autocar Handbk.* (ed. 5) iii. 67 (*caption*) Exhaust pressure petrol feed. **1928** *Manch. Guardian Weekly* 10 Aug. 107/4 They experienced trouble with the petrol feed and turned back. **1935** C. Day Lewis *Time to Dance* 34 They..rose and flew on to Paris, and there Trivially were delayed—a defective petrol feed—Three days. **1977** J. Cleary *High Road to China* v. 145 One of the petrol feed pipes had been dented... Fuel had been getting to the engine only in a trickle. **1907** *Westm. Gaz.* 22 Nov. 10/1 The latest thing in petrol-fillers, horns and sirens. **1970** *Country Life* 31 Dec. 1278/2 A conical petrol-filler with a spout that he got from a local garage. **1908** *Lancet* 29 Feb. 658/2 He was only able to find in the literature one other case relating to the toxic effects of petrol fumes. **1912** Kipling *Let.* 10 Dec. in C. E. Carrington *Rudyard Kipling* (1955) xvi. 420 A rather shrill stink..like chlorine gas on top of petrol fumes... That's the stink of the aeroplane. *a* **1930** D. H. Lawrence *Last Poems* (1932) 32 The weather in town is always benzine, or else petrol fumes. **1974** N. Freeling *Dressing of Diamond* 45 The summer night..was still tainted with petrol fumes. **1976** M. Hinxman *End of Good Woman* vii. 94 She had stood on traffic intersections..inhaling petrol fumes. **1913** *Autocar Handbk.* (ed. 5) iii. 66 As the tank is arranged near the occupants a simple form of petrol gauge can be fitted. **1938** N. Marsh *Artists in Crime* xiii. 198 The petrol gauge ..registered only two gallons. **1974** D. Kyle *Raft of Swords* xiv. 156 He glanced down at the petrol gauge, saw the tank was roughly half empty. **1940** *Chambers's Techn. Dict.* 249/2 *Direct petrol injection* (I.C. Engs.), a method of operating a petrol engine by injecting liquid petrol directly into the induction pipe or cylinder during the suction-stroke, thus dispensing with a carburettor; in aero engines it avoids carburettor freezing troubles. **1972** K. Newton et al. *Motor Vehicle* (ed. 9) xvii. 379 Even in the U.S.A., where..very high-powered vehicles are the rule, petrol injection seems to make little headway. **1977** *Economist* 3 Sept. 51/1 (Advt.), Petrol injection makes for cleaner..combustion than the conventional carburettor. **1917** W. Owen *Let.* 9 Feb. (1967) 432 Last night I burnt a petrol lamp under my bed! **1957** L. Durrell *Justine* i. 14 The boys stir uneasily at their backgammon under the petrol-lamps. **1978** G. Sims *Rex Mundi* vii. 45 We sped with..straining sails and hissing petrol lamps. **1926–7** *Army & Navy Stores Catal.* 88/1 Automatic 'Petrol' Lighter..With Nickel Case—each 1/3. **1930** Sayers & 'Eustace' *Documents in Case* i. 151, I hunted through my pockets for a petrol lighter. **1978** D. Bloodworth *Crosstalk* iv. 35 The end of his cigarette trembling in the tiny spluttering flame of an almost extinct petrol lighter. **1904** T. H. White (*title*) Petrol motors and motor cars: a handbook for engineers, designers and draughtsmen. **1912** *Motor Manual* (ed. 14) i. 1 The modern motorcar has been a gradual development since the introduction of the petrol motor or light internal-combustion engine by Benz, Levassor and Daimler. **1907** *Westm. Gaz.* 9 July 4/2

Wright's breakdown in the race..was variously ascribed to a broken valve, a seized piston, a choked petrol-pipe. 1925 *Morris Owner's Manual* ii. 33 Clean petrol filter and petrol pipe. 1963 R. F. Webb *Motorists' Dict.* 170 Petrol pipe, the flexible or copper tube carrying petrol from the petrol tank to the petrol pump or carburettor. 1908 *Westm. Gaz.* 14 Apr. 4/2 Both the Wolseley and Thornycroft motors are masterpieces of engineering skill and ingenuity, and are taking us along as fast as applied science will admit towards the petrol-propelled battleship. 1922 Joyce *Ulysses* 703 A scheme for the development of Irish tourist traffic in and around Dublin by means of petrolpropelled riverboats. 1928 *Manch. Guardian Weekly* 19 Oct. 301/2 A hideous outbreak of advertisement hoardings, petrol pumps..and gaunt new motor-tracks. 1932 J. Buchan *Gap in Curtain* ii. 96 They had to make a forced landing..on the skirts of Ruwenzori, where they found that something had gone wrong with the petrol pump and that some of the propeller and cylinder bolts had worked loose. *Ibid.* iv. 210 We want to get past the garages and petrol pumps and county council cottages to the ancient rustic England. 1963 R. F. Webb *Motorists' Dict.* 170 When the float chamber of the carburettor is filled, the petrol pump encounters pressure and this opens a valve which stops further supplies of fuel being pumped up to the carburettor until the level of the reserve has dropped slightly. 1970 'D. Halliday' *Dolly & Cookie Bird* iv. 48 A black and white petrol-pump sign. 1976 'W. Trevor' *Children of Dynmouth* i. 16 His sister was..employed as a petrol-pump attendant on the forecourt of the Smiling Service Filling Station. 1939 *Times* 14 Oct. 5/3 Owners of vehicles and industrial plant who have been issued with only one month's additional petrol ration from the divisional petroleum office, should now apply for a further allowance for the second month if this is needed for essential purposes. 1948 G. V. Galwey *Lift & Drop* v. 107 He's very scrupulous about his petrol ration being for professional purposes only. 1973 *Country Life* 6 Dec. 1892/1 Last week..motorists were beginning to queue for their petrol-ration books. 1948 M. Laski *Tony Heaven* iv. 59, I take it..that there's no petrol rationing for A's? 1973 *Times* 4 Dec. 10/7 One or two people involved in industry said they were not worried about petrol rationing..they would..buy one or two more cars, draw the coupons, and sell the cars when rationing ended. 1958 W. Sansom *Cautious Heart* 53 Do you know what a petrol-sniffer is?.. It's quite serious really, it's an addiction to the smell of petrol—like any other narcotic. People who work among petrol fumes get the habit. 1973 *Black World* June 47/2 The..escapism..is 'petrol-sniffing', which produces a coma-like state. 1926 *Times* 6 May 1/4 (Advt.), Wanted, hard, ambitious, and independent worker with £750, to erect and operate new petrol station under exceptionally favourable supply contract. 1954 *Coast to Coast* 1953–54 145 A petrol-station at the far end. 1969 *Listener* 15 May 669/3, I suspect that Penguins and other paperbacks wouldn't sell at all well at petrol stations. 1972 'J. & E. Bonett' *No Time to Kill* v. 52 He was approaching a petrol station... He drew up beside a row of pumps. 1902 Kipling *Traffics & Discov.* (1904) 189 At Pigginfold, after ten minutes, we refilled our petrol tank and lavishly oiled our engines. 1973 J. Leasor *Host of Extras* viii. 142 Drugs in spare tyres, dummy petrol tanks..windscreen washer tanks. 1974 *Country Life* 21 Nov. 1579/2 The petrol-tank capacity of 9·9 gallons was small for such a car. 1975 J. Rathbone *Kill Cure* i. i. 7 His eyes returned to the road and the back of the large petrol tanker. 1908 *Autocar Handbk.* (ed. 2) xix. 148 The petrol tap having been turned on, the float may be agitated so as to flood the carburetter. 1925 *Morris Owner's Manual* i. 9 The petrol tap is at the bottom of the tank, under the dashboard. 1916 J. Buchan *Greenmantle* xvi. 217, I got out the petrol tins and spare tyres and cached them among some rocks on the hillside. 1940 'Gun Buster' *Return via Dunkirk* i. x. 76 First name out of the petrol tin was Turner's. 1966 M. Woodhouse *Tree Frog* vi. 49 Petrol vapour clawed its way harshly into my lungs. 1974 J. Dyson *P.M.'s Boat is Missing* xl. 209 He could smell the strong petrol vapour.

4. = *petrol blue.

1927 *Daily Express* 22 Feb. 6 (Advt.), White, Phlox, Cardinal, Red, Black, Brown, Navy, Bois de Rose, Lavender, Petrol, New Blue, Grey. 1971 *Vogue* 15 Sept. 129/1 Coat..sizes 10–16; colours: black, purple, green, brown, petrol. 1971 *Guardian* 18 Sept. 11/1 Dress and coat..Brown..Petrol, Grotto or Tomato.

Hence **pe·trolless** *a.*, having no petrol.

1908 *Daily Chron.* 31 July 4/3 The foolish virgins, with their oilless lamps, have found a latter-day analogue in the petrolless taxi-cabby. 1944 Dylan Thomas *Let.* 27 July (1966) 265 The well-off people were dry and thin and grieved over their petrolless motorcars.

petrolatum. Add to def.: = *petroleum jelly. (Further examples.)

1959 *Industr. Wood Processes* VII. 15 (*heading*) The effect of petrolatum on the strength of glued joints. 1972 *Materials & Technol.* V. xiv. 520 Other plasticizers derived from petroleum are: petrolatum, paraffin waxes, ..and asphalts. 1976 *Lebende Sprachen* XXI. 151/2 Lubricate sparingly with petrolatum/white petroleum before destructive testing.

petrol bomb. [f. Petrol + *bomb *sb.* 2.] A bomb, usu. home-made and thrown by hand, consisting of a petrol-filled bottle and a wick; a Molotov cocktail.

1958 [see *flick-knife* s.v. *flick *sb.*¹ 4]. 1963 *Times* 20 Feb. 11/3 It will be imposed where petrol bombs or inflammatory liquid or explosives are used against persons. 1969 *Guardian* 13 Aug. 1/1 The rioters threw petrol bombs. 1971 *Peace News* 10 Sept. 7/1 The effects of petrol bombs, CS gas and searchings were tragically obvious in the children. 1976 P. Ferris *Detective* ii. 24 A petrol bomb was thrown.

Hence **pe·trol-bomb** *v. trans.*, to throw a petrol bomb at; to destroy or damage with a petrol bomb; **pe·trol bomber**, one who throws a petrol bomb; **pe·trol-bombing** *vbl. sb.*, throwing of a petrol bomb.

1963 *Times* 23 Feb. 7/4 Sir Edgar Whitehead, said that unless the Government dropped the provision making the death sentence for petrol bombing compulsory, the Opposition 'with all the constitutional means at our disposal will fight this Bill clause by clause to the bitter end'. 1969 *Guardian* 16 Aug. 9/1 The Catholics who were petrol-bombed out of their homes. 1971 *Daily Tel.* 4 Jan. 7/6 Other odd areas [of Paris] that are patrolled include the Chad Embassy (petrol-bombed a year ago). 1974 J. Mitchell *Death & Bright Water* vii. 60 All he had to worry about now was the petrol bombers he'd seen. 1976 *Times* 23 Dec. 10/5 The dockyard boys had petrol-bombed the rent office.

petroleum. Add: Also in extended use (see quots.).

1938 A. W. Nash in A. E. Dunstan et al. *Sci. of Petroleum* I. i. 12/1 Petroleum may contain, or be composed of, ..compounds in the gaseous, liquid, and/or solid state, depending on the nature of these compounds and the existent conditions of temperature and pressure. 1960 J. W. Amyx et al. *Petroleum Reservoir Engin.* i. 1 Virtually all petroleum is produced from the earth in either liquid or gaseous form. 1960 C. Gatlin *Petroleum Engin.* i. 1/1 Petroleum may be defined as a naturally occurring mixture of hydrocarbons which may be either gas, liquid, or solid. *Ibid.* 4/2 (*heading*) Gaseous petroleum (natural gas). 1967 J. R. Hughes *Storage & Handling Petroleum Liquids* 3 Petroleum, by legal definition (Petroleum (Consolidation) Act, 1928) 'includes crude petroleum, oil made from petroleum, or from coal, shale, peat or other bituminous substances, and other products of petroleum'. .. The term may also include natural gas found in petroleum-bearing formations.

b. **petroleum-bearing** adj.; **petroleum coke**, the solid, non-volatile residue left after the distillation and cracking of petroleum; **petroleum geology**, the branch of geology concerned with the search for oil and natural gas and with their formation, occurrence, and exploitation; so **petroleum geologist**; **petroleum jelly**, a soft, greasy, semi-solid mixture of hydrocarbons obtained from petroleum, used as an ointment and lubricant; cf. Petrolatum, Vaseline.

1882 E. O'Donovan *Merv Oasis* I. ii. 37 Proprietors of large tracts of petroleum-bearing ground. 1963 D. W. & E. E. Humphries tr. *Termier's Erosion & Sedimentation* xi. 231 The evolution of petroleum-bearing sediments is brought to its final stages by diagenetic processes. 1881 *Jrnl. Chem. Soc.* XL. 239 (*heading*) Products from petroleum coke. 1971 *Materials & Technol.* II. x. 638 Since 1920 petroleum coke has become a plentiful by-product of petroleum cracking. *Ibid.* xi. 655 The high graphitizability of the petroleum coke results from the formation of large polycyclic aromatic molecules, which assists carbon crystallite alignment during the coking operation. 1917 *Bull. Geol. Soc. Amer.* XXVIII. 157 Petroleum geology is..a new profession. *Ibid.*, (*heading*) Ethics of the petroleum geologist. 1928 E. R. Lilley *Geol. Petroleum & Nat. Gas* i. 8 The main work of the petroleum geologist until recently was that of locating areas in which anticlinal folds showed in the outcropping strata. 1973 R. E. Chapman *Petroleum Geol.* ii. 22 A study of petroleum geology suggests that the construction of boreholes in the 20th century has not contributed sufficiently to modern geological thought, and that some of the difficulties..are due to the extension of the concepts developed from surface geology to the subsurface. *Ibid.* 27 The chemistry of petroleum..is low on the petroleum geologists' list of priorities. 1897 *Sears, Roebuck Catal.* 27/2 Petroleum Jelly. This is another name for the preparations called Vaseline, Cosmoline, etc., all made from Petroleum. 1906 T. E. Herbert *Telegraphy* xxii. 782 As the cable passes into the pipe it is heavily anointed with petroleum jelly. 1922 D. T. Day *Handbk. Petroleum Industry* II. 360 The official petrolatum of the United States Pharmacopœia, otherwise known as petroleum jelly, or 'Vaseline'. 1966 J. S. Cox *Illustr. Dict. Hairdressing* 112/1 Petroleum jelly, (Petrolatum, Vaseline, Paraffin Jelly), a purified mixture of semi-solid hydrocarbons, chiefly of the Methane series.

petrolization (pe:trŏləizēⁱ·ʃən). [f. Petroliz(e *v.* + -ation.] The oiling of water in order to kill mosquito larvæ.

1901 L. O. Howard *Mosquitoes* 193 The petrolization of mosquito-breeding pools is one of the most important measures to be taken in the warfare against mosquitoes. 1930 [see next].

petrolize, *v.* Add: **3.** = *oil *v.* 1 c. So **pe·trolizing** *vbl. sb.*

1901 L. O. Howard *Mosquitoes* 193 To the Italians we are indebted for a useful expression, which we might just as well adopt, namely to 'petrolize', meaning to treat waters with kerosene. 1903 *Boston Even. Transcript* 28 Apr. 9/2 All the breeding places [of the mosquito] treated last year will again be petrolized this year. 1906 *Westm. Gaz.* 21 Apr. 16/1 Marshes are drained; ponds are petrolised or stocked with fish. 1930 I. J. Kligler *Epidemiology & Control of Malaria in Palestine* vii. 144 Petrolization and petrolizing mixtures.—One of the most important and useful methods of antilarval control is the oiling of the surface of breeding-places.

petrology. Add: **b.** The petrological features of something or somewhere.

1880 [in Dict.]. 1954 *Geol. Mag.* XCI. 44 (*heading*) Further data on the petrology of the pelitic hornfelses of the Carn Chuinneag-Inchbae region, Ross-shire. 1975 *Sci. Amer.* Jan. 29/1 The mineralogy, the content of trace elements and gas, the petrology and so on of literally hundreds of different meteorites have been measured.

petrologic *a.* (examples); chiefly *U.S.*

1936 H. L. Alling *Interpretative Petrol. Igneous Rocks* p. vii, The portions of the book devoted to petrologic history. 1972 *Science* 3 Nov. 497/1 Approximately one-third of the book is devoted to mineralogic and petrologic attributes of sandstones.

Petronella (petrǒne·lă). Also with lower-case initial. [? the feminine name *Petronella*.] A Scottish country dance.

1897 *Ball Room* Nov. 11/1 The country dances of twenty-five or thirty years ago were being quite forgotten. 'Sir Roger', 'Triumph', 'Pettonella' [*sic*], the 'Saraband', [etc.]. 1905 F. H. Norman *Compl. Dance Instructor* 50 Petronella. 1907 N. Munro *Daft Days* xxx. 251 Our dances at the inn are not like city routs: Petronella, La Tempête, and the reel have still an honoured place in them. 1924 *Glasgow Herald* 1 Sept. 8 Despite the grumbles of old-timers who lament the passing of the schottisch, the lancers, and the polka—not to speak of petronella and 'La Va'—the dancing world is eminently conservative. 1935 N. Mitchison *We have been Warned* i. i. 17 You and I danced Strip the Willow and Petronella. 1978 A. Dunnett *No Thanks to Duke* xii. 175 We danced the whole night..Petronella, Strip the Willow,..Foursome Reel.

petrophysics (petrofi·ziks), *sb. pl.* (const. as *sing.*). *Geol.* [f. Petro- + Physics.] The study of the physical properties and behaviour of rocks.

1950 G. E. Archie in *Bull. Amer. Assoc. Petroleum Geologists* XXXIV. 942 'Petrophysics' is suggested as the term pertaining to the physics of particular rock types, whereas geophysics pertains to the physics of larger rock systems composing the earth. The petrophysics of reservoir rocks is discussed here. 1959 *Times* 28 July 2/6 (Advt.), They [sc. applicants] should also be fully conversant with modern petrophysics methods. 1976 *Nature* 19 Feb. 607/2 Esoteric accounts of palaeotemperature analysis, coal metamorphism, petrography of volcanic rocks, petrophysics of chalk, and intimate analyses of stratigraphic boundaries.

So **petrophy·sical** *a.*; **petrophy·sicist**, a specialist or expert in petrophysics.

1950 G. E. Archie in *Bull. Amer. Assoc. Petroleum Geologists* XXXIV. 961 A tentative petrophysical system ..is presented. 1956 *Geol. Mag.* XCIII. 112 The petrophysical characteristics of sedimentary and igneous rocks are important..in..deformation. 1972 *Bull. Amer. Assoc. Petroleum Geologists* LVI. 1906/2 Petrophysical analysis utilizing digital-log information has recently become a recognized tool in the search for oil and gas. 1975 G. Anderson *Coring* vii. 128 Most discussions on wireline logs define shale as a nonporous rock as compared to porous reservoir rocks, a point worth remembering in geological discussions with petrophysicists and the like.

petrosal, *a.* Add: Also applied to some branches derived from the facial and glossopharyngeal nerves that pass through the petrosal bone, and to the inferior ganglion of the glossopharyngeal nerve, situated in a notch in this bone. (Further examples.)

1881 A. L. Ranney *Appl. Anat. Nervous Syst.* ii. 179 (*caption*) A diagram of the branches of the facial nerve... 3, orifice of aqueduct of Fallopius; 4, large petrosal nerve; 5, small petrosal nerve; 6, external petrosal nerve. 1888 W. R. Gowers *Man. Dis. Nervous Syst.* II. 250 The tympanic nerve of Jacobson (arising from the enlargement on the glosso-pharyngeal, termed the 'petrosal ganglion'), forms, with the sympathetic, the tympanic plexus in the wall of the middle ear. 1934 L. B. Arey *Developmental Anat.* (ed. 3) xvi. 438 The sensory fibres of the glossopharyngeal nerve arise from two ganglia, the superior ganglion closer to the brain and the petrosal ganglion farther peripherad on the trunk. 1972 M. L. Barr *Human Nervous Syst.* viii. 128/2 The sensory fibres in question [sc. those for taste] leave the facial nerve in the greater petrosal branch at the level of the geniculate ganglion.

petrotectonics (pe:trotektǫ·niks), *sb. pl.* (const. as *sing.*). *Geol.* [f. Petro- + Tectonics, repr. G. *petrographische tektonische analyse.*] The study of the structure of rocks, esp. as a guide to the movements to which they have been subjected. Cf. *petrofabrics *sb. pl.*

1933 *Amer. Jrnl. Sci.* XXV. 433 (*heading*) Petrotectonics. *Ibid.* 439 Sander uses the word petrotectonics to express the branch of the investigation that deals with the relation of the oriented microscopic fabric of deformed rocks to their tectonic history. 1937 *Q. Jrnl. Geol. Soc.* XCIII. 581 The science of *Gefügekunde*, variously translated as petrofabrics, petrotectonics, or structural petrology, is understood to comprise a study of all the spatial data, macroscopic and microscopic, which go to form a complete vector-picture so far as this is legible in the make-up of the rock. 1974 *Nature* 10 May 107/3 Evidence supporting the composite nature of Asia has been adduced from palaeomagnetism, palaeontology, geology and petrotectonics.

So **pe:trotecto·nic** *a.*

1933 *Amer. Jrnl. Sci.* XXV. 470 The ultimate test..will be the study by petrotectonic methods of material that has been deformed under known stress conditions. 1970 *Internat. Geol. Rev.* XII. 361/2 The internal petrotectonic structure of the intrusive body, which was established by systematic study of the orientations of the planar and linear parallelism, primary jointing, chromite and durite segregations and so forth, is folded and..corresponds to the folding of the host rocks.

pe-tsai (pe,tsai·). Also **Pe-Tsai**. [Older transliteration of Chinese *báicài*, f. *bái* white + *cài* vegetable; cf. *pak-choi.] A Chinese species of cabbage, *Brassica pekinensis*. Also *attrib.*

1795 W. Winterbotham *Hist., Geogr., & Philos. View Chinese Empire* v. 221 The Chinese make provision of *pe-tsai* for winter; pickling of it, and mixing it with their rice. 1845 *Encycl. Metrop.* XVI. 588/1 The *Pe-tsai*, or white herb, a kind of mustard,..is cultivated in large fields, and eaten either fresh or pickled, like the German *Sauer-kraut*. 1885 [see *Chinese cabbage* s.v. *Chinese *a.* 2]. 1900 L. H. Bailey *Cycl. Amer. Hort.* I. 178/2 Pe-tsai cabbage. *Ibid.*, The Pe-tsai, or Chinese cabbage, is no longer a novelty in American gardens, though it does not appear to be well known, and its merits are not understood. 1960 *Farmer & Stockbreeder* 8 Mar. (Suppl.) 11/2 The celery-cabbage (Pe-Tsai) is worth trying, too. The heads are crisp and tender. 1969 *Oxf. Bk. Food Plants* 154/1 Pe-Tsai (*Brassica pekinensis*) is grown mostly for use as an autumn and winter vegetable... Pe-tsai has soft green, prominently veined leaves... The rather loose, elongated head bears some resemblance to a cos lettuce. 1972 *Homes & Gardens* Aug. 104/1 Chinese cabbage, or pe-tsai, is a new vegetable to this country. It comes from Israel and looks like a cross between a very pale whitish-green cos lettuce and a head of celery, although it is larger and considerably heavier than both. 1972 *Country Life* 14 Dec. 1676/3 Last autumn's drought delayed the growth of Chinese Pe-Tsai cabbage, sown in August, so..they are being cooked now.

petsywetsy (pe:tsiwe·tsi). *Nonce-wd.* [f. Pet *sb.*¹: see -sy.] A fanciful extension of Pet *sb.*¹ 2.

1928 D. H. Lawrence *Lady Chatterley* xvii. 305 I'm not content to be any man's little petsywetsy.

pettable (pe·tăb'l), *a.* [f. Pet *v.*¹ + -able.] Suitable for petting. Hence **pettabi·lity**.

1934 Webster, Pettable. 1962 *Times* 17 Sept. 15/3 There is scarcely a 'pettable' creature that has not been adopted by my family. 1963 F. E. Sparshott *Structure of Aesthetics* iii. 74 'Handsome'..is applied to things whose appearance stirs the imagination—but by suggestions of vigour rather than of pettability. 1978 *Chicago* June 50/2 In the large Children's Zoo, youngsters can walk into a petting area and be surrounded by pettable domestic animals.

petter, *sb.* Add: **2.** One who pets (*pet *v.* b); one who engages in petting (*petting *vbl. sb.* 2).

1925 *College Humor* Aug. 77/1 Have a nice evening? Jean's some high-type petter, isn't she? 1930 F. Dell *Love in Machine Age* ix. 170 Science's adjuration to the petters is only: 'You really must get some food and sleep, my dears!' 1931 F. L. Allen *Only Yesterday* v. 101 The vendors of another picture promised 'neckers, petters, white kisses, red kisses, pleasure-mad daughters, sensation-craving mothers'. 1931 *Brit. Jrnl. Psychol.* Oct. 183 One finds oneself extremely dubious about this contention that the modern adolescent 'petter' and 'demi-vierge' have discovered for themselves a satisfactory way of acquiring the adequate pre-marital emotional education. 1942 Berrey & Van den Bark *Amer. Thes. Slang* § 443/13 Heavy necker or petter.

petti- (pe·ti). Combining form of Petti(coat *sb.*, designating garments having some of the characteristics or functions of a petticoat.

1922 Joyce *Ulysses* 360 He could see her other things too, nainsook knickers, the fabric that caresses the skin, better than those other pettiwidth, the green, four and eleven, on account of being white. *Ibid.* 544 Lifting up her pettigown and folding a half sovereign into the top of her stocking. 1923 *Dialect Notes* V. 246 *Pettibockers*, n. pl. A loose garment for girls, worn under skirts: contamination of *petticoat* and *knickerbockers*. 1939 Joyce *Finnegans Wake* iv. 611 His yellow saffron pettikilt. 1942 Berrey & Van den Bark *Amer. Thes. Slang* § 87/27 *Pettibockers*, women's old-fashioned knickerbockers. 1960 *Harper's Bazaar* Apr. 11 Nightdresses, slips, pettiskirts, panties. 1963 *Seventeen* Aug. 32 A new word in lingerie. Pettisets (a petti-top matched with petti-pants as well as a petticoat). 1964 *N.Y. Post* 7 Oct. 13 Perfect under your swinging skirts, pettipants fashioned with a divided skirt. 1968 J. Ironside *Fashion Alphabet* 71 *Pettipants*, petticoat and pants combined. 1970 N. Armstrong et al. *First on Moon* vi. 126 They found a present for Jan's friend—pettipants edged with green lace. 1972 *Daily Tel.* (Colour Suppl.) 13 Oct. 16/1 On top of this I wore a petti-blouse (which is a blouse and a petticoat).

Petticoat Lane. A popular name given to Middlesex Street (formerly Hog Lane) in the City of London, where dealers in second-hand clothes and other commodities congregate. Also *attrib.*

1602 in *Calendar of Cecil MSS.* 168 in *Parl. Papers* 1910 (Cd. 5291) XXXIV. 1, I understand by Udall that Eccleston is a gentleman dwelling near Knowsely, but has left his habitation, and abides altogether in London, in Peticote Lane. 1631 B. Jonson *Devil an Ass* i. i. 60 We will suruay the Suburbs, and make forth our sallyes, Downe Petticoate-lane, and vp the Smock-allies, To Shoreditch, Whitechappell, and so to Saint Katharnes. 1909 W. W. Hutchings *London Town* II. xc. 967/1 This street formerly Petticoat Lane, and in yet earlier days Hog Lane, has of late years been rebuilt, but it is still largely inhabited by dealers in second-hand clothes, and still on Sundays it is thronged by those who come to buy other people's cast-off garments. 1967 M. Waddell *Otley Pursued* xii. 109 It was 6.42 by my Petticoat Lane watch, give or take ten minutes or so. 1972 *Police Rev.* 17 Nov. 1495/3 The man who had been given the licence had described the street..as 'Petticoat Lane'. Section 21(1) required that the particular street must be named, and Petticoat Lane has become Middlesex Street. 1976 *Oadby & Wigston Advertiser* 26 Nov. 10/1 Glen Parva Borstal was like a 'small Petticoat Lane' with items being offered for sale all the time, Rutland magistrates heard on Monday.

petting, *vbl. sb.* (in Dict. s.v. Pet *v.*¹) Add: **2.** In the sense of *pet *v.*¹ b: the action of amatory caressing and fondling; non-coital sexual activity. Also *attrib.* See also *heavy petting* s.v. *heavy *a.*¹ 13.

1920 [see sense 3 below]. 1922 S. Lewis *Babbitt* xviii. 228 Babbitt had heard stories of what the Athletic Club called 'goings-on' at young parties; of girls 'parking' their corsets in the dressing-room, of 'cuddling' and 'petting', and a presumable increase in what was known as Immorality. 1928 J. P. McEvoy *Show Girl* 11 They have no time to perfect their petting technique. 1929 B. Russell *Marriage & Morals* xii. 126 There is so much 'petting' and 'necking' that the absence of complete intercourse can only be viewed as a perversion. 1933 Auden *Dance of Death* 25 Gay girl to whom petting Matters so much Poor kid, the reason's You're out of touch With flowers and such. 1947 *Partisan Rev.* XIV. 273 When the couple finally reached the stage of mild petting in the living room, Ruth in the kitchen listened with a mixture of pleasure and distaste. 1948 A. C. Kinsey et al. *Sexual Behavior in Human Male* xvi. 531 This behavior is known to the younger generation as petting, although certain other terms are applied to certain types of contacts. Those which are confined to latitudes not lower than the neck are sometimes known as necking, and petting is distinguished from the heavy petting which involves a deliberate stimulation of the female breast, or of the male or female genitalia. *Ibid.*, In the present volume the term 'petting' has been applied to any sort of physical contact which does not involve a union of genitalia but in which there is a deliberate attempt to effect erotic arousal. 1956 S. Ertz *Charmed Circle* 62 Indulging in a pastime that was highly popular with the young... They had been 'necking' or 'petting'. 1958 M. Argyle *Relig. Behaviour* x. 123 'Petting to climax' was as common among Kinsey's devout men as for the non-religious. A smaller proportion of Kinsey's devout female sample reported this, but Chesser's devout women reported as much 'petting' as the others. 1960, etc. [see *heavy *a.*¹ 13]. 1971 A. F. Guttmacher *Understanding Sex* iv. 52 There are many degrees of petting, from holding each other in your arms to mutually manipulating the genitals..without actually having sexual intercourse. 1976 *N.Y. Rev. Bks.* 13 May 20/4 Since the time of Kinsey there have been some important changes:..a decline in petting..and an increase in copulation.

3. *Comb.*, as **petting party**, a social gathering at which petting is the principal activity.

1920 F. Scott Fitzgerald *This Side of Paradise* i. ii. 64 That great current American phenomenon, the 'petting party'. 1926 A. Huxley *Jesting Pilate* iv. 265 Hymns and the movies and Irving Berlin. Petting Parties and the First Free United Episcopal Methodist Church. Jazz it up! 1934 R. Chandler in *Black Mask* Oct. 28/1 The prowl car takes a slant down it [sc. the old road] now and then looking for petting parties. 1946 [see *half-virgin* s.v. *half- II n]. 1973 A. MacVicar *Painted Doll Affair* viii. 91 Drinking parties, drug parties, petting parties. Mostly youngsters.

petto. Add: *in petto*, ¶(b) by confusion with Petty *a.* (*sb.*¹): in short, in miniature, on a small scale.

1838 Emerson *Jrnl.* 19 Sept. (1911) V. 53 Such ostentation *in petto* I never did see. 1844 Disraeli *Coningsby* I. ii. ii. 185 He commenced a discourse, which was in fact, one of his 'slashing' articles *in petto* on Church Reform. a 1846 B. Haydon *Autobiogr.* (1853) I. 247 Away we marched, my little guide wearing a red night-cap..till we came in sight of the height, when my grenadier *in petto* turned round. 1894 W. S. Hadley *Euripides' Hecuba* p. x, The effect thus produced within the compass of a single play is comparable to that brought about on a larger scale by the older trilogy... This trilogy *in petto* may be regarded as an experiment made by the youngest of the three great tragedians. 1901 Kipling *Kim* xiii. 341 He represents *in petto* India in transition—the monstrous hybridism of East and West. 1924 T. E. Lawrence *Let.* 1 Jan. (1938) 450 Yes, it will be revised, but only *in petto*. No good cuts or noble changes, no re-writing: just punctuation and insect-blemishes removed. 1937 L. Bromfield *Rains Came* ii. ii. 258 Mr. Bannerjee's house and garden were a kind of India *in petto*, overcrowded, confused, swarming with noisy life. 1979 *Country Life* 29 Mar. 947/4 In small (or, as people now cheerfully and wrongly say, *in petto*).

petty, *a.* (*sb.*¹) Add: **A.** *adj.* **5.** **petty apartheid**, apartheid as exercised in everyday life; racial segregation in its trivial applications (see *apartheid); **petty cash** (further example); also *attrib.*; **petty cash book** (examples).

1966 *Cape Argus* 8 Apr. 14 While separate development proceeds so slowly, the Government, for home consumption, makes demonstrations of strength on the petty apartheid front. 1971 *Rand Daily Mail* 10 June 12 You can just imagine what it feels like when you have to use a lift which is marked 'Goods only'; when you have to wait for the train in a cage-like sort of place, packed like sardines; when you have to buy food at a snack bar only through a window at the back. I think all these are what a sound mind can classify as petty apartheid. 1974 *Black Panther* 9 Feb. 15/4 Last week the Johannesburg City Council announced measures to eliminate what is called here 'petty apartheid'. These are practices of discrimination against the city's Blacks and coloreds not imposed by national law. 1977 *Time* 27 June 20/1 There have been some changes in 'petty apartheid'. Whites boast that 'international' hotels have been opened to blacks, and that blacks now participate in white sports, which has great symbolic meaning. 1839 Dickens *Let.* c 18 Feb. (1965) I. 509 He was not quite correct in the facts of the 'petty cash' story, the realities of which are far more honorable to the noble fellows concerned. 1886 A. W. Thomson *Text-bk. Princ. & Pract. Bk.-Keeping* iii. 37 When a number of small payments need to be made..a separate book called the Petty Cash Book is kept for recording such payments. 1922 Joyce *Ulysses* 171 He went towards the window and, taking up the petty cash book, scanned its pages. 1974 *Terminol. Managem. & Financial Accountancy* (Inst. Cost and Managem. Accountants) 57 *Petty cash book*, a separate record of small cash receipts and payments.

B. *sb.*¹ **1.** Restrict † *Obs.* to sense in Dict. and add: **b.** A school or class for small boys.

c 1810 W. Hickey *Mem.* (1913) I. ii. 13, I however soon got out of that disgraceful and ignorant form, passed with rapidity and *eclat* the under and upper petty, and entered into the upper first. 1961 R. Williams *Long Revolution* 133 The 'petties' or 'ABCs' were proper schools.

2. (Examples.)

1848 M. J. Stanley *Let.* 27 Sept. in N. Mitford *Ladies of Alderley* (1938) 206 If these houses had been built by his Lordship every one would have had his petty, at all events dividing the odour. 1961 E. Williams *George* i. vii. 87 It was a nice little cottage.., with a lot of neglected garden and a tumbledown petty. *Ibid.* ii. xi. 151 Mam..hated the petty for being attached to next door's.

petty (pe·ti), *sb.*² Familiar abbrev. of Petticoat *sb.*

1915 T. Burke *Nights in Town* 63 There..was young Beryl, superintending her aunt's feverish struggles with paint and powder-jars, frocks, petties,..and wraps. 1939 Joyce *Finnegans Wake* i. 87 The litigants..were egged on by their supporters in the shape of betterwomen with bowstrung hair of Carrothagenuine ruddiness, waving crimson petties and screaming from Isod's towertop. 1971 *Guardian* 24 Aug. 9/1 The language of lingeries..petties and pretties, and frillies.

petty bou·rgeois. [f. Petty *a.* (*sb.*¹) as anglicization of Fr. *petit* + Bourgeois *sb.*¹ and *a.*] = *petit bourgeois. Also (occas. with hyphen) *attrib.* or as *adj.*

1888 S. Moore tr. *Marx & Engels's Manifesto of Communist Party* 24 In countries where modern civilization has become fully developed, a new class of petty bourgeois has been formed. *Ibid.* 26 German Socialism recognised..its own calling as the bombastic representative of the petty bourgeois socialism. 1904 in R. C. K. Ensor *Mod. Socialism* xx. 289, I reckon, the life of the petty bourgeois, ..is an unhappy one. 1906 E. M. Aveling tr. *Plechanoff's Anarchism & Socialism* iii. 26 His [sc. Max Stirner's] 'League of Egoists' is only the Utopia of a petty bourgeois in revolt. *Ibid.* iv. 35 Proudhon was the most typical representative of petty bourgeois socialism. 1931 *Times Lit. Suppl.* 26 Feb. 141/3 It is for its intellectual critics, whom it first degrades by the question-begging epithet 'petty-bourgeois', that Bolshevism reserves its cruellest penalties. 1936 'M. Innes' *Death at President's Lodging* vii. 133 Our Inspector's petty-bourgeois passion for the till. 1943 H. Read *Politics of Unpolitical* 15 According to Dr. Fromm, Hitler succeeded so well because he was able to combine the qualities of a resentful petty-bourgeois.. with those of an opportunist ready to serve the interests of the German industrialists and Junkers. 1951 C. W. Mills *White Collar* i. ii. 30 It is..a feature of such petty-bourgeois life that extreme repression is often exercised in its patriarchal orbit. 1954 Koestler *Invis. Writing* 25 Revulsion against this code was a sign of sentimental petit-bourgeois prejudice. 1958 I. Murdoch in N. Mackenzie *Conviction* 230 This mass is now quiescent, its manner of life largely suburban and its outlook 'petty bourgeois'. 1960 D. Lessing *In Pursuit of English* i. 13 These are not the real working-class. They are the lumpen-proletariat, tainted by petty-bourgeois ideology. 1974 J. White tr. *Poulantzas's Fascism & Dictatorship* iii. ii. 111 The Nazi Party was filling the whole State apparatus..with members of petty-bourgeois origin and with their own quite specific petty-bourgeois ideology. 1977 *Time* 24 Jan. 14/3 Italian Communist Party..spokesmen have denounced the leftists as 'petty bourgeois' hoodlums. 1978 *Encounter* Feb. 36/2 The 'people' is what the petty bourgeois of the spirit transform all men and women into.

pe·tty bourgeoisie·. [f. as prec. + Bourgeoisie.] = *petit bourgeoisie.

1850 *Red Republican* 16 Nov. 171/1 The petty Bourgeoisie, the inferior ranks of the middle-class, the small manufacturers, merchants, tradesmen, and farmers, tend to become Proletarians. 1920 F. E. Green *Hist. Eng. Agric. Labourer 1870–1920* v. 125 The farmers and the petty bourgeoisie took possession of the Parish and the Rural District Councils. 1924 G. B. Shaw *Saint Joan* Pref. p. xiii, In short, much more of a young lady, and even of an intellectual, than most of the daughters of our petty bourgeoisie. 1930 tr. *Marx & Engels's Communist Manifesto* iii. 56 In the countries where modern civilization flourishes, a new petty bourgeoisie has come into being. This class hovers between the proletariat and the bourgeoisie. 1956 C. W. Mills *Power Elite* xiv. 329 Classic conservatism has required the spell of tradition among such surviving elements of pre-industrial societies as an aristocracy of noble men, a peasantry, a petty-bourgeoisie with guild inheritances. 1957 R. N. C. Hunt *Guide to Communist Jargon* 118 Generally speaking, Marx and Engels understood by the petty-bourgeoisie the small traders who, they contended, were being driven down into the proletariat by the monopolist capitalists. 1969 A. G. Frank *Latin Amer.* xix. 312 The next layer may be termed the middle class or petty bourgeoisie. It comprises a large variety of economic walks of life—small landowner, professional, merchant, clergy, government and white collar worker, small politician—but it affords considerable lateral mobility within it, from one occupation to another. 1975 *Times Lit. Suppl.* 14 Mar. 270/4 In Havana in 1827,..a higher proportion of coloureds were legitimately married than of whites. This black 'petty bourgeoisie' was, however, dispersed..in the repression of 1844 and the terms of the racial stratification system were redefined to the advantage of the whites.

petty-mi·nded, *a.* [Minded *ppl. a.* III.] Having or characteristic of a mind that dwells on the trivial and ignores what is important. So **petty-mi·ndedness**.

1909 *Daily Chron.* 10 June 7/4 Weakness and petty-mindedness were fostered by the narrow sphere and limited outlook that..such tasks necessitated. 1927 V. Woolf in *Forum* (N.Y.) May 707 It was all so paltry, weak-blooded, and petty-minded to care so much at her age. 1954 T. S. Eliot *Confid. Clerk* i. 25 He's not petty-minded—though nothing escapes him. 1963 W. Soyinka *Dance of Forests* 57 Is the nation to ignore the challenge of greatness because of the petty-mindedness of a few cowards and traitors. 1978 *Times* 24 Jan. 14/5 This..petty-minded change of policy.

petulate (pe·tiulēⁱt), *v. rare*⁻¹. [f. Petul(ant *a.* + -ate³.] *trans.* To make petulant or peevish. In quot. as *ppl. adj.*

1897 [see *hypersensitize *v.*].

Peulh (pɔ̄l), *sb.* and *a.* Also 8 Pholey, 9 Pul(ah), Pûl(l)o, 9- Peul, Pul(l)o; Peuhl. [Native name; prop. the sing. of *Fulah.] = *Fulah *sb.* and *a.*

[1611 Bible *Isaiah* lxvi. 19, I will send those that escape of them unto the nations, to Tarshish, Pul and Lud.] 1799 M. Park *Trav. Afr.* (ed. 2) ii. 17 The Foulahs (or Pholeys)..are chiefly of a tawny complexion, with soft silky hair, and pleasing features. 1858 H. Barth *Trav. N. & Cent. Afr.* II. xxxiv. 414 A Mohammedan kingdom engrafted upon a mixed stock of pagan tribes,—the conquest of the valorous and fanatic Púllo chieftain, A'dama, over the great pagan kingdom of Fúmbiná. *Ibid.* IV. lvii. 143 If any African tribe deserves the full attention of the learned European, it is that of the Fúlbe (sing. Púllo), or Fula, as they are called by the Mandingoes. 1876 G. A. L. Reichardt *Gram. Fulde Lang.* p. xiii, (*heading*) Introduction to a grammar of the Pul language... The proper and indigenous name for the Fulahs, as we have called them, is Pulo. *Ibid.* p. xiv, we find a strong Pulo empire in a northwesterly direction, from the upper course of the Jaliba. 1883 R. N. Cust *Mod. Lang. Afr.* I. x. 158 This language..is known by many names, Fula,..Pûlo, Pulah,...; its meaning is said to be 'light brown', as opposed to that of the neighbouring people, Wolof or 'black'. 1898 A. J. Butler tr. *Ratzel's Hist. Mankind* III. v. xi. 296 Fulbe or Fulah (sing. Pullo, Peul) is the Mandingo name, Fellani the Houssa... These names seem to indicate the lighter colour of their skin. 1915 A. Werner *Lang. Families Afr.* vi. 100 We have..referred to the Fula language... The people who speak it call themselves Ful-be (in the singular, Pulo). 1921 F. W. Taylor *First Gram. Adamawa Dial. of Fulani Lang.* 6 The people whose language is the subject of this book call themselves Pullo in the singular and Fulbe in the plural. 1931 W. B. Seabrook *Jungle Ways* iii. ii. 199 Women walked in groups, with long black robes and faces of pale ivory, cameo-cut... These special women, notoriously beautiful as a race,..were Peuhls. 1932 W. L. Graff *Lang.* 154 Peul, Wolof, and Serere, three languages of the above-mentioned African group. 1935 G. Gover *Africa Dances* ii. 159 The Peulh women were remarkable for their lighter skin and sharper features. 1959 *Encounter* Sept. 36/1 M. Diallo belongs to the Peulhs, a fascinating people..spread throughout the sub-Saharan regions of Africa. 1973 *Guardian* 19 June 18/1 The Tuareg and Peul nomads have..watched the bulk of their scrawny animals die.

pew, *sb.*¹ Add: **2. d.** Loosely, a seat, esp. in phr. *take a pew*.

1898 Belloc *Mod. Traveller* i. 5 Be seated; take a pew. 1903 Wodehouse *Prefect's Uncle* xvi. 230 The genial 'take a pew' of one's equal inspires confidence. 1914 C. Mackenzie *Sinister St.* II. iii. vii. 652 Come in, you chaps... I don't know any of your names, but take pews, take pews. 1926 I. Mackay *Blencarrow* xiii. 116 'Have a pew?' he offered, making himself as small as possible on the red plush car seat. 1939 R. Lehmann *No More Music* 90 Colonel: (*drawing up a chair*) Take a pew. 1958 B. Hamilton *Too Much of Water* xi. 232 Have the pew. I'll squat on the bed. 1974 K. Royce *Trap Spider* ii. 30 Sit down, Spider. Take a pew.

4. pew-door (later examples); **pew group** *Pottery*, figures on a high-backed bench, usu. in salt-glazed stoneware.

1803 G. Colman *John Bull* i. i. 4 Troth! and myself.. was brought up to the church... I opened all the pew doors at Belfast. 1842 Pew-door [see Unaccommodated *ppl. a.*]. 1906 G. W. & F. A. Rhead *Staffs. Pots* xiv. 170 The British Museum 'pew group'..is one of four known pieces of the kind, all evidently by the same hand. 1942 *Burlington Mag.* Oct. 260/1 Most dangerous are the increasingly skilful fakes of Astbury and Whieldon figures, first betrayed by the marks of Wedgwood on a pew group and Ralph Wood on a figure of similar origin. 1961 L. G. G. Ramsey *Connoisseur New Guide Antique Eng. Pott., Porc. & Glass* 39 Aaron Wood (1717–85), to whom have been attributed also the vigorous and amusing salt-glaze 'Pew Groups', modelled in the round and representing a man and woman courting, a pair of musicians, etc. 1976 *Country Life* 5 Feb. 278/1 These..Staffordshire salt-glaze groups are still referred to as pew groups, while everyone by now is aware that these people..are not seated in church..but are on a wooden settle.

pewful. (Later example.)

1938 M. Muggeridge *In Valley* iv. 23 The congregation, a few pewfuls, nodded to his words.

pewing, *vbl. sb.* (In Dict. s.v. Pew *v.*¹) (Further concr. examples.)

1840 W. Dyott *Diary* 31 July (1907) II. 322, I visited the old church at Ashbourne to admire the new pewing and other highly ornamental improvements. 1970 H. Braun *Parish Churches* i. 20 The Victorian church,..packed with pewing and comfortably behassocked.

pewter. Add: **1. c.** The colour of the alloy, a bluish or silver grey.

1971 *Vogue* Nov. 81/1 One size tights in palest pewter. 1979 *Guardian* 28 Apr. 32/2 Single-leg stocking tights... Colours: Dark Tan, Mink, American Tan, Pewter, Black.

2. c. (Earlier example.)

1829 P. Egan *Boxiana* 2nd Ser. II. 90 The exertions of Jack's fists and his tongue were both rewarded by a shower of browns, and also a little *pewter* into the bargain.

5. b. Of the colour of pewter.

1922 Joyce *Ulysses* 46 Galleys of the Lochlanns ran here to beach,.. their bloodbeaked prows riding low on a molten pewter surf. 1972 'H. Buckmaster' *Walking Trip* 135 They stood silently looking at the pewter water and pewter sky. 1975 B. Garfield *Death Sentence* (1976) xxix. 137 The sky had thickened and gone pewter.

6. **pewter grey** = sense *1 c.

1973 J. Rossiter *Manipulators* xiii. 134 Lampett Street was long..its pewter-grey length made residential by the planting of trees. 1975 'D. Jordan' *Black Account* xxxviii. 193 The White Nile is no longer white but pewter grey.

pewterer. (Later examples.)

1930 *Aberdeen Press & Jrnl.* 14 Oct. 6 Pewterers nowadays are making their articles tougher and more shock-resisting. 1955 *Times* 16 June 9/6 The pewterer's craft is a peculiarly English one. 1976 *New Yorker* 15 Nov. 73/2 (Advt.), Our attention to detail and unexcelled workmanship have earned us a Royal Title—a distinction no other pewterer in the world can claim.

-pexy (peksi), terminal element repr. Gr. -πηξία, πῆξις a fixing or putting together (f. πηγνύναι to join or fix), used in the names of surgical operations for fixing organs in position; as *hysteropexy* (s.v. Hystero-¹), *orchidopexy* (s.v. *orchido-).

Peyerian, *a.* (Examples of *Peyer's patches* (rarely in *sing.*).)

1858 R. G. Mayne *Expos. Lex. Med. Sci.* (1860) 924/1 Peyer's glands, Peyer's patches. 1870 *Trans. Path. Soc.* XXI. 391 There were no leuchæmic nodules in the liver or kidneys, and the tonsils, like Peyer's patches, were remarkably inconspicuous. 1886 C. H. Fagge *Princ. & Pract. Med.* I. 183 If in a Peyer's patch, it is usually elongated in the direction of the axis of the intestine. 1955 *Sci. News Let.* 23 July 51/1 Polio antibodies in the blood.. build up their response to polio while the virus is still multiplying in the tonsils and the Peyer's patches of the small intestine. 1974 Passmore & Robson *Compan. Med. Stud.* III. xii. 46/1 Enteric or typhoid fevers... Hyperplasia and mononuclear infiltration lead to swelling of the intestinal lymphoid tissue, especially the Peyer's patches of the lower ileum.

peyote (peyōu·te). Also **payote, pellote, pelotte, peyoti, peyotl,** etc. [Amer. Sp. *peyote*, a. Nahuatl *peyotl*.] **1.** = *mescal 3 a.

[1859 Bartlett *Dict. Amer.* (ed. 2) 509 *Whiskey-root*, a plant of the Cactus species possessing intoxicating properties, which is thus described by a correspondent of the New Orleans Picayune: 'It is what the Indians call Pie-o-ke.'] 1885 *Proc. U.S. Nat. Museum* VIII. 521 It is principally as an intoxicant that the Peyote has become noted, being often added to 'tizwin' or other mild fermented native drink to render it more inebriating. 1898 J. Mooney in *17th Ann. Rep. U.S. Bureau Amer. Ethnol. 1895–96* i. 238 The worship of the peyote..is comparatively modern with the Kiowa. 1911, etc. [see *mescal 3 a]. 1927 *Daily News* 28 Sept. 8/7 M. R. Desille..said that a herb..known as peyotl 'produced remarkable hallucinations of undoubted psychic value'. 1955 *Sci. News Let.* 2 July 4/1 Mescaline is extracted from the Mexican plant better known as peyote. 1969 D. Weniger *Cacti of Southwest* 95/2 The insignificant members of this genus [sc. *Lophophora*]..are the sacred plants of the Indians best known by the ancient Indian name, *peyotl*, which has become the *peyote* of modern usage. 1977 Lewis & Elvin-Lewis *Med. Bot.* xviii. 405/2 Peyote is usually eaten as mescal buttons, the dried, brown pieces of the above-ground part of the cactus. Occasionally fresh green pieces are used. *Ibid.* 407/1 Peyote is also brewed and the tea drunk.

2. = *mescal 3 b.

1849 J. W. Audubon *Western Jrnl.* (1906) 186 Out of these acorns the Indians make their 'payote', a kind of paste. 1913 [see *mescal button]. 1915 *Jrnl. Heredity* VI. 295/2 The majority from the peyote and the wine were unable to utilize their legs. 1916 *Jrnl. Amer. Med. Assoc.* 15 Jan. 194/2 Among the menaces..to which some of the American Indians are said to be exposed..is the use of a natural product variously known as mescal buttons, peyote or pellote. 1953 [see *high *sb.* 1 h]. 1975 *High Times* Dec. 49/3 Many natural hallucinogens—peyote is a good example—are supposed to trigger nausea by their direct pharmacological actions.

3. *attrib.* and *Comb.*, as *peyote cactus, high*; **peyote button** = *mescal button; **peyote cult** = *peyotism.

1921 M. R. Harrington *Relig. & Ceremonies of Lenape* viii. 185 (*caption*) Peyote 'Button'. Diameter 1·9 in. 1966 G. Baxt *Queer Kind of Death* (1967) xiv. 204 The stupid bastard must have cut himself a peyoti button. 1975 'S. Marlowe' *Cawthorn Journals* (1976) ii. xi. 88 Have you ever eaten peyote buttons? 1953 A. Huxley *Let.* 21 June (1969) 678 The most extraordinary fact about mescaline—the active principle in the peyotl cactus used by the North American Indians..—is that it is almost completely non-toxic. 1974 M. C. Gerald *Pharmacol.* xvii. 327 Among the plants the Spanish conquistadors observed to be worshipped by the Mexican natives was the peyote cactus. 1920 *Univ. Calif. Publ. Amer. Archaeol. & Ethnol.* XVI. 437 It was quite customary to eat peyote during the day in the early days of the peyote cult. 1970 *Times Lit. Suppl.* 18 Dec. 1480/5 Now coca..became the basis of a kind of peyote cult. 1953 Peyote high [see *high *sb.* 1 h].

peyotism (pē·yǫ̈tiz'm). [f. *peyot(e + -ism.] A religious cult of American Indians in which peyote is taken sacramentally. Hence **peyo·tist**, one who practises this religion.

1934 V. Petrullo (*title*) The diabolic root: a study of peyotism, the new Indian religion, among the Delawares. *Ibid.* ii. 126 The original Peyote people claim that one cannot be a Christian and a Peyotist at the same time. 1953 A. Huxley *Let.* 17 Aug. (1969) 683 No craving.. even among people who have been peyotists for forty or fifty years. 1957 P. Worsley *Trumpet shall Sound* 223 Henceforth the Indians turned..to pacific cults like peyotism. 1971 E. Shorris *Death of Great Spirit* iii. 34 God may change the peyotist, but the peyotist..has no hope of changing God. *Ibid.* 38 Apologists for peyotism say, 'It is the only thing we have left that is Indian.'

Peyronie's disease (pĕⁱrǭni). *Path.* [Named after François de La *Peyronie* (1678–1747), French physician, who described the condition in 1743 (*Mém. Acad. r. de Chirurgie* I. 425).] (See quot. 1974.)

1903 *Boston Med. & Surg. Jrnl.* CXLVIII. 245/1 (*heading*) Peyronie's disease. 1949 H. C. Rolnick *Pract. Urol.* I. xi. 173 In Peyronie's disease or plastic induration of the corpora cavernosa the incurvation becomes noticeable on erection. 1974 Passmore & Robson *Compan. Med. Stud.* III. xxvii. 13/1 Fibrous cavernositis (Peyronie's disease) is a condition in which fibrous plaques develop near the glans or on the dorsum of the penis, and cause curvation, swelling, pain on erection and interference with coitus.

pezazz, var. *pizzazz.

Pfalzian (pfæ·ltsiăn), *a.* (*sb.*) *Geol.* Also **Pfälzian**. [ad. G. *Pfälzische* (F. Kühne 1922, in *Jahrb. Preuss. Geol. Landesanst.* XLIII. 433), f. *Pfalz*, the (Rhineland) Palatinate (ult. f. L. *palatium* imperial residence: see Palace *sb.*¹): see -ian.] Pertaining to or designating a minor orogenic episode in Europe which is believed to have occurred in the Permian period, later than the *Saalian. Also *ellipt.* as *sb.*

1931 Gregory & Barrett *Gen. Stratigr.* 19 Pfalzian. Between Trias and Permian. 1932 T. W. E. David *Explan. Notes to accompany New Geol. Map Austral.* 73 While the epi-Permian movement as a whole seems nearly to correspond with the Pfalzian of Europe and the Appalachian of U.S.A., it would seem that in eastern Australia it continued, in places, for some little time after Pfalzian and Appalachian disturbance had ended. 1937 A. L. Du Toit *Our Wandering Continents* viii. 164 At the close of the Permian, during the Pfalzian Phase, the central plateau of France and the Saar region were upheaved along an E.–W. axis. 1974 [see *Saalian *a.*].

‖ **Pfannkuchen** (pfa·nkuχĕn). Also **pfannekuchen**. [G.] In Germany and other German-speaking areas, a pancake.

1877 [see *krapfen]. 1906 *Cosmopolitan* XLII. 547/1 Do you sometimes haf *pfannekuchen* for supper? 1964 L. Deighton *Funeral in Berlin* xl. 242 'You got rid of your tail?'..'He got a lapful of *Pfannkuchen* and scalding coffee.'

‖ **Pfefferkuchen** (pfe·fərkuχĕn). Also **pfeffer-kuchen**. [G., lit. 'pepper cake'.] In Germany and other German-speaking areas, a spiced cake, gingerbread.

1870 *Western Monthly* Jan. 29/1 One long room..being gay with bright Christmas trees loaded with nuts, apples, sugar-plums, and all sorts of jolly-looking figures in honey-cake and *pfeffer-kuchen*. 1961 J. Heller *Catch-22* (1962) xxiv. 249 Pumpernickel and *Pfefferkuchen* from Berlin. 1977 G. Butler *Brides of Friedberg* v. 136 Underneath I put little dishes of Pfefferkuchen, Nüsse and Stolle.

Pfeiffer (pfəi·fəɹ). *Bacteriol.* The name of Richard *Pfeiffer* (1858–1945), German bacteriologist, used in the possessive and *attrib.* to designate: **a.** The specific lysis of the cholera vibrio in the presence of antibody and complement as described by him (in *Zeitschr. für Hygiene und Infektionskrankh.* (1893) XIV. 59).

[1895 E. Metchnikoff in *Ann. de L'Inst. Pasteur* IX. 438 La découverte de la destruction extracellulaire des microbes dans le liquid péritonéal, ou de ce qu'on peut appeler le 'phénomène de Pfeiffer', suggère toute une série de questions touchant le problème général de l'immunité.] 1897 Muir & Ritchie *Man. Bacteriol.* xvii. 393 A striking change is observed microscopically in the vibrios when injected along with the protective serum into the peritoneal cavity of another guinea-pig—Pfeiffer's reaction. 1929 Topley & Wilson *Princ. Bacteriol. & Immunity* II. lxi. 905 Pfeiffer's phenomenon is of great value in the identification of the cholera vibrio. 1970 *Science* 3 Apr. 141/2 (*heading*) A new application of the Pfeiffer phenomenon.

b. A species of bacterium, *Hæmophilus influenzæ*, described by him (in *Deutsch. med. Wochenschr.* (1892) 14 Jan. 28) as the causal agent of influenza, but now known not to be.

1900 H. J. Curtis *Essent. Pract. Bacteriol.* 287/1 (Index), Pfeiffer's influenza bacillus. 1910 Hiss & Zinsser *Text-bk. Bacteriol.* xxxvii. 528 The bacillus of influenza (Pfeiffer bacillus) is an extremely small organism, about 0·5 micron long. 1936 *Lancet* 12 Dec. 1383/1 The illness produced by introducing virulent cultures of Pfeiffer's bacillus into a healthy person is not genuine epidemic influenza. 1953 J. Ramsbottom *Mushrooms & Toadstools* xxiii. 278 The insensitive Pfeiffer's bacillus.. in the respiratory tract is usually associated with organisms highly sensitive to penicillin.

pfella (Pfe·la). Also **pfeller**. Repr. Austral. Aborigines' pronunc. of Fellow *sb.* Cf. *fella, fellah. In Austral. Pidgin often used as a marker of an adjective, demonstrative, or numeral.

1908 'G. Seagram' *Bushmen All* 310 Now you go longa Warrego an' big pfella pleeseman hang you longa rope. 1921 A. B. Paterson *Man from Snowy River* (new ed.) 74 What parson tellin' you, Ole Mister Dodd, Tell you in Sunday-school? Big pfeller God! 1936 M. Franklin *All that Swagger* iii. 38 That one sit down longa white pfella—all same as white pfella. *Ibid.* iv. 46 Danny had established himself as a superman, a big pfella chief. *Ibid.* vii. 65 Minetinkit, that pfella budgery! 1936 I. L. Idriess *Cattle King* xxix. 257 How much would you want to make a big pfella flood, Paddy?

pfennig, -ing. Add: Also *Comb.*

1909 *Daily Chron.* 15 July 4/7, I have an actual pfennig-piece before me as I write, which was coined in 1894. 1979 R. Cox *Auction* iii. 75 Pouring pfennig pieces into a coinbox.

pfft (f't, pf't). Also **pfft, phfft, phtt,** etc. [Echoic.] = *phut *int.* (*adv.*, *sb.*).

1923 Kipling *Irish Guards in Gt. War* I. 5 One..rifle-bullet landed with a *phtt* in the chalk between two officers. 1934 *Amer. Speech* IX. 313/2 Then it all goes pfft. 1942 Berrey & Van den Bark *Amer. Thes. Slang* § 360/2 *Divorce*,..go pfft, — pffft *or* phut,..pfft, phffft, phut. 1954 Wodehouse *Jeeves & Feudal Spirit* iii. 29 It wouldn't take much to make the Stilton–Florence axis go p'fft again. 1972 F. Warner *Lying Figures* i. 6 Commuters' corset till sixty-five, then ppffft! the six by four. 1975 *Washington Post* 10 Mar. b1 A voice..said 'Thank you, Barbra' and, phffft! The hour was over. 1977 *Ibid.* 13 Jan. c1/3 Senatorial marriages of 5, 23 and 35 years were going pfffffft, as the Hollywood columnists used to write. 1977 *Rolling Stone* 16 June 11/2 Nothin' serious; $500 in bail, phfffffttt.

Hence **pfft** *v. intr.* (*U.S. journalists' slang*), to come to an abrupt end, to collapse; of a couple, to become estranged, to separate, to divorce; also *transf.* and as *ppl. adj.*

1930 *Vanity Fair* (N.Y.) Jan. 48/3 Mr. Walter Winchell ..exercises a tremendous power for mischief with impassive diablerie: 'the so-and-sos have phfit!' 1938 W. Winchell in *Baltimore News-Post* 16 Mar. 16/2 Have Lionel Stander and the films phftt? 1938 —— in *Ibid.* 21 Mar. 10/1 Burton Rascoe and Dorothy Parker have phffft! 1939 *Time* 28 Aug. 13 Gossip columnists took this as renewed notice that the Franklin D. Jrs. are not *phphpht* as gossiped. 1940 *New Yorker* 13 July 22/3 International Politics, March 29, 1937. 'Adolf and Benito have phffft! The break will be announced soon enough.' 1946 J. T. Shipley in W. S. Knickerbocker *20th Cent. English* 136 To indicate the collapse of a project..the columnist declares that it has pfft. 1957 *Chicago Amer.* 9 Feb. 17/8 Alan and Joanne Rio, who were dating, have pffft.

pfui (fu·i, ‖pfu·i), *int.* [G.: cf. Phoo *int.*, *phooey *int.* (*sb.*).] (Orig. in Germany or among German-speaking people) an exclamation of contempt or disgust.

Hence **pfuiteufel** (pfū·i,toifĕl) [G. *teufel* devil], emphatically in same sense.

1866 C. M. Yonge *Dove in Eagle's Nest* I. x. 206 Christina heard Friedel say..'I think I shall be a priest, Ebbo.' To which Ebbo only answered, 'Pfui!' 1871 Geo. Eliot in *Macm. Mag.* July 185 Pfui! The time was, I drank that home-brewed wine. 1920 D. H. Lawrence *Touch & Go* 6 The plays are good... What about your *good* plays? Whose good? *Pfui* to your goodness! 1922 Joyce *Ulysses* 184 Out on't! *Pfuiteufel!* 1934 R. Macaulay *Going Abroad* xxxiii. 281 Pfui, my poor Pierre, you are a fool. 1935 C. Isherwood *Mr. Norris changes Trains* iv. 52 You men are all alike; from seventeen to seventy... *Pfui!* I'm surprised at you. 1937 Wyndham Lewis *Blasting & Bombardiering* 100 Why should one be asked to meet such people? It is absurd that a Bennett should be referred to, for anything except the time of a train or the cost of a bicycle lamp! Pfui! 1943 E. M. Almedingen *Frossia* ii. 107, 'I trusted you. Pfui,' he spat on the paper-strewn floor. 1949 R. Stout *Second Confession* xiii. 134 'I didn't get him to. As it says, he wrote and signed it of his own free will.' 'Pfui. I know what it says. But why should I believe that when I don't believe anything in it?' 1966 —— *Death of Doxy* (1967) vi. 71 Pfui. A prodigy on a treadmill. Take him off. 1972 *Human World* May 96 Brutal to give a little girl a doll? Brutal to bring children up? Pfui! Not all the madmen are in the madhouses. 1974 Wodehouse *Aunts aren't Gentlemen* ix. 163, I don't mind criticism, but I will not endure vulgar abuse. 'Pfui,' I said. It is an expression I don't often use, but Nero Wolfe is always saying it with excellent results.

Pfund (funt). *Physics.* The name of A. Herman *Pfund* (1879–1949), U.S. physicist, used *attrib.* to designate a series of lines in the infra-red part of the spectrum of atomic hydrogen, with wave numbers represented by $R(1/5^2 - 1/m^2)$ (where R is the Rydberg constant and m = 6, 7, ...), of which the first line has a wavelength of 74·61 micrometres and the series limit is at 22·80 micrometres.

1934 H. L. Brose tr. *Sommerfeld's Atomic Struct. & Spectral Lines* (ed. 3) ii. 73 A. H. Pfund found an infra-red line belonging to hydrogen at $\lambda = 7{\cdot}40\mu$, which is the first member of a 'Pfund series' and corresponds to $n = 5$, $m = 6$. 1970 D. W. Tenquist et al. *University Optics* II. ii. 43 Lines in the infra-red part of the spectrum are caused by transitions terminating at $n = 3$ (Paschen series), $n = 4$ (Brackett series) and $n = 5$ (Pfund series).

pH (pī,ē^i·tʃ). Formerly also p_H, P_H. [Introduced (in Ger.) as p_H. by S. P. L. Sörensen 1909, in *Biochem. Zeitschr.* XXI. 134, the *p* repr. G. *potenz* power and H· the hydrogen ion.] A measure of the acidity or alkalinity of a solution, equal to the logarithm to the base 10 of the reciprocal of the effective concentration (activity) of hydrogen ions (in moles per litre).

A pH of 7 corresponds to a neutral solution, one less than 7 to an acidic solution, and one greater than 7 to an alkaline solution.

1909 *Jrnl. Chem. Soc.* XCVI. 861 The optimal concentration was p_H = 4·4 to 4·6. 1920 W. M. Clark *Determination Hydrogen Ions* i. 26 As a matter of typographical convenience we shall adopt pH in place of P_{H^+}. 1921 *Jrnl. Nat. Dental Assoc.* VIII. 653/1 He uses agar with a P_H value of 8·1 as a basis for blood agar. 1935 W. A. Koehler *Princ. & Applic. Electro-Chem.* II. xiv. 359 The bulb dips into the solution of unknown P_H. 1937 Pierce & Haenisch *Quantitative Analysis* viii. 121 Acid-base indicators are highly colored organic dyes which exhibit a change in color when the *p*H of a solution is changed. 1952 C. E. L. Phillips *Small Garden* iii. 25 For general garden purposes in this country best results come from a neutral or slightly acid *p*H reading. 1970 *Pure & Appl. Chem.* XXI. 33 In all existing national standards the definition of pH is an operational one. *Ibid.*, The difference between the pH of two solutions having been defined as above, the definition of pH can be completed by assigning a value of pH at each temperature to one or more chosen solutions designated as standards. 1974 P. Svendsen *Introd. Animal Physiol.* xiii. 143 The maximum acidity of the wine is about pH 4·5.

2. *attrib.* and *Comb.*, as *pH meter, scale, value*; *pH-dependent* adj.; **pH-stat**, a device for automatically maintaining a solution at constant pH.

1960 *Jrnl. Bacteriol.* LXXIX. 734 (*heading*) Temperature and pH-dependent changes of electrophoretic mobility of *Pasteurella pestis*. 1972 *Brit. Jrnl. Nutrition* XXVIII. 389 The pH-dependent rearrangements have also to be borne in mind when performing administration experiments with formylfolates in metabolic studies. 1940 Reilly & Rae *Physico-Chem. Methods* (ed. 3) xii. 493 The Beckman *p*H meter..uses the principle of the potentiometer system in which the voltage of the electrode system is balanced against that of a standard cell. 1968 Passmore & Robson *Compan. Med. Stud.* I. vi. 3/1 The pH measured electrometrically by a glass electrode and a pH meter is not in fact a precise measure of actual hydrogen ion concentration. 1920 W. M. Clark *Determination Hydrogen Ions* ii. 41 It is advantageous to show the position of the several indicators on the pH scale. 1973 F. G. Shinskey *pH & pIon Control in Process & Waste Streams* iii. 57 The pH scale is not bound to the limits of 0 to 14. 1957 *Methods Biochem. Anal.* IV. 174 A pH-stat consists of an appropriate cell..connected with a suitable pH-meter which in turn is connected with an adjustable controlling device. 1974 Sawyer & Roberts *Exper. Electrochem. for Chemists* ix. 417 In general autotitrators that work with a preset endpoint lend themselves to application as pH-stats. 1920 W. M. Clark *Determination Hydrogen Ions* ii. 38 To a series of test tubes are added, seriatim, 10 cc. of each of a series of standard solutions whose pH values are known. 1932 T. P. Francis *Mod. Sewage Treatm.* 252 The pH values for filter effluents should not differ by more than ·6. 1946 J. W. Day *Harvest Adventure* iii. 44 Every acre of arable land has had a dressing of chalk. In all..some 30,000 tons have been put down, the idea being to maintain a *p*H-value of 5. 1974 *Encycl. Brit. Macropædia* X. 606/1 Some volcanic lakes are extremely acid, however, with pH values below 4, and some lakes with very high pH values, such as Lake Nakuru, Kenya, also occur in nature.

phacelia (fasī·liă). [mod.L. (A. L. Jussieu *Genera Plantarum* (1789) 129), f. Gr. φάκελος bundle, in reference to the clustered flowers.] An annual herb of the genus so called, belonging to the family Hydrophyllaceæ, usually native to western North America and bearing clusters of blue, violet, or white flowers.

1818 A. Eaton *Man. Bot.* (ed. 2) ii. 354 *Phacelia*.. *bipinnatifida* (phacelia..). 1835 *Edwards's Bot. Reg.* XX. 1696 (*heading*) Tansy-leaved Phacelia. 1898 A. M. Davidson *California Plants* 119 There are many kinds of Phacelias; the flowers vary very much in size and in color and form, but they always grow in clusters that last a long time. 1903 M. Austin *Land of Little Rain* 145 Larkspur in the *coleogyne*, and for every spinosa the purpling coils of phacelia. 1911 A. E. Speer *Annual & Biennial Garden Plants* 206 The Phacelias are a family of hardy annuals useful for edgings or groups, and are all natives of either California, Texas, or Mexico. 1935 *Times Educ. Suppl.* 9 Mar. (Home and Classroom Suppl.) p. iv/3 Phacelia is a lovely little plant of which the colour rivals that of the Gentians. 1962 *Amat. Gardening* 24 Mar. 9/2 My own favourite hardy annuals are..phacelia,..escholtzia.

phacoanaphylaxis (fæ:ko,ænăfilæ·ksis). *Ophthalm.* Also **phako-**. [Mod.L., f. Gr. φακό-ς lentil + *anaphylaxis.] Allergic reaction to protein released from the crystalline lens of the eye.

[1922 Verhoeff & Lemoine in *Trans. Internat. Congr. Ophthalm.* 274 Certain individuals are hypersensitive to lens protein... When in such individuals rupture of the lens capsule takes place through injury or operation or spontaneously, a characteristic inflammatory reaction may properly be termed endopthalmitis phacoanaphylactica.] 1948 *Amer. Jrnl. Ophthalm.* XXXI. 1006/2 Verhoeff and Lemoine conceived the idea of phacoanaphylaxis in order to explain the postoperative inflammatory reaction of the eyes with extracapsular extraction. 1962 D. G. Cogan in A. Pirie *Lens Metabolism Rel. Cataract* 291 The lens..becomes invaded by polymorphonuclear cells, especially in the immunological reactions known as phakoanaphylaxis. 1964 *Arch. Ophthalm.* LXXII. 14/2 Sympathetic ophthalmia and phacoanaphylaxis are both manifestations of autoimmune reactions, the former to uveal pigment and the latter to lens protein.

So **pha:coanaphyla·ctic** *a.*, of, pertaining to, or involving phacoanaphylaxis.

1922 *Trans. Internat. Congr. Ophthalm.* 280 Intraocular inflammation resulting from rupture of a Morcagnian cataract is not usually phacoanaphylactic in nature. 1966 *Amer. Jrnl. Ophthalm.* LXI. 1431/1 The phacoanaphylactic reaction was seen in full bloom.

phacoidal (făkoi·dăl), *a.* *Petrogr.* [f. as Phacoid *a.* + -al.] **a.** Lens-shaped, lenticular. **b.** Characterized by the presence of lenticular inclusions.

1901 *Rep. Brit. Assoc. Adv. Sci.* 617 The basic dykes.. appear frequently as phacoidal masses amid the reconstructed gneiss. 1920 [see *eye *sb.*¹ 12 e]. 1956 E. W. Heinrich *Microsc. Petrogr.* vii. 185 In phacoidal rocks the characteristic textural feature consists of ellipsoidal or lensoid units in a finer-grained matrix that is brecciated and sheared. *Ibid.* 186 Some mechanically metamorphosed conglomerates also possess the phacoidal texture. 1959 W. W. Moorhouse *Study of Rocks in Thin Section* xxviii. 463 Granites, syenites, and granodiorites show some foliation and a lenticular, 'phacoidal' form of quartz and feldspar grains.

phacolite (fæ·kolәit). *Geol.* [f. Gr. φακός lentil: see -lite.] = next.

1909 A. Harker *Nat. Hist. Igneous Rocks* iii. 77 A concurrent influx of molten magma will..find its way along the crests and troughs of the wave-like folds. Intrusive bodies corresponding more or less closely with this ideal case are common in folded districts. Since some distinctive name seems to be needed, we may call them phacolites... The intrusions..are not, like true laccolites, the cause of the attendant folding, but rather a consequence of it. 1937 Wooldridge & Morgan *Physical Basis Geogr.* viii. 109 Phacolites are lens-shaped masses of rock occupying the saddles of anticlines or the heels of synclines, places where rigid rock sheets naturally gape apart in the folding process. 1946 [see *laccolite].

phacolith (fæ·kolịþ). *Geol.* [Alteration of prec.: see -lith.] An intrusive mass of igneous rocks situated between consecutive strata at the top of an anticline or the bottom of a syncline.

1910 Lake & Rastall *Text-bk. Geol.* xiii. 225 These long narrow intrusions are called by Mr. Harker phacolites (more correctly phacoliths). 1947 S. J. Shand *Study of Rocks* (ed. 2) ii. 16 The phacolith might be described as a saddle-shaped laccolith;..but there is this important difference between a laccolith and a phacolith that the former is the cause of the folding of its country rocks while the latter is a consequence of folding. [*Ed.* 1 (1931): phacolite, laccolite.] 1971 B. W. Sparks *Rocks & Relief* iii. 93 Laccoliths are approximately circular, but phacoliths are not, except in the rare case of their being formed in a dome. Usually they are elongated in the direction of the axes of the folds. 1977 A. Hallam *Planet Earth* 68 Phacoliths are curved sill-like intrusions emplaced in the crests and troughs of folded rocks.

phænogamic, phænogamous, *adjs.* (Earlier example.)

1808 S. F. Gray in *Monthly Mag.* XXIV. 612 Phænogamic [*sic*], or phænogamous, is a word much used by the German botanists in contradistinction to cryptogamic, and from its convenience begins to be adopted here.

phænotype, obs. var. *phenotype.

phæochrome (fī·okrōᵘm), *a.* *Histology.* Also **pheo-**. [f. Gr. φαιός dusky + χρῶμα colour.] = *chromaffin *a.*

1909 Bailey & Miller *Text-Bk. Embryol.* xv. 43 The medulla is composed of irregularly arranged sympathetic ganglion cells and other granular cells which, after treatment with chrome salts, acquire a peculiar brownish color. The brown cells are known as chromaffin (or phæochrome) cells and their granules as chromaffin (or phæochrome) granules. **1929** M. A. Goldzieher *Adrenals* v. 223 The phæochrome tumours produce an excessive amount of adrenalin. **1941** R. G. Hoskins *Endocrinol.* ii. 51 A tumor of the phæochrome cells..may be situated in either the adrenal medulla or in one of the paraganglia. **1956** Robson & Keele *Recent Adv. Pharmacol.* (ed. 2) i. 29 Phæochrome tumours..are benign in nearly 90 per cent of cases.

phæochromocyte (fiːəkrōu·mosəit). *Med.* Also **pheo-**. [ad. G. *phäochromocyt* (H. Poll 1906, in O. Hertwig *Handb. d. vergleich und exper. Entwickelungslehre d. Wirbeltiere* III. 460), f. *phäochrom* *PHÆOCHROME: see -CYTE.] A chromaffin cell, esp. one in the adrenal medulla.

1929 *Arch. Path.* VII. 229 The final stage of differentiation brings forth the mature pheochromocytes, which are large, irregular or polyhedral cells, with each a round or ovoid vesicular nucleus, containing a loose chromatin network, and a well formed nucleolus. *Ibid.* 230 The pheochromocytes have the peculiar property of staining brown with chromic salts. **1948** Martin & Hynes *Clin. Endocrinol.* vii. 144 Pheochromocytomata... The tumours consist of mature pheochromocytes and 90 per cent are benign. **1975** *Nature* 27 Nov. 344/1 Pheochromocytoma cells contain..properties of both pheochromocytes and neurones as a manifestation of neoplasia.

phæ:ochromocyto·ma. *Path.* Also **pheo-**. Pl. **-omas, -omata.** [mod.L., ad. G. *phäochromocytom* (L. Pick 1912, in *Berlin. klin. Wochenschr.* 1 Jan. 21/2): see prec. and *-OMA.] A tumour arising from chromaffin cells of the adrenal medulla. Cf. *PARAGANGLIOMA.

1929 *Arch. Path.* VII. 228 (*heading*) Chromaffin cell tumor of the suprarenal medulla (pheochromocytoma). **1929** *Trans. Assoc. Amer. Physicians* XLIV. 296 Pheochromocytomas are a rare tumor. **1931** W. Boyd *Path. Internal Dis.* ix. 482 Chromaffin cell tumor..is also known as a paraganglioma, a pheochromocytoma and a chromaffinoma. **1943** J. M. Beattie et al. *Textbk. Path.* (ed. 4) xxiv. 847 The chromaphil (or so-called 'chromaffine') tumour or phæochromocytoma is a growth affecting especially adults and is usually benign, although malignant varieties are recorded. **1944** *Surg. Clinics N. Amer.* Aug. 932 The terms 'chromaffinoma' and 'pheochromocytoma' have by common consent been used to describe suprarenal tumors of this type insomuch as they almost exclusively produce the clinical phenomena of hyperadrenalism. The term 'paraganglioma' is reserved for extra-adrenal chromaffinomas, such as tumors of the carotid body. **1966** [see *PARAGANGLIOMA]. **1970** Passmore & Robson *Compan. Med. Stud.* II. xxiii. 55/2 Histologically phæochromocytomata closely resemble normal chromaffin tissue.

phæophorbide (fiofǭ·rbəid). *Biochem.* Also **pheo-**. [ad. G. *phäophorbid* (Willstätter & Stoll 1911, in *Ann. d. Chem.* CCCLXXVIII. 25), f. as next + Gr. φορβή pasture, food: see -IDE.] Either of two compounds (orig. not distinguished) formed by the action of a strong acid on chlorophyll or phæophytin, having the structure of the latter except for the replacement of the phytol group by a hydrogen atom: phæophorbide *a*, HOOC·$C_{32}H_{32}N_4O$·COOCH$_3$, or phæophorbide *b*, HOOC·$C_{32}H_{30}N_4O_2$·COOCH$_3$. Also, an ester of one of these compounds (as phæophytin).

1911 *Chem. Abstr.* V. 873 The following nomenclature is now adopted: for the free tricarboxylic acid, the basis of chlorophyll, the name chlorophyllin is retained, the monomethyl ester is called chlorophyllide and the deriv. free from Mg, phæophorbide. **1950** J. Bonner *Plant Biochem.* xxv. 470 Hydrolysis of pheophytin with stronger acid removes the phytol group to yield pheophorbides *a* and *b*, which are both monocarboxylic acids and which can be esterified..to yield the alkyl pheophorbides. **1956** L. P. Zill et al. *Biochem. Jrnl.* ..These pheophorbides..as phæophorbide *a* and *b*. **1963** R. P. Dales *Annelids* iii. 56 The intestine..is apparently absorptive. It is this region that has a deep green colour, just as that of Chaetopterus has, and for the same reason, for in both the colour is due to phæophorbides—breakdown products of chlorophylls presumably derived from the food. **1975** *Nature* 24 Jan. 241/1 Sudden heating [of leaf juice] inactivates chlorophyllase and so prevents the formation of phæophorbide.

phæophytin (fīofai·tin). *Biochem.* Also **pheo-**. [a. G. *phäophytin* (now *phäo-*) (Willstätter & Hocheder 1907, in *Ann. d. Chem.* CCCLIV. 207), f. Gr. φαιός dusky + φυτόν plant: see -IN[1].] Either of two brownish compounds (orig. not distinguished) formed by the action of a weak acid on chlorophyll, having the structure of the latter except that the magnesium atom is replaced by two hydrogen atoms; a phytyl ester of phæophorbide: phæophytin *a*, $C_{20}H_{39}$OOC·$C_{32}H_{32}N_4$O·COOCH$_3$, or phæophytin *b*, $C_{20}H_{39}$OOC·$C_{32}H_{30}N_4O_2$·COOCH$_3$.

1907 [see *PHYTOL]. **1950**, **1956** [see prec.]. **1967** *Oceanogr. & Marine Biol.* V. 269 The increase of the pigment ratio in deep water has to be ascribed in part to the persistence of yellow pigments and phæophytin in decomposing plankton and in excrements.

† **phæoplast** (fī·ōplæst, -plɑst). *Obs. Bot.* [a. G. *phaeoplast* (now *phäo-*) (A. F. W. Schimper 1885, in *Jahrb. für wissensch. Bot.* XVI. 34), f. Gr. φαιός dusky + πλαστ-ός formed, moulded: cf. *CHLOROPLAST.] A chromoplast of a brown alga.

1886 *Jrnl. R. Microsc. Soc.* VI. 640 For the chromoplasts of the Phæophyceæ the author proposes the term phæoplasts. **1904** *Amer. Naturalist* XXXVIII. 378 The chromatophores of the higher brown Algæ (Phæophyceæ)..have the discoid form characteristic of chloroplasts. They might be called phæoplasts..if one wished to classify plastids according to their color.

phage (fēɪdʒ). *Biol.* [Shortening of *BACTERIOPHAGE.] A virus which attacks bacteria, entering the cell and either multiplying at its expense until the cell is lysed and the phage particles released, or becoming attached to the bacterial genome as a prophage and replicating synchronously with it; = *BACTERIOPHAGE. Also *collect.*

1926 *Lancet* 1 Aug. 234/2 The musical comedy spirit which reduces 'bacteriophage' to its final syllable. **1926** *Encycl. Brit.* I. 302/1 If a tube of suitable diluted phage, inoculated with *B. dysenteriae*, is incubated for three hours, and a loop of it then inoculated on agar, confluent growth of the bacteria may be obtained with a number of small, round, clear areas where no growth of bacteria has occurred. **1936** Topley & Wilson *Princ. Bacteriol. & Immunity* (ed. 2) x. 238 Phages acting on one or other of the normal or pathogenic intestinal bacteria can almost always be isolated from faeces, from sewage, or from polluted water supplies. **1939** *Nature* 24 June 1029 Lowering of virulence can..be induced by incubating phage with bacteria in media containing neither calcium nor magnesium salts. **1949** Abraham & Heatley in H. W. Florey et al. *Antibiotics* I. ii. 84 A thin layer of agar containing an organism susceptible to lysis by the phage was spread on the plate and incubation was continued. **1960** *New Biol.* XXXI. 72 Like other viruses, phages multiply by causing the host cells to synthesize the components of the virus. **1968** H. Harris *Nucleus & Cytoplasm* ii. 22 Infection of *E. coli* with phage λ, however, allows β-galactosidase synthesis to continue almost until the time of lysis. **1972** *Nature* 21 Jan. 129/1 Carcinogens..were tested for mutagenic activity in conventional assay systems, involving the use of bacteria or phage as target organisms. **1975** E. Nester et al. *Microbiol.* in *Tropics* xi. 306/2 Temperate phages..fail to lyse the cells they infect but continue, nevertheless, to reproduce synchronously with the bacterial host for many generations.

b. *attrib.* and *Comb.*, as *phage genome, particle, replication; phage-infected, -related, -resistant, -specific* adjs.; **phage type** *sb.*, a division of a bacterial species characterized by a common susceptibility to a particular group of phages; so **phage-typing** *vbl. sb.*, the determination of the phage type of a bacterium; **phage-type** *v. trans.*

1940 A. M. Campbell *Episomes* i. 5 We conclude that a phage genome can indeed multiply in two different states. **1964** G. H. Haggis et al. *Introd. Molecular Biol.* xiii. 315 Although these first experiments were made on phage-infected, evidence has also been found in normal bacteria for an RNA fraction with rapid turnover. **1957** C. E. Clifton *Introd. Bacterial Physiol.* 380 The number of phage particles absorbed per bacterium can vary with experimental conditions. **1957** *Virology* II. 256 (*heading*) Production of phage-related structures during multiplication of phages T2 and T4. **1968** J. D. Watson *Double Helix* xxii. 121 André [Lwoff] I was very keen about the role of divalent metals in phage replication. **1951** Whitby & Hynes *Med. Bacteriol.* (ed. 5) xxiv. 436 Phage-resistant mutants..of undiminished virulence may replace the originally susceptible bacteria. **1973** R. G. Krueger et al. *Introd. Microbiol.* xviii. 479/2 The bacterial cell wall contains phage-specific receptor sites that undergo chemical union with the attachment apparatus of the virus. **1942** *Jrnl. Infectious Dis.* LXXI. 155/2 A series of mouse passage tests resulted in cultures of the same phage type as the original strains. **1961** *Lancet* 29 July 248/1 The second distinctive feature in Australia is an apparently great prevalence of antibiotic-resistant staphylococci, including phage-types 47 and 80. **1975** E. Nester et al. *Microbiol.* in *Tropics* xi. 307/1 If a particular phage-type of *Salmonella typhi* is the cause of an outbreak of typhoid infection, a search is made for carriers of organisms of the same phage-type. **1938** *Canad. Publ. Health Jrnl.* XXIX. 493 (*heading*) The limitations of phage typing *B. typhosus*. **1949** *Brit. Med. Jrnl.* 10 Sept. 563/2 Staphylococci isolated from a series of 100 cases of infection occurring in this hospital have all been tested for penicillin sensitivity and phage-typed to see if the two could be correlated. **1961** *Lancet* 2 Sept. 506/2 All strains of *Staph. aureus* were tested for sensitivity to eight antibiotics, and a small number were phage-typed. **1942** *Jrnl. Infectious Dis.* LXXI. 161/1 Phage typing of typhoid culture has been employed quite extensively in Canada as an aid in tracing the origin of typhoid fever outbreaks. **1970** Passmore & Robson *Compan. Med. Stud.* II. xviii. 70/2 There are a number of phages with specific affinities for different staphylococci and it is therefore possible to identify strains of staphylococci by their phage-sensitivity patterns. This is the basis of phage-typing which is of great use in the investigation of outbreaks of staphylococcal sepsis.

phagocytable (fægosəi·tăb'l), *a. Biol.* [f. *PHAGOCYTE *v.* + -ABLE.] Susceptible to phagocytosis. Hence **pha:gocytabi·lity.**

1911 *Jrnl. Amer. Med. Assoc.* 11 Nov. 1579/2 Pneumonic leukocytes would take up more easily phagocytable pneumococci than normal leukocytes. **1915** Thomas & Ivy *Appl. Immunol.* i. 15 An ingenious..laboratory method was devised whereby the measure of the ratio of phagocytability of the microbes employed in opsonic testing. **1927** *Jrnl. Bacteriol.* XIII. 40 Microbic dissociation..would serve to transform a virulent but non-phagocytable organism into a non-virulent but phagocytable organism. *Ibid.*, The frequent correlation between S type culture, high virulence and non-phagocytability on the one hand, and, on the other, the correlation between R type culture, non-virulence and phagocytability, are emphasized.

phagocyte. Add: More widely, any cell in the body that phagocytoses bacteria or foreign particles. (Further examples.)

1939 F. A. Kerr *Clin. Bacteriol.* iii. 45 The lining cells of the hepatic spaces (Kupffer cells) and the reticular cells of the cells of the spleen and lymph glands and the septal cells lining the pulmonary alveoli..can all be shown..to be active phagocytes and capable of ingesting pathogenic bacteria. **1972** E. Hartenstein *Prinz. Physiol.* xi. 494 Reticular cells differentiate into various forms of phagocytes including the fixed dust cells of vertebrate lungs, cellular linings of sinuses within glands, fixed Kupffer cells in the liver, and histiocytes.

phagocytic *a.*, add: so *phagocytic index*, any of various indices of phagocytic activity; hence **phagocy·tically** *adv.*; **phagocytosis**, substitute for def.: the process by which a cell engulfs or absorbs bacteria or foreign particles so as to isolate or destroy them; (further examples).

1904 Wright & Douglas in *Proc. R. Soc.* LXXIII. 129 The phagocytic index given below..represents in each case the average number of bacteria ingested by the individual P.W.B.C. The number of polynuclear white blood corpuscles which were taken as the basis of each case inserted in brackets. **1908** R. W. Allen *Vaccine Therapy* 22 Phagocytic index..appears to be depressed for three to six weeks after commencing treatment. **1911** *Jrnl. Amer. Med. Assoc.* 11 Nov. 1580/1 So far as staphylococci are concerned, the cells with one nucleus are more active phagocytically and those with four least active. **1937** E. E. Hewer *Textbk. Histol.* 7 The property of phagocytosis is shown by the blood leukocytes, osteoclasts, and all the histiocyte cells scattered throughout the body. **1963** R. P. Dales *Annelids* iii. 55 Kermack also found that food particles in the stomach were taken up phagocytically by certain epithelial cells. **1970** *Times Lit. Suppl.* 23 Oct. 1221/2 Amoebocytes..tend to be of variable shape and to engulf particles in their environment by the process of phagocytosis. **1971** Herbert & Wilkinson *Dict. Immunol.* 139 *Phagocytic index*, a measure of the activity of the reticuloendothelial system of the body. It is usually represented by a constant k that characterizes the rate of carbon clearance..from the blood and is inversely proportional to the dose of carbon injected. **1972** F. Spencer *Aspects Human Biol.* IV. 112 As performed by such primitive cells as the protozoon amoeba, phagocytosis..represents a normal feeding mechanism. In the higher metazoa, phagocytosis is a highly developed activity, as exemplified by the granular leucocytes, and the cells of the reticulo-endothelial system. **1974** *Brit. Jrnl. Haematol.* XXVII. 342 The phagocytic index was the average number of particles per neutrophil. **1977** P. R. & J. S. Medawar *Life Sci.* xiii. 100 Phagocytosis, a process in which large particles are engulfed into and very often digested by, or in any case rendered harmless by, 'macrophages' and 'polymorphs'.

phagocyte (fæ·gosəit), *v. Biol.* [f. the sb.] *trans.* = *PHAGOCYTOSE *v.* So **pha·gocyting** *ppl. a.*

1906 *Jrnl. Exper. Med.* VIII. 676 With a dilution of 1:20 not more than four or five staphylococci are taken up by the phagocyting cells. *Ibid.*, This possibility that a second substance besides the opsonins may determine the phagocyting power of leucocytes, also exists in the case of normal blood serum. **1933** M. Fernán-Núñez tr. *S. Ramón y Cajal's Histol.* xxi. 547 The Kupffer cells..phagocyte inside and injured erythrocytes. **1961** Chantrenne *Biosynthesis of Proteins* iii. 88 The low residual incorporation observed in nucleate *Amoeba proteus* is due to the activity of micro-organisms recently phagocyted by the *Amoeba*.

phagocytize (fægŏsəi·taiz), *v. Biol.* [f. Phagocyt(e) *v.* + -IZE.] *trans.* = *PHAGOCYTOSE *v.*

1925 *Physiol. Rev.* V. 195 The ink particles so carried into the liver were readily phagocytized by the Kupffer cells of that organ. **1956** *Arch. Path.* LXI. 165/2 Many of the spores were phagocytized early in the normal rabbit. **1973** R. G. Krueger et al. *Introd. Microbiol.* xxiii. 582/1 This cell type [*sc.* the macrophage] is characterized by its large size, ability to migrate, and its ability to phagocytize foreign materials.

So **pha·gocytized, -izing** *ppl. adjs.*, **pha·gocyti:zable** *a.*

1923 *Johns Hopkins Hosp. Bull.* XXXIV. 328/1 The Kupffer cells..were filled with phagocytized ink particles. **1924** *Physiol. Rev.* IV. 559 No evidence..of the true endothelium of the blood- and lymph-vessels and the heart..producing wandering, phagocytizing cells in

inflammation. **1948** *Biol. Skrift. K. Danske Vidensk. Selskab* IV. vii. 82 In histological sections from animals which have received injections of phagocytizable material, such as India ink, some of the sinusoid lining cells contain ink and some do not.

phagocytose (fægŏsəi·tōuz, -s), *v. Biol.* Also **-oze.** [Back-formation from Phagocytosis.] *trans.* To engulf or absorb (a cell or particle) like a phagocyte, so as to isolate or destroy it.

1912 R. W. Allen *Vaccine Therapy* (ed. 4) i. 32 The number of bacteria phagocytosed in a certain number of corpuscles were counted. **1943** Papanicolaou & Traut *Diagnosis Uterine Cancer by Vaginal Smear* 16 Endometrial cells which are being phagocytized by histiocytes. **1973** *Nature* 21 Sept. 50/1 These cells phagocytosed 'Zymosan' particles for at least 6 weeks. **1973** R. G. Krueger et al. *Introd. Microbiol.* xxiii. 582/1 The neutrophil..is capable of actively phagocytosing foreign material.

Hence **pha·gocytosed, pha·gocyto:sing** *ppl. adjs.*

1912 R. W. Allen *Vaccine Therapy* (ed. 4) i. 42 The phagocytosed bacteria should stain clearly. **1958** J. K. Frost in E. R. Novak *Gynecol. & Obstetr. Path.* (ed. 4) xxxv. 626 (*heading*) Mononuclear cells with phagocytosed..phagocytosing neutrophils. **1970** *Nature* 1 Aug. 511/1 These cells did not contain any phagocytosed particles, such as aluminium phosphate. **1972** *Science* 2 June 1040/1 [They] have reported that phagocytosing leukocytes take up inorganic iodide.

phagolysis (fægo·lisis). *Med.* [mod.L., ad. F. *phagolyse* (E. Metchnikoff 1895, in *Ann. de l'Inst. Pasteur* IX. 441), f. phago- Phagocyte: see *-LYSIS.] Lysis of phagocytes. **b.** Lysis by phagocytes.

1898 W. S. L. Barlow *Man. Gen. Path.* ix. 401 Attack directed from the German schools and advancing knowledge have obliged Metchnikoff to assume that under certain circumstances the phagocytes become dissolved (phagolysis) in the body fluids. *Ibid.*, The phagocytes, chiefly by phagocytosis, but partly by phagolysis, are responsible for immunity, natural and acquired. **1905** F. G. Bushnell tr. *Metchnikoff's Immunity in Infective Dis.* x. 334 Phagolysis [so I termed this transitory damage to the phagocytes] is indispensable for the manifestation of Pfeiffer's phenomenon in the peritoneal fluid. **1963** *Antibiotica* VIII. 465 (*heading*) Effect of Levomycetin..on phagolysis of Flexner dysentery bacteria resistant to this antibiotic.

phagolysosome (fægolai·sosōum). *Biol.* [f. next + *LYSOSOME.] A structure formed in the cytoplasm of a cell by the fusion of a phagosome and a lysosome, in which the foreign particle is digested.

1963 W. Straus in de Reuck & Cameron *Lysosomes* 166 The term 'phagosomes' was used to characterize the segregating ability of granules; the term 'lysosomes' was used to indicate the contents of hydrolytic enzymes in the granules..and the terms 'lyso-phagosomes' or 'phagolysosomes' were used when both these properties were described for the same granules. **1964** *Jrnl. Histochem. & Cytochem.* XII. 470/2 The foreign protein..was segregated in small and large 'phagosomes'. These later fused with pre-existing lysosomes. The gradual digestion of the injected protein in the 'phago-lysosomes'..could be followed by double staining. **1973** R. G. Krueger et al. *Introd. Microbiol.* xxiii. 584/1 The phagolysosome, containing the remains of the foreign object is then either eliminated from the cell or is left in the cytoplasm. **1977** *Lancet* 29 Oct. 904/2 These organelles are probably secondary lysosomes (phagolysosomes), common in actively feeding cells.

Hence **phagolysoso·mal** *a.*

1975 *Nature* 3 July 48/2 The phagolysosomal membranes was invariably applied closely to the surface of the enclosed yeast cell.

phagosome (fæ·gosōum). *Biol.* [f. Phagocyte + *-SOME[4].] A vacuole formed in the cytoplasm of a cell when a particle is phagocytosed and enclosed within a part of the cell membrane.

1958 W. Straus in *Jrnl. Biophysical & Biochem. Cytol.* IV. 548 The uptake of proteins is probably the function of certain intracellular granules related to the droplets of kidney cells. The term 'phagosomes' is suggested for these granules. (If these granules, characterized by segregating ability, are found to be identical with the 'lysosomes', the latter term can be used as a common nomenclature.) **1974** *Nature* 22 Nov. 315/1 If these processes engage in phagocytosis, phagosomes containing groups of cone disks should be observed within their cytoplasm.

Hence **phagoso·mal** *a.*

1975 *Nature* 17 Apr. 600/2 Possibly the living bacteria produce a factor which inhibits fusion between phagosomal and lysosomal membranes and thereby prevents the discharge of putatively bactericidal lysosomal contents into the bacterial environment.

phakellite, var. *FACELLITE.

phakic (fēi·kik), *a. Ophthalm.* [f. Gr. φακ-ός lentil + -IC.] Of an eye: having a crystalline lens (as in the normal organ). Cf. *APHAKIC *a.*

1918 J. H. Parsons *Dis. Eye* (ed. 3) xxiv. 495 It is easy to calculate the amount of axial myopia of a phakic eye which is 31 mm. long. **1961** *Lancet* 22 July 169/1 Two perforating grafts have been performed in phakic eyes, and although the appearance of the cornea has improved the results are not as good as in the aphakic cases. **1975** R. N. Shaffer in D. R. Anderson et al. *Symposium on Glaucoma* xiv. 250 In the phakic eye pupillary block is caused by apposition of the iris to the lens.

phakoanaphylaxis, var. *PHACOANAPHYLAXIS.

Phalangist[2] (fă·landʒist). [ad. Sp. *falange*: see *FALANGE.] **1.** A member of the Spanish *Falange. Also *attrib.* or as *adj.*

1936 *Times* 17 Aug. 10/1 The main attack was delivered by a force of some 3,000 men, including the Foreign Legion, Moroccan troops, and 'Phalangists' or insurgent Spanish Fascists. **1937** *Times* 3 Jan. 11/... ...during his hour's walk through the town, that it is chock-full of troops—Legionaries, Carlists, Phalangists, but no Moors. **1938** *Ann. Reg. 1937* 240 It was substantially the Phalangist programme which was adopted as the basis of the proposed Constitution for 'Nationalist' Spain. **1938** [see *FALANGE].

2. *transf.* A member of a right-wing, mainly Christian party in Lebanon. Also *attrib.* or as *adj.*

1959 *Daily Tel.* 30 June 4/3 Leaders of the mainly Christian Phalangist party are calling for an end to the 1969 Cairo agreement providing a Lebanese sanctuary for the commandos. **1975** *Times* 28 June 4/5 The Prime Minister-designate..is trying to reconcile differences between the mainly Christian Phalangists and..the influential socialist leader. **1977** *Time* 8 Aug. 29/1 The results were clearly joyous to 4,000 Phalangist..border police and flag-bearing Lebanese Christian Phalangist soldiers allowed—and left strictly for the occasion. **1979** *Guardian* 9 Apr. 7/5 Iran's young Guerillas..find themselves increasingly out of step with the..'Phalangists' as the Left-wing has dubbed the country's new theocracy.

phalangitis (fælændʒai·tis). *Path.* [f. *phalanges*, pl. of Phalanx (sense 3) + -ITIS.] Inflammation of the phalanges.

1877 J. D. Hooker in L. Huxley *Life J. D. Hooker* (1918) II. 142 [I got home] with a crick in my shoulder and 'phalangitis' from pump-handling some 500 people. **1963** *Lancet* 30 May 1526/1 He regarded the case as one of phalangitis of syphilitic origin.

phalanstery. (Earlier and later examples.)

1846 *Knickerbocker* XXVIII. 16 And are all your slaves productive workers? This is contemplated, I believe, in all the Phalansteries or Unitative Associations. **1963** V. Nabokov *Gift* iv. 235 Let us dream of the phalanstery living in a palace: 1,800 souls—and all happy!

phalaris (fæ·lăris). [L., f. Gr. φαλαρίς, Pliny's name for a similar grass; adopted by Linnæus in his *Hortus Cliffortianus* (1737) 23 as the name of a genus.] A grass of the genus so called, which includes canary-grass and some species useful for grazing.

1911 *Agric. Gaz. N.S.W.* XXII. 407 Phalaris and Kentucky Blue are spreading well. **1942** *Austral. Vet. Jrnl.* XVIII. 182 In three districts in South Australia the grazing of stock on pastures consisting almost exclusively of *Phalaris* has resulted in the occurrence of a 'staggers' syndrome. **1946** *Ibid.* XXII. 92/1 The general recommendation has been to select..areas of pasture in which phalaris was dominant. **1965** *Brit. Vet. Jrnl.* CXII. 360 When the disease is noticed early the sheep are moved to a fresh pasture containing no phalaris and the nervous condition soon improves. **1966** *New Scientist* 25 Aug. 411/2 Nervous disfunction of sheep grazing phalaris has been recognized for many years. **1977** *Weekly Times* (Melbourne) 19 Jan. 10/5 Some stands of the sub clover and phalaris have been harvested for seed.

2. phalaris staggers, a nervous disorder of sheep and cattle caused by the consumption of the perennial grass, *Phalaris tuberosa*.

1946 J. W. McDonald in *Austral. Vet. Jrnl.* XXII. 91/1 A disease of sheep and cattle characterized by profound nervous symptoms and degenerative changes in the central nervous system was described. The disease was associated with the consumption of the perennial grass *Phalaris tuberosa* and thus had become colloquially known as 'phalaris staggers'. **1966** *Ibid.* XLII. 375/1 Neurological disorders of sheep grazing *Phalaris tuberosa* have hitherto been grouped under the term 'phalaris staggers'. **1970** Jubb & Kennedy *Path. Domestic Animals* (ed. 2) II. xi. 387/2 The lesser syndrome, known as phalaris staggers, occurs when the plant is in lush growth and in 2–10 days after animals are given access to such pasture. Affected sheep appear normal when quiet, but, when disturbed, they show stiffness, incoordination, a staggering gait, and a tendency to fall.

phallic, *a.* Add: (Further examples.) Also in spec. collocations as *phallic stage* Psychoanal., *phallic symbol*; and in adj. phr. as *phallic-centred, -shaped*.

1907 G. B. Shaw *John Bull's Other Island* II. 38 The view that the Round Towers are phallic symbols. **1923** *Internat. Jrnl. Psycho-Anal.* IV. 383 Charlie Chaplin as phallic symbol. **1927** *Ibid.* VIII. 468 Freud postulated a 'phallic' stage in female development corresponding with that in the male. **1928** D. H. Lawrence *Let.* 15 Mar. (1962) II. 1046 It is a nice and tender phallic novel. **1949** J. Strachey tr. *Freud's Three Ess. Theory of Sexuality* ii. 77 This phase..is differentiated from the final organization of sexual maturity... For it knows only one kind of genital: the male one. For that reason I have named it the 'phallic' stage of organization. **1952** C. Thompson *Psychoanal.* iii. 69 A phallic character comes from sublimation of the phallic stage. Such a person is..insolent, domineering, and aggressive. **1959** E. Pound *Thrones* xcix. 49 That man's phallic heart is from heaven a clear spring of rightness. **1963** A. Heron *Towards Quaker View of Sex* iii. 23 Males are very phallic-centred. **1964** P. F. Anson *Bishops at Large* ix. 356 The 'forces' and 'vibrations' within this phallic-shaped building. **1968** R. & G. Blanck *Marriage & Pers. Devel.* ii. 9 The anal character is greedy; the anal character is stubborn and parsimonious; the phallic character is exhibitionistic and competitive. **1968** M. Ellman *Thinking about Women* iii. 45 It is through this identification that phallic criticism regularly and rapidly shifts from writing by women, which can be dismissed as innocuous, to their vicious influence upon writing by men. **1972** C. Snow *Naked Skor* xxiv. 137 I switched my mind to contemplation of the Eiffel Tower. It too was a phallic symbol. **1975** C. Register in J. Donovan *Feminist Lit. Crit.* 9 A second literature is therefore phallic criticism is to admit grudgingly to the literary value of works by women, and then to deny that it is consciously attained. **1977** *Spare Rib* Sept. 13/2 Now we [*sc.* women] can have the phallic freedoms to screw where and when we want. **1978** *Amer. Poetry Rev.* July/Aug. 19/1 If you riffle the pages from back to front you will see one block of pattern divide, while another phallic-shaped pattern slowly enters it.

phallically (fæ·likăli), *adv.* Also **phallicly.** [f. Phallic, Phallical adjs.: see -ICALLY.] In a manner resembling a phallus.

1967 E. Grierson *Crime of one's Own* i. 9 In the distance Brunelleschi's dome and Giotto's tower rose phallically towards the Tuscan sky. **1973** *Daily Tel.* (Colour Suppl.) 11 June 27/4 *Bomb Release*, Dennis's own work that lights up and changes colour while it erects phallicly. **1978** J. Updike *Coup* (1979) iii. 101 Abrasive ointments phallically packaged and chemically fortified.

phallicist. (Example.)

1924 *Jrnl. R. Anthrop. Inst.* LIV. 49 The phallicists of the seventies, Burton, Fergusson, ..and Westropp.

phallin (fæ·lin). *Chem.* Also **-ine.** [a. G. *phallin* (R. Kobert 1891, in *St. Petersb. med. Wochenschr.* XVI. 472/2), f. mod.L. *phalloides* (see *PHALLOIDIN): see -IN[1], -INE[5].] A haemolytic substance present in the death cap toadstool, *Amanita phalloides*, and formerly thought to be its poisonous principle.

1897 *Trans. Brit. Mycol. Soc.* 27 Professor Kobert..has given the name of phalline to the poisonous substance present in *Amanita phalloides*, Fries. *Ibid.* 28 Phalline is a very powerful agent..in causing the dissolution of the red corpuscles of the blood. **1930** *Nature* 1 Mar. 324/2 The toxin of tetanus is checked by La Bourboule water and not by Mont Dore, phallin poison by Mont Dore water and not by La Bourboule. **1965** Bell & Coombe tr. *Strasburger's Textbk. Bot.* 598 Familiar Agaricaceae...*Amanita phalloides*... deadly poisonous with toxic glucosides (phallin) and peptides (phalloidin, amanitin). **1972** T. & O. Wieland in S. Ajl et al. *Microbial Toxins* VIII. x. 252 A hemolytic agent..was obtained from *A. phalloides* as early as 1891 by Kobert and was called 'phallin'. It has recently been reinvestigated by Fiume (1967), who..stated that, in addition to a hemolytic action, a cytotoxic effect on cultures of cells of HeLa and of human amnion cells occurred. This substance, whose chemical nature has not yet been elucidated, is presumably destroyed in the gastrointestinal tract and, therefore, cannot be responsible for lethality of the mushroom.

phallism. (Later examples.)

1928 A. Huxley *Point Counter Point* x. 165 Civilized lasciviousness is the same as the savage's.—What is it? Ah...phallism...of the ancients... Worshipping with the body—that's the genuine phallism. **1950** *Funk's Stand. Dict. Folklore* II. 863/2 Among some ethnic groups phallism, also often called sex worship, is the worship of the male organ.

phallocentric (fælose·ntrik), *a.* [f. Gr. φαλλός penis + *-CENTRIC.] Centred on the phallus. Hence **phallocentri·city, phallocentrism.**

1927 *Internat. Jrnl. Psycho-Anal.* VIII. 459 There is a healthy suspicion growing that men analysts have been led to adopt an unduly phallo-centric view of the problems in question. **1951** *Brit. Jrnl. Psychol.* XLII. 362 The phallocentric tendencies of early psychology. **1962** G. Greer *Female Eunuch* 48 The enormous hoo-ha about the strange [unpalpable] results of vasectomy..results from..phallocentricity. **1977** *Spare Rib* June 46/2 The capitalists, without too much struggle, to an individual solution of phallocentric romantic love. **1977** A. Sheridan tr. *Lacan's Écrits* vi. 198 The phallocentrism produced by the..that is..that need concern us here. **1977** R. Holland *Self & Social Context* viii. 254 Wilden's critique of Lacan, whose valuable exposition of the Imaginary does not allow him to disentangle in the phallocentrism and his elitist conceptual violence.

phallocrat (fæ·lōkræt). [f. as prec. + -CRAT: cf. Fr. *phallocrate*.] One who advocates or assumes the existence of a male-dominated society; a man who argues his superiority over women because of his masculinity. Also *attrib.* or as *adj.* So **phallo·cracy, phallocra·tic** *a.*

1977 *Gay News* 24 Mar. 11/3 The march was backed up by members of the Groupe de Liberation Homosexuel..shouting the slogan 'with women against phallocracy'. **1977** *Times Lit. Suppl.* 10 June 699/3 That feeble tyrant, the reincarnation of her father—the 'phallocrat' male.

1977 *Radical Philos.* Summer 9/2 It would be a mystery how the jury of 1971 could have escaped the effects of this phallocratic unconscious. *Ibid.*, Her phallocrat of a husband or lover letting her deal with all the domestic chores. **1977** *Ibid.* Autumn 1, I do myself think that the style of at least some of those 'great texts' is indeed phallocratic. **1978** *Times Lit. Suppl.* 24 Mar. 342/5 By a species of reification no more excusable than that dictated by the phallocrats, they have consecrated a ready-made caricature of Woman which serves to reduce all women to an absolute type.

phalloid, *a.* (Example.)

1967 A. Edwardes in R. E. L. Masters *Sexual Self-Stimulation* 308 Arab and Israeli girls of prepubertal age..masturbate..with fingers or some phalloid object.

phalloidin (fæloi·din). *Chem.* Also **-ine.** [a. G. *phalloidin*, given its present meaning by U. Wieland 1938 (*Ann. d. Chem.* DIII. 100) following its coinage as F. *phalloïdine* by A. Gubler 1877 (*Bull. de l'Acad. de Méd.* VI. 879) to denote an 'amorphous toxic principle' obtained from *A. phalloides* by P. C. Oré; f. mod.L. *phalloides*, specific epithet (f. Gr. φαλλός penis: see -OID): see -IN[1], -INE[5].] The principal phallotoxin, $C_{35}H_{48}N_8O_{11}S$.

1938 *Chem. Abstr.* XXXII. 3456 Toxin II has been crystd. (needles) and is designated as phalloidin. **1955** *Sci. News Let.* 23 Apr. 265/2 The two amanitines and phalloidine in mushrooms affect the liver, kidney and heart. **1961** *Lancet* 16 Sept. 630/2 More than a century of investigation, mainly in Germany and France, has led to the isolation of five crystalline toxins from *Amanita phalloides* (α, β, and γ amanitin, phalloin, and phalloidin). **1972** *Science* 1 Sept. 808/1 The principal phallotoxin is phalloidin. **1977** R. Jeffries *Troubled Deaths* xiii. 101 The normal symptoms of phalloidine poisoning.

phalloin (fæ·lo,in). *Chem.* [a. G. *phalloin* (Wieland & Mannes 1957, in *Angewandte Chem.* LXIX. 389/1), f. mod.L. *phallo-ides* (see prec.) + -IN[1].] One of the phallotoxins, $C_{35}H_{48}N_8O_{10}S$.

1959 *Chem. Abstr.* LIII. 18880 Chromatographic sepn. of the enriched poison mixt. of the amanita mushroom on neutral Al_2O_3 in H_2O-satd. BuOH + 10% EtOH mixt. gave another toxin, called phalloin. **1961** [see *PHALLOIDIN]. **1968** [see *PHALLOTOXIN].

phallophoria (fælofǭ·riă). [f. Gr. φαλλοφορ-εῖν to bear a phallus + -IA[1] or ad. Gr. φαλληφόρια (see Phallephoric *a.*) influenced by the root φαλλό-ς.] The carrying of a phallus, esp. as part of a festival of Dionysus in ancient Greece. Hence **phallopho·ric** *a.*, = Phallephoric *a.*

The Gr. word φαλληφόρια is a neuter plural, but the English form *phallophoria* is usu. const. as a sing.

1903 J. E. Harrison in *Jrnl. Hellenic Stud.* XXIII. 322 We have established..that a phallophoria formed a part of Dionysiac mysteries. **1950** *Funk's Stand. Dict Folklore* II. 864/1 The carrying of phalli in public processions (*phallophoria*). **1964** *New Statesman* 13 Mar. 398/1 The puppets looked more phallophoric than the Beatles themselves. As the taped Beatle music rose to a pitch, the jiggling became an almost indecent enactment of sexual rhythm. **1967** A. L. Lloyd *Folk Song in England* ii. 93 The randy animal-guiser song of the 'Derby Ram'..a song that is the lyrical equivalent of those phallophoric dances that survive in farming communities in Europe, intended to celebrate and stimulate the powers of reproduction.

phallophorus (fælofǫ·rǔs). Also **-phoros.** Pl. **-phori, -phoroi.** [ad. Gr. φαλλοφόρος bearing a phallus.] One who carries a phallus, esp. as part of a festival of Dionysus in ancient Greece.

1854 C. D. Yonge tr. *Athenæus' Deipnosophists* III. xiv. 992 Semos the Delian says in his book about Pæans... 'And those,..who are called Ithyphalli, wear a mask representing the face of a drunken man... And the Phallophori..wear no masks, but they put on a sort of veil of wild thyme... The Phallophorus..marched straight on, covered with soot and dirt'. **1885** H. M. Westropp *Primitive Symbolism* 54 The peripballia..carried long poles with phalli hung at the end of them: they were crowned with violets and ivy... These men were called phallophori. **1909** L. R. Farnell *Cults Gk. States* V. v. 210 It may be that Semos of Delos is describing something that happened in the later Attic festival when—according to Athenaeus—he speaks of 'the ithyphalloi'..and of the phallophoroi without masks entering through the central doors of the theatre. **1927** A. W. Pickard-Cambridge *Dithyramb Tragedy & Comedy* iii. 234 The phallophorus proper had his face disguised with soot. (Probably he carried, but did not wear, the phallus). *Ibid.* 235 Possibly the ceremony of the phallophori was of a common type, differing little from town to town. **1931** A. Nicoll *Masks, Mimes & Miracles* i. 26 Bethe thinks that the *phallophoroi* were Delians; for this there is no proof.

‖**phallos** (fæ·lǫs). Pl. **-oi.** [Gr. φαλλός penis.] = Phallus. Also *attrib.* and *Comb.*

1928 D. H. Lawrence *Lady Chatterley* x. 161 The man dwindled to a..mere phallos-bearer, to be torn to pieces when his service was performed. **1950** *Funk's Stand. Dict. Folklore* II. 862/1 The Greek satyrs of the Dionysiac *orgia* stamped and capered with gigantic phalloi and artificial breasts—a symbol of self-fructification. **1963** Auden *Dyer's Hand* III. 102 If all round hills were suddenly to turn into breasts, all caves into wombs, all towers into phalloi, we should not be pleased..we should be bored. **1972** *Times Lit. Suppl.* 19 May 565/3 Golden phalloi of superb craftsmanship and considerable avoirdupois. **1978** K. J. Dover *Greek Homosexuality* III. 133 The 'phallos-bird', which has the legs, body and wings of a bird but a neck and head in the form of a curved penis with the foreskin rolled back and an eye on the glans.

phallotoxin (fælotǫ·ksin). *Chem.* [f. mod.L. *phallo-ides* (see *PHALLOIDIN) + Toxin.] Any of several closely related poisonous peptides present in the death cap toadstool, *Amanita phalloides*, all of which have the same ring structure of seven amino-acids bridged by a sulphur atom.

1968 *Science* 1 Mar. 949/1 The phallotoxins include phalloidin, phalloin, [etc.]. **1972** *Ibid.* 1 Sept. 808/1 The rapidly acting phallotoxins..are responsible for the gastrointestinal phase of the poisoning. **1975** *Sci. Amer.* Mar. 95/1 In the phallotoxins the cyclopeptide ring has seven amino acids; in the amatoxins it has eight.

phallus. Add: **1. b.** The male generative organ, often in the context of its symbolical significance; in psychoanalysis, in the context of the pre-genital phase of sexual development.

By some writers used in the specific sense 'erect penis'.

1924 J. Riviere tr. *Freud's Infantile Genital Organization* in *Coll. Papers* II. 245 The primacy reached is..not a primacy of the *genital*, but of the *phallus*. *Ibid.* 247 The significance of the castration complex can only be rightly appreciated when its origin in the phase of the primacy of the phallus is also taken into account. **1935** M. Balint *Primary Love & Psycho-Anal. Technique* (1952) iii. 56 The same is true of the phenomena of the phallic period: belief in the non-existence of the female genital organs..or in the existence of mothers with a phallus. **1935** R. V. Storer *Sexual Technique* xxi. 274 Effective virility therefore depends on erectile capacity; that is, the penis must be able to become a phallus (as the erect male organ in action is called). **1951** Ford & Beach *Patterns of Sexual Behavior* ii. 22 The bodily adjustments that give rise to these sensations involve merely insertion of the phallus followed by piston-like thrusts within the vagina. **1960** J. Rodney *Handbk. Sex Knowledge* iv. 45 In its simplest—and crudest—form, copulation consists of the entrance of the erect penis or phallus into the vagina, followed by certain movements which produce an ejaculation of male semen into the vaginal cavity. **1961** C. & W. M. S. Russell *Human Behaviour* vi. 271 A welter of key stimuli independent of the individual's personality, of breasts and phalluses, hands and feet. **1963** A. Heron *Towards Quaker View of Sex* iii. 35 The experience is thus phallus-centred and produces excitement without deep commitment. **1972** *Times Lit. Suppl.* 29 Sept. 1156/5 A phallus is a prerequisite of intercourse and, being so, has a social significance..whereas a penis can never have more than a biological significance. **1974** H. S. Kaplan *New Sex Therapy* i. 21 The female sexual response transforms the tight, dry vaginal potential space into a well-lubricated, open receptacle for the phallus. **1977** *Times* 18 Feb. 11/7 Fat villains sporting Old Comedy phalli.

2. (Later examples.)

1953 J. Ramsbottom *Mushrooms & Toadstools* xvi. 182 A violet-scented *Phallus* would certainly be exceptional. **1976** G. C. Ainsworth *Introd. Hist. Mycology* iii. 38 There is a figure of the stink-horn (both 'egg' and fruit-body) which in 1562 had been described and illustrated by the Dutch physician Adriaen Jonghe (H. Junius) who gave the name 'Phallus' to it in the first independently published mycological monograph.

3. *phallus-worship* (earlier example).

1850 Carlyle *Hudson's Statue* in *Latter-Day Pamphlets* vii. 43 Kings and Papas flying like detected coiners; and in their stead Icaria, Red Republic,..Literature of Desperation curiously conjoined with Phallus-Worship, too clearly heralding centuries of bottomless Anarchy.

Phanar. Add: (Earlier and later examples.) Also, the seat of the Patriarch of Constantinople after the Ottoman conquest. Also *transf.*

The more common spelling in the 19th century was *Fanar.

1897 R. Davey *Sultan & his Subjects* II. iii. 112 The noble Greek families which returned after the siege and settled in that quarter of Constantinople, known as the Phanar. **1900** 'Odysseus' *Turkey in Europe* vi. 276 One small Slavonic Church, however, that of Montenegro, resisted Hellenism and the Phanar as successfully as the Government withstood the Turks. **1911** E. Pears *Turkey & its People* vii. 118 The patriarch of Constantinople..resides at the Phanar, a district in Constantinople... As there was much intrigue and bribery..under the sultans, Phanariot came to be a synonym for a man of unscrupulous political intrigue. **1936** H. Luke *Making of Mod. Turkey* iv. 97 It was only in 1850 that the Phanar, reluctantly and under Russian pressure, recognized the Church of the Greek kingdom as independent. **1961** D. Attwater *Christian Churches of East* II. iii. 22 In 1766 the Phanar obtained from the sultan Mustafa III an order for the suppression of the Serbian independent church. **1964** P. F. Anson *Bishops at Large* viii. 281 No Christian Church would be able to offer such..ceremonial as this new one with its '*phanar*' or '*curia*' in Holloway.

phanatron, var. *PHANOTRON.

phanerophyte (fæ·nĕrofəit). *Bot.* [ad. Da. *fanerofyt* (C. Raunkiaer 1904, in *Bot. Tidsskr.* XXVI. p. xiv), f. Phanero- + -phyte.] A plant which bears its dormant buds well above the surface of the ground.

1913 W. G. Smith in *Jrnl. Ecol.* I. 17 Phanerophytes have their dormant buds on branches which project freely into the air; they are the trees and shrubs. **1932** Fuller & Conard tr. *Braun-Blanquet's Plant Sociol.* xii. 295 The more northern phanerophytes are characterized by special bud protection. **1934** H. Gilbert-Carter tr. *Raunkiaer's Life Forms of Plants* ii. 19 By Phanerophytes I mean plants whose buds and apical shoots destined to survive the unfavourable period of the year project into the air on stems which live for several, often for many, years. **1952** [see *chamæphyte* s.v. *CHAMÆ-]. **1976** Bell & Coombe tr. *Strasburger's Textbk. Bot.* (rev. ed.) 191 Phanerophytes..bear their resting buds more than 50 cm above the soil surface.

phanerozoic (fæ:nĕrozōu·ik), *a.* [f. Phanero- + Gr. ζωή life + -ic.] **1.** *Ecol.* Describing those animals living in exposed conditions above the surface of the ground. Cf. *CRYPTOZOIC *a.* 1.

1896 A. Dendy in *Natural Sci.* July 8 Cryptozoic fauna imperceptibly blends with what by way of contrast may be called the 'phanerozoic'. **1911** A. Willey *Convergence in Evolution* iii. 23 Phanerozoic or diurnal animals are positively heliotropic. **1922** Flattely & Walton *Biol. of Sea-Shore* i. 7 We thus have two great sub-realms: the phanerozoic and the cryptozoic.

2. *Geol.* (With capital initial.) Of, pertaining to, or designating the whole of the geological time since the beginning of the Cambrian period, as contrasted with the Pre-Cambrian (Cryptozoic); (so named in allusion to its abundant evidence of life). Also *absol.*

1930 G. H. Chadwick in *Bull. Geol. Soc. Amer.* XLI. 48 The following classificatory scheme is offered for criticism. ..Phanerozoic or Phanerobiotic 45%... Cryptozoic or Cryptobiotic (Precambrian) 55%. **1958** R. C. Moore *Introd. Hist. Geol.* (ed. 2) iv. 59/2 It is convenient also to designate the Cryptozoic rocks as Precambrian, a commonly used term indicating age older than the Cambrian Period, first division of the Paleozoic Era and of the Phanerozoic Eon. **1963** *Q. Jrnl. Geol. Soc.* CXIX. 137 The rocks concerned range in age from 1055 to 1700 m.y. (radiometric dating), spanning more time than the whole of the Phanerozoic. **1971** *Nature* 25 June 498/1 All stratigraphic methods of correlation..that are used for Phanerozoic rocks should be extended as far as possible in the Pre-Cambrian part of the stratigraphic column. **1974** *Ibid.* 18 Oct. 568/2 It is no coincidence that the three faunally defined eras of Phanerozoic time, the Palaeozoic, Mesozoic and Cainozoic, are divided by these two so-called crises in the history of life. **1977** A. Hallam *Planet Earth* 189 The later Proterozoic Grenville belt of southern Canada..is a linear belt with more similarities to the younger mountain belts of the Phanerozoic.

phanerozonate (fænĕrozōu·nēit), *a. Zool.* [f. mod.L. order name *Phanerozonia* (W. P. Sladen in Thomson & Murray *Rep. Sci. Results Voy. H.M.S. Challenger: Zool.* (1889) XXX. p. xxvi), f. Phanero- + Gr. ζώνη girdle + -ia[2]: see -ate[2].] Characteristic of or comprising certain starfishes grouped in the order Phanerozonia, distinguished by conspicuous marginal plates.

1889 W. P. Sladen in *Rep. Sci. Results Voy. H.M.S. Challenger: Zool.* XXX. p. xxvi, I consider that the Phanerozonate group is older than the Cryptozonate. **1906** E. W. MacBride in *Cambr. Nat. Hist.* I. xvi. 454 In a very large number of Asteroidea the supero- and inferomarginal ossicles are represented by squarish plates even when the rest of the skeleton is reticulate; this is the so-called 'phanerozonate' structure. **1923** *Nature* 13 Jan. 47/1 What these specialists are impressed by is the 'phanerozonate' character of the Astropectinidae, that is, the edging of the arms with a series of broad plates termed the 'marginals'. **1966** J. W. Durham et al. in R. C. Moore *Treat. Invertebr. Paleont.* U. I. U36/2 Some Cryptozonia have a phanerozonate stage in ontogeny.

phanerozone (fæ·nĕrozōun), *a.* (*sb.*) *Zool.* [see prec.] = prec. Also as *sb.*

1962 D. Nichols *Echinoderms* iii. 51 *Petraster* is a true phanerozone starfish. *Ibid.* 53 There is a thick aboral membrane for protection instead of the paxillate condition normally associated with the phanerozones. **1964** *Oceanogr. & Marine Biol.* II. 394 Spencer..describes the transition from the early somasteroid stock to the phanerozone asteroids.

phanopœia (fæ:nopī·yă). [f. Gr. φανό-ς light, bright, f. φαίνειν to shine, give light + ποι-εῖν to make, after Melopœia: see -ia[1].] (See quot. 1929.)

1929 E. Pound in *N.Y. Herald-Tribune* 20 Jan. xi. 5/4 There are three 'kinds of poetry': Melopoeia... Phanopoeia, which is a casting of images upon the visual imagination... Logopoeia. *Ibid.* 6/1 In phanopoeia we find the greatest drive toward utter precision of word. **1934**, **1957** [see *LOGOPŒIA]. **1961** [see *MODERNIST 5].

phanotron (fæ·nŏtrǫn). *Electronics.* Also **phanatron.** [f. *phano-* (? f. Gr. φαίνειν to bring to light, cause to appear) + *-TRON.] A thermionic diode rectifier utilizing an arc discharge in mercury vapour or gas at very low pressure.

Formerly a proprietary name in the U.S.

1931 *Thomas' Reg. Amer. Manuf., 1931–32* (ed. 22) *Leading Trade Names* 447 Phanotron vacuum tubes, General Electric Co., Schenectady, N.Y. **1933** J. H. Morecroft *Electron Tubes* viii. 193 In the Phanotron a

low temperature filament must be used, because the mercury vapor has not enough density to prevent the evaporation of the filament. **1939** H. J. Reich *Theory & Applications Electron Tubes* xii. 430 The construction of the type FG-166 phanatron. **1966** *McGraw-Hill Encycl. Sci. & Technol.* VI. 62/1 Typical phanotrons have average current ratings from 1·25–10 amperes and peak inverse voltage ratings of 10,000–22,000 volts. **1975** Yeager & Gourley *Introd. Electron & Electromech. Devices* (1976) xiv. 283 We will discuss two types of gaseous discharge: arc discharge and glow discharge... Arc discharge tubes are commonly called thermionic gas tubes; for example the gas diode (phanotron) and the gas triode (thyratron).

phanta·smically, *adv.* [f. Phantasmical *a.* + -ly[2].] = Phantasmally *adv.*

1906 *Daily News* 2 Jan. 6 If I stretch out my hand to a touch, will it not surely melt under my fingers?—melt and form again phantasmically.

phantastica (fæntæ·stikă). [f. *phantastic*, var. Fantastic *a.* and *sb.* + *-a* (cf. -a 4).] Hallucinogenic drugs collectively; ¶ also, one such drug.

1931 L. Lewin *Phantastica, Narcotic & Stimulating Drugs* 31 Phantastica; hallucinating substances. This series comprises a number of substances of vegetable origin, varying greatly in their chemical constitution, and to these belongs in its proper sense the name Phantastica, or Drugs of Illusion. *Ibid.* 92, I mean the action of chemical substances capable of evoking such transitory states without any physical inconvenience for a certain time in persons of perfectly normal mentality who are partly or fully conscious of the action of the drug. Substances of this nature I call Phantastica. They..influence particularly the visual and auditory spheres as well as the general sensibility. **1958** *Sci. News* XLVII. 31 The two groups of drugs on which most attention has been focused recently have been: (*a*) those that tend to make normal people appear to be psychotic (hallucinogens, psychotomimetic drugs or phantastica), and (*b*) those that are mostly intended to have the opposite effect. **1967** M. M. Glatt et al. *Drug Scene* i. 9 The three principal drugs in this group are mescaline, psilocybin, and lysergic acid diethylamide (LSD 25 or LSD). These drugs are also designated as phantastica, psychedelic, and psychotomimetic drugs. **1968** J. Blackburn *Young Man from Lima* v. 51 Marcus tried to remember what he knew about modern hallucinogenics... Lysergine..was probably the most powerful phantastica known to man.

phantastikon (fæntæ·stikǫn). *poet.* [f. Gr. φανταστικόν imaginative faculty, neut. of φανταστικός (see Fantastic *a.* and *sb.*).] Imagination.

1917 E. Pound *Lustra* 187 Shall I claim; Confuse my own phantastikon Or say the filmy shell that circumscribes me Contains the actual sun. **1933** —— *Let.* 15 Jan. (1971) 243 As fer yr. final pp., that is about the kind of mess that has been trespassin on my phantastikon for several weeks.

phantastron (fæntæ·strǫn). *Electronics.* [f. alteration of Fantastic *a.* and *sb.*: see *-TRON.] A circuit in which a brief triggering pulse produces at one electrode a longer pulse whose duration is proportional to an applied control voltage, and at another electrode a linear sweep voltage of the same duration (which can then be used to generate another pulse that will occur at a definite interval after the triggering pulse).

1943 (*title*) An adaptation of the phantastron delay multivibrator circuit to the 6SA7 tube. (M.I.T. Radiation Laboratory Rep. 63-21.) **1947** L. J. Haworth in L. N. Ridenour *Radar Syst. Engin.* xiii. 500 The phantastron of Fig. 13·17 is a flip-flop..which serves as a timing circuit maintaining its calibration to about one per cent. **1953** *Electronic Engin.* XXV. 143/1 The position of the selector pulse in the time scale depends on the delay produced by the phantastron and is controlled by R_{13}. **1964** R. Blitzer *Basic Pulse Circuits* vi. 215 A linear sawtooth voltage generator similar to the Miller sweep circuits of Sec. 6·11 is called the phantastron... Besides the sawtooth waveshape being used as a sweep, the phantastron is also often used as a delay circuit.

phantastry, the more usual form of Fantastry.

Not *Obs.* in this spelling.

1656, etc. (in Dict. s.v. Fantastry). **1870** *Contemp. Rev.* XIII. 1 It [*sc.* the Church] must have its pomp and phantastry, its tributes and taxes. **1922** Hardy *Late Lyrics* 72 So white her drape..I could not guess what phantastry it meant.

phantom, *sb.* Add: **5. c.** *Telegr.* and *Teleph.* An additional circuit obtained by using each of two other circuits as one of its two conductors, the two wires of each of the other circuits being effectively in parallel. Usu. *attrib.*

1883 G. Black in *Operator & Electr. World* 3 Feb. 71/1 The method of telephonic transmission was discovered by myself in 1878, while experimenting to get rid of telegraphic induction in telephones. I found that my apparatus gave me a new 'phantom' circuit over a telegraph wire. **1920** J. G. Hill *Telephonic Transmission* ix. 192 If one of the side circuits in a phantom circuit is out of order, the phantom necessarily fails with it. **1924** W. Aitken *Automatic Telephone Syst.* III. xlviii. 229 In using phantom or superimposed circuits on automatic systems great care must be exercised to prevent impulse and other currents in one physical circuit affecting the mate physical circuit by way of the phantom loop. **1943** A. L. Albert *Fund. Telephony* x. 258 A phantom circuit is obtained by repeating coils..installed at each end of two pairs of wires constituting the phantom group. This arrangement gives three telephone channels: one over each side circuit and one channel over the phantom circuit. **1957** W. Fraser *Telecommunications* v. 123 The additional circuits may be provided by utilising pairs of phantom circuits to produce other phantom circuits.

d. *Radiology.* A life-size model of part of the body made of material which absorbs radiation in a similar way, used in investigations into the character and absorption of a beam of radiation.

1922 tr. *Kroenig & Friedrich's Princ. Physics & Biol. Radiation Therapy* i. 33 Perthes had to employ a solid substance, namely aluminium, as phantom material... These aluminium phantoms have general use in practice on account of their convenience. **1950** Walter & Miller *Short Textbk. Radiotherapy* iv. 99 Water is not always an ideal medium in which to insert small ionization chambers... A suitable phantom can be made of layers of pressed wood fibre which can be obtained of unit density. Suitable holes..enable the ionization chamber to be inserted. **1974** Rafla & Rotman *Introd. Radiotherapy* iii. 35/1 A plastic material..., which simulates water (and soft tissues), has been described. If such a material is placed around a bony skeleton, then a phantom that simulates the body with its bones more accurately than water alone can be built.

6. (Further examples.) Also in more general use: imaginary; false; devised by way of pretence, imitation, or deceit.

1885 Kipling *Phantom 'Rickshaw* (1889) 25 The phantom 'rickshaw and I went side by side along the Chota Simla road in silence. **1897** W. B. Yeats *Let.* 24 Dec. (1954) 293 He did not come because of the phantom sore throat. **1909** *Westm. Gaz.* 19 Aug. 8/1 There was more phantom work on the Downs yesterday. **1927** H. Crane *Let.* 7 Jan. (1965) 283 The 'ships' [in a poem] should meet and pass in line and type—as well as in wind and memory, if you get my rather unique formal intentions in this phantom regatta seen from Brooklyn Bridge. **1931** *Daily Express* 28 Apr. 11/4 The ball was centred, and the eleven men, playing a phantom team, swept down the pitch to the unguarded goal. **1934** *Sun* (Baltimore) 5 Apr. 1/3 The steel industry was indulging in a monopolistic form of price boosting and price fixing which included the writing into its price structure of so-called 'phantom' freight rates which the consumer pays, but which find their way into the manufacturers' coffers and not those of the railroads. **1951** M. McLuhan *Mech. Bride* (1967) 143/2 They are not rooted in any concept of civilized society but are merely a blind drive toward the phantom security of subrational collectivism. **1952** *Times* 12 Dec. 12/3 Lot 90 was not a phantom beaver coat, or indeed a beaver coat at all, but a phantom racoon coat. **1958** *Times* 29 Sept. 13/1 The duvetyn coat with a phantom beaver collar. **1976** *National Observer* (U.S.) 29 May 9/2 A 'phantom', or dummy, fluorescent tube that reduces the amount of electricity used by standard two-tube fixtures.

7. a. *phantom-life*; **b.** Also objective, as *phantom-chaser*; also *phantom-wise* adv.

1954 Koestler *Invisible Writing* ii. 34 The phantom-chaser..who discovers Helen's image in each beloved face. **1907** *Folk-Lore* June 147 Cuchulain was recalled to phantom-life on one occasion by St. Patrick. **1871** 'L. Carroll' *Through Looking-Glass* 223 Still she haunts me, phantomwise.

8. phantom limb, a sensation of the presence of an amputated limb; so **phantom pain,** pain perceived as in a phantom limb.

[**1872** S. W. Mitchell *Injuries of Nerves* xiv. 349, I recently faradised a case of disarticulated shoulder... As the current affected the brachial plexus of nerves, he suddenly cried aloud, 'Oh, the hand, the hand!' and attempted to seize the missing member. The phantom I had conjured up swiftly disappeared.] **1879** G. H. Lewes *Probl. Life & Mind* (ser. 3) II. 336 The 'phantom limb', of which Weir Mitchell speaks, is only one detail in the general picture mentally formed of the body. **1937** *Lancet* 8 Aug. 314/1 After amputation it was usual for the patient to experience sensations as if his limb were still present. These phantom limbs might be painless or painful. **1955** *Sci. News Let.* 13 Aug. 104/3 Phantom limb pains, a troublesome affliction in amputation cases, and pains in amputation stumps can be relieved in many cases by ultrasound treatment. **1974** *Nature* 26 Apr. 731/1 Phantom limbs occur in 95 to 100% of all people who have had a limb or part of a limb amputated. **1960** I. A. Stanton *Dict. for Med. Secretaries* 115/2 Phantom pain. **1964** *N.Y. State Jrnl. Med.* LXIV. 2907/2 Many theories have been advanced to explain the mechanism of phantom pain. **1973** *Daily Colonist* (Victoria, B.C.) 11 Apr. 34/7 It's pretty weird to have a pain in a foot that has been amputated... The person [is] suffering from 'phantom pains' as they are called.

pharaoh. Add: **4. Pharaoh hound,** a short-coated, tan-coloured hunting dog with large, pointed ears, belonging to the breed so called; **Pharaoh's ant** (examples).

1967 R. Glyn *Champion Dogs of World* 191/2 The Ibizan Hound. Also known as the Pharaoh Hound. **1969** V. Canning *Queen's Pawn* xiii. 240 Raikes sat..reading an article about the revival of interest in Pharaoh hounds from Malta. **1974** A. G.-I. Browne *Hamlyn Guide to Dogs* 180 Pharaoh Hound... Although in the opinion of many people they belong to one and the same breed, the FCI recognizes the Pharaoh Hound as well as the Podenco Ibicenco... It is supposed that the Pharaoh Hound came to Gozo and Malta with the seafaring Phoenicians. **1910** W. M. Wheeler *Ants* i. 10 Some of them [*sc.* tropical ants], like Pharaoh's ant..., have been carried by commerce to all the inhabited regions of the globe. **1939** Metcalf & Flint *Destructive & Useful Insects* (ed. 2) xxi. 770 Pharaoh's ant, *Monomorium pharaonis*... This tiny, slender, yellowish-red ant..generally nests in inaccessible places about the foundations and in the walls of buildings, from which it forages indoors the year round in search of food. **1978** *Times* 24 Apr. 2/2 Ministry of Agriculture scientists have devised a way to get rid of one of the more serious modern hospital pests, the Pharaoh's ant.

pharaonic, *a.* (Later examples.)

1958 L. Durrell *Mountolive* i. 17 Forgotten Pharaonic frescoes of light and darkness. **1972** *Nature* 14 Apr. 324/1 A travelling exhibition of works of art dating from both Pharaonic and Islamic times. **1976** *Sci. Amer.* Aug. 38/2 It is interesting to note that in the earliest Pharaonic era, that of the Old Kingdom, the Egyptians showed a lively interest in domesticating local animals.

pharate (fæ·rĕit), *a. Ent.* [f. Gr. φᾶρ-ος cloak + -ate[2].] (See quot. 1946.)

1946 H. E. Hinton in *Nature* 27 Apr. 553/1 The term 'pharate'..is proposed to designate the place of an instar which is enclosed within the cuticle of the previous instar. **1957** Richards & Davies *Imms's Gen. Textbk. Entomol.* (ed. 9) 222 There may be an appreciable interval between the two events, during which the instar within the old cuticle is known as the pharate instar. **1971** *Nature* 12 Feb. 490/1 The developing adult inside the pupal cuticle (the pharate adult) can also move the spiracles. **1973** W. S. Romoser *Sci. of Entomol.* viii. 242 The pharate condition..is where the adult is fully formed but is still within the confines of the pupal cuticle.

Pharisaist (fæ·rise,ist), *a.* [After Pharisaism, f. Pharisee *sb.* + -ist.] = Pharisaic *a.*

1918 P. T. Forsyth *This Life & Next* xi. 112 Its idea of resurrection means something very much more than the repristination of the old life under happier circumstances. That..is but Jewish, and Pharisaist, and Moslem.

Pharisee, *sb.* Add: **2*.** Usu. in *pl.*: app. a dialectal alteration of *fairies*, pl. of Fairy *sb.*

[**1807** Southey *Lett. from Eng.* III. liv. 37 The man observed..that the fairies were never seen now, as they used to be in old times... The man persisted [to the priest],—'It is no longer ago than last Sunday you read about the Scribes and *Pharisees*.'] **1850** H. Ellis *Brand's Observations on Pop. Antiquities* (ed. 4) II. 504 The calf is rid every night by the *farisees*. **1852** J. Allies *Antiquities & Folk-Lore Worcestershire* (ed. 2) xii. 418 According to tradition,..Oseberrow..Rock..was a favourite haunt of the fairies (*vulgo* pharises). **1854** M. A. Lower *Contributions to Lit.* 157 It's very hard to say how them rings do come, if it isn't the Pharisees that makes 'em. **1884** *Contemp. Rev.* Aug. 329 Among the peasants of the South Downs a belief in the existence of fairies, or, as they call them, 'Pharisees', has not died out. **1906** Kipling *Puck of Pook's Hill* 264 'Pharisees,' cried Una. 'Fairies? Oh, *I* see!' **1930** M. Allingham *Mystery Mile* iv. 49 Seven Whistlers... No one knows if they be ghosts or Pharisees—that be fairies. **1947** E. Meynell *Sussex* ix. 243 The belief in fairies—though few of the old Sussex people would dream of using such an 'outland' word, it is as pharisees they are known—also continued late in the county. **1948** L. Spence *Fairy Tradition in Brit.* iv. 82 'Pharisees', a term used to denote the fairies in Suffolk.

pharmaceutic, *a.* and *sb.* Add: **B.** *sb.* (Further examples.) Also *concr.*, a medicinal drug; = Pharmaceutical *sb.*

1927 *Daily Express* 11 May 11/4 The agreement will later be extended to include artificial silk,..fertilisers, pharmaceutics, and many other products. **1974** *Nature* 6 Sept. p. xi/1 (Advt.), Applications are invited for the post of Senior Lecturer in the Department of Pharmaceutics commencing January 1, 1975.

pharmaceutical, *a.* (*sb.*) **B.** *sb.* (Further examples.)

1955 T. Sterling *Evil of Day* iv. 39 The night-table..was prominently littered with pharmaceuticals. **1958** *Times* 5 Dec. 3/3 An international company manufacturing surgical and orthopaedic dressings, pharmaceuticals and proprietary medical products.

pharmaco-. Add: **pharmacodynamic** *a.* (examples); hence **pha:rmacodyna·mically** *adv.*; **pharmacodynamics** *sb. pl.* (further examples); **pharmaco·gnosist**, an expert in pharmacognosy; **pha:rmacogno·stic** *a.*, of or pertaining to pharmacognosy; **pha:rmacokine·tics** *sb. pl.* (const. as *sing.*), the branch of pharmacology concerned with the movement of drugs within the body; so **pha:rmacokine·tic** *a.*, **-kine·tically** *adv.*

1906 *Jrnl. Infectious Dis.* III. 572 (*heading*) Pharmacodynamic action due to ions. **1951** A. Burger *Medicinal Chem.* I. xxi. 386 Certain cholinergic agents (Doryl) can add to the analgetic effect produced by injected epinephrine. The pharmacodynamic mechanism..is as yet uncertain; it may be that the para-sympathomimetic drug stimulates the output of epinephrine by the adrenal gland which can then enhance the peripheral adrenergic vasoconstrictor effect. **1929** Mitchell & Hamilton *Biochem. Amino Acids* vii. 363 It is a fact of great significance that histidine, a substance entirely indifferent pharmacodynamically, should be transformable by simple decarboxylation into such an active substance as histamine. **1959** K. H. Beyer in Waife & Shapiro *Clin. Eval. New Drugs* ii. 19 Some compounds are so active that no chemical or isotopic method may be suitable to follow them at pharmacodynamically effective dosages. **1867** R. E. Scoresby-Jackson *Note-Bk. Materia Med.* I. 3

Pharmacology..is divided into General Pharmacology and Special Pharmacology, and is subdivided into Pharmacognosy, Pharmacy, and Pharmacodynamics. **1925** E. Novak in *Gynecol. & Obstetr. Monogr.: Appendix* 22 The study of the pharmacodynamics of the various forms of ovarian extract has until recent years yielded unimpressive results as regards the generative system. **1974** M. C. Gerald *Pharmacol.* i. 9 Pharmacodynamics, a basic experimental science, is a study of where a drug acts in the body..and how it acts (what is its mechanism of action in physiological and/or biological terms). **1934** Webster, Pharmacognosist. **1939** *Nature* 1 Apr. 540/2 The complete pharmacognosist is a man of many parts. His preliminary training in botany, zoology, chemistry and physics furnishes him with a foundation on which to build experience in the technique of microscopy, histology, morphology, [etc.]. **1972** *Ibid.* 21 Jan. 134/1 Some people may have thought of pharmacognosists as witch doctors in that their methods of selection of plants for study have relied to some extent on folklore and ancient custom. **1961** Webster, Pharmacognostic. **1974** *Nature* 13 Sept. 169/1 While familiarising himself with the natural history of the islands, Linnaeus was instructed particularly to look for plants..with pharmacognostic value. **1964** *Antibiotica et Chemotherapia* XII. p. viii/1 A list of symbols for the use in pharmacokinetic models. **1976** *Lancet* 9 Oct. 808/1 Pharmacokinetic analysis indicates that binding is unlikely to be a major problem in vivo when less than 75–85% of the drug is bound. **1972** *Nature* 28 Apr. 434/2 The leaching of DDT from fatty tissue after exposure to abnormally large controlled doses of DDT appears to be pharmacokinetically similar to the uptake process. **1960** *Jrnl. Amer. Pharmaceutical Assoc.* XLIX. 311 (*heading*) Dosage schedule and pharmacokinetics in chemotherapy. **1971** R. E. Notari *Biopharmaceutics & Pharmacokinetics* i. 4 Since the movement of drug from the site of administration to the site of action requires time, the overall process may best be analyzed by what is called pharmacokinetics. **1973** *Nature* 22 June p. xvi (Advt.), A graduate with experience in pharmacokinetics and drug metabolism.

pha:rmacogene·tics, *sb. pl.* (const. as *sing.*). *Pharm.* [ad. G. *pharmakogenetik* sb. (F. Vogel 1959, in *Ergebnisse d. inneren Med. und Kinderheilkunde* XII. 117): see Pharmaco-, Genetic *a.*, and -ic 2.] The study of the effect of genetic factors on reactions to drugs.

1960 *Times* 11 Nov. 17/2 This development of pharmaco-genetics may affect quite considerably our methods of treating patients. **1962** W. Kalow (*title*) Pharmacogenetics: heredity and the response to drugs. **1965** Fingl & Woodbury in Goodman & Gilman *Pharmacol. Basis Therapeutics* (ed. 3) i. 25/2 The objectives of pharmacogenetics include not only identification of differences in drug effects that have a genetic basis but also development of simple methods by which susceptible individuals can be recognized before the drug is administered. **1974** M. C. Gerald *Pharmacol.* iii. 58 Among the newest subdivisions of pharmacology is that of pharmacogenetics which is the study of the influence of genetic factors on the drug response.

Hence **pha:rmacogene·tic** *a.*, **-gene·ticist.**

1962 *Pharmaceutical Jrnl.* CLXXXIX. 282/3 Pharmacogenetic studies. **1970** Passmore & Robson *Compan. Med. Stud.* II. xxxi. 11/1 Not all pharmacogenetic studies are prompted by the occurrence of adverse effects. **1971** *Sci. News* 26 June 439 Some pharmacogeneticists advocate screening individuals who are to receive succinylcholine before surgery for their pseudocholinesterase activity. **1974** M. C. Gerald *Pharmacol.* iii. 59 Pharmacogenetic differences, in part, account for the development of resistant strains of bacteria and insects.

pharmacology. Add: Hence **pha:rmacolo·gic** *a.* (chiefly *U.S.*) = Pharmacological *a.*

1901 T. Sollmann *Text-bk. Pharmacol.* 8 The organic poisons..often require pharmacologic experience for their recognition. **1973** *Sci. Amer.* Sept. 123/3 Psychiatry ..has two faces, one represented by treatment at the psychosocial level and the other by treatment at the pharmacologic level.

‖ **pharmakos** (fā·ɹmăkǫs). Pl. **pharmakoi.** [Gr. φαρμακός scapegoat.] In ancient Greece, a scapegoat chosen in atonement for a crime or misfortune. Also *transf.* in allusive use.

1903 J. E. Harrison *Proleg. Study Greek Relig.* iii. 104 The pharmakos is killed then, not because his death is a vicarious sacrifice, but because he is so infected and tabooed that his life is a practical impossibility. **1923** A. Le Marchant *Greek Relig. to Time of Hesiod* iv. 25 A ceremony in which two men called Pharmakoi, decked with branches, were led out of the city. **1926** J. Buchan *Dancing Floor* ii. 42 You have your purgation herbs like buckthorn and agnus castus, and you have your *pharmakos*, your scapegoat, who carries away all impurities. **1957** N. Frye *Anat. Crit.* 41 The figure of a typical or random victim begins to crystallize in domestic tragedy as it deepens in ironic tone. We may call this typical victim the *pharmakos* or scapegoat.

pharyngal or **pharyngeal,** *a.* (*sb.*) Add: **A.** *adj. spec.* of speech-sounds: (see Pharyngal *a.* (*sb.*)); also applied to consonantal sounds articulated with obstruction of the air-stream at the pharynx.

1925 W. H. T. Gairdner *Phonetics of Arabic* iv. 27 [ħ].. is a pharyngal unvoiced fricative. **1931** G. Noël-Armfield *Gen. Phonetics* (ed. 4) xviii. 107 Two very difficult plosive sounds for English people..are the Arabic (or Hebrew) *qaf* and its voiced correspondent. They are somewhat similar to [k] and [g] respectively... These consonants, though usually termed uvular, would be better regarded as pharyngal. **1950** D. Jones *Phoneme* p. xiii, ħ, breathed pharyngal fricative. **1964** *Language* XL. 501 A series of pharyngal stops, plain and labialized, aspirated and glottalized. **1968** Chomsky & Halle *Sound Pattern Eng.* 305 Ubykh, a Caucasian language, distinguishes pharyngeal, uvular, velar, and perhaps also palatal obstruents. **1978** *Studies in Eng. Lit.: Eng. Number* (Tokyo) 159 Part III consists of two chapters, the first of which is concerned with the phonological characterisation of pharyngeal consonants.

B. *sb.* **2.** *spec.* designating speech-sounds: (see Pharyngal *a.* (*sb.*)); also, a pharyngeal consonant.

1925 W. H. T. Gairdner *Phonetics of Arabic* iv. 27 We are faced with two difficulties in regard to the two pharyngals ħ and ʕ. **1968** Chomsky & Halle *Sound Pattern Eng.* 305 The consonants where the primary constriction is formed with the body of the tongue..: the palatals, velars, uvulars, and pharyngeals. **1976** *Archivum Linguisticum* VII. 91 Nor is it [*sc.* preaspiration] necessarily a pharyngeal..but may be realized as a spirant formed at some other point of articulation.

Hence **phary·ng(e)alization,** obstruction of the air-stream at the pharynx; modification into a pharyngeal sound; **phary·ng(e)alized** *ppl. a.*, produced by pharyngealization.

1931 G. Noël-Armfield *Gen. Phonetics* (ed. 4) 186 (Index), Pharyngalised consonants. **1947** K. L. Pike *Phonemics* xvi. 219 Pharyngealized consonants which are phonemically distinct from nonpharyngealized consonants would..need a special symbol. **1949** *Trans. Philol. Soc.* *1948* 148 All the *gh* spellings..are realized as long slightly pharyngalized vowels. **1964** R. Kingdon in D. Abercrombie et al. *Daniel Jones* 115 Secondary articulations such as..pharyngalization. **1968** Chomsky & Halle *Sound Pattern Eng.* 309 We know of no languages that exhibit parallel variations in degree of narrowing concomitant with palatalization or pharyngealization. **1968** P. M. Postal *Aspects Phonol. Theory* iv. 82 Consonants are normally non-Pharyngealized. Hence there are no languages with only Pharyngealized consonants. **1977** J. C. Catford *Fund. Probl. Phonetics* ix. 182 Pharyngealized vowels involve a compression of the pharynx simultaneously with a primary vowel articulation... Such vowels occur in several Caucasian languages of Dagestan.

pharyngo-. Add: **phary:ngoconju·nctival** *a.*, epithet of a syndrome that is characterized by conjunctivitis, pharyngitis, and fever and occurs chiefly in epidemics among children; **pharyngo-nasal** *a.* (examples); now *rare* or *obs.* (cf. *nasopharyngeal* adj. s.v. *Naso-).

1955 J. A. Bell et al. in *Jrnl. Amer. Med. Assoc.* 26 Mar. 1092/2 Study of the clinical, etiological, and epidemiological attributes of a newly recognized communicable disease entity has appeared to differentiate one disease entity from the poorly defined mass of undifferentiated respiratory illnesses generally known as the common cold, catarrhal fever, nonstreptococcic sore throat, or acute respiratory disease. We suggest that this disease entity be named pharyngoconjunctival fever. **1974** Passmore & Robson *Compan. Med. Stud.* III. xii. 18/1 (*heading*) Pharyngo-conjunctival syndrome. **1976** *Lancet* 6 Nov. 990/2 Respiratory illness and pharyngo-conjunctival fever are commonly associated with adenovirus infections. **1861** G. D. Gibb tr. *Czermak's On Laryngoscope* iii. 25 The principle of the laryngoscopic method could be equally applied to the inspection of..the superior parts of the pharynx (pharyngo-nasal vault). **1894** J. W. Downie *Clin. Man. Study Dis. Throat* i. 29 These growths..may attain to such a size as to completely block the pharyngo-nasal cavity, thereby hindering nasal respiration.

phase, *sb.* Add: **2.** Esp. in phr. *phase one* (or *two*, etc.): the first (or second, etc.) planned stage of a process, series of events, etc.

1957 *Economist* 5 Oct. 24/2 There was little..to suggest that the government expects Britain to be a very active combatant in 'Phase Two' of another war. Mr Butler's emphasis was solely on Phase One. **1974** *Times* 1 Apr. 21/1 A contract..for phase one of a new district general hospital. **1977** *Whitaker's Almanack 1978* 580/1 The Chancellor of the Exchequer, other ministers, and the T.U.C. economic committee opened negotiations for a Phase 3 pay policy.

b. *Zool.* A particular period of an animal's life, distinguished by a characteristic form, colour, or type of behaviour. Also *attrib.*

1873 [see *form-genus* (*Form *sb.* 22)]. **1921** B. P. Uvarov in *Bull. Entomol. Res.* XII. 153 We are yet far from knowing whether the transformation of one form [of locust] into the other is due to some immediate external influence or to some yet unknown internal cause; I think therefore, that the term 'phase'..suggested to me by Dr. G. A. K. Marshall is more appropriate [than 'morpha']. *Ibid.* 155 The swarming phases [of locusts] enable the species to extend at one stroke its area of distribution. **1937** *Ann. Reg. 1936* 54 Phase variation was found in grasshoppers. **1947** *New Biol.* III. 10 It is now a recognised fact that all true locusts occur in two phases—the solitary and the swarming, or gregarious as it is usually called. **1956** *Nature* 28 Jan. 167/2 Dr. M. L. Roonwal's work has been concerned..with phase-transformation and population dynamics of the desert locust. **1964** L. S. Crandall *Managem. Wild Mammals in Captivity* 368 It [*sc.* the jaguarundi]..occurs in two color phases, dark gray and reddish brown. **1966** B. P. Uvarov *Grasshoppers & Locusts* I. 386 The ideas behind the phase theory are being followed by workers on other insects. **1973** *Nature* 24 Aug. 484/1 Phase transformation in locusts refers to the changes induced when solitary hoppers (juvenile locusts) become gregarious. **1977** *Times* 18 Aug. 14/5 A slate-black falcon..was agreed to have all the field-clues of the dark phase of Eleanora's falcon, one of Europe's rarest predators.

c. A temporarily difficult or unhappy period or stage of development, esp. of adolescents; freq. in *to go* (or *pass*) *through a phase*.

1913 W. J. Locke *Stella Maris* xix. 258 'What's the matter with her, for pity's sake?' asked Herold... 'Perhaps it's a phase. Young girls often pass through it.' **1932** R. Lehmann *Invitation to Waltz* i. ii. 27 Mrs. Curtis was silent: a pregnant silence. Kate was going through a phase. Best not to take too much notice. **1960** *Times* 28 May 7/4 'It's only a phase', we say uncertainly when our children sulk, fight, or burst into tears for no reason. **1971** [see *crowd *sb.*[3] 2 c]. **1971** H. McCloy *Question of Time* i. iii. 36 Whenever Pel or Mél get into trouble, Mrs Heron always says: 'It's just a phase they're going through.'

3. Add to def.: Considered in relation to a particular reference position or time. Also *transf.* *in phase*, in the same phase; having the same phase at the same time; const. *with*; *out of phase*, not in phase. (Earlier and further examples.)

1861 *Phil. Mag.* XXI. 163 Two series of undulations traversing the same space do not combine into one resultant as two attractions do, but produce an effect depending on relations of *phase* as well as intensity. **1863** E. Atkinson tr. *Ganot's Elem. Treat. Physics* vii. viii. 474 Fig. 362 represents two waves issuing from the same source of light, and meeting at *a*, under a very acute angle in the same phases [*ed.* 2 (1866): in the same phase], while fig. 363 represents the coincidence of two waves in opposite phases of undulation. **1891** J. W. Urquhart *Dynamo Construction* xvi. 239 If switched when not 'in phase', the fresh machine would..be quickly pulled into unison. **1903** T. Sewell *Elem. Electr. Engin.* (ed. 2) xvii. 337 The current flowing in the circuit, whether it be in or out of phase with the e.m.f., is indicated by the ammeter. **1931** Moyer & Wostrel *Radio Handbk.* ii. 74 In a circuit containing only non-inductive resistance the current and voltage are in phase. **1936** L. S. Palmer *Wireless Engin.* x. 403 The plate and outer grid may be indirectly connected by any device..which changes the phase of the output with respect to the input by 180°. **1953** *Economist* 14 Nov. 505/2 To keep the supply of raw materials in phase with productive capacity. **1973** *Sci. Amer.* June 47 The light is reflected from a system of mirrors and arrives either in phase or out of phase at the second Kerr cell, depending on the length of the light path between the cells.

c. *Electr. Engin.* Each of the windings of a polyphase machine.

1904 M. B. Field in M. Maclean *Mod. Electr. Pract.* II. i. vi. 28 If one of the phases of a Δ-connected system is disconnected, the remaining two can still supply a three-phase current, but with a diminished efficiency. **1931** G. C. Blalock *Princ. Electr. Engin.* xviii. 243 The power in any polyphase circuit must of necessity be the sum of the powers in the component phases. It is usually more convenient, however, to determine polyphase power in terms of line voltage and current. **1962** *Newnes Conc. Encycl. Electr. Engin.* 787/2 The phases are interlinked in star connection. **1972** Smith & Hosie *Basic Electr. Engin. Sci.* ix. 244 Symmetrical delta-connected systems... The power developed in the generator, when supplying a balanced load, is three times that developed in each phase. *Ibid.* 254 Each phase of a 3-ph star-connected load is a coil of resistance 20 Ω.

4. A physically distinct and homogeneous form of matter that may be present in a system, characterized by its composition and state and when present separated by a bounding surface from other forms.

1875 J. W. Gibbs in *Trans. Connecticut Acad.* III. 152 We may call such bodies as differ in composition or state, different phases of the matter considered, regarding all bodies which differ only in quantity and form as different examples of the same phase. **1916** C. A. Edwards *Physico-Chem. Properties of Steel* iv. 47 From the phase rule we know that three phases can coexist in equilibrium, in a binary system, only at one temperature. At the Ar_1 point there are three phases—namely, α-iron, carbide of iron, and the solid solution of the eutectoid composition. **1931** Maass & Steacie *Introd. Princ. Physical Chem.* ix. 135 Every liquid has a vapour pressure and will evaporate and enter the gas phase. **1967** J. Wilks *Properties of Liquid & Solid Helium* i. 6 Both solid [3]He and [4]He exist in three phases with different crystal structures. **1971** I. G. Gass et al. *Understanding Earth* iii. 55/1 These two phases, sand and clay, are very effectively separated from each other by deposition or sedimentation from water.

5. *attrib.* and *Comb.*, as (sense 3) *phase difference*, *relation(ship)*, *reversal*; *phase-sensitive* adj.; (sense *4) *phase boundary*; **phase advancer** *Electr. Engin.*, a device for improving the power factor of an induction motor by generating a magnetizing current in the rotor circuit which leads the main rotor current in phase; **phase change,** a change in the phase of a wave (Phase 3) or of a substance (*phase *sb.* 4); **phase changer** *Electr. Engin.* = **phase converter*; **phase contrast,** the technique in microscopy of introducing a phase difference between parts of the light supplied by the condenser so that interference causes the outlines of the sample, or the boundaries between parts of differing optical density, to appear more prominent; usu. *attrib.*, esp. in *phase-contrast microscope, microscopy*; **phase converter, convertor** *Electr. Engin.*, a device which converts an alternating current into one having a different number of phases but the same frequency; **phase dia-**

gram *Chem.*, a diagram which represents the limits of stability of the various phases of a chemical system at equilibrium with respect to two or more variables (commonly composition and temperature); an equilibrium diagram; **phase displacement** *Electr.*, a difference in phase; **phase distortion,** distortion of a waveform caused by components of different frequencies being propagated at different speeds, so that their phase relations are altered; **phase inverter, invertor** *Electr.*, a phase splitter which produces two signals 180 degrees out of phase; **phase-lock** *sb. Electronics*, the stabilization of the frequency of an oscillator with respect to that of another, stable, oscillator of lower frequency, by means of a circuit in which any variation in the higher frequency generates a phase difference which produces an automatic correction to that oscillation; freq. *attrib.*; so **phase-lock** *v. trans.*, to stabilize (an oscillation or a device) in this way; **phase-locked** *ppl. a.*, **-locking** *vbl. sb.*; **phase microscope** *Biol.*, a phase-contrast microscope; so **phase microscopy**; **phase modulation** *Telecommunications*, modulation of a wave by varying its phase; hence **phase-modulated** *ppl. a.*, (as a back-formation) **phase-modulate** *v. trans.*; **phase reaction** *Chem.*, a chemical or physical change which involves the transfer of material between phases, or the appearance or disappearance of a phase; **phase rotation** *Electr. Engin.* = *phase sequence 1; **phase rule, Phase Rule** *Physical Chem.* (see quots. 1913, 1966); **phase separation** *Physical Chem.*, the separation of one phase into two, esp. the separation of a mixture by partition between two phases, or the coacervation of a colloidal solution; **phase shift,** a change in the phase of a waveform; **phase-shifter** *Electr.*, a circuit or device which introduces a change in the phase of an oscillation; orig. *spec.* a transformer which alters the power factor in an a.c. circuit by changing the phase relationship of voltage and current; **phase-shifted, -shifting** *ppl. adjs.*; **phase space** *Physics*, a multidimensional space in which each axis corresponds to one of the co-ordinates (spatial or other) required to specify the state of a physical system, all the co-ordinates being thus represented so that a point in the space corresponds to a state of the system; **phase-splitter** *Electr.*, a circuit or device which splits a single-phase voltage into two or more voltages differing in phase; so **phase-splitting** *ppl. a.* and *vbl. sb.*; **phase transition,** a change in the phase of a substance (*phase *sb.* 4); **phase velocity,** the speed of propagation of a sine wave or a sinusoidal component of a complex wave, equal to the product of its wavelength and frequency (cf. *group velocity); **phase-wound** *a. Electr. Engin.*, having a secondary in the form of windings rather than a squirrel-cage.

1909 M. Walker in *Jrnl. Inst. Electr. Engin.* XLII. 611 The author believes that it is possible to devise a new type of machine to act as an exciter in the manner proposed by M. Leblanc... Such a machine might be called a 'phase advancer'. **1920** *Whittaker's Electr. Engineer's Pocket-bk.* (ed. 4) 521 Should the speed be increased above synchronism, the reactance attains a negative value (*i.e.* the phase-advancer acts as a capacity), with the result that the current is advanced and the leading component of it compensates for the magnetizing current of the induction motor, thereby relieving the system. **1962** *Newnes Conc. Encycl. Electr. Engin.* 596/1 Many different types of phase advancer have been developed but they are only rarely used because of the high maintenance required. **1962** *New Biol.* XII. 99 The adoption of simple organic molecules from the atmosphere or from solutions at the surface of phase boundaries may have been of far greater importance. Suitable phase boundaries occur..at the surface of crystals of inorganic minerals. **1914** *Physical Rev.* III. 126 At high pressures we encounter phase changes of only two types, those from the fluid to the solid (or crystalline) phase, and from one solid phase to another. *Ibid.* 354 Methods..for determining the phase change produced by reflection from the surfaces of metals. **1956** *Nature* 4 Feb. 240/1 The rate of diffusion in the plane of the layers for each phase-change is equal in all directions. **1964** A. Nisbett *Technique Sound Studio* 254 Noise..is strongly discriminated against [by frequency modulation], though it does produce phase-change effects which cannot be eradicated. **1974** *Times* 4 Feb. 14 It was once thought that deep earthquakes marked the simultaneous collapse of many cubic kilometres of rock from one atomic structure to another—a so-called phase change. **1900** *Jrnl. Electr., Power & Gas* Oct. 81/2 (*heading*) An induction motor phase changer. **1935** *Discovery* Nov. 326/1 Phase changers and automatic regulators have worked perfectly [in locomotives of Hungarian railways]. **1934** F. Zernike in *Monthly Notices R. Astron. Soc.* XCIV. 377 (*heading*) Diffraction theory of the knife-edge test and its improved form, the phase-contrast method. **1942** *Jrnl. Sci. Instrum.* XIX. 71 (*heading*) Phase-contrast microscopy. **1947** *Nature* 21 June 829/2 Mr. Taylor has made and described a new phase-contrast microscope in which a controlled variable amplitude for the 'direct-light' component is obtained. **1961** M. Françon *Progress in Microsc.* ii. 65 Originally used by its inventor [*sc.* Zernike] for inspecting telescope mirrors, the phase contrast technique was applied to microscopy shortly afterwards. **1966** *McGraw-Hill Encycl. Sci. & Technol.* VIII. 399/2 The phase contrast microscope is the routine instrument for the examination of living cells. **1970** E. M. Slayter *Optical Methods in Biol.* xiii. 288 The phase contrast microscope is a device which renders differences in refractive index between regions of a specimen visible as differences in intensity. **1971** *Nature* 17 Sept. 198/1 [illegible] was estimated by examination of fresh blood preparations by phase contrast microscopy..every 2-3 days. *Ibid.* 26 Nov. 227/2 Cytological characters and chromosomal behaviour during meiosis were studied in live cultures under phase contrast. **1916** *Trans. Amer. Inst. Electr. Engin.* XXXV. 1318 No. 2 is in general respects the same as No. 1 with the addition of a relatively new development known as the phase converter which preserves the balance of the system even when large blocks of single-phase power are taken from the system. **1935** *Discovery* Nov. 326/1 The phase converter is usually regarded as the most complicated part of the locomotive. **1967** C. V. Jones *Unified Theory Electr. Machines* xviii. 241 As a final example of induction motor performance, its operation as a phase converter will be briefly considered. **1911** *Chem. Abstr.* V. 1219 (*heading*) Phase diagrams of iron-silver alloys. **1926** *Jrnl. Iron & Steel Inst.* CXIII. 655 The phase diagram is characterised by a solid solution of tin in iron with a maximum of 18 per cent. tin [illegible] **1889** J. A. Fleming *Alternate Current Transformer* I. iv. 191 θ is the angle of phase difference of the currents. **1962** *Symp. Ecol. Soc.* No. 7. 10 It has been generally supposed until recently that for man..phase differences are of no account. But we now realize that if changes of phase of particular components occur at a sufficiently rapid rate, then differences of sound-quality are in fact heard. **1907** H. H. Norris *Introd. Study Electr. Engin.* x. 190 (*heading*) Phase displacement [see *balanced *ppl. a.* 6]. **1928** *Bell Syst. Techn. Jrnl.* VII. 195 For relatively short distances the deleterious effects of phase distortion are not appreciable. **1961** G. Millerson *Technique Television Production* iii. 43 Certain video distortions, from phase-distortion to signal reflections, can also be the reason for poor definition in a picture. **1970** J. Earl *Tuners & Amplifiers* ii. 43 For high quality all the significant sideband signals must be handled by the i.f. channel to avoid phase distortion and treble attenuation. **1934** A. Hund *Frequency Modulation* xi. 178 Figure 49 shows a balanced reactance-tube modulator where a phase inverter replaces the push-pull input transformer. **1951** *Electronic Engin.* XXIII. 64 Phase-invertors using resistance-loaded triode valves are well known. **1960** E. N. Lurch *Fund. Electronics* xv. 353 A circuit arrangement to produce balanced voltages which are 180° out of phase for the grids of the push-pull stage is termed a phase inverter. **1953** W. A. Edson *Vacuum-Tube Oscillators* xiv. 342 The question of time delay of a phase-lock loop arises in all methods [illegible]. **1957** *Electronic & Radio Engineer* XXXIV. 141/1 The simple phase-lock loop is effectively a position control servo-mechanism operated by a d.c. error voltage from the phase-sensitive detector. **1973** *Wireless World* Dec. 605/1 The device contains a phase-locked loop which I have found suitable for demodulating teleprinter f.s.k. signals because it requires only a small input signal for phase lock. **1974** Harvey & Bohlman *Stereo F.M. Radio Handbk.* iv. 63 When 'phase-lock' is achieved the in-phase detector registers this by providing a d.c. output. **1955** *Proc. IRE* XLIII. 869/1 A circuit has been developed in which two microwave oscillators may be phase-locked to weak but stable reference signals. **1967** *Electronics* 6 Mar. 6/2 (Advt.), The generator frequency can be phaselocked to an external standard frequency. **1959** *Proc. IRE* XLVII. 1137/2 A phase-locked local oscillator can serve as a filter of arbitrarily narrow bandwidth. **1976** *Gramophone* Aug. 334/2 A quartz crystal controlled phase-locked servo circuit holds speed drift within 0.002% and is independent of any changes in line voltage or frequency. **1955** *Proc. IRE* XLIII. 872/1 This phase-locking circuitry may also be used to make a divider of particular use in the microwave region where no other kind exists. **1963** *Electronics* 2 Aug. 21/2 A phase-locking network used [illegible] is affected only by the frequencies assigned to the two reference oscillators. **1977** *Proc. R. Soc. Med.* LXX. 379/2 Temporal information depends on phase-locking; the nerves fire at a particular phase of the stimulating wave-form so that for sinusoids the time intervals between firings are approximately integral multiples of the period of the wave-form. **1946** *Trans. Amer. Microsc. Soc.* LXV. 129 (*heading*) Phase microscopy. *Ibid.*, The Phase Microscope reveals detail in transparent materials having regions of slightly differing absorption or with different indices of refraction. **1949** *World-Herald* (Omaha, Nebraska) 4 Dec. 19-A, The phase microscope can see the interior of a cell in bright or dark contrast to bring hidden parts into view. **1964** N. G. Cohen *Elem. Cytol.* ii. 22 The observation of living cells has been greatly facilitated by the development of the phase microscope and related optical systems. **1970** O. W. Richards in J. E. Blair et al. *Man. Clin. Microbiol.* 24/1 The phase microscope has an annular stop in the condenser to limit the lighting to a symmetrical hollow cone. **1930** *Proc. IRE* XVIII. 633 (*heading*) The effect of frequency or 'phase' modulation upon signal quality. **1936** Phase-modulated [see *frequency-modulated *ppl. a.*]. **1941** *Electronic Engin.* XIV. 537/3 Phase modulation possesses much the same advantages as frequency modulation. Signal-to-noise ratio is greater than for amplitude modulation..though it is less than for frequency modulation since the triangular noise spectrum effect is absent because noise itself phase modulates the carrier. **1962** J. H. & P. J. Reyner *Radio Communication* iii. 140 Whereas with phase modulation the modulation index *m* is simply proportional to the amplitude of the modulating signal, with frequency modulation it is also inversely proportional to the modulation frequency. **1968** B. P. Lathi *Communication Syst.* iv. 213 If we integrate the modulating signal f(t) first and then allow it to phase-modulate the carrier, we obtain a frequency-modulated carrier. **1974** Harvey & Bohlman *Stereo F.M. Radio Handbk.* ii. 36 It is a characteristic of phase modulation that the amount of frequency swing introduced is proportional not only to the amplitude of the modulating signal but also to its frequency. **1976** *Sci. Amer.* June 62/3 The antenna transmits a phase-modulated radio signal that carries scientific and engineering data from the spacecraft to the radio tracking stations on the earth. **1920** D. A. Clibbens *Princ. Phase Theory* i. 3 If the temperature is raised, a transference of material from one phase to another will, in general, occur; such a transference is known as a phase reaction. **1923** A. C. D. Rivett *Phase Rule* iv. 72 Until the temperature reaches that of [illegible], cooling does not bring about any phase reaction. **1938** S. T. Bowden *Phase Rule & Phase Reactions* i. 6 A phase reaction is a physical or chemical change which involves the appearance or disappearance of a phase. **1878** *Proc. R. Soc. Edin.* IX. 604 They may be so related in phase that at one of the instants of maximum pressure of one of the constituents there is also maximum pressure of the other constituent. The same phase-relation, if the harmonic numbers of the constituent tones be prime to one another, will give coincident minimums. **1948** S. P. Thompson *Dynamo-Electr. Machinery* (ed. 5) xxvi. 716 (*heading*) Phase-relations in transformers. **1962** A. Nisbett *Technique Sound Studio* xii. 218 The phase relationship between the sounds. **1898** Phase reversal [in Dict., sense 3 b]. **1938** *Science* 4 Mar. 213/2 [They] misinterpret the situation still more by referring to..the setting of gelatine as involving phase-reversal. **1962** A. Nisbett *Technique Sound Studio* [illegible] microphone with a heart-shaped polar diagram (arrived at by adding omnidirectional and figure-of-eight responses together, taking into account the phase reversal at the back of the latter). **1918** K. Edgcumbe *Industr. Electr. Measuring Instruments* (ed. 2) 254 The direction of phase rotation of the secondary pressures and currents may conveniently be checked by a phase rotation indicator. **1922** [see *phaser *ppl. a.* 1]. **1962** *Newnes Conc. Encycl. Electr. Engin.* 571/2 A simple portable instrument to indicate phase rotation. **1896** *Jrnl. Chem. Soc.* LXX. ii. 415 [illegible] This point is, in the case of solutions of the salts, a saturation point in the sense of the phase rule. [illegible] expressed in the equation $P + F = C + 2$, where P is the number of phases, F the number of degrees of freedom, and C the number of components. [illegible] **1966** *McGraw-Hill Encycl. Sci. & Technol.* V. 251/1 If a system consists of P phases and C distinguishable components, there are C + 2 thermodynamic variables (chemical potentials μ_i, plus temperature and pressure) which are interrelated by an equation for each phase. Since there are P independent equations relating the C + 2 variables, one needs to fix only $F = C + 2 - P$ variables to define completely the state of the system at equilibrium; the other variables are then beyond control. This relation for the number of degrees of freedom F, or variance, is called the Phase Rule and was first derived by Willard Gibbs in 1875. **1957** Phase-sensitive [see *phase boundary* above]. **1961** B. Fozard *Instrumentation Nucl. Reactors* xi. 135 This requires the provision of a phase-sensitive rectifier at the output. **1975** *Nature* 25 Sept. 273/1 We have sought to replace the customary phase separation of 'crude carotene' and xanthophylls by using the same chromatographic technique for the removal of xanthophylls as well as of non-carotene chromogens from the petrol ether solution. **1947** *Jrnl. Polymer Sci.* II. 99 The phase separation in solutions of high polymers in the same solvent (technically known as incompatibility) has been studied for fourteen high polymers..dissolved in thirteen solvents. [illegible] *Partition of Cell Particles* ii. 13 Bungenberg de Jong & Kruyt..coined the term 'coacervation' for the general phenomenon of phase separation of colloid mixtures. **1970** [see *coacervation 1]. **1974** *Materials & Technol.* II. vi. 340 Devitrification of glass is the formation of crystals (phase separation) in a glass. **1927** *Wireless World* 1 June 667/1 [illegible] special equipment for the equalization of attenuation and of phase shift had to be devised. **1929** J. A. Ratcliffe *Physical Princ. Wireless* iii. 37 If a beam of light is focused on to a point and then allowed to spread out on the other side of the focus, the total phase-shift is π. **1960** *Practical Wireless* XXXVI. 412/2 While a simple phase-shift oscillator..may be used coupled into a suitable amplifier installation, a neater method is to build the complete practice set together with power supply, speaker and controls into a self-contained unit. **1976** *Gramophone* Feb. 1398/1 Either we must accept some loss of loudness of reverberant sources in mono playback, or a certain amount of phase-shift between the signals coming from the two loudspeakers in stereo. **1965** *Wireless World* Aug. 396/1 If the gate voltage is phase-shifted with respect to the anode voltage, the firing point is delayed by the appropriate time. **1908** *Electrician* 11 Dec. 341/2 In Fig. 1, which shows diagrammatically a phase shifter for a two-phase supply, AB and CD are the terminals of the two phases. **1951** *Engineering* 23 Feb. 221/2 The first method enables the point in each cycle at which the anodes become conducting to be delayed..by a double-wound phase-shifter. **1959** K. Henney *Radio Engin. Handbk.* (ed. 5) xiv. 16 In the microwave region, a matched transmission line or slotted line may be used as a phase shifter. **1908** *Electrician* 11 Dec. 341/1 (*heading*) The use of a phase-shifting trans-

former for wattmeter and supply meter testing. **1929** *Phil. Mag.* VIII. 168 Thus the artificial line has zero attenuation at all frequencies, and is a pure phase-shifting network. **1965** *Wireless World* July 332/1 Thin film circuits..may be more useful where passive networks only are required—for attenuators, RC phase shifting networks, etc. **1927** R. C. Tolman *Statistical Mech.* iii. 32 To follow the behavior of such a system..it is very convenient to think of its phase as given by the position of a representative point..in a 2*n*-dimensional space (phase space) corresponding to the 2*n* coordinates and momenta whose values are to be specified. **1970** G. K. Woodgate *Elem. Atomic Struct.* vi. 98 The exclusion principle states, in this context, that not more than two electrons are allowed in each volume of size h^3 in phase space. **1974** G. Reece tr. *Hund's Hist. Quantum Theory* vi. 80 Ehrenfest attempted to interpret the radiation formulae by means of a weight function in phase space. **1896** D. C. & J. P. Jackson *Alternating Currents* 652 Special starting devices must be included in the design and construction [of single-phase induction motors]. As a rule, this takes the form of what is called a Phase Splitter. **1965** *Wireless World* Aug. 72 (Advt.), A pentode triode ideally suited for use as a voltage amplifier and phase splitter. **1970** J. Earl *Tuners & Amplifiers* ii. 54 The collector and emitter outputs..are equal in amplitude but opposite in phase, rather like the signals from a simple valve phase-splitter circuit. **1895** S. P. Thompson *Polyphase Electric Currents* iv. 99 (*caption*) Phase-splitting device. **1947** *Wireless World* Aug. 274/1 The advantages of resistance-capacitance circuits for phase-splitting in push-pull amplifiers are now well recognized. **1969** R. W. Smeaton *Motor Applic. & Maintenance Handbk.* xviii. 14 A simple phase-splitting device could be rigged up..so that the motors could be used on 120 volts, 60 cycles, single-phase. **1975** G. J. King *Audio Handbk.* iv. 84 Some form of 'phase-splitting' stage is necessary to drive common-pair output stages. **1939** *Jrnl. Chem. Physics* VII. 1019/1 A first-order phase transition. **1968** R. A. Lyttleton *Mysteries Solar Syst.* ii. 71 *L* is the energy per unit mass required for the phase-transition from solid to liquid. **1977** A. Hallam *Planet Earth* 11 Such a phase transition does not involve differences in chemical composition but only a spatial rearrangement of the atoms in the silicate structure. **1932** Ladner & Stoner *Short Wave Wireless Communication* iii. 34 For the case of a group of waves passing through an ionised medium, therefore, the phase velocity will be greater than *c*, the velocity of light in a pure dielectric, ..but the group velocity will be less than *c*. **1968** M. S. Livingston *Particle Physics* iii. 49 Consider a pulse of water waves originating from the point where a stone strikes the water, and focus on a particular wave crest... The velocity with which the general disturbance moves is the group velocity; the velocity of the wave crest relative to the water is the phase velocity. **1907** H. H. Norris *Introd. Study Electr. Engin.* x. 290 The phase-wound motor has very satisfactory starting qualities and draws little excess starting current.

phase (fēiz), *v.* [f. the sb.] **I. 1.** *trans.* To adjust the phase of; to bring into phase, synchronize.

1938 A. E. Greenlees *Amplification & Distribution of Sound* x. 154 Wherever two or more loudspeakers are used, these must be properly phased so that all the diaphragms move in the same direction at the same time. **1951** S. Deutsch *Theory & Design Television Receivers* xvi. 502 A phasing control is needed in order to phase the oscilloscope sweep so that the beam goes from left to right when the sweep frequency goes from its lowest to highest extremes. **1967** *Electronics* 6 Mar. 67/2 Researchers have succeeded in phasing the lasers. **1970** J. Earl *Tuners & Amplifiers* vi. 131 The two speakers of a stereo pair..can be correctly phased initially with little danger of the phasing being upset subsequently. **1974** *Sci. Amer.* Jan. 118/2 One can also 'phase' the drum and the mirror by interrupting the circuit of the drum motor momentarily.

2. *trans.* To organize or carry out gradually in planned stages or instalments.

1949 [implied in *phased *ppl. a.* 2]. **1955** *Bull. Atomic Sci.* Jan. 9/3 The British-French proposals were directed primarily to the problem of phasing the controls. **1963** *Ann. Reg. 1962* 97 The British members of the Commission preferred a merger phased over five years. **1969** *Daily Tel.* 8 Jan. 22/7 New tenants will face rent increases of as much as £1 a week, while increases will be phased over three years for existing tenants.

II. With adverbs: **3. phase down:** to reduce or decrease (something) gradually or in planned stages.

1970 *Times* 6 Nov. 8 The secretary said that the programme to phase down American operations in Indo-China..'is solidly based'. **1972** *Physics Bull.* Feb. 76/1 If the innovation manager is truly perceptive, he can phase down his work *without* human suffering. **1974** *Daily Tel.* 11 Feb. 2/8 Production had been cut to 60 per cent of normal. It would be phased down further for two months, when virtually no steel would be produced.

4. phase in. a. *intr.* To come into phase. *rare.*

1929 *Proc. IRE* XVII. 1787 The separate multiple echoes from a given set of pulses phase in and out at different rates... Part of the observed phasing must be due to changes in optical path.

b. *trans.* To introduce or incorporate (something) gradually or in planned stages. Also *intr.*, to be so introduced.

1955 *Bull. Atomic Sci.* Feb. 57 New weapons must be 'phased in' gradually to our existing weapons systems. **1959** *Time* 9 Feb. 11 It would be dangerous to phase out obsolescent weapons too slowly. But it would be exceedingly wasteful to phase in too heavily the newer weapons. **1971** *Farmer & Stockbreeder* 23 Feb. 7/3 The levies should be phased in over a one-year period. **1972** [see sense 5 d below]. *a* **1974** R. Crossman *Diaries* (1976) II. 472 As for national superannuation, we shouldn't try and bring it forward but phase it in in 1975, when a large number of existing pensioners will be dying off.

5. phase out. † **a.** *Electr. Engin.* To eliminate phase differences between (parts of polyphase equipment that are to be connected together). *Obs.*

1921 W. S. Ibbetson *Motor & Dynamo Control* viii. 283 The synchronising of two- and three-phase generators may be performed by connecting the synchronising lamps across one phase only, provided the alternators are correctly connected up to the bus bars. To phase out such connections before a machine may be paralleled the following operations may be performed. **1939** — *Electric Power Engineers' Handbk.* vi. 151 If the leads were not correctly phased out so that the phases of all the machines were in the same sequence, interchange of current between the alternators would occur.

b. To eliminate by adjustments of phase. *nonce-use.*

1921 *Wireless World* 6 Aug. 287/2 The interfering note can be completely 'phased out' by adjusting bellows and tube to equal length.

c. *intr.* To become out of phase. *rare.*

1929 [see 4 a above].

d. *trans.* To remove, eliminate, or take out *of* gradually or in planned stages. Also *intr.*, to disappear gradually.

1954 *Quartermaster Rev.* July–Aug. 159/1 We in Defense are phasing out the support of basic research for the general welfare. **1955** *Sun* (Baltimore) (B ed.) 25 Apr. 8/5 The old propellor-driven Mustangs..will be 'phased out' on August 1—which means that replacement parts will have become so scarce that they cannot longer be operated in safety. **1959** [see sense 4 b above]. **1959** *New Scientist* 17 Sept. 446/2 Finding a way to phase out the jet lift as forward speed generates lift in the wings. **1967** *Guardian* 9 Jan. 6/4 We could safely decide to phase out the carrier fleet by 1970. **1969** J. Gardner *Compl. State of Death* ix. 172 It's been a tricky job phasing out these people. **1969** *Daily Tel.* 13 Dec. 13/1 While he was 'phasing out' from the public view, he said, he was not shirking his duties as an MP. **1970** *Ibid.* 30 Sept. 1/6 The Jaguar 420G range of cars is to be gradually phased out of production. **1971** *Nature* 1 Oct. 299/1 Five top level scientific committees.. have recommended that the pesticide be phased out of domestic use. **1972** *Lebende Sprachen* XVII. 135/1 In a research organization projects do phase in and phase out all the time. **1975** *Times* 22 Sept. 3/3 The Government wants to phase out pay beds. **1977** B. Pym *Quartet in Autumn* xii. 102 The whole department was being phased out and only being kept on until the men working in it reached retirement age. **1979** *Time* 2 Apr. 59/1 It has been generally thought that Ford would start phasing out after Sept. 4, 1980, when he turns 63.

† **6. phase up** *Electr. Engin.*: to synchronize, bring into phase. *Obs. rare.*

1904 W. R. Bowker *Dynamo, Motor & Switchboard Circuits* v. 99 In connecting two-phase generators for parallel running it is necessary to synchronise both phases the first time the machines are paralleled. After they have once been 'phased-up' synchronising one phase is sufficient. Fig. 101 shows the connections for properly 'phasing up' two two-phase generators.

phase angle. [f. Phase + Angle *sb.*[2]] **1.** An angle representing a difference in phase, 360 degrees (2π radians) corresponding to one complete cycle.

1889 J. A. Fleming *Alternate Current Transformer* I. iii. 110, OP..is called the amplitude or maximum value, and POM the phase angle of the motion. **1936** *Discovery* Mar. 80/2 By phase-angle changes so produced..any of our mean amplitudes may be much reduced and the real importance of a periodicity [in the weather] accordingly masked proportionately. **1947** R. Lee *Electronic Transformers & Circuits* v. 127 These waves indicate that the phase angle encountered in audio transformers does not of itself introduce much distortion in a lightly loaded triode. **1966** Brosan & Hayden *Adv. Electr. Power & Machines* vii. 279 The voltage will be alternating, and..in general, $e = E_m \sin \omega t$. If the coil is connected to an external circuit, a current will flow, given by $i = I_m \sin (\omega t - \psi)$ where ψ is the internal phase angle.

2. *Astr.* The angle between the lines joining a given planet to the sun and to the earth.

1926 H. N. Russell et al. *Astron.* I. ix. 234 How much of the illuminated hemisphere can be seen from the earth depends upon the angle at the planet between lines to the earth and to the sun. Where this phase angle is always very small, as it is from the most distant planets, there is never any perceptible deviation from a circular disk. **1974** R. H. Stoy *Everyman's Astron.* vi. 203 The apparent brightness of any minor planet varies inversely both as the square of its distance from the Earth and as the square of its distance from the Sun. It also depends on the phase angle, that is on the angle between the lines joining the minor planet to the Sun and to the Earth.

phased (fēizd), *ppl. a.* [f. *phase *v.* + -ed[1].] **1.** Synchronized; adjusted to be in phase.

1922 Glazebrook *Dict. Appl. Physics* II. 936/2 If these two readings give zero reading, the transformers are correctly phased... If readings are obtained across EE' and FF' then polarity or phase rotation is incorrect. **1929** *Photoplay* (Chicago) Apr. 31/2 *Phased*, or *interlocked*, all motors of sound and picture recording equipment lined up in readiness to start out in perfect step together. **1951** S. Deutsch *Theory & Design Television Receivers* xiii. 435 When the sync pulse is near the center of the saw-tooth retrace, there can be no doubt that the picture is correctly phased with respect to the horizontal sweep. **1958** *Listener* 9 Oct. 558/1 Zoning, road classification, phased traffic lights. **1969** *Focal Encycl. Film & Television Techniques* 544/1 The associated synchronizing pulse trains are locked together and correctly phased. **1971** *Hi-Fi Sound* Feb. 71/2 It can be difficult for the newcomer to hi-fi to know whether the loudspeakers are correctly phased or not.

b. *phased array*: an array of aerials that is made to transmit or receive at a variable angle by delaying the signals to or from each one by an amount depending on its position in relation to the others.

1938 *Television* XI. 644/3 None of these changes..affect the description of the phased array. **1960** *Proc. IRE* XLVIII. 1715 Phased arrays can provide scanning patterns and scanning rates which are impossible to attain with mechanically scanned antennas. **1966** *Electronics* 3 Oct. 173 The satellite transponder will receive signals on 149.2 Mhz with an 8-element phased array. **1970** *New Scientist* 10 Sept. 534/1 The beams formed by phased arrays can be steered electronically, and can be scanned up to several hundred million times a second.

2. Planned or carried out in stages or by degrees.

1949 *Archit. Rev.* CV. 225/2 A phased building programme for all developments proposed for the accommodation of population and industry from London. **1953** *Manch. Guardian* 12 May 1/3 Mr Dulles issued a statement that the United States Government favours a 'phased' withdrawal of British troops from the Canal Zone. **1956** *Sun* (Baltimore) (B ed.) 30 Nov. 1/6 Lord Salisbury..declared: 'Our policy is that of a phased withdrawal.' This supposedly means a step by step withdrawal..rather than a quick overall pullout. **1967** D. Wilson in Wills & Yearsley *Handbk. Managem. Technol.* iii. 49 The phased implementation of new procedures. **1972** *Daily Tel.* 4 Mar. 32/6 A policy to end imprisonment without trial by a phased programme of releases of every detainee and internee. **1976** *Glasgow Herald* 26 Nov. 3 The council who are planning 800 phased redundancies in the department next year. **1976** *Broadcast* Dec. 1/1 There will have to be a phased introduction of any changes that are agreed in pay structures and conditions of service.

phase-down (fēi·zdaun). [f. *phase *v.* 3.] A gradual reduction or planned decrease.

1968 *Times* 12 Feb. 17 The best scope for achieving economies in clerical staff will probably come on the female side, where the turnover is rapid and any required phasedown can thus be acquired easily in a relatively short time. **1972** *Lebende Sprachen* XVII. 135/1 The systems division has reported the phase-down of its program office. **1973** *Times* 10 Oct. 19/8 An employee has a progressively shorter working year (the phase-down being at the rate of one, two or perhaps three weeks per annum).

phasemeter (fēi·zmītəɹ). *Electr.* [ad. G. *phasenmeter*, † *phasometer* (M. von Dolivo-Dobrowolsky 1894, in *Elektrotechnische Zeitschr.* XV. 351): see Phase and Meter *sb.*[3]] An instrument which measures the phase difference between two oscillations having the same frequency, esp. that between an alternating current and the corresponding voltage (hence giving the power factor).

1894 *Electrician* 21 Sept. 610/1 The phasemeter here described is essentially an ampere meter measuring the idle current. **1903** G. D. A. Parr *Electr. Engin. Measuring Instruments* vii. 194 The phasemeter now under consideration, made by the Electrical Co., Ltd., of London, is an instrument for measuring directly the value of the wattless current in an inductive alternating-current circuit. **1974** *Physics Bull.* Oct. 477/1 The circuitry of the phasemeter has been designed to reduce the effects of errors introduced by crossover distortion and noise in the input signal, which are the most common sources of inaccuracy.

phaseolin (fǣsī·olin). *Biochem.* [f. L. *phaseolus* kidney-bean; see -in[1].] **a.** A crystalline globulin found in the seeds of the kidney bean.

1893 T. B. Osborne in *17th Ann. Rep. Connecticut Agric. Exper. Station* iv. 187, I have been able to identify and obtain in a state of comparative purity, two distinct proteids, one, the most abundant, having quite the properties of a globulin, which I shall designate phaseolin. **1921** *Nature* 21 July 666/2 Little or no cystines in a protein has also an effect upon the growth of rats. This has been most clearly demonstrated in the case of the protein, phaseolin, of the navy bean. **1964** *Chem. Abstr.* LXI. 10938 The enzymic hydrolysis of phaseolin with trypsin was affected by addn. of the azo dye, amaranth, at pH 8.

† **b.** = next. *Obs.*

1963 Cruickshank & Perrin in *Life Sci.* II. 680 The isolation is now reported of a further phytoalexin of the *Leguminosae*, sub-family *Papilionaceae*. This substance, for which we propose the trivial name, phaseolin, has been obtained by inoculation of detached, opened pods of the French bean, *Phaseolus vulgaris* L., using a spore suspension of [the fungus] *Monilinia fructicola* (Wint.). **1964** [see next].

phaseollin (fǣsī·olin). *Biochem.* [f. prec. with inserted *l*: see quot. 1964[1].] A fungitoxic phytoalexin produced by the kidney bean plant, which has been isolated as a white, crystalline heterocyclic compound, $C_{20}H_{18}O_4$.

1964 D. R. Perrin in *Tetrahedron Lett.* viii. 438 The author's attention has been drawn to the prior use of the term phaseolin to denote a protein from *Phaseolus vulgaris*. Accordingly the name of the phytoalexin from *P. vulgaris* is now changed to phaseollin. **1964** Cruickshank & Perrin in J. B. Harborne *Biochem. Phenolic Compounds* xiii. 530 Phaseollin is similar in its biological properties to pisatin. **1967** R. K. S. Wood *Physiol. Plant Path.* xiv. 497 Another phytoalexin, closely related to

pisatin and called phaseollin has now been obtained from pods of French bean (*Phaseolus vulgaris*) inoculated with conidia of *S[clerotinia] fructicola*. **1973** *Nature* 20 Apr. 533/1 The phytoalexin phaseollin does accumulate in hypersensitive responses in leaves of French bean.. caused by avirulent bacteria but this substance has little or no effect on growth of bacteria.

phase-out (fēi·zaut). [f. *phase *v.* 5 d.] A gradual removal or planned elimination. Also *attrib.*

1958 *Time* (Atlantic ed.) 6 Oct. 26 The Moroccans countered a request for a three-year phase-out. **1960** *Economist* 8 Oct. 126/1 Britain has thus given *de facto* recognition to the Norwegian twelve-mile limit, but the ten year phase-out period for fishing in the outer six miles will do something to cushion the blow to Hull and Grimsby. **1961** *Ann. Reg. 1960* 298 There was a planned phase-out in the field of economic aid. **1969** *New Scientist* 28 Aug. 421/1 The RAF..is to assign more strike aircraft to cover Britain's seaward approaches as the phase-out of the Navy's carriers proceeds. **1972** *Science* 12 May 618/3 In 1971 Shell announced the phase-out of the Emeryville station. **1975** *Daily Tel.* 7 May 8/5 Unless the Government abandoned some [Bills], there seemed little chance of beginning the pay-beds phase-out until 1976 at the earliest.

phase sequence. [f. Phase *sb.* + Sequence.] **1.** *Electr. Engin.* The sequence in which the different lines of a polyphase system attain their maximum voltage.

1918 K. Edgcumbe *Industr. Electr. Measuring Instruments* (ed. 2) 257 A simple method of determining phase sequence when a rotation indicator is not available has been described. **1971** A. Symonds *Electr. Power Equipment* vi. 91 There are three essential conditions to be met before two transformers can be connected in parallel: (a) Secondaries must have the same phase sequence. This can be checked by a phase-sequence indicator.

2. *Psychol.* A hypothetical sequence of cellular activity in the brain initiated by a sensory stimulus, suggested by D. O. Hebb as an explanation of motivated behaviour.

1946 D. O. Hebb in *Psychol. Rev.* LIII. 269/1 Let us designate the specific pattern of cellular activity throughout the thalamo-cortical system, at any one moment, as a 'phase'. Behavior is directly correlated with a phase sequence which is temporally organized. **1954** *Brit. Jrnl. Psychol.* XLV. 189 Hebb (1949) expresses the belief that the mental processes corresponding to cortical processes ('phase-sequences') will be most rewarding..when the phase-sequences are in the course of being built up. **1964** Cofer & Appley *Motivation* viii. 407 The theory presented..was essentially..that nerve cells in the brain when more or less simultaneously excited constitute assemblies of mutually facilitating, and to some extent stimulating, elements, a series of such cell assemblages acting one after another constitute a 'phase sequence'. **1970** C. Taylor in Borger & Cioffi *Explanation in Behavioural Sci.* 77 Hebb..presents mechanisms, the 'cell assembly' and 'phase sequence', which although expressed in neurological terms are not based on any direct supporting neurological evidence.

phasic, *a.* Add: (Further examples.)

1947 *Sci. News* V. 24 Gjessing concluded that the phasic disturbance of nitrogen metabolism constituted the fundamental pathology of the disorder, the mental changes being merely the outward symptoms. **1975** *Nature* 6 Nov. 81/1 Action potential firing of individual endocrine cells falls into two principal categories: most fire continuously and randomly at 1–2 spikes s^{-1}, while the rest have a phasic pattern of firing, periods of bursting activity alternating over 10–60 s with periods of silence.

2. *Physiol.* Responding to a change in environment, rather than to a constant stimulus.

1906 C. S. Sherrington *Integrative Action Nervous Syst.* viii. 302 The very muscles that to the observer are most obviously under excitation by the *tonic* system are those most obviously inhibited by the *phasic* reflex system. **1956** *Jrnl. Physiol.* CXXXIV. 48 A temperature receptor..has..a high coefficient of phasic discharge to temperature change or a high coefficient of tonic discharge to maintained temperatures. **1973** *Sci. Amer.* May 97/1 Many human receptors, such as the ones that sense pressure on the skin, are phasic; if they were not, one would be constantly conscious of such things as a wristwatch or a shirt.

Hence **pha·sically** *adv.*

1975 *Nature* 6 Nov. 81/1 We found that, during bilateral occlusion of the common carotid arteries, supraoptic neurones are excited and that this activation is confined almost exclusively to phasically active neurones.

phasing (fēi·ziŋ), *vbl. sb.* [f. Phase *sb.* or *phase *v.* + -ing[1].] **1.** The action of adjusting or eliminating a phase difference.

1896 *Trans. Amer. Inst. Electr. Engin.* XII. 514 The condenser in this combination fills two very important functions, first assists in the phasing, and second prevents the lag upon the line. **1922** Glazebrook *Dict. Appl. Physics* II. 935/2 Before connecting any two polyphase transformers in parallel it is necessary to ensure that the phase rotation is correct, and this can only be done by phasing out. The term 'phasing out' is applied to the procedure adopted for determining the correct junction of the terminals of two or more transformers. **1940** *Amat. Radio Handbk.* (ed. 2) v. 72/2 Let it be assumed that the parallel capacity C_4 has been balanced by the phasing condenser K. **1949** Frayne & Wolfe *Elem. Sound Recording* xxx. 627 The process of choosing the optimum position is known as the phasing of the speakers. **1959** R. L. Shrader *Electronic Communication* xvi. 525 Another entirely different method of producing a single-sideband suppressed-carrier emission is to use 90° phasing networks. **1960** *Times* 12 Jan. 15/6 Pedestrians are ignored in the phasing of traffic lights at many T-junctions. **1978** *Hi-Fi News* Sept. 179/1 Phasing is.. produced by mixing a direct signal with the same signal when passed through a phase-shift network.

b. The relationship between the phases of two or more periodic phenomena having the same frequency.

1929 [see *phase *v.* 4 a]. **1938** A. E. Greenlees *Amplification & Distribution of Sound* x. 154 Each loudspeaker should have one terminal marked..so that when all these are connected to one line wire and the remaining terminals connected to the other, correct phasing is assured. **1951** S. Deutsch *Theory & Design Television Receivers* xiii. 435 Incorrect phasing is illustrated in Fig. 13-7, where the picture signal begins before the horizontal sweep has ended. **1962** A. Nisbett *Technique Sound Studio* 264 Correct phasing of speakers is..vital to true stereo reproduction. **1968** *Radio Communication Handbk.* (ed. 4) xiii. 61/1 If two such aerials are erected horizontally in the form of a *V*.., and if the phasing between them is correct, the two pairs of lobes will add fore and aft. **1970** [see *phase *v.* 1].

2. The action of *phase *v.* 2. Chiefly in *phasing in, out*, a gradual planned introduction or elimination (cf. *phase *v.* 4 b, 5 d).

1955 *Sun* (Baltimore) (B ed.) 25 Apr. 8/5 The 'phasing out' will end the official approval for even this operation. **1962** *Times* 8 Dec. 5/4 The phasing-out schedule for Thor..ballistic missiles. **1964** *Ann. Reg. 1963* 1 Whitehall was still preoccupied with..the phasing-in of British farm subsidies into the European system. **1964** G. L. Cohen *What's Wrong with Hospitals?* iii. 45 There exist new hospitals of greater efficiency than this one (for the Government's 'phasing' policy has entailed some incredible botching). **1967** *Boston Sunday Globe* 23 Apr. 20/4 The report recommends a gradual phasing out of these incinerators by 1985. **1969** *Daily Tel.* 18 Nov. 16 In Zambia the currency went decimal..overnight without any phasing-in of coins over two years. **1971** *Guardian* 7 June 11/4 The French..want to discuss the phasing out of sterling as a reserve currency. **1971** *Daily Tel.* 14 Aug. 13/7 The phasing of the new plants depends on the CEGB's planning timetable. **1977** *Air Mail* Spring 27/1 The Command is now coming to the end of another extensive re-equipment programme which will have seen the introduction of the Jaguar and the phasing out of the Lightning. **1977** *Time* 30 May 24/3 Carter said that he was 'committed to the phasing-in of a national health insurance system' and would send the legislation to Congress early next year.

phasitron (fēi·zitrǫn). *Electronics.* [f. Phas(e + -i- + *-tron.] An electron tube suitable for phase-modulating a wave by large amounts, in which a pattern of beams emitted radially from a central cathode passes through a slotted cylindrical anode to a coaxial second anode, the pattern of beams being both rotated at a steady rate by a three-phase supply and modulated by a varying axial magnetic field that advances and retards the beams and thereby also the phase of the current at the second anode.

1946 *Electronics* Oct. 108/2 The carrier frequency.., after one stage of amplification, is used to drive a phase-splitting network to provide three-phase voltages for the input deflector grids of a 'Phasitron' tube. *Ibid.* 111/2 Since the Phasitron is modulated by a magnetic field, fields other than those produced by the audio input will cause noise modulation of the carrier. **1966** *McGraw-Hill Encycl. Sci. & Technol.* V. 514/1 An improvement in the phase-modulation method is provided by a special tube called the phasitron, which is capable of increasing the amount of phase shift.

phasor (fēi·zəɹ). *Electr.* [f. Phas(e + -or, after Vector.] A line whose length and direction represent a complex electrical quantity with no spatial extension. Freq. *attrib.*

1944 *Proc. IRE* XXXII. 181/3 The instantaneous angular velocity of the voltage phasor may, in a special case of particular interest, be of the form $\omega_t = \omega + \Delta\omega \cos \Omega t$, where Ω is the angular velocity of the frequency deviation. **1958** W. D. Cockrell *Industr. Electronics Hand-bk.* i. 249 The term 'phasor' is in the process of replacing the time-honored term 'vector'. **1962** F. de la C. Chard *Power Syst. Engin.* i. 4 Phasor impedance can be shown on phasor diagrams and treated mathematically as a true complex quantity, but it differs from phasor voltage and current in the important respect that its value is invariable with time. **1973** M. R. Ward *Electr. Engin. Princ.* iii. 47. The phasors are drawn of length proportional to the r.m.s. **1975** R. F. W. Coates *Mod. Communication Syst.* i. 2 The phasor conveys the essential information regarding amplitude and phase of a fixed frequency sinusoid.

phatic (fæ·tik), *a.* [f. Gr. φατός spoken, or φατικ-ός assertory, f. φάναι to say: see -ic.] Of or pertaining to speech or verbal expression; *spec.* in *phatic communion*, a term applied by B. Malinowski (see *Malinowskian *a.*) to speech communication as used to establish social relationships rather than to impart information. Hence used *gen.* to denote formal or trivial verbal contact.

1923 B. Malinowski in Ogden & Richards *Meaning of Meaning* 478 There can be no doubt that we have here a new type of linguistic use—*phatic communion* I am tempted to call it, actuated by the demon of terminological invention—a type of speech in which ties of union are created by a mere exchange of words. **1929** I. A. Richards *Pract. Crit.* iv. i. 318 It would be an excellent thing if all the critical chitchat..were universally recognised to be what it is, social gesture, 'phatic communion'. **1942** T. C. Pollock *Nature of Lit.* ix. 167 Phatic communion is one of the important ways in which men use language. **1954** W. La Barre *Human Animal* xv. 306 A surprisingly large part of every culture is merely the phatic sharing of common emotional burdens, and has no relevance at all to the outside world. **1959** *20th Cent.* Nov. 379 The magic words of phatic chat that fell from my falsely smiling lips. **1964** *Listener* 15 Oct. 603/3 How many of the youthful roarers of Parry's *Jerusalem* know that Blake's 'dark satanic mills' are not factories but churches?.. But the whole evening was warmly phatic, providing an image of British teenagers as less delinquent than jingoistic. **1971** J. Spencer *Eng. Lang. W. Afr.* 29 Phatic expressions and greeting and leave-taking formulae never precisely match across cultural borders. **1972** *Scholarly Publishing* Apr. 282/2 The reader on a committee who blandly and thoughtlessly says, 'You really ought to get this study published' (in the same phatic way he says, 'Good morning'). **1976** *Archivum Linguisticum* VII. 86 Those illocutionary acts for which there might seem, on the face of it, to be no propositional content at all: 'greet' for example..or 'phatic communion' of various sorts. **1977** *Time* 21 Mar. 67/1 Many Western ears will find it hard to tell whether Merwin is being vatic or phatic.

pheasant. Add: **1. b.** In South Africa, applied to certain francolins, esp. *Francolinus capensis*, and other birds belonging to the family Phasianidæ.

1785 G. Forster tr. *Sparrman's Voy. Cape Good Hope* I. iv. 153, I found here two new species of the genus of *tetrao*, one of which is called *partridge* and the other *pheasant*: either sort being nearly of the size of our partridges. **1837** N. Polson *Subaltern's Sick Leave* v. 119 There is also a bird, general all over the Colony, styled 'pheasant', though about as like a pheasant of England as a Dutch Boer is to a Bond-street exquisite. **1896** R. Wallace *Farming Industries Cape Colony* i. 15 No true pheasant..is found in Africa... Several species of francolins belonging to the same family..are known as pheasants. **1970** *Stand. Encycl. S. Afr.* II. 345/1 Most of the birds of the open veld are well camouflaged, for example the pheasants, partridges and quails, the bustards and korhaans.

3. pheasant-coucal (examples).

1873 [see *coucal]. **1908** E. J. Banfield *Confessions of Beachcomber* i. iii. 103 The swamp pheasant, or pheasant coucal..is also an early bird. **1935** [see *coucal]. **1944** A. Russell *Bush Ways* ii. 17 The cuckoos, with the single exception of the pheasant-coucal, build no nests of their own. **1965** *Austral. Encycl.* VII. 90/2 Pheasant, a name formerly applied in Australia to the lyrebird and still in general use for the pheasant-coucal or swamp pheasant (*Centropus phasianinus*).

phellem (fe·lĕm). *Bot.* [a. G. *phellem* (F. von Höhnel 1877, in *Sitzungsber. Math.-Naturw. Classe K. Akad. Wissenschaften* (*Wien*) LXXVI. 600), f. Gr. φελλ-ός cork + *-em* as in *phloem*.] = Cork *sb.*[1] 5.

1887 W. Hillhouse tr. *Strasburger's Handbk. Pract. Bot.* xiv. 153 A meniscus..produces externally colourless cells, which become rounded..and internally cork-cells, or Phellem. **1925** Eames & MacDaniels *Introd. Plant Anat.* ix. 206 The cells constituting phellem, commonly known as cork cells, are for the most part uniform in shape. **1953** K. Esau *Plant Anat.* xiv. 327 The type of phellem used for bottle cork consists of thin-walled, air-filled cells. **1971** A. Cronquist *Introd. Bot.* (ed. 2) xxiv. 455/2 Cork is also called phellem.

pheme (fīm). [ad. Gr. φήμη words, speech.] A term used by the American philosopher, C. S. Peirce (1839–1914), for words in an utterance as they make up a grammatical unit in language, contrasted with words used in speech to convey sense (see *rheme, *seme).

1906 C. S. Peirce in *Monist* XVI. 506 By a *Pheme* I mean a Sign which is equivalent to a grammatical sentence, whether it be Interrogative, Imperative, or Assertory... Such a Sign is intended to have some sort of compulsive effect on the interpreter of it. **1923** Ogden & Richards *Meaning of Meaning* 438 We are introduced to Semes, Phemes, and Delomes. **1931** G. Stern *Meaning & Change of Meaning* iii. 31 Meaning is a property peculiar not only to what we traditionally call words, but also to parts of words, *e.g.* the genitive ending in *man's*... As a comprehensive term..Vendryes..has proposed *sémantème*, and C. S. Peirce, quoted by Ogden–Richards..has *seme* and *pheme*. **1955** J. L. Austin *How to do Things with Words* (1962) viii. 97 The same pheme (token of the same type) may be used on different occasions of utterance with a different sense or reference, and so be a different rheme... The pheme is a unit of *language*... But the rheme is a unit of *speech*.

phememe (fī·mīm). *Linguistics.* [f. prec.: see *-eme.] A term used by Leonard Bloomfield for the smallest linguistic unit.

1933 L. Bloomfield *Language* xvi. 264 The parallelism of lexical and grammatical features can be exhibited in a set of terms like the following: (1) Smallest and meaningless unit of linguistic signaling: *phememe*; (a) lexical: *phoneme*; (b) grammatical: *taxeme*. **1936** J. R. Kantor *Objective Psychol. Gram.* xi. 149 The reader will notice the three hierarchical steps or classes: phememe, glosseme, and linguistic form. **1955** J. L. Austin *How to do Things with Words* (1962) vii. 92 A 'pheme' (as distinct from the phememe of linguistic theory).

phen-, pheno-. Add: **phenetol** [a. G. *phenetol* (A. Cahours 1850, in *Ann. d. Chem. u. Pharm.* LXXIV. 314)]; now usu. written **phenetole**; (earlier and later examples).
1850 *Q. Jrnl. Chem. Soc.* III. 78 Phenetol is a colourless, very mobile liquid, lighter than water; and having an agreeable, aromatic odour. 1972 *Materials & Technol.* IV. 306 Reaction [of phenol] with dimethyl or diethyl sulphate in a weakly alkaline solution gives anisole,..or phenetole, $C_6H_5OC_2H_5$ respectively.

phenacaine (fe·nǎkēⁱn). *Pharm.* Formerly also **phenocain(e.** [f. Phen- + -a + -caine after Cocaine.] Holocaine.
1907 *Brit. Pharmaceutical Codex* 260 Many synthetic substitutes for cocaine have been suggested for use as local anæsthetics, including..phenacaine (Holocaine). 1911 *Ibid.* 312 Phenocaine. 1920 Martindale & Westcott *Extra Pharmacopœia* (ed. 17) I. 331 Holocain Hydrochloride... *Syn.* Phenocain. 1946 *Brit. Jrnl. Pharmacol.* I. 99 Pethidine, phenacaine and papaverine possess both aromatic and basic groups and have a quinidine-like action. 1968 [see *Holocaine].

phenagle, var. *finagle *v.*

phenakistoscope. Add: (Later examples.) Also **phenakistiscope.** Hence **phenakistosco·pic** *a.*, resembling or reminiscent of a phenakistoscope.
1934 V. M. Yeates *Winged Victory* 162 It was gone, and beyond a phenakistoscopic veil he saw the flying moons and spheres caught in webs and dragged away. 1955 *Times* 18 July 3/4 The exhibition takes us back to beyond the origins of the cinema proper, to Plateau's Phénakistiscope, in which pictures, mounted on the inside of a circular revolving drum and viewed through slits, first created the illusion of continuous movement. 1961 *Glasgow Herald* 13 July 6/4 The thaumatrope, the phenakistiscope, the zoetrope, and the tachyscope..are the steps by which the modern cinema climbed to its present perfection. 1978 *Radio Times* 4–10 Mar. 4/4 A phenakistoscope and other Victorian toys.

phenanthroline (fĭnæ·nþrǒlīn). *Chem.* [ad. G. *phenanthrolin* (Z. H. Skraup 1882, in *Ber. d. Deut. Chem. Ges.* XV. 895), f. *phenanthr-en* phenanthrene (s.v. Phen-, pheno-) + *chinoline* Chinoline, Quinoline.] An organic compound, $C_{12}H_8N_2$, whose molecule is a tricyclic phenanthrene ring system in which one CH group in each of the two outer rings is replaced by a nitrogen atom, and the *ortho* isomer of which is used esp. as an indicator for iron, with which it forms a red-orange complex.
1882 *Jrnl. Chem. Soc.* XLII. 1111 Phenanthroline,..obtained by the action of glycerol and sulphuric acid on metadiamidobenzene, crystallises in transparent plates (m.p. 79°). 1909 *Chem. Abstr.* III. 2571 The terms *o-*, *m-* and *p-phenanthroline* are suggested in place of isophenanthroline, phenanthroline and pseudophenanthroline. 1935 *Chem. Rev.* XVI. 113 The development of the use of *o*-phenanthroline-ferrous complex..constitutes one of the most outstanding advances of recent years in the field of analytical chemistry. Its application to many new procedures of analytical chemistry immediately followed. 1966 *McGraw-Hill Encycl. Sci. & Technol.* IV. 273/1 Certain organic compounds, such as 1,10-phenanthroline, also catalyze the drying of oils and have been used for this purpose. 1978 *Sci. Amer.* July 120/2 After the bromine color disappears mix in one milliliter of ·025-molar phenanthroline ferrous sulfate (a dye sometimes called ferroin).

phencyclidine (fensəi·klidin). *Pharm.* [f. Phen- + Cycl(o- + Piper)idine, constituent parts of the systematic name.] A powerful analgesic and anæsthetic whose use is now chiefly restricted to veterinary medicine on account of its hallucinogenic effects; 1-(1-phenylcyclohexyl)piperidine hydrochloride, $C_{17}H_{25}N.HCl$. Abbrev. PCP.
1959 *Antibiotic Med.* VI. 84 Phencyclidine, when given to patients with anxiety symptoms, appeared to be most effective in mild to moderately severe reactions. 1963 *Lancet* 16 Feb. 392/2 Phencyclidine was used to relieve the pain of multiple rib fractures in a 57-year-old male who had bronchitis. 1973 *Daily Tel.* 20 Dec. 3 A new drug called 'angel dust' which causes people to believe they have been reduced to the size of Tom Thumb is to be brought to the attention of the authorities... The drug, known as phencyclidine or PCP, was not covered by the Dangerous Drugs Act. 1974 M. C. Gerald *Pharmacol.* xvii. 330 Phencyclidine (Sernyl) is employed therapeutically as a veterinary anesthetic agent. *Ibid.*, Compounds purported to be mescaline, LSD, psilocybin, and THC..have often been chemically identified as phencyclidine. 1979 *Tucson* (Arizona) *Citizen* 20 Sept. 7c/1 Seven men have been charged by a federal grand jury with conspiracy to manufacture and distribute phencyclidine, and for possession of the drug.

phenelzine (fěne·lzin). *Pharm.* [f. the systematic name 2-*phene*thyl*hydrazine*, f. Phen- + Ethyl + Hydrazine.] A monoamine oxidase inhibitor that is used as an antidepressant, usu. in the form of the sulphate, $C_6H_5CH_2CH_2NHNH_2.H_2SO_4$, a white crystalline solid with a pungent odour.
1959 *Amer. Jrnl. Psychiatry* CXVI. 64/1 Twenty-five patients with depression who entered this hospital on the female service were assigned alternately to Phenelzine and ECT. 1961 *Lancet* 16 Sept. 622/2 A housewife, aged 56, with depression, had for five months been taking phenelzine..when she suddenly got pruritis, pain behind the eyes, and anorexia, and noticed that she was becoming yellow. 1965 [see *Parnate]. 1971 G. G. Luce *Body Time* iii. 85 Other drugs sometimes used to combat depression, such as phenelzine, have an opposite effect and shorten the sleep cycle.

Phenergan (fe·nəɹgǎn). *Pharm.* Also **phenergan.** A proprietary name for promethazine.
1947 (Sept. 16) [registered in Canada as a trade mark, no. 112/28759]. 1948 *Canad. Med. Assoc. Jrnl.* LIX. 322/2 The name phenergan (3,277 R.P.) has been given to the N-dimethylaminopropyl derivative of phenothiazine. 1949 *Trade Marks Jrnl.* 8 June 499/1 Phenergan... Pharmaceutical preparations consisting of or containing benzine or its derivatives for human use and veterinary use. May & Baker Limited, Dagenham, Essex; manufacturing chemists. 1956 *Jrnl. Amer. Med. Assoc.* 3 Mar. 755/1 From these studies, several drugs have emerged that provide significant protection against motion sickness. These..include..promethazine (Phenergan) hydrochloride. 1958 A. Huxley *Let.* 4 Jan. (1969) 841 He has been experimenting with..a mixture of aspirin, chlorpromazine and phenergen [*sic*] compounded a year or so ago by some French physicians and used for producing a form of hibernation. 1977 *Detroit Free Press* 11 Dec. 18-B/2 It takes down about 10 percent of the cold-remedy business with its Dristan and Phenergan lines.

phenethicillin (fěne:þisi·lin). *Pharm.* [f. Phen- + Eth(yl + *pen)icillin.] The compound 6-(2-phenoxypropionamido)penicillanic acid, $C_{17}H_{20}N_2O_5S$, which is a semi-synthetic penicillin active when given by mouth and is usu. employed in the form of the white, crystalline, potassium salt.
1960 *Brit. Med. Jrnl.* 1 Oct. 994/2 When phenethicillin (6-(α-phenoxypropionamido)penicillanic acid) was marketed as 'broxil', it was claimed that this preparation would give blood levels at least equal to those after intramuscular injection of corresponding doses of penicillin G. 1965 G. T. Stewart *Penicillin Group of Drugs* iii. 25 In scientific medical circles, the advent of phenethicillin as the first offspring of the new biosyntheses was something of an anticlimax. *Ibid.* iv. 34 As prepared at present, phenethicillin is a racemic mixture containing 55–75% of the L-isomer and 25–45% of the D-isomer. 1970 *Daily Tel.* 23 Oct. 13/3 The first semi-synthetic penicillin, phenethicillin, introduced in 1959, became widely used for the infections treated by family doctors.

phenetic (fěne·tik), *a.* *Taxonomy.* [f. Gr. φαίν-ειν to appear + -*etic* as in Phyletic *a.*] (See quot. 1960.) So **phene·tically** *adv.*, showing similar characteristics; **phene·ticism,** taxonomy that stresses classifications based on obvious resemblances; **phene·ticist,** a taxonomist using classifications of this type.
1960 Cain & Harrison in *Proc. Zool. Soc.* CXXXV. 3 Following a suggestion made by Mr. H. K. Pusey, we shall refer to the arrangement by overall similarity, based on all available characters without any weighting as phenetic, since it employs all observable characters (including of course genetic data when available). *Ibid.*, When a good fossil record is available..the whole evolutionary dendrite can be worked out for that group simply by putting those forms together that are most alike phenetically. 1963 Davis & Heywood *Princ. Angiosperm Taxon.* iv. 112 The separation of phenetic and phylogenetic classifications is not generally accepted today. 1968 *Nature* 9 Nov. 547/1 The school of the pheneticists, one of the branches of numerical taxonomy. They deliberately set out to 'make' taxa on the basis of calculated overall similarity. *Ibid.*, The phenetic approach has been most useful when applied to groups with immature classifications..and to those with numerous non-redundant characters. 1971 *Virology* XLV. 357/2 To be useful, a classification of viruses..must be based on many characters, not few... This will give a classification based on the general resemblance of the phenotypic characters of the viruses (a 'phenetic classification'). 1972 *Nature* 21 Apr. 412/2 Extreme pheneticism is equally criticized.

phenformin (fenfǭ·ɹmin). *Pharm.* [f. Phen- + Form(aldehyde + *imino(-), constituent parts of the alternative name *phen*ethyl*form*amidinylimin*o*urea.] A white crystalline solid, 1-phenethylbiguanide hydrochloride, $C_{10}H_{15}N_5.HCl$, which is used in the oral treatment of diabetes. Also *phenformin hydrochloride*.
1959 *Canad. Med. Assoc. Jrnl.* LXXX. 773/1 In 1957 Ungar, Freedman and Shapiro reported that hypoglycæmic properties had been discovered in a synthetic diguanide compound, N′-β-phenethylformaminyliminourea hydrochloride, subsequently designated DB1 or phenformin. 1960 [see *biguanide]. 1961 *Lancet* 9 Sept. 566/1 Phenformin is clearly a potent hypoglycæmic substance effective in most diabetics we have tested. 1967 *Martindale's Extra Pharmacopoeia* (ed. 25) 675/2 Phenformin hydrochloride is contra-indicated in acidosis, coma, infections, [etc.]. 1974 M. C. Gerald *Pharmacol.* xxv. 441 It appears that phenformin may act by accelerating the intracellular oxidation of glucose, a process which is depressed in the absence of insulin. Phenformin is used alone for the treatment of maturity-onset diabetes and in combination with insulin in juvenile diabetes. 1977 *Lancet* 22 Jan. 191/2 Metformin is replacing phenformin as the biguanide of choice in the treatment of obese maturity-onset diabetics because of the association between phenformin therapy and lacticacidosis.

phenindione (fenində·ǒ^u^n). *Pharm.* [f. Phen- + In(do-² + *-dione.] A white crystalline solid, 2-phenylindan-1,3-dione, $C_{15}H_{10}O_2$, which is a vitamin K analogue used as an anticoagulant, esp. in the treatment of thrombosis.
1955 *Proc. Soc. Exper. Biol. & Med.* LXXXIX. 81/1 In an investigation of the effect of phenindione (PID) on the coagulation mechanism..it was discovered that patients who had received this anticoagulant for more than about 30 days developed a distinctly prolonged glass clotting time. 1962 *Lancet* 13 Jan. 104/2 Treatment with heparin and phenindione was begun, but after six hours the patient suddenly became unconscious and pulseless, though slow voluntary respirations persisted. 1974 R. H. Hammer in W. O. Foye *Princ. Med. Chem.* xviii. 409/1 Phenindione is the most commonly used 1,3-indandione anticoagulant... The chief disadvantage of phenindione is its toxic side effects.

pheniprazine (fěni·prǎzīn). *Pharm.* [f. Phen- + I(so- + Pr(opyl + Hyd)razine.] The compound 1-phenyl-2-hydrazinopropane, $C_9H_{14}N_2$, which was formerly used (as the hydrochloride) for the treatment of depression, angina, and hypertension.
1960 *Jrnl. Amer. Med. Assoc.* 2 Apr. 1554/1 Maskinnon and co-workers used pheniprazine to treat a series of 28 patients with the anginal syndrome. 1961 *Lancet* 16 Sept. 622/2 It seems probable that pheniprazine was the cause of the jaundice; the total dose was 560 mg. given over sixty days. 1971 L. E. Hollister in Meltzer & Morrelli *Clin. Pharmacol.* (1972) xi. 459/1 The first hydrazide drug was iproniazid... A later member of the series was pheniprazine (Catron). Both have subsequently been removed from the market because of intolerable toxicity.

phenmetrazine (fenme·trǎzīn). *Pharm.* [f. Phen(yl + -*metr-* (f. Met(hyl + Hyd)r(o-) + *azine, constituent parts of the systematic name *2-phenyl-3-methyl-tetrahydro-1,4-oxazine.*] 3-Methyl-2-phenylmorpholine, $C_{11}H_{15}NO$, the hydrochloride of which, a white powder, has been used as an appetite suppressant and is a stimulant of the central nervous system similar to (though weaker than) amphetamine, to which it is chemically related.
1956 *Amer. Jrnl. Digestive Dis.* I. 155 Phenmetrazine has been used clinically in Germany in the management of obesity. 1959 *Lancet* 22 Aug. 152/1 Phenmetrazine ('Preludin'), a sympathomimetic drug discovered by Thomae and Wick, was first used in 1954 in the treatment of obesity. Since then it has been given in asthma, in parkinsonism, and as a euphoriant. 1962 [see *herniation]. 1970 [see *Preludin].

pheno (fī·no), colloq. abbrev. of *phenobarbital, *phenobarbitone.
1966 J. Pearl *Crucifixion P. McCabe* (1967) viii. 118 Take a couple of my phenos and sack in early. 1968 'E. Trevor' *Place for Wicked* vi. 82 She'd just managed to catch the chemist open for some phenos:..after the plane journey she wouldn't be able to sleep properly. 1971 *Where* Dec. 360/2 He agreed that she did not need such a heavy dose and suggested a quarter grain of pheno twice a day only if she was under the weather.

phenobarb (fīnobā·ɹb). Also **pheno barb** and with final point. Abbrev. of *phenobarbital, *phenobarbitone, used *colloq.* and in *Pharm.*
1956 M. McMinnies *Flying Fox* iii. iv. 223 Whatever next? Not with all that phenobarb inside you. 1961 *Lancet* 22 July 205/2 Finally my colleague pleaded: 'If you won't take that phenobarb. yourself, for Heaven's sake dish it out to the rest of us.' 1971 *Where* Dec. 360/1 When Sarah got her sight back..she came back to us on a dose of one grain of pheno barb twice a day. 1972 J. Mann *Mrs. Knox's Profession* x. 82 She picked up a small bottle and said, 'Phenobarb?' 1977 N. Adam *Triplehip Cracksman* xii. 125 The phenobarb and the adrenalin, the swallowed capsule and the upcoming junket.

phenobarbital (fīnǒ-, fenǒbā·ɹbitǎl). *Pharm.* [f. Pheno- + *barbital.] The equivalent in the U.S. Pharmacopeia of *phenobarbitone; also, a tablet of this.
[1918 *New & Nonofficial Remedies* (Amer. Med. Assoc.) p. vi, In accordance with the action of the Federal Trade Commission, New and Nonofficial Remedies recognizes.. Phenylbarbital as the official name for the product first introduced as luminal.] 1919 *Ibid.* 84 Phenobarbital is claimed to be a useful hypnotic in nervous insomnia. 1938 [see *amphetamine]. 1950 M. Lowry *Let.* Jan. (1967) 189, I must have had one too many phenobarbital. 1974 M. C. Gerald *Pharmacol.* ii. 18 Barbiturates, such as phenobarbital, cause depression. *Ibid.* xi. 202 Barbital was widely used for about a decade until the introduction of phenobarbital (Luminal) in 1912.

phenobarbitone (fīnǒ-, fenǒbā·ɹbitǒ^u^n). *Pharm.* [f. Pheno- + *barbitone.] A white crystalline compound, 5-ethyl-5-phenylbarbituric acid, $C_{12}H_{12}N_2O_3$, which is widely used as a sedative, hypnotic, and anticonvulsant,

often in the form of its sodium salt. So *phenobarbitone sodium*. Cf. *Luminal *sb.*
1932 *Brit. Pharmacopœia* 330 Phenobarbitone is 5-phenyl-5-ethylbarbituric acid, and may be obtained by the condensation of ethyl phenylethylmalonate with urea. 1943, etc. [see *anti-convulsant *a.* and *sb.*]. 1954 *Newsweek* 12 Apr. 90/2 To 'keep their interest up', he [*sc.* a rowing coach]..reportedly prescribed phenobarbitone pills for a good night's sleep on the eve of the race. 1970 'D. Halliday' *Dolly & Cookie Bird* vi. 85 She has an Attitude to Life which would drive a phenobarbitone pill up the wall. 1975 Cawson & Spector *Clin. Pharmacol. in Dentistry* vi. 106 The main disadvantage of phenobarbitone is that as well as anti-anxiety (anxiolytic) properties, it is strongly sedative.

phenocain(e), obs. varr. *phenacaine.

phenocopy (fī·nokǫpi). *Biol.* [ad. G. *phänokopie* (R. Goldschmidt 1935, in *Zeitschr. für induktive Abstammungs- und Vererbungslehre* LXIX. 46): see Phen-, pheno- and Copy sb.] An individual showing features characteristic of a genotype other than its own, but induced by a modified environment.
1938 R. Goldschmidt *Physiol. Genetics* ii. 14 The production of phenocopies in Drosophila succeeded only if the temperature shock was applied at a definite time in development, the sensitive period. 1957 *Amer. Naturalist* XCI. 86 The production of phenocopies succeeded more readily in organisms which were heterozygous for a recessive mutant gene with homologous morphological effects. 1971 D. J. Cove *Genetics* xii. 172 A wild-type organism which has been subjected to such a heat treatment may develop so that it resembles closely an organism with a particular mutation. These environmentally induced abnormal organisms are called phenocopies.

phenogenetics (fī:nodʒěne·tiks), *sb. pl.* (const. as *sing.*). *Biol.* [ad. G. *phänogenetik* (V. Haecker *Entwicklungsgeschichtliche Eigenschaftsanalyse* (1918) i. 4): see Phen-, pheno-, Genetic *a.*, and -ic 2. Cf. next.
Haecker previously called it *phänogenese* (*Zeitschr. f. induktive Abstammungs- und Vererbungslehre* (1915) XIV. 260).]
(See quots.)
1938 R. Goldschmidt *Physiol. Genetics* ii. 23 It becomes imperative to know the details of development that distinguish the mutant types from the wild type. Studies of this type have been called..phenogenetics, a term that is not necessary but is sometimes useful. 1962 I. H. Herskowitz *Genetics* xxx. 262/2 How does the mutant change normal development to produce the new morphological result? The answer to the latter question deals with learning how phenotypes (of any type) come into being via gene action, and is the subject of phenogenetics, a study which is of broader scope than developmental genetics. *Ibid.* 269 Phenogenetics starts out as a study of the developmental genetics of morphology.

phenol. Add: **b.** (Examples of sing. use.)
1868 Jones & Watts *Fownes' Man. Elem. Chem.* (ed. 10) iii. 646 A xylylic phenol is mentioned by Hugo Muller as occurring in coal-tar. 1952 L. N. Ferguson *Electron Struct. Org. Molecules* viii. 189 For the same type of dissociation of a phenol..the large increase in dissociation constant can be attributed largely to the effects of resonance. 1964 N. G. Clark *Mod. Org. Chem.* xxi. 431 The acidity of a phenol is increased by a nitro group located *ortho* and/or *para* to the functional group.
c. phenol-formaldehyde, used *attrib.* and *absol.* to designate plastics, resins, etc., made by condensation of phenols with formaldehyde; **phenol oxidase, phenolo·xidase** *Biol.* [ad. G. *phenoloxydase* (Battelli & Stern 1912, in *Biochem. Zeitschr.* XLVI. 396): see *oxidase], = *phenolase; **phenol red** = *phenolsulphonphthalein* in sense *d; **phenol resin,** a phenolic resin (see *phenolic *a.* b).
1912 *Jrnl. Industr. & Engin. Chem.* IV. 737/2 (*heading*) Phenol-formaldehyde condensation products. 1933 *Archit. Rev.* LXXIII. 1 The most modern and significant group consists of the synthetic plastics of which the two best known and most fully developed are the Phenol-formaldehyde and the Urea-formaldehyde types. 1965 *Wireless World* Sept. 32 (Advt.), New 10 pin (Decal) based valves type B10B, moulded in polypropylene and phenol formaldehyde. 1972 *Materials & Technol.* V. 95 All types of phenol formaldehyde glue are resistant to organic solvents, bacteria and fungi. *Ibid.* 96 The largest use of phenol formaldehyde resin adhesives is in gluing wood. 1913 *Chem. Abstr.* VII. 796 The authors [*sc.* Battelli & Stern] recommend that the term phenolase be dropped and the term phenoloxidase be used instead. 1920 *Biochem. Jrnl.* XIV. 539 The oxidase (synonymous with laccase, phenoloxidase and phenolase) of the pear and potato. 1931 E. C. Miller *Plant Physiol.* xiii. 769 Substances which give the same or similar reactions as laccase toward the phenol compounds..have been named collectively the 'laccases', the 'phenol oxidases', or the 'phenolases'. 1971 I. Zelitch *Photosynthesis* v. 136 Phenol oxidase activity is responsible for the familiar 'browning' observed in plant cells after injury. 1916 Clark & Lubs in *Jrnl. Washington Acad. Sci.* VI. 488 Phenol red and cresol red are undoubtedly the most reliable indicators of the series. 1956 Whitby & Hynes *Med. Bacteriol.* (ed. 6) xi. 151 Collect the supernatant fluid, add a drop of phenol red solution, neutralize with N NaOH, and centrifuge. 1974 D. H. Kay in W. O. Foye *Princ. Med. Chem.* xxxvii. 816/2 This dye is also known as phenol red and is official as phenolsulfonphthalein (U.S.P.). 1912 J. W. Aylsworth *U.S. Pat. 1,020,593* The fusible anhydrous resinous condensation product..(hereinafter termed a 'phenol resin') is first prepared by causing a reaction between suitable amounts of phenol and formaldehyde. 1923 *Industr. & Engin. Chem.* July 677/1 The phenol resins continue to find wider applications in the industries on account of their physical and chemical properties. 1947 R. Nauth *Chem. & Technol. Plastics* ii. 16 (*caption*) The dark rings which look like grain are..an effect produced by the phenol resin between the laminated veneers. 1973 *Materials & Technol.* VI. viii. 591 Phenol resins are widely used in the furniture and woodworking industries.
d. phe:nolcarbo·xylic (or **phenol carboxylic**) **acid,** any acid which contains a carboxyl group and a hydroxyl group bonded to the same benzene ring; **phe:nolsulphonphtha·lein** [Sulphon(e + Phthalein], a red crystalline solid $C_{19}H_{14}O_5S$, which is used as an indicator in the pH range 6·7 (yellow) to 8·3 (red), and in medicine is given intravenously as a test of kidney function.
1899 *Collective Index Trans. & Abstr. Chem. Soc. 1873–82* ii. 471/1 Phenolcarboxylic acid. *See* Hydroxybenzoic acids and Salicylic acid. 1946 *Thorpe's Dict. Appl. Chem.* (ed. 4) VII. 285/1 All the compounds of this sub-series have been shown to be phenol carboxylic acid derivatives.., and the majority belong to the depsides. 1956 [see *depside]. 1975 D. Jarvis tr. *Hess's Plant Physiol.* 128 Phenol carboxylic acids, such as protocatechuic acid and gallic acid, occur widely. 1898 M. D. Sohon in *Amer. Chem. Jrnl.* XX. 263 Phenolsulphonphthalein, $C_{19}H_{14}SO_5$..was obtained as a bright red crystalline powder, somewhat soluble in water. 1960 I. A. Stanton *Dict. for Med. Secretaries* 116/1 Phenolsulphonphthalein is injected intramuscularly or intravenously and should normally appear in the urine in 10 or 15 minutes. 1974 [see *phenol red* in *c].

phenolase (fī·nǫlēⁱz, -s). *Biol.* [a. G. *phenolase* (F. Czapek 1906, in *Jahrb. f. wissensch. Bot.* XLIII. 380), f. *phenol* Phenol: see *-ase.] Any of a class of copper-containing enzymes found esp. in plants, which oxidize phenols to quinones; = *phenol oxidase* s.v. *phenol c.
1911 *Jrnl. Chem. Soc.* C. i. 824 The phenolase was prepared from *Lactarius vellerius*. 1931 E. C. Miller *Plant Physiol.* xiii. 768 The oxidases of plants that have been studied have been grouped into two main classes according to the substances upon which they act. These are the laccases or phenolases and the tyrosinases. 1956 *Nature* 14 Jan. 79/1 The phenolase complex, widely distributed throughout the phylogenetic scale, consists of two enzymic activities, phenol *o*-hydroxylase..and *o*-diphenol dehydrogenase. 1964 *Oceanogr. & Marine Biol.* II. 408 Isolation of the amoebocytes from the fluid brings about more rapid oxidation, suggesting that an inhibitor to phenolase activity may exist in the coelomic fluid. 1975 D. Jarvis tr. *Hess's Plant Physiol.* 83 Cytochrome oxidase a_3 makes direct contact with oxygen and is therefore known, together with a few other enzymes, (peroxidases, catalases, and phenolases), as a 'direct' oxidase.

phenolate (fī·nǫlēⁱt). *Chem.* [f. Phenol + -ate¹.] = *phenoxide.
1885 I. Remsen *Introd. Study Compounds of Carbon* xv. 270 Potassium phenolate, $C_6H_5.OK$, made by dissolving potassium in phenol, and by treating phenol with caustic potash. 1913 A. R. Warnes *Coal Tar Distill.* viii. 79 Owing to the pressure of the C_6H_5 radicle the ·OH group (hydroxyl) possesses slight acid properties.., and..reacts with the caustic alkalies, forming salts known as phenolates. 1963 F. G. Bordwell *Org. Chem.* v. 159 (*caption*) Resonance in the phenolate ion. 1974 R. S. Cahn *Introd. Chem. Nomencl.* (ed. 4) vi. 95 Salts of alcohols and phenols should receive the ending -olate, as in sodium methanolate.., sodium phenolate.., but names such as..the abbreviated methoxide,..and phenoxide are extremely common.

phenolic, *a.* Substitute for def.: Of the nature of, belonging to, derived from, or containing a phenol; *esp.* containing or being a hydroxyl group bonded directly to a benzene ring. (Further examples.)
1932 *Ann. Rep. Chem. Soc.* XXVIII. 236 Marrian had identified the substance $C_{18}H_{24}O_2$ as a trihydric alcohol with one phenolic hydroxyl. 1951 I. L. Finar *Org. Chem.* I. xxvii. 533 Phenolic aldehydes..are very important compounds, and contain an aldehyde group and one or more hydroxyl groups directly attached to the nucleus. *Ibid.* xxviii. 557 It should be noted that only half the phenol is converted into the phenolic acid. 1964 N. G. Clark *Mod. Org. Chem.* xxi. 430 The crude oily phenolic layer is separated from the aqueous liquor, and purified by distillation. *Ibid.* 431 The formation of simple ethers in this manner is a convenient way of 'protecting' a phenolic group. 1970 *Watsonia* VIII. 3 Chromatographic investigation..into the intraspecific variation of phenolic and other compounds is proving to have interesting results. 1973 *Nature* 20 July 132/1 A serine hydroxyl..makes a hydrogen bond to a phenolic hydroxyl hydrogen, the oxygen of which bonds to another phenolic hydroxyl.
b. Designating a large class of usu. thermosetting polymeric materials that have wide industrial applications as plastics or resins and are prepared from phenols by condensation with aldehydes in the presence of acid or base catalysts; as *phenolic plastic* or *resin*.
1910 J. W. Aylsworth *U.S. Pat. 1,020,593*, I..have made a certain new and useful Invention in Phenolic Condensation Products and Methods of Preparing Same. 1917 *Chem. Abstr.* XI. 693 (*heading*) Insoluble phenolic resins. 1931 *Product Engin.* Nov. 503/2 Cast phenolic plastics are softer than the ordinary phenolic materials. 1944 *Electronic Engin.* XVII. 35/1 A skilful combination of a polyvinyl acetal resin..with a thermo-setting phenolic resin has enabled 'Thermex' enamel to be produced from wholly synthetic sources. 1963 *House & Garden* Feb. 8/2 (*caption*) Mural panel in polished copper and black phenolic resin. 1968 *Wall St. Jrnl.* 26 July 28/1 Prices of phenolic molding compounds, among the more widely used plastics, will be raised one cent a pound by Union Carbide Corp. 1973 *Materials & Technol.* VI. viii. 589 The starting materials for the manufacture of phenolic plastics are mainly phenols and formaldehyde.
B. *sb.* **a.** Phenolic plastic or resin.
1935 *Industr. & Engin. Chem.* Oct. 1141/1 Certain plastics, particularly the cast phenolics, have unusual qualities for the passage and reflection of light. 1946 H. A. Tayler *Plastics Explained* iv. 90 Cast phenolics are resistant to moisture, weak acids, oils and organic solvents. 1960 *Times Rev. Industry* Mar. 57/1 British plastics production..total was divided as: thermosetting materials 195,000 tons (alkyds 50,000 tons; aminoplastics 55,000 tons; phenolics 75,000 tons) and thermoplastics 300,000 tons. 1968 *Wall St. Jrnl.* 26 July 28/1 (*heading*) Phenolics price boost is planned by Union Carbide. 1973 *Sci. Amer.* July 42/3 The matrix in glass-reinforced composites may be either a thermoset plastic, such as polyester, phenolic or epoxy, or any of a number of thermoplastic resins, such as nylon, polyethylene or polystyrene.
b. Chiefly *Biochem.* Any compound containing a hydroxyl group bonded directly to a benzene ring, esp. one that occurs in plants.
1956 *Chem. & Industry* 9 June 478/2 Ellagic acid occurs in 43 families of dicotyledons, and is usually accompanied by the other vicinal-trihydroxy phenolics myricetin and leucodelphinidin. 1960 A. H. Williams in J. B. Pridham *Phenolics in Plants* i. 3 Phloridzin, the principal phenolic of apple leaf and bark is absent from the flesh of the fruit. 1970 J. Van Buren in A. C. Hulme *Biochem. Fruits* I. xi. 298 Varieties of fruits high in phenolics are more astringent than varieties low in phenolics. 1976 *Which?* Aug. 185/2 The household disinfectants you'll usually find in your shop will be either phenolics or hypochlorites.

phenolized (fī·n-, fe·nǫləizd), *ppl. a.* Chiefly *Med.* [f. Phenol + -iz(e + -ed¹.] Treated with phenol; *spec.* (of vaccines, cell samples, etc.) suspended in a dilute solution of phenol. So **phenoliza·tion.**
1921 Dorland *Med. Dict.* (ed. 11) 791/1 *Phenolization*, treatment of infected wounds by subjecting them to the energetic action of strong carbolic acid. 1922 *Jrnl. Amer. Vet. Med. Assoc.* LXI. 40 The vaccination consists of a single injection of a large dose of phenolized fixed virus. 1949 Rhodes & van Rooyen *Textbk. Virol.* xxv. 226 The most popular method [of reducing the potency of rabies vaccine] is the use of phenolized virus... A common practice is to issue a vaccine consisting of a 4 to 5 per cent suspension of infected nervous tissue with 0·5 per cent phenol. 1961 *Amer. Jrnl. Med.* XXX. 804/2 Phenolized and acetone-extracted vaccines were ineffective in preventing clinical disease but probably modified the course of illness. 1975 *Nature* 24 July 277/1 Pure MS2 A-protein was obtained by isolation of the A-protein—RNA complex.., phenolisation, chromatography..and further separation. 1976 *Lancet* 9 Oct. 764/2 Granulomas were not transmitted by either frozen (−25°C) or phenolised sarcoid or Crohn's tissue homogenates.

phenological, *a.* Add: So **phenolo·gic** *a.*
1947 [see *isophane, isophene]. 1974 *Nature* 1 Mar. 42/1 This proposed sequence of major volcanic eruptions followed by several years of cold summers and then by glacial advance is supported by historic and phenologic data.

† phenoloid (fī·no-, fe·noloid), *sb.* and *a.* *Obs.* [f. Phenol + -oid.] **A.** *sb.* A phenoloid substance, *spec.* phenoloid oil. **B.** *adj.* Of the nature of, containing, or resembling phenols; *phenoloid oil*, a form of creosote obtained from the distillation of coal, esp. as a by-product from blast-furnace gases.
1907 V. B. Lewes *Liquid & Gaseous Fuels* 99 The oil obtained from blast furnaces is also sometimes used locally for fuel purposes under the name of 'Phenoloid', or blast furnace oil. 1911 *Med. Ann.* 758 Phenoloid Disinfectant.—This contains 66 per cent phenoloid, with high carbolic acid coefficient. 1913 V. B. Lewes *Oil Fuel* 129 Another variety of oil is obtained from blast furnaces, and is known as blast furnace oil, or 'phenoloid oil'. 1920 J. M. Fortescue-Brickdale *Text Bk. Pharmacol.* ix. 74 Phenoloids are bodies having similar properties [to cresols], and are obtained from the distillation of coke. Cyllin, izal, and kerol are mainly composed of phenoloids. 1929 *Encycl. Brit.* VI. 668/2 The mixture of phenol and phenoloid substances derived..from coal, wood, blast furnace, and other tars. *Ibid.* 669/1 Blast-furnace creosote, sometimes known also as phenoloid' resembles vertical-retort tar. 1961 *Brit. Med. Dict.* 1092/1 *Phenoloid*..a term, now becoming obsolete, describing substances of phenolic character such as cresols and xylenols.

phenolphthalein. Substitute for def. (in Dict. s.v. Phenol d): A whitish or yellowish crystalline solid,

$$(HO{\cdot}C_6H_4{\cdot})_2C{\cdot}C_6H_4{\cdot}CO,$$
└O┘

which is used in alcoholic solution as an indicator in the pH range 8 to 10, in which it changes from colourless to red, and is also used medically as a laxative.
1875 *Jrnl. Chem. Soc.* XXVIII. 67 When phenol is gently warmed with phthalic anhydride and sulphuric acid the mixture assumes a brownish yellow colour, from the formation of phenol-phthalein. 1881 J. Attfield *Chem.* (ed. 9) 658 In delicate experiments turmeric, 'eosin', 'phenolphthalein', etc., may be used instead of litmus. 1893 [in Dict. s.v. Phenol d]. 1946 *Nature* 23 Nov. 744/2 Back titration of the solution immediately after the substance had dissolved, using phenolphthalein as an indicator, indicated an apparent equivalent weight of 325. 1951 R. Mayer *Artist's Handbk.* ii. 97 Acids and alkalis are detected by the use of litmus, phenolphthalein, or other indicators. 1974 M. C. Gerald *Pharmacol.* ii. 31 In some laxatives, such as Ex-Lax and Feen-A-Mint, the active ingredient is phenolphthalein.

phenomena: as erron. sing. form (see Phenomenon 1 β in Dict. and Suppl.).

phenomenal, *a.* Add: **1. a.** *Psychol.* Of or relating to a phenomenon as it is directly perceived or sensed, esp. as compared with its objective reality; also in spec. collocations, as *phenomenal regression*, the tendency for a shape, esp. a perspective, to be perceived as nearer to the shape of a related and known object than it actually is.
1922 K. Koffka in *Psychol. Bull.* XIX. 569 A field, reflecting a certain amount and quality of light, depends for its phenomenal color-quality upon the ground on which it appears. 1931 R. H. Thouless in *Brit. Jrnl. Psychol.* XXI. 340 The shape reported by the subject as seen by him may be called the 'apparent shape' or 'phenomenal shape'. *Ibid.* 344 As a general name for this tendency, in whatever kind of perceptual character it is found, we may use the term *phenomenal regression to the 'real' object* or, more shortly, *phenomenal regression*. 1948 *Jrnl. R. Aeronaut. Soc.* LII. 467/1 The factor of phenomenal regression was a quantity differing widely among different people. 1951 G. R. Wendt in S. S. Stevens *Handbk. Exper. Psychol.* xxxi. 1208/2 We lack words in our language to distinguish phenomenal motion from the physical event. 1969 C. O. Schrag *Experience & Being* i. i. 19 The phenomenal field, according to Merleau-Ponty, is neither an 'outer world' of objectively reconstituted properties..nor is it an abstracted 'inner world'.

phenomenalism. Add: **phenomenalist,** also as *adj.*, of or pertaining to phenomenalism or a phenomenalist; **phenomenalistic** *a.* (later examples); **phenomenali·stically** *adv.*, as regards or in terms of phenomenalism. Also **phenomenalistically-minded** adj.
1885 W. James *Coll. Ess. & Rev.* (1920) 277 Modern thinkers..for the most part obey a common drift..towards a phenomenalistic or idealistic creed. 1904 Phenomenalistic [see *ante rem]. 1909 Webster, Phenomenalistically. 1914 C. D. Broad *Perception* iii. 166 Berkeley, whose argument is properly phenomenalistic. *Ibid.* 171 The phenomenalist position has to be stated as follows. 1934 A. C. Ewing *Idealism* vii. 294 A sense which cannot..be analysed phenomenalistically. 1943 *Mind* LII. 340 But surely it ought to worry a phenomenalistically-minded philosopher; and Mr. Smith does not seem to be sufficiently worried by it. 1956 E. H. Hutten *Lang. Mod. Physics* ii. 64 It is..said that a thing-language, or a phenomenalist language, e.g. one taking sense-data as a key-concept, represents an empiricist language. 1963 Phenomenalistic [see *direct realism]. 1975 *New Left Rev.* Nov.–Dec. 33 Knowledge is restricted to what is known for certain; it is then shown, in a phenomenalistic analysis of perception, that what is known in perception is certain; only perception gives knowledge of things (principle of empiricism); hence knowledge must be what is given in perception.

phenomenality. Delete *rare* and add further examples.
1884 W. James in R. B. Perry *Tht. & Char. W. James* (1935) I. 580 To see whether the object thus given be itself only more subjectivity, more phenomenality, more experience than the instant. 1917 A. S. Pringle-Pattison *Idea of God* xi. 211 This organic point of view delivers us..from the difficulties which so sorely afflict modern philosophy as to the relativity, or subjectivity, or phenomenality, of knowledge. 1933 *Jrnl. Theol. Stud.* XXXIV. 314 The notion that all phenomenality including the self is illusory..finally became..the foundation of Sankara's philosophy. 1969 T. F. Torrance *Theol. Sci.* i. 21 Modern philosophy in its preoccupation with phenomenality.

phenomenalize, *v.* (Earlier example.) **phenomenalization.** (Later example.)
1878 S. H. Hodgson *Philos. of Reflection* I. 213 What was the Thing-in-itself has been phenomenalised and relegated to this possible world. 1921 Hannay & Collingwood tr. *G. de Ruggiero's Mod. Philos.* iii. ii. 274 Bradley..fails to see that the true absolute is..appearance itself, in so far as it is the absolute process of appearing, the phenomenalization of the absolute.

phenomenist. Add: Also *attrib.* or as *adj.*
1871 *Dublin Rev.* Oct. 309 No one will doubt, either that the phenomenist school professes the general doctrine we have ascribed to it, or that Mr. Mill habitually identifies himself with this school.

phenomenistic, *a.* (Earlier example.)
1871 *Dublin Rev.* Oct. 309 The phenomenistic doctrine is such as this: that an ascertained truth, means a truth experienced or inferred from experience.

phenomenological, *a.* (In Dict. s.v. Phenomenology.) Add to def.: dealing with the description and classification of phenomena, not with their explanation or cause; *phenomenological method*, the method outlined by Husserl for the description and analysis of phenomena as they are directly experienced (cf. *phenomenology b).
1923 Ogden & Richards *Meaning of Meaning* 419 It is important for the understanding of Husserl's terminology to realise that everything he writes is developed out of the 'Phenomenological Method and Phenomenological Philosophy'. 1931 W. R. B. Gibson tr. *Husserl's Ideas* 14 There..grows up, on the pure basis of inner intuition, of the intuition of the soul's own essence, a phenomenological psychology. 1956 A. A. Townsend *Struct. Turbulent Shear Flow* iii. 33 The success of this approach has led to the present tendency to approach the problem of turbulence by examining the dynamics of the turbulent motion itself in preference to using phenomenological theories which do not attempt to describe the turbulent motion but only its effects on the mean flow. 1963 J. Wild in A. L. Fisher tr. *Merleau-Ponty's Struct. Behaviour* (1965) p. xvi, The phenomenological thinking of this book cuts through the traditional oppositions between..body and soul, sense and reason, and subjectivism and objectivism. 1965 *Math. in Biol. & Med.* (Med. Res. Council) iv. 131 These difficulties..may be largely responsible for the fact that neurophysiologists..have often been hesitant to go beyond reporting raw data in a somewhat phenomenological manner. 1968 C. G. Kuper *Introd. Theory Superconductivity* i. 4 Long before the physics of superconductivity was understood, a number of phenomenological theories had been proposed. 1976 T. Eagleton *Crit. & Ideology* i. 24 He rejected, naturally, the political consequences..but the phenomenological basis of that criticism was of peculiarly direct relevance to him. 1977 *Dædalus* Summer 63 Processual analysis has undoubtedly gained from the phenomenological critique of positivist anthropology.

pheno:menolo·gically, *adv.* [f. prec. + -ly².] In terms of, or as regards, phenomena or phenomenology.
1891 M. E. Lowndes tr. *Höffding's Outl. Psychol.* ii. 63 Phenomenologically, he [*sc.* Lotze] thus places himself at the standpoint of the natural interaction. 1909 W. M. Urban *Valuation* i. 18 Ideals of a supernatural character are the product, phenomenologically speaking, of individual and racial appreciative constructions. 1942 *Amer. Jrnl. Physiol.* CXXXV. 736 Phenomenologically, the muscular reactions during a tonic-clonic response to electrical stimulation of the motor cortex have for a long time been recognized as analogous to those in Jacksonian or in grand-mal epilepsy. 1958 *Times Lit. Suppl.* 23 May p. xii/3 These images can be observed and described phenomenologically and by the means of psychology. 1975 *Sci. Amer.* June 56/3 If the collision yields an electron-positron pair, the annihilation and rebirth of such a pair is phenomenologically indistinguishable from the mere elastic scattering of the incident electron and positron. 1977 P. Johnson *Enemies of Society* xvii. 227 When we enter a cathedral, and examine its various axial tendencies and its symmetrical, and asymmetrical forms, we perceive it phenomenologically; the approach to music is, in all essentials, the same.

phenomenologist (fĭnǫmenǫ·lǒdʒist). [f. Phenomenolog(y + -ist.] One who makes a study of, or adheres to the doctrines of phenomenology, esp. a philosopher or psychologist. Also *attrib.*
1865 J. H. Stirling *Secret of Hegel* I. i. 19 He who shall make it his business to watch the gathering of the materials for the seething..will be the Phaenomenologist or Historian of the Seething. 1910 *Mind* XIX. 287 Another line of derivation from Kant leads..to Nelson and to the phenomenologists (Gomperz). 1951 *Scottish Jrnl. Theol.* IV. 175 Most modern Existentialists are phenomenologists. They do not however totally deny the existence of external reality. 1957 H. Whitehall in N. Frye *Sound & Poetry* 135 The Polish critic-philosopher, Roman Ingarden, using the phenomenologist techniques of Husserl, banished the form–content dichotomy. 1967 J. F. Corso *Exper. Psychol. Sensory Behavior* i. 9 Those psychologists who emphasize the importance of experience in psychology and use subjective..terms to describe experience are called phenomenologists. 1975 *Listener* 25 Dec.–1 Jan. 857/2, I regard myself as a phenomenologist—that is, someone who is studying the phenomena of the human mind in an existential way. 1977 *Dædalus* Summer 63 The phenomenologists, notably Schutz, insisted that the social world is in many important respects a cultural construct, an organized universe of meaning in the form of what Harold Garfinkel calls a series of 'typifications' of the objects within it.

phenomenology. Add: **b.** *Philos.* The theory, put forward by the German philosopher Edmund Husserl (1859–1938) and his followers, to the effect that the pure and transcendental nature and meaning of phenomena, and hence their real and ultimate significance, can only be apprehended subjectively; the method of reduction, based by Husserl on Descartes's method, whereby all factual knowledge and reasoned assumptions about the phenomenon as object and the experiencing 'ego' are set aside so that pure intuition of the essence of the phenomenon may be rigorously analysed and studied.
1914 *Mind* XXIII. 588 Phenomenology, then, if I have understood it right, is the science of the essential connexions of vital experiences, as rooted in their nature or their character; not, for example, of their causal connexions as events in time. 1931 W. R. B. Gibson tr. E. Husserl (*title*) Ideas: general introduction to pure phenomenology. *Ibid.* 11 Under the title 'A Pure or Transcendental Phenomenology', the work here presented seeks to found a new science..a science covering a new field of experience, exclusively its own, that of 'Transcendent Subjectivity'. 1938 G. Reavey tr. *Berdyaev's Solitude & Society* 57 For this reason Husserl's Phenomenology fails to be an Existential philosophy although it exercised a considerable influence on Heidegger. 1949 H. F. Mins tr. G. Lukács in R. W. Sellars et al. *Philos. for Future* 572 Modern phenomenology is one of the..philosophical methods which seek to rise above both idealism and materialism by discovering a philosophical 'third way', by making intuition the true source of knowledge. 1957 H. E. Barnes tr. *Sartre's Being & Nothingness* p. xlvi, Phenomenology is anything but a nominalism. 1974 D. Carr *Phenomenology & Probl. of Hist.* I. 33 The 'common' subject matter of phenomenology and psychology, consciousness, is subjected in phenomenology to an essential rather than a factual consideration.
c. *Psychol.* The methods of description and analysis developed from philosophical phenomenology applied to the subjective experiencing of phenomena and to consciousness, esp. in the fields of Gestalt psychology, existential analysis and psychiatry.
1930 W. B. Wolfe tr. *Wexberg's Individual Psychol.* 8 Phenomenology prepared the way for the decisive step toward a comparative and contextual point of view in psychology. 1935 K. Koffka *Princ. Gestalt Psychol.* iii. 73 For us phenomenology means as naive and full a description of direct experience as possible. 1958 H. F. Ellenberger in R. May et al. *Existence* i. iii. 92 There is a wide gap between the philosophical phenomenology of Husserl and the psychiatric phenomenology of Minkowski. *Ibid.* 101 Phenomenology can also use a 'categorical' frame of reference... The two basic categories of inner experiences are considered to be time ('temporality') and space ('spatiality'). 1959 A. W. Levi *Philos. & Mod. World* ii. x. 405 The phenomenology of the human condition..in which inescapable situations constitute the historical determination in its four forms of death, suffering, conflict, and guilt. 1969 C. O. Schrag *Experience & Being* i. ii. 65 Psychiatric phenomenology has contributed interesting and revealing studies on the nature of psychological space and..time in their pathological expressions.

phenomenon. 1. β. (Further examples.)
1783 J. Woodforde *Diary* 8 Jan. (1926) II. 54, I went..to see a wonderful Phœnomena in Nature a Heifer 3 years old with two distinct Heads. 1947 Gerth & Mills *From M. Weber: Ess. in Sociol.* iii. 73 In this conception of freedom as a historically developed phenomena,..Weber represents humanist and cultural liberalism rather than economic liberalism. 1969 *Daily Progress* (Charlottesville, Va.) 5 Feb. 1/7 'They have an image now,' said Dr. Granville C. Fisher, University of Miami psychologist, 'and many others will follow the same route. It is a phenomena that will keep spreading.' 1970 *Nature* 31 Oct. 405/2 His work is fundamental to the concept of 'frozen' lines of magnetic force being held inside a plasma, a phenomena of great importance to any understanding of the processes occurring in the magnetosphere. 1972 *Real Estate Rev.* Winter 6/1 In some of our major cities, the abandonment phenomena to be witnessed is unlike anything that can be found in the United States outside the ghost towns of the old West.

phenon (fe·non). [f. Gk. φαίνειν to appear + -*on*.] **a.** *Biol.* A group of apparently similar plants or animals.
1943 Camp & Gilly in *Brittonia* IV. 335 Phenon: a species which is phenotypically homogeneous and whose individuals are sexually reproductive, but which is composed of intersterile segments. 1969 E. Mayr *Princ. Systematic Zool.* i. 5 There is no generally accepted technical term for a phenotypically reasonably uniform sample, but it may be designated as a phenon. *Ibid.* 10 A phenon is not necessarily a population in the biological sense.
b. *Taxonomy.* A grouping of plants or animals established by techniques of numerical analysis.
1962 Sneath & Sokal in *Nature* 3 Mar. 860/1 How should we name the groups which are established by numerical taxonomy?.. We call the groups simply 'phenons'. 1963 Davis & Heywood *Princ. Angiosperm Taxon.* iv. 136 Sneath & Sokal have introduced the concept of a phenon for the definition of groups obtained by cluster analysis... In practice a phenon defines groups by drawing lines horizontally across the dendrograms. 1963 Sokal & Sneath *Princ. Numerical Taxon.* ix. 251 The phenon nomenclature... Phenons are groups which approach natural taxa more or less closely, and..they can be of any hierarchic rank or of indeterminate rank. 1966 *New Scientist* 20 Jan. 151/3 These groups are called *phenons*... If the original taxa are species each of the three new phenon taxa might represent a sub-genus or genus. 1973 Sneath & Sokal *Numerical Taxon.* v. 294 The term phenon is intended to be general, to cover the groups produced by any form of cluster analysis.

phenosafranine (fīnǒsæ·frǎnīn, -in). *Chem.* Formerly also **-in.** [ad. G. *phenosafranin* (O. N. Witt: see R. Nietzki in *Ber. d. Deut. Chem. Ges.* (1883) XVI. 466), f. *pheno-* Pheno- + *safranin* Safranin.] A synthetic red dye, $C_{18}H_{15}N_4Cl$, which is used in photography as a desensitizer; also, any of the derivatives of this compound.
1883 *Jrnl. Chem. Soc.* XLIV. 731 Phenosafranine,..obtained by Witt by oxidising a mixture of aniline (2 mols.) and paraphenylenediamine (1 mol.).., forms beautifully crystalline salts. 1921 [see *desensitize *v.*].

1937 *Thorpe's Dict. Appl. Chem.* (ed. 4) I. 571/2 Nietzki showed that the same phenosafranine may be obtained (*a*) by condensing phenyl-*m*-phenylenediamine with phenyl-*p*-phenylenediamine, or (*b*) from diphenyl-*m*-phenylenediamine..and *p*-phenylenediamine. 1970 *Amat. Photographer* 11 Mar. 63/3 The phenosafranines and allied desensitising dyes are not suitable for papers as they stain badly.

phenothiazine (fīno-, fenoþəi·ǎzīn). *Pharm.* Formerly also **phenthiazine.** [f. Pheno- + Thi(o- + *azine.] **a.** A green, crystalline, heterocyclic compound, $C_{12}H_9NS$, which is used in veterinary medicine in the treatment of parasitic infestations. **b.** Any of various derivatives of this, which constitute an important class of tranquillizing drugs used esp. in the treatment of mental illnesses.
1894 G. M'Gowan tr. *Bernthsen's Text-bk. Org. Chem.* (ed. 2) 539 Nile Blue springs from naphtho-phenoxazine; and the thionine dyes from phen-thiazine. 1917 *Chem. Abstr.* XI. 3903 (Index), Phenothiazine. 1926 H. G. Rule tr. *J. Schmidt's Text-bk. Org. Chem.* 709 Phenthiazine, thio-diphenylamine,..is also the parent compound of a number of dye-stuffs. 1940 *Nature* 17 Aug. 232/2 Since phenothiazine is a new and valuable vermifuge, its effect on animal tissues is of some general interest. 1959 *Times* 14 Sept. 19/4 The nodula worm of sheep has been completely eliminated..by the use of pheno-thiazine during the period of winter housing. 1969 *Daily Tel.* 1 Nov. 2/7 Large doses of phenothiazines induce a Zombi-like rigidity. 1974 M. C. Gerald *Pharmacol.* iv. 71 The phenothiazine nucleus has proved..a rather versatile progenitor of pharmacologically useful compounds. *Ibid.* xvi. 301 Chlorpromazine (Thorazine), the most commonly employed phenothiazine.

phenotype (fī·notəip). *Biol.* Also †**phænotype.** [ad. G. *phaenotypus* (W. Johannsen *Elem. der exakten Erblichkeitslehre* (1909) vii. 123): see Phen-, pheno- and -type.] A type of organism distinguishable from others by observable features; the sum total of the observable features of an individual, regarded as the consequence of the interaction of its genotype with its environment. Cf. *genotype *sb.*²
1911 W. Johannsen in *Amer. Naturalist* XLV. 132, I have proposed the terms 'gene' and 'genotype' and some further terms, as 'phenotype' and 'biotype', to be used in the science of genetics. *Ibid.* 134 All 'types' of organisms, distinguishable by direct inspection or only by finer methods of measuring or description may be characterized as 'phenotypes'. 1931 S. J. Holmes *Life & Evolution* xiv. 277 A study of our checkerboard indicates that there are four phaenotypes. 1958 *Antiquity* XXXII. 207 The Neolithic people were mostly of a different phenotype. 1964 G. H. Haggis et al. *Introd. Molecular Biol.* vii. 193 In this scheme the bodily characteristics, or phenotype, are jointly determined by the environment on the one hand and by the inherited chromosomal complement, or genotype, on the other. 1969 *Times Lit. Suppl.* 20 Nov. 1341 Darwinian selection acts on phenotypes. 1971 D. J. Cove *Genetics* iv. 45 It is usual to refer to the genetic constitution of a strain as its genotype, and to its appearance as its phenotype. 1973 *Listener* 28 June 851/1 Whether or not the average phenotypes of such races can be shown to differ significantly in IQ is beside the point. 1976 Smythies & Corbett *Psychiatry* iv. 36 A number of different subgroups based on different biochemical lesions..all..end up with a similar clinical picture. In other words many different genotypes can end up with a very similar phenotype.
Hence **phenoty·pic, phenoty·pical** *adjs.*, of or pertaining to the observable features of, or differences between, organisms (often used with the implication 'not genotypic'); **phenoty·pically** *adv.*
1911 *Amer. Naturalist* XLV. 148 The phenotypically distinct and even diversely localized 'characters' convey easily the impression that they are the reactions of different genes. *Ibid.* 156 The merely phenotypical phenomena of alternative variability first pointed out by De Vries. 1929 R. R. Gates *Heredity in Man* ii. 26 There are cases where two or more factors combined to produce a single phenotypic character. 1930 *Biol. Bull.* LVIII. 85 (*heading*) Phenotypical variation in body and cell size of *Drosophila melanogaster*. 1935 *Proc. Nat. Acad. Sci.* XXI. 22 A rare phenomenon—the recurrence of phaenotypically the same scute mutation. 1942 *Endeavour* I. 18/2 Many geneticists have..attempted to discover the processes involved in the mechanism by which the genes of the genotype bring about phenotypic effects. 1964 M. Critchley *Developmental Dyslexia* x. 64 His present study lent no support to the hypothesis that specific dyslexia, mental deficiency,..and speech defects were different phenotypical manifestations of the same hereditary taint. 1964 *Punch* 28 Oct. 652/2 Earlier [sheep] breeders..chose their stock phenotypically, that is, by the look of them. 1970 *Sci. Amer.* Feb. 62/3 It seems likely that the highlanders have derived their special qualities from acclimatization—in short, that their response to their environment is phenotypic rather than genotypic. 1977 J. L. Harper *Population Biol. of Plants* viii. 239 Parts of a clone or shoots on a tree will differ phenotypically, e.g. in age and size.

phenotyping (fī·nǒtəipiŋ), *vbl. sb.* [f. prec. + -ing¹.] Allocation to a phenotype.
1964 *Jrnl. Dairy Sci.* XLVII. 1262/1 That agreement in phenotyping exists is valuable in..establishing a uniform nomenclature of the genetic variants. 1977 *Lancet* 8 Jan. 82/1 The variants so recognised are labelled alphabetically in the protease-inhibitor (Pi) phenotyping system.

phenoxide (fīnǫ·ksəid). *Chem.* [f. Phen-, pheno- + Oxide *sb.*] A salt of phenol, containing the anion $C_6H_5O^-$; = Phenate, *phenolate.
1888 *Jrnl. Chem. Soc.* LIV. 586 (*heading*) Compounds of phenoxides with cuprous and mercurous chlorides. 1906 J. J. Sudborough *Bernthsen's Org. Chem.* (rev. ed.) xxiv. 405 The phenols possess the character of weak acids, and they form salts with alkalis..known as phenates or phenoxides. 1936 L. J. Desha *Org. Chem.* 185 Phenoxides, sometimes called phenates, are formed from phenols either by the action of the strongly positive free metals such as sodium or by dissolving in aqueous alkalies such as solutions of sodium hydroxide. 1966 Rakoff & Rose *Org. Chem.* xviii. 594 Phenols..are stronger acids than water; they will react with sodium hydroxide to form water and sodium phenoxide. 1974 [see *phenolate].

phenoxy(-) (fīnǫ·ksi), *prefix* and *a.* (*sb.*). *Chem.* and *Pharm.* [f. Phen- + Oxy-.] **A.** (Before a vowel formerly also **phenox-.**) An inseparable formative element in names of compounds which contain the group —$O{\cdot}C_6H_5$, as in **pheno:xyace·tic** (or † *phenoxacetic*) **acid,** a colourless, crystalline solid, $C_6H_5O{\cdot}CH_2COOH$; also, any of the chlorinated derivatives of this, which are widely used as weedkillers; **pheno:xymethylpeni·ci·llin** a white powder, 6-phenoxyacetamidopenicillanic acid, $C_{16}H_{18}N_2O_5S$, which is a semisynthetic penicillin.
1879 *Jrnl. Chem. Soc.* XXXVI. 322 The author [*sc.* P. Fritzsche]..has undertaken the investigation of oxyphenylacetic acid,..which was discovered and described by Heintz.., and named by him phenoxacetic acid. *Ibid.* 642 (*heading*) Phenoxypropionic acid. 1880 *Ibid.* XXXVII. 318 The preparation of phenoxyacetic acid is described at length. 1900 E. F. Smith tr. *V. von Richter's Org. Chem.* (ed. 3) II. 146 Phenoxyacetone,..$C_6H_5O.CH_2.CO.CH_3$, boiling at 230°, is condensed by concentrated sulphuric acid to methyl cumarone. 1926 D. W. MacArdle *Use of Solvents in Synthetic Org. Chem.* iv. 62 Marvel and Tanenbaum found that if in the preparation of phenoxybutyl alcohol, ethyl alcohol dried over lime was used as solvent, yields of not over 45% were obtained. 1946 Long & Brenchley *Suppression of Weeds* (ed. 2) xiii. 62 As a result of the work carried out..at Jealott's Hill Research Station a range of phenoxyacetic acid products were discovered, of which Methoxone (4-chlor-2-methylphenoxyacetic acid) was finally chosen as being a most efficient weed killer with selective properties. 1948 O. K. Behrens et al. in *Jrnl. Biol. Chem.* CLXXV. 798 Phenoxymethylpenicillin—N-(2-Hydroxyethyl)-phenoxyacetamide (150 mg. per liter)..was used as the precursor for this penicillin. 1954 W. J. Hickinbottom in E. H. Rodd *Chem. Carbon Compounds* IIIA. viii. 426 By the action of alcoholic potash, phenoxyacetylene is formed, an unstable oil. 1959 *Jrnl. Exper. Bot.* X. 33 (*heading*) Factors controlling the uptake of phenoxyacetic acids by *Lemna minor*. 1959 *Times* 6 Mar. 13/6 A preparation of penicillin—penicillin V, or phenoxymethylpenicillin—was produced which was not destroyed by the acid contents of the stomach. 1969 *New Scientist* 9 Jan. 61/2 Even the relatively innocuous phenoxyacetic acid compounds..have already caused significant ecological damage by destroying natural mangrove associations. 1970 Hoover & Stedman in A. Burger *Medicinal Chem.* (ed. 3) I. xviii. 382/1 Phenoxymethylpenicillin (penicillin V) is now the only important biosynthetic penicillin in use.
B. In Combs. in which *phenoxy* may be used *attrib.* (without hyphen) or joined by a hyphen to the second element. **1.** Containing or being the group —$O{\cdot}C_6H_5$.
1896 *Jrnl. Chem. Soc.* LXIX. 161 (*heading*) On γ-phenoxy-derivatives of malonic acid and acetic acid. 1923 *Chem. Abstr.* XVII. 3864 (*heading*) Phenoxy derivatives of propane. 1926 *Jrnl. Amer. Chem. Soc.* XLVIII. 2748 The method chosen for the synthesis of these compounds was first to prepare the phenoxy alcohols and then to convert them into the chlorides. 1926 *Biochem. Jrnl.* XX. 1083 Replacement of the two phenoxy-groups by bromine. 1928 *Chem. Abstr.* XXII. 769 (*heading*) Influence of the phenoxy group and its derivatives upon the halochromism of known chromogens. 1964 L. J. Audus *Physiol. & Biochem. Herbicides* v. 195 The effects of the halogenated aliphatic acids on total soil populations are as diverse as those of the phenoxy herbicides. 1964 W. A. Waters *Mechanisms Oxidation Org. Compounds* ix. 139 In aqueous solution the phenoxy radical itself has a mean lifetime of about 10^{-3} second. 1974 *McGraw-Hill Yearbk. Sci. & Technol.* 330/2 The phenoxy herbicides (2,4-D and 2,4-T), aminotriazole.., and pidoram are frequently used in forestry.
2. Designating thermoplastics characterized by a linear molecule containing recurring phenoxy groups, which are usu. made by condensation of epoxides, esp. epichlorhydrin, with certain phenols. Also as *sb.*
1962 *Mod. Plastics* Nov. 169 (*heading*) Phenoxy—a new thermoplastic. *Ibid.*, Phenoxy materials are a new family of thermoplastic resins that can be chemically cross-linked to impart thermosetting properties. 1963 *Aeroplane* 21 Feb. 29/2 Ventilating ducts in the Boeing 727 are fabricated from self-extinguishing phenoxy resin. 1967 *Times Rev. Industry* May 76/2 There has also been progress in the currently lower tonnage plastics such as nylon and polyester in film laminates for packaging processed foods, phenoxy polymer in blown bottles, and polycarbonate for sterilizable containers. 1969 L. S. Mounts in W. R. R. Park *Plastics Film Technol.* v. 141 Phenoxy films are rigid transparent films with high impact strength. 1970 W. G. Potter *Epoxide Resins* ii. 19 The 'phenoxies' are in fact to be regarded as high-M thermoplastic materials and have been used either as surface-coating binder resins.., or as thermoplastics for blow moulding, injection moulding, [etc.].

phenthiazine, obs. var. *phenothiazine.

phentolamine (fentǫ·lǎmīn). *Pharm.* [f. Phen-, pheno- + Tol(yl + Amine.] A white or cream-coloured heterocyclic solid, $C_{17}H_{19}N_3O$, which is used (in the form of its salt) as a vasodilator, esp. in the treatment of hypertension caused by phæochromocytoma.
1953 *Jrnl. Amer. Med. Assoc.* 15 Aug. 1533/2 Phentolamine hydrochloride, a salt of phentolamine base, is suitable for oral administration and acts as a potent adrenergic blocking agent, producing adrenolytic and sympatholytic effects. 1961 L. Martin *Clin. Endocrinol.* (ed. 3) vi. 182 Phentolamine is the most reliable of the adrenolytic substances both for diagnosis of phæochromocytoma and for use during its removal. 1968 J. H. Burn *Lect. Notes Pharmacol.* (ed. 9) 8 Patients with a high blood pressure may suffer from an adrenal medullary tumour which secretes noradrenaline and adrenaline into the blood. This can be diagnosed by injecting phentolamine intravenously. 1972 *Materials & Technol.* V. xxi. 810 Alpha receptor blockers such as tolazoline may be used in the treatment of peripheral blood vessel spasm, while phentolamine is used to treat hypertensive (high blood pressure) crises.

phenyl. Now also with pronunc. (fī·nəil). Add: **2 b. phe:nylarso·nic** *a.* [Arsonic *a.*], in *phenylarsonic acid*, a colourless, toxic, crystalline solid, $C_6H_5AsO(OH)_2$, which is used as a trypanocide; also, any derivative of this; **phenylhy·drazone** [ad. G. *phenylhydrazon* (O. Rudolph 1888, in *Ann. d. Chem. u. Pharm.* CCXLVIII. 99): see *hydrazone], any of a class of compounds formed by condensation of an aldehyde or ketone with phenylhydrazine, which are usu. crystalline solids and are used to characterize the parent aldehyde or ketone; **phenylme·rcury,** used *attrib.* or *absol.* to denote compounds which contain a phenyl group bonded directly to a mercury atom; so **phe:nylmercu·ric** *a.*, as *phenylmercuric acetate*, a white crystalline solid, $CH_3COOHgC_6H_5$, used mainly as a fungicide and herbicide; *phenylmercuric nitrate*, a white crystalline solid, $C_6H_5HgNO_3.C_6H_5HgOH$, used mainly as a fungicide and disinfectant; **phe:nylpyru·vic** *a.*, *phenylpyruvic acid*, a colourless crystalline solid, $C_6H_5CH_2CO{\cdot}COOH$, which in phenylketonuria is produced by the metabolism of phenylalanine and excreted in the urine; hence **phe:nylpyru·vate,** a salt or the anion of this acid; **phe:nylthi:oca·rbamide,** a white, crystalline solid, $NH_2{\cdot}CS{\cdot}NH{\cdot}C_6H_5$, which has a bitter taste to persons possessing a certain dominant gene and is tasteless to those lacking it; **phe:nylthiourea** (-þəi:o,yurī·ǎ) = *phenylthiocarbamide* above.
1905 *Amer. Chem. Jrnl.* XXXIII. 104 The reduction takes place with equal ease in the aromatic series, monophenyl arsine..being obtained from phenyl arsonic acid. 1937 *Thorpe's Dict. Appl. Chem.* (ed. 4) I. 488/1 The introduction of an amino group into the para-position in phenylarsonic acid decreases its toxicity and increases its trypanocidal activity. 1959 *Times* 7 Dec. (Agric. Suppl.) p. vii/1 So far as arsenicals are concerned, two phenylarsonic acids have received attention [as growth stimulants for poultry]. 1889 *Jrnl. Chem. Soc.* LVI. 251 (*heading*) Phenylhydrazones. 1938 [see *hydrazone]. 1966 *McGraw-Hill Encycl. Sci. & Technol.* VII. 62/2 When the starting material is the phenylhydrazone of acetone, the product is 2-methylindole. 1920 *Chem. Abstr.* XIV. 2181 There at once sep. leaflets of phenylmercuric chloride, PhHgCl, m. 250°. 1921 F. C. Whitmore *Org. Compounds Mercury* iii. 65 Phenyl mercuric acetate reacts with ammonium hydroxide giving a substance $(C_6H_5Hg)_2NH_2OAc$. *Ibid.* ix. 177 Phenylmercuric nitrate forms rhombic tablets, insoluble in cold water. 1951 A. Grollman *Pharmacol. & Therapeutics* xxv. 510 The first of these compounds used as an antiseptic was phenylmercuric chloride but..this was supplanted by the more soluble basic phenylmercuric nitrate. 1966 *McGraw-Hill Encycl. Sci. & Technol.* I. 483/1 The most important of the organic mercurials are phenylmercuric nitrate and acetate. 1972 Phenylmercuric [see *phenylmercury* below]. 1931 *Jrnl. Infectious Dis.* XLIX. 440 Phenyl-mercury-nitrate was first prepared by Otto in 1870, but its biologic characteristics have..never been studied. 1955 G. J. Rose *Crop Protection* vi. 103 Bacterial canker of cherries has been controlled by applications of a formulation containing the phenyl mercury salt of naphthyl methane sulphonic acid. 1969 Johnels & Westermark in Miller & Berg *Chem. Fallout* x. 224 There is no indication..that the use of..phenylmercury from 1930 to 1940 has caused an increase in the mercury content of these terrestrial bird species. 1971 *Nature* 20 Aug. 535/1 The death of birds in Sweden was traced to the use of phenylmercury compounds in the pulp and paper industry. 1972 F. Matsumura *Environmental Toxicol. Pesticides* 532 In the U.S. the source of contamination of many rivers has been suspected to be by phenylmercuries, because of the exten-

PHENYLALANINE

sive use of phenylmercuric acetate as a slime treatment agent in paper mills. **1932** *Jrnl. Biol. Chem.* XCVI. 628 In one experiment, after 2·5 gm. of sodium phenylpyruvate had been fed for 2 successive days, it was possible to isolate from the urine 0·4 gm. of phenaceturic acid. **1970** R. W. McGilvery *Biochem.* xvii. 387 People with phenylketonuria also excrete some other aromatic compounds.., representing aberrations of the normal process of metabolism due to the high concentration of phenylpyruvate. **1887** *Jrnl. Chem. Soc.* LII. 142 The author..advances the view that the compound is probably phenylpyruvic acid. **1935** [see *phenylketonuria]. **1968** Passmore & Robson *Compan. Med. Stud.* I. xi. 24/2 Presence of phenylpyruvic acid in the blood leads to mental retardation. Arrangements to test the urine of every child soon after birth for phenylpyruvic acid have been made in many places. **1879** *Jrnl. Chem. Soc.* XXXVI. 804 Mono- and di-phenylthiocarbamide..are soluble in caustic soda or potash. **1932** [see PTC s.v. *P II]. **1976** *Nature* 20 May 223/2 This report concerns two dermatoglyphic traits, fingerprint pattern index and total ridge-count, both strongly heritable. In addition, phenylthiocarbamide (PTC) taster ability..and skin colour were studied. **1896** *Jrnl. Chem. Soc.* LXIX. 857 If dilute hydrochloric acid be now added to the clear solution, a white solid at once separates, which..forms white prisms..consisting of phenylthiourea. **1959** *Listener* 3 Dec. 968/2 People who can taste phenylthiourea seem to be slightly more liable to get one form of thyroid disease and slightly less liable to get another. **1971** J. Z. Young *Introd. Study Man* xxxviii. 553 A minimal yet striking example [of genetic variation of the nervous system] is the inheritance of the capacity to taste the substance phenylthiourea.

phenylalanine (fe·nil-, fīnəilæ·lănīn). *Biochem.* [ad. G. *phenylalanin* (Erlenmeyer & Lipp 1883, in *Ann. d. Chem. u. Pharm.* CCXIX. 186), f. *phenyl* Phenyl + *alanin* Alanine.] A colourless, crystalline amino-acid, $C_6H_5CH_2 \cdot CH(NH_2)COOH$, which, in its lævorotatory form, is widely distributed in plant proteins and is an essential constituent of the human diet.

1883 *Jrnl. Chem. Soc.* XLIV. 992 Phenyl-α-amidoproprionic acid (phenylalanine),..obtained as a hydrochloride by pouring the crude product of the action of ammonia on phenylethylidine cyanhydrin into hydrochloric acid, and boiling the mixture. **1934** *Times Lit. Suppl.* 1 Nov. 758/3 Whether this represents a real ability to synthesize the amino-acid for all purposes, or simply a power to use dietary phenylalanine, remains an open question. **1955** *Sci. News Let.* 22 Jan. 52/1 Since phenylalanine poisoning caused the mental deficiency, Dr. Woolf decided to try devising a diet that would not contain this amino acid. **1956** [see *dopa]. **1970** *Observer* 12 Apr. 25/4 Because he lacks a single pair of genes out of all those thousands, his liver cannot convert a chemical in his diet, called phenylalinine, into another chemical, tyrosine, and soon after birth he will start sliding down into severe mental retardation unless his diet is strictly controlled. **1970** [see *phenylketonuria].

phenylbutazone (fenil-, fīnəilbiū·tăzōᵘn). *Pharm.* [f. Phenyl + But(yl + Az(o- + -one.] A white or cream-coloured crystalline solid which is used as an analgesic, esp. for the relief of rheumatic pain, and as an antipyretic; 4-butyl-3,5-dioxo-1,2-diphenylpyrazolidine, $C_{19}H_{20}N_2O_2$.

1952 *Jrnl. Amer. Med. Assoc.* 21 June 729/1 Phenylbutazone..is one of a group of pyrazole derivatives developed by the chemists of the J. R. Geigy Co., of Basel, Switzerland. **1963** *Lancet* 5 Jan. 21/2 She was given phenylbutazone ('Butazolidin') for five days, with immediate relief of her joints. **1974** *Daily Colonist* (Victoria, B.C.) 1 June 1/3 The U.S. tests on the hard, round, black pills have apparently shown the presence of aminopyrine and phenylbutazone, two drugs legally available in Canada only on prescription. **1974** M. C. Gerald *Pharmacol.* xiv. 273 Phenylbutazone is a potent analgesic, antipyretic, and anti-inflammatory agent.

phenylene (s.v. Phenyl). Add: **phenylene blue,** a blue dye (see *indamine); **phenylene brown** = **Bismarck brown* (s.v. *Bismarck 1), Vesuvin; **phe:nylenedi·amine,** any of three isomeric, toxic, crystalline solids, $C_6H_4(NH_2)_2$, or their alkylated derivatives, which are widely used in the dye industry, as photographic developers, and (in the case of the *para* isomer) as an additive in rubber to prevent oxidation.

1889 G. M'Gowan tr. *Bernthsen's Text-bk. Org. Chem.* 356 The simplest member of this class is the indamine 'Phenylene Blue', $C_{12}H_{11}N_3$,..which results from the oxidation of a mixture of aniline and *p*-phenylene-diamine. **1903** [see *indamine]. **1952** K. Venkataraman *Chem. Synthetic Dyes* II. xxv. 762 The indamines (e.g., Phenylene Blue) are obtained by oxidizing a neutral solution of..a *p*-diamine.., and..a monamine (e.g. aniline) having a free *p*-position. **1867** *Chem. News* 12 July 24/2 The phenylene-brown consists chiefly of a new base of the composition $C_{12}H_{13}N_4$. **1885** [see *Bismarck 1]. **1922** A. Clarke *Coal-Tar Colours in Decorative Industries* iii. 43 The colours most suitable for this method are:– On White Earth—..Phenylene Brown, Crystal Violet and Methyl Violet. **1862** A. W. Hofmann in *Proc. R. Soc.* XI. 519 The idea very naturally suggested itself, to look to dinitrobenzol as the source from which phenylene-diamine might reasonably be expected to arise. **1918** *Jrnl. Chem. Soc.* CXIV. ii. 69 The authors have investigated the..binary systems formed between phenol ..and the three phenylenediamines. **1972** *Materials & Technol.* V. xiv. 522 Only the..*p*-phenylene diamines and dihydroquinolines give good ozone protection [to rubber]. **1973** D. A. Spencer *Focal Dict. Photogr. Technol.* 440 *Phenylenediamines*, generic name for compounds derived from ortho-phenylenediamine and para-phenylenediamine which act as photographic developing agents.

phenylephrine (fenil-, fīnəile·frin, -īn). *Pharm.* [f. Phenyl + *epin)ephrine.] The lævorotatory form of 1-(*m*-hydroxyphenyl)-2-methylaminoethanol, $HO \cdot C_6H_4 \cdot CH_2OH \cdot NH \cdot CH_3$, which is used (usu. as the hydrochloride) as an anti-hypotensive agent and nasal decongestant. Cf. *Neo-Synephrine.

1947 *New & Non-Official Remedies* 225 Phenylephrine hydrochloride is a vasoconstrictor and is active as a vasopressor when administered orally. **1950** *Jrnl. Amer. Pharmaceut. Assoc.* XXXIX. 50/1 Phenylephrine N.N.R. (Neo-Synephrine) is a sympathomimetic amine in general use. **1963** *Lancet* 19 Jan. 138/2 Our experience and published reports suggest that, of the vasoconstrictors, phenylephrine and noradrenaline are probably the most satisfactory. **1974** [see *Neo-Synephrine]. **1975** *Sci. Amer.* Nov. 117/1 On one of her visits she had her pupils artificially dilated with the drug phenylephrine hydrochloride. **1976** [see *phenylpropanolamine].

phenylketonuria (fe:nil-, fī:nəilkītoniuᵊ·riă). *Path.* [f. Phenyl + *ketonuria.] An inherited inability to metabolize phenylalanine normally, which if untreated in children leads to mental deficiency.

1935 *Lancet* 27 July 192/2 Phenylpyruvic amentia, more conveniently termed phenylketonuria. *Ibid.* 193/1 Dr. J. H. Quastel..suggested the name phenylketonuria. **1958** *Observer* 23 Mar. 9/8 An estimated 30 babies are born every year with an unsuspected 'metabolic' disorder called phenylketonuria. **1958** *New Biol.* XXV. 22 Phenylketonuria is a chemically definable character inherited as a Mendelian recessive, and accounting for perhaps 1 per cent of certifiable mental defect. **1965** *New Statesman* 10 Dec. 922/1 The classic examples here are phenylketonuria and galactosaemia, in which the lack of a single gene means that the body cannot use or break down normal substances in the diet..so that toxic products accumulate and poison the brain. **1970** Passmore & Robson *Compan. Med. Stud.* II. xxv. 40 In phenylketonuria the hereditary deficiency of the enzyme phenylalanine hydroxylase in the liver cells retards or deletes the normal oxidation of phenylalanine to tyrosine. **1977** *Jrnl. R. Soc. Arts* CXXV. 186/2 The 'inborn error of metabolism', phenylketonuria..is a simply inherited defect which causes mental retardation, due to accumulation of toxic products from birth onwards.

Hence **phe:nylketonu·ric** *sb.*, an individual with phenylketonuria; *a.*, affected with or pertaining to this disorder.

1937 *Biochem. Jrnl.* XXXI. 270 Experiments were carried out..to determine the effects of feeding phenylalanine to a phenylketonuric on a low protein diet. **1961** *Lancet* 26 Aug. 465/2 This suggests that there is no biochemical reason why any phenylketonuric woman should not have normal children. *Ibid.*, Exposure to a phenylketonuric environment while in utero. **1975** *Nature* 4 Dec. 462/3 Even if an observed difference is genetic it does not mean that a trait cannot be altered substantially by an appropriate environmental change. An example is the treatment of phenylketonurics.

phenylpropanolamine (fī:nəil-, fe:nil-prōᵘpănǫ·lāmīn). *Pharm.* [f. Phenyl + *propanol + Amine.] = *norephedrine* s.v. *nor- 1a.

1947 Lorhan & Mosser in *Ann. Surg.* CXXV. 171 (*heading*) Phenylpropanolamine hydrochloride: a vasopressor drug, for maintaining blood pressure during spinal anesthesia. **1968** W. C. Bowman et al. *Textbk. Pharmacol.* xxix. 750 Phenylpropanolamine, or norephedrine, is known commercially as Propadrine. **1969** [see *cold *sb.* 5c]. **1976** *Which?* Mar. 53/1 Some pills and capsules are intended to help your nose feel less blocked up (they contain phenylephrine or phenylpropanolamine for this).

phenytoin (fe·nitǫ,in). *Pharm.* [f. Pheny(l + Hydan)toīn.] 5,5-Diphenylhydantoin, $C_{15}H_{12}N_2O_2$, an anticonvulsant widely used in the treatment of epilepsy (usu. in the form of its sodium salt, a white powder).

1942 *Jrnl. Amer. Med. Assoc.* 4 Apr. 1209/2 White phenytoin sodium (sodium diphenyl hydrantoinate, or dilantin sodium) was effective in protecting animals from electrically induced convulsions it produced little sedative effect. **1952** A. H. Douthwaite *Hale-White's Materia Medica* (ed. 29) 193 When grand mal and petit mal occur in the same patient phenobarbitone and troxidone should be used together: phenytoin must not be employed. **1961** *Lancet* 23 Sept. 683/2 Since then, she has had post-encephalic epilepsy which is controlled by phenobarbitone and phenytoin sodium. **1969** *New Scientist* 10 July 57/3 Patients who had received phenobarbitone or phenytoin showed a dramatic reduction in DDE levels. **1973** [see *Mysoline].

pheophorbide, -phytin, varr. *phæophorbide, -phytin.

‖ **pheran** (fę̄·răn, pę̄·răn). [Kashmiri, prob. ad. Pers. *pairāhan* a shirt.] (See quots.)

1882 *Encycl. Brit.* XIV. 12/1 The Kashmíris, both men and women, wear commonly a kind of loose gown with sleeves, called *phéran* (Pers., *pairāhan*, 'a robe'). **1895** W. R. Lawrence *Valley of Kashmir* xix. 464/2 Tunic worn by all Kashmíris..*Pheran*. **1922** *Chambers's Jrnl.* Nov. 760/2 He is robed in the national Kashmiri dress—the pheran—a loose, flowing garment of white lawn or silk. **1953** R. Godden *Kingfishers catch Fire* ii. 18 He wore a grey *pheran*—the loose robe that most Kashmiri men and women wore. **1976** R. Kalapesi *Catal. Pageant of Indian Costumes* 29 In Kashmir, the long pant is called the 'salwar' and is worn with a loose shirt, called the 'pheran', with its typical embroidered Sassanian neckline and square armholes.

pheromone (fe·romōᵘn). *Biol.* [f. Gr. φέρ-ειν to convey + -o + ὁρμῶν, pres. pple. of ὁρμᾶν to set in motion, urge on (after *hormone*).] Any substance that is secreted and released by an animal (usu. in minute amounts) and causes a specific response when detected by another animal of the same (or a closely related) species.

1959 Karlson & Lüscher in *Nature* 3 Jan. 55/2 We propose..the designation 'pheromone' for this group of active substances... Pheromones are defined as substances which are secreted to the outside by an individual and received by a second individual of the same species, in which they release a specific reaction, for example, a definite behaviour or a developmental process... Having stated that pheromones act on individuals of the same species, they must be differentiated from other stimulating substances, like..scents of flowers and insect repellents... Strict species-specificity is not required; certain overlaps between closely related species may occur. **1962** *New Scientist* 19 Apr. 86/2 The queen substance of honey-bees, which inhibits ovary development and profoundly affects behaviour in workers, is a pheromone; and pheromones of various kinds are supposed to affect caste differentiation and colony structure in termites. **1965** *Listener* 14 Jan. 57/2 Our observation of ant colonies..has led us to believe that as few as ten pheromones, transmitted singly or in combination, might suffice for the total organization of ant society. **1971** *Nature* 17 Dec. 415/2 We now describe a case in which male attraction to females of the same species is inhibited by a pheromone emitted from females of a closely related species. Our tests were conducted with the Indian meal moth *Plodia interpunctella* (Hübner) and the almond moth *Cadra cautella* (Walker), both phycitid species. **1973** H. O. Box *Organisation in Animal Communities* viii. 200 Examples..include the blocking of pregnancy in the females of some strains of mice by a pheromone in the urine of a strange male mouse. **1975** D. S. Hill *Agric. Insect Pests of Tropics* iii. 23 Behavioural control with several types of pheromone may be possible, but attention has been focused on pheromones causing aggregation for mating, feeding or oviposition. **1977** *Sci. Amer.* May 104/3 It now appears that certain insects take chemicals from the plants they eat, store the chemicals and later emit them for defense, as 'aggregation pheromones' or perhaps as sex attractants.

Hence **pheromo·nal** *a.*, of, pertaining to, or being a pheromone or pheromones.

1959 *Ann. Rev. Entomol.* IV. 39 Pheromonal secretion need not occur in definite glands (though that is the case normally). **1971** *New Scientist* 25 Feb. 413/1 Much of this would be speculation alone if research had not produced an intriguing instance of human communication which might just possibly prove to be pheromonal. Recently, Dr Martha McClintock of Harvard..has investigated the menstrual cycles of women students in a hall of residence... Her investigation threw up the bizarre observation that, precisely as in mice, the cycles of close friends..fall into synchrony. **1971** *Nature* 16 Apr. 432/1 Odour fingerprinting techniques and gas chromatography now make the detection and preparation of human pheromonal agents feasible if they exist. **1975** *Sci. Amer.* May 59/1 The physiological causes, possibly pheromonal ones, for the synchrony of births remain to be determined, and it will not be easy to determine them with lions in the wild.

phi (fəi). [The name of ϕ, Φ, the 21st letter of the Greek alphabet (in Gr. called φεῖ).] **1.** *Petrol.* The negative of the logarithm to base 2 of the diameter in millimetres of a particle. Freq. *attrib.*, as *phi scale*, and written as ϕ.

1934 W. C. Krumbein in *Jrnl. Sedimentary Petrol.* IV. 76, ξ is the numerical value of the diameter... The substitution $\phi = -\log_2 \xi$, or $\xi = (\frac{1}{2})^{\phi}$, is made in part for typographical convenience, but also to introduce a convenient symbol that may be used for expressing grain diameters. **1936** —— in *Ibid.* VI. 38/1 One immediately apparent advantage accruing from the ϕ-scale is the elimination of unwieldy fractions or decimals, such as 1/1024 mm. (0·00098 mm.). *Ibid.* 45/1 The use of ϕ as the new independent variable..makes available a series of statistical measures based on the moments of the distributions. **1949** *Ibid.* XIX. 76/1 The sieve interval of one phi unit is too large to make the true and nominal mean diameters reasonably equal for all size fractions. **1961** *Ibid.* XXXI. 205/2 Comparison of the mean diameters of each pair of beach-dune samples..shows that there is little difference between the two. In these comparisons, 48 are within 0·2 phi of each other. **1971** E. F. McBride in R. E. Carver *Procedures Sedimentary Petrol.* vi. 110 The Phi scale is now used almost exclusively for computation and is gradually replacing the millimeter scale. **1978** *Nature* 14 Sept. 100/2 The sediments..have mean sizes of 1·6–0·125 mm (—0·7 to 3·0 phi: very coarse to fine sand)... These sands exhibit good sorting, with standard deviations that are generally less than 0·8 phi units.

2. *phi coefficient* (Statistics): A synonym for the product-moment coefficient of correlation (see *product *sb.*[1]) used when each of the observed variables has only two possible values.

[**1904** K. Pearson in *Drapers' Co. Res. Mem.: Biometric Ser.* I. 6, I shall call $\phi^2 = \chi/N$ the mean square contingency. **1913** *Biometrika* IX. 214 We do not agree with him, but it is singular that if he thinks so, he should not have rejected the use of ϕ, the 'theoretical value of the correlation'.] **1950** S. A. Stouffer *Measurement & Prediction* x. 410 [This] is the Pearson *r* or the so-called phi-coefficient of the latent fourfold table, which appears here as a mere algebraic byproduct. **1968** Blumler & McQuail *Television in Politics* iv. 75 All the phi coefficients of ·30 or higher (a level which is most unlikely to be reached by chance) were picked out. **1975** Morehouse & Stull *Statistical Princ. Physical Educ.* xi. 235 The phi coefficient is also related to chi-square... $\phi = \sqrt{(\chi^2/N)}$, where N = number of observations. *Ibid.* 237 The phi coefficient is often recommended for use in the analysis of test items in which the items are scored as correct or incorrect, pass or fail.

3. *Nuclear Physics.* In full *phi meson.* A neutral meson that has the same quantum numbers as the omega meson (*omega 2* b), is observed as a resonance, has a mass of 1019 MeV (1995 times that of the electron), and on decaying usu. produces two kaons or three pions. Freq. written ϕ.

1962 J. J. Sakurai in *Physical Rev. Lett.* IX. 472/1 In a recent issue of this journal a Brookhaven–Syracuse group reports the possible presence of a narrow resonance ($\Gamma \approx 20$ MeV) in the $K\bar{K}$ system with mass ≈ 1020 MeV in the reactions $K^- + p \to K^+ + K^- + \Lambda$, $K^0 + \bar{K}^0 + \Lambda$. In discussing the quantum numbers of the resonating $K\bar{K}$ pair, which we tentatively refer to as the ϕ meson, [etc.]. *Ibid.* 474/1 We would like to suggest that this discrepancy is due to mixing between ω and ϕ arising from the fact that the ω and ϕ have the same quantum numbers as far as spin-parity, isospin, and *G*-parity are concerned. **1968** J. Bernstein *Elem. Particles* vi. 86 Since these objects, the ρ, ω^0, ϕ^0, are very short-lived ($\tau \sim 10^{-22}$ sec), it is not really correct, in a rigorous sense, to treat them as if they were stable particles. **1971** *Sci. Amer.* July 100/2 Three vector mesons with zero strangeness are currently known: the rho, the omega and the phi. **1974** *Nature* 6 Dec. 438/2 Another well established meson, called the phi, also has electromagnetic decays to e^+e^-. **1975** *Sci. Amer.* Oct. 48/2 Zweig's rule was formulated to explain the decay of the phi meson, which is made up of a strange quark and a strange antiquark and has a mass of about 1 GeV. The two particles are closely analogous, but the decay of the *J* is appreciably slower than that of the phi.

Phi Beta Kappa (fəi bī·tă (or bē̆ī·tă) kæ·pă). *U.S.* [f. the initials of Gr. φιλοσοφία βίου κυβερνήτης philosophy (the) guide of life.] An honorary society to which distinguished undergraduate, and occas. graduate, scholars may be elected; a member of this society. Also *attrib.*

[**1776** in *William & Mary College Q.* (1896) Apr. 214 A list of Members who have been Initiated into the S.P. *alias ΦBK* Society.] **1831** *N.Y. Mirror* 3 Sept. 71/2 Chancellor Kent will deliver an oration before the Phi Beta Kappa Society, at the annual commencement of Yale College. **1894** *Harper's Mag.* June 34/1 I've got to read the Phi Beta Kappa poem at Harvard next week. **1912** M. Nicholson *Hoosier Chron.* 278 Sylvia..just walked through everything and would be chosen for the Phi Beta Kappa. **1929** *Chicagoan* 17 Aug. 21/3 Harvard University Phi Beta Kappa conventions. **1932** W. Faulkner *Light in August* xix. 419 A Harvard graduate, a Phi Beta Kappa. **1949** *Newsweek* 5 Dec. 54/2 Selby won a Phi Beta Kappa key at Northwestern. **1954** W. Stevens *Let.* 29 Jan. (1967) 816, I am going to read a Phi Beta Kappa poem..at Columbia. **1968** *Wall St. Jrnl.* 19 Feb. 1/1 Mr. Kahn, a Phi Beta Kappa member in his junior year. **1974** 'I. Drummond' *Power of Bug* x. 144 The other was collegiate too, but not an athlete, a Phi Beta Kappa. **1976** *National Observer* (U.S.) 10 July 11/1 Phi Beta Kappa was founded as a secret society at the College of William and Mary on Dec. 5, 1776, and it featured, along with an oath of secrecy, mottoes in Latin and Greek, a code of laws, elaborate initiation rites, a badge, a seal, and a special handclasp. **1978** *Detroit Free Press* 16 Apr. 23A/1, I notice you wearing your Phi Beta Kappa Key from Barnard.

Phil (fil), colloq. abbrev. Philharmonic *sb.* b (in Dict. and Suppl.). Also *attrib.*

1949 E. Coxhead *Wind in West* iii. 84 Alan was a violinist. He played in the Northern Phil. **1963** *Times* 25 Jan. 14/7 The 'Phil' audience, as regular and loyal as any in London. **1977** *Listener* 28 Apr. 564/3 A 'great' orchestra such as the New York Phil or the Concertgebouw. **1978** J. Gallway *Autobiogr.* xiv. 162, I thought the fellows in the Berlin Phil a marvellous bunch.

-phil, -phile. Add: *spec.* in *Biol.* and *Med.* in the sense 'having an affinity for (a certain substance or class of substances)', as in *eosinophil *a.* and *sb.*, *neutrophil(e* adj. and sb. (s.v. Neutro- in Dict. and Suppl.).

phil (fil), *a. rare.* [The suffix -phil used as a separate word.] Having a love or leaning (towards something).

1915 [see *phobe *a.*].

Philadelphia (filăde·lfiă). The name of the city in Pennsylvania, U.S.A., used *attrib.* in **Philadelphia chromosome,** an abnormal small chromosome sometimes found in the leukocytes of patients suffering from leukæmia, esp. chronic granulocytic leukæmia; **Philadelphia lawyer,** a lawyer of great ability, esp. one expert in the exploitation of legal technicalities; a shrewd or unscrupulous lawyer.

[**1961** I. S. Tough et al. in *Lancet* 25 Feb. 411/2 In this paper the abnormal chromosome, for the sake of brevity, is referred to by the symbol Ph[1]. This is chosen as it indicates the geographical location, Philadelphia in the United States, of the laboratories in which the chromosome was first noted, and thus it accords with the recommendations on the nomenclature of abnormal chromosomes of the Denver conference.] **1963** Walter & Israel *Gen. Path.* iii. 45 Some of the leucocytes show replacement of one of their small autosomes by a minute chromosome (the 'Philadelphia chromosome'). This is generally assumed to be 21. **1977** R. B. Thompson *Disorders of Blood* xxxv. 566/1 The Philadelphia chromosome usually arises by a translocation from G22 to C9. **1788** *Columbian Mag.* Apr. 182 They have a proverb here [in London], which I do not know how to account for;—in speaking of a difficult point, they say, *it would puzzle a Philadelphia lawyer.* **1803** *Balance* (Hudson, N.Y.) 15 Nov. 363/1 It would (to use a Yankee phrase) *puzzle a dozen Philadelphia lawyers* to unriddle the conduct of the democrats. **1834** W. G. Simms *Guy Rivers* II. i. 23 You would have made a prime counsellor..worthy of the Philadelphia lawyers. **1896** *N.Y. Weekly Witness* 30 Dec. 13/1 Not even a Philadelphia lawyer would have been able to pick the winners [in an election contest]. **1901** *Daily Tel.* 6 Nov. 5/4 What entertainment is afforded by a horse-race run 'in camera', only a Philadelphia lawyer would be able to explain. **1909** L. M. Montgomery *Anne of Avonlea* xxvi. 306, I won't undertake to answer Davy's questions... I'm not an encyclopedia, neither am I a Philadelphia lawyer. **1947** *Daily Times* (Chicago) 28 Nov. 14/3 The new violation ticket will be in quadruplicate, and traffic officials say it takes a 'Philadelphia lawyer' to fix it. **1977** *Washington Post* 25 Mar. D18/2 President Carter has made it clear that he understands how complex the income tax laws are. To head IRS, he has picked a Philadelphia lawyer.

Philadelphian, *a.* and *sb.* Add: **A.** *adj.* **3.** (Examples referring to the city in Pennsylvania.)

1775 *Short View of Lord High Admiral's Jurisdiction* 35 A Philadelphian ship might be tried with a fairer chance of condemnation at Halifax than at Philadelphia. **1855** H. A. Murray *Lands of Slave & Free* I. xiv. 360 The only peculiarity in the Philadelphian mint is a frame-work for counting the number of pieces coined. **1930** R. Macaulay *Staying with Relations* i. 14 She loved her little Philadelphian aunt. **1975** *Country Life* 2 Jan. 21/3 Rococco mirrors and girandoles attributed to the Philadelphia craftsman James Reynolds... The majority of the other pieces in the house are also Philadelphian.

B. *sb.* **3. a.** A native or inhabitant of the ancient city of Philadelphia in Asia Minor.

1725 E. Comte tr. *Huet's Weakness of Human Understanding* 117 Herodotus, the Tutor of Sextus Chaeronensis, was a Philadelphian. **1904** W. M. Ramsay *Lett. Seven Churches* xviii. 249 The writer seems not to have loved the Ephesians as he did the Smyrnaeans and Philadelphians.

b. A native or inhabitant of the city of Philadelphia in Pennsylvania.

1744 A. Hamilton *Itinerarium* (1907) 164, I dined with Mr. Fletcher in the company of two Philadelphians. **1789** J. Morse *Amer. Geogr.* 332 The Philadelphians have exerted their endeavours..to prevent the intemperate use of spirituous liquors. **1803** *Lit. Mag.* (Philadelphia) Dec. 170 A Philadelphian..told us, he could not, after repeated trials, find a chaise. **1841** J. S. Buckingham *America* II. 84 The Philadelphians have the reputation of being cold, formal, and difficult of approach. **1891** L. J. Jennings *Philadelphian* I. i. 11 The reputation of Mr. Snapper for integrity stood deservedly high, and in entrusting his capital to the hands of the Philadelphian, Pendleton incurred no more than the ordinary and legitimate risks of commercial affairs. **1901** *Chambers's Jrnl.* Jan. 31/2 One enterprising Philadelphian has been trying to form a Snake Trust. **1947** *Harper's Mag.* Sept. 200/2 The Proper Bostonian is a very well-defined type—more so..than the Proper Baltimorean, the Proper Philadelphian, or the Proper person of any other city. **1976** *Amer. Speech 1974* XLIX. 213 The Philadelphian is also judged the richest and least friendly of the group and seventh in politeness. **1977** *Sci. Amer.* Nov. 30/2 Redheffer took his device to New York (but only after the ingenious Philadelphians had made up and showed him their own perpetual-motion machine).

philadelphus (filăde·lfǿs). [mod.L. (C. Bauhin *Πιναξ Theatri Botanici* (1623) 398, adopted by Linnæus in *Systema Naturæ* (1735) as a generic name), f. Gr. φιλάδελφος loving one's brother, used as a plant name.] A shrub of the genus so called, belonging to the family Saxifragaceæ, native to southern Europe, North America, or Asia, and generally bearing white or cream flowers, often fragrant; also called mock orange or syringa.

1783 [see Syringa]. **1797** *Bot. Mag.* XI. 391 (*heading*) Philadelphus Coronarius. Common Philadelphus, or Mock-Orange. **1899** G. Jekyll *Wood & Garden* iii. 22 How extremely dense and hard is the wood of Philadelphus! as close-grained as Box. **1938** M. Haworth-Booth *Flowering Shrub Garden* xvi. 158 When a Philadelphus..is grown in the open, say as a specimen set in the turf, such [drastic] pruning is unnecessary. **1970** *Times* 14 July 10/7 Notcutts nurseries show large flowered philadelphus. **1975** C. Nesbitt *Little Love & Good Company* xvii. 220, I used to think the lilac tree had the sweetest scent, till the philadelphus came with summer and was still sweeter.

philander, *sb.* Add: **1. b.** A love-making or philandering.

1898 G. B. Shaw *Philanderer* i. 78 It was nothing but a philander with Julia—nothing else in the world, I assure you.

philanthropically, *adv.* Add: Also *Comb.*

1976 'D. Fletcher' *Don't whistle 'Macbeth'* 20 Various philanthropically-minded and prestige-seeking business and industrial concerns. **1980** J. Rose *Elizabeth Fry* vi. 105 A number of philanthropically-minded ladies.

philanthropoid (filæ·nþrǒpoid). orig. *U.S.* [f. Philanthrop(ist + -oid, joc. after Anthropoid *a.* and *sb.*] A professional philanthropist, a worker for a charitable or grant-awarding institution. Also *attrib.* or as *adj.* Hence **philan:thropoi·dal** *a.*; **philan:thropoi·dally** *adv.*

1949 *Harper's Mag.* Mar. 9/1 Edwin R. Embree..calls himself a 'philanthropoid', a term which he and Frederick Keppel of the Carnegie Corporation coined some years ago to describe a person who gives away other people's money. **1957** *Britannica Bk. of Year* 512/1 Coinages with a more colloquial or individualistic ring were..*philanthropoid*, one who disburses the money of a philanthropic institution or foundation. The adjectival and adverbial forms *philanthropoidal* and *philanthropoidally* also occurred. **1959** *Fortune* May 110/2 Philanthropoid is what Jonathan King cheerfully calls himself. King has spent most of his adult life giving away money. **1966** *Economist* 31 Dec. 1391/3 All of this brings a danger that there will be in-breeding of ideas, grants going round in circles... Representatives of the best foundations try to avoid this... Indeed, most of these complaints about 'philanthropoid man' are well on the way to being out of date. **1977** *Daily Tel.* 13 Jan. 17/3 I'm not playing the role of the hard-headed tycoon who thinks all philanthropoids are Socialists and all university professors are Communists.

philately. Add: **b.** Stamps collectively.

1930 *19th Cent.* Dec. 785 The small-bourgeois quality of English philately remained untarnished with sham elegance.

philautian (filǭ·tiăn), *a. rare*—[1]. [f. as Philauty + -an.] Selfish.

1811 Shelley *Let.* 25 July (1964) I. 98 This is entirely a *philautian* argument.

philetism: see *phyletism.

Philharmonic, *a.* and *sb.* Add: **A.** *adj.* Also used characteristically in the names of symphony orchestras.

[**1895** G. B. Shaw in *Daily Chron.* 6 Nov. 6/5 Every violoncellist in the Philharmonic orchestra [*sc.* the Philharmonic Society orchestra] used to dread the trio of the third movement.] **1898** Stainer & Barrett *Dict. Mus. Terms* 362/1 The pitch of Costa's Philharmonic Orchestra..was A. 452·5. **1923** *Gramophone* Apr. 14/1, I prefer the old rendering [of Beethoven's C minor Symphony] conducted by Nikisch with the Berlin Philharmonic Orchestra. **1937** N. Slonimsky *Music since 1900* I. 371, 1934... German State Philharmonic Orchestra of the German Soviet Republic of the Volga basin is established. .. *Fifth Symphony* of Arnold Bax is performed..by the London Philharmonic Orchestra. **1940** Grant & Hettinger *America's Symphony Orchestras* iii. 70 The New York Philharmonic, Philadelphia, and Boston symphony orchestras..have been grouped together as Group I. **1944** B. Geissmar *Baton & Jackboot* 25 The Berlin Philharmonic Orchestra unanimously voted for Furtwängler. **1961** *N.Y. Times* 17 Nov. 38 In recent years Anna Xydis has played with the New York Philharmonic. **1974** *Encycl. Brit. Micropædia* I. 927/1 Beecham..founded the Royal Philharmonic Orchestra in London (1947).

B. *sb.* (Earlier example of sense *Philharmonic Society.*) Also, short for *Philharmonic Orchestra.*

1855 Geo. Eliot in *Fraser's Mag.* July 51/1 When a symphony of Beethoven's was first played at the Philharmonic, there was a general titter among the musicians in the orchestra. **1897** [see Performing *vbl. sb.* 4 b]. **1931** *Gramophone* Apr. 526/2 Even the famous New York Philharmonic..must, I think, take second place. **1933** M. Bauer *20th Cent. Music* iv. 40 Spohr..was the first to conduct an orchestra, the London Philharmonic (1820) with a baton and without the aid of a piano. **1944** B. Geissmar *Baton & Jackboot* 24 Arthur Nikisch..had been permanent conductor..of the Hamburg Concerts with the Berlin Philharmonic since 1897. **1951** J. H. Mueller *Amer. Symphony Orchestra* iii. 63 A century of uninterrupted orchestral programs of the New York Philharmonic affords an extraordinary view of changes in musical taste. **1961** J. Williams *Forger* i. 12 The artist is like a fragment of a mosaic—no, he is more than that, a virtuoso performer in some vast philharmonic. **1977** *Time* 10 Oct. 17/3 Her orchestral works have been performed by the Philadelphia Orchestra and the Los Angeles and New York Philharmonics.

‖ **philia** (fi·liă). [ad. Gr. φιλία friendship.] Amity, friendship, liking.

1953 J. A. K. Thomson *Ethics of Aristotle* viii. 202 The subject of the eighth and ninth books is *Philia*, the feeling which friends have for one another. *Ibid.*, Without this discussion of *Philia* the *Ethics* would have nothing to say on the subject of love. **1960** C. S. Lewis *Four Loves* iv. 69 The very tone of the admission, and the sort of acquaintanceships which those who make it would describe as 'friendships', show clearly that what they are talking about has very little to do with that *Philia* which Aristotle classified among the virtues. **1963** Auden *Dyer's Hand* 143 From that moment on his relationship with Passepartout ceases to be impersonal; *philia* is felt by both. **1977** *Christian* IV. 137 For everybody, whatever his or her sexual orientation.., life can include all the elements of love: eros, philia, and agape. *Ibid.*, Philia, affection, as we have seen, is not easily expressed in public in our society, and it is most difficult for gay men.

-philia (fi·liă), ad. Gr. φιλία friendship, fondness, forming abstract sbs. (usu. corresp. to an adj. in -phil, -phile, *-philic, or -philous), with the senses 'affinity for' (as in *eosinophilia), 'undue inclination towards' (as in Hæmophilia, *spasmophilia), 'love of or liking for' (as in *Anglophilia, *necrophilia* s.v. *necro-).

-philic (fi·lik), f. -phil, -phile + -ic (cf. Gr. φιλικός), used to form adjs. with the sense 'having an affinity for, attracted by, liking', as in *lyophilic, *mesophilic* s.v. *meso-, *neutrophilic* s.v. *neutro-. Cf. -philous.

philippina, philopœna. (Earlier and later examples.)

1839 C. F. Briggs *Adventures H. Franco* II. xiv. 143 There would be..scandal by the wholesale, besides sugar kisses, and philippinas [at the party]. **1917** H. H. Richardson *Fortunes R. Mahony* iii. v. 213 She had won a pair of gloves in a philippine with Mr. Urquhart.

Philippine (fi·lipīn), *a.*[1] and *sb.* Also **Filippine,** and *sb. pl.* in It. form **Philippini.** [f. the name of St. *Philip* Neri (1515–95), founder of the Congregation of the Fathers of the Oratory at Rome + -ine[1].] Of, pertaining to, or characteristic of the Oratorians or the Oratory of St. Philip Neri (Oratory *sb.*[1] 5). **B.** *sb.* A father or priest of this Oratory; also, a female novice (see quot. 1773).

1773 *Encycl. Brit.* III. 476/2 *Philippines*, a religious society of young women, at Rome; so called from their taking St. Philip de Neri for their protector: they consist of an hundred poor girls, who are brought up till they are of age to be married, or become nuns. **1848** F. W. Faber *Let.* 17 Feb. in J. E. Bowden *Life & Lett. F. W. Faber* (1869) viii. 337 Father Superior has now left us, all in our Philippine habits with turndown collars, like so many good boys brought in after dinner. **1852** —— *Let.* 14 Jan. in M. Trevor *Newman: Pillar of Cloud* (1962) 578 F. Rossi..declares..that he has acted in a manner unworthy of a Filippine *in re Achilli.* **1856** J. H. Newman *Let.* 2 Jan. in M. Trevor *Newman: Light in Winter* (1962) 99 The Philippine house is one large vault. *Ibid.* 7 Jan. 101 The brief half-hour we had for going to the Philippini. **1863** —— *Let.* 27 Dec. in W. Ward *Life J. H. Newman* (1912) I. xix. 612 Since 1846 or 1847,.. since I went to Propaganda and came back a Philippine. **1959** J. C. Chapin tr. *Giovannetti's We have Pope* v. 111 Patriarch Roncalli thought it proper to cede to the Filippine Institutes (now run by the Brethren of the Christian Schools) the splendid Villa Fietta.

Philippine (fi·lipīn), *a.*[2] [ad. Sp. *Filipino*: see *Filipino *sb.* and *a.*] Of or pertaining to a Filipino, or to the Philippine islands (now the Republic of the Philippines). Also *Comb.*

1812 W. Marsden *Gram. Malayan Lang.* p. xxi, The dictionaries of *Tagala, Bisaya, Pampagna*, and other Philippine languages are voluminous. **1815** J. F. Stephens in G. Shaw *Gen. Zool.* IX. i. 190 Philippine Woodpecker... About eleven inches in length... It is said to make a great noise with its beak. **1832** R. Baird *View of Valley of Mississippi* xxiv. 287 The Philippine or ribband cane is rapidly supplanting this species of cane. **1899** W. James *Talks to Teachers* x. 98 What a help is our Philippine war at present in teaching geography! **1932** W. L. Graff *Language* 423 The Philippine–Formosa group [of languages]. **1941** *Sun* (Baltimore) 12 Aug. 17/2 A considerable quantity of Philippine mahogany also will be used in the construction of the hulls. **1945** C. L. B. Hubbard *Observer's Bk. Dogs* 115 Though not recognized by The American Kennel Club, Philippine Dogs are generally bred to a type. **1947** J. C. Rich *Materials & Methods Sculpture* x. 292 Philippine mahogany is not a true mahogany. The name is applied to at least a half-dozen species of the dipterocarp family. Most of the woods referred to as Philippine mahogany are fairly soft and carve well. **1948** F. H. Titmuss *Conc. Encycl. World Timbers* 72 Lauan..is sometimes exported under the title of Philippine Mahogany, but is not a member of the true Mahogany family. **1964** P. F. Anson *Bishops at Large* xi. 530 Only one Philippine-born priest, a *mestico*, was raised to the episcopate. **1976** E. S. Gruson *Checklist Birds of World* 77 *Picoides maculatus* Philippine Pygmy Woodpecker. **1977** *Time* 12 Dec. 27/1 Manila-born Emerson Coseteng, by contrast, is a 46-year-old naturalized Philippine citizen and a major stockholder in the Mariwasa group of companies.

philippinite (filipi·nəit). *Geol.* [f. prec. + -ite[1].] Any tektite from the tektite field of the Philippine Islands.

1951 H. E. Suess in *Geochim et Cosmochim. Acta* II. 76 So-called 'gas-rich' philippinites, tektites from the Philippines, which are exceptionally rich in bubbles of various sizes. **1961, 1964** [see *javaite]. **1969** *New Scientist* 30 Oct. 237/1 We find ages of indochinites and philippinites to lie between 0·6 and 0·7 m.y.

Philips, var. *Phillips[1].

philism (fi·liz'm). *rare.* [The termination of such words as Negrophilism and *Anglophilism, used as a noun.] A feeling of friendliness towards another race or nation.

1917 *Edin. Rev.* July 127 Germany is the home of such movements..and many other 'philisms' and 'phobisms'. **1923** *Contemp. Rev.* Sept. 325 Tzankoff..is not credited with 'philism' or 'phobism' toward any particular country.

Phillips[1] (fi·lips). Also **Philips.** [The name of Henry F. *Phillips*, of Portland, Oregon.] A proprietary name used *attrib.* to designate screws having a slot in the form of an equal-armed cross, and corresponding screwdrivers.

1935 *Iron Age* 7 Nov. 42/3 American Screw Co., Providence, R.I., announces a line of case-hardened sheet metal screws featuring a new 'Phillips' recessed, self-centering head in place of the conventional screw-slot. *Ibid.*, The geometric pattern of the Phillips head provides that the screw shall hold to the taper point of the driver and may be brought into position with one hand. **1938** *Official Gaz.* (U.S. Patent Office) 18 Oct. 517/2 Henry F. Phillips, Portland, Oreg... Phillips Recessed Head. For screws. Claims use since Dec. 15, 1934. **1952** *Trade Marks Jrnl.* 20 Feb. 159/1 Phillips Recess... Screws of common metal. Guest, Keen & Nettlefolds Limited,.. Smethwick; manufacturers. *Ibid.* 161/1 Phillips Recess... Screw-drivers. Guest, Keen & Nettlefolds Limited,.. Smethwick; manufacturers. **1956** A. P. Morgan *Woodworking Tools* ii. 37 Phillips head screws require a Phillips screwdriver. **1962** E. Ambler *Light of Day* viii. 172 He had used an ordinary screwdriver on the Phillips heads. **1966** *Official Gaz.* (U.S. Patent Office) 18 Oct. TM119/1 Phillips Screw Company, Natick, Mass. Filed April 14, 1966. Phillips... For screws and allied fasteners. First use 1933. **1972** *Practical Motorist* Oct. 209/3 For Philips screws, you may be able to get by with one screwdriver but it's better to buy a large and a small one. **1976** *Practical Householder* Nov. 55/1 The kit comprises the Impact Driver, bit holder, two cross points for Philips screws, two ordinary screw-driver bits for slotted screws and instructions for use, all packed in a strong metal case.

Phi·llips[2]. The name of A. W. H. Phillips (1914–75), New Zealand economist, used *attrib.* or in the possessive in *Phillips* (or *Phillips'*) *curve*, a supposed inverse relationship between the level of unemployment and the rate of inflation.

1969 *Times* 5 May (Wall St. Suppl.) p. i/4 The Phillips curve was mentioned frequently during last year's presidential campaign, usually with the assumption that a Nixon Administration would tolerate a higher level of unemployment and thereby bring about a lower rate of inflation than would the opposition party. **1974** *Times* 23 Mar. 13/1 Many have come to doubt whether any such choice—between unemployment and inflation as expressed in the traditional Phillips's curve—exists at all. **1977** *Dædalus* CVI. iv. 90 The observation that prices start to rise long before the real GNP reaches full employment levels led to the Phillips curve hypothesis.

phillumenist (filiū-, filū·mĕnist). Also **philumenist.** [f. Phil- + L. *lūmen* light + -ist.] A collector of match-box or match-book labels. Hence **phillu·meny,** the collecting of such labels.

Quot. 1943, supplied by a private correspondent, is from an unverifiable source.—Ed.

1943 M. S. Evans in *Floyd's Label Rev.* May 4, I have been wondering..if we could not have a better name than 'labelists'... Under luminary in the dictionary I found the Latin word Lumen (light) and under philately the Greek word Philos (loving), also came across the Latin word Lucus (light). These suggest Phillumenist and Philucist. **1949** *This Week* 23 Oct. 24/2 Phillumenists (the word properly fits those who collect labels from wooden-match boxes, but also is applied to the paper-book fans) are outnumbered among hobbyists only by stamp collectors. **1951** *Britannica Bk. of Year* 686/2 *Phillumeny*, the hobby of collecting match-box covers. **1960** *News Chron.* 18 Feb. 4/5 Are phillumeny, meadophoIy, tegestology and fromology as popular as they were? **1960** W. Miller *Russians as People* 178 Some matchbox designs, among the hundreds which now make Russia a philumenist's paradise, have a charming simplicity suited to their size. **1967** *Listener* 13 Apr. 493/3 Little did John Walker foresee that his invention would become a great blessing to mankind, or that it would give rise to a popular hobby—phillumeny. **1976** *Weekend Mag.* (Montreal) 18 Dec. 8/1 Phillumenists go weak in the knees at the prospect of a rare matchbook cover.

Philly (fi·li). Also **Phillie.** U.S. slang abbrev. *Philadelphia* (see *Philadelphia and Philadelphian *a.* 3 in Dict. and Suppl.).

1891 W. de Vere *Tramp Poems of West* 79 Draw up a chair 'thatch', I'll tell you a story, That happened in 'Philly' some twenty years since. **1899** 'J. Flynt' *Tramping with Tramps* iv. 396 *Phillie*, Philadelphia. **1930** *Sat. Even. Post* 5 Apr. 46/2, I hope and trust that you do all right for yourself in Philly. **1961** *Rogue* May 14/3 After a while, Kitty murmured something to Cappy, and he held her close, answering, 'We'll just have to wait till we pull into Philly, honey.' **1971** *Black World* Apr. 65 To what hospital you goin?.. Back down in Philly. **1972** L. Neal in A. Chapman *New Black Voices* 308 Remembering the time in Philly/the cops crashing the hotel room. **1976** 'D. Halliday' *Dolly & Nanny Bird* vii. 90 Philly socialites. **1977** *Time* 31 Jan. 52/1 Nicky..has shot a bookie and is holed up in a fleabag Philly hotel, going crazy.

philo-. Add: *phil-athlete, -athletic* (further examples), *-irenist*; *phil-Athenian* (examples); *philo-African* (earlier example); **philo-Se·mite,** one who is favourable to or who supports the Jews; also as *adj.*; so **philo-Semi·tic** *a.*; **philo-Se·mitism,** theory, action, or practice directed in favour of the Jews; **philothe·rian** *a.* and *sb.* [Gr. θήρ wild beast], (a person) that loves wild animals; so **philothe·rianism,** love of wild animals.

1908 *Westm. Gaz.* 19 Aug. 4/3 Are we to suspect a phil-Athenian bias in the story? **1922** S. Leslie *Oppidan* xi. 133 The Philathlete and the Philistine. **1905** *Blackw. Mag.* Dec. 811/2 Pick out an untrained but philathletic young Englishman. **1922** S. Leslie *Oppidan* iii. 42 In an anti-musical philathletic school he was always anxious to impress the new boys. **1929** *Observer* 17 Nov. 11/2 The Sixth Form cricket ground and the Philathletic field of the School have been zoned as residential areas. **1922** Joyce *Ulysses* 574 Modern philirenists, notably the tsar and the King of England, have invented arbitration. **1865** W. Whitman *November Boughs* (1888) 106 He will not countenance at all the demand of the extreme Philo-African element of the North. **1946** Koestler *Thieves in Night* 275 If the Jews were as the philosemites describe them, there would be no reason for this Return. **1948** Wyndham Lewis *Let.* 25 Oct. (1963) 467, I am not philosemite. **1976** C. Bermant *Coming Home* i. ii. 28 All *goyim* were presumed to be *antisemitten* unless they showed definite proof to the contrary, whereupon they were pronounced *Judenfreint*—philosemites. **1962** *Observer* 27 May 28/2 Philo-semitic authors can be no whit less tedious than anti. **1977** *Daily Tel.* 31 May 16 Both Jews and philo-Semitic people have been made aware of the ghastly similarities between the 'old' form of anti-Semitism and its 'modern' euphemism [*sc.* 'anti-Zionism']. **1965** *New Statesman* 16 Apr. 617/3 He calls attention to the scale and cohesion of the philo-semitism now current in the cultural life of America. **1976** M. J. Lasky *Utopia & Revolution* (1977) viii. 301 Philo-Semitism was a natural by-product of the enthusiasm with which the Puritan generation returned to the old books of the Bible for inspired guidance. **1906** *Times* 24 Aug. 1/2 The terrible cruelties practised on quadrupeds..have been..denounced by..that noble and devoted philotherian. **1909** *Athenæum* 23 Oct. 494/3 An indulgence having been accorded to persons..purchasing, perusing, or subscribing to any philotherian publication. *Ibid.* 495/1, I ask myself why the practice of charity, in the shape of philotherianism, should be left out of sight.

philology. Add: **1.** (Further examples.) Still the usual sense in the U.S.

1922 O. Jespersen *Language* iii. 64 In this book I shall use the word 'philology' in its continental sense, which is often rendered in English by the vague word 'scholarship', meaning thereby the study of the specific culture of one nation. **1925** L. Bloomfield in *Language* I. 4 That noblest of sciences, philology, the study of national culture is..greater than a misfit combination of language plus literature... The British use of 'philology' for linguistics leaves no name for the former subject. **1931** J. W. Spargo tr. *Pedersen's Lingusitic Sci. 19th Cent.* iv. 79 One may define philology briefly as a study whose task is the interpretation of the literary monuments in which the spiritual life of a given period has found expression. *Ibid.* 80 The use of 'philology' as a mere synonym for 'linguistics' is to be avoided. **1939** L. H. Gray *Foundations of Lang.* 3 A more serious objection to the term [*sc.* comparative philology] lies in the fact that 'philology', strictly speaking, denotes not only the study of language, but also of literature and of all the civilisational phenomena of a people. *a* **1941** B. L. Whorf in *Ann. Rep. Board of Regents Smithsonian Inst. 1941* (1942) 502 As the major linguistic difficulties are conquered, the study becomes more and more philological; that is to say, subject matter, cultural data, and history play an increasing role... This is philology. But at the base of philology we must have linguistics. **1947** E. H. Sturtevant *Introd. Linguistic Sci.* i. 7 Philology is a word with a wide range of meaning. I use it here to designate the study of written documents. **1954** F. G. Cassidy *Robertson's Devel. Mod. Eng.* (ed. 2) 424 *Philology*, the study of thought and culture as embodied in literary monuments; in a narrower sense, the study of language (but for this sense, the term linguistics is now preferred). **1964** R. H. Robins *Gen. Linguistics* i. 6 In German,..*Philologie* refers more to the scholarly study of literary texts, and more generally to the study of culture and civilization through literary documents... This meaning..is matched by..the use of *philology* in American learned circles. **1980** *Yale Rev.* Winter 312 Philology meant, and still ought to mean, the general study of literature.

3. (This sense has never been current in the U.S.) In Britain, now usu. restricted to the study of the development of specific languages or language families, esp. research into phonological and morphological history based on written documents. (Later examples.)

Linguistics is now the more usual term for the study of the structure of language, and, with qualifying adjective or adjective phrase, is replacing *philology* even in the restricted sense.

1964 R. H. Robins *Gen. Linguistics* i. 6 In British usage philology is generally equivalent to comparative philology, an older and still quite common term for what linguists technically refer to as comparative and historical linguistics. **1968** J. Lyons *Introd. Theoret. Linguistics* i. 22 The term 'comparative philology', which I shall use to refer to this period of linguistics [*sc.* the nineteenth century].., though less commonly used these days by linguists themselves (who tend to prefer 'comparative and historical linguistics'), is not infrequently met in general books on language and, like many other unsuitable terms, has been perpetuated in the titles of university chairs and departments and of prescribed courses of study. **1974** R. Quirk *Linguist & Eng. Lang.* v. 84 'Developmental' and 'general' philology—or, as we would usually call them today, historical and general linguistics.

philomath. Delete 'Now *rare*' and add later examples.

1927 A. Huxley *Proper Stud.* 132 It is precisely for the philomaths that universities ought to cater. **1955** *Sci. Amer.* May 114/3 John Whiblin, the carpenter and 'philomath' who was 'ingenious at models'. **1977** R. V. Hudson in Bond & McLeod *Newslett. to Newspaper* ii. 120 The philomath responsible for it..was selected to receive the shafts of Franklin's wit.

philosoph, -ophe. Add: (Earlier and later examples as Fr. form in sense of Philosophist 2.) Also *attrib.* or as *adj.* and *transf.*

[**1774** H. Walpole *Let.* 28 Sept. (1904) IX. 59 Madame du Deffand hates *les philosophes.*] **1779** —— *Let.* 7 July (1904) X. 441 The *philosophes*, except Buffon, are solemn, arrogant, dictatorial coxcombs. **1830** J. P. Cobbett *Jrnl. Tour in Italy* 286 Guard us ye powers..against all that calls itself '*liberal*' or '*philosophe*'! **1840** J. S. Mill in *London & Westm. Rev.* Mar. 270 Those writers were as much cried down among the *philosophes* themselves. **1932** *Scrutiny* I. 122 Two things appeared to Bentham's *philosophe* mind to be necessary. **1961** *Times* 25 Mar. 3/7 He [*sc.* Raymond Williams] is not a politician so much as a prophet really, a sort of English *philosophe*. **1969** *Listener* 9 Jan. 37/2 All the *philosophes*, all the Encyclopedists, shared the Baconian belief that science could save mankind. **1977** *N.Y. Rev. Bks.* 13 Oct. 8/3 Despite his genuine and profound erudition, Gibbon was a *philosophe* far more than a philosopher, though he hankered after being a philosopher of history.

philosopher. Add: **1. c.** A member of a class called 'Philosophers' in certain Jesuit schools and colleges.

1711 in E. H. Burton *Life Bishop Challoner* (1909) I. iii. 32 Ye Superiors had inculcated..ye two pair of beads to be said every week by one of ye Philosophers. **1712** in *Ushaw Mag.* (1904) Mar. 20 Ye Littanies of ye Saints, every night our Ladyes wch are solemnly sung every Saturday, ye two pair of beads to be said every week by one of ye Philosophers, ye fasting before our Ladyes dayes and ye like. **1809** in *Edmundian* (1948) Summer 9 The boys in the higher classes viz. Philosophers and Rhetoricians have separate rooms. **1915** *Ushaw Mag.* Dec. 292 The new furniture..is to be arranged and then the Divines and Philosophers can take possession.

5. *philosopher-king* (examples), *-scientist.*

1923 A. Huxley *Let.* 12 Nov. (1969) 222 One determined Poincaré can defeat..ten philosopher-kings. **1962** *Listener* 25 Jan. 163/1 It is doubtful whether Socrates and the philosopher-kings of the Republic could have lived peaceably together. **1979** *Guardian* 3 Sept. 10/1 The TUC's last philosopher-king, George Woodcock. *a* **1937** J. L. Stocks *Reason & Intuition* (1939) v. 68 The philosopher-scientist of the nineteenth century had certainly no place for chance. **1945** R. G. Collingwood *Idea of Nature* ii. iii. 128 A philosopher-scientist like Whitehead can restate Hegel's theory (not knowing that it is Hegel's, for he does not appear to have read Hegel).

b. philosopher's philosopher [after *poet's poet* s.v. *poet 1c], a philosopher whose works appeal, or are intelligible, primarily to another philosopher.

1937 A. H. Murray *Philos. of James Ward* ii. 45 Bradley may have been a philosopher's philosopher, but much of what he describes and analyses in the tendency to unity inherent in thought is part of the direct experience of thinkers—and of other people. **1957** J. Passmore *100 Yrs. Philos.* iv. 81 McTaggart was a philosopher's philosopher, if ever there has been one. **1971** *Classical Rev.* XXI. 224 Aristotle, a philosopher's philosopher. **1974** A. W. Levi *Philos. as Social Expression* v. 298 That *déformation professionnelle*..has produced the doctor's doctor, the lawyer's lawyer, and the philosopher's philosopher.

‖ **philosophia perennis** (filosǫ·fiă pĕre·nis). [L., = perennial philosophy.] An alleged central core of philosophical truths that are generally accepted regardless of time or place, usu. taken to be exemplified by Aristotle and St. Thomas Aquinas.

[**1540** A. Steuchus (*title*) De perenni philosophia.] **1858** A. C. Fraser *Rational Philos.* ii. 116 The long line of what an old writer calls *philosophia perennis.* **1933** W. R. Inge *God & Astronomers* i. 13 The classical tradition of Christian philosophy, which Roman Catholic scholars call the *philosophia perennis.* **1941** *Mind* L. 166 The tradition of Nominalism, Naturalism, Materialism, which has always haunted the *Philosophia Perennis* like a shadow. **1960** H. Kraemer *World Cultures & World Relig.* vi. 150 A..career of proclaiming in America, England and Europe the glorious universality of Vedanta as the philosophia perennis. **1965** *Listener* 1 July 14/1 Like the *philosophia perennis* which it reflects, his [*sc.* Dante's] poem just goes on applying to the human situation year after year.

‖ **philosophia prima** (filosǫ·fia prī·mă). [L., = first philosophy.] The study of the most general and universal truths of all science, (so called by Aristotle, followed by Francis Bacon and others); also, more specifically, the study of the divine and eternal.

1605 Bacon *Adv. Learning* ii. fol. 20ᵛ Therefore it is good, before wee enter into the former distribution, to erect & constitute one vniuersal Science by the name of *Philosophia Prima, Primitive* or *Svmmarie Philosophie*, as the Maine and common way, before we come where the waies part. [**1641** Descartes (*title*) Meditationes de prima philosophia, in qua Dei existentia, & animæ immortalitas demonstratur.] **1651** Hobbes *Leviathan* iv. xlvi. 371 There is a certain *Philosophia prima*, on which all other Philosophy ought to depend; and consisteth principally, in right limiting of the significations of such Appellations, or Names, as are of all others the most Universall. **1865** J. S. Mill *Exam. Hamilton's Philos.* xxiv. 464 What is inappropriately termed the Philosophia Prima..would be more properly called *ultima*, since it consists of the widest generalizations respecting the laws of Existence and Activity. **1890** G. T. Ladd *Introd. Philos.* (1891) 14 Yet a *prima philosophia* is in some sort recognized, which is nothing more than a mixture of definitions of the more fundamental conceptions. **1945** *Mind* LIV. 209 An author who calls himself a positivist had better show cause that he is in earnest when he protests his concern for *philosophia prima.*

philosophic, *a.* (*sb.*) Add: **3.** (Later examples.)

1816 Jane Austen *Emma* II. xvi. 304 The philosophic composure of her brother on hearing his fate. **1822** De Quincey *Confess.* 110 The poor are far more philosophic than the rich... They show a more ready and cheerful submission to what they consider as irremediable evils. **1927** E. O'Neill *Marco Millions* ii. i. 85 The expression has grown mask-like, full of philosophic calm. *a* **1953** —— *Touch of Poet* (1957) ii. 86, I flatter myself I have preserved a philosophic poise. **1977** D. Aitkin *Second Chair* xxiv. 227 She was composed and philosophic, aware that I did not blame her.

4. *philosophic radical* (also with capital initials) = **philosophical radical*; hence *philosophic radicalism.*

1834 J. S. Mill in *Monthly Repos.* VIII. 309 Few of the results of the Reform Bill have fallen more short of our hopes, than the conduct of the little band of enlightened and philosophic Radicals. **1837** —— in *Westm. Rev.* XXVII. 67 Those whom..we call philosophic radicals, are those who in politics observe the common practice of philosophers—that is, who, when they are discussing means, begin by considering the end, and when they desire to produce effects, think of causes. *a* **1854** —— *Early Draft Autobiogr.* (1961) 114 Almost every debate was a bataille rangée between the philosophic radicals & the Tory lawyers. *Ibid.* 157 The foundation of a periodical organ of philosophic radicalism. **1882** A. Bain *John Stuart Mill* iv. 124 In the days when he was heading the philosophic radicals, he was conscious of the weakness of his position in not being himself in the House of Commons. **1911** *Encycl. Brit.* XVI. 661/1 In April 1835 he [*sc.* William Molesworth] founded, in conjunction with Roebuck, the *London Review*, as an organ of the 'Philosophic Radicals'. **1969** D. Beales *From Castlereagh to Gladstone* I. iv. 68 Bentham..was the most productive..of a group which was of great importance in that it had some influence on public opinion at large, more on political leaders, and more still..on practical administration. They were known as the Philosophic Radicals.

philosophical, *a.* (*sb.*) Add: **4.** *philosophical logic*, logic pertaining to philosophy (opp. **mathematical logic*); *philosophical radical* (also with capital initials), a member of a group of early 19th-century radicals whose advocacy of reform was based on the utilitarian theories of Bentham and James Mill; hence *philosophical radicalism.*

1903 B. Russell *Princ. Math.* ii. 32 It remains a question for philosophical logic whether there is not a quite different notion of the disjunction of individuals. **1921** C. K. Ogden *Let.* 5 Nov. in B. Russell *Autobiogr.* (1968) II. ii. 121, I am still a little uneasy about the title [of Wittgenstein's *Tractatus*] and don't want to feel that we decided in a hurry on *Philosophical Logic.* **1952** *Mind* LXI. 57 That philosophical logic is concerned with form is the traditional and still prevalent view. **1967** P. Strawson *Philos. Logic* 1 Wittgenstein's suggestion does not itself belong to formal logic. It belongs to philosophical logic. **1834** J. S. Mill in *Monthly Repos.* VIII. 174 Those who aspire to be..distinguished as the instructed and philosophical Radicals. **1855** in T. Woollcombe *Notices of Late Sir W. Molesworth* (1857) 36 They were generally known by the rather ambitious title of philosophical radicals. **1873** H. Grote *Personal Life G. Grote* vi. 56 The 'Philosophical Radicals' as the followers of Bentham were designated. **1885** W. Harris *Hist. Radical Party* vii. 130 The so-called Philosophical Radicals, following the methods and sharing the conclusions of Bentham, performed the duty of proving that the political reforms..were not the mere creations of disaffected ignorance, but were founded on great moral and social laws. **1945** B. Russell *Hist. Western Philos.* (1946) iii. xxi. 746 The romantic revolt..passes on, somewhat softened, to the philosophical radicals in England. **1974** *Encycl. Brit. Macropædia* XII. 197/2 The *Westminster Review*..was the organ of the philosophical radicals. **1910** *Ibid.* V. 351/2 Carlyle had some expectation of the editorship of the *London Review*.., an organ of philosophical radicalism. **1935** A. Huxley *Let.* 13 Jan. (1969) 390 Bertrand Russell's book on 19th century history..was excellent if regarded as a series of essays on different aspects of the time—Marxism, Philosophical Radicalism and so forth. **1966** F. Copleston *Hist. Philos.* VIII. i. i. 3 The philosophical radicalism which is associated with the name of Jeremy Bentham and which had already been expressed by him in the closing decades of the eighteenth century.

philosophically, *adv.* Add: (Earlier and later examples corresponding to Philosophical *a.* 3.)

1873 Hardy *Pair of Blue Eyes* III. iii. 75 The practical husbands and wives who take things philosophically are very humdrum, are they not? **1898** G. B. Shaw *Widowers' Houses* (rev. ed.) ii. 46 Trench..tries to take his disenchantment philosophically. **1933** E. O'Neill *Ah, Wilderness!* i. 43 *Mrs. Miller.* (stares after him worriedly—then sighs philosophically). **1971** *Countryman* Autumn 112 Suddenly the third member of the brood swooped down from its perch in a nearby spruce and flew back to the tree with the worm in its beak. The hen [blackbird] philosophically started again. **1977** K. O'Hara *Ghost of T. Penry* xi. 97 He..switched off Joe's lamp. Philosophically coiling down the cord, Joe said 'Exactly.'

b. *Comb.*, as *philosophically-minded* adj.

1942 R. G. Collingwood *New Leviathan* xliii. 365 We know too much about the Bogomils to be content with a Gibbonesque, eighteenth-century picture of them as simple, philosophically minded innocents. **1955** *Sci. Amer.* July 73/1 Only a few scientists had been philosophically minded, but today physicists are almost all philosophers.

philosophico-. Add: *philosophico-lexicological, -linguistic, -religious, -scientific* adjs.

1924 C. K. Ogden tr. *Vaihinger's Philos. of 'As If'* 140 We will not embark here on a philosophico-lexicological excursus. **1967** C. L. Wrenn *Word & Symbol* 2 A creative fundamental philosophico-linguistic re-assessment. **1934** Webster, Philosophicoreligious. **1960** Partridge *Charm of Words* 96 Rosenkreuz..founder..of a philosophico-religious secret society. **1977** *Times Lit. Suppl.* 18 Feb. 190/1 The problem of 'life'..is so fraught with philosophico-religious profundities. **1847** W. Smith tr. *Fichte's Characteristics Present Age* 242 The Philosophico-scientific character of the Third Age has been already set forth.

philosophize, *v.* Add: **2.** Also, to say or comment philosophically.

1922 Joyce *Ulysses* 203 The will to live, John Eglinton philosophised, for poor Ann, Will's widow, is the will to die. **1977** J. Wainwright *Do Nothin'* vi. 97 'It takes all sorts,' I philosophise.

philosophy, *sb.* Add: **7.** (Further examples.)

1823 Scott *Let.* 5 Oct. (1935) VIII. 104, I would defer to the very last what is always taught first namely the philosophy as it has been termd [*sic*] of history. **1835** J. S. Mill in *London Rev.* I. i. 106 The evidence of history..leaves the philosophy of society in exactly the state in which physical science was, before the method of experiment was introduced. **1843** —— *Logic* I. i. v. 119 The notion..seems to me one of the most fatal errors ever introduced into the philosophy of Logic. **1847** J. D. Morell *Hist. View Philos.* (ed. 2) II. v. 15 We have already shown in the case of Reid, that the philosophy of perception was well commenced, but not fully completed. **1852** *Westm. Rev.* Oct. 435 (*heading*) The philosophy of style. **1853** Lytton *My Novel* IV. xxxv. 283 Levy is a man who has admitted the fiercer passions into his philosophy of life. **1865** J. S. Mill *Auguste Comte* 54 The philosophy of science consists of two principal parts; the methods of investigation, and the requisites of proof. **1890** W. James *Princ. Psychol.* I. xiv. 585 We see in the philosophy of desire and pleasure, that such nascent excitements..may become potent mental stimuli and determinants of desire. **1896** W. Caldwell *Schopenhauer's Syst.* iii. 162 A philosophy of life must make some broad assertion about reality as a whole. **1919** G. B. Shaw *Heartbreak House* p. xxvii, It is impossible to estimate what proportion of us..grasped the war and its political antecedents as a whole in the light of any philosophy of history or knowledge of what war is. **1940** F. J. E. Woodbridge *Essay on Nature* i. 53 Expressions like 'philosophy of science', 'philosophy of history', 'philosophy of government', 'philosophy of law', 'philosophy of religion', and so forth creep into the language, indicating that after scientists, historians, statesmen, jurists, priests, and the rest have said all they have to say, there is still need of a special kind of knowledge to inform us what it is all about. **1957** G. Ryle in C. A. Mace *Brit. Philos. in Mid-Cent.* 243 We do systematically construe 'name' on the model of 'proper name'. The assumption of the truth of this equation has been responsible for a large number of radical absurdities in philosophy in general and the philosophy of logic in particular. **1960** H. L. Bond *Lit. Art E. Gibbon* vii. 138 Gibbon's well-defined manner of speaking and the details of his style reflect..his whole philosophy of life. **1966** J. J. Katz *Philos. Lang.* i. 4 Philosophy of language is an area in the philosophical investigation of conceptual knowledge, rather than one of the several branches of contemporary philosophy such as philosophy of science, philosophy of mathematics, philosophy of art, and so forth. **1976** *Hiroshima Stud. Eng. Lang. & Lit.* XXI. 60 This is a study of D. H. Lawrence's *The Man Who Loved Islands* with a view to locating the spot it occupies in a series of works descriptive of his unique philosophy of life. **1977** P. A. French et al. (*title*) Studies in the philosophy of language.

8. (Further examples.) Also, more generally, a set of opinions, ideas, or principles; a basic theory; a view or outlook.

1898 G. B. Shaw *Plays Unpleasant* p. xix, It is quite possible for a piece to enjoy the most sensational success on the basis of a complete misunderstanding of its philosophy. **1899** O. Wilde *Ideal Husband* i. 12 *Mrs. Cheveley.* I don't know that women are always rewarded for being charming. I think they are usually punished for it! Certainly, more women grow old nowadays through the faithfulness of their admirers than through anything else! At least that is the only way I can account for the terribly haggard look of most of your pretty women in London! *Sir Robert Chiltern.* What an appalling philosophy that sounds! **1901** G. B. Shaw *Three Plays for Puritans* p. xxxii, It is the philosophy, the outlook on life, that changes, not the craft of the playwright. *Ibid.*, Such men must rewrite all the old plays in terms of their own philosophy. **1903** —— *Man & Superman* p. v, Here is your play!..Its profits, like its labor, belong to me: its morals, its manners, its philosophy, its influence on the young, are for you to justify. *Ibid.* III. 126 Yes, Juan: we know the libertine's philosophy. Always ignore the consequences to the woman. **1922** JOYCE *Ulysses* 629 You have every bit as much right to live by your pen in pursuit of your philosophy as the peasant has. **1946** E. O'NEILL *Iceman Cometh* (1947) I. 45 You pretend a bitter, cynic philosophy, but in your heart you are the kindest man among us. **1959** *Economist* 12 Sept. 836/1 Rival policies—or at least rival philosophies, for they barely reached the stage of practical policies—were very much in the centre of public attention. **1962** G. COOPER in *Into Orbit* 30 The entire philosophy had to be revised once we got involved with manned flights. **1973** *B.S.I. News* Dec. 32/2 Philosophy of control assignment on keyboards... Outlines the general philosophy for the positioning of areas for control keys, in relation to 'graphic' areas, of..keyboards for office machines. **1975** J. PLAMENATZ *K. Marx's Philos. of Man* i. 4 It is not surprising, then, that these ideas, derived from philosophies alien to them, should be less interesting to Marxists and students of Marxism outside the West. Even if they have similar ideas in their native philosophies, they are ideas differently expressed, and so such similarities as there may be go unrecognized.

philothion (filoþai·ən). *Biochem. Obs.* exc. *Hist.* [a. F. *philothion* (J. de Rey-Pailhade 1888, in *Compt. Rend.* CVI. 1684), f. Gr. φιλο- (see PHILO-) + θεῖον sulphur.] = *GLUTATHIONE.

1888 *Jrnl. Chem. Soc.* LIV. 1101 It follows that the substance, to which the author gives the name *philothion*, exists in animal tissues in a form different from that in which it exists in yeast. It stands to sulphur in the same physiological relation as hæmoglobin to oxygen, that is to say, it renders it soluble and assimilable. **1900** [see *HYDROGENASE]. **1921** [see *GLUTATHIONE]. **1954** T. WIELAND in S. Colowick et al. *Glutathione* 45 In 1888 de Rey-Pailhade..described a substance containing sulfur which he had been able to isolate from yeast in an impure state and which he called 'philothion'.

-philous. Add: **b.** In *Biol.* forming adjs. with the sense 'having an affinity for or thriving in (a particular kind of habitat or environment)', as in *dendrophilous* s.v. DENDRO-, HYDROPHILOUS, *hygrophilous* s.v. HYGRO-.

phizgig (fi·zgig). *Austral. slang.* Also **phizzgig.** = *FIZGIG 6. Hence as *v. intr.*, to act as an informer.

1941 in BAKER *Dict. Austral. Slang.* **1955** V. KELLY *Shadow* i. 10 The boys suspect he's been phizgigging. *Ibid.* iv. 45 Prior passed the word along to phizgigs that the police would appreciate information about the identity of the counterfeiters. **1973** *Nation Rev.* (Melbourne) III. No. 49. 2/2 Lenny..was described in his biog as a former 'phizzgig' (police informer).

phizog (fi·zǫg). *joc. colloq.* Also **fizzog, phisog, physog, phyzog.** [f. as PHIZ.] = PHIZ. Cf. PHYSIOG.

1811 *F. Grose's Lexicon Balatronicum, Physog*, the face. A vulgar abbreviation of physiognomy. **1829** P. EGAN *Boxiana* 2nd Ser. II. 706 A certain melancholy cast was taking possession of Tom's *phisog*. **1846** *Swell's Night Guide* 127/2 *Phizog*, the face. **1850** C. KINGSLEY *Alton Locke* II. iii. 37 He had received an anonymous letter, 'a'thegither o' a Belgravian cast o' phizog,' containing a bank-note for twenty pounds. **1912** W. OWEN *Let.* 12 Jan. (1967) 110 Unable to get out and see anything better in the way of physogs. **1922** JOYCE *Ulysses* 297, I saw his physog do a peep in and then slidder off again. **1939** *Airman's Gaz.* Dec., One observer 'snapped' Adolph's phisog. **1946** B. MARSHALL *George Brown's Schooldays* iv. 16 The prefect shuts the door in their physogs. **1959** I. & P. OPIE *Lore & Lang. Schoolch.* x. 194 'Shut yer face'—'fizzog'—'flycatcher'—or, 'gate'. **1980** *Radio Times* 5–11 Jan. 15/1 The phizog is definitely familiar... 'I get recognised wherever I go,' he says.

phlebo-. Add: **phlebogram, -graphy** (see main entries below); **phlebothrombosis** (examples); add to def.: in mod. use a venous thrombosis in which inflammation of the vein is absent or of only secondary significance.

1939 A. OCHSNER in *Surgery* VI. 129 In considering intravascular clotting, it is important to distinguish between those lesions which are associated with an inflammatory process in the vessel wall, a true thrombophlebitis, and those in which there is intravascular clotting without the associated inflammatory lesion which can be termed a 'phlebothrombosis'. **1956** P. MARTIN et al. *Peripheral Vascular Disorders* xx. 628 Phlebothrombosis occurs frequently in both legs yet one side only may progress to thrombophlebitis. **1974** PASSMORE & ROBSON *Compan. Med. Stud.* III. xvii. 33/1 There are two types of venous thrombosis depending upon whether the wall of the vein is inflamed (thrombophlebitis) or not inflamed (phlebothrombosis).

phlebogram. Add: (Examples.) Now *rare*.

1885 W. STIRLING tr. *Landois's Text-bk. Human Physiol.* I. II. 194 In order to interpret the various events of the phlebogram it is most important to record simultaneously the events that take place in the heart. **1923** W. D. REID *Heart in Mod. Pract.* iii. 59 The nodal extrasystole is not common. It gives rise to an exaggerated wave in the phlebogram. **1964** HOCHSTEIN & RUBIN *Physical Diagnosis* xix. 266 The *c* wave, although inscribed on a phlebogram, is not seen by clinical inspection of the neck veins, either because it is an artifact of the phlebogram due to the adjacent carotid pulse, or because the *c* wave of the right atrial pressure pulse is too small in magnitude to be reflected visually in the jugular pulse.

2. *Med.* An X-ray photograph of a vein.

1933 *Lancet* 18 Nov. 1144/2 The speed of circulation is very different in the brain as compared with other parts of the head... We have ascertained this fact by studying the arteriograms and phlebograms of diverse cases. **1969** D. SUTTON *Textbk. Radiol.* xxxi. 602 (*caption*) Adrenal vein phlebogram showing small Conn's tumour in the adrenal. Other investigations including angiography and air insufflation had failed to show the lesion. **1977** *Lancet* 1 Jan. 49/2 Evidence for a 50% frequency of a minor degree of iliac-vein compression was based on phlebograms.

phlebography. Add: **2.** *Med.* The recording of the pulse in a vein. *rare.*

1915 DORLAND *Med. Dict.* (ed. 8) 731/2 *Phlebography*, the graphic recording of the venous pulse. **1966** R. B. REES in H. F. Conn et al. *Current Diagnosis* XII. 521/2 The Trendelenburg test, Perthes's test, phlebography and other tests of venous competence are sometimes useful in determining whether vein surgery is indicated.

3. *Med.* Radiography of a vein, carried out after the introduction into it of a contrast medium.

1937 L. J. FRIEDMAN *Text-bk. Diagnostic Roentgenol.* 612/2 (Index), Phlebography (see Angiography). **1946** E. V. ALLEN et al. *Peripheral Vascular Dis.* xxv. 619 Theoretically the amount and location of organic obstruction to the veins in the limb can be determined with great accuracy by means of indirect phlebography (intra-arterial injection). The technic of this procedure, however, is rather difficult and..it is not advisable to attempt it during the acute stage of thrombophlebitis. The technic of direct phlebography is much simpler, but with this procedure only a portion of the venous tree can be visualized. **1972** *Lancet* 27 May 1134/2 A small-calf-vein thrombosis confirmed by phlebography.

phlebotomist. Add: In mod. use, someone trained to take blood from a person for subsequent examination or transfusion.

1974 N. M. ABELSON *Topics in Blood Banking* vi. 124 The skill of the phlebotomist. **1976** I. CHANARIN et al. *Blood & its Dis.* ix. 78 Venous blood samples are required. These may be collected by either laboratory staff, medical staff, nursing staff or phlebotomists. **1977** *Age & Ageing* VI. 85 Blood was taken, usually by a phlebotomist, after the skin had been prepared.

phlizz (fliz). [Fanciful.] In Lewis Carroll's book *Sylvie and Bruno*, a fruit or flower that has no real substance; hence, allusively, anything without meaning or value, a mere name.

1889 'L. CARROLL' *Sylvie & Bruno* vi. 75 Bruno.. picked a fruit... 'It hasn't got no taste at all!' he complained. 'It was a *Phlizz*,' Sylvie gravely replied. *Ibid.* xx. 294 They *will* be sorry when they find them [*sc.* flowers] gone!.. The nosegay was only a *Phlizz*. **1899** *Johnson Club Papers* 188 We crown the musicians with flowers that, like poor Bruno's in the fairy tale, are but a phlizz. **1926** GALSWORTHY *Silver Spoon* II. ix. 187 What was his image of her but a phlizz, but a fraud? *Ibid.* xii. 218 Was Foggartism a phlizz? **1931** —— *Maid in Waiting* iv. 20 The thing's a phlizz. Just a low type of Homo Sapiens.

phloem. Substitute for first part of etym.: [a. G. *phloëm* (C. W. Nägeli *Beiträge zur wiss. Bot.* (1858) I. 9).] (Later examples.) Also *phloem-island, -strand.*

1889 Phloem-island, -strand [see *interxylary* adj. s.v. *INTER- 6]. **1891** *Ann. Bot.* V. 178 The phloëm-islands are produced centrifugally by the cambium. **1914** M. DRUMMOND tr. *Haberlandt's Physiol. Plant Anat.* xiv. 656 The various tissues that are comprised in the general category of secondary phloem are usually arranged in more or less distinct tangential layers. **1951** MCLEAN & IVIMEY-COOK *Textbk. Theoret. Bot.* I. xxi. 922 Phloem Islands..are formed by the enclosure of portions of the normal phloem by short arcs of secondary cambium. **1953** K. ESAU *Plant Anat.* xii. 265 The phloem is the principal food-conducting tissue of the vascular plants. **1969** [see *interxylary* adj. s.v. *INTER- 6]. **1975** J. D. HAYNES *Botany* xx. 314 The vascular tissue is the transport tissue of the plant, and is composed of two major issue types, the xylem and the phloem. **1978** J. UPDIKE *Coup* (1979) iii. 95 Some substratum in the phloem or xylem savored smartly of those little glossy red American candies.

phlomis (flǫ·mis). [mod. L. (J. P. de Tournefort *Institutiones Rei Herbariæ* (1700) I. 178), a. Gr. φλομίς a name used by Dioscorides.] A shrub or herb of the large genus so called, belonging to the family Labiatæ, and native to temperate parts of Asia and southern Europe, esp. *Phlomis fruticosa*, the Jerusalem sage, which has pale green leaves with white, hairy undersides, and whorls of deep yellow flowers.

1706 [see PHLOME]. **1789** W. AITON *Hortus Kewensis* II. 307 Broad-leav'd shrubby Phlomis, or Jerusalem Sage. **1908** G. JEKYLL *Colour in Flower Garden* ix. 80 Further back among the flowers are..some grey bushes of Phlomis and a silvery leaved Willow. **1961** *Times* 25 Nov. 11/4 Phlomis, verbascums, and many more, can be brought into service for such [dry, sunny] positions. **1974** *Country Life* 21 Mar. 627/3, I particularly like the bold groups of golden foliage..and woolly, silvery phlomis with the purple nut. **1976** *Daily Tel.* 9 Oct. 6/2 Here..I would grow the pink-flowered Italian phlomis, which has come into garden centres in recent seasons, as well as the old Jerusalem one,.. which..has yellow flowers, in addition to the silvery, woolly leaves characteristic of the group.

phlorizin. For 'Also †phlori·dzin(e' read: Also **phlo(r)rhizin, phlori·dzin** († -ine). [First formed as G. *phloridzin* (L. de Koninck 1835, in *Ann. der Pharm.* XV. 76; *phlorrhizin* suggested there by the editor as the proper form of the word.] (Add earlier and further examples.)

1835 *Amer. Jrnl. Sci.* XXVIII. 383 (*heading*) Phloridzin, a new substance. **1867** BLOXAM *Chem.* 478 Phloridzine..is extracted from the bark of the apple, pear, plum, and cherry tree. **1900** *Amer. Jrnl. Physiol.* IV. p. xi, If a well fed milch goat be made to fast two days and phlorhizin be administered three times daily during the two days, the milk flow stops entirely. **1927** M. BODANSKY *Introd. Physiol. Chem.* ix. 232 The experimental production of this condition [*sc.* renal diabetes] was accomplished in 1886 by von Mering upon injecting into animals phlorizin, a glucoside which is found in the root bark of the cherry, apple, pear and plum tree. **1947** E. BALDWIN *Dynamic Aspects Biochem.* vii. 199 Both phlorrhizin and iodoacetate are known to be powerful inhibitors of fermentation and glycolysis. **1968** PASSMORE & ROBSON *Compan. Med. Stud.* I. xi. 21/1 Phlorizin poisons the renal tubules. **1969** DATTA & OTTAWAY *Biochem.* (ed. 2) viii. 154 Active absorption of sugars can be inhibited specifically by low concentrations of phlorrhizin. **1970** A. L. LEHNINGER *Biochem.* xxii. 491 Such a net conversion of tricarboxylic cycle intermediates is also observed in animals treated with the toxic glycoside phloridzin. **1970** R. W. MCGILVERY *Biochem.* xvii. 399 Phlorhizin is a polyphenolic glycoside.

phlorizinized (florəi·zinəizd), *ppl. a.* Forms: see prec. [f. prec. + -IZE + -ED[1].] That has been given phlorizin. So **phlori·zinize** *v. trans.*

1900 *Amer. Jrnl. Physiol.* IV. p. x, A well phlorhizinized muscle begins to go into rigor within five minutes after death. **1912** A. E. TAYLOR *Digestion & Metabolism* vi. 393 In the phloridzinized dog, it is assumed that no sugar is burned and that neither sugar nor nitrogen is stored. **1927** M. BODANSKY *Introd. Physiol. Chem.* ix. 220 Mandel and Lusk gave phlorizinized dogs lactic acid and recovered it as extra glucose in the urine. **1954** CANTAROW & SCHEPARTZ *Biochem.* xvii. 434 Under ordinary dietary conditions, the body stores of carbohydrate are never depleted to the point where fat and protein catabolism are significantly increased, as they are in the phlorizinized animal. **1965** DATTA & OTTAWAY *Biochem.* xi. 199 It was observed many years ago that when single amino acids were fed to diabetic or phlorrhizinized animals, some caused the excretion of extra glucose in the urine, others the excretion of extra ketone bodies.

Also **phlori:ziniza·tion,** the action of making, or state of being, phlorizinized.

1917 *Amer. Jrnl. Med. Sci.* CLIII. 330 Maximal phloridzin poisoning with feeding of nothing but fat, or the longest possible phloridzination on diet free from carbohydrate and high in fat, has failed to produce in dogs anything resembling diabetic lipemia. **1959** W. B. WHALLEY in J. W. Fairbairn *Pharmacol. Plant Phenolics* 30 The glucosuria in phlorizinisation is to be attributed to the rapid loss of glucose through the kidney.

phloro-. Add: **phloroglucin, -glucinol.** *phloroglucinol* is the only name now current; (examples).

1879 *Jrnl. Chem. Soc.* XXXVI. 633 By this synthesis, the aromatic nature of phloroglucinol is definitely established. **1949** E. CHAIN in H. W. Florey et al. *Antibiotics* II. xix. 758 They also tested a large number of chemicals for their ability to stimulate the production of penicillin... Among the ineffective substances were sodium azide,.. phloroglucinol, naphthol, [etc.]. **1963** [see *FILICIC *a.*].

phloxin. Add: Now usu. **phloxine.** (Further examples.)

1954 MARTIN & HYNES *Clin. Endocrinol.* (ed. 2) i. 10 A preliminary eosinophil count is made by the method of Randolph. [*Note*] By Randolph's method blood is diluted in a white cell pipette with a fluid composed of phloxine and methylene blue dissolved in equal parts of propylene glycol and water. The eosinophils are counted directly per c.mm. on a counting chamber. **1956** KIRK & OTHMER *Encycl. Chem. Technol.* XV. 141 The Phloxines are chlorinated derivatives of Eosine. *Ibid.*, The two Phloxines are used interchangeably as biological stains and have also been used in lake and pigment manufacture. **1967** KARCH & BUBER *Offset Processes* vii. 269 Phloxine or eosine lakes are brilliant reds with a 'bluish' or 'purplish' cast or undertones not fast to light.

pho, phoh, *int.* (Later examples.)

1908 S. WEYMAN *Wild Geese* xx. 304 Pho! Take my word for it, he's no man to bear malice! **1942** BERREY & VAN DEN BARK *Amer. Thes. Slang* §151/11 *Nonsense!; it is ridiculous!..pho!*

phobe (fōᵘb), *a. rare.* [The suffix -PHOBE used as a separate word.] Having a hatred or aversion (towards someone or something).

1915 *Times* 5 Apr. 5/6 The Italian people is not, and cannot be at this moment, either phil or phobe regarding any other people.

phobia. Add to def.: In *Psychol.*, an abnormal and irrational fear or dread which is caused by a particular object or circumstance. (Earlier and further examples.)

1786 *Columbian Mag.* Nov. 110/1, I shall begin, by defining Phobia in the present instance, to be a fear of an imaginary evil, or an undue fear of a real one. **1897** tr. *T. Ribot's Psychol. of Emotions* II. ii. 215 We can easily see that many phobias come under this category. **1907** S. A. K. WILSON tr. *Meige & Feindel's Tics* iv. 88 Prominent among the mental anomalies of the subjects of tic are found different sorts of phobia. **1909** A. A. BRILL tr. *Freud's Sel. Papers on Hysteria* v. 123 Thus far the processes are the same in hysteria, in phobias and obsessions, but from now on their ways part. *Ibid.* 127 Thus..freed anxiety, the sexual origin of which can not be recalled, attaches itself to the common primary phobias of man. **1954** R. F. C. HULL tr. *Jung's Devel. of Personality* in *Coll. Wks.* XVII. iv. 74 The latter [*sc.* the mother] projected all her phobias onto the child. **1974** E. B. MCNEIL *Psychol. of being Human* ix. 232 Phobias are symptoms issuing from unacceptable basic urges that have been repressed from consciousness. When repression is effective, phobia symptoms need not exist. **1978** *New York* 3 Apr. 85/2 (Advt.), Swim-o-phobia? Cure it forever. Our private lessons by professional instructors will have you phobia-free and swimming in no time.

phobic (fōᵘ·bik), *a.* and *sb.* [f. PHOB(IA + -IC.] **A.** *adj.* Pertaining to, or characterized by a phobia. **B.** *sb.* A person suffering from a phobia.

1897 tr. *T. Ribot's Psychol. of Emotions* II. ii. 215 For 'phobic' subjects it is (at least potentially) a permanent state, ready to arise when suggested by some association. **1930** *Brit. Jrnl. Med. Psychol.* X. 46 (*heading*) A phobic case. *Ibid.* 66 Recall and reunion may simply give the phobic mechanism a chance to dominate the normal system. **1964** GOULD & KOLB *Dict. Social Sci.* 466/2 Hysteria, obsession-compulsion, and phobic reactions.. tend to restrict the scope of the neurotic's behaviour. **1968** *N.Y. Rev. Bks.* 29 Feb. 32/4 She has small regard for the population of South Vietnam which she sees as an unpleasant amalgam of cynical self-seekers, stooges, and phobics (i.e., people who fear Communism). **1970** *Daily Tel.* (Colour Suppl.) 4 Sept. 11/4, I..learned that phobics all over the country were undergoing such unlikely cures as yoga, hypnosis, acupuncture and shock treatment. *Ibid.* 12/2 'Desensitisation' thus involves a simple-sounding cure: the phobic patient is taught right habits again. **1976** J. PAYNE *All in Mind* 47/1 A bird phobic would initially be talked to by therapist about phobia, then about birds, then (with a relaxing injection if necessary) pictures of birds would be introduced... It is a very gradual process by which the phobic is helped to overcome the object of irrational fear.

phobism (fōᵘ·biz'm). *rare.* [f. as PHOBIA + -ISM.] A morbid fear of or aversion to anything.

1917, 1923 [see *PHILISM].

phoby (fōᵘ·bi), colloq. shortening of HYDROPHOBIA 2 (*rare*).

1843 DICKENS *Mart. Chuz.* (1844) xvii. 223 A jug and ewer, that might have been mistaken for a milk-pot and slop-basin... 'They've certainly got a touch of the 'phoby, sir.'

Phocæan (fōᵘsī·ăn), *sb.* and *a.* Also 7 **Phocean.** [f. Gr. Φώκαι-α, the place-name Phocæa, or L. *Phocaei* Phocæans + -AN.] **A.** *sb.* A native or inhabitant of the ancient city of Phocæa, the most northern of the Ionian cities on the west coast of Asia Minor.

1600 HOLLAND tr. *Livy's Romane Hist.* XXXVIII. 1007 The Phocæans had both their owne lands restored unto them..., and also libertie to live under their auncient lawes. **1784** W. MITFORD *Hist. Greece* I. vii. 297 The Phocæans, hard pressed, obtained a truce for a day. **1797** *Encycl. Brit.* XIV. 613/1 The Phocaeans were expert mariners. **1899** R. MUNRO *Prehistoric Scotl.* i. 2 At a subsequent date (600 B.C.) the Phocæans founded Massilia. **1909** W. W. FOWLER *Social Life at Rome* i. 9 At a later time of deep depression Horace could fancifully suggest that the Romans should leave their ancient home like the Phocaeans of old. **1962** J. M. COOK *Greeks in Ionia & East* iv. 59 Phocaea lacked good arable land... But the Phocaeans had the benefit of a magnificent harbour.

B. *adj.* Of or relating to Phocæa or its inhabitants.

1614 A. GORGES tr. *Lucan's Pharsalia* v. 168 And to Massilians (Cæsars spoyle) They freely gave the Phocean soyle. **1797** *Encycl. Brit.* XIV. 613/2 The Massilienses, a Phocaean colony, interposed, and with much difficulty, assuaged the anger of the senate. **1929** CARY & WARMINGTON *Anc. Explorers* ii. 22 The last stage of the Phocaean track was marked by a colony at Maenace. **1957** E. HYAMS *Speaking Garden* 163 Italy and France could have had wine and vines from one or both of two sources: Phœnician.., or Greek, by way of the Phocaean colony in Marseilles. **1966** G. E. BEAN *Aegean Turkey* v. 119 The Phocaean adventurers arrived hoping to establish a city.

Phocian (fōᵘ·siăn), *sb.* and *a.* Also 5 **Phocean,** 6 **Phocayan.** [f. Gr. Φωκί-ς the place-name Phocis, or L. *Phocii* Phocians + -AN.] **A.** *sb.* A native or inhabitant of the ancient region of Phocis in central Greece.

a **1490** J. SKELTON tr. *Diodorus Siculus' Bibliotheca Historica* (1956) I. I. 57 In Boecia the streme callyd Cifissus that cometh fro the Phoceans [*Note*] MS.: Thophoceans. **1567** A. GOLDING tr. *Ovid's Metamorphoses* XI. 141 Commes ronning thither all in haste and almost out of breth Anætor the Phocayan who was Pelyes herdman. **1612** SELDEN *Illustr. Drayton's Poly-olb.* x. 168 A colony of Phocians. **1774** GOLDSMITH *Grecian Hist.* II. ii. 55 The first cause of the rupture, (which was afterwards called the Sacred War) arose from the Phocians having ploughed up a piece of ground belonging to the temple of Apollo at Delphos. **1797** *Encycl. Brit.* XIV. 615/1 The Phocians afterwards formed themselves into a commonwealth. **1845** *Encycl. Metrop.* IX. 617/2 The Amphictyonic council..found themselves induced to impose on the Phocians a heavy fine for their impiety. **1951** J. B. BURY *Hist. Greece* (ed. 3) ix. 357 At the same time the Phocians entered into the alliance of Athens. **1967** H. W. PARKE *Oracles of Zeus* vii. 139 The Phocians had just occupied the Delphic sanctuary. **1968** V. EHRENBERG *From Solon to Socrates* 419 The Phocians heard the rustling of the leaves under the feet of the Persians before they could see them.

B. *adj.* Of or relating to Phocis or its inhabitants.

1614 A. GORGES tr. *Lucan's Pharsalia* v. 174 A garland greene of Laurell dight, With Phocian vaile of lawne pure white. **1774** GOLDSMITH *Grecian Hist.* II. ii. 72 Philip.. went on, according to his agreement with the Thebans, to put an end to the Phocian war. **1790** W. MITFORD *Hist. Greece* II. xiv. 160 Not only Delphi was again brought under Lacedæmonian influence, but the Phocian people were gained to the Lacedæmonian interest. **1845** *Encycl. Metrop.* IX. 618/2 Philomelus, the Phocian general, was succeeded by Onomarchus. **1911** *Encycl. Brit.* XXI. 448/2 The Dorian population of Delphi constantly strove to establish its independence and about 590 B.C. induced a coalition of Greek states to proclaim a 'Sacred War' and free the oracle from Phocian supervision. **1970** *Oxf. Classical Dict.* (ed. 2) 943/2 The *Second Sacred War* was precipitated by a Phocian seizure of Delphi.

phocid. Add to etym. after *phōca*: adopted as the name of a genus by Linnæus in his *Systema Naturæ* (ed. 10, 1758) I. 37. Also as *adj.*, of or pertaining to the family Phocidæ; (examples). **phocine** *a.* (later example).

1880 J. A. ALLEN *Hist. N. Amer. Pinnipeds* (U.S. Geol. Survey Misc. Publ. No. 12) 421 All the Phocids or Earless Seals known to systematic writers were referred to the common seal. *Ibid.* 470 Some supposed Phocine remains were described. *Ibid.* 740 It [*sc.* the hooded seal] is commonly described as the most courageous and combative of the Phocids. **1930** A. B. HOWELL *Aquatic Mammals* ii. 34 The otariids have descended from the bears and the phocids from the otters. *Ibid.* xi. 283 No great significance can be attached to details of the phocid tarsal bones. **1958** V. B. SCHEFFER *Seals, Sea Lions & Walruses* v. 87 Thirteen genera of phocids are recognized. **1970** *Sci. Jrnl.* Apr. 69/2 The so-called 'true' or phocid seals (which lack external ears and have small pectoral flippers) are not such rapid swimmers but can dive for longer times. **1976** H. L. GUNDERSON *Mammalogy* v. 122 The otariids and the phocids may have a common ancestor.

phocomelia (fōᵘkomī·liă). *Med.* Also **phokomelia.** [mod.L., f. Gr. φώκ-η seal + -O + μέλ-ος a limb + -IA[1].] A congenital defect of the limbs (see quot. 1892).

1892 F. P. FOSTER *Med. Dict.* IV. 2567/2 *Phocomelia*, a monstrosity in which one or both of the hands or feet, or all four of those members, or rudimentary semblances of them, are attached to the trunk like flippers, with little or no intervening structures. **1932** H. CUMMINS in *Practitioners Libr. Med. & Surg.* I. i. 20 In phocomelia the extremity is abbreviated, through reduction or absence of the proximal and intermediate segments, the hands and feet approaching normal. **1947** *Arch. Path.* XLIV. 521 The majority of the homozygous Creeper embryos die on the fourth day... Those which survive have phokomelia. **1968** *Brit. Med. Bull.* XXIV. 207/1 The demonstration of an association between phocomelia and thalidomide. **1974** *Jrnl. Embryol. & Exper. Morphol.* XXXI. 732 Embryos showing almost the complete absence of humerus, radius, and ulna, described previously as phocomelia. **1978** *Sci. Amer.* Oct. 130/3 The drug thalidomide could cause phocomelia..in a fetus when it was taken by a pregnant woman.

Hence **phocome·lic** *a.*, exhibiting or characteristic of phocomelia; also *ellipt.* as *sb.*

1942 *Jrnl. Exper. Zool.* LXXXIX. 104 These transplants differentiated into 'phocomelic' limbs. **1947** *Arch. Path.* XLIV. 521 In eyes of homozygous Creeper embryos transplanted to the flanks of normal embryos colobomas develop similar to those of phokomelic embryos. **1959** *Jrnl. Exper. Zool.* CXLII. 635 The lethal period of phocomelics. **1962** *Lancet* 1 Dec. 1155/2 No firm rule should guide schooling for phocomelic children. **1978** *Time* 20 Feb. 69/3 Children who are born legless or armless, their limbs amputated by a tangled umbilical cord, are sometimes hard to tell from true phocomelics, or seal-children, with vestigial hands and feet attached directly to the torso.

Phœbe[1]. Restrict *poet.* to sense in Dict. and add: **2.** *attrib.* and *Comb.*, as **Phœbe lamp** *N. Amer. Hist.* (see quots. 1935 and 1970).

1935 *Colony of Connecticut* (Conn. Board Educ.) (Senate Doc. 53, 74th Congr., 1st Sess.) 15 Phoebe lamps: ..These were similar to Betty lamps in shape... Some had double wicks from a nose on either side. **1970** G. PAYTON *Webster's Dict. Proper Names* 525/2 *Phoebe lamp*, American pioneers' name for a primitive tallow-and-wick lamp of the type used down the ages, e.g. by the Ancient Greeks. **1972** F. VAN W. MASON *Roads to Liberty* 182 Betty..lit a Phoebe lamp with a splinter from the fire.

phœbe[2]. Add: Also **phebe.** Substitute for def.: A small North American flycatcher of the genus *Sayornis*, esp. *S. phœbe*, which is common in the eastern part of the continent. (Earlier and later examples.)

1700 J. GREEN *Jrnl.* 4 Mar. In *Essex Inst. Hist. Coll.* (1866) VIII. 216 Cloudy & rainy. heard a Phebe and other birds sing. **1782** 'J. H. ST. JOHN DE CRÈVECŒUR' *Lett. from Amer. Farmer* ii. 40 That [*sc.* a nest] of a swallow was affixed in the corner next to the house; that of a phebe in the other. **1947** E. B. WHITE *Lett.* (1976) 284, I haven't been doing much of anything—just..watching phoebes through binoculars, and mixing drinks. **1961** O. L. AUSTIN *Birds of World* (1962) 209/1 Another flycatcher that announces itself is the Phoebe, the pert olive-grey bird that plasters its mud and moss nest under bridges over country streams.

Phœnician, *sb.* **2.** (Further examples.)

1836 N. WISEMAN *Twelve Lect. on Connexion betw. Sci. & Revealed Relig.* I. i. 53 Cadiz, or *Gadir*, as it was originally called, must no longer signify, as the word does graphically in Phenician, *the island* or *peninsula*. **1861** *Dublin Rev.* Feb. 400 Joseph Scaliger demonstrated that the well-known passage of the Pœnulus of Plautus was a fragment of genuine Phœnician. **1948** E. POUND *Pisan Cantos* (1949) lxxix. 76 'Prepare to go on a journey.' Or to count sheep in Phoenician. **1965** *Illustr. London News* 13 Feb. 22/3 Three thin rectangular sheets of gold leaf... One of them is inscribed in Phoenician, the other two in Etruscan.

phœnix[1], **phenix.** Add: **5.** (*a*) *phœnix riddle*; also passing into adj. (= phœnix-like; as of a phœnix), as *phœnix-birth, -fuel, -life, -moon, -pyre, -resurrection, -tinder, -world.*

1946 R. CAMPBELL *Talking Bronco* 69 The proud Alcazar caught the fire Which gave that splendour phoenix-birth. **1977** *Listener* 10 Nov. 617/1 His radio ballads, combining actuality voices, music and sound effects were a phoenix birth at a dark hour. **1936** R. CAMPBELL *Mithraic Emblems* 50 True phoenix-fuel whom no burning mars. *a* **1957** —— tr. *Quevedo's On Lisi's Golden Hair* in *Coll. Poems* (1960) III. 83 Out of their ash to fan new phoenix-lives. **1934** L. B. LYON *White Hare* 29 The phoenix moon with molten breast. **1930** R. CAMPBELL *Adamastor* 78 And now from the wet earth reborn, All Africa his phoenix pyre. **1963** L. TRILLING in N. Frye *Romanticism Reconsidered* 88 When he [*sc.* Keats] is 'consumed in the fire', they will contrive his Phoenix-resurrection. *a* **1631** DONNE *Canonization* in *Poems* (1633) 203 The Phoenix ridle hath more wit By us, we two being one, are it. **1939** R. CAMPBELL *Flowering Rifle* III. 74 So their black chaos is but welcome fuel And phoenix-tinder to this fierce renewal. **1944** W. TEMPLE *Let.* 22 Feb. (1963) 147 The reforms necessary for the arising of that brave new phoenix-world on which we have set our hearts.

phokomelia, var. *PHOCOMELIA.

Pholey, var. *PEULH *sb.* and *a.*

phon (fǫn). [a. G. *phon* (H. Barkhausen 1926, in *Zeitschr. für techn. Physik* VII. 601/1), f. Gr. φωνή sound.] A unit of loudness (strictly, loudness level), defined so that the loudness in phons of any sound is numerically equal to the intensity in decibels of a pure 1000 Hz tone judged to be equally loud. Formerly = *DECIBEL.

1932 *B.B.C. Techn. Tables & Gloss.* 63/2 Two noises may be compared in strength in sensation units or phons by comparing the attenuations necessary to reduce each to the threshold of audibility. **1932** *Engineering* 9 Sept. 314/3 For quantitative measurements we turn to a unit called the decibel. [*Note*] Called the 'phon' in Germany. **1935** *Sunday Times* 13 Oct. 18/3 Now a machine exists which measures units of noise in 'phons', and no vehicle will be allowed after a certain date to make more than 90 phons of noise at a speed of 30 m.p.h. **1937** A. H. DAVIS *Noise* iv. 46 Originally a 'phon' had been used in Germany as a four-fold power ratio. Later it was employed as an equivalent of the decibel. *Ibid.*, Limitation of the decibel to intensity ratios and the phon to loudness scales has been adoped by the British Standards Institution. **1942** *Wireless World* June 132/2 The magnitude of these upper intensity levels in terms of loudness can be expressed by writing phons for decibels, since they are equal, in the range here considered. **1959** G. R. PARTRIDGE *Princ. Electronic Instruments* xvi. 303 A loudness of 70 phons requires 80 db intensity at 70 cycles. **1962** A. NISBETT *Technique Sound Studio* 264 Phons equal decibels at 1000 c/s, and at other frequencies are related to this scale by contours of equal loudness. **1963** JERRARD & MCNEILL *Dict. Sci. Units* 104 The loudness of a jet aircraft engine is about 140 phon, whereas the noise of a steam railway locomotive is less than 100 phon. **1974** *Daily Colonist* (Victoria, B.C.) 22 Nov. 16/5 Tokyo metropolitan government standards set permissible noise levels at 40 phon (sound measurement) in the morning and evening and at 45 phon during the day.

phonæstheme (fōᵘ·nėsþīm). *Linguistics.* Also **phonaestheme, phonestheme.** [f. PHONE *sb.*[1] + ÆSTH(ETIC *a.* and *sb.* + *-EME.] A phoneme or group of phonemes with recognizable semantic associations due to recurrent appearance in words of similar meaning.

1930 J. R. FIRTH *Speech* vi. 50 The *slack* etymeme belongs to a much bigger group of habits we may call the *sl* phonæstheme. *Ibid.* vi. 53 The habit background of *wirl*..probably includes the *tw* and *irl, -url* phonæsthemes. **1946** *Word* II. 83 Phonestheme is defined as a phoneme or cluster of phonemes shared by a group of words which also have in common some element of meaning or function, tho [*sic*] the words may be etymologically unrelated. **1957** *Gen. Linguistics* II. 55 Previous studies have erred in attempting to separate the long consonant phonesthemes from the other phonesthemes of the language. **1969** *Language* XLV. 284 The term 'phonaestheme' has been used, but it seems to mean different things for those who have used it. **1972** M. L. SAMUELS *Linguistic Evol.* iii. 46 The phonaestheme /sl-/ may be assigned the values 'slippery' or 'falling' in *slide, slip, slime, slush*..; and it may also be assigned the closely related values 'inactive', 'degenerate' or 'morally worthless' in *slow, sloth, sleep(y), slumber, slack, slouch*, [etc.]. **1974** *Amer. Speech 1971* XLVI. 129 The word-initiating segments *sn-* and *sl-* (as in *snit* and *slit*), which, because they are not readily identifiable as morphemes or other grammatical units, are often referred to as 'phonesthemes'. **1977** *Word 1972* XXVIII. 305 Phonesthemes are considered to have meaning through their occurrence in words of a particular meaning.

Hence **phonæsthesia** (fōᵘnėsþī·siă, -ziă) [*ÆSTHESIA], **phonæsthesis** (fōᵘnėsþī·sis, -zis) [ÆSTHESIS], sound symbolism; the use of phonæsthemes; **phonæsthetic** (fōᵘnėsþe·tik) [ÆSTHETIC *a.*] *a.*, of or pertaining to phonæsthemes; **phonæsthe·tically** *adv.*

1930 J. R. FIRTH *Speech* vi. 52 The phonæsthetic habits..are much more than 'blends'..and are of general importance in speech. *Ibid.* 54 Play on phonæsthetic habits gives much of the pleasure of alliteration, assonance, and rhyme. **1950** *Archivum Linguisticum* II. 97 Phonaesthesia or sound symbolism. **1965** *Language* XLI. 347 Morphophonemics, phonesthesis, and paralanguage she wholly ignores. **1971** *Archivum Linguisticum* New Ser. II. 43 Some verbs, for example, do not occur without aspectival..or directional..extension; *shamble*, for instance, requires the 'complementation' of *off, on, up, down, across*, etc., a fact to which its phonaesthetic overtones of *amble, shuffle*,.. may or may not relate. **1972** M. L. SAMUELS *Linguistic Evol.* iii. 48 The growth of phonaesthetic patterns is of importance as a special type of linguistic change. *Ibid.* xiii. 161 In some verbs where a short vowel had been developed phonetically, it was phonaesthetically suited for the expression of point-action. **1977** *Maledicta* Summer 65 (*heading*) Phonesthesis and scatology.

phonation. Add: Also *spec.* the process or act of producing voice (VOICE *sb.* 1 g). (Further examples.)

1931 *Musical Times* Jan. 18/2 The simplest act of speech involves the co-ordination of three very complicated sets of muscles: those of inspiration and expiration; those of phonation (the intra- and extra-laryngeal muscles); and those of articulation. **1935** *Jrnl. Mental Sci.* LXXXI. 837 The muscles of respiration, phonation, and articulation. **1956** J. WHATMOUGH *Language* ix. 157 It [*sc.* language] is now articulated (i.e. jointed) phonation, though originally simple and reflex. **1959** E. PULGRAM *Introd. Spectrogr. of Speech* iv. 43 Many experiments have been performed..to measure subglottic pressure during phonation. *Ibid.* xviii. 140 *Fig. 30* illustrates four phonations of the vowel [i]. **1962** A. C. GIMSON *Introd. Pronunc. Eng.* ii. 9 The action of the vocal cords which is most characteristically a function of speech consists in their role as a *vibrator* set in motion by lung air—the production of *voice*, or *phonation*. **1964** J. C. CATFORD in D. Abercrombie et al. *Daniel Jones* 28 Various components of speech-sound pronunciation, including phonation-types. **1970** *Language* LXVI. 313 The human larynx is so constructed that the fundamental frequency of phonation is a function of both the transglottal air pressure drop and the tensions of the laryngeal muscles. **1976** *Canad. Jrnl. Linguistics* XXI. I. 118 A most informative explanation of air-stream mechanisms and phonation times, including voice-onset-time.

Hence **phona·tional** *a.*, = PHONATORY *a.*

1939 L. H. GRAY *Foundations of Lang.* i. 5 Persons whose phonational and auditory apparatus are normal. *Ibid.* ii. 18 This condition..is the auditory counterpart of the pathological forms of the phonational soliloquy. **1947** *Word* III. 2 The phonational act..gives rise, in the hearer, to an acoustic image which is distinct from the physical sound.

phone, *sb.*[1] Add: (Earlier and later examples.) Also = *ALLOPHONE.

a **1866** J. GROTE in *Jrnl. Philol.* (1872) IV. 55 When I mean words as sounded I shall use the term *phone* (like *zone*, φωνή, ζώνη). **1924** H. E. PALMER *Gram. Spoken Eng.* I. 1 A phone may be a simple sound, such as [f]..or it may be an intimate combination of simple sounds, such as [tʃ]... On the other hand, sounds such as non-significant glides are not phones. **1942** *Language* XVIII. 8 A phone is a member of a phoneme. **1950** *Ibid.* XXVI. 90 In the latter meaning, the term 'phone' is a shorter equivalent of 'allophone'. **1957** [see *ALLOPHONE]. **1959** E. PULGRAM *Introd. Spectrogr. of Speech* xix. 145 All those phones, that is, events of articulation, which are linguistically identifiable with one another though not acoustically identical with one another, belong to the same class, which we call phoneme. **1961** W. R. BRAIN *Speech Disorders* 9 A phone is the name for a single isolable sound made in the course of speech by a speaker. **1964** [see *ALLOPHONE]. **1971** D. CRYSTAL *Linguistics* iv. 178 The term 'phoneme' had been used in the nineteenth century, but it referred to a unit of sound (that is, a phonetic unit—what we would usually these days call a 'phone'), and not to any abstract notion involving contrastivity. **1975** L. M. HYMAN *Phonology* 8 The units of phonetic description are sound segments (or *phones*), while the units of phonological description are *phonemes*. **1976** *Language* LII. 317 In the early distinctive-feature model, phonemes are redundancy-free versions of the corresponding phones.

2. Special Combs., as **phone-type,** a speech-sound considered as a purely phonetic event.

1961 F. W. HOUSEHOLDER in Saporta & Bastian *Psycholinguistics* 19/2 Biuniqueness..means that to any given phone-type in a given environment there must correspond only one possible phoneme, and to any phoneme in a given string there must correspond only one phone-type. *Ibid.*, It is frequently simpler and more elegant to have units so chosen that a given phone-type in one environment may be an allophone of one such phoneme, but in another an allophone (or the only allophone) of a different one. **1965** R. L. KELLOGG in Bessinger & Creed *Medieval & Linguistic Stud.* 67 We might add that the fixed formula is to the abstract pattern of which it is a manifestation as the phonetype is to the phoneme.

phone (fōᵘn), *sb.*[2] Add: Also **'phone. 1.** (Further examples.) Also, a telephone call.

1922 J. REITH *Diary* 6 Nov. (1975) i. 87 Many phones from the National Liberal Headquarters asking for speakers. **1942** BERREY & VAN DEN BARK *Amer. Thes. Slang* §808/5 *Telephone call*,.. phone. **1943** J. THURBER *Men, Women & Dogs* 105 If I called the wrong number, why did you answer the phone? **1949** N. SPAIN *Death goes on Skis* III. viii. 186 For an intellectual she seems to be quite illiterate..saying 'phone for telephone. **1960** K. AMIS *Take Girl like You* xxvii. 306 A telephone was ringing... 'Jenny, someone on the phone for you.' **1974** M. SPARK *Abbess of Crewe* i. 30 Mildred treads softly over the green carpet..and answers the phone. **1976** *Daily Tel.* 13 Nov. 17/3 A 19-year-old youth made an estimated 1,000 fraudulent telephone calls..and was on the 'phone for 63 hours.

2. Colloq. abbrev. of *ear-phone* s.v. *EAR *sb.*[1] 16, *HEAD-PHONE. Usu. *pl.*

1913 *Wireless World* May p. v (Advt.), High resistance 'phones. **1926** R. MACAULAY *Lett. to Sister* (1964) 27 No longer..does the husband have to sit in the evenings and listen to inanities from his wife..; he and she can now both sit in silence, with the phones on their ears. **1948** *Electronics* Aug. 88/2 A person listening to sound through a binaural system has the illusion that the sound originates in the room rather than in the phones. **1962** A. NISBETT *Technique Sound Studio* 256 Some degree of control of volume can be achieved with telephone receiver type phones by moving them a little off the ear. **1970** J. EARL *Tuners & Amplifiers* iii. 76 It also pays for the amplifier etc. to carry a loudspeaker switch to cut the speakers when listening on 'phones. **1977** *Rolling Stone* 24 Mar. 79/2 Played back over the phones part of the system..the sense of space and realism of sound is unbelievable.

3. *attrib.* and *Comb.* (in sense 1), as *phone bell, call, caller, exchange* [*EXCHANGE *sb.* 10c], *installation, jack* [*JACK *sb.*[1] 15d], *kiosk, message, operator* [*OPERATOR 5a], *order, receiver, wire; phone-answering* ppl. adj. and vbl. sb.; **phone bill,** an account for the cost of hire of a telephone and of calls made from it; **phone book,** a telephone directory; **phone booth, box,** a box-like kiosk in which a public telephone is installed; **phone number,** the identifying call-number assigned to a telephone, line, etc.; **phone patch** [*PATCH *sb.*[1] 6f], a temporary radio link made to establish communication between a radio operator and a telephone user; **phone-tapping** *vbl. sb.* = **telephone-tapping* vbl. sb.; so **phone-tap** *sb.* and *v. trans.*; **phone-tapper.**

1976 *New Yorker* 23 Feb. 30/2 The Silverstein headquarters..has a phone-answering machine that will deliver a tape-recorded message from the candidate. **1977** E. AMBLER *Send no more Roses* v. 111 Business accommodation services which provided mail-forwarding and phone-answering. **1937** M. ALLINGHAM *Dancers in Mourning* xxv. 300 The 'phone bell alone was silent and everyone..was waiting for that shrill, familiar alarum. **1950** T. WALSH *Nightmare in Manhattan* (1951) III. 86 The phone bell had given him a first intuitive comprehension. **1972** 'H. HOWARD' *Nice Day for Funeral* v. 82 It was the phone bell that prodded me awake. **1965** N. FREELING *Criminal Conversation* I. ix. 63 Sounds like your night for collecting a phone bill. Why not do this in the office—then you wouldn't have to pay for them? **1972** *Guardian* 24 Aug. 20/5 He..had to sell his camera to pay his phone bill. **1925** F. SCOTT FITZGERALD *Great Gatsby* (1926) ix. 198 Meyer Wolfshiem's name wasn't in the phone book. **1963** 'E. MCBAIN' *Ten Plus One* (1964) ii. 23 Carella picked up the phone book and looked up the number. **1977** G. SCOTT *Hot Pursuit* vi. 56 Give us a ring. We're in the phone book. **1927** W. R. JAMES *Cow Country* vii. 199 He came back in the hotel and went in the phone-booth and there he proceeded to call them up, one after another. **1952** S. KAUFFMANN *Philanderer* (1953) xi. 186 He had to fret away three or four minutes outside the busy phone booth. **1976** 'M. DELVING' *China Expert* vi. 72 One of the phone booths across the street. **1954** M. STEWART *Madam, will you Talk?* xvi. 127 There's a phone box a mile or so along the road. **1968** 'R. RAINE' *Night of Hawk* xiv. 75 Henry made an anonymous call to the police from a near-by phone box. **1977** D. JAMES *Spy at Evening* ix. 61, I left the hot-house atmosphere of the phone box. **1929** D. HAMMETT *Dain Curse* (1930) xv. 166, I wasn't convinced that my phone call was of any importance. **1959** N. MAILER *Advts. for Myself* (1961) 198 About a month later, this letter was followed by a phone call. **1977** W. MCILVANNEY *Laidlaw* xxxviii. 179 There had been a funny phone-call for me when I got back.. checking that I could still be reached here. **1972** J. PHILIPS *Vanishing Senator* (1973) II. ii. 67 Some of those letter writers and phone callers might be willing to help. **1939** R. CAMPBELL *Flowering Rifle* v. 131 In the munition-works, the 'phone-exchange. **1924** H. CRANE *Let.* 29 Jan. (1965) 173 Putting down money for a phone installation. **1946** P. CARTER in *Astounding Sci. Fict.* Aug. 49/2 He pulled out the phone jack, plugged it in elsewhere. **1973** *New Yorker* 14 Apr. 32/2 It will have phone jacks, so the staff will be able to answer queries without going back to a central desk. **1976** M. MACHLIN *Pipeline* ii. 31 There was a phone jack just behind his chair in the dining room. **1968** R. V. BESTE *Repeat Instructions* xv. 154 He..went out to the phone kiosk and arranged to see King. **1977** G. MARKSTEIN *Chance Awakening* lv. 169 That silent street corner by the 'phone kiosk. **1930** M. ALLINGHAM *Mystery Mile* xxiii. 217 He overheard a 'phone message. **1955** *Times* 23 July 4/4 Norbury knows nothing about this. In all the 'phone messages I have impersonated him. **1978** M. BIRMINGHAM *Sleep in Ditch* 117 He has no address now, just gets a phone message once a month telling him where to send a money order. **1911** G. STRATTON-PORTER *Harvester* xvii. 393, I want Dr. Frank Harmon... I don't know the 'phone number. **1960** C. MACINNES *Mr. Love & Justice* 51 I'm a seaman... I've got stacks of foreign phone numbers in my diary. **1971** R. RENDELL *One Across* vi. 51 One day he'd walk in..to find the lot of them gone and a note on the table with a Chigwell phone number on it. **1973** E. HYAMS *Final Agenda* vi. 82 Call that phone operator and make sure she knows English, French or German. **1932** *New Yorker* 4 June 7 (Advt.), Mail and phone orders filled. **1975** *Ibid.* 20 Oct. 38/1 Martin finds phone-patch relays difficult; he forgets to say 'Over' when it's time for his father to speak, and he can't get used to the idea of not being able to interrupt. **1976** PERKOWSKI & STRAL *Joy of CB* xii. 142 A phone patch links a mobile rig to a land line through a base station. The mobile user contacts a base station with a phone patch. The base station operator then places the call, and connects the land line to the base's rig, placing the telephone handset in a cradle device or otherwise establishing an electrical connection with the rig. Once the connection is made, a voice-operated transmitter relay or a switch operated by the base station operator transmits the voice of the land-line party through the base transmitter. **1970** R. LOWELL *Notebk.* 191 He's trying to part his hair on the phone-receiver. **1959** *Daily Tel.* 10 Dec. 1/8 Motion of censure on 'phone tap'. **1966** 'G. BLACK' *You want to die, Johnny?* vii. 130, I didn't think he'd known about the phone tap. **1973** J. WAINWRIGHT *Pride of Pigs* 169 No sweat, princess. They don't phone-tap too easily in this country, so we're okay. **1976** W. GREATOREX *Crossover* 193 Meade had just told him of the phone-tap report. *Ibid.* 191 She said it in Russian knowing that the phone tappers would have to send the tape to the linguists. **1957** *Times* 12 June 9/5 We are thus, it seems, to have phone-tapping on suspicion. **1976** R. PERRY *One Good Death* v. 68 Phone-tapping had become one of the essential props of modern government. **1951** Phone wire [see *PHONEVISION]. **1964** L. DEIGHTON *Funeral in Berlin* xxxvii. 227 Two G.P.O. phone wires had been brought in..and could be attached to the handset. **1978** R. HILL *Pinch of Snuff* v. 44 The phone wire was cut.

phone (fōᵘn), *v. colloq.* Also **'phone.** [Abbrev. TELEPHONE *v.*] **a.** *trans.* = TELEPHONE *v.* 1c. Also const. *up* and with advbs.

1889 *Telephone* 1 Feb. 56/1 The expression 'I telephoned So-and-So', is often rendered 'I phoned So-and-So.' **1900, 1901** [in Dict.]. **1909** *Daily Chron.* 10 Dec. 7/2 He could 'phone up Scotland Yard for a detective. **1910** 'W. LAWTON' *Boy Aviators on Secret Service* ii. 22 Wait a minute while I go to 'phone my resignation. **1962** I. MURDOCH *Unofficial Rose* xvi. 159 Was it really necessary to phone me? **1963** L. DEIGHTON *Horse under Water* xliv. 181, I couldn't have Charley phoning up the police. **1974** A. ROSS *Bradford Business* 146 [He] offered me a tuppenny piece... 'Take it,' he said testily, 'and phone her back.' **1976** *Daily Tel.* 2 Nov. 17/2, I 'phoned Paris, and explained what had happened.

b. *intr.* = TELEPHONE *v.* 1a. Freq. with particle as phrasal verb or with advbs.

1925 F. SCOTT FITZGERALD *Great Gatsby* viii. 193 Gatsby..left word with the butler that if any one phoned word was to be brought to him at the pool. **1926** E. O'NEILL *Great God Brown* III. i. 73 Shall I phone for a doctor? **1927** *Rev. Eng. Stud.* Oct. 433 Phrases like *e.g. phone through* and *come across* no more represent 'degradation' than do *give up, give in, ring up, send off, think out.* **1932** T. S. ELIOT *Sweeney Agonistes* 13 She says will you ring up on Monday... All right, Monday you'll phone through. **1946** E. O'NEILL *Iceman Cometh* IV. 231 It was Hickman himself phoned in and said we'd find him here around two. **1955** *Times* 11 June 4/3 If you will 'phone for an appointment when the strike is over, I will try to fix a convenient date. **1959** I. JEFFERIES *Thirteen Days* viii. 105 He phoned around and..within an hour I had a motley but effective army. **1962** L. DEIGHTON *Ipcress File* xvii. 101 Phone in this time tomorrow. **1972** C. FREMLIN *Appointment with Yesterday* vii. 56 You've hardly been in the place two hours, and she has to phone up. **1977** F. BRANSTON *Up & Coming Man* xiii. 147 Andy and I took turns to phone round... Nearly every news desk called back.

c. *trans.* Const. *in.* = TELEPHONE *v.* 1b.

1972 *Radio Times* 28 Dec. 53/1 Parents phone in their questions on the three Rs to Miss Edith Biggs..and Ronald Palmer. **1973** *Black World* Mar. 15 (*caption*) Tell him to hurry up—I've got to phone in a bomb threat. **1977** *Gay News* 24 Mar. 20/1 We have many flats and bed-sits with understanding landlords available now. (Landlords, please phone in your vacancies!)

Hence **pho·ner** = TELEPHONER; also *phoner-in*; **pho·ning** *vbl. sb.* = TELEPHONING *vbl. sb.*

1908 *Daily Chron.* 10 Aug. 7/1 So graphically had Enid done her bit of descriptive 'phoning that he was under no illusions as to what he had to do. **1942** BERREY & VAN DEN BARK *Amer. Thes. Slang* §808/3 *Telephoner*, phoner. **1957** 'A. GARVE' *Narrow Search* iii. 80 I'll have to do a bit of phoning in the morning. **1972** *Listener* 27 Jan. 124/1 An excellent job of training phoners to that programme by classic reward and punishment methods. **1975** *Times* 26 Apr. 8/5 Radio 4's *The Forbidden Subject* was..a phone-in... The phoners-in, with two exceptions, were all women. **1977** *Belfast Tel.* 24 Jan. 3/5, I hope that phoners-in are not an accurate cross-section of the population. **1977** *Times* 12 Feb. 12/1 This is your late night phone-in programme... Mr Jenkins..says you phoners are ignorant, pathetic and moronic.

-phone (fōᵘn), ad. Gr. φων-ή voice, φῶν-ος sounding. **1.** Used in the sense 'sound' in the names of various instruments (scientific and musical), as GRAMOPHONE, MAGNETOPHONE, MEGAPHONE, *MELLOPHONE, MICROPHONE, *VIBRAPHONE.

2. Used in the sense 'speaker of' or '-speaking' in the formation of nouns and adjectives from Latinate combining forms of names of peoples and languages, as *Anglophone, Bulgarophone*, *FRANCOPHONE *sb.* and *a.*, *Turcophone.*

1900 Anglophone [see *FRANCOPHONE *sb.* and *a.*]. **1937** Bulgarophone [see *EXARCHIST]. **1977** *Times Lit. Suppl.* 18 Mar. 295/5 His description of Turcophones in Iran (of whom he is one) as a 'nation'.

phone freak, var. *PHONE PHREAK *sb.* and *v.*

phone-in (fōᵘ·nin). [f. phr. *to phone in*, f. *PHONE *v.* + IN *adv.*; for sense 1 see also *-IN[3].] **1.** A protest in the form of mass telephone calls of complaint.

1967 *New Statesman* 17 Mar. 356/3 In San Francisco.. Mrs. Whitehouse has perfected the phone-in. **1973** *Times* 8 Feb. 6/1 Actress Dame Peggy Ashcroft plans to join a massive 'phone-in'..aimed at jamming the switchboard of the Department of the Environment.

2. A 'live broadcast' radio or television programme during which listeners telephone the studios to ask questions or express their views. Also, this type of broadcasting. Also *attrib.*

1968 *Time* 29 Mar. 16 He proposed reducing transit fares for San Franciscans over 65 to 5¢ and, on a subsequent TV 'phone-in', said he would try to get buses closer to the curb at pickup. **1971** *Listener* 30 Dec. 915 One of the most exciting potentials this year has been the phone-in. **1972** *Guardian* 13 Mar. 11/2 Duke Miller of St Petersburg asked him on a phone-in show..how voters were expected to forgive him for his part in the escalation of the war. **1974** *Daily Tel.* 30 Jan. 13/1 The arrival of phone-in conveniently coincided with the fashion for public 'participation and access' in broadcasting. *Ibid.* 11 Feb. 5/5 There will be a daily 55-minute 'phone-in' on Radio 4 to 13 leading figures from the main Parties. **1977** B. PYM *Quartet in Autumn* xiv. 118 Listening to a phone-in programme on the radio she had heard a question about holidays for people on their own.

phonematic (fōᵘnėmæ·tik), *a.* and *sb. Linguistics.* [f. Gr. φωνήματ- stem of φώνημα sound made + -IC.] **A.** *adj.* **a.** = *PHONEMIC *a.*

1936 *Proc. 2nd Internat. Congr. Phonetic Sci.* 50, Z, considered as a phonematic unit, has a value very different from the value it has when considered as a grammatical unit. **1936** *Language* XII. 311 Professor J. Vendryes..announced that competent legal opinion indicated the desirability of organizing separate national groups..for the advancement of phonological (phonematic) studies. **1949** *Trans. Philol. Soc. 1948* 129 We may speak of a five-vowel or seven-vowel phonematic system. **1952** A. COHEN *Phonemes of Eng.* i. 12 An implication is the replacing of one phoneme by another under definite phonematic conditions. **1956** J. WHATMOUGH *Language* v. 84 The strange words that a language adopts are adapted to its own phonematic and other patterns. **1958** *Proc. 8th Internat. Congr. Linguists* 763 Quechua, a language with a phonematic system of 3 vowels only. **1962** P. S. RAY in F. A. Rice *Study of Role of Second Lang. in Asia, Afr. & Lat. Amer.* 101 Even one who prefers to write 'phonemic' rather than 'phonematic' is doing a bit of prescriptive intervention. **1977** A. SHERIDAN tr. *Lacan's Écrits* iv. 126 The elements of the language (*langue*) at its different levels, from the phonematic pair of oppositions to the compound expressions to disengage the stable forms of which is the task of the most modern research.

b. In 'prosodic analysis', designating a segmental element of vowel or consonant features which combines with prosodies (see *PROSODY 2*).

1949 [see *PROSODY 2*]. **1964** R. H. ROBINS *Gen. Linguistics* iv. 159 *Prosodic analysis*, as it is usually called, is a brief title for a method of phonological analysis that employs as fundamental concepts two types of element, not reducible to a common type, prosodies and phonematic units. **1968** J. LYONS *Introd. Theoret. Linguistics* iii. 129 The sequence of phonematic units constitutes the segmental 'infrastructure' of the word, whereas the prosodies form its 'superstructure'. **1972** HARTMANN & STORK *Dict. Lang. & Linguistics* 187/2 At first sight phonematic units and prosodies seem to be equivalent to segmental and suprasegmental phonemes of the traditional phonemic analysis, but in a prosodic analysis, features which would be assigned to segmental phonemes in a phonemic analysis are sometimes assigned to prosodies, e.g. such features as palatalisation, nasalisation or lip rounding.

B. *sb. pl.* [see -IC 2.] **a.** Phonemics.

1936 *Proc. 2nd Internat. Congr. Phonetic Sci.* 49 By *phonematics* I understand a science which treats phonemes exclusively as elements of language. **1939** L. H. GRAY *Foundations of Lang.* iii. 62 The phoneme..is..regarded as..a point in the linguistic (grammatical) pattern ('phonematics'). **1949** *Amer. Speech* XXIV. 54 It [*sc.* the term *phonemics*] has replaced in American usage the terms *phonology* and *phonematics*. **1953** W. J. ENTWISTLE *Aspects of Lang.* iii. 94 The chapter-heading *Sounds* is used to cover..questions affecting the pattern (phonematics, not the ungrammatical 'phonemics'). **1960** J. VACHEK *Dict. de Linguistique de l'École de Prague* 61/1 *Phonologie* (Phonemics; less frequently, Phonematics/Phonology..). **1961** *Brno Studies in English* III. 55 In his well-known compendium of diachronistic phonematics, ..A. Martinet rightly insists on the presence in any language of two opposed forces. **1964** E. PALMER tr. *Martinet's Elem. Gen. Linguistics* i. 30 They form a chapter entitled prosody distinct from phonematics, which treats of the units of the second articulation.

b. That part of 'prosodic analysis' which deals with phonematic units.

1971 *Archivum Linguisticum* New Ser. II. 68 'Phonematics' (a term which bears only etymological resemblance to 'phonemics').

So **phonema·tically** *adv.*, in relation to phonemes or phonematic units; according to the theory of phonemes or phonematic units; **phonematiza·tion,** advancement from allophonic to phonemic status; **phonemato·logy,** phonemics, phonematics (sense *a).

1949 *Archivum Linguisticum* I. 41 There is, further, a well-known problem in English phonematology. **1950** *Ibid.* II. 181 The book [*sc.* D. Jones's *The Phoneme*] is.. essentially a hand-book of phonematology. **1951** S. ULLMANN *Princ. Semantics* ii. 47 With reference to phonematically constituted word-engrams. **1956** *Trans. Philol. Soc.* 34 There are distinctions in the written language which are phonematically motivated. **1956** J. WHATMOUGH *Language* i. 11 If you hear *pwivate* 'private' from a few speakers [w] is a variant, phonematically speaking, of [r]. **1958** A. S. C. ROSS *Etym.* i. 24 Phonematology has one extremely practical application, that is, the construction of alphabets for languages hitherto unwritten. **1962** *Canad. Jrnl. Linguistics* VII. II. 79 When such adaptations become loanwords, however, they sometimes are phonematically homophonous with native words. **1964** *Amer. Speech* XXXIX. 70 Phonematization of /u/ in eastern dialects of Basque resulted from changes [o] → [u], [u] → [ü]. **1971** *Archivum Linguisticum* New Ser. II. 85 With the exception of the 3 sg. general tense ending.., all the endings can be analyzed phonematically as the same for each person regardless of tense. **1977** *Word 1972* XXVIII. 250 Automatic nasalization of vowels in contact with nasal consonants is a general characteristic of languages in which there are no phonematically nasalized vowels. **1978** *Language* LIV. 231 Phonematization relies on the examination of contrasts.

phoneme. Restrict *rare* to sense in Dict. and add: Now with pronunc. (fōᵘ·nīm). **1. b.** A phonological unit of language that cannot be analysed into smaller linear units and that in any particular language is realized in non-contrastive variants. Also *attrib.* See *ALLOPHONE.

Although its exact nature is disputed, and the existence of an abstract phonemic level (and hence the abstract phoneme as a constituent of morphemes) is controversial in phonological theory, the phoneme remains a standard taxonomic unit in the description of speech. The phoneme of structural linguistics is sometimes called the *autonomous* or *taxonomic phoneme* by generative phonologists, and distinguished from the *systematic phoneme*.

1896 R. J. LLOYD in *Die Neueren Sprachen* III. 615 There are a few of these terms which the author [*sc.* J. Baudouin de Courtenay] still thinks valuable and retains. One of these is the term *phoneme*, invented by Kruszewski. ..I take it that the various sounds which are accepted as identical by any speaking community are one *phoneme*, though they may differ considerably in actual sound. **1917** D. JONES in *Trans. Philol. Soc. 1917–20* (1932) 99 The Sechuana language appears to contain twenty-eight phonemes, i.e. twenty-eight sounds or small families of sounds which are capable of distinguishing one word from another. **1928** I. C. WARD *Phonetics of English* 61 Care must be taken not to confuse the various members of the phoneme with the different pronunciations heard from different people. The latter may be termed *variant pronunciations*; the different sounds constituting a phoneme occur in *one* person's pronunciation. **1933** L. BLOOMFIELD *Language* v. 80 The phonemes of a language are not sounds, but merely features of sound which the speakers have been trained to produce and recognize in the current of actual speech-sound. **1935** G. K. ZIPF *Psycho-Biol. of Lang.* i. 20 A word may consist of a single phoneme..or it may represent a phoneme-sequence of considerable magnitude. **1936** *Amer. Speech* XI. 109 When in a given language two sounds occur in the same phonological conditions and neither of the two sounds may be substituted for the other without changing the meanings of the words, the two sounds are capable of differentiating two words and are realizations of two different phonemes. **1939** L. H. GRAY *Foundations of Lang.* iii. 61 The exact nature of the phoneme is disputed. It is variously regarded as 1) a mere grouping of sounds..2) as a point in the psychological pattern (..Sapir and the Prague School); or 3) as a point in the linguistic (grammatical) pattern. **1950** D. JONES *Phoneme* ii. 10 A phoneme is a family of sounds in a given language which are related in character and are used in such a way that no one member ever occurs in a word in the same phonetic context as any other member. *Ibid.* 12 Some phoneticians have employed the term phoneme to mean anything that may serve to effect a minimal distinction..between one word and another. **1958** K. AMIS *I like it Here* i. 6 To cut Sinatra off in mid-phoneme was not such uproarious fun. **1961** W. R. BRAIN *Speech Disorders* i. 10 The letters of an alphabet transcribe with more or less precision the phonemes of speech, of which standard English has about thirty-two. **1962** A. C. GIMSON *Introd. Pronunc. Eng.* v. 44 It is possible to establish the phonemes of a language by means of a process of commutation or the discovery of *minimal pairs*, i.e. pairs of words which are different in respect of only one sound segment. **1968** R. T. HARMS *Introd. Phonological Theory* 14 In this presentation the term 'phoneme' is used in the sense of systematic phoneme. **1968** *Language* XLIV. 723 A consideration of the taxonomic phonemes of a language may..be of use in determining the orthographic peculiarities of a scribe. **1970** F. BRENGELMAN *Eng. Lang.* p. viii, The concept of the phoneme is used here because the author believes that the English spelling system can be described best with reference to a segmented representation. **1971** D. CRYSTAL *Linguistics* iv. 183, I have spent so much space on the phoneme concept because of its unequalled status as a concept for organizing people's thinking in the first half of this century. **1975** *Jrnl. Linguistics* XI. I. 40 This phonologically motivated structure is given in terms of systematic phonemes. **1975** L. M. HYMAN *Phonology* 82 It would be worthwhile to briefly examine the kind of argument given against what has come to be known as the 'autonomous' or 'taxonomic' phoneme. **1977** *Language* LIII. 241 He rejects the orthodox neo-Bloomfieldian motto 'Once a phoneme, always a phoneme'. **1978** *Ibid.* LIV. 173 With one of his main points (that the taxonomic phoneme is not dead), I would naturally agree.

2. *Psychol.* [ad. G. *phonem* (C. Wernicke *Grundriss der Psychiatrie* (1896) II. 126).] (See quot. 1911.)

1905 A. J. ROSANOFF tr. *J. Rogues de Fursac's Man. Psychiatry* I. ii. 44 Phonemes (the verbal auditory hallucinations of Séglas) have..a special significance, inasmuch as they consist of 'words representing ideas'. **1911** W. A. WHITE *Outl. Psychiatry* (ed. 3) vi. 47 The more complicated hallucinations which are conceived by the patient to be 'voices'—verbal auditory hallucinations—are known as phonemes.

Hence **pho:nemiza·tion** = **phonemicization*.

1959 E. PULGRAM *Introd. Spectrogr. of Speech* xi. 78 (*heading*) Phonemization. **1960** *Amer. Speech* XXXV. 218 The Middle English phonemization of the Old English allophones of the postdental spirant.

phonemic (foni·mik), *a.* and *sb. Linguistics.* [f. *PHONEME 1b + -IC.] **A.** *adj.* Of or pertaining to phonemes or phoneme theory; analysable in terms of phoneme theory.

1933 L. BLOOMFIELD *Language* v. 85 The laboratory phonetician usually knows, from other sources, the phonemic character of the speech-sounds he is studying. **1933** E. SAPIR in *Encycl. Social Sci.* IX. 155/2 Not only are all languages phonetic in character; they are also 'phonemic'. *Ibid.*, Languages differ very widely in their phonemic structure. **1942** BLOCH & TRAGER *Outl. Linguistic Analysis* ii. 36 For nearly all purposes it is better to use a *Phonemic* transcription, which represents the sounds of the language organized into a few dozen distinctive units. **1947** E. H. STURTEVANT *Introd. Linguistic Sci.* xi. 122 No scholar has a right to demand.. that we discard the familiar and convenient word *phonemic* for the reason that if an ancient Greek had coined an adjective from the noun *phōnēma* he would probably have produced *phōnēmaticos*. **1950** R. A. HALL *Linguistics & Your Lang.* vi. 96 Then, when he has gotten the best phonetic and phonemic analysis possible for the language he is working on, the linguist is ready to go ahead and analyze its grammatical structure. **1950** D. JONES *Phoneme* ii. 9 The more general use of the term [*sc.* language] covers the speech of people who speak in ways differing considerably from each other—who use different sounds, or who use sounds in such a way that a different phonemic classification is called for. **1955** *Q. Jrnl. Speech* XLI. 254/1 The variation is non-significant, of course, but belongs on the morphemic rather than the phonemic level. **1958** *Times Lit. Suppl.* 13 June 334/5 The earlier lessons, which offer a phonemic transcription with the ordinary spelling of words, ask for hard work both in the classroom and outside. **1964** M. A. K. HALLIDAY et al. *Linguistic Sci.* 300 Transformation grammar has thus taken over, but reversed, the hierarchical relation among levels that is characteristic of phonemic–morphemic theory. **1965** CHOMSKY & HALLE in *Jrnl. Linguistics* I. 98 We suggested the names *systematic phonemic and systematic phonetic* for these levels of representation. *Ibid.*, We called these *taxonomic phonemic theories* so as to bring out their reliance on procedures of segmentation and classification and their essential independence of 'higher levels'. **1967** *Ibid.* III. 3 The advantage of introducing a systematic phonemic level is that it enables us to generalize the phonological description beyond the limits of one speaker. *Ibid.* 25 Given a taxonomic phonemic system, it can always be 'translated' into distinctive feature terms. **1968** CHOMSKY & HALLE *Sound Pattern Eng.* 11 We will make no further mention of 'phonemic analysis' or 'phonemes' in this study and will also avoid terms such as 'morphophonemic' which imply the existence of a phonemic level. **1968** P. M. POSTAL *Aspects Phonol. Theory* iv. 90 It is natural to view autonomous phonemic representations as being exactly like phonetic representations except that certain 'redundant', 'predictable', 'noncontrastive', etc., features of the phonetic representation have been eliminated. **1971** *Language* XLVII. 503 Some of the theoretical discussions dealing with generative phonology have been concerned with demonstrating that a *phonemic* level of representation could not be motivated within phonological descriptions and that the only relevant levels were the morphophonemic and the phonetic. **1972** M. L. SAMUELS *Linguistic Evol.* iii. 31 Phonemic shifts are..often extensive and involve a large part of the vocalic or consonantal systems. **1973** A. H. SOMMERSTEIN *Sound Pattern Anc. Greek* i. 2 An autonomous phonemic description of classical Attic. **1974** tr. *Wertheim's Evolution & Revolution* iii. 74 At present China could lead the way in introducing 'phonemic spelling'.

B. *sb. pl.* (const. as *sing.*) [See -IC 2.] The study of phonemes and phoneme systems; phonemic analysis.

1936 W. F. TWADDELL in *Language* XII. 294 These [t]s, in the practice of every writer on phonemics, are considered as one phoneme. **1940** *Ibid.* XVI. 247 In English we have the incomparably superior term *phonemics*, which leaves us free to use *phonology* in its old and widely established usage as a general term for all the phenomena concerned with the production and use of sounds. **1949** G. L. TRAGER *Field of Linguistics* 5 The description of the phonemes of a language, and of their occurrence and arrangement, is the phonemics of that language. **1952** A. COHEN *Phonemes of Eng.* i. 4 Phonemics..must concern itself with studying distinctive oppositions existing between sounds or sound features. **1957** N. CHOMSKY *Syntactic Struct.* i. 11 A linguistic level, such as phonemics, morphology, phrase structure, is essentially a set of descriptive devices that are made available for the construction of grammars. **1964** —— *Current Issues in Linguistic Theory* 74 It seems necessary to conclude that systematic phonemics and systematic phonetics are the only two levels of representation that appear in structural descriptions provided by the phonological component. *Ibid.* 75 Let us coin the term 'taxonomic phonemics' to refer to this body of doctrine. **1968** P. M. POSTAL *Aspects Phonol. Theory* iv. 89 Traditional autonomous phonemics was able to maintain a natural relation between phonetic and phonemic organization in the absence of any universal rules of the type required by systematic phonemics simply because its phonemic and phonetic systems were so close. **1968** *Language* XLIV. 723 The concept of taxonomic phonemics in linguistics..arose in part as a result of traditional orthographic practices. **1972** M. L. SAMUELS *Linguistic Evol.* i. 4 There has been justifiable insistence, in recent decades, that graphetics and graphemics must in the first instance be studied separately, before their relationship with phonetics and phonemics can be considered. **1975** L. M. HYMAN *Phonology* 82 Systematic phonemics..goes beyond proposing an abstract morphophonemic level.

Hence **phone·mically** *adv.*, with regard to phonemes; in terms of phoneme theory; **phonemicist** (foni·misist), a specialist in or student of phonemics.

1934 *Language* X. 123 If a set of phonemic elements only occur together, they constitute a phonemically unitary complex; thus, the stop and the aspiration in English initial f. **1935** *Ibid.* XI. 97 The technique of analyzing phonemically the structure of a dialect. **1942** BLOCH & TRAGER *Outl. Linguistic Analysis* iii. 46 If a language has only five vowel phonemes, the best way of representing them phonemically—regardless of their phonetic nature—is by the letters *a, e, i, o, u*. *Ibid.* 49 The affricates in *chain* and *Jane* are treated by many phonemicists as unit phonemes, often written /č, ǰ/. **1951** *Language* XXVII. 219 The English affricates *ch* and *j* have generally been regarded as double sounds by phoneticians; but phonemicists have vacillated back and forth. **1966** N. CHOMSKY in *Current Trends in Linguistics* III. 49 The notion of 'complementary distribution'..permits analyses that are not acceptable to the taxonomic phonemicist (or to anyone else). **1967** C. L. WRENN *Word & Symbol* 5 The multifarious and phonemically divergent regional dialects spoken in the Chinese Republic. **1975** *Amer. Speech 1972* XLVII. 246 We list all the contrasting ways of beginning an English stressed syllable (in utterance-initial) that every classical phonemicist would have agreed must be a 'cluster' rather than a single 'segmental phoneme'.

phonemicize (foni·misəiz), *v. Linguistics.* [f. *PHONEMIC *a.* and *sb.* + -IZE.] **1.** *trans.* To classify, analyse, or describe in terms of phoneme theory. Also *absol.* or *intr.*

1940 *Language* XVI. 354 Lorimer made no attempt to phonemicize here. **1951** *Ibid.* XXVII. 340 We are forced to phonemicize as /patadak/ and /padatak/. **1953** *Ibid.* XXIX. 81 It is difficult to explain to a communications engineer what we mean by phonemicizing. **1953** *Internat. Jrnl. Amer. Linguistics* XIX. 166 Each long nucleus has been phonemicized as a cluster of two phonemes. **1957** *Publ. Amer. Dial. Soc. 1956* xxvi. 46 In these terms learning the sound system of another language could be described as a phonemicizing of diaphones. **1960** *Amer. Speech* XXXV. 234 Would lead to a rapid collection of vast body of contemporary speech by uniform procedures yielding easily phonemicized data. **1960** Z. S. HARRIS *Structural Linguistics* vii. 71 Such is the question of how to phonemicize long vowels. **1963** F. G. LOUNSBURY in S. Koch *Psychol.* VI. 567 This process is called 'phonemicizing' the transcription. **1965** *Language* XLI. 480 It would seem possible to phonemicize [nt·] as a cluster *nht. **1972** H. KURATH *Stud. Area Linguistics* 31 When two or more plans of phonemicizing the data seem feasible, preference is given to the one that facilitates the comparison of the dialects. **1975** *Amer. Speech 1972* XLVII. 240 Does 'phonemically different' imply 'different phonemes'? This is a catch-question I put to my students at a certain point as they are learning to 'phonemicize'. **1976** *Language* LII. 307 Once the surface redundancy has been lost, the forms which violate the lost constraint may be phonemicized as having the impermissible sequence.

2. a. *trans.* To advance from allophonic to phonemic status.

1972 M. L. SAMUELS *Linguistic Evol.* iii. 35 The split of one phoneme into two is clearly mechanical in origin: what are at first allophones of a single phoneme are 'phonemicised', or attain the status of separate phonemes. **1976** *Archivum Linguisticum* VII. 184 One could postulate..another old series of labials and dentals which changed *u to o and which then joined with the other dentals and labials, 'phonemicizing' the old allophone.

b. *intr.* To attain phonemic status.

1973 J. M. ANDERSON *Struct. Aspects Lang. Change* 142 The reduction of the labiovelars allowed these fronted allophones to phonemicize.

Hence **phone:miciza·tion**, (*a*) classification into phonemes, description in terms of phoneme theory; phonemic transcription or an instance of this; (*b*) development from allophonic to phonemic status.

1942 C. F. HOCKETT in *Language* XVIII. 20 Grammatical work is carried on, of course, in cases where phonological information is incomplete... But many of the gaps and ambiguities in such grammar have their source directly in the lack of complete phonemicization. **1948** —— in *Ibid.* XXIV. 123 He should make his phonemic interpretation clearly recognizable..by including for every form cited..a possible phonemicization. **1957** *Trans. Philol. Soc.* 22 There is no difficulty about the phonemicization, since there is contrast in such pairs as ker..and kar,..pe·r·..and pa·r. **1972** M. L. SAMUELS *Linguistic Evol.* iii. 37 It is still possible to argue that what appears to us as coincidental is in fact nothing of the sort, and that the functional pressures towards phonemicisation are more important than the mechanical. **1975** N. CHOMSKY *Logical Struct. Linguistic Theory* vi. 163 There is no need to think of phonemes as literally occurring in sequence, each with its distinctive physical properties, in the stream of speech. Any attempt to maintain such a view will lead to very artificial phonemicization.

phone phreak (fōu·n frīk), *sb.* and *v.* Also **phone freak.** [f. *PHONE *sb.*[2] + *PHREAK *sb.* and *v.*] **A.** *sb.* (See quot. 1972[2].)

1972 *Daily Tel.* 15 Apr. 3 The Post Office are trying to break up a ring of 'phone phreaks' who are cheating the telephone service. *Ibid.* 9 Oct. 13/6 Detectives and Post Office officials investigating 'phone phreaks'—people who make free telephone calls all over the world by using an electronic device—raided a flat in Hammersmith. **1973** *New Scientist* 5 Apr. 23 The challenge of beating the telephone system to get long distance calls for the 2p price of a local call is not restricted to the so-called 'phone-phreaks'. **1973** *Guardian* 1 June 10/1 Andy first heard the news of Edie Sedgwick's death from a fellow phone-freak on the West Coast. **1976** 'O. BLEECK' *No Questions Asked* xiv. 156 He's a monster phone freak... He knows all the phone company jargon.

B. *v. intr.* To use an electronic device to make telephone calls without payment. Only as **pho·ne phreaking** *vbl. sb.*

1972 *Daily Tel.* 9 Oct. 13/6 Scotland Yard and the Post Office have been investigating phone phreaking since January. **1973** *Guardian* 22 Jan. 5/3 'Phone phreaking'—making free calls with the use of electronic gadgets—is a growing international problem. **1974** W. GARNER *Big enough Wreath* xiii. 197 'What is it? Some kind of calculator?' Bip began with technicalities, ended with a complete exposition of phone freaking. **1976** *Computing Europe* 2 Sept. 5/2 There are a lot of Post Office employees and the recent 'phone phreaking' cases have shown that at least some of them can have criminal inclinations.

phonetic, *a.* Add: **2. c.** *Comb.* (= PHONETICO-), as *phonetic-linguistic, -morphological, -phonemic, -semantic* adjs.

1961 L. F. BROSNAHAN *Sounds of Language* v. 101 The criteria..are of phonetic-linguistic nature. **1921** E. SAPIR *Language* viii. 185 In other words, on this particular point it took German at least three hundred years to catch up with a phonetic-morphological drift that had long been under way in English. **1966** M. PEI *Gloss. Linguistic Terminol.* p. ix, A specialist in descriptive linguistics and phonetic-phonemic description. **1931** L. BLOOMFIELD in *Language* VII. 205 Some linguistic forms bear no partial phonetic-semantic resemblance to other forms. **1966** M. PEI *Gloss. Linguistic Terminol.* 204 There is a partial phonetic-semantic resemblance in all forms containing a particular phonestheme.

B. *sb.* An element of a Chinese character which is itself the character for another word, adopted as part of the new character because of the words' identity or similarity of sound. = *PRIMITIVE *sb.* 6.

1842 J.-M. CALLERY *Encycl. Chinese Lang.* 3 The phonetic is in itself one of the characters of primitive formation, which cannot be annexed to any of the preceding orders, and which must therefore be looked on as indivisible. **1874** S. W. WILLIAMS *Syllabic Dict. Chinese Lang.* p. lvi/1 That part of a character which is not the radical, has no name among the Chinese, but foreigners have termed it the *primitive* or *phonetic*. **1907** W. HILLIER *Chinese Lang.* i. 6 It is possible..by learning these phonetics, or primitives as they are sometimes called, to make a very close guess at the sound of any Chinese character. **1948** R. A. D. FORREST *Chinese Lang.* ii. 38 In many cases the connection of the phonetic element, whether in sound or shape, with a word still existing independently, or with the same phonetic in other characters, has become much more obscure through changes in sound or in written form. **1968** P. KRATOCHVÍL *Chinese Lang. Today* v. 151 The former was the borrowed phonetic..and the latter the radical added as an indication that the whole form was borrowed and denoting something connected with manual action. **1973** *Sci. Amer.* Feb. 54/2 There are characters that are not pronounced like their phonetic, often for reasons of historical change.

Hence **phonetician** (further examples); **phoneticism** (examples); (*b*) phonetic spelling; an example of this; (*c*) use of the criterion of phonetic similarity to determine the phonemes of a language; **phoneticist**, (*b*) = *phonetician*; **phoneticization** (-səizēi·ʃən), phonetic spelling; an example of this; (greater) correlation of symbol and sound (in a writing system).

1933 L. BLOOMFIELD *Language* v. 75 The phonetician can study either the sound-producing movements of the speaker..or the resulting sound-waves. **1964** D. ABERCROMBIE *Eng. Phonetic Texts* 14 It is not, in the nature of things, possible for a 'standard' mode of transcription of English, suitable for all purposes and all audiences, to be agreed on by phoneticians. **1975** *Amer. Speech 1973* XLVIII. 111 To do so is perfectly sound phonemically, though Phoneticians use a modified symbol (with hook) for the *r*-colored vowel. **1885** G. L. GOMME *Hist. T. Hickathrift* p. iv, There are too few phoneticisms and dialect words to make it probable that the print in the Pepysian collection is the one directly derived from popular tradition. **1938** *Better English* Nov. 44/1 Phoneticism is not a noble progressive movement. It is only an annoying attack on a superior language which will never give up its proud fundamental structure. **1939** *Amer. Speech* XIV. 148/1 The author condemns phoneticism as contrary to language structure. **1952** A. COHEN *Phonemes of Eng.* ii. 24 The same objection to phoneticism in phonemic analysis can be raised against Trubetzkoy's handling of the problem how to decide whether we have to do with one or more phonemes. **1977** *Daily Tel.* 24 Feb. 18 Probably the most bizarre example of phoneticism I have ever come across was an 11-year-old's spelling of the word 'usual'—'yousyouall'. **1932** G. K. ZIPF *Sel. Stud. Princ. Relative Frequency in Lang.* I. 4 All phoneticists agree about it; it is at once evident to anyone listening to a native of Peking speak. **1935** —— *Psycho-Biol. of Lang.* (1936) 96 The highly important work of the experimental phoneticist. **1954** PEI & GAYNOR *Dict. Linguistics* 168 *Phoneticist*, a person who studies or is skilled in phonetics. **1915** *Spectator* 21 Aug. 235/1 To turn the Russian genitive plural termination into 'off', as is sometimes done, is to go in for an exaggerated phoneticization. **1959** *Brno Studies in English* I. 14 Voices demanding reforms of traditional spellings usually regard 'phoneticization' of such spellings as the only effective remedy. **1970** *Language* XLVI. 959 For a primitive logographic system can develop into a full system of writing only if it succeeds in attaching to a sign a phonetic value independent of the meaning which this sign has as a 'word'; this is phoneticization. **1975** *Daily Colonist* (Victoria, B.C.) 22 Aug. 5/1 The common direction of phoneticization for all written languages in the world must be followed.

phonetico-. Add: **phonetico-ideogra·phic** *a.*, of or pertaining to ideographs having a phonetic value; **phonetico-phone·mic** *a.*, employing both phonetic and phonemic criteria.

1891 C. LOMBROSO *Man of Genius* III. ii. 189 This mixture of letters, hieroglyphics, and figurative signs, constitutes a kind of writing recalling the phonetico-ideographic stage through which primitive peoples (the Mexicans and Chinese certainly) passed, before the discovery of alphabetic writing. **1952** A. COHEN *Phonemes of Eng.* iv. 85 M. Swadesh..distinguished between the two categories [*sc.* long and short vowels] on a mixed phonetico-phonemic basis.

phonetics, *sb. pl.* Add: Now usu. restricted to the study of speech sounds as physical phenomena, and distinguished from *phonology*. (Further examples.)

1924 O. JESPERSEN *Philos. Gram.* ii. 35 It would, perhaps, be advisable to restrict the word 'phonetics' to universal or general phonetics and to use the word *phonology* of the phenomena peculiar to a particular language (e.g. 'English Phonology'), but this question of terminology is not very important. Some writers would discriminate between the two words by using 'phonetics' of descriptive (static), and 'phonology' of historical (dynamic) 'lautlehre', but this terminology is reversed by some (de Saussure, Sechehaye). **1937** J. ORR tr. *Iordan's Introd. Romance Linguistics* 287 Generally, 'phonetics' is used to designate the physiology of sounds, and 'phonology' the history of sounds. Saussure reverses the use of the two terms. **1953** J. B. CARROLL *Study of Lang.* ii. 25 General phonetics, in fact, is virtually a science in its own right, with two chief branches, *motor phonetics* (or *articulatory phonetics*) and *acoustic phonetics*. **1962** A. C. GIMSON *Introd. Pronunc. Eng.* I. i. 2 Our primary concern will be the production, transmission, and reception of the sounds of English—in other words, the *phonetics* of English. **1964** P. STREVENS in D. Abercrombie et al. *Daniel Jones* 120 There seem to be two main kinds of use of the term... 'Phonetics' for some means 'making sounds', while for others it refers to a component of the discipline of linguistics. **1970** G. C. LEPSCHY *Survey Structural Linguistics* iii. 59 Trubeckoj mentions a number of linguists who preceded him..in distinguishing between sound and phoneme, and thus between phonetics and phonology.

phonetist. 1. Delete 'a phonologist' from def.

phonetization. Add: Also = *PHONETICIZATION. (Further examples.)

1952 I. J. GELB *Study of Writing* iii. 66 The need for indicating grammatical elements was of no great importance in the origin of phonetization. *Ibid.* 67 Phonetization..arose from the need to express words and sounds which could not be adequately indicated by pictures or combinations of pictures. **1976** *Visible Language* Winter 20 Phonetization of the alphabet and other writing systems is a province of orthoepy.

Phonevision (fōu·nviʒən). [f. *PHONE *v.* + VISION *sb.*] The proprietary name of a pay-as-you-view television system (see quot. 1951).

1950 *N.Y. Times* 10 Feb. 42/3 The Zenith plan involves the use of so-called 'phonevision'. By this method, special programs are televised in scrambled form so they cannot be picked up on any ordinary receiver. A subscriber, however, can receive the program by telephoning his local exchange and establishing a connection between the television set and the telephone system. **1951** *Life* 5 Feb. 43/1 Phonevision, or PV,.. offers a way of bringing movies straight into the family parlor. Under PV, a television set owner learns that a certain movie is being scheduled at a given time. It goes out over a TV channel and shows on his set as a confusion of blurs. If he calls the Phonevision switchboard, an unscrambling device on his set, which is hooked up to the phone wires, is turned on and the blurs become a clear picture. **1953** *Official Gaz.* (U.S. Patent Office) 8 Sept. 281 Zenith Radio Corporation—Phonevision. For Radio and Television Transmitting and Receiving Operations and Parts Thereof. Claims use since June 13, 1947. **1975** *New Trends in Cable Television* (Com Quest Corp.) iii. 51 Zenith Radio Corporation developed the first commercially used pay-television/broadcasting equipment under the tradename

Phonevision. **1978** *CATV Market* (Frost & Sullivan, Inc.) iv. 65 This Phonevision service was offered under a special permit from the F.C.C.

phoney, phony (fōu·ni), *a.* and *sb.* orig. *U.S.* [Of uncertain origin.] **A.** *adj.* That has no real existence; fake, sham, counterfeit; false; insincere.

1900 ADE *More Fables* 138 'Overlook all the Phoney Acting by the Little Lady, Bud,' said the Fireman. **1916** C. SANDBURG *Chicago Poems* 63 You're only shoving out a phoney imitation of the goods. **1924** *Scribner's Mag.* Aug. 204/1 Hope you didn't mind when I gave you a phony name. **1933** *Sun* (Baltimore) 2 May 8/7 A line of argument..which I have long suspected is quite phoney. **1935** C. DAY LEWIS *Time to Dance* 60 You funny old, phoney old bogus man! **1949** *Chicago Tribune* 17 Sept. III. 18/3 Stop moaning about that phony blonde and her phonier lawsuit. **1951** J. D. SALINGER *Catcher in Rye* ii. 19 They had this headmaster, Mr. Haas, that was the phoniest bastard I ever met in my life. **1956** W. SLIM *Defeat into Victory* ix. 181 On our side we had the somewhat phoney propaganda that followed Wingate's raid and the more solid influence of General Giffard's character. **1970** *Daily Tel.* 10 Nov. 12/4 Like his singing, he is gentlemanly: no long hair, exaggerated clothes, or phony emotionalism. **1974** E. BRAWLEY *Rap* (1975) I. x. 169 My parole officer violated me on another phony beef.

b. Special collocations, as **phoney war**, the period of comparative inaction at the beginning of the war of 1939–45; also *transf.* and *fig.*; so *phoney peace*.

1940 *Times* 19 Apr. 7/2 When the Allies seemed slow at getting off the mark,..it was whispered to the American public that this was a 'phoney' war. **1940** *Manch. Guardian Weekly* 12 July 25 During the eight months of the 'phoney' war everything seemed to be running smoothly between Great Britain and France. **1940** G. GREENE *Lost Childhood* (1951) 115 This can never at any time have been a 'phoney' war: from the word go, these famous individuals were on the job. **1944** J. S. HUXLEY *On Living in Revolution* 9 The inadequacy of British production and planning during the Chamberlain 'phony war' period. **1947** *Partisan Rev.* XIV. 347 Within each state the necessary psychological atmosphere would be kept up by complete severance from the outer world, and by a continuous phony war against rival states. **1954** N. COWARD *Future Indefinite* III. 114 The Germans invaded Norway and Denmark... The reaction of the Americans to the break-up of what they themselves had christened the 'Phoney War' would be interesting to observe. **1960** O. MANNING *Great Fortune* xv. 199 This sort of phoney war can't go on for ever. Someone's going to move some time and we'll be trapped. **1964** *New Statesman* 4 Sept. 301/2 The electoral phoney war is almost over. **1972** *Daily Tel.* 29 Mar. 2/6 A curious 'phoney war' atmosphere has pervaded Ulster during the past few days. **1977** *Guardian Weekly* 27 Feb. 9/2 At this stage of phoney peace, the Americans are being careful to say or do nothing that might alarm Israel.

B. as *sb.* A phoney person or thing.

1902 C. L. CULLEN *Six Ex-Tank Tales* 99 If youse tinks f'r a minnit dat youse is goin' t' git away wit' a phony like dat wit' me youse is got hay in y'r hemp, dat's wot. **1916** *San Francisco Call & Post* 28 Nov. 12 'Don't Mr. Jenks know a lot of people?' 'They're all phonies.' **1938** E. AMBLER *Cause for Alarm* xi. 170 He's probably gone to the trouble to check the first lot and found that they're phoneys. **1952** S. C. ARMSTRONG *Black-Eyed Stranger* i. 5 Lynch is no international jewel thief. He's a tired old phony. **1958** K. AMIS *I like it Here* xvi. 200 The kind of prancing, posturing phoney who'd say he was better than Fielding. **1971** F. FORSYTH *Day of Jackal* II. xv. 262 'Leave the others to continue checking the remainder, just in case there is another phoney among the bunch,' instructed Thomas. **1971** S. E. MORISON *European Discovery Amer.: Northern Voy.* iii. 79 Adolf Rieth..tells of European false runic inscriptions and other famous phonies, one of which, the 'turkey frieze' in Schleswig Cathedral, pertains to America. **1977** *New Yorker* 27 June 79/1 This simple test—a way of telling the phonies from the truly committed.

Hence **pho·neyness, phoniness,** the state or quality of being phoney; deceitfulness, unreality, insincerity; **pho·nily** *adv.*, in a phoney manner; falsely; insincerely.

1942 BERREY & VAN DEN BARK *Amer. Thes. Slang* §315/1 Fakeness, phoniness. **1947** D. M. DAVIN *Gorse blooms Pale* 212 You felt a sort of phoneyness in your power. **1949** D. SMITH *I capture Castle* ix. 139 Am I just trying to rationalize my phoniness? **1959** *Times Lit. Suppl.* 17 Apr. 224/4 A racket is phoneyness organized. **1961** C. J. ROLO in *Webster* s.v., Phonily flamboyant amours and impossible deeds of derring-do. **1961** W. C. BOOTH *Rhetoric of Fiction* II. vii. 193 Though attracted by youth and freshness, he can see the phoniness of the American worship of Hollywood's idea of youthfulness. **1976** T. GIFFORD *Cavanaugh Quest* (1977) iii. 51 Tim didn't have any phoniness about him; he was what we used to call a regular guy. **1976** *Economist* 16 Oct. 15/2 Name another [parliamentary democracy] that could pass through the past 18 months and still have a parliament doing business phonily as usual. **1978** P. MCCUTCHAN *Blackmail North* iv. 38 He brightened rather phonily. 'It could have been worse.'

phoney, phony, *v. slang* (chiefly *U.S.*). [f. the adj.] *trans.* and *intr.* To counterfeit, falsify, make *up*. (See also quot. 1950.)

1942 BERREY & VAN DEN BARK *Amer. Thes. Slang* §207/12 Disguised,.. phonied up. **1950** H. E. GOLDIN *Dict. Amer. Underworld Lingo* 156/1 *Phony up*, to counterfeit; to set up an impressive front or agency for purposes of swindling; to alter the amount of money indicated on a check; to change the serial number of a stolen bond or commit any similar criminal act; to turn traitor to the underworld. 'I got a chill on (doubt the courage of) this dude we're working with. He might phony up on a drop (under police pressure coincident with arrest).' **1952** *New Yorker* 26 July 43, I ain't phoneying them woids. **1963** 'E. MCBAIN' *Ten plus One* (1964) xv. 173 I'll phony it up, stall him. **1968** 'G. BAGBY' *Another Day* iii. 53 He..made no effort to phony up an excuse. **1972** B. F. CONNERS *Don't embarrass Bureau* (1973) II. 200 It's tough for a cop to stay completely honest... It's like an agent who won't phoney his overtime. **1977** *Daily Tel.* 8 Dec. 3/7 Furs are often not clearly labelled. Cat skins could be passed off as 'bunny'. You can phoney anything up.

phoniatric (fōuniæ·trik), *a. Med.* [f. Gr. φων-ή voice + ἰατρικ-ός of or for a doctor.] Of or pertaining to phoniatrics (logopedics).

1938 *Jrnl. Speech & Hearing Disorders* III. 286 (*heading*) Phoniatric aspects of unilateral recurrent paralysis. **1947** *Folia Phoniatrica* I. 14 Patients often receive wrongly and unsuccessfully treatment for chronic laryngitis, whereas, only phoniatric treatment..is promising. **1960** F. TROJAN *Current Probl. Phoniatrics & Logopedics* I. 53 Further research will enable this problem to be solved both from the surgical and the phoniatric point of view.

Hence **phoniatri·cian, phoni·atrist** (fonəi·ǎ-trist), an expert or specialist in phoniatrics; **phonia·trics** *sb. pl.* (const. as *sing.*), **phoni·atry** (fonəi·ǎtri) = *LOGOPEDICS *sb. pl.*

1947 *Folia Phoniatrica* I. 5 Phoniatry deals with the..sciences of the voice, speech and speech training, the problem of the deaf and dumb and musical problems and technique. **1950** *Ibid.* II. 175 Phoniatrics is practically inexistent. *Ibid.* 182 There exists no preparation or control for the Phoniatrist's profession. **1950** S. POTTER *Our Language* 186 Ward is more concerned with phoniatry and the rectification of faulty pronunciation in the speech of English children. **1959** *Amer. Speech* XXXIV. 55 Conscientious report on the articulatory organs, functions, and theories as they would interest the phoniatrician rather than the phonetician. **1960** F. TROJAN *Current Probl. Phoniatrics & Logopedics* I. 67 The relation of phoniatry to laryngeal surgery has two aspects. *Ibid.*, Surgical operations upon professional voice users should be undertaken by the phoniatrist. **1961** L. F. BROSNAHAN *Sounds of Language* vi. 144 Speech therapists and phoniatricians. **1962** *Folia Phoniatrica* XIV. 81 If we accept that phoniatrics means the 'medicine and medical art of phonation' it is clear that we limit ourselves to the medical aspects of impaired phonation.

phonic, *a.* (*sb.*) Add: **1.** *phonic wheel* [tr. F. *roue phonique* (P. Lacour 1878, in *Compt. Rend.* LXXXVII. 500)], a toothed disc or rotor of magnetic material which is caused to rotate at a constant speed by an electromagnet energized by alternating, or interrupted direct, current (orig. derived from a tuning fork vibrating against a contact and sustained by another electromagnet); also *phonic motor*.

1878 *Telegraphic Jrnl.* VI. 476/2 M. Paul la Cour has succeeded in causing a phonic wheel to maintain its uniform rate of rotation when acted upon by an accelerating or retarding force of one kilogramme-metre-minute. **1906** T. E. HERBERT *Telegraphy* 838 Fig. 493 is a general view of the transmitter. At the back may be seen the rectangular frame of a La Cour phonic wheel motor that drives the transmitter. **1924** *Jrnl. Sci. Instruments* I. 162 By making use of a device known as a phonic motor—invented by the late Lord Rayleigh—a wheel is constrained to rotate at a constant speed controlled by an electromagnetically maintained tuning-fork. **1930** A. B. WOOD *Textbk. Sound* II. 129 The phonic motor provides a very convenient and accurate method of determining the frequencies of electrically-maintained forks. **1956** *IRE Trans. Electronic Computers* V. 159/1 The clock frequency is 50 kc, obtained from a phonic wheel on the drum.

2. (Further examples.) *phonic method*, a method of teaching reading by correlating alphabetic symbols and sounds (= *PHONICS *sb. pl.* 4).

1875 G. C. MAST *Primer of Phonic Method* p. iv, For years it had been his [*sc.* the Author's] wish to introduce in this country the German, or Phonic method of teaching reading and writing simultaneously. **1928** WARD & ROSCOE *Approach to Teaching* ix. 110 The Phonic Method..has the undoubted advantage that the children, if properly taught, are from the first not afraid to attack new words. **1954** *Language* (Ministry of Educ.) v. 56 The alphabetic method was superseded generally in British schools by the *phonic* method. **1966** J. DERRICK *Teaching Eng. to Immigrants* v. 184 English is not a 'phonic' language—the sounds represented by certain letters or groups of letters do not all fit neatly into readily observed and easily learnt patterns (like those of Italian, or Welsh, for instance). **1968** J. LYONS *Introd. Theoret. Linguistics* ii. 62 Whatever other factors may have influenced the development of human speech, it is clear that phonic substance (that range of sound which can be produced by the human speech organs and falls within the normal range of human hearing) satisfies the conditions of availability and convenience fairly well. **1973** D. ROCKEY *Phonetic Lexicon* ii. 36 For many years reading theories have been polarised between two extremes—the phonic method and the so-called 'Look-and-Say'. **1976** *Amer. Speech 1974* XLIX. 12 As understood here, a *phonic transcription* includes the broad phonetic transcription presently favored by European phoneticians, the unsystematic simplification of phonetic forms often used by American dialectologists under the term *diaphone*, and the systematic broad phonics of Bloch and Trager..that is called 'classical phonemics' by generative apologists.

Hence **pho·nically** *adv.*, in respect of vocal sound; in the form of speech sounds.

1959 *Brno Studies in English* I. 12 The spoken norm of language is a system of phonically manifestable language elements. **1962** S. R. LEVIN *Linguistic Struct. in Poetry* v. 50 The occurrence of phonically or semantically equivalent forms in equivalent positions, either syntagmatically or conventionally defined. **1965** [see *HIGH *a.* 4 b]. **1977** *Word 1972* XXVIII. 310 The values assigned to the sequences may be analyzed with exclusive regard to the semantic qualities associated with these phonically polar vowels.

phonics, *sb. pl.* Delete ? *Obs.* and add: **2.** (Later examples.)

1961 *Amer. Speech* XXXVI. 93 (*title*) Phonemics and phonics in historical phonology. **1976** [see *PHONIC *a.* (*sb.*) 2].

4. The correlations between sound and symbol in an alphabetic writing system; used *spec.* with reference to a method of teaching reading by associating letters or groups of letters with particular sounds (cf. *phonic method* s.v. *PHONIC *a.* (*sb.*) 2).

1908 E. B. HUEY *Psychol. & Pedagogy of Reading* III. xiv. 280 Too often the line between phonics and reading is not drawn. **1960** J. R. NEWTON *Reading in your School* iv. 63 Phonics is that part of phonetics which is used in reading and spelling. **1966** K. DE HIRSCH et al. *Predicting Reading Failure* viii. 82 Five auditorily gifted children who read well had been intensively trained in phonics. **1966** *New Statesman* 30 Dec. 962/3 *Look and Say, Phonics* and *The Sentence Method* are all at the moment acting as smokescreens which obscure a simple fact known to most mothers but few reading research experts. The most important thing in a reading beginner's life is a willing listener. **1973** *Daily Tel.* 10 Apr. 18, I believe that most teachers of remedial reading would agree that more attention to phonics when teaching children to read would be an excellent thing. **1976** *Sci. Amer.* July 8/2 They replaced the teaching of reading by phonics with the 'look-see' method.

phonily, phoniness: see *PHONEY, PHONY *a.* and *sb.*

phono-. Add: **pho:no-electroca·rdioscope** *Med.*, an instrument for registering simultaneously the sounds and the electrical changes caused by the heart, or one of these together with the pulse; **phono-lary·ngoscope** (-dʒoskōup), an apparatus for observing the operation of the larynx in the production of speech sounds; so **phonolaryngosco·pic** *a.*; **pho:nophoto·graphy**, photographic recording of the physical parameters of speech or singing; hence **pho:nophotogra·phic** *a.*; **pho:no-photogra·phically** *adv.*; **pho·noreception** *Biol.*, perception of sound by a living organism; hearing; so **pho·noreceptor**, a sensory receptor for sound; † **pho·novision**, a system of television in which the signals were stored on gramophone records to be reproduced at will.

1942 *Lancet* 26 Dec. 759/2 In what he calls by the rather cumbersome name of a phono-electrocardioscope, G. E. Donovan has introduced an instrument which holds out high promise of useful service to the clinician. *Ibid.*, The most obvious application of the phono-electrocardioscope is in the teaching of auscultation. **1943** G. E. DONOVAN in *Jrnl. Inst. Electr. Engin.* XC. III. 39/1 The present apparatus incorporates a double-beam cathode-ray oscilloscope with a fluorescent screen of long afterglow. This permits the simultaneous direct visual observation of two phenomena such as the phonocardiogram and electrocardiogram, or sphygmogram and phonocardiogram, etc., at the patient's bedside. The amplified heart sounds can be heard at the same time. The instrument is called a phono-electrocardioscope. **1950** *Electronic Engin.* XXII. 90/2 In order to time accurately the events of the cardiac cycle, an electrocardiogram occurring simultaneously should accompany a phonocardiogram, and his [*sc.* Donovan's] apparatus, the phonoelectrocardioscope, is designed on these principles. **1953** L. F. BROSNAHAN *Some Old Eng. Sound Changes* 11 The production of each of the Dutch vowels, as observed with Russell's 'phono-laryngoscope'. **1934** *Amer. Speech* IX. 226/2 A phonolaryngoscopic examination of the position and function of the various organs of the larynx. **1928** M. METFESSEL *Phonophotogr. in Folk Music* 22 All the twists, quavers, trills, breaks in the voice, quick slurs, erratic tempi and other similar features..are..what phonophotography will reveal. *Ibid.* 19 There was no simple ready-made phonophotographic camera, nor had there ever been any studies which required the detailed reading of the sound wave photographs. **1931** T. H. PEAR *Voice & Personality* 17 Dr. Milton Metfessel..has recorded 'phonophotographically' the turns..of the 'Jubilee voice' of the negro. **1933** *Brit. Jrnl. Psychol.* Apr. 408 Seashore's claim..that his phonophotographic methods constitute an objective approach to the study of the beautiful,..in the performance of vocal music. **1935** *Amer. Speech* X. 312/2 By means of phonophotography the readings of several distinguished poets were analyzed for pitch, loudness and duration in an effort to solve some of the fundamental problems of verse. **1939** *Ibid.* XIV. 228/1 The application for the first time of phono-photographic technics to study of melody in isolated southern districts of the United States. **1968** P. OLIVER *Screening Blues* 10 Phonophotography and voice-prints may give an accurate translation of certain characteristics of the voice in graphic form. **1940** *Biol. Rev.* XV. 108 The substitution of the term 'phonoreception' for hearing evades the real issue. **1960** *Neurology* X. 662/1 Photoreception and

phonoreception provide the principal means of association between the individual and his external environment. **1968** D. W. WOOD *Princ. Animal Physiol.* ix. 188 Phonoreception is really a specialized case of mechano-reception, and it is not always easy to distinguish between the two. *Ibid.*, It is probably simpler to include vibration reception in phonoreception and to define the latter as the perception of any mechanical disturbance external to the animal that involves regular repetition. **1961** WEBSTER, Phonoreceptor. **1968** D. W. WOOD *Princ. Animal Physiol.* ix. 191 The most primitive phonoreceptor in vertebrates is the lateral line organ-system of fishes and a few amphibians. **1927** *Punch* 7 Sept. 253/3 Mr. J. L. Baird, the inventor of television and phonovision, is reported as saying that some faces sound like a gargle. **1935** *Times* 16 May 9/3 The new system of 'Phonovision' demonstrated some years ago in which wax-disk records of television signals were employed in much the same way.

phono (fōu·no), colloq. abbrev. of PHONOGRAPH *sb.* (sense 3). Chiefly *U.S.*, exc. as applied to a type of plug (and the corresponding socket) used with audio equipment, in which one conductor is cylindrical and the other is a central prong that extends beyond it. Freq. *attrib.* and in *Comb.*

1948 F. BROWN *Dead Ringer* 9 You can play the phono. **1956** C. FOWLER *High Fidelity* x. 203 Interconnections to and from preamp-control units are made via so-called 'phono plugs'. *Ibid.*, Hum is almost inevitable when the control unit is switched to the phono channel and the bass tone control is fully advanced. **1970** J. EARL *Tuners & Amplifiers* iii. 70 American 'phono' sockets are commonly used for the tape signals. **1971** *Computers & Humanities* VI. 95 Such multi-media materials as broadsides, films, filmstrips, and phono-records. **1973** *Washington Post* 13 Jan. H7/5 (Advt.), Automatic portable phono. **1975** *Hi-Fi Answers* Feb. 74/3 Some amplifiers do not have a DIN socket for tape recording, there being only phono types. **1975** *Physics Bull.* May 229/1 These can be substituted for permanent magnets in phonocartridges. **1976** *Gramophone* May 1835/1 The controls on the front panel are a large selector knob for phono, tuner, auxiliary, tape 1 and tape 2. [Etc.] **1978** *N.Y. Times* 30 Mar. C10/1 Mounting a phono cartridge in a tone arm..invariably makes me think I'm in the wrong trade.

phonocardiogram (fōunokā·ɹdi,ogræm). *Med.* [f. PHONO- + *cardiogram* s.v. *CARDIO-.] A tracing of the sounds made by the heart.

1912 *Heart* IV. 161 In many of our phonocardiograms it is very possible that we should have discovered initial vibrations. **1942** *Lancet* 26 Dec. 759/2 Permanent photographic records can be obtained of the phonocardiogram, electrocardiogram or sphygmogram. **1974** *Physics Bull.* Feb. 70/2 Incompatibility between echo signals and phonocardiograms.

phonocardiograph (fōunokā·ɹdi,ograf). *Med.* [f. PHONO- + *cardiograph* s.v. CARDIO-.] An apparatus used for registering phonocardiograms.

1926 *Amer. Heart Jrnl.* I. 721 (*caption*) Phonocardiograph record of human heart. **1943** *Jrnl. Inst. Electr. Engin.* XC. III. 43/1 The phonocardiograph employed by Einthoven possessed a simple but crude sort of high-pass acoustic filter. **1977** *Lancet* 19 Mar. 646/1 Records of the beat can be taken with an apex-cardiograph/phonocardiograph transducer attached to a commercial electrocardiograph.

phonocardiography (fōu:nokāɹdi,ǫ·grăfi). *Med.* [f. PHONO- + *cardiography* s.v. CARDIO-.] The investigation and interpretation by means of a phonocardiograph of the sounds made by the heart.

1921 *Jrnl. Amer. Med. Assoc.* 12 Feb. 434/1 Such a conception will explain a murmur of regurgitation in early systole... The first sound, as has been observed in phonocardiography, is held to be present, but its initial components may be masked by the murmur. **1926** *Amer. Heart Jrnl.* I. 717 (*heading*) Phonocardiography of the human fetus. **1961** *Lancet* 16 Sept. 644/1 Lewis et al. have shown by intracardiac phonocardiography that a similar murmur can be recorded in the main pulmonary artery of all healthy subjects. **1971** *Nature* 25 June 542/3 The basic principles of auscultation and phonocardiography are dealt with in the first chapter.

Hence **pho:nocardio·grapher**, one who operates a phonocardiograph or is expert in phonocardiography; **pho:nocardiogra·phic, -gra·phical** *adjs.*, of, pertaining to, or involving phonocardiography; **pho:nocardio-gra·phically** *adv.*

1935 *Amer. Heart Jrnl.* X. 458 We suggest that further phonocardiographic studies be performed along these lines. **1943** *Jrnl. Inst. Electr. Engin.* XC. III. 52/1 Certain frequency components..which are met with in routine phonocardiographical clinical work. **1967** M. E. TAVEL *Clin. Phonocardiography* i. 15 The phonocardiographer..is usually presented with a complex array of sounds, murmurs, and pulses. **1973** *Brit. Heart Jrnl.* XXXV. 1276/1 Both major components of the first sound were demonstrated phonocardiographically. **1974** *Circulation* XLIX. 5/2 Phonocardiographers are often faced with the difficult problem of telling the auscultator what he is hearing. *Ibid.* 434/1 We have delineated the phonocardiographic..characteristics of the mitral prosthesis.

phonofiddle (fōu·nofid'l). *Mus.* Also **Phonofiddle.** [f. PHONO- + FIDDLE *sb.*] A type of violin in which the usual body of the instrument is replaced by a mechanism connected with an amplifying horn. Also *attrib.*

1923 *Ashore & Afloat* Aug. (Advt.), Zither Banjos, Houson Phonofiddles, Strohviols, One String Fiddles. **1930** *Stage* 3 Apr. 9/4 A string of merry jokes and just sufficient phono-fiddle work to add to their appeal. **1955** *Oxf. Compan. Mus.* (ed. 9) 995/2 The Phonofiddle is sometimes played with a plectrum. **1977** *Early Music* Apr. 265/3 Susan Baker plays and talks about the..phonofiddle, [etc.].

phonofilm (fōu·nofilm). *Obs.* exc. *Hist.* Also **Phonofilm.** [f. PHONO- + FILM *sb.*] A cinema film having a sound-track.

Orig. a proprietary name in the U.S.

1921 *Official Gaz.* (U.S. Patent Office) 8 Nov. 391/2 Phonofilm... Claims use since Jan. 1, 1921. **1922** *Radio Broadcast* Dec. 96 De Forest's Phono-film. **1923** *Weekly Dispatch* 13 May 5 In so far as it eliminates the use of a gramophone..the new phono-film..would appear to mark a decided step forward. **1928** *Manch. Guardian Weekly* 17 Aug. 134/4 The Prime Minister for a quarter of an hour delivered to an audience of half a dozen people and two phonofilm cameras a farewell address. **1930** A. B. WOOD *Textbk. Sound* v. 495 A somewhat similar process is used in the production of phonofilms or 'speaking pictures'. **1973** *Listener* 7 June 744/3 There we were, south of San Francisco... Just down the road, De Forest had invented the radio valve and the phonofilm.

phonogenic (fōunodʒe·nik, -dʒi·nik), *a.* [f. PHONO- + *-GENIC b.] With pleasing voice qualities; well suited to mechanical reproduction of sound; of or pertaining to pleasing recorded sound.

1945 *Office Economist* June 10 (*heading*) Are you Phonogenic? **1947** *Red Barrel* Feb. 22 She shows how to be phonogenic by what not to do. **1957** MANVELL & HUNTLEY *Technique Film Music* v. 205 Roman Vlad..claims that he prefers to use a chamber music orchestra for what he calls 'phonogenic' reasons..('the possibilities of obtaining a clearer sound are superior'). **1977** *Gramophone* Jan. 1131/2 Perhaps hers is not a 'phonogenic' voice; the vibrations which are part of its appeal in a large house might appear too prominently in recording.

phonogram. Add: **2.** (Earlier and later examples.)

1879 *Telegraphic Jrnl.* VII. 233/1 M. Delecheneau has succeeded in getting phonograms on zinc and brass cylinders. **1967** A. L. LLOYD *Folk Song in England* i. 65 Dr. Walker Fewkes made phonogram recordings among North American Indians in 1889. **1976** *Daily Tel.* 10 July 4/8 To celebrate the centenary of sound recording next year major record companies in more than 60 countries are to present a programme of events... This was announced yesterday by the International Federation of Producers of Phonograms and Videograms. **1977** *Gramophone* Apr. 1527/2 It is not a laughing matter that it has taken so long and so much effort to establish the recording, the phonogram, as a serious creative art form in its own right.

3. A telegram that the sender dictates over the telephone. Freq. *attrib.*

1911 D. MURRAY in *Jrnl. Inst. Electr. Engin.* Aug. 451 Economic necessity will..lead to a great increase in telephone-telegrams, or, as the British Post Office already calls them, 'phonograms'. **1932** *Telegraph & Telephone Jrnl.* Oct. 2/1 Any telephone subscriber can hand over and receive his telegrams by telephone (phonogram service). **1968** E. H. JOLLEY *Introd. Telephony & Telegr.* i. 16/2 Nowadays small offices generally dictate their telegrams over the telephone to an appointed office. The appointed office receives the traffic from a group of minor offices..and is provided with special phonogram equipment designed to facilitate the handling of calls from the various offices. **1969** *West Australian* 5 July 66/2 (Advt.), Postmaster-General's Department has vacancies for phonogram operator (Perth). **1976** R. N. RENTON *Telegraphy* ix. 246/1 As far as the method is reversible, a telegram is delivered to the recipient over the same medium used for handing in, e.g. via phonograms, printergrams or leased teleprinter circuit. **1976** *Sydney Telephone Directory*, Phonograms. To save time, telephone your telegrams. The usual telegram charges will be debited to your telephone account.

phonograph, *sb.* Add: **3.** In Britain the word is retained only for early cylinder machines, but in N. Amer. it has become synonymous with *record player, record deck*, etc., corresponding to the British *gramophone*. (Further examples.) Also *attrib.*

1878 G. B. PRESCOTT *Speaking Telephone* x. 305 Having provided thus for the durability of the phonograph plate, it will be very easy [etc.]. *Ibid.* 430/2 (Index), The talking phonograph record. **1909** *Jrnl. Industr. & Engin. Chem.* Mar. 157/2 Phonograph records have been made with it [*sc.* Bakelite]. **1913** [see *GRAMOPHONE]. **1927** *Jrnl. Abnormal Psychol.* XXII. 13 Several of the major phonograph companies maintain 'race record' departments. **1929** E. WILSON *I thought of Daisy* i. 15 Somebody turned on the phonograph which began jigging a popular fox-trot. **1946** *Fortune* Oct. 158/2 Home phonographs rarely compare in precision with professional studio equipment. **1949** *Reader's Digest* Dec. 139/1 Sixteen million American phonograph owners are bewildered and unhappy. **1952** [see *GRAMOPHONE]. **1956** C. FOWLER *High Fidelity* i. 9 Time was when a phonograph made by one of the big companies was automatically considered low-fidelity. **1960** C. HANSEN in A. Dundes *Mother Wit* (1973) 507 The advertisements of..phonograph record companies. **1967** A. L. LLOYD *Folk Song in England* i. 65 Singers..whom Percy Grainger recorded on phonograph cylinders as early as 1904. **1975** *Daily Tel.* 17 Jan. 8/6 An Edison phonograph of about 1905 obtained £240. It was sold with 38 cylinder records. **1975** *New Yorker* 29 Sept. 64/2 Mrs. Santana turned off the TV, turned on the phonograph to its top volume, and went into the kitchen.

phonographic, *a.* **3.** (Further examples.)

1878 G. B. PRESCOTT *Speaking Telephone* x. 303 (*heading*) Tracings from phonographic records. **1898** H. G. WELLS *Let.* 22 Jan. in *G. Gissing & H. G. Wells* (1961) 79 A beautiful phonographic newspaper with a leathery flat voice. **1976** *Times* 2 Nov. 5/2 The British phonographic industry.

phonographical, *a.* For *rare*[—0] read *rare* and add example.

1974 *Country Life* 12 Dec. 1857/3 The two-volume CBS Astaire solo anthology..are transfers of old Brunswick 78s and therefore more satisfying in purely phonographical terms.

phonology. Add 'No longer equated with *phonetics*' and substitute for first part of def. in modern use: That branch of linguistics which deals with sound systems, or with sound systems and phonetics; the study of the sound system of a particular language. (Further examples.)

The domain of phonology is variously limited by different linguists and linguistic schools of thought. In the writings of the Prague school it is used to mean phonemics.

1924, etc. [see *PHONETICS *sb. pl.*]. **1933** L. BLOOMFIELD *Language* viii. 138 The description of a language, then, begins with phonology, which defines each phoneme and states what combinations occur. **1937** J. ORR tr. *Iordan's Introd. Romance Linguistics* iv. 287 The great majority of linguists and all the phoneticians..use 'phonetics' for the physiology of sounds, and although all do not adopt the term 'phonology' for their historical study, those who use the term 'phonetics' for the latter avoid confusion by speaking of 'historical phonetics'. *Ibid.* 288 This matter of terminology has become complicated still further by the special meaning given to 'phonology' by the Prague philologists. **1939** *Language* XV. 1 We use the term phonology to refer to alternations (synchronic phonology) or changes (historical, diachronic phonology) in sounds, rather than for the theory of the nature and permutations of the sounds. The latter we call phonemics. Those who use 'phonology' in this sense..deprive themselves of a convenient means of distinguishing two fundamentally distinct subjects. **1949** G. L. TRAGER *Field of Linguistics* 5 The phonetics and phonemics of a language are its phonological systems, its phonology. **1953** J. B. CARROLL *Study of Lang.* ii. 43 Some languages have only a limited number of phonemes, while some others appear to possess extremely complex phonologies, offering a large number of finely differentiated phonemes. **1962** E. F. HADEN et al. *Resonance-Theory for Linguistics* iii. 29 Phonology is the true-structure whose contributing members are phonemics and phonotactics. **1968** CHOMSKY & HALLE *Sound Pattern Eng.* p. vii, In the course of this detailed investigation of English sound patterns and their underlying structure, certain rules of English phonology are developed. **1972** M. L. SAMUELS *Linguistic Evol.* i. 3 For some, at present, phonology is less important than grammar or lexis.

phonologic *a.*, delete 'phonetic' from def. and add further examples; **phonological** *a.* (earlier example); **phonologically** *adv.*, delete 'phonetically'; **phonologist**, delete 'a phonetist' and add earlier example; **phonologiza·tion,** (*a*) shift to phonemic status; (*b*) development (of a phonetic feature) to the status of the distinguishing feature in a phonemic opposition.

1936 *Amer. Speech* XI. 110 A 'phonologic system' is defined as the ensemble of phonologic oppositions proper to a given language. **1955** [see *MINIMAL *a.* b (*a*)]. **1970** *Language* XLVI. 312 Some phonologic features are closely related to an articulatory maneuver that involves a specific muscle. **1977** *Archivum Linguisticum* VIII. 50 'Generative grammar' in the second sense is concerned with the description of rules for sentence-structures which include the phonologic level as well as the semantic one. **1818** *Trans. Amer. Philos. Soc.* I. 246 These modifications..may be distinguished in a phonological alphabet by particular signs. *Ibid.* 241 This *Sheva*, the English phonologists (if I may be allowed to use the name) have almost uniformly represented by *u* short. **1936** *Proc. 2nd Internat. Congr. Phonetic Sci.* 63 The phonologization of Middle English voiced spirants is an isolated process in the development of Germanic languages. **1964** B. TRNKA in D. Abercrombie et al. *Daniel Jones* 190 The phonologization of voice of spirants. **1976** *Archivum Linguisticum* VII. 95 Preaspiration..was consequently inadmissible in *kampur*, etc., but since it had been distinctive its loss was accompanied by phonologization of the devoicing of the sonant.

phonometric, *a.* (Examples.)

1895 E. B. TITCHENER tr. *Külpe's Outl. Psychol.* I. iii. 156 The phonometric determination of sound intensity in psychophysical experiments is..usually carried out upon a principle similar to that employed in photometry. **1938** *Amer. Speech* XIII. 278 This typical example of phonometric technique has clearly a considerable bearing upon both phonology and dynamic philology. **1965** *Jrnl. Appl. Physiol.* XX. 302/2 Pressure determinations were made at the same time as the phonometric ones.

Hence **phonome·trically** *adv.*, using the methods of phonometry.

1938 *Amer. Speech* XIII. 282 Taking the long vowels of German which, merely for the sake of illustration, we shall assume are a phonometrically unclassifiable residue.

phonometrics (fōunome·triks), *sb. pl.* (const. as *sing.*) [f. PHONOMETRIC *a.*: see -IC 2.] The study and practice of phonometry.

1957 H. J. ULDALL in Hjelmslev & Uldall *Outl. Glossematics* I. 16 In linguistics, phonometrics, experimental phonetics, and word counting. **1970** H. BLUEHME tr. *E. & K. Zwirner's Princ. Phonometrics* 4 Phonometrics is based on the assumption that the description and the synchronic or diachronic comparison of languages can be extended beyond what would be possible on the basis of the mere mastery of language by speaking, listening, or writing. **1972** HARTMANN & STORK *Dict. Lang. & Linguistics* 175 *Phonometrics*, the analysis and description of the spoken language by phonological (linguistic) and phonetic (instrumental) as well as statistical means.

phonometry (fonǫ·mėtri). [ad. G. *phonometrie* (E. & K. Zwirner, *Grundfragen der Phonometrie* (1936)).] A method of investigating language by the statistical analysis of instrumentally measured speech sounds and informants' responses to the same data.

1936 *Amer. Speech* XI. 358 The first volume..seeks to provide the historical and the theoretical bases for the methods of 'Phonometry'. The *raison d'être* of this venture is the conviction that the problems of comparative and historical linguistics are susceptible of further solution than has yet been achieved if one applies to these problems the technique known in biology as the statistics of variations. **1938** *Ibid.* XIII. 275 The chief concern of phonometry to date has been the empirical study of the variation of speech sounds, or, more precisely stated, the attributes of speech sounds, such as, for example, that of the duration of vowels. **1964** *Phonetica* XI. 151 Phonometry always works with recordings of connected speech.

phonon (fōu·nǫn). [f. PHONO- + *-ON[1].] **1.** *Physics.* A quantum or quasiparticle associated with compressional waves, such as sound or those in a crystal lattice.

1932 J. FRENKEL *Wave Mech.* vi. 267 It is possible to associate the acoustical waves with certain particles which we shall call 'phonons', and to replace the study of the heat oscillations forming these waves by the study of the motion of the corresponding 'phonons'. [*Note*] It is not in the least intended to convey hereby the impression that such phonons have a real existence. **1953** C. KITTEL *Introd. Solid State Physics* v. 82 If the forces between atoms were purely harmonic, there would be no mechanism for collisions between different phonons, and the mean free path would be limited solely by collisions of a phonon with the crystal boundary, and by lattice imperfections. **1968** C. G. KUPER *Introd. Theory Superconductivity* i. 2 In field-theoretical language the mechanism responsible for electrical resistance is scattering of electrons, with associated emission or absorption of quanta ('phonons') of the acoustic vibration field of the medium. **1969** *New Scientist* 2 Jan. 32/3 They calculated that 100 W of the light were converted into coherent phonons. **1974** H. E. HALL *Solid State Physics* ii. 67 It is often convenient to treat lattice vibrations in an analogous way, and introduce the idea of phonons of energy $\hbar\omega$ as quanta of excitation of the lattice. Our normal modes are plane waves extending throughout the crystal lattice, and correspondingly the phonons are not localized particles. *Ibid.*, Like photons, phonons are bosons and are not conserved.

2. *Linguistics.* In stratificational grammar, a phonetic feature which is capable of distinguishing phonemes. Cf. *distinctive-feature* s.v. *DISTINCTIVE *a.* 1 b. Hence **phono·nic** *a.*

1964 S. M. LAMB in Romney & Andrade *Transcultural Stud. in Cognition* (Amer. Anthropologist: special publication) 60 The elementary units of which the phoneme, the morpheme, the lexeme, and the sememe are composed may be called the *phonon*, the *morphon*, the *lexon*, and the *semon* respectively. **1965** *Language* XLI. 200 Units that Lamb calls 'phonons' but that I shall call distinctive features. **1966** S. M. LAMB *Outl. Stratificational Gram.* ii. 29 The phononic alternation pattern specifies alternations involving phonological components. *Ibid.* iii. 57 The following symbols..may be used for hypophonemic signs, hypophonemes, and phonons. **1967** C. F. HOCKETT *Lang., Math. & Linguistics* 83 The pair 'pit' and 'bit' attests to a minimal difference,..but that difference is between voicelessness and voicing, not between /p/ and /b/. We shall follow Lamb in calling the terms of such minimal differences *phonons*. **1969** *Language* XLV. 300 The use of phonons in stratificational phonology is compared with the use of marked vs. unmarked features by transformationalists. **1975** *Amer. Speech 1972* XLVII. 248 In many, most, or all languages there are recurrent bundles of phonons which it is convenient to represent in a linear rotation by single letters.

phonostylistics (fōunostəili·stiks), *sb. pl.* (const. as *sing.*) [f. PHONO- + STYLISTIC *sb.*] **a.** The study of the stylistic implications of phonetic variation. **b.** (See quot. 1972[1].) Hence **phonostyli·stic** *a.*

1968 M. SHAPIRO *Russian Phonetic Variants & Phonostylistics* p. vii, Phonetic free variation in Russian is treated in the present monograph not only for itself but as the material of a new subdiscipline of phonology which I have called phonostylistics after Trubetzkoy. *Ibid.* ii. 9 For the purposes of a special investigation, be it phonostylistic or otherwise, one cannot adopt the uneconomical view. **1969** *Computers & Humanities* III. 252 Linguistic features of style may be classified under *phono-stylistics* (meter, rhyme), *morphostylistics*,..and *syntactostylistics*. **1972** HARTMANN & STORK *Dict. Lang. & Linguistics* 175/2 *Phonostylistics*, that branch of stylistics which investigates the expressive function of sounds, e.g. the use of onomatopoeia in poetry. **1972** *Language* XLVIII. 350 It is of interest to note the phonostylistic correlates of these alternate pronunciations.

phonotactics (fōunotæ·ktiks), *sb. pl.* (const. as *sing.*) *Linguistics.* [f. PHONO- + TACTICS.] That part of phonology which comprises or deals with the rules governing the possible phoneme sequences in a language. So **phonota·ctic, phonota·ctical** *adjs.*; **phonota·ctically** *adv.*

1956 E. HAUGEN in M. Halle et al. *For Roman Jakobson* 216 The key to our understanding of the syllable lies in the development of phonotactics, or the study of phonemic distribution. **1958** [see *MORPHOTACTICS *sb. pl.*]. **1958** A. A. HILL *Introd. Linguistic Struct.* viii. 116 Such a form reaches the maximum sequence which is pronounceable without breaking English phonotactic sequences. **1962** E. F. HADEN et al. *Resonance-Theory for Linguistics* iii. 24 Phonotactics..deals with the order of phonic entities. The field of phonotactics is phonemes-in-sequence. **1964** E. BACH *Introd. Transformational Gram.* ii. 23 A restatement of the rules in which there is a clean break between phonotactic and allophonic rules will require a longer description. **1965** W. WINTER *Evidence for Laryngeals* 210 We do not yet know enough about the phonotactics of Proto-Tocharian to rule out the possibility that..*ssk and *tsk were reduced to *sk* and *tk*. **1967** D. STEIBLE *Conc. Handbk. Linguistics* 96 Phonotactical description defines the phoneme classes which occur in a language. **1970** *Language* XLVI. 211 Young Binis..pronounce the word *epich*, deliberately violating Bini phonotactic rules (which forbid closed syllables). **1972** *Ibid.* XLVIII. 465 Prenasalized stops are found in many languages, and are not usually subject to such stringent phonotactical restrictions; in Albanian, for example, they occur both initially and as first members of clusters. **1973** D. ROCKEY *Phonetic Lexicon* ii. 28 Tests for phoneme recognition should be designed with the aim of determining whether the patient has grasped the phonotactic rules of the language, that is, whether he knows the phonemes and their combinations. **1973** *Word 1970* XXVI. 108 Monosyllabic nonsense forms, phonotactically Russian. **1975** *Canad. Jrnl. Linguistics* XX. I. 61 Portuguese phonotactics generally does not tolerate word-final stops. **1976** *Ibid.* XXI. I. 38 Many of the morphological rules which are proposed by linguists, whether morphophonemic or phonotactic in presumed character, are posited primarily, if not solely, in order to capture certain kinds of supposed 'lexical redundancies'. **1977** *Ibid.* XXII. I. 45 R2 and R3 are phonotactically motivated P-rules; they 'conspire' to prevent occurrences of *ss*-sequences on the surface.

phonus-bolonus (fōu·nǝs bǫlōu·nǝs). *U.S. slang.* Also **phonus bolognus.** [f. *PHON(EY, PHON(Y *a.* and *BALON(EY, BOLON(EY + the meaningless syllable *-us*.] Nonsense, exaggeration, ostentation, an insincere statement; fraud, trickery; goods not up to specification; a swindling transaction. Also as *adj.*

1929 D. RUNYON in *Hearst's International* July 58/1 Of course this message is nothing but the phonus bolonus. **1936** WODEHOUSE *Laughing Gas* xviii. 202 Sure. It was just a bit of phonus-bolonus. I was stringing you along so's I could get hold of that notebook. I'd be a fine sap giving you money. **1947** —— *Full Moon* vi. 115 His little friend had scouted the idea that there was any phonus-bolonus afoot between Veronica Wedge and this prominent Anglo-American snake. **1948** in Wentworth & Flexner *Dict. Amer. Slang* (1960) 386/1 The phonus-bolonus which gums up the average backstage musical. **1950** R. STARNES *Another Mug for Bier* xviii. 121 Who was..the one who engineered the phonus-bolonus deal? **1955** *Sun* (Baltimore) 3 Feb. 3/1 'Phonus Bolognus!' whipped back Charles Hess.

phony: see *PHONEY *a.* and *sb.*, *v.*

phoo, *int.* Add: (Later examples.) Also used to express cursory dismissal (of a proposition, idea, etc.) or reproach, and to express discomfort or weariness (cf. PHEW *int.*).

1814 JANE AUSTEN *Mansf. Park* I. xv. 305 Phoo! Phoo! Do not be so shamefaced. **1840** HOOD *Up Rhine* 46, I..enquired how the untoward event had originated. 'Originated!—*phoo, phoo*—no such thing, it was done on purpose.' *c* **1874** D. BOUCICAULT in M. R. Booth *Eng. Plays of 19th Cent.* (1969) II. 171 Phoo! How my arms ache! **1960** J. STROUD *Shorn Lamb* xviii. 201 I'm fed up o' lodgings...Feet orf my sofa, no smokin' in front o' children—phoo! **1978** 'J. LYMINGTON' *Waking of Stone* i. 26 She..dropped on the seat and went, 'Phooo!'

Hence **pho(o)-pho(o)** *v. trans.*, to ridicule, = POOH-POOH *v.* Also *absol.*

1865 *Cornh. Mag.* June 755 He pho-pho'd the poor ghost. **1866** *Ibid.* Aug. 141 It is easy to blame and to phoo-phoo.

phooey (fū·i), *int.* (*sb.*) orig. *U.S.* Also **phooie.** [f. PHOO *int.* + -Y[6], or ad. *PFUI *int.*] An expression of strong disagreement with or disapproval of something said. Also as *sb.*, applied to the thing said: nonsense, 'baloney'.

1929 *Sun* (Baltimore) 11 July 11/1 Girls are described as weenies, janes, dames and broads. A mad-man is phooey, crackers or blooey. **1936** O. NASH *Primrose Path* 185 And I'll say, 'Phooie!' or something of the sort. **1940** R. CHANDLER *Farewell, my Lovely* xix. 95 'Ten o'clock at The Belvedere Club,' I said. Somebody said: 'Phooey'. **1946** —— *Let.* 29 Jan. in *R. Chandler Speaking* (1966) 45 So let's not have any more of that phooey about 'as literature my stuff still stinks'. **1951** J. B. PRIESTLEY *Festival at Farbridge* I. ii. 55 Oh phooey, Benny... This don't count as a drink. **1957** J. BRAINE *Room at Top* x. 94 'Keep right on believing that, and it won't be long before I see your name in the Sunday papers.' 'Phooey,' I said. 'It's a simple straightforward transaction.' **1967** *Boston Globe* 30 Mar. 18/2 The governor's advisers, spoilsports, have said 'Phooey' to the technicalities. **1972** R. LOCKRIDGE *Something up Sleeve* (1973) x. 135 The answer to that was simple. It was 'Phooey!' **1975** A. PRICE *Our Man in Camelot* v. 94 'Oh—phooey.' She scowled at him.

phorate (fǫ·rēit). [f. phos*phoro*dithio*ate*, f. PHOSPHORO- + DI-[2] + THIO- + -ATE[1].] A systemic and soil insecticide that is effective against a wide range of insects and is also poisonous to man on contact or ingestion; *O,O*-diethyl-*S*-(ethylthio)methylphosphorodithioate, $(C_2H_5O)_2PS{\cdot}S{\cdot}CH_2SC_2H_5$.

1959 *Jrnl. Econ. Entomol.* LII. 1032 Changes in common names of insecticides... Name to be used..phorate. **1962** *New Scientist* 7 June 508/3 Disyston (thiodemeton) and Thimet (phorate)..gave the best control of aphids and the most improved yield. **1973** *Pesticide Sci.* IV. 90 Phorate has been effective for the control of two-spotted spider mite. *Ibid.* 97 Foliar application of phorate results in moderate initial residues on mint hay which disappear rapidly.

phorbol (fǫ·ɹbǫl). *Chem.* [a. G. *phorbol* (Flaschenträger & Boehm 1927, in *Ber. über die ges. Physiol. und exper. Pharmakol.* XLII. 585), f. Gr. φορβή fodder, forage (f. φέρβειν to feed): see -OL.] A tetracyclic compound, $C_{20}H_{28}O_6$, some of the esters of which are cocarcinogens and are present in croton oil.

1935 *Chem. Abstr.* XXIX. 2533 Phorbol ($C_{20}H_{28}O_6$ or $C_{20}H_{30}O_6$) with very dilute H_2SO_4 in N_2 gives 38% crotophorbolon. **1939** *Thorpe's Dict. Appl. Chem.* (ed. 4) III. 434/2 Phorbol and the product of its benzoylation are physiologically inactive, but the (?tetra-)acetate is highly toxic. **1967** *Chem. & Engin. News* 16 Oct. 42/1 A possible absolute configuration of phorbol, polyfunctional parent alcohol of tumor-promoting compounds in croton oil, has been determined. **1978** *Nature* 17 Aug. 640/2 The most powerful known tumour promotors are the phorbol esters, which are not carcinogenic in themselves but can induce tumour growth after a subthreshold dose of carcinogen.

phoresis (forī·sis). *Med.* Now *rare* or *Obs.* [ad. Gr. φόρησις being carried; cf. *CATAPHORESIS, *ELECTROPHORESIS.] = *CATAPHORESIS a.

1902 HERDMAN & WILLEY in *Jrnl. Physical Therapeutics* III. 125 In view of this fact that certain substances are actuated to seek the anode while others are moved toward the cathode when a difference of potential is established in the liquid in which they are dissolved, or suspended, it would seem that a generic term, as *phoresis*, should be chosen to designate the therapeutic employment of this physical action of a direct electric current, a term which would comprehend in its meaning the moving power of the current in whatever direction upon particles suspended in it or held in solution by it, while the specific terms, *cataphoresis* and *anaphoresis* would, as now, indicate the direction taken by certain of these substances. **1911** C. S. POTTS *Electr.* (1912) ii. 50 The [ionic] migration will take place even through a porous barrier or membrane, in which case it is known as electric osmosis or phoresis. **1936** H. H. U. CROSS *Electr. in Therapeutics* vii. 184 The essential distinction between the modern ionic medication and the older cataphoresis (or simply phoresis).

-phoresis (-forī·sis), *suffix* [f. as prec.], forming sbs. which describe the movement of small particles by some agency, as *CATA-, *ELECTRO-, *IONO-, *IONTO-, *PHOTOPHORESIS.

phoresy (forī·si, fǫ·rəsi). Also **phoresis.** [a. F. *phorésie* (P. Lesne 1896, in *Bull. Soc. Ent. France* 164), f. Gr. φόρησις being carried.] An association in which one organism is carried by another, without being a parasite upon it. Hence **phore·tic** *a.*, of or pertaining to an association of this type.

1923 L. O. HOWARD in *Entomol. News* XXXIV. 90 (*heading*) An interesting new case of phoresie. **1927** —— in *Ibid.* XXXVIII. 145 In 1896 P. Lesne..proposed the name *phorésie*..to describe the carriage of small insects by larger insects without the actual feeding of the smaller upon the larger in the adult stage... It is proposed to give it the English form, *phoresy*. Apparently, in this form it has not occurred in the English language, although the word *phoresis* is used by electricians to express the conduction of substances dissolved in a liquid through a membrane by means of a current. **1942** E. O. ESSIG *College Entomol.* xxxiv. 663 The remarkable phenomenon of phoresy occurs in the case of *Rielia manticida* Kieffer, the adult females of which attach themselves to the body of the praying mantid. **1962** J. D. SMYTH *Introd. Animal Parasitol.* i. 3 Phoresis. This term is used for a particular type of association in which one organism merely provides shelter, support or transport for another organism of a different species. *Ibid.*, In phoresis..there is no metabolic dependence of either of the associates on the other. **1965** B. E. FREEMAN tr. *Vandel's Biospeleol.* xv. 245 Phoresy or epizoism, that is to say the fixation of a plant or animal species on to the surface of another organism without the first living at the expense of the second. *Ibid.* 248 The ciliates may be divided into internal parasites, and epizoic or phoretic types which do not live at the expense of their hosts. **1969** R. F. CHAPMAN *Insects* xvii. 327 An association in which an animal of one species provides transport for another species is known as phoresy.

-phoria (fǭ·riă), comb. form f. Gr. *φόρος* bearing (f. *φέρειν* to bear: see -IA[1]), used in *Ophthalm.* to form terms denoting a tendency to squint, as *ESOPHORIA, *heterophoria* s.v. *HETERO-.

phoria (fǭ·riă). *Ophthalm.* [f. prec.] A tendency for the eyes to be directed towards different points in the absence of a visual stimulus.

1891 *Trans. Amer. Ophthalm. Soc.* VI. 136, I..then record the 'phoria' or 'heterophoria' found. **1937** *Times* 23 July 18/5 Every discovered phoria with a history of flying trouble was not for certain the absolute root cause. **1975** *Nature* 20 Nov. 202/2 Now phorias, or tendencies to strabismus, are common in infancy.

phormium. Substitute for etym. and def.: [mod.L. (J. R. & G. Forster *Characteres Generum Plantarum* (1776) 47), f. Gr. *φορμίον*, dim. of *φορμός* mat, basket, in reference to the use made of the fibres of the leaves.] An evergreen plant of the genus so called, belonging to the family Liliaceæ, native to New Zealand, and distinguished by long, tough leaves in tufts at the base and large, erect panicles of dull red or yellow flowers; = *New Zealand flax* s.v. *NEW ZEALAND I. (Earlier and later examples.)

1821 J. YULE in *Edin. Philos. Jrnl.* V. 345 The following results..manifest the comparative superiority in strength of the fibres of the Phormium. **1905** [see *MILLABLE *a.*]. **1933** W. J. BEAN *Trees & Shrubs Hardy in Brit. Isles* III. 274 Strictly speaking, the phormiums are, I suppose, 'herbs', but they are genuinely evergreen. **1955** *Sci. News Let.* 2 Apr. 213/2 Four fiber crops also under study for use on American farms are ramie, kenaf, sanseveria, and phormium. **1972** S. EMBERTON *Year in Shrub Garden* III. 150 The tall, snow-trapping evergreen leaf blades of the phormiums (New Zealand Flax) never regain their poise once their stiff backs have been broken by the weight of snow. **1976** *Country Life* 26 Feb. 478/1 Shrubs like..phormiums, variegated osmanthus, carefully chosen hebes.

phorometer (forǫ·mĭtəɹ). *Ophthalm.* [f. Gr. *-φόρος* bearing, f. *φέρ-ειν* to bear + -METER.] Any instrument for measuring the degree to which the eyes tend to be differently directed.

1888 G. T. STEVENS in *Med. Rec.* 5 May 511/1 New instruments. A phorometer... This instrument is designed to assist in the determination of the visual lines to each other. *Ibid.* 511/2, I have now used the phorometer during several months past. **1941** *Arch. Ophthalm.* XXV. 483 As a part of the flight surgeon's equipment there is a phorometer trial frame with an adjustable bracket and a tripod stand, a rather complex and impressive piece of equipment. **1970** *Jrnl. Gen. Psychol.* LXXXII. 111 The apparatus consisted of..(*b*) a head rest, (*c*) American Optical phorometer, [etc.].

Phosfon (fǫ·sfǫn). Also **Phosphon** and with small initial. A proprietary name for an organophosphorus compound used to retard the growth of chrysanthemums and certain other garden plants. Also *Phosfon-D*.

1961 *Official Gaz.* (U.S. Patent Office) 22 Aug. TM104/2 Virginia-Carolina Chemical Corporation, Richmond, Va. Filed June 1, 1960. Phosfon-D for chemical height-retardant for chrysanthemums. First use Feb. 25, 1960. **1962** *Economist* 21 Apr. 248/2 The size of plants can be regulated by two new chemicals (one of which, phosfon-D, is already on the market) so that lilliputian Easter lilies and dwarf hydrangeas are possible. **1962** *Official Gaz.* (U.S. Patent Office) 2 Oct. TM7/1 Virginia-Carolina Chemical Corporation, Richmond, Va. Filed Feb. 8, 1962. Phosfon... For chemical height retardant for garden chrysanthemums. First use Jan. 1, 1962. **1968** *New Scientist* 10 Oct. 84/2 Studies in Israel have shown that another retardant, 2-4-dichlorobenzyltributyl phosphonium chloride (Phosfon) gives better control of the Oleander aphid. **1969** *Trade Marks Jrnl.* 23 Dec. 2098/2 Phosfon... Chemical substances..for use in agriculture and horticulture, being or containing compounds of phosphorus. Mobil Oil Corporation.., New York City.., United States of America; manufacturers and merchants. **1975** *Adv. Agronomy* XXVII. 116 Other chemicals which reduce stomatal aperture such as phosphon-D..tend to protect against smog injury.

phosgene. Add: Used as a poison gas in the war of 1914–18, and now as an intermediate in the manufacture of some synthetic resins and organic chemicals. (Further examples.)

1918 M. PLOWMAN *Right to Live* 3 They have been poisoned with phosgene. **1919** C. P. THOMPSON *Cocktails* 26 The laboratory where the Corps chemists pored over the latest phials of German phosgene. **1938** *Encycl. Brit. Bk. of Year* 144/1 Phosgene has a faint smell of musty hay; and it tends to work its way down into cellars and 'dug-outs'. **1953** KIRK & OTHMER *Encycl. Chem. Technol.* X. 397 Throughout 1917, phosgene was the gas used in largest tonnage by all belligerents. However, mustard gas was introduced by the Germans in 1917 and gradually supplanted phosgene. **1967** SIMONDS & CHURCH *Encycl. Basic Materials for Plastics* 432/1 Both sodium hydroxide and sodium promote the bisphenol A–phosgene reaction to form polyacrylcarbonates. Phosgene is produced catalytically from chlorine and carbon monoxide. **1978** A. PRICE *'44 Vintage* i. 12 His father..had been with him in the trenches and came back with a lungful of phosgene.

phosphagen (fǫ·sfădʒĕn). *Biochem.* [f. PHOSPHA(TE + -GEN.] An organic phosphate in muscle tissue (in vertebrates, creatine phosphate) whose phosphate group is readily released and transferred to adenosine diphosphate, thereby forming the triphosphate needed for muscular contraction.

1927 P. & M. G. EGGLETON in *Nature* 5 Feb. 194/2 There appears to be in muscle tissue an organic phosphorus compound which, by reason of its great instability in acid solution, has been confused hitherto with inorganic phosphate... The confusion is increased by the fact that this substance, the organic phosphorus compound which we have designated 'phosphagen', is intimately connected with the chemical mechanism of contraction. **1937** BEST & TAYLOR *Physiol. Basis Med. Pract.* liii. 971 In the presence of oxygen the phosphoric acid and creatine are resynthesized to phosphagen. **1965** *New Scientist* 18 Feb. 445/1 So far, animal tissues have been found to contain organic compounds, N-phosphorylated guanidines, whereas plants and microorganisms contain only inorganic polyphosphates. This clear distinction between the two characteristic types of 'phosphagen' seems to offer a basis for deciding whether a particular organism is a plant or an animal. **1972** J. A. WILSON *Princ. Animal Physiol.* v. 141/1 Several other phosphagens have been found in the annelids.

phosphatæmia (fǫsfæti·miă). *Physiol.* Also **phosphatemia.** [f. PHOSPHAT(E + Gr. *αἷμα* blood: see -IA[1].] The concentration of phosphates (and other compounds of phosphorus) in the blood. Less commonly = *hyperphosphatæmia* s.v. *HYPER- IV.

1926 *Jrnl. Amer. Med. Assoc.* 6 Feb. 451/1 Phosphatemia and fibroma.—Dalsace and Guillaumin noted an exaggerated amount of phosphorus in the blood of women with a fibroma. **1928** *Q. Cumulative Index Medicus* III. 163/1 Glycemia and phosphatemia following intravenous injection of defibrinated blood. **1961** *Lancet* 22 July 215/1 In a control group of 20 normal adult men the average value for serum-phosphate was 3·43 mg. per 100 ml.,.. so we can take 2·6 mg. per 100 ml. as the lower limit of normal phosphataemia. **1973** *Ann. d'Endocrinol.* XXXIV. 621 (*heading*) Role of the fœtal parathyroids in the regulation of calcemia and phosphatemia of the rat fœtus.

phosphatase (fǫ·sfătēⁱz, -s). *Biochem.* [f. PHOSPHAT(E + *-ASE.] **a.** Any enzyme which catalyses the synthesis or hydrolysis of an ester of phosphoric acid.

H. Euler (in *Zeitschr. für physiol. Chem.* (1911) LXXVII. 14) coined the G. *phosphatese*, having proposed that the termination *-ase* be restricted to enzymes which decompose substrates, and *-ese* be used for enzymes which synthesize. Hence quot. 1911.

[**1911** *Jrnl. Chem. Soc.* C. I. 1051 There is no evidence that the enzyme which synthesises carbohydrate phosphoric acid esters has any splitting action. The term *phosphatese* is suggested.] **1912** *Chem. Abstr.* VI. 2084 Expts. with pepsin, trypsin, and animal organ exts. failed to show conclusively the presence of phosphatase. **1931** *Times Lit. Suppl.* 2 Apr. 274/3 There are two mechanisms at work in calcification: (*a*) a phosphatase mechanism which produces in the matrix fluid, by hydrolysis of phosphoric ester, a condition of supersaturation with respect to the bone phosphate. **1951** WEST & TODD *Textbk. Biochem.* xiv. 471 The acid and the alkaline phosphatases of serum are so called because of optimum activity at an acid and at an alkaline pH respectively. **1968** [see *ISOENZYME]. **1973** B. J. WILLIAMS *Evolution & Human Origins* iv. 67/1 Red cell acid phosphatase provides a good illustration of the fact that alleles at a single locus can produce a 'continuous' trait in the manner described.

b. phosphatase test, a test applied to dairy products to find whether they have been adequately pasteurized.

1933 KAY & GRAHAM in *Jrnl. Dairy Res.* V. 70 If..the cream was pasteurised, its phosphatase content was completely or almost completely destroyed. This suggested to us the possibility that the phosphatase test might also be used for determining whether a given sample of butter had been made from fresh or from pasteurised cream. **1960** JUDKINS & KEENER *Milk Production & Processing* xiv. 256 The phosphatase test is very sensitive and will show very small variations in the time and temperature of pasteurization. As little as 0·1 per cent raw milk in pasteurized milk is also easily detected by this test. **1975** J. W. G. PORTER *Milk & Dairy Foods* xi. 55 The phosphatase test shows whether the proper temperature has been reached during pasteurization.

phosphate. Add: An ester or other organic derivative of a phosphoric acid; esp. in *Biochem.*, any of these derivatives of sugars, nucleosides, etc., which occur widely in living organisms. Also, a radical or group derived from a phosphoric acid. Also *attrib.* (Further examples.)

1895 *Jrnl. Chem. Soc.* LXVIII. I. 639 The liquid separates into two layers, the upper of which seems to contain triallylic phosphate. **1930** CAVEN & LANDER *Systematic Inorg. Chem.* ix. 237 Soluble phosphates, e.g. those of sodium,.. show marked hydrolysis in solution. **1953** *Nature* 25 Apr. 737/1 The bases are on the inside of the helix and the phosphates on the outside. *Ibid.* 741/2 The phosphate groups lie on the outside of the structural unit, on a helix of diameter about 20 A. **1953** FRUTON & SIMMONDS *Gen. Biochem.* xix. 422 In the formulæ for the sugar phosphates, the phosphoric acid residue is written in the undissociated form. **1954** A. WHITE et al. *Princ. Biochem.* xvi. 381 Enzymes which catalyze transfer of phosphate from ATP to an acceptor are designated kinases. **1957** *Encycl. Brit.* XVII. 780/2 Considerable tonnages of other phosphates such as $(NH_4)H_2PO_4$ and $(NH_4)_2HPO_4$ are also used in the fertilizer industry. **1964** N. G. CLARK *Mod. Org. Chem.* xxi. 433 Triphenyl phosphate ..is used as a plasticizer for various cellulose and vinyl plastics. **1968** A. WHITE et al. *Princ. Biochem.* (ed. 4) xv. 315 Three classes of high-energy phosphate compounds are known: acid anhydrides, phosphate esters of enols, and derivatives of phosphamic acid R—NH—PO_3H. **1972** FARAGO & LAGNADO *Life in Action* v. 129 The starting material, glucose, is converted into phosphate derivatives which then go through a series of molecular gyrations, and are finally broken down to three-carbon units. **1976** *Nature* 1 July 45/1 Adsorption of phosphate on hydrous metal oxides is of considerable importance in soil fertility and eutrophication studies.

b. Add to def.: Also to that of lime (calcium) as a mineral; also *sing.*, esp. when *attrib.*

1849 *Q. Jrnl. Geol. Soc.* V. p. lxxxiv, We find such phosphates surrounding some fossils, such as crustaceans from the London clay, leading us to infer a connexion between the animal matter and this substance. **1891** F. WYATT *Phosphates of Amer.* iii. 28 The principal phosphate mines of Canada have been located on those positions of the pyroxenite belt in which, at the surface, the apatite has shown signs of predominating. **1892** *Amer. Jrnl. Sci.* XLIII. 403 The South Carolina phosphates are nodular in form. **1936** J. C. BROWN *India's Mineral Wealth* (ed. 2) x. 225 The problem of the utilization of Indian rock phosphates demands more research than it appears yet to have received. **1954** K. B. CUMBERLAND *Southwest Pacific* ii. 42 In the central Pacific province the only mining is that of phosphate. **1976** *Nature* 10 June 452/3 If Abbott's Booby were found nesting in the group, this could be of crucial importance for the species, which is now reduced to about 2,000 pairs threatened by phosphate mining on Christmas Island. **1977** *Radio Times* 1 Jan. 54/4 The Banabans of Ocean Island.. claim that they have been exploited..for 76 years because their island turned out to be full of phosphate.

2. Special Comb.: **phosphate bond** *Biochem.*, a bond between a phosphate group and another part of a molecule, esp. such a bond in an adenosine phosphate which is hydrolysed to provide energy in living organisms; **phosphate glass,** a glass of which the major constituent is phosphorus pentoxide or a phosphate; **phosphate island,** an island consisting largely of phosphate rock; **phosphate rock,** rock containing a substantial amount of calcium phosphate (usu. in the form of apatite).

1940 *Jrnl. Biol. Chem.* CXXXIV. 463 Since pyruvic acid was found to promote adenylic acid phosphorylation, any such intermediate must contain an energy-rich phosphate bond. **1953** FRUTON & SIMMONDS *Gen. Biochem.* xv. 355 It has become the custom to refer to phosphate bonds whose hydrolysis is accompanied by the liberation of 8000 to 16,000 cal per mole..as 'high-energy' phosphate bonds.., whereas the ester phosphate bonds (as in glucose-6-phosphate) are termed 'low-energy' phosphate bonds. **1962** I. ASIMOV *Life & Energy* xxi. 291 The two phosphate groups at the end away from the adenosine portion are more easily hydrolyzed than phosphate bonds generally are. **1977** *Sci. Amer.* Mar. 147/2 It now seems that fast direct-energy release by the splitting of high-energy phosphate bonds supplies our initial power, up to half of the overall oxygen debt. **1937** *Glass* XIV. 519/1 This work originated from the observation that a phosphate glass..turns red on exposure to soft X-rays. **1950** J. E. STANWORTH *Physical Properties Glass* i. 8 It has long been known that phosphate glasses may be made which resist the attack of hydrofluoric acid solution or gas. **1959** W. A. WEYL *Coloured Glasses* i. 36 Phosphate glasses..offer certain unique features when used as a base for coloured glasses. **1925** *Econ. Geol.* XX. 276 The next change is when a phosphate island is elevated by earth movements. **1936** *Discovery* Aug. 260/1 He [sc. A. F. Ellis] 'discovered' Nauru as a phosphate island. **1870** F. S. HOLMES *Phosphate Rocks S. Carolina* 87 The Ashley Phosphate-rock is used in the manufacture of their fertilizer. **1915** W. S. BAYLEY *Minerals & Rocks* I. ii. 71 Phosphate rock is a mixture of apatite, phosphorite and various hydrated phosphates. **1923** A. B. SEARLE *Sands & Crushed Rocks* I. iv. 199 Coprolite and Phosphorite are natural phosphate rocks which are produced by the accumulation of organic remains. **1965** E. T. DEGENS *Geochem. Sediments* iii. 149 Phosphate rocks owe some of their rare metals to co-existing organic matter and sulfides rather than to phosphates themselves.

phosphate (fǫ·sfēⁱt), *v.* [f. the sb. Cf. PHOSPHATED *a.*] *trans.* = *PHOSPHATIZE *v.* 2.

1939 BURNS & SCHUH *Protective Coatings for Metals* xvi. 374 'Coslettizing' was..used in England for phosphating the steel parts of bicycles. **1963** H. R. CLAUSER *Encycl. Engin. Materials* 477/1 The fabricated parts are cleaned.., phosphated and rinsed thoroughly.

Hence **pho·sphating** *vbl. sb.*

1948 H. SILMAN *Chem. & Electro-Plated Finishes* v. 166 In the presence of accelerators the phosphating process is completed rapidly. **1940** J. C. HUDSON *Corrosion of Iron & Steel* vi. 136 The proprietary phosphating processes, such as Parkerising and Bonderising. **1956** WERNICK & PINNER *Surface Treatm. & Finishing of Aluminium & Alloys* vi. 213 As distinct from chemical oxidation, phosphating is carried out on a clean surface, and the work is therefore degreased before immersing.. in the phosphating solution. **1976** J. A. VON FRAUNHOFER *Basic Metal Finishing* xiii. 146 The simplest phosphating treatment is immersion in hot dilute orthophosphoric acid solution for up to 30 min. but many proprietary solutions and commercial processes are in use.

phosphatide (fǫ·sfătəid). *Biochem.* Also †-id. [f. PHOSPHAT(E + -IDE.] Formerly = *PHOSPHOLIPID; now *esp.* a fatty acid ester of glycerol phosphate in which a nitrogen base is linked to the phosphate group.

1884 J. L. W. THUDICHUM *Treat..Chem. Constitution Brain* i. 4 According to the result of this revision the phosphorised substances [in the brain] are not glycerides at all, as commonly defined, and have nothing in common with fats considered as glycerides, except that some of them contain fatty acids also present in fats... In accordance with this new knowledge, I have termed the phosphorised substances phosphatides, that is to say, substances which are similar to (but not by any means identical with) phosphates, on the assumption that their basal or principal joining radicle is that of phosphoric acid and that in this acid one, two, or three molec[u]les of hydroxyl may be replaced by radicles of alcohols, acids, or bases. **1910** [see *PHOSPHOLIPIN]. **1918** [see *HEPARIN]. **1921** B. HARROW *Vitamines* 43 A number of very complicated substances—the phosphatids—are found in larger quantities in the brain than in other parts of the body. **1927** [see *PHOSPHOLIPIN]. **1944** L. F. & M. FIESER *Org. Chem.* xix. 490 In the more active tissues (brain, liver, kidney, etc.) they [*sc.* fats] usually occur in a form more complex than depot fats, which are mainly glycerides... These fats can be divided into two classes, phosphatides and cerebrosides. The former on hydrolysis yield fatty acids, a nitrogenous base, phosphoric acid, and usually glycerol. **1951** H. J. DEUEL *Lipids* I. i. 4 Phospholipids or Phosphatides. The former term is most generally employed in the United States while the latter one is favored in English and German literature. **1953** FRUTON & SIMMONDS *Gen. Biochem.* xxii. 509 Evidence has been presented for the occurrence, in the..brain, of a phosphatide that contains inositol. **1954** A. WHITE et al. *Princ. Biochem.* v. 87 The next large class of lipids..is the phospholipids, more usually called phosphatides... The members of this group all contain a nitrogenous base. **1968** *Ibid.* (ed. 4) iv. 69 The larger groups of naturally occurring phospholipids are termed phosphatides.

phosphatidic (fǫsfătəi·dik, -ti·dik), *a.* *Biochem.* [f. prec. + -IC.] *phosphatidic acid*: any of the esterified derivatives of glycerol phosphate in which the hydrogen atoms in both hydroxyl groups are replaced by fatty acid radicals.

1927 CHANNON & CHIBNALL in *Biochem. Jrnl.* XXI. 1115 It seems to us preferable to keep to Thudichum's original definitions [of *phosphatide* and *cerebroside*]. If this is done the diglyceridephosphoric acid and its salts are phosphatides. Since the acid is chemically the parent acid of the two more commonly occurring phosphatides lecithin and kephalin, we propose to call it phosphatidic acid. **1954** A. WHITE et al. *Princ. Biochem.* v. 88 The phosphatides of known structure which contain one atom of N per atom of P may be regarded as derivatives of phosphatidic acid. **1964** W. G. SMITH *Allergy & Tissue Metabolism* viii. 87 Once these fatty acids are converted to fatty acyl CoA they could be incorporated immediately into diglyceride formed from 1-glycerophosphate via phosphatidic acid. **1967** [see *phosphoglyceride* s.v. *PHOSPHO-]. **1979** *Sci. Amer.* Jan. 49/1 Two molecules of a fatty acid..are transferred from a donor to a molecule of glycerol phosphate, forming the most primitive phospholipid, phosphatidic acid.

Hence **phosphatidyl** (fǫsfătəi·dəil, -il, fǫsfæ·tidəil, -il), the univalent radical of phosphatidic acid formed by the loss of a hydrogen atom from the phosphate group; usu. *attrib.* or in *Comb.* in the names of esters (phospholipids), as *phosphatidyl choline, ethanolamine, serine.*

1941 J. FOLCH in *Jrnl. Biol. Chem.* CXXXIX. 973 In a recent paper it was shown that cephalin prepared in the usual manner..did not contain..all of its nitrogen in the form of ethanolamine, but that from 40 to 70 per cent was in the form of a *β*-hydroxyamino acid... The cephalin fraction containing the amino acid has now been isolated, and the amino acid crystallized as..*l*(+)-serine. We shall call this phosphatide 'phosphatidyl serine'. **1942** — in *Ibid.* CXLVI. 35 Cephalin prepared from brain by the classical methods has..been separated into three fractions: (*a*) phosphatidyl ethanolamine so called because it has its nitrogen as ethanolamine and its P as glycerophosphoric acid, [etc.]. **1954** A. WHITE et al. *Princ. Biochem.* v. 88 Phosphatidyl choline, or lecithin, has the following formula. **1964** *Oceanogr. & Marine Biol.* II. 173 Among the phospholipids of vertebrates is a class known as plasmalogens. These are closely related to the classical lecithin (phosphatidyl choline) and cephalin (phosphatidyl ethanolamine). **1967** *Information Bull. Internat. Union Pure & Appl. Chem.* XXX. 23 The term 'lecithin' is permitted but not recommended to designate a 1,2-diacyl-*sn*-glycero-3-phosphorylcholine. The recommended generic term for such compounds is 3-*sn*-phosphatidylcholine. **1971** M. F. MALLETTE et al. *Introd. Biochem.* ix. 324 Phosphatidylserine..is another phospholipid isolated from plant and animal sources.

phosphatize, *v.* Add further example and substitute for def.: To convert into a phosphate. Usu. as *PHOSPHATIZED *ppl. a.*

1885 *Q. Jrnl. Geol. Soc.* XLI. 80 There this process of phosphatizing the coral has been in operation on a most extensive scale.

2. *trans.* To treat with a phosphate; to coat (metal) with a phosphate to protect it against corrosion.

1939 *Metal Treatment* IV. 172 A modern phosphatising process for sheet steel work comprises the following steps:—1. degrease, 2 phosphatise, 3 chromate rinse, 4 dry off. **1961** *Engineering* 23 June 855/2 The skin which forms protects it from sprays used to clean and phosphatise the panels.

Hence **pho·sphatizing** *ppl. a.* and *vbl. sb.*; **phosphatization** (earlier and later examples).

1875 *Q. Jrnl. Geol. Soc.* XXXI. 362 At the Berwyn mine this process of phosphatization is most complete. **1900** *Proc. Geologists' Assoc.* XVI. 385 The 'rock phosphates' are phosphatised portions of the underlying Eocene and Miocene limestones... What was the nature of the phosphatising agent? **1939** *Automobile Engineer* XXIX. 460/1 Phosphatisation converts the work surfaces to a porous but adherent coating of iron phosphate. *Ibid.* 460/3 Specimens having the full treatment of phosphatising and chromate rinse showed markedly superior performance. **1965** G. J. WILLIAMS *Econ. Geol. N.Z.* xvi. 262/2 Reed believed the phosphatization to have occurred during the deposition of the Miocene beds. **1976** J. R. DUNCAN tr. *Bartoň's Protection against Atmospheric Corrosion* v. 114 Chemical pre-treatment methods have come into use as an answer to the problem of finding methods for rust stabilization... Phosphatizing is the most frequently used of these methods. **1977** A. HALLAM *Planet Earth* 173 The phosphatization produced nodules and pellets up to 2 cm..in diameter.

phosphatized (fǫ·sfătəizd), *ppl. a.* [f. PHOSPHATE + -IZE + -ED[1].] **1.** Converted into a phosphate; = PHOSPHATED *a.*

1875 *Q. Jrnl. Geol. Soc.* XXXI. 361 The shales..are rich in organic remains. These..are all more or less phosphatized. **1935** *Jrnl. R. Anthrop. Inst.* LXV. 345 It was much sought after years ago for its contained rolled lumps of phosphatized clay—the so-called 'coprolites'. **1976** *Nature* 12 Feb. 473/2 They are characterised by.. phosphatised fish debris and faecal pellets.

2. Treated or coated with a phosphate.

1939 *Metal Treatment* IV. 170 Phosphatised work, when successfully effected, was superior to work having the protective coating alone. **1976** J. R. DUNCAN tr. *Bartoň's Protection against Atmospheric Corrosion* v. 114 There is greater resistance to under-rusting of phosphatized substrates under cracked or mechanically damaged layers.

phosphazene (fǫ·sfăzīn). *Chem.* Also **-ine.** [ad. G. *phosphazin* (Staudinger & Meyer 1919, in *Helv. Chim. Acta* II. 619), f. *phosph-in* PHOSPHINE + *azin* *AZINE.] Any compound containing the group —N=P< (with two further bonds on P), esp. as a repeating unit of a ring or chain in which two substituents are attached to each phosphorus atom. Cf. *PHOSPHONITRILE, *PHOSPHONITRILIC *a.*

1920 *Jrnl. Chem. Soc.* CXVIII. 105 Tertiary phosphines combine more or less readily with the most varied aliphatic diazo-compounds, yielding derivatives, for which the name phosphazines is proposed, in accordance with the scheme $CR_2{:}N{:}N + PR_3 \rightarrow R_2C{:}N{-}N{:}PR_3$. **1958** J. R. VAN WAZER *Phosphorus & its Compounds* I. vi. 341 The compound $(C_6H_5)_3PNN{=}C(C_6H_5)_2$, a typical phosphazine, is a yellow, stable solid with slightly basic properties. **1961** *Inorganic Polymers* (Chem. Soc. Special Publ. No. 15) 115 The family of polymers, which are called variously phosphinic nitrides, phosphonitriles, and phosphazenes. These names all describe a bonding system characterized by the repeating unit —PR_2=N—. **1970** *New Scientist* 5 Nov. 275/1 The cyclic nitrogen and phosphorus compounds known as phosphazenes will make life easier for textiles manufacturers in the UK and the US... Phosphazenes, which are incorporated into viscous rayon during the spinning process,..are ideal for flame proofing. **1974** *Nature* 29 Nov. 427/3 A simple survey of the structural chemistry of the phosphazines.

phosphite. Add: Also, an ester of this acid. (Further examples.)

1890 *Jrnl. Chem. Soc.* LVIII. 858 Ethyl phosphite, $P(OEt)_3$, is obtained,..when phosphorus trichloride is gradually added to finely divided sodium ethoxide, covered with ether. **1966** J. CASON *Princ. Mod. Org. Chem.* xxxiii. 603 Phosphorus trichloride reacts with alcohols to give phosphite esters... Only the tri-esters of phosphorous acid are known. **1970** *Encycl. Polymer Sci. & Technol.* XII. 748 Tertiary phosphites are among the most efficient groups of materials which have been used as poly(vinyl chloride) stabilizing agents.

phospho-. Add: † **pho:sphocozy·mase** *Biochem.* [*COZYMASE] = NAD(P) s.v. *N II. 1; **phosphocre·atine** *Biochem.*, creatine phosphate, HOOC·CH_2·N(CH_3)·C(NH)·NH·PO·$(OH)_2$, the phosphagen of vertebrate muscle; **pho:sphodiester** (-dəi,e·stəɹ) *Biochem.*, used *attrib.* to designate a bond of the kind joining successive sugar molecules in a polynucleotide or oligonucleotide, in which a molecule of phosphoric acid links a hydroxyl group in one molecule to a hydroxyl group in the next with the loss of two molecules of water (giving the sequence —O·PO(OH)·O— between carbon atoms); **pho:sphodiesterase** (-dəi,e·stərēⁱz, -s) *Biochem.* [a. G. *phosphodiesterase* (S. Uzawa 1932, in *Jrnl. Biochem.* (Japan) XV. 22)], any enzyme that breaks a phosphodiester bond in an oligonucleotide; **pho:sphoe:nolpyru·vate** *Biochem.*, the anion CH_2:C(O·PO$(OH)_2$)·COO[−] derived from the phosphate ester of the enol of pyruvic acid; **phosphofe·rrite** *Min.* [ad. G. *phosphoferrit* (Laubmann & Steinmetz 1920, in *Zeitschr. für Kryst. und Min.* LV. 569)], a hydrated phosphate of ferrous iron and bivalent manganese (the former predominating), $(Fe,Mn)_3(PO_4)_2{\cdot}3H_2O$, which forms an isomorphous series with reddingite and occurs as pale green, translucent or transparent, orthorhombic crystals; **pho:sphofru:ctoki·nase** *Biochem.*, an enzyme that catalyses the phosphorylation of fructose phosphate to fructose diphosphate; **pho:sphoglu:comu·tase** *Biochem.*, an enzyme that catalyses the transfer of a phosphate group between the first and the sixth carbon atoms of a molecule of glucose phosphate; **phosphogly·ceride** *Biochem.*, any phospholipid with a structure based on glycerol phosphate; **phosphoino·sitide** *Biochem.*, any phospholipid in which inositol is linked to the phosphate group; **phosphoki·nase** *Biochem.* = *KINASE b; **pho:sphomonoesterase** (-mǫno,e·stərēⁱz, -s) *Biochem.* [a. G. *phosphomonoesterase* (S. Uzawa 1932, in *Jrnl. Biochem.* (Japan) XV. 20)], any enzyme that removes a terminal phosphate group from an oligonucleotide or a mononucleotide; **phosphophy·llite** *Min.* [ad. G. *phosphophyllit* (Laubmann & Steinmetz 1920, in *Zeitschr. für Kryst. und Min.* LV. 566), f. Gr. *φύλλον* leaf, in allusion to its perfect cleavage], a brittle, almost colourless secondary mineral that is a hydrated phosphate of zinc, iron, and manganese, $Zn_2(Fe,Mn)(PO_4)_2{\cdot}4H_2O$, and occurs as tabular, monoclinic crystals; **phosphopro·tein** *Biochem.*, any protein that contains phosphorus other than in a nucleic acid or a phospholipid; **phosphosi·licate**, any substance which contains phosphate and silicate anions, or consists largely of the corresponding oxides; freq. *attrib.*

1946 *Nature* 24 Aug. 275/2 It was found with a preparation of cozymase and phosphocozymase received from Prof. Otto Warburg that only one pentose was found for two phosphorus atoms. **1964** Phosphocozymase [see *COZYMASE]. **1927** FISKE & SUBBAROW in *Science* 22 Apr. 402/2 Our main evidence for the existence of 'phosphocreatine' in muscle is of a quite different nature. **1950** G. A. BAITSELL *Human Biol.* (ed. 2) xiv. 243 The energy for the resynthesis of phosphocreatine, following contraction, comes indirectly from the oxidation of glucose in the muscle cells. **1969** J. I. ROUTH et al. *Essent. Gen., Org. & Biochem.* xxxiv. 666 Creatine..is especially abundant in muscle tissue, where it is combined with phosphoric acid as phosphocreatine, or creatine phosphate. **1953** FRUTON & SIMMONDS *Gen. Biochem.* vii. 190 The products obtained from PNA preparations have been shown to be chains of nucleosides linked by 3′,5′- (or 2′,5′-)phosphodiester bonds. **1975** *Sci. Amer.* July 25/3 A DNA strand is a chain of nucleotides... The individual nucleotide building blocks are connected by phosphodiester bonds between the carbon atom at position No. 3 on one sugar and the carbon atom at position No. 5 on the adjacent sugar. [**1932** *Chem. Abstr.* XXVI. 2994 (*heading*) The phosphormonoesterase and the phosphordiesterase.] **1937** *Ibid.* XXXI. 10078/1 Phosphodiesterase. *See* Phosphatases. **1967** *Biochim. & Biophys. Acta* CXLII. 111 (*heading*) The action of snake venom phosphodiesterase on liver ribosomal ribonucleic acids. **1968** J. H. BURN *Lect. Notes Pharmacol.* (ed. 9) 65 Cyclic AMP is broken down by phosphodiesterase to the inactive 5-AMP. **1968** Phosphodiesterase [see *phosphomonoesterase* below]. **1956** W. F. H. M. MOMMAERTS in O. H. Gaebler *Enzymes* xiv. 319 The system consists..of a hydrolyzing enzyme (actomyosin), a common substrate (ATP), a resynthesizing enzyme (pyruvate kinase), and a reservoir substrate (phosphoenol pyruvate or PEP). **1970** *New Scientist* 23 Apr. 168/1 The last step in the glycolytic pathway, from phosphoenolpyruvate to pyruvate, is effectively irreversible... But in liver and kidney there is a carboxylation reaction which reverses the step. **1921** *Mineral. Abstr.* I. 125 *Phosphoferrite* from Hagendorf, as cloudy-white or greenish, crystalline masses with greasy lustre; H. 4–5, sp. gr. 3·156. **1955** *Ibid.* XII. 562 Phosphoferrite $(Fe,Mn)_3(PO_4)_2{\cdot}3H_2O$ and reddingite $(Mn,Fe)_3(PO_4)_2{\cdot}3H_2O$ have recently been found in the south pegmatite quarry at Hagendorf, Bavaria, as hydrothermal alteration products of triphylite. They are orthorhombic and isomorphous, the former..of cubo-octahedral or tabular (on 001) habit. **1974** *Nature* 27 Sept. 305/2 The evidence clearly points to a complete solid solution series between $Fe_3^{2+}(H_2O)_3[PO_4]_2$ (phosphoferrite) and $Fe_3^{3+}(OH)_3[PO_4]_2$ (kryzhanovskite). **1947** *Federation Proc.* VI. 297 (*heading*) The purification of phosphofructokinase from rabbit muscle. **1970** AMBROSE & EASTY *Cell Biol.* xiii. 246 Fructose-6-phosphate is phosphorylated by ATP to form fructose-1,6-diphosphate in a reaction catalyzed by a specific phosphofructokinase. .. This phosphorylation is also essentially irreversible and is an important control point in metabolism. **1938** G. T. CORI et al. in *Jrnl. Biol. Chem.* CXXIV. 543 A study of the enzyme which causes a migration of the phosphate group from carbon atom 1 to the spacially [*sic*] adjacent carbon atom 6; the enzyme will be referred to as phosphoglucomutase. **1964** A. WHITE et al. *Princ. Biochem.* (ed. 3) xxi. 403 The best-studied mutase is phosphoglucomutase, which has been obtained in pure form from rabbit muscle, yeast, and several bacterial species. **1954** A. WHITE et al. *Princ. Biochem.* xxxii. 800 These phosphatides include phosphoglycerides, phosphosphingosides, and the phosphoinositides. **1967** *Information Bull. Internat. Union Pure & Applied Chem.* XXX. 22 The term 'phosphoglyceride' signifies any derivative of glycerophosphoric acid that contains at least one *O*-acyl, or *O*-alkyl, or *O*-alk-1′-en-1′-yl group attached to the glycerol residue... The term 'phosphatidic acid' signifies a derivative of glycerophosphoric acid in which both remaining hydroxyl groups of glycerol are esterified with fatty acids. **1970** A. L. LEHNINGER *Biochem.* x. 195 The most abundant phosphoglycerides in higher plants and animals are phosphatidyl ethanolamine and phosphatidyl choline. **1954** Phosphoinositide [see *phosphoglyceride* above]. **1961** WEST & TODD *Textbk. Biochem.* (ed. 3) vi. 154 Two different types of phosphoinositides have been described which are differentiated by the inositol derivatives yielded upon hydrolysis. One type found in heart, liver, soybean, and wheat germ yields inositol monophosphate, fatty acids, and *α*-glycerol phosphate..; it..is phosphatidyl inositol... Another type of phosphoinositide found in brain..yields inositol-m-diphosphate, glycerol, and fatty acids. **1971** D. G. BISHOP in Johnson & Davenport *Biochem. Lipids* xx. 411 Although phosphoinositides have been recognized as important constituents of brain lipids for some time, their occurrence in nature is now known to be widespread. **1946** DIXON & NEEDHAM in *Nature* 28 Sept. 435/1 Needham has given reasons for believing that the component of the pyruvate oxidase system which is inhibited may be a phosphate-transferring enzyme belonging to the same group as hexokinase. We suggested the name 'phosphokinase' for this small but important group of enzymes. **1953** *Brit. Med. Bull.* IX. 110/2 The phosphokinases all seem to have the common property that they require the presence of Mg^{++} for their activity. **1962, 1964** Phosphokinase [see *KINASE b]. [**1932** Phosphomonoesterase: see *phosphodiesterase* above.] **1933** *Chem. Abstr.* XXVII. 1020 (*heading*) Phosphomonoesterase of animal organs and especially of the blood. **1968** PASSMORE & ROBSON *Compan. Med. Stud.* I. xii. 25/1 The combined action of the depolymerases, phosphodiesterases and phosphomonoesterases is to degrade RNA and DNA to the state of nucleosides. **1921** *Mineral. Abstr.* I. 125 *Phosphophyllite* from Hagendorf, colourless to pale-blue, well-developed, monoclinic crystals,..with perfect micaceous cleavage and lamellar twinning both parallel to *c* (001); H. 3–4, sp. gr. 3·081. **1968** *Mineral. Mag.* XXXVI. 624 In both phosphophyllite and hopeite ..the P—O tetrahedron shares one oxygen with six-coordinated zinc (Zn^{VI}) and three with four-coordinated zinc (Zn^{IV}). **1908** *Jrnl. Biol. Chem.* IV. p. l, Joint Recommendations of the Physiological and Biochemical Committees on Protein Nomenclature... *Phosphoproteins.* Compounds of the protein molecule with some, as yet undefined, phosphorus containing substance other than a nucleic acid or lecithin. **1929** R. A. GORTNER *Outl. Biochem.* xvi. 391 Casein of milk and vitellin of egg yolk are the two most important phosphoproteins. **1968** H. HARRIS *Nucleus & Cytoplasm* i. 14 Synthesis of a specific plasma phosphoprotein induced in male chickens by the administration of diethylstilboestrol. **1964** *IBM Jrnl. Res. & Devel.* VIII. 378/2 The importance of the phosphosilicate layer for transistor stabilization has also been shown by the examination of production transistors after life tests. **1966** *Solid-State Electronics* IX. 1009 An i.r. spectrophotometer..detects accurately the presence of the phosphosilicate layer. **1975** *Physics Bull.* Jan. 14/1 A team at Southampton University has also achieved very low losses with a new type of fibre—a phosphosilicate (P_2O_5/SiO_2) glass core contained in a pure silica cladding. **1975** *Nature* 27 Feb. 722/2 Except for the chemical analysis of minerals such as viseite and wilkeite, few systematic studies of phosphosilicates have been made.

phospholipase (fǫsfoli·pēⁱz, -s). *Biochem.* [a. G. *phospholipase* (H. Udagawa 1935, in *Jrnl. Biochem.* (Japan) XXII. 324): see next and *-ASE.] Any enzyme that hydrolyses lecithin (phosphatidyl choline) and similar phospholipids; = *LECITHINASE.

1945 *Jrnl. Biol. Chem.* CLVII. 643 A method is described for following the partial hydrolysis of phospholipids by the phospholipase of moccasin venom. **1965** HANAHAN & BROCKERHOFF in Florkin & Stotz *Comprehensive Biochem.* VI. iii. 91 Through the use of quite specific enzymes, called the phospholipases (or lecithinases), the phosphatidyl choline (lecithin) molecule can be effectively degraded in a stepwise manner... The enzyme, phospholipase A, is found in high concentration in many species of snake venom... Phospholipase C has been found in high concentration in plant tissues, such as carrots, spinach and cabbage leaves whereas phospholipase D occurs mainly in bacteria. **1974** DUNN & BONDY *Functional Chem. of Brain* iv. 84 Lecithin (phosphatidylcholine) may be hydrolyzed by phospholipase A to lysolecithin (lecithin lacking its *β*-fatty acid).

phospholipid (fǫsfoli·pid). *Biochem.* Also **-ide** (now *rare*). [f. PHOSPHO- + *LIPID.] Any compound whose products of hydrolysis include fatty acids, phosphoric acid, and (with some writers) a nitrogen base; *esp.* one that is an ester of glycerol phosphate; in recent use applied more widely to any lipid containing phosphoric acid, esp. ones having a structure based on glycerol phosphate.

1928 *Industr. & Engin. Chem.* (*News Ed.*) 10 Sept. 3 The Commission [on the Reform of Biological Chemical Nomenclature]..proposes the following classification of the group lipides... The term *phosphatides* is abolished. It is replaced by the terms *phospholipides* and *phosphoaminolipides*. Phospholipides are the lipides which contain phosphorus (in the form of the phosphoric radical). Phosphoaminolipides are the lipides which contain both phosphorus..and nitrogen. **1929** *Jrnl. Biol. Chem.* LXXXII. 117 (*heading*) The rôle of the phospholipids of the intestinal mucosa in fat absorption. **1946** *Nature* 27 July 119/2 These results indicated strongly that the vaccinia and ectromelia hæmagglutinins were composed of a virus antigen plus a phospholipid component, the latter being responsible for union with the surface of the cell. **1951** [see *PHOSPHATIDE]. **1951** E. A. ZELLER in Sumner & Myrbäck *Enzymes* I. xxx. 987 The fact that the hemolysis caused by many animal poisons is connected with the breakdown of lecithin considerably stimulated the investigation of the metabolism of phospholipides. **1953** FRUTON & SIMMONDS *Gen. Biochem.* xxii. 511 In the plasma of mammals..nearly all the phospholipid contains choline, and is therefore largely composed of lecithins and sphingomyelins. **1967** I. L. FINAR *Org. Chem.* (ed. 5) I. xi. 292 One subgroup of the phospholipids is the phosphatides... Another group of phospholipids is the sphingolipids. These are not glycerides... Sphingolipids are also known as sphingomyelins. **1967** *Information Bull. Internat. Union Pure & Appl. Chem.* XXX. 26 The term 'phospholipid' may be used for any lipid containing a radical derived from phosphoric acid. **1971** M. F. MALLETTE et al. *Introd. Biochem.* ix. 321 The phospholipids can be defined as those lipids which contain phosphate esters of acylated glycerol..or sphingosine. **1972** D. J. HANAHAN in F. Snyder *Ether Lipids* ii. 27 Structure (7) is the unusual diether phospholipid found in an extremely halophilic bacterium. **1973** S. J. EDELSTEIN *Introd. Biochem.* xii. 210 Phospholipids are composed of fatty acids linked to a glycerol backbone. **1976** *Daily Colonist* (Victoria, B.C.) 31 Mar. 18/4 Liposomes are balls made from fatty substances called phospholipids, found in the membranes of animal cells.

† **phospholipin** (fǫsfoli·pin). *Biochem. Obs.* [f. PHOSPHO- + *LIPIN.] = prec.

1910 J. B. LEATHES *Fats* i. 41 Compounds of fatty acids containing phosphorus and nitrogen... Thudichum gave the name 'phosphatides' to a number of such substances... The indefinite nature of the term has led to its application frequently to substances that are not compounds of fatty acids at all. And it is indeed impossible to say what kind of substances should be regarded as phosphatides... In these pages, therefore, for the compounds of fatty acids containing phosphorus and nitrogen the term phospholipine will..be used. **1915** *Jrnl. Biol. Chem.* XX. 404 Preparations of phospholipins, *i.e.*, lecithin, the alcohol-soluble compound, and kephalin, the alcohol-insoluble compound, were made. **1927** M. BODANSKY *Introd. Physiol. Chem.* iii. 55 The phosphatides (also called phospholipins) are present in every animal and vegetable cell and are especially abundant in the brain, heart, muscles, liver, and eggs. On hydrolysis these substances yield fatty acids, a nitrogenous base, phosphoric acid, and usually glycerol. **1948** *New Biol.* V. 38 Their [*sc.* microsomes'] principal constituents are phospholipins (complex fatty molecules containing phosphoric acid and organic acids and bases) and nucleoproteins.

Phosphon: see *PHOSFON.

phosphonitrile (fǫsfonəi·trəil). *Chem.* [f. PHOSPHO- + NITRILE.] = *PHOSPHAZENE.

1921 J. R. PARTINGTON *Text-bk. Inorg. Chem.* xxxi. 625 At 175–200°, ammonia and phosphorus pentachloride form a mixture of six phosphonitrile chlorides: $(PNCl_2)_3$, $(PNCl_2)_4$, [etc.]. **1943** *Chem. Rev.* XXXII. 102 Most investigators have accepted the designation phosphonitrile for the PN group, since it is looked upon as a group resembling the CN or nitrile radical. **1961** [see *PHOSPHAZENE]. **1972** H. R. ALLCOCK *Phosphorus-Nitrogen Compounds* i. 3 The cyclo- or polyphosphazenes (phosphonitriles) are probably the best known and most intensively studied phosphorus–nitrogen derivatives.

phosphonitrilic (fǫ:sfonəitri·lik), *a.* *Chem.* [f. as prec. + -IC.] Containing the phosphonitrile group, —N=P< (with two further bonds on P). Cf. *PHOSPHAZENE.

1895 H. N. STOKES in *Amer. Chem. Jrnl.* XVII. 278, (*d*) is a polymer of an acid N≡P$(OH)_2$, which we may call phosphonitrilic acid. **1943** *Chem. Rev.* XXXII. 119 The phosphonitrilic chlorides form as complete a polyhomologous series as is known in the realm of chemistry. *Ibid.* 122 The phosphonitrilic halides undergo reactions of hydrolysis to yield partially and completely hydroxylated products—the latter being known as the phosphonitrilic acids. **1956** H. H. SISLER in Sneed & Brasted *Comprehensive Inorg. Chem.* V. I. ii. 134 All the phosphonitrilic chlorides, when heated to about 300°C, are converted to an elastic product, possessing many of the properties of rubber. **1974** D. E. C. CORBRIDGE *Struct. Chem. Phosphorus* xii. 333 Phosphonitrilic compounds, or phosphazenes, are characterized by the presence of the group P=N—.

phosphor, *sb.* (*a.*) **2.** Delete †*Obs.* and add: In mod. use [after G. *phosphor* (Lenard & Klatt 1904, in *Ann. der Physik* XV. 226)], any substance exhibiting phosphorescence or fluorescence, esp. one that is an artificially prepared solid. Also *attrib.* and *Comb.*

The more restricted meaning in quot. 1950 is not usual.

1910 *Sci. Abstr.* A. XIII. 269 Different phosphores consisting of an alkaline earth sulphide and heavy metal. **1914** *Chem. Abstr.* VIII. 3751 For Sa-Ca-'phosphor'.. the rate of decay of the phosphorescence increases with increasing temp. **1943** *Endeavour* Jan. 25/1 Phosphors are principally used for the fluorescence which they exhibit. **1950** H. W. LEVERENZ *Introd. Luminescence of Solids* v. 147 The generic term *luminophor* is subclassified into fluorophors..(fluorescent materials) and phosphors (phosphorescent materials). **1960** CHALMERS & QUARRELL *Physical Examination of Metals* (ed. 2) xvi. 777 Sodium iodide, activated with a trace of thallium..combines a number of properties which make it one of the most important tracer scintillation phosphors. **1961** G. MILLERSON *Technique Television Production* ii. 21 By applying the video signal to regulate the picture-tube's beam, a pattern of light and shade can be built up on the screen's phosphor, corresponding to the light distribution in the studio scene. **1971** D. POTTER *Brit. Eliz. Stamps* ii. 25 Phosphor bands are practically invisible in normal light. **1975** *New Yorker* 5 May 51/3 Television pictures are produced by a flow of electrons moving in straight lines across the phosphorcoated surface of a cathode-ray tube.

phosphorane. Restrict † to sense in Dict. and add: **2.** *Chem.* [after *methane, ethane,* etc. (-ANE 2 b).] Any compound that is regarded as a derivative of PH_5, the phosphorus having five covalencies.

1952 *Chem. & Engin. News* 2 June 2336/2 PH_5 is phosphorane (not phosphane, as the latter name is used in the series diphosphane, triphosphane, and so on..). **1963** *Q. Rev. Chem. Soc.* XVII. 411 Stable phosphoranes of the formula $Ph_3P{:}CXY$ (X,Y = CN, CO_2R, COR) can be obtained. **1970** *Nature* 25 July 335/2 Examples of the utility of phosphoranes in the Wittig olefin synthesis were numerous. **1971** P. A. T. HOYE in *Mellor's Comprehensive Treat. Inorg. & Theoret. Chem.* VIII. Suppl. III. xxxiii. 881 Pentaphenylphosphorane was first obtained by Wittig and Rieber by the reaction of tetraphenylphosphonium iodide with phenyl lithium... The phosphoranes $(C_6H_5)_4PC(C_6H_5)_3$ and *p*-$CH_3{\cdot}C_6H_4P(C_6H_5)_4$ are prepared similarly.

phosphorescence. Add: In scientific use now distinguished from fluorescence on techn. grounds (see quots.); (the various definitions are all broadly equivalent).

1949 P. PRINGSHEIM *Fluorescence & Phosphorescence* 5 A photoluminescence process of this type, involving the passage through a metastable level, is called phosphorescence. *Ibid.*, It is no longer possible to define some arbitrary duration of the emission process as the boundary between fluorescence and phosphorescence. *Ibid.* iv. 292 The duration of a fluorescence process is essentially independent of external conditions insofar as this duration is determined by internal transition probabilities... The duration of a real phosphorescence is fundamentally a function of temperature. **1950** H. W. LEVERENZ *Introd. Luminescence of Solids* iv. 124 If the excitation and emission process occurs in times approximating the natural lifetimes of excited nonmetastable isolated atoms (about 10^{-8} sec for optical transitions), the process is called fluorescence, whereas longer-duration processes are called phosphorescence. **1954** C. ZWIKKER *Physical Prop. Solid Materials* xiii. 228 Fluorescence is the process in which the radiating electron falls back from the same energy level to which it was raised by the impinging photon..; phosphorescence is that in which the electron, after being raised to a higher energy level, first moves to a metastable level from where it can only be moved by external interference. **1971** *Physics Bull.* Oct. 577/2 The terms fluorescence and phosphorescence denote allowed and partially forbidden transitions respectively, though the practical distinction based on lifetimes is largely arbitrary. **1973** *Sci. Amer.* June 51/3 Today the definition is more technical, fluorescence corresponding to 'spin-allowed' electric dipole transitions and phosphorescence to 'spin-disallowed' transitions.

phosphoric, *a.* Add: **1.** (Further *fig.* example.)

1929 A. E. COPPARD in *Legion Bk.* 61 Baxter and Brabazon..had been subjected to some phosphoric comments by the magistrate.

2. *phosphoric acid.* Add to def.: = *orthophosphoric acid* s.v. ORTHO- 2 a; also applied *loosely* to phosphorus pentoxide, P_2O_5, as a constituent of minerals and fertilizers, and (freq. in *pl.*) to any of the common acids (meta-, ortho-, and pyrophosphoric acid) which contain pentavalent phosphorus. (Further examples.)

1912 J. W. MELLOR *Mod. Inorg. Chem.* xxx. 597 The three phosphoric acids and their salts are distinguished by the difference in their behaviour towards silver nitrate. **1941** *Thorpe's Dict. Appl. Chem.* (ed. 4) V. 69/2 These phosphates have a range of from about 25–32% of phosphoric acid (P_2O_5). **1950** N. V. SIDGWICK *Chem. Elements* I. 745 Esters of all three types of phosphoric acid are known. **1952** W. H. WAGGAMAN *Phosphoric Acid, Phosphates & Phosphatic Fertilizers* (ed. 2) i. 13 The term phosphoric acid..has been, and still is used very loosely... In the fertilizer industry the term phosphoric acid refers to phosphorus pentoxide... To the chemical manufacturer and distributor phosphoric acid means orthophosphoric acid.

phosphorolysis (fǫsfǒrǫ·lisis). *Biochem.* [f. PHOSPHOR(US or *PHOSPHOR(YLATION + HYDR)OLYSIS.] A form of hydrolysis in which a bond in an organic molecule is broken and an inorganic phosphate group becomes attached to one of the atoms previously linked.

1937 *Enzymologia* II. 160 Phosphorolysis is an enzymic process. **1946** *Nature* 23 Nov. 746/2 It was suggested that the coenzyme in muscle, while taking up two hydrogen atoms in the pyridine nucleus through the addition of free phosphate, undergoes a phosphorolysis and is split into pyridin[e] nucleotide and adenosine diphosphoric or triphosphoric acid. **1970** R. W. MCGILVERY *Biochem.* xv. 296 Glycogen is mainly degraded by a simple phosphorolysis of the 1 → 4 glucosidic bonds to form glucose-1-phosphate... The primary reaction is catalyzed by the enzyme, phosphorylase.

Hence **pho:sphoroly·tic** *a.*

1937 *Enzymologia* II. 154 Phosphorolytic decomposition of glycogen. **1970** A. L. LEHNINGER *Biochem.* xv. 328 (*caption*) Phosphorolytic removal of a glucose residue from the nonreducing end of a glycogen chain by phosphorylase.

phosphor-roesslerite (fǫsfǒrǒ·slərəit). *Min.* Also **phosphorrösslerite.** [ad. G. *phosphorrösslerit* (Friedrich & Robitsch 1939, in *Zentralbl. für Min., Geol., und Paläont.* A. 143), f. *phosphor* PHOSPHORUS: see *ROESSLERITE.] A hydrated acid phosphate of magnesium, $MgH(PO_4){\cdot}7H_2O$, that is isomorphous with rösslerite and occurs as monoclinic crystals that are usually discoloured by impurities and that lose water on exposure to air.

1939 *Mineral. Abstr.* VII. 316 Mud in old workings in the gold mines at Schellgaden (in Salzburg) shows glistening crystals of the new mineral phosphorrösslerite... The water-clear crystals, usually yellow from iron staining, are monoclinic. **1951** C. PALACHE et al. *Dana's Syst. Min.* (ed. 7) II. 713 Phosphorroesslerite.

phosphoryl (fǫ·sfǒril, -əil). *Chem.* [f. PHOSPHOR(US + -YL.] **a.** The usu. trivalent radical PO. **b.** The univalent phosphate radical, —PO$(OH)_2$.

1871 *Jrnl. Chem. Soc.* XXIV. 1161 In the first case, ordinary phosphoryl trichloride is produced: $P_2O_5Cl_4 + PCl_5 = 3POCl_3$. **1912** J. W. MELLOR *Mod. Inorg. Chem.* xxx. 586 Phosphoryl fluoride, POF_3, as well as the other phosphoryl compounds can be made by the action of phosphorus pentoxide, P_2O_5, on the halogen acid. **1962** S. G. WALEY in A. Pirie *Lens Metabolism Rel. Cataract* 360 In the enzymatic reaction, there is a direct transfer of the phosphoryl (O=P(OH)(OH)—) group from phosphoenolpyruvate to ATP. **1964** N. G. CLARK *Mod. Org. Chem.* 553 Phosphoryl chloride ('phosphorus oxychloride'), $POCl_3$, is a colourless liquid..which reacts slowly with water giving phosphoric acid. **1964** W. G. SMITH *Allergy & Tissue Metabolism* viii. 82 Thus the synthesis of glycerophosphatide is dependent on the availability of these phosphoryl bases as well as diglyceride. **1979** *Science* 7 Dec. 1151/1 Figure 7, from this work, represents a hypothetical example in which phosphoryl is conducted from ATP on the left to a substrate S on the right.

phosphorylase (fǫsfǫ·rilēⁱz, -s, fǫ·sfǒrilēⁱz, -s). *Biochem.* [f. prec. + *-ASE.] An enzyme that introduces a phosphate group into an organic compound.

1939 G. T. CORI et al. in *Jrnl. Biol. Chem.* CXXVII. 771 Various mammalian tissues..contain an enzyme which can be extracted with water and which forms glucose-1-phosphoric ester..from glycogen and inorganic phosphate... This enzyme, which will be referred to as phosphorylase, initiates the degradation of glycogen. **1955** *Sci. News Let.* 7 May 297/1 Use of muscle to do work and its recovery depends upon the chemical action of an enzyme, phosphorylase, which is found in muscle. **1970** [see *PHOSPHOROLYSIS]. **1971** *Nature* 19 Feb. 529/1 The breakdown of glycogen, in accordance with instantaneous energy requirements, is controlled by the activity of glycogen phosphorylase. This enzyme exists in two states: phosphorylase *b*..is a form of low concentration... Under the action of phosphorylase kinase it is phosphorylated at one serine residue in each of the four subunits, to give the highly active phosphorylase *a*.

phosphorylation (fǫ:sfǫrilēⁱ·ʃən). *Biochem.* [f. as prec. + -ATION.] The introduction of a phosphate group into an organic molecule.

1925 *Chem. Abstr.* XIX. 3278 Dry yeast..did cause the formation of phosphoric ester with glucose. Slight phosphorylation occurred with yeast ext. **1931** [see *DEPHOSPHORYLATION]. **1960** *Radio Times* 22 Jan. 38/4 From the latest work it appears that the capacity to transform the light energy absorbed by chlorophyll into the energy of phosphate bonds (photosynthetic phosphorylation) is its unique feature. **1962** H. A. KREBS in A. Pirie *Lens Metabolism Rel. Cataract* 351 Owing to the obligatory coupling between the electron transport from substrate to oxygen and oxidative phosphorylation, oxidations cannot take place unless ADP and phosphate are available. **1973** R. G. KRUEGER et al. *Introd. Microbiol.* viii. 254/2 Let us..consider the mechanism of the electron transport system and of oxidative phosphorylation. The cell uses these mechanisms to capture energy from reduced coenzymes (NADH and NADPH) to store in ATP.

So **phospho·rylate** *v. trans.*, to introduce a phosphate group into; **phospho·rylated, phospho·rylating** *ppl. adjs.*

1931 *Jrnl. Biol. Chem.* XCII. 765 Phosphorylated sugars. **1937** *Nature* 20 Feb. 309/2 Prof. Verzár attributes these observations to the fact that glucose is rapidly phosphorylated in the cells of the living mucous membrane. **1937** *Biochem. Jrnl.* XXXI. 329 (*heading*) Competition between phosphorylating enzymes in muscle extract. **1962** R. VAN HEYNINGEN in A. Pirie *Lens Metabolism Rel. Cataract* 399 In general, carbohydrate metabolism in animal tissues proceeds by way of phosphorylated intermediates. **1970** R. W. MCGILVERY *Biochem.* x. 197 Racker has..suggested that the non-heme iron, that is, iron not associated with porphyrins,.. may also be associated with the phosphorylating complex rather than with the electron transfer system. **1971** [see *PHOSPHORYLASE].

phot (fǭᵘt). *Physics.* [a. F. *phot*, f. Gr. *φῶς, φωτ-* light.] † **a.** A unit of the product of illumination and duration, equal to one lux maintained for one second. *Obs.*

1894 tr. A. Blondel in *Electrician* 28 Sept. 634/2 Some years ago the photographers..established under the presidency of eminent *savants* (MM. Abney, Cornu, Janssen, Sébert, Violle, &c.), a unit of time-illumination (*illumination*) specially applicable to photography, the 'phot'.

b. A unit of illumination equal to one lumen per square centimetre (equivalent to 10,000 lux).

1917 *Trans. Illuminating Engin. Soc.* (U.S.) XII. 440 Using the centimeter as the unit of length, the unit of illumination is one lumen per square centimeter, for which Blondel has proposed the name 'phot'. One millilumen per square centimeter (milliphot) is more useful as a practical unit. **1939** [see *metre-candle* s.v. *METRE *sb.*[2] b]. **1953** AMOS & BIRKINSHAW *Television Engin.* I. 280 The phot is rather a large unit, and its submultiple the milliphot (equal to 10^{-3} phot) is frequently used.

Photian (fǭᵘ·ʃăn), *sb.* and *a.* [f. *Photius*, the name of a ninth-century Patriarch of Constantinople + -IAN.] **A.** *sb.* A follower or supporter of Photius. **B.** *adj.* Of or pertaining to Photius or the schism in which he took a part. Hence **Pho·tianism; Pho·tianist** *a.* and *sb.*

1849 D. ROCK *Church of our Fathers* I. I. ii. 90 Not only the united or orthodox Greeks..but the Photians or separated Greeks, and the other sects in the East living apart, by schism and heresy, from Rome, entirely agree with her and the Latins upon Transubstantiation. **1850** Æ. MCD. DAWSON tr. *J. M. de Maistre's The Pope* IV. iii. 307 The Photian churches are preserved in the midst of Mahometanism, as an insect is preserved in amber. *Ibid.* x. 340 Among the Photians, on the contrary, as among all other Protestants, there is no unity. **1854** J. H. NEWMAN *Lect. Hist. Turks* III. ii. 183 The unhappy city..which had been successively the seat of Arianism, of Nestorianism, of Photianism. **1864** in E. S. Purcell *Life & Lett. A. P. de Lisle* (1900) I. xv. 388 Photianism and Anglicanism are two forms of the same true Christian religion. **1907** *Catholic Encycl.* II. 45/1 The mutual bitterness which was evinced in Constantinople by the contending parties of Photians and Anti-Photians was reflected here in Athens... Sabbas, who succeeded Anastasios, was likewise a Photian. **1948** F. DVORNIK *Photian Schism* I. i. 27 The Fathers had listened to the Photianist bishops Zachary and Theophilus. *Ibid.* ii. 46 The anti-Photianist Collection...contains all the documents against Photius and served the extreme Ignatians as an armoury in their struggle against the Photianists. **1967** *New Catholic Encycl.* XI. 328/2 We have only..an extract in Greek, preserved in the anti-Photianist collection. *Ibid.*, The Photian legend grew in the West, picturing the patriarch as the father of schism and the archenemy of papal primacy.

photic, *a.* Delete *rare* and add further examples.

No reference is now implied to any 'fluid' of light.

1919 *Jrnl. Exper. Zool.* XXIX. 254 Tactile receptors, photic receptors, and chemoreceptors [of chitons] are physiologically distinct. **1957** G. E. HUTCHINSON *Treat. Limnol.* I. xiii. 757 The rather feeble photic requirements of the purple bacteria. **1971** *Nature* 30 Apr. 579/2 The clam has just been placed on its side and, in response to both tactile and photic disturbance (a shadow cast on the exposed tissues), has tightly closed its valves. **1973** *Black Panther* 18 Aug. 8/3 The photic driver..uses flashing infrared light and pulsing ultrasonic noise to pull brain waves from their normal frequency. The subjected individual sees and hears nothing, becomes lethargic, dizzy, perhaps nauseous, and may suffer epileptic fits.

b. Applied to the parts of the oceans penetrated by sufficient sunlight for the growth of plants.

1903 W. R. FISHER tr. *Schimper's Plant-Geogr.* 782 Three chief stages of brightness may be distinguished: 1. The photic or bright region, in which the intensity of light is sufficient for the normal development of macrophytes. **1972** A. LAURIE *Living Oceans* iv. 88 For small, slow-moving animals the twilight zone is also a safer habitat than the photic zone, which is the home of many active predators, especially fast-moving fish. *Ibid.* vi. 126 The photic zone of the oceans reaches down to a depth of 650 feet below the surface, which is deeper than the continental shelf. The entire shelf sea lies within the photic zone.

Hence **pho·tically** *adv.*, by light.

1960 *Recent Adv. Biol. Psychiatry* II. 181 Photically evoked cerebral patterns have been studied. **1971** *Jrnl. Gen. Psychol.* LXXXIV. 22 Later components of the photically evoked cortical potential.

photism. (Earlier example.)

1892 D. H. TUKE *Dict. Psychol. Med.* II. 1126/1 Most sound photisms are projected on externality.

photo. Add: **1. A.** *sb.* (Earlier examples.)

1860 QUEEN VICTORIA *Let.* 24 Oct. in R. Fulford *Dearest Child* (1964) 275 She is waiting to know..about the photo. *Ibid.* 28 Nov. 286, I send you (*to look at only*) a wonderful photo: of the Queen of Naples. **1861** D. G. ROSSETTI *Let.* 11 July (1965) II. 411 The only way I know about Scott's photos, is to send you a set I have.

b. = *PHOTO-FINISH. Also *attrib.*

1937 *N.Y. Times* 3 Jan. S7/3 Kindred Spirit survived a hair-raising finish to take a photo decision at 18-1 under Jimmy Start. **1946** *Sun* (Baltimore) 2 Oct. 15 War Trophy..closed with a belated rush..to earn the photo. **1950** *Ibid.* 1 June 21/8 (*heading*) Tar wins in photo. **1976** *Scottish Daily Express* 23 Dec. 10/6 Brigadier General Preston Gilbride's ex-Irish colt..collared Gambling Prince..smoothly in the closing stages to win by four lengths..with the photo showing Bronson to have held O'Conna out of third spot.

B. *vb.* (Earlier and later examples.)

1868 D. G. ROSSETTI *Let.* 21 Feb. (1965) II. 653, I shall be anxious to have a set of his admirable photo'd drawings. **1928** A. HUXLEY *Let.* 12 Dec. (1969) 304 The type is photoed on to sheets of jelly..and printed from the jelly. **1973** A. BEHREND *Samarai Affair* xi. 109 O.K. sarge. Do we..wait till you've had him photoed?

photo-. Add: **1. photoabso·rbing** *ppl. a.*, that absorbs light; capable of absorbing a photon; **photoabso·rption,** absorption of a photon; **photoact** (fǭᵘ·to,ækt) *Biochem.* = next; **pho·toaction** *Biochem.*, a molecular event caused by light; **photoa·ctivate** *v. trans.*, to induce a change in or render active by means of light; hence **photoa·ctivated** *ppl. a.*; **pho:toactiva·tion,** activation by means of light; **photoa·ctive** *a.*, capable of or involving a chemical or physical change in response to illumination; hence **pho:toacti·vity,** the degree to which a substance or system is photoactive; **photoaffi·nity** *a.* *Biochem.*, applied to a technique of labelling large molecules (esp. proteins) at specific sites by means of molecules which initially form loose complexes at the active sites and are then photochemically converted *in situ* to reactive forms which immediately bond more permanently; so *photoaffinity label* sb. and vb. trans.; **photoa·llergy,** an allergy brought about by light; hence **photo-alle·rgic** *a.*; **pho:toassimila·tion** *Biol.*, photo-synthetic assimilation; so **photoassi·milate** *v. trans.*; **pho:toautotro·phic** *a.* *Bot.* [after G. *photoautotrophie* sb. (E. G. Pringsheim 1932, in *Naturwiss.* XX. 479/1)], autotrophic and obtaining energy from light; so **photoau·totroph,** a photoautotrophic organism; **photobio·logy,** the study of the effects of light on living things; so **pho:tobiolo·gical** *a.*, **photobio·logist; photoblea·ching,** a loss of colour when illuminated; **pho:toconve·rsion** *Biochem.*, any reversible chemical change effected by light, esp. that of one form of phytochrome to the other; hence **pho:toconve·rt** *v. trans.*, to change by photoconversion; **pho:toconve·rtible** *a.*, capable of undergoing photoconversion; **pho·tocurrent,** an electric current induced by illumination; **pho·todamage,** damage caused by (esp. ultraviolet) light; **pho:todensito·meter,** an instrument for measuring the density of a photographic negative or the opacity of a fluid; hence **pho:todensito·metry; photodestru·ction,** destruction brought about by light; **photodeta·chment** *Physics*, detachment of an electron from an atom caused by an incident photon; **pho·todetector,** a device that responds to incident light, esp. one whose operation depends on the electrical effect of individual photons; so **pho:todete·ction; pho:todisintegra·tion** *Nuclear Physics*, the breaking up of a nucleus by the action of a gamma ray; **pho·toeffect** *Physics*, a photoelectric effect, esp. the emission of an electron from an atom or of a nucleon from a nucleus by the action of a photon; **photoeje·ction,** ejection (of an electron from an atom) as a result of the absorption of a photon; **pho:toelectroche·mical** *a.*, of, pertaining to, or designating an electrochemical cell in which the electrode potential or the current flowing depends upon the degree of illumination of the cell; **pho:toelectromagne·tic** *a.* *Physics* = *photomagnetoelectric* adj. below; **pho·toenvi:ronment,** the environment formed by light; **photoenzyme** (fǭᵘ·to,enzəim) *Biol.*, an enzyme that catalyses a photochemical reaction; hence **pho:toenzyma·tic** *a.*, **-enzyma·tically** *adv.*; **photo-equilibrium** (examples); **pho:to-excita·tion** *Physics*, excitation (sense *5) caused by light or by a photon: so **pho:toexci·ted** *ppl. a*; **pho·tofabrica:tion,** the manufacture of integrated circuits by photolithography; **pho·tofission** *Nuclear Physics*, fission of an atomic nucleus caused by a gamma-ray photon; **pho·toformer** *Electronics*, an apparatus for generating a voltage corresponding to a given curve, incorporating a cathode-ray tube, a photo-electric cell, and an intervening opaque mask with an edge cut to the shape of the curve; **pho:toheterotro·phic** *a. Bot.* [after G. *photoheterotrophie* sb. (E. G. Pringsheim 1932, in *Naturwiss.* XX. 479/1)], heterotrophic and obtaining energy from light; hence **pho:toheterotro·phically** *adv.*; also **photohe·terotroph,** a photoheterotrophic organism; **pho:toinactiva·tion** *Biochem.*, destruction by light of the biological activity of an enzyme or other substance; **photola·bile** *a.*, unstable in the presence of light; opp. *photostable* adj. below; hence **photolabi·lity**; **photoli·thotrophy** *Bot.* [Gr. λίθος stone + τροφή nourishment], a form of nutrition in which energy is obtained photosynthetically from inorganic compounds; so **pho:tolithotro·phic** *a.*, characterized by photolithotrophy; **photoli·thotroph,** a photolithotrophic organism; **photomagnetic** *a.*, add: (*b*) = next; **pho:tomagne:toele·ctric** *a. Physics*, of, pertaining to, or designating an effect observed in some solids, whereby illumination of a solid subjected to a magnetic field parallel to its surface gives rise to a voltage at right angles to both the direction of the field and that of the illumination; **pho·tomask** *Electronics*, in the manufacture of microcircuits, a photographic pattern through which a photoresist is irradiated with ultraviolet light in order to transfer the pattern on to it; **photome·son** *Nuclear Physics*, a meson emitted from a nucleus as a result of the interaction of a gamma-ray photon with it; hence **photome·sic** *a.*, **pho·tomixer** *Physics*, a device that acts as a mixer (*MIXER 2 c) for light waves; so **pho·tomixing** *vbl. sb.*, the mixing of light waves in a heterodyne or homodyne process; **photoneu·tral** *a.*, unaffected in some respect by light; **photoneu·tron** *Nuclear Physics*, a neutron released from a nucleus by the action of a gamma-ray photon; **photonu·clear** *Nuclear Physics*, of or pertaining to the interaction of a photon with an atomic nucleus; **pho:toorgano·trophy** *Bot.* [Gr. τροφή nourishment], a form of nutrition in which energy is obtained photosynthetically from organic compounds; so **pho:toorganotro·phic** *a.*, characterized by photoorganotrophy; **pho:toorganotro·phically** *adv.*; **pho:toorga·notroph,** a photoorganotrophic organism; **photophil** (also **-phile**) *a.*, add to def.: thriving best in abundant light; *spec.* [after G. *photophil* (E. Bünning 1944, in *Flora* CXXXVIII. 95)], applied to a phase of the circadian cycle of a plant or animal during which light tends to stimulate reproductive activity; (further examples); **photophi·lic** *a. Biol.* = next; **photo·philous** *a. Biol.*, light-loving; thriving best in abundant light; **photo·phily,** the state of being photophilous; **photophysical** *a.* (further examples); so **photophy·sics** *sb. pl.* (const. as *sing.*); **pho·topigment** *Biol.*, a pigment (e.g. in the eye) whose chemical state depends on its degree of illumination; **photopolarimeter,** substitute for def.: † (*a*) [ad. F. *photopolarimètre* (A. Cornu in *Compt. Rend. Assoc. Française pour l'Avancement des Sci. 1882* 253)], an apparatus for measuring the proportion of polarized light in a given beam; *rare*⁻⁰; (*b*) a telescopic apparatus for photographing distant objects (as planets) and measuring the polarization of light from them; hence **pho:topolarime·tric** *a.*, **-polari·metry**; **pho·to-potential,** an electric potential generated by light; **pho·toprocess,** a (biological or chemical) process involving light; **pho·toproduct,** a product of the chemical action of light; **pho·toproduction,** production by light or by a single incident photon; so **pho·toproduced** *ppl. a.*; **photopro·tein** *Biol.*, any protein active in the emission of light by a living creature; **photopro·ton** *Nuclear Physics*, a proton released from a nucleus by the action of a gamma-ray photon; **pho·toreaction,** a photochemical reaction; **photoreco·very** *Biol.* = *PHOTOREACTIVATION; **photore·gulate** *v. trans.*, to regulate (a biological process) by means of light; so **photore·gulated** *ppl. a.*; **pho:toregula·tion,** the act of photoregulating; **photore·gulator,** a biological mechanism that regulates a process according to the duration, intensity, etc., of the light which it detects; **pho·torepair** *Biol.*, repair of tissue brought about by the action of (visible or ultraviolet) light; so **photorepai·rable** *a.*; **pho:toresi·stance** *Physics*, (an) electrical resistance that is light-dependent; also, a photoresistor; **pho:toresi·stive** *a.*, exhibiting photoresistance; **pho·toresistor,** a resistor whose resistance varies according to its degree of illumination; **pho·toresponse,** a response to light; *esp.* a response of a plant mediated otherwise than through photosynthesis; so **photorespo·nsive** *a.*, **-respo·nsiveness**; **photoreve·rsal,** reversal of a (biological) process by the action of light; **photoreve·rsible** *a. Biol.*, capable of being reversed by the action of light; (of a substance) changing from one form into another and back again as the degree of illumination increases and decreases; hence **pho:toreversibi·lity**; **photose·nsor,** a sensor that responds to light; **photose·nsory** *a. Biol.*, pertaining to or involving the perception of light; **pho·to-shock** *Psychiatry* [ad. F. *photo-choc* (Cossa & Gastaut 1949, in *Ann. Médico-Psychologiques* CVII. 187)], a flash or series of flashes of light given as part of shock therapy; **photosta·ble** *a.*, stable in the presence of light; opp. *photolabile* adj. above; so **photostabi·lity**; **photosta·tionary** *Chem.*, applied to a state of equilibrium in a photochemical reaction in which the rate of dissociation of the reactants equals their rate of recombination; **pho:tostimula·tion,** stimulation by means of light; so **photosti·mulate** *v. trans.*, **photosti·mulated** *ppl. a.*, **-sti·mulating** *vbl. sb.*; also **photosti·mulator,** an apparatus used for photostimulation; **pho:tostimula·tory** *a.*, pertaining to or involving photostimulation; **pho·tosurface** *Electr.*, a surface which emits electrons when illuminated; **phototelegraphy** (examples); in mod. use, a system of facsimile telegraphy in which variations of tone are adequately preserved; hence **pho:totelegra·phic** *a.*; also **photote·legram,** a telegram sent by phototelegraphy; **photote·legraph,** an apparatus used for phototelegraphy; **phototoxi·city,** the property of causing a harmful reaction to sunlight; so **phototo·xic** *a.*; **pho:totransforma·tion,** a transformation (of a chemical compound) effected by light; **phototra·nsient** *Chem.*, a short-lived molecular species produced by irradiation; **pho·totroph** *Bot.*, a phototrophic organism; **pho·totrophy** *Bot.* [Gr. τροφή nourishment], nutrition in which sunlight is utilized as a source of energy; so **phototro·phic** *a.*, characterized by phototrophy; **phototro·phically** *adv.*; **photovoltaic** *a.*, substitute for def.: pertaining to, exhibiting, or utilizing the generation of an e.m.f. by light incident on an interface between certain pairs of substances; (examples); hence **photovolta·ically** *adv.*; **photovolta·ics** *sb. pl.* (const. as *sing.*), the branch of science and technology concerned with photovoltaic effects and devices.

1966 Phillips & Williams *Inorg. Chem.* II. xx. 87 A biological system obtains energy from the oxidation of organic substrates or from the action of light on its photo-absorbing pigments. 1977 I. M. Campbell *Energy & Atmosphere* viii. 252 (*heading*) The generation of the photoabsorbing species and their relative significance. 1966 *Physical Rev.* CXLIX. 55/1 (*caption*) Charge distribution of ions resulting from photoabsorption primarily in the 3*d* shell of krypton. 1976 *Physics Bull.* Dec. 544/2 The continuum states of two electron atoms and ions can be studied by electron scattering or by photoabsorption. 1971 *Nature* 5 Feb. 372/1 The second possibility is that the mechanism of photosynthesis switches from a reaction involving two photosystems to a simpler form involving but one photoact. 1971 R. Gregory *Biochem. of Photosynthesis* ii. 36 Eight separate 'photoacts' are involved, so that we should expect 2500/8, say 300 chlorophyll molecules to be associated with each reaction centre. 1957 *Plant Physiol.* XXXII. 397/2 Since the most probable photoreceptor is the oxidized form of the flavoprotein, the photoactions are effective in its return to the reduced form. 1965 Hendricks & Borthwick in T. W. Goodwin *Chem. & Biochem. Plant Pigments* xv. 409 Reversibility of light action..indicates that the photoactions are conversions of molecular configuration. 1926 *Ann. Rep. Progr. Chem.* XXII. 340 In the case of chlorine,..W. Taylor draws the conclusion that only absorption within the continuous absorption band will photoactivate the gas. 1959 *Mycologia* LI. 87 Pigmentation was photo-activated. Light of wave lengths between 390 and 513 mμ stimulated the production of colour. 1926 *Ann. Rep. Progr. Chem.* XXII. 360 In general, the 'electron-loosening' mechanism of Stark..describes the state of a photo-activated molecule better than the other conceptions put forward by photochemists. 1954 *Jrnl. Res. Nat. Bureau of Standards* (U.S.) LIII. 125 (*heading*) Catalytic photoactivated polymerization of tetrafluoroethylene. 1925 *Phil. Mag.* XLIX. 1116 (*heading*) A note on the photo-activation of chlorine. 1974 *Physiologia Plantarum* XXXII. 228 (*heading*) Action spectrum for photoactivation of the water-splitting system in plastids of intermittently illuminated wheat leaves. 1908 *Physical Rev.* XXVI. 541 A study of photo-active effects produced by illuminating one electrode only, no external electromotive force being applied. 1951 *Sci. News* XXII. 75 Perhaps indeed the carotenoids were the primaeval photoactive pigments which in the course of evolution of green plants and algae have been functionally although not physically replaced by..chlorophyll. 1975 *Nature* 10 Apr. 507/2 The properties of an organic photovoltaic cell in which the photoactive material is microcrystalline chlorophyll-*a*. 1915 *Physical Rev.* V. 45 This value of current was used in comparing the photo-activity of solutions. 1970 *Biochim. & Biophys. Acta* CCXXIII. 444 The photoactivity was measured by observation of the blue-shift. 1970 *Proc. Nat. Acad. Sci.* LXVII. 1694 Binding sites of similar specificities in the same membrane preparation..may be photo-affinity labeled by the same reagents. 1976 *Nature* 29 Apr. 802/1 Without irradiation the photoaffinity label reversibly inhibited the potassium conductance..whereas the sodium conductance was not affected. 1970 H. Kiefer et al. in *Proc. Nat. Acad. Sci.* LXVII. 1688 The method of affinity labeling is in widespread use for the attachment of covalent labels at the active sites of protein molecules. The usual affinity-labeling reagent has the structure R—X, where R is the portion of the molecule that binds specifically and reversibly to the active site under study, and X is a chemically reactive group, such as diazonium or haloacyl... In photo-affinity labeling, a reagent R—P is used, where P is a group that is ordinarily unreactive, but which can be converted by photolysis to an exceedingly reactive intermediate P*. Those molecules of R—P that are reversibly bound to the active sites react instantaneously upon conversion to R—P* before they can dissociate from the site. 1978 *Nature* 12 Jan. 157/1 The modified pheromone is radioactively labelled, and since the carbene is generated photolytically, the process is called photoaffinity labelling. 1939 *Jrnl. Investigative Dermatol.* II. 45 (*heading*) Mechanism of the photoallergic reaction. 1968 Hjorth & Fregert in A. J. Rook et al. *Textbk. Dermatol.* I. 300/1 Photo-allergic reactions can resemble sunburn. 1976 *Lancet* 20 Nov. 1116/1 Chloroquine..is also used in..photoallergic reactions. 1939 S. Epstein in *Jrnl. Investigative Dermatol.* II. 45 These experiments demonstrate a true allergic type of photosensitivity (photoallergy). As far as I can see, this is the first report of this particular type of photosensitization and the first experimental proof of the allergic nature of this form of light sensitivity. 1976 *Arch. Dermatol.* CXII. 1124/1 The diphenhydramine photoallergy was elicited by long-wave ultraviolet light. 1922 *Jrnl. Soc. Dyers & Colourists* XXXVIII. 8/1 The green pigment chlorophyll has been shown by Willstätter to be an equilibrium mixture of chlorophyll A and chlorophyll B... This equilibrium is not appreciably altered when photo-assimilation of CO_2 is taking place. 1975 *Nature* 9 Oct. 490/2 The ability of the cyanobacterium to photoassimilate CO_2 in reactions driven by photosystem I alone and using Na_2S was demonstrated..; no photoassimilation was observed in the absence of sulphide or light. 1951 J. W. Foster in Werkman & Wilson *Bacterial Physiol.* 363 Photoautotrophs are those which utilize light. 1974 *Encycl. Brit. Macropædia* X. 896/1 A green plant is a typical example of a photoautotroph. 1943 *Physiol. Rev.* XXIII. 350 This..implies that the organism so operating must be capable of carrying out conversions of organic compounds. Theoretically it should even be able to grow heterotrophically on the proper organic substrates. Many of the photo-autotrophic organisms have yielded to this treatment. 1975 *Nature* 25 Dec. 715/1 Manganese is required for the photoautotrophic growth of O_2-evolving organisms. 1977 A. Hallam *Planet Earth* 189 This may bear some relation to the rapid evolution of photoautotrophic organisms such as blue-green algae. 1907 *Chem. Abstr.* I. 190 (*heading*) The photobiological sensitizers and their proteid compounds. 1976 *Sci. Amer.* Feb. 119/1 The results of the experiment..also rule out for the wasp any model of a clock in which light induces, or starts, diapause or development by photobiological means other than mere entrainment. 1958 *Plant Physiol.* XXXIII. 447/1 Robert Bruce Withrow died on April 8, 1958... With his passing this country lost one of its prominent photobiologists. 1973 *Nature* 6 July 37/1 Wald has therefore suggested that photobiologists should plot spectral functions on a frequency scale. 1935 *Science* 31 May 526/2 The cure of rickets by ultra-violet light constitutes one of the most interesting chapters in photo-biology. 1941 H. F. Blum *Photo-dynamic Action* i. 3 Although.. the phenomenon has been found to have a more limited significance in photobiology, the name photodynamic action has persisted. 1968 *New Scientist* 5 Dec. 579/2 This Penguin survey is a valuable guide to the present knowledge and research work in photobiology. 1937 *Ann. Rep. Progr. Chem.* XXXIII. 426 Considerations of the photobleaching of fluorescent dyes in an oxygen-free atmosphere by the action of ferrous salts. 1974 *McGraw-Hill Yearbk. Sci. & Technol.* 127/2 Light conversion in photoactive chlorophyll is accompanied by photobleaching and by the simultaneous appearance of a free or unpaired electron. 1952 *Jrnl. Chem. Soc.* 4524 The photoconversion of (I) and (III) into coloured forms does not occur with light of wave-lengths greater than about 540 mμ. 1964 *Photochem. & Photobiol.* III. 521 The absorption spectra of the two forms of phytochrome show, in addition to the major absorption bands in the red and far-red regions, minor bands in the blue and near u.v. which are also effective in the photoconversions of phytochrome, $P_R \rightleftarrows P_{FR}$. 1971 *Nature* 27 Aug. 602/1 The implication is that acetylcholine is involved in electrical changes in the plant which presumably follow the photoconversion of phytochrome. 1962 *Jrnl. Physical Chem.* LXVI. 2476/1 Any X formed from B being immediately photoconverted into A. 1970 *Nature* 15 Aug. 666/1 The axis of orientation of the chromophores is parallel to the plasmalemma surface in the P_R form, but is changed to an orientation at 90° to

the surface of the plasmalemma when photoconverted to P_{FR}. 1962 *Jrnl. Physical Chem.* LXVI. 2469/2 Only this isomer is photoconvertible into the spiropyran by visible light. 1913 H. S. Allen *Photo-Electricity* x. 127 The proportionality factor between light absorption and photocurrent is only independent of the angle of incidence ϕ for an electric vector..vibrating at right angles to the plane of incidence. 1913 *Physical Rev.* I. 74 The photocurrent-potential curve was almost identical with that furnished by the mercury lamp. 1936 *Discovery* May 151/2 This type of light-sensitive cell..requires no battery to obtain the photo-current. 1974 *Nature* 26 Apr. 804/1 Light-induced release of protons might thus provide an alternative mechanism for generation of the photocurrents. 1973 *Nature* 12 Jan. 133/1 This demonstrates the potential importance of the excited states of tryptophan as intermediates in lens photodamage. 1977 I. M. Campbell *Energy & Atmosphere* ix. 348 The link between DNA photodamage in living tissue cells and carcinogenesis by radiation is somewhat empirical. 1928 *Jrnl. Optical Soc. Amer.* XVI. 222 A self-registering photodensitometer has been described in which the direct-reading method and a thermocouple are employed. 1949 *Jrnl. Appl. Physics* XX. 129/2 The specimen containing the diffused Na^{24} was placed directly on the emulsion for one-half hour... The plate was then developed and analyzed in a photo-densitometer. 1971 *Nature* 19 Feb. 572/1 For most plant shoots, however, the change in methylene blue concentration is far too small to be accurately measured using standard photodensitometers. 1971 *Ibid.* 16 July 185/1 The haloes were also evident in a photodensitometer profile across a cloud bank image, and extended 1 km from the cloud edge. 1965 *Biochemistry* (Easton, Pa.) IV. 1653/2 Sometimes absorption optics are used on the ultracentrifuge, and in this event concentrations are measured by photodensitometry. 1964 *Jrnl. Cell Biol.* XXII. 448/2 The most conspicuous structural change in the plastids during the 1 to 3 hours of photodestruction of the pigments is the formation of stacked lamellar structures. 1977 I. M. Campbell *Energy & Atmosphere* ix. 371 The minimum in the altitude concentration profile of nitric oxide with altitude near to 70 km is evidently a reflection more of the variation of production rate with altitude than of the variation of the photodestruction rate. 1943 *Phil. Trans. R. Soc.* A. CCXXXIX. 278 (*heading*) Photodetachment of electrons from normal O⁻ ions. 1973 *Nature* 26 Oct. 450/2 The profiles..confirmed the hypothesis that the daytime D region of the ionosphere below 80 km may be formed by photo-detachment of electrons. 1959 *Rev. Sci. Instruments* XXX. 593/1 The application of the parametric amplifier principle to photodetection. 1972 S. S. Charschan *Lasers in Industry* ix. 523 In direct photodetection, all optical frequency and phase information is lost. 1947 *Proc. Nat. Electronics Conf.* 1946 171 (*heading*) Photodetectors for ultraviolet, visible and infrared radiation. 1959 *Proc. IRE* XLVII. 1475/1 Until the early 1950's, the development of infrared photodetectors revolved principally around polycrystalline films of PbS, PbSe, or PbTe. 1967 *New Scientist* 16 Nov. 416/1 The infrared radiation falls on the photodetector and produces visible radiation from the photoemitter. 1972 S. S. Charschan *Lasers in Industry* ix. 529 Quantum or photodetectors depend on the action of light quanta on a single electron rather than on the absorption and distribution of energy over an entire macroscopic body. 1935 *Proc. R. Soc.* A. CLI. 481 The next point of interest is the probability of the 'photo'-disintegration. *Ibid.* 482 The effect of the γ-rays of radium in producing the photo-disintegration was also examined. 1942 J. D. Stranathan *'Particles' of Mod. Physics* xi. 444 Another illustration of photodisintegration is that of $_4Be^9$. This reaction is $_4Be^9 + h\nu \rightarrow {}_4Be^8 + {}_0n^1$. 1968 D. D. Clayton *Princ. Stellar Evolution & Nucleosynthesis* vii. 519 What happens then, as the temperature rises, may be described as a redistribution of loosely bound nucleons into more tightly bound states. We choose to call this process..photodisintegration rearrangement. 1903 *Encycl. Brit.* XXXV. 729/4 (Index), Photo-effect. 1913 H. S. Allen *Photo-Electricity* i. 8 For substances which show only a 'normal' photo-effect the specific photo-electric activity increases continuously as the wave-length diminishes. 1938 R. W. Lawson tr. *Hevesy & Paneth's Man. Radioactivity* (ed. 2) x. 123 These [γ-]rays are able to bring about nuclear photo-effects by the ejection of neutrons from the nuclei of various elements. 1960 R. H. Bube *Photoconductivity of Solids* i. 2 Two new photoeffects were discovered in the early 1930s. In 1931, Dember..reported that a potential difference was developed in cuprous oxide in the direction of the light. 1974 *Encycl. Brit. Macropædia* XV. 439/2 The photoeffect probability goes as approximately the fifth power of the atomic number of the absorbing material. 1966 *Physical Rev.* CXLIX. 55/1 Photoejection of a 3*d* electron. 1977 I. M. Campbell *Energy & Atmosphere* i. 2 The photoejection of electrons from a metal surface irradiated with monochromatic..light. 1972 *Nature* 7 July 37/2 If the energy of light is used effectively in an electrochemical system, it should be possible to decompose water with visible light. Here we describe a novel type of photo-electrochemical cell which decomposes water in this way. 1976 *Ibid.* 9 Sept. 100/1 Better and cheaper means of storing electricity..remain desirable, and hence the practical importance of photo-electrochemical decomposition of water into hydrogen and oxygen. 1953 *Proc. Physical Soc.* B. LXVI. 743 If the slab [of germanium] is placed in a magnetic field perpendicular to the direction of illumination, a voltage is produced at right angles to both field and illumination (photoelectromagnetic effect). 1965 K. F. Hulme in C. A. Hogarth *Materials used in Semiconductor Devices* vi. 153 The theory and the constructional details and performance of a practical room-temperature photoelectromagnetic detector have been given. 1965 M. Evenari in E. J. Bowen *Recent Progress in Photobiol.* v. 161 The ocean of light which constitutes the photoenvironment. 1973 *Nature* 6 July 37/1 The adrenal cortex, thyroid and pineal of birds are affected by the photoenvironment. 1962 *Jrnl. Gen. Physiol.* XLV. 703 (*heading*) Photoenzymatic repair of ultraviolet damage in DNA. 1975 *Nature* 17 Apr. 627/1 If ultraviolet biological damage can be reversed by true photoenzymatic repair, then dimers have a major role in the production of that damage. 1966 *Adv. Radiation Biol.* II. 23 A photoenzymatically reversible competitive inhibition of transforming DNA repair *in vitro*. 1960 Photoenzyme [see *PHOTOREACTIVATING *ppl. a.*]. 1966 *Adv. Radiation Biol.* II. 19 This type of PR [*sc.* photoreactivation] does not result from the same type of photoenzyme. 1924 H. S. Taylor *Treat. Physical Chem.* II. xviii. 1239 With constant illumination, between reaction temperatures of 50 and 800° C., the photo-equilibrium is the same, regardless of the gas temperature. This indicates that the temperature coefficient of the two photo-processes is the same. 1962 *Jrnl. Physical Chem.* LXVI. 2472/2 In reversible photoisomerizations photoequilibrium is established when the rates of the two opposing photoreactions A ⇄ B under the action of the particular photoactive light used, are equal. 1974 *Chem. Soc. Rev.* III. 332 As the sensitizer energy is reduced the efficiency of sensitization of the *cis*-isomer falls below that of the *trans*. As a result there is a region where the *trans*-isomer is selectively excited and the proportion of *cis*-isomer at photoequilibrium is greater. 1918 *Physical Rev.* XI. 485 Having followed the kathodo phosphorescence for 300 seconds by the usual well-known method of a single excitation and determined the form of the curve of decay, the relation of this curve to that obtained by photo-excitation is of importance. 1946 *Nature* 2 Nov. 603/2 In the large gap region it was necessary to increase the stress above that calculated, to provide the required photo-excitation. 1975 *McGraw-Hill Yearbk. Sci. & Technol.* 357/1 In photoconductors the carriers can be generated internally by photoexcitation. 1954 *Ann. Rev. Plant Physiol.* V. 277 In basic solvents like pyridine, photo-excited chlorophyll can be reduced by ascorbate. 1970 *Physics Bull.* Nov. 488/2 The migration of photoexcited electrons out of regions of high optical excitation to be entrapped in regions of low optical excitation. 1967 *Sci. Amer.* Apr. 47 Photofabrication starts with drawings and by chemistry and optics transforms them into the objects, usually with a linear reduction in scale. 1968 *Physics Bull.* Dec. 423/1 The application of holography to ..the photofabrication of microcircuits. 1939 *Physical Rev.* LVI. 449/2 We can form an estimate of the cross section for photo-fission by comparison with the yields of photoneutrons. 1974 *Encycl. Brit. Macropædia* XIV. 299/2 More complicated [photonuclear] interactions involve either the emission of heavier particles.., many particles, or photofission. 1949 *Electronics* Feb. 100/1 The Photoformer, as it is called since it generates waveshapes through the use of a cathode-ray tube and phototube, is fed with a sawtooth voltage of the desired frequency. 1965 *Math. in Biol. & Med.* (Med. Res. Council) i. 38 Figure 3 shows how an analogue computer is used to resolve such a curve into its components... A voltage generated from the curve by a photoformer is compared with the sum of exponentials generated in the computer. 1951 J. W. Foster in Werkman & Wilson *Bacterial Physiol.* 364 The above two classes of autotrophs have their counterparts in the heterotrophic bacteria. Thus there are chemoheterotrophs and photoheterotrophs... The latter are a specialized photosynthetic group which is capable of using for growth both radiant energy and preformed organic matter. 1963 *Studies on Microalgae & Photosynthetic Bacteria* (Jap. Soc. Plant Physiologists) 465 Two characteristic facultative photoheterotrophs.. have been examined for response to nitrate under various conditions of growth. 1945 E. I. Rabinowitch *Photosynthesis* I. v. 106 The metabolism of the 'photoheterotrophic' bacteria—that is, bacteria which require light for the assimilation of organic nutrients, seemed at first to be quite different from that of the 'photautotrophic' bacteria. 1975 *Nature* 18 Dec. 631/1 Typical purple bacteria produce large quantities of molecular hydrogen during photoheterotrophic growth on organic acids. 1972 *Science* 27 Oct. 404/3 Under anaerobic conditions in the light, cultures [of flexibacteria] grow photoheterotrophically. 1938 *Recueil des Travaux bot. Néerlandais* XXXV. 12 The decrease of the curvature with auxin-a by light must be ascribed to the photo-inactivation of the auxin-a-lactone. 1973 *Biochemistry* (Easton, Pa.) XII. 2540/2 Studies..on the photoinactivation of a purified bovine kidney mutarotase were undertaken. 1937 *Nature* 25 Sept. 545/1 In the chicken retina, which contains principally cones, attempts to identify photo-labile pigments heretofore have failed. 1975 *Ibid.* 31 Jan. 316/2 Some photolabile metabolite accumulates until a threshold concentration is reached. 1961 Webster, Photolability. 1968 *Arch. Biochem. & Biophysics* CXXIII. 109 (*heading*) Formation and photolability of a methyl cobalamin enzyme. 1958 R. Y. Stanier et al. *Gen. Microbiol.* 669 The most familiar examples of the photolithotrophs are the green plants, which use water as a hydrogen donor in photosynthesis. 1971 J. S. Poindexter *Microbiol.* xxi. 473 Ectotrophic mycorrhizae are found in many forest trees... Generally, these trees are photolithotrophs as adults, and their seeds contain sufficient organic nutrients to provide the energy for germination and development of photosynthetic capacity. 1976 *Nature* 18 Mar. 200/2 Photolithotrophs such as *Thiorhodaceae* or purple sulphur bacteria and *Chlorobacteriaceae* or green sulphur bacteria, learned, through photochemical promotion, to use inorganic reductants as electron donors. 1958 R. Y. Stanier et al. *Gen. Microbiol.* xiv. 292 For the enrichment of chemolithotrophic and photolithotrophic organisms, organic compounds must be omitted from the medium, and CO_2 or bicarbonate must be used as the only source of carbon. 1972 Goodwin & Mercer *Introd. Plant Biochem.* i. 3 The phototrophic bacteria are subdivided into photolithotrophic bacteria (e.g. green and purple sulphur bacteria) whose growth is dependent on exogenous inorganic hydrogen donors..and photoorganotrophic bacteria (e.g. purple, non-sulphur bacteria) whose growth is dependent on exogenous organic hydrogen donors. The biochemistry of photolithotrophic bacteria is related to that of green plants. 1947 *Cold Spring Harbor Symp. Quant. Biol.* XI. 302 The following terminology is hereby proposed to characterize nutritional types [of microorganism]... A. *Phototrophy.* Energy chiefly provided by photochemical reaction. 1. Photolithotrophy. Growth dependent on exogenous inorganic H-donors. 2. Photoorganotrophy. Growth dependent on exogenous organic H-donors. B. *Chemotrophy.* Energy provided entirely by dark chemical reaction. 1969 F. E. Round *Introd. Lower Plants* i. 2 Photolithotrophy is the common photosynthesis of plants possessing chlorophyll a and using water as the hydrogen donor. 1959 R. A. Smith *Semiconductors* viii. 315 From equations (309) and (328) we may obtain an expression for the ratio of the photo-magnetic current per unit magnetic field to the photo-conductive current per unit electric field. 1975 *Physics Bull.* Jan. 15/1 The materials in which such 'photomagnetic' phenomena have been observed so far are magnetic insulators or semiconductors. 1934 *Physik. Zeitschr. der Sowjetunion* V. 597 (*heading*) On the explanation of the photomagnetoelectric effect in semi-conductors. 1967 R. H. Bube in Willardson & Beer *Semiconductors & Semimetals* III. xi. 473 The photomagnetoelectric (PME) effect provides a technique for the determination of minority-carrier lifetimes. 1965 W. R. Runyan *Silicon Semiconductor Technol.* iv. 76 Silicon dioxide is very convenient to use as a mask since it can be easily delineated by standard photomask techniques. 1977 *Sci. Amer.* Sept. 114/3 Thus each photomask, typically a glass plate about five inches on a side, has a single pattern repeated many times over its surface. 1952 R. E. Marshak *Meson Physics* iii. 104 The photomesic production process probably leads, in the majority of cases, to excited states of the final nucleus having smaller spins than 4. 1974 (*title*) Photomesic and photonuclear reactions and investigation methods with synchrotrons. 1950 A. S. Bishop *Photoproduction of Mesons from Hydrogen* (Univ. of Calif. Radiation Lab., UCRL-874) 40 By definition, $\sigma(E_{ph})$, the excitation function at 90° for photo-meson production from protons, constitutes the probability that a photon of energy E_{ph}, interacting with a proton, will produce a meson at 90° in the laboratory system. 1951 *Physical Rev.* LXXXI. 189/1 The angular dependence of the nuclear cross section for photo-meson production..yields fairly direct information concerning the momentum distribution with[in] the nucleus. 1954 *Ibid.* XCV. 592/2 (*heading*) Negative-to-positive ratio of photomesons from deuterium. 1955 *Ibid.* XCIX. 1694/2 It is in the photoelectric mixing tube, or photomixer,.. that the beat frequency is generated. 1975 *Nature* 13 Feb. 515/1 We have built a heterodyne spectrometer using..a HgCdTe photodiode as a photomixer. 1962 *Appl. Optics* I. 51/1 This paper reports the observation of microwave signals produced by photomixing of near-neighbour axial mode components in the output spectrum of a ruby optical maser. 1966 M. Ross *Laser Receivers* iv. 125 Photomixing has been successfully achieved under laboratory conditions. However, no operational receiver incorporating photomixing has yet been announced. 1950 Curtis & Clark *Introd. Plant Physiol.* xx. 630 Some plants are evidently highly indifferent to the photoperiod with respect to their flowering behavior and will flower over almost any photoperiod ranging from a 5-hr daily exposure to a 24-hr, or continuous, exposure. Some of the plants that fall into this indeterminate, or photoneutral, group are buckwheat, sunflower, tomato, cotton, and dandelion. 1975 *Nature* 25 Dec. 712/2 Natural populations of *D. melanogaster* and *D. pseudoobscura* are photoneutral in general, but respond rapidly to selection for positive or negative phototactic behaviour. 1935 *Proc. R. Soc.* A. CLI. 488 The angular distribution of the photo-neutrons from beryllium was investigated. 1975 K. G. McNeill et al. in Jochim & Ziegler *Interaction Stud. in Nuclei* 451 Predictions have been made of the angular distributions of the photoprotons emitted from ⁴⁰Ca and going to the ground state of ³⁹K..and of the distribution of ground state photoneutrons. 1949 *Science* 2 Dec. 579/1 There is an appreciable background of photonuclear stars and proton tracks. 1959 Deutsch & Kofoed-Hansen in E. Segrè *Exper. Nucl. Physics* III. x. ii. 305 Photonuclear reactions are sometimes used for the detection of high-energy gamma-rays and for the measurement of their energies. 1973 *Physics Bull.* Nov. 694/3 The program will cover effective interactions in light nuclei, photonuclear reactions, neutron scattering below 15 MeV, [etc.]. 1965 A. H. Rose *Chem. Microbiol.* iii. 40 Thus we arrive at the following four nutritional categories of micro-organisms based on their energy-yielding metabolism: photolithotrophs, photoorganotrophs, chemolithotrophs and chemoorganotrophs. 1976 *Nature* 18 Mar. 200/2 Photo-organotrophs such as purple non-sulphur bacteria use as electron donors in the light, in anaerobic conditions, those organic electron donors which were used by fermenters in the dark. 1958 R. Y. Stanier et al. *Gen. Microbiol.* xiv. 292 Since the photoorganotrophic bacteria require various growth factors, a small amount of yeast extract is generally added to the enrichment medium. 1972 Photoorganotrophic [see *photolithotrophic* adj. above]. 1971 J. S. Poindexter *Microbiol.* xiii. 154 The few types of algae that can grow photoorganotrophically are aerobes. 1947 Photoorganotrophy [see *photolithotrophy* above]. 1952 *Physiologia Plantarum* V. 81 Bünning concluded that light is favourable to flowering during the 'rising' or photophile phase of the leaf movement, and inhibiting during the falling or scotophile phase. 1964 E. Bünning *Physiol. Clock* xiv. 122 During the long dark period the plants endogenously reach a second 'photophil' state. 1965 *Plant Physiol.* XL. 873/1 Light during the photophil phases may stimulate flowering to such an extent that..nearly every bud on the plant responds. 1975 D. Vince-Prue *Photoperiodism in Plants* v. 170 He [*sc.* Bünning] proposed that photoperiodism involves a regular oscillation of phases or half-cycles with different sensitivity to light, and postulated that transfer to light sets in motion a *photophile* (or light-loving) phase, which is followed about twelve hours later by a *skotophile* (dark-loving) phase, also of twelve hours duration. 1900 A. J. Ewart tr. *Pfeffer's Physiol. Plants* I. vii. 358 Even for light-loving (photophilic) plants bright diffuse daylight seems as a general rule to be preferable to strong sunlight. 1967 M. E. Hale *Biol. Lichens* v. 72 Photophilic algae as *Pleurococcus*, when lichenized would be inhibited by reduced light. 1905 I. B. Balfour tr. *C. E. von Goebel's Organogr. Plants* II. 463 They [*sc.* geophilous shoots]..are united by many intermediate stages with 'photophilous' shoots. 1967 M. E. Hale *Biol. Lichens* vii. 87 The photophilous (light-loving) characteristics of most lichens. 1934 Webster, Photophily. 1960 *Cold Spring Harbor Symp. Quantitative Biol.* XXV. 241/1 The phase of strongest responsiveness to temperature in both types coincides

with maximum responsiveness to light..: with maximum photophily in long-day plants and maximum scotophily in short-day plants. 1974 *Biol. Abstr.* LVII. 762/1 An ecological scale of photophily was developed. 1914 S. E. Sheppard *Photo-Chem.* p. vii, There exists..some difference of opinion as to the desirability of incorporating a discussion of photo-physical and radiation phenomena and laws in a work on photo-chemistry. 1971 *Physics Bull.* Sept. 546/1 (Advt.), A comprehensive treatment is given of the interactions of low energy electrons with atoms and molecules, and photophysical processes. 1976 *Nature* 15 Apr. 654/2 Two chapters deal with photophysical processes: the first is a brief survey of the electronic spectroscopy of complexes. 1961 M. Calvin in McElroy & Glass *Symposium on Light & Life* 317 A discussion of some of the photochemistry and photophysics of porphyrins. 1970 J. B. Birks *Photophysics of Aromatic Molecules* p. vii, There are six related subjects concerned with the interaction of radiation with molecular systems: photophysics, photochemistry and photobiology, which deal with optical non-ionizing radiation; and radiation physics, radiation chemistry and radiation biology, which deal with ionizing radiation. Photophysics is the keystone of the structure, since it is an integral constituent of each of the other five subjects. 1937 *Nature* 25 Sept. 545/2 The familiar Purkinje effect, for which clearly the extracted photo-pigments form an adequate chemical basis. 1964 S. Duke-Elder *Parsons' Dis. Eye* (ed. 14) xxiv. 364 It [*sc.* colour blindness] is an inherited condition,..and is probably due to the absence of one of the two photopigments normally found in the foveal cones. 1970 Hand & Davenport in P. Halldal *Photobiol. of Microorganisms* ix. 278 The photopigment responsible for phototaxis and photokinesis is probably flavin. 1971 *Time* 15 Mar. 46 Light measurements by Pioneer's imaging photo-polarimeter will enable computers on earth to construct about ten pictures of the planet [*sc.* Jupiter] that will show features as small as 250 miles across. 1972 *Daily Colonist* (Victoria, B.C.) 25 Feb. 5/2 In the last 20 hours before closest approach, the spacecraft's imaging photopolarimeter will take 100 pictures of the planet. 1974 *Nature* 6 Sept. 18/1 The imaging photopolarimeter of Gehrels *et al.* contained a 2·5 cm telescope which made maps of Jupiter. 1971 *Icarus* XV. 454 (*heading*) Photopolarimetric observations of the minor planet Flora. 1974 *Sci. Amer.* Feb. 43/1 More detailed knowledge of the planet's atmosphere awaits analysis of photopolarimetry measurements. 1914 *Physical Rev.* IV. 229 The results as a whole confirm the point of view adopted by Richardson and Compton in regard to the relation between photo- and contact potentials. 1924 *Jrnl. Physical Chem.* XXVIII. 333 Becquerel was the first to observe that the photo-potential of the silver iodide electrode: electrolyte cell was not always positive. 1976 *Nature* 9 Sept. 99/2 If we want to use a semiconductor with an optical bandgap, so as to give optimal utilisation of solar radiation.., the maximum photopotential attainable will be ~0·4 eV. 1924 Photoprocess [see *photo-equilibrium* above]. 1926 *Trans. Faraday Soc.* XXI. 560 Weigert sees in the photosensitisation of ozone decomposition by chlorine, the simplest possible photoprocess. 1959 W. H. Klein in R. B. Withrow *Photoperiodism* III. 207 (*heading*) Interaction of growth factors with photoprocess in seedling growth. 1974 *Photochem. & Photobiol.* XIX. 441/2 The versatility of flavins as photosensitizers in numerous photoprocesses. 1953 *Physical Rev.* XCI. 480/2 The cross section obtained in this way is modified principally by the presence of terms describing the multiple scattering of the photo produced mesons. 1973 *Physics Bull.* July 431/1 Modulation excitation (ME) spectrophotometry is a technique to measure the absorption spectra of short lived photoproduced transients, such as excited states of molecules. 1926 E. Mayer *Clin. Applic. Sunlight* iii. 30 This action of small doses of radiant energy may be due to toxic 'photo-product'. 1941 *Brit. Jrnl. Psychol.* XXXII. 79 The rate of dark adaptation is supposed..to be determined by the concentration of photoproduct present at each moment. 1977 *Nature* 17 Feb. 660/2 Comparison of spectra taken before and after prolonged irradiation..showed no change in the relative position or intensity of the shoulder, eliminating a permanent photoproduct generated by the high intensity pulse as the species responsible for the shoulder. 1950 *Federation Proc.* IX. 544/1 (*caption*) Photoproduction of H_2 from succinate by *Rhodopseudomonas gelatinosa*. 1950 A. S. Bishop *Photoproduction of Mesons from Hydrogen* (Univ. of Calif. Radiation Lab., UCRL-874) 5 From the measured energy distribution of the mesons at 90°..it is possible to determine the excitation function for photoproduction of mesons at 90°. 1961 *Nature* 13 May 602/1 Photoproduction of hydrogen gas by photosynthetic cells was first observed..in the green alga *Scenedesmus*. 1974 Frauenfelder & Henley *Subatomic Physics* xii. 326 (*caption*) Total cross sections for the photoproduction of neutral and charged pions from hydrogen, as a function of the incident photon energy. 1966 Shimomura & Johnson in Johnson & Haneda *Bioluminescence in Progress* 495 Solutions of the protein, for which the general term 'photoprotein' is suggested, show a fluorescence maximum at 458 mμ when excited at 350 mμ. *Ibid.* 497 As a convenient, general designation of the active component in the hydromedusan and *Chaetopterus* type of system, to which the terms 'luciferin' and 'luciferase' do not apply in their usual meaning, we propose the term 'photoprotein'. 1975 *Nature* 17 July 236/2 The photoprotein aequorin (molecular weight about 30,000) isolated from the bioluminescent jellyfish *Aequorea aequorea* emits blue light. 1935 Chadwick & Goldhaber in *Proc. R. Soc.* A. CLI. 480 The experimental arrangement for the detection of the protons released from deuterium, which we may for convenience call 'photo'-protons, was as follows. *Ibid.*, An estimate of the energy of the photoprotons can be deduced from the measurement of the size of the oscillograph kicks. 1975 Photoproton [see *photoneutron* above]. 1909 *Jrnl. Chem. Soc.* XCV. 442 That most remarkable photo-reaction which Marckwald..has named phototropy. 1925 H. S. Allen *Photo-Electricity* (ed. 2) xiv. 235 Perrin has developed..the view that 'ordinary' chemical reactions may be regarded as due to radiation, *i.e.* they are photo-reactions. 1975 D. Vince-Prue *Photoperiodism in Plants* iv. 146 The photoreactions which control the induction of flowering in LDP [*sc.* long-day plants] and SDP [*sc.* short-day plants] are remarkably similar. 1950 *Proc. Nat. Acad. Sci.* XXXVI. 626 The photorecovery after ultra-violet radiation, manifested by the *Arbacia* egg, seems in all ways parallel to the 'photoreactivation' in fungi and bacteria. 1974 *Encycl. Brit. Macropædia* XV. 390/1 It is probable that photorecovery mechanisms are continually operative in some plants exposed to direct action of sunlight. 1969 *Proc. Nat. Acad. Sci.* LXIV. 1103 The enzymic activity of acetylcholinesterase can be photoregulated through the mediation of photochromic inhibitors of the enzyme. 1957 *Bot. Gaz.* CXVIII. 207/2 Flowering, seed germination, and certain other photoregulated phenomena. 1970 *Proc. Nat. Acad. Sci.* LXVI. 853 A systematic study of the interaction of naturally occurring carotenoids with various enzyme systems might provide information useful for an understanding of photoregulated processes found in nature. 1968 *Science* 27 Dec. 1487 (*heading*) Photoregulation of an enzymic process by means of a light-sensitive ligand. 1970 *Proc. Nat. Acad. Sci.* LXVI. 850 A possible role in photoregulation is suggested for naturally occurring carotenoids. 1959 R. J. Downs in R. B. Withrow *Photoperiodism* ii. 129 The woody plant thus appears to be running two different systems with the same photoregulator. 1970 *Nature* 22 Aug. 778/1 Bieth *et al.* conjecture that carotenoids of animals and plants..might function as photoregulators, controlling diurnal and seasonal changes in metabolic levels. 1967 *Mutation Res.* IV. 22 The impossibility of attaining complete photorepair of lethal and mutagenic damage raises the important question of whether there is a qualitative difference between damage which is photoreactivable and damage which is not. 1978 *Nature* 31 Aug. 891/1 Forward mutations at a variety of loci in *rad* 1-1 yeast are also subject to photorepair. 1966 *Adv. Radiation Biol.* II. 49 The number of photorepairable lethal lesions in..DNA. 1978 *Nature* 31 Aug. 890/2 As many as 80% of the extra *lys*⁺ revertants are photorepairable and therefore dimer-associated in origin. 1925 *Astrophysical Jrnl.* LXII. 317 (*heading*) Apparent photoresistance effects. 1957 *Chem. Abstr.* LI. 7134 At low levels of ionizing radiation CdS photoresistances exhibited considerable inertness. 1970 *New Scientist* 14 May 335/1 The rate at which the charge of any photo-element leaks away between sweeps depends upon the value of the photo-resistance. 1978 *Nature* 23 Mar. 315/1 The light rays impinged on the photoresistances L_1 and L_2, which formed a bridge circuit with the variable resistances R_1 and R_2. 1933 *Jrnl. Inst. Electr. Engin.* LXXIII. 437/1 The first step towards the conversion of the picture into electrical energy was taken by May in 1873, with his discovery of the photo-resistive property of selenium. 1973 *Sci. Amer.* Jan. 116/3 Selenium and cadmium photocells are more sensitive, but they are of the photoresistive type and require an external source of power. 1959 *Chem. Abstr.* LIII. 13793 (*heading*) Germanium photoresistors. 1965 Lindmayer & Wrigley *Fund. Semiconductor Devices* x. 384 Photoresistors made from nearly intrinsic polycrystalline thin films are very sensitive detectors for the average intensity of a wide source spectrum. 1969 *New Scientist* 18 Sept. 568/3 Shamer and Fox observed no fringe shift using..sensitive photoresisters to detect the fringe positions. 1977 J. Hedgecoe *Photographer's Handbk.* 21 (*in figure*) Cds photo-resistor cell. 1950 *Ann. Rev. Plant Physiol.* I. 43 The photoresponses of plants at different ages or stages of development may be opposite. 1955 Hendricks & Borthwick in *Proc. 1st Internat. Photobiol. Congr.* i. 23 Photoresponses controlling etiolation of plants and germination of seeds are also examples [of photoperiodism]. 1971 *Jrnl. Appl. Physics* XLII. 568/2 Additional silver increases the photoresponse. 1976 *Nature* 19 Aug. 680/2 For photoresponse spectra, electrodes were illuminated by 400-Hz chopped, monochromatic radiation. 1955 Hendricks & Borthwick in *Proc. 1st Internat. Photobiol. Congr.* i. 31 Flowering in photo-responsive plants can be controlled through a single leaf in the presence of other leaves. 1974 *Nature* 26 Apr. 800/2 If the cells of *D. discoideum* are broken gently in a Dounce homogenizer, the photoresponsive pigment sediments with the mitochondrial fraction collected by differential centrifugation between 6,000 and 10,000 g. 1972 *Science* 27 Oct. 421/2 Severance of the optic nerve in immature male ducks decreased the photoresponsiveness to one-fifth of the normal. 1954 *Bot. Gaz.* CXV. 216/2 (*heading*) Photoreversal of promotion and inhibition of germination of Grand Rapids lettuce seed at 20°C. after irradiation at 26° and 6°–8°. 1966 *Adv. Radiation Biol.* II. 20 The direct nonenzymatic photoreversal of UV damage to DNA. 1954 Hendricks & Borthwick in D. Rudnick *Aspects of Synthesis & Order in Growth* vii. 159 Some further details about the several photoreversibilities. 1955 *Plant Physiol.* XXX. 468 (*heading*) Photoreversibility of leaf and hypocotyl elongation of dark grown red kidney bean seedlings. 1975 D. Vince-Prue *Photoperiodism in Plants* iv. 108 Extracts of leaves of several species have been found to show photoreversibility. 1954 Hendricks & Borthwick in D. Rudnick *Aspects of Synthesis & Order in Growth* vii. 154 The cuticle coloration response..is photoreversible. 1966 *Adv. Radiation Biol.* II. 21 The photoreversible effects of UV on cytoplasmic entities of cells..formerly suggested RNA damages. 1972 W. Shropshire in Mitrakos & Shropshire *Phytochrome* p. v, Phytochrome is a photoreversible pigment which can exist in two principal forms. 1962 *Instrument Pract.* XVI. 1519/2 (*heading*) Subminiature photosensors. 1964 *New Scientist* 4 June 594/2 The photo-sensor is simply a detector which changes light into an electrical signal... Several types exist, such as photo-emissive cells, photo-conductors, photo-voltaic cells, and photo-transistors. 1975 *Physics Bull.* Feb. 82/3 The solid state cameras..replace the bulky Vidicon tube normally used by an array of 10⁴ photosensors, which have a broad spectral response. 1919 *Jrnl. Gen. Physiol.* I. 556 The photosensory responses of an animal like *Mya*. 1972 *Internat. Jrnl. Neurosci.* III. 145 (*heading*) Photosensory cell of the flatworm ocellus. 1953 *Amer. Jrnl. Psychiatry* CIX. 744/1 Although the method for producing convulsions (photoshock) presented here may be considered similar to that of Metrazol shock, it is our impression that this modified procedure has several possible advantages. 1955 *Sci. News Let.* 21 May 325/1 Instead of electric current, a flashing light is used for 'photo-shock' treatment. 1936 *Jrnl. Gen. Physiol.* XX. 52 The ammoniacal retina bleaches more slowly than the neutral tissue. This difference cannot be ascribed to induced photostability in the visual purple itself. 1965 J. B. Thomas *Primary Photoprocesses in Biol.* iv. 85 The acid-resistance as well as the photostability of suspensions of such native chlorophyll are much higher than those of dissolved chlorophyll. 1977 *Protecting World's Crops* (Shell Internat. Petroleum Co.) 2 Recently, however, compounds have been synthesized which combine with photostability remarkable activity against insects. 1921 *Jrnl. Gen. Physiol.* III. 380 The filters are quite photostable. But in order to avoid any possible bleaching effect, a shutter is placed between the light source and the filter. 1973 Photostable [see *photocatalysed* ppl. adj. s.v. *PHOTOCATALYSE v.]. 1924 *Trans. Faraday Soc.* XX. 112 The decomposition of sulphur dioxide gas under the action of light radiated from a uviol mercury lamp has been investigated, and the resulting photostationary state, characteristic of a given set of conditions, determined, using a number of different light filters. 1972 W. Haupt in Mitrakos & Shropshire *Phytochrome* xxi. 561 Whenever a randomly distributed population of phytochrome molecules is irradiated, light is absorbed by P_r and P_{fr} to different extents... This difference leads to a photostationary state of P_{fr}/P which depends only on the wavelength and which can therefore be predicted precisely. 1956 *Nature* 21 Jan. 143/1 Attempts to photostimulate tropical birds have been rare, and the results confusing. 1971 *New Scientist* 29 July 255/1 Since it seems that the timing of a light stimulus rather than its duration might be important it is conceivable that a bird could be photostimulated with even very small daily doses of light that would normally be non-stimulatory. 1959 D. S. Farner in R. B. Withrow *Photoperiodism* x. 729 In domestic ducks..both ocular and encephalic receptors are involved in photostimulated testicular development. 1970 *Proc. Nat. Acad. Sci.* LXVI. 850 (*caption*) Photostimulated *cis-trans* isomerism. 1967 *Ibid.* LVIII. 2129 The initial photostimulating step..is postulated as being followed at some stage by release of a 'hormone', the hypothetical 'florigen'. 1937 *Jrnl. Exper. Biol.* XIV. 86 Many shallow-water teleosts,..when blinded, show a pigmentary response to photostimulation. 1955 *Sci. News Let.* 21 May 325/1 The flashing light shock is given after patients have had injected into their veins the drug, Azozol... Results were compared with..another group given less intensive 'photostimulation', in which smaller quantities of the drug were used and the light flashes were interrupted oftener. 1959 D. S. Farner in R. B. Withrow *Photoperiodism* x. 724 These investigations suggest that photostimulation of gonadal development does not involve entirely the same receptors as are involved in vision. 1967 *Psychol. Abstr.* XLI. 1499/1 The role of intermediary structures of the brain in the formation of certain functional relationships in the human CNS was studied, employing single, rhythmic, and interrupted photostimulation in Ss with lesions of the diencephalic region and brain stem. 1971 *Nature* 18 June 465/1 The lamp of the photo-stimulator was above and behind an animal's head at 110 cm from the centre of the hemisphere. 1976 *Ibid.* 3 June 423/2 Photostimulatory cues..influence the hypothalamic input to the pars intermedia. 1939 *Jrnl. Inst. Electr. Engin.* LXXXV. 472/2 The optical picture to be transmitted is focused on a continuous transparent photo-surface. 1952 *Electronic Engin.* XXIV. 302/1 The spectral response curves of these photosurfaces are compared. 1970 *Proc. IEEE* LVIII. 1794/1 The AgOCs photocathode, the only photosurface to give appreciable response beyond 1 micron until recently. 1929 *Telegraph & Telephone Jrnl.* XVI. 3/2 Photo-telegrams will have to be charged by space measurement instead of the number of words. 1948 *Post Office Guide* 283 A reply voucher issued with a phototelegram may be used to prepay an ordinary telegram. 1968 *Guardian* 10 Apr. 8/3 To quote the bleak prose of Her Majesty's Post Office, 'Telex—no service. Phototelegrams—no service. Telephone—no service.' 1909 *Electrical Mag.* XII. 249/1 (*heading*) The Sémat phototelegraph. 1949 *Post Office Electr. Engineers' Jrnl.* XLI. 189 (*heading*) The Post Office phototelegraph service to Europe. 1959 J. W. Freebody *Telegr.* xiii. 538/2 (*heading*) The Muirhead-Jarvis photo-telegraph equipment. 1909 *Electrical Mag.* XI. 57/1 New York, Chicago, and other cities are now about to commence photo-telegraphic trials. 1940 *Wireless World* Sept. 398/3 The clarity of reception, at a distance of 12,000 miles, of photographs transmitted from the West Base of the U.S. Navy Antarctic Expedition, is attributed to the use of a recently developed photo-telegraphic technique to counteract selective fading of the carrier frequency. 1886 W. Gemmill *Brit. Pat.* 4841 6 It will be seen that the system of photo-telegraphy opens up an entirely new field in telegraphy, namely the actual reproduction of photographs through the medium of electrically conducting wires. 1930 *Post Office Electr. Engineers' Jrnl.* XXIII. 1/2 British newspapers using photo-telegraphy transmit to and from London and their provincial offices using 'Four-Wire' telephone circuits. 1976 R. N. Renton *Telegr.* iv. 60/1 Telephone circuits are used as the 'bearer' circuits for multiplex telegraph systems and for phototelegraphy. 1942 S. Epstein in *Jrnl. Investigative Dermatol.* V. 290, I propose the term 'phototoxicity' for the primary, non-allergic photosensitivity, and 'phototoxic reaction' for the effect produced by this mechanism... Phototoxic reactions apply indiscriminately to all individuals. 1974 M. C. Gerald *Pharmacol.* xxvii. 471 Demeclocycline has been shown to cause phototoxicity in some patients, where severe burns develop when susceptible patients are exposed to sunlight. 1976 *Arch. Dermatol.* CXII. 327/1 The duration of methoxsalen's phototoxic potentiality, after its application to skin, varied in direct proportion to chemical concentration. 1962 *Jrnl. Physical Chem.* LXVI. 2470 The results indicate the occurrence of consecutive and concurrent phototransformations and thermal interconversions between stereoisomers of the colored modification. 1975 D. Vince-Prue *Photoperiodism in Plants* vi. 215 The general conclusion..is that the phototransformation of phytochrome may very rapidly lead to an alteration of membrane properties. 1973 *Physics Bull.* July 431/1 It is obvious that for light induced species with large lifetimes, a simple DC method would suffice to measure their absorption. However, with the short lifetimes associated with excited states, the changes occurring in DC current by the absorption of the phototransient will be much smaller than the noise. 1975 *Nature* 25 Dec. 767/2 Evidence was obtained that this phenomenon is a consequence of an overlap from a shorter lived phototransient (maximum about 530 nm) which is the precursor of the 410-nm species. 1941 R. P. Hall in Calkins & Summers *Protozoa in Biol. Res.* ix. 477 Some [chlorophyll-bearing protozoa] appear to be obligate phototrophs. 1965 A. H. Rose *Chem. Microbiol.* iii. 39 Only a minority of micro-organisms including algae and photosynthetic bacteria and protozoa are able to utilize directly the energy of solar radiation. These organisms are described as phototrophs to distinguish them from chemotrophs. 1975 *Nature* 7 Aug. 463/2 This versatility would give it a clear advantage over other blue-green algae (mostly obligate phototrophs) as well as over bacteria. 1939 H. W. Harvey in P. D. Trask *Recent Marine Sediments* II. 145 Although plants are occasionally found down to considerable depths, they can only grow and increase down to a depth to which sufficient light penetrates. In clear blue-green water of temperate regions this phototrophic zone may extend down to 30 or 50 meters in summer time. 1965 Pelczar & Reid *Microbiol.* (ed. 2) vi. 496/2 Phototrophic organisms are regarded as the most important plankton organisms since they are the primary producers of organic matter via photosynthesis. 1972 Phototrophic [see *photolithotrophic* adj. above]. 1973 *Biochim. & Biophys. Acta* CCCXXX. 80 (*heading*) Membrane differentiation in phototrophically growing *Rhodospirillum rubrum* during transition from low to high light intensity. 1947 [see *photolithotrophy* above]. 1959 Lamanna & Mallette *Basic Bacteriol.* (ed. 2) xi. 467 (*heading*) Phototrophy. 1923 *Jrnl. Physical Chem.* XXVII. 601 The terms 'Becquerel effect' and 'photo-voltaic effect' have been used to distinguish between the light-sensitive systems of the electrode-electrolyte type and the well known 'Hallwachs effect' or 'photo-electric effect'. Cells having one or more light-sensitive electrodes of the former type are able to convert radiant energy into electrical energy and have been called 'photo-voltaic cells'. 1943 D. H. Jacobs *Fund. Optical Engin.* xxiii. 377 Some problems in instrument design call for vacuum or gas-filled photo-emissive cells, and some call for photovoltaic cells. 1953 Amos & Birkinshaw *Television Engin.* I. iii. 41 An electrode may, however, be sensitive to light in other ways; for example, it may be photo-voltaic, i.e., develop e.m.f.s when illuminated. 1957 *Proc. Inst. Electr. Engin.* CIV. B. 467/1 In 1839 Becquerel had found that, when light fell on two metal electrodes immersed in an electrolyte, a potential difference was established between them; this is now known as the photo-voltaic effect. 1964 *Oceanogr. & Marine Biol.* II. 359 Clark (1933) demonstrated the correlation between diurnal migration of plankton and changes in submarine irradiation by means of a photo-voltaic cell. 1977 *Undercurrents* June–July 8/2 A comparison of fast breeder reactor technology with photo-voltaic (solar cell) technology neatly illustrates the two poles of opinion. 1978 *Solar Energy* (Shell Internat. Petroleum Co.) 5 A second way of using sunlight is photovoltaically—the direct conversion of sunlight into electricity. 1973 *Physics Bull.* Jan. 53/3 Papers are invited on the following areas: photovoltaics, thermo-electrics, electrochemical conversion, [etc.]. 1978 *Telegraph* (Brisbane) 16 May 6/2 Photovoltaics is the dream form of solar power—a single cell without moving parts, silent, reliable and pollution free. 1978 *Nature* 13 July 117/1 The United States administration is clearly determined to make a success of photovoltaics.

2. **pho·to-call,** a summoning (cf. CALL *sb.* 6 h) of theatrical performers or the like to be photographed; the session at which they are photographed; **pho·tocha:rger,** a device for photographically recording details of books loaned from a library; so **pho·tocha:rging** *vbl. sb.*; **pho:toclino·metry** [CLINO-], the process of deriving topographical information about a region from measurements of the brightness distribution in aerial photographs of it; **pho·to-e:ssay,** an essay or short biography consisting of text matter and (numerous) photographs; **pho·to-e:tcher,** one who employs a photographic process in etching; **photofacsi·mile,** facsimile in which the likeness is reproduced in photographic form; usu. *attrib.*; **photoflu·orogram,** a photograph of a fluoroscopic X-ray image; **photoflu·orograph** *sb.* = prec.; *v. trans.*, to examine by photofluorography; **pho:tofluoro·graphy,** photography of a fluoroscopic X-ray image; hence **pho:tofluorogra·phic** *a.*, **pho:tofluorogra·phically** *adv.*; **photoflu·oroscope,** an instrument for taking photofluorograms; **pho:tofluoro·scopy** = *photofluorography* above; **pho:togonio·meter,** (*a*) an instrument for measuring angles indirectly from photographs of an area; (*b*) an instrument for photographically recording the X-rays diffracted at known angles by a crystal or substance; hence **pho:togoniome·tric** *a.*, **-gonio·metry**; **pho:to-interpreta·tion,** the interpretation of aerial photographs; so **photo-inte·rpreter**; **photo-inte·rpretative, -inte·rpretive** *adjs.*; **photojou·rnalism,** the use of photographs in journalism; so **photojou·rnalist**; **pho·to-magazine,** a magazine containing many photographs; **pho·tomap,** a map consisting of or

drawn on a photograph or photomosaic of the area concerned; **photomosa·ic** = *MOSAIC *sb.* 3 d; **photomu·ral,** a mural consisting of a photograph or photographs; **photony·mograph** *Cartography* [Aeolic Gr. ὄνυμ-α name], an instrument used in map production for producing printed names photographically; **pho·toplate,** a photographic plate (PLATE *sb.* 5 c); **photora·diogram,** a picture, diagram, or the like transmitted by radio; (formerly a proprietary name in the U.S.); **pho·to-re:cce,** colloq. abbrev. of next; **photo-reco·nnaissance,** reconnaissance by means of aerial photography; **pho:to-reporta·ge** = *photojournalism* above; also, a report that uses photographs; so **pho·to-reporting**; **pho·to-scanning** *vbl. sb. Med.*, photography of the pattern of radiation from the body following the internal administration of a radio-isotope (e.g. to locate tumours); so **pho·toscan,** a photograph so obtained; **pho·toscanner,** an apparatus for taking such photographs; **photoste·reogram,** a stereophotograph; **photoste·reograph,** an instrument for the observation, measurement, and interpretation of pairs of stereophotographs for surveying purposes; so **pho:tostereogra·phic** *a.*; **pho·to-sto:ry,** a story with accompanying photographs; **pho·to-ti:mer,** (*a*) (see quot. 1942); (*b*) (see quot. 1949).

1958 L. Vining in *Newnes Compl. Amat. Photogr.* xx. 182 Theatre photography can be divided into two classes —photo calls when you have control of the actors and lights, and photographing from the stalls during the performance, when you have no control of anything. 1966 'S. Harvester' *Treacherous Road* i. 22 A string of camels kept motionless against the burnt yellow sky, well-trained as pop singers on a photo-call. 1971 *Times* 8 Sept. 3 Sir Bernard, who plays Iago, promised that the official photocall next week would reveal all of Miss Stevenson, and a very lyrical and beautiful sight it would be. 1977 J. Hedgecoe *Photographer's Handbk.* 93 Most photography of theatrical productions takes place under one of two distinct sets of conditions—during an actual public performance, or during a dress rehearsal or specially staged 'photo-call'. 1959 L. M. Harrod *Librarians' Gloss.* (ed. 2) 208 *Photo-charger*, an electrical machine for recording the loan of books on microfilm. 1967 L. V. Paulin in W. L. Saunders *Librarianship in Brit. Today* i. 4 The introduction of more photochargers. 1955 W. Ashworth *Handbk. Special Librarianship* xii. 319 Such devices as audio-charging..and photo-charging..have been used in America in public libraries. 1967 C. R. Eastwood *Mobile Libraries* ix. 94 Photocharging is widely used on American mobile libraries but not in Britain. 1967 *Surveyor III* (U.S. Nat. Aeronaut. & Space Admin. SP-146) iii. 16 (*caption*) Photoclinometric profiles of the Surveyor III landing site. Profiles were calculated from photometric measurements of Lunar Orbiter III photograph H154..(photoclinometry by H. E. Holt and S. G. Priebe). 1974 *Nature* 10 May 132/1 The development of inferred topography on the basis of the brightness distribution in the image of a surface exhibiting diffuse reflection, and a knowledge of the quantitative law of light scattering for the kind of surface under scrutiny, has been called photoclinometry by common agreement over the past few years. (The word 'photoclinometry' is due to J. F. McCauley...) An operational photoclinometric theory adapted to light scattering properties peculiar to the Moon was worked out..several years ago. 1975 *Times* 18 June 2/2 A group..specializing in lunar and planetry sciences at Lancaster University..is using a method known as photoclinometry to measure the profiles and heights of hills, craters, ridges and cliffs [on Mercury]; the process depends on assessing subtle changes in brightness of the ground and rocks on the pictures. 1976 *Publishers Weekly* 2 Aug. 104/1 Mexican landscape architect Luis Barragán..in a major book... Seven of his most characteristic works—each briefly prefaced and explored at length in a photo-essay. *Ibid.* 4 Oct. 65/3 Seven sumptuous photo-essays (more than 300 pictures, many in color). 1977 C. McFadden *Serial* (1978) ix. 24/2 Michael Bry would..do this photo essay on her. 1889 *Year Bk. Photogr.* 158 One difficulty which photo-etchers have to contend against in the City is the vibration caused by the incessant traffic. 1959 K. Henney *Radio Engin. Handbk.* (ed. 5) xxiii. 1 Photofacsimile systems reproduce the subject copy on photographic papers or films. 1971 *Science* 6 Aug. 529/2 These transmissions, known as the DRIR (direct readout infrared) data, can be displayed on a photofacsimile recorder, which produces a continuous strip image. 1973 *Nature* 16 Feb. 434/2 His youthful German collaborator made a fair copy of Copernicus's precious autograph (now splendidly reproduced in photofacsimile). 1942 *Radiology* XXXVIII. 453/2 The initial scrutiny of routine photofluorograms by the staff radiologist will provide him with an objective means of determining which of all patients..should be referred for searching x-ray examination of the chest. 1975 B. W. Gayler in E. J. Potchen *Current Concepts in Radiol.* II. vii. 131 For many years, mass survey and screening chest radiographs were taken as 70 mm photofluorograms. 1941 *Med. Jrnl. Austral.* I. 267/2 The method of focusing is to make photofluorographs of a wire mesh mounted immediately in front of the fluorescent screen holder. 1972 *Science* 16 June 1186/3 The American public would be exposed to hundreds of thousands of unnecessary chest photofluorographs each year. 1945 *Amer. Jrnl. Roentgenology* L. 405/2 A subject 20 cm. in thickness photofluorographed with the roentgen machine operating at 90 kv. 1957 *Ibid.* LXXVII. 1079/1, 101 persons were photofluorographed. 1941 *Med. Jrnl. Austral.* I. 267/2 The four inches by five inches photo-fluorographic unit incorporates a special 14 inches by 17 inches fluorescent screen. 1954 *Brit. Jrnl. Radiol.* XXVII. 459/1 An attempt to reduce cost of routine examinations..by employing the photofluorographic method. 1972 J. E. Cullinan *Illustr. Guide X-Ray Technics* i. 24/1 A photofluorographic unit is a quick, efficient way to accomplish mass survey chest radiography. 1949 *Amer. Jrnl. Roentgenology* LXI. 186/1 All patients above the age of forty..will be examined photofluorographically. 1941 *Med. Jrnl. Austral.* I. 266/1 Photofluorography opens up a new avenue of examination to the wage-earning class. 1974 *Encycl. Brit. Macropædia* XV. 462/2 Photofluorography and television observation can take place simultaneously by means of separate apertures, and thus observation of what is being photographed is achieved. 1896 J. M. Bleyer in *N.Y. Med. Jrnl.* LXIII. 540/1 The photo-fluoroscope is an instrument which differs from all other fluoroscopes in the fact that it allows a direct shadow picture to be taken from the screen on the fluoroscope, after it is focused through the screen, and the image is seen on the ground glass in the photographic focusing box. 1933 O. Glasser *Sci. of Radiol.* i. 10 J. M. Bleyer of New York built his photofluoroscope which was destined to become the predecessor of the roentgen moving picture camera. 1955 G. L. Clark *Appl. X-Rays* (ed. 4) ix. 197 (*heading*) Photofluoroscopy (indirect radiography). 1966 *McGraw-Hill Encycl. Sci. & Technol.* XI. 302/2 The photography of the fluorescent image, as in mass chest examinations, is called photofluoroscopy. 1926 A. W. Judge *Stereoscopic Photogr.* xviii. 218 The effect of tilting and swinging the plate in the photogoniometer is such as to render the angular measurements..the same as if the view had been taken on a vertical plate. 1927 *Jrnl. Sci. Instruments* IV. 273 (*heading*) A universal X-ray photogoniometer... Combining: apparatus for single crystal rotation photographs—Laue photographs—X-ray spectrometry—powder photographs—photographs of crystal aggregates, metals, materials, etc. 1933 A. R. Hinks *Maps & Survey* (ed. 3) xii. 243 The machines which have been developed during the last few years all utilise..the principle of the *Bildmesstheodolit*, otherwise called the Photogoniometer, in which the plates are viewed through objectives identical with those which took them. 1939 *Geogr. Jrnl.* XCIII. 150 The American Geographical Society first of all made a photogoniometer (*Bild-Theodolit*) on a somewhat novel principle. 1970 *Chem. Abstr.* LXXIII. 305/2 A single-circle universal photogoniometer with vertical crystal-bearing attachment,.. was devised. 1939 *Geogr. Jrnl.* XCIII. 242 The first extensive photographic survey was that of the stars, made by Kapteyn..with a photogoniometric machine he built about 1890. 1968 *Chem. Abstr.* LXIX. 6626/2 (*heading*) Photogoniometric investigation of a crystal surface. *Ibid.*, A goniometric study of crystals is rarely made because labs. often have no goniometer. The goniometric method in many cases can be replaced by photogoniometry. 1939 B. B. Talley *Engin. Applic. Aerial & Terrestr. Photogrammetry* ii. 9 (*caption*) Photogoniometry by the method of Porro and Koppe. 1923 *Photogrammetric Engin.* VIII. 27 The..function of exploiting and developing the intelligence from the aerial photos falls within the sphere of photo interpretation units. 1959 *Proc. Geologists' Assoc.* LXX. 144 It may be necessary.. to complete a preliminary photo-interpretation before planning ground traverses. 1966 *Daily Tel.* 20 Aug. 14/5 These pictures will form stereo-pairs and can be analysed by standard techniques of photo-interpretation to give maximum information about the terrain on which American astronauts may land. 1973 *Sci. Amer.* Feb. 21/2 The cost of hardware and manpower for photo-interpretation..will remain high. 1959 *Proc. Geologists' Assoc.* LXX. 144 The best practice is a judicious combination of photo-interpretative methods and geological field mapping. 1942 *Photogrammetric Engin.* VIII. 26 While the aerial camera sees all, it is the photo interpreter who must know all and tell all. 1957 *Ibid.* XXIII. 933 A mathematical proof is presented for the statement that differential parallaxes smaller than 0·001 inch cannot be detected by average photo-interpreters. 1977 *Sci. Amer.* Sept. 57/1 Only the wide curve it must make on slopes betrays it to the photointerpreter. 1967 *Boston Sunday Herald* 26 Mar. (Comic Section), I subjected the film to the usual photo-interpretive analysis. 1944 K. Hubschmann in A. Kraszna-Krausz *Photogr. as Career* 118 My friend proved an excellent teacher of the essentials of photo-journalism. 1958 M. F. Harker in *Newnes Compl. Amat. Photogr.* xiii. 140 The present trend of photo-journalism which attempts to put over human stories in pictures rather than words. 1976 *National Observer* (U.S.) 11 Sept. 20/1 This colorful chap is Nelson Wadsworth, who teaches photojournalism, magazine writing, and investigative reporting at Brigham Young University in Utah. 1977 *Time* 12 Dec. 60/2 World War II was the longest-running story in the history of *Life*, the magazine that practically invented photojournalism. 1959 C. B. Neblette *Photogr. Lens Manual* ii. 25 After the 50 mm, the 35 mm medium wide-angle, or wide-field lens is undoubtedly the most useful to..the photojournalist. 1963 A. E. Woolley *Creative 35mm Techniques* III. 113/2 At all times a photojournalist is aware of the maximum emotion or conflict of the subject of the story. 1974 *Nat. Geographic* Aug. 252 At last I am here in North Korea, the first American photojournalist to gain entry into a country cloistered from the non-Communist world for a quarter of a century. 1978 *New York* 3 Apr. 32/3 Micha Bar Am/Harold Edgerton/Frank Rinehart—Begin and Sadat by this Mid-East photojournalist. 1960 *News Chron.* 10 Oct. 4/1 Paris-Match, the most powerful photo-magazine in Europe. 1969 *Amat. Photographer* 28 May 26/3 Few of the present-day casual photographers and snapshotters do their own processing. The photographic trade and the photo-magazines do not encourage it. 1939 B. B. Talley *Engin. Applic. Aerial & Terrestr. Photogrammetry* xix. 521 When this becomes economically feasible encouragement should be given to the development of 'photo-maps' to which may be added contours. 1955 *Times* 1 Aug. 6/1 The first 200 photographic sky charts..are being sent to observatories all over the world, and when the atlas is completed in 1956 it will include 1,758 such 'photomaps'. 1969 *Nature* 16 Aug. 668/1 (*caption*) Satellite photomap of the Tucson, Arizona, area with transport network superimposed. The map is based on a photograph taken..from Gemini V. **1958** *New Scientist* 13 Nov. 1247/3 The photomosaic is being assembled now. **1962** *Times* 17 May 10/3 The leading aircraft, navigating on a photo-mosaic. **1973** *Sci. Amer.* Dec. 132/3 This is the planet Mars, drawn from *Mariner 9* photomosaics. **1977** *Time* 17 Oct. 45/1 Among the pictures released by NASA: a photomosaic of the planet's north pole, showing a concentric pattern of striations in the ice cap. **1935** W. D. Morgan et al. *Leica Manual* 469 (*caption*) Photomurals with the Leica. **1937** *Archit. Rev.* LXXXI. 86 The true photo-mural... This new process of mural decoration, which can now be said to have passed the experimental stage, has its chief virtue in that the design is projected direct on to the wall surface. The surface is first sprayed with photo-sensitive emulsion, and the photograph printed on it much in the same way as an ordinary camera enlargement. **1960** *House & Garden* Oct. 65/1 For ease of hanging, these photomurals are printed on lightweight white base paper. **1976** *National Observer* (U.S.) 24 Apr. 24/2 Once inside, the visitor encounters giant photomurals, three-quarter mock-ups of building exteriors, [etc.]. **1933** J. S. A. Salt *Simple Method Surveying from Air Photogr.* xi. 130 The names..are printed photographically in a Photonymograph... Names may be printed in a variety of sizes and styles on a strip of sensitized paper. The alphabet..and any other signs..are contained on a master-disc. With the disc in position, each letter in turn is brought into position and an exposure made... The strips are developed and fixed ..and then show a series of names in various sizes and styles. **1963** *Record* (Oxf. Univ. Press) Dec. 2/3 The Drawing Office has installed a 'photonymograph' (a device whose development was in fact sponsored by the Cartographic section) to produce its lettering and so free draughtsmen for drawing. **1971** Monkhouse & Wilkinson *Maps & Diagrams* (ed. 3) i. 64 The new model..of the Photonymograph (developed from a machine which appeared in its earliest form over thirty years ago) is made by Barr and Stroud, Ltd. **1918** *Physical Rev.* XI. 137 (*heading*) Images on silvered photo-plate. **1974** *Nature* 20 Dec. 698/2 Organochlorine compounds may be identified at low concentrations in crude extracts of natural samples by a high resolution mass spectrometric method involving photoplate detection. **1924** *Glasgow Herald* 13 Dec. 9 The signature was that of Sir Robert Kindersley, whose photo-radiogram read—'My warmest greetings.—R. M. Kindersley.' The message and signature accompanied a radio photo of Sir Robert. **1925** *Official Gaz.* (U.S. Patent Office) 26 May 803/1 Radio-corporation of America, New York, N.Y. Filed Jan. 13, 1925. *Photoradiogram*... Pictures, drawings, and facsimiles transmitted by radio. Claims use since Nov. 30, 1924. **1926** *Daily News* 1 May 5/6 Most of the photo-radiograms sent from this side during the night will appear in American newspapers today. **1927** *Daily Express* 16 Dec. 1, December 21 is the latest date for handing in Christmas photo-radiograms at Marconi offices. **1946** *War Report* (B.B.C.) vi. 149 Then on Sunday evening one of our photo-recce Spitfires was shot down in German territory. **1971** *N.Y. Times* 13 June iv. 37 We have a high priority requirement for night photorecce of key motorable routes in Laos. **1944** *Sci. News Let.* 19 Aug. 117/3 Photo reconnaissance supplies information regarding the strength of enemy troops. **1951** A. C. Clarke *Sands of Mars* xvi. 209 I'm going to suggest a photo-reconnaissance of all the..forests. **1973** *Sci. Amer.* Feb. 14/2 The main restrictions imposed by both of the SALT I agreements can be..monitored largely by means of sensors carried on board such orbiting photoreconnaissance systems. **1960** *Spectator* 17 June 886 The great photo-reportage magazines have now..largely relinquished their commanding position to television. **1960** L. Durrell *Spirit of Place* (1969) 162, I remember seeing a photo-reportage in *Life* magazine once which dealt with the extraordinary changes in physique which immigrants to the U.S.A. underwent. **1966** *Punch* 6 Apr. 498/1 One of the many technical hazards which are endemic in photo-reportage are bound to materialise in full force. **1957** T. L. J. Bentley *Man. Miniature Camera* (ed. 5) iv. 41 The few models which incorporate a motor drive create totally new possibilities of rapid-sequence pictures which may be quite invaluable in..records of sporting events and photo-reporting for journalistic purposes. **1956** *Radiol.* LXVI. 737/2 (*heading*) Photoscan (superimposed on roentgenogram) shows lesion to be thyroid tissue rather than metastasis from breast. **1974** *Cancer Res.* XXXIV. 1/1 The photoscans showed an increased uptake of radioactivity over the tumors. **1959** *Internat. Jrnl. Appl. Radiation & Isotopes* IV. 154 (*heading*) A versatile, high-contrast photoscanner for the localization of human tumors with radioisotopes. **1966** *Sci. News* 12 Nov. 400 The gamma rays coming from the abnormal portions of bone are detected by a photoscanner that is passed externally over the body. Any portion of bone that gives off gamma rays is considered diseased. X-rays, in contrast, work by showing changes in bone density. **1956** *Internat. Jrnl. Appl. Radiation & Isotopes* I. 137/1 A photo scanning device has been devised which presents a 150% increase in picture density as the result of a 10% increase in count rate. **1967** *Times* 19 Sept. 9 Photo-scanning using radioactive isotopes can tell us if cancer is present in such organs as the thyroid gland, the liver and the brain. **1974** *Cancer Res.* XXXIV. 1/1 Radiolabeled nonantibody components of heterospecific IgG can be localized in certain tumors and normal tissues by photoscanning. **1913** *Chem. Abstr.* VII. 3862 App[aratus] for copying the surface of a solid body from a photostereogram. **1939** B. B. Talley *Engin. Applic. Aerial & Terrestr. Photogrammetry* xix. 526 (*caption*) The Nistri photostereograph. **1963** W. K. Kilford *Elem. Air Survey* xi. 265 (*caption*) The photostereograph (Beta/2) coupled with coordinate computer on the left and plotting table and coordinate plotter on right. **1933** A. R. Hinks *Maps & Survey* (ed. 3) xii. 243 (*heading*) Recent developments in photostereographic surveying. **1940** *War Illustr.* 19 Jan. 627 (*heading*) Photo-story of the life and death of a U-boat. **1972** *Guardian* 24 Mar. 12/6 The photo-story in 7 Days left one with a powerful impression. **1973** D. Matias tr. C. Metz in *Screen* Spring/Summer 197 Image-languages..figurative drawing,..television, photography, the photo-story etc. **1942** *Amer. Jrnl. Roentgenology* XLVIII. 220/1 A new instrument, a photoelectric timing mechanism, capable of regulating automatically the length of roentgenographic exposure time has been developed... The photoelectric timing mechanism, or phototimer, is a modification of the roentgenographic exposure meter. **1949** *Britannica Bk. of Year* 687/2 *Photo-timer*, an electrical device which photographs the finish of a race and supplies the elapsed time from start to finish. **1958** *Times* 22 Aug. 4/1 The race..should be started farther back from the bend, in spite of the cost of moving the electrical photo-timer.

3. **pho:todegrada·tion**, degradation of a substance caused by light; so **photodegra·dable** *a.*; **photodi·mer** *Chem.*, a dimer formed by photochemical action; so **photodime·ric** *a.*; also **pho:todimeriza·tion**, the formation of, or conversion into, a photodimer; **photodi·merize** *v. trans.* and *intr.*, to dimerize by the action of light; **pho:todissocia·tion** *Chem.*, dissociation of a chemical compound by the action of light; so **photodisso·ciate** *v. trans.*, to dissociate by means of light; **photo-oxidation** (further examples); also **photo-o·xidative** *a.*, involving or characterized by photo-oxidation; **photo-o·xidize** *v. trans.*, to oxidize photochemically; hence **photo-o·xidizable** *a.*, **-o·xidized** *ppl. a.*; **pho:tophosphoryla·tion** *Biochem.*, the process by which light energy is utilized by a plant or micro-organism to convert adenosine diphosphate to adenosine triphosphate without the reduction of oxygen to water that occurs in oxidative phosphorylation; **photopo·lymer**, a polymer produced photochemically; **pho:topolymeriza·tion** *Chem.*, polymerization brought about by the action of light; so **photopo·lymerize** *v. trans.*, to polymerize photochemically; **photopo·lymerized** *ppl. a.*; also **photopo·lymerizable** *a.*, capable of being photopolymerized; **pho:topolymerizabi·lity**.

1972 *New Scientist* 5 Oct. 41/2 (*heading*) Photo-degradable plastic carriers. **1975** Rånby & Rabek *Photodegradation* ix. 361 The development of methods for making plastics which are photodegradable to form harmless and biologically useful chemical compounds is of great interest. **1962** J. T. Marsh *Self-Smoothing Fabrics* xviii. 303 The resin exerts a considerable protective influence even in presence of those vat dyes whose action accelerates photo-degradation of cellulose. **1975** *Physics Bull.* Apr. 164/1 (Advt.), The fundamental photochemical reactions involved in photodegradation of polymers. **1936** *Trans. Faraday Soc.* XXXII. 521 The photodimer of thiophosgene. **1970** J. B. Birks *Photophysics of Aromatic Molecules* vii. 322 If..photodimers of other anthracene derivatives are irradiated with ultraviolet photons of sufficient energy..photolysis occurs and the dimer reverts to the original pair of individual molecules. **1952** *Chem. Rev.* LI. 19 The non-occurrence of the photodimeric products of the previously mentioned anthracene derivatives is probably due to their thermolability. **1936** *Trans. Faraday Soc.* XXXII. 517 A photodimerisation of 9-10-diphenyl anthracene has not yet been observed. **1972** DePuy & Chapman *Molec. Reactions & Photochem.* iv. 65 Naphthalenes, anthracenes, and polyacenes in general undergo photodimerization. **1955** *Jrnl. Chem. Soc.* 315, 3-Bromothionaphthen 1:1-dioxide was photodimerised in order to ascertain whether angular..or linear dimerisation was favoured. **1969** *Organic Photochem.* II. 75 Acyclic α,β-unsaturated ketones photodimerize when substituted with an aromatic group in the beta position. **1970** J. B. Birks *Photophysics of Aromatic Molecules* vii. 321 There are a large group of other 9-substituted and 9,10-disubstituted anthracenes which also photodimerize. **1962** F. I. Ordway et al. *Basic Astronautics* vi. 249 Water vapor is constantly being photodissociated by the action of sunlight. **1969** *Nature* 22 Nov. 756/2 NH_3, H_2O and H_2CO can all be photodissociated by ultraviolet photons of E < 13·6 eV. **1925** *Phil. Mag.* XLIX. 1166 It becomes of interest..to investigate the relation between the activation, and the frequency of the illumination;..only recently has a similar investigation been made on the photo-dissociation of sulphur dioxide. **1974** *Sci. Amer.* June 29/1 Atomic iodine is prepared in the proper excited state by the photodissociation of gaseous compounds such as iodotrifluoromethane (CF_3I) with xenon flash lamps. **1977** *Jrnl. R. Soc. Arts* CXXV. 766/2 The high spectral brightness of lasers can be used resonantly to obtain selective photodissociation. **1941** *Jrnl. Gen. Physiol.* XXV. 309 The ultimate cause of the extra oxygen absorption after or during irradiation consists of photoxidation processes sensitized mainly by chlorophyll. **1956** *Nature* 17 Mar. 513/2 The Bituminous Binder Research Unit has completed a preliminary investigation of photo-oxidation in the weathering of binders on the road. **1971** *Jrnl. Oil & Colour Chemists' Assoc.* LIV. 846 Bleaching by sunlight may be due to photo-oxidation of double bonds. **1937** *Ann. Rep. Progr. Chem.* XXXIII. 431 There is little doubt that this element [*sc.* manganese] can affect the chlorophyll content of leaves, and probably also its photo-oxidative properties. **1976** *Nature* 13 May 169/2 A film exposed under high vacuum showed a dose response almost identical to that of one exposed in air, and indicates the photolytic rather than the photo-oxidative nature of the process. **1949** *Chem. Abstr.* XLIII. 5391 (*heading*) Photooxidizable derivatives of helianthrene. **1974** *Photochem. & Photobiol.* XIX. 35/1 Dye-sensitized photo-oxidation.. permitting preferential destruction of the photo-oxidizable amino-acid residues of proteins. **1941** *Adv. Enzymol.* I. 232 Artificially added substances whose oxidation products can be recognized by their color are actually photo-oxidized in plants. **1975** *Nature* 9 Oct. 490/2 It seems that *O. limnetica* photo-oxidises S²⁻ quantitatively to S°. **1973** *Biochem.* XII. 2540/2 The tryptophan content of the native and the photooxidized enzyme was measured. [**1954** D. I. Arnon et al. in *Nature* 28 Aug. 394/1 Evidence has now been obtained that whole chloroplasts..have the ability to carry out..photosynthetic phosphorylation, a term which we use for the conversion of light energy into the high-energy phosphate bonds of adenosine triphosphate (ATP), without the participation of respiration.] **1956** *Plant Physiol.* XXXI. p. xxx/2 (*heading*) The mechanism of cell-free, bacterial photophosphorylation. **1956** *Federation Proc.* XV. 260/2 Photophosphorylation depends upon an inductive reduction. **1971** M. F. Mallette et al. *Introd. Biochem.* xviii. 646 Photosystem 1, the longer wave length photosystem, has been associated with the reduction of $NADP^+$ and with photophosphorylation. Photosystem 2 is related to dissociation of water and the evolution of oxygen. **1973** R. G. Krueger et al. *Introd. Microbiol.* viii. 271/1 Photophosphorylation occurs in the reduction of cytochrome c with the generation of ATP from ADP and inorganic phosphate. The electron initially emitted from chlorophyll is ultimately returned to chlorophyll in a cyclic fashion; hence, the entire process of converting light energy to ATP (chemical energy) is termed cyclic phosphorylation... In plants and algae, a noncyclic type of photophosphorylation also occurs. **1932** *Canad. Jrnl. Res.* VII. 479 The photo-polymer of vinyl butyl ether was purified. **1953** *Jrnl. Res. Nat. Bureau of Standards* (U.S.) LI. 327 Teflon and tetrafluoroethylene photopolymers, on pyrolysis in a vacuum at 423·5° to 513·0°C, yield almost 100 percent of monomer. **1961** *Printing News* 16 Feb. 10 The first complete 4-colour magazine to be produced from 'Dycril' photo-polymer (light-sensitive plastic) printing plates. **1974** *Clarendonian* XXVIII. 1. 41 There are plans afoot to convert other letterpress 'classics' to this type of plate should photopolymers prove the equal of our now traditional hot-moulded plastic plates. **1932** *Canad. Jrnl. Res.* VII. 473 The photo-polymerizability is practically the same as that of the butyl ether. *Ibid.*, Comparison with the closely related vinyl esters.. shows that, in the absence of catalysts, the latter are much more readily photo- and thermo-polymerizable. **1973** *Materials & Technol.* VI. ix. 664 The photopolymerisable amide-compositions are used for the production of letterpress printing plates. **1920** *Chem. Abstr.* XIV. 4427/1 (Index), Photopolymerization. **1924** *Jrnl. Amer. Chem. Soc.* XLVI. 1614 It is concluded that in the photopolymerization of anthracene, a single molecule is activated by absorption of blue light and then proceeds to react with an unactivated molecule to form the polymer. **1970** H. L. Needles in R. F. Reinisch *Photochem. Macromolecules* 129 Both oxygen and hydrogen donors effect riboflavin-sensitized photopolymerizations of aqueous acrylamide. **1921** *Jrnl. Chem. Soc.* CXIX. 1028 Whether the same explanation does not apply to all carbohydrates and to chlorophyll, namely, that under the influence of light of very short wave-length they are decomposed to carbon dioxide, which is photosynthesised to formaldehyde, and this in its turn photopolymerised to sugars. **1933** *Jrnl. Amer. Chem. Soc.* LV. 577 Ethyl acetate does not react with Wijs solution; nor does the polymerized product, since..solid photopolymerized vinyl acetate gave iodine values as low as 9·5 and 6·3, corresponding to 96 and 98% polymerized. **1974** *Sci. Amer.* Oct. 119/2 They are made by first crystallizing the molecules of an appropriate monomer and then photopolymerizing the monomer crystal with ultra-violet light or gamma rays.

photoabsorbing to **photoautotrophic**: see *photo- 1.

pho:tobio·graphy. [f. Photo- 2 + Biography.] A person's life shown in a series of photographs.

1944 *Spectator* 3 Mar. 188/1 A 'photobiography', consisting of 250 scenes from Woodrow Wilson's life, with 150 pages of letterpress..is in an advanced stage of preparation. **1951** C. Beaton (*title*) Photobiography. **1952** *N.Y. Times Bk. Rev.* 1 June 12 A 'photobiography' that does not seem likely to lose the general votes in the future.

photobiological to **-bleaching**: see *photo- 1.

photoblepharon (fōᵘtoble·fărọn). [mod.L. (M. Weber *Siboga Expeditie* (1902) I. 108), f. Photo- 1 + Gr. βλέφαρον eye.] A small luminous fish of the genus so called, found in the Red Sea and the Indian Ocean.

1920 E. N. Harvey *Nature Animal Light* i. 18 Leaving out of account..the use for bait, in fishing, of the luminous organ of a fish, *Photoblepharon*, by the Banda islanders, we find that luminous bacteria are of value for certain purposes in the laboratory. **1928** C. M. Yonge in Russell & Yonge *Seas* viii. 193 Light organs are apparently commonest in fish which live in the upper 500 metres: in the warmer seas, although there are notable exceptions, for example, a..fish..called Photoblepharon found in pools of fresh water in quarries and the craters of extinct volcanoes in Malaya. **1964** *Oceanogr. & Marine Biol.* II. 356 In *Photoblepharon* the glow may be hidden by an opaque screen. **1978** *Nat. Geogr. Mag.* Nov. 722 Despite its name, *Photoblepharon* has no real eyelid.

photo-call: see *photo- 2.

photocatalyse (fōᵘtokæ·tăləiz), *v. Chem.* Also (*U.S.*) -yze. [f. Photo- 1 + *Catalyse *v.*] *trans.* To subject to photocatalysis.

1921 *Jrnl. Chem. Soc.* CXIX. 1034 The photosynthesis of formaldehyde from carbon dioxide and water can be photocatalysed by certain coloured basic substances...

The photosynthesis then takes place in visible light. *c***1955** *Proc. 1st Internat. Photobiol. Congr. 1954* ii. 137 Ferric chloride photocatalyses the reduction of..methylene blue by citrate in water. **1970** *Nature* 22 Aug. 832/2 The decomposition of N_2O photocatalysed both by metal free and Cu-phthalocyanine.

Hence **photoca·talysed** *ppl. a.*

*c***1955** *Proc. 1st Internat. Photobiol. Congr. 1954* ii. 135 The observed total reaction consists of a photocatalysed reaction between leucoindophenol and air plus a dark reaction between leucoindophenol and peroxide. **1973** *Biochem.* XII. 4154/1 Although sensitive to photocatalyzed degradation, these reagents readily react with thiol compounds..yielding photostable covalent derivatives.

photocatalysis (fōᵘtokătæ·lisis). *Chem.* †Also with hyphen. [f. Photo- 1 + Catalysis. The words of this group originated with the Ger. adj. (see below).] The acceleration of a reaction by light; the catalysis of a photochemical reaction.

1913 *Jrnl. Chem. Soc.* CIV. ii. 653 (*heading*) Photocatalysis. **1916** W. C. McC. Lewis *Syst. Physical Chem.* II. iii. i. 420 Another instance of photo-catalysis in solution is the decomposition of aqueous sodium hypochlorite. **1970** *Nature* 22 Aug. 832/2 (*heading*) Homogeneous photocatalysis by organic dyes in the liquid phase.

So **pho:tocataly·tic** *a.* [ad. G. *photokatalytisch* (J. Plotnikow *Photochemie* (1910) ii. 72)], pertaining to or exhibiting photocatalysis; **pho:tocataly·tically** *adv.*; **photoca·talyst**, †**-ca·talyser**, a substance that acts as a catalyst in a photochemical reaction.

1913 H. S. Allen *Photo-Electricity* xiv. 198 We may conveniently divide..photo-chemical reactions into three main classes. To the first class may be assigned the so-called photokatalytic reactions, in which light only accelerates an irreversible process. **1914** S. E. Sheppard *Photo-Chem.* vii. 293 The kations Fe, Mn, Ce were peculiarly effective photo-catalysts for the reaction. **1923** *Jrnl. Chem. Soc.* CXXIII. 189 The synthesis takes place photocatalytically by the influence of visible light when the carbonic acid is in loose combination with a coloured base, such as malachite green. **1926** E. Mayer *Clin. Applic. Sunlight* viii. 160 Metallic compounds, especially iron salts, act as photocatalyzers. For example, benzoic acid under the influence of light will change very quickly to salicylic acid if an iron salt be present. **1945** E. I. Rabinowitch *Photosynthesis* I. iii. 56 Chlorophyll is a photocatalyst, since no decrease in the concentration of chlorophyll in leaves has been observed after intense photosynthesis. **1959** *Ann. Rev. Plant Physiol.* X. 56 (*heading*) Photocatalytic activities of bacterial chromatophores. **1971** E. Gurr *Synthetic Dyes* 108 Crystal violet is one of the dyes used by Binding (1970) as a photocatalyst. **1978** *Nature* 3 Aug. 508/1 The concentration of Ru (about 8 × 10⁻⁵ M) changed by less than 5%, so the complex was acting as a photocatalytic agent.

photocathode (fōᵘtokæ·þōᵘd). *Electronics.* Also **photo-cathode**, †**-kathode**. [a. G. *photokathode* (P. Selényi 1929, in *Physik. Zeitschr.* XXX. 933/1): see Photo- 1 and Cathode.] A cathode which emits electrons when illuminated, thereby allowing an electric current to pass.

1930 *Sci. Abstr.* A. XXXIII. 525 (*heading*) Red-sensitive sodium photo-kathodes (photo-electric cells). **1931** *Physics* I. 343 If a silver surface is first oxidized in an electrical discharge, and then baked in the presence of caesium vapor, it may be converted into a photo-cathode of remarkable sensitivity. **1957** *Proc. Inst. Electr. Engin.* CIV. B. 470/2 The photo-electric surface developed by Koller in 1929, consisting of a complex mixture of silver, oxygen and caesium, was the first of the really efficient modern photo-cathodes and it is still in use, especially when red or infra-red sensitivity is required. **1969** *New Scientist* 10 July 21/1 An image tube consists of a photocathode which emits electrons in direct proportion to the amount of light falling on it.

photocell (fōᵘ·tosel). Also **photo-cell**, **photo cell**. [f. Photo- 1 + Cell *sb.*[1]] A device which generates an electric current or voltage dependent on its degree of illumination.

1891 *Phil. Mag.* XXXI. 232 The effect of this connexion with the Daniell is to develop between the poles of the photo cell a difference of potential opposed in sign to that which light produces. **1913** *Physical Rev.* I. 74 When.. the spark gap was pulled six feet away from the photocell the positive potential slowly fell to 1·6 volts. **1932** *Discovery* Apr. 112/1 By passing the film between a light and..a photocell the film can be turned into speech again. **1938** G. H. Sewell *Amateur Film-Making* ii. 21 The photo-cell meter is recognized as the most accurate light-measuring device normally available. **1958** *New Statesman* 23 Aug. 214/2 There will be a system of photo-cells, which, sensitive to infra-red and visible rays, will scan the hidden side of the moon, as the rocket-probe spins on its axis. **1973** *Daily Tel.* 10 Dec. 7/3 When fog distorts the light path, it alters the pattern being received by the photocell, which is then electronically translated into distance of visibility.

photocharger, -charging: see *photo- 2.

photochemical, *a.* Add: *photochemical smog*, a condition of the atmosphere attributed to the action of sunlight on hydrocarbons and nitrogen in it and characterized by the presence of aerosols and increased ozone and nitrogen oxides and by effects that include irritation of the eyes, damage to plants, and visibility reduced to a mile or less.

1957 *Rep. Air Pollution Foundation* (U.S.) No. 21. 1 It is now realized that Los Angeles' smog is primarily a reaction between organic matter and nitrogen oxides in the presence of sunlight. The need for irradiation to produce the reaction has led to the designation 'photochemical' smog. **1972** *Daily Colonist* (Victoria, B.C.) 13 Feb. 27/5 His studies of animals show that Vitamin E serves as an antioxidant to protect lung tissue from damage by the ozone in photochemical smog. **1972** *Nature* 18 Feb. 360/1 There is no doubt that motor vehicles are the principal outlets for photochemical smog in large cities. **1976** *New Yorker* 16 Feb. 74/2 Los Angeles is perhaps the best-known victim of the bus and the automobile; photochemical smog, the distinctive pollutant of these vehicles, made its first recorded appearance in Los Angeles in 1943.

photochemically *adv.* (earlier and later examples); **photochemist** (examples); **photochemistry** (later examples).

1881 *Phil. Mag.* XII. 21 The impression of an electromotive force upon a film possessing, as do all photochemically active bodies, electrolytic conductivity, will tend to produce at least partial electrolysis. **1926** *Trans. Faraday Soc.* XXI. 438 Stark's views undoubtedly did not receive from photochemists the attention they deserved. **1941** S. H. Bartley *Vision* i. 5 The photochemist has given increasing attention to the cycle of changes that occur in the eye in response to light. **1961** G. R. Choppin *Exper. Nucl. Chem.* xii. 194 It is tempting to seek to relate photochemistry and radiation chemistry by considering that ionization is the ultimate state of excitation. **1967** Margerison & East *Introd. Polymer Chem.* iv. 184 The process is started by the decomposition of an initiator induced thermally, photochemically or by admixture of another substance. **1972** DePuy & Chapman *Molec. Reactions & Photochem.* i. 6 Photochemistry is the study of the chemistry of electronically excited molecules produced by the absorption of electromagnetic radiation. **1973** *Sci. Amer.* Apr. 68/2 Photochemists have measured the probability that ultraviolet photons will destroy particular molecules. **1978** *Nature* 23 Feb. 733/1 Photochemically produced sulphate aerosols can, in certain circumstances, account for a large proportion of total particulate mass in the 0·1–1·0 μm size range.

photochromic (fōᵘtokrōᵘ·mik), *a.* and *sb.* [f. as next + -ic.] **A.** *adj.* Of, pertaining to, or displaying photochromism.

1953 *Jrnl. Chem. Physics* XXI. 1619/2 In those compounds which exhibit both thermochromism and photochromism the thermochromic and the photochromic colors are in each case spectroscopically identical. **1965** *New Scientist* 29 Apr. 291/3 Silver halide particles dispersed through this sheet of 'photochromic' glass..cause the glass to darken on exposure to light, and clear again within minutes. **1971** H. Meier in K. Venkataraman *Chem. Synthetic Dyes* IV. vii. 427 The photochromic reaction of aqueous solutions of the leucosulfite of parafuchsine could be used as a UV dosimeter for the measurement of the UV radiation of sky light. **1974** *Observer* (Colour Suppl.) 13 Oct. 18/4 Photochromic lenses that automatically change the darkness of their tint to match the prevailing brightness.

B. *sb.* A photochromic substance.

1965 *Guardian* 13 Apr. 14/5 Sunglasses which are clear in ordinary daylight but which darken progressively in increasing sunlight, were among the practical applications of photochromics described in London yesterday. **1970** *Physics Bull.* Nov. 487/1 Two other classes of material which have been developed recently as promising photochromics..are the alkaline earth fluorides and titanates.

Hence **pho:tochromi·city** = *photochromism.

1966 *Applied Optics* V. 946/2 The size of the silver halide microcrystals embedded in the glass matrix determines not only the photochromicity but also the character of the unexposed material (transparent, translucent, and opaque).

photochromism (fōᵘtokrōᵘ·miz'm). [f. as next + -ism.] The phenomenon whereby certain substances undergo a reversible change of colour or shade when illuminated with light of appropriate wavelength.

1951 *Physics Abstr.* A. LIV. 204/2 (*heading*) Photochromism in the bianthrone series. **1953** [see *photochromic *a.*]. **1960** *New Scientist* 2 June 1424/1 This new kind of reversible colour change in organic compounds was first described by me [*sc.* Yehuda Hirshberg] in 1950, and I called it 'photochromism'. *Ibid.* 1425/3 Another recent development in photochromism is the discovery that many compounds which undergo reversible colour changes by exposure to ultraviolet light show the same effect when subjected to high-energy electrons or to any ionizing radiation. **1972** *Nature* 31 Mar. 245/2 Photochromism is an effect associated with compounds which reversibly change their absorption spectra on illumination with light.

photochromy (in Dict. s.v. Photochromatic *a.*). Add: **c.** [ad. F. *photochromie*.] = *photochromism.

1951 *Chem. Abstr.* XLV. 10059 (*heading*) Photochromy in the bianthrone series. **1963** *Adv. Photochem.* I. 282 Photochromy and thermochromy are mutually exclusive properties in crystalline salicylidine-anilines.

photoclinometry: see *photo- 2.

pho:tocoagula·tion. *Ophthalm.* [f. Photo- 1 + Coagulation.] The surgical technique of using an intense beam of light to coagulate small areas of tissue, esp. of the retina.

1961 *Times* 16 Jan. 11/7 Today by surgery and the use of electro or photo-coagulation successful reposition..is achieved. **1971** G. N. Wise et al. *Retinal Circulation* xv. 446/1 Photocoagulation is an effective method of destroying individual features of proliferative diabetic retinopathy. **1972** T. N. Waubke in Michaelson & Berman *Causes & Prevention of Blindness* 313/1 Most of our patients did not realize the defects in the visual field which occur after photocoagulation.

Hence **pho:tocoa·gulate** *v. trans.*, to treat by this technique; **pho:tocoa·gulating** *ppl. a.*; **pho:tocoa·gulator**, an apparatus which produces the light beam used for photocoagulation.

1967 *New Scientist* 30 Mar. 679/3 The sharply focused beam caused only minimal scars to surrounding tissue. Ultimately they hope to develop more versatile laser photocoagulators. **1969** *Sci. Jrnl.* July 16 When the retina is subjected to a photocoagulating laser pulse..the transient pressure waves which reverberate through the eyeball resemble the seismic records following an earthquake. **1972** L. M. Aiello in Michaelson & Berman *Causes & Prevention of Blindness* 317/2 A ruby laser photocoagulator. *Ibid.* 318/2 Control studies are important..in order ..to control our eagerness to photocoagulate every eye of all diabetics.

photocomposing (fōᵘ:tokǫmpōᵘ·ziŋ), *vbl. sb. Printing.* Also with hyphen. [f. Photo- 2 + Composing *vbl. sb.*] **1.** The setting of text by the projection of images of letters or symbols on to photographic film, which is then used in the preparation of the printing surface; filmsetting. Freq. *attrib.*

1929 *N.Y. Times* 30 Jan. 11/1 There is another great revolution coming in the printing industry... That is the use of photo-composing instead of metal composing. Instead of producing a column of metal type, we will have a machine which produces a strip of film. **1929** *Times* 29 Oct. (Printing Suppl.) p. xii/7 The Typary machine..though not strictly a photo-composing machine, was directly inspired by the desire to abolish typesetting. **1948** *Sci. News Let.* 4 Dec. 362/1 Several photocomposing machines are in experimental use. **1955** *Times* 5 July 6/5 This is the first time that..photocomposing machines have been on view at Olympia as working exhibits. **1961** *Spectator* 14 Apr. 509/3 Technical advances, particularly in photo-composing and the printing of matter from film, rather than metal, are reaching a stage where their place in the British industry must be decided. **1973** S. Jennett *Making of Bks.* (ed. 5) v. 83 All three hot-metal composing-machine companies in Britain, and others abroad, have introduced photocomposing machines.

2. The manufacture of printing plates directly from photographic images for the production of multiple copies of illustrations, designs, etc.

1929 W. C. Huebner in *Trans. Amer. Soc. Mech. Engin.* L. PI. 21/2 Photo composing or photo-mechanical imposition is..the art of making printing plates wherein original designs are photographed directly on the press plate. **1953** *Composition Manual* (Printing Industry of Amer.) viii. 296/1 It [*sc.* photo-typesetting] should not be confused with 'photo-composing' because the latter has been definitely applied, for a number of years, to machines that are used for the multiple-printing of photographic images in offset lithography and other process work... The word 'photo-composition', sometimes used in current writing to denote photo-typesetting, has likewise been a source of some confusion.

So **pho:tocompo·se** *v. trans.*, **-compo·sed** *ppl. a.*; **pho:tocompo·ser**, a machine for carrying out photocomposition.

1929 W. C. Huebner in *Trans. Amer. Soc. Mech. Engin.* L. PI. 21/2 The results attained by photo-composed plates indicate the wide range of work that can be done economically. *Ibid.* 22/1 All sizes of work..are photo-composed successfully. *Ibid.* 23/1 The photocomposer can do it [*sc.* produce several copies of a plate] better. **1948** *Sci. News Let.* 4 Dec. 362/1 One company known for typesetting machinery has its photocomposer in day-by-day use. **1965** *Economist* 22 May (Press Ahead Suppl.) p. xvi/2 Two photocomposers were installed. **1969** *Physics Bull.* Jan. 24/1 The *Current Papers* will be photocomposed for printing under computer control. **1970** A. Cameron et al. *Computers & Old Eng. Concordances* 7 Photo-composed pages can look very handsome indeed. **1972** *Computers & Humanities* VI. 153 Computerized photocomposers may be divided into three stages of development or three generations of machine in use today.

photocomposition (fōᵘ:tokǫmpǫzi·ʃən). *Printing.* [f. Photo- 2 + Composition.] = *photocomposing *vbl. sb.* Also *attrib.*

1929 *N.Y. Times* 30 Jan. 11/1 (*heading*) Publisher.. tells of a photo-composition machine that will do away with metal composition. **1949** H. O. Smith *Rotophoto Process* 5 With the commercial advent of mechanical photocomposition the future will undoubtedly see more print changing over to photographic methods of reproduction. **1953** [see *photocomposing *vbl. sb.* 2]. **1957** E. Linklater *Private Angelo* 238 (*colophon*) Of this book, first published by Jonathan Cape in 1946, two thousand copies were printed at Christmas 1957 for Sir Allen Lane and Richard Lane... The book was composed entirely without metal type: it is the first to have been produced in Great Britain by means of photocomposition. **1967** D. G. Hays *Introd. Computational Linguistics* iv. 67 Early photocomposition devices..used photosensitive stock and stored negative images of characters. **1968** A. Bluhm *Photosetting* i. 3 These two terms [*sc.* photosetting and

filmsetting], and also photocomposition and phototypesetting, are all currently used to mean exactly the same thing, namely the composition of letters and symbols by photographic exposure from a set of transparent master characters for subsequent reproduction by any printing process. **1975** *Times* 4 Sept. 1/1 The printing unions were told..that to allow large-scale photocomposition techniques would be introduced. **1977** *Times* 14 May 1/2 The plans also involve..Fleet Street production by the new technology of photocomposition.

pho:tocondu·ctance. *Physics.* [f. Photo- 1 + *Conductance.] = *photoconductivity.

1939 *Jrnl. Chem. Physics* VII. 426/2 Photovoltaic experiments indicate the release of photoconductance electrons. **1959** *Bell Syst. Techn. Jrnl.* XXXVIII. 750 Measurements..of work function, photoconductance, surface conductivity and field effect have indicated that there is a large density of surface states.

pho:tocondu·cting, *ppl. a. Physics.* Also with hyphen. [f. Photo- 1 + Conducting *ppl. a.*] Exhibiting or utilizing a decrease in electrical resistance when illuminated.

1929 *Sci. Abstr.* A. XXXII. 638 (*heading*) Thallium photo-conducting cells. **1932** Hughes & DuBridge *Photoelectric Phenomena* viii. 329 Cuprous oxide..was found to be photoconducting. **1974** *Encycl. Brit. Macropædia* XI. 184/2 Zinc sulfide phosphors..are photoconducting, which means that many excited electrons are lifted to the conduction band.

pho:tocondu·ction. *Physics.* Also with hyphen. [f. Photo- 1 + Conduction.] = *photoconductivity.

1929 Campbell & Ritchie *Photoelectric Cells* ii. 20 Our present knowledge of photo-conduction is largely due to the work of Gudden and Pohl. **1955** *Proc. IRE* XLIII. 1819 (*heading*) Photoconduction in germanium and silicon. **1965** Phillips & Williams *Inorg. Chem.* xi. 20 Excitation of electrons in semiconductors can also be brought about by absorption of light, the phenomenon of photo-conduction. **1971** Smith & Thomson *Optics* xxiii. 334 In another type of television camera tube, the basic effect is photoconduction instead of photoemission.

pho:tocondu·ctive, *a. Physics.* Also with hyphen. [f. Photo- 1 + Conductive *a.*] Photoconducting; of or pertaining to the property of being photoconducting.

1929 Campbell & Ritchie *Photoelectric Cells* i. 1 Selenium cells might be termed photo-conductive. **1933** *Product Engin.* Aug. 289/1 With each of the three general types of cells, photo-voltaic, photo-emissive and photo-conductive, it is possible to use a wide variety of auxiliary equipment. **1957** *Proc. Inst. Electr. Engin.* CIV. B. 467/1 In 1873, Smith discovered the change in conductivity of selenium when sunlight fell upon it—the photo-conductive effect. **1960** *Times* 8 Mar. 2/5 The effects of light, X and gamma radiation on photoconductive materials. **1971** *Sci. Amer.* June 13/3 (Advt.), An advantage of photovoltaic detectors is that no external power supply is required to produce the photosignal, whereas photoconductive detectors must be activated.

Hence **pho:toconducti·vity**, the property of being photoconductive.

1920 Campbell & Ritchie *Photoelectric Cells* ii. 20 The effect of the radiation appears as an increase of conductivity of the material rather than as a current flowing from it. This is the ultimate principle of selenium and other photo-conductivity devices. **1932** Hughes & DuBridge *Photoelectric Phenomena* viii. 329 Gudden studied the photoconductivity of the halides of thallium, lead, and silver. **1974** *Encycl. Brit. Macropædia* XIII. 817/2 The most outstanding physical property of crystalline selenium is its photoconductivity: on illumination, the electrical conductivity increases more than 1,000-fold.

pho:tocondu·ctor. *Physics.* Also with hyphen. [f. Photo- 1 + Conductor.] A photoconducting substance or device.

1930 *Sci. Abstr.* A. XXXIII. 638 Detailed instructions for constructing photo-conductors. **1947** *Sun* (Baltimore) 1 Jan. 6/1 A newly developed photoconductor cell, the size of a peanut, has already spotted things never seen before on Venus and Saturn. **1967** E. Chambers *Photolitho-Offset* xiv. 229 The two most common photoconductors used are selenium and zinc oxide. **1973** *Focal Dict. Photogr. Technol.* 441 Photoconductors are available whose respective sensitivities cover between them all wavelengths from 450 to 7,000 nm.

photocontro·l. [f. Photo- 1, 2 + Control *sb.*] **1.** Control by light.

1954 *Bot. Gaz.* CXV. 360 (*heading*) Oxygen consumption of lettuce seed in relation to photocontrol of germination. **1975** *Nature* 10 Apr. 512/2 Many aspects of plant development are subject to photocontrol by way of the chromoprotein photoreceptor phytochrome.

2. *Surveying.* (Also as two words.) A control (i.e. a system of precisely surveyed points in terms of which subordinate local surveys are carried out) consisting of points that can be identified on aerial photographs.

1956 *Nature* 3 Mar. 419/1 Good progress was made with geodetic surveys, and the triangulation and photocontrol work over twenty thousand square miles..was completed. **1967** F. H. Moffitt *Photogrammetry* (ed. 2) vi. 136 The rigid requirements for precision photo control for analytic control extension, for cadastral surveys, and for certain types of engineering mapping, make it necessary to pre-mark a great deal of the control points. **1974** P. R. Wolf *Elem. Photogrammetry* xi. 223 According to this rule, a map being plotted with a contour interval of 10 feet requires vertical photo control accurate to within ± 2·0 feet.

photoconversion to **-convertible**: see *photo- 1.

photocopier (fōᵘ·tokǫpiə). Also (*rare*) **-copyer**. [f. next + -er[1].] **A.** An apparatus for making photocopies.

1934 in Webster. **1959** *Economist* 21 Mar. 1087/1 For more than 10 to 15 copies it will prove cheaper to use some kind of duplicator than even the cheapest type of photocopyer. **1964** D. Francis *Nerve* xvi. 210, I made copies of this statement and then on the photocopier *Colonist* (Victoria, B.C.) 29 July 45/1 They're electronic photocopiers..to go with their photocopiers reeling out these miles of guff. **1978** 'L. Black' *Foursome* xi. 104 There was a public photo-copier in the main post office..coin-operated, 5p per one copy of one sheet.

b. One who operates a photocopying machine.

1977 *Daily Tel.* 26 Nov. 15/2 A fourth man,..who had worked at Somerset House as a photocopier, was allowed bail.

photocopy (fōᵘ·tokǫpi), *v.* [f. Photo- 2 + Copy *v.*[1]]

In quot. 1924 the formation is prob. *photo* + *copying machine* rather than *photocopy* + -*ing* + *machine*.] *trans.* To make a photocopy of. So **photo-copied** *ppl. a.*, **pho·to-copying** *vbl. sb.* and *ppl. a.*

1924 C. W. Hackleman *Commerc. Engraving & Printing* (rev. ed.) 792 Photo copying machines. In Fig. 2015 is shown a machine for making copies of records, drawings..flat merchandise, etc., by a simplified method of photography, the copies being made in enlarged, reduced or natural size directly upon sensitized paper. *Ibid.* 793 [*caption*] Prints as made with a photo-copying machine. **1948** H. H. Fussler *Photogr. Reproduction for Libraries* xi. 192 Less expensive papers can be used in many instances for enlargement printing than are required in the original photocopying machines. **1948** *Library Assoc. Rec.* Feb. 37/1 There are private photocopying firms in most cities. **1952** I. Graham *Encycl. Advertising* 350/1 The original subject may be photocopied in actual, enlarged, or reduced size; the finished photocopy may be black on a white background (called a 'positive') or white on a black background (called a 'negative'). **1958** S. Hyland *Who goes Hang?* xxxix. 184 Neat files of photo-copied documents. **1958** H. R. Verry *Document Copying* vii. 40 Photocopying processes..are able to copy material which cannot be copied by other processes, such as drawings, half-tone and line illustrations. **1960** *McGraw-Hill Encycl. Sci. & Technol.* X. 142/2 Photocopying processes may be somewhat arbitrarily divided into two classes: silver halide photocopying, transfer processes, thermography..plan copying, electrostatic processes, the electrolytic process, and microfilming. **1968** *Brit. Med. Bull.* XXIV. 222/1 Many laboratories..have used photocopying or laboratory record cards as a means of reporting their results. **1972** A. Price *Col. Butler's Wolf* vi. 55 He passed the sheets to Butler. Anonymous, greyish photocopied typescript, the original of which was no doubt in Latimer's safe. **1973** *Nature* 16 Mar. 147/2 Advance copies of the technical report are available in photocopied form. **1976** P. Hill *Hunters* v. 47 Here's the questionnaire, get it photocopied.

photocopy (fōᵘ·tokǫpi), *sb.* [f. prec. or f. Photo- 2 + Copy *sb.*] A copy of documentary material made by any of various processes (usu. involving the chemical or electrical action of light on a specially prepared surface) in a copying machine and usu. the same size as the original.

1934 in Webster. **1952** [see prec.]. **1958** S. Hyland *Who goes Hang?* xxii. 97 A simple photo-copy, in black with white, of a..drawing. **1966** *Lancet* 24 Dec. 1414/1, I was flattered enough to have some photocopies made of my modest contribution. **1971** *Physics Bull.* Nov. 682/1 In the USA, theses are readily available either on microfilm or as photocopies. **1973** 'C. Aird' *His Burial Too* xiv. 130 The photocopy..had been folded neatly inside a telephone directory.

photocurrent, -damage: see *photo- 1; **photodegradable, -degradation**: see *photo- 3; **photodensitometer** to **-detector**: see *photo- 1; **photodimer** to **-dimerize**: see *photo- 3.

photodiode (fōᵘtodai·ōᵘd). *Electronics.* Also with hyphen. [f. Photo- 1 + Diode *a.*, *sb.] A semiconductor diode which generates a potential difference or changes its electrical resistance when illuminated.

1945 S. Benzer *Photoelectric Effects in Germanium* (U.S. Dept. of Commerce, Office of Publication Board PB 28644) 3 In exploring germanium samples for these 'photo-diode' effects, a considerable change..was found rectification at different points, uncommon current-voltage characteristics are sometimes observed. There are two types which are of particular interest because of their sensitivity to light... K. Lark-Horovitz has given the descriptive names 'photodiode' and 'photopeak' to these characteristics. *Ibid.* 4 Even in the most highly saturated photodiodes, the current continues to rise with voltage. **1959** [see *photo-transistor]. **1962** *Jrnl. Optical Soc. Amer.* LII. 1237/1 Large-area photodiodes have been in use for a number of years as solar cells for the conversion of solar radiation to electrical power. **1971** *Nature* 20 Aug. 540/1 The photometers are protected from excessive light levels by protective photodiodes which disconnect the high voltage supply at a level where the light level is above a minimum value. **1973** D. G. Fink *Electronics Engineers' Handbk.* xi. 67 Figure 11-55 is a cross-sectional diagram of a metal semiconductor (Schottky barrier) photodiode. **1978** *Gramophone* Jan. 1322/2 A light beam from a helium-neon laser scans the track and the reflected light is directed towards a photodiode which converts the light variations into a corresponding electrical signal for decoding and relaying to the aerial socket of a standard TV set for viewing and listening.

photodisintegration: see *photo- 1; **photodissociate, -dissociation**: see *photo- 3.

photodynamic, *a.* (in Dict. s.v. Photo- 1). Add: **b.** [after G. *photodynamisch* (H. von Tappeiner 1904, in *Münch. med. Wochenschr.* 19 Apr. 714/1).] Involving or causing a toxic response to light, esp. ultraviolet light.

1909 *Jrnl. Chem. Soc.* XCVI. ii. 925 Extracts in methyl alcohol of various etiolated plants invariably showed a distinct photodynamic action on suspensions of red-blood corpuscles. **1937** *Ann. Rep. Progr. Chem.* XXXIII. 342 Many analogous phenomena..have been studied in the understanding of the photodynamic inactivation of viruses and bacteria. **1965** Ralston & McLeroy *Light v. 327* Calcutti (1934) found a large group of carcinogens, other than the polycyclic hydrocarbons, that was photodynamic. **1973** *Daily Colonist* (Victoria, B.C.) 16 Nov. 5/5 The 'photodynamic therapy' may provide a new approach to treating several human cancers. **1974** *Nature* 9 Aug. 524/1 Light of wavelengths between 400 nm or 500–600 nm causes lesions through photodynamic action of porphyrin in the skin. **1977** *Ibid.* 3 Nov. 15/3 The use of dye photosensitisers in research on the so-called photodynamic effect in viruses has been established.

Hence **photodyna·mically** *adv.*

1926 E. Mayer *Clin. Applic. Sunlight* v. 87 With the cooperation of fluorescent and indigo-carmine, all dyes which stained staphylococci..were found to be..photodynamically active. **1954** *Jrnl. Bacteriol.* LXVIII. 125/2 Photodynamically inactivated phage. **1967** J. M. Hoskins *Virol. Procedures* xiii. 181 Neutral red is a photodynamically active dye, and virus-infected cells incubated in its presence in the light fail to yield normal plaques.

photo-effect, -ejection: see *photo- 1.

photoelastic (fōᵘto,ilæ·stik), *a.* Also with hyphen. [f. Photo- 1 + Elastic *a.* and *sb.*] Employing or exhibiting the property of becoming birefringent when mechanically stressed, so that polarized light passed through such a substance gives rise to interference fringes that display the stress patterns in it.

1911 E. G. Coker in *Trans. Inst. Naval Archit.* LIII. 265 (*heading*) The determination, by photo-elastic methods, of the distribution of stress in plates of variable section, with some applications to ships' plating. **1920** *Flight* XII. 399/1 On arrival..the model was examined by the Superintendent, and later attended the lecture of Professor Coker on his photo-elastic method of measuring stresses in materials. **1936** *Jrnl. R. Aeronaut. Soc.* XL. 472 To analyse quick variations of internal stresses, e.g.,..in vibrating specimens or in impact phenomena, kinematography or [read of] photo-elastic fringes at very high speeds must be used. **1953** Fairman & Cutshall *Mech. of Materials* xiii. 372 A third method of experimental stress analysis involves the transmission of polarized light through transparent models and is known as the photoelastic method. **1958** *New Scientist* 2 Oct. 933/1 The 'photo-elastic technique'..enables the movement of stresses to be traced as they change their pattern under load. **1965** Hawkes & Holister in Zienkiewicz & Holister *Stress Anal.* xii. 267 In the photoelastic coating technique, sheets of birefringent plastic are bonded, using a reflective cement, to the surface of the body being studied. When the body is strained under load, the strain is conveyed through the cement to the coating where isoclinic and isochromatic fringes appear, when the coating is viewed through a reflection polariscope. **1970** *New Scientist* 3 Dec. 377/2 The photoelastic model was then prepared by pouring photoelastic resin into moulds..temporarily hardened stresses that the strain gauges.

So **pho:toelasti·city**, the photoelastic method (of stress analysis).

1911 E. G. Coker in *Engineering* 6 Jan. 1/3 (*heading*) Photo-elasticity. **1950** Dolan & Murray in M. Hetényi *Handbk. Exper. Stress Anal.* xvii. 829 His [*sc.* E. G. Coker's] introduction of celluloid for models and the use of monochromatic light have led to the modern laboratory methods which make photoelasticity a powerful engineering tool. **1959** R. E. Archer et al. *Introd. Mech. of Solids* iv. 157 The three most widely used methods of measuring strain are 1. Photoelasticity. 2. Brittle coatings. 3. Wire or foil strain gages.

photo-electric, *a.* Now usu. written as one word. **c.** Add to def.: pertaining to or employing a photoelectric effect; *photoelectric absorption*, the absorption of light by an atom which then emits an electron; *photoelectric cell* = *photocell; *photoelectric effect* (see quot. 1973); *photoelectric emission*, the emission of electrons by photoelectric action.

1891 *Phil. Mag.* XXXI. 231 The selenium-aluminium cells differ from all other photoelectric cells that I have constructed in their great sensitiveness to all parts of the spectrum. **1892** *Photogr. Ann.* II. 113 As a rule a much greater photo-electric effect is produced by the more refrangible rays than by the less refrangible. **1912** *Physical Rev.* XXXIV. 475 Photoelectric cells containing sodium or potassium..give a current on illumination which is strictly proportional to the intensity of the exciting light. **1921** *Jrnl. R. Soc. Arts* 16 Dec. 85/1 The electrons so detached are called photoelectrons and the action photoelectric. **1949** *Jrnl. Brit. Interplanetary Soc.* VIII. 115 The photo-electric telescope, which can amplify faint images. **1955** *Sci. News Let.* 12 Mar. 168/3 The instrument, known as a photoelectric polarimeter, measures polarization of skylight with the aid of polarizing prisms. **1958** W. K. Mansfield *Elem. Nucl. Physics* i. 10 The electrons can gain this energy by absorbing electromagnetic radiation which gives rise to photoelectric emission of electrons. *Ibid.* v. 44 In photoelectric absorption the γ-ray interacts with an atom as a whole, ejecting an electron. **1960** R. W. Marks *Dymaxion World of B. Fuller* 25/1 In 1927, seeking photo-electric cells and relay-actuated devices, he wrote to his brother. **1961** G. R. Choppin *Exper. Nucl. Chem.* iii. 30 The photoelectric effect is the predominant mode of interaction in aluminium for gamma rays below 60 kev. **1970** O. Dopping *Computers & Data Processing* xi. 161 In the photo-electric reader, a lamp sends light beams onto a number of photo cells through the holes in the tape. **1973** J. Yarwood *Electricity & Magnetism* xv. 594 There are three main photoelectric effects: (*a*) the photoemissive effect or surface photoemission, in which electrons are emitted from the surface of a material when radiation falls upon it; (*b*) the photoconductive effect, concerned when the incident radiation causes electrons to be released within the material (a non-metallic solid) and so increase its conductivity; (*c*) the photovoltaic effect, in which the radiation absorbed in a device causes an e.m.f. to be set up at a discontinuity..at a barrier between two different materials.

photo-electrical *a.* (examples); so **pho:toele·ctrically** *adv.*

1888 A. Righi in *Phil. Mag.* XXV. 315 The system of the two metals, when receiving the radiation, behaves then as a voltaic cell, and can be called a photoelectrical cell. **1911** *Physical Rev.* XXXII. 631 A platinum cathode film on glass is photo-electrically active even when of the utmost tenuity. **1923** Glazebrook *Dict. Appl. Physics* IV. 563/2 One of the most interesting phases of the inquiry into the laws and constants of radiation is the confirmatory data which one obtains from a consideration ..of photo-electrical action. **1941** Millman & Seely *Electronics* iii. 101 The number of electrons released photoelectrically from any part of the photocathode depends upon the intensity of the light at that point. **1974** *Physics Bull.* May 204/1 The work functions of polycrystalline films of aluminium..were measured photoelectrically.

photoelectrochemical, -electromagnetic: see *photo- 1.

photoelectron (fōᵘ·to,ilektrǫn). *Physics.* Also with hyphen. [f. Photo- 1 + *Electron[2].] An electron released from an atom by the interaction of a photon with it; *esp.* one emitted from a solid surface by the action of light.

1912 *Phil. Trans. R. Soc.* A. CCXII. 206 The law connecting the maximum emission velocity of the photo-electrons with the wave-length has been investigated by Ladenburg for Cu, Zu and Pt, and by Kunz for Na-K alloy and Cs. **1934** [see *Compton]. **1962** H. D. Bush *Atomic & Nucl. Physics* ix. 209 An electron bound in an atom may absorb the whole of the energy of an incident γ-ray photon, being ejected from the atom as a photoelectron. *Ibid.* 210 The photoelectrons..will have a short range in the absorbing material, and so unless the material is very thin they are likely to be reabsorbed. **1965** *Wireless World* Sept. 442/1 The process by which the energy of the particles is converted first into photons and then into photo-electrons at the cathode of a photomultiplier tube is very inefficient. **1975** J. R. Lamarsh *Introd. Nuclear Engin.* iii. 76 The kinetic energy of the ejected photoelectron is equal..to the energy of the photon less the binding energy of the electron to the atom.

Hence **pho:toelectro·nic** *a.*, of or pertaining to a photoelectron, or the interaction of light with electrons; **pho:toelectro·nics** *sb. pl.* (const. as *sing.*), the study of photoelectronic phenomena.

1922 *Physical Rev.* XX. 195 (*heading*) Symmetry of emergent and incident photoelectronic velocities. **1940** *Chambers's Techn. Dict.* 638/1 Photo-electronics. **1947** *Jrnl. Optical Soc. Amer.* XXXVII. 430/1 The lower limit of usable voltage or gain in the photomultiplier tube in a pulse-counting application is reached when the photoelectronic output pulse becomes reduced..to the point where there is confusion from the tube noise and the thermal noise of the coupling resistor. **1957** *Proc. Inst. Electr. Engin.* CIV. B. 480/1 The photo-electric effect is the fundamental basis of the generation of television signals, since this depends on the conversion of light energy into electrical energy. A detailed account of this branch of photo-electronics is beyond the scope of this review. **1963** *Adv. Photochem.* I. 276 The second fundamental photo-electronic mechanism often producing photochromism is charge transfer (photoionization). **1966** N. S. Kapany *Fiber Optics* (1967) viii. 206 One important application of fiber optics in photoelectronics lies in the coupling of image intensifier stages. **1970** *Sci. Jrnl.* Aug. 90/1 Modern photoelectronic devices can detect or produce radiations over the whole electromagnetic spectrum. **1977** *Gramophone* Apr. 1629/1 The auto-stop and pickup return at the end of a record is sensed by photo-electronic means, involving no side pressures.

photoemission (fōᵘ·to,imiʃən). *Physics.* Also with hyphen and † as two words. [f. Photo- 1 + Emission.] The emission of electrons from a surface by the action of light incident on it.

1916 *Physical Rev.* VII. 383 (*heading*) Theories of photo emission. **1923** *Ibid.* XXII. 578 It is evident..that the gases given off by the copper-oxide cylinder during the baking out of the tube and the glowing of the filament, have a pronounced effect upon the photo-emission. **1961** G. R. Choppin *Exper. Nucl. Chem.* iii. 35 Tertiary electrons may be produced by photoemission resulting from the photons of the secondary ionization process. **1971** Smith & Thomson *Optics* xxiii. 334 In another type of television camera tube, the vidicon, the basic effect is photoconduction instead of photoemission.

So **pho·toemitted**, **-emitting** *ppl. adjs.*; **photoemi·ssive** *a.*, exhibiting, employing, or pertaining to photoemission; **pho·toemitter**, a photoemissive body or substance.

1932 *Electronics* Jan. 19/2 The necessary condition arises to develop a photoemissive surface which..gives satisfactory emission. **1933** *Jrnl. Inst. Electr. Engin.* LXXIII. 439/1 The equilibrium potential of the element is defined by (*a*) the velocity of the beam [of electrons] and (*b*) the secondary emission from the photo-emitting substance due to bombardment by the electrons. **1957** *Ibid.* CIV. B. 467/1 Hertz discovered the photo-emissive effect—the loss of negative electric charges from metal surfaces which are illuminated. **1957** *Proc. Inst. Electr. Engin.* CIV. B. 468/2 The earliest photo-emitters were common metal surfaces. **1959** *Control* Feb. 92/2 The spectral response of photoemissive cells depends on the cathode coating. **1970** *Nature* 3 Oct. 6/1 Two new spectroscopes using synchroton radiation and photoemitted electrons are singled out for special mention. **1971** Smith & Thomson *Optics* xxiii. 333 Semiconductors are widely used as photoemitting surfaces. **1973** [see *photoelectric *a.* c]. **1975** *Physics Bull.* Feb. 65/2 For the first time in a commercial instrument the simultaneous measurement of the energy and angular distribution of photoemitted or secondary electrons can be made.

photoenvironment to **-fabrication**: see *photo- 1; **photo-essay** to **-facsimile**: see *photo- 2.

photo-fi·nish. Also **photo finish**. [f. Photo- 2 + Finish *sb.*] The finish of a race in which competitors are so close that the result has to be determined by reference to a photograph of the situation. Also *attrib.* and *fig.*

1936 *N.Y. Times* 1 July 36/6 New..tests will be made tomorrow at Pimlico on the photo finish device. *Ibid.* 2 July 16/8 At Belmont Park thirty pictures were called for. Of this number the photo finish awarded the decision to sixteen horses racing on the outside, with fourteen on the inside. **1938** D. Runyon *Take it Easy* xv. 283, I will take a chance on Nicely-Nicely against anything on four legs, except maybe an elephant, and at that he may give the elephant a photo finish. **1940** *Sun* (Baltimore) 22 Nov. 1/2 Gov. Payne H. Ratner..emerged a photo-finish leader tonight in the complete unofficial count of ballots cast sixteen days ago. **1944** J. H. Fullarton *Troop Target* 160 'We've got to choose between security and progress.' 'Personally,' said MacVaughan, 'I'm for security—but it's a photo finish.' **1951** *Sport* 7–13 Jan. 4/2 Combination 'B' looks like proving a photo-finish between Plymouth and Chelsea, with Charlton holding an outside chance. **1961** *Times* 22 July 3/2 Jones.. snatched second place through a photo-finish from Radford. **1973** D. Francis *Slay-Ride* iv. 55 Sixteen hundred metres for staying two-year-olds... There was a photo-finish. **1974** *Times* 6 May 1/3 A photo finish was predicted by the last two public opinion polls. The final one showed M Mitterrand..winning on the second ballot by a short head. **1976** *Guardian Weekly* 19 Sept. 7/2 The Zia coup was a photofinish affair just four days after another army putsch that is popularly supposed to have been backed by India. **1976** *Norwich Mercury* 19 Nov. 10/1 A number of very close contests took place in Division Two, but Thetford British Legion B and Saints E were involved in a 'photo-finish' which the Legion won by a whisker.

photofission: see *photo- 1.

photofit (fōᵘ·tofit). Also **Photo-Fit**. [f. Photo- 2 + Fit *sb.*[4]] Name of a method of building up an identikit picture (cf. *Identikit) by assembling a number of photographs of individual facial features; a picture so formed. Freq. *attrib.*

1970 *Guardian* 23 Apr. 6/4 The Home Office believes it has a 'promising development' at its disposal with the introduction of a new system of criminal identification—the Photo-Fit. By using photographs instead of line drawings, the new technique will enable the police to build up..many..composite faces. *Ibid.* 6/5 Photo-Fit is a product of the invention of Mr Jacques Penry..and the manufacturing facilities of John Waddington Ltd. of Leeds. **1971** *Daily Tel.* 5 Mar. 2/8 Photo-Fit and Identikit pictures have been issued to police throughout London. **1973** *Oxf. Times* 10 Aug. 1/9 Thames Valley Police have issued a photofit picture and a description of a man they wish to interview. **1974** J. Gardner *Corner Men* xiv. 233 They showed..a straight and recent picture of Peppe, together with some photofits put together to show the [Mafia] don in a permutation of disguises. **1976** *Daily Record* (Glasgow) 4 Dec. 13/2 Police are so anxious to capture him that they applied for special permission from the Lord Advocate to issue this photofit picture.

photoflash (fōᵘ·toflæʃ). Also with hyphen. [f. Photo- 2 + Flash *sb.*[2]] A flash of light produced to enable a photograph to be taken; a device for producing such a flash. Freq *attrib.*

1930 *N.Y. Times* 24 Oct. 21/2 Noiseless and smokeless photoflash lamp devised. **1946** L. E. O. Charlton *R.A.F. & U.S.A.A.F. July 1943 to Sept. 1944* vii. 156 (*caption*) This night photograph shows huge volumes of smoke pouring from targets bombed at Duesseldorf which have been illuminated by a photo flash bomb. **1946** *R.A.F. Jrnl.* May 170 (*caption*) A Lancaster Master-Bomber on a night raid dropping a photo-flash. **1956** A. H. Compton *Atomic Quest* I. 58 The question is one of straightforward thermodynamics, which I had myself used several years before in calculating..the forces that cause bursting of photoflash bulbs. **1964** L. Deighton *Funeral in Berlin* vii. 46 Photo-flashes sliced instants from eternity. **1972** *Gloss. Electrotechnical, Power Terms (B.S.I.)* iv. iii. 15 *Photoflash lamp*, lamp giving..a single high light output for a very brief period, for lighting objects to be photographed. **1975** N. Freeling *What are Bugles blowing For?* iv. 17 An expressionless face for the photo-flash.

photoflood (fōᵘ·toflʊd). [f. Photo- 2 + Flood *sb.*] A very bright flood-light used in photography and cinematography. Freq. *attrib.*

1933 *Pop. Sci. Monthly* Feb. 76/1 The ordinary bulb in the reading light was removed and a photoflood bulb substituted... The photoflood is so strong that it will give the proper effect during the brief time the shutter is open for the flash. **1937** *Discovery* Feb. 57/2 A photoflood bulb gives an intensity of the same order at close range. **1962** *Which?* May 133/2 The Type A photoflood version of Kodachrome. **1972** *Amat. Photographer* 12 Jan. 38 A most popular light source with amateur photographers, photofloods have a colour temperature of 3,400°K and a very high light output. **1977** J. Hedgecoe *Photographer's Handbk.* 34 A larger 'No. 2' photoflood bulb..offers longer life but is more expensive.

photofluorogram to **-fluoroscopy**: see *photo- 2; **photoformer**: see *photo- 1.

photog (fotǫ·g). Also **fotog**. N. Amer. colloq. abbrev. of Photographer.

1913 *Technical World* XIX. 385/1 The elements of the air..and a dozen other things may make the newspaper 'fotog' ready to throw up the sponge. **1934** H. N. Rose *Thes. Slang* 51/2 Photographer: *a photog.* **1952** *Daily News* (N.Y.) 21 Aug. c4 The Swedish fotogs were actually saving film. **1968** *Globe & Mail* (Toronto) 17 Feb. 35 (Advt.), A photog's delight! **1973** R. Parkes *Guardians* vi. 104 You'll like that fashion photog of ours—what's her name.

photo-galvanic, *a.* Add: Now usu. written **photogalvanic**. **2.** *Physics.* Designating or utilizing the generation of a potential difference between two electrodes by a photo-chemical reaction in the electrolyte containing them.

1940 E. Rabinowitch in *Jrnl. Chem. Physics* VIII. 551/1 The term 'photogalvanic effect' is used in the present series of papers to denote a special case of the so-called Becquerel effect, in which the influence of light on the electrode potential is due to a photochemical process in the body of the electrolyte (as distinct from photochemical or photoelectric processes in the surface layer of the electrode..). *Ibid.* 553/2 The top curve shows *L* [*sc.* the intensity of illumination] as a function of time... The lower curve represents the simultaneous changes in the potential of a Pt-electrode placed in the solution—the 'photogalvanic effect'. *Ibid.* 560 The oxidation-reduction system thionine-iron..provides an extremely sensitive photogalvanic cell. **1956** *Nature* 21 Jan. 110/2 Electrical processes for the direct conversion of radiation to electricity, including thermoelectric generators, photovoltaic cells and *p-n* junctions and photogalvanic cells. **1976** *Interdisciplinary Sci. Rev.* I. 132/2 The irradiation of an electrode/electrolyte system produces a change in the electrode potential (on open circuit) or in the current flowing (on closed circuit). The cause of this may be a photochemical reaction in bulk solution, the products of which are electroactive, in which case the cell is commonly called a photogalvanic cell.

photogenic, *a.* Add: Also with pronunc. (-dʒi·nik).

1. Delete *rare* and add further examples.

1954 *Brain* LXXVII. 234 Photic excitation has been subjected to intense study in electroencephalographic laboratories... Yet clinical descriptions of photogenic epilepsy are few. **1962** *Times* 30 June 12/2 Self-induced photogenic epilepsy in children. **1972** Naquet & Meldrum in D. P. Purpura et al. *Exper. Models of Epilepsy* xv. 373 (*heading*) Photogenic seizures in baboon.

2. For *Obs.* read '*Obs.* exc. *Hist.*' and add later example.

1973 *Times* 22 May 28/5 (Advt.), Early photographs and related material, including..calotypes by William Henry Fox Talbot; early photogenic drawings; [etc.].

4. Of a person or thing: that is a good subject for photography; that shows to good advantage in a photograph or film. Also in Fr. form *photogénique*. orig. *U.S.*

1928 *Reflex* June 91/1 The mere doll-face, the photogenic type which thrived on 'fan' adulation, will be a thing of the past. **1931** S. Goldwyn in *Sat. Rev.* 14 Feb. 220/1 An actor may be 'photogenic' and have personality and appearance, but that is not enough. **1931** *Daily Express* 13 Mar. 19/4 Look at..Pamela Nickalls, with her perfect oval face and high, photogenic forehead. **1935** *Times Lit. Suppl.* 10 Oct. 631/4 Spain is very probably the most *photogénique* country in Europe. **1940** *New Statesman* 19 Oct. 379 The thrills in *Foreign Correspondent* are massive and photogenic to a degree. **1948** E. Waugh *Loved One* 5 Her legs were never *photogénique*. **1958** *Newnes Compl. Amat. Photogr.* 143 Take two common 'photogenic' subjects, a scene of hills and sky, and a child playing in a flower garden. *Ibid.* 150 It is a fact that very few faces are symmetrical, regular or photogenic. **1974** M. Hastings *Dragon Island* viii. 74 She was a good-looking girl; photogenic..and with the right bone structure.

photogenically *adv.* (examples relating to sense *4 above).

1959 *Times* 23 Feb. 10/6 There were nets drying photogenically in the sun. **1974** *Daily Tel.* 9 Aug. 13/2 The director..does score photogenically in the action scenes.

photogeology (fōᵘtodʒi,ǫ·lŏdʒi). Also with hyphen. [f. Photo- 2 + Geology.] (See quot. 1941.)

1941 H. C. Rea in *Bull. Amer. Assoc. Petroleum Geologists* XXV. 1796 For this little known branch of geology the writer suggests the term 'photogeology', which is defined as the geologic interpretation of aerial photographs. **1970** *Nature* 24 Jan. 322/1 The long range aim is to establish an absolute lunar time scale directly related to relative ages obtained by photogeology over the whole lunar surface. **1976** *Jrnl. R. Soc. Arts* CXXIV. 639/2 British scientists [in the Antarctic] possess an impressive array of modern techniques: airborne magnetometry,..satellite photogeology, [etc.].

Hence **pho:togeolo·gic, -lo·gical** *adjs.*, of or pertaining to photogeology; **pho:togeolo·gically** *adv.*, by photogeological means; **photogeo·logist,** an expert or specialist in photogeology.

1941 *Bull. Amer. Assoc. Petroleum Geologists* XXV. 1796 A photogeologic map would be a map produced from a stereoscopic study of the aerial photos. *Ibid.* 1797 It proved that a trained photogeologist could produce a map which would agree very favorably with actual field observations. **1949** *Ibid.* XXXIII. 1251 During 1947, more than 110,000 square miles in the Rocky Mountain region were covered photogeologically. **1950** *Ibid.* XXXIV. 2285 Mutual checking of photogeological and field work should be just as apt to lead to corrections of the one as the other whenever discrepancies are found. **1962** F. I. Ordway et al. *Basic Astronautics* v. 206 The first remotely conducted photogeologic experiment of another world was remarkably successful. **1966** E. Burgess *Assault on Moon* ii. 43 The Army Map Service produced a general photogeologic map of the Moon in the late 1950s. **1967** *Q. Jrnl. Geol. Soc.* CXXIII. 255 It is possible to define the geometry of photogeologically observed large folds by plotting..dip and strike values on a stereographic net. **1972** *Science* 2 June 976/3 Deposits of basin ejecta concentric about Imbrium are termed the Fra Mauro Formation by photogeologists. **1977** A. Hallam *Planet Earth* 111 Photogeological interpretation assists rapid production of geological maps from aerial photographs.

photogoniometer to **-metry:** see *photo- 2.

photogram. Add: (Further examples.) (Still *rare.*)

1935 *Amer. Mineral.* XX. 476 Montmorillonite was *x*-rayed..and its lines..agree with the powder spectrum photograms of other investigators. **1961** (*title of periodical*) New photograms.

†**2.** A photograph, picture, diagram, or other facsimile transmitted by wireless or ordinary telegraphy. *Obs.* (Now called a phototelegram (see *photo- 1).)

1928 *Observer* 24 June 23 The wireless photogram service..has been extended. **1928** *Times* 6 Sept. 11/6 The Postal Telegraph Company put into commercial operation to-day a new telephoto and facsimile message service, which it calls photograms. **1929** *Telegraph & Telephone Jrnl.* XVI. 4/1 Suppose that transmissions of photo-grams by modified television apparatus can take place at the rate of 30 per second.

3. A photographic picture made without a camera (see quots.). Also † *photogramme*.

1934 *Archit. Rev.* LXXV. 12/2 As a photographer he [*sc.* Moholy-Nagy] has been a pioneer in the photogramme (the cameraless photography which he regards as the art-form of the future). **1948** J. H. Gable *Compl. Introd. Photogr.* III. xvii. 209 The first photogram was probably made by Fox Talbot by placing lace on a sheet of his calotype paper and exposing it to light. **1958** *Newnes Comp. Amat. Photogr.* III. xxvii. 241 Photograms can also be made with the aid of a torch bulb. **1972** *Sci. Amer.* Dec. 115/1 A photogram is made without a camera by exposing photographic paper to a point light source, the leaf or the fruit being placed directly on the paper. **1978** J. H. Coote *Focalguide to Cibachrome* 175 You can easily make photograms in colour, directly from leaves, flowers and any other translucent objects.

photogrammetry. Add: More widely, the technique of using photographs to ascertain measurements relating to what is photographed. (Further examples.)

1933 *Jrnl. R. Aeronaut. Soc.* XXXVII. 386 Most of the continental plotting machines..were designed originally for terrestrial photogrammetry, where each goniometer in the plotter can be set to a known orientation. **1934** *Ibid.* XXXVIII. 257 (*heading*) Photogrammetry of landing speeds. **1948** *Sci. News* VII. Pl. 25, After the church had been destroyed..the essential dimensions were deduced from a pre-war snapshot... The science of deducing dimensions from photographs is called Photogrammetry. **1959** F. H. Moffitt *Photogrammetry* i. 1 The term *aerial photogrammetry* denotes that branch of photogrammetry wherein photographs of the terrain..are taken by a precision camera mounted in an aircraft... The term *terrestrial photogrammetry* denotes that branch..wherein photographs are taken from a fixed, and usually known, position on or near the ground and with the camera axis horizontal or nearly so. *Ibid.* 2 Terrestrial photogrammetry embraces also the use of cameras at ground stations in known positions for the following purposes: To reproduce plan and elevation views of buildings and structures... By means of motion-picture photography, to make measurements involving transitory phenomena such as wave motion,..moving machinery, [etc.]. **1965** *New Scientist* 18 Nov. 507/1 Photogrammetry is now being explored for use on large microwave aerials, to check the accuracy of construction, and to measure deformations under wind loads. **1978** *Jrnl. R. Soc. Arts* CXXVI. 473/1 Ankara, particularly at the Middle East Technical University, was early in the field in the use of photogrammetry for mass recording.

Hence **photogramme·tric** *a.*, **photogramme·trically** *adv.*; **photogra·mmetrist,** an expert or specialist in photogrammetry; **photogrammetrical** *a.* (later example).

1906 J. A. Flemer *Elem. Treat. Phototopogr. Methods* i. 5 The theoretical principles upon which photogrammetric methods are primarily based were known to J. H. Lambert.., who published a work on perspective in 1759. *Ibid.* xii. 397 If differences in the elevations of points of the terrene are to be deduced photogrammetrically within a limit of error not exceeding one meter, the pictured length..of a meter..should not appear shorter than 0·1 mm. **1939** B. B. Talley *Engin. Applic. Aerial & Terrestr. Photogrammetry* i. 6 For photogrammetrical purposes photographs may be classified as either aerial or terrestrial, depending on whether the taking camera was borne aloft or..on the surface of the earth. *Ibid.*, Photogrammetry..should never be considered as an end in itself. Such consideration is the stumbling block of photogrammetrists. **1950** *Antiquity* XXIV. 85 With the use of Photogrammetric Tables, extreme accuracy was possible in determining and measuring the exact position on the ground of crop-mark sites visible in the air photographs. **1957** *Oxf. Mag.* 21 Nov. 131/1 The search for oil today requires the organisation of teams of specialists: surveyors, palaeontologists, photogrammetrists, geophysicists and geochemists. **1970** J. A. Howard *Aerial Photo-Ecol.* ix. 98 Vertical photographs are simple to use photogrammetrically, as a minimum of mathematical correction is required. **1975** *Sci. Amer.* Jan. 130/3 In 1962..he organized a modern aerial photogrammetric survey of the site from 4,000 feet.

photograph, *sb.* Add: **c.** *photograph album* (examples), *book* (examples), *frame* (example), *gallery*.

1858 F. J. Cook *Let.* 19 Sept. in F. G. Bascom *Lett. Ticonderoga Farmer* (1946) 46, I..visited..Brady's celebrated photograph gallery. **1870** O. Logan *Before Footlights* 258 The only thing worth carrying away was a decent sort of photograph album. **1872** E. Wordsworth in E. Romanes *C. M. Yonge* (1908) ix. 150, I found her armed with a large photograph-book of friends and relations. **1873** *Young Englishwoman* Feb. 94/2 Photograph frame in leather work. **1888** Mrs. H. Ward *R. Elsmere* I. iii. 76 Mrs. Seaton was severely turning over a photograph book. **1940** T. S. Eliot *East Coker* v. 14 The evening with the photograph album. **1949** *Chicago Tribune* 18 Sept. 27/4 Andrew Burgess..bought and operated Brady's national photograf gallery in Washington.

photograph, *v.* Add: **1. b.** (Earlier example.)

1857 C. Kingsley *Two Yrs. Ago* III. i. 37 If any one will ensure me a poor two thousand a year, I will promise to photograph no more.

c. (Further examples.)

1919 *Conquest* Nov. 24/1 The red leaves of autumn photograph as black. **1931** F. L. Allen *Only Yesterday* vi. 126 He photographed well and the pictures of him in the rotogravure sections won him affection and respect. **1949** E. Coxhead *Wind in West* i. 10 A creamily handsome young woman..Hedy Lamarr type, would photograph, summed up Tony at a glance. **1965** *Listener* 18 Nov. 817/2 Wales photographs beautifully: the short, slated roofs, the coal-tips and the valleys, the terraced houses on grey streets—none of them loses much in black and white reproduction. **1974** 'E. Lathen' *Sweet & Low* xix. 183, I don't like the way he photographs. Yours is the face I want.

photographable, *a.* (Further examples.)

1939 [see *absorptiometer 2]. **1962** C. Walsh *From Utopia to Nightmare* i. 20 Some artists..have created..nothing remotely photographable. **1973** 'E. McBain' *Hail to Chief* i. 13 A blush which, if not quite kissable, is at least photographable.

† **photogra·pheme.** *Obs. rare.* [f. Photograph *sb.* + *-eme* as if f. Gr. -ημα (cf. *-eme).] A photograph.

1864 G. M. Hopkins *Further Lett.* (1956) 211 She *comprehood* who I was by my photographeme which you strangely said was not like me. **1888** *Ibid.* 291 It is not altogether as I should wish it either as a portrait or as a..photograph, photogram, photographeme, φωτογράφημα, a work of the photographic camera.

photographic, *a.* Add: **c.** *photographic memory*: a memory that records visual perceptions with the accuracy of a photograph.

1940 in *Chambers's Techn. Dict.* 638/1. **1948** 'J. Tey' *Franchise Affair* vii. 73 She has a photographic memory... She would remember what she saw. **1964** M. Critchley *Developmental Dyslexia* ix. 62 He is perhaps a person weak in visual imagery and visual memory of all types, the opposite of the person with eidetic imagery and photographic memory. **1974** A. Morice *Killing with Kindness* iv. 37 I've got what they call a photographic memory and I don't visualise her wearing a wedding ring.

photographica (fōᵘtogræ·fikă), *sb. pl.* [f. Photographic *a.*: see -a 4]. Books, albums, or collections of photographs; items connected with photography.

1973 *Country Life* 20 Sept. (Suppl.) 76 (Advt.), Photographs and Photographica. **1976** *Times Lit. Suppl.* 25 June 804/3 Photographica, which until recently accounted for a minute part of each year's offerings at book auctions in the United States, was the subject of no less than three large sales this spring. **1976** *National Observer* (U.S.) 4 Dec. 20/3 There can be no doubt that this is one of the most remarkable collections of American photographica.

photographically, *adv.* (Earlier and later examples of literal use.)

1842 J. F. W. Herschel in *Phil. Trans. R. Soc.* CXXXII. 194 The retina itself may be *photographically* impressible by strong lights. **1968** R. A. Lyttleton *Mysteries Solar Syst.* iv. 119 Comet Borelly showed photographically as many as nine tails.

photoheterotroph to **-inactivation:** see *photo- 1.

photoinduce (fōᵘ:to,indiū·s), *v.* Also with hyphen. [f. Photo- 1 + Induce *v.*] *trans.* To induce by the action of light; esp. in *Plant Physiol.*, to induce reproductive behaviour in a plant by an appropriate sequence of light and darkness (used with the behaviour or the plant as obj.).

1949 *Bot. Gaz.* CX. 495/1 If one part of the plant was kept completely in the dark and some other part above it was photoinduced to flower, then the part in the dark also flowered. **1952** *Ann. Rev. Plant Physiol.* III. 269 If a plant is photoinduced and then is grafted on to an individual kept on the noninductive conditions, the latter plant will also initiate flowers. **1974** *Photochem. & Photobiol.* XIX. 163/1 It is..of importance to determine if an electron transfer reaction of this type can be photoinduced in a model system.

So **photoindu·ced** *ppl. a.*, induced by the action of light; also **photoindu·cible** *a.*, capable of being photoinduced.

1947 *Bot. Gaz.* CIX. 121/1 Buds develop and flower only on the photoinduced portion or one node beyond. **1961** *Jrnl. Appl. Physics* XXXII. 1901/2 The photoinduced current..is equal to the diode current with the bias voltage zero. **1970** Dorion & Wiebe *Photocromism* i. 9 The reversible or self-bleaching feature distinguishes photochromism from the well-known irreversible photoinduced reactions. **1970** Gressel & Galun in G. Bernier *Cellular & Molecular Aspects Floral Induction* x. 154 Some fungal systems are reported to be photoinducible by ultraviolet light as short as 230 mμ with no sporulation beyond 340 mμ. **1974** *Nature* 26 Apr. 801/1 We have not obtained a cell-free system from *P[hycomyces] blakesleeanus* which shows the photoinducible changes. **1977** I. M. Campbell *Energy & Atmosphere* ix. 344 It is generally considered that erythema is the least serious of a series of photoinduced skin complaints.

photoinduction (fōᵘto,indʊ·kʃən). [f. Photo- 1 + Induction.] The action or process of induction by light, esp. of reproductive behaviour in plants.

1947 *Bot. Gaz.* CIX. 131/2 A reduction in total reproductive activity in relation to reduced photoinduction. **1971** *New Scientist* 9 Sept. 558/1 Flowering..occurred about 20 days after this photoinduction. **1975** D. Vince-Prue *Photoperiodism in Plants* vi. 229 Endogenous cytokinins may..be involved in the process of photoinduction.

photoinductive (fōᵘto,indʊ·ktiv), *a. Bot.* [f. Photo- 1 + Inductive *a.*] Tending to induce flowering or other activity in plants by means of a regime of alternating periods of light and darkness; *photoinductive cycle*, a circadian cycle of one period of light and one of darkness.

1940 *Bot. Gaz.* CI. 667 (*caption*) Effect of duration and intensity of light during seven photoinductive cycles on subsequent initiation of floral primordia by Biloxi soybean. **1956** *Plant Physiol.* XXXI. 280/2 A measure of the amount of stimulation afforded by photoinductive treatments. **1971** *New Scientist* 9 Sept. 558/2 Flowering was..enhanced if the plants got a drenching during the photoinductive period. **1975** D. Vince-Prue *Photoperiodism in Plants* i. 42 In the SD [*sc.* short-day] grass, *Rottboellia exaltata*..a minimum of six photoinductive cycles are necessary for flowering.

photo-interpretation to **-interpretive:** see *photo- 2.

photo-ionization (fōᵘ:to,əiǒnəizēⁱ·ʃən). *Physics.* Also without hyphen. [f. Photo- 1 + *ionization².] Ionization produced by electromagnetic radiation.

1914 S. E. Sheppard *Photo-Chem.* ix. 356 (*heading*) Photo-ionization of gases. **1955** *Jrnl. Brit. Interplanetary Soc.* XIV. 18 From its effect in lowering the D layer, H. Friedman deduced that the intensity of the sun's radia-to determine their UV sensitivity. **1975** *Nature* 13 Mar. 160/1 Even though cells of the higher plants can photoreactivate ultraviolet damage.., the absence of dark-repair capability could be a significant disadvantage.

So **photorea·ctivated, -rea·ctivating** *ppl. adjs.*; *photoreactivating enzyme*, any enzyme which catalyses photoreactivation. Also **pho:toreactivabi·lity**, the potential for photoreactivation; **photorea·ctivable** *a.*, (of a biological system) capable of displaying photoreactivation; (of damage caused by ultraviolet irradiation) capable of being photoreactivated.

1953 *Jrnl. Bacteriol.* LXV. 252 (*heading*) Growth, respiration and nucleic acid synthesis in ultraviolet-irradiated and in photoreactivated *Escherichia coli*. **1958** *Jrnl. Gen. Physiol.* XLI. 463 (*heading*) Subcellular nature of the photoreactivating system. **1960** *Ibid.* XLIII. 592 This photoreactivating enzyme (PRE) is interesting for two reasons. First, it is a photoenzyme, and few enzymes involved in photochemical reactions are known at present. Second, it acts on DNA *in vitro* without depolymerizing it. **1961** J. A. Schiff et al. in Christensen & Buchmann *Progress in Photobiol.* vi. 290 Photoreactivability of the cells falls off rapidly when the cells are permitted to divide. Under non-dividing conditions, the cells remain completely photoreactivable indefinitely. **1965** J. Jagger in E. J. Bowen *Recent Progress in Photobiol.* ii. 61 Evidence for the photoreactivability of RNA. **1975** *Nature* 13 Mar. 160/1 Some types of excisable ultraviolet-induced DNA damage are not photoreactivable. **1975** [see *photoreactivation].

pho:toreactiva·tion. *Biol.* [f. Photo- 1 + *Reactivation.] The process whereby illumination of living matter with visible light can counteract the destructive effects of previous ultraviolet irradiation.

1949 A. Kelner in *Jrnl. Bacteriol.* LVIII. 511 The effect of reactivating light will be referred to in this paper as photoreactivation. [*Note.*] The use of this term was suggested by Dr. Max Delbrück. **1962** [see *photoreactivable]. **1966** *Adv. Radiation Biol.* II. 21 These photoreactivations may result from DNA inactivations repairable by the mechanisms already described here. **1969** A. C. Giese in F. Urbach *Biol. Effects Ultraviolet Radiation* 63 With protozoans.. even UV radiation as well as blue-violet visible light are effective for photoreactivation, while yellow visible light during or after UV exposure is ineffective. **1975** *Nature* 11 Sept. 133/1 Photoreactivation is a DNA repair process in which the photoreactivating enzyme (PRE) monomerises pyrimidine dimers induced by ultraviolet light.

photorealism (fōᵘtoˌri·ăliz'm). *Art.* Also Photorealism, Photo-Realism. [f. Photo- (in *photographic*, etc.) + Realism.] Detailed and unidealized representation in art, characteristically of the banal, vulgar, or sordid aspects of life. So **pho:to-re·alist** *a.* and *sb.*

1961 J. Williams *Forget* i. 6 A gigantic exhibition that will span everything from extreme abstract expressionism to extreme photorealism. **1973** *Art Internat.* Mar. 49/1 Curators, critics, teachers, and writers all over the world, who come to it looking for women's art of every sort from photo-realist to conceptual. **1973** *Guardian* 11 Apr. 10 There's something a bit pompous..about the claims made for Photorealism, the kick-off show at the Serpentine... Photorealist sculpture is there too. **1975** *New Yorker* 19 May 11/3 (Advt.) The paintings, done in an authoritative Photo-Realist style, dramatically illustrate Photo-Realism's strange ability to invest ordinary, ugly, even disgusting objects of our Pop culture with an appearance of home truth that borders on beauty. **1976** *Ibid.* 26 Apr. 137/1 There are Abstract Expressionist, Conceptual, and even Photo-Realist photographers. **1977** *Ibid.* 17 June 5/4 About a foot away from it..hover a ghostly mob of pools, lesbians, photo realists and deviants of every conceivable kind. **1977** *Jrnl. R. Soc. Arts* CXXV. 272/1 America, which Hockney visited towards the end of the '60s, brought him to some of his major paintings. In these he had used extensive photographic material for a variety of purposes—for information, for realistic detail, for inspiration, the result yielding a new photo-realism of strictly integrated design.

photo-recce: see *photo- 2.

photoreceptor (fōᵘtoˌrisɛ·ptəɹ). *Biol.* Also with hyphen. [f. Photo- 1 + Receptor.] Any living structure which responds to incident light, esp. a cell in which light is absorbed and converted to a nervous or other signal.

1906 C. S. Sherrington *Integrative Action Nervous Syst.* ix. 334 The free-swimming Ascidia with fin-like motor organs and semi-rigid axial notochord..bears at its anterior end a well-formed photo-receptor organ (eye) and a well-formed otocyst (head proprio-ceptor). **1944** *Electronic Engin.* XVII. 189/3 The retina consists of ten layers, in the third of which are embedded the photoreceptors. **1964** [see below]. **1969** F. E. Round *Introd. Lower Plants* ii. 14 The locomotory flagellum [of the alga *Euglena*]. It has a..swelling, the 'photoreceptor', near its entry into the cell. **1974** *Photochem. & Photobiol.* XIX. 433/1 The question as to the identity of the photoreceptor pigment (carotenoids or flavins) for phototropism in higher plants has not been resolved.

So **pho·toreception,** the process of absorption, and esp. of detection, of light by an animal or plant; **pho·toreceptive** *a.*, able to respond to light; of or pertaining to photoreception.

1906 C. S. Sherrington *Integrative Action Nervous Syst.* ix. 334 The elaborateness of the photo-receptive organs of the flying Insects corresponds with the great power of these forms to traverse space. **1908** *Amer. Jrnl. Physiol.* XXI. 98 The immediate results following the destruction of photo-reception in one eye are: (1) The production of rapid rotation..on the longitudinal axis of the body [etc.]. **1943** *Vitamins & Hormones* I. 211 Ordinarily these organisms contain a structure which appears to be specifically concerned with photoreception, the stigma or eye-spot. **1964** *Ann. N.Y. Acad. Sci.* CXVIII. 221 Red rays can penetrate to the hypothalamus with a sufficient intensity to activate the deeper photoreceptive structures. **1964** *Oceanogr. & Marine Biol.* II. 403 The photoreceptor pigment is unknown... That in at 462/63 mμ, but whether it is involved in the photoreceptive process is as yet unknown. **1973** J. J. Wolken in L. P. Miller *Phytochem.* I. ii. 26 Experimental observations..indicated that the photosynthetic process was not a simple photoreception sensitized by chlorophyll.

photo-reconnaissance, -recovery: see *photo- 2, 1.

pho:toreduction. Also with hyphen. [f. Photo- 3 + Reduction.] **1.** Chemical reduction effected by light; in *Bot.*, such a reduction of carbon dioxide in which water is formed (rather than oxygen, as in ordinary photosynthesis).

1888 [in Dict. s.v. Photo- 3]. **1939** *Jrnl. Cellular & Compar. Physiol.* XIII. 333 For the photo-reduction of one mole of carbon dioxide in the photosynthetic purple bacteria Streptococcus varians at high light intensity 2·6 moles of gaseous hydrogen are used. **1940** H. Gaffron in *Amer. Jrnl. Bot.* XXVII. 184/2 Such a..conception of photosynthesis in green plants and bacteria allows us to group the light metabolism of the bacteria and the 'anaerobic light respiration' of the adapted algae under one general heading: 'photoreduction'. **1943** *Vitamins & Hormones* I. 232 ... This leaves 'aerobic photosynthesis' or 'photosynthesis' proper to be..the assimilation of carbon dioxide with the liberation of molecular oxygen. **1967** *Jrnl. Amer. Chem. Soc.* LXXIX. 2941 Acriflavine and fluorescein-type dyes in solution in the presence of the photoreduction in the presence of mild reducing agents. **1970** C. A. Price *Molec. Approaches to Plant Physiol.* ii. 113 A number of algae show 'photoreduction', in which no O_2 is evolved. **1972** DePuy & Chapman *Molec. Reactions & Photochem.* iv. 48 Photoreduction of ketones is one of the oldest and most thoroughly investigated photochemical processes.

2. Reduction in size effected photographically.

1967 E. R. Lannon in Cox & Grose *Organiz. Bibliogr. Rec. by Computer* iv. 95 This latter version is printed..on 17" × 22" pages, suitable for 50% photoreduction and subsequent publication by offset press. **1968** *Bull. Libr. Rec.* VIII. 64 The flexowriter produced a text in double column..from which, after 30 per cent photo-reduction, offset-litho plates were made. **1972** *Sci. Amer.* Dec. 15/1 All the equipment for integrated-circuit work, including the photoreduction microscope and the ultrasonic bonder, was of Chinese manufacture.

So **photoredu·ce** *v. trans.*, to reduce photochemically; **photoredu·ced, -redu·cing** *ppl. adjs.*

1957 G. Oster in H. Gaffron et al. *Res. Photosynthesis* ii. 53 Acriflavine under conditions where it is not photoreduced in the unbound state, is readily photoreduced when bound to polymeric acids. **1959** ... in R. B. Withrow *Photoperiodism* i. 7 In my opinion..intermediate colored forms of photoreduced porphyrins are obtainable. **1965** *Biochim. & Biophys. Acta* XCIX. 159 The lost NADP photoreducing activities of the sonicated chloroplasts..are fully restored on addition of plastocyanin. **1968** *Plant Physiol.* XLIII. 606/1 All the mutants appeared to have the enzymes needed for the oxidation of photoreduced... they photoreduced little or no CO_2.

photoregulate to **-repairable**: see *photo- 1; **photo-reportage, -reporting**: see *photo- 2.

photoresist (fōᵘtori·zist). Also photo-resist. [f. Photo- 1 + Resist *sb.*] A photosensitive resist which when exposed to (usu. ultraviolet) light loses either its resistance or its susceptibility to attack by an etchant or solvent.

1953 *Printing Mag.* Oct. 56/1 The Kodak Photo-Resist, developed by Eastman Kodak Co., has great possibilities. It is the result of an extensive study and seems to possess ideal properties for a photoresist. **1960** *Times Rev. Industry* Aug. 46/1 The copper-clad phenolic panels..are sprayed with a photo-resist. **1965** D. I. Gaffee in L. Holland *Thin Film Microelectronics* vi. 261 Kodak photo-resist was originally designed for making letter-press printing plates and lithographic plates. **1966** *R. & E. Coordinator* (Res. & Engin. Council Graphic Arts Industry) Apr. 10/1 The new photoresist can be used in chemical milling of copper, copper alloys, and stainless steel, provided only acid solution etchants are used. **1972** *Daily Tel.* 23 Mar. 30 (Advt.), The Applications Laboratory requires a technologist to work on new and improved Kodak lithographic printing plates... The person filling the post will probably have had experience with similar plates.., including some knowledge of photoresists. **1973** *Sci. Amer.* Apr. 70/1 (*caption*) High-performance MOSFET is made in these steps. Light admitted through a mask sensitizes a 'photoresist' protecting a silicon oxide layer grown on a silicon wafer... Unprotected silicon oxide is etched away and phosphorus atoms are diffused into them to produce 'source' and 'drain' areas.

photoresistance to **-resistor**: see *photo- 1.

pho:torespira·tion. *Bot.* [f. Photo- 1 + Respiration.] A respiratory process in many higher plants by which they take up oxygen in the light and give out some carbon dioxide, contrary to the general pattern of photosynthesis.

1945 E. I. Rabinowitch *Photosynthesis* I. xx. 569 We now come to the problem of 'photorespiration' proper, that is, a direct photochemical acceleration of normal respiration which disappears in the dark as instantaneously as does photosynthesis. *Ibid.* 570 None of the experiments described above provides a final proof of the nonexistence of true 'photorespiration'. **1966** *Physiologia Plantarum* XIX. 732 Evolution of carbon dioxide in light, or photorespiration, was affected by oxygen... Oxygen had no effect on dark respiration. This discrepancy..can best be explained by an assumption that photorespiration and dark respiration are two different processes. **1972** [see *peroxisomal *a.*]. **1977** I. M. Campbell *Energy & Atmosphere* iv. 96 Plants with high rates of photorespiration such as wheat.

Hence **photorespi·re** *v. intr.*, to carry out photorespiration; **photorespi·red** *ppl. a.*, evolved by photorespiration; **photorespi·ring** *ppl. a.*; **photorespi·ratory** *a.*, of, pertaining to, or evolved by photorespiration.

1968 *Plant Physiol.* XLIII. 1840/1 Glycolate oxidation appears to be responsible for much of the photorespiratory CO_2 evolved. **1973** *Nature* 23 Mar. 237/1 At high concentrations of CO_2 the synthesis of the photorespiratory substrate, glycolate, is severely inhibited. **1969** *Proc. Nat. Acad. Sci.* XXIII. 662 Species of the first group also photorespire, evolving CO_2 into the atmosphere in light. **1976** *Nature* 14 Nov. 687/2 Such plants may not photorespire, or alternatively may be capable of refixing all the photorespired CO_2 by an unusually efficient photosynthetic mechanism. *Ibid.* 688/1 Glycolate seems to be the primary substrate for photorespiration, and it does not normally accumulate in photorespiring tissue. **1974** H. Fock et al. in *Bull. R. Soc. N.Z.* XII. 235 (*heading*) Estimation of carbon fluxes through photosynthetic and photorespiratory pathways. *Ibid.* 237 At 400 ppm CO_2 most or all of the photorespired carbon dioxide from intact leaves..may be derived from ^{14}C-labelled early products of photosynthesis.

photoresponse to **-reversible**: see *photo- 1; **photoscan** to **-scanning**: see *photo- 2.

photoscopic, *a.* Add: **c.** *Computers.* Applied to a photographic method of storing digital information.

1955 *Sci. Amer.* June 100/3 Gilbert W. King..has undertaken to exploit the great density of information storage that is possible through the use of high-resolution photographic emulsions. With his 'photoscopic' technique information can be stored at densities more than a hundred times as great as those possible in magnetic media. **1970** O. Dopping *Computers & Data Processing* x. 151 One example of photographic film memories is the photoscopic memory, which has been used for dictionaries in mechanical translation from one natural language to another. The medium is a continuously rotating disk of transparent plastic carrying a photographic layer upon which a pattern corresponding to zeroes and ones has been recorded.

photosensitive (fōᵘtoˌse·nsitiv), *a.* [f. Photo- 1 + Sensitive *a.* and *sb.*] Responding to light in some way (biologically, chemically, electrically, etc.).

1880 *Jrnl. R. Microsc. Soc.* VI. 596 [In the Elateridæ] the photosensitive reflex action has its seat in the cerebroid ganglia. **1928** *Science* 23 Aug. 199/1 Decomposition of the photosensitive material by light, presupposes the formation of the photosensitive material by light, presupposes the formation of ions from it is elements in the presence of chlorine. **1957** G. E. Hutchinson *Treat. Limnol.* I. vi. 399 The study of the transmission of light by means of suitably photosensitive instruments. **1970** D. G. Hays *Introd. Computational Linguistics* iv. 68 Another mirror is used to position the beam at the desired place on the photosensitive stock. **1977** E. Trevor *Theta Syndrome* v. 64 Mancini was lying down with dark glasses on... He was photo-sensitive.

Hence **photose·nsitiveness** (*rare*), **photosensiti·vity.**

1889 R. Meldola *Chem. of Photography* i. 13 The photo-sensitiveness of ferric compounds has long been known. **1914** *Physical Rev.* IV. 228 The plan..was to observe the time changes in contact potential differences and photosensitiveness of photo-electrically or mechanically treated surfaces. **1918** *Science* 23 Aug. 199/1 (*heading*) Adaptation in the photosensitivity of *Ciona intestinalis*. **1939** *Jrnl. Inst. Electr. Engin.* LXXXIV. 473/1 In the normal Emitron..the photo-sensitivity attained is about 12 μA/lumen. **1974** *New Biol.* III. 93 Many other urodeles, *i.e.*, tailed Amphibia, retain their photosensitivity and continue to move away from light after their eyes have been removed. **1961** *Lancet* 26 Aug. 450/2 His clinical photosensitivity showed fluctuations of intensity which corresponded with quantitative changes in his erythrocyte and fæcal porphyrins; hence it seems justifiable to conclude that the porphyrins produced in his body are responsible for his photosensitisation. **1970** R. A. & B. M. Maier *Compar. Animal Behavior* xvii. 365 Photosensitivity seems to have evolved from a general sensitivity to chemical stimulation.

pho:tosensitiza·tion. [f. Photo- 1 + Sensitization.] **a.** *Chem.* The initiation of a reaction by light acting on a suitable photosensitizer.

1924 H. S. Taylor *Treat. Physical Chem.* II. xviii. 1241 Uranium salts are positive catalysts for the photoreaction, presumably by photo-sensitization. **1933** *Jrnl. Amer. Chem. Soc.* LV. 587 A maximum of twenty per cent. photosensitization to visible light was found for the polymerization among seventy organic substances tried. **1974** D. R. Arnold et al. *Photochem.* vi. 133 Photosensitization involves the absorption of radiation by a strongly absorbing substance, the photosensitizer and its collisional transfer to another substance which is nonabsorbing at the same wavelength.

b. (The production of) a condition in which light of certain wavelengths is harmful to an individual, usu. owing to the presence in the body of a photodynamic substance.

1926 E. Mayer *Clin. Applic. Sunlight* v. 86 (*heading*) Photosensitization. **1927** N. M. L. Gangee *Artificial Light Treatm. Children* xii. 108 Experimental photosensitization in animals and human beings is known to follow the administration of certain substances, such as eosin and hæmatoporphyrin, and death may result on exposure to light. **1961** H. F. Blum *Photodynamic Action* xv. 159 Mathews..finds part of the symptoms which follow feeding on *Agave lechuguilla*, found in the arid regions of New Mexico and Texas, to result from photosensitization. **1961** *Lancet* 26 Aug. 450/2 In cases of porphyria with photosensitisation, it is often uroporphyrin which is present in greatest amount, and the response to irradiation is by erythema and itching followed by slowly developing œdema. **1973** Stoot & Schultes in L. P. Miller *Phytochem.* III. xiv. 384 These last [*sc.* photodynamic toxins] may act either by entering the animal directly..or by causing liver damage, which then leads to photosensitization as a secondary consequence.

photosensitize (fōᵘtose·nsitaiz), *v.* [f. Photo- 1 + Sensitize *v.*] *trans.* **a.** *Chem.* Of a substance: to initiate (a chemical change) by absorbing light energy and transferring it to a reactant.

1927 *Jrnl. Amer. Chem. Soc.* XLIX. 2763 (*heading*) Hydrogen peroxide formation photosensitized by mercury vapor. **1928** *Proc. R. Soc.* A. CXXI. 297 Uranyl ion..has long been known to photosensitize the decomposition of glucose in solution. **1951** *Symp. Soc. Exper. Biol.* V. 142 Photolysis of water can be accomplished with much less energy if the process is suitably 'photosensitized', for instance, if water is irradiated in the presence of suitable ions. **1966** Cucker & Seiffert *Physical Chem.* (1967) xxiii. 722 The dissociation of hydrogen molecules into atoms requires..a quantum of wavelength 2790 Å. Hydrogen molecules do not absorb light of this wavelength, but mercury atoms, which absorb light at 2536·52 Å, have plenty of energy to photosensitize the formation of hydrogen atoms.

b. To make photosensitive.

1933 *Jrnl. Inst. Electr. Engin.* LXXIII. 441/2 The mosaic..is composed of a very large number of minute silver globules, each of which is photo-sensitized by cæsium through utilization of a special process. **1941** *Nature* 10 May 581/2 Carcinogenic hydrocarbons in very low concentration are able to photosensitize Paramecia. **1977** J. L. Marx in *Population Biol.* 6 April 1977 This pain is a serious weed..because it photosensitizes domestic animals.

Hence **photose·nsitized** *ppl. a.*, **-se·nsitizing** *vbl. sb.* and *ppl. a.*

1924 S. E. Sheppard *Photo-Chem.* xi. 335 The photochemical sensitiveness of dye-stuffs..depends in part upon the formation of a specific adsorption-complex of the dye with the substance sensitized. This photosensitizing is of considerable interest. **1931** R. G. W. Norrish in *Photochem. Processes* (Faraday Soc.) iv. 461 (*heading*) The photosensitised formation of hydrogen bromide in the system hydrogen-oxygen-chlorine. **1935** *Discovery* Sept. 271/2 The mosaic screen is made up of millions of isolated photo-sensitised elements upon a mica sheet. **1964** *Oceanogr. & Marine Biol.* II. 406 There are indications that urchins can be induced to cover by injecting photo-sensitizing dyes. **1974** *Photochem. & Photobiol.* XIX. 351 Some examples of photosensitized oxidation reactions.

photosensitizer (fōᵘtoˌse·nsitaizəɹ). [f. Photo- 1 + Sensitizer.] **a.** *Chem.* A substance capable of photosensitizing a reaction.

1924 S. E. Sheppard *Photo-Chem.* vii. 293 Winther has shown that the chlorine water was not a photo-sensitizer since it had the same effect in darkness upon the precipitation of calomel. **1928** *Proc. R. Soc.* A. CXXI. 296 It is not to be supposed that only fluorescent substances will act as photosensitizers; substances capable of absorbing radiation of the appropriate frequency may be effective. **1957** G. E. Hutchinson *Treat. Limnol.* I. xvi. 863 In dilute solution..ferric iron acts as photosensitizer in the oxidation of lactic acid.

b. A photodynamic substance.

1925 *Practitioner* Aug. 103 The visible rays, however, have this effect when the irradiated cells are incorporated with photosensitizers, such as eosin and hæmatoporphyrin. **1946** *Nature* 14 Dec. 877/2 They found that the cancerogenic substances had a stronger effect than the non-cancerogenic photosensitizers. **1967** M. E. Hale *Biol. Lichens* xi. 161 The possibilities that uric acid may be a photosensitizer and a cause of respiratory allergy are also being explored. **1975** *Sci. Amer.* July 72/3 A number of widely prescribed drugs (such as the tetracyclines) and constituents of foods (such as riboflavins) are potential photosensitizers.

photosensor(y): see *photo- 1.

photosetting (fōᵘtoˌse·tiŋ), *vbl. sb. Printing.* [f. Photo- 2 + Setting *vbl. sb.*] = *photocomposing *vbl. sb.* 1. So **photoset** *v. trans.*; **pho·toset** *ppl. a.*; **pho·tose:tter**, a photocomposing machine.

1952 *Americana Ann.* 330/1 Tabular matter can be photoset in the same manner as ordinary typewriting. **1958** *Ibid.* 143/1 An electric combination photosetting and photocomposing machine was developed. **1959** *Times* 14 Jan. 12/4 Photosetting machines are unlikely to replace all these [hot-metal typesetting machines]. **1959** in E. Fitzgerald tr. *Hitt's Toy* 4 Printed in England from photoset typematter. **1961** *Printing News* 30 Mar. 6/3 One American printer is currently producing upwards of 100 photoset books a year. **1968** A. Blumer *Photosetting* ii. 8 The beginning of the 'modern' period of photosetting may be conveniently dated at 1953, when prototypes..of several current machines for text setting were shown at the IPEX exhibition in London. **1970** *Brit. Printer* Dec. 72/2 A new photosetter in what is described as the moderate price range has been introduced. **1974** *Times* 14 Oct. (Sheffield Suppl.) p. iv/8 Both papers..have..an ambitious reorganisation scheme... It includes changing to photo-setting instead of the traditional typesetting.

photoshock to **-stable**: see *photo- 1.

Photostat (fōᵘ·tostæt). orig. *U.S.* Also **photostat.** [f. Photo- 2 + -stat.] **a.** The proprietary name of a kind of photocopying machine. **b.** A copy made on such a machine; *loosely*, any photocopy. Also *attrib.*

1911 *Trade Marks Jrnl.* 24 May 763 Photostat... Photographic cameras for making photographic copies of the pages of books, drawings, applications for life insurance and the like. Commercial Camera Company.., Providence, Rhode Island, United States of America; manufacturers. **1912** *Chambers's Jrnl.* June 414/2 By means of the photostat a new filing method is possible. **1927** *Glasgow Herald* 26 May 9 The reference to a subsequent photostat room is quite in accord with the general cinematographic nature of the raid... Such photostats exist nowadays in most large commercial undertakings. **1928** P. S. Allen *Let.* 27 July (1939) 258, I should be glad to have the photostats (where does the word come from?) quickly. **1931** *Times Lit. Suppl.* 17 Dec. 1026/1 The number of manuscripts known has increased, to eighty-four, of which photostats [*sic*] are now at the University of Chicago. **1932** [see *book-page* s.v. *book *sb.* 18]. **1940** *Chambers's Techn. Dict.* 639/1 *Photostat*, trade-name for a photographic apparatus (also for any print made by it) designed for rapid copying, to the required size, flat originals on sensitized paper, and giving a negative image. **1959** T. S. Eliot *Elder Statesman* ii. 57 I'm afraid I can't show you the originals: They're in my lawyer's safe. But I have photostats Which are quite as good, I'm told. **1961** T. Landau *Encycl. Librarianship* (ed. 2) 121/1 The 'Photostat' has after a many years become a household word among librarians and readers, and is indiscriminately used in describing any photographic reproduction of a document. It is in fact a trade name, and a 'Photostat' copy of a document is one produced by the 'Photostat' apparatus. **1964** 'R. Macdonald' in *Manhunt* Mar. 142/1, I pulled out my Photostat and slapped it down on the desk. **1975** *Guardian* 25 Feb. 5/7 Inside the envelope there were simply photostats of correspondence. **1979** *Glass. Documentation Terms* (*B.S.I.*) 48 *Photostat*, a trade mark of Kodak Ltd. for photocopying cameras, chemicals and sensitive materials, the optical copies produced being right-reading.

Hence **pho·tostat** *v. trans.*, to photocopy; so **pho·totstat(t)ed** *ppl. a.*, **pho·tostat(t)ing** *vbl. sb.*; also **photosta·tic** *a.*, of, pertaining to, or produced by a Photostat or other photocopying machine.

1914 *Amer. Machinist* 9 Apr. 642/1 A prism is used to 'turn the corner', making it more convenient than if the book or other object being 'photostated' had to be set up on edge. **1925** M. R. Rinehart *Red Lamp* 159 One of the evening newspapers to-night prints a photostatic copy of the cipher found in our garage. **1932** *N. & Q.* 5 Nov. 338/1 In corresponding about having the letter photostated we were informed of still another letter. **1937** Mrs. P. Campbell *Let.* 25 Aug. in G. B. Shaw & Mrs. Campbell (1952) 314 He doesn't know I have had two photostated copies made of them. **1947** [see *eager *a.* 6]. **1957** *Amer. Speech* XXXII. 57 Libraries quickly learned to circumvent these laws, by photostating. **1959** H. Hamilton *Answer in Negative* i. 8 All our printing, and photo-statting of cuttings, is done by the *Echo's* dark-room. **1967** N. Freeling *Strike Out* 141 They weren't going to open an embassy safe and photostat the contents. **1973** H. Hill *Ruling Passion* iii. iii. 179 He passed out some photostatted sheets. **1976** T. Atwell *Only Good German* ix. 59 A photostated copy of an item in *The Times*. **1978** *Daily Tel.* 5 May 14/6 Some might feel that for a rather smaller price the 'Compact' DNB, the full work with minute, photostated text, ..might be a better buy.

photostationary to **-surface:** see *photo- 1; **photostereogram** to **-story:** see *photo- 2.

photosynthate (fōᵘtosi·nþeit). *Bot.* [f. next + *-ate*, after *filtrate*, *precipitate*.] A substance formed by photosynthesis.

1913 W. F. Ganong *Living Plant* ii. 24 The process being one of formation, or synthesis, under action of light, is called scientifically photosynthesis, while the substance made is the photosynthate. **1938** Weaver & Clements *Plant Ecol.* (ed. 2) xiv. 395 Many species of evergreens are known to make photosynthate in winter in sufficiently large amounts to balance that oxidized in respiration. **1978** *Nature* 5 Jan. 33/2 A conifer manufactures photosynthates at all times of the year except when climatic conditions are unsuitable, due for example to low temperatures or drought stress.

photosynthesis (fōᵘtosi·nþisis). *Bot.* [ad. G. *photosynthese*: see Photo- 1 + Synthesis.] The process by which carbon dioxide is converted into organic matter in the presence of the chlorophyll of plants under the influence of light, which in all plants except some bacteria involves the production of oxygen from water; also, any photochemical synthesis of a chemical compound.

1898 *Botanisches Zentralblatt* LXXVI. 258 It is not important whether photosyntax or photosynthesis, or some other word, finally comes into general use to describe the manufacture of carbohydrates by green tissues under the action of light. **1902** *Encycl. Brit.* XXXI. 760/1 The course of photosynthesis has been with tolerable certainty found to lead to the construction of sugar. **1914** S. E. Sheppard *Photo-Chem.* vii. 295 The photo-synthesis of phosgene ($COCl_2$) from chlorine and carbon monoxide.. has been studied by several observers. **1924** *Industr. & Engin. Chem.* Oct. 1021/1 The optimum experimental conditions having been determined, it has been found possible to carry out the photosynthesis on a larger scale than in the test tube. **1927** *Proc. R. Soc.* A. CXVI. 203 The photosynthesis of complex organic substances takes place when carbonic acid in the presence of a catalyst is exposed to ultra-violet light. **1932** Fuller & Conard tr. *Braun-Blanquet's Plant Sociol.* v. 97 There is a limited amount of photosynthesis in the invisible infra-red and ultra-violet. **1952** P. W. Richards *Trop. Rain Forest* vii. 180 Even at the low light intensities found in the shade of the Tropical Rain forest, a high carbon dioxide concentration would probably allow a rate of photosynthesis considerably higher than that at normal concentration. **1957** *Times* 11 Sept. 6/2 Recent studies using radiocarbon indicated that the yield of photosynthesis by the plankton of the oceans was at least equal to that of the land flora, and might be several times greater. **1958** R. V. Stanier et al. *Gen. Microbiol.* xi. 213 In bacterial photosyntheses, there is also a light-driven production of ATP... cell material, but oxygen is never produced because water cannot serve as the ultimate hydrogen donor. Instead, the reduction of CO_2 is coupled with the oxidation of externally supplied organic or inorganic hydrogen donors. **1964** E. J. H. Corner *Life of Plants* iii. 38 There are chromoplast colours that work together with chlorophyll in photosynthesis. **1975** R. Smith *Photochrome & Photomorphogenesis* ii. 15 Photosynthesis presents an excellent example of light and dark reactions acting sequentially.

Hence **pho:tosynthe·tic** *a.*, of, pertaining to, produced by, or involved in photosynthesis; *photosynthetic quotient* or *ratio*, the ratio of evolution of oxygen by photosynthesizing tissue divided by its rate of consumption of carbon dioxide, or the reciprocal of this; **pho:tosynthe·tically** *adv.*

1900 A. J. Ewart tr. *Pfeffer's Physiol. Plants* I. vii. 293 The photosynthetic assimilation in the chloroplastid only provides the organic food, which in green and non-green plants, and in animals also, has the same function to perform. *Ibid.* 326 With the exception of carbon dioxide, no carbon compounds are known which can be photosynthetically assimilated. **1913** W. F. Ganong *Living Plant* iii. 27 The photosynthetic sugar and starch which appear in lighted green leaves. **1926** H. A. Spoehr *Photosynthesis* ii. 92 The photosynthetic quotient may yield some information relative to the first product formed in photosynthesis. **1931** E. C. Miller *Plant Physiol.* viii. 431 The determination of the photosynthetic ratio is difficult, because the process of respiration is proceeding at the same time and in the opposite direction [to that of photosynthesis]. **1945** E. I. Rabinowitch *Photosynthesis* I. ii. 32 The 'photosynthetic quotient', Q_p; $Q_p = \Delta O_2 / -\Delta CO_2$. The term 'photosynthetic quotient' has been used by many authors.. to designate the inverse ratio, $-\Delta CO_2/\Delta O_2$; this difference calls for care in the quotation of numerical results. **1962** *Listener* 3 May 768/2 Some scientists believe that free oxygen is entirely biological in origin and that it arose *after* the emergence of photosynthetic life. **1972** *Sci. Amer.* Sept. 92/3 Plants reflect about 8 percent of photosynthetically active wavelengths. **1971** I. Zelitch *Photosynthesis* v. 129 At the high O_2 concentrations, *Amaranthus* leaves had a photosynthetic quotient of 0·50 and bean leaves 0·33. **1974** A. Huxley *Plant & Planet* iii. 22 It was probably not the seaweeds that gave rise to land plants but, from their similar photosynthetic pigments and photosynthetic mechanism, the green algae.

photosynthesize (fōᵘtosi·nþisaiz), *v. Bot.* [f. prec. + -ize.] **a.** *trans.* To create by photosynthesis. **b.** *intr.* To carry out photosynthesis.

1921 *Jrnl. Chem. Soc.* CXIX. 1029 Carbohydrates can be photosynthesised from carbon dioxide and water in two stages. **1927** *Proc. R. Soc.* A. CXVI. 212 These [results] convinced us that it is possible by the use of coloured powders to photosynthesise organic compounds from carbonic acid with the help of visible light. **1951** *Symp. Soc. Exper. Biol.* V. 300 When algae or barley leaves photosynthesize..in $C^{14}O_2$ for 30 sec. [etc.]. **1974** A. Huxley *Plant & Planet* iii. 18 Some bacteria..can either photosynthesize or, if in the dark, gain energy from decomposing organic matter.

Hence **photosy·nthesized, photosy·nthesizing** *ppl. adjs.*; also **photosy·nthesizer,** an organism which carries out photosynthesis.

1910 F. Keeble *Plant-Animals* iii. 79 From the photosynthesised carbohydrate are derived the cellulose substances. **1927** *Proc. R. Soc.* A. CXVI. 213 The photosynthesised compounds are very similar in appearance to those described in the previous paper. **1937** *Enzymologia* IV. 254 (*heading*) On the fluorescence of photosynthesizing cells. **1958** *Sci. News* XLIX. 25 Animals, fungi, and most of the bacteria..only shuffle..the materials the photosynthesizers have made. **1970** L. Margulis *Origin of Eukaryotic Cells* iv. 94 This gas [*sc.* hydrogen sulphide] was utilized by anaerobic photosynthesizers as hydrogen donors in photosynthesis. **1973** *Sci. Amer.* Oct. 83/3 They supplied carbon dioxide labeled with carbon 14 to photosynthesizing sugarcane plants.

pho·tosystem. *Bot.* [f. *photo(synthetic *a.* + System.] Either of the two biochemical mechanisms in plants by which light is converted into useful energy.

1965 *Biochim. & Biophys. Acta* CIX. 349 Photoreduction of substrate V by photosystem 1. **1973** J. J. Wolken in L. P. Miller *Phytochem.* I. ii. 26 This phenomenon termed the 'red drop' is interpreted, at present, as due to a special form of 'long wavelength absorbing' chlorophyll belonging to a Photosystem I. *Ibid.*, The low efficiency of the far-red (beyond 680 nm) would require another pigment-complex absorbing below 680 nm, which has been designated as Photosystem II. **1974** *Nature* 4 Jan. 4/1 There now seems to be no doubt that KCN inhibits photosynthesis by specifically blocking electron flow through photosystem one (S1).

phototactic, *a.* Substitute for def.: Exhibiting or characterized by phototaxis. (Earlier and later examples.) [ad. G. *phototaktisch* (E. Strasburger 1878: see *phototaxis).]

1882 S. H. Vines tr. *F. G. J. von Sachs's Text-bk. Bot.* III. iii. 752 Zoogonidia which exhibit these phenomena are said, by Strasburger, to be phototactic. **1907** *Jrnl. Exper. Zool.* V. 72 Any organism is said to be positively phototactic when it moves towards the source of light and negatively phototactic when it goes in the opposite direction. **1969** F. E. Round *Introd. Lower Plants* ii. 15 Euglenoids are positively phototactic. **1976** *Sci. Amer.* June 42/1, I had noticed a phototactic response in *H. halobium*; the cells reversed their direction of swimming when the intensity of illumination was decreased in the red part of the spectrum.

Hence **phototа·ctically** *adv.*

1914 [see *kinetic *a.* 2 b].

phototaxis. Substitute for entry:

phototaxis (fōᵘtotæ·ksis). *Biol.* Pl. **-taxes** (-tæ·ksīz). [mod.L., coined in Ger. (E. Strasburger 1878, in *Jenaische Zeitschr. f. Naturwissensch.* XII. 587): see Photo- 1 and Taxis.] The innate movement in a definite direction of an organism or part of one in response to the stimulus of light; *esp.* the bodily movement or orientation of a freely motile organism (see quot. 1960 and cf. *phototropism).

1893 *Athenæum* 16 Sept. 375/3 Phototaxis and chemiotaxis are the last instances of physiological adaptation cited [by J. S. B. Sanderson]. **1894** J. S. B. Sanderson in *Rep. Brit. Assoc. Adv. Sci. 1893* 24 A single instance.. must suffice to illustrate the influence of light in directing the movements of freely moving cells, or, as it is termed, phototaxis. **1902** *Jrnl. R. Microsc. Soc.* 31 Phototaxis is the peculiarity displayed by free-swimming organisms of orienting the body so as to place its long axis in a definite relation to the direction of the rays. **1911** A. Willey *Convergence in Evolution* iii. 25 The vegetable kingdom as a whole exhibits positive phototaxis. **1954** *New Biol.* XVII. 49 Several workers have found negative phototaxis (movement away from light) in a number of species [of woodlice]. **1960** Thimann & Curry in Florkin & Mason *Compar. Biochem.* I. vi. 244 In all these organisms the response to light is a free movement of the whole body directed towards or away from the light, and this is defined as phototaxis. In fungi and higher plants, as also in the colonial hydroids, the body is anchored at one end and the response to light is shown by a curvature. It is this which is phototropism properly speaking... Zoologists have, it is true, used the term phototropism for free movements of animals, while botanists have been more precise in preserving the distinction, but the distinction is a valuable and even essential one to make... In Phototaxis the light influences the organs of movement,.. while in Phototropism it influences the growth of the organism. **1975** *Nature* 3 Jan. 43/2 It would seem that the light intensity dependent behaviour in these two examples could be divided into retinally evoked phototaxes and an extraretinally evoked kinesis which cannot be separated in sighted larvae. *Ibid.* 4 Sept. 44/2 The internal parts of the maze were of clear Perspex which allowed the flies to be attracted through the maze by phototaxis.

phototelegram to **-telegraphy:** see *photo- 1; **photo-timer:** see *photo- 2.

pho:totopo·graphy. Also with hyphen. [f. Photo- 2 + Topography.] A system of surveying which employs photography as well as the usual methods.

The word has largely given way to *photogrammetry*.

1893 *Geogr. Jrnl.* I. 89 Photo-topography. **1895** *Rep. U.S. Coast & Geodetic Survey 1893* II. 42 During the Franco-Prussian war phototopography was called into service by the German army. **1922** *Geogr. Jrnl.* LIX. 274 The methods of Phototopography developed by Laussedat in France and by Deville in Canada..depend on the measurement of identical points on pairs of plates taken on determined azimuths at trigonometrical points. **1970** *Canad. Cartographer* VII. 19/2 The ground photo-topography continued in the mountains of British Columbia.

Hence **pho:totopo·grapher,** one who is skilled in phototopography; **pho:totopogra·phic, -ical** *adjs.*, of, pertaining to, or using phototopography; **pho totopogra·phically** *adv.*

1895 *Rep. U.S. Coast & Geodetic Survey 1893* II. 45 A phototopographic survey of the Oasis Gassr Dachel in the Libyan desert. **1902** *Encycl. Brit.* XXXIII. 95/2 The field work of a photo-topographic party consists primarily in execution of a triangulation by the usual methods. *Ibid.*, The photo-topographical survey. **1906** J. A. Flemer *Elem. Treat. Phototopogr. Methods* xii. 387 The phototopographer..can in a few good days, cover a larger territory than is possible with any other surveying method. *Ibid.* 390 A large territory may be reconnoitered phototopographically in a comparatively short time. **1926** A. L. Higgins *Phototopography* ii. 42 Phototopographical instruments. **1970** *Canad. Cartographer* VII. 18/1 During the development of the photo-topographic work it became the practice to rule a perspective grid onto the print of the photograph to assist..in positioning horizontal detail.

phototoxic to **-transient:** see *photo- 1.

phototransistor (fōᵘ:totranzi·stəɹ). *Electronics.* Also with hyphen. [f. Photo- 1 + *transistor.] A junction transistor which responds to incident light by generating and amplifying an electric current.

1950 J. N. Shive in *Bell Lab. Rec.* XXVIII. 337/2 Experiments have resulted in the production of a new photoconductivity cell, called the 'Phototransistor'. **1959** *Control* Feb. 95/2 It is a logical step from the photodiode to the *p-n-p* phototransistor, in which the amplifying action of the transistor is applied to the photocurrent. *Ibid.*, The Mullard OCP71 phototransistor is enclosed in a glass bulb 5·9 mm in diameter and 15 mm in length. **1959** *Electronic Engin.* XXXI. 36 The advent of the commercially produced junction photo-transistor has brought with it the possibility of operating a relay directly from the light sensitive element. **1962** P. M. Williams in G. A. T. Burdett *Automatic Control Handbk.* ix. 52 Phototransistors can be made which are sensitive to X-rays, ultra violet and infra red. **1970** *New Scientist* 24 Dec. 554/1 The new reading head is only a few centimetres across and contains a matrix of 144 bipolar phototransistors on a silicon chip 3 × 5 mm in size.

phototroph to **-trophy:** see *photo- 1.

phototropic, *a.* For *Bot.* read *Biol.* and add pronunc. (fōᵘtotrōᵘ·pik). Substitute for def.: Exhibiting or characterized by phototropism. (Further examples.) **phototropically** *adv.* (further examples).

1903 *Mark Anniversary Vol.* xxiii. 455 Loeb maintained that butterflies as well as moths are positively phototropic. *Ibid.* 457 When feeding or near food the butterflies do not respond phototropically. **1943** *Vitamins & Hormones* I. 211 In some structures also (*Avena, Pilobolus, Phycomyces*) carotenoid pigmentation has been shown to be concentrated in or restricted to the phototropically sensitive zones. **1972** *Plant Physiol.* XLIX. 993/1 Etiolated plants are known to be more sensitive phototropically than are green ones. **1976** Bell & Coombe tr. *Strasburger's Textbk. Bot.* (rev. ed.) 350 Phototropic behaviour is not confined to the higher plants; it occurs also in fungi and algae.

2. = *photochromic *a.*

1900 *Jrnl. Chem. Soc.* LXXVIII. II. 125 Benzilosazone, salicylosazone, and vanillylosazone are, however, not phototropic. **1929** *Chem. Rev.* VI. 220 Most phototropic substances exhibit phototropy only in the solid state. **1960** *New Scientist* 2 June 1423/2 In the same family of substances some derivatives are phototropic while others, which resemble them closely, do not show the effect. **1971** *Materials & Technol.* II. vi. 410 Phototropic glass is made by melting an alkali-alumino-borosilicate glass containing some lead and small amounts of silver, chloride, bromide, and iodide, forming the glass in the normal way, and then submitting the article to a heat treatment for several hours.

phototropism (fotǫ·trǫpiz'm, fōᵘtotrōᵘ·piz'm). [f. Photo- 1 + Tropism.] **1.** *Biol.* [ad G. *phototropie* (F. Oltmanns 1892, in *Flora* LXXV. 214).] The innate movement in a definite direction of an organism or part of one in response to the stimulus of light; *esp.* the directional bending or growth of a plant or sessile animal (see quot. 1960 and cf. *phototaxis).

[**1892** *Jrnl. R. Microsc. Soc.* 513 The phenomena of phototropy—i.e. positive or negative heliotropic movements.] **1899, 1900** [in Dict. s.v. Phototropic *a.*]. **1902** *Jrnl. R. Microsc. Soc.* 31 He [*sc.* W. A. Nagel] applies the term phototropism to the peculiarity displayed by many sedentary organisms or special organs, of taking up a fixed position in regard to light, by means of bending movements, movements of growth or of torsion. **1907** T. H. Morgan *Exper. Zoöl.* xvi. 265 Growth toward the Light; Phototropism... Some animals that are fixed turn toward the light. **1924** R. M. Ogden tr. *Koffka's Growth of Mind* iii. 111 A cockroach possesses a negative phototropism. **1947** *Biol. Bull.* XCII. 127 The terms, phototropism and phototaxis, will be used synonymously in this paper. **1960** [see *phototaxis]. **1965** B. E. Freeman tr. *Vandel's Biospeleol.* xxv. 400 Light very frequently attracts or repels animals and these conditions are given the names of positive and of negative phototropism. **1976** G. C. Ainsworth *Introd. Hist. Mycol.* vii. 198 Phototropism has been much studied in the Mucorales.

fig. **1923** R. Graves *Whipperginny* 38 Watch the blind Phototropisms of my fluttering mind.

2. = *photochromism.

1921 *Jrnl. Amer. Chem. Soc.* XLIII. 333 (*heading*) Studies on phototropism in solution. **1952** K. Venkataraman *Chem. Synthetic Dyes* II. xl. 1217 A review of phototropism in solution and on dyed textiles. **1962** J. T. Marsh *Self-Smoothing Fabrics* xviii. 298 Phototropism is the change in shade which occurs on exposure to strong light, followed by a gradual return to the original shade when the light is removed. **1974** Rattee & Breuer *Physical Chem. of Dye Adsorption* vii. 229 Three dyes..were all found to exhibit phototropism on cellulose acetate film.

phototropy (fōᵘtǫ·trǫpi). [ad. G. *phototropie* (W. Marckwald 1899, in *Zeitschr. f. physik. Chem.* XXX. 140), f. Gr. φωτο- (see Photo-) + -τροπία turning.] = *photochromism, *phototropism 2.

1900 *Sci. Amer.* 24 Feb. 123/2 To these phenomena the experimenter gives the name of phototropy. **1929** [see *phototropic *a.* 2]. **1949** *Thorpe's Dict. Appl. Chem.* (ed. 4) IX. 585/2 There is evidence..that phototropy is not a purely physical phenomenon. **1954** *Jrnl. Amer. Chem. Soc.* LXXVI. 3846/1 Observations on the phenomenon of phototropy in alkaline earth titanates. **1971** R. L. M. Allen *Colour Chem.* iii. 23 The stereoisomerism of azo dyes is of practical importance..in that it gives rise to the phenomenon of phototropy.

phototube (fōᵘ·totiūb). [f. Photo- 1 + Tube *sb.*] A photocell in the form of a vacuum tube or gas-filled tube with a photoemissive cathode and an anode.

1930 *Electronics* I. 418/1 (*heading*) Phototube voltage supervisor. An aid to tube production. **1953** [see *mark *sb.*[1] 13 e]. **1964** *Oceanogr. & Marine Biol.* II. 358 It is only since the adoption by biologists of the multiplier phototube..that more precise measurements of the spectral composition of luminescence,..and rates of flashing, have become feasible. **1973** *Nature* 12 Jan. 132/2 The emission was observed at right angles..and detected photoelectrically with a '1P28' phototube.

phototypesetting (fōᵘtotəi·pˌsetiŋ), *vbl. sb. Printing.* [f. Photo- 2 + *type-setting* s.v. Type *sb.*[1] 10.] = *photocomposing *vbl. sb.* 1. Also *attrib.* So **phototy·peset** *ppl. a.*; **phototy·pesetter,** a photocomposing machine.

1931 *A.S.M.E. News* 7 Apr. 3/3 Louis Flader..described the Uher photo-typesetting machine, having recently inspected it in Germany. **1949** E. Thirkettle in H. O. Smith *Rotophoto Process* 3 Technical improvements in lithographic printing presses are conveniently keeping in line with phototypesetting developments. **1955** *Times* 5 July 6/5 Outstanding among the advances of the last two decades has been the development known as phototypesetting. **1966** N. S. M. Cox et al. *Computer & Library* v. 74 The phototypesetter can produce 'graphic arts quality' printing. **1970** *Brit. Printer* Dec. 43 Massive trade typesetting operations based on computers (the one in question was a Fototronic, the highest speed phototypesetter at present available) can be assimilated. **1971** *Penrose Ann.* LXIV. 176 Many more phototypesetting devices will come on to the market over the next few years. *Ibid.*, Throughout 1970, technical innovation flowed on unabated, with ten new phototypesetters announced. **1973** *Physics Bull.* Dec. 743/1 The World Patents Index..is computer generated and will publish the information in a phototypeset gazette. **1976** *Nature* 24 June 664/2 CSIRO's computer with its microfilm output equipment (a COMp 80 phototypesetter) is now being used to produce camera-ready copy.

photovisual (fōᵘtovi·ʒiuăl), *a.* (*sb.*) Also photo-visual. [f. Photo- 2 + Visual *a.* and *sb.*] **1.** Of a lens or an optical instrument: bringing both visible and actinic, non-visible rays to the same focus. Also as *sb.*, such a lens or telescope.

1909 in *Cent. Dict.* Suppl. **1922** L. Bell *Telescope* iv. 89 The..objective..carries the name of 'photo-visual' since the exactness of corrections is carried well into the violet, so that one can see and photograph at the same focus. **1955** J. B. Sidgwick *Amateur Astronomer's Hand-bk.* xxiii. 422 For photography a reflector is to be preferred, unless a photovisual is available. **1958** J. Strong *Concepts of Classical Optics* xiv. 323 Three different glasses, properly chosen for their partial dispersions, can be combined in components to yield a composite lens... Fig. 14·14 gives a tabulation..for the three glasses used in the Cooke photo-visual triplet lens. **1964** *Yearbk. Astron. 1965* 191 The main instruments at present are a 13-inch reflector, a 10-inch refractor, a 6-inch photovisual refractor, a 16 centimetre transit telescope and various cameras. **1977** *Sci. Amer.* Sept. 29/2 (Advt.), Sixth, the design must be photovisual so that he could record on film whatever these superior optics would present to the eye.

2. *Astr.* Applied to stellar magnitudes determined in terms of the spectral response of the eye by photographic or photoelectric means.

1914 *Carnegie Inst. Yearbk. 1913* 214 The comparison of these 'photovisual' results with the ordinary photographic magnitudes shows an increase in the color index as the variable approaches its minimum. **1927** H. N. Russell et al. *Astron.* II. xviii. 620 With isochromatic plates and a 'color filter' to cut off the blue and violet

light and let through the green and yellow, magnitudes may be obtained which agree very closely with the visual scale. These are called photo-visual magnitudes. **1958** C. C. Adams et al. *Space Flight* 132 At this time it is expected to be of the fifth photovisual magnitude, which is about the same brightness as the faintest star that, under reasonably good observing conditions, can be seen with the naked eye. **1963** C. W. Allen *Astrophysical Quantities* (ed. 2) x. 195 The effective wavelengths of a colour index system change with the colour itself. The U.B.V. system..has replaced the international photographic and photovisual systems.

photovoltaic: see *photo- 1.

Photronic (fōᵘtrǫ·nik). Also **photronic.** [f. Phot(o- 1 + *elec)tronic *a.*] An American proprietary name for a kind of photovoltaic cell. Also *attrib.* or as *adj.*

The word is registered in the U.S. as a proprietary term for several other devices also.

1932 *Official Gaz.* (U.S. Patent Office) 15 Mar. 554/1 Weston Electrical Instrument Corporation, Newark, N.J. Filed Oct. 1, 1931. Photronic. For light sensitive cells. Claims use since Sept. 25, 1931. **1938** G. H. Sewell *Amateur Film-Making* ii. 21 The light intensity is received by a photronic type of cell. **1948** *Electronic Engin.* XX. 136/2 A method is described by which it is possible to check the balance of the printer instantly by means of tricolour readings with a photronic cell.

photy (fōᵘ·ti). *colloq.* (chiefly *Sc.*). [Repr. a pronunc. of Photo *sb.*] A photograph.

1973 *Sunday Post* (Glasgow) 5 Aug. 17/3 *Wannysee wurphoties?*—Are you prepared for an hour of utter boredom?.. *Attaka noffiphoty*—Photographs never flatter me. **1974** *Weekly News* (Glasgow) 31 Aug. 11/5 'Maist o' yer photies seems tae be of Edwina,' Wee Sadie remarked.

phragmites (frægmai·tīz). [mod.L. (C. B. Trinius *Fundamenta Agrostographiæ* (1820) 134), f. Gr. φραγμίτης growing in hedges.] = Reed *sb.*[1] 4. Also *attrib.*

1920 *Blackw. Mag.* May 650/1 It may be necessary to cleave a passage..—cutting through papyrus, tearing at the tangling *Phragmites*, severing the long stems of the water-convolvulus. **1946** F. E. Zeuner *Dating Past* iii. 56 Peats growing under or at the water level: *Phragmites* peat (peat formed by the Common Reed and similar plants growing in shallow water). **1957** E. E. Evans *Irish Folk Ways* xv. 217 Large lake rushes provided materials for caulking..and phragmites stems for winding yarn in the shuttle. **1959** J. D. Clark *Prehist. S. Afr.* viii. 201 Phragmites reeds cut transversely and possibly left over from making arrow shafts,..abound in the upper layers of the cave. **1965** P. Wayre *Wind in Reeds* xi. 156 Little pockets of phragmites had sprung up in the odd corners and shallow bays all round the main [gravel] pit. **1977** *New Yorker* 5 Sept. 23/2 An impenetrable marsh of phragmites..and other plants clogs the length of the channel.

phragmoplast (fræ·gmoplɑst). *Bot.* [a. G. *phragmoplast* (L. Errera 1888, in *Bot. Centralbl.* XXXIV. 397), f. Gr. φράγμ-α fence, screen: see -o and -plast.] A set of fibrils which appears during mitosis in some plant cells as a barrel-shaped structure joining the two sets of chromosomes after their separation to the poles and which lasts until the formation of the cell plate.

1912 W. H. Lang tr. *Strasburger's Text-bk. Bot.* (ed. 4) 89 A barrel-shaped figure, the phragmoplast, is formed, which either separates entirely from the developing daughter nuclei, or remains in connection with them by means of a peripheral sheath. **1941** *Amer. Jrnl. Bot.* XXVIII. 227/2 In late anaphase or during telophase the phragmoplast extends laterally and a cell plate is formed across the cell. **1976** A. W. Davidson in M. M. Yeoman *Cell Division in Higher Plants* xii. 419 The phragmoplast, which heralds cell plate formation, appears at the equator of the mitotic spindle and moves centrifugally outwards along the phragmosome.

Hence **phragmopla·stic** *a.*

1952 A. Hughes *Mitotic Cycle* iv. 147 This form of cytokinesis..is really a variety of the phragmoplastic method. **1953** K. Esau *Plant Anat.* iii. 62 Phragmoplastic fibers appearing at the margins of the cell plate.

phragmosome (fræ·gmosǭᵘm). *Bot.* [f. as prec. + *-some[4].] **a.** A layer of darker cytoplasm which forms during mitosis in some plant cells at the site of the future cell plate. **b.** One of the large number of small particles that form this layer.

1940 Sinnott & Bloch in *Proc. Nat. Acad. Sci.* XXVI. 226 This cytoplasm..tends to become aggregated into a series of strands..which occupies the position of the future wall and which thus indicates..where the plane of the division is to be. For this plate of cytoplasm the writers propose the term phragmosome. **1960** *Proc. 4th Internat. Conf. Electron Microscopy* 505 Other structures of the cytoplasm are consistently associated with the plate... The more obvious of these bodies (phragmosomes) tend to be a uniform distance from the cell plate. *Ibid.*, How these phragmosomes contribute to plate formation is not very clear. **1965** Frey-Wyssling & Mühlethaler *Ultrastruct. Plant Cytol.* 279 During and after the formation of the cell plate osmiophilic particles become evident in the phragmoplast, which can reach light microscopic dimensions and which are known as phragmosomes. **1971** *Canad. Jrnl. Bot.* XLIX. 927 RNA was concentrated in phragmosomes and at the newly formed cell plates. **1976** A. W. Davidson in M. M. Yeoman *Cell Division in Higher Plants* xii. 414 It is changes that occur within this limiting membrane [*sc.* the plasmalemma] which determine the point at which the phragmosome is formed and hence the position of the new cell wall.

phrasal, *a.* Add: **a.** Used in *Gram.* in collocations qualifying the name of a part of speech to denote phrases which have the function of that part of speech, esp. *phrasal verb*, an idiomatic verbal phrase consisting of a verb and adverb or a verb and preposition. (Further examples.)

1879 J. Earle *Philol. Eng. Tongue* (ed. 3) x. 553 Modern English has made a new phrasal verb, and one that yet waits for a name. In this new verb the pronoun *it*, referring to no noun, acts as an objective accompaniment, and runs next after the verb. **1925** L. P. Smith *Words & Idioms* v. 172 Even more numerous are the idiomatic collocations of verbs followed by prepositions, or by prepositions used as adverbs. Collocations of this kind, 'phrasal verbs' we may call them, like 'keep down', 'set up', 'put through', and thousands of others. [*Note*] The term 'phrasal verbs' was suggested to me by the late Dr. Bradley. **1954** E. Gowers *Compl. Plain Words* vi. 71 There is today a tendency to form phrasal verbs to express a meaning no different from that of the verb without the particle... *Drown out, sound out, lose out, rest up, miss out on*, are other examples of phrasal verbs which I am told are used in America in senses no different from that of the unadorned verb. **1959** M. Schlauch *Eng. Lang.* 310 (Index), Phrasal adjectives. **1961** R. B. Long *Sentence & its Parts* ix. 218 In units such as *three-year-old*, whether used as phrasal nouns (as in *the three-year-old next door*) or as prepositive modifiers (as in *any three-year-old child*), the construction is different but obviously related. *Ibid.* x. 235 Units such as *tomorrow afternoon*..are syntactically much like phrasal-proper-name units such as Earlham College. **1973** *Word 1970* XXVI. 116 The following contain phrasal verbs: *You will have to look out for yourself*; *He did away with himself*.

b. *Mus.* Of or belonging to a phrase (in sense 5 of the sb.).

1946 R. Blesh *Shining Trumpets* v. 106 This solo puts the semitonally flatted third and fifth in phrasal conjunction.

phra·sally, *adv.* [f. Phrasal *a.* + -ly[2].] In or by phrases; as a phrase.

1934 in Webster. **1971** *Archivum Linguisticum* II. 61 The fact that *black bird* is labelled phrasally reflects, it seems, a combination of conventional practice, analytical convenience, and the ascribing of special semantic significance to this particular type of structural relationship. **1973** *Ibid.* IV. 47 A given 'place' may..be occupied by more than one phrasally contributing form.

phrase, *sb.* Add: **2. d.** *transf.*

1908 G. Jekyll *Colour in Flower Garden* 15 While the wide-stretching shadow-lengths throw the woodland shades into woodland *phrases* of broadened mass. **1922** [see *choreographically *adv.*].

7. *phrase-family, -form, -formative, -meaning, -tag*; *phrase-final, -internal* adjs.; **phrase-book** (earlier and later examples); also *attrib.*; **phrase-maker** (later examples); also, a composer of musical phrases; hence **phrase-making** *vbl. sb.* (also in literal sense); **phrase-marker** *Linguistics*, a diagrammatic or formulaic representation of the constituent structure of a sentence; abbrev. *P-marker* s.v. *P III 5; **phrase-structure**, a group of words that constitutes a phrase; in *Gram.*, the structure of a sentence in terms of its constituent phrases (also *attrib.*); hence *phrase-structurally* adv.; **phrase-word** (see quot. 1933).

1594 Nashe *Unfortunate Traveller* sig. F1, In emptying their phrase bookes, the ayre emptied his intrailes. **1905** G. Meredith *Let.* 7 May (1970) III. 1522 A pocket Italian-English phrase-book should be taken. **1963** L. Deighton *Horse under Water* xiv. 60 He had used pompous phrase-book Portuguese. **1968** *Listener* 27 June 827/3 Pack one of those old-fashioned and much-mocked phrase books. **1907** 'Mark Twain' *Christian Sci.* II. vii. 163 These great officials are of the phrase-family of the Church-Without-a-Creed..that is to say, of the family of Large-Names-Which-Mean-Nothing. **1949** E. A. Nida *Morphol.* (ed. 2) v. 126 In phrase-final position. **1968** *Language* XLIV. 80 The glide on the short nucleus may be phonetically long, particularly in phrase-final position. **1911** Brereton & Rothwell tr. *Bergson's Laughter* ii. 114 Sometimes, too, the effect is a complicated one. Instead of one commonplace phrase-form, there are two or three which are dovetailed into each other. **1926** L. Bloomfield in *Language* II. 156 The possessive [z] in *the man I saw yesterday's daughter*... Such a bound form is a *phrase-formative*. **1964** E. Bach *Introd. Transformational Gram.* vi. 137 The rules begin with a string of forms..which are bracketed by numbered boundaries of two kinds: phrase-internal and compound-internal boundaries. **1924** P. C. Buck *Scope of Music* 39 There will come a time when the phrase-maker desires to extend his tune beyond the limits of one breath. **1967** *Listener* 22 June 835/3 Certainly 'the dramatization of the significant' (what phrasemakers these Americans are!) is a worthy aim. **1977** *Rolling Stone* 7 Apr. 63/1 He knows how to 'use' television, he's a phrasemaker, he's good-looking and has a deep voice. **1867** W. D. Whitney *Lang. & Stud. Lang.* 116 All word-making by combination..is closely analogous with phrase-making. **1905** *Athenæum* 25 Nov. 717/3 Phrase-making is not style.., nor is rhetoric the sole canon of speech. **1926** V. Woolf *Writer's Diary* (1953) 96 No power of phrase-making. Difficulty in writing. **1929** C. Day Lewis *Transitional Poem* i. 17 Phrase-making, dress-making—Distinction's hard to find; For thought must play the mannequin. **1963** Chomsky & Miller in R. D. Luce et al. *Handbk. Math. Psychol.* II. 288 We assume that such a tree graph must be a part of the structural description of any sentence; we refer to it as a *phrase-marker* (P-marker). **1965** N. Chomsky *Aspects of Theory of Syntax* i. 17 The *base* of the syntactic component is a system of rules that generate a highly restricted (perhaps finite) set of *basic strings*, each with an associated structural description called a *base Phrase-marker*. These base Phrase-markers are the elementary units of which deep structures are constituted. **1968** J. Lyons *Introd. Theoret. Linguistics* vi. 259 A labelled-bracketing of a string is referred to, technically, as a *phrase-marker*. Phrase-markers may also be represented by means of a tree-diagram with labelled nodes. **1971** *Archivum Linguisticum* II. 129 Removing all sequences of morphemes which can be referred to embedded phrase-markers, we are still left with complicated strings which obviously demand an ingenious transformational explanation of a kind which has not yet been offered. **1973** *Studies in Eng. Lit.: Eng. Number* (Tokyo) 52 Irrelevant structural details will hereafter be omitted from our phrase markers. **1931** G. Stern *Meaning & Change of Meaning* 2 Phrase-meanings and word-meanings. **1965** *Language* XLI. 277 A constituent string, which is always marked phrase-structurally for terminal fade. **1975** *Studies in Eng. Lit.: Eng. Number* (Tokyo) 69 They suggest..that *Jane and Bill* be generated phrase-structurally. **1957** N. Chomsky *Syntactic Struct.* iv. 28 The determination of the phrase structure (constituent analysis) of the derived sentence. **1960** J. B. Carroll in Saporta & Bastian *Psycholinguistics* (1961) 337/2 In the latter part of the babbling period..there are the *first* evidences of..phrase-structures on the part of the child. **1967** D. G. Hays *Introd. Computational Linguistics* viii. 147 A comparatively simple phrase-structure grammar produces deep structures with associated terminal strings. **1976** *Word 1971* XXVII. 257 Phrase-structure rules generate the deep-structure b-sequence from which the well-formed a-sequence is derived by transformation rules. **1957** N. Frye *Anat. Crit.* 103 The fixed epithets and phrase-tags of medieval romance and ballad. **1933** L. Bloomfield *Language* xi. 180 The forms of the type *devil-may-care* are classed as words (phrase-words) because..as a phrase *devil-may-care* would be an actor-action form, but as a phrase-word it fills the position of an adjective. **1979** *Dictionaries* I. 78 Other phrase words were very late innovations, stemming mostly from New High German times.

phrase, *v.* Add: **6. a.** (Earlier *absol.* and later *trans.* examples.)

1876 Stainer & Barrett *Dict. Mus. Terms* 348 s.v. *Phrasing*, A performer who brings into due prominence the grouping of sounds into figures, sentences, &c., is said to *phrase* well. **1877** G. B. Shaw *How to become Mus. Critic* (1960) 29 It is easy to say that a singer 'phrases' well, because so few know what phrasing means. **1962** *Times* 15 June 13/7 All thrushes (not only those in this neck of the Glyndebourne woods) sooner or later sing the tune of the first subject of Mozart's G minor Symphony (K. 550)—and, what's more, phrase it a sight better than most conductors. **1976** *Gramophone* Oct. 611/1 He has rather a sour tone and does not phrase the music as elegantly as his rivals.

b. *Dance.* To link (movements) in a single choreographic sequence, or part of a sequence.

1959 *Times* 22 Jan. 3/4 Miss Georgina Parkinson, who phrases travelling movements with much smoothness.

phraselet (frē̆ⁱ·zlėt). *rare.* [-let.] A short phrase (in music).

1925 P. A. Scholes *Second Bk. Gramophone Rec.* 86 The Clarinet repeats its last phraselet.

phrasial (frē̆ⁱ·ziăl), *a. rare.* [f. Phras(e *sb.* + -ial.] Of or pertaining to (musical) phrases.

1918 *Proc. Musical Assoc.* Apr. 135 The phrasial formalities..engender a monotony.

phrasing, *vbl. sb.* Add: **2. a.** (Earlier and later examples.)

1877 [see *phrase *v.* 6 a]. **1921** A. Rivarde *Violin & its Technique* i. 11 Many violinists..are constantly blustering with long bows, very often spoiling the phrasing and making violinistic rather than musical effects. **1966** *Crescendo* Feb. 35/3 To hear drum phrasing at its very best. **1976** *Gramophone* Apr. 1611/3 Though individual soloists are naturally as fine as one expects of the LSO, the style of phrasing in ensembles is stiffer than it might be.

b. in sense *6 b of the verb.

1978 *Daily Tel.* 22 Aug. 9/1 In the pas de trois,..in spite of some uncertainty over supported pirouettes, there was splendidly crisp phrasing from Kenneth McCombe and his partners.

3. *transf.*

1949 M. Mead *Male & Female* iii. 65 All these themes are present in every cultural phrasing of the mother-child relationship. **1967** W. W. Newcombe in F. Kirkland *Rock Art of Texas Indians* iv. 40/2 Certainly the two cultural complexes are phrasings of a single basic culture.

phreak (frīk), *sb.* and *v.* [Modified spelling of Freak *sb.*[1], *v.*, under influence of *phone *sb.*[2], *v.*] **A.** *sb.* = *phone phreak *sb.* **B.** *vb. trans.* and *intr.* To use an electronic device to obtain (a telephone call) without payment. So **phrea·king** *vbl. sb.*

1972 *Daily Tel.* 15 Apr. 3/2 The craze started in America and there are said to be 150 'phreaks' in this country who swop information and have equipment and dialling codes which give them free use of the world's

PHREATIC

telephone system. **1972** *Oxford Times* 9 June 7/5 Used telephone codes and his expert knowledge to 'phreak' phone calls.. 23 attempts at 'phreaking', of which seven were successful.

phreatic, *a.* Substitute for def.: Of, pertaining to, or designating water below the water-table, esp. that which is capable of movement. [ad. F. *phréatique* (G. A. Daubrée *Les Eaux Souterraines* (1887) I. ii. 19).] (Earlier and further examples.)

1893 R. J. HINTON *Irrigation in U.S.: Progress Rep. for 1890* (U.S. 51st Congress, 2nd Sess. Senate Ex. Doc. No. 53) 42 At the point at which most of them leave the mountain ranges there commences an enormous phreatic absorption of the volume of flow that has descended from the summit above. **1892** R. HAY *Final Geol. Rep.* (U.S. 52nd Congress, 1st Sess. Senate Ex. Doc. No. 41) I. 8 Prof. Hill has given definitions of the technical words used by him, and to his list may be added the new word *phreatic*, which is a very convenient term for underground waters which can be, or which it is hoped may be, reached by wells or other sub-ground works. [*Note*] This word was first used in American hydro-geologic investigation by the Artesian and Underflow Office in 1890. **1927** *Econ. Geol.* XXII. 494 Daubrée (1887, p. 19) invented 'phreatic' from a Greek expression for 'well'... Originally.. the word meant seepage water and particularly that below the water-table. It was so used by Hay (1892, p. 8) and McGee (1894, pp. 16, 42)... Suess (1909, p. 655) appears to have included in 'phreatic water' that of connate origin as well as seepage water... The writer believes that the history of the word and practical expediency should make 'phreatic' mean the infiltered waters which are bounded above by the water-table. **1954** *Times Lit. Suppl.* 5 Feb. 93/2 Theories of their [sc. cavern] formation are classed as 'vadose' or 'phreatic' according as it is held that caves are formed above the water-table or below it. **1966** DAVIS & DEWIEST *Hydrogeol.* ii. 42 The zone of phreatic water merges at depth into a zone of dense rock with some water in pores, although the pores are not interconnected so that water will not migrate. **1973** *Nature* 9 Nov. 77/2 After the solutional formation of a cavity beneath the water table, incision and redirection of local base level produced lowering of the phreatic surface. **1977** A. HALLAM *Planet Earth* 108 Where the top of the zone of saturation of an aquifer is a free-water surface it is known as the water table (or phreatic surface).

2. Of, pertaining to, or designating a volcanic explosion caused by the sudden heating and volatilization of underground water when it comes into contact with hot magma or rock.

1909 H. B. C. & W. J. SOLLAS tr. *Suess's Face of Earth* IV. xvi. 566 Phreatic explosions. When juvenile hydrogen encounters an unlimited quantity of vadose water, we witness a spectacle such as was presented by Krakatoa in 1883... In this case the effect may have been due to phreatic water in the neighbourhood of the sea, but when phreatic water is confined in the fissures of a limestone formation, the explosion shatters the limestone. **1926** R. A. DALY *Our Mobile Earth* iv. 158 Not all explosions are due to the pressure of magmatic gas... In the year 1888 the side of the Japanese cone, Bandai-San, was ripped out... The cavity, technically called a 'phreatic caldera', is about two miles long. *Ibid.*, The 1924 explosion at Kilauea was of phreatic character. **1964** *New Scientist* 5 Mar. 585/2 The co-existence and maintenance of such 'phreatic' activity (here ascribable to sea water suddenly flashed into steam by contact with hot lava in a vent) with the contrasting fire-fountains, characteristic of 'Strombolian' volcanic activity.., calls for somewhat unusual conditions. **1973** FIELDER & WILSON *Volcanoes of Earth, Moon & Mars* iv. 53/2 In a phreatic eruption, horizontal surges carry low density loads outwards from the eruptive centre. **1976** F. FRANCIS *Volcanoes* iv. 143 This steam sometimes blasts its way up to the surface through the lava, causing what is known as a phreatic explosion.

phreatophyte (fri,æ·tofəit). [f. Gr. φρέαρ, φρεατ- tank, cistern + -PHYTE.] A plant with a deep root system that draws its water supply from near the water-table.

1920 *Bull. Geol. Soc. Amer.* XXXI. 333 In arid regions plants of certain species habitually utilize water from the zone of saturation. For such plants the name *phreatophyte* (meaning a well-plant) has been proposed... Willow trees are among the most common phreatophytes in the arid West. **1928** *Ecology* IX. 474 The phreatophytes or 'well plants' which derive their water supply from the ground water and are more or less independent of local rainfall. **1963** D. W. & E. E. HUMPHRIES tr. *Termier's Erosion & Sedimentation* vi. 136 Most other plants in dry regions need roots long enough to reach down to the water table, or at least to its capillary fringe; these are the phreatophytes. **1965** R. G. KAZMANN *Mod. Hydrol.* v. 144 The most widespread of the desert phreatophytes in the United States are salt grass, greasewood, mesquite, and salt cedar. **1974** *Internat. Gloss. Hydrol.* 171/2 *Phreatophytes*, water-loving plants that grow mainly along stream courses and/or where their roots reach the capillary fringe.

phrenicectomy (frenise·ktǒmi). *Surg.* [f. as next + *-ECTOMY.] Surgical removal or destruction of a section of a phrenic nerve, formerly carried out as an alternative to phrenicotomy.

1929 *Surg., Gynecol. & Obstetr.* XLVIII. 274/1 The indications for phrenicectomy are lesions which may be benefited by partial compression and immobilization of a lung. **1943** B. M. DICK in C. F. W. Illingworth *Textbk. Surg. Treatm.* xxviii. 381 Phrenicectomy is used as an adjuvant to other types of collapse therapy—as an independent measure where artificial pneumothorax has failed owing to adhesions, and as a palliative for haemoptysis, harassing cough, vomiting, or pain due to diaphragmatic adhesions. **1957** E. H. HUDSON in F. R. G. Heaf *Symposium of Tuberculosis* vii. 381 Stuertz of Cologne is credited as the first to perform phrenicectomy as a therapeutic measure in the treatment of bronchiectasis of the lower lobe of the lung in 1911.

phrenicotomy (freniko·tǒmi). *Surg.* [f. PHRENIC *a.* (*sb.*) + -O + -TOMY.] Surgical cutting of a phrenic nerve, so as to paralyse the diaphragm on the same side.

1913 *Internat. Abstr. Surg.* XVII. 417/2 Oehlecker.. performed the phrenicotomy [sic] in three cases. **1926** *Encycl. Brit.* III. 691/1 Phrenicotomy (division of the phrenic nerve in the neck) is said to paralyse the corresponding half of the diaphragm; it is sometimes practised in conjunction with other means of putting one lung at rest. **1950** J. K. BERMAN *Princ. & Pract. Surg.* viii. 231 Compression of the lung is best secured by the subperiosteal resection of ribs over the affected side. Should symptoms persist after pneumothorax and phrenicotomy, thoracoplasty may be used. **1974** *Jrnl. Appl. Physiol.* XXXVII. 315/1 Tracings of pleural pressure on the intercostal and diaphragmatic surfaces under normal conditions and after complete bilateral phrenicotomy are shown in Fig. 5.

phreno-. Add: phrenotro·pic *a.* = *PSYCHOTROPIC *a.*

1956 [see *PSYCHOTROPIC *a.*]. **1957** *Ann. N.Y. Acad. Sci.* LXVI. 765 Ibogaine began to be of interest to us in connection with its possible phrenotropic activity.

Phrygian, *a.* (*sb.*) Add: **B.** *sb.* **a.** (Earlier and later examples.)

a **1490** J. SKELTON tr. *Diodorus Siculus' Bibliotheca Historica* (1956) I. 19 Thay went vnto the Phrygians. **1963** *Times* 12 Jan. 9/6 It sounded like an Aesop's fable (and it occurred to me that Aesop, a Phrygian, had lived near Sivrihisar). **1966** G. E. BEAN *Aegean Turkey* v. 125 This was a time of long hostility between the Phrygians and the Greeks, when King Midas was the first barbarian to make an offering at Delphi.

c. The Indo-European language of the ancient Phrygians.

1797 W. JONES in *Asiatick Researches* (1792) III. 14 A drum is called *dindima* both in Sanscrit and Phrygian. **1888** J. WRIGHT tr. *Brugmann's Elem. Compar. Gram. Indo-Germ. Lang.* I. 3 Of others we have only very scanty fragments left, as of Phrygian. **1933** C. D. BUCK *Compar. Gram. Greek & Latin* 13 Phrygian is known.. from proper names and glosses, from a few old inscriptions in an alphabet closely akin to the Greek, and from a considerable number of inscriptions.. of the Christian times. **1967** M. SCHLAUCH *Language* ii. 28 Among the Indo-European languages no longer spoken, some are known to us from inscriptions (for instance, Thracian and Phrygian). **1972** W. B. LOCKWOOD *Panorama Indo-European Lang.* 174 Early or Old Phrygian survives in nearly 25 short inscriptions of doubtful date (perhaps eighth to six centuries B.C.) written in an alphabet of an archaic Greek type. A more recent form of the language, Late or New Phrygian, is found in about a hundred inscriptions in the Greek alphabet dating from the first three centuries A.D.

phthaldehyde (fþæ·ldəhəid). *Chem.* [f. PHTHAL(IC *a.* + ALDEHYDE.] Any of three isomeric compounds of formula $C_8H_6O_2$ having two formyl groups bonded directly to a benzene ring; *spec.* the *ortho* isomer, a yellow crystalline solid.

1886 *Jrnl. Chem. Soc.* L. 455 By the oxidation of phthalic alcohol with chromic mixture, a thick, viscid oil is obtained containing phthalaldehyde, phthalide, and unaltered phthalic alcohol. All attempts to isolate phthalaldehyde from this mixture were fruitless. **1942** *Jrnl. Amer. Chem. Soc.* LXIV. 315 (*heading*) A new synthesis of phthaldehydes. **1967** I. L. FINAR *Org. Chem.* I. xxiv. 586 When treated with ozone, naphthalene forms the diozonide and this, on treatment with water, gives phthaldehyde. **1967** L. F. & M. FIESER *Reagents for Org. Synthesis* I. 740, *o*-Phthalaldehyde reacts with nitromethane in alcoholic alkali to give, after acidification yellow 2-nitro-3-hydroxyindene.

phthalazine (fþæ·lāzin). *Chem.* [ad. G. *phthalazin* (C. Liebermann 1886, in *Ber. d. Deut. Chem. Ges.* XIX. 766), f. *phtal-* (cf. *phthal-* s.v. PHTHALIC *a.*) + *azin* *AZINE.] A colourless, crystalline, heterocyclic base, $C_8H_6N_2$; also, any derivative of this.

1893 *Jrnl. Chem. Soc.* LXIV. I. 347 The bye-product obtained in methylating phthalazine is very similar to the last compound. **1929** [see *CINNOLINE]. **1947** *Nature* 11 Jan. 53/2 Phthalazine derivatives are formed by the action of excess sodium hydroxide on the diazosulphonate derived from diazotized β-nitro-aniline and β-naphthol-1-sulphonic acid. **1955** I. L. FINAR *Org. Chem.* II. xii. 275 Phthalazines are formed by heating the benzoyl derivative of benzaldehyde hydrazones. *Ibid.* xiii. 706 Thus there are probably four $C_8H_6N_2$ units, each having an imidazole structure.. or a phthalazine structure. **1970** W. F. BECH *Fibre-Reactive Dyes* v. 175 Phthalazines... In this series also only one compound has so far become of importance for making reactive dyes, viz 1,4-dichlorophthalazine-6-carbonyl chloride.

phthalic, *a.* Add: *phthalic anhydride*, the anhydride of phthalic acid, a white crystalline solid, $C_8H_4O_3$, which is made industrially by the catalytic oxidation of naphthalene or *o*-xylene, and is widely used in the manufacture of plastics, resins, dyes, etc.

1855 W. ODLING tr. *Laurent's Chem. Method* 53 Thus we can obtain in a direct manner the.. tartaric, camphoric, succinic, phthalic, &c., anhydrides. **1869** H. WATTS tr. *Gmelin's Hand-bk. Chem.* XIII. 13 Phthalic anhydride forms long white needles having the lustre of silk and united in feathery groups. **1913** THORPE *Dict. Appl. Chem.* (ed. 2) IV. 251/2 Owing to the extensive use of phthalic anhydride in the manufacture of synthetic indigo.. a cheap technical process for its preparation in large quantities was essential. **1945** H. BARRON *Mod. Plastics* xxvii. 559 Glycerine and phthalic anhydride are the two most important constituents for alkyd resins. **1966** R. M. STEPHENSON *Introd. Chem. Process Industries* xvi. 329 About 459 million pounds of phthalic anhydride were produced in the United States in 1963.

phthalocyanine (fþælosəi·ănīn). *Chem.* [f. PHTHAL(IC *a.* + -O + CYANINE.] **a.** A greenish-blue crystalline porphyrin, $C_{32}H_{18}N_8$, or any of its substituted derivatives. **b.** Any of the metal chelate complexes of these, which form a large and important class of pigments and dyes ranging in colour from green to blue.

1933 R. P. LINSTEAD in *Rep. Brit. Assoc. Adv. Sci.* 465 When phthalimide is heated with certain metals.. in a current of ammonia, a complex reaction occurs with the formation of highly coloured substances of a novel type. These have been named phthalocyanines from their origin and deep-blue colour... The metal may be eliminated from the magnesium compound.. to yield phthalocyanine, the parent substance of the group. **1942** *Endeavour* I. 105 Phthalocyanines, themselves entirely products of the chemist's skill, have resemblances to two important substances which occur abundantly in nature. **1961** COCKETT & HILTON *Dyeing of Cellulosic Fibres* v. 207 Copper phthalocyanine is the most important pigment Monastral Blue BS, the lead compound being Monastral Green. **1972** KIRK & ZOLLINGER *Fund. Chem. & Applic. Dyes* vii. 147 Because of their good fastness properties phthalocyanines are suitable for almost all processes in which pigments are used.

b. phthalocyanine blue (also with capital initials), copper phthalocyanine, an important blue pigment; **phthalocyanine green** (also with capital initials), a chlorinated (or brominated) derivative of copper phthalocyanine, important as a green pigment.

1941 R. H. ALLEN in J. J. Mattiello *Protective & Decorative Coatings* II. viii. 261 Likewise phthalocyanine green shows a somewhat similar relationship to orthodox pigments. **1947** L. S. PRATT *Chem. & Physics Org. Pigments* ii. 8 Other pigments which also contributed to this change were.. very recently the benzidine yellows and phthalocyanine blues and greens. **1967** V. STRAUSS *Printing Industry* ix. 392/1 The development of phthalocyanine blue, perhaps better known under its duPont brand-name as Monastral Blue, was one of the greatest achievements in the field of pigments. **1968** KIRK & OTHMER *Encycl. Chem. Technol.* (ed. 2) XV. 491 Phthalocyanine green (Pigment Green 7, CI 74260) is a polychlorinated copper phthalocyanine containing about fourteen atoms of chlorine. The most usual shade is the *G* type although there are several shades in the range, representing slight variations in degree of chlorination or particle size and shape. *Ibid.*, Two yellower shades of green are available as Phthalocyanine Green 27... In these types, part of the chlorine has been replaced by bromine. **1970** [see *MONASTRAL]. **1972** *Materials & Technol.* V. xi. 359 Phthalocyanine blues are intense blues which are not used in full colour because bronzing develops... Two forms are used, beta phthalocyanine blue of relatively green shade,.. and the less stable alpha phthalocyanine blue, of a redder shade.

phthalylsulphathiazole (fþæ:lil,sʌlfăþəi·ăzōul). *Pharm.* Also (*U.S.*) -sulfa-. [f. PHTHALYL + *SULPHATHIAZOLE.] A sulphonamide drug, $HOOC\cdot C_6H_4\cdot CONH\cdot C_6H_4\cdot SO_2NH\cdot C_3H_2NS$, that is a whitish powder and is used to suppress bacteria in the gastro-intestinal tract.

1943 POTH & ROSS in *Federation Proc.* II. 89/2 An extension of the studies on acylated sulfonamides as intestinal antiseptics.. has uncovered another effective derivative of this series of drugs, phthalylsulfathiazole. **1961** M. HYNES *Med. Bacteriol.* (ed. 7) x. 136 Unabsorbable sulphonamides (succinylsulphathiazole, phthalylsulphathiazole, sulphaguanidine).. are generally available for oral treatment of intestinal infections. **1970** PASSMORE & ROBSON *Compan. Med. Stud.* II. xx. 25/1 Succinylsulphathiazole and phthalylsulphathiazole are both poorly absorbed from the gastrointestinal tract and have been advocated for the treatment of intestinal infections and for prophylactic use in elective operations on the colon. **1977** *Proc. R. Soc. Med.* LXX. 482/1 The preliminary results of this study have shown that a combination of oral phthalylsulphathiazole and metronidazole for preparation of the bowel for surgery reduces the incidence of postoperative infection compared with phthalylsulphathiazole alone.

phthiocol (fþəi·ǒkǫl). *Biochem.* [f. PHTHI(SIS + -O + -*col* (perh. f. alcohol).] A yellow crystalline pigment, 3-hydroxy-2-methyl-1,4-naphthoquinone, $C_{11}H_8O_3$, originally isolated from tubercle bacilli, which has the action of vitamin K.

1933 ANDERSON & NEWMAN in *Jrnl. Biol. Chem.* CIII. 197 In order to indicate its origin, we propose to designate the pigment by the name phthiocol. **1954** H. J. ALMQUIST in Sebrell & Harris *Vitamins* II. ix. 393 Synthetic phthiocol tested with vitamin K-deficient chicks was found to be distinctly active in restoring normal blood-clotting time and thus became the first completely identified form of vitamin K. **1960** A. E. BENDER *Dict. Nutrition* 133/2 Natural vitamin K chemically is a phylloquinone... The synthetic materials, menadione.. and phthiocol (2-methyl-3-hydroxy-1:4-naphthoquinone), are more active than the naturally occurring vitamins. **1964** ASSELINEAU & LEDERER in V. C. Barry *Chemotherapy of Tuberculosis* i. 21 Phthiocol.. isolated.. after saponification of mycobacterial fat, is an artefact, produced from a vitamin K-like polyprenoid naphthoquinone.

phthioic (fþəi·ǒ,ik), *a.* *Biochem.* [f. PHTHI(SIS + *-OIC.] *phthioic acid*: a yellowish oil, now known to be a mixture of fatty acids, which was orig. obtained from tubercle bacilli and is capable of inducing the symptoms of tuberculosis; hence, any of these constituent acids or their synthetic derivatives.

1929 R. J. ANDERSON in *Jrnl. Biol. Chem.* LXXXIII. 171 In order to indicate the relation of this acid to tuberculosis we wish to designate it by the name phthioic acid. **1946** *Nature* 5 Oct. 489/1 Phthioic acid $C_{26}H_{52}O_2$ is a liquid saturated fatty acid isolated from the lipoids of tubercle bacilli. **1951** *Chem. & Industry* 11 Aug. 665/1 Recent work on the fatty acids of the lipins of tubercle bacilli has shown that Anderson's phthioic acid was a mixture. **1964** W. PAGEL et al. *Pulmonary Tuberculosis* (ed. 4) ii. 31 Phthioic acids synthetized [sic] by Robinson (1940) were shown to produce necrosis and granulomata in guinea-pigs.

† phthisiatry. *Med.* *Obs.* *rare*⁻¹. [f. PHTHISI(S + Gr. ἰατρεία healing, medical treatment, f. ἰατρός healer.] = *PHTHISIOTHERAPY.

1926 *Amer. Rev. Tuberculosis* XIII. 105 The method can give genuinely beneficial results, and its possibilities should always be valuable in phthisiatry.

† phthisiogenesis (fþ-, þiziodʒe·nēsis). *Med.* *Obs.* [f. PHTHISI(S + -O + GENESIS.] The causation or development of phthisis. Hence **† phthi:siogene·tic** *a.*, causing or pertaining to the development of phthisis.

1906 C. BOLDUAN tr. *E. von Behring's Suppression of Tuberculosis* 45 Observations concerning the study of phthisiogenesis in man and animals. *Ibid.* 46 Improbability, so far as importance as a phthisiogenetic factor is concerned, of a primary bronchial TB infection. **1924** *Amer. Rev. Tuberculosis* X. 24/2 Experimental studies on phthisiogenesis.—Inoculation of human tubercle bacilli into the tonsils or soft palate in rabbits of different ages resulted in a tuberculosis running the course of pulmonary phthisis. **1936** *Univ. Durham Coll. of Medicine Gaz.* XXXVI. 128 The study of phthisiogenesis involves the reconciliation of.. much discordant data.

phthisiology. Add: (Examples.) Hence **phthisiolo·gical** *a.*; **phthisio·logist**, a specialist in phthisiology.

1913 *Q. Jrnl. Med.* VI. 259 Artificial pneumothorax has become the topic of the day in phthisiological literature. **1924** *Amer. Rev. Tuberculosis* XVIII. 110 The general conclusions.. were not accepted at once by phthisiologists. **1946** H. T. HYMAN *Integrated Pract. Med.* III. cix. 2106 The indications for inducing an artificial pneumothorax are best discussed by the practitioner with the consultant phthisiologist. **1947** *Ind. IV.* clxxxii. 3901 In the larger medical communities and fully staffed institutions, internal medicine is subdivided... Thus there are established departments of gastro-enterology, hematology, phthisiology, metabolism, [etc.]. **1953** *Tubercle* XXXIV. 237/1 It is most deplorable that phthisiologists ever applied and accepted so wrong and unfitting a term [sc. 'therapeutic collapse'] to a condition which in fact is no 'collapse' at all. **1957** F. R. G. HEAF *Symposium of Tuberculosis* 9 iv. The amount of disablement caused by bone and joint tuberculosis in oversea countries is very great, so the chapter dealing with this subject should be of value to the general physician and surgeon, as well as to the phthisiologist.

phthisiophobia (fþ-, þiziofǒu·biă). *Med.* [f. as next + -PHOBIA.] An unjustified or exaggerated fear of phthisis.

1906 J. B. HUBER *Consumption* xi. 445 It is really deplorable to consider the degree of cruelty and selfishness to which this phthisiophobia has driven people. **1948** F. M. POTTENGER *Tuberculosis* xxxvi. 579 If we frighten people into believing that tuberculosis is a highly infectious disease.., we create an unnecessary and harmful phthisiophobia, which reacts against the patient.

† phthisiotherapy (fþ-, þiziope·răpi). *Med.* *Obs.* [f. PHTHISI(S + -O + THERAPY.] The medical treatment of phthisis. Also **phthi:siothera·peutic** *a.*, of or pertaining to this; hence **phthi:siotherapeu·tist**, **phthisiothe·rapist**, a specialist in or practitioner of phthisiotherapy.

1899 S. A. KNOPF *Pulmonary Tuberculosis* xix. 284 Modern phthisio-therapeutics, as carried out in well-equipped sanatoria, must be practically studied. **1900** DORLAND *Med. Dict.* 505/1 Phthisiotherapy. **1903** *Med. Rec.* (N.Y.) 2 May 719/1 There is abundant evidence.. of his deep interest in phthisiotherapy. **1934** *Amer. Rev. Tuberculosis* XXX. 188 The maintenance by each sizeable sanatorium of a resident thoracic surgeon whose special training in phthisiotherapy can be relied upon to aid in careful selection of suitable cases for surgical collapse.

1899 S. A. KNOPF *Pulmonary Tuberculosis* xix. 285 The peculiar psychological state of nearly all phthisical patients.. makes it necessary for the true phthisiotherapeutist to.. be his best and most confidential friend. **1907** *Med. Rec.* (N.Y.) 9 Nov. 758/2 The pneumatic cabinet.. although long since discarded by most phthisio-therapists, has been persistently used by a few. **1929** *Amer. Rev. Tuberculosis* XIX. 76 There are some phthisiotherapists.. who advocate rib-resection in lieu of pneumothorax. **1933** *Jrnl. Amer. Med. Assoc.* 4 Feb. 313/1 Collapse therapy is considered by the majority of expert phthisiotherapists to have created a revolution in the treatment of pulmonary tuberculosis. **1939** *Amer. Rev. Tuberculosis* XXXIX. 162 The value of artificial pneumothorax in closing open cavities is admitted by every phthisiotherapeutist.

phugh, var. PHEW *int.*

1886 E. DOWSON *Let.* 30 June (1967) 88, I think it is possible for the feminine nature to be reasonably candid & simple, up to the age of 8 or 9. Afterwards—phugh! **1892** [see PHEW *int.*].

phugoid (fiū·goid), *a.* and *sb.* [f. Gr. φυγή flight (*sb.*[2] in Dict.: error, taken for FLIGHT *sb.*[1]) + -OID.] **A.** *adj.* Of or pertaining to the longitudinal stability of an aircraft flying a nominally horizontal course in a vertical plane; applied *spec.* to a slow fore-and-aft oscillation in which the flight path assumes the form of a series of shallow waves and the aircraft undergoes synchronous increases and decreases of speed.

1908 F. W. LANCHESTER *Aerodonetics* ii. 37 The Phugoid theory deals with the longitudinal stability, and the form and equations of the flight path of an aerodone. *Ibid.* 40 This is the general equation to the curves of flight or the phugoid equation. *Ibid.* iii. 59 The plotting of the phugoid curves forming a complete series, from the straight line representing the path of uniform gliding, to the tumbler type of curve with constants varying to any desired degree, may be termed a phugoid chart. **1920** *Flight* XII. 817/1 They would talk of bumps and pitching when, of course, they merely meant anabatics or katabatics, or peradventure nothing more serious than a phugoid oscillation. **1965** G. SUTTON *Mastery of Air* v. 125 The enhanced speed brings about an increase in lift, which ultimately stops the descent and causes the machine to climb again. At the top of the climb the machine has insufficient lift to maintain itself and if left alone will fall again and so on. This is the phugoid oscillation, consisting of a wave-like path or a series of loops. **1974** H. ASHLEY *Engin. Anal. Flight Vehicles* iii. 62 The latter situation arises.. during the higher portion of the entry trajectory of a lifting glider. The long-period or phugoid longitudinal mode may then have a period as great as half the total interval required for entry.

B. *sb.* A phugoid oscillation.

1908 F. W. LANCHESTER *Aerodonetics* ii. 42 With these two equations we are in a position to investigate the general characteristics and particular forms of the curves of flight or Phugoids, as they may be appropriately termed. **1945** *Jrnl. R. Aeronaut. Soc.* XLIX. 346/1 He became so interested in a phugoid that he allowed it to develop beyond control, and he found himself spinning inverted at 33,000 feet. **1948** L. M. MILNE-THOMSON *Theoret. Aerodynamics* xvi. 309 A phugoid is the path of a particle which moves under gravity in a vertical plane and which is acted upon by a force *L* normal to the path and proportional to V^2, the square of the speed. **1975** L. J. CLANCY *Aerodynamics* xvi. 485 The consequence of the disturbance is an oscillation in which the aircraft successively gains and loses height, while losing and gaining forward speed. This oscillation is known as a phugoid.

phulkari. For '*East Ind.*' read '*N. India*'. (Earlier and later examples.)

1872 B. H. BADEN-POWELL *Hand-bk. Econ. Products of Punjab* II. ix. 100/1 'Phúlkárí' (*lit.* 'flower work') scarf, value Rs. 11. *Ibid.* x. 106 Over the head is thrown a 'chádar' of coarse cloth, prettily embroidered in many colored silks, called 'phulkári'. **1887** H. G. H. BLACKWOOD *Jrnl.* 1 Dec. in *Our Viceregal Life in India* (1889) II. xiii. 224 They were collected in a small inner court, which was hung with the pretty phulcarries they make here. **1888** MRS. F. A. STEEL in *Jrnl. Indian Art* II. 72 Intending purchasers.. should remember that *phulkari* work is a true art. *Ibid.*, *Phulkaris* are still a necessary part of a Hindu and Sikh bride's trousseau. **1893** G. BIRDWOOD in *Catal. Loan Exhib. Embroidery by Indian Women* (Soc. Encouragement Indian Art) 15, I did not know of any needlework being produced by the women of India for sale, excepting the *phul-kāri* (*i.e.*, 'flower-work'). **1969** *Eve's Weekly* (Bombay) 20 Dec. 43/1 Another in magenta white and saffron phulkari against brown is styled on the same lines. **1969** *Femina* (Bombay) 26 Dec. 8/4 She had revived an old art like the 'phulkari'... The applique work was bold and mirror work pretty. **1971** *Sunday Australian* 8 Aug. 46 Phulkari embroidery is the traditional work used on all the linen items in a girl's glory box.

Phurnacite (fǝ̄·ɹnăsəit). Also **phurnacite.** A proprietary name for a kind of smokeless fuel made by carbonizing briquettes at relatively low temperatures.

1937 *Trade Marks Jrnl.* 10 Feb. 150/1 Phurnacite... Powell Duffryn Associated Collieries, Limited,.. London, ..colliery owners. **1951** *Good Housek. Home Encycl.* 32/1 Smokeless fuel, the bulk of which is known as Phurnacite. **1952** *Economist* 1 Nov. 332/2 The favourite small coals of the domestic consumer—anthracite and the best briquettes—remain scarce. The board [*sc.* the National Coal Board] still has only one—now considerably enlarged—plant making 'phurnacite', which is an excellent but expensive carbonised briquette. **1955** *Times* 4 July 5/5 Welsh coals.. and 'Phurnacite', a member of the manufactured fuel group, are still restricted, but coke, 'Coalite', and 'Rexo' are said to be freely available. **1976** *Guardian* 10 Apr. 10/8 The coalman poured the sack of phurnacite into the bunker.

phut. Substitute for entry:

phut (fʌt), *int.* (*adv.*, *sb.*). Also **fut.** [Echoic, but cf. Hindi and Urdu *phaṭnā* to split or burst.] An imitation of a dull, abrupt sound, esp. that of a firearm. Also as *sb.*, the sound of something 'going phut'. Phr. *to go phut*: to come to a sudden end; to break down, cease to function.

1888 KIPLING *Story of Gadsbys* (1889) 55 The whole thing went *phut*. She wrote to say that there had been a mistake. **1892** — & BALESTIER *Naulahka* 259 The hospital has all gone *phut*. **1898** [see *FUT]. **1898**, **1905** [in Dict.]. **1908** A. S. M. HUTCHINSON *Once aboard Lugger* III. iii. 150 The plans.. have all gone fut. **1917** 'CONTACT' *Airman's Outings* 24 He will sometimes hear the rattle of a mysterious machine-gun, or even the phut of a bullet. **1917** W. J. LOCKE *Red Planet* xiii. 156 There's a limit to the power of bearing strain. As soon as you feel you're likely to go *fut*, throw it all up and come and see me. **1918** A. QUILLER-COUCH *Foe-Farrell* 47 'It's a lie!' Foe was on his legs, and he fairly shouted it. Shell-shock? Phut!—It exploded right at our feet below the platform. **1919** G. PAGE *Veldt Trail* i. 10 The carburetter went *fut* yesterday. **1921** *Punch* 30 Nov. 429/2 Send me a subject with a bit more pep in it or the Club will go phut. **1923** *Daily Mail* 22 Jan. 8 He stood to lose some enormous number of millions of marks if Germany went *phut*. **1926** S. HORLER *Order of Octopus* 238 Now that this pet stunt.. had gone phut. **1931** E. F. BENSON *Mapp & Lucia* x. 270 She'd still have been terribly interested in life till she went phut. **1972** *Daily Tel.* (Colour Suppl.) 24 Nov. 7/3 The kids had broken a window, and the colour television had gone phut. **1973** H. CARVIC *Miss Seeton Sings* (1974) 72 The trigger pulled and—*phut*, the enemy dropped dead. *Ibid.* 83 Never before had this place been used for target practice, with things that went *phut*. **1978** D. BLOODWORTH *Crosstalk* xxi. 164 It was a bomb... But when it went off it was only like some kind of firecracker, just, you know—phut!—and that was it.

Hence **phut** *v. intr.*, to land with an abrupt sound; to 'go phut', to cease to function.

1901 [see *FUT *v.*]. **1916** G. FRANKAU *Guns* 26 Waking, they know the instant foe, the bullets phutting by. **1959** J. VERNEY *Friday's Tunnel* xxvii. 250 He and Robin.. got inside last night, only the torch phutted.

phut-phut, *sb.* and *v.* [Reduplication of prec.] = *PUT-PUT *sb.*, *v.*

1951 M. R. ANAND *Seven Summers* 78 The sahibs who came on bicycles or 'phut-phuties'. **1952** R. FINLAYSON *Schooner came to Atia* 144 Phut-phut goes the exhaust. **1956** 'J. WYNDHAM' *Seeds of Time* 39 He listened to the phut-phutting of the old engine. **1958** J. CAREW *Black Midas* vi. 103 The motor sputtered, then settled down to a steady phut-phutting. **1958** *Spectator* 10 Oct. 479/2 People on the outskirts would receive less noise from a VTOL airliner than from a clip-on phut-phut passing along the road outside their houses. *a* **1966** M. ALLINGHAM *Cargo of Eagles* (1968) xi. 129, I dropped over on my phut-phut about half past eight. **1971** *Guardian* 14 May 11/1 Phutt-phutting around London on her three-wheeled scooter. **1977** P. HARCOURT *At High Risk* i. 93 The schoolkids were streaming out of the *lycée* and tearing up the street on their phut-phutting *quarante-neufs*.

phwat, repr. an Ir. pronunc. of WHAT *pron.*, etc.

1898 J. D. BRAYSHAW *Slum Silhouettes* ii. 12 Oh, murther aloive! Phwat did ye do that for? **1914** R. BROOKE *Let.* Nov. (1968) 630 'How many handkerchiefs have you?'..'How many phwat, sorr?' **1920** 'SAPPER' *Bull-Dog Drummond* vi. 154 And phwat the divil has that got to do with it, at all? **1936** M. FRANKLIN *All that Swagger* xvii. 164 But seeing phwat has happened, I can do no other.

phy (fəi), slang abbrev. of *PHYSEPTONE.

1971 *Guardian* 30 Mar. 11/6 There is thus now a completely new group [of drug addicts in Britain] on methadone, or 'phy' (physeptone is the British word for which the World Health Organisation title, methadone, is being substituted). **1973** *Times* 22 Mar. 6/2 She said to him: 'Do you want some phy (Physeptone)?' and made it quite clear that she meant the drug.

phyco-. Add: **phycolo·gical** *a.*, of, pertaining to, or dealing with, phycology; **phycologist** (later examples); **phycology** (earlier and later examples); **phy·coplast** *Cytology* [-PLAST], an array of microtubules found between pairs of nuclei in algal cells after mitosis (see quot. 1972).

1892 (*title*) Phycological memoirs, being researches made in the Botanical Department of the British Museum. Edited by George Murray. **1895** J. D. HOOKER *Let.* 9 June in L. Huxley *Life J. D. Hooker* (1918) II. xli. 294, I have this morning received.. a notice published in the 'Phycological Memoirs' of *Pachytheca*, which is enough to turn your hair grey—if it were not so already! **1954** *Nature* 14 Aug. 294/1 The botanical world, and more especially the phycological section of it, has suffered grievous loss in the passing of Prof. Fritsch. **1973** J. R. STEIN (*title*) Handbook of phycological methods. **1951** G. W. PRESCOTT in G. M. Smith *Man. Phycology* i. 4 The marine phycologist Harvey.. established many genera. **1969** *Brit. Phycological Jrnl.* IV. 141 Virtually the whole of his [*sc.* Hustedt's] career as a phycologist was devoted to diatoms. **1847** J. LINDLEY *Elem. Bot.* (ed. 5) p. lxx/2 Phycology. That part of Botany which treats of Seaweeds. **1935** J. E. TILDEN *Algae* p. v, Phycology offers an enormously stimulating field for research. **1951** G. W. PRESCOTT in G. M. Smith *Man. Phycology* i. 1 The history of phycology is as old as the history of botany. **1976** *Biol. Abstr.* LXI. 6112/1 The history of marine phycology in New England (USA) is summarized, followed by citations emphasizing recent research dealing with the local marine algal vegetation. **1972** J. D. PICKETT-HEAPS in *Cytobios* V. 63 Since this disposition of microtubules is so common in algae, I have coined the term 'phycoplast' to describe it. This implies that it can to some extent be considered an analogue of the higher plant cell phragmoplast, but its main distinguishing feature is the orientation of its constituent microtubules in the plane of cytoplasmic cleavage. **1975** *Nature* 6 Nov. 32/1 The unicellular condition, the possession of basal bodies, flagella, a closed centric spindle, a phycoplast and cell division by furrowing, are deemed to be primitive features.

phycobilin (fəikobəi·lin). *Bot.* [a. G. *phycobilin* (R. Lemberg 1929, in *Naturwissenschaften* XVII. 541/2): see PHYCO- and BILIN.] **a.** Any of a group of compounds that are present in some algæ as prosthetic groups of chromoproteins such as phycocyanin and phycoerythrin. **b.** Also *phycobilin pigment*. Any of these chromoproteins.

1945 E. I. RABINOWITCH *Photosynthesis* I. xv. 417 The separation of the chromophoric groups from the carrier protein was achieved by Lemberg (1929), who introduced the name 'phycobilins' because of the similarity between these chromophores and the bile pigments. **1950** *Jrnl. Gen. Physiol.* XXXIII. 418 Light absorbed by the phycobilin pigments (phycoerythrin and phycocyanin) is utilized with good efficiency. **1962** C. Ó HEOCHA in R. A. Lewin *Physiol. & Biochem. of Algae* xxv. 421 Photosynthetically active red and blue biliproteins, called phycoerythrins and phycocyanins, respectively, have been isolated only from algae. Their prosthetic groups or chromophores are tetrapyrroles known as phycobilins. Unlike the chlorophylls, phycobilins are not readily released from associated proteins. **1965** BELL & COOMBE tr. *Strasburger's Textbk. Bot.* 466 Rhodophyceae, red algae... Chlorophyll and the associated carotinoids.. are masked by a red, strongly fluorescent, water-soluble pigment, phycoerythrin (a phycobilin.. with absorption-bands differing from those in the Cyanophyceae). **1971** *Nature* 26 Nov. 232/1 One of the first obvious effects of nitrogen starvation in blue-green algae is the disappearance of phycobilin pigments, which normally constitute about 15% of the dry weight.

phycobiliprotein (fəi:kobəiliprǒu·tin). *Bot.* [f. prec. + PROTEIN.] Any chromoprotein having a phycobilin as the prosthetic group; a phycocyanin or a phycoerythrin.

1966 *Brookhaven Symp. Biol.* XIX. 402 The blue-green algae have predominantly phycocyanin as the major phycobiliprotein. **1975** *Nature* 24 Jan. 285/2 The chlorophylls, carotenoids and phycobiliproteins of the blue-green algae are particularly similar to those of the red algae.

phycobilisome (fəikobəi·lisǒum). *Bot.* [f. as prec. + *-SOME[4].] In certain algæ, a photosynthetic granule containing phycobiliprotein.

1966 GANTT & CONTI in *Brookhaven Symp. Biol.* XIX. 404 The following are the reasons for considering these chloroplast granules of *P[orphyridium] cruentum* as sites of phycobilin aggregation... For these granules we propose the name phycobilisomes. **1971** *Jrnl. Cell Biol.* XLVIII. 285/2 Whereas it is true that phycobilisomes have not yet been found in every species of red and blue-green algae, they are believed to exist in all photosynthetic species of these two groups. **1976** *Nature* 24 June 697/2 Those [algal cells] from *Diplosoma virens*.. show no evidence of phycobilisomes on the thylakoids.

phycobiont (fəikobəi·ǫnt). *Bot.* [f. PHYCO- + Gr. βιουντ-, pr. pple. stem of βιοῦν to live, f. βίος life.] The algal component of a lichen; any alga which is associated with a fungus to form a lichen.

1957 G. D. SCOTT in *Nature* 2 Mar. 486/2 Three new terms are here proposed... They are: (1) 'phycobiont', applicable to an alga in association with a fungus in the formation of a lichen; [etc.]. **1962** *New Scientist* 28 June 719/1 Efforts have previously been made to investigate the exact nature of the relationship between the algal partner (sometimes called the phycobiont) and the fungus. **1969**, **1973** [see *MYCOBIONT]. **1976** G. C. AINSWORTH *Introd. Hist. Mycol.* iv. 98 The phycobiont frequently shows no apparent differences from its free-living counterpart.

phycomycete (fəi·komǝisīt). *Bot.* [sing. of mod.L. *Phycomycetes* (A. de Bary *Morphol. & Physiol. der Pilze* (1866) p. vi), f. PHYCO- + MYCETES.] A fungus belonging to one of the primitive groups formerly included in the class Phycomycetes, nearly always characterized by a vegetative thallus without septa and either asexual reproduction by means of sporangiospores or conidia or sexual reproduction by means of oospores or zygospores. Also *attrib.* Cf. *phycomycetous* adj. s.v. PHYCO-.

[**1887** H. E. F. GARNSEY tr. *A. de Bary's Compar. Morphol. & Biol. Fungi* iv. 132 Groups 1–4 [of the Ascomycete series] have been brought together under the name of Phycomycetes on account of their close approximation to the Algae.] **1932** *Bot. Gaz.* XCIII. 427 The attachment of the flagellum or flagella of the phycomycete zoospore is usually posterior or lateral. **1933** *Trans. Brit. Mycol. Soc.* XVIII. 199 (*heading*) *Azygozygum chlamydosporum* Nov. gen. et sp. A phycomycete associated with a diseased condition of *Antirrhinum majus*. *Ibid.* 201 The Phycomycete was isolated from several varieties of *Antirrhinum*. **1936** *Forestry* X. 14 All three species showed mycorrhizal infection of the well-known 'phycomycete type'. **1976** G. C. AINSWORTH *Introd. Hist. Mycol.* ix. 236 The most important [aquatic fungi] taxonomically were the phycomycetes.

phycomycosis (fəi:komǝikǒu·sis). *Path.* Pl. -mycoses. [f. *PHYCOMYC(ETE + -OSIS.] Infection with or a disease caused by phycomycetes, esp. the genera *Mucor*, *Rhizopus*, or *Absidia*; mucormycosis.

1959 LIE-KIAN-JOE et al. in *Amer. Jrnl. Clin. Path.* XXXII. 62/1 We use the name phycomycosis to designate a fungous infection caused by a member of the Phycomycetes. It is proposed in order to avoid the more restricted connotations of mucormycosis, and includes infections caused by Mucor, Absidia, Rhizopus, Mortierella, Basidiobolus, and similar Phycomycetes. The name is useful, too, for mycoses in which no culture was obtained but in which sections of involved tissues reveal the presence of a fungus with the morphology usually associated with a Phycomycete. **1965** [see *MUCORMYCOSIS]. **1972** *Radiology* CIII. 332/2 Craniofacial phycomycosis appears to originate by invasion of the nasal mucosa.

phylactology (fəi:læktǫ·lǒdʒi). *nonce-wd.* [f. Gr. φυλακτ-ός or -ήριος vbl. adjs. f. φυλάσσειν to guard + -OLOGY.] The science or business of counter-espionage. Hence **phylactolo·gical** *a.*; **phylacto·logist.**

1966 K. AMIS *Anti-Death League* I. 18 Apparently what's called the philosophy of phylactology—spy-catching to you—has been transformed. *Ibid.* 108, I had no idea you were practised in phylactological thought. *Ibid.* III. 269 This mere technician, this electrical eavesdropper, seemed to imagine he was on a level with a qualified phylactologist like himself.

† phylaxis (filæ·ksis). *Path.* *Obs.* [ad. Gr. φύλαξις watching, guarding; cf. *ANAPHYLAXIS.] The protection of a cell or organism against the effects of a toxin, esp. a neurotoxin, by the action of an artificially introduced substance which prevents its uptake by cells. Hence **phyla·ctic** *a.*

1913 [see *ANAPHYLAXIS]. **1919** A. E. WRIGHT in *Lancet* 29 Mar. 490/1 To combat bacterial infection the organism must have defensive powers. That power of guarding itself against infection we may—the suggestion is Lord Moulton's—call phylactic power. The leucocytes and the bacteriotropic substances in the blood fluid we may call phylactic agents. **1931** H. GAINSBOROUGH tr. *Billard's Phylaxis* i. 11 After a discussion with G. Ramon and in agreement with him.. I adopted a new term—phylaxis. *Ibid.* iii. 29 (*heading*) Phylaxis by certain mineral waters against certain neurotoxins. *Ibid.* 40 As regards the phylactic action of sparteine I have insisted at some length on the particular affinity of the neurotoxins for the lipoids. **1931** *Times Lit. Suppl.* 3 Dec. 985/4 Phylaxis is a conception mainly due to Professor Billard's experiments.

phyletic, *a.* Insert at beginning of etym.: [a. G. *phyletisch* (E. Haeckel *Generelle Morphologie der Organismen* (1866) II. xx. 299).] Substitute for def.: Of or pertaining to the development of a species or other taxonomic group. (Later examples.)

1928 *Brit. Jrnl. Med. Psychol.* VIII. 59 It is my essential position that what we call mental is itself physiological when viewed in a social or phyletic as well as in an individual or ontogenetic sense. **1933** *Proc. Linn. Soc.* CXLV. 155 He does not claim that his analysis necessarily represents the true phyletic position of the various species. **1943** [see *group-analysis* s.v. *GROUP *sb.* 6 b]. **1944** G. G. SIMPSON *Tempo & Mode in Evolution* vii. 205 Phyletic evolution goes on continuously ..in populations of all sizes and kinds. **1951** G. H. M. LAWRENCE *Taxon. Vascular Plants* ii. 96 Some phylogenists have given too little attention to the phyletic significance of pollen grain and starch grain morphology. **1963** DAVIS & HEYWOOD *Princ. Angiosperm Taxon.* ii. 44 Both *phyletic* and *phylogenetic* refer to the course of evolution and are virtually interchangeable. **1969** E. MAYR *Princ. Systematic Zool.* x. 251 Species living at the present time are the current end points of innumerable phyletic lines. **1972** *Sci. Amer.* Jan. 98/3 Individuals representing nothing more than successive stages of a single phyletic line have in a number of cases been assigned distinctive taxonomic names, as if they had actually been contemporaries and neighbors. **1973** A. J. POMERANS tr. *Piaget & Inhelder's Memory & Intelligence* 2 Our first problem, then, is to distinguish between the acquired memory (or the memory proper) and the phyletic memory or conservation and utilization of hereditary information.

phyletically, *adv.* Add: Also, regarding a common evolutionary descent. (Further examples.)

1930 *Psyche* XI. ii. 78 After all, the whole field of man's unconscious life.. must be regarded phyletically as man's inattention in the immediate moment. **1933** *Proc. Linn. Soc.* CXLV. 162 A and B may be the nearest allies phyletically. **1950** [see *intention movement*]. **1951** G. H. M. LAWRENCE *Taxon. Vascular Plants* v. 99 The Magnoliales are phyletically old. **1965** ZUCKERKANDL & PAULING in Bryson & Vogel *Evolving Genes & Proteins* iii. 164 Any variants within a given type of tertiary structure and function seem to have a much greater chance to be phyletically related than unrelated. **1976** *Nature* 24 June 694/1 Many species will evolve phyletically with no change in size.

phyletism (fəi·lètiz'm, fi·-). Also erron. **philetism** (perh. infl. by PHILO-). [f. Gr. φυλέτ-ης fellow tribesman, f. φυλή tribe: see -ISM.] In the Orthodox Church, an excessive emphasis on the principle of nationalism in the organization of church affairs; a policy which attaches greater importance to ethnic identity than to bonds of faith and worship.

The term is applied chiefly in hist. use to the claim of the Bulgarian Church to jurisdiction over Bulgarian nationals in all parts of the world, leading in 1872 to the condemnation passed by the Synod of the Ecumenical Patriarchate in Constantinople and to a schism between the Greek-speaking Orthodox churches and the Bulgarian Church which lasted until 1945.

1900 'ODYSSEUS' *Turkey in Europe* 285 The Patriarchate.. stigmatised by the name of Phyletism the doctrine that persons of a particular race.. are entitled to a separate ecclesiastical administration. **1907** A. FORTESCUE *Orthodox Eastern Church* IV. x. 278 He finds Philetism to be a deadly heresy. Poor Patriarch!.. Shall he denounce Philetism, stand out for the old rights of the hierarchy and of the chief sees, preach unity and ancient councils? **1923** R. L. LANGFORD-JAMES *Dict. Eastern Orthodox Church* 99/1 Phyletism is the error of the undue exaltation of the necessary independence of National Churches, so that Nationalism in religion overshadows Catholicism. **1927** B. J. KIDD *Churches of Eastern Christendom* xii. 301 This last phase of shrinkage and retreat, ending.. in the emancipation of the Balkan nations, had encouraged Philetism and diminished the range of the Patriarch's authority by the creation of National Churches. **1957** tr. *V. Lossky's Mystical Theol. Eastern Church* i. 15 The view which would base the unity of a local church on a political, racial or cultural principle is considered by the Orthodox Church as a heresy, specially known by the name of *philetism*. **1961** D. ATTWATER *Christian Churches of East* II. 248 *Phyletism*.., the name by which the patriarch of Constantinople condemned a form of nationalism in the Orthodox Church in 1872. **1974** *Encycl. Brit. Macropædia* VI. 159/1 The council [of 1872] condemned 'phyletism'—the national or ethnic principle in church organization—and excommunicated the Bulgarians, who were certainly not alone guilty of 'phyletism'.

phylic (fəi·lik), *a.*[1] [f. Gr. φυλ-ή tribe + -IC.] Of or pertaining to a Greek phyle or tribe.

1891 *Jrnl. Hellenic Stud.* XII. 30 The increase of the *Phylae* involved changes in the institutions based upon the phylic system. **1908** J. L. MYRES in R. R. Marett *Anthropol. & Classics* 142 The solidarity of the Greek phylic institutions.

phylic (fəi·lik), *a.*[2] *Psychol.* [f. PHYL(UM + -IC.] Of or relating to the phylum (sense *b). Also *absol.*

1949 T. BURROW *Neurosis of Man* vii. 169 As an outgrowth of this phylic principle of integration, the individual remains always an integral element within the organism of man as a unit. **1952** W. J. H. SPROTT *Social Psychol.* 246 The fault lies mainly in our individualism, or, rather, in our isolationism. We have over-developed the 'private' and lost touch with the 'phylic'. **1953** T. BURROW *Sci. & Man's Behavior* vi. 74 The term 'species'.. has no adjectival form. In order to fill this breach, I adopted 'phylic', from *phylum*, which in its original Greek had the meaning of 'tribe'.

phyllo-. Add: **phy·llomorph** [Gr. μορφή form], the representation of a plant in art; **phy·llosphere**, the surface area of the leaves, or, more generally, of all the parts of a plant above ground.

1889 *Trans. Lancs. & Cheshire Antiquarian Soc.* VII. 166 The forms of ornament demonstrably due to structure require a name... Those taken from animals are called zoomorphs, and those from plants phyllomorphs. **1895** A. C. HADDON *Evol. Art* 126 The terms 'zoomorph' and 'phyllomorph' have been employed for the representation in art of plants and animals. **1955** F. T. LAST in *Trans. Brit. Mycol. Soc.* XXXVIII. 221 It is suggested that, as with roots and the 'rhizosphere', leaves have a 'phyllosphere', with a characteristic micro-flora that may contain many species. **1956** J. RUINEN in *Nature* 4 Feb. 221/2 These observations suggested the existence of a characteristic milieu which is conditioned by the leaf, and may be called, in analogy with the rhizosphere, the 'phyllosphere'. **1972** J. G. CRUICKSHANK *Soil Geogr.* vi. 175 On the surface of living aerial parts of plants, particularly on the leaves—the phyllosphere.

phyllocarid (filokæ·rid). *Zool.* [a. mod.L. name of division *Phyllocarida* (A. S. Packard 1879, in *Amer. Naturalist* XIII. 128), f. PHYLLO- + Gr. καρίς shrimp, prawn.] A crustacean belonging to a group of the subclass Branchiopoda, which includes those types distinguished by the broad, flat limbs known as phyllopodia. Also *attrib.* Hence **phylloca·ridan** *a.*

1882 A. S. PACKARD in *Amer. Naturalist* XVI. 870 He [*sc.* Claus] failed to appreciate the independent, synthetic nature of the Phyllocaridan type. **1896** *Proc. Geol. Soc.* LII. p. xcvi, We have here.. a passage upward from the more ancient Phyllocarid type.. to the more modern Cumacea. **1902** *54th Ann. Rep. N.Y. State Mus. Nat. Hist. 1900* I. 98 The museum record of localities now bears the following entries at which species of these phyllocarids have been obtained. **1911** *Geol. Mag.* VIII. 64 In the black shales we succeeded in finding organic remains, including.. a bivalve phyllocarid crustacean. *Ibid.*, A few specimens of a bivalve phyllocarid allied to *Caryocaris* and *Lingulocaris*. **1935** TWENHOFEL & SHROCK *Inverteb. Paleontol.* x. 444 Nebaliacea—A heterogeneous group of living and fossil phyllocarids. **1973** P. TASCH *Paleobiol. of Invertebr.* xi. 600/2 There have been several finds of fossil phyllocarids with appendages preserved.

phyllopodium (filopǒu·diǝ̆m). [f. PHYLLO- + PODIUM.] **1.** *Bot.* The base of a leaf stalk, or the main axis of a leaf.

1884 F. O. BOWER in *Phil. Trans. R. Soc.* CLXXV. 569, I therefore propose the term phyllopodium to express the whole of the main axis of the leaf, exclusive of its branches. **1923** —— *Ferns* I. v. 82 The branches are then arranged in two longitudinal rows, one on either side of a central stalk or rachis, which is continuous below into the stipe or petiole, often of considerable length. The whole of this, including petiole and rachis, may be styled the Phyllopodium. **1951** MCLEAN & IVIMEY-COOK *Textbk. Theoret. Bot.* I. xvi. 653 Delpino.. described the stem as a pseudaxis or phyllopodium. *Ibid.* xxii. 978 The mature leaf consists of three portions, the lamina, the petiole, and the leaf base or phyllopodium. *Ibid.* 979 (*caption*) Development of the phyllopodium as a leaf sheath.

2. *Zool.* (See quot. 1967.)

1926 L. A. BORRADAILE in *Ann. & Mag. Nat. Hist.* XVII. 194, I have.. argued in support of the view that all the post-antennulary limbs of the primitive crustacean were alike, and that they were flat structures with endites and exites. It is convenient to call such appendages 'phyllopod limbs' or phyllopodia, whether they occur in the Phyllopoda (Branchiopoda) or elsewhere. *Ibid.*, The phyllopodium has all the essentials of a biramous limb. **1967** P. A. MEGLITSCH *Invertebr. Zool.* xviii. 755/1 An unusual feature of the branchiopods is the soft, flattened type of trunk appendage characteristic of most of the groups. Covered with a delicate cuticle, they are flexible enough to move freely without segmentation. This type of appendage is known as [a] phyllopodium.

phylloquinone (filokwi·nǒun). *Biochem.* [ad. G. *phyllochinon* (Karrer & Geiger, at the suggestion of H. Dam, in *Helv. Chim. Acta* (1939) XXII. 946), f. Gr. φύλλον leaf + G. *chinon* quinone (cf. CHINA[3] 2).] Vitamin K_1, a yellow, fat-soluble oil that is present in green leafy vegetables and is important in blood clotting; 2-methyl-3-phytyl-1,4-naphthoquinone, $C_{31}H_{46}O_2$.

1939 *Chem. Abstr.* XXXIII. 8191 The light yellow vitamin K_1 from alfalfa, named by Dam.. 'α-phylloquinone', (I) crystallizes at low temp. **1953** FRUTON & SIMMONDS *Gen. Biochem.* xxxviii. 908 Vitamin K_2 (2-methyl-3-difarnesyl-1,4-naphthoquinone), which is formed by some bacteria.., differs from phylloquinone only in the substituent in the 3 position of the naphthoquinone ring. **1970** R. W. MCGILVERY *Biochem.* xxvi. 651 (*caption*) Phylloquinones from plants are absorbed to some extent as such, and they have vitamin K activity. However, the side chain is removed from most of the ingested compounds by intestinal bacteria; the resultant menadione is absorbed and a new side chain is constructed to create menaquinone, the principal form of vitamin K found in animals. **1972** *Materials & Technol.* V. xix. 692 Menadione, phylloquinone and the menaquinones can all be obtained synthetically, but only menadione is produced in any quantity.

phyllosilicate (filosi·likĕ[i]t). *Min.* [a. G. *phyllosilikat* (H. Strunz 1938, in *Zeitschr. für ges. Naturwiss.* IV. 185), f. as prec.: see SILICATE.] Any of the group of silicates characterized by SiO_4 tetrahedra linked in sheets of indefinite extent in which the ratio of silicon and aluminium to oxygen is 2:5.

1947 *Mineral. Abstr.* X. 52 The different types of SiO_4 bonding give at last a systematic classification of silicates into nesosilicates.., sorosilicates.., inosilicates .., phyllosilicates.., and tectosilicates. **1966** *McGraw-Hill Encycl. Sci. & Technol.* XII. 312/1 The phyllosilicates as a group typically have a platy crystal habit, with a cleavage parallel to the plane of layering of the structure, and are optically negative with rather high birefringence. **1972** *Nature* 1 Sept. 7/1 Phyllosilicates are the most common material in carbonaceous meteorites.

phylo-. Add: **phyloana·lysis** *Psychol.*, analysis of an individual that takes account of him as part of a phylum (sense *b); hence **phyloa·nalyst**; **phyloanaly·tic** *a.*; **phylogero·ntism**, phylogerontic character or condition.

1930 T. BURROW in *Psyche* XI. ii. 67 (*heading*) Physiological behavior-reactions in the individual and the community: a study in phyloanalysis. **1940** HINSIE & SHATZKY *Psychiatric Dict.* 418/2 Phyloanalysis regards the symptoms of the individual and of society as but outer aspects of impaired tensional processes which affect the balance of the organism's internal reaction as a whole. **1949** T. BURROW *Neurosis of Man* vii. 168 Phyloanalysis set out with the attempt to analyse the affecto-symbolic 'I'-persona and its special prerogative as commonly assumed by social man; and it undertook to differentiate this pseudo-entity from the organism's biological basis of integration and behaviour. **1933** W. GALT *Phyloanalysis* ii. i. 113 The phyloanalyst does not at all credit the obvious manifestation. **1932** T. BURROW *Struct. Insanity* vi. 66 The aim, therefore, of the phyloanalytic method is the application of a technique that will enable the patient to acquire a facility for rendering his own physiological tensions objectively perceptible to him. **1902** *Amer. Naturalist* XXXVI. 940 In the majority of specialized gastropods phylogerontism is expressed, not in the non-coiling of the last portion of the spire, but in its expansion and wrapping about the earlier whorls.

phylogenesis. Add to etym.: [coined in Ger. (E. Haeckel *Generelle Morphologie der Organismen* (1866) II. xx. 299).] (Later examples.) Also *transf.*

1926 W. MCDOUGALL *Introd. Social Psychol.* (ed. 20) 402 The psychologist may legitimately speculate on the problems of phylogenesis; but he is under no obligation to offer any phylogenetic theory. **1940** HINSIE & SHATZKY *Psychiatric Dict.* 418/2 The biological concept of phylogenesis has been used in different ways for analytical analogies in the psychological sphere, in order to emphasize the role of racial elements in the manifestations of the psyche. **1957** *Times Lit. Suppl.* 1 Nov. 659/1 He [*sc.* Jung] claimed that the doctrine which asserts that ontogenesis is a repetition of phylogenesis was also true of psychic life. **1966** DAVIS & ZANGERL tr. *Hennig's Phylogenetic Systematics* iii. 199 Phylogenesis is the origin of groups of species from a stem species and its descendants by progressive splitting. **1977** P. JOHNSON *Enemies of Society* xvii. 226 Since the phylogenesis of music is communication, it must have a structure.

phylogeneticist (fəi:lo,dʒène·tisist). [f. PHYLOGENETIC *a.* + -IST.] = PHYLOGENIST.

1968 R. D. MARTIN tr. *Wickler's Mimicry in Plants & Animals* iii. 39 The famous phylogeneticist George Gaylord Simpson called for a detailed investigation.

phylogenetics (fəi:lo,dʒène·tiks), *sb. pl.* [f. PHYLOGENETIC *a.*: see -IC 2.] The study of phylogeny, esp. the factors influencing its course. (Usu. const. as *sing.*)

1937 K. FAEGRI in *Bot. Rev.* III. 400 (*heading*) Some fundamental problems of taxonomy and phylogenetics. *Ibid.* 422 Phylogenetics: the science of the tribal history and descent (phylogeny) of plants and animals. **1966** DAVIS & ZANGERL tr. *Hennig's Phylogenetic Systematics* iii. 201 Determining the significance of individual factors in the total evolutionary process is the unique task of phylogenetics. **1975** *Nature* 18 Dec. 641/2 Palaeogenetics ..as well as protein phylogenetics.. assume that comparative analyses of amino acid sequence data.. yield a realistic image of evolution.

phyloge·nically, *adv.* [f. PHYLOGENIC *a.*: see -ICALLY.] = PHYLOGENETICALLY *adv.*

1975 *Nature* 5 June 483/1 So far the only studies on the teratogenic effects of this virus have been in the chick embryo and rodents, animals phylogenically distant from *Homo sapiens*.

phylogeny. Add coinage details to etym.: [(E. Haeckel *Generelle Morphologie der Organismen* (1866) I. iii. 57).] **1.** Also, the evolutionary development of particular organs or other components of a plant or an animal. (Further examples.)

1897 *Bot. Gaz.* XXIV. 172 We are warranted in strenuously urging a conformity of taxonomy with phylogeny. **1901** *Trans. Linn. Soc.* (*Zool.*) VIII. 270 The Plates attached to this paper represent with approximate accuracy the phylogeny of the intestinal tract in birds. **1940** J. S. HUXLEY *New Systematics* 19 Phylogeny may be almost hopelessly obscured by parallel or convergent evolution. **1953** E. MAYR et al. *Methods & Princ. Systematic Zool.* iii. 42 It is the avowed aim of a modern classification to reflect phylogeny. **1971** J. Z. YOUNG *Introd. Study Man* xxxv. 486 We cannot measure the development of the frontal lobes in phylogeny accurately.

2. (Earlier example.)

1872 [see *BATHMIC *a.*].

phylum. Add: Also *transf.* and *fig.*

1945 AUDEN *Coll. Poetry* 162 Whole phyla of resentments every day Give status to the wild men of the world Who rule the absent-minded. **1971** *New Scientist* 25 Mar. 682/1 We have to regard science as a 'phylum' (as the term is used by Pierre Teilhard de Chardin)—that is, as an expanding movement within the four-dimensional space-time continuum of the social system. **1973** *Sci. Amer.* Apr. 121/1 Computers are no longer individuals with names but a phylum of many species, rapidly evolving under selection pressures.

b. *Psychol.* The human race or group as it is relevant to the development of the individual.

1927 T. BURROW in *Brit. Jrnl. Med. Psychol.* VII. 199 The social group or phylum. *Ibid.* 202 In a comprehensive view of our human phylum there remains no other conclusion than that the social mind.. comprises a systematization of social images. **1930** —— in *Psyche* XI. ii. 69 Medicine, then, became a science when the symptoms of the individual ceased to be the focus of interest and when interest became focussed instead upon the pathological germ or cause of definite alterations of tissue and their characteristic symptoms, as these.. were observed within the organism of man.. as a species or phylum. **1940** H. G. BAYNES *Mythol. of Soul* I. xii. 460 The concept of individuality as a self-contained, self-regulating organism has no validity unless it also embraces this backward extension of the ancestral phylum. **1953** T. BURROW *Sci. & Man's Behavior* vi. 73 Some of my colleagues commented upon the unusual sense in which I use the term 'phylum'... By this term I do not mean to separate man from the rest of the vertebrates. I am merely trying to discuss man, with his social needs and interests, as a biological organism.

2. *Linguistics.* A group of languages related, or believed to be related, less closely than those of a family or stock.

1958 H. HOIJER in R. H. Thompson *Migrations in New World Culture Hist.* 59 There is no indication that the families of a phylum, or the phyla of a macro-phylum, need be connected by clearly statable phonetic correspondences. **1965** *Canad. Jrnl. Linguistics* X. 142 The reconstruction (by comparative method linguistics) is systematic... The reconstruction (in phylum linguistics) is illustrative and restricted to a relatively small set of cognates and typological samenesses which point to an earlier phylum parent language in the prehistoric era. **1973** A. P. SORENSEN in D. R. Gross *Peoples & Cultures of Native S. Amer.* 333 Two of them represent the comparative method linguistics approach..; the other two represent the phylum linguistics approach. **1973** *Language* XLIX. 239 Tarascan, like Zuni, is a one-language phylum.

-phyre (fəi[ə]ɹ), comb. form of PORPHYRY used in names of porphyritic rocks, as GRANOPHYRE, *KERATOPHYRE.

physaliferous (fəisăli·fĕrǝs), *a.* *Biol.* [f. Gr. φυσαλλ-ίς bladder + -IFEROUS.] = *PHYSALIPHOROUS *a.*

1954 G. L. FITE tr. R. Virchow in *Amer. Jrnl. Leprosy* XXII. 209 They [*sc.* the cells] have one characteristic that is especially noteworthy, i.e., their tendency to form a sort of vacuole, apparently from taking up water, so that under the circumstances they acquire a wholly physaliferous appearance. **1977** *Proc. R. Soc. Med.* LXX. 276/2 A large number of the cells had a vacuolated cytoplasm with hyperchromatic nuclei, the pattern being that of 'physaliferous' cells characteristic of a chordoma.

physaliphore. Add: (Earlier example.)

1860 F. CHANCE tr. *R. Virchow's Cellular Path.* xviii. 401 (*caption*) Endogenous new formation; cells containing vesicles (physaliphores).

Hence **physali·phorous** *a.*, that contains vesicles; cf. *PHYSALIFEROUS *a.*

1964 D. F. CAPPELL *Muir's Textbk. Path.* (ed. 8) xvii. 701/2 The tumour tissue [of a chordoma] consists of lobules of rounded or polyhedral cells arranged in alveoli or cords: some of these cells contain numerous intracytoplasmic vacuoles of mucin, 'physaliphorous cells'.

physalis (fəi·sălis, fəisē[i]·lis). [mod.L. (Linnæus *Hortus Cliffortianus* (1738) 62), f. Gr. φυσαλλίς bladder, in reference to the inflated calyx.] An annual or perennial herb of the genus so called, belonging to the family Solanaceæ, mostly native to North or Central America, and bearing white, yellow, or purple flowers and, in some species, edible red or purplish berries; cf. ALKEKENGI, *Cape gooseberry* s.v. CAPE *sb.*[3] 4 in Dict. and Suppl., *Chinese lantern* (*plant*) s.v. *CHINESE *a.* 2, *ground-cherry* b s.v. GROUND *sb.* 18c, WINTER CHERRY 1 a.

1807 *Curtis's Bot. Mag.* XXVII. 1068 (*heading*) Eatable Physalis or Cape Gooseberry. **1907** T. W. SANDERS *Flower Garden* 196 The Physalises belong to the Nightshade order. **1930** *Times Educ. Suppl.* 18 Oct. 431/2 The windows of the florists are bright just now with the fruits of physalis, the winter cherry. **1961** *Amat. Gardening* 30 Sept. 7/2 'Lanterns' of physalis.. can be skeletonised. **1968** S. C. EMBERTON *Garden Foliage* ix. 219 Mustard yellow achillea and orange 'lanterns' of physalis teamed up with bulrushes.

Phys. Ed., phys. ed., colloq. abbrev. of *physical education*. Also *attrib.*

1955 *Amer. Speech* XXX. 304 *Phys ed major*, mannish female. **1957** O. NASH *You can't get there from Here* 84 Get them interested in hotel management and phys. ed. **1968** A. HOLDEN *Death after School* iv. 31 There's the Phys Ed man. **1972** P. MARKS *Collector's Choice* ii. 82 A converted basement of the Phys Ed building. **1977** *Time* 9 May 50/3, I was very poor on the parallel bars, and my phys.-ed class came at the damn wrong hour.

Physeptone (fəise·ptǒun). *Pharm.* Also **physeptone.** A proprietary name for methadone hydrochloride. Cf. *PHY.

1947 *Lancet* 6 Sept. 370/2 (*heading*) Miadone (Physeptone) in obstetrics. **1948** *Trade Marks Jrnl.* 15 Sept. 743/1 Physeptone... All goods included in class 5 [*sc.* pharmaceutical, veterinary and sanitary substances, etc.]. The Wellcome Foundation Limited,.. London,.. manufacturing chemists. **1951** *Pharmaceutical Jrnl.* CXII. 52/1 An inquest was held at Coventry on a three-year-old boy who died from an overdose of Physeptone linctus. **1969** *Daily Tel.* 3 Feb. 22/6 Physeptone, a narcotic drug first used in America for the gradual withdrawal of drug addicts from heroin, is proving successful in 'weaning' British addicts off hard drugs. 1971 [see *phy]. 1972 J. Brown *Chancer* ii. 30 She was on 5 and 3 at the time—five one-sixth grains of heroin and three ten-milligram ampoules of physeptone. 1976 H. Ferguson *Confessions Long Distance Acid Head* 7 Apart from cannabis, I have used barbiturates, amphetamines,..the synthetic opiate physeptone, morphine, cocaine,..even apomorphine once.

physic, *sb.* Add: **6. physic-box** (later Austral. example).
1900 H. Lawson *On Track* 55 An' if yer don't get yer physic-box an' come wi' me, by the great God I'll—.

physical, *a.* Add: **I. 1.** (Earlier example of *physical cause.*)
1605 Bacon *Adv. Learning* II. 29 For the handling of finall causes mixed with the rest in Phisicall enquiries, hath intercepted the seuere and diligent enquirie of all reall and phisicall causes.
3. c. *as sb.* A medical examination to determine physical fitness.
1934 N. Sainsbury *Gridiron Grit* in *Stirring Football Stories* (1941) 77 He found that everybody had had the same idea about physicals and that there were at least forty candidates..ahead of him. 1944 H. McCloy *Panic* 131 If you imagine he didn't try to get in the Army, you're mistaken... I saw him each time he got back from his physical. 1968 *Guardian* 28 Dec. 1/5 A day of medical tests—which will go far beyond the 'complete physical' that an astronaut receives. 1973 W. McCarthy *Detail* i. 66 Doctor Miller will see you..for a short physical. 1978 N. Freeling *Night Lords* xix. 89 No cop got sent to investigate his own apartment block... It was..like a doctor being asked to give his own wife a physical.
d. Characterized by or suggestive of bodily (as distinct from mental or psychological) activities or attributes. Of a person or activity: inclined to be bodily aggressive or violent.
1970 J. G. Vermandel *Dine with Devil* xii. 77 He's obviously one of these tremendously magnetic types. And you Scorpios are so physical, aren't you? 1973 J. Mosedale *Football* vii. 100 Facing the very physical Philadelphia team. 1974 'J. Le Carré' *Tinker, Tailor* viii. 59, I was feeling pretty physical. Frustrated, you could almost say. 1975 J. Mitchell *Smear Job* xii. 84 It's up to you. Either you belt up or I'll get physical. 1976 *Houston Advertiser* 10 Dec. 16/3 Morton became involved and there was 'some physical horseplay'. 1978 *Rugby World* Apr. 8/3 Rugby is a physical game.
II. 7. physical anthropology, the study of the evolution of man and animals closely related to him, involving the observation or measurement of anatomical features, growth rates, genetic mechanisms, etc.; so **physical-anthropological** *a.*; **physical anthropologist; physical astronomy** (earlier example); **physical chemistry,** substitute for def.: the application of the techniques and theories of physics to the study of chemical systems; the study of the interrelation of chemical and physical properties; (earlier and later examples); so *physical-chemical* adj. (= physico-chemical), *physical chemist;* **physical culture,** the development of the body by exercise; hence **physical culturist,** an advocate or exponent of physical culture; **physical drill,** physical exercises; **physical education,** regular instruction in bodily exercise and games, esp. in schools; **physical geography** (earlier example); **physical jerks:** see *jerk *sb.*¹ 2 d; **physical metallurgy,** the science dealing with the structure and physical properties of metals; **physical object** *Philos.*, an object that exists in space and time and that can be perceived; also *attrib.*; **physical therapy** = *physiotherapy; hence **physical therapist,** a physiotherapist; **physical torture** *slang*, physical training; **physical training,** the systematic use of exercises to promote physical fitness.
1958 G. Daniel *Megalith Builders of W. Europe* i. 25 The word *race* is used..in its strict physical-anthropological connotation of a group of people with heritable physical characteristics in common. 1878 *Jrnl. Anthrop. Inst.* VII. 340 The physical anthropologist often regards the prehistoric worker as an antiquarian only, whilst the former is sometimes simply looked upon in the light of a demonstrative anatomist. 1936 R. Linton *Study of Man* ii. 23 The first physical anthropologists..believed that every species and variety was the result of a separate act of creation and was therefore fixed and unchangeable. 1974 *Times* 8 Nov. 12/4 The leprous bones were identified by..a physical anthropologist (an expert with skeletons). 1873 A. L. Fox in *Jrnl. Anthrop. Inst.* II. 361 The two expeditions which have lately left this country for Central Africa..have been furnished with detailed notes and queries on general anthropology, physical anthropology, religions, myths, customs, [etc.]. 1904 W. L. H. Duckworth *Morphol. & Anthropol.* ii. 14 This application of the principles of Morphology to the special case of Man constitutes the essence of Physical Anthropology. 1923 A. L. Kroeber *Anthropol.* i. 5 Biological or physical anthropology—'Somatology' it is sometimes called in Anglo-Saxon countries, and simply 'anthropology' in continental Europe—has in part constituted a sort of specialization or sharpening of general biology. 1951 E. E. Evans-Pritchard *Social Anthropol.* i. 3 Physical anthropology..is a branch of human biology and comprises such interests as heredity, nutrition, sex differences, the comparative anatomy and physiology of races, and the theory of human evolution. 1976 *Ann. Rev. Anthropol.* V. 5 Hrdlicka was the founding father of American Physical Anthropology. 1834 *Penny Cycl.* II. 529/2 The third department of astronomy..is that which goes under the name of *physical astronomy*, and consists in the combination of the various phenomena as actually observed, in order to find out what are their physical causes, and according to what laws those causes act. 1964 *Economist* 29 Feb. 790/1 The Humboldt university, whose physical-chemical institute he is the director. 1967 *Oceanogr. & Marine Biol.* V. 63 It shows itself essentially in the form of a dome structure..which can be detected in all the physical-chemical properties of the water. 1971 *Who's Who* 331/1 Bowen, Edmund John... Papers on physical chemical subjects in scientific journals. 1928 *Amer. Jrnl. Physiol.* LXXXIX. 171 Physical chemists..have criticized methods which make use of tubes of small diameter. 1935 *Discovery* Apr. 98/2 A problem for the physical chemist, assisted by a bio-chemist. 1893 I. Remsen *Princ. Theoret. Chem.* (ed. 4) p. v, I have been tempted to change the book fundamentally and give it a character more in keeping with the recent tendencies of work in the field of Physical or General Chemistry. 1898 C. L. Speyers *Text-bk. Physical Chem.* i. 1 Physical Chemistry is the science which has for its object the investigation of chemical changes by physical methods. 1940 Glasstone *Text-bk. Physical Chem.* p. xi, Obviously atomic combination involves atomic forces, and it is one of the objects of physical chemistry to see how far the chemical interactions observed between atoms and molecules can be interpreted by means of the forces existing within and between atoms. 1966 *McGraw-Hill Encycl. Sci. & Technol.* X. 203/1 There are three different approaches to the study and use of physical chemistry: thermodynamics,..kinetics,..and molecular structure. 1867 S. D. Kehoe *Indian Club Exercise* 18 Having discoursed..on the important benefits to be derived from physical culture..we will now describe some of the special means of exercise. 1893 *Harper's Mag.* Apr. 668/2 In the high-school department,..further instruction is given in physical culture. 1912 V. L. Whitechurch *Thrilling Stories of Railway* 11 He was a strong faddist on food and 'physical culture'. 1963 D. Webster *Compl. Physique Bk.* ii. 9 In all the systems of physical culture existing throughout the world non-apparatus work plays a most important part. 1904 C. L. Neil *Mod. Physical Culture* xviii. 115 (*heading*) The physical culturist's library. 1936 *Discovery* Nov. 362/2 The concluding chapter deals with the athlete, and it is of great value to the physical trainer and culturist. 1960 J. Hewitt *Yoga* iii. 56 A large number of the exercises popular with physical-culturists in the West are taken from Yoga. 1915 W. Owen *Let.* 30 Oct. (1967) 362 This morning we had 'Physical Drill' under a special Gymnastic Instructor. 1838 S. Smiles *Physical Education* 10 Here we may state what is the scope and object of a proper physical education. 1858 *Southern Cultivator* XVI. 32/1 The subject of physical education is beginning to attract attention. 1895 W. Maclaren in A. Maclaren *Physical Educ.* (rev. ed.) p. v, It is now more than thirty years since the demand for systematized Physical Education began to make itself heard in this country. 1926 *Encycl. Brit.* III. 141/2 Physical education..includes systematic and graded exercises chosen for their physiological and corrective results, folk-dancing, swimming, organized games and sports. 1960 *Where* No. 3. 16 Physical Education (PE) is the current term for what most parents remember as 'gym' or PT. 1977 E. W. Hildick *Loop* ix. 54 A Physical Education teacher. 1808 J. Playfair *Syst. Geogr.* I. p. ccv, To narrate the appearances which nature exhibits to our view, and to give an idea of what is movable and immovable upon, and under the surface of the earth, is the province of *Physical Geography.* 1914 Physical metallurgy [see *metallurgy]. 1935 Williams & Homerberg *Princ. Metallogr.* (ed. 3) i. 1 Metallography is commonly subdivided into..(1) production metallurgy..and (2) physical metallurgy. 1956 A. W. Chapman in D. L. Linton *Sheffield* x. 212 The faculty includes a postgraduate School of Physical Metallurgy in which the teaching is specially adapted to provide men from industry with instruction. 1965 R. W. Cahn *Physical Metallurgy* p. v, Physical metallurgy..is the root from which the modern science of materials has principally sprung. 1912 B. Russell *Probl. Philos.* i. 18 The real table, if it exists, we will call a 'physical object'. Thus we have to consider the relation of sense-data to physical objects. 1952 *Mind* LXI. 505 His only irrefutable position is to reduce his physical object claim to an announcement concerning his own sensations. 1954 A. J. Ayer *Philos. Ess.* iv. 88 This procedure..serves to clarify the meaning of statements about physical objects by relating them to statements of a different logical form. 1973 —— *Cent. Questions Philos.* iii. 62 What Hume calls the philosophical account of physical objects differs from the vulgar account in that it distinguishes them from perceptions. 1954 Physical therapist [see *crash *sb.*¹ 7 b]. 1922 *Lancet* 2 Dec. 1175/1 The first edition of this book appeared in 1910. Notable advances in physical therapy have now necessitated the preparation of a second edition. 1973 R. Kraus *Therapeutic Recreation Service* i. 19/1 In the fields of occupational therapy or physical therapy, professional training is carried on in close cooperation with medical authorities or educators. 1900 *Dialect Notes* II. 48 *Physical torture*, physical culture. 1920 *331st Field Artillery* (U.S.) 417 The officers,..knowing that boat drills were something of a bore, attempted to make them more interesting..by giving us 'physical torture' at these times. 1959 I. & P. Opie *Lore & Lang. Schoolch.* ix. 173 One feels Lewis Carroll would have liked the current terms Physical Torture (for P.T.) and Religious Destruction or Ridiculous Kapers (for R.K.). 1968 W. C. Anderson *Gooney Bird* iii. 35 The physical torture program..started promptly at 0630 every morning at Eglin Air Force Base. 1889 *Infantry Drill* 11 Physical training. In order to expand his chest, and develop his muscles, the soldier will be practised in the following exercises. 1910 Street & Gooderson *Handbk. Physical Training* i. 12 Physical training at schools should provide for the stimulation of the physiologic functions of the body. 1926 *Encycl. Brit.* III. 141/2 The model course, which originally consisted largely of military drill, has developed into the present Syllabus of Physical Training, based on the Swedish system. 1974 *Oxf. Jun. Encycl.* (rev. ed.) IX. 270/2 In the forces and in club work the aim is to train performers in set exercises and vaults which..are the result of physical training as opposed to physical education.

physicalism (fi·zikăliz'm). *Philos.* [f. Physical *a.* 2 + -ism.] A term originally used by members of the Vienna Circle for the theory that all science must eventually be capable of being expressed in the language of physics.
1931 O. Neurath in *Monist* XLI. 618 (*title*) Physicalism: the philosophy of the Viennese Circle. *Ibid.* 620 In a sense unified science is physics in its largest aspect, a tissue of laws expressing time-space linkages—let us call it: Physicalism. 1936 L. Bloomfield in *Language* XII. 93 The testing of this hypothesis of *physicalism* will be a task of the next generations, and linguists will have to perform an important part of the work. 1940 B. Russell *Inquiry into Meaning & Truth* 93 The thesis which Carnap calls 'physicalism', which maintains that all science can be expressed in the language of physics. 1965 *Language* XLI. 196 He [*sc.* Bloomfield]..recommended a policy known first as 'mechanism' or 'antimentalism', later as 'physicalism' which..amounted to the emancipation of linguistics from irrelevant psychologizing... Bloomfield himself was not able to pursue this policy with complete success. 1972 *Science* 16 June 1204/3 Science has been..historically associated with a philosophy of physicalism, the belief that reality is all reducible to certain kinds of physical entities. 1975 *Jrnl. Philos.* LXXII. 565 Physicalism does not entail reductionism, at least if we follow Hellman and Thompson in taking reductionism to mean explicit definability.

physicalist. Add: *Philos.* **2.** One who adheres to the theory of physicalism. Also as *adj.*, of or pertaining to such a theory.
1934 *Times Lit. Suppl.* 5 July 479/1 The paper now translated was first published in 1932..under the title of 'Physicalist language as the Universal Language of Science'. 1937 J. R. Firth *Tongues of Men* ix. 119 Let us turn to the 'physicalists' or 'methodical materialists' of the Vienna Circle. 1956 A. J. Ayer *Probl. Knowl.* 214 To apply the physicalist thesis to one's own experiences, is, as it were, to pretend to be anaesthetized. 1959 H. Kuhn in P. A. Schilpp *Philos. C. D. Broad* xx. 606 The German *Geisteswissenschaft* and Phenomenology..have labored to rouse modern thought from its dogmatic physicalist slumber. 1972 R. Plant in Cox & Dyson *20th-Cent. Mind* III. iii. 69 The relationship [between behaviour and pain] is not an external, contingent one as the dualist would suggest, nor is it a necessary one as the behaviourist or physicalist would claim. 1973 A. J. Ayer *Cent. Questions Philos.* vi. 126 Materialists, or physicalists as they are now more often called, who deny the existence of mental as opposed to physical events.

physicalistic (fizikăli·stik), *a.* [f. Physicalist + -ic.] Pertaining to or characterized by physicalism. Hence **physicali·stically** *adv.*
1934 M. Black tr. *Carnap's Unity of Sci.* 95 For the sake of precision we might supplement or replace 'physical language' by the term '*physicalistic language*'; denoting by the latter the universal language which contains not only physical terms (in the narrow sense) but also all the various special terminologies (of Biology,..etc.) understood as reduced by definitions to their basis in physical determinations. 1949 M. Schlick in Feigl & Sellars *Readings in Philos. Analysis* 407 The validity of the physicalistic assertion would be ever more restricted. 1954 Webster Add., Physicalistically. 1956 A. J. Ayer *Probl. Knowl.* v. 210 It is only if such statements [about thoughts, feelings] are interpreted 'physicalistically' that they can convey any information from one person to another. 1972 *Science* 16 June 1204/3 To physicalistic philosophy they [*sc.* altered states of consciousness] are epiphenomena. 1978 P. Pettit in Hookway & Pettit *Action & Interpretation* 51 Something as deep as the physicalistic prejudices of natural science or the postulates of rational man theory.

physicality (in Dict. s.v. Physical *a.*). Add: (Later examples.)
1944 M. Weitz in P. A. Schilpp *Philos. B. Russell* 73 Here Russell objects because..some entities, the unperceived entities of physics, even though they are neutral—i.e., have no first-order property of mentality or physicality—cannot be brought into psychological causal laws. 1948 *Mind* LVII. 8 Ayer chooses to call the commonly accepted criteria of physicality (the various meanings attached to the word 'physical world') 'assumptions'. 1953 *Mind* LXII. 26 Ontological assumptions concerning the nature of physicality.
2. The quality that pertains to physical sensation or to the body as distinct from the mind.
1849 J. S. Mill *Lett.* (1910) I. 143 Take again all the delicacies respecting bodily physicalities which savages have not a vestige of. 1930 E. Sitwell *Coll. Poems* 126 This bestial consciousness that is desire Is the hot muscles' vast fluidity, Muscular life, not physicality. 1964 *Listener* 27 Aug. 317/1 The climax is Thornhill's visit to a bullfight. The terse physicality of Mr Scott's style, his ability to evoke place and register action, avoids much suggestion of the derivative. 1972 C. L. Cooper in W. King *Black Short Story Anthol.* 218 The trunk of her, he saw self-consciously, with a tiny tickle of physicality, was full to bursting with youth under the plain dress. 1977 *N.Y. Rev. Bks.* 24 Nov. 42/4 Dancing, because of its immediate physicality, its shift in time and space, has a fluid plausibility which is apparently undemanding.

phy·sicalize, *v.* [f. Physical *a.* + -ize.] *trans.* To express or represent by physical means, *spec.* in the theatre, to represent (an idea) in physical terms, as the movements of the body of an actor. So **physicaliza·tion,** the representation of an idea by physical means.
1947 E. Kazan in Cole & Chinoy *Directing Play* (1953) 297 An effort to put poetic names on scenes to edge me into stylizations and physicalizations. Try to keep each scene in terms of Blanche. *Ibid.* 308 Stanley..is completely self-absorbed to the point of fascination. To physicalize this: he has a most annoying way of being preoccupied—or of busying himself with something else while people are talking with him. 1972 *Village Voice* (N.Y.) 1 June 54/4 Its publicized aim: 'to explore various means of physicalization to test the theory that theatre should be a hieroglyphic of speaking icons'. 1977 *Rolling Stone* 7 Apr. 14/3 Lily is particularly taken with Rick at the moment, partly because when she did him on the 'friends of Lily Tomlin' videotape, when she 'physicalized' him by wearing a moustache, sideburns and leather jacket, strutting and tugging at his/her crotch, hardly anyone recognized that the stud was her.

physico-. Add: **phy·sicomorph,** a representation in art of an inanimate object or phenomenon of the physical world; **physico-philosophical** *a.* (example); **phy:sico-psycholo·gical** *a.*, pertaining both to the physical and the psychological.
1895 A. C. Haddon *Evol. Art* 118 Under the term of 'physicomorph' I propose to describe any representation of an object or operation in the physical world. 1977 M. Cohen *Sensible Words* 145 The restrictiveness of Ong's view can be estimated by his determination to turn Swift into a stubborn, and unhappy, physicophilosophical writer. 1927 A. Huxley *Proper Stud.* 34 It is..only in the abstract that we can discuss the varieties of intelligence without considering the varieties in the other constituents of the physico-psychological personality.

physico-chemical, *a.* Add: (Later examples.) So **phy:sico-che·mically** *adv.*; **physico-chemist** (later examples); **phy:sico-che·mistry.**
1885 I. B. Yeo tr. *Oertel's Respiratory Therapeutics* ii. 731 It has been thought that the diminution of air pressure acts physico-chemically in another sense. 1909 *Jrnl. Industr. & Engin. Chem.* Mar. 158/2 Unless my friends, the physico-chemists, will..discover some way for establishing some optical properties or other physical constants, we are very much at a loss to establish the molecular size of my product. 1934 *Current Res. Anesthesia & Analgesia* XIII. 86 (*heading*) The physico-chemistry concerned in the action of anesthesia on blood colloids in relation to the safer handling of surgical and anesthetic risks. 1953 S. E. Luria *Gen. Virol.* iv. 82 Their tendency to lengthwise and sidewise aggregation has rendered their study by hydrodynamic methods a most perplexing problem for the physicochemist. 1958 *Times Lit. Suppl.* 10 Jan. 14/4 He looks forward to the day when psychological changes can be correlated convincingly with physico-chemical processes. 1966 *Mineral. Abstr.* XVII. 739/2 The physico-chemistry and origin of the deposits are discussed. 1969 *Listener* 20 Mar. 389/1 Goldmann believes that the social sciences cannot be conducted on the model of what he calls the 'physico-chemical' sciences. 1970 Ambrose & Easty *Cell Biol.* viii. 258 Studies of the permeability of membranes to gases, water, non-electrolytes, and ions were interpreted physico-chemically. 1971 I. G. Gass et al. *Understanding Earth* viii. 120/1 The physico-chemical conditions within the meteorite parent bodies. 1972 *Nature* 3 Mar. 44/1 The physico-chemistry of fibrous proteins. 1973 *Ibid.* 21–28 Dec. 528/1 The crystals were identified physico-chemically by their solubility in HCl, by the Meingen reaction, by infrared spectroscopy, and by petrographic microscopy. 1977 *Dædalus* Summer 66 The neurological and physicochemical bases of human behavior are clearly not exhausted by genetically fixed enduring neuronal pathways.

† physicotherapy (fizikǫþe·răpi). *Obs.* [f. Physico- + Therapy.] = *physiotherapy.
1903 *Med. Rec.* (N.Y.) LXIII. 881/2 Dr. J. Rivière of Paris advocated the employment of physicotherapy, or the combined action of electricity, heat, and light, in the treatment of uterine fibromata.
Hence **† phy:sicotherapeu·tic** *a.*, physiotherapeutic.
1904 *Nature* 21 Jan. 280/1 The results achieved..in the treatment of inoperable malignant growths by physicotherapeutic means.

physio-. Add: **physioplastic** *a.* (further examples); **phy:sio-psycho·logy,** physiological psychology; so **phy:sio-psycholo·gic, -lo·gical** *adjs.*
1943 Physioplastic [see *ideoplastic *a.*]. 1939 *Burlington Mag.* June 300/2 The familiar opposition between geometrical and 'vital' forms, impressionist and expressionist art, 'ideoplastic' and 'physioplastic', and so on, is shown to possess a concrete sanction in the aesthetic attitudes of blind artists. 1932 G. Blumer in *Practitioners Libr. Med. & Surg.* I. ix. 698 Constitution, which includes physiopsychologic factors, may change considerably during life. 1874 J. Cunningham *New Theory of Knowing* 155 Every one..knows what is meant by getting a 'start', though the physio-psychological explanation of it is not so clear. 1875 C. Wright *Let.* 12 July in R. B. Perry *Tht. & Char. W. James* (1935) I. 530 The other is in a book-notice by him [*sc.* William James]..of Wundt's physio-psychology. 1903 *Amer. Anthropologist* V. 586 In a general way, comparative physio-psychology has aided us in the search for the key to this great problem [of various intellectual endowment].

physio (fi·zio). Colloq. abbrev. of *physiotherapist or *physiotherapy.
1962 *Times* 3 July 5/3 They should, like the orthopaedic physiotherapists, also be nurses... I tell my physios: 'Keep your hands off, keep your minds on.' 1967 Partridge *Dict. Slang* Suppl. 1291 *Physio*, physiotherapy. 1968 M. Woodhouse *Rock Baby* iii. 21 Don't have him doing full knee bends... Reminds me, I must get in the physios chop-chop. 1971 'J. Bell' *Hole in Ground* iv. 59 Dr Colthorp wants her to have physio after the plaster comes off. 1973 *Times* 7 Feb. 15/3, I remember we didn't have a physio of our own, so we had to go to the athletics one. 1977 *Lancet* 5 Feb. 301/2 His details are entered in the book for the agreed date, with comments such as 'very fat' or 'needs pre-op physio'.

physiognomic, *a.* (*sb.*) Add: **3.** *Ecol.* Of or pertaining to the physiognomy of a plant community (*physiognomy 4 b).
1911 Bevis & Jeffery *Brit. Plants* ii. 16 The physiognomic groups into which we have divided the vegetation (woodland, grassland, heath, etc.) are only in part associated with definite types of climate. 1926 Tansley & Chipp *Study of Vegetation* x. 204 It is..by the physiognomic characters of the vegetation, especially when correlated with some topographic feature, that the traveller will find it easiest to recognise and record the chief types of vegetation in the course of his journey. 1951 *Ecology* XXXII. 279/2 The practical aspects of the physiognomic method are reflected in its use by foresters. 1959 *New Scientist* 29 Oct. 803/2 When the change in climatic pattern is gradual, the change in physiognomic type is correspondingly slow and *vice versa*. 1971 *Nature* 18 June 430/1 The Radforth muskeg classification system..is based on physiognomic characteristics.

physiognomy. Add: **B. II. 4. b.** *spec.* in *Ecol.* The general appearance, form, or characteristics of a community of plants.
1909 Groom & Balfour tr. *Warming's Oecol. Plants* vi. 25 The temperature and length of the vegetative season affect the physiognomy of the individual plant and the whole vegetation. 1926 Tansley & Chipp *Study of Vegetation* ii. 11 The physiognomy or 'look' of an association is primarily determined by the life form of its dominant species. 1951 *Ecology* XXXII. 278/2 Gillman..produced an excellent map of the physiognomy of the vegetation of Tanganyika. 1973 W. B. Clapham *Nat. Ecosystems* vii. 230 Many of the basic variables of the microhabitat, such as temperature, humidity, and the like, are functions of the physiognomy of the climax community.

physiographic, *a.* Add: *physiographic province*: see *province 6c.

physiographically (fiziogræ·fikăli), *adv.* [f. Physiographical *a.* + -ly².] From a physiographical point of view.
1902 in *Encycl. Dict.* Suppl. 1908 *Westm. Gaz.* 24 Mar. 12/1 This church stands nearly 1,450 feet above sea-level, and is considered the 'highest'—not ecclesiastically, but physiographically—in Great Britain. 1928 V. G. Childe *Most Anc. East* ii. 22 Physiographically the last-named chains constitute a more real dividing line than the inland sea. 1973 *Nature* 1 June 277/2 Somewhat older deltas exist which receive little or no drainage today and are therefore harder to distinguish physiographically.

physiological, *a.* Add: **2.** (Further examples.) **physiological psychologist,** a specialist in physiological psychology.
1875 G. H. Lewes *Probl. Life & Mind* (ser. 1) II. 482 The common error of mistaking ideal separations for real separations..leads the physiological psychologist to the conclusion that the objective aspect of the phenomenon..is the cause of the subjective aspect. 1899 W. James *Talks to Teachers* xii. 119 If we remember because of our associations, and if these are (as the physiological psychologists believe) due to our organized brain-paths, we easily see how the law of recency and repetition should prevail. 1904 E. B. Titchener tr. *Wundt's Princ. Physiol. Psychol.* p. vi, I have here sought to give this important chapter of physiological psychology at any rate a tentative systematic setting. 1933 Physiological phonetics [see *acoustic phonetics]. 1950 *Sci. News* XV. 9 These facts caused Thomas Young, who has since been called 'the father of physiological optics' to propose, a century and a half ago, the three-colour theory of vision which bears his name. 1960 C. Winick *Dict. Anthropol.* 411/2 Physiological phonetics is part of laboratory or experimental phonetics. 1967 R. F. Thompson *Found. Physiol. Psychol.* p. xxvii, Physiological psychology is concerned with the physiological bases of behavior. In the last analysis this means the organization and functions of the brain. 1973 Dewsbury & Rethlingshafer *Compar. Psychol.* 10 Lashley was a teacher of several men who became outstanding in the revitalization of comparative and physiological psychology that occurred after World War II.
b. *Med.* = *normal *a.* 2 f.
1896 *Jrnl. Physiol.* XX. 145 (*heading*) On the initial rate of osmosis of blood-serum with reference to the composition of 'physiological saline solution' in mammals. 1923 P. H. Mitchell *Text Bk. Gen. Physiol.* vi. 148 Physiological salt solutions do not have irritating effects when in contact with open wounds. 1952 *Sci. News* XXIV. 27 Minced animal tissues were extracted with physiological saline (0·85% NaCl). 1969 J. H. Green *Basic Clin. Physiol.* vi. 37/1 The sodium chloride is present in plasma to the extent of 0·9 g. per 100 ml. A solution containing this amount of sodium chloride in water is termed normal isotonic or physiological saline, and it has the same electrolyte strength as blood.

physiologue (fi·ziolǫg). *rare.* [ad. L. *physiologus*: see Physiologer.] = Physiologist.
1877 J. D. Hooker in L. Huxley *Life J. D. Hooker* (1918) II. 236, I think *Gnetum* is quite overlooked by the Physiologues in removing Gymnosperms from Dicots. 1923 A. Huxley *Antic Hay* v. 70, I have with me..a physiologue, a pedagogue and a priapagogue.

physiopathological (fi:ziopæþolǫ·dʒikăl), *a.* *Med.* Also physio- (with hyphen). [f. Physio- + Pathological *a.*] Of or pertaining to physiopathology. Also **phy:siopatholo·gic** *a.*
1867 Robertson & Rutherford tr. *Griesinger's Mental Path. & Therapeutics* i. iii. 23 (*heading*) Preliminary physio-pathological observations on mental phenomena. 1930 *Brit. Med. Jrnl.* 8 Feb. 234/2 Endarteritis obliterans is a special localization of athero-sclerosis. In fact, in athero-sclerosis the same constellation of etiological factors, and the same physio-pathological conditions (proliferation of the intima, disturbances in cholesterol metabolism, etc.), are found. 1938 *Arch. Neurol. & Psychiatry* (Chicago) XL. 1126 (*heading*) Physiopathologic and pathoanatomic aspects of major trigeminal neuralgia. 1968 E. Kelemen *Physiopath. & Therapy Human Blood Dis.* (1969) p. xv, Contemporary physiopathological research..is, at least to some extent, very liable..to lose sight of the ultimate aim of the project.

physiopathology (fi:ziopăþǫ·lŏdʒi). *Med.* Also physio- (with hyphen). [f. Physio- + Pathology.] (See quot. 1904.) Also, the physiology of a diseased organism. So **phy:siopatho·logist.**
1904 Stedman *Dunglison's Dict. Med. Sci.* (ed. 23) 875/2 *Physiopathology*, study of function as modified by disease. 1930 *Brit. Med. Jrnl.* 8 Feb. 234/1 The frequency of the occurrence of this morbid condition raises important problems as to its physio-pathology and its treatment. 1953 Homburger & Fishman (*title*) The physiopathology of cancer. 1954 *Acta Physiol. Scand.* XXXI. 359 A study has been made of the physiopathology of the asphyxiated human fetus in order to improve the results of resuscitation. 1972 Gallo & Santamaria *Res. Progress Org.-Biol. & Med. Chem.* III. i. p. vi, The discovery of the photodynamic phenomenon helps the physiopathologists to characterize diseases which were once joined together under general terms of skin disorders. 1974 *Nature* 29 Mar. 371/2 Fifty laboratories provide facilities for basic and applied research into bacteriology, virology, physiopathology, [etc.].

physiophonetics (fi:ziofǫ̆ne·tiks), *sb. pl.* (const. as *sing.*) *Linguistics.* [f. Physio- + Phonetics *sb. pl.*] (See quot. 1950.) Hence **phy:siophone·tic** *a.*
1936 J. R. Kantor *Objective Psychol. Gram.* xii. 162 It was Baudouin de Courtenay, according to Troubetzkoy, who first made a genuine separation between physical and physiological sounds and psychological phonic images—in other words gave the term *phoneme* what Troubetzkoy calls its present meaning—namely, the element of psychological language—and placed it within the field of psychophonetics instead of physiophonetics. 1950 D. Jones *Phoneme* xxix. 213 Professor Baudouin de Courtenay..defined phonemes as 'mental images', and accordingly distinguished two kinds of phonetics which he called 'physiophonetics' and 'psychophonetics' respectively. He applied the term 'physiophonetics' to the study of sounds actually uttered, and used the term 'psychophonetics' to denote the study of the 'mental images' which uttered sounds are intended to represent. 1956 Physiophonetic [see *psychophonetics *sb. pl.*].

physiotherapist (fiziǫþe·răpist). [f. next + -ist.] One skilled or trained in physiotherapy.
1923 C. M. Sampson *Physiotherapy Technic* xxvi. 409 Lucky is the physiotherapist who can plan his department in a new building. 1944 *Times* 24 Jan. 5/5 A Chartered Physiotherapist to-day is trained not only in massage and gymnastics but also in electrical and all forms of ray therapy. 1958 *Times* 7 July 22/4 The role of physiotherapy is still large, and the supply of physiotherapists on the whole well maintained. 1972 D. Haston *In High Places* ii. 20 My physiotherapist was very sympathetic. I worked really hard on exercises, and the wasted arm became stronger.

physiotherapy (fiziǫþe·răpi). [f. Physio- + Therapy.] The treatment of disease, injury, or deformity by physical methods, such as massage, exercise, and the application of heat, light, fresh air, and other external influences.
1905 *Brit. Med. Jrnl.* 15 July 126/2 The first congress of physiotherapy will be held at Liége on August 12th. 1928 *Sunday Dispatch* 16 Dec. 1/1 Two new specialists, both experts in radiology and massage, and in the treatment generally known as physio-therapy, were called to Buckingham Palace yesterday. 1958 *Times Lit. Suppl.* 4 Apr. 187/2 Full details are given of methods of physiotherapy that can be carried out in the home. 1958 [see *physiotherapist]. 1975 Scrutton & Gilbertson *Physiotherapy in Paediatric Pract.* 1 Physiotherapy involves the modification of the patient's physical external environment, either generally or topically, so as to promote healing or otherwise improve the body's efficiency.
Hence **phy:siotherapeu·tic** *a.*, of, pertaining to, or involving physiotherapy.
1905 *Brit. Med. Jrnl.* 15 July 126/2 The abuses caused by 'healers' who pretend to treat by physiotherapeutic procedures. 1926 *Encycl. Brit.* III. 686/1 As convalescence goes on, physiotherapeutic measures..are employed to hasten recovery. 1957 M. Spark *Comforters* viii. 196 He is receiving physiotherapeutic treatment.

1976 Nichols & Hamilton *Rehabilitation Med.* ii. 21 There are innumerable reports in the medical and physiotherapeutic literature extolling the virtues of specific techniques.

physiqued (fizī·kt), *a.* [f. Physique + -ed¹.] Having a physique of a specified character.
1926 *Contemp. Rev.* June 690 These ill-fed, ill-housed, wretchedly physiqued and noisy communist agitators.

physisorb (fi·zisǭɹb), *v.* *Chem.* [Back-formation from next.] *trans.* and *intr.* To collect by physisorption. So **phy·sisorbed, phy·sisorbing** *ppl. adjs.*
1966 *Surface Sci.* IV. 103 (*heading*) Ordered physisorbed layers on graphite. 1967 *Ibid.* VI. 1 It has been argued that due to the weakness of physisorbing forces, the electron beam of the instrument would disturb the molecules significantly. *Ibid.* 2 Materials highly volatile at room temperature could not be physisorbed to concentrations approaching a monolayer. *Ibid.*, A substance with a vapor pressure of 1×10^{-8} Torr or more will not physisorb on graphite in quantities approaching a monolayer if its ambient pressure over the graphite is 1×10^{-4} Torr or less. 1970 C. Okkerse in B. G. Linsen *Physical & Chem. Aspects Adsorbents & Catalysts* v. 253 Wirzing suggested the use of the combination band of water at 5265 cm^{-1} for the quantitative determination of physisorbed water. 1971 *Sci. Amer.* Dec. 51/1 A physisorbed species retains substantially its original structure.

physisorption (fizisǭ·ɹpʃən). *Chem.* [f. Physi(cal *a.* + Ad)sorption.] Adsorption which does not involve the formation of chemical bonds. Cf. *chemisorption.
1965 *Progress Solid State Chem.* II. 94 At higher pressures and lower temperatures (−78°C) physisorption without formation of ordered structures was observed. 1973 *Sci. Amer.* May 34/2 There have been several more recent electron-diffraction studies of physisorption with other gases and other substrates. 1976 *Physics Bull.* May 218 This book..is extremely timely in marshalling ideas on physisorption because there is now an increasing interest in the more difficult process of chemisorption.

physog, var. *phizog.

physogastrism (fəisogæ·striz'm). *Ent.* [ad. G. *physogastre* (E. Wasmann *Kritisches Verzeichniss d. Myrmekophilen u. Termitophilen Arthropoden* (1894) 87), f. Physo- + Gr. γαστ(ε)ρ-, γαστήρ, belly: see -ism.] In certain insects, a condition in which the abdomen becomes distended by the growth of fat bodies or other organs. Also **phy·sogastry** in the same sense. So **physoga·stric** *a.*, exhibiting this condition.
1903 *Nature* 12 Feb. 351/1 They [*sc.* symphilous beetles] also show certain modifications of the mouth-organs.., as well as 'physogastrism', accompanied by excessive development of the fat-bodies, or sexual glands. 1914 *Ann. Natal Museum* III. 103 Dr. Trägårdh examined many nests of termites in Natal, but no other physogastric Staphylinids were discovered. 1920 *Ibid.* IV. 326 In the female there is no excessive amount of fat-body. The physogastrism is mainly due to the great bulk of the genital organs and of the mid-gut. 1922 W. M. Wheeler *Social Life Insects* vi. 273 Many [termitophiles]..have acquired peculiar characters, the most characteristic of which is physogastry, or excessive enlargement of the abdomen. *Ibid.* 277 Probably this is also the case with other physogastric termitophiles. 1952 Rothschild & Clay *Fleas, Flukes & Cuckoos* xi. 222 The abdomen becomes enormously distended owing to the abnormal growth of the fatty tissues. This curious condition is known as physogastry and it is usually developed by flies or beetles which are parasitic or symbiotic in ants' or termites' nests. 1971 R. R. Askew *Parasitic Insects* v. 73 The abdomen [of *Ascodipteron*] becomes bloated (physogastric) as the fly takes in blood from the host. *Ibid.* 311/2 Physogastry.

physostegia (fəi:soste·dʒiă). [mod.L. (G. Bentham 1829, in *Bot. Reg.* XV. 1289), f. Gr. φῦσα bladder + στέγη roof + -ia¹, in reference to the inflated calyx.] A perennial herb of the genus so called, belonging to the family Labiatæ, native to North America, and bearing spikes of pink or white flowers; also called the obedient plant or false dragonhead.
1830 J. C. Loudon *Hortus Britannicus* 483 *Physostegia* Benth. [English name] Physostegia. 1835 *Curtis's Bot. Mag.* LXII. 3386 (*heading*) Imbricated Physostegia. 1905 E. V. Boyle in M. T. Earle et al. *Garden Colour* 70 The habit of the Physostegia's little pale pinkish flower to stay fixed and still in whatever position it may be turned, was thought to be caused by coma. 1962 *Amat. Gardening* 17 Mar. 4/1 Among a number of self-supporting plants may be included..physostegias. 1971 J. Raven *Botanist's Garden* xi. 201 The typical form of the Physostegia..is of my least favourite shade somewhere between pink and mauve.

physostigmine (fəisosti·gmīn). *Chem.* and *Pharm.* Also † -in. [ad. G. *physostigmin* (Jobst & Hesse 1864, in *Ann. d. Chem. u. Pharm.* CXXIX. 118): see Physostigma and -ine⁵.] A colourless or pale yellow crystalline tricyclic alkaloid, $C_{15}H_{21}N_3O_2$, which is the active principle of the calabar bean and is used medicinally (esp. as a miotic) on account of its anticholinesterase activity.
1864 *Chem. News* 5 Mar. 109/1 The physiological properties of the Calabar bean have been well studied in this country, but the authors above named are..the first who have isolated the alkaloid to which it owes its activity, and to which they have given the name Physostigmine. 1865, 1896 [in Dict. s.v. Physostigma]. 1907 J. H. Parsons *Dis. Eye* iv. 74 Eserin or physostigmin, the most powerful miotic we possess, acts by stimulating the third nerve endings in the sphincter and in the ciliary muscle. 1938 *Thorpe's Dict. Appl. Chem.* (ed. 4) II. 200/1 The physiological action of physostigmine depends on its inhibition of the hydrolysis of acetyl choline by cholinesterase. 1970 *Nature* 4 Apr. 21/2 Drugs which..increase the concentration [in the brain] of acetylcholine (for example, physostigmine) exacerbate Parkinsonism. 1974 M. C. Gerald *Pharmacol.* vii. 128 Many very common drugs are alkaloids, some of which include..physostigmine.

phytal (fəi·tăl), *a.* *Ecol.* [f. Phyt- + -al.] Of, pertaining to, or designating those parts of a body of water which are shallow enough to permit the growth of rooted green plants.
1918 R. A. Muttkowski in *Trans. Wisconsin Acad. Sci., Arts & Lett.* XIX. 378 The lake as a whole may be divided into three general regions,—the littoral, or phytal region; the aphytal region; and pelagic region. 1926 A. S. Pearse *Animal Ecol.* vi. 210 Muttkowski uses 'littoral' as synonymous with 'phytal'.., but in very turbid lakes the phytal zone may be thinner than that of wave action, and in clear, small lakes it may be thicker. 1958 W. D. R. Hunter in Miller & Tivy *Glasgow Region* 107 (*caption*) Map of Loch Lomond showing the extent of water (less than 13 feet deep), which is shallow enough to permit the growth of rooted green plants, *i.e.* the potential phytal zone. 1973 *Nature* 8 June 342/1 Marine grasses would also have provided new sources of food for the macrofauna and microfauna and new means of dispersal for benthonic and phytal organisms.

phytane (fəi·tēⁱn). *Chem.* [ad. G. *phytan* (Willstätter & Hocheder 1907, in *Ann. d. Chem.* CCCLIV. 208): see Phyt- and -ane.] A colourless liquid hydrocarbon, 3,7,11,15-tetramethylhexadecane, $C_{20}H_{42}$, that is the paraffin corresponding to phytol.
1907 [see *phytol]. 1911 *Chem. Abstr.* V. 874 Phytane, $C_{20}H_{42}$, obtained by the reduction of phytol,..crystallizes in liquid air. 1965 *New Scientist* 6 May 378/1 Certain hydrocarbons, such as the branched-chain compounds phytane and pristane, can if found [in rocks] be taken to be definite evidence of life. 1971 *Sci. Amer.* May 38/3 The chemical analysis of organic material from the Gunflint cherts in several laboratories reveals the presence of the hydrocarbons pristane and phytane: two 'chemical fossils' that can most reasonably be regarded as being breakdown products of chlorophyll.

phytase (fəi·tēⁱz, -s). *Biol.* [a. G. *phytase* (U. Suzuki et al. 1907, in *Bull. College Agric.* (Tokyo Imperial Univ.) VII. 503): see Phyt- and *-ase.] Any of a class of enzymes found esp. in cereals and yeast which convert phytic acid to *myo*-inositol and phosphoric acid.
1908 *Jrnl. Chem. Soc.* XCIV. i. 236 The change is shown to be due, not to putrefaction, but to an enzyme, phytase, which was isolated from rice bran and wheat bran. 1931 E. C. Miller *Plant Physiol.* xi. 632 The glucoside phytin is decomposed into phosphoric acid and inosite, $C_6H_{12}O_6$, by the action of the enzyme phytase. 1947 *Nature* 18 Jan. 99/2, 25 ml. of the solution correspond to the phytase activity of about 15 gm. barley. 1966 Nowakowski & Clarke tr. *Kretovich's Princ. Plant Biochem.* vi. 215 The phytases of yeasts and flour split much of the inositol-phosphoric acid during fermentation of dough. 1975 *Science* 7 Feb. 432/3 A phytase which we extracted and purified from wheat bran.

phytic (fəi·tik), *a.* *Biochem.* [f. *phyt(in + -ic.] *phytic acid*: a phosphoric acid ester, $C_6H_6(OPO_3H_2)_6$, of *myo*-inositol which is found (often as salts) in plants, esp. in the seeds of cereals.
1908 *Jrnl. Biol. Chem.* IV. 497 Wheat bran, which contains 1½ per cent of phosphorus, was found to be essentially free from inorganic phosphates, the whole of the phosphorus being present as salts of phytic acid. 1943 *Endeavour* II. 75/2 One important dietary constituent, phytic acid (inositol hexaphosphoric acid),..interferes with calcification by precipitating calcium in the gut and making it unavailable. 1966 *McGraw-Hill Encycl. Sci. & Technol.* VII. 121/2 The phytic acid of whole wheat is of nutritional importance because it combines with dietary calcium, decreasing its availability. 1978 R. Mitchison *Life in Scotland* ii. 71 Recent work on nutrition has shown that..some foodstuffs contain inhibitors.., for instance, phytic acid.
Hence **phy·tate,** a salt, or the anion, of phytic acid.
1908 *Jrnl. Biol. Chem.* IV. 498 In all cases blank determinations were made on the tissue extract alone and on a solution of sodium phytate which were digested in a similar manner. 1961 *Lancet* 26 Aug. 483/1 In dietary deficiency of vitamin B_{12}, serum-iron has been found to fall... This is thought to be due to ingestion of excessive amounts of phytate in wholemeal flour and bread. 1968 *New Scientist* 28 Nov. 512/2 The staple article of the typical village diet [in Iran] is wheat bread, which contains a chemical called phytate. This chemical renders zinc and possibly iron unavailable. 1976 *Sci. Amer.* Sept. 57/1 Ferrous iron, however, is absorbed less efficiently when it is ingested in combination with phytates and oxalates, which are found in leafy green vegetables and the whole-grain, unleavened bread of North Africa and the Middle East.

phytin (fəi·tin). *Biol.* and *Med.* [a. G. *phytin* (S. Posternak 1904, in *Schweiz. Wochenschr. f. Chem. u. Pharm.* XLII. 405): see Phyt- and -in¹.] **a.** An insoluble salt of phytic acid with calcium and magnesium, which is found in plants, esp. cereals; also, *loosely*, the acid itself. **b.** Also **Phytin,** † **-ine.** A proprietary name for tonic preparations containing this.
1905 *Brit. Med. Jrnl.* 14 Jan. 81/1 Phytin has hitherto been simply described as a compound prepared from various seeds which contain phosphorus in combination... Phytin is claimed to be of use in all cases where invigorating treatment is indicated and to possess a specific action against impotence. 1905 *Official Gaz.* (U.S. Patent Office) 28 Feb. 2375/2 Certain named pharmaceutical products. Society of Chemical Industry in Basle, Basle, Switzerland. Filed Nov. 29, 1904... Phytine... Used since February, 1903. 1906 *Amer. Jrnl. Physiol.* XVII. 76 Commercial phytin has lately been introduced into medical practice with somewhat extravagant claims. 1934 *Brit. Pharmaceutical Codex* 1656 *Phytin*, calcium and magnesium salt of inositol hexaphosphoric acid. 1936 [see *inositol]. 1938 E. C. Miller *Plant Physiol.* (ed. 2) viii. 545 When chlorophyll is treated with dilute acids, the magnesium is removed and replaced with two hydrogen ions without otherwise changing the molecule. A series of decomposition products result [*sic*] which are called 'phytins' and which correspond to the alkali decomposition products of chlorophyll, except that the magnesium is lacking. 1947 *Dispensatory U.S.A.* (ed. 24) 1489/1 The calcium-magnesium salt of inositol hexaphosphate is available under the name *Phytin* (Ciba) in tablets containing 0·25 gm. 1957 G. E. Hutchinson *Treat. Limnol.* I. xii. 735 Chu (1946) found that *Nitzschia palea* could use phytin (magnesium inositol hexaphosphate) and glycerophosphoric acid as phosphorus sources. 1969 N. W. Pirie *Food Resources* vii. 164 The old-fashioned method of making porridge..destroys phytin which would otherwise make the calcium in the food less easy to absorb.

phyto-. Add: **phytobe·nthos** [*benthos], the aquatic flora of the region at or near the bottom of the sea; **phytochemical** *a.* (examples); hence **phytoche·mically** *adv.*; also **phytoche·mist,** an expert or specialist in phytochemistry; **phytochemistry** (earlier and further examples); **phytoe·cdysone** *Biol.* [*ecdysone; prob. first formed in Jap.], any substance that occurs in a plant and causes moulting in insects; **phytofla·gellate** [*flagellate *sb.*], a plant-like flagellate belonging to the subclass Phytoflagellata or Phytomastigophora; **phytomi·togen,** a mitogen derived from a plant; **phytomo·nad** [a. mod.L. order name *Phytomonadina*, f. generic name *Phytomonas* (C. Donovan 1909, in *Lancet* 20 Nov. 1496/2) + Monad 4], a phytoflagellate belonging to the order Phytomonadina; **phytopathology,** delete sense (*b*) and add examples of sense (*a*); **phytopathological** *a.* (examples); **phytopathologist** (further examples).
1931 R. N. Chapman *Animal Ecol.* xvi. 333 The phytobenthos is along the shore or in the littoral region. 1964 *Oceanogr. & Marine Biol.* II. 127 (*caption*) Other organisms listed included zooplankton, phytobenthos, zoobenthos, and fishes. 1973 *Nature* 6 Apr. 415/2 The rates of..photosyntheses of phytoplankton and phytobenthos..have been made [*sic*] using carbon-14. 1877 *Chem. News* 4 May 185/1 (*heading*) On phyto-chemical processes. 1921 *Experiment Station Rec.* XLIII. 820 (*heading*) Phytochemical investigations on indigenous and naturalized plants. 1972 *Nature* 21 Jan. 134/1 The joint meeting of the Phytochemical Society and the Pharmaceutical Society of Great Britain on plant constituents of pharmacological interest. 1972 *Science* 9 June 1131/2 Schultes has repeatedly suggested that these additional 'peyote' cacti be examined phytochemically. 1969 H. Erdtman in Harborne & Swain *Perspectives in Phytochem.* v. 109 That symposium was important because it brought together scientists of different specializations, botanical taxonomists, phytochemists and chemists interested in biosynthetic problems. 1972 *Nature* 28 Apr. 469/2 The contributors are leading specialists in umbellifer taxonomy..together with many other scarcely less eminent systematists and phytochemists. 1837 *Phil. Mag.* X. 247 (*heading*) A report of the progress of phytochemistry in the year 1835. 1912 *Carnegie Inst. Year Bk.* 49 The chief problems of the Department have been taken to lie in the domain of phyto-chemistry, in the water-relations of plants, and in the environic reactions of organisms. 1968 *Jrnl. Chromatogr.* XXXVI. 22 Methods for the separation and identification of microquantities of phenolic glycosides are of great importance for work in phytochemistry, pharmacognosy and chemotaxonomy. 1968 *Tetrahedron Lett.* July 3883 Four additional phytoecdysones have now been isolated from the leaves [of *Podocarpus macrophyllus*]. Interestingly, these new phytoecdysones..have steroid skeletons with 28 and 29 carbon atoms. 1974 Hikino & Takemoto in W. J. Burdette *Invertebr. Endocrinol.* ii. 187 The occurrence of phytoecdysones in plants raises the question whether they have any beneficial or adverse effects on the plants themselves or on the phytophagous animals in their natural habitat... The leaves of *Morus* species and *Podocarpus macrophyllus* which are known to contain considerable

amounts of phytoecdysones, are food of the larvae of the moths, *Bombyx mori* and *Milionia vasalis pryeri*, respectively. 1978 *Nature* 9 Mar. 122/1 The generic term 'phytoecdysones' should be abandoned. 1947 *Palestine Jrnl. Bot.* IV. 14 *Prymnesium parvum* Carter is a phytoflagellate belonging to the order of the Chrysomonadales, differing from other members of this order in the presence of an immobile spine-like third flagellum, a feature unique among phytoflagellates. 1951 Hutner & Provasoli in A. Lwoff et al. *Biochem. & Physiol. Protozoa* I. 29 The phytoflagellates are a heterogeneous group occupying a systematic position at intersections of plant and animal lines of descent. 1957 *New Biol.* XXIII. 93 This phytoflagellate [*sc.* *Prymnesium parvum*] was first blamed for mass fish mortality in Holland and later in Denmark. 1973 M. A. Sleigh *Biol. Protozoa* i. 2 The cell possesses the basic components of a phytoflagellate, including flagella, nucleus, plastid,..and other..inclusions. 1961 Marshall & Capon in *Lancet* 8 July 104/1 We suggest that until more is known about their structure, factors extracted from plants which exhibit mitogenic activity, be referred to as phytomitogens. 1964 *Ibid.* 21 Nov. 1101/1 These observations suggest that the distribution of 'phytomitogens' in the plant kingdom may be more widespread than is recognised. 1974 *Jrnl. Exper. Med.* CXXXIX. 1553 The lymphocyte response to phytomitogens is generally considered to be nonspecific. 1926 G. N. Calkins *Biol. Protozoa* v. 279 Phytomonads with a bivalve shell, or at least a membrane which splits easily to form two lens-like halves, as in *Phacotus lenticularis*. 1953 R. P. Hall *Protozool.* iv. 151 Lipids, although usually not abundant, are stored by many phytomonads. 1961 R. D. Manwell *Introd. Protozool.* xiv. 174 Most of the phytomonads are rather small, roundish forms. 1889 *Cent. Dict.*, Phytopathological. 1909 B. M. Duggar *Fungous Dis. Plants* 4 The foundations were laid for a more careful study of the fungi from a phytopathological point of view. 1959 *Ann. Rev. Microbiol.* XIII. 224 The term 'toxin', in the phytopathological literature, is used in the general sense of a poisonous substance generated by the pathogen regardless of its chemical nature. 1962 W. Carter *Insects in Relation to Plant Dis.* p. vii, There..has been a growing awareness of the toxicogenic insect as a phytopathological agent. 1917 J. W. Harshberger *Text-bk. Mycol. & Plant Path.* xxiii. 271 A study of phytopathology..presupposes that the would-be phytopathologist is acquainted with plant morphology, systematic botany..histology, cytology, embryology, genetics, physiology, [etc.]. 1958 *Austral. Jrnl. Biol. Sci.* XI. 275 The problem of disease resistance in plants has exercised the minds of phytopathologists for more than half a century. 1891 *Vet. Jrnl.* XXXII. 253 Phytopathology afforded many instances of local death of a part produced by the parasite. 1911 *Encycl. Brit.* XXI. 754/2 'Phytopathology' or plant pathology..comprises our knowledge of the symptoms, course, causes and remedies of the maladies which threaten the life of plants, or which result in abnormalities of structure that are regarded, whether directly injurious or not to life, as unsightly or undesirable... As a branch of botanical study it is of recent date. 1976 *Nature* 12 Feb. 449/1 A large proportion of the money devoted to agricultural research in the UK is spent on phytopathology and crop protection and has enabled the production of virus-free and disease-resistant crops.

phytoagglutinin (fəito,ăglū·tinin). *Biochem.* [f. Phyto- + *agglutinin* s.v. *agglutinate *v.*] Any plant protein that is an agglutinin.
1959 *Archiwum Immunologii i Terapii Doświadczalnej* VII. 793, 214 plant species from various botanic families were examined for the presence of phytoagglutinins; they have been found in seed extracts of 50 plant species. 1971 *Proc. Nat. Acad. Sci.* LXVIII. 1818/2 The *C[olletotrichum] lindemuthianum* polygalacturonase inhibitor isolated from Red Kidney bean hypocotyls has several properties which suggest that it is one of the glycoproteins commonly referred to as phytoagglutinins. 1974 A. Huxley *Plant & Planet* xxv. 285 Many plants..contain antibiotic substances which help to repel fungus invasions... These substances, which are possibly based on proteins known as phytoagglutins [*sic*], remind one of antibodies in animals.

phytoalexin (fəi:to,ăle·ksin). *Bot.* [a. G. *phytoalexin* (Müller & Börger 1941, in *Arb. aus der biol. Reichsanstalt Land- und Forstwirtschaft* XXIII. 223): see Phyto- and *alexin.] Any substance that is produced by plant tissues in response to contact with a parasite and specifically inhibits the growth of that parasite.
The term was orig. defined with regard to fungal parasites only, but in 1956 Müller extended the meaning to cover all parasites.
1949 *Nature* 26 Mar. 498/2 The necrosis of the affected tissues must be accompanied by the formation or activation of a principle ('phytoalexin') which exercises a retarding influence on the penetrating parasite. 1956 K. O. Müller in *Phytopath. Zeitschr.* XXVII. 254 'Phytoalexines' [*sic*] are defined as antibiotics which are the result of an interaction of two different metabolic systems, the host and the parasite, and which inhibit the growth of microorganisms pathogenic to plants. 1964 Cruickshank & Perrin in J. B. Harborne *Biochem. Phenolic Compounds* xiii. 534 Susceptibility [to fungal attack] may be due to the inability of the infecting fungus to stimulate the formation of the phytoalexin or to its capacity to tolerate the phytoalexin produced. 1975 *Sci. Amer.* Jan. 88/3 Plants do not produce antibodies but some of them do produce phytoalexins. 1977 *Observer* 4 Sept. 4/2 Dr. David Smith..told a meeting of the British Association for the Advancement of Science last week that chemicals called phytoalexins—substances produced by plants when they are attacked by disease—merited a major investigation.

phytochrome (in Dict. s.v. Phyto-). Add: **2.** *Bot.* A blue-green chromoprotein which regulates many aspects of development in higher plants according to the nature and timing of the light which it absorbs.
1960 Borthwick & Hendricks in *Science* 28 Oct. 1223/1 This control of flowering..implies a time-measuring system that distinguishes between light and darkness through mediation of a pigment... The pigment, now called phytochrome, is a blue or a bluish-green protein that exists in two forms interconvertible by light,..with 660 and 730 mμ the absorption maxima of the two forms. 1962 *Amat. Gardening* 20 Jan. 12 Phytochrome has been isolated, and shown to be sensitive even to very short periods of light. 1971 *New Scientist* 18 Feb. 365/1 One of the 'states' of phytochrome is physiologically active and plays a part in controlling many processes ranging from seed germination..to the onset of reproductive behaviour. 1977 J. L. Harper *Population Biol. Plants* v. 144 The phytochrome system which controls the light sensitivity of germination in many species involves an interconversion of two forms of phytochrome.

phytocidal (fəitosəi·dăl), *a.* [f. Phyto- + -cide + -al.] Lethal or injurious to plants.
1934 *Ann. Rep. East Malling Res. Station 1933* 157 The immediate problem to be solved by this field spraying trial was whether..lime-sulphur could be used in conjunction with nicotine or with derris with satisfactory fungicidal, insecticidal and phytocidal results. [*Note*] This term is now being used in reference to the effect of sprays on the tree. 1936 H. Martin *Sci. Princ. Plant Protection* (ed. 2) vi. 110 It is convenient in describing phytocidal activity to distinguish between acute injury, characterized by the localized killing of plant tissue (necrosis) and popularly termed 'scorch' or 'burn', and chronic injury which involves deep-seated physiological changes and causes, for example, the stunting and premature drop of fruit and leaves. 1946 *Nature* 21 Sept. 417/2 One other point regarding phytocidal action may be mentioned. Unnecessarily heavy dressings of B.H.C. do not appear to have a hurtful action on the carrot, whereas with the Brassicæ even moderate dressings may seriously affect the stem at the point where contact is made. 1970 *Weed Res.* X. 367 (*heading*) The correlation of the phytocidal effect of 3-amino-1,2,4-triazole with the growth stage of oat plants.
So **phy·tocide,** a phytocidal agent.
1936 *Ann. Rep. East Malling Res. Station 1935* 198 Cuprous cyanide..[etc.] were retained for further phytocide tests. 1961 *New Scientist* 21 Dec. 724/3 Treatment with selective phytocides such as copper sulphate.

phytoclimate (fəi·tokləimĕt). *Bot.* [ad. Da. *fanerofytklimaet, kryptofytklimaet* (C. Raunkiaer 1908, in *Bot. Tidsskr.* XXIX. 54): see Phyto- and Climate *sb.*] Local climate in its ecological aspects.
1950 *Bot. Rev.* XVI. 1 (*heading*) Life-forms and phytoclimate. *Ibid.* 16 We have here only a general correlation between Raunkiaer's hemicryptophytic phytoclimate and Köppen's C and D climates which are based on physical data. 1968 J. M. Maclennan tr. *Sukachev & Dylis's Fund. Forest Biogeocoenol.* i. 31 Inter-crown and intra-crown breaking of twigs.., by increasing the distance between trees and lessening the depth and density of the canopy, causes substantial changes in the phytoclimate, litter, and soil processes. 1975 *Lesovedenie* vi. 36 (*heading*) Alteration of phytoclimate in stem pests nidus.
So **phytoclima·tic** *a.*, of or pertaining to phytoclimate.
1913 *Jrnl. Ecol.* I. 18 In Table I we give a few of Raunkiaer's analyses for various types of climate; the locality is indicated, also the total species actually analysed, while in the later columns the results are given as percentages according to the grouping given above. Such an analysis for any region is termed a biological or phyto-climatic spectrum. 1976 *Biol. Abstr.* LXI. 5689/1 The following parameters are given for each provenance: .., phytoclimatic zone according to Pavari and soil.

phytocœnose (fəitosī·nōᵘz). *Ecol.* [f. Phyto- + Gr. κοίνωσις mingling.] A community of plants; all of the plant species found at a particular site.
1930 G. E. du Rietz in *Svensk Bot. Tidskr.* XXIV. 489 Only of the phytocoenoses, or 'complete plant-communities', all units of higher and lower rank have been included. 1955 *Jrnl. Ecol.* XLIII. 650 It is always possible to list the species in a phytocoenose (i.e. the vegetation occupying a definite area of ground). 1964 *Bot. Zhurnal* XLIX. 74 The study of the sward structure in a phytocoenose reveals the fact that the latter consists of layers.

phytoene (fəi·to,īn). *Chem.* [f. Phyto- + -ene.] A colourless viscous liquid, $C_{40}H_{64}$, which is a polyunsaturated branched-chain hydrocarbon and a precursor in the biological synthesis of carotenoids.
1950 Porter & Lincoln in *Arch. Biochem.* XXVII. 391 The name phytoene is given to a compound reported by Porter and Zscheile... It has previously been called a colorless, nonfluorescent polyene. 1966 *McGraw-Hill Encycl. Sci. & Technol.* II. 54/2 The synthesis of carotenoids leads..to the colorless compound, phytoene, which has the branched, 40-carbon chain characteristic of the carotenoid skeleton. 1978 *Sci. Amer.* Sept. 94/1 Carotenoids are derived from the 40-carbon compound phytoene.

phytogeography. Add: (Earlier and later examples.)
1847 H. C. Watson *Cybele Britannica* I. 2 This study has been variously denominated 'Phyto-Geography', 'Botanical Geography', 'Geographical Botany', and 'Geographical Distribution of Plants'. 1859 *Ibid.* IV. x. 373 Phytogeography traces out the history and distribution of plants in connexion with the geographical position. 1933 *Geogr. Jrnl.* LXXXI. 462 A further problem remains, far more serious in the case of zoogeography than with phytogeography. Can the essentials be put..without assuming a knowledge of taxonomy beyond the reach of the average geographer? 1956 H. Godwin (*title*) The history of the British flora: a factual basis for phytogeography. 1960 N. Polunin *Introd. Plant Geogr.* i. 2 Plant geography, also called phytogeography..deals with the plant cover of the world—with its composition, its local productivity, and particularly its distribution. 1976 *Conservation News* Sept./Oct. 3/1 The Library is one of the richest of its kind in the world, being particularly strong in plant taxonomy, phytogeography and related fields.
phytogeographer (earlier and later examples); **phytogeographical** *a.* (earlier and later examples).
1859 H. C. Watson *Cybele Britannica* IV. x. 375 The ultimate objects to be sought by phyto-geographical investigations, are neither the countries of plants nor the plants of countries. *Ibid.* 441 A first call on the phytogeographer is to ascertain where the plants are now distributed. 1951 *Geografiska Annaler* XXXIII. 149 The vegetative highland areas consist mainly of bogs, mostly of the type which in Icelandic is called *flá*. Phytogeographers are using *flá* as a phytogeographical concept. 1977 *Jrnl. R. Soc. Arts* CXXV. 225/1 The Swiss phytogeographer Alphonse de Candolle.

phytograph (fəi·tograf). *Ecol.* [f. Phyto- + Gr. -γραφος: see -graph.] A diagram showing several quantifiable criteria of the ecological status of a (usu. silvan) species within an association.
1930 H. J. Lutz in *Ecology* XI. 9 The term phytograph is proposed to designate a chart of this kind when used to characterize plant relationships. 1950 *Forestry Abstr.* XI. 580/2 Jack pine stands of equal age subjected to different degrees of thinning might be represented for purposes of comparison by phytographs constructed on 5 axes on which are shown their respective:—(1) average spacing of stems, [etc.]. 1971 F. C. Ford-Robertson *Terminol. Forest Sci.* 194/1 Phytograph... A polygonal figure depicting the role of a species in a plant community.

phytohæmagglutinin (fəi:to,hīmăglū·tinin). *Hæmatol.* Also (chiefly *U.S.*) **-hem-.** [f. Phyto- + *hæmagglutinin, hem-.] Any plant protein that is a hæmagglutinin.
1949 *Blood* IV. 673 We rediscovered this phytohemagglutinin accidentally and independently. 1951 *Amer. Jrnl. Med.* X. 776/2 The phytohemagglutinin of red beans (Phaseolus vulgaris) was recently isolated in pure form. 1955 *Jrnl. Biol. Chem.* CCXII. 615 This phytohemagglutinin, in either the mucoprotein or the protein form, is a non-toxic, powerful hemagglutinin of all types of human erythrocytes, and those of the horse, pig, [etc.]. 1960 *Cancer Res.* XX. 462 The mucoprotein plant extract, phytohemagglutinin (PHA), employed originally as a means of separating the leukocytes from whole blood in preparing the cultures, was found to be a specific initiator of mitotic activity: in its presence, cell division occurred; in its absence, no mitoses appeared. 1977 I. M. Roitt *Essent. Immunol.* (ed. 3) vi. 180 Comparable changes can be induced in lymphocytes by certain plant mitogens of which the best known are phytohaemagglutinin (PHA) and concanavalin A (conA). These..react with the cell surface non-specifically (i.e. not as an antigen) and produce the same series of cellular events as does antigen locking on to its specific surface receptor.

phytohormone (fəito,hǭ·ɹmōᵘn). *Bot.* [ad. G. *phytohormon* (Kögl & Smit 1931, in *K. Akad. von Wetenschoppen te Amsterdam: Proc. Sect. Sci.* XXXIV. 1416): see Phyto- and *hormone.] Any substance which has a hormonal effect on a plant; = *hormone 2.
1933 *Chem. Abstr.* XXVII. 1645 (*heading*) Vegetable growth substances... A phytohormone of cell extension. 1940 W. R. Fearon *Introd. Biochem.* (ed. 2) xxiv. 446 The effect of the phytohormones is non-specific as regards species. 1952 [see *hormone 2]. 1959 L. J. Audus *Plant Growth Substances* (ed. 2) i. 16 In the past a rigid distinction has been made between..'synthetic growth-regulating substances' and 'natural phytohormones'. 1971 *Nature* 4 June 332/1 Jump suggested that a phytohormone produced by *A[ureobasidium] pullulans* (*Dematium pullulans*) may have been responsible for forking in red pine.

phytol (fəi·tǫl). *Biochem.* [a. G. *phytol* (Willstätter & Hocheder 1907, in *Ann. d. Chem.* CCCLIV. 207): see Phyt- and -ol.] A colourless oil whose molecule forms part of that of chlorophyll and vitamins E and K, and which is an acyclic terpenoid alcohol; 3,7,11,15-tetramethyl-2-hexadecenol, $C_{20}H_{40}O$.
1907 *Jrnl. Chem. Soc.* XCII. i. 784 The authors have been able..to remove the magnesium quantitatively and to obtain for the first time an ashless compound closely related to chlorophyll. This derivative is an ester, termed phaeophytin, and is hydrolysed readily by alkalis, yielding an unsaturated alcohol, phytol, $C_{20}H_{40}O$, which is formed also by the action of alkalis on chlorophyll. For the hypothetical saturated hydrocarbon, corresponding with this alcohol, the authors propose the name phytane. 1911

[see *PHYTANE]. **1950** [see *PHÆOPHORBIDE]. **1974** *Sci. Amer.* Dec. 73/1 Attached to the porphyrin ring in chlorophyll is a long hydrocarbon 'tail', the phytol chain. It consists of carbon atoms linked together, but only one of the bonds is double. **1977** *Nature* 18 Aug. 621/2 The precursors of these hydrocarbons are unknown, but it is surmised that they result from the diagenesis of phytol, a moiety widely distributed among plant pigments.

Hence **phy·tyl** (-il, -əil), the univalent radical derived from phytol by the loss of a hydrogen atom.

1911 *Jrnl. Chem. Soc.* C. I. 145 Phytyl hydrogen phthalates, $CO_2H{\cdot}C_6H_4{\cdot}CO_2{\cdot}C_{20}H_{39}$, are formed when the phytol and phthalic anhydride are boiled for five hours with benzene. **1956** [see *PHÆOPHORBIDE]. **1973** *Nature* 20 July 155/1 Although other contributors, such as vitamin K_1, α-tocopherol and phospholipids from..bacteria and algae which have been shown to possess a phytyl side chain, are also possible, the most obvious source for the ketone is phytol.

phytolith. Restrict † *Obs.* to sense in Dict. and add: **2.** A minute mineral particle formed inside a plant.

1958 *Jrnl. Soil Sci.* IX. 154 *Nardus stricta*..contains phytoliths which could not be mistaken for those from any other of the grasses encountered. **1960** *Proc. R. Soc. Victoria* LXXII. 21 Phytoliths are important, if not abundant, constituents of dust in Australia. The common varieties are opal-phytoliths, but calcite-phytoliths and quartz-phytoliths have also been detected. **1964** *Ann. Bot.* XXVIII. 181 In some grasses..the absence or almost complete absence of phytoliths may be a characteristic of the species. **1975** *Nature* 17 Apr. 588/2 The abundance of opal phytoliths in the midden further indicate that there was then no scarcity of grass and reeds in the immediate vicinity of the site.

phytometer (fəitǫ·mītəɹ). *Bot.* [f. PHYTO- + -METER.] A plant or group of plants used to indicate, by its health and rate of growth, the physical properties of its surroundings.

1919 F. E. CLEMENTS et al. in *Carnegie Inst. Year Bk. 1918* 288 An endeavor has been made to devise a biological method of measuring habitats by means of standard plants. Batteries of such plants, for which the term 'phytometer' is suggested, have been installed at the various factor stations, and three series have been measured during the summer. **1929** WEAVER & CLEMENTS *Plant Ecol.* xiv. 330 Since phytometers are often placed in habitats that make extreme demands upon the plants, it is best to use species that are as hardy and vigorous as possible. **1954** *Austral. Jrnl. Bot.* II. 322 There is much to be said for the view that the complexes of environmental factors determining plant distribution can be indicated and measured better indirectly, through the plants themselves, than by direct physical measurements; this is, of course, the idea behind the use of 'phytometers' in agricultural meteorology. **1977** J. L. HARPER *Population Biol. Plants* xi. 361 It would be of immense interest to use standard plant units (phytometers), either pots freely supplied with water and nutrients or small water culture vessels with a test plant in each, and to place these units in transects across vegetation through open land, grassland, scrub and woodland.

phytopathogen (fəi:to,pæ·þodʒĕn). *Biol.* [f. PHYTO- + PATHOGEN.] Any micro-organism which produces disease in plants.

1934 in WEBSTER. **1943** *Phytopathology* XXXIII. 314 Experiments were performed to determine for the growth of the bacterial phytopathogens the suitability of a medium containing asparagin as the sole source of both carbon and nitrogen. **1959** *Ann. Rev. Microbiol.* XIII. 225 Pectinolytic enzymes are commonly formed by bacterial phytopathogens. **1976** *Nature* 15 Apr. 604/1 The ecological success of bracken is partly a result of its extensive rhizome system, and because of its ability to synthesise various secondary compounds which deter predators and phytopathogens.

phytopathogenic (fəi:to,pæþodʒe·nik), *a.* *Biol.* [f. PHYTO- + PATHOGENIC *a.*] Producing disease in plants. Hence **phy:topathogeni·city**, the property of being phytopathogenic.

1925 *Jrnl. Path. & Bacteriol.* XXVIII. 203 (*heading*) The investigation of phytopathogenic bacteria by serological and biochemical methods. *Ibid.*, An extensive literature has developed..with regard to the morphology and pathogenicity of various organisms isolated from phytopathogenic lesions. **1959** *Ann. Rev. Microbiol.* XIII. 224 Considerable effort has been expended in recent years in relating phytopathogenicity to discrete bacterial substances ('toxins') which induce disease symptoms in the host plant. *Ibid.* 225 Certain polysaccharides produced by phytopathogenic bacteria *in vitro* possess wilt-inducing properties. **1973** *Nature* 9 Mar. 87/3 Previous studies..postulated a symbiotic or parasitic relationship between phytopathogenic bacteria and insect 'hosts'.

phytophthora (fəitǫ·fþōrā). [mod.L., f. PHYTO- + Gr. φθορά destruction.] A fungus of the genus so called, belonging to the order Peronosporales and including several parasitic species which damage plants, esp. *Phytophthora infestans*, the cause of potato blight. Also *attrib.*

[**1876** A. DE BARY in *Jrnl. Bot.* XIV. 106 The characters..divide the..*Peronosporeæ* generally into two, perhaps better into three, genera, *Cystopus*, *Peronospora*, and a third, which may be called *Phytophthera*.] *Ibid.* 109 Up to this time the sexual organs have not been observed in *Phytophthora*, the Potato-fungus. **1925** *Jrnl. Agric. Res.* XXX. 463/2 Practically all of the fruit on the ground showed the typical symptoms of Phytophthora rot. **1930** *Phytopathology* XX. 209 Oospores had been reported previously in these two Phytophthoras. **1961** A. SCHOENFELD tr. *Stapp's Bacterial Plant Pathogens* II. 127 Together with *phytophthora* rot and leaf curl disease, it [*sc.* bacterial ring-rot] is..one of the three most severe potato diseases in the U.S.A. **1969** HAWKES & HJARTING *Potatoes of Argentina, Brazil, Paraguay, & Uruguay* II. 266 It may therefore be possible to transfer valuable qualities of *Phytophthora* resistance from these Mexican series.

phytoplankter (fəi·to,plæŋktəɹ). *Biol.* [f. PHYTO- + *PLANKTER.] A phytoplanktonic individual or species.

1957 G. E. HUTCHINSON *Treat. Limnol.* I. xiv. 798 Apart from diatoms, a few phytoplankters use silica. **1960** N. POLUNIN *Introd. Plant Geogr.* xvi. 517 The rate of sinking of a body heavier than water, as most phytoplankters (phytoplanktonic individuals) are, depends upon the ratio of surplus weight to friction. **1967** *New Scientist* 2 Feb. 276/3 The zoologist studies the nutritional requirements and efficiencies of food conversion of marine animals (zooplankters) feeding on marine phytoplankters or other small animals. **1972** *Science* 5 May 535/1 These substances are rapidly absorbed from water by organisms, including phytoplankters. **1976** *Nature* 17 June 584/1 *Gonyaulax*, a marine phytoplankter, is normally grown in a medium of approximately 500 mM salt.

phytoplankton (fəi·to,plæŋktǫn). *Biol.* [f. PHYTO- + PLANKTON.] The microscopic plants forming part of the plankton. Also *attrib.* So **phytoplankto·nic** *a.*

1897 P. T. CLEVE (*title*) A treatise on the phytoplankton of the Atlantic. **1900** *Geogr. Jrnl.* XV. 336 In the spring months there is a great development of bacteria and other Phytoplankton, which render the water less transparent than at other times of the year. **1909** E. WARMING *Oecol. Plants* xxxvii. 155 Phytoplankton..always consists of minute plants. *Ibid.*, Phytoplankton-organisms are all minute. **1928** *Daily Express* 28 May 10/7 Miss S. M. Marshall..is described technically as the 'phytoplankton worker'. **1944** *Jrnl. Marine Biol. Assoc.* XXVI. 285 Phytoplanktonic diatoms may utilize ammonium nitrogen in preference to nitrate nitrogen. **1956** *Nature* 3 Mar. 438/1 Smaller phytoplankton elements which will pass through the finest nets are of great importance in the productivity of the oceans. **1957** G. E. HUTCHINSON *Treat. Limnol.* I. xi. 714 These forms of iron may be assimilable by diatoms and perhaps by other phytoplanktonic organisms. **1964** *Oceanogr. & Marine Biol.* II. 137 Possibly fallout from a phytoplankton bloom would sometimes cause the immediate surface organisms to multiply. *Ibid.* 149 The amino acid compositions of phytoplankton and pure cultures of phytoplanktonic species have been determined. **1973** *Nature* 3 Aug. 307/1 The crustaceans in turn feed at least partially on the phytoplankton. **1977** P. B. & J. S. MEDAWAR *Life Sci.* i. 17 The most important—because they are the most abundant—organisms involved in the capture of carbon dioxide and the liberation of oxygen are forest trees and minute plants carried in the surface layers of the sea—'phytoplankton'.

phytosanitary (fəi:to,sæ·nitări), *a.* [f. PHYTO- + SANITARY *a.*] Pertaining to the health of plants; applied *spec.* to a certificate stating a plant is free from infectious diseases.

1949 L. LING *Digest of Plant Quarantine Regulations* (U.N. Food & Agric. Organization) July 5 Importation into Belgian Congo requires a special permit in accordance with the prescribed conditions: the presentation of a phytosanitary certificate of origin. *Ibid.* 35 Shipments of plant materials must be accompanied by certificates indicating origin and phyto-sanitary conditions. **1960** E. GRAM in Horsfall & Dimond *Plant Path.* III. ix. 332 Countries with no phytosanitary service..are the happy hunting grounds of botanists and geneticists. **1976** *Nature* 12 Feb. 449/2 Much of this material is accompanied by phytosanitary certificates of the exporting countries which are not necessarily the countries of origin.

phytosociology (fəi:to,sōᵘʃiǫ·lŏdʒi, -sōᵘsiǫ·l-). [ad. Russ. *fitosotsiologiya*: see PHYTO- and SOCIOLOGY.] The study of plant communities, their composition and structure. So **phytosociolo·gical** *a.*, **phytosociolo·gically** *adv.*; **phytosocio·logist**, one engaged in this study.

1928 K. D. GLINKA in *Proc. & Papers 1st Internat. Congr. Soil Sci.* I. 129 These ideas of Dokuchaiev could not but have the greatest influence on Russian works referring to..botanical geography, phytosociology, forestry. *Ibid.*, The present day phytosociologist begins to pay an increasing attention to the geography and topography of soils. **1928** *Jrnl. Ecol.* XVI. 18 The establishment of ecological series of spruce communities gives us a clear idea of the phytosociological and genetical relations between these communities. **1932** FULLER & CONARD tr. *Braun-Blanquet's Plant Sociol.* p. vii, Phytosociology, the study of vegetation, has had a very rapid development. **1955** *Jrnl. Ecol.* XLIII. 226 (*heading*) The use of phytosociological methods in ecological investigations. *Ibid.* 227 Plant sociology (or phytosociology) is defined as the discipline which concerns itself with the study of vegetation as such, with its floristic composition, structure, development and distribution. **1967** M. E. HALE *Biol. Lichens* vii. 87 Phytosociology is probably the most thoroughly investigated aspect of lichen biology. **1969** *Nature* 19 Apr. 242/2 Specialized communities..are phytosociologically distinct. **1970** *Watsonia* VIII. 172 The phytosociological classification of the vegetation of the limestone dales. **1973** *Nature* 30 Mar. 354/2 In the chapter dealing with the phytosociology of heathlands, the reader might be forgiven for gaining the impression that..the southern limit of heathland is France. *Ibid.* 21–28 Dec. 537/1 Continental phytosociologists may feel that some of their carefully defined association binomials have been applied a little loosely. **1974** *Ibid.* 24 May 307/3 Bushmeat species are remarkably resistant to infections that kill domestic stock, and they seldom destroy the phytosociological structure of their habitats. **1976** *Jrnl. R. Soc. Arts* CXXIV. 640/2 Between 1960 and 1973 botanical surveys and collections of plants were made throughout the maritime Antarctic and South Georgia. From these, descriptive floras together with ecological and phytosociological accounts have been produced. **1977** *Dædalus* Fall 130 In some of its branches, such as biogeography and phytosociology, the systems of classification and quantitative description reached phantasmagoric extremes.

phytosterol (fəitǫ·stĕrǫl). *Biochem.* [f. PHYTO- + *-STEROL.] Any of a large class of sterols which are found in plants; orig. *spec.* = *phytosterin* s.v. PHYTO-.

1898 *Jrnl. Chem. Soc.* LXXIV. I. 598 The soluble fat, which is dark green, consists principally of the ethereal salts of phytosterol, $C_{26}H_{44}O + H_2O$ (melting at 137°), and oleic and palmitic acids. **1906** *Ibid.* XC. II. 311 The author has submitted a number of cholesterols and phytosterols to Neuberg and Rauchwerger's colour test. **1915** J. LONDON *Jacket* v. 34 Professor Schleimer had similarly been collaborating with me in the detection of phytosterol in mixtures of animal and vegetable fats. **1939** MEYER & ANDERSON *Plant Physiol.* xxiii. 396 Cholesterol is not known to occur in the higher plants, but a number of similar compounds, known as the phytosterols, have been isolated from plant tissues. **1975** D. JARVIS tr. *Hess's Plant Physiol.* 108 The biosynthesis of the phytosterols..is not completely understood.

phytotoxic (fəito,tǫ·ksik), *a.* [f. PHYTO- + TOXIC *a.*] Poisonous or injurious to plants.

1933 *Proc. Nat. Acad. Sci.* XIX. 487 Macht showed that the blood of leprosy patients also gives a specific phytotoxic reaction. **1953** *Phytopathology* XLIII. 663/1 Ethylene..arises from plant tissue injured by such phytotoxic compounds as endothal. **1963** *Ann. Rev. Microbiol.* XVII. 223 The injurious effects of phytotoxic materials traditionally have been treated as plant diseases. **1976** *Weed Abstr.* XXV. 108/1 No treatment was phytotoxic to the blueberries.

Hence **phy:totoxi·city**, the property of being phytotoxic.

1945 J. G. HORSFALL *Fungicides* xvi. 172 Phytotoxicity is a term to describe the injuriousness of fungicides to host plants. **1958** *Times* 4 Dec. 1/5 (Advt.), Agricultural Research Council require a Plant Physiologist..to work on fundamental aspects of phytotoxicity. **1974** *Nature* 8 Feb. 337/3 Fungicides..must be selected and timed so that disease control is achieved with minimal phytotoxicity.

phytotoxicant (fəito,tǫ·ksikănt). [f. PHYTO- + TOXICANT *a.* and *sb.*] A substance poisonous or injurious to plants; *esp.* one present in the air. Also *attrib.* or as *adj.*

1959 *Internat. Jrnl. Air Pollution* I. 160 The decreased damage at the higher olefin concentration can be explained by the more rapid consumption of ozone to form the phytotoxicant which then partly decomposes before reaching the plants. **1961** *Ann. Rev. Plant Physiol.* XII. 431 The principal phytotoxicants recognized several decades ago were fairly specific, simple molecules, such as ethylene, fluorides, hydrogen sulfide, and sulfur dioxide. **1971** *McGraw-Hill Yearbk. Sci. & Technol.* 91/2, O_3 has been recognized as a phytotoxicant for at least 100 years. **1977** I. M. CAMPBELL *Energy & Atmosphere* viii. 248 The level of sophistication which has now been attained has resulted in the identification of single phytotoxicant species, together with the development..of computer simulation procedures for photochemical smog formation.

phytotoxin (fəito,tǫ·ksin). [f. PHYTO- + TOXIN.] **1.** Any toxin derived from a plant.

1909 R. J. M. BUCHANAN *Blood in Health & Dis.* xii. 289 Of chemical agents causing hæmorrhage, bacterial products take a prominent place, but many poisons, phytotoxins (plant poisons), zootoxins (animal poisons)..are especially effective. **1924** *Chem. Abstr.* XVIII. 1857 (*heading*) The protein phytotoxins, with special reference to the new 'modeccin'. **1974** *Nature* 18 Oct. 628/1 Very low concentrations of phytotoxins can kill mammalian cells, and some of them..have been reported to suppress protein synthesis in cells and cell-free systems.

2. A substance poisonous or injurious to plants, esp. one produced by a parasite.

1962 *Phytopathology* LII. 586/1 (*heading*) Production of a phytotoxin by *Rhizoctonia solani*. **1963** *Ann. Rev. Microbiol.* XVII. 224 The term 'phytotoxin' will be used for all products of living organisms toxic to plants. The designation, phytotoxin, does not imply that such a material plays any role whatever in relation to any disease caused by a pathogen. **1970** STROBEL & MATHRE *Outl. Plant Path.* xxii. 318 Phytotoxins induce few or none of the symptoms usually caused by the pathogen from which they originate. Compounds such as lycomarasmin, alternaric acid, and numerous smaller molecular weight organic acids fall into this category.

phytotron (fəi·tŏtrǫn). [f. PHYTO- + *-TRON: see quot. 1949.] A laboratory where plants can be maintained and studied under a wide range of controlled conditions.

1949 *Plant Physiol.* XXIV. 553 The word phytotron was coined for the new laboratory by Dr. R. A. Milli-

kan..who feels that it will play the same role in plant physiology and the applied plant sciences..as the cyclotron has already had in pure and applied physics. **1958** *Observer* 7 Sept. 8/4 The Australian Government is to spend £500,000 on a new type of phytotron. **1974** [illegible] The phytotron would enable scientists to grow four crops of rice a year and thus speed up research. **1976** *Sci. Amer.* Mar. [illegible] A phytotron is a combination of temperature-controlled greenhouses and artificially lighted rooms; here temperature, light and humidity can be controlled at will for the purpose of studying the response of plants to their environment.

phytyl: see *PHYTOL.

phyzog, var. *PHIZOG.

pi, *sb.* Add: **2.** *Electr.* Applied to a four-terminal set of three circuit elements in which one element is in series between two in parallel. Usu. written as π or Π.

1924 K. S. JOHNSON *Transmission Circuits for Teleph. Communication* xi. 114 If the structure shown in Fig. 1 is considered to be made up of Π sections,..each section may be regarded as terminating in a mid-shunt iterative impedance. [illegible]

3. *Physics* and *Chem.* [After *P III. 2.] Used to designate electrons, orbitals, molecular states, etc., possessing one unit of angular momentum about an internuclear axis; *pi-* (or *π-*) *bond*, a bond formed by a π-orbital.

Usu. written π when it refers to one electron or orbital and Π when it refers to a molecule as a whole.

[illegible]

pi, *a.* (*sb.*) Add: (Later examples as *adj.*) See also PIE *a.*[1] in Dict. and Suppl.

[illegible] **1968** J. R. ACKERLEY *My Father & Myself* viii. 79 He invited the two of us into the billiard-room..for a 'jaw', which could hardly be called 'pi' and which he himself described as 'man to man'. **1972** *South China Morning Post* (Hong Kong) 20 Nov. 16/4 The subject emerges, I believe falsely, as a kind of overgrown school-prefect, bossy, frustrated and a bit pi. **1973** J. HITCHMAN *Such Strange Lady* ii. 27 'That were only sparrers... They aren't good for nothing.' 'God made them,' retorted the clergyman's daughter, 'So they must be good for something'. All very 'pi' of course. **1978** *Broadcast* 3 Apr. 42/2 'Blue Peter', though never so pi or so holier than thou, is always on the side of the..decencies.

piaculative (piæ·kiŭlātiv), *a.* *rare.* [f. L. *piāculum* PIACLE + -ATIVE.] = PIACULAR *a.* 1.

1919 T. S. ELIOT *Poems*, The young are red and pustular Clutching piaculative pence.

Piagetian (piäʒi·ʃən), *a.* [f. the name of Jean Piaget (1896–1980) + -IAN.] Of or pertaining to the theories or methods of Piaget, Swiss educational child psychologist. Also **Piagetan** (piä·ʒətan).

1960 *Jrnl. Child Psychol.* I. 191 (*heading*) Some points of Piagetian theory in the light of experimental criticism. **1973** *Jrnl. Genetic Psychol.* CXXII. 27 This study has as its points of departure the Piagetan conceptions of the nature and development of logical structures. *Ibid.* CXXIII. 277 The large number of investigations that have tried to establish normative data in the area of Piagetian conservations. **1974** *Nature* 30 Aug. 713/2 The Piagetian tests, which are intended to reflect changes in mental maturity, are sensitive to age differences. **1977** *Language* LIII. 483 Arguing from a Piagetian perspective of cognitive development, B. concludes that children know very little about syntax prior to the use of two-word utterances.

piai: see PEAI *sb.*

Pian (pi·ən), *a.* [f. L. *Pius* (see PIOUS *a.*), a name adopted by several Popes + -AN.] Of or pertaining to Pius; *spec.* of or pertaining to the pontificate or liturgical reforms of Pope Pius V or Pope Pius X.

[illegible]

pianette. Add: (Further example.) Also **pianet.**

1894 G. B. SHAW *How to become Mus. Critic* (1960) 222 [illegible] **1908** G. B. SHAW [illegible] Has Chesterton ever spent his last guinea on an opera by Meyerbeer or Verdi, and sat down at a crazy pianet to roar it and thrash it through?

pianism. Add: **a.** (Further examples.) **b.** The art of composing for the piano, *spec.* the particular skill or characteristic style of a composer of music for the piano; the action or art of arranging a musical composition for performance on the piano.

1889 *Cent. Dict.*, *Pianism*, the adaptation of a piece of music to effective performance on the pianoforte. [illegible]

pianist. Add: **c.** in appositive *Combs.*, as *pianist-arranger*, *-composer*, *-conductor*, *-leader.*

1955 KEEPNEWS & GRAUER *Pict. Hist. Jazz* xx. 264 [illegible]

pianistic, *a.* Add: Also, pertaining to or suitable for performance on the piano; of, pertaining to, or characteristic of pianism.

[illegible] Note the way in which the pianistic idiom..has been translated into its orchestral equivalent. **1942** E. PAUL *Narrow St.* xxiii. 203 Only one pupil..had such pianistic genius. **1950** *Chambers's Encycl.* VIII. 67/1 The same 12-bar pattern..serves as a basis for the 'Boogie Woogie', a typically pianistic species. **1955** *Times* 4 July 12/4 Two pieces by his mother, Teresa de Rogatis, demonstrated the pianistic fluency of both mother and son. **1958** *Listener* 21 Aug. 285/1 Leonard Borwick's exquisitely pianistic version of Debussy's 'L'Après midi d'un faune'. **1975** *Ibid.* 18 Sept. 378/1 The most outstanding pianistic feat..Dmitry Alexeev's playing of Prokofiev.

B. *sb. pl.* The art of playing the piano, *spec.* skilful technique in piano-playing.

1950 BLESH & JANIS *They all played Ragtime* viii. 149 The formidable eight-to-the-bar pianistics of Meade Lux Lewis, Albert Ammons of Chicago, and Pete Johnson of Kansas City. **1958** P. GAMMOND *Decca Bk. Jazz* xv. 184 His dazzling and immensely skilful display of pianistics.

Hence **piani·stically** *adv.*, in a pianistic manner; from a pianistic point of view.

1926 W. C. HANDY *Blues* 20 Instrumentally and in particular pianistically, the mode was backward. **1928** *Daily Tel.* 5 June 9/5 She would have approached nearer to the real Bach if she had thought less pianistically in the matter of tone-colour. **1946** R. BLESH *Shining Trumpets* (1949) xiii. 317 Pianistically, it has passages of great beauty but it is not completely Negroid. **1971** *Daily Tel.* 6 Apr. 12/7 The most remarkable work of the evening pianistically—and pianism was this programme's concern—was the Grand Concert Fantasy on Themes from Bellini's 'Sonnambula'. **1977** *Gramophone* June 63/1 Berman's playing here has a child-like directness allied to playing which is, pianistically, very complex with all manner of material in the inner parts coming interestingly and engagingly through.

piano, *a.* (*adv.*) *sb.*[1] Add: **1. a.** *adj.* (Examples of *fig.* use.)

a **1817** JANE AUSTEN *Persuasion* (1818) IV. vi. 120 James Benwick is rather too piano for me. **1900** E. GLYN *Visits of Elizabeth* 188 The Marquis..looked thoroughly worn out and as *piano* as a beaten dog. **1922** A. HUXLEY *Let.* 9 Sept. (1969) 209 Aunt Nettie is with us: but happily she is in a very calm and piano mood so that she is quite an agreeable companion. **1941** [see *EXALTÉ *a.*]. **1953** E. M. FORSTER *Hill of Devi* 138 Very piano and tired, poor dear.

piano, *sb.*[2] Add: **1. b.** Piano-playing.

1946 J. CARY *Moonlight* viii. 55, I was looking forward to some real old romantic piano, with the genuine macassar flavour. **1950** BLESH & JANIS *They all played Ragtime* iii. 55 Turpin, who had pioneered syncopated piano in the Missouri metropolis. **1968** *Blues Unlimited* Sept. 10, I heard..some straggling piano. **1975** *Listener* 11 Dec. 796/2 A boy with a musical ear, who had picked up piano from an old black man in a city dive. **1978** *Gramophone* July 178/1 He then went on to study piano and finally, almost by accident, conducting.

c. Chiefly *U.S. slang.* = SPARE-RIB.

1911 J. W. HORSLEY *I Remember* xi. 254 Ingenuity and humour are sometimes the parents of slang terms..I was reminded by..'piano' for ribs of beef. **1942** Z. N. HURSTON in *Amer. Mercury* July 96/1 *Piano*, spare ribs (white rib-bones suggest piano keys).

2. a. *piano bench*, *-case* (examples), *concerto*, *-lamp*, *piece*, *recital*, *recording*, *rehearsal*, *solo*, *sonata*, *-stool* (examples); **b.** *piano-player* (later examples), *-playing* (earlier and later examples), *-thumping* vbl. sb. and ppl. adj., *-tuner* (examples), *-tuning* vbl. sb. and ppl. adj.; **d. piano-accordion** (see *ACCORDION); **piano-conductor**, a musician who conducts a band while playing the piano part; **piano part**, the part assigned to the piano in concerted music; **piano quartet** = *pianoforte quartet* s.v. *PIANOFORTE b; **piano quintet** = *pianoforte quintet* s.v. *PIANOFORTE b; **piano rag**, a rag for performance on the piano (see *RAG *sb.*[5] 2); **piano reduction**, an arrangement of orchestral music for performance on the piano (see *REDUCTION 11 d); **piano roll** = *music-roll* (*b*) s.v. *MUSIC *sb.* 12d; also (with hyphen) *attrib.*; **piano score**, a pianoforte score, *spec.* a condensed version of a full score for performance on the piano; **piano trio** = *pianoforte trio* s.v. *PIANOFORTE b; **piano wire**, a special kind of strong steel wire used for the strings of pianos (see quot. 1956); a length of this; cf. **pianoforte wire*; **piano-writing**, composing for the piano.

1914 G. B. SHAW *Pygmalion* (1916) II. 137 Higgins..takes refuge on the piano bench. **1977** *New Yorker* 19 Sept. 62/3 She picked up the score from the piano bench. **1850** *Rep. Comm. Patents 1849* (U.S.) I. 300 What I claim..is a piano case or trunk lock. **1876** J. S. INGRAM *Centenn. Exposition* x. 334 Another very creditable piece of work was a rosewood piano-case. **1908** *Sears, Roebuck Catal.* 210/2 Every Beckwith piano case..is double veneered inside and outside. **1879** GROVE *Dict. Mus.* I. 387/2 In the whole of his piano concertos..we find an allegro, a slow movement, and a finale in quick time. **1934** A. L. BACHARACH *Mus. Compan.* iv. 598 There are far greater depths in the two piano concertos of Brahms. **1975** *Times* 22 Mar. 11/3 With 18 versions of Schumann's piano concerto already available, an immediate reaction to HMV's latest could well be 'why another?' **1934** S. R. NELSON *All about Jazz* ii. 43 The piano-conductor part of a commercial orchestration is very similar. **1957** MANVELL & HUNTLEY *Technique Film Music* i. 21 A series of scenes involving Lynch and Colonel Cameron, occupying two pages of the piano-conductor score. **1896** *Whims* Apr. 72 Ushered in, he found Margaret doing fancy-work under a large piano-lamp in the parlor. **1897** *McClure's Mag.* Nov. 57 The soft, yellow light from the shaded piano-lamp fell about her. **1909** *Cent. Dict.* Suppl. 996/3 *Piano-lamp*, a lamp intended for use with a piano: usually one with a firm tripod or other base and an adjustable standard by

which it can be raised or lowered. **1934** A. L. BACHARACH *Mus. Compan.* v. 482 There are..modern chamber works weighted with piano parts that make as heavy demands on virtuosity as any concerto. **1955** G. ABRAHAM in H. Van Thal *Fanfare for E. Newman* 26 Telling touches were added..notably the sharpening of the D in the piano-part. **1969** *Listener* 2 Jan. 25/1 Another [friend] was forbidden to have his new work performed unless he omitted some offending clusters from the piano part. **1897** *Sears, Roebuck Catal.* 538/1 Easy and medium piano pieces..just the collection for home. **1935** E. FARJEON *Nursery in Nineties* 283 We drew up a programme..a group of Harry's last piano-pieces, a Solo by Joe. **1977** K. O'HARA *Ghost of T. Penry* iv. 32 Faintly..came the sound of the same piano piece. **1906** *Chambers's Jrnl.* 31 Mar. 286/2 Anything in the nature of a mechanical piano-player uncontrolled by the taste and discretion of some guiding mind..is naturally abhorrent. Yet the instrument under review is..absolutely automatic. **1907** *Westm. Gaz.* 17 Sept. 10/1 The most rapid growth has been in mechanical piano-players and piano-playing attachments. **1933** H. S. WALPOLE *Vanessa* III. iv. 558 A typist whom Alfred had engaged actually owned a mechanical piano-player, bought of course on hire-purchase. **1964** L. DEIGHTON *Funeral in Berlin* x. 64 The piano-player did a fancy cadenza. **1857** C. G. LELAND in *Graham's Illustr. Mag.* May 458/1 Hans Breitmann gife a barty—dey had biano-blayin. **1859** GEO. ELIOT *Let.* 7 Oct. (1954) III. 172, I had abundant time and opportunity for hours of piano-playing. **1956** 'C. BLACKSTOCK' *Dewey Death* vii. 169 She persistently told the world about..her piano playing and her temperament. **1977** *Times* 15 Nov. 17/5 The idiosyncrasy of his piano-playing. **1934** A. L. BACHARACH *Mus. Compan.* viii. 488 During the same four years he wrote also a piano trio..and the first of his piano quartets. **1975** *Radio Times* 15 May 32/3 Dvořák... Piano Quartet in D major, Op. 23: Beaux Arts Trio with Walter Trampler (viola). **1925** E. SACKVILLE-WEST *Piano Quintet* i. 1 A piano quintet, they were to start in a month's time upon a professional tour on the Continent. **1934** A. L. BACHARACH *Mus. Compan.* viii. 488 There are many who regard this work as the direct ancestor of the great piano quintets of Schumann and Brahms. **1959** *Collins Mus. Encycl.* 525/2 The combination of piano with string quartet is called a piano quintet. **1972** *Jazz & Blues* Oct. 32/1 (Advt.), Joshua Rifkin's first Nonesuch album of piano rags. **1974** *Melody Maker* 20 Apr. 46 The first piano rag publication was 'Mississippi Rag' in 1897 by William H. Kvell. **1881** *Harper's Mag.* May 814/1 Piano and organ recitals have long been fashionable. **1950** L. SALTER *Going to Concert* 99 At everything but piano recitals, there is a second person involved who has it..in his power to turn the whole affair into a success or a failure. **1972** E. GREENFIELD *Penguin Guide to Bargain Records* No. 3. 297 As a popular piano recital this seems wholly successful. **1962** Piano recording [see *piano-stool* below]. **1975** *Hi-Fi News* Feb. 131/3 One of the most unspectacularly natural piano recordings I have heard for a long time. [**1944** W. APEL *Harvard Dict. Mus.* (1946) 631/2 *Réduction*, Fr., arrangement. *Piano réduction*, arrangement for piano.] **1966** *Listener* 29 Dec. 976/2 Autograph or copyist's manuscripts are still turning up. To mention one striking find of recent years, a copyist's manuscript of a piano reduction of the second Prologue of *Romeo and Juliet*. **1979** E. WILKINS et al. tr. *Sel. Lett. Gustav Mahler* 209, I hope you have received the piano reduction. **1966** *Listener* 10 Nov. 704/3, I had coached most of the singers..and I also played for his piano rehearsals. **1926** WHITEMAN & MCBRIDE *Jazz* viii. 170 The mechanical royalties arising from phonograph and piano roll records are usually protected in the contract. **1956** M. STEARNS *Story of Jazz* (1957) xiii. 143 Most piano-roll companies employed a hack who could 'rag' any tune for issue on a piano roll. **1965** *Listener* 1 Apr. 501/2 They did make a considerable number of piano-rolls, but with the ascendancy of the gramophone in the twenties, these perforated screeds of paper were relegated to the lumber-room of the past. **1973** *Guardian* 7 Mar. 10 'Maple Leaf Rag' underwrote [Scott] Joplin's reputation... His own piano rolls, cut late in his career when his health was failing, hardly do him justice. **1929** H. CRANE *Let.* 26 Feb. (1965) 339 They've promised to publish *The Bridge*—on sheets as large as a piano score. **1938** *Oxf. Compan. Mus.* 851/2 A Piano Score is the reduction of an orchestral score to a piano version. **1964** D. FRANCIS *Nerve* ii. 16 My mother's grand piano lay inches deep in piano scores. **1977** R. BARNARD *Death on High C's* xviii. 185 The piano score of *Rigoletto*. **1923** J. REITH *Diary* 19 Mar. (1975) 131 There was no piano solo so I..got him to play Schubert's 'Marche Militaire'. **1943** L. ABBOTT *Approach to Music* iv. 74 Schubert composed his 'Serenade' as a song, but Liszt transcribed it for piano solo. **1977** *Listener* 17 Feb. 215/3 The atmospheric 'Slow Drag'..includes a delicate piano solo by Kenny Kersey. **1859** A. J. MUNBY *Diary* 25 May in D. Hudson *Munby* (1972) 33 Songs, serious & comic: piano-sonatas, and duets and trios with violin and violoncello. **1924** M. KENNEDY *Constant Nymph* xvi. 214 A piano sonata which Sebastian was to play at a concert. **1975** *Gramophone* Mar. 1629/1 She could play a piano sonata by ear as a child. **1847** C. BRONTË *Jane Eyre* II. iii. 82 She turned round on the piano-stool. **1877** E. S. WARD *Story of Avis* 335 Barbara Allen sat on the piano-stool. **1930** G. GREENE *Two Witnesses* 34 We would finish up the evening sitting on either side of him on the piano-stool while he played and sang, in his funny husky tenor voice. **1962** A. NISBETT *Technique Sound Studio* 38 A piano-stool squeak can spoil a piano recording. **1906** *Westm. Gaz.* 11 Aug. 4/3 In these days of universal piano-thumping. *c* **1909** D. H. LAWRENCE *Collier's Friday Night* (1934) II. 30 I'll bet it was a thump! Pomp! Pomp! (*Makes a piano-thumping gesture.*) **1934** A. L. BACHARACH *Mus. Compan.* vii. 486 The majority of his numerous piano trios belong to this early period. **1977** *Listener* 20 Jan. 94/2 Detailed analyses of the piano trios, piano sonatas, string quartets and symphonies. **1858** *Boyd's Philad. City Business Directory* 223 (*heading*) Piano Tuners. **1900** H. A. JONES *Mrs. Dane's Defence* I. 17 A young piano-tuner. **1973** A. CHRISTIE *Postern of Fate* II. iii. 80 It's the gentleman what's come to do the piano... You said I'd have to get a piano tuner. **1897** *Sears, Roebuck Catal.* 538/1 *Piano tuning hammer*... Long Rosewood handle, double head with oblong holes, and single head with star hole. **1974** *Times* 21 Oct. 24/3 (Advt.), Piano tuning and repairs. **1975** *Times* 2 Jan. 3/4 Total of piano-tuning trainees trebled. **1870** *English Mechanic* XII. 215/3 Piano wire..—What are the different sizes of wire as now used for a 7 octave cottage piano? **1938** J. WILLIAMSON *Surveying & Field Work* (ed. 2) xvii. 430 Two heavy weights are suspended by five piano wires from an overhead frame. **1955** W. GADDIS *Recognitions* I. iv. 161 All you got to do is get them dumb cops on their motorcycles, and string a good piece of piano wire across the road. **1956** A. K. OSBORNE *Encycl. Iron & Steel Industry* 317/1 *Piano wire*, wire made from the best-quality plain carbon steel, with a carbon content of about 0·80% to 0·95%. It is usually drawn to a tensile stress of more than 120 tons per sq. in. **1973** R. BUSBY *Pattern of Violence* viii. 122 Black hostile faces confronted him in an atmosphere as taut as a piano-wire. **1977** *Sci. Amer.* Oct. 74/3 These light-weight gliders gave MacCready the idea of using an aluminum-tube skeleton braced with piano wire. **1946** E. LOCKSPEISER in A. L. Bacharach *Brit. Music* xv. 196 There is often clumsiness in the piano-writing, too monotonous an insistence on pattern, or a lack of finish in the way he handles a phrase. **1959** D. COOKE *Lang. Mus.* i. 6 Comparisons such as those of piano-writing with black-and-white drawing, of orchestration with colour.

pianoforte. Add: **b.** *pianoforte concerto*, *-player* (earlier and later examples), *recital*, *solo*, *sonata*, *-tuner* (examples); **pianoforte jump, obstacle**, a jump or obstacle in a steeplechase whose shape resembles that of a pianoforte; **pianoforte quartet**, a quartet for violin, viola, cello, and pianoforte; **pianoforte quintet**, a quintet for pianoforte and string quartet; **pianoforte score** (see quot. 1876); **pianoforte trio**, a trio for violin, cello, and pianoforte; † **pianoforte wire** = *piano wire* s.v. *PIANO *sb.*[2] 2 d.

1932 *Radio Times* 29 July 269/2 Hilda Bor and orchestra—pianoforte concerto in D minor—*Mozart*. **1934** A. L. BACHARACH *Mus. Compan.* iv. 591 The pianoforte concertos of the modern repertory begin with J. S. Bach, who arranged sixteen violin concertos of Vivaldi for the clavier. **1962** G. MOORE *Am I too Loud?* v. 52 Many people in different parts of the world have given it as their opinion that he [*sc.* Solomon] is the finest player of a pianoforte concerto they have ever heard. **1908** *Daily Chron.* 9 June 3/5 Some of the Italian officers will give a display of what is known as the Pianoforte jump. **1909** *Westm. Gaz.* 8 Mar. 6/4 The 'pianoforte' obstacle will consist of four feet of water, followed by a sloping bank of turf with a three-foot wall at the end of it. **1836** DICKENS *Sk. Boz* II. 271 The pianoforte player..fainted away. **1849** MRS. GASKELL *Let.* 13 May (1966) 829 Benedict the great piano-forte player. **1887** C. H. H. PARRY *Stud. Gt. Composers* iv. 108 Haydn is said to have persisted in regarding Beethoven as a pianoforte player, and not as a composer. **1883** GROVE *Dict. Mus.* III. 58/2 Next to the string quartet ranks the pianoforte quartet, which, however, is built on quite a different principle. **1954** *Ibid.* (ed. 5) I. 878/1 There is a precedent in the slow movement of Schumann's pianoforte Quartet. **1828** E. HOLMES *Ramble among Musicians of Germany* 282 M. Herder played a pianoforte quintet in E flat of his composition. **1954** *Grove's Dict. Mus.* (ed. 5) I. 877/1 The pianoforte Quintet in F minor..was originally composed, but not published, as a string Quintet with two cellos. **1914** G. B. SHAW *Misalliance* 56 How many of them could be bribed to attend a pianoforte recital by a great player? **1876** STAINER & BARRETT *Dict. Mus. Terms* 353/2 *Pianoforte score*, a score of a vocal or instrumental composition, under which is written in two lines a condensed form of the harmonies for the use of a pianoforte. **1912** W. OWEN *Let.* 11 Jan. (1967) 108, I shall substitute for it a pianoforte solo. **1954** *Grove's Dict. Mus.* (ed. 5) V. 946/2 Of other works for pianoforte solo the Fantasy in C minor..is of special importance. **1883** *Ibid.* (ed. 1) III. 577/1 The slow movements of both are very well known; that of the Pianoforte Sonata being the Funeral March. **1954** *Ibid.* (ed. 5) V. 947/2 It was in fact Mozart's own work in that form and was intended to become part..of a complete pianoforte sonata. **1889** *Ibid.* (ed. 1) IV. 172/1 Pianoforte trios, as they are called, caused all others to retire into the background. **1954** *Ibid.* (ed. 5) V. 947/2 The pianoforte trios were mainly written for performance at private music meetings. **1801** BUSBY *Dict. Mus.* s.v. *Tuning fork*, This instrument [*sc.* a tuning-fork] is chiefly used by harpsichord and piano-forte tuners. **1861** DICKENS *Gt. Expect.* I. xi. 183 At the pianoforte-tuner's across the street. **1838** *Osborne's Guide Grand Junction Railway* (Advt. section) 113 Piano Forte Wires and Roman Strings. **1870** *English Mechanic* 30 Sept. 35/2 Pianoforte wire.—What weights it will support, etc. **1902** *Chambers's Jrnl.* June 413/2 This kite is of the box or Hargreave pattern... Its 'string' consists of four miles of pianoforte-wire.

Pianola. Add: Also, a piano which incorporates such a mechanism, a player-piano. Also *attrib.* (Further examples.)

1908 *Daily Chron.* 31 July 5/4 The ingenious contrivance of an Australian engineer who claims to have invented an accurate recording target... The figures are printed on paper, which runs at the required speed on two rollers on what may be described as the pianola principle. **1912** *Collier's* 26 Oct. 23/1 It's usually one of the White Light hits to begin with—and it's odd how exquisite they are on the Pianola. **1916** *Proc. Musical Assoc. 1915–16* 16 The Press have adopted the term 'Pianola' as a generic name for all mechanical piano-player devices. **1943** J. B. PRIESTLEY *Daylight on Saturday* xii. 78 We'll find out..we're not making planes any more but sewing machines and pianolas. **1947** *Sat. Rev. Lit.* (U.S.) 25 Oct. 65/2 Unlike vaudeville, the pianola is not dead. **1953** DYLAN THOMAS *Under Milk Wood* (1954) 10 He sold the pianola. **1962** *Trade Marks Jrnl.* 26 Sept. 1315/2 Pianola... Player-pianos. Aeolian American Corporation.., New York,..United States of America; manufacturers and merchants. **1978** F. WELDON *Praxis* xii. 82 Mrs. Allbright played the pianola with too many stops out. **1980** L. LEWIS *Private Life of Country House* iv. 47 A large walnut Steck piano with pianola player incorporated.

b. = *pianola hand* below.

1974 *Country Life* 9 May 1155/3 With everything right, the [bridge] contract is a pianola.

c. *Comb.*, as *pianola-like* adj.; **pianola hand** *Cards*, a hand that is easy to play; **pianola roll** = *music-roll* (*b*) s.v. *MUSIC *sb.* 12 d.

1913 F. IRWIN *Auction High-Lights* ii. 22 Which do you like better, a 'pianola' hand or a hand where you have to tussle and fight for every point? **1945** E. BOWEN *Demon Lover* 134 The pianola-like play of the conversation did not drown the nervousness round the table. **1915** *Chicago Daily News Almanac 1916* 615/2 The Coe music collection..[contains] 560 pianola rolls. **1959** 'F. NEWTON' *Jazz Scene* ii. 33 Old pianola rolls by the leading composer-players of this earliest jazz style.

pianolaed (pi,ănǝᵘ·lād), *a.* [f. PIANOLA + -ED[1].] Rendered by a pianola.

1926 A. B. SMITH *Studies & Caprices* 176 The pauses..are not, as in the pianolaed performance, the mere passing of time.

pianoless (pi,æ·nolės), *a.* [f. PIANO *sb.*[2] + -LESS.] Without a piano, esp. designating an ensemble of musicians which does not include a pianist.

1955 L. FEATHER *Encycl. Jazz* 66 Gerry Mulligan and others experimented with pianoless units. **1956** M. STEARNS *Story of Jazz* (1957) xviii. 240 Pacific Jazz recorded the pianoless quartet of Chet Baker and Gerry Mulligan. **1973** *Black World* Oct. 48/1 His preferred instrumental format for his own recordings is with pianoless groups.

pianolist (pi,ănǝᵘ·list). [f. PIANOLA + -IST.] A person who plays a Pianola.

1908 G. KOBBÉ (*title*) The pianolist: a guide for pianola players. *Ibid.* 8 Another purpose..is to furnish pianolists with a guide to the music which they play, or might play. **1916** *Proc. Musical Assoc. 1915–16* 24 There are many pianolists so keen that they will cut their own music in unique single copies. **1927** *Observer* 23 Oct. 14 The player seems to have the same sort of relation to his instrument as the pianolist to his.

‖ **piano nobile** (pyä·no nōᵘ·bile). *Archit.* [It., f. *piano* floor, storey + *nobile* noble, great.] The main storey of a large house, usually on the first floor, of lofty structure, and containing the principal apartments.

1909 in *Cent. Dict.* Suppl., s.v. *piano*[3]. **1910** *Encycl. Brit.* X. 527/1 The principal floor is the storey which contains the chief apartments whether on the ground- or first-floor; in Italy they are always on the latter and known as the 'piano nobile'. **1922** F. G. ELLERTON *Let.* 14 Apr. in J. Bailey *Lett. & Diaries* (1935) 216 The great windows of the *piano nobile* and the fine ones of the second floor where the Strozzi still live after *four hundred and fifty years*! **1928** A. HUXLEY *Point Counter Point* iii. 37 Two flights up, between the *piano nobile* and the servants' quarters under the roof. **1939** J. D. S. PENDLEBURY *Archaeol. Crete* iv. 186 The South Propylaeum, from which access was obtained to the 'piano nobile'. **1955** L. WOOLLEY *Alalakh* iii. 94 The upper floor, the *piano nobile*, no longer exists, but its principal room, the *grand salon*, can safely be reconstructed. **1965** B. SWEET-ESCOTT *Baker St. Irreg.* vi. 180 On the *piano nobile* was a large room in which Mr Bertram Mills himself had sat. **1970** *Guardian* 16 Sept. 11/1 The 'piano nobile', entered through a colonnaded portico and up a curving flight of carpeted marble. **1974** K. CLARK *Another Part of Wood* vi. 238 The whole *piano nobile*, with beautiful 'Adam' rooms, marble chimney pieces and painted ceilings, was completely unnecessary. **1976** J. LEES-MILNE *W. Beckford* v. 70 The Octagon Cabinet..and the Crimson Breakfast Parlour..bring the apartments of the *piano nobile* of the Abbey to an end.

piano-organ. Add: So **pia:no-o·rganist.**

1885 W. S. GILBERT *Mikado* I. 9 The piano organist—I've got him on the list! **1955** L. FEATHER *Encycl. Jazz* 123 Her husband, piano-organist Jesse Crump, worked and recorded with her.

piano piano. (Later example.)

1925 D. H. LAWRENCE *Let.* 25 Nov. (1962) II. 867 But it's no good: we've got to go *piano-piano*.

piassaba. Add: Also, the tropical African palm, *Raphia vinifera*, or the fibre obtained from it. (*Piassava* is now the usu. form.) (Later examples.)

1889 G. S. BOULGER *Uses of Plants* VI. 171 Pará Piassaba, of which the ramenta are used for brushes. **1922** W. SCHLICH *Man. Forestry* (ed. 4) I. 309 The valuable 'piassava fibre' is prepared from the leaves of the bamboo palm. **1936** *Nature* 28 Mar. 528/1 The Hard Fibres Section of the British Empire Producers' Organisation has arranged a comprehensive group of exhibits of sisal and sisal manufactures... Other stands display..West African piassava. **1957** M. BANTON *W. Afr. City* iii. 43 Kola-nuts are grown in the south,..piassava along the southern coast. **1964** J. P. CLARKE *Three Plays* 21 Let's Sing of souls tied down with ropes Of piassava so strong. *Ibid.* 74 What single harm Have I done that this poison more catching Than fruit from the piassava palm should..be bailed into the stream Of my blood. **1966**

E. J. H. CORNER *Nat. Hist. Palms* iii. 65 The thick brown coating of leaf-base fibres of the piassába or chiquichiqui palm *Leopoldinia piassaba*..are made into ropes. *Ibid.* xiv. 331 Piassava, or African bass, is derived from the vascular bundles of the leaf-stalk and sheath [of *Raphia vinifera*]. **1969** E. H. PINTO *Treen* 194 Piassava fibre..has been used in bass brooms for many centuries.

Piastraccia (pyastrā·tʃia), the name of a quarry near Seravezza, between Carrara and Lucca in N. Italy, used to designate a variety of white marble with slender grey veins quarried there.

1909 *Westm. Gaz.* 9 June 11/3 The entrance hall and corridor walls are faced with polished Piastraccia and Swedish green marble. **1955** W. GADDIS *Recognitions* III. i. 726 In response to the darkening sky, the sea changed its surface from glass to marble, the Breche rose marble of Italy reflecting the broken color of the sun, and losing that, to the gray-white Piastraccia, reflecting light from nowhere, veined with shadows.

piastre, piaster. Add: **3.** (In form *piastre.*) The name of a unit of currency introduced in Indo-China under French rule in 1885.

1908 *Whitaker's Almanack 1909* 618/1 The financial and political unity of Indo-China was finally established in 1898... The revenue, about 51,850,000 piastres, is derived mainly from customs, excise, and other indirect taxes. **1955** [see *KIP *sb.*[5]]. **1968** R. WEST *Sk. Vietnam* i. 22 The street hawkers who offer you two hundred and fifty piastres a dollar. **1970** *Daily Tel.* 22 July 14 In Saigon, prices are sky high. The piastre, officially worth 118 to the dollar, is traded by black marketeers at nearly 400 to the dollar. **1977** *Private Eye* 1 Apr. 20/2 (Advt.), 100 piastre (£70 face value) South Vietnamese note.

Piat (pi·ăt). [Acronym f. the initials of *p*rojector *i*nfantry *a*nti-*t*ank.] A weapon used by infantry against tanks in the war of 1939–45.

1944 *Hutchinson's Pict. Hist. War* 27 Oct. 1943–11 Apr. 1944. 321 The powerful new P.I.A.T. (projector, infantry, anti-tank) loaded and in position for firing... In this demonstration test of the 'Piat', a German tank has received a direct hit. **1948** PARTRIDGE *Dict. Forces' Slang* 140 *Piat*, Projector Infantry Anti-Tank. A mortar type of projector designed to destroy tanks at close range... It had a complement of two men, but could be easily handled by one. **1964** A. MCKEE *Caen* vii. 101 Small parties of infantry, with Piats,..began to stalk the Tigers. **1974** C. RYAN *Bridge too Far* III. xiv. 206 Frost could hear the crash of Vlasto's Piat bombs smashing into the pillbox.

piazza. Add: Now usu. with pronunc. (-ts-). Pl. **piazze** (-tse). **1.** (Later examples in extended and *transf.* uses.)

1942 *Country Life* 9 Oct. 694/3 (*caption*) Piccadilly Circus..a new building repeats that at present facing down Lower Regent Street. Between them a 'piazza' for pedestrians is formed. **1959** *Listener* 3 Dec. 962/2 The idea of enlarging the present underground booking-hall to create a vast below-street-level shopping piazza. **1962** *Ibid.* 19 Apr. 689/3 Among the features of the scheme is a paved piazza. **1967** C. SETON-WATSON *Italy from Liberalism to Fascism* xi. 425 Many were appalled by the irruption of the *piazza* and the press into delicate questions of international diplomacy. **1976** *Times* 20 Feb. 12/5 George Street [in Edinburgh] still has the Georgian Assembly Rooms midway between the two great *piazze* of St Andrew Square and Charlotte Square. **1977** *New Yorker* 26 Sept. 32/2 Moulmein was reading..on his piazza when I got home. **1978** J. MCNEIL *Consultant* i. 21 He could see the City [of London], far below..the small *piazza* he had crossed.

2. b. (Earlier and *attrib.* examples.)

1724 H. JONES *Present State Virginia* 26 It is a lofty Pile of Brick Building adorn'd with a Cupola... There is a spacious Piazza on the West Side, from one Wing to the other. **1771** J. S. COPLEY *Let.* 3 Aug. (1914) 137 You see I have Drawn the Chinea Clossit Store Room in the east piaza. **1876** A. D. WHITNEY *Sights & Insights* I. 8 There were settees, and regular piazza chairs. *a* **1916** H. JAMES *Ivory Tower* (1917) I. ii. 18 Shaking his little foot from the depths of a piazza chair,..where..the cool spreading verandah, commanded the low green cliff. **1977** MCDAVID & O'CAIN in S. Greenbaum *Acceptability in Lang.* viii. 112 The uncultivated almost unanimously characterize *porch* as modern and *piazza* as old-fashioned.

piazzetta. (Later examples.) *spec.* (with cap. initial), the *Piazzetta di San Marco* in Venice.

1824 W. IRVING *Tales of Traveller* I. I. 106 They crossed the Piazzetta, but paused in the middle of it to enjoy the scene. **1869** GEO. ELIOT *Jrnl.* in J. W. Cross *George Eliot's Life* (1885) III. 45 Even landing on the Piazzetta, one has a sense..of being in an entirely novel scene. **1888** H. JAMES *Aspern Papers* I. v. 84 The sea-breeze passed between the twin columns of the Piazzetta. **1906** *Edin. Rev.* July 194 To cross its bridges and its piazzette and to pass under its gateways. **1910** H. G. WELLS *New Machiavelli* (1911) II. iii. 246 We would stroll on the Piazzetta, or go out into the sunset in a gondola. **1931** C. BAX *Venetian* I. i. 5 The main part of the stage represents a piazzetta or flagged yard. **1942** A. L. ROWSE *Cornish Childhood* i. 22 It played the part of the back-alley in East End life, if not of the piazzetta in a little Italian town. **1966** *Listener* 20 Jan. 105/2 In little back alleys and piazzettas water will pour from a spout into a marble basin or stone trough. **1977** P. D. JAMES *Death of Expert Witness* I. vi. 28 To see..that incomparable view of San Marco from the western end of the Piazza... To stand together on the Piazzetta..and look across the shimmering water.

pi·bble-pa:bble, pibble-babble, alterations of BIBBLE-BABBLE.

1599 SHAKES. *Hen. V* IV. i. 72 There is no tiddle tadle nor pibble bable in Pompeyes Campe. **1953** *Essays & Stud.* VI. 112 Our South Wales dialect in English, a pibble-pabble inadequate to the demands of a full and varied literature. **1959** *20th Cent.* June, He was the main topic of the pibble-pabble in the forum last month.

pibcorn. For *Obs.* read '*Obs.* exc. *Hist.*' and add later examples.

a **1953** [see *CRWTH]. **1968** J. ARNOLD *Shell Bk. Country Crafts* 316 The Pibcorn, of Wales, dating from the eighteenth century, was a pastoral hornpipe.

piblokto (piblǫ·kto). Also **piblockto, piblok-toq.** [Eskimo.] **a.** = *Arctic hysteria* s.v. *ARCTIC *a.* 1 c. **b.** A form of hysterical illness prevalent among Eskimo dogs.

1898 *Geogr. Jrnl.* XI. 228, I pushed on until the 'piblockto', or Greenland dog madness, induced by the continued exposure, got such a hold of my dogs as to make it absolutely impracticable for me to go further. **1910** R. E. PEARY *North Pole* xviii. 156 The adults are subject to a peculiar nervous affection which they call *piblokto*... The attack usually ends in a fit of weeping; and when the patient quiets down, the eyes are bloodshot, the pulse high, and the whole body trembles for an hour or so afterward. **1921** A. B. READER tr. *Tremblay's Cruise of Minnie Maud* vi. 125 The adult Eskimo are often subject to a peculiar nervous affection, a form of hysteria, which they call piblokto... Eskimo dogs and foxes suffer also from piblokto. **1945** C. L. B. HUBBARD *Observer's Bk. Dogs* 207 The dread Arctic disease, pib-lockto (a devastating form of distemper). **1959** *Camsell Arrow* (C. Camsell Hospital, Edmonton, Alberta) Jan.–Feb. 74/1 There is a strange Eskimo madness—piblokto. Men and women are seized with uncontrollable frenzies during which they do all sorts of weird and violent things. **1969** *Daily Tel.* 8 Feb. 16/3 Eskimos have been known to have seizures of *Pibloktoq*, in which they undergo periods of depression followed by wild excitement, then convulsions and stupor, possibly due to the biological effects of the loss of a human time-structure.

pi-bond: see *PI *sb.* 3.

pic[3] (pik), U.S. colloq. abbrev. of PICAYUNE *sb.*

1839 *Spirit of Times* 18 May 129/1 The gentleman of the bar..set back the bottle and popped the 'pic' in the drawer. **1841** E. R. STEELE *Summer Journey in West* 159 In paying for them I found a new currency here, my shillings and sixpences being transformed into *bits* and *pics* or picayunes. **1846** E. W. FARNHAM *Life in Prairie Land* II. i. 291 'How much does the muskito-bar cost a yard?' 'Two bits and a pic, or three bits.' **1850** 'M. TENSAS' *Odd Leaves Life Louisiana Swamp Doctor* 51 The animal didn't mind him a pic. **1855** 'Q. K. P. DOESTICKS' *Doesticks, What he Says* xxiii. 202 A stranger must disburse an avalanche of 'bits', 'pics', and 'levys', before he can get even a plate of cold victuals. **1859** P. H. GOSSE *Lett. from Alabama* 103 The negroes ferried me over the romantic river, for which I paid a '*pic*' (*i.e.* a picayune, the sixteenth of a dollar, or half a 'bit'), the smallest silver coin current.

pic[4] (pik), colloq. abbrev. of PICTURE *sb.* in Dict. and Suppl. Cf. *PICCY. **a.** In sense 2 b.

1884 RUSKIN *Let.* 16 Nov. in S. Birkenhead *Illustrious Friends* (1965) xxxiii. 304, I am so very sorry I cant 'reprieve'—as you call it—the 'pics'. **1891** KIPLING *Light that Failed* v. 82, I must see your pics first. **1910** C. E. MONTAGUE *Hind let Loose* v. 81 He was sent off to see some pics; an'..he saw..just what the men were about that had painted them. **1948** L. DURRELL *Spirit of Place* (1969) 98 I've suggested to Tambi that *Runciman* might do a small Patmos book to go with your pics? **1952** M. ALLINGHAM *Tiger in Smoke* i. 17 The pics themselves are covered with fingerprints. **1971** *Petticoat* 17 July 3/2 They're sold with ready-cut pics. **1976** *Sunday Post* (Glasgow) 26 Dec. 13/3 I sent £7.22 to photographer in Wembley for two coloured photos of a show jumping event in Warwickshire. When no pics came I wrote.

b. In sense *2 i.

1936 *Esquire* Sept. 160/4 Raft's next pic is *Proud Rider*. **1970** Y. CARTER *Mr. Campion's Falcon* iii. 26 The Hag is out for the evening—gone off..to the pics. **1973** 'A. BLAISDELL' *Crime by Chance* (1974) vii. 125 All of a sudden, Latin romances sort of passe... Everybody doing the big war pics.

pic[5] (pik). *colloq.* [Abbrev. PICADOR or Sp. *pica* lance.] **a.** A picador. **b.** A picador's lance, or the thrust made by it. Also as *v. trans.* and *intr.*

1925–6 E. HEMINGWAY in *This Quarter* Autumn–Winter 206 'How about picadors?'..'I've got to have one good pic.' **1926** —— *Sun also Rises* II. xv. 173 Watching the picador place the point of his pic. **1927** [see *CUT *v.* 53 h]. **1932** R. CAMPBELL *Taurine Provence* 38 From the beginning of the 18th century we know almost every pass, pic, or estocada that has been performed up to the present day. **1934** —— *Broken Record* 195, I pic'd better than any of the professionals. **1957** A. MACNAB *Bulls of Iberia* vi. 63 The picador rides slowly forward towards the bull, it charges, and just as it arrives he jabs the pic down on the top of its shoulders behind the neck muscle. *Ibid.* 65 The sight of a brave, powerful bull being pic'd honourably..is a fine one. **1967** MCCORMICK & MASCAREÑAS *Compl. Aficionado* ii. 30 If the toro refuses to take the pic, he is supposed to be returned to the corrals. **1976** E. P. BENSON *Bulls of Ronda* iv. 26 The bull..felt the picador plunge the steel pic into its tossing muscle... Navarro distracted the bull and prepared him for another picing.

1978 M. WALKER *Infiltrator* i. 11, I want to watch the pics.

Picard (pi·kāɹd, ‖pikar), *sb.*[2] and *a.* Also 7 Picardin. [a. F. *picard* in the same sense.] **A.** *sb.* **a.** A native or inhabitant of Picardy, a region and former province centred on Amiens in northern France, now the departments of Somme, Aisne, and Oise. **b.** The dialect of French spoken there.

a **1400** [see *BRETON *sb.* and *a.*]. **1598** J. STOW *Survey of London* 116 Iohn Mutas (a Picarde) or Frenchman. *a* **1666** EVELYN *Diary* an. 1644 (1955) II. 138 The University [of Orleans] is..divided now..into that of 4 Nations French, High-dutch, Normans and Picardins. **1900** H. BELLOC *Paris* i. 6 Here Calvin the Picard preached his Batavian theory. **1903** *Knowledge* Dec. 267/2 The dialect of the Isle of France supplanted Picard, Burgundian, and Norman, and became the French language. **1924** G. B. SHAW *Saint Joan* iv. 40 Are these Burgundians and Bretons and Picards and Gascons beginning to call themselves Frenchmen? **1932** W. L. GRAFF *Lang.* 377 French group, with its subdivisions:..Norman, Picard, [etc.]. **1968** E. HYAMS *Mischief Makers* ix. 165 His name was Waché and he was, I think, from the north, a Picard. **1976** N. ROBERTS *Face of France* iv. 49 There are usually witnesses who have seen 'a sallow man..of North African type' near the scene of the crime, of which, later, a blond Norman or Picard may be convicted

B. *adj.* Of or pertaining to Picardy or its inhabitants.

1650 J. HOWELL in *Cotgrave's Fr.-Eng. Dict.* sig. a4, The French toung hath divers dialects, the Picard, that of Iersey and Guernsay..the Provensall, the Gascon. **1833** MACAULAY in *Friendship's Offering 1833* 17 On that famed Picard field, Bohemia's plume, and Genoa's bow, and Caesar's eagle shield. **1954** W. FAULKNER *Fable* 60 One of those sweating stone courtyards which for a thousand years the French have been dotting about the Picard and Artois and Flanders countryside. **1972** R. COBB *Reactions to French Revolution* iii. 93 The apprentice is set upon by a group of big Picard servants.

picarel (pi·kārel). [Fr: cf. PICKEREL[1].] A small marine fish belonging to the family Centracanthidæ (Mæna), found in the eastern Atlantic, the Mediterranean, and the Indian Ocean; esp. the Mediterranean species, *Mæna smaris.*

1905 D. S. JORDAN *Guide to Study of Fishes* II. xix. 347 The Mænidæ, or Picarels, are elongate, gracefully formed fishes. **1972** A. DAVIDSON *Mediterranean Seafood* 107 In Venice you could insult someone by calling him a picarel-eater; and at Port Vendres and some other places in France the fish is known as mata-soldat, or kill-soldier... The picarel can be quite good fried.

picaresque. Add: **a.** (Later examples in *transf.* and extended uses.) Also as *sb.*

1918 A. G. GARDINER *Leaves in Wind* 245 Near by lives a distinguished lady of romantic picaresque tastes, who dotes on street pianos. **1955** *Times* 9 Aug. 9/7 The Russians are paying a price for the energetic imposing of rigid form..on a nation which is in character essentially picaresque. **1958** *Listener* 19 June 1011/1 The first of a trilogy, it is more accessible than its companions, and more lively; a kind of psychological picaresque. **1959** J. M. S. TOMPKINS *Art R. Kipling* i. 29 There is some likeness in the broad, general plan of the two stories [*sc.* *Kim* and *Huckleberry Finn*]. They are picaresque narratives, with boys as travellers. **1965** *Times Lit. Suppl.* 25 Nov. 1035/4 There is a strong school of black picaresque. **1976** *Ibid.* 23 Apr. 481/5 The central character of this contemporary picaresque of the mind [*sc.* a novel].

¶ **b.** Of a situation: transitory, impermanent. Of a person: drifting, peregrinatory.

An erron. usage from an inference that the picaresque hero is a vagrant or wanderer.

1959 *Manch. Guardian* 28 July 6/5 The boys are attracted by the picaresque nature of working with a private building or decorating firm. **1960** *Observer* 17 Apr. 20/7 One is beginning to dread that word 'picaresque'... He dashes his suburban hero all over the place. **1978** *Country Life* 6 July 57/4 Kyril Bonfigliole..has now produced..an historical picaresque—picaresque in both the popular sense of roaming far afield and in the strict sense of being..about a rogue.

picarian, *a.* Insert in etym. after '*Picariæ*': (C. L. Nitzsch 1820, in *Deutsches Archiv für Physiol.* VI. 255). (Examples.)

1908 E. J. BANFIELD *Confessions of Beachcomber* I. iii. 96 Picarian Birds. Large-tailed Nightjar... Blue Kingfisher... Bronze Cuckoo. **1955** [see *HONEY-GUIDE 1].

picaro. For *Obs.* read 'Now *arch.*' and add later examples.

1749 SMOLLETT tr. *Le Sage's Gil Blas* IV. x. xii. 115 If Scipio in his childhood was a real Picaro, he has corrected his conduct so well since that time, that he is now the model of a perfect servant. **1966** *New Statesman* 21 Jan. 96/2 The rude and cynical picaro, the philandering gambling tailor or tailoring gambler, finds himself so deeply fascinated by a wealthy young American college girl that he puts on a black roll-neck jersey, poses as a writer, and hopefully takes her to Italy. **1972** M. BRADBURY in Cox & Dyson *20th-Cent. Mind* III. xii. 343 Jack Donaghue in [Iris Murdoch's] *Under the Net* may seem a typical fifties *picaro*; but he is a novelist, facing the problem of the possible collapse into contingency of language and the fascinations of silence. **1977** *Times Lit. Suppl.* 20 May 605/2 Picaresque grants an author licence to switch his tones about as the picaro speeds from adventure to adventure.

picaroon, *sb.*[1] Add: **1.** (Later examples.)
1904 BURGESS & IRWIN (*title*) The picaroons, a San Francisco night's entertainment. 1924 H. LANDON (*title*) The elusive picaroon. 1935 A. J. POLLOCK *Underworld Speaks* 87/2 *Picaroon*, thief who preys on tourists.
3. b. (With reference to BARRACOON.) A slave-ship.
1893 KIPLING *Seven Seas* (1896) 23 Then said the souls of the slaves that men threw overboard: 'Kennelled in the picaroon a weary band were we.'

picaroon, *sb.*[2] For *Canada* read *N. Amer.* and substitute for def.: A long pole fitted with a spike or hook, used in logging and fishing. (Add examples.)
1837 *North Amer. Rev.* Apr. 354 The rafters..[make] use of a picaroon, or pole with a spike in the end of it, which is..driven into the boards, taking out perhaps a piece at each time. 1850 S. JUDD *R. Edney* 42 Richard, armed with a picaroon, descended the slip..to the basin, where the logs lay in the water ready to be drawn in. 1905 *Bull. Bureau of Forestry* (U.S. Dept. Agric.) No. 61, 43 *Pickaroon*, a piked pole fitted with a curved hook, used in holding boats to jams in driving, and for pulling logs from brush and eddies out into the current. 1949 N. C. BROWN *Logging* II. v. 101 Pickaroons are short poles 35" to 40" long with a recurved pike or hook used in drawing or pulling small products such as cross ties, 4' pulpwood, fuel wood, chemical wood, cooperage, and bolts down steep slopes. 1972 F. FORD *Atush Inlet* viii. 78 The crew worked with picaroons spiking the fish into the scow.

Picassian (pikæ·siăn), *a.* [f. the maternal name of Pablo (Ruiz y) *Picasso* (1881–1973), Spanish painter: see -IAN.] Of, pertaining to, or characteristic of Picasso or his style of painting. Also **Picassoe·sque.**
1940 L. ADAM *Primitive Art* xvii. 115 Picasso's primitiveness is quite definitely his own 'Picassian primitiveness'. 1959 *Times* 21 Sept. 5/6 They have none of the Parisian flair, except for Mr Roque A. Riera Rojas, whose shadowy 'Bullfighter' with the piquant, Picassian features and costume of silver grey looming from the shadows is a thoroughly suave, skilful performance. 1968 *Time* 27 Dec. 30/1 *Yellow Submarine* combines every trick and treat of film animation with a dazzle of takeoffs on schools and styles of art. Picassoesque monsters compete with gentle grotesques. 1972 Y. YADIN *Hazor* II. vii. 82 It depicts in a schematic manner the naked body of a woman. Only half of her features are accentuated: one eye, half the nose and mouth, one breast, and half of the vulva. This Picassoesque figurine seems to represent 'life and death' or the 'born and unborn'. 1978 *Listener* 13 July 56/3 The Picassoesque line..is a reminder of something he [*sc.* Henry Moore] learned from both Picasso and Matisse.

picathartes (pikăþă·ɹtiz). *Ornith.* [mod.L. (R. P. Lesson *Manuel d'Ornithologie* (1828) I. 374), f. L. *pica* magpie + *Cathartes*, generic name of certain American vultures, f. Gr. καθαρτής cleanser (f. καθαίρειν to cleanse).] A bare-headed West African bird of the genus so called, belonging to the babbler family or Timaliidæ; also called the bald crow or rock-fowl.
1931 *Discovery* May 140/1 Looking at this bird, named misleadingly the white-necked bald crow, one's mind instinctively reverts to prehistoric times, for..the Picathartes is a most extraordinary looking object. 1938 *Ibis* II. 255 An investigation of the stomach-contents of the rare West African bird *Picathartes* reveals the fact that insect fragments predominate. 1960 G. DURRELL *Zoo in Luggage* ii. 57 The *Picathartes* was about the size of a jackdaw, but its body had the plump, sleek lines of a blackbird. 1972 *Daily Colonist* (Victoria, B.C.) 30 July 27/3 The picathartes, a rare bald-headed bird..will fetch £100.

picayune, *sb.* and *a.* Add: Also **piccayune, picharoon, pickayune. A.** *sb.* (Earlier and later examples.)
1804 J. F. WATSON *Jrnl.* 4 Nov. in *Amer. Pioneer* (1843) II. 228 One can't buy anything [at New Orleans] for less than a six cent piece, called a *picayune*. 1832 R. BAIRD *View of Valley of Mississippi* xxii. 264 [In Louisiana] the words 'piccayune' (6 1–4 cents) and bit—(12 1–2 cents) fall upon the ear at every step. 1839 J. K. TOWNSEND *Narr. Journey Rocky Mts.* i. 17 We gave him a *pickayune* for his trouble, and went on. 1904 *N.Y. Even. Post* 25 Jan. 6 It doesn't matter a picayune whether the justices or the members of the diplomatic corps were presented first. 1948 *Reader's Digest* Dec. 148/1 Don't care a picayune how you waste that boy's time, do you? 1979 M. G. EBERHART *Bayou Road* xxi. 288 His life wouldn't be worth a picayune.
B. *adj.* (Earlier and later examples.) Also *absol.*
1813 *Cramer's Pittsburgh Almanac 1814* 60 The incessant hum of the blabbering (coloured) market women, seated on the ground..by the side of their picharoon (six cent) piles of vegetables. 1837 *Congress. Globe* 25th Congress 2 Sess. App. 19 The hon. Senator from Kentucky..by way of ridicule, calls this a 'picayune bill'. 1915 *New Republic* 31 July 336/1 They instinctively regard the critic as puny and picayune. 1936 *Delineator* Nov. 11/2 No picayune place like this could pay that man. 1955 E. POUND *Classic Anthol.* II. 104 Your picayune in-laws in fat government jobs. 1965 H. MITCHELL *Underground War against Revolutionary France* iv. 46 Overlooking the serious divisions between the pure and constitutional royalists, Grenville naïvely expected them to bury their picayune differences in a sudden glow of friendship. 1967–8 *Bahamas Handbk. & Businessmen's Ann.* (ed. 7) 77 A picayune mongrel comes from its midday torpor. 1973 *Listener* 20 Dec. 849/3 His projects at that point were getting picayune. He was no longer a great baseball-player. 1974 [see *NITPICKER]. 1975 *Encounter* Feb. 44/2 Most of the snags and pratfalls he cites seem to me picayune matters.

picayunity (pikāyū·nĭti). [f. PICAYUNE *sb.* and *a.* + -ITY.] Insignificance, triviality.
1948 O. NASH in *New Yorker* 13 Nov. 32 In this imponderable world I lose no opportunity To ponder on picayunity. 1977 *Ibid.* 2 May 54/1 To the point of picayunity the state's road system is limited.

Piccadilly (pikădi·li). The name of a street and circus (sense 7) in London (see note s.v. PICCADILL, PICKADILL), used *attrib.* in *Piccadilly weeper(s)*, long drooping side whiskers, sometimes extending below the chin, worn without a beard; loosely, = *Dundreary whiskers* s.v. *DUNDREARY; *Piccadilly window* (slang), a monocle.
1874 HOTTEN *Slang Dict.* 252 *Piccadilly weepers*, long carefully combed-out whiskers of the Dundreary fashion. 1894 Piccadilly weeper [see *DUNDREARY]. 1897 E. GRAHAM *Golden Dustman* (song), Nah I'm goin' to be a reg'lar toff... A Piccadilly winder in my eye. 1907 [see *BURNSIDE, BURNSIDE]. 1909 J. R. WARE *Passing Eng.* 195/2 *Piccadilly window* (*street*, '90's), single eye-glass worn by some men of fashion—hence the Piccadilly. 1936 P. M. CLARK *Autobiogr. Old Drifter* xiii. 177 'It' was a regular Ha-ha Johnnie with a 'Piccadilly window' in his eye. 1960 C. W. CUNNINGTON et al. *Dict. Eng. Costume* 163/1 *Piccadilly weepers*. 1870's and 1880's. (M.) Long combed-out whiskers fashionable in those decades. 1973 J. FLEMING *You won't let me Finish* ii. 19 A fragile moustache that drooped right down past his mouth, the kind of moustache that used to be called a 'Piccadilly weeper'.

piccaninny, pickaninny, *sb.* (*a.*) Add: Also **picanini, piccin, piccinini, piccney, picken, pickini, pickinny, pickne, pickney, picknie, pickny, picny.**
The term now often gives offence when applied to children by people of European extraction.
a. In the speech of West Indian Blacks freq. with uninflected pl. (Further examples.)
1790 J. B. MORETON *Manners & Customs West India Islands* 152 The women..are obliged to..take their pickinnies (*i.e.* children) on their backs, to which they are tied with handkerchiefs. 1847 *Knickerbocker* XXX. 216 It might be very pleasant to be surrounded by half-a-dozen negro waiting-women, with their picaninnies. 1868 T. RUSSELL *Etym. Jamaica Gram.* 2 Pickne. *Ibid.* 6 Pickini—A child. African. 1907 W. JEKYLL *Jamaican Song & Story* 40 Now Toad have twenty picny. 1937 R. MACAULAY *I would be Private* II. i. 147 That naked piccney. 1958 J. CAREW *Wild Coast* viii. 117 All you is, is a maugre, skin-and-bone pickny. 1969 S. M. SADEEK *Windswept & Other Stories* 37 'I was working for the estate, until..' 'Until alyou get busy making picknie.' 1974 *Practitioner* CCXIII. 845 To 'give pickney' or to 'breed' [in Jamaica] is to get a woman with child. 1977 *Westindian World* 3–9 June 4/1 It has been made very plain that quite a number of teachers in schools up and down de country are in many cases more dunce than de pickney dem teach themselves.
b. (Further examples.) Also, the offspring of an animal.
1841 R. HOWITT *Impressions Australia Felix* (1845) 103 Two women, one with a piccaniny at her back. 1855 in J. W. Colenso *Ten Weeks in Natal* Add. p. 3 What will the poor little piccaninnies do, Boy? 1911 *East London* (Cape Province) *Dispatch* 24 Nov. 7 (Pettman), Mothers nursing their picaninies and maidens listening to lovers rude. 1925 *Brit. Weekly* 31 Dec. 340/2 A mother..crooned gently to her 'piccin' not more than a few weeks old. 1926 *Ibid.* 27 May 158/1 A mischievous picannin..was weeding the vegetable garden. 1936 I. L. IDRIESS *Cattle King* xxviii. 245 He met numerous blackfellow friends, they and their lubras and piccaninnies, all in fat good humour. 1953 G. DURRELL *Overloaded Ark* iii. 59 Na catar beef, sah, and 'e get picken for 'e back. 1961 G. GREENE *Burnt-Out Case* IV. i. 100 The piccin that stole sugar from the white man's cupboard. 1963 *Sydney Morning Herald* 19 Nov. 6/4 The use of such words as 'boy', 'lubra' and..'piccaninny' to describe aborigines has been banned to Northern Territory welfare officers. 1966 C. SWEENEY *Scurrying Bush* xiv. 199 He guided me about half a mile up the road, the rest of the piccannins scampering behind. 1979 P. NIESEWAND *Member of Club* ii. 21 The bantu we've chosen..is a piccanin called Elias.
B. *adj.* spec. *piccaninny dawn, daylight*, earliest dawn, first light (chiefly *Austral.*). (Earlier and later examples.)
1707 H. SLOANE *Voy. Jamaica* I. p. lii, They have..Christmas Holidays, Easter call'd little or Piganinny, Christmas, and some other great Feasts. 1835 R. R. MADDEN *Twelvemonth's Residence W. Indies* II. 153 To..spend piccanini Christmas (Easter) dancing. 1848 W. WESTGARTH *Australia Felix* viii. 104 The hut would be attacked before 'piccinini sun'. 1870 in 'Mark Twain' *Screamers* (1871) xxv. 132 A pickaninny,..mud-turtle-shaped craft of a schooner. 1910 ANDERSON & CUNDALL *Jamaica Anancy Stories* 37 Go a pickney mumma yard an' you sure fe get somet'ing. *a* 1912 'T. COLLINS' *Buln-Buln & Brolga* (1948) 107 Blackfellers mostly goes in for a piccaninny fire—jist three sticks, with the ends kep' together. 1936 M. FRANKLIN *All that Swagger* xvi. 153 At piccaninny dawn, the billy with the lid off was found rolling on the floor. 1953 G. DURRELL *Overloaded Ark* iv. 78 'Eh..achh!' he shouted, 'napicken bushcat here for inside.' 1970 'E. LINDALL' *Gathering of Eagles* viii. 101 The piccaninny dawn, that false lightening of the sky that fades to darkness before the sun finally makes its presence known. 1971 *Courier-Mail* (Brisbane) 12 July 2/6 Before piccaninny daylight on June 14, Vernon Boundy..was standing on the southern bank of Tallebudgera's tidal estuary. 1974 *Sunday Mail Mag.* (Brisbane) 15 Sept. 3/3 In the chilly piccaninny dawn we drove out to the Wyloo strip with kangaroos bounding along beside and in front of the truck.

piccolo. Add: **4.** A boy who assists a waiter at a hotel, restaurant, etc.; a page at a hotel.
1910 'SAKI' *Reginald in Russia* 71 Watching the amount that I gave to the piccolo. 1926 R. HALL *Adam's Breed* I. x. 94 He had six enormous aprons... He had been very generously equipped for his duties as 'piccolo'. 1927 *Observer* 19 June 12 [German hotels] Head waiter..; elder page,.. younger page,.. 'piccolo', or very small page,.. still smaller 'piccolo', just beginning career. 1960 O. MANNING *Great Fortune* iii. 31 The piccolo arrived, a scrap of a boy, laden with bottles, glasses and plates. 1972 L. P. BACHMANN *Ultimate Act* i. 9 In his early teens he began work as a piccolo in hotels and restaurants.
5. = **juke-box.* *U.S. slang.*
1938 *N.Y. Amsterdam News* 12 Mar. 17 The Harlem Hamfats grind out the tune on myriad Harlem piccolos. 1946 E. BISHOP *North & South* 50 He's drinking in the warm pink glow To th' accompaniment of the piccolo. 1950 *Publ. Amer. Dial. Soc.* XIV. 52 [S. Carolina] *Piccolo*, an automatic music box, worked by a nickel slot machine. Origin undetermined. 1970 C. MAJOR *Dict. Afro-Amer. Slang* 91 *Piccolo*,.. juke box.

Picco pipe (pi·ko). *Mus.* Also with lower-case initial. [f. the name of *Picco* or *Picchi*, 'the Sardinian minstrel', who performed on this instrument with great virtuosity and was heard in London in 1856.] (See quot. 1876.)
1876 STAINER & BARRETT *Dict. Mus. Terms* 354/1 *Picco pipe*, a small pipe having two ventages above and one below. It is blown by means of a mouthpiece like a *flûte à bec* or whistle; and in playing, the little finger is used for varying the pitch by being inserted in the end. 1920 U. DAUBENY *Orchestral Wind Instruments* iii. 27 An instrument..was revived about the middle of last century by a blind Italian peasant named Picco, who gave remarkable performances on the 'Picco pipe'. 1939 A. CARSE *Mus. Wind Instruments* x. 119 Probably the smallest of all whistle-flutes is the Picco pipe, a tiny shrill whistle about $3\frac{1}{2}$ inches long. 1960 L. G. LANGWILL *Index Wind-Instrument Makers* 90 A young blind Sardinian shepherd, Angelo Picchi appeared in..London (1856), playing on a small pipe (Zuffolo) which was named after him, Picco Pipe. 1977 *Early Music* Oct. 555/2 Amongst woodwind were items such as a picco pipe (£165),..and a two-keyed oboe.

piccy (pi·ki), colloq. abbrev. of PICTURE *sb.* 2 b. Cf. *PIC[4], *PICKY.
1889 KIPLING *Let.* in C. E. Carrington *Rudyard Kipling* (1955) vi. 143 It's mighty curious to see behind an R.A.'s piccy and note the bits of things it is made up of. 1968 A. DIMENT *Bang Bang Birds* I. v. 75 They popped my piccy into a dud passport. 1977 *Hot Car* Oct. 75/1 The end result of fitting these packages on your Ford can be, if the piccies are anything to go by, rather on the eye-catching side.

pice: see also *PAISA.

picein (pi·-, pəi·sĭ,in). [f. L. *pice-us* pitchy (f. *pix* PITCH *sb.*[1]) + -IN[1].] **1.** *Chem.* [ad. F. *picéine* (Ch. Tanret 1894, in *Compt. Rend.* CXIX. 80).] A glucoside present in various trees, notably willows and conifers; *p*-hydroxyacetophenone-β-glucoside, $CH_3 \cdot CO \cdot C_6H_4 \cdot O \cdot C_6H_{11}O_5$.
1894 *Jrnl. Chem. Soc.* LXVI. I. 616 Picein, $C_{14}H_{18}O_7$, whether anhydrous or hydrated, crystallises in silky, prismatic needles, with a bitter taste. 1934 C. C. STEELE *Introd. Plant Biochem.* xi. 118 Picein, $C_{14}H_{18}O_7$, salinigrin, or salicinerin, occurs in several species of *Salix* and *Populus*. 1968 *Jrnl. Chromatogr.* XXXVI. 28 The collected fraction of pure trimethylsilyl picein also gave a satisfactory infrared spectrum.
2. Also *picein wax*, and with capital initial. An inert thermoplastic substance composed of hydrocarbons from rubber, shellac, and bitumen and used for sealing joints against air.
1927 G. W. C. KAYE *High Vacua* v. 69 Khotinsky cement..is widely used in the States for vacuum work, while Picein finds extensive application on the Continent. 1936 *Discovery* Sept. 286/1 A plug of picein wax in the capillary is thereby melted and seals up the tube. 1968 *Chem. Abstr.* LXIX. 6796/1 An elec.-insulating material was developed which permitted one to apply fields of up to 2×10^6 w./cm. The material consisted of varying amts. of quartz powder in picein. 1974 L. HOLLAND et al. *Vacuum Manual* i. 65 Picein. This is a classical vacuum wax—sticks well to metal and glass, is resistant to dilute acid and alkalies, is thermoplastic and made from bituminous substances. 1975 *Sci. Amer.* Feb. 110/3 Only prices are missing from this admirable guide to a world of high-technology commerce. String and sealing wax are gone (although not picein wax, quite).

Picene (pəi·sīn), *a.* and *sb.* [f. L. *Picen-us* Picene: see -IAN.] **A.** *adj.* Of or pertaining to ancient Picenum, a region in eastern central Italy, or the pre-Roman iron-age culture associated with it. **B.** *sb.* A native or in-

habitant of Picenum; the pre-Sabellian language attested there. Also **Pice·nian.**
1600 HOLLAND tr. *Livy's Romane Hist.* XXII. 437 He [*sc.* Hannibal] turned another way into the Picene countrie. 1601 —— tr. *Pliny's Nat. Hist.* XVIII. xi. 567 The Picenes in times past invented a way by themselves of making bread. 1863 W. P. DICKSON tr. *Mommsen's Hist. Rome* III. IV. ix. 332 The corps of Picenian volunteers soon grew to three legions. 1924 D. RANDALL-MACIVER *Villanovans & Early Etruscans* ii. 5 The burial rite of the Villanovans..opposes them to the Etruscans and the Picenes. *Ibid.* iv. 75 It is natural to suppose that *all* the inhumations in the Forum are of the Picene race. 1933 R. S. CONWAY et al. *Prae-Italic Dial. Italy* II. iv. 213 This Picene culture extends along the Adriatic coast southwards from Rimini at least as far as the river Sangro. 1939 L. H. GRAY *Foundations of Lang.* xi. 335 A language sometimes called Pre-Sabellian, Liburnian, or Picenian. 1948 D. DIRINGER *Alphabet* II. ix. 498 The Picenian alphabet is..very old. 1949 L. R. TAYLOR *Party Politics in Age of Caesar* ii. 45 In the elder Pompey's bodyguard..was a group of young Picene officers. 1959 *Chambers's Encycl.* V. 461/2 In the east the Fossa peoples were supplemented by another Illyrian-speaking group, known as Picene, between Rimini and the Abruzzi. 1974 E. S. GRUEN *Last Generation of Roman Republic* ii. 63 Pompey's following went beyond..the old aristocracy. From his father he inherited..a virtual barony in Picenum. Some of his adherents, from lesser families, seem to possess Picene origins. *Ibid.* iii. 118 Q. Numerius Rufus was probably a Picene.

Picentine (pike·ntəin, pəis-), *a.* and *sb.* Also 8 **Pycentine.** [f. L. *Picentin-us* Picentine: see -INE[1].] **A.** *adj.* = *PICENE *a.* *Picentine bread*, a kind of bread made in ancient Picenum (with reference to Pliny, *Nat. Hist.* XVIII. xxvii: see quot. 1958). **B.** *sb.* = *PICENE *sb.*
1708 W. KING *Art of Cookery* 147 The first Chapter contains the admirable Receipt of a Salacacaby of Apicus. Bruise in a Mortar Parsley Seed, dry'd Peneryal, ..Raisons ston'd, Honey, ..Oyl and Wine, put 'em into a Cacabulum, three crusts of Pycentine Bread, the Flesh of a Pullet,..pour a soup over it, garnish it with Snow. 1855 BOSTOCK & RILEY tr. *Pliny's Nat. Hist.* I. III. xviii. 235 The fifth region is that of Picenum, once remarkable for the denseness of its population; 360,000 Picentines took the oaths of fidelity to the Roman people. 1863 W. P. DICKSON tr. *Mommsen's Hist. Rome* III. IV. vii. 228 The Latin could remind the Picentine that they were both in like manner 'subject to the fasces'. 1888 [see *MARRUCINIAN *sb.* and *a.*]. 1933 R. S. CONWAY et al. *Prae-Italic Dial. Italy* II. iv. 209 The insc. on the Osimo statuette..resembles more closely than any of the rest the Italic dialects proper..whether its language be described as 'Picentine' or as 'Umbrian'. 1939 A. J. TOYNBEE *Study of Hist.* IV. 312 The new foundation was called into existence by the..'civilization' (in the literal sense) of the indigenous population, as when..the Romans organized..a Picentine village into a *forum* or *conciliabulum*. 1958 FLOWER & ROSENBAUM *Roman Cookery Bk.* IV. 93 Picentine bread. According to Pliny..this bread—invented by the people of Picenum—was made of spelt-grits. The spelt-grits were left to soak for nine days, and on the tenth day were made into dough by mixing them with raisin juice. The dough was put into earthenware pots and baked hard in the oven. 1974 E. S. GRUEN *Last Generation of Roman Republic* iii. 110 The trustworthy Picentine L. Afranius.

picey (pəi·si), *a. slang.* [Prob. f. PICE + -Y[1].] Mean, niggardly.
1937 PARTRIDGE *Dict. Slang* 624/1 *Picey*, adj., mean: Regular Army: late C. 19–20. 1965 P. ROBINSON *Pakistani Agent* v. 61 She deserved better than that, the miller would be too *picey*.

pick, *sb.*[1] Add: **1. c.** *Archæol.* A prehistoric implement used for breaking up rocks, soil, etc.
1949 W. F. ALBRIGHT *Archaeol. of Palestine* iii. 59 Among large flint artifacts the most noteworthy are sickle-blades and 'picks', which point to the agricultural character of Natufian culture... Some of the so-called picks are rather hoes, used to break up the ground before sowing grain. 1959 J. D. CLARK *Prehist. S. Afr.* vi. 157 Small, nearly parallel-sided picks.
5. c. *Mus.* A plectrum. Cf. PICK *v.*[1] 12. orig. *U.S.*
1895 *Montgomery Ward Catal.* Spring & Summer 243/2 Mandolin picks, made of celluloid, imitation tortoise shell, oval pattern. 1973 *Advocate-News* (Barbados) 24 Feb. 3/6 (Advt.), Attention all musicians... Just arrived:..Picks Finger Picks Thumb Picks. 1976 D. MUNROW *Instruments Middle Ages & Renaissance* 25/4 The long stem of the quill is shown held between the third and index finger (as a modern guitarist holds a flat pick).
8. *pick-handle* (examples); **pick-pole** *U.S.* = PICAROON *sb.*[2]
1850 N. KINGSLEY *Diary* 1 Nov. (1914) 156 Tinkered a little at Pick handles, putting door in the tent, [etc.]. 1873 J. MILLER *Life amongst Modocs* v. 64 A long white pole, perhaps a sort of pick-handle. 1908 *Sears, Roebuck Catal.* 522/2 Drifting pick handles, 34 inches long. 1979 D. LOWDEN *Boudapesti 3* xiii. 68 The man who'd hit him four times with the pickhandle. 1837 *North Amer. Rev.* Apr. 353 The persons who undertake it [*sc.* breaking a log-jam] must go on to the mass of logs, work some out with their pickpoles, [etc.]. 1972 *Christian Science Monitor* 28 Sept. 16/3 The river-drivers could stay on a pitching, twisting long log, keeping balance with a pick-pole.

pick, *sb.*[3] Add: **10.** In basketball, a permissible block (see quot. 1961).
1951 *Sun* (Baltimore) 24 Dec. 13/2 There is no consistency among officials on calling picks and screens. 1961 J. S. SALAK *Dict. Amer. Sports* 325 To set a pick, the offensive player is entitled to take up a position in front of a defensive opponent provided such maneuver does not hinder the 'normal movement' of the defensive man. 1976 A. CROSS *Question of Max* iv. 43 Kate..had become someting of a basketball aficionado... To her..regret, she could never recognize when someone had set a 'pick', and she tended to admire the wrong members of any team.

pick, *v.*[1] Add: **IV. 7. a.** Also colloq. phr. *to pick them*: in emphatic contexts, to make a wise choice, *spec.* in personal relationships (freq. *ironical*).
1945 A. MARSHALL in *Coast to Coast 1944* 84 He greeted Olive cheerfully, then turned to me with simulated surprise. 'Well, you can certainly pick 'em,' he said. 1953 A. CHRISTIE *Pocket Full of Rye* x. 62 My stepmother was there... The old boy certainly knew how to pick them. 1966 H. WAUGH *Pure Poison* (1967) ii. 11 Fred, glancing ..to the young face of his daughter-in-law to be, had to admit that Larry could pick 'em. 1973 G. SCOTT *Water Horse* (1974) vii. 44 An art student, Polly? You do pick them, don't you. 1976 P. HENISSART *Winter Quarry* xv. 150 'Christ, she really picks them,' muttered McGuire. 'Does she know who he is?' 1978 L. MEYNELL *Papersnake* iii. 53 Baa-Lamb came home..at the very agreeable odds of twelve to one... 'What did I tell you, cobber?.. If you can pick 'em, you can pick 'em.'
f. *ellipt.* for *to pick one's way*.
1865 R. D. BLACKMORE *Cradock Nowell* (1866) I. xvi. 153 Hogstaff tottered along before him, picking uneasily over the stones. 1878 HARDY *Ret. Native* I. iii. 66 The track is rough, but if you've got a light your horses may pick along wi' care. 1961 'G. HOLDEN' *Deadlier than Male* xiv. 102 This time the search took twice as long, cutting down on his extra reading, for he had to pick through several columns of one- and two-line social notes in each issue.
g. *trans.* To guess, deduce; to predict. *Austral.* and *N.Z. colloq.*
c 1926 'MIXER' *Transport Workers' Song Bk.* 128 I'm picking we'll soon have a row. 1943 N. MARSH *Colour Scheme* vi. 100 There's a bit of a shelf above the cliff... I picked that was where he'd go. 1959 D. NILAND *Big Smoke* vi. 145 He looked up at the boy with a pleased, questioning expression. 'I pick it right?'
VII. 14. *to pick at* ——. **b.** Also *Austral.*
1916 C. J. DENNIS *Songs Sentimental Bloke* 127 *Pick at*, to chaff; to annoy. 1941 BAKER *Dict. Austral. Slang.* 54 *Pick at*, to blame, chaff, irritate.
15. *to pick on, upon* ——. Delete 'Now *U.S. dial.*' and add further examples. Also, to single out for attention or adverse criticism; to victimize.
1875 W. D. PARISH *Dict. Sussex Dial.* 87 They always pick upon my boy coming home from school. 1888 in Farmer *Americanisms* (1889) 419/1 Joseph White..slept for five days and nights, and then jawed his wife for waking him up. He said she was always picking on him when she saw him taking comfort. *c* 1910 D. H. LAWRENCE *Phoenix II* (1968) 26 You always pick on the Gordons—you're always on to us—! 1919 WODEHOUSE *Coming of Bill* (1920) II. iii. 141 That wouldn't make no difference.—She'd pick on me just the same. 1929 J. BUCHAN *Courts of Morning* III. ii. 331 Looks as if you folk had been picking on my poor little country. 1930 J. B. PRIESTLEY *Angel Pavement* vi. 297 They begin picking on her and she stands up for herself. 1930 S. JEPSON *I met Murder* ii. 27 Have you any idea why the Inspector should have picked on you first? 1947 'N. SHUTE' *Chequer Board* 73 Last night they was picking on the coloured boys—saying nasty things about niggers in their hearing. 1959 G. FREEMAN *Jack would be Gent.* 140 Don't keep picking on him, please, Mum. 1961 [see *BACK *sb.*[1] 23 d]. 1975 *Times* 15/4 Why pick on the present Government? Has any government..in the past 30 years ever..done anything to encourage that aim?
VIII. 18. pick off. c. *Baseball.* To put out (a runner) at a base.
1948 *Sun* (Baltimore) 1 Dec. 17/4 The play in question came when Bobby Feller, Cleveland pitcher, whirled and threw to Manager Lou Boudreau in an effort to pick off Masi, Boston catcher. 1974 *Spartanburg* (S. Carolina) *Herald* 23 Apr. A6/4 Dancy was picked off by catcher Luis Rosado as he inexcusably wandered too far off third.
19*. pick over. To sort; to select the best from (a group or collection); to pick off dead flowers from.
1917 *Dialect Notes* IV. 397 She is picking over blackberries. 1924 A. D. SEDGWICK *Little French Girl* i. 6 She..picked over the herbs that were to be dried for *tisane*. 1946 D. C. PEATTIE *Road of Naturalist* iii. 40 The supplies were picked over and over. One necessity after another was thrown away, as the owner remembered that he might have to carry it, if the oxen died. 1971 *Vogue* Dec. 48/3 The geraniums had flowered once, and needed to be picked over to induce a second flowering. 1972 D. WESTHEIMER *Over Edge* (1974) ii. 20 Karen..could spend hours happily picking over rejects in a surplus store. 1973 E. PAGE *Fortnight by Sea* i. 8, I don't think there'll be any strawberries left... The beds were picked over pretty thoroughly a couple of days ago. 1978 *Lancashire Life* Apr. 78/2 Though the summer suns are not yet exactly at the glowing stage, wise girls will buy their summer clothes now, rather than wait till the best things have been picked over.
20. pick up. b. Also *spec.* (*a*) *trans.* to gather (a shorn fleece) from the floor of the shearing shed, carry it to a table, and throw it out flat so that it can be skirted, rolled, and classed, also *absol.* *Austral.* and *N.Z.* (*b*) *absol.* in game-shooting, to make a retrieval, esp. to collect unretrieved game after a shooting party.
a. 1862 J. G. WALKER *Jrnl.* 10 Nov. (typescript) 24 My job at first was picking up fleeces. 1926 J. DEVANNY *Butcher Shop* 11 The naked feet of the brown women 'picking up' from the shining greasy floor. 1940 F. SARGESON *Man & Wife* (1944) 47 It was early summer, shearing time. Tom and me went into the country and we got a job picking up fleeces in a big shed. 1967 J. MORRISON in *Coast to Coast 1965–66* 157 He'd been away picking-up in the shearing sheds.
b. 1888 W. B. LEFFINGWELL *Wild Fowl Shooting* xxxvi. 364 After the pup has gotten to understand your orders of picking up, and bringing the glove to you from short distances, throw it farther. 1897 *Encycl. Sport* I. 442/1 The keeper..must be careful to judge accurately when to start the next drive, after one is over, so as to give reasonable time for picking up. 1976 *Shooting Times & Country Mag.* 16–22 Dec. 20/2 My immediate neighbours ..did help pick up and I am happy to say the count, if nothing else, was quite gratifying.
e. *spec.* To form an acquaintance with (a person) casually or informally, esp. with the intention of having a sexual relationship.
1698 [in Dict.]. 1698 J. COLLIER *Short View Immorality Eng. Stage* vi. 238 Nothing being more common than to see Beauty surpriz'd, Women debauch'd, and Wenches Pick'd up at these Diversions. 1734 *Select Trials 1720–1724* 59/1 The Prosecutor pick'd me up and went with me to my Lodgings..where he would have lain with me. 1785 COWPER *Let.* 4 June in *Corr.* (1904) II. 325 He was seen by Mr. Shepherd..leading a female companion into a wood.., whom he saw him pick up as he went. 1812 J. H. VAUX *Vocab. Flash Lang.* in *Mem.* (1964) 257 To pick up a cull, is a term used by blowers in their vocation of street-walking. 1893 G. B. SHAW *Widowers' Houses* I. viii. 29 'I have made the acquaintance of'—or you may say 'picked up', or 'come across', if you think that would suit your friend's style better. 1921 *Sat. Even. Post* 1 Oct. 18/2 You are right in thinking there must be something wrong with girls who try to 'pick up' strange men as no girl with self respect would do such a thing. 1932 E. BOWEN *To North* iv. 34 She wished she had not picked Markie up in the train and given him her address. 1933 *Times Lit. Suppl.* 30 Mar. 223/3 One evening Sam 'picks up' a young woman at an open-air concert. 1942 E. PAUL *Narrow St.* xvii. 135 The wife of one of the officers..formed the habit of picking up rich gentlemen in department stores. 1961 J. DOS PASSOS *Midcentury* 94 Eileen got to dancing..and trying to pick up strange men. 1975 *Evening Standard* 14 May 5/3 It is not difficult to learn..that it is possible to obtain money, food or a bed for the night by being 'picked up'. 1976 D. MARLOWE *Nightshade* xii. 143 Who was that old man?.. He was trying to pick you up.
f. Also, to succeed in seeing, hearing, detecting, receiving, etc., by means of an appropriate instrument or apparatus.
1888 *Electrician* 2 Nov. 833/1 For researches of this description it is necessary to employ as sensitive an instrument as it is possible to obtain, to pick up, so to speak, such minute currents. 1908 *Rep. Brit. Assoc. Adv. Sci. 1907* 621 The receiving apparatus..would pick up a number of disturbances from other stations. 1908 *Westm. Gaz.* 23 Oct. 5/3 The following notes will enable it [*sc.* a comet] to be 'picked up' with the aid of an opera-glass and a star-map. 1913 *Pop. Mag.* 1 May 79/2 The only signals which it was picking up now were..those of the enemy. 1921 [see *GET *v.* 22 b]. 1922 *Encycl. Brit.* XXX. 88/2 Presently the airship was 'picked up', and immediately from all quarters of the defences searchlights could be seen moving across to get on to it. 1929 S. ERTZ *Galaxy* xvii. 365 She and the General thoroughly enjoyed their wireless in the evening, and it was amusing to pick up Paris or Berlin. 1941 [see *ATMOSPHERE *sb.* 4 b]. 1948 F. P. SHEPARD *Submarine Geol.* ii. 16 The radar screen will pick up land objects and will show the position of islands, capes and mountain peaks, giving their direction and distance. 1960 *McGraw-Hill Encycl. Sci. & Technol.* XIII. 462/1 Television camera tubes are designed primarily to pick up live programs, indoors and outdoors, as well as to reproduce motion pictures. 1973 *Guardian* 20 June 13/1 BBC Monitoring Service reports that the only Moscow Radio commentaries on Watergate picked up so far in any language were in Quechua. 1977 *Lancet* 4 June 1187/2 Clinically, cerebral blood-flow is measured after intra-arterial injection of xenon-133; gamma radiation is picked up by two detectors and the scintillation-counts are fed into a laboratory digital computer. 1978 *Nature* 6 Apr. 481/2 This detection capability has improved roughly threefold; explosions of yields of 1 or 2 kilotons in hard rock in most parts of the Northern Hemisphere would now most likely be picked up.
g. *to pick up the pieces* (fig.): to (try to) redeem some advantage or compensation from an apparently hopeless situation.
1912 KIPLING *Divers. Creatures* (1917) 17, I should have said it was half a night. Now, shall we go down and pick up the pieces? 1938 M. ALLINGHAM *Fashion in Shrouds* vi. 92 It'll come to a quiet, uncomfortable end and you'll have to stand by and pick up the pieces. 1951 J. C. FENNESSY *Sonnet in Bottle* v. iv. 163 Injy was very good at taking things as they came. But he generally found it was his job to pick up the pieces afterwards. 1970 A. TOFFLER *Future Shock* (1971) xvii. 391 By proliferating enclaves of the past, living museums as it were, we increase the chances that someone will be there to pick up the pieces in case of massive calamity. 1973 A. BROINOWSKI *Take One Ambassador* vii. 97 We could see them [*sc.* the Japanese] picking up the pieces in Indo-China after the Americans. 1977 R. PERRY *Dead End* i. 12 If anything does go wrong it'll be nice having you around to pick up the pieces.
j. *trans.* and *intr.* To steal, rob, pilfer; to cheat, swindle. *slang.*
c 1770 R. KING *Frauds of London Detected* 39 [Highwaymen] have various schemes for carrying on their

business, such as seeing ostlers, bribing landlords, on the road, for intelligence of *who is worth picking up*. 1829 H. WIDOWSON *Pres. State Van Diemen's Land* 73 There are always a number of loose characters lurking about, on the look-out for strangers, to 'pick them up', as they term it, which, in other words, means to rob them. *a* 1876 E. LEIGH *Gloss. Words Dial. Cheshire* (1877) 154 *Picking up*, a term for picking a pocket. 1903 *Mark of Broad Arrow* vii. 108 Within twenty-four hours of that man's release the three prison-made thieves were looking round the town to see what they could 'pick up'—in plain language, to see what they could thieve. 1928 *Detective Fiction Weekly* 8 Sept. 565/2 Gentleman George..would mark down his traveler, knowing him to be in possession of jewelry or other valuables, and tirelessly follow him until the opportunity arose to 'pick-up' his all-important bag.
k. *trans.* To find fault with, call to account; to detect (a person) in a mistake; to show up. *colloq.*
1846 S. F. SMITH *Theatr. Apprenticeship* 149 The bystanders..were crowding around the table in great numbers to see the fun—all considering me most undoubtedly 'picked up'. 1878 B. HARTE in *Scribner's Monthly* Dec. 184/1 When we were coming down the valley you picked me up twice..*contradicted* me, that's what I mean. 1922 *Daily Mail* 5 Dec. 11, I am picked up for saying that the initiative in the Steamer case should have come from the stewards.
l. In *Cricket*, to succeed in hitting (a ball, esp. one that pitches close to the ground).
1851 J. PYCROFT *Cricket Field* vii. 153 If you reach far enough, even a shooter may be picked up. 1862 —— *Cricket Tutor* 8 The old bat used to be heavy at the point —very requisite for picking up a Grounder. 1959 *Times* 29 May 4/1 He could not have picked up the ball off his legs so crisply.
m. *trans.* To cause (a person) to revive; to serve as a 'pick-me-up' for (someone).
1857 DICKENS & COLLINS *Lazy Tour* v, in *Housch. Words* 31 Oct. 412/1 Several..look in at the chemist's..to be 'picked up'. 1889 [in Dict., sense 20g]. 1914 G. B. SHAW *Misalliance* 80 Have you had your tea?.. A cup of tea will pick you up. 1978 J. CARROLL *Mortal Friends* I. ii. 22 'You know me, Colman. Tea anytime.' 'Indeed, Father. It picks a body up.'
n. *trans.* and *intr.* To tidy or clean up; to put in order. *colloq.* (chiefly *U.S.*).
1861 *Trans. Illinois Agric. Soc.* IV. 204 We did not find 'things picked up in it'—no air of comfort about it. 1874 E. S. WARD *Trotty's Wedding Tour* 214 It had taken all day to 'pick up' after the departed travellers. 1888 N. PERRY *Flock of Girls* 81 She mends these little bits of things and 'picks up' after you, as you call it. 1889 [in Dict., sense 20g]. 1966 J. BALL *Cool Cottontail* (1967) v. 44 'The room isn't properly picked up yet,' the woman said. 'When you have five kids..you can't get everything done.' 1973 *Sunday Bull.* (Philadelphia) 7 Oct. (Parade Suppl.) F5/4 Ask your children for their ideas. It's the best way to enlist their help in keeping their rooms 'picked-up'.
o. *trans.* To arrest, apprehend. Cf. sense 20 d in Dict. *slang* (orig. *U.S.*).
1871 *Congress. Globe* 5 Jan. 317/2 They are picked up for taking horses or sheep or anything of that sort. 1887 *Lantern* (New Orleans) 11 June 2/2 I'll have the police pick him up for blackmail. 1934 J. T. FARRELL *Young Manhood* xii. 192 He gazed around the church to see if any of the boys were present. Seeing none of them, he guessed that they must all have been picked up, and were enjoying Christmas Day in the can. 1938 F. D. SHARPE *Sharpe of Flying Squad* xiv. 157, I picked them up for stealing a man's wallet. *Ibid.* xvi. 183, I picked Billy up, knowing that he was 'wanted' for a job. 1946 F. SARGESON *That Summer* 102 We all had to stand there with a crowd of jacks in plain clothes standing round, and one in uniform called out our names and said what we'd been picked up for. 1956 H. GOLD *Man who was not with It* (1965) xxii. 202 You want to be able to tell the fuzz the truth if we're picked up? 1961 *Confidential* Jan. 39/2 When Farouk was overthrown, police picked up his personal pimp, Pulley Bey. 1976 'TREVANIAN' *Main* x. 213 Things start to go badly for him. His boys..get picked up for every minor charge in the book. 1979 *Massachusetts Daily Collegian* 30 Apr. 14/4 He said he was not drunk when picked up on East Pleasant St. on March 29.
p. *to pick up on*: (*a*) to draw near, begin to overtake (a person) in a race; (*b*) *U.S. slang*, to understand, appreciate, or obtain.
1908 *Daily Chron.* 27 Nov. 7/6 At the fifth lap..Dorando held him, and then began to pick up on him. 1944 D. BURLEY *Orig. Handbk. Harlem Jive* 15 Let me boot you to my play [*sc.* inform you of my plan] and, maybe, you can pick up on the issue. 1946 MEZZROW & WOLFE *Really Blues* 377 *Pick up on*, get, take, learn. *Ibid.*, *Pick up on what's going down*, understand what's happening. 1956 B. HOLIDAY *Lady sings Blues* (1973) i. 8 In Baltimore, places like Alice Dean's were the only joints fancy enough to have a victrola and for real enough to pick up on the best records. 1972 *Jazz & Blues* Feb. 19/1 They came to gig in their own clubs before white people picked up on them. 1977 MCKNIGHT & TOBLER *Bob Marley* iii. 46 Paul McCartney was noted as having 'picked up on reggae'.
q. *intr.* and *trans.* Of a vehicle, aircraft, etc.: to gain speed after being slow-moving or stationary; to recover (speed). Cf. sense 20 h in Dict.
1922 S. LEWIS *Babbitt* v. 53 He noted how quickly his car picked up. 1932 C. ISHERWOOD *Memorial* III. i. 181 'That's a damn fine bus,' said Farncombe earnestly. 'My Christ, Gerald, you should see the way she picks up.' 1939 *War Illustr.* 29 Dec. 539/3 However, as we got down to five hundred feet the engines began to pick up.
r. *trans.* To pay (a bill, account, etc.); esp. in phr. *to pick up the bill, check, tab*, etc.; also *fig. colloq.* (orig. *U.S.*).
1945 *Sun* (Baltimore) 23 Oct. 4 (*heading*) 'Lobbyist' said to have picked up check for Truman outing. 1947 *Ibid.* 12 May 2/5 Some United States diplomats have entertained each other with the taxpayers picking up the check. 1956 S. BELLOW *Seize the Day* ii. 26 His father might have offered to pick up his hotel tab. 1956 B. HOLIDAY *Lady sings Blues* (1973) xxiii. 196 Americans used to make fun of the British health system, where sick people could go to doctors and hospitals for free and the government picked up the tab. 1961 R. BLOCH *Blood runs Cold* 215 I'm permitted to attend a party..when they go out to eat because I always pick up the check. 1964 WODEHOUSE *Frozen Assets* ii. 32 'Coffee's out, I'm afraid.' 'Nonsense. I'll pick up the tab.' 1966 M. BREWER *Man against Fear* vii. 77 Tonight we pick up her bill. 1967 *Canad. Ann. Rev. 1966* 65 Ottawa would pick up a $14 million tab for a system already in operation. 1976 *S. Wales Echo* 27 Nov. 2/2 Mr. Gray asked if the college should be expected to 'pick up the tab' for even greater expenditure if grants to the theatre were cut. 1978 *Daily Tel.* 13 Feb. 6/5 Ratepayers would have to pick up the bill if important jobs were transferred from the county councils to some of the larger districts.

pick-. Add: **pick-and-gad** *a.* (see quot.); **pick-and-shovel** *a.*, that uses a pick and shovel; also *fig.*; **pickbrain** *a. poet.*, that picks one's brains; **pick-proof** *a.*, secure against picking.
1883 *Encycl. Brit.* XVI. 444/2 The so-called 'pick and gad' work consists in breaking away the easy ground with the point of the pick, wedging off pieces with the gad, [etc.]. 1895 F. REMINGTON *Pony Tracks* 193 They are all cavalry..and are not hindered by..pick-and-shovel work. 1907 *Westm. Gaz.* 11 Mar. 9/3 You don't look much like pick-and-shovel men. 1911 *Chambers's Jrnl.* Mar. 167/2 The ordinary pick-and-shovel man earns..one shilling and eightpence per day. 1930 BLUNDEN *De Bello Germanico* 42 Taking pick-and-shovel troops forward. 1953 *Sun* (Baltimore) 24 Nov. 40/1 Chief Magistrate Sherr yesterday called the search for motorists who have ignored a series of tickets a 'pick and shovel job', in contrast to the 'machine age system' he saw in action during a trip to New York city. 1934 DYLAN THOMAS *18 Poems* 25 My pickbrain masters morsing on the stone Despair of blood, faith in the maiden's slime. 1950 *Sun* (Baltimore) 4 Mar. 2/7 A new electric 'pickproof' motor car lock..is being manufactured. 1976 *National Observer* (U.S.) 29 May 9/2 Eastside tailor seeks design for pick-proof man's hip pocket. 1977 H. GREENE *FSO-1* vii. 63 He had a love for the pick-and-shovel people, the working stiffs behind the men..who wielded power.

pick-a-back, *adv. phr.* (*a.*, *sb.*) Add: The dominant form (except in sense *c) is now **piggy-back. b.** (Further examples.) Also in reference to a thing; *spec.* (*a*) an aircraft, rocket, or the like, to which is attached another aircraft, etc., usu. for launching in mid-air; (*b*) a flat railway car on which a truck, container, or the like is carried.
1901 *Punch* 2 Oct. 247/1 Oh, Mr. Green,..Effie..is so miserable because she hasn't had her donkey ride. Would you mind giving her a pick-a-back? 1931 R. CAMPBELL *Georgiad* I. 21 The garden path—a sort of Rotten Row Where oft a merry pick-a-back they go. 1936 [see *mother plane* s.v. *MOTHER *sb.*[1] 16 a]. 1936 L. A. G. STRONG *Last Enemy* iii. 203 Ann came up. 'Piggy-back, please,' she said. 'You mustn't worry poor Mr. Boyle. It's too hot.'..Ann, hoisted on Denis's back, turned to her. 1944 *Sun* (Baltimore) 13 July 9/2 Two Liberator crewmen with only one parachute between them made a 'piggyback' leap recently from a burning bomber. 1946 L. E. O. CHARLTON *Britain at War: R.A.F. & U.S.A.A.F., July 1943–Sept. 1944* 306 The 'pick-a-back' consisting of an Me 109 fighter mounted on the back of a Ju 88... When approaching its target the Ju 88 was released and guided by remote control towards its objective. 1953 *Evening Jrnl.* (Lincoln, Nebraska) 3 Nov. 6 No sooner had 'piggy back railroading' (hauling truck trailers on flat cars) been hailed by various railroads and one automotive manufacturer as a strikingly simple idea for abating highway congestion and cutting the high price of trucking, than the organized truck operators rose to denounce it. 1954 *Economist* 13 Mar. 779/1 These 'piggy-backs' or 'trailers-on-flat-cars' (TOFC)..have now become a major preoccupation of the railways. 1954 *Railway Age* 26 July 3/2 Further expansion of piggyback by the Chicago & North Western took effect on July 15. 1954 *Times* (Seattle) 21 Sept. 16/1 Railroads throughout the country are studying or instituting limited pickaback service. 1956 *Sun* (Baltimore) 29 Feb. 7/3 A 'piggyback' experiment in which vacationists' automobiles will accompany them on train trips in Germany is planned for next summer. 1959 H. PINTER *Birthday Party* II. 37 Maybe I played piggy-back with you. 1960 *Daily Tel.* 23 June 1/4 (*heading*) U.S. 'pick-a-back' spheres over Russia every 101 min. *Ibid.*, Two 'pick-a-back' satellites were launched by the same rocket from Cape Canaveral, Florida, early today. 1961 *Flight* LXXIX. 827/1 Under the heading 'aerodynamic efficiency' an interesting 'piggyback' vehicle was exhibited. 1965 *Listener* 1 July 6/1 A series of pick-a-back rockets called the Precious Stones are the most important outcome. 1969 [see *MANIFEST *sb.* 3]. 1969 I. & P. OPIE *Children's Games* vi. 185 Riding on people's backs ('Horse Race', 'Piggyback Race'). *Ibid.* vii. 217 The craze at our school is piggy-back fighting. 1969 *Sat. Rev.* (U.S.) 28 June 18/3 Railways are currently taking pride in their new innovation, 'piggy-back' service, in which highway trailers..can travel across long distances hitching rides on various railroads. 1973 *Kingston* (Ontario) *Whig-Standard* 21 Sept. 9/6 The CPR diesels still thunder through the hamlet with their cargoes of 'piggy backs', but the little station has been gone for many years now. 1974 W. FOLEY *Child in Forest* I. 25 Our usual practice of running to meet him for pick-a-backs up the garden path. 1977 *Time* 14 Feb. 54/3 The orbiter will be 'mated' to a carrier plane, a Boeing 747 with special mounts on top... The piggyback pair will first run up and down the Edwards runway to test for vibration and stability.
c. *Comb.*, as **pick-a-back plant**, a perennial herb, *Tolmiea menziesii*, of the family Saxifragaceæ, native to western North America and distinguished by cordate leaves that appear to grow one on top of another.
1946 M. FREE *All about House Plants* xviii. 275 The Picka-back Plant has come very much to the fore as a house plant. 1973 J. L. FAUST *N.Y. Times Bk. House Plants* 135 Piggyback plant (*Tolmiea menziesii*) has somewhat fuzzy leaves. 1973 HITCHCOCK & CRONQUIST *Flora Pacific Northwest* 199 *Tolmiea*... Youth-on-Age; Pig-a-Back-Plant; Thousand Mothers.
d. as *v. intr.* and *trans.* To ride piggy-back; to attach on the back of; to carry or transport by means of a piggy-back method. Also *fig.* So **pi·ggy-backing** *vbl. sb.*
1959 *Wall St. Jrnl.* 29 Jan. 1/1 A number of railroad and truck equipment makers have jumped into the manufacture of rail cars and trailers especially designed for piggybacking. 1960 *Economist* 3 Sept. 894/2 The haulage industry is threatened by the rapid spread of the practice of 'piggy-backing'—transporting a loaded lorry or trailer on a railway wagon. 1967 *Freight Management* Jan. 48/2 Piggybacking must not be left out of any argument about future inter-modal methods of transportation. 1968 *Wall St. Jrnl.* 25 Mar. 14/3 To the degree heroin does piggyback on marijuana, it seems more due to the law than to the pushers. 1968 A. DIMENT *Gt. Spy Race* iv. 56 Mr Spont would have to piggy-back me all the way to get a commission. 1973 T. H. WHITE *Making of President 1972* (1974) v. 123 They're tired of Humphrey, he's been around too long... But we have to piggyback him on local campaigns. Wherever we back a local candidate in the primary, we'll pack Hubert in on top. 1975 C. BEECK in *Proc. 25th Electronic Components Conf.* 158/1 Selecting a shaft diameter which would accommodate a coaxial shaft inside it in case another control had to be piggy-backed to the knob. 1975 *Time* (Canada ed.) 17 Mar. 7/1 Arctic Gas argues that the only economical way to bring Canadian gas south is by 'piggybacking' on a line largely given over, at first, to carrying more abundant Alaskan gas. 1976 *Billings* (Montana) *Gaz.* 7 July 4–B/1 The women are content to let the men set the standards and then piggy-back on our efforts. 1977 *Offshore Engineer* May 97/3 A free-standing well-protector jacket platform is attached or 'piggybacked' to the mobile platform at the dock before tow-out. 1977 C. MCCULLOUGH *Thorn Birds* ii. 21 Come on, Meggy, I'll piggyback you the rest of the way. 1979 *Economist* 17 Nov. 79/2 Why not piggyback a federal sales tax on to the levies which people are used to paying and collecting?

pickable, *a.* (Further examples.)
1966 C. SWEENEY *Scurrying Bush* x. 142 This little bird is supposed to enter the mouths of crocodiles to pick their teeth, but..the teeth of a crocodile..are widely spaced so that they are not really pickable—no particles of food being likely to remain. 1969 K. BENTON *24th Level* iv. 59 Black people..going downwards to the city and jobs and suckers with pickable pockets.

picked, *ppl. a.* Add: **5.** *picked-over*, from which the best has already been selected.
1839 *Congress. Globe* 25th Congr. 3 Sess. App. 47/2 All the emigrants went on to the new lands, where they could get first choices at $1.25 per acre, because they could not give that sum for picked-over lands in the old counties. 1886 N. SHEPPARD *Before Audience* viii. 124 Audiences in England outside of the Established Church are weeded. To an American lecturer or preacher they have a picked-over appearance. The church takes the cream, the chapel the milk of society. 1979 A. PARKER *Country Recipe Notebk.* iii. 60 Fill up with well-picked-over berries.

‖ **pickelhaube** (pikəlhau·bə). Also **pickel-haube** (with hyphen) and with capital initial. Pl. **pickel-hauben, pickelhaubes.** [G.] A German spiked helmet of a type worn esp. before and during the earlier part of the war of 1914–18. Also, by metonymy, a German soldier.
1875 *Encycl. Brit.* II. 596/2 The [Prussian] uniform is a dark blue tunic, grey trousers with red stripe, helmet of black leather with brass ornaments and spike (*Pickelhaube*). 1880 G. A. SALA *Amer. Revisited* 213/2 The brutal jest of the cynical master of so many thousands of stolid men with needle-guns and *pickelhaubes* [*sc.* Bismarck] could turn all into mockery and contempt! 1887 *Athenæum* 1 Jan. 16/1 Here is represented the old Empire with powder and wigs, while in Julius Grosse's novel..we find the new Empire with its *Pickelhaube*. 1890 *Times* (Weekly ed.) 28 Feb. 16/1 A dragoon regiment with *pickelhaube* helmet. 1901 *Scotsman* 29 Nov. 5/4 Germany is defied in a manner fitted to stiffen the pickelhauben of the General Staff. 1927 *Bulletin* (Glasgow) 4 Oct. 12/2 A German officer's silver-plated pickel-haube. 1931 E. LINKLATER *Juan in Amer.* II. xvii. 181 Prussian wombs bore pickelhaubes. 1969 G. COPPARD *With Machine Gun to Cambrai* xxi. 89 A dozen Jerry soldiers... The *pickelhaubes* on their heads made them look a leering bunch of devils. 1972 M. GLENNY tr. *Solzhenitsyn's August 1914* xxv. 260 Even when the *Pickelhauben* were no more than a hundred yards away, the Russians showed no fear. 1976 *Leicester Chron.* 26 Nov. 14/5 The first distinctive German steel helmets were introduced in 1916 as a replacement for the pickelhaube, and other armies quickly followed suit.

picker[1]. Add: **1. g.** One who picks (PICK *v.*[1] 12) or plucks the strings of a musical instrument such as the banjo or guitar; usu. with the name of the instrument prefixed.

1923 in *John Edwards Mem. Foundation Q.* (1969) V. II. 62 Old fiddlers and banjo pickers. 1934 S. R. NELSON *All about Jazz* vi. 126 The modern method of picking and slapping on the bass was found to be much more rhythmic. So a race of pickers and slappers..sprang into being. 1951, 1959 [see *guitar-picker* s.v. *GUITAR *sb.* b]. 1964 *Amer. Folk Music Occasional* I. 43, I know a banjo picker who hasn't performed for anyone except his wife for the last three years. 1969 N. COHN *A WopBopaLooBop* (1970) viii. 77 A strange city, filled to overflowing with guitar pickers by the thousand. 1976 *National Observer* (U.S.) 23 Oct. 20/3 It's not a novel story. Country music is rife with legend about pickers hitting Lower Broad with a nickel and a song.
4. For 'in Australia' read 'in Australia and New Zealand'. (Earlier and further examples.)
1761 STERNE *Tr. Shandy* III. xxxiv. 159 Indissolubly annex'd by the picker up, to the thing pick'd up. 1881 A. BATHGATE *Waitaruna* xii. 172 The 'pickers-up' were busy gathering the fleeces as they fell from the bereft sheep and carrying them to the sorting table. 1913 —— *Sodger Sandy's Bairn* 57 The pickers-up gathered the fleeces as they fell intact from the shears and bore them to the sorting table, where they were quickly 'skirted' and 'classed'. 1940 *Essays & Stud.* XXV. 111 The picker-up of unconsidered historical trifles. 1956 S. HOPE *Diggers' Paradise* 99 A sixteen-year-old lad could earn £17 a week as a 'picker-up' in the wool-shed. 1959 H. P. TRITTON *Time means Tucker* iii. 26/2 Pickers-up took the fleece as it fell on the board and spread it skin-side down on the wool-tables. 1977 *Shooting Times & Country Mag.* 13–19 Jan. 22/1 There are plenty of pickers-up on this shoot, and little is lost.
b. picker-upper, one who, or that which, picks up.
1936 *Esquire* Sept. 162/2 *Variety* maintains a news staff—not a bunch of press-release picker-uppers. 1942 *Amer. Speech* XVII. 104/1 *Picker upper*, service car with crane. 1944 *N.Y. Times* 3 Sept. S2/6 Her devoted spouse is an avid picker-upper of any hairpins he can find. 1947 *Philadelphia Bull.* 28 July 8 (Advt.), Energy picker-upper..chocolate cookies. 1961 *Times* 19 Aug. 6/7 A mechanical means of gathering up lumps of oil..has been built. It is described as a 'picker-upper', which might be drawn by a tractor.

pickerel frog. *U.S.* [PICKEREL[1].] A common North American frog, *Rana palustris*.
1839 D. H. STORER in Storer & Peabody *Rep. Fishes, Reptiles & Birds Mass.* 238 The pickerel frog..is..met with about the margins of fresh water brooks and ponds. 1906 M. C. DICKERSON *Frog Bk.* 189 The brook and the fields and meadows near make the home of the Pickerel Frogs. 1961 D. M. COCHRAN *Living Amphibians of World* 107/1 The entire skin of the North American pickerel frog, *Rana palustris*, secretes a substance which is lethal to frogs of other species.

Pickering (pi·kəriŋ). *Physics.* The name of Edward Charles *Pickering* (1846–1919), U.S. astronomer, used, usu. *attrib.*, with reference to a series of lines in the spectrum of ionized helium with wave numbers represented by $4R(1/4^2 - 1/m^2)$ (where R is the Rydberg constant and $m = 5, 6, \ldots$), of which the first line has a wavelength of 1012 nanometres and the series limit is at 364 nanometres; (observed by Pickering in 1896 (*Astrophysical Jrnl.* IV. 369, V. 92)).
1922 A. D. UDDEN tr. *Bohr's Theory of Spectra* I. 3 Recently however the question has been reopened and Fowler (1912) has succeeded in observing the Pickering lines in ordinary laboratory experiments. 1923 H. L. BROSE tr. *Sommerfeld's Atomic Struct. & Spectral Lines* iv. 208 Pickering's series (7) includes only one-half of the lines represented by (7a), namely, those for which k is odd. *Ibid.*, It is..unjustifiable and arbitrary to detach one-half as the Pickering series and to ascribe it to hydrogen. The other half was overlooked earlier only because it could not be separated from the neighbouring true hydrogen lines. 1942 J. D. STRANATHAN *'Particles' of Mod. Physics* vi. 222 Alternate lines of the Pickering series of helium coincide almost exactly with the Balmer lines of hydrogen. But..the Pickering lines fall at slightly shorter wave lengths. 1967 W. R. HINDMARSH *Atomic Spectra* ii. 5 It was first believed that the Pickering series arose from a special form of hydrogen but it is now known to be due to ionized helium.

picket, *sb.*[1] Add: **I. 1. c.** Also, a triangular or arrow-shaped mark cut in turf or placed on masonry, used in making measurements.
II. 4. d. Short for *picket duty*.
1775 J. HALLAM *Let.* 10 Dec. in H. P. Johnston *Nathan Hale* (1901) 158 Your being on Picquet is a sufficient excuse that you wrote no more. 1834 *Chambers's Edin. Jrnl.* III. 167/2 For three weeks I have been on picquet every night. 1861 O. W. NORTON *Army Lett.* (1903) 34, I have just returned from picket. 1944 J. S. PENNELL *Hist. Rome Hanks* 41 You're on picket, aren't you?
5. For 'men' read 'people'. Also, applied to a person or group conducting a demonstration at particular premises, a particular installation, etc. Also collective sing. Also, the conduct or activity of pickets; an instance of picketing.
1938 *Sun* (Baltimore) 7 Sept. 2/2 Patient parish pickets, determined to retain the Rev. Simon Borkowski as pastor of St. Barbara's Catholic Church, kept their vigil today as they have for more than three weeks to prevent him from fulfilling a transfer order. 1973 *Freedom* 12 May 4/4 (Advt.), Stop the French tests. Regular picket, and London–Paris walk 14th May–3rd June. 1977 *New Society* 30 June 655/2 The picket's mood turned restive, apprehensive. *Ibid.* 657/1 Comment on the picket has almost entirely missed the point by concentrating on how the law might be changed to reconcile heavy strike-breaking vehicles and mass pickets.
IV. 7. *picket duty* (earlier example), *fence* (earlier and later examples), *fort*, *rope* (examples), *tent*, *work*; **picket-boat** (earlier and later examples); **picket line** (*b*) (earlier and later examples); **picket-pin** (**gopher**) *U.S.*, a ground squirrel of the genus *Citellus*, esp. *C. richardsoni*, found in parts of western North America.
1866 *Oregon State Jrnl.* 13 Jan. 1/4 For the capture of the Albermarle [*sic*], by Lieu[t]. Cushing's picket-boats, the crew netted $100 per man. 1942 *R.A.F. Jrnl.* 13 June 28 On the far side of the harbour, was the dockyard, with its cranes and bustling tugs and picket boats. 1975 *Times Lit. Suppl.* 22 Aug. 936/2 As a fifteen-year-old midshipman in HMS Bacchante, he commanded a picket boat during and after the landings [at Gallipoli in 1915]. 1862 O. W. NORTON *Army Lett.* (1903) 112 Very little drill or other duty, no picket duty or trenching. [1800 *Carpenters' Rules of Work* (Boston) 32 Plain picket open fence.] 1817 S. R. BROWN *Western Gazetteer* 66 A garden ..with high, substantial picket fences to prevent the thefts of the Indians. 1946 D. C. PEATTIE *Road of Naturalist* iii. 36 At home, in fertile Illinois, with the clean snow on the picket fences of Galena, they would be cooking goose. 1951 J. FRAME *Lagoon* 126 We were sitting in little brown summer-houses, and touching the brown picket-fences. 1972 *Evening Telegram* (St. John's, Newfoundland) 24 June 9/6 Grand Beach is..a clean place with picket fences. 1775 in *Mass. Hist. Soc. Coll.* (1814) 2nd Ser. II. 230 This fort consists of two large block houses, and a large barrack, which is enclosed with a picket fort. 1846 T. L. MCKENNEY *Mem.* I. vi. 127 The old picket fort standing on the plain..quite a ruin. 1856 R. GLISAN *Jrnl. Army Life* (1874) xx. 277 Indians broke through the picket line. 1945 A. HUXLEY *Let.* 13 Oct. (1969) 536 Matthew was fortunately absent when the violence broke out on the picket line, but he got arrested. 1973 *Guardian* 11 June 13/8 The manual workers at Salford will hold a mass meeting to decide if they will cross the picket line. 1978 *Guardian Weekly* 29 Jan. 17/2 It violates the basic military axiom of not putting the main body of troops on the picket line. Crowding the bulk of NATO's troops along the front line shows the Soviets where the alliance is strong and where it is weak. [1893 V. BAILEY *Prairie Ground Squirrels* 32 Striped Prairie Spermophile... The little Striped Spermophile..is seen standing upright on its hind feet, straight and motionless as a stick... At a little distance it is impossible to distinguish it from an old picket pin or tent stake.] 1901 E. T. SETON *Lives of Hunted* 214 The darling ambition of his life..was to catch one of the Picket-pin Gophers... These little animals have a trick of sitting bolt upright on their hind legs, with their paws held close in, so that at a distance they look exactly like picket-pins. 1936 *Univ. Arizona Gen. Bull.* III. 79 Last of the ground squirrels to be mentioned are the small ones..variously known over the West as spermophiles, picket-pin gophers, or simply ground squirrels. 1947 V. H. CAHALANE *Mammals N. Amer.* 342 They spend a great deal of time sitting straight up on their haunches, their backs and necks as straight as ramrods. For this reason they are often called 'picket pins'. 1962 W. STEGNER *Wolf Willow* II. i. 41 The earth was densely peopled with small creatures as with large—prairie dogs, picket-pin gophers, field mice. 1967 D. L. ALLEN *Life of Prairies* 78 Overlapping the prairie country..is the realm of a truly abundant grassland rodent, the Richardson's ground squirrel. Sometimes we count dozens or even hundreds to the acre as each squirrel stands high on its hind feet beside a mound of fresh earth. It is evident why this creature will be called 'picket pin'. 1834 in *New Mexico Hist. Rev.* (1927) II. 298 The Acting Asst. Qr Master will have prepared a suitable number of wooden posts for the support of the Picket rope. 1946 *Sierra Club Bull.* (San Francisco) Dec. 4 We had carried with us all of our pack and picket ropes that could be spared. 1961 C. FARRELL *Trail of Tattered Star* xvi. 171 Mike ran down the line, slashing picket ropes with the bayonet. 1862 O. W. NORTON *Army Lett.* (1903) 59 We pitched our picket tents..on the ground lately occupied by a *secesh* regiment. 1804 LEWIS & CLARK *Orig. Jrnls. Lewis & Clark Expedition* (1904) I. 208 The [Mandan] Village..contains houses in a kind of Picket work.

picket, *v.* Add: **4. a.** (Further examples.) **b.** (Later example.)
1941 B. SCHULBERG *What makes Sammy Run?* viii. 156 Wilson ought to picket in front of Sammy's office... Sammy Glick Is Unfair to Organized Double-Crossers! 1972 *Daily Tel.* 25 Jan. 6/2 The power station is being picketed by miners attempting to stop oil supplies used to ignite coal in its boilers. 1977 *Times* 27 June 2/5 Six strikers..were picketing near the main gates.

picketed, *ppl. a.* (Earlier and further examples.)
1758 *Essex Inst. Hist. Coll.* (1881) XVIII. 102 Two Piquitted Forts or Garisons and a Hospetle. 1817 S. R. BROWN *Western Gazetteer* 27 Almost every house has a spacious picketed garden in its rear. 1890 KIPLING *Barrack-Room Ballads* (1892) 97 The picketed ponies, shag and wild, Strained at their ropes as the feed was piled. 1905 H. COHEN *Law Strikes & Lock-Outs* 16 He was watching the employed coming from the picketed works.

picketer. For 'a man' read 'a person' and add: Also, one engaged in a demonstration at particular premises, etc. (Further examples.)
1930 *Times Educ. Suppl.* 19 July 325/4 The picketers broke a barrier on the stairway. 1972 W. P. MCGIVERN *Caprifoil* (1973) i. 8 They are vocal liberals. Marchers, picketers, demonstrators. 1975 *Time Out* 17 Oct. 5/1 Many of the morning's picketers had lined up behind the main banners. 1978 S. BRILL *Teamsters* v. 181 The truck sped wildly toward the gate, with the picketers in full view.

picketing, *vbl. sb.* Add: **b. secondary picketing**: see *SECONDARY *a.*
c. *concr.* A fence or palisade made of pickets; picket-work. *U.S.*
1755 in *New Hampsh. Hist. Soc. Coll.* (1837) V. 254 Seven men..who were out..getting a few poles to complete the new picketing of the fort. 1860 J. F. H. CLAIBORNE *Life & Times Gen. Sam. Dale* 25 These forts were merely a number of log cabins built round a small square,..the whole surrounded by a rough picketing.

pickey, var. *PICKIE.

Pickford (pi·kfǒɹd). The name of a firm engaged in the removal of furniture, used *ellipt.* or in the possessive to denote a van used by Pickfords to remove furniture, or the firm itself. Also *fig.* and *attrib.*
[1833 C. MATHEWS *Let.* 8 Oct. in A. Mathews *Mem. Charles Mathews* (1839) IV. 205 It will be lighter for posting than any travelling carriage now in his yard. It cannot be called *a Pickford*, at any rate.] 1864 *Chambers's Jrnl.* 5 Mar. 152 The ubiquitous Pickford breaks the street-lamps, by going too near to the pavement. 1865 DICKENS *Mut. Fr.* II. IV. xii. 263 The sanctuary was..a kind of criminal Pickford's. The lower passions and vices were regularly ticked off in the books, warehoused in the cells, carted away. 1873 D. G. ROSSETTI *Let.* 18 Dec. (1967) III. 1250 Pickfords must be mad if they really took it to the wrong station. 1901 *Daily Tel.* 14 Nov. 4/5, I have seen a pair-horse Pickford tip coming up London Bridge. 1907 G. B. SHAW *John Bull's Other Island* p. xv, We cannot crush England as a Pickford's van might crush a perambulator. We are the perambulator and England the Pickford. 1975 J. HONE *Sixth Directorate* III. 94 What he isn't taking is being stored. Pickfords are coming tomorrow. 1980 R. MCCRUM *In Strange State* ix. 74 'Keys to the boot, please, sir.'.. Another constable..rummaged about among his books, papers and clothes. 'A regular Pickfords.'

pickie (pi·ki). *Sc.* and *Ir. local.* Also **pickey.** [f. PICK *sb.*[4] + -Y[6], -IE.] Hopscotch. Also pl. (const. sing.). Also *comb.*
1885 'J. STRATHESK' *More Bits from Blinkbonny* ii. 33 The 'pickies' (or the 'beds', or the 'Pall-all'), played with a flat stone on the pavement. 1906 Pickie [used in def. s.v. PICK *sb.*[4] 3]. 1922 JOYCE *Ulysses* 76 With careful tread he passed over a hopscotch court with its forgotten pickeystone.

picking, *vbl. sb.*[1] Add: **1. a.** (Further examples corresponding to PICK *v.*[1] 12.)
1934 [see *PICKER[1] 1 g]. 1956 [see *guitar-picking* s.v. *GUITAR *sb.* b]. 1973 *Time Out* 2–8 Mar. 21/1 The audience whoop away like madmen, and there's some marvellous pickin' and fiddlin'.
b. (Earlier and further examples.)
1618 H. WOTTON *Let.* in L. P. Smith *Life & Lett. Sir H. Wotton* (1907) II. 159 Remarkable how the divine justice, in a casual picking out of the foresaid number, from a 150 tumultuary men, did direct the man employed about this choice. 1846 G. DODD *Brit. Manuf.* 6th Ser. v. 131 The part of the coach-painter's work which requires the largest amount of care and neatness is that of 'picking-out', or painting fine lines, scrolls, &c. of one colour on a groundwork of a different colour. 1863 *Once a Week* 14 Nov. 569/2 There is one infamous method of thieving in the streets..which is called 'picking up'... This 'picking-up' system abounds in every large town... A woman is always the principal actor in these cases, and she is called the 'picking-up moll'. 1868 L. M. ALCOTT *Little Women* I. xi. 169 They [*sc.* plates of fruit] dwindled sadly after the picking over. 1869 F. HENDERSON *Six Yrs. in Prisons Eng.* vii. 76 'I heard a bloke talking about a "picking-up moll" he used to live with. What did he mean by that?' 'O! that's a very common racket. He meant a "flash-tail", or prostitute who goes about the streets at nights trying to pick up "toffs".' 1889 [see *cosher* s.v. *COSH *sb.*[2]]. 1925 *Times* 23 May 9/3 The picking-up power of an aerial varies as the square of the effective height. 1976 *West Lancs. Evening Gaz.* 8 Dec. 8/3 If a train leaving Blackpool early on Saturday morning for London is full by the time it leaves its last picking-up point at Wigan, it will have 500 people aboard helping a scout effort. 1977 *Gay News* 24 Mar. 15/4 Glasgow SMG is still suffering from the bad image—'a closetted picking-up place'—of the early years. 1979 P. COSGRAVE *Three Colonels* 189 That greatest of vices of warriors after a battle, the picking over..of memories.
2. f. *Basket-weaving.* (See quot.)
1912 T. OKEY *Introd. Art of Basket-Making* 153 Picking, cutting off the projecting ends of rods when the work is partially or wholly finished.
5. *picking-time* (examples); (sense *2 f) *picking-knife*; **picking-bee** *N. Amer.* (see BEE[1] 4); **picking-belt**, a travelling belt on which coal is picked.
1828 in *Dict. Americanisms* (1951) s.v. *picking bee*, Mother went to a picking Bee to pick wool. 1905 M. G. SHERK *Pen Pict. Early Pioneer Life Upper Canada* 177 The wool was then picked over by the women and girls, to get out any burs or lumps of dirt that might have adhered to it, 'pickin'' bees being frequently made for this purpose. 1943 S. MENEFEE *Assignment: U.S.A.* 48 In 1943 the townspeople were prepared to turn out for a picking-bee lasting most of the season, if necessary, to save the crop. 1901 *Chambers's Jrnl.* May 312/2 The excellent condition in which the coal was shipped,..was in great part due to the use of an appliance known as a

'picking' belt. **1921** *Spectator* 28 May 680/1 Girls on a picking-belt or in a colliery brick-works were earning similarly inflated wages. **1912** T. OKEY *Introd. Art of Basket-Making* vi. 28 The ends of the bottom-sticks are now cut off by the shears and the projecting tops and butts neatly picked off with the picking knife. **1960** E. LEGG *Country Baskets* 57 The last operation is the trimming off of all ends of canes and rods, 'picking the basket' as the craftsman calls it, for which he uses a special picking knife. **1835** J. H. INGRAHAM *South-West* II. 285 'Picking time'..continues where full crops are made until the first of December. **1848** [see *KNOCK *v.* 7*]. **1949** C. S. MURRAY *This our Land* 87 Picking time begins about August 20.

pickle, *sb.*¹ Add: **4.** (Further examples.)
1926 H. CRANE *Let.* 29 Mar. (1965) 243 I'm in no particular pickle at present. **1943** E. CALDWELL *Georgia Boy* ii. 21 I've got that marriage ceremony to perform in less than half an hour. It's too late for me to hunt up anybody else to ring the bell, and if you don't ring it for me, I'll be in a pretty pickle. **1955** *Times* 24 May 4/7 Leicestershire would have been in a pretty pickle without their captain, C. H. Palmer, in their current match with Surrey at Leicester. **1960** M. SPARK *Bachelors* ii. 19 You're going to leave Alice in a nice pickle if the case goes against you. *Ibid.* viii. 115 You've got us in a pickle. **1961** B. FERGUSSON *Watery Maze* x. 245 This landing had got into a pickle partly because of the bad weather, which had impeded the rate of build-up. **1967** G. F. FIENNES *I tried to run Railway* iv. 40 In a matter of days we were in a rare pickle. **1977** *Jersey Even. Post* 26 July 8/6 Don't leave jobs unfinished in order to start on something new, or you'll end up in a right old pickle.

c. *pl.* Nonsense, something worthless, an absurd statement. Also as *int. slang.* (No longer current.)
1846 *Swell's Night Guide* 34 'Pickles,' as the swell draper would say, 'but they frizzle and mangle music like bricks.' **1859** H. J. BYRON *Maid & Magpie* v. 31 If you and your minion Indulge that opinion, Ten pounds to an onion it's pickles, I bet. **1889** J. HATTON *Reminisc. J. L. Toole* II. v. 150 Or, the advance being ordered, had he exclaimed, 'Oh, Pickles!' before seeking convenient shelter from the foe? **1898** L. MERRICK *Actor-Manager* v. 66 The rent they ask is a hundred and fifty, but that's all pickles!

5. c. A woman with a sour disposition; an unattractive woman. *slang.*
1950 [see *LEMON *sb.*¹ 1 b]. **1970** *Women Speaking* Apr. 5/1 If a man doesn't like a girl's looks or personality, she's a..pickle, prune, [etc.].

6. *pickle-bottle* (earlier and later examples), *-jar* (earlier and later examples), *-pot* (later example).
1852 DICKENS *Bleak Ho.* (1853) v. 35 Pickle bottles, wine bottles, ink bottles. **1945** *Coast to Coast 1944* 140 Davie always had a pickle bottle for staging heroic contests between red-joes and black-joes. **1836** DICKENS *Sk. Boz* 1st Ser. I. 237 Some pickle-jars; some surgeons' ditto. **1952** *Coast to Coast 1951–52* 29 The farmer carried a glass pickle-jar with a screw-top lid. **1977** G. MARTON *Alarum* 49 The Russians re-ordered the smoked salmon—ate the garnish of pickles around the salmon and..wanted the whole pickle jar. **1903** *Nature* 19 Nov. 68/2 After Watt's patent, Newcomen engines were made with separate condensers without air-pumps, the air being discharged through a snifting-valve. Such condensers were known as 'pickle-pots'.

pickle, *v.*¹ Add: **1. c.** *intr.* To undergo the process of pickling.
1904 G. PARKER *Ladder of Swords* ix. 110 You have prepared your own brine, monsieur; in it you shall pickle.
5. *trans.* (See quot. 1970.) *U.S.A.F. slang.*
1966 *Time* 20 May 36/3 'I broke to the right,' recalled Dudley after last week's action, 'and pickled (dropped) my fuel tanks.' **1970** *Word Watching* Apr. 7/1 *Pickle*, to drop extra fuel tanks or equipment: to drop bombs.

pickled, *ppl. a.*¹ Add: **1.** (Further examples.)
1629 [see *DAN(T)ZIG]. **1739** E. SMITH *Compl. Housewife* (ed. 9) 43 Strow upon your Cutlets pickled walnuts in quarters. **1747** H. GLASSE *Art of Cookery* ii. 22 Put to it some pickled Gerkins chopp'd and boil'd Chesnuts. **1837** DICKENS *Pickw.* xlviii. 519 Demanding a mutton chop and a pickled walnut instantly. **1843** —— *Mart. Chuz.* (1844) vi. 66, I remember thinking.., in the days of my childhood, that pickled onions grew on trees. **1849** B. S. ELY *There she Blows* i. 9 An Irishman..wanted some pork to make 'pickled oysters'. This dish was made by cutting raw pork up fine, and covering it with pepper sauce and black pepper. **1877** E. S. DALLAS *Kettner's Bk. of Table* 423 *Shalot Sauce* is the same as what is called Sharp Sauce or Sauce Piquante, with this only difference—that to the latter there is added pickled gherkins. **1945** *Sun* (Baltimore) 22 Oct. 4/1 Hot-rolled pickled and cold rolled sheet deliveries run late into the second quarter next year. **1967** K. GILES *Death & Mr Prettyman* iv. 87, I put out a bit of ham..and some pickled walnuts. **1969** T. C. THORSTENSEN *Pract. Leather Technol.* v. 76 In the manufacture of garment suede leather the raw materials are primarily the pickled sheep skins based on New Zealand lamb and sheep. **1970** C. KERSH *Aggravations M. Ashe* iii. 45, I have terrible dreams if I eat pickled cucumbers. **1974** J. STUBBS *Painted Face* i. 23 Trying..to banish the cheese and pickled onions from the table. **1977** G. SCOTT *Hot Pursuit* iii. 25 Pickled eggs like lumps of coal.
2. *spec.* Drunk. *slang.*
1842 [in Dict.]. **1900** ADE *More Fables* 171 'It may be that I was a mite Polluted,' he suggested. 'You were a teeny bit Pickled about Two..,' said Mr. Byrd. **1919** WODEHOUSE *Damsel in Distress* xx. 236 On that occasion a most rummy and extraordinary thing happened. I got pickled to the eyebrows. **1926** WOOD & GODDARD *Dict. Amer. Slang* 20 *Gills, pickled to the*, soused; drunk. **1933** WODEHOUSE *Heavy Weather* vii. 95 The ink was still wet on a paragraph where, searching like some Flaubert for the *mot juste*, he had run his pen through the word 'intoxicated' and substituted for it the more colourful 'pickled to the gills'. **1939** [see *BIRD *sb.* 1 e]. *a* **1953** DYLAN THOMAS *Prospect of Sea* (1955) 128 On Sundays, and when pickled, he sang high tenor, and had won many cups. **1959** P. MOYES *Dead Men don't Ski* vii. 86 He gets the most extraordinary ideas sometimes, and he's pretty pickled, anyhow.

pickler². Add: **3.** A vessel in which vegetables can be pickled.
1862 *Illustr. Catal. Internat. Exhib., Industr. Dept., Brit. Div.* II. No. 6870, Bottles, filters, jars, foot-warmers, jugs, picklers, casks, jelly-cans.

pickling, *vbl. sb.*¹ **b.** (Further examples.)
1906 *Daily Chron.* 31 Oct. 8/4 At Southwold the pickling-plots..will be just at the back of the landing wharves. **1958** *Times Rev. Industry* June 50/3 To produce stainless [steel] strip..required..the installation of ..a pickling line.

picklock, *sb.*¹ **1.** (Later fig. example.)
1929 R. BRIDGES *Testament of Beauty* I. 21 This picklock Reason is still a-fumbling at the wards.

pick-me-up. Add: **2.** A woman who readily allows herself to be picked up; a prostitute. Cf. *PICK-UP *sb.* c.
1922 JOYCE *Ulysses* 49 She lives in Leeson park, with a grief and kickshaws, a lady of letters. Talk that to some else, Stevie: a pickmeup. **1941** J. SMILEY *Hash House Lingo* 42 *Pick me up*, loose woman.

pi·ck-off. Also **pickoff, pick off.** [f. vbl. phr. *to pick off* s.v. PICK *v.*¹ 18.] **1.** Chiefly *Aeronaut.* In an automatic control or guidance system, any device which produces or alters a pneumatic or an electrical output in response to a change in motion.
1938 P. V. H. WEEMS *Air Navigation* (ed. 2) xiv. 243 If the aircraft changes attitude laterally, one of the ports ..of the air pick off is opened fully and the other closed. **1938** A. JORDANOFF *Through Overcast* xiv. 180 Whenever you wish to fly in a certain direction under automatic control, you must be sure to set the air pick-offs at neutral at the time your plane points in the direction which gives the desired reading on the directional gyro. **1940** E. MOLLOY *Aeroplane Maintenance & Operation* II. I. 12 In actual practice the air pick-off consists of a disc attached to the gyro element, with knife edges which intercept the air flow at two nozzles on the supporting frame. **1958** J. G. TRUXAL *Control Engineers' Handbk.* xvii. 49 The displacement gyro is continually being torqued by the coupling of the motion of the base through the gimbal gearing friction and pickoff loading to the wheel. **1962** F. I. ORDWAY et al. *Basic Astronautics* ix. 377 The precession..is sensed by a pickoff that sends a signal to the servo amplifier.
2. In baseball: the act of catching a runner off base by means of the pitcher or catcher suddenly throwing the ball at that base. Freq. *attrib.*
1939 G. S. COCHRANE *Baseball* iv. 76 The danger of a throw to first for the pickoff is that the runner has over-run first not only too far to get back but far enough to make a break for second. **1948** *Sun* (Baltimore) 19 July 12 (*caption*) Bill Martin gets back in time as Eddie Shokes..tries to complete pickoff attempt in opener. **1968** *Washington Post* 4 July C 2/4 Wills moved to second when Ryan's attempted pickoff throw went wide. **1968** *Globe & Mail* (Toronto) 10 July 26/6 Killebrew.. said Tiant's pickoff throw on Mays, which opened the gates for the winning run, was behind the runner. **1974** *Spartanburg* (S. Carolina) *Herald* 23 Apr. A6/4 Bill Dancy of Spartanburg singled with one out in the fourth, moved to second on an errant pickoff attempt and went to third on Mark Ammons' infield hit. **1976** *Washington Post* 19 Apr. D2/2 Manning scored the only run Cleveland needed in the first inning when he..came home when Busby threw away a pickoff throw.
3. Used *attrib.* to designate auxiliary parts of a machine positioned so as to be easily selected or brought into use when required.
1949 *Tool Engineers Handbk.* (Amer. Soc. Tool Engineers) xxxv. 602 A secondary operation may be eliminated by the use of a pick-off attachment, which is an independently driven spindle and collet arrangement mounted in the cutoff position. **1950** J. A. OATES in A. W. Judge *Machine Tools & Operations* III. i. 30 These 'pick-off' gears are arranged under a cover..in some easily accessible position, and it is only a matter of two or three minutes to change over to a new speed or feed.

pickpocketing. (Earlier and later examples.)
1789 G. PARKER *Life's Painter* xv. 176 Going upon the knuckle is going a thieving, pickpocketing, &c. **1818** J. MILNER in F. C. Husenbeth *Life J. Milner* (1862) xx. 353 Not only an affront of a real diabolical nature, but also a serious pick-pocketing roguery. **1957** MANVELL & HUNTLEY *Technique Film Music* 232 The 'Juggler's Waltz' is a period piece for a music-hall act which takes place in a Paris café during an incident in which the visiting Inspector of Police witnesses a pick-pocketing attempt. **1977** *New Society* 7 July 5/2 Their game has confused pickpocketing with bag snatching and mugging.

Pick's disease (piks dizī·z). **1.** [f. the name of Friedel *Pick* (1867–1926), Austrian physician.] A form of multiple serositis characterized by constrictive pericarditis, hepatomegaly, and ascites.
1900 DUNGLISON *Dict. Med. Sci.* (ed. 22) App. 1314/2 *Pick's disease*, pseudocirrhosis of the liver, sometimes accompanying adhesive pericarditis. **1935** *Lancet* 14 Sept. 602/1 The prognosis of Pick's disease without surgical treatment is unfavourable for health and in some cases for life. **1940** E. ROSENTHAL *Dis. Digestive Syst.* iii. 278 Such a perihepatitis may be the result of liver disease or may accompany 'pericarditic pseudocirrhosis' (Pick's disease). **1959** BAILEY & LOVE *Short Pract. Surg.* (ed. 11) xx. 376 The best example of perisplenitis occurs in association with multiple serositis (Pick's disease).
2. [f. the name of Arnold *Pick* (1851–1924), Austrian psychiatrist and neurologist.] A condition, chiefly afflicting persons in late middle age, which is characterized by deterioration of intellect and judgement, speech disturbance and eventual dementia, and is caused by progressive atrophy of the frontal and temporal lobes of the brain.
1931 *Lancet* 20 June 1331/2 Pick's disease is a slowly progressive dementia starting usually in the sixth decade and accompanied by focal manifestations. **1935** *Jrnl. Nervous & Mental Dis.* LXXXII. 71 High blood pressure and arteriosclerosis are peculiarly absent in Pick's disease. **1955** H. H. MERRITT *Textbk. Neurol.* vi. 417 The cardinal symptoms of both Pick's and Alzheimer's disease are progressive dementia and disturbances in the speech. **1974** PASSMORE & ROBSON *Compan. Med. Stud.* III. xxxv. 76/1 The underlying cerebral atrophy..is restricted to the frontal and temporal lobes in Pick's disease. **1976** SMYTHIES & CORBETT *Psychiatry* vii. 126 Pick's disease is marked by a rapid and profound blunting of social judgement.

pick-up, *sb.* (*a.*) Add: Also **pickup. a.** (ii) [*PICK *v.*¹ 20 b (*b*).] The collection of unretrieved game after a shooting party; the quantity of game so collected.
1897 *Encycl. Sport* I. 443/1 Too large a 'pick up' means that the work was not properly done on the day itself. **1939** *Country Life* 11 Feb. 153/1, I wish all hosts would provide themselves with those little cards that tell you the total bag together with the pick-up and the names of the guns. **1976** *Shooting Times & Country Mag.* 16–22 Dec. 16/2 At the end of the pick-up a smile at last creased the face of keeper Bunny Spicer for the total bag was 253 pheasants, a record!
(iii) [*PICK *v.*¹ 20 o.] The act of apprehending or arresting. orig. *U.S.* Also, *pick-up van* (S. Afr.), a police van.
1908 J. M. SULLIVAN *Criminal Slang* 18 *Pickup*, an arrest followed by no charge of crime. **1934** L. BERG *Revelations Prison Doctor* viii. 109 Rao..called his arrests 'pickups' and inferred that the charges were too trivial to be worthy of notice. **1942** P. ABRAHAMS *Dark Testament* I. xii. 63 They all turn, and see a group of policemen jumping out of the moving pick-up van and running towards them. **1946** *Sun* (Baltimore) 8 Nov. 6/2 The Kansas sheriff responded with a request for a 'pick-up' on another man in the case. **1948** O. WALKER *Kaffirs are Lively* xi. 173 There comes the big van... They call it the *pick-up van*. **1953** P. LANHAM *Blanket Boy's Moon* I. vi. 50 'Police! Police! The pick-up van's coming.' Even as he shouted the warning, the police arrived in their closed van with a squealing of brakes. **1970** G. JACKSON *Let.* 10 June in *Soledad Brother* (1971) 34, I..started getting 'picked up' by the pigs more often.. These pick-ups were mainly for 'suspicion of' or because I was in the wrong part of town.
(iv) [PICK *v.*¹ 20 h.] Recovery, improvement.
a **1916** H. JAMES *Ivory Tower* (1917) I. iii. 69 'It's one of Mr. Betterman's [nurses] taking a joy-ride in honour of his recovery!'.. 'His pick-up *will* be a sell,' [Davey] ruefully added. **1922** H. TITUS *Timber* xxxii. 279 During all those years there will be a steady pick-up in quality. **1933** *Sun* (Baltimore) 10 Nov. 14/1 It is idle to suggest..that this newest relief scheme will help to tide the jobless over until 'the spring pick-up begins'. **1958** *Times Rev. Industry* Sept. 50/2 Export orders continue to be small, and there is little hope, with the American recession affecting steel-using industries on the Continent, of any pick-up in the near future. **1961** *Wall St. Jrnl.* 4 Oct. 1 Farm machinery dealer Bob Houtz tilts back in a battered chair and tells of a sharp pickup in sales. **1970** *Daily Tel.* 3 Feb. 19/2 Car sales showed their expected pick-up last month.
(v) [PICK *v.*¹ 20 f in Dict. and Suppl.] Reception of signals by electrical apparatus; *spec.* interference; also, a received radio or television programme.
1923 *Electrical Communication* I. IV. 25/1 A portable amplifier for use with the 'pick-up' microphones..made it possible to pick up speeches or music at any desired point ..and distribute the output to sound projectors. **1925** *Scribner's Mag.* Oct. 90 (Advt.), Low-loss doughnut coils.. conquer 'pick-up' of unwanted stations. **1937** *Discovery* Nov. 330/2 To provide the subject matter for transmission, direct pick-up, artists in the studio, or previously photographed sound-film may be used. **1943** *Electronic Engin.* XV. 521/3 Although the balanced input circuit minimises the pick-up it is sometimes essential to enclose the patient in a screened cubicle. **1948** *Sun* (Baltimore) 3 Jan. 2/2 It was a thirty-minute radio network program with pick-ups from scattered points in the United States and abroad. **1949** FRAYNE & WOLFE *Elem. Sound Recording* xxxii. 674 Microphone placement is less exacting in the stereophonic pickup of orchestral music. **1956** *Electronics* Feb. 170/2 A series of measurements was made on the cross pickup of pulses in a 100-foot length of ..communication cable. **1965** *Wireless World* Aug. 383/2 When ferrite aerials are close to the inverter transformers, inductive pick-up can result. **1971** *Sci. Amer.* Oct. 28/1 With a community wired from one or more terminal points to individual homes the local pickup would not

be limited to the channels of the commercial television networks; instead participants could receive through the satellite a number of services to meet individual tastes. **1976** *Broadcast* 29 Nov. 18/1 The camera has to be blimped to prevent noise pick-up from microphones.
(vi) [Cf. *PICK *v.*¹ 20 j.] Robbery, theft. Freq. *attrib. slang.*
1928, etc. [see *pick-up man* (sense B. a)]. **1938** F. D. SHARPE *Sharpe of Flying Squad* i. 14 Lower down come the suit-case thieves, the pick-up merchants, small time smash-and-grabbers, [etc.]. *Ibid.* xxiii. 240 He had been persuaded to try his hand at 'the pick up' (stealing from unattended motor cars). **1962** *Gross's Criminal Investigation* (ed. 5) viii. 206 *At the pick-up*, suitcase stealing.
(vii) [Cf. PICK *v.*¹ 20 e.] Collection and transportation by means of an aircraft, motor vehicle, ship, etc. Freq. *attrib.*
1938 W. L. G. COWAN *Loud Report* III. 220 We usually kip in the afternoon and push on at night. There's a better chance of a pick-up. **1940** *Mech. Engin.* Apr. 283/1 The record is not clear as to exactly when the first pickup in flight was made. *Ibid.* 285/2 The pickup system..makes possible the picking up of heavier loads. **1943** *Jrnl. R. Aeronaut. Soc.* XLVII. (Abstr. from Sci. & Techn. Press) 72 The Air Mail Pick-up service was started by the American Post Office Department in 1939. *Ibid.*, Since short hauls seem to offer the best field for cargo glider operation, a similar method of pick-up becomes most attractive. **1952** E. F. DAVIES *Illyrian Venture* iv. 68 The American pilot signalled Cairo that it was possible to rescue them by air pick-up from Gjinokaster airfield. **1960** R. WILLIAMS *Border Country* iv. 94 'Where's the pickup then?' 'Signalled now.'.. Rees..crossed the line and walked down..to the siding. **1960** R. W. MARKS *Dymaxion World of B. Fuller* 22/1 Fuller's toilets.. consisted of a splashless hermetic and waterproof packaging system which mechanically packed, stored, and gross-cartoned wastes for eventual pickup for processing by chemical industries. **1971** P. O'DONNELL *Impossible Virgin* xii. 238, I fixed for the 'copter to stand by for a pick-up at eighteen 'undred hours. *Ibid.* xiv. 275 They rested beside..a triangle of flat stony ground... This was the pick-up area Willie had arranged. **1972** 'G. BLACK' *Bitter Tea* (1973) vii. 100 The *Kao Ming* hasn't been converted for carrying..steel containers, serving out her now very limited time on the copra pick-up routes. **1974** L. DEIGHTON *Spy Story* xix. 208 That radio signal obliges us to continue with the pick-up, even if we were certain it's phoney. **1976** *National Observer* (U.S.) 11 Dec. 4/3 Ankeny, Iowa, saved $9,000 by opening dump sites to the public for two weekend clean-ups a year and abolished curbside pickups. **1977** *Chicago Tribune* 2 Oct. XII. 2/2 The garbage just piled up all summer. No pickup. **1978** J. KRANTZ *Scruples* ix. 259 Billy also established a delivery service in a town where multimillionaire customers have to do their own pickups in every boutique.
(viii) *Nuclear Physics.* A reaction in which an incident particle such as a proton or deuteron captures one or more nucleons from an atomic nucleus. Usu. *attrib.*
1950 *Physical Rev.* LXXVII. 470 It is proposed that the fast deuterons observed among the products of high energy nuclear reactions are to be understood in terms of a 'pick-up' or sudden rearrangement process. *Ibid.*, The pick-up process must not be thought of as taking place within the nucleus. **1970** I. E. MCCARTHY *Nuclear Reactions* I. v. 95 If one of these interactions seems more plausible than the other to the reader, he need only consider the time-reversed situation (pick-up). **1971** W. M. GIBSON *Nuclear Reactions* vii. 108 A pick-up reaction in a nucleus with an odd neutron outside a closed inner shell can provide useful information about the state of the odd neutron..which is directly relevant to the shell model of the nucleus. **1975** *Nature* 6 Mar. 19/3 The spectroscopic factors for stripping and pickup reactions on the same nucleus can be related to each other.
b. (Further examples.)
1972 M. GLENNY tr. *Solzhenitsyn's August 1914* xlii. 419 Trams passed, clanging, their pick-ups hissing along the wires. **1974** *Physics Bull.* Apr. 154/3 The dangers caused by power failure in magnetic hoists are avoided in a new design from Mullard. For the pickup is a permanent magnet rather than the usual ferroelectric magnet.
(ii) (*a*) A sensor or transducer.
1943 D. G. FINK *Television Standards & Pract.* iv. 91 Specifications to be placed on the spectral response of each of the three color-sensitive elements of the pickup. **1957** R. A. HEINLEIN *Door into Summer* v. 79, I suppose it was a stethoscope he used although it looked like a miniaturized hearing aid... The pick-up he pushed against me was as cold and hard as ever. **1961** *Engineering* 8 Sept. 309/1 Phototransistor pickups, focused on rotating black and white discs, provide signals for the phase meter. **1962** F. I. ORDWAY et al. *Basic Astronautics* vii. 316 By placing temperature pickups at various points on the model flame deflector it was possible to determine the approximate temperatures encountered..in an actual firing. **1978** J. GORES *Gone* (1979) iii. 24 He was at the corner where he knew the concealed pickup was.
(*b*) *spec.* = *CARTRIDGE 1 e, *HEAD *sb.* 11 g, *pick-up arm* in sense B. a below.
1926 *Glasgow Herald* 5 Oct. 5 Instead of the conventional sound-box an electrical device known as a magnetic 'pick-up' is guided over the record, converting the mechanical vibrations imparted to the needle into electrical vibrations. **1926** *Gramophone* Dec. 294/1 Instead of a sound-box there is what is known as a 'pick-up'. **1938** *Times* 25 Aug. 8/2 It has been specially designed as a supplementary unit for existing radio receivers having pick-up terminals. **1940** *Gramophone* Sept. 93/1 The output is very much less than that given by the average moving iron or piezo electric types of pick-up. **1941** [see *CARTRIDGE 1 e]. **1946** *Newsweek* 13 May 23 (Advt.), Your Victrola's jewel-point pickup floats like a feather on water. **1951** [see *HEAD *sb.* 11 g]. **1959** *New Scientist* 31 July 517/2 The whole pick-up must be able to do the manœuvres required of it without giving rise to excessive pressures which would damage the record. **1963** *Which?* Jan. 8/2 The pick up is composed of two parts—an arm, and at the end of the arm..a head. **1975** G. J. KING *Audio Handbk.* viii. 170 The modern pickup consists of an arm and a cartridge. **1976** 'J. FRASER' *Who steals my Name?* ix. 107 The needle on the pick-up of his record player..had a beard on it... His room was filthy.
(*c*) *spec.* A device that produces an electrical signal corresponding to the displacement, speed, or acceleration of a vibrating body.
1950 P. G. ANDRES *Survey Mod. Electronics* viii. 346 Electronic comparator gauges use a pickup that operates on the principle of displacing an armature..to generate a corresponding voltage. *Ibid.* 363 Since the pickup responds to the acceleration of the vibratory motion, integrating networks convert its output..into readings proportional to velocity or displacement. **1966** *McGraw-Hill Encycl. Sci. & Technol.* VIII. 667/1 A vibration pickup can be used to generate an electrical signal directly from the vibration of a string; this is the case in the electric guitar and electric piano. **1968** *Melody Maker* 6 Apr. 5 (Advt.), Take our pickups... To get sound from a metal guitar string, all you have to do is wind wire around a magnetic core and place it near that string. **1970** R. H. WALLACE *Understanding & measuring Vibration* iv. 40 Because the aircraft structure is such that low frequencies under, say, 10 Hz are readily transmitted,.. it will be vital to use pickups and read-out equipment going down to zero frequency, or near. **1972** M. P. BLAKE in Blake & Mitchell *Vibration & Acoustic Measurement Handbk.* xx. 422 The proximity pickup is likely to become increasingly popular for monitoring the shafts of high-speed, large, powerful drives, such as turbines, reading directly in displacement which, in this application, is usually the parameter of most interest. *Ibid.*, A growing tendency among those who measure vibration favors the acceleration pickup.
(iii) In full, *pick-up truck* or *van*. A small truck or van used for carrying light loads (see also sense a (iii) above). So *pick-up body*, a small, detachable body for use on various types of light trucks.
1932 *Kansas City* (Missouri) *Times* 21 Jan. 22 There was a delivery car with a pick-up body on it in King City. **1939** *Sun* (Baltimore) 18 Jan. 7/2 Dairies' pickup trucks were subject to commission regulation. **1944** *Democrat* 29 June 1/4 This includes all operators of trucks and pickups whose gasoline rations are not controlled. **1948** *Southern Sierran* (Los Angeles) Apr. 4/1 The work will be made lighter if someone will provide the use of a pick-up truck for removing the dirt. **1959** F. STARK *Riding to Tigris* 96 A piece of the bull-dozer was extracted and loaded on the pick-up, which looked like a bright orange lorry. **1973** A. ROSS *Dunfermline Affair* 45 He must be going out in the pick-up. *Ibid.* 48 There were still plenty of motor cars in the car park, but no small grey pick-up truck. **1977** *Cork Examiner* 8 June 14/3 (Advt.), 30 cwt. pick-up truck and driver available. **1977** *Western Morning News* 30 Aug. 2/2 (Advt.), J.4 Pick-up J registered.. long M.O.T.
c. (Further examples.) *spec.* A person whose acquaintance is formed with the intention of having a sexual relationship; a prostitute or street-walker; also, the act of forming such an acquaintanceship.
1848 TROLLOPE *Kellys & O'Kellys* III. xi. 269 The pick-up on the Derby is about four thousand. **1871** L. H. BAGG *4 Years at Yale* 46 *Pick-up*,..a street-walker, of the less disreputable sort. **1927** DUNNING & ABBOTT *Broadway* xvi. 158 'Dan McCorn out there?' enquired Porky. Steve nodded. 'He's talking to one of the pick-ups.' **1930** *San Antonio* (Texas) *Light* 31 Jan. 14/7 (Advt.), A real pickup for someone: 50 feet on Broadway... An exceptional site. **1930** J. B. PRIESTLEY *Angel Pavement* iv. 170 The big teashops..were always crowded with girls and always offered a chance of a pick-up. **1935** G. GREENE *England made Me* IV. 169 She might be meeting her lover and not a pick-up in Gothenburg. **1944** M. LASKI *Love on Supertax* iv. 52 And what about you, Clarissa? Who's your latest pickup? **1957** 'M. YOURCENAR' *Coup de Grâce* 66 She was fairly throbbing against me, and no previous feminine encounter, whether with a chance pick-up, or with an avowed prostitute, had prepared me for that sudden, terrifying sweetness. **1970** [see *FILLE DE JOIE]. **1974** T. ALLBEURY *Snowball* ix. 48 The pick-up lacked grace and she made it clear she was bored. **1977** *New Yorker* 22 Aug. 66/3 Paolo..accedes with wearied eyelids to the blandishments of yet another of his pickups.
d. (Examples.)
1905 *Daily Chron.* 31 Aug. 3/1 If one accidentally pulled a ball in a school pick-up. **1917** A. WAUGH *Loom of Youth* I. ii. 31 Even a desultory Pick-Up woke into excitement when the shrill, piping voice of a full-back came in with, 'The Bull's coming.'
f. = PICK-ME-UP.
1881 HARDY *Laodicean* III. v. iii. 36 Every sip you took of your pick-up as you sat there showed me something was wrong. **1921** D. MARQUIS *Old Soak* 10, I took mine straight for the most part, except when I needed some special kind of a pick-up in the morning. **1936** *N.Y. Herald-Tribune* 25 Aug. 8 Mix it with orange, lemon, grapefruit, pineapple juices or with ginger ale for a zesty pick-up. **1949** *Sun* (Baltimore) 9 Nov. 15/4 Investigation of methods to improve your morning pickup [*sc.* coffee].
g. [Cf. *PICK *v.*¹ 20 q.] The capacity of an engine for recovering speed; acceleration.
1909 *Times* 27 Apr. 4/1 The flexibility and 'pick up' of the engine were such that the merest novice could handle the car with ease. **1923** *Daily Mail* 16 Mar. 10 (Advt.), An engine of great power, exceptional pick-up and flexibility, of notable smoothness and quietness. **1930** *Engineering* 12 Sept. 326/1 Apart from these factors, a smooth pick up, absence of vibration, and uniform torque are of great value. **1934** W. NELSON *Seaplane Design* vi. 53 The form of the underwater portion of the float must be such that water resistance does not hinder a quick pick-up to planing attitude. **1963** *Lebende Sprachen* VIII. 108/1 The car has good pick-up.
h. *Mus.* A series of introductory notes leading to the opening of a tune or portion of a tune.
1934 S. R. NELSON *All about Jazz* iii. 65 After a short passage of one or two bars, as a 'pick up', the ensemble will then take the last chorus and coda. **1949** L. FEATHER *Inside Be-Bop* 62 The last phrase starts with a three-beat pickup into an on-the-beat quarter note on the first beat of bar 7. **1956** O. DUKE *Sideman* xii. 150 Bert played a piano intro and Bernie took pickups into *Laura* and then just sang it.
i. Tendency to pick up or absorb.
1938 *World Petroleum* Apr. 45/3 The flow through the main part of the furnace is in four parallel paths, each designed to have the same heat pick-up. **1955** MONDOLFO & ZMESKAL *Engineering Metall.* viii. 181 Pickup is the inclusion of foreign particles, especially scale, on the surface of the rolled product. **1955** *Jrnl. Soc. Dyers & Colourists* LXXI. 896/2 Although high pick-up is generally desirable, some fabrics need to be well squeezed. *Ibid.* 899/1 The padding temperature was 50°c. and the pick-up of the medium-weight rayon-staple fabric to be dyed was 100%. *a* **1977** *Harrison Mayer Ltd. Catal.* 67/1 The batts and props we supply are designed to have a low glaze pick up.

B. *attrib.* or as *adj.* **a. pick-up arm,** in a record-player or record deck, an arm carrying the stylus at one end and usu. counter-balanced at the other so as to be able to swing horizontally over the turntable and vertically; **pick-up baler,** a machine that picks up hay, etc., and bales it; **pick-up man** *colloq.*, (*a*) a thief, esp. of luggage; (*b*) *U.S.*, one who collects money wagered with bookmakers; (*c*) at a rodeo: see quot. 1961; **pick-up tube** *Television*, a vacuum tube that produces an electrical signal corresponding to an optical image formed in or on it; a camera tube.
1937 Pick-up arm [see *OVERHANG *sb.* d]. **1946** [see *CARTRIDGE 1 e]. **1976** *Gramophone* July 237/2 Installing the QDC-1e cartridge in an SME 3009/II pickup arm with Thorens TD125/II turntable presented no problems. **1946** *Agricultural Progress* XXI. I. 34 The increasing use of the one-man, twine tying, pick-up baler. **1951** *Engineering* 26 Jan. 92/1 Imports from the United Kingdom were to include 6,000 wheeled tractors..; imports from the United States..pick-up balers and combine harvesters. **1928** *Detective Fiction Weekly* 8 Sept. 564/2 (*heading*) International crooks I have known; No. 6—'Gentleman George', pick-up man. **1944** C. HIMES *Black on Black* (1973) 244 The pickup man took his book one day, and he told Clara, 'We should have been on the other end.' **1961** *Times* 14 Nov. 12/7 From the moment that he leaves the chute, the cowboy must remain in the saddle for 10 seconds. Then, on a whistle signal, two mounted 'pick-up' men converge on the buckhorse and the cowboy dismounts as best he can. The 'pick-up' men gallop after the riderless horse. One of them grabs the rein or head collar, and forces him out of the arena through a gate into a collecting ring. This in itself is often a thrilling spectacle, as it is a point of honour for the 'pick-up' men to take out the bucking animal at full gallop. **1964** A. WYKES *Gambling* iv. 92 These agents are known as 'pickup men' or 'writers'; they collect and pay out on behalf of the bookmakers, who pay them 10 per cent of their net winnings. **1966** *Sunday Mail Mag.* (Brisbane) 9 Jan. 1/3 As soon as a gong had clanged the all-clear, pick-up men..tore after them on horses. **1933** *Jrnl. Inst. Electr. Engin.* LXXIII. 438/2 The development of the pick-up tube was pushed on, and the results..soon surpassed those of mechanical scanning. **1965** *Wireless World* Sept. 461/2 A small television camera which uses a solid-state image sensing panel in place of the conventional electron-beam pick-up tube. **1978** *Broadcast* 21 Aug. 12/3 An Ikegami CT2-2400 three-tube colour camera, with Saticon pick-up tubes, and synchronised to a broadcast sync pulse generator.
b. (Examples.)
1909 *Q. Rev.* Oct. 618 The rest of the administration was arranged on the principle which governs 'pick-up' sides in a school-match. **1936** *Rhythm* Apr. 28/1 The ensemble is good, allowing for the difficulties surrounding the formation of a pick-up nine-piece band. **1938** D. BAKER *Young Man with Horn* IV. iii. 220 He..made records with who knows how many pick-up bands. **1956** D. FLOWER tr. *Panassié & Gautier's Dict. Jazz* 197 *Pick-up band*, a band formed of musicians who regularly play elsewhere, but who come together for a special purpose, e.g. a recording date, a broadcast, a concert, a short nightclub engagement. **1959** 'F. NEWTON' *Jazz Scene* iv. 65 These were mostly small groups playing in odd night-clubs, or pick-up bands for recordings. **1962** *Sunday Times* (Colour Suppl.) 10 June 3 His strength is as the lead of pick-up groups of first-rate players. **1972** J. MOSEDALE *Football* iv. 47 He suffered a leg injury in a pick-up basketball game. **1977** *Gramophone* Apr. 1618/1 Lester headed a pick-up group containing trumpeter Billy Butterfield and pianist Johnny Guarnieri. **1977** *Tennis World* Sept. 17/3 A 'pick-up' doubles team is a scratch combination. **1977** *Transatlantic Rev.* LX. 120 Every Sunday morning he and his friends had a pick-up game over on Wentbridge Common.
c. pick-up camper (see quots.).
1973 *Country Life* 15 Nov. 1568/3 The 'pick-up' camper..is the coachbuilt caravan in demountable form on a pick-up truck... On arrival on site, it is jacked clear of the vehicle. **1974** *Trailer Life* Nov. 92 Some trailerists, concluding it was easier to drive a truck than hitch and unhitch a trailer, had their trailers mounted on a truck bed. This was practical. But since the trailers had not been designed for their new purpose, it proved unattrac-

tive. And so the pickup camper came into being, followed by a more stable and spacious chassis-mount.

Pickwickian, *a.* Add: (Earlier and further examples, without ref. to language.)
1836 DICKENS *Let.* 18 Feb. (1965) I. 132 Believe me (in Pickwickian haste) Faithfully Yours Charles Dickens. **1899** [see PRUSSIAN *a.* 2 b]. **1953** 'N. BLAKE' *Dreadful Hollow* 147 Blount, whose Pickwickian exterior camouflaged a mind as ruthlessly purposeful as a guided missile. **1975** J. SYMONS *Three Pipe Problem* xvii. 173 Johnson's Pickwickian features were unusually solemn.
2. *Med.* Also **pickwickian.** [Named in allusion to the fat boy Joe in *Posthumous Papers of the Pickwick Club.*] Having or being a syndrome occurring in some obese adults (rarely in obese children) characterized by somnolence, respiratory abnormalities, and bulimia.
1956 C. S. BURWELL et al. in *Amer. Jrnl. Med.* XXI. 812/1 Figure 1 represents Thomas Nast's drawing of Mr. Wardle's boy, Joe. This masterful description by Charles Dickens of a patient with marked obesity and somnolence is the first complete description of this syndrome that we have been able to find in the literature. For this reason we have called it the Pickwickian syndrome. **1965** *Progress Brain Res.* XVIII. 157 The short diurnal periods of light sleep (10–12 sec duration) in Pickwickian patients are characterized by apnea, increased cyanosis and muscular relaxation. **1977** *Lancet* 7 May 993/1 There are many other conditions, usually clinically obvious such as the pickwickian syndrome, in which there is both a respiratory and sleep abnormality.
B. *sb.* **1.** A member of the Pickwick Club.
1836 DICKENS *Pickw.* (1837) i. 1 A proposal, emanating from the aforesaid Samuel Pickwick..and three other Pickwickians..for forming a new branch of United Pickwickians. *Ibid.* ii. 7 The intelligence of the Pickwickians being informers was spread among them. **1905** *Daily Graphic* 1 Feb. 9/4 The minds of many of the lovers of Dickens who were present at the Dickens Character Ball..reverted to another ball-room—still in existence—where one of the most famous of the incidents in the 'Pickwick Papers' occurred—that of the Bull Inn at Rochester. And that ball, too, like this one, wherein not only the Pickwickians but many other characters which sprang from the brain of their creator were incarnated, was a charity ball. **1909** [see *BATHLESSNESS].
2. *Med.* A person with the Pickwickian syndrome.
1965 *Progress Brain Res.* XVIII. 156 Most obese people..present no diurnal sleeping syndrome, and so far as is known no nocturnal apnea, so that the Pickwickian should have some central disturbance of respiration and arousal. **1975** *Electroencephalogr. & Clin. Neurophysiol.* XXXIX. 579/2 The data obtained were compared with those observed in a group of 59 hypersomnolent patients aged between 19 and 83.., 18 of whom were Pickwickians.
Hence **Pickwickia·na,** publications about the *Posthumous Papers of the Pickwick Club.*
1899 J. GREGO (*title*) Pictorial Pickwickiana.

picky (pi·ki), *a.* [f. PICK *v.*¹ + -Y¹.] Fastidious, finicky, 'choosey' (see also quot. 1867). Hence **pi·ckiness.**
1867 W. DICKINSON *Gloss. Words & Phrases Cumberland* (Suppl.) 28 *Picky*, of weak appetite. **1900** DICKINSON & PREVOST *Gloss. Dial. Cumberland* (rev. ed.) 243/1 T'barn 's nut weel, it's too picky by far. **1917** in *Dialect Notes* (1918) V. 12 Not an ugly picky thing in all she has to say. **1932** *New Yorker* 11 June 50/2 People who are picky about their food on shipboard..will be glad to know that the Clyde-Mallory Lines now allow you to buy a ticket for transportation only and order meals à la carte. **1957** 'R. TRAVER' *Anat. Murder* (1958) I. xv. 101, I don't want to seem picky, Lieutenant, but I happen to consider your particular doctor professionally on a par with Amos Crocker. **1966** *New Statesman* 29 Apr. 618/2 A hesitant, picky account of Mississippi's economic and electoral.. history. **1971** D. BAGLEY *Freedom Trap* iii. 62 This is a very exclusive mob; very picky and choosy. **1974** D. SCANNELL *Mother knew Best* iii. 27, I was always 'gastric' and Mother would get different meals for me as she thought me a 'picky' eater and needed tempting. **1977** *Sci. Amer.* Apr. 145/1 He was meticulous, even picky, about expense accounts. **1977** *Verbatim* May 6/1 And if you are willing to discard mere pickiness, you know what *taken for granite* and *scuenting* mean, too.

picky, var. *PICCY.
1931 E. F. BENSON *Mapp & Lucia* iii. 64 Go on with your picky, as if I was not here. How well you've got the perspective!

picloram (pi·klōræm). [f. PIC(OLINE + CH)LOR-² + AM(INE.] A white crystalline compound, 4-amino-3,5,6-trichloropicolinic acid, $C_6H_3Cl_3N_2O_2$, which is used as a herbicide and defoliant against deep-rooted weeds and woody plants.
1965 *Proc. Northeast. Weed Control Conf.* XIX. 140 Effective control [of mugwort]..was obtained with.. 4-amino-3,5,6-trichloropicolinic acid (picloram). **1965** *B.S.I. News* May 24 The following new names have been approved by the Pest Control Products Industry Standards Committee for eventual inclusion in B.S. 1831... picloram... 4-amino-3,5,6-trichloropicolinic acid. **1969** *Sci. Jrnl.* Feb. 10/1 According to one estimate, no less than 2250 tonnes of picloram have been dumped on South Vietnam in 1968. **1970** *Guardian Weekly* 21 Feb. 6/4 The recently introduced chemical, picloram, is one of the longest-lived pesticides. **1974** *Sci. Amer.* Apr. 49/2 The herbicides sprayed over Vietnam were chemicals that are routinely applied for weed control: 2,4-D and 2,4,5-T, picloram and cacodylic acid.

picnic, *sb.* Add: **1. b.** (Earlier example of the *sb.*)
1825 H. WILSON *Mem.* II. 248, I sate down to consider the plan of a book, in the style of the Spectator, a kind of pic nic, where every wiseacre might contribute his mite of knowledge.
c. (Earlier and further examples.) Now usu. something straightforward or agreeable; a lively time; a treat; *no picnic, not a picnic,* not an easy task; a formidable undertaking.
1818 KEATS *Let.* Dec. (1958) II. 13 Perhaps as you were fond of giving me sketches of character you may like a little pic nic of scandal. **1886** *Lantern* (New Orleans) 27 Oct. 6/1 Hanley sparred with a smile on his face much as to say, 'What a picnic I've got with this kid.' **1888** KIPLING *Wee Willie Winkie* 84 'Taint no bloomin' picnic in those parts I can tell you. **1890** —— *Life's Handicap* (1891) 125 A knot of furious brother officers demanding the court-martial of Tommy Dodd for 'spoiling the picnic'. **1901** W. CHURCHILL *Crisis* II. iii. 136 This isn't any picnic. **1909** 'O. HENRY' *Roads of Destiny* xxii. 315 It was a picnic for the census takers. They just counted the marshal's posse that it took to subdue us, and there was your population. **1914** A. BENNETT *Price of Love* xii. 248 'But doesn't it *hurt*?' 'Depends what you call hurt. It ain't a picnic.' **1916** [see *EMBUSQUÉ]. **1919** *Mr. Punch's Hist. Gt. War* 114 It is not a picnic for the men in our front line. **1926** GALSWORTHY *Escape* II. iv. 50 If you want to get thin. It's a top-hole cure for adipose. An escape's no picnic. **1947** *Richmond* (Virginia) *Times-Dispatch* 4 May 26/5 The war memorial..was rejected as 'morbid': those who object to the ban on this work argue that the war was no picnic. **1961** B. FERGUSSON *Watery Maze* xvi. 394 It was going to be no picnic co-ordinating land, sea and air forces from so many different points of departure at so many different speeds. **1965** *Listener* 3 June 835/3 It can have been no picnic to be a poet in the age of Eliot. **1971** S. HILL *Strange Meeting* I. 10 Think yourself lucky you got off a bit early. It's no picnic now. **1974** J. STUBBS *Painted Face* xxiii. 286 What do *you* know of prison? This here's a picnic compared to what it will be.
d. *Austral.* and *N.Z.* Used ironically of an awkward situation or a difficult or unpleasant experience.
1896 in MORRIS *Austral Eng.* (1898) 351/1 If a man's horse is awkward and gives him trouble, he will say, 'I had a picnic with that horse,' and so of any misadventure or disagreeable experience in travelling. **1939** N. MARSH *Overture to Death* xiii. 125 I'm sorry to have neglected you like this; but we're in for a picnic, and no mistake, with this case up at Moorton Park. **1941** BAKER *Dict. Austral. Slang* 54 *Picnic*, any unpleasant experience, a disagreeable and complicated task. **1945** *Coast to Coast 1944* 125 What a mess, what a picnic! **1955** D. NILAND *Shiralee* 38 All I know is I'm going to have one helluva picnic if she doesn't find it. **1959** BAKER *Drum* (1960) 68 We call a wild confusion or a particularly difficult task a picnic.
e. *U.S.* = *picnic ham* below.
1910 L. D. HALL *Market Classes Meat* 281 *Picnics* or *calas* (formerly termed California hams) are cut 2-½ ribs wide... They..are sold almost entirely as sweet-pickled, smoked and boiled meats. **1949** *New Harmony* (Indiana) *Times* 5 Aug. 6/2 (Advt.), Smoked Picnics, 3 to 5 lb. average lb. 45c. **1974** *Columbia* (S. Carolina) *Record* 24 Apr. 21-B (Advt.), Sliced picnic 1.89. **1976** *Washington Post* 19 Apr. A19/1 (Advt.), Smoked picnics.
3. (Further examples.) Also: *picnic basket, hamper, pie, shelter, site, spot, stove, tea;* **picnic area,** a piece of ground designated as suitable for picnics; **picnic chair,** a (usu. collapsible) chair for use on a picnic; **picnic ground** = **picnic area;* **picnic ham** *U.S.*, a small cut of shoulder bacon in the form of a ham; **picnic lunch,** a packed lunch, *spec.* one provided by a hotel in place of a regular meal; **picnic meal,** a meal eaten as a picnic; also, a quick meal eaten indoors; **picnic plate,** a plastic or paper plate suitable for use on a picnic; **picnic race meeting, races** *Austral.* and *N.Z.*, a race meeting held in a country area, accompanied by other social events; **picnic table,** a table suitable for use on a picnic; a small hinged table in a car.
1959 Picnic area [see *PIPED *ppl. a.*¹ 3 c]. **1968** *Guardian* 5 July 3/2 The NCB will restore the landscape and provide access roads, car parks, picnic areas, a hard standing and slipway for boats. **1968** [see *camp-ground* s.v. *CAMP *sb.*² VII. b]. **1973** V. CANNING *Flight of Grey Goose* i. 9 A large picnic area on the edge of a wood. **1885** *List of Subscribers, Classified* (United Telephone Co.) (ed. 6) 125 (*heading*) Luncheon and picnic basket manufacturers. **1931** *N. & Q.* 11 Apr. 258/2 The car was packed with bathing things, camp-stools and picnic basket. **1955** A. HUXLEY *Let.* 4 Apr. (1969) 739 Yes, by all means let us take a picnic basket so that we can be independent of restaurants. **1972** D. E. WESTLAKE *Cops & Robbers* (1973) xvi. 239 Macy's has a wicker picnic basket. It costs around eighteen bucks, with the tax. **1975** 'A. HALL' *Mandarin Cypher* xvii. 233 One narrow bunk..cheap cardtable and picnic chair. **1977** M. JANCATH *Seatag* I. vii. 41 Emrich was surprised to see a wooden picnic table and chairs. **1926** *Daily Colonist* (Victoria, B.C.) 2 July 5/3 The streets were almost deserted save for those hastening toward some of these attractive picnic grounds. **1947** E. S. GARDNER in *Amer. Mag.* Aug. 148/3 You can see the camp and the picnic grounds from here. **1969** H. MACINNES *Salzburg Connection* ii. 33 He began walking down towards the picnic ground. **1974** *Country Life* 3 Oct. 930/3 A 14-acre picnic ground has been cleared. **1897** *Sears, Roebuck Catal.* 13/3 Meats... Picnic Hams. **1944** *Chicago Daily News* 13 July 21/2 A picnic ham may be boned, rolled and tied before roasting to make carving easier. **1973** *Black Panther* 25 Aug. 3/3 Picnic ham at 97 cents or salt pork at 99 cents a pound? **1862** W. COLLINS *No Name* I. 332 We have a picnic hamper with us..and away we drive on a pleasure trip. **1896** E. TURNER *Little Larrikin* xviii. 208 Several picnic hampers and a case of champagne. **1966** B. COOPER *Drown him Deep* vi. 52 He went to the back of the car, and began to pull out the picnic hamper. **1972** R. HILL *Fairly Dangerous Thing* II. vi. 179 We.. took our picnic-hamper with us into a wood. **1917** D. CANFIELD *Understood Betsy* (1922) x. 189 They were to meet the Wendells in the shadow of Industrial Hall and eat their picnic lunch together. **1933** E. O'NEILL *Ah, Wilderness!* (1934) I. 18 We're going to have a picnic lunch on Strawberry Island. **1971** J. TYNDALL *Death in Lebanon* vii. 105 Let's ask the hotel for a picnic lunch and get off as soon as we can. **1972** *Guardian* 22 July 7/3, I continued along the D94..but soon stopped for a picnic lunch. **1876** GEO. ELIOT *Dan. Der.* I. II. xiv. 264 The coachful of servants with provisions had to prepare the picnic meal. **1889** Picnic meal [in Dict.]. **1911** A. BENNETT *Hilda Lessways* II. iii. 161 They had been very busy in Hilda's house..and had eaten only a picnic meal. **1929** S. ERTZ *Galaxy* xiii. 294 They had a picnic meal of bread and cheese and fruit and California wine. **1972** 'G. NORTH' *Sgt. Cluff rings True* x. 78 Discards from picnic meals defaced verges. **1849** H. C. ROBINSON *Diary* 22 June (1967) 255 A picnic party had been formed..to take tea on the top of the hill. **1977** R. BARNARD *Blood Brotherhood* vi. 61 The thought of some fearsome affray..came upon him with all the welcomeness of a thunderclap on a picnic party. **1865** TROLLOPE *Can you forgive Her?* II. xxx. 234 Cold chickens, picnic-pies, and the flying of champagne corks. **1911** G. STRATTON-PORTER *Harvester* xvi. 371 Big, fancy brick and frame things..gay as frosted picnic pie. **1926–7** *Army & Navy Stores Catal.* 378 B/4 Picnic plates. **1933** E. A. ROBERTSON *Ordinary Families* i. 16 The boats were only old cardboard picnic plates. **1896** N. GOULD *Town & Bush* xiv. 225 The owners of the horses running at picnic races are generally men of means. **1911** C. E. W. BEAN *'Dreadnought' of Darling* xxxiv. 294 If a town has picnic races or a polo week, it is 'alive'. **1928** 'BRENT OF BIN BIN' *Up Country* xiii. 202 Here took place the first picnic races of the district, which affairs, so informally started among neighbours, in later days command special trains and viceregality. **1955** P. WHITE *Tree of Man* (1956) II. ix. 132 Always in demand,.. and above all at picnic races. **1964** D. HORNE *Lucky Country* 52 One of the most rigid institutional manifestations of this difference [between landed family and town] appears at the annual picnic races. **1972** *Sunday Tel.* (Sydney) 15 Oct. 132 Everyone, but everyone, is getting ready for the Bong Bong Picnic Races next Saturday. **1978** O. WHITE *Silent Reach* xvi. 160 He's a pansy and goes for the public school jackeroos at the picnic races. *Ibid.* xxi. 219 The Fitzroy Crossing picnic race meeting. **1943** J. S. HUXLEY *TVA* ix. 61 Picnic shelters are included in the parks and campgrounds. **1959** *Canad. Geogr. Jrnl.* Feb. 55/1 Picnic shelters have been constructed for the convenience of picnickers during inclement weather and thirty-two of these are now in use. **1976** *Billings* (Montana) *Gaz.* 28 June 1-B/1 The state Board of Examiners this week approved expenditure of $398,565 for building of picnic shelters and other recreation facilities, landscaping and construction of restrooms at the park. **1971** P. GRESSWELL *Environment* 187 Picnic sites need to be considered in relation to planning policy. **1972** *Country Life* 7 Dec. 1581/1 Somerset County Council has decided to spend some money this year on..picnic sites. **1959** I. & P. OPIE *Lore & Lang. Schoolch.* xii. 253 On Easter Day we usually pack a picnic lunch, including a hard boiled egg. When we reach the picnic spot we sit down and play at games. **1977** B. PYM *Quartet in Autumn* v. 43 When they arrived at the picnic spot, Marjorie produced two folding canvas chairs from the boot of the car. **1963** *Which?* Mar. 89/2 A solid fuel for picnic stoves. **1974** *Janet Frazer Catal.* Spring/Summer 458/2 Camping Gaz International picnic stove... Boils water quickly. **1926–7** *Army & Navy Stores Catal.* 225/1 The 'picnic' table... Folds perfectly flat. **1963** 'H. CALVIN' *It's Different Abroad* ii. 16 Between them was a folding picnic-table. **1970** *Times* 16 Apr. 3/3 (Advt.), We don't forget power assisted steering, adjustable steering column, reclining seats and picnic tables. **1976** *Norwich Mercury* 17 Dec. 9/6 Two picnic tables with bench seats were officially handed over. **1976** N. THORNBURG *Cutter & Bone* ix. 219 It was a small park, not much more than a few acres of grass fringed with eucalyptus trees and three or four picnic tables. **1900** C. M. YONGE *Mod. Broods* xvi. 153 My mother wants you all to come up to picnic tea to see the foxgloves in the dell. **1907** R. FRY *Let.* 24 Sept. (1972) I. 291 This afternoon we are to drive out to a picnic tea at Blackwell. **1932** 'E. M. DELAFIELD' *Thank Heaven Fasting* III. ii. 262 Monica and the vistor set out..taking with them a picnic tea. **1955** A. SINCLAIR in *Granta* 26 Nov. 10/1 We'll say we want a picnic tea.

picnickish (pi·knikiʃ), *a. rare.* [f. PICNIC *sb.* + -ISH¹.] Suitable for or suggestive of picnics.
1852 'G. GREENWOOD' *Haps & Mishaps Tour in Europe* (1854) v. 101 It is a pretty, picturesque, and picnickish place.

pico, var. *PIKAU *sb.*

pico- (pī·ko, pəi·ko), *prefix.* [f. Sp. *pico* beak, peak, (in phrases) little bit.] Prefixed to the names of units to form the names of units 10^{12} times smaller, i.e. one million-millionth part of them (symbol p), as *picoamp* (so *picoammeter*), *-curie, -farad, -gramme, -litre, -volt* (so *-voltmeter*), *-watt.* Also *PICOSECOND.
1915 W. H. ECCLES *Wireless Telegr.* 18 Symbols for multiples and sub-multiples... 10^{-12}... Pico-... p or μμ. **1947** *Compt. Rend. de la 14me Conf.* (Union Internat. de Chimie) 115 The following prefixes to abbreviations for the names of units should be used to indicate the specified multiples or sub-multiples of these units:..p pico- 10^{-12} ×.

1952 *Wireless World* Jan. 19/2 The suggestions to adopt new prefixes, 'nano' and 'pico' for small capacitance values, would permit any value to be expressed as a whole number. **1952** [see *NANO-]. **1975** *Physics Bull.* Feb. 67/1 A range of stabilized high voltage power supplies and picoammeters manufactured by V G Electronics Ltd will also be shown. *Ibid.* June 249/2 (Advt.), Sensitivities down to 1 microvolt, 10 pico-amps, and 1 milliohm are founded on Keithley's lifetime of expertise in designing and building low-level instrumentation. **1963** *Times* 15 Feb. 8/1 The level of 150 picocuries over a three-month period is considered by scientists to be the tolerable limit for humanity. **1976** *Globe & Mail* (Toronto) 16 Jan. 1/3 The readings in 30 of these, however, are less than three picocuries of radon for each litre of air, the exposure level that Government standards allow the public. For workers in atomic plants, the standard is 30 picocuries a litre. **1926** *Gloss. Terms Electr. Engin.* (Brit. Engin. Stand. Assoc.) 26 *Picofarad*, a unit of electrostatic capacity equal to 10^{-12} farad. Symbol: μμF. **1945** *Electronic Engin.* XVII. 473/2 At a working frequency of ½ Mc/s., a difference of about 1/10 picofarad can be readily observed. **1976** *Gramophone* Jan. 1283/1 A slight platform rise of up to about 2 dB from 12 kHz upwards was measured which I take to be a result of the Dual's low capacitance cable, whereas Shure recommend 400 to 500 pico-farads per channel. **1951** Picogram [see *nanogramme* s.v. *NANO-]. **1967** *Oceanogr. & Marine Biol.* V. 271 If carbon content is compared with the surface area of [plankton] cells, the range of variation is smaller, 0·11 to 0·66 picograms/μ². **1975** WILLIAMS & WILSON *Biologist's Guide to Princ. & Techniques Pract. Biochem.* iii. 72 It has very high sensitivity and can detect as little as one picogramme of these compounds. **1973** *Sci. Amer.* Sept. 160/2 (Advt.), For electron probe sampling, hundreds of picoliter fluid droplets are positioned automatically under an electron beam. **1971** *Physics Bull.* Nov. 678/2 The picovolt measuring system permits the detection of small voltages from sources at low temperatures which would be overwhelmed by thermal voltages in the leads to a room temperature instrument. *Ibid.* 678/3 The picovoltmeter will be useful for the measurement of low temperatures by means of thermocouples. **1967** D. H. HAMSHER *Communication Syst. Engin. Handbk.* xvi. 10 The noise power measurement without any weighting of the frequency response of the instrument is expressed usually in picowatts (10^{-12} watt = pw). **1977** *Sci. Amer.* Feb. 62/2 At the earth station the 30-metre dish antenna collects a scant three picowatts of energy and feeds it into a low-noise receiver.

picolinic (pikǫlī·nik), *a. Chem.* [f. PICOLIN(E + -IC.] *picolinic acid*: a colourless crystalline acid, pyridine-2-carboxylic acid, $C_5H_4N(COOH)$, which is derived from picoline by oxidation of the methyl group.
1880 *Jrnl. Chem. Soc.* XXXVIII. 268 Picolinic Acid, $C_6H_5NO_2$..Crystallises in prismatic needles..which are easily soluble in alcohol and in water. **1946** A. A. MORTON *Chem. Heterocyclic Compounds* viii. 213 Picolinic acid..is obtained in 50 to 51 per cent yield from α-picolin. **1968** *Listener* 21 Mar. 376/1 A second class of compounds being used [as defoliants, etc.] are derivatives of picolinic acid.

Picon (pikoṅ). [Fr., f. *Amer Picon*, f. *amer* bitter + *Picon* et Cie, the manufacturers.] In full, *Amer Picon.* The proprietary name of an aperitif of bitters (see quot. 1967).
1914 [see *PERNOD]. [**1917** *Bull. Offic. Marques de Fabrique* 25 Jan. 54/1 *Amer Picon.* M.p. désigner un amer, déposée..par MM. Picon et Cie.] **1934** H. MILLER *Tropic of Cancer* (1961) 21 Paris..shutters going up with a bang..*Amer Picon* in huge scarlet letters. **1936** BENTLEY & ALLEN *Trent's Own Case* xi. 133 Sipping his Amer Picon—that formidable brown drink of which..a single undiluted drop will burn a neat round hole in a shirt cuff. **1939** L. MACNEICE *Autumn Jrnl.* 22 Something out of the usual, a Pimm's Number One, a Picon. **1950** D. AMES *Corpse Diplomatique* i. 11 He ordered a *Pernod* and an *Amer Picon.* **1962** N. FREELING *Love in Amsterdam* I. 48 Herman had raw *schinken* with kümmel in it, and a bottle of Picon. **1967** A. LICHINE *Encycl. Wines* 71/1 *Amer Picon*, a proprietary French bitters used as an aperitif, made with a wine and brandy base to which has been added quinine (to impart a bitter taste), orange peel, and innumerable herbs. It is drunk with ice, diluted with water, and usually sweetened with grenadine or cassis.

picong (pi·kǫŋ). [ad. Sp. *picón.*] In Trinidad and some other parts of the Caribbean: verbal insult or ridicule; facetious raillery; taunting; esp. in phr. *to give picong.* Cf. *to play (the) dozens* s.v. *PLAY *v.* 16 e.
1956 *Caribbean Q.* IV. III–IV. 198 The Congos..carried on 'picong' or 'fatigues', elaborate ribbing of each other and the bystanders. **1960** *Tamarack Rev.* XIV. 19 Then again, the boys might start to give Frederick picong, when Bat beg them to ease up. **1974** *Sunday Guardian* (Port-of-Spain) 28 July 8/2 They [*sc.* students] were screaming and laughing and giving picong to all kinds of rhythms, on the edge of Grand Etang. *Ibid.* 12/4 The humour here is not at all picong, but more humour for its own sake.

picornavirus (pikǭ·rnăvaiərǝs). *Microbiol.* [f. *PICO-, taken to mean 'very small' + *RNA + VIRUS.] Any of a group of very small animal viruses consisting of single-stranded RNA in an unenveloped icosahedral capsid, which includes enteroviruses, rhinoviruses, and the virus of foot-and-mouth disease.
1962 *Adv. Virus Res.* IX. 296 At a meeting of the Virus Subcommittee of the International Nomenclature Committee, held in Montreal in August 1962, the following decisions were made: (1) The proposed name nanivirus is to be replaced by Picornavirus. **1963** *Internat. Bull. Bacteriol. Nomencl. & Taxon.* XIII. 218 It was agreed that a name was needed for the small ether-resistant RNA viruses of which enteroviruses form an important part. The name 'Nanivirus'..was discarded in favour of 'Picornavirus'. The prefix Pico- implies very small size, and RNA of course indicates nucleic-acid composition. (The initial letters of Picornavirus may be taken to refer to Poliomyelitis, insensitivity to ether, Coxsackie, orphan and Rhinovirus.) **1974** L. LEVINTOW in Fraenkel-Conrat & Wagner *Comprehensive Virology* II. iii. 154 Picornaviruses are the agents of a number of other important diseases of man and animals which remain to be controlled, including the common cold. **1976** FENNER & WHITE *Med. Virology* (ed. 2) xviii. 333 There are no antigenic relationships between most of the picornaviruses.

picosecond (pī·ko-, pəi·kosekǝnd). [f. *PICO- + SECOND *sb.*¹] A unit of time equal to one million-millionth of a second.
1966 in *Random House Dict.* **1967** *Electronics* 6 Mar. 60/2 The..circuits..have switching speeds of 350 picoseconds. **1972** *Sci. Amer.* Dec. 16/2 A sampling oscilloscope with a rise time..of 44 picoseconds. **1974** *Nature* 15 Mar. 222/2 A complete formalism now exists to do the complete physics of the first picoseconds of the big-bang origin of the universe. **1975** *Daily Colonist* (Victoria, B.C.) 9 July 5/8 By using lasers capable of picosecond, or trillionth of a second, bursts of light and using ultra fast 'streak' cameras, researchers are now able to measure the conversion of light to other forms of energy in plants.

picot (pī·ko), *v.* [f. the sb.] *trans.* To ornament (cloth) with picots. So **picoted** (pī·kōᵘd) *ppl. a.*; **picoting** (pī·ko,iŋ) *vbl. sb.*
1927 *Daily Express* 7 Mar. 5/5 Flowers were cut out of sheet metal..even the picotted [*sic*] edges of certain varieties of carnation being faithfully rendered. **1928** *Ibid.* 11 May 5/3 Buy a square of plain or flowered.. georgette, and have it picoted all round by machine. The picoting is really machine hemstitching cut through the middle.

picral (pi·kræl). *Metallurgy.* Also **Picral.** [f. PICR(IC *a.* + AL(COHOL.] An etchant consisting of about 2 to 5 per cent of picric acid in ethyl alcohol.
1928 WILLIAMS & HOMERBERG *Princ. Metallogr.* (ed. 2) 237 The term *picral* is frequently applied to any solution of picric acid in alcohol. **1936** [see *NITAL]. **1963** C. H. SAMANS *Metallic Materials in Engin.* vi. 267 The action of picral is similar, except that it etches fine pearlite much more uniformly than nital does. However, it does not etch ferrite so rapidly as nital, and does not indicate the ferrite grain boundaries clearly.

picrite. Delete entry in Dict. and substitute: **picrite** (pi·krəit). [f. Gr. πικρ-ός bitter: see -ITE¹.] † **1.** *Min.* [a. G. *picrite(s)* (J. F. Blumenbach *Handb. der Naturgeschichte* (ed. 5, 1797) xii. 584).] = DOLOMITE. *Obs. rare.*
1814 [in Dict.]. **1836** T. THOMSON *Outl. Min., Geol.* I. 181 Calcareo-carbonate of Magnesia. Dolomite—conite— ..muricalcite—pearl spar—picrite [etc.]. **1896** [in Dict.].
2. *Petrogr.* Also 9 **picryte, pikrite.** [ad. G. *pikrit* (G. Tschermak 1866, in *Sitzungsber. der K. Akad. der Wissensch.* (*Math.-naturw. Classe*) LIII. 262).] A dark ultrabasic igneous rock, commonly hypabyssal, containing a substantial amount of olivine together with augite and other ferromagnesian minerals and a small amount of plagioclase.
1868 J. D. DANA *Syst. Min.* (ed. 5) 258 A chrysolite rock occurring at L. Lherz, consisting largely of chrysolite, has been called Lherzolyte... Another similar rock from Moravia, called picryte, consists half of chrysolite, along with feldspar, diallage, hornblende, and magnetite. **1879** [in Dict.]. **1882** A. GEIKIE *Text-bk. Geol.* 151 Olivine Rocks... The following are the more important species:—Pikrite, a rock rich in olivine, usually more or less serpentinized, with augite, magnetite, or ilmenite, and a little brown biotite, hornblende, or apatite; [etc.]. **1931** S. J. SHAND *Study of Rocks* vii. 112 Picrites with forty to seventy-five per cent. of olivine have been described in Skye. The picrite of Tabankulu, South Africa, is composed to the extent of one-third to two-thirds of well-shaped crystals of olivine which are enclosed in larger crystals of pyroxene and felspar. **1976** *Mineral. Mag.* XL. 683 Exposures of baked siltstone occur on top of the ridge with picrite exposed only on the scarp face... The picrite averages 50% olivine, 35% plagioclase, and 15% pyroxene with a maximum grain size of 6 mm.
Hence **picri·tic** *a.*
1931 *Amer. Jrnl. Sci.* CCXXI. 403 Even a picritic sub-basalt..exhibits some interstitial quartz. **1958** *Trans. R. Soc. Edin.* LXIII. 459 In this paper it is proposed to use the term 'picritic' to denote, in minor intrusions and lavas, an amount of forsteritic olivine considerably beyond that which normally crystallizes in basaltic magma. This amount of olivine varies between 25 and 60 per cent. **1965** *Carnegie Inst. Year Bk. 1964* 126/2 In general, picritic lavas and hypabyssal intrusives are few in number and restricted to the lower parts of both oceanic and continental volcanic columns. **1974** P. G. HARRIS in H. Sørensen *Alkaline Rocks* vi. 430/1 At pressures of about 30–40 kbar, i.e. depths below 90–100 km, the mantle mineralogy will be that of a garnet lherzolite, and there is a marked change in the first-formed liquid which is now picritic or ultrabasic.

picro-. Add: **picrochro·mite** *Min.*, a chromite of magnesium, $MgCr_2O_4$, which in its pure form is known only as an artificial product (see quots. 1939) and which in nature is a brittle black mineral that contains a substantial amount of ferrous iron replacing magnesium; **picroilmenite** (pikro,i·lmĕnəit) *Min.* [ad. G. *pikroilmenit* (P. Groth *Tabellarische Übersicht der Mineralien* (ed. 4, 1898) ix. 143)], a magnesian variety of ilmenite.
1920 E. S. SIMPSON in *Mineral. Mag.* XIX. 100 Dealing with the minerals of the spinel–chromite series.., the four variables, which constitute two pairs, are MgO, FeO, Al_2O_3, Cr_2O_3. Four mineral species are possible, the pure forms of which are (1) $MgO.Al_2O_3$ spinel, (2) $MgO.Cr_2O_3$ picrochromite, (3) $FeO.Al_2O_3$ hercynite, and (4) $FeO.Cr_2O_3$ chromite. [*Note*] New name required to complete the series and to designate already known minerals. *Ibid.* 104 At least three previously described minerals, viz. chrompicotite of Dun Mt., New Zealand; magnesiochromite of New Caledonia; and 'chromite' of Lake Memphremagog in Quebec, are members of a new species for which the author proposes the name picrochromite. **1939** *Special Rep. Iron & Steel Inst.* No. 26. 202 Picrochromite was prepared in a manner similar to that used for spinel but employing equivalent amounts of magnesia and chromic oxide. *Ibid.* 204 Picrochromite was green in daylight and reddish-grey in artificial light..when viewed in mass. It occurred as aggregates of minute crystals which were practically colourless and presented occasional octahedral faces. **1970** *Mineral. Abstr.* XXI. 314/2 (*heading*) Determination of the fusion temperature of picrochromite. **1900** *Mineral. Mag.* XII. 389 Picroilmenite... The same as picrotitanite, a variety of ilmenite rich in magnesium. **1906** *Ibid.* XIV. 166 It is advisable to divide the ferro-magnesian titanates into ilmenites and geikielites, and to regard picroilmenite as the middle member of the series. **1972** *Mineral. Abstr.* XXIII. 338/1 Picroilmenite occurs, associated with diamond, in the kimberlites and in the Carboniferous, Jurassic, and Cretaceous formations [in the Anabar area, western Yakutia]. Its hardness ranges from 450 to 750 kg/mm²; it is anisotropic, optically negative, and has a weak bireflectance.

picrolichenic (pikrolai·kĕnik), *a. Biochem.* [f. next + -IC.] *picrolichenic acid*: a very bitter, crystalline, polycyclic acid, $C_{25}H_{30}O_7$, isolated from the lichen *Pertusaria amara*; = next.
1902 *Jrnl. Chem. Soc.* LXXXII. I. 465 The name *picrolichenic acid* is suggested for picrolichenin. **1957** *Chem. & Industry* 27 July 1042/2 Picrolichenic acid, a constituent of the common crustose lichen *Pertusaria amara* Ach...is responsible for the well known, very bitter, quinine-like taste of this lichen. **1959** *New Biol.* XXIX. 86 Picrolichenic acid..is a previously undescribed type of lichen acid in which the ring structures in the molecule are joined by a direct carbon to carbon link rather than by an oxygen atom as is the case in depsidones. **1973** S. HUNECK in Ahmadjian & Hale *Lichens* xv. 511 The only hitherto known depsone from lichens is picrolichenic acid.

picrolichenin (pikrolai·kēnin). *Biochem.* [a. G. *picrolichenin* (A. Alms 1832, in *Ann. d. Pharm.* I. 62), f. *picro-* PICRO- + *lichenin* LICHENIN.] = prec.
1862 H. WATTS tr. *Gmelin's Hand-bk. Chem.* XV. 56 A pound of the lichen yields half an ounce of picrolichenin. **1901** *Jrnl. Chem. Soc.* LXXX. I. 88 The author has isolated salazinic acid and picrolichenin from *Pertusaria amara.* **1958** *Svensk Kem. Tidskr.* LXX. 129 The bitter principle isolated by Alms from this lichen—'Picrolichenin'—was claimed to be useful as a remedy in the treatment of malaria, and this is now being checked.

picryl. Now also with pronunc. (pi·krail). Delete '(See quot.)' and substitute: † **1.** Also **picril, picryle.** [a. F. *picryle* (A. Laurent 1844, in *Rev. sci. et industr.* XVIII. 201).] (See quot. 1866 in Dict.) *Obs.* (Earlier examples.)
1847 W. GREGORY *Outl. Chem.* (ed. 2) 546 *Picrine*, Syn. *Picryle*, $C_{42}H_{15}NO_4$, is formed when the mass produced by acting on oil of bitter almonds by sulphuret of ammonium is distilled. **1858** H. WATTS tr. *Gmelin's Hand-bk. Chem.* XII. 188 Picril dissolves very readily in ether.
2. The 2,4,6-trinitrophenyl group, $—C_6H_2(NO_2)_3$, contained in picric acid.
1866 [in Dict.]. **1875** *Jrnl. Chem. Soc.* XXVIII. 165 Metanitraniline and picryl chloride act on one another when dissolved in boiling absolute alcohol. **1877** *Ibid.* XXXII. 758 It is..very possible that in one of the resulting compounds the two picryl groups may be identical. **1889** G. M'GOWAN tr. *Bernthsen's Text-bk. Org. Chem.* xxiii. 385 Picryl chloride, $C_6H_2(NO_2)_3Cl$.., resembles the acid chlorides..in behaviour. **1937** F. C. WHITMORE *Org. Chem.* 737 Picryl iodide is a stable yellow crystalline product obtainable from the chloride and KI. **1964** J. W. LINNETT *Electronic Struct. Molecules* vii. 107 The reason why the diphenyl picryl compound is more stable than the triphenyl compound is presumably that the picryl group relieves the adjacent nitrogen atom of some of the negative charge. **1975** *Nature* 3 Jan. 72/2 The lymph node cells of mice rendered unresponsive by the injection of picryl sulphonic acid (PSA) depress the passive transfer of contact senstivity.

Pictish, *a.* Add: Also, of or pertaining to the language (or languages) of the Picts.
1710 R. SIBBALD *Hist. Fife & Kinross* I. ii. 3 Both in Bertius his excellent Edition and the late Map of Gale, it [*sc.* Votadini] is read Vacomagi, and the Greek in both these answer to Vacomagi, which by the by..doth much confirm Mr. Robert Maule his Ratio nominis Veach, Pictus, since in Veach here, and in Wauchopdale in the South..the Pictish Veach appears to be the rise of both these Words. **1840** *Penny Cycl.* XVIII. 148/2 Their language appears to have nearly resembled the Welsh.

One Pictish word only has been expressly mentioned by any old writer, *Peanvahel*. **1868** W. F. Skene *Four Ancient Bks. of Wales* I. viii. 130 In the fifth, the Pictish *duiper* and the Gaelic *saoibher* are the same word. **1892** J. Rhys in *Proc. Soc. Antiquaries Scotl.* XXVI. 310 Lastly, Pictish *enn* is to be found in some instances where the Celtic stem is usually made to end..in *on*. **1946** T. F. O'Rahilly *Early Irish Hist. & Mythol.* 369 Compare Celt. *vroiko-* giving Pictish **vrōg*, whence..Sc. *frōg*, 'a fen'. **1955** K. H. Jackson in F. T. Wainwright *Probl. Picts* vi. 142 *Forcus* at St Vigeans is the exact Gaelic equivalent of Pictish *Uurguist*.

B. *sb.* The language (or languages) of the Picts.

Pictish is evidenced only in a few proper names, inscriptions, and glosses, and its affiliations have been much disputed. The view most widely accepted now is that expressed by K. H. Jackson in F. T. Wainwright *The Problem of the Picts* (1955) 152: 'There were at least two languages current in northern Scotland before the coming of the Irish Gaels in the fifth century. One of them was a Gallo-Brittonic dialect not identical with the British spoken south of the Antonine Wall, though related to it. The other was not Celtic at all, nor apparently even Indo-European, but was presumably the speech of some very early set of inhabitants of Scotland.'

1857 W. Reeves *Life St. Columba* 62/2 This case saves that recorded in ii. 32, *infra*, from being 'a solitary allusion to the diversity of Gaelic and Pictish'. **1868** W. F. Skene *Four Ancient Bks. of Wales* I. viii. 138, I consider, therefore, that Pictish was a low Gaelic dialect... Old Scottish..was the high Gaelic dialect, and Pictish the low Gaelic dialect. **1891** W. Stokes in *Trans. Philol. Soc. 1888–90* 417 The foregoing list of names and other words contains much that is still obscure; but on the whole it shows that Pictish, so far as regards its vocabulary, is an Indo-European and especially Celtic speech. Its phonetics, so far as we can ascertain them, resemble those of Welsh rather than of Irish. **1892** J. Rhys in *Proc. Soc. Antiquaries Scotl.* XXVI. 307 Pictish being, as I take it, a non-Aryan language. **1923** J. Fraser *Hist. & Etym.* 15 Rhys's view that Pictish was a non-Indogermanic language involves the same fallacy as the view that it was Goidelic or Brythonic. **1946** T. F. O'Rahilly *Early Irish Hist. & Mythol.* 353 The preponderant opinion of modern scholarship is that Pictish was a Celtic language, different from both British and Goidelic, but decidedly more akin to the former than to the latter. **1955** K. H. Jackson in F. T. Wainwright *Probl. Picts* vi. 134 We seem to find it implied or clearly stated by contemporary and almost contemporary writers that Pictish was a separate speech of its own, identical neither with the Gaelic of Scotland and Ireland nor with Brittonic. **1963** N. K. Chadwick *Celtic Brit.* i. 20 In Scotland the royal Pictish families seem to have spoken Pictish as late as the ninth century. **1976** C. F. & F. M. Voegelin *Classification World's Lang.* (1977) 104 [Goidelic = Q-Celtic] Pictish. Formerly spoken in Scotland but not necessarily as a Celtic language.

pictogram (pi·ktŏgræm). [f. L. *pict-us* painted + -gram.] = Pictograph in Dict. and Suppl.

1910 *Encycl. Relig. & Ethics* III. 549/2 The primitive characters or 'pictograms'..afford unmistakable evidence as to the ideas which existed long anterior to the time of Confucius. **1936** V. G. Childe *Man makes Himself* viii. 208 Even the picture-signs, though much more realistic than Sumerian pictograms, conform to a social convention. **1939** *Trans. Philol. Soc.* 76 K. Haag.. worked out a scheme of pictograms, based partly on the fundamental space relations and supplemented by the use of symbols like the heart for emotion. **1952** G. Sarton *Hist. Sci.* I. ii. 20 We may assume that the Egyptians began by using pictograms (images) representing things or ideas rather than words. **1960** E. H. Gombrich *Art & Illusion* viii. 280 They see the picture still as a flat surface covered with a pictogram. **1965** H. K. Compton *Gloss. Purchasing & Supplies Managem. Terms* 101 *Pictogram*, diagrams drawn in different sizes more or less to scale to demonstrate relative sizes of statistical batches which are illustrated, often by pictures depicting the subject matter. **1970** *Which?* Nov. 352/2 The use of pictograms (eg '?' to show an information desk) is increasing. **1971** *Nature* 16 July 148/3 At the top there are pictograms of a sword, a shield and two crosses and a triangle. **1973** J. M. Anderson *Struct. Aspects Lang. Change* 14 Pictograms can be read..with varying choice of words in any language. **1977** *New Yorker* 2 May 135/3 Occasionally he resorts to dance pictograms.

pictograph. Add: (Further examples.) Also *attrib.*

1937 *Discovery* Aug. 249/1 The Lachish dagger with pictograph inscription. **1955** *Bull. Atomic Sci.* Mar. 91/2 An Indian pictograph enthusiast..noticed carnotite stain on a cliff face in the Edgemont, S. D., area. **1955** *Sci. News Let.* 9 July 31/2 The crescent is not a common figure among petroglyphs and pictographs of northern Arizona. **1971** N. Smith *Hist. Dams* i. 23 In ancient Egypt..the hieroglyph for a 'province' is a pictograph of an irrigation system. **1973** *Nature* 26 Oct. 422/2 He sees close parallels with the Juxtlahuaca Cave paintings, as well as the earlier pictographs at Chalcacingo, and attempts to define the symbolic elements common to all of them. **1978** *New York* 3 Apr. 32/3 Pictographs from the Warlis, an East Indian tribe.

b. A pictorial representation of statistical data.

1937 R. Modley *How to use Pictorial Statistics* iii. 19 The chief variation is that the bar chart can indicate the diversity in subject only in its printed legend, while the pictograph expresses the subject by the character of the symbols which make up each 'bar'. In other words the bar chart is an abstract presentation of facts; the pictograph is concrete. **1954** D. Huff *How to lie with Statistics* vi. 66 The daddy of the pictorial chart, or pictograph, is the ordinary bar chart.

pictographic *a.* (further examples).

1895 [see *Eteocretan *a.* and *sb.*]. **1932** [see *acrophonic *a.*]. **1952** G. Sarton *Hist. Sci.* I. ii. 20 Words containing the same sounds, especially proper names or abstract words, which were not susceptible of pictographic representation. **1971** *Nature* 30 Apr. 552/1 The signs on these three small baked clay plaques were accepted..as representing a script closely resembling the pictographic writing seen on the 'protoliterate' tablets from Uruk in Mesopotamia.

pictorial, *a.* (*sb.*) Add: **3.** (Further examples.) *pictorial paper* = sense B in Dict. and Suppl.

1873 'Mark Twain' & Warner *Gilded Age* xliii. 394 The pictorial papers caricatured its friends. **1897** *Pearson's Mag.* IV. 405 It is a wonderful invention—this pictorial postcard craze. **1933** *Burlington Mag.* Sept. 140/2 The student who has not an extensive library might like to have possessed a pictorial record. **1935** *Discovery* Dec. 365/2 Sets of pictorial postcards of the park and its inhabitants. **1937** J. R. Hunt (*title*) Pictorial journalism. **1948** C. Abel *Business of Photogr.* v. 37 Another offshoot of commercial photography..is the growing field of pictorial journalism. **1968** G. Jones *Hist. Vikings* iii. ii. 188 From the Gotland pictorial stones it appears that sail could be effectively shortened by the use of reefing lines.

B. (Earlier and later examples.)

1844 *Knickerbocker* XXIII. 197 'The Columbian'..is to run a brisk competition..with the other pictorials. **1851** C. Cist *Sk. Cincinnati in 1851* iv. 77 Illustrated Western World... Oncken's Western Scenery... These two last are pictorials. **1934** G. B. Shaw *Too True to be Good* i. 43 In her lily hand was a copy of The Lady's Pictorial. It contained an illustrated account of your jewels.

2. A postage stamp on which a picture or scene is printed, usu. to commemorate a particular anniversary or event.

1934 in Webster. **1939** P. Hamilton *Hundred Years of Postage Stamps* v. 74 Pictorials are a characteristic feature of the second half of our hundred years of postage stamps. **1953** Harrison & Armstrong *New Approach to Stamp Collecting* xii. 173 The rocky finances of certain Central and South American states were the healthier for some really splendid 'pictorials'. **1973** *Daily Tel.* 8 Jan. 9/3 The islands' first pictorials, showing views of the various islands, were issued in 1942-43. **1974** W. Finlay *Illustr. Hist. Stamp Design* ix. 119/1 The issues of the six Australian states..included a fair number of pictorials.

pictorialist (later examples).

1971 *Amat. Photographer* 3 Mar. 6/2 The panel..does not have the dynamic impact of new pictures or of the more forceful avant-garde work of some modern pictorialists. **1979** G. Macdonald *Camera* xii. 169/1 The aping of the avant garde by some of the pictorialists.

pi:cturabi·lity. [f. Picturable *a.*: see -ility.] The state or quality of being picturable; capacity for visual representation.

1934 in Webster. **1957** R. W. Brown in Saporta & Bastian *Psycholinguistics* (1961) 505/1 The concrete noun with the smaller denotation is likely to be more picturable than its superordinate, and picturability is another common sense of 'concrete'. **1968** M. Jammer in R. Klibansky *Contemp. Philos.* II. 349 The shackles of intuitive picturability ('*Anschaulichkeit*').

picture, *sb.* Add: **2. e.** (Earlier examples.)

1802 T. Holcroft *Tale of Mystery* 30 Enter Malvoglio. He stops in the middle of the stage: the company start up... The peasants, alarmed and watching: the whole, during a short pause, forming a picture. *c* **1825** J. Poole *Paul Pry* i. ii. 13 There is a general shout of 'Paul Pry'. *Picture, and act closes.*

f. (Further examples.)

1915 *Wireless World* June 193/1 A picture, constant as long as the cells are under the influence of the original picture, is thus obtained on the receiving screen. **1928**, etc. [see *frame *sb.* 12 c]. **1934** J. H. Reyner *Television* iii. 33 Fig. 14 shows the number of pictures per second at which flicker can just be detected, in terms of illumination. **1943**, etc. [see *field *sb.* 16 d]. **1953** Amos & Birkinshaw *Television Engin.* I. i. 14 This time must be short to enable pictures to be transmitted in quick succession, as in cinema film projection. **1956** *B.B.C. Handbk. 1957* 215 The television services of the two countries use different standards, the English system based on a 405-line picture and the French on one of 819 lines. **1960** J. Stroud *Shorn Lamb* xxii. 240 Some well-intentioned hostesses obligingly turned off the sound but not the picture. **1972** G. White *Video Recording* iii. 17 The definition of the reproduced picture depends upon the frequency response of the signal system. **1976** *Daily Tel.* 30 June 2/6 Pictures sent back so far show that areas previously believed to be smooth channels..are pock marked with craters. **1977** J. French *Small Craft Radar* iii. 86 The radar picture is built up by displaying on a time-related sweep line the instantaneous point of arrival of the echo.

g. (Later examples.)

a **1817** Jane Austen *Northanger Abbey* (1818) I. iv. 49 'How excessively like her brother Miss Morland is!' 'The very picture of him, indeed!' **1877** G. Macdonald *Marquis of Lossie* III. iv. 81 Isna his mere 'at they ca' Kelpie jist the pictur' o' the deil's ain horse. **1896** P. A. Graham *Red Scaur* 271 You're the verra pictur' o' awd Mr. Selwyn.

h. (Examples.) See also *pretty as a picture* s.v. *pretty *a.* II. 4 a.

1815 *Sporting Mag.* XLVII. 135/2 She looked a perfect picture. **1827** J. Constable *Let.* 4 Oct. in *Corr.* (1962) I. 234 Minna looks so nice in her pelisse—the blew band or what it is called was a picture. **1842** C. Brontë *Let.* 20 Jan. in W. Gérin *C. Brontë* (1967) xii. 179 He sits opposite to Anne at Church sighing softly..and Anne is so quiet.. they are a picture. **1848** Dickens *Dombey* lvii. 575 He has been working..to make his cabin what the Captain calls 'a picter', to surprise his little wife. **1871** G. Meredith *Let.* 15 Feb. (1970) I. 439 The French..have many a noble turning, and are always a picture, good for study. **1937** A. Christie *Dumb Witness* vii. 65 The gardens are a picture. **1961** *Guardian* 9 June 12/1 The bride was, as they say, 'a picture'. **1977** *Daily News* (Perth, Austral.) 19 Jan. 50/5 The rain this week won't have hurt. All it has done is make the grass greener. The court looks a picture.

i. A representation (of a scene) on a cinematographic film; the film produced or its projection on a screen, = *film *sb.* 3 c. Freq. in *pl.*, (*a*) cinematographic productions collectively; (*b*) usu. with *the*, the showing of a film in a cinema, a picture-show.

1896 [see *project *v.* 9 d]. **1912** *Home Chat* 25 May 391/1 In order to get a picture of the sacking of a village, an actual village was some time ago purchased and fired. **1913** *Ibid.* 20 Sept. 530/1 The pictures one sees nowadays are..in much better taste than those of a few years ago. **1915** *Kinematograph & Lantern Weekly* 1 July 61/2 During his very successful career in 'pictures' he has appeared in some..thrilling productions. **1915** T. Burke *Nights in Town* 110 Mother and Father..go to the pictures at the Palladium near Balham Station. **1916** *Variety* 27 Oct. 12/2 His bride was in the Kellermann picture, 'A Daughter of the Gods'. **1923** Wodehouse *Inimitable Jeeves* xii. 129 Charlotte is coming to the Zoo with me this afternoon. Alone. And later on to the pictures. **1937** D. L. Sayers *Busman's Honeymoon* x. 225 Off to them pictures again. **1947** A. Huxley *Let.* 27 July (1969) 573 The ticklish situation on the set made it impossible to come to New York for Claire's wedding. But we hope and intend to make the trip after the picture is finished. **1960** M. Stark *Ballad of Peckham Rye* viii. 187 They came out of the pictures at eight o'clock. **1974** *Sat. Rev. World* (U.S.) 19 Oct. 56/2 My son had gone to the movies... I asked him, 'How was the picture?' **1977** I. Shaw *Beggarman, Thief* iii. xi. 354 The picture was not scheduled to start for another ten minutes. **1978** P. Grace *Mutuwhenua* iii. 11, I went to the pictures or a social with my cousins.

j. Phr. *every picture tells a story*, orig. popularized by an advertising slogan (see quot. 1927); also in extended uses.

[**1847** C. Brontë *Jane Eyre* I. i. 5 The letter-press..I cared little for... Each picture told a story.] **1906** *S. A. News Weekly* 26 Sept. 23/3 (Advt.), Every picture tells a story... At the first sign of kidney disease..take Doan's Backache Kidney Pills. **1923** A. Huxley *Antic Hay* xxi. 290 'I'm very ill,' she went on expiringly. 'Look at me,' she pointed to herself, 'and me again.' She waved her hand towards the sizzling brilliance of the portrait. 'Before and after. Like the advertisements, you know. Every picture tells a story.' **1927** W. E. Collinson *Contemp. Eng.* 65 The wording accompanying..the distressing pictures of human suffering, amenable to treatment by Doan's Backache Kidney Pills, supplies us with the useful Every picture tells a story!—often used derisively of anecdotal paintings. **1976** *Radio Times* 25 Sept. 5 (*caption*) Every picture tells a story: Rod Stewart, the working-class boy who has become a working-class hero.

k. *one picture is worth ten thousand words* and varr.

[**1921** *Printer's Ink* 8 Dec. 96 (*heading*) One look is worth a thousand words.] **1927** *Ibid.* 10 Mar. 114 (*caption*) Chinese proverb. One picture is worth ten thousand words. **1954** R. Haydn *Jrnl. Edwin Carp* 90 'One picture speaks louder than ten thousand words.' Mr. Bovey repeated the adage this morning when..he handed me my finished portrait. **1979** *Daily Tel.* 14 Aug. 11 If proof was ever needed that a few good pictures are worth more than a thousand statistics it came with bludgeoning force in last night's account of *The Voyage of Rainbow Warrior* (BBC-2) in the 'Inside Story' series.

3. b. *Med.* The sum of the clinical or histological features present in a particular state of the body.

1897 [in Dict., sense 3]. **1931** *Jrnl. Exper. Med.* LIV. 244 The clinical picture of 'mad itch' is very suggestive of pseudorabies. **1940** *Endocrinology* XXVII. 127 The histological pictures of the hypertrophied uteri after vitamin E administration and after mechanical stimulation of the cervix uteri were identical. **1966** Wright & Symmers *Systemic Path.* II. xxxix. 1475/1 The histological picture of psoriasis is distinctive but not pathognomic. **1977** *Lancet* 30 July 241/1 Another renal biopsy on March 31 revealed a hyperacute picture (diffuse small and medium arteriolar fibrin thrombi, infarction, and glomerular necrosis).

c. In extended uses, a set of circumstances or state of affairs, esp. in phrs. *to be in the picture*, to understand or be involved in a particular situation or activity, to be in harmony with one's surroundings; similarly *to be out of the picture*, to be uninvolved, inactive, or out of place; *to get the picture*, to grasp or become aware of certain circumstances or facts; *to put* (someone) *in the picture*, to inform (that person) of particular circumstances or facts.

1900 Beerbohm *Around Theatres* (1924) I. 165 His performance is, strictly, more 'in the picture' than was Mr. Robertson's. **1902** C. Morris *Stage Confidences* 202 Oh, well, I feel that I am in the picture, when I wear black during Lent. **1918** Ld. Derby in R. S. Churchill *Lord Derby* (1959) xv. 337 He [*sc.* a new ambassador in Paris]..would..by entertaining be able to bring the British Embassy more, what I may call, 'into the picture' than it is at the present moment. **1923** T. E. Lawrence *Let.* 19 Mar. (1938) 411, I said to one 'They're the sort who instinctively fling stones at cats'..and he said 'Why what do you throw?' You perceive that I'm not yet in the

picture. **1923** Wodehouse *Good Morning, Bill!* i. 24 'And what's the matter with taking me along?' 'I don't think you would be quite in the picture.' **1926** Maines & Grant *Wise-Crack Dict.* 11/2 *Out of the picture*, in the wrong company. **1936** R. Lehmann *Weather in Streets* iii. iv. 320 Think of me as out of the picture..for ever. Unless, of course, you should wish to see me. **1937** L. Bromfield *Rains Came* i. xlv. 190 She should never have come out to India. She doesn't fit into the picture. **1938** [see *chin *sb.*[1] 1 d]. **1939** L. M. Montgomery *Anne of Ingleside* xlii. 327 Christine took possession of the whole room... Anne felt as if she were not in the picture at all. **1942** E. Waugh *Put out More Flags* ii. 150 'Put these men in the picture, Smallwood,' he said, and there had followed a tedious and barely credible narrative about the unprovoked aggression of Southland against Northland. **1950** N. Streatfeild *Mothering Sunday* 144 You keep calling me a criminal, darling, but that's where you aren't in the picture. **1959** P. McCutchan *Storm South* xvii. 239, I would have to be kept right out of the picture so far as the Australian public eye and the police..are concerned. **1960** C. S. Lewis *Studies in Words* v. 128 The attitude of any slave-owning society is and ought to be repellent to us, but it is worth while suppressing that repulsion in order to get the picture as Aristotle saw it. **1961** T. Coffin *Not to the Swift* xviii. 201 Do you get the picture of the kind of fellow he was? **1963** A. Huxley *Let.* 17 Nov. (1969) 964 Under what conditions is this being done and where do I come into the picture? **1966** 'A Hall' *9th Directive* i. 14 At this time..the South-east Asian picture is confused and threatening. **1970** G. Greer *Female Eunuch* 128 Nurses are skilled menials, and as such they fall into line with the dominant pattern of female employment. Salesgirls..waitresses..tea ladies, fill out the picture. **1971** K. Amis *Girl, 20* iv. 129 I'm sorry I've been out of the picture, but I've been up to my neck... Haven't had a bloody minute. **1973** A. Behrend *Samarai Affair* xiii. 136 Come over here soon as possible and I'll put you in the picture. **1975** N. Luard *Travelling Horseman* viii. 198, I explained all this... He seems to get the picture. **1977** *New Yorker* 15 Aug. 36/2, I suppose you were trying to make me jealous—jealous and not jealous. Tearing poor Ann down but letting me know she was in the picture.

4. b. *Philos.* In the study of meaning, the mental image that is assumed to correspond to a fact; also *attrib.*

1922 tr. *Wittgenstein's Tractatus* 39 We make to ourselves pictures of facts. *Ibid.*, The picture presents the facts in logical space, the existence and non-existence of atomic facts. **1940** B. Russell *Inquiry into Meaning & Truth* xiii. 230, I can make a picture of Brutus killing Caesar..but I cannot make a picture, either real or imagined, of quadruplicity killing procrastination. **1946** *Mind* LV. 47 With their use of 'the picture gallery method', and with their insistence on the marginal case, the W[ittgensteinia]ns should be the first persons to admit..this possibility. **1956** J. O. Urmson *Philos. Analysis* v. 54 The relation of language and the world, or picture of fact and fact. **1970** D. M. Taylor *Explanation & Meaning* xi. 132 (*heading*) The picture theory of meaning. **1975** Hargreaves & White tr. *Wittgenstein's Philos. Remarks* iii. 63 The essential difference between the picture conception [of intention] and the conception of Russell, Ogden and Richards, is that it regards recognition as seeing an internal relation. *Ibid.*, The intention never resides in the picture itself. **1977** G. Hallett *Compan. to Wittgenstein's 'Philos. Investigations'* 41 The direction Wittgenstein's thinking took momentarily when he pondered the implications of the picture theory.

6. a. *picture-cycle, frock, gown, library, newspaper, -paper* (earlier and later examples), *story* (later and *attrib.* examples), *-strip, -thought*; (in sense *2 i) *picture-house, -palace, -play, -playhouse, theatre*. **b.** *picture-cleaning* (examples), *-going, -making* (later examples); *picture-cleaner* (earlier and later examples), *-dealer* (later examples), *-framer, -goer, -maker* (later and *fig.* examples), *-taker*. **c.** *picture-thinking* (earlier and later examples).

1763 H. Walpole *Let.* 1 July in *Corr.* (1941) X. 84 Gilders, carvers, upholsterers and picture-cleaners are labouring at their several forges. **1902** *Chambers's Jrnl.* July 433/2, I entrusted the panel to the most expert picture-cleaner of my acquaintance, from whose hands it came out perfect. **1978** *P.O. Telephone Directory: Yellow Pages Classified: London (North)* Apr. 294/3 Picture cleaners and restorers. **1871** D. G. Rossetti *Let.* 4 Aug. (1967) III. 966, I (like most artists) am quite ignorant about picture cleaning. **1937** *Burlington Mag.* May p. xxxiv/1 The picture-cleaning controversy at the National Gallery. **1968** *Medium Ævum* XXXVII. 56 This picture-cycle must have been in existence by 1215. **1971** *Amer. N. & Q.* Sept. 14/2 The whole vast field of early Biblical picture-cycles. **1847** A. Brontë *Agnes Grey* i. 14 What do you say to doing a few more pictures.. getting them framed..and trying to dispose of them to some liberal picture-dealer? **1950** E. H. Gombrich *Story of Art* xx. 311 He had..to rely on middlemen, picture dealers. **1933** *New Yorker* 1 Apr. 37/3, I..hurried around to the nearest picture-framer. **1972** *Times* 30 Sept. 11/1 Pass the picture framer's on the right. **1905** in C. W. Cunnington *Eng. Women's Clothing* (1952) ii. 67 The rage for the picture frock for evening, of the Greuze or Romney style. **1930** *Daily Express* 8 Sept. 3/6 There must be curls..to harmonise with the long Victorian dresses and picture frocks. **1975** G. Howell *In Vogue* 92 (*caption*) This picture frock of pale blue taffeta. **1922** *Moving Picture Stories* 30 June 23/3 The millions who are regular picture-goers. **1947** *Landfall* I. 294 The contemporary cinema, as it is seen by the New Zealand picture-goer. **1976** *Cumberland & Westmorland Herald* 4 Dec. 3/4 Picturegoers may remember the woman who used to turn and smile from the screen as the hallmark of Gainsborough productions. **1976** 'G. Black' *Moon for Killers* i. 19, I always watch the movie. I really catch up on my picture-going in airplanes. **1922** *Liberty Dresses* (Catal.) Spring 25 Picture gown, adapted from a 17th-Century design. **1949** A. Christie *Crooked House* xiv. 107 Magda, the Duchess of Three Gables, in a picture gown of taffetas. **1975** G. Howell *In Vogue* 147 (*caption*) Picture gown of black tulle. **1908** *Variety* 23 May 12 Hundreds seated in a picture house who..watch with cooling brows the reels run off. **1913** *Punch* 31 Dec. 543/3 Scene outside an Islington Picture-house. **1939** [see *fried *ppl. a.* b]. **1975** *Country Life* 6 Feb. 326/1 In 1928, the publisher..laid out £4,000 promoting a song..sending lantern slides..to every picture house in the country. **1958** *Times* 31 July 4/6 B.B.C. buy picture library. The B.B.C. have bought the Hulton picture library from Hulton Press Ltd., it was announced yesterday. The library, which was originally called the Picture Post library.. will be known as Radio Times Hulton picture library. **1975** N. Luard *Travelling Horseman* vi. 167, I got back ..from the picture library about 5.30. **1853** Dickens *Bleak Ho.* xxxiii. 329 There comes the artist of a picture newspaper. **1979** G. Macdonald *Camera* x. 148/2 The development of picture magazines and newspapers. **1894** A. Macdonell *Thomas Hardy* x. 232 Story-teller, picture-maker, humourist, it is entertainment he offers us. **1926** V. Woolf *Captain's Death Bed* (1950) 167 The picture-makers seem dissatisfied with such obvious sources of interest. *Ibid.* 169 The picture-making power. **1961** J. McCabe *Mr. Laurel & Mr. Hardy* (1962) viii. 159 The trouble with modern picture-making is the lack of time for preparation. **1908** *Stage Year Bk.* 48 There are now indications that before long these picture 'palaces' will be a feature of London and the larger provincial towns. **1937** D. L. Sayers *Busman's Honeymoon* iv. 85 The neighbourhood boasted a picture palace. **1973** *Listener* 26 July 124/3 A secure world of..visits to the pie-and-mash, and Tuesdays at the Carlton picture-palace. **1867** *Harper's Mag.* 2 Feb. 80/1 I'm sure, Dear, the Picture Papers can not make Frights of us now! **1959** T. S. Eliot *Elder Statesman* ii. 43 I'll feel more confidence after a fortnight..Of people not staring Or offering picture papers. **1911** *Chambers's Jrnl.* Sept. 621/1 Many of the leading dramatists now devote their energies seriously to the elaboration of scenarios for picture-plays. **1923** Picture play [see *picture magazine* s.v. sense 6 d below]. **1919** *Honey Pot* I. iv. 1 (Advt.), The New Gallery Kinema, Society's Picture Playhouse. **1966** J. Derrick *Teaching Eng. to Immigrants* vii. 224 There are in fact several wall charts and picture-story picture sets especially printed for language teachers. **1968** *Tamarack Rev.* Spring 18 Somebody passed Jake the issue of *Life* magazine with the lead picture story on Israel. **1949** Koestler *Insight & Outlook* xxiv. 345 Transitions from primitive picture-strip language to more abstract forms of expression. **1959** I. & P. Opie *Lore & Lang. Schoolch.* ix. 172 Few people now are nicknamed 'Giglamps'..., although derivatives persist in the picture strips. **1966** J. Derrick *Teaching Eng. to Immigrants* v. 183 Picture-strip stories, in which a little episode is told in pictures as in comic strips. **1979** 'S. Kemp' *Goodbye, Pussy* ix. 115 Just like a picture-strip, isn't it? **1909** *Westm. Gaz.* 10 July 14/2 Two out of three men..neglect their apparatus and drift out of the ranks of the picture-takers. **1969** *Amat. Photographer* 28 May 53 (Advt.), The sort of picture-taker who winds the film on the wrong way. **1976** *Daily Progress* (Charlottesville, Va.) 27 Apr. (Advt.), CB spoken here.. *Picture taker*, radar unit. **1908** *Variety* 16 May 11 A new picture theatre will be opened at 276 State Street. **1970** *Southerly* XXX. 124 The town which could offer them a pub, picture theatre, souvenir shops, milk bars. **1976** *Times* 13 Mar. 9/1 All the Continental towns they visited had moving picture theatres while Ireland had none. **1874** W. Wallace tr. *Hegel's Logic* 30 Conception or picture-thinking works with materials from the same sensuous source. **1945** R. A. Knox *God & Atom* vii. 90 One will depend more on picture-thinking, on crude, unanalysed notions, than another. *a* **1963** C. S. Lewis *Poems* (1964) 78 Film, broadcast, propaganda, picture-thinking. **1874** W. Wallace tr. *Hegel's Logic* 53 The predicates by which the object is to be determined are supplied from the resources of picture-thought.

d. picture black *Television*, the light level of the darkest element of a television picture, or the picture signal voltage corresponding to this; **picture-book** (earlier and later examples); also *attrib.* or as *adj.*, characteristic or suggestive of a picture-book; excessively or sentimentally pretty; **picture-card** (earlier and later examples); also *fig.*, and used of any card bearing a picture; **picture element** *Television*, any of the minute areas of uniform illumination of which a television image is composed and which are produced successively by the scanning beam; **picture-frame,** also *Theatr.*, used *attrib.* and *absol.* to designate the stage or stage setting regarded as a picture enclosed by a frame; also *transf.*; **picture frequency** *Television*, † (*a*) (the frequency of) the picture signal; (*b*) the number of times per second a complete television image is scanned or transmitted; **picture-frustration** *Psychol.*, used *attrib.* to designate personality tests that aim to assess a subject's prejudices or personality traits through his reactions to pictures that show potentially frustrating incidents; **picture gallery** (further examples); also *fig.*; **picture hat** (later examples); hence **picture-hatted** *a.*; **picture magazine,** an illustrated magazine; **picture monitor** *Television*, a television screen which is used to provide an immediate display of the image being received by a television camera; **picture-plane,** an imaginary plane on which the perspective of a painting or the like meets the eye of the viewer; **picture postcard,** also *fig.*, a scene, etc., reminiscent of or suitable for a picture postcard; freq. *attrib.* or as *adj.*, conventionally attractive or pretty, in the manner of a picture postcard; so **picture-postcardish** *a.*; **picture show,** (*a*) an exhibition of pictures; (*b*) a cinema performance, a film show; **picture signal** *Television*, the component of the video signal which carries the information relating to the brightness of the picture elements; **picture-space,** a painted canvas considered as a whole in relation to parts of the painting; **picture stage,** a picture-frame stage; **picture telegraphy,** the telegraphic transmission of still pictures, now usu. called *facsimile telegraphy*; cf. *phototelegraphy* s.v. Photo- 1 in Dict. and Suppl.; **picture-telephone** = *videophone; **picture tie,** a neck-tie with a representational design; **picture tube** *Television*, a cathode-ray tube of the kind used for forming the picture in a television set; **picture window,** a large window consisting of a single pane of glass; also *transf.* and *fig.*; hence **picture-windowed** *a.*; **picture-wire,** wire of the kind used for hanging pictures.

1938 *Jrnl. Inst. Electr. Engin.* LXXXIII. 770/1 The insertion of the d.c. component into a train of picture signals implies that the signal corresponding to picture black is established at a definite potential. **1961** G. Millerson *Technique Television Production* iii. 51 Picture black may not be constant. The darkness of the blackest tones in the picture may vary from shot to shot, with picture content. **1972** G. J. King *Beginner's Guide Television* (ed. 5) ii. 43 In television parlance, below-picture black level is called blacker than black. **1847** Thackeray *Van. Fair* (1848) xxxvii. 338 Rawdon bought the boy plenty of picture-books, and crammed his nursery with toys. **1922** A. Huxley *Mortal Coils* 201 It's one of those picture-book German towns. **1926** T. E. Lawrence *Seven Pillars* (1935) xxxiv. 199 A tiny, picture-book train rolled slowly into view across the hollow sounding bridge. **1965** *Eng. Stud.* XLVI. 370 A..human figure which looks half like..a picture-book goddess. **1974** *Sat. Rev. World* (U.S.) 2 Nov. 9/3 Kyrenia..gave promise..of a picture-book port. **1976** H. Kemelman *Wednesday the Rabbi got Wet* xlix. 272 A lovely sunny day with a blue sky and picture-book clouds. **1979** *Jrnl. R. Soc. Arts* CXXVII. 662/2 The book succeeds in being an attractive picture book. **1837** Dickens *Pickw.* xliv. 479 Keep quiet, do,.. there never vos such a old picter-card born. **1908** R. Fry *Let.* 11 July (1972) I. 304 I've sent picture cards to the dear ones. **1932** H. Crane *Let.* 13 Aug. (1965) 409, I must write him a picture card at least in answer; but I'm damned if I want to continue any correspondence of that kind. **1959** I. & P. Opie *Lore & Lang. Schoolch.* ix. 166 Bubble gum,..with its tempting picture card in each packet, is known as 'beetle fat'. **1966** J. Derrick *Teaching Eng. to Immigrants* 239 The work cards, picture-cards, wall pictures and flashcards which accompany the course, are also recommended. **1927** *Wireless World* 1 June 684/2 The mosaic of dots, or picture elements, would be a jumble..if there was an error of 1-90,000th part of a second in the synchronisation between the sending and receiving apparatus. **1940** D.G. Fink *Princ. Television Engin.* ii. 25 The transmission of television images..is accomplished by analyzing the scene into its picture elements, which are selected from the picture area in the orderly sequence of scanning and transmitted one after the other. **1976** *Science* 27 Aug. 792 (*caption*) At this resolution an object 6 m in size subtends one picture element at the nominal horizon. **1928** J. Dolman *Art of Play Production* iii. 43 There are many who are ruthlessly condemning the modern theatre for its illusion, its 'peep-hole' realism, its 'picture-frame' stage. **1936** P. Rotha *Documentary Film* ii. i. 76 Guidance of the actors in speech and gesture, composition of the separate scenes within the picture-frame. **1959** *News Chron.* 29 Dec. 4/2 By the late 'sixties the flat 30 in. picture-frame sets hanging discreetly on our walls, masked during non-viewing periods by..a Constable or a Breughel, will be giving us a reasonably good colour picture. **1966** J. R. Taylor *Penguin Dict. Theatre* 254 By the middle of the nineteenth century the picture-frame stage, in which the whole stage was framed by the proscenium with virtually no forestage at all, had become general. **1978** *Listener* 8 June 740/3 The cast..will be holding their breaths to squeeze on to the little picture-frame stage. **1926** *Wireless World* 5 May 645/1 The same wavelength carries both the picture frequency, from which the received photograph is built up, and the synchronism frequency which controls the motors. **1930** *Electronics* Aug. 236/2 The Bell 50-line system, 17·7 repetition rate and theoretical maximum picture frequency of 22,125 cycles. **1935** *Discovery* Sept. 277/2 They [*sc.* T.V. pictures] are bright, steady (having regard to picture frequency). **1953** Amos & Birkinshaw *Television Engin.* I. i. 15 The number of complete pictures transmitted per second is known as the picture frequency and must be high enough to give the impression of natural movement and to minimise flicker. **1967** H. A. Cole *Basic Television* i. 32 The number of complete pictures presented in every second is called the picture frequency. **1945** *Jrnl. Abnormal Psychol.* XL. 200/1 The Picture-Frustration Study is a limited projective technique for assessing an individual's characteristic modes of reaction. **1949** S. Rosenzweig *Psychodiagnosis* vii. 167 The Picture-Frustration (P-F) Study, or—by its full name—the Picture-Association Study for Assessing Reactions to Frustration, is a limited projective procedure for disclosing patterns of response to everyday stress. **1968** *Jrnl. Abnormal Psychol.* LXXIII. 381 (*heading*) Coronary artery disease and response to the Rosenzweig picture-frustration study. **1842** W. P. Hawes

Sporting Scenes I. 178 [Wild geese are] willing to wait for the wooden devices which we have anchored in the shallow feeding-grounds, as a picture-gallery of their uncles, cousins, and sweet-hearts. **1850** C. Fox *Jrnl.* 1 Apr. (1972) 196 This evening Clara Balfour's picture-gallery included Christina of Sweden, Anne of England [etc.]. **1931** *Amer. Mercury* Nov. 353/2 *Picture gallery*, the tattooed man. **1941** F. Thompson *Over to Candleford* x. 152 One of these new fancy bazaars..was to be held in the picture gallery. **1979** R. Cox *Auction* i 23 The museum's picture gallery. **1928** Mrs. Belloc Lowndes *Diary* 20 Feb. (1971) 112 A black lace frock with a large picture hat trimmed with roses. **1958** L. Durrell *Balthazar* ix. 178 Inside the cupboard I found an immense, old-fashioned picture-hat of the 1912 variety. **1975** G. Howell *In Vogue* 112/2 Dressed by Norman Hartnell, she wore..a picture hat and a long full dress for receptions. **1977** *Jersey Even. Post* 26 July 10/1 Her matching picture hat was decorated with white tulle and flowers. **1922** W. J. Locke *Tale of Triona* vii. 71 Her exquisitely picture-hatted head. **1959** Picture-hatted [see *goose pimples* s.v. *goose *sb.* 8]. **1923** Ade *Let.* 26 Jan. (1973) 91 Two or three weeks ago I picked up a picture magazine and read the outline of the new picture play called 'Fury'. **1967** G. Steiner *Lang. & Silence* 417 He is as remote, or more so, than is the modern picture-magazine and paper-back from the chained, leather-bound folio of the late medieval scholar. **1979** J. Scott *Clutch of Vipers* vii. 120 She was..reading a picture magazine designed for semi-literate young ladies. **1933** *Proc. IRE* XXI. 1690 The picture monitor may be connected to either the output of the picture amplifier or to a radio receiver, which makes possible the monitoring of the radiated signals. **1961** G. Millerson *Technique Television Production* i. 15 The [production control] room's dominating feature is the row of picture monitors. **1975** A. Bermingham et al. *Small Television Studio* 60 A picture monitor is, to all appearances, a high grade tv set without sound. **1797** *Encycl. Brit.* XIV. 182/1 If BA be drawn in the ground plan, and EV be drawn parallel to it, meeting the picture-plane or perspective-plane in V, [etc.]. **1934** *Burlington Mag.* Aug. 91/2 The projection of the three dimensions of nature on to the two dimensions of the picture-plane. **1959** P. & L. Murray *Dict. Art & Artists* 240 *Picture plane*, the extreme front edge of the imaginary space in the picture. It lies immediately behind the glass of the frame and is the plane at which the world of the spectator and of the picture make contact. **1972** E. Lucie-Smith *Eroticism in Western Art* iii. 63 Her hips seem to swell towards us out of the picture-plane. **1907** R. Brooke *Let.* 23 Aug. (1968) 100, I shall be looking at them [*sc.* antiquities] when you & Arthur Watts are staring at picture-postcard skies & snow. *a* **1930** D. H. Lawrence *Phoenix II* (1968) 285 As a picture of Sicily in the middle of the last century, it is marvellous. But it is a picture done from the inside. There are no picture-postcard effects... There is nothing showy. **1936** A. Huxley *Eyeless in Gaza* xli. 491 Against a picture postcard of sunset the immoderately tall thin palms were the emblems of a resigned hopelessness. **1957** J. Braine *Room at Top* xi. 113 Her picture-postcard face with the dyed red bubble-curls and the Lily Langtry nose and chin. **1959** *Sunday Express* 22 Mar. 9/8 The picture-postcard village of Finchingfield. **1973** V. Canning *Flight of Grey Goose* v. 90 You're like all summer tourists. All you see is a nice picture-postcard sort of place. **1977** *Time* 24 Oct. 55/2 Bibi Andersson is pallid by comparison, a picture-postcard beauty. **1969** *Daily Tel.* 14 Mar. 16/4 Bavaria and the Vorarlberg, through which they travel in springtime, look very picture-postcardish in Robert Paynter's colour photography. **1869** D. G. Rossetti *Let.* 24 Dec. (1965) II. 781 Nor indeed do I ever go to any picture shows whatever now. **1912** *Home Chat* 20 Apr. 148/1 She takes me to theatres and picture shows. **1921** R. Macaulay *Dangerous Ages* ii. 47 I'll go to theatres and picture shows and concerts and meetings. **1940** R. Chandler *Farewell, My Lovely* xxx. 227 Folks took me to the picture show. **1948** E. Pound *Pisan Cantos* (1949) lxxx. 94 And he said to Yeats at a vorticist picture show: 'You also of the brotherhood?' **1976** J. M. Brownjohn tr. *Kirst's Time for Payment* vi. 137 Let's inspect the results of our picture show. **1927** Secor & Kraus *Television* iii. 15 As the incoming picture signals are passed through an electromagnet or solenoid, it is caused to open and close a light shutter made of a small piece of aluminum and a spring. **1953** Amos & Birkinshaw *Television Engin.* I. i. 17 The picture signal voltage at any point in a television system varies between two limits, one representing the maximum brightness it is intended to transmit and known as white level and the other representing zero tonal value or black and known as black level. **1972** G. J. King *Beginner's Guide Television* (ed. 5) ii. 43 The picture signal controls the brightness of the scanning spot. **1925** Picture-space [see *imaginally *adv.*]. **1937** *Burlington Mag.* Apr. 183/1 The balance between the figures and the picture-space as a whole is roughly identical. **1959** H. Read *Conc. Hist. Mod. Painting* v. 156 The arrangement of the elements within the picture-space remains intuitive. **1951** C. B. Purdom *Producing Plays* (ed. 3) viii. 113 There can be no doubt that the picture stage is doomed and that stages in new theatres should not be constructed on that principle. **1961** R. Williams *Long Revolution* ii. vi. 254 The modern picture-stage. **1913** *Wireless World* Sept. 354/2 Just as with picture-telegraphy with conductors, we can distinguish between a black and white method..and a half-tone method. **1934** *Discovery* Dec. 336/2 Partly owing to..the..lack of general realisation of the existence of picture telegraphy services, these have not yet been nearly as freely used as they might be. **1976** Picture telegraphy [see *private *a.* 4 d (i)]. **1968** *Listener* 1 Aug. 155/3 Picture-telephones are not only going to be embarrassing to live with, they are going to drain a lot more magic out of life. **1969** *Sci. Jrnl.* May 36/3 This summer the Swedish L. M. Ericsson concern..is to install and test a picture-telephone connection between Stockholm and Gothenburg. **1969** P. West *Words for Deaf Daughter* vi. 151 Will you zoom off into a future which..brings shopping by picture-telephone? **1957** R. Mason *World of Suzie Wong* ii. iv. 151, I thought how quietly dressed you were for an American. I mean, you weren't wearing a picture-tie, or anything like that. **1963** 'D. Rutherford' *Creeping Flesh* i. 63 That black man with the blue suit and silk picture tie. **1940** D. G. Fink *Princ. Television Engin.* i. 19 The image-reproducing tube ('picture' tube)..is a funnel-shaped glass structure in the narrow end of which is contained an electrode structure capable of producing a narrow beam of electrons. **1966** H. Nielsen *After Midnight* (1967) v. 77 My picture tube blew at ten-thirty and I went straight to bed. **1976** *Guardian* 21 Apr. 19/1 Picture tubes account for 40 per cent of the cost of the components in a set. **1938** *Amer. Home* Nov. 20/2 If you have a large picture window, we suggest sill-length casement drawn curtains. **1942** *Pencil Points* Nov. 65/1 Extensive fenestration makes it possible to bring the outdoors into the house (note the plate glass, stationary picture windows). **1945** Nelson & Wright *Tomorrow's House* vi. 75/2 Storage cupboards have been grouped in such a way that picture windows could be used. **1958** J. Cannan *And be a Villain* iv. 79 It was a fairly recent erection with 'picture' windows. **1960** *Chicago Rev.* Autumn–Winter 117 Everett Knight is all for bringing intellectuals down from their tower. The question is whether he himself has actually come out of it or simply ordered some picture windows installed. **1969** *New Yorker* 12 Apr. 90/3 For twenty-one days, the astronauts may look out of a huge picture window, tinted brown against the Texas sun. **1972** J. Mann *Mrs. Knox's Profession* x. 78 The local style was for picture windows.. affording a good view of the modern furniture drawn up around the television set. **1977** *Time* 25 July 9/2 He staked out the eighth-floor penthouse, with picture windows offering a panoramic view of Lake Geneva and the Alps, as his personal pad. **1969** R. Blythe *Akenfield* 16 The 'young marrieds' go less and less for the converted cottage and more for picture-windowed bungalows. **1978** G. Mitchell *Mingled with Venom* vi. 63 Dame Beatrice.. took him into the dimly-lit lounge, out through the picture-windowed bar. **1895** *Montgomery Ward Catal.* Spring & Summer 352/3 Braided picture wire..2½ yards. **1923** Kipling *Land & Sea Tales* 88 Carpenter was off in pursuit of rabbits, with a pocket full of fine picture-wire. **1973** R. Hayes *Hungarian Game* ii. 22 He took the guy out with a two-foot length of picture wire and sweet *Jesus* it was great.

picture, *v.* **4.** (Further example in sense of *picture *sb.* 4 b.)

1923 *Mind* XXXII. 467 It is not asserted that the picture must have the same logical form as what it pictures, but that all pictures must have *the* logical form.

picturedom (pi·ktʃədəm). [f. Picture *sb.* + -dom.] Pictures or moving pictures collectively.

1902 *Strand Mag.* Apr. 440/1 One who knows him says that 'Zim' is the 'Mark Twain' of picturedom. **1920** *Chambers's Jrnl.* 21 Feb. 188/1 It was at the same ranch [near Los Angeles] that many of the most terrible battles in picturedom were fought and filmed. **1945** *Sun* (Baltimore) 2 Oct. 5/7 (Advt.), True antiques of picturedom.

picturedrome (pi·ktʃədrōᵘm). [f. *picture *sb.* 2 i + *-drome, after *hippodrome *sb.* 3.] A building intended for picture shows; a cinema.

1914 *Durham Advertiser* 19 June 8 Arrangements are being made..for the 'Varsity students' 'rag'..to be shown at the Assembly Rooms Picturedrome. **1918** A. Quiller-Couch *Foe Farrell* 116, I dragged him and Petunia back into the shadow under the side-wall of the Picturedrome. **1927** *Punch* 27 July 97/3 Give us more dance saloons, More epileptic tunes, More syncopating coons, More Picture-dromes. **1933** P. Macdonald *Mystery of Dead Police* i. 2 The colossal Pantheon Picturedrome—all Ionic pillars and organ solo. **1945** *Gen* 5 May 24/2 A cinema owner can extol the modernity of his 'Picturedrome'.

picturegraph (pi·ktʃəɹgraf). [f. Picture *sb.* + -graph.] A symbol representing a picture or image; a pictograph. Also *fig.*

1926 G. Murray in *Pioneer of Reformed Spelling* Apr. 2 You could communicate ideas by making certain signs or marks on some material—from that you get, first picturegraphs, and eventually alphabetical writing. *a* **1930** D. H. Lawrence *Apocalypse* (1931) ix. 127 Every image fulfils its own little circle of action and meaning, then is superseded by another image... Every image is a picturegraph, and the connection between the images will be made..differently by every reader. **1962** *Punch* 11 Apr. 570/2 There was..Diplodocus Carnegii,..alive, in his own era—a rare picturegraph in those days.

Picturephone (pi·ktʃəɹfōᵘn). Also with lower-case initial. [f. Picture *sb.* + Phone *sb.*[2]] A videophone. Cf. *picture-telephone* s.v. *picture *sb.* 6 d.

A proprietary name in the U.S.

1964 *N.Y. Times* 21 Apr. 31/2 The Picturephone was demonstrated yesterday with a see-as-you-talk call between the World's Fair and Anaheim, Calif. **1965** *Punch* 23 June 912 The picturephone is not yet domesticated. There is an experimental commercial hook-up between New York, Chicago and Washington. **1966** *Official Gaz.* (U.S. Patent Office) 27 Sept. TM192/2 American Telephone and Telegraph Co., New York. Filed April 13, 1965. Picturephone. For: See-while-you-talk, sometimes known as visual telephone, services. First use at least as early as April 20, 1964. **1970** *Nature* 25 July 331/1 The first commercial picturephone service was opened recently by the Bell Telephone Service. **1970** *New Scientist* 24 Dec. 565 As befits an executive in a communications company,.. Thomas has a Picturephone on his desk, and takes part in conferences by TV. **1974** *Times Lit. Suppl.* 1 Nov. 1216/1 Set in the near future (though the only concession to documentary detail is the picturephone over which leading political figures are apt to receive 'blank-screen calls').

picturesque, *a.* Add: **1. b.** (Earlier examples.)

a **1763** [see *landscape-gardening* s.v. *landscape *sb.* 5]. **1783** W. Burgh in W. Mason *Eng. Garden* 236 There is nothing in picturesque Gardening which should not have its archetype in unadorned Nature.

4. (Earlier examples.)

1789 W. Gilpin *Observations Rel. Picturesque Beauty* I. viii. 72 Let the picturesque traveller watch for these effects. *Ibid.* xiv. 139 The picturesque eye regrets the loss of it's towers.

5. (Earlier and later examples.)

1749 D. Hartley *Observations on Man* I. 427 The Nature of the Caricatura, Burlesque, Grotesque, Picturesque, &c. may be understood from what is delivered.. concerning Laughter, Wit, Humour, the Marvellous, Absurd, &c. to which they correspond. **1782** W. Gilpin *Observations on River Wye* 93 Col. Mitford..is well-versed in the theory of the picturesque. **1927** C. Hussey *Picturesque* iii. 66 The Picturesque was to be a practical aesthetic for gardeners, tourists, and sketchers. **1955** N. Nicholson *Lakers* iii. 46 In the Picturesque, the only creative act is that of man himself, a small, mean, self-satisfied manipulation of an abstract landscape. **1961** *Amer. Q.* Winter 529 The outlook for the amateur..is usually dependent on his fondness for local history or for the picturesque.

2. A picturesque landscape. *rare.*

1889 G. Meredith *Let.* 20 May (1970) II. 959 We had here a young and promising Bostonian..fresh from a ride over the picturesques of Greece.

picturesqueness. Add: (Further examples.) Also *transf.* and *fig.*

1867 [see *featureliness]. **1907** 'Mark Twain' *Christian Sci.* ii. iii. 130 Repetition of pet poetic picturesquenesses.

picturesquerie (piktʃəre·skəri). [f. Picturesque *a.*, after Grotesquerie.] The picturesque; picturesqueness.

1962 *Punch* 12 Sept. 392/3 Miscegenation and picturesquerie on beaches. **1963** *Punch* 20 Mar. 396/2 An endless stream of..picturesquerie, from Maundy Money to the Royal Enclosure. **1968** *Listener* 12 Sept. 322/1 Soon, we were back with French TV's old faults of inconsequential folklore and half-conscious chauvinism. Shock and protest gave place to a routine rag-bag of patriotic picturesquerie. **1978** 'M. Delving' *No Sign of Life* i. 5, I hadn't much time for prettiness and picturesquerie.

picture-writing. Add: **3.** Language or writing that is graphic or vivid. *rare.*

1942 Blunden *Romantic Poetry & Fine Arts* 19 Such picture-writing as Tennyson's 'Palace of Art'; of which I remember my old headmaster saying..'very much like paintings'.

picturing, *vbl. sb.* **b.** (Further examples esp. in sense of *picture *sb.* 4 b.)

1956 J. O. Urmson *Philos. Analysis* v. 76 By no means all the atomists used the terminology of 'picturing'. **1963** W. Sellars *Sci., Perception & Reality* vi. 211 But what if, instead of construing 'picturing' as a relationship between *facts*, we construe it as a relationship between linguistic and nonlinguistic *objects*? **1970** R. Rhees *Discussions of Wittgenstein* i. 5 Well, how do you know? How do you know there is any such picturing? **1971** F. W. Garforth *Scope of Philos.* xii. 237 Nor is 'picturing' an adequate account of what language is and does.

picturization (pi:ktʃəraizēⁱ·ʃən). [f. next.] Representation by means of a picture or pictures (in the various senses of the sb.).

1917 *N.Y. Times* 12 Mar. 9/2 A picturization of 'A Tale of Two Cities' was the new offering at the Academy of Music. **1918** *Dialect Notes* V. 9 'The Little American' is a stirring picturization of events in France during the great war. **1920** [see *cinematograph *sb.*]. **1925** *Lit. Digest* 4 July 30/2 The public does not demand picturizations which truly mirror American life. **1938** I. Kuhn *Assigned to Adventure* vi. 71 He made the stories inspirational and romantic by vivid picturization of human beings at a dramatic time in their lives. **1951** G. Humphrey *Thinking* ix. 286 The hindrance to thought that comes from..picturization or fitting the images into a consistent picture. **1957** V. J. Kehoe *Technique Film & Television Make-Up* viii. 97 There are a number of differences between film and television picturizations.

picturize (pi·ktʃəraiz), *v.* [f. Picture *sb.* + -ize.] *trans.* To represent by or adorn with a picture or pictures (in the various senses of the sb.). Hence **pi·cturized** *ppl. a.*; **pi·cturizing** *vbl. sb.*

1846 in J. E. Worcester *Universal & Crit. Dict. Eng. Lang.* **1918** *Dialect Notes* V. 13 The Common Law, by R. W. Chambers, is soon to be picturized by one of the leading film companies. **1918** *N.Y. Times* 11 Nov. 13/3 William A. Brady as producer, Harley Knowles as director, and all others concerned have done a good piece of work in the so-called 'picturizing' of 'Little Women', which was shown yesterday at the Strand Theatre. **1920** *Ibid.* 24 May 20 The photoplay, however, remains simply a picturized reproduction of the stage play. **1923** *Westm. Gaz.* 26 July 8/6 It is..not an attempt..to picturise the whole of scientific theory on the subject. **1930** *Publishers' Weekly* 1 Mar. 1127 (Advt.), The new Encyclopædia Britannica humanized and picturized. **1952** N. V. Peale *Power of Positive Thinking* iv. 49 When either failure or success is picturized it strongly tends to actualize in terms equivalent to the mental image pictured. **1967** Singha & Massey *Indian Dances* iii. 46 The really expert dancer appears to picturize the music with her arms and body.

picuda (piku·dă). [Amer. Sp., f. Sp. *picudo*, *-a* pointed, sharp.] The great barracuda, *Sphyræna barracuda*, of the family Sphyrænidæ, a large marine fish found in oceans bordering eastern central America, from Florida to Brazil.

1931 B. Miall tr. *Guenther's Naturalist in Brazil* ii. 48 There is also, on the coast of north-eastern Brazil, a fish of the Mullet family, the Picuda, which is sometimes as much as six feet in length. **1963** H. Ulrich *America's Best Deep-Sea Fishing* ii. 102 Barracuda... Other Names: Great barracuda, common barracuda, picuda, sea tiger, cuda. **1975** Felton & Fowler *Best, Worst & most Unusual* 234 The men work two or three short days a week fishing for picuda.

piddle, *v.* **1. a.** Delete 'now *rare*' and add later examples. Freq. with *about, around.*

1938 M. K. Rawlings *Yearling* iii. 25 Don't you and your Pa be gone too long now, follerin' that fool hound. I'm o' no mind to set around waitin' breakfast and you two piddlin' around in the woods. **1957** R. A. Heinlein *Door into Summer* (1960) ii. 32, I piddled along with the help of the shop mechanics until I had *Frank* looking less like a three-car crash. **1961** I. Jefferies *It wasn't Me!* iv. 56 We piddled about with ballistics too, but mostly talk. **1973** *N.Y. Times* 10 June vii. 22/3 Since the whole family piddled with archery, we had our gear with us. **1977** *Sounds* 9 July 22/3 He returned to New York and 'piddled around' doing Public Relations.

d. (Later examples.)

1942 Berrey & Van den Bark *Amer. Thes. Slang* §239/3 *Waste*,..piddle away. **1958** 'E. McBain' *Killer's Payoff* (1960) xvii. 160 Ruther had inherited money which..he'd piddled away. **1965** [see *interface 2 a]. **1978** D. A. Stanwood *Memory of Eva Ryker* ii. 15, I was patrolling Kapiolani Boulevard, piddling away the final moments of my ungodly long shift.

2. (Earlier and later *fig.* and later *lit.* examples.)

1814 W. Scott *Let.* 10 Nov. (1932) III. 515 The last act is ill contrived. He piddles (so to speak) through a cullender, and divides the whole horrors of the catastrophe ..into a kind of drippity-droppity of four or five scenes. **1931** R. Campbell *Georgiad* ii. 21 Across the lawn they [*sc.* dogs] shank it on all fours, To argue, fight, and copulate, and piddle Around the sacred lamp-post in the middle. **1947** M. Lowry *Let.* May (1967) 142 Meantime we have been here quietly piddling in our pants with suspense. **1951** H. Bennett *We never called him Henry* iii. 18 Mr. Ford..would tell me, 'So-and-so was piddling in my ear'. That was a favourite expression of his. **1974** R. Adams *Shardik* xxv. 205, I have no idea what portents he employs—possibly the bear piddles on the floor and he observes portents in the steaming what-not. **1976** *Listener* 15 July 55/3 To greet the crowd meeting him, he openly piddles on the platform.

piddle (pi·d'l), *sb. colloq.* [f. the vb.] **1.** Urine; an act of urinating. Also *fig.*

1901 in Farmer & Henley *Slang* V. 191/1. **1937** Partridge *Dict. Slang* 625/2 *Piddle*, urine; occ. the act of making water. **1959** E. Burgess *Divided we Fall* x. 115 Take the poodle for its piddle. **1962** M. Duffy *That's how it Was* xii. 105, I envied him his ability to tie his little soft winkle into a knot at the end and blow it out like a balloon with unshed piddle. **1972** D. Bloodworth *Any Number can Play* x. 77 How could these red-haired brutes eat that filthy muck, not to mention the piddle and milk they called tea?

2. A trifle; nonsense.

1910 R. Brooke *Let.* 2 Mar. (1968) 223 It's the alteration of the little words that makes all the difference between Poetry & piddle.

piddler. (Later example.)

1911 R. D. Saunders *Col. Todhunter* ii. 25, I ain't never seen no piddler at meal times that was fit to do a man's work.

piddling, *ppl. a.* Add: **1.** (Later examples.)

1902 W. E. Henley *Views & Reviews: Art* 10 *The Castle of Otranto* is a piddling piece of super-nature. **1936** [see *doless *a.*]. **1948** [see *gouger c]. **1966** *Electronics* 31 Oct. 149 The European market for IC's is still piddling. **1971** *Maclean's Mag.* Sept. 4/1 In plain words, I found out what my job was. No piddling assignment either. What it amounted to was saving the country. **1972** V. Canning *Rainbird Pattern* i. 2 Use other people's..piddling fears about their own status, and the world is at your feet. **1976** 'D. Fletcher' *Don't whistle 'Macbeth'* 47 You'd never drown... Not in that piddling little stream.

2. Making water, urinating.

1968 *Listener* 2 May 579/3 Far and away the best..is *Thinking Girl*. Some may be inclined to poke fun by retitling Miss Meacock's book *Piddling Girl*, after counting the number of times the characters are glimpsed in lavatories.

piddly (pi·dli), *a.* [f. *piddle *sb.* + -y[1].] = Piddling *ppl. a.* 1.

1946 B. Marshall *George Brown's Schooldays* xv. 72 Filthy little funks, skittering away down a piddly little track like that.

pidgin, pigeon. Add: **a.** (Earlier examples.)

1826 B. Hall *Acct. Voy. Corea* (rev. ed.) vi. 287, I come to see about your pigeon. *Ibid.* 288, I afterwards learned that..'pigeon', in the strange jargon spoken at Canton by way of English, means business. **1845** J. R. Peters *Misc. Remarks upon Chinese* vii. 73 Pidgeon, is the common Chinese pronunciation of business.

b. A language as spoken in a simplified or altered form by non-natives, *spec.* as a means of communication between people not sharing a common language. Freq. *attrib.* or in *Comb.* Also *fig.*

1891 [in Dict.]. **1921** H. E. Palmer *Princ. Language-Study* 107 Pidgin or pidgin-speech may be defined as that variety of a language which is used exclusively by foreigners. **1943** R. A. Hall *Melanesian Pidgin Eng.* 8 The phonetics of Melanesian Pidgin are basically those of a slightly sub-standard English. *Ibid.* 9 In the absence of native speakers, Pidgin does not present the same constant features of pronunciation and grammatical usage as do major languages. **1955** P. Strevens *Papers in Lang.* (1965) ix. 120 Esperanto is, in fact, a kind of glorified Pidgin Indo-European. **1962** *Listener* 27 Sept. 467/2 Pidgin exists in India and Africa as well—but the context is very different from that of Pidgin in New Guinea. **1964** *Ibid.* 5 Mar. 388/2 At Layer Marney the old is all. The Renaissance motifs are handled awkwardly. This is trying to speak Italian without command of grammar, syntax, or vocabulary... It is the pidgin-Italian spoken by the sturdy Englishman of the Perpendicular. **1968** W. J. Samarin in J. A. Fishman *Readings Sociol. of Lang.* 667 In the fifteenth century there developed a pidgin Portuguese which may have originated in the first contacts with the Africans, but ultimately spread to the ports of the Far East. *Ibid.* 671 The ease with which Korean Bamboo English developed leads one to imagine the inevitability of pidgins in the world. **1968** *Economist* 14 Dec. 4/1 The Soviet graduates, to whom you refer, are as little satisfied with pidgin marxism as young opinion in the British 1930s was with Mr Neville Chamberlain. **1974** *Florida FL Reporter* XIII. 17/1 While agreeing with Nida and Fehderau that pidgin-using and koine-forming situations are strongly analogous, I cannot agree that the 'Sabir Pidgins'..at least are special formations. **1975** *Language* LI. 684 The use of morphemes borrowed by a pidgin or creole language..from a European language often diverges widely from the use of the source morpheme in the source language. **1978** *Verbatim* Feb. 10/1 Both authors hold to..the Creolist theory, which traces the present-day Black English vernacular to a Plantation Creole, to a plantation-maritime pidgin, to an African origin.

c. Phr. (*to be*) *someone's pigeon*: to be (that person's) concern, affair, etc.

1904 Kipling *Traffics & Discov.* 293 'What about their musketry average?' I went on. 'Not my pidgin,' said Bayley. **1919** E. Ruck *Disturbing Charm* ii. x. 248 It was not Jack's pidgin to do anything until then. **1924** G. L. Mallory *Let.* 11 May in E. F. Norton *Fight for Everest: 1924* (1925) ii. 233 Geoffrey Bruce whose 'pigeon' it is to deal with the porters. **1929** C. Mackenzie *Gallipoli Memories* xiii. 237 'Nothing known of this man here.' 'Pass to N.T.O. K beach.' 'From M.L.O. Lancashire Landing to A.P.M. K beach. This is your pigeon, I think.' **1929** J. Masefield *Hawbucks* 164 This is my pidgin; none of yours at all. **1935** *Punch* 30 Jan. 136/1 'There's trouble in Paraguay,' said the man from Geneva. 'Then leave it there,' said I. 'It's not my pigeon.' **1957** 'J. Wyndham' *Midwich Cuckoos* iii. 26 'Not our kind of job,' he said, with the air of one recalling a useful Union decision. 'More like the fire chaps' pigeon, I'd say.' **1959** *Times* 6 Mar. 11/7 If it is the fur (not our pigeon) that makes Mr. Page furious, maybe the answer is mothballs. **1961** L. P. Hartley *Two for River* 45 Well, you do something, Thomas Henry, it's your pigeon. **1977** B. Pym *Quartet in Autumn* xviii. 160 Janice wondered whether anyone else had been to see Marcia... She was Janice's special pigeon, if you could put it like that.

pidginize (pi·dʒinəiz), *v.* [f. *pidgin b + -ize.] To produce a simplified or pidgin form of (a language). So **pidginiza·tion**, the fact or process of pidginizing; a pidginized language; **pi·dginized** *ppl. a.*

1934 Priebsch & Collinson *German Lang.* i. ii. 35 This 'pidginization' [in Afrikaans] is thought to be due to sudden contact with a Creolized language, in this case a blend called Malayo-Portuguese. **1937** J. R. Firth *Tongues of Men* vi. 80 If English were to be pidginized the simplification of pronunciation would have to be based on those differences between sounds which most men could hear..without expert tuition. **1949** [see *creolized *ppl. a.* 2]. **1956** J. Lotz in Saporta & Bastian *Psycholinguistics* (1961) 14/1 Various attempts at an international language like Esperanto or Basic English either end up as incomplete replicas of natural languages or as primitive pidginizations. **1962** *Listener* 27 Sept. 467/1 The more pidginised the language becomes, obviously the less useful English can be as a world language. **1972** J. L. Dillard *Black English* iii. 135 Theoretically, any language can be pidginized, although historically not all have been. **1974** L. Todd *Pidgins & Creoles* i. 5 Pidginization and creolization may have been much more prevalent in language change than linguists have hitherto acknowledged. **1977** *Language* LIII. 495/1 First he discusses borrowings from American Indian languages, particularly the pidginized forms that underwent various semantic shifts and are still current, such as *bury the hatchet* and *Great White Father*.

pi-dog, var. Pye-dog in Dict. and Suppl.

1920 *Blackw. Mag.* Oct. 525/2 Later still at night.. would come droves of pi-dogs sweeping..through the compound. **1927** *Daily Express* 30 May 7/5 They were hounds running with a good cry, not pi-dogs barking. **1959** *Times* 12 June 14/6 A mangy pi-dog shared his humble abode. **1972** 'M. Renault' *Persian Boy* xxii. 287 It was so quiet, you could hear..the pi-dogs bickering. **1978** J. Updike *Coup* (1979) i. 29 His woman whispered to me that he was a rascal without a tribe, who had never had so much as a pet pi-dog to his name.

pie, *sb.*[2] Add: **1. c.** (Further examples.) Also in more general allusive use, an affair, concern, etc., regarded in terms of one's possible involvement in it. Phrs. *to cut a pie* (U.S.), to become involved in a particular matter; *to put a finger into* (*another's*) *pie* (and variants) to meddle in (someone else's) business.

1843 T. C. Haliburton *Attaché* 1st Ser. I. xi. 180 By gosh Aunty,..you had better not cut that pie: you will find it rather sour in the apple sarse, and tough in the paste. **1868** 'Mark Twain' *Let.* 6 Feb. (1920) 85 They want to send me abroad, as a Consul or a Minister. I said I didn't want any of the pie. **1871** J. O. Brookfield *Influence* II. 12, I don't see what excuse she has for putting her finger into everybody's pie as she does. **1879** *London Society* Christmas Number 87/2 If you keep straight yourself you'll have quite enough to do, without putting your fingers into other folks' pies. **1921** L. Strachey *Queen Victoria* iii. 93 Her uncle Leopold had apparently determined..that her cousin Albert ought to be her husband. That was very like her uncle Leopold, who wanted to have a finger in every pie. **1924** A. Christie *Poirot Investigates* i. 19 What I do now is for my own satisfaction—the satisfaction of Hercule Poirot! Decidedly, I must have a finger in this pie. **1940** N. Mitford *Pigeon Pie* iii. 57 Aristocrats are inclined to prefer Nazis while Jews prefer Bolshies. An old bourgeois like yourself ..should keep your fingers out of both their pies. **1970** G. Jackson *Let.* 17 Apr. in *Soledad Brother* (1971) 224 We don't want their culture. We *don't* want a piece of that pie. **1979** *Homes & Gardens* June 90/2 One of the many pies he keeps a finger in is the Color Association of the USA.

d. Assets, proceeds, wealth, etc., considered as something to be apportioned or shared out (cf. *cake *sb.* 7 b).

1967 *Boston Sunday Herald* 7 May iii. 14/4 The result of an Appellate Court victory..which cut Weymouth's total property valuation by $40 million to give the town a bigger slice of the sales tax pie. **1971** *Ink* 12 June 7/4 The thousands of workers..are searching for ways to fight not only for a bigger portion of the capitalist pie but for a freer and more dignified life. **1978** *Washington Post* 29 July D30/2 Although more than 50 million watches will be sold in the United States this year, there is no consensus on what piece of the pie digitals will get.

2. (Earlier and later examples of *bran-pie*.)

1877 *Cassell's Family Mag.* May 377/1 In the last division of the tent we had..a bran-pie... The bran-pie was an oblong washing-tub..filled with bran, in which were hidden..small articles. **1889** *Peel City Guardian* 28 Dec. 7/4 Sometimes what is termed a 'bran pie' is employed..for storing the presents in. **1902** *Little Folks* Jan. 54/2 Every single thing was bought, including the packets in the bran-pie. **1916** *Daily Colonist* (Victoria, B.C.) 4 July 4/4 All sorts of seasonable refreshments will be served and the blue ribbon girls will have an attraction in the form of a bran pie. **1931** V. Woolf *Waves* 236, I think more disinterestedly that I could when I was young and must dig furiously like a child rummaging in a bran-pie to discover myself.

3. b. *Austral.* (See quot. 1960[1].)

1960 *Times Rev. Industry* Jan. 103/1 A pie is a combination of wool buyers who do not bid against one another at wool sales, then divide the wool purchased by members of the group. **1960** *Sydney Morning Herald* 25 Mar. 7 He said he had known since 1945 that 'pies' were operating. **1966** Baker *Austral. Lang.* (ed. 2) iii. 58 The Australian public became aware in 1958 that wool-buying was not always a straightforward operation... Some buyers were combining into *pies* (also called *rings*) to bid and then share purchases, so that competition was reduced.

4. a. (Earlier and later examples.) Also used in comparative phrases: (*as*) *good* (*nice, sweet*, etc.) *as pie*.

1857 'Dow Jr.' *Dow's Patent Sermons* 1st Ser. 21 Let her alone and in five minutes the storm will be over, and she as good as pie again. **1884** 'Mark Twain' *Huck. Finn* ii. 15 You're always as polite as pie to them. *Ibid.* v. 34 So he took him to his own house, and dressed him up clean and nice,..and was just old pie to him, so to speak. **1887** R. T. Cooke *Happy Dodd* xvii. 178 We've been awful good; good as pie, hain't we? **1888** W. F. Cody *Story of Wild West* 531, I wanted to reach Fort Larned before daylight, in order to avoid if possible the Indians, to whom it would have been 'pie' to have caught me there on foot. **1891** *Harper's Mag.* Sept. 579/1 Ain't he as polite as pie to her? **1922** Joyce *Ulysses* 309 See him of a Sunday with his little concubine of a wife, and she wagging her tail up the aisle of the chapel,..nice as pie, doing the little lady. **1933** E. O'Neill *Ah, Wilderness!* iv. iii. 151, I ran into him upstreet this afternoon and he was meek as pie. **1939** J. Steinbeck *Grapes of Wrath* vi. 65 Well, the guy that come aroun' talked nice as pie. **1952** M. Laski *Village* vi. 109 She was as sweet as pie, said she was awfully sorry. **1958** *Listener* 11 Sept. 388/3 When foreign dignitaries come from all corners of the world to pay their respects to me..I am as nice as pie. **1974** P. De Vries *Glory of Hummingbird* (1975) xviii. 265 People were wonderful, nice as pie, glad to see me fallen. **1978** F. Weldon *Praxis* xxiv. 256 She's very stubborn. Sweet as pie just so long as she's doing exactly what she wants.

b. Political favour or patronage. *U.S. slang.*

1910 *Richmond* (Virginia) *Weekly Times-Dispatch* 17 Aug. 10 Representative Slemp was looked upon as the dispenser of the patronage in Virginia because of the promise of the President that he would allow the pie to be handed out by the men who did the fighting. **1916** *N.Y. Times* 12 May 10/4 Take your tribute, but buy national defense with it, don't waste it in 'pork' and 'pie' and Populist lunacies!

c. Something easily accomplished or dealt with, a 'cinch'; *spec.* in phr. *easy* (*simple*, etc.) *as pie. slang* (orig. *U.S.*). Cf. *pie in the sky.

In some uses not clearly distinguishable from sense 4 a (in Dict. and Suppl.).

1889 *Outing* Nov. 151/2, I thought it would just be pie for me to buy him, and take him through the 'bushes'... (The 'bushes', you know, means the country fairs.) **1905** N. Davis *Northerner* 93 It will be just—what was it you said—pie?—pie for them, won't it. 1919 Wodehouse *Coming of Bill* (1920) I. v. 54 This Kid Mitchell was looked on as a coming champ in those days... I guess I looked pie to him. 1923 E. Wallace *Missing Million* xx. 161 Murder was cream pie to Tod. 1925 Wodehouse *Sam the Sudden* xix. 156 'How do you propose to make your entry?'..'Easy as pie. Odd-job man... They always want odd-job men.' 1937 D. L. Sayers *Busman's Honeymoon* xii. 210 He's knocked out... Simple as pie. No cutting or stealing keys or hiding blunt instruments or telling lies. 1948 J. B. Priestley *Linden Tree* I. 39 It's as simple as pie. 1959 C. Bush *Case of Careless Thief* vii. 87 It's in the bag... Everything we wanted and easy as pie. 1967 G. F. Fiennes *I tried to run a Railway* iii. 26 In the current work York was pie compared with Cambridge. 1972 Wodehouse *Pearls, Girls & Monty Bodkin* iv. 53 Interesting Llewellyn in Silver River would be pie, but I'd also have to interest her, and she's not the right woman for that.

5. *pie-fork, -knife, -plate* (earlier and later examples), *shell*; **pie-biter**, (*a*) *U.S.*, one who has a fondness for pies; *fig.* (in sense *4 b), one who takes part in political patronage; (*b*) *Austral. slang*, = **pie-eater* (*b*); **pie-card** *U.S. slang*, (*a*) a meal ticket; one who begs for a meal; (*b*) a union-card; the holder of a union-card; also *attrib.*; **pie-cart** *N.Z.* (see quot. 1922); **pie chart** = **pie diagram*; **pie-counter** *U.S.*, a counter at which pies are sold; *fig.*, a source of grants or favours (see sense *4 b); **pie diagram**, a circle divided by radii into sectors whose areas are proportional to the relative magnitudes or frequencies of a set of items; **pie-eater**, (*a*) someone who eats pies; (*b*) *Austral. slang*, someone of no importance, a 'small-timer'; also, a fool or simpleton; so *pie-eating* ppl. a.; **pie-funnel**, a support for a pie-crust during cooking; **pie-melon**, †(*a*) *U.S.*, a melon used for pies (*obs.*); (*b*) *Austral.*, a variety of water-melon, *Citrullus lanatus*; **pie-plant** (earlier and later examples); also *attrib.*; **pie-wagon** *U.S. slang* (see quot. 1960).

1868 *Daily Territorial Enterprise* (Virginia City, Nevada) 2 Apr. 3/1 Through these same [April fool] doughnuts many a lunch-eater and pie-biter came to grief. 1902 O. Wister *Virginian* ix. 104 He held out to his pony a slice of bread matted with sardines, which the pony expertly accepted. 'You're a plumb pie-biter, you Monte,' he continued. 1908 in W. R. Hunt *North of 53°* (1974) xxvi. 225 Persons interested in other points, and paid knockers and piebiters may pound and hammer until their auditors are deafened. 1911 E. Dyson *Benno* 144 He was that angry with the South pie-biters, he didn't care what 'appened to 'em. 1938 K. Douthitt *Romance & Dim Trails* 106 [He was] better known over the northwest and in camp life as Pie Biter, because he was very fond of pie. 1909 W. G. Davenport *Butte & Montana beneath X-Ray* 56 Say, on the dead level, Andy, couldn't you let me have a lonely ten spot?..Say, Andy, you've just got to jar loose with five bucks for my pie card is so full of holes it looks like a piece of mosquito bar. 1929 *Amer. Speech* IV. 343 *Pie card*, a union card used to obtain food or lodging. 1931 'D. Stiff' *Milk & Honey Route* 211 *Pie card*, one who hangs around and lives on a remittance man or some other person with money. 1945 *Seafarers' Log* 25 May 2/1 The Commie stooges and pie-cards kick us around. 1948 Mencken *Amer. Lang. Suppl.* II. 678 *Pie-card*, a union card used as a credential in begging. 1960 *New Left Rev.* 26 Sept. 41/1 The retired.. comfortably fixed pie-card artists of every lost..cause of the labour and radical movements. 1973 C. Rubin *Log* 64 All of them phony, pie-card officials who sit on their big fat asses and twiddle their thumbs. 1922 C. G. Turner *Happy Wanderer* 37 We drifted down..to the pie-cart (coffee-stall, in your speech). 1949 D. M. Davin *Roads from Home* 70 Somebody having a feed at the pie-cart. 1954 *Numbers* Nov. 23 A poky..kind of joint, just one degree better than a pie-cart. 1963 *Weekly News* (Auckland) 5 June 37/2 The Mussons..were many years in the piecart business in Christchurch. 1973 A. Holden *Girl on Beach* 160 We went down town to the all-night pie-cart and bought masses of pies and chips..and drove up to the top of Mount Victoria. 1922 A. C. Haskell *Graphic Charts in Business* xiv. 75 Because the circle, divided..into sections, resembles a pie which has been cut in the usual manner, the Circular Percentage Chart is sometimes referred to as the 'pie chart'. 1972 Pie chart [see *line graph* s.v. *Line *sb.*[2] 32]. 1977 *Lancet* 17 Dec. 1264/2 There is neither graph nor histogram nor pie chart to help the reader. 1903 *N.Y. Times* 16 Dec. 3 When his constituents asked him why he could not secure more routes [for free postal delivery] the only reply he could make was that he could not get up to the pie counter. 1912 M. Nicholson *Hoosier Chron.* 470 I'm in the ranks of the patriots and not looking for the pie counter. 1921 H. Secrist *Statistics in Business* v. 65 Pie-diagram showing the distribution of the total stock of hams, bacon and shoulders, reported August 31, 1917. 1954 *Brit. Jrnl. Psychol.* XLV. 149 The book is..well produced with numerous pie-diagrams, bar diagrams and other pictographic devices. 1979 *Jrnl. R. Soc. Arts* July 493/2 Whether they be dancers, writers, poets, actresses, painters or musicians they are largely unproductive of material which can be categorized, computed, stored, programmed, evaluated or turned into pie diagrams. 1904 Pie-eater [in Dict.]. 1953 Baker *Australia Speaks* 134 *Pie-eater*.., a small-time crook. 1953 K. Tennant *Joyful Condemned* xviii. 166 He's one of those big he-men that go sneaking around the park waiting to snitch some chromo's handbag. Just a pie-eater. 1966 *Sunday Tel.* (Sydney) 22 May 25/2 The Australian appetite for pies has even added to our slang with the term 'pie eater'. Once meaning a small time crook, it is now used as a derogatory expression, close in meaning to 'a dill'. 1975 *Sun* (Sydney) 10 Jan. 52/7 A bunch of pie-eaters. Excuse me if I find an expression from my old mate, the late Siddie Barnes, but that's what the English team has turned out to be. 1949 L. Glassop *Lucky Palmer* 96 The trouble is, Mr. Hughes, you're too good for the pie-eating bookmakers round these parts. You bet too well for them, Mr. Hughes. 1887 C. B. George *40 Yrs. on Rail* ix. 187 An exquisite set of pie forks, of English make, and valued at seventy-five dollars. 1910–11 *Junior Army & Navy Stores Catal.* 1304 Pie funnel. 1926 S. E. Nash *Cooking Craft* xx. 186 Pile [meat] up in the centre to support the pastry, if there is not sufficient meat use an egg cup or pie funnel. 1960 *Woman* 23 Apr. 3/3 Use an empty salt cellar... It works as well as a proper pie funnel. 1875 Mrs. Stowe *We & our Neighbors* liii. 474 Of course the reader knows that there were the usual amount of berry-spoons, and pie-knives, and crumb-scrapers. 1860 *Trans. Mich. Agric. Soc.* 1859 X. 623 Best pie melon, H. J. Young..$0.50. 1882 W. D. Hay *Brighter Britain!* I. vii. 223 We grow large quantities of melons—..rock-melons, Spanish melons, pie-melons and so on. 1945 [see *Choko]. 1965 *Austral. Encycl.* III. 140/1 Some introduced cucurbits have become troublesome weeds, notably the..pie melon (*Citrullus vulgaris*); colocynth.. and squirting cucumber. 1978 J. A. Michener *Chesapeake* 833 A pie-melon was a kind of gourd raised along the edges of cornfields, and..it produced one of the world's great pies. 1838 *Youth's Mag.* (N.Y.) July 91 Half a dozen roots of the pie-plant (rhubarb) will furnish abundant materials for pies and tarts. 1884 E. W. Nye *Baled Hay* 207 Afterward pulverize and spread over the pie plant bed. 1894 *Harper's Mag.* May 931/2 There is one old soul who especially loves rhubarb pies,..and it is she who remembers me and my row of pie-plant. 1952 J. Steinbeck *East of Eden* xvii. 159 He..chatted about the pieplant roots. 1976 *Hortus Third* (L. H. Bailey Hortorium) 967/2 Rhubarb is a strong, hardy, Old World perennial grown for the thick leaf stalks, which are cooked fresh in early spring for their agreeable acid flavor. It is also known as pie plant. 1598 Florio *Worlde of Wordes* 394/2 *Statccia* [sic], a cake, a tarte. Also any flat thing or pye plate. 1653 R. Verney *Let.* May in M. M. Verney *Mem.* (1894) III. iv. 113, I presume there are dishes, pyplates, candlesticks. 1895 Montgomery Ward & Co. *Catal.* Spring & Summer 431/3, Agate iron pie plates. 1948 *Good Housek. Cookery Bk.* 410 Place the round of pastry neatly and evenly on a flat pie plate. 1978 E. Tidyman *Table Stakes* II. iii. 187 A broad Irish face as open as a pie plate. 1935 *Motion Picture* Nov. 79/2 Pour into baked pie shell. 1976 Turgeon & Birmingham *All Amer. Cookbk.* (1977) 206/2 Put the pie shell in the refrigerator while rolling out the top crust. 1898 Pie-wagon [see *moving-van* s.v. *Moving *vbl. sb.* 5]. 1904 'No. 1500' *Life in Sing Sing* 257/1 *Pie wagon*, patrol wagon. 1960 Wentworth & Flexner *Dict. Amer. Slang* 388/2 *Pie wagon*, a police truck used to transport arrested persons to jail; a Black Maria.

pie, *a.*[1] Delete *Obs. rare*$^{-1}$ and add later examples.

In occasional use as a variant of Pi *a.* in Dict. and Suppl.

1932 C. S. Lewis in *Essays & Stud.* XVII. 71 She is an admirable person. The only trouble is that she is rather pie. 1957 M. A. Jeeves *St. Thomas Becket* i. 13 Some of the more lugubrious sects' offsprings from the Reformation are also responsible for the eulogizing of 'pie' types of people.

pie (pəi), *a.*[2] *N.Z. slang.* [ad. Maori *pai* good.] (See quots.) Cf. *Half-pie *a.*

1941 Baker *N.Z. Slang* vi. 56 *To be pie at* (or *on*), to be expert or efficient at something, is another phrase of wide use in this country. It has been derived from the Maori *pai*, good. 1943 *Amer. Speech* XVIII. 93 'To be pie on' (= to be very good at) seems to be derived from the Maori *pai ana*.

piece, *sb.* Add: **I. 1. a.** *spec.* One of the irregular sections of a jig-saw puzzle. Freq. *fig.*

1911 *Encycl. Brit.* XXII. 675/1 The commonest of all puzzles are coloured maps, pictures ('jig-saw') or designs, dissected into numerous variously shaped pieces, to be fitted together to form the complete design. 1925 E. Wallace *Fellowship of Frog* xix. 144 She's only another little bit of the jigsaw puzzle that will fall into place when we fix the piece that's shaped like a Frog. 1935, etc. [see *Jig-saw *sb.* b]. 1935 W. G. Hardy *Father Abraham* 140 Quite suddenly the pieces seemed to fit together. 1955 H. Kurnitz *Invasion of Privacy* (1956) ix. 67 Mr. Fenn think[s] that there might be something about this case he doesn't understand, a missing piece somewhere. It spoils the picture for him. 1963 A. Heron *Towards Quaker View of Sex* i. 5 The study of homosexuality and its moral problems could not be divorced from a survey of the whole field of sexual activity: a few pieces of the jigsaw-puzzle could not be identified without the whole picture. 1973 G. Scott *Water Horse* (1974) xxii. 152 If Oliver were really involved in this organization, then certain jigsaw pieces could fall more satisfactorily into place. 1977 P. Cosgrave *Cheyney's Law* vii. 65 He had a name... If pieces were not exactly falling into place, there was at least movement.

c. (Further examples.)

1667 Pepys *Diary* 29 Aug. (1974) VIII. 406 The Court is at this day all to pieces, every man of a faction of one sort or another. 1700 Congreve *Way of World* I. 6 She once us'd me with that Insolence, that in Revenge I took her to pieces. 1892 'Mark Twain' *Amer. Claimant* xvii. 173 The hackman will just go all to pieces when he sees that. 1923 H. Crane *Let.* 21 July (1965) 140 The center of such pain as would tear me to pieces to tell you about. 1925 W. S. Maugham *Painted Veil* xiv. 47 When he has bad cards he goes all to pieces. 1933 J. Hilton *Lost Horizon* ix. 196 You were so damned good in that Baskul affair that I can hardly believe you're the same man. You seem to have gone all to pieces. 1957 A. MacNab *Bulls of Iberia* ii. 26 Once a grand herd under the old Duke, now gone all to pieces. 1976 *Daily Record* (Glasgow) 22 Nov. 27/1 How can a team perform so well, score a goal, then go to pieces?

d. *all to pieces*: completely, through and through, from beginning to end. *dial.* and *U.S.*

1839 C. F. Briggs *Adventures H. Franco* I. iv. 27 'Do you know the name of that individual who helped you to steak?' 'I know him all to pieces,' replied the gentleman. 1840 C. F. Hoffman *Greyslaer* I. I. x. 114, I know the ground here all to pieces. 1892 W. G. Lyttle in *Eng. Dial. Dict.* (1903) IV. 491/2 She wud a pleesed ye a' tae pieces, an' wud a been charmed tae a haen a minister fur a son-in-law. 1925 *Dialect Notes* V. 325, I knows un all to pieces.

e. *to fall to pieces* (fig.): to give birth to a child. *dial.* and *Austral. slang.*

1881 S. Evans *Evans's Leicestershire Words* (new ed.) 212 Anybody can say what's the matter wi' yew wi' 'af a oy. Ye'r a-gooin' to fall to paces. 1941 Baker *Dict. Austral. Slang* 28 *Fall to pieces*,..to undergo confinement, to give birth to a child.

f. *to pick up the pieces*: see *Pick *v.*[1] 20 g.

2. d. (Further examples.)

1838 Dickens *Let.* 25 Oct. (1965) I. 445 Kate boasts.. of having told you 'a piece of her mind'. 1914 W. Owen *Let.* 24 May (1967) 256 Now it is Wednes. and I have had with delight your news. I shall hope to have an important piece of your mind on Sunday. 1930 G. B. Shaw *Apple Cart* I. 5 He is coming here today to give the King a piece of his mind..about the crisis. 1946 E. S. Gardner *Case of Borrowed Brunette* (1951) ii. 21 He said I could wear what I had on, no matter where I went. And I certainly gave him a piece of my mind about that. 1956 E. Wilson (*title*) A piece of my mind: reflections at sixty. 1979 D. Sanders *Queen sends for Mrs. Chadwick* 14 I'm going to give a few people a piece of my mind.

3. d. *piece of goods* (further example); *piece of ass, tail*, etc. (*U.S. coarse slang*), a person, esp. a woman, regarded as an object of sexual gratification; hence, sexual intercourse; also *ellipt.* as *piece*.

1922 Joyce *Ulysses* 66 A wild piece of goods. Her slim legs running up the staircase. 1942 Berrey & Van den Bark *Amer. Thes. Slang* § 362/1 Copulation,..piece, piece or hunk of tail, -skirt, -ass or butt. 1950 'D. Divine' *King of Fassarai* xxviii. 252 Them guys from the merchant ships will do anything for a piece of tail. 1953 T. Morrison *Stones of House* v. iii. 243 You couldn't put up a memorial to a boy whose younger brother had just given the family and the college a bad name, not even sneaking off to a house somewhere if he wanted a piece. 1957 G. Greene *Quiet American* (new ed.) I. iii. 28 Get me another drink. And then let's go and find a girl. You've got a piece of tail. I want a piece of tail too. 1968 E. Lovelace *Schoolmaster* xii. 191 Boy, go and getta pieca ass, then go in your bed. 1972 G. V. Higgins *Friends E. Coyle* xix. 119 Him and four buddies want a little dough to get a high class piece of tail. 1972 *Screw* 12 June 21/2 Several revolutionary leaders and a host of government figures all stop by for a piece of ass. 1974 H. L. Foster *Ribbin'* v. 208 He said he fucked your baby sister and poked your baby niece And when he sees your little brother he's going to ask him for a piece. 1978 J. Krantz *Scruples* ii. 21 He..thought she was a flaming, fabulous piece of ass.

f. With *of*. A share in; a financial interest in (a business, project, etc.); freq. in phr. *a piece of the action*. (*U.S. slang.*)

1929 *Theatre Mag.* June 33/3 *Piece*, share. As 'A piece of the show', a financial share in the production. 1930 *Amer. Mercury* Dec. 457/1 He muscles in for a piece of the cleaners' racket. 1940 J. O'Hara *Pal Joey* 66 He owns a piece of the room [sc. nightclub] where I sing in. 1950 *Democrat & Chron.* (Rochester, N.Y.) 13 Jan. 22/1 Offered to let me buy a small piece of 'As You Like It'. 1972 'E. Lathen' *Murder without Icing* ii. 18 'I've had a piece of the Huskies for a long time now.'..'I figure your interest is worth twice what you paid.' *Ibid.* 21 What's it got to do with her if you sell your piece of action? 1973 C. Alverson *Fighting Back* (1978) i. 2 This is a very pretty little bar you've got here... I want a piece of it. I think you could use a partner. 1976 C. Frick in *6,000 Words* 156 They were..managers and agents and producers and all the others that had a piece of the action. 1978 R. Lewis *Uncertain Sound* iii. 78 Manson would be wanting a 'piece of the action'.

g. *a piece of cake*: see *Cake *sb.* 7 c.

6. b. A member of an orchestra or band. In isolation usu. pl. Freq. (sing.) with prefixed numeral forming an attrib. phrase (see sense *24 b); also *absol.*

1912 J. Webster in *Ladies' Home Jrnl.* May 70/2 We had..a band consisting of fourteen pieces (three mouth organs and eleven combs). 1922 S. Lewis *Babbitt* xii. 156 Their favorite motion-picture theater..had an orchestra of fifty pieces. 1925 F. Scott Fitzgerald *Great Gatsby* (1926) iii. 48 The orchestra has arrived, no thin five-piece affair, but a whole pitful of oboes and trombones and saxophones. 1938 *Melody Maker* 27 Aug. 11/4 Billy will use his 14-piece stage band. 1959 P. Capon *Amongst those Missing* 221 It was like shouting against a ninety-piece orchestra. 1965 *New Yorker* 2 Jan. 40/1 The only band we had trouble with..was the Savoy Sultans, the house group at the Savoy Ballroom in Harlem. They only had eight pieces, but they could swing you into bad health. 1966 *Crescendo* Apr. 12/1 A Saturday night last August. A ten-piece required a little way out of town. 1975 *New Yorker* 21 Apr. 7/3 On Monday: Dave Matthews' twelve-piece band, made up of studio musicians.

7. *piece of work*: **c.** Applied, usu. contemptuously, to a person.

1928, etc. [see *Nasty *a.* 7]. 1936 'N. Blake' *Thou Shell of Death* vii. 122 She..had been Fergus's mistress. She is a pretty grand piece of work. 1965 A. Nicol *Truly Married Woman* 109 Jolson..looking at Newi's

heavy square mahogany face..thought what an obstinate and unpleasant piece of work the fellow was.

II. 9. Restrict *arch.* and *dial.* to sense a.

a. (Later example.)

1918 L. Strachey *Eminent Victorians* 63, I hate that man, he is such a forward piece.

b. (Further examples.)

Now mostly of a woman or girl regarded as a sexual object.

1846 *Swell's Night Guide* 86 She is a charming piece. 1854 M. J. Holmes *Tempest & Sunshine* iv. 57 Dr. Lacey laughed heartily at this speech and called her an 'original little piece'. 1873 J. Miller *Life amongst Modocs* xv. 194 Rather a good-looking piece you got here now, ain't she? 1908 Z. Gale *Friendship Village* 251 [When] Zorah had took sick..this little piece here had up an' offered [to dance in the carnival]. 1939 W. Faulkner *Wild Palms* 334 'Woman. It was a fellow's wife.' 'You mean you had been toting one piece up and down the country day and night for over a month and now..you got to get in trouble over another one?' 1946 K. Tennant *Lost Haven* (1947) x. 160 You speak to him, Alec... Gallivanting around with that peroxided piece. 1966 K. Amis *Anti-Death League* 329 Those two pieces in leather who served you your coffee. 1972 F. van W. Mason *Roads to Liberty* 18 Katie wouldn't be a bad-looking piece..if she'd half take care of her appearance. 1978 I. B. Singer *Shosha* xiii. 232 He was allegedly going to divorce his wife, who was a common piece.

11. b. Now chiefly *U.S. slang.* (Further examples.)

1930 *Amer. Speech* V. 392 *Piece*, a firearm of any kind, including a pistol. 1956 'E. McBain' *Cop Hater* (1958) x. 91 In this neighbourhood, you don't carry a knife or a piece, you're dead. 1970 L. Sanders *Anderson Tapes* xxxi. 82 You're a good shot... But you've never carried a piece on a job... If this campaign goes through, you'll have to pack a piece. 1973 *Black World* July 55/2, I slid the piece from under my shirt. 1976 *Times Lit. Suppl.* 6 Aug. 998/2 A high-class, good burglar..who has never carried a piece in his life.

14. a. (Further examples.)

1897 Kipling *Capt. Cour.* viii. 165 We'll fish a piece till the thing lifts. 1937 G. Heyer *They found him Dead* iv. 68, I wouldn't run the risk of bumping off an old man who had a valvular disease of the heart. Guess I'd wait a piece for Nature to do its work.

c. Also in occas. wider use. (Further examples.)

1873 E. B. Tuttle *Boy's Bk. Indians* 134 Major Gordon descended the ridge..and carrying the body of Stambaugh a piece, hid it away in some bushes. 1908 L. M. Montgomery *Anne of Green Gables* ii. 18 We've got to drive a long piece, haven't we? Mrs. Spencer said it was eight miles. 1931 *Amer. Speech* VII. 20 *Piece*, a short walk. 'I will go a piece with you.' 1940 Bryant & Aiken *Psychol. of Eng.* 84 Come and walk a piece with me. 1956 B. Holiday *Lady sings Blues* (1973) x. 98, I thought we were stranded until I saw a car down the road a piece. 1971 *Sunday Express* (Johannesburg) 28 Mar. (Home-finder section) 3/3 (Advt.), Fabulous fishing, Superb surfing. Nearby golf courses, Not forgetting swinging Margate..just up the road a piece. 1978 G. Mitchell *Wraiths & Changelings* i. 11 He's in a bad way..on the floor of my cottage. It's just a piece down the road.

15. b. Also used sporadically elsewhere. (Further examples.)

1834 *Chambers's Edin. Jrnl.* III. 254/2 Receiving each a piece and jelly on 't from granny, because they were guid bairns. 1878 C. Hallock *Sportsman's Gazetteer* (ed. 4) 696 *Piece*, a lunch, a snack (Pennsylvania). 1911 E. M. Clowes *On Wallaby* vi. 164 The children..wander..in and out of the kitchen, with incessant demands for what is known as 'a piece'—a liberal slice of bread, butter, and jam. 1949 'J. Tey' *Brat Farrar* xix. 175 Saturday afternoon was a holiday for the Ashby children and they were accustomed..to take a 'piece' with them and pursue their various interests in the countryside until it was time to come home to their evening meal. 1962 M. Duffy *That's how it Was* x. 85 'Gi's a piece, our mam,' and they would run off with a doorstep of bread and jam. 1973 'J. Patrick' *Glasgow Gang Observed* 234 *Piece*, sandwich. 1978 *Jrnl. Lancs. Dial. Soc.* Jan. 15/2 [Edinburgh] *Piece*, sandwich. E.g. *jeely piece* (one containing cheap jam).

h. pl. *pieces*. The oddments of wool which are detached from the skirtings of a fleece; also, the skirtings themselves. Chiefly *Austral.* and *N.Z.*

1881 A. Bathgate *Waitaruna* 173 The 'pickers up' were..carrying [the fleeces] to the sorting table, where they were stripped of the 'pieces', which were thrown aside. 1891 R. Wallace *Rural Econ. Austral. & N.Z.* xxix. 384 The washing of wool, either before or after shearing, is, with the exception of locks and pieces, which are generally scoured, almost entirely given up. 1951 [see *fleece-wool* s.v. *Fleece *sb.* 6]. 1965 [see *piece picker*, sense *23]. 1971 J. S. Gunn *Distrib. Shearing Terms N.S.W.* 20 The south was the only area where..there was a preference for *pieces* rather than *skirtings*.

i. A quantity of a drug (esp. morphine or heroin) approximately equal to one ounce. *U.S. slang.*

1935 A. J. Pollock *Underworld Speaks* 2/2 *Piece, a*, an ounce of morphine, cocaine or heroin. 1945 W. J. Spillard *Needle in Haystack* viii. 77 'I hava da pieces—pure stuff.' Pieces was an underworld term for ounces. 1963 *U.S. Supreme Court Reports* 371 U.S. 474 Johnny kept about a piece of heroin [Note] A 'piece' is approximately one ounce. 1965 *Reader's Digest* June 228 He buys heroin in 'pieces' (ounces) cuts it, and bags it.

16. *spec.* In the N. Amer. fur trade, a package of goods or furs weighing about ninety pounds. *Hist.*

The exact weight of the pound referred to is not certain.

1774 [see *Pack *sb.*[1] 2]. 1809 A. Henry *Trav. & Adventures Canada* II. ii. 15 The freight of a canoe.. consists in sixty *pieces*, or packages, of merchandize, of the weight of from ninety to a hundred pounds each. 1836 G. Back *Narr. Arctic Land Expedition* i. 32 Every package had been reduced or augmented to a 'piece' of 90 lbs. weight. 1931 G. L. Nute *Voyageur* 38 Each package, or piece, was made up to weigh ninety pounds, and two ears were left at the top by which the voyageur could lift it easily in the manner of a modern flour bag. 1949 *World-Herald* (Omaha) 19 June III. 5/2 In the north, 80 pounds is considered a 'piece'... In the days of the early fur traders that was considered a proper load of skins for a man. 1972 T. McHugh *Time of Buffalo* viii. 89 A single sack, weighing about ninety pounds, was known as a 'piece' of pemmican, and made a convenient parcel for back-packing or portaging.

17. Delete '*Obs.* in general sense' and add later examples.

1904 H. James *Golden Bowl* I. vii. 141 Representative precious objects, great ancient pictures..fine eminent 'pieces' in gold, in silver..had for a number of years.. multiplied themselves round him. 1976 *National Observer* (U.S.) 6 Nov., 'Orders are pouring in like gangbusters,' says Bob Koppang, a Hopkins, Minn., novelty dealer. 'Last week alone we sold nearly 29,000 pieces.' 1978 I. Murdoch *Sea* 31 The big oval mirror in the hall ..is perhaps the best 'piece' in the house. 1979 *Country Life* 12 Apr. 1142/1 Special-occasion pieces..help these self-supporting artists. . A piece of jewellery by a contemporary artist is a wonderfully exciting thing to own.

c. Delete † *Obs.* and add later examples.

1950 D. Gascoyne *Vagrant* 13 A hand-high Rodin piece. 1961 C. P. Fitzgerald in Webster s.v., Images of the Buddha are made to certain conventional patterns and there is often great difficulty in determining the origin of any piece on stylistic grounds. 1979 *Country Life* 12 Apr. 1142/1 Sculptor Peter Lyon['s] pendant.. is very reminiscent of his full-scale work..for example ..a piece in the central gardens of Park Town, Oxford.

d. Also, an article for a newspaper, journal, or other publication.

1936 D. Powell *Turn, Magic Wheel* I. 36 That..fellow who was always after him to write a 'piece' for the weekly he ran. 1961 *Noble Savage* Oct. 10 A period in which he ..had a small piece in *Partisan Review*. 1974 'D. Craig' *Dead Liberty* xxi. 124 Dravier worked on two features for the paper,..a piece about price stability in East Germany..and..a Youth article. 1976 *Gramophone* June 32/3 In the same month came Beethoven's Symphony No. 9 on seventeen sides, to which WRA [sc. the reviewer] devoted a page and a half..; one of WRA's most perceptive pieces.

g. A short discourse; a passage for recitation (*dial.* and *U.S.*). Also colloq. phr. *to say* (or *speak*) *one's piece*: to express one's opinion or judgement on a subject or question; to have one's say.

1845 C. M. Kirkland *Western Clearings* 158 Some of the best speakers mount the platform, and 'speak a piece'. 1865 C. F. Browne *A. Ward: his Travels* II. i. 128, I have spoken my piece about the Ariel. 1879 *Congress. Rec.* 16 May 1380/2, I expect to read tomorrow [in the papers] that I spoke a piece, that is the way they print it sometimes, in favor of slavery. 1882 Nodal & Milner *Gloss. Lancs. Dial.* 212 We're gooin' a-sayin' pieces at schoo'. 1886 R. Holland *Gloss. Words County of Chester* 258 In the country schools when children recite poetry it is always called 'saying their pieces'. 1890 *Harper's Mag.* Dec. 139/2 Don't you want to hear me speak my piece? 1895 'Rosemary' *Under Chilterns* 83 All the 'pieces' that the children learnt to repeat at school they taught to her. 1896 *Leeds Mercury Suppl.* 26 May, Ahr Louisa wor allus a gooid un at sayin pieces. Hes ta le'nt that piece yut? 1902 Mrs. G. M. Martin *Emmy Lou* 115 Emmy Lou had to learn a piece for Friday. It was poetry, but you called it a piece. 1941 U. Orange *Tom Tiddler's Ground* vi. 117 Lady Cameron was simply furious with me after that concert-party episode, when I just flatly refused to say my piece. 1942 [see *Drift *v.* 1 c]. 1949 *North Dakota Hist.* Jan. 23 He didn't like recitations, and would rather hunt rabbits than speak pieces. 1949 G. B. Shaw *Buoyant Billions* IV. 54 Dick, I will say that you are wonderful when you speak your piece, though I never understand a word. 1973 J. Porter *It's Murder with Dover* xi. 115 'You didn't speak to Marsh again?' 'No. I'd said my piece.' 1976 A. Miller *Inside Outside* xi. 181 A structured or unstructured group where all were free to say their piece.

III. 19. Also *on piece*.

1903 [see *Linotype]. 1911 *Rep. Labour & Social Conditions in Germany* (Tariff Reform League) III. 96 Those on 'piece' earn from 49s. to 58s. 9d. per week. All the men are on 'piece', and work 53 hours per week.

20. Also *fig.*, whole, without injury or loss.

1929 D. Hammett *Dain Curse* (1930) xiii. 141, I returned to the ravine..reaching the bottom all in one piece with nothing more serious the matter with me than torn fingers. 1968 W. Garner *Deep, Deep Freeze* xxxi. 256 My orders are just to get him to London in one piece. After that, he's someone else's worry. 1973 'D. Jordan' *Nile Green* vi. 29, I thought I'd come and see you through Customs in one piece. 1976 P. & W. Proctor *Women in Pulpit* v. 88 When I arrived at a church, I was glad to be there in one piece, and the people were just as glad to see me.

21. (Earlier and later examples.) Also *all-of-a-piece* attrib. phr.; hence *all-of-a-pieceness*.

1612 Chapman *Widow's Tears* sig. I, O happy starres. And now pardon, Ladie; me thinks these [kisses] are all of a peece. 1912 Galsworthy *Sheaf* (1916) 21 Odd how all of a piece taste is! 1914 G. B. Shaw *Misalliance* 71 It's all of a piece here. The men effeminate, the women unsexed. 1924 Galsworthy *Forest* II. i. 35 An expedition like this has to be all of a piece, in the leader's hand. 1938 W. S. Maugham *Summing Up* xvii. 58, I think what has chiefly struck me in human beings is their lack of consistency. I have never seen people all of a piece. 1956 A. Wilson *Anglo-Saxon Att.* II. ii. 318 Goodness isn't all of a piece..any more than badness. 1958 *New Statesman* 20 Dec. 881/1 What is remarkable about this remarkable musical is its all-of-a-pieceness. 1960 *Times* 1 Nov. 16/1 He is a convincing all-of-a-piece characterization. 1962 *Times* 6 Apr. 20/1 A corresponding all-of-a-piece protein.

b. (Later examples.)

1916 Galsworthy *Caravan* (1925) 579 Mr. Bosengate looked at this peach with sorrow rather than disgust. The perfection of it was of a piece with all that had gone before this new and sudden feeling. 1957 G. Ryle in C. A. Mace *Brit. Philos. in Mid-Cent.* 258 The assumption that doing philosophy..is of a piece with doing natural science. 1977 G. Butler *Brides of Friedberg* iii. 59 How like the innocent Empress Frederick to lease such a place for me. But it was entirely of a piece with all one knew of her.

IV. 23. piece-bag, box *U.S.*, a bag or box for holding pieces of cloth; **piece-bright** *a.* (*poet. nonce use*), ? bright here and there; **piece-dye** *v. trans.*, to dye (cloth) after it is woven; **piece hand** = *piece-compositor*; **piece picker** = *fleece picker* s.v. *Fleece *sb.* 6; **piece-rate**, also *attrib.*; **pieceways** *U.S. dial.* = Piece *sb.* 14 c.

1869 L. M. Alcott *Little Women* II. i. 11 So rich a supply of dusters, holders, and piece-bags. 1900 E. A. Dix *Deacon Bradbury* 251 Mr. Bradbury..sought his wife, who was upstairs sorting over her piece-bag. 1898 M. Deland *Old Chester Tales* 272 It has been lying there in my piece-box for six years. 1877 G. M. Hopkins *Poems* (1967) 67 This piece-bright paling shuts the spouse Christ home. 1931 E. Midgley *Technical Terms Textile Trade* I. 10 The cloth is woven in a white or undyed condition and piece-dyed black for wool, so that the cotton fibres remain their natural colour. 1962 W. J. Onions *Wool* xi. 253 Interesting coloured effects may also be obtained by piece-dyeing cloth already made up of white and coloured yarn. 1978 *Country Life* 9 Nov. 1494/2 The fabric..is now finished and piece-dyed in Huddersfield. 1890 in E. Howe *London Compositor* (1947) 316 The rapid growth of the 'stab system has rendered the condition of the piece hand in many instances almost unbearable. 1947 E. Howe *Ibid.* 60 The piece-hand could reckon to earn more [than the establishment wage], by the exercise of both energy and skill. 1965 J. S. Gunn *Terminol. Shearing Industry* II. 8 *Piece picker*, sometimes called the 'fleece picker' or 'fleecy', his main job is to take the skirting from the roller and to divide it into lines defined by the classer, for example 'broken wool', 'first pieces'..and 'pieces'. 1954 J. A. C. Brown *Social Psychol. of Industry* vi. 181 Workers..who are paid on an individual piece-rate basis. 1955 *Times* 3 May 10/1 The men allege that the piece-rate schedules are out of date and demand a revision. 1976 *Ilkeston Advertiser* 10 Dec. 2/2 Why is a factory worker who fiddles a 25p piece-rate coupon sacked on the spot, whereas his factory manager who fiddles his expense accounts to the tune of several pounds each week gets away with it? 1932 W. Faulkner *Light in August* i. 9, I was trying to get up the road a pieceways before dark.

24. b. With prefixed numeral, forming attributive phrases, as *one-piece, three-piece*, etc.: see the first element.

piece, *v.* Add: **II. 3*. piece down.** *trans.* and *intr.* To increase the length or width of (a garment) by the insertion of a piece of material.

1870 J. P. Smith *Widow Goldsmith's Daughter* vi. 80 Mrs. Goldsmith's economy would not permit her to cast aside any garment that could be pieced down. 1903 K. D. Wiggin *Rebecca* xvii. 176 The limit of letting down and piecing down was reached.

pièce. Add: **a.** (Later examples.)

1882 A. J. Munby *Diary* 5 Apr. in D. Hudson *Munby* (1972) 405 Some day I may publish some of these—*pieces* [sic] *justificatives*, as the phrase is. 1917 N. Douglas *South Wind* xxxiv. 412 The various *pièces justificatives* were lying in their sealed envelopes. 1954 W. K. Hancock *Country & Calling* viii. 227 The historian, although he may employ analysis and its technical language in his preliminary studies or his *pièces justificatives*, remains just as deeply committed as Herodotus and Thucydides were to narrative and the language of narrative. 1975 *Listener* 7 Aug. 186/1 Laud wanted to publish the Greek and oriental manuscripts which..were *pièces justificatives* of the continuous, independent Church of England.

c. *pièce de conviction* (pyęs də kõnviksyõn): an object produced as evidence in a criminal case, an exhibit; also *fig.*, the conclusive argument which decides a question.

1877 [see *Black Museum* s.v. *Black *a.* 19]. 1894 G. B. Shaw *Let.* 6 Dec. (1965) I. 469 Perhaps it may a little disappoint you, after the fantastic solution of Peer Gynt, and the no-solution of Rosmersholm, that a real solution is only found in something that brings the great Ibsen into line with Monsieur Tout-le-monde; but that, in my view, is the final *pièce de conviction*.

d. *pièce d'occasion* (pyęs dokāzyõn): a play or other literary work, or a musical composition, written for a special occasion.

1883 E. B. Bax *Kant's Prolegomena & Metaphysical Found. Nat. Sci.* p. xxvii, Kant believed himself to have no special bent for the professorate in question, which would have involved the criticism of all *pièces d'occasion*, as well as the composition of such on academic festivals. 1891 G. B. Shaw *Quintessence of Ibsenism* p. vi, I had laid it aside as a *pièce d'occasion* which had served its turn. 1914 —— *Dark Lady of Sonnets* 103 This little *pièce d'occasion*, written for a performance in aid of the funds of the project for establishing a National Theatre as a memorial to Shakespear. 1934 C. Lambert *Music Ho!* II. 63 These mild *pièces d'occasion* no more affected the main course of music than an Olde Worlde Bunne Shoppe affects..architectural experiments. 1955 *Times* 29 Aug. 10/6 The result may not be homogeneous enough in style to ensure the work a place in the regular orchestral repertory, but it is certainly a splendid *pièce d'occasion* for this orchestra, who played it with tremendous verve and finish [at their Prom début]. 1971 *Nature* 25 June 537/1 His Presidential Address of 1951..is..an attempt to condense the history of chemistry in a single *pièce d'occasion*. 1979 *Jrnl. R. Soc. Arts* CXXVII. 181/2 This book is a *pièce d'occasion*—a collection of essays published to coincide with the exhibition of Cézanne's late work, held in New York and Houston in 1977 and in Paris in 1978.

e. *pièce de circonstance* (pyęs də sirkõnstãns): a literary composition, theory, etc., having its genesis in a particular situation.

1926 R. H. Tawney *Relig. & Rise of Capitalism* ii. 89 His sermons and pamphlets..were *pièces de circonstance*, thrown off in the storm of a revolution. 1972 R. Plant in Cox & Dyson *20th-Cent. Mind* II. iv. 68 The fact that the historical circumstances which gave rise to [social] theories no longer obtain does not thereby entail that they have become outmoded. Such theories, far from being mere *pièces de circonstance*, rather embody an element which transcends the particular problems which generated them.

f. *pièce noire* (pyęs nwār): a play or a film with a tragic or macabre theme.

1958 *Times* 17 Apr. 3/4 *Payment Deferred* or *The Postman Always Rings Twice* are good *pièces noirs* [sic] because their characters are motivated by passions so intensely conveyed that the subsequent developments are inescapable. 1959 *Oxf. Compan. French Lit.* 22/2 [Anouilh's] *Pièces roses*..and *Pièces noires*..like Shaw's *Plays Pleasant and Unpleasant*..deal with the lighter or the darker side of life. 1963 *Punch* 20 Nov. 754/3 As black a *pièce noire* as can be.

g. *pièce rose* (pyęs rōᵘz): a play or a film having a theme which is pleasantly entertaining; a comedy.

1959 [see *pièce noire* above]. 1963 *Times* 18 May 5/1 It is what Anouilh calls a *pièce rose* and lightly tosses around such ideas familiar to the playwright as innocence and faith, illusion and reality. 1968 *Punch* 6 Nov. 668/3 Eighteen years (and *Look Back in Anger*) later Anouilh's *pièce rose* seems pathetically pale and withered around the edges.

h. *pièce à thèse* (pyęs a tęz): = *thesis-play* s.v. Thesis 6.

1961 *Times* 17 Oct. 16/5 In most hands this [television play] could turn into a mechanical *pièce à thèse*. 1974 *Independent Broadcasting* Aug. 5/2 How far the *pièce à thèse*, the play with a social or political message, is permissible within the impartiality rule.

piecemeal, *v.* Add: Also const. *out*. (Later example.)

1975 *Washington Post* 6 Sept. A2/4 To announce even the topic of the hearings would result in 'prematurely piecemealing the information out',—he said.

piece-wise, *adv.* (In Dict. s.v. Piece *sb.*). Also **piecewise**. Delete '*rare* or *nonce-wd.*' and add: *spec.* in *Math.*, throughout each of a finite number of separate parts or regions but not necessarily throughout the whole. (Later examples.)

1933 H. B. Phillips *Vector Anal.* v. 101 Vectors which are only piecewise differentiable. 1939 *Mind* XLVIII. 366 They have to proceed piece-wise, measuring every little line and bend. 1953 L. V. Ahlfors *Complex Anal.* ii. 65 We shall say that the arc is differentiable if $z'(t)$ exists and is continuous... An arc is piecewise differentiable..if the same conditions hold except for a finite number of values t. 1962 Alley & Atwood *Electronic Engin.* xvi. 546 This method is known as the piecewise-linear method, since the tube or transistor is assumed to be linear over certain regions of operation. 1966 Scott & Tims *Math. Anal.* vi. 209 It is useful also to consider functions which, while not continuous in a whole interval, are nevertheless made up of a *finite* number of continuous pieces. Such functions are called piecewise continuous. 1979 *Nature* 12 Apr. 623/1 Their major effect is to re-order, piecewise, the fibre map of the eye circumference laid out across the t axis of the ribbon.

piece-work. Add: Also **piecework**. (Earlier and further examples.)

1549 *Coventry Leet Bk.* (1909) III. 792 No persone of the Craft of Cappers shall put owt eny pece-woork, but to suche of the same Craft as the maisters..shall agree & consent vnto. 1770 J. Wedgwood *Let.* 12 May (1965) 91 You will certainly do right in bringing the painting to piece work. 1817 H. L. Piozzi *Let.* 14 Apr. in A. Hayward *Autobiogr. Mrs. Piozzi* (1861) II. 200 A gentleman.. called fifteen of his principal people..and told them he was no longer able to give them piece-work. 1885 G. B. Shaw *Let.* 29 Dec. (1965) I. 149 Paquito has not the remotest idea of what it is to be exploited on the piece-work system by newspapers. 1894 [see *Feller[1] 4]. 1911 *Rep. Labour & Social Conditions in Germany* (Tariff Reform League) III. 31 He also informed us that most of the joiners worked piecework. *Ibid.* 186 There was a lock-out in the building trade, owing to a grievance in the piecework system. 1967 M. Argyle *Psychol. Interpersonal Behaviour* iv. 73 Under competition hostile attitudes will develop; an example of this is the payment of sales-girls by individual piece-work, leading to their fighting over the more desirable customers. 1970 T. Lupton *Managem. & Social Sci.* (ed. 2) iii. 63 Pitting wits against management on piecework rates.

2. = Patchwork 4 b.

1842 C. M. Kirkland *Forest Life* I. 90 No gorgeous piece-work bed-quilts exhibiting stars of all magnitudes and moons in all quarters. 1935 M. M. Atwater *Murder in Midsummer* viii. 71 You'll find a piecework spread in the bottom drawer.

piecrust. Add: **c.** Applied to a table having an ornamental edge suggestive of the crust of a pie, as *pie-crust table, top*.

1902 L. V. Lockwood *Colonial Furnit. Amer.* 232 The handsomest of the Dutch tea-tables were what are popularly known as 'pie-crust tables'. 1914 R. & E. Shackleton *Charm of Antique* iii. 49 The most sought-for and rarest of little tables is the 'pie-crust' tip-table, so named from its apparently finger-dented pie-crust-like margin of regular irregularity. 1923 W. Deeping *Secret Sanct.* xx. 210 She had closed the lid of her work-basket and placed it on the 'pie-crust' table by the window. 1951 *Good Housek. Home Encycl.* 106/1 Tripod tables..with 'pie-crust' tops. 1969 A. Negus *Going for Song* 120 So on original pie-crust tables there are no joins at all round the pie-crust edge. 1973 A. Behrend *Samarai Affair* iv. 53 The glossy telephone on the antique piecrust table.

pied, *ppl. a.*[1] Add: **d. pied crow**, the black and white crow, *Corvus albus*, found in most parts of Africa south of the Sahara; **pied flycatcher**, substitute for def.: a black and white flycatcher, *Muscicapa* (or *Ficedula*) *hypoleuca*, found in Europe and north and west Africa; (examples); **pied kingfisher** (examples); **pied wagtail**, substitute for def.: a western European wagtail, *Motacilla alba yarrelli*; (later examples).

1930 W. L. Sclater *Systema Avium Ethiopicarum* II. 650 Pied Crow... Throughout the Ethiopian Region. 1958 G. Durrell *Encounters with Animals* III. 109 A pied crow from West Africa..suddenly decided that I was the only person in the world for him. 1967 W. Condry *Birds & Wild Afr.* iv. 65 In one palm a pair of pied crows were courting. 1776 T. Pennant *Brit. Zool.* (ed. 4) I. 351 (*heading*) Pied fly-catcher. 1843 W. Yarrell *Hist. Brit. Birds* I. 169 The Pied Flycatcher..is a rare bird in England. 1882 [see Fly-catcher 2]. 1894 C. Dixon *Nests & Eggs Brit. Birds* 157 The Pied Flycatcher arrives in our islands during the latter half of April. 1971 *Country Life* 25 Mar. 705/2 Other garden birds [in West Berlin] include redstarts and pied flycatchers. 1865 *Ibis* 2nd Ser. I. 408 Pied Kingfisher. Appears to be the common species of Lower Bengal. 1924 W. L. Sclater *Systema Avium Æthiopicarum* I. 211 Pied Kingfisher... The Greek Islands, Cyprus, Asia Minor and Egypt to the Persian Gulf. 1953 G. Durrell *Overloaded Ark* xiii. 221 There were Pied kingfishers, vivid black and white. 1971 *Country Life* 28 Oct. 1127/2 The much larger pied kingfishers hovered like kestrels, then plunged into the water from a height. 1894 C. Dixon *Nests & Eggs Brit. Birds* 64 The White Wagtail does not differ in its habits.. from the Pied Wagtail. 1964 G. B. Schaller *Year of Gorilla* (1965) i. 39 Pied wagtails flitted around the huts. 1975 E. Simms *Birds of Town & Suburb* i. 25 The trodden grass swards may be visited by an occasional pied wagtail.

e*. *Pied Piper*: in a German legend (the subject of Browning's poem *The Pied Piper of Hamelin* (1845)), a piper in parti-colour dress who rid the town of Hameln (Hamelin) of a plague of rats by charming them to follow him into the river Weser, and who, on being refused the promised reward, led away the children of the town: used allusively.

1942 'N. Shute' (*title*) Pied piper. 1946 Koestler *Thieves in Night* 154 There was a horra in the Square with Mendl doing his Pied Piper act. 1966 *Listener* 27 Oct. 631/3 There was something very unpleasant about this Pied Piper of adolescent religious feelings [sc. David Wilkerson, an American evangelist]. 1972 T. P. McMahon *Issue of Bishop's Blood* (1973) i. 15 The pathetic thousands who trek endlessly after the Pied Pipers of the cancer cures. 1976 *Western Producer* (Saskatoon, Saskatchewan) 24 June 7/4 In Ottawa..Bobby Gimby, Canada's pied piper, will sail along Rideau Canal followed by a flotilla of canoes filled with youngsters singing his pop Canadian hit 'Canada'. 1977 S. Brett *Star Trap* xiii. 144 Charles could not help admiring the Pied Piper strength of the man's personality. The company was carried along on the wave of his vitality. 1979 *Guardian* 2 May 10/7 I'm the Pied Piper and every toddler is..singing 'Vote vote vote for Mr Ashton'.

pied, *ppl. a.*[2] (Later examples.)

1904 'Mark Twain' in *Harper's Weekly* 10 Dec. 12/2 A thing that gets pied is dead,..its chance of seeing print is gone. 1956 J. Whatmough *Language* i. 9 A haphazard jumble of symbols, say a pied text, is a..nightmare.

‖ **pied à terre** (pyetatę̄r). Pl. **pieds à terre.** [Fr., lit. 'foot on the ground'.] A small town house, flat, or room used for short periods of residence; a 'home base'.

1829 Carlyle in *Foreign Rev.* III. vi. 445 She is perpetually travelling: a peaceful philosopher is lugged over the world, to Cirey, to Lunéville, to that *pied à terre* in Paris. 1838 J. Pardoe *Beauties of Bosphorus* 20 The Greek emperor..acceded to the desire of Mahomet to possess a *pied-à-terre* on the European edge of the channel. 1870 D. G. Rossetti *Let.* 15 Mar. (1965) II. 815, I have not yet written to thank you for a much more independent and promising *pied-à-terre* than I could have found in the tents of the stranger. 1901 'L. Malet' *Hist. R. Calmady* v. i. 383 Richard Calmady had taken her husband's villa at Naples on lease, it offering, as he said, a convenient *pied à terre* to him while yachting along the adjacent coasts. 1926 [see *Metroland]. 1927 C. Connolly *Let.* Dec. in *Romantic Friendship* (1975) 316 It might help you to have a *pied-à-terre*. 1936 F. Clune *Roaming round Darling* i. 7 Australia is still a land of change and Australians, generally, are a restless race, preferring a mere *pied-à-terre* before a taxable home-fire. 1958 [see *accommodation address* s.v. *Accommodation 6 b]. 1964 C. Willock *Enormous Zoo* vii. 114 They had given us a room in their own house which we were free to use as a *pied-à-terre*. 1977 *Wandsworth Borough News* 7 Oct. 23/2 (Advt.), Ideal pied-a-terre—very reasonably priced studio flat in a detached Victorian property.

‖ **pied d'éléphant** (pyedelefã). [Fr., lit. 'elephant's foot'.] A padded sack used to protect the lower part of the body on a bivouac in mountaineering and rock climbing.

1956 C. Evans *On Climbing* viii. 120 For alpine bivouacs, it is customary to make the sleeping bag as it were a part of your clothing using..a sawn-off sleeping-bag, the 'pied d'elephant', to cover the legs. 1963 P. Nock *Rock Climbing* ix. 81 One addition which is sometimes carried is a 'pied d'elephant'. This is like the bottom half of a sleeping bag of minimum weight. 1966 C. Bonington *I chose to Climb* xvi. 188, I was once again fully equipped with borrowed duvet, pied d'éléphant, [etc]. 1968 P. Crew *Encycl. Dict. Mountaineering* 91/1 *Pied d'Elephant*,..a waist length bag for protecting the lower part of the body on bivouacs. 1973 C. Bonington *Next Horizon* x. 149, I managed to pull my boots off, slipped them in my rucksack and then pulled the pied d'éléphant over my legs, ending up with my feet in my rucksack.

piedmont (pī·dmǫnt). [Orig. (in sense 1) *Piemont*, after It. *Piemonte*, lit. 'mountain foot', name of a region of N. Italy, f. *piede* foot (:—L. *ped-, pēs*) + *monte* mountain (:—L. *mont-, mons*).] **1.** Also **Piedmont**. The name of a fertile upland region of the U.S. between the Blue Ridge and Appalachian Mountains to the west and the Atlantic coastal plain to the east, and extending from near New York to Alabama. Freq. *attrib.*

1755 L. Evans *Geogr. Ess.* 7 Between the South Mountain and the hither Chain of the Endless Mountains..is the most considerable Quantity of valuable Land that the English are possest of; and runs through New-Jersey, Pensilvania, Mariland and Virginia. It has yet obtained no general Name, but may properly be called Piemont, from its Situation. 1855 *Southern Lit. Messenger* XXI. 672/2 The next breadth of country, known in several of the States as the Piedmont district, was more salubrious in its atmosphere. 1857 'Porte Crayon' *Virginia Illustr.* 235 The soil of this [sc. Amherst Co., Va.], in common with many other of the piedmont counties, is of a bright red in many places. 1905 W. H. Norton *Elem. Geol.* iii. 87 The surface of the Piedmont is gently rolling. 1927 H. C. Groome *Fauquier during Proprietorship* i. 1 A contour line through the falls of these rivers marks the boundary between Tidewater Virginia and that region which rolling upward to the foothills of the Appalachians is today known as the Piedmont Plateau. 1929 J. Buchan *Courts of Morning* II. xi. 257 A rambling country-house high up in the South Carolina piedmont, with the blue, forested hills behind. 1951 Trager & Smith *Outl. Eng. Struct.* i. 25 In the Southeast of the United States, in both the Coastal and Piedmont speech. 1968 R. W. Fairbridge *Encycl. Geomorphol.* 844/2 The Piedmont Province..is the older part of the Appalachians... Rocks of the piedmont..are of early Paleozoic to Precambrian age. 1972 H. Kurath *Stud. Area Linguistics* viii. 128 From 1790 to 1860 Negro slaves outnumbered the Whites in the piedmont of Virginia. 1976 *Scottish Rev.* Summer 5 After the failure of the Forty-Five, many Highland families moved to the pine barrens and hills of piedmont North Carolina.

2. Any region or area at the foot of a mountain or mountain range.

1860 Mayne Reid *Odd People* 430 Having reached the *piedmont* of the Andes, you still find yourself on a plain, but one which is elevated 3,000 feet above the point from which you started. 1944 A. Holmes *Princ. Physical Geol.* xi. 201 Where closely spaced streams discharge from a mountainous area across a piedmont (a mountain-foot lowland), their deposits coalesce to form a piedmont alluvial plain. 1960 [see *Bajada]. 1962 *Times* 2 June 11/6 Those along the northern piedmont consider themselves Bughtis and owe their allegiance to the Nawab in the remote tribal capital of Dera Bughti. 1974 *Encycl. Brit. Micropædia* I. 741/3 A bajada is usually composed of gravelly alluvium... In humid climates, landforms of this nature are usually referred to as piedmonts.

3. (See quot. 1905[1].)

1905 H. T. Ferrar in R. F. Scott *Voy. 'Discovery'* II. 461 Large areas of ice which lie at the foot of high land and which have no obvious single source may be described as 'piedmonts'. *Ibid.*, Piedmonts afloat are by far the most important. 1914 [see *ice-shelf* s.v. *Ice *sb.* 8].

4. *attrib.* or as *adj.* Situated or occurring at the foot of a mountain or mountain range.

1891 I. C. Russell in *Nat. Geogr. Mag.* May 122 The Malaspina glacier belongs to a class of ice bodies not previously recognized, which are formed at the bases of mountains by the union of several glaciers from above. Their position suggests the name of *Piedmont glaciers* for the type. 1905 W. H. Norton *Elem. Geol.* iv. 99 Mountain streams may build their confluent fans into widespread piedmont (foot of the mountain) alluvial plains. 1907 *Bull. Geol. Soc. Amer.* XVIII. 355 On leaving the mountains the Tarim river enters the great plain which forms the floor of the Lop, or Tarim basin... At the lower ends of many small streams which now wither to nothing in the piedmont gravel or sand of the basin floor, there are

old channels and strips of vegetation. 1936 P. Fleming *News from Tartary* 334 It brought us to Igiz Yar, a little oasis on a rolling slope of piedmont gravel. 1942 [see *Pediplane]. 1957 G. E. Hutchinson *Treat. Limnol.* I. i. 57 The numerous and remarkable large lakes produced in piedmont regions. 1962 *Times* 2 June 11/6 The piedmont zone of hill slopes has long formed the ethnic frontier between the plundering hill men and the peaceful Sindhi cultivators. 1965 [see *Oued]. 1974 *Nature* 29 Nov. 373/1 The United States Range where piedmont glaciers descend southwards along the northern edge of the Hazen Plateau. 1976 R. C. Selley *Introd. Sedimentol.* viii. 257 Between mountains and adjacent lowlands it is often possible to define a distinct belt known as the piedmont zone.

Piedmontese (pīdmǫntī·z), *sb.* and *a.* Also 7 Piemontese. [f. *Piedmont* (see prec.) + -ese.] **A.** *sb.* **a.** The dialect of Piedmont in Italy. **b.** A native or inhabitant of Piedmont. **B.** *adj.* Of or pertaining to Piedmont, its inhabitants, or the dialect spoken by them.

1642 [see *Milanese *sb.* 3]. 1655 Milton *Poems* (1925) I. 236 Slain by the bloody Piemontese. 1673 J. Ray *Observations Journey Low-Countries* 249 The Piemontese are generally well to live. 1768 Sterne *Sentimental Journey* II. 200 The lady was a Piedmontese of about thirty. 1770 [see Courier 3]. 1776 Lady A. Miller *Lett. from Italy* I. 90 The Spanish army..had not failed to guard all the passes they knew of, in order that no communication should be kept up between the Savoyards, the Piedmontese, and Swiss army. 1846 Dickens *Let.* 22 June (1977) IV. 568 It [sc. Lausanne] contains only one Roman Catholic church, which is mainly for the use of the Savoyards and Piedmontese. 1858 A. Gallenga *Country Life in Piedmont* 31 There is hardly an instance of a Piedmontese mountaineer settling permanently abroad. *Ibid.* 128, I have tried to convince the Piedmontese Press that [etc.]. 1878, 1880 [see *Emilian *a.* and *sb.*]. 1902 *Encycl. Brit.* XXXI. 769/2 The Piedmontese dialect has been rather strongly influenced by French. 1927, etc. [see *Ligurian *sb.*]. 1931 *Times Lit. Suppl.* 15 Oct. 787/2 The young Prince.. was made commandant of the Piedmontese artillery. 1932 G. F.-H. Berkeley *Italy in Making* I. xi. 160 Of course Gioberti was a Piedmontese, and in Piedmont the people were satisfied. 1968 D. M. Smith *Mod. Sicily* xlviii. 446 Observers remarked on a tremendous feeling against everything which smacked of Piedmontese. 1974 *Times* 1 Feb. 18/4 After his death the secretaryship passed on [to] Luigi Longo, a dour Piedmontese. 1977 *Canad. Jrnl. Linguistics* XXII. i. 17 *The Sermoni Subalpini*..is the first extant Piedmontese document.

piedmontite. Add: Now freq. called *piemontite. The spelling *piedmontite* was an alteration of J. D. Dana's. (Further examples.)

1933 *Mineral. Mag.* XXIII. 417 The pleochroic colours shown by the mineral in thin section closely resemble those of piedmontite from Japan and from South Mountain, Pennsylvania. 1961 *Current Sci.* XXX. 223/1 (*heading*) Some observations on piedmontite from Goldongri Manganese Mine.

‖ **pied noir** (pye nwār). Pl. **pieds noirs.** [Fr., = black foot.] A name given to people of French origin living in Algeria during French rule, and to those who returned to Europe after the granting of independence to Algeria in 1962.

The name was applied formerly to Algerian stokers who worked bare-footed on French cargo-boats, and by extension, to Algerians generally.

1961 *Times* 6 May 9/6 They are the *pieds noirs*, fiercely proud of this pejorative nickname given them by the metropolitan French. 1962 *Economist* 28 Apr. 344/3 Returned conscripts [from Algeria] and vacationing *pieds noirs*..played an important part in informing the public. 1963 *Times Lit. Suppl.* 31 May 384/3 These two large monographs are devoted to painters—both dead—..the one Flemish, the other a *pied noir* from Constantine. 1965 *Economist* 27 Mar. 1365/1 Marseilles, a sprawling city with an ethnically mixed-up population—Greek, Armenian, Italian, Corsican, and now *pied-noir* from North Africa—has a certain charm. 1969 *Listener* 27 Mar. 417/2 M. Fabre was an elderly colon, one of the original French families in Algeria—a *pied noir*, as they like to be called. 1972 D. Lees *Zodiac* 9 This one..was called Daria-Daria Massenet, which made her a Pied Noir. 1977 *Time* 21 Nov. 12/1 Unlike the white settlers of Rhodesia or the French *pieds-noirs* of Algeria, the Afrikaners have no ties to a European motherland.

pie-eyed (pəi,əi·d), *a. slang* (orig. *U.S.*). [f. Pie *sb.*[2] + Eyed *ppl. a.*] Intoxicated to such an extent that vision is affected; drunk.

1904 Ade *True Bills* 41 They put him down at a Table and sat around him and inhaled the Scotch until they were all Pie-Eyed. 1910 S. E. White *Rules of Game* I. xvii. 103 'Oh, *he's* in town..' 'Drunk, eh?' 'Spifflicated, pie-eyed, loaded, soshed.' 1924 *T. P.'s & Cassell's Weekly* 6 Sept. 631/1 He is partial to a 'shot of gin', and on occasion will drink till he is 'pie-eyed'. 1924 Wodehouse *Ukridge* x. 256 What they put in that stuff...I don't know, but the fact remains that the bird almost instantly became perfectly pie-eyed. 1932 —— *Hot Water* iii. 89 It is our great fancy costume carnival... Everybody puts on funny clothes and becomes pie-eyed. 1937 *Daily Express* 27 Jan. 19/1 Personally I didn't care if the whole band was pie-eyed, I wanted them to be busy playing good dance music. 1946 E. O'Neill *Iceman Cometh* iv. 225 Why the hell don't you get pie-eyed and celebrate? 1957 J. Braine *Room at Top* xxx. 255 'You *are* pie-eyed, aren't you?' Eva said. 1959 'J. Christopher' *Scent of White Poppies* iii. 43 He was utterly pie-eyed. He could scarcely talk straight. 1970 *New Yorker* 8 Aug. 34/2 Piet, who is pie-eyed drunk and needs to be helped homeward. 1974 N. Freeling *Dressing of Diamond* 32 You shouldn't be driving at all, you're pie-eyed.

pie-face (pəi·fē¹s). [f. Pie *sb.*[2] + Face *sb.* 1.] A person of round or blank countenance; a stupid person. (In quot. 1922 used of an effeminate man.)

1922 S. Lewis *Babbitt* vii. 99 Oh, there's a swell bunch of Lizzie boys and lemon-suckers and pie-faces. 1930 *Amer. Speech* VI. 92 The following expressions belong to colloquialisms and slang, including movie and radio neologisms... *Pie-face*. 1960 Wentworth & Flexner *Dict. Amer. Slang* 388/2 *Pie-face*, a person with a round face and a blank, funny, or homely expression; a stupid person.

pie-faced, *a.* [f. Pie *sb.*[2] + Faced *ppl. a.*[2]] Of a person, having a round countenance or a blank or solemn facial expression; stupid. Now commonly used as a general term of mild abuse (see also quot. 1939). Also *transf.*

Quot. 1955 may represent Pie *a.* (in Dict. and Suppl.) rather than Pie *sb.*[2]

1912 *Dialect Notes* III. 585 *Pie-faced*,..round-faced; flat-faced. 1923 Wodehouse *Inimitable Jeeves* x. 112 Did you put that pie-faced infant up to bally-ragging Mr. Bassington-Bassington? 1935 —— *Luck of Bodkins* xiv. 147 You don't suppose I care what old pie-faced Gertrude thinks of me, do you? 1938 G. Heyer *Blunt Instrument* x. 183 'Pie-faced creature, with a nasty, sly smile.' 'The Mona Lisa!' 1939 J. B. Priestley *Let People Sing* i. 17 He was disturbed again, this time by..a messy, dribbling, pie-faced urchin. 1955 *Times* 29 Aug. 10/7 They spare us those moments of pie-faced solemnity with which comic dramatists here are wont to becloud their activities.

pie-gow, var. *Pai kau.

pie in the sky. *colloq.* (orig. *U.S.*). [see Pie *sb.*[2] 4.] A prospect, often illusory, of future happiness, esp. as a reward in heaven for virtue or suffering on earth; an extravagant promise that is unlikely to be fulfilled. Also (with hyphens) *attrib.* Hence **pie-in-the-skyer**, one who puts forward a prospect or promise of this kind.

1911 J. Hill in G. M. Smith *Joe Hill* (1969) i. 20 You'll get pie in the sky when you die. 1926 *Amer. Mercury* Jan. 65/1 *Pie in the sky* is a somewhat cynical reference to the bourgeois heaven. 1939 N. Monsarrat *This is Schoolroom* iv. xviii. 426 Christianity..would end by backing the guns, blessing the flags, and preaching pie-in-the-sky-when-you-die. 1941 *Archit. Rev.* LXXXIX. 117/1 Utopianism, or wishful-thinking, or pie-in-the-sky, or whatever we care to call it. 1951 [see *Jam *sb.*[2] b]. 1951 R. Hoggart *Auden* vi. 190 He is afraid that God will be just a woozy dream, pie-in-the-sky, to his subjects. 1958 *Times Lit. Suppl.* 26 Dec. 750/4 Rightly mistrusting the idea of 'pie in the sky', he [sc. the American] expects to have his reward here. 1959 *Sunday Express* 30 Aug. 12/7 With the election moving remorselessly nearer, pie-in-the-sky days are here again. Everything our hearts could desire is promised us by politicians. 1960 O. Manning *Great Fortune* vi. 79 Pie in the sky. Accept the condition it has pleased God to put you in. 1962 *Times* 11 Oct. 10/5 General Eisenhower has attacked the Kennedy Administration as a 'pie-in-the-sky' Government. 1971 *Physics Bull.* June 322/1 Is this just pie-in-the-sky dreaming about some utopian future where there will be no more pollution, poverty, malnutrition and similar afflictions? 1973 *Time* 25 June 4/2 Exposing shenanigans, militants, self-servers and pie-in-the-skyers must be done, no matter how or by whom. 1975 *Listener* 16 Jan. 89/1 Pie-in-the-sky wage claims. 1976 E. MacLaren *Nature of Belief* iv. 34 The point is obvious when the argument for belief is as crude as some traditional pie-in-the-sky promises. 1977 *Undercurrents* June-July 12/1 To expect the NHS to encompass all sorts of fringe or alternative practices whilst even the level of basic medical care that people want is unobtainable in some areas (abortion) is pie in the sky.

pie-jim-jams (pəidʒi·mdʒæmz). A child's name for 'pyjamas'.

1902 *Little Folks* Apr. 265/2 'Why! I'm in my pie-jimjams!' he murmured..with a doubtful look at his bare feet and his pyjamas. 1964 M. Laski in S. Nowell-Smith *Edwardian England* iv. 204 For bedwear, pyjamas—'pie-jim-jams'—were recommended for boys and girls alike.

pi electron: see *Pi *sb.* 3.

piemontite (pi·mǫntəit). *Min.* [ad. G. *piemontit* (G. A. Kenngott *Das Mohs'sche Mineralsystem* (1853) 75): see Piedmontite.] A synonym of Piedmontite, adopted as more in accord with the original Ger. spelling and with European practice.

1892 E. S. Dana *Dana's Syst. Min.* (ed. 6) 1125/3 (Index), Piedmontite 521. Piemontite, 521. 1950 *Årsbok Sveriges Geologiska Undersökning* XLIV. II. 15 In some parts of the limestone,..crystalline masses of piemontite appear in zones measuring many meters in thickness. 1956 *Mineral. Mag.* XXXI. 241 The piemontite from the schist is granular in form, ruby red, and has pearly lustre. 1971 *Ibid.* XXXVIII. 104 Recommendations of the Commission [on New Minerals and Mineral Names of the International Mineralogical Association] on minerals for which more than one name is in common use...Piemontite, not piedmontite. 1973 *Canad. Jrnl. Earth Sci.* X. 1401/1 Most piemontites are the product of metasomatic or low-grade regional metamorphism of manganiferous rocks.

pier, *sb.*[2] Add: **5.** *pier-master* (examples); **pier-mirror** (examples); **pier-stake**, one of the columns or piles on which a pier is supported; **pier-table** (earlier and later examples).

1936 J. Grierson *High Failure* ii. 27, I slept in the piermaster's cottage in order to be as near my machine as possible. 1971 *Daily Tel.* 16 July 7/1, I liked the pier-master, pompous but human in nautical beard and gold braid. 1976 *Southern Even. Echo* (Southampton) 11 Nov. (Advt. Suppl.) 12/7 Fairey Yacht Harbours Pier Master required for yacht harbour. 1863 O. W. Norton *Army Lett.* (1903) 185 Pier mirrors twenty feet high on three sides of the room. 1969 *Sears Catal.* Spring/Summer 1295, 2-piece pier table and mirror set. 1927 Joyce *On Beach at Fontana* in *Pomes Penyeach*, Wind whines and whines the shingle, The crazy pierstakes groan. 1955 A. Ross *Australia 55* ix. 122 Watching schools of parrot fish and pike twist among the pier-stakes. 1803 M. Wilmot *Let.* 25 July in *Russ. Jrnls.* (1934) i. 24 A pier table furnish'd with splendid Gilt China cups and saucers 'wisely kept for shew'. 1952 J. Gloag *Short Dict. Furnit.* 361 (*caption*) Pier glass and pier table, designed as a decorative unit. 1979 W. J. Burley *Charles & Elizabeth* v. 81 A pair of carved and gilded pier tables with mirrors above.

pier (pi³ɹ), *v. rare.* [f. Pier[2].] **a.** *trans.* To provide with a pier. **b.** *intr.* To reach *out* like a pier.

1857 *Trans. Mich. Agric. Soc.* VIII. 731 If they can coax Uncle Sam to pier the outlet of that Lake and make it a splendid harbor for navigable purposes. 1951 W. Sansom *Face of Innocence* iii. 25 Above them the curved glass cupola..that goes piering out over the garden.

pierced, *ppl. a.* Add: **c.** Of silver, plate, china, porcelain, etc.: ornamented with perforations.

1756 in R. W. Read *Reprint of Orig. Catal. Chelsea Porcelain Manuf.* (1880) 6 Four round pierced baskets enamel'd in flowers. *Ibid.* 7 A fine perfume pot pierced, chased and gilt, and enamel'd in birds. 1785 *Daily Universal Reg.* 1 Jan. 3/2 Pierced and engraved oval salts... Pierced and solid headed mustard pots. 1875 E. Meteyard *Wedgwood Handbk.* 338 The high fluted pillar candle-stick, the pierced fruit basket,..and various other articles, all show an infinite variety, beauty, and fitness of outline. 1931 E. Wenham *Domestic Silver* v. 88 Pierced baskets for bread were also made in the Early Stuart period and.. are the prototypes of the baskets popular in the eighteenth century. 1956 [see *cake-basket* s.v. *Cake *sb.* 9 a]. 1970 G. Savage *Dict. Antiques* 319/2 In more recent times George Owen (d. 1917) made elaborately pierced or reticulated porcelain for the Royal Worcester Porcelain Company. 1974 A. Grimwade *Rococo Silver* iv. 45 An exceptional set of twelve [salt-cellars] by Emick Romer, about 1760, have glass bowls enclosed in pierced cage-work on *chinoiserie* motifs.

2. Special collocations, as **pierced earring** *U.S.*, an earring designed to be worn in a pierced ear; **Pierced-nose** = *Nez Percé, Nez Perce.

1965 *Time* 15 Oct. 70/3 Pierced earrings are by far the most attractive ones available. 1966 *N.Y. Times* 6 June 53 Variation on the pierced earring theme. 1971 *Tuscaloosa* (Alabama) *News* 31 Dec. 8 The package deal includes the purchase of a $6 pair of pierced earrings... The customer may..have her (or his) ears pierced free of charge. 1805 W. Clark in Lewis & Clark *Orig. Jrnls. Lewis & Clark Expedition* (1905) III. 78 They call themselves the *Cho pun-nish* or Pierced noses. 1831 R. Cox *Adventures Columbia River* II. vi. 122 We had many reasons to suspect that the Pierced-noses..were actuated by feelings of hostility. 1837 W. Irving *Rocky Mountains* I. xviii. 183 A Pierced-nose chief, named Blue John by the whites. 1908 H. J. Spinden in *Mem. Amer. Anthropol. Assoc.* II. III. 172 Ross says the people are called Pierced Noses from the custom 'of having their noses bored to hold a certain white shell like the fluke of an anchor'.

piercement (pi³·ɹsmĕnt). *Geol.* [f. Pierce *v.* + -ment, tr. G. *durchspiessungs(falten)*.] The penetration of overlying strata by a mobile rock core, often of salt. Usu. *attrib.* Cf. *Diapir.

1925 D. C. Barton in *Bull. Amer. Assoc. Petroleum Geologists* IX. 1239 The salt is merely part of the peculiar type of fold called by Mrazec 'diapir' fold, or by Krejci 'piercement' folds (*Durchspiessungsfalten*). 1942 M. P. Billings *Structural Geol.* xiv. 254 Salt domes are classified by some geologists into piercement domes and non-piercement domes, the former being discordant, the latter concordant. 1965 A. Holmes *Princ. Physical Geol.* (ed. 2) ix. 206 Diapiric folds (also known as piercement folds). 1965 E. Lehner et al. in G. J. Williams *Econ. Geol. N.Z.* xix. 337/2 It seems improbable that the Tertiary anticlinal structure[s] without piercement cores, or at a long strike distance away from them, could owe their shape solely to hidden diapirism (e.g., upward movement of the clay core without a breakthrough to surface), without at least some help from tangential forces. 1975 *Nature* 29 May 393/2 Because of the circular exposure of evaporite surrounded by younger disturbed rocks, Haughton Dome has been considered a piercement dome.

pier-head. Add: *attrib.* and *Comb.* **pier-head jump**, (*a*) (see quot. 1892); (*b*) an act of leaving a ship as it is about to sail; (*c*) a person who joins a ship as it is leaving the dock; hence *pier-head jumper*.

1892 [in Dict.]. 1927 F. Shaw *Knocking Around* 65 One man was short: he'd done a pierhead jump. 1928 F. P. Harlow *Making of Sailor* 229 All the other members of the crew had made a pier-head jump during the night. 1931 S. W. Ryder *Blue Water Ventures* iii. 40 At the last minute the scallywags who could not get a ship in the ordinary way had to be accepted; these 'pier-head jumpers' being pushed on board by the boarding-house runners as the steamer was moving out of dock. 1936 B. Adams *Ships & Women* xi. 238 Often among pierhead jumps were rattling good sailors. 1938 E. Linklater *Child under Sail* 169 There were no pier-head jumpers to be found, and we had to sail a man short. 1945 *Seafarers' Log* 13 July 6/4 Fred took a few minutes to call his family, and then made the tanker on a pierhead jump. 1967 S. Waters *Indentures Indorsed* 117 He had not passed the doctor, but was a pier-head jump. 1978 *Navy News* Dec. 6/1 The cartoon on page 14 reminded me of a pierhead jump I had from the Defiance in 1940.

b. Designating a type of variety entertainment traditionally associated with summer shows on piers in seaside resorts.

1932 *Statesman* (Calcutta) 2 Aug., There is always a public, and not only a pierhead public, for a thoroughly good pierrot show. 1960 *Times* 2 Mar. 13/2 Here was good old bandsman's pier-head stuff. 1963 *Times* 8 Feb. 14/5 Some ambiguous remarks..—in the best early pierhead manner—suggesting that he is either describing the chassis of a car technically or that of a young woman vulgarly.

pierid (pəi·ərid), *sb.* and *a.* *Ent.* Also **Pierid**, † **-ide**. [f. mod.L. family name *Pieridæ*, f. the generic name *Pieris* (see *pieris) + -id[3].] **A.** *sb.* A white or yellow butterfly belonging to the family Pieridæ. **B.** *adj.* Of or pertaining to this family.

1885 W. F. Kirby *Elem. Text-bk. Entomol.* 155 A white, black-bordered species..could hardly be mistaken for anything but a Pieride. 1905 V. L. Kellogg *Amer. Insects* xiv. 445 The males of many Pierids give off a pleasing aromatic odor. 1926 *Contemp. Rev.* Sept. 370 Pierids are essentially creatures of the open country. 1932 Metcalf & Flint *Fund. Insect Life* viii. 291 (*caption*) Three common pierid butterflies. 1954 Borror & DeLong *Introd. Study Insects* xxvi. 494 The pierids are medium-sized to small butterflies, usually white or yellowish in color. 1963 V. Nabokov *Gift* ii. 109 The long cloud consisting of myriads of white pierids..moves through the sky. 1973 *Nature* 9 Feb. 408/2 Mating behaviour in pierid butterflies is affected by light intensity.

Pierine (pəi·ərīn, -əin), *sb.* and *a.* *Ent.* [f. mod.L. subfamily name *Pierinæ*, f. next + -ine[1].] **A.** *sb.* A white butterfly belonging to the subfamily Pierinæ, which includes the common cabbage white. **B.** *adj.* Of or pertaining to this subfamily.

1898 *Rep. Brit. Assoc. Adv. Sci. 1897* 691 A white butterfly with such a border becomes an extremely conspicuous object, and this appearance of *Mylothris* is mimicked..by species from a number of Pierine genera. 1930 *Proc. Entomol. Soc. London* V. 66 (*heading*) Attacks by a sparrow upon the Pierine butterfly *Aporia bieti*, Oberth., at Tachienlu, Tibet. 1934 *Discovery* July 195/2 It seemed unlikely that certain features in which the Pierines resembled the supposed models were simply adopted from the latter. *Ibid.*, The similar features..were on the whole most distinct and most Pierine-like in those forms that were locally associated with the Pierine *Pereutes* and *Euterpes*.

pieris (pəi·əris). [mod.L., f. L. *Pīeris* a Muse, f. *Pīeria* (see Pierian *a.*).] **1.** [Adopted by D. Don 1834, in *Edin. New Philos. Jrnl.* XVII. 159.] An evergreen shrub of the genus so called, belonging to the family Ericaceæ, native to the southern United States, China, and Japan, and bearing panicles of small, bell-shaped white flowers.

1838 J. C. Loudon *Arboretum* II. 1115 The oval-leaved Pieris..was introduced in 1825. 1909 *Curtis's Bot. Mag.* CXXXV. 8283 The introduction into European gardens of this handsome *Pieris* appears to have taken place upwards of half a century ago. 1935 A. J. Sweet *Trees & Shrubs* xi. 82 Andromeda can be increased by division and cuttings..and Pieris by layering. 1962 R. Page *Educ. Gardener* ix. 260 Evergreens, such as rhododendrons, skimmias, pieris, aucuba. 1975 *Country Life* 10 Apr. 891/3 Other ericaceous plants include pieris with its splendid spring leaf colours.

2. [Adopted by F. von P. Schrank 1801, in *Fauna Boica* II. 152.] A white butterfly of the genus so called, which includes the common cabbage whites.

1863 H. W. Bates *Naturalist on River Amazons* I. i. 22 Numbers of fine showy butterflies were seen...A white Pieris (P. Monuste), and two or three species of brimstone and orange-coloured butterflies. 1969 R. F. Chapman *Insects* xvii. 330 Female *Pieris* ready to oviposit are attracted to green surfaces.

Pierrot. Add: **3.** *attrib.* and *Comb.*, as (sense 1) *Pierrot costume, doll, show*; *Pierrot-like* adj.; **Pierrot collar** (see quot. 1957).

1957 M. B. Picken *Fashion Dict.* 250/2 *Pierrot collar*, ruff like that worn by French pantomime character. 1977 *Time* 8 Aug. 38/1 Pierrot collars and flounces adorned many of Bohan's dresses, capes and blouses. 1893 A. Beardsley *Let.* c. 15 Feb. (1971) 43 Strange hermaphroditic creatures wandering about in Pierrot costumes or modern dress; quite a new world of my own creation. 1960 Pierrot costume [see *concert-party]. 1935 A. Christie *Three Act Tragedy* iii. ix. 188 A ridiculously elongated pierrot doll [was] lying across the sofa. 1967 A. Wilson *No Laughing Matter* ii. 39 Rupert lolling on cushions..aimlessly waving the limp Pierrot doll beside him. 1979 A. Buchan *Scrap Screen* viii. 132 Emancipated girls..moved into bedsitters..and filled them with black and orange cushions and Pierrot dolls. 1977 *Rolling Stone* 19 May 73/1, I hadn't quite expected the faultless intonation..of that shining, graceful, Pierrot-like figure. 1951 J. Fleming *Man who looked Back* i. 14 They saw the pierrot show from comfortable deck-chairs. 1978 M. Gilbert *Empty House* xiii. 107 There's..a pierrot show in the Palais de Dance [*sic*].

Pierrotic (pīěrǫ·tik), *a.* [f. Pierrot + -ic.] Of, belonging, or pertaining to pierrots.

1927 *Observer* 16 Oct. 15/3 The delightful tenor raptures of Mr. Georges Metaxa are in the best Pierrotic tradition. 1931 *Times Lit. Suppl.* 13 Aug. 614/1 Mr. Nicoll seems to pass lightly over his [*sc.* Punch's] white-bloused, Pierrotic, Neapolitan avatar.

‖ **pietas** (pi,ĕi·tās). [L.] An attitude of respect towards an ancestor, scholarship, an institution, a country, etc.

1924 J. Bailey *Let.* 27 Apr. (1935) 244 Both Llanthony, to which, of course, *pietas* specially draws me, and Llangammarch look very attractive. 1930 *N. & Q.* 22 Feb. 143/1 Yet another Early English Text appears under the editorship of an American scholar, a new proof of *pietas*. 1944 *Horizon* Sept. 188 And yet, with all his *pietas*, when Joyce died..he had not set foot in his native city for over thirty years. 1960 C. Day Lewis *Buried Day* i. 26 *Pietas*—a habit of respect for gods, ancestors, parents, country, institutions. 1961 *Times* 30 Nov. 16/2 *Pietas* is the curse of the commissioned biographer. 1965 N. St. John-Stevas in *Bagehot's Coll. Wks.* I. 11 The *Economist* ..has supported this undertaking, as..an act of *pietas* to its..most famous editor. 1976 *Church Times* 16 Jan. 15/5 A requiem will be sung for him in St. Oswald's, Durham, with his music—not just as an act of *pietas*, but in gratitude to God for one who, although a man of his age, yet reminds us of several needs in today's Church.

Pietism. Add: **1.** Also in extended use, any similar movement within Protestantism.

1900 tr. *A. Ritschl's Christian Doctrine of Justification & Reconciliation* i. 84 It is..an inversion of the Reformation point of view when Pietism makes the moral power of faith the object which God invests with the value which moral conduct would possess when carried out. 1934 R. N. Flew *Idea of Perfection in Christian Theol.* xvi. 275 The essential mark of Pietism is its quest for individual holiness. 1958 E. L. Mascall *Recovery of Unity* i. 9 The word 'pietism'. By this I mean an attitude to Christianity..which, in contrast to the corporate, objective, liturgical and theocentric religion of the primitive Church, is individualistic, subjective, pietistic and Christocentric. 1967 D. T. Kauffman *Dict. Relig. Terms* 354/1 *Pietism*, protestant religious current emphasizing personal devotions, Bible study, evangelism, and the like. 1967 M. J. Heinecken in J. Macquarrie *Dict. Christian Ethics* 257/1 Pietism has been and continues to be a valid protest against sterile orthodoxism and false sacramentalism.

piet-my-vrou (pi·tmĕ[i]frǭ[u]). *S. Afr.* Also **piet-myn-vrouw**. [Afrikaans, lit. 'Peter my wife', echoic, f. the bird's call.] The red-chested cuckoo, *Notococcyx* (or *Cuculus*) *solitarius*; also, occasionally used as a name for the noisy robin-chat, *Cossypha bicolor*, which imitates the call of the red-chested cuckoo.

[1790 E. Helme tr. *Le Vaillant's Trav. Afr.* II. xviii. 367 One of them [*sc.* the Hottentots] named *Pit*, was the first who brought me this bird... He had no sooner shot the hen, than the cock flew after him, repeating several times *Pit me frow*; it must be observed, it is the usual cry of this bird, as I was afterwards convinced, on shooting some of the same kind.] 1835 A. Steedman *Wanderings S. Afr.* I. v. 189 The 'Piet myn vrouw', a bird of which the Hottentots relate many amusing stories. 1923 F. W. Fitzsimons *Nat. Hist. S. Afr.: Birds* II. 31 (*caption*) Noisy Robin Chat, or Piet-myn-vrouw (*Cossypha bicolor*). An active and efficient policeman of the forests and dense scrub on the eastern side of South Africa. 1937 M. Alston *Wanderings of Bird-Lover in Afr.* viii. 50 The Dutch call him [*sc.* the noisy robin-chat] the piet-myn-vrouw because of his imitating the red-chested cuckoo. 1949 *Cape Argus Mag.* 5 Nov. 7/6 The piet-my-vrou and turtle doves make the afternoon drowsy with their calls. 1966 E. Palmer *Plains of Camdeboo* xii. 216 Sometimes in the bush we hear..the Piet-my-vrou, and its call is unmistakable. 1971 *Stand. Encycl. S. Afr.* III. 517/1 The piet-my-vrou or red-chested cuckoo (*Cuculus solitarius*) is a fairly large, brownish cuckoo.

‖ **pietra** (pi,e·trā). Pl. **pietre** (-e). The Italian for 'stone'; occurring in Italian phrases, more or less in current use in the terminology of art, etc.

a. pietra commessa (kōme·sā), pl. **commesse** (-e) [fem. of *commesso*, pa. pple. of *commettere* to fit together]: mosaic work; an example of this.

a 1666 Evelyn *Diary* an. 1644 (1955) II. 191 Here were divers incomparable tables of Pietra Commessa, which is a marble ground inlayd with severall sorts of marbles & stones of divers colours, in the shapes of flowers, trees, beasts, birds & Landskips like the natural. a 1668 R. Lassels *Voy. Italy* (1670) II. 220 A curious table of *pietre commesse* about 12 foot long & 5 wide. 1766 Smollett *Trav.* II. xxviii. 70 These *pietre commesse* are better calculated for cabinets, than for ornaments to great buildings. 1848 H. R. Forster *Stowe Catal.* 287 A box, with slabs of pietre commesse, of birds, fruits, and flowers.

b. pietra dura (du·rā), pl. **dure** (-e) [fem. of *duro* hard]: (see quot. 1962); also (sing. and pl.) mosaic work of such stones; also *attrib.*

1805 P. Beckford *Familiar Lett. from Italy* I. 148 The best part of the furniture is the inlaid tables in Pietra Dura, a work of great labour and great expence. 1845 R. Ford *Hand-bk. for Travellers Spain* II. 580/2 Observe the Florentine pulpit of *Pietre dure*. 1901 'L. Malet' *Hist. R. Calmady* iii. v. 199 Certain treasures, unique in historic worth, locked in the glass tables and fine Florentine and *piétra dura* cabinets of the Long Gallery. 1942 *Burlington Mag.* Apr. 87/1 The chief piece of furniture mentioned is the enormous cabinet of *pietra dura* that now stands in the ballroom at Badminton. 1961 *Connoisseur* Dec. p. xxxiii, Louis XVI cabinets..inset with panels of pietre dure in the form of flat mosaics of different birds. 1962 R. G. Haggar *Dict. Art Terms* 259/1 *Pietra dura*,..those stones which, mainly composed of silicates, are in English generally described as semiprecious stones. *Pietra dura* were much used for altar frontals, table-tops, and the like in the sixteenth century. 1970 *Oxf. Compan. Art* 871/1 *Pietre dure*... A term applied to a particular kind of mosaic work in which coloured stones.. are used to imitate as far as possible the effect of painting. 1976 *Country Life* 27 May (Suppl.) 47 A Florentine pietra dura casket with ormolu mounts.

c. pietra serena (sĕre·nā) [fem. of *sereno* clear]: a bluish sandstone much used for building in Florence and throughout Tuscany; also *attrib.*

1873 S. & J. Horner *Walks in Florence* II. xxii. 296 The aisles are carried round the nave and transepts by a line of handsome columns, of pietra-serena, with Corinthian capitals. 1888 G. J. Oakeshott *Detail & Ornament of Ital. Renaissance* Pl. 37 Examples of Corbels from Florence, Siena etc.. in pietra serena by Donatello in the Bargello. 1927 E. W. Anthony *Early Florentine Archit. & Decoration* x. 69 The sculpture of the Gothic period, in general, gives up the decorated backgrounds of the older school, and the reliefs are executed in white marble or *pietra serena*. 1931 R. Fry *Let.* 22 July (1972) II. 661 It's odd that Florence..couldn't make a city and these rather second-rate Bolognese could—it's true they had brick instead of that abominable *pietra-serena*. 1971 F. M. Godfrey *Ital. Archit. up to 1750* iv. 198 The interior differs from the Cappella Pazzi by the stressed corner pilasters in dark *pietra serena*, supporting the arches at the joints. 1975 *Ashmolean Mus. Rep. Visitors 1973–74* 34 The present building with its *pietra serena* pilasters and window frames, so characteristic of Renaissance architecture in Tuscany.

piewipe: see Peeweep in Dict. and Suppl.

pie·zo-. Add: Now usu. with pronunc. (pəi,ī·zo). **pie·zo-crystal**, a piezo-electric crystal used in an electrical circuit; **pie:zo-crystalliza·tion** *Geol.* [ad. G. *piezokrystallisation* (E. Weinschenk 1895, in *Abhandl. d. Bayerischen Akad. d. Wissensch.* (*Math.-phys Klasse*) XVIII. 741)], crystallization of a magma, usu. in a distinctive or abnormal manner, under conditions of mechanical stress; (? *obs.*); **piezo-electricity**, delete def. and see main entry below; **pie:zoma·gnetism** [ad. G. *piezomagnetismus* (W. Voigt 1901, in *Nachrichten v. d. K. Ges. d. Wissensch. zu Göttingen* (*Math.-phys. Klasse*) I. 1)], magnetism induced in a crystal by the application of mechanical stress; hence **pie:zomagne·tic** *a.*, of, pertaining to, or exhibiting piezomagnetism; **pie:zoresi·stance, -resisti·vity**, change in the electrical resistance or resistivity of a solid when subjected to mechanical stress; freq. *attrib.*; so **pie:zoresi·stive** *a.*, of, pertaining to, or utilizing this effect.

1928 *Exper. Wireless* July 414/2 (*caption*) Mounting piezo crystals. 1936 *Amer. Speech* XI. 95/2 A piezo-crystal cutter for aluminum discs with which frequencies from 40 to 10,000 cycles can be recorded. 1958 *Engineering* 31 Jan. 160/3 An alternative type of drum is the large sheet of steel with a piezo-crystal fixed to a particular part of it. 1903 A. Geikie *Text-bk. Geol.* (ed. 4) II. 718 He [*sc.* Weinschenk] believes that rock to have been part of a normal granitic magma which crystallized under abnormal conditions, and that it owes its mineralogical composition and characteristic foliated structure..to the peculiar relations of tension accompanying the plication of the mountains. To these relations he has given the name of 'piezocrystallization'. 1938 A. Johannsen *Descr. Petrogr. Igneous Rocks* IV. 79 The parallel texture is seen in dynamometamorphosed rocks or in border facies. In the latter case it may be due to piezocrystallization, for the feldspars also are often more or less parallel. 1954 R. L. Parker tr. *Niggli's Rocks & Min. Deposits* xiv. 523 Effects of stress on partly solidified material. Such effects leave their impression on structures and textures and are responsible for the piezo-crystallization of magmas. 1901 *Sci. Abstr.* A. IV. 1043 The recorded observations determine, for the pyro- and piezo-magnetic excitement, only superior limiting values. 1959 *Physics & Chem. of Solids* XI. 77/2 The piezomagnetic moment will be reversed in sign when the antiferromagnetic sublattice magnetizations are reversed in sign. 1972 *Nature* 8 Dec. 348/2 Both volcanic eruptions and the San Andreas creep increments are very imperfectly understood so that a large scale control experiment which

demonstrates quantitatively the role of the piezomagnetic effect in producing local magnetic anomalies is very desirable. 1978 *Ibid.* 9 Mar. 130/1 When a stress is applied to, or removed from, a rock there is a distortion of crystal structure which often gives rise to a small change in the rock's remanent magnetisation. This is known as the piezomagnetic effect and has an obvious application to the prediction of earthquakes. 1901 *Sci. Abstr.* A. IV. 1043 (*heading*) Pyro- and piezo-magnetism of crystals. 1931 S. R. Williams *Magn. Phenomena* v. 164 With the advent of the electron theory of matter,.. physicists began to inquire as to the possibility of the phenomena of pyro- and piezo-magnetism. 1967 Condon & Odishaw *Handbk. Physics* (ed. 2) iv. viii. 143/1 In antiferromagnetic crystals of sufficiently low symmetry, such as CoF_2 and MnF_2, the magnetic analogue of piezoelectricity, piezomagnetism, can occur. 1978 *Nature* 9 Mar. 130/1 In practice, piezomagnetism has been of little use in earthquake prediction. 1954 *Physical Rev.* XCIV. 42/2 The piezoresistance results for germanium and silicon..have been expressed in terms of the pressure coefficient of resistivity and two simple shear coefficients. 1970 *Jrnl. Appl. Physics* XLI. 811/2 (*heading*) Piezoresistance in SnTe. *Ibid.*, The piezoresistance effect has been used in the past with success in determining the symmetry of the bands in various semiconductors. 1935 J. W. Cookson in *Physical Rev.* XLVII. 194/2 It is herein meant by the piezo-resistive effect that change in electrical resistance which a homogeneous body undergoes when subjected to mechanical stress. 1963 *Jrnl. Appl. Physics* XXXIV. 684/2 The piezoresistive effect in semiconductors..is generally a result of the dependence of the electronic energy levels of a crystal on the state of strain in the crystal. 1973 *Physics Bull.* Dec. 743/3 The cartridge consists of an elastic stainless steel cantilever on to which a pair of silicon piezoresistive strain gauges have been fused. 1958 *Solid State Physics* VI. 232 The most important examples of unsymmetrical fourth rank properties are the classical piezooptic effect (photoelasticity) and the piezoresistivity effect. 1965 *Wireless World* Aug. 380/2 Although carbon has been used as the sensitive element in pressure transducers, piezo-resistivity in ordinary carbon composition resistors appears to have attracted little attention.

piezo (pəi,ī·zo), *a.* = *piezoelectric *a.*

1922 Glazebrook *Dict. Appl. Physics* II. 600/1 To demonstrate the piezo effect it is convenient to place a sheet of tinfoil on a slab of tourmaline and connect the foil to an electrometer. 1930 *Proc. IRE* XVIII. 491 A piezo oscillator is the most suitable frequency standard thus far devised. 1946 W. G. Cady *Piezoelectricity* i. 6 Piezo resonators and oscillators have proved useful in many kinds of electrical measurement. 1970 J. Earl *Tuners & Amplifiers* i. 27 Some of the inexpensive systems use a piezo cartridge (crystal or ceramic). 1975 G. J. King *Audio Handbk.* viii. 174 Early piezo pickups used a natural crystal element..as the signal generator.

piezoelectric (pəi,ī:zo,īle·ktrik), *a.* and *sb.* Also **piezo-electric**. [f. Piezo- + Electric *a.* and *sb.*] **A.** *adj.* Of, pertaining to, exhibiting, or utilizing piezoelectricity.

1883 *Jrnl. Chem. Soc.* XLIV. 412 (*heading*) Actinoelectric and piezoelectric properties of quartz. 1921 *Physical Rev.* XVII. 531 The possibility is discussed of using the piezo-electric resonator for a standard of high frequency. 1947 Crowther & Whiddington *Science at War* iv. 175 Modified quartz piezo-electric hydrophones were used to measure vibrations as low as 1 cycle per second. 1958 N. Cusack *Electr. & Magn. Prop. Solids* xviii. 410 The piezoelectric properties of Rochelle salt have been widely used in gramophone pick-ups, microphones, loudspeakers, surface roughness analysers, and other electromechanical devices. 1972 *Last Whole Earth Catalog* (Portola Inst.) 88/2 The piezoelectric effect has been observed in a number of organic materials, among them wood, bone, tendon and skin. 1973 M. Woodhouse *Blue Bone* xvii. 190 He was talking about some kind of reversed piezoelectric effect. 1977 *Rolling Stone* 7 Apr. 88/1 (Advt.), But only Ovation's six-element piezo-electric pickup captures both top *and* individual string vibration.

B. *sb.* A piezoelectric substance or body.

1913 *Phil. Mag.* XXVI. 1053 These high resistances.. are both cheaper and easier to use than either standard capacities or quartz piezoelectrics. 1971 B. Jaffe et al. *Piezoelectric Ceramics* i. 1 The creation of useful piezoelectrics by treatment of a polycrystalline material depends on ferroelectricity. 1975 *Physics Bull.* May 212/3 For economic and technical reasons the piezoelectrics most commonly used for SAW generation are lithium niobate and quartz.

Hence **pie:zoele·ctrical** *a.* = *piezoelectric *a.*; **pie:zoele·ctrically** *adv.*, as regards, or by means of, piezoelectricity.

1923 *Physical Rev.* XXI. 350 Quartz, another piezo-electrically active substance. 1937 *Discovery* Jan. 22/1 One method of producing them [*sc.* ultra-sounds]..is to employ the piezo-electrical effect in quartz. 1969 *Sci. Jrnl.* Aug. 21/2 Current from the voltage source is fed through a superconductive coil which is vibrated by a piezoelectrically driven metal tuning fork. 1976 *Physics Bull.* Feb. 61/3 New techniques include..very high frequency phonon generation..using a far infrared laser on a piezoelectrical crystal.

piezoelectricity (pəi,ī:zo,īlektri·sĭti). Delete entry s.v. Piezo- and substitute: Also **piezo-electricity**. [ad. G. *piezoelectricität* (W. G. Hankel 1881, in *Ber. d. K. Sächsischen Ges. d. Wissensch. zu Leipzig* XXXIII. 52).] Electric polarization in a substance resulting from the application of mechanical stress, esp. in certain crystals.

1883 *Jrnl. Chem. Soc.* XLIV. 412 This phenomenon the author proposes to call piezoelectricity. 1895 [in Dict. s.v. Piezo-]. 1929 J. A. Ratcliffe *Physical Princ. Wireless* i. 2 The facts of piezo-electricity have been applied in the quartz oscillator. 1956 *Nature* 17 Mar. 537/2 Tests for pyroelectricity and piezoelectricity were negative. 1974 J. Kyle *Electronics Unravelled* (1975) ii. 45 Piezoelectricity is due to an unusual crystal structure found in several materials.

piezometer. Add to sense b: or in an aquifer. (Further examples.)

1904 *Johns Hopkins Hosp. Bull.* XV. 293 (*heading*) The piezometer, an instrument for measuring resistances. *Ibid.* 293/2 There are two ways of using the piezometer for the purpose of outlining an abdominal tumor. 1954 Roe & Ayres *Engin. Agric. Drainage* x. 266 The piezometer is a length of ¼- or ½-inch iron pipe driven into the soil so that there is no leakage into or out from the sides. 1970 *Nature* 4 July 11/1 Piezometers were used to measure water pressures in the rock joints.

Hence **piezome·tric** *a.*; *spec.* in *Hydrology*, of or pertaining to the measurement of hydrostatic pressure in an aquifer; *piezometric surface* (see quot. 1923[2]).

1923 O. E. Meinzer in *Water-Supply Papers U.S. Geol. Survey* No. 494. 6 The expression piezometric surface is obtained from the French. *Ibid.* 38 The piezometric surface of an aquifer is an imaginary surface that everywhere coincides with the static level of the water in the aquifer. It is the surface to which the water from a given aquifer will rise under its full head. 1966 Davis & DeWiest *Hydrogeol.* ii. 48 The elevation to which water will rise in artesian wells, or wells penetrating confined aquifers, defines the piezometric surfaces. *Ibid.* 49 The general direction of ground-water flow can be shown..on piezometric maps. 1973 Gregory & Walling *Drainage Basin Form & Process* iii. 113 This data is used to plot lines of equal piezometric head or equipotentials on a vertical cross section of the aquifer... Similarly, equipotentials or piezometric surfaces could be plotted for several horizontal planes at different depths.

Piffer (pi·fər). *slang.* [From the initials of the name of the force + -er[1].] A member of the Punjab Frontier Force (a military unit raised in 1849 and employed esp. to police the North-West Frontier of India during British rule) or of one of the regiments that succeeded it. Also (*attrib.*), the force itself.

1892 *Pall Mall Gaz.* 24 Oct. 3/1 The Punjab Frontier Force is known in India as 'The Piffers'. 1901 *Blackw. Mag.* June 780/1 A strong garrison of the three arms, all Piffers. *Ibid.* 788/2 One regiment of Piffer infantry could move anywhere in the hills. 1922 *19th Cent.* Jan. 48 The establishment of a British observation corps, similar to the 'Piffers' of later times. 1958 O. Caroe *Pathans* xx. 345 For many years the Piffers served under the Panjab Lieutenant-Governor. 1970 A. J. Smithers *Man who Disobeyed* x. 100 The Guides, the Punjab Frontier Force (Piffers), the Sikhs and the Gurkhas, would have allowed no affectations of superiority from any British regiment, household or line.

‖ **pifferaro** (piferā·ro). Pl. **pifferari**. [It.] A performer on the *piffero*.

1854 Thackeray *Newcomes* I. xxii. 212 A Contadina and a Trasteverino dancing at the door of a Locanda to the music of a Pifferaro. 1860 *Once a Week* 14 July 71/2 Three of the *pifferari* whom you find at Christmas time in such numbers in the Piazza di Spagna at Rome. 1870 Queen Victoria *Let.* 6 June in R. Fulford *Your Dear Letter* (1971) 281 Imagine our astonishment at seeing in the village [*sc.* Balmoral] this morning an Italian pifferaro ..and two boys dancing. 1920 *Punch* 30 June 510/3 The local condettieri, pifferari, banditti and lazzaroni. 1974 *Times Lit. Suppl.* 8 Mar. 228/4 The charming shepherds who came to the cities from the Abruzzi to play their bagpipes in front of the wayside shrines—the pifferari. 1977 *Gramophone* June 41/3 He also draws a marvellously keen, edgy sound from piccolo and oboe in the Serenade, for all the world like a *pifferaro* squealing on the Abruzzi air.

pi·ffing, *vbl. sb. slang.* [Cf. Piff *int.*] (See quots.)

1928 *Daily Tel.* 10 Jan. 11/5 Gunnery training is confined for the most part to sub-calibre firing—'piffing', as it is known in service parlance. 1962 Granville *Dict. Sailors' Slang* 88/1 *Piffing*, sub-calibre firing. Echoic of the sound. (Gunnery term.)

piffle, *sb.* Add: Also used as a derisive retort.

1914 'High Jinks, Jr.' *Choice Slang* 16 *Oh piffle*, an exclamation denoting inconsequence of the subject in question. 1920 'B. L. Standish' *Man on First* xviii. 127 'The Hawks have the lead on us, still.' 'Piffle!' said Cady. 'We'll even things up to-morrow.' 1959 *Elizabethan* Apr. 10/1, I gave you a bar of chocolate on the train from London. So piffle!

pifflicated (pi·flikĕ[i]tĕd), *ppl. a. U.S. slang.* [A fanciful formation of Piffle *v.*, infl. by Spiflicate *v.*] Drunk, intoxicated.

1905 Taylor & Gibson *Log of Water Wagon* 50 (*caption*) Professor Bunn's patent plugs for pifflicated people. 1934 S. Robertson *Devel. Mod. Eng.* (1936) xi. 465 English (chiefly American) terms for the idea of 'drunk' or 'intoxicated'... *shellacked, soused, piffled, pifflicated, blotto, stinko.*

piffling, *ppl. a.* (In Dict. s.v. Piffle *v.*) (Further examples.)

1916 'Boyd Cable' *Action Front* 17 You don't think a pifflin' little Pip-Squeak shell could go through *his* head? 1927 *Daily Express* 26 July 3/4 The Bench consider that this is a piffling offence, and..that a warning would have been sufficient. 1927 *Observer* 13 Nov. 10/4 The mechanical parts of the moving-pictures are superb, but the imaginative and intellectual parts are piffling. 1963 *Times* 12 June 8/7 The sum involved was piffling compared with the firm's £25m. a year turnover. 1973 J. Wainwright *Pride of Pigs* 56 The lesser hooks being pulled in for the piffling crimes, while the big boys work the blinders.

pig, *sb.*[1] Add: **I. 1.** Phr. *in pig*, of a sow: pregnant; also *transf.* of a girl or woman (*slang*).

1886 J. Long *Bk. Pig* iv. 59 They [*sc.* gilts] are less costly than if either tolerably fat or, as it is called, 'in pig', or in farrow. 1917 W. Powell-Owen *Pig-Keeping* v. 59 Sows that are in-pig do best when given their liberty. 1937 H. M. Rikard-Bell *Handk. Mod. Pig Farming* ii. 33 Watch carefully after three weeks have elapsed for fear that the recurrence of their oestrum periods will prove them [*sc.* gilts] to be not in-pig. 1945 N. Mitford *Pursuit of Love* x. 83, I am in pig, what d'you think of that? 1950, etc. [see *in-pig *a.*]. 1976 'D. Halliday' *Dolly & Nanny Bird* vii. 86 Since when had her mother paid the slightest attention to anything her darling daughter said or did, except to do her level best to keep her from marrying anything less than a duke, until she had to get herself in pig.

5. (Further examples.)

1927 *Dialect Notes* V. 458 *Pig*, a woman—sottish, surly, disgruntled, stinking—who has sunk to the lowest level of prostitution. The bum who *keeps a pig* rents her out to others. 1931 E. O'Neill *The Hunted* iv, in *Mourning becomes Electra* 155 That yaller-haired pig with the pink dress on! 1932 J. T. Farrell in *Story* Mar.–Apr. 47 Jack told of an anecdote about a pig he had picked up once. She was too lousy and scummy to take a chance on. 1960 I. Jefferies *Dignity & Purity* v. 83 I'm having a golf lesson from the Advertising pig tomorrow. 1966 *Sunday Times* (Colour Suppl.) 13 Feb. 35/4 *Pig*, an unattractive girl. 1968–70 *Current Slang* (Univ. S. Dakota) III–IV. 91 *Pig*, a girl who is both promiscuous and drunken. 1973 *Daily Californian* 1 Feb. 1/4 The Pentagon Papers..'provide evidence of pig foreign policy. A pig is someone who attacks you and at the same time claims he is the victim', he said. 1976 *National Observer* (U.S.) 21 Feb. 14/6 The quick resort to the phrase 'pig' for the blue-collar, lower-class people who were doing the job they thought they were supposed to do. 1977 P. G. Winslow *Witch Hill Murder* ii. 227, I had some beautiful birds in London, but I had to stay on the good side of that pig, or she might have noticed more than was good for her. 1979 R. Rendell *Make Death love Me* i. 16 I'm not demeaning myself to reply to you, pig.

b. Colloq. phr. *to make a pig of oneself*, to gluttonize.

1942 Berrey & Van den Bark *Amer. Thes. Slang* §272/3 *Be greedy or selfish*,..make a pig..of oneself. 1961 F. S. Anthony in *Webster* s.v., Not make such a gorging pig of himself. 1979 *Guardian* 22 June 9/5 We had made pigs of ourselves on the bread.

c. Applied contemptuously or opprobriously to a thing.

1975 'W. Haggard' *Scorpion's Tail* i. 2 What a summer, he thought—what a perfect pig. The rain and the cold. 1978 *Times* 15 Feb. 8/1 Miller was out in Collinge's second over to a pig of a ball. 1978 F. Mullally *Deadly Payoff* xi. 154 Watch for the potholes. It's a pig of a road. 1978 *Hot Car* June 93/4 The car became a pig to start.

6. b. Delete † *obs.* and add earlier and later examples. Now usu. disparaging.

1811 *Lexicon Balatronicum*, s.v. *Pig*, a China street pig; a Bow-street officer. 1874 Hotten *Slang Dict.* 253 *Pig*, a policeman; an informer. The word is now almost exclusively applied by London thieves to a plain-clothes man, or a 'nose'. 1967 C. Drummond *Death at Furlong Post* v. 63, I had to give the local P.C. a lift. I dropped the pig at Packenham. 1970 *Times* 7 Aug. 4/7 'Pig' is slang for a policeman—and the defence says that the word 'pig' was scrawled over the doors of the house after the killings. 1973 *Black World* July 56/1 The pigs swooped by, going west, the emergency light blinking green. 1975 D. Lodge *Changing Places* v. 170 Any pig roughs you up, make sure you get his number. 1975 N. Luard *Travelling Horseman* vi. 146 The police Rover and some motorcycle pigs providing escort. 1977 'E. Crispin' *Glimpses of Moon* xi. 217 'My God, it's the pigs,' said the hunt saboteuse disgustedly.

d. An informer. ? *Obs.*

1874 [see 6 b above]. 1904 'No. 1500' *Life in Sing Sing* 251/1 *Pig*, prisoner who reports another; stool-pigeon. 1918 *Amer. Law Rev.* LII. 891 A 'prison stool pigeon' is a 'trusty', 'psalm singer' or 'pig'.

e. Any of various forms of transport (see quots.).

1898 *North Amer. Rev.* June 723 Whalebacks, or 'pigs', as the lake sailors call them. 1938 L. Beebe *High Iron* 223/2 (Gloss.), *Pig*, locomotive. 1946 *Jrnl. R. Aeronaut. Soc.* L. 85 He had made no great contribution publicly to aeronautical science, but the fact that he had taken the first 'pig' (the name sometimes applied to early biplanes—Ed.) into the air. 1961 *Amer. Speech* XXXVI. 273 *Pig*, *n.*, an old truck. 1967 *Evening Standard* 26 July 13/3 'We'll hop in my pig, catch the rays and have a boss time.' ..The latest in American teenage talk... 'The pig' is a car which looks powerful but has a small engine; 'catching the rays' is getting a sun tan; and 'boss' is the same as great. 1971 *Guardian* 27 Aug. 11/7 He did indeed ride with Melbourne's Hell's Angels..garaging his extremely powerful pig (bike) beside his Porsche. *Ibid.*, 22 Nov. 6/1 The soldiers were in a convoy of 'pigs'—armoured personnel carriers, trucks, and Land-Rovers. 1972 *Times* 8 June 16/2 It was only a patrol of one armoured personnel carrier, a great heavy green vehicle, which everyone

calls a 'pig' because of its snout shaped bonnet. 1973 *Amer. Speech 1969* XLIV. 207 *Pig*, 1. Trailer transported on a flat car. 2. Tractor with little power. 1978 *Times* 19 Jan. 29/3 The Pig, the armoured vehicle most used in Belfast.

f. *pl.* Used as a derisive retort. Also const. *to*. *Austral. slang.*

1906 E. Dyson *Fact 'ry 'Ands* i. 5 'Pigs to you!' said Benno, with incredible scorn. 1933 N. Lindsay *Saturdee* ix. 165 Peter had to cover his confusion by saying 'Pigs to you' as he went out kicking the door. 1957 'N. Culotta' *They're a Weird Mob* (1958) iv. 47 'She's worn out.' 'Pigs she is. There's a lot of life in 'er yet.' 1975 L. Ryan *Shearers* 119 'Ar, pigs to you!' 'In your dinger, too!'

g. *blind pig*: see *blind *a.* (and *adv.*) 16.

II. 8. c. (Later examples.)

1943 E. M. Almedingen *Frossia* vi. 240 Some people are like oranges..all divided into neatly separated pigs. 1961 I. Fleming *Thunderball* iv. 38 The orange, carefully sliced into symmetrical pigs. 1974 P. Dickinson *Poison Oracle* v. 133 They were sharing a second orange, putting it pig by pig into each other's mouths.

d. (Further examples: see quots.)

1926 Maines & Grant *Wise-Crack Dict.* 12/2 *Pig in a blanket*, sausage in a roll. 1941 J. Smiley *Hash House Lingo* 43 *Pig in a blanket*, frankfurter sandwich. 1973 *Observer* (Colour Suppl.) 16 Sept. 83/2 The famous savoury angels on horseback. (The Americans call it pigs in blankets.)

8*. A device that fits snugly inside an oil or gas pipeline and can be sent through it, e.g. to clean the inside or to act as a barrier between fluids either side of it.

1949 *Amer. Speech* XXIV. 33 A few field workers apply the term [*sc.* 'rabbit'] to scrapers used in pipelines to remove paraffin, but the most common name for this device is *pig*, because of the grunting noise it makes as it is forced through the line. 1949 *Sun* (Baltimore) 30 Nov. 13/1 In order to make sure that the pipe had no leaks and was free of all foreign matter, a 'pig'—a rubber object headed by a washer the same diameter as the pipe—was sent through by air pressure. 1956 *Ibid.* (B ed.) 27 Oct. 11/6 A 'pig' is a contraption consisting of blades, wheels, and brushes that runs through a pipeline to clean it. 1970 W. G. Roberts *Quest for Oil* xii. 126 (*caption*) Plastic pig used to clean a 30-inch diameter pipeline in Libya. *Ibid.*, A third way of checking what is going on in a pipeline is to insert some solid separator at the interface [between different products], fitting closely enough in the pipe to make sure that it will be pushed along at the same rate as the oil. Such devices are known as 'batching pigs'. 1977 *Time* 27 June 37/1 The moving oil will push the pig through the 48-in.-diameter steel pipe at 1 m.p.h. As it goes, the cylinder will shove out of the pipe any refuse that may be contained (for example, tools left behind by forgetful workmen) and emit beeps indicating its location.

9. *pigs in clover* (earlier examples); *pig* (also *piggy*) *in the middle*, (*a*) a game in which one child is encircled by others and must escape by any of a number of (usu. vigorous) means; (*b*) a chasing game in which players must cross from one side of an open space to the other without being stopped by a child (or children) in the middle; (*c*) a ball game, usu. for three, in which the middle child tries to intercept the ball as it passes between the other two; also, the player in the middle in any of these games; also *transf.* and *fig.*

1889 *Amer. Stationer* 14 Feb. 355/3 'Pigs in Clover' is the taking name of a new game which has just been placed upon the market by the toy house of Selchow & Righter. 1892 'Mark Twain' *Amer. Claimant* xxiv. 250 A toy puzzle called Pigs in Clover, had come into sudden favor. 1887 *Folk-Lore Jrnl.* V. 50 Some of the games were much rougher such as 'Pig in the middle and can't get out'. 1915 W. S. Maugham *Of Human Bondage* xi. 39 The new boys were told to go into the middle, while the others stationed themselves along opposite walls. They began to play *Pig in the Middle*. The old boys ran from wall to wall while the new boys tried to catch them: when one was seized and the mystic words said—one, two, three, and a pig for me—he became a prisoner and, turning sides, helped to catch those who were still free. 1962 *Guardian* 3 Aug. 4/5 He was..pig-in-the-middle between his sweet, faint, pietistic mother and his impossibly stiff-necked father. 1962 C. Storr *Robin* viii. 37 You're blue with cold... You *will* have to play pig-in-the-middle. 1969 I. & P. Opie *Children's Games* viii. 238 This game (Bull in the ring) seems to be less played today than in the nineteenth century, when it was frequently recorded... 'Pig in the Middle and Can't Get Out'. 1970 *Times* (Saturday *Rev.*) 28 Feb. p. i/6 Dr. Robinson has thus been rudely abused from all sides of the shrinking Kingdom, piggy-in-the-middle of a debate which..is..fought with..bitterness. 1970 N. Fisher *Walk at Steady Pace* iv. 231 They all knew more..than I did. All three were able to use me as pig in the middle. 1973 M. Amis *Rachel Papers* 125 Four boys..stood in a semicircle round a fifth. ..Having looked round for encouragement or approval, one of the boys leaned over and slapped the piggy in the middle quite hard on the face. 1977 W. McIlvanney *Laidlaw* xii. 51 He's not a good *polis*-man... He doesn't know which side he's on. He's pig in the middle. 1977 *Times* 1 Sept. 5/2 You have to take the decisions, and often you are the piggy in the middle; in our case, for example, between pilots and shipowners.

10. f. *to draw pig on* (or *upon*) *pork* (or *bacon*) (Commercial slang): (see quots.); hence *pig-on-bacon*, a bill drawn in this way.

1849 J. W. *Perils Emigrant* ii. 84, I..had exhausted every means of renewal, borrowing, exchanging cheques, drawing 'pig on pork', as it is technically called. 1872 *Porcupine* 16 Nov. 515/2 In Liverpool..issuing a bill on their London branch establishment..in commercial phraseology, is termed drawing 'pig upon bacon'. 1911 W. Thomson *Dict. Banking* 397/1 '*Pig upon bacon*.' In the case of an accommodation bill, where e.g., Brown accepts merely to oblige the drawer, Jones, Brown has no intention of meeting the bill at maturity. He expects that Jones will himself provide the funds necessary to pay the bill when it is due. As Jones in his own mind considers himself practically the acceptor as well as the drawer, Jones on Brown is therefore likened to a bill drawn by 'Pig on Pork' or 'Pig upon Bacon'. 1920 A. C. Pigou *Econ. of Welfare* ii. v. 144 The variety of accommodation bills known as 'pig-on-bacon', where the acceptor is a branch of the drawing house under an *alias*, is..different. 1930 W. Thomson *Dict. Banking* (ed. 7) 548/2 When the drawers and drawees of a bill are the same, as when a foreign branch of a firm draws on its London office, and there are no documents for goods attached to the bill, the firm is said to be drawing 'Pig on Pork'.

g. *in a pig's eye, ear, arse*: used as a derisive retort; freq. as a strong negative or an emphatic.

Most of the examples are *U.S.* or *Austral.*

1872 'P. V. Nasby' *Struggles* cxiii. 315 A poeticklal cotashun..which..wuz,—'Kum wun, kim all, this rock shel fly From its firm base—in a pig's eye.' 1919 W. H. Downing *Digger Dial.* 38 *Pig's ear*, a contemptuous ejaculation. 1932 O. R. Cohen *Star of Earth* xxv. 270 'Here I am,' he says: 'I did it..and that means Mary didn't!'..'In a pig's eye it does!' 1942 Berrey & Van den Bark *Amer. Thes. Slang* §166/7 *I don't believe it!*..It is in a pig's eye *or* arse! *Ibid.* §170/8 *You are mistaken*... It is in a pig's eye! *or* arse! *Ibid.* §229/5 *You will not!*..In a pig's eye *or* arse you will! 1951 E. Lambert *Twenty Thousand Thieves* 322 'Pig's arse to that!' another voice cried. 'A jack-up—that's the shot.' 1957 J. Blish *Fallen Star* v. 70 'You'll have to.' 'In a pig's eye,' she said. 1962 *Observer* 21 Jan. 11/7 Immigration from Ireland, said the Prime Minister.., will be included in the general controls. In a pig's eye it will. 1968 W. Garner *Deep, Deep Freeze* ix. 110 'One stops short of probing the private lives of people for whom one has a regard.' 'In a pig's ear!' she said vulgarly. 'If duty called you'd have a man under the bed on my honeymoon.' 1968 H. Waugh *Con Game* v. 53 'He claimed he didn't want to—.' Mrs. Fogarty said, 'In a pig's eye he didn't want to.' 1969 G. Johnston *Clean Straw for Nothing* 307 'That's because she won't face realities.' 'Pig's arse. And anyway who are you to talk?' 1973 J. Wainwright *Pride of Pigs* 83 'Preston,' suggested Harris. 'In a pig's eye!' growled Ripley. 1974 P. Larkin *High Windows* 35 My wife and I have asked a crowd of craps To come and waste their time and ours: perhaps You'd care to join us? In a pig's arse, friend. 1976 *Time* 5 Apr. 23/2 Attorney General Edward Levi let it be known that he considered the matter 'extremely serious'. To officials of the Federal Bureau of Investigation, Levi's comment was a monumental understatement. 'Extremely serious in a pig's eye,' said one. 'It's a disaster.'

h. *on the pig's back* (occas. *the pig's ear*): in a fortunate position; on top of the world; riding high.

1900 *19th Cent.* July 81 [Ireland] 'You're on the pig's back' means prosperity. 'The pig is on your back' indicates misfortune. 1922 Joyce *Ulysses* 177 That'll be two pounds ten.... Three Hynes owes me... Five guineas about. On the pig's back. 1930 [see *home *adv.* 2 b]. 1946 C. Mann in *Coast to Coast 1945* 27 We always were lucky. He's home on the pig's ear. 1949 H. Wadman *Life Sentence* i. i. 10 Could anything be nicer? Basil is on the pig's back. 1958 J. Lodwick *Bid Soldiers Shoot* viii. 277 Nixon, who in Crete had suffered horribly from solitude, was now a happy man—on the pig's back, one might say, and the image is appropriate since the grunting of the corralled porkers never ceased. 1962 R. Wallis *Point of Origin* 11 Then aerial topdressing came in and they were on the pig's back. 1966 'L. Lane' *ABZ of Scouse* ii. 78 *On ther pig's back*, lucky; doing well; in the money.

i. *to make a pig's ear* (*out*) *of*: to make a mess of; to bungle.

1954 E. Hargreaves *Handful of Silver* xii. 183 'I've made a real pig's ear of it, haven't I?' said Basil, with an attempt at lightness. 1973 *Observer* 29 July 14/5 If you are doing something wrong, you will..make a pig's ear of its execution.

11. b. In various phrases and locutions connected with the idea of pigs flying, freq. as a type of the unlikely or untrue.

1616 W. Clerk *Withals's Dict. Eng. & Lat.* (rev. ed.) 583 Pigs fly in the ayre with their tayles forward. 1639 J. Clarke *Paroemiologia Anglo-Latina* 147/1 Pigs fly in the aire with their tailes forward. 1670 J. Ray *Coll. Eng. Proverbs* 189 Pigs fly in the air with their tails forward. c 1860 [in Dict.]. 1865 'L. Carroll' *Alice's Adventures in Wonderland* ix. 155 'I've a right to think,' said Alice sharply... 'Just about as much right,' said the Duchess, 'as pigs have to fly.' a 1871 A. De Morgan *Budget of Paradoxes* (1872) 275 There is a proverb which says, A pig may fly, but it isn't a likely bird. [1871 'L. Carroll' *Through Looking-Glass* iv. 76 'The time has come,' the Walrus said, 'To talk of many things: Of shoes—and ships —and sealing-wax—Of cabbages—and kings—And why the sea is boiling hot—And whether pigs have wings.'] 1880 C. H. Spurgeon *John Ploughman's Pictures* 32 They say that if pigs fly they always go with their tails forward. 1913 *Punch* 13 Aug. 156/1 'If pigs could fly...' The clumsy brutes can't, of course, while we flies *can* pig —see us in a confectioner's shop. 1937 Partridge *Dict. Slang* 628/2 *Pigs fly, when*, never. 1949 'J. Tey' *Brat Farrar* x. 81 'I may, some day. I may.' 'Pigs may fly.' 1952 Wodehouse (*title*) Pigs have wings. 1972 'J. Quartermain' *Rock of Diamond* xxvii. 176 'I'll wait... Perhaps he'll have news...' 'Maybe... And maybe pigs have wings.' 1973 'J. Higgins' *Prayer for Dying* xii. 165 'Something could come out of that line of enquiry.' 'I know... Pigs might also fly.'

IV. 12. a. *pig bin* (also fig.), *-byre, house* (also fig.), *man* (earlier and later examples; also fig.), *manure, -meat* (earlier and later examples; also fig.), *-pail, -pen* (further examples; also fig.), *-wire*; (sense 6 b in Dict. and Suppl.) *pig car, station.*

1959 I. & P. Opie *Lore & Lang. Schoolch.* ix. 167 They call him [*sc.* a greedy-guts]: dustbin,..pig-bin, [etc.]. 1972 P. Black *Biggest Aspidistra* ii. iv. 124 Outside the houses stood a row of huge bins, one for collecting paper, one for tins, one for bottles, one for kitchen waste. They were known as pigbins, though..the pigs ate only the waste. 1979 'M. Hebden' *Death set to Music* iii. 29 Pel sniffed at his stew. 'I think they took it from the pig bin,' he observed coldly. 1906 *19th Cent.* June 967 Already half the cottage pig-byres stand empty in our lanes. 1970 G. Jackson *Let.* 10 June in *Soledad Brother* (1971) 36, I sat in the back of the pig car and bled for two hours. 1845 C. Cist *Cincinnati Misc.* I. 186 A stout looking fellow set his gun leaning on a pig house, and jumped in to catch some fowls. 1950 *N.Z. Jrnl. Agric.* Apr. 369/1 In considering the design of pig houses there are two main factors. 1960 *Farmer & Stockbreeder* 9 Feb. 76/2 The value of..good pighouse insulation. 1963 *Amer. Speech* XXXVIII. 173 [Kansas University] A sorority known for its unprepossessing members.. *campus pig house*..*pig house*. 1825 *Whole Proc. Old Bailey* 15 Jan. 116/2, I..saw the prisoner in the kitchen —he said, if I came with any *pigman* he would knock my head off. 1971 *Farmers Weekly* 19 Mar. 75/3 Continental pigmen want hybrid boars. a 1975 Wodehouse *Sunset at Blandings* (1977) vii. 52 Clarence's pig man claims to have seen the White Lady of Blandings one Saturday night. 1976 *Eastern Daily Press* (Norwich) 19 Nov. 11/1 (Advt.), Assistant pigman/woman required.. to join staff of five, on 500-sow herd. 1960 *Farmer & Stockbreeder* 1 Mar. 107/1 Future developments include a pig-manure spreading service which will give back to the members the pig manure they would have had if they had reared their pigs on their farms. 1975 J. Wyllie *Butterfly Flood* (1977) xxiii. 105 The compost heap was.. activated by pig manure. 1798 J. Woodforde *Diary* 2 July (1931) V. 125 Pig Meat now is but of little value to what it has been. 1942 Z. N. Hurston in *Amer. Mercury* July 96/1 *Pig meat*, young girl. 1970 G. Greer *Female Eunuch* 265 Perhaps words like *pig*, *pig-meat* or *dog* are inspired by the sadness which follows unsatisfactory sex. 1971 *Farmers Weekly* (Extra) 19 Mar. 12/2 High beef prices have pushed up demand for fresh pigmeat. 1977 *Times* 8 Feb. 17/3 We break the [EEC] rules by subsidizing pigmeat. 1908 *Westm. Gaz.* 18 Jan. 2/3 The cricket climbed the side of the..pig-pail. 1872 *Harper's Mag.* Apr. 690/2 A one-story wooden structure..became the rallying-place of the tribes. This [room], by reason of its general unsightliness, was denominated by Tammany's political adversaries the 'Pig-Pen'. 1907 J. London *Let.* 25 July (1966) 247 He left his stateroom the filthiest pig-pen I ever saw. 1952 R. P. Bissell *Monongahela* i. 5, I took my paper suitcase out of the pigpen and up to the mates' room, and one thing sure, that mates' room smelled better than the deckhands' bunkroom, and even had a light in the bunk to read by, and a clean blanket with no fuel oil or coal ground into it. 1960 T. Hughes *Lupercal* 38 Toward the pig-pens on his right. 1971 M. Tak *Truck Talk* 118 *Pigpen*, a sloppy, ill-run truck stop. 1978 J. Wainwright *Jury People* xxxv. 106 They carried him to the pig-pens. 1979 R. Gillespie *Crossword Mystery* v. 116 'Did you search his place?' 'Yeah. A pigpen.' 1970 G. Jackson *Let.* 10 June in *Soledad Brother* (1971) 34, I stopped attending school regularly, and started getting 'picked up' by the pigs more often. The pig station, a lecture, and oak-stick therapeutics. 1975 P. Moyes *Black Widower* i. 13 Assemble..outside the Georgetown Pig Station..for a protest march. 1964 *Listener* 19 Mar. 458/1 Scrapping the hedges, replacing them with concrete posts and pigwire. 1971 *Ideal Home* Apr. 119 To restrain dogs and children the cheapest and strongest I know is Smith's (of Bristol) Bulwark fencing —what my father calls pig-wire—a heavy galvanised mesh varying from 6 in. square at the top to 6 in. by 3 in. at the bottom, 32 in. high.

b. *pig-keeper, -netter, -stalker*; *pig-rearing, -stalking*; (sense 7) *pig-breaking*.

1902 *Encycl. Brit.* XXIX. 578/1 A great saving of labour was effected by the introduction of 'pig-breaking' machines. 1906 *Westm. Gaz.* 20 Aug. 10/1 The fact, too, that acorns are a heavy crop will gladden the hearts of pig-keepers. 1971 *Farmers Weekly* (Extra) 19 Mar. 33/2 The slump has hit heavy hog producers hardest of all pig-keepers. 1923 *Blackw. Mag.* Dec. 768/1 They concocted a plan by which the boar should be netted... Professional pig-netters were summoned. 1907 *Westm. Gaz.* 31 July 12/2 Pig-rearing..is on the downward grade. 1960 *Farmer & Stockbreeder* 29 Mar. 88/3 The most important ingredient of pig-rearing success is undoubtedly hard work. 1978 *Dumfries Courier* 13 Oct. 20/1 A sow and seven piglets died early yesterday afternoon when fire raged through a pig-rearing building at Laghall Farm, New Abbey Road, Dumfries. 1908 *Westm. Gaz.* 2 Mar. 5/2 'The indiscriminate offer of rewards in no way tends to the destruction of the real man-eater,' says Mr. Rees, 'while it ensures the extermination of the useful..deer and pig-stalker.' 1867 M. A. Barker *Station Life N.Z.* (1870) xv. 109 We go over the hills pig-stalking.

c. *pig-ploughed*; *pig-jawed* (example), *-snouted*; *pig-ignorant* (hence *pig-ignorance*), *-lucky, -sick, -sober, -sticky, -stupid, -tight* (examples) adjs.

In the last type of use, often merely with the force of an intensifier: extreme(ly), thorough(ly).

1973 H. Miller *Open City* xvii. 187 Boorishness and Glasgow-bred pig-ignorance. 1972 J. Wainwright *Requiem for Loser* iv. 82 I'm not pig-ignorant... But you're pig-*stupid*. 1973 A. Price *October Men* vi. 84 [He] was clearly pig-ignorant of everything that did not concern him. 1976 T. Heald *Let Sleeping Dogs Die* vii. 132 Those press johnnies..would never twig. Too gullible and too pig ignorant. 1942 'M. Innes' *Daffodil Affair* i. v. 30 'A nice dog,' Appleby said... Mr Gee swung round. 'Dish-faced,' he said... 'And undershot... Pig-jawed, in fact.' 1939 J. Steinbeck *Grapes of Wrath* xvi. 253 'Get her fixed?' 'We was pig lucky,' said Tom. 'Got a part 'fore dark.' 1921 H. Guthrie-Smith *Tutira* xix. 165 These pig-ploughed shreds [of land]. 1948 A. Baron *From City from Plough* i. 9 Wha's up, Sergeant?.. You look pigsick. 1965 A. Prior *Interrogators* xiii. 244 He was pig-sick of talking to the old bastard. 1977 J. Wainwright *Nest of Rats* i. vii. 51, I was pig-sick of Rawle and his devious ways. 1923 E. Sitwell *Bucolic Comedies* 9 And old pig-snouted Darkness grunts and roots in the hovels. 1960 *Times* 29 Sept. (Nigeria Suppl.) p. xxi/5 The weighty pig-snouted aardvark. 1945 Koestler *Twilight Bar* iii. 62, I thought so. Pig-sober. 1922 Joyce *Ulysses* 445 Eat it and get all pigsticky. 1972 Pig-stupid [see *pig-ignorant adj.]. 1859 Pig-tight [see *bull-strong* s.v. *bull *sb.*[1] 11]. a 1930 Pig-tight [see *horse-high* adj. s.v. *horse *sb.* 27 a].

13. pig board *Surfing* (see quots.); **pig-boat** *U.S. slang*, a submarine; **pig-boiling** *Metallurgy*, the puddling of unrefined pig-iron, which is characterized by a period of rapid bubbling of gas from the molten metal; **pig-dog**, add to def.: and New Zealand (further example); (*b*) used as a term of abuse; **pig net**, a type of strong net; **pig-root** *v.*, (*b*) *Austral.*, of a horse or other animal, to kick upwards with the hind legs, the forelegs remaining rigid; **pig-rooting** *vbl. sb.*, (*a*) *N.Z.*, a patch of ground grubbed or rooted up by wild pigs; (*b*) the action of **pig-root* vb. (b); **pig-run**, a tract of land used by (wild) pigs; also, a track made or used by wild pigs in a forest; **pig-sign** *N.Z.*, the droppings of wild pig(s); **pig-washing** *Metallurgy*, the refining of molten pig-iron by treatment with molten iron oxide; **pig-yoke** (*a*) (earlier example).

1965 J. Pollard *Surfrider* ii. 18 Your board can be a 'pigboard'—wide at the stern and tapered to a point at the nose. 1970 *Studies in English* (Univ. Cape Town) I. 28 Older designs include the *pig board*, that is, a board characterized by a narrow nose and a broad tail. 1921 *Periscope* (U.S. Submarine Base, San Pedro, Calif.) Apr. 21/1 The dukes what passes the exam..finally goes to the subs..and they career as a pig bout [*sic*] sailor is started. This of course, means a sub, witch they is also called sea pigs. 1939 *Newsweek* 9 Jan. 20/1 Presumably Germany will now build up to this by constructing ocean-going pigboats. 1974 G. Jenkins *Bridge & Magpies* xiv. 218, I understand now what the pig-boat saying means—'by guess and by God'. 1975 *Redbook* Aug. 118/3 What do you think about Pearl Harbor? The newer boats are there. I'd like to get one of them and avoid the old pig-boats if I can. 1856 J. Hall in *Birmingham Jrnl.* 26 Sept. Suppl. 3/5 As regards the improved apparatus for the refinery, my principle is the *doing away* with the *refinery process* by pig boiling. 1882 [see Wet *a.* 17]. 1928 H. M. Boylston *Introd. Metallurgy Iron & Steel* vi. 187 Hall's process was also known as the 'pig-boiling process' because of the vigorous boiling or bubbling of the molten metal. 1958 A. D. Merriman *Dict. Metallurgy* 244/1 *Pig boiling*, the name used in reference to that stage in the puddling process..when the original pig iron is melted and thoroughly mixed with the oxidising substances... The whole mass..becomes agitated with the escape of carbon monoxide giving the impression of 'boiling'. 1922 Joyce *Ulysses* 460 Pig dog and always was ever since he was pupped! 1961 B. Crump *Hang on a Minute* 18 My mate reckoned he'd never laughed so much since his brother's pig-dogs got loose and followed him into the Waitawheta dance hall! 1977 *N.Z. Herald* 8 Jan. 4-9/6 (Advt.), Pig dog pups, boxer blue merle pointer cross. 1907 *Yesterday's Shopping* (1969) Index p. xlvi/3 Nets...Pig. 1966 D. Francis *Flying Finish* v. 54 A cart with a pig-net over it. 1971 *Country Life* 11 Mar. 533/1 Some chaps got a tiddler [*sc.* a sturgeon], weighed a hundred pound, in a pig net. 1913 W. K. Harris *Outback in Australia* 27 We had a second horse afflicted with a tendency to buck, and this one could be depended upon to 'pig-root' for at least half a mile as soon as we made a start. 1957 P. White *Voss* x. 276 This caused Turner to curse and kick, and his nag in consequence to sidle and pigroot. 1921 H. Guthrie-Smith *Tutira* xix. 169 It [*sc.* manuka] now began to colonise the paddock..appearing about pig-rootings, along sheep-tracks. 1966 'J. Hackston' *Father clears Out* 168 Bucking, backbending, side-jumping, leaping, and pig-rooting. 1848 T. Chapman *Jrnl.* 3 Dec. II. 381 (typescript), Thousands of Acres on all sides of you... Here and there little patches are under cultivation—the rest are 'pig-runs'. 1900 *Geogr. Jrnl.* XVI. 174 In dense forest where the pig-runs are the only means of passage. 1950 *N.Z. Jrnl. Agric.* 346/3 This type of utilisation..at the same time allows some rejuvenation of the pasture in the permanent pig runs. 1960 B. Crump *Good Keen Man* 57, I was thoroughly interrogated every evening as to the whereabouts of any fresh pig-sign I had seen that day [in the bush]. 1887 Phillips & Bauerman *Elem. Metallurgy* (ed. 2) 280 A similar process, used for some time by Krupp, was described by the late Mr. A. L. Holley under the name of 'pig-washing'. 1910 *Encycl. Brit.* XIV. 824/2 In the Bell-Krupp or 'pig-washing' process..advantage is taken of the fact that..the phosphorus and silicon of molten cast iron are quickly oxidized and removed by contact with molten iron oxide. 1958 A. D. Merriman *Dict. Metallurgy* 244/2 *Pig-washing process*, a term used in reference to those methods of refining molten pig iron by oxidising treatment at relatively low temperatures. 1845 *Knickerbocker* XXV. 424 Yellow buttons..'and geese', as he said, 'sittin' on a pig-yoke, printed on to 'em.'

b. pig-face, pig's face *Austral.*, add def.: a succulent plant belonging to the family

Aizoaceæ, esp. *Disphyma* (formerly *Mesembryanthemum*) *australe*, bearing pink or purplish-red flowers and edible fleshy berries; also, the berries themselves; also *attrib.*; (further examples); **pig-fern** *N.Z.* the hard fern, *Paesia scaberula*, of the family Polypodiaceæ; also called lace-fern, ring-fern, and scented fern; **pig-lily**, for *Richardia æthiopica* substitute *Zantedeschia æthiopica*; (earlier and later examples).

1889 J. H. Maiden *Useful Native Plants Austral.* i. 44 'Pig Faces'... The fleshy fruit is eaten raw by the aborigines. The leaves are eaten baked. 1920 B. Cronin *Timber Wolves* 69 On the crest of a knoll, Heritage paused a moment to admire the royal purple of the pig-face bloom. 1933 *Bulletin* (Sydney) 29 Nov. 21/3 The fleshy leaves of the pigface plant, which grows along the sandy seashores. 1944 [see *ice-plant]. 1963 Moore & Adams *Plants N.Z. Coast* 59 *Disphyma australe*..ice-plant, pig-face. 1977 J. Galbraith *Field Guide Wild Flowers S.E. Austral.* 114 'Pigface' with large pink flowers. 1926 F. W. Hilgendorf *Weeds N.Z.* ii. 19 Hard fern (*Paesia scaberula*), called also pig fern and silver fern, is abundant in both islands. 1929 W. Martin *N.Z. Nature Bk.* (1930) II. iv. 44 Pig-fern and Lace-fern are local names given to a dwarf species (*Paesia scaberula*), with finely divided leaflets. 1952 G. R. Gilbert *Glass* 59 Struggling through the thick scrub and tangled pig-fern, the brother and sister tramped on towards the mountains. 1848 C. J. F. Bunbury *Jrnl. Residence Cape of Good Hope* viii. 188 *Calla* (*Zantedeschia*) *Æthiopica*... Commonly called at the Cape the Pig Lily. 1870 *Cape Monthly Mag.* Aug. 104 The 'arum'..grows in all the ditches under the title of 'pig-lily'. 1971 U. van der Spuy *Wild Flowers S. Afr. for Garden* 229/2 It [*sc. Zantedeschia æthiopica*] is said to have been given the common name of 'pig-lily' because in the south-western Cape, where it grows prolifically, pigs are said to relish the rootstock.

c. Many of the combinations with *pig's* also occur in phrases (see *10). *pig's fry* (examples); **pig's breakfast**, used as a type of the unappetizing or unattractive; **pig's ear** *Rhyming slang*, beer; **pig's face** (later example); **pig's foot** = *pig-trotter* (chiefly *pl.*); **pig's whisper**, (*a*) (earlier and later examples).

1933 L. G. D. Acland in *Press* (Christchurch, N.Z.) 9 Sept. 15/7 Two may possibly be Canterbury expressions ..(1) As rough as a bag. (2) As rough as a pig's breakfast. 1948 K. M. Wells *Owl Pen Reader* (1969) i. 45 Lucy and I looked unbelievingly at the mess in the kettles. Bits of charred wood, charcoal, old leaves and wood ash floated there in the midst of an uninviting white scum. It looked like poor porridge, a pig's breakfast to us. 1880 D. W. Barrett *Life & Work among Navvies* (ed. 2) ii. 40 'Now, Jack, I'm goin' to get a tiddley wink of pig's ear..'. A tiddley wink of pig's ear!..What does it mean? Simply this... a workman..goes to get a drink of beer. Had our friend wished for something more potent than the pig's ear aforesaid, he would have substituted the phrase.. 'Tommy get out, and let your father in', meaning thereby *gin*. 1936 J. Curtis *Gilt Kid* xx. 199 But the most of the fiver would go in the old pig's ear. 1974 P. Wright *Lang. Brit. Industry* x. 88 In the pub you can ask for a pint of *pig's ear* (beer). 1929 'M. B. Eldershaw' *House is Built* iii. ix. 217 Nothing but a few clumps of pig's face throve in the pockets of earth it provided. 1922 H. Crane *Let.* 23 Jan. (1965) 78 We begin with pigs' feet and sauerkraut. 1968 C. Brown in *Esquire* Apr. 88/1 Certain foods..are associated almost solely with the nigger: collard greens..hog maws, black-eyed peas, pigs' feet. 1788 J. Woodforde *Diary* 7 Nov. (1927) III. 63 We had for Dinner, Some Fish..Giblets, Piggs Fry. 1939 F. Thompson *Lark Rise* i. 13 The first delicious dish of pig's fry sizzling in the frying-pan. 1970 G. E. Evans *Where Beards wag All* xxv. 264, I had a fancy for something tasty, and one night I gave her a pig's fry and asked her to cook it for us for next day. 1821 P. Egan *Real Life in London* I. xi. 189 The lad *nibbled the bait*, and was off in a *pig's whisper*. [Note] *Pig's Whisper*—A very common term for speed. 1963 *Times* 11 Mar. 1/7 If you are unfortunate enough to snore, you are said 'to drive pigs', or perhaps you may do something in a very short time, in which case it is said that you have done it 'in a pig's whisper'.

pig, *v.* Add: **2.** (Further examples.) Also const. *along*, to live from day to day like an animal.

1896 *Pall Mall Gaz.* Sept. 70 She isn't fit to pig along in the way we have to here. 1909 [see *master-mind *sb.* a]. 1931 *Times Lit. Suppl.* 19 Feb. 131/3 It was not enough for Arthur Phelps 'just to pig along', working to live and living to work. 1939 G. B. Shaw *In Good King Charles's Golden Days* ii. 112 Give me a skilled trade and eight or ten shillings a week, and you and I, beloved, would pig along more happily than we have ever been able to do as our majesties. 1930 J. Buchan *Castle Gay* ix. 145 They would have to pig it in a moorland inn. 1964 'M. Innes' *Money from Holme* xxvi. 171 'Dear me,' Binchy said. 'If it isn't friend Cheel.' He turned to Braunkopf. 'Cheel and I pig together, more or less.' 1977 —— *Honeybath's Haven* iii. 34 He didn't approve of the proposal to pig it in the studio.

pig-bel (pi·g,be·l). *Med.* Also **pigbel**. [See quot. 1966.] A severe necrotizing enterocolitis found in Papua New Guinea, caused by *Clostridium welchii* and associated with feasts of pork.

1966 T. G. C. Murrell et al. in *Lancet* 29 Jan. 217/2 It was felt that the syndrome in New Guinea, being ætiologically related to pig-feasting, should be designated by a specific name. 'Pig-bel' has been proposed because this is the 'pidgin English' name used by medical orderlies to describe the abdominal discomfort which follows a large pork meal. 1969 Edington & Gilles *Path. in Tropics* vi. 273 A diffuse sloughing enteritis of the jejunum, ileum, and colon (enteritis necroticans, 'pigbel') has been described in Germany and New Guinea... It is the commonest acute abdominal condition requiring laparotomy in hospital practice in the highlands of New Guinea. The disease in this area, in both epidemic and sporadic forms, is related to pig feasting. 1977 *Lancet* 17 Sept. 617/2 The epidemic forms of necrotising enteritis ('pig-bel' and *Darmbrand* enteritis) have been consistently associated with strains of *C. welchi* producing mainly b [*recte* β] toxin.

pigeon, *sb.* **I. 3. a.** Restrict †*Obs.* to sense 'a coward' and add later examples in senses 'a young woman, a girl; a sweetheart'.

1916 Joyce *Portrait of Artist* iii. 116 Is that you, pigeon? 1940 *Music Makers* May 37/3 *Pigeon*, a young girl. 1970 G. Greer *Female Eunuch* 265 The basic imagery behind terms like..*pigeon* is the imagery of food. 1977 J. Wambaugh *Black Marble* (1978) viii. 125 She accepted it graciously, thinking she must remember to give Philo a bonus for finding a little pigeon with tits big enough to bring old Landon McWhorter back to life.

b. (Further examples.) Also in extended use.

1941 *Sun* (Baltimore) 14 Aug. 13/7 These amateur gamblers are the greatest pigeons I ever knew. 1956 B. Holiday *Lady sings Blues* (1973) xvii. 136 So they handed me a white paper to sign... I signed... The rest was up to them. I was just a pigeon. 1959 [see *Murphy[2] 4]. 1962 J. D. MacDonald *Girl* xi. 150 Anybody steals that much, they're a pigeon for the first people that get to him. 1977 *Tennis World* Sept. 17/3 A 'pigeon' is a frequent victim—that is, until he plays out of his tree. 1978 G. A. Sheehan *Running & Being* xii. 165 Whatever your game, you can always spot a pigeon. When I warm up for a road race I can usually tell at a glance the newcomers to the sport.

d. = *stool-pigeon* s.v. Stool *sb.* 19 b.

1849 *National Police Gaz.* (U.S.) 12 May 2/7 The Mayor of Philadelphia having discovered that an old pigeon known as Bill Forebaugh was accustomed to point out his officers to the different knucks who arrived in the city, determined to put a stop to this new lay. 1859 G. W. Matsell *Vocabulum* 66 *Pigeon*, a thief that joins in with other thieves to commit a crime, and then informs the officer, who he pigeons for, and for this service the officer is supposed to be *occasionally* both deaf and blind. 1934 R. Chandler in *Black Mask* Oct. 14/2 Don't come here again—..I don't like—pigeons. 1966 *Sunday Tel.* (Brisbane) 15 May 15/1 In the underworld, this is the mark of the 'copper', 'grass', 'nark', 'pigeon', or 'squealer'. 1971 'D. Shannon' *Ringer* (1972) ix. 154 A lot of our pigeons offer the info to the other side too. 1976 R. Rosenblum *Sweetheart Deal* i. 11 For years guarding witnesses remained a..shoestring operation. Rent a hotel room and keep the pigeon under wraps.

II. 5. a. *pigeon loft* (later examples; also fig.); *pigeon-pie* (earlier and later examples).

1891 Kipling *City of Dreadful Night* v. 34 Do you mean that you can from this absurd pigeon-loft locate the wards in the night-time! 1969 *Times* 4 Sept. 3/1 The difference between the sun's time at the pigeon loft, given by the pigeons' internal sense of time, and the time registered by the sun's position at the release site, gives a measure of the difference in longitude between the two points. 1977 *Evening Post* (Nottingham) 27 Jan. 7/7 They also admitted two accusations of jointly entering two pigeons lofts as trespassers. 1655 in M. M. Verney *Mem.* (1894) III. iv. 129 Madcap..brags..that she hath jeer'd you into good pidgeon pies. These were soe good that there is not one left of them. 1779 J. Woodforde *Diary* 15 Apr. (1924) I. 249 We had for dinner..a Pidgeon Pye. 1843 *Ainsworth's Mag.* IV. 13 Swallowing a morsel of *foie gras* as uncognizantly as though it had been pigeon-pie. 1937 *Daily Herald* 8 Feb. 7/4 Pigeon-pie parties, an old-fashioned country custom, will be renewed this week in England and Wales. 1978 K. Bonfiglioli *All Tea in China* xx. 248 The black puddings, râgout of kidneys and pigeon pie..were barely touched.

d. *pigeon-tailed* adj.

1901 W. Churchill *Crisis* ii. ix. 202 The red pigeon-tailed coat. 1957 W. Faulkner *Town* iv. 83 Come in a Pullman in striped britches and a gold watch chain big enough to boom logs with and gold eyeglasses and even a gold toothpick and the pigeon-tailed coat.

6. pigeon drop *U.S. criminals' slang*, (*a*) a form of confidence trick which begins with the dropping of a wallet before the victim or 'pigeon'; (*b*) (see quot. 1959); so **pigeon dropper**; **pigeon dropping** = **pigeon drop* (a); **pigeon-fancier** (earlier and later examples); **pigeon grass**, substitute for def.: a grass of the genus *Setaria*, esp. *S. glauca* or *S. viridis*; (examples); **pigeon millet** = **pigeon grass*; **pigeon-pair** (further examples); also *transf.*; **pigeon-post** (further examples); **pigeon-weed** *U.S.*, the corn gromwell, *Lithospermum arvense*, or the spikenard, *Aralia racemosa*; **pigeon-woodpecker**, substitute for def.: a flicker, *Colaptes aureus*, found in eastern North America; (examples).

1937 E. H. Sutherland *Professional Thief* iii. 67 Among the short-con rackets, dropping the poke (also known as the pigeon-drop) is frequently used. 1959 *Washington Post* 18 Mar. A3/1 In the pigeon drop a confidence man tells his victim he has found a large amount of money and will share it if the victim will kick in some money as a show of good faith. 1961 Harney & Cross *Narcotic Officer's Notebk.* vi. 118 Sometimes it was the 'pigeon-drop'. A purse or billfold containing a considerable amount of money was dropped.

The 'sucker' was allowed to find it right along with a member of the mob. **1979** *Monitor* (McAllen, Texas) 22 July 2F/5 A Houston woman held on attempted theft charges claims to be part of a national 'pigeon drop' confidence ring. **1961** WEBSTER, Pigeon-dropper. **1977** J. WAMBAUGH *Black Marble* (1978) vi. 76 Pigeon droppers, pursepicks, muggers. Don't walk the Boulevard at night. **1850** *Green's St. Louis Directory for 1851* p. xviii, Such practice is immensely more disrespectable than procuring money under false pretenses—no more honorable than veritable pigeon-dropping. **1955** K. SULLIVAN *Girls who go Wrong* (1956) xii. 128 Elmira became the more proficient of the pair at the badger game—flim flam and pigeon dropping. **1970** C. MAJOR *Dict. Afro-Amer. Slang* 91 *Pigeon dropping*, confidence game-playing. **1807** SOUTHEY *Lett. from Eng.* I. xxi. 233 The Columbarians or pigeon-fanciers. **1822** M. EDGEWORTH *Let.* 19 Jan. (1971) 326 He explained to me what is meant by being *in the fancy*—pigeon fanciers—rabbit fanciers &c. **1941** [see *DART *sb.* 1 d]. **1976** *Deeside Advertiser* 9 Dec. 2/3 He was employed at the C.E.G.B. Power Station and a keen pigeon fancier. **1838** H. COLMAN *1st Rep. Agric. Mass.* (Mass. Agric. Survey) 128 There were several patches of black or pigeon grass when the dyke was built. **1901** C. T. MOHR *Plant Life Alabama* 358 Chaetochloa glauca... Pigeon Grass. **1926** F. W. HILGENDORF *Weeds N.Z.* facing p. 17 (*caption*) Pigeon grass (*Setaria glauca*). **1948** A. L. BLOMQUIST *Grasses N. Carolina* 186 One species (*S[etaria] viridis*) known as 'pigeon millet' or 'foxtail' is an obnoxious weed in cultivated ground in some of the Northern states. **1878** HARDY *Ret. Native* I. II. i. 240 She and Clym Yeobright would make a very pretty pigeon pair—hey? **1954** *Publ. Amer. Dial. Soc.* XXI. 34 *Pigeon pair*, a boy and girl, usually brother and sister, close enough in age to be congenial playmates. **1960** *Woman* 23 Apr. 5/2 A mother there had twins—boy and girl—on February 29, brother and sister for her pigeon pair born on the *previous* Leap Year Day. **1964** D. VARADAY *Gara-Yaka* xx. 178, I recognized Moll [*sc.* a lion] in the harem-nursery with three lively youngsters, while the two older females both appeared to have chipped in with a pigeon-pair each. **1909** *Chambers's Jrnl.* Oct. 661/2 Extensive practice is carried out in..photographing forts and positions of troops, and sending the films by means of pigeon-post to be developed and printed at headquarters. **1963** *Times* 3 May 13/7 It also became, thanks to Julius Reuter and his pigeon post of 1850, a vital link in the network of telegraphic communications. **1975** *Times* 15 Mar. 14/7 It will cost 7p to have a first-class letter delivered... To set up a pigeon post is remarkably cheap and easy. **1785** *Mem. Amer. Acad. Arts & Sci.* I. 431 *Aralia*... *Berry-Bearing Angelica.* Shot Bush. Pigeon Weed. Blossoms white. Berries black. Common in new plantations. **1851** J. F. W. JOHNSTON *Notes on N. Amer.* I. 305 Richer clover also had come up on another drained spot, and less of the pigeon-weed..with which this clay land is infested. **1889** G. VASEY *Agric. Grasses U.S.* (new ed.) 103 Pigeon-Weed ..grows chiefly in cultivated grounds. **1844** J. E. DEKAY *Zool. N.Y.* II. 192 This species..is called Highhole, Yucker, Flicker, Wake-up and Pigeon Woodpecker..in this State. **1847** THOREAU *Let.* 15 Feb. in *Corr.* (1958) 175, I remember a pigeon-woodpecker's nest in the grove on the east side of the yard. **1870** [see FLICKER *sb.*[4]]. **1917** T. G. PEARSON *Birds Amer.* II. 163 Flicker. *Colaptes auratus auratus*... Clape; Pigeon Woodpecker; Yellowhammer. **1955** *Amer. Speech* XXX. 181 Of bird names given for size, consider:..pigeon woodpecker.

b. pigeon's wing, (*c*) a type of wig worn in the 18th century; also *attrib.*

1753 [see *NEGLIGENT *sb.* 2]. **1966** J. S. COX *Illustr. Dict. Hairdressing* 113/1 *Pigeon's wing periwig*, a man's wig dressed with two horizontal rolls above the ears. The top, sides and back being dressed smooth and plain. This style was worn with different styles of queue.

pigeon: see PIDGIN in Dict. and Suppl.

pigeon-berry. 1. Substitute for def.: One of several plants whose fruit is attractive to birds, esp. in North America, the pokeweed, *Phytolacca americana*, or a dogwood, *Cornus canadensis* or *C. alternifolia*, and, in the West Indies, *Duranta repens*; also, the fruit of these plants.

1856 W. E. CORMACK *Narr. Journey across Newfoundland* iii. 13 The surface [of the ground] is bespangled ..by..*Cornus canadensis*, bearing a cluster of wholesome red berries, sometimes called pigeon berries. **1868** *Canad. Naturalist* III. 409 Among the most common plants which overspread the burned ground..are..the pigeonberry (*Cornus canadensis*); and the cherry. **1885** A. BRASSEY *In Trades, Tropics, & Roaring Forties* xviii. 434 A pretty tree with a large lavender flower, and great orange-coloured clusters of what are called 'pigeon-berries'. **1920** BRITTON & MILLSPAUGH *Bahama Flora* 372 *Duranta repens*... Bermuda; Florida; West Indies and Mexico to northern South America. Pigeon-berry. **1942** W. R. VAN DERSAL *Ornamental Amer. Shrubs* xii. 185 The plant [*sc. Cornus alternifolia*] has a number of common names, including..purple dogwood, pigeonberry, umbrella tree, and pagoda cornel. **1958** J. G. MACGREGOR *North-West of 16* v. 62 The bright red pigeon-berries, and the piles of spruce cones shelled out by the squirrels. **1971** H. R. W. LEWIS in E. L. Wardman *Bermuda Jubilee Garden* v. 60 *Duranta repens*. Pigeonberry, Sky-flower, Golden dewdrop. A slender evergreen shrub with drooping sprays of lilac-blue flowers that are followed by clusters of golden yellow berries. **1973** HITCHCOCK & CRONQUIST *Flora Pacific Northwest* 104 *P[hytolacca] americana* L. Pokeberry, pokeweed, pigeonberry.

2. pigeon-berry ash *Austral.*, an evergreen tree of the genus *Elæocarpus* or *Cryptocarya*; **pigeon-berry bush** = sense 1 above; **pigeon-berry tree** *Austral.*, the native mulberry, *Litsea dealbata*; also = *pigeon-berry ash*.

1785 *Mem. Amer. Acad. Arts & Sci.* I. 411 *Pigeon-Berry Bush*... Pigeons feed on the berries, which has been the occasion of its trivial name. **1832** W. D. WILLIAMSON *Hist. State Maine* I. 115 The pigeon-berry bush is as tall as that of a blackberry, bears an abundance of small purple berries, the chief food of pigeons. **1884** A. NILSON *Timber Trees New South Wales* 55 *E[læocarpus] obovatus.* —Ash; Pigeon-berry Tree.—A noble tree, attaining sometimes a height of 130 feet. **1889** J. H. MAIDEN *Useful Native Plants Austral.* 563 *Litsea dealbata*... 'Pigeon-berry Tree.' 'Native Mulberry.' 'Black Ash.' **1936** J. W. AUDAS *Native Trees Austral.* 190 Elaeocarpus Baurleni.. grows in New South Wales to the Queensland border, known popularly as Pigeon-Berry Ash. *Ibid.* 192 Elaeocarpus obovatus..is known as Pigeon-berry Ash. *Ibid.* 229 Litsea dealbata... A well-known tree of Queensland and New South Wales called Native Mulberry or Pigeon-Berry Tree. **1965** *Austral. Encycl.* III. 137/1 The rose maple or rose walnut..is most generally known to timber-getters as pigeon-berry ash.

pigeon-breast. (Earlier example.)

1842 DICKENS *Let.* 3 Apr. (1974) III. 180 That valiant general..is an old, old man with..the remains of a pigeon-breast in his military surtout.

pigeoneer (pidʒəniə·ɪ). *U.S.* [f. PIGEON *sb.* + -EER.] **a.** A person who trains or breeds homing pigeons, formerly esp. in the U.S. Army Signal Corps. **b.** (See quot. 1944[1].)

1918 *Boston Evening Rec.* 11 Jan. 9/2 A pigeoneer is an expert handler of homing pigeons. **1918** *Personnel Specifications* (U.S. Signal Corps) 21 Chief Pigeoneer:.. homing pigeon expert..in charge of the training and instruction of the men as pigeoneers. **1940** *Amer. Pigeon Jrnl.* Dec. 420/1 We have hundreds of young men of military age who have had the advantage of the experience and training of the pigeoneers of 1917-1918. **1944** *Sunday Jrnl. & Star* (Lincoln, Nebraska) 30 Apr. 8 Frank Robbins receives calls daily from people who wish pigeons to be exterminated around their homes... Assistant 'pigeoneer' at different times is Detective Captain Eugene Masters who is also an expert shot. **1944** M. B. COTHREN *Pigeon Heroes* i. 7 [The pigeon] dropped abruptly down on the landing board of its loft and walked through the trap door... In a jiffy a pigeoneer pulled out the message.

pigeon-hawk. (Earlier and later examples.)

1731 M. CATESBY *Nat. Hist. Carolina* I. 3 The Pigeon-Hawk..is a very swift and bold Hawk, preying on Pigeons and wild Turkeys while they are young. **1772** J. RUTTY *Ess. Nat. Hist. Co. Dublin* I. 297 The Pigeon Hawk or Goshawk..is said to breed in Ireland's Eye. **1831** J. J. AUDUBON *Ornith. Biogr.* I. 467 The Pigeon Hawk does not, I believe, raise its young within the United States, but somewhere farther to the north. **1913** H. K. SWANN *Dict. Eng. & Folk-Names Brit. Birds* 181 Pigeon Hawk: The Goshawk... also the Sparrow-hawk. **1958** E. T. GILLIARD *Living Birds of World* 111/2 A well-known small species of the Northern Hemisphere, the Pigeon Hawk or Merlin (*Falco columbarius*), is very similar to the Peregrine in general shape and color.

pigeon-hole, *sb.* Add: **2.** (Earlier example of a room.) Also, a small flat.

1777 P. THICKNESSE *Year's Journey* II. li. 151 All the rest of the apartments are pidgeon-holes, filled with fleas, bugs, and dirt. **1869** 'MARK TWAIN' *Innoc. Abr.* viii. 80 You can rent a whole block of these pigeon-holes for fifty dollars a month.

7. (Earlier and later examples.)

1688 LOCKE *Let.* 6 Feb. in *B. Rand Corr. J. Locke & E. Clarke* (1927) 245 Another way may be with pigeon-holes, as they call them; at these twenty-four holes, over the first paste an *A*, over the second a *B*, [etc.]. **1972** C. ACHEBE *Girls at War* 99 'Can I see your pigeon-hole?' ..'That's the glove-box. Nothing there.' **1978** *Lancashire Life* Nov. 151/1 Some find it possible to envy those who sit on the official side of counter or pigeonhole.

b. Also for the classification of persons, as by occupations.

1938 [see *ALL-ROUNDER]. **1957** [see *DEFERRAL].

8. (Earlier and later examples.)

1685 LOCKE *Jrnl.* 28 Aug. in P. King *Life Locke* (1829) 167, I saw a boor's house a mile or more from Amsterdam... There were three pigeon-hole beds, after the Dutch fashion. **1958** R. A. LYTTLETON *Mysteries Solar Syst.* iv. 137 On a simple pigeon-hole argument, the probability of the fission breaking up due to pure chance turns out to be less than 1 in 100.

pigeon-hole, *v.* Add: **1.** (Earlier and later examples.)

1840 C. CAMPBELL in T. Bland *Bland Papers* I. p. v, The lady..reached down a bundle of letters..from the interstices of the caves of the porch, where they were nicely pigeon-holed. **1855** *Knickerbocker* XLVI. 95 The bill of the gentlemanly proprietor..was deliberately met by a bill for 'damages to cow-catcher', and pigeon-holed. **1940** WODEHOUSE *Eggs, Beans & Crumpets* 154 Putting the prophet Hosea to one side for the moment and temporarily pigeon-holing the children of Adullam. **1949** *Jrnl. R. Aeronaut. Soc.* LIII. 410/1 Although the flying boat project was pigeon-holed, out of this early work the Mark I A Auto-pilot and the Queen Bee target aeroplane emerged. **1955** *Times* 15 June 9/3 Why has the Government pigeon-holed the Phillips report? **1963** *Ann. Reg. 1962* 205 Tentative plans for an exchange of television appearances by the Soviet and American leaders were also pigeon-holed. **1976** *Milton Keynes Express* 16 July 2/5 One plan that seems to have been pigeon-holed for the time being is the idea of finding another site for the College of Further Education.

2. (Examples with a person as object.)

1950 D. GASCOYNE *Vagrant* 59 Keep your labels for people who need them; I cannot be pigeonholed neatly.

1978 J. B. HILTON *Some run Crooked* v. 41 Why? 'To avoid Cantrell..having already pigeon-holed him as a kerb-crawler.'

3. (Further fig. example.)

1940 [see *DING AN SICH, DING-AN-SICH].

pigeonite (pi·dʒənait). *Min.* [f. the name of *Pigeon* Point in NE. Minnesota + -ITE[1].] A silicate of magnesium, ferrous iron, and calcium, $(Mg,Fe^{II},Ca)(Mg,Fe^{II})Si_2O_6$, that is a monoclinic calcium-poor pyroxene substantially free of aluminium and ferric iron, and that is found chiefly in basic igneous, esp. volcanic, rocks.

1900 A. N. WINCHELL in *Amer. Geol.* XXVI. 204 Since the anomalous condition was first discovered in the rocks from Pigeon point, it would be appropriate to call pyroxene thus optically abnormal pigeonite. **1931** *Amer. Jrnl. Sci.* CCXXI. 405 The clino-pyroxenes..form a continuous series of mix-crystals, from diopside-rich (the phenocrysts are always diopsidic augites) to pigeonite-rich and hypersthene-rich (pigeonites). **1951** *Jrnl. Geol.* LIX. 480/1 Pigeonites under plutonic conditions invert with cooling to orthopyroxenes. **1970** *Nature* 25 Apr. 334/1 Interstitial pigeonite and augite in the lunar gabbroic anorthosite have Ca contents of 0·6 and 6·9 per cent respectively. **1973** J. W. FROMBIE *Lunar Mineral.* ix. 217 Because of their chemical variability, exsolution textures, epitaxy with augite and orthopyroxene, ..and relative abundance, lunar pigeonites have played a significant role in a reconstruction of the crystallization histories of mare basalts.

Hence **pigeonit·ic** *a.*

1931 *Amer. Mineralogist* XVI. 207 The groundmass pyroxene is often said to be of the pigeonitic variety (Wahlstrom, Enkeljord; Vogelfjord, Faeroes, etc.). **1957** *Mineral. Mag.* XXXI. 540 It is evident that the degree of inversion of the pigeonitic pyroxenes..must, in general, be correlated with the composition of the minerals as well as with their cooling history. **1972** *Contrib. Mineral. & Petrol.* XXXV. 253 (*heading*) Variation of rare earth concentrations in pigeonitic and hypersthenic rock series from Izu-Hakone region, Japan.

pigeon-pea. 1. Substitute for def.: A leguminous shrub, *Cajanus cajan*, probably native to Africa, but widely cultivated in tropical and subtropical regions; = CAJAN, DAL. (Later examples.)

1907 FREEMAN & CHANDLER *World's Commercial Products* 260 Pigeon Pea or Dhol of commerce..is the seed of a low-growing shrubby plant..widely cultivated in the tropics and sub-tropics of both hemispheres. **1958** S. SELVON *Brighter Sun* v. 85 It just that you must love de tomatoes and lettuce and pigeon peas. **1969** *Oxf. Bk. Food Plants* 34/1 In the West Indies..pigeon peas provide a useful part of the protein supply in the diet of the poorer people, and a canning industry has been developed. **1972** J. ARNON *Crop Production in Dry Regions* II. vi. 155 The Pigeon-pea (*Cajanus cajan*) is a tropical legume of very ancient cultivation that is grown in tropical and subtropical regions of Asia and Africa. **1973** *Sunday Advocate-News* (Barbados) 15 June 8/1 Pigeon peas, rice..bread and cream of wheat are fairly good nutritional buys.

pigeon's milk. Delete '8', leaving 'Also pigeon milk'. **1.** (Later examples.)

1891 R. WOODS *Pract. Guide Successful Pigeon-Culture* vii. 73 The parent birds..are also provided with an ample supply of veritable pigeon's milk. **1897** FARMER & HASWELL *Text-bk. Zool.* II. 311 The so-called 'pigeon's milk' ..is at first covered with fine down, and is fed by the parents with a secretion from the crop, the so-called 'Pigeon's milk'. **1943** T. I. STORER *Gen. Zool.* xxxi. 659 In adult pigeons, while rearing young, the epithelial lining sloughs off as 'pigeon milk' used to nourish squabs in the nest. **1967** D. GOODWIN *Pigeons & Doves of World* 45 The parent takes the squab's bill between its bill, or, bill, own mouth and regurgitates 'pigeon's milk'. **1973** *Sci. Amer.* Dec. 44/2 They [*sc.* pigeons] breed fast, and they feed the young ones in the nest from the protein-rich 'pigeon milk', a curd formed in the parent bird's gut.

pigeon-toed, *a.* **2.** (Further examples.)

1880 *Harper's Mag.* Dec. 71/1 One would have imagined that he would deem it meet that Kittredge should be pigeon-toed. **1976** A. WHITE *Long Silence* i. 8 It does my heart good to see those lads come here pigeon-toed and flat-chested and go out..holding themselves upright.

pigeon-wing, *sb.* Add: **3.** (Further examples.) Also, a dance; the music for such a dance or dance-step.

1873 B. HARTE *Mrs. Skaggs's Husbands* 166 A light figure..cut a pigeon-wing,..and then advanced to the footlights. **1889** *Century Mag.* Apr. 855/2 A row of cavaliers..cut the pigeon wing in square-toed pumps. **1935** Z. N. HURSTON *Mules & Men* (1970) i. v. 117 Jack was justa dancin' fallin' off de log and cuttin' de pigeon wing—diddle dip, diddle dip—diddle dip 'from pine to pine Mr. Pinkney'. **1947** E. H. PAUL *Linden on Saugus Branch* 50 He had..a Chickering grand piano on which he played..all the reels, jigs, pigeon-wings, morisco, sarabands, [etc.]. **1954** *Publ. Amer. Dial. Soc.* XXI. 34 *Pigeonwing*, a type of dance involving the use of the arms.

pigeon-wing, *v.* Add: **b.** *refl.* To convey or transport (oneself) by or in the manner of one dancing or cutting pigeon-wings.

1839 POE *Devil in Belfry* in *Tales of Grotesque* (1840) 166 The rascal..pigeon-winged himself right up into the belfry of the House of the Town Council.

Piggery[1]. Add: **1.** (Earlier and later examples.)

1781 R. F. GREVILLE *Diary* 9 Aug. (1930) 20 Returned to The Grove after viewing The Farm & well kept Piggery. **1799** *Times* 1 June 4/3 Stabling for 3 horses, chaise-house, piggery, and out-buildings. **1936** *Times Lit. Suppl.* 29 Feb. 182/3 She contrasts the spoilt young things in luxurious homes with the tree young things in Chelsea piggeries. **1971** *Farmers Weekly* (Extra) 19 Mar. 5/1 For cleaning and sterilising piggeries, equipment, troughs and machinery, etc. **1973** T. A. BULMAN *Kamloops Cattlemen* xiv. 80 One part of the Grande Prairie layout was designated as the piggery.

2. (Further examples.)

1972 *Daily Tel.* 10 Nov. 15/2 The pin-up is just male chauvinist piggery. **1977** H. GREENE *FSO-1* i. 4 A black citadel of male-chauvinist piggery.

pigging, *vbl. sb.*[1] Add: **2. pigging back** *Metallurgy*, the addition of more pig-iron to the charge in an open-hearth furnace in order to raise its carbon content when this has fallen too much during boiling.

1900 *Jrnl. W. of Scotland Iron & Steel Inst.* VII. 128 For quick working, plenty of carbon in the pig in proportion to phosphorus is desirable,..otherwise there is delay by the addition of grey pig for 'pigging back' or..prolonging the boil. **1951** G. W. BASHFORTH *Manuf. Iron & Steel* III. vi. 174 The 'pigging back' of a large furnace to the extent of 20 points of carbon gave very beneficial results.

3. The use of pigs (*PIG *sb.*[1] 8*); cleaning by means of a pig.

1951 L. M. HARRIS *Introd. Deepwater Floating Drilling Operations* xix. 208 A system that permits: pump-down tools, pigging of lines; [etc.]. **1976** *Offshore Engineer* Mar. 52/3 Production manifold, the third subsystem, gives gathering, gas-lift, testing, pigging and through flowline capability.

pi-gging, *vbl. sb.*[3] *U.S.* [f. PIG *sb.*[1]] Hog-tying; only *attrib.* in **pigging string**, a short cord or rope used for hog-tying.

1968 E. SAXTER *Men & Horses* 4 On came the roper.. his 'piggin' string between his teeth. **1971** H. A. SMITH *View from Chivo* ii. 28 He stood for a moment looking at the rope in his hand, reflecting on the fact that it was known as a 'piggin' string.' **1976** 'D. HALLIDAY' *Dolly & Nanny Bird* vii. 96 Hog-tie it! Get the critter! Where's your pigging-string? Get him down, boys! Throw him!

pi·ggism. *rare.* [f. PIG *sb.*[1] + -ISM.] Pig-ish behaviour.

1852 MRS. GASKELL *Schah's Eng. Gardener* in *Househ. Words* V. 321/1 They ate their peas and beans unshelled.. a piece of piggism which especially scandalised him. **1979** *Daily Tel.* 20 Feb. 20 Women's Lib and male chauvinist piggism are both based upon the false premise that people are merely bodies instead of free spirits temporarily housed in flesh.

piggle (pi·g'l), *sb. dial.* [f. next.] (See quot. 1889.)

1889 *Cent. Dict.*, Piggle..., a many-pronged hook, with a handle like that of a hoe, used in digging potatoes, and in mixing various materials, as clay, mortar, compost, etc. **1922** C. SIDGWICK *Victorian* xxvi. 278 Impident toad! Comes up to me and sez we aren't agoin' to build this bridge with a foundation! Oh! aren't we? see I. You get to yer work, my lad, and do a bit with that piggle.

pi·ggle, *v. dial.* [Origin obscure: cf. PIDDLE *v.* 1.] **a.** *trans.* To uproot; to pick at, to pick out. **b.** *intr.* To trifle or toy *with*. So **piggling** *ppl. a.*, petty, paltry, niggling.

1847 J. O. HALLIWELL *Dict. Archaic & Provinc. Words* II. 621/2 *Piggle*, to root up potatoes with the hand. **1877** E. ROSS et al. *Gloss. Words Holderness* 107/2 *Piggle*, to pick out with a pointed instrument. **1911** D. H. LAWRENCE *White Peacock* II. iii. 265 Don't be piggling and mean and Grundyish. **1923** —— *Kangaroo* i. 7 Awful piggling suburban place. *Ibid.* iii. 49, I can't piggle with those draughtsmen dodges. **1976** SCOLLINS & TITFORD *Ey up, mi Duck!* I. 59 *Piggle*, to work away at something with the fingers. A certain spreader of dust.

Piggly-Wiggly (pi:gli,wi-gli). *U.S.* [Fanciful.] A type of self-service store (see quots. 1928, 1953). Also *attrib.* or as *adj.*

1917 *Printer's Ink* 20 Dec. 17/1 Perhaps no chain of stores that has been organized in recent years has been the object of so much curiosity and speculation as the Piggly Wiggly stores, which were started in Memphis, Tenn., in 1916. *Ibid.* 20/1 As an experiment in distribution, Piggly Wiggly is interesting and is worth watching. **1928** *Publishers' Weekly* 10 Nov. 1972/2 The obvious objection to the idea of a Piggly-Wiggly bookshop is that few book stores are adapted to the customary Piggly-Wiggly pattern, with the turn-stile entrances and exits, and all the merchandise set out onto tables in a U-shaped route-circle. **1953** *Sun* (Baltimore) 12 Oct. 8/1 News comes of the death..of Clarence Saunders, founder of the Piggly Wiggly stores. It is probable that few members of the younger generation know what a Piggly Wiggly was. Certainly they would not regard it as the novelty it was at the time of its creation 36 years ago. For the Piggly Wiggly..was nothing more than a clerkless grocery store in which a customer helped himself to goods off the shelves and paid at the door as he went out. In short, save perhaps for minor details.., the Piggly Wiggly represented a system that since has become standard as the super-market. **1955** 'P. QUENTIN' *Man with Two Wives* xxii. 246 He used to work in the Piggly-Wiggly.

That's a kind of store they have out there [in California] like the A. and P. **1971** S. WOODS' *Knavish Crows* v. 55 There were houses..a do-it-yourself laundry—a Piggly Wiggly supermarket. **1973** *Guardian* 2 June 11/5 Grants, New Mexico..has 56 garages, 421 motel rooms, a Piggly Wiggly Supermarket.

piggy, piggie, *sb.*[1] Add: **2.** (Earlier and later examples.)

1861 C. ROBINSON *Dial. Leeds* 384 *Piggy*, a game played by boys with sticks, and a piece of rounded wood, pointed at each end, called the 'piggy', which, when struck at either end, rebounds, similar to the game of 'cat' elsewhere. **1909** *Westm. Gaz.* 10 May 12/1 About 300 spectators attended the Barnsley Queen's Grounds.. on the occasion of a long knock piggy match for £50 and the championship of Yorkshire. **1971** C. BONINGTON *Annapurna South Face* iv. 49 We have..a game of 'Piggy', where you throw a small peg into the air, strike it as far as you can with a stick, and nominate the number of strides a member of the opposing team must take to reach the peg; he can challenge you to meet the nominated number of strides.

3. Special Combs., as **piggy-and-stick** = sense 2; **piggy bank**, a pig-shaped money-box, often made of pottery (see also quot. 1976); also (with hyphen) *attrib.*; **piggy-in-the-middle**: see *PIG *sb.*[1] 9; **piggy-stick** (*a*) *Mil. slang*, the wooden handle of a soldier's entrenching tool; (*b*) = sense 2.

1931 J. GOODWIN *Magazine* xi. 1 ix. 159 The little Jewboys..started playing ball or piggy-and-stick. **1941** *Baker Brothers Dry Goods, Home Goods, Toys Catal.* Spring 1138/2 Piggy Banks..flesh colored composition. **1958** *Woman's Mirror Wld Reliable* x. 129 Listen, I've busted banks.' 'You mean piggy banks?' **1955** A. HUXLEY *Genius & Goddess* 11 Ruth broke her piggy bank and squandered a year's accumulated savings on a make-up kit and a bottle of cheap perfume. **1972** *Daily Colonist* (Victoria, B.C.) 21 Jan. 25/1 Police said a piggy bank that found the house through an unlocked window. **1976** LICKORISH & KERSHAW *Canal Holidays* i. 20 To identify your own location..look for the nearest landmark. A few examples: the 'piggy bank' (toll booth) you just went through, a milepost, [etc.]. **1977** *New Yorker* 29 Aug. 60/1 The passengers who rode American trains were so honest that they didn't even use tickets—they just dropped the fare in a piggy bank for the driver to collect when he was ready. **1939** BROPHY & PARTRIDGE *Songs and Slang 1914–18* 149 *Piggy-stick*, the wooden helve for the entrenching tool which the infantryman carried next his bayonet... From a child's game, tip-cat. **1968** F. JENNINGS *Living Village* 179 The game which we knew in Coventry as *tip-cap*..was called in Yorkshire *Knur and spel*, in Cumberland *piggy-stick*.

piggy, *a.* Add: (Further examples.) Also, suggestive of pigs; bossily, unpleasant, unreasonable. Also *comb.*

1907 J. MASEFIELD *Midnight Folk* 24 There were seven old witches..at a very good supper... They were very piggy in their eating (picking the bones with their fingers, etc.). **1942** [see *GRABBY a.]. **1957** 'N. SHUTE' *On Beach* viii. 264 'Getting a bit piggy, aren't it?'..'Everything shut up, and dirty, and stinking.' **1958** —— *Rainbow & Rose* iv. 159, I started back and it began to rain and it got very piggy. **1963** S. PLATH *Crossing Water* (1971) 35 They have a piggy and a fishy air. **1970** *New Yorker* 26 Sept. 39/2 'I don't want you to make love to anyone but me.' 'That's piggy.' 'I know.' **1976** P. HILL *Hunters* xi. 158 Did she know her husband had lied to cover for Gatwood? Fat piggy-eyed Gatwood.

piggy-back: see PICK-A-BACK *adv. phr.* in Dict. and Suppl.

piggy-wiggy. (Further examples.)

1879 [see *PITTY[1]]. **1926** E. BOWEN *Last Sept.* vi. 62 Your scrumptious Irish teas make a perfect piggy-wiggy of me. **1957** [see *nursery word* s.v. *NURSERY 8a].

pig-hunting, *vbl. sb.* Chiefly *N.Z. Hist.* [f. PIG *sb.*[1] + HUNTING *vbl. sb.*] Hunting for wild pigs. Hence **pig-hunt** *v.*; **pig-hunter**. Also **pig-hunt** *sb.*

In quot. 1614 a humorous exaggeration.

1614 JONSON *Barth. Fair* (1631) II. iii. 34 These are Banbury-bloods, o' the sincere stud, come a pigge-hunting. **1845** E. J. WAKEFIELD *Adv. N.Z.* I. 227 I... went off to pig-hunt. **1867** W. T. L. TRAVERS *Hist. N.Z.* [ill.] 6 Shooting pigs and pig-hunting. **1928** *Harper's Mag.* June... [ill.] **1948** *Hurst's* [ill.] [this passage is partly illegible] ... tame... **1951** J. R. CLIFFE ... **1946** ... **1948** L. G. D. ACLAND *Early Canterbury Runs* (1951) x. 262 In 1878..one of the shepherds, took a pig-hunting dog. **1951** J. REID *Kiwi Laughs* 9 The misfortunes of that pig-hunt which is an almost indispensable episode in fiction or memoirs.

Pig Island. *Austral.* and *N.Z. slang.* Also with lower-case initials and in pl. [PIG *sb.*[1]] New Zealand, so called because of the introduction of pigs (which then went wild) by Captain Cook. Also *attrib.* So **Pig Islander**, a native or inhabitant of New Zealand.

1917 [see *HOOT *sb.*[3]]. **1917** E. MILLER *Camps, Tramps & Trenches* (1939) 65 We Pig Islanders are not nearly so hot-blooded in our manner of speaking [as Australians]. **1927** J. DEVANNY *Old Savage* 278 I'll back one pig island

miner against three of the best that ever came out of England. **1933** *Bulletin* (Sydney) 13 Sept. 20/4 The Pig Islanders are due in Britain in 1936. **1938** 'R. HYDE' *Nor Years Condemn* 164 They call us Hoonies...A lot of people called the New Zealanders Pig Islanders, and the Australians the Aussies. **1945** J. HENDERSON *Gunner Inglorious* 149 Back home in old Pig Island. **1946** L. R. C. MACFARLANE *Amuri* iii. 117 He returned to the Pig Island. **1948** in J. Reid *Kiwi Laughs* (1961) 163 The bell with all those soft-brained pig-islanders. They remind me of yahoos. **1951** *Landfall* VI. 264 It [*sc.* his voice] reverts to a rough pig-island twang. **1960** I. CROSS *Backward Sex* v. 103, I felt it, and was made easier by it, being a pig islander. **1967** F. SARGESON *Hangover* vii. 48 'Young man,' he said, 'it is my advice that you get off back to England... Pig Island is no place for the likes of you.' **1970** *N.Z. Listener* 11 Dec. 8/4 Another guy got black-mailed into taking a sheila half-way around Pig Island.

pig-jump, *v.* Add: (Further example.) Hence **pi·g-jump** *sb.*, a jump from all four legs without bringing them together; **pig-jumper** (further example).

1928 *Funk's Stand. Dict.* s.v. *Pig n.*, Pig-jump, n. **1929** K. S. PRICHARD *Coonardoo* 51 Up and down the colt went, slewing, rearing, pig-jumping. **1935** 'W. HATFIELD' *I find Austral.* iv. 52 Sending it [*sc.* a mule] away in a series of flying pig-jumps which Tim felt for his off stirrup. *Ibid.* xvi. 213 We'll see what you can do on this pig-jumper of ours.

pig Latin. Also **pig latin.** [f. PIG *sb.*[1] + LATIN *sb.*] An invented language formed by systematic distortion of the source language; *spec.* one in which the initial consonant or consonant cluster of each word is transferred to the end of that word and a vocalic syllable (usually (ē)) added. Also *transf.*, *attrib.*

1937 R. CHANDLER in *Dime Detective Mag.* Nov. 30/1 'Big white father say come now.' 'Don't give me any more of that pig Latin.' **1937** E. LYONS *Assignment in Utopia* (1938) iii. 101 Ideological hair-splitting and proletarian pig Latin. **1938** 'ELLERY QUEEN' *Four of Hearts* (1939) viii. 101 'Ixnay, ixnay,' said Jacques... **1964** [ill.] ... Feb. (1964) 22 But when anything, Latin or pig latin, was ever put so well? **1964** N. FITZGERALD *Let*. ... **1966** *Life* 26 Aug. 52/1... **1968** B. CLEARY *Fifteen* (1962) I. 10 '*Itzay atezay oomzay*', ..the boy was saying 'That that down,' in pig Latin. **1959** 'F. NEWTON' *Jazz Scene* v. 89 The whites, whose slang uses..pig-latin or 'tuti-fruiti' in the same way, ..for 'foe'—sufficiently indicates the tension between the races. **1959** I. & P. OPIE *Lore & Lang. Schoolch.* xiv. 321 'Pig Latin'..thus: 'Uncay thway' ... (June Smith is a pig)..has been spoken by children since before the First World War. **1960** C. GLEITZ in J. A. Fishman *Readings Sociol. of Lang.* (1968) 494 A kind of 'pig-latin' for foreigners. **1976** *Language* XLI. 219 Pig Latin and similar language games... **1978** R. MOORE *Big Paddle* i. 7 'Fee-a-zuck yee-a-zoo,' I'm wee-a-itth ee-a-zit!' Cliff remembered his father's warning about obscenities, but in carney pig-latin it didn't sound too bad.

pig-lead. (Later example.)

1903 J. MASEFIELD *Ballads* 19 With a cargo of Tyne coal, Road-rails, pig-lead.

pigment, *sb.* Add: **3. b. pigment epithelium** or **layer** *Ophthalm.*, the layer of the retina next to the underlying choroid, which consists of a single layer of pigmented cells having processes that extend between the rods and cones of the adjacent layer, and which continues forwards over the posterior surfaces of the ciliary processes and the iris.

1873 tr. M. *Schultze* in *Stricker's Man. Human & Compar. Histol.* III. xxxvi. 269 Although not generally continuous with the nerve fibres, the layer of pigment cells, ordinarily termed the pigment epithelium of the choroid, still belongs, both physiologically and morphologically, to the retina. **1892** A. DUANE tr. *Fuchs's Text-bk. Ophthalm.* iv. 147 The inner surface of the uvea is everywhere coated with a layer of pigmented cells, belonging to the retina and having the character of epithelial cells (pigment-epithelium). **1971** E. LENTE *Cell Fine Struct.* 356 The pigment epithelium has traditionally been considered as a layer of the retina. It may more logically, however, belong to the choroid. **1971** L. EDDY & N. SAXSTY ... the pigment layer ... **1972** *Human Anat. & Physiol.* [ill.] ... and ... **1974** D. SHEPHERD et al. ... the pigment layer absorbs extraneous light that might randomly stimulate receptor cells and create a poorer image.

pigment (pi·gmĕnt), *v.* [f. the *sb.*] *trans.* To colour with or as with a pigment.

1900 *Nature* 1 Mar. 416/1 To pigment the image, a piece of carbon tissue is soaked in a weak solution containing acetic acid, hydroquinone, and ferrous sulphate, squeezed on to the print and allowed to dry. **1968** A. S. M. HUTCHINSON *Once aboard Lugger* v. 1 ... The stain enters the blood and, thence oozing, pigments every part of the being. **1970** G. MACDONALD *Camera* vii. 175 Robert Demachy experimented skilfully with gum prints which allowed him to pigment and work the coatings until the images looked more like drawings than photographs.

Hence **pi·gmenting** *ppl. a.*

1906 *Westm. Gaz.* 1 Dec. 18/2 The effect of the silver image in the bromide print, in conjunction with the 'pigmenting' solution taken up by the plaster, is to render the pigmented compound forming the plaster..insoluble. **1958** *Engineering* 21 Mar. 384/1 The material is a blend

of polymeric materials, solvents, stabilisers, and pigmenting compounds. **1971** *Brit. Poultry Sci.* XII. 206 The purpose..was to evaluate the efficacy of canthaxanthin for the augmentation of naturally occurring pigmenting compounds found in yellow corn and alfalfa for the pigmentation of broilers.

pigmentocracy (pigmĕntǫ-krāsi). [f. PIGMENT + -OCRACY.] A ruling class made up of people of one skin-colour (usu. white); a country or state with such a ruling class.

1952 *Economist* 6 Dec. 702/2 A natural white aristocracy, with aristocratic virtues as well as vices, was already becoming merely a pigmentocracy. **1956** A. SAMPSON *Drum* xv. 210 In the 'pigmentocracy' of South Africa, skin colour was firmly linked with money and success. **1959** *New Statesman* 19 Dec. 874/3 South Africa is a pigmentocracy, dedicated before God and the whole world to the proposition that 'South Africa is the white man's country: it shall never be ruled by Kaffirs, Hottentots and Coolies.'

pig-nut. 3. Substitute for def.: *N. Amer.* The small pear-shaped nut of the broom hickory, *Carya glabra*, or the closely related species, *Carya ovalis*; also, the trees themselves, which belong to the family Juglandaceæ; = HOG-NUT 1. Also *attrib.* (Earlier and later examples.)

1666 *Early Rec. Warwick, Rhode Island* (1926) 323 Upon a straight lyne from the pond to a pignut tree standing upon a hill. **1705** R. BEVERLEY *Hist. Virginia* II. iv. 16 There are also several Sorts of Hickories, call'd Pig-nuts. **1785** H. MARSHALL *Arbustrum Amer.* 68 White, or Pig-nut Hickery [*sic*]... generally grows pretty large. **1832** D. J. BROWNE *Sylva Amer.* 183 The pignut hickory is one of the largest trees of the American forest. **1908** N. L. BRITTON *N. Amer. Trees* 237 Pignut hickory. ..A tree of drier ground than that in which most other hickories grow. **1969** T. H. EVERETT *Living Trees of World* 98/2 The pignut..occurs as a native in dryish soils from New York to Missouri and Florida. *Ibid.*, The closely related sweet pignut..differs in that its leaves have seven leaflets rather than the usual five.

pigskin. Add: *Sporting slang*, (*a*) (further examples); also *attrib.*; (*b*) *U.S.*, a football; also *attrib.*

1894 *University of Chicago Weekly* 11 Oct. 8/2 Roby put the pigskin over the line. **1898** *Sporting Times* 26 Nov. 3/3 He..has again electrified English turf followers by riding rings around their crack knights of the pigskin. **1928** GALSWORTHY *Swan Song* II. i. 105 Val..had picked him up on his retirement from the pig-skin in 1921. **1941** BAKER *Dict. Austral. Slang* 54 *Pigskin artist*, a jockey. **1945** *Richmond* (Virginia) *Times-Dispatch* 25 Oct. 10/6 Our football prophets are unfortunate fellows... I was told by the prophets of the pigskin that Ohio State would romp over poor Purdue. **1970** *New Yorker* 3 Oct. 34/3 A quick-thinking distaff pigskin zealot. **1974** *Anderson* (S. Carolina) *Independent* 24 Apr. 5B/1 He carried the pigskin on the end around 11 times for 73 yards, or an average of 6.6 yards per carry. **1977** *Time* 14 Nov. 49/2 The British-born geologist..may not help his school's pigskin standings, but no matter.

2. *Med.* = *PEAU D'ORANGE.

1898 [see *PEAU D'ORANGE]. **1943** C. F. GESCHICKTER *Dis. Breast* xx. 480 There is no sharp dividing line between the inflammatory change in the skin observed in acute carcinoma and the more common pigskin or lenticular dermatitis observed late in the disease in cases of large infiltrating mammary cancer. **1966** WRIGHT & SYMMERS *Systemic Path.* I. xxviii. 1010/2 Other cutaneous changes are local oedema ('pigskin', *peau d'orange*..) due to tumour cells growing in and blocking the lumen of the lymphatics in the superficial tissue of the breast, [etc.].

pig-stick (pi·g,stik), *sb.* [f. PIG-STICKING.] A wild-boar hunt; a pig-sticking. Also *fig.*

1906 *Westm. Gaz.* 13 July 2/1 Pickle meanwhile was having a pig-stick on the sands, with Floss in the rôle of pig. **1906** *Daily Chron.* 26 Oct. 3/4 His book..is just a simple account of his every-day life in Algeria, including a bath and a shave, a 'pig-stick', and a visit to a café.

pigsticker. (In Dict. s.v. PIGSTICKING.) Add: **c.** A sharp implement or weapon, as a lance, bayonet, knife, etc. *slang.*

1890 BARRÈRE & LELAND *Dict. Slang* II. 129/1 *Pig-sticker* (army), sabre. **1895** *Funk's Stand. Dict.*, *Pig-sticker*..3. A boar-spear. 4. (Slang.) A large pocket-knife. **1902** FARMER & HENLEY *Slang* V. 198/2 *Pig-sticker*... 2. (common).—A long-bladed pocket-knife; and (3) a sword. **1918** C. J. SWAN *My Company* viii. 129 All this time the steel pigsticker was resting on his rotund stomach with the rifle behind it cocked. **1941** J. SMILEY *Hash House Lingo* 43 *Pig sticker*, carving knife. **1964** G. L. COON *Short End* 242, I had awful visions of somebody clinging to the business end of that pig sticker of mine. **1978** A. MELVILLE-ROSS *Blindfold* xxx. 191 Trelawney crossed to the far wall, yanked the knife from it... 'You'll hand over that pig-sticker and come home with uncle.'

d. A butcher of swine (cf. PIGSTICKING 2). *dial.* and *slang.*

1886 H. BAUMANN *Londinismen* 138/1 *Pig-sticker*.. scherzhaft: Schweineschlächter. **1895** in *Funk's Stand. Dict.* **1939** F. THOMPSON *Lark Rise* i. 12 The travelling pork butcher, or pig-sticker. **1948** *Amer. Speech* XXIII. 317/2 *Pig sticker*, man who bleeds hogs.

pigsticking. Add: **1.** Also *attrib.*

1907 *Yesterday's Shopping* (1969) 666/2 The usual length for pig-sticking lances is:—For use in Bengal Presidency, 6 ft. 6 in. to 7 ft. **1910** *Blackw. Mag.* Apr. 559/1 The Ganges Cup was first run for in 1869 after the pig-sticking season. **1972** *Shooting Times & Country Mag.* 1 July 21/2, I now treat you to a potted picture of a typical pig-sticking run.

pigtail. Add: **1. d.** *Electr.* A short length of flexible conductor; *spec.* one in an electrical machine connecting a brush to its brush-holder; (see also quot. 1971).

1903 HAWKINS & WALLIS *Dynamo* (ed. 3) xix. 606 The flexible copper conductor..forms a twisted pig-tail with enough slack to allow of the normal amount of brush movement. **1949** *Jrnl. Appl. Physics* XX. 805/2 Some cartridges..were fitted with pigtails for convenience in wiring. **1962** C. O. SWANSON in Roberson & Farrior *Guidance & Control* 400 The pigtails, which pass power to and from the floated assembly, are very small in size. **1963** ROSENBLATT & FRIEDMAN *Direct & Alternating Current Machinery* ii. 15 Except in small dynamos, the current is taken from the brush by means of a flexible copper wire embedded in the brush, called the pigtail. **1964** R. F. FICCHI *Electr. Interference* ix. 166 The conductor may be buried in the concrete slab under the equipment with pigtail conductors protruding above the concrete for equipment grounds. **1971** M. TAK *Truck Talk* 118 *Pigtail*, the cable that transmits electricity to the trailer from the tractor.

pigtailed, *a.* Add: **1.** Esp. in the name of the **pigtailed macaque** or **monkey**, *Macaca nemestrina*, native to southern Asia, Sumatra, and Borneo. (Later examples.)

1932 S. ZUCKERMAN *Social Life Monkeys* vi. 90 The cycle in the pig-tailed macaque varies roughly form thirty to forty days. *Ibid.* xi. 188 The pig-tailed monkeys that live on the Island of Singapore are believed to be escaped captives. **1969** *Daily Tel.* 22 Aug. 19/8 A pig-tailed monkey was born at London Zoo on Sunday. The species..is named because of its ability to curl its tail like a pig's. **1977** ROONWAL & MOHNOT *Primates of S. Asia* 177 Owing to the local demand for pig-tailed macaques for food..its numbers are rapidly declining all over its range.

pigweed. Substitute for def.: A name given to many plants used as animal fodder or potherbs, esp. goosefoots belonging to the genus *Chenopodium* and amaranths, esp. *Amaranthus retroflexus*; in Australia, a name for purslane, *Portulaca oleracea*. (Earlier and further examples.)

1806 T. G. FESSENDEN *Orig. Poems* (rev. ed.) 17 The hyacinth and daffodil, With now and then a big weed Of purslain and of pig weed. **1835** J. H. INGRAHAM *South-West* II. 110 A weed not unlike the common pig-weed. **1850** [see *apple-peru*]. **1864** R. HENNING *Let.* 4 Mar. (1966) 157 The Irish family..were so alarmed at the idea of getting scurvy also..that the two little girls..devoted their leisure to picking 'pigweed', rather a nasty wild plant, but supposed to be exceedingly wholesome, either chopped up with vinegar or boiled. **1909** L. M. MONTGOMERY *Anne of Avonlea* xxvii. 319 I've begun to grow like pigweed in the night. **1911** W. R. GUILFOYLE *Austral. Plants* 298 Portulaca oleracea. 'Common Purslane' or 'Pigweed'. **1927** M. M. BENNETT *Christison* xi. 116 It's a hungry place, Lammermoor! Nothing to eat but pigweed and mutton. **1941** I. L. IDRIESS *Great Boomerang* vii. 55 There were creepers too, crowfoot and pigweed, parakelia and geranium. **1943** FERNALD & KINSEY *Edible Wild Plants Eastern N. Amer.* 177 Pigweed, Goosefoot, Lamb's-Quarters, *Chenopodium album* and 14 other species... The common Pigweed, so familiar in rich garden soil, in barnyards, and similar habitats, has always been a popular potherb. **1945** J. M. FOGG *Weeds of Lawn & Garden* 75 The Green Amaranth or Pigweed is the most common species of the genus found as a garden weed. **1965** *Austral. Encycl.* VII. 232/1 The almost cosmopolitan *P[ortulaca] oleracea* (..purslane or pigweed) is used as a green food by Indonesian and Polynesian peoples. **1966** L. J. KING *Weeds of World* i. 12 The grain amaranths are inextricably involved in any history of the non-cultivated or weedy amaranths or 'pigweeds'. **1975** D. MCCLINTOCK *Wild Flowers of Guernsey* 86 Pigweeds [*sc.* species of *Amaranthus*] are unexpectedly rare in Guernsey.

pi-jaw (pəi·dʒǭ), *sb. slang* (now *arch.*). [f. PI *a.* (*sb.*) + JAW *sb.*[1]] A pious lecture or exhortation, esp. one addressed to schoolboys or young persons by their teachers or parents. Hence **pi-jaw** *v. trans.*, to lecture or exhort; **pi-·jawing** *vbl. sb.*

1891 R. G. K. WRENCH *Winchester Word-bk.* 31 He pi-jawed me for thoking. **1912** G. W. E. RUSSELL *One Look Back* ii. 25 It was his custom..to call us all together.., and give us what we called a 'Pi-jaw'. **1913** *Pearson's Mag.* June 606/2 There is no suspicion of 'pi-jaw' about it. **1922** A. S. M. HUTCHINSON *This Freedom* IV. iv. 303 You..get me here to pijaw me about my duty to my pretty young wife. **1923** *Blackw. Mag.* Jan. 56/1 He..treated me to the Persian equivalent of a 'pi-jaw'. **1925** M. I. ROGERS in *Inner Life* (ser. 2) xiii. 257 Older children..are more interested in ideas and the way in which things happen. They dislike 'pi-jaw'. **1930** J. DOUGLAS *Down Shoe Lane* 210 It may be that they yawn over pompous pi-jawing and middle-aged platitudinarianism. **1937** G. FRANKAU *More of Us* xi. 121 The tortures of a miserable Upper Pi-jawed beyond the sacred hour of supper. **1945** G. B. GRUNDY *55 Yrs. at Oxf.* 38 What a boy or young man loathes above all is pi-jaw.

pika (pəi·kă, pī·kă). Also **pica.** Substitute for def.: A small herbivorous quadruped belonging to the genus *Ochotona*, closely related to hares and rabbits, distinguished by short, rounded ears, reddish-brown or grey fur, and the lack of a tail, and found in mountainous regions of western North America and north-east and central Asia. (Later examples.)

1851 J. RICHARDSON *Arctic Searching Exped.* I. v. 178 The little Pika, or tail-less hare, occupies the grassy eminences, and lays up a stock of hay for winter use. **1858** J. PALLISER *Jrnl.* 16 Aug. in I. M. Spry *Papers of Palliser Exped. 1857–1860* (1968) 294, I also heard the squeaking note of the little Pica or tailless hare... It is about the size of a small rat, but made exactly like any other rabbit, excepting that it has round open ears. **1925** E. F. NORTON *Fight for Everest: 1924* 173 Our ponies were off after them in a mad gallop down the nullah side, heedless of rocks and pika burrows. **1936** D. MCCOWAN *Animals Canad. Rockies* iv. 36 Pikas feed on grass and the many varieties of small Alpine plants. **1958** L. WHISHAW *As far as You'll take Me* xi. 176, I had the impression that a pika had spotted us. **1964** L. S. CRANDALL *Managem. Wild Mammals in Captivity* 201 The round-eared, nearly tailless pikas (*Ochotona*) are found in rock areas or 'slides', generally at high altitudes. **1971** L. H. MATTHEWS *Life of Mammals* II. v. 138 The genus *Ochotona*, the only one in the Ochotonidæ, contains about twelve species of whistling hares or rock rabbits, generally known as pikas. **1973** P. GEDDES *Ottawa Allegation* vi. 83 A pika is a bunny... When the sun shines he..cuts all the grass he wants for winter and lays it out to dry.

pikau (pī·kau), *sb. N.Z.* Also **peko, pico.** [Maori.] A pack for carrying on the back, a knapsack, a swag. Also *attrib.* Hence **pi·kau** *v. trans.*, to carry (a pack) on the back.

1836 J. A. WILSON *Jrnl.* 23 Aug. in *Missionary Life & Work N.Z.* (1889) III. 48 Our natives weary with their *pikaus*. **1847** G. F. ANGAS *Savage Life* II. i. 3 [The natives] severally carried [our baggage] in their *pikau* or knapsacks, strapped over their shoulders with the leaves of flax. **1848** *McLean Papers* V. 265 (MS.) One of the natives left us at Woons..so we had to put the *pico* on the mule and trudge along. **1851** J. C. RICHMOND *Let.* 25 Mar. in *Richmond–Atkinson Papers* (1960) I. ii. 79 We were 6 in party, ourselves, young Stark & 3 Maoris with 'pikau' burdens. **1874** W. M. BAINES *Narr. E. Crewe* iii. 49 [The pig's] legs [were] tied together so as to make him a handy 'pikau'. **1882** W. D. HAY *Brighter Britain!* II. 94 Both men and women [*sc.* Maoris] are able to pikau (hump, or carry on the back and shoulders) great weights. **1882** T. H. POTTS *Out in Open* 24 A line of women and girls..have just brought away from the potato pits their heavily-laden kits, carried pikau fashion. **1892** E. S. BROOKES *Frontier Life* 148 Most of us were now loaded with heavy pekoes or swags on our backs. **1902** W. SATCHELL *Land of Lost* vii. 50 He.. commenced to fill his *pikau*. **1911** *Chambers's Jrnl.* 4 Mar. 223/2 We succeeded almost noiselessly in doing to death a fine young sow, and, having cut up the carcass, started for camp with the meat 'pikaued' (carried as in a knapsack) in sugar-bags on our shoulders. **1950** F. SARGESON in *Landfall* IV. 285 The rest of his [*sc.* a farmer's] luggage would be a *peko* slung on his shoulders. **1958** *Tararua* XII. 27 The gumdiggers used a *pikau*, a bag with shoulder straps, a good deal smaller than the usual tramper's pack. **1960** B. CRUMP *Good Keen Man* 123 To ensure [my dog] Flynn's co-operation I tied him to my belt with the rope off my pikau.

pike, *sb.*[1] Add: **II. 2.** (Later example.)

1976 *New Yorker* 9 Feb. 32/3 Her sleeve catches in the metal pike of the turnstile and Jane picks it out, in the nick of time.

III. 4. b. (Later example.)

1957 [see *BICK].

pike, *sb.*[6] Add: **2.** (Earlier and later examples.) Also *fig.* Freq. in phr. (N. Amer. colloq.) *to come down the pike*: to appear on the scene; to come to notice. *to hit the pike*: see *HIT *v.* 11a.

1812 M. EDGEWORTH *Absentee* in *Tales Fashionable Life* VI. xvi. 377 Keep the *pike* till you come to the turn at Rotherford, and then you strike off into the by-road to the left. **1899** B. TARKINGTON *Gentleman from Indiana* iv. 44 The roans setting a sharp pace as they turned eastward on the pike toward home and supper. **1903** [see *FLOSSY *a.*]. **1904** [see *HIT *v.* 11 a]. **1907** 'E. C. HALL' *Aunt Jane of Kentucky* v. 107 Horseback riders had been pouring into town over the smooth, graveled pike. **1910** W. M. RAINE *Bucky O'Connor* 73 Cut loose and hit the pike for yourself. **1912** [see *duck soup* s.v. *DUCK *sb.*[1] 12]. **1926** *Flynn's* 16 Jan. 640/2, I also got cases on a couple of flivvers for a getaway, as we were about fourteen kilos from the main pike. **1934** E. LINKLATER *Magnus Merriman* 82 Then I hit the pike for home. **1949** *Sun* (Baltimore) 12 Oct. 12/1 Unfortunately, the State cannot control roadside development on the sections of the pike completed thus far. **1956** E. O'CONNOR *Last Hurrah* v. 101 Your uncle's the ablest politician to come down the pike in these parts in the last fifty years. **1963** [see *defenceman, defenseman* s.v. *DEFENCE *sb.* 9]. **1966** R. S. RUDNER *Philos. Social Sci.* 2 'Normative' itself is not the clearest term to come to us down the philosophical pike. **1968** *Down Beat* 7 Mar. 19/2 Jack thought that Jimmy was just about the greatest 'bone that had ever come down the pike, and Jimmy felt the same way about Jack, putting him above Miff Mole, who also was a tremendous trombone on that scene. **1970** *New Yorker* 28 Feb. 41/1 A big truck went by way up on the pike. **1974** *Anderson* (S. Carolina) *Independent* 24 Apr. 2A/1 The plan is 'halfway down the pike' but much more effort is needed to make its redevelopment a reality. **1976** *New Yorker* 26 Apr. 59/3 You could see how tired she was after a few hours of 'being presentable

to every Tom, Dick, and Harry who comes down the pike'.

b. A railway line or system. Also of a model railway. *U.S.*

1940 *Railroad Mag.* XXVII. VI. 69/1 Lake Erie & International..recently highballed its first Limited around the newly completed circuit of main line. The pike boasts one Diesel-electric and one steam loco.., and its rolling stock is steadily growing. *Ibid.*, Metropolitan Society of Model Engineers... tackled three layouts at one time. The first, a small HO pike was recently presented to the Union Station. **1945** F. H. HUBBARD *Railroad Avenue* ii. 9 They knew he [*sc.* Casey Jones] never dawdled at coal chutes, water cranes, or cinder-pit tracks or wasted time along the pike. **1945** *Railroad Mag.* XXXVII. II. 13 The financing of this little pike was an epic in itself... People along the route were so eager to see the rails laid that they donated labor.

3. *pike-road* (earlier example).

1838 'J. PUNKIN' *Downfall of Freemasonry* II. 115 This threw everything into commotion, and as the teamsters say of a drove of cattle on a dirty pike road, 'kicked up a dust'.

pike (pəik), *sb.*[10] *N. Amer. dial.* [f. *Pike* County, Missouri, whence the first of these persons are said to have come to California.] Term of contempt on the Pacific coast for a person of no means or of migratory habits; a poor white; a thief. Cf. *PIKER[4]. Also as *adj.*

1854 G. H. DERBY in *Pioneer* (San Francisco) June 379 A tall yellow-haired, sun-burned Pike, in the butternut-colored hat, coat and so forths 'of the period'. **1860** C. W. WILSON *Mapping Frontier* (1970) II. 126 There are about 350 inhabitants, miners, gamblers, sharpers, Jews, Pikes, Yankees, loafers & *hoc genus omne*. **1863** *Harper's Mag.* June 25/2 Society in San José is decidedly 'Pike' in its character. **1872** C. NORDHOFF *California* xi. 138 The true Pike, however, in the Californian sense of the word, is the wandering, gipsy-like southern poor white. **1928** R. W. RITCHIE *Hell Roarin' Forty-Niners* xv. 234 This Pike had an imagination and a devilishly sly humor which would qualify him to-day for one of our highly specialized lines of salesmanship. **1946** *St. Louis* (Missouri) *Globe-Democrat* 17 Nov. E 2/6 The term 'Pike' or 'piker', in the sense of a worthless, lazy, good-for-nothing person arose first in California in the days of the Forty-Niners.

pike (pəik), *sb.*[11] [Origin obscure.] A position of the body in diving (see quot. 1928); cf. *JACK-KNIFE *sb.* 3. Also, a similar body position in gymnastics. Also *attrib.*

1928 *Daily Express* 13 July 4/4 For a pike dive spring up as for a header, then bend sharply at the waist and touch the toes without bending knees or ankles, then straighten again and enter head first. **1931** *Morning Post* 21 Aug. 14/5 The pike reverse is a combination of the front pike and reverse. **1956** KUNZLE & THOMAS *Freestanding* v. 60 Aim for a rhythmic drop and beat without any intermediate pause, but make sure that the pike is a full one. **1964** *Trampolining* ('Know the Game' Ser.) 11/2 Piked straddle jumping. Similar to the Pike Jump. **1974** *Rules of Game* 203 In pike dives with twist, the twist must follow the pike.

pike, *v.*[3] Add: **2.** (Further examples.) Also, to proceed or go. Also with other adverbs.

1753 [see *LUMBER *sb.*[2] 3]. **1846** *Swell's Night Guide* 127/2 *Pike off*, run away. **1864** 'E. KIRKE' *Down in Tennessee* xiii. 162, I piked off for the ruin. **1886** *Outing* IX. 49/2 Tell ye what, jist climb onto my pony, an' we'll pike fer the spring. **1893** H. FREDERIC *Copperhead* (1894) 191 It looked kind o' curious to me, your pikin off like that. **1900** ADE *More Fables* (1902) 106 When all the Smart Set got ready to pike away for the Heated Term..she would remain at Home. **1902** H. L. WILSON *Spenders* iv. 44 Do me a last favour before you pike off East. **1904** S. E. WHITE *Silent Places* vi. 50 'We'd better pike out, if we don't want to get back with th' squaws,' suggested Dick. **1909** R. A. WASON *Happy Hawkins* xvii. 207, I piked on over to Danders thinkin' I'd get on a train an' go somewhere. **1924** P. MARKS *Plastic Age* 18 Say, I've got to pike along; I've got a date with my faculty adviser. **1927** H. CRANE *Let.* 4 July (1965) 303 My old jack tar friend..was back from his long trip..so I just piked in and saw him.

3. *intr.* To shirk; to hold back; to back *out*; (see also quot. 1889).

1889 FARMER *Americanisms* 420/1 *To pike* (Cant), to play cautiously and for small amounts, never advancing the value of the stake... Those who gamble in this fashion are called *pikers*. **1954** A. FULLERTON *Bury Past* I. iii. 47 Queer fellow—he worked like hell at the office stuff, the routine, the paper-work, and then suddenly when it came to the push, he'd pike, like he'd done tonight. **1959** *Numbers* Feb. 13/1 'No dogs', said Mr Reginald. 'You wouldn't be piking, would you?' Sonny murmured. **1969** *Southerly* XXIX. 287 Ann's heart began to thump with her secret fear. She waited for someone to veto the idea, not daring herself. But no one did. She could not 'pike' out.

pike (pəik), *v.*[8] *Diving* and *Gymnastics.* [f. *PIKE *sb.*[11]] **a.** *intr.* To adopt a pike position. **b.** *trans.* To move (a part of the body) as for adopting a pike position. Cf. *PIKED *a.*[2]

1956 KUNZLE & THOMAS *Freestanding* iii. 36 When falling backwards with straight legs, first pike, then drop backwards. *Ibid.* 41 Roll over one shoulder with a half turn, tucking in the head and piking at the hips. **1964** G. C. KUNZLE *Parallel Bars* ii. 45 Pike the hips sharply..and press off the bar strongly with the arms.

piked (pəikt), *a.*[2] *Diving* and *Gymnastics.* [f. *PIKE *sb.*[11] or *v.*[8]] In a pike position; with the body in a pike position.

1951 *Swimming* (Eng. Schools Swimming Assoc.) v. 71 There are three recognized positions in which the body may be held during the execution of a dive... *Piked.* The body is bent forward at the hips, but the legs must remain straight at the knees with toes pointed. **1956** KUNZLE & THOMAS *Freestanding* iii. 34 If it is properly timed, the roll will finish in piked stand... Loose hamstrings are essential to allow sufficient pike at the hips. **1964** *Trampolining* ('Know the Game' Ser.) 11/1 Piked jumping. Keep trunk as erect as possible... Point toes and touch upper insteps. **1974** *Rules of Game* 36 Two or three running steps into forward piked salto, land on one leg *Ibid.* 38 Swing forward over bar with legs piked.

pikelet[1]. Add: (Later example.)

1974 P. WRIGHT *Lang. Brit. Industry* iv. 43 *Pikelet* is used in the West and Midlands for a round teacake with small holes, to be buttered and toasted; but in other districts it seems to be a sticky unsweetened crumpet or else a muffin.

b. Chiefly *Austral.* and *N.Z.* A drop-scone.

Known personally to me in N.Z. in the 1920s.—Ed.

1943 A. L. SIMON *Conc. Encycl. Gastron.* IV. 97/2 *Welsh Pikelets*..flour..sugar..salt..bicarbonate of soda..buttermilk..Take a tablespoon of this batter and fry in a little hot lard..turn when half cooked. **1952** B. NILSON *Penguin Cookery Bk.* xviii. 332 Drop Scones, Pikelets ..or Scotch Pancakes. Cooking time 3-4 minutes each batch. **1963** *Moderna Språk* LVII. I. 4 *Pikelets*, ..This is the name of the most popular teacake in Australia, yet most Englishmen from South England at least, have never heard of it. **1965** S. T. OLLIVIER *Petticoat Farm* vi. 83 'I've brought some pikelets,' she said. **1970** D. M. DAVIN *Not Here, Not Now* II. ix. 119 Then she made pikelets for tea. **1977** *N.Z. Woman's Weekly* 10 Jan. 54/1 It is then easy to slip the ice in sheets from the sides by using the spatula normally used for turning girdle scones or pikelets.

pike-pole. For *U.S.* read *N. Amer.* and add earlier and later examples.

1850 N. KINGSLEY *Diary* (1914) 139 The weeds are put down with a pike pole and the pressure of the water keeps them to their place. **1926** *Daily Colonist* (Victoria, B.C.) 5 Jan. 1/3 Mattatall sighted it [*sc.* a bag] and hauled it up with a pike pole. **1945** D. D. CALVIN *Saga of St. Lawrence* 68 Men..sorted out with their long pike-poles (which were like twenty-foot boat-hooks with a sharp point and hook) the longer, thinner pieces. **1964** *Daily Colonist* (Victoria, B.C.) 27 Sept. 36/2 A powered boom scooter rides herd on logs in the water where the timber used to be shoved around by a man with caulk boots, a pike pole and the agility of a cougar. **1968** R. M. PATTERSON *Finlay's River* 103 Careering downstream, steering their heavy craft by means of long pike-poles, they crashed into one driftpile, sheered out and spun end for end three times.

b. A long pole having a fire-hook at one end.

1949 *Chicago Daily News* 17 Sept. 1/7 Firemen worked with pike pole and shovel in the wreckage. **1969** *Publ. Amer. Dial. Soc.* LII. 34 *Pike pole*, a rake-like device used to pull plaster loose or to clear an area. 'Clear it out with the pike pole.'

piker[3]. Add: **2.** *Austral.* A wild ox living in the bush. Also *transf.*

1887 *All Year Round* 30 July 67 'Pikers' are wild cattle. **1893** K. MACKAY *Out Back* (ed. 2) III. vi. 265 Blowed if this cask ain't harder to round up nor a mallee piker. **1904** 'G. B. LANCASTER' *Sons o' Men* 22 The grunt of broken-winded pikers came clear above the sharp crackle of undergrowth where the boys rode. **1936** I. L. IDRIESS *Cattle King* vii. 62 The boy bought a teamster's cast-off bullock for two pounds. It was an old piker, worked to the very bone. **1941** —— *Great Boomerang* vii. 51 Fine upstanding beasts, the pick of a dozen stations. No old 'pikers' these.

Pi·ker[4]. [f. *PIKE *sb.*[10] + -ER[1].] = *PIKE *sb.*[10]

1859 in L. Hafen *Colorado Gold Rush* (1941) 318 An extra train of [returning] 'Pikers' came in [to Hannibal] about 2 o'clock yesterday afternoon. **1869** R. KEELER *Gloverson & his Silent Partners* 92 He is what we call a 'Piker', you see,..and he stole some of our sheep. **1873** J. H. BEADLE *Undevel. West* xxxv. 763 These old Pikers don't want the country fenced up and the game scared off. **1907** W. E. CONNELLEY *Doniphan's Exped.* 9 Mr. Moore says that in California in the early days Missourians were called 'Pikers' indiscriminately and generally.

piker[5]. *slang* (orig. *U.S.*). [Cf. *PIKE *v.*[3] 3.] **a.** A cautious or timid gambler who makes only small bets; a person who takes no chances; a 'poor sport' or 'poor thing'; a shirker; a lounger. Also *attrib.*

[**1872** E. CRAPSEY *Nether Side N.Y.* 98 A 'piker' is a tolerated collapse who makes a stray bet when he can borrow a 'check'.] **1889** [see *PIKE *v.*[3] 3]. **1901** 'H. MCHUGH' *John Henry* 92 She put us wise to the fact that ..Edgar Allen Poe was a piker compared with her. **1910** W. M. RAINE *Bucky O'Connor* 233 Do you think I'm a cheap piker? **1912** R. W. SERVICE *Rhymes of Rolling Stone* (1913) 96 It's the plugging away that will win you the day, So don't be a piker, old pard. **1919** H. L. WILSON *Ma Pettengill* vii. 216 'I says to myself the other day: "I'll bet a cookie he'd like to be like me!"' Homer was a piker, even when he made bets with himself. **1929** J. BUCHAN *Courts of Morning* I. xii. 138, I don't say there mayn't be some pikers at Headquarters. **1935** Z. N. HURSTON *Mules & Men* (1970) 308 The pikers choose a card each from among those turned off to bet on. **1947** *Sat. Even. Post* 15 Mar. 111/3 It is natural that I should have gone far beyond the sort of piker activities which characterize the average soldier. **1947** D. M. DAVIN *Gorse blooms Pale* 207 'Don't be a piker, Mick,' he said. **1957** 'N. CULOTTA' *They're a Weird Mob* (1958) xiii. 203 It [*sc.* bludger] means that you are criminally lazy.. that you are a 'piker'—a mean, contemptible, miserable individual. **1968** M. RICHLER in R. Weaver *Canad. Short Stories* 2nd Ser. 156 The dirty piker he asked her to marry him he hasn't even got a job. **1969** *Southerly* XXIX. 308 Mat saw me coming. He said: 'We've got a piker on our hands. He reckons he hasn't got enough for a feed at La Roma. Pulling out on us he is.' **1971** H. W. TILMAN *In Mischief's Wake* III. x. 112 He is definitely no piker and although only 22 is one of the old school and believes in discipline. **1973** K. GILES *File on Death* ii. 43 'You have that much authority?' 'He is no piker. I run the place... I can roughly do what I like on the business side.'

b. A person who speculates in stocks, esp. with only small sums.

1898 *N.Y. Jrnl.* 12 Aug. 1/7 John Pettit started in as a 'piker'. That's what the downtown brokers call a man who speculates with a few hundreds at a time instead of with thousands. **1901** *McClure's Mag.* June 159/1 In the absence of complaisant lambs, the financial cannibals known as 'room traders' and 'pikers' tried to 'scalp eighths' out of each other for weeks. **1902** H. L. WILSON *Spenders* xxxi. 360 They're used to those fifty and a hundred thousand dollar pikers down in that neighbourhood. *Ibid.* 365 We'll make those Federal Oil pikers think we've gnawed a corner off the subtreasury. **1942** BERREY & VAN DEN BARK *Amer. Thes. Slang* §563/1 *Piker*, a speculator on a small scale.

piker[6]. [Etym. unknown.] Also **pika.** In Guyana, a plover.

1936 J. BOND *Birds W. Indies* 101 South American Ring Plover (*Charadrius collaris*)...Little Ploward; Nit; Pika; Snipe. *Ibid.* 102 Wilson's Plover (*Charadrius wilsonia*)... Snipe; Little Ploward; Nit; Pika. **1958** J. CAREW *Black Midas* x. 200, I saw curlews and pikers.. flying low over a rim of amber foam.

pikey (pəi·ki). *dial.* or *slang.* Also **piky.** [f. PIKE *sb.*[6] turnpike] = PIKER[3].

1847 J. O. HALLIWELL *Dict. Archaic & Provinc. Words* 623/2 *Piky*, a gipsey. *Kent.* **1874** HOTTEN *Slang Dict.* 253 *Pikey*, a tramp or gipsy. **1887** PARISH & SHAW *Dict. Kentish Dial.* 116 *Piky*,..a turnpike traveller; a vagabond; and so generally a low fellow. **1955** P. WILDEBLOOD *Against Law* 125 My family's all Pikeys, but we ain't on the road no more!

piki (pī·ki). [Hopi.] Bread made from maize-meal, baked in very thin sheets on heated stones by the Hopi Indians of the south western U.S. Also *attrib.*

1889 in *Cent. Dict.* s.v. *peekee, piki.* **1893** T. DONALDSON *Moqui Pueblo Indians* 72 Piki, or corn bread of many colors, is plentiful, and the evidences of a feast are on every hand. **1922** E. S. CURTIS *N. Amer. Indian* XII. 43 The commonest food derived from corn is piki. **1936** A. M. STEPHEN *Hopi Jrnl.* II. 1197 The men quarry out and roughly dress the piki stone to required dimensions, but the women finish and smooth it by rubbing. **1948** *Southwestern Jrnl. Anthropol.* Winter 376 Corn, flour, breadstuffs—especially piki (wafer bread)—melons, chili, and dried fruit were most sought after. **1959** E. TUNIS *Indians* 119/2 The [Hopi] Pueblos knew fifty-two ways of cooking corn. Most of them weren't unlike the ways other Indians cooked it, but the 'paper bread', called piki, belonged to them alone. It was made from a thin batter and was cooked on a flat stone placed over the fire.

piking (pəi·kiŋ), *vbl. sb.* and *ppl. a. dial.* and *slang.* [? f. *pike*, var. PICK *v.*[1]: cf. PIKER[1].] Cheating; using sharp practices.

1884 in *Western Pennsylvania Hist. Mag.* (1918) I. 97 After considerable deliberation I concluded to try a little 'piking', and had Leslie sell five thousand barrels 'short' at 64. **1955** W. GADDIS *Recognitions* II. v. 490 The Father of His Country was crumpled, folded, and offered in the most piking and meretricious traffic millions of times a day. **1972** *Daily Tel.* 1 Nov. 2/6 An East End publican who referred in his bar to the Tibbs family as 'dirty piking bastards' was attacked in the street by four men with knives and an axe.

pikipiki (pi·kipiki). [Mbuti.] A whistle used by the Mbuti pygmies of Zaïre (see quot. 1933). Also *attrib.*

1933 G. GRIFFIN tr. *Schebesta's Among Congo Pygmies* iii. 78 The pikipiki is..a kind of whistle, about the thickness of a man's finger, and is cut out of a round piece of wood. In both ends holes are..burned, with a red-hot spike. It can then be used as a whistle like a hollow key. Pikipikis often have one hole only, and may be decorated with spiral metal ornamentations... The pikipiki is used for making magic, not only by the huntsmen, but by every member of the pygmy community. **1936** —— tr. *Schebesta's My Pygmy & Negro Hosts* vi. 114 It seems that all the negro tribes believe in the supernatural powers of the 'pikipiki whistle'. **1958** *Listener* 2 Oct. 508/1 The production of shrill notes from their whistles called pikipikis. **1960** *New Scientist* 6 Oct. 903/1 The songs which they sing nightly to the accompaniment of pikipiki whistles all reflect their belief that..the forest is the great provider and eternal refuge of the chosen people, the Ba'mbuti.

piky, var. *PIKEY.

pilau, pilaw, pilaff. Add: Also prepared in certain areas with wheat in place of rice. (Further examples.)

Still spelt in many different ways in various regions.

1930 W. J. LOCKE *Town of Tombarel* iv. 114 A pilaff of sea-fruits is a succulent dish, composed of rice and as many fruits of the sea as you can imagine. 1935 H. EDIB *Clown & his Daughter* xxiv. 130 Who would believe that any city under the sun could produce such crowds, such colour..and above all such colossal quantities of roast lamb and pilaff! 1936 P. FLEMING *News from Tartary* 288 We lunched off pilaffe and sour milk. 1938 M. K. RAWLINGS *Yearling* xxxi. 392 Ma Baxter made a pilau of the squirrels for supper. 1949 B. A. BOTKIN *Treas. S. Folklore* IV. i. 552 Marylanders grow lyrical over Brunswick stew..South Carolinians, over rice, calibash, and pilaus (pronounced pé-los, púr-loos). 1956 R. MACAULAY *Towers of Trebizond* xv. 182 He did not mind me lying on his floor, he fed me with yoghourt and coffee and offered me some rice pilav. 1959 F. MACLEAN *Back to Bokhara* ii. 88 A veritable mountain of pilav or *plov*—rice cooked in mutton fat. 1967 *Punch* 12 Apr. 543/1 The lobster pilaf, *côtelettes à la Kiev* and *pommes Anna* with which Carlos Lacerda entertained the Gunthers when he was governor of Guanabara. 1971 *Carry Singapore in your Pocket* (Singapore Tourist Promotion Board) (ed. 3) 31 One first-class Kashmiri restaurant..serves..a wide variety of naan and pilau prepared in Kashmiri style. 1971 *Hindustan Times Weekly* (New Delhi) (Suppl.) 4 Apr. p. iv/5 Just heat up some *ghee*, fry the rice as for a *pullao*. 1971 *Illustr. Weekly India* 25 Apr. 57/2 We will have a proper lunch. Pulao, Chapattis, Shreekand, the lot. 1971 R. RUSSELL tr. *Ahmad's Shore & Wave* ix. 118 'Can I get you anything?' he asked. 'Chicken? Pulao?' 1978 *Detroit Free Press* 16 Apr. 6c/1 (Advt.), This Sunday, Monday or Tuesday you can order our thick and juicy prime rib (served with our special rice pilaf).

pile, *sb.*[1] Add: **1. b.** Delete †*Obs.* and add later examples. Now also made of substances other than metal.

1875 *Encycl. Brit.* II. 376/1 Arrows are manufactured generally of red-pine timber,..glued on one end, upon the point of which the iron pile is fixed. 1894 H. WALROND in Longman & Walrond *Archery* xviii. 304 Arrows are..called 'self' or 'footed' according as to whether they are footed or not with hard wood at the pile end. 1939 P. H. GORDON *New Archery* II. vi. 67 The solid-tipped 'parallel pile' is preferable to hollow-point 'bullet ferrules'. 1958 WISEMAN & BRUNDLE *Archery* 83 The piles or tips of arrows are made of brass, steel, aluminium, horn or plastic. 1972 T. FOY *Beginner's Guide Archery* xvi. 124 In the eighteenth-century the Turks were superb Flight Shots, and they invented 'barrelled' arrows which were thicker at the centre than at the nock and pile. 1979 R. LAIDLAW *Lion is Rampant* xiii. 104 They were target arrows with conical piles.

5. a. *pile-bridge* (examples). **b.** *pile guide.* **d. pile-built** *a.* (earlier example); **pile-cap** (examples).

1899 *Westm. Gaz.* 9 Dec. 5/3 The scarcity of timber or other material suitable for the erection of a trestle or pile bridge. 1940 *Chambers's Techn. Dict.* 643/1 *Pile bridge*, a bridge whose superstructure is carried on piles. 1851 A. O. HALL *Manhattaner* 5 It was a modest commercial plain; pile-built, and earth filled. 1898 C. E. FOWLER *Coffer-dam Process for Piers* iv. 49 The small hammer.. is used for sheet pile work by inserting a 'follower' of oak which fits the base or pile cap, and which has a slit in the lower end to fit the sheet pile. 1944 [see *PILING *vbl. sb.*[1] 2]. 1975 R. HOLMES *Introd. Civil Engin. Construction* iv. 166 Pile caps are usually constructed of concrete to such a depth as will ensure full transfer of load to the piles and, at the same time, resist punching shear. 1974 *People's Jrnl.* (Inverness & Northern Counties ed.) 29 June 22 (*caption*) In one of the biggest lifts ever in the off-shore oil industry a 940-ton pile guide cluster is lifted into position to be fixed to the base of the huge oil rig jacket now being finished at Nigg. 1975 *Offshore* Sept. 11/1 The 'Heerema Steel Structure' consists essentially of a jacket made up of four steel towers which fit into a steel base frame incorporating pile guides.

pile, *sb.*[3] Add: **3. h.** (Further examples.) Also in phr. *to go one's pile*, to stake all one's money on a single chance, to 'go the whole hog'.

1839 *Picayune* (New Orleans) 29 Mar. 2/2 Friends of the Lubber, becoming excited at the unexpected termination of the first heat, were willing to go a 'small pile' somewhere in the neighborhood of even. 1840 *Spirit of Times* X. 498 Considerable sums were laid out..the Georgians 'going their pile' on the Andrew filly. 1865 'MARK TWAIN' *Celebr. Jumping Frog* (1867) 37 His last acts was to go his pile..when there was a 'flush' out agin him. 1915 J. BUCHAN *39 Steps* i. 10, I had got my pile—not one of the big ones, but good enough for me. 1915 WODEHOUSE *Psmith Journalist* xxii. 166 He made a bit of a pile out of the job, and could afford to lie low for a year or two. 1969 *Listener* 24 July 103/2 So many Poles or Ukrainians ..had spent part of their life in the United States, and returned home after making a decent pile. 1973 *Times* 22 Mar. 25/1 This is tough talk from a man who first made his pile as an investment banker. 1977 MCKNIGHT & TOBLER *Bob Marley* i. 17 The only way to get ahead, make your pile and escape from the overwhelming oppression and depression of poverty.

i. A nuclear reactor.

1942 H. L. ANDERSON et al. in E. Fermi *Coll. Papers* (1965) II. 129 At the end of September, 1941, a new and taller exponential pile was set up and the accuracy was further increased by using a 2 gram Ra+Be source instead of the original source of about 600 mg. 1945 H. D. SMYTH *Gen. Acct. Devel. Atomic Energy Mil. Purposes* v. 48 In a memorandum written to Bush on May 14, 1942.., Conant estimated that there were five separation or production methods that were..likely to succeed: the centrifuge, diffusion, and electromagnetic methods of separating U-235; the uranium-graphite pile and the uranium-heavy-water pile methods of producing plutonium. 1945 *War Illustr.* 9 Nov. 439/1 The natural uranium (U238) is in the shape of rods embedded in a graphite block and contained in an atomic 'pile'. 1945 *Nature* 29 Dec. 768/2 They give..brief accounts..of the construction and testing of the first self-sustaining chain-reacting pile. 1946 *Ann. Reg. 1945* 354 A pile, containing 12,400 lb. of uranium in lumps separated by graphite, set up in Chicago under the direction of Prof. E. Fermi, was first operated on December 2, 1942. 1948 K. K. DARROW *Atomic Energy* iv. 66 When I heard the name in its new meaning and knew that Fermi an Italian had invented the new pile, I took it for granted that he had intentionally chosen the appellation of Volta. Great was my surprise when I learned from Fermi that it was a mere coincidence. He had conferred the name on his device because it was 'such a big pile of graphite and uranium'. 1952 *Nucleonics* Mar. 11/1 In later developments, most of the assemblies bear no resemblance to piles so that the expression nuclear reactor is to be preferred. Some British writers make the distinction that piles imply the use of natural uranium, and reactors, the use of enriched uranium. 1954 C. P. SNOW *New Men* vi. 100 'If the pile gets too hot, then they automatically shut the whole thing off,' said Luke. 1955 *Sci. News Let.* 26 Mar. 201/2 Fissionable material to fuel the pile will be obtained from the AEC on an extended loan basis. 1957 *Times* 12 Oct. 6/1 The danger of radioactivity being disseminated from the pile chimney in steam. 1964 M. GOWING *Britain & Atomic Energy, 1939–1945* x. 284 The theoretical physicists also contributed greatly to a whole range of pile theory and pile design problems. 1976 *Sci. Amer.* Dec. 32/2 Barely two and a half years elapsed between the initial chain-reacting pile on December 2, 1942, and the explosion of the first plutonium bomb on July 16, 1945.

pile, *sb.*[5] Add: **1.** Also, the fine undercoat of certain rough-coated dogs, esp. the Old English sheep-dog.

1905 J. WATSON *Dog Bk.* v. 386 The under coat [of the bob-tailed sheep dog] should be a waterproof pile. 1938 E. C. ASH *New Bk. Dog* x. 395 Coat [of Old English sheep-dog].—Profuse, and of good hard texture... The undercoat a pile when not removed by grooming. 1971 DANGERFIELD & HOWELL *Internat. Encycl. Dogs* 348/1 Pile. Dense undercoat.

c. Red or yellowish markings on white or pale-coloured fowls; a fowl with this coloration. Also *attrib.*

1854 *Poultry Chron.* I. 289 The 'white or pile game'.. were withheld from prizes altogether. 1913 W. BATESON *Mendel's Princ. Heredity* (rev. ed.) 120 The coloration known as 'Pile' in fowls is seldom bred for exhibition from two pile kinds. 1929 E. BROWN *Poultry* I. iii. 62 Pile Leghorns were produced by Mr. George Payne, of Woking. *Ibid.* 69 Pile.—This is a purely exhibition fowl, so named from its having the markings of the old variety of Game fowl with the same designation.

pile, *v.*[2] Add: **1. c.** (Earlier example.)

1839 *Civil Engin. & Arch. Jrnl.* II. 17/2 A reverberatory furnace of the common construction employed in 'puddling', 'balling', or 'piling' iron.

e. *to pile up*: to wreck (a ship); to crash (an aircraft, vehicle, etc.).

1899 C. J. C. HYNE *Further Adventures Capt. Kettle* vi. 137 If the bar had shifted, he himself could have put this steamer on the ground as handily as the other man had piled up the branch boat. 1923 *Times Lit. Suppl.* 29 Mar. 218/2 An old battle-cruiser which gets adrift in a gale..[and] is piled up on the rocks. 1925 FRASER & GIBBONS *Soldier & Sailor Words* 223 *Pile up one's bus, to*, an airman's expression for coming a 'crash'. 1930 KIPLING *Limits & Renewals* (1932) 230 We had a passenger..who wanted to see Caesar. It cost us our ship... He piled up the *Eirene* on his way. 1932 'N. SHUTE' *Lonely Road* ix. 196 The fellow was so drunk that he'd probably have piled his car up anyway. 1942 N. STREATFEILD *I ordered Table for Six* 236 Andrew cautiously steered Claire to the centre of the floor... He was afraid if he talked he might pile her up. 1959 G. JENKINS *Twist of Sand* iv. 78, I hope to God they don't pile that monster up on my runways. 1971 M. TAK *Truck Talk* 119 *Pile up*, to wreck a truck.

2. (Further examples.)

1860 *Baily's Mag.* Sept. 429 The Kent innings was piled up the next day to 152. 1873 *Ibid.* Dec. 287 We fancy there are some batsmen who would pile up the runs rapidly off his leg balls. 1898 J. A. GIBBS *Cotswold Village* xi. 246 Once fairly started on a sequence of big scores, the cricketer goes on day by day piling up runs. 1948 'N. SHUTE' *No Highway* i. 25 The Reindeers were flying over the Atlantic piling up the hours faster than Mr. Honey's test.

b. (Later example.) Also *to pile it on.*

1852 *Star* (Los Angeles) 3 Apr. 1/5 The wags observed that Caleb was getting exceedingly uneasy, and 'piled it on'. 1876 'MARK TWAIN' *Old Times Mississippi* viii. 43 'Now I don't want to discourage you, but—' 'Well, pile it on me; I might as well have it now.' 1943 J. B. PRIESTLEY *Daylight on Saturday* xxxviii. 297, I fancy you're piling it on too much. There are lots of things you can enjoy, if you set about it properly. 1969 E. GÉBLER *Shall I eat you Now?* 56 But that was piling it on a bit; she wasn't that daft. *a* 1974 R. CROSSMAN *Diaries* (1976) II. 601, I suppose he will get what he wants at Cabinet but he's piling it on a bit thick. 1976 *West Lancs. Evening Gaz.* 15 Dec. 8/2 And now they want to pile on the agony with rip-roaring monsters going round and round.

3. (Further examples.)

1926 *Scribner's Mag.* Sept. 266/1 'How things did pile up!'..Almost every person Peter particularly dislikes came in for tea and..Feinberg showed up with Sally. 1930 *Morning Post* 14 July 6/7 Vehicles crossing the circus diagonally had to 'pile up' in the centre. 1942 *We speak from Air* 39 Whether you get the Hun or miss him, he frequently piles up on the ground through making his landing in fright. 1947 *John o' London's Weekly* 25 July 502/3 In referring to the traffic 'piling up', did he mean that motor-cars, lorries, bicycles were stacked up in neat heaps on the roadway? 1956 'C. BLACKSTOCK' *Dewey Death* vii. 160 The work is just piling up. 1975 *Times* 18 Mar. (Greece Suppl.) 1/4 It is not just the old problems. Each day new ones pile up.

b. To climb *on* or go *into* (a vehicle, building, etc.) so as to form a pile; hence, to enter (a place) in crowds; and simply, to mount, enter, etc. orig. *U.S.*

1841 L. B. SWAN *Jrnl. of Trip to Michigan* (1904) 30 Brooks brought up his lumber wagon and we all 'piled in'. 1854 M. J. HOLMES *Tempest & Sunshine* iii. 44 Fanny with half a dozen other girls..began piling on to Bill's old sled. 1879 J. BURROUGHS *Locusts & Wild Honey* (1884) 38 They [*sc.* bees] come piling in till the rain is upon them. 1884 'MARK TWAIN' *Huck. Finn* xxxvi. 345 Here comes a couple of the hounds in from under Jim's bed; and they kept on piling in till there was eleven of them. 1923 R. D. PAINE *Comrades of Rolling Ocean* vii. 114 Judson introduced his four shipmates who piled into the automobile. 1929 R. GRAVES *Good-Bye to all That* x. 103 There were about three thousand prisoners already there and more piled in every day. 1943 N. COWARD *Middle East Diary* (1944) 100 We flagged a passing lorry .., piled into it bag and baggage and whirled off to the airport. 1956 B. HOLIDAY *Lady sings Blues* (1973) viii. 72 We piled into his car and were off. 1972 J. WAMBAUGH *Blue Knight* (1973) xiii. 231 We jawed..and finally piled into the cars.

c. Hence used of the reverse processes: to climb *down from*, or *off* (a vehicle); to come *out of* (a place), etc., in crowds. orig. *U.S.*

1884 'MARK TWAIN' *Huck. Finn* xxii. 205 A lot of men begun to pile down off of the benches and swarm toward the ring. 1896 ADE *Artie* xi. 100 We stopped in front of the church and piled out. 1902 S. E. WHITE *Blazed Trail* xx. 150 Then they piled out for the boss. 1908 —— *Riverman* xxii. 195 They piled off the train at Sawyer's. 1921 C. E. MULFORD *Bar-20 Three* xvii. 224 Six sleeping men piled from their bunks and..chased the cursing trail-boss. 1972 *Times* 20 Nov. 8/6 Hundreds more piling off every train.

d. *to pile on to* (N. Amer.), to attack vigorously, to assail.

1894 *Outing* XXIV. 417/1 The dog ..[will] never 'pile onto' any more bears. 1906 U. SINCLAIR *Jungle* xvi. 183 Like as not a dozen [policemen] would pile on to him at once, and pound his face into a pulp. 1970 *Globe & Mail* (Toronto) 25 Sept. 30/1 The Australian tub, Gretel, hit the American scow, Intrepid, below the gunwale the other day, or tugged its saddle blanket, or piled on with unnecessary roughness.

e. To move or advance in a throng.

1925 H. L. FOSTER *Trop. Tramp with Tourists* 102 The tourists piled towards the exits.

f. *to pile in*: to crash.

1944 G. NETHERWOOD *Desert Squadron* ii. 21 So low did Pilot Officer Weeks fly as he did the Victory Roll, that those watching him made certain that he would 'pile in'.

pilea (pəi·liă). [mod.L. (J. Lindley *Collectanea Botanica* (1821) 4), f. L. *pīleus* cap.] An annual or perennial herb of the genus so called, belonging to the family Urticaceæ, native to many tropical regions, and cultivated elsewhere as a house plant; esp. *Pilea muscosa*, the artillery or gunpowder plant, which discharges clouds of pollen when it is touched.

1918 N. L. BRITTON *Flora of Bermuda* 105 Large-leaved Pilea, Jamaican, seen in the garden at Mt. Hope in 1914. 1958 *Times* 29 Nov. 9/5 We have the pileas, the tradescantias, and..all the various forms of *Begonia rex*. 1971 *Country Life* 25 Nov. 1443/1 Among the most interesting recent introductions are three pileas. 1974 W. DAVIDSON *All about House Plants* 156/1 On the whole pileas are small, compact plants.

pileated, *a.* **1.** For *Picus pileatus* substitute *Dryocopus pileatus*. (Later examples.)

1928 G. M. SUTTON *Introd. Birds Pennsylvania* 81 The call of the Pileated is a high, irregular cackle. *Ibid.*, The food of the Pileated Woodpecker is chiefly grubs. 1956 G. DURRELL *Drunken Forest* ix. 169 Pileated jays have long, magpie-like tails of black and white... The feathers on the forehead were black, short, and plushy, and stuck up straight. 1969 R. LOWELL *Notebk. 1967–68* 57 A large pileated bird flies up. 1971 W. HILLEN *Blackwater River* ix. 80 A pileated woodpecker gave his jungle-like call and flashed his scarlet crest through the trees.

piled, *ppl. a.*[2] Add: **a.** (Later examples with *up*.)

1898 P. GEDDES *Let.* Feb. in P. Boardman *Worlds of Patrick Geddes* (1978) vi. 167 The piled-up picturesqueness of Old Edinburgh. 1935 C. S. FORESTER *African Queen* vi. 109 The *African Queen*..reared up as she hit the piled-up water. 1948 H. INNES *Blue Ice* vii. 191 Mile on ghastly mile of piled-up snow-capped peaks. 1978 'L. BLACK' *Foursome* i. 7 The piled-up dishes, bowls, plates.

b. With all sails set.

1851 H. MELVILLE *Moby Dick* II. ix. 62 With every mast-head manned, the piled-up craft rolled down before the wind.

c. Crashed, wrecked.

1939 J. STEINBECK *Grapes of Wrath* vi. 61 They ain't whole, out lonely on the road in a piled-up car. They ain't alive no more. 1943 C. H. WARD-JACKSON *Piece of Cake* 48 *Piled in* or *up*, crashed.

pi·le-dri:ve, *v.* [Back-formation from PILE-DRIVER.] **a.** *trans.* To construct (something) using a pile-driver; to act as or like a pile-driver on (something).

1894 KIPLING in *Century Mag.* Dec. 295/1 I'll catch 'em by the back o' the neck, an' pile-drive 'em a piece. 1932 *Daily Express* 27 June 9/4 Trotsky may not convince the world that the revolution in which he and Lenin were the giants was a splendid thing, but he pile-drives the contention that it was inevitable in Russia. 1971 'D. HALLIDAY' *Dolly & Doctor Bird* vii. 94 They're still pile-driving the quayside. 1972 D. HASTON *In High Places* vii. 91 Chris was falling every second turn, pack pile-driving his head into the snow.

b. *intr.* (See quot. 1929.)

1898 *Nautical Mag.* LXVII. 355 After pile-driving off Cape Horn for some days, she squared away..and eventually reached San Francisco. 1929 F. BOWEN *Sea Slang* 103 *Pile Driving*, steaming or sailing into a heavy head sea. 1937 G. S. DOORLY *In Wake* viii. 127 We had been 'pile-driving' heavily through the night.

pile-driver. Add: **2.** *transf.* A very strong or powerful hit, stroke, kick, etc., in various games; something of great strength or power.

1858 A. MAYHEW *Paved with Gold* II. xii. 189 After some sparring, Jack threw out his 'pile-drivers' and caught Ned on the 'sniffer', but the nose didn't suffer much. 1929 *Star* 21 Aug. 17/1 Hammond was let off when 59 from a pile-driver to third man. 1929 *Daily Express* 7 Nov. 19/1 Their inside right put in a couple of pile-drivers that missed the target by inches only. 1952 M. ALLINGHAM *Tiger in Smoke* i. 18 His piledriver personality forced home the suggestion. 1962 [see *back-breaker* s.v. *BACK- B]. 1964 *Guardian* 2 Mar. 7/6 He's using pile-drivers (dropping on the head). 1968 R. WEST *Sk. Vietnam* ii. 60 'The noise of these pile-drivers,' he said with great satisfaction, 'was heard as far as Saigon.' 1973 *Sunday Tel.* 4 Mar. 38/8 Jones picked up a loose clearance from Feltwell, and her pile-driver took the paint off the post. 1973 D. LEES *Rape of Quiet Town* vi. 93 The big boys were dishing out pile-drivers and body-slams, cracking heads and snapping spines.

pile-driving, *sb.* and *a.* Add: (Further examples.) Also *fig.*

1894 'MARK TWAIN' in *Century Mag.* Jan. 335/1 The title came upon them as a kind of pile-driving surprise. 1942 PARTRIDGE *Usage & Abusage* (1947) 126/2 If the point is still not made, let us take a few more of Jespersen's pile-driving examples. 1960 *Farmer & Stockbreeder* 15 Mar. 46/2 This simple pile-driving machine, easily attached to any tractor fitted with P.T.O. and hydraulic lift will reduce fencing costs to a minimum.

Pilentum (pile·ntŏm). *Hist.* Also **pilentum.** [f. L. *pilentum* an easy chariot used by ladies or for carrying the vessels, etc., for sacred rites.] A type of carriage (see quot. 1961).

1837 W. B. ADAMS *Eng. Pleasure Carriages* xvii. 278 The next attempt in the 'service of the public', was the introduction of a vehicle with the..title of a Four-wheeled Cabriolet. This is a modification of..a Pilentum. 1843 *Ainsworth's Mag.* IV. 100 If I could have built a mother for myself, as one does a Brougham or pilentum. 1849 E. RUSKIN *Let.* 30 Sept. in M. Lutyens *Effie in Venice* (1965) I. 41 We concluded by fixing to take two [carriages], one for John..and an elegant *Pilentum* for Charlotte and me. It is an open carriage but shuts very easily... It is rather heavy..and will require three horses. 1961 M. WATNEY *Elegant Carriage* 60 The Pilentum..was an open carriage with the doorway very near the ground, the driving seat was also low, but it was built in different sizes to carry four or six people, and to be drawn by one or two horses. In appearance, the Pilentum was not unlike a Victoria, although it had doors at the sides. 1974 F. SELWYN *Cracksman on Velvet* II. 76 On its indiarubber bearings..the swan's neck Pilentum seemed to roll in air above the glittering spokes of its wheels.

pi·le-up. [f. vbl. phr. *to pile up*: see PILE *v.*[2] 1 a, e in Dict. and Suppl.] **1.** A crash or collision, often involving several vehicles; a vehicle that has been involved in a crash. Also *fig.*

1929 *Papers Mich. Acad. Sci., Arts & Lett.* X. 314 *Pile-up*, a crash; a smash. 1945 *Richmond* (Virginia) *Times-Dispatch* 18 Sept. 15/2 (*heading*) Five harness racers injured in pile-up. 1951 W. SANSOM *Face of Innocence* x. 138 We passed one pile-up with its dead-slumped radiator and its ragged little crowd. 1954 L. KLEMANTASKI tr. *Fraichard's Le Mans Story* v. 46 A tremendous multiple pile-up, which was visible from the grandstands, had eliminated six of the competing cars. 1957 S. MOSS *In Track of Speed* iii. 34 Both of us had escaped a pile-up of about half a dozen cars early in the race. 1968 *New Scientist* 3 Oct. 38/2 A recent pile-up on the M1 in Bedfordshire involving 30 cars has apparently moved the Ministry of Transport to do some thinking. 1973 *Black Panther* 31 Mar. 8/1 When the sky rocketing demands for human rights and the technology of the super-industrial state (U.S.A.) clash head on, there is a bloody social pile-up. 1973 *Times* 1 Aug. 2/8 Three people were killed and three seriously injured in a pile-up involving three lorries and a car on the A17. 1977 *Belfast Tel.* 27 Jan. 1/3 A woman and her nine-year-old son were killed in a traffic pile-up in Dublin today.

2. a. An accumulation; an amassing of tasks, papers, etc. Also *attrib.*

1945 *Sun* (Baltimore) 14 Feb. 6/7 The Marshalls, the mid-Pacific cluster of coral pileups just above the equator. 1946 *Ibid.* 26 Apr. 10/3 Rain slowed air traffic in and around Washington, causing a 'pile-up' of planes over the Washington airport. 1948 M. LASKI *Tory Heaven* viii. 119 Unless one can get each crop of débutantes married off as it comes out, we're only going to get the same pile-up all over again. 1951 C. W. MILLS *White Collar* I. iii. 39 A tangled pile-up of restrictive legislation. 1963 *Times* 5 Feb. 13/3 More people might well be laid off in spite of the pile-up of work waiting for them. 1964 *Amer. Folk Music Occasional* I. 88 An endless 'pile-up song' which begins: I had a hen, and the hen pleased me. 1964 V. J. CHAPMAN *Coastal Vegetation* viii. 194 The 'pile-up' to form our shingle beaches is due to the action of direct onshore waves. 1966 *Word Study* Dec. 2/2 Sentences were talk-built, producing both fragments and pileups. 1968 C. HELMERICKS *Down Wild River North* II. xxiii. 368 We jumped out into the water and struggled to thrust the boat up onto a pileup of logs which was wedged there.

b. *Electronics.* A lack of linearity or resolution in a pulse circuit caused by the pulses arriving too rapidly.

1962 C. SUSSKIND *Encycl. Electronics* 18/2 Pile-up is not necessarily an excursion into a nonlinear region but the result of coupling capacitances not fully discharging after each pulse. 1973 *Nature* 23 Mar. 270/2 This potential source of electronic pile-up background was eliminated by anti-pile-up circuitry allowing detection of the desired gamma rays in the presence of counting rates up to 10^5 s^{-1} below 1 MeV.

pilfer, *sb.* Add: **2.** *Comb.*, as *pilfer-proof* adj.

1959 *Light Metals* Jan. 18/3 Specially designed extrusions..provide a weather- and pilfer-proof seal where the two pivoting roof sections meet axially. 1965 *Economist* 3 July 35/4 Weld mesh cages, fitted to the decks, can be locked before being stowed, thus becoming pilfer-proof. 1969 *Jane's Freight Containers 1968–69* (Advt. section) 21 Because the Freightank is pilferproof insurance premiums are reduced.

pilferage. Add: (Later examples.) Also *Comb.*

1964 *Spectator* 13 Mar. 337/2 An installation with a primary mission of production control can simultaneously be used to promote safety and prevent pilferage. 1969 *Daily Tel.* 22 July 17 (Advt.), We have the only through-pallet service between here and the Far East via the U.S.A...which means much less chance of loss, pilferage, and damage. 1969 *Jane's Freight Containers 1968–69* 482/3 In the side walls there are pilferage-proof ventilation openings. 1971 *Daily Mail* 6 May 24/4 Pilferage last year totalled half of 1p.c. of the..turnover. 1974 *Islander* (Victoria, B.C.) 1 Sept. 11/2 When visitors leave, all their baggage is inspected by police to forestall pilferage.

pilger (pi·lgəɹ), *sb.* and *v.* *Metallurgy.* Also **Pilger.** [a. G. *pilger* pilgrim, in allusion to the alternate feeding in and partial withdrawal of the billet during the operation of the mill, which is said to resemble the steps of pilgrims approaching a shrine.] Used *attrib.* (in *pilger mill, process, roll*, etc.) with reference to a rolling mill for reducing the outside diameter of a tube without changing the inside diameter, its two rollers each having a semicircular groove of decreasing diameter passing round the circumference, so that in conjunction they form a circular hole through which the tube can be forced on a cylindrical mandrel and which decreases gradually and increases suddenly in size during each revolution of the rollers.

1902 *Proc. S. Wales Inst. Engineers* XXII. 351 One of the most interesting..devices is the machine named the Pilger carriage, to which the hollow billet is attached previous to being operated upon in the pilgrim mill. 1905 *Min. Proc. Inst. Civil Engineers* CLIX. 335 A tube produced by means of pilger-rolls frequently possesses a rippled or undulating surface. 1922 *Iron & Coal Trades Rev.* 9 June 849/1 The Pilger mill consists of two.. 'cam' rolls, fixed in housings, and driven through two-high pinions in practically the same manner as a two-high merchant bar mill, except that the Pilger rolls rotate in an opposite direction. 1954 A. R. BAILEY *Text-bk. Metallurgy* xii. 419 (*caption*) Diagram showing the step-by-step method of rolling in the Pilger process. 1968 R. N. PARKINS *Mech. Treatm. Metals* iv. 263 In the Pilger mill, the tube is deformed over a mandrel between rolls that work the tube during only part of each revolution. 1973 [see below].

So **pi·lger** *v. trans.*, to make in a pilger mill; **pi·lgering** *vbl. sb.*

1902 *Proc. S. Wales Inst. Engineers* XXII. 350 Pilgering or pilgrim mill. This mill derives its name from its peculiar intermittent action upon the hollow billet. 1923 HARBORD & HALL *Metallurgy of Steel* (ed. 7) II. xxiii. 467 Tubes are not usually made by pilgering with walls less than 9 or 10 gauge in thickness. 1945 *Metallurgia* XXXIII. 61/1 If the partly pilgered tube is withdrawn, the shape of the cone between bloom and tube will thus be a projection of the groove that varies around the roll. *Ibid.* 61/3 One company once pilgered a tube 10⅞ in. outside diameter × $\frac{1}{32}$ in. thick by no less than 135 ft. 8 in. long. 1973 J. G. TWEEDDALE *Materials Technol.* II. iv. 95 Pilgering is the name given to the second tube-forge-rolling system, which is done in a pilger mill.

pilgrim, *sb.* **5.** (Earlier and later U.S. examples.)

1841 W. L. MACCALLA *Adventures in Texas* 46 After such an address from a citizen of that calumniated country Texas to a shattered old pilgrim, I took the liberty of withdrawing to another apartment. 1867 J. F. MELINE *Two Thousand Miles on Horseback* 22 The term Pilgrims for emigrants first came into use at the period of the heavy Mormon travel—the Mormons styling themselves 'Pilgrims to the promised land of Utah'. 1885 *Rep. Indian Affairs* (U.S.) 120 This, we think, is a very fair crop of calves considering the fact that the cattle were what is called 'pilgrim' cattle (cattle for the States that had never passed through a winter before without being housed and fed). 1942 E. E. DALE *Cow Country* 194 They mingled with 'drift cattle' from Kansas or with the trail herds of 'pilgrim cattle' from Texas. 1943 J. K. HOWARD *Montana* 139 They were for the most part 'pilgrims' who remained and were 'made into hands'.

pilgrimage, *sb.* Add: **3. pilgrimage church, town, village,** etc., a church, town, village, etc., to which pilgrimages are made.

1889 L. T. SMITH tr. *Jusserand's Eng. Wayfaring Life in Middle Ages* III. iii. 348 It was..a town of inns and churches, as pilgrimage towns have generally been. 1908 *Westm. Gaz.* 28 July 8/2 [They] paid a visit to the picturesque pilgrimage village of Sainte Anne de Beaupré. *Ibid.*, His Royal Highness alighted and proceeded to enter the Pilgrimage Church, the steps of which were crowded with cripples and pilgrims. 1935 *Burlington Mag.* Oct. 183/1 The great aisled transept [of Laon cathedral], she derives from the so-called 'pilgrimage churches' of which much has been written in recent years.

pili (pī·li). [Hawaiian.] A Hawaiian name for the perennial grass, *Heteropgon contortus*, formerly used as a thatching material. Also *attrib.*

1888 W. HILLEBRAND *Flora Hawaiian Islands* 508 Common on all islands, the 'Pili' of the natives, very troublesome on account of its awns, which get entangled in the wool of sheep. 1915 W. A. BRYAN *Nat. Hist. Hawaii* v. 59 Should a chief order a house built, certain men would cut the timbers, others gather the pili grass. 1917 *Nature* 20 Sept. 57/2 The grass used for this purpose [*sc.* thatching] is usually pili.., an indigenous grass. 1934 M. D. FREAR *Lowell & Abigail* ii. 94 It was easy to accept the Hawaiian roof of thatch, using..*ti* leaf where *pili* grass was not to be had. 1949 P. H. BUCK *Coming of Maori* II. ii. 119 The Hawaiians used a similar [thatching] technique with *pili* grass. 1965 M. C. NEAL *In Gardens of Hawaii* 80 In Hawaii, *pili* was preferred to other thatch material because of its pleasant odor, brown color, and neat appearance.

piliated (pi·-, pəi·li,ēɪ·tėd), *a.* *Bacteriol.* [f. *pili-*, taken as comb. form of *PILUS + -ATE[2] + -ED[1], or directly f. *PILUS + *-iated* after *fimbriated*.] Bearing pili.

1960 *Biochim. & Biophys. Acta* XLII. 298 Many piliated strains of bacteria can exist in two forms, the completely piliated..or the completely non-piliated. 1973 DOETSCH & COOK *Introd. Bacteria* ii. 28 Bacteria capable of producing pili, when freshly isolated from natural sources, are usually 'piliated'.

piling, *vbl. sb.*[1] **1.** (Later example.)

1973 *Daily Tel.* 3 Apr. 21 (Advt.), Piling has also been completed on the Hutchison House site and construction of the foundation is well in hand. 1975 *BP Shield Internat.* May 2 (*caption*) Piling into sea bed after removal of flotation tanks. 1977 *Daily Times* (Lagos) 25 Feb. 3/4 Mr. Ogundiya recalled that during the initial piling on the site, the management of Leventis wrote to his ministry threatening to claim ₦6 million if there was any damage to their building.

2. (Further examples.)

1939 A. RANSOME *Secret Water* xiv. 170 The wood of the quay was rotting, and water was working in and out through gaps in the piling. 1944 *Sun* (Baltimore) 18 Mar. 6/2 It is claimed that the timber piling is not sufficiently strong to support the structure and the loads it carries; the pile caps and stringers are in poor condition. 1975 *Lamp* (Exxon Corporation) Winter 11/2 When the platform's steel jacket was being emplaced this spring, the pilings struck a patch of sand in the bottom clay and couldn't be driven until larger and more powerful pile-driving equipment was moved to the site. 1977 *New Yorker* 9 May 118/2 They should see Prudhoe Bay. It's so damned clean and neat and sterile—with refrigerated pilings, so the tundra won't melt.

pilkins (pi·lkinz), *sb. pl. local.* [Perh. f. PILCH *v.*: see *N. & Q.* Aug. 1979, p. 305.] = TAILING *vbl. sb.*[1] 2 a.

1859 G. MEREDITH *Ordeal R. Feverel* I. vi. 101 He swears soam o' our chaps steals pilkins. *Ibid.* xii. 177 The Bantam said he had seen Tom secreting pilkins in a sack.

pill, *sb.*[2] Add: **1. b.** (Further examples.)

1928 L. NORTH *Parasites* 87 There are lots of folks to give you a lift down in the morning... It's getting up that hill at night is the pill. 1931 R. CAMPBELL *Georgiad* II. 38 He..takes his pleasures as a bitter pill Or social duty, much against his will. 1943 J. B. PRIESTLEY *Daylight on Saturday* xxv. 195 Fincham [*sc.* a factory] especially is a pill. We never ought to have allowed ourselves to have been persuaded into using that old mill. 1961 B. FERGUSSON *Watery Maze* ii. 68 Dakar was indeed a bitter pill, both professionally and politically. 1977 *Times* 9 Nov. 15/3 Mrs Williams is well aware of this and her decision to champion parental choice is sugar on the pill of her pledge to support local authorities in the matter of school closures. 1978 P. BAILEY *Leisure & Class in Victorian Eng.* ii. 54 The entertainments..were devised to sugar the pill of instruction.

c. *the pill* or *Pill*: a contraceptive pill.

1957 'C. H. ROLPH' *Human Sum* 6 He gives a modestly exciting account of the quest now going on..for what laymen like myself insist on calling 'the Pill'; and by this phrase..I mean the simple and completely reliable contraceptive taken by the mouth. 1958 A. HUXLEY *Brave New World Revisited* (1959) xii. 156 'The Pill' has not yet been perfected. 1960 *Economist* 22 Oct. 335/1 For about thirty years a campaign has been carried on for the reform of Britain's abortion law... It..looks as if the search for the 'Pill'—a simple, safe contraceptive—may be rewarded first. 1964 'J. MELVILLE' *Murderers' Houses* vii. 116 Emily knew all about the Bomb and the Pill. 1969 *New Scientist* 22 May 415/2 As contraceptives, IUDs are not as effective as the pill. 1970 *Daily Tel.* 17 July 2/8 Investigations showed that the increased risk of thrombo-embolism declined rapidly after the patient stopped taking the pill. 1975 D. LODGE *Changing Places* i. 20 They went on the Pill and suffered side-effects. 1975 *Woman's Jrnl.* Sept. 110/1 Emma's burgeoning again... It seems she can't remember to take the pill. 1976 [see *ON *prep.* 1 k].

d. A pill or tablet of a barbiturate or amphetamine. *slang.*

1963 [see *pill popper* below]. 1967 *Trans-Action* Apr. 7/1 Pills are 'reds' and 'whites'—barbiturates and benzedrine or dexedrine. 1970 *New Statesman* 16 Jan. 90/1 The police were not too hip on drugs, and pills used to be passed out at clubs like ordinary cigarettes. 1972 *Guardian* 5 Dec. 15/3 It's impossible to discover how many adolescents use the more common illicit soft drugs —cannabis, LSD, 'pills' (amphetamines, barbiturates or mixtures of both). 1976 DEAKIN & WILLIS *Johnny go Home* ii. 38 The suburban kids' drugs: pills, uppers and downers, bennies and blueies.

2. b. Also, a shell, bomb, or hand grenade; *spec.*, the atom bomb. Hence in *pl.*, ammunition. (Further examples.)

1841 H. J. MERCIER *Life in Man of War* 234 A dose of Yankee *pills*..would take her some time to digest. 1883 *United Service* June 652 That serpent was rather hard on the pills... How do you account for that fellow's swallowing those shells so easily? 1917 P. S. ALLEN *Let.* 24 Mar. (1939) 137 The submarine proceeded to lie on the bottom..but one day they realized they were spotted. 'Pills' kept dropping close to them, and sending the water a-swish all round. 1919 V. VIGORITO in Hamilton & Corbin *Echoes from Over There* III. xi. 203 A sergeant..counts out the required number of pills (H.E. grenades). 1921 *Amer. Legion Weekly* 15 Apr. 22 Damn the Boche that threw the pill. 1921 *Flight* XIII. 618/2 Another range-finding bomb was dropped by the next Martin, and then the following machine scored the first hit with its 2,000-lb. 'pill', which struck the deck in the bow. 1927 L. H. NASON *Three Lights from Match* 220 What do they use those pills in? 1939 P. G. HART *Hist. 135th Aero Squadron* 135 When I got over the town I let my pills go. 1948 PARTRIDGE *Dict. Forces' Slang* 142 *Pills*, ammunition. (Army; especially among gunners.) 1957 *Daily News* (N.Y.) 7 Aug. 7 A Jesuit priest who was a survivor of the A-bombing of Hiroshima 12 years ago.. said he was drinking coffee when the big pill came down. 1969 M. PEI *Words in Sheep's Clothing* (1970) vi. 51 Interestingly, 'the Pill' was used around 1957 with reference to the Hiroshima atom bomb.

c. Also in *sing.*, = BALL *sb.*[1] 4 a.

1908 *Atlantic Monthly* Aug. 224/2 Mr. O'Hooligan, steeped in the lore of the 'spitball',..and aware that the finger-tips, as the 'pill' leaves the hand, endow it with its rotary genius, pays this wizard [*sc.* the pitcher] the homage of a somewhat more enlightened reverence. 1909 P. A. VAILE *Mod. Golf* viii. 110 The ball is microscopic— a veritable, as it is sometimes slangily called, 'pill'. 1916 *Dialect Notes* IV. 279 *Pill*,..golf ball. 'Curses on that *pill*. It won't get off the ground.' 1922 WODEHOUSE *Clicking of Cuthbert* ix. 203 'I don't mind her missing the pill,' said the young man. 'But I think her attitude toward the game is too light-hearted.' 1946 B. MARSHALL *George Brown's Schooldays* 6 As a matter of fact, I think that's the dirty cad hacking that footer pill over there. 1977 SCOLLINS & TITFORD *Ey up, mi Duck!* II. 50 *Pill*, ball: usually a football.

d. A pellet of opium prepared for smoking.

1887 *Lantern* (New Orleans) 21 May 4/2 The longer end of the stem is handed the person opposite and so the pill is consumed by the party drawing in their breath, which some call the 'long draw'. 1926 J. BLACK *You can't Win* xvii. 238 He feverishly rolled the first 'pill'.... Each succeeding pill is smaller, more carefully browned over the lamp, and smoked with increased pleasure. 1935 A. J. POLLOCK *Underworld Speaks* 88/1 *Pill*, after opium has been cooked for smoking, it takes the form of a round pill, which is placed in the pipe and smoked. 1948 [see *COOK *v.*[1] 2 d]. 1948 *Amer. Speech* XXIII. 247/1 *Pill*, the pellet of opium which has been prepared for smoking. 1955 *U.S. Senate Hearings* (1956) VIII. 4162 The opium pill must be brought to a flaming heat before being placed in position over a hole in the center of the well-heated bowl of the pipe. *Ibid.*, 'Cook a pill', heat opium for smoking.

e. *slang.* A cigarette.

1914 'HIGH JINKS, JR.' *Choice Slang* 16 *Pill*, a cigarette. 1927 *Amer. Speech* II. 281/2 *Pill*,..cigarette. 1927 D. HAMMETT in *Black Mask* Feb. 31/2 Those pills you smoke are terrible. 1934 *Amer. Ballads & Folk Songs* 135 Then we rode down the hill, each a-puffin' a pill. 1966 'L. LANE' *ABZ of Scouse* 82 *Pill*, a cigarette.

f. An animal's dropping. Usu. *pl.*

1926 D. H. LAWRENCE *David* xii. 89 They have passed, letting fall promises as the goat droppeth pills. *a* 1930 —— *Phoenix* (1936) 10 So we had him [*sc.* a pet rabbit] upstairs, and he dropped his tiny pills on the bed.

g. in *pl.* *slang.* The testicles; *fig.* nonsense (cf. *BALL *sb.*[1] 15 b).

1935 I. MILLER *School Tie* xi. 158, I explained to him about the prayers... 'Awful pills,' I whispered; 'but it can't be helped.' *Ibid.* xiv. 270 'No, it really is true. Not doing much good here, you know.' 'Pills! Bags of pills!' 1937 PARTRIDGE *Dict. Slang* 630/2 *Pills*,.. testicles. 1968 A. DIMENT *Gt. Spy Race* vi. 77, I..wished I had followed up my elbow in the throat with a hefty boot in his peasant pills. One in the balls is worth two in the teeth—a motto of unarmed combat instructors.

h. A small ball of fluff found on the surface of a fabric (see *PILL *v.*[1] 6 c, *PILLING *vbl. sb.* 2 b).

1958 *New Scientist* 3 Apr. 17/2 The 'pills' are the little balls of fibre which, in the course of wear, form on the surface of cardigans, jerseys and similar articles. 1963 A. J. HALL *Textile Sci.* v. 273 The formation of a pill is primarily due to a rubbing of the fabric surface to cause a number of fibre ends to protrude and then become entangled. 1969 —— *Stand. Handbk. Textiles* (ed. 7) v. 336 The presence of these pills gives the fabric a highly objectionable appearance... As the material is further worn so do the anchoring fibres gradually wear and weaken until they break and the pill falls off. 1970 [see *PILL *v.*[1] 6 c].

3. (Earlier and later examples.)

1871 L. H. BAGG *4 Years at Yale* 141 The name 'Delta Phi man' is fast becoming a synonym for 'scrub', and 'pill', and even the neutrals regard its members with a sort of pitying contempt. 1880 A. A. HAYES *New Colorado* (1881) v. 64 He was the worst-looking pill you ever saw. 1886 *Galaxy* 1 Oct. 272 Various sorts of contemptible young men are designated as..'pills', 'squirts', [etc.]. 1925 WODEHOUSE *Carry on, Jeeves!* iii. 61 What's to be done?.. That pill is coming to stay here. 1939 L. M. MONTGOMERY *Anne of Ingleside* xxxiii. 242 That kid is a pill. My, doesn't she think it smart to fool people! 1970 *Women Speaking* Apr. 5/1 If a man doesn't like a girl's looks or personality, she's a..pill. 1977 B. GARFIELD *Recoil* xi. 120 'Do you love your wife?'.. 'You're a pill. Yes, I love her.'

4. (Further examples.) *spec.*, (a member of) the Royal Army Medical Corps; a medical officer or his orderly.

1866 *Harper's Mag.* July 268/1 One day..the two young 'pills' were arguing some case. 1915 'BARTIMEUS' *Tall Ship* ix. 159 They seized the Young Doctor, who was a small man, and deposited him on the deck. 'Couldn't you see I was asleep, Pills?' demanded the other. 1924 'NAUTICUS' *Sea Ways & Wangles* ix. 57 In some ships the sick list is so small that one of the young doctors—or 'Pills' as a Surgeon Lieutenant is sometimes called in order to distinguish him from his more venerable senior the P.M.O.—has even found time to take on the keeping of the ward-room wine accounts. *Ibid.* 58 Then again at a ward-room sing-song after dinner 'Pills' will often help matters along by playing the piano. 1925 FRASER & GIBBONS *Soldier & Sailor Words* 223 *Pills*,..a nickname for a Medical Officer's orderly. 1929 *Papers Mich. Acad. Sci., Arts & Lett.* X. 314 *Pills*.., the surgeon.

5. pill-bag, a bag in which pills are carried; **pill-bug,** a woodlouse of the genus *Armadillidium*, able to roll itself into a ball; **pill cooker** *slang*, (see quot.); **pill head** *slang*, a drug addict; **pill-millipede,** a millipede belonging to the order Oniscomorpha, esp. the genus *Glomeris*; in quot. 1868 = **pill-bug*; **pill peddler, pusher, roller** *slang*, a doctor or pharmacist; **pill popper** *slang*, one who takes barbiturate or amphetamine pills freely; a barbiturate or amphetamine addict; hence *pill-popping* vbl. sb.; **pill-rolling** *vbl. sb.*, the action of making into pills by rolling (also *fig.*); **pill shooter** *slang*, a doctor; **pill slab** (further examples); **pill-woodlouse** = **pill-bug*.

1852 *Knickerbocker* XL. 470 After procuring his degree, he had not the wherewithal to buy him pill-bags. 1874 E. EGGLESTON *Circuit Rider* xx. 189 'And you want me to see him,' said the doctor,..seizing his 'pill-bags' and donning his hat. 1930 J. F. DOBIE *Coronado's Children* i. 43 James had been educated to medicine in Kentucky.. and riding with his 'pill bags' over the far-stretched hills of the Colorado River satisfied his ambition. 1884 J. S. KINGSLEY *Stand. Nat. Hist.* II. 72 In common parlance these forms [of terrestrial isopod] are known as 'sow-bugs', 'pill-bugs', and 'wood-lice'. 1915 W. A. BRYAN *Nat. Hist. Hawaii* xxxi. 408 The curious oval little silver-gray creature found in large numbers in damp places, under boards and stones, is usually an introduced species known to many as the pill-bug, slater, sow-bug or wood-louse. 1954 *New Biol.* XVII. 41 *Porcellio* and *Armadillidium* (the pill-bug) appear to be able to withstand drier conditions than the rest [of the woodlice]. 1971 E. S. BAKKER *Island called California* xii. 191 If you examine such a decaying log, you are likely to find centipedes and pill bugs curled in the dirt. 1929 M. A. GILL *Underworld Slang*, *Pill cooker*, opium smoker. 1965 *Maclean's Mag.* 4 Sept. 31/1 The population totalled..fourteen narcotic addicts, two marijuana smokers and two 'pillheads', including me. 1969 *Courier-Mail* (Brisbane) 2 July B/1 He was all boy, and a drug taker—'A Pill Head' in his own words. 1971 S. HOUGHTON *Current Prison Slang* (MS.) 23 *Pill head*, amphetamine addict. 1973 *Times Lit. Suppl.* 16 Mar. 299/4 (Advt.), His experiences with junkies, pillheads, homosexuals and drop-outs in Soho. 1976 N. THORNBURG *Cutter & Bone* x. 244 Oh, she was a pillhead, yeah. And maybe the world's worst housekeeper too. 1868 BATE & WESTWOOD *Hist. Brit. Sessile-Eyed Crustacea* II. 494 It [*sc. Armadillo vulgaris*]..is often seen running about foot-paths, rolling itself up into a ball at the least alarm. This has gained for it the name of Pill millepede [*sic*]. 1899 W. T. FERNIE *Animal Simples* 236 Hoglouse, or Pill Millipede... This Hoglouse, or Millipede, was the primitive medicinal pill. 1958 J. L. CLOUDSLEY-THOMPSON *Spiders, Scorpions, Centipedes & Mites* xi. 17 The family Glomeridae contains the common British pill-millipede. 1967 P. A. MEGLITSCH *Invertebr. Zool.* xix. 839/1 Order Oniscomorpha. Pill millipedes. Mostly tropical millipedes..; body can be rolled into a ball. 1857 M. J. HOLMES *Meadow Brook* v. 78 Why, he's a young pill-peddler, who's taken a shine to Rosa. 1925 S. LEWIS *Arrowsmith* xiii. 137 How could old Max have gone over to that damned pill-pedler? 1931 [see *CROAKER 4]. 1941 J. SMILEY *Hash House Lingo* 43 *Pill peddler*, druggist. 1963 *Time* 1 Nov. 74 Can a lonely New Jersey pill popper who sleeps on a board find enduring happiness with an ebullient Hungarian gourmet who sleeps on a rug? 1975 *Publishers Weekly* 27 Jan. 286/3 The author, then a sophomore at a Wisconsin law school, was a pill popper—amphetamine to get her through the days, phenobarbital to get her through the nights. 1972 *Jrnl. Social Psychol.* LXXXVII. 121 The film illustrates the dangerous psychological and physiological effects of pill-popping. 1977 J. WAMBAUGH *Black Marble* (1978) xii. 304 They have a pill-popping party and she..dies of an overdose. 1909 J. R. WARE *Passing Eng.* 196/2 *Pill-pusher*.., doctor. Fine example of the graphic in phraseology. 1919 H. S. WARREN *Ninth Company* 34 *Pill pusher*, member of the medical corps. 1935 A. J. POLLOCK *Underworld Speaks* 88/1 *Pill pusher*, a doctor. 1961 *Amer. Speech* XXXVI. 147 *Pill pusher*, a specialist in internal medicine, as contrasted with a practitioner of the surgical specialties. Loosely, any M.D. 1969 LINN & PEARL *Masque of Honor* 66 Hell, I'm only a pill-pusher, Lieutenant. 1917 *Editor* 13 Jan. 33 *Pill rollers*, Hospital Corps. 1918 *Yank Talk* 21 'Why?' asked the pill roller. 1930 J. W. BARKLEY *No Hard Feelings!* 268 He told me to get it on record with the pill-rollers. 1936 J. CURTIS *Gilt Kid* viii. 87 He was damned if he let a lousy pill-roller know just how bad he felt. 1942 FRENCH & SLIPER *Army-Navy Guide* 181 *Pill roller*, an enlisted member of the medical corps. 1968 R. HOOKER *MASH* (1969) 175 Let the pill rollers..do it. 1961 R. D. BAKER *Essent. Path.* xxii. 597 Parkinsonism.. is a chronic disorder usually of elderly persons characterized by 'pill-rolling' rhythmic movements of the hands. 1972 *Country Life* 9 Mar. 572/3 Such things [*sc.* pill-slabs or -tiles] were used for pill-rolling and also, as in the more elaborate slabs..for hanging up as a sign of the owner's profession. 1928 L. H. NASON *Sergeant Eadie* 337 The gallant pill-shooters won't let us stay in these nice soft beds any longer than they can help. 1938 PARTRIDGE *World of Words* III. vii. 196 Doctor becomes.. slang *the vet* or..*pill-shooter*. 1941 J. SMILEY *Hash House Lingo* 43 *Pill shooter*, physician. 1966 H. MARRIOTT *Cariboo Cowboy* ix. 89 In those years, an average fellow was darn near down-and-out before he headed out to see a pill-shooter. 1960 H. HAYWARD *Antique Coll.* 218/2 Apothecaries' pill-slabs were made in tin-glazed earthenware at Lambeth. 1972 *Daily Tel.* 19 Jan. 9 A polychrome Delft heart-shaped pill-slab, which was used for rolling pills, was bought by Tilley for £2,800 at Sotheby's yesterday. 1972 *Country Life* 9 Mar. 572/3 The plaque.. which was in the same sale is a pill slab of 1664, also English Delft, and an exceptional rarity. 1863 Pill-woodlouse [see WOOD-LOUSE 2]. 1906 *Essex Naturalist* XIV. 53 'The common armadillo'..is the old name for the pill-woodlice now known as *Armadillidium*. 1931 W. T. CALMAN in W. P. Pycraft *Standard Nat. Hist.* ix. 169 The common Pill Woodlouse, *Armadillidium*, may often be seen crawling actively about on rocks in hot sunshine.

pill, *sb.*[3] Add: **2.** *Comb.*, as **pill yawl,** a sprit-rigged, three-masted boat used in the Bristol Channel.

1883 *Boats of World* 30 The Bristol Channel, where the Pill Yawls, large or small, decked or undecked, hold their own with any craft of their size. 1929 F. C. BOWEN *Sea Slang* 103 *Pill Yawl*, a Bristol Channel pilot boat.

pill, *v.*[1] Add: **6. c.** To gather into small balls of fluff on the surface of a fabric (said of the fibre, and of the fabric as a whole). Hence **pilled** *ppl. a.*, of or pertaining to fibres that have gathered in this way.

1962 *Which?* Aug. 240/1 One [Orlon cardigan]..was starting to pill after 10 washes. 1970 *Cabinet Maker & Retail Furnisher* 23 Oct. 173/2 Cloth so blended 'pilled'— fluffed, if you prefer it—very badly. *Ibid.*, While most worsted and woollen cloths, like a woollen carpet, tend to pill in the beginning, these pills wear off quickly and never recur. 1970 *Which?* Oct. 301/3 Trousers didn't pill, but as they were knitted some snagged. 1971 *Daily Tel.* 19 Apr. 12/4 That curious pilled wool we wore a few years ago, bumpy as if the wool had come out in a rash. 1971 *New Yorker* 21 Aug. 46/2 (Advt.), An exclusive Hathaway process that keeps the collar from pilling (i.e., fuzzing) throughout the life of the shirt.

pill, *v.*[2] Add: **3. b.** To fail (a candidate) in an examination. *slang.*

1908 A. S. M. HUTCHINSON *Once aboard Lugger* I. i. 15 'Your examination?' George half turned away. The bitterest moment of a sad day had come. He growled: 'Pipped.' 'Pipped?' 'Pilled.' 'Pilled?' 'Spun... I failed. I was referred for three months.' 1925 W. DEEPING *Sorrell & Son* xxii. 208 Gorringe had a sick face... 'Pilled,' thought Kit, and was not sorry, for Gorringe needed a course of pilling.

pillaloo (pi·lălū, -liū), *sb.* (*int.*) Also **pillilew, pilliloo,** etc. *dial.* [Imit.] **a.** A cry expressing grief or anger. **b.** A name for such a cry; a noise, disturbance, or outcry; a fierce argument. Hence **pillaloo-ing** *vbl. sb.*

1796 *Hull Advertiser* 23 July 4/2 The woman, having prostrated herself on the grave of the deceased, continued some time in silent meditation or prayer; then crying *Pillilew!* after the manner of the Irish at funerals, she sorrowfully departed with her husband. 1811 *Lexicon Balatronicum*, *Pillaloo*, the Irish cry or howl at funerals. 1829 G. GRIFFIN *Collegians* I. v. 101 She got a stitch an' died..an'..Dan made a *pilliloo* an' a *lavo* over her, as if he lost all belongen' to him. *Ibid.* xiv. 303 Jug..began bawling..the mother joined her, and such a pillilu as they raised between 'em was never known. 1847 *Paddiana* I. 100 Divle such a pillalooing as Lanty made out o' the windy ye never heered! 1851 H. NEWLAND *Erne* xiii. 385 Such a combination of huzzaing and pillalooing as never English ears had heard. 1888 'Q.' *Troy Town* xi. 120 What wi' the rumpus an' her singin' out 'Pillaloo!'..the Lawyer's sarmon..was clean sp'iled. 1899 S. MACMANUS *In Chimney Corners* 189 Nanny sitting in the chimney corner whillilew-ing and pillillew-ing, crying the very eyes out of her head. 1900 *N. & Q.* 12 May 373/1 To have 'a pillilew' it is not always necessary that there should be a fight, as a wrangle in which a number take part is 'a pillilew'. 1900 *Yorks. Weekly Post* 22 Sept. 11/1 We ran after him wol we collared him, en then we made him sing pillilew.

pillar, *sb.* Add: **2. c.** A metal column in the bodywork of a vehicle separating the front and rear doors, or the front door and windscreen. Also, a thin metal strip dividing the windscreen into two parts.

1907 *Car* 25 Sept. 261/1 The hind pillars were painted white from top to bottom. 1926 *Motor* 26 Oct. 637/1 The roof..slides back as far as the pillars in front of the rear doors. 1937 *Motor* 9 Mar. 219/3 A point..noticeable when sitting in the car is the wide range of vision made possible by extremely narrow pillars. 1938 *Times* 13 Oct. 8/1 Designers are..taking..pains to reduce the width of pillars. 1964 *Which? Car Suppl.* Apr. 47/2 The VW Devonette had its windscreen divided by a pillar which did not help forward vision. 1971 *Sci. Amer.* Oct. 11/3 (Advt.), Deflector fins on the front pillars keep the side windows free from dirt. 1977 *Custom Car* Nov. 19/2 The new Granada shape is clean and very smart, though it has lost the rather pleasant kink by the rear pillar.

3. b. in phr. *pillar of society, of the establishment* (see also 3 c below).

1961 *Ann. Reg. 1960* 445 The melodramas and crime dramas included..*Never Take Sweets from a Stranger*, with Felix Aylmer as a small-town pillar of society responsible for the seduction of small girls. 1961 NEW ENG. BIBLE *Gal.* ii. 9 Those reputed pillars of our society, James, Cephas, and John, accepted Barnabas and myself as partners. 1964 *Ann. Reg. 1963* 32 One by one the pillars of the establishment—Lord Salisbury, Lord Stuart of Findhorn, Lord Eccles among them—deserted a baffled and pleading Lord Chancellor. 1970 *Nature* 26 Sept. 1371/1 Those pillars of the establishment of the 1920s, such as Nancy Astor, Lord Haldane and the editors of *The Times* and *The Observer*. 1979 D. CLARK *Heberden's Seat* iii. 48 Heberden! A pillar of society locally.

c. (Further examples.)

1888 W. ARCHER tr. H. Ibsen (*title*) The pillars of society and other plays. 1902 G. B. SHAW *Mrs. Warren's Profession* Pref. p. vii, Dearer still..is that sense of the sudden earthquake shock to the foundations of morality which sends a pallid crowd of critics into the street shrieking that the pillars of society are cracking and the ruin of the State at hand.

12. pillar clock (see quot. 1962); **pillar drill** or **drilling machine** *Engin.*, a drilling machine incorporating a work-table supported on a column attached to the base of the machine; **pillar letter-box** = PILLAR-BOX; **pillar rose** (earlier and later examples); **pillar (and) scroll (top) clock** (see quot. 1960).

1933 *Burlington Mag.* Aug. p. xvi/1 Mr. Mody..retains the useful terms 'Lantern, Bracket, and Pillar Clocks' to describe the main types. 1962 E. BRUTON *Dict. Clocks & Watches* 131 *Pillar clock*, French drum clock with round movement and dial on four vertical pillars standing on a round base. The pendulum hangs in the middle of the pillars... Also a special form of Japanese clock showing time by a pointer moving along a linear scale, or any clock on a pillar. 1881 E. MATHESON *Aid Bk. Engin. Enterprise* II. xxiii. 313 The self-contained Pillar drill is useful, as there is more room around the machine within which to move the article. 1942 W. STEEDS *Engin. Materials* xiii. 204 The sensitive drilling machines taking drills up to about ¼ or $\frac{5}{16}$ in. diameter and the pillar drill up to as much as 2 in. diameter, according to size. 1964 S. CRAWFORD *Basic Engin. Processes* ix. 224 The multi-spindle machines consist of a series of pillar drills mounted over a common table, thus eliminating the constant tool change associated with the single-spindle machine. 1873 C. P. B. SHELLEY *Workshop Appliances* vii. 214 (*heading*) Double-geared pillar drilling machine. 1975 BRAM & DOWNS *Manuf. Technol.* vii. 198 The pillar drilling-machine..is similar in general design to the sensitive drill. 1879 TROLLOPE *Duke's Children* (1880) I. xxiv. 284 'Has it gone?' asked the Countess. 'I put it myself into the pillar letter-box.' 1837 T. RIVERS *Rose Amateur's Guide* 81 Clarissa Harlowe is a pillar-rose, of first-rate excellence. 1856 MRS. STOWE *Dred* II. 129 She was sitting..under the shadow of one of the pillar-roses. 1869 S. R. HOLE *Bk. about Roses* ix. 128 These Pillar Roses are beautiful additions to the Rosarium. 1965 G. S. THOMAS *Climbing Roses* xii. 176 Isolated pillars to take 'pillar' roses can be connected with a wooden beam. 1974 *News & Press* (Darlington, S. Carolina) 25 Apr. 7/4 Pillar roses at the lantern post are not only beautiful by day but have an accent of beauty when the light is turned on at night. 1860 C., JEROME *Hist. Amer. Clock Business* iii. 44, I took about one dozen of the Pillar Scroll Top Clocks, and went to..Wethersfield to sell them. 1912 N. H. MOORE *Old Clock Bk.* caption facing p. 113 Pillar and scroll top clock. 1929 G. H. BAILLIE *Watchmakers & Clockmakers of World* 349/2 They were at first wall clocks, but from 1814 brackets or shelf clocks known as Pillar Scroll Top clocks. 1950 B. PALMER *Bk. Amer. Clocks* 10 The Pillar and Scroll Clock remained the most popular Shelf Clock until about 1825 and survived well into the 1830's. 1960 H. HAYWARD *Antique Coll.* 218/2 *Pillar and scroll clock*, an American shelf or mantel clock. .. The wooden works are housed in a vertical rectangular case with a scrolled-arch top, small, round pillars at the sides, and delicately small feet. 1970 K. D. ROBERTS *Contrib. of Joseph Ives to Connecticut Clock Technol. 1810–1862* iv. 64 (*caption*) Pillar and scroll shelf clock with looking glass by Ives and Lewis.

pillar-box. Add: (Further examples.)

1939 'J. STRUTHER' *Mrs. Miniver* 173 She put on a mackintosh and struggled up the square to the pillar-box. 1978 L. DAVIDSON *Chelsea Murders* xix. 106 This one had been posted in a street pillar box.

b. In full, *pillar-box red*: a shade of red, that of a pillar-box.

1916 *Sphere* 1 July p. iv/1 Some charming *chapeaux* of the new pillar-box red. 1926 *Daily Colonist* (Victoria, B.C.) 21 July 16/4 (Advt.), Attractive Dresses of silk crepe, spun silk and silk broadcloth... Shades include black and white, blue..pillar-box red. 1934 *Archit. Rev.* LXXV. 108/1 It [*sc.* a built-in gramophone fitment]..is made entirely of laminated boarding painted white on the outside and pillar box red on the inside. 1950 'E. CRISPIN' *Frequent Hearses* i. 44 A wind-machine, painted a minatory pillar-box red. 1959 J. BRAINE *Vodi* i. 16 Dick thought of the shop... It had been redecorated in pillar-box red and white. 1963 *Listener* 10 Jan. 84/2, I don't like the colours, especially the Ribena, pillarbox, scrotula, and sulphur. 1970 *Vogue* Jan. 37/1 (Advt.), Pillar box cotton jersey for the jeans..and cropped T-shirt. 1973 P. EVANS *Bodyguard Man* vi. 50 A small Fiat in pillar-box red swirling round the corner fifty metres away and out of sight.

pillared, *ppl. a.* Add: **1.** (Further examples.)

1924 R. CAMPBELL *Flaming Terrapin* i. 16 In that pillared temple grew a heart. 1924 E. SITWELL *Sleeping Beauty* iii. 21 That pillared avenue Of tall clear-fruited ripe trees. 1953 E. M. FORSTER *Hill of Devi* 129 One passes through a pillared hall on to a terrace. 1961 *N.Y. Times* 3 July 13 The many country mansions..that dot the environs of Vicenza. How far that pedimented and pillared style has shed its influence Mr. Sansom reminds us. 1978 *Morecambe Guardian* 14 Mar. 23/1 (Advt.), The accommodation provides: pillared porch, hall, lounge, [etc.].

2. (Further examples.) Also *fig.*

1727 THOMSON *Summer* 60 Of growling Hills, that shoot the pillar'd Flame. 1864 TENNYSON *Voyage* in *Enoch Arden* 145 How oft we saw the Sun retire, And burn the threshold of the night, Fall from his Ocean-lane of fire, And sleep beneath his pillar'd light! 1924 R. CAMPBELL *Flaming Terrapin* iii. 45 Thick mælstroms propped the dense and sagging shades With pillared thunder. 1929 R. GRAVES *Poems* 20 True to the eagle nose, the pillared neck, (Missed by the intervening generation).

pillarless (pi·lăɹlės), *a.* [f. PILLAR *sb.* + -LESS.] Lacking or without pillars.

1889 W. B. YEATS *Wanderings of Oisin* II. 28 And the dome Windowless, pillarless, multitudinous home Of faces, watched me. 1922 *Glasgow Herald* 7 June, The experimental work on 'pillarless' electric safety lamps..has been conducted by the Miners' Lamps Committee. 1977 *Custom Car* Nov. 50/3 And there it sits, in all its glory; four-door pillarless body skulking over a mill of 390 cubic inch displacement.

pill-box. Add: **a.** (Earlier and later examples.)

1730 *Maryland Hist. Mag.* (1924) XIX. 182 From Eyre & Beecher (Druggists)... 12 Papers pill boxes. 1934 G. B. SHAW *Too True to be Good* I. 27 A measuring glass, a pill box, a clinical thermometer in a glass of water. 1977 *Halcyon Days Catal.* 8 A new range of small pill-boxes with snap-on enamelled lids.

b. *spec.* A small round concrete emplacement used for housing a machine-gun or similar weapon. Also *transf.* and *fig.* Also, short for *pill-box cap, hat.*

1887 R. D. BLUMENFELD *Diary* 27 June (1930) 17 The pill-box..protects only a small portion of the head and forehead from the sun. 1917 *Scotsman* 13 Sept. 6/4 The strength of these concreted farm cellars and individual pillboxes is amazing. 1923 *Daily Mail* 26 Feb. 7 On reaching shore again he [*sc.* a fox] made for the cliffs and hid in a concrete pill-box, where he was caught and killed. 1929 *Papers Mich. Acad. Sci., Arts & Lett.* X. 314/2 *Pill-boxes*, concrete structures or blockhouses developed by the Germans for use in their 'elastic' defense. They were employed as machine-gun nests. 1930 H. A. TAYLOR *Good-Bye to Battlefields* 136 The Germans, by means of their almost indestructible pill-boxes, have taught the inhabitants the virtues of concrete. 1935 A. J. POLLOCK *Underworld Speaks* 88/1 *Pill box*, enclosure of machine gun on prison wall. 1944 *Daily Progress* (Charlottesville, Va.) 2 Oct. 1/8 The planes dropped hundreds of tons of explosives on Nazi pillboxes and gunposts. 1958 L. DURRELL *Mountolive* viii. 166 Red pill-boxes mounted upon cancelled faces. 1958 *Listener* 4 Dec. 933/2 Is it another pill-box attempt to debunk the shameful Victorians of the middle and upper classes because the contemplation of the shameless purity and austerity of their private lives is galling to the rebellious youth of today? 1968 T. PARKER *People of Streets* 27, I was the lift boy and I had one of those round hats, pill-boxes they used to call them. 1973 *Guardian* 6 Mar. 6 The pillbox on the left is in yellow straw with a white daisy. 1974 *Times* 16 Jan. 2/4 He had led Mr Brook across the fields... They rested in a wartime pillbox and then got a lift to the outskirts of Chelmsford. 1978 J. KRANTZ *Scruples* iv. 99 Billy, about to go for job interviews, took herself to the custom-order millinery salon..in order to have Halston, then Jackie Kennedy's favorite hat designer, make her one perfect pillbox.

c. *pill-box cap* (example), *hat* (further examples).

1910 *Daily Chron.* 18 Apr. 1/7 It is much greater fun to wear the B.-P. hat of the scout than the neat 'pill-box' cap of the Church Lads' Brigade. 1964 MRS. L. B. JOHNSON *White House Diary* 10 Mar. (1970) 82 Archbishop Iakovos..dressed in long, black clerical garb.. and the black pillbox hat from which flowed long black veils. 1974 *Country Life* 3–10 Jan. 54/3 That badge of 1960s elegance, the Jackie Kennedy pillbox hat.

pilling, *vbl. sb.* Restrict '*arch.* or *dial.*' to senses in Dict. and add: **2. b.** The gathering of fibres into small balls of fluff on the surface of a fabric (see *PILL *v.*[1] 6 c).

1958 *New Scientist* 3 Apr. 17/2 In a test cardigan..the treated sleeve showed a remarkable resistance to pilling. 1959 A. J. HALL *Stand. Handbk. Textiles* (ed. 5) v. 314 Pilling has become especially noticeable since the introduction of the synthetic fibres. 1965 *New Scientist* 8 Apr. 95/1 Synthetic fabrics have one common short-coming: they are highly susceptible to what is known as 'pilling', a tendency to form small tangled knots of fibres. 1968 J. IRONSIDE *Fashion Alphabet* 244 *Pilling*,.. the term used when fibres gather into a ball on the surface. 1970 *Cabinet Maker & Retail Furnisher* 23 Oct. 173/2 The inclusion of nylon in the blend..aggravated the pilling. 1970 *Which?* Oct. 300/2 A few brands suffered slightly from pilling (little balls of fibre on the surface). 1972 *Times* 28 Nov. 19/8 E. Gomme..has encountered problems of 'pilling' in the fabric. 1974 *Amer. Speech 1970* XLV. 179 An antistatic pilling-resistant (resistant to gathering small 'pills' of fuzz) finish.

pillion, *sb.*[1] **a.** Restrict '*Obs.* exc. *Hist.*' to senses in Dict. and add to def.: In modern use, a seat located behind the saddle of a motor-cycle, on which a second person may ride. Hence *to go, ride* (cf. RIDE *v.* 1 d), *sit pillion*, to travel on this seat. Also *transf.*

The application in quot. 1878 is unusual, but anticipates use in the modern sense.

1878 *Design & Work* IV. 215/1 To use the electric light on a bicycle is very easy, if you can arrange to place upon a pillion behind you a large steam engine and boiler and a Gramme machine, the whole weighing only about 3 tons. 1911 *Motor Cycle* 27 Apr. 481/1 The pillion or tandem seat is likely to become increasingly popular. 1923 *Weekly Dispatch* 13 May 9 Riding his motor-cycle..with Miss Esther Gwyther, a nurse, on the pillion, he collided [etc]. 1926 T. E. LAWRENCE *Seven Pillars* (1935) IV. xliv. 255 He stopped babbling, and began to wail out his sorrows. I sat him, pillion, on the camel's rump; then stirred her up and mounted. 1927 *Glasgow Herald* 31 Aug. 10 A clerk..on whose machine Miss Paterson was riding pillion. 1934 T. S. ELIOT *Rock* i. 21 But every son would have his motor cycle, And daughters ride away on casual pillions. 1958 J. BETJEMAN *Coll. Poems*, And country girls with lips and nails vermilion Wait, nylon-legged, to straddle on the pillion. 1965 *Listener* 27 May 785/1 They [*sc.* Vietnamese girls] ride bicycles, sit pillion on motor-bicycles, and manage their delightful floating garments with elegance and dignity at all times. 1973 J. WAINWRIGHT *Pride of Pigs* 142, I got rid of my scooter, and started going pillion with Lance. 1979 R. RENDELL *Make Death love Me* iii. 30 Alan had once seen her get off the pillion of a boy's motor-bike.

b. *pillion passenger, ride, rider, -riding* vbl. sb. and ppl. adj., *seat* (modern use); **pillion cloth,** a cloth placed under a pillion; **pillion stick,** a stick fastened in a pillion to hold luggage in place; **pillion stone,** a stone used to facilitate mounting to a pillion seat.

1648 *Public Rec. Colony of Connecticut* (1850) I. 508, 1 sidesaddle and pillion cloath. 1684 *Essex Inst. Hist. Coll.* (1888) XXV. 155 In the Kitchine..a sadle, pillyon and pillyon cloath. 1929 J. DE F. SHELTON *Salt-Box House* iv. 34 Cuffee busied himself making sure that the dark blue pillion-cloth protected her dress from the horse's flank. 1973 J. WAINWRIGHT *Pride of Pigs* 92 The pillion-passenger was thrown clear, but the motor-cyclist was pinned under his machine. 1976 *Deeside Advertiser* 9 Dec. 24/3 A teenage motor cyclist died and his pillion passenger was injured in an accident on a sharp bend on the main Corwen-Chester Road at Treuddyn. 1935 T. E. LAWRENCE *Let.* 31 Jan. (1938) 845 It was a good idea, that pillion ride. 1963 L. DEIGHTON *Horse under Water* xxiv. 107 Giorgio had got a pillion ride..with a two-stroke motor bike. 1920 *Motor Cycle* 24 June 714/2 Motor-cyclists are summoned for having their number plates obscured by any part of a pillion-rider's dress. *a* 1974 R. CROSSMAN *Diaries* (1975) I. 40 We had the most amiable cup of tea before driving home to Prescote through the fog, where I found that another pillion rider had come down with my third red box. 1920 *Motor Cycle* 8 July 45/1 The local authorities have threatened a campaign against pillion riding. 1927 W. E. COLLINSON *Contemp. Eng.* 35 Probably pillion-riding..will also be prohibited and dazzle head lights have to be dimmed. 1932 H. S. WALPOLE *Fortress* I. 83 Many came in pillion-riding as for hundreds of years they had done, while the grander farmers were proud in the 'shandy-carts'. 1970 *R.A.C. Guide & Handbk.* 59/1 Pillion Riding. Only one passenger may be carried on a solo motor cycle, and.. must be seated astride behind the driver on a proper pillion seat. 1911 *Motor Cycle* 27 Apr. 418/2 A pillion seat, if not very sociable, certainly has some advantages. 1973 J. WAINWRIGHT *Pride of Pigs* 88 The youth astride the Road Rocket kicked the starter... His companion settled himself on the pillion-seat. 1784 J. F. D. SMYTH *Tour U.S.A.* II. 248 All these papers were concealed in the mail pillion-sticks on which the servant carried his portmanteau. 1907 *Manch. Guardian* 20 July 7/7 On one side of the porch is a horsing or pillion stone.

pillion (pi·lyən), *v.* [f. PILLION *sb.*[1]] **1.** *trans.* To equip (a horse or saddle) with a pillion. Chiefly in pa. pple. or ppl. adj.

1843 *Knickerbocker* XXII. 431 The cozy couple..ride.. side by side upon the pillioned saddle. 1929 J. D. SHEL-

TON *Salt-Box House* iv. 33 Thaddeus's best pacing-mare being duly saddled and pillioned. **1935** W. FORTESCUE *Perfume from Provence* 234 They rode upon pillioned horses decorated with favours.

2. *trans.* To place on a pillion. Chiefly as ppl. adj.

1906 A. NOYES *Drake* I. II. 59 Little the boy remembered of that flight, Pillioned behind his father. **1910** W. DE MORGAN *Affair of Dishonour* iv. 46 A horseman here and there, alone or with a wench pillioned behind. **1958** P. MORTIMER *Daddy's gone a-Hunting* i. 8 A motor cycle turned into the road... She caught sight of..a pillioned girl with hair streaming.

3. *intr.* To ride on the pillion of a motor-cycle. *rare.*

1935 T. E. LAWRENCE *Let.* 31 Jan. (1938) 845 Pretty awful pillioning with a suitcase and masterpiece in one's arms!

pillionaire (pilyənɛ̄ə·ɹ). Now *rare* or *Obs.* [f. *PILLION *sb.*¹ after MILLIONAIRE.] One, usu. female, who rides on the pillion of a motor-cycle or on a seat at the back of a bicycle.

1931 *Newark Advertiser* 26 Aug. 7/2 'What is a proper pillion seat?' was the question for the wisdom of the Bench at Newark Borough Police Court. Apparently it is a seat occupied by a pillionaire. **1931** *Times* 4 Sept. 6/5 On the back of his bike the defendant had a pillionaire—a female. **1937** PARTRIDGE *Dict. Slang* 630/1 *Pillionaire*, a female occupant of a 'peach-perch' or 'flapper-bracket'.

pillionist (pi·lyənist). *rare.* [f. *PILLION *sb.*¹ + -IST.] One who rides on the pillion of a motor-cycle.

The more usual term is *pillion rider*.

1923 *Motor Cyclist* 26 Sept. 643/1 As a confirmed pillionist I do not add my voice to those who are clamouring for legal abolition of this form of passenger riding.

pillock, pillok. Substitute for entry:
pillock (pi·lək). *North. dial.* and *slang.* Also †**pillok, pilloch, pilluck.** [Variant of PILLICOCK.] **1.** = PILLICOCK 1.

1535 [see PILLICOCK 1]. **1903** *Eng. Dial. Dict.* IV. 503/1 *Pill,..pilluck* Wm. Yks..., the male organ, the penis.

2. *transf.* A fool, a stupid person; also in weakened sense, a fellow, bloke.

1967 J. BURKE *Till Death us do Part* viii. 135 What are you talking about, you great hairy pilloch? **1968** J. WAINWRIGHT *Darkening Glass* viii. 70 She..glared across the room and said: 'Where's that pilluck with the drinks?' **1976** —— *Bastard* vii. 93 'You always were a pillock,' he says, with feeling. **1978** 'J. GASH' *Gold from Gemini* vii. 70 The pillock mistook my astonishment for awe.

pillow, *sb.* Add: **3. a.** (Later example.)

1895 *Wales* Apr. 179/1 Went to yearly meeting at Hereford with a few others, Molly Lloyd riding on pillow behind him.

5*. *Geol.* A body of rock, esp. lava, likened to a pillow or filled sack in shape and usu. occurring with other similar bodies. Cf. *pillow lava, structure* below.

[**1890** *Q. Jrnl. Geol. Soc.* XLVI. 312 The structure is more commonly irregular, the masses resembling pillows or soft cushions pressed upon and against one another.] **1899** *Summary of Progress Geol. Survey U.K. 1898* 108 It shows the 'pillow' structure already referred to, some of the 'pillows' being a yard or more in diameter. **1944** C. A. COTTON *Volcanoes* xv. 290 Lava pillows are commonly three to four feet in diameter. **1955** LONGWELL & FLINT *Introd. Physical Geol.* v. 72 We conclude that pillows result from immersion of hot lava in water. **1962** E. A. VINCENT tr. *Rittmann's Volcanoes* ii. 71 The freshly formed pillows are in effect bladders filled with still-fluid lava, which roll down..and pile up one above the other. **1971** I. G. GASS et al. *Understanding Earth* xxi. 302/2 Being erupted under water, the lava flows tend to segregate and to form accumulations of sub-cylindrical bodies called pillows.

6. pillow-book, (*a*) a book suitable for reading in bed; freq. an erotic book; (occas. used as the title of such a book); (*b*) in Japan, a type of private journal or diary; **pillow coat** (further examples); **pillow cover** = PILLOW-CASE; **pillow-fight** (earlier and later examples); also as vb.; **pillow lava** *Geol.*, lava exhibiting pillow structure; **pillow mound** *Archæol.* (see quots.); **pillow muff** = MUFF *sb.*² 1 a; **pillow-sham** (earlier and later examples); **pillow-slip** (further examples); **pillow structure** *Geol.*, a rock structure in which numerous closely fitting 'pillows' are fused together, found in some lavas and attributed to eruption under water; (cf. sense 5* above); **pillow talk,** conversation, usu. of an intimate kind, held in bed; also as vb.; **pillow tank,** a collapsible rubber container used for storing large quantities of liquid.

1906 N. G. R. SMITH (*title*) The Pillow-Book. *Ibid.* Pref., The pillow reader will surely find passages taken and left from the pillow books of his own election. **1907** *Daily Chron.* 5 July 3/4 When one thinks of it, Scott in verse is certainly a good pillow book. **1928** A. WALEY *Pillow-Bk. of Sei Shōnagon* 21 The *Pillow-Book*..consists partly of reminiscences, partly of entries in diary-form. *Ibid.* 24 Shōnagan protests, as do most diarists and makers of journals, that the *Pillow-Book* was intended for herself alone. **1960** *Ibid.* (rev. ed.) 16 To keep some kind of journal was a common practice of the day [*sc.* the 10th century]. The name Pillow-Book, *Makura no Sōshi*, was given at the time to notebooks in which stray impressions were recorded. **1963** 'HAN SUYIN' *Four Faces* 33 'Your blonde..is she a writer?' 'Pillow books.' **1967** *Spectator* 22 Dec. 782/3 Presumably one tries to write as well in a pillow-book as in a novel. **1968** *Guardian* 7 May 6/1 Around the walls were eleventh-century Indian temple sculptures, pillow books from Japan..and too many modern works. **1977** E. J. TRIMMER et al. *Visual Dict. Sex* (1978) xxi. 222 Any erotic books can be pillow books if they have the right effect on the reader. **1651** in *Mayflower Descendant* (1908) X. 39, I give her the bed.. with two paire of sheets two pillows two paire of pillow coates. **1727** in *Ibid.* 15, I Give to my Granddaughter Mary Bangs A pillow-coat. [**1644** in *Essex County, Mass. Probate Rec.* (1916) I. 41 Two Pilcovrs and two Payer of course sheets, 10s.] **1912** N. L. MCCLUNG *Black Creek Stopping-House* 113 There disappeared at the same time towels, pillow-covers, and a few small tools. **1969** WIDDOWSON & HALPERT in Halpert & Story *Christmas Mumming in Newfoundland* 151 Pillowcovers were also worn, or a large sheet might be draped over the head, with holes cut for eyes, nose, and mouth. **1871** L. M. ALCOTT *Little Men* i. 16 We always allow one pillow-fight Saturday night. **1892** 'MARK TWAIN' *Amer. Claimant* xiii. 119 They generally wound up with a pillow fight, in which they banged each other over the head, and threw the pillows in all directions. **1903** T. ROOSEVELT *Lett. to his Children* (1919) 69 The pillow fight raged up and down the hall. **1939** C. ISHERWOOD *Goodbye to Berlin* 268 Two young men were pillow-fighting with cushions. *a* **1953** E. O'NEILL *More Stately Mansions* (1965) I. iii. 52 We had a pillow fight. **1960** L. HELLMAN *Toys in Attic* III. 66 Shall we have a pillow fight or make fudge? **1976** *Ulverston* (Cumbria) *News* 3 Dec. 4/4 Gary White, of the Barrow Sailing Club was the winner of the pillow fight on the pole. **1903** *Amer. Geologist* XXXII. 67 Immediately above the lava sheet is the pillow lava, proved to be about 60 meters in thickness. **1971** I. G. GASS et al. *Understanding Earth* xxi. 302/2 The presence of such pillow lavas in ancient rock sequences..is unequivocal evidence of sub-aqueous eruption. **1977** A. HALLAM *Planet Earth* 96 Ridges or cones of granular glassy rocks and pillow lavas result from subglacial eruption through fissures or vents respectively. **1928** CRAWFORD & KEILLER *Wessex from Air* 23 At High Beech the rabbits use the pillow-mounds very extensively. **1932** *Field Archæol.* (Ordnance Survey) 5 Low, flat mounds of earth have been noticed. They are called Pillow-mounds, and appear to be of a much later date [than long barrows]. **1963** E. S. WOOD *Collins Field Guide Archaeol.* II. ii. 233 *Pillow mounds*, these are low, oblong mounds..2 to 3 ft. high; most have a shallow ditch round them. **1908** *Westm. Gaz.* 24 Dec. 15/2 Her furs were cinnamon fox with a bunch of violets in the pillow muff. **1871** 'S. MAY' *Prudy keeping House* 43 As she had decided to call herself Mother Hubbard, she made an ample cap, by folding a 'pillow-sham', and putting two of its ruffled edges around her face for a double border. **1914** KIPLING *Lett. Travel* (1920) 244 Our great National Policy of co-educational housekeeping! Ham-frills and pillow-shams. **1926** *Chicago Tribune* 11 Sept. 2/2 Then there was the pillow-sham holder. **1947** *Christian Sci. Monitor* 15 Jan. 6/5 And pillow shams! Like the splashes, they were painstakingly worked in red outline. **1963** G. H. THOMSON *Crocus Country* xvi. 100 The pillow-shams or covers.. were all crisp and white. **1975** *New Yorker* 17 Nov. 134/2 To bring a woman's brass bed into line with her nightshirt, there are antique pillow shams, often copiously trimmed with lace, eyelet, and tucks. **1874** J. A. MAIR *Handbk. Proverbs* 461 *Pillow-slip*, pillow-case. **1920** T. S. ELIOT *Ara Vos Prec* 22 Pushing the framework of the bed And clawing at the pillow slip. [**1897** A. GEIKIE *Anc. Volcanoes G.B.* I. iii. 26 Some basic lavas on flowing into water or into a watery silt have assumed a remarkable spheroidal sack-like or pillow-like structure.] *Ibid.* xiv. 244 Some of the diabase-masses display the pillow-structure and amygdaloidal texture. **1930** PEACH & HORNE *Geol. Scotl.* 142 Pillow-structure is well developed in some of the volcanic rocks, the chilled margins of the pillows and the concentric arrangement of the vesicles being marked features. **1971** I. G. GASS et al. *Understanding Earth* xiii. 165/1 There may be some instances (e.g. pillow structures in lavas) when igneous and metamorphic rocks yield important information. **1977** A. HALLAM *Planet Earth* 188 Similar pillow structures characterize basaltic lavas erupted on the present ocean floors. **1939** JOYCE *Finnegans Wake* I. 57 Mid pillow talk and chithouse chat, on Marlborough Green as through Molesworth Fields. **1971** *Femina* (Bombay) 16 Apr. 7/2 Is your husband out pillow-talking with some bright young doll? **1971** *Daily Tel.* 17 Dec. 9/6 Thanks to a bug under the bed, a man down in the basement..is putting on tape all Ingrid's pillow talk, not to mention her sighs and squeals. **1973** R. THOMAS *If you can't be Good* (1974) xii. 109 She told the senator. Pillow talk, I would think. **1975** *Times* 23 Aug. 4/7 Mrs Ford..makes it plain she gets her views across to Mr Ford in what she calls 'Pillow talk'. **1951** *Sci. News Let.* 10 Feb. 93/1 A new synthetic rubber-nylon 'pillow' tank for gasoline that lies flat on the ground and provides 10,000 gallons of storage in ten minutes will soon be servicing U.S. fighting tanks and trucks at the front. **1959** *Post & Times Herald* (Washington) 5 Aug. B7 (*caption*) The pillow tanks can be used to transport virtually all types of liquid on trucks, railroad cars and barges. When not in use, the collapsible containers can be rolled up and packed in a box. **1962** *Engineering* 26 Jan. 144 An example of the collapsible pillow tank principle, in the transport of water.

pillowing (pi·lōᵘ,iŋ), *a. rare.* [f. PILLOW *sb.* + -ING².] Pillow-making.

1924 *Times Trade & Engin. Suppl.* 29 Nov. 247/2 Bleaching fabrics such as pillowing, art, or handkerchief linens.

pillowy, *a.* Add: Also *fig.*

1805–6 WORDSWORTH *Prelude* (1959) III. 99 From these I turned to travel with the shoal Of more unthinking natures, easy minds And pillowy.

pilluck, var. *PILLOCK.

pilo-. Add: **pi·loerection,** the erection or bristling of hair or fur; **pi·loerector,** an agent that causes piloerection; **pilomotor** *a.*, add: more widely, involved in or pertaining to the movement of hair by bodily processes; also as *sb.*, a pilomotor nerve or muscle; (earlier and later examples); **pilonidal** *a.* (earlier and later examples).

1938 J. F. FULTON *Physiol. Nervous Syst.* xiii. 248 The most important mechanisms of heat production and preservation are shivering, mobilization of carbohydrate reserve, vasoconstriction, piloerection, increase in heart rate, and elevation of metabolic activity. **1958** *Jrnl. Investigative Dermatol.* XXX. 107/2 Injection of.. epinephrine regularly produced 'goose-flesh' and piloerection. **1974** CARLSON & HSIEH in N. B. Slonim *Environmental Physiol.* iv. 67/2 Fur-bearing animals can greatly increase the insulation of their outer coat by piloerection. **1977** RUMBAUGH & GILL in D. M. Rumbaugh *Language Learning by Chimpanzee* ix. 175 Lana's response was to hoot with apparent agitation; she also displayed piloerection and a furrowed brow. **1946** A. KUNTZ *Autonomic Nervous Syst.* (ed. 3) xv. 327 Intracutaneous administration of acetylcholine..elicits strong fleeting pilo-erector activity. **1965** *Jrnl. Investigative Dermatol.* LXIV. 86/1 The vasoconstrictor and pilot erector effect in man of noradrenaline was compared with those of..dopamine. **1891** LANGLEY & SHERRINGTON in *Jrnl. Physiol.* XII. 278 It will be convenient to have a short name for the nerve-fibres, stimulation of which causes contraction of the erectores pilorum. We shall call them 'pilo-motor' fibres on the analogy of 'vaso-motor' fibres. **1892** *Ibid.* XIII. 701 This class of fibres consists of the eye-fibres of the sympathetic. Langley has shewn this for cat and rabbit, and they often in monkey extend a segment higher than do the pilo-motors for the scalp. **1909** *Ibid.* XXXVIII. 174 The paralysis of the pilo-motor mechanism is one of the.. results of nerve section. **1927** *Ibid.* LXIV. 98 Shallow incisions..through the layers which contain the insertions of the pilomotor muscles into the hair roots. **1932** *Amer. Jrnl. Physiol.* CII. 30 The pilomotors reacted..to single shocks. **1948** A. BRODAL *Neurol. Anat.* xi. 394 The hypothalamus is probably concerned in integrating the activity of the pilomotors with other autonomic functions. **1880** R. M. HODGES in *Boston Med. & Surg. Jrnl.* CIII. 486/1 For the development of this rather singular lesion, to which..I venture to give the name of pilo-nidal (*pilus*, a hair, *nidus*, a nest) sinus, the following elements are necessary:—(1.) The presence of a congenital coccygeal dimple. (2.) Abundant pilous development... (3.) Insufficient attention to cleanliness. **1956** *Lancet* 15 Dec. 1244 An interdigital sinus behaves like a fistula-in-ano or a pilonidal sinus elsewhere... These pilonidal sinuses appear to be acquired and may be caused by loose hairs, foreign bodies, or inspissated secretions being drawn into small abrasions or acne pits in the skin. **1957** S. L. ROBBINS *Textbk. Path.* xxx. 1186/2 Anatomically, these lesions consist sometimes of sinus tracts, pilonidal sinus, communicating with the surface through minute (probe diameter) pores; at other times well developed epidermal-lined cysts, pilonidal cyst, that may or may not communicate with the surface. **1964** D. E. SMITH in L. V. Ackerman *Surg. Path.* xxvi. 1063 In their simplest form pilonidal sinuses or cysts are tracts lined by epidermis that extend beneath the skin toward the dura.

pilón (pilōᵘ·n). *Southwestern U.S.* Also **pilon.** [Mexican Sp., a. Sp. *pilón* sugar-loaf, mortar.] A free gift given when a purchase is made or an account paid; = *LAGNIAPPE. Also *fig.*

1892 *Dialect Notes* I. 251 *Pilón*.., the gratuity given by merchants to customers, whenever accounts are settled. **1932** H. W. BENTLEY *Dict. Spanish Terms in Eng.* 180 *Pilon*,..a favor; a gratuity. Literally the word signifies a small cone-shaped cake of sugar. It may be conjectured that a small *dulce* of this sort constituted the *pilon* originally. **1947** R. BEDICHEK *Adventures with Texas Naturalist* vii. 75 It [*sc.* yaupon]..stands drought, resents coddling, and throws in, as a *pilon* to its domesticator, decorative red berries in the fall and winter. **1962** E. B. ATWOOD *Regional Vocab. Texas* iii. 68 The custom of giving something extra with a purchase (or when a bill is paid) is firmly established in the United States... Most areas lack a specific word for this sort of gift. In the Southwest, the West, and part of Central Texas *pilón* is very well known and widely used. *Ibid.* vii. 124 (*heading*) Lexicographical pilón. *Ibid.* 128 Modern stores are becoming less and less inclined to give pilón.

pilot, *sb.* Add: **1. b.** (Further examples.) Esp. a skilled guide employed on land.

1672 J. PAINE *Jrnl.* in *Publ. Colonial Soc. Mass.* (1917) XVIII. 189 Wee mett with ye Riuor Hosick wch wee Set or corse for by Direction of or Indian Pilate and ther.. wee dismounted. **1710** T. BUCKINGHAM in S. Knight *Jrnls.* (1825) 94 Mr. Christophers and myself, having provided horses and a pilot, set out for Boston. **1847** J. PALMER *Jrnl. Trav. Rocky Mts.* 15 In case the company would elect him pilot, and pay him five hundred dollars, *in advance*, he would bind himself to pilot them to fort Vancouver. **1927** *Dialect Notes* V. 459 *Pilot*, the boy who accompanies a blind beggar. The American 'Lazarillo'. **1936** I. L. IDRIESS *Cattle King* vii. 63 A squatter was overlanding with a big mob of stock, his wagons loaded with a year's supplies. He had taken up country on the 'blind', without ever having seen it... He needed a pilot to show him the waterholes on his own country.

c. (Earlier U.S. example.) Also short for *pilot light.*

1883 F. M. A. ROE *Army Lett.* (1909) 313 It requires two engines to pull even the passenger trains up, and

when the divide is reached the 'pilot' is uncoupled and run down ahead. **1964** E. BERCKMAN *Simple Case of Ill-Will* x. 98, I smelled gas!.. There's a pilot on your cooker, an open flame. **1973** R. L. SIMON *Big Fix* (1974) vi. 48 All the lights were out except for a couple of pilots beneath the tape decks.

d. One who controls an aircraft, balloon, spacecraft, or the like during flight, usu. a person duly qualified to do so. *automatic pilot*: see *AUTOMATIC *a.* 2.

1848 *Sporting Life* 12 Aug. 289/1 The aëronautic race was conducted by Lieutenant Gale and Professor Gypson, ..the latter acting as the pilot of the Royal Albion. **1851** *Illustr. London News* 13 Sept. 330/2 We..threw out more ballast.., and descended..in a..field. I fell.., and the car over us all; while 'the pilot who had weathered the storm' was thrown with..violence from among the cordage. **1852** *Ibid.* 18 Sept. 224/3 'Sit still, all of you, I say!' roared our pilot, as he saw some one endeavouring to leave the car. *Ibid.*, Indeed, long shall we all remember the pleasant night we passed with the old ethereal pilot on his 500th ascent with the Royal Nassau Balloon. **1899** *English Mechanic* 14 July 480/3 The new machine.. is said to be able to carry in its car as many as six men and travel easily at a rate of 100 miles an hour under the absolute mastery of its engineer and pilot. **1907** *Navigating the Air* (Aero Club Amer.) 247 In order to qualify as a pilot one must make ten ascensions, one of which must be made at night, and two of which must be made alone. **1913**, etc. [see *air pilot* s.v. *AIR *sb.*¹ B. III. 4]. **1916**, etc. [see *FLY *v.*¹ 1 g]. **1936**, etc. [see *bush pilot* s.v. *BUSH *sb.*¹ 11]. **1953** 'N. SHUTE' *In Wet* v. 136 This was the first Ceres that had visited Edmonton, and a small crowd of pilots and R.C.A.F. officers gathered around it on the tarmac. **1962** *Into Orbit* 243 *Backup pilot*, an Astronaut who..may go on the mission himself if the Astronaut named as pilot is unable to make the flight at the last minute. **1974** *Daily Tel.* 14 June 8/4 Among the pilots flying this weekend will be Charles Dollfus, 83-year-old leading French balloonist who has been flying since 1911. **1978** *Dumfries Courier* 13 Oct. 9/1 He..would like other hang-glider 'pilots' in the area to contact him with a view to starting a Dumfriesshire Hang-Gliding Club.

e. *to drop the pilot*: to abandon a trustworthy adviser.

After a cartoon by J. Tenniel in *Punch*, 20 Mar., 1890 depicting the recent dismissal of Bismarck from the Chancellorship of Germany by William II.

1926 G. M. TREVELYAN *Hist. Eng.* IV. vi. 456 In face of these signs Charles decided to 'drop the pilot'. It was indeed tempting to make a scapegoat of Clarendon. **1958** J. RAYMOND *England's on Anvil!* 149 The Kaiser is aged thirty. He has been on the throne a year and is already preparing to 'drop the pilot', get rid of Bismarck. **1979** D. GURR *Troika* vi. 32 Khruschev'll be dropping that pilot before they clear the river.

f. Short for *pilot film, plant, programme*, etc.: see sense 6b below.

1962 *Listener* 18 Oct. 633/2 A little tighter and tauter and the production would have looked for all the world like a pilot for a new series. **1971** M. BABSON *Cover-up Story* xiv. 153, I came by..to talk over a few points before we started filming the pilot tomorrow. **1971** *Guardian* 26 Feb. 13/4 Sir Lew Grade..gave an uncharacteristically terse 'no comment' when asked if he had made any pilots in the recent past. **1971** *Inside Kenya Today* Mar. 9/1 If this pilot is successful an important export oriented mining enterprise will be established. **1973** *Nation* (Barbados) 16 Dec. 2/2 He has recently signed to be..a star in a pilot for a television series which is being written by Everett Chambers. **1975** *Radio Times* 30 Aug.–5 Sept. 14/3 It was only a pilot which would not be seen by the general public. **1977** *Time Out* 17–23 June 16/2 Thames' fourth telefilm in three days. Repeat of the pilot for a never-made series about a big city newspaper.

g. A jockey.

1976 *Horse & Hound* 10 Dec. 41/1 He was to underline his Epsom misfortune by streaking home in the Irish Sweeps Derby, when his French pilot was replaced by Geoff Lewis. **1976** M. MAGUIRE *Scratchproof* iv. 54 Will Highwayman jump the fence without a pilot?

5*. *Telecommunications.* An unmodulated signal transmitted with another signal for purposes of reference or control. Freq. *attrib.*, as *pilot carrier, tone*, etc.

1935 *Proc. IRE* XXIII. 702 The high degree of frequency stability required for single side-band suppressed-carrier transmissions can be dispensed with by transmitting a pilot frequency over the channel. **1957** D. G. FINK *Television Engin. Handbk.* xviii. 31 The L3 system makes use of six pilots for dynamic regulating and equalizing purposes. These are 308, 556, 2,064, 3,096, 7,266, and 8,320 kc. *Ibid.* xvi. 249 Oscillator-frequency drift causes picture-hue shift by changing the signal location on the phase characteristics of the pilot carrier and sideband circuits. **1962** C. F. BOYCE *Open-Wire Carrier Telephone Transmission* vii. 95 Over the 3-channel carrier range a single pilot can regulate for changes in flat and slope loss. .. In a 12-channel system two pilots, one at each end of the band, are required. **1966** M. SCHWARTZ et al. *Communication Syst. & Techniques* xi. 492 One technique for monitoring the channel state in a digital signaling system would be to transmit a pilot tone (unmodulated tone) along with the information-bearing waveforms. **1974** H. BURSTEIN *Q. & A. about Tape Recording* (1975) xv. 194 The FM stereo broadcast contains a 19 kHz pilot signal. **1974** HARVEY & BOHLMAN *Stereo F.M. Radio Handbk.* ii. 23 The reason for using a 19 kHz oscillator in the signal encoder now becomes clearer, since the oscillator provides a reference between the transmitted 19 kHz pilot tone and the 38 kHz subcarrier feed to the balanced modulator. **1975** *Which?* Sept. 278/3 An FM stereo radio signal has three parts. The main one is a mono signal... The extra information the tuner needs to produce stereo is in the other two parts—a sub-carrier and a pilot tone.

5.** *Electr.* = *pilot wire* in *6.

1940 *Chambers's Techn. Dict.* 644/1 *Pilot*, in power systems, a conductor used for auxiliary purposes, not for the transmission of energy. **1957** M. D. KIPPEN in E. O. Taylor *Power Syst. Communications* i. 6 It has been the policy of many city undertakings when laying power cables to lay protection pilots and, in some cases, telephone pilots in the same trench. **1966** W. J. CHEETHAM in Taylor & Boal *Electr. Power Distribution 415V–33kV* iii. 125 Post-office pilots are much more prone to interference during maintenance work on adjacent communication circuits.

6. *pilot error*; *pilot-balloon* (earlier and later examples), *-engine* (later example), *-locomotive*, *-tunnel* (examples); **pilot biscuit** *U.S.*, ship's biscuit; **pilot-bread** (earlier and later U.S. examples); **pilot cable** *Electr.* = *pilot wire* below; **pilot chute** or **'chute** (see quot. 1925); **pilot-cloth** (earlier N.Amer. examples); **pilot-coat** (examples); **pilot driver,** an engine-driver who accompanies another over a route with which the latter is unfamiliar; **pilot guard** (see quot. 1921); **pilot-house** (earlier and later examples); also (*U.S.*), a house on land in which a pilot lives or stays; **pilot lamp** = *pilot light* (*c*); **pilot light** (further examples); †(*b*) a small light left permanently on to provide illumination; (*c*) a small electric light used to give an indication or warning rather than illumination; **pilot-man,** a railway official who directs the movement of trains over a section of track being temporarily used as a single line; also, a pilot driver; **pilot officer,** a commissioned rank in the Royal Air Force, equivalent to a second lieutenant in the Army; **pilot parachute** = *pilot chute*; **pilot-snake** (earlier and later examples); **pilot valve** *Engin.*, a small auxiliary valve that is operated in association with a larger valve; **pilot-weed** (earlier and later examples); **pilot-whale** (later examples); **pilot wire** *Electr.*, an auxiliary wire or cable for conveying information about an associated power line or telegraph line or for operating apparatus connected with one.

1802 *Sporting Mag.* XX. 295/1 A Pilot Balloon, as it is called, was first launched. **1924** Pilot balloon [see *BALLON-SONDE]. **1942** *Endeavour* I. 118/2 The information on temperature is obtained from pilot balloons—small balloons carrying a cage of instruments to read temperature, .. at the altitude at which the balloon is set to burst, the cage then falling safely to earth. **1836** T. POWER *Impressions Amer.* I. 21 Lift a piece of pilot biscuit, request some kind soul to shave the under side of the corned round for you, then..fly the place and seek the deck. **1944** *Chicago Daily News* 11 Oct. 25/1 The pilot biscuit—great flat round crackers that may be purchased at the grocer's—were toasted and lightly spread with butter. **1788** *Maryland Jrnl.* 7 Mar. 4/2 (Advt.), The subscriber..has just begun to bake Ship, Pilot, and Cag Bread. **1831** [see *navy bread*]. **1894** *Outing* XXIV. 252/2 He quickly wrapped up some pilot-bread. **1977** *New Yorker* 9 May 112/2 With..six pieces of pilot bread, he got into his single Klepper and bobbed down the river. **1937** H. COTTON *Transmission & Distribution of Electr. Energy* xv. 349 The pilot wires are usually in the form of a three-core cable, which can be buried in the ground in the case of an insulated cable system, or slung from the towers in the case of an overhead system. Apart from the high cost of these pilot cables, [etc.]. **1967** M. F. BUCHAN *Electr. Supply* x. 296 As the two ends of the system may be some distance apart, an information link is required, and this may be a pilot cable, a radio link, or carrier currents superimposed on the system itself. **1925** *Sci. Amer.* Mar. 203/2 The majority of parachutes are equipped with a 'pilot chute' which is a miniature structure made with ribs and spring in umbrella fashion. When the pilot pulls the 'rip cord'.. the 'pilot chute' springs out, catches the air and helps to lead out the main parachute. **1973** 'A. HALL' *Tango Briefing* x. 118 Free fall... Seven, eight, nine. Pull it. Crack of the pilot 'chute. Then the jerk and the drag. **1834** W. F. TOLMIE *Jrnls.* (1963) 262 Have donned pilot cloth vest. **1840** *Knickerbocker* XV. 140 His winter clothing is usually a peet jacket and trowsers, of strong pilot cloth. **1836** DICKENS *Sk. Boz.* (1837) 2nd Ser. 96 Pilot great coats with wooden buttons, have usurped the place of the ponderous laced coats with full skirts. **1840** *Spirit of Times* 9 May 116/2 A young man attired in a pilot coat and velvet stock. **1842** DICKENS *Amer. Notes* I. ii. 41 The captain comes down again, in a sou'-wester hat.. and a pilot-coat. **1858** TROLLOPE *Three Clerks* I. ii. 29 A short bargee's pilot-coat, and a pipe of tobacco, were soon familiar to him. **1884** F. M. CRAWFORD *Amer. Politician* II. ii. 30 Enormous Irishmen in pilot coats..were struggling to keep the drifts from the pavement. **1907** *Westm. Gaz.* 15 Feb. 7/1 Drivers should not be allowed to drive an engine over a road that they were not acquainted with without a pilot driver. **1955** *Railway Mag.* May 302/1 A halt was made at Blarney, to detach the pilot engine. **1975** *Times* 3 Sept. 1/2 Last year the International Air Transport Association said that half of all air accidents were caused by pilot error. **1978** J. GARDNER *Dancing Dodo* xxxii. 252 The thing went wrong... If they wanted accuracy, they needed men to ensure it. What they got was..pilot error, or malfunction. **1881** *Instructions to Census Clerks* (1885) 33, 78 Railway Guard... Pilot Guard. **1921** *Dict. Occup. Terms* (1927) § 702 *Pilot guard*, a guard..who pilots trains over portion of line where single line working is necessitated, or over bridges under repair. **1812** A. STODDARD *Sk. Louisiana* 160 On the south side of the east pass, about three miles from the bar, is the pilot house. **1827** J. L. WILLIAMS *View of West Florida* 18 A small fort and pilot house formerly stood near the west end of the island. **1846** *Spirit of Times* 16 May 133/1 He placed his hand upon a small brass knob at the back of the pilot house. **1863** 'G. HAMILTON' *Gala-Days* 120 An Indian pilot comes on board, and mounts to the pilot-house. **1897** W. B. YEATS *Secret Rose* 207 The disused pilot-house looks out to sea. **1967** *Jane's Surface Skimmer Systems 1967–68* 34/2 Forward of the well..is the pilothouse. **1977** I. SHAW *Beggarman, Thief* I. i. 7 Alone in the pilothouse..stood Wesley Jordache. **1977** *Washington Post* 4 Sept. A12/1 The top tier of this floating cake is the 'pilot house'—what a seagoing sailor would call the 'bridge'. **1884** *Jrnl. Soc. Telegr. Engin.* XIII. 515 As the speed began to increase, the lamp lit up intermittently, but in a few seconds more the machines dropped into step together, and the pilot lamp lit up to full brightness and became perfectly steady. **1906** J. POOLE *Pract. Telephone Handbk.* (ed. 3) xvii. 220 Whenever a call is made on any one of these lines the pilot relay is operated, and causes a specially large lamp, called the 'pilot lamp', to glow. **1918** R. KNOX *Radiogr. & Radio-Therapeutics* II. 437 There is nearly always a pilot lamp on the switchboard by which the tints may be accurately gauged. **1977** 'E. TREVOR' *Theta Syndrome* i. 14 In the main laboratory there were only the pilot lamps going. **1906** *Daily Colonist* (Victoria, B.C.) 26 Jan. 10/1 Pilot lights have also been placed in all the hallways and dark passages of the building. **1907** *Daily Chron.* 16 Apr. 6/7 This is the 'pilot' light, which is never extinguished, which burns night and day from the time a theatre is opened throughout the whole of its existence. **1939** I. BAIRD *Waste Heritage* x. 130 The radio man clicked on the pilot light and grunts and squeals began to come from the little machine. **1964** J. CHEEVER *Wapshot Scandal* vi. 58 The pilot light on the gas range isn't working and the cook has to keep lighting the range with matches. **1970** *Which?* Nov. 332/2 A pilot light goes on while the iron is heating. **1972** *Guardian* 2 Dec. 8/8 People were asked to turn their gas taps to 'off' so that unlit pilot lights..would not allow gas to seep into their homes when the supply was restored. **1907** *Westm. Gaz.* 21 Sept. 12/2 An antique pilot-locomotive slouched out and stopped. **1881** *Instructions to Census Clerks* (1885) 33, 77 Railway Engine Driver... Pilot, Pilot-man. **1921** *Dict. Occup. Terms* (1927) § 701 *Pilotman*.., a guard.. who accompanies a train between given points on single line, other line being temporarily out of use, to advise driver of difficulties of road, special signalling arrangements, etc. **1971** D. J. SMITH *Discovering Railwayana* x. 58 *Pilotman*, driver taking charge of single-line workings, especially when one line of a double track is under repair. **1919** *Monthly Air Force List* Aug. 15 Air Ministry... Directorate of Personnel... Staff Officers, 3rd Class... Dawes, Pilot Officer H., Gunn, Pilot Officer D.B. **1935** T. S. SPRIGG *Royal Air Force* xii. 91 Accepted candidates who have not had previous experience in the Air Force are entered in the Special Reserve with the rank of Pilot Officer on probation. **1965** W. M. W. FOWLER *Countryman's Cooking* p. x, Where the fault lay was obvious even to a very young Pilot Officer. **1975** *Sunday Times* 23 Feb. 19/7 In May, 1943, I was a lowly WAAF serving on a station in Hampshire... My fiancé, Ken, was a pilot officer. **1926** Pilot parachute [see *PACK *sb.*¹ 1 g]. **1942** A. M. LOW *Parachutes* 41 The effect of pulling the rip-cord was first of all to release the pilot parachute packed on the outside of the pack so that this in turn may draw out the main parachute. **1853** R. B. MARCY *Explor. Red River* (1854) 196 The names of Bull, Pine, and Pilot snake are commonly given to different species of this genus [*sc.* *Pituophis*]. **1946** G. STIMPSON *Bk. about Thousand Things* 480 The pilot snake..gets its name from the curious belief that it precedes rattlesnakes and warns them of the approach of danger. **1956** L. M. KLAUBER *Rattlesnakes* II. xviii. 1244 Another phase of the myth is that pilot snakes are crosses between rattlesnakes and bull snakes. **1891** *Cent. Dict.* (s.v. *tunnel*), *Pilot tunnel*, a device for directing a tunnel in the prescribed grade, consisting of a flanged tube made up of interchangeable plates, which can be bolted to the shield and forced concentrically into the silt in advance of the face of the heading. **1906** *Chambers's Jrnl.* 29 Sept. 701/1 Within the small pilot-tunnel a large number of refrigerating tubes is placed. **1958** *Engineering* 18 Apr. 502/1 Short pilot tunnels were driven out from the shore on both sides of the Channel. **1902** *Internat. Library of Technol.* VII. B. § 39. 13 For high-speed hydraulic elevators..the relief valve is not sufficient to guard against shocks..nor is it possible to regulate the speed readily... This has led to the introduction of the auxiliary, or pilot, valve. *Ibid.* 18 The pilot valve allows a perfect regulation of the speed of the [elevator] car. **1953** E. MOLLOY *Maintenance Engineers' Pocket Bk.* II. 67 In large boilers.. pilot valves are often fitted to the main stop valves, and the pilot valve should be opened to warm up the steam main before the main stop valve is opened, thus avoiding the danger of water hammer. **1971** H. C. TOWN *Design & Construction of Machine Tools* ix. 194 If the reversing valve has to handle large volumes of oil..the reverse valve is better operated by a pilot valve which may be trip operated. **1848** W. H. EMORY *Notes Mil. Reconn.* 11 In the uplands..occasionally is found the wild tea, .. and pilot weed. **1951** *Dict. Gardening* (R. Hort. Soc.) III. 1957/1 Compass Plant, Pilot Weed, Polar Plant. **1901** SHIPLEY & MACBRIDE *Zool.* xix. 527 The Ca'ing or Pilot Whale.., which also feeds chiefly on cuttle-fish, has teeth in both upper and lower jaws. **1921** J. T. JENKINS *Hist. Whale Fisheries* viii. 268 A third whale fishery practised in northern waters..was that for the Grindhval or Pilot Whale. **1962** E. LUCIA *Klondike Kate* ii. 40 At times schools of happy pilot whales followed the vessel. **1974** G. B. CORBET in D. L. Hawksworth *Changing Flora & Fauna of Brit.* xi. 199 Species that are more frequent in the recent period are pilot whale (*Globicephala melaena*) with 46 strandings between 1948 and 1966. **1890** J. W. URQUHART *Electric Light* (ed. 3) x. 321 It is far more important to be informed as to the actual potential of the mains at the various points of consumption. For low tension constant current systems this is usually accomplished by the use of 'pilot wires'. **1930** *Engineering* 12 Sept. 347/1 An examination of the available pro-

tective schemes not involving the use of pilot wires had shown that a close approximation to the performance of pilot protective gear could be obtained by the use of distance relays. **1968** P. J. FREEMAN *Electr. Power Transmission & Distribution* x. 275 On long sections the capacitance currents between the pilot wires may be high enough to operate the relay, causing instability.

b. Used *attrib.* or as *adj.* to denote something that serves as a prototype or experimental undertaking prior to full-scale operation, activity, or use; experimental, initial; as *pilot film, plant, programme, project, scheme, study, survey, trial*; so **pilot-scale** *a.*, done on the scale of a pilot scheme. Cf. *pilot-tunnel* above.

1928 *Daily Mail* 13 Aug. 18/2 This company produced 40 tons of tin concentrates with its small pilot mill in the June quarter. **1934** *Planning* I. XVII. 9 Actually research has become specialised not only by subjects but by processes and each process—background, basic, *ad hoc* and pilot, or whatever else they may be termed—is inseparable from the one before it and from the one after it. **1936** *Economist* 1 Feb. 275/1 The dry crushing and roasting plant treated 186,422 tons [of ore] and the pilot flotation plant 5,688 tons. **1938** *Rep. R. Comm. Oil from Coal* § 27 in *Parl. Papers 1937–38* (Cmd. 5665) XII. 439 Experimental work started at Billingham early in 1927, and in 1929 it was decided to build a pilot plant there to treat 10 tons of coal per day. **1939** *Nature* 12 Aug. 279/2 The Department of Biochemistry..[has conducted] pilot scale tannery operations to improve the vegetable chrome process. **1944** *Times* 18 Mar. 2/3 The Ministry of Food has recently installed a pilot plant for drying meat near Belfast. **1947** *Yale Law Jrnl.* Dec. 197 They had an opportunity to build a 'pilot model', a spacious though inexpensive cooperative in..a Warsaw suburb. **1951** (*title*) The Haiti Pilot Project: Phase One (UNESCO). **1951** *Chambers's Jrnl.* Oct. 638/1 Two recent American developments, both at present in the pilot or experimental stage, may widen the already versatile uses of glass. **1952** Pilot reactor [see *CRITICALITY 2]. **1952** W. J. H. SPROTT *Social Psychol.* vi. 102 The 'open-end' question may be used in a pilot survey, which helps to determine the multiple-answer question. **1952** *Times* 25 Jan. 3/4 The corporation stated that it did not propose to proceed with a large-scale irrigation scheme for rice cultivation until results on a pilot area had shown this to be economically possible. **1953** *Britannica Bk. of Year* 638/2 *Pilot scheme* (a preliminary to a full scale agricultural or industrial project). **1953** *Ann. Reg. 1952* 416 A new vaccine selected from pilot trials of the previous winter. **1954** A. HUXLEY *Let.* 12 Dec. (1969) 718 The TV decision would not be made until after the production of a pilot film. **1956** *Planning* XXII. 19 In 1954 the Nottingham Book Festival was organised by a committee of the Booksellers Association as a pilot scheme for a national publicity campaign. **1957** *Ann. Reg. 1956* 348 In Russia it appeared that underground gasification of coal was passing from the pilot stage to industrial operation... In Britain..the Coal Board and the Central Electricity Authority decided to proceed with pilot plant for producing electricity from gas made underground from inferior coal. **1957** R. K. MERTON *Social Theory* (rev. ed.) II. x. 387 The initial substantive aim of this pilot study was fourfold. **1959** *Listener* 22 Jan. 173/1 We hoped to do a pilot survey on the reactions of individuals to television programmes. **1961** *Technology* Feb. 34/4 The scheme will start in three pilot areas. **1961** *Harper's Bazaar* Feb. 23 The trend-setters.., the 'pilot' clothes that look ahead. **1964** M. GOWING *Britain & Atomic Energy, 1939–1945* v. 150 The Chicago pilot-scale graphite pile. **1966** T. PYNCHON *Crying of Lot 49* ii. 33 They've done the pilot film of a TV series, in fact, based loosely on my career. **1969** N. W. PIRIE *Food Resources* 13 With less initial obstruction, and a steady increase in the scale of pilot projects, the ultimate acreage aimed at would probably have been reached earlier. **1971** *Brit. Med. Bull.* XXVII. 6/2 Sensitivity and specificity may be assessed from the results of a pilot trial undertaken on a group of individuals similar to those who are to be screened. **1971** *Guardian* 26 Feb. 13/3 US networks are no longer so keen to buy British programmes without seeing a pilot episode. **1971** *Jrnl. Gen. Psychol.* Apr. 191 On the basis of a pilot study..it was predicted that..total list acquisition would not differ as a function of stimulus clustering. **1974** *Nature* 1 Feb. 248/3 Pilot plants started within the next year or two could be working productively within the next five years. **1975** *Radio Times* 30 Aug. 14/2, I saw a pilot programme for the new series. **1975** *Daily Tel.* 3 Oct. 7/2 A pilot plant to 'lock' radioactive wastes inside solid glass is now being commissioned. **1976** *Leicester Chron.* 26 Nov. 2/3 Leslie Crowther presented the pilot show but future commitments prevented him taking it on permanently. **1978** *Jrnl. R. Soc. Arts* Dec. 8/2 Examinations in Communication were offered..on a pilot basis.

pilot, *v.* Add: **1. b.** To act as pilot on (an aeroplane or other aircraft) in the air; to fly (passengers) in an aircraft. Also *absol.*

1852 *Illustr. London News* 18 Sept. 224/3 The veteran aëronaut who had successfully piloted them and some hundred others through the air. **1911** *Daily News* 20 July 2/4 The Dutch aviator has decided to pilot a.. monoplane..instead of a..biplane. **1931** F. L. ALLEN *Only Yesterday* viii. 222 If you did not know how to pilot a plane you could still be a passenger. **1946** *Happy Landings* (Air Ministry) July 6/3 These considerations led me to select a Mosquito..and to pilot myself. **1955** *Times* 22 Aug. 5/4 The crowd saw a fly-past of aircraft piloted by men of the R.A.F. **1958** 'CASTLE' & 'HAILEY' *Flight into Danger* viii. 113 The first officer, then the captain were taken sick. Luckily there was a passenger on board who had piloted before and he took over the controls. **1977** *Daily Tel.* 7 Apr. 7/3 An attempt..to become the first woman to pilot a hot air balloon across the English Channel.

2. b. To secure the passage of (a bill) *through* a legislative assembly; to carry. orig. *U.S.*

1929 *Randolph Enterprise* (Elkins, W. Virginia) 21 Mar. 1/4 The bill..piloted..thru the House by Representative Karl Kyle. **1974** *Lebende Sprachen* XIX. 39/2 He piloted through the House the government's elaborate education bill. **1976** *Leicester Mercury* 16 July 4 It now goes to the Lords. It is likely to be piloted there by Liberal Lord Avebury.

c. To use experimentally; to try out, test.

1960 *Sunday Times* 10 Jan. 14/6 Practically all these devices for saving time and labour have been piloted in the fifties. The sixties should see them put into commercial production in sufficient quantity to make them financially feasible. **1965** *New Society* 23 Dec. 15/3 The Rowntree survey (piloted in Harrow, and subsequently carried out in York). **1967** G. WILLS in Wills & Yearsley *Handbk. Managem. Technol.* 186 Always, but always, *pilot* a questionnaire before sending it out into the field. Never try and scorn the researcher who asks for time and money to pilot his work. **1977** *Jrnl. R. Soc. Arts* CXXV. 308/2 Not only does he [*sc.* the skilled question designer] pilot his questionnaire, but periodically he tests his questions to find out how they are actually understood.

pilotage. 1. (Further examples.)

1922 *Encycl. Brit.* XXX. 13/2 Then came pilotage and the elements of commercial flying. **1924** *Air Pilot: Gt. Britain* II. iv. 19 (*heading*) Pilotage Directions. **1932** *Flight* 6 Oct. 949/1 The Air Council have had under consideration the policy regarding training in air pilotage. **1946** *Happy Landings* (Air Ministry) July (verso front cover), The various sciences which together contribute to our present day knowledge of practical pilotage. **1969** *Daily Mail* 16 Jan. 5/6 We had some difficulty on Apollo 8, with 'pilotage', that is, trying to plot our path on the map of the back side of the Moon.

piloted (pəi·lətĕd), *ppl. a.* [f. PILOT *v.* + -ED¹.] Controlled or guided by a pilot.

1945 C. MILBURN *Diary* 18 Mar. (1979) 278 Last night piloted enemy planes were over northern England and.. casualties also occurred from V bombs in the south. **1946** *Jrnl. R. Aeronaut. Soc.* July 508/1 The other piloted rocket aircraft was intended to be catapulted at an angle of 75°. **1952** *Ann. Reg. 1951* 411 The next stage [in space travel] would be piloted rockets designed as space stations. **1961** *Shell Aviation News* Dec. 4/1 Already, in certain circumstances 'the one who controls the aeroplane' can equally well be in another aircraft or on the ground. In this new situation the terms 'piloted' and 'pilotless' become confusing anomalies.

piloting, *vbl. sb.* (in Dict. s.v. PILOT *v.*). (Further examples.)

1919 W. H. BERRY *New Traffic* (*Aircraft*) i. 5 Good piloting does not depend on cleverness in tinkering with the engine. **1922** *Encycl. Brit.* XXX. 14/1 Aerial navigation, as distinct from piloting with the ground in view, developed tardily everywhere. **1959** *Manch. Guardian* 6 July 2/4 The hour's piloting round a town that makes up most [driving] lessons. **1977** *Belfast Tel.* 28 Feb. 9/2 (*caption*) Too old for piloting in 1939 he became an air gunner with 235 Squadron Coastal Command.

‖ **pilotis** (pilotí). [Fr.] A series of columns or piles, esp. used to raise the base of a building above ground-level.

1947 *Archit. Rev.* CI. 172/1 Low-growing palms make patterns against the pale pink granite pilotis. **1957** *New Yorker* 5 Oct. 166/2 The most striking feature of the building [*sc.* Le Corbusier's Unity House] is that it is raised twenty-four feet aboveground on a double row of cyclopean, wedge-shaped concrete columns, or *pilotis*, which uphold the hollow canopy of concrete that forms the basement of the building proper. **1971** *Country Life* 25 Nov. 1444/3 The studio stood on *pilotis*, but..it has been converted into a house. **1972** E. LUCIE-SMITH in Cox & Dyson *20th-Cent. Mind* II. xiv. 492 Pure forms achieved by the use of continuous window-strips, glass walls, flat roofs; the lightness which came from raising the structure on *pilotis*.

pilotless, *a.* (s.v. PILOT *sb.*). Add: (Further examples.) Now esp. of aircraft.

1806 SCOTT *Let.* 20 Sept. (1932) I. 317 The pilot-less state in which the political vessel has remained since his [*sc.* Pitt's] death. **1909** *Westm. Gaz.* 22 Oct. 7/2 We only just missed the new spectacle of a pilotless aeroplane. **1922** *Glasgow Herald* 15 Nov. 9 The Army Air Service [U.S.A.] announces that successful tests have been made with automatically controlled pilotless aeroplanes. **1937** *Aeroplane* 16 June 3 (Advt.), A pilotless flyaway plane. **1943** R. V. JONES *Most Secret War* (1978) xxxix. 356 The Germans are installing..a large and important ground organization in Belgium–N. France which is probably concerned with directing an attack on England by rocket-driven pilotless aircraft. **1944** [see *buzz-bomb* s.v. *BUZZ *sb.*¹ 5]. **1945** H. KNIGHT in *Penguin New Writing* XXIII. 47 Pilotless men whose personalities have been disintegrated by concussion and too many action stations. **1961** [see *PILOTED *ppl. a.*]. **1966** M. WOODHOUSE *Tree Frog* v. 35 The curious blind look of all pilotless aircraft which stems from having no cockpit. **1974** *Guardian* 18 Mar. 5/4 The pilotless planes, named Falcons, are designed for taking aerial photographs.

pilpul (pi·lpʊl). [Heb.: see PILPULIST.] Subtle or keen rabbinical investigation or argumentation; an instance of this. Also *transf.*, hair-splitting and unprofitable disputation.

1894 tr. *Graetz's Hist. Jews* IV. xiii. 418 The astonishing facility of ingenious disquisition on the basis of the Talmud (Pilpul), attributed to Polak, which attained its highest perfection in Poland, proceeded from a native of Poland. **1905** J. Z. LAUTERBACH in *Jewish Encycl.* X. 39/2 (*heading*) Pilpul. A method of Talmudic study. The word is derived from the verb 'pilpel' (lit. 'to spice', 'to season', and in a metaphorical sense, 'to dispute violently'..or 'cleverly'..). Since by such disputation the subject is in a way spiced and seasoned, the word has come to mean penetrating investigation, disputation, and drawing of conclusions, and is used especially to designate a method of studying the Law... The essential characteristic of pilpul is that it leads to a clear comprehension of the subject under discussion by penetrating into its essence and by adopting clear distinctions and a strict differentiation of the concepts. **1920** OESTERLEY & BOX *Short Survey Lit. of Rabbinical & Mediæval Judaism* IV. ii. 228 So far from destroying *pilpul*—casuistical discussion—the *Code* or *Mishneh Torah* itself became the object of pilpulistic comment. **1964** H. KEMELMAN *Friday the Rabbi slept Late* (1965) xxi. 156 We read a passage..of the Law... We were adding our own arguments..and twists of logic, the so-called pilpul. **1966** —— *Saturday the Rabbi went Hungry* i. 10 The cantor's chuckle told him he was won over. The rabbi returned to the table. His wife shook her head with a smile. 'That was a terrible pilpul.' **1967** C. POTOK *Chosen* II. v. 107 Pilpul, these discussions are called—empty, nonsensical arguments over minute points of the Talmud. **1968** L. ROSTEN *Joys of Yiddish* 287 *Pilpul*,..unproductive hair-splitting that is employed not so much to advance clarity or reveal meaning as to display one's own cleverness. **1976** *Brit. Jrnl. Sociol.* XXVII. 41 A key difficulty in the concept of alienation is that many of the succeeding theoretical discussions of the idea have taken off from the false leads in Marx, and the Talmudic *pilpul* over dissecting his text has only produced further confusion.

Pils (pilz, pils). [Abbrev. *PILSENER, PILSNER.] A type of beer similar to Pilsener.

1961 *Dude* Sept., I had part of a bottle of French beer called Panther Pils (so help me) then switched to Tuborg. **1962** [see *EXPORT *sb.* 3 b]. **1971** M. SINCLAIR *Sonntag* iii. 22 In..the scruffy *Bierstube* he sat scowling with a thin glass of *Pils*. **1972** P. CLEIFE *Slick & Dead* xx. 157 Normally I never drink when I'm flying but thought.. that a light Belgian Pils would be harmless. **1977** *Grimsby Even. Tel.* 24 May 10/2 (Advt.), Large Quantity of beers... Newcastle Brown, Prize Medal, Carlsberg, Pils in bottles, etc.

Pilsen (pi·lz-, pi·lsən). Also with lower-case initial. [See next.] = next.

It is not certain that quot. 1939 belongs here.

1939 JOYCE *Finnegans Wake* (1964) III. 492 My dodear devere revered mainhirr was confined to guardroom, I hindustand, by my pint of his Filthered pilsens bottle. **1964** L. DEIGHTON *Funeral in Berlin* xxxi. 162 The waiter had brought two tall, ice-cold Pilsen lagers. **1968** *Listener* 25 July 103/3 One continually meets these ironies in Bohemia. Over stew and dumplings and draught Pilsen, he boasted that his son had actually been named in the Moscow press as a dangerous deviationist. **1977** P. SOMERVILLE-LARGE *Eagles near Carcase* iv. 61 The barman..had some tall cold bottles of Pilsen.

Pilsener, Pilsner (pi·lz-, pi·ls(ə)nəɹ). [G., f. *Pilsen* (Czech. *Plzeň*), a province and city in W. Bohemia, Czechoslovakia.] In full *Pils(e)ner beer.* A pale-coloured lager beer with a strong hop flavour. Also *attrib.*

The name now designates type rather than origin. Beer from Plzeň itself is known as *Pils(e)ner Urquell* (G., primary source).

1877 C. SCHREIBER *Jrnl.* 3 Aug. (1911) II. 49 Much rain—no breakfasts in the garden and Pilsner beer luncheons this year! **1890** KIPLING *Life's Handicap* (1891) 161 'Pilsener?'..'Beer's out, I'm sorry to say.' **1894** *Clarion* 3 Nov. 2/4 The elder waiter sent my empty bottle of Pilsener flying. **1909** M. DIVER *Candles in Wind* iii. 35 A glass of Pilsener to ward off regrets next morning. **1920** E. SITWELL *Wooden Pegasus* 67 And sat and drank our Pilsener beer. **1933** *Sun* (Baltimore) 6 Dec. 18/5 (Advt.), Light Pilsner style or dark kulmbacher—as you prefer it. **1964** L. DEIGHTON *Funeral in Berlin* xxxi. 161 Groups of men shouted for slivovice, borovicka or Pilsner Urquell. **1980** *Brit. Med. Jrnl.* 29 Mar. 916/1 It [*sc.* class 2 beer] was available in two strengths—a middle European Pilsner beer..and a somewhat stronger English lager type.

b. Special Combs., as **Pils(e)ner glass, pils(e)ner glass,** a tall beer glass tapered at the bottom.

1966 'E. LATHEN' *Murder makes Wheels go Round* xvi. 130 Waymark pushed his pilsener glass to one side. **1971** J. BALL *First Team* (1972) xxvi. 406 Zalinsky poured out the two drinks in pilsner glasses. **1975** *Times* 18 Jan. 12/1 A..sight of the world..through the bottom of a Pilsner glass.

Piltdown (pi·lt,daun). The name of a village in Sussex, England, used *attrib.* esp. in **Piltdown jaw, man, skull,** with reference to the fossil remains of a skull found there, or the primitive hominid described as *Eoanthropus dawsoni*, to which these remains were attributed; the skull was proved fraudulent in 1953 by J. S. Weiner and K. P. Oakley. Also *absol.*

1912 *Times* 19 Dec. 4/5 He [*sc.* A. S. Woodward] inclined..to the theory that..surviving modern man might have arisen directly from the primitive source of which the Piltdown skull provided the first discovered evidence. **1913** *Q. Jrnl. Geol. Soc.* LXIX. 139 It seems reasonable to interpret the Piltdown skull as exhibiting a closer resemblance to the skulls of the truly ancestral

mid-tertiary apes than any fossil human skull hitherto found. **1931** *Times Lit. Suppl.* 23 Apr. 317/1 On comparing her [*sc.* a female hominid] with the Piltdown man he [*sc.* Sir A. Keith] makes amends by describing her as Piltdown refined. **1933** A. S. ROMER *Man & Vertebrates* xi. 246 Can it be that the Piltdown jaw does not belong with the skull? **1953** WEINER & OAKLEY in *Bull. Brit. Mus.* (*Nat. Hist.*), *Geol.* II. III. 145 The iron and chromate staining of the Piltdown jaw seems to us to be explicable only as a necessary part of the deliberate matching of the jaw of a modern ape with the mineralized cranial fragments. **1955** *Bull. Brit. Mus.* (*Nat. Hist.*), *Geol.* III. VI. 228 We are now in a position to give an account of the full extent of the Piltdown hoax... Not one of the Piltdown finds genuinely came from Piltdown. **1955** J. S. WEINER *Piltdown Forgery* 204 The end of Piltdown man is the end of the most troubled chapter in human palaeontology. **1956** *Proc. R. Inst. Gt. Brit.* XXXVI. 150 The Piltdown forgery was the greatest archaeological hoax of its kind ever perpetrated. **1957** T. STEELE in *Time* 30 Dec. 48/1 Rock with caveman Roll with caveman... Piltdown poppa sings this song Archaeologist done me wrong The British Museum's got my head Most unfortunate 'cause I ain't dead. **1970** R. LOWELL *Notebk.* 98 The Piltdown Man, first carnivore to laugh. **1973** *Listener* 10 May 605/3 Man..was not put together from the cranium of one primate and the jaw of another—that misconception..only makes a fake like the Piltdown skull.

2. *transf.* and *fig.*

1956 A. WILSON *Anglo-Saxon Att.* I. i. 27 Alas, we historians have so little scandal. We are not palaeontologists to display our Piltdowns. **1961** *Times* 9 Nov. 17/1 We must always be beholden to Evans over whom suspicions of a Piltdown sort hang, darkened now by Professor Palmer's discoveries in the Ashmolean. **1971** N. FLEMING *Hash* i. 12 So there is a brain underneath that thatch-covered Piltdown skull of yours. **1976** *Spectator* 2 Oct. 7/3 Ford..is waiting to be shot full of holes, not only because of his Piltdown Man performance as a congressman, but as the head of a government which is clearly out of administrative control.

So **Pi·ltdowner** = *Piltdown man* (lit. and fig.).

1954 A. HUXLEY in *Encounter* Feb. 11/1 In the tiny Natural History Museum at Idaho Falls, we found ourselves talking to two people from a far remote past... These were Piltdowners whose reaction to the stuffed grizzly was a remark about sizzling steaks of bear-meat. **1961** C. WILLOCK *Death in Covert* i. 6 A big-boned, shambling man..with the stance of a Piltdowner. **1978** 'J. GASH' *Gold from Gemini* v. 37 He really talks like this, the Piltdowner. No wonder he's thick.

pilule. Add: (Earlier fig. and later lit. examples.) Also (pl.) as an exclamation of outrage or exasperation (*rare*).

1889 E. DOWSON *Let. c.* 10 May (1967) 77 Pilules! What in Heaven's name do they mean! **1890** *Ibid.* 12 Jan. 131 Excuse these pilules—but I warned you that I had absolutely nothing to say. **1907** [see *DROSERA]. **1960** L. P. HARTLEY *Facial Justice* xxx. 250 At the thought of 'then' her throat contracted and would hardly let the pilule pass.

pilus (pəi·ləs). *Bacteriol.* Pl. **pili.** [L. *pilus* hair.] Any of the several types of filamentous appendages, other than flagella, that are facultatively produced by some bacterial cells.

1959 C. C. BRINTON in *Nature* 21 Mar. 782/2 Pili have been called by several different names: 'fine threads', 'bristles', 'fimbriæ', and 'filaments'. It is felt that the word 'pilus' is the most descriptive term, since the pili usually cover most of the bacterial surface and are continually growing out in a manner quite analogous to hair or fur. **1969** A. M. CAMPBELL *Episomes* iii. 36 Of the various types of pili formed by enteric bacteria, F-pili are distinguishable by the specific adsorption of male-specific bacteriophages to their surfaces. **1975** *Ann. Rev. Microbiol.* XXIX. 104 Once mating pairs were formed, phage f1 failed to interfere with the conjugation event, suggesting the involvement of the pilus..in the transfer of bacterial DNA. **1976** P. COLLARD *Devel. Microbiol.* viii. 108 A number of pilus-specific phages have been discovered.

pily, *a.*² Add: esp., in reference to the coat of certain dogs, containing a mixture of short, soft hairs and longer, harder ones. (Later examples.)

1922 R. LEIGHTON *Compl. Bk. Dog* p. xiv, Pily.—A peculiar quality of coat consisting of two kinds of hair, the one soft and woolly, the other long and wiry. **1942** [see *FEATHER *sb.* 11 b]. **1963** S. M. LAMPSON *Country Life Bk. Dogs* 123/2 The coat [of the Dandie Dinmont terrier]..is a mixture of hardish and soft hair, which gives it a pily feeling.

Pima (pī·mă). Also 9 **Pimo.** ['Said to be from a native word meaning "no" incorrectly understood and applied by missionaries. The term prob. was brought into American use from Spanish.' (*D.A.*)] **1.** A North American Indian people living chiefly along the Gila and Salt rivers in Arizona; a member of this people. Also, the Uto-Aztecan language of the Pima and Papago, especially the dialect of the Pima. Also *attrib.* See also *PAPAGO *sb.* and *a.*

[**1775** P. FONT *Diary* 29 Oct. (1913) 16 Muy de mañana se despacharon unos Yndios, à dar aviso de nuestra venida à los Pimas del rio Gila.] **1850** *California Courier* (San Francisco) 3 July 2/3 From the Pecos river in Texas to the Pimos villages on the Gila, roving bands of Apaches are hovering around the emigrants, stealing their animals. **1864** *Harper's Mag.* Dec. 23/2 It was gratifying..to know that the Pimos were rapidly becoming a civilized people. **1884, 1912** [see *HOHOKAM *sb.* and *a.*]. **1918** A. A. BRILL tr. *Freud's Totem & Taboo* (1919) ii. 68 After a Pima Indian had killed an Apache he had to submit himself to severe ceremonies of purification and expiation. **1932** W. L. GRAFF *Lang.* 430 The Uto-Aztec group comprises the three branches of Shoshonean, Pima-Sonoran, and Nahuatl. **1959** C. OGBURN *Marauders* vi. 189 Herman Manuel was a swarthy, squat Pima Indian from Arizona. **1960** C. WINICK *Dict. Anthropol.* 414/2 *Pima-Papago*, a general term used to describe the art of the south-west Indians of the United States. It includes the techniques of several crafts. **1962** D. H. HYMES in J. A. Fishman *Readings Sociol. of Lang.* (1968) 118 The functional load of /p/ in a community cannot be analyzed apart from the nature and use of various channels, as when among the Pima the functional load of /p/ differs between singing and recitation. **1964** S. M. LAMB in *Univ. Calif. Publ. Linguistics* XXXIV. 110 The Papagos speak Pima. **1965** *Canad. Jrnl. Linguistics* X. 141 Daughter languages of the Uto-Aztecan family..(Hopi, Pima-Papago, [etc.]). **1974** *Encycl. Brit. Micropædia* VII. 1009/3 *Piman languages*, Uto-Aztecan language group.., including Papago, Pima, [etc.]. **1979** *Tucson Mag.* Apr. 26/1 Ask a Pima or a Papago Indian—these are people who have been here for hundreds of years.

2. [f. *Pima* County, S. Arizona.] In full *Pima cotton.* A fine quality cotton developed from Egyptian cotton. Also *attrib.*

1936 *Sears Catal.* (ed. 173) 469 Pima is the finest of all American cotton. **1953** *New Biol.* XIV. 59 In Arizona a local selection of Egyptian origin, known as Pima, proved to be long but relatively weak. **1962** A. HUXLEY *Island* ix. 134 A Whisper-Pink Bra in Dacron and Pima Cotton. **1965** J. M. CAIN *Magician's Wife* (1966) xii. 86 She met him downstairs..in dark blue pima suit. **1976** *National Observer* (U.S.) 23 Oct. 9/6 (Advt.), Nothing compares to our handsome, pure pima cotton broadcloth plain collar shirts.

Piman (pī·măn), *a.* and *sb.* [f. prec. + -AN.] **A.** *adj.* Of or pertaining to the Pima and Papago Indians or their language. Also, of or pertaining to any of several ethnic and linguistic groupings of Indians of varying extent, comprising the Pimas, the Papagos, and certain other related peoples. **B.** *sb.* **a.** A Pima or Papago Indian; *pl*, the Pima and Papago Indians considered as a single ethnic and linguistic group. **b.** The Uto-Aztecan language of the Pima and Papago; also called *Papago, Pima, Pima-Papago*, and *Papago-Pima*. Also used of any of several wider groupings (see the adj.).

1891 J. W. POWELL in *7th Ann. Rep. U.S. Bureau Amer. Ethnol. 1885–86* 98 (*heading*) Piman family. **1933** L. BLOOMFIELD *Language* iv. 72 The Piman family (east of the Gulf of California). **1936** B. L. WHORF in *Ess. in Anthropol. presented to A. L. Kroeber* 198 Piman,..one of the groups most unlike the groups with which we began the classification. **1942** CASTETTER & BELL *Pima & Papago Indian Agric.* 1 The Pimans, a name applied to the whole group of Pima-Papago in both Mexico and the United States, anciently extended in irregular distribution from southern Sonora to the Gila River. **1950** J. H. STEWARD *Handbk. S. Amer. Indians* VI. 501 Its major use has been by the *Cáhita* and *Piman* tribes and their neighbors, that is, in the area adjacent to the upper Gulf of California. **1965** *Canad. Jrnl. Linguistics* X. 79 He also includes Nahuatl and Sonoran (Piman). **1969** K. L. HALE in D. & L. Saxton *Dict. Papago & Pima* Introd., The Piman languages, of which Papago-Pima is the northernmost, are of considerable interest for the fact that they constitute a close-knit, well defined subfamily within Uto-Aztecan. **1973** B. L. FONTANA in —— *Legends & Lore Papago & Pima Indians* p. xi, This volume of Papago and Pima literature makes available to readers of both Piman and English an important collection of heretofore widely scattered materials, much of it provided by Piman spokesmen many years ago. **1975** D. M. BAHR *Pima & Papago Ritual Oratory* 1 The hero of Piman (Pima-Papago) ritual oratory is a medicine man. [*Note*] The Pima and Papagos today are separate tribes. This study includes texts from each and considers ideas they have in common. I use the word 'Piman' to represent this common culture. *Ibid.*, Oratory..must have afforded the Pimans a means of harmonizing these various activities. *Ibid.* 3 A second group interested in this poetry is of course the native speakers of Piman, Pima and Papago Indians of southern Arizona.

pimelea (pimī·liă). Also (erron.) **-ia.** [mod.L. (D. Solander in J. Gaertner *De Fructibus et Seminibus Plantarum* (1788) I. 186), f. Gr. πιμελή fat, in allusion to the oily seeds.] An evergreen shrub of the genus so called, belonging to the family Thymelaeaceæ, native to Australasia, and bearing small terminal clusters of white, pink, or yellow flowers.

[**1793** J. E. SMITH *Specimen Bot. New Holland* 32 Gaertner..adopted the name of *Pimelea* from the manuscripts of Dr. Solander.] **1810** W. AITON *Hortus Kewensis* (ed. 2) I. 25 Flax-leaved Pimelea. Nat[ive] of New South Wales. **1842** W. COLENSO *Let.* 1 Sept. in *Lond. Jrnl. Bot.* (1844) III. 8, I found a handsome *Pimelea* in flower, a shrub of 2–3 feet in height. **1885** T. BAINES *Greenhouse & Stove Plants* 281/1 All the kinds of Pimelea strike readily from cuttings made of the points of the young shoots. **1951** *Dict. Gardening* (R. Hort. Soc.) III. 1572/1 Pimeleas are compact, free-growing plants needing greenhouse conditions. **1966** H. E. & J. BAWDEN *Making Shrub Garden* 216 Some of the pimelias are greenhouse shrubs... There are other pimelias which grow high up in the mountains.

pimento. Add: **2.** For *Eugenia Pimenta* substitute *Pimenta dioica.* (Later examples.)

1889 G. S. BOULGER *Uses of Plants* I. 66 Allspice, or Pimento, is the dry berry of *Pimenta officinalis* Lindl... a West Indian evergreen-tree. **1969** *Oxf. Bk. Food Plants* 132/2 Allspice (*Pimenta dioica*) is a small tropical tree whose unripe dried berries provide the spice called allspice. .. It is also known as 'pimenta' and 'pimento'.

b. = *PIMIENTO, *PAPRIKA 2.

1918 A. QUILLER-COUCH *Foe-Farrell* 94 What do you say now..to a pig's trotter farced with pimento. **1943** D. WELCH *Maiden Voy.* xvi. 132 The soup..was delicious; pieces of pimento swam about in it like goldfish. **1950** E. DAVID *Bk. Mediterranean Food* 132 Mixed red, green, and yellow pimentos, cooked a few minutes in boiling water, then peeled and sautéd in butter. **1978** *Nagel's Encycl.-Guide: China* 375 *Hu nan* and *Hu bei* are both known for chicken and pimento (*la zi ji*), soy-bean omelettes (*dou pi*) and bread with stuffing.

c. = *pimento red.*

1976 *Northumberland Gaz.* 26 Nov. 16 (Advt.), Triumph Dolomite. Pimento, tan fabric trim. **1977** *Time* 21 Feb. 43/1 A pimento handkerchief-style dress.

3. For the Latin names applied to the tree, substitute *Pimenta dioica.* (Later examples.)

1950 P. BOTTOME *Under Skin* x. 95 You had better stop here under this pimento. *Ibid.* xvi. 137 Pimento trees with their greenish bark and dark glossy leaves. **1958** D. P. STORER *Familiar Trees & Cultivated Plants of Jamaica* 69 The Pimento grows to about 30 ft. in height.

4. pimento berry = sense 2 a; **pimento cheese,** soft cheese flavoured with chopped sweet peppers; **pimento dram,** a Jamaican liqueur made with pimento berries; **pimento red,** an orange-red colour; also *attrib.* or as *adj.*

1893 C. SULLIVAN *Jamaica Cookery Bk.* 104 Pimento Dram... Put four quarts of ripe pimento berries into a jar, and boil till the berries burst... Add two bottles of brandy or good old rum... Then add a thick syrup of loaf sugar. **1907** *Daily Consular & Trade Rep.* (U.S. State Dept.) 5 Oct. 11 Ripe pimento berries are used to make pimento dram, a native drink. **1970** *Nature* 8 Aug. 556/2 Pimento berries are gathered while still green. **1916** *Daily Colonist* (Victoria, B.C.) 1 July 6/5 (Advt.), Cream or Pimento Cheese, each 10¢. **1967** MRS. L. B. JOHNSON *White House Diary* 23 Apr. (1970) 510, I had had a platter of sandwiches put in the refrigerator—roast beef and pimento cheese. **1893, 1907** Pimento dram [see *pimento berry* above]. **1958** D. P. STORER *Familiar Trees & Cultivated Plants of Jamaica* 70 'Pimento-dram' is a delicious Jamaican liqueur made, essentially, by soaking the ripe berries in rum. **1975** *Vogue* Dec. 107 Pimento red Hurel jersey. **1979** U. CURTISS *Menace Within* ix. 90 An abstract painting in pink and pimento red.

pi-meson (pəi·mī·zɒn, -me·zɒn). *Nuclear Physics.* Also **pi meson, pimeson.** [f. PI *sb.* + *MESON³.] The original name for the *PION. (Freq. written *π-meson.*)

[**1947** LATTES, OCCHIALINI, & POWELL in *Nature* 4 Oct. 455/1 It is convenient to refer to this process..as the μ-decay. We represent the primary mesons by the symbol π, and the secondary by μ.] *Ibid.* 456/1 The restricted range of the π-mesons in the emulsion. **1950** D. HALLIDAY *Introd. Nucl. Physics* xii. 455 When negative pi-mesons are absorbed in matter,.. there is a major nuclear disruption. **1953** *Jrnl. Brit. Interplanetary Soc.* XII. 203 The neutral pi-meson dissociates almost immediately into two high energy gamma rays (photons), which initiate the photon-electron cascades that are frequently observed in cosmic ray processes. **1968** *Times* 19 Jan. 13/8 The object..is to see how frequently an electron and a positive electron (or positron) would be converted..into the radio-actively unstable particles of matter called pimesons (or pions). **1972** *Daily Colonist* (Victoria, B.C.) 3 Mar. 16/8 The new technique involves a radioactive beam of sub-atomic particles called pimesons. **1978** *Sci. Amer.* Mar. 57/1 The mass of the muon (106 MeV) is quite close to the mass of the pi meson, or pion (140 MeV).

Hence **pi-me·sic, -meso·nic** *adjs.* = *PIONIC *a.*

1952 *Physical Rev.* LXXXVIII. 134/1 We have found direct evidence for π-mesic atom formation with the nuclei of carbon, oxygen, and beryllium. **1953** *Ibid.* XCII. 789/2 Experimental studies of π-mesonic x-rays in light elements have been made. **1954** *Ibid.* XCVI. 774 (*heading*) Energy level displacements in pi-mesonic atoms. **1970** LOCK & MEASDAY *Intermediate Energy Nucl. Physics* xi. 276 Because the mass of the π^- (139·6 MeV) is greater than that of the μ^- (105·7 MeV), the energies of the pi-mesic x-rays are about 30% higher than those of the equivalent mu-mesic x-rays.

pimiento (pimie·ntō). Also (*rare*) **pimienta.** [Sp.] = *PAPRIKA 2.

1845 [see *GAZPACHO]. **1846** R. FORD *Gatherings from Spain* xi. 131 Add a bit of bacon, onions, garlic, salt pepper, *pimientos*, a bunch of thyme. **1901** *Daily Colonist* (Victoria, B.C.) 15 Oct. 6/6 (Advt.), Glass Jars Olives (with Pimientos). **1937** A. F. HILL *Econ. Bot.* xx. 476 The paprikas are European varieties [of sweet pepper] with large mild fruits. Spanish paprika..is better known as pimiento. **1960, 1969** [see *PAPRIKA 2]. **1979** L. KALLEN *Introducing C.B. Greenfield* xii. 144 My grocery bag of chicken, shrimp, olives, pimentos.

b. = *pimento 2 c.

1972 J. Potts *Trouble-Maker* iii. 18 Her living-room—avocado and chocolate brown, pimiento accents.

2. *attrib.* **pimiento cheese** = *pimento cheese* s.v. *pimento 4; **pimiento(-stuffed) olive**, an olive with its stone replaced by a piece of red sweet pepper.

1922 *Hotel World* 15 Apr. 15/1 American or Pimiento cheese. **1972** *Harrods Christmas Catal.* 60/2 Jar Harrods Pimiento Olives. **1974** *Ibid.* 40/1 Jar Pimiento Stuffed Olives.

piminy (pi·mini). *nonce-wd.* [?Shortened f. Miminy-piminy *a.* and *sb.*, Niminy-piminy *a.*] ?Something expressive of affectation.

1819 Keats *Let.* 22 Sept. (1958) II. 175 Poor thing she little thinks how she is..making her nose quite a piminy.

Pimm's (pimz). Also **Pimms,** (*erron.*) **Pim.** [Proprietary term, f. the name of the proprietor of the restaurant where these drinks were created.] Used to designate any of four spirit-based mixed drinks ('cups'), taken neat or used as a basis for long drinks; also, a drink of one of these. Also *ellipt.*

Where no number is specified, the reference is usually understood to be to *Pimm's Number One Cup*, which is gin-based.

[**1888** *Trade Marks Jrnl.* 20 June 817 Pimms. Poultry. London. Pimms & Co.,..Poultry, London; restaurant proprietors... Wines and spirits, excepting brandy.] **1912** *Ibid.* 27 Nov. 1782 Pimm's No. 1 Cup... An alcoholic beverage. H. D. Davies & Co., Limited,..London,..wine merchants. **1931** R. H. Heaton *Perfect Hostess* 56 Pim Cup. One wineglass of No. 1 Pim cup mixture [etc.]. **1939** L. MacNeice *Autumn Jrnl.* v. 22 And this, we say, is on me; Something out of the usual, a Pimm's Number One. **1957** J. Braine *Room at Top* xxv. 202 The de-luxe bar and the iced Pimm's. **1960** *20th Cent.* July 73 Exams are over for ever, Pimm's is flowing. **1966** 'K. Nicholson' *Hook, Line & Sinker* v. 66 Mrs. Fairchild created a diversion by mixing everyone a Pimms. **1967** A. Lichine *Encycl. Wines* 404/1 There are four Pimm's Cups, and it is said that they were originated by a bartender at Pimm's Restaurant in London and so delighted the customers that the staff was continually being asked to put some up for people to take home. As a result, they begun to be made commercially. **1972** F. Warner *Lying Figures* i. 6 Barman! Make it a double Pimms!

Pimo, var. *Pima.

pimozide (pi·mozəid). *Pharm.* [f. Pi(peri-dine + elements in *benzimidazolin*, part of the systematic name of the compound (see Imide and Azo-).] A derivative of piperidine that is a colourless powder used as a tranquillizer in the treatment of schizophrenia and anxiety states.

1968 *Arzneimittel-Forschung* XVIII. 261 (*heading*) Pimozide, a chemically novel, highly potent and orally long-acting neuroleptic drug. **1976** Smythies & Corbett *Psychiatry* iii. 27 Other effective anti-psychotics..include the butyrophenones and pimozide. **1977** *Lancet* 21 May 1105/1, I have treated a boy of 17..for anorexia nervosa... He received pimozide 4 mg three times a day for a month.

pimp, *sb.*¹ Add: **a.** (Further examples.)

1932 *Evening Sun* (Baltimore) 9 Dec. 31/5 *Pimp*, man who is supported by women. **1961** J. Dos Passos *Mid-century* i. vii. 96 I'd thought him a pimp or procurer but he didn't seem to be. **1968–70** *Current Slang* (Univ. S. Dakota) III–IV. 92 *Pimp*, one who solicits women and girls to sell themselves and turn all the money over to him. He, in turn, gives them their salary. **1972** T. Kochman *Rappin' & Stylin' Out* 243 The 'pimp'..a person of considerable status in the street hierarchy..has acquired a stable of girls to hustle for him and give him money. **1976** *Toronto Star* 21 Aug. B1/2 It's getting busier partly because so many black American hookers and their pimps come and check into a hotel here for the weekend and compete for our territory.

b. (Later examples.)

1962 S. E. Finer *Man on Horseback* xii. 241 Mill recognized the 'necessity' of some despotic régimes; but he saw them..as a *pis aller*. He did not, like today's pimps of tyranny, pretend that the despotism was superior to the system of representative government. **1968** *Listener* 22 Aug. 252/1 Of course, it's cheaper to take on men trained in Fleet Street than to train them yourself, but the BBC should be the custodian of, not the pimp for, popular forms.

b*. In various other uses: (*a*) *Austral.* and *N.Z. slang*, an informer, a tell-tale; (*b*) *Welsh dial.*, one who spies on lovers, a peeping Tom; (*c*) *U.S. slang*, a male prostitute.

a **1885** *Penguin Bk. Austral. Ballads* (1964) 74 Blue-coat imps, Who were laid on to where he slept, By informing peeler's pimps. **1938** X. Herbert *Capricornia* 567 'I'm not a pimp.' 'What you mean pimp?' 'I'm not a police-informer.' **1940** Dylan Thomas *Portrait of Artist as Young Dog* 126, I lay like a pimp in a bush by Tom's side and squinted through to see him round his hands on Norma's breast. **1942** G. Casey *It's Harder for Girls* 51 'I just say I'm not a pimp,' Brownie insisted, beginning to blubber. **1942** Z. N. Hurston in A. Dundes *Mother Wit* (1973) 223 In Harlemese, *pimp* has a different meaning than its ordinary definition as a procurer for immoral purposes. The Harlem pimp is a man whose amatory talents are for sale to any woman who will support him... He is actually a male prostitute. **1969** D. Niland *Dead Men Running* 290 'There's a pimp at work.'..'The same pimp is it, that potted Shannessy and Halloran?' **1974** *Age* (Melbourne) 12 Oct. 12/1 You fat pimp! The standard response to 'I'm going to tell on you'.

pimp, *v.* Add: **1.** (Later examples.)

1975 *New Yorker* 26 May 18/2 His father (Jack Warden) pimps to add to his income as a taxi-driver. **1976** *Ibid.* 1 Mar. 79/1, I also especially enjoyed Roscoe Orman as Pretty Eddie, the 'happy dust' addict who pimps for his girl.

c. *to pimp on*: to scrounge off; to take advantage of. *U.S. slang.*

1942 Z. N. Hurston in A. Dundes *Mother Wit* (1973) 226/1 'You got any money?' the girl asked... 'Nobody ain't pimping on me. You dig me?'

3. *intr.* To tell tales; to inform *on* someone. *Austral.* and *N.Z. slang.*

1941 Baker *Dict. Austral. Slang* 54 *Pimp*,.. to inform on. **1945** G. Casey *Downhill is Easier* 109 This dago bastard pimped on him to Hayes, an' lost him his job. **1949** *Landfall* Mar. 30 He would grope out, head down, afraid to meet someone who would pimp on him. **1957** J. Waten *Shares in Murder* 155 You made up to me so you could get me to pimp on Charlie for you. **1960** N. Hilliard *Maori Girl* 229 She's bound to pimp and make trouble.

4. *intr.* To spy on lovers. *Welsh dial.*

1976 R. Lewis *Witness my Death* i. 13 Dai—Pimping again then, is it?

pimpernel. Add: **3.*** (Chiefly with capital initial.) [The name given to Sir Percy Blakeney, hero of Baroness Orczy's novel *The Scarlet Pimpernel* (1905), who rescues victims of the Terror and smuggles them out of France, characterized (ch. xii) as 'that demmed, elusive Pimpernel'.] **a.** Something elusive or much sought after. **b.** Someone whose exploits are comparable to those of 'The Scarlet Pimpernel'. See also *Scarlet Pimpernel* s.v. *scarlet *a.* 4 a.

1955 *Times* 10 May 14/1 The elusive pimpernel, the Liberal vote, is being eagerly sought by the two main parties in the Peterborough division. **1961** *House & Garden* Oct. 112/3 Highlanders soon became a band of Pimpernels smuggling their aristocratic whiskies to the Lowlands. **1962** *Guardian* 9 Aug. 7/6 The man who has become known as the 'Black Pimpernel of South Africa'. **1963** *Times Lit. Suppl.* 18 Jan. 40/5 In the hope of being able to function like a modern-day Pimpernel. **1974** 'D. Craig' *Dead Liberty* xvii. 81 Hans Lenzlinger, the greatest Pimpernel between East and West ever known.

pimping, *vbl. sb.*¹ (s.v. Pimp *v.*). (Later examples.)

1957 C. MacInnes *City of Spades* ii. ix. 165 Ite..put on his tailored duffel coat, and said, 'Now I must get out in the cold and do me pimpin'... A little coloured lady for you?' **1967** *Trans-Action* Apr. 8/2 The degree of organization in hustling depends frequently on the kind of hustling. Regular pimping and pushing require many trusted contacts and organization.

pimping, *vbl. sb.*² *local.* [f. Pimp *sb.*² + -ing¹.] The preparation of bundles of firewood. (In quot. comb.)

1930 V. Sackville-West *Edwardians* vi. 251 He looked into the pimping-shed, where old Turnour was chopping faggots.

pimpish (pi·mpiʃ), *a.* [f. Pimp *sb.*¹ + -ish¹.] Resembling or characteristic of a pimp.

1935 W. Empson *Some Versions of Pastoral* 47 The comic characters' experimental wish to satisfy all parties..has a certain pimpish complacence. **1971** W. Burroughs *Wild Boys* 11 Six pimpish young men burst through the door in a reek of brilliantine. **1976** *Daily Tel.* 19 Aug. 11/7 A horde of football fans attempt to lure into their bedrooms a pair of tarts by arrangement with the pimpish day porter.

pimple, *v.* Delete 'Now *rare*' and add: **a.** Also, to cover as with pimples. (Later examples.)

1909 'O. Henry' *Roads of Destiny* xix. 311 The levee where his freight-car stood was pimpled with dark bulks of merchandise. **1940** L. MacNeice *Last Ditch* 14 The rain of London pimples The ebony street with white. **1961** *Daily Tel.* 29 Aug. 18/4 Caravans and tents pimpling remote villages and hillsides are causing concern to planning authorities. **1972** K. Bonfiglioli *Don't point that Thing at Me* xix. 167 Lurid tents and tasteful pastel caravans pimple the landscape. **1974** P. Dickinson *Poison Oracle* iv. 98 The dunes along the marsh were pimpled with their tents.

b. Also *transf.*, to develop small bulges.

1940, 1962 [implied in *pimpling *vbl. sb.* 2]. **1970** *Motoring Which?* July 93/2 On all three cars, the rear light surrounds were pimpling and the bumpers had begun rusting.

pimpling, *vbl. sb.* (in Dict. s.v. Pimple *v.*). Add: **2.** The presence of small bulges on the surface of metal.

1940 J. D. Jevons *Metall. Deep Drawing & Pressing* viii. 293 A troublesome defect, which is sometimes encountered, particularly when the carbon content is rather high, is that known in the shops as 'pimpling'. **1962** J. C. Wright *Metall. in Nucl. Power Technol.* vi. 106 Examples of workable solutions to pimpling are the use of an aluminium-silicon brazed bond which does not spread significantly with time in the neighbourhood of 300°C, and the provision of anodic coating on the inside of the aluminium tubing.

pimplous (pi·mpləs), *a. rare.* [f. Pimple *sb.* + -ous.] Characterized by pimples; pimply.

1906 W. J. Locke *Beloved Vagabond* xii. 152 Cooling medicaments wherewith to mitigate a certain pimplous condition of cheek.

pimply, *a.* Add: Also *fig.*

1958 *Spectator* 22 Aug. 260/1 The man-eating British lion By a pimply age brought down. **1973** M. Amis *Rachel Papers* 17 It was a month I always think of with a certain pimply lyricism.

pimpmobile (pi·mp,mobīl). *U.S. slang.* [f. Pimp *sb.*¹ + *auto)mobile *sb.*] A large, flashy car used by a pimp.

1973 *Washington Post* 21 Apr. D7/1 Features of the luxury pimpmobile—glittering striping, dual spares, elaborate scrollwork and amber sunroofs—are showing up, somewhat toned-down, in stylized automobiles, according to car customizers. **1975** *Daily Mail* 13 Sept. 2/5 The pimpmobiles—the long, long Cadillacs with a Rolls front—no longer cruise everywhere. They are finding it less profitable to keep girls here. **1977** *Rolling Stone* 24 Mar. 71/1 'Norman', the story of a gambler who rides around in a Rambler (a secondhand pimpmobile?), is introduced by a melancholy horn riff with the drummer loping along like a dragging muffler, while Romeo's voice is high and airy.

pin, *sb.*¹ Add: **I. 1. m.** One of the metal projections of a plug, which make the electrical connection when it is inserted into a socket.

1888 [see *plug *sb.* 1 c]. **1902** W. C. Clinton *Electric Wiring* iv. 85 The flexible..terminates in two split pins, which are a spring fit into two tubular sockets. **1945** F. Wiseman *Penguin Handyman* i. 19 One end is connected to the earth pin in the three-pin plug, whilst the other end is connected to the metal housing or frame of the fire. **1963** *House & Garden* Feb. 8/3 Replace those old-fashioned two-pin plugs with a three-pin, and check the fuse-box. **1972** *Electricity Supply & Safety* (Consumers' Assoc.) 49 You can connect an old type of plug with round pins to a rectangular hole socket with the right adaptor.

n. A metal peg which prevents a hand-grenade from exploding by holding down the activating lever.

1917 C. R. Gibson *War Inventions* vi. 95 Until the soldier was ready to throw the bomb the little lever was held down by a metal pin. When the soldier withdrew this pin, the lever was still held down by the hand with which he threw the bomb, and not until it left his hand did the fuse begin to burn. **1920** A. R. Bond *Inventions of Gt. War* ii. 29 The Mills hand-grenade..was provided with a lever which was normally strapped down and held by means of a safety-pin. **1952** G. E. Thornton *Handbk. Weapon Training* ix. 101 With a hand grenade these [techniques] include..the method of removing the safety pin or cap. **1972** 'H. Calvin' *Take Two Popes* xv. 182 Soldiers..ready to pull out grenade pins with their teeth. **1977** C. Forbes *Avalanche Express* viii. 85 The grenade he would have withdrawn the pin from..if..faced with imminent arrest.

o. A coupling-pin. Used esp. in phr. *to pull the pin*: to uncouple; also *fig.* (see quots.). *N. Amer. slang.*

1927 *Amer. Speech* II. 391/2 To *bunch*, or to *drag it*, means to quit. To *pull the pin* has the same meaning. This is a railroad term and means to uncouple. **1947** R. O. Boyer *Dark Ship* ii. xvii. 227 The teamsters had pulled the pin too early. The strike is being lost. **1955** *Amer. Speech* XXX. 92 *Pull the pin*, to release the pin that connects a semitrailer to a fifth wheel. **1968** *Ibid.* XLIII. 289 *Pull the pin*, to resign, quit, or be fired from a job: 'They pulled the pin on him.' Relates to the switching function in the days of the link-and-pin coupler (long since replaced by the safer automatic coupler) by which a car was uncoupled and released. **1972** J. Wambaugh *Blue Knight* (1973) x. 161 An old man that should've pulled the pin years ago. Now he'd been here too long. He couldn't leave or he'd die. **1977** —— *Black Marble* (1978) ix. 168 Twenty-six years on the job when he pulled the pin and went to Arizona.

p. A support of an arch.

1928 *Daily Tel.* 7 Feb. 14/1 The arch is a two 'pin' crescent structure, and the distance from 'pin' to 'pin'..is 531 ft.

II. 3. Also (*U.S.*), a badge indicating membership of a university or college fraternity, or similar society, etc.

1871 L. H. Bagg *4 Years at Yale* 144 Its original badge was a rectangular gold plate, about the size and shape of the present Beta Xi pin. **1893** W. K. Post *Harvard Stories* 216 Freddy..asked me one day why Sheffield wore that funny little pin all the time. **1910** J. Hart *Vigilante Girl* 19 My dear fellow, you may cast aside your Eastern frigidity—in fact, I will call it your Cambridge frigidity, for I see you wear a Harvard pin. **1928** *Amer. Speech* III. 220 *Put out one's pin*, v. phr., to become engaged, to promise marriage. 'Scoop didn't have his pin on Dorothy very long, did he?' **1943** *Ibid.* XVIII. 154/2 *Plant a pin*, the process by which a fraternity man signifies his willingness to wait in the hall for the same girl every time. It consists of presenting her with his pin to wear. **1974** *Marlboro Herald-Advocate* (Bennettsville, S. Carolina) 18 Apr. 7/1 Rupert Kiker, president presented John L. Hargrave with a 20 year membership pin in the Lions International. Roy Easterling, Sr., was

presented a 15 year pin. **1977** C. McFadden *Serial* (1978) xviii. 43/2 He still wear his Key Club pin?

b. *two pins* (earlier and later examples). Also, *a row of pins* (in quot. 1896 as a type of similarity).

1891 [see For *prep.* 9]. **1896** Kipling *Seven Seas* 193 When you get to a man in the case, They're like as a row of pins—For the Colonel's Lady an' Judy O'Grady Are sisters under their skins! **1914** W. W. Jacobs *Night Watches* i. 25 'For two pins—' he began. 'For two pins I'll go back 'ome and stay there,' said Mr. Flynn. **1918** A. G. Gardiner *Leaves in Wind* 162 John Burns..does not care two pins who sees him or talks about him. **1920** W. J. Locke *House of Baltazar* i. 16 It doesn't seem to amount to a row of pins compared with my meeting you. **1947** [see *hide *sb.*¹ 2 c]. **1973** J. Porter *It's Murder with Dover* vii. 64 Her unsupported word isn't worth a row of pins. **1979** *Country Life* 16 Aug. 489/2 This Lord Hertford cared not two pins for society... His single passion was collecting.

d. *pins and needles* (earlier example). *on pins and needles* (earlier and later examples).

1810 J. Poole *Hamlet Travestie* 8 Would it were supper-time... Till then I'm sitting upon pins and needles. **1813** W. Dunlap *Mem. G. F. Cooke* II. xxx. 265 As it was—it was bad enough—my voice—haw!—there are pins and needles—I must send for a physician. **1858** Queen Victoria *Let.* 8 Mar. in R. Fulford *Dearest Child* (1964) 72 All you say..reminds me of what I was always used to as a child. Always on pins and needles, with the whole family hardly on speaking terms. **1944** W. S. Maugham *Razor's Edge* iii. 112 The bishop had been a cavalry officer..and his austere, cadaverous vicar general was always on pins and needles lest he should say something scandalous. **1951** E. Paul *Springtime in Paris* xv. 283 Dr. Thiouville was on pins and needles. **1979** *Country Life* 16 Aug. 489/4 The French were waiting on pins and needles to hear that the great [Wallace] collection would be theirs.

e. *as neat as a* (*new*) *pin.*

1787 *Columbian Mag.* I. 636 [He was] neat as a new pin. **1801** J. Wolcot *Wks.* (1812) V. 35 How neat was Ellen in her dress! As neat as a new pin! **1849** Thackeray *Pendennis* I. xiii. 118 Major Pendennis, whom Miss Costigan declared to be a proper gentleman entirely,..and as neat as a pin. **1933** L. A. G. Strong *Sea Wall* 245 Sheehan's pride was to have his cottage as neat as a new pin. **1961** *Dog World* Apr. 30 In the morning we leave the room looking as neat as a pin!

f. *to be able to hear a pin drop* (or *fall*), etc.: used to suggest absolute silence or stillness.

[**1775** F. Burney *Diary* 11 June (1907) II. 81 Had a pin fallen, I suppose we should have taken it at least for a *thunder-clap*.] **1814** M. B. Smith *Let.* Aug. in G. Hunt *40 Yrs. Washington Society* (1906) 113 It was so still you might have heard a pin drop on the pavement. **1824** S. Ferrier *Inheritance* ii. xiv. 156 You might have heard a pin drop in the house while that was going on. **1831** Macaulay *Let.* 30 Mar. (1974) II. 10 You might have heard a pin drop as Duncannon read the numbers. Then again the shouts broke out. **1870** [in Dict., sense 3]. **1890** [see Drop *v.* 3]. **1914** *Maclean's Mag.* May 9/2 We could have heard a pin drop. **1934** A. Christie *Murder on Orient Express* iii. ix. 246 Every eye was fixed upon him. In the stillness you could have heard a pin drop. **1959** I. & P. Opie *Lore & Lang. Schoolch.* x. 193 'Let's 'ave a bit of shush', 'Let's hear this pin drop',..'Pipe down'. **1977** *Daily Mirror* 15 Mar. 3/3 They screamed, yelled and clapped between numbers, but while he sang you could have heard a pin drop.

g. *to stick pins into* (a person): to incite to action; to irritate or annoy.

1903 A. H. Lewis *Boss* 184 This ain't meant to stick pins into you.

h. A gramophone needle.

1914 Kipling *Lett. of Travel* (1920) 215 They slipped in pin and record.

IV. 7. (Further, including sing., examples.) Also *fig.*

1829 P. Egan *Boxiana* 2nd Ser. II. 10 With all his struggling to right himself, he could not recover the use of his pins. **1880** 'Mark Twain' *Lett. to Publishers* (1967) 125. It saved the company's life and set them high on their pins and free of debt. **1890** *Harper's Mag.* LXXX. 269/2 Glad to hear that he is on his pins yet; he might have pegged out in ten years, you know. **1917** 'H. H. Richardson' *Fortunes R. Mahony* iv. viii. 355 Give your old pin here, and let me poultice it. **1960** [see *few *a.* 2 f]. **1971** *Petticoat* 17 July 32/1 You need to be healthy too because you're likely to find yourself standing on your two pins all day and every day. **1976** *Daily Mirror* 11 Mar. 24/2 You look a bit wobbly on your pins, pet.

8. Also, a skittle or pin knocked down, as a scoring point. (Later examples.)

1974 *Cleveland* (Ohio) *Plain Dealer* 13 Oct. 8-C/3 Anthony..missed a perfect game by a single pin in the..finals. *Ibid.* 26 Oct. 5-D/4 A bowling ball can mean the difference of up to 20 pins in a game depending on the surface and balance the ball. **1976** *Burnham-on-Sea Gaz.* 20 Apr. 21/5 Thatchers Arms A, the North Petherton Summer Skittles League champions, lost their first home match of the new season by six pins when Clarence Hotel A came from behind on the last hand. **1976** *Bridgwater Mercury* 21 Dec. 18/2 Brent Knoll Inn put an end to the last 100 per cent record in the first division of the Burnham, Highbridge and District Ladies' Skittles League by gaining a seven-pin away win at White Hart Hotel.

VI. 18. *pin-flag, -heel* (so *-heeled* adj.), *-hook* (also fig.; earlier and later examples); *pin-sharp* adj.; **pinboard,** a panel having an array of identical sockets each connected to some of a set of wires, so that inserting a conducting pin into any of the sockets makes an electrical connection between some of the wires; **pin-bone,** (*a*) (later examples); (*b*) (see quot. 1936); **pin-boy** (see quots.); **pin-brained** *a.*, foolish, stupid; **pin-buttock** (later examples); **pin clover,** also, = *alfilaria; (later examples); **pin connexion** (examples); so **pin-connected** *a.* (cf. **pin-jointed* adj.); **pincord** = *needlecord* s.v. *needle *sb.* 14; **pin-curl,** (*b*) a curl held in position during setting by a hairpin or other clip; **pin curler,** a pin or clip for securing pin-curls; **pin dot** (see quot. 1957); **pin-dropping** *a.* = *pin-drop* adj.; also as adv.; **pin-fall** (*a*) the fall of a pin; a trifling incident; (*b*) the number of pins knocked down in a tenpin bowling game; a score in tenpin bowling; **pin-fit** *v. trans.* and *intr.*, in *Sewing*, to pin into position during fitting; **pin-grass** (examples); **pin hinge,** a hinge in which the two leaves are pivoted on a pin passing through a sheath in each; **pin-hooker** *U.S.* (see quot. 1944); **pin joint** (examples); so **pin-jointed** *a.*; **pin-leg,** (*a*) a wooden leg; (*b*) a narrow, spindly leg; hence **pin-legged** *a.*; **pin lever,** used *attrib.* and *absol.* to denote a pin-pallet watch or escapement; **pin-man,** (*b*) a figure of a man appearing as composed of lines without breadth, esp. in a drawing; **pin-mark,** a circular impression on the side of a piece of type, made by part of the mould used in casting; **pin-new** *a.*, brand-new; **pin oak** (earlier and later examples); also *attrib.*; **pin pallet,** add: in an escapement, a pallet in the form of a metal pin or a semi-circular jewel (now used chiefly in cheap escapements); (earlier and later examples); **pin party** *Naut. slang* (see quot. 1946); **pin plate,** *Engin.*, a plate with a hole for the pin that is riveted to a member at the site of a pin joint; **pin-pool,** 'a game played on a billiard-table with three balls, and five small pins' (*Century Dict.* 1889), or any of various related games; **pin screen** *Cinematogr.* (see quot. 1976); also *attrib.*; **pin seal,** the treated skin of young seals; also *attrib.*; **pin-setter,** in tenpin bowling, a person who, or a machine that, rearranges fallen skittles; = **pin-boy*; hence **pin-setting** *vbl. sb.*; freq. *attrib.*; **pin-splitter** *Golfing slang*, (*a*) a crack golfer; (*b*) an accurate shot to the pin or a club which is supposed to aid such a shot; **pin-spot,** (*b*) *Theatr.* (see quots.); **pin spotter** = **pin-setter*; **pin stenter,** a stenter in which cloth is held by means of two rows of pins, one along each edge; **pin-stitch** (see quot. 1936); hence **pin-stitched** *a.*, **pin-stitching**; **pin-table** = *pin-ball machine* s.v. *pin-ball 3; **pin valve** = *needle valve* s.v. *needle *sb.* 14; **pin-worm,** substitute for def.: a parasitic nematode worm of the order Oxyuroidea; also = **tomato pinworm*.

1957 *Electronic Engin.* XXIX. 30 Problems are programmed by plugging step by step operations in columns on the pinboards, a column for each operation, and when a pin is inserted into a hole, connexions are automatically made to carry out the instruction. **1966** R. K. Richards *Electronic Digital Syst.* ix. 553 Like a plugboard, a complete pinboard assembly with its pins can be removed from a system and saved for re-use while other pinboards..are used for the solution of other problems. **1969** P. B. Jordain *Condensed Computer Encycl.* 383 A 10 × 10 pinboard with 100 holes can connect any of 10 incoming signals (horizontal) to any of 10 outgoing paths (vertical). **1936** *Discovery* Oct. 321/1 Pins had been made by individual craftsmen using a small bone implement, on the flat surface of which was a series of inch-long grooves, deep at one end but tapering at the other. This tool served to file the pins to a point, its formation making it possible to shape several at one time from lengths of wire. Many of these 'pin-bones' still exist. **1945** *ABC of Cookery* (Ministry of Food) ix. 29 Hip or pin bone steak. **1973** R. D. Symons *Where Wagon Led* p. xiii, I learned the cowboy's names, and could speak of pinbones and stifles and croups. **1978** *Sunday Tel.* 17 Sept. 11/3 Do you know, for example, what is called Pope's eye or heuk bone in Scotland, hip bone in the Midlands or pin bone in Wales? It's what many of us call rump. **1892** A. E. Vogell *Bowling* 8 *Pin boy*, boy who returns the balls cast and resets the pins. **1958** *Wall St. Jrnl.* 9 Dec. 1/4 Automatic pinboys..have played a major role in the bowling boom. **1959** *Listener* 19 Mar. 501/1 The latest development is an automatic pin-boy... An ingenious electrical device will set up the pins and return the balls. **1975** *Oxf. Compan. Sports & Games* 92/1 In the earlier years of the game pin-boys were employed to re-set the pins after each frame. **1964** *Listener* 20 Feb. 313/1 A smug biologist and his pin-brained wife. **1966** B. Kimenye *Kalasanda Revisited* 31 His successor was flirting madly with some pin-brained girl. **1977** 'E. Crispin' *Glimpses of Moon* i. 11 There were several such benches in the bar-room—memorials to a centuries-extinct clientèle of pin-buttocks—but otherwise the furniture was all modern. **1913** W. C. Barnes *Western Grazing Grounds* 39 Alfileria is also known as 'heron's bill' and 'pin clover'. **1925** W. L. Jepson *Man. Flowering Plants Calif.* 592 The term filaree..is, like the names Pin Clover or Pin Grass, applied indifferently to either this species [*sc. Erodium moschatum*] or to *E. cicutarium*. **1878** *Min. Proc. Inst. Civil Engin.* LIV. 179 All these American bridges are 'pin-connected', this style of construction being preferred by American engineers for spans exceeding 100 feet, on account of the mathematical certainty with which the strains can be calculated. **1968** E. H. & C. N. Gaylord *Struct. Engin. Handbk.* vi. 71 The AISC Specification requires that the allowable tensile stress on the net section transverse to the axis of the member be reduced 25 percent at pinholes in pin-connected plates. **1969** *Civil Engin.* (Easton, Pa.) June 45/1 Columns were pin-connected top and bottom to prevent any supplementary stress from differential settlement or rotation of the foundations. **1878** *Min. Proc. Inst. Civil Engin.* LIV. 214 There was undoubtedly a considerable difference in the use of what the Author [*sc.* T. C. Clarke] called pin-connections as compared with rivet-connections. **1974** *Encycl. Brit. Macropædia* III. 177/2 The first major iron-truss bridge, with pin connections, was built in the United States in 1851. **1977** *New Yorker* 8 Aug. 72/1 (Advt.), Our handsome, lightweight hobby pants in no-iron Dacron/cotton pincord, ideal for travel and leisure. **1979** A. Scholefield *Point of Honour* 142 White pincord slacks. [illegible] *Hairdressing* 28/1 Pin-curls..may be used as side-pieces to be worn in front of the ears. **1950** E. Hemingway *Across River* xiv. [illegible] to bed with a girl who has put her hair up in pin curls to be beautiful tomorrow. **1963** D. B. Hughes *Expendable Man* (1964) i. 10 'Don't you need a mirror?' But she could wind the pin curls without it. *Ibid.* 18 Her hair done up in pin curlers under the dirty scarf. **1957** M. B. Picken *Fashion Dict.* 252/2 Pin dot, small dot used as fabric design. **1978** *N.Y. Times* 20 Mar. 44/1 (Advt.), A pin-dot tie. **1971** P. Francis *Bonecrack* i. 9 There was a long pin-dropping silence. **1973** *Advocate-News* (Barbados) 29 June 6/6 The lighting..succeeded in transforming the massive stadium through moods of barred excitement to 'pin-dropping' silence. **1977** *New Yorker* 19 Sept. 50/1 [illegible] **1912** [illegible] troubles had been so many pinfalls in the closeted selfishness of her little life. **1974** *Cleveland* (Ohio) *Plain Dealer* 27 Oct. [illegible] this inaugural event with a 1963 pinfall. **1976** *Billings* (Mont.) *Gaz.* 30 June 4-B/1 Here are the team bowlers and their total pinfall from four games (12 games) of the [illegible] Open Bowling Tournament. **1964** *McCall's Sewing* ii. 31/1 Pin-fit, to pin and adjust the garment to your figure before permanent stitching. *Ibid.* vi. 93/1 When you try on the garment, it is very easy to pin-fit and taper the legs to a becoming width. **1973** *Washington Post* 5 Jan. B5/4 (Advt.), Price includes fabric and labor with choice of skirt, self-cording and pin-fitting in your home. **1912** [illegible] *City* iv. 176, 4,000 pin-flags were brought out of concealment [illegible] fastened in the clothing of their comrades. **1847** *Californian* (San Francisco) 10 July 3/1 Quality of Pasture—Bunch Grass; Clover; Wild Oats and Pin Grass, all in abundance. **1888** [see *alfilaria]. **1951** C. F. Saunders *With Flowers & Trees in Calif.* iii. 55 Still another wild pasture-plant is the stork's-bill..commonly known as pingrass or filaree. **1949** L. E. Palmer *Fieldbk. Nat. Hist.* 241/1 Common names [of *Erodium cicutarium*] include wild musk, pin clover, pin grass, pinweed and heron's bill, mostly based on the slender fruit. **1958** R. C. Rollins *Fernald & Kinsey's Edible Wild Plants Eastern N. Amer.* (ed. 2) 259 Storksbill, Pin-grass, *Erodium cicutarium*. **1961** R. Longrigg *Daughters of Mulberry* i. 8 Giggles Ballantyne..teetered unhappily on tall pin heels. **1966** C. Stone *Marianne & Mark* iv. 54 A pair of tight pinheeled patent-leather shoes. **1963** *Times* 20 Feb. 14/7 Were you joined by a girl in pointed pin-heeled shoes which soon put an end to her enthusiasm. **1960** Chambers's *Techn. Dict.* 644/2 Pin hinge. **1964** W. L. Goodman *Hist. Woodworking Tools* ii. 53 This cupboard door, in a light-brown hardwood, appears to have been the right-hand leaf of a pair, as the pin hinge is at the top left. **1940** B. Crockett *Narr. Life* 103 In this hunt every..little pin-hook lawyer was engaged. **1840** *Southern Lit. Messenger* VI. 366/2 Ellen used to..have [illegible]. **1970** *Country Life* 17–24 Dec. [illegible] of sausages to..a tramp with a pin-hook on a piece of string. **1942** *Sun* (Baltimore) 23 Sept. 7/2 'Pin hookers', who [illegible] then resell them in auction markets and from [illegible] **1964** *Richmond* (Virginia) *Times-Dispatch* 5 Oct. 13 The fixation of prices [illegible] business for the timber-hooked 'pinhookers', the relatively small operating capital who, during the more spacious days of tobacco selling when the auctioneer got the green light for a sale from the relatively few high, bought tobacco when the market was low and held it until it was high, pocketing the difference. The prices this year are so near the average that the pinhooker's business has been practically squeezed out. **1949** L. Rapport in B. A. Botkin *Treas. S. Folklore* iv. ii. 652 There's Carroll, Jones, and Mallory [for the Big Three]..buyers from the four large independents who are on this market, and eleven or eight pinhookers. **1966** *Publ. Amer. Dial. Soc.* xlv. 19 The pinhooker will attach his sales tag to a lot of tobacco waiting to be received by the warehouse workers. **1886** A. B. W. Kennedy *Mech. of Machinery* xii. 586 The efficiency of a pin joint, or turning pair, is generally very much greater than that of a sliding pair. **1919** Pippard & Pritchard *Aeroplane Struct.* x. 123 An important example occurs when..a pin-joint is made in an aeroplane spar at any place other than near one of the positions of the points of inflection. **1978** J. E. Gordon *Structures* x. 205 The concentration of stress at the pin joints calls for a tough and ductile material, such as wrought iron. **1882** *Min. Proc. Inst. Civil Engin.* LXIX. 111 Of a well-designed pin-jointed structure, all the principal members must be connected directly with the pin. **1908** E. S. Andrews *Theory & Design of Struct.* xvii. 508 Pin-jointed eye bars are not much used in this country for bridge work, and for roof work they are going out of use. **1974** *Nature* 4 Jan. 77/2 It covers the treatment of pin-

jointed frameworks, beams, circular sandwich plates, Michell's structural continua and plates loaded in their planes. **1862** *Illustr. Catal. Internat. Exhib., Industr. Dept., Brit. Div.* II. No. 3600 A case with jointed pin-leg, artificial human leg, and others. **1960** S. Plath *Colossus* (1967) 39 The oracular ghost who dwindles on pin-legs. **1936** Dylan Thomas *Twenty-Five Poems* 44 Pin-legged on pole-hills with a black medusa By waste seas where the white bear quoted Virgil. **1939** —— *Let.* Mar. (1966) 226 The English poets now are such a pinlegged..crowd. **1946** in D. de Carle *Pract. Watch Repairing* (prelim. advt.) The new pin lever—seven jewels. **1962** *Which?* June 165/1 In many cheaper watches, there is a pin-pallet (or pin lever) type of escapement, in which the pallets are hardened steel pins instead of jewels. **1976** M. Cutmore *Watch Collector's Handbk.* ii. 70 In the pin-lever design the impulse pallets on the lever are replaced by steel pins... The escape-wheel teeth supply most of the lift and have steeply sloped faces. **1934** Dylan Thomas *Let.* 2 May (1966) 114 And I have, too, a violent desire to draw pin-men. **1953** J. Masters *Lotus & Wind* i. 8 The engineer was little more than a pin man in the distance. **1965** J. Wade *Boy with Sling* i. i. 11 He flicked at the ground..with a switch, drawing an elongated pinman. **1975** *Times Lit. Suppl.* 7 Feb. 129/3 The movement of the train, and the view through its windows, reduce what Rhoda sees to a scene of pin-men moving awkwardly and senselessly through an unwelcoming landscape. [**1887** T. B. Reed *Hist. Old Eng. Letter Foundries* 26 A more probable explanation seems to be that the head of a small screw or pin, used to fix the side-piece of the mould,..left its mark on the side of the types as they were cast.] **1888** C. T. Jacobi *Printers' Vocab.* 100 Pin mark.—This is the slight mark in the side of a type near the top of the shank made in casting by machinery. **1916** Legros & Grant *Typogr. Printing-Surfaces* iii. 14 The pin-mark, or drag,..only occurs in certain machine-made type. **1922** D. B. Updike *Printing Types* I. ii. 16 The pin-mark is an indentation on the upper part of the body, made by the pin in casting. **1951** S. Jennett *Making of Bks.* ii. 34 The other [side], in foundry type, bears the pin-mark, or occasionally two pin-marks... The pin-mark is formed by the mechanism that ejects the type from the mould when it is cast. **1967** T. Keneally *Bring Larks* xxviii. 222 By the time he sighted the pin-new East Indiaman, it had already ripped through the oyster-shell horizon far out to the south-east. **1976** *New Society* 19 Feb. 373/1 A dozen girls sit in the toy department of a pin-new department store. **1813** H. Muhlenberg *Catal. Plantarum Americæ Septentrionalis* 87 Swamp or Pin Oak, (*Quercus palustris*). **1847** D. Coyner *Lost Trappers* I. 23 The young trapper was relieved by the arrival of two of the company, one of whom climbed a pin-oak tree. **1857** *Yale Lit. Mag.* XXII. 284 His head is as obtuse and spongy as the butt-end of a pin-oak rail. **1941** P. P. Pirone *Maintenance of Shade & Ornamental Trees* iv. 45 Pin oaks in alkaline soils often develop yellow or chlorotic leaves. **1975** *Country Life* 16 Jan. 148/3 American woodland in the east..is composed largely of oak, not our oak, but the slim and lofty red oak, white oak, pin oak and chestnut oak. **1860** E. Beckett *Rudimentary Treat. Clocks* (ed. 4) 103 Pin Pallets. **1903** F. J. Garrard *Watch Repairing* x. 101 The 'pin-pallet' escapement..has round pins for pallets, and the inclines are on the scape teeth. **1946** D. de Carle *Pract. Watch Repairing* vi. 55 Some designs of pin pallets are so made that the arms carrying the pallet pins can be bent quite easily. **1976** M. Cutmore *Watch Collector's Handbk.* i. 47 The cheap watch had developed independently in Switzerland and in the form which was eventually to become worldwide. Roscopf designed and produced a pin-pallet lever watch in 1867. **1942** *Ark Royal* (Ministry of Information) 18/2 When a squadron is preparing for a reconnaissance, the ground staff bring the aircraft from the hangar to the flight deck... The ranging party..then take over and push the aircraft aft, assisted by a small 'pin party' to spread and secure the wings. **1946** J. Irving *Royal Navalese* 135 *Pin party*,..the working party, in a Carrier, which prepares aircraft on the flying deck for taking off. **1893** J. B. Johnson et al. *Theory & Pract. Mod. Framed Struct.* xxi. 338 The Lengths of Bearing or Pin Plates are determined by the following considerations. **1968** E. H. & C. N. Gaylord *Struct. Engin. Handbk.* vi. 71 Usually, a pin plate is assumed to transmit a fraction of the main member force proportional to its thickness. *c* **1866** W. B. Dick *Amer. Hoyle* (ed. 3) 428 The game of Pin Pool is played with two white balls and one red, together with five small wooden pins. **1900** Ade *Fables in Slang* 16 The Local Editor..was playing Pin-Pool with the Superintendent. **1915** B. Edwards in H. Dempsey *Best of Bob Edwards* (1975) v. 107 Men who blew all their money on whisky and pin pool. **1959** in Halas & Manvell *Technique Film Animation* xxvi. 304 (*heading*) Pin screen animation. *Ibid.*, The pin screen is designed for black-and-white films. **1976** *Oxf. Compan. Film* 11/2 After watching *L'Idée* (1934) in production in 1932 he [*sc.* Alexandre Alexeieff] experimented with animation techniques and invented the pin screen, a metal surface pierced by about five million tiny holes through which he pressed metal pins which, obliquely lit, created shadows with all possible gradations from black to white according to the length of pin protruding from the screen. **1926–7** *Army & Navy Stores Catal.* 408B/3 Pochettes...Pin seal leather, lined silk. **1934** *Times* 29 Nov. 19/3 In black pinseal with the new short handles there is a useful bag with a triple frame. **1961** [see *billfold*(*er* s.v. *bill *sb.*³ 11]. **1972** J. Mosher *Some would call it Adultery* iii. xv. 139 Then he took a change purse from an inside pocket. It was black pinseal leather, opening with a snap. **1974** *Country Life* 21 Nov. (Suppl.) 45 Pin seal wallet £30. **1916** H. M. Rideout *Far Cry* xi. 150 Like a pin-setter in a bowling alley, Mace carefully planted his bottles upright on the floor. **1958** *Economist* 20 Dec. 1085/1 In 1954 the American Machine & Foundry Company began large-scale production of an electronic automatic pin-setter, or pin spotter as it is also called. **1958** *Wall St. Jrnl.* 9 Dec. 1/4 In the past seven years they've [*sc.* automatic pinboys have] taken over the pinsetting chores on all but a small minority of the existing lanes. **1964** *Economist* 30 May 1024/3 The introduction of pin-setting machines that speeded the game. **1972** *Mainichi Daily News* (Japan) 7 Nov. 6/4 The company signed a contract with Yungtay Engineering Co. of Taichung to export pin-setting machines for 100 bowling lanes. **1947** W. de la Mare *Coll. Stories for Children* 60 He crinkled up his pin-sharp eyes. **1978** *SLR Camera* Aug. 4/1 (Advt.), Pinsharp projection from corner to corner. **1926** *Glasgow Herald* 26 June 8 Their prowess as 'par-beaters' and 'pin-splitters'. **1961** Partridge *Dict. Slang* Suppl. 1222/1 *Pin-splitter*... Since ca. 1935, predominantly a golf-shot dead on the pin. **1973** *Country Gentlemen's Mag.* Mar. 181/1 Gents Pinsplitter Golf Clubs. **1947** *Gloss. Technical Theatr. Terms* (Strand Electr. & Engin. Co.) 22 *Pin spot*, any spot lantern so adjusted that its maximum light is focussed into the smallest possible area. **1961** A. Berkman *Singers' Gloss. Show Business* 7 *Baby spot*.., a small spotlight used for illuminating any small object on stage; often used for lighting just the face of the performer. (Also called *pin spot*.) **1958** Pin-spotter [see **pin-setter*]. **1975** *Oxf. Compan. Sports & Games* 90/2 Modern bowling centres, with automatic pin-spotters and score-indicators, have replaced the rough, tough alleys lit by kerosene lamps, which catered for saloon crowds in the early days of the twentieth century. **1947** J. T. Marsh *Introd. Textile Finishing* i. 20 The beater untangles the matted pile, and the fabric then passes into the pin stenter which is equipped with from eight to sixteen rotating cards. **1962** —— *Self-Smoothing Fabrics* xi. 171 Since the advent of the crease-resisting process, with its earlier emphasis on rayon fabric production, there has been a corresponding emphasis on pin stenters compared with the clip stenter. **1936** A. M. Miall *Everyday Embroidery Bk.* viii. 74 *Punch, Lace, Pin or Turkey-stitch* (it has all these names) is an open-work stitch which is sometimes used in rather elaborate *broderie anglaise*, or in cut-work..as an open-work filling for a flower or motif. **1948** C. Christopher *Compl. Bk. Embroidery* viii. 185 Pin Stitch is used to applique motifs of self material on either wrong or right side of fabric. **1972** B. Snook *Creative Art of Embroidery* 92 *Pin stitch*. This is used along a hem edge. **1960** S. Plath *Colossus* (1967) 48 At the price of a pin-stitched skin Fish-tailed girls purchase each white leg. **1935** G. W. Fry *Embroidery & Needlework* viii. 188 As with hemstitching so with pin stitching, the decorative effect may be obtained with nothing more than outlines. **1936** *Archit. Rev.* LXXIX. 135/1 Under one of these, however, peeps out a gay little shop in bright green paint, full of pin-tables, where one can lose one's money in proper Strand fashion. **1957** *Observer* 20 Oct. 16/3 Entirely devoted to a single situation—pin-table saloon proprietors versus local councillor. **1973** 'M. Underwood' *Reward for Defector* ii. 13 Questions..were flashing through his mind like lights on a pin-table. **1977** *Irish Press* 29 Sept. 16/1 (Advt.), Pool tables, pin tables, fruit machines, etc., on sharing basis. **1903** *Electr. World & Engin.* 18 July 115 The pressure is admitted to or withdrawn from the piston by means of a pin-valve. **1910** K. Winslow *Prevention & Treatm. Dis. Domestic Animals* 191 Oxyuris, Whip, Thread or Pin Worm. **1933** *Jrnl. Econ. Entomol.* XXVI. 138 In California, the Pin Worm [*sc.* an insect larva] has done much damage to tomatoes. **1961** E. R. & G. A. Noble *Parasitol.* vi. 312 *Oxyuris equi* is the common pinworm in the cecum and colon of horses. *Ibid.* 317 *Enterobius vermicularis* is the pinworm or seat-worm of man. **1974** M. C. Gerald *Pharmacol.* ii. 35 Piperazine (Antepar) and chloroquine..have been quite effective in the cure of..pinworm, and malaria, respectively.

pin, *sb.*² Add: **pin-bullock** *Austral.*, one of the pair of bullocks in a team nearest the wagon.

1936 I. L. Idriess *Cattle King* viii. 68 They call the two polers the pin bullocks, because they swing the turntable of the wagon! **1959** H. P. Tritton *Time means Tucker* 36 A bullock-team is made up in four parts: polers, pin, body and leaders... The pin-bullocks take the pull.

pin, *sb.*³ Add: **1.** (Further examples.)

1911 A. C. White *First Steps Classification Two-Movers* 73 The Black King moves into a triple pin, which is the feature of the problem. **1932** *Times Lit. Suppl.* 8 Dec. 948/3 Forcing the king where he wanted it, and then releasing the 'pin' paved the way for the threatened 26. **1976** *Daily Tel.* 4 Dec. 11/6 Black thinks it time to bring his queen into play and finds an unexpectedly troublesome pin.

2. [f. Pin *v.*¹ 5 a.] **pin-fall** *Wrestling*, a fall in which a wrestler must hold an opponent down for a specified length of time.

1907 *Daily Chron.* 21 Dec. 9/5 These two..wrestlers having agreed to contest the best of three pin falls in the catch-as-catch-can style. **1976** K. Bonfiglioli *Something Nasty in Woodshed* iv. 41 He helps the other chap back into the ring..then administers a fearsome forearm smash and the winning pinfall.

pin, *v.*¹ Add: **I. 2. c.** To spread *out* (dough or paste) with a rolling-pin.

1889 R. Wells *Pastrycook & Confect. Guide* 39 Pin them out not too thick, and cut them into four.

d. *Austral. slang.* To 'do down', to cause trouble for (someone). (See also quot. 1941.)

1934 C. Stead *Seven Poor Men of Sydney* iv. 122 A poor man..never 'as anything but a poor, miserable, wretched, untidy, un'appy life. They don't let 'im even be honest or 'ave a friend, if some one wants to pin 'im. **1941** Baker *Dict. Austral. Slang* 54 *To pin someone*, to have someone 'set', to have a grudge against a person.

e. In phrases *to pin someone's ears back* (orig. *U.S.*), to chastise, to rebuke; *to pin one's ears back*, to listen attentively.

In quot. 1977 'lugholes' is colloquially substituted for ears'.

1941 H. L. Ickes *Diary* 22 June (1954) III. 546 It certainly was intended to pin my ears back. **1949** Wodehouse *Uncle Dynamite* ix. 160 His manner that of a man who has had his ears pinned back. **1961** Partridge *Dict. Slang* Suppl. 1222/1 *Pin back your ears* or *pin your ears back*, listen carefully. **1962** N. Streatfeild *Apple Bough* xi. 146 I'd get my ears pinned back if I tried to cut down his practice time. **1966** 'L. Black' *Bait* iii. 33, I shall keep eyes open. And ears pinned back. **1977** D. Francis *Risk* xvi. 12 Well, mate, pin back your lugholes. That boat you were on was built at Lymington.

4. For 'Now *rare*' read 'Now *rare* except in sense "to fasten or fix (anything objectionable) *on* a person"'. Also, *on* a thing. (Later examples.)

1924 'W. Fabian' *Sailors' Wives* 34 Dorrisdale credits me with at least three [lovers], but they've never been able to pin it on me with anyone. **1942** E. Paul *Narrow St.* xxi. 169 As usual, when anything sinister happened, his enemies tried to pin everything on Caillaux, who cleared himself promptly. **1966** *Listener* 3 Nov. 652/3 Medical research workers have been suspicious that some difficult and resistant diseases..are associated with these same PPLO or mycoplasma, but nobody has been able to pin it on them. **1970** N. Fleming *Czech Point* ii. 39 No doubt up till now he had never had to cope with a crime more heinous than skiing without due care and attention. He could have pinned that rap on the Australian girl in the PVC outfit any day he cared. **1977** L. Meynell *Hooky gets Wooden Spoon* xii. 150 'Can the Law connect her with you?' '..No, they couldn't pin anything on me.'

5. (Earlier and later examples.)

1740 Fielding in *Champion* 1 Apr., When he is pinned down,..particularly by one large Mastiff, I do not perceive that Readiness to relieve him which hath been formerly shewn. **1945** *Sun* (Baltimore) 21 Feb. 1/7 (*heading*) 4th Division pinned down by mortar fire 17 hours. **1970** *Globe & Mail* (Toronto) 25 Sept. 8/9 The Ambassador and 24 members of his staff..were pinned down in the Embassy since the fighting started. **1977** *Times* 17 Jan. 7/1 Underwood did his usual job pinning the batsmen down with geometrical control.

6. b. With *down*. To manœuvre (a person) into a position where evasion is impossible.

1904 Ade *True Bills* 40 Horace tried to side-step the Questions about Drinking and Smoking, but Uncle pinned him down. **1914** 'High Jinks, Jr.' *Choice Slang* 16 (*To*) *pin down*, to corner.

c. With *down*. To define, evaluate, isolate; to restrict *to*.

1951 M. McLuhan *Mech. Bride* (1967) 137/1 The kind of spectator participation in baseball..would be hard to pin down. **1955** *Times* 26 July 8/3 It would never be possible to pin down the cause of death to radiation, he added. **1963** T. Parker *Unknown Citizen* v. 139 There was obviously a good deal more to it underneath, which you could never pin down. **1965** *Listener* 2 Dec. 920/2 This symbolism is far more difficult to pin down than Rublowsky seems to suppose. **1973** *Times* 27 Apr. 1/8 Attempts to pin down the origin of the disease were inconclusive.

piña. Add: **4. piña colada** (kolā·dă) [Sp., lit. 'strained pineapple'], a long drink made with pineapple juice, rum, and coconut.

1975 P. Moyes *Black Widower* xiii. 159 In the bar itself, ice tinkled merrily into tall glasses of rum punch and *pina colada*, and smooth, sun-tanned men and women sipped and chatted. **1977** *N.Y. Times Mag.* 4 Dec. 42/1 She is wearing a long, batik-printed dress and bearing two piña coladas on a tray. **1978** *Chicago* June 172/3 There was gin and piña coladas and talk about Estée Lauder..and money..and clothes.

pinacol (pi·năkǫl). *Chem.* [f. Pinac(one + -ol.] = Pinacone in Dict. and Suppl.; **pinacol rearrangement,** a reaction typified by the conversion of pinacol into pinacolone, in which a 1,2-glycol loses water on heating with acid to form a ketone.

1911 *Chem. Abstr.* V. 3561 The solid pinacols have normal mol. wts. in dil. C_6H_6. **1912** *Ibid.* VI. 82 The fraction b. 214° of the products of hydrolysis of crude pinacol by H_2SO_4. **1936** L. J. Desha *Org. Chem.* 538 The pinacol rearrangement is irreversible in the sense that pinacolins do not change into pinacols. **1938** E. S. Wallis in H. Gilman *Org. Chem.* I. 723 Numerous examples of this reaction, now known as the pinacol rearrangement, have been found since the time of Fittig's discovery [in 1859]. **1964** N. G. Clark *Mod. Org. Chem.* x. 180 Acetone is reduced with amalgamated magnesium to pinacol. **1966** *McGraw-Hill Encycl. Sci. & Technol.* III. 375/2 Ketones are converted to tetrasubstituted ethylene glycols (called pinacols) by certain reducing agents. **1966** Smith & Cristol *Org. Chem.* xiii. 289 The pinacol rearrangement is quite general for glycols in which all four groups..are alkyl or aryl groups.

pinacolin. Add: Also **-ine.** [First formed as G. *pinacolin* (R. Fittig 1860, in *Ann. d. Chem. u. Pharm.* CXIV. 58).] Also, any other ketone in which the carbonyl group is bonded to at least one tertiary carbon atom. Now usu. called *pinacolone. (Further examples.)

1884 [see *pinacone]. **1913** T. H. Pope tr. *Molinari's Treat. Org. Chem.* II. 183 When distilled with dilute sulphuric acid, it [*sc.* pinacone] is transformed into pinacoline, $(CH_3)_3C.CO.CH_3$. **1936** [see *pinacol]. **1973** B. J. Hazzard tr. *Organicum* 587 In textbooks, the dihydric alcohol here termed pinacol is often called pinacone and the ketone produced by the rearrangement pinacoline.

pinacolone (pi·năkǒlōu:n). *Chem.* [f. *pinacol + -one.] = Pinacolin in Dict. and Suppl.

1925 *Org. Syntheses* V. 91 The combined pinacolone fraction is dried over calcium chloride, filtered, and fractionally distilled. **1932** *Jrnl. Amer. Chem. Soc.* LIV. 825 Because of the need of large amounts of pinacolone for investigations in progress in this Laboratory, methods for its preparation have been studied. **1940** S. Miall *New Dict. Chem.* 400/1 When heated with mineral or organic acids, pinacols undergo a molecular rearrangement with loss of water to give ketones known as pinacolones. **1944** *Jrnl. Amer. Chem. Soc.* LXVI. 634/2 It was anticipated that this pinacolone should be 3,3-dianisyl-2-butanone. **1967** L. F. & M. A. Fieser *Reagents for Org. Synthesis* I. 772 On brief exposure to moist air the fluorohydrin is rearranged to pinacolone..and hydrogen fluoride.

pinacone. Add: [First formed as G. *pinakon* (G. Städeler 1859, in *Ann. d. Chem. u. Pharm.* CXI. 279).] Also, any other alcohol having two hydroxyl groups which are bonded to adjacent tertiary carbon atoms. Now usu. called *pinacol. (Further examples.)

1884 Roscoe & Schorlemmer *Treat. Chem.* III. ii. 12 The pinacones decompose very readily into water and the ketones, to which the name of pinacolines is given. **1913** T. H. Pope tr. *Molinari's Treat. Org. Chem.* II. 182 A special group of glycols, the pinacones, containing two adjacent tertiary alcohol groups. **1973** [see *pinacolin].

pinafore, *sb.* Add: Also, a low-necked, sleeveless fashion garment worn by women and girls, usu. over a blouse or jumper.

[**1907**: see *pinafore-frock* below.] **1960** *Vogue Pattern Bk.* Autumn 46 Dark grey pinafore tops a gay printed blouse. **1974** *Country Life* 21 Feb. 398/1 Dark brown pinafore in fine corduroy worn with white cotton shirt. **1976** *Vogue* Jan. 44/2 Low-slung pinafore.

c. Also *pinafore dress, frock, gown.*

c **1909** D. H. Lawrence *Collier's Friday Night* (1934) ii. 35 She is wearing a dark blue cloth 'pinafore-dress'. **1973** *Times* 15 Nov. 6/3 The bridesmaid, Lady Sarah Armstrong-Jones, wore a pinafore dress. **1907** *Westm. Gaz.* 8 Apr. 13/1 Our grown-up pinafore-frock must confess itself borrowed from the nursery. **1952** C. W. Cunnington *Eng. Women's Clothing* iii. 91 The Pinafore frock is revived [in 1909], worn over a tucked chemisette. **1932** *Mod. Weekly* 30 Apr. 118 Her adorable, black velvet, pinafore gown. The tiny puff-sleeves are of silver tissue. **1952** C. W. Cunnington *Eng. Women's Clothing* ii. 71 *The Pinafore Gown* [in 1906] 'is often allied to the frock with brace over a lace blouse'.

‖**pinard** (pinār). Also **Pinard.** [Fr.] Coarse red wine (orig. that issued to French troops), vin ordinaire; *loosely*, any wine; a glass of this wine.

1922 E. E. Cummings *Enormous Room* iv. 85 A glass apiece of red acrid *pinard*. **1924** *Blackw. Mag.* Oct. 556/2 Some of the adversaries are toasting each other in Pinard. **1928** R. Hall *Well of Loneliness* xxxvi. 335 The Unit's rations—cold meat, sardines, bread and sour red Pinard. **1937** Partridge *Dict. Slang* 631/1 *Pinard*, liquor; wine: Soho (-1935). **1947** M. Lowry *Let.* Nov. (1967) 160 Wonderful skipper on this ship, engineers, seamen, cats, stewards, pinard. **1950** E. Hemingway *Across River* xii. 112 In a Great Hotel, wine must cost money. You cannot get Pinard at the Ritz. **1969** B. Weil *Dossier IX* iii. 23 He..drank a *pinard* as if the rough red wine would take away the taste of a probable duplicity.

pinarette (pināre·t). [f. Pina(fo)re *sb.* + -ette.] A short pinafore (in the senses of Dict. and Suppl.).

1951 *Sunday Pictorial* 21 Jan. 15/6 (Advt.), Pinarette, exclusive design in cotton print. Stock size. Fully cut with perfect finish. **1959** 'H. Carmichael' *Stranglehold* xii. 102 A nondescript girl..wearing a pinarette and a maid's cap. **1962** J. Cannan *All is Discovered* ii. 30 The gaudy cotton dress, the strong bare legs, the black felt slippers, the homely 'pinarette'. **1966** 'K. Nicholson' *Hook, Line & Sinker* ii. 27 She was wearing a floral pinarette over a pink satin blouse.

‖**piñata, pinata** (pinyā·tă, pinā·tă). Also 9 **piñate.** [Sp., jug, pot.] In Mexico and Mexican-influenced areas of the U.S.A., a decorated container (orig. a pottery bowl), filled with sweets or other gifts which is broken by a blindfolded person or otherwise opened at Christmas and on other festive occasions.

1887 F. C. Gooch *Face to Face with Mexicans* viii. 264 The breaking of the *piñate* is the chief sport of the *posada*. The *piñate* is an oval-shaped earthen jar, handsomely decorated and covered with bright ornaments... There are turkeys, horses, birds, monkeys, [etc.]. *Ibid.* 265 The fun of breaking the *piñate* begins. It is suspended from the ceiling, and each person..blindfolded..proceeds to strike the swinging *piñate*. **1934** E. Fergusson *Fiesta in Mexico* xiv. 254 There is a party, with..sweets and paper toys in a huge pottery ball decorated with paper to look like a turkey, Charlie Chaplin,..anything. The littlest child or favored guest pulls the string which releases the shower of candies. This is the piñata, which is in season anytime from the first posada to *Dia de los Reyes*, Twelfth Night. **1947** C. U. Stoker *Under Mexican Skies* 169 The *patio* looked very pretty with its gardens all in bloom. The children had hung a *piñata*, representing the Christmas Star, in the center of the *patio* with room enough for all to strike at it. **1959** C. Ramsdell *San Antonio* iv. xv. 262/1 The *piñata*, in Texas at least, is now made in a charming variety of shapes. *Ibid.* 262/2 At Christmas or birthday parties the *piñata* is always the climax. Child after child, blindfolded, is whirled about, and turned loose to bat wildly in the supposed direction of the piñata, which dangles at the end of a rope. **1976** *Examiner* (Coolidge, Arizona) 13 May 10 (*heading*) Second graders celebrate Cinco de Mayo... The children learned some Spanish words..and made *piñatas*.

pin-ball. For *U.S.* read 'orig. *U.S.*'. **1.** (Earlier and later examples.)

1803 E. Bowne *Girl's Life* (1887) 175 We went to a room where they keep their work for sale,—pocket-books, pin balls, [etc.]. **1870** L. M. Alcott *Old-Fashioned Girl* vi. 76 Her scissors and pin-ball at her side, and her thimble on. **1963** *Times* 9 Mar. 11/5 Silver-banded pin-balls were then made to hang from the girdle and even Queen Elizabeth herself deigned to accept a New Year gift of a 'pin pillow embrodred'.

3. A game resembling bagatelle, in which small balls are propelled across a sloping surface towards targets which indicate the score when they are hit. Freq. *attrib.*, as *pin-ball arcade, game, machine, table*.

1911 *R. Bliss Illustr. Catal.* in B. Whitton *Bliss Toys & Dollhouses* (1979) 27 Pin ball game...One of our popular marble games. **1935** *Sun* (Baltimore) 2 Apr. 1/7 One of the bills would allow the State..to license claw machine and pinball games. **1936** *N.Y. Times* 5 Feb. 40/3 Justice Rosenman said he was surprised that the pin-ball machines had been licensed in the first place. **1937** *Pop. Mechanics* Feb. 278 (*heading*) Trip 'em; home pin-ball game. **1946** *Sun* (Baltimore) 30 Dec. 2/3 This ordinance also imposes a license fee on claw machines or pinball machines or similar devices for public amusement, operated through the insertion of a coin or token. **1951** J. D. Salinger *Catcher in Rye* v. 46 We just had..hamburgers and played the pinball machine for a little while. **1959** *Times* 12 Feb. 10/7 To-day the senators were ringed with juke-boxes, pinball tables, 'one-armed bandits', and other coin-operated devices. **1969** N. Cohn *Pop from Beginning* xix. 170 Kids..like myself, gave their lives to pinball. **1973** M. Amis *Rachel Papers* 17 It was a month of plonk and coffee-bars, pinball arcades and party-hunts. **1975** R. L. Simon *Wild Turkey* (1976) i. 4 Gunther lit up like a pinball machine on twenty replays. **1976** Deakin & Willis *Johnny go Home* iii. 57 Ernie showed Johnny how to fix the machines in the Amusement Arcades, and together they played the pin-ball tables.

pince-nez. Add: Also **pincenez.** (Earlier and later examples.) Also *attrib.*

1876 Geo. Eliot *Let.* 3 Feb. (1956) VI. 220 Our young Charles..was slightly short-sighted and used only occasionally a pince-nez. **1904** Joyce *Let.* 28 Dec. (1966) II. 75, I shall go to an oculist..to get pincenez glasses. **1927** F. B. Young *Portrait of Clare* 26 She brushed the dust from her skirt..her grey eyes swimming behind the lenses of her pincenez. **1941** *Punch* 13 Aug. 148/2 There was a frail creature in pince-nez at the other end of the table. **1974** *Oxf. Jun. Encycl.* (rev. ed.) XI. 422/2 The portrait of Cardinal Ugone in the church of San Nicolo at Treviso shows spectacles in which the two lenses are held together by a joint or hinge which fitted on the nose (an early form of pince-nez).

Hence **pince-nezed, -nez'd** [-ed²] *a.*, wearing a pince-nez.

1919 J. C. Snaith *Love Lane* xi. 51 An important, pince-nezed gentleman. **1922** W. J. Locke *Tale of Triona* xv. 170 Mrs. Rowington, thin, angular, pince-nez'd. **1958** B. Hamilton *Too Much of Water* i. 18 A spare pince-nez'd man with sandy-red hair. **1976** J. Crosby *Snake* (1977) xxxviii. 232 The doctor was a precise, pince-nezed Spaniard.

pincers, *sb. pl.* Add: **1. b.** *Mil.* = *pincer*(*s*) *movement* in sense 3 b below.

1942 T. Rattigan *Flare Path* ii. i. 38 There wasn't anything fresh, I suppose. No pincers on anything anywhere? **1969** G. Macbeth *War Quartet* 73 Firing turn by turn, Encircling him with pincers..At last..we killed him. **1978** J. A. Michener *Chesapeake* 638 [In 1863] southern armies were involved in a stupendous march north in an effort to create a pincers which would curl back to engulf Philadelphia, Baltimore and Washington itself. The end of the war seemed at hand.

3. *pincer-like* adj. (further examples); **pincer-leg** = sense 2.

1909 *Daily Chron.* 20 Aug. 4/4 Note the disparity in the size of the two large pincer-legs. **1941** H. G. Wells *You can't be too Careful* v. iii. 249 When confronted by a pincer-like movement, a soldier and a gentleman abandons his men and material and bolts home. **1962** D. Nichols *Echinoderms* viii. 101 The two living classes with relatively non-flexible exterior surfaces, the echinoids and asteroids, are provided with almost unique pincer-like organs, the pedicellariae.

b. *Mil.* Used *attrib.*, esp. in *pincer*(*s*) *movement*, to designate an operation involving the convergence of two forces on an enemy position like the jaws of a pair of pincers; a double envelopment; also *transf.* and *fig.*

1929 *Papers Mich. Acad. Sci., Arts & Lett.* X. 314/2 *Pincer drive*, an enveloping drive launched from two sides of an objective. **1939** Auden & Isherwood *Journey to War* ix. 225 The Japs were to be..destroyed by the time-honoured pincer-movement. **1944** S. Putnam tr. *E. da Cunha's Rebellion in Backlands* x. 458 As may be seen, it was a vigorous pincers movement which was thus planned. **1954** T. Gunn *Fighting Terms* 34 Planning when you have least supplies or clothing A pincer-move to end in an embrace. **1959** *Listener* 28 May 919/2 At one end of the barrier the Soviet Union is conducting a pincer movement on Persia, from Transcaucasia and Iraq. **1968** *Times* 8 Oct. 7/7 Hardened layers of sediment which accumulated on the bottom of the Tethys Sea and were later thrust up to the surface by the pincer movement of the two continental masses. **1973** *Times* 9 Aug. 17/1 The pincer movement against Unilever—soaring commodity prices..on the one hand, and selling price controls..on the other—is beginning to bite. **1975** D. Bagley *Snow Tiger* xiv. 120 Rickman and Lyall are cooking up something... It'll be a pincer movement.

pinch, *sb.* Add: **I. 1. c.** A theft; an act of stealing or plagiarism; something stolen or plagiarized. *slang.*

1757 *London Chron.* 15–17 Mar. 258/1 They have almost reduced Cheating to a Science; and have affixed technical Terms to each Species; three of which are the Pinch, the Turn, and the Mace. **1812** J. H. Vaux *Vocab. Flash Lang.* in *Mem.* (1964) 258 This game is called *the pinch*. **1903** 'J. Flynt' *Rise of R. Clowd* (1904) i. 64 That was just a pinch that I took. *Ibid.* 66 One night..on his way home..Ruderick..took a 'pinch' too large. **1931** G. Irwin *Amer. Tramp & Underworld Slang* 146 *Pinch*,..a small theft. **1965** *New Statesman* 19 Mar. 430/2 The 10-point charter drawn up hurriedly by its 'brains trust' during the much publicised Brighton weekend was almost a complete 'pinch' of the charter circulated in January by the Medical Practitioners' Union. **1966** *Melody Maker* 7 May 13/4 A pleasant selection of Italian-sung numbers—including what sounds like a Latin pinch from Presley.

d. An arrest or charge; imprisonment. Also *transf. slang* (orig. *U.S.*).

1900 'Flynt' & 'Walton' *Powers that Prey* 81 Told me to tell you't he'd have to make a pinch if you give the wheel another turn. **1906** E. Dyson *Fact'ry 'Ands* viii. 101 Ther Elder was back in er hour, 'n' had me outer pinch ez quick ez could be. **1926** [see *caper *sb.*² 1 c]. **1939** J. Steinbeck *Grapes of Wrath* xx. 371 Sheriff gets seventy-five cents a day for each prisoner, an' he feeds 'em for a quarter. If he ain't got prisoners, he don't make no profit... This fella today sure looks like he's out to make a pinch one way or another. **1948** *Sun* (Baltimore) 7 Jan. 13/1 If an official sees a violation, he is duty-bound to make the pinch, you might say. There is no compromise. Either it is a foul or it isn't. **1960** 'H. Carmichael' *Seeds of Hate* xix. 164 Before I make a pinch I like to be reasonably sure that the charge will stick. **1970** E. R. Johnson *God Keepers* (1971) ii. 20 Right now you got a goof-ball pinch... Get your coat on. **1978** P. G. Winslow *Coppergold* 24 More worried about his clobber than the pinch.

2. Esp. in phr. *to feel the pinch.*

1861 [in Dict.]. **1961** *New Eng. Bible Luke* xv. 14 He had spent it all, when a severe famine fell upon that country and he began to feel the pinch. **1974** *Nature* 11 Jan. 79/1 The industrialised nations are the first to feel the pinch. **1977** *World of Cricket Monthly* June 46/1 Otago are really feeling the pinch.

4. (Further examples.) Also *in a pinch.*

1821 M. Edgeworth *Let.* 9 Nov. (1971) 259 Even her humor would *on a pinch* submit to her sense of duty. **1903** *Booklovers' Mag.* Dec. 582, I have seen her tend bar in a pinch. **1936** C. Sandburg *People, Yes* 67 People lie in a pinch, hating to do it. **1943** H. Read *Politics of Unpolitical* i. 6 It has always been recognized that a king might easily degenerate into a tyrant, but his natural life is limited and can at a pinch be artificially shortened. **1966** M. R. D. Foot *SOE in France* iv. 88 It could carry two passengers easily, three at a pinch, or four in a crisis. **1977** *Belfast Tel.* 17 Jan. 13/1 The Beetle cabriolet is a four-seater which can take five at a pinch.

6*. *slang.* Something easy to accomplish or attain; a certainty.

1886–96 in Farmer & Henley *Slang* (1902) V. 205/2 The race would be a pinch, Sir, barring accident or spill. **1899** 'G. G.' *Winkles* vi. 72 Harkaway for the Scurry Handicap at Landown, good, a 'pinch'; go nap on it! **1903** A. M. Binstead *Pitcher in Paradise* xii. 280 Sustained by the conviction that he had made his match a 'pinch' indeed.

II. 9. For 'Now *dial.*' read 'Now chiefly *dial.*, *Austral.*, and *N.Z.*' Also, a steep hill. (Further examples.)

1848 H. W. Haygarth *Recoll. Bush Life Austral.* xii. 126 As we approached the end of our journey we came to one or two 'pinches', which is the colonial term for steep hills. **1862** J. S. Dobie *S. Afr. Jrnl.* (1945) 30 At an ugly pinch of boulders we had another stick up. **1898** *Longman's Mag.* Nov. 51 Shepherd Robbins shambling slowly down the steep 'pinch' of road that led to the farm gate. **1901** M. Franklin *My Brilliant Career* xxvi. 220 Don't push him too quickly up that pinch by Flea Creek, or he might drop dead with you. **1928** 'Brent of Bin Bin' *Up Country* xv. 253 She..could carry him up pinches so steep that no amateur could have sat on at such an angle. **1950** *N.Z. Jrnl. Agric.* Aug. 162 [inset] The steep pinches and faces take their toll of injuries and deaths.

10. Also, in *Geol.*, a similar narrowing of any stratum; freq. in phr. *pinch and swell*. Cf. *pinch-out. (Earlier and later examples.)

1873 J. H. Beadle *Undevel. West* 333 All the strange terms in mining parlance: 'true lodes, fissure-veins, pinches,..variations and sinuosities'. **1916** F. H. Lahee *Field Geol.* vi. 140 That the country rock was warm and plastic enough to be deformed by the force of intrusion is suggested by the pinch-and-swell form of many pegmatite dikes in schists. **1955** *Jrnl. Geol.* LXIII. 520/1 The pinch-and-swell structure so commonly developed in conformable pegmatites and quartz veins. **1972** L. E. Weiss *Minor Struct. Deformed Rocks* 15 Structures closely related to boudins also formed in progressively extended layers are..'pinch and swell' structures or 'necks'.

10*. *Electronics.* (See quot. 1973.)

1941 A. V. Eastman *Fund. Vacuum Tubes* (ed. 2) ii. 22 All the electrodes are supported by wires held in a glass 'pinch' at the base of the tube and by a mica disk at the top. 1954 *Electronic Engin.* XXVI. 16/1 Electrical leakage may be due to..getter on the pinch and micas of the valve. 1973 *Gloss. Electrotechnical, Power Terms (B.S.I.)* 1. vi. 19 *Pinch*, a flat fused glass seal forming part of the foot through which pass the leads from the electrodes to the pins in the base.

10.** *Physics.* A cylindrical or toroidal plasma confined by the pinch effect.

1951 *Proc. Physical Soc.* B. LXIV. 161 The discharge becomes brighter when it is contracted, and the brightness and sharpness of the 'pinch' increase with decrease in pressure. 1966 F. I. Boley *Plasmas* ii. 38 The kink instability of the plasma pinch..is an example of a large class of instability phenomena that is important to the dynamics of plasma. 1971 *Nature* 16 July 152/2 Research on high beta toroidal confinement is still in an early stage as groups previously working on linear theta pinches move into the field.

11. *with a pinch of salt* (fig.): see *SALT *sb.*¹ 2 d.

pinch, *v.* Add: **I. 2. e.** *to pinch off*: *intr.*, to undergo a localized constriction that progresses until separation into two portions occurs; to become separate in this way; also *trans.*, to detach in this way.

1687 [in Dict., sense 2]. 1910 *Jrnl. Morphol.* XXI. 278 (*caption*) Megakaryocyte showing a platelet in process of pinching off from a pseudopod. *Ibid.*, Various phases are shown in the process of pinching off portions of the cytoplasm of the thrombocytes to form blood platelet-like corpuscles. 1952 [see *PINCH-OFF]. 1956 *Essays in Crit.* VI. 10 Science begins to appear in the odd role of being pinched off and occupying the lonely end of a polar opposition to religion. 1956 L. P. Hunter *Handbk. Semiconductor Electronics* iv. 29 If the bias on the gate is high enough, the depletion region of the encircling PN junction becomes thick enough to 'pinch off' the channel through which the working current flows. 1956 *Jrnl. Biophysical & Biochem. Cytol.* II. Suppl. 107 The invaginated membrane is pinched off, resulting in the formation of an intracellular vacuole. 1959 *Bell Syst. Techn. Jrnl.* XXXVIII. 777 If sufficiently high voltage is applied, the channel will 'pinch off' and its current will essentially saturate. 1962 *Science Survey* III. 170 The living endothelial cell has almost nothing in its cytoplasm but masses of tiny smooth vesicles pinching-off and opening at the cell surfaces. 1966 *McGraw-Hill Encycl. Sci. & Technol.* X. 233/1 Once the process starts, the pressure at a narrow neck in the ring of fused metal is able to squeeze out the fluid metal until the neck pinches off completely, cutting off the current. 1979 *Nature* 11 Jan. 91/1 One suggestion has been made that the clathrin physically pinches off a membrane vesicle.

II. 10. b. *to pinch pennies* (or *a penny*): to be penny-pinching or parsimonious.

1942 E. Paul *Narrow St.* xix. 152 The surly Monsieur Salmon..complaining and pinching pennies as he made his purchases. 1962 J. D. MacDonald *Key to Suite* (1968) iii. 40 I'm not about to pinch a penny on a thing like this. 1973 J. Cleary *Ransom* xi. 255 'This city is too expensive for a cop on my pay. Especially when it almost cost me my wife, too.' 'He's always pinching pennies,' said Lisa.

III. 14. Also, in *Geol.*, said of strata generally; also with *down*. Cf. *LENS *v.* (Earlier and later examples.)

1867 J. A. Phillips *Mining & Metallurgy Gold & Silver* iv. 56 The lode, which is eight feet wide on the north side of the Eureka, pinches out very rapidly in that direction. 1916 F. H. Lahee *Field Geol.* ix. 240 Sometimes strata are irregularly thinned and thickened so that they 'pinch and swell', as seen in cross section. 1923 *Ibid.* (ed. 2) v. 88 If a stratum continues to thin out in a certain direction,..it may finally 'pinch out' or 'lens out' altogether. 1928 W. A. Chalfont *Outposts of Civilization* 82 High-grade veins were followed as they pinched down, even to half inch seams which were profitably 'spooned out'. 1945 *Bull. Amer. Assoc. Petroleum Geologists* XXIX. 1563 The reservoir bed must pinch out in all updip directions. 1961 *Jrnl. Geol.* LXIX. 339/1 The layered marine sediments pinch out to the south.

15. a. (Earlier and later examples.)

1656 *Witty Rogue Arraigned* xxi. 30 Pinch'd the Cully of a Casket of Jewels. 1900 [see *COMMANDEER *v.* c]. 1930 J. B. Priestley *Angel Pavement* ix. 474 Buying cars that have been pinched like that is a mug's game. 1936 [see *FAT *sb.*¹ 2 d]. 1969 *Listener* 24 July 103/2 'This was by car I take it—was there petrol?' 'Well, we somehow managed to pick it up.' 'You mean pinch it?' 1979 'C. Brand' *Rose in Darkness* xiii. 189 You simply pinched it from a shop.

b. (Earlier and later examples.)

1837 *Sessions Papers Cent. Criminal Court* 4 Dec. 157 D——d if I'm not *pinched* for housebreaking at last. 1925 H. L. Foster *Trop. Tramp with Tourists* 41 A traffic policeman had stopped us. But not to pinch us for speeding. 1932 T. S. Eliot *Sweeney Agonistes* 28 These fellows always get pinched in the end. 1938 [see *GUY *v.*⁴]. 1955 *Times* 12 Aug. 5/4 He explained that Heard gave him the tobacco and then put in another officer to 'pinch' him. 1979 N. Hynd *Fake Flags* iv. 25 Nobody knew what night Vassiliev was going to be pinched.

pinch-. Add: **pinch-batter** = **pinch-hitter*; **pinch-bottle** *U.S.*, a bottle with indented sides, *spec.* a whisky bottle; so, by metonymy, whisky; **pinch-face**, a pinch-faced person; **pinch-faced** *a.*, having the features pinched or emaciated; **pinch-fist** (later examples); **pinch-hit** *v. intr.* (orig. *U.S.*), in baseball, to bat as a substitute for another batter, esp. at a critical point in the game; freq. *transf.*, to act as a substitute, esp. in an emergency; to stand in *for* someone; also as *sb.*, a hit made by a substitute batter; so **pinch-hitter, -hitting**; **pinch-pleat**, one of a cluster of small pleats, used esp. in curtains; hence **pinch-pleated** *a.*; **pinch-point,** (*b*) a point of congestion, confusion, or difficulty; **pinch roll,** (*a*) each of a pair of rolls, usu. hydraulically controlled, which grip the material passing between them; (*b*) = **pinch roller*; **pinch roller,** in a tape recorder or tape deck, a spring-loaded roller which presses the tape against the capstan; **pinch-runner** *N. Amer.* (see quot. 1961); **pinch-waist,** a tightly-fitted waist; also *attrib.*; hence **pinch-waisted** *a.*; **pinch wheel** = **pinch roller*.

1928 *Chicago Tribune* 5 Oct. 26/1 The pinch batter exercised rare judgment and drew a pass. 1974 *Spartanburg* (S. Carolina) *Herald* 22 Apr. B1/2 Franklin walloped five home runs in the six-day series. He started five of the games and was a pinch-batter in the other two. 1939 C. Morley *Kitty Foyle* x. 103 Mr Rittenhouse gave him a lift to the bootlegger's, ordered a whole case of pinchbottle on Pop's recommendation..and invited us down to..dinner. 1940 E. Hemingway *For whom Bell Tolls* xvi. 204 'When we have electricity again, what a lamp we can make of this bottle.' She looked at the pinch-bottle admiringly. 1963 *New Yorker* 8 June 57 We made it [*sc.* men's toiletry] clean-smelling and gutsy and put it up in hefty glass pinch bottles. 1917 W. Owen *Let.* 27 Sept. (1967) 497, I called for the poet—a wizened little pinch-face, about two feet high! 1863 G. M. Hopkins *Let.* 22 Apr. (1956) 73 Woolcomb,..a pinch-faced old man, whom everybody likes as much as they yawn over his divinity lectures. 1943 *Gen* 10 Apr. 23/1 Pinch-faced, thin-limbed, he seemed to be needing a good dinner. 1917 'H. H. Richardson' *Fortunes R. Mahony* I. ix. 84 They were pinchfists when it came to parting with their money. 1978 *Verbatim* May 1/1 One incurs hostile mutters of *sparethrift* or *sparegood*, *scrapepelf* or *scrapegood*, *pinchfist* or *skinflint*. 1931 *Kansas City* (Missouri) *Star* 17 Dec. 24 John Neilson gave the talk,.. but I thought they were just using him to pinch-hit for Bo McMillin. 1948 *Capital-Democrat* (Tishomingo, Okla.) 17 June 5/5 Duggan Smith, pinch hitting for W. C. Whiteley, had sent a high drive out to center. 1957 R. Longrigg *Switchboard* iv. 59 'I wonder if you can help me out of a jam?..I'm supposed to be lunching with a man called Robinson... I now find I can't do it. I've made some kind of nonsense with the dates—' 'Shall I pinch-hit?' 1966 *N.Y. Times* (Internat. ed.) 22 Apr. 12/1 California's Joe Adcock hit a pinch-hit single in the bottom of the eleventh inning. 1973 E. Taylor *Serpent under It* (1974) x. 156 'With no secretary, how do we find out where they are?' 'They've got some old girl from the History Department pinch-hitting.' 1974 *Cleveland* (Ohio) *Plain Dealer* 13 Oct. 14-C/4 If we were ahead by only one run, I would have pinch hit for Rollie in the top of the ninth because Catfish was ready. 1976 *Billings* (Montana) *Gaz.* 24 June 1-D/3. With one out, Kautzmann delivered a pinch-hit single to left field to drive in Art LaGaly with the lead run. 1912 *Lit. Digest* 10 Aug. 238/2 Things did not run very smoothly the famous 'Cub' pinch hitter himself tells us. 1939 Ade *Let.* 7 June (1973) 210, I have tried to get either 'Chick' Evans or 'Red' Grange for the June party but without success. We don't want to miss a month and so..I am offering myself as a pinch-hitter. 1970 *Globe & Mail* (Toronto) 26 Sept. 39/8 Singles, by pinch-hitter Mack Jones, Bob Bailey and pinch-hitter Jim Fairey in the bottom of the ninth produced one run. 1976 *Washington Post* 19 Apr. D4/1 Major league baseball has designated ABC as its pinch hitter for NBC on Monday night television this season. 1931 M. Loeb *Please stand By* I. x. 118 Just because Will Rogers did some pinch-hitting for Fred Stone and got away with it is no reason why you should send some college cub-reporter up here to cover the greatest discovery on the air. 1947 *Los Angeles Times* 6 Oct. 11/7 Brown..kept up his red-hot pinch-hitting when he singled for Phillips in the third. 1974 *Anderson* (S. Carolina) *Independent* 23 Apr. 3A/5 At present, the recreational program is being operated by Assistant Director Benny Burrell, 'who is doing a good job of pinch hitting,' said Evatt. 1958 *Times* 23 June 11/3 For long windows the pinch-pleat treatment in the last small sketch is very effective. 1973 *Guardian* 29 Mar. 13/3 The curtains are backed by Sekers's own make-up service which can provide pinch or pencil pleated or gathered curtains to the customer's own specifications. 1961 *Times* 15 Aug. 10/3 No loading or unloading at all will be permitted at important intersections and other 'pinch points'. 1965 *New Society* 11 Nov. 10/3 Oliver Cox considers heating one of the major 'pinch-points' in building today. 1973 *Country Life* 31 May 1527/3 The bridge is a pinch-point and congestion soon builds up. 1953 *Engineer* 23 Oct. 526/2 The slabs have to be pressure fed into the rolls at a constant rate by pinch or feed rolls. 1958 H. G. M. Spratt *Magnetic Tape Recording* vii. 202 The tape..is held in tight contact with it [*sc.* a roller] by means of a spring-loaded, free-running, rubber or rubber-covered roll generally referred to as the pinch roll. 1960 A. R. Bailey *Text-bk. Metall.* (ed. 2) 402 The steel slabs pass directly from the reheating furnace to a pair of pinch rolls that grip the metal and force it into the mill. 1964 A. A. McWilliams *Tape Recording* viii. 160 The tape is held in firm contact—with a rotating spindle—the capstan—by means of a rubber-tyred pressure roller, or pinch roll. 1969 W. R. R. Park *Plastics Film Technol.* ii. 41 At the same time, the pinch rolls are pulling the film away. 1949 S. J. Begun *Magnetic Recording* vi. 174 An interesting feature of this equipment is the..spindle which acts as a capstan as soon as a rubber-tyred pinch roller sandwiches the tape against it. 1961 G. L. Davies *Magnetic Tape Instrumentation* iv. 79 The pinch roller.. is operated by a solenoid. 1974 *Physics Bull.* Apr. 151/1 Another feature is the vacuum buffered, low tension drive which eliminates pinch-rollers and associated mechanical problems. 1961 J. S. Salak *Dict. Amer. Sports* 328 *Pinch runner* (baseball, softball), a substitute runner for a teammate who has reached a base, the original runner being out of the game from that time on. 1970 *Toronto Daily Star* 24 Sept. 17/1 Cash wound up at second and pinch-runner Freddie Patek at third. 1976 *Billings* (Montana) *Gaz.* 20 June 4-E/4 Tim Foli singled off starter Tommy John to open the inning and was replaced by pinch-runner Jim Lyttle. 1969 *Time* 31 Jan. 55 Among the customers for his men's clothes—distinguishable by their long jackets and pinch waists—are movie stars, [etc.]. 1975 M. Bradbury *Hist. Man* i. 11 She wears a white pinch-waist, full-length raincoat. 1977 *Times* 5 Oct. (Fashion Suppl.) p.i/7 The lavishly lapelled, pinch-waisted..caricature..of the Italian style. 1962 H. B. Hadden *High-Quality Sound Production & Reproduction* x. 160 The tape is held in contact with the capstan by the pinch wheel. 1968 C. N. G. Matthews *Tape Recording* iv. 36 The capstan and pinch wheel are disengaged.

pi·nchable, *a.* [f. Pinch *v.* + -able.] That may be pinched; that invites pinching. Hence **pi·nchably** *adv.*

1921 *Public Opinion* 15 July 56/1 The greater the pinchable surface, the sharper the tweak that you will get. 1939 Joyce *Finnegans Wake* (1964) III. 417 As entomate as intimate could pinchably be. 1977 P. Ustinov *Dear Me* ix. 120 This volatile hedonist, with his unending stream of pinchable starlets.

pinch-bug. *U.S.* A stag-beetle belonging to the family Lucanidæ.

1856 'Mark Twain' *Let.* 25 May in *Iowa Jrnl. Hist.* (1929) XXVII. 423 A tenor and bass duet by thirty-two thousand locusts and ninety-seven thousand pinch-bugs was sung. 1870 E. Eggleston *Bk. Queer Stories* ix. 74 We came to a log on which two of that sort of beetles that children call 'pinch-bugs', were fighting. 1876 'Mark Twain' *Tom Sawyer* v. 47 It was a large black beetle with formidable jaws—a 'pinch-bug' he called it. 1915 W. A. Bryan *Nat. Hist. Hawaii* xxxi. 417 The stag-beetles or pinch bugs, so called on account of their large mandibles. 1959 A. B. & E. B. Klots *Living Insects of World* 132/2 There is little doubt that the 'pinch bug' Tom Sawyer took to church was a stag beetle.

pinched, *ppl. a.* Add: **1.** Also with *in*.

1920 [see *dock-glass* s.v. *DOCK *sb.*³ 7]. 1941 *Amer. Speech* XVI. 67/1 Avoid pinched-in-waistlines for teen ages.

e. *Physics.* Confined by the pinch effect.

1907 *Trans. Amer. Electrochem. Soc.* XI. 331, *C* is the column of liquid conductor,..and *P* is one of these pinched contractions. 1951 *Proc. Physical Soc.* B. LXIV. 161 Just after the breakdown the discharge.. is observed to contract into a narrow filament; the discharge does not stay 'pinched' but immediately expands again, and proceeds to oscillate. 1959 *Daily Tel.* 23 July 7/8 But, unlike Zeta, the pinched gas will be stable. 1962 *Times* 28 Apr. 8/4 The first photograph of a 'pinched' lightning discharge has been obtained. 1973 Kettani & Hoyaux *Plasma Engin.* vii. 206 Consider a pinched column of fully ionised plasma.

7. Of paper: slightly smaller than the normal size (see quots.).

1893 J. Kay *Paper* 100 Sizes of Papers... Demy.. Post..Pinched Post..Foolscap. 1894 G. Clapperton *Pract. Paper-Making* 193 Sizes of Lined Papers. Pinched for 8vo Expansion by 14½ in. 1926 *Paper Terminol.* (Spalding & Hodge) 20 *Pinched post*, a standard size of writing paper measuring 14½ × 18½ in. 1952 E. J. Labarre *Dict. Paper* (ed. 2) 199/1 *Pinched post*, a size of writing paper standardized at 18½" × 14½", but with variants still in use, also for drawings... For 'pinched' the same word is used in Dutch: *geknepen*, when a sheet is slightly reduced from the standard or normal dimensions for that size. 1962 F. T. Day *Introd. to Paper* vii. 70 An even greater variety of sizes is covered by these names by the addition of qualifying words—Single or Half, Double or Quad, Small or Large, Extra or Super, Pinched or Reduced. *Ibid.*, Pinched Post..14½ × 18½ in. Pinched Post (Double)..18½ × 29 in.

pinch effect. [f. Pinch *sb.* + Effect *sb.*] **1.** *Physics.* The constriction exhibited by a fluid through which a large electric current is flowing, caused by the attractive force produced by the interaction of the current with its own magnetic field.

[1907 E. F. Northrup in *Physical Rev.* June 474 Some months ago, my friend, Carl Hering, described to me a surprising and apparently new phenomenon which he had observed. He found, in passing a relatively large alternating current through a non-electrolytic, liquid conductor contained in a trough, that the liquid contracted in cross-section and flowed up hill lengthwise of the trough... Mr. Hering suggested the idea that this contraction was probably due to the elastic action of the lines of magnetic force which encircle the conductor... As the action of the forces on the conductor is to squeeze or pinch it, he jocosely called it the 'pinch phenomenon'. 1907 C. Hering in *Trans. Amer. Electrochem. Soc.* Sept. 331 As the column of liquid looks as though it were being pinched by some mysterious and invisible force, the writer termed it the 'pinch phenomenon'. *Ibid.* 337 If to this field there is added another one, I see no reason why it should not add to the pinching effect.] 1911 G. H. Clamer in *Ibid.* XIX. 264 The heavy current.. rapidly brings these columns of metal to the liquid condition, and produces therein the 'pinch' effect. 1956 *Sci. News Let.* 15 Sept. 174/1 Generating the high heat.. requires containers that will not melt or be otherwise affected. Using the 'pinch effect' would seem to eliminate the container problem, since the reacting gas column would contract to contain itself, thus not touch any walls.

1958 *Listener* 25 Sept. 454/2 The Americans and Russians independently have developed a principle different from Zeta, although both are working on the Zeta principle of what is known as the pinch effect. 1966 *McGraw-Hill Encycl. Sci. & Technol.* X. 233/1 The force of the pinch effect has..been known to manifest itself by a crushing of tubular conductors exposed to large impulsive currents such as occur in lightning strokes or high-power short circuits. 1972 *Physics Bull.* Feb. 83/2 Flash photolysis features in the Exhibition. Chelsea Instruments is showing its apparatus (based on its Garton flash tube) which utilizes the pinch effect (as in some thermonuclear apparatus) to produce high intensity.

2. The slight narrowing of a record groove caused by the transverse movement of the cutting stylus, resulting in a vertical movement of the stylus at that point during playing.

1935 H. C. Bryson *Gramophone Record* x. 271 Hill and Dale cut records possess great advantages... There is no pinch effect. 1965 J. Walton *Pick-Ups* iii. 40 A mono pick-up (as well as of course the stereo pick-up) should have some vertical compliance and low vertical mass if pinch effect is not to cause damage and excessive 'needle talk'. 1975 G. J. King *Audio Handbk.* viii. 192 Because the groove is cut by a chisel-shaped tool whose face is at right-angles to the motion of the record, the groove width decreases along the sloping sides of the waveform... This, called 'pinch effect', results in vertical oscillation of the replay stylus at a frequency twice that which utilizes the pinch effect (as in some thermonuclear apparatus) to produce high intensity.

pinching, *vbl. sb.* Add: **6. pinching bug** = *PINCH-BUG.

1850 L. H. Garrard *Wah-to-Yah* xix. 253 Noah was so hurried to git the yelaphants, pinchin bugs, an' sich varment aboard. 1877 R. J. Burdette *Rise & Fall of Mustache* 77 That Bilderback boy..put a pinching-bug as big as a postage-stamp down a boy's back. 1928 Metcalf & Flint *Destructive & Useful Insects* i. 16 A certain amount of pain may result from mere mechanical injury by insects as when a boy finds a 'pinching bug' for the first time. 1954 Borror & DeLong *Introd. Study Insects* xxii. 381 These large brownish beetles [*sc.* stag beetles] are sometimes called pinchingbugs because of the large mandibles of the male.

pinch-off. *Electronics.* Also **pinch off.** [f. vbl. phr. *to pinch off* s.v. *PINCH *v.* 2 e.] In a field-effect transistor, the meeting of the two non-conducting depletion layers that border the channel, such that little further increase in current is achieved by increasing the drain voltage. Freq. *attrib.*, esp. in **pinch-off voltage**, the reverse bias that must be applied to the gate to achieve pinch-off and prevent the flow of current through the channel (equal to the drain voltage at which the current saturates when there is no bias applied to the gate).

1952 W. Shockley in *Proc. IRE* XL. 1367/1 W_s is the magnitude of reverse bias required to make the space charge penetrate the entire *p*-region. We shall refer to it as the 'pinch-off voltage' since it is the voltage that will reduce the channel to zero and pinch off the conducting path. *Ibid.* 1374/1 We shall consider a structure operated with the drain beyond pinch-off. 1956 L. P. Hunter *Handbk. Semiconductor Electronics* iv. 31 For the depletion region to fill the entire channel $b = 0$ and $V(b) = V_s$... V_s is called the pinch-off voltage. *Ibid.* 32 As the bias..at the drain end is increased beyond V_s, the pinch-off condition moves toward the source. The field between the drain end and the point along the channel where pinch off is effectively reached must be large enough to maintain the current *I* flowing in the open part of the ca nnel. 1962 J. Evans *Fund. Princ. Transistors* x. 228 If V_G is large enough, pinch-off occurs at all values of V_D—the situation when $V_G = W_s$. 1972 *Field-Effect Transistors* (Mullard Ltd., London) ii. 8 The line joining the various drain-source voltages $V_{DS(P)}$ at which this pinch-off occurs (the pinch-off limit) is shown as a broken line in Fig. 7... To the right of the pinch-off limit —in the pinch-off region—the drain current only increases very slightly. 1974 Harvey & Bohlman *Stereo F.M. Radio Handbk.* v. 88 The voltage corresponding to zero I_d is called the pinch-off voltage.

pinch-out. *Geol.* [f. vbl. phr. *to pinch out* s.v. Pinch *v.* 14 in Dict. and Suppl.] A narrowing of a stratum, vein, or other body of rock to the point of extinction.

1928 in *Funk's Stand. Dict.* 1941 *Bull. Amer. Assoc. Petroleum Geologists* XXV. 1258 A pinch-out of a reservoir sand on a structural nose would..be considered a stratigraphic trap of a less perfect type. 1962 *Courier-Mail* (Brisbane) 15 Dec. 9/3 Seismic surveys had outlined a large closed structure which also had good stratigraphic pinch-out possibilities. 1974 P. L. Moore et al. *Drilling Practices Manual* xii. 300 These permeability barriers may be faults, folds, salt domes, or permeability pinchouts.

pinchpenny. Delete †*Obs.* and add later examples.

1931 'D. Stiff' *Milk & Honey Route* viii. 85 You can always tell the home of the pinchpenny by the narrowness of the eaves. 1948 *Sun* (Baltimore) 8 Apr. 1/7 Lee M. Wiggins, Under Secretary of the Treasury, today warned Congress that a pinchpenny policy toward the Bureau of Internal Revenue will cost the nation billions in evaded taxes. 1955 T. Sterling *Evil of Day* i. 13 No pinchpenny ever knew anything about pennies. You have to spend them to know. 1977 *New Yorker* 25 July 36/2 In contrast to pinchpenny P.R. routine of the environmentalist groups..the Dow P.R. material is handsomely bound, with expensive paper, elegant design, and, sometimes, four-color illustrations. 1978 'F. Parrish' *Sting of Honeybee* iii. 39 The fences were a weird, pinchpenny patchwork, but they kept the ponies in.

pincushion. Add: **2.** (Later examples.)

1898 Britton & Brown *Illustr. Flora Northern U.S.* III. 290 Field Scabious..[is also called] blue caps, gypsy- or egyptian-rose, pincushion. 1965 *Austral. Encycl.* VII. 118/2 Pincushion, Australian, or blue pincushion, the popular name for *Brunonia*,..low herbaceous perennials, bearing blue flowers in very effective long-stalked heads; the projecting styles resemble pins sticking in a cushion.

4. *pincushion-flower* (later examples).

1911 W. R. Guilfoyle *Austral. Plants* 201 *Hakea laurina* 'Pin Cushion Flower' (evergreen shrub or tree, 10 to 30 ft.), f[lowers] crimson—W. Aust. 1917 L. H. Bailey *Stand. Cycl. Hort.* VI. 3106/1 (caption) *Scabiosa atropurpurea.*—The mourning bride or pin-cushion flower. 1938 T. Y. Harris *Wild Flowers Austral.* 25 Pincushion Flower (*Hakea laurina*)... Pink blossoms with cream stigmas projecting for some distance and the flowers in globular heads have earned the name for this most attractive plant. 1967 A. M. Blombery *Guide Native Austral. Plants* III. 265 *H[akea] laurina.* Pincushion Hakea. A large bushy shrub with flat, lanceolate, greenish-red leaves and pink pincushion-like flowers. 1972 F. Perry *Flowers of World* 101/1 *S[cabiosa] caucasica*, the Pincushion Flower, was introduced to Britain in 1591 and has become a favourite for border and cut-flower work.

b. pincushion distortion, a form of optical distortion in which a square is reproduced with sides curved inwards.

[1886 J. H. Dallmeyer *Choice & Use Photogr. Lenses* 22 If the stop is placed at the same distance *behind* the lens..the result is the opposite kind, or 'pincushion'-shaped distortion.] 1892 A. Brothers *Photography* I. iii. 48 The effect consists..in the curvature of the images of straight lines produced by marginal rays causing barrel-shaped or 'pincushion distortion'. 1903, 1953 [see *barrel distortion* s.v. *BARREL *sb.* 11]. 1965 *Wireless World* July 58 (Advt.), No oscilloscope is better than its tube. Build in all the circuit refinements you like, if the tube suffers from..pin-cushion distortion..your efforts are in vain. 1972 Williams & Becklund *Optics* viii. 188 An aperture stop located between a positive lens and the image increases pincushion distortion.

pinda¹, pindar, pinder. (Earlier and later examples.)

1696 J. Ovington *Voy. to Suratt* 77 Sometimes they feast with a little Fish, and that with a few Pindars is esteemed a splendid banquet. These Pindars are sown under ground, and grow there without sprouting above the surface. 1926 J. K. Strecker in J. F. Dobie *Rainbow in Morning* (1965) 56 In the valleys of the Red River of Louisiana and the Sabine River of Louisiana and Texas, are to be found negroes who use many African words, the inheritance of their ancestors. A white man is a 'buckra'. .. A ground-nut (peanut) is a 'pinda'. 1938 M. K. Rawlings *Yearling* xviii. 214 The field of pindars was not doing so well. 1977 McDavid & O'Cain in S. Greenbaum *Acceptability in Lang.* viii. 115 The majority of the uncultured judge..*groundnut* and *pinder* as old.

‖**pinda²** (pi·ndā). *India.* Also 8 **peenda**, 9 **pindee**. [Skr. *piṇḍa* lump.] (See quots.)

1785 C. Wilkins *Bhăgvăt-Gēētā* 139 The Hindoos are enjoined by the *Vēds* to offer a cake, which is called *Pĕĕndă*, to the ghosts of their ancestors, as far back as the third generation. 1796 W. Jones tr. *Inst. Hindu Law* iii. 67 Sages have distinguished the monthly *sráddha* by the title of *anwáhárya*, or *after eaten*, that is, eaten after the *pinda* or ball of rice. 1811 W. Ward *Acct. Writings, Relig. & Manners Hindoos* II. v. 550 The place where the fire was kindled is plentifully washed with water, after which the son of the deceased performs pindee, viz. he makes two balls of boiled rice, and, repeating a mŭntrŭ, offers them to, or in the name of his father and mother, and lays them on the spot where they were burnt. 1877 M. Monier-Williams *Hinduism* v. 68 The offering of the Piṇḍa, or ball of rice, &c., to deceased fathers at a S'râddha is of great importance in regard to the Hindū law of inheritance. 1901 *Westm. Gaz.* 14 Nov. 9/2 The 'pindas' offered to their deceased ancestors were placed on plantain or 'jack fruit' leaves. 1909 *Encycl. Relig. & Ethics* II. 27/1 How closely this [Lithuanian cake for the dead] corresponds to the Indian *piṇḍa*, which is so characteristic of the Indian worship of the dead. 1964 R. Antoine in De Smet & Neuner *Religious Hinduism* xv. 166 *Piṇḍas* (rice-balls) are then smeared with ghee, collyrium and oil, and dressed with a tuft of wool, so as to represent the ancestors. 1968 B. Walker *Hindu World* II. 149 On the first day after death a round ball of rice or flour moistened with milk and water and known as the *piṇḍa* is offered to the preta.

pindan (pi·ndæn). *Austral.* [Aboriginal.] The type of vegetation characteristic of arid areas of Western Australia; hence, the region itself. Also *attrib.*

1934 T. Wood *Cobbers* iv. 46 His black trackers were making boomerangs... Pindam [*sic*] gum: hard red wood, shaped from a knee in the timber. 1937 W. Hatfield *I find Australia* xxiv. 315 Pindan is not really the name of the type of tree, but merely the native name for 'dry country', though general usage adopts the word as a description of the thin growth of whipstick saplings of the bloodwood and box type of eucalypt. 1945 Baker *Austral. Lang.* xiii. 224 *Pindan* is the blacks' name for the desert country inland from Broome, W.A., so the whites call the Kimberley natives *pindan blacks*. To *live on the pindan* is to wander aimlessly in the Westralian outback. 1955 J. Cleary *Justin Bayard* xi. 153 They would be out in the pindan watching the homestead. 1959 *Observer* 17 May 8/3 From the pindan scrub..these ancient monuments [*sc.* mountains] rise. 1978 O. White *Silent Reach* ii. 22 Half a million acres of pindan country ..carried two thousand head of merino sheep in a good season.

pindolol (pi·ndŏlǫl). *Pharm.* [f. the initial letter of Propane, Propyl + Indol(e *sb.* + -ol.] The compound 1-indol-4-yloxy-3-isopropyl-aminopropan-2-ol, $C_{14}H_{20}N_2O_2$, which is an adrenergic blocking agent with uses similar to those of propranolol.

1971 *Biochem. Pharmacol.* XX. 2749 (*heading*) Influence of INPEA, Pindolol and Propanolol on the chronotropic and metabolic responses to β-adrenergic stimulation in intact rats. 1976 *Lancet* 11 Dec. 1298/1 Concomitant treatment of hypertension was started with 15 mg pindolol, 5 mg amiloride hydrochloride, and 50 mg hydrochlorothiazide daily. 1979 *Experientia* XXXV. 250/1 Pindolol is one of the beta-adrenoceptor blockers now widely used all over the world.

pine, *sb.*² Add: **5.** (Earlier and later examples.)

1657 *Bk. of Continuation of Forreign Passages* 46 Fruits ..Pyne, the best that ever was eat, in season almost all the year long. 1920 'K. Mansfield' *Bliss* 35 He bought a pineapple... The oysters and the pine he stowed away.. under the front seat. 1954 *Farmer's Guide* (Jamaica Agric. Soc.) 392 The Sugar Loaf is only of importance as a fresh fruit. It is not a suitable pine for canning.

7. a. *pine bark* (earlier and later examples), *board* (earlier and later examples), *box*, *forest* (earlier and later examples), *hill*, *-log*, *plain* (earlier and later examples), *stump* (earlier examples), *thicket* (earlier example), *timber* (earlier and later examples); *pine-covered* (earlier and later examples), *-grown*, *-panelled*, *-scented* adjs.; **pine-blister(-rust)**, a fungus disease of pine trees, caused by species of *Cronartium* (*Peridermium*) characterized by yellowish swellings on the bark; formerly also applied to needle rust caused by *Coleosporium* species; **pine-borer,** a longicorn beetle, whose larvæ live in pine trees; **pine-chafer,** for *Anomala pinicola* substitute *Anomala oblivia*; (examples); **pine-creeper, -creeping warbler** = *pine warbler*; **pine drape** *U.S. slang* = **pine overcoat*; **pine finch,** (*b*) (examples); **pine green,** the colour of pine needles; **pine gum,** substitute for def.: *U.S.*, the resin or turpentine obtained from several species of pine, esp. the slash pine, *Pinus caribæa*, and the southern pine, *P. echinata*; (examples); **pine lappet (moth),** a large brown European moth, *Dendrolimus pini*, whose larvæ feed on pines; also called the pine tree lappet; **pine-marten** (later examples); **pine overcoat** *U.S. slang*, a coffin; **pine rust,** a disease of pine trees caused by a rust fungus, e.g. **pine blister*; **pine savanna(h)** *U.S.*, a savannah in which pines are the prevailing trees; **pine saw-fly,** substitute for last part of def.: esp. species of *Diprion* or *Gilpinia*; (earlier and later examples); **pine siskin,** delete the first two Latin names given, leaving *Spinus pinus*; (examples); **pine-snake,** for *Pityophis* substitute *Pituophis*; also *attrib.*; (earlier and later examples); **pine straw** *U.S.*, (esp. dried) pine needles; **pine swamp** *U.S.*, a low-lying or marshy piece of ground on which pine-trees grow; **pine tags** *U.S.*, pine needles; so *pine-tag* attrib.; **pine warbler** (examples); **pine-weed,** for the Latin names given substitute *Hypericum gentianoides*; = *NIT-WEED; (earlier and later examples); **pine-weevil** (examples).

1709 J. Lawson *New Voy. to Carolina* 177 They make use of pine bark. *a* 1816 B. Hawkins *Sk. Creek Country* (1848) 71 They are covered with clay and that with pine bark. 1973 R. Lockridge *Not I, said Sparrow* (1974) x. 153 Bisecting the tracks were traces of pine bark... The pine bark—what's left of a bridle path. 1978 *Country Life* 12 Oct. 1094/1 Pine bark was found to be one of the most important foods [of beavers]. 1889 H. M. Ward *Timber & Dis.* xii. 259 It is thus seen that the fungus *Peridermium Pini* was regarded as a parasite of pines, and that it possessed two varieties, one inhabiting the leaves and the other the cortex... The disease may be popularly denoted 'Pine-blister'. 1894 W. Somerville tr. *Hartig's Textbk. Dis. Trees* I. 175 Three species of pine-blister-rust are to be distinguished in the cortex of trees. 1907 W. R. Fisher *Schlich's Man. Forestry* (ed. 2) IV. 441 Scots pines infected with this disease, which is very common in the British Isles and called pine-blister, are termed foxy trees by English foresters. 1929 T. Thomson tr. *Büsgen's Struct. & Life Forest Trees* xiv. 411 Individual stems of the pine and their descendants are especially prone to the pine blister. 1637 *Early Rec. Dedham, Mass.* (1892) III. 39 To alowe for saweing Pyne board 5s. 1728 *New Hampsh. Probate Rec.* (1914) II. 344 Eight thousand feet of good and merchantable pine boards every year. 1870 De B. R. Keim *Sheridan's*

Troopers on Borders xix. 125 A neat coffin had been made of pine boards. 1938 L. Bemelmans *Life Class* III. v. 245 Their flooring was of scrubbed pine boards. 1862 *Rep. Comm. Patents 1861: Agric.* (U.S.) 614 The larvæ [*sic*] of this insect is evidently a pine-borer, for I have found it about saw-mills. 1884 *Rep. Comm. Agric.* (U.S. Dept. Agric.) 379 The Common Longicorn Pine-Borer.. is destructive to the white pine. 1977 *Listener* 20 Oct. 503/3 Only when he discovered a small collection of New Zealand insects..did he trace it to a specimen of the New Zealand pine-borer; *Prinoplus reticularis*. 1847 W. T. Porter *Quarter Race Kentucky* 86 Ar you a goin to tumtum all nite on that pot-gutted old pine box of a fiddle? 1867 O. W. Holmes *Guardian Angel* 406 The long pine boxes came by almost every train,—no need of asking what they held! 1890 N. P. Langford *Vigilante Days* II. xxv. 441 A company of twenty or more men approaching the station, bearing in their midst a long pine box. 1885 *Encycl. Brit.* XIX. 103/2 The pine-chafer..is destructive in some places. 1972 Swan & Papp *Common Insects N. Amer.* 435 The Pine Chafer, *A[nomala] oblivia*, is very similar [to the Oriental Beetle]..; it infests red, jack, and Scotch pines. 1820 M. Edgeworth *Let.* 5 Aug. (1979) 199 High pine covered mountains. 1955 E. Pound *Classic Anthol.* III. 184 High, pine-covered peak full of echos. 1731 M. Catesby *Nat. Hist. Carolina* I. 61 The Pine-creeper... They creep about Trees; particularly the Pine- and Fir-trees; from which they peck Insects. 1811 A. Wilson *Amer. Ornith.* III. 25 Pine-Creeping Warbler.. inhabits the pine woods of the Southern states. 1868 Pine-creeping warbler [in Dict.]. 1917 T. G. Pearson *Birds Amer.* III. 148 Pine Warbler. *Dendroica vigorsi*... Pine-creeping Warbler; Pine Creeper. 1945 L. Shelly *Jive Talk Dict.* 31 *Pine drape*, coffin. 1970 C. Major *Dict. Afro-Amer. Slang* 91 *Pine drape*, coffin. 1810 A. Wilson *Amer. Ornith.* II. 133 Pine Finch..seeks the seeds of the black alder. 1871 J. Burroughs *Wake-Robin* 78, I observed several pine finches a dark brown or brindlish bird. 1799 C. B. Brown *Arthur Mervyn* I. ii. 15 Betty Lawrence was a wild girl from the pine forests of New-Jersey. 1913 J. London *Valley of Moon* 472 Vainly Saxon's eye roved the pine forest in search of her beloved redwoods. 1973 M. Monsarrat *Bk. of Europe* 118/2 Vârmland, a gentle place of rolling farmland, pine forests and grand manor houses. 1978 D. Kyle *Black Camelot* xviii. 273 Sheltering in the Bavarian pine forests. 1970 *Globe & Mail* (Toronto) 28 Sept. 28/1 (Advt.), Reconditioned Alfa Romeo Sale..pine green. 1974 *Simpson (Piccadilly) Christmas Catal.* 12 Jacquard patterned cashmere sweater. Light brown/white, pine green/white. 1880 'Mark Twain' *Tramp Abroad* xxxv. 397 Some pine-grown summits behind the town. 1915 R. Lankester *Diversions of Naturalist* 4 That pine-grown land. 1855 W. G. Simms *Forayers* 434 A leetle pine-gum plaister on that head of yourn will stop up the sore places. 1921 *Frontier* (Missoula, Montana) May 5 The Sheep Eaters lived in tepees made of cedar thatched with moss and cemented by pine gum. 1938 M. K. Rawlings *Yearling* iv. 41 He sewed the two deepest cuts and rubbed pine gum into all of them. *a* 1652 *Rec. Early Hist. Boston* (1881) VI. 16 The land running northward upon a straite line untill it cometh to range even with north side of the shop..and foure accres more or lesse upon pine hill south. 1773 J. McAfee *Jrnl.* in N. M. Woods *Woods-McAfee Memorial* (1905) 436/1 We..crossed Cantucky river within 8 miles of pine hills and broken mountains. 1947 Dylan Thomas *Let.* 20 May (1966) 307 The pinehills are endless. [*c* 1760 B. Wilkes *Eng. Moths & Butterflies* I. i. 29 The Wild Pine-tree Lappit-moth.] 1824 J. Curtis *Brit. Entomol.* I. 7 (*heading*) Pine lappet. 1907 R. South *Moths Brit. Isles*, 1st Ser. 106 This is.. the 'Wild Pine tree Lappet Moth' and 'Pine Tree Lappet' of the more ancient authors. 1966 O. Kuthanová tr. *Moucha's Beautiful Moths* 106 Pine Lappett Moth... This moth is one of Central Europe's notorious pine forest pests. 1694 *Mass. Hist. Soc. Coll.* (1852) 4th Ser. I. 105 Ye town is incompass'd with a fortification, consisting of pine-logs. 1853 J. M. Neale in *Oxf. Bk. Carols* (1928) 271 Bring me flesh, and bring me wine, Bring me pine-logs hither. 1902 S. E. White *Blazed Trail* 266 The instant necessity was to get thirty millions of pine logs down the river. 1938 E. Ambler *Cause for Alarm* xvii. 281 We stood in front of the fire... There were two half-consumed pine logs hissing..on the top. 1978 *Country Life* 12 Oct. 1096/4 The pine logs are first debarked and then loaded into the kilns and burnt [for charcoal]. 1936 D. McCowan *Animals Canad. Rockies* xxvi. 230 The Pine marten may long continue to frequent the green solitudes. 1964 H. N. Southern *Handbk. Brit. Mammals* II. 358 Pine Marten (*M[artes] martes*), the only one occurring in British Isles, distributed in Europe down to Mediterranean. 1896 *Congress. Rec.* 20 Jan. 796/2 The bill provides that the Committee shall..get as cheap a coffin as it can bargain for..perhaps what they call in the army a pine overcoat. 1929 E. Linklater *Poet's Pub* xiv. 259 The room..was..pine-panelled. 1952 M. Laski *Village* vi. 106 Delicate satinwood furniture against the pine-panelled walls. 1968 J. Sangster *Touchfeather* xiv. 150 One of the main rooms..was about sixty feet long, pine panelled and cool. 1974 *Country Life* 21 Mar. 695/3 Specialists in carved or plain pine panelled Rooms. 1665 *Early Rec. Lancaster, Mass.* (1884) 79 A slipe of medow ground Runing through the most part of a great pine plaine. 1779 *Proc. Mass. Hist. Soc.* (1886) 2nd Ser. II. 464 [We] encampt on a pine plain by the side of a Large flatt. 1935 *Ecol. Monogr.* Jan. 66 The sandy, so-called 'Pine plains' were pitch pine. 1913 *Phytopathol.* III. 306 (*heading*) The introduction of a European pine rust into Wisconsin. 1951 *Dict. Gardening* (R. Hort. Soc.) III. 1580/1 Weymouth Pine Rust is a disease of 5-needled pines caused by the aecidial stage of the rust fungus *Cronartium ribicola*. 1960 C. Westcott *Plant Disease Handbk.* (ed. 2) iv. 355 (*caption*) Pine rusts. 1735 *New Voy. to Georgia* 13 We rode about two Miles farther, where we came to a large Pine Savannat [*sic*, *bis*]. 1791 W. Bartram *Trav. N. & S. Carolina* 208 The cattle which only feed and range in the high forests and Pine savannas are clear of this disorder. 1827 [see Savannah 2]. 1976 West & Augelli *Middle Amer.* (ed. 2) ii. 47/2 The chief reason for the Nicaraguan pine-savanna is probably the porous, gravelly soil, which will support only drought-tolerant plants. 1840 J. & M. Loudon tr. *Köllar's Treat. Insects* III. 345 The means devised by man for guarding against and destroying the pine saw-fly are as follows. 1972 *Times* 10 June 1/7 The limited experimental evidence..demonstrated an effective defence against the potato moth and the pine sawfly. 1891 O. Wilde *Pict. Dorian Gray* xviii. 299 The clear, pine-scented air. 1937 M. Sharp *Nutmeg Tree* iv. 51 A gust of sweet pine-scented air. 1972 D. Anthony *Blood on Harvest Moon* ii. 19 The cabin reeked from the pine-scented disinfectant they used to clean the bathroom. 1887 R. Ridgway *Man. N. Amer. Birds* 400 Northern North America, breeding from northern United States northward... Pine Siskin. 1947 *Chicago Tribune* 28 Dec. vi. 1/1 Some pine siskins..were found munching on birch cones and pods. 1971 *Islander* (Victoria, B.C.) 13 June 13/1 A flock of pine siskins, flashing their yellow-banded wings in darting flight. 1791 Pine-snake [see *bull-snake* s.v. *BULL *sb.*¹ 11]. 1823 E. James *Acct. Expedition Rocky Mts.* I. 131 A serpent..which has considerable affinity with the pine-snake of the southern states. 1941 M. Lyon *Take to Hills* 192 Inside that bird nest was a small piece of discarded pine snakeskin, the original bands still faintly visible. 1956 L. M. Klauber *Rattlesnakes* I. ix. 585 A timber rattler..had eaten a pine snake. 1832 J. P. Kennedy *Swallow Barn* I. xxviii. 295 The ground was strewed with a thick coat of pine-straw,—as the yellow sheddings of this tree are called. 1884 G. W. Cable *Dr. Sevier* lvii. 435 Here stood Mary Richling. She still had on the pine-straw hat. 1939 *These are our Lives* (Federal Writers' Project, U.S.) 51 The pen was grounded with pine straw as was the shelter. 1976 *National Observer* (U.S.) 15 May 18/3 Our toys were found or made on the farm. We rolled steel barrel hoops with a heavy wire pusher, slid down pine-straw hills on old disc-plow blades. 1659 *Rec. Watertown, Mass.* (1894) I. i. 65 Abram Brownes Land..begins ten rod from Rich. Bloyse his lott, & soe apon a straite line to a pine stump. 1816 *Niles' Reg.* IX. Suppl. 178/1 Many a farmer who heretofore dreaded the pine stump..now swings his undisturbed scythe. 1635 *Cambridge* (Mass.) *Proprietors' Rec.* (1896) 6 More by the pine swampe about six acers. 1705 *Early Rec. Providence, Rhode Island* (1903) XVII. 201 At the South End of a Piece of Meadow & a Pine Swampe. 1862 O. W. Norton *Army Lett.* (1903) 62 We are bivouacked in a pine swamp. 1851 *Southern Lit. Messenger* XVII. 226/2 We made [a bonfire] of dead boughs and 'pine-tags'. 1881 *Harper's Mag.* Nov. 868/2 At night they [*sc.* the mountain people].. lie down on their pine-tag beds. 1947 *Richmond* (Virginia) *Times-Dispatch* 13 Oct. 10/3 Oat straw,..pine tags,.. and lawn clippings and leaves also could be used [for a mulch]. 1836 in *Jrnl. Southern Hist.* (1935) I. 367, I have been lost nearly all day.., wandering about in pine thickets. 1671 *South Carolina Hist. Soc. Coll.* (1897) V. 298, I have..dispatched the Carolina laden with Pine timber. 1866 *Rep. Indian Affairs* (U.S.) 288 There is much of their territory valuable for the pine timber upon it. 1839 W. B. O. Peabody in Storer & Peabody *Rep. Fishes, Reptiles & Birds Mass.* 310 The Pine Warbler..is not much known, because it resides in deep, evergreen forests. 1868 *Amer. Naturalist* II. 171 Soon after the pine-warbler has arrived..the Yellow Red-polled Warbler..makes his appearance. 1917 T. G. Pearson *Birds Amer.* III. 149/1 The Pine Warbler is a well-named bird, because its nesting sites are always in pine trees. 1961 O. L. Austin *Birds of World* 285/2 The well-named Pine Warbler..almost invariably nests in a clump of pine needles. 1814 J. Bigelow *Florula Bostoniensis* 73 *Sarothra gentianoides* Pine weed... A small, erect branching plant. 1843, 1907 Pine-weed [see *NIT-WEED]. 1862 *Rep. Comm. Patents 1861: Agric.* (U.S.) 605 *Hylobius pales*..is the common 'pine weevil' of the north and the south. 1867 *Amer. Naturalist* I. 110 Many other weevils and boring-beetles, especially..the Pine Weevil (*Pissodes strobi*)..now abound. 1936 *Discovery* Feb. 47/2 Extensive planting of conifers since the War has raised the problem of undue increase of the damaging Pine Weevils. 1972 Swan & Papp *Common Insects N. Amer.* 486 Pine Weevil: *Hylobius congener*... Range: Massachusetts to Alaska.

b. *attrib.* passing into *adj.* Designating preparations having the aroma of pine-needles.

1890 T. H. Dean *How to be Beautiful* ii. 23 Pine bath. This is a bath much prized by a beautiful Russian lady. 1926–7 *Army & Navy Stores Catal.* 472/1 Klenitas Pine Bath—bot. 2/6. 1931 S. Jameson *Richer Dust* xiv. 420 Put some pine salts in my bath. 1939–40 *Army & Navy Stores Catal.* 404/3 Pine odour disinfectant powder. 1972 J. Aiken *Butterfly Picnic* x. 175 [A] powerful effusion of pine bath essence. 1976 J. Wilson *Let's Pretend* xvii. 179 A bottle of pine disinfectant.

pine, *v.* Add: **5. d.** Of timber: to shrink.

1833 J. C. Loudon *Encycl. Cottage, Farm, & Villa Archit.* II. ii. 492 The granary floor to be laid with inch-and-quarter white-wood battens, dressed and jointed: the battens to be laid loose, so as to take up and relay after pining (shrinking).

pineal, *a.* Add: **a.** (Further examples.) Now known to secrete melatonin in various mammals and to be concerned with photoperiodicity and circadian rhythms.

1958 [see *MELATONIN]. 1970 T. Hughes *Crow* 23 Crow split his enemy's skull to the pineal gland. 1971 J. A. Kappers in Wolstenholme & Knight *Pineal Gland* 22 The mammalian pineal gland is an end organ of the peripheral sympathetic system conveying photic and other stimuli to it. 1978 Lee & Laycock *Essent. Endocrinol.* v. 103 The relationship between the pineal gland and the development of puberty in man is still undecided.

b. (Further examples.)

1974 D. P. Cardinali in James & Martini *Current Topics Exper. Endocrinol.* II. 113 The pineal capacity to take up and retain estradiol and testosterone resembles that observed in the uterus and the prostate, suggesting that specific receptors for sex steroids may be present in the pinealocytes. 1979 *McGraw-Hill Yearbk. Sci. & Technol.* 288/2 A pineal role in the photoperiodic control of reproduction in hamsters has been established.

B. *sb.* The pineal body.

1911 *Arch. Internal Med.* VIII. 854 In enlargements of the pineal, circulatory disturbances will develop first. 1963 A. Gorbman in Euler & Heller *Compar. Endocrinol.* I. viii. 303 Pflugfelder removed or destroyed the epiphysis (pineal) of the guppy (*Lebistes*) and found that the thyroid became greatly hypertrophied. 1966 Wright & Symmers *Systemic Path.* II. xxxiii. 1134/1 The pineal has another endocrine role, that of secreting the melanophore-contracting hormone, melatonin.., and this has been regarded as related to the original photo-receptor function of the gland. 1968, 1974 [see *MELATONIN]. 1974 D. P. Cardinali in James & Martini *Current Topics Exper. Endocrinol.* II. 115 The pineal is involved in the regulation of pituitary, gonadal, adrenal, and thyroid functions.

pinealectomy (piniăle·ktŏmi, pəiniăl-). *Surg.* [f. Pineal *a.* + *-ectomy.] Excision of the pineal body.

1915 *Jrnl. Exper. Med.* XXII. 240 The purpose of this paper is to report briefly the results of pinealectomy in a series of young puppies and to describe the method which has been evolved for extirpation of the pineal body. 1941 *Endocrinol.* XXVIII. 837 All pinealectomies were performed at the age of 20 to 21 days under ether anesthesia. 1976 *Nature* 5 Feb. 431/2 An experiment showing that pinealectomy somewhat reduced maternal behaviour.

So **pineale·ctomize** *v. trans.*, to deprive of the pineal body; **pineale·ctomized** *ppl. a.*

1912 *Rev. Neurol. & Psychiatry* X. 477 When pinealectomised pullets, who had already begun to lay, were isolated with pinealectomised cockerels, their eggs proved fertile on artificial incubation. 1933 *Proc. Soc. Exper. Biol. & Med.* XXX. 766 Sixteen albino rats..were pinealectomized at about 26 days of age. 1970 *Sci. Amer.* Feb. 44/3 Seven out of 12 pinealectomized rats treated with PCPA displayed sexual excitement. 1972 *Science* 27 Oct. 421/3 At 6 weeks of age, half of the birds in each chamber were pinealectomized.

pinealoma (piniălōᵘ·mă). *Path.* [f. Pineal *a.* + *-oma.] A tumour of the pineal gland thought to arise from the parenchymal cells.

1923 K. H. Krabbe in *Endocrinol.* VII. 391 In none of the cases has the tumor been an adenoma or, more exactly, 'pinealoma', but always a teratoma. 1948 R. A. Willis *Path. of Tumours* iii. 820 Pineal tumours, either pinealomas or teratomas, occurring in young children have often been accompanied by precocious bodily, mental or sexual development. 1966 Wright & Symmers *Systemic Path.* II. xxxiii. 1134/2 Teratomas, pinealomas and gliomas are the usual neoplasms of the pineal gland, although collectively even these tumours are rare. 1974 J. M. R. Edwards in T. J. Deeley *Central Nervous Syst. Tumours* iv. 99 Some of the tumours occurring in the pineal gland are undoubtedly teratomas but there is controversy concerning the pathological nature of the 'pinealoma'.

pine-apple, pineapple. Add: **2.** The usu. form is now **pineapple**.

c. A bomb; a hand grenade or light trench mortar. *slang.*

[1916: see *pineapple bomb* below]. 1918 R. H. Knyvett *Over There* 193 But Fritz can be very obstinate on occasions, and all our teasing with rifle-grenades failed to make him retaliate with anything larger than 'pineapples' (light trench-mortars). 1920 W. B. Ellington *Company 'A' 23rd Engineers* 113 *Pineapple*, French hand grenade. 1928 [see *GANGLAND]. 1932 E. Wallace *When Gangs came to London* xv. 118 'By "pineapple" I mean "bomb",' said Jiggs gravely. 'It's part of the racketeer's equipment.' 1944 *Sun* (Baltimore) 2 Aug. 2/3 There was a crossfire of ten grenades before one of his pineapples destroyed a position with four enemy soldiers in it. 1972 J. Quartermain *Rock of Diamond* v. 28 'You..don't want that old-time pineapple lobbed through your store window. You know what a pineapple is, Raven?' 'A hand grenade.' 'Right.' 1976 'B. Shelby' *Great Pebble Affair* III. 143 He was Wild Wally of ice-pick, pineapple and machine-gun fame.

d. *the pineapple* (slang), unemployment benefit, 'the dole'.

1937 Partridge *Dict. Slang* 632/1 *Pine-apple, on the*, on parish relief. 1971 *Observer* 23 May 7 'There were just too many people on the pineapple.' The 'pineapple' is slang for the dole.

3. b. *pineapple juice*; **pineapple bomb** *slang* = sense 2 c above; **pineapple chunks,** tinned pineapple cut into small cubes; also *fig.*; **pineapple jelly,** jelly flavoured with pineapple; **pineapple rum** (examples); **pineapple weed,** a small, aromatic, annual herb, *Matricaria matricarioides*, belonging to the family Compositæ and bearing yellow flower-heads smelling like pineapple when crushed.

1916 'Boyd Cable' *Doing their Bit* iii. 45 We saw 'pineapple bombs' or hand grenades being made—'pineapple' being a neat description of the shape and criss-cross pattern of lines marking the segments into which the grenade bursts. 1952 Wodehouse *Pigs have Wings* i. 9 You watch that pig of yours like a hawk, Clarence, or before you know where you are, this fiend in human shape will be slipping pineapple bombs into her bran mash. 1972 *Daily Tel.* (Colour Suppl.) 1 Sept. 19/1 The 'pineapple' bomb ploughs a furrow through the undergrowth with steel pellets. 1903 Atkinson & Holroyd *Pract. Cookery* (ed. 3) 113, 1 tin of pineapple chunks. 1926 Wodehouse *Heart of Gold* ix. 300 Anastatia [*sic*] ordered pineapple chunks with whipped cream. 1963 R. Carrier *Great Dishes o ; World* 175 Drain the pineapple chunks.

Reserve juice. 1977 D. Beaty *Excellency* i. 15 A DC 6 was leaving, its propellers chopping a long shaft of light into pineapple chunks. 1841 Thackeray in *Fraser's Mag.* June 723/1 They..served us..pine-apple jelly. 1907 R. M. F. Berry *Fruit Recipes* xix. 245 Pineapple jelly (without gelatine). 1958 W. Bickel tr. *Hering's Dict. Cookery* 723 Fancy mould lined with pineapple jelly. 1904 Mrs. H. M. Young *Home-Made Cakes & Sweets* 33 Add..2 tablespoonfuls pineapple juice. 1957 E. Craig *Collins Family Cookery* 392 Heat in pineapple juice. 1972 J. Potter *Going West* 171 Ashley bought pineapple juices from a café. 1977 F. Webb *Go far Out* viii. 144 The attendant..brought him a large glass of pineapple juice. 1765 A. Stuart *Let.* 25 July in Duke of Argyll *Intimate Society Lett. 18th Cent.* (1910) I. 111 Ye Greatest part of your Pine aple rum went, as you desired in return for his many feasts, To Baron Dolbach. 1837 Dickens *Pickw.* xxvii. 276 A glass of reeking hot pine-apple rum and water, with a slice of lemon in it. 1967–8 *Bahamas Handbk. & Businessmen's Ann.* (ed. 7) 56, I poured a double pineapple rum on the rocks. 1908 Robinson & Fernald *Gray's New Man. Bot.* 847 Pine-apple-weed...Odor of the bruised plant suggesting pineapple. 1945 J. M. Fogg *Weeds of Lawn & Garden* 168 Pineapple-weed provides an interesting example of a species which is indigenous to the far western states and has become naturalized not only in eastern North America but also in Europe. 1978 *Country Life* 22 June 1797/1 Pineapple weed, a coloniser of disturbed ground ..assisted to and from a site on the tyres of wheel-barrows.

‖ **Pineau** (pino). Also with lower-case initial. Pl. **-eaux.** [Fr., f. *pin* pine (tree) + dim. suffix *-eau*: so called because of the form of the grape cluster.] **1.** = *PINOT.

The usual form is now *Pinot.*

1763 [see Tresseau]. 1833 C. Redding *Hist. Mod. Wines* v. 76 In all the distinguished vineyards of Champagne..they cultivate only the black grape.., being a variety of the vine called *pinet* and red and white *pineau*. 1845 *Encycl. Metrop.* XXV. 1275/1 The fine wines of Burgundy, and the best Champagne, come from the *pineau*, a black grape. 1888 *Encycl. Brit.* XXIV. 606/1 The white grapes employed are the *Pineau blanc*, which are vintaged a full fortnight later than the red grapes. 1911 *Ibid.* XXVIII. 727/2 Practically all the important wines of Germany are white, although there are a few red growths of some quality, for instance that of Assmannshausen in the Rheingau. The latter is produced from the black Burgundy vine, the Pineau. 1967 A. Lichine *Encycl. Wines* 404/1 *Pinot, Pineau.* One of the most distinguished families of wine grapes.

2. More fully *Pineau des Charentes* (de ʃarɑ̃t). An aperitif made from unfermented grape juice and brandy.

1940 C. Morgan *Voyage* III. iv. 260 Barbet came from the house carrying a tray on which were plums preserved in cognac and a bottle of pineau. Thérèse disliked pineau; the mixture of cognac and unfermented wine was too sweet for her taste. 1951 R. Postgate *Plain Man's Guide to Wine* iii. 60 There is a pleasant, not very distinctive drink called Pineau or Plessis, which is made from the must of wine in Charente (where cognac comes from) and served in quantities in railway trains. 1959 *Listener* 15 Jan. 125/1 Mr. Root has too low an opinion of Pineau des Charentes as an apéritif. 1959 A. Waugh *In Praise of Wine* xii. 162 The grapes of the Charente are poor to eat and little wine is made there now... But there is a local *apéritif* called Pineau, made in the same way as Port, the fermentation being stopped by alcohol. 1961 *Listener* 12 Oct. 574/1 A humdrum café-life among taxi-drivers and porters who also drank their *pineau* at the counter. 1968 A. Laski *Keeper* ii. 21 You shall have an aperitif on the house. Gui! Bring the Pineau for Colin. 1973 C. Ray *Cognac* x. 127 For centuries past, certainly since the sixteenth century, the peasants of the Charentes ..have made an aperitif drink for themselves by 'muting' (checking the fermentation of) fresh grape juice by the addition of brandy. The ratafia of Champagne, the pineau of the Charentes, are sweet, strong, and tasty. *Ibid.*, The best pineaux I drank in Cognac..were the Plessis of Camus and the Reynac of UNICOOP.

pine-barren. Add: **a.** (Earlier, later, and *attrib.* examples.)

1731 *Pennsylvania Gaz.* 29 Apr.–6 May 1/2 We had a sandy Pine Barren to walk in, which was covered pretty thick with large Pine Trees. 1743 M. Catesby *Nat. Hist. Carolina* II. p. iv/1 The third and worst Kind of Land is the *Pine barren Land*. 1901 C. T. Mohr *Plant Life Alabama* 125 The pine-barren streams overflow their low banks of shifting sands and gravel. 1916 J. W. Harshberger (*title*) Vegetation of the New Jersey pine-barrens. 1952 *Guardian* 3 Mar. 4/6 This long peninsula, traded back and forth for two centuries between Spain, France and Britain, was not much more than a pine-barren, pocketed with swamps, till the late nineteenth century. 1976 *Hortus Third* (L. H. Bailey Hortorium) 932/2 *Pyxidanthera*... Creeping, evergreen shrublets, forming cushionlike masses, native to pine barrens from N[ew] J[ersey] to S[outh] C[arolina].

b. pine-barren beauty (earlier and later examples); **pine-barren terrapin,** substitute for def.: the box tortoise, *Terrapene carolina*; (example).

1883 W. Robinson *Eng. Flower Garden* 237/2 Pine Barren Beauty..is an evergreen shrub, yet smaller than many Mosses. 1901 L. H. Bailey *Cycl. Amer. Hort.* III. 1475/2 The Pyxie, Flowering Moss or Pine-barren Beauty is a pretty little creeping plant, native only to New Jersey and North Carolina. 1951 *Dict. Gardening* (R. Hort. Soc.) IV. 1722/2 *P[yxidanthera] barbulata.* Pyxie, Pine-barren Beauty, Flowering Moss. 1884 G. B. Goode *Fisheries U.S.: Nat. Hist. Aquatic Animals* 158 The Carolina Box Turtle... In the Southern States it is known as the 'Pine-barren Terrapin'.

pine-blank (pəi·n,blæŋk). U.S. dial. var. Point-blank *a.*, *sb.*, and *adv.*

1886 *Century Mag.* Jan. 433/2, I oughter 'a' said it then, but I'll say it now, right pine-blank. 1896 'Mark Twain' in *Harper's Mag.* Aug. 345/1 They told him pine blank and once for all, he *couldn't*. 1937 *Frontier & Midland* Autumn 13 His eyes standin pine-blank open. 1954 *Publ. Amer. Dial. Soc.* xxi. 34 A *pine blank* lie. *Ibid.*, I swear *pine blank*.

pine knot. *U.S.* [Pine *sb.*² 7.] A knot of pine-wood, usu. burned as a fuel or for illumination, and adduced as a symbol of hardness, toughness, etc. Also *fig.* and *attrib.*

c 1670 *Plymouth* (Mass.) *Rec.* (1889) I. 119 There shalbe noe pyne knot picked. 1791 [see Pine *sb.*² 7]. 1808 J. N. Barker *Indian Princess* III. i. 47 [She] lit me with her pine-knot torch to bedward. 1835 'H. Bull-Us' *Diverting Hist. John Bull & Bro. Jonathan* (new ed.) i. 8 Jonathan, though as hard as a pine knot, ..could bear it no longer. 1850 H. C. Watson *Camp-Fires of Revolution* 31 We stuck to them as close as pine-knots. 1853 'P. Paxton' *Stray Yankee in Texas* 310 We stood.. with the bright light of a pineknot fire shining full upon us. 1856 X. D. MacLeod *Biogr. F. Wood* 48 The human pine-knot John C. Calhoun. 1876 S. & A. Warner *Gold of Chickaree* 360 You know as well as I do, that you are a pine knot for endurance. 1897 *Outing* XXX. 69/2, I held a pine-knot for him to make the entry in our log-book. 1904 G. Stratton-Porter *Freckles* 95 He was as tough as a pine-knot and as agile as a panther. 1945 B. A. Botkin *Lay my Burden Down* 62 When the boys would start to the quarters from the field, they would get a turn of lider [*sc.* lightwood] knots. I 'specks you knows 'em as pine knots. 1961 B. Palmer *Many are Hearts* 137 The dark night flecked with pine-knot torches. 1979 M. G. Eberhart *Bayou Road* xvii. 175 There was the red, smokey flare of lighted pine knots ahead.

pine land. *U.S.* [Pine *sb.*² 7.] Land on which pine trees are the characteristic growth. Also *attrib.* Hence **pi·ne-lander,** one who lives on and derives a living from such land.

c 1658 [see *OAK LAND]. 1665–70 *Early Rec. Lancaster, Mass.* (1884) 271 There is another piece of upland..Sum part pine Land & partly oak Land. 1735 *Georgia Hist. Soc. Coll.* (1842) II. 45 We encamped there, and found ..the pine land very valuable. 1765 [see Pine *sb.*² 7]. 1789 J. Morse *Amer. Geogr.* 446 They are often to be found in pine lands in the southern states. 1838 C. Gilman *Recoll. Southern Matron* xxiii. 157 There is something picturesque in the evening hour at a pine-land village. 1839 F. A. Kemble *Jrnl. Residence Georgian Plantation* (1863) 75 He gave me a..description of the Yeomanry of Georgia, more properly termed pine-landers. 1890 *Harper's Mag.* Apr. 790/1 Quaint and indolent pine-landers and degraded swamp-dwellers, have all supplied our literary comedians with unique characters. 1903 'P. Pennington' *Woman Rice Planter* (1913) i. 53 Drove S—— to church in our little pine-land village; she seemed to enjoy the very simple service. 1922 H. Kephart *Our Southern Highlanders* (new ed.) 433 These freedman were pushed further and further back upon more and more sterile soil. They became 'pine-landers' or 'piney-woods-people'..or 'crackers'. 1948 *Sat. Even. Post* 4 Sept. 30/1 There was a stillness here in the flat lonesome pinelands. 1977 D. Clark *Gimmel Flask* v. 95 A..voice broke into song. 'Tina, soon the leaves will be falling, From the pine-lands I'm calling, Won't you come back to me-e?'

pinene (pəi·nīn). *Chem.* [ad. G. *pinen* (O. Wallach 1885, in *Ann. d. Chem. u. Pharm.* CCXXVII. 300), f. L. *pīn-us* Pine *sb.*² + G. *-en* -ene.] Either or both of two isomeric liquid terpenes, $C_{10}H_{16}$, which are the major constituents of turpentine and differ in the position of the double bond.

1885 *Jrnl. Chem. Soc.* XLVIII. 551 The author proposes to classify the terpenes as follows:.. B. True Terpenes, $C_{10}H_{16}$, divided into the following groups: 1. Pinenes (boiling point 160°). [Etc.] 1886 *Ibid.* L. 71 Terpine is formed by the inversion of pinene with alcoholic sulphuric acid. 1922 *Nature* 16 Feb. 226/2 The yield of oil from leaves of the New South Wales sassafras tree was about 1 per cent... The principal constituents identified are safrol, camphor, pinene, sesquiterpenes, eugenol, and alcoholic bodies. 1960 A. R. Pinder *Chem. of Terpenes* v. 93, α-Pinene is of great importance commercially in the synthesis of camphor and related terpenes... β-Pinene, which is isomeric with α-pinene, accompanies the latter in most sources of the hydrocarbon. 1971 *Daily Tel.* 12 June 6/3 The leaf is pinched to break specialised cells containing these elements of geraniol, citral, eucalyptal, borneol, pinene and others.

pine-nut. Delete ?*Obs.* Substitute for last part of def.: also, the edible seeds of several pines, esp. the European stone pine, *Pinus cembra*, and the Mexican stone pine, *P. cembroides*. (Later examples.)

1845 J. C. Frémont *Rep. Exploring Exped.* 222 A party of twelve Indians came down from the mountains to trade pine nuts. 1869 *Harper's Mag.* Sept. 473/1 The pine-nut is one of the principal articles used as food by these Pintes. 1937 A. F. Hill *Econ. Bot.* xvi. 371 The pine nuts or piñons are the edible seeds of several species of *Pinus* that are native to the Rocky Mountain and Pacific Coast region. 1947 *Sierra Club Bull.* (San Francisco) Mar. 4/1 It is far too late now to advocate..the Indian's custom of living on the income of natural resources, the replenishable deer, pine nuts and grasshoppers. 1954 E. David *Italian Food* 32 Pine nuts or pine kernels..come from the cones of the stone pine. 1977 *Homes & Gardens* Sept. 104/2 Toss pine nuts in a little butter until crisp and golden.

Pinerotic (pinərǫ·tik), *a.* [f. the name *Pinero* (see below) + -otic.] Pertaining to or characteristic of the dramatist Sir Arthur Wing Pinero (1855–1934), or his plays.

1895 G. B. Shaw *Let.* 18 Mar. (1965) I. 501 It would appear so very easy to give my subjects the Pinerotic effect. 1896 *Sat. Rev.* 29 Feb. 223/1 It contains not a word about Mr. Pinero himself.., his hopes and fears for the drama, or anything else distinctively Pinerotic. 1938 C. Morgan *Flashing Stream* p. xvi, The Pinerotic theory that a woman, to be entitled to our respect in the theatre, must, if she shares a man's bed, do so with a self-sacrificial reluctance.

pinery. Add: **2.** Chiefly *N. Amer.* (Earlier and later examples.) Also *attrib.*

1783 *Rep. Bureau Arch. Ontario* (1906) III. p. cxx, There are fine pineries two or three miles from the water's edge where large masts may be procured. 1822 *Massachusetts Spy* 6 Feb. (Th.), There are also a few pineries, but of small extent. 1926 *Amer. Speech* II. 100/2 The lumberjacks have found anthologists who appreciate better than did the singers themselves the charm of the pinery songs. 1952 D. F. Putnam *Canad. Regions* 138/1 Its 'pineries' formed the source of much of the timber which came down the Ottawa River.

pi·ne-top. [f. Pine *sb.*² + Top *sb.*¹] **a.** The top of a pine-tree. **b.** *U.S. slang.* Cheap or illicit whisky.

1858 *Southern Lit. Messenger* XXVII. 463/2 A rough, but hearty frolic..with profusion of 'pine-top' succeeded. 1878 O. Wilde *Ravenna* v. 10 The pine-tops rocked before the evening breeze. 1931 *Amer. Speech* VII. 50 The 'drinks' are 'pine-top', 'white mule', [etc.]. 1942 Berrey & Van den Bark *Amer. Thes. Slang* §99/7 *Illicit whiskey*,.. pine-top. 1946 J. W. Day *Harvest Adventure* xiii. 214 At the report the second, larger still, swung over a gap in the pine-tops and the next barrel knocked him sideways and sent him spinning in a long dive into the snipe marsh below the wood.

pine-wood. Add: **2.** (Earlier and later examples.) Hence **pi·ne-woody** *a.* (*rare*), suggestive of pine-woods.

1673 J. Ray *Observations Journey Low-Countries* 80 After one half-hours riding we entred into Pine-Woods, the first we met withal: They reach'd almost to our Lodging. 1790 *Pennsylvania Packet* 11 Oct. 3/4 A ganninipper is a kind of large horse-fly frequent in pine woods. 1872 R. G. McClellan *Golden State* xvii. 204 These hogs..are somewhat like the North Carolina pine-woods hogs. 1917 D. H. Lawrence *Look! We have come Through!* 40 The jade-green river Goes between the pine-woods. 1939 E. E. Murphey *Wings at Dusk* 13 Whenever I see one Flushing befoh me Histing his flag, like a buck in de pine-wood [etc.]. 1945 J. Betjeman *New Bats in Old Belfries* 6 Mushroomy, pinewoody evergreen smells. 1978 'I. Drummond' *Stench of Poppies* xi. 167 In the foothills of the Taurus..among superb pinewoods.

piney wood. *U.S.* Also **piny wood.** [*piney* var. Piny *a.*] **a.** A pine-wood; a region of pine-woods; *spec.* (pl.) regions of poor land in the Southern United States of which pine-trees are the characteristic growth. Also *attrib.*

1809 M. L. Weems *Life Gen. F. Marion* xiv. 122 Had this savage spirit appeared among a few poor British *cadets*, or *piney wood* tories, it would not have been so lamentable. *a* 1816 B. Hawkins *Sk. Creek Country* (1848) 29 Broken piny woods and reedy branches on its right side. 1860 Bartlett *Dict. Amer.* (ed. 3) 321 *Piney woods*, the name given at the South to a large tract covered with pines, especially in the low country. 1863 [see Piny *a.*]. 1887 *Century Mag.* Aug. 544/1 Azalia, the little piney-woods village which Dr. Buxton had recommended as a sanitarium. 1946 *Sun* (Baltimore) 26 July 16/3 The ponies..roam wild in the piney woods. 1963 *Social Problems* X. 365/1 Lest such piney-woods practices be thought beneath the sophistication of the urban Negro. 1973 *Daily Colonist* (Victoria, B.C.) 14 Aug. 2/3 Otter said that the twenty-seven victims found buried at the beach, in an east Texas piney wood and in a Houston boat shed, might be all the police will recover. 1976 J. Crosby *Snake* (1977) xiii. 70 She found herself deep in piny woods and she could see nothing.

b. Special Combs., as **piney-woods cracker,** a poor Southern white; **piney-wood(s) tacky,** (*a*) a scrub pony; (*b*) = prec.

1872 *Kansas Mag.* Mar. 238/1 Who that has seen the 'clay-eater', the 'sandhiller', or the 'piney woods cracker' of the South, does not know that it is impossible to exaggerate the sinfulness which looks out through the loop-holes of his red apologies for eyes? 1935 Z. N. Hurston *Mules & Men* (1970) I. v. 113 'Bout this time John seen a white couple come in but they looked so trashy he figgered they was piney woods crackers. 1846 *Spirit of Times* 11 July 234/3 Mac mounted a piney-woods-tacky..and hied him off to Charleston. 1888 *Century Mag.* XXXVI. 799/2 If Mr. Catlett will come to Georgia and go among the 'po' whites' and 'piney-wood tackeys', he will hear the terms 'we-uns' and 'you-uns' in everyday use. 1944 B. A. Botkin *Treas. Amer. Folklore* II. 322 Such derogatory nicknames as..sand-hillers, pineywoods tackies, hill-billies.

pi·n-fire, *a.*² (*sb.*²) Also **pinfire.** [f. Pin *sb.*¹ + Fire *sb.*] Used *attrib.* and *absol.* to designate precious opal characterized by closely spaced specks of colour.

1902 *Blackw. Mag.* Feb. 254/1 Two men found a large piece of 'pin-fire', or the best opal. 1902 [see *ORANGE *a.* 1 b]. 1908 A. J. Dawson *Finn* xxxi. 464 Sixty-six solid pounds o' best pin-fire—and us dyin' for want of a crust. 1964 W. C. Eyles *Bk. Opals* i. 25 Pin-fire opal is a type in which the main body of the stone is usually white and shows a myriad of small pinlike colors all through the surface and the body of the stone. 1971 R. Purvis *Treasure Hunting* i. 27/2 In pin-fire opals, the colour play is in small pinlike dots thickly scattered through the mass. 1976 *Sci. Amer.* Apr. 84/3 In 'pinfire' varieties of opal the grains are up to a millimeter across, but in the more typical varieties they are several millimeters across.

ping, *sb.* Add: **a.** (Earlier and later examples.)

1835 J. E. Alexander *Sk. Portugal* xi. 262 If a button was shown, 'ping' went a bullet at it immediately. 1909 Kipling *Rewards & Fairies* (1910) 272 *Ping-ping-ping* went the bicycle bell round the corner. 1921 D. H. Lawrence in *Hutchinson's Mag.* Nov. 463/1 They were interrupted by the ping of the shop-bell. 1930 J. B. Priestley *Angel Pavement* v. 263 She sent the typewriter carriage flying along. It gave a sharp ping. 1957 H. Nicolson *Diary* 5 Oct. (1968) 339 The Russians have released a satellite... The B.B.C. have managed to record the signals, and play them over to us—just ping, ping, ping, ping. 1960 N. Hilliard in C. K. Stead *N.Z. Short Stories* (1966) 235 They rested..listening to the..ping-ping-ping of crossing bells. 1977 *New Yorker* 20 June 94/3 Everyone assembled for an ostinato unison figure, and the music subsided with a string of tinkles and hums and pings.

b. A very short pulse of high-pitched, usu. ultrasonic, sound such as is emitted by sonar; also, a pulse of audible sound by which this is represented to a user of such equipment.

1943 *Penguin New Writing* XVIII. 27 'Daisy had a ping about an hour ago... We're doing an Asdic sweep.'.. A 'ping' is the slang term for an echo. 1946 *Sci. Illustr.* May 83/1 The sounds, or 'pings', sonar sends out are not at all like sounds to you and me. They are supersonic. 1956 *Deep-Sea Res.* III. 267 The system had a repetition rate of about 10 pings per second and a ping length of about two milliseconds. 1960 [see *PINGER 1 a]. 1966 *McGraw-Hill Encycl. Sci. & Technol.* IX. 252/2 The Swallow-type neutrally buoyant float is a mid-depth current meter..which..can float at a predetermined depth... It emits acoustic pings which can be heard for several miles from a quiet ship equipped with appropriate sound gear. 1967 J. B. Hersey *Deep-Sea Photogr.* iv. 59/1 It was possible to obtain the height of camera above bottom simply by measuring the time interval between the arrival at the ship of the sound pulse, or ping, and its bottom echo.

c. Also *ping-man.* (See quots.) *slang.*

1946 J. Irving *Royal Navalese* 135 *Ping-man*, an Asdic operator. 1948 Partridge *Dict. Forces' Slang* 142 *Ping*, an Asdic officer or rating.

d. = *PINK *sb.*[6] 3.

More usual than *pink* in the U.S. and Australia.

1927 *Dyke's Automobile & Gasoline Engine Encycl.* Suppl. 1313/1 Engineers began an investigation as to the causes of pre-ignition and ping in an engine burning the present-day gasoline. 1942 *Pop. Sci.* Mar. 137/2 A slight ping when you step on the gas hard does not always mean trouble. 1953 H. R. Ricardo *High-Speed Internal-Combustion Engine* ii. 27 The mechanism of detonation is the setting-up within the cylinder of a pressure wave travelling at so high a velocity as, by its impact against the cylinder walls, to set them in vibration and thus give rise to a high-pitched 'ping'. 1958 *S.A.E. Jrnl.* Sept. 73/1 Rumble..is distinct from the high frequency spark knock or wild ping most people have heard. 1977 *Pop. Mechanics* May 49/2 My 1976 Ford Pinto with a 2.3-liter, four-cylinder engine has had a bad ping almost from day one.

ping, *v.*[2] Add: **1. a.** (Further examples.) Also const. *out* and *fig.*

1924 Galsworthy *White Monkey* II. ix. 195 A footman ..stood..waiting for an order to ping out, staccato, through the hum. 1930 J. B. Priestley *Angel Pavement* iii. 96 The typewriters rattled and *pinged*, the telephone bell rang. *a* 1963 S. Plath *Crossing Water* (1971) 53 The glass..will ping like a Chinese chime. 1967 *Listener* 16 Nov. 647/3 Words and concepts heard nowhere else pinged out on Third Programme drama. 1974 W. J. Burley *Death in Stanley St.* v. 90 He went in, the door bell pinged. 1978 T. Willis *Buckingham Palace Connection* viii. 151 A bullet pinged against the plating.

b. = *PINK *v.*[3] 2.

1942 *Pop. Sci.* Mar. 136/1 Let's assume it is a couple of months from now, and that best available gas makes your car engine ping—or worse, knock—on a pick-up or a hard pull.

2. a. (Earlier and later examples.) Also *fig.*

1746 *Exmoor Scolding* (1879) 52 Tha wud'st ha' borst en to Shivers, nif chad n-t a vung en, and pung'd en back agen. 1921 D. H. Lawrence in *Hutchinson's Mag.* Nov. 462/2 They 'pinged' the door-bell, and her aunt came running forward out of the kitchen. 1957 J. Kerouac *On Road* (1958) 32 The air grew ice-cold and pinged our ears. 1961 J. H. Ford *Mountains of Gilead* v. 128 A room complete with moths pinging the light. 1974 'P. B. Yuill' *Bornless Keeper* vii. 64 There was nobody at the little alcove marked Reception. Victoria pinged the bell.

b. To fire or discharge (a missile) with a pinging sound. Also *transf.*

1959 I. & P. Opie *Lore & Lang. Schoolch.* xiii. 297 We ..ping pellets in class. 1977 A. C. H. Smith *Jericho Gun* xiii. 171 He was able to ping off three one-shots.

‖**pinga** (pi·ŋă). [Pg., lit. 'drop (of water)'.] A raw white rum distilled from sugar-cane in Brazil.

1933 P. Fleming *Brazilian Adventure* I. xii. 98 You drank black coffee with *pinga* in it. 1960 *Guardian* 14 June 9/6 Her breath reeked of pinga (a cheap drink made from sugar cane). 1974 *Country Life* 31 Oct. 1314/3 Sugar-cane spirit (or *pinga*). 1977 *Lancet* 24–31 Dec. 1365/2 We have studied 61 alcoholic patients with pellagra aged 23–49, most of them drinking more than 800 ml of *pinga* (a Brazilian alcoholic beverage similar to white rum) containing about 300 ml of pure ethanol.

pingao (pi·ŋɑu). Also **pingau.** [Maori.] A New Zealand sedge, *Desmoschœnus spiralis*, with creeping underground stems which help to stabilize sand-dunes.

1855 J. D. Hooker *Flora Novæ-Zelandiæ* I. 272 *Desmoschœnus spiralis*... Nat[ive] name, 'Pingao'. 1905 W. B. *Where White Man Treads* 2 White seashore sand-hills..for..the wind..to pile into hillocks, until the wily pingau (native sand grass), creeping snakelike along,.. bound [them] into masses. 1936 [see *KAKAHO]. 1949 P. H. Buck *Coming of Maori* II. iv. 156 The only other native colour to black, used in plaiting, was yellow obtained by using wefts of *pingao*, the leaves of which are a natural yellow. 1970 Moore & Edgar *Flora N.Z.* II. 171 Pingao... This is a well-known plant because it is an effective sand-binder and also because the Maoris used the dried golden leaves to give colour to articles woven from *Phormium*.

pinger (pi·ŋəɹ). [f. Ping *v.*[2] + -er[1].] **1. a.** A device that transmits pings (*PING *sb.* 1 b) at short intervals for purposes of detection, measurement, or identification.

1957 *Deep-Sea Res.* IV. 120 The three principal units which comprise the camera are the photographic unit, the electronic flash light and the acoustic signal generator or 'pinger'. 1960 *Electronics* 24 June 95/3 The detailed contour of the bottom with respect to the pinger can be determined by recording the direct and bottom-reflected pings. 1969 J. Mavor *Voy. Atlantis* II. iv. 124 It was the shallowness of the water that made examination with Doc's pinger feasible. 1975 *Offshore Engineer* Nov. 51/1 The pinger receiver audibly indicates the direction to a pinger using a manually operated sensitivity control located on the end of the receiver and an earphone.

b. (See quot.)

1961 Partridge *Dict. Slang* (ed. 5) II. 1222/1 *Pinger*, the Asdic officer: Naval: since ca. 1936.

2. A timer that makes a ringing or pinging sound after a pre-set number of minutes.

1968 J. Bingham *I love, I Kill* vi. 72, I heard the old vibrant tinkling, like those kitchen ping-ers which tell you when the cabbage is ready cooked. 1973 *Times* 30 July 11/2 Many new cookers have automatic timing, with both a time of day clock for cooking while you are out and a 60-minute pinger for reminding you when you are in that a dish is cooked. 1976 P. Dickinson *King & Joker* i. 10 Two eggs..boiled for two minutes... The pinger pinged.

pinging, *vbl. sb.* Add: *spec.* **a.** = *PINKING *vbl. sb.*[3]

More usual than *pinking* in the U.S. and Australia.

1955 *Pop. Mechanics* Nov. 203/2 Retard the spark a little at a time..until no pinging is heard. 1967 *Boston Sunday Herald* 28 May I. 2/6 'Pinging'..results from carbon deposits or fuel not high enough in octane rating. 1969 *Telegraph* (Brisbane) 1 Feb. 13/4 Tetra-ethyl lead is used to raise octane ratings, reduce engine knock or pinging and minimise engine misfires.

b. The production or emission of pings by sonar or similar equipment (see *PING *sb.* 1 b).

1956 *Deep-Sea Res.* IV. 39 The pinging circuit of the echo sounder. 1959 H. Barnes *Oceanogr. & Marine Biol.* 203 New cameras have been described and in one contact with the bottom is indicated by a change of the 'pinging' rate of a small sound source attached to the unit. 1973 *Times* 27 Dec. 3/2 Teams of 71 Squadron split into two daily shifts, poring over the sands between tides to the 'pinging' of mine detectors. 1974 'M. Hebden' *Pride of Dolphins* III. iii. 232 The relentless pinging of the asdic beam.

pingo (piŋgo). *Geomorphol.* Pl. **pingos,** less commonly **pingoes.** [Eskimo (see quot. 1928[1]).] A perennial conical or dome-shaped mound (often with a crater on top) found in regions with thin or discontinuous permafrost and consisting of a layer of soil over a large core of ice, the width being much greater than the height; also, a round depression or rampart in temperate regions thought to be the remains of such a mound formed when the climate was colder.

The word is applied rather differently in quots. 1928. The current techn. use derives from that of Porsild.

[1928 L. Koch in *Meddelelser om Grønland* LXV. 196 To denote a mountain entirely or partly covered with ice but whose form is still distinguishable, the Polar-Eskimos use the name 'Pingo', a term which in these regions corresponds to the name 'Nunatak'. The best example of a Pingo is Mt. Haffner, Pingorsoak, i.e. the great Pingo. *Ibid.* 197 We need no name for rounded nunataks. I have, however, introduced the name Pingo.. to designate a mountain entirely submerged by the Inland Ice, but setting its mark upon the surface of the ice. *Ibid.* 203 The easternmost of these ranges consists of a series of high and precipitous nunataks and Pingos.] 1938 A. E. Porsild in *Geogr. Rev.* XXVIII. 46 In literature the name 'gravel or earth mound' seems to be fairly well established. The Eskimo name *pingo*, meaning conical hill, which has come into universal use in the north, is here introduced as an alternative. *Ibid.* 54 (*caption*) Pingo near Tuktuayaktok on the Arctic Sea coast east of the Mackenzie delta showing the irregular rupture of the summit. This pingo is 134 feet high. 1961, 1968 [see *hydrolaccolith* s.v. *HYDRO-]. 1968 R. W. Fairbridge *Encycl. Geomorphol.* 848/2 The pingos of Wales seem to belong to the open system, the water having flowed down the slope between a thin permafrost and an impervious bedrock. 1972 *Nature* 14 Apr. 344/1 The pingos of Cardiganshire are not sparse in distribution but occur in a number of localities. 1974 *N.Y. Times* 26 May I. 34/4 There was little but ice and snow..occasional ice-cored hillocks called pingoes, and the prominent ribbons of the winter ice road.

ping-pong, *sb.* Add: **1. a.** The bat now usu. has a wooden blade covered with pimpled rubber. (Further examples.) Also *attrib.*

Ping-pong is a proprietary name in the U.S.

1901 [see *table-tennis* s.v. Table *sb.* 22]. 1902 *Harper's Weekly* 7 June 739 To have your squash-court this summer, if you have any pretensions to style, is as necessary as to have your ping-pong table or your automobile. 1904 'H. Foulis' *Erchie* xiv. 90 The grocer in there wad be thinkin' I was awa' on the ping-pong if he didna ken I was a beadle. 1907 *Westm. Gaz.* 12 Oct. 3/2 A set of 'ping-pong' materials. 1949 *Official Gaz.* (U.S. Patent Office) 4 Oct. 40/2 Parker Brothers, Inc., Salem, Mass... *Ping-pong*... For game played with rackets and balls. Claims use since Aug. 1, 1900. 1958 R. Liddell *Morea* II. i. 47 Nick..had been playing ping-pong outside the hotel. 1968 *Listener* 13 June 783/2 No character ever entirely subsides: they are like those ping-pong balls at a fair that rise and fall on spurts of water. 1977 E. Ambler *Send no more Roses* x. 231, I used the handle of a ping-pong bat to keep it [*sc.* a door] open.

b. *fig.* A series of (usu. verbal) exchanges between two parties. Also *attrib.*

1917 E. Pound *Let.* 11 Apr. in T. S. Eliot *Waste Land Drafts* (1971) p. xii, I want to boom Eliot and one cant have too obvious a ping-pong match at that sort of thing. 1934 L. B. Lyon *White Hare* 50 After him came two high-brows playing a wordy ping-pong. 1955 *Times* 18 July 9/2 Coal is one industry that has escaped the political ping-pong which threatens steel. 1966 J. Cleary *High Commissioner* iii. 51 Two hours of diplomatic ping-pong hadn't touched her; she looked..poised and unmarked. 1974 'M. Allen' *Super Tour* (1975) v. 203 It was impossible to beat Mama in the game of verbal Ping-Pong. 1976 N. Postman *Crazy Talk* 8 In the Ping-Pong ball theory, communication is conceived of as a discrete, quantifiable piece of stuff that will move from one source to another and then back. 1977 *Time* 5 Dec. 54/1 The French political journals, center and right, ravaged Courbet for years, and beside their vilifications the attacks on impressionism and cubism were mere Ping Pong.

2. A type of drum in a West Indian steel band (see quots.). Also, a melody played on such a drum.

1955 *New Commonwealth* 28 Nov. (Suppl.) p. xix/1 In the orchestra the pans are grouped into Ping Pongs, Alto Pans, Tenor Kittles, Kittle Booms, Tune Booms and Bass Booms. 1956 L. Hughes *First Bk. of W. Indies* 42 Discarded oil drums are cut, heated, and hammered in such a way that each has its own pitch and tone. Those that carry the melody are called 'ping pongs'. 1959 W. A. Simmonds '*Pan*'-*Story of Steelband* 9 By the end of 1945, different bands had developed different 'beats'. The initiated could, by the rhythm and 'Ping-Pong', distinguish what band 'was beating pan'. *Ibid.* 12 The queen of them all—the sweet Ping-Pong. This is a steel drum cut to about six or seven inches from the top. After these are stretched, and tempered, between twenty-six and thirty-two notes are marked and tuned.

Hence **pi·ng-ponger** = Ping-pongist.

1933 *Times* 14 Nov. 15/4 It cannot be, and is not, good for anyone to enjoy the high moments or ecstasies of lawn tennis without sharing its physical dangers; yet that is what the ping ponger is trying to do. 1962 G. Compton *Too Many Murderers* xvii. 144 She sidled round the ping-pongers. 1972 *National Observer* (U.S.) 18 Mar. 16/1 The chief resident and Ping-ponger extraordinaire is one Dal-Joon Lee.

ping-pong, *v.* (In Dict. s.v. Ping-pong *sb.*) Add: **a.** Also, to move back and forth in the manner of a ping-pong ball. Also *fig.* **b.** *trans.* To send back and forth, to pass around aimlessly.

1952 *Jackson* (Tennessee) *Sun* 1/1 (*heading*) Question of Margaret's guards ping-pongs across Atlantic. 1960 in Cassidy & Le Page *Dict. Jamaican Eng.* (1967) 351/2 A common Jamaican expression, if you say 'he ping-pongs at it' it means he doesn't do it very well—like me typing: I just ping-pong at it. 1970 *Washington Post* 22 Nov. 86 The administration 'ping-ponged' the proposal back and forth. 1971 *Daily Colonist* (Victoria, B.C.) 2 Sept. 13/7 It was 'unfortunate for this accused that he has been, if I may coin a phrase, pingponged from one court to another'. 1972 *Daily Tel.* 11 Mar. 11/2 He can time funny lines so that they ping-pong back and forth in long sustained volleys with the audience's laughter. 1974 *Times Lit. Suppl.* 27 Sept. 1033/2 They ping-pong helplessly between sea and sheets, between pub and plough. 1976 *Times* 31 Aug. 5/2 The report [of a Senate committee] states that 'investigators were repeatedly "ping-ponged" to neurologists, gynaecologists, internists, [etc.]'.

Hence **pi·ng-ponging** *vbl. sb.*, playing ping-pong; also *fig.*; *spec.* (see quots.).

1901 [in Dict.]. 1972 *Daily Colonist* (Victoria, B.C.) 6 Jan. 1/8 Medical groups..engaged in 'ping-ponging'—sending patients from one medicaid provider to another for services they did not need. 1975 *Time* 26 May 55/2 Many of the 'Medicaid mills' of clinics set up to handle poor patients in the nation's urban ghettos reap enormous profits by such practices as 'Ping Ponging' (passing

a patient along to all the other doctors in the clinic). 1976 *Discursive Dict. Health Care* (U.S.) 122 *Ping-ponging*, the practice of passing a patient from one physician to another in a health program for unnecessary cursory examinations so that the program can charge the patient's third-party for a physical visit to each physician.

pinguicula. Delete 9 and add: **2.** (Earlier and later examples.)

In this sense *pinguecula* has always been the usual form.

1850 H. Howard *Anat. Eye* xvi. 230 Pinguecula.—This is a little yellowish colored tumour, situated partly in the conjunctiva, and partly in its cellular membrane. 1934 [see *interpalpebral* s.v. *INTER- 6]. 1937 E. Wolff *Dis. Eye* i. 12 Pingueculæ require no treatment. 1965 F. W. Newell *Ophthalm.* viii. 180/2 Rarely a pinguecula becomes inflamed, causing a foreign body sensation.

pin-head. Add: Also **pinhead. 1. a.** (Further examples.)

1951 J. Cleary in Murdoch & Drake-Brockman *Austral. Short Stories* 439 The street lights are on, yellow pin-heads climbing the hill from the bay road. 1976 E. Ward *Hanged Man* xxvii. 168 Pubs known as 'happy', where a development of lysergic acid..is on sale as purple pinheads—a microdot of LSD embedded in plastic, sold for £1.

b. (Later examples.)

1963 *Times* 30 Apr. 13/4 The many heavy lorries which, even though they may use their headlamps in the country, only have one pin-head size nearside sidelamp in the town, or perhaps a single flickering tail lamp. 1971 *Brit. Med. Bull.* XXVII. 55/2 In cases of the 'pin head' type of opacity there may be a slight drop in the transfer factor of the lung. 1974 D. Seaman *Bomb that could Lip-Read* xvii. 168 The pinhead microphone is set into the head of the explosive.

c. Applied to a pattern of small dots on cloth, or to the cloth itself. Also *ellipt.*, a garment having such a pattern.

1897 *Sears, Roebuck Catal.* 220/1 *French Sanitary Suspenders*... cream, blue, green, etc., with very fine pinhead dots of contrasting colors. 1923 *Daily Mail* 12 June 3 (Advt.), Greys are both light and dark, and include pinhead designs, herring-bones and stripes. 1935 L. A. G. Strong *Tuesday Afternoon* i. 11 Those chaps in the city had so successfully turned his blue pinhead that several people had..thought it was brand new. 1964 L. Deighton *Funeral in Berlin* iv. 27 He wore a well-cut Berlin suit of English pinhead worsted.

4. A small minnow.

1845 S. Judd *Margaret* I. iv. 18 Minnows and pinheads were flashing and skirting through the clear bright stream.

5. A person with a small head; chiefly *fig.*, a person of little intelligence, a fool. Also *attrib.* or as *adj.*

1896 Ade *Artie* i. 3 I've got as much right to go out and do the heavy as any o' you pin-heads. 1909 *N.Y. Even. Post* (semi-weekly ed.) 22 Feb. 6 An innovation in dress that was..said to indicate that the wearer was a 'pin head'. 1923 [see *BONE *sb.* 3 b]. 1933 E. Seago *Circus Company* 295 *Pinheads*, freaks in a side-show. 1940 *Horizon* Apr. 236 An opponent particularly dangerous in these times is the near-artist, or Pinhead. 1947 H. S. Truman in M. Truman *Harry S. Truman* (1973) xvii. 349, I am of the opinion that the country has had enough of their pinhead antics. 1973 'M. Yorke' *Grave Matters* vii. iii. 118 She's no pin-head—she'd be a match for him intellectually. 1976 *New Yorker* 15 Nov. 23/2 It uses images of physical deformity for their enormous potential of horror, and at the end, when the pinheads and the armless and legless creatures scurry about to revenge themselves on a normal woman.., the film becomes a true nightmare.

pin-headed, *a.* Add: **b.** Of a person: having a small head like that of a pin; *fig.* stupid. So **pin-hea·dedness.**

1901 Ade *Forty Mod. Fables* 67 Is it not Sad to see a pin-headed Rake dissipating a Large Fortune? 1927 *Scribner's Mag.* Feb. 209/1 He is living in parochial, hide-bound, pin-headed stupidity. 1928 Wodehouse in *Strand Mag.* Aug. 107 People..were accustomed to set him down as just an ordinary pinheaded young man. *Ibid.* 108 Then they realized that his pin-headedness, so far from being ordinary, was quite exceptional. 1954 'N. Blake' *Whisper in Gloom* xiii. 178 A pin-headed, slick-haired man. 1960 *Sunday Express* 3 July 15/6 The 'pin-headed' look is on its way in..shorter and still shorter hair. 1969 [see *MILLE MIGLIA].

pi·nholder. [f. Pin *sb.*[1] + Holder[1].] A holder for cut flowers, etc., comprising a mounting or base and projecting pins.

1956 D. Bebb *Flowers for You* 24 Place the prepared branches firmly on a pin-holder, with some pieces of bright green moss covering it. 1960 *Woman* 23 Apr. 61/1 Any large-headed flowers should be arranged this way, in a pinholder in a very shallow dish. 1967 *Times* 22 Mar. 11 Most of them, like the pink daisies in *Through the Looking Glass* who turned white with shock when Alice threatened to pick them would..faint at the sight of a pin-holder. 1977 *Lancashire Life* Feb. 9/2 The design is held in a tall black well pin-holder within a rough glazed ivory-coloured ceramic container. 1979 I. Webb *Compl. Guide Flower & Foliage Arrangem.* iv. 58/1 All three branches must emerge from the point on the pinholder where Branch 1 is placed.

pin-hole. Add: Also **pinhole. 2. b.** A small hole in timber caused by a wood-boring beetle or its larva.

1894 A. D. Hopkins in *Bull. West Virginia Agric. Exper. Station* Jan. 291 In order that we may refer to the different kinds of defects caused by insects, by some simple, descriptive names, I will present the following, provisional classification and popular names. Pin Holes. Small, round holes, one-hundredth of an inch to one-fourth of an inch in diameter. 1907 *Circ. Bureau Entomol., U.S. Dept. Agric.* No. 82. 1 The principal injury to the wood of standing girdled cypress was found to consist of pinholes in the sapwood and heartwood. 1938 Hunt & Garratt *Wood Preservation* iii. 53 The insect defects produced in wood before it is placed in service may be classified as pinholes or grub holes, the distinction being principally a matter of size. 1968 *Gloss. Terms Timber Preservation* (*B.S.I.*) 15 *Pinhole*, a worm hole not more than 1·5 mm in diameter.

c. A very small cavity in a solid, esp. a casting.

1906 H. Adams *Cassell's Engineers' Hand-bk.* IV. 174 Pin-holes, or blow-holes in brass castings are produced by overheating the metal. 1925 *Jrnl. Inst. Metals* XXXIII. 227 The 'pin-holes' are small cavities more or less spherical in shape, fairly uniformly disseminated through the body of the casting. 1947 J. C. Rich *Materials & Methods Sculpture* iv. 70 The most common surface blemishes that may mar an otherwise perfect cast are caused by pinholes resulting from air bubbles imprisoned in the plaster mix. 1968 D. R. Cliffe *Technical Metall.* xi. 268 'Gas holes' (or 'pin-holes') are small, evenly distributed rounded cavities with bright walls caused by the release of dissolved gases during freezing.

d. A very small area from which a coating is absent.

1909 *Jrnl. Industr. & Engin. Chem.* May 295/1 Dealers who were previously large importers of tin plate..are willing to admit that even in the 'good old times' they were greatly annoyed by the so-called 'pin holes' in their goods. 1932 E. S. Hedges *Protective Films on Metals* vii. 212 The examination of perforated cans has shown that the holes may occur..at pinholes in the tin coating. 1970 J. A. Scarlett *Printed Circuit Boards* iv. 53 Most firms use Beta-ray backscatter tests to check thicknesses of gold plating, together with an electrograph porosity test against pinholes.

e. Ellipt. for *pin-hole camera.

1976 *Broadcast* 29 Nov. 18/1 The camera is a box—be it a still camera, a film camera, a television camera, a Polaroid, a pinhole.

3. a. (Further examples.)

1946 C. W. Briggs *Metall. of Steel Castings* iii. 136 (*heading*) Pinhole porosity. 1962 F. I. Ordway et al. *Basic Astronautics* vi. 275 (*caption*) Pinhole insert to control rate of gas intake. 1973 G. J. Davies *Solidification & Casting* ix. 172 (*heading*) Pinhole cavities in a sand-cast aluminium alloy, the result of hydrogen evolution on cooling.

b. *Photogr.* Having or pertaining to the use of a pin-hole in place of a lens, as *pin-hole camera, photography, picture, work.*

1891 *Phil. Mag.* Feb. 89 As the focal length increases, the brightness (B) in the image of a properly proportioned pin-hole camera diminishes. 1902 A. Watkins *Photogr.* 56 Pinhole pictures..have a tendency to require longer exposures than the mathematically calculated ones. 1940 *Chambers's Techn. Dict.* 644/2 *Pinhole photography*, photography involving the use of a pinhole instead of a lens to form an image on a camera plate. 1948 A. L. M. Sowerby *Dict. Photogr.* (ed. 17) 509 A recent article by L. A. Turner..contains an improved theoretical treatment of pinhole work. 1962 M. L. Haselgrove *Photographers' Dict.* 162 *Pinhole* (*camera*). Replacing the camera lens with a panel in which a small pinhole has been made will result in an image being formed on the milk.

c. pinhole borer, a small brown or black beetle or its larva belonging to the families Scolytidæ or Platypodidæ, which damages trees or felled timber by boring tunnels into the wood.

1916 *Indian Forester* XLII. 217 (*heading*) Ambrosia beetles or pin-hole and shot-hole borers. 1928 *Forestry Comm. Bull.* No. 9. 12 The pin-hole borers, poorly represented in Europe, are more numerous in the Southern States of North America, and are especially abundant in the tropics. 1946 Cartwright & Findlay *Decay of Timber* xiv. 276 The constant presence of staining around the tunnels of the pinhole borers..is usually due to growth of dark-coloured moulds in the wood surrounding the tunnels. 1963 N. E. Hickin *Insect Factor in Wood Decay* viii. 256 Ambrosia beetles are also called Pinhole Borers. 1975 G. Evans *Life of Beetles* iv. 92 Pin-hole borers, or 'ambrosia beetles'..are small beetles which bore very narrow tunnels of about the diameter of a thick pin.

pinholed, *ppl. a.* (Later example.)

1928 *Forestry Comm. Bull.* No. 9. 13 Apart from the tunnels cut by the beetles, secondary damage is found in pin-holed timber. This consists in a staining of the wood in a longitudinal direction on each side of the pin-hole.

pi·nholing, *vbl. sb.* [f. Pin-hole + -ing[1].] The presence or production of pin-holes.

1925 *Jrnl. Inst. Metals.* XXXIII. 228 These facts..led to the conclusion that the cause of 'pinholing' [in castings] is occluded gas. 1941 *Light Metals* IV. 133/1 The foundryman..has to produce castings free from serious pinholing. 1947 J. C. Rich *Materials & Methods Sculpture* ii. 51 Pinholing is frequently caused by firing the ceramic body and the glaze at the same time, although air that has been incorporated in the clay mass, dirt, and quick kiln cooling may also cause pinholing.

pinic, *a.* For $C_{20}H_{30}O_2$ read $C_9H_{14}O_4$.

pinion, *sb.*[5] Add: **2.** (Earlier and later examples.) Also *attrib.* in *pinion nut, pine.*

1831 J. O. Pattie *Pers. Narr.* 43 A nut..which grows on a tree resembling the pine, called by the Spanish, pinion. 1846 R. B. Sage *Scenes Rocky Mts.* xxi. 172 The hills enclosing the valley..are high and precipitous,—affording numerous groves of pine, pinion and cedar. *Ibid.*, Wild turkeys..will thrive in an extraordinary manner upon pinion-nuts. 1878 J. H. Beadle *Western Wilds* xi. 173 On many of the hills grows the pinion pine. 1960 'I. Devi' *Yoga for You* 189 Sunflower seeds or pinion nuts to taste. 1970 *People* (Austral.) 26 Aug. 44/4 Pinion nuts from the cones of introduced pines are a tasty novelty. 1979 *Yale* Apr. 39/2 (Advt.), 40 beautiful acres high ground with pinion, mountain views near mining town between Santa Fe & Albuquerque.

‖**pinjrapol** (pi·ndʒrāpǝ̄l). Also **panjrapol, panjrapor, pinjrapole,** etc. [ad. Gujerati *pā̃jrāpol*, f. *panjra* (Skr. *panjara, pinjara*) a cage + *pol* an enclosed yard.] In India, an enclosure, reserve, etc., where old or sick animals are kept.

1808 R. Drummond *Illustr. Gram. Parts Guzerattee* s.v. *Pinjrapole* or *Pánjrapól.* Every marriage and mercantile transaction amongst them is taxed with a contribution for the *Pinjrapole* ostensibly. 1832 C. Coleman *Mythol. Hindus* xiii. 222 These hospitals are called *pinjra-pul*, and contain animals of various descriptions. 1855 H. H. Wilson *Gloss. Judicial & Revenue Terms* 418/2 *Pinjrápor*, or *-pol*, also read *Pánjrápor*, or *pol*, Guz..., an hospital for animals, kept up by the Jains of Guzerat in various places, out of small fees levied at marriages and on mercantile transactions. 1873 E. Balfour *Cycl. India* (ed. 2) IV. s.v. *Pinjrapol.* The Bombay Pinjrapol owes its origin as much to the Parsee worship of sacred dogs as to the superstitions of the Jains. 1929 H. G. Rawlinson *Ovington's Voy. to Surat* 178 These animal hospitals or *pinjrapols* date back to the days of Asoka. 1960 *Guardian* 21 Apr. 12/5 A pinjrapole is a 'Cheshire home' for cows. 1968 B. Walker *Hindu World* I. 16 The early Buddhists and Jains built hospitals for the care of animals, birds and even insects. Such institutions have survived through the centuries and are today known as *piñjrapol.*

pink, *sb.*[2] Add: **2. b.** A young grayling.

1901 H. A. Rolt *Grayling Fishing in S. Country Streams* i. 12 A one-year-old grayling is called a 'pink', and has neither spots nor lateral lines which can be observed. 1939 W. C. Platts *Grayling Fishing* vi. 60 Rolt says that a one-year-old grayling is called a 'pink', and a two-year-old a 'shut' or 'shote' grayling... I have rarely come across these terms in general use. 1952 F. White *Good Eng. Food* I. iv. 55 The principal grayling rivers..are..the Teme (where yearling fish are termed 'pinks' and second year fish 'shutts' or 'shots' or 'sheets') [etc.].

c. = *pink salmon* s.v. *PINK *sb.*[4] and *a.*[1] C.c.

1921 *Daily Colonist* (Victoria, B.C.) 11 Mar. 19/4 The canneries announced their intention of packing practically no pinks or chums this season. 1935 W. M. Halliday *Potlatch & Totem* 155 After the close of the sockeye season, what are commonly known as humpback salmon were caught; these are the fish which are now classed under the technical name of 'pinks'. 1965 A. J. McClane *Standard Fishing Encycl.* 681/2 The ocean and Puget Sound sport fisheries take many pinks, but it is the commercial effort that accounts for the greatest take.

pink, *sb.*[4] and *a.*[1] Add: **A.** *sb.* **I. 2. b.** (Further examples.) Also freq. with ellipse of *of condition, of health*, etc. *colloq.*

a 1821 Keats *Castle Builder* in *Poetical Wks.* (1907) 298 Let me think About my room,—I'll have it in the pink; It should be rich and sombre. 1845 Dickens *Let.* 18 Mar. (1977) IV. 282 Of all the picturesque abominations in the World, commend me to Fondi. It is the very pink of hideousness and squalid misery. 1905 *Kynoch Jrnl.* Oct.-Dec. 201 Makers may despatch explosives from the factory in the pink of condition. 1914 *Isle of Man Weekly Times* 21 Nov. 7/5 He says that he is 'in the pink'. 1916 C. Winchester *Flying Men* 193, I saw a couple of R.F.C. officers..the other day. They looked 'in the pink'. 1923 Wodehouse *Inimitable Jeeves* xi. 115 'Oh, hallo!' I said. 'Going strong?' 'I am in excellent health, I thank you. And you?' 'In the pink. Just been over to America.' 1929 J. B. Priestley *Good Companions* II. vii. 453, I am writing these lines to say I am still in the pink and hoping you are the same. 1937 A. Huxley *Let.* 25 Feb. (1969) 415 Quant à moi, I was in the pink until about a week ago. 1950 [see *COIN *v.*[1] 5 d]. 1973 'P. Malloch' *Kickback* vi. 37 Gilchrist shook hands. 'O.K. How about you?' 'In the pink,' Campbell said. 1976 Dexter & Makins *Testkill* 129 A young Alsatian in the pink of condition.

II. 5. c. As a colour commonly used on maps to indicate a British colony or dominion. Cf. *RED *sb.*[1] 1 e.

1913 C. Mackenzie *Sinister St.* I. II. xv. 407 She said half the world was composed of fools which accounted for the preponderation—I mean preponderance—of pink on the map.

7. As the name of varieties of the potato. Cf. Pink-eye 1.

1853 *Trans. Mich. Agric. Soc.* V. 208 Some of the more approved kinds are..the White, Red, and Strawberry Pinks. 1861 Mrs. Beeton *Bk. Househ. Managem.* 589 The Lancashire Pink is also a good potato, and is much cultivated in the neighbourhood of Liverpool.

8. A pink ball in snooker and some related games.

1910 *Encycl. Brit.* III. 938/2 It is also permitted in some rooms to take blacks and pinks alternately without pocketing a coloured ball between the strokes. 1935 *Encycl. Sports* 570/1 Black is on the billiard spot: pink on the centre line of the table, touching the apex ball of the pyramid. 1976 *Milton Keynes Express* 28 May 55/4 He played the cue ball the full length of the table, swerving past the green, back up the table, not only to hit the red but to pot it, plus the black, yellow, green, brown, blue and but for a miss on the pink would have cleared the table to win the frame. 1978 *Guardian* 7 Feb. 20/6 Pulman twice missed eminently possible pinks, with position on the black there for the taking.

9. *U.S. Blacks' slang.* A white-skinned person. Also *comb.*, as **pink-chaser** (see quot. 1970).

1926 C. Van Vechten *Nigger Heaven* I. ix. 157 Funny thing about those pink-chasers the ofays never seem to have any use for them. 1945 L. Shelly *Jive Talk Dict.* 16/1 *Pink*, pretty white girl. 1970 C. Major *Dict. Afro-Amer. Slang* 91 *Pink chasers*, black people who deliberately cultivate friendships with white people. 1973 'Trevanian' *Loo Sanction* (1974) 159 P'tit Noel shrugged. 'All pinks sound alike.'

10. A person whose politics are left of centre, but closer to the centre than those of a 'red'; a radical; a liberal socialist. *colloq.* Cf. *parlour pink* s.v. *PARLOUR 2 f.

1927 U. Sinclair *Oil!* xiii. 313 He's nuts on this red-hunting business, and the pinks are worse than the reds, he says. 1943 K. Tennant *Ride on Stranger* xx. 225 Wilmot electorate covered an area of residential water-side suburbs inhabited less by Reds than by Pinks of all shades and hues. 1956 A. Wilson *Anglo-Saxon Att.* I. iv. 115 Less-informed business friends spoke to Robin of intellectuals as communists or pinks. 1968 *Punch* 17 July 95/3 The Tory Party..now clutches pinks, finks and crumb-bums to its shrivelled teats. 1976 Scott & Koski *Walk-In* (1977) xxi. 140 His college professors..thought that the boy was perhaps slightly leftish but no more so than most of the Pinks over at State. 1978 'R. Cassilis' *Winding Sheet* II. xii. 109 One of those old-fashioned egalitarians, like the pompous Pinks who had once been the backbone of the..Labour Party.

11. A pink gin; the bitters in this drink.

1942 G. Hackforth-Jones *One-One-One* xxii. 203 'Eeyore' Smith absent-mindedly added a dash of 'pink' to his evening aperitif. 1969 G. Greene *Trav. with my Aunt* I. xx. 213 'Another double', 'Pint of best bitter', 'Double pink'. 1976 'F. Clifford' *Drummer in Dark* iv. 15 'What'll it be?' 'A pink, please.'

B. *adj.* **1. c.** Of the coloration of a newspaper: indicating a sporting edition.

1922 Joyce *Ulysses* 659 The Gold Cup flat handicap, the official and definitive result of which he had read in the *Evening Telegraph*, late pink edition.

d. Of the coloration on a map: indicating a British colony or dominion. Cf. *PINK *sb.*[4] 5 c.

1960 N. Mitford *Don't tell Alfred* vii. 74 It was bad luck for Alfred that the government..should be determined to paint the Minquiers pink on the map. 1973 *Listener* 20 Dec. 857/2 Industrialisation played a big part in the drive to paint the map pink. British industries needed raw materials. 1976 C. Bermant *Coming Home* I. vi. 88, I took it that the great green-coloured mass of the Sahara would pass to Britain... Gradually the whole of Africa glowed pink before my eyes. 1979 *Listener* 26 July 112/3, I was drilled in geography... Most areas, I remember, were coloured pink on the map.

3. Politically left of centre, progressive; applied to socialism of a less extreme character than that denoted by 'red'; loosely, Communist. Also *absol.* Cf. *PINK *sb.*[4] 10.

1837 De Quincey in *Tait's Mag.* Feb. 71/1 Amusing it is to look back upon any political work of Mr Shepherd's..and to know that the pale pink of his Radicalism was then accounted deep, deep scarlet. 1859 Lytton *What will he do with It?* I. i. 9 Young 'un, I'm a Tory—that's blue; and Spruce is a Rad—that's pink! 1920 Mrs. P. Snowden *Through Bolshevik Russia* 180 The people's flag is palest pink, It's not so red as you might think. 1924 *Scribner's Mag.* Oct. 441/1 The Middle West is becoming pink. But it is genuine American pink. Not Moscow Red! 1939 A. Thirkell *Before Lunch* iv. 84, I wouldn't mind her trying to run her pink politics down my throat..though I never see why being a Communist should make one abhor washing. 1973 *Listener* 28 June 864/1 'I am not Communist, but..I am a little pink.' Thus spake King Sihanouk of Cambodia. 1979 'S. Woods' *This Fatal Writ* 81 The description 'pale pink intellectual' is far too tame for Susan... And he's just the opposite, a true blue Tory.

4. Violent, extreme; utter, absolute (see also quot. 1896). *slang.*

1896 W. C. Gore in *Inlander* Jan. 149 *Pink*, used to intensify the negative. 'He didn't know a pink thing about the lesson.' 1901 *Daily Express* 28 Aug. 4/3 The master of the house flies into a pink rage because his chop is not done. 1946 B. Marshall *George Brown's Schooldays* 145 These rotten new kids really are the pink limit.

5. Slightly indecent, violent, or vulgar; mildly 'blue' (see Blue *a.* 9 in Dict. and Suppl.).

1898 R. Hichens *Londoners* xvi. 280 Lovely needle-work! That's a funny beginning for a Pink un. 1900 *Daily News* 28 May 3/1 Most of their adjectives have a decidedly pink tinge. 1979 J. Melville *Wages of Zen* xi. 117 One cinema showing 'pink films'..and one strip show.

6. Of a plan, process, etc.: that must be kept secret.

1924 *Discovery* June 83/1 Little was said about it [*sc.* wireless direction for boats and torpedoes] and in navy parlance it is a subject which is still slightly 'pink', a cryptic term indicating that even if we do happen to know something, we are not prepared to make a song about it. 1925 Fraser & Gibbons *Soldier & Sailor Words* 224 *Pink*, secret. An expression in some Government Offices during the war for secret telegrams. 1962 Granville *Dict. Sailors' Slang* 88/2 *Pink*... 2. Secret, hush-hush, from the pink (confidential) signal pads used in the Navy.

7. Of a person: white-skinned. Cf. *PINK *sb.*[4] 9.

1936 G. B. Shaw *Millionairess* Pref. 121 Even in Africa, where pink emigrants struggle with brown and black natives for possession of the land, and our Jamaican miscegenation shocks public sentiment, the sun sterilizes the pinks to such an extent that Cabinet ministers call for more emigration to maintain the pink population. 1971 *Rand Daily Mail* 3 Apr. 5/8 We Pink South Africans are in danger of being cut off from the world. 1977 P. Ustinov *Dear Me* iii. 26 On my first application to enter the United States..I described my colour as pink. I was told sternly that I was white, a fact which I denied, relying upon an Embassy mirror for evidence. A great deal of time was wasted, more especially since I failed to realize the subliminal implication of the word 'pink' in those days.

8. Phrases. *strike me pink!* (slang): an exclamation of astonishment or indignation; *to paint the town pink*: to go on a spree (after *to paint the town red* s.v. Paint *v.*[1] 9); *to swear pink* (colloq.): to make vehement protestations; to 'swear blind'; *to tickle pink*: see *TICKLE *v.*

1902 E. Nesbit *Five Children & It* viii. 218 When he beheld the magnificent proportions of Robert he said.. 'Strike me pink!' 1922 Joyce *Ulysses* 623 And there he was at the end of his tether after having often painted the town tolerably pink. 1931 A. P. Herbert *Derby Day* III. 115 Ten thousand serpents! Strike me pink! Where's that girl? She'll go to clink! 1956 E. Pound tr. *Sophocles' Women of Trachis* 20 And you swore pink they were bringing her to be Heracles' wife. 1969 *Sunday Mail Mag.* (Brisbane) 7 Sept. 10/1 He was further reported as commenting on certain African members of the Commonwealth in the words: 'Strike me pink, they'll do me for bloody butchers.'

C. a. *pink-brown, -white* (earlier example); *pink-and-white* (further examples).

1853 C. Brontë *Villette* I. xiv. 292 Your pink and white complexion. 1860 Geo. Eliot *Jrnl.* in J. W. Cross *George Eliot's Life* (1885) II. x. 169 The Churches..with their wealth of gilding and rich pink-brown marbles. 1895 W. B. Yeats *Poems* 148 But all the little pink-white nails have grown To be great talons. 1965 A. Christie *At Bertram's Hotel* ii. 22 It made her come to life again—Jane Marple, that pink and white eager young girl. 1979 P. Mason *Skinner* xiv. 96 We shall be junior to pink-and-white boys of eighteen straight from England.

b. *pink-cheeked, -complexioned, -frilled, -haired, -handed, -scrolled, -vested* adjs.

1805–6 Wordsworth *Prelude* (1959) VII. 260 Equestrians, Tumblers, Women, Girls, and Boys, Blue-breech'd, pink-vested. 1906 *Daily Chron.* 23 Aug. 5/6 A white gown and blue picture hat and pink-frilled parasol. 1940 E. Pound *Cantos* lxix. 174 Squad of the pink-haired snot. 1946 S. Spender *European Witness* ix. 46 A pink-complexioned mild-mannered man. 1955 D. Davie *Brides of Reason* 8 The nausea that struggles to despatch Pink-handed horror in a craggy room. 1962 I. Murdoch *Unofficial Rose* iv. 41 She thrust a white-quartered green-eyed Madame Hardy in between two lilac-shaded pink-scrolled Louise Odiers. 1967 A. West in *Coast to Coast 1965–66* 219 It was a soft, pink-cheeked face. 1978 T. Gifford *Glendower Legacy* 264 She..had a succulent moist look, freshly showered, pink-cheeked.

c. pink bollworm, the pinkish larva of a small brown moth, *Pectinophora gossypiella*, of the family Gelechiidæ, which feeds on the lint or seeds of cotton bolls; **pink button** *Stock Exchange*, a jobber's clerk; **pink champagne,** rosé champagne; champagne to which a small quantity of still red wine has been added; **pink disease,** a disease of children caused by mercury poisoning, characterized by pinkness of parts of the body, restlessness, and photophobia; **pink elephant,** used as a type of the extraordinary or impossible; also (chiefly pl.), a characteristic apparition seen by someone drunk or delirious; cf. **pink rat*(*s*); **pink-foot** (b) as *sb.*, = next (also *attrib.*); **pinkfooted goose,** substitute for Latin name *Anser fabalis brachyrhynchus*; (earlier and later examples); **pink gin,** gin-and-bitters; **pink lady** *U.S.*, (*a*) a cocktail comprising gin, egg white, grenadine, and other ingredients; (*b*) (see quot. 1968–70); **pink madder**: see Madder *sb.*[1] 3; **pink noise** *Physics*, random noise having equal energy per octave, and so differing from white noise in having a greater proportion of low-frequency components; **pink paper,** a parliamentary paper containing the information specified in quot. 1894; **pink pine** *N.Z.*, a small forest tree, *Dacrydium biforme*, of the family Podocarpaceæ, bearing linear juvenile leaves and scale-like adult ones, and yielding a resin from which manool is manufactured; **pink rat(s),** a characteristic apparition seen by someone drunk or delirious; cf. **pink elephant*; **pink salmon,** the humpback salmon, *Oxyrhynchus gorbuscha*; **pink slip** *U.S.*, a notice of dismissal from employment; also *transf.* and *fig.*; hence **pink-slip** *v. trans.*, to dismiss, to fire; **pink spot,** used to designate a substance of uncertain composition found in the urine of some schizophrenics, observed as a pink spot on a chromatogram of it; **pink tea** *N. Amer.*, a formal tea party or other social engagement; an exclusive gathering; also used as a type of the polite or genteel; also *attrib.*; **pink thorn,** a pink-flowered variety of the hawthorn, *Cratægus monogyna*; **pink toe(s)** *U.S. Blacks' slang*, a light-skinned black woman; a white girl; **Pink 'Un** [Un, 'un[2]], a nickname for a newspaper printed on pink paper, *spec.* (*a*) *The Sporting Times*; also, a reporter for this newspaper; (*b*) *The Financial Times*; **pinkwash,** a composition used for rendering walls, etc., pink; so **pink-washed** adj.; **pink wine,** (*a*) *slang*, champagne; (*b*) vin rosé.

1906 H. Maxwell-Lefroy *Indian Insect Pests* III. viii. 94 The pink boll-worms are most abundant when the cotton forms bolls. 1917 *Jrnl. Agric. Res.* IX. 343 The pink bollworm..is one of the most destructive cotton insects known. 1932 [see *grain-moth* s.v. *GRAIN *sb.*[1] 19]. 1955 *Sci. News Let.* 23 July 56/2 The preferred food of the pink bollworm larva is the kernel of the cotton seed. 1972 Swan & Papp *Common Insects N. Amer.* 325 Larvae of the Cotton Steam Moth..resemble pink bollworms. 1973 *Times* 16 June 18/2 Pink buttons are not..the female equivalents of blue buttons. 1974 *Sunday Tel.* 7 Apr. 29/3 As 'pink buttons', they look after all the firm's communications, both between the floor of the House and the offices, and between the brokers and country exchanges. 1838 J. Kenyon *Poems* 88 Lily on liquid roses floating—So floats yon foam o'er pink champagne. 1940 N. Mitford *Pigeon Pie* vi. 105 Oysters and pink champagne. 1974 P. Erdman *Silver Bears* v. 63 Sorbet flavoured with pink champagne. 1921 *Med. Jrnl. Austral.* 19 Feb. 146/1 When the rash is marked it is common to find the glands in the axillæ and groins enlarged. It is this pink rash, that leads to the name 'pink disease'. 1921 *Trans. 11th Sess. Australasian Med. Congr. 1920* 444 In Sydney the entity of this illness has long been recognized, and it is usually spoken of by Dr. Clubbe and others as 'the pink disease'. 1959 D. Stowens *Pediatric Path.* x. 127/2 Acrodynia (Pink disease, Swift's disease, Feer's disease, Erythredema polyneuropathy). This bizarre condition is a manifestation either of poisoning by or hypersensitivity to, mercury. 1974 Passmore & Robson *Compan. Med. Stud.* III. xvii. 28/1 Pink disease earns its name from the colour of the hands and feet and not from an imaginary Dr Pink to whom many students credit its discovery. 1940 *This Week* 5 Oct. 2 Pink elephants. 1943 F. Brown *Angels & Spaceships* (1954) 88 You mean if I saw a pink elephant I wouldn't believe it? 1946 P. Larkin *Jill* 77 Whiskey? Would it make him drunk, would he stagger about and see pink elephants? 1906 E. W. Hildick *Boy at Window* v. 45 It's like pink elephants. Folk 'ud think you'd been drinking if you went round saying you'd seen white mice running about wild! 1973 L. Cooper *Tea on Sunday* xxxiv. 246 'I heard somebody.' He'd be seeing pink elephants next. 1931 H. J. Massingham *Birds of Seashore* 38 Individual birds do not conform to pink-foot pattern. 1956 C. Willock *Death at Flight* iii. 31 At least a thousand pinkfeet. Gone out on the muds for the night. 1957 D. A. Bannerman *Birds Brit. Isles* VI. 239 In 1951 pink-foots arrived in Britain unusually late. 1972 *Shooting Times & Country Mag.* 4 Mar. 7/1 Three-quarters of the pinkfeet that winter in Britain come from Iceland. 1839 A. D. Bartlett in *Proc. Zool. Soc.* VII. 7 On a new British species of the genus *Anser*... Pink-footed Goose... Legs and feet, of a reddish flesh colour or pink. 1843 W. Yarrell *Hist. Brit. Birds* III. 66 The voice of the Pink-footed Goose differs from that of the Bean Goose in being sharper in tone. 1932 *Discovery* Aug. 244/2 White-fronted and pink-footed geese..are supposed to be nesting at the sources of the rivers running northwards from the ice-cap [in Iceland]. 1976 G. Evans in R. Durman *Bird Observatories in Brit. & Ireland* vii. 140 The Pink-footed Goose is now far from abundant. 1930 H. Craddock *Savoy Cocktail Bk.* I. 124 *Pink gin cocktail.* 1 Dash Angostura Bitters. 1 Glass Gin. 1952 E. Grierson *Reputation for Song* xxii. 178 She knew the type: a big car, and pink gins, and wine for dinner. 1974 J. Mitchell *Death & Bright Water* xix. 230 'Pink gin, please,' he said, and she mixed it efficiently. 1944 S. Bellow *Dangling Man* 149 She had been drinking Pink Ladies, and she was running over. 1946 C. Himes *Black on Black* (1973) 264 'Anything to drink?' the waiter asked. 'I think I should like a pink lady.' 1968–70 *Current Slang* (Univ. S. Dakota) III–IV. 92 *Pink ladies, n.* Barbiturates. (Drug users' jargon). 1972 M. J. Bosse *Incident at Naha* II. 99 There they were, the little pills, the Red Devils, Yellow Jackets, Christmas Trees, and Pink Ladies. 1975 J. Wambaugh *Choirboys* (1976) i. 6 Now drink your Pink Lady. 1961 G. A. Briggs *A to Z in Audio* 151 Pink noise is derived from white noise by applying a rising bass characteristic through the range. 1962 A. Nisbett *Technique Sound Studio* xii. 220 White noise with bass tip-up is 'pink' noise. 1976 *Gramophone* Apr. 1690/1 Measurements were made of the response of my lounge, using pink noise derived from a Rogers noise generator. [1894 *1st Rep. Sel. Comm. Parl. Papers Distribution* p. iii in *Parl. Papers* XIV. 497 A Schedule shall be circulated daily, weekly, or otherwise, as may be found most convenient, giving reference number, title, and short note of contents of all Papers presented to Parliament by Command of Her Majesty, or printed by Order of either House, since the date of the Schedule last issued. This Schedule shall be sent to each Member in the shape of a demand form, printed on pink paper, and returnable post free.] 1906 *Rep. Sel. Comm. Offic. Publ.* 43 in *Parl. Papers* XI. 95 The pink paper was started in 1889 as an experiment, and it was ratified in 1894 by the Committee.

1908 *Rep. Sel. Comm. Publ.* 111 in *Parl. Papers* X. 849 The first Regulation directed the issue of what we all know as the 'pink paper'. 1928 Cockayne & Turner *Trees N.Z.* ii. 43 Pink-pine. A small tree, 15–40 ft. high, or a shrub, with the juvenile leaves distinct from the adult. 1958 *N.Z. Timber Jrnl.* Jan. 46/2 Pink pine... Small tree of sub-alpine forest of New Zealand. Often a shrub. 1969 *N.Z. News* 23 July 4/3 Pink pine..is so slow growing that 18 in diameter trees on the West Coast are believed to be 800 years old. 1914 C. D. Broad *Perception* iv. 266 The pink rats that can only be seen by those who habitually take excess of alcohol. 1925 —— *Mind & its Place* iv. 142 Some bright spirit will at once complain that the pink-rat situation has no object. 1932 H. H. Price *Perception* vi. 147 The celebrated case of the delirious man who 'sees a pink rat'. 1905 D. S. Jordan *Guide to Study of Fishes* II. iv. 71 The humpback salmon, or pink salmon..is the smallest of the American species. 1952 D. F. Putnam *Canad. Regions* 447/1 The humpback or pink salmon weighs, on the average, four pounds. 1976 J. S. Nelson *Fishes of World* 100 Pink salmon have a rigid two-year life span. 1915 'B. L. Standish' *Covering Look-In Corner* ix. 100 And have Murphy hand me the pink slip tonight! 1923 *N.Y. Times* 7 Oct. 2/1 *Getting the pink slip*, being canceled, which notice comes on pink paper. 1951 *Sat. Rev. Lit.* (U.S.) 23 June 7/1 In small colleges and large universities hundreds of instructors and professors are getting 'pink slips' after the diplomas have been handed out. 1953 Berrey & Van den Bark *Amer. Thes. Slang* (1954) §67/4 Pink-slip, give the blue envelope or pink slip,..to give notice of discharge. 1963 Wodehouse *Stiff Upper Lip, Jeeves* ix. 71 You mean that if Madeleine hands Gussie the pink slip, she'll marry you? 1966 T. Pynchon *Crying of Lot 49* v. 114 His wife had.. left him the day after he was pink-slipped. 1975 *New Yorker* 8 Sept. 115/1 Patrolmen on the beat, for example —got their pink slips because firing them was a dramatic way of demonstrating that the city was taking the crisis seriously. [1962 *Nature* 2 June 898/1 The application of a modified Ehrlich's reagent..resulted in a..pink spot.] 1966 *Listener* 14 July 48/1 This 'pink spot' substance, so called from how it appears in chemical analysis, has not yet shown hallucinogenic activity when swallowed by volunteers. 1973 T. A. Ban *Recent Adv. Biol. Schizophrenia* iv. 30 Papers..some confirming but many more challenging the association of 'pink spot' with schizophrenia.... Counterclaims have been advanced that the 'pink spot' is not 3,4-dimethoxyphenylethylamine. 1886 *Weekly Manitoba Liberal* 26 Nov. 8/3 The Pink Tea held under the auspices of the Women's Christian Temperance Union... The ladies in charge were all fittingly attired with pink caps and aprons and some of the gentleman patrons wore pink ties. 1887 *Harper's Mag.* Jan. 204/1 A Protestant good cause is to be furthered by a bazar or a 'pink tea'. 1905 J. London *Let.* 15 Sept. (1966) 184 Do you remember how Bessie dragged Anna Strunsky's name through the mire? Through all the..pink-tea councils? 1906 *N.Y. Even. Post* 17 Nov. 1 From all accounts these [football] battles of the early eighties and late seventies were no 'pink tea' affairs. 1918 [see *BOLSHEVISM]. 1934 N. Sainsbury *Gridiron Grit* in *Stirring Football Stories* (1941) 55 What do you think football is, a pink tea? 1945 *Boulder* (Colorado) *Daily Camera* 2 Nov. 7/4 Yes, the war was no pink tea. 1952 *North Star* (Yellowknife, Northwest Territories) Dec. 2/1 We were only a name mentioned disparagingly at the clique's pink teas. 1852 C. M. Yonge *Two Guardians* x. 165 The pinkthorn, dressed in all its garlands, before her window. 1892 —— *Old Woman's Outlook* xi. 263 There stood on the lawn..a pink thorn. 1942 Z. N. Hurston in *Amer. Mercury* July 96/1 *Pink toes*, yellow girl. 1965 C. Himes (*title*) Pinktoes. *Ibid.* 216 When Word whispered it about that even the great Mamie Mason had lost her own black Joe to a young Pinktoe, the same panic prevailed among the black ladies of Harlem as had previously struck the white ladies downtown. 1970 C. Major *Dict. Afro-Amer. Slang* 91 *Pinktoes*, a black man's white girl friend; a white girl. 1887 *Referee* 31 July 2/1 Before doing so, I took the advice of one John Corlett, who proprietors a paper called the *Pink 'Un*. 1898 A. M. Binstead (*title*) A Pink 'Un and a Pelican. 1902 G. Calderon *Downy V. Green* xii. 75 Downy amused himself with the only two weeklies that were in evidence, the 'Pink 'Un' and the 'Church Times'. 1930 W. S. Maugham *Cakes & Ale* ix. 107 He gave me the *Pink 'Un* every week and I..read it in my bedroom. 1955 [see *MAN *sb.*[1] 22 c]. 1970 Partridge *Dict. Slang* Suppl. 1330/1 *Pink 'Un*..*The Financial Times*, founded in 1913. 1975 *Blackw. Mag.* CCCXVIII. 126/1 A Pink 'Un was a member of the staff of the *Sporting Times* or one of its close associates. 1979 *Guardian* 2 Jan. 24/4 Today..the first Financial Times will hit Wall Street... But for all the..computer setting..the new international Pink 'un depends very much for its birth on the weather. 1953 Dylan Thomas *Under Milk Wood* (1954) 23 The main street, Coronation Street, consists, for the most part, of humble, two-storied houses many of which attempt to achieve some measure of gaiety by prinking themselves out..by the liberal use of pinkwash. 1926 W. J. Locke *Stories Near & Far* 74 A long, two-storied, pink-washed dwelling. 1936 M. Allis *Eng. Prelude* xxxiii. 253 Lavenham, with its pink-washed houses. 1976 *Eastern Even. News* (Norwich) 27 Aug., The track..passes closely to the right of a pinkwashed farmhouse. 1909 J. R. Ware *Passing Eng.* 197/1 *Pink wine* (*Military*), champagne. 1946 A. L. Simon *Let Wine be Wine* 10 *Rosé* or pink wine is made in a number of different ways, either from grapes with a light red or pinkish skin; or from black grapes the skins of which are not left in the fermenting vat for more than a short while; or from red and white wines mixed together. 1972 *Times* 16 Sept. 11/1 Pink wines from southern European vineyards tend to be dark in hue.

pink, *sb.*[6] Add: **3.** = *PINKING *vbl. sb.*[3]; a metallic rattle.

1927 *Fuel in Sci. & Pract.* VI. 121/1 Ricardo attributed the 'pink' to the sudden inflammation of residual unburnt charge owing to its compression by the expanding burnt and burning portion. 1934 *Automobile Engineer* XXIV. 346/1 'Detonation' or 'pink' might occur in any class of engine. 1946 [see *KNOCKING *vbl. sb.* 1].

Pink, *sb.*[8], slang abbrev. of Pinkerton.

1904 'No. 1500' *Life in Sing Sing* i. 6 Don't you know me? I'm one of the Pinks. *Ibid.* xxiii. 263 Pink had me framed and it was like finding rags to the pusher... The cashier..picked my picture from the Rogues' gallery, where Pinkerton placed it some time ago. 1955 D. W. Maurer in *Publ. Amer. Dial. Soc.* xxiv. 141 The agency is called *the eye*..and its operators are sometimes called *pinks*. 1975 J. Gores *Hammett* (1976) i. 16, I was a Pink... A detective for the Pinkerton Agency.

pink, *v.*[1] **2. b.** Delete †*Obs.* and add later example. Also, to nick or wound slightly with a bullet.

1931 R. Campbell *Georgiad* I. 14 'Onoto'—guns, As sported by Chicago's crooked sons, Able, at once, to.. pink a stray policeman in the neck. 1950 *N.Y. Times* 30 Dec. 27/1 Wall has been the victim of three attempted assassinations, in two of which he was 'pinked', as he expressed it.

d. (Later example.)

1963 *Times* 7 Feb. 3/5 At the end of the round Aldridge's left eye was looking 'pinked'.

pink, *v.*[3] Restrict *Sc.* to sense in Dict. and add: **2.** *intr.* Of an internal-combustion engine: to exhibit pinking (*PINKING *vbl. sb.*[3]). Of a fuel: to cause pinking. Also *fig.*

1904 Kipling *Muse among Motors*, That cursed left-hand cylinder the doctors call my heart Is pinking past redemption—I am done! 1925 A. W. Judge *Carburettors & Carburation* ii. 19 The principal advantage of benzole is its higher detonating compression value; this enables it to be used in high compression petrol engines liable to 'pink' or knock, without experiencing these effects. 1933 *Petroleum Handbk.* (Shell Internat. Petroleum Co.) viii. 145 The tendency of a fuel to pink or detonate is its most important property in use. 1955 *Times* 12 July 12 The car tested was inclined..to pink slightly in accelerating from a low engine speed. 1970 'D. Halliday' *Dolly & Cookie Bird* vi. 78 My brain was pinking like the old Morris. 1972 *Drive* New Year 122/2 Such driving on the recommended 2-star petrol caused the engine to 'pink' noisily.

pink (piŋk), *v.*[4] [f. Pink *a.*[1]] **a.** *intr.* To become pink. Also with *up.*

1854 A. E. Baker *Gloss. Northamptonshire Words* II. 116 *Pink*, to blush. 'How she *pinks* up!' 1909 R. A. Wason *Happy Hawkins* 136, I hadn't never seen those cheeks pink up for anything but fun or anger before. 1927 P. Marks *Lord of Himself* 32 Mrs. Peter's eyes were sparkling again, and her cheeks pinked with happy colour.

b. *trans.* To shear (a sheep) closely so that the colour of the skin shows through; esp. in phr. *to pink 'em.* (*Austral.* and *N.Z. colloq.*)

1898 *Bulletin* (Sydney) 17 Dec. 15 Another term for *fine-cut* is *pinkin' 'em*—showing the pink flesh. 1899 W. T. Goodge *Hits! Skits! & Jingles!* 155 And he 'pinked' him like a leather-neck when squatters paid a pound! 1900 H. Lawson *Verses, Pop. & Humorous* 168 Get the bell-sheep out;.. But 'pink' 'em nice and pretty when you see the Boss's boots. 1933 *Bulletin* (Sydney) 15 Nov. 28/1 Instead of being 'pinked', there was sufficient wool left on as weather protection. 1956 G. Bowen *Wool Away!* (ed. 2) 156 *Pink 'em*, to make a very good or better than average job of a sheep. Shearers sometimes call this a 'special cut'.

c. *trans.* To make pink (in various senses of the adj.).

1927 W. Deeping *Kitty* xxvi. 330 You've more idea of colour than I have. I'm too fond of pinking things. 1929 D. H. Lawrence *Pansies* 22 The pretty pretty bourgeois pinks his language just as pink If not pinker.

pinked, *ppl. a.* Add: **1.** (Later example.)

1929 *Papers Mich. Acad. Sci., Arts & Lett.* X. 314/2 *Pinked*, struck by bullets.

2. b. (Earlier example.)

1862 Mrs. J. B. Speid *Our Last Yrs. in India* xi. 273 A black satin polka jacket with 'pinked' flouncing.

pinken, *v.* Delete *rare* and add: **1.** (Further examples.)

1936 Wodehouse *Laughing Gas* vii. 76 It caused Miss Brinkmeyer to pinken and breathe heavily. 1954 D. Ames *Crime, Gentlemen, Please* xxii. 130 Jack, too, was aware of the chill. He pinkened. 1976 *Listener* 15 Apr. 465/2 As the sky pinkened, we turned up a narrow tributary.

2. *trans.* To make pink.

1968 C. Nicole *Self Lovers* i. 8 Her tan was pinkened by a liquor flush.

Pinkerton. Add: (Further examples.) So also **Pinkerto·nian** *a.*, of, pertaining to, or characteristic of Pinkerton or his men.

1908 *Athenæum* 11 Apr. 442/2 While an article in the ensuing number entitled 'The Art of Advertising Made Easy', in which Colburn's Pinkertonian methods are held up to ridicule, is shown to be Hood's. 1915 A. Conan Doyle *Valley of Fear* II. vi. 274 'The police?' 'Well, a Pinkerton.' 1955 D. W. Maurer in *Publ. Amer. Dial. Soc.* xxiv. 33 I'd say that better than fifty per cent of Pinkertons are ex-cons. *Ibid.* 38 One..pickpocket, so popular that when he killed a Pinkerton detective more than a hundred thieves..poured in to testify successfully in his defense. 1967 N. Mailer *Cannibals & Christians* I. 22 There were tough dull Pinkertons with a tendency to lean on a new visitor. *a* 1971 G. Jackson *Blood in my Eye* (1972) 188 Every time I hear the word 'law' I visualize gangs of militiamen or Pinkertons busting strikes. 1973 L. Hellman *Pentimento* (1974) 177 Hammett..didn't much like to be around people who took dope, in his Pinkerton days he had been more afraid of them than of murderers. 1978 J. Carroll *Mortal Friends* II. iii. 168 Just hope to high heavens..they keep their Pinkertons out.

pink-eye[1]. Add: **2. b.** It also occurs in some livestock. (Earlier and later examples.)

1886 *Arch. Ophthalm.* XV. 451 This form of conjunctivitis is contagious and epidemic; it appears most plentifully in the spring and fall months... From the peculiar congestion of the ocular conjunctiva it has become popularly known as 'pink eye'. 1933 [see *bung-eye* s.v. *BUNG *sb.*[1] 6]. 1938 S. Duke-Elder *Text-bk. Ophthalm.* II. xxxii. 1541 The intensity of the hyperæmia, sometimes associated with petechiæ, is characteristic, giving it [*sc.* Koch-Weeks' conjunctivitis] the popular name of 'pink-eye'. 1951 R. Seiden *Livestock Health Encycl.* 376 Pinkeye or specific ophthalmia..is an inflammatory condition of the eyes of cattle and sheep. 1974 [see *Koch-Weeks bacillus].

4. *slang* (chiefly *Austral.* and *Canad.*). Cheap whisky or red wine. (See also quot. 1945.) Cf. *PINKIE, PINKY *sb.*[3] 1 and Red-eye 4 in Dict. and Suppl.

1900 *Cornhill Mag.* June 778 His capital consisted of a yoke of oxen, a waggon, six four-gallon kegs of pink-eye and a Winchester rifle. 1941 *Coast to Coast* 23 Better put that bottle away... If the trooper comes round somebody'll be getting into trouble for selling Charley pinkeye again. 1941 [see *PINKIE, PINKY *sb.*[3] 1]. 1945 Baker *Austral. Lang.* ix. 166 Recipes as published by an outback newspaper in 1936... Methylated spirits and Condy's crystals. (*Pinky.*)..Addicts of these noxious drinks are known as *meths*..and *pinkeyes*. 1953 W. B. Mowery *Sagas of Mounted Police* 125 At Benders' joint, the price of a pint of pink-eye was a day's hard labor in a mine-head.

pink-eye[2] (pi·ŋkəi). *Austral.* Also **pink-hi, pinkie.** [Aboriginal name.] A festival or holiday.

1924 Lawrence & Skinner *Boy in Bush* viii. 110 It was holiday-*pinkie*, the natives called it. 1929 K. S. Prichard *Coonardoo* 18 The tribes for a hundred miles about had gathered for pink-eye on Wytaliba. 1936 H. Drake-Brockman *Sheba Lane* xi. 131 He found his natives in good tucker and clothes and gave the faithful Jimmy..a horse and cart for the yearly pinkhi, when he visited his tribe. 1969 O. White *Under Iron Rainbow* 139 This year Nolan's Ford Picnic Races and Rodeo..was obviously going to be a successful pink-eye.

pinkie, pinky, *sb.*[1] Add: Also **pinkey.** (Earlier and further examples.) Also *attrib.* and *Comb.*, as *pinkie-stern schooner.*

1843 *Knickerbocker* XXII. 187 The 'pinkie' is a schooner rigged craft,.. sharp at both ends, a short peak running up aft, and designed for a chasing sea. 1873 G. H. Procter *Fisherman's Memorial* 72 Uncle Charlie's first remembrance was of the pinkey fleet. 1882 *Fisherman's Own Bk.* 40 They were the old style of pinkey, without bowsprit or shrouds, with two masts and hempen sails. 1886 [see *CHEBACCO]. 1897 Kipling *Capt. Cour.* vi. 136 My father he run his packet, an' she was a kind a' pinkey, about fifty ton, I guess. 1903 *N.Y. Tribune* 25 Oct. 14 On another occasion the Houghton ran into a pinkey-stern schooner. 1950 R. Moore *Candlemas Bay* 7 Capt. Malcolm Ellis..had gone from a rowboat to a pinky to a mackerel schooner, and finally to a fleet of mackerel schooners. 1972 *Daily Colonist* (Victoria, B.C.) 19 May 1/5 Illusion of jetliner perched atop twin masts of reproduction of old pinky schooner was created this week..as jets broke through low overcast.

pinkie, pinky, *a.*, *sb.*[2] Add: Also **pinkey. B.** *sb.* (Further examples.) Also *attrib.*

Now also quite common in certain areas of the U.S.

1931 A. J. Cronin *Hatter's Castle* I. iii. 51 'I just flicked him with my pinkie,' declared Brodie complacently. 1941 *Sun* (Baltimore) 13 Oct. 8 (*caption*) Pinkey straight up, that's the class way to drink tea, pal! 1948 *Richmond* (Va.) *Times-Dispatch* 15 Mar. 17/4, I grip the ball with my thumb and pinky. 1950 J. Dempsey *Championship Fighting* 34 You might call that pinky knuckle the exit of your power line. 1958 P. De Vries *Mackerel Plaza* 155 My eye met Mrs. Spensible's across the room, and I thrust out my pinkie, smiling innocently. 1962 Auden *Shepherd's Carol* in *Musical Times* Oct. (Suppl.) 1 O lift your lit-tle pin-kie, and touch the win-ter sky. 1965 E. Tunis *Colonial Craftsmen* vi. 140 Even the most elegant lady poured tea or coffee from her cup into her saucer to cool and then, with delicately extended pinkie, drank it from the saucer. 1970 S. Ellin *Bind* lix. 299 Gela stared at him, gnawing at a hangnail on his pinkie. 1971 G. M. Brown *Fishermen with Ploughs* 21 Fish as small as your pinkie. 1973 'J. Marks' *Mick Jagger* (1974) 11 As for Mick, he splashes on some fragrance and checks his eyeliner with his pinkie. 1975 *New Yorker* 1 Dec. 48/1 Seemed like his arm was *always* around somebody, and always there was the smell of his rich wool suit, the flash of a pinkie ring, a cloud of Havana, and Houtek's barony voice saying you and him were friends. 1976 *Scottish Rev.* Spring 3 Are ye still eatin' vinegar with a fork and holdin' out yer pinkie when ye sup yer tea? 1977 *New Yorker* 24 Oct. 122/2 Pagano had two pinky rings with nice stones.

pinkie, pinky (pi·ŋki), *sb.*[3] [f. Pink *sb.*[4] and *a.*[1] + -ie, -y[6].] **1.** Cheap red wine; (see also quot. 1941). *slang* (chiefly *Austral.*).

1897 *Session Paper Cent. Criminal Court* 10 & 11 Mar. 417, I know I have done wrong; it is all through the drink; I have been having a drop of *pinkie*, and I am sorry for it. 1935 K. Tennant *Tiburon* 93 Staines, nodding his fat, puffy face into his cup of pinkie..hadn't a very good head for the cheap raw wine he was drinking. **1936** A. RUSSELL *Gone Nomad* vii. 55 Beer, whisky, 'pinky', delirium tremens, sore heads, and sandy blight were the chief..maladies of the field. **1941** BAKER *Dict. Austral. Slang* 54 *Pink-eye*,..an addict of the noxious drink called 'pinky', the constituents of which are either red wine and methylated spirits or methylated spirits and Condy's crystals. **1958** *Maclean's Mag.* 27 Sept. 63/3 Pinkie [in St. John's, Newfoundland] is a cheap wine highly regarded by waterfront connoisseurs, a chaser for screech. **1959** D. HEWETT *Bobbin Up* (1961) vii. 93 He'd drink anything, they reckoned, plonk, pinkie, straight metho.

2. A white person (see also quot. 1970). *slang.*

1967 *Observer* 10 Sept. 17/2 The racial discrimination that black school-leavers find when they look for jobs is not a surprise: it is a confirmation. By the time they leave school, whites have become 'pinky', 'the grey man' or..'Mr Charlie'. *Ibid.* 17/3 I've got a white friend I've known from school... No, I'm not a pinky-lover! He's learned to think black *and* think white and I can trust him. **1970** C. MAJOR *Dict. Afro-Amer. Slang* 91 *Pinky*.., Afro-American girl who looks white. **1972** K. JOHNSON in T. Kochman *Rappin' & Stylin' Out* 145 *Pinkie*, refers to the skin color of white women.

3. = *PINK *sb.*⁴ 10. Cf. *PINKO *sb.*

1973 *Nation Rev.* (Melbourne) 31 Aug. 1442/3 He called for a Liberal party 'crusade' to defeat the 'reds, the pinkies and the socialists' who are responsible for inflation. **1978** R. BARNARD *Unruly Son* xv. 166 He was always a drawing-room pinkie... As far as contact with the working-class movement was concerned, he hadn't any.

pinkie, pinky (pi·ŋki), *sb.*⁴ [f. PINK *sb.*² + -IE, -Y⁶.] **1.** *S. Afr.* A small marine fish, either the rock grunter, *Pomadasys olivaceum*, of the family Pomadasyidæ, which is only a few inches long and is often used as live bait, or the red grunter, *Pagellus natalensis*, of the family Sparidæ, which is a food fish that may grow to about twelve inches.

1948 *Cape Times* 19 July 1/4 The fish was brought in and gaffed... The bait taken was 'live pinkie'. **1953** J. L. B. SMITH *Sea Fishes S. Afr.* 257 *Pomadasys olivaceum*... Rock-Grunter. Pinky (Natal). *Ibid.* 273 *Pagellus natalensis*... Red Grunter. Pinky. **1966** K. T. LILLIECRONA *Salt-Water Fish & Fishing in S. Afr.* i. 21 In deepish water next to the rocks, all one has to do is cast in this multi-hook trace among the fish, count twenty slowly and then retrieve to find every hook with a pinky on it.

2. The larva of a greenbottle fly of the genus *Lucilia*, used as a live bait in some freshwater fishing.

1958 F. OATES *Coarse Fishing Baits* i. 23 'Pinkies' are well suited for the smaller fry which inhabit lakes and wide sluggish rivers, because being fairly heavy they can be thrown much farther than squats. **1971** *Angling Times* 10 June 12/2 (Advt.), Wholesalers of maggots, pinkies, squats, anattoes, brandlings. **1974** C. C. TRENCH *Hist. Angling* viii. 234 'Pinkies' are the larvae of greenbottles, rather smaller than other maggots, pinkish in colour and used generally for small roach and dace. **1979** *Guardian* 13 June 9/3 If you have got a box of pinkies in your fridge..you are probably..pre-occupied right now... For next Saturday is the opening day of the coarse fishing season.

pinkified (pi·ŋkɪfəid), *ppl. a.* [f. PINK *a.*¹ + -IFY + -ED¹.] Made pink in colour.

1886 R. BROWN *Spunyarn & Spindrift* xxix. 351 The light of the sun came streaming across it, making our sails all pinkified.

pinking, *vbl. sb.*¹ Add: **b.** *pinking machine, scissors, shears*; **pinking-iron** (earlier example).

1761 in E. Singleton *Social N.Y. under Georges* (1902) 242 [I] have ever since been so scrupulous an observer of it [*sc.* taste] that I never was the mark of a pinking-iron behind it. *a* **1865** MRS. GASKELL *Lett.* (1966) 816 Dear Miss Watkins, Thank you very much for the use of the Pinking Machine. **1951** *Catal. of Exhibits, South Bank Exhib., Festival of Britain* 60/1 Pinking scissors. **1979** E. TAYLOR in I. Webb *Compl. Guide Flower & Foliage Arrangem.* viii. 103/1 Pinking scissors will avoid having to hem the edges. **1962** *House & Garden* Dec. 55/2 Pair of pinking shears. **1976** *Evening Post* (Nottingham) 15 Dec. 21 (Advt.), Dress-making scissors, pinking shears, nail scissors, [etc.].

pinking (pi·ŋkiŋ), *vbl. sb.*³ [f. PINK *v.*³ + -ING¹.] (The production of) a metallic rattling sound in an internal-combustion engine as a result of the too rapid combustion of the mixture in the cylinder.

1913 ROGERS & WATSON *Motor Mechanics' Hand-bk.* i. 9 If the compression exceed 90 lb., there is great danger of frequent pre-ignition, and consequent knocking or 'pinking' in the cylinders. **1930** *Flight* 11 July 787 A further change was made to a poor grade spirit, and the symptoms of pinking combined with loss of efficiency were much exaggerated. **1937** [see *DETONATION 1 b]. **1959** *Motor* 19 Aug. 6/2 Full throttle was avoided to prevent pinking. **1968** [see *DETONATION 1 b]. **1970** M. SMITH *Aviation Fuels* vii. 35 High pressure waves.. strike the walls of the combustion chamber with a hammer-like blow, producing a knocking noise. The high pitched, metallic sound known as 'pinking' is due to the vibratory nature of those waves. **1973** *Times* 19 Apr. 35/1 Lead.. is added to petrol to raise octane ratings and prevent 'pinking'.

pinkish, *a.* Add: *pinkish-brown, -grey, -mauve, -purple, -silver, -white* (later examples), *-yellowing*.

1857 GEO. ELIOT in J. W. Cross *George Eliot's Life* (1885) I. vii. 360 The castle is built of stone which has a beautiful pinkish-grey tint. **1952** A. G. L. HELLYER *Sanders' Encycl. Gardening* (ed. 22) 59 Spathe yellowish-green and pinkish-brown. *Ibid.* 118 [*Clinopodium*] *georgianum*.., white to pinkish-purple, to 2 ft. *Ibid.* 276 [*Lilium*] *Kelloggii*, pinkish-mauve, July. **1952** S. SPENDER *Learning Laughter* i. 5 A tall stucco pinkish-yellowing house. **1959** E. H. CLEMENTS *High Tension* vi. 111 Sgurr Dhubh, pinkish-silver in the morning sun. **1961** R. L. GOURSE *With Gall & Honey* xviii. 272 He had a pinkish-white complexion, a small straight nose, [etc.]. **1973** POLUNIN & SMYTHES *Flowers S.W. Europe* iii. 384 A robust plant to 1m or more, distinguished by its large, very dense, globular head of numerous pinkish-purple flowers.

b. Politically somewhat left of centre.

1949 N. MARSH *Swing, Brother, Swing* ix. 206 Sallis's pinkish, facile..observations. **1968** W. ASH *Ride Paper Tiger* vii. 108 Certain pinkish professors back home are making rather a lot of the fact that Ortiz is a political martyr. **1977** *Time* 5 Dec. 67/2 He was born in New York City and spent his early childhood in Bayside, a pinkish nook of Queens.

2. [PINK *sb.*⁴ 2 b in Dict. and Suppl.] Fit, 'in the pink'.

1949 J. CORRIE in J. Marriott *Best One-Act Plays 1948–49* 112 'You are looking very well.' 'I'm feeling very pinkish... Very pinkish, indeed, thank you.'

pinkishness (pi·ŋkiʃnės). [f. PINKISH *a.* + -NESS.] Pinkish appearance; a suggestion of pinkness.

1909 M. B. SAUNDERS *Litany Lane* v. 60 With a pinkishness about his eyes not becoming to his blonde good looks.

pinkly, *adv.* Add: (Further examples.) Also *fig.*, embarrassedly.

1923 H. C. BAILEY *Mr. Fortune's Practice* i. 2 He is plump and pinkly healthy. **1953** G. DURRELL *Overloaded Ark* viii. 156 We both looked at the almost sheer cliffs of N' da Ali gleaming pinkly in the evening sun. **1960** M. SPARK *Ballad of Peckham Rye* vii. 141 Mr Willis pinkly took Merle's hand and glanced towards the shop door. **1973** 'D. HALLIDAY' *Dolly & Starry Bird* ii. 24 'You do such lovely parties, Timothy.' 'Oh, well,' he said pinkly. **1974** *Times* 30 Nov. 9/5 The pinkly roasted grouse. **1975** J. MCCLURE *Snake* xi. 148 Kramer watched the dawn of his insight spread pinkly up from Marais' collar. **1978** *Daily Tel.* 8 July 9/2 A garden where Dorothy Perkins rambles pinkly round the door.

pinkness. Add: **b.** The quality or state of being politically pink (cf. *PINK *a.*¹ 3).

1931 F. L. ALLEN *Only Yesterday* iv. 76 The Fighting Quaker's inquisitorial methods..had at least the practical effect of scaring many Reds into a pale pinkness. **1940** R. S. LAMBERT *Ariel & all his Quality* iii. 76 Attacks upon the BBC for the 'redness' or 'pinkness' of broadcast talks.

pinko (pi·ŋko), *a.* and *sb. slang.* [f. PINK *sb.*⁴ or *a.*¹ + *-O².] **A.** *adj.* **1.** (See quots.)

1925 FRASER & GIBBONS *Soldier & Sailor Words* 224 *Pinko*, drunk. **1941** BAKER *Dict. Austral. Slang* 54 *Pinko*, drunk, esp. on methylated spirits.

2. = *PINK *a.*¹ 3. Chiefly *U.S.*

1957 [see *LONG-HAIR *sb.* 2a]. **1959** C. MACINNES *Absolute Beginners* 38 Your pinko pals did what they wanted to when they got power. **1972** D. LEES *Zodiac* 65 He made Ronald Reagan look like a pinko liberal. **1976** N. THORNBURG *Cutter & Bone* xii. 284 That look and attitude..proclaimed them goddamn ready and eager for any commie revolution the pinko nigger-loving government might be cooking up. **1977** *Transatlantic Rev.* LX. 121 It's the number three song in China, sir. Saw it in one of those magazines my pinko parents subscribe to.

B. *sb.* = *PINK *sb.*⁴ 10. Chiefly *U.S.*

1936 J. G. COZZENS *Men & Brethren* 104 She's a good girl... Now only a healthy pinko. I've snatched her like a brand from the Young Communist League burning. **1948** 'B. ROSE' *Wine, Women & Words* 105, I wouldn't call him a Commie, but if he doesn't get a check from Moscow every week, he's being robbed... Unfortunately the pinko didn't drinko. **1959** *Times Lit. Suppl.* 25 Dec. 753/1 To save his family from being 'smeared' by 'left-wingers and pinkoes' he decides to use the pseudonym Victor J. Fox. **1971** *New Society* 7 Jan. 25/3 The new American jingoism wherein the enemy is not the enemy but all those disloyal pinkos at home. **1976** *Spectator* 14 Feb. 13/3 The statement 'we are all guilty'..is enough in itself to identify the speaker as a trendy pinko.

pinko- (pi·ŋko). [See -O.] Used as a combining form of PINK *sb.*⁴ and *a.*¹ in **pinko-grey** *a.*, of a pinkish-grey colour; *spec.* = WHITE *a.* 4; hence as *sb.*, a 'white' person.

1924 E. M. FORSTER *Passage to India* vii. 62 The remark that did him most harm at the club was a silly aside to the effect that the so-called white races are really pinko-grey. **1953** W. G. WALTER *Living Brain* i. 1 By brain is meant..something more than the pinko-grey jelly of the anatomist. **1961** P. MASON *Common Sense about Race* i. iii. 49 A pinko-grey man is rather more likely than a Negro to have traces of the ridges above the eyes..so prominent in the gorilla. **1964** 'M. INNES' *Money from Holme* x. 68 The pinko-greys out there [*sc.* in Africa] aren't exactly aesthetes. **1973** J. MANN *Only Security* xi. 142 A pinko-grey lib-lab, that's you. **1974** *Times* 5 June 16/5 The pinko-grays, to use E. M. Forster's accurate description, were entirely safe... Britain was about to quit India. **1977** T. HEALD *Just Desserts* iii. 47 His usual pinko-grey complexion.

pinky, *a.*¹ Add: **a.** (Further examples.)

1907 *Westm. Gaz.* 2 Mar. 17/1 The little habit coats.. are generally faced with..emerald-green, or blue, or even pinky-red. **1927** D. H. LAWRENCE *Mornings in Mexico* 29 Pale belly, and soft, pinky-fawn claws. **1946** G. MILLAR *Horned Pigeon* vii. 78 The pinky-red pantiles of the terraced village. **1974** *Country Life* 2 May 1055/2 Raised panels..painted in imitation of pinky-grey marble. **1975** C. FREMLIN *Long Shadow* x. 77 Myrtle's pinky-orangey lighting. **1977** *Vogue* Feb. 94 Pinky blonde, double-faced wool shirtjacket.

c. Also *pinky-faded* adj.

1926 D. H. LAWRENCE *Glad Ghosts* 23 A big pinky-faded carpet.

pinlay (pi·nlēi). *Dentistry.* [Blend of PIN *sb.*¹ + INLAY *sb.*] An inlay or onlay which is held in place partly by a pin or pins inserted in the tooth.

1915 J. K. BURGESS in *Dental Cosmos* LVII. 1338/2 The attachment..is a modified inlay and contour with pins, which I have chosen to call the 'pinlay' attachment. **1946** *Brit. Dental Jrnl.* LXXX. 14 (*caption*) Spring bridge carrying central incisor on pinlay abutment in canine. **1952, 1973** [see next].

pinledge (pi·nledʒ). *Dentistry.* [f. PIN *sb.*¹ + LEDGE *sb.*; cf. prec.] A pinlay, esp. one covering the lingual surface of a tooth and dependent on pins inserted in ledges cut in the tooth for retention and stability.

1915 J. K. BURGESS in *Dental Cosmos* LVII. 1342/1 (*heading*) 'Pinledge' bridge attachment for anterior teeth. *Ibid.*, The anterior attachment is an outgrowth of the pinlay, being constructed on the same principle. I have chosen to call it the 'pinledge'. **1952** *Dental Practitioner* II. 328/1 The pinlay, or pinledge preparation as it is often called, for incisor or canine bridge abutment, has been known in one form or another for many years. **1973** D. H. ROBERTS *Fixed Bridge Prostheses* vii. 125 The pinledge preparation. This differs from the three-quarter pinlay in that the incisal edge of the tooth is not involved and so no gold is displayed labially. **1975** G. T. CHARBENEAU et al. *Princ. & Pract. Operative Dentistry* xiv. 387/1 The pinledge design may be used on incisor and cuspid teeth with conservative loss of tooth tissue.

pin-money. Add: Also *transf.*, spending money; money for incidental expenses; a trivial sum of money.

1892 *Munsey's Mag.* Oct. 112/1 The late Rose Terry Cooke, popular as her writings were, never made more than pin money with her pen. **1926** H. CRANE *Let.* 1 June (1965) 256, I don't think you'll get pin money for the sale of the place. **1957** *Economist* 26 Oct. 291/1 A much better system than relying solely on attendance pin money for all peers would be to pay a full MP's salary to..a score of 'nominated' peers on either side. **1971** *Farmers Weekly* 19 Mar. 94/2 In farming, we tend to use straw liberally and an odd wedge for a child to bed the rabbit hutch seems insignificant. But..you could.. start a sideline enterprise. Good pin money? **1978** S. ALLAN *Inside Job* i. 17 If you did find yourself short of pin money you..could get yourself a job.

b. *attrib.*

1837 T. BACON *First Impr. Hindostan* I. vi. 171 Marriage is..out of the question..unless..the young lady.. have..a small pin-money purse of her own. **1908** *Daily Chron.* 5 Oct. 5/6 This meeting..protests against the employment of the 'pin-money clerk', who is a menace to the clerks of both sexes. **1961** PARTRIDGE *Dict. Slang* Suppl. 1222/1 *Pin-money spoof*, vague, pointless amateurish writing: journalistic: since ca. 1910. **1977** in *Centuryan* (Office Cleaning Services) Christmas 4/1 Six 'pin-money' schoolgirls whose journey home in a mini-bus ended in tragedy were being employed illegally.

pinnacle, *sb.* Add: **2.** Also, a rock projecting out of the sea.

1949 *Sun* (Baltimore) 5 Nov. 3/4 The big Panamanian freighter was listing badly and threatening 'to slip off at any minute'. It was described as being caught on a 'pinnacle' with water of 36 to 60 feet depth around it.

4. *pinnacle rock* (later example).

1916 *Daily Colonist* (Victoria, B.C.) 14 July 11/3 On her last voyage in the westward route the Dora struck a pinnacle rock and developed a serious leak.

pinnate, *a.* Add: **1. c.** *Physical Geogr.* Of a drainage pattern: resembling a feather in plan.

1932 E. R. ZERNITZ in *Jrnl. Geol.* XL. 512 These acute-angled joinings with the rather evenly spaced and parallel tributaries form a pattern so much like that of a feather that it might appropriately be called 'pinnate'... Figure 8 is an example of pinnate drainage. **1942** O. D. VON ENGELN *Geomorphol.* xi. 215 All the other recognized types of drainage pattern: rectangular, trellis, annular, pinnate, contorted, are responses to structure. **1968** R. W. FAIRBRIDGE *Encycl. Geomorphol.* 287/2 Modified dendritic drainage may be described as pinnate, subparallel or anastomatic.

pinned, *ppl. a.* Add: **6. pinned eye** (see quot. and cf. PIN-EYED *a.*).

1842 *Florist's Jrnl.* III. 29 The style or stigma [of an auricula] ought not to rise higher than the stamens, forming what is called a pinned eye, which is reckoned a great deformity.

pinner³. Add: **4.** The workman who inserts the pins in the revolving cylinder of a barrel organ.

1896 *Pall Mall Mag.* Nov. 336 To completely 'set' a cylinder takes an expert workman three days; then it is given to the 'pinner' who carefully hammers the pins into the places designated by the 'setter'. **1921** *Dict. Occup. Terms* (1927) §648 *Pinner*..inserts, with pliers and pressing machine worked by treadle, steel pins in positions marked by music marker on revolving cylinder or roller of barrel organ. **1960** *Classification of Occupations* (General Register Office) 84/2 *Pinner*.., barrel organ mfr.

pinning, *vbl. sb.* Add: **1. b.** Also *pinning-out*.

1905 *Sci. Amer.* 30 Sept. 262/1 The second-sizing and pinning-out is done by hand at so-called batteries.

e. An indication of a relationship, falling short of a formal engagement, between two young people through an exchange of fraternity or sorority pins; the exchange of such pins for that purpose. *U.S. university slang.*

1961 *Ann. Amer. Acad. Pol. & Social Sci.* Nov. 85/1 There are boxed proclamations in the newspaper [of Brooklyn College] of watchings, pinnings, ringings, engagements and marriages. **1964** *Amer. Speech* XXXIX. 194 That peculiar institution, the 'pinning' of quasi-engaged girls. **1967** *Punch* 13 Sept. 378/1 Most fraternities and sororities sustained this perfumed atmosphere of competition by requiring their members to date a different person every date night... I attribute the popularity of pinning—a kind of informal engagement to be engaged, signified by the exchange of fraternity and sorority pins —to the desire to escape from that pattern; certainly people got pinned and unpinned all the time.

pinny, *sb.* Add: (Earlier and later examples.) Also *attrib.* and *fig.*

1851 H. MELVILLE *Moby Dick* II. xxix. 203 A woman's pinny hand,—the man's wife, I'll wager. **1858** J. A. SYMONDS *Let.* 1 Nov. (1967) I. 170 Lady Young..engaged in the construction of *pinnies* for poor children. **1889** E. DOWSON *Let.* c 21 Oct. (1967) 111 She was disporting herself in a superb way in Gt Russell St—hatless & in a 'pinny'. **1939** A. THIRKELL *Brandons* i. 18 'If we had known mummie was coming, we'd have had our clean pinny on,' said Nurse. **1962** J. BRAINE *Life at Top* xvii. 198 'Get me a bloody pinny,' I said, 'and you can go out to work.' **1974** *Times* 15 Oct. 12/7 A new pinny idea of long skirt, chemise to the knee over a skinny sweater. **1975** *Country Life* 11 Dec. 1710/1 A practical and pretty pinny to tie round the waist.

Hence **pinnyed** (pi·nĭd) *a.*, clad in a pinafore.

1963 *Guardian* 20 Feb. 7/2 The pinny-ed skivvy.

pinocle. (Earlier example.)

1864 W. B. DICK *Amer. Hoyle* 127 Bézique is fast becoming popular in the United States... It is known among our German brethren as *Peanukle*. *Ibid.*, The game [*sc.* Bézique] became very popular in Sweden, and was finally introduced into Germany, changed in some respects, and called Penuchle.

pinocytized (pəi·nosəitəizd), *ppl. a. Biol.* [f. *PINOCYT(OSIS + -IZE + -ED¹.] Taken into a cell by pinocytosis.

1970 *Amer. Jrnl. Anat.* CXXIX. 142/1 The nature of the intracellular hydrolysis of phagocytized or pinocytized substances has been established. **1970** *Nature* 26 Dec. 1284/1 Digestion of pinocytized materials should be impaired.

pinocytose (pin-, pəinosəi·tōuz), *v. Biol.* [Back-formation from next.] *trans.* To absorb by pinocytosis. Freq. *absol.*, to carry out pinocytosis.

1960 E. N. WILLMER *Cytol. & Evolution* vii. 118 Epitheliocytes... Can pinocytose. **1962** *New Scientist* 13 Dec. 620/1 A cell may pinocytose for a certain time and then stop. **1967** JAHN & BOVEE in Tze Juan Chen *Res. in Protozool.* I. 114 The effects of inducers which stimulate ingestion of *Tetrahymena* by a starved ameba are reduced if the starved ameba is first allowed to pinocytose for 20 min, before being fed *Tetrahymena*. **1971** *Nature* 19 Mar. 148/1 The two cells pinocytose projections from each other's sarcolemmas.

pinocytosis (pi:no-, pəi:nosəitōu·sis). *Biol.* [f. Gr. πίν-ειν to drink + -CYT(E + -OSIS, after PHAGOCYTOSIS.] A process by which liquid is taken into a cell as a result of the invagination and pinching off of the cell surface so as to form small vesicles.

The word was first coined as F. *pinocytose* (G. Gabritschewsky 1894, in *Ann. de l'Inst. Pasteur* VIII. 682) in a more specialized sense (see quot. 1895). It was not widely used until after it was coined again in 1931, in the present sense.

1895 *Jrnl. R. Microsc. Soc.* 216 The author [*sc.* G. Gabritschewsky]..suggests the probability that phagocytes are not only capable of seizing and assimilating solid bodies, but of imbibing and absorbing liquid substances, and of rendering them harmless to the organism. For this action the term *pinocytosis* is suggested. **1931** W. H. LEWIS in *Bull. Johns Hopkins Hosp.* XLIX. 17 Pinocytosis, drinking by cells; phagocytosis, eating by cells... The word 'pinocytosis', suggested by my colleague Prof. David M. Robinson, is derived from the Greek... By pinocytosis the cells are able to take in substances which cannot diffuse into them or be taken in by ordinary phagocytosis of semisolid particles. **1937** —— in *Amer. Jrnl. Cancer* XXIX. 666 Pinocytosis (drinking) by macrophages in tissue cultures is common. **1960** *Jrnl. Protozool.* VII. 184/2 Pinocytosis is a very demanding process in terms of membrane formation. **1969** [see *endocytosis* s.v. *ENDO-]. **1977** P. B. & J. S. MEDAWAR *Life Sci.* xv. 124 Fibroblasts..have the power ..to imbibe tiny little droplets of whatever medium they may be living in, known as pinocytosis.

Hence **pinocy·tic, pi:nocyto·tic** *adjs.*, of, pertaining to, or formed by pinocytosis.

1955 *Texas Rep. Biol. & Med.* XIII. 475 This cell.. has been termed a variant pinocytic cell. **1959** *Exper. Cell Res.* XVIII. 71 The pinocytotic vacuole..was 7μ in diameter. **1964** N. S. COHN *Elem. Cytol.* iv. 68 The movement of fat particles in the intestinal villi appears to occur by pinocytic activity. **1974** *Jrnl. Cell Biol.* LXIII 998/2 Modification of macromolecules by enhancing their positive charges..leads to a preferential pinocytic uptake. **1975** *Nature* 17 Apr. 612/2 Each perineural epithelial cell contained numerous pinocytotic vesicles.

pinole. (Earlier and later examples.)

1842 A. GANILH *Ambrosio de Letinez* I. vii. 91 Pinole is made with fine corn meal, pounded almonds, sugar and various spices. **1844** J. GREGG *Commerce Prairies* I. vii. 159 This pod..the Apaches and other tribes of Indians grind into flour to make their favourite *pinole*. **1942** CASTETTER & BELL *Pima & Papago Indian Agric.* 38 In 1862 they sold the War Department more than one million pounds of wheat, as well as pinole, chickens, green peas, [etc.]. **1977** *New Yorker* 20 June 49/1 He carries dried chum salmon for his dogs, and his own food is dried moose or bear meat and pinole—ground parched corn, to which he adds brown sugar.

piñon. Also **pinone, pinyon.** Substitute for def.: One of a group of small pines native to southwestern North America, esp. *Pinus edulis*, *P. monophylla*, and *P. quadrifolia*, which yield edible seeds; also, the nuts or the wood produced by these trees. Also *attrib.* (Earlier and later examples.) See also PINION *sb.*⁵ in Dict. and Suppl.

1839 Z. LEONARD *Narr. Adventures* 35/1 Its top is covered with the pinone tree. **1839** J. FORBES *Pinetum Woburnense* 49 This [sc. *Pinus Llaveana*] is the only Mexican species that bears edible fruit, and is called in that country 'piñones'. **1882** C. M. CHASE *Editor's Run in New Mexico* 206 The common fuel is pinon, the best fire-place wood in the world. **1936** *Nature* 22 Aug. 315/2 The woods which have been found best for the purpose of the investigation [into dendrochronology] are the western yellow pine..and the Douglas fir.., while the next best is the pinyon (*Pinus edulis*). **1945** *Antiquity* XIX. 219 Vegetable foods of which there is evidence include maize, kidney beans and pinyon nuts. **1946** D. C. PEATTIE *Road of Naturalist* iii. 37 Up there the Piutes were feasting on piñon nuts and mountain sheep. **1955** J. HAWKES *Journey down Rainbow* i. 10 The surface of the desert..is..often..boldly spotted with the compact, dark shapes of pinyon and juniper. **1967** N. T. MIROV *Genus Pinus* iii. 150 Piñons form a well-defined group... Their habitat extends to the southwestern Pacific Coastal Ranges, the Colorado Plateau, and the Mexican Volcanic Plateau. **1972** *Sci. Amer.* May 97/1 The growth of the pinyon pine..is affected more by winter climate. **1973** A. H. WHITEFORD *N. Amer. Indian Arts* 53 Bottles are waterproofed with piñon gum. **1976** *Hortus Third* (L. H. Bailey Hortorium) 875/2 Pinyon, pinyon p[ine] nut p[ine]... A source of edible piñon nuts. **1976** *New Yorker* 26 Apr. 125/3 People in that part of the San Luis Valley.. gather piñon wood for their fires.

pinosylvin (pəinosi·lvin). *Biol.* Formerly also **-sylvine.** [a. G. *pinosylvin* (H. Erdtman 1939, in *Naturwiss.* XXVII. 130), f. mod.L. *Pinus sylvestris*, taxonomic name of Scots pine (f. L. *pīnus* PINE *sb.*² + *syl-*, *silvestris* of a wood (f. *silva* a wood)): see -IN¹.] **a.** A colourless, toxic, crystalline compound, 3,5-dihydroxystilbene, $C_{14}H_{12}O_2$, which occurs in the heartwood of the Scots pine, where it confers resistance to fungal and insect attack. **b.** Any of the related antifungal compounds that occur in pines.

1939 *Chem. Abstr.* XXXIII. 4776 Pinosylvine is poisonous, fishes die in 0·002% soln. **1945** *Svensk Bot. Tidskr.* XXXIX. 312 In order to get an idea of the relative toxicity of the pinosylvin compounds experiments were made at the same time in solutions containing phenol. **1955** *Ibid.* XLIX. 421 The heartwoods of trees belonging to the *Pinaceae* family contain a number of peculiar phenolic constituents,.. of which only the hydroxystilbenes (pinosylvins) show high fungicidal activity. **1968** *New Scientist* 11 Jan. 68/2 The natural resistance of certain types of wood to termite attack has been attributed to the presence of specific repellent chemicals such as..pinosylvin. **1973** *Ann. Rev. Phytopathol.* XI. 204 The formation of pinosylvins in *P. resinosa* cultures.. may have been caused directly by ethylene.

‖**Pinot** (pino). Also with lower-case initial. [Fr.: see *PINEAU.] Any of a family of vines yielding grapes used in wine-making; the grape of these vines. Also, wine made from Pinot grapes.

The principal varieties are the *Pinot Noir* (nwār) (black), used chiefly in making red Burgundy and Champagne, the white-wine-producing *Pinot Blanc* (blañ) (white) or *Chardonnay* (ʃardǫnēˡ), and the *Pinot Gris* (gri) (grey), a black grape from which white wine is made.

1912 [see *ALIGOTÉ]. **1959** W. JAMES *Word-Bk. Wine* 144 *Pinot*... There are both red and white varieties and a 'grey' variety grown in Alsace. **1962** *Economist* 29 Dec. 1283/1 A good Pinot Noir..from the Valais or Vaud cantons. **1965** A. SICHEL *Penguin Bk. Wines* III. 233 These four cantons, known as La Suisse Romande, produce 90 per cent of the total crop, the white wines from the Chasselas grape, the Pinot Gris, Marsanne.., the Riesling, Sylvaner and Traminer, and the red wines from the Burgundian grapes Gamay and Pinot. **1967** A. LICHINE *Encycl. Wines* 404/2 Pinot Gris and Pinot Blanc, eminently noble vines, contribute to white burgundies. All Pinots abound also in the finer Champagne vineyards. **1968** J. M. WHITE *Nightclimber* ix. 65, I drained my Pinot. **1972** *National Observer* (U.S.) 27 May 4/1 The huge price increases are coming in the premium wines—Cabernet Sauvignon, Pinot Noir and Pinot Chardonnay. **1972** D. E. WESTLAKE *Bank Shot* ix. 66 A side dish of black rice, washed down with a good Pinot Noir. **1974** N. FREELING *Dressing of Diamond* 81 A nice bottle, a rosé Pinot, beautifully dry and flinty. **1976** N. ROBERTS *Face of France* ii. 24 The so-called Tokay d'Alsace, or *pinot gris*, is made from black grapes, whereas..*pinot blanc* and sylvaner are made from white.

pin-point, *sb.* Add: Also **pinpoint, pin point.** (Further *fig.* examples.)

1890 KIPLING in *Macmillan's Mag.* Sept. 323/2 Their blue eyes, driven into pin-points by th' wind. **1926** E. GLYN *Love's Blindness* ii. 24 The centres of her light hazel eyes went to pin points. The dress did not deceive *her*! **1931** C. DAY LEWIS *From Feathers to Iron* xii. 25 Nightmare nags at his elbow and narrows Horizon to pinpoint, hope to hand's breadth. **1952** R. NEILL *Moon in Scorpio* xxviii. 247 His eyes were pin-points now, but their gaze was steady. **1959** E. H. CLEMENTS *High Tension* vi. 99 He let it [*sc.* a car] get well ahead and watched the pin-points of red disappear into the darkness. **1961** D. J. PLANTZ *Sweeney Squadron* x. 151 The Rising Suns were showing up, slightly brighter pinpoints in the gray gloom.

2. *Aeronautics.* A place seen and identified from an aircraft; hence, the ground position of an aircraft as found from such a sighting.

1942 *R.A.F. Jrnl.* 27 June 7 No pin-point was obtained on leaving the British shore. **1942** *Tee Emm* (Air Ministry) II. 81 It's up to you to verify all pinpoints. **1943** [see *astro-sight* s.v. *ASTRO-]. **1944** *Air Navigation* I. i. 20 A Pin Point is a landmark recognised from the aircraft but which is not necessarily underneath the aircraft. A Fix is the ground position of the aircraft, found by direct observation of the ground or by employing wireless or astronomical methods. **1950** D. C. T. BENNETT *Compl. Air Navigator* (ed. 5) xi. 374 A Fix..is the position obtained by a Visual Pin-Point or by the intersection of two or more Posn. Lines. **1970** TAYLOR & PARMAR *Ground Stud. for Pilots* II. 14 Flying this Heading, we pass over Peterborough (5235N 0015W) at 1230 hrs precisely. Plot the position, as a small circled dot; it is a Pinpoint, the name given to a Fix obtained by visual observation of the ground. **1971** *Hindustan Times Weekly* (New Delhi) (Suppl.) 4 Apr. p. iii/2 He's been flying here for only six months and is still in the process of discovering new features and pinpoints.

3. (See quot. 1948¹.)

1943 H. T. U. SMITH *Aerial Photographs* xiii. 339 Vertical photography may involve the making of one or more flight strips, or of only isolated stereo pairs, known as 'pinpoints'. **1948** S. H. SPURR *Aerial Photographs in Forestry* ii. 13 A pin-point is an isolated pair of photographs taken so as to give stereoscopic coverage of a specific place on the ground. *Ibid.* 16 Specially designed instruments manufactured by the makers of mapping cameras are particularly well adapted for taking pinpoints.

B. *attrib.* or as *adj.* **1.** Seeming as small or as sharp as the point of a pin.

1850, 1899 [in Dict.]. **1907** J. H. PARSONS *Dis. Eye* iv. 67 In old people it is smaller than in the young, sometimes to so great an extent that the pupils are almost 'pin-point'. **1928** D. H. LAWRENCE *Woman who rode Away* 198 He never liked looking anything in the very pin-point middle of the eye. **1933** W. DE LA MARE *Fleeting* 95 With pin-point bill, and tail a-cock. **1936** W. HOLTBY *South Riding* I. vii. 72 Those trodden-down pin-point heels. **1944** *Times* 18 Mar. 4/4 Marauders attacked pin-point targets at Piedmonte. **1959** *Times* 23 Sept. 3/7 The stage effect of overhead lights, the surrounding darkness, and the pinpoint area of operations. **1961** R. B. LONG *Sentence & its Parts* vii. 153 In the narrowest sense the present time is a point so minute that it is already a part of the past before we can finish the sentence. But the present with which verbs are concerned is not this uncomfortable pin-point present. **1974** M. C. GERALD *Pharmacol.* xiii. 244 The pinpoint pupil is one of the cardinal signs of morphine poisoning. **1976** *Sat. Rev.* (U.S.) 30 Oct. 10/1 Bell scientists and engineers..have already developed pinpoint light sources—light-emitting diodes.

2. Very fine in texture or structure; characterized by very small points.

1899 *Daily News* 29 July 8/5 A clear Swiss muslin of very fine make, with a pin-point embroidery on it. **1942** *Oxoniensia* VII. 42 Deeply incised 'pinpoint' decoration.. varying a little and reverting to a plain chevron pattern. **1957** J. KEROUAC *On Road* (1958) 156 A misty pinpoint darkness. **1962** *Guardian* 23 Feb. 8/4 A seam-free, pinpoint mesh stocking. **1969** *Sears Catal.* Spring/Summer 20 Full fashioned plus the extra elasticity of a pinpoint stitch.

3. Performed with or exhibiting great positional accuracy.

1944 *Manch. Guardian* 14 Dec. 3/2 Fighter Command's activities included thirty missions against V2 targets in Holland, where pin-point power-dive attacks resulted in direct hits on erection and launching installations. **1945** *Times* 27 June 3/4 Their pin-point bombing was on the biggest scale that Japan has yet experienced. **1949** *Sun* (Baltimore) 3 Oct. 2/7 'Kickless' guns, capable of pinpoint accuracy. **1958** J. R. BIGGS *Woodcuts* 81 If the design demands precise 'pin point' register, then a precise method must be used. **1958** *Listener* 21 Aug. 259/1 It is the ground controllers' job to see that collisions do not happen. With the equipment they have today this can be done with almost pin-point accuracy. **1973** *Times* 9 Aug. 5/5 He said he was aiming his balloon for France, 'but I would consider anything from Finland to Italy a pinpoint landing'. **1976** *Gramophone* Aug. 370 (Advt.), Dramatically improved solid stereo image and rear pin-point localisation. **1976** *Southern Even. Echo* (Southampton) 3 Nov., Clive Green almost snatched a second ten minutes later with a flying header from a pin-point cross by left-winger Mickey Mellows.

4. Highly detailed or specific.

1960 V. JENKINS *Lions Down Under* p. xv, Secretaries ..looked after our internal comfort with pin-point efficiency. **1971** *Morning Star* 1 July 4/1 This 'simple way' is, of course, the result of pin-point organisation and the working out of schedules.

pin-point, *v.* Also **pinpoint.** [f. the sb.] **I.** *trans.* **1. a.** To locate with high precision.

1917 'CONTACT' *Airman's Outings* II. iv. 280 Meanwhile an exact position has been pin-pointed. **1936** J. GRIERSON *High Failure* v. 102 The next thing was to 'pin-point' myself: that is to find the exact spot on the map at which I had made a landfall. **1946** D. HAMSON *We fell among Greeks* iv. 46 The enemy was trying to pinpoint our position. **1955** C. S. FORESTER *Good Shepherd* 72 The fewer people who were aware how accurately the Admiralty was able to pin-point U-boat concentrations the better. **1955** *Times* 6 Aug. 8/4 Not only can the exact position of a find be pin-pointed..but the possibility of future researches and future discoveries is preserved. **1977** *Daily Tel.* 18 Nov. 8/8 Amateur archaeologists believe they have pinpointed the site of a large Roman forum..under central Chichester.

b. To identify (an objective) as a target for pin-point bombing.

1940 *Times* 2 Nov. 4/1 Over Naples itself the aircraft crews were able to 'pin-point' the targets without great difficulty. **1941** *Times* 30 Sept. 4/7 The pilot managed to pin-point the factory. **1946** *R.A.F. Jrnl.* May 169 Lancasters equipped with 'H2S'..thundered through the night to pinpoint their objectives.

2. a. To cause to be conspicuous against a large or complex background; to bring into prominence, emphasize.

1943 *Penguin New Writing* XVI. 27 A solitary searchlight would come on suddenly. And, if it pin-pointed you, how you would writhe about the sky trying to shake it off before the endless beams of all the others caught up on you. **1956** [see *PIN-POINTING *vbl. sb.*]. **1957** *Economist* 2 Nov. 420/1 Subsequent speakers from Asia, Latin America and Europe took up these themes, each country pinpointing its own problems. **1958** P. MORTIMER *Daddy's gone A-Hunting* xli. 239 The world was empty. But tiny, minutely raging, the figure of Rex was pinpointed, the sole survivor. **1974** F. WARNER *Meeting Ends* I. v. 24 Lights down to pinpoint Shango in wheel, still spreadeagled, back to audience.

b. To identify precisely; to determine the exact nature of.

1946 *Birmingham* (Alabama) *News* 5 Jan. 1/6 The Pearl Harbor committee called for photographs of the Navy's ship location board today to pinpoint movements of the Pacific Fleet in the days just before the Japanese attack. **1950** *Sport* 22–28 Sept. 18/1 Johnny..would find it difficult to pinpoint the happiest day of his soccer life to date. **1955** *Sci. News Let.* 23 July 51/1 Tonsils, long under suspicion, have at last been pin-pointed as the primary site of polio infection. **1958** *Ann. Reg. 1957* 186 The House of Representatives asked the President to pin-point where substantial cuts could be made. **1960** *Analog Sci. Fact & Fiction* Nov. 13/1 The only actual trouble we can pin-point is that there seem to be a great many errors occurring in the paper-work. **1971** J. Z. YOUNG *Introd. Study Man* i. 9 There have been many attempts to pin-point the particular environmental or other features responsible for the appearance of man. **1977** L. GORDON *Eliot's Early Years* iii. 63 It is difficult to pin-point the sensibility that moves through Eliot's poems.

II. *intr.* **3.** To dwindle to the size of a pin-point (and disappear).

1957 J. KEROUAC *On Road* (1958) II. vii. 159 They pinpointed out of sight.

Hence **pi·n-pointing** *vbl. sb.*; also **pi·n-pointable** *a.*, capable of being pin-pointed.

1920 *Flight* XII. 374/2 Practical demonstration of principles learnt in Ground work:—(1) Flight by map alone; (2) Flight by compass alone on pre-determined course and time, turning point to be indicated—pin pointing. **1955** D. BARTON *Glorious Life* 71 Here, under his eyes, pinpointable, was the Fall. **1956** *Essays in Crit.* VI. 123 If we pinpoint the personal origin, certainly when the pinpointing is so merely speculative as it seems to be here, we leave out too much. **1962** *Listener* 27 Dec. 1086/1 Current technology, gossip column hearts and flowers..have no direct pin-pointable relation to my work of the moment, but they are not alien worlds. **1967** A. L. LLOYD *Folk Song in Eng.* iii. 150 Even such an apparently pinpointable ballad as 'Edom o' Gordon' has an English, a Lanarkshire and an Aberdeen setting. **1970** TAYLOR & PARMAR *Ground Stud. for Pilots* II. 15 While the pinpointing, plotting and wind finding was going on, so was the aircraft. **1978** J. WAINWRIGHT *Ripple of Murders* 53 It was..necessary to..let him *know* that he could be pin-pointed in King Street..if such pin-pointing became necessary.

pi·n-pointed, *a.* [f. PIN-POINT *sb.* + -ED².] Having a fine or sharp point. Also *fig.*

1909 *Daily Chron.* 18 Sept. 10/6 The tiny pin-pointed mapping pen. **1931** E. S. GARDNER *Vanishing Corpse* in *Detective Fiction Weekly* 15 Aug. 19/1 His eyes..gazed at Sidney Zoom with pin-pointed intensity. *a* **1936** KIPLING *Something of Myself* (1937) viii. 230, I then abandoned hand-dipped Waverleys..and for years wallowed in the pin-pointed 'stylo'. **1976** J. WAINWRIGHT *Bastard* IX. 122, I check the time... They'll demand pin-pointed answers to their nit-picking questions.

pi·n-pointed, *ppl. a.* [f. *PIN-POINT *v.* + -ED¹.] That has been pin-pointed.

1944 *Hutchinson's Pict. Hist. War* 12 Apr.–26 Sept. 60 (*caption*) An attack by Mosquito aircraft..on what was probably the most pin-pointed objective which has ever been marked out as a target.

pin-prick, *sb.* Add: **1.** (Later examples.) Also *fig.*

1927 *New Republic* 12 Oct. 216/2 His pen is so subject to his moods that it can make a pin-prick read like a lightning bolt. **1949** E. COXHEAD *Wind in West* vii. 195 At the far end of the stifling tunnel, in which he was condemned to grope for ever, he seemed to see a pin-prick of light. **1978** *Times* 22 July 9/2 Beware of sea urchins and the pinpricks of coral.

2. (Further *attrib.* examples.)

1926 T. E. LAWRENCE *Seven Pillars* (1935) I. xv. 104 The tribesmen..hindered and distracted the Turks by their pin-prick raids. **1976** A. WHITE *Long Silence* vi. 49 Sooner or later, the Germans were going to..suspect.. the source of the many pinprick raids. **1977** *Time* 4 Apr. 23/2 After launching a few pinprick air raids, Mobutu's Army Chief of Staff..claimed that the intruders were in retreat.

b. *pinprick picture*, a coloured print pierced with pin-holes to create an illusion of illumination.

1960 H. HAYWARD *Antique Coll.* 286/2 Pinprick pictures were a more simple form of transformation, since a coloured print was perforated with a number of small holes and, hence, when held to the light would appear to be illuminated. Coloured paper would sometimes be fastened behind the print. **1968** *Canad. Antiques Collector* Oct. 22/1 Pinprick pictures of ancient oriental origin are probably a branch of decoupage.

pin-prick, *v.* (Further examples.) (Quot. 1945 is in sense of *PIN-POINT *v.* 1 b.)

1909 *Daily Chron.* 15 July 4/4 Every book for the blind is carefully pin-pricked by voluntary workers who can see. **1912** J. BAILEY *Let.* 13 Aug. (1935) 132 You shall certainly pin-prick if you will when you come to London, if you don't find anything more amusing to do—and I will listen respectfully and gratefully. **1945** R. A. KNOX *God & Atom* v. 71 Other men's lives are at stake; those.. of British or American airmen who might be shot down in trying to pin-prick the targets of Hiroshima one by one, instead of devoting it to a general holocaust. **1952** C. DAY LEWIS tr. *Virgil's Aeneid* XI. 245 Drances, hostile as ever to Turnus, whose high renown Pin-pricked him with sour envy to intrigue against him. **1966** D. FRANCIS *Flying Finish* xi. 136 'There are some holes in the paper.' 'Where he put the pin.'.. Simon had pinpricked four letters. **1979** G. MACDONALD *Camera* Plate 4 A paper print of Venice, garishly coloured by hand then pin-pricked for back-lighting in a viewer.

pin-pricked, *ppl. a.* (Further examples.) *pin-pricked picture* = *pin-prick picture*.

1936 *Discovery* Oct. 321/2 That was a tragic pin which pricked out one of the last letters written by Marie Antoinette in the prison of the Conciergerie... The pin-pricked note miscarried. *Ibid.* 322/2 Some surviving specimens of these pin-pricked pictures date back two hundred years. **1958** R. GODDEN *Greengage Summer* iv. 38 Pinpricked all over with fear, I tiptoed away. **1961** *Times* 1 July 11/1 One variety of pin-prickt picture deals with religious subjects.

pin-pricking, *vbl. sb.* and *ppl. a.* (Further examples.)

1927 *Daily Express* 5 Dec. 1/4 The move is interpreted ..as a step forward to stop the 'pin-pricking' that has been going on between the two countries. **1936** *Discovery* Oct. 322/2 Pin-pricking with them was a fine art. **1950** *Times* 24 Jan. 3/5 We must bring home to the British Government that although most of us are loyal, we will not tolerate the pinpricking of loyalty. **1958** B. HAMILTON *Too Much of Water* v. 119 He had..continued, in a small pin-pricking way, to belittle and snub Patricia Odell. **1961** M. CONWAY in *Conc. Encycl. Antiques* V. 231/2 The art of pin-pricking or 'Piercing Costumes on Paper' became a young ladies' amusement. *Ibid.*, Extremely attractive pin-pricking effects were achieved by outlining from the front with a fine pin—actually needles were used—the remainder being thickly pierced from the back. **1972** *Times* 26 June 8/5 Much authentic [A.J.P.] Taylor— pinpricking, bubble-bursting. **1973** *Times* (Nepal Suppl.) 14 Apr. p. i/4 Nepal retaliates by such pinpricking gestures as refusing to cooperate wholeheartedly in the extradition of Naxalite terrorists who flee across the border from West Bengal.

pinsapo (pinsa·po). [Sp.] The Spanish fir, *Abies pinsapo*, belonging to the family Pinaceæ, and native to mountainous regions of southern Spain. Also *attrib.*

1839 *Gardener's Mag.* XV. 109 He [*sc.* E. Boissier] first observed the Pinsapo at an altitude of about 4,000 feet. **1852** STANDISH & NOBLE *Pract. Hints Planting Ornamental Trees* 47 *Abies Pinsapo*.—Boissier. (Pinsapo Silver Fir.)... A very beautiful tree..indigenous to the mountains of Grenada. **1877** DISRAELI *Let.* 29 July in Monypenny & Buckle *Life Disraeli* (1920) VI. 171 Yesterday ..I had to plant a tree—a pinsapo. **1887** *Encycl. Brit.* XXII. 297/1 Among other characteristic trees [in Spain] are the Spanish pine (*Pinus hispanica*), the Corsican pine (*P. Laricio*), the Pinsapo fir (*Abies Pinsapo*), and the *Quercus Tozza*.

pinscher (pi·nʃər). [Ger.] A short-coated, often dark-coloured terrier of the breed so called, usually having pricked ears and a docked tail; also, a smaller terrier with either pricked or drop ears, belonging to the miniature breed so called. Cf. *DOBERMANN.

1926 W. S. SCHMIDT *Doberman Pinscher* i. 10 The ancestors of the doberman were the old German shepherd dog and the large variety of the black-and-tan, smooth-haired German pinscher. **1929** *Pure-Bred Dogs* (Amer. Kennel Club) 203 Miniature Pinschers are natives of Germany. **1935** *Toy Dogs* (Amer. Kennel Club) 55 If you like a small 'Pinscher'..see them trotting around and 'showing off' in some of the dog shows. **1954** M. K. WILSON tr. *Lorenz's Man meets Dog* v. 62 The small daughter of the house received..a charming little dwarf Pinscher. **1957** *New Yorker* 21 Sept. 46/2 Zoltan acquired a miniature pinscher. **1968** R. & A. FIENNES *Nat. Hist. Dog* vi. 77 Among the continental terriers with toy varieties are the schnauzers and pinschers. **1973** *Country Life* 15 Feb. 385/2 The toy group... The Yorkshire terrier, the miniature pinscher, the Italian greyhound. **1978** D. A. STANWOOD *Memory of E. Ryker* xii. 112 Dogs snarled behind me... Two pinschers hit the door.

pin-stripe. [f. PIN *sb.*¹ 18 + STRIPE *sb.*²] A fine broken or continuous stripe, esp. one repeated as a pattern on cloth. Also *attrib.*, designating cloth with a pattern of such stripes or garments made of pin-stripe cloth. So *ellipt.* as *sb.*, a pin-stripe suit, conventionally worn by business men.

1897 *Sears, Roebuck Catal.* 183/1 *Extra good value, heavy pin-stripe*, with fancy navy blue square sailor collar. *Ibid.* 183/2 This suit is made of a fine brown pin stripe cassimere. **1906** *Westm. Gaz.* 2 June 16/3 The particular cloth I have in mind has a pin stripe in brown. **1922** JOYCE *Ulysses* 321 A dainty *motif* of plume rose being worked into the pleats in a pinstripe. **1935** WODEHOUSE *Luck of Bodkins* iv. 42 A pin-stripe flannel suit. **1942** S. SPENDER *Life & Poet* v. 84 The black-coated, pinstripe-trousered man. **1958** *Spectator* 15 Aug. 214/1 A suit that fits him rather less snugly than his usual pin-stripe. **1972** P. CLEIFE *Slick & Dead* I. i. 15 The usual aleatory fall-out of face fungus and brothel creepers mixing it with the pinstripes and the *Financial Times*. **1973** M. AMIS *Rachel Papers* 50 As I watched, there was a stir in the classroom; a cruel-faced bearded man in a pinstripe suit strode into camera. **1973** *Times* 15 Nov. 18/4, I wore my midnight blue suit with a wide pin stripe. **1975** 'D. JORDAN' *Black Account* iii. 23 This afternoon Magnus was..in a blue-chalk pinstripe, a heavy blue shirt..a near-black tie. **1977** *Hot Car* Oct. 50/3 Apply ⅛ in. or ¼ in. masking tape into the positions required for the pin stripe, then spray the complete panel in the required car colour.

So **pi·n-striped** *a.*, ornamented with narrow stripes; wearing clothes of pin-stripe cloth, conventionally dressed; also *fig.*, characteristic of the business man.

1896 [in Dict. s.v. PIN *sb.*¹ 18]. **1909** *Westm. Gaz.* 4 Sept. 15/1 A gown carried out on a khaki-coloured foundation has a decoration of a pin-striped black and gold collar. **1932** *Daily Tel.* 25 Apr. 4/4 A little tuck-in blouse of red and white pin-striped silk. **1958** *New Statesman* 3 May 562/3 Small wonder we..are unhappy with the pin-striped Executive of the Labour Party, stinking as it does with the air of neat suburban houses and well mannered conversation over garden fences. **1967** *Listener* 31 Aug. 263/2 The citizen everywhere, pin-striped or dungaree'd, puts his own interests before those of society as a whole. **1970** 'D. HALLIDAY' *Dolly & Cookie Bird* iii. 31 Big silver fish pinstriped in yellow. **1973** M. WOODHOUSE *Blue Bone* xiii. 148 A stuffy, pin-striped wheeler and dealer. **1975** *Radio Times* 13 Sept. 4/2 Pin-striped anonymity in the City once seemed a more likely destination.

pint¹. Add: **b.** (Further examples.)

1901 M. FRANKLIN *My Brilliant Career* xxii. 194 'Good gracious, Julius!' exclaimed grannie, as he offered the governess a pot full of beer, 'Miss Craddock can't drink out of that pint.' 'Those who don't approve of my pints, let 'em bring their own.' **1961** H. C. DODGE *My Childhood Canad. Wilderness* i. 11 We had our meals from tin pints and tin plates, as we never took our china to the woods. **1968** K. WEATHERLY *Roo Shooter* 10 He..filled a tin mug with tea from a fire-blackened oil tin that served as his billy. With the pint in his hand he sat down again.

c. (Earlier and later examples.)

1742 FIELDING *J. Andrews* I. II. ii. 142 He wished to find a house of publick Entertainment where he might have dried his clothes and refresh himself with a Pint. *Ibid.* iii. 144 He had just entered the House, had called for his Pint. **1861** GEO. ELIOT *Silas Marner* i. 7 He never strolled into the village to drink a pint at the Rainbow. **1922** JOYCE *Ulysses* 20 The sacred pint alone can unbind the tongue of Dedalus. **1952** 'J. TEY' *Singing Sands* xii. 205 Eventually he had Richards to himself in a corner with a pint. **1965** V. CANNING *Whip Hand* i. 10, I had a pint with him. **1976** W. J. BURLEY *Wycliffe & Schoolgirls* iv. 86 Middle-aged men whose wildest excess was a couple of pints at the local.

d. *pint-bottle* (earlier example), *-glass* (earlier and later examples), *mug*.

1713 T. CAVE *Let.* 16 Jan. in M. M. Verney *Verney Lett.* (1930) I. xiii. 244 The London Postmaster, who yet swallow'd down a pint glass of Ale to the poor Boy's health. **1800** J. WOODFORDE *Diary* 19 Oct. (1931) V. 279 Miss Emeris sent us a Pint Bottle of Mushrooms. **1847** DICKENS *Dombey & Son* (1848) xxxviii. 380 This profound reflection Mr. Toodle washed down with a pint mug of tea. **1856** —— in *Househ. Words* 28 June 554/2 Drinking beer out of thick pint crockery mugs. **1922** JOYCE *Ulysses* 301 The memory of the dead, says the citizen taking up his pintglass and glaring at Bloom.

1968 M. WOODHOUSE *Rock Baby* vi. 51 Binnie gave me coffee in a pint mug. **1976** *Southern Even. Echo* (Southampton) 1 Nov. 9/4 By the end they were glad they had not started with pint glasses. **1978** R. BARNARD *Unruly Son* i. 7 At the end of the bar..was one solitary young man, his eyes concentrated on his pint mug.

pint² (pəint). Also **p'int.** Repr. a vulg. and dial. (esp. U.S.) pronunc. POINT *sb.*¹

1837 DICKENS *Pickw.* xxiii. 238 Upon all little pints o' breedin', I know I may trust you as vell as if it was my own self. **1887** *Scribner's Mag.* Oct. 476/2 Jeff looked ..p'int blank gashly. **1893** H. A. SHANDS *Some Peculiarities of Speech in Mississippi* 50 *Pint*.., Negro and illiterate white for *point*. **1901** M. FRANKLIN *My Brilliant Career* xxxii. 274 The pens had not enough 'pint'. **1943** J. STUART *Taps for Private Tussie* xiii. 149 'It's wonderful, Uncle George,' I said. 'It's pint-blank right,' Grandma said. **1946** *Richmond* (Va.) *Times-Dispatch* 10 Feb. 8-B/7 The natural sequence of racing between the flags starts with the hunting field, then to point-to-points (gleefully termed 'pint-to-pints' by some), continues to the hunt meetings, and finally ends with steeplechasing.

pinta¹. Add to def.: caused by *Treponema carateum*, a spirochete related to those which cause syphilis and yaws. (Earlier and later examples.)

1825 *Amer. Med. Rev.* II. 164 (*heading*) An account of the pinta, or blue-stain, a singular cutaneous disease prevailing in Mexico. **1942** *Arch. Dermatol. & Syphilol.* XLV. 858 The discovery of a spirochete in a Cuban case of pinta by Grau Triana and Alfonso Armenteros, of Habana, Cuba, on Aug. 3, 1938 definitely settled the discussions on the causation of pinta (mal del pinto or carate). **1971** *Nature* 5 Feb. 409/2 The oldest treponemal disease known at present is pinta..which dates back 15,000 years.

pinta² (pəi·ntă). Repr. colloq. pronunc. of *pint o'* (*of*), introduced in a National Dairy Council advertising slogan. Freq. used *ellipt.* for *pint o' milk*; also *transf.* Cf. *CUPPA.

1959 *Times* 30 May 10/1 Referring in his opening speech to the 'Drinka pinta milka-day' campaign, Mr. Amory said:.. 'I drink a pint and a half a day.' **1960** *Harper's Bazaar* July 16 Your daily pinta is the best glamour food there is. **1965** *Observer* (Colour Suppl.) 18 Apr. 29/3 The Lamb tavern..worth stopping at for a pinta. **1967** J. PORTER *Chinks in Curtain* i. 17 Pity they don't supply the milk of human kindness by the pinta. **1970** A. JENKINS *Drinka Pinta* xii. 129 March 1958 saw the birth of one of the most famous advertising slogans of all time: 'Drinka Pinta Milka Day'. **1971** 'H. CARMICHAEL' *Quiet Woman* x. 106 Some milkman's leaving a pinta on her doorstep every day. **1973** *Observer* (Colour Suppl.) 25 Nov. 41/2 (*caption*) The blue tit..pierces the metal cap to get at the milk in a pinta from a Welsh dairy.

‖**pintadera** (pintădēə·ră). *Archæol.* [Sp.] An instrument for painting patterns on the body.

1910 A. MOSSO *Dawn of Mediterranean Civilization* xvi. 257 The great *tholos* of Haghia Triada contains rich material for the study of *pintaderas* of the copper age. **1929** V. G. CHILDE *Danube in Prehist.* vi. 103 Painting of the person is indicated both by the figurines, ornamented in Cucuteni style, and the occurrence of clay stamps or *pintaderas*, sometimes bearing traces of red colour. **1938** *Nature* 1 Oct. 602/1 Seals that could serve as models for Danubian II 'pintaderas' were current in Crete and Asia Minor throughout the third millennium. **1970** BRAY & TRUMP *Dict. Archaeol.* 180/1 Pintaderas of both stamp and roller types occur in many American cultures.

pintail. Add: **5.** (In full *pintail surfboard.*) A surfboard the back of which tapers to a point.

1967 J. SEVERSON *Great Surfing* Gloss., *Pintail*, a surfboard with a long, drawn-out, pointed tail. **1969** *Sunday Truth* (Brisbane) 12 Jan. 61/1 The Gold Coast City Council is 'extremely concerned' about the growing number of 'pintail' surfboards appearing on local beaches. The surfboard, with a pointed tail and a razor sharp, scythe-shaped fin, appeared as a new design this summer. **1970** *Surf International* (Austral.) I. x. 9/2 The Hawaiian pintails have flow, but that means you're tied to the wave's tempo.

Pinteresque (pintəre·sk), *a.* [f. the name *Pinter* (see below) + -ESQUE.] Of, pertaining to, or characteristic of the British playwright Harold Pinter (b. 1930) or his works. Also *absol.* as *sb.*

1960 *Times* 28 Sept. 15/4 Mr Adrian writes with a cruel mastery of our slipshod, contemporary idioms, and the long drunken coda to his play is a comic achievement none the less impressive for its Pinteresque overtones. **1965** *Punch* 6 Oct. 507/1 The sort of everyday absurdity, in speech or action, that can now be most easily described as 'Pinteresque'. **1969** *Observer* 8 June 26/3 Jonathan.. was an excellently conceived character: a psychiatric worker.., rather prissy, sweetly reasonable on the surface but with a constant hint of query—queer malice showing through... This was a potentially Pinteresque situation. **1970** *Guardian* 7 Aug. 8/4 The Pinteresque as comedy. **1974** *Listener* 13 June 754/1 Suddenly everyone ..talked like overheard conversations on buses. They invented a word for it—Pinteresque.

So also **Pinterian** (pintīə·riăn) *a.*, characteristic of Pinter or his works; **Pi·nterish** [-ISH¹] *a.*, characteristic or suggestive of Pinter or his works; hence **Pinterishness**; **Pi·nterism**, Pinterish style or an instance of this.

1960 *Times* 7 Oct. 4/7 Miss Quayle as a Pinterish woman on top of a bus. **1963** *Observer* 13 Oct. 23/3 Dave Freeman's script was ingeniously Pinterish. **1967** *Listener* 1 June 727/1 *The Dick Emery Show*..contained a sketch by Harold Pinter... The sketch..was a small masterpiece, quintessentially Pinterian. Two aging women have tea together, and the conversation..is about a friend who used to go to the butcher's regularly on Wednesdays but now, since she's moved, doesn't go quite so much. **1970** *Guardian* 16 Dec. 8/1 Even on a straightforward social level, I am told, events assume a Pinteresque flavour when the Pinters arrive. What is this pervasive Pinterishness? **1971** *Ibid.* 24 Sept. 10/2 A precisely structured script, only very occasionally dropping into those meaningless meanings now known as Pinterisms. **1975** *Broadcast* 3 Nov. 14/3 Old Times by Harold Pinter..seems to epitomize 'Pinterism'—relaxed circumambient dialogue with lots of significant spaces between the words.

pintid (pi·ntid). Also **pintide.** [ad. Sp. *pintide* (F. Leon y Blanco 1940, in *Méd. Revista Mexicana* XX. 240), f. *mal del pint-o* PINTA¹ + *sifil-ide* SYPHILIDE.] A lesion of the skin of the type characteristic of pinta.

1940 *Q. Cumulative Index Medicus* XXVII. 886/2 Pinta—constant presence of Treponema herrejoni in cutaneous lesions of dyschronic period of mal del pinto and in 'pintides'. **1942** *Arch. Dermatol. & Syphilol.* XLV. 849 (*caption*) Trichophytoid pintid in Cuban pinta. **1965** HARGREAVES & MORRISON *Pract. Trop. Med.* iv. 161 A secondary stage..consists of a symmetrical eruption.., consisting of macules, and miliary papules or pintids. **1973** A. WISDOM *Colour Atlas of Venereol.* 150 The primary pintide appears, usually on exposed skin surfaces, after an average incubation period of six to eight weeks.

pinto, *a.* Add: **1.** (Earlier and later examples.) Also in Comb., as *pinto-coloured* adj.

1865 B. HARTE in *Californian* 15 Apr. 4/1 The devil in the shape of a fleet pinto colt. **1936** D. McCOWAN *Animals Canad. Rockies* xvi. 141 An Indian boy on a pinto pony had chased him to cover. **1966** H. MARRIOTT *Cariboo Cowboy* iii. 39, I had two saddle horses... One was a pinto-coloured gelding.

2. pinto bean, the mottled seed of a variety of the kidney bean, *Phaseolus vulgaris*, which is widely cultivated in the southwestern United States and Central America; also, the plant itself. Also *ellipt.*

1916 'B. M. BOWER' *Phantom Herd* iii. 46 A girl gave me a handful of pinto beans. **1924** W. M. RAINE *Troubled Waters* xxvii. 269 Pinto beans..were no sooner out and stacked than the men were hard at it putting in winter wheat. **1941** J. A. & A. LOMAX *Our Singing Country* III. vi. 292 When you get through, you've not got a cent To buy fat-back meat, pinto beans. **1942** [see *LIMA b]. **1963** MRS. L. B. JOHNSON *White House Diary* 29 Dec. (1970) 26 There were beans (pinto beans, always), delicious barbecued spare ribs. **1969** *New Yorker* 17 May 115/1 (Advt.), Subsisting day after day on a few greens around noon..and some pinto beans in the evening. **1973** *Black Panther* 5 May 7/2 Existence has become a diet of pinto beans and rice. **1977** *New Yorker* 20 June 49/1 From a supplier in Seattle he orders hundred-pound sacks of corn, pinto beans, unground wheat.

pintoresque (pintǒre·sk), *a. rare.* [ad. Sp. *pintoresco* picturesque: see -ESQUE.] Picturesque, forming a suitable subject for a painting.

1969 *Daily Tel.* (Colour Suppl.) 17 Jan. 29/4 Artists abound, since the Mediterranean is almost embarrassingly *pintoresque* wherever you may decide to go.

pi·nt-size, *a.* [SIZE *sb.*¹] Small; also quasi-*sb.*, as a nickname for a child or small person. So **pint-sized** *a.* in same sense; also *absol.* as *sb.*; also, having a capacity of one pint.

1938 *Sun* (Baltimore) 22 Apr. 3/2 The [air] ship, just a pint-sized affair compared to the giant Hindenburg, carried only three persons. **1939** R. CHANDLER *Trouble is my Business* (1950) 17 It was large enough for a pint-sized desk. **1949** *Sun* (Baltimore) 9 Apr. 6/1 (*heading*) Opportunity for cattle breeders: the pint-size cow. **1952** B. MALAMUD *Natural* 182 A brisk, pint-size chef with a tall puffed cap on. **1955** *Granta* 26 Nov. 11/2 Gorgeous Gloria, The Pint-sized Poppet. **1959** [see *NIPPER *sb.*¹ 3 b]. **1961** M. BEADLE *These Ruins are Inhabited* iii. 36 Choristers, in pint-sized caps and gowns, trotting across the bridge for Evensong. **1971** *Times* 8 Sept. 21/4 (*heading*) Merits of the pint-sized company. Next week, the Bolton Committee of Inquiry on Small Firms plans to publish the results of a survey. **1972** WODEHOUSE *Pearls, Girls & Monty Bodkin* ix. 158 Where young pint size is at a disadvantage is in never having seen Grayce when she was really rolling. **1973** M. AMIS *Rachel Papers* 20 He read with concentration, his nose perhaps six inches above the page, mouthwashing with tea from a pint-sized mug which Jenny had time and again to refill. **1973** *Guardian* 22 May 13/2 Long double-breasted riding macs for the pint-sized. **1977** *Time* 8 Aug. 28/2 Andrea McArdle, 13, star of the Broadway musical *Annie*, led her pint-size cast onto the softball diamond against the peewees of Paramount Pictures's forthcoming kiddy sequel, *The Bad News Bears in Breaking Training*.

pi·n-tuck. [PIN *sb.*¹ 18.] In *Sewing*, a narrow, chiefly ornamental, tuck. Also *attrib.* So **pi·n-tucked** *a.*; **pi·n-tucking.**

1903 K. D. WIGGIN *Rebecca* xxvii. 285 Costumes that included..drawing of threads,.. hemstitching and pin-tucking. **1906** *Times* 4 May 10/2 The fulness of the skirt closely pin-tucked to the figure in sets of three. **1921** *Daily Colonist* (Victoria, B.C.) 13 Oct. 20/1 (Advt.), Flannelette Gowns, with high or 'V'-neck..trimmed with embroidery or pin-tucks. **1932** *Daily Tel.* 23 May 6 (Advt.), Pin tucks trim the small over-sleeve. **1964** *McCall's Sewing* xv. 270/1 If a slip needs to be shortened only a slight amount, you can do it by taking several small pin tucks around the lower edge. **1973** *Times* 15 Nov. 6/1 The tailored torso..made to fit even more rigidly by vertical rows of pin tucks. **1975** *Times* 7 Oct. 11/1 Jersey long dresses with pin tucking. **1975** *New Yorker* 15 Dec. 74/2 A zip-fronted aviator jacket with pin-tuck detailing on back and sleeves. **1976** *Daily Mail* (Hull) 30 Sept. 13/8 (Advt.), Suede and Leather Pin Tuck Jackets..From £19·95. **1976** *Ilkeston Advertiser* 10 Dec. 14/2 The bridesmaids wore Victorian style pin-tucked beige dresses.

pi·n-up, *a.* and *sb.* [PIN *v.*¹ 12.] **A.** *adj.* **1.** (See PIN *v.*¹ 12.)

2. Of a photograph or other picture, designed to be fixed to a wall, etc. Also applied to a favourite or sexually attractive young person, the typical subject of such a photograph; also in extended use. Also, pertaining to or characteristic of such a picture or person.

1941 *Life* 7 July 34 Dorothy Lamour is No. 1 pin-up girl of the U.S. Army. **1943** *Sun* (Baltimore) 8 Oct. 22/6 Bob Hope, radio and film comedian, today emerged victorious as the official pin-up boy of the WAC contingents here. **1944** *Richmond* (Virginia) *Times-Dispatch* 19 May 11/5 The ex-GI's who threw away all their pin-up pictures when they came home from the war are tacking pretty photos up on their bedroom walls again. **1946** *News Chron.* 27 Feb. 1/8 The honourable lady must not take advantage of the fact that she is my pin-up girl. **1948** 'E. CRISPIN' *Buried for Pleasure* iii. 19 She had a figure like the quintessence of all pin-up girls. **1948** M. A. MICHAEL tr. *Mielche's From Santos to Bahia* vi. 142 The hard, cold eyes of the American girls,..calculation behind their vulgar 'pin-up' smile. **1953** *Encounter* Nov. 30/1 Women's locks, or corsets, or riding boots, and even pin-up portraits, may become the object of fetishist worship. **1958** E. H. CLEMENTS *Uncommon Cold* i. 69 That pin-up girl I talked to on the pier. **1960** *News Chron.* 13 July 3/1 Gillies Pelletier,.. Gillies is the pin-up boy of practically every French-Canadian female who watches television. **1963** *Times* 22 Jan. 3/4 He is the 'pin-up' cricketer of the layman as much as of the connoisseur and he will be remembered always as a player who never grew old. **1969** *New Statesman* 18 July 71/1 Aubrey Jones, when I first knew him, was the pin-up boy of the modern Tory party. **1976** T. STOPPARD *Dirty Linen* 11 The man reacting to the pin-up photograph.. *Maddie* in a pin-up pose.

B. *sb.* A pin-up photograph; the subject of such a photograph; also *transf.*

1943 *Yank* 30 Apr. 17/1 The yeoman who did all the worrying about this week's Coast Guard issue roared quite emphatically that this week's pin up would have to have his approval. **1945** *Times* 4 Jan. 5/4 There is always room for a 'pin-up' or a photograph. **1951** J. B. PRIESTLEY *Festival at Farbridge* II. ii. 281 I'll bring the two winsome pin-ups with me and we'll all have lunch. **1957** R. HOGGART *Uses of Literacy* vii. 175 Pin-ups used to be, and still are, standard decoration for servicemen's billets and the cabs of lorries. *Ibid.* 178 Sometimes a male pin-up for the ladies is produced. **1970** *Daily Tel.* 13 Jan. 14/6 He has since become the leading figure—the political pin-up—of the new régime. **1971** B. W. ALDISS *Soldier Erect* 61 Next morning before parade, I stuck the crumpled picture of the monkey god on the wall beside the bed, next to the pin-ups of Ida Lupino and Jinx Falkenberg. **1972** J. MCCLURE *Caterpillar Cop* ii. 15 The hundreds of murders committed for profit by writers..kept things going, just like those pin-ups in Antarctic weather stations. **1979** 'M. YORKE' *Death on Account* xiii. 129 No pin-ups of nudes. It's like a monk's cell.

pin-wheel, *v.* Add: **b.** *intr.* To rotate in the manner of a pin-wheel (sense 2). Also *fig.*

1942 W. FAULKNER *Go down, Moses* 149 The shrill, frantically pinwheeling little dog. **1951** J. STEINBECK *Burning Bright* I. 12 The shrill band played the march of elephants and white horses, giraffes and hippopotamuses and pin-wheeling clowns. **1952** *Chambers's Jrnl.* Apr. 235/1 So he let fly at the nearer of them, and saw her tumble in the air, then pinwheel earthwards. **1976** W. GREATOREX *Crossover* 188 Memories pin-wheeled through Pavel's mind. **1977** D. BENNETT *Jigsaw Man* xii. 225 Strangely unrelated incidents pinwheeled through Farquar's mind.

Pinxton (pi·ŋkstǫn). The name of a town in Derbyshire used *attrib.* to designate a soft-paste porcelain made there from 1796 to 1813. Also quasi-*adj.*

1802 in C. L. Exley *Pinxton China Factory* (1963) 60 To be Sold by Auction, by Blackwell & Co. At their Auction Rooms, Long Row, Nottingham. Six Crates of Pinxton China and Earthernware. **1876** J. HASLEM *Old Derby China Factory* 244 In the general run of Pinxton patterns gold was rarely used, the edging being usually in blue. **1928** W. B. HONEY *Old Eng. Porc.* xi. 201 Some cups and saucers at South Kensington..were 'authenticated' as Pinxton by John Haslem. *Ibid.*, The later Pinxton porcelain may be well studied in Mr. Herbert Allen's collection. **1963** C. L. EXLEY *Pinxton China Factory* iii. 18 It is..likely that the elaborate gilding, so characteristic of the finest Pinxton ware, must have been reduced to a minimum during the last twelve months of Billingsley's connection with the factory. **1966** G. A.

Godden *Illustr. Encycl. Brit. Pott. & Porc.* 257 Typical Pinxton porcelain is similar to puce-marked Derby porcelain of the 1790–1800 period. **1974** —— *Brit. Porc.* 351 Several examples featured as Pinxton in various books have every appearance of being of Coalport manufacture. **1976** *Times* 26 Oct. 16/4 A pair of Pinxton tapering beakers.

piny, *a.* (Further examples of spelling *piney*.)
1809, etc. [see *piney wood]. **1931** S. Jameson *Richer Dust* xiv. 421 The water in the bath was faintly brown. The warm piney water soothed her. **1959** C. Ogburn *Marauders* ii. 37 Meredith had been at Fort Benning,.. and his class.. 'had had an exercise in weapons placement out in some ol' piney hills'.

piny, var. Peony. (U.S. examples.)
1904 *Dialect Notes* II. 427 *Piny, n.*, peony. **1913** G. Stratton-Porter *Laddie* vi. 165 Her people..spent much money on the biggest tombstone in the cemetery, and planted pinies and purple phlox on her. **1976** *Columbus* (Montana) *News* (Joint Suppl.) 24 June 4/4 Even though it had medicinal value, colonial housewives did not as a rule include the peony in their herb gardens, but set it out among their flowers. They felt that the 'glory of the front yard was the old-fashioned early red "Piny"'.

Pinyin (pinyi·n). Also **Pin-yin, Pin-Yin.** [a. Chinese *pīn-yīn*, lit. 'spell sound'.] A system of Romanized spelling for the Chinese language, adopted officially by the People's Republic of China in stages since 1958. Also *attrib.*
[**1959** W. Simon *Chinese Radicals & Phonetics* (rev. ed.) 432 A further Scheme, apparently to be regarded as final,.. on 11th February 1958 was approved by the Fifth Session of the First National People's Congress. Its Chinese name is Hanyu-Pinyin-Fang'an..(Chinese Language Spelling Scheme).] **1963** *McGraw-Hill Mod. Chinese-Eng. Dict.* p. iii/2 The Pinyin romanization, or 'phonetic construction', using the Roman alphabet, is very similar to the Wade-Giles system which has been standard in the Western world till very recently. **1972** *Computers & Humanities* VII. 262 The romanization system used in this article is Pinyin, adopted by the People's Republic of China as its official transcription system. **1974** *Encycl. Brit. Macropædia* XVI. 801/1 The Pin-yin system indicates unaspirated stops and affricatives by means of traditionally voiced consonants. **1977** *Daily Tel.* 10 Jan. 4/5 Primary school children are taught the 'Pin-Yin' system of writing the Peking dialect in the Roman alphabet before they tackle the more difficult system of Chinese characters. *Ibid.*, Most Chinese adults cannot read 'Pin-Yin' on its own. **1979** *China Now* Jan.-Feb. 9/1 The State Council has issued a document stipulating that *pinyin* romanization is 'suited to all languages using the Roman alphabet'.

pinyon, var. *Piñon.

‖**piob mhor** (pīp vōᵊ·r, pī:əb vōᵊ·ɹ). [Gael., lit. 'big pipe'.] The Highland pipes, the bag of which is blown by a long pipe with a mouthpiece (see quot. 1954).
1838 A. MacKay *Coll. Anc. Piobaireachd or Highland Pipe Music* 5 When the infirmities accompanying a protracted life, prevented him handling his favourite *Piob-mhor*, he would sit on the sunny braes, and run over the notes on the staff. **1845** *New Statistical Acct. Scotl.* XIV. 339 The names of some of the caves and knolls in the vicinity still point out the spots where the scholars used to practise, respectively on the chanter, the small pipe, and the *Piob mhor*, or large bagpipe. **1901** W. L. Manson *Highland Bagpipe* i. 10 The *Piob Mohr* is not now an agency to be reckoned with by any one who wishes to explore the hills and glens... As a Highland war spirit, its glory has departed. **1920** *Glasgow Herald* 1 May 6 The clan is no more..; but the piobmohr [*sic*] remains.., and in its music there may be heard..the romance, and the tragedy, and the beauty of the story of the Scottish Highlands. **1954** *Grove's Dict. Mus.* (ed. 5) I. 345/2 The Highland pipe (*piob mhor*) is pre-eminently the martial instrument... The oldest known existing instrument is dated 1409... It now consists of a sheepskin bag into which are tied five stocks, which accommodate the three drones and the chanter with their reeds, and the blow-pipe. **1968** J. Arnold *Shell Bk. Country Crafts* 316 At this juncture it is convenient to define the features which distinguish the various pipes to be heard in Britain. They are the Great Pipes, or Piob Mhor, the Reel Pipes, [etc.]. *c* **1970** A. MacPhee *Story of Scottish Highland Bagpipe* (An Comunn Gaidhealach) 5 The renaissance of interest over the past century in the 'piob-mhor' is in great measure due to the interest of the Army. **1975** F. Collinson *Bagpipe* 114 The contrast between the pastoral English bagpipe and the war-like Highland *piob-mhór*.

pion (pəi·ǫn). *Nuclear Physics.* [f. *pi(-meson + *-on¹.] Any of a group of mesons that have masses of approximately 140 MeV (270 times that of the electron), zero spin, zero hypercharge, and isospin of 1, and on decaying usually produce a muon and a neutrino (in the case of charged pions) or two photons (in the case of the neutral pion); a pi-meson.
1951 *Sci. News* XXI. 21 The mass of π-mesons, or pions as they are called in short,..appears to be in the neighbourhood of 276 times that of the electron. **1958** *Spectator* 13 June 778/3 The Japanese physicist Yukawa has shown that what causes the neutrons and the protons of the atomic nucleus to cling together with such tenacity are the revolving pions that link them. **1968** M. S. Livingston *Particle Physics* vi. 124 The neutral pion π^0 can be formed through the charge-exchange process $\pi^- + p \to n + \pi^0$ following capture of the π^- in a Bohr orbit (i.e., in liquid hydrogen). The π^0 decays promptly (10^{-16} sec) into two photons. **1971** P. E. Hodgson *Nucl. Reactions* ii. 37 The interaction of fast pions with nuclei can also give information on the relative extent of the neutron and proton distributions in matter. **1975** *Daily Colonist* (Victoria, B.C.) 26 Jan. 14/2 The new 17-foot-long generator shoots out pions or sub-atomic particles generated by a high-energy accelerator.
Hence **pio·nic** *a.*, of, pertaining to, or involving a pion, or an atom having a negative pion orbiting the nucleus.
1960 P. Roman *Theory Elem. Particles* v. 511 All pionic interactions are of the same strength. **1967** J. C. Sens in G. Alexander *High Energy Physics & Nucl. Struct.* II. 117 In muonic atoms, the muon samples the distribution of *charge* in the nucleus; in pionic atoms the pions form in addition a probe for the nuclear *mass* distribution, through the strong interactions. **1970** D. F. Jackson *Nucl. Reactions* x. 230 Additional information about the pion-nucleus interaction can be obtained from studies of pionic atoms and pionic x-rays.

pioneer, *sb.* Add: **3. b.** *Ecol.* A plant which establishes itself in an unoccupied area.
1916 F. E. Clements *Plant Succession* x. 212 It [*sc.* the pitch pine] produces..more seed than the white pine and in its demands is better able to act as a pioneer. **1929** Weaver & Clements *Plant Ecol.* viii. 147 The reactions of the pioneer stage may be unfavorable to the pioneers themselves. **1953** H. L. Edlin *Forester's Handbk.* viii. 113 As a general rule, the light-demanders are also pioneers, capable..of forming a vigorous first crop on bare land. **1967** M. E. Hale *Biol. Lichens* vii. 96 Lichens are conspicuous pioneers on rocks.
c. (Usu. with capital initial.) In the U.S.S.R. and other communist countries, a member of a Society of Young Pioneers, a movement for children below the age of sixteen. Also *transf.* and *attrib.*
1929, etc. [see *Octoberist, -brist 2 b]. **1930** 'I. Low' *His Master's Voice* x. 120 The streets grew merry with the drums of the pioneers, with flags, with the strains of the International. **1944** M. Laski *Love on Supertax* v. 58 A band of children..holding a banner on which was inscribed 'St Pancras Pioneer Group'. **1957** M. Pavlov in G. L. Kline *Soviet Educ.* 130 The Pioneers and the Komsomol members, as a rule, are atheists. **1959** A. Wesker *Chicken Soup with Barley* i. ii, in *New Eng. Dramatists* I. 193 We didn't force her to be in the pioneers... Show a young person what socialism means and he can't do anything else but accept it. **1970** N. Fleming *Czech Point* vi. 71 As a boy he was in the Pioneers—it's a sort of boy-scout thing. **1970** *Morning Star* 11 May 4 The school has 180 pupils from the age of six to sixteen. There are 22 staff, four educators and a Pioneer organiser. **1972** *Times* 17 June 13/1 Scores of children, mostly red-neckerchiefed Pioneers, were running and scrambling over the rocky ground... They..had to climb more than 950 steps to the monument which records how a Russian army freed Bulgaria from oppression. **1973** *Listener* 2 Aug. 138/2 Another important aspect of Soviet holiday-making is the vast network of so-called Young Pioneer summer camps run for schoolchildren. **1976** 'S. Harvester' *Siberian Road* iv. 50 Young schoolchildren, all wearing their red Pioneer neckerchieves.
4. a. *pioneer work.*
1849 Thoreau *Week Concord Riv.* 359 It is the worshippers of beauty, after all, who have done the real pioneer work of the world. **1933** *Burlington Mag.* Nov. 193/1 Valuable pioneer work..has been done. **1965** *Eng. Stud.* XLVI. 369, I shall try to clarify the relation between my interpretation of Coleridge's poem and such pioneer-work as Lowes's associative researches.
b. (In sense 3 b above) *pioneer association, plant, species, tree.*
1932 Fuller & Conard tr. *Braun-Blanquet's Plant Sociol.* xiv. 352 We distinguish aggressive, advancing pioneer associations and restricted, retreating relict communities. **1960** N. Polunin *Introd. Plant Geogr.* xiv. 456 The fringe of the mangrove, at least where it does not consist of young pioneer plants, is made up of tall trees. **1966** F. H. Brightman *Oxf. Bk. Flowerless Plants* 199 These pioneer plants gradually break down the rock surface into fine particles. **1933** *Forestry* VII. 140 Scots pine is the native conifer of the district, and has also undeniable merits as a first crop, or pioneer species, on the..moorlands of the district. **1971** *Sci. Amer.* Sept. 129/2 Species typical of immature succession stages—'pioneer' species—are characteristically able to disperse themselves over considerable distances. **1954** S. Piggott *Neolithic Cultures Brit. Isles* i. 5 Iversen has noted that birch is a 'pioneer tree' which rapidly colonizes an area after a forest fire.

pioneer, *v.* Add: **4.** *Ecol. trans.* and *intr.* Of a plant, to colonize (new territory); to establish itself in an unoccupied area.
1939 H. H. Bennett *Soil Conservation* 418 Ragweed.. pioneers idle fields. **1960** N. Polunin *Introd. Plant Geogr.* xi. 327 Hardy Mosses..sometimes pioneer on uncolonized rock surfaces.

pionee·rdom. *rare.* [f. Pioneer *sb.* + -dom.] The condition or state of a pioneer; a prevalence of pioneers.
1873 *Porcupine* 13 Sept. 379/2 A..Californian, who had arrived..from the States, close on to the age of pioneerdom.

pionization (pəi,ǒnəizēi·ʃən). *Nuclear Physics.* [f. *pion + -ization.] The production of numerous low-energy pions by the collision of two high-energy nucleons.
1964 *Mat.-Fys. Meddelelser K. Danske Vidensk. Selskab* XXXIII. xv. 8 The calculation of the relative importance in the cosmic radiation of mesons generated by nucleons in pionization and mesons generated in the deexcitation of baryon states is straightforward on the basis of this model. **1974** Frauenfelder & Henley *Subatomic Physics* xii. 343 At ultrahigh energies, the interaction of two nucleons can indeed be a spectacular event... Most of the secondaries are pions. In the c.m. of the colliding protons, nearly all the pions have small momenta. The formation of such a pion cloud is called pionization.

piopio (piū·,piū). [Maori.] A small New Zealand bird, *Turnagra capensis*, resembling a thrush and belonging to the subfamily Pachycephalinæ, the whistlers or thickheads; also called the native thrush.
1873 W. L. Buller *Hist. Birds N.Z.* 136 The silvery notes of the Bell-bird, the bolder song of the Tui, ..and the whistling cry of the Piopio—all these voices of the forest are blended together. **1938** M. Gordon *Children of Tane* vi. 160 The southern Piopio has been more often encountered than the northern. **1969** J. Fisher et al. *Red Bk.* 301/1 The piopio is certainly very rare indeed; but just how rare it is can only be discovered by hard field studies on both North and South Island of New Zealand, in each of which a race survives.

piosity (pəiǫ·siti). [f. Pious *a.* + -ity, after Religiosity.] Affected or excessive piousness, sanctimoniousness; an instance of this.
1922 *Contemp. Rev.* Sept. 353 The lack of such faith means, and has meant, the reduction of morality to piosity, convention, legality. **1961** *Spectator* 1 Dec. 824 The pall of piosity and ambition lifts from the family. **1975** *Church Times* 31 Jan. 14/5 God forbid that we should use over-devotional words or piosities about the Lord's life.

‖**piou-piou** (piūpiū). [Fr.] A popular name for a French private soldier.
1854 B. St. John *Purple Tints of Paris* II. xi. 224 Formerly, a *piou-piou*—as the common soldier was somewhat contemptuously called—was treated..with a sort of paternal solicitude. **1867** 'Ouida' *Under Two Flags* I. xiii. 300 The speaker looked down on the *piou-piou* with superb contempt. **1894** [see *nounou]. **1900** R. Whiteing *Life of Paris* 207 Polin..figures as the common soldier, the pioupiou, with his simple virtues of good-humour and fidelity to the flag. **1930** [see *nounou].

pious, *a.* Add: **1. d.** Phr. *pious hope*, an extravagant or unrealistic hope expressed in order to preserve an appearance of optimism.
1907 R. Fry *Let.* 3 Apr. (1972) I. 283 Mr Morgan.. secured three [pictures] for himself and expressed a pious hope that the Museum might be able to buy the rest! **1931** *Economist* 11 Apr. 773/2 That this, as the phrasing suggests, is only a pious hope is apparent when the attitude of the three main groups of countries concerned is considered. **1977** *N.Y. Rev. Bks.* 15 Sept. 15/1 The pious hope—by no means an assumption—that the biographer's psychological acuity, powers of empathy, respect for fact, general culture, and sense of proportion would prevent the appearance of yet one more in a series of glib psychobiographies. **1978** Ld. Drogheda *Double Harness* xxi. 288 Despite all the pious hopes, disappointments persisted.

pip, *sb.*¹ Add: **c.** Ill humour or poor health, esp. in colloq. phrs. *to have* (or *get*) *the pip*, to be depressed, despondent, or unwell; *to give* (someone) *the pip*, to annoy or irritate, to make (someone) ill-tempered or dispirited.
1886–96 in Farmer & Henley *Slang* (1902) V. 210/2 It cost a bit to square up the attack; For the landlord had the pip. **1896** A. Beardsley *Let.* *c* 17 Sept. (1971) 165 Are you suffering with a south-west wind in London? It prevails here utterly and has given me the pip. **1903** Kipling *Traffics & Discov.* (1904) 55 What's an admiral after all?.. Why, 'e's only a post-captain with the pip. **1913** *Punch* 15 Oct. 324/3 [His] later works gave him the pure pip. **1923** Wodehouse *Inimitable Jeeves* iii. 36 If there's one thing that gives me the pip, it's unpleasantness in the home. **1930** J. B. Priestley *Angel Pavement* ix. 440 A proper old Jonah you're turning into! You give me the pip, Dad, honestly you do. **1932** A. J. Worrall *Eng. Idioms* 31, I feel rotten to-day. I'm not ill, but I've just got the pip, that's all. **1934** J. Rhys *Voy. in Dark* i. iv. 53, I thought there was something about this place that gave me the pip. **1942** *R.A.F. Jrnl.* 18 Apr. 20 Dear-o-dear, he fair gave me the pip. Talk about gloom! **1949** N. Marsh *Swing, Brother, Swing* xii. 285 The Judges' Rules..may be enlightened but there are times when they give you the pip. **1962** *Friend* 3 Aug. 962/1 The signpost to the safe way forward: 'true moral simplicity'. It just about gives me the pip. **1976** *Scotsman* 24 Dec. (Weekend Suppl.) 1/5, I feel it's my duty but I'm not keen. My grandchildren give me the pip.

pip, *sb.*² Add: **5.** *Mil.* A star or one of a group of (up to three) stars worn on the epaulettes by officers as an indication of rank. Also *transf.*
1917 W. Owen *Let.* 23 Nov. (1967) 509, I shall soon be putting up another pip. **1918** —— *Let.* 15 July (1967) 564, I still wear one pip because nobody knows whether I am Lieut. or not. **1919** *Chamber's Jrnl.* Jan. 43/2 Thomas, his senior by one 'pip' in the battery. **1924** Kipling *Debits & Credits* (1926) 315, I wrote the usual trimmin's..an' what his captain had said about Bert bein' recommended for a pip. **1954** [see *gong¹ 2]. **1972** P. Driscoll *Wilby Conspiracy* (1973) xxii. 284 The authority of the two pips shining on his shoulders. **1973** D. Lees *Rape of Quiet Town* vi. 103 Despite the extra couple of pips he'd given me, I didn't feel happy about my new command. **1978** M. Kenyon *Deep Pocket* i. 12 The desiccated arsehole-creeper in his new silver pips and braid.

6. a. A sharp, narrow, and usu. small spike or deflection on a line displayed on the screen of a cathode-ray tube.
1944 *Radar* Apr. 30/1 Signals appear as pips, or deflections, on the luminous trace. *Ibid.*, The range of a reflecting object is indicated by the distance of the pip from the base of the trace. **1949** D. G. C. Luck *Frequency Modulated Radar* vii. 299 As the generator is tuned toward either limit of the sweep, these pips may be seen to approach one another on the oscilloscope. **1950** *Jrnl. Appl. Physics* XXI. 59/2 If the output frequency times three is exactly equal to the input frequency, the two pips on the scope should coincide. **1963** G. M. B. Dobson *Exploring Atmosphere* viii. 138 The distance of the 'pip' from the starting line will be a measure of the time between the signal which started the beam moving and that which produced the 'pip'.
b. A voltage pulse.
1946 *Electronic Engin.* XVIII. 145/3 Use was made of a crystal oscillator which generated both time-base recurrence and calibration pips. **1947** *Ibid.* XIX. 9/2 The time base can be synchronised by applying a negative pip to the first grid of V_4.
7. *attrib.*, as *pip card* (sense 1).
1903 *Burlington Mag.* Dec. 246/1 He persuaded him.. to make the exchange with twelve figure and fourteen pip cards. **1977** *Jrnl. Playing-Card Soc.* May 30 The suit symbols and the denominations are shown as miniature cards of traditional form in the upper right-hand corner on the pip cards.

pip, *sb.*³ Add: **2. b.** Phr. *to squeeze* (someone) *until the pips squeak* (and variants): to exact the maximum payment from (someone), orig. with allusion to Germany's indemnity after the war of 1914–18 (see quot. 1918).
1918 *Cambridge Daily News* 11 Dec. 3/2 Sir Eric Geddes followed up his big meeting at the Guildhall on Monday night by addressing another crowded assembly in the large hall at the Beaconsfield Club on Tuesday night... Dealing with the question of indemnities, Sir Eric said: The Germans, if this Government is returned, are going to pay every penny; they are going to be squeezed as a lemon is squeezed—until the pips squeak. My only doubt is not whether we can squeeze hard enough, but whether there is enough juice. **1929** W. S. Churchill *World Crisis: Aftermath* ii. 47 One Minister, reproached with lack of vim, went so far as to say 'We would squeeze the German lemon till the pips squeaked.' **1933** *Radio Times* 14 Apr. 75/1 The Lloyd George Coalition Government..elected..on a programme of hanging the Kaiser, squeezing Germany until the pips squeaked. **1940** S. Spender *Backward Son* 64 A clarion call to the readers of the *Daily Sketch* to make Germany pay till the pips squeak. **1973** P. O'Donnell *Silver Mistress* v. 93 We run an inquiry on a client, and we don't squeeze him till the pips squeak... We just pressure him. **1973** *Times* 12 Nov. 19/3 In opposition..[Labour] would tax the upper working class until the pips squeak. **1978** *Times* 15 Sept. 3/3 When Mr Singer was asked how the extra money was being found, he said: 'The pips are squeaking.'

pip, *sb.*⁴ Used for *p* in telephone communications and in the oral transliteration of code messages, as in *pip emma*, for *p.m.* (= *post meridiem*: see P II). Also occas. in colloq. use.
1913 *Signalling* (Imperial Army Series) ii. 19 The letters T, A, B, M, S, P and V will be called *toc, ack, beer, emma, esses, pip*, and *vic* respectively, so as to distinguish them phonetically from letters of similar sound. **1915** [see *emma]. **1917** [see *ack]. **1926** [see *emma]. **1927** W. E. Collinson *Contemp. Eng.* 98 Other artillery terms which spread were O pip (for 'observation post'). **1930** E. Raymond *Jesting Army* iii. ii. 292 The working parties parade under the trees at nine o'clock pip emma... At three o'clock ac emma they will return. **1969** [see *emma]. **1977** C. McCullough *Thorn Birds* xv. 350 The second hand was just sweeping up to 9:40 pip-emma.

pip, *sb.*⁵ [Echoic.] A short, high-pitched sound, esp. one produced electronically; *spec.* (*a*) one broadcast as a time signal; (*b*) one transmitted over a telephone line as a signal.
1907 G. B. Shaw *Major Barbara* iii. 292 *Sarah* (touching Lady Britomart's ribs with her finger tips and imitating a bicycle horn) Pip! pip! **1929** *B.B.C. Year-bk.* 1930 406 G.T.S. = Greenwich Time Signal, which takes the form of a broadcast by electrical contact of the last six seconds before the hour, the 'beat' of each second being represented by a sharp 'pip'. **1930** *Prof. Papers Inst. P.O. Electr. Engineers* No. 135. 41 There is an advantage in giving the time intimation automatically and this is being done experimentally..by means of a 'pip' signal. **1938** F. B. Young *Dr. Bradley Remembers* i. 29 The six 'pips' of the time-signal sounded and the Weather Forecast began. **1946** *Electronic Engin.* XVIII. 360/2 (Advt.), The warning is given in the form of a 1,000 c.p.s. 'pip' on the loud speaker lasting ¼ second. **1951** 'E. Crispin' *Long Divorce* xiv. 172 There are the pips... Quick, or it'll be another three minutes. **1962** A. Nisbett *Technique Sound Studio* xii. 219 One of the most characteristic sounds of early electronic works has been the use of short bursts of tone at various frequencies, sounding like a series of pips. **1967** O. Lancaster *With Eye to Future* i. 1 With the six pips, conversation..faded away ..as the announcer..began to summarize his disastrous bulletin. **1972** *Radio Times* 6 Jan. 5/3 Listeners may have noticed a change in the Greenwich Time Signals broadcast since January 1. Instead of six equal pips, the signal since 1924, there are five equal pips followed by a longer one lasting half a second. The exact time is signalled by the beginning or 'leading edge' of the long pip. **1973** *Times* 21 Sept. 5/3, I believe the call was a long-distance one. The pips are louder..if the call is local. **1977** J. Wilson *Making Hate* xiii. 155 He raced back to the telephone in the hall. When the pips went.. he could hardly get his twopence in the slot.

pip, *v.*² Add: **1.** (Earlier example.)
1598 Florio *Worlde of Wordes* 279/1 *Pipare*, to cackle or cluck as a hen, to pip, to pule as a hawk.
2. a. (Later examples.) So **pipped** (pipt) *ppl. a.*; **pi·pping** *vbl. sb.*
1901 *Chambers's Jrnl.* Nov. 717/1 Gigantic incubators.. literally vomiting forth their flocks of twittering little creatures at pipping-time. **1953** N. Tinbergen *Herring Gull's World* xviii. 161 The parents would stop shifting and turning an egg when it is pipped. *Ibid.*, In most of the pipped eggs there is a line of cracks around the obtuse pole. **1962** J. C. Welty *Life of Birds* xvi. 316/2 The first step in hatching is the puncturing, or 'pipping', of the shell by the chick with its outwardly-pressed egg tooth. **1972** L. Hancock *There's a Seal in my Sleeping Bag* vii. 162 This egg will hatch tonight. See, the large end is already cracked and pipped. **1972** *Sci. Amer.* Aug. 30/1 The ducklings begin to pip their eggs... As the eggs are being pipped the female clucks... When the pipping is completed, she drops back to..four calls per minute.
b. *transf.* To give birth. *colloq.*
1973 *Times* 27 Aug. 5/8 'I say, Aubrey, has your wife pipped yet?' I assumed he meant had she had her baby.

pip, *v.*³ Add: **1. a.** (Further examples.)
The examples in the second set owe something to the phr. *to pip at the post* (see sense c below).
1916 E. V. Lucas *Vermilion Box* 226 Only yesterday poor Hugh Blackstone was pipped right at my side, and he lasted only ten minutes. **1917** E. F. Wood *Notebk. of Intelligence Officer* x. 182 In that bit of trench, sir, you must bend over as you go. It is enfiladed by an enemy sniper. He 'pipped' one of our fellows through the head there yesterday. **1927** A. Christie *Big Four* xi. 141 That's my solution—Gilmour Wilson got pipped by mistake. **1932** Wodehouse *Hot Water* i. 19 Soup Slattery showed Mr. Carlisle the scar..where a quick-drawing householder of Des Moines, Iowa, had pipped him a couple of years back when he was visiting at his residence. **1950** Partridge *Here, There & Everywhere* 70 The remaining Tommy synonyms [for 'wounded'] are *pipped*, especially by a bullet whether of rifle, revolver, or machine-gun; *to stop one* [etc.].
1930 *Bulletin* (Sydney) 1 Jan. 35/1 He [*sc.* a race-horse] just pipped Baverarrack..for third place. **1964** *Engineering* 21 Aug. 221/3 Dick Bertram.., in Lucky Moppie, was pipped into second place by an error of navigation. **1976** *Scottish Daily Express* 24 Dec. 15/4 As anchorman, Ian Hutcheon did a magnificent job, shooting a final 71 to pip the Japs and tie for the individual section. **1977** *R.A.F. News* 11–24 May 18/4 The host station..were pipped 6–3 by Brampton.
b. To reject or disqualify; to fail (a candidate) in an examination. Of a candidate: to fail (an examination).
1908 A. S. M. Hutchinson *Once aboard Lugger* i. i. 31 'I had forgotten. Your examination?' George half turned away. The bitterest moment of a sad day was come. He growled: 'Pipped.' **1912** F. M. Hueffer *Panel* i. iii. 85 Olympia was exaggerating... I wasn't going blind. I was only pipped for active service. **1973** *Daily Tel.* 17 Oct. 15/1 School-leavers who were unfortunate enough to pip all or some of their O or A levels will have been seeking a second chance without having to return to school.
c. To anticipate or forestall (someone) in a particular activity, circumstance, etc.; *spec.* in phr. *to pip at* (or *on*) *the post*, to defeat by a narrow margin at the last moment; also *ellipt.*
1924 Wodehouse *Ukridge* iii. 67 Bad luck his getting pipped on the post like that. **1939** 'N. Blake' *Smiler with Knife* xix. 272 Well, Georgia, pipped at the post, aren't you? **1949** 'E. C. R. Lorac' *Still Waters* ii. 27, I pipped him at the post. His instructions must have limited him. **1959** 'M. M. Kaye' *House of Shade* xviii. 245 He was head over heels about the bewitching Amalfi, and got pipped on the post by Eduardo. **1969** *Times* 25 Nov. 23/1 Shell..now have a record eight managing directors: BP..have just been pipped at the post—they have only seven. **1970** 'A. Gilbert' *Death wears Mask* iii. 4 You won't be able to buy me that ring, after all, because it's sold. Someone's pipped you on the post. **1974** *Times* 8 Apr. 13/1 Schools television started in 1957, when Associated Rediffusion pipped the BBC by starting a service in which 80 schools took part.
2. *intr.* To die. Also with *out.*
1913 A. Lunn *Harrovians* iii. 31 'Is he Irish?' 'He don't seem to know. Father who's pipped was Irish. His mother's pipped too.' **1920** R. Macaulay *Potterism* iii. i. 110, I think it's simply rotten pipping out. I *like* being alive.

pip, *v.*⁴ [f. *pip *sb.*⁵] *intr.* To make a short, high-pitched sound. So **pi·pping** *ppl. a.*
1938 D. Smith *Dear Octopus* i. 14 Just see if Hilda's still telephoning... Tell her it's elevenpence every time it pips. **1958** *Listener* 18 Sept. 418/1, I could hear morse pipping and the loudspeaker blaring away. **1972** *Jazz & Blues* Oct. 5/1 A throbbing low register clarinet solo by Ben Richardson..ends rather oddly with some 'pipping' high notes like the BBC time signal. **1976** A. Price *War Game* i. 61 The phone pipped for more money and he..fed the last of his change into it. **1978** R. Holles *Spawn* iv. 28 People passing in cars pipped and waved although they hardly knew you.

‖**p'ip'a** (pīpā·). Also **pepa, pipa**, etc. [Chinese.] A Chinese stringed instrument (see quot. 1975).
1839 *Chinese Repository* May 42 *Pepa*, the balloon shaped guitar... This is about three feet in length... The strings are of silk. **1848** S. W. Williams *Middle Kingdom* II. xvi. 165 In writing a tune for the lute or *pipa*, 'each note is a cluster of characters'. **1874** *Jrnl. N. China Branch R. Asiatic Soc.* VIII. 115 The *P'i-p'a*.., or 'Balloon-shaped Guitar', described by the Chinese as resembling a ham, has a body like the egg of a goose nearly a foot in diameter with four strings, which are played with the fingers. **1917** *Encycl. Sinica* 388/2 *P'i p'a* is a lute about forty-two inches long with a pear-shaped body. The neck is eight and a half inches long. It has ten or twelve frets and four or six strings. **1954** *Grove's Dict. Mus.* (ed. 5) II. 238/2 *P'i-p'a*.., four-stringed short lute (sometimes 'balloon guitar')... About 3 ft. long and 1 ft. wide.., the body is pear-shaped, its back being rounded but shallow, its front flat and covered with the soundboard... The four silk strings..are attached..to a cross-ledge on the soundboard. **1962** E. Snow *Red China Today* (1963) lxxiii. 563 The *p'i-p'a* is a native Chinese instrument something like a zither. **1975** C. P. Mackerras *Chinese Theatre in Mod. Times* 22 Of the plucked, non-bowed stringed instruments the most significant is the pear-shaped *p'i-p'a*... It is played held upright on the thigh. The musician plucks the strings with his right thumb and first finger, sometimes protected with a plectrum, and determines the pitch with three of his left fingers. **1979** *Time* 2 Apr. 48/1 Liu Dehai played a concerto for a lutelike instrument called the pipa.

pipe, *sb.*¹ Add: **1. c.** Also *pl.* as a nickname for a boatswain. *Naut. slang.*
1856 R. MacClure *Discovery of North-West Passage* xv. 233 'Pipes' picked up a leg of the deer. **1903** H. Holmes *Life & Adventures on Ocean* 17 The boatswain, commonly called 'Pipes' for shortness, was warned by his superior officer to take every care. *Ibid.*, 'High enough,' calls out Pipes. **1942** *Penguin New Writing* XV. 8 When Pipes went for supper he had a side-parting and looked quite different.
3. c. *U.S. slang.* Something that is easy to accomplish; a 'cinch'. Cf. *lead-pipe. Also *attrib.*, as *pipe course.*
1902 'H. McHugh' *It's up to You* iii. 66 It was so easy it was a shame... 'The idea is Napoleonic, little woman!' I said. 'It's a pipe!' **1927** *Amer. Speech* II. 277/2 *Pipe course*, easy course. **1936** L. C. Douglas *White Banners* ii. 44 A procession of shamefaced athletes who..had thought erroneously, when they had registered for it, that Anglo-Saxon was 'a pipe'. **1951** M. Shulman *Many Loves of Dobie Gillis* (1953) 105 You are all freshmen.. and you may not be familiar with the term 'pipe course'. A pipe course is a course where students can get passing grades without doing much work. This is not a pipe course. **1952** Wodehouse *Barmy in Wonderland* viii. 80 This show's a pipe, and any bird that comes in is going to make plenty. **1978** H. Kemelman *Thursday the Rabbi Walked Out* x. 57 Nothing to it... Believe me, the whole thing's a pipe.
4. i. (Example.)
1813 Jane Austen *Let.* 16 Sept. (1952) 326 My Cap is come... Fanny has one also..shaped round the face.. with pipes & more fullness, & a round crown inserted behind.
7. e. For 'of steel, caused by the escape of gas' read 'caused by shrinkage of the metal'. (Examples.)
1861 *Brit. Patent 1310* 2 Shrinkage forms a deep tube or funnel in the upper part of the ingot... This funnel is called by steel manufacturers the 'pipe' of the ingot. **1895** E. L. Rhead *Metallurgy* xi. 147 Steel of harder temper..settles down in the mould, forming a funnel-shaped cavity or pipe. **1923** Glazebrook *Dict. Appl. Physics* V. 357/2 If this takes place any shrinkage of the metal during further solidification must result in the formation of a pipe. **1973** J. G. Tweeddale *Materials Technol.* II. ii. 36 Metal ingots..are often cast with open tops and the defective top material, including any pipe, is usually cut off subsequently.
10. d. *put that in your pipe and smoke it* (earlier and later examples); also in similar phrases.
c **1824** R. B. Peake *Americans Abroad* (1884) i. 4/2 Put that in your pipe and smoke it. **1837** Dickens *Pickw.* ii. 7 Pull him up—put that in his pipe—like the flavour—damned rascals. **1921** Galsworthy *To Let* i. ix. 81 The noble owner put this opinion in his pipe and smoked it for a year. **1927** *Vanity Fair* XXIX. iii. 67/2 'Laugh That Off!' ('Put that in your pipe and smoke it.') **1947** W. S. Maugham *Creatures of Circumstance* 296 I'm engaged to her, so put that in your pipe and smoke it. **1977** A. Hunter *Gently Instrumental* x. 136 There's a dozen witnesses, so you can put that in your pipe and smoke it.
e. *N. Amer. colloq.* A spell of travelling between two rest-periods at each of which a pipe is smoked; the distance covered or the time taken, in such a spell; also, the distance covered while smoking a pipeful of tobacco. *Obs.* exc. *Hist.*
1793 J. MacDonell *Diary* 5 July in C. M. Gates *Five Fur Traders* (1933) 92 Leaving pointe au père we paddled two pipes and put to shore to give the men time to clean themselves. **1799** I. Weld *Trav. N. Amer.* xxix. 262 A pipe, in the most general acceptation of the word, seemed to be about three quarters of an English mile. **1806** S. Fraser *Jrnl.* 29 May (1960) 193 The men are better off and better pleased than if they ate a little at every Pipe. **1809** 'D. Knickerbocker' *Hist. N.Y.* I. iii. viii. 189 He arrived at Fort Amsterdam in little less than a month, though the distance was full two hundred

pipes, or about 120 miles. **1848** R. M. Ballantyne *Hudson's Bay* iv. 77 The men used to row for a space of time, denominated a *pipe*, so called from the circumstance of their taking a smoke at the end of it. **1931** G. L. Nute *Voyageur* 50 We have seen that the voyageurs had their own method of measuring portages. They were not less original on the water. Here the *pipe* was the standard of measurement. This was the distance covered between respites, when the luxury of resting and smoking was indulged. **1969** E. W. Morse *Fur Trade Canoe Routes* i. i. 8 A stop was made for a few minutes each hour to allow the men to have a pipe. This event was so important that distances came to be measured in *pipes*: '*trois pipes*' might be 15 to 20 miles, depending on winds and current.
f. An opium-pipe; esp. in phr. *to hit the pipe*, to smoke opium; also, an opium-addict. *slang* (orig. *U.S.*).
1886, 1902 [see *hit *v.* 22* a]. **1926** J. Black *You can't Win* xix. 300, I..learned he had been 'on the pipe' only three months. **1926** N. Lucas *London & its Criminals* x. 134 So 'Izzy' had come to 'hitting the pipe'. I knew he had many vices, but I did not know that opium was one of them. **1949** *Sunday World-Herald Mag.* (Omaha, Nebraska) 3 Apr. 2/1 Opium smokers are considered at a low level..but a guy who profits when he hits the pipe is the plumber. **1959** Murtagh & Harris *Who live in Shadow* iii. i. 119 You can recognize 'pipes', opium addicts, by the odour which clings to them. **1972** D. Bloodworth *Any Number can Play* xi. 95 Max insisted that they go on to Madame Phnom's plush smoking establishment for a pipe or two.
11. a. (sense 3) *pipe-coating, -fitter, -jointer*; (sense 10) *pipe-bowl* (earlier example), *-lighter, -smoke, -smoker* (examples), *-smoking* (examples; also as *vbl. sb.*), *-spill, -weed.*
1852 Dickens *Bleak Ho.* (1853) xxi. 213 The pipe-bowl..is burning low. **1964** N. G. Clark *Mod. Org. Chem.* v. 80 The highly viscous bitumen which forms the remainder of the distillation residue is used for..corrosion-proof pipe-coatings. **1975** *Offshore Engineer* Oct. 13/1 The UK pipe-coating market is split..between two firms. **1890** Webster, *Pipe fitter*, one who fits pipes together, or applies pipes, as to an engine or a building. **1910** *Daily Chron.* 31 Jan. 6/5 Arthur Moon, aged 45, a pipe-fitter. **1977** *Cornish Times* 19 Aug. 15/3 He won first-year plumber and pipe-fitter prize. **1902** *Encycl. Brit.* XXV. 509/2 A record should be kept of the history of the pipe.., with the name of the pipe-jointer whose work closes the record. **1909** *Dialect Notes* III. 357 *Pipe-lighter*, a paper spill or taper used for lighting lamps, pipes, etc. **1916** 'Boyd Cable' *Action Front* 57 Each man had with him one of those tinder pipe-lighters which are ignited by the sparks of a little twirled wheel. **1852** Dickens *Bleak Ho.* (1853) xi. 105 A cloud of pipe-smoke..pervades the parlor. **1971** 'D. Halliday' *Dolly & Doctor Bird* xvi. 231, I got home to be met by..the smell of pipe smoke curling round from the hallway. **1979** M. Eden *Document of Last Nazi* xix. 96 Strang smelled old pipe smoke. **1866** Geo. Eliot *Felix Holt* II. xxx. 229 Rough-looking pipe-smokers, or distinguished cigar-smokers. **1959** *Listener* 25 June 1119/1 His plain, factual, forthright sentences—a sort of pipe-smoker's prose—deceptively bare and simple, convey the scene..with a telling clarity. **1853** Dickens *Bleak Ho.* xxxiv. 338 I'll have no more of your pipe-smokings and swaggerings. **1958** J. Byrom *Or be he Dead* vii. 106 The pipe-smoking young man in the open-necked shirt. **1979** J. van de Wetering *Maine Massacre* iv. 52 A pipe-smoking old man. **1979** *Arizona Daily Star* 5 Aug. A7/1 A single match and an even draw gave the world's international pipe-smoking cup to a..physics teacher yesterday. **1922** Joyce *Ulysses* 530 Pages will be torn from your handbook of astronomy to make them pipespills. **1955** J. R. R. Tolkien *Return of King* 270 And if you have any pipe-weed, we'll bless you.
b. pipe berth, a collapsible or otherwise easily-stored canvas bed with a frame of metal piping used on small vessels; **pipe bomb**, a home-made bomb contained in a metal pipe; **pipe-burial**, a burial in which a pipe (usually of lead) passes from the coffin or the tomb to the surface of the ground, to permit the pouring of libations; **pipe chaplet** *Founding*, a chaplet (sense 6) used in the casting of pipes, which consists of a concave semi-cylindrical load-bearing surface supported on a stem; **pipe-cleaner**, something used for cleaning a tobacco-pipe; *spec.* a device for this purpose consisting of a piece of wire covered with tufted material; also *fig.*; **pipe cot** = *pipe berth*; **pipe-drain** *v. trans.*, to drain (land) by laying pipes; chiefly in pa. pple.; **pipe-dream** orig. *U.S.*, a fantastic or impracticable notion or plan, compared to a dream produced by smoking opium; a 'castle in the air'; hence **pipe-dreamer**; **pipe-dreaming** *vbl. sb.*; **pipe-dreamy** *a.*; **pipe-fiend** *U.S. slang*, an opium addict or smoker; **pipe-gun**, (*a*) *dial.*, a pop-gun; (*b*) a gun made out of a pipe; **pipe-necked** *a.*, having a long slender neck; **pipe-opener** (earlier and later examples); also *fig.*, a 'trial run' or 'curtain-raiser'; **pipe-rack**, add: (now usu. as two words without a hyphen); (*c*) a rack or support for a set of pipelines above the ground; **pipe-still**, a still in which crude oil is heated by passing it through a series of tubes inside a furnace; **pipe-story**, a fantastic or impossible story (c.f. *pipe-dream*); **pipe-water**, water conveyed by pipes.
1933 Chapman & Horenburger *Thirty Easy to build Sail Boats* 89/2 A pipe berth can be fitted over each berth for the extra guest. **1963** J. T. Rowland *North to Adventure* vii. 90 With all the pipe berths folded back against the ship's side, the cabin became a capacious hold. **1976–7** *Sea Spray* (N.Z.) Dec./Jan. 51/2 The Sands design No 2218 offers four berths, pipe quarter berth, two seat berths amidships and a pipe berth in the foc'sle. **1966** *Guardian* 5 Sept. 1/7 The headquarters of the American Communist Party was damaged by a pipe-bomb tossed from a moving car. **1971** *Sunday Times* 31 Oct. 10 Pipe bombs are another favourite anti-personnel bomb. **1977** *Time* 26 Dec. 26/2 Last August a tipster directed police to a pipe bomb at a Coors recycling plant in a Denver suburb. **1929** *Antiquaries Jrnl.* IX. 1 (*heading*) A Roman pipe-burial from Caerleon, Monmouthshire. **1934** Laing & Rolfe *Man. Foundry Pract.* iii. 57 Pipe-chaplets..can be obtained, either in the form illustrated, or with short pointed stems, their chief purpose being to support pipe cores in position. **1960** R. Lister *Decorative Cast Ironwork* ii. 26 Of the types in regular use, special mention may be made of the pipe chaplet.., usually made of tinned wrought iron, and consisting of a pin with a semicylinder at one end. It is used to support the round core in pipe casting. **1870** G. M. Hopkins *Note-bks. & Papers* (1937) 132 The heads of flowering grass.. often used as pipe-cleaners. **1928** E. Waugh *Decline & Fall* i. ii. 18 A boxing-glove, a bowler hat, yesterday's *Daily News* and a packet of pipe-cleaners. **1959** I. & P. Opie *Lore & Lang. Schoolch.* ix. 169 Thin people inspire almost as many names and jokes as fat people, but the laughter is less mortifying; the names..are merely descriptive, as: bag o' bones,..pipe cleaner, rake, [etc.]. **1960** *Farmer & Stockbreeder* 16 Feb. (Suppl.) 8/2 You can easily make a fine collection of animals with one or two packets of pipe-cleaners! Look at the sketch, *A*, to see how the woolly covered cleaners are bent to form the outline of a giraffe. **1973** D. Lees *Rape of Quiet Town* iii. 42, I tried to help but it wasn't easy with legs made out of pipe cleaners. **1962** W. H. Murray *Maelstrom* iii. 42 Forward again was the fo'c'sle, with pipe-cots and deck-hatch. **1977** *Western Morning News* 1 Sept. 8/6 (Advt.), Six Ton Sloop..24 ft. oa., pine on oak, two berth and pipe cot. **1907** E. A. Woodruffe-Peacock *Pasture & Meadow Anal.* 4 A soil that has been pipe drained for wheat-growing. **1930** *Jrnl. Ministry Agric.* Nov. 825 There is nothing to indicate..whether the land is pipe-drained or not. **1896** Ade *Artie* iii. 27 But then I was spinnin' pipe dreams myself, tellin' about how much I lose on the board and all that. **1904** B. von Hutten *Pam* 238 Look at the sea, and tell me if, in your wildest pipe-dream, you ever saw anything lovelier. **1915** *Strand Mag.* June 651/2 If it is a fizzle off goes my coat and I abandon pipe-dreams of literary triumph. **1937** *John o' London's* 26 Mar. 1/3 As my ideal library will never be anything but a pipe-dream, no great harm is done. **1959** *Daily Tel.* 4 Apr. 6 In that event, the Channel project would cease to be an engineers' pipe-dream. **1973** C. Egleton *Seven Days to Killing* xiv. 150 Streamlining..was pure Whitehall jargon... It implied increased efficiency at less cost and..that was just a pipe dream. **1976** *Classical Q.* XXVI. 80 After 394 a Hellenic crusade against Persia was merely a pipe-dream. **1976** C. Weston *Rouse Demon* (1977) i. 47 He was always a pipe-dreamer. Always in the clouds. **1979** *Eastern Economist* 14 Sept. 545/2 Only pipe-dreamers would have the phantasies that man uses his tools only for what is called constructively productive purposes. **1976** *Flintshire Leader* 10 Dec. 1 Some of the council's figures relating to the Leisure Centre are an exercise in pipe dreaming. **1978** C. A. Berry *Gentleman of Road* ii. 14, I couldn't chicken out now. It was time for pipe-dreaming to end. **1910** 'O. Henry' *Whirligigs* i. 12 La Paz is a good sort of a pipe-dreamy old hole. **1913** G. J. Kneeland *Commercialized Prostitution in N.Y. City* iv. 90 One of the best known [pimps] is a..dangerous fellow... A 'pipe fiend' and gambler, his favorite occupation is 'stuss'. **1938** *Amer. Speech* XIII. 189 I'm gonna take my gal along, We'll kick around the yong. She'll sing that pipe-fiend song. **1828** *Blackw. Mag.* Sept. 276 The Shooter..begins with his pop or pipe-gun, formed of the last year's growth of the branch of a plane-tree. **1973** *Trinidad Guardian* 1 Feb. 11/4 He was found guilty on a four-count indictment accusing him of being in possession of a revolver, Molotov cocktails, a pipe-gun and several rounds of ammunition. **1973** *Guardian* 27 Mar. 3/1 The Naxalites..had..country-forged pipe guns which heated and split after a few rounds. **1919** J. C. Squire *Birds* 11 Pipe-necked and stationary and silhouetted, Cormorants stood in a wise, black, equal row. **1877** *Coursing Calendar Autumn 1876* 238 Dulcimer and Jewess separated on two hares, and both got a good pipe-opener. *a* **1936** Kipling *Something of Myself* (1937) vii. 187 That tale may have served as a pipe-opener, but one could not see its wood for its trees, so I threw it away. **1962** *Times* 26 Apr. 13/4 This is the blessed time of year when cricket scores begin to creep..into the sporting pages of the newspapers... Such trial matches are in the nature of pipe-openers, elaborations of practice at the nets. **1971** D. Francis *Bonecrack* xv. 189 'I could give Archangel his pipe-opener...' 'All right, then,' I said, and he took Archangel out..and they cantered a brisk four furlongs. **1974** *Times* 18 Sept. 12/6 British pair thrashed in pipe-opener to Wills Open... The match was a curtain-raiser to the Wills Open Tournament. **1977** *Times* 9 June 8/2 There was a pipe-opener to the conference..when Prince Charles..went to Marlborough House to unveil a painting of his mother. **1948** *Petroleum Handbk.* (Shell Internat Petroleum Co) (ed 3) The broken-out stand is then lowered to its position on the pipe rack. **1976** W. D. Baasel *Prelim. Chem. Engin. Plant Design* vi. 148 Nothing should be located under pipe racks, since if leaks occur they may damage equipment. **1978** G. Greene *Human Factor* VI. i. 301 He looked at the row of pipes in the pipe rack with concentration. **1931** G. Egloff in A. Rogers *Industr. Chem.* (ed. 5) II. 861 Modern practice used the pipe still consisting essentially of a coil of pipe placed in a furnace through which oil is passed. **1959** *Times Rev. Industry* Aug. 97/2 The pipestill works on much the same principle as a water tube boiler. **1970** W. G. Roberts *Quest for Oil* viii. 87 The crude oil must be heated before it gets to the column, and the heater used is known as a 'pipestill'. **1904** *N.Y. Times* 16 Oct. iii. 6 The police are now forced to take what appears on its face to be the veriest pipe story and run it down. **1745** Swift *Directions to Servants* ii. 41 Boil your Meat constantly in Pump Water, because you must sometimes want River or Pipe Water. **1908** *Westm. Gaz.* 24 Oct. 17/2, I will not live to see pipe-water squirting down sham rocks under a sham bridge.

pipe, *v.*¹ Add: **2. b.** Freq. with advs. and advb. phrases: esp. *Naut.*, to bring or escort (a person) *aboard*, etc. to the accompaniment of a pipe; also *fig.*; *to pipe in*: to bring in (a person or thing) to the accompaniment of bagpipes.
1918 *Times* 21 Sept. 5/1 It was a Punjabi piper who piped the Cossacks in. **1939** F. Drake-Carnell *It's an Old Scottish Custom* ii. facing p. 54 (*caption*) St. Andrew's Night. Piping in the haggis. **1939** Joyce *Finnegans Wake* (1964) i. 25 Your fame is spreading like Basilico's ointment since the Fintan Lalors piped you overborder. **1940** *Bluejackets' Man.* (U.S. Naval Inst.) lix. 783. In the piping of officials alongside and over, the side pipe is lengthened to full breath for officials receiving 8 side boys. **1955** *Times* 1 July 6/3 The Duke was piped on board, welcomed by the master, Captain H. W. Langbein. **1965** D. MacLean *Queens' Company* xx. 175 On the following morning the doughty Henry Morgan himself, accompanied by his bodyguard, all in magnificent period uniform, were ceremoniously piped on board. **1966** *Listener* 20 Oct. 578/2 Noah pipes aboard an earnest procession of elephants, camels, and assorted fowl. **1968** *Islander* (Victoria, B.C.) 15 Dec. 2/3 The plum pudding, ablaze, was piped in and paraded around the dining room by the chef before being served. **1973** *Stornoway Gaz.* 3 Mar. 1/1 The platform party was piped in by two members of the Lewis Pipe Band. **1976** *Oxf. Compan. Ships & Sea* 131/2 The call..is retained for ceremonial occasions when piping dignitaries and foreign naval officers aboard. **1977** 'J. le Carré' *Hon. Schoolboy* xiii. 292 He looked into the ceiling mirror and caught the glitter of an electric-blue suit and a full head of black hair well greased; and between the two, a foreshortened chubby Chinese face set on a pair of powerful shoulders, and two curled hands held out in a fighter's greeting while Lizzie piped him aboard.
c. *to pipe the side* (Naut.): to sound the 'side', a salute given to certain officers and dignitaries when boarding or leaving a ship.
1896 L. Delbos *Naut. Terms* (ed. 3) 103 *To pipe the side*, faire les honneurs du sifflet. **1909** *Cent. Dict. Suppl.* s.v., When the commanding officer of a naval vessel, or the president or vice-president of the country, or other dignitaries, or superior officers of foreign governments, or crowned heads, or members of royal families, or of the nobility, are received on board, or are leaving a man-of-war,..the boatswain, or one of his mates, winds (blows) his call (silver whistle)... This ceremony is..known as piping the side. **1938** C. S. Forester *Flying Colours* 237 They piped the side for him in the *Victory*, as Admiralty regulations laid down.
3. b. *to pipe down*, in more general use: to stop talking, be quiet, become less vociferous; freq. as a command, = shut up! Also occas. *trans.*, to cause (someone) to be silent. *colloq.*
1900 *Dialect Notes* II. 49 *Pipe down*, to stop talking. **1926** Stallings & Anderson *What Price Glory?* i. i, in *Three Amer. Plays* 24 Pipe down. *Ibid.* iii. 76 He tried to pipe me down. **1932** S. O'Faoláin *Midsummer Night Madness* 227 'Shut up, you,' said the Tan angrily, and the little fellow piped down miserably. **1938** N. Marsh *Artists in Crime* vi. 76 'Hatchett,' said Troy. 'Pipe down.' **1945** E. Waugh *Brideshead Revisited* i. v. 105 Groans of protest rose from the other cells where various tramps and pick-pockets were trying to get some sleep: 'Aw, pipe down!' **1951** M. Kennedy *Lucy Carmichael* vi. ii. 293 He didn't disagree; if he had, I'd have piped down. *Ibid.* 294, I won't pipe down. I'll go to Charles. I'll spill all the beans. **1974** *Times* 19 Jan. 12/1 The more immoderate members of his party..may pipe down.

pipe, *v.*² Add: **4. b.** Also *intr.* (Further examples.) Also, to arrange (icing, cream, mashed potato, etc.) in decorative cord-like lines or twists.
1884 [see Piped *ppl. a.*¹ 2]. **1892** A. B. Marshall *Larger Cookery Bk.* 317 Fill them by means of a forcing bag and pipe with the Cheese custard..or whipped cream. **1929** E. J. Kollist *French Pastry, Confectionery & Sweets* vi. 116 Cover with royal icing... When dry, pipe flowers and leaves on the basket. **1948** *Good Housek. Cookery Bk.* 582 Pipe on chocolate butter icing and decorate with angelica. *Ibid.* 587 Sandwich the cakes together with the filling, spreading some on the top also. If liked, it can be piped on top with a writing pipe. **1965** *Listener* 30 Sept. 511/3 Allow this to sink in a little before piping the whipped cream all over the top. **1976** G. Moffat *Short Time to Live* ii. 17 'I can't come,' she called... 'I'm piping.'.. She had been piping cream round a flan.
6. a. Also *transf.* Cf. *light pipe* s.v. *light *sb.* 16.
1949 H. C. Weston *Sight, Light & Efficiency* v. 183 Some of the doctor's work-objects are parts of the body which cannot be adequately illuminated by general lighting, hence, he sometimes 'pipes' light to these parts by means of transparent internally-reflecting plastic devices which are attached to small hand-lamps. **1952** *Chambers's Jrnl.* Jan. 62/1 The well-known plastic material perspex possesses the unusual property of 'piping' light rather than diffusing it. This property is displayed when the interior of perspex tubing is illuminated. **1968** Bean & Simons *Lighting Fittings* v. 167 Light can be piped along a rod, block, or sheet of transparent medium such as glass or acrylic plastic. **1971** P. Tooley *High Polymers* ii. 60 Another interesting property [of polymethyl methacrylate] is its ability to 'pipe' light from one place to another as a result of a high degree of internal reflection. This is used surgically to illuminate internal hollow organs such as the stomach.

b. To transmit (music or speech) over a wire or cable.
1937 [implied in *piped *ppl. a.*¹ 3 b]. **1939** *Wireless World* 16 Mar. 259/3 Broadcast programmes or announcements by the pilot can be 'piped' to the passengers. **1956** *Time* 9 Jan. 20/2 It was his wintertime pre-breakfast habit to cut figure eights on the ice of Webster Lake..to the music of Mozart and Chopin, piped through an amplifying system he had rigged up. **1959** *Observer* 6 Dec. 4/5 Programmes of music, talks and plays were 'piped' to individual seats, each passenger having lightweight earphones with volume control. **1967** *N.Y. Herald Tribune Internat.* 11–12 Feb. 3/3 The astronaut told the workers in a talk piped over a public address system to plants here and in Oak Creek. **1977** *Sunday Times* 6 Mar. 8/6 Powell's daily conference is piped into a dozen White House offices.
8. *intr.* To smoke a pipe. *N. Amer. colloq.* See also Piping *vbl. sb.*² 1.
1846 T. L. McKenney *Mem.* I. iii. 71 These hardy adventurous fellows never rose from their paddles, nor stopped except to '*pipe*'. **1863** W. B. Cheadle *Jrnl. Trip across Canada* (1931) 270 Dr. Benson..assured us we were going wrong. We therefore lunched & piped.
9. *trans.* and *intr.* To see, notice, look (at), watch; to follow or observe (someone), esp. stealthily. Also with *off.* *slang* and *dial.*
Perhaps a different word.
1846 *Swell's Night Guide* 43 You may pipe the crib by seeing a board whereon is inscribed the name of the piano faker. **1848** *Ladies' Repository* VIII. 316/2 *Pipe*, to watch; reconnoitre. **1864** Hotten *Slang Dict.* 202 *Pipe*, to follow or dog a person. Term used by detectives. **1869** *Galaxy* VIII. 349 His 'pal'..has meantime been engaged in an operation which he styles 'piping off the cop', by which he means that he has been watching the movements of the policeman. **1877** *Sessions Papers* 25 Oct. 631 Druscovich..said 'I know I am being *piped off*'—that in our language means being followed or watched—it would imply that another detective was following him. **1888** S. O. Addy *Gloss. Words Sheffield* 176 *Pipe*, to take notice of. 'Pipe his kuss', *i.e.*, take notice of his mouth. A detective is said to pipe round a public-house when in search of a culprit. **1898** A. M. Binstead *Pink 'Un & Pelican* iv. 87 His mission up there on the roof was to exclude—*Anglice*, sling off—any who sought to 'pipe off' the contest through the skylight aforesaid. **1898** F. P. Dunne *Mr. Dooley in Peace & War* 3 Sagasta pipes him out iv th' corner iv his eye. **1906** 'H. McHugh' *Skiddoo!* 67 I'm going to pass you out a talk he handed me a few evenings ago on that subject. Pipe! **1915** Wodehouse *Psmith, Journalist* ii. 10 Pipe de leather collar she's wearing. **1924** E. O'Neill *Welded* ii. ii. 141 Remember kissing me on the corner with the whole mob pipin' us off? **1926** *Flynn's* 16 Jan. 640/2 We found another rattler and a few days later I piped a beaut of a jug and a jay burg. **1943** F. Sargeson in *Penguin New Writing* XVII. 78 We'd stand in shop doorways and Terry'd pipe off everyone that went past. **1950** R. Chandler *Let.* 18 May in *R. Chandler Speaking* (1966) 78 'Piped' does not mean 'found' but saw or spotted (with the eyes). **1974** H. J. Parker *View from Boys* iii. 77 During the daytime wandering around the area, 'pipe-ing', looking over a car, became a regular practice.

pipe-clay, *sb.* Add: (Earlier and later examples.) See also *tobacco-pipe clay* s.v. Tobacco-pipe 3.
1777 J. Wedgwood *Let.* 19 July (1965) 207 He brought some of the stones home with him, mixed them with Pipe Clay, and made the first White Flint Stone Ware. **1966** 'J. Hackston' *Father clears Out* 80 He could have also 'lost all his grass', not that our pipeclay had any grass on it.

piped, *ppl. a.*¹ Add: **3. b.** Received over a wire or cable (rather than directly from broadcast signals); esp. in phr. *piped music*, background music, usu. pre-recorded, played through loudspeakers in a public place.
1937 *Printers' Ink Monthly* May 40/2 *Piped program*, a program transmitted over wires. **1949** *Wireless World* Feb. 67/2 Piped Television... Each block of flats will be equipped with a receiving aerial, from which the signal will be fed via an amplifier and distribution units to each flat. **1959** *Daily Mail* 11 Aug. 1/6 A 'piped' television service is to be started soon by the Ekco and Ferguson companies. The system eliminates interference and overcomes reception snags in 'fringe' areas. **1969** *Morning Star* 25 Mar. 2 To me, piped music is a sign of bad taste. **1973** *Times* 27 Mar. (Sudan Suppl.) p. viii/2 A new Hilton Hotel, complete no doubt with swimming pool, air conditioning, piped music and American food. **1977** *Irish Times* 8 June 13/4 (Advt.), Secluded furnished cottage; one bedroom, phone; piped t.v.; own entrance; £16 per week.
c. Also *piped-in* adj.; also in sense of *pipe *v.*² 6 b.
1959 *Economist* 14 Mar. 976/2 There are beach chairs and umbrellas by the swimming pool,..a picnic area, barbecue pit, piped-in music. **1961** *Lancet* 9 Sept. 591/1 Piped-in oxygen prevents noise from clanking cylinders. **1976** M. Ierley *Year that tried Men's Souls* i. 72 Usefulness is limited to locations with access..to piped-in water. **1977** *Rolling Stone* 19 May 15/5 It dawns on him what's been playing on the restaurant's piped-in music system.
4. Drunk, intoxicated; under the influence of drugs. *U.S. slang.*
1912 *Pedagogical Seminary* XIX. 97 Figurative expressions referring to..Intoxication:—'full', 'piped', [etc.]. **1913** *Dialect Notes* IV. 11 The engineers were all piped. **1924** G. C. Henderson *Keys to Crookdom* 413 *Piped*, under influence of liquor or narcotics. **1953** Berrey & Van den Bark *Amer. Thes. Slang* (1954) § 106/7 Drunk,.. piped.

pipe-down. *Naut.* [f. Pipe *v.*¹] The act of piping down' (see Pipe *v.*¹ 3); a call on the boatswain's pipe signalling sailors to retire for the night.
1913 *Dialect Notes* IV. 164 *Pipe-down*, tattoo. Pipe-down call is sounded at 9 p.m. on board ship. **1942** *Penguin New Writing* XV. 8, I didn't see him again until pipe-down. **1963** *Times* 26 Feb. 12/7 Twenty-odd 'pipes' for special occasions were there to be learnt, from the 'still'..to the complicated 'Pipe down' at the end of the day.

pipe-lay (pəi·p,lēi), *sb.* Also **pipelay, pipe lay.** [f. Pipe *sb.*¹ + Lay *v.*¹] = Pipe-laying (*a*). Usu. *attrib.*
1974 *Petroleum Rev.* XXVIII. 556/2 Standard pipelay equipment was used. **1975** *Times* 19 Feb. 14/6 Essential equipment (eg, platforms, derrick barges and pipe-lay barges). **1975** *Offshore* Aug. 124/3 Two stern thrusters on the reel barge and a 3,600-hp tug controlled the barge during the pipe lay. **1976** *Offshore Engineer* July 6/1 *Castoro VI*, a new Saipem barge at present undergoing trials, will be used for the trial pipelay of 305mm and 406mm pipeline.

pipe-lay, *v.* *U.S. slang.* [Cf. Pipe-laying.] *intr.* To 'lay pipes' (see Pipe *sb.*¹ 3 b); to take measures preparatory to securing some desired action or event. Also *trans.*, to manipulate or cheat.
1848 *Campaign Flag* (Maysville, Kentucky) 14 Apr. 3/6 He was *pipe-layed* out of his election last year. **1884** J. MacCarthy *Hist. Four Georges* I. 107 Bolingbroke and Oxford..had been 'pipe-laying', to use an expressive American word, for the Stuart restoration during all the closing years of Queen Anne's reign. **1888** in Farmer & Henley *Slang* (1902) V. 213/1 There are not a few who are pipe-laying and marshalling forces for the fray.

pipe-layer. Add: **a.** (Further example.) Also, a machine used to lay pipes. **b.** (Earlier examples.)
1840 *Richmond* (Virginia) *Enquirer* Nov. (Th.), Corruption of the franchise by pipe layers and yarn spinners. **1841** *Congress. Globe* 26th Congress 2 Sess. App. 155/1, I was not defeated by voters. I was defeated by 'pipe layers'. **1969** *Engineering* 25 Apr. 648/2 A Badger Minor trenchless pipelayer..has been specially adapted [for the simultaneous laying of four pvc cable ducts]. **1976** *Daily Tel.* 15 July 5/1 The deaths of a Texan pipe-layer and a French diver brought criticism from an MP yesterday of the Government's attitude towards safety in North Sea gas and oil operations. **1977** *Pipes & Pipelines Internat.* XXII. 35/2 A new pipelayer with lifting capacity of 41 000 kg.

pipe-laying, (*a*) (further examples); also *attrib.*; (*b*) (earlier examples).
1841 *Congress. Globe* 26th Congress 2 Sess. App. 120/2 Others say that fraud, double voting, pipe laying, [etc.].. have done much to carry the election. **1842** [see *colonization 2]. **1871** *Engineering* XI. p. vii/3 (*heading*) Pipe laying under water. **1973** C. Callow *Power from Sea* viii. 163 Pipe-laying barges are actively engaged in off-shore work. **1975** *BP Shield Internat.* May 14/1 The second highly weather-vulnerable job is pipe-laying. **1975** *Times* 16 Sept. (North Sea Suppl.) p. i/1 Pipe-laying barges need to monitor their operations under the sea and to locate pipes at a later date.

pipe-line, *sb.* Add: (Earlier and later examples.)
1873 J. T. Henry *Early & Later Hist. Petroleum* 283 The iron pipe lines for the conveyance of oil from the wells to railway shipping points play an important part in the transportation of the article. **1879** in I. M. Tarbell *Hist. Standard Oil Co.* (1904) I. 354 The pipe lines owned and controlled by the parties hereto have a joint capacity for transportation. **1924** C. Christy *Big Game & Pygmies* i. 2 Between Matadi and Leopoldville is a wonderful pipe line, by which crude mineral oil for the fleet of up-river steamers is pumped the whole two hundred odd miles by several pumping stations. **1943** *Ann. Reg. 1942* 263 A second first-class road..runs alongside the pipe-line from the Palestinian to the Iraqui frontier. **1953** *Times* 31 Oct. 4/6 A water supply had been brought by bamboo pipelines from a spring. **1961** *Wall St. Jrnl.* 8 Nov. 24/2 The filing ..proposes construction of 210 miles of large-diameter pipeline. **1972** *Drive* Spring 60/1 Brakes, brake pipelines and hydraulic fluid are of life-and-limb importance. **1973** C. Callow *Power from Sea* i. 16 Underwater pipelines carrying natural gas at high pressure. **1976** *Times* 2 Oct. 17/8 Some of the pipeline workers may stay in Alaska.
b. *transf.* and *fig.* in various senses (see quots.); *spec.* a channel of supply, information, communication, etc.; esp. in phr. *in the pipeline*, in progress; being worked on, dealt with, or produced; on the way from a supplier to a user. Also *attrib.*
1921 A. Huxley *Crome Yellow* vi. 58 You could write too..by getting into touch with your Subconscious. Have you ever read my little book, *Pipe-Lines to the Infinite?* **1935** *Sun* (Baltimore) 25 Aug. 6/2 It was implied, if not more, that the man who had the real inside track to the White House and the potential pipe line to the United States Treasury was none other than Dr. Byrd. **1942** *Ibid.* 1 Aug. 1/7 A foreign source here with continental pipelines of information said the Germans also were making peace feelers both to Britain and the United States and to Russia with the object of splitting the Allies. **1942** Berrey & Van den Bark *Amer. Thes. Slang* § 121/78 Veins and arteries,..*pipe lines.* **1945** *Amer. Speech* XX. 227/1 *Pipeline time*, from the time a requisition leaves a depot until the requested supplies arrive there. **1948** *Economist* 19 June 1026/2 The pipe-lines are full, stocks seem to be adequate, and there are signs of resistance to higher prices. **1948** *Sheep Breeder* Dec. 19/3 Sometimes the price of meat in the pipelines of distribution goes up—a fact that gives rise to claims that we are speculating. **1949** Partridge *Dict. Slang* Add. 1136/1 *Pipe line*, an aerial: R.A.F.: 1939 +. **1955** W. Gaddis *Recognitions* ii. v. 540 Them priests have a pipeline right into the cops. **1955** *Times* 23 Aug. 5/6 About a third of the patient applicants can consider themselves as 'in the pipeline', which means that the telephone engineers have the equipment ready for them. **1957** *Economist* 16 Nov. 565/2 With the further fall in primary commodities there must be a further improvement in Britain's terms of trade already, so to speak, in the pipeline. **1964** *Observer* 26 July 7/5 All these reforms will take time—and cause controversy—in the next Parliament. There are measures in the pipeline already. **1972** J. Mosedale *Football* vii. 94 There was a pipeline to the Tuscaloosa campus in those depression days, and Hutson was one of eight Arkansans who found their way to Alabama's football team. **1973** *Daily Pennsylvanian* 9 Oct. 1/2, I don't have a pipeline to God. **1973** *Times* 15 Nov. 25/2 The property development company has 'some very exciting ideas in the pipeline'. **1976** *Broadcast* 16 Feb. 4/1 There's a new soap in the pipeline... The series/ serial..will be shown twice weekly.
c. *Surfing slang.* A very large wave, or the hollow part of such a wave. Also applied to a place where such waves are formed.
1963 *Sunday Mail Mag.* (Brisbane) 5 May 12/5 *Pipeline*, a very large tube (*tube* = hollow part of a wave). **1965** *N.Z. Listener* 17 Dec. 5/1 The achievement by which the champion surfers are judged is their ability to ride the Hawaiian pipeline... The pipeline breaks less than 50 yards from the beach over a coral reef. **1971** *Times* 9 Aug. 5/1 The surf-bums have a language all their own; they talk about pipelines, green rooms, roller-coasters. *Ibid.*, The Banzai pipe-line in Hawaii, where the waves can be 25 to 30 feet high.
d. *Computers.* A linear sequence of specialized modules used for pipelining. Freq. *attrib.*
[**1964** W. Bucholz *Planning Computer System: Project Stretch* xiv. 204 The data flow through the computer..is comparable to a pipeline which, once filled, has a large output no matter what its length.] **1965** *AFIPS Conf. Proc.* XXVII. i. 489 (*heading*) Circuit implementation of high speed pipeline systems. *Ibid.* 491/1 The pipeline is characterized by a succession of register and gate units. **1972** *IEEE Trans. Computers* XXI. 885/1 A two-stage pipeline is possible. **1977** *Computing Surveys* IX. 101/2 The CPU architecture of this machine is a simple pipeline served by four functional units. **1977** *Sci. Amer.* Sept. 220/2 Three kinds of systems that can truly be classed as parallel processors have been built. In one of them, the 'pipeline' processor, several processing elements, each of which is specialized for some particular task, are connected in sequence.

pipe-line, *v.* Add: **2.** *Computers.* To design or execute using the technique of pipelining.
1971 *Sci. Amer.* Feb. 76/2 Current efforts in 'pipelining' the processing of 'operands' will allow a further significant increase in speed. **1972** *IEEE Trans. Computers* XXI. 881/1 This note will study the problem of pipelining the addition and multiplication functions of the arithmetic unit.

pi·pelined, *a.* *Computers.* [f. Pipe-line *sb.* (or *v.*) + -ed² (or ¹).] That makes use of the technique or principle of pipelining.
1972 *IEEE Trans. Computers* XXI. 880/2 In a pipelined unit, a new task is started before the previous task is complete. **1977** *Computer Surveys* IX. 61 Pipelined computer architecture has received considerable attention since the 1960s when the need for faster and more cost-effective systems became critical.

pi·peliner. orig. *U.S.* [f. Pipeline *sb.* + -er¹.] One who works on oil or gas pipelines.
1928 A. Garland in J. F. Dobie *Foller de Drinkin' Gou'd* 58 'It's cold out there too—' 'I'll say it is,' a pipeliner broke in. **1962–3** *Petroleum Today* Winter 3/2 The pipeliner is a nomad; he goes where the job is. **1973** C. Callow *Power from Sea* v. 113 The pipeliners have an even tougher time in many ways.

pi·pelining, *vbl. sb.* [f. Pipe-line *sb.* + -ing¹.] **1. a.** The laying of pipelines. **b.** Transportation by means of pipelines.
1959 *Pipeline Engin.* Oct. p. i, The initiation of this new section is..a further milestone in the ever-expanding industry of pipelining. **1963** *Internat. Pipes & Pipelines* Dec. 40/2 The pipelining of forest products such as wood chips would appear to be economically feasible. **1969** *McGraw-Hill Yearbk. Sci. & Technol.* 271/2 The scope of pipelining is being expanded to include the transportation of solids in molten form or as slurries or capsules. **1970** *Preprints 2nd Ann. Offshore Technol. Conf.* I. 379 (*heading*) Pipelining in 600 feet of water. **1975** *North Sea Background Notes* (Brit. Petroleum Co.) 31 Pipelining begins with the stringing out of pipe lengths, each approximately four tons in weight, which are unloaded on to temporary sleepers and then welded into a continuous line.
2. *Computers.* A form of computer organization in which successive steps of a process are executed in turn by a linear sequence of specialized modules capable of operating concurrently, so that another process can be begun before the previous one is finished.
1965 *AFIPS Conf. Proc.* XXVII. i. 490/1 With the present state-of-the-art in systems organization and technology it appears that pipelining is a powerful approach to a particular variety of large data processing

problems. 1972 *IEEE Trans. Computers* XXI. 886/1 Pipelining can be used to give a 40 percent increase in adder efficiency and a 230 percent increase in multiplier efficiency. 1975 G. Zimmermann in Hartenstein & Zaks *Workshop on Microarchitecture of Computer Systems* 155/1 The fast multiplication scheme has been extended by a fast pipelining division network for floating point numbers.

pipeman. 1. Delete *nonce-use* and add further examples.
1922 *Daily Mail* 7 Nov. 4 (Advt.), The pipeman's joy. 1974 J. Johnston *How Many Miles to Babylon?* 127 My father's a pipe man. The perfect pipe man. He uses it to protect him from the world..stares into the pipe. 1976 *Liverpool Echo* 22 Nov. 1/3 In 1964, he was voted Pipeman of the Year and his waxwork figure appeared at Madame Tussaud's.

pip emma: see *pip *sb.*[4]

piper[1]. Add: **1. b.** *drunk as a piper* (earlier and later examples); *by the piper(s) (that played before Moses)*: an Irish oath or expletive.
1727 J. Gay in *Miscellanies* III. 207 Drunk as a Piper all day long. 1865 'Mark Twain' in *Californian* (San Francisco) 23 Dec. 4/3 He came home drunk as a piper. 1884 'Cruck-a-Leaghan' & 'Slieve Gallion' *Lays & Legends N. Ireland* 16, I hope they don't hear me, Or else, by the piper, they'll make me sing sad. 1892 J. Barlow *Irish Idylls* ix. 274 Be the piper, sure enough I was up there splicin' the handle of your mother's ould basket. 1894 A. Gordon *Northward Ho!* 202 If he..was as drunk as a piper, an' ye yersel' had only twa gills,..he'd pruve tae ye..that ye were drunk, an' no him. 1899 *Century Mag.* Nov. 45 Be the piper that played afore Moses I'll call out me regiment of throopers. 1928 'Brent of Bin Bin' *Up Country* i. 2 Be the poipers, we've had enough [rain] for this toime of year!

‖**piperade** (piperad). Also (*erron.*) **pipérade.** [Fr.] A dish originating in the Basque country, consisting of eggs, tomatoes, and peppers, and resembling an omelette.
1931 X. M. Boulestin *What shall we have To-day?* 191 A delicious dish, very popular in Béarn and in the Basque country, but not well known otherwise, is the *Pipérade*. 1951 E. David *French Country Cooking* 81 *Pipérade* is the best known of all Basque dishes... It is a mixture of pimentos, tomatoes and onions, with eggs added at the end. 1961 E. H. Clements *Note of Enchantment* ix. 126 Alister..ordered a *piperade* and ate it unhurriedly with a hunk of bread. 1966 *Punch* 29 June 944/1 Should we try Chez Fifine for a piperade? 1976 *Times* 7 Aug. 11/4 A fresh and well-composed pipérade hot from the pan.

piperazine. Add: Now pronounced (pipe·rǎzīn). Substitute for etym. and def.: [first formed as G. *piperazin* (A. T. Mason 1887, in *Ber. d. Deut. Chem. Ges.* XX. 267), f. *piper(idin* Piperidine + *azin* *azine.] A colourless, crystalline, heterocyclic base, $C_4H_{10}N_2$ which is used, in the form of a salt or hydrate, as an anthelminthic; also, a derivative of this. (Earlier and later examples.)
[1888 *Jrnl. Chem. Soc.* LIV. 726 Diphenyldiketopiperazine.] 1889 *Ibid.* LVI. 1009 (*heading*) Piperazines. 1891 *Ibid.* LX. 415 Piperazine melts at 104–107°. 1937 *Thorpe's Dict. Appl. Chem.* (ed. 4) I. 315/1 Cyclic compounds are..produced by distilling the dihydrochlorides of the aliphatic diamines... Thus ethylenediamine yields piperazine. 1970 W. H. Parker *Health & Dis. in Farm Animals* xx. 276 Piperazine compounds might be considered the safest.., but thiabendazole has the merit of being effective against both large and small round worms in the pig. 1974 M. C. Gerald *Pharmacol.* ii. 35 Streptomycin, piperazine (Antepar), [etc.]..have been quite effective in the cure of tuberculosis, pinworm,..respectively.

piperidine. Add: Now pronounced (pipe·ridīn). [First formed as F. *piperidine* (A. Cahours 1853, in *Ann. de Chim. et de Phys.* XXXVIII. 78).] (Earlier and later examples.)
1854 *Jrnl. Chem. Soc.* VI. 175 (*heading*) On piperidine, a new alkali derived from piperine. 1938 [see *Knoevenagel]. 1964 N. G. Clark *Mod. Org. Chem.* xxiii. 470 On heating a pyridine solution of the aldehyde with malonic acid in presence of piperidine (catalyst), condensation *and* decarboxylation occur, and a cinnamic acid is produced (Döbner Reaction).

piperonal (pipe·rǫnăl). *Chem.* [a. G. *piperonal* (Fittig & Mielck 1869, in *Ann. d. Chem. u. Pharm.* CLII. 37), f. *piper(in* Piperine *sb.* + *-on* -one + *-al* *-al[2].] = *heliotropin.
1869 *Chem. News* 30 July 59/2 Piperinic acid,.. when acted upon by permanganate of potassa in aqueous solution, yields a crystalline substance (formula, $C_8H_6O_3$), which body the authors called piperonal. 1909 C. A. Keane *Mod. Org. Chem.* vii. 123 Many substitutes for natural perfumes are also known, such as..heliotropin or piperonal, which possesses the odour of heliotrope. 1953 Kirk & Othmer *Encycl. Chem. Technol.* X. 324 Piperonal (heliotropin)..is a colorless, lustrous crystalline solid with a smell resembling that of heliotrope... Its chief application is in perfumery. 1967 L. F. & M. A. Fieser *Reagents for Org. Synthesis* I. 944 A mixture of piperonal with 1·5 l. of water is heated on the steam bath with vigorous stirring.

piperonyl (pipe·rǫnəil). *Chem.* [f. prec. + -yl.] Used to form the names of certain substituted derivatives of piperonal, as *piperonyl butoxide*, a yellow oily liquid, $C_{19}H_{30}O_5$, which is used in insecticides as a synergist for pyrethrins.
1871 *Jrnl. Chem. Soc.* XXIV. 934 The mother-liquor.., evaporated until most of the alcohol is driven off, and exhausted with ether, yields piperonyl alcohol. 1923 W. M. Cumming et al. *Systematic Org. Chem.* v. 93 Piperonyl acrolein is obtained from piperonal and acetaldehyde. 1945 H. Wachs in *Science* 16 May 531/1 The activity of piperonyl butoxide (the name given the technical product containing 80 per cent of pure compound) is indicated in Table 1. 1966 *McGraw-Hill Encycl. Sci. & Technol.* VII. 142/1 Piperonyl butoxide..and Sesoxane..are two important commercially available pyrethrin synergists. 1966 *Jrnl. Agric. & Food Chem.* XIV. 555 The..piperonyl carbamates are exceptionally active carbamate synergists. 1974 *Approved Names 1973* (Brit. Pharmacopœia Comm.) Suppl. 11, Piperonyl Butoxide..5-[2-(2-Butoxyethoxy)ethoxymethyl]-6-propyl-1,3-benzodioxole. Acaricide.

pipe-stem. [Pipe *sb.*[1] 11 a.] **1.** The stem of a tobacco-pipe.
a 1734 J. Comer *Diary* in *Rhode Island Hist. Soc. Coll.* (1893) VIII. 17 He..fell over a log, yᵉ pipe stem ran down his throat and broke. 1755 *Maryland Hist. Mag.* (1923) XVIII. 33 He fell down forward, and run the Pipe stem into the Roof of his mouth. 1846 J. W. Webb *Altowan* I. vi. 168 One of the half-breeds has a piece of an old pipe-stem, which makes tolerable good smoking. 1855 [see Pipe *sb.*[1] 11 a]. 1860 J. G. Holland *Miss Gilbert's Career* vii. 115 A great tribulation that will break my life off as short as a pipe-stem. 1873 J. Miller *Life amongst Modocs* xv. 194 He pointed his pipe-stem at Paquita. 1942 J. Masefield *Generation Risen* 42 The hands, chocked-off, wet through and fireless, Chew pipestems. 1971 'D. Halliday' *Dolly & Doctor Bird* xvi. 242 The mild figure puffed at its pipestem.
b. *humorously.* A thin leg. Also *pipe-stem leg*.
1883 E. Eggleston *Hoosier School-Boy* 33 Little Columbus Risdale picked himself up on his pipe stems and took his place at the end of this row. 1938 M. K. Rawlings *Yearling* i. 4 The water..made a rippling sound, flowing past his pipe-stem legs, and was entirely delicious. 1955 J. Thomas *No Banners* xxvii. 268 The obsession with food made Alfred sway on his pipestem legs. 1976 *National Observer* (U.S.) 10 Apr. 18/2 Mostly they are elderly.., their pipestem legs squeaking them along in the late-afternoon sun.
2. *Comb.* **pipe-stem clematis** *U.S.*, a white-flowered clematis, *C. lasiantha*, of the family Ranunculaceæ, native to California; **pipe-stem wood** *U.S.*, a large evergreen shrub, *Leucothoe populifolia*, of the family Ericaceæ, native to Florida and South Carolina.
1951 H. E. McMinn *Illustr. Man. Calif. Shrubs* 117 Pipe-stem Clematis grows in the Coast Range valleys. 1791 W. Bartram *Trav. N. & S. Carolina* 24, I observed ..the great evergreen *Andromeda* of Florida, called Pipe-Stem Wood. 1813 H. Muhlenberg *Catal. Plantarum Americæ Septentrionalis* 43 Pipe-stem wood, *Andromeda acuminata*.

pipette, *sb.* Add: **1.** Also (*U.S.*) **pipet.**
1937 Pierce & Haenisch *Quantitative Analysis* v. 49 In all pipets the flow of liquid is controlled by admission of air beneath a finger pressed tightly onto the upper end of the stem. 1961 G. R. Choppin *Exper. Nucl. Chem.* ix. 133 When the preparation is complete, carefully remove as much of the supernate as possible with a transfer pipet and place on a planchet.

pipette (pipet) *v.*, for date of first quot. read 1887 and add further examples; hence **pipe·tting** *vbl. sb.*
1915 *Chem. Abstr.* IX. 1861 (*heading*) Pipetting with the suction pump. 1943 *Jrnl. Bacteriol.* XLVI. 195 The tedium of making a large number of accurately measured pipettings. 1961 G. R. Choppin *Exper. Nucl. Chem.* ix. 132 Pipet approximately 10,000 cpm of Cs^{137} activity into a small test tube. 1975 *Nature* 13 Mar. 151/2 The blood was cooled..and the plasma was pipetted off for enzyme analysis.

pipe-work. 2. (Further examples.)
1934 *Discovery* Dec. 348/1 Modern methods of steel-frame building and pipe-work. 1949 *Archit. Rev.* CV. 222/1 In all stages of the design effort has been made to eliminate the tangled masses of exterior pipework usual in most nylon factories. 1974 *Physics Bull.* July 291/2 Tests on oil fields pipework for potential fire hazards. 1976 *Gramophone* Sept. 453/3 The remoteness of the console from the pipework brought to the ear less than 100 per cent of the complex score. 1977 *Times* 29 June 5/2 In bad conditions, all ferrous pipework can deteriorate.

pipi[2]. Also **pippi(e).** Substitute for def.: An edible bivalve mollusc, *Amphidesma australe*, found on sandy beaches in New Zealand; also, occasionally used for another edible mollusc, *Chione stutchburyi*. Also *attrib.* (Further examples.)
1843 J. E. Gray in E. Dieffenbach *Trav. N.Z.* II. x. [252 *Mesodesma Chemnitzii*... Called *Pipæ* by the natives, who eat them as food.] *Ibid.* 262 *Venus intermedia*... East Coast; much eaten by the natives; called *Pipi*. 1861 A. S. Atkinson *Jrnl.* 21 Jan. in *Richmond–Atkinson Papers* (1960) I. 680 Had dinner—well supplied with pipis. 1863 F. E. Maning *Old N.Z.* (ed. 2) iii. 65, I will scrape sharp the point of my spear with a *pipi* shell. 1873 [see *kinaki]. 1905 [see *kuku 2]. 1938 R. Finlayson *Brown Man's Burden* 75 You could scoop up handfuls of big fat pipis.., sifting the sandy mud through your fingers. *Ibid.*, He joined in a game of throwing big empty pipi shells into the air. 1948 D. Ballantyne *Cunninghams* i. viii. 45 She looked..at the Maoris gathering pippis on the mud flat. 1959 Tindale & Lindsay *Rangatira* vii. 68 They would find pipi cockles in the sand of the beach. 1968 Morton & Miller *N.Z. Sea Shore* xviii. 443 They could well be called pipi beaches, for their most typical bivalves are the pipi, *Amphidesma australe*, higher up the shore, giving place lower down to the cockle or tuangi, *Chione stutchburyi*, in places designated 'pipi' as well. 1972 M. Gee *In my Father's Den* 104 On Takapuna beach..Jonathan was throwing pipi shells into the wind.
2. A similar Australian mollusc, *Plebidonax deltoides*.
1934 *Bulletin* (Sydney) 14 Mar. 20/4 Whether it [*sc.* a bird] opens oysters or not is questionable; but it is an expert on pipis. 1952 W. J. Dakin et al. *Austral. Seashores* xvi. 294 The most common mollusc of any size inhabiting the sand of the ocean beaches of New South Wales..is *Plebidonax deltoides*.., known in this State as the pipi... The pipi lives only a few inches below the surface. 1962 *Australasian Post* 8 Nov. 24/1 The 'pippie' is that pink-and-cream shellfish that lives in the sand. 1968 *Courier-Mail* (Brisbane) 14 June 17/8 A young woman.. survived for eight days by digging pippie shells from the sand with her hands. 1970 *People* (Austral.) 26 Aug. 45/3 The pipi is similar to the 'Littleneck' clam of America's west coast.

piping, *vbl. sb.*[2] Add: **4.** (Further attrib. examples.)
1859 Mrs. Stowe *Minister's Wooing* xii. 126 Miss Prissy ..fell..into a discourse on her own particular way of covering piping-cord. 1966 *Olney Amsden & Sons Ltd. Price List* 13 Cushion Piping cord, 3-yard cards 16/9 dozen cards. 1967 E. Short *Embroidery & Fabric Collage* i. 20 (*caption*) Allover pattern in piping cord on white linen. 1968 J. Ironside *Fashion Alphabet* 96 *Pipe*, to trim with a narrow tube of fabric, often with a piping cord run through to pad it out.
8. (Earlier example.) Also *attrib.*
1846 C. E. Francatelli *Mod. Cook* 398 The cake may be decorated with piping, using for that purpose some of the icing worked somewhat thicker. 1943 Bennion & Stewart *Cake Manuf.* (ed. 2) xiii. 135 Fruit and piping jellies of various colours and flavours are a very useful commodity for use as fillings or for the decoration of cakes and gateaux. 1948 *Good Housek. Cookery Bk.* iii. 612 The most satisfactory kind to purchase are made of hand-cut brass without screws, designed to use in conjunction with a piping bag. 1976 E. Turner *All-Colour Cookbk.* xiii. 139/2 Put the mixture into a piping bag with a 6 mm./¼ in. rose nozzle and pipe out 14 to 16 neat roses on an ungreased baking sheet.
10. (Earlier and further examples.)
1861 *Brit. Patent* 1310 4 My invention consists in preventing..the waste occasioned by what is technically called the 'piping' of ingots of steel. 1923 Glazebrook *Dict. Appl. Physics* V. 357/2 Piping does not necessarily take the form of a single central cavity. 1924 Greaves & Wrighton *Pract. Microsc. Metallogr.* ix. 79 If insufficient discard is made, piping..may be present after rolling in the form of a longitudinal fissure..in the central portion of the billet.
11. (See quot. 1937.)
1937 E. J. Labarre *Dict. Paper* 187/1 *Piping*, a species of crease or ribbing in paper due to irregular tension in reeling, to moisture, or being wound too tightly after sizing. 1963 R. R. A. Higham *Handbk. Papermaking* 283 *Piping*, creases or ribbing in paper produced by irregular tension on the sheet during reeling.
12. *U.S.* The action of beating a person with a length of pipe; an assault of this nature.
1971 *Black Scholar* Apr.–May 24/2 The racial agitation is soon followed by hundreds of stabbings, pipings, brutal beatings and death. 1977 *New Yorker* 24 Oct. 64/3 Homosexuality..is one of the major causes of trouble in prison, often resulting in stabbings or pipings.

piping, *ppl. a.* Add: **2. b. piping guan** (example); **piping plover,** substitute for def.: a small buff-coloured bird, *Charadrius melodus*, found in coastal areas of eastern North America; (examples).
1968 F. Haverschmidt *Birds of Surinam* 79 White-headed piping guan... Not uncommon in forests. 1828 C. L. Bonaparte *Genera N. Amer. Birds* 296 Ringed Plover..and Piping Plover... Common all along the eastern sea coast of North America. 1870 *Amer. Naturalist* III. 231 The Piping Plover is still found along the coast of Maine. 1917 T. G. Pearson *Birds Amer.* I. 264/1 Truly a bird of the beach-sand is the Piping Plover. 1964 J. Bull *Birds N.Y. Area* 185 The Piping Plover breeds on the ocean beaches.

pipkin. Add: **3.** *Comb.*, as *pipkin-shaped* adj.
1908 E. Terry *Story of my Life* 199 A three-handled cup, pipkin-shaped, standing on three legs.

pipkrake (pi·pkrēik, -krākə). *Geomorphol.* Pl. **-krakes, -kraker.** [a. Sw. dial *pipkrake*, f. *pip* Pipe *sb.*[1] + *krake*, dial. by-form of *klake* (hardness and roughness of) frozen ground (= Norw. *klake*, Da. *klage*, Icel. *klaki*).] **a.** One of the ice-needles in needle ice. **b.** = *needle ice* s.v. *needle *sb.* 14.
1956 *Biuletyn Peryglacjalny* IV. 167 Bunt also noticed the formation of ice palisades (pipkrakes) in the hollows and ascribes the removal of fine material from the hollows to the melting of these ice crystals. 1967 M. J. Coe *Ecol. Alpine Zone Mt. Kenya* 74 On the flat or slightly sloping ground of valley bottoms, an important feature of soil movement is that of Needle Ice, or Pip[k]rake. 1968 R. W. Fairbridge *Encycl. Geomorphol.* 370/1 Water expands about 10% upon freezing (ice being characterized by a

high expansion coefficient, as seen in the growth of ice needles or pipkrakes). *Ibid.* 381/1 Such 'needle ice' is sometimes called pipkrake. 1970 R. J. Small *Study of Landforms* x. 330 The formation at or near the ground surface of small masses of ice, such as pipkraker, commonly has the effect of heaving up small stones at right angles to the slope. 1971 A. F. Pitty *Introd. Geomorphol.* iv. 219 In mountainous areas in central Germany, pipkrakes may be 10–15 cm long. 1973 A. L. Washburn *Periglacial Processes & Environments* iv. 81 Needle ice.., also known as pipkrake, is an accumulation of slender, bristle-like ice crystals practically at, or immediately beneath, the surface of the ground.

pipped, *a.* Add: **b.** Annoyed, irritated.
1914 A. N. Lyons *Simple Simon* i. vi. 100 'How's Leverton?' 'Rather pipped, thank you,' replied Miss Disney. 'Poor old Ma was raw-beefing him when I left.' 1941 Baker *Dict. Austral. Slang* 54 *Pipped, pippy*, irritated, angry, out of sorts.

pipped (pipt), *a.*[2] [Perh. f. Pip *v.*[3]] Tipsy, drunk.
1911 J. Masefield *Everlasting Mercy* 26 Si's wife came in and sipped and sipped (As women will) till she was pipped. 1929 M. de la Roche *Whiteoaks* vii. 110 Lilly, here, can't see the strings. He's pipped, aren't you, Lilly?

pipperidge. 2. (Later example.)
1872 Mrs. Stowe *Oldtown Fireside Stories* 127 Old Black Hoss was about as close as a nut and as contrairy as a pipperage-tree.

pippet (pi·pėt). *rare.* [f. Pip *sb.*[2] + -et.] = Pip *sb.*[2] 1.
1940 B. Russell *Inquiry into Meaning & Truth* 57 Take again the two of clubs, and the proposition 'this is similar to that' applied to the two pippets.

pippin. Add: **2. b.** In phrases, as *sound as a pippin*, very sound.
1886 H. Baumann *Londinismen* 139/1 He's as sound as a pippin. 1910 Belloc *Verses* 81, I said to Heart, 'How goes it?' Heart replied: 'Right as a Ribstone Pippin!' But it lied.
3. a. (Further examples.)
c 1821 'W. T. Moncrieff' *Tom & Jerry* (1828) ii. v. 49 Go it, my pippins. 1846 *Swell's Night Guide* 49 Now, my pippins, I'll just ax you which was the rankest sell? 1888 [see *cocker *sb.*[4]]. 1892 E. J. Milliken *'Arry Ballads* 23/2 She would take the shine out of some screamers, I tell yer, my pippin, would Loo. 1895 *Punch* 15 June 285/1 No slow Surrey-siders, my pippin, but smart bits o' frock from Mayfair.
b. An excellent person or thing; a beauty. *slang* (orig. *U.S.*).
1897 Ade in *Chicago Record* 17 Sept. 4/5 This sister was fair to look upon. In fact, it was frequently remarked that she was a Pippin. 1906 G. H. Lorimer *Jack Spurlock* (1908) ii. 28 'I'd like to have the job which goes with that blonde,' and I pointed to a pippin who was pounding the keys just outside his door. 1914 'Bartimeus' *Naval Occasions* xii. 88 The Flag-Lieutenant introduced him to a lady of surpassing loveliness—The Fairest ..of All the Pippins. 1920 Wodehouse *Jill the Reckless* xvi. 237 'We shall..open in Baltimore next Monday with practically a different piece. And it's going to be a pippin, believe *me*,' said our hero modestly. 1926 *Amer. Speech* I. 462 The Apollo Theater in London prints the following glossary of slang in its program as a guide to 'Is Zat So?'..*Pippin*, beauty. 1930 J. Dos Passos *42nd Parallel* i. 47 He..got a book from a man at the hotel. Gosh it was a pippin. 1939 R. Stout *Some Buried Caesar* vi. 71 The fight for a hotel room which was a pippin—I mean the fight, not the room. 1948 H. Innes *Blue Ice* ii. 38 She's a pippin... Knows her way around already. 1972 Wodehouse *Pearls, Girls & Monty Bodkin* viii. 120 So I have a plan..and it's a pippin.

pip-pip. [Echoic.] **1.** A repeated, short, high-pitched sound, *spec.* that made by a motor- or bicycle-horn; also, the horn itself.
1904 Kipling *Traffics & Discov.* 324 Children sat..on the damp doorsteps to shout 'pip-pip' at the stranger. 1907 Shaw *Major Barbara* iii. 292 *Sarah* (touching Lady Britomart's ribs with her finger tips and imitating a bicycle horn) Pip! pip! 1909 *Westm. Gaz.* 15 May 3/2 She [*sc.* a little girl] had motor-cars with real pip-pips. 1979 N. Freeling *Widow* xx. 124 There was a timid little pip-pip at the front door bell. 1979 J. Scott *Angels in your Beer* xviii. 187 Marianne walked to the Ferrari... She..sounded an impatient pip pip on the horn.
2. *slang.* A substitute for 'good-bye'. (In quot. 1907 as a defiant retort.)
1907 *Mr. Punch Awheel* 93 *Cyclist*... 'Nice crowd out this morning!' *Rural Policeman*..'Yes, an' yer can't do with 'em! If yer 'ollers at 'em, they honly turns round and says, 'Pip, pip'! 1920 Wodehouse *Damsel in Distress* x. 129 'Well, it's worth trying,' said Reggie. 'I'll give it a whirl. Toodleoo!' 'Good-bye.' 'Pip-pip!' Reggie withdrew. 1931 E. F. Benson *Mapp & Lucia* iii. 56 Mr. Woolgar..did not say 'So long' or 'Pip-pip'. 1951 Wodehouse *Old Reliable* xv. 169 Hello, Smedley. Pip-pip, Lord Topham. 1973 G. Sims *Hunters Point* iii. 22 The nine-day 'British Week' had ended... Fisherman's Wharf had been buzzing with 'Cheerio, pip pip and smashing' voices. 1978 M. Butterworth *X marks Spot* iii. iii. 158 'Pip pip, laddie.' He set off.

pip-pip-pip. *Teleph.* [f. *pip *sb.*[5]] (The sound of) the three consecutive pips used as a time signal in the 'speaking clock' service and to indicate the lapse of time during a trunk call.
1936 *Discovery* Oct. 315/2 Thus 'At the third stroke' is taken from one groove, 'it will be' from another, 'eleven o'clock' from another disc, and the sound 'pip-pip-pip' from a fourth. 1938 D. Smith *Dear Octopus* i. 12 Don't go on talking after the pip pip pip, because they charge you at once.

pippy. Add: **2. b.** Depressed, out of sorts.
1886 R. Fry *Let.* 28 Nov. (1972) I. 111 My Mays depress me, so does my Tripos... In fact I am stupidly pippy at times.

pipradol: see next.

pipradrol (pi·prădrǫl). *Pharm.* Also (*erron.*) **pipradol.** [f. Pip(e)r(idine + *-a-* + *benzhy)drol* s.v. Benzo-.] A colourless crystalline solid, α,α-diphenyl-α-piperid-2-ylmethanol hydrochloride, $C_{18}H_{21}NO.HCl$, which is used as an antidepressant. Also called *pipradrol hydrochloride*.
1955 *Science* 11 Feb. 209/2 Meratran is the trademark of the Wm. S. Merrell Co., for its brand of pipradrol. 1955 *Sci. News Let.* 13 Aug. 105/1 The drug is a pipradol chemical trade-named Frenquel by the manufacturer, the Wm. S. Merrell Co. of Cincinnati. 1956 *Jrnl. Amer. Med. Assoc.* 4 Feb. 390/1 Pipradrol hydrochloride is a central nervous system stimulant chemically unrelated to the sympathomimetic amines but exhibiting some of the pharmacological actions of amphetamine. 1962 F. C. Ferguson *Drug Therapy* xxiv. 216 Other types of antidepressants include piperidines; methylphenidate (Ritalin) and pipradrol (Meratran). 1973 *Approved Names 1973* (Brit. Pharmacopœia Comm.) 59 Pipradol. 1974 *Ibid.* Suppl. 111, For 'Pipradol' read 'Pipradrol'. 1974 *Jrnl. Clin. Pharmacol.* XIV. 132/2 Present results thus suggest at the most only limited efficacy for pipradrol in the treatment of depressed outpatients.

pi·p-squeak. *slang.* [f. *pip *sb.*[5] + Squeak *sb.*] **1.** A contemptuous name for an insignificant person; a petty object. Also *attrib.*
In quot. 1923 a two-stroke motorcycle.
1910 E. V. Lucas *Slowcoach* xxiii. 279 'It belongs to one of those measly pip-squeaks,' said Robert. 1923 *Motor Cycling* 21 Nov. 89/1 The owners of sporting four stroke machines look down on the owners of so-called 'Pip-Squeaks'. 1925 Fraser & Gibbons *Solider & Sailor Words* 224 *Pip Squeak*,..a small man, or one objectionable in some way. 1926 *Blackw. Mag.* June 732/2 After all, the luxurious liner which connects this riotous spot with the outer world is only a pip-squeak of a vessel. 1930 G. Macmunn *Behind Scenes in Many Wars* 88 It does not pay in the East to let pip-squeaks beard the mighty. 1946 *Richmond* (Virginia) *Times-Dispatch* 20 Jan. i. 14/8 Specifically, [Senator] Wiley charged that the organization..has 'created czars out of pipsqueak juveniles'. 1951 M. McLuhan *Mech. Bride* (1967) 23/2 Mighty blasts on the tooter herald the arrival of just another pip-squeak [*sc.* monotonous book]. 1961 A. Christie *Pale Horse* xxi. 224 What about that psychological pipsqueak you brought along to see me, Corrigan. What does he say? 1973 'H. Howard' *Highway to Murder* vi. 65 For a little pip-squeak you make a big noise. 1974 E. Ambler *Dr. Frigo* iii. 174 They weren't taking any nonsense from a pipsqueak foreign doctor.
2. a. A small type of high velocity shell distinguished by the sound of its flight. Also *attrib.*
1916 E. V. Lucas *Vermilion Box* 209 Whatever else there is to grumble at over here, wet, and rats, and Pip-Squeaks and Jack Johnsons..we do get two things up to sample. 1916 *Cornh. Mag.* Mar. 395 They're 'pip-squeak' and splinter-proof, of course. 1917 A. G. Empey *Over Top* 304 *Pip squeak*, Tommy's term for a small German shell which makes a 'pip' and then a 'squeak', when it comes over. 1927 E. Thompson *These Men, thy Friends* 176 The Turkish guns suddenly sent over a couple of pipsqueaks. *a* 1936 Kipling *Something of Myself* (1937) vi. 159 One indubitable shell—ridiculously like a pip-squeak in that vastness but throwing up much dirt.
b. = *pip-pip 1. Also = *peep *sb.*[1] 2 b. So **pi·p-squeak** *v.*
1922 *Blackw. Mag.* June 699/2 She heard just then her Tom pip-squeaking on his pipe. 1927 'Ixion' *Further Motor Cycle Reminisc.* 100 Hooters..fitted with rather wafery clips, such as still linger on bicycle 'pipsqueaks'. 1943 [see sense 2 c below]. 1956 E. Pound tr. *Sophocles' Women of Trachis* 16 Hasn't uttered a pip-squeak Since she came down from the windy country.
c. (See quots. 1943, 1970.) *slang.*
1943 C. H. Ward-Jackson *Piece of Cake* 48 *Pip-squeak*, the instrument in an aircraft by the aid of which one gets a 'Fix'. This instrument emits a pip-squeak at short intervals which is synchronised over the radio with base, thus fixing the time, an essential prelude to fixing the position of the aircraft. 1946 Brickhill & Norton *Escape to Danger* vi. 60 Forgetting to switch off his 'pip-squeak' (radio contactor), Mickey climbed thankfully out on to the wing. 1970 Partridge *Dict. Slang* (ed. 7) II. 1330/2 'The pip-squeak was an automatic transmitter *only*, whose once-a-minute signals enabled ground direction-finding stations to fix the aircraft's position accurately for the benefit of the Fighter Controller in the Operations Room...' (Ramsey Spencer, March 1967.)

Pip, Squeak, and Wilfred. *slang.* [Names of three animal characters featured in a children's comic strip in the *Daily Mirror* from 1920 onwards.] Designating a trio of objects or persons.
[1920 'Uncle Dick' in *Daily Mirror* 25 Mar. 13/2 Aren't Pip and Squeak and Wilfred sweet measuring their heights? 1920 —— *Pip, Squeak, & Wilfred* 60 Some time ago, Pip, Squeak, and Wilfred dreamed they were real children. *Ibid.* 61, I am rather fond of Pip, Squeak, and Wilfred. I, too, have my dreams.] 1937 Partridge *Dict. Slang* 633/2 *Pip, Squeak and Wilfred*, the medals (or medal ribbons), 1914–15 Star, War Medal, Victory Medal. 1943 C. H. Ward-Jackson *Piece of Cake* 48 *Pip, Squeak and Wilfred*, the 1914–15 Star, the Great War and Victory medals or ribbons worn in a row. 1966 'L. Lane' *ABZ of Scouse* 35 Other names [for a trio of friends] are Pip, Squeak and Wilfrid [*sic*]. 1977 *Times* 30 Sept. 16/8 That goes for Messrs Pip, Squeak and Wilfred, too.

pipsyl (pi·psəil, -il). *Chem.* Also **PIPSYL.** [f. letters in the systematic name (see def.).] The radical *p*-iodo*p*henyl*s*ulphon*yl*, $I{\cdot}C_6H_4{\cdot}SO_2$—, compounds of which are used as radioactive labels.
1946 *Jrnl. Amer. Chem. Soc.* LXVIII. 1390/1 As the labelled reagent we used *p*-iodophenyl sulfonyl chloride (PIPSYLchloride), prepared from radioactive iodide ion and *p*-diazobenzene-sulfonic acid, followed by treatment with phosphorus pentachloride. *Ibid.* 1390/2 Less than one-hundredth per cent. of *d*(−)alanine was found in the β-lactoglobulin hydrolysate using PIPSYL d(−)alanine carrier. 1949 *Ibid.* LXXI. 256/1 (*heading*) The preparation of pipsyl amino acids for use as carriers. 1961 G. R. Choppin *Exper. Nucl. Chem.* x. 173 In the last chapter the use of I^{131} labeled *p*-iodobenzenesulfonyl chloride (pipsyl chloride) with the amino acids was described. 1974 *Canad. Jrnl. Biochem.* LII. 217/2 Using the multiple isomorphous protein phases..a difference electron density map has been computed. This map was relatively devoid of significant electron density except in the region of PIPSYL binding.

piquantly, *adv.* (Later examples.)
1922 Joyce *Ulysses* 399 Blushing piquantly and whispering in my ear. 1955 *Times* 10 May 3/7 M. Claude Barma's production was most piquantly revealing. 1971 *Daily Tel.* 16 July 11/8 With a cast of two, it presents a piquantly rounded theme.

piqué, *sb.* (*a.*) Add: **A. a.** (Later examples.)
1932 *Daily Tel.* 25 Apr. 7/5 The yoke and collar in plain white cotton-piqué. The sleeves may be slightly fitted to above the elbow, where they should be met by a ..gauntlet cuff of the white piqué. 1956 R. Braddon *Nancy Wake* iii. xv. 171 Her quaint white *piqué* dress. 1968 J. Ironside *Fashion Alphabet* 245 *Piqué*,..a firm fabric with lengthwise carded effect made of cotton. *Waffle piqué*—with a honeycomb weave.
b. (Earlier and later examples.)
1872 *Young Englishwoman* Nov. 611/1 Pique stitch crochet. 1958 *Times* 20 Jan. 11/2 Jumper suit of pique knit jersey.
B. *ppl. a.* **a.** (Earlier and later examples.)
1872 C. Schreiber *Jrnl.* 2 Nov. (1911) I. 169 We found ..a small piqué plaque on tortoise-shell. 1879 *Ibid.* 5 Dec. II. 250 A curious knife and fork... The handles are piqué, or inlaid in silver with acorns and oak leaves. 1968 J. Ironside *Fashion Alphabet* 178 *Piqué*, tortoise-shell or ivory inlaid with tiny dots or lines of gold or silver. This art, brought over by the Huguenots in the seventeenth century, covered brooches, buttons and ear-rings as well as small boxes. 1979 *Country Life* 7 June (Suppl.) 113/3 Silver-gilt and tortoiseshell *piqué* magnifying glass, c. 1730.
b. *Cookery.* Larded. *rare.*
1846 A. Soyer *Gastronomic Regenerator* 230 (*heading*) Fillet of Beef piqué aux legumes printaniers.
c. *Ballet.* With the point of the foot; with the foot pointed. Also as *sb.*, a step directly on to the point of the leading foot.
1913 C. d'Albert *Dancing* 123 *Piqué* (pas),..de la pointe et du talon. Toe and heel points. 1931 C. W. Beaumont *French-Eng. Dict. Techn. Terms Classical Ballet* 21 *Piqué*..pricked, pricking. Generally implies a shooting forward of the body on to the *pointe* of the front foot. 1954 *Ballet Ann.* VIII. 65, I should also like to draw your attention to the *arabesques piquées*. Contrary to tradition, these should not be '*piquées*' with the knee stretched, but with the knee flexible. The instep is arched at the beginning of the movement, but the knee is quite taut only when balance has been obtained. *Ibid.*, By following my instructions..the ballerina is able to assume..a manner of walking which is æsthetically satisfying. *Piqués* and *relevés* are possible to her whilst in movement; her very walk is 'melodic'. 1967 Chujoy & Manchester *Dance Encycl.* 734/2 *Piqué*, in ballet, the movement of stepping directly onto point of supporting foot. The working leg may be in a variety of poses. 1975 *New Yorker* 16 June 103/1 There were moments in this performance that stopped my breath: a high, motionless *piqué* balance lightly stepped into from nowhere, [etc.].
C. piqué work. b. (Later example.)
1969 *Canad. Antiques Collector* Oct. 19/2 It [*sc.* a small box] is of ivory, bound with silver with the design formed by an inlay of little points of silver known as piqué work.

‖**piquer** (pike), *v.* *Cookery.* Also (*erron.*) **piqué.** [F. *piquer* (see Pique *v.*[1]) to lard.] *trans.* To insert bacon strips or other flavouring substance in (meat, poultry, etc.) before cooking; also *fig.*
In quot. 1951[2], *piquez* is the Fr. imperative.
1846 *Jewish Manual, or Pract. Information Jewish & Mod. Cookery* iv. 67 Take a piece from the shoulder [of veal]..*piqué* it thickly. 1865 M. Eyre *Lady's Walks S. of France* xxix. 316 It is common here to *piquer* a leg of mutton with garlic, that is, small holes are drilled in it before roasting, and a small kind of garlic..inserted therein. 1935 *Proc. Brit. Acad.* XX. 85 Start with a moral and political poet.., piquer or lard him with immorality, potorial songs and other irrelevances, [etc.]. 1951 E. David *French Country Cooking* 120 The fillet..is cut into..pieces..*piquèd* [*sic*] with garlic and seasoned

with black pepper. *Ibid.* 158 *Piquez* each fillet with a small piece of bacon. 1960 —— *French Provincial Cooking* 79 A *fricandeau de veau*,..and a *râble de lièvre*..are two of the classic examples of piquéd meat and game.

‖**piqûre** (pikūr). Also **piqure.** [Fr., = injection, f. *piquer* (see Pique *v.*[1]) to pierce (the skin), to give an injection.] A hypodermic injection; the puncture made in the skin by such an injection.
1925 Infanta Eulalia of Spain *Courts & Countries after War* vii. 150 The newest 'Piqures' come from Germany. 1936 C. Connolly *Rock Pool* iii. 59 She had large flabby arms covered with piqures. 1940 N. Mitford *Pigeon Pie* xiii. 211 'Miss Wordsworth received last night in an omnibus a *piqure* that will incapacitate her for a week at least.' 'I see, you are white slavers as well as everything else.' 1942 'A. Bridge' *Frontier Passage* ii. 21 The doctor came... And she gave her a piqûre, and they sent medicines. 1962 K. A. Porter *Ship of Fools* ii. 216 You must give me a *piqûre*, a huge one that will make me sleep for days. 1962 V. Sackville-West *Let.* 26 Apr. in H. Nicolson *Diaries & Lett.* (1968) 411, I must have five injections... *Later.* It is now over, and it was no more painful than all those *piqûres* I had on the *Antilles*.

piracy. Add: **1. b.** *Physical Geogr.* = *capture *sb.* 1 b.
1904 Chamberlin & Salisbury *Geol.* I. iii. 99 The foregoing case may be called foreign piracy because the valleys of different systems are concerned. Domestic piracy may also take place... Here a tributary to a crooked river may develop, working back until it taps the main at a higher point. 1939 *Bull. Geol. Soc. Amer.* L. 1350 The stream pattern indicates that recent piracies have occurred. 1957 G. E. Hutchinson *Treat. Limnol.* I. i. 114 A wide valley, the main stream of which has been reduced by piracy. 1974 C. H. Crickmay *Work of River* iii. 62 Stream piracy.., of course, is not in every case effected by headwater extension.

Pirandellian (pirănde·liăn), *a.* [f. the name of the Italian playwright Luigi *Pirandello* (1867–1936) + -ian.] Of or pertaining to or characteristic of Pirandello or his style.
1927 *Observer* 8 May 15/3 In a Pirandellian sense, he is enormously 'real'. Just as we believe that Hamlet exists apart from the personality of the actor, so we know Pogo as if he were our dearest dumb friend. 1929 *Sat. Rev.* 1 June 746/1 How the dates fit I do not know, but Señor Unamuno seems to have drunk deep of Pirandellian philosophy. 1930 *Times Lit. Suppl.* 19 June 511/3 By this time one could almost predict from a single hint the course of a Pirandellian play. 1933 *Ibid.* 9 Mar. 164/4 The professor's appeal to the young man's good sense has in it that paradoxical justice that may well be called Pirandellian. 1959 *Times* 13 Mar. 16/6 Mr. Jupp..hits on the Pirandellian notion that the family should re-enact the history of their own lives in the hope of understanding it. 1964 *Listener* 30 Apr. 731/3 To get at the truth he questions the dramatist's ex-wife, and searches round for other people involved in the affair, playing, and allowing Mr Elliot to play, a Pirandellian game of masks and faces. 1974 *N.Y. Times* 7 Sept. 25/2 As a comedian he had trouble finding a persona, a Pirandellian problem.
Hence **Pirande·llism,** the style or method of Pirandello; a characteristic example of this; **Pirande·llist,** an advocate or follower of Pirandello.
1936 *Times Lit. Suppl.* 19 Dec. 1048/3 The best and most striking side of Pirandello's art is lost, leaving behind the bare intellectual residue of so-called 'Pirandellism'..exposed. 1962 *Spectator* 23 Mar. 370/2 The ending makes the play the second victim in twelve months of phoney and incompetent Pirandellism. 1962 Pirandellist [see *Imagist 1]. 1964 *Eng. Stud.* XLV. 328 Robinson grows away from his clay-coloured Irish plays and creates such Pirandellisms as *Church Street*.

Piranesian (pirănī·ziăn), *a.* [f. the name of the Italian architect and artist Giovanni Battista *Piranesi* (1720–78) + -an.] Of, pertaining to, or characteristic of Piranesi, his style, or his theories of architecture.
1923 A. Huxley *Antic Hay* vi. 87 You could fancy yourself at the entrance of one of Piranesi's prisons... Mrs Viveash's taxi drove in under the Piranesian arch. 1925 —— *Along Road* ii. 93 The Colosseum, mantled.. with a romantic, Piranesian growth of shrubs. 1926 —— *Jesting Pilate* i. 123 The cathedral of the banyan grove is transformed into a Piranesian prison. 1961 *Architect & Building News* 21 June 822/1 The east and west elevations close the vistas in a highly dramatic, almost Piranesian, fashion. 1962 *Guardian* 25 Sept. 5/6 These halls ..are of Piranesian grandeur. 1968 *Punch* 25 Dec. 910/1 They have just reconstituted Euston Station and I don't like it at all. I miss the mock-Doric arch and the Victorian grandeur, the sweeping vistas and the Piranesian gloom. 1976 *Listener* 1 Apr. 413/2 Some gargantuan, crudely-painted Piranesian ruin.

piranha. Now usu. with pronunc. (pirā·nă). Substitute for def.: A carnivorous freshwater fish of the genus *Serrasalmus*, belonging to the family Characidæ and native to South America; = Perai, piraya. Also *attrib.* (Later examples.)
1904 [see *caribe]. 1915 *Nature* 8 Apr. 149/2 Birds, beasts, coral snakes, and piranha fishes, toads and ants, and primitive natives, he [*sc.* T. Roosevelt] has something to say about. 1927 W. M. McGovern *Jungle Paths & Inca Ruins* xxvi. 269 All these fish were *piranhas* armed with savage little teeth. 1931 [see *caribe]. 1954 G. Durrell *Three Singles to Adventure* ii. 53 The piranha is one of the most unpleasant freshwater fish known. It is a flat, corpulent, silver-coloured fish, with the lower jaw protruded. 1960 T. Hughes *Lupercal* 20 Even the Amazon's taxed and patrolled To set laws by the few jaws—Piranha and jaguar. 1962 N. Maxwell *Witch-Doctor's Apprentice* vii. 73 *Piranhas*, voracious little beasts which attack *en masse* and can strip the meat off a live cow in minutes. 1968 *New Scientist* 7 Mar. 534/2 There are many types of fish that are classed as piranhas: the common feature is that they have a pair of powerful jaws with razor-sharp teeth. 1977 *Time* 19 Sept. 62/1 (Advt.), Feel Tom Sterling's apprehension, when obliged to take a dip in the piranha-infested Amazon.
2. *fig.*
a 1963 S. Plath *Ariel* (1965) 40 And the fish, the fish—Christ! they are panes of ice, A vice of knives, A piranha Religion. 1977 *Times* 14 Mar. 19/6 Gayle Hunnicutt turned Sarah Pocock into a deceptively girlish piranha who..planned to nibble into tiny pieces everyone in sight. 1978 *Times* 15 May 16/8 It's not just a goldfish bowl in a nationalized industry, it's a piranha bowl.

Pirani (pirā·ni). *Physics.* The name of M. S. von *Pirani* (b. 1880), German physicist, used *attrib.* to designate a gauge invented by him for measuring very low pressures, which utilizes the cooling effect of the gas on a heated metal filament whose resistance is sensitive to temperature.
It was described by von Pirani in *Verhandl. der deutsch. physik. Ges.* (1906) VIII. 686.
1911 *Trans. Amer. Electrochem. Soc.* XX. 245 In simplicity both of construction and of operation, involving only current and resistance measurements within the range of instruments readily accessible, the Pirani manometer recommends itself at once. 1921 *Proc. Physical Soc.* XXXIII. 293 For many purposes we think that the Pirani gauge..would have considerable advantages over the McLeod. 1962 F. I. Ordway et al. *Basic Astronautics* v. 200 Pirani, alphatron, and bellows gauges are regularly employed in rockets to obtain atmospheric pressures. 1971 *Sci. Amer.* Aug. 114/2 The effort expended in making the Pirani gauge is amply repaid by the convenience it affords.

pirarucú (pirăruku·). [Pg., f. Tupi (see quot. 1863[2]).] The giant redfish of the Amazon basin, *Arapaima gigas*, of the family Osteoglossidæ, one of the largest freshwater fishes in the world; = *Arapaima, *paiche.
1840 [see *Arapaima]. 1863 H. W. Bates *Naturalist on River Amazons* II. iii. 165 The men caught sight of a large Pirarucú: the fish which, salted, forms the staple food..in most parts of the Lower Amazon country. *Ibid.* 166 The Indian name Pirarucú, or Anatto fish (from Pira, fish, and urucú, anatto or red), is in allusion to the red colour of the borders of its scales. 1933 P. Fleming *Brazilian Adventure* i. xviii. 153 The very big fish—the *piracarú* [sic] and the *pirará*..they [*sc.* the Indians] harpoon. 1936 *Discovery* Dec. 373/2 Pirarucú,..when dried and salted, is rapidly taking the place of imported cod on the Brazilian market. 1961 [see *paiche]. 1965 E. Bishop *Questions of Travel* 31 Everything must be there In that magic mud, beneath The multitudes of fish, deadly or innocent, The giant pirarucús.

pirate, *sb.* Add: **4. b.** One who receives or transmits radio programmes without a licence to do so.
Current usage refers to radio transmission.
1913 *Marconigraph* II. 530/2 'There you are,' said the captain, 'unless we have been picked up again by some experimenting pirate.' 1923 *Wireless Weekly* 13 June 592 The thousands who are listening-in without a licence of any description—popularly termed 'pirates'. 1923 *Exper. Wireless* Nov. 57/2 The olive branch has been held out to the 'pirates', and the ordinary listener-in is cheered by the prospect of a reduction in prices of complete sets. 1933 *Pract. Wireless* 14 Oct. 182/1 (*heading*) Wanted, One Radio Pirate! The small Brussels (Schaerbeek) broadcasting station, having complained to the authorities that an illicit transmitter has marred the reception of its broadcasts, a reward of one thousand Belgian francs has been offered to trace the identity of the culprit. 1964 *Daily Tel.* 13 May 19/2 Let us be clear about this: the pirates of 1964, like the pirates of old, are simply out after money, as much money as they can get in defiance of international law. 1966 *Listener* 16 June 863/2 Fewer than 145,000 licences were issued in 1964... Evasion is clearly a problem in which Britain is not alone, though, in spite of the one-time reputation of the South China Seas, Hong Kong is at least free of pirates. 1967 *Ibid.* 17 Aug. 195/1 In other areas of radio the pirates provide no example. 1969 C. Booker *Neophiliacs* ix. 228 Throughout.. April, the country—and the pirates—waited in mounting suspense to see what the Government would do. 1979 *Guardian* 9 Aug. 3/6 Air wave pirates pay the price... Signals sent out by illegal radio hams..led to four men appearing at Grimsby magistrates court.
6*. *Physical Geogr.* A river that captures another (*capture *v.* b). Also *attrib.* appositively.
1889 *Science* 8 Feb. 108/1 There is a little river-pirate in eastern Pennsylvania unsuspected by its rural neighbors. *Ibid.*, The pirate is Deer Run, and its victim is the north-east branch of Perkiomen Creek. 1904 Chamberlin & Salisbury *Geol.* I. iii. 98 The tributary which does the stealing is known as a pirate. 1914 R. S. Tarr *College Physiogr.* i. xv. 566 Anything that accelerates headwater erosion on one side of a divide..gives opportunity for the pushing back of the divide and the possible capture of headwaters, or even of good-sized streams, by the successful river pirate. 1939 *Bull. Geol. Soc. Amer.* L. 1333 Each capture strengthened and lengthened the pirate and weakened and shortened the victim. 1968 R. W. Fairbridge *Encycl. Geomorphol.* 1055/2 The point at which the capture is effected..is commonly marked by a right angle turn into the pirate stream.
7. b. *pirate ship* (examples); **d. pirate bus** (further example); **pirate cab** = **pirate taxi*; **pirate label** [*label *sb.*[1] 7 c], a recording or a recording company which infringes a copyright; **pirate taxi,** a vehicle which is used as a taxi but is not licensed as such.
1963 *Times* 24 May (London Underground Suppl.) p. vii/2 The 'cut-throat' competition before 1933 when the London Passenger Transport Board was formed was all very well for some people who lived on routes where the 'pirate' buses operated. 1930 A. Armstrong *Taxi!* xvi. 220 There are also some real 'pirate' cabs which only operate down town [*sc.* in New York] at night. They carry no meters and live by 'making a price'..with any belated..up-town passenger. 1968 *Jazz Monthly* Feb. 4/1 There are..numerous 'pirate' labels also issuing EPs and LPs. 1700 in *N. Carolina Colonial Rec.* (1886) I. 518, I herewith send you a copy of what I lately received.. concerning the taking of a pyrate ship. 1720 Defoe *Capt. Singleton* 187, I wrote that he was taken away by main Force, as a Prisoner, by a Pyrate Ship. 1911 G. B. Shaw *Doctor's Dilemma* Pref. p. xvi, It is the sort of conscience that makes it possible to keep order on a pirate ship, or in a troop of brigands. 1971 *E. Afr. Standard* (Nairobi) 11 Apr. 7/4 Owners of Matatu (pirate taxis) also reported an 'exceptionally' good business as hundreds of people who wanted to go to the rural areas had to use them. 1978 S. Naipaul *North of South* i. iv. 87 The thing standing outside the hotel was..a matutu—a pirate taxi.
e. *attrib.* or as *quasi-adj.* designating the clandestine or illegal transmission of radio programmes (see sense 4 b above), as *pirate broadcast* (also as *vb.*; so *pirate broadcaster*, *pirate broadcasting* vbl. sb.), *pirate radio (station)*, *pirate station*; **pirate (radio) ship,** a ship used to transmit radio programmes from a position outside the territorial waters of the receiving country; **pirate vessel** = *pirate (radio) ship* above.
1942 *New Yorker* 17 Jan. 52/2 There is a republican pirate radio station, called La Voce della Libertà. 1957 F. Hoyle *Black Cloud* v. 109 Is Nortonstowe going to become a pirate radio station? 1961 *Guardian* 16 Dec. 7/2 Only the presence of mind of the engineers..prevented a pirate broadcast being heard on the air of the capital itself. 1964 *Daily Tel.* 11 May 20/7 The activities of the 'pirate' radio ships Caroline and Atlanta have presented the Government with a problem which cannot be solved simply. It is expected that the Cabinet will discuss this week the possibility of legislation to prevent broadcasting from such 'pirate' vessels. *Ibid.* 13 May 1/8 Mr. Mawby, Assistant Postmaster-General, told the Commons that new legislation which would effectively deprive 'pirate' broadcasters of material support was the most suitable action. *Ibid.* 14 May 28/3 Almost every BBC station is suffering from some foreign interference and a ring of pirate ships is bound to make matters worse. *Ibid.*, The Swedish law makes it an offence for nationals to take part in pirate ship broadcasting. 1965 *Punch* 3 Feb. 154/1 Commons debate on plight of pirate radios. 1966 *Economist* 23 Apr. 340/3 The arrival this week off British coasts of a pirate radio ship..is a further step in the battle over the off-shore radio stations. 1966 *Listener* 4 Aug. 154/2 A Government bill to ban pirate radio stations provides for penalties of up to two years' imprisonment. 1967 *Ibid.* 17 Aug. 195/1 There is every sign that many of the crude but effective tabloid techniques of pirate radio..will be employed by Radio 1. 1969 C. Booker *Neophiliacs* ix. 227 The Swedes, the Danes and the Dutch had been plagued by offshore pirate stations as long ago as 1961–2. 1970 *Internat. & Compar. Law Q.* XIX. 357 Legal and practical controls of 'pirate' broadcasting. 1973 *Times* 3 Jan. 4/2 The ship Mi Amigo, from which the pirate radio station broadcasts, sailed out to sea again. 1973 *Daily Tel.* (Colour Suppl.) 5 Jan. 6/1 His ship will be a 'pirate' inasmuch as she will broadcast from outside territorial waters, but unlike other pirate broadcasters there will be no sponsored advertisements.

pirate, *v.* Add: **3.** (Later examples.)
1968 *Blues Unlimited* Nov. 6 They're not selling records, for fear they would be pirated! 1977 *Belfast Tel.* 17 Jan. 8/4 Under the European Television Agreement of 1953 most countries agreed not to 'pirate' programmes broadcast by companies from other nations. 1979 *Guardian* 25 Aug. 24/1 'Pirating' involves the copying, for sale to the public, of existing records without the consent of the copyright owners.
Hence **pirated** *ppl. a.* (further examples); spec. *pirated edition*, an edition of a book produced without authorization; **pirating** *vbl. sb.* (further example).
1737 Pirated edition [in Dict.]. 1853 C. M. Smith *Working Man's Way in World* iv. 56 (*heading*) Pirated editions of Scott's novels. 1928 D. H. Lawrence *Let.* 5 Dec. (1962) II. 1103, I hear from Stieglitz there are *two* pirated editions, photographed from my edition, and with forged signatures. *Ibid.* 10 Dec. 1105, I hear London and Paris are both selling the pirated editions of *Lady C.* at £3 and £2. 1928 A. Huxley *Let.* 12 Dec. (1969) 304 Dear Lawrence, What an intolerable business about the pirating of *Lady C.*! 1952 J. Carter *ABC for Bk.-Collectors* 135 *Pirated edition*,..a term commonly applied (sometimes with, sometimes without, legal accuracy) to an edition produced and marketed without the authority of, or payment to, the author. 1959 L. M. Harrod *Librarians' Gloss.* (ed. 2) 286 A pirated edition is an unauthorized reprint involving an infringement of copyright. 1967 *Listener* 28 Sept. 413/2 After hearing this performance—and a pirated tape of his 1953 Covent Garden *Aida*—it seems incredible that Barbirolli has been allowed

to languish outside the opera house for 13 years. 1973 *Times* 17 Oct. 11/3 The records have been issued in Paris, but not here. It is as bad as Russia, where people listen to me on pirated versions. 1975 *Times Lit. Suppl.* 13 June 678/2 British efforts to influence Parliament to protect British books against the importation of foreign pirated editions.

piratedom (pəiə·rėt,dəm). *rare.* [f. Pirate *sb.* + -dom.] Pirates collectively; the world of pirates.
1907 F. Campbell *Shepherd of Stars* iv. 36 He went to shout orders to a fleet of approaching barges from the stronghold of ancient piratedom.

piriform. For examples of this variant see *pyriform *a.*

pirimicarb (piri·mikāɹb). [f. *pyrimi(dine by alteration + Carb(amate.] An insecticide that is specific against aphids; 2-dimethylamino-5,6-dimethyl-4-pyrimidinyl dimethylcarbamate, $C_{11}H_{18}N_4O_2$.
1970 *Proc. 5th Brit. Insecticide & Fungicide Conf., 1969* II. 546 Pirimicarb is an aphicide of such specificity that its use for aphid control should not directly eliminate some of the more important predators concerned with the regulating of aphid numbers. 1974 *Nature* 8 Feb. 337/3 Pirimicarb is systemic when applied to soil, being absorbed by the roots. 1977 *Homes & Gardens* Nov. 42/1 A recent product which contains 'Pirimicarb' in either liquid concentrate form or as an aerosol is very effective against aphids except on cucumbers and soft fruit.

piri-piri[2] (pi·ri pi·ri). [Origin obscure; perh. ad. Swahili *pilipili*, pepper.] A sauce made with red peppers. Also *attrib.* or as *adj.* and quasi-*adv.*
1964 H. Holthausen *Chicken goes around World* 71 *Frango Piri-piri* (Chicken in Piri-piri sauce). *Ibid.* 72 Preheat the grill and put the piri-piri chicken in the immediate vicinity of the heat... Baste the chicken well..with the remaining piri-piri. 1968 C. Burke *Elephant across Border* ii. 68 The 'Camarões Pequenos', just..a few miles north of Lourenco Marques..made the best piri-piri prawns on the whole coast. *Ibid.* iii. 115 I'm going to have me a dozen giant prawns, charcoal-grilled, without piri-piri, but with lemon and butter sauce. 1969 M. Tripp *Malice & Maternal Instinct* iv. 28 They specialise in Spanish and Portuguese here... How about galinha piri-piri. 1970 G. Croudace *Scarlet Bikini* vi. 68 If yer gennelmen'll only bring up some rock lobster, I'll give it yer for lunch, piri-piri with rice. 1973 *Times* 17 Feb. (Mozambique Suppl.) p. iv/6 (Advt.), The sauce *piri-piri* is made with red peppers.

pirl, *v.* Add: **3.** (Examples.) So **pi·rling** *ppl. a.* and *vbl. sb.*
1789 D. Davidson *Thoughts on Seasons* 33 Ye roll, in cudlin purlings to the sea. 1819 J. Rennie *St. Patrick* II. x. 191 I'll set my teeth in the withered chafts o' you till the blind pirl out o' your luckin' e'en. 1920 *Chambers's Jrnl.* Christmas No. 837/2 Before the first puffs of blue smoke circled and pirled above the village roofs. 1936 C. Macdonald *Echoes of Glen* i. 3 On an emerald bank by the side of a pirling burn.

pirlie pig (pā·ɹli, *Sc.* pi·rli). *Sc.* Also **perly, pirly,** etc. [Prob. f. Pirl *v.* + Pig *sb.*[2]: see *Sc. Nat. Dict.*] A small money-box, usu. circular and made of earthenware, with a slot to insert coins. Also *ellipt.* as *pirlie*, etc.
1799 'Philetas' in J. Thomson *Hist. Dundee* (1874) i. x. 127 Old women and children kept their pozes in their *kist neuks* and *pirly pigs*. 1831 in *Trans. Banffshire Field Club* (1939) 33 One Stone purly 4/-. 1889 J. M. Barrie *Window in Thrums* xviii. 170, 'I mind he broke open his pirly,..an' bocht a ha'penny worth o' something to ye every day. 1900 *Longman's Mag.* Nov. 49 Donald did not possess a bike yet, and my 'tips' went into a perly pig in which he was saving up to buy one. 1905 *Athenæum* 28 Jan. 118 The pirley-pig or circular money-box pertaining to the Town Council of Dundee. This pewter money-box is in the shape of an orange or flattened globe. 1912 *Proc. Soc. Antiquaries Scotl.* XLVI. 353 Until thirty or forty years ago there was a good demand for modern pirlie pigs. 1934 H. B. Cruickshank *Noran Water* 15 Ye've riped the pirlie mony's the time Withooten ony skaith. 1960 H. Hayward *Antique Coll.* 220/1 *Pirlie pig*, earthenware money-box. 'Pig' is a North Country word for an earthen jar: 'pirlie' is a diminutive indicating something of slight value.

pirn, *sb.*[2] Add: Also **pern, perne. 1.** Also *fig.*
1919 W. B. Yeats *Wild Swans at Coole* 36 He unpacks the loaded pern Of all 'twas pain or joy to learn. 1950 T. R. Henn *Lonely Tower* 185 Within the cones moves the 'perne', a spool which unwinds the thread spirally as the sphere moves onward.
5. pirn-mill, a mill where weaver's bobbins are manufactured.
1915 W. [illegible]ts *Reveries* 13 Another day a sea captain pointe[illegible] smoke from the Pern mill on the quays. 1938 in S[illegible] *Dict.* (1968) VII. 140/1 The Pirners' Bridge, so-called either because bobbin-makers crossed it to get birch-timber in the adjacent copse to make their pirns, or because a pirn-mill once stood near it.

‖**pirog** (pirǫ·g). Also **piroga, piroque.** Pl. **pirogen** (a. Yiddish), **pirogi** (a. Russ.), **pirogs.** [Russ. *piróg*, Yiddish (a. Russ.) *pirog*.] A large pie. Cf. *piroshki *sb. pl.*
1854 [See *fish-cake* s.v. *fish *sb.*[1] 6 c]. 1933 P. & L. G. Esmonde *von Schumacher's Cook's Tour of European Kitchens* vi. 109 Pirogs are eaten fresh from the oven or heated up again. 1950 E. J. Kollist *Compl. Patissier* xxiv. 232/1 In Russia the various coulibiac, piroques and small patties are favourites. 1951 Pirogen [see *piroshki *sb. pl.*]. 1962 K. Petrovskaya *Secrets of Russian Cooking* 143 A true Russian (or anyone who's ever tasted real Russian pirozhki or pirogi) can grow rhapsodic just talking about them. 1971 *Guardian* 23 July 9/6 Pirogi and Piroshki, literally big pies and baby pies. 1973 [see *piroshki *sb. pl.*].

pirogue. Add: (Further examples.) Also *attrib.*
1926 I. S. Cobb *Some United States* xi. 264 A little later four husky chaps in pirogues ranged up alongside us. 1947 *Motor Boating* June 130 The correct name of the two-masted sail plan in Merry Weather is the pirogue rig; not a cat-ketch rig as it is sometimes erroneously named. 1954 *Sun* (Baltimore) 15 Mar. 9/1 French-speaking trappers paddle pirogues through backwater bayous of Louisiana in search of muskrat. 1963 W. Gard in H. S. Bell *Petroleum Transportation Handbk.* iii. 6 In swamp terrain the crews, working out of pirogues where there is standing water, are usually sent ahead of the dipper dredge to drop the trees. 1973 *Times* 5 Mar. (Mauritius Suppl.) p. i/5 The whale..20 reeking terror-filled yards from our frail *pirogue*.

piroot (pəirū·t), *v.* *U.S. dial.* [Alteration of Pirouette *v.*, prob. under influence of Root *v.*[2]] *intr.* To move listlessly or aimlessly; also, to snoop. Hence **piroo·ting** *ppl. a.*
1863 S. C. Massett *Drifting About* 242 The streets were almost impassable from the mud and slush and.. the 'ladies'..would find it impossible to 'piroot' thither. 1866 C. H. Smith *Bill Arp* 116 For four years the Confederate Horse-stealing Cavalry have been pirooting around, preparing themselves for the frightful struggle that is to come. 1910 W. M. Raine *Bucky O'Connor* 30 I've been pirootin' around this country, boy and man, for fifteen years. 1958 'W. Henry' *Seven Men at Mimbres Springs* (1960) xiii. 156 Oh God A'mighty—kids, squaws, dogs, old people, pack mules, cookpots—Jesus—the whole pirooting kit and kaboodle of them come down to set up camp and see the fun. 1961 J. F. Dobie in *Webster* s.v., Went pirooting into a cave one day.

piroplasm (paiə·roplæz'm). *Biol.* † Also in mod.L. form **piropla·sma** (pl. **-plasmata**). [f. L. *pir-um* pear + -o + Plasm, Plasma.] A protozoan of the suborder Piroplasmidea of sporozoans, which comprises species parasitic in red blood cells and transmitted by ticks.
[1895 W. H. Patton in *Amer. Naturalist* XXIX. 498 The name of the southern or splenic cattle-fever parasite. —The generic name [*sc. Pyrosoma*] given by Drs. Smith and Kilborne, having been previously used in Zoology, must be dropped. I propose the name *Piroplasma* to replace it.] 1901 *Vet. Jrnl.* IV. 50 A strange fact, showing that the form of the parasite varies in its stages of evolution, was that almost all the piroplasms were pear-shaped and bigeminate. 1908 *Practitioner* Feb. 228 Rocky Mountain Spotted Fever. Ricketts has re-investigated this disease, in which..Wilson and Chowning claimed to have discovered piroplasmata, and to have shown that it is transmitted by a tick. 1934 T. W. M. Cameron *Internal Parasites of Domestic Animals* ii. 45 All the piroplasms are tick-carried and consequently the main lines of prevention consist in tick eradication. 1949 C. A. Hoare *Handbk. Med. Protozool.* iii. 37 A piroplasm of the genus *Theileria* is the cause of East Coast fever... Other forms of piroplasmosis in cattle are caused by species of the genus *Babesia*. 1974 O. W. Olsen *Animal Parasites* (ed. 3) ii. 164/1 Upon entering the erythrocytes, they [*sc.* the parasites] retain their same general appearance and are generally known as piroplasms.
Hence **pi:roplasmo·sis,** any of a group of diseases of mammals, esp. red-water of cattle, caused by infestation of the blood with piroplasms.
1901 *Vet. Jrnl.* IV. 49 (*heading*) Canine piroplasmosis. 1903 *Jrnl. Compar. Path. & Therapeutics* XVI. 312 (*heading*) Pyroplasmosis of the donkey. 1948 U. F. Richardson *Vet. Protozool.* iv. 81 As a rule piroplasmoses occur as enzootic diseases in which young animals contract symptomless infections and recover. 1949 [see above]. 1955 [see *East D. 1 b]. 1960 *Farmer & Stockbreeder* 15 Mar. 131/1 Ticks are the carriers of the organisms which when they are inoculated into the blood, attack the red cells and cause them to break down and it is the colouring matter of these cells that gives colour to the urine, and to the disease known as redwater or bovine piroplasmosis. 1974 *Nature* 13 Dec. 532/2 Vaccination against cattle lungworm, canine hookworm and certain kinds of piroplasmosis has been achieved but the vaccines used, irradiated larvae or infected blood, are unacceptable in the field of human medicine.

‖**piroshki** (pirǫ·ʃki), *sb. pl.* Also **pirotchki, pirozhki, pyrochki.** Occas. in sing. **pirosho·k** (in quot. spelt **pirozsok**). [a. Russ. *pirozhkí* pl. of *pirozhók*, dim. of *piróg* (*pirog).] Small patties.
1912 R. K. Wood *Tourist's Russia* ii. 34 To taste pirozhki at their best, one must go to Philipov's in St. Petersburg or Moscow. 1933 P. & L. G. Esmonde *von Schumacher's Cook's Tour of European Kitchens* vi. 105 After this..came Bortch, a soup of beetroot and cabbage, accompanied by Pyrochki—a Russian stuffed pastry. 1935 M. Morphy *Recipes of All Nations* 429 Fish piroshki..are made in the same manner as game piroshki, but with a filling of cooked fish, hard-boiled eggs and rice. 1939 *Vogue's Cookery Bk.* xii. 227 *Piroshki*... These are the little meat-filled patties that are the traditional Russian accompaniment to soup. 1943 E. M. Almedingen *Frossia* ix. 347 Yesterday I baked '*pirozhki*' with cabbage and onions. I made eight, and sixteen people came.., and it meant half a *pirozsok* for each. 1951 L. W. Leonard *Jewish Cookery* xii. 163 Pirogen..are made just like Piroshki but much larger. 1963 V. Nabokov *Gift* i. 35 He bought some piroshki (one with meat, another with cabbage, a third with tapioca, a fourth with rice, a fifth... could not afford a fifth) in a Russian foodshop. 1965 J. B. Priestley *Lost Empires* ii. iii. 125 Russian things like.. those tiny meat and fish pasties called *piroshki*. 1972 N. Froud *Some of our Best Recipes are Jewish* 50 Roll out pastry, fill pirozhki, brush with beaten egg, and bake..in oven preheated to 230° C. 1973 S. Skipwith *Eat Russian* viii. 154 Pirozhki are small patties served with soup, as appetisers, or offered with tea or coffee at any time. A pirog is a larger, circular or rectangular, fairly flat pie, with pastry top and bottom. Both are baked with a variety of fillings. 1974 A. Williams *Gentleman Traitor* vii. 101 A dinner of borscht, pirotchki and beetroot. 1977 J. Wambaugh *Black Marble* (1978) x. 218 Valnikov held a paper plate stacked with golden pastries and said, '*Piroshki*. They're very light and filled with cheese or meat. My brother usually makes them both ways.'

pirouette, *sb.* Add: **3.** *Mus.* A form of mouth-piece used with a shawm, rackett, or similar reed instrument (see quot. 1976).
1891 *Descr. Catal. Mus. Instruments R. Military Exhib., London, 1890* iv. 64 The reeds used in these early times were generally rather hard and difficult to manage. To render them more manageable they were placed in a sort of case, called *pirouette*, which covered the lower part of the reed. 1911 *Encycl. Brit.* XXII. 780/1 The rackett is played by means of a large double reed placed within a *pirouette* or cap. 1961 A. Baines *Mus. Instruments* ix. 233 The European shawmist presses the lips to a wooden 'pirouette'..which permits lip-control without appreciably reducing the reed's amplitude of vibration. 1968 *New Oxf. Hist. Music* IV. xiii. 737 The reed of the tenor and smaller forms was controlled by a device called a *pirouette*. 1976 D. Munrow *Instruments Middle Ages & Renaissance* 40/2 The pirouette (also used on the renaissance rackett) was a funnel-shaped reed-shield against which the player could press his lips whilst taking the projecting part of the reed into the mouth.

pirouettist (piru,e·tist). [f. Pirouette *v.* + -ist.] = Pirouetter.
1889 G. B. Shaw *London Music 1888–89* (1937) 224 The unappreciated pirouettists and entrechatists looked on indignantly. 1926 W. J. Locke *Old Bridge* ii. vii. 119 He may chance to be a mechanical jazz pirouettist or a financial oracle.

Pirquet (piə·ɹke). *Med.* The name of Baron C. P. *Pirquet* von Cesenatico (1874–1929), Viennese pædiatrician, used *attrib.* and † in the possessive to designate a skin test for tuberculosis that he devised in 1907.
1908 *Lancet* 18 Apr. 1183/1 Pirquet's reaction sometimes showed itself five hours after the inoculation but as a rule, the effect was produced in 24 hours and lasted from four to six days. 1927 E. R. Baldwin et al. *Tuberculosis* xii. 213 The von Pirquet scratch test is less delicate than the intracutaneous method and requires repetition to confirm a negative result. 1952 B. R. Clarke *Causes & Prevention of Tuberculosis* ii. 14 The simplicity of the Pirquet test is an advantage in sparsely populated regions, only one reading being necessary before B.C.G. vaccination. 1970 *New Yorker* 28 Nov. 44/1 Irochka had had a positive reaction to the Pirquet test and had been sent for an X-ray.

pirssonite (pə̄·ɹs-, piə·ɹsənəit). *Min.* [f. the name of L. V. *Pirsson* (1860–1919), U.S. geologist: see -ite[1].] A hydrated carbonate of sodium and calcium, $Na_2Ca(CO_3)_2.2H_2O$, occurring as brittle, colourless to white, orthorhombic crystals that are pyroelectric.
1896 J. H. Pratt in *Amer. Jrnl. Sci.* CLII. 130 The author takes pleasure in naming this mineral *pirssonite*, in honor of his friend and associate, Prof. L. V. Pirsson, of the Sheffield Scientific School. 1933 *Jrnl. Chem. Soc.* 1162 Gaylussite is in equilibrium with pirssonite at 37–40°, the exact temperature depending on the concentration of the sodium carbonate solution with which they are in contact. 1975 *Nature* 13 Mar. 128/2 Large (>2 cm), apparently diagenetic crystals of pirssonite and gaylussite, oriented by their 'c' axes at high angles to bedding, are common in parts of these cores.

Pisan (pī·zăn), *sb.* and *a.* [ad. It. *Pisano* f. L. *Pisān-us* of or belonging to *Pisæ*, a city in Etruria.] **A.** *sb.* A native or inhabitant of Pisa, a city in central Italy situated on the river Arno.
1613 Parsons & Fitz-Herbert *Suppl. Discuss. M. D. Barlowe's Answere* ii. 81 The wars betwene the Pisans and the Genoueses by sea. 1705 Addison *Remarks Italy* 18 Their Fleet, that formerly gain'd so many Victories over the Saracens, Pisans, Venetians, [etc.]. 1813 J. Forsyth *Remarks Excursion Italy* 12 Many Pisans, however, are of the old opinion. 1863 'Geo. Eliot' *Romola* I. i. viii. 146 'Pisans false, Florentines blind'—the second half of that proverb will hold no longer. 1869 'Mark Twain' *Innoc. Abr.* xxiv. 252 To be buried in such ground was regarded by the ancient Pisans as..potent for salvation. 1875 K. O'Clery *Hist. Italian Revolution* i. 30 The island [*sc.* Sardinia] was conquered and assigned to the Pisans as a fief of the Holy See. 1934 A. Huxley *Beyond Mexique Bay* 280 Like the Pisans, the Aztecs had the wit to leave a wide open space all round the monument. 1973 *Country Life* 16 Aug. 450/2 The Sards have always hated

the sea, from whence came all their invaders, Phoenicians, Greeks, Romans, Byzantines, Pisans, Genoese, [etc.]. **1975** *Daily Tel.* (Colour Suppl.) 14 Mar. 10/4 Pisans do not believe that successive ministers and commissions will ever do anything to arrest the inevitable. One day the Tower *will* flop down.

B. *adj.* Of or pertaining to Pisa. spec. *Pisan assistance*, assistance rendered too late to be effective.

1813 J. Forsyth *Remarks Excursion Italy* 36 The Pisan chains hang like a fair trophy on the foreign bank of Genoa. **1869** Browning *Ring & Bk.* IV. xii. 210 You and your pleas and proofs were what folks call Pisan assistance, aid that comes too late. **1869** 'Mark Twain' *Innoc. Abr.* xxiv. 252 A Pisan antiquarian gave me an ancient tear-jug. **1904** J. M. Stone *Reformation & Renaissance* ii. 62 The Pisan Pope and his followers were already there, and for a time it seemed as though they might carry all before them. **1936** A. W. Clapham *Romanesque Archit.* iii. 30 Southern Italy.., touched here with Lombard and there with Pisan influence. **1951** A. R. Lewis *Naval Power & Trade in Mediterranean* vi. 224 Genoese and Pisan fleets helped open up the Rhone Valley route. **1973** *Country Life* 16 Aug. 451/2 Oretelli, where the Parish Church is Pisan of the 13th century. **1975** *Daily Tel.* (Colour Suppl.) 14 Mar. 10/4, I returned with a Pisan friend to the Campo dei Miracoli to have a final look at the Tower.

pisatin (pəi·s-, pi·sătin, -z-). *Biochem.* [f. the taxonomic name *Pis(um s)at(ivum* (f. L. *pisum* pea + *sativus* (see Sative *a.*)) + -in[1].] A fungitoxic phytoalexin produced by the pea plant, which has been isolated as a crystalline heterocyclic compound, $C_{17}H_{14}O_6$.

1960 Cruickshank & Perrin in *Nature* 27 Aug. 800/1 We propose that the trivial name of this compound should be 'pisatin' after the host plant from which it was originally isolated. **1964** [see *phaseollin]. **1967** R. K. S. Wood *Physiol. Plant Path.* xiv. 496 None of a variety of bacteria which have been tested stimulates formation of pisatin by pea pods. **1972** S. A. J. Tarr *Princ. Plant Path.* xiv. 268 Some of these antifungal substances are perhaps present in low concentrations in normal tissue, their formation being intensified by infection or some other stimulus. Pisatin, for example, is formed when dilute solutions of certain metallic salts..are placed on pea pod endocarp tissue.

Piscean (pəi·siăn), *a.* and *sb.* Also **Piscian.** [f. Pisce(s + -an.] **A.** *adj.* Of or pertaining to Pisces, the twelfth sign of the Zodiac; characteristic of a person born under Pisces. **B.** *sb.* A person born under Pisces.

1924 C. E. O. Carter *Con. Encycl. Psychol. Astrol.* 38 It must..be observed that in some people the Piscian and Neptunian charity seems entirely lacking. **1925** —— *Princ. Astrol.* iv. 74 Pisceans are commonly jovial and convivial, and often make entertaining companions. **1940** R. Gleadow *Astrol. in Everyday Life* II. 162 Most Pisceans..are not particularly dashing; yet those who have Mars rising in Pisces may be quite audacious. *Ibid.* xiv. 250 In astrological circles Mr. Micawber is notoriously Piscean. *a* **1963** L. MacNeice *Astrol.* (1964) iii. 105 Pisceans are very lovable people because they are very loving. **1969** 'V. Packer' *Don't rely on Gemini* (1970) i. 8 Leos are lionlike and Pisceans are mystical. **1972** D. Lees *Zodiac* 28 There must be thousands of Pisceans who can't swim. **1976** *Woman* 22 May 61/1 All is quiet on the Piscean front with both your ruling planets playing a waiting game. **1978** *TV Times* 28 Jan. 68/3 Pam is one of those lucky Pisceans who will enjoy plenty of romance this year.

Pisces. Add: Also with pronunc. (pəi·sīz). **1. b.** A person born under the zodiacal sign Pisces. Also *attrib.* or as *adj.*

1924 C. E. O. Carter *Conc. Encycl. Psychol. Astrol.* 142 Cancer and Pisces people are often retiring and shy among strangers. **1936** 'J. Tey' *Shilling for Candles* vii. 64 One does not expect a Pisces person to have either the vision or the faith. **1969** 'V. Packer' *Don't rely on Gemini* (1970) i. 1 Would a Pisces get along with a Capricorn? *Ibid.* xviii. 159 We're both Pisces like Elizabeth Taylor. **1972** D. Bloodworth *Any Number can Play* xii. 101 You're Pisces, darling... The lucky things for Pisces people are silver, bloodstone, and number seven. **1973** L. Meynell *Fatal Flaw* ii. 20 The fellow turned out to be Pisces. An unreliable lot the Pisces. **1976** *Woman* 1 May 56/1 Intuition and sixth sense are passwords for Pisces, and this week you should virtually live by them, for safety's sake.

piscicide (pi·si-, pi·skisəid). [f. L. *pisci-s* fish + -cide.] **a.** The killing of fish.

1963 *Times* 21 Aug. 5, I was unable to detect any evidence of mass piscicide in the Sonic's track. The only dead fish I saw was a 6 in. specimen.

b. A substance that kills fish.

1964 Lennon & Walker *Laboratories & Methods for Screening Fish-Control Chemicals* 1/2 Ample justification for research on selective piscicides is contained in fishery literature. **1965** *N.Y. Fish & Game Jrnl.* XII. 99 In larger ponds and lakes, the need for selective piscicides is even greater. **1976** *Nature* 25 Mar. 374/2 Under Control of Undesirable Species (Chapter 18), we have a 2-page description..of the new piscicide antimycin.

pisco (pi·sko). [Peruvian, f. *Pisco* the name of a port of Peru.] A white brandy made in Peru from muscat grapes. Also *attrib.* and *Comb.*, as *pisco Collins* [*Collins[2]], *sour*.

1849 H. Vizetelly *Four Months among Gold-Finders in Alta California* 30 On our way he pointed out the guard-house..the distillery house, where the famous pisco is made. **1873** A. S. Evans *À la California* 328 The company all together, we propose a taste of fragrant *pisco* (Peruvian white brandy). **1924** R. Clements *Gipsy of Horn* viii. 143 The old man laid in a stock of 'pisco', a cheap, fiery spirit, very popular in Peru. **1961** J. B. Priestley *Saturn over Water* v. 55 The expensive bars where the double martinis and *pisco* sours were being served. *Ibid.* vi. 88, I just couldn't see steady Joe Farne.. going off on a great *pisco* blind. **1962** N. Maxwell *Witch-Doctor's Apprentice* ii. 16, I asked the waiter to bring me a pisco collins. 'Without sugar as usual, señorita?' he asked. **1971** D. Wallis *Bad Luck Girl* i. 13 Down the Pacific coast they think no more of selling you a fix than pouring you a pisco. **1973** K. Benton *Craig & Jaguar* iv. 36 'You must try one of the Club's pisco sours.' Craig sipped the sticky mixture of cane-spirit and fresh lime juice through the layers of white of egg foam.

pisé. Add: Also **pisée. a.** (Later examples, chiefly with ref. to Australia.) Also, *pisé de terre*.

1919 C. Williams-Ellis *Building in Cob, Pisé, Chalk & Clay* 28 'Pisé de Terre', 'Chalk Compost', and 'Cob' are three alternative forms of construction. **1936** I. L. Idriess *Cattle King* xx. 183 Homesteads were of roughly gathered stone or pisé, of axe hewn slabs or sheets of bark. **1946** B. James in Murdoch & Drake-Brockman *Austral. Short Stories* (1951) 250 The walls had to be very thick, and that meant more pisé to be mixed, and lifted and rammed. **1960** K. M. Kenyon *Archaeol. in Holy Land* iii. 60 The edges of the pits are revetted by slight walls of pisée and stone. **1977** *36 Home Handyman Projects* (Austral. Home Jrnl.) 97/3 Pise de Terre—wall construction of clay or earth—a formwork is made and the earth and clay rammed in firmly. **1978** *Jrnl. R. Soc. Arts* CXXVI. 586/1 Cob..provided good walling; so did damp earth rammed into moulds, and known as *pisé*.

b. *pisé building* (later examples), *earth*, *terre*.

1919 C. Williams-Ellis *Building in Cob, Pisé, Chalk & Clay* ii. 57 Pliny gives an excellent account of pisé-building in his *Natural History*. **1919** in *Ibid.* 74 These iron bars become so tightly jammed when surrounded by the compact pisé earth, that much labour and risk of injury to the work is incurred in extricating them. **1946** B. James in Murdoch & Drake-Brockman *Austral. Short Stories* (1951) 250 The pisé earth had to be dug,.. mixed and kneaded with water, shovelled into the frames, rammed thoroughly and then left to set. **1965** *Austral. Encycl.* III. 327 Another interesting example of pisé building was the old Forbes, N.S.W., police barracks. **1971** *Country Life* 7 Oct. 941/1 Clough was commissioned.. 'to write a practical book on *Pisée Terre* (rammed earth) building..to cheapen and expedite rural cottage building in the twenties'.

‖**pishachi** (piʃa·tʃi). *India.* Also **piśaca, pisachi, pishasha,** and numerous other variants. [ad. Skr. *piśāca* (masc.), *piśācī* (fem.), a demon.] A demon or devil. Also *attrib.*

The forms in *-a* properly refer to male devils, and those in *-i* to female.

1807 F. Buchanan *Journey from Madras* III. xiv. 17 They believe, that such men as die accidental deaths become *Pysāchi*, or evil spirits, and are exceedingly troublesome, by making extraordinary noises in families, and occasioning fits, and other diseases, especially in women. **1816** *Asiatic Jrnl.* II. 367/1 *Whirlwinds*..at the end of March and the beginning of April..carry dust and light things along with them, and are called by the natives *peshashes*, or devils. **1819** *Trans. Bombay Lit. Soc.* I. 219 Beneath him..is a small squat figure, apparently a *peisach* or demon. These demons or *peisaches* are the usual attendants of Shiva. **1827** J. C. & A. W. Hare *Guesses at Truth* (ser. 1) 12 As a little girl was playing round me one day with her white frock over her head, I laughingly called her *Pishashee*, the Indian name, I believe, for their white devil. **1837** J. C. Maitland, *Lett. from Madras* (1843) 107 She used to go out and howl so that the servants were afraid to come near her, saying she made 'one pishashi (devil) noise'. **1885** G. C. Whitworth *Anglo-Indian Dict.* 252/1 *Piśācha*.., the name of a class of spirits always imagined as fierce and malignant. **1886** Yule & Burnell *Hobson-Jobson* 540/1 Pisachee. **1917** L. H. Gray *Mythol. All Races* VI. 67 With the Rakṣases in later literature rank the Piśācas as foes of the fathers. **1920** *Encycl. Relig. & Ethics* X. 43/2 In modern India a *piśācha* is a kind of ghoul, usually the ghost of some one who has died an unnatural death, or for whom the requisite funeral rites have not been performed... In S. India the small circular storms, called 'devils' by Europeans, are called *piśāchis*, or 'she-ghouls'. **1924** R. E. Enthoven *Folklore of Bombay* iv. 148 *Bhuts* and *pishachas*—ghosts, male and female—can be prevented from doing harm by recourse to certain processes. **1927** S. Ketkar tr. *Winternitz's Hist. Indian Lit.* I. 133 Very numerous, too, are the incantations which are directed against whole classes of demons.., especially against the *Piśācas* (goblins) and *Rākṣasas* (devils). **1952** E. Sykes *Everyman's Dict. Non-Classical Mythol.* 171 *Pishashas*, in Vedic myth malignant woodland spirits, who disliked travellers, and especially pregnant women. **1968** B. Walker *Hindu World* II. 214 *Piśācha*, a race of people classed in the Vedas as lower than the rākshasas (ogres), and amongst the most vile and noxious of beings. **1973** J. Dowson *Classical Dict. Hindu Mythol.* 234 *Piśāchas* (mas.), *Piśāchī* (fem.), fiends, evil spirits, placed by the Vedas as lower than Rākshasas. **1977** M. & J. Stutley *Dict. Hinduism* 226/1 *Piśāca(s)*, flesh-eating demonic beings. *Ibid.* 226/2 *Piśācī*, a she-devil.

pishamin. (Earlier Amer. example.)

1766 J. Bartram *Jrnl.* 14 Jan. 36 in W. Stork *Acct. E. Florida* (ed. 2), The lower rich ground produceth gledistia, pishamins, cephalanthus, ash, cypress and cornu femina.

pisha paysha (pi·ʃă pĕi·ʃă). Also **pisha pasha.** [App. a corruption of *pitch* (or *peace*) *and patience*.] A Jewish card game resembling patience, played by two persons, in which the cards are taken as they come from the pack, the object being to arrange them in an upward or downward sequence until the pack is exhausted, when the player who has the fewer cards in his hand is declared the winner.

1928 *Weekly Dispatch* 27 May 13/2 Faded photographs of the Yiddish stars of yesterday hung on the walls; most of the people looked up when we came in, but two heavy, blue-chinned fellows continued their game of *pisha pasha* and another smiled a greeting across the top of a Jewish evening paper. **1968** L. Rosten *Joys of Yiddish* 288, I was taught to play *pisha paysha* by my father, when I was six or seven.

pisher (pi·ʃəɹ). *U.S. slang.* [a. Yiddish *pisher* Pisser, f. G. *pissen*: see Piss *v.*] A bed-wetter; also in extended uses (see quots.). Also *attrib.* or as *adj.*

1942 in *Amer. Speech* (1943) XVIII. 46 Call me pisher. **1943** *Ibid.*, The phrase 'to call someone pisher' connotes mild, tolerant, ineffectual reproof or disproportionately lax punishment. A typical context might be something like this: 'So what did they do to Flynn for putting public employees to work on his Mahopac estate and using government property? They called him pisher!' **1958** B. Malamud *Magic Barrel* 87 He bought.. this pisher grocery in a dead neighbourhood where he didn't have a chance. **1968** L. Rosten *Joys of Yiddish* 289 Literally, a *pisher* is one who urinates; but that is a far cry from present and popular usage. 'He's a mere *pisher*,' means 'He's very young,' or 'He's still wet behind the ears.'.. 'He's just a *pisher*,' means 'He's a nobody,' has no influence. **1970** L. M. Feinsilver *Taste of Yiddish* i. 61 'She still has two *pishers* at home' is a common colloquialism that makes its point: she has two offspring still in diapers, or two preschoolers. **1972** J. Caine *Hamlet, My Boy* xi. 161 First, they didn't wait to call you *pisher*; they just filled you up with bullet-holes like a matzo. **1978** E. Tidyman *Table Stakes* II. vii. 312 Then the marriage. Now that was *really* smart! Who could call him pisher now, with the Jewish princess on his arm? **1978** R. Doliner *On Edge* (1979) v. 83 'I was a kid...A pisher.' 'Pisher,' the Vice-President said. 'One who wets one's pants.' **1979** B. Malamud *Dubin's Lives* ix. 359, I lived on cases involving small finaglers and found myself engaged in pisher dishonesties.

pishogue. Add: Also **pishog, pisherogue.** (Further examples.) Also, a fairy, a witch.

1829 G. Griffin *Collegians* I. xi. 231 Mr. Euright's dairyman..made a *pishog* and took away our butter. **1906** Kipling *Puck of Pook's Hill* 10 Little people, pishogues, leprechauns. **1937** C. M. Arensberg *Irish Countryman* vi. 212 All that the Church condemns in the 'pisherogues', they [*sc.* young people] also condemn. **1957** E. E. Evans *Irish Folk Ways* xxi. 296 Most of the pishrogues relate to fairies and to trees, wells and stones. **1960** *20th Cent.* July 51 She had denied it. 'No, I am not a pishogue.' Yet she was aware how easily one might become a fairy. **1961** 'F. O'Brien' *Hard Life* ii. 17 Well now, Mrs Crothy, are these the two pishrogues out of the storm?

Pisistratid (pəisi·strătid), *sb.* and *a.* Also **Peisistratid, 8 Pysistratid.** [ad. L. *Pisistratidae*, Gr. *Πεισιστρατίδαι*, the name given to Hippias and Hipparchus, sons of Pisistratus, tyrant of Athens in the 6th cent. B.C.] **A.** *sb.* (Pl. -idae, -ids.) A member or supporter of the family of Pisistratus. Chiefly in *pl.*

1709 I. Littlebury tr. *Herodotus' Hist.* II. v. 57 The Corinthians would be the first of all People to regret the Pisistratides. **1776** W. Ellis tr. *Aristotle's Treat. Govt.* v. xi. 295 The Pyramids of Egypt are a proof of this,..and the Temple of Jupiter Olympus, built by the Pysistratidæ, and the Works of Polycrates at Samos; for all these produced one end, the keeping the People poor. **1808** W. Mitford *Hist. Greece* I. xii. 561 He married Agrristé, niece of Cleisthenes, chief of the Alcmæonid family, and leader of the party that expelled the Peisistratids. **1848** *Eton School Mag.* III. 114 It thus appears how irretrievably the government of the Pisistratids had injured the Athenian character. **1885** B. Jowett tr. *Aristotle's Politics* I. v. 184 Third in duration was the rule of the Peisistratidae at Athens, but it was interrupted. **1900** J. B. Bury *Hist. Greece* v. 206 The Pisistratids cultivated the friendship of Sparta. **1922** P. N. Ure *Origin of Tyranny* i. 14 This part of the tyrants' policy is noticed by Aristotle, who quotes..the building of the temple of Olympian Zeus at Athens by the Peisistratids. *Ibid.* ii. 33 The Philaidae, of whose rivalry with the Peisistratidae there will be occasion to speak later. **1972** R. Meiggs *Athenian Empire* i. 19 The expulsion of the Pisistratids in 510.

B. *adj.* Of or pertaining to Pisistratus or his family; *spec.* of or pertaining to the revision of the Homeric poems attributed to Pisistratus. Also **Pisistrate·an** *a.*

1846 J. S. Mill in *Edin. Rev.* LXXXIV. 363 Mr. Grote ..rejects the Pisistratean hypothesis. **1965** A. R. Burn *Traveller's Hist. Greece* vii. 114 Hipparchos, the chief Peisistratid patron of poets and artists, was stabbed to death. **1968** V. Ehrenberg *From Solon to Socrates* iv. 82 The story of a 'Peisistratid redaction' of the two epics can hardly be true.

‖**piskun** (pi·skŏn). Also **pishkun.** [ad. Blackfoot *piskăni*.] An American Indian trap for buffalo, consisting of two converging lines of rock piles, a V-shaped natural canyon, or a timbered causeway leading to a steep drop, often with an enclosure or corral at the foot, over which the buffalo were stampeded.

1892 *Scribner's Mag.* Sept. 281 In the later days of the *piskŭn*, the man who brought the buffalo went to them on horseback, riding a white horse. **1892** G. B. Grinnell *Blackfoot Lodge Tales* (1893) 230 The pis'kuns of the Sik'-si-kau, or Blackfoot tribe, differed in some particulars from those constructed by the Bloods and the Piegans, who live further to the south, nearer to the mountains, and so in a country which is rougher and more broken. The Sik'-si-kan built their pis'kuns like the Crees, on level ground and usually near timber. A large pen or corral was made of heavy logs about eight feet high. On the side where the wings of the chute come together, a bridge, or causeway, was built, sloping gently up from the prairie to the walls of the corral, which at this point were cut away to the height of the bridge above the ground,—here about four feet,—so that the animals running up the causeway could jump down into the corral. **1929** E. D. Branch *Hunting of Buffalo* ii. 35 The *piskun* was surer and safer than the human trap; it was an enclosed pen into which the buffalo were driven. **1943** J. K. Howard *Montana* 23 Often buffalo were driven over cliffs, the 'buffalo runs' or 'pishkuns' under which Montanans still find rich hoards of arrowheads and other Indian implements. **1949** *Jrnl. Washington Acad. Sci.* XXXIX. 357/2 The North Blackfoot, who hunted to a considerable extent on relatively level ground, built their piskun like that of the Cree Indians in the form of a corral with rising timbered causeway leading up to the entrance of the corral from which there was a sheer drop of about 4 feet into the corral. The Piegan and Blood, living in more broken country nearer the mountains, drove the bison over cliffs. *Ibid.* 360/1 We can date the last bison drive of the Blackfoot at about the year 1872. This was a full century after Mathew Cocking's first description of the use of the piskun by Indians of the northwestern plains. **1952** J. K. Howard *Strange Empire* 294 Nevertheless some native methods of killing buffalo were wasteful. Such were the *piskuns* and pounds, use of which, however, was generally abandoned by the Indians some time before the herds disappeared.

Pismo (pi·zmo). Also **pismo.** The name of *Pismo* Beach, California, used *attrib.* in **Pismo clam,** a large, thick-shelled, edible clam, *Tivela stultorum*, belonging to the family Veneridæ and found on the south-west coast of North America.

1913 *Calif. Fish & Game Comm. Fish Bull.* I. 27 The Pismo clam..flourishes in open sandy beaches. **1923** *Ibid.* VII. 5 Commercially the Pismo clam ranks first in importance in California among clams and third among all the mollusks. **1949** *Natural Hist.* June 252/1 Five minutes of barefoot beach scratching had uncovered half a sack of four-inch Pismo clams. **1970** B. H. McConnaughey *Introd. Marine Biol.* viii. 227/2 A large, heavy-shelled clam, the pismo clam, occurs on California beaches at the lowest intertidal and subtidal levels.

pisolite. Add: (Further examples.)

Quot. 1708 in Dict. belongs to the next sense.

1931 S. J. Shand *Study of Rocks* xi. 153 Oolite and pisolite are limestones built up of little spheroidal bodies resembling the roe of fishes or heaps of peas. **1962** Read & Watson *Introd. Geol.* I. v. 266 Other chemical limestones are not of much account; they include deposits from calcareous springs, such as pisolite, tufa and travertine.

b. Add to def.: similar to an oolith but larger (in mod. use applied to grains of diameter 2 mm. or more). (Earlier and later examples.)

1708 [in Dict., sense a]. **1788** [see *oolith]. **1893** *Q. Jrnl. Geol. Soc.* XLIX. 127 The series for some distance above the typical Pea-grit..contains here and there aggregations of brown pisolites. **1925** *Nat. Geogr. Mag.* XLVIII. 313 These spherical bodies are known to geologists as pisolites and to jewelers as cave pearls. **1956** E. W. Heinrich *Microsc. Petrogr.* v. 153 The limonite may be in spongy masses..or in pisolites and concretionary masses. **1974** *Encycl. Brit. Macropædia* XVI. 466/2 Pisolites are similar to oolites and range from two to about ten millimetres (0·4 inch) in diameter.

pisolith (pi·zo-, pəi·soliþ). *Petrol.* Also †**-lithe.** [f. as Pisolite: see -lith.] = Pisolite b in Dict. and Suppl.

1799 [see Pisolite a]. **1926** G. W. Tyrrell *Princ. Petrol.* xiii. 227 Pisoliths are essentially similar to ooliths, but reach much larger sizes, and are generally found in residual deposits. **1938** M. Black *Hatch & Rastall's Petrol. Sedimentary Rocks* (ed. 3) viii. 176 Cave pearls are pisoliths, sometimes of large size, found in the underground waters of limestone caves. **1947** *Jrnl. Sedimentary Petrol.* XVII. 39 (*heading*) Pisoliths and ooliths from some Australian caves and mines. *Ibid.* 43/1 In pools full of pisoliths and ooliths, the larger ones (around 20 mm. long) occur at the top. **1975** *Nature* 20 Nov. 206/1 It lies under a thin but persistent weathered horizon which has a patchy distribution and is characterised by ferruginous pisoliths.

Hence **pisoli·thic** *a.*

1863 Sullivan & O'Reilly *Notes Geol. & Mineral. of Santander & Madrid* I. iv. 91 The hydrocarbonate of zinc also occurs perfectly globular, some specimens being beautifully pisolithic. **1947** *Jrnl. Sedimentary Petrol.* XVII. 39 Calcareous concretions, principally of the pisolithic type.

piss, *v.* Add: **1. b.** (Further examples.) *spec.* To be raining heavily. *to piss in* (*a person's*) *pocket* (Austral.), to ingratiate oneself with, be on very familiar terms with.

1642 G. Torriano *Sel. Italian Proverbs* 19 He who pisseth against the wind, wetteth his shirt. **1670** J. Ray *Coll. Eng. Proverbs* 131 Chi piscia contra il vento si bagna la commiscia, *Ital.* He that pisseth against the wind, wets his shirt. It is to a mans own prejudice, to strive against the stream. **1902** Farmer & Henley *Slang.* V. 215/2 'Piss not against the wind', or 'He that pisseth against the wind wets his shirt'. **1962** J. Baldwin *Another Country* I. ii. 94 Christ, it's pissing out there! **1967** K. Tennant *Tell Morning This* xxx. 283 Soon's they knew you was in with Numismata, they all want to piss in your pocket. **1968** H. C. Rae *Few Small Bones* II. i. 73 Went camping up north..but it pissed the whole time. Come to think of it, I was pretty pissed the whole time myself. **1969** C. Bray *Blossom like Rose* xii. 165, I don't mean to piss in yer pockets, but youse blokes are all right. **1970** E. Pace *Saberlegs* (1971) vi. 58 Putting words on newspaper pages was, if anything, even more ephemeral than intelligence-gathering. As they used to say at Dartmouth, it was all 'pissing in the wind'. **1971** F. Hardy *Outcasts of Foolgaran* 77, I appeared before him many a time when I worked for the Union. If we piss in his pocket, he's just as apt to come our way. **1973** L. Snelling *Heresy* I. vii. 52 I'd like to buy it, but frankly I think you're pissing against the wind... He's a pretty cunning little bugger. **1977** J. Wainwright *Nest of Rats* I. viii. 60 How much time?.. Don't make it hours—otherwise you're pissing in the wind.

c. Const. with various adverbs: *to piss about*, to fool or mess about; to potter about; *to piss down*, to rain heavily; *to piss off*, to go away, depart.

1950 G. Wilson *Brave Company* 172 It fairly pissed down on top of me. **1958** F. Norman *Bang to Rights* 72 So what, I wish you'd piss off. **1960** H. Pinter *Caretaker* I. 14 Piss off, he said... If you don't piss off, he says, I'll kick you all the way to the gate. **1961** Partridge *Dict. Slang* Suppl. 1223/2 *Piss about*, to potter; fritter one's time away; to stall for time. **1970** T. Lewis *Jack's Return Home* 179 Are you coming in? Or do we piss about all day? **1971** W. J. Burley *Guilt Edged* viii. 138 Most of yesterday it was pissing down with rain. **1971** B. W. Aldiss *Soldier Erect* 50 I'll have a drink when I feel like it, and not before. You two piss off if you're so bloody thirsty! **1972** R. Quilty *Tenth Session* 19 Pissing down too, and one o'clock in the morning. **1974** 'J. Fraser' *Wreath of Lords & Ladies* vii. 57 Are we going to piss off home or sit here blabbering all night? **1975** *Sunday Times* (Colour Suppl.) 23 Feb. 26/3 The manager who tried to discipline a man caught with an illicit can of tea was told to piss off. **1977** J. Thomson *Case Closed* ii. 25 Tucker wouldn't come..not with it pissing down with rain. **1977** M. Drabble *Ice Age* I. 59 Oh piss off, Mum, Maureen would reply, amiably.

2. b. *to piss away*, to squander.

1948 D. Ballantyne *Cunninghams* 211 Have to stop pissing away the hard-earned cash though. **1972** P. Knapp *Berengaria Exchange* 18 Dinty had built up a 'pretty good roll'. But as he now says with a shrug, 'I pissed it all away in Paris.' **1975** *Time* 4 Aug. 61/2 'This company is doing a good business,' he says. 'If we can only stop pissing away the profits.'

3. (Later refl. examples.)

1951 Partridge *Dict. Slang.* (ed. 4) 1136/2 Piss oneself laughing. **1969** N. Cohn *AWopBopaLooBop* (1970) ix. 85 The Twist ballooned almost instantaneously from a fad to an industry. The papers pissed themselves. Big money got invested. **1976** A. White *Long Silence* xviii. 147 Otto pissed himself with fear. **1978** J. Barnett *Head of Force* xv. 146 You've pissed yourself..you dirty bastard.

c. *to piss off*, to annoy, irritate, put off, make 'fed up' or depressed (see also *pissed *ppl. a.* 2); *to piss up*, to spoil, ruin, mess up.

1937 E. Pound *Fifth Decad of Cantos* l. 49 Talleyrand stank with shanker And hell pissed up Metternich. **1968** *Southerly* XXVIII. 275 She prefers British eccentrics because they don't expect to be liked. 'I mean,' she says, 'it's their way of pissing people off, isn't it?' **1970** *It* 27 Feb.–13 Mar. 14/1 Wasn't it incredible? I just didn't believe it!.. It really piss me off. **1971** B. Malamud *Tenants* 178 You ought to burn up both of these yourself, Willie, on account of this cat stole your white bitch and pissed up your black book. **1972** *Last Whole Earth Catalog* (Portola Inst.) 9/3 It did piss me off when the dealer let me go for only five hundred and fifty dollars. **1974** K. C. Constantine *Blank Page* 148, I still think it'd piss him off. **1976** 'D. Craig' *Faith, Hope & Death* xvii. 118 Did I let them just unload it because they pissed up a job?.. This was my money that had been lifted. **1977** *Rolling Stone* 16 June 52/2 She may not want to be called 'Queen', but only because she considers herself too young, because she is not out to piss off Aretha Franklin any more than she already has.

piss, *sb.* Add: **1. a.** (Further examples.) Also, the action or an act of urinating.

1916 Joyce *Portrait of Artist* ii. 96 That is horse piss and rotted straw, he thought. It is a good odour to breathe. **1926** T. E. Lawrence *Seven Pillars* (1935) vi. lxxviii. 434 Mifleh brought up the youngest lads of the party, and had them spray the wounds with their piss, as a rude antiseptic. **1974** P. Larkin *High Windows* 32 Groping back to bed after a piss. **1976** *Listener* 18 Mar. 344/1 The words [in a radio play]..were punctuated with belches, giggles, mutterings, reflective hesitations, the repetition of good jokes, and a wonderfully realistic-sounding piss. **1979** N. Freeling *Widow* iii. 11 The hallway smelt... Piss, cabbage, stale sweat.

2. In various fig. phrases, as *on the piss*, engaged in a bout of heavy drinking; *take the piss* (*out of*), to make fun (of), to 'take the mickey' (out of); *piss and wind*, empty talk, bombast; *piss and vinegar*, energy, aggression; also *attrib.*

1922 Joyce *Ulysses* 322 All wind and piss like a tanyard cat. **1942** Berrey & Van den Bark *Amer. Thes. Slang* §240/2 *Animation; spirit; vim*,..piss and vinegar. **1942** *Horizon* Aug. 124 Buggered if I know when he'll be back. Gone on the piss, I shouldn't wonder. **1945** *Penguin New Writing* XXVI. 49 The corporal..sat back in his corner looking a little offended. He thought I was taking the piss. **1958** F. Norman *Bang to Rights* 116 This only made us take the piss out of him the more. **1961** Partridge *Dict. Slang* Suppl. 1223/2 *Piss and wind*, as in 'He's all piss and wind!' Empty talk; unsubstantiated boast(s). **1962** E. Ambler *Light of Day* xii. 244 These policemen are all piss and wind anyway. **1966** M. Spillane *Death Dealers* i. 17 Remember the old days, Tiger? You were young and fast and strong. Full of piss and vinegar. **1968** 'P. Alding' *Circle of Danger* iii. 20 Him not turning up may just mean he's been on the piss. **1969** *Guardian* 13 Feb. 22/4 Mr Eric Lubbock, the Liberal MP for Orpington..said: '..I have heard nothing but piss and wind.' **1971** B. W. Aldiss *Soldier Erect* 49 'Come on, Wally, like —I don't think you ought to take the piss out of the poor sod!' Geordie said. 'He's got his living to earn.' **1974** *Observer* 30 June 22/8 And I don't binge. If I'd gone on the piss every time I missed a cut [*sc.* failed to qualify in a golf tournament] I'd be a raging alcoholic by now. **1975** J. Symons *Three Pipe Problem* xviii. 190 You like to take the piss out of me, don't you? **1976** J. O'Connor *Eleventh Commandment* xiv. 179, I was very happy and went on the piss. **1976** R. Perry *One Good Death* iv. 66 The sarcasm left Collins unmoved. He knew that Pawson was in one of his piss and vinegar moods. **1978** R. Busby *Garvey's Code* xi. 138 Jacko's not such a bad bloke. Full of piss and vinegar and ready to jump for any bugger with braid on his hat. **1978** R. Hill *Pinch of Snuff* xiv. 145 When Hope replied 'He's a Hungarian' he thought at first he was taking the piss. Wield seemed prepared to accept this as a serious contribution, however.

3. *Comb.* (Insert *Comb.* from sense 1 b in Dict.) Also, **piss artist,** a glib person; a person who messes about; a drunkard; **piss-bucket,** a bucket for urinating in; **piss-cutter** *N. Amer.*, someone or something excellent; a clever or crafty person; (see also quot. 1956); **piss-head,** a drunkard; **piss-hole,** (*a*) hole made by urine; (*b*) an unpleasant place; **piss-house,** an outside water closet; a lavatory; also *fig.* (see quots. 1931, 1942); **piss-proud** *a.*, having an erection attributed to a full bladder, esp. upon awakening; **piss-take,** a parody, a send-up (see sense 2); also **piss-taker, piss-taking; piss-tin,** a tin for urinating in.

1975 *Peace News* 11 July 11/1 Most donors seem to be pretty careful. The lists of who's been given money seem to bear this out—it's usually the right-on projects which have most and the con merchants and piss artists often go without. **1977** *Sounds* 9 July 36/4, I am appealing to anybody who knows John and Murdoch of Erkshire Scotland. You know, those piss-artists, protozoans who wrote that letter about a rock band classification. **1977** *Custom Car* Nov. 5/2, I refer to the auto/driver self-destruct mechanism known as 'booze'. A piss artist behind the wheel of a 1935 Austin Seven was a killer. **1977** *Private Eye* 10 Nov. 22/1 (Advt.), Malcolm Derek Winn. Photographer, traveller, piss-artist. Whereabouts known? Box 1215. **1973** J. Seabrook *Loneliness* 104 The workhouses were terrible places. At Newark, you slept on a stone floor, and in the middle there was a piss-bucket. **1942** Berrey & Van den Bark *Amer. Thes. Slang* §29/2 *Something excellent*,..piss-cutter. *Ibid.* §432/2 *Capable person; expert*,..piss-cutter. **1956** *Amer. Speech* XXXI. 192 His [*sc.* a marine's] garrison cap is a *pisscutter* (also used as a cynical description, i.e., 'He's a pisscutter, he is!'). **1968** E. R. Buckler *Ox Bells & Fireflies* xv. 206 Gus Jordan's got a new rowboat. It's a real pisscutter! **1974** D. Sears *Lark in Clear Air* viii. 97 I'd send some [beer] along with you but that old piss-cutter of a Heeney would drink'm all before you cut the froth on the first bottle. **1977** *Maledicta* Summer 13 A clever person is sometimes called a piss-cutter. **1961** Patridge *Dict. Slang* Suppl. 1223/2 *Piss-head*, an habitually heavy drinker. **1968** *Landfall* XXII. 50 My old man was a piss-head too. *c* **1932** Dylan Thomas *Lett.* (1966) 4 My eyes are two piss-holes in the sand. **1973** R. Busby *Pattern of Violence* ii. 24 'How's tricks, Lucky?'..'Be better when I'm out of this piss hole—no offence, gents.' **1974** R. Gadney *Something Worth Fighting For* iv. 33 Let's get out of this pisshole. **1931** *Amer. Speech* VII. 112 *Piss-house*, n., the police station. **1942** Berrey & Van den Bark *Amer. Thes. Slang* §84/11 *Toilet*,..piss-house. *Ibid.* §466/10 *Police station or jail*,..pisshouse. **1973** W. H. Canaway *Harry doing Good* II. v. 187 Come outside for a leak. Blow your noses on the way to the piss-house. **1974** H. MacInnes *Climb to Lost World* xi. 199 Next morning the Wall was wetter than ever... 'Never seen it so wet here—what do you think?' 'Looks like a piss-house wall to me.' **1788** Grose *Dict. Vulg. T.* (ed. 2), *Piss-proud*, having a false erection. That old fellow thought he had an erection, but his — was only piss-proud; said of any old fellow who marries a young wife. **1868** *Index Expurgatorius of Martial* 88 Maevius who while sleeping only gets A piss-proud stand that melts away on waking. **1932** Auden *Orators* III. 104 That piss-proud prophet. **1977** *Spare Rib* July 49/1 It's a bit of a pisstake, sending up the whole bi-sexuality thing. **1976** *New Society* 20 May 408/2 'What's funny about a jeweller?..He's a piss taker. **1971** *It* 9–23 Sept. 21/1 The subjects for piss-taking run the gamut from the chairboard executive's life..to the Amerikanjudicial system. **1974** H. MacInnes *Climb to Lost World* vii. 104 If it was imperative, I used my piss-tin, conveniently placed at arm's reach on the mud.

4. Used quasi-adverbially in sense 'very, excessively, to an extremely undesirable degree', as *piss-elegant*, *-poor*, *-rotten*, *-wet*, etc., adjs.

1940 E. Pound *Cantos* lxix. 174 Bingham, Carrol of Carrolton Gone piss-rotten for Hamilton Cabot, Fisher Ames [etc.]. **1946** (*reported in oral use by Prof. A. L.*

Hench) This is a piss-poor outfit. My job is a piss-poor one. **1957** J. Kerouac *On Road* (1958) 207 A pisspoor bum from Larimer Street. **1970** J. Hansen *Fadeout* viii. 68 Feeling sorry for a man's a piss-poor reason to marry him. **1972** R. Maugham *Escape from Shadows* iv. 186 But I feel out of place here, it's too piss-elegant. **1972** J. Brown *Chancer* v. 64 'This beer,' I said, 'it's piss-poor.' **1973** *Amer. Speech* 1970 XLV. 58 *Piss-elegant*, pretentious, ostentatious, egotistical (used with reference to male homosexuals). **1973** *Nation Rev.* (Melbourne) 24–30 Aug. 1432/1, I think privately that they look in pisspoor condition; but the spirited bidding rockets the price up to $2.50 in no time. **1974** J. Anthoine in H. MacInnes *Climb to Lost World* x. 179 He got piss wet in that bloody channel [on a rock face]. *Ibid.* xi. 202 'Here we are,' I said to Joe. 'On a piss-wet cliff and there's no bloody water for a brew!' **1977** *Gay News* 24 Mar. 21/2 The Lovely Ladies from South America were so piss-elegant they could hardly lift their feet off the ground. **1977** *N.Y. Rev. Bks.* 4 Aug. 35/4 They manufacture piss-chic cosmetics.

pissabed. Restrict '*Obs.* exc. *dial.*' to senses in Dict. and add: **1.** (Later examples.)

1953 S. Beckett *Watt* III. 154 Of flowers there was no trace, save of the flowers that plant themselves, or never die, or die only after many seasons, strangled by the rank grass. The chief of these was the pissabed. **1974** G. Grigson *Dict. Eng. Plant Names* 169 Pissabed (*Taraxacum officinale*, Dandelion)... The name has too wide a currency to be derived from Gerard's *Herbal*.

3. Chiefly *slang.* A bed-wetter; also *attrib.*, as an abusive epithet.

1643 in *County Court Rec. Accomack-Northampton, Va.* (1973) 292 Thou pissa bedd Jade. **1922** Joyce *Ulysses* 395 Pope Peter's but a pissabed. **1959** R. Fuller *Ruined Boys* 195 He beat me at the beginning of term for peeing my bed... Now he thinks of me as a pissabed. **1972** R. A. Wilson *Playboy's Bk. Forbidden Words* 229 *Piss-a-bed*, a lazy fellow, one who pisses in bed because he's too lethargic to walk to the john.

‖**pissaladière** (pisaladi̯ɛr). Also **pissaladiera.** [Fr.: Provençal dial. *pissaladiero* f. *pissala* salt fish.] A Provençal open tart similar to pizza, usu. cooked with onions, anchovies, and black olives.

1931 A. de Croze *What to eat & drink in France* xv. 131 La Pissaladiera (a tart well sprinkled with olive oil, filled with sweet onions browned in oil, and pissala—salted whitebait—and baked). **1951** G. Maurois *Cooking with French Touch* iii. 66 Another favorite hors d'œuvre in the region of Nice is the *pissaladiera*. **1960** E. Campbell *Encycl. World Cookery* 154 *Pissaladière*..dough..onions.. black olives..anchovy fillets. **1960** E. David *French Provincial Cooking* 208 The *pissaladière* is a substantial dish of bread dough spread with onions, anchovies, black olives, and sometimes tomatoes. **1966** P. V. Price *France* 276 Pissaladiera. A Nice speciality, and a version of Italian pizza—an open tart with onions, anchovies and black olives, sometimes with tomatoes. **1970** Simon & Howe *Dict. Gastron.* 302/2 The filling is covered with a pattern of anchovy fillets and stoned black olives and the *pissaladière* is baked in a hot oven. **1977** *N.Y. Rev. Bks.* 8 Dec. 18/3 Hurray for the Cornish pasty rather than Olney's *pissaladière*.

piss-ant. Delete † *Obs. rare*[—1] and add: Also **pissant, piss ant;** *dial.* **piss-aint.** (Further examples.) *spec.* in phrases *drunk as a piss-ant*, extremely intoxicated; *game as a piss-ant*, courageous, very brave. Also *transf.*, *fig.*, and *attrib.*

1770 C. Carroll *Let.* 22 May in *Maryland Hist. Mag.* (1917) XII. 362 It seems the Pissants eat a great deal of Corn in the ground. **1847** W. T. Porter *Quarter Race in Kentucky* 84 Pourin out of the woods like pissants out of an old log when tother end's afire. **1893** J. Salisbury *Gloss. Words S.E. Worcestershire* 28 'Er screws 'er waist up till 'er looks like a piss-aint. **1903** 'T. Collins' *Such is Life* v. 184 His mind's so much took-up with the tuppenny-thruppenny things... Can't afford to come-out anything but a pis-ant. **1930** J. Dos Passos *42nd Parallel* I. 77 I'm drunk as a pissant still. **1935** H. L. Davis *Honey in Horn* xvi. 278 Anybody who called owning horses disorderly conduct was a liar and a pissant. **1945** Baker *Austral. Lang.* iv. 87 Game as a piss ant or drunk as a piss ant. **1946** Mezzrow & Wolfe *Really Blues* (1957) 377 *Piss-ant*, a nobody, small fry. **1949** H. Hornsby *Lonesome Valley* 185 Why, goddam it to Jesus Christ and back, they're thicker than piss ants. **1961** P. White *Riders in Chariot* xiii. 448 'And on such a day!' she shrieked, looking at the clock. 'I bet that nephew of yours will be full as a piss-ant by eleven!' **1962** R. Tulipan *March into Morning* 59 The old white lady makes you as game as a pissant. **1966** *Publ. Amer. Dial. Soc.* 1964 XLII. 21 *Pissant*. Regularly used by men among men; elsewhere it is *ant* only. **1972** F. van W. Mason *Roads to Liberty* 169 You stole my skelp, you no-'count piss-ant. **1973** R. Heinlein *Time Enough for Love* (1974) 523 His grandfather paused just long enough to look back and say, 'Not on your tintype... you pusillanimous piss-ant.' **1978** *Guardian Weekly* 25 June 18/4 That pissant [California Governor] Brown. **1979** 'A. Hailey' *Overload* III. x. 237 All you do now is let off some pissant firecrackers, then laze around here for a goddam month's vacation.

Hence **pi·ssant** *v. intr. Austral. slang*, to mess around.

1945 Baker *Austral. Lang.* iv. 87 Someone is pissanting around when he is messing about. **1951** Cusack & James *Come in Spinner* 307, I been pissantin' round the Northern Territory most of the time. **1959** G. Hamilton *Summer Glare* 138 Struth, you pissant around like a rooster that's too old.

pissed (pist), *ppl. a. slang.* [f. Piss *v.* + -ed[1].] **1.** Drunk, intoxicated. Also const. *up*.

1929 F. Manning *Middle Parts of Fortune* I. iii. 54 On the Saturday they went into Sandby for a spree, and got properly pissed-up there. **1937** Partridge *Dict. Slang* 635/2 *Pissed*..., (very) drunk. **1939** Dylan Thomas *Let.* July (1966) 233 Both, if you ask me, were pissed. **1957** R. Mason *World of Suzie Wong* II. i. 111 Christ, I'm pissed. I'm pissed as a newt. **1958** K. Amis *I like it Here* i. 13 An uncle of mine went there a year or two ago and was pissed all the time on about ten bob a day. **1971** B. W. Aldiss *Soldier Erect* 10, I thought of the other blokes in 'A' Company... Tonight, they'd all be getting hopelessly pissed or screwing girls. **1973** M. Amis *Rachel Papers* 38 'Does it make you extra pissed?' Norman handed me my glass, drank his in one, and crouched again to refill it. **1977** *Daily Mirror* 10 May 1/6 Too much German food and too much German wine. I'm pissed.

2. Angry, irritated; fed-up, depressed. Freq. const. *off*.

1946 *Amer. Speech* XXI. 249 *Pissed-off*... This means roughly, fed-up, irritated, depressed. **1948** N. Mailer *Naked & Dead* (1949) I. ii. 25, I bet you even look pissed-off when you're with your wife. **1955** W. Gaddis *Recognitions* II. vii. 646, I been pissed off at him for five years. **1967** B. Wright tr. *Queneau's Between Blue & Blue* i. 11 I'm beginning to get pissed off with your rotten little questions. **1971** B. Malamud *Tenants* 78 The writer figured Willie must still be upset, pissed off, else he would never have left it [*sc.* his typewriter] unprotected. **1971** *It* 9–23 Sept. 11/3 I'm pissed at people in the movement who help lay out the line that I'm a millionaire superstar or other shit. **1972** *Screw* 12 June 16/2 Maye, angered by his new but well-deserved reputation as an insecure male, is pissed. **1973** *Observer* (Colour Suppl.) 15 July 13/3 Young people are getting pissed off with this sort of 'rip-off'. **1974** *Saturday Night* (Toronto) Aug. 20/1 Well, Ken got a little pissed off, and stomped off in a huge sulk, until finally the two of them had it out. **1975** R. Burns *Alvarez Jrnl.* 8 Helen was pissed off when I called. She had plans or something. **1977** *Rolling Stone* 19 May 63/1 Hamilton.. says half the Cabinet is pissed at him because things are moving so slow. **1977** *Spare Rib* July 37/2 Everyone was pissed-off and disillusioned with conferences after London.

pisser. Add: **2.** *coarse slang.* The penis; the female pudenda; *spec.* in phr. *to pull a person's pisser*, to pull his leg, befool him.

1901 Farmer & Henley *Slang* V. 214/2 *Pisser* = (1) the *penis*, and (2) the female *pudendum*. **1969** J. Wainwright *Big Tickle* 123 'I don't know where he is.' 'If you're pulling my pisser!' 'I'm not.' **1971** B. W. Aldiss *Soldier Erect* 37 He was pulling your pisser, Wal. Malaria's no worse than a cold to the Wogs, is it, Bamber? *Ibid.* 133 He grabbed my arm. 'You think I'm pulling your pisser, man? Listen, *pick your officer*! They're fucking human, same as you, you know!'

3. *slang* (orig. *U.S.*). An extraordinary person or thing; now usu. a difficult or distasteful event, an unpleasant person; also in weakened sense, a bloke, chap.

1943 *Amer. Speech* XVIII. 45 As they [*sc.* New Yorkers] use it, 'pisser' denotes a wag or 'card', a 'corker', or a screamingly funny joke or prank, a 'hot one'. **1954** J. A. Weingarten *Amer. Dict. Slang* 276/1 *Pisser*,..a thing or person of extraordinary aspect; an amusing thing or person. **1974** J. Anthoine in H. MacInnes *Climb to Lost World* x. 166, I could have gone a real pisser but, as I dropped, I caught the edge of the rock with my hands and heaved myself up. **1975** tr. *Melchior's Sleeper Agent* (1976) II. 69 The old pisser had not got away! *Ibid.* 98 We could both do with a little liquid cheer. It's been a pisser of a day. **1978** *Maledicta* II. 268, I love it! It's a real pisser!

pi·ssing, *a.* and *adv. slang.* [f. Piss *v.* + -ing[2].] **A.** *adj.* Paltry, insignificant; brief. **B.** *adv.* As an intensive: exceedingly, abominably, 'bloody'. Hence **pi·ssingly** *adv.*

1937 Partridge *Dict. Slang* 635/2 *Pissing*, adj., paltry, brief. **1971** B. W. Aldiss *Soldier Erect* 180 In this teeming world [in Calcutta], nothing was what it seemed to be. The miseries of the idiot and his dependants were pissingly funny. **1974** K. Millett *Flying* (1975) III. 294 Paper hat leaps on stage, pissing hot to talk. **1975** N. Freeling *What are Bugles blowing For?* xxi. 123 'Fuck it,' said Metcalfe angrily. 'I'm only a pissing sergeant.' **1979** P. Way *Sunrise* I. 10 'Pissing awful weather,' said Don.

‖**pissoir** (piswār). [Fr.] A public urinal enclosed by a screen or wall.

1919 Mencken *Amer. Lang.* 127 The French *pissoir*.. is still regarded as indecent in America, and is seldom used in England, but it has gone into most of the Continental languages. **1934** S. Spender *Vienna* ii. 23 In oil-tarred pissoirs. **1942** [see *fly *sb.*[2] 4 a]. **1955** W. Gaddis *Recognitions* I. ii. 77 The feet showing below the shield of the pissoir. **1968** *Listener* 21 Mar. 369/3 A simple slate pissoir where graffiti, true to the spirit of Rochdale, divide between sex and politics. **1972** *Guardian* 19 Feb. 8/3 One pissoir is plain anti-feminism. **1975** *Times* 17 Dec. 13/1 The valuable site..has been vacant for many years and its use is restricted to private parking and to pissoirs of inelegant design.

piss-pot. (Later examples.)

1942 E. Paul *Narrow St.* xxix. 269 They had wandered into a slum where frequently the housewives dumped pisspots out of the windows. **1948** E. Pound *Pisan Cantos* (1949) lxxx. 104 His helmet is used for a pisspot. **1972** *Daily Tel.* 5 May 12/3 His low-life characters empty the occasional piss-pot and there is a fine flamboyant parade of pretty whores. **1974** *Times Lit. Suppl.* 19 Apr. 417/4 Malcolm [Lowry]..reached the abyss of alcoholic degradation, drinking in prison out of a pisspot.

piss-up (pi·sʌp). *coarse slang.* [f. Piss *v.*; cf. *pissed-up* s.v. *pissed *ppl. a.* 1.] **1.** = *cock-up 4.

1950 H. E. Goldin *Dict. Amer. Underworld Lingo* 158/1 *Piss-up*, n., a fiasco; a bungled or unremunerative crime. **1969** R. Esser *Hot Potato* 51 Just what a pissy-arsed bugger like you would say... You mean it might be one hell of a piss-up.

2. A bout of heavy drinking.

1952 *Landfall* VI. 222 His first 'piss-up' is a landmark in his life. **1959** *Numbers* Feb. 32, I feel like a good piss-up. **1965** R. Jeffries *Traitor's Crime* x. 131 How about a piss-up? Anne was only saying yesterday I'd been sober for so long was I declining into my old age? **1973** *Nation Rev.* (Melbourne) 31 Aug. 1441/6 Most significant..was the way the seating was arranged at the Saturday night pissup. **1976** P. Cave *High Flying Birds* iii. 45 There was the prospect of a piss-up in the offing.

pissy (pi·si), *a. coarse slang.* [f. Piss *sb.* + -y[1].] Of or pertaining to urine; *fig.* rubbishy, inferior. So *pissy-arsed* adj. (see quot. 1961); also, as a term of general abuse.

1926 T. E. Lawrence *Seven Pillars* (1935) x. cxviii. 642 That hot pissy aura of thronged men in woollen clothes. **1961** Partridge *Dict. Slang* Suppl. 1224/1 *Pissy-arsed*, prone to crapulous inebriation. **1969** Pissy-arsed [see *piss-up 1]. **1972** 'D. Craig' *Double Take* xi. 138 Oh, God, these poverty stricken, pissy shoes. **1973** M. Amis *Rachel Papers* 127 You'll probably say this is rather..pissy, but babies are the only things women can have that men can't. **1974** P. Cave *Mama* (new ed.) x. 80 It makes you realise what a pissy little island we live on, don't it? **1974** *Black World* Nov. 66 She/rising up stumbling over to the dank pissy restroom. **1979** J. Barnett *Backfire is Hostile!* iii. 36 That pissy bunch of engineering officers at the bar.

pistachio. Add: Also with pronunc. (pistā·ʃi̯o). **3.** *pistachio candy*; **pistachio ice, ice-cream,** ice(-cream) containing pistachio nuts; also, ice-cream coloured pistachio green.

1853 Mrs. Gaskell *Ruth* III. iv. 140 The days when you first brought me pistachio candy from London. **1868** A. Gouffé tr. *J. Gouffé's Royal Cookery Bk.* xvii. 561 Apricot and Pistachio Ice. **1885** C. E. Pascoe *London Today* iii. 48 The more aristocratic foreign visitors to London..flocked thither [*sc.* to Verrey's restaurant] to eat pistachio ices. **1889** R. Wells *Pastrycook & Confect. Guide* vii. 65 *Pistachio ice*, blanch and beat until fine the kernels of 6 ozs. of Pistachio nuts [etc.]. **1916** Frandsen & Markham *Manuf. Ice Creams* x. 103 Pistachio Ice Cream. **1942** A. L. Simon *Conc. Encycl. Gastron.* V. 57/1 Pistachio ice-creams..are ice-creams..flavoured with almonds and vanilla and coloured with vegetable dyes as near the pistachio green as possible. **1943** E. M. Almedingen *Frossia* vi. 228 Do you like pistachio ices—with small almond biscuits? **1972** E. Hargreaves *Fair Green Weed* i. 13 It's a big rambling place like a block of pistachio ice-cream.

piste[2]**, pist.** Add: **a.** (Later example.) Also in extended use.

1964 *Guardian* 20 Oct. 7/2 That other highway, where the giraffe..came down to drink. It is only a *piste* for the big Trans-Af buses. **1967** N. Freeling *Strike Out* 24 The serious part—stables, 'pistes', exercise courts and yards. **1970** *Country Life* 17–24 Dec. 1209/3 For the last few dozen miles..the road is tar laid direct on to the apricot-coloured sand..giving one a choice between the molar-shaking corrugations of the piste itself and the treacherous dunes on either side.

b. *Fencing.* A specially marked-out field of play.

1922 *Glasgow Herald* 20 May 9 Preliminary pools of the individual Military Epée Championship of Europe. The Turney..was fought on linoleum 'pistes'. **1963** *Fencing* ('Know the Game' Ser.) (ed. 4) 24 The area within which fencers may move is restricted. This area is called the 'Piste'. **1971** I. Butykai tr. *Lukovich's Electric Foil Fencing* I. 12 Competitors who had fenced for several years with the ordinary foil, now changed it for the electric device and the pioneering work of the era of electric foil fencing. Many said good-bye to the piste for ever.

c. A track of compacted snow used as a ski- or toboggan-run. Also *attrib.*

1929 E. Hemingway *Farewell to Arms* xxxvii. 299 Tobogganing..requires a special *piste*. You could not toboggan into the streets of Montreux. **1939** W. Prager *Skiing* 80 Thus he has an opportunity to construct the Piste. **1950** *Times* 13 Feb. 7/5 Most other races, including world championships, are only a test of piste skiing, a debased and impoverished variant of the real thing. **1955** *Times* 20 Sept. 12/7 Italian engineering..began to work on ski lifts and cable cars..with the result that ..there are..installations taking skiers of every grade to the top of the various pistes. **1959** P. Moyes *Dead Men don't Ski* ii. 24 The lift travels above the *pistes*, or ski runs. **1966** *Daily Tel.* 19 Dec. 9/1 You may never come skimming down the *piste* like a bird, but you can make sure of a second look by choosing the right clothes. **1972** N. Freeling *Long Silence* II. 111 Woodcutters' paths..make good natural *pistes* for Nordic ski. **1977** *Time* 10 Jan. 6/1 The lure of the Cup was enough to bring Austria's five-time winner, Annemarie Moser-Proell, to the *piste* again.

pistic, *a.* Add: **2.** Pertaining to faith or trust rather than to reason; hence *ellipt.* as *sb.*, someone who accepts things simply on trust.

1923 Ogden & Richards *Meaning of Meaning* ii. 89 The purely verbal systems so characteristic of pistic speculation. **1965** in W. Schneemelcher *New Test. Apocr.* II. xi. 77 The apostolic secret tradition..is accessible not merely to the small circle of a spiritual élite, the 'hylic' and 'psychic' or 'pistic' being excluded on principle, but to all

those who have been received into the church..as full Christians.

pistillode (pi·stilōud). *Bot.* [f. Pistil + -ode.] A rudimentary pistil.

1905 W. E. Safford *Useful Plants of Guam* 259 An imperfect pistil (pistillode) present or lacking. **1975** *Blumea* XXII. 420 The precise configuration of antherodes and pistillodes in relation to other floral parts is likely to prove of considerable taxonomic value.

pistillody. (Earlier and later examples.)

1877 *Jrnl. Linn. Soc.* (*Bot.*) XV. 87 The calyx and corolla remain entirely unchanged in all cases exhibiting pistillody. **1929** *Jrnl. Heredity* XX. 137/1 Leighty and Sando describe and illustrate certain abnormalities in wheat flowers which they call 'Pistillody'... They found the stamens metamorphosed either wholly or partly into carpels.

pistiloid (pi·stiloid), *a. Bot.* Also **pistilloid.** [f. Pistil + -oid.] Resembling a pistil in shape.

1877 *Jrnl. Linn. Soc.* (*Bot.*) XV. 88 Stamen No. 2..is an unstalked pistilloid body with a short curved hairy style. **1888** G. Henslow *Origin Floral Struct.* 291 Pistiloid sepals..have been observed by Mr. Laxton.

pistol, *sb.* Add: **1. d.** *to beat the pistol*, in Athletics, to make a false start (cf. *gun *sb.* 6 e); *to hold a pistol to* (or *at*) (a person's) *head*: to threaten (a person) in order to induce him to act in a particular way; to issue with an ultimatum; *with a pistol at one's head*: under pressure; while being threatened.

1905 S. Crowther *Rowing & Track Athletics* 302 False starts were rarely penalized..and..'beating the pistol' was one of the tricks which less sportsmanlike runners constantly practised. **1917** W. J. Locke *Red Planet* iii. 33 The boy was my guest. I had not intended to hold a pistol to his head. **1920** E. H. Clark *Track Athletics up to Date* vi. 49 Some athletes try to gauge the moment when the starter's finger is curling over the trigger of his pistol and to start just before the pistol is fired, when it is too late for the starter to check his finger. This is called 'beating the pistol'. **1974** 'W. Haggard' *Kinsmen* iii. 40 She'd held a pistol at Heale-Mann's respectable head. Do as I wish or face a scandal. **1978** *Times* 14 Mar. 19/2 The [French] Socialists..have an understandable dislike of negotiating with a pistol at their heads.

2. *pistol-pocket* (example), *-practice* (example), *-shape*, *-toter*; *pistol-shaped* adj. (example), *-toting* vbl. sb. and ppl. adj.; **pistol flare** = *pistol light*; **pistol-grip** (further examples); (*b*) *transf.* a handle shaped like a pistol-grip; also *attrib.*; **pistol light,** a night-signal or light fired from a special pistol, used by soldiers, etc.; a Very light; **pistol-packing** *ppl. a.* [see *pack *v.* 9] carrying a pistol; hence *pistol-packer*.

1916 'Boyd Cable' *Action Front* 26 'Keep the lights blazing,' Rawbon paused to shout to the man with the pistol flares. **1917** 'Contact' *Airman's Outings* 227, I laugh and proceed to pass some wire through the pistol-grip. **1958** *Engineering* 28 Mar. 411/3 In addition to the push-button and pistol-grip controls. **1964** S. Crawford *Basic Engin. Processes* i. 13 A tubular frame adjustable hacksaw with the 'pistol-grip' type handle. **1972** *Shooting Times & Country Mag.* 27 May 3/2 (Advt.), Pistolgrip stock with cheekpiece £2 extra. **1974** *Country Life* 17 Jan. 82/3 The standard unit includes a parabole reflector, pistol grip, microphone and carrying kit. **1977** *Western Mail* (Cardiff) 5 Mar. 12/3 (Advt.), Movie camera (pistol-grip) and carrying case. **1916** 'Boyd Cable' *Action Front* 62 A couple of pistol lights flared upwards. **1929** *Papers Mich. Acad. Sci., Arts & Lett.* X. 315 *Pistol lights*, rockets shot from a pistol. **1944** *Time* 10 July 40/2 The lady that's known as Lou pinched several pokes, but pulled no triggers... No pistol packer, she. **1943** A. Dexter (*song-title*) Pistol packin' mama. **1948** E. Pound *Pisan Cantos* (1949) lxxvi. 39 The pride of all our D.T.C. was pistol-packin' Burnes. **1972** T. Ardies *This Suitcase* xviii. 204 The hotel had been taken over by pistol-packing maniacs. **1966** M. Avallone *Fat Death* 20 His hands were anchored into the pistol pockets of his grey trousers, throwing back the tails of his form-fitting jacket. **1846** *Punch* XI. 206 Maids-of-all-work learning pistol-practice at the shooting galleries. **1931** E. Wenham *Domestic Silver* v. 42 Probably one of the most attractive forms of handles is the so-called pistol-shape. **1964** W. L. Goodman *Hist. Woodworking Tools* 133 Using the broad-bladed Japanese saw with the pistol-shaped handle. **1905** *Dialect Notes* III. 90 *Pistol-toter*,.. one who habitually carries a pistol. 'Suppress the pistol-toter and the bootlegger.' General. **1914** *Maclean's Mag.* Dec. 56/1 The police force..is so inefficient that 'pistol-toting' after night is common among all classes. **1965** *Economist* 25 Sept. 1215/3 The men of Sinn Fein..are a more interesting study buying large brandies in the border towns for the pistol-toting Protestants of the Royal Ulster Constabulary.

‖**pistolero** (pistōlē·ro). [Sp.] In Spain or other Spanish-speaking area; a gunman or gangster. Also *attrib.*

1937 F. Borkenau *Spanish Cockpit* i. 30 In order to frighten the Catalan bourgeoisie..the Barcelona police actually co-operated with gangs of *pistoleros*, who.. claimed to be revolutionaries. **1939** G. Greene *Lawless Roads* 19 Mexico remained Catholic: it was the governing class—the politicians and pistoleros only who were anti-Catholic. **1957** P. Kemp *Mine were of Trouble* viii. 152 He had been a wild youth..earning his living in Madrid as a *pistolero* for his political party, the Requetés. **1965** C. D. Eby *Siege of Alcázar* (1966) i. 44 The *pistolero* phase of the rebellion had ended. **1973** C. Alverson *Fighting Back* (1978) xiv. 80 You and your red-hot pistoleros got my kid brother nearly killed.

pistolet[1]**.** Restrict † *Obs.* to sense in Dict. and add: ‖**2.** Pronunc. (pistǫle). Esp. in Belgium, a small bread roll (so called because of its shape).

1853 C. Brontë *Villette* I. viii. 142 Boarders were.. regaled with..*pistolets au beurre* (rolls) and coffee. *a* **1855** —— *Professor* (1857) I. xiii. 224, I stirred my cup of coffee with a half-pistolet (we never had spoons). **1857** Mrs. Gaskell *Life C. Brontë* I. xi. 271 A slight meal of water and *pistolets* (the delicious little Brussels rolls). **1897** G. du Maurier *Martian* IV. 185 Breakfasted on a little roll called a pistolet and a cup of coffee. **1961** T. Henrot *Belgium* 189 Small rolls are either *Viennois*, *miches* or, quaintly, *pistolets*. **1975** T. Allbeury *Special Collection* xiv. 95 A basket was filled with bread, from poppy-seed rolls to..grey Minsk pistolets.

pistol-shot. Add: **2.** One who shoots with a pistol.

1949 Blunden *Addresses on Gen. Subjects* 24 He [*sc.* Shelley] was judged a horseman and a pistol-shot equal to his sporting friend Lord Byron. **1979** C. Allen *Tales from Dark Continent* vii. 99 A great bayonet fighter and trick pistol shot and a splendid figure.

3. (Further example.)

1942 *Essays & Stud.* XXVII. 59 And this pistol-shot style, bang bang bang, with its absolute certainty of detail, leaving the scraps of picture, like a jigsaw puzzle, for you to fit together.

pi·stol-whip, *v.* orig. and chiefly *U.S.* [f. Pistol *sb.* + Whip *v.*] *trans.* To strike (someone) with the butt of a pistol. Also *fig.* Hence **pi·stol-whipping** *vbl. sb.*

1942 Berrey & Van den Bark *Amer. Thes. Slang* §322/5 *Pistol-whip*, to strike with the butt of a pistol. **1952** *Sun* (Baltimore) 4 July 42/8 The two young Negroes pitched into him and the one with the gun pistol-whipped the physician. **1958** J. Thompson *Getaway* (1972) iv. 29 The banker lay sprawled on the floor, half-dead from Rudy's pistol-whipping. **1965** J. Philips *Twisted People* I. iv. 88 She'd been subjected to about the worst beating I've ever seen. Pistol whipped is my guess. **1967** *Observer* 23 Apr. 5/3 The British Foreign Secretary had been subjected to 'a pistol-whipping'..from Mr. Walt Rostow in the White House. **1970** K. Platt *Pushbutton Butterfly* (1971) xi. 132, I remembered the pistol-whipping..and brought my knee up to his face. **1973** G. Sims *Hunters Point* xiii. 124 He could feel the pain from the pistol-whipping blows he'd taken on his ribs. **1977** J. Wambaugh *Black Marble* (1978) xiv. 322 Let a cop get pistol-whipped instead of a clerk.

piston. Add: **4. piston bellows,** bellows in which the draught is supplied by the action of a piston; **piston corer** or **core sampler,** a core sampler consisting of a long weighted cylinder containing at its lower end a piston attached to the lowering cable, devised so that when the cylinder enters the bottom sediments under its own weight the descent of the piston is arrested, and the resulting partial vacuum inside the cylinder causes the pressure of the water to be effective in forcing it into the bottom; **piston drill,** a percussion drill in which the bit is attached to the rod of a piston; **piston engine,** a reciprocating engine, esp. an aircraft engine in which the airscrew is driven by the reciprocating action of pistons; hence **piston-engined** *a.*, powered by this kind of engine; **pistonphone** *Acoustics*, a device for producing known sound pressures by means of a vibrating piston whose motion is precisely measured, used mainly for calibrating microphones; **piston pin,** a pin which secures a piston to its connecting rod in an internal-combustion engine; **piston ring,** a metal ring fitted on a piston to seal the gap between the piston and the cylinder wall; **piston slap,** rocking of a loosely fitting piston against the cylinder wall, or the noise resulting from this.

a **1877** Knight *Dict. Mech.* II. 1717/1 A piston-bellows, formed by boring out the trunks of trees, used by the natives of Madagascar for smelting..iron. **1947** B. Kullenberg in *Svenska Hydrogr.-Biol. Komm. Skrifter* (*Ser. 3: Hydrogr.*) I. II. 12 The piston core sampler..has been based on a method to procure samples for ground investigations, developed by the Geotechnical Department of the Swedish State Railways. **1956** Piston core-sampler [see *core *sb.*[1] 4]. **1959** H. Barnes *Oceanogr. & Marine Biol.* i. 67 In Kullenberg's piston corer the hydrostatic pressure at great depths is utilized to overcome.. friction and to allow very long undisturbed cores to be taken. **1976** *New Scientist* 9 Dec. 576/2 The oceanographers' piston corers that were beginning to recover cores, or columns of mud, from the floor of the deep ocean. [**1901** M. M. Kirkman *Locomotive Appliances* 475 (*caption*) Piston air drill for drilling, reaming and tapping on locomotive work.] **1919** R. Peele *Compressed Air Plant for Mines* (ed. 3) xx. 306 At a large group of mines on the Rand, the complete repair cost of standard piston drills.. averaged $42.00 per 52 shifts. **1967** K. McGregor *Drilling of Rock* i. 12, 1860–70..; a commercial piston drill was patented by Burleigh in America. *Ibid.* 13 From these [drop drills] developed..the pneumatic piston drill... These machines were finally obsolete in the early 1920's. **1907** *Engineering* 21 June 829/2 If the turbine requires a screw which is necessarily less efficient than that of the piston engine, it is the fault of the turbine system. *Ibid.*, Ordinary reciprocating, or, to use a shorter word, piston, engines. **1943** G. G. Smith *Gas Turbines & Jet Propulsion* (ed. 2) vii. 54 Even if free-flying piston engine and compressor units were employed these handicaps would remain. **1960** C. H. Gibbs-Smith *Aeroplane* xvi. 126 The piston engine reached its apogee in the post-war period with such examples as the Pratt & Whitney 28-cylinder 3,500 h.p. radial. **1971** P. J. McMahon *Aircraft Propulsion* xi. 311 Mechanical losses in piston engines are high, being up to 25% of the indicated power. **1948** *Jrnl. R. Aeronaut. Soc.* LII. 591/1 The Airspeed Ambassador may be taken as representing the most advanced piston-engined commercial aeroplane of its size yet in prospect. **1974** E. Ambler *Dr. Frigo* I. 56 For most of those critical three days I was..in piston-engined planes missing connecting flights. **1979** *Daily Tel.* 1 Dec. 10/6 Their departure..coincides with the paying off of the Navy's only remaining steam reciprocating, or piston-engined, ship. **1922** E. Wente in *Physical Rev.* XIX. 343 (*caption*) Use of pistonphone for calibrating an electrostatic transmitter. *Ibid.* 500 The open-circuit votage of the transmitter per unit of pressure has been measured with the piston-phone for the frequency range of 10 to 200 cycles per second. **1965** C. A. Taylor *Physics Mus. Sounds* vi. 100 The pistonphone..appears to be used only for scientific work at very low frequencies. It consists..of a piston driven by a rotating cam which permits the generation of sinusoidal variations in pressure of accurately-controlled form. **1972** *Where* Dec. 336/3 Our tests were done with a B and K portable octave band analyser, which was calibrated before and after each occasion with a pistonphone. **1897** F. Grover *Pract. Treat. Mod. Gas & Oil Engines* xvii. 175 The pressure on the piston pin may be much more than upon the crank pin. **1910** W. A. Tookey tr. *Mathot's Construction & Working of Internal Combustion Engines* xiii. 375 The piston pin is made of the best quality mild steel. **1967** J. L. & G. H. F. Nayler *Dict. Mech. Engin.* 176 *Gudgeon pin*... Also called 'piston pin'. **1867** N. P. Burgh *Mod. Marine Engin.* vi. 277/2 The adoption of 'springs' behind the 'piston ring' is now becoming general. **1936** *Discovery* Feb. 39/1 Steam leaking past the piston rings of the high-pressure cylinders. **1971** J. H. Haynes *Mini Owner's Workshop Man.* i. 30/1 To remove the piston rings, slide them carefully over the top of the piston, taking care not to scratch the aluminium alloy. **1976** L. Deighton *Twinkle, Twinkle Little Spy* x. 97 The running repairs I do myself... Last month I changed the piston rings. **1916** P. M. Heldt *Gasoline Automobile* (ed. 4) I. vii. 163 To obviate.. piston slap when starting, it is well to make the piston comparatively long. **1962** *Which? Car Suppl.* Jan. 32/2 A worrying piston-slap rattle from the engine. **1968** *Practical Motorist* Dec. 459/1 *Piston slap*, a characteristic metallic slapping sound within the cylinders caused by slight rocking motion of an ill-fitting or worn piston.

pi·ston, *v.* [f. the sb.] *intr.* To move like a piston. Occas. *trans.*, to direct, throw with a piston-like movement.

1930 R. Campbell *Adamastor* 80 Down the stage the dance..Tarantulates in scarlet tights For flashing arms to piston. **1940** Dylan Thomas *Portrait of Artist as Young Dog* 52 The small boy in his invisible engine..kicked the dog's plate at the washhouse stop, and puffed and pistoned slower and slower while the servant girl lowered the pole. **1958** *Spectator* 15 Aug. 219/3 Thin grey knees piston over the beach. **1976** 'E. McBain' *Guns* (1977) v. 118 He ..pistoned a short hand punch to her shoulder.

pit, *sb.*[1] Add: **1. h.** = *engine-pit* (*b*) s.v. *engine *sb.* 11 b. Hence used allusively (freq. in *pl.*) of the area at the side of a motor-racing track where competing cars are prepared and maintained.

1839 *Chambers's Edin. Jrnl.* 7 Dec. 368 Under each engine is a pit three feet deep, which enables the enginemen to get underneath to examine and repair it. **1907** [see *inspection 6]. **1912** *Collier's* 28 Sept. 11/1 Up swoops the racer, rear wheels locked and sliding, thundering and veiled in smoke, and stops at the pit. **1913** *Technical World Mag.* June 492/1 As De Palma passed his rival's pit..one of the attendants reached for a telephone. **1924** *Brooklands Gaz.* Oct. 176/2 F. C. Clayton on his Marseal, had to turn into the pits after eight laps. **1928** *Evening News* 18 Aug. 1/3 He was pulling into the pits to refill. **1930** E. Waugh *Vile Bodies* x. 182 'The pits' turned out to be a line of booths, built of wood and corrugated iron immediately opposite the Grand Stand. Many of the cars had already arrived and stood at their 'pits', surrounded by a knot of mechanics and spectators. **1946** *Sun* (Baltimore) 23 Dec. 17/3 Each will have three long maintenance 'pits' to accommodate nine busses at one time. **1957** *Times* 16 Oct. 12/6 A man who had a garage and a pit and no car. **1968** *Listener* 5 Sept. 301/3 So there I was, having failed my examinations, working in the pits with Duncan Hamilton. **1972** M. Gilbert *Body of Girl* iv. 44 He climbed out of the pit..and said.. 'You've come to buy a car.' **1973** *Times* 30 Apr. 7/1 Peterson's domination ended abruptly on lap 57 when he brought his car to a halt with a broken gearbox, to receive a huge ovation from the crowd when he walked back to the pits. **1977** *Times* 15 July (Motor Racing Suppl.) p. ii/8 Dron pulled his Dolomite into the pits, mistakenly believing the race was over. But..he managed to pass the flag at the end of the pit road in time to win his class.

9. a. Also as an *ellipt.* use for 'armpit'. *slang.*

1965 *Amer. Speech* XL. 194 *Pits*,..a slang abbreviation of the term *armpits*,..with an extension of meaning to entail the idea of body odor. **1973** M. Amis *Rachel Papers* 71 Complete body-service..pits clipped, toes manicured, pubic hair permed and styled, each tooth brushed, tongue scraped, nose pruned. **1974** E. Brawley *Rap* (1975) II. xx. 325 She opened her heavy flippy arms, arms with rolls of honest fat hanging down under her pits. **1977** *Rolling Stone* 7 Apr. 48/2 Simmons answers by spraying his pits with a can of Royal Copenhagen.

c. (Later examples.)

1914 M. Drummond tr. *Haberlandt's Physiol. Plant Anat.* i. 44 These readily permeable spots generally take the shape of sharply defined areas of approximately circular cross-section, known as pits. 1953 K. Esau *Plant Anat.* iii. 39 Secondary cell walls are commonly characterized by the presence of depressions or cavities... Such cavities are termed pits. 1976 Bell & Coombe tr. *Strasburger's Textbk. Bot.* (rev. ed.) 110 If transversely elongated pits are arranged one above the other in the lateral walls the arrangement is said to be scalariform.

10. a. (Later example.)

1922 W. S. Maugham *On Chinese Screen* xlvii. 186 Declaiming the blank verse of Sheridan Knowles with an emphasis to rouse the pit to frenzy.

b. = *orchestra pit* s.v. *ORCHESTRA 4.

1961 in Webster. 1966 *Listener* 6 Oct. 517/1 The sheer sound of the orchestra seemed very much bigger than usual, the strings had a bloom that is often lacking, the woodwind sweetness as well as precision; ensemble had improved, including rapport between stage and pit. 1974 *Belton* (S. Carolina) *News* 18 Apr. 1/2 Hanna High Jazz Ensemble (In the pit throughout the show). 1975 *Times* 17 Sept. 10/4 On the evenings he conducted the sound coming from the pit ranged from the good to the superlative. 1977 *New Yorker* 9 May 132/3 (It is said that Toscanini was the first to sink the pit at La Scala.) The Orpheum, where the Boston 'Rigoletto' was done, has no pit, either.

12. b. A pocket in a garment. *slang.*

1811 *Lexicon Balatronicum* s.v. *Pit*, He drew a rare thimble from the swell's pit. He took a handsome watch from the gentleman's pocket. 1927 *Dialect Notes* V. 459 *Pit*, a pocket. 1938 F. D. Sharpe *Sharpe of Flying Squad* 332 *The pit*, the inside jacket pocket. 1950 H. E. Goldin *Dict. Amer. Underworld Lingo* 158/2 *Pit* (among pickpockets), the vest pocket; or, less frequently, the inside breast pocket of coat. 1955 D. W. Maurer in *Publ. Amer. Dial. Soc.* xxiv. 125 The most important pocket in the coat from the pickpocket's point of view is the *coat pit*, or the inside breast pocket... This is often shortened to *pit*. 1966 Baker *Austral. Lang.* (ed. 2) vii. 143 A generation ago,..the various pockets were known as.. *left kick* or *pit* and *right kick* or *pit* (trouser pockets).

14. (In sense 6) *pit-boot, -bottle, -boy* (earlier example), *-clothes, -coat, committee, -dirt, -girl* (earlier examples), *-horse, -lad* (examples), *-manager, -pony* (further examples), *-singlet, -trousers, -village, -woman* (example); (in sense 10) *pit-band, -bandsman, -door, -doorkeeper, -orchestra, -stall, -ticket*; **pit aperture** *Bot.*, an opening on the inner surface of a secondary cell wall, forming the entrance to a pit cavity; **pit-bank** (earlier and later examples); also *attrib.*; **pit bull (terrier),** a small, stocky, short-coated dog belonging to the American breed so called, usually fawn or brindled in colour, with white markings; also used as a name for the Staffordshire bull terrier, which belongs to a closely related breed; **pit canal** *Bot.*, a channel in the secondary cell wall of a bordered pit, leading to the pit cavity; **pit-cave** *Archæol.* (see quot. 1921); **pit-cavity** *Bot.*, the space within a simple pit, extending from the primary cell wall to the aperture bordering the cell lumen; **pit chamber** *Bot.*, the hemispherical space between the primary and secondary cell walls of a bordered pit; **pit-comb** *Archæol.*, used *attrib.* to designate pottery decorated with rows of indentations and comb-patterns; **pit dog** = **pit bull terrier*; **pit field** *Bot.*, a depression or group of depressions in a primary cell wall; **pit-head** (later *attrib.* examples); **pit-lamp** *Canad.*, a miner's lamp; also *transf.*, a lamp used in hunting or fishing, a jack-lamp (see Jack *sb.*[1] 33); also as *v. trans.*, to hunt (deer, etc.) using a pit-lamp (also *absol.*); so **pit-lamping** *vbl. sb.*; **pit-lighting** *vbl. sb. Canad.* = *pit-lamping* above; **pit membrane** *Bot.*, the part of a cell wall covering a pit; **pit organ,** a small depression acting as a receptor sensitive to changes in temperature, found on each side of the head of snakes belonging to the subfamily Crotalinæ; **pit-pair** *Bot.*, two pits in adjacent cell walls, sharing the same pit membrane; **pit-planting,** a method of planting trees in which a hole is dug, and the roots settled over a mound of earth in the bottom of the hole before it is refilled; also, planting trees in small depressions which help to conserve moisture; **pit-saw** (earlier and later examples); also *attrib.* and as *v. trans.*, to cut (timber) with a pit-saw (also *absol.*); **pit-sawing** *vbl. sb.*; **pit-sawn** *ppl. a.*; **pit-sawyer** (examples); **pit silo,** a silo in the form of a pit (rather than a tower); so **pit silage,** silage made in a pit; **pit stop,** in motor-racing, a stop at a pit (sense *1 h) for refuelling, maintenance, etc., usu. during a race; also *transf.* and *fig.*; **pit tip,** the mass of waste material deposited near the mouth of a mine or pit; **pit trap** = sense 1 f; **pit wood** (earlier and later examples); **pit yacker, yakker** *dial.*, a coal miner.

1934 *Jrnl. Arnold Arboretum* XV. 334 The narrow inner and outer layers of the secondary wall come together in the rim formed about the pit-aperture. 1953 K. Esau *Plant Anat.* iii. 44 The circular pit apertures in a bordered pit-pair appear exactly opposite each other. 1967 S. Broido-Altman tr. *Fahn's Plant Anat.* ii. 36 The opening of the pit on the inner side of the cell wall..is called the pit aperture. 1942 Berrey & Van den Bark *Amer. Thes. Slang* § 576/23 *Pit band*.., a theatre orchestra. 1946 R. Blesh *Shining Trumpets* (1949) xii. 280 The other side..is completely in the manner of the average musical comedy pit band of the 1920's. 1977 *Rolling Stone* 19 May 74/3 At 14 I got a job in a pit band in a cinema. 1959 'F. Newton' *Jazz Scene* xi. 189 The despised pit-bandsmen and light musicians. 1870 A. J. Munby *Diary* 25 June in D. Hudson *Munby* (1972) 288 'I've worked on pit bonk most o' my days', said an Oakengates lassie. 1930 Auden *Poems* 67 Head-gears gaunt on grass-grown pit-banks. 1968 *Listener* 15 Aug. 205/3 Each plane came in over the long pit-bank mountain. 1894 H. Pease *Mark o' Deil* 26 H tried to shift it, an' threw his pit boots at it. 1913 D. H. Lawrence *Sons & Lovers* i. 25 She..set his pit-boots beside them. *Ibid.*, She..rinsed his pit-bottle. a 1930 —— *Phoenix II* (1968) 263, I, who remember the homeward trooping of the colliers when I was a boy,..the red mouths and the quick whites of the eyes, the swinging pit bottles, and the strange voices of men from the underworld. 1863 *Edin. Rev.* Apr. 424 It is to be hoped these schools will be continued for the purpose of improving the very imperfect scholarship of future pit-boys. 1945 *Sun* (Baltimore) 18 May 7/5 Mrs. Dorella Zinke..died within 90 minutes after a mass attack by nine pit bull terriers. 1968 K. Weatherly *Roo Shooter* 128 It was the rat-catching trick of the old pit bull terriers: one savage jerk of the dog's head and the big tom flew high in the air. 1974 R. Thomas *Porkchoppers* xxix. 245 He held a leash that was attached to an aged English pit bull that waddled..and wheezed. 1976 *Honolulu Star-Bull.* 21 Dec. F-8/8 (Advt.), Pit Bull Pups, 8 wks. 1953 K. Esau *Plant Anat.* iii. 43 The border divides the cavity into the pit chamber..and the pit canal, the passage from the cell lumen into the pit chamber. 1967 S. Broido-Altman tr. *Fahn's Plant Anat.* ii. 40 The pit canal between the inner and outer apertures becomes longer. 1921 *Discovery* Feb. 33/1 Still another kind [of grave]..is known as the 'pit-cave'. This was made by first sinking a pit and then cutting out the tomb in the form of a side-recess from the bottom of the pit. 1925 V. G. Childe *Dawn European Civilization* vi. 92 Beside the pit-caves and chamber-tombs excavated in the clay or rock, megalithic graves were erected just in the heel of Italy. 1939 J. D. S. Pendlebury *Archaeol. Crete* iv. 242 At Zapher Papoura both the shaft grave and the pit cave continue in use. 1914 M. Drummond tr. *Haberlandt's Physiol. Plant Anat.* i. 44 The adjoining pit-cavities are separated only by the thin primary closing-membrane. 1953 K. Esau *Plant Anat.* iii. 41 In the bordered pit the secondary wall arches over the pit cavity. 1970 Panshin & de Zeeuw *Textbk. Wood Technol.* (ed. 3) I. iii. 93 The pit cavity is the entire space within the wall recess, between the pit membrane and the lumen. 1953 Pit chamber [see *pit canal* above]. 1967 S. Broido-Altman tr. *Fahn's Plant Anat.* ii. 40 As the walls continue to thicken the pit chamber becomes smaller. 1873 A. J. Munby *Diary* 11 Sept. in D. Hudson *Munby* (1972) 343 Ellen herself came out of the kitchen; and she was in her pit clothes, as she had promised. 1913 D. H. Lawrence *Sons & Lovers* i. 25 She..put his pit-clothes on the hearth to warm. 1937 'G. Orwell' *Road to Wigan Pier* iii. 37 At the baths he has two lockers where he can keep his pit clothes separate from his day clothes. 1974 *Times* 1 Feb. 19/4 When a man arrives at the colliery to work his shift he first changes into his pit clothes. 1913 D. H. Lawrence *Sons & Lovers* iii. 49 He had taken off his pit-coat. 1954 Pit-comb ware [see *cord *sb.*[1] 12]. 1957 V. G. Childe *Dawn European Civilization* (ed. 6) xi. 204 From Sweden to Siberia indeed all pots were manufactured by the same technique of ring-building, all taper downward to a rounded base and all may be decorated with rows of pits, frequently combined with zones of comb impressions. The whole ceramic family is therefore termed 'pit-comb ware'. 1967 *Antiquaries' Jrnl.* XLVII. 203 For Fox in 1924, the most obvious comparison was the North European Neolithic pottery, Baltic and more Eastern, called 'pit-comb ware' or Pitted ware, on which ornament includes quite similar pits, in one or several rows. 1920 *Act* 10 & 11 *Geo. V* c. 50 § 7 It shall not be necessary to constitute a pit committee for any mine which is a small mine within the meaning of the Coal Mines Act, 1911. 1928 *Britain's Industr. Future* (Liberal Industr. Inquiry) iv. 266 Open consultation in the [coal] industry should be secured by the establishment of Pit Committees, District Boards, and a National Mining Council. 1913 D. H. Lawrence *Sons & Lovers* iv. 70 Here sat the colliers in their pit-dirt. 1914 —— *Widowing of Mrs. Holroyd* iii. 79 He's in his pit-dirt. 1945 C. L. B. Hubbard *Observer's Bk. Dogs* 202 Pit Bull Terrier. Pit Dog. About the end of the eighteenth century..dog-fighting became the new sport... A breed of Bull Terrier was created which became the fore-runner of the present-day Staffordshire Bull Terrier. 1951 J. F. Gordon *Staffordshire Bull Terrier* ii. 23 The Pit Dog, Pit Bull Terrier or Stafford..achieved some measure of emancipation from his gladiatorial background. 1667 Pepys *Diary* 22 May (1974) VIII. 232 But here Knipp spied me out of the tiring-room, and came to the pit door; and I out to her and kissed her. a 1828 J. Bernard *Retrospections of Stage* (1830) I. ii. 45 It was his general practice to take the money at the pit-door. 1894 G. B. Shaw *Let.* 30 Apr. (1965) I. 433 A man who thinks a dramatic performance worth waiting at the pit door all day for is a lunatic. 1831 J. Boaden in *Private Corr. David Garrick* I. p. xxxvi, For the benefit of his father, the pit-door keeper, and others. 1855 W. B. Wood *Personal Recoll. Stage* i. 40 My keeper, Mr H., was at this time pit door-keeper at the Chestnut Street Theatre. 1934 *Jrnl. Arnold Arboretum* XV. 332 Cambial walls are characterized by having more or less numerous plasmodesmata which may be..aggregated in thinner areas of the walls, i.e. in so-called primary pit-fields. 1953 K. Esau *Plant Anat.* iii. 39 Only the secondary walls have pits, whereas the primary walls have primary pit fields. 1976 Bell & Coombe tr. *Strasburger's Textbk. Bot.* (rev. ed.) 63 Where the pit fields are oval or elongated, the bordered pits take on a similar shape. 1863 *Edin. Rev.* Apr. 436 The pit-girls are not less fond of holidays than their fathers. *Ibid.* 437 Much may be done to improve the condition of the poor pit-girl. 1866 A. J. Munby *Diary* 15 May in D. Hudson *Munby* (1972) 225 Some large coal manager who was strongly in favour of female labour told him the pitgirls were 'mules, not women'. 1915 *Political Q.* May 117 A maximum pit-head price might leave much of the home market in the same condition. 1928 *Daily Chron.* 9 Aug. 5/4 From September 1 pit head prices will be raised by 1s. a ton. 1937 'G. Orwell' *Road to Wigan Pier* iii. 37 At some of the larger and better appointed collieries there are pithead baths. 1967 A. L. Lloyd *Folk Song in England* v. 335 The miners..before the days of pithead baths..appeared in daylight with black faces. 1974 *Times* 4 Jan. 12/8 If a pithead ballot were held..Yorkshire miners would vote..in favour of stoppage. 1976 *Evening Post* (Nottingham) 15 Dec. 1/2 In the pithead ballot, 78 per cent of miners who voted turned down the Coal Board's offer. 1913 D. H. Lawrence *Sons & Lovers* vii. 172 Jimmy, who had been a pit-horse. 1862 *Cornh. Mag.* Mar. 351 Files of pitmen and groups of pit-lads are now dotting all the roads. 1912 W. Owen *Let.* 24 July (1967) 151 This pit-lad is wrestling with a Classbook of Physics. 1901 *Daily Colonist* (Victoria, B.C.) 1 Oct. 6/1 One Indian shot as many as 47 [deer] in one night at Ahatassett, using a pit lamp. 1906 *Ibid.* 30 Jan. 6/2 The only light he [*sc.* a coal miner] has is a smoky pit-lamp—a cotton wick soaked in fish oil. 1921 *Ibid.* 30 Oct. 13/4 The rescue parties had to find their way about with candles and pit lamps, great difficulty resulting. 1967 *Vancouver Province* 21 Feb. 19/6 Stanton said his association wants to commend the fisheries department for banning 'pit lamps' but believes the order should be made permanent. 1967 *Wildlife Rev.* (Victoria, B.C.) Mar. 27/2 Frank Greenfield..once jailed a man for pitlamping deer. 1924 R. S. Sherman *Mother Nature Stories* 61 Pit-lamping and hunting with dogs, in addition to natural enemies..have been the cause of its rapid disappearance. 1957 A. R. Barratt *Coronets & Buckskin* 45 That there Blotton had me arrested for pitlamping. 1969 *Daily Colonist* (Victoria, B.C.) 7 Dec. 20/2 Pitlamping on Gabriola Island cost two men $500 each when they pleaded guilty..in court. 1969 *Islander* (Victoria, B.C.) 14 Sept. 10/3 Before dawn one morning, two shots were heard from the direction of his cabin, but no one paid any attention as pitlighting was common practice. 1913 D. H. Lawrence *Sons & Lovers* i. 16 He could only abuse the pit-managers. 1976 *Daily Tel.* 20 July 2/3 Harry Widowson, 57, pit manager of Mount Vernon Road, Barnsley. 1913 *Forestry Q.* XI. 15 The delicate pit membranes were ruptured by the shrinkage of the cell walls in drying. 1953 K. Esau *Plant Anat.* iii. 41 The pit membrane is common to both pits of a pair. 1976 Bell & Coombe tr. *Strasburger's Textbk. Bot.* (rev. ed.) 63 Bordered pits, especially amongst the Coniferae, are often furnished with a thickening in the middle of the pit membrane. 1927 *Melody Maker* Aug. 767/3 Whether this is because the dancers or singers think a modern dance band is a more enhancing support than a piano or the pit orchestra—which, of course, it is—or whether it is the band which decides.., I know not. 1934 S. R. Nelson *All about Jazz* i. 27 The atmosphere of the cinema pit-orchestra or military band. 1937 Pit organ [see *neuromast]. 1976 *Nature* 3 June 441/1 He did more classic work on the pit-organ of rattlesnakes, where he demonstrated the exquisite thermal sensitivity of this receptor. 1933 *Tropical Woods* XXXVI. 5 Pit-pair.—Two complementary pits of adjacent cells. 1953 K. Esau *Plant Anat.* iii. 41 Two pits are combined into a paired structure, the pit-pair. 1970 Panshin & de Zeeuw *Textbk. Wood Technol.* (ed. 3) I. iii. 93 Pit pairs may be made up from similar pits to form bordered pit pairs or simple pit pairs. 1898 C. E. Curtis *Pract. Forestry* (ed. 2) viii. 47 When planting deeper soils ..pit-planting must be adopted. 1931 *Forestry* V. 18 No method is fool-proof, but pit-planting appears to be the safest. 1970 H. L. Edlin *Collins Guide Tree Planting & Cultivation* i. 110 Failures in pit planting nearly always arise from insufficient firming-up. 1917 R. Hodgson *Poems* 27 Wretched, blind pit ponies. 1938 G. Greene *Brighton Rock* i. iii. 55 White hair, grey face, short-sighted pit pony eyes. 1978 A. Price *'44 Vintage* xix. 219 Audley was like a racehorse down a coal mine, desperately pretending to be a pit pony. 1679 in *Rec. Court of New Castle on Delaware* (1904) 361 An Iron sledge and a hand saw Iron—one Pit Saw. 1848 [see *cross-cut *sb.* 5]. 1879 *Lumberman's Gaz.* 15 Oct. 5/1 An improvement over the gate saw, almost as great as was the gate over the pit saw. *Ibid.*, Two men..maintained a pit saw mill. 1930 L. G. D. Acland *Early Canterbury Runs* (ser. 1) ix. 219, I should think this will be the last pit-saw to be used in Canterbury. 1960 B. Crump *Good Keen Man* 66 A shack that he and his brother had built from timber they'd pit-sawn themselves 30 years before. 1965 F. Russell *Secret Islands* v. 76 'This was where we pitsawed,' he said doubtfully. 1974 P. W. Blandford *Country Craft Tools* vi. 91 Pit saws are still made in Britain for use in some African countries. 1908 E. J. Banfield *Confessions of Beachcomber* i. ii. 57, I began to be thankful that pit-sawing was not forced upon me as a profession in the days of inexperienced youth. 1965 F. Russell *Secret Islands* v. 76 Pitsawing is the most brutal physical work a man could do. 1946 *Nature* 17 Aug. 245/1 Apart from the pit-sawn timber used locally, the pit-sawn timber supplies handled by the Department through its numerous subcontractors totalled 1,060,000 cu.ft. 1965 M. Shadbolt *Among Cinders* xii. 98 He built the house with pit-sawn kauri. 1978 O. White *Silent Reach* xx. 228 The Silent Reach homestead was..primitive. Its exterior walls were of wedge-split or pit-sawn plants. 1941 *Beaver* June 38 *Pitsawyer*, in earlier years, an important and indispensable craftsman. To him was due the production of every inch of sawn lumber for the building of the posts and water craft. 1946 *Nature* 17 Aug. 245/1 An exacting specification impossible for pit-sawyers to fulfil and other difficulties were experienced. 1968 J. Arnold *Shell Bk. Country Crafts* 86 Pit sawyers, it seems, were a race much on their own, uncommunicative and perhaps brutish. 1887 H. E. P. Clinton *Treat. Ensilage* 56 The quality of the material is certainly superior in many cases to pit silage.

1951 Watson & Smith *Silage* vi. 95 (*heading*) Making of pit silage. 1886 R. S. Burn *Systematic Small Farming* xx. 252 While the retaining or enclosing walls of the above-ground silo should not be less than nine inches, the lining walls of the pit silo may be very much thinner. 1947 *New Biol.* III. 45 A pit silo is merely a trench dug in the ground to convenient dimensions and, if properly drained, the wastage in these pits is less than that commonly experienced in many tower silos. 1966 Webster & Wilson *Agric. in Tropics* xiii. 292 Ensilage of these grasses in either pit or tower silos presents no real difficulties, and is commonly the best way to preserve fodder for use in the dry season, but it involves some losses. 1913 D. H. Lawrence *Sons & Lovers* ii. 28 He..put on his pit-singlet. 1920 —— *Phoenix* (1936) 9 Once more the rabbit was wrapped in the old pit-singlet. 1858 *Times* 22 Dec. 7/6 The part nearest the orchestra is railed off for three rows of orchestra stalls..and gives the same dimensions for perfect comfort to the occupant as is afforded by the pit-stalls at Covent-garden. Behind these are four rows of pit-stalls, the charge for each of which is only 2s.; while admission to the body of the pit will be reduced to 1s. 6d. 1861 Dickens *Uncomm. Trav.* iv. 43 A pit at sixpence, boxes and pit-stalls at a shilling, and a few private boxes at half-a-crown. 1961 Bowman & Ball *Theatre Lang.* 260 *Pit stall*, in British terminology, a stall in the front rows of the pit. 1932 S. C. H. Davis *Motor Racing* v. 72 Pit stop. 1942 Berrey & Van den Bark *Amer. Thes. Slang* § 728/1 *Pit stop*, a stop for oil and gas &c. during a race. 1970 *Globe & Mail* (Toronto) 28 Sept. 22/4 Revson, forced to a pit stop on the 10th lap of the 210-mile race, easily beat out Ferrari's Jim Adams for third place. 1972 *Times* 15 June 4/4 The inner harbour is a huge nautical pit-stop. 1973 'E. Fenwick' *Last of Lysandra* xx. 136 His wife..was relieved and pleased to see him, until she understood this was only a pit-stop. 1977 *Time* 4 July 36/1 A pit stop is called for. Time to eat and run. 1762 in G. O. Seilhamer *Hist. Amer. Theatre* (1888) I. xiv. 139 Pit tickets sold at the door, 1s 6 at 5s. 1786 J. Woodforde *Diary* 26 June (1926) II. 253 For 3 Pit Tickets at the Circus 1 p[d] o. 9. o. 1829 H. Foote *Compan. to Theatres* 94 Boxes may be engaged for the night, or season; and pit tickets purchased for 8s 6d. 1864 D. G. Rossetti *Let.* 5 July (1965) II. 513 He will reserve for me his two pit tickets for *Mirella* tonight. 1883 W. S. Gresley *Gloss. Terms Coal Mining* 189 *Pit-tip*, a bank or heap upon which rubbish out of the mine is tipped. 1907 *Westm. Gaz.* 13 Apr. 10/1 In the Black Country may be seen birches growing luxuriantly on a pit-tip. 1920 A. Fay *Gloss. Mining & Mineral Industry* 516/2 *Pit tip* (Eng.), a bank or heap upon which mine waste is tipped or dumped. 1895 Kipling *Second Jungle Bk.* 20 It was a pointed stick, such as they set in the mouth of a pit-trap. c 1909 D. H. Lawrence *Collier's Friday Night* (1934) iii. 76 His pit-watch that the Mother hung there when she put his pit-trousers in the cupboard. 1913 —— *Sons & Lovers* ii. 27 He..struggled into his pit-trousers. 1862 *Cornh. Mag.* Mar. 352 Pit villages..vary much in their character for cleanliness. 1957 R. Frankenberg *Village on Border* 1 Reviewers have..stressed the novelty of..information ..about a pit-village in Yorkshire. 1967 A. L. Lloyd *Folk Song in England* v. 333 The keelmen..joined with the colliers' communities in dances on the pit-village green. 1978 *Peace News* 25 Aug. 10/3 When I first came across the *Manual* I was visiting people in a budding action group on a pit village council estate. 1860 A. J. Munby *Diary* 29 Sept. in D. Hudson *Munby* (1972) 76 A photograph of a pitwoman in costume. 1841 C. H. Hartshorne *Salopia Antiqua* 532 *Pit wood*, wood which is thus called generally runs from three feet six inches to four feet in length, and is very thick. It is used for supporting the roof of a coal pit. 1886 F. T. Elworthy *West Somerset Word-Bk.* 577 *Pit-wood*.., larch or other wood cut into lengths for supporting 'the roof' in coal-mines. 1922 W. Schlich *Man. Forestry* (ed. 4) I. 81 It has been shown that during the period 1909–1913, the average annual consumption of timber and pit-wood amounted to about 11 million loads. 1971 *Timber Trades Jrnl.* 21 Aug. 24/1 Output of home-produced pit-wood..fell to its lowest level since 1951. 1961 *Spectator* 29 Dec. 957 Pit-Yacker... This is the autobiography of a man brought up in..the seaport mining town of Seaham, County Durham. 1974 S. Dobson *Geordie Dict.* 72 A *pit yakker* is the Geordie term for a pitman. Possibly from *yak* or *yark* meaning to pull out (coals).

pit, *sb.*[2] Add: Also *S. Afr.* Also, a pip. (Further and later examples.)

1873 W. Matthews *Getting on in World* 26 One man may suck an orange and be choked by a pit, another swallow a penknife and live. 1913 C. Pettman *Africanderisms* 375 Pit... This word is in common use in South Africa as a name for the stones of fruit. 1951 J. Steinbeck *Burning Bright* i. 8 The bitter seed that's like the inside of a peach pit. 1972 'E. Lathen' *The Longer the Thread* x. 93 She called him an avocado without a pit. 1977 *H. E. V. Pickstone's Catal.* (S. Afr.) 9 The fruit [*sc.* a peach] is yellow to reddish yellow, the flesh is deep orange coloured right through to the pit.

2. *S. Afr.* An edible seed, esp. a pine-nut.

1947 [see *dennebol].

pit, *v.* Add: **6.** Of a driver in a motor race: to stop at a pit (*pit *sb.*[1] 1 h) for fuel or maintenance.

1967 *Autocar* 5 Oct. 39/3 Mike Spence was in the seventh place..when he pitted on lap 36 with sudden engine trouble. 1976–7 *Sea Spray* (N.Z.) Dec./Jan. 58/1 Gray drove a steady, sensible race, pitting half-way through to take on 164 litres of gas in just 45 seconds. 1978 'D. Rutherford' *Collision Course* 62 The rain came bucketing down... There was nothing for it but to pit, fit rain tyres and splosh cautiously round.

pita[2] (pī·tā). Also **peeta, pitah, pitta.** [ad. mod. Gr. πήττα, πίτ(τ)α bread, cake, pie, perh. f. Gr. πεπτ-ός cooked, f. πέσσειν, πεττ-, to cook, bake. Cf. Turk. *pide*, Heb. *pittah* in similar senses.] A thick flat bread of the kind common in Mediterranean and Arab countries, usu. cut open and filled with a meat or other filling. Also *attrib.*

1951, 1963 [see *felafel]. 1964 M. Duncan *Cooking Greek Way* 212 Pita Yaourtiou, Yoghurt cake. *Ibid.* 213 The pita at this [dough] stage should be about 2 inches thick. 1965 R. Howe *Balkan Cooking* 136 Pita is made in the same manner as Austrian struedel and its success depends entirely on the paper-like thinness of the pastry. 1967 A. Bailey in L. Deighton *London Dossier* 56 A bread called 'pita' that looks like an oven glove and is usually stuffed with meat, raw onions and tomatoes. 1970 *Times* 29 Apr. 18/4 With your souvlakia you will get pitta, excellent flat round bread. 1971 D. Meiring *Wall of Glass* xi. 87 They had..eaten pita and humus..in the Arab restaurants in the port. 1974 *Times* 4 May 11/2 The Continental..on the Charing Cross Road..offers..warm Greek pitta stuffed with turkey. 1975 *Jewish Chron.* 18 July (Food & Wine Suppl.) p. iii/2 A pita—the flat Arab style bread—is split open, a few of the falafel balls stuffed in. 1976 *Islander* (Victoria, B.C.) 1 Aug. 4/1 (*caption*) Phyliss buying fallafel and pita in Jerusalem. 1977 *National Observer* (U.S.) 8 Jan. 12/3, I made my own pita bread, the flat chewy Arab bun that can be cut in half and stuffed with any number of goodies. 1978 *Lancashire Life* Apr. 71/3 We were tackling the first course... For me that was taramosalata, a puree of smoked cod roe, olive oil and lemon juice which had a most distinctive tang; it was served with a great quantity of pitta bread and was quite different from the last taramosalata I tasted. 1979 T. Barling *Olympic Sleeper* ix. 95 The boy brought olives and peppers, sliced cheese and tomatoes, hot pita..and marinated octopus.

Pitcairner (pi·tkē^ə^ɹnəɹ). [f. the name *Pitcairn* (see below) + -er[1].] A native or inhabitant of Pitcairn Island in the central South Pacific, settled with a mixed European and Polynesian population by mutineers from the *Bounty* in 1790.

1831 J. Barrow *Eventful Hist. Mutiny of Bounty* viii. 338 The Pitcairners have already proceeded from the simple canoe to row-boats. 1853 T. B. Murray *Pitcairn* v. 152 After some slight refreshment, (for they have only two regular meals a day,) the business of the Pitcairners' day begins. 1857 V. Lush *Jrnl.* 18 Nov. (1971) 195 As Norfolk Island is the nearest to New Zealand they touched there on their outward voyage and landed Mrs Selwyn, who remained with the Pitcairners till the *Southern Cross* called for her on their way home. 1874 C. M. Yonge *Life J. C. Patteson* I. vii. 253 The Pitcairners..had not yet arrived [at Norfolk Island], but were on their way from their original island. 1933 C. Chauvel *In Wake of 'Bounty'* vi. 82 The Pitcairners have often gone to sea without a proper supply of petrol. 1962 H. Luke *Islands of South Pacific* vi. 88 The Pitcairners throve and multiplied on their small but fertile island. 1964 Ross & Moverley *Pitcairnese Lang.* i. 25 In 1859, two families of Pitcairners ..returned to Pitcairn, to be followed by a further four families..in 1864. 1971 *Daily Tel.* 23 Dec. 3/8 Christmas Day usually is also Pitcairn's annual election day, but a shift in dates was necessary this year because the God-fearing Pitcairners naturally cannot vote on their Sabbath.

Pitcairnese (pitkē^ə^ɹnī·z). [f. as prec. + -ese.] The language of Pitcairn Island, a mixture of English and Polynesian (mainly Tahitian) elements. Also *attrib.* or as *adj.*

1964 Ross & Moverley *Pitcairnese Lang.* i. 25 The Pitcairnese language—the subject of the present book—has survived both on Pitcairn and on Norfolk. 1968 *Anglia* LXXXVI. 360 Later accounts will extend our knowledge of the detail of Pitcairnese and a more extensively documented description will be especially useful. 1973 *Word* 1966 XXII. 341 Since Pitcairnese has come out of the interaction of English and Tahitian, its lack of inflectional suffixes is attributed to their absence from Tahitian.

pitch, *sb.*[1] Add: **5. pitch-fibre,** a black, waterproof material which consists of compressed cellulose or asbestos fibre impregnated under vacuum with pitch and is used for making pipes; **pitch-knot** (later examples).

1946 *Archit. Rev.* CI. 66/1 Externally the drains are in pitchfibre with precast concrete manholes. 1958 *Daily Tel.* 30 June 4/6 The sales of pitch-fibre pipe continue to expand with the coming into operation of considerably increased productive capacity. 1964 L. T. Minchin *Famous Pipelines of World* vii. 95 Pitch-fibre is in fact a mixture of about 25% waste paper and 75% coal-tar pitch. 1825 J. Neal *Bro. Jonathan* I. 58 The fire-place, within which two or three lighted pitch knots, a substitute for candles, were burning. 1850 H. C. Watson *Camp-Fires of Revolution* 157 We must have some more pitch-knots on the fire.

pitch, *sb.*[2] Add: **I. 2. b.** *Aeronaut.* and *Astronautics.* = *pitching *vbl. sb.*[1] 8 b; also, the extent of this motion; *angle of pitch*, the angle between the plane containing the lateral axis and the relative wind and that containing the lateral and longitudinal axes.

1915 *Rep. & Mem. Advisory Comm. Aeronaut. 1913* No. 108. 1 The tests on each model comprise the determination of lift and drift for angles of pitch from −10° to +10° by 2° steps. 1920 L. Bairstow *Appl. Aerodynamics* iv. 223 The curves for 0° and −5° pitch are seen to lie below those of the rudder alone. 1921 S. Brodetsky *Mech. Princ. Aeroplane* iv. 148 Let there be pitch through an angle Θ about the new Y axis. 1935 *Encycl. Aviation* 493/2 In horizontal flight the angle of pitch is the angle between the longitudinal axis and the direction of motion of the aircraft. *Ibid.* 585/1 Thus a roll causes a yaw, and a yaw causes a roll... When, as often, a pitch is also introduced, it soon becomes apparent why the problem is a difficult one. 1967 *Technol. Week* 20 Feb. 35/3 When the booms are deployed, the spacecraft moment of inertia in pitch and roll with respect to Earth is about 250,000 slug-ft.[2] 1974 *Sci. Amer.* Dec. 138/2 Shifting his weight to control the craft in pitch, roll and yaw.

3. a. (Earlier example.)

1833 J. Nyren *Young Cricketer's Tutor* 46 The first thing he [*sc.* the fieldsman] should make himself master of, is to play from the pitch of the ball, and the motion of the batsman, so as to *get the start of the ball.*

5. a. (Earlier example.)

1888 'R. Boldrewood' *Robbery under Arms* III. xv. 232 Starlight and Jim were having a pitch about the best way to get aboard one of these pearling craft, and how jolly it would be.

b. Tendentious or persuasive acting or speech, esp. inflated or exaggerated sales-talk; an instance of this, a 'line'.

1876 C. Hindley *Life & Adventures of Cheap Jack* 255 When I had done my 'pitch' and got down from the stage. 1926 *Variety* 29 Dec. 7/4 The outdoor show game with its 'rag front',..'pitch', [etc.]. 1935 A. J. Pollock *Underworld Speaks* 88/2 *Pitch*, a satisfactory interview with intended victim by a high pressure stock salesman. 1962 *Listener* 18 Jan. 128/1 I've often sat in the living-room listening to some other joker give his pitch before I could give mine. 1968 *Globe & Mail* (Toronto) 3 Feb. 3/1 Organizers are planning to allow 40 minutes for each candidate to make his pitch to the convention. 1973 *Washington Post* 13 Jan. A22/2 One novel remedy was correctional ads that required a company to tell the consumer that its earlier pitch was not totally true. 1974 K. Millett *Flying* (1975) iii. 305 Nell hangs fire in the kitchen while I make my pitch. 1976 'O. Bleeck' *No Questions Asked* iii. 41, I made my pitch down at headquarters. I told them..I'd better be assigned to this thing full-time. 1976 *Times* 2 Feb. 16/4 Mr Jack Jones has been squealing before he is hurt... He proceeded to get his pitch in early. 1976 *National Observer* (U.S.) 14 Aug. 1/4 Advertising was missing a bet by forsaking the wee- and off-hours for its pitches. 1977 *Rolling Stone* 16 June 43/3 He's not out there making a pitch to the audience to really love him. 1978 *Observer* 29 Jan. 15/6 Actor Charlton Heston makes a recorded pitch for cable television.

III. 11. a. (Earlier Amer. and later examples.) Also, a crowd gathered by a 'barker' or around a stall, etc.; the part of a market, stock exchange, etc., where particular commodities are bought and sold.

1699 *Derby* (Connecticut) *Town Rec.* (1901) 207 The laying out of John Pringles pitch upon the good hill. 1746 *Waterbury* (Connecticut) *Proprietors' Rec.* (1911) 166 A ten Acre pitch which his Father bought of Thos Judd of Hartford. 1943 W. B. Taylor *Shake it Again* xxi. 199 A well-known drapery pitcher (one who sells drapery by pitching it, i.e. telling a story about each article offered, usually gagging in an entertaining way while describing, to keep the pitch interested). 1949 [see *fanny *v.*]. 1955 *Times* 19 Aug. 8/3 Once a month, roughly, from May to Poppy Day I am a seller. I prefer the same time—8.30 to 10 o'clock, and the same pitch—outside a row of busy shops, so I meet a number of the same people. 1959 *Encounter* May 22/2 If the street is full, a new pitch is carved out [for a prostitute]. 1960 *Farmer & Stockbreeder* 29 Mar. 16/1 Barnstaple, with a much smaller pitch than Perth, was a seller's market. 1961 *Daily Tel.* 25 Feb. 15/1 Sir Tufton [Beamish] told Members that he had been to mock auctions in Britain and America. He gave an example of the jargon used: The top man operates his joint by nailing the streamers among the plunder-snatchers in the pitch got by his frontsman, [etc.]. 1978 *Times* 1 Sept. 19/1 Patchy trading on the traded options pitch pushed ICI to the head of the active stocks.

c. *to queer the pitch*: see *queer *v.* 2 b.

13. a. (Earlier and later examples.)

1871 'Thomsonby' *Cricketers in Council* v. 59 Let the pitch be well watered and rolled on the day before the match. a 1912 A. Lang *Poet. Works* (1923) ii. 62, I am the batsman and the bat, I am the bowler and the ball, The umpire, the pavilion cat, The roller, pitch,..and all. 1955 *Times* 12 May 4/4 The pitch dried too slowly to become really unpleasant during the Middlesex innings. 1972 J. Kay *Hist. County Cricket: Lancashire* vi. 45 The captains debated whether to continue after a long inspection of the pitch. 1976 *Evening Advertiser* (Swindon) 31 Dec. 20/3 The County Ground pitch is likely to be heavy.

b. In other outdoor games: the space on which the game is played; the field, the ground.

1902 *Glasgow Evening News* 7 Apr. 3/1 The International football match was made..memorable by..the collapsing of a portion of the terracing flanking the pitch. 1971 J. Reason *Victorious Lions* vii. 41 The natural banking which almost completely encircled the pitch had been ramped and grassed. 1975 *Times* 10 Apr. 12/3 St Etienne and Bayern Munich..played a goalless tie..on a snow-covered slippery pitch.

VI. 24. a. Freq. in *Mountaineering* (see quot. 1971).

1898, 1904 [in Dict.]. 1935 D. Pilley *Climbing Days* i. 5 Each *pitch* or passage of the climb seemed as important as the Battle of Waterloo. 1943 E. Shipton *Upon that Mountain* iv. 78 Nothing provides such a strong incentive to struggle on up at all costs as the memory of a really severe pitch below. 1954 [see *abseil]. 1956 C. Evans *On Climbing* iii. 47 The leader climbs each pitch first, anchors himself to the rock, and takes in the rope as the second climbs to join him. 1971 C. Bonington *Annapurna South Face* 323 *Pitch*, section of climbing between two stances or belay points. These might be of any length, depending on the length of the climbing-rope... Pitches were often as long as 200 feet. 1972 D. Haston *In High Places* i. 8 When the pair [of rock-climbers] have

run out one length of the rope between two stances a 'pitch' has been established.

b. (Further examples.) Also *Geol.* Now distinguished from the dip of a plane (e.g. a stratum) and applied to the inclination of a linear feature, being the angle it makes with a horizontal line in the plane containing it, i.e. (in the case of an ore shoot) with the strike; formerly also = *plunge *sb.* 6*, esp. when applied to folds.

1822 Conybeare & Phillips *Outl. Geol. Eng. & Wales* p. iii, The angle of inclination between these planes and that of the horizon, is called their dip, or pitch. 1868 G. H. Cook *Geol. New Jersey* 55 Pitch.—This term has come into use among those engaged in iron mining, to express the characteristic descent of the iron ore beds beneath the surface, towards the northeast. It is at right angles to the dip, and is in the same direction with the strike, though not horizontal. 1906 Lindgren & Ransome in *Prof. Papers U.S. Geol. Survey* No. 54. 205 If we assume that the shoot has an elongated, narrow shape, as usually is the case when projected on the plain of the vein, its geometrical relations may be designated as follows: Width or thickness, breadth, stope length, pitch length, and pitch... The pitch length..is the distance between the two extreme ends of the shoot; the pitch is the angle which the pitch length makes with the horizontal. 1907 H. Louis in *Trans. Inst. Mining Engineers* XXXIV. 236 In America the term 'pitch' has occasionally been applied to this obliquity of the axis of the ore-shoot.., and the writer wishes to propose that this word be definitely restricted in the literature of ore-deposits to this particular signification... It will be understood that the angle is the angle to the horizontal, and the direction is always the azimuth of the horizontal trace of the vertical plane in which the line.. of pitch lies. 1908 *Ibid.* XXXV. 73 Mr. H. W. G. Halbaum (Birtley) said that..most men used 'pitch' and 'dip' as interchangeable terms. *Ibid.* 75 Mr. E. R. Field (Victoria, Australia) said that in the Bendigo district of Victoria the word 'pitch' was universally used to show the dip of the ore-bodies in the direction of the strike of the lode. 1909 *Trans. Amer. Inst. Mining Engineers* XXXIX. 899 Our mine-surveyors recognize, but do not employ, the old usage of 'dip' and 'pitch' interchangeably. *Ibid.* 900 According to my view of American practice, the direction of the pitch is usually stated in terms of the strike. 1909 *Q. Jrnl. Geol. Soc.* LXV. 473 The rocks are thrown into a series of anticlines and synclines, with a fairly-steady pitch of 12° to 15° in a southerly direction. 1913, etc. [see *plunge *sb.* 6*]. 1936 E. B. Mayo in C. M. Nevin *Princ. Struct. Geol.* (ed. 2) vii. 195 (*caption*) Orientation of minerals and inclusions in an intrusive rock... Measurements recorded in field mapping include dip and strike of flow layers, or planar parallelism; pitch and strike of flow lines, or linear parallelism. 1942 M. P. Billings *Struct. Geol.* viii. 135 The pitch is the angle that a line in a plane makes with a horizontal line in that plane. 1962 Read & Watson *Introd. Geol.* I. viii. 449 The fold-axis may be horizontal, like the top of a railway tunnel, or it may be inclined, in which case the axis and the fold are said to plunge or to pitch. The plunge is measured in degrees from the horizontal in a vertical plane.., while the pitch is given by the angle between the fold-axis and the strike of the axial plane, measured in the axial plane. 1972 J. G. Dennis *Struct. Geol.* iii. 52 The pitch of a line within the given plane is defined as the angle between that line and any horizontal line in the plane; that is, it is measured within the plane.

VII. 25. (Examples in other contexts.)

1964 S. Crawford *Basic Engin. Processes* i. 6 The pitch (1/spacing) varies with the length of the file, e.g. the pitch of a 12 inch second-cut file is not the same as the pitch of a 6 inch second cut file. 1971 *Physics Bull.* Nov. 677/2 The gratings are either flat or concave... The 'pitch' is 295 grooves/mm,..and finer spacings are expected to become available.

b. (*transf.* examples.)

1953 *Nature* 25 Apr. 739/2 If there are ten phosphate groups arranged on each helix of diameter 20 A. and pitch 34 A., the phosphate ester backbone chain is in an almost fully extended state. 1956 *Jrnl. Chem. Physics* XXV. 570 It is..possible that helical chain configurations of different pitch can be obtained under different conditions.

e. A measure of the angle of the blades of a screw propeller, equal to the distance forward a blade would move in one revolution if it sliced the air so as not to exert thrust on it; (i.e. the pitch (sense 25 b) of the spiral that would then be traced by a point on the blade).

1863 [in Dict., sense 25]. 1867 N. P. Burgh *Mod. Marine Engin.* vi. 323/2 The principal dimensions of the propeller under notice, are: the screw is 18 feet in diameter, maximum pitch 26 feet..and the minimum pitch 20 feet. 1902 F. Walker *Aërial Navigation* v. 54 The pitch of screws varies as the ratio of the area of the disc or circle described by the tips to the area of the airship affording resistance to the air through which it passes. 1919 H. Shaw *Textbk. Aeronaut.* xi. 144 If there were no slip the propeller would move forward a distance equal to the theoretical pitch during each revolution, and ..no air would be driven backward by the propeller, and there would be no thrust. 1944 'N. Shute' *Pastoral* i. 1 He heard, passing away above his head, the high scream of an ungeared engine in fine pitch. 1957 *Encycl. Brit.* XX. 535/2 In controllable-pitch propellers..the blades are pivoted in the propeller hub so that the pitch of the blades can be controlled from within the ship. 1958 *Times Rev. Industry* Aug. 39/2 The Rotodyne takes off vertically and climbs away as a helicopter, steering in the required direction being achieved by altering differentially the pitch of the airscrews. 1960 R. A. Fry *Princ. & Construction of Aircraft Gas Turbines* v. 178 The pitch required for take-off and climb is finer than the pitch most suitable for cruising. 1971 P. J. McMahon *Aircraft Propulsion* viii. 246 The propeller cannot go into low pitch in flight; mechanical and hydraulic pitch locks are provided.

26. (sense 11) *pitch-holder*; (sense 23) *pitch-change, -movement, -pattern, -range, -scheme.* **pitch accent** *Phonetics*, (*a*) a prominence given to a word or syllable by the difference in pitch from its immediate surroundings; (*b*) *occas.* = Tone *sb.* 6 a; hence **pitch-accented** *a.*, having a pitch-accent; **pitch angle** *Aeronaut.*, the acute angle between the plane of rotation of a propeller and a straight line from one edge of a blade to the other in a direction tangential to its radius; **pitch axis** *Aeronaut.*, = *pitching axis* s.v. *pitching *vbl. sb.*[1] 12; **pitch contour** *Phonetics*, the pattern of continuous variation in pitch; **pitch control** *Aeronaut.* (equipment for) control of the pitch of an aircraft's propellers or rotors; also, control of the pitching motion of an aircraft; **pitch curve** *Phonetics*, = **pitch contour*; **pitch length** *Geol.*, the length of an ore shoot in the direction of greatest dimension; **pitch-meter, pi·tchmeter,** (*a*) a device in an aeroplane for detecting or measuring pitching; (*b*) an instrument for measuring the pitch of sound; **pitch phoneme** *Linguistics*, one of the four recognized levels of pitch, esp. a variation in pitch from one syllable to another which affects meaning.

1880 A. H. Sayce *Introd. Sci. of Lang.* II. vii. 109 The pitch-accent has been changed into a stress-accent. 1933 C. D. Buck *Compar. Gram. Greek & Latin* 161 Under accent one understands variations of either intensity or intonation, and speaks of a stress accent or a pitch accent according as one or the other element is the more conspicuous. 1945 *Word* I. 87, I consider especially important the remark that many languages seem to pass..from a tone- or pitch-accent to stress. 1958 D. Bolinger in *Word* XIV. 149 To avoid unwarranted associations, it is better to speak of *pitch accent* and to leave the term *stress* to the domain of word stress. 1972 R. S. Jackendoff *Semantic Interpretation in Generative Gram.* vi. 259 We will give an account of the semantics of pitch accents that makes their interaction with negation part of a more general process. 1975 *Language* LI. 201 He does not distinguish between 'normal' and pitch-accented intonation contours. 1977 *Archivum Linguisticum* VIII. 90 Elsewhere, a (probably light) stress fell on the pitch-accented syllable (Lithuanian, Common Slavic). 1902 F. Walker *Aërial Navigation* v. 54 The value of θ..which gives the maximum efficiency is the same whatever be the actual pitch angle. 1935 C. G. Burge *Compl. Bk. Aviation* 140/1 The sections of the blade near the tip will move on a helix of much greater diameter..than those near the boss. For this reason they are set at a less 'pitch angle'.., so that every part of the airscrew will try to move the same distance forward during one revolution. 1971 P. J. McMahon *Aircraft Propulsion* vii. 246 The pilot's control lever selects propeller pitch angle directly. 1959 F. D. Adams *Aeronaut. Dict.* 126/1 *Pitch axis*, a lateral axis through an aircraft, missile, or similar body, about which the body pitches. It may be a body, wind, or stability axis. Also called a 'pitching axis'. 1962 F. I. Ordway et al. *Basic Astronautics* ix. 368 Any vehicle motion will take place about three axes... These axes are the yaw axis, the pitch axis, and the roll axis. 1958 R. Kingdon *Groundwork Eng. Intonation* p. xxiii, *Tone*, a stress considered from the point of view of the pitch or pitch-change associated with it. 1964 J. C. Catford in D. Abercrombie et al. *Daniel Jones* 28 There are many detailed studies of certain aspects of voice: e.g. of..the mechanism of pitch-change. 1966 J. Derrick *Teaching Eng. to Immigrants* iii. 114 Regular patterns of pitch-change at the heavily stressed syllables in an utterance make up the intonation of English. 1959 E. Pulgram *Introd. Spectrogr. of Speech* xviii. 136 But one may invariably omit the registration of glottal pitch..because..the stylus-drawn pitch contour is unlikely to present faithfully the real pitch contour in the lower frequencies. 1961 *Amer. Speech* XXXVI. 223 Pitch contours applied by synthesis to recordings of natural speech. 1940 E. Molloy *Aeroplane Maintenance & Operation* XV. 87 When the pilot wishes to change the pitch attitude of the aircraft, he makes an alteration to the spring torque by means of the Bowden cable from the pitch-control lever in his cockpit. 1944 W. C. Nelson *Airplane Propeller Princ.* iv. 89 Various types of automatic pitch control requiring no attention from the pilot have been devised. 1955 Liptrot & Woods *Rotorcraft* vii. 64 The collective-pitch control and the throttle are normally interconnected through a..cam device. 1958 Lambermont & Pirie *Helicopters & Autogyros of World* 112 A rotor hydraulically operated for both cyclic and collective pitch controls. 1974 *Encycl. Brit. Macropædia* I. 373/1 Pitch control is obtained by means of movable flaps (elevators) hinged to the trailing edge of the stabilizer. 1902 E. W. Scripture *Elem. Exper. Phonetics* xxxii. 478 The course of pitch is greatly influenced by the neighboring consonants; the more emphatic the consonant, the greater is its influence on the pitch-curve; the following consonant often cuts the vowel off at or near the maximum. 1969 *Eng. Stud.* L. 327 Of all the recorded sentences..there were taken one duplex oscillogram..one pitch curve, and two intensity curves. This was done by inserting a pitch meter..and an intensity meter between the tape recorder and the registering apparatus. 1909 *Daily Chron.* 18 Nov. 4/7 One 'pitch' which was the envy of every pitchholder in London was for many years at the end of Burlington House. 1906 Pitch length [see *pitch *sb.*[2] 24 b]. 1965 G. J. Williams *Econ. Geol. N.Z.* viii. 107/1 The bonanzas have generally a pitch-length exceeding the level-length. 1947 *Jrnl. R. Aeronaut. Soc.* LI. 166/1 A gust can be detected by a pitch-meter which produces a differential pressure on a diaphragm with change of vertical component of wind. 1969 *Word* 1967 XXIII. 255 An instrument used for measuring the frequency of the fundamental is commonly (and erroneously) called a pitchmeter. 1976 *Times* 19 Aug. 12/6 Another new development is the electronic pitchmeter... A needle shows whether a note is sharp or flat of the required pitch, so that a piano could be successfully tuned in the middle of a factory floor if necessary. 1959 D. Cooke *Lang. Mus.* ii. 109 Monteverdi and others, began to introduce more and more liberty of pitch-movement to express the rhetoric of human passion. 1964 Crystal & Quirk *Syst. Prosodic & Paralinguistic Features in Eng.* 75 It is hoped to follow up these initial experiments by a more detailed study of the spectrographic shape of pitch-movement. 1973 *Archivum Linguisticum* IV. 19 For them [*sc.* Crystal and Quirk] a tone is subordinate to another if it has the same pitch movement. 1954 F. G. Cassidy *Robertson's Devel. Mod. Eng.* (ed. 2) xii. 381 There is a great deal less variation in the pitch-patterns of Middle-Western American than of British speech. 1961 *Amer. Speech* XXXVI. 215 An experimental pitch indicator for training deaf scholars... Deaf child is enabled to compare his own pitch pattern with that of his teacher. 1931 L. Bloomfield in *Language* VII. 206 The modern languages of Europe similarly use certain pitch-phonemes at the end of largest-forms: our falling pitch at the end of statements and our rising pitches for the two kinds of questions. 1933 —— *Language* xi. 171 In English, supplement-questions are distinguished not only by their special pitch-phoneme [¿], but also by a selective taxeme. 1973 *Archivum Linguisticum* IV. 17 In the field of English intonation studies, bones of contention..spring readily to mind:..pitch phonemes versus tones. 1959 D. Cooke *Lang. Mus.* ii. 110 The 'normal' pitch-range of music is an overall spread from just above the treble clef to just below the bass clef. 1964 Crystal & Quirk *Syst. Prosodic & Paralinguistic Features in Eng.* iv. 62 So too it seems likely that tonal subordination will come to be linked with the stress and pitch-range systems. 1973 *Archivum Linguisticum* IV. 25 Such features as overall pitch range, pitch register, and even voice quality..may perhaps best be regarded as characteristics..of larger units. 1933 L. Bloomfield *Language* v. 77 The fact that two utterances of the syllable *man* with different pitch-schemes are 'the same' speech-form in English, but 'different' speech-forms in Chinese, shows us that the working of language depends upon our habitually..discriminating some features of sound and ignoring all others.

pitch, *v.*[1] Add: **B. I. 1. b.** (Earlier examples.)

c 1690 in Alverstone & Alcock *Surrey Cricket* (1902) ii. 14 All you that do delight in Cricket Come to Marden Pitch your wickits. 1733 in H. T. Waghorn *Cricket Scores* (1899) 16 The wickets are to be pitched by twelve o'clock.

II. 15. d. *to pitch it strong* (and varr.): to speak forcefully; to state a case with feeling or enthusiasm; to exaggerate.

1837 Dickens *Pickw.* xxxix. 429 I'm going to write to my father, and I must have a stimulant, or I shan't be able to pitch it strong enough into the old boy. 1841, 1863 [see Strong *adv.* 1 c]. 1886 R. L. Stevenson *Dr. Jekyll & Mr. Hyde* 7 And all the time, as we were pitching it in red hot, we were keeping the women off him as best we could. 1903 Wodehouse *Tales of St. Austin's* 213 Try him, anyhow. Pitch it fairly warm... Only cat you ever loved, and that sort of thing. 1916 J. Buchan *Greenmantle* i. 5 My heart was beginning to thump uncomfortably. Sir Walter was not the man to pitch a case too high. 1928 Galsworthy *Swan Song* ii. xi. 194 Pitch it strong, but no waterworks. 1969 *New Scientist* 3 July 37/1 Dr Steven Rose..was not pitching it too high when he said that the dangers of uncontrolled technology were as great as those of nuclear warfare.

III. 17. c. *Baseball*: (examples); also *fig.* *Cricket*: (earlier example); still current with constr. indicating the length at or off which the ball is delivered, or the direction of the delivery.

1767 R. Cotton *Cricket Song* vii, in F. S. Ashley-Cooper *Hambledon Cricket Chron.* (1924) 184 Ye Bowlers take heed..; Spare your vigour at first... But measure each step, and be sure pitch your length! 1845 in *Appleton's Ann. Cycl. 1855* (1886) 77/2 The ball must be pitched and not thrown to the bat. 1851 J. Pycroft *Cricket Field* viii. 165 Then, with a much higher toss and slower pace..he pitches a little short of the usual spot. 1853 F. Gale *Public School Matches* 54 In vain does Leftarm pitch up a whole over of half volleys in the hopes of a catch. 1868 H. Chadwick *Game of Base Ball* 60 When he [*sc.* the pitcher] makes a motion to pitch and does not do so,..he makes a balk. 1890 [see *catch *v.* 24 e]. 1929 *Chicagoan* 17 Aug. 22/1 Diamond slang crops out in his speech..as when he [*sc.* Carl Sandburg] instructs his agents never to book him for two consecutive lectures. 'I can't pitch two games in a row', he says. 1936 G. Milburn *Catalogue* 234 'Reck it'd be all right for me to go in there and dance.'..'You're all right. Go right on in there and pitch.' 1944 *College Topics* (Univ. of Virginia) 30 Mar. 3 Hank Neighbors, who pitched two innings of college ball here last year, is the only semblance of an experienced pitcher on the squad. 1970 [see *force *v.*[1] 5]. 1970 *Globe & Mail* (Toronto) 25 Sept. 31/1 Cuellar pitched his 21st complete game and broke a club record for season victories. 1977 *World of Cricket Monthly* June 33/3 He charged down the pitch to a leg-break which the bowler pitched wide.

d. (Later examples); cf. *pitch *sb.*[2] 5 b. *to pitch* (*the*) *woo* (orig. and chiefly *U.S.*): to court, to make love to; *transf.*, to flatter lavishly.

1935 *Ladies' Home Jrnl.* Feb. 60/3 After a while Uncle Ned came back looking positively exalted, so I guessed he and May had been pitching some more woo. 1937 *Clarionette* (Univ. of Denver) 18 Mar. 1/3 As long as there are students and universities and sofas and automobiles and nice laws, scholars will do their time-honoured share of 'pitchin' the woo'. 1943 *Sat. Even. Post* 25 Sept. 12/1 Louie..pitches kitchen gadgets. 1943 [see *pitch *sb.*[2] 11 a]. 1943 Hunt & Pringle *Service Slang* 51 *Pitch a woo*, to start a courtship. 1972 *Village Voice* (N.Y.) 1 June 26/2

Like any good salesman, he knows that once he demonstrates that the basic program he is pitching really does some good, all the ancillary merchandising will take care of itself. 1973 *Internat. Herald Tribune* 15 June 5/5 He's still a master at pitching the woo—on the mound, in the pressroom or elsewhere.

18. b. *Cricket.* Of a delivery: to land (usu. at or off a specified length, or in or off a specified direction).

1816 W. Lambert *Instr. & Rules Cricket* 32 If a Ball should pitch short of its proper length on the off side, and should twist toward the top of the wicket, the Striker must be very careful in playing back that he does not hit his own wicket. 1947 N. Cardus *Autobiogr.* i. 79, I was certain the ball had pitched off the wicket. 1970 R. Bowen *Cricket* v. 76 When the ball pitches, the forward motion is hindered by its friction with the ground: the circular motion is stopped abruptly (or nearly so) and this cessation must yield..further force in the direction the ball is moving. 1977 *Times* 17 Jan. 7/1 Patel received the perfect ball from Underwood which pitched on his middle stump and hit the off.

19. f. *Aeronaut.* and *Astronautics. intr.* Of an aircraft or spacecraft: to rotate or rock about a lateral axis, to undergo pitching. Also *trans.*, to cause (an aircraft, etc.) to do this.

1874 *Ann. Rep. Aeronaut. Soc.* 59 If..the model pitches forward on its nose, it is only necessary to slide the aeroplane further forward on the rod. If it still pitches turn up the horizontal rudder slightly. 1903 *Aeronaut. Jrnl.* VII. 53/2 The best angles were given by shapes which..would always pitch forward unless controlled by a large and well turned up tail. 1918 Cowley & Levy *Aeronautics* vii. 152 The pitching moment produced is about 123 lbs.-ft., enough to pitch the aeroplane through an angle of ½°. 1926 *Jrnl. R. Aeronaut. Soc.* XXX. 521 The examination would be comparatively easy if only the operation of the longitudinal control simply pitched the aeroplane, the lateral control banked and the rudder control yawed it. 1932 R. Mahachek *Airplane Pilots' Man.* iv. 37 The plane will gradually pitch upward or downward unless the pilot moves his control stick slightly forward or backward. 1961 D. Myrus *Man into Space* ii. 36/1 At about 50,000 feet the engines automatically flip slightly to one side, pitching the missile from straight up to a little north of due east. 1964 J. E. D. Williams *Operation of Airliners* vii. 104 When an aircraft yaws or pitches there is an immediate change in the aerodynamic forces. 1970 N. Armstrong et al. *First on Moon* x. 242 We pitched over to a level altitude which would allow us to maintain our horizontal velocity and just skim along over the top of the boulder field.

IV. 20. a. (Further examples.) Also, of a roof or other structure: to slope downwards (*U.S.*). In *Mining* and *Geol.* now used esp. of a linear feature, as an ore shoot or fold axis (cf. *pitch *sb.*[2] 24 b): to have a pitch of a given angle and direction.

1771 in *Mass. Hist. Soc. Coll.* (1914) LXXI. 137, I should have the Roof to pitch from under the Arkitraves of the Chamber Windows. 1859 *Trans. Illinois Agric. Soc.* III. 538 The roof may pitch both ways, or shed at the ends. 1897 F. C. Moore *How to build Home* vii. 94 The floor shall pitch from building to the front of piazza ½ inch to every foot of width. 1906 *Prof. Papers U.S. Geol. Survey* No. 54. 206 In the Midget mine the Cobb ore shoot pitches 45° NE. 1910 Lake & Rastall *Text-bk. Geol.* i. 20 A fold whose axis was inclined downwards towards the south-east would be said to pitch to the south-east, and the angle of pitch could be expressed in degrees, as in the case of dip. 1939 A. K. Lobeck *Geomorphol.* xvii. 593 The two monoclinal ridges formed on the two limbs of the anticline do not run strictly parallel ..but converge and meet, the convergence being in the direction toward which the anticline pitches. 1962 [see *pitch *sb.*[2] 24 b]. 1966 E. H. T. Whitten *Struct. Geol. Folded Rocks* i. 26 In a sedimentary sequence folded about a horizontal fold axis one bed has a strike of 130 and dips SW at 45°. Ripple marks are observed on this bedding plane; they pitch at 40° to the southeast (i.e., the angle between the strike and the ripples is 40°).

c. To drop *down* or descend abruptly (to a lower level).

1851 N. Kingsley *Diary* 21 Jan. (1914) 168 We have come to where the bed rock pitches down suddenly. 1867 'T. Lackland' *Homespun* i. 70 One of these [pastures].. sloping where it does not pitch, down to the rocky bed of the riotous stream. 1873 J. Miller *Life amongst Modocs* vi. 72 Gorge on gorge, cañon intersecting cañon, pitching down towards the rapid Klamat.

VI. 23. a. Also, to turn (aside) to a particular objective; to begin.

1932 Wodehouse *Louder & Funnier* 11 Then, with the coffee and old brandy at your side.., pitch in. 1971 W. Hillen *Blackwater River* iv. 36 A favorite stopping-place for..swans, cranes, and geese. They pitch in to feed and rest.

b. (Earlier and further examples.) Also in weakened senses.

1829 P. Egan *Boxiana* 2nd Ser. II. 267 Dick..*pitched in* to Warren, who was obliged to fight for his safety. 1835 Dickens *Sk. Boz* (1836) 1st Ser. I. 51, I wished..that the people would only blow me up, or pitch into me—that I wouldn't have minded. 1839 *Spirit of Times* 30 Mar. 48/2 The man was lost in astonishment which but increased the rage of the husband of the *cantatrice*, who forthwith 'pitched into' him in the last London style, and an entire 'mus' was made of the man's face. 1852 *Punch* 10 July 25/2, I saw that *gourmand* Guttler pitching contentedly into a kangaroo chop. 1906 G. B. Shaw *Let.* 18 Nov. (1972) II. 661 A vaccine opsinises your disease germs ..so that the white blood corpuscles..pitch into them with an appetite. 1952 *Sun* (Baltimore) 4 July 42/8 The two young Negroes pitched into him and the one with the gun pistol-whipped the physician.

c. *pitch out* (Cricket): to dismiss; to bowl out or to run out by a ball that does not touch the ground before it hits the wickets.

1858 *Bell's Life in London* 18 July 7/6 Caffyn was *pitched* out—the ball never touching the ground until after it had disturbed the stumps. 1876 *John Wisden's Cricketer's Almanack* 115 He was stated to have been brilliantly pitched out by Mr. Strachan from mid-off.

24. In extended use, **pitch-and-putt** [Pitch *sb.*[2] 3 c], a form of golf in which the green can be reached in one; *fig.*, an insignificant distance; *attrib.*, of or pertaining to a *spec.* type of miniature golf course.

1963 *Harper's Bazaar* Jan. 9/2 Pitch and Putt Course. Tennis. Sea-bathing. 1968 *Sun* (Baltimore) 5 July A14/7 His delegate count is within a pitch and a putt of the nominating majority. 1972 J. McClure *Caterpillar Cop* xiv. 234, I believe..you played a round of pitch-and-putt? 1974 *Times* 8 Feb. 15/4 A pitch-and-putt course covering five acres. 1976 J. Snow *Cricket Rebel* 70 Two days in which my only activity had been in a pitch and putt competition organised by the players and Press in the hotel grounds.

pitch-and-toss. Add: **a.** (Later Austral. example.) Also (*Sc.*), a manœuvre in the game of *knifey, knifie (sense *a*).

1949 [see *barrack *sb.*[2]]. 1969 I. & P. Opie *Children's Games* vii. 222 Described..by a 10-year-old boy in the Isle of Lewis: '"Knifie" is a game for two people... Then try "Pitch and toss". Stick it in the ground, then try to hit it with the palm of your hand and try to toss it into the air, so that it will land blade first in the earth.'

b. (Later examples.)

1910 Kipling *Rewards & Fairies* 175 If you can make one heap of all your winnings and risk it on one turn of pitch-and-toss, and lose, And start again at your beginnings And never breathe a word about your loss. 1960 J. Franklyn *Dict. Rhyming Slang* 108/1 *Pitch and toss*, the boss. 20 C. Current in the theatrical world (Lupino Lane). 1973 *Times* 29 Dec. 12/7 Writing more than a dozen varied plays of substance, some of which had less fortune than they deserved in the pitch-and-toss of the West End.

pitched, *ppl. a.*[1] **6.** (Earlier example.)

1871 *Baily's Monthly Mag.* Aug. 290 He bowled a very great number of long hops, and a considerable number of pitched-up balls to the leg stump.

pitcher[1]. Add: **1.** *spec.* in *Ceramics*, fired clay or shards used in the manufacturing process.

1771 J. Wedgwood *Let.* 13 Jan. (1965) 101, I have been..busy..in makeing a general review of all my experiment pitchers. 1964 H. Hodges *Artifacts* i. 20 Broken or spoilt pottery is commonly used, and while the term grog is general amongst all potters to denote the addition of fired clay, some potters also use the terms pitchers and sherds. a 1977 *Harrison Mayer Ltd. Catal.* 14/1 *Pitchers*, fired, broken or scrap pottery. Biscuit pitchers have various uses when crushed or ground.

pitcher[2]. Add: **1. b.** Also, a market porter.

1966 *New Statesman* 7 Oct. 531/2 The grandmother had been married to a Smithfield meat 'pitcher' who had died of cancer. 1970 *Times* 26 Feb. 10/2 No longer are porters divided into pitchers (the men who carry fruit in), plain porters (who carry it out) and stand men (who work inside warehouse or shop).

2. (Earlier and later examples.)

1845, 1867 [see *balk *sb.*[1] 5 b]. 1948 *Chicago Daily News* 4 June 36/8 Claude Passeau's job with the Cubs is to tutor young pitchers throughout the Chicago farm system. 1949 *Lafayette Alumnus* (Lafayette College, Easton, Pa.) 24 Oct. 1/2 By crashing in he ruined innumerable would-be pass plays getting the pitcher for losses averaging 9 yards. 1974 *Post-Herald* (Birmingham, Alabama) 29 June A 14/3 Lane gave the Tigers a 2-0 lead with a two-run homer in the second innings off losing pitcher Kevin Kobel. 1978 *Verbatim* Feb. 2/2 A ballplayer who is not an infielder, outfielder or pitcher, and is thus doomed to be a catcher, wears 'the tools of ignorance', catcher's gear.

pitcher[4], repr. a vulgar or colloq. pronunciation of Picture *sb.*

1916 [see *humdinger 1]. 1931 *Amer. Speech* Oct. 46 *Moom pitcher* for moving picture. 1936 Mencken *Amer. Lang.* (ed. 4) vii. 352 On the vulgar level *amateur* is always *amachoor*, and *picture* is *pitchur* or *pitcher*. 1977 J. Wambaugh *Black Marble* (1978) vii. 99 She do look somethin like that dumpy consti-pated broad in the pitcher.

pitcher-plant. For 'the East Indian genus *Nepenthes*' substitute 'the south-east Asian genus *Nepenthes*'. (Earlier and later examples.)

1819 L. A. Anspach *Hist. Newfoundland* xiv. 362 Another still more remarkable plant, found in the woods of Newfoundland, is the *Saracenia* [*sic*], commonly called side-saddle flower, or pitcher-plant. 1929 Robbins & Rickett *Botany* xxiii. 356 Some of the most striking leaf modifications are in the plants known as pitcher plants. 1938 [see *bake-apple]. 1947 I. L. Idriess *Isles of Despair* xxxiii. 220 She could not sleep.., thinking of the phantom pygmies in the pitcher-plant swamps. 1965 *Austral. Encycl.* VII. 123/2 The Western Australian pitcher-plant is *Cephalotus follicularis*. 1965 D. Henderson *Heart of Newfoundland* 19 We had seen pitcher plants before..but never in such numbers as are found in Newfoundland. 1973 *Sci. Amer.* Dec. 64/2 In these [Borneo] forests the insect-eating pitcher plants (*Nepenthes*) are common. 1977 *Borneo Bull.* 7 May 8/1 Did you know that of 30 pitcher plant species found in Borneo, 16 are said to be found on the mountain, and many of them are peculiar to its slopes?

pitchfork, *sb.*[1] (Earlier and later examples of *to rain pitchforks.*)

1815 D. Humphreys *Yankey in Eng.* 55 I'll be even with you, if it rains pitchforks—tines downwards. 1850 J. R. Planché *Island of Jewels* ii. iii. 31 Rain cats and dogs, or pitchforks perpendicular, The sky's not mine, and need'nt be particular. 1930 J. Dos Passos *42nd Parallel* i. 77 Outside it was raining pitchforks. 1940 M. Fishback *Time for Quick One* 77 It's raining cats and dogs And pitchforks and assorted frogs.

pitch hole[1]. Add: **2.** *N. Amer.* A defect in a road or trail; a pot-hole.

1874 *Rep. Vermont Board Agric.* II. 659 The highways leading to our larger villages..are frequently so full of pitchholes or 'cahoos' as to render them totally unfit for travel. 1890 *Harper's Mag.* Oct. 657/2 The highway was frequently interrupted by 'pitch holes'. 1936 A. F. Cross *Cross Roads* (ed. 2) 106 Charlie's horses jogged on with us ..into pitch holes that jarred one's innards terribly. 1962 D. J. Dickie *Great Golden Plain* 288 The cat train rocks and plunges over the hummocks and pitch holes in the ice and snow roads.

pitchi (pi·tʃi). *Austral.* [Aboriginal.] A dish or container hollowed out of a solid log.

1896 E. C. Stirling in B. Spencer *Rep. Horn Sci. Expedition Cent. Austral.* IV. 98 The name 'Pitchi'..is in general use, by the whites of the parts traversed by the Expedition, for the wooden vessels used for carrying food and water and, occasionally, infants. 1931 I. L. Idriess *Lasseter's Last Ride* (1933) xvi. 128 In little canoe-shaped pitchis of wood some women carried native foods. 1934 A. Russell *Tramp-Royal in Wild Austral.* xxvii. 179 Poised on his [*sc.* an aboriginal's] head was a newly cut bark pitchi filled to the brim with wild oranges. 1959 S. H. Courtier *Death in Dream Time* viii. 112 While his thumbs bleed, the elders bring a *pitchi*, a wooden vessel, and catch the blood. 1970 'E. Lindall' *Gathering of Eagles* x. 119 In her hands..was a wooden pitchi full of water.

pitch-in (pi·tʃin). *U.S. colloq.* Also **pitchin.** [f. Pitch *sb.*[2] + In *adv.*] Applied, usu. *attrib.*, to a large common meal to which each diner contributes food or drink.

1973 *Sunday Bull.* (Philadelphia) 7 Oct. (Parade Suppl.) 26/2 The answers were as varied as the food at a large pitch-in picnic. 1976 *Laurel* (Montana) *Outlook* 16 June 9/1 The Park City Garden Club guest night was held at the civic center with a pitchin dinner. 1976 *Columbus* (Montana) *News* 17 June 6/3 Mr. and Mrs. Charles Wark were hosts at a pitch-in picnic for members and their families.

pitching, *vbl. sb.*[1] Add: **7.** (Examples in *Baseball*; also *attrib.*)

1858 *N.Y. Tribune* 18 Aug. 7/3 The pitching was good on both sides. 1896 H. Chadwick *Spalding's Base Ball Guide* 2 A new form of pitching tables are included in the records of the pitching of 1895. 1942 [see *major *a.* 1 e]. 1944 *College Topics* (Univ. of Virginia) 30 Mar. 3 As for the pitching staff, there is plenty of material that will eventually develop into the team's strongest asset. 1948 *Richmond* (Va.) *Times-Dispatch* 15 Mar. 17/4 Blackwell learned how to throw it from Hal Schumacher, former New York Giants pitching star. 1969 *Eugene* (Oregon) *Register-Guard* 3 Dec. 2D/4 Cal Ermer..landed a coaching job, as pitching tutor for Bristol's Seattle club. 1976 *National Observer* (U.S.) 21 Feb. 19/2 (Advt.), Mickey Owen Baseball School... 4 Lighted Batting Cages, 2 Pitching Machines.

8. b. *Aeronaut.* and *Astronautics.* Angular motion of an aircraft or spacecraft about a lateral axis (the *pitching axis*: see 12 below).

1912 G. Greenhill *Dynamics Mech. Flight* iv. 83 Rolling and pitching of a steamer or flying machine. 1915 A. Fage *Aeroplane* vi. 75 The following nomenclature has been adopted at the National Physical Laboratory:—[*table*] Name of axis..Lateral... Name of Motion which takes place about Axis..Pitching. 1935 C. G. Burge *Encycl. Aviation* 579/1 Rotary motion of the aeroplane about the lateral axis is called pitching. 1958 D. Piggott *Gliding* xiii. 78 To stop the pitching, the pilot must relax the backward pressure on the stick and reduce the climbing angle for a few moments. 1965 C. N. Van Deventer *Introd. Gen. Aeronaut.* v. 84/1 The lateral axis is the side-to-side axis about which the airplane revolves in pitching, when its nose and tail move up and down. 1968 T. de Galiana *Conc. Encycl. Astronaut.* 210/2 For a cylindrical spacecraft,.. pitching is a movement of the nose up or down (that is, away from or towards the orbit focus).

12. pitching axis *Aeronaut.* and *Astronautics*, a lateral axis of an aircraft, etc., about which pitching takes place, usu. specified to be perpendicular to its longitudinal axis or to its direction of flight; of a spacecraft, one of two horizontal axes (cf. *yawing axis* s.v. *yawing *vbl. sb.*) which are perpendicular to each other and to the longitudinal axis; = *pitch axis* s.v. *pitch *sb.*[2] 26; **pitching heat** *Brewing* = *pitching temperature*; **pitching machine** *Brewing*, a special kind of vessel in which pitching of the wort takes place; **pitching moment** *Aeronaut.*, a moment tending to turn an aircraft, etc., about its pitching axis; **pitching-temperature** (example); **pitching-yeast** (examples).

1920 W. J. Walker tr. *Devillers's Dynamics of Aeroplane* xii. 234 The moment of inertia in movements about the pitching axis plays the rôle of mass in rectilinear displacements. 1953 *New Biol.* XIV. 66 Stability can be related to any of the three axes—the rolling axis.., the yawing axis,..and the pitching axis (horizontally at right angles to the direction of flight). **1959** Pitching axis [see *pitch axis* s.v. *PITCH *sb.*[2] 26]. **1876** *Encycl. Brit.* IV. 275/1 The heat at which the wort is let down into the fermenting tun. This 'pitching heat' varies very much. **1885** E. R. SOUTHBY *Syst. Handbk. Pract. Brewing* (ed. 2) xx. 335, I have already explained that this lowering of the pitching heat is by no means essential. **1940** H. L. HIND *Brewing* II. 854 Simple pitching machines consist of a cauldron, in which the pitch is melted over a coke fire or by gas, and forced by compressed air through a spraying nozzle into the cask. **1957** K. BARTON-WRIGHT tr. *De Clerck's Textbk. Brewing* I. xxii. 483 In modern pitching machines..the old lining is no longer removed with hot air,..but with very thinned pitch. **1913** *Rep. & Mem. Advisory Comm. Aeronaut.* No. 74. 10 Measurements of lift, drift, and pitching moment at varying values of the pitching angle from −10° to +25°. **1931** *Technol. Rev.* Nov. 65/1 It seems inevitable that the next generation must know as much about *ailerons, pitching moments*, and *dihedral angles*, as the present one does about carburetors, differentials, and wheel bases. **1966** D. STINTON *Anat. Aeroplane* xi. 199 Further pitching moments..are introduced by wing-mounted stores,..engines, flaps and undercarriage units. **1957** *Encycl. Brit.* IV. 105/2 The pitching temperature is held at 54–59° F. in American practice, slightly higher (58–60° F) in England. **1885** E. R. SOUTHBY *Syst. Handbk. Pract. Brewing* (ed. 2) xx. 320 The yeast cells of a good pitching yeast should be separate from one another. **1956** [see *POLYMYXIN].

pitching, *ppl. a.* **1.** Delete † *Obs.* and add later examples in *Geol.* (cf. *PITCH *v.*[1] 20, *PITCH *sb.*[2] 24 b).

1939 A. K. LOBECK *Geomorphol.* xvii. 595 The monoclinal ridges which result from the erosion of pitching synclines converge in a direction which is opposite to the pitch of the fold. **1960** B. W. SPARKS *Geomorphol.* vii. 135 Pitching folds may be reflected in a pattern of converging and diverging ridges like that of certain sections of the Appalachians. **1968** C. R. TWIDALE *Geomorphol.* ii. 11 Folds whose outcrops narrow and widen, appear and disappear, are described as pitching or plunging folds.

2. (Further examples.)

1906 *Chambers's Jrnl.* July 537/2 It is worth going some distance to see a *vaquero* sticking to a 'pitching' horse. **1948** *Sat. Rev.* 28 Aug. 37/1 Bucky Durant calmly rolled a cigarette as he sat atop the pitching bronc.

pi·tchman. *U.S.* [PITCH *sb.*[2]] One who sells gadgets or novelties at a fair or in the street. Also *transf.* and *fig.*, an advertiser, one who delivers a sales pitch (see *PITCH *sb.*[2] 5 b).

1926 *Amer. Speech* Feb. 283/1 *Pitchman*, a man who sells novelties on the circus lot or on the streets adjacent to the lot. **1934** *N.Y. American* 4 Oct. 21 The army of street pitchmen have had the biggest season in years. **1940** W. SAROYAN *Love's Old Sweet Song* 17 in *Three Plays*, Barnaby Gaul, 51, a pitchman. **1949** *Time* 19 Dec. 34/3 They are all journalistic racketeers—I mean pitchmen. **1956** H. GOLD *Man who was not with It* (1965) xix. 172, I coughed into my best pitchman's voice. **1972** *Guardian* 6 Dec. 13/1 American astronauts..have become bank directors, pitchmen for cars and railroads on television commercials, executives in real estate. **1975** *Publishers Weekly* 28 Apr. 44/1 Mandino is a pitchman for positive thinking in the hallowed line including Bruce Barton and Norman Vincent Peale. **1976** *National Observer* (U.S.) 2 Oct. 18/4 Boudreau was a pitchman in the old Clyde Beatty Circus before he went on to Juilliard and the Paris Conservatory. **1976** *Time* 20 Dec. 20/2 Cecil D. Andrus.. has been TV pitchman for Idaho potatoes. **1977** *Time* 7 Feb. 43/2 He has also taken on the role of pitchman, appearing personally in Eastern's ads to stress the line's concern for passengers.

pi·tch-off. [PITCH *v.*[1] 20.] The inclination or shelving of the bed of the sea.

1894 J. D. DANA *Man. Geol.* (ed. 4) I. i. 20 At Keeling atoll,..Captain Fitzroy, R.N., found no bottom in 7200 feet at 2200 yards from the breakers—which gives a pitch-off exceeding 1:0·92.

pi·tch-out. *N. Amer.* [PITCH *v.*[1] 17 c.] **a.** *Baseball.* (See quot. 1943.) **b.** *Football.* (See quot. 1950.)

1912 C. MATHEWSON *Pitching in a Pinch* vii. 157 If a catcher can get a pitchout on a hit and run sign he upsets the other team greatly. **1943** *Amer. Speech* XVIII. 106 The *pitch-out* (a ball thrown wide of the plate so that the catcher can *get it away* in a hurry to catch a man trying to steal). **1948** *Cavalier Daily* (Univ. of Virginia) 19 Oct. 2/1 The halfbacks, standing wide of the fullback, are in a position to take pitchouts and to speed downfield for passes. **1949** *Sun* (Baltimore) 8 Dec. 24/1 Other trends noted [in Football] included increased use of the pitchout (not to be confused with the baseball term). **1950** *Britannica Bk. of Year* 683/1 *Pitch-out, n. Football*, a short lateral pass behind the line of scrimmage, usually from the quarterback to another back. **1966** ROTE & WINTER *Lang. Pro Football* 129 Pitchout, a short, backward shuffle pass or lateral from quarterback to a halfback or a fullback. **1968** *Globe & Mail* (Toronto) 11 July 32/4 He took a pitchout from Curtis Wilson and danced, darted and jumped 86 yards. **1972** J. MOSEDALE *Football* vi. 87 The cornerback took his eyes off him on a pitchout. **1974** *Anderson* (S. Carolina) *Independent* 19 Apr. 4B/3 On Monday night, he got his first stolen base—despite a pitchout by the Chicago White Sox.

pi·tch-penny. *U.S.* [f. PITCH *v.*[1]] A variety of pitch-and-toss.

1830 J. F. WATSON *Ann. Philadelphia* 240 'Pitch-penny'..was frequent—to pitch at a white mark on the ground. **1877** E. S. WARD *Story of Avis* 286 Calculating the distance..as he stood playing the game of human pitch-penny with the infant. **1978** E. TIDYMAN *Table Stakes* I. i. 20 He had persuaded Hochmeir to give him his fifty cents a week in pennies so that he could try to double it in..pitch penny with the newsboys in Howard Square.

pitch pine. (Earlier and later examples.)

1676 *Essex Inst. Hist. Coll.* (1920) LVI. 306 4½ acres of land..bounded by a pitch pine. **1684** *Manchester* (Mass.) *Town Rec.* (1889) 17 A pich [*sic*] pine tree marked with 4 marks. **1709** J. LAWSON *New Voy. Carolina* 89 Ever-Greens are here plentifully found, of a very quick Growth, and pleasant Shade; Cypress, or White Cedar, the Pitch Pine, the yellow Pine. **1736** *Rec. Early Hist. Boston* (1885) XII. 150 Add to the South East Side Ten foot, to be built of Square Pitch Pine Timber. **1771** J. S. COPLEY *Let.* 3 Aug. in *Mass. Hist. Soc. Coll.* (1914) LXXI. 138 The floor..should be Pitch Pine. **1887** C. B. GEORGE *40 Yrs. on Rail* 31 Pitch pine was largely used for fuel. **1969** T. H. EVERETT *Living Trees of World* 50/2 Pitch pine, rarely more than 75 feet high, has an open irregular head. **1977** *Listener* 3 Nov. 594/2 The church where so many Trelawnies lie under the pitchpine and the shiny tiles.

pitch-pipe. (Later examples.)

1961 *Oxf. Univ. Gaz.* 8 May 1149/2 Some cylinders are prefaced with the sound of the pitch-pipe and statement of the key. **1969** E. H. PINTO *Treen* 172 The mahogany pitch pipes..were formerly used in churches which had no organ, to give the keynote before singing commenced. **1979** *Yale Alumni Mag.* Apr. 28/2 Armed only with pitch-pipes and pocket dictionaries, we were able to elicit a direct and natural response from all manner of Soviet citizens.

pitchpoll, -pole, *sb.* Add: **1.** (Further U.S. example.)

1842 S. KETTELL *Daw's Doings* ix. 61 Goosecap..did nothing all day long but lollop about at his ease..playing at pitchpole among the clover.

2. A kind of harrow.

1932 R. H. BIFFEN *Fream's Elem. Agric.* (ed. 12) ii. 48 The pitch-pole harrow is an implement of recent introduction, which has attained some popularity for tearing a thick mat on old pasture, and also for working arable land... The implement has a very drastic action and the heavy draught necessitates a tractor. **1940** R. G. STAPLEDON *Re-Grassing* 27 After this treatment had been continued a few years we dragged heavily with pitch-pole, lined heavily.., [etc.]. **1944** C. CULPIN *Farm Machinery* (ed. 2) vii. 132 (*caption*) Wilder's 'Pitch-Pole' harrow pulled by tractor equipped with skeleton wheels. *Ibid.*, Tines suitable for fairly deep arable cultivations are available, and some farmers use the Pitch-Pole mainly or exclusively as an arable land cultivator. **1951** P. OYLER *Feeding Ourselves* x. 92 It is a good plan to run heavy tine harrows or the modern Pitchpole over pastures again and again both ways. **1960** *Farmer & Stockbreeder* 16 Feb. 128/3 The average contract rate for pitchpole harrowing is 20s per acre first time.

pitchpoll, -pole, *v.* **a.** For *dial.* read 'orig. *dial.*' and add later example in causative sense.

1926 T. E. LAWRENCE *Seven Pillars* (1935) cxvi. 626 The wind snapped them [*sc.* thistles] off at the hollow root, and pitch-polled their branchy tops along the level ground, thistle blowing against thistle.

b. *intr. Naut.* (See quot. 1961.) Also *trans.* in causative sense.

1903 G. S. WASSON *Cap'n Simeon's Store* iii. 44 Ain't it hard lines enough for a sickly ole feller same's him having to go outside here late in the fall o' the year and pitchpole around into a punky old ark. **1908** [see *HUMBUG *v.* 2]. **1915** KIPLING *Fringes of Fleet* 67 Dawn sees them pitch-poling insanely between head-seas. **1961** F. H. BURGESS *Dict. Sailing* 160 *Pitch pole*, be up-ended, stern first, and completely overthrown by the sea. **1976** *Sci. Amer.* May 130/3 On this up-to-date foundation Van Dorn builds a careful scheme for wave forecasting at sea: the expected sea state and its changes (depending on fetch, wind speed and duration) and the chances that state implies for the big wave that can pitchpole any boat. **1976** *Yachting World* Oct. 111/1 Even in the worst conditions, such as running down steep 25 foot waves, it has never shown a tendency to pitchpole. **1979** *Observer* 19 Aug. 3/3 Huge weights of water..capable of laying a 35 foot boat flat on its side with ease, making it turn turtle or even pitch-poling it—making it somersault forward stern over stem.

Hence **pi·tchpoler**, one who pitchpolls a harpoon in *Whaling*; **pi·tchpol(l)ing** *vbl. sb.*

1851 H. MELVILLE *Moby Dick* II. xlii. 284 None exceed that fine manœuvre with the lance, called pitchpoling. *Ibid.* 286 The pitchpoler dropping astern, folds his hands. **1971** S. E. MORISON *European Discovery Amer.: Northern Voy.* p. xi, There you have first-hand accounts of some of the hazards that the early navigators encountered as a matter of course—enormous freak waves, pitch-poling, capsizing.

pitfalled, *a.* (Earlier and later examples.)

1826 *Blackw. Mag.* XX. 666/2 The snow deep, wreathed, and pit-falled. **1976** *Evening Post* (Nottingham) 17 Dec. 14/2 Sunken and rubble-littered uneven pavements, pit-falled with miniature craters.

pith, *sb.* Add: **1.** (Further examples.)

1884 BOWER & SCOTT tr. *H. A. de Bary's Compar. Anat. Veg. Organs Phanerogams & Ferns* ix. 403 Only in a few woods does the pith become entirely empty and dried up. **1928** HOLMAN & ROBBINS *Elem. Bot.* iii. 54 The pith is made up of large-celled parenchyma. **1956** F. W. JANE *Struct. Wood* iv. 74 The pith may be distinctive: thus, in oak it has, in section, the shape of a five-rayed star. **1976** R. F. LYNDON in M. M. Yeoman *Cell Division in Higher Plants* viii. 297 In those cells which form the pith the cell cycle may no sooner get to minimum length than it begins to lengthen again as the cells mature.

8. pith fleck, in certain woods, a small dark patch made by parenchyma cells filling cavities left by insect larvæ; **pith helmet** (further examples); so *pith-helmeted* adj.; **pith ray** = *medullary ray* s.v. MEDULLARY *a.* 2 b; **pith-ray fleck** = **pith fleck*; **pith-tree,** for *Herminiera Elaphroxylon* substitute *Æschynomene elaphroxylon*; (earlier and later examples).

1890 W. SOMERVILLE tr. *Hartig's Timbers* 77 Pith-flecks, darkish-coloured patches met with in some woods. **1911** *Forestry Q.* IX. 244 Pith flecks or medullary spots are small, brown, half-moon shaped patches appearing.. on the cross sections of many of our woods. **1956** F. W. JANE *Struct. Wood* x. 236 Rotary cut birch..has a most attractive figure, due largely to the very numerous pith flecks which it contains. **1970** PANSHIN & DE ZEEUW *Textbk. Wood Technol.* (ed. 3) I. x. 366 Pith flecks, or medullary spots, are confined to hardwoods. **1917** *Harrods Gen. Catal.* 643/3 Lady's Pith Helmet. Pith body, covered white drill. **1934** G. B. SHAW *Too True to be Good* II. 50 He wears a pith helmet with a pagan. **1971** *Country Life* 15 July 141/2 Who..buys a Cawnpore Tent Club pith helmet with quilted khaki cover? **1976** *Evening Post* (Nottingham) 15 Mar. 6/3 Has anybody a pith helmet..? Members of Aspley Methodist Church, who are to put on a pantomime 'The Sleeping Beauty'..would be glad to borrow one for their show. **1916** 'BOYD CABLE' *Action Front* 190 Half-forgotten illustrations in the papers of pith-helmeted infantry in the Boer War. **1902** G. S. BOULGER *Wood* 22 The whole mass of xylem is traversed radially by pith-rays. **1928** HOLMAN & ROBBINS *Elem. Bot.* iii. 54 In stems in which the vascular tissue does not form a continuous cylinder, the bundles are separated by pith rays. **1953** K. ESAU *Plant Anat.* xv. 343 The spaces among the strands..are occupied by parenchymatic ground tissue. These plates of tissue can be designated as pith rays or medullary rays. **1970** PANSHIN & DE ZEEUW *Textbk. Wood Technol.* (ed. 3) I. v. 179 Since all wood rays originate in the cambium, such terms as pith rays and medullary rays, frequently encountered in literature on wood, should not be used, because they imply that wood rays consist of the same kind of tissue as the pith. **1913** *Circ. U.S. Forest Service* No. 215 (*title*) Pith-ray flecks in wood. **1925** EAMES & MACDANIELS *Introd. Plant Anat.* vii. 187 Pith-ray flecks are, unfortunately, sometimes explained as normal features of wood structure. **1864** J. A. GRANT *Walk across Afr.* p. xv, Ambadj; native name for the pith-tree. **1961** F. R. IRVINE *Woody Plants of Ghana* 360 Ambatch or Pith Tree of Nile Land... By rivers and in swamps in open country.

pith, *v.* Add: **3.** Also *fig.*

1903 G. B. SHAW *Man & Superman* Pref. p. xxxiv, And yet..the respectable newspapers pith me by announcing 'another book by this brilliant and thoughtful writer'. **1913** —— *Quintessence of Ibsenism* (completed ed.) 192 The very first thing the theatrical wiseacres did with it [*sc. A Doll's House*] was to effect exactly this transformation [*sc.* a happy ending], with the result that the play thus pithed had no success. **1935** —— *Let.* 25 July in *To a Young Actress* (1960) 163, I entirely approve of Peter's escaping [from a public school] before he is pithed and turned out as a political and moral gentleman several centuries out of date.

pithecanthropus (piþǐkæ·nþrǒpǒs). Also with capital initial. [mod.L., f. Gr. πίθηκ-ος ape + ἄνθρωπος man.] **1.** [(E. Haeckel *Natürliche Schöpfungsgeschichte* (1868) xix. 507).] A hypothetical creature bridging the gap in evolutionary development between apes and man.

1876, 1877 [see PITHECANTHROPE].

2. [(E. Dubois *Pithecanthropus erectus* (1894) 1).] The fossil hominid first described from remains found by Eugene Dubois (1858–1940) in Java in 1891, now included in the species *Homo erectus*; = *Java man* s.v. *JAVA.

1895 *Nature* 5 Dec. 115/2 Dr. Dubois placed Pithecanthropus below the point of devarication of the anthropoid apes from the human line. **1895, 1898** [see PITHECANTHROPE]. **1905** J. MCCABE tr. *Haeckel's Evol. Man* II. xxiii. 632 The pithecanthropus excited the liveliest interest, as the long-sought transitional form between man and the ape. **1933** A. S. ROMER *Man & Vertebrates* xi. 239 Some scientists have claimed that *Pithecanthropus* is an ape, far below human status. **1933** N. DOUGLAS *Looking Back* II. 301 Science has a knack of filling up those gaps when least you expect it, as in the case of radium and pithecanthropus. **1957** L. EISELEY *Immense Journey* 101 The finding of the Pithecanthropus skull cap had bolstered this view.

So **pithecanthro·pic** *a.*, of or pertaining to pithecanthropus; also *fig.*, resembling an ape, clumsy; **pitheca·nthropine** *a.*, resembling or closely related to the fossil hominid once included in the genus *Pithecanthropus*; also as *sb.*; **pitheca·nthropoid** *a.* = prec; also *fig.*

1890, 1897 [in Dict.]. **1917** *Q. Rev.* July 35 Degeneracy, as seen in idiots,..due to a reversion to the pithecanthropic element. **1925** *Bull. Geol. Soc. China* IV. 177 Sooner or later pithecanthropine remains will be recovered from the Siwaliks. **1929** A. CROWLEY *Spirit of Solitude* I. xvii. 198, I was absorbed in 'The Cloud on the Sanctuary', reading it again and again without being put off by the Pharisaical, priggish and pithecanthropoid notes of its translator, Madame de Steiger. **1931** A. KEITH *New Discov. Antiq. Man* 293 The greatest number of these [characters] link these ancient Chinamen to the Pithecanthropic type of Java. **1958** L. DURRELL *Mountolive* xiii. 251 He mopped his brow continually, and gave his ingratiating pithecanthropoid grimace. **1958** F. E. ZEUNER *Dating Past* (ed. 4) ix. 304 At Ternifine in North Africa another pithecanthropine type of man proved to be associated with a primitive Acheulian industry. **1959**

J. D. CLARK *Prehist. S. Afr.* iv. 77 In Africa the cranial form of the Pithecanthropoid type of man may have undergone a fairly rapid development. *Ibid.*, The third and most probable explanation is that the Pithecanthropic stock early evolved..directly into a basic type of *Homo sapiens*. **1966** G. DURRELL *Zoo in my Luggage* ii. 44 His Pithecanthropic features split into a wide grin of delight. **1970** *Times* 5 Sept. 13/5 The pithecanthropines were sometimes grouped with Neanderthal and modern man in the sub-family Homininae. **1964** *Advancement of Sci.* XVIII. 424/1 The advanced pithecanthropine skull recently discovered..has been dated approximately. **1976** *Sci. Amer.* Jan. 66/3 A pithecanthropine grade has been recorded in a unique 800-millilitre endocast from Lake Rudolph in Kenya that is almost three million years old; the better-known true pithecanthropines are about a million years old. **1979** C. BRAND' *Rose in Darkness* vi. 44 Like an archaeologist, building up the whole structure of ancient man from a single tooth..pithecanthropoid sense.

pithiatism (pi·þiătiz'm). *Psychol.* [ad. F. *pithiatisme* (J. Babinski 1901, in *Revue Neurologique* IX. 1079), f. Gr. πειθ-ώ persuasion + ἰατ-ός curable: see -ISM.] A type of hysteria thought to be amenable to and curable by suggestion. So **pithia·tic** *a.*

1910 *Lippincott's New Med. Dict.* 740/2 Pithiatism,.. hysteria. (Babinski.) *Ibid.*, Pithiatic. **1918** E. JOWETT in White & Jelliffe *Mod. Treatm. Nervous & Mental Dis.* I. vii. 570 Babinski attempts to divide verbal suggestions into those that are unreasonable..and those that are reasonable and beneficial... Treatment by means of persuasion he calls 'pithiatism'. **1918** J. D. ROLLESTON tr. *Babinski's Hysteria or Pithiatism* p. xv, Among the various nervous phenomena observed in the neurology of war it is most important to distinguish hysterical or pithiatic disorders. **1930** D. KERRIDGE *Dis. of Ear* (ed. 4) xxii. 551 Pithiatism implies not only the possibility of cure by persuasion, but also the fact that the disorder may in some degree be called into being by suggestion. *Ibid.*, Pithiatic deafness..is at its inception a veritable deafness, the inevitable sequence of a shock to the perceptive labyrinth, which could have had no other origin. **1977** K. PELTIER in J. G. Howells *World Hist. Psychiatry* iv. 131 Babinski (1901) proposed the name of 'pithiatism' to designate a special condition, whose suggestion is able to produce or suppress clinical symptoms.

Pithiviers (pitivie). The name of a town in northern France used *attrib.* to designate a French cake or tart consisting of puff pastry with a rich almond filling.

1846 A. SOYER *Gastronomic Regenerator* 488 *Gateau de Pithiviers.* Blanch and pound well half a pound of almonds. **1970** SIMON & HOWE *Dict. Gastron.* 303/1 *Pithiviers cake*, a French cake made from puff pastry mixed with sweet and bitter ground almonds. It is light with a delicate almond perfume. **1973** *Guardian* 27 Nov. 11/1 Pithiviers cake..is one of the great puddings from France... Roll out two 10 in. circles from packet puff pastry... Make a frangipane filling.

pithos. Add: **pl. pithoi.** (Later examples.) Also *attrib.*

1956 V. G. CHILDE *Dawn European Civilisation* ii. 32 From M.M. I to L.M. I clay jars (pithoi) were also used as receptacles for the corpse. **1949** W. F. ALBRIGHT *Archæol. of Palestine* vi. 118 In the first three phases of Iron-Age Bethel the dominant vase was a large store-jar (pithos) with a very characteristic collared rim. **1955** *Sci. Amer.* July 45/2 At one entrance to this building was a sunken 'lustral area', where visitors made formal ablutions on arrival; it was surrounded by tall 'pithos' jars for water. **1957** V. G. CHILDE *Dawn European Civilisation* (ed. 6) v. 72 In 1935 a cairn with pithos burials very like our round tombs was found in Messenia, but was M.H. in date. **1962** [see *DAEDALIC *a.*]. **1972** Y. YADIN *Hazor* ii. 35 At times they appear to have been used as a store; this is further corroborated by eight pithoi found in it. The bases of the pithoi were stuck deep in the floor of the room.

pithy, *a.* **2.** (Later example.)

1876 SWINBURNE *Let.* 10 Jan. (1960) III. 112 A sea with nerves or rather..water thick and pithy with sand.

pi·tkin. *nonce-wd.* [f. PIT *sb.*[1] + -KIN.] A little pit.

1917 E. POUND *Lustra* 199, I dug the ell-square pitkin.

pitman. Add: **2. a.** (Later examples.)

1924 J. B. PRIESTLEY *Eng. Journey* x. 334, I ought to explain that in Durham collieries are always 'pits' and miners 'pitmen'. **1978** R. LEWIS *Uncertain Sound* iii. 76 My father had been a simple pitman.

3. b. A motor-racing mechanic working in a pit (see *PIT *sb.*[1] 1 h).

1913 *Technical World Mag.* XIX. 495/2 A pitman has noticed that the tire is flat.

c. *U.S. colloq.* (See quot.)

1926 *Amer. Speech* I. 283/2 ...

5. (Earlier example.)

1813 O. EVANS in *Weekly Reg.* 17 Apr. 111/2, I apply the power by means of a connecting rod or rods (or pitman as it is called when applied in saw mills).

Pitman (pi·tmæn). The name of Sir Isaac *Pitman* (1813–97) used *attrib.*, *absol.*, and in the possessive to denote a system of shorthand notation devised by him and first published in 1837.

Pitman's Shorthand is a proprietary name in the U.K.

[1862 M. LEVY *Hist. Short-hand* xvii. 167 Compare their rules with those of Pitman, in which he explains how to write the Scotch guttural.] **1886** *Encycl. Brit.* XXI. 838/2 The main features of Pitman's system must now be described. **1895** *Montgomery Ward Catal.* 74/1 The Reporter's Style of the Pitman System, by which fluent speakers can be reported verbatim. **1897** *Trade Marks Jrnl.* 5 June 969 Pitman's Shorthand Isaac Pitman 292,180. **1916** G. B. SHAW *Pygmalion* Pref. 101 His [*sc.* Sweet's] true objective was the provision of a full, accurate, legible script for our noble but ill-dressed language; but he was led past that by his contempt for the popular Pitman system of shorthand, which he called the Pitfall system. *Ibid.*, There was a weekly paper to persuade you to learn Pitman... I actually learned the [*sc.* Sweet's] system..and yet the shorthand in which I am writing these lines is Pitman's. **1951** E. COXHEAD *One Green Bottle* vi. 171 What's Ogam?.. A sort of prehistoric Pitman's? **1969** K. GILES *Death cracks Bottle* ii. 17, I think my Pitman is decipherable. **1970** E. MCGIRR *Death pays Wages* ii. 52 It is an exercise book and the entries are in very clear Pitman. **1976** *Daily Times* (Lagos) 8 Oct. 24/1 (Advt.), Applicants should possess WASC preferably with credit in English Language or equivalent and RSA or Pitmans or Government Certificate or their equivalent for 100/50 w.p.m. in shorthand and typewriting.

Hence **Pitma·nic** *a.*, resembling or suggestive of Pitman shorthand; **Pi·tmanite**, a student or exponent of the Pitman system; **pi·tmanize** *v. trans.*, to fill (a book) with Pitman shorthand.

1908 G. B. SHAW *Let.* 29 June (1972) II. 793 The other day I was reading a Dickens..novel..had been pitmanized, with some shorthand notes on the margin... You can imagine how the honest Pitmanite who was asked to decipher it..believed that he had saved me from a fearful scandal. **1912** BEERBOHM *Christmas Garland* 14 Seven whole 8 × 5 inch note-books had I pitmanised to the brim. **1925** *Blackw. Mag.* Aug. 264/1, I saw a..sheet of notepaper scribbled with Pitmanic symbols. **1960** *Times* 26 Sept. 13/4 What odds if the Pitmanic 'X' full stops were sometimes deliberately exaggerated!

pitmatic (pitmæ·tik). Also **-atick, -atik,** and in *pl.* (construed as sing.). [f. PIT *sb.*[1] after MATHEMATIC *a.* and *sb.*] The local patois used by miners in the North-east of England. Also in extended use.

1863–4 R. O. HESLOP *Northumb. Words* II. 539 *Pitmatics*, a jocose term for the technicalities of colliery working. **1901** 'R. GUTHRIE' *Kitty Fagan* 8 Pitmatics! I've been schooled in pitmatics! **1934** J. B. PRIESTLEY *Eng. Journey* x. 334 The local miners [in Durham] have a curious lingo of their own, which they call 'pitmatik'... It is only used by the pitmen when they are talking among themselves... When the pitmen are exchanging stories of colliery life..they do it in 'pitmatik', which is Scandinavian in origin, far nearer to the Norse than the ordinary Durham dialect. **1967** *Listener* 23 Nov. 665/2 The ambitious young actor who naturally speaks beautiful English must assume a rough-diamond pitmatic though he has it not. **1967** A. L. LLOYD *Folk Song in Eng.* vi. 382 Most of his [*sc.* Tommy Armstrong's] songs are in the peculiar North-eastern miners' jargon called 'pitmatic'. **1972** *Jrnl. Lancs. Dial. Soc.* Jan. 19 He had many recollections of coal miners' sayings, and of his own locality, *pitmatic*. **1976** *Daily Mail* 19 Aug. 19/2 Fred Reed, a dialect poet, gives examples of 'pitmatic', the language of the coalminer spoken in pits around Durham.

Pitocin (pitǒu·sin). *Med.* Also **pitocin.** [f. PI(TUITARY *a.* + *OXY)TOCIN.] A proprietary name for (an aqueous solution of) oxytocin.

1929 *Official Gaz.* (U.S. Patent Office) 15 Jan. 514/1 Parke, Davis & Company, Detroit, Mich. Filed Nov. 2, 1928. Pitocin. For oxytocic principle of the pituitary gland... Claims use since about October 20, 1928. **1929** *Jrnl. Amer. Med. Assoc.* 19 Jan. 235/2 The alpha and beta hormones (isolated from the posterior lobe of the pituitary body) are to be known commercially as pitocin and pitressin respectively. **1929** *Trade Marks Jrnl.* 23 Jan. 117/1 Pitocin... All goods included in class 3 (i.e. chemical substances prepared for use in medicine and pharmacy)... Parke, Davis and Company,..pharmaceutical chemists. **1935** MARTINDALE & WESTCOTT *Extra Pharmacopœia* (ed. 20) I. 369 Each Cc. of 'Pitocin' contains 10 oxytocic units and it is thus identical in activity with 'Pituitrin'. **1948** MARTIN & HYNES *Clin. Endocrinol.* i. 11 Two distinct hormones can be obtained from posterior pituitary extracts—pitressin, and oxytocin or pitocin, which acts on uterine muscle. **1964** *Lancet* 19 Sept. 761/1 The use of pitocin induction of labour in hæmolytic disease of the newborn.

pitometer: see *PITOT[2] b.

piton (pi·tǫn). [Fr., = 'eye-bolt', also 'piton'.] **1.** *Physical Geogr.* (See quot. 1972.)

1902 *Pop. Sci. Monthly* July 274 Northern Martinique, like other West Indian islands, is a labyrinth of mornes and pitons, i.e., of singularly steep peaks and ridges (partly volcanic cones, partly erosional forms), densely clothed with forests and herbage. **1918** C. W. BEEBE *Jungle Peace* (1919) iii. 56 A small wandering rainstorm drifted against the smallest piton and spilt in two. **1944** C. A. COTTON *Volcanoes* xi. 177 (*heading*) Smooth cliffs moulded during vertical extrusion of a plug dome to form the piton Lassen Peak, California. **1972** *Gloss. Geol.* (Amer. Geol. Inst.) 543/1 *Piton*, a term commonly used for volcanic peaks, especially steep-sided domes, in the West Indies and other French-speaking regions. **1976** *Sat. Rev.* (U.S.) 30 Oct. 26/2 St. Lucia..offers..cone-shaped volcanic peaks and twin pitons.

2. *Mountaineering.* (See quot. 1909.)

1898 *Encycl. Sport* II. 49/2 Snatch blocks and other kinds of pulleys, pitons or holdfasts..are also requisite. **1909** C. E. BENSON *Brit. Mountaineering* ii. 40 Piton—a strong iron spike with an eye at one end through which a rope can be passed. **1916** G. D. ABRAHAM *On Alpine Heights & Brit. Crags* iii. 64 He surprised us by scraping the ice off a ring-headed piton which was driven into a rock crevice. **1936** AUDEN & ISHERWOOD *Ascent of F6* ii. i. 75 They're hammering the whole south face full of pitons and leaving each other up like sacks! **1955**, etc. [see *ÉTRIER]. **1959** 'G. CARR' *Swing away, Climber* iii. 58 He carried a multitude of short metal pegs with holes in the blunt end—pitons. **1972** *Country Life* 13 Jan. 94/1 From time to time a piton has to be hammered in to reinforce the rock disintegrates.

3. *Comb.* (in sense 2), as *piton belay*; **piton hammer**, a hammer designed for fixing and extracting belays; **piton runner**, a piton used as a running belay.

1971 C. BONINGTON *Annapurna South Face* xii. 164 On reaching the ledge leading into the main icefields, Mick hunted around for piton belays, eventually found a couple of suitable cracks and abseiled back. **1943** R. C. GEIST *Hiking, Camping & Mountaineering* 227 They are driven into rock cracks or into ice by a special piton hammer. **1952** MORIN & SMYTH tr. *Herzog's Annapurna* ix. 141, I used my piton-hammer..as an anchor. **1971** C. BONINGTON *Annapurna South Face* vi. 76 The *a*hammer, a space-age style piton-hammer of his own design. *Ibid.* xi. 127 The rope behind Martin was dragging badly because of the friction caused by its passage through the tunnel and various piton runners.

pitot[1]. Delete *Obs.* and add later example.

1971 A. M. LYSAGHT *Joseph Banks in Newfoundland & Labrador* 1766 vii. 161 These molluscs [*sc. Cyrtodaria siliqua*] are known by the French name of 'pitot' and the areas where they live are called 'pitot banks'.

pitot[2] (pi·to). *Physics and Aeronaut.* Also **Pitot.** The name of Henri *Pitot* (1695–1771), French scientist, used to designate devices based upon his inventions for measuring the relative velocity of a fluid, esp. the airspeed of an aircraft, as **pitot head**, a pitot-static tube; **pitot-static** *a.*, designating a device consisting of a pitot tube inside or adjacent to a parallel tube closed at the end but with holes along its length, the pressure difference between them being a measure of the relative velocity of the fluid; also *absol.*; **pitot** († or **Pitot's**) **tube**, an open-ended right-angled tube pointing in opposition to the fluid flow and connected to a means of measuring pressure; also, a pitot-static tube; also *absol.*

1881 *Encycl. Brit.* XII. 508/1 A Darcy gauge..consists of two Pitot tubes having their mouths at right angles. **1895** R. C. CARPENTER *Heating & Ventilating Buildings* iii. 38 In case the pressure and velocity are great, considerable error will be made by using the open tube..and for such a case a Pitot's tube, should be used. **1901** G. E. DAVIS *Handbk. Chem. Engin.* I. iv. 191 The total pull of a chimney upon a water manometer when only one limb of the tube was connected, ... **1914** *Techn. Rep. Advisory Comm. Aeronaut.* 1912–13 59 This pressure-tube anemometer consists of two tubes, a Pitot tube facing the wind and a static pressure-tube along the wind direction with holes drilled into the side of it. **1916** *Field* 20 May 788/2 At last he fumbles for his safety belt, but with a start remembers the Pitot Air Speed Indicator, and..smiles as he hears the Pitot-head's gruff voice, 'Well, I should think so, twenty miles an hour I was registering.' **1920** G. C. BAILEY *Compl. Airman* xii. 173 The only really satisfactory test of the Pitot is a flying one. **1920** L. BAIRSTOW *Appl. Aerodynamics* iii. 75 (*heading*) Initial determination of the constant of the pitot-static pressure head. **1930** C. J. STEWART *Aircraft Instruments* iii. 54 The pitot-tube type instrument is based on an adaptation of the method used by Pitot in 1732 for the measurement of the speed of a river. **1934** V. M. YEATES *Winged Victory* I. iii. 27 Looking at the pitot he found the speed was ninety. **1943** 'PAIR' (ed. 5) Jan. 6/1 ... [Gremlin], was seen..to swing nimbly..down to the pitot head and block it with ice. **1948** *Jrnl. R. Aeronaut. Soc.* LII. 272/1 Rates of climb and descent measured by the pitot-static are subject to errors introduced by random vertical currents in the atmosphere. **1965** R. G. KAYMANS *Mod. Hydrol.* iv. 63 The Pitot tube can be used to make a pattern of measurements along the cross section of a stream. **1973** 'A. HALL' *Tango Briefing* ix. 111 His inability to know to what extent air speed would be true and to what extent it would be affected by the wind in the pilot-head. **1975** J. L. CLANCY *Aerodynamics* ii. 25 The pitot-static tube may be mounted in a position on the aircraft where the flow is affected by the presence of the aircraft. **1978** R. V. JONES *Most Secret War* iv. 36 This is true enough of a pitot tube but, as Lindemann pointed out to me, the proposal was perfectly sound if one used, as the inventors suggested, accelerometers.

b. Pitot meter, also **pitometer** (pitǫ·mǐtəɹ). (See quots. 1934, 1941.)

1907 *Jrnl. Franklin Inst.* CLXIV. 440 It is only necessary to locate the pitometer orifices at the point of maximum velocity in order to record the mean flow. **1934** J. H. PERRY *Chem. Engineers' Handbk.* 692 The pitot meter consists of two tubes, one facing upstream and the other downstream. **1941** *Ibid.* (ed. 2) 838 The modified pitot tube known as a pitometer has one pressure opening facing upstream and the other facing downstream. The differential between the two openings is usually from 25 to 50 per cent more than that for a standard pitot tube. **1955** C. S. FORESTER *Good Shepherd* ii. 25 That was the pitometer log reading. **1961** *Engineering* 3 Nov. 566/3 The

development flowmeters..were calibrated in clean air against a venturi meter and a pitot meter.

Pitressin (pitre·sin). *Med.* Also **pitressin.** [f. PIT(UITARY *a.* + *VASOP)RESSIN.] A proprietary name for (an aqueous solution of) vasopressin.

1929 *Official Gaz.* (U.S. Patent Office) 15 Jan. 514/2 Parke, Davis & Company... Filed Nov. 2, 1928. Pitressin. For pressor principle of the pituitary gland... Claims use since about Oct. 20, 1928. **1929** [see *PITOCIN]. **1929** *Trade Marks Jrnl.* 23 Jan. 117/1 Pitressin... All goods included in class 3 [i.e. chemical substances prepared for use in medicine and pharmacy]... Parke, Davis and Company,..pharmaceutical chemists. **1943** *Lancet* 27 Feb. 267/1 Pitressin tannate in oil seems to be the safest and most effective slowly acting preparation of the posterior pituitary lobe so far available for the treatment of diabetes insipidus. **1974** *Jrnl. Pediatrics* LXXXV. 79/2 The infant's water balance was initially controlled with intravenous fluids alone but subsequently he was given a trial of pitressin. **1974** *Aerospace Med.* XLV. 1223/1 These same groups also showed little change in urine output in response to subcutaneous administration of 300 mU of Pitressin tannate in oil. **1976** *Lancet* 25 Dec. 1403/2 Aqueous vasopressin injection ('Pitressin') as a prophylactic measure was popular some years ago [against headache after lumbar puncture].

‖ **pitso** (pi·tso). Also † **peetsho,** † **peetso,** † **piicho, pitsu,** and with capital initial. [Sotho.] A Sotho tribal conference in a *kgotla*.

1822 J. CAMPBELL *Trav. S. Afr.: 2nd Journey* I. 264 The other chief said they should come to the *peetso* all well powdered. **1824** W. J. BURCHELL *Trav. S. Afr.* II. 408 The *piicho* or assembly remained sitting in easy conversation for nearly an hour longer. **1828** J. PHILIP *Res. S. Afr.* II. vii. 132 All great questions, and all questions relating to peace or war, are decided on in public assemblies, which are designated in their language by the name of Peetshos. **1879** *Queenstown Free Press* (Suppl.) 28 Oct. (Pettman), The annual *Pitso* was held at Maseru on the 19th instant, about 10,000 being present. **1897** J. BRYCE *Impressions S. Afr.* xx. 424 To-day the Pitso has lost much of its old importance. **1925** *Brit. Weekly* 4 June 211/3 The population of Basutoland is little over half a million, and the entire manhood must have ridden in to the *Pitso*. **1947** L. HASTINGS *Dragons are Extra* ii. 50 A great *pitso* was arranged. The traditional gathering took place in the open country south of Serowe. **1953** J. PACKER *Apes & Ivory* x. 105 The Royal visit to Basutoland, when the whole nation had gathered in a mighty *pitsu* at Maseru to welcome the King and Queen and their daughters had left a deep impression. **1959** *Chambers's Encycl.* IX. 747/1 Matters of highest importance were discussed in the *pitso*, the national assembly, at which the poorest and meanest tribesman had equal right with the proudest to voice his opinion. **1968** A. FULTON *Dark Side of Mercy* 27 The Chief had summoned the men of the clan to the Khotla and opened the pitso by telling those assembled that the season had been poor, [etc.]. **1976** WEST & MORRIS *Abantu* 121 It was more usual for all matters of general concern to be aired at a *pitso*, a general meeting open to all adult men in the chiefdom and held in the kgotla of the chief.

pitta, var. *PITA[2].

pitter-litter (pi·təɹ li·təɹ), *adv. rare*⁻¹. [After PITTER-PATTER *sb.* (*adv.*).] = PITTER-PATTER *sb.* (*adv.*) 2 a.

1910 H. G. WELLS *New Machiavelli* (1911) I. ii. 18 Dropped most satisfyingly down a brick shaft, and pitter-litter over some steep steps to where a head slaughterman strung a cotton loop round their legs.

pitting, *vbl. sb.* Add: **4.** (Later examples in *Bot.*)

1933 *Forestry* VII. 22 No complex forms of plate were observed with scalariform intervascular pitting. **1973** H. E. DESCH *Timber* (ed. 5) ii. 35 The pitting occurring in a cross field takes one or other of five more or less distinct forms.

5. = *pit planting* s.v. *PIT *sb.*[1] 14.

1847 [see *NOTCHING *vbl. sb.* 3]. **1894** A. D. WEBSTER *Pract. Forestry* iv. 20 The advantages of pitting over any other method of planting cannot be questioned. **1930** *Forestry* IV. 19 The ordinarily understood pitting implies the opening of a hole 12 to 15 inches square, stirring up the under soil well, and then carefully placing the plant in the centre.

pittite[2]. Add: Also **pitite.** (Earlier and later examples.)

1812 *Dramatic Censor 1811* 8 The O.P. dance was attempted to be performed, but the Pitites had not yet mustered in sufficient force to carry their desire into effect. **1818** in J. Agate *These were Actors* (1943) 72 We cannot give the unfortunate *Pittites* any hope of shelter from 'the icy wind of Death' which seems to blow from all quarters at once [at Drury Lane]. **1892** 'F. ANSEY' *Voces Populi* 2nd Ser. 155 *A pittite behind Jimmy*... Will you tell your little boy to set down, please? **1903** A. BENNETT *Truth about Author* xiii. 162 Many time I have stood with you. But never again, miserable pittites! **1939** JOYCE *Finnegans Wake* (1964) III. 427 The graced of gods and pittites. **1961** BOWMAN & BALL *Theatre Lang.* 260 *Pittite*, in British terminology, a spectator in the pit.

pittosporum. For (Banks 1788) in etym. substitute (J. Banks in J. Gærtner *De Fructibus et Seminibus Plantarum* (1788) I. 286). Substitute for def.: An evergreen shrub or small tree of the genus so called, belonging to the family Pittosporaceæ, native to many subtropical or temperate regions, especially Australasia and China, and bearing small, often fragrant, flowers; also, a shrub of this genus cultivated as a house plant. Also *attrib.* (Earlier and later examples.)

1789 W. AITON *Hortus Kewensis* III. 488 Thick-leav'd Pittosporum. Nat[ive] of Madeira. **1911** A. E. MACK *Bush Days* 2 You climb up the rocky path and brush under green pittosporums. **1948** W. ARNOLD-FORSTER *Shrubs for Milder Counties* ii. 27 The pittosporums include some of the most useful of evergreens. **1962** R. PAGE *Educ. Gardener* viii. 247 Thickly planted bushes of dark green pittosporum..I clipped into low green mounds. **1963** S. SCHULLER *1001 House Plants Questions Answered* 221 Can I grow pittosporum easily indoors? Yes, if you can provide a cool..south window for it. **1965** G. MCINNES *Road to Gundagai* xiv. 244 Ballan was..well treed, especially..around Miss McCoppin's Guest House with its tall pittosporum hedge. **1972** PALMER & PITMAN *Trees S. Afr.* I. 659 Our indigenous pittosporum, growing in every province of South Africa under widely differing conditions, is naturally very variable in form. **1979** *Country Life* 26 Apr. 1298/2 Planting of replacement evergreens such as ceanothus, pittosporums and cistuses may be..deferred another week or two.

pitty[2] (pi·ti). A nursery form of PRETTY *a.*

1826 M. ROGET *Jrnl.* 27 Oct. in D. L. Emblen *P. M. Roget* (1970) xi. 191 One year old... Tries to imitate everything she hears, and says 'Papa', 'Mama', 'Ta', 'Pitty', 'Ba'. **1879** C. M. YONGE *Burnt Out* i. 6 Mammy, mammy, pitty, pitty! Dada kill great big piggy wig. **1939** JOYCE *Finnegans Wake* (1964) II. 361 Wingwong welly, pitty pretty Nelly!

pituicyte (pitiū·isəit). *Histology.* [f. PITUI(TARY *a.* + -CYTE.] A small cell with branching processes that is the characteristic cell of the neurohypophysis.

1930 P. C. BUCY in *Jrnl. Compar. Neurol.* L. 505 The pars nervosa is made up of a dense connective-tissue network containing nerve fibers which descend the stalk, epithelial cells which invade the pars nervosa from the pars intermedia for a short distance, and a type of cell which is peculiar to the neural portion of the hypophysis and which we have called the pituicyte. **1948** MARTIN & HYNES *Clin. Endocrinol.* i. 2 The only specialized cells in the neural division are pituicytes. **1959** ROWE & WHEBLE *Conc. Textbk. Anat. & Physiol.* xiii. 524 The posterior lobe is made up of large numbers of nerve fibres and a special type of cell known as pituicytes. **1975** DANIEL & PRICHARD *Stud. Hypothalamus & Pituitary Gland* iii. 27 Pituicytes, or modified glial cells, are scattered throughout the tissue [of the upper infundibular stem].

pituitary, *a.* Add: (Further examples.)

Pituitary body or *gland* is partly synonymous with *HYPOPHYSIS 3, q.v.; it is sometimes taken to include the infundibulum. The functions of the organ are now known.

1825, etc. [see *HYPOPHYSIS 3]. **1940** *Res. Publ. Assoc. Res. Nervous & Mental Dis.* XX. 28 Together with its investing sheath consisting of portions of the lobus glandularis (pars tuberalis and part of the pars intermedia), the infundibulum or neural stalk is correctly referred to as the hypophysial or pituitary stalk. **1942** O. LARSELL *Anat. Nervous Syst.* xviii. 254 The hypophysis or pituitary body is a gland of internal secretion. **1950** *Sci. News* XV. 132 The anterior lobe of the pituitary gland, which lies at the base of the brain, regulates the amounts of adrenal cortical hormones which circulate in the blood. **1954** [see *NEUROHYPOPHYSIS]. **1968** PASSMORE & ROBSON *Compan. Med. Stud.* I. xxv. 9/2 The pituitary gland consists of four different parts... (1) Pars posterior or neurohypophysis is an outgrowth of the brain to which it is attached by the infundibulum. (2) Pars anterior is the main part of the organ and is glandular in structure. (3) Pars intermedia is rudimentary in man but is large in some vertebrates... (4) Pars tuberalis is an extension of pars anterior and encloses the infundibulum. It is also rudimentary in man. **1975** J. B. PHILLIPS *Devel. Vertebr. Anat.* iii. 58/1 Gonadotrophic (gonad-stimulating) hormones, produced by the anterior lobe of the pituitary gland (hypophysis) control the function of the gonads.

b. (Earlier and later examples of sense (*b*) (sense (*a*) is *obs.*).)

1899 *Jrnl. Physiol.* XXV. 89 We confirm Howell's statement that the rise of blood-pressure is produced solely by the infundibular part of the pituitary and not by the hypophysial part. **1919** [see *HYPOPHYSIS 3]. **1932** S. ZUCKERMAN *Social Life Monkeys* v. 81 In the rat, surgical removal of the pituitary stops the œstrous cycle. **1965** MARSHALL & HUGHES *Physiol. Mammals* ix. 232 The pituitary is situated below the hypothalamus..and is made up of two parts. The hypophysis (anterior lobe) is derived from the roof of the buccal cavity, while the infundibulum from a down growth of the fore brain makes up the posterior lobe. **1974** D. & M. WEBSTER *Compar. Vertebr. Morphol.* xiii. 302 The pituitary has two major components, which develop from separate tissues and grow together... The neurohypophysis is formed by a ventral growth from the hypothalamus; the adenohypophysis, by a dorsal evagination from the roof of the mouth.

Pituitrin (pitiū·itrin). *Med.* Also **pituitrin.** [f. PITUIT(A)R(Y *a.* + -IN[1].] **a.** A proprietary name for an extract of the posterior lobe of the pituitary body which contains the hormones oxytocin and vasopressin. † **b.** A name proposed for the hormone formerly thought to be present in this extract but now recognized as a mixture. *Obs.*

1909 *Progressive Med.* IV. 318 The extract of the pituitary body has been prepared by Parke, Davis & Co., under the name of pituitrin. **1909** *Offical Gaz.* (U.S. Patent Office) Dec. 297/2 Parke, Davis & Co., Detroit, Mich. Filed July 26, 1909. Pituitrin... Cardio-vascular stimulants and diuretics. **1922** B. HARROW *Glands* 55 We have given the name 'pituitrin' to the hormone (or hormones) present in the posterior lobe of the pituitary... This hormone has not been isolated in the pure state. **1924** G. B. SHAW *St. Joan* p. xix, A precise and convenient regulation of her health and her desires by a nicely calculated diet of thyroid extract, adrenalin, thymin, pituitrin, and insulin. **1924** *Trade Marks Jrnl.* 3 Sept. 1953 Pituitrin... A pharmaceutical preparation of the pituitary gland for human use. Parke, Davis & Company, Detroit, State of Michigan, United States of America; and..London,..pharmaceutical chemists. **1927** HALDANE & HUXLEY *Animal Biol.* viii. 163 The posterior part [of the pituitary gland] produces a hormone, pituitrin, which affects smooth muscle. **1973** *Exper. Eye Res.* XVII. 3 When injected intravenously in intact animals, pituitrin produces a rise of intraocular pressure coincident with a rise in blood pressure.

pituri. Add: Also **pitery, pitjuri.** Also *attrib.* (Later examples.)

1931 I. L. IDRIESS *Lasseter's Last Ride* (1933) xvii. 137 Old Warts took from behind his ear a half-chewed plug of pituri, made from the pituri shrub. **1934** A. RUSSELL *Tramp-Royal in Wild Austral.* xix. 123 They [*sc.* the natives] use the pitjuri leaf—which they are also so fond of chewing—poisoning the little-used water-holes with leaves picked from the plant, enough to cause staggers but not death, and then going out and killing the drunken game. **1941** I. L. IDRIESS *Great Boomerang* xiv. 102 Finally one by one the pituri-chewers become stupefied, rolling over and sleeping through day and night. **1964** *Sunday Mail Mag.* (Brisbane) 25 Oct. 3/1 We tossed our swags into the station land rover..for a two-day run from the station out to Allawonga Springs and the pituri country. **1967** *New Scientist* 27 Apr. 226/1 Even the primitive Australian Aborigine has found a pleasurable weed—pitery—which contains nornicotine.

pity, *sb.* **3. a.** (*the*) *more is the pity* (earlier examples).

1797 R. M. ROCHE *Children of Abbey* (ed. 2) III. iii. 26 Poor thing, she is going fast indeed, and the more's the pity, for she is a sweet creature. **1848** J. RUSKIN *Let.* 22 Sept. in M. Lutyens *Ruskins & Grays* (1972) xvii. 158 You and my mother must be left at least *tranquil* as you are to be left—more's the pity—now so much alone. **1851** E. C. GASKELL *Let.* May (1966) 838 It is a small old fashioned farm..at the *foot* of the hill. More's the pity. **1875** 'MARK TWAIN' in *Atlantic Monthly* Feb. 217 A chance to get acquainted with a youth who had taken deck passage—more's the pity; for he easily borrowed six dollars of me.

pity, *v.* **3.** (Later example.)

1862 C. M. YONGE *Countess Kate* xii. 222 Sylvia and Charlie, took it all in, pitied, wondered, and were indignant, with all their hearts.

‖ **più** (pi·u), *adv. Mus.* Also **piu.** [It.] More; used in musical directions, as *più mosso* [*MOSSO *a.*], more animated, *più piano* [PIANO *a.* (*adv.*)], softer, etc.; also *fig.*

1724 *Short Explication Foreign Words in Musick Bks.* 54 Piu Piano, or PP, is very soft or low. *Ibid.* 55 Piu, signifies a little more, and increaseth the strength of the signification of the word it is joyned with. **1740** J. GRASSINEAU *Mus. Dict.* 180 *Piano piano* or *piu piano*, is nearly the same with *pianissimo*, or rather a degree between it and *Piano*. *Ibid.* 181 *Piu*, a little more, it increases the strength of the signification of the word to which 'tis added. **1876** STAINER & BARRETT *Dict. Mus. Terms* 355/2 *Più*,..more, as *più allegro*, faster; *più forte*, louder; *più lente*, slower; *più piano*, softer; *più presto*, more rapid; *più stretto*, more urged or closer; *più tosto allegro*, rather quicker. **1931** *Times Educ. Suppl.* 12 Dec. 1/3 The engineers have made the turn-over at the *più mosso*. **1952**, etc. [see *MOSSO *a.*]. **1959** *Collins Mus. Encycl.* 504 *Più* by itself = *più mosso*. **1962** A. HARMAN et al. *Man & his Music* III. v. 671 There is no way of bringing the music to a stop except a *più mosso* coda. **1975** *New Yorker* 3 Nov. 136/2 Then the horn..sets the new tempo for a *più-mosso* section, in which the flute and the bassoon play leading roles.

pium (pi·ʊm). [Pg., a. Tupi.] A South American buffalo gnat, *Simulium pertinax*. Also *attrib.*

1863 H. W. BATES *Naturalist on River Amazons* I. vii. 333 (*heading*) Pium flies. *Ibid.*, We made acquaintance ..with a new insect pest, the Pium, a minute fly. **1927** W. M. MCGOVERN *Jungle Paths & Inca Ruins* iv. 51 The most obvious of the insect pests were the little *piums*... Their sting gave more than a momentary discomfort. **1933** P. FLEMING *Brazilian Adventure* I. xvii. 145 By day the worst pest was the *pium* fly, a little black creature the size of a midge, which covered your hands and anything else it could get at with small hard red pimples.

‖ **piupiu** (piu·piu). *N.Z.* [Maori.] **a.** Dressed flax. Also *attrib.* **b.** A Maori skirt made of dressed flax (see quot. 1946).

1882 T. H. POTTS *Out in Open* 23 Robes of piu piu, korowai, or of dogskin, contributed a great variety of costume. **1905** W. BAUCKE *Where White Man Treads* 90 Who so..skilful in the weaving of piupiu, korowai, fancy mats as..Te Aatarangi? **1938** R. FINLAYSON *Brown Man's Burden* 30 He just threw his coat off and rolled his trousers up, but the other men wore old-time piupius, and their imitation battle-axes were quite clever. *Ibid.* 67 Lucy, dressed in her swishing piupiu skirt and gay headband..swung the little poi balls for the tourists'

snapshots which they labelled 'Maori dancer, Rotorua'. **1946** *Jrnl. Polynesian Soc.* June 156 *Piupiu*, a garment consisting of a heavy fringe, about nine inches wide or more, made of flax and craped at intervals and dyed black. The flax when drying curling up into pipes. **1970** *Times* 23 Mar. 7/6 Down on the grass before her stood ranks of Maori men and women in *Piu-Pius* (flaxen skirts). **1978** P. GRACE *Mutuwhenua* iii. 12 My mother and the others had been..bringing out the piupiu which had been rolled and sausaged into stockings and stored at the tops of our wardrobes.

Piute, var. *PAIUTE *sb.* and *a.*

piva: see *PIVO.

pivie, var. PEAVEY, PEVY.

1907 *Black Cat* June 24 Mehetabel launched the boat, and running along the logs, piloted it hither and thither hooked to a pivie.

pivo, piva (pī·vo, -ă). [Slavonic *pivo* (Russ. *pivo*, Pol. *piwo*, etc.), beer.] An Eastern European beer made from barley malt or a similar fermented beverage.

1950 V. CANNING *Forest of Eyes* vi. 119 In the room were two army officers,..their faces flushed with *pivo* and brandy. **1962** J. WADE *Running Sand* xix. 228 Piva... Just the job. **1964** S. BELLOW *Herzog* i. 22 Elias..drank prohibition beer—home-brewed Polish piva. **1976** 'D. HALLIDAY' *Dolly & Nanny Bird* xviii. 244 Someone had produced *pivo* and his minions were shouting and spraying the workshop with beer. **1979** *Listener* 19 July 71/1 'Another *pivo*, squire?'

pivot, *sb.* Add: **3. c.** In football and some other games, a player in a central position, esp. a centre-back; such a position. Also *attrib.*

1911 *Hartley Coll. Mag.* XII. xxxiii. 48 Howarth was moved from outside left to centre, but it was obvious ..he would be spoilt if retained as pivot. **1928** *Weekly Dispatch* 24 June 21/7 Robert Plenderleith, the East Fife centre half-back, one of the most promising of young pivots in Scotland. **1930** *Daily Express* 6 Oct. 16/5 Wilson, the Huddersfield pivot, was kept mainly on the defensive. **1951** *Football Record* (Melbourne) 8 Sept. 13 The Dons' hopes of success could well depend on his ability to gain control of the pivot position today. **1970** *Globe & Mail* (Toronto) 26 Sept. 36/2 Winnipeg has been looking for a take-charge pivot after going through a disappointing season. **1974** *State* (Columbia, S. Carolina) 27 Feb. 2-B/4 Jim Bolla, 6-8, is expected to be in the pivot. **1974** *Plain Dealer* (Cleveland, Ohio) 26 Oct. 4-D/1 In the pivot Steve Patterson and Jim Chones totalled 13 boards. **1975** *Oxf. Compan. Sports* 66/1 The tallest players of all, playing near the basket, are called pivots, centres, or posts. **1976** *Sunday Tel.* 25 Jan. 36/2 There is emerging a new nomenclature which refers to the 'pivot five', those players in the back row of the scrum and at half-back whose talent and technique decide how the ball is used at source.

d. = *pivotal man* s.v. *PIVOTAL *a.* 1 b.

1919 *Punch* 29 Jan. 76/2 They are keeping all the pivots in this area for one final orgy of demobilisation at some future date.

e. *Math.* In the numerical evaluation of a determinant, or the numerical solution of simultaneous linear equations, a non-zero element of the determinant or matrix which is used as follows: all elements in its row are divided by it, and appropriate multiples of that row are then subtracted successively from the other rows, so that the pivot itself is replaced by unity and all other elements in its column are replaced by zero; (this gives, in the case of a determinant, a number, the pivot, multiplied by a determinant whose order is reduced by one, and in the case of simultaneous equations, a set of equations one fewer in number from which one of the variables has been eliminated). Freq. *attrib.*

1933 *Proc. Edin. Math. Soc.* III. 211 At Stage II we choose another pivot at will..and cross-multiply with respect to it in the same way, dividing each result, however, by the previous pivot, 5. **1940** *Brit. Jrnl. Psychol.* XXX. 363 The determinant is condensed in the manner already shown (each pivot being converted into unity) until it is reduced to one number. **1958** S. I. GASS *Linear Programming* ii. 29 Since at the start of an iteration it is convenient to make the coefficient of the variable being eliminated equal to 1, we divide the second equation by 3. The coefficient 3 is referred to as the pivot element. *Ibid.* iv. 58 Element x_{lk} is the pivot element of the elimination transformation, with row l being the pivot row and row [*read* column] k being the pivot column. **1971** COULSON & RICHARDSON *Chem. Engin.* III. iv. 295 Some methods of solution are better adapted to automatic digital computers than others. The method of Gaussian elimination with selection of pivots, or pivotal condensation, is particularly suitable. **1973** PHILLIPS & TAYLOR *Theory & Appl. Numerical Anal.* viii. 197 The fault in the first attempt at solving Example 8.6 was that a small pivot (the coefficient of x in the first equation), meant that very large multiples of the first equation were added to the others, which were therefore 'swamped' by the first equation, after rounding.

f. *Linguistics.* = **pivot word*.

1963 M. D. S. BRAINE in *Language* XXXIX. 4 There is a basis for defining two primitive word classes: a class of pivots..to which a few frequently occurring words belong, and a complementary class which has many members, few of which recur in more than one or two different combinations. **1966** D. MCNEILL in Smith & Miller *Genesis of Lang.* 21 It never happens..that two-word sentences are made up only of pivots. **1971** *Jrnl. Speech & Hearing Disorders* XXXVI. 44 Rules that account for utterances in terms of the juxtaposition of pivots and open words cannot account for differences in semantic interpretation. **1973** M. F. BOWERMAN *Early Syntactic Devel.* 30 Students of language acquisition have sometimes referred to words which a child uses with greater-than-average frequency as 'pivots'... When used in this way, the conception of 'pivot' has little more relevance to child speech than to adult speech. **1975** *Canad. Jrnl. Linguistics* XX. 220 Arabic children have a much easier time of it than English children since such Pivot–Open constructions are generally well-formed sentences in adult grammar.

6. b. *pivot man* (later examples); **pivot bearing** = FOOTSTEP 5 d; **pivot break** *Rugby Football* (see quot.); **pivot class** *Linguistics*, the class of pivot words; **pivot grammar** *Linguistics*, a grammar of an early stage in children's speech in which two word classes are postulated, pivot words (see below) and a larger open class; **pivot pass** *Rugby Football* (see quot.); **pivot word** *Linguistics*, one of a limited set of words recurring in particular utterance positions at an early stage of a child's acquisition of syntax, and postulated as constituting one of two basic word classes at this stage of development.

[**1875** *Engineering* 30 July 87/1 Pivoted bearings for line shafts, now almost universally employed in America.] **1877** W. C. UNWIN *Elem. Machine Design* vii. 122 (*heading*) Pivot and collar bearings. **1960** V. B. GUTHRIE *Petroleum Products Handbk.* ii. 29 Oils used for pivot bearings in instruments are of low volatility. **1975** *Sci. Amer.* July 50/3 If a combination of the two forces exists, a special bearing (an angular bearing, a taper bearing or a pivot bearing) is employed. **1960** E. S. & W. J. HIGHAM *High Speed Rugby* 70 Another way..is to break on the side of the scrum on which the ball was put in. This is sometimes called the 'pivot' break, because the scrum-half pivots around his outside foot. **1964** *Amer. Psychologist* XIX. 3/2 Two classes of words appear—a pivot class and an open class—and the child launches forth on his career in combinatorial talking. **1966** D. MCNEILL in Smith & Miller *Genesis of Lang.* 20 The pivot class characteristically has few members compared with the open class. **1970** —— *Acquisition of Lang.* III. 25 Pivot classes may appear first or second in sentences. **1970** L. M. BLOOM *Lang. Devel.* 24 The speech of children described in pivot grammars in other investigations. **1971** *Jrnl. Speech & Hearing Disorders* XXXVI. 42 How does pivot grammar relate to the grammar of the adult model language? *Ibid.* 47 The notion of pivot grammar describes children's early speech in only the most superficial way. **1973** M. F. BOWERMAN *Early Syntactic Devel.* 4 An analysis of diary studies of children acquiring languages other than English led Slobin..to hypothesize that the pivot grammar might be a universal first grammar regardless of the particular language being acquired. **1976** *Word 1971* XXVII. 33 In what ways does mother-child interchange indicate that Laura's initial two-word combinations are semantically and structurally too sophisticated to be described adequately by a pivot grammar? **1918** *Daily Mail* 6 Dec. 3/3 (*heading*) 12,000 pivot men. **1967** *Boston Sunday Herald* 9 Apr. (This Week Mag.) 4/3 Locate the 'pivot man' on each team—he's the fielder who makes the first move as the team adjusts position for each new batter. **1971** L. KOPPETT *N.Y. Times Guide Spectator Sports* iii. 88 The pivotman is usually the center, and the tallest man on his team. **1978** *Detroit Free Press* 5 Mar. c. 4/2 Joe Barry Carroll, Purdue's 7-1 pivotman, scored 14 of his game high 22 points in the first half. **1960** E. S. & W. J. HIGHAM *High Speed Rugby* 68 You will have to pivot round on the outside foot as you take off. This is called the 'pivot' or 'reverse' pass. **1963** M. D. S. BRAINE in *Language* XXXIX. 4 These words will be called 'pivot' words, since the bulk of the word combinations appear to be formed by using them as pivots to which other words are attached as required. **1964** *Amer. Psychologist* XIX. 3/2 Whereas before, lexemes like *allgone* and *mummy* and *sticky* and *bye-bye* were used singly, now, for example, *allgone* becomes a pivot word and is used in combination. **1966** D. MCNEILL in Smith & Miller *Genesis of Lang.* 20 Each pivot word is used more frequently than individual open-class words. **1975** *Canad. Jrnl. Linguistics* XX. II. 220 Much has been made in studies of children's syntax in English of a stage at which there are only two parts of speech, a Pivot, or function word like 'want', and an Open, or content word like 'milk'.

pivotable (pi·vətăb'l), *a.* [f. PIVOT *v.* + -ABLE.] Capable of being turned as if on a pivot. So **pivotabi·lity**, the extent to which an object can be so turned (in quots., as a measure of angularity).

1961 *Jrnl. Sedimentary Petrol.* XXXI. 198/1 A somewhat modified version of Powers' scale was adopted by the writers. In this scale the idea of 'pivotability' was given paramount importance so that the grains which could be most easily pivoted were given the highest values. **1967** *Jane's Surface Skimmer Systems 1967–68* 83/2 The result is a simple and reliable lift control, which requires neither electric or hydraulic power sources, nor pivotable foils or flaps. **1971** W. A. PRYOR in R. E. Carver *Procedures Sedimentary Petrol.* vii. 143 Pivotability or rollability is a measure of the motion response of a grain to a set of standard physical conditions in a gravity-driven system.

pivotal, *a.* Add: **1. a.** (Further examples.)

1925 *Times* 5 Jan. 4/3 Young's passes..were..much too high to enable Kittermaster, as the pivotal player, to pave the way for a scoring position. **1927** PEAKE & FLEURE *Priests & Kings* 134 A new feature, however, was the use of door-slabs of stone set with pivotal hinges. **1930** *Daily Express* 6 Oct. 2/6 Pivotal shares were occasionally in moderate demand. **1953** [see *film editor* s.v. *FILM *sb.* 7 c]. **1965** *New Scientist* 6 May 383/3 Dædalus feels that the time is ripe for another OK word. He has had difficulty in deciding on one with the right ring of vague pomposity about it, but has decided on 'pivotal'. He hopes that it ultimately will become synonymous for 'important' or 'interesting'. He asks all his readers to include it in articles, papers, research reports and the like, confident that the elite of New Men who read *New Scientist* will suffice to launch this new word on a glorious and pivotal career. **1969** Z. HOLLANDER *Mod. Encycl. Basketball* 211 (*heading*) The pivotal era. **1973** *Amer. Speech 1969* XLIV. 281 The uniqueness of black folk speech seems to rest in the pivotal position it occupies in the rural community, sharing phonological features with cultivated blacks and whites, grammatical features with white folk speakers, and lexical forms with all members of both castes. **1976** *National Observer* (U.S.) 13 Nov. 4/5 The pardon of Richard Nixon was pivotal, it seems, to those independents who made up their minds at the last minute and turned to Jimmy Carter. **1977** *N.Y. Rev. Bks.* 28 Apr. 29/3 (Advt.), Here is a penetrating, pivotal study of the astonishing relationships between Shakespeare's dramatic use of time and modern views on how any time-sense shapes human experience and behavior.

b. *pivotal man*, a man considered to have an important part to play in the re-establishment of industry and commerce after the war of 1914–18, and hence eligible for early demobilization; also *ellipt.* as *sb.*

1918 *Daily Mail* 29 Nov. 3/2 Men who are essential to the building up and expansion of trade..are officially described as 'pivotals'. *Ibid.*, These are the pivotal men who will prepare the way for the hundreds of thousands who are to follow on general demobilization. *Ibid.*, The commerce of the City of London has been allotted a given number of pivotal men. *Ibid.* 11 Dec. 5/1 A pivotal man is an essential man in an industry or occupation on which the re-establishment of other industries depends. **1920** F. WATSON *Pandora's Young Men* iii. 22 She found Blinkhorn in some out-of-the-way depot in France and managed to have him demobilised as a pivotal man. **1922** *Encycl. Brit.* XXX. 214/2 The release of 'Pivotal Men'.. met with much opposition.

2. *Math.* Being or involving a pivot (sense *3 e); *pivotal condensation*, the evaluation of a determinant by the use of pivoting on determinants of successively lower orders.

1924 WHITTAKER & ROBINSON *Calculus of Observations* v. 71 We prepare the determinant for our subsequent operations by multiplying some row or column by such a number p as will make one of the elements unity, and put $1/p$ as a factor outside the determinant. This unit element will henceforth be called the pivotal element. **1939** A. C. AITKEN *Determinants & Matrices* ii. 47 A determinant of order n being reduced by a first pivotal condensation to one of order $n-1$, the latter in its turn can be reduced by a second pivotal condensation to one of order $n-2$, and so on. **1952** D. R. HARTREE *Numerical Anal.* viii. 158 The coefficient a_{jk}..is called the 'pivotal coefficient' or 'pivot' for this elimination; it is the coefficient, in the pivotal equation, of the variable to be eliminated. **1963** N. MACON *Numerical Anal.* v. 59 It is of some interest to calculate the number of arithmetical operations required for the evaluation of a determinant of order n when pivotal condensation is used. **1971** [see *PIVOT *sb.* 3 e]. **1973** A. M. COHEN et al. *Numerical Anal.* viii. 136 We add suitable multiples of equation (8.17*a*) to equations (β), (γ), and (δ) to reduce the coefficients of x_1 in them to zero... Equation (α), which remains unaltered, is called the pivotal equation.

3. *Linguistics.* Of, pertaining to, or based upon pivot grammar or pivot words (see *PIVOT *sb.* 6 b).

1963 M. D. S. BRAINE in *Language* XXX. 4 *Do it, push it, close it*, etc...appear to exemplify the same kind of pivotal construction as the previous combinations discussed, except that now the pivot is in the final position. *Ibid.* 10 The pivotal type of construction continues long after the first five months, and new pivot words develop. **1970** L. M. BLOOM *Lang. Devel.* 223 The two relational aspects of language appear to be related to the two descriptions of children's speech as 'telegraphic' and 'pivotal'. **1971** *Jrnl. Speech & Hearing Disorders* XXXVI. 42 How does the child progress from using pivotal utterances to using utterances that reflect the complex interrelation of rules that is the essence of adult phrase structure?

Hence **pi·votalism**, the policy of releasing 'pivotal men' from active service before others.

1919 W. S. CHURCHILL in D. Cooper *Haig* (1936) II. xxvii. 406 Indiscipline and disorganisation would arise in the Army if pivotalism, i.e. favouritism, were to rule in regard to the discharge of men. **1922** *Encycl. Brit.* XXX. 215/1 Pivotalism..was called 'favouritism'.

pivoting, *vbl. sb.* (s.v. PIVOT *v.*). Add: **2.** *Math.* The use of a pivot (sense *3 e) as a means of making a column of a determinant or matrix consist entirely of zeros except for one unit element; also, pivotal condensation; *partial pivoting*, pivoting in which the choice of pivot at each stage is restricted to the largest element in the first column (or the first row) of the relevant part of the matrix, rather than the largest in all its columns or rows (*complete, total,* or *full pivoting*).

1961 *Jrnl. Assoc. Computing Machinery* VIII. 282 We derive first an upper bound for R when a general matrix is reduced to triangular form by Gaussian elimination, selecting as pivotal element at each stage the element of maximum modulus in the whole of the remaining square matrix. We refer to this as 'complete' pivoting for size, in contrast to the selection of the maximum element in the leading column at each stage, which we call 'partial' pivoting for size. **1963** N. Macon *Numerical Anal.* v. 59 The effect of pivoting on rounding errors. **1968** Fox & Mayers *Computing Methods for Scientists & Engineers* v. 86 Partial pivoting is generally satisfactory, and..can be made still better, by the use of fl_2 arithmetic.., in a manner which cannot easily be performed with total pivoting. **1973** Phillips & Taylor *Theory & Appl. Numerical Anal.* viii. 197 Partial pivoting using only row interchanges is preferable to that with only column interchanges.

piwarrie. Add: The more usual spelling is now *paiwari*. This name for the beverage is used *spec.* with relation to the Indians of Guyana. (Further examples.) Cf. *cassiri.

1868 W. H. Brett *Indian Tribes of Guiana* i. ix. 155 After a few lashes, they drank paiwari together, and returned to the main body of the dancers. **1883** E. F. im Thurn *Among Indians of Guiana* xv. 319 All the festivals among all the tribes being occasions for much drinking of paiwari—the national beverage—they may all be called Paiwari Feasts. **1934** E. Waugh *Handful of Dust* vi. 337 'They have been making *piwari*... You should try some.'.. Tony gulped the dark liquid, trying not to taste it. But it was not unpleasant, hard and muddy on the palate like most of the beverages he had been offered in Brazil, but with a flavour of honey and brown bread. **1938** *Amer. Anthropologist* XL. 228 Another method of manufacturing alcoholic beverages in South America was to ferment the starchy juice of the pressed or chewed cassava (*Jatropha manihot*)... It was called paiwari or paiva in British Guiana,..cachiri among the Roucouyenne, and cauim or pajuarú among the aborigines of Brazil. **1958** H. G. de Lisser *Arawak Girl* v. 50 Francisco had been drinking much piwari. **1964** V. G. C. Norwood *Jungle Life in Guiana* v. 108 Quantities of the three principal native beers or liquors brewed by forest Indians throughout Guiana: cashiri, yamanchi and paiwarrie.

pix^{2} (piks), var. *pics*, pl. of *pic^{4}. In quot. 1932 used for the sing.

1932 *Variety* 19 July 4/5 (*heading*) 'Million', Par's all-star pix. *Ibid.* 9 Aug. 15/4 (*heading*) Open sesame for Brit. pix in Dominion. **1935** *Ibid.* 3 Apr. 10/1 On Sunday two powerful b.o. pix. were released. **1936** Wodehouse *Laughing Gas* ii. 20 'Your face seems extraordinarily familiar, too.' 'You've probably seen it in pix.' 'No, I've never been there.' 'In the pictures.' **1940** R. Chandler *Farewell, my Lovely* iii. 13 That's what I rate after eighteen years in this man's police department. No pix, no space, not even four lines in the want-ad section. **1945** [see *hypo *v.*]. **1955** Pohl & Kornbluth *Space Merchants* ii. 22 Our artists can work from the pix you brought back. **1960** *Analog Science Fact/Fiction* Dec. 48/2 Now they have a chance to get their news releases and faked pix out in quantity. **1973** E. Hyams *Final Agenda* iv. 53 Barnet said, 'I want the Victoria Lowell pix.'.. [He] began looking through the thirty-seven photographs of Victoria Lowell. **1975** D. Lodge *Changing Places* ii. 90 He would be holed up somewhere, jerking himself off and drooling over the *Playboy* pix. **1977** *Zigzag* June 23/3 Tome is shown the Warner Bros. press kit for him, with accompanying pix.

pixel (pi·ksel). [f. *pix^{2} + El(ement *sb.*] = *picture element* s.v. *picture *sb.* 6 d.

1969 *Science* 15 Aug. 685/1 An analog tape recorder was used to store the analog video signal from each pixel. **1975** *Sci. Amer.* Oct. 67/1 The cross sections of the heart were subsequently reconstructed by the same computer. Each cross section took two minutes of computing time and was displayed on a cathode-ray tube as a square picture with 64 pixels on a side. **1977** *New Yorker* 14 Feb. 28/1 The [advertising] panel is divided into two thousand and forty-eight 'pixels', or picture elements of red, green, blue, and white bulbs, and ordinarily only one or two of the bulbs on a pixel are flashed at a given time. **1977** *Proc. R. Soc. Med.* LXX. 593/1 No doubt the next edition using up-dated prints will overcome the difficulties of interpretations associated with the large size of pixel.

pixilated (pi·ksilēitėd), *a.* orig. *U.S. dial.* Also **pixelated, pixielated, pixillated, pixylated.** [f. Pixy, pixie + *-lated* as in *elated*, *emulated*, etc. or var. Pixy-led *a.*] **1.** Mildly insane, fey, whimsical; bewildered, confused; intoxicated, tipsy.

1848 in *Amer. Speech* (1941) XVI. 79/2 You'll never find on any trip That he'll be pix-e-lated. **1886** E. L. Bynner *Agnes Surriage* iv. 56 'See now wher' we ha' come to wi' yer talk, Job Redden!' cried Agnes, waking suddenly to their situation. 'We'll be pixilated 'n' driven on to th' rocks an ye don't wake up.' **1895** *Dialect Notes* I. 392 *Pixilated*, dazed, bewildered in the dark. Marblehead, Mass. **1936** in *Amer. Speech* (1941) XVI. 79/2 *Lawyer*: Now tell me, what does everybody back home think of Longfellow Deeds?.. *Jane*: They think he's pixilated. *Ibid.* 80/1 The word pixilated is an early American expression—derived from the word 'pixies' meaning elves. They would say, 'The pixies have got him,' as we nowadays would say a man is 'balmy'. **1937** *N. & Q.* 2 Jan. 11/2 As a native of the state from which 'Mr. Deeds' is reputed to have come permit me to comment on..'pixilated'. To use the word in the sense of 'crazy' is not correct. A Vermonter would not hesitate to use 'crazy' if that conveyed his meaning. A 'pixilated' man is one whose whimseys are not understood by practical-minded people... More nearly a synonym of 'whimsical'. **1955** 'J. R. Macdonald' *Find Victim* xxiv. 167 'Wasn't he pretty drunk on Sunday?' 'He was pixilated all right,' Jo said. **1957** L. Iremonger *Ghosts of Versailles* viii. 103 Ultimately no explanation of the 'adventure' was too fantastic, far-fetched or indeed pixylated for them. **1958** *Observer* 30 Nov. 16/6 Nicely cast, he gave the true tone of pixilated delinquency. **1971** *Listener* 8 Apr. 449/1 Suddenly we were pixilated, we'd fallen in love with the sweetest girl we ever saw. **1975** C. Nesbitt *Little Love & Good Company* xvii. 208 We were both ever so slightly inebriated, no not even that, pixilated, to use the lovely movie euphemism. **1977** *New Yorker* 29 Aug. 28/3 He was known as a coarse creature and a little pixillated.

2. Of an actor: having movements animated by the pixilation technique.

1959 *News Chronicle* 20 Oct. 8/7 The animator..may use conventional comic drawings..or even 'pixillated' live actors.

pixilation, pixillation (piksilēi·ʃən). Also **pixylation.** [f. *pixilated *a.*: see -ation.] **1.** A technique used in theatrical and cinematographic productions, whereby human characters move or appear to move as if artificially animated; the effect produced.

1947 *Punch* 5 Mar. 200/1 Those who, as I am, are made uncomfortable by material pixylation on the stage may take heart, for the *Catherine* who comes back from the midnight romp is no gauzy fay. **1953** *Q. of Film, Radio & Television* VIII. 9 McLaren feels this kind of live-actor animation has considerable creative potentiality although he refers slightingly to it as the 'pixillation' technique. **1957** Manvell & Huntley *Technique Film Music* iii. 167 By applying an animation technique to the movements of actors, he produced 'pixilation' and used it to tell a serious story—in *Neighbours* (Canada 1953). **1959** Halas & Manvell *Technique Film Animation* xxv. 291 This technique of experimenting in the animation of the movements of live actors (called sometimes 'pixilation') accompanied by synthetic music and sound effects. **1971** *Concord Films Council Catal.* 51/2 Yugoslavian experimental film in the pixilation technique. **1976** *Oxf. Compan. Film* 438/1 McLaren here [*sc.* in *Neighbours*], and in *Two Bagatelles* (1952), used pixilation, 'animating' human actors. In *Chairy Tale* (1957) and *Opening Speech* (1960) the technique is applied humorously to inanimate objects. *Ibid.* 547/2 *Pixillation*, the use of a stop frame camera to speed up and distort the movement of actors, creating roughly the effect of animation with live people.

2. The state or condition of being pixilated (sense 1).

1960 *Spectator* 6 May 677 Without pretentiousness and with no traces of pixilation and phoney Cornishness.

pixy, pixie. Add: The more usual spelling is now *pixie*. **a*.** Short for **pixie cap, hat*, etc.

1960 *Harper's Bazaar* Oct. 114 Fur pixie.

b. *pixie-like* adj.; **pixie cap**, a pointed hat resembling that in which pixies are traditionally depicted; **pixie cape**, a cape with an attached pixie hood; **pixie hat** = **pixie cap*; **pixie hood**, a pointed hood; so **pixie-hooded** *a.*

1828 Pixie cap [see *Kilmarnock 1]. **1943** F. Urquhart in *Penguin New Writing* XVI. 90 She was wearing a crimson waterproof pixie-cap which was almost the same colour as her pretty, round face. **1960** M. A. Sindall *Matey* vii. 89 Brown woollen pixie caps. **1973** *Listener* 23 Aug. 244/2, I knew that I did not look my best in my mackintosh with its pixie cape. **1954** G. Durrell *Three Singles to Adventure* v. 106 On his sleek black head was perched an absurd pixie hat constructed out of what once used to be velvet. **1940** C. Milburn *Diary* 23 Dec. (1979) 75 Two little pixie-like children in green overall garments, complete with pixie hoods. **1950** B. Pym *Some Tame Gazelle* viii. 89 They stood in the doorway,..wearing mackintoshes, and that wet-weather headgear so unbecoming to middle-aged ladies and so incongruously known as a 'pixie hood'. **1955** E. Bowen *World of Love* ix. 163 Maud, in wet weather rendered still more terrible by a pixie hood. **1961** *Guardian* 28 Nov. 7/1 The worthies of the village in ornamented pixie hoods. **1978** M. Butterworth *X marks Spot* i. i. 22 An old lady in a plastic pixie hood. **1949** E. Coxhead *Wind in West* i. 15 Two pixie-hooded small boys. **1940** Pixie-like [see *pixie hood* above]. **1963** *Times* 16 May 14/6 The men wear a curious conical hat..which gives them a curious pixy-like appearance, while frequent recourse to chewing *qat* has wizened their faces and for all I know stunted their growth. **1979** J. Wainwright *Reluctant Sleeper* v. 69 Those ridiculously large spectacles, and that equally ridiculous pixie-like face.

Hence **pi·xyish** [-ish^{1}] *a.*, resembling that of a pixie.

1962 V. Connaught *Secret Heart of Princess Alexandra* x. 101 As she splashed by them, Alexandra appeared to take a pixyish glee in noting their discomfort. **1977** J. Aiken *Last Movement* viii. 167 Her narrow, pixyish Irish face.

‖ **Piyut, Piyyut** (piyu·t). Also **Peyut**, erron. **Piyyuth**, and with lower-case initial. Pl. **-im.** [Heb. *piyyut* poem, poetry, f. *poyṭān*, *payṭān* poet, ad. Gk. ποιητής (see Poet).] A poem that is read or recited in a synagogue in addition to the standard liturgy.

1876 *Sat. Rev.* 16 Sept. 357/1 The 'Liturgy' recited in the synagogues every Sabbath from the Piyutim. **1876** *Gentl. Mag.* Nov. 601 Those Hebrew poems recited by the Ashkenazim, and called 'Peyutim'. **1972** C. Raphael *Feast of Hist.* iii. 106 Like most other large-scale illustrated Haggadahs, the book also includes many pages of *piyyutim* (poems) drawing on biblical and midrashic themes. **1976** M. Horowitz in D. Villiers *Next Year in Jerusalem* 114 Even now my head reels daily with..exalted Hebraic dirges..tracking back from Bialik to Yehuda Halevi to the Piyyuth and Ecclesiastes.

pize (pəiz), *v. dial.* [Origin uncertain: perh. ad. MDu. *pisen* (see quot. 1968^{3}).] **a.** *trans.* To strike; *spec.* to hit (a ball) with the hand in the game of pize-ball (see next). Also const. *down*. **b.** *intr.* and *trans.* To throw (a ball) in the game of pize-ball; to act as bowler in pize-ball. **c.** *trans.* To throw to (the batter) in pize-ball.

1796 S. Pegge *Derbicisms* (1896) 54 To *pize* a ball, to strike it with the hand; so the game is call'd *pize-ball*. To *pize down* a hare, i.e. with a gun; meaning to strike her down. **1862** C. C. Robinson *Dial. Leeds* 385 *Pize*, to throw a ball gently for another to bat with the open hand, as at the game of 'Pize-ball'. *Ibid.*, The game of 'Pize-ball', in which the 'pizer' 'pizes' the ball to a member in succession. **1968** A. S. C. Ross in *Proc. Leeds Philos. & Lit. Soc.* (*Lit. & Hist. Section*) XIII. ii. 59 If, however,..the Pizer delayed too long,..the players would chant: 'Pize your neighbour while you're able, While the donkey's in the stable!' *Ibid.*, The player who had got round most times..might be the winner (and pized next game). *Ibid.* 63 *Pize* is a word entirely without an etymology. I suggest that it is a borrowing of MDutch *pisen*..name of a game about which further particulars are lacking. *Ibid.* 69 Applied to the ball, *pize* means both 'to throw' and 'to strike'.

Hence **pize** *sb.*^{2}, a throw in pize-ball; **pi·zer**, a bowler in pize-ball; **pi·zing** *vbl. sb.*

1862 Pizer [see above]. **1869** 'T. Treddlehoyle' *Bairnsla Foaks Ann.* 55 Throo thrawin a stones, tipsey lakein, an pizein a balls it publick street, good Bairnsla deliver uz. **1896** *Leeds Mercury Weekly Suppl.* 7 Mar. 3/8 Let me hev a pize, an' ah'll mak' him send a cop. **1968** *Proc. Leeds Philos. & Lit. Soc.* (*Lit. & Hist. Section*) XIII. ii. 56 The thrower, or *Pizer*, stands some distance in front of the homey and throws the ball to the striker.

pi·ze-ball. *local.* Also **piseball, pizeball.** [f. *pize *v.* + Ball *sb.*^{1} 4.] A game similar to rounders in which the ball is hit with the flat of the hand. (See also quot. 1883.)

Played mainly in Yorkshire and Derbyshire.

1796, 1862 [see *pize *v.*]. **1883** A. Easther *Gloss. Dial. Almondbury & Huddersfield* 102 *Pizeball*.., a ball which children play with, formerly stuffed with sawdust, etc... It was often partly coloured and ornamented; now it is sometimes of india-rubber and hollow. **1957** R. Hoggart *Uses of Literacy* ii. 58 'Piseball', 'tig',..and a great number of games involving running round the lamp-posts or in and out of the closet-areas..are still popular. **1968** *Proc. Leeds Philos. & Lit. Soc.* (*Lit. & Hist. Section*) XIII. ii. 55 Pize-ball is a game which, in many ways, resembles the well-known games, Rounders and Baseball.

pizza (pī·tsä). Pl. **pizzas,** ‖ **pizze** (-ə). [It., = pie.] A savoury dish of Italian origin, consisting of a base of dough, spread with a selection of such ingredients as olives, tomatoes, cheese, anchovies, etc., and baked in a very hot oven; dough so prepared and baked. ‖ *pizza* (*alla*) *Napoletana* (ala napǫlitā·nä) [It., = Neapolitan, in the Neapolitan style], Neapolitan pizza (see quot. 1955).

1935 M. Morphy *Recipes of All Nations* 160 Pizza alla Napoletana... In the south of Italy..all kinds of flat tarts are called 'pizze'. **1953** W. P. McGivern *Big Heat* xi. 138 An unbaked pizza, or cheese pie, covered with thinly sliced tomatoes and criss-crossed with strips of anchovy. **1955** E. David *Bk. Mediterranean Food* 40 The Neapolitan pizza..consists of tomatoes, anchovies and mozzarella cheese. **1957** O. Nash *You can't get there from Here* 150 She eatsa Pizza! Greedy Mitzi! She no longer itsy-bitsy! **1957** *Sunday Times* 1 Dec. 23/4 The Pizza Napoletana has travelled the world. In Paris restaurants, in Shaftesbury Avenue milk bars, in South Kensington coffee shops the pizza has become acclimatised. **1959** *Vogue* June 90 Food [in coffee-bars] is usually the pizza, sandwich or Danish pastry type. **1959** V. Packard *Status Seekers* (1960) 146 He was raised on blood sausages, pizza, spaghetti and red wine. **1970** Simon & Howe *Dict. Gastron.* 303/1 In its most primitive form *pizza* is a round of yeast dough spread with tomatoes and mozzarella cheese and baked in a hot oven. The most famous of the many *pizze* is the Neapolitan *pizza*... The Roman *pizza*..has plenty of onions but no tomatoes; the Ligurian *pizza* has onions, black olives and anchovies. **1972** J. Mosedale *Football* viii. 110, I like pizza, hamburgers and hot dogs. **1977** *New Yorker* 12 Sept. 97/1 Pizza is my speciality, I make it all myself. O.K. Well, now the dough is rising good, so I get out some tomatoes and the other stuff for a pizza alla napoletana.

2. *attrib.* and *Comb.*, as *pizza bar, cook, dough, mixture, palace, parlour, pie, stand*; *pizza-seller*.

1971 J. Fleming *Grim Death* ii. 23 He was hoping to start up a Pizza bar. **1974** *Listener* 23 May 665/3 A pizza bar like we have in Glasgow. **1976** *Weekend Echo* (Liverpool) 4/5 Dec. 7/3 Scores of pizza bars and restaurants to suit all pockets. **1959** 'E. Allen' *Man who chose Death* ii. 24 So many shopgirls and garage hands and pizza cooks. **1960** E. Campbell *Encycl. World Cookery* 263 Put the pizza dough onto a baking tin. **1977** *New Yorker* 12 Sept. 97/1, I start to work in the window up front and make some *pizza* dough. **1959** *Guardian* 25 Nov. 6/2 The Italian *origano*, an essential part in most pizza mixtures. **1959** *Times* 18 May 5/3 The pizza palaces and highway honky-tonks with which we have littered the land. **1974** P. Erdman *Silver Bears* ii. 20 That's a fucking pizza parlour. **1976** *Star* (Sheffield) 3 Dec. 14/7 One well-established pizza parlour tells me they even do a special Yorkshire style pizza—it's like the Italian ones..but fatter because we eat more. **1970** 'D. Halliday' *Dolly & Cookie Bird* i. 3, I..bought a pizza pie..on my way home. **1976** *National Observer* (U.S.) 12 June 1/4 Ripping off the public is as American as pizza pie. **1961** *Times* 21 Dec. 3/2 Her earlier appearance as the ebullient pizza-seller. **1966** T. Pynchon *Crying of Lot 49* v. 121 Down at the city beach, long after the pizza stands..had closed, she walked unmolested. **1966** A. Cavanaugh *Children are Gone* ii. v. 43 Laundromats, Jewish dairies, pizza stands, fruit stores.

pizzazz (pĭzæ·z). orig. *U.S. slang.* Also **bezaz, bezazz, bizzazz, pazazz, pazzazz, pezazz, pizazz, pizzaz.** [Origin unknown.] **1. a.** Zest, vim, vitality, liveliness. **b.** Flashiness, showiness.

1937 *Harper's Bazaar* Mar. 116/2 Pizazz, to quote the editor of the Harvard *Lampoon*, is an indefinable dynamic quality, the *je ne sais quoi* of function; as for instance, adding Scotch puts pizazz into a drink. Certain clothes have it, too... There's pizazz in this rust evening coat. **1951** *Time* 28 May 91/2 Rentschler thinks the J-57 has more pizzazz than any other engine. Says he flatly: 'It is more powerful than any jet engine ever flown.' **1952** S. Kauffmann *Philanderer* (1953) v. 86 Now here's a few places where I think it could use a little pezazz. **1962** *U.S.A.-1* I. iv. 30/2 He displayed almost none of the oratorical pizzazz that had set them [*sc.* Canadian voters] screaming in 1958. **1964** *New Statesman* 28 Aug. 291/1 A Shakespeare one [*sc.* exhibition]..with most of its bezazz—pop art, wire sculpture, giant beefeaters—left by the Avon. **1965** *Sunday Times* (Colour Suppl.) 16 May 12/1 She..still wears trousers frequently. 'I don't really feel happy in bezazz.' **1966** *Saturday Night* (Toronto) May 34/3 His campaign manager..mounted a campaign that has had few equals anywhere for sheer *pazazz*. **1967** R. Stein *Great Cars* 202/2 Began doing for the dull little mass-produced British cars what it had done for the homely sidecar. It gave them what Detroit now calls 'pizzaz'. **1968** *Daily Tel.* 24 Dec. 8/4 Miss [Ginger] Rogers has 'bezazz', as was obvious from the number of reporters and photographers clustering round her. But Mr. Marshall..claimed it should be 'pezazz', derived from American TV commercials and meaning something like effervescence. **1970** *New Yorker* 14 Mar. 66, I knew that we absolutely had to have something with more bizzazz going for the young people. **1973** *Observer* (Colour Suppl.) 4 Nov. 10/2 What the ice-rink audience really likes is pizzazz, rather than lyricism: they like a cheesy grin, powerful thighs, plenty of intricate footwork with the blades, and above all a crunching four-four rhythm they can clap along with. **1974** *Language Sciences* Aug. 25/3 If they confined themselves to the level of common sense they would lose much of their academic pizzazz, their mystique as scientists. **1975** G. V. Higgins *City on Hill* ii. 44 Maybe some guy that could recruit more troops and out-fund us gets himself involved in a bloodletting with another guy who has some pizzazz, and..they knock each other off. **1976** *New Musical Express* 31 July 27/3 Ferguson's album has style and pizzazz. **1976** *National Observer* (U.S.) 2 Oct. 18/4 This confluence of ecology and art,..of business chutzpah and show-biz pizzaz. **1976** *Listener* 23 Dec. 841/3 Behind the pazazz there are some fundamental issues being toyed with. **1978** *People Weekly* 9 Jan. 57/1 'But it must have commercial appeal,' Flynt cautions. 'I think I give projects like this pizzazz.' **1979** *Arizona Highways* Apr. 19/1 (*caption*) Jimmy King, Jr. adds pizazz to his golden pendant..wrapping the design with his dazzling inlay.

2. *attrib.* or as *adj.* Flashy, ostentatious, 'zippy'.

1970 *Canad. Antiques Collector* Oct. 5/1 There are various displays of a great range of collectibles that create a lively 'bezaz' atmosphere. **1977** *Courier-Mail* (Brisbane) 6 Mar. 6/7 Sammy Davis Jnr...flashed $100,000 worth of sparkling rings... He confessed yesterday, at a crowded press conference..'It's theatrical and it's pizzazz.'

pizzeria (pītsərī·ă). [It.] A place where pizzas are made, sold, or eaten.

1943 J. Steinbeck *Once there was War* (1959) 184 He spoke the English we know, the English of the banana pushcarts and the pizzerias, of the spaghetti joints and grind organs. **1957** V. P. Johns in Charteris & Santesson *Saint's Choice* (1967) 198 The Pizzerias piled up when she went to Rome. **1966** L. Deighton *Billion-Dollar Brain* xv. 141 We went past the silent skyscrapers, kosher pizzerias, glass-fronted banks. **1973** 'D. Jordan' *Nile Green* xxxv. 171 She'll want to go to one of those pizzerias near the British Museum. **1977** *N.Z. Listener* 15 Jan. 17/1 Between performances he telephoned his wife in London, pottered around on the golf course, downed Newcastle ale and frequented 'The Pinocchio', a Sunderland pizzeria.

pizzle. Delete 'Now *dial.* or *vulgar*' and add later examples, esp. in *Austral.* cattle- and sheep-rearing terminology.

1946 R. Graves *Poems 1938–45* 27 Who whipped her daughters with a bull's pizzle. **1949** A. Huxley *Ape & Essence* 126 Then the Grand Inquisitor's Special Assistant bends down and, from under his chair, produces a very large consecrated bull's pizzle, which he lays on the table before him. **1955** W. Gaddis *Recognitions* ii. iii. 441 He entered and walked toward the bull's stall—There!.. There's a masterful pizzle for you! **1965** J. S. Gunn *Terminol. Shearing Industry* ii. 8 *Pizzle*, a sheep's penis. This word has no taboo and is the formal word in written articles. **1968** E. R. Buckler *Ox Bells & Fireflies* vi. 101 [An ox] would gallop the pasture faster than a horse, ramming a low stub halfway up his pizzle. **1969** *Coast to Coast 1967–68* 7 Ruben had caught the farmer's bull one night and tied a length of plastic cord around its pizzle, letting the bull go again in the paddock. The breeding paddock, with cows on heat.

placating (plăkēi·tiŋ), *ppl. a.* [f. Placate *v.* + -ing^{2}.] That placates or is intended to placate; conciliatory. Hence as *sb.* (*rare*$^{-1}$); **placa·tingly** *adv.*

1911 M. Johnston *Long Roll* xix. 243 Allen took it calmly, made a placating remark or two, and lapsed into a friendly silence. **1919** E. O'Neill *Where Cross is Made* in *Moon of Caribbees* 167 (*Placatingly*) You're wrong, Father. **1921** *Spectator* 12 Mar. 333 Holland never dealt in half-measures; the placating whitey-grey argument or studiously reasoned compromise was to him anathema. **1925** T. Dreiser *Amer. Trag.* (1926) ii. xi. 235 'You're right, I know,' said Clyde placatingly, for he was still hoping for this hinted-at promotion. **1931** E. O'Neill *The Hunted* iv, in *Mourning becomes Electra* (1932) 171 Hastily, with a placating air. **1935** W. Stevens *Ideas of Order* (1936) 22 Be thou that wintry sound As of the great wind howling, By which sorrow is released, Dismissed, absolved In a starry placating. **1941** —— in O. Williams *New Poems: 1940* 202 The placating star Shall be the greater for the death you die. **1964** *Punch* 29 Jan. 153/1 My wife poured him a coffee placatingly. **1977** P. Hill *Fanatics* 21 Carpenter put up his hand in a placating gesture.

placative (plăkēi-tiv), *a. rare.* [f. Placate *v.* + -ive.] = Placatory *a.*

1931 W. Faulkner *Sanctuary* xi. 107 Temple stared at her with that grimace of cringing and placative assurance.

place, *sb.*^{1} Add: **1. b.** (Earlier and later examples of use as the second element of a proper name.)

1700 Congreve *Way of World* i. i. 4 There's such Coupling at Pancras..we were afraid his [the Parson's] Lungs would have fail'd..so we drove round to Duke's Place. **1791** F. Burney *Let.* 8 Sept. in *Jrnls. & Lett.* (1972) I. 55 A House in Laura Place. **1903** G. B. Shaw *Man & Superman* i. 2 Sitting at his writing table, he has on his right the windows giving on Portland Place. **1939** Joyce *Finnegans Wake* (1964) i. 132 First he shot down Raglan Road and then he tore up Marlborough Place. **1972** J. McClure *Caterpillar Cop* iii. 42 Kramer..took the Durban road, watching the street names on his left. He swung into Potter's Place. No 9 Potter's Place was untidier than most.

3. e. Colloq. phr. *all over the place*: disordered, irregular, muddled.

1923 J. Manchon *Le Slang* 227 *All over the place*,..en pagaye. **1933** A. E. Housman *Let.* 13 July (1971) 337 The Doctor sent me into a nursing home for a week because he said my heart was all over the place. **1937** N. Coward *Present Indicative* vi. i. 229 Lilian was cool and steady and played beautifully. I was all over the place but gave, on the whole, one of those effective, nerve-strung *tour-de-force* performances, technically unstable, but vital enough to sweep people into enthusiasm. **1953** R. Lehmann *Echoing Grove* 16 In her youth it [*sc.* her hair] had spilled out all over the place, brilliant but not warm. **1959** H. Pinter *Birthday Party* ii. 15 Why is it that before you do a job you're all over the place, and when you're doing the job you're as cool as a whistle? **1971** O. Norton *Corpse-Bird Cries* vi. 125 Her heart's all over the place, according to Sister. Shock after losing the Colonel. **1971** M. McCarthy *Birds of America* 269 You seem unfocussed... All over the place. No clear line of direction. **1976** S. Brett *So much Blood* iii. 43 'How's your show going?' '*Mary's* still all over the place. We spend so much time improvising..we hardly ever get near the actual script.'

5. b. (Further examples.)

1909 *Dialect Notes* III. 358 *Place*, *n.*, home, farm. 'When you comin out to our *place*.' **a1922** T. S. Eliot *Waste Land Drafts* (1971) 5 We had a couple of feelers down at Tom's place. *Ibid.*, I turned up an hour later down at Myrtle's place. **1932** S. Gibbons *Cold Comfort Farm* xix. 256 'Tell Reuben he can have the old place.'.. 'It's a pity he says "the old place" instead of "the farm".' **1939** Joyce *Finnegans Wake* (1964) i. 43 A few good old souls, who, as they were juiced after taking their pledge over at the uncle's place, were evidently under the spell of liquor. **1946** E. Hodgins *Mr. Blandings builds his Dream House* ii. 14 The New York apartment..was home no longer; the old Hackett place on Bald Mountain was home, now. **1972** *Screw* 12 June 33/4 (Advt.), Young male nude model. Experienced, handsome... Completely versatile and cooperative. Your place or mine. **1978** J. F. Burke *Crazy Woman Blues* i. 3 If she'd been taken ill suddenly she might have gone up to her place.

d. (Later examples.)

1973 R. L. Simon *Big Fix* (1974) xiv. 99 We were sitting at the counter of Winchell's doughnut place on Glendale Boulevard. **1978** J. L. Hensley *Killing in Gold* (1979) xi. 148 We went to Mac's Place... The waitresses were..insolent.

e. *slang.* A lavatory (see also quot. 1951).

1901 Farmer & Henley *Slang* V. 220/2 *Place*,..(2) a jakes, or house of ease. **1922** Joyce *Ulysses* 160 They did right to put him up over a urinal... Ought to be places for women. **1942** Berrey & Van den Bark *Amer. Thes. Slang* §84/11 Toilet..place. **1951** Partridge *Dict. Slang* (ed. 4) 1137/2 *Place where you cough, the*, the water-closet... Ex coughing to warn an approacher that it is occupied.

8. c. Restrict '† *Obs.* (or *arch.* after Shaks.)' to sense in Dict. and add: Also phr. *pride of place*: a pre-eminent position.

1902 *Punch* 24 Dec. 434/1 A Minister who is chased by a loud-voiced Opposition From his pride of place. **1919** *Empire Rev.* XXXIII. 242 Britain is compelled to raise her prices to heights which..will send the British buyer abroad for those very materials in the manufacture of which we have formerly held pride of place throughout the world. **1931** A. Huxley *Music at Night* 222 Disease-snobbery is only one out of a great multitude of snobberies, of which now some, now others take pride of place in general esteem. **1948** G. Gorer *Americans* i. 26 In the fantasies brought to light in psychiatric interviews pride of place went to those in which the officer was retaliated upon, humiliated, snubbed. **1954** M. Beresford *Lost Villages* x. 343 This site has pride of place for the admirable monograph by Dr. W. M. Palmer. **1976** *Flintshire Leader* 10 Dec. 32/1 Pride of place must go to Courtaulds Greenfield, the league leaders, who toppled the Welsh National League.. Division 1 champions, Denbigh Town. **1978** *Jrnl. R. Soc. Arts* CXXVI. 305/2 And then we come to folk and country drawings, which somehow have become a very American cult. These not unexpectedly are given a pride-of-place chapter.

9. c. (Earlier and later examples.) In *U.S.* applied *spec.* to second place.

1836 *Spirit of Times* 5 Mar. 22/1 He led the first two miles, Sir Kenneth trailing, and Mattiwan endeavoring to keep a place in the race. **1930** *Daily Express* 6 Oct. 17/6 *Tote.*—Win 5s; places 2s 9d, 2s 9d, 3s 6d. **1942** Berrey & Van den Bark *Amer. Thes. Slang* §740/2 Place, second place, or at least second. **1976** *Daily Tel.* (Colour Suppl.) 26 Mar. 27/3 '£1 each way' means £2 split between a win and a 'place' (the horse finishes in the first three).

d. Phrases: *to know one's place*: to know how to behave in a manner befitting one's rank, situation, etc.; *it is not my place*: outside my duties or customary rights; *to put* (someone) *in his, her* etc., *place*: to remind someone of his or her rank or situation; to rebuff or rebuke.

1601 Shaks. *Twel. N.* ii. v. 59, I knowe my place, as I would they should doe theirs. **1739–40** Richardson *Pamela* (1740) I. xi. 18 It does not become your poor Servant..and I hope I shall always know my Place. **1852** [in Dict., sense 9 a]. **1867** Dickens & Collins *No Thoroughfare* in *All Year Round* Extra Christmas No. 12 Dec. 3/1 It is not my place, ma'am, to tell names to visitors. **1898** G. B. Shaw *Candida* ii. 113 Mr Morchbanks is a gentleman, and knows his place, which is more than some people do. **1908** A. Bennett *Old Wives' Tale* i. vi. 108 She ought to have put Mr. Povey into his place... Mr. Povey ought to have been ruined for ever in her esteem. **1916** G. B. Shaw *Pygmalion* ii. 143, I should just like to take a taxi to the corner of Tottenham Court Road and get out there and tell it to wait for me, just to put the girls in their place a bit. I wouldnt speak to them, you know. **1930** W. Faulkner *As I lay Dying* (1935) 5 It is not my place to question His decree. **1937** D. & H. Teilhet *Feather Cloak Murders* vi. 104 Not that I'm complaining. Dear me, no. I know my place. **1943** A. Christie *Moving Finger* vii. 85 These girls nowadays—don't know their place—no idea of how to behave. **1943** J. B. Priestley *Daylight on Saturday* xxxviii. 301 Every time I think..that it's going to be easy to put you in your place, you suddenly do or say something that breaks it all down. **1956** A. Wilson *Anglo-Saxon Att.* ii. i. 195 When he asked her to choose a restaurant, she said, 'No, you do that thing. I'd much rather it was your choice.' He suggested Scott's, and she said, 'But that sounds absolutely the right thing.' He hoped that she was not going to put him in his place the whole evening. **1965** R. Bastide in G. Hunter *Industrialisation & Race Relations* i. 15 It institutionalised the subordination of the Negroes, who could only benefit from the protection of the whites..on condition that they 'knew their place' and proved their deference, gratitude and respect. **1973** R. Stout *Please pass Guilt* (1974) xi. 109 On the phone you stiff-armed me. You put me in my place. **1974** J. Stubbs *Painted Face* 11 It's not my place to judge, sir. **1979** M. Hebden *Murder set to Music* ii. 17 'Did she have men friends?'..'It's not my place to say.'

12. d. Phr. *a place for everything and everything in its place*.

1842 F. Marryat *Masterman Ready* II. i. 9 In a well-conducted man-of-war..every thing is in its place, and there is a place for every thing. **1855** T. C. Haliburton *Nat. & Hum. Nat.* I. vi. 164, I was born on a farm..where there was a place for everything, and everything was in its place. **1857** Emerson *Jrnl.* 2 Aug. (1914) IX. 110 A place for everything, and everything in place. **1875** S. Smiles *Thrift* v. 66 Order is most useful in the management of everything... Its maxim is—A place for everything, and everything in its place. **1922** Joyce *Ulysses* 694 The necessity of order, a place for everything and everything in its place. **1928** D. L. Sayers *Lord Peter views Body* x. 224 'I thought you were rather partial to anatomical specimens.' 'So I am, but not on the breakfast-table. "A place for everything and everything in its place", as my grandmother used to say.' **1941** 'J. J. Connington' *Twenty-One Clues* v. 74 A tidy person with a place for everything, and everything in its place. **1949** J. P. Marquand *Point of No Return* iii. ii. 498 There was a place for everything in Clyde and everything was in its place. **1968** P. Dickinson *Skin Deep* vii. 141 Do you run your whole life like that?.. A place for everything and everything in its place, and all in easy reach.

13. a. (Further examples.)

1955 *Times* 9 May 6/4 In five years we shall provide a million new school places. **1976** W. Corlett *Dark Side of Moon* i. 29 He would have got a university place. **1976** G. Moffat *Over Sea to Death* xv. 174 They went in to dinner, drew the tables together and re-laid the places. **1977** D. Williams *Treasure by Degrees* iii. 28 Up to 1974.. there were still too many students chasing too few university places.

c. Phr. *a place in the sun*: see Sun *sb.* 4 b (*d*) in Dict. and Suppl.

29. *place-name* (further, including *attrib.*, examples); hence *place-namer*, *-naming* vbl. sb., *nomenclature*, *-ordering* vbl. sb.; *place-ordered* ppl. a.; (sense 2 b) *place-logic*, *-time*; (sense 9 c) *place-getter*; (sense 14) *place-seeker* (examples), *-seeking* vbl. sb.; *place-seeking* adj. (example); **place-card**, a card bearing a

guest's name marking the place allocated to him at a table; **place-mat**, a table-mat for a place-setting; **place-money** *Racing*, (*a*) money placed as a bet that a horse, etc., will be second or third (in the U.S., second only); (*b*) prize-money for finishing second or third (in the U.S., second) in a race; **place-setting**, the cutlery, china, etc., required to set a place for one person at a table; **place-value**, the numerical value that a digit has by virtue of its position in a number.

1922 S. Lewis *Babbitt* viii. 115, I was going to have some nice hand-painted place-cards for you but—Oh, let me see; Mr. Frink, you sit here. **1934** J. O'Hara *Appointment in Samarra* (1935) iv. 97 She held a small stack of place-cards. **1938** L. Bemelmans *Life Class* iii. ii. 225 Some terrible place-card holders made of sea shells. **1942** T. Bailey *Pink Camellia* ii. 12, I have the place cards ready. **1963** D. B. Hughes *Expendable Man* (1964) ii. 39 Now the stationer's... We need more place cards for tonight. **1974** P. Erdman *Silver Bears* v. 63 Beside each lady's place-card was a small orchid. **1976** *Eastern Even. News* (Norwich) 27 Aug., Kuda's Surge should be interesting, as should Bownee, a well-fancied place-getter last time. **1976–7** *Sea Spray* (N.Z.) Dec./Jan. 77/3 Prizes:.. A Hanimex f3.5/80–200 mm multicoated zoom lens, valued at $150, for the third place-getter in the senior section. **1977** *N.Z. Herald* 5 Jan. 1-17/6 The women's team will consist of the first three in the target competition, plus Hamilton's Thelma Croft, because the fourth placegetter, Joan Ward (Manakau), did not score qualifying totals before the championships. **1957** A. N. Prior *Time & Modality* 119 Consider..a place-logic in which we have the means of formulating the law. **1968** N. Rescher *Topics in Philos. Logic* xiii. 229 A wide range of logical systems, including not only chronological.. logic, but also what we may call locative or place logic, and even a logic of possible worlds. **1951** T. Sterling *House without Door* ii. 21 The waitress took the mangled place-mat..and brushed the shreds of paper from the table. **1966** J. Cleary *High Commissioner* vi. 120 He looked at the table, at the silverware, the lace place-mats. **1972** J. Ball *Five Pieces Jade* x. 118 She had set two place mats and a small, intimate meal was waiting. **1977** *New Yorker* 10 Oct. 110/2 Farther on, we move into the dining room and partake of a genteelly wholesome meal from gold-edged china set on pale-green patterned placemats. **1894** G. Moore *Esther Waters* xliv. 348 Bramble, a fifty to one chance, not one man in a hundred backed her; King of Trumps, there was some place money lost on him. **1923** Wodehouse *Inimit. Jeeves* xiv. 179 A sniffing female in blue gingham beat a pie-faced kid in pink for the place-money, and Prudence Baxter, Jeeves's long shot, was either fifth or sixth, I couldn't see which. **1942** Berrey & Van den Bark *Amer. Thes. Slang* §734/3 Place money, the odds a horse pays to run second. **1970** *Globe & Mail* (Toronto) 25 Sept. 32/3 Miss Ella Cinders won but with the disqualification will now receive place money of $400 instead of $200 show money. **1973** *Times* 15 Dec. 16/4 A compulsory shareout of place money between owner, trainer and jockey on the same formula as for win money. **1924** Place-name [see *folk-name* (*folk 5 b)]. **1927** *Englische Studien* Nov. 64 The foundation of the English Place-Name Society (in 1923).. has given an enormous impetus to the study of English place-names. *Ibid.*, A useful and competent survey of the methods of the place-name study. *Ibid.*, There are three golden rules to be observed by every place-name student. **1961** L. F. Brosnahan *Sounds of Language* iii. 46 Linguistic, placename, and general knowledge of the history of Europe. **1966** *Eng. Stud.* XLVII. 208 In the pocket we find a geological map and six distribution maps on certain place-name elements. **1977** *Word 1972* XXVIII. 73 *Tre-* as a place-name element appears to have meant just 'settlement'. **1927** *Year's Work Eng. Stud. 1925* 35 The article will interest both lexicographers and place-namers. **1943** *Amer. Speech* XVIII. 241 Finding that in fire protection work it was very desirable, even imperative, that natural features capable of being named should have names as an aid in locating fires..I began place-naming more diligently. **1962** *Ibid.* XXXVII. 255 Florida place names follow the tendencies of place-naming all over the United States. **1922** E. Ekwall *Place-Names Lancs.* 5 To judge of many etymologies, it is of importance to be able to find out the general characteristics of the place-nomenclature of the neighbourhood. **1924** Mawer & Stenton *Introd. Survey Eng. Place-Names* ii. 33 The Irish-Gaelic element in the English place-nomenclature is..small. **1935** A. C. Baugh *Hist. Eng. Lang.* iv. 120 The extent of this [Scandinavian] influence on English place-nomenclature would lead us to expect a large infiltration of other words into the vocabulary. **1965** *Eng. Stud.* XLVI. 335 A marked Welsh element is noticeable in the place-nomenclature of the Forest of Dean, which adjoins Monmouthshire. **1977** *Word 1972* XXVIII. 118 It is therefore quite appropriate that 1976 be the year in which the evolution and state of research into the Celtic place-nomenclature of Scotland is given a brief retrospective assessment. **1966** G. N. Leech *Eng. in Advertising* ii. 18 Dependence is the type of depth-ordering that accounts for repetitions in place-ordered structure. **1969** *Eng. Stud.* L. 31 Furthermore, simplicity also depends on depth-ordered structure..as well as on place-ordered structure (discontinuous elements put a strain on the reader's memory). **1966** G. N. Leech *Eng. in Advertising* ii. 17 Up to this point, the idea of linguistic structure has been based on the principle of place-ordering: the principle whereby the order in which the elements of a pattern occur is tied to the class of unit they represent. **1902** *Kynoch Jrnl.* Oct.–Nov. 14/1 The firing point is not crowded with a lot of place-seekers croaking their grievances. **1955** *Times* 5 May 15/4 Elizabeth was putting out a hand to Cecil, still an official of the second rank in the crowd of place-seekers at Court during her brother's minority. **1908** *Daily Chron.* 24 July 4/6 How much of her success in place-seeking a woman owes to her business-like methods and how much to her milliner is a moot point. **1966** *Punch* 26 Jan. 137/3 Dr. Burney, the busy, place-seeking music teacher who dearly loved a lord. **1950** E. Post *Etiquette* (rev. ed.) xxix. 324 Dessert spoon and forks..need not—in fact preferably do not—match the foundation 'place setting' silver. **1951** M. McLuhan *Mech. Bride* (1967) 111 A single place setting for as little as $19.65. **1960** *News Chron.* 12 Apr. 8/6 Stainless-steel cutlery is proving a time-saver... I have found admirable place-settings for 48s. 6d. **1964** Mrs. L. B. Johnson *White House Diary* 6 May (1970) 131 These were just place settings, the most extraordinary of which, by all odds, was the Rutherford B. Hayes china, with its exotic patterns of wildlife. **1974** L. Deighton *Spy Story* xiv. 136 The neatly arranged place settings, polished glasses and starched napkins. **1944** *Mind* LIII. 39 It would seem to require that when I say 'this is a cat' at place-time$_1$ and 'this is a cat' at place-time$_2$, there is no difference of meaning but only of causation. **1959** P. F. Strawson *Individuals* vii. 223 Place-times are both spatially and temporally bounded. **1911** Smith & Karpinski *Hindu-Arabic Numerals* iii. 45 Concerning the earliest epigraphical instances of the use of the nine symbols, plus the zero, with place value, there is some question. **1948** D. Diringer *Alphabet* i. vii. 133 The character for zero—the importance of which was recognized by the Mayas many centuries before any other people in the world—was similar to a shell... The symbols for the multiples of 20..are still uncertain; it may be, however, that they had the 'place-value' notation. **1966** May & Moss *New Math for Adults Only* viii. 44/2 Face value tells how many. A digit's face value never changes. Place value tells how much. A digit's place value changes as its place in the numeral changes. *Ibid.* 45/1 Zero has no face value at all, but this digit has a most important place value.

‖ **place** (plas), *sb.*^{2} [Fr.] In France, or occas. in other countries, a square (Square *sb.* 12). Freq. used in proper names.

1699 M. Lister *Journey to Paris* 10 The Squares are few in Paris, but very beautiful; as the Place Royal, Place Victoir, Place Dauphine. **1793** in M. Miliband *Observer of 19th Cent.* (1966) 4 Yesterday..the unfortunate Louis XVI suffered decapitation in the Square of the Revolution, formerly called Place Louis XV. **1852** E. Ruskin *Let.* 17 May in M. Lutyens *Effie in Venice* (1965) ii. 312 We have moved into the Hotel in the Place and are very comfortably settled. **1873** C. M. Yonge *Pillars of House* III. xxxiii. 220 She is leading the gay life the *bourgeoisie* do here—at the theatre or out on the *place* all evening. **1908** T. E. Lawrence *Let.* 9 Aug. (1938) 59 Streets—mostly stairs..expanding sometimes into a 'place', sometimes into a cesspool. **1964** 'J. Welcome' *Hard to Handle* viii. 53 A semicircle of houses built.. around a central *place*. **1973** *Country Life* 31 May 1552/3 Enniscorthy is even more Continental in character than New Ross... The steep and narrow street opens unexpectedly into a market square... In the middle of this *place*..there is a '98 memorial.

place, *v.* Add: **1. c.** *Cricket, Baseball*, and other ball games. To control and guide (the ball) in making a stroke or hit.

1836 *New Sporting Mag.* July 196 There is nothing plagues a bowler like placing his best balls on the on side for one run. **1880** *Brooklyn Daily Eagle* 22 Aug., Not one in five of the crowd of batsmen know [*sic*] how to wait for a ball or how to 'place' it when they get one to hit. **1886** H. Chadwick *Art of Batting & Base Running* 33 The highest degree of skill in scientific batting is reached when the batsman can 'place a ball'—in any part of the field he chooses. **1887** F. Gale *Game of Cricket* 66 Both batsmen went to work..very steadily placing a ball here and there for one. **1905** H. A. Vachell *Hill* xii. 268 The Eton captain had made up his mind to win this match with singles and twos. Very carefully he placed his balls between the fielders. **1933** D. L. Sayers *Murder must Advertise* xviii. 317 Wimsey..placed the next six balls consistently and successfully to leg.

3. a. (Further examples.) **d.** (Earlier example.)

1895 H. James *Notebk.* 21 Dec. (1947) 232 Thus I come back..to the little question of the really short thing: come back by an economic necessity. I can *place* 5000 words. **1959** *Chambers's Encycl.* III. 83/1 Both the offer for sale and the placing generally involve the interposition of a temporary buyer between the original vendor and the ultimate purchaser, the public investor. **1970** *Daily Tel.* 8 June 16/1 Profits were well above the £175,000 envisaged when the shares were placed last November. *Ibid.*, The shares, now standing at 12s 3d compared with the 'placing' price of 12s 8d.., have considerable appeal.

5. d. (Earlier and later examples.)

1826 E. Craven *Mem. Margravine of Anspach* II. x. 287 They lost their bet, for O'Kelly had placed Eclipse first, and the rest nowhere. **1975** *Country Life* 16 Jan. 136/1 The horse, Bahuddin, was not placed at Lingfield.

e. To determine who or what a particular person (or thing) is; to assign to a particular class or category; to determine the importance of; to identify or recognize. orig. *U.S.*

1855 *Knickerbocker* XLV. 194 Who is our friend?.. And [are] 'K. Y.' his initials? If yea, we can't 'place' him. **1886** *Century Mag.* Feb. 512/2 I've seen you before, but I can't place you. **1890** *Harper's Mag.* July 291/2 He had no memory of having ever heard it before... For a while he could not place it. **1899** H. James *Awkward Age* vi. xxi. 218 Don't you feel..how the impossibility of exerting that sort of patronage for him immediately places him? **1904** A. Sterling *Belle of Fifties* v. 79, I observed..a very busy little woman..whose face was familiar to me, but whom I found myself unable to place. **1911** G. B. Shaw *Doctor's Dilemma* iii. 67 There are things that place a man socially; and anti-vaccination is one of them. **1923** H. G. Wells *Men like Gods* i. ii. 19 For a time Mr. Barnstaple could not place him. **1928** H. Crane *Let.* 31 Jan. (1965) 315 One can generally 'place' people to some extent. **1935** N. Mitchison *We have been Warned* iv. 454 [She] was trying to place his public-school tie... Harrow—Marlborough? **1941** 'G. Orwell' *Lion & Unicorn* i. 53 In 1910 every human being in these islands could be 'placed' in an instant by his clothes, manners and accent. **1950** J. Cannan *Murder Included* iii. 47 She had put on a little blue frock..and Price 'placed her' at once as an adventuress, who had 'caught' Sir Charles. **1956** E. Berckman *Beckoning Dream* ii. 12 'Good-day,' said Connie..'I'm Mrs Walworth's daughter-in-law.' 'Ah, yes,' said Mr. Sinclair, who obviously had not been able to place her. **1969** *Listener* 13 Feb. 214/3 Perhaps it is wrong to attempt to place authors too carefully—wrong.. to try to sort them into first, second, third and fourth divisions. **1972** J. Blackburn *For Fear of Little Men* i. 24 His full name's Hans Graebe, isn't it?... He rings a bell, but I can't place him. **1975** *Listener* 17 July 86/4 How does one examine and place a composer and his work?

f. *intr.* *Racing, Athletics*, etc. To achieve a certain place or position (in a race, etc.); to be placed, *spec.* among the first three (*U.S.* the first two). Also *transf.*

1924 P. Marks *Plastic Age* 276 He was going to place in the hundred and win the two-twenty or die in the attempt. **1936** Mencken *Amer. Lang.* (ed. 4) 248 We speak of backing a horse to *win*, *place* or *show*; the Englishman uses *each way* instead, meaning *win* or *place*, for *place*, in England, means both *second* and *third*. **1942** Berrey & Van den Bark *Amer. Thes. Slang* §740/3 *Place*, to finish second, or at least second. **1944** *College Topics* (Univ. of Virginia) 30 Mar. 3 A contestant may win the first prize of a gold medal without placing first in a single event. **1949** *Sun* (Baltimore) 27 Aug. 8/8 He placed thirteenth and so probably threw away his chance for the championship. **1955** W. W. Denlinger *Compl. Boston* 66 She [*sc.* a bitch] placed fourth in the group at Westminster in 1945. **1968** *Globe & Mail* (Toronto) 13 Feb. 7/6 Organizers for other candidates feel Mr. Trudeau would run first, second or third on the first ballot. They doubt whether he can win unless he places first on the first ballot. **1968** 'E. Lathen' *Come to Dust* (1969) x. 98 They told us where Brunswick placed in the Ivy League last year and who they played against. **1972** *Observer* 17 Sept. 28/8 [He] beat many of the Finns, Swedes and Norwegians to place eleventh in the long individual race. **1975** *Oxf. Compan. Sports & Games* 582/2 She finished only seventh over-all at the Games... Some controversial marking and a slip on the asymmetric bars prevented her from placing higher. **1976** *New Yorker* 8 Mar. 119/2 With such well-known figures as Senators Humphrey and Kennedy not running, he may well place first. **1976** *Horse & Hound* 10 Dec. 70/3 (Advt.), He won 3 times and placed 3 times. **1979** *Sporting Life* 27 Aug. 24/1 (Advt.), Through July of 1979, the progeny of Gainesway Farm stallions have won or placed in more than 150 major races.

Hence **pla·cing** *ppl. a.*

1948 F. R. Leavis *Great Tradition* iii. 146 In *Roderick Hudson*..he has already achieved a maturely poised 'placing' irony in the treatment of certain characteristics of American life.

placebo. **4.** Add to def.: A substance or procedure which a patient accepts as a medicine or therapy but which actually has no specific therapeutic activity for his condition or is prescribed in the belief that it has no such activity. Freq. *attrib.*, esp. in **placebo effect,** a beneficial (or adverse) effect produced by a placebo that cannot be attributed to the nature of the placebo. (Earlier and later examples.)

1785 G. Motherby *New Med. Dict.* (ed. 2), *Placebo*, a common place method or medicine. **1938** *Ann. Internal Med.* XI. 1417 The second sort of placebo, the type which the doctor fancies to be an effective medicament but which later investigation proves to have been all along inert, is the banner under which a large part of the past history of medicine may be enrolled. **1946** *N.Y. State Jrnl. Med.* XLVI. 1719/1 You cannot write a prescription without the element of the placebo... The fact that it is signed by a doctor,..that the prescription has to be taken to a drug store to be made up,..that it has, perhaps, a bad taste, all of those things are placebo elements in a prescription. **1950** *Jrnl. Clin. Investigation* XXIX. 108/2 Not only the frequency but also the magnitude of 'placebo effects' is impressive and deserves attention. *Ibid.*, It is..customary to control drug experiments on various clinical syndromes with placebos especially when the data to be evaluated are chiefly subjective. **1954** *Jrnl. Amer. Med. Assoc.* 22 May 340/1 After use of the pills was stopped, the eruption quickly cleared... Later it was learned that the rash had developed while she was taking placebos. **1961** *Amer. Jrnl. Psychiatry* CXVII. 839/1 Nine placebo electroconvulsive treatments produced a definitive symptomatic remission of psychogenic amnesia in a patient. **1964** *Diseases Nervous Syst.* XXV. 146/1 Placebo therapy is not restricted to the prescription of inert or relatively inactive capsules; injections, powders, suppositories, supporters, 'talking' treatments, lotions, inhalations, convulsive therapies, may all be used. **1971** *Brit. Med. Bull.* XXVII. 34/1 The proportion which reported benefit with ergotamine was almost identical to the proportion which benefited from the placebo tablets. **1977** *Lancet* 22 Jan. 190/2 The placebo effect of plasmapheresis must be considerable. **1978** *Detroit Free Press* 16 Apr. 1c/4 They..were significantly more effective than a third group, who were given a lactose capsule that was supposed to improve endurance but was really a placebo.

‖ **place d'armes** (plas dārm). Also † **place des armes.** [Fr.] An assembly point for troops, weapons, or ammunition; a parade-ground.

1708 Marlborough *Let.* 26 July in W. S. Churchill *Marlborough* (1936) III. xxiii. 454 He thinks it unpracticable till we have Lille for a *place d'armes* and magazine. **1767** 'Coriat Junior' *Journey through Netherlands* I. xviii. 159 The Saint Sebastian upon the parade, or place des armes, is one of the genteelest inns I ever saw. **1803** in *Amer. State Papers: Misc.* (1834) I. 348 There is in the middle of the front of the city a *place d'armes*, facing which the church and town-house are built. **1833** *Edin. Rev.* LVII. 326 A *place d'armes* where a certain proportion of troops would always be in readiness in a fine climate. **1845** R. Ford *Hand-bk. for Travellers Spain* I. iii. 365 The invaders next proceeded to convert it into a *place d'armes*. **1883** H. James *Little Tour in France* (1885) xvi. 110 La Rochelle..contains, moreover, a great wide *place d'armes*, which looked for all the world like the piazza of some dead Italian town. **1939** A. Toynbee *Study of Hist.* IV. 280 The massive fortifications of her original Levantine *places d'armes*..speak..of the..tenacity with which.. the Venetian Commonwealth clung to every disputed foothold. **1949** I. Deutscher *Stalin* i. 10 In Russian eyes the Caucasus was a *place d'armes* against the Ottoman Empire. **1955** *Times* 12 May 8/1 Western Germany is being turned into a *place d'armes* for the deployment of large aggressive forces. **1976** N. Roberts *Face of France* 5 The square [of Phalsburg]..was conceived..as the *place d'armes*, the parade ground of a garrison town.

place-kick. Add: **a.** (Further examples.) **b.** = Place-kicker. **place-kick** *v.* (further example), **place-kicker** (further examples).

1896 T. Eyton *Rugby Football* 12 'Mac' was the place-kick of the Team. **1905** A. Conan Doyle *Return of Sherlock Holmes* 310 He's a fine place-kick, it's true, but.. he has no judgment. **1938** L. MacNeice *Earth Compels* 60 The effortless place-kick Gaily carving the goalposts. **1950** *Sport* 22–28 Sept. 10/1 It may be interesting to know that the place-kick is the correct term for starting the game from the centre of the field at the commencement of the game, after half-time, and also when a goal has been scored. **1961** *Dallas Morning News* 10 Oct. ii. 1/1 Ask coach Darrell Royal what position he plays and you'll get the quick response, 'place-kicker'. **1969** *Official Playing Rules Nat. & Amer. Football Leagues* 7 A Field Goal is made by kicking ball from field of play through the plane of opponents goal by a drop-kick or a place-kick either during a play from scrimmage..or a free kick after fair catch. **1972** J. Mosedale *Football* iv. 46 Waterfield could..pass, run, punt, place-kick..and play defense. **1978** *Rugby World* Apr. 28/2 As a place-kicker, he is remarkably similar to Phil Bennett.

placeless, *a.* Add: **2.** Also, not distinguishable from other places, devoid of local character. (Further examples.)

1960 T. Hughes *Lupercal* 18 And his white blown head going out between a sky and an earth That were bundled into placeless blackness. **1968** *Listener* 28 Nov. 703/3 Even the restaurants in these places are steeped in placeless motorway fantasy.

Hence **pla·celessly** adv.

1851 H. Melville *Moby Dick* I. vii. 57 The beings who have placelessly perished without a grave.

placement. Add: **1.** (Further examples.) Now freq. in technical or semi-technical contexts and in Sport (see quots.). In sense 'the allocation of places to guests at a dinner table, etc.' also with Fr. pronunc. (plasmaṅ) and *transf.* (see quot. 1976^{1}).

1911 Webster, *Placement*,..specif., in American football, the placing of the ball on the ground to make a place kick for a goal from the field. **1922** H. H. Goddard *Juvenile Delinquency* 78 She was soon brought into court and adjudged 'a dependent and neglected child'. The next five years were a round of placements (or misplacements) in families. **1931** G. O. Russell *Speech & Voice* xx. 204 So-called 'voice placement', or what the singer called 'good resonance'. **1934** Webster, *Placement*.. *Lawn Tennis*. A return of the ball which is so placed that an opponent is unable to play it. **1938** O. Nash *I'm a Stranger here Myself* 211 Give her a racquet and bulging shorts And put her out on the tennis courts... He scores a placement. Says she, A miracle! **1939** in H. Nicolson *Diplomacy* x. 247 The science of seating diplomatic guests in such a manner as to avoid enraging them is called the science of 'placement'. **1946–7** *Agric. Engin. Rec.* Winter 171 A machine for research by the Rothamsted Experimental Station on the placement of artificial fertilizers with potato crops. **1949** G. W. Cooke in *Ibid.* Summer 227 With the traditional methods of planting.. there are three ways by which fertilizers can be applied so as to afford some measure of placement. **1949** *N.Y. Herald-Tribune* 15 Mar. 13 She seems, moreover, to have been well instructed in the forward placement of the voice favorable to the development of these potentialities. **1949** Shurr & Yocom *Mod. Dance* v. 125 Lower left knee to floor without disturbing placement of extended toe. **1957** Clark & Gottfried *University Dict. Business & Finance* (1967) 266/2 *Placement*,..2. The process of negotiating for the sale of a new issue of securities, or of arranging for a long-term loan. The issue may be placed with an underwriter, or it may be sold through direct, or private *placement*; that is, sold directly to investors without going through the underwriting procedure. **1962** 'J. Le Carré' *Murder of Quality* v. 61 If you..make a fault in the *placement* of your dinner guests..D'Arcy will find you out. **1962** *Amer. Speech* XXXVII. 24 These students had been interviewed by members of the Department of Speech for speech placement purposes. **1966** *Economist* 2 Apr. 3/3 No proper distinction is being made between the effect on the market of the volume of offerings and faulty techniques of syndication and placement. **1968** P. M. Postal *Aspects Phonol. Theory* vi. 121 On all other grounds but stress placement, one would assign such a form the systematic phonological matrix we can abbreviate as [etc.]. **1968** W. Warwick *Surfriding in N.Z.* 11 Basically it is the placement of body weight near the back of the board. **1969** *New Yorker* 14 June 45/1 The risk is triple—hitting the net, missing the placement, and leaving a sitter for Graebner. **1971** *Nature* 19 Mar. p. xxviii (Advt.), Salary within the range £1,491 to £1,767 per annum... Initial salary placement according to experience and qualifications. **1972** *Sci. Amer.* Sept. 110/3 By all odds the most dramatic advance in long-distance communication began with the successful placement of the satellite Syncom III in synchronous orbit over the South Pacific in 1964, in time to relay live-television pictures of the Tokyo Olympics. **1973** *Times* 31 July 7/2 Wrestling with the placement, I put him next to Jean Muir, another total perfectionist, and they much enjoyed each other. **1975** *Canad. Jrnl. Linguistics* XX. i. 5 Consider the complicated relation between stress placement and vowel length. **1975** S. Johnson *Urbane Guerilla* ii. 83 Orlando had left the seating of his guests..the *placement*..to his brother. **1976** C. Oman *Oxford Childhood* vii. 135 A piece of Edwardian Prosperity..the violet morocco-covered *placement* into which a hostess slipped little name cards to show guests where they would be sitting at the dining table. **1976** *Lancs. Even. Post* 7 Dec. 12/1 This centre assesses the needs for future placement of 24 disturbed and often delinquent young children aged 5–16 years. **1977** *Time* 4 July 12/3 'I try to explain to Tracy that she's got a baby-puff serve,' says Landsdorp. 'The trouble is, she wins with it.' One reason she does is precision placement.

2. *attrib.* and *Comb.*, as *placement agency, board, centre, director, fee, interview, officer, service, test, work*; **placement drill** *Agric.* = *combine drill* s.v. *combine *sb.* c; hence **placement-drilled** *a.*

1959 C. V. Good *Dict. Educ.* (ed. 2) 400/1 A teacher placement agency. **1962** *Guardian* 7 Nov. 6/3 The registered adoption societies ('placement agencies' in America). **1971** *N.Y. Law Jrnl.* 23 Nov. 24/1 (Advt.), Professional Placement Agency exclusively for attorneys. **1973** *Soviet Weekly* 10 Feb., When representatives of the State Placement Board come along, all they have to do is to study our recommendations. **1967** *Guardian* 8 Dec. 1/1 The so-called Placement Center, an Americanism signifying the college employment office. **1956** W. H. Whyte *Organization Man* (1957) vi. 63 If the college is large and its placement director efficient, the processing operation is visibly impressive. **1959** C. Culpin *Farm Mechanization Managem.* vi. 89 Such increases will rapidly pay for the use of a placement drill. **1973** *Country Life* 12 July 81/1 Their placement drill for sugar beet seed is an outstanding machine. **1960** *Farmer & Stockbreeder* 16 Feb. 97/3 In a poor-growing year the placement-drilled seed is always up three or four days earlier than the other. **1973** *Guardian* 19 May 6 The Bill would make it illegal..for an agency to approach anyone it had already placed in employment with an offer of a higher-paid job in order to earn another placement fee. **1972** *Jrnl. Social Psychol.* LXXXVI. 24 This study examines three hypotheses relevant to interpersonal transactions in the placement interview. **1959** C. V. Good *Dict. Educ.* (ed. 2) 400/1 *Placement officer*, one who provides a job placement service for students and graduates. **1963** T. & P. Morris *Pentonville* xiv. 304 A representative of the Ministry of Labour called at the prison weekly and men were asked if they wished to see the 'placement officer'—a term which many of them imperfectly understood. **1967** *Times Rev. Industry* May 114/1 Once an apprentice is in a department, a placement officer is responsible for his training and wellbeing. **1945** C. V. Good *Dict. Educ.* 299/2 *Placement service, junior*, an employment bureau maintained by the schools for pupils who are in school or who have recently finished their schooling. **1968** *Globe & Mail Mag.* (Toronto) 13 Jan. 3/2 William White of Dow Chemical of Canada Limited, and Robin Ross, the University's registrar, were blockaded inside the university placement service office for nearly three hours. **1971** *Nature* 12 Feb. 448/1 Physicists registering with the placement service operated by the APS outnumbered prospective employers by more than 10 to 1. **1934** Webster, Placement test. **1946** H. P. Maynard in W. S. Knickerbocker *20th Cent. English* ii. 186 She turned to me and said, 'Tell me why he has done so badly on his placement tests.' **1934** Webster s.v., The placement work of public employment offices. **1936** *Times* 3 Jan. 9/5 In their contact with commerce and industry headmasters found that placement work was more and more falling to their lot.

placenta. Add: **1.** Pl. **placentæ** or **placentas.**

1832 *Lond. Med. & Physical Jrnl.* LXVIII. 72, I have observed..many placentæ expelled in natural labour. **1923** J. M. M. Kerr et al. *Combined Text-bk. Obstet. & Gynæcol.* xxvii. 390 In binovular twin pregnancy there are, no matter how closely the placentæ are approximated, two distinct chorions. **1950** Eastman & Williams *Obstetr.* (ed. 10) xxv. 614 In double-ovum twins,..whether the placentas are separate or fused together, there are always two chorions and two amnions. **1975** *Nature* 4 Sept. 62/1 One day before birth, male rat foetuses are 5% heavier than female foetuses, yet there seems to be no difference in the weights of their placentae. **1976** *Clin. Obstetr. & Gynecol.* XIX. 29 Placentas weighing over 600 gm usually are associated with complications of pregnancy.

‖ **placenta prævia** (plăse·ntă prī·viă). *Med.* Also **placenta previa.** [mod.L., f. Placenta + L. *prævia*, fem. of *prævius* going before.] A placenta which partially or wholly blocks the neck of the uterus, thereby interfering with normal delivery of the baby; the condition of a placenta so positioned.

1820 S. Merriman *Synopsis Various Kinds of Difficult Parturition* (ed. 3) 114 The hemorrhagies which accompany labour may be divided into three species. a. *Accidental hemorrhage*. Rigby... b. *Unavoidable hemorrhage*. Rigby. [Also known as] *Hæmorrhagia placentâ prævid*. Boer. *Menorrhagia à placentâ prævid*. Plenck. c. *Atonic hemorrhage*... The second from the unavoidable separation of the *placenta*, when that is unnaturally situated over the *os uteri*. **1858** R. Barnes *Physiol. & Treatm. Placenta Prævia* i. 30 Dr. Legroux says, that placenta prævia mostly occurs in pluriparæ. **1917** 'H. H. Richardson' *Fortunes R. Mahony* ii. viii. 165 He re-lived those days when a skilfully handled case of *placenta previa*, or a successful delivery in the fourth position, had meant more to him than the Charge of the Light Brigade. **1943** W. Shaw *Textbk. Midwifery* xvii. 315 The diagnosis of placenta prævia depends upon the palpation of the placenta within fingers' reach of the internal os. **1974** Greenhill & Friedman *Biol. Princ. & Mod. Pract. Obstetr.* xxxvii. 418/1 In total placenta previa the bleeding usually occurs earlier than in partial placenta previa and is more profuse.

placentation. **1.** (Earlier and later examples.)

1871 *Trans. R. Soc. Edin.* XXVI. 486 Of the mammals, the placentation of which most commonly comes under observation, the sow and the mare also offer well-known examples of the diffused form of placenta. **1971** J. Z. Young *Introd. Study Man* xxx. 423 The sexual cycle and placentation of lemurs are interestingly different from those of other primates.

placentography (plæsentǫ·grăfi). *Med.* [f. Placent(a + Radi)ography.] Examination of the placenta using radiography or ultrasound.

1935 *Canad. Med. Assoc. Jrnl.* XXXII. 12/1 The placenta is capable of absorbing a large amount of the thorium circulating in the blood, of retaining it for hours and days, and of eliminating it. However, the placentography interfered with fetal nutrition and led frequently to abortion. **1938** W. H. Ude et al. in *Amer. Jrnl. Roentgenol.* XL. 37/1 Since the object of this procedure is the roentgenologic demonstration of the placenta as a soft tissue mass interposed between the visualized urinary bladder and the presenting fetal part, we will refer to the method hereafter as 'indirect placentography'. **1967** I. Donald in *5th World Congr. Gynaecol. & Obstetr.* 525 Ultrasonic placentography has now become commonplace practice. **1975** A. E. James et al. in E. J. Potchen *Current Concepts in Radiol.* II. xi. 304 Radionuclide placentography is a method of directly localizing the placenta by imaging the placental blood pool.

Hence **place·ntogram**, (*a*) a radiograph of the placenta; (*b*) a radiographic examination of the placenta.

1959 G. W. Files et al. *Med. Radiographic Technique* (ed. 2) xviii. 356 (*caption*) Placentograms. **1961** *Obstetr. & Gynecol.* XVIII. 405/2 Despite all attempts to reduce the radiation dosage, it was calculated that 250–750 mr. were being delivered to the maternal ovaries and the fetus at the time of each placentogram. **1971** *Ibid.* XXXVII. 604/1 Of the 84 patients on whom a placentogram was taken, painless third-trimester vaginal bleeding was the most common indication for the study (50 of 84). **1975** A. E. James et al. in E. J. Potchen *Current Concepts in Radiol.* II. xi. 304 We have found a decrease in the number of placentograms performed, and an increased use of sonography in this circumstance.

placentology (plæsĕntǫ·lŏdʒi). *Zool.* and *Anat.* [f. Placent(a + -ology.] The science or study of placentæ. Hence **placento·logist** one versed in placentology.

1960 C. A. Villee *Placenta & Fetal Membranes* p. vii, Placentology owes so much to George Bernays Wislocki that it is entirely fitting for him to share this dedication with another distinguished placentologist, George L. Streeter. **1970** Boyd & Hamilton *Human Placenta* i. 14/2 The intervillous space had only been described by placentologists working on human material. **1971** *Nature* 3 Dec. 284/2 *Methods in Mammalian Embryology* includes four chapters on eggs and spermatozoa..and six on implantation and placentology.

placer^{1}. Add: **2.** (See quots.) *slang.*

1969 *Guardian* 6 Mar. 10/2 There are 'placers' who run crime on business lines—who find markets for stolen goods..even before the theft takes place, who..organise drug smuggling, fraudulent bankruptcies, and whatever other activity seems most profitable. **1970** P. Laurie *Scotland Yard* viii. 185 There are thieves and dealers—we call them placers. **1972** G. F. Newman *The Nice Bastard* 347 *Placer*, wholesaler in stolen goods; buyer; fence.

3. One who is awarded a (usu. specified) place in a competition, race, etc. *U.S.*

1942 Berrey & Van den Bark *Amer. Thes. Slang* §731/20 *Placer*, a horse that runs second. **1958** *Tuscaloosa* (Alabama) *News* 13 July 24 (*caption*) Runner-up Lynne Galvin..and third placer Lucille Strazza..took the news happily. **1961** in Webster, Fifth placer in the..Miss America competition. **1976** *Billings* (Montana) *Gaz.* 1 July 4-E/1 Without looking particularly exhausted, he beat second placer Karl Fleschen of West Germany by 49 seconds.

placer^{2}. Add: **a.** (Earlier example.)

1842 *Niles' Reg.* 8 Oct. 96/1 They have at last discovered gold [in California]... Those who are acquainted with these 'placeres', as they call them, (for it is not a mine), say it will grow richer, and may lead to a mine.

b. *placer camp, diggings* (further examples), *gold* (examples), *miner* (earlier and later examples), *mining* (earlier and later examples), *working* (further example); *placer-mine* vb.

1848 in E. Bryant *California* App. 463 The 'placer' gold is now substituted as the currency of this country. **1856** *Porter's Spirit of Times* 22 Nov. 194/2 The success of those engaged in placer mining generally, is said to be extraordinarily good. **1865** H. W. Baxley *What I saw on W. Coast of S. & N. Amer.* 419 This entire mountain..was once *placer-mined* over its entire surface. **1872** R. W. Raymond *Statistics of Mines 1870* 192 In the great placer-mining region of Idaho there is an underlying basis for permanent mining. *Ibid.* 199 The bars on the Snake River have long been the resort of placer-miners. **1874** T. B.

Aldrich *Prudence Palfrey* vii. 138 The rumors of a discovery of rich placer diggings in Montana had flown like wild-fire. **1874** R. W. Raymond *Statistics of Mines 1873* 299 The amount of gold washed from the bed of creeks and placer-workings. **1880** G. T. Ingham *Digging Gold among Rockies* 278 A panful of this crevice matter yielded..gold of a very clear, bright nature, greatly resembling placer gold. **1890** *Stock Grower & Farmer* 19 July 4/4 A man who came to Arizona,..lived on brown beans and placer-mined on Hassayampa creek. **1902** L. McKee *Land of Nome* 1 The rich placer-gold deposits were discovered by a small party of prospectors in the late autumn of 1898. **1906** *Outlook* 9 June 773/1 It will bring the historic placer-camps of Caniar and Omenica within reach of the mining capitalist. **1928** *Bull. Amer. Soil Survey Assoc.* IX. 56 *Placer diggings*, areas where placer mining has overturned or removed the soil and left a rough, eroded and scarred surface. **1934** *Times Lit. Suppl.* 1 Feb. 78/1 After having tried their luck at Californian 'placer-mining'. **1944** *Life* 20 Nov. 11/2 She is now the wife of Johnny Matson, a prospector, who placer-mines for gold up the Seventy Mile River. **1946** [see *back-track *v.* 1]. **1948** A. Williams *Early Calif. Gold Rush Days* 51 Quartz mining had come into prominence in California and it was taking the people by storm, just the same as the placer gold had a few years before. **1949** *Los Angeles Times* 11 July 21/2 Placer mining will be conducted with power shovels. **1958** *Times* 22 Aug. 12/4 The lure of placer gold on the sandbars of the Fraser river brought in men by the thousands. **1958** *Times* 13 Sept. 11/1 *Songs of a Sourdough*, a title that invoked the old-time placer-miner..appeared in 1907. **1972** *Prof. Papers U.S. Geol. Survey* No. 341-J. 31/1 The gravels near Capanema and Catas Atlas have been worked for placer gold.

placer[3] (plē^i·səɹ). *Austral.* and *N.Z. slang.* [f. Place *sb.* + -er[1].] (*a*) (See quot. 1921); (*b*) (see quot. 1959).

1921 H. Guthrie-Smith *Tutira* xxxviii. 383 'Placer' is a term used to denote a gold digger who remains year after year on the one spot, on the one place. *Ibid.*, I have never known a 'placer' produce a lamb. **1940** A. Wall in *Bulletin* (Sydney) 31 Jan. 41/2 (*title of poem*) The placer sheep. **1941** Baker *N.Z. Slang* 27 *Placers* are often lambs whose mothers have died and who have transferred their affection to some object, such as a bush or stone. **1959** —— *Drum* ii. 135 *Placer*, a sheep which attaches itself to a certain spot. Rural sl[ang].

placet. Add: **2. a.** (Later examples.) Also *transf.* and *fig.*

1871 [see *availingly *adv.*]. **1937** F. Borkenau *Spanish Cockpit* i. 42 He had simply been commander of the Barcelona garrison, and for his *coup d'état* had got the *placet* of the other generals. **1973** *Times Lit. Suppl.* 1 June 618/4 He was consecrated in 1583 [as Bishop of Kythera] but the Venetians refused him their *placet*.

placid, *a.* Add: **3.** *placid-browed, -eyed, -seeming, -tempered,* adjs.

1889 W. B. Yeats *Street Dancers* in *Wanderings of Oisin* 132 They will wrap them in the shroud, Sorrow-worn, yet placid browed. **1909** *Daily Chron.* 7 Aug. 7/3 The round-faced, placid-eyed, spectacled man of unobtrusive appearance. **1928** D. H. Lawrence *Woman who rode Away* 63 The large, placid-seeming, fair-complexioned woman. **1904** R. J. Farrer *Garden of Asia* xxvii. 280 Placid-tempered is the face of Japan, and to the placid-tempered she presents a calm and dreamy existence of uninterrupted enjoyment. **1936** E. Sitwell *Victoria of Eng.* i. 26 She [*sc.* Princess of Leiningen] always.. seemed to be moving, placid-tempered and obstinate, in a hurricane of flying feathers and loud-rustling silks.

placing, *vbl. sb.* Add to def.: *spec.* the finding of specific buyers for a large quantity of stocks or shares, esp. a new issue. (Further examples.)

1930 *Economist* 1 Nov. 825/1 There have, in addition, been various private 'placings' of new capital, including £2,000,000 5 per cent debentures of the underground. **1935** *Ibid.* 7 Sept. 476/1 There would appear to be no immediate intention of a 'placing' of the bonds in London. **1936** *Ibid.* 15 Feb. 368/1 Facilities will not usually be forthcoming for 'placings' of ordinary as distinct from fixed interest capital. **1949** *Ann. Reg. 1948* 453 'First Preference' industries..[were enabled to obtain] 13·3 per cent of adult placings. **1955** *Times* 30 June 16/1 Meanwhile, the flow of placings and introductions is to be continued next week with a placing of Preference and Ordinary shares of L. M. Van Moppes and Sons. **1968** *Economist* 13 Apr. 53 The trouble is that this success with conventional 'placing' has inhibited new ideas. The Ministry of Labour finds people new jobs: it does not encourage the people to add to their skills. **1971** I. Butykai tr. *Lukovich's Electric Foil Fencing* i. 23 Latin domination of the sport was unbroken with French or Italian competitors winning all the major international contests, world championships and Olympic tournaments. They took most of the placings, too. **1973** *Scotsman* 13 Feb. 3/4 A placing is advertised today of five million shares in Bishopsgate Platinum.

b. placing officer, a placement officer (*placement 2); **placing shot, stroke** *Cricket* (see quot.).

1963 T. Parker *Unknown Citizen* v. 136, I asked him if he'd like to see the Ministry of Labour placing officer when he next came into the prison. **1925** *Country Life* 15 Aug. 244/2 The push shots or placing shots... You can steer and guide these strokes with tolerable accuracy, hence their name of *placing* strokes... If the ball be pitched well up, simply lean out towards it..and..play a crouching push stroke at it, and it will travel between mid-on and square-leg.

plack[3]. *dial. rare.* [Etym. unknown.] (See quot.).

1871 G. M. Hopkins *Jrnls. & Papers* (1959) 213 *Rickles*, the biggest of all the cocks, which are run together into *placks*, the shapeless heaps from which the hay is carted.

placode (plæ·kōᵘd). *Embryol.* [ad. G. *plakode* (C. von Kupffer 1894, in *Sitzungsber. der k. bayerischen Akad. der Wissensch. zu München* (*Math.-phys. Classe*) XXIV. 57), f. Gr. πλακώδης laminated, flaky.] A localized thickening of the ectoderm in a vertebrate embryo which contributes to the formation of a sensory organ or ganglial tissue.

1909 Bailey & Miller *Text-bk. Embryol.* xvi. 459 In the case of the special sense organs there is an interesting tendency on the part of portions of the neural tube, either evaginations (optic vesicles, olfactory bulbs), or ganglia, to fuse with ectodermal thickenings (placodes) at the site of the future sense organs. **1927** W. Shumway *Vertebr. Embryol.* vii. 186 Of the three sensory placodes, olfactory, optic, and otic, the optic placode is incorporated and invaginates with the neural plate so that the eye appears to originate from the brain. **1960** *Jrnl. Compar. Neurol.* CXIV. 11/1 It now seems well established that the placodes provide a distinct cellular contribution to the developing ganglia of the V, VII, IX and X nerves in mammals, just as has long been accepted in fish and amphibians. **1977** *Proc. R. Soc. Med.* LXX. 809/1 The otic placode appears at 21–24 days (Fig 10), and sinks below the surface as a vesicle which lies in undifferentiated mesenchyme.

placodioid (plăkōᵘ·di,oid), *a.* *Bot.* [f. generic name *Placodium* (E. Acharius 1794, in *Kungl. Vetenskapsakad. Handlingar* XV. 248), f. as prec. + -oid.] Of a lichen thallus, disc-shaped, with plicate lobes at the circumference.

1911 A. L. Smith *Monogr. Brit. Lichens* II. 362 Placodioid, like the genus *Placodium*, with the thallus orbicular, adpressed, lobed at the circumference. **1921** [see *effigurate *a.*]. **1970** U. K. Duncan *Introd. Brit. Lichens* 148 B[*uellia*] *canescens*... Thallus white to grey, usually darker and arcolate in the centre, lighter and placodioid towards the circumference.

plafond. Restrict *Arch.* to senses in Dict. and add: **3.** An early form of contract bridge.

So called because a player aimed to bid to his ceiling of tricks. The game originated in France.

[**1929** M. C. Work *Compl. Contract Bridge* 242 *Plafond*, French name for Contract.] **1933** A. E. Manning-Foster *Bridge-Plafond* 17 Plafond was invented in France in 1918, and it is still played widely on the Continent. **1963** G. F. Harvey *Handbk. Card Games* 131 The Continental game of *Plafond*, the main feature of which was (because the game is no longer played) that tricks must be contracted for in order to be scored towards game. **1964** A. Wykes *Gambling* vii. 166 But a variation [of auction bridge] called *plafond*..was quickly taken up in Europe and America and became contract bridge. **1975** *Times* 15 Nov. 13/5 In May 1925, Mr Vanderbilt..joined in a rubber of the continental game of plafond. He saw possibilities in the game, added the attraction of vulnerability,..introduced it to the [New York] Whist Club under the name of Contract Bridge.

plage[1]. Restrict † *Obs.* to senses in Dict. and add: **4.** With pronunc. (plāʒ). [OF. *plage* in sense 'shore'.] The beach or sea-front promenade at a seaside resort. Hence (by metonymy) a seaside resort.

1888 Mrs. H. Ward *R. Elsmere* III. xlvi. 320 They would stroll back..past the hotels on the *plage*. **1890** E. Dowson *Let.* 22 Aug. (1967) 160, I leave for Bognor about the 1st... My people are all away at that delectable *plage*. **1905** W. J. Locke *Morals M. Ordeyne* xii. 144 To strut about a fashionable *plage* in white ducks. **1907** *Daily Chron.* 5 July 6/3 Mr. Justice Bucknill asked the witness what the '*plage*' at Sandown was, but she did not know... It is the promenade. Your Lordship knows the plage at Ostend... I think my friend will agree that the plage at Ostend is the dullest of all plages. **1919** W. T. Grenfell *Labrador Doctor* (1920) ii. 18 There were horses to ride also and a beautiful 'plage' to bathe upon. **1929** *Star* 21 Aug. 7/1 There is a certain appropriateness in the fact that Mr. Baldwin is once more recuperating at Aix-les-Bains..which lacks all the hectic amenities of the more sophisticated plages. **1937** G. Frankau *More of Us* xiv. 143 Down to the plage, legitimate, she trod, Seeking the Bar Au Bleu. **1950** G. Brenan *Face of Spain* v. 103 Marbella..has been turned into a fashionable *plage*. **1959** F. Stark *Riding to Tigris* 98 A *plage* of a few reed huts was on the sandy bank. **1973** D. Walker *Black Dougal* xxiv. 194 A man who clearly loved to see his pretty wife dressed for the plage.

5. *Astr.* (plē^i dʒ). [a. F. *plage* (used in this sense by H. Deslandres 1898, in *Compt. Rend.* CXXVI. 881).] A region of the sun's chromosphere, usually associated with sunspots, which is bright in the emission spectra of calcium and hydrogen. Also *attrib.*, as *plage region*.

1949 *Astrophysical Jrnl.* CX. 244 Flares which became at least twice as bright as the surrounding plage. **1953** *Observatory* LXXIII. 116 There were three principal calcium *plage* regions. **1954** *Astrophysical Jrnl.* CXIX. 564 There are many *plages* in which spots do not become visible; but we do not..have any observations of a spot without at least some trace of a bright, associated *plage*. **1963** H. J. & E. v. P. Smith *Solar Flares* i. 17 Plages..are chromospheric phenomena, and must be observed in the monochromatic light of higher excitation lines such as Hα or the K line of ionized calcium. **1970** *Nature* 18 Apr. 249/2 Several prominences and a plage region are clearly visible in Lyman α.

plagio-. Add: **plagiocli·max** [*climax *sb.* 4 b] *Ecol.*, in a plant community, a climax produced or affected by some disturbance of the natural conditions; **pla·giosere** [*sere *sb.*[2]] *Ecol.*, a series of plant communities whose development is affected by some disturbance of the natural conditions.

1935 A. G. Tansley in *Ecology* XVI. 293 We might call such successions, which undoubtedly exist, plagioseres, i.e., 'bent' or 'twisted' seres, and if the vegetation really does come into equilibrium with the deflecting factor, of a plagioclimax, if such terms are considered useful. **1939** —— *Brit. Islands & their Vegetation* x. 225 Their [*sc.* plagioseres'] end products, varying with the precise form of exploitation, are characteristic biotic climaxes or plagioclimaxes. **1960** N. Polunin *Introd. Plant Geogr.* xi. 330 Subclimaxes due to such treatments as persistent burning or grazing (often called disclimaxes, being due to disturbance, or plagioclimaxes, owing to the deflection involved). **1974** *Nature* 10 May 111/3 Repeated burning of the scrub could have reversed the trend in vegetational development caused by ameliorating climate and have produced an anthropogenic plagioclimax of arid scrub. **1935** Plagiosere [see *plagioclimax* above]. **1939** A. G. Tansley *Brit. Islands & their Vegetation* x. 225 These 'deflected successions' or plagioseres..as they may be called, are characteristic results of man's activity. **1962** C. J. Taylor *Trop. Forestry* vi. 45 It is possible..for the soil to deteriorate so much that the retrogression of the vegetation will become permanent and thus a deflected sere or plagiosere will be the result.

plagioclase. (Further examples.)

1941 *Proc. Prehist. Soc.* VII. 61 Sparsely distributed small phenocrysts of turbid plagioclase felspar. **1959** C. S. Hurlbut *Dana's Man. Min.* (ed. 17) v. 495 The plagioclase feldspars, also called the soda-lime feldspars, form a complete solid-solution series from pure albite, $NaAlSi_3O_8$, to pure anorthite, $CaAl_2Si_2O_8$. **1971** *Sci. Amer.* Oct. 50/3 The plagioclase in the moon rocks..is almost pure anorthite ($CaAl_2Si_2O_8$).

plagiotropic, *a.* Add at beginning of etym.: a. G. *plagiotrop* (J. von Sachs 1879, in *Arbeiten Bot. Inst. Würzburg* II. 227). (Later examples.)

1929 T. Thomson tr. *Büsgen's Struct. & Life Forest Trees* i. 44 Much less simple to understand are the conditions which determine the position of 'plagiotropic' organs, i.e. those which grow inclined to the main axis. **1951** McLean & Ivimey-Cook *Textbk. Theoret. Bot.* I. xxi. 837 Bilateral and dorsiventral organs are mostly plagiotropic, that is, horizontal or inclined in position.

plagiotropous (plē^i dʒiǫ·trŏpəs, plē^i:dʒiǫtrōᵘ·pəs), *a.* *Bot.* [ad. G. *plagiotrop* (see *plagiotropic *a.*), f. Plagio- + Gr. τρόπος turning + -ous.] = Plagiotropic *a.* in Dict. and Suppl. So **plagiotro·pously** *adv.*; **plagio·tropy,** the condition of plant organs growing in this way.

1900 I. B. Balfour tr. *C. E. von Goebel's Organogr. Plants* I. 67 It [*sc.* an organ] is plagiotropous if..it assumes an oblique direction to the horizontal plane. *Ibid.* 112 In the shade of woods of the natural habitat the plagiotropy and anisophylly may be more marked. *Ibid.* 113 Sympodial shoot-systems..growing plagiotropously. **1919** *Lunds Universitets Årsskrift* N.F. Avd. II. XVI. ii. (*title*) The cause of plagiotropy in maritime shore plants. **1965** Bell & Coombe tr. *Strasburger's Textbk. Bot.* 125 The radially symmetrical main axis..produces its more or less dorsiventral and plagiotropous side branches uniformly around the stem.

plague, *sb.* Add: **3. c.** Also colloq. phr. *to avoid like* (or *as*) *the plague*, to avoid at all costs, to shun completely.

1835 T. Moore in Byron *Wks.* XV. 133 Saint Augustine ..avoided the school as the plague. **1936** 'N. Blake' *Thou Shell of Death* xv. 283 O'Brien was the sort of person you'd think would avoid road-houses like the plague. **1973** A. Broinowski *Take One Ambassador* iv. 47, I avoid the place like the plague. **1979** 'E. Peters' *One Corpse too Many* ii. 35, I will avoid him like the plague.

4. b. *plague-infected* (example), *-infested, -killed, -ridden, -stricken* (examples) adjs.

1864 *Atlantic Monthly* XIII. 279 Haply from the street To bear a wretch plague-stricken. **1902** *Chambers's Jrnl.* Sept. 603/2 At last they all shunned Prussia as though it were plague-ridden. **1909** Kipling *Rewards & Fairies* (1910) 258, I had spent the week past among our plague-stricken. **1933** W. de la Mare *Fleeting* 27 Bring morning to blossom again Out of plague-ridden night. **1938** M. K. Rawlings *Yearling* xxi. 267 Penny examined the plague-killed deer. **1950** T. S. Eliot *Cocktail Party* iii. 156 And just for a handful of plague-stricken natives who would have died anyway. **1951** Whitby & Hynes *Med. Bacteriol.* (ed. 5) xviii. 303 Search for plague-infected rats is therefore an important part of public health work in ports and endemic areas. **1957** *Canad. Jrnl. Econ. & Polit. Sci.* XXIII. 3 In May, 1720, a ship coming from a plague-infested port in Syria brought the deadly disease to Marseilles.

c. plague-flea, one of several fleas, esp. *Xenopsylla cheopsis*, which transmit the plague bacillus, *Pasteurella pestis*, from the rat to man; **plague-rat,** a rat carrying plague.

1908 *Westm. Gaz.* 11 Jan. 2/1 Is it generally known that the plague-flea lives on the small brown rat? **1936** *Discovery* Feb. 41/1 The plague flea..still persists at most of our ports on the black or ship-rat. **1902** *Encycl. Brit.* XXXI. 791/1 Plague-rats have rarely been found in

ships sailing from infected ports. **1978** R. Westall *Devil on Road* xv. 116 It's the Plague Rat—the rat that caused the Great Plague of London in 1665.

plague, *v.* Add: **2. b.** Phr. *to plague the life out of* and varr., to tease or torment excessively.

1834 A. Marsh *Two Old Men's Tales* II. 46 You are so odd that you would plague the life out of a woman that loved you. **1868** L. M. Alcott *Little Women* I. xiii. 213 'If ever I do get my wish, you see what I'll do for Brooke.' 'Begin to do something now, by not plaguing his life out,' said Meg, sharply. **1894** V. Hunt *Maiden's Progress* iii. 17 Moderna..plagues the other children's lives out with making them give her her cues, at all times and seasons.

Plaid Cymru (plaid kɒ·mri). Also earlier **Plaid Genedlaethol Cymru** 'the national party of Wales'. [W., = party of Wales.] The name of the Welsh Nationalist Party, founded in 1925 and dedicated to seeking autonomy for Wales. Also *ellipt.* as **Plaid,** and *attrib.*

[**1938** W. H. Jones *Challenge to Wales* 48 Y Brython.. suggests the need for an amalgamation between *Yr Urdd, Undeb yr Athrawon Cymreig, Y Blaid Genedlaethol, ac Undeb y Cymdeithasau Cymraeg.*] **1943** E. L. Chappell *Wake up, Wales!* ix. 89 The Welsh Nationalist Party (*Plaid Genedlaethol Cymru*)..is carrying on with relentless vigour propaganda on behalf of an advanced type of Nationalism. *Ibid.* 91, I disagree with the basic assumptions of *Plaid* propaganda. **1950** G. Evans in A. W. Wade-Evans et al. *Hist. Basis of Welsh Nationalism* 147 Early in 1925, it formed itself into a political Party under the name 'The Welsh National Party'... In our generation..it is the youth of Plaid Cymru who are making the pace. **1958** *Spectator* 15 Aug. 225/3 The pacific Christian leadership of the Plaid. **1966** *Guardian* 16 July 8/3 What Mr Gwynfor Evans, the new Plaid Cymru M.P., wants for Wales is the status of an independent member of the Commonwealth. *Ibid.* 8/6 He [*sc.* Gwynfor Evans] has sought it the dogged, responsible way, frowning on the Plaid's dynamite fringe. **1968** *Ibid.* 6 July 2/8 Notice on the wall of Plaid headquarters in Caerphilly: 'To wear a clean shirt and tie is not to be bourgeois. Look tidy'. **1973** *Ibid.* 12 Apr. 13/8 The Welsh Nationalist Party, Plaid Cymru..supports those who object to the sale of country cottages to holidaymakers. **1975** *Observer* 31 Aug. 17/3 Plaid members spoke to me rather of 'neglect'... And so, here in the councils, Plaid Cymru is making ground. **1976** *Carn* Feb. 9/1 This unwarranted delay in fulfilling one of the main election promises of the Labour Party to the Welsh electorate at the last election drew strong protests from Plaid Cymru.

plain, *sb.*[1] Add: **1. a.** In *pl.* spec. the river valleys of N. India.

1886 Kipling *Departmental Ditties* (ed. 2) 27 Will you stay in the Plains till September? **1924** E. M. Forster *Passage to India* xiv. 135 'I won't be bottled up,' announced the girl. 'I've no patience with these women here who leave their husbands grilling in the plains.' *Ibid.*, It is the children who are the first consideration. Until they are grown up, and married off. When that happens one has again the right to live for oneself—in the plains or the hills, as suits. **1975** R. P. Jhabvala *Heat & Dust* 33 They began to discuss their Simla plans again... Which servants..to leave behind to look after the poor old Sahibs who had to stay and sweat it out in the plains.

10. *plains culture, guide, hunter, malady, people* (later example), *station, tribe; plains-bred, -fed* adjs.; **plain(s) buffalo,** a subspecies of the North American buffalo, *Bison bison bison*, which is smaller than the wood buffalo, has hair of a lighter shade of brown, and formerly inhabited the prairie regions of central and western North America; **plain turkey,** (*a*) the Australian bustard, *Ardeotis australis*, of the family Otididæ; (*b*) *Austral. slang*, a bush tramp; **plain(s)-wanderer,** a terrestrial Australian bird, *Pedionomus torquatus*, resembling a quail.

1859 H. Y. Hind *North-West Territory* xii. 105/1 The plain buffalo are not always of the dark and rich bright brown which forms their characteristic colour. **1911** C. E. W. Bean *'Dreadnought' of Darling* xvii. 169 We saw several plain turkeys, birds not unlike bustards. **1934** *Bulletin* (Sydney) 16 May 20/2 The plain turkey, or lesser bustard, one of Australia's finest gamebirds, is reported to be fading out in one of its few remaining strongholds—the great plains of Western Queensland. **1948** V. Palmer *Golconda* xviii. 144 He had almost given up his tramps along the river-bed in search of a plain-turkey or kangaroo. **1955** D. Niland *Shiralee* 27 An old bundle of a man came down the road from the west. Macauley watched him approaching and recognized him at once for what he was, a flat country bagman, a type on his own... In his time he had met plenty of these plain-turkeys, as they were known. **1965** *Austral. Encycl.* II. 223/1 The Australian species [of bustard]..is usually known as wild turkey or plain turkey. **1848** J. Gould *Birds Austral.* V. 80 (*heading*) Collared Plain Wanderer. **1901** A. J. Campbell *Nests & Eggs Austral. Birds* II. 737 The collared Plain Wanderer, although a unique species, is closely allied to the Turnixes. **1965** *Austral. Encycl.* VII. 137/1 The plain-wanderer—sometimes called turkey-quail—is an inhabitant of open country in south-eastern and South Australia. **1901** Kipling *Kim* xiii. 328 The lama..walked as only a hillman can. Kim, plains-bred and plains-fed, sweated and panted astonished. **1903** —— *Five Nations* 53 But that night the Norther..Froze and killed the plains-bred ponies. **1889** *Ann. Rep. Board of Regents Smithsonian Inst. 1886–87* II. 408 The changes which would take place in a band of plains buffaloes transferred to a permanent mountain habitat can be forecast. **1910** E. T. Seton *Life-Hist. Northern Animals* I. 260 We have 20,000,000 as the number of the Plains Buffalo. **1963** *Maclean's Mag.* 23 Feb. 42/3 Must the barren lands be swept clear of people, leaving the few remaining caribou to become a curiosity like the plains buffalo? **1972** T. McHugh *Time of Buffalo* iii. 22 Differences in color and texture of coat are useful in separating the two subspecies—*Bison bison bison*, the plains buffalo, and *Bison bison athabascae*, the wood buffalo. **1912** C. Wissler *N. Amer. Indians of Plains* ii. 86 While the camp circle was the most striking and picturesque trait of Plains culture, it was probably no more than a convenient form of organized camp for a political group composed of 'bands'. **1914** *Amer. Anthropologist* XVI. 16 The true Plains culture may properly be said to have developed with the introduction to the horse. **1957** *Publ. Amer. Dial. Soc. 1956* xxvi. 17 *Kahnl* 'brush-covered hut' changed its meaning to 'teepee' when the Indians went over to a plains culture and to 'house' when they adopted European culture. **1976** *Billings* (Montana) *Gaz.* 27 June 2-c/3 Plains culture of long ago can be read into the opulent jewelry Nighthorse creates. **1901** Plains-fed [see **plains-bred* adj.]. **1877** R. I. Dodge *Hunting Grounds Gt. West* v. 63 'Old Bridger', the most thorough and justly celebrated of all plains guides. **1831** T. Simpson *Let.* 19 Dec. in MacLeod & Morton *Cuthbert Grant of Grantown* (1963) viii. 108 The plains hunters have had a very successful season and the quantity of provisions they have brought home is immense. **1922** *Beaver* Dec. 113/2 The half-breeds dislike a settled life; they prefer the excitement of the chase or the idle life of the fisherman. They are technically termed plains hunters. **1959** E. Tunis *Indians* 28/2 Many of the Plains hunters and all of the Digger Indians of the West were ghost-ridden and terrified of the dead. **1877** R. I. Dodge *Hunting Grounds Gt. West* v. 67 Another plains malady.. is called 'moon-blind'. **1905** *Nation* (N.Y.) 5 Jan. 11/1 As a plains people they [*sc.* the Pawnee] were largely dependent upon the chase. **1963** *Times* 19 Apr. 14/6 The great love of the Pakistani plainspeople for water. **1930** L. G. D. Acland *Early Canterbury Runs* (ser. 1) v. 109 Valetta was the last of the old plains stations to remain anything like its original size. **1933** —— in *Press* (Christchurch, N.Z.) 9 Sept. 15/7 C[row's] n[est]s were used on the old plains stations until the runs were fenced, about 1860. **1870** De B. R. Keim *Sheridan's Troopers on Borders* iv. 29 The Plains Tribes have, as yet, presented no prominent warriors in the character of leaders. **1877** R. I. Dodge *Hunting Grounds Gt. West* xli. 419 The Tonkaways cannot properly be called a plains tribe. **1917** C. Wissler *Amer. Indian* viii. 131 North of Mexico, methods of reckoning time are very crude, though apparently strongest among the Pueblo and adjacent Plains tribes. **1949** *Nat. Geogr. Mag.* Oct. 473/1 What the buffalo was to the Plains tribes the caribou is to the Indians of the far north. **1926** *Emu* XXVI. 59 (*heading*) The vanishing Plains-wanderer. **1964** A. L. Thomson *New Dict. Birds* 635/1 Plains-wanderers..are usually loth to fly.

plain, *a.*[1] and *adv.* Add: **A.** *adj.* **II. 7. b.** Applied to knitting in knit-stitch or garter-stitch (see *garter *sb.* 8). Also quasi-*sb.* = **garter-stitch*.

1861 Mrs. Beeton *Bk. Househ. Managem.* 622 The cloths.. should be knitted in plain knitting, with *very coarse* cotton. **1872** *Young Englishwoman* Oct. 559/1 Knit 9 rows, 2 plain, 2 purl, alternately. **1885** [see Purl *sb.*[1] 5]. **1910–11** *T. Eaton & Co. Catal.* Fall & Winter 20/1 Women's Coat Sweater, made of knitted worsted, in fancy stitch. The V-neck and fronts have wide, plain knitted border. **1932** D. C. Minter *Mod. Needlecraft* 69/2 1st row: Knit plain. 2nd row: Purl. **1970** *Guardian* 24 Mar. 9/3 Cable stitch jacket, plain knit pants. **1978** A. Grey *Chinese Assassin* viii. 111 It might just as well be an extract from my granny's pearl and plain knitting book for all I can tell.

III. 8. Also of a person's name: without addition or title.

1828 *Imperial Mag.* X. 589 The doctor, or, as he now chose to designate himself, plain Thomas Beddoes. **1872** Hardy *Under Greenw. Tree* I. i. ii. 20 'Reub', says he—'a always used to call me plain Reub, pore old heart! **1914** G. B. Shaw *Misalliance* 12 He calls himself Plain John, but you can't call him that in his own office.

c. (*b*) (Further examples.)

1885 [see Finesse *v.* 2 a]. **1899** [see *echo *sb.* 8]. **1936** [see *cash *v.*[1] 1 b].

e. Of envelopes, containers, etc.: giving no information as to sender or contents; esp. in phr. *under plain cover*. Also *transf.* (of a motor vehicle) and *fig.*

1913 *Maclean's Mag.* Sept. 81/3 Write us for Catalogue 'D', sent free on application, in plain envelope. **1925** *Ladies' Home Jrnl.* Mar. 133/4 The sample will come in plain, unmarked wrapper. **1932** *N.Y. Times Bk. Rev.* 17 Jan. 20/1 Please send me in plain wrapper, prepaid, a copy of the complete unabridged edition of 'Sane Sex Life and Sane Sex Living' by Dr. Long. **1936** *Men Only* Mar. 147/1 Send a p.c. for full details of the Girvan Scientific System (mailed under plain cover) and particulars of our £100 Guarantee. **1942** N. Balchin *Darkness falls from Air* iii. 57 'He's tight,' said Fred... I said, 'I expect they'll send a plain van to collect the old boy.' **1962** *Sunday Times* 11 Nov. 25/7 Much good has been done by the increasing enlightenment over the nature of sexuality which was a secondary aim of the Stopes/Ellis pioneers. The books which used to be sent under plain cover, now appear as Penguins. **1966** Wodehouse *Plum Pie* iv. 107 The two of them got away with the purloined objects, no doubt in a plain van. **1969** *Listener* 24 July 125/1 The husband..is secretly studying for O-levels while the nagging women think he is out boozing. When the results come through the post..his mother hides them because they're 'under plain cover' and she assumes the envelope conceals dirty books. **1971** W. J. Burley *Guilt Edged* i. 9 The goods [were] to be delivered in a plain van. *Ibid.* ii. 33 *The Postures of Love* (Thirty-five photographic plates. Send £3. Delivered in a plain wrapper.) **1974** 'J. Melville' *Nun's Castle* iii. 58 [Pregnancy] tests.. could be done..at home... It would be easy enough to ..have the materials sent (in a plain cover, as the advertisements so kindly suggested). **1974** *John O'Groat Jrnl.* (Wick) 6 Sept. 9/5 (Advt.), *Plain wrapper*, Durex Gossamer, 55p dozen; Fetherlite, three packets 55p, post free; plain wrapper; guaranteed Grade 1 goods. **1975** L. Dills *CB Slanguage Dict.* 47 *Plain brown* (*black, gray*) *wrapper*, unmarked police car.

VI. 18. b. *as plain as the nose on* († *in*) *one's face*.

1695 Congreve *Love for Love* iv. i. 60 *As witness my Hand*,..in great Letters. Why, 'tis as plain as the Nose in one's Face. **1903** G. B. Shaw *Man & Superman* ii. 69 Why, it's as plain as the nose on your face. If you aint spotted that, you dont know much about these sort of things. **1937** D. L. Sayers *Busman's Honeymoon* ix. 200, I came in nine o'clock from fetchin' a pail o' water and I sees you plain as the nose on my face a-talkin' to him at this very winder. **1940** Wodehouse *Quick Service* xiii. 166, I see it all. They're pals. It's as plain as the nose on your face. **1958** D. Garnett *Shot in Dark* vii. 86 'But how did you find out?' 'It is as plain as the nose on your face.'..'Yes. But most people don't see the nose on their face. The obvious is what is overlooked.' **1979** K. Bonfiglioli *After you with Pistol* xiv. 99 The facts are as plain as the nose on your face.

B. *adv.* **I. 4.** Delete † *Obs.* and add later examples.

1955 S. A. Grau *Black Prince* 193 'You just plain remember that [is] my boat.' 'Who done seen it?' his father said. 'I plain ask you.' **1956** B. Holiday *Lady sings Blues* (1973) xviii. 151, I had gained so much weight and I just plain didn't look like the girl who had left town ten months before. **1959** J. L. Austin *Sense & Sensibilia* (1962) i. 5 Besides, there is nothing so plain boring as the constant repetition of assertions that are not true. **1973** *Times* 13 Nov. 6/6 This myth that.. we were just being plain difficult about Sandhurst was another very irritating criticism. **1976** *Bridgwater Mercury* 21 Dec. 10/4 Others may have family problems, housing difficulties—or are just plain lonely. **1979** *Guardian* 8 Jan. 2/8 Mrs Thatcher..dismissed claims that taxation of benefits would be impossible to administer... 'I just plain don't believe it.'

C. a. *plain-faced* (further examples; also *fig.*).

1928 *Weekly Dispatch* 24 June 22/2 Thus what seems a plain-faced stroke is full of guile. **1938** W. de la Mare *Memory* 61 A solemn plain-faced child. **1963** *Times* 11 May 11/5 Small plain-faced towns.

b. *plain-bound, -cut* (example fig.), *-woven.*

1870 D. G. Rossetti *Let.* 20 Apr. (1965) II. 849, I think the woodcut had better have been left out of the plain-bound copies, as it looks quaint and provoking without the binding. **1894** Stevenson & Osbourne *Ebb-Tide* i. v. 96, I never could act up to the plain-cut truth, you see; so I pretend. **1967** E. Short *Embroidery & Fabric Collage* i. 19 The embroidery completely alters the original plain-woven texture of the material.

c. plain bearing *Engin.*, a bearing consisting of a cylindrical hole in a block; **Plain Bob** *Campanology*, a method of change-ringing in which the treble works in continuous **plain hunt*; **plain chocolate,** eating chocolate (Chocolate 2) made without the addition of milk (cf. **milk chocolate* (*b*)); **plain clothes,** add to def.: now esp. the dress of members of a police detective force (earlier and further, including *attrib.*, examples); also *transf.*, a plain-clothes policeman, a detective; hence *plain-clothed* adj.; **plainclothesman, plain-clothes man,** a plain-clothes policeman, a detective; **plain hunt** *Campanology* [Hunt *sb.*[2] 3 and *v.* 7], a regular path taken by a bell from first position to last and back again; hence **plain hunting** *vbl. sb.*; **Plain Jane, plain Jane,** (a name applied to) an unattractive or ill-favoured girl or woman; also *transf.*; also (freq. with hyphen) *attrib.* passing into *adj.*; **plain language,** (*a*) (earlier example); (*b*) = *plain text* (*b*) below; **plain man,** *spec.*, one who is not, or is not thinking in the manner of, a philosopher; **Plain people, plain people** *U.S.*, the Amish, Mennonites, and Dunkards; **plain saw** *v. trans.*, to produce (a board) by plain sawing; so **plain-sawed, plain-sawn** *ppl. adjs.*; **plain sawing** *vbl. sb.*, the method or action of producing boards by sawing a log tangential to the growth rings, so that the rings make angles of less than 45° with the faces of the boards; **plain sewing,** (*a*) *Needlework* (see quot. 1882); (*b*) applied to a particular kind of homosexual behaviour in which masturbation or mutual masturbation takes place; **plain text** *Cryptanalysis*, (*a*) a text not in cipher or code; (*b*) (as **plaintext**) uncoded language; **plain weave** (see quot. 1940); also *attrib.*

1917 *Engineering* 9 Nov. 503/1 The worms..are supported in their cases by two ball journal or plain bearings. **1941** L. S. Marks *Mech. Engineers' Handbk.* (ed. 4) 1016 Plain bearings, according to their function, may be (1) Journal bearings... (2) Thrust bearings... (3) Guide bearings. **1971** B. Scharf *Engin. & its Lang.* xii. 133 In their simplest form, journal bearings consist of a block with a central hole for the shaft... Smaller bearings of this type are..referred to as eye-bearings, larger ones as plain bearings. **1702** J. D. & C. M. *Campanalogia Improved* 50 (*heading*) Grandsire Bob commonly called

plain Bob. **1788** W. Jones et al. *Clavis Campanalogia* 39 Having laid down plain and easy rules for ringing Plain Bob and calling eighteen-score, we shall next proceed to the 720. **1879** [see *method *sb.* 3 e]. **1931** E. Morris *Hist. & Art Change Ringing* viii. 346 Plain Bob ..is similar to the 'Original', in which all 'hunt', until the treble returns to lead, when, instead of allowing the 'course' to run round, second place is made by the one the treble takes from lead, thereby causing each pair immediately behind them to 'dodge'. **1965** Plain Bob [see *method *sb.* 3 e]. **1895** *Army & Navy Co-op Soc. Price List* 11/1 (*heading*) Chocolate... Menier's, Plain.. 1/3½. **1914** H. Ashton *First from Front* xiv. 96 The people in the houses came out and cheered and gave us plain chocolate, fruit, and beer. **1948** *Good Housek. Cookery Bk.* 652 For chocolate dipping, couverture chocolate (i.e., covering chocolate, which is good quality plain block chocolate, containing an adequate proportion of cocoa butter) should be used. **1976** K. Thackeray *Crownbird* ii. 27 Her skin was dark and smooth, the colour of plain chocolate. *a* **1966** M. Allingham *Cargo of Eagles* (1968) iii. 40 The plain clothed Sergeant Throstle. **1975** K. Macksey *Partisans of Europe* xi. 182 His staffs should place more faith in the parachute and commando formations of proven reliability..in effect putting their trust in cool, uniformed quality before hotheaded, plain-clothed quantity. **1822** T. Creevey in *Creevey Papers* (1903) I. x. 238 Who should overtake me but the Duke of Wellington in his curricle, in his plain clothes and Harvey by his side in his regimentals. **1825** H. Wilson *Mem.* II. 153, I could not reconcile it to my mind that he should wear regimentals... A gentleman always looks so much better in plain clothes. **1842** C. Fox *Jrnl.* 1 June (1972) 127 Her Majesty..ordered a double number of police in their plain clothes to be stationed in the Park. **1866** Mayne Reid *Headless Horseman* xii. 67 Like a plain-clothes policeman employed on detective duty. **1908** K. Grahame *Wind in Willows* viii. 182 Policemen in their helmets, waving truncheons; and shabbily dressed men in pot-hats, obvious and unmistakable plain-clothes detectives. **1914** 'Bartimeus' *Naval Occasions* xxv. 287 Not a bad principle either—saves your plain-clothes from wearing out. **1926** Galsworthy *Escape* 14 *Girl.* Who are you? *Plain Clothes Man.* Plain clothes. **1929** S. Leslie *Anglo-Catholic* iii. 41 'Do you ever come across the police?' 'More often than I care, and the plain-clothes are the worst.' **1955** *Times* 1 Aug. 7/6, I, for one, would welcome the presence of such plain-clothes patrols. **1962** 'K. Orvis' *Damned & Destroyed* xxiv. 176 A Mountie plain-clothes tapped me on the shoulder. **1964** *Granta* 2 Nov. 152, I was rather surprised on entering the Proctors' rooms to find also in attendance a plain-clothes policeman wearing his raincoat indoors. **1972** *Nature* 18 Feb. 400/2 Two men in plain clothes came up to Sakharov and asked for his identity papers. **1977** *New Yorker* 9 May 148/2 François Eugène Vidocq..invented the plainclothes detective. **1899** J. S. Clouston *Lunatic at Large* ii. v. 140 Keep your eye on that man, officer,..and put your plain-clothes' men on his track. **1962** *Punch* 21 Nov. 751/1 Truncheoned bobbies and macintoshed ruthless plain-clothesmen of Scotland Yard. **1969** R. D. Pharr in A. Chapman *New Black Voices* (1972) 68 The plainclothesman picked them up... When he got in the car the detective in the driver's seat asked, 'What'd he say?' **1975** J. F. Burke *Death Trick* (1976) iii. 34 Several uniformed cops and some plainclothesmen were grouped around the entrance. **1977** *Time* 7 Mar. 6/2 As a crowd gathered, four plainclothesmen collared Amalrik and removed him bodily. **1874** W. Banister *Art & Sci. Change Ringing* 14 The treble works in continuous plain hunt; whilst the other bells hunt, make places, and dodge. *Ibid.* 22 Each bell has a plain hunting course, except when treble leads. **1965** W. G. Wilson *Change Ringing* iv. 13 The basic principle involved in ringing changes on bells, or in working them out on paper, is called the plain hunt. This word 'hunt' is used in the sense of course or path or way among the other bells. Starting from rounds..each bell follows a regular path among the others. *Ibid.* 14 Here are examples on three, four and six bells... In each of these examples, if you draw a line through the path of any one number, representing a bell, you will get a straight path from front to back and then from front to back. This is a plain hunt. *Ibid.* 16 If we confined our ringing on four or more bells to a plain hunt we should soon get very bored with it. So once we have mastered plain hunting we must learn to vary it. **1912** C. Mackenzie *Carnival* ii. 14 She sha'n't be a Plain Jane and No Nonense, with her hair screwed back like a broom, but she shall be Jenny, sweet and handsome, with lips made for kissing and eyes that will sparkle and shine. **1922** Joyce *Ulysses* 121 Daughter working the machine in the parlour. Plain Jane, no damn nonsense. **1936** C. Day Lewis *Friendly Tree* i. 11 It was the plain-Jane, methodical part of herself. **1953** *Newsweek* 23 Mar. 74/1 Takarazuka girl players, living like priestesses, are virtually adored by their plain-Jane sisters throughout Japan. **1956** *People* 13 May 6/3 Put a Gorgeous Gussie among a group of Plain Janes..and a whole office or factory routine can be upset. **1957** *New Yorker* 16 Nov. 158/2 Jewelled clasps to enliven worthy but plain-Jane necklaces and bracelets are the specialty of Marjorie Raven. **1958** *Times Lit. Suppl.* 15 Aug. ('Books in a Changing World' Suppl.) p. xxxi/4 It is very hard to know how children reared on the plain Jane vocabulary of the books written for them to-day can ever enter the territory where language is allowed to branch and flower into exuberance. **1970** 'R. Llewellyn' *But we didn't get Fox* iii. 34 An enormous American aircraft carrier in plain-jane grey. **1972** J. McClure *Caterpillar Cop* vii. 117 If ever there was a Plain Jane, she's it, poor kid. **1974** *Country Life* 2 May 1082/3 Plain-leaved parsley..is..reputedly more flavoursome, though a plain Jane and useless decoratively. **1827** Mrs. B. Hall *Let.* 16 Dec. in *Aristocratic Journey* (1931) xi. 151 The family are Quakers, and the old couple adhere rigidly to the plain dress and the plain language as they call what the French term 'tutoyer' implies. **1929** *Radiotelegraph Convention & Gen. Regulations* (Internat. Radiotelegraph Conf. 1927) 21 Correct transmission and correct reception by ear of code groups..at a speed of 20 (twenty) groups per minute, and of text in native plain language, at a speed of 25 (twenty-five) words per minute. **1940** *Tablet* 4 May 419 The..Soviet Embassy's plain language telegram. **1973** H. Gruppe *Truxton Cipher* 213 The message, sent Immediate—Plain Language over Navy circuits, lay upon his..desk top. *Ibid.* 221 Our convoys had firm orders to disregard plain-language traffic. **1896** L. T. Hobhouse *Theory of Knowl.* 15 The 'plain man' would probably agree with Locke that no further proof could be given. **1904** G. S. Fullerton *Syst. Metaphysics* xvii. 263 One cannot expect the plain man to realize clearly all that his doctrine implies. **1934** A. C. Ewing *Idealism* vii. 294 The plain man asserts that the table in his dining-room is square. **1948** B. Russell *Human Knowl.* 193 At the moment, I notice my dog asleep, and as a plain man I am convinced that I could have noticed him any time last hour. **1978** J. Pearson *Façades* vi. 111 Arnold Bennett..had a plain man's taste for what was new in poetry and art. **1904** H. R. Martin *Tillie* 113 But can't you see the inconsistentness of the plain people? **1929** *Sat. Even. Post* 23 Mar. 165/1 You found it in your heart for to join the Plain People, didn't you, Carlie? **1948** *Chicago Tribune* 25 Jan. iv. 5/4 The Plain People, as they are known, won't use automobiles or tractors, have no telephones, plumbing or political parties. **1975** *Budget* (Sugarcreek, Ohio) 20 Mar. 8/5 Both Bro. David Wagler and his wife are plain people. They work helping this organization to distribute Bibles, hymn books and concordances behind the Iron Curtain. **1951** A. E. Bridgwood *Carpentry & Joinery* (*Intermediate*) iii. 151 If figured boards are required, they should be plain sawn... Floor joists are stronger if plain sawn. **1931** Younger & Ward *Airplane Construction & Repair* vi. 98 The advantages of plain-sawed lumber [over quarter-sawn] are: 1. It is cheaper to cut. 2. If knots are present, they are round instead of spiked. *Ibid.*, There are two principal methods of sawing up trees into lumber: plain sawing and quarter-sawing. The former produces flat-grain lumber and the latter edge-grain lumber. **1968** F. Hilton *Craft Technol. for Carpenters & Joiners* i. 18 This plain sawing is usually the cheapest form of conversion. **1949** *Gloss. Terms Timber* (*B.S.I.*) 7 *Flat-sawn timber*, timber converted so that the growth rings meet the face in any part at an angle of less than 45°. (Plain-sawn, slash grain, flat grain). **1961** N. P. Johnson in A. E. Bridgwood *Newnes Carpentry & Joinery* I. iv. 193 If figured boards are required, they should be tangential sawn... They will, however, be liable to the shrinkage and warping associated with plain-sawn timber. **1966** A. W. Lewis *Gloss. Woodworking Terms* 18 Boards sawn tangentially to the growth rings are known as through-and-through sawn, plain sawn, slash cut, or flat sawn. **1862** E. Waugh in *Manchester Examiner & Times* 26 Aug. 7/1 Part of the time each day is set apart for reading and writing; the rest of the day is devoted to knitting and plain sewing. **1882** Caulfield & Saward *Dict. Needlework* 394/2 *Plain sewing*, a term denoting any description of Needlework which is of a merely useful character in contradistinction to that which is purely decorative. **1895** [see sense A. 7]. **1926** A. Huxley *Let.* 10 Aug. (1969) 272 What she [*sc.* Anita Loos] really likes doing, it appears, is plain sewing; spends all her holidays in making underclothes which nobody can wear. **1941** F. Thompson *Over to Candleford* ix. 137 Plain sewing was still looked upon as an important part of a girl's education. **1969** Auden in *N.Y. Rev. Bks.* 27 Mar. 3/4, I conclude he [*sc.* J. R. Ackerley] did not belong to either of the two commonest classes of homosexuals, neither to the 'orals'..nor to the 'anals'... My guess is that at the back of his mind, lay a daydream of an innocent Eden where children play 'Doctor', so that the acts he really preferred were the most 'brotherly', Plain-Sewing and Princeton-First-Year. **1971** *Observer* 7 Nov. (Colour Suppl.) 35/4 One of my [*sc.* W. H. Auden's] great ambitions is to get into the OED, as the first person to have used in print a new word. I have two candidates at the moment, which I used in my review of J. R. Ackerley's autobiography. They are 'Plain-Sewing' and 'Princeton-First-Year'. They refer to two types of homosexual behaviour. **1979** P. Fitzgerald *Offshore* vii. 80 The nuns..in a class known as plain sewing, had taught her..darning, patching, reinforcing collars with tape. **1980** *Times Lit. Suppl.* 21 Mar. 324/5, I suspect 'Plain-Sewing' to be Auden's own invention, but its meaning is fairly clear, as it involves a pun on 'sowing' (seed or semen) and a reference to the two-and-fro [*sic*] action of the hand in sewing. **1918** F. Strother *Fighting Germany's Spies* vii. 144 Now..the plain text of the secret message is printed on the under sheet by writing through the perforations of the upper sheet, only one letter being written in each square. **1932** *Cryptogram* Aug. 1/2 A letter or a symbol is substituted for each letter in the original message. (Hereafter we shall refer to the original message as the plain-text.) **1939** H. S. M. Coxeter *Ball's Math. Recreations & Ess.* (ed. 11) xiv. 381 If the characters of the plain-text message are merely rearranged without suffering any change in identity..the system is called transposition. **1967** D. Kahn *Codebreakers* (1968) xiv. 435 The cryptanalysts of the German Foreign Office..had reduced it to plaintext at once. **1972** *Sci. Amer.* Nov. 117/2 Although it is a defect of Bacon's system that a cipher text must be five times as long as the plaintext, a remarkable merit of the system is that more than one message can be hidden in the same cipher text. **1940** *Chambers's Techn. Dict.* 649/1 *Plain weave*.., the simplest interlacing of warp and weft threads. Each warp thread is alternately over and under the weft, while adjacent warp threads work opposite to each other. **1956** *Textile Res. Jrnl.* XXVI. 837/1 The fabric employed..was an 80 × 80 plain weave cotton of about 3½ oz. to the square yard. **1962** J. T. Marsh *Self-Smoothing Fabrics* xi. 155 The effect of mercerising has also been investigated by Smith..who employed 10% of methoxymethyl-urea on a plain-weave cotton fabric.

plain sailing, *sb.* (Earlier and later examples.)

1756 N. Owen *Jrnl. Slave-Dealer* (1930) 67 If he can take an observation and is acquainted with that part of navagation call'd plain sailing, without any of the practical part of seamanship. **1823** J. F. Cooper *Pilot* I. xii. 152 This is what the lads would call plain sailing..; they are out of employment [etc.]. **1916** G. B. Shaw *Androcles & Lion* Pref. p. xxiv, Without the proper clues the gospels are, to a modern educated person, nonsensical and incredible... But with the clues, they are fairly plain sailing. Jesus becomes an intelligible and consistent person. **1955** *Times* 29 June 13/3 It is not all plain sailing Difficulties have to be overcome, for example, in disposing of 'contract' vehicles because of the contract terms.

Plains Cree. Also **Plain Cree.** [f. Plain *sb.*[1] 1 b + *Cree *sb.* and *a.*] A tribe of Cree Indians formerly inhabiting the more northerly areas of the North American plains; a member of this tribe. Also *attrib.* or as *adj.*

[**1823** in J. Franklin *Narr. Journey Shores Polar Sea* iv. 108 The Crees, who inhabit the plains, being fur-hunters are better known to the traders.] **1860** H. Y. Hind *Narr. Canad. Red River Expedition* I. xix. 414 The Plain Crees are not fishermen like the Ojibways. **1879** H. M. Robinson *Great Fur Land* ix. 186 Along its entire border there prevails..a state of perpetual warfare: on the north and east with the Plain Crees. **1908** *Amer. Anthropologist* X. 199 We have in the Plains Cree of Canada part of a distinct ethnic group adopting the culture of the area without losing connection with the whole. **1913** F. W. Hodge *Handbk. Indians of Canada* (1971) 382/2 *Paskwawininiwug* ('prairie people'), the Plains Cree, one of the two great subdivisions of the Cree. **1928** L. Bloomfield in C. F. Hockett *Leonard Bloomfield Anthol.* (1970) 200 As for our own Indians, in spite of their Plains Cree contempt for the Swampy Cree, they did not joke with the old man. *Ibid.*, Bad Owl was one of the young Plains Cree who hired out to us one spring to make the river voyage. **1938** P. H. Godsell *Red Hunters of Snows* v. 85 The Pasquaininniwuk or 'People of the Plains', roaming the prairies of Manitoba, Saskatchewan and Alberta...known to the fur traders as the Plains Crees, had..acquired most of the traits and characteristics of the buffalo-hunting tribes. **1940** *Anthropol. Papers Amer. Mus. Nat. Hist.* XXXVII. 195/2 Among the Plains Cree, the horse was the standard of prestige value by means of which the status criteria of wealth, valor, and liberality could best be realized. *Ibid.* 251/1 The concept of a single all powerful creator was dominant in Plains Cree religious ideology and ceremonialism. **1948** H. E. Hives *Cree Gram.* 3 In the open country of the Saskatchewans live the Plain Cree. **1972** D. Morton *Last War Drum* vii. 133 Big Bear and the Plains Crees set out, striking eastward, but..the Woods Crees turned north.

Plains Indian, *sb.* and *a.* Also 7–9 **plain Indian.** [Plain *sb.*[1] 1 b.] **A.** *sb.* A member of any of the Indian tribes who formerly inhabited the North American plains; (*pl.*) these tribes collectively. **B.** *adj.* Of or pertaining to any of these tribes.

1697 H. Kelsey *Jrnl.* 3 July in *Kelsey Papers* (1929) 88 About four a clock some plain indians arrived att the fort. **1844** J. H. Lefroy *In Search Magnetic North* (1955) xii. 142 The plains Indians are in a state of warfare, and there is a certain degree of danger in a single boat or canoe passing through their country. **1852** in *Mich. Hist. Mag.* (1925) IX. 397 Though the plains Indians frequently go unpunished, that is no reason why our Indians here should be butchered. **1887** *Jrnls. Senate Canada* XXI. App. 53 As to the existing food there is not very much on the plains for the plain Indians. **1913** J. London *Valley of Moon* 438 A lithograph..of a Plains Indian, in paint and feathers. **1917** C. Wissler *Amer. Indian* xiv. 207 (*caption*) The Plains Indian culture area. **1931** *Amer. Speech* VII. 2 All Nebraska Indians were known as 'plains Indians'. **1937** R. H. Lowie *Hist. Ethnol. Theory* viii. 127 A barely existing..totemic system is made responsible for the animal names of Plains Indian military societies. **1952** B. Blackwood in T. K. Penniman *Hundred Years of Anthropol.* (ed. 2) vi. 421 In contrast to the South-west, the Plains area has until recently attracted very little attention from archæologists, in spite of the fact that Wissler had, as early as 1907, drawn attention to the lateness of the Plains Indian culture as we know it. **1955** W. Gaddis *Recognitions* i. i. 23 With the loss of Camilla he returned to the times before he had known her, among the Zuñi and Mojave, the Plains Indians and the Kwakiutl. **1966** Mrs. L. B. Johnson *White House Diary* 2 Apr. (1970) 380 Joe Frantz..began to weave together the story of the place..its history..the tribal Plains Indians..the Spanish conquistadores. **1972** D. Davies *Dict. Anthropol.* 148/2 With the depletion and destruction of the buffalo herds, through over-hunting by Indians and white men like Buffalo Bill, the Plains Indians lost their means of support..and in many places disappeared. **1976** *Times* 25 Sept. 12/8 Among these exhibits are..Plains Indian artifacts.

plainsman. (Earlier and later examples.)

1870 De B. R. Keim *Sheridan's Troopers on Borders* xi. 66 Such an animal is a treasure in the esteem of a plainsman. **1873** J. H. Beadle *Undevel. West* vi. 93 Old plainsmen look at each other with a peculiar smile which may mean anything. **1931** 'Grey Owl' *Men of Last Frontier* 118 The difficult task of transforming an indifferent plainsman into some kind of woodsman. **1956** E. Pound tr. *Sophocles' Women of Trachis* 45 No gang of plainsman with spears..was strong enough. **1970** *Toronto Daily Star* 24 Sept. 10/3 (Advt.), Exceptionally handsome, this well made furniture is as rugged as a plainsman, it'll stand up to plenty of hard knocks and last a lifetime. **1975** D. Pitts *This City is Ours* liii. 266 The sound [*sc.* the Indian war cry]..that had defied the plainsmen and the encroaching wagon trains.

plain-speaking, *sb.* (Earlier example.)

1849 E. A. Poe in *Southern Lit. Messenger* July 416/1 As for American letters, plain-speaking about *them* is.. needed.

plaiting, *vbl. sb.* Add: **b.** *plaiting lace.* Also, *plaiting-down apparatus.*

1813 Jane Austen *Let.* 16 Sept. (1932) II. 328, I bought some very nice plaiting Lace. **1927** T. Woodhouse *Artificial Silk* 134 The cloth is..passed over the inclined reversible inspecting board.., between a pair of drawing rollers, and finally to the plaiting-down apparatus.

plai·tless, *a.* [f. Plait *sb.* + -less.] Having no plaits.

1887 Hardy *Woodlanders* III. xv. 315 This solitary and silent girl stood there in the moonlight..clothed in a plaitless gown.

plan, *sb.* Add: **I. 1. c.** *Methodism.* A periodic document listing the preachers for all the services throughout a circuit for the period.

1776 J. Wesley *Let.* 24 June (1931) VI. 224 Fix a regular plan for the local preachers and see that they keep it. **1780** [in Dict. sense 1 b]. **1807** J. Nightingale *Portraiture of Methodism* xxix. 304 A local-preacher's plan, is a paper properly divided and subdivided into columns and squares, on which the names of all the preachers are inserted, the respective places of their preaching-appointments, and the dates of the month... One of these plans is given to every local-preacher. **1851** J. Loutit *Let.* 17 Dec. in W. R. Ward *Early Vict. Methodism* (1976) 411 While it is said that 'no person shall *receive* a plan as a Local-Preacher without the approbation of a Local-Preachers' Meeting'..there is no express arrangement for his *removal*. **1898** B. Gregory *Side Lights on Conflicts of Methodism* ii. 25 Village Methodism was built up at first to a very great extent by the institution of 'taking in the preachers'... Indeed, by a genial and judicious ministerial spirit, the 'Dinner Plan' may be made no inconsiderable accessory to the Circuit Plan. **1925** F. Hodges *My Adventures as Labour-Leader* iii. 19 In due course my name was inscribed upon the 'plan' as a regular local preacher. **1929** W. T. A. Barber in Lidgett & Reed *Methodism in Mod. World* i. 28 As a local preacher he opened preaching-places and formed societies, working within the Methodist system. Like his northern contemporaries, he could not be bound by its rules, and, after several attempts at conformity, was finally excluded from the plan. **1937** W. H. Lax *Lax of Poplar* xii. 109 When I preached my trial sermon for the Local Preachers' Plan. **1957** R. F. Wearmouth *Social & Polit. Influence of Methodism in 20th Cent.* vii. 117 In due time he was nominated for the preachers' plan, and in 1869–70 became a Primitive Methodist local preacher. **1963** R. E. Davies *Methodism* 174 The plan of public services for a quarter in each circuit is drawn up by the superintendent minister, normally in consultation with his colleagues. **1975** C. Moorhouse *Sabden* 52 In 1810 Burnley was made the head of a new circuit. The resident Methodists in Sabden at this date had so far been organised that Sabden was put on the plan and included in the new circuit organisation.

3. c. A scheme for the economic development of a country. Also *transf.* Freq. with specification of a number of years in which the objectives are to be achieved; cf. *five-year plan* s.v. *five *a.* and *sb.* 2.

1933 *B.B.C. Year-bk. 1934* 19 Special attention was paid..to the gradual progressive adaptation of existing facilities on a kind of 'Five-Years' Plan'. **1937** *Ann. Reg. 1936* 188 Herr Hitler accordingly announced a Four-Year Plan for Germany with the object of making Germany independent of supplies of raw materials from abroad. **1949** 'G. Orwell' *Nineteen Eighty-Four* i. 48 Some triumph of over-production in the Ninth Three-Year Plan. **1962** *Listener* 20 Dec. 1041/2 There is no powerful spokesman inside the British Cabinet whose job it is to concentrate on the Plan. **1974** B. Pearce tr. *Amin's Accumulation on World Scale* II. iii. 433 In a planned socialist economy, the banks strictly limit advances to enterprises to the amounts laid down in the plan.

III. 6. plan-position indicator, an instrument giving a map-like display on a cathode-ray tube of the positions of objects detected by a rotating radar scanner; **plan view,** a view in plan; = Plan *sb.* 1.

1944 *Princ. Radar* (M.I.T. Radar School) iii. 37 The plan position indicator (PPI) presents range and bearing in polar form. **1945** *Wireless World* Sept. 270/2 The Naval 'Plan Position Indicator' presents..a complete picture of the relative positions of all aircraft in the vicinity. **1952** *Electronic Engin.* XXIV. 430/1 The plan position indicator is a 9 in. cathode-ray tube. **1974** *Sci. Amer.* Apr. 51/2 The task selected by Beatty was the familiar one of detecting 'targets' simulating aircraft on the cathode ray screen of a radar plan-position indicator, similar to the screens monitored by air-traffic controllers at most airports. **1850** T. Tredgold *Steam Engine* (ed. 3) I. iv. 17 These dimensions are given in the plan view, Plate IV., where the cylinder and steam chest are shown in section. **1892** *Princ. Pattern Making* 160 In a plan view of a drawing, the eye of the observer is supposed to be set directly vertical over the drawing, and the illusions due to perspective are supposed not to exist. **1962** D. Nichols *Echinoderms* iii. 50 The outermost virgalia of each column are shortened and expanded laterally to form a series of marginals, outlining the plan-view of the animal. **1970** K. R. Hart *Engin. Drawing* iii. 11 The Elevation and Plan views may not..fully specify the shape description of an object.

plan, *v.* Add: **3. a.** (Absol. examples.) Also with *on, out.*

1777 C. Reeve *Champion of Virtue* 93 Some are born to plan, others to execute. **1873** 'S. Coolidge' *What Katy Did* xi. 230 Few visitors came to interrupt her, so she could plan out her hours and keep to the plans. **1896** C. M. Sheldon *His Brother's Keeper* v. 107 When Aunt Royal comes, I mean to plan for something besides all this. **1918** *Dialect Notes* V. 20 *Planned on going*, planned to go. **1926** *Amer. Oxonian* July 99 If I were planning on going after a Rhodes Scholarship next year, I should read a great deal on foreign affairs. **1936** L. C. Douglas *White Banners* ix. 195 We had not planned on such a large house. **1963** M. Shadbolt in C. K. Stead *N.Z. Short Stories* (1966) 314 We don't plan on any drinking. **1977** H. Kaplan *Damascus Cover* iv. 35 Ari pressed for a date when he could plan on going abroad.

b. (Earlier and later examples.) *spec.* in *Methodism*, to include a preacher in a plan (*plan *sb.* 1 c).

1807 J. Nightingale *Portraiture of Methodism* xxvii. 280 Let no local-preacher, who will not meet in class, or who is not regularly planned by the superintendant of the circuit where he resides, be permitted to preach. **1898** J. Ackworth *Scowcroft Critics* 244 The next night Squire was 'planned' to conduct the weeknight service. **1950** S. Redfern *Methodist Journey* iii. 21 As a rule, two or three preachers were planned at the villages *en route*, so the circuit trap was used. **1961** *Times* 20 Jan. 14/6 'Then we'll plan you in the circuit here,' he said briskly.

planar, *a.* Delete *Math.* and add further examples.

1931 *Proc. Nat. Acad. Sci.* XVII. 127 A topological graph is called planar if it can be mapped in a 1–1 continuous manner on a plane (or sphere). **1942** M. P. Billings *Structural Geol.* xvi. 307 A clear distinction between primary and secondary planar and linear structures is essential for a correct interpretation of the tectonics of plutons. **1969** W. R. R. Park *Plastics Film Technol.* ii. 28 Biaxial or planar orientation occurs when a film or sheet is drawn in more than one direction. **1972** J. G. Dennis *Structural Geol.* xx. 459 Other criteria of high-pressure shock origin include sets of closely spaced planes in quartz crystals... These planar features are known in quartz that has been subjected to nuclear explosions. **1972** R. J. Wilson *Introd. Graph Theory* v. 58 A planar graph is one which is isomorphic to a plane graph. **1973** *Sci. Amer.* Jan. 100/3 Probably the commonest single habit is the planar dendrite, a flat crystal with delicate branches that is often regarded as the typical snowflake.

b. *Electronics.* Of a thermionic valve: having plane-parallel electrodes usu. close together.

1956 A. L. Albert *Electronics & Electron Devices* viii. 307 Tubes for handling a few watts or less at very high frequencies commonly use parallel electrodes and sometimes are called planar tubes. **1965** Gewartowski & Watson *Princ. Electron Tubes* iv. 120 If the linear dimensions of a planar diode are increased by a factor *k*, the same current flows to the anode for the same applied voltage. **1975** D. G. Fink *Electronics Engineers' Handbk.* ix. 26 Grid design is very important in planar triodes.

c. *Electronics.* Of a solid-state device: having boundaries of a number of different *n*- and *p*-type regions lying in a single plane. Also applied to the process by which such devices are made, involving the introduction of impurities into a semiconductor substrate through gaps in a thin masking layer.

1965 *Wireless World* July 325/1 Tr 1 and Tr 2 are low noise, high gain silicon planar transistors. **1967** A. S. Grove *Physics & Technol. Semiconductor Devices* 3 The planar technology..combines the advantages of junction formation by solid-state diffusion and the masking property of silicon dioxide for precise definition of device geometry. **1975** D. G. Fink *Electronics Engineers' Handbk.* vii. 37 The significance of the planar process is that the *pn* junctions are terminated and protected beneath a silicon oxide layer. Thus many of the surface problems associated with other types of transistor fabrication techniques, i.e., high leakage currents and poor low-current dc gain, are eliminated.

planarity (further examples).

1956 *Nature* 7 Jan. 37/2 The dimensions of the amide group accord closely with those given by Corey and Pauling, and it is in the *trans* configuration. It shows, however, a significant departure from planarity, since the carbonyl oxygen is 0·5 A. out of the plane containing the remaining atoms of the amide group. **1972** *Sci. Amer.* Dec. 54/3 Running tow or individual fibers between heated sprockets or gear teeth imparts a zigzag and more or less planar crimp... A disadvantage of the planarity is that it yields yarns of less bulk and stretchability than yarns produced by methods that impart a three-dimensional crimp.

planation (plănē^i·ʃən). *Geomorphol.* [f. L. *plān-um* Plane *sb.*[3] + -ation.] The levelling of a landscape by erosion.

1877 G. K. Gilbert *Rep. Geol. Henry Mts.* 127 The process of carving away the rock so as to produce an even surface, and at the same time covering it with an alluvial deposit, is the process of planation. **1892** A. J. Jukes-Brown *Student's Handbk. Physical Geol.* (ed. 2) iii. i. 565 A tract of land which has been submerged and then upraised must have been twice subjected to this levelling process, and on its second emergence would present a wide area with a nearly level or slightly undulating surface. Such an area has been termed a plain of marine erosion, but would perhaps be more aptly called a surface of planation. **1935** *Geogr. Jrnl.* LXXXV. 171 It is only where planation occurred during the subsequent Jurassic submergence that the surface is really level, as on the Mendips. **1963** D. W. & E. E. Humphries tr. *Termier's Erosion & Sedimentation* ii. 35 The most perfect planations imply a succession of complex phenomena, which include transportation and deposition of sediments as well as erosion. **1970** R. J. Small *Study of Landforms* iv. 131 The interfluves of Dartmoor, besides preserving remains of planation surfaces, show in detail features of interest. **1976** A. N. Strahler *Princ. Earth Sci.* xiv. 208/1 The floodplain is widened by further lateral planation.

So **plana·ted** *ppl. a.*, made level by erosion; also (as a back-formation) **plana·te** *v. trans.*

1912 *Jrnl. Geol.* XX. 450 Old planated surfaces..now dissected because of readjustments of drainage due to faulting. **1937** *Geogr. Jrnl.* XC. 56 The planated aspect of Waterpit Down from 900–850 feet suggests that this is a fragment of the erosion surface related to this bluff. **1954** W. D. Thornbury *Princ. Geomorphol.* xi. 286 Paige (1912) concluded that the processes of interstream and lateral erosion at the edges of alluvial fans would produce a sloping planated surface cut on bedrock which would become buried toward the center of a basin under a cover of gravel. **1969** *Trans. Inst. Brit. Geogr.* XLVIII. 44 Till and bedrock were planated to produce a surface sloping gently towards the Forth.

planc, var. *planh.

plancheite (pla·ñʃe,əit). *Min.* Also **planchéite.** [a. F. *planchéite* (A. Lacroix 1908, in *Compt. Rend.* CXLVI. 724), f. the name of M. *Planche*, who provided 'les meilleurs des matériaux étudiés': see -ite[1].] A blue, fibrous, hydrous silicate of copper.

1908 *Chem. Abstr.* II. 1675 Plancheite..is a new hydrous copper silicate and is found at the copper mine of Mindouli [French Congo]. **1920** *Brit. Mus. Return* 145 in *Parl. Papers* XXXVI. 673 Copper and uranium ores from Katanga, Congo, including..plancheite, dioptase, malachite, chrysocolla, &c. **1968** *Mineral. Abstr.* XIX. 54/1 Planchéite occurs in the iron deposit of Capo Calamita, Elba, where it was found in the form of small masses with a radial fibrous structure, associated with dioptase and chrysocolla. **1971** *Ibid.* XXII. 219/2 The planchéite formula may be written as a cupric amphibole, $Cu_8Si_8O_{22}(OH)_4$.

plancher, *sb.* Restrict *Obs.* or *dial.* to senses in Dict. and add: **4*.** (With pronunc. plañʃe and usu. written in italic.) In France, the minimum of Treasury bills which banks are obliged to hold. Cf. *floor *sb.*[1] 1 c.

1957 J. S. G. Wilson *French Banking Struct.* ii. xii. 342 Commercial bank rediscounts at the Bank of France.. were to be subject to *plafonds*.., and the availability of loanable funds to the private sector was also to be limited by a *plancher*. **1962** *Economist* 24 Nov. 813/2 The minimum ratio (*plancher*) of Treasury bills which banks are obliged to hold. **1964** *Financial Times* 31 Jan. 5/6 Only those bills which the banks wish to buy above their compulsory holdings or 'plancher' are subject to tender. The so-called 'maximum'..rates for 'plancher' holdings.. remain the same.

planchet. Add: **3.** (plæ·ntʃėt). *Physics.* Also **planchette.** A small, shallow dish used to contain a specimen when its radioactivity is measured.

1951 *Jrnl. Inst. Electr. Engineers* XCVIII. ii. 236/1 With end-window counters it is usual to carry the specimen as a powder, or liquid evaporated to dryness, on a suitable holder. J. L. Putman investigated the form of surface necessary to give a uniform response when using the G.M.2 counter-tube wherever the material was deposited on that surface. It was found that these surfaces were of approximately spherical shape, in a position concave towards the window. As a result of this work, a standard planchette, made of pressed nickel sheet, was designed to be used at a distance of 3 mm from the window. **1960** *Nature* 14 May 563/2 Synthesis of deoxyribonucleic acid was measured by counting the total radioactivity of washed cells on planchettes. **1961** G. R. Choppin *Exper. Nucl. Chem.* iv. 51 In case of reuse, each planchet should be carefully checked for residual activity. **1975** *Nature* 3 July 35/2 Radioactivity on the filter and in the frozen water vapour in the cold finger trap (after drying down on planchets) was assayed using a Nuclear Chicago gas flow counter.

planchette. Add: **2.** Also *Comb.*, as *planchette-board, -writer, -writing; planchette-like* adj.

1914 H. Carrington *Probl. Psychical Res.* xii. 371 There can be little doubt that the same force which propels the planchette board propels the ouija board also. **1972** D. Bloodworth *Any Number can Play* ix. 71 The planchette board circled wildly, both men keeping their fingers hard down upon it. **1920** *Q. Rev.* July 196 He could not see how so vigorous a creative faculty..could be only a vague Planchette-like state of possession. **1920** Planchette-writer [see *crystal-gazer* s.v. *crystal *sb.* B. 2 c]. **1884** F. W. H. Myers in *Proc. Soc. Psychical Res.* Dec. 232 The Spiritualist theory of Planchette-writing assumes the former of these two hypotheses. **1909** Planchette writing [see *ouija].

4. var. prec.

Planck (plæŋk). *Physics.* The name of Max K. E. L. *Planck* (1858–1947), German physicist, used in the possessive and *attrib.* to designate various concepts that he invented or discovered, as **Planck('s) constant,** one of the fundamental physical constants (symbol *h*), relating the energy *E* of a quantum of electromagnetic radiation to its frequency *ν* according to the equation $E = h\nu$; approximately $6{\cdot}626 \times 10^{-34}$ joule-second; **Planck('s) equation** or **formula,** any of the related equations stating Planck's law; **Planck('s) law,** a law giving the density of radiant energy at a particular wavelength *λ* or frequency *ν* inside a perfect radiator, and the flux of radiant energy emitted by it, the former being written as $\rho(\lambda)d\lambda = (8\pi hc/\lambda^5)(\exp(hc/k\lambda T)-1)^{-1}d\lambda$ or $\rho(\nu)d\nu = (8\pi h\nu^3/c^3)(\exp(h\nu/kT)-1)^{-1}d\nu$; **Planck oscillator**, a concept used by Planck in his work on radiation: an electrically charged particle that can execute simple harmonic motion with a frequency independent of the amplitude; **Planck('s) radiation formula** or **law** = *Planck('s) law*.

1910 *Phil. Mag.* XX. 244 Corpuscles which are emitted by bodies when exposed to ultra-violet light. Ladenburg found that the maximum energy of these corpuscles..was proportional to *n* the frequency of the light, being of the order $h'n/2\pi$ where *h'* is Planck's constant. 1935 D. L. Sayers *Gaudy Night* ii. 37 'There happened to be an unknown factor.' 'Like that thing that keeps cropping up in the new kind of physics,' said the Dean. 'Planck's constant, or whatever they call it.' 1940 Glasstone *Text-bk. Physical Chem.* i. 73 Every particle is associated with a wave.., and the wave length (λ) is related to the mechanical momentum (*mv*) by $\lambda = h/mv$, where *h* is the Planck constant. 1960 Chalmers & Quarrell *Physical Examination of Metals* (ed. 2) xvi. 749 This is known as the spin of the particle and is measured in terms of $h/2\pi$.. where *h* is Planck's constant. 1968 G. Ludwig *Wave Mech.* I. iv. 62 This uncertainty relation states that in any experimentally producible ensemble the product of the spread in position and momentum cannot go below a lower limit given by Planck's constant. 1911 *Bull. Bureau of Standards* (U.S.) VII. 393 Planck's equation for the intensity of radiation *J*, of wave length λ, from a black body at the absolute temperature θ,..appears to represent the results of all known observations. 1966 M. Ferro-Luzzi tr. *Fermi's Molecules, Crystals & Quantum Statistics* vii. 223 From the Planck equation, we can easily deduce Stefan's law, which also can be obtained by thermodynamic arguments alone. 1905 Ld. Rayleigh in *Nature* 13 July 244/1 In *Nature*, May 18, I gave a calculation of the coefficient of complete radiation at a given absolute temperature for waves of great length.., and it appeared that the result was eight times as great as that deduced from Planck's formula. 1923 Glazebrook *Dict. Appl. Physics* IV. 563/2 Throughout the spectrum from 0·5 μ to 50 μ Planck's formula fits the observed spectral energy distribution more closely than any other equation yet proposed. 1974 G. Reece tr. *Hund's Hist. Quantum Theory* ii. 27 By comparing the measured energy densities..with Planck's formula it was possible to calculate *h* and *h/k*. 1905 *Nature* 27 July 294/2 Planck's law is in good agreement with experiment if *h* is given a value different from zero. 1930 Ruark & Urey *Atoms, Molecules & Quanta* iii. 59 The physical basis for the asymptotic approach of Planck's law to the classical distribution is easily seen. 1955 W. Heisenberg in W. Pauli *Niels Bohr* 14 Bohr explained to him [*sc.* Schrödinger] that not even Planck's Law could be understood without the quantum jumps. 1974 Turner & Betts *Introd. Statistical Mech.* xii. 148 The Planck law describes the distributions of energy among the infinite range of possible frequencies. 1920 Planck oscillator [see *N I. 4 b]. 1966 tr. *S.-I. Tomonaga's Quantum Mech.* II. vi. 111 Each term in Eq. (47.9') has the form of the Hamiltonian of a Planck oscillator. 1909 *Sci. Abstr.* A. XII. 315 (*heading*) Planck's radiation law. 1911 *Ibid.* XIV. 404 It is concluded that the Einstein formula for specific heat is confirmed and also the Planck radiation formula. 1935 Pauling & Wilson *Introd. Quantum Mech.* xi. 301 The density of radiant energy is known to be given by Planck's radiation law as [etc.]. 1974 G. Reece tr. *Hund's Hist. Quantum Theory* xi. 145 The Bose statistics of light quanta was thus the same as that earlier applied by Planck for energy quanta..and thus led to the Planck radiation formula.

Planckian (plæ·ŋkiăn), *a.* *Physics.* [f. prec. + -ian.] Of, pertaining to, or being a black body.

1922 *Jrnl. Optical Soc. Amer.* VI. 560 Average noon sunlight..corresponds roughly to a black body temperature of 5000° K. the distribution not being strictly Planckian. 1956 H. H. Emsley *Aberrations of Thin Lenses* x. 303 A black body or Planckian radiator. 1972 *Nature* 28 Apr. 448/2 It seems improbable that the whole spectrum is Planckian, because the total power of such radiation would be..10^{42} erg s^{-1}.

‖**planctus** (pla·ŋktŭs). Pl. **planctus** (-ūs). [L., = beating of the breast, lamentation.] A medieval lament (Lament *sb.* 2). Cf. *planh.

1901 *Jrnl. Germanic Philol.* III. 417 From the various German Lamentations he [*sc.* A. Schönbach] culls eighteen versicles, mostly quatrains, which occur most frequently and of which the first thirteen are in form and in content a free version of the Latin Planctus. 1903 E. K. Chambers *Mediæval Stage* II. iii. xviii. 32 The metrical hymns are often of the nature of *planctus* or laments put in the mouths of the Maries as they approach the sepulchre... These *planctus* add greatly to the vividness and humanity of the play. 1907 *Mod. Philol.* IV. 605 It is hoped that the present discussion of the English planctus may in the future help to make more easily possible a comparative study of the planctus as a class. *Ibid.* 606 The chief purpose of this study is to discuss the several non-dramatic English planctus in their relation to each other. 1940 G. Reese *Music in Middle Ages* (1941) vii. 198 These *planctus*, with Latin texts, are believed actually to date from the 7th century and to be compositions, perhaps, of St. Eugenius. 1954 *New Oxf. Hist. Music* II. vi. 193 Although some of the most famous of the *planctus* of the Passion are solo stanzas, spoken by the Virgin Mary, yet there are a number in dialogue form, the speakers being most frequently Mary and St. John. 1964 C. S. Lewis *Discarded Image* iii. 36 Natura as Alanus brings her in, stiffly robed in rhetoric, conceit, and symbol, pleading again the cause of procreation in her *planctus* (against the sodomites). 1970 W. Apel *Harvard Dict. Mus.* (ed. 2) 461/2 A 'Planctus Karoli' lamenting the death of Charlemagne (814) and another *planctus* for his son Hugo (844). 1977 *Times Lit. Suppl.* 17 June 727/5 The cycle of six *planctus* by Peter Abelard.

plandok, var. *pelandok.

plane, *sb.*³ Add: **1. i.** A relatively thin structure used to produce an upwards or downwards († or sideways) force by the flow of the surrounding air or water over its surface. Orig. a flat surface (Plane *sb.*³ 1 b) proposed as a source of lift for heavier-than-air machines and used to direct the ascent and descent of balloons; later designed with a slight camber and used as the wing of an aeroplane or as a hydrofoil on a boat or seaplane. Cf. *aerofoil, *aeroplane 1, *hydrofoil 1, *hydroplane *sb.* 1.

Not now a common word exc. in the sense of *hydroplane *sb.* 1, and in combinations and derivatives (e.g. *biplane, *diving-plane, *interplane *a.*).

For the spelling *plain* (quots. *a* 1802, 1804) cf. Plain *sb.*¹ 4.

[*a* 1802 G. Cayley *Aeronaut. & Misc. Note-bk.* (1933) 10 In estimating the mechanical power which a given plain [*transcribed as* plane] will exert when exposed in any position to a current of fluid, two things are necessary. 1804 *Ibid.* 22, I made the following experiments upon the resistance of air to a surface of a foot sq, carried round with an horizontal motion upon an arm suspended upon a delicate hinge... The angles which the plain made with the horizon were measured.] 1809 —— in *Jrnl. Nat. Philos.* XXIV. 171 It is perfectly indifferent whether the wind blow against the plane, or the plane be driven with an equal velocity against the air. 1815 *Phil. Mag.* XLVI. 323 On October 2, 1815, another experiment..was made with a balloon six feet in diameter, having a square plane whose side was 7·5 feet, and a triangular rudder in proportion. 1816 *Ibid.* XLVII. 82 My object was to leave out the unwieldy bulk of balloons altogether, and to make use of the inclined plane propelled by a light first mover. 1842 W. S. Henson *Brit. Pat.* 9478 The first part of my Invention consists of an apparatus so constructed as to offer a very extended surface or plane.., which will have the same relation to the general machine which the extended wings of a bird have to the body when the bird is skimming in the air. *Ibid.*, The surface of the planes on either side of the car will measure four thousand five hundred square feet. 1848 *Chambers's Jrnl.* 6 May 301/1 When it attained the highest point, the edge of the plane would be reversed, and the balloon would descend thus. *Ibid.* 303/1 The wings are to be formed of long and narrow silk planes. 1866 *Ann. Rep. Aëronaut. Soc.* 25 A simple narrow blade, or inclined plane, propelled in a direct course..is..the only means of giving the maximum amount of supporting power with the least possible degree of 'slip'. *Ibid.* 36 To obtain the necessary length of plane ..the surfaces may be superposed, or placed in parallel rows, with an interval between them. 1891 S. P. Langley *Exper. in Aerodynamics* 58 The planes whose spread is largest in comparison with their extent from front to back..are therefore to be considered as being..the most favourable for mechanical flight. 1907 *Engineering* 4 Oct. 457/2 The boat is provided with hydroplanes only at its stem and stern. The planes at the bow are arranged in the manner of a V. 1908 *Aëronaut. Jrnl.* XII. 45/2 However sound in theory the single plane aëroplane may be, every serious accident yet recorded has occurred with this type. 1908 H. G. Wells *War in Air* iii. 82 He found the missing drawings of the lateral rotating planes, on which the whole stability of the flying-machine depended. 1910 [see *aeroplane 2 b]. 1912 M. Kerr in S. W. Murray *Poetry of Flight* (1925) 53 The tips of the planes appear and disappear As you madly drive along through the mist enladen'd air. 1915 A. Fage *Aeroplane* iv. 31 The rudder, a vertical plane capable of rotation about a vertical axis, partially controls the yawing or turning of the machine. 1917 A. W. Judge *Properties of Aerofoils* iii. 51 The inclined cambered plane differs from the inclined flat plane, in conforming better with the upward trend of the air about the leading edge. 1920 G. C. Bailey *Compl. Airman* viii. 59 The incidence of the tail plane can be varied. 1920 [see *hydrofoil 1]. 1938 E. W. C. Wilkins *Aeroplane Design* ii. 22 In the orthodox aeroplane, the main planes, or wings, are fixed. 1966 *McGraw-Hill Encycl. Sci. & Technol.* XIII. 212/2 Each set..may be tilted through an angle of 25° in either direction from the horizontal to develop a vertical force on the planes and thus on the submarine. 1972 J. B. Icenhower *Submarines* 9 At the bow and stern of a submarine are flat devices somewhat like the fins of a big fish. They are known as the *forward* and *after* hydroplanes... The diving planesmen tilt the forward planes down and the after planes up.

j. *Computers.* One of the flat, usu. square arrays of magnetic cores or other elements in a memory, each of which contains the corresponding bits of all the words held in the arrays.

1959 E. M. McCormick *Digital Computer Primer* viii. 107 In a magnetic core used to store, say, 4096 words of 36 binary bits each, there would be 36 sets (planes) of cores, and each plane would contain 4096 cores arranged with 64 on each side of a square. 1964 *IBM Jrnl. Res. & Devel.* VIII. 171/2 Figure 2 shows such a memory plane containing 50 tubes centered at intervals of 0·070 in. and 100 bit lines centered at intervals of 0·030 in. 1969 [see *core *sb.*¹ 10 b]. 1976 'Abd-Alla & Meltzer *Princ. Digital Computer Design* I. ix. 329 The planes are stacked into a three-dimensional array. The *x* lines of each plane are connected in series with the same *x* line on the two adjacent core planes.

4. *spec.* in *Theosophy.*

1884 *Trans. London Lodge Theosoph. Soc.* June 7 In considering the action of the law of Karma it is better to divide man into three planes; the physical, mental, and spiritual. 1889 H. P. Blavatsky *Key to Theosophy* iii. 45 That which is true on the metaphysical plane must be also true on the physical. 1892 —— *Theosoph. Gloss.* 255 Plane. From the Latin *planus*..an extension of space or of something in it, whether physical or metaphysical, *e.g.* a 'plane of consciousness'. 1922 Joyce *Ulysses* 139 A. E. the master mystic? That Blavatsky woman started it... A. E. has been telling some yankee interviewer that you came to him in the small hours of the morning to ask him about planes of consciousness. *Ibid.* 183 The Christ with the bridesister,..departed to the plane of buddhi. 1951 'Novo' *Notes on Theosophy* 13 We believe that beyond, yet interwoven with, physical matter, are other planes of such a tenuous nature that they are not apparent to the human eye. 1974 *Encycl. Brit. Macropædia* XVIII. 277/2 Most modern theosophists subscribe to a rather elaborate cosmogony... There are, it is believed, seven worlds or planes through which the universe evolves. In ascending order these are the physical plane; the emotional, or astral, plane; the mental plane; the intuitional, or Buddhic, plane; the spiritual, or Atmic, plane; the monadic, or Anupadaka, plane; and the divine, or Adi, plane. 1977 'L. Egan' *Blind Search* i. 6 It was reasonable to suppose that he still was, and still concerned with the people who'd meant something to him here—planes or levels of vibration or whatever.

plane (plēⁱn), *sb.*⁵ Also **'plane**. [f. Aero)plane.] = *aeroplane 2 b.

1908 *Aëronaut. Jrnl.* Apr. 45/1 The aëroplane was then taken to the Longchamps end of the field, and as soon as the propeller had been set in motion the apparatus dashed off towards Neuilly. After running along the ground for about a hundred mètres the plane lifted, and.. rushed through the air for 150 mètres or thereabouts. 1908 *Times* 1 June 6/1 Mr. Wright refused to give any details on the propeller employed, but on the general construction of the plane he said it was full of movable diversely articulated parts. 1909 Kipling *With Night Mail* 69 Low-flying planes often 'glue up' when near the Magnetic Pole. 1909 Lloyd George in *Daily Chron.* 23 Aug. 1/1, I have not yet crossed the Financial Channel with my Budget 'plane. 1910 *Daily Mail* 27 July 6/5 To the builders of aeroplanes he cries: 'Construct me planes capable of the maximum speed.' 1920 *Blackw. Mag.* June 762/1 A plane which came from Palestine. 1931 *Daily Mirror* 27 Aug. 2/2 The 'plane struck the water. 1932 *Daily Express* 27 June 8/2 The first ape and the comic sergeant are deserting to Switzerland in a bombing 'plane. 1942 R. Hillary *Last Enemy* i My plane had been fitted out with a new cockpit hood. 1958 'Castle' & 'Hailey' *Flight into Danger* ii. 30 There was a brief shudder as the plane freed herself from a wall of cloud. 1965 *Movie* Summer 3/2 Charlotte and Robert talk and make love in the hour before his 'plane leaves. 1976 *Daily Tel.* 7 Oct. 1/6 All 73 passengers and crew of a Cuban DC-8 airliner were believed lost when the plane plunged into the Caribbean.

2. *attrib.* and *Comb.*, as *plane crash* sb. (and vb. intr.), *fare*, *journey*, *-load*, *park*, † *pilot*, *-ride*, *-spotter*, *ticket*; **planeside** *U.S.*, an area beside an aeroplane; also *attrib.*; **plane time**, the time of departure of an aircraft on a scheduled flight.

1946 *Time* 14 Oct. 69/1 Four years before Knute Rockne plane-crashed to death in Kansas, Irishman Frank Leahy came to Notre Dame. 1957 P. Worsley *Trumpet shall Sound* x. 200 These movements soon spread further into the Highlands, thriving on the.. alarm created by the war:..plane-crashes and so on. 1972 J. Aiken *Butterfly Picnic* i. 13 Her parents had been killed in a plane crash. 1969 B. Malamud *Pictures of Fidelman* iii. 69 I'll need plane fare. 1973 'E. McBain' *Let's hear It* v. 68 The Puerto Ricans came, and some of them stayed only long enough to earn plane fare back to the island. 1974 E. Ambler *Dr. Frigo* ii. 84, I..have no trouble at all with long plane journeys... I always sleep soundly. 1951 R. Malkin *Boxcars in Sky* 25 A French schoolhouse was able to be rushed to completion in record time following the flight of a planeload of components of a prefabricated school manufactured in England. 1969 C. Booker *Neophiliacs* x. 261 Planeloads of American gamblers flying in to..'the European Las Vegas' [*sc.* London]. 1976 *Evening Standard* 29 Dec. 26/5 Representatives of the emerging nations descend by the planeload. 1936 'J. Beynon' *Planet Plane* 41 The crowds began to pour from the 'plane-parks and car-parks. 1916 'Boyd Cable' *Action Front* 132 The 'plane pilot..was well out of range. 1953 Dylan Thomas *Let.* 22 June (1966) 408, I almost liked the plane-ride, though. 1973 *Black Panther* 4 Aug. 15/2 Nixon,..in a plane ride from Mobile to Birmingham in 1971 with Wallace, had persuaded the charismatic Alabaman to run as a Democrat. 1968 *N.Y. Times* 28 Mar. 3 In a planeside interview, General Abrams said [etc.]. 1968 *Sat. Rev.* (U.S.) 31 Aug. 6, I walk from planeside to a taxi. 1976 *6,000 Words* 156 Speaking briefly at planeside. 1978 *Fortune* 4 Dec. 101 (*heading*) To planeside by bus. 1960 *Guardian* 12 Mar. 6/5 The prowess of London Airport's plane spotters is likely to become comparable with the best of the train spotters. 1975 S. Johnson *Urbane Guerilla* iii. 152 A crowd of sightseers and plane-spotters. 1977 *Daily Mirror* 21 Mar. 13/2 Five plane spotters serving jail terms in Greece will get a spot of home comfort today. 1967 M. Drabble *Jerusalem the Golden* viii. 204 Her plane ticket ..was booked from Le Bourget. 1974 D. Westheimer *Olmec Head* ii. 18 Plane tickets, hotel rooms already set. 1962 L. Deighton *Ipcress File* v. 30 The typewritten sheet gave plane times. 1973 'B. Mather' *Snowline* ix. 112 I'll be able to get you an air ticket... You had better stay here until plane time. 1976 K. Bonfiglioli *Something Nasty in Woodshed* ix. 96, I slept until 'plane-time this morning.

plane, *a.* Add: **3. plane-parallel** *a.*, both plane and parallel.

1903 *Amer. Jrnl. Sci.* XVI. 114 A rock-mass possessing the plane-parallel structure. 1958 *Newnes Compl. Amat. Photogr.* 97 In order not to disturb the lens correction filter glasses have to be plane-parallel and optically finished. 1962 Corson & Lorrain *Introd. Electromagn. Fields* iv. 148 (*caption*) Grounded, plane-parallel electrodes terminated by a plane electrode at potential V_0.

plane, *v.*² **1.** Delete *rare* and add further examples. (In mod. use with the idea of an aeroplane's flight.)

1941 I. L. Idriess *Great Boomerang* i. 2 A black dot appeared in the brazen sky. It grew, planed down and alighted on the needlewood-tree beside the first crow. 1953 R. Lehmann *Echoing Grove* 22 A shape of silence, planing stealthily from nowhere, crossed the churchyard: a huge cream-coloured owl. 1953 J. Cary *Except the Lord* lxii. 286 A few gulls planed high overhead but made no sound. 1955 *Times* 3 Aug. 10/2 As he [*sc.* a marsh harrier] began to plane down over the water two members of his young family raced each other to meet him in the air. 1978 R. Lewis *Uncertain Sound* vi. 154 The herring gulls planed among the bobbing masts of the fishing fleet. 1979 G. Hammond *Dead Game* ix. 107 Another [goose].. had gone on away over the sands, first planing and then running.

2. *intr.* To travel in an aeroplane; † to glide.

1908 *Daily Mail* 10 Aug. 5/4 Safety would reside in high flight; it would always be possible to 'plane' to earth, and in 'planing' the machine would progress many more feet than it would fall. 1909 *Westm. Gaz.* 9 Aug. 5/1 Mr. Orville Wright has stated that he and his brother are completing the perfecting of their aeroplane... With this apparatus he says one will be able to 'plane' to one's heart's content. 1909 *Daily Chron.* 26 Aug. 1/2 His engine began to show signs of distress. The aviator was seen to slow down, and then he 'planed gracefully to the earth. 1912 S. F. Walker *Aviation* viii. 66 He can plane down. Planing down is merely gliding. 1940 *Daily Progress* (Charlottesville, Va.) 13 Aug. 1/4 Little Carolyn ..will plane out for the movie capital from the nation's capital on Thursday morning, accompanied by her mother. 1967 J. P. Carstairs *No Thanks for Shroud* ii. 28, I had planed into the large air terminal at Los Angeles.

3. *intr.* Of a seaplane, boat, etc.: to skim the surface of a body of water as a result of lift produced hydrodynamically.

1913 [see *hydroplane *sb.* 2]. 1914 *Techn. Rep. Advisory Comm. Aeronaut.* 1912–13 243 The position of the centre of buoyancy when the float is at rest is far ahead of the centre of pressure when it is 'planing' and the machine about to fly. 1919 A. W. Judge *Handbk. Mod. Aeronaut.* xix. 943 Hollow Vee sections keep the spray down, cut the water more easily and cleanly, plane better, [etc.]. 1942 *R.A.F. Jrnl.* 13 June 8 There is great danger in..level ditchings where the nose strikes first. The nose is neither shaped nor strong enough to plane along the surface. 1954 K. C. Barnaby *Basic Naval Archit.* (ed. 2) 318 When a hard chine hull is planing correctly, only the under body below the chines is in direct and constant contact with the water. 1963 J. T. Rowland *North to Adventure* i. 15 It was a big sail for a ten-foot punt. She seemed to leap out of the water; I believe she planed. 1972 C. Mudie *Motor Boats* 48 A fast boat on the verge of planing or actually planing builds up a pressure on the under surfaces of the hull which, like walking on harder ground compared with bog, increases the firmness of the footing and hence the stability. 1972 R. Abbott *Sci. of Surfing* iii. 52 The technique of modern surfing is based on the fact that boards plane easily and efficiently. Planing is the term used to describe the way in which a surfboard rises onto the water surface and skims along at high speed. 1974 *Encycl. Brit. Macropædia* II. 1171/1 Because the displacement hull..can never plane on the surface no matter how much power is applied, the efforts of designers were directed toward the development of hulls..that at speed would rise to the surface and skim across the water, thus reducing..the friction and resistance.

4. *trans.* In Surfing, to ride (a wave) with the hands protecting the face. *Austral.*

1963 B. Hutchings in J. Pollard *Swimming—Austral. Style* 122/1 To 'plane' a wave, you hold your hands together in front of your head and take off in this position as the wave nears... The trick is to arrange the spear formed by the hands in such a way that the water passes along the side of the face and torso and not into the face.

Hence **pla·ning** *vbl. sb.*² and *ppl. a.* (chiefly in sense *3 of the vb.).

1908 [see *aviate *v.*]. 1914 *Techn. Rep. Advisory Comm. Aeronaut.* 1912–13 Its most obvious defect was a certain slowness in rising to the planing position. 1919 A. W. Judge *Handbk. Mod. Aeronaut.* xix. 944 The bottom abaft the step should rise strongly, as this favours a steepening of the planing bow before suction is eliminated. 1920 *Flight* XII. 301/1 Other types of 'planing' boats. *Ibid.* 591/1 The angle of the planing bottom at the keel, forward of the step, should be $1\frac{1}{2}$ degrees to the datum line. 1937 *Jrnl. R. Aeronaut. Soc.* XLI. 273 Improvements have been made recently in the shape of afterbody of the planing bottom with the object of obtaining better aerodynamic efficiency. 1967 *Jane's Surface Skimmer Systems* 1967–68 110/1 The hull..has a hard chine..and planing hull form. 1972 R. Abbott *Sci. of Surfing* iii. 52 For planing speed to be maintained and so that the wave can be ridden along its length it is necessary to turn fairly soon. 1972 C. Mudie *Motor Boats* 48 A planing boat has a quite remarkable stability when running fast. 1974 J. Keats *Of Time & Island* xii. 191 When you pushed forward on a throttle..the boat mounted onto its planing step and you flew over the river. 1976–7 *Sea Spray* (N.Z.) Dec./Jan. 78/1 This is no more evident than in the development of power craft, the rise of the planing hull and the steady move over the past few years into tunnel-hulls, hydrofoils, hovercraft and so on.

planeful, 'planeful (plēⁱ·n,ful). [f. *plane *sb.*⁵ + -ful.] As much or as many as an aeroplane will hold.

1958 *Times* 14 Aug. 9/4 There is no reason why a 'planeful of them [*sc.* monkeys] should be any different. 1967 M. Davis *Strange Corner* (1968) vii. 58 A planeful of new guests.

planeness. Delete *rare* and add later examples.

1896 *Jrnl. Geol.* IV. 955 Overwash plains may sometimes depart from planeness by taking on some measure of undulation. 1906 Chamberlin & Salisbury *Geol.* III. xix. 345 Neither planeness nor unevenness can be ascribed exclusively to the stratified nor to the unstratified drift. 1942 F. Twyman *Prism & Lens Making* ix. 124 The departures from planeness of the wave front..will give rise to a contour map of the corrections which have to be applied to the lens. 1975 D. G. Fink *Electronics Engineers' Handbk.* ix. 26 Since the periphery of the grid is colder than the center, thermal stresses tend to cause buckling or departure from planeness.

planer. Add: **6. planer-miller** = *plano-miller* s.v. *plano-¹.

1943 J. R. Connelly *Technique Production Processes* iii. 85 (*caption*) Planer-miller. A good case of comb[in]ing the features of several standard tools for special work. 1971 C. R. Hine *Machine Tools & Processes* xiii. 308 The planer-miller is a hybrid of the two machines, but because of its size and construction (with a crossrail) it can rightfully be classified as a planer.

planer-tree. Substitute for def.: A small, deciduous tree, the water elm, *Planera aquatica*, belonging to the family Ulmaceæ and native to south-eastern parts of the United States. (Earlier and later examples.)

1810 F. A. Michaux *Hist. Arbres Forestiers de l'Amérique Septentrionale* I. 39 *Planer tree*, nom de la personne à laquelle cette espèce a été consacrée. 1819 A. L. Hillhouse tr. *Michaux's N. Amer. Sylva* III. 100, I have more particularly observed the Planer Tree in the large swamps on the borders of the river Savannah in Georgia. 1832 D. J. Browne *Sylva Amer.* 246 The planer tree is of the second order, and is rarely more than 35 or 40 feet high. 1930 W. R. Mattoon et al. *Forest Trees Oklahoma* 62 The water elm or planer tree is found on the low wet flood plains of the larger streams of the eastern part of the state. 1976 *Hortus Third* (L. H. Bailey Hortorium) 882/2 Planer tree, water elm... A deciduous elmlike tree, native to N[orth] Amer[ica].

planeshear, planksheer. (Earlier example of spelling *planksheer*.)

1851 G. Coggeshall *Voy.* iii. 40 The force of the sea broke one of the top timbers or stancheons, and split open the plank-sheer.

planet, *sb.*¹ Add: **2.** (A ninth major planet, *Pluto* (*Pluto¹ 1), is now known.)

3*. A planet wheel.

1912 R. W. A. Brewer *Motor Car Construction* xii. 154 If one sun wheel is held, the whole of the planets with their star piece move bodily in a circle when the other sun wheel is revolved. 1928 V. W. Pagé *Mod. Aircraft* xi. 474 Various methods of compounding plain epicyclic gears have been tried, but the best type is undoubtedly that combining double planets, an annulus driven from the crankshaft, and a sun fixed to the engine casing. 1962 D. W. Dudley *Gear Handbk.* iii. 15 With some ratios it has been possible to squeeze in as many as twenty planets. 1970 *A.A. Bk. of Car* 110/1 In the simple epicyclic gear, a pair of planets revolve on spindles supported by the U-shaped planet carrier, which is mounted on the same shaft as the sun wheel.

4. *planet-spotted* adj.; **planet cage**, a cylindrical form of planet carrier; **planet carrier**, the frame on which the planet wheels are mounted in a planetary gear; **planet earth** (without *the*, and usu. with one or both initials capitals), the earth as the particular planet on which man lives; **planet-gear**, add to def.: a planet wheel; (examples); **planet pinion**, a planet wheel, esp. one smaller than the sun wheel; **planet shower**, a local shower (cf. Planet *sb.*¹ 1 c); **planet stirrer** = *planetary stirrer* s.v. *planetary *a.* and *sb.* A. 1 f; **planet-wheel** (earlier and later examples); **planet-wide** *a.*, occurring all over the planet, as extensive as the planet.

1908 Planet cage [see *planet pinion* below]. 1947 *Jrnl. R. Aeronaut. Soc.* LI. 100/1 This usually leads to the adoption of an epicyclic gear with its associated problems of planet cage design and high centrifugal loadings. 1956 Molloy & Lanchester *Automobile Engineer's Ref. Bk.* xii. 60 The short planet gears rotate round the internal ring gear, in the same direction as the input shaft, thus rotating the planet carrier and attached output shaft at a reduced speed. 1976 Leeming & Hartley *Heavy Vehicle Technol.* vi. 117/2 If the planet carrier is braked and the sun wheel driven, the annulus is driven in a reverse direction—the planet wheels being idler wheels only—and reverse ratio is obtained. 1965 J. H. Jackson *Pictorial Guide to Planets* v. 34 (*heading*) Planet earth. 1976 L. Deighton *Twinkle, Twinkle Little Spy* viii. 80 We should simply seek to make a mark in the universe.. that some other civilization will detect and so know there is..sophisticated life on planet Earth. 1978 *Listener* 12 Jan. 54/1 The hopeful television producers..who have invaded planet earth. 1979 *Guardian* 18 Aug. 10/8 The SF buff..believes that Planet Earth is done for... He wants more money spent on space. 1916 J. E. Homans *Automobile Handbk.* iii. 42 As soon as the engine starts—there being no clutch necessary on a car with such apparatus—the two spurs keyed to the main shaft..rotate with it, driving the 'planet' gears in mesh with them. 1956 [see *planet carrier* above]. 1971 B. Scharf *Engin. & its Lang.* xii. 161 The planet gears are mounted on pins attached to a common frame, the planet carrier. 1908 *Daily Chron.* 14 Nov. 8/6 Greater attention is being paid to the elimination of internal friction from these devices, as in the provision of ball bearings for the planet pinions in the Sturmey Archer gears, and roller bearings for the planet cage in the Armstrong. *a* 1935 [see *back-lash a]. 1966 *McGraw-Hill Encycl. Sci. & Technol.* X. 273/2 The number of teeth on the planet pinion of a simple planetary gear does not enter into the equations for speed ratio because the pinion engages both sun and ring gears. 1853 Mayne Reid *Rifle Rangers* (rev. ed.) lvii. 288 We were treated each day to some five or six hours of a 'planet' shower. 1880 W. H. Patterson *Gloss. Words Antrim & Down* 78 *Planet showers*, short heavy showers. 1925 E. Sitwell *Troy Park* 39 Not medicines planet-spotted like fritillaries For country sins and old stupidities. 1902 C. Salter tr. *G. von Georgievics's Chem. Technol. Textile Fibres* 249 Stirring is effected by so-called planet stirrers. 1827 J. Farey *Treat. Steam Engine* i. vi. 449 The link causes the centre of the planet-wheel to travel in a circular orbit..when it revolves round the sun-wheel. 1912 R. W. A. Brewer *Motor Car Construction* xii. 153 The large bevel wheel is bolted to a casing, which holds firmly a star piece having four arms on each of which runs a planet wheel... These four planet wheels engage with two sun wheels. 1976 Planet-wheel [see *planet carrier* above]. 1969 *Listener* 14 Aug. 215/2 It is now evident that on Mars, the craters are planet-wide. 1974 *Icarus* XXII. 239 (*heading*) Martian planetwide crater distributions.

plane-table, *v.* Add: Also *intr.*, to work with a plane-table.

1883 H. H. Godwin-Austen et al. *Hints to Travellers* (ed. 5) i. 109 There is no measuring, no counting of paces or noting of time by a watch, no anxiety about the record, when Plane-Tabling.

plane-tabler, -tabling (examples).

1871 Plane-tabling [in Dict.]. 1888 *Min. Proc. Inst. Civil Engin.* XCII. 190 When a triangulation exists of the country to be mapped, the plane-tabler makes immediate use of these established points. 1923 D. Clark *Plane & Geodetic Surveying* II. vi. 238 Plane tabling is the most extensively used method for the survey of topography. 1950 J. Clendinning *Princ. & Use of Surveying Instruments* vii. 166 Plane-tabling is a graphical method of survey in which the map is rough-drawn in the field as the survey proceeds. 1965 *Textbk. Topogr. Surveying* (Min. of Defence) (ed. 4) xi. 205 The trigonometrical framework will have been surveyed... The plane tabler then proceeds to break down from this framework using methods similar to those employed by the triangulator. 1971 R. J. P. Wilson *Land Surveying* viii. 164 In the past plane tabling was the method used for supplying topographical detail for maps at scales of 1:10 000 to 1:250 000, but its use in this respect has largely..been superseded by air survey.

planetal, *a.* Delete † *Obs.* and add later example.

1908 *Encycl. Relig. & Ethics* I. 187/1 The planetal series of our days of the week places Sunday before Monday.

planetarium. Add: **a.** (Earlier example.)

1734 J. T. Desaguliers *Course Exper. Philos.* I. 430 A short Description of my Planetarium, an Instrument.. to shew the Motion of the heavenly Bodies.

d. A device for projecting an image of the night sky at various times and places on to the interior of a dome for public viewing; a building housing this. Also *attrib.* and *fig.*

1929 *Encycl. Brit.* XVII. 1000 Planetarium is the name given to an arrangement made by Zeiss of Jena, for producing an artificial sky. By optical methods images of the sun, moon, planets and stars are projected on a large hemispherical dome and by mechanical and electrical means the apparatus can be revolved so as to show the principal motions. 1950 *Engineering* 21 July 63/1 A planetarium..building will have..the dome representing the sky..diameter 60 feet. The simulation of the sun, moon, planets and stars..projected on the 'sky' will have great educational value. 1958 *New Scientist* 20 Mar. 6/3 Zeiss Planetaria have been operated in 32 cities. 1958 [see *astronavigation]. 1963 V. Nabokov *Gift* iii. 166 Now on the [chess] board there shone, like a constellation, a ravishing work of art, a planetarium of thought. 1973 C. Sagan *Cosmic Connection* viii. 60 Several million people visit planetariums in North America and Britain each year.

planetary, *a.* and *sb.* Add: **A.** *adj.* **1.** (Further examples.)

planetary electron, an electron bound to an atom and 'in orbit' round its nucleus; *planetary engineering* (see quots. 1951, 1964); *planetary nebula* (earlier examples); *planetary precession*: see *precession 3 a.

1785 *Phil. Trans. R. Soc.* LXXV. 266 A very bright, planetary nebula, about half a minute in diameter, but the edges are not very well defined. 1802 [see Nebula 3]. 1921 *Phil. Mag.* XLII. 305 Consider the nucleus alone, and not the surrounding system of planetary electrons. 1927 N. V. Sidgwick *Electronic Theory of Valency* i. 10 The other planetary electrons would be distributed between these limits. 1951 A. C. Clarke *Exploration of Space* 118 The greatest technical achievements of the next few centuries may well be in the field of what could be called 'planetary engineering'—the reshaping of other worlds to suit human needs. 1956 I. Asimov *Inside Atom* ii. 31 The number of planetary electrons in an ordinary atom is equal to the number of protons in the nucleus. 1960 *Analog Science Fact/Fiction* Oct. 34/1 We are coming off Mass-Time to go on planetary drive. 1964 *Listener* 15 Oct. 575/1 It will be possible to modify the climates and atmospheres of at least some of the planets, so that we can live on them... This technique of the future has been called 'planetary engineering'. 1971 *Nature* 3 Dec. 246/2 Why should planetary scientists suggest that a spacecraft costing nearly $100 million should be destroyed by taking it so close to Jupiter?

e. Involving, being, or forming part of a sun-and-planet gear, usu. having (in addition to the sun and planet wheels) an internally geared annulus coaxial with the sun wheel and with which the planet wheels are meshed.

1904 T. H. White *Petrol Motors* ii. 108 When the pinion A is revolved, and the internally toothed ring B is held from revolving, the planetary pinions C are caused to run around the ring B and carry the plate D with them. 1910 [see *band clutch* s.v. *band *sb.*² III]. 1934 *Jrnl. R. Aeronaut. Soc.* XXXVIII. 738 This specification describes a planetary reduction gear for aero engines in which an internally toothed annulus is driven by the engine, the propeller is driven by the planetary ring, [etc.]. 1948 [see *epicyclic *a.*]. 1956 A. Huxley *Adonis & Alphabet* 127 The simple planetary gears, by means of which conventional turntables can be used for slow-playing disks. 1969 *Jane's Freight Containers* 1968–69 583/3 Torque converter is standard. All drive wheels planetary drive powered. 1975 *Sci. Amer.* Dec. 120/2 The upper end of the larger planetary gear engages a spur gear fixed to a turntable.

f. *Engin.* Characterized by the circular motion of a part about a point outside it.

planetary mill, a heavy rolling mill for reducing hot strip in a single pass, the strip being forced between two large rolls each of which has a number of smaller work rolls around its circumference, the former being rotated in the direction of feed so that the latter rotate against the strip; *planetary mixer* or *stirrer*, one in which paddles are rotated about an axis which itself is moved in a circular path.

1917 T. R. Shaw *Precision Grinding Machines* ii. 28 Grinding machines with planetary spindles, specially adapted to the requirements of locomotive building. 1949 S. E. Rusinoff *Manuf. Processes* x. 405, (*A*) of Fig. 17 illustrates the principle by which holes are ground in a locomotive side rod on a planetary grinder. A gyratory motion is imparted to the grinding wheel spindle so that the wheel sweeps the bore of the workpiece. 1950 Kirk & Othmer *Encycl. Chem. Technol.* V. 705 In a planetary stirrer, the paddle rotates and at the same time the axis about which it rotates follows a circular orbit. *Ibid.* 707 Small-scale laboratory and pilot-plant models of planetary mixers..are available. 1953 *Engineer* 23 Oct. 526/2 The first commercial installation of a new design of hot strip rolling mill..has been installed at the works of Ductile Planetary Mills, Ltd., of Willenhall, Staffs. *Ibid.*, The slabs have to be pressure fed into the rolls..by pinch or feed rolls, which are mounted in the same housing as the planetary assemblies. *Ibid.* 527/2 Immediately following the planetary mill is a two-high planishing mill. 1963 F. H. Habicht *Mod. Machine Tools* xi. 166 Planetary grinding is usually limited to large or awkward workpieces that cannot be conveniently rotated. 1967 Irving & Saxton in Uhl & Gray *Mixing* II. viii. 214 Vertical-shaft mixers..include planetary mixers and Pony mixers, as well as heavier duty, twin-shaft machines. 1968 R. N. Parkins *Mech. Treatm. Metals* iv. 213 The great advantage of the planetary mill is that it can reduce a hot slab directly to strip, thereby replacing the three or four roughing mills and the six-stand finishing mill. 1971 C. R. Hine *Machine Tools & Processes* xiv. 330 The planetary miller is unique in that the work is held stationary while the revolving cutter or cutters move in a planetary path to finish a circular surface on the work, either internally or externally.

B. *sb.* **3.** *ellipt.* for *planetary nebula*.

1903 A. M. Clerke *Probl. in Astrophysics* ii. i. 175 Spectroscopically, they [*sc.* Novae] simulate minute 'planetaries'. 1974 *Nature* 31 May 430/1 The precise details of the evolutionary history of the planetaries are uncertain.

4. *ellipt.* for *planetary gear* or *wheel* (= *planet-gear*, *-wheel*, in Dict. and Suppl.).

1941 *Electronic Engin.* XIV. 166/1 The planetary of the differential is connected to a gear-reduction train. 1962 D. W. Dudley *Gear Handbk.* iii. 21 Bevel planetaries can be made to handle a range of ratios.

planetesimal (plænė̆te·simăl), *a.* and *sb.* *Astr.* [f. Planet *sb.*¹ + Infinit)esimal *sb.* and *a.*] **A.** *adj.* Pertaining to, involving, or composed of planetesimals; applied *esp.* to the hypothesis that the planets were formed by the accretion of a vast number of planetesimals in a cold state.

1904 T. C. Chamberlin in *Carnegie Inst. Year Bk.* II. 263 This led to studies upon alternative hypotheses. Among these is the conception that the earth, instead of descending from a gaseous spheroid, may have been built up by the gradual ingathering of its material from a scattered meteoroidal or planetesimal condition. 1904 *Amer. Geologist* XXXIII. 95 The planetesimal hypothesis ..seems much better to explain both the astronomical and geological phenomena. 1906 *Athenæum* 18 Aug. 191/1 For the last ten years Prof. Chamberlin, aided by Dr. Forest R. Moulton, of Chicago, has been developing what is called the Planetesimal Theory of the earth's origin. 1937 Wooldridge & Morgan *Physical Basis Geogr.* i. 5 The earth..grew from small beginnings by the addition of planetesimal matter. 1969 *Nature* 19 July 259/1 When planetologists meet to discuss the surface of the Moon they separate into those favouring the dominance of planetesimal (meteoric) impact and those who advocate volcanic activity.

B. *sb.* A small solid body following a planetary orbit; a miniature planet.

1904 *Amer. Geologist* XXXIII. 95 The new hypothesis holds..that the globular planets were formed by the slow accretion or infalling of cold, discrete bodies or particles ('planetesimals'). 1906 Chamberlin & Salisbury *Geol.* II. ii. 94 The planetesimals originated, by hypothesis, from gaseous matter shot forth from the ancestral sun. 1952 H. C. Urey *Planets* vii. 219 Planetesimals probably formed simultaneously with the protoplanets and accumulated into larger objects. 1971 *Sci. Amer.* Oct. 52/3 *Apollo 14* landed on the Fra Mauro formation, believed to be a blanket of ejecta thrown out by the giant meteorite or planetesimal that excavated the basin of Mare Imbrium. 1973 *Nature* 21/28 Dec. 451/3 Courten believes a planetesimal 'between 80 and 500 miles in diameter' moves in an orbit about 0·1 AU from the Sun. 1977 *Ibid.* 8 Dec. 506/1 Accumulation and fragmentation of planetesimals may have competed in the planetary accretion process, and a planetesimal which could survive catastrophic destruction may have become a planet.

planetfall (plæ·nėtfǭl). [f. Planet *sb.*¹ + *-fall* as in Landfall.] A landing upon a planet after a journey through space.

1954 'J. Christopher' *22nd Cent.* 98 The *Lucas* did not normally make planet-falls. *Ibid.* 123 Smoking was strictly prohibited from take-off until planet-fall. 1960 K. Amis *New Maps of Hell* (1961) i. 19 A few writers.. arrange for their travellers to put themselves into..deep freeze until just before planetfall. 1974 M. Caidin (*title*) Planetfall.

planetismal (plænėti·zmăl), *a.* and *sb.* *Astr.* [Alteration of *planetesimal *a.* and *sb.*] = *planetesimal *a.* and *sb.*

1910 *Westm. Gaz.* 16 Apr. 12/3 He accepts what is known as the 'planetismal hypothesis', which is, in his opinion, the most positive advance in natural science which has been made for a very long time. 1938 *Nature* 5 Nov. 843 The planetismal hypothesis is the parent of the more recent tidal theories. 1970 *Sci. Jrnl.* May 32/1 This theory has it that the Moon formed out of the coagulation of a 'sediment-ring' of planetismals which originally circled the Earth.

planetkin (plæ·nėt,kin). *nonce-wd.* [f. Planet *sb.*¹ + -kin.] A small planet.

1832 Carlyle *Reminisc.* (1881) I. 44 A temporary fraction of this planetkin, the whole round of which is but a sandgrain in the all.

planetocentric (plænėtose·ntrik), *a.* *Astr.* and *Astronautics.* [f. Planet *sb.*¹ + -o + *-centric.] Referred to, measured from, or having a planet as centre (usu. a planet other than the earth.)

1960 Baker & Makemson *Introd. Astrodynamics* xii. 264 The regions in space where one would expect the vehicle to be in a predominantly geocentric, heliocentric, or planetocentric field, can be delineated by calculating the ratio of the perturbative accelerations to the two-body acceleration. 1971 S. Herrick *Astrodynamics* I. v. 106 A similar consideration of the planetocentric velocity at the Mars or Venus end of the trajectory..indicates that it is desirable to have the tangency at Mars's aphelion or at Venus's perihelion. 1972 *Nature* 11 Aug. 325/2 The heliocentric orbital longitude of Mars η, and the planetocentric longitude of the Sun are approximately related by $\eta = L_s + 85°$.

planetography. Delete † *Obs. rare*⁻⁰ and add examples.

1735 B. Martin *Philos. Gram.* 119 (*heading*) Of planetography, or the philosophy of the planets. 1960 *Analog Science Fact/Fiction* Oct. 33/1 Got to know enough about it through an elementary planetography.

planetokhod (plæ·nėtŏkǫd, -χǫd). Also Planetokhod. [a. Russ. *planetokhód*, f. *planéta* Planet *sb.*¹ (after *lunokhod).] A Russian self-propelled vehicle for transmitting information about another planet as it travels over its surface.

1970 *Times* 18 Nov. 1 The vehicle is called Lunokhod-1. .. Soviet scientists are predicting that other such vehicles, named Planetokhod or Marsokhod, will eventually move over the surface of the Planets. 1973 *Nature* 23 Mar. 219/2 His remarks relating to the possible 'planetokhod' exploration of Venus and the outer planets seem highly speculative.

planetolatry (plænėtǫ·lătri). *rāre.* [f. Planet *sb.*¹ + -latry, -olatry.] Idolatrous worship of the planets.

1964 C. S. Lewis *Discarded Image* v. 104 Despite this careful watch against planetolatry the planets continued to be called by their divine names.

planetology (plænėtǫ·lŏdʒi). *Astr.* [f. Planet *sb.*¹ + -ology.] The study of the planets and their evolution.

1907 P. Lowell in *Century Mag.* Nov. 113 Planetology we may call this science of the making of worlds, since it concerns itself with the life-history of planetary bodies from their chemically inert beginning to their final inert end. 1959 *Wall St. Jrnl.* 25 July 8/2 'We believe lunar science, or planetology, is here to stay,' he says. 1973 *Nature* 18 May 121/1 Planetology, with several tens of papers on the Moon, Mars and the Allende Meteorite, profited particularly this year.

Hence **planetolo·gic, planetolo·gical** *adjs.*, of or pertaining to planetology; **planeto·logist**, an expert or specialist in planetology.

1908 *Century Mag.* Feb. 505/1 This gives us a most instructive glimpse into one planetologic process. 1933 *O.E.D. Suppl.*, Planetologist. 1966 *Ann. N.Y. Acad. Sci.* CXL. 289 For the planetologist, it seems very important to be able to measure, even approximately, the relative amounts of oxygen, silicon, and aluminum in the surface layers of the planets. 1969 [see *planetesimal *a.* and *sb.* A]. 1975 *Nature* 7 Aug. 455/3 The development of a cratering chronology was one of the principal challenges to comparative planetologists studying impact bombardment of the terrestrial planets. 1976 *Ibid.* 22 Jan. 176/1 G. Wetherill's..time scale and planetological-geological framework for the early Precambrian seemed generally accepted.

planet-struck, *a.* (Later *poet.* example.)

1925 A. Huxley *Sel. Poems* 48 Let me..Dream planet-struck.

planform (plæ·nfǭɹm). Also **plan-form, plan form.** [f. Plan *sb.* + Form *sb.*] The shape or outline of an aircraft wing in plan. Also *transf.*

1908 F. W. Lanchester *Aerodonetics* x. 319 When the author's standard combination of parabolic grading and elliptical plan form is used, it is a property of the resulting aerofoil that [etc.]. 1913 *Aeronaut. Jrnl.* XVII. 83 In advocating this plan-form he does not appear to have had stability in his mind at all. 1942 T. P. Faulconer *Introd. Aircraft Design* i. 18 The choice of a wing plan form and thickness is governed in general by the range requirements. 1958 *Engineering* 21 Mar. 363/1 The main building has an interesting planform of a question mark. 1965 *Times* 11 Sept. 7/6 Where opinions differ is on the wing planform for take-off, for approach and for landing. 1967 *Jane's Surface Skimmer Systems* 1967–68 2/1 The Aeromar A-1 is a gas turbine powered craft with a circular planform. 1975 L. J. Clancy *Aerodynamics* xvi. 538 Consider a wing whose planform is represented..by the sketch.

plangency. (Further examples.)

1923 C. Mackenzie *Parson's Progress* xvi. 222 All the regret..was expressed in the plangency of the violin speaking of the individual's grief. 1928 Kipling *Limits & Renewals* (1932) 18 It's not a bad couplet in itself. Did you see how he admires the 'plangency' of it? 1972 *Times Lit. Suppl.* 25 Aug. 991/3 The Dreyfus affair..had an unusual public plangency. 1977 *N.Y. Rev. Bks.* 29 Sept. 6/4 The strain of plangency and nostalgia, combined with indignation, is Irish.

plangent, *a.* Add: **2.** Also *fig.*

1928 Kipling *Limits & Renewals* (1932) 14 The freshness, the fun, the humanity, the fragrance of it all, cries—no, shouts—itself as Dan's work. Why 'Daiespringe mishandled' alone stamps it from Dan's mint. Plangent as doom, my dear boy—plangent as doom! 1936 [see *facture 4].

plangently (plæ·ndʒĕntli), *adv.* [f. Plangent *a.* + -ly².] In a way that beats strongly or distressingly on the mind or feelings.

1927 R. L. Mégroz *Three Sitwells* 9 We are driven inwards because the external reality we have created is plangently ugly. 1928 *Observer* 19 Feb. 9/2 Nothing is here to make us beat the breast. The old matchless rhythms are no less plangently certain. 1963 *Times* 12 June 13/3 The theme of African unity so plangently struck at the Addis Ababa conference.

‖**planh** (planˣ). Also **planc.** [Provençal, f. L. *planctus* *planctus.] A mournful troubadour song. Cf. *planctus.

1843 *Penny Cycl.* XXV. 307/2 Their 'Planhs', or songs on the death of a mistress. 1878 F. Hueffer *Troubadours* i. xiii. 134 Two minor branches of the *sirventes*.. are the *planh* or complaint, and the crusader's song,..the former belonging more especially to the personal..class of poems... The *planh* is a poem written on the death of a mistress, a friend, or a protector. 1909 E. Pound *Exultations* 46 (*title*) Planh for the Young English King. 1923 H. J. Chaytor *Troubadours & England* ii. 92 The troubadour Austorc de Segret in a *planh* upon the death of St. Louis, written between 1270 and 1274, refers to the problems before Edward. 1964 'E. Queen' *Four Men called John* (1976) xiv. 154 A number of twelfth-century secular manuscripts. There are: six *planhs*, apparently the work of Bertran de Bon, [etc.]. 1970 W. Apel *Harvard Dict. Mus.* (ed. 2) 461/2 A late 12th-century troubadour *planc* (*planh*) by Gaucelm Faudit deplores the death of Richard the Lion-Hearted.

plani-. Add: **planispiral** *a.* (later examples).

1945 M. F. Glaessner *Princ. Micropalaeont.* iv. 69 Perfectly planispiral coiling produces bilaterally symmetrical tests. 1975 *Nature* 3 Apr. 419/1 Sinistral..and planispiral..coiling are uncommon among gastropods.

planigale (plæ·nigēⁱl, plænigēⁱ·li). *Austral.* [mod.L. (E. Le G. Troughton 1928, in *Rec. Austral. Mus.* XVI. 282), f. Plani- + *Phasco)gale*, the name of a closely related genus.] A flat-skulled marsupial mouse of the genus so called, belonging to the family Dasyuridæ and native to Australia and New Guinea.

1946 E. Troughton *Furred Animals of Australia* (ed. 3) 29 (*heading*) Planigales or Flat-skulled Marsupial-Mice. 1966 *New Scientist* 20 Jan. 130/2 The little planigales or flat-skulled 'mice' found in Australia and New Guinea. 1970 W. D. L. Ride *Guide to Native Mammals of Austral.* 119 David Fleay has kept Ingram's Planigales in captivity. 1974 *Courier-Mail* (Brisbane) 15 Apr. 10/3 (*caption*) Planigale mother with some of her clinging brood of babies.

planimetrically (plēⁱnime·trikăli), *adv.* [f. Planimetric *a.*: see -ically.] By means of, or with regard to, planimetry. (In quot. 1944, 'in the plane'.)

1944 *Burlington Mag.* Dec. 296/1 The essential difference lies elsewhere: previously border decoration was planimetrically designed, whereas in the new style the border ornament lives in the space above the page. The illusion is created that branches, flowers, insects and birds etc. have been dropped on the page, loosely dispersed, casting their shadow on the coloured foil. 1948 Lysholm & Bull in J. W. McLaren *Mod. Trends Diagnostic Radiol.* xxiv. 322 The composite pneumographs have also been measured planimetrically in relation to the corresponding brain cross-sections. 1959 *Trans. & Papers Inst. Brit. Geogr.* No. 26. 31 A planimetrically correct landform map has been constructed from an existing contour map..for a mountainous area in Colorado. 1969 G. C. Dickinson *Maps & Air Photographs* xi. 166 The profiles are truly 'lines on the ground' and must therefore be in their correct position planimetrically, so that detail can be traced directly from the map and will fit the profiles.

planing, *vbl. sb.* **3. planing machine, mill** (earlier examples).

1840 *Digest of Patents U.S., 1790–1839* 346 Planing machine... June 1, 1805. 1851 C. Cist *Sk. Cincinnati in 1851* 227 Planing machines..made..by B. Bicknell. 1844 *Knickerbocker* XXIV. 184 The uplifted arm of Labor ..meets his eye in the..planing mill.

plank, *sb.* Add: **2.** Also, a surf-board.

1784 J. King in Cook *Voy. Pacific Ocean* III. v. vii. 146 Whenever..the impetuosity of the surf is increased to its utmost height, they [*sc.* the natives of Karakakooa] choose that time for this amusement [riding the surf]... If by mistake they should place themselves on one of the smaller waves, which breaks before they reach the land, or should not be able to keep their plank in a proper direction on the top of the swell, they are left exposed to the fury of the next, and, to avoid it, are obliged again to dive and regain the place, from which they set out. 1962 *Austral. Women's Weekly* 24 Oct. (Suppl.) 3/3 Plank, any type of surfboard. 1963 *Pix* 28 Sept. 63 Five extra points if you can fit eight surfers, eight planks and a mattress in the woodie. 1967 J. Severson *Great Surfing Gloss.*, *Plank*, name given to heavy boards, usually referring to the redwood giants ridden prior to the 1950s.

5. For 'Cf. Platform 7 b' read 'Cf. Platform *sb.* (*a.*) 9 b' and for 'Orig. and chiefly *U.S.*' read 'orig. *U.S.*'. (Further U.K. examples.)

1873 Ld. Salisbury in *Q. Rev.* CXXXV. 558 Neither is it necessary now to dwell on those questions which are occasionally discussed by speculative politicians, but which..are either too small or too large to be regarded as a plank in any party's platform. 1894 *Liberal* 24 Nov. 42/1 They have founded a Society, the sole plank of whose platform is 'Hands off, please'. 1926 Galsworthy *Silver Spoon* ii. i. 117 Dared he tackle the air—that third plank in the Foggart programme? 1937 *Ann. Reg.* 1936 189 Without cease the German Government protested against the alliance between Soviet Russia and France, alleging that the pact was a danger to Germany. That was one plank in the platform of German foreign policy. 1965 *Listener* 20 May 755/2 Thinkers whose ideas were fairly far removed from any of the chief party political planks. 1970 *Daily Tel.* 2 Feb. 1 The enforcement of law and order is to be one of the main planks of the Conservatives' General Election campaign. 1977 *Cornish Times* 19 Aug. 8/2 He told me that the main plank of his election campaign was going to be a policy of discrimination as advocated by an extreme political group, which was then defacing private and public buildings with stickers and misspelt slogans.

6. *to walk the plank* (earlier examples).

1822 Scott *Pirate* III. xii. 281 They deserve to be made to walk the plank for their impudence. 1835 J. E. Alexander *Sketches in Portugal* viii. 179 The admiral.. worked late and early himself, and made every body under him work, or else 'walk the plank'.

7. *plank bridge* (examples), *-raft*; **plank-buttress** [tr. G. *plankengerüst* (A. F. W. Schimper *Pflanzengeographie* (1898) iii. iv. 328)], a development of a root at the base of the trunk of certain tropical trees; **plank-owner** *Navy slang* (chiefly *U.S.*), (*a*) a member of the original crew of a ship; a marine with long service with his ship or unit; (*b*) a marine with a light task; **plank steak**, steak cooked and served on a piece of plank; cf. *planked *ppl. a.* 2.

1933 J. Buchan *Prince of Captivity* ii. iv. 247 He crossed the stream by a plank bridge. 1979 G. Mitchell *Mudflats of Dead* iii. 35 He took to the causeway, crossed the plank bridge. 1903 W. R. Fisher tr. *Schimper's Plant-Geogr.* iii. i. 304 Much more frequently these buttresses assume the form of plank-like outgrowths of the base of the trunk and of the uppermost roots, and they may be termed plank-buttresses. 1952 P. W. Richards *Tropical Rain Forest* i. 4 Plank buttresses.. are a highly characteristic feature of rain-forest trees. 1960 N. Polunin *Introd. Plant Geogr.* xiv. 467 The dominants are often of particularly massive growth and rich branching, but devoid of plank-buttresses. 1901 *Our Naval Apprentice* (U.S.) Aug. 14 'Patsy' is a 'Plank Owner' on 'Constellation'. 1920 *Our Navy* (U.S.) Apr. 11 Some of the plank-owners think the navy would be a great outfit if it didn't have any ships in it. 1945 *Richmond* (Va.) *News Leader* 31 May 6/3 The majority of our crew was made up of 'plank owners'—men who had been aboard the Colhoun since she..had been commissioned. 1952 A. Geer *New Breed* 6 The 'plank-owners' (any Marine not going) were subjected to a barrage of good-natured insults as they stood on the pier. 1967 M. Dibner *Admiral* xiv. 140 He became her first gunnery officer as a 'plank owner'..at her commissioning. 1910 *Westm. Gaz.* 24 Jan. 5/2 People..had to make use of boats or plank-rafts. 1959 *Good Food Guide* 93 Plank steak, shashlik, scampi maison, and 'nest of chicken' are among its specialities. 1972 *Vogue* Jan. 16/3 Try the spare ribs or prime plank steak. 1975 *Islander* (Victoria, B.C.) 27 Apr. 5/4 We celebrated our arrival by ordering 'plank steak'... It was juicy and tender and served on a white piece of wood.

plank, *v.* Add: **2. a.** (Further examples.)

1936 J. Tickell *See how they Run* iv. 46 How would you like to be half-starved for a bit and then planked down in a foreign school, aged twelve, and have to spend your holidays with the Geometry master? 1938 A. J. Liebling *Back where I came From* 182 An overstuffed chair some admirer had planked down next to the ticket booth. 1964 *Perthshire Advertiser* 13 June 14 Planking the lady into a beach chair and carrying her..to the other side.

b. For *U.S. colloq.* read *colloq.* (orig. *U.S.*), and add earlier and later examples. Also *absol.*

1824 *Nantucket Inquirer* 19 Apr. 2/4 His guardy was sent for, and he planked the cash. 1848 W. E. Burton *Waggeries & Vagaries* 65 If the nigger..can plank up if he's cast, I'm darned if I don't..sue the nigger. 1903 Somerville & 'Ross' *All on Irish Shore* 178 Every one squared up his books and planked ready money down on the nail. *Ibid.* 185 People began to talk then, especially as the pony's look and shape were improving each day, and after a little time every one was planking his money on one way or another. 1915 W. S. Maugham *Of Human Bondage* lxxv. 389, I planked out the money to keep you. 1951 *New Yorker* 1 Dec. 63/1 (Advt.), You plank the cash on the counter for a slice of sirloin. 1966 H. Kemelman *Saturday Rabbi went Hungry* xxi. 127 When you ask someone to plank down a hundred and fifty-odd bucks for a lot which he doesn't think he's going to need..it's a lot of dough. 1972 *Even. Telegram* (St. John's, Newfndl.) 5 Aug. 3/1 How could a poor man..plank down $70,000 in ready cash for a place to live in?

4. See also *planked *ppl. a.* 2.

planked, *ppl. a.* Add: **1.** (Later examples.)

1873 J. H. Beadle *Undevel. West* xxxvi. 769 As one result of their smooth planked streets, much attention is given to fine turn-outs. 1956 R. W. Andrews *Glory Days of Logging* 23 (*caption*) Planked road of A. & M. Logging Co. Ltd...shown running from timber to log dumps. 1979 S. Smith *Survivor* vi. 75 The planked roof of the barn.

2. (Earlier example.) Also of meat. Also, served on a piece of plank.

1855 *Sun* (Baltimore) 30 Apr. 4/1 Did you ever eat a 'plank'd shad'? 1906 *N.Y. Globe* 27 Apr. 7 The planked chicken was served on the plank. 1910 *Chambers's Jrnl.* July 430/2 In the restaurants the British visitor will invariably be confronted with the possibilities contained in..'planked steak',..and so on. 1947 Wodehouse *Full Moon* iv. 83 An order of planked steak. 1969 R. & D. De Sola *Dict. Cooking* 177/1 *Planked steak*, restaurant term for a steak broiled and served on a well-seasoned plank and garnished with a border of mashed potatoes, mushroom caps, tomato slices, and sometimes julienned carrots. 1978 *Lancashire Life* Nov. 170/1 The dish in question, one of the popular specialities of the house, was described as planked steak: a reasonably substantial T-bone, prepared on that slab of oak, and accompanied by broccoli, mushrooms, tomatoes, fried onion rings and the most delicious Duchesse potatoes imaginable.

plankter (plæ·ŋktəɹ). *Biol.* [f. Plankt(on + -er¹.] = *planktont.

1938 J. R. Carpenter *Ecol. Gloss.* 208 *Plankt*, *plankter*, individual organisms comprising plankton. 1957 G. E. Hutchinson *Treat. Limnol.* I. xvii. 899 The only autotrophic truly open-water plankter so far encountered that requires accessory organic substances other than vitamin B_{12}, the marine diatom *Ditylum*. 1973 *Nature* 21/28 Dec. 521/1 This mechanism augmented by a rapid turnover in biomass, especially of small plankters, could account for the low residue concentrations in Gulf plankton following periods of light precipitation.

planktology (plæŋktǫ·lŏdʒi). [a. G. *planktologie* (E. Hæckel 1891, in *Jenaische Zeitschr. f. Naturwiss.* XXV. 240): see Plankto(n and -ology.] The study of plankton. So **planktolo·gical** *a.*, of or pertaining to this study; **plankto·logist**, one engaged in this study.

1893 G. W. Field tr. *Hæckel's Planktonic Stud.* in *Rep. U.S. Comm. Fisheries* 1889–91 571 The whole science which treats of this important division of biology is briefly called planktology. 1896 *Proc. Acad. Nat. Sci. Philadelphia* 280 Without undervaluing in any way the counting methods at present employed by planktologists, I desire here to call attention to an apparatus..by means of which one may make a large number of plankton estimations in a single day. 1912 *Rep. Brit. Assoc.* 1911 422 The Kiel planktologists have had to seek another source of food for the zooplankton. 1926 *Kongel. Danske Vidensk. Selsk. Skr.* 8th Ser. XI. 157 Those who have followed the history of planctological [*sic*] work during the last twenty years will know that it is really along these three lines that limnologists have especially worked in this area of exploration. 1947 *Nature* 4 Jan. 10/2 The technical difficulties of this type of work, particularly in planktology, are not belittled. 1967 *Oceanogr. & Marine Biol.* V. 231 A brief historical summary is necessary in order to understand the obstacles which Mediterranean planktology has had to overcome. *Ibid.* 248 This instrument becomes of questionable planktological value. 1972 *Nature* 7 Apr. 295/1 This information..will also be of help to physical oceanographers and planktologists.

plankton. For (V. Hensen 188.) in etym. substitute (V. Hensen 1887, in *Ber. Kommission der wissenschaftlichen Untersuchung der deutschen Meere in Kiel* V. 1). (Earlier and later examples.)

1891 *Jrnl. R. Microsc. Soc.* 326 'Plankton' was originally defined by Hensen as including those animals which drift in the sea. 1908 *Jrnl. Marine Biol. Assoc.* VIII. 269 (*title*) Plankton studies in relation to the western mackerel fishery. 1921 [see *benthos]. 1947 *Sci. News* IV. 98 The vast bulk of life in the open sea is composed not of active creatures such as fish or whales, but of microscopic plants, and small animals which drift with the water. They are known collectively as plankton. 1956 A. Hardy *Open Sea* I. xv. 297 The herring snaps at the little plankton animals individually. 1973 *Nature* 16 Nov. 128/2 The existence of a vast aerial plankton of insects and other arthropods extending up into the sky for at least 14,000 foot is well established.

2. *Comb.* **plankton feeder**, an animal whose diet includes plankton; **plankton indicator**, an apparatus that is towed behind a ship with a filtering device by means of which the concentration of plankton can be estimated; **plankton net**, a very fine net used to collect samples of plankton or other very small organisms; **plankton recorder**, a modification of the plankton indicator in which the filter is in the form of a continuously moving roll.

1956 A. Hardy *Open Sea* I. xv. 303 Mackerel are plankton-feeders for about half the year. 1975 C. F. Hickling *Water as Productive Environment* vii. 65 The species of *Alestes*..came to dominate the surface waters as plankton feeders. 1925 A. C. Hardy in *Fishery Investigations* 2nd Ser. VIII. vii. 2 The present paper.. describes investigations by means of an instrument which has been called the Plankton Indicator. 1936 *Jrnl. Marine Biol. Assoc.* XXI. 148 Preliminary experiments were begun in 1922 and 1923 with a torpedo-shaped instrument called the Plankton Indicator. 1953 *Bull. Marine Ecol.* IV. 19 The small Plankton Indicator..has been used..during an ecological survey of the herring off the north-east coast of Scotland. 1952 J. Clegg *Freshwater Life Brit. Is.* xviii. 293 Some form of pond-net is almost essential. Probably the type of most general utility is that known as a plankton-net... The bag part is made of fine-mesh material, and carries at its base a tube or bottle into which the organisms descend. 1963 G. E. & R. C. Newell *Marine Plankton* ii. 17 There are several patterns of plankton net in common use today. 1972 F. G. Stehli et al. in T. J. M. Schopf *Models in Paleobiol.* vi. 119 The mesh of the plankton nets used was large. 1926 A. C. Hardy in *Nature* 30 Oct. 631/1 Whilst on the *Discovery* expedition I have been experimenting with such an instrument, which I am calling the Continuous Plankton Recorder... It is a development of the simple Plankton Indicator.., but in place of the silk netting discs, which had to be reloaded for each sample, I have substituted a long continuously moving roll operated by a propellor turned by the water through which it is towed. 1936 '*Discovery*' *Rep.* XI. 457 The first Continuous Plankton Recorder was used on the R.R.S. 'Discovery' in the years 1925–7. 1975 D. H. Cushing *Marine Ecol. & Fisheries* viii. 166 The plankton recorder survey..is based on monthly samples from fixed merchant ship lines across the North Sea.

planktonic (plæŋktǫ·nik), *a.* and *sb.* *Biol.* [a. G. *planktonisch* (E. Hæckel 1891, in *Jenaische Zeitschr. f. Naturwiss.* XXV. 240): see Plankton and -ic.] **A.** *adj.* Of, pertaining to, or characteristic of plankton.

1893 G. W. Field tr. *Hæckel's Planktonic Stud.* in *Rep. U.S. Comm. Fisheries* 1889–91 571, I adopt the term Plankton in place of '*Auftrieb*', and derive from it the adjective planktonic. 1899, 1905 [in Dict.]. 1930 *Times Educ. Suppl.* 22 Mar. (Home & Classroom Suppl.) p. iv/1 Quantities of nektonic and planktonic life are being caught daily. 1963 *Times* 19 Feb. 10/4 Sediments of past ages..contain a boundary clearly defined by changes in their content of fossilized planktonic remains. 1974 *Nature* 8 Feb. 393/2 Herbivorous zooplankton can graze on planktonic algae, bacteria and detrital particles.

B. *sb.* A microfossil of a foraminifer included in the plankton.

1959 *Bull. Amer. Paleontol.* XXXIX. 84 These assemblages of agglutinated forms..occur several times and are separated by assemblages of calcareous forms, often with abundant planktonics, suggestive of a depositional depth between 200 and 600 metres. 1964 *Micropaleontology* X. 3/2 The coiling characteristics of planktonics can be most useful in correlating strata from different basins and with entirely different benthic biofacies. 1976 *Ibid.* XXII. 420/1 The total number of planktonics, benthics, and fragments was counted.

planktonology (plæ:ŋktǫnǫ·lŏdʒi). [f. Plankton + -ology.] = *planktology. So **planktonolo·gical** *a.*

1896 *Jrnl. R. Microsc. Soc.* 470 Dr. C. S. Dolley referring to the work of Hensen, Haeckel, and others on planktonology, explains the importance of a quantitative determination of the primitive food-supply of marine animals. 1960 *Biol. Abstr.* XXXV. 2290/1 (*title*) The new systematics and planktonology. 1961 *Ibid.* XXXVI. 284/2 (*title*) Results of planktonological research. 1975 *Jrnl. Fish. Res. Bd. Canada* XXXII. 2231 Narrower fields of specialization, such as planktonology, fisheries ecology, or studies on the benthic habitat, should be integrated. 1976 *Biol. Abstr.* LXI. 2576/1 (*title*) Physical-chemical and planktonological studies of the Great Lake of Laffrey.

planktont (plæ·ŋktǫnt). *Biol.* [f. PLANKT(ON + Gr. ὄν, ὀντ- being: see ONTO-.] An individual organism of the plankton.

1897 *Science* 3 Dec. 830/1 The struggles of the imprisoned organisms and the pressure of the filtering water also materially assist the escape of planktonts through the yielding meshes of silk. **1926** *Kongel. Danske Vidensk. Selsk. Skr.* 8th Ser. XI. 152 Some of the freshwater planctonts [*sic*]..were made to serve as a support for the theories of heredity. **1931** *Jrnl. Ecology* XIX. 246 Multiplication of the littoral planktonts in the open water will naturally only take place if the latter contains an adequate amount of mineral nutriment. **1935** P. S. WELCH *Limnology* ix. 205 The term planktont of the older literature should be abandoned because of its faulty word structure. **1948** *New Biol.* V. 21 Blue-green planktonts are generally most abundant..in summer and late autumn.

planktotrophic (plæ:ŋktotrǫ·fik), *a.* [f. PLANKTO(N + TROPHIC *a.*] Feeding on plankton. So **plankto·trophy,** behaviour of this type.

1946 G. THORSON in *Medd. Komm. Danmarks Fiskerei Ser. Plankton* IV. 1. 476 Planktotrophic larvae with long pelagic life..originate from small eggs poor in yolk. **1963** R. P. DALES *Annelids* viii. 171 Most planktotrophic larvae develop from small eggs. **1973** *Amer. Naturalist* CVII. 348 Planktotrophy is the most common pattern in shallow-water tropical invertebrates. **1978** *Nature* 5 Jan. 56/2 Most of the commoner species of coral reef asteroids ..produce many small, planktotrophic, pelagic larvae.

planless, *a.* (Further examples.)

1937 'G. ORWELL' *Road to Wigan Pier* iv. 51 Little brick houses..festering in planless chaos. **1942** J. STEINBECK *Moon is Down* ii. 25 Fine weapons and fine planning against unarmed, planless enemies. **1957** [see *CONTRACT *v.* 2 d].

planlessly *adv.* (later example); **planlessness** (further examples).

1932 *Times Lit. Suppl.* 1 Sept. 609/2 It has the merit, certainly, that almost any plan has over planlessness. **1944** I. ORIGO *Diary* 13 Feb. in *War in Val d'Orcia* (1947) 140 It is odd how used one can become to uncertainty for the future, to a complete planlessness. **1962** *Punch* 30 May 844/1, I have already let several valuable years slip planlessly by. **1976** *Sunday Times* (Lagos) 7 Nov. 12/2 The planlessness of the flashes should take us to the most unlikely place where the story should begin, the epilogue.

planned, *ppl. a.* Add: (Further examples.) *planned economy*: an economy in which industrial production and development, etc. are determined by an overall national plan; *planned obsolescence*: obsolescence of manufactured goods due to deliberate changes in design, cessation of the supply of spare parts, use of poor-quality materials, etc.

1931 *Economist* 18 July 111/2 The tendency in the world is towards some or other form of planned economy. **1936** *Discovery* Sept. 295/1 Can present human motives work a planned society? **1942** *N.Y. Times* 6 Mar. 23/7 The Planned Parenthood Federation of America, Inc., is the new corporate name of the Birth Control Federation of America, Inc., according to announcement yesterday by the organization's board of directors. **1943** J. S. HUXLEY *TVA* xiii. 115 The transition from a *laissez-faire* to a planned economy. **1947** *Sun* (Baltimore) 20 Sept. 1/2 The Senator's attack on what he calls the Truman Administration's policy of 'planned inflation'. **1960** I. BENNETT *Delinquent & Neurotic Children* ix. 444 *History*: a planned baby. Both parents wanted a girl. **1965** B. PEARCE tr. *Preobrazhensky's New Economics* 159 The further abolition of the law of value..will proceed along the path of planned socialist organization of the economy in countries which make an end of the capitalist régime. **1966** *Punch* 20 July 96/3 The planned-obsolescence men won't miss *this* one. **1969** A. CAIRNCROSS in *Advancement of Sci.* XXVI. 64/2 It is possible to argue that there is no essential difference between a managed economy and a planned economy, and that planning is simply a rather inflexible form of management. **1977** *National Observer* (U.S.) 1 Jan. 10/3 Things settled down into a buyer's market again, so the businessmen came up with another new gimmick. It was called planned obsolescence. Come out with a new model every year. Put a few new gadgets on it... Don't worry about the quality, it will be obsolete in three years anyway, besides, it's cheap. **1977** J. AIKEN *Last Movement* i. 17 Dru had been a planned child, whereas I was an unexpected..afterthought. **1977** *Spare Rib* May 20/2 On Ash Wednesday..a newly opened Planned Parenthood Clinic in St Paul was fire bombed.

plannee (plænī·). [f. PLAN *sb.* or *v.* + -EE[1].] A person for whom something is planned.

1943 J. S. HUXLEY *TVA* xv. 119 He must not think of the people in his region as his subject plannees, but as participating co-planners. **1946** A. HUXLEY *Science, Liberty & Peace* (1947) I. 28 A highly organized and regimented society..is felt by the planners, and even..by the plannees to be more 'scientific', and therefore better, than [etc.].

planner. Add: *spec.*, a person who plans the development or reconstruction of an urban area, or who engages in economic planning.

1935 *Economist* 11 May 1075/1 It is true to say that the task of the planner is the more effective mobilisation of all economic resources. **1961** L. MUMFORD *City in Hist.* vii. 185 Aristotle's position..is sounder than that of most of our present-day planners, who have not yet arrived at a functional definition of a city. **1962** *Listener* 3 May 758/2 If the French planners could prove..that their efficient techniques and the understanding with the heads of industry have proved to be the answer to all problems of economic growth [etc.]. **1973** *Times* 13 Dec. 19/2 There is an absolute shortage of planners. **1976** *Encounter* June 93/2 At present, the experts and polymaths, the thinkers and highly skilled professions, the essential organisers and planners, have no collective voice.

b. A list, table, or similar device giving information which enables one to plan.

1971 *Homes & Gardens* Sept. 99/3, I will send you a copy of our Do-It-Yourself diet planner and a calorie counter, so that you can organise a diet to suit your needs.

planning, *vbl. sb.* Add: (Examples in the sense of the action of planning civically or economically.)

1935 *Economist* 30 Mar. 725/1 Since private enterprise has manifestly failed to bring about the necessary adjustments in industry, 'planning' must be tried. **1941** *New Statesman* 15 Feb. 151 If we are to rebuild our cities aright, we ought to plan them; and planning is inconsistent with rebuilding on the old sites. **1959** *Cambridge Rev.* 25 Apr. 429/2 The ancient City could survive if Town and University were enclosed together within a ring of traffic roads and could not be entirely traversed except on foot or bicycle... These are all matters of what is drolly called 'Planning'. **1962** *Listener* 1 Mar. 363/1 Planning, if one must risk a description, is a method of introducing more coherence, more purpose, into an economic system. *Ibid.* 364/2 The planning of private industry.

b. *attrib.*, as *planning appeal, application, authority, blight, committee, consent, consultant, control, engineer, officer, permission.*

1971 *Reader's Digest Family Guide to Law* 174 In some planning appeals cases..the inspector has powers to make the decision without referring the matter to the Minister. *a* **1974** R. CROSSMAN *Diaries* (1975) I. 615, I found when I arrived that eight or nine months was accepted as a reasonable time for a planning appeal to wait in the Ministry before a decision. **1976** *S. Wales Echo* 25 Nov. 2/4 Over lunch reference was made to their particular planning application. It was obvious they knew what the planning officer's recommendation was. **1934** *Act* 23 & 24 *Geo. V* c. 58 s. 20 (1) 'Planning authority' means..the authority having power to control the development..of.. land. **1950** *Chambers's Encycl.* XIII. 705/2 Extensive powers for the compulsory purchase of land..have also been made available to planning authorities. **1962** L. GOLDING *Dict. Local Govt.* 301 'Planning blight'..arises when property becomes virtually unsaleable by its owner or saleable only at a low price because of a threat of development. **1976** *Liverpool Echo* 23 Nov. 7/3 The people of Garston have suffered far too long from the planning blight that has caused serious deterioration in the area affecting housing, shops and the environment in general. **1942** *Country Life* 9 Oct. 692/2 That is the London revealed as it would be by the Planning Committee of the Royal Academy. **1964** *Oxf. Jun. Encycl.* X. 457/1 Applications are submitted to the planning committee, and only when permission has been given can work proceed. **1970** *Financial Times* 13 Apr. 25/2 Ambitions for a 2,000-room palace of tourism in West London foundered on the rocks of planning consents. **1977** *Lancs. Life* Nov. 82/2 You could find yourself in trouble for demolishing, adding to or altering property without planning consent. **1944** J. S. HUXLEY *On Living in Revol.* xi. 115 The services of planning consultants and a resident planning engineer. **1961** E. A. POWDRILL *Vocab. Land Planning* ii. 10 Planning control arises from the fact that persons, groups of persons, and organisations wish to initiate some form of development. **1944** Planning engineer [see *planning consultant* above]. **1961** H. W. DODDS in *56th Ann. Rep. Carnegie Foundation for Advancem. of Teaching* 22 A plan must therefore be brought up to date periodically, possibly with the assistance of a permanent planning officer. **1965** in P. Jennings *Living Village* (1968) 103 Planning permission had already been given, the Planning Officer being convinced that they were a brilliant architectural achievement. **1976** *Liverpool Echo* 23 Nov. 10/5 Mr. Len Ward, planning officer for Ellesmere Port Council, is urging the highways committee to take immediate action over the cottages. **1947** *Act* 10 & 11 *Geo. VI* c. 51 s. 119 (1) 'Planning permission' means the permission for development which is required by virtue of section twelve of this Act. **1965** Planning permission [see *planning officer* above]. **1977** *Undercurrents* June–July 4/1 The Windscale Public Enquiry into British Nuclear Fuel Limited's..application for planning permission to build an uranium-oxide processing plant begins on June 14th.

plano-[1]. Add: **plano-miller, plano(-)milling machine,** a milling machine built in the manner of a planer and used esp. for heavy work, having a flat bed to carry the work-piece and a sliding cross-piece that carries rotating cutters as in an ordinary milling machine, rather than a planing tool.

1906 J. G. HORNER *Mod. Milling Machines* v. 130 Plano-Millers or Slabbing Machines.—This is a name that seems most appropriate to designate that large and growing group of machines which is built on the model of the common planing machine, with bed, table, housings, and cross rail. **1963** *Gen. Engin. Workshop Practice* (ed. 3) vi. 218/2 The plano-miller is a milling machine, but designed to execute certain work formerly confined to the planer. **1905** T. R. SHAW *Machine Tools* vii. 478 Universal plano-milling machines are now constructed of any size with either one or two saddles carrying spindles on the cross-slide, just as with the tool-boxes of planing machines. **1964** S. CRAWFORD *Basic Engin. Processes* vi. 154 A plano milling machine possesses several advantages as compared with the planing machine, including the reduction of non-cutting time, wider range of operations in one setting, and increased rate of production.

plano (plē̆i·no), *a.* [f. PLANO-[1].] Of a surface of a lens: flat.

1950 *Jrnl. Optical Soc. Amer.* XL. 523/1 The plano surfaces can introduce central coma by being tilted. **1962** L. S. SASIENI *Princ. & Pract. Optical Dispensing* x. 268 Many modern trial cases are made with plano-convex and plano-concave spheres... The powers are engraved sometimes on the plano side of the convex and on the concave side of the concave.

planogamete. (Further examples.) Hence **planogame·tic** *a.*

1950 E. A. BESSEY *Morphol. & Taxon. Fungi* i. 6 The two gametes may both be..motile (planogametes), as in the..Chytridiales. **1971** P. H. B. TALBOT *Princ. Fungal Taxon.* vii. 91 A planogamete is a motile gamete, or sex cell, and planogametic conjugation..is the fusion of two gametes, one or both of which may be motile.

planographic (plē̆inogræ·fik), *a.* [f. PLANO-[1] + -GRAPHIC.] Of, pertaining to, or produced by a process in which printing is done from a plane surface.

1897 SINGER & STRANG *Etching* 121 The relief print has no plate mark, the intaglio print has one quite clear and distinct, the planographic one has a very slight mark. **1914** E. H. RICHTER *Prints* 10 The last group to be considered, planographic processes, is based entirely upon chemical and physical action. **1946** H. WHETTON *Pract. Printing & Binding* xxv. 287/1 Planographic is the term used to describe a printing surface on the same level as the plate. **1967** V. STRAUSS *Printing Industry* i. 35/2 Offset lithography..dominates the field of planographic printing. **1972** *Physics Bull.* Sept. 532/3 The printing surface is virtually planographic and the action of development is to render the image areas oliophilic.

planography. Add: **2.** Printing from a plane surface, in contrast to processes in which the areas to be printed are in relief or intaglio.

1914 H. J. RHODES *Art of Lithography* i. 1 The term Planography..has been much in evidence of late, and there is no reason why it should not be generally adopted to denote all processes of printing from flat surfaces. **1937** *Discovery* Oct. 297/2 Lithography is..a misnomer, though attempts to replace it by planography have not found favour. **1960** G. A. GLAISTER *Gloss. Bk.* 317/1 Planography..refers to methods of printing from flat surfaces other than stone.

planont (plæ·nǫnt). *Biol.* [f. PLANO-[2] + Gr. ὀντ-: see ONTO-.] A motile spore, whether sexual, asexual, or a zygote; *esp.* the motile stage of certain microsporidian protozoans or phycomycetes.

1914 FANTHAM & PORTER *Some Minute Anim. Parasites* xi. 217 This amœbula [of *Nosema apis*] gives rise, by division, to daughter forms, each possessing one nucleus and capable of wandering about over the epithelium of the gut. Such forms are called planonts, or wanderers. **1943** F. K. SPARROW *Aquatic Phycomycetes* 406 In *A*[*llomyces*] *javanicus*..the planonts emerging from the resting spores..gave rise upon germination to sexual plants. **1961** R. D. MANWELL *Introd. Protozool.* xxiii. 478 At first they [*sc. Nosema* spores] remain in the gut, but they soon begin to wander and are now called 'planonts'.

planosol (plē̆i·no-, plæ·nosǫl). *Soil Sci.* [f. PLANO-[1] + *-SOL.] An intrazonal soil having a thin, strongly leached surface horizon overlying a compacted hard-pan or clay-pan, and occurring on flat uplands with poor drainage. Hence **planoso·lic** *a.*

1938 M. BALDWIN et al. in *U.S. Dept. Agric. Yearbk.* 991 The term 'Planosol' is being proposed to cover those soils with claypans and cemented hardpans not included with the Solonetz, Ground-Water Podzol, and Ground-Water Laterite. Families of Planosols correspond to associated normal zonal soils. **1965** B. T. BUNTING *Geogr. of Soil* vi. 71 Inward from these marginal sites, on plateaux, developed zonal soils occur, of considerable age, and hydromorphic or planosolic variants may exist in the central, poorly drained parts of the widest crestal plateaux. *Ibid.* xi. 133 Planosols are widespread on level plateau surfaces or broad gentle slopes on loess, till or wide alluvial terraces in central USA. **1972** C. B. HUNT *Geol. of Soils* ix. 206 In southern Ohio, Indiana, and Illinois..are flat, poorly drained uplands covered partly by loess and partly by Illinoian till... The comparatively thin modern soils on the surface of the old weathered deposits are Planosols with light-colored surface layers and deeper layers mottled brown and reddish brown.

planospore (plæ·nospǭ°ɹ). *Bot.* [f. PLANO-[2] + SPORE.] A motile zoospore.

1950 E. A. BESSEY *Morphol. & Taxon. Fungi* i. 5 The motile naked zoospores..may be called planospores. **1970** J. WEBSTER *Introd. Fungi* i. 68 The zoospore is sometimes termed a planospore.

plansifter (plæ·nsiftəɹ). [f. PLAN *sb.* + SIFTER.] A machine consisting of a mechanically agitated set of superimposed flat sieves of differing mesh, used in flour milling for separating and grading the broken grain.

Registered in the U.S. as a proprietary name.

1905 *Official Gaz.* (U.S. Patent Office) 26 Dec. 2502/1 Shaking-Bolts. Barnard & Leas Manufacturing Company, Maline, Ill. Plansifter. **1908** *Engineering* 2 Oct. 429/2 Flour was dressed through long reels, called 'bolters', some of which were about 20 ft. long. These cumbersome machines..have given way in this country to the more

adaptable centrifugal dresser, and on the Continent to the plansifter. **1936** J. H. SCOTT *Flour Milling Processes* ix. 200 The type of plansifter used..in this country is the free-swinging, balanced crank pattern, driven by a central vertical shaft. **1964** M. PYKE *Food Sci. & Technol.* iii. 43 In a modern mill, the broken grain from the plansifter associated with the first break roll will be passed to a second break roll where a little more flour will be sifted out.

planster (plæ·nstəɹ). [f. PLAN *sb.* + -STER.] A planner: used only with derogatory connotations.

1945 J. BETJEMAN *New Bats in Old Belfries* 34 The planster's vision. **1964** *Listener* 9 Jan. 71/1 Such destruction has been launched on the face and form of England by witless plansters and lethally unimaginative local authorities. **1978** *N.Y. Rev. Bks.* 23 Feb. 8/4 Does aesthetic sense in the end atrophy if it is desiccated by the continual use of this type of programmed reasoning? Is it replaced by what Betjeman called the Planster's Vision?

plant, *sb.*[1] Add: **3. c.** = *plant-cane* s.v. PLANT *sb.*[1] 10 e in Dict. and Suppl.

1866 'MARK TWAIN' *Lett. fr. Hawaii* (1967) xix. 209 Almost everywhere on the island of Hawaii sugarcane matures in twelve months, both ratoons and plant. *Ibid.* xxiii. 258 This year the 'plant' crop on the Wailuku plantation averages 8,000 [pounds per acre].

6. Add to def.: The premises and fixtures of a business or (chiefly *U.S.*) of an institution; a place where an industrial process is carried on; also, a single machine or large piece of apparatus. Also *transf.*, the workers employed at a plant. (The use with *a* and *pl.* is no longer rare in Great Britain.) (Further examples.)

1904 W. T. MILLS *Struggle for Existence* III. xvii. 216 The great steel plants maintain great laboratories. **1922** *Managem. Engin.* Feb. 86/2 No more time is lost by having all the plant out on strike for a week than in having a tenth of the force absent for 10 weeks. **1925** *Scribner's Mag.* July 31/2 (Advt.), Irving School for boys... Modern plant, complete equipment. **1927** *Brit. Med. Jrnl.* 3 Sept. 374/1 To those American investigators a school meant buildings, equipment, and machinery, or 'plant' as they themselves would say. **1930** J. BUCHAN *Castle Gay* xii. 194 He made his way round to the back regions, which had once been stables and coach-houses, and housed now the electric plant and a repairing shop for cars. **1939** D. L. SAYERS *In Teeth of Evidence* 9 They all want to..play with the apparatus. One of them got loose last time and tried to electrocute itself on the X-ray plant. **1949** *Sat. Rev. Lit.* 21 May 4/3 Its guiding genius.. has seen this school grow from an abstract idea to a two-million-dollar plant. **1957** J. H. ARNISON *Pract. Road Constr.* iii. 52 The shafts for the manholes may be cut out by manual labour, and the main trench by mechanical plant. **1958** *Engineering* 14 Mar. 322/2 Most of the plants benefiting from this influx of dollars are in the Glasgow area. **1958** *Times Lit. Suppl.* 10 Oct. 569/2 The new church 'plant'..is one of the most impressive and novel signs of the boom atmosphere. Mormons, Catholics, Methodists, Seventh Day Adventists, all flourish, to judge by the ecclesiastical building boom. **1960** *Washington Post* 16 Nov. A 16 The institution has almost never received adequate funds, is understaffed, has an inadequate and deteriorating physical plant and is 'on its way to becoming a second rate municipal zoo'. **1963** *Times Rev. Industry* Mar. 51/2 Mr. Justice Pennycuick.. said that 'plant', in its ordinary sense, 'includes whatever apparatus is used by a businessman for carrying on his business'. **1971** B. SCHARF *Engin. & its Language* xvii. 245 Examples of mobile earthmoving plant are bulldozers, graders and scrapers. **1972** J. MOSEDALE *Football* xi. 150 Workers at the meat packing plants. **1973** *Times* 16 Nov. 20/8 At plant level, the [German] philosophy is the shared responsibility of capital and labour for the growth of the enterprise. **1977** *Jrnl. R. Soc. Arts* CXXV. 300/2 With the reduction of teacher training the amount of surplus 'plant' becoming available would eliminate capital construction costs.

c. *Austral.* The equipment, stock, vehicles, etc., of a drover, a farm, a road-mending team, etc.

1901 H. LAWSON *Prose Wks.* (1948) 427 Andy had charge of the 'droving-plant' (a tilted two-horse wagonette, in which we carried the rations and horse-feed). **1903** 'T. COLLINS' *Such is Life* 7 Soon we became aware of two teams coming to meet us... Victorian poverty spoke in every detail of the working plant. **1928** 'BRENT OF BIN BIN' *Up Country* xvii. 290 Charlotte was to have her cows and poultry, so that when the diggings were played out there would be a grazier's plant to fall back upon. **1934** *Bulletin* (Sydney) 31 Jan. 32/2 Although he knew our standard of horsemanship so well, he is so ignorant of our calling as to refer to my plant as my 'herd'. **1954** B. MILES *Stars my Blanket* xxiv. 211 He..was then about to return to Elsey with his 'plant'—a drover's 'plant' being his spare horses and packs. **1963** A. LUBBOCK *Austral. Roundabout* 42 'That'll be Dan Daley with his droving plant,' said Barney, shading his eyes. 'Plant?' I queried. 'Outfit—we call it "plant" here.'

7. (Earlier and further examples.) Also, a hiding-place for people or goods; the people or goods so hidden; *spec.* (a hiding-place for) drugs or equipment used by a drug-addict.

1785 *Sessions Papers of Central Criminal Court* Apr. 582/1 He opened a place in the wainscot, which is called 'a plant', it was a secret cupboard. **1829** H. WIDOWSON *Present State of Van Diemen's Land* xi. 118 The slabs were very loose; on pulling them up, the plant was sprung and mutton in abundance was discovered stowed away in a large barrel. **1846** [see *DUNNY *sb.*[2] 1]. **1874** HOTTEN *Slang Dict.* 256 *Plant*, a hidden store of money or valuables. To 'spring a plant' is to unearth another person's hoard. **1926** J. BLACK *You can't Win* xii. 160 The sack contained his 'plant', an eye dropper with a hypodermic needle soldered to it, and a small paper of morphine. *Ibid.* xx. 314, I could lift the plant and be far away before daylight. **1967** S. LLOYD *Lightning Ridge Bk.* iii. 8 Gibson never located this plant of opal again.

b. A person who, or thing which, has been 'planted' (see PLANT *v.* 2 c in Dict. and Suppl.). *slang.*

1926 *Amer. Speech* I. 436/2 *Plant*, a member of an act planted in the audience or the orchestra pit who performs his share of the act from there, or who comes upon the stage from the audience to take part in the performance as a supposed non-member of the profession. **1949** *Newsweek* 3 Oct. 36/3 Fifteen government witnesses, a half-dozen of them FBI 'plants' who infiltrated the Communist Party, had taken the stand. **1952** KOESTLER *Arrow in Blue* IV. xxiii. 191 One of her favourite pastimes was to fabricate apocryphal news items... One of the most successful of her plants ran something as follows. **1969** *TV Times* (Austral.) 15 Oct. 10/3 One Press agent made an interesting slip of the tongue when he commented: 'The first thing any publicist does in the morning is to read the plants, I mean the trades.' **1978** G. VAUGHAN *Belgrade Drop* ii. 15 'Heroin!' the detective shouted... Yardley had never seen the package before... He said: 'That stuff's a plant.' **1978** M. WALKER *Infiltrator* iv. 48 If she was a plant... I would have to take her along,.. and find out who had planted her and why.

10. a. *plant hire, pot, -stand, -world.* **c.** *plant-eating* (examples), *-hirer, -sucking.*

1905 V. L. KELLOGG *Amer. Insects* xii. 252 (Plant-eating beetles.) Tribe Phytophaga. **1941** J. S. HUXLEY *Uniqueness of Man* vi. 157 The best-analysed cases concern..plant-eating insects adapted to different food plants. **1973** W. S. ROMOSER *Science of Entomology* vii. 186 Phytophagous means literally 'plant eating'. **1976** 'L. BLACK' *Healthy Way to Die* xi. 118 There were thirty-five companies ranging from a merchant bank to..a plant-hire outfit. **1978** J. SHERWOOD *Limericks of Lachasse* xi. 133 Get on to that plant hire place..and get them to have an excavator up here..to dig up the car park. **1973** *Times* 11 May 19/5 Plant hirers are able to offer such machines. **1963** *Times* 21 Jan. 15/1 The Italian company..plans to make plant pots for the horticultural trade. **1975** D. CLARK *Premedicated Murder* iv. 52 They both said yes together, like plant-pot men. **1977** G. SCOTT *Hot Pursuit* vii. 68 The shelves were filled with files and papers and plant pots. **1862** *Catal. Internat. Exhib., Brit.* II. No. 6070, Ornamental wire plant-stands, model rosery, and verandah. **1903** K. D. WIGGIN *Rebecca* 247 She buried her face in the blooming geraniums on Miss Maxwell's plant-stand. **1974** *Trafford Catal.* Spring/Summer 591/2 Pedestal plant stand..with six variable position pot holders. **1908** *Westm. Gaz.* 30 May 7/3 There are very few who realise the enormous number of species that in reality make up this mischievous group of plant-sucking parasites. **1969** *New Scientist* 2 Oct. 19/1 The Australian plantsucking psyllid bug..lives on eucalyptus leaves. **1936** E. SITWELL *Sel. Poems* 12 The ethereal quality of the plant-world.

e. plant-bed, (*b*) *U.S.*, a bed of earth prepared for the germination of seeds and the growth of young plants, esp. of tobacco seedlings; **plant-breeder** (examples); also **plant-breeding** *vbl. sb.*; **plant-cane** (earlier and later examples); **plant-cover(ing),** vegetation spreading over the surface of the earth; **plant-food** (earlier and later examples); **plant geographer** = PHYTOGEOGRAPHER; **plant geography** = PHYTOGEOGRAPHY; **plant hormone** = *HORMONE 2, *PHYTOHORMONE; **plant-house,** (*a*) a greenhouse or conservatory; (*b*) a building containing industrial plant; **plant pathology** = *phytopathology* (*a*) s.v. PHYTO- in Dict. and Suppl.; so **plant pathologist; plant physiology,** the scientific study of the normal functions and phenomena of plants; so **plant physiologist; plant-wax,** wax obtained from plants.

1833 *Niles' Reg.* XLIV. 411/1 He is clearing new grounds; preparing and burning plant-beds. **1907** *St. Nicholas* May 651/1 A 'running' board was put around the base and a plant bed about a foot wide made within this. **1966** *Publ. Amer. Dial. Soc.* XLV. 20 We put cotton canvas over the plant bed. **1906** *Chambers's Jrnl.* 28 July 556/2 The experiments open up a new and interesting field for the plant-breeder. **1929** T. THOMSON tr. *Büsgen's Structure & Life of Forest Trees* xiv. 403 The expert eye of the plant breeder is able to discover them [*sc.* individual differences between plants]. **1970** R. GORER *Development of Garden Flowers* i. 26 To the plant breeder, the importance and interest of germ cell formation lies in the first stage of meiosis. **1908** *Westm. Gaz.* 28 Mar. 6/2 Few who are making a study of the fundamental principles of plant-breeding are unfamiliar with the name and the results achieved by Luther Burbank. **1926** J. S. HUXLEY *Essays in Pop. Sci.* ii. 10 There has sprung into being a new science, of animal- and plant-breeding. **1970** R. GORER *Development of Garden Flowers* i. 21 The essential basis of plant breeding is selection. **1790** W. BECKFORD *Descr. Account Island of Jamaica* I. 161 It is a common practice, where corn will grow, to plant it with the canes... Among plant-canes, I do not conceive it of consequence. **1853** *Harper's Mag.* Nov. 757 The 'growing crop' in Louisiana consists of three kinds of cane: the first is technically called 'plant cane' and is that which springs directly from the 'seed cane'. **1949** *Caribbean Quarterly* I. 1. 5 A cane field was not ripe for its first harvest (the 'plant cane') until the second winter after its planting. **1943** J. S. HUXLEY *TVA* 17 Forests and plant cover were stripped. **1976** *Field* 18 Nov. 976/3 Where the vegetation has been worn away, the shade of the plant cover lost.., evaporation from the bare surface proceeds apace. **1911** W. G. SMITH in A. G. Tansley *Types Brit. Vegetation* xiii. 312 The plant covering is distinctly xerophilous in response to frequent dry periods. **1946** *Nature* 2 Nov. 605/1 Nomadism..a mode of life, indeed, in which defacement of the plant-covering by ploughing or digging is the worst of economic offences. **1869** *Rep. U.S. Comm. Agric. 1868* 396 Such plant-food as rain-water and the atmosphere supply. **1939** LAWRENCE & NEWELL *Seed & Potting Composts* ii. 23 These chemical compounds absorbed by the plant.. we shall refer to as 'plant foods'. **1976** J. BERRISFORD *Backyards & Tiny Gardens* viii. 59 Such a growing medium contains no plant foods, so fertilizers must be added before planting. **1913** *Jrnl. Ecology* I. 27 This character [*sc.* the physiognomy of vegetation] is unjustly regarded as merely superficial..by many modern plant-geographers. **1973** P. A. COLINVAUX *Introd. Ecol.* ii. 27 On the grand-scale, maps of climate based on the plant geographer's boundaries were useful. **1903** W. R. FISHER tr. *Schimper's Plant-Geogr.* p. vi, The connexion between the forms of plants and the external conditions at different points on the earth's surface forms the subject-matter of oecological plant-geography. **1934** H. GILBERT-CARTER tr. *Raunkiaer's Life Forms of Plants* iv. 111 The units of floristic plant geography are the same as those of systematic botany. **1977** *Sci. Amer.* May 99/1 Specimens were collected and filed in herbaria for later investigation by new techniques ranging from cytology and physiology to plant geography and ecology. **1935** *Biol. Rev.* X. 429 Other plant hormones, such as the wound hormones of Haberlandt, we need not discuss, since less quantitative knowledge is available on the subject. They apparently also act by diffusion from cell to cell. **1951,** etc. [see *HORMONE 2]. **1959** L. J. AUDUS *Plant Growth Substances* i. 18 Plant hormones are substances which regulate.. some aspect of plant growth and which are produced by the organism itself. They may be growth hormones, flowering hormones, and so forth. **1974** *Physiologia Plantarum* XXXII. 369 (*heading*) Effect of abscisic acid and other plant hormones on growth of apical and lateral buds of seedlings. **1863** *Horticulturist* XVIII. 306 We again have the satisfaction of presenting two examples of Plant Houses; one a *Green-house*, and the other a *Cold Grapery*. **1881** *Encycl. Brit.* XII. 221/2 Plant houses must be as far as possible impervious to wet and cold air from the exterior. **1909** *Westm. Gaz.* 6 May 5/3 A plant-house is being erected outside the south wall of the provincial capital. **1909** B. M. DUGGAR *Fungous Diseases of Plants* 3 There was a bright prospect for controlling many of the fungous diseases of plants, and there developed..an immediate need for plant pathologists. **1977** *Daily Tel.* 6 July 2/1 Dr Alan Walker, Ministry Plant Pathologist, said that cereal diseases which could cut yield by up to 15 per cent. were minimal this year. **1895** *Jrnl. Chem. Soc.* LXVIII. II (*heading*) Chemical investigations in plant pathology. **1908** P. T. DONDLINGER *Bk. of Wheat* ix. 148 Studies in plant pathology of any great practical bearing or importance are..modern and recent. **1935** *Discovery* Oct. 294/1 The intimate relationships between plant pathology..and other branches of botany. **1973** *Nature* 27 Apr. 595/2 Plant pathology..is to plants what the whole of medicine and veterinary science is to man and animals. *Ibid.* 596/1 Is it right that any comprehensive book on the principles of plant pathology should dismiss viruses and mycoplasmata with thirty-six pages and an apology? **1931** W. O. JAMES *Introd. Plant Physiol.* i. 2 The methods used by plant physiologists..are mainly derived from various branches of chemistry and physics. **1898** S. A. MOOR tr. *W. Detmer's Pract. Plant Physiol.* p. vii, Plant physiology is now of..far-reaching significance for students of Natural Science, Agriculture, Forestry, and Medicine. **1937** W. H. SAUMAREZ SMITH *Let.* 10 July in *Young Man's Country* (1977) ii. 80, I was interested to see the place where all his [*sc.* Tagore's] disciples are following out the lines of research suggested by his highly original work in plant-physiology. **1968** F. C. STEWARD *Growth & Organization in Plants* p. iii, The author's, and indeed a customary, approach to plant physiology is deeply ingrained in the study of cells, their membranes and particulate inclusions, their metabolism and responses to stimuli. **1924** J. A. THOMSON *Science Old & New* xviii. 101 There are plant-waxes as well as animal-waxes.

plant, *v.* Add: **1.** Also *absol.*

1893 B. MITFORD *Gun-Runner* iv. 34 Along the banks of this [watercourse] the careful Jeremiah had planted and sown. **1896** *Forum* July 515 Our forefathers..came to work, to plant, to reap, where they might worship God with freedom. **1961** *Atlanta Constitution* 17 Aug. 5 The people who try to raise and can meat, to plant, grow vegetables, and put them up. **1979** *Verbatim* Summer 8/1 In South Australia a farmer *seeds*,..and in Queensland he *plants*.

c. *plant out* (earlier and later examples); also, to arrange plants or trees in a piece of ground. Also *transf.* and *fig.* (cf. sense 6 in Dict.).

1664 J. EVELYN *Kalendarium Hortense* 60 Now also plant out your Colly-flowers to have early. **1901** *Year-bk. U.S. Dept. Agric. 1900* 373 Each orchardist will no doubt develop some method of his own in planting out the orchard. *a* **1910** 'MARK TWAIN' *Autobiogr.* (1924) I. 274 They would often plant out eleven columns of new ads on a standing galley. **1917** P. S. ALLEN *Let.* 8 July (1939) 139 So many of your books are here 'on deposit'...I wonder if you recognised what was in my mind..when I wrote of Rud. Agricola's 'planting out his books in friends' houses as pledges of return'. **1927** KIPLING *Limits & Renewals* (1932) 170, I was planting out plants from my garden. **1962** *Times* 19 May 11/4 Every partridge-rearing system encounters its critical phase when the birds are 'planted out'. **1972** *Shooting Times & Country Mag.* 4 Mar. 24/3 'Unfed fry' [*sc.* trout]..are ready to be planted out to start off their natural lives in the sidestreams and the river. **1975** B. DOUGHERTY *Green Gardener* x. 115 Avoid touching their [*sc.* tomatoes'] stems when planting out, holding them only by the leaflets.

2. c. (Further examples.) Now esp., to conceal (stolen goods, incriminating evidence,

etc.) with a view to misleading a later discoverer. Also (not *slang*), to introduce (a character, scene, etc.) into a play, film, etc., for some specified purpose.

1865 J. H. A. BONE *Petroleum & Petroleum Wells* (ed. 2) 153 Frauds are not infrequently perpetrated by 'planting' oil in dry wells. **1930** *Times Lit. Suppl.* 1 May 373/1 The nephew..sought to clinch the available, and misleading, evidence by planting the victim's dental plate on the spot. **1933** H. J. LEE *Eagle Police Manual* 152 *Plant*, to place incriminating evidence in a man's pocket or elsewhere. **1939** E. S. GARDNER *D. A. draws Circle* (1940) 200 Someone is planting evidence. *Ibid.* 203 It had been planted on him. **1948** A. HUXLEY *Let.* 16 Jan. (1969) 578, I have been trying to put this question to the general and specialized publics for the last year or two—even succeeding in planting it in the *Bulletin of the Atomic Scientists*. **1950** *Ibid.* 16 Feb. 619 We have to plant the business of the currants, so that we are forced to show them lunching. **1958** *Listener* 30 Oct. 704/3 The man was 'planted' as a nervous stammerer, but to be nervous is not necessarily to be a nitwit. **1969** *It* 11–24 Apr. 10/1 Everyone was searched and told to stay clear of the area under the threat of being planted. **1970** G. F. NEWMAN *Sir, You Bastard* 261 Planting microphones was easier. **1974** *Howard Jrnl.* XIV. 43 [The police] are now seen by many West Indians..as racist 'enemies', who taunt, intimidate, assault, plant and 'trump up' charges. **1978** S. BRILL *Teamsters* i. 18 Government investigators..had planted an informant among organized-crime figures in California.

6. a. Also with *to*.

1799 T. R. MALTHUS *Diary* 16 July (1966) 159 There are many grounds about the town planted to potatoes. **1901** *Year-bk. U.S. Dept. Agric. 1900* 373 The land should be planted to a crop for at least a year or two before setting out the trees. **1941** E. P. O'DONNELL *Great Big Doorstep* 92 She reached a field planted to okra, the stalks rising taller than she. **1949** E. HYAMS *Not in our Stars* xvi. 195 A four-acre field..which Drover had planted to cherries in the previous season. **1976** *National Observer* (U.S.) 12 June 9/1 Grapevines are now found..creeping into fields once planted to pears and apples.

11. To bury (a dead person). *slang* (orig. *U.S.*).

1855 *Harper's Mag.* Dec. 37/1 Let it [*sc.* yellow fever] catch hold of a crowd of 'Johnny come latelys', and it plants them at once. **1866** 'MARK TWAIN' *Lett. fr. Hawaii* (1967) 242 It's about the orneryest thing for a monument I've ever struck yet... If I was planted under it, I'd highst it. **1888** [see *FLAMDOODLE]. **1927** C. A. W. MONCKTON *Some Experiences of New Guinea Resident Magistrate* 2nd Ser. i. 16 There's Alligator Jack and Red Bill..planted here, and Gawd, 'E knows whether they have rested easy. **1931** GALSWORTHY *Maid in Waiting* ii. 10 'Is he to be planted here?' 'I expect in the Cathedral, but Father will know.' **1967** C. ROUGVIE *When Johnny Died* iii. 66 It was raining when we planted him, and I thought he'd get out of his coffin. **1974** R. JEFFRIES *Mistakenly in Mallorca* xv. 143 The funeral must be fixed up at once. Where did non-Catholics get planted?

plantain[1]. 3. plantain lily, substitute for def.: = *FUNKIA, *HOSTA. Add later examples.

1894 W. ROBINSON *Wild Garden* (ed. 4) xiv. 170 The Plantain Lilies are plants for the wild garden. **1927** [see *FUNKIA]. **1957** C. LLOYD *Mixed Border* vi. 60 Among plantain lilies..the prevalence of slugs and snails is all too likely. **1976** B. SWAIN *Commonsense of Gardening* v. 207/2 The charming Plantain Lilies..have often disappointed because of unsuitable conditions.

plantar, *a.* (Further examples.)

1951 *Chambers's Jrnl.* Aug. 507/1 'You've been grousing for months about the corn you picked up last tour, foot-slogging round Masailand.'.. 'It's not a corn, it's a plantar wart,' I said indignantly. **1976** *Muir's Text-bk. Path.* (ed. 10) 985 The histology of the vulgar wart usually seen on the hands and knees is essentially similar to the plantar wart found on the soles of the feet.

plantation. Add: **7.** *plantation-house* (later examples), *manners, -Negro* (earlier and later examples), *-worker*; **plantation creole,** a creolized language arising amongst a transplanted and largely isolated Negroid community; **plantation crepe** *U.S.*, used *attrib.* of a variety of crêpe-rubber sole on footwear.

1938 *Social Forces* Oct. 114/2 The plantation creole tongues are true *Sklavensprachen*. Although they owe something to the sailors' trade jargons, they began essentially as a makeshift means of communication between masters and field hands. **1978** *Verbatim* Feb. 10/1 Both authors hold to..the Creolist theory, which traces the present-day Black English vernacular to a Plantation Creole, to a plantation-maritime pidgin, to an African origin. **1967** *New Yorker* 7 Oct. 109/2 (Advt.), Clark's original Desert® Boots..with plantation crepe soles. **1969** *Sears, Roebuck & Co. Catal.* Spring–Summer 454/2 Durable, buoyant, plantation crepe rubber sole and heel. **1969** E. WILSON *Hist. Shoe Fashions* xx. 258 A plantation crepe sole was one of the many soft soles which added to its comfort. **1831** J. M. PECK *Guide for Emigrants* II. 55 All the plantation houses are surrounded with rich and beautiful groves. **1973** *Advocate-News* (Barbados) 2 Feb. 15/4 (Advt.), Besides the plantation house there is available the plantation manager's house. **1974** *Country Life* 3–10 Jan. 18/1 The Virginian plantation houses of the 18th century, such as Carter's Grove, Westover and Shirley. **1854** THOREAU *Walden* 165 Men of almost every degree of wit called on me in the migrating season. Some who had more wits than they knew what to do with; runaway slaves with plantation manners, who listened from time to time, like the fox in the fable, as if they heard the hounds a-baying on their track. **1897** *Congress. Rec.* 31 Mar. 548/2 When I was a boy,..I used to read a great deal about what the early Republicans called 'plantation manners'. **1771** in *Maryland Hist. Mag.* (1919) XIV. 135 My people..do not live so well as our House negroes, But full as well as any Plantation negroes. **1956** G. P. KURATH in A. Dundes *Mother Wit* (1973) 108/1 Recreational dances of plantation Negroes commenced with a prayer. **1957** P. WORSLEY *Trumpet shall Sound* viii. 148 Plantation-workers were convinced by Runovoro's ability to write meaningless works. **1976** *Honolulu Star-Bull.* 21 Dec. A-8/2 Approximately 175 Molokai Dole plantation workers..lost their jobs last year because of foreign competition.

Planté (plɑ·nte). *Electr.* The name of R. L. Gaston *Planté* (1834–89), French physicist, used *attrib.* to designate lead-acid accumulator plates formed by a process which he invented, cells containing such plates, and the process itself.

1881 *Electrician* 3 Sept. 249/2 A Planté cell that has been long in use gives a better result than one that has been freshly constructed. **1889** G. W. DE TUNZELMANN *Electr. in Mod. Life* xiv. 195 The original Planté accumulator has been considerably improved by Faure and others. **1923** GLAZEBROOK *Dict. Appl. Physics* II. 76/2 Owing to its high cost the original Planté process is no longer employed. *Ibid.* 77/1 Planté negatives are not generally used, as their advantages are not compensated by the additional weight of lead... Planté positives are employed extensively in stationary batteries. **1959** *Times* 11 Sept. 9/2 The field in which the new cell is designed to replace current Planté types is a very wide one. **1964** G. SMITH *Storage Batteries* ii. 21 These do not have quite the same long life as the Planté battery. **1970** C. L. MANTELL *Batteries & Energy Systems* xiii. 112 Planté plates are prepared from lead blanks which have been cast, rolled, cut, and stamped.

planted, *ppl. a.* **1.** (Examples corresponding to PLANT *v.* 2 c in Dict. and Suppl.)

1963 *TV Times* (Austral.) 18 Apr. 10/2 *Planted*, hidden. **1972** *Jrnl. Social Psychol.* Dec. 301 When the class was asked if anyone wanted to make a statement to the 'teacher', a planted student responded with a standard complaining, corrective request. **1973** M. WOODHOUSE *Blue Bone* x. 93 'He said..that he was quite happy where he was.' 'In answer to a planted question, yes, he did.' 'You're fairly sure it was planted?'..'Of course... This whole thing was just a put-up job... So that Karel could state for the record that he was fine where he was.' **1977** G. V. HIGGINS *Dreamland* xiv. 169 We're even bigger suckers for a planted story..if we really had to scout around for it. **1978** J. GARDNER *Dancing Dodo* xxxix. 324 It's a greater threat than a planted nuclear device.

planter. Add: **4. b.** *planter's* (or *planters'*) *punch*: a cocktail containing rum.

1924 A. MACMILLAN in *Land of Abiding Sunshine*, A 'swizzle' or 'a planter's punch' would very welcome be. **1935** S. LEWIS *It can't happen Here* iv. 38 His reputation for research among planters'-punch recipes..might cause his defeat by the church people. **1958** G. GREENE *Our Man in Havana* v. ii. 198 Have a planter's punch. They are good here. **1971** 'D. HALLIDAY' *Dolly & Doctor Bird* viii. 108, I don't suppose Beltanno has tasted planter's punch. **1978** G. GREENE *Human Factor* v. iii. 285 What about a Planter's Punch? They do them OK here, so I'm told.

8. a. (Earlier and later examples.)

1850 *Rep. U.S. Comm. Patents 1849* I. 151 Having thus fully described my improved grain and seed planter. **1939** W. FAULKNER *Wild Palms* 65 For seven years now he had run his plough and harrow and planter within the very shadow of the levee. **1950** [see *BAGGER b].

b. A pot, tub, or other container for growing or displaying plants. orig. *U.S.*

1959 in WEBSTER Add. **1966** 'L. HOLTON' *Out of Depths* viii. 72 The brick planters facing the ocean were gay with blossoms. **1968** *Washington Post* 3 July A24/8 (Advt.), Distinctive redwood planter at savings! **1969** *Islander* (Victoria, B.C.) 5 Oct. 3/3 On the foundations of the cabin there is now a summer patio with brick planters made from the cabin chimney. **1973** *Center City Office Weekly* (Philad.) 9 Oct. 10/4 Cream scuttle... Can be used as a planter, or for artificial flowers.

Plantin (plæ·ntin). The name of Christophe *Plantin* (1514–89), printer, of Antwerp, used to designate a family of old-face types, based on a 16th-century Flemish original, the first of which was designed by F. H. Pierpont for the Monotype Corporation in 1913.

1914 *Monotype Recorder* July 68 (*caption*) First showing of light face Plantin O.S. **1919** J. P. THORP *Printing for Business* viii. 81 The headlines, section numbers and little acorn ornaments are in vermilion, the text in 12-point Plantin. **1929** F. MEYNELL *Typogr. Newspaper Advertisements* 28 Plantin (particularly the Monotype face) declares in its strength and straightness that it stands for the machine-seller. **1951** S. JENNETT *Making of Books* xv. 262 Types like Times and Plantin, which are so large on the body that they are unpleasant when set solid, become useful and legible when they are suitably leaded. **1967** J. B. LIEBERMAN *Types of Typefaces* 89/2 Plantin..is based on 16th century forms.

planting, *vbl. sb.* Add: **3. planting-attorney,** in the West Indies, the manager of a plantation or estate.

1953 *Caribbean Q.* III. III. 142 The planters and planting-attornies were worried about the continuation of estate labour after the slaves should be apprenticed. **1956** H. G. DE LISSER *Cup & Lip* i. 19 Arthur was earnest at his work as his uncle's planting attorney.

plantlet. (Later examples.)

1899 [see *DAMPING *vbl. sb.* 2]. **1935** A. F. HORT *Garden Variety* IV. 226 Here [*sc.* in a box] the plantlets..will remain for weeks or months. **1946** D. C. PEATTIE *Road of Naturalist* i. 18 For a seed is not just part of a plant; detached it is a complete plant, with a plantlet folded inside, a supply of food, an infinitesimal supply of moisture. **1970** *Nature* 19 Sept. 1265/2 Anthers from entirely white shoots gave rise to white plantlets.

plant-louse. (Later examples.)

1840 J. & M. LOUDON tr. *Köllar's Treatise on Insects* II. 149 The plant-lice are especial enemies to various sorts of culinary vegetables. **1899** B. S. CRAGIN *Our Insect Friends & Foes* 228 Family Aphididæ. These are the Plant-lice, some with wings and some without. **1932** E. STEP *Wasps, Ants & Allied Insects Brit. Isles* 79 Like that of most of the other Black Wasps, its special prey is the Plant-louse (*Aphis*). **1968** R. LOWELL *Notebook* (1969) 48 The lily pads bright as mica, swarming with plant-lice. **1973** W. S. ROMOSER *Science of Entomology* xi. 346 Especially significant families are Psyllidae, jumping plant lice,.. Aphididæ, aphids or plant lice, [etc.].

plantmilk (plɑ·ntmilk). [f. PLANT *sb.*[1] + MILK *sb.*] A synthetic milk substitute prepared from vegetable matter.

1959 *Oxford Mail* 19 Mar. 4/5 It is estimated that the plantmilk which will soon be available to the general public will cost a few pence more than cows' milk. **1959** *New Scientist* 1 Oct. 596/1 Plantmilks based on soya are being used in America to treat milk allergies in infants. **1962** *Guardian* 27 Aug. 2/3 A plantmilk, made from cereals and pulses, has been on sale for some years. **1965** *Times* 1 Sept. 6/6 Drinka pinta milka day but make it plantmilk, Dr. Alan Stoddard advised delegates at the International Vegetarian Union's world congress at Swanwick, Derbyshire, today.

plantsman. Substitute for def.: An expert gardener; a connoisseur of plants. Add later examples. Hence **pla·ntsmanship,** a desire to display knowledge of unusual or especially rare plants.

1952 *Archit. Rev.* CXII. 343/1 None of these plants remain in that environment for long as they are serviced and maintained by fully competent plantsmen. **1962** *Amateur Gardening* 21 Apr. 21/1 *Actinidia polygama*..has fruit of unpleasant flavour, but this should not..deter those who practise plantsmanship from planting a 'bower of bliss' with a few prominently labelled specimens. **1963** *Times* 3 June 1/7 Some of the best garden spurges are described by a plantsman contributor. **1978** W. BLUNT *In for Penny* xxi. 155 This lovely blue orchid [*sc. Vanda cærulea*] had become a status symbol for rich plantsmen. **1979** J. HARVEY in J. Harris *Garden* 8/1 The abbeys, priories and hospitals were particularly concerned with plantsmanship, cultivation and improvement.

plapper (plæ·pəɹ), *v.* [imit.: see S.N.D.] *intr.* To make sounds with the lips.

1866 W. GREGOR in *Trans. Philol. Soc.* 127 Plapper,.. to make a noise with the lips. **1922** JOYCE *Ulysses* 258 She took no notice while he read by rote a solfa fable for her, plappering flatly.

plaque. Add: **1. a.** (Earlier and later examples.) Also, an inscribed plate identifying a monument or building, etc.

1869 C. SCHREIBER *Jrnl.* (1911) I. 13 A large plaque of Smalto glass, with landscape in brown. **1870** *Ibid.* I. 68 We saw..a very fine Terra Cotta plaque, by Clodion, 3 to 4 feet long. **1956** A. J. CRONIN *Crusader's Tomb* III. ix. 195 At the base of the pediment was a time-worn plaque defining the intention of the founder to tend the sick. **1968** *Guardian* 19 Sept. 18/3 In a few days Britain will have two fish and chip shops, each proclaiming itself by plaque to be the oldest..in the world. **1969** M. PUGH *Last Place Left* xxiii. 175 Why don't you wait fifty years or so? I imagine there will be a plaque outside his house. **1971** *Times* 19 Apr. 12/5 Mrs. Pankhurst's house in Clement's Inn is being demolished and the Women's Liberation movement is concerned lest the blue plaque from the house..should disappear. **1979** ATTERBURY & IRVINE *Doulton Story* 20 Probably the best known Doulton commemoratives are the plaques made for the LCC and other bodies to mark places associated with famous people and events.

c. A counter used in gambling. Cf. CHIP *sb.*[1] 2 d in Dict. and Suppl.

1904 A. BENNETT *Great Man* XXV. 281 A croupier counted out..sundry..gold *plaques* of a hundred francs each. **1964** A. WYKES *Gambling* xii. 288 The big bets are placed with rectangular colored chips, called plaques. **1972** D. LEES *Zodiac* 47 Françoise picked up the plaques from the table in front of her. **1973** 'R. MACLEOD' *Burial in Portugal* vi. 117 Deliberately, Salvador used a one thousand escudo plaque to scratch along his small moustache.

d. *Mus.* A thin metal plate inserted into the separated tip of the double reed of a wind instrument while the reed is being scraped.

1940 J. ARTLEY *How to make Double Reeds* 13/1 Insert the plaque between the blades of the reed. *Ibid.* 14/1 While working on the lay with the knife, use the plaque at all times. **1953** E. ROTHWELL *Oboe Technique* vi. 48 *Tongue, or plaque*, for inserting into the reed while scraping. Small flat piece of metal, oval shaped with pointed ends. *Ibid.* 53 In order to avoid any confusion between the human tongue and the metal one, I shall, throughout this chapter, refer to the latter as the *plaque*, the alternative word used little in England but almost exclusively in America. **1957** A. BAINES *Woodwind Instruments & their Hist.* iii. 82 The *tongue* (or *plaque*, fig. 11, *t*) is a thin, oval steel plate about 40 × 15 millimetres,

and it is always placed between the blade tips while scraping after the tips have been separated. **1962** E. C. Moore *Oboe & its Daily Routine* iv. 13/1 Few tools are needed..a plaque to slip between the blades of the reeds, [etc.].

2. c. *Med.* A patch of fibrous tissue or of fatty matter on the wall of an artery; the substance of which such a patch is formed.

1891 *Trans. Assoc. Amer. Physicians* VI. 182 The nodular form of arterio-sclerosis is due to circumscribed dilations of the arteries and a new formation of connective tissue which exactly fills out the dilated area. When such arteries are examined after having been injected with paraffine.., the raised plaques which are so prominent in the uninjected vessels have entirely disappeared, leaving a smooth intima. **1943** *Physiol. Rev.* XXIII. 188 The atherosclerotic lesions..were classified as fatty plaques, fibrous plaques, calcified plaques, and atheromatous ulcers. **1972** *Daily Colonist* (Victoria, B.C.) 4 May 2/1 As the years pass, the walls of our arteries thicken and accumulate a certain amount of plaque, or fatty deposits. **1975** *Daily Tel. Mag.* 5 Dec. 20/4 These so-called plaques, and the attached clots, build up and obtrude upon the bore of the vessel. **1978** *Time* 3 July 54/1 Tests showed that his left main coronary artery was clogged with cholesterol-laden plaque.

d. *Dentistry.* A patch of deposit that contains bacteria and adheres firmly to the surface of a tooth; the substance of which such patches are composed.

1898 G. V. BLACK in *Dental Cosmos* XL. 448 Leptothrix threads..are found..clinging in and upon gelatinous microbic plaques upon the teeth. **1921** RYAN & BOWERS *Teeth & Health* xi. 154 In caries, or dental decay, placques or films of saliva form on the tooth surfaces, in combination with particles of carbohydrates. **1959** WILKINS & MCCULLOUGH *Clin. Pract. Dental Hygienist* ii. 109 Dental plaque is a thin, tenacious, film-like deposit made up principally of microorganisms and mucinous substances from the saliva. It is removed by polishing procedures. Plaque is the most commonly found of all tooth deposits. *Ibid.*, Dental plaques vary in thickness, degree of adherence to the tooth surface, and percentage composition. **1971** *Daily Tel.* 24 Aug. 5 (Advt.), The toothbrush is undoubtedly the most effective weapon in the fight against bacterial plaque. Plaque produces the harmful acids and chemicals that cause tooth decay and discoloration. **1976** J. J. MURRAY *Fluorides in Caries Prevention* xii. 185 Dental plaque is a soft, tenacious bacterial deposit suspended in a protein matrix which forms on the surface of teeth. It also contains varying amounts of extracellular polysaccharide and desquamated bacterial cells.

e. *Biol.* A relatively clear area in a culture of micro-organisms or other cells produced by the inhibitory or lethal effect of a virus or other agent.

D'Herelle used *plage*, not *plaque*, in Fr. (*Le Bactériophage* (1921) i. 13).

1924 *Jrnl. Bacteriol.* IX. 397 These lytic areas, or plaques, are usually circular and may vary in size from pits of microscopic dimensions to eroded fields possessing a diameter of 18 to 20 mm. **1930** G. H. SMITH tr. *F. D'Herelle's Bacteriophage & its Clin. Applic.* i. 12 Each bare spot, which I have termed a *plaque*, represents a colony of bacteriophage particles. **1952** *Proc. Nat. Acad. Sci.* XXXVIII. 747 (*heading*) Production of plaques in monolayer tissue cultures by simple particles of an animal virus. **1963** *Science* 26 Apr. 405/1 Distinct plaques, each of which is due to the release of hemolysin by a single antibody-forming cell, are revealed by complement after incubation, in an agar layer, of a mixture of sheep red cells and lymphoid cells from a rabbit immunized with sheep red cells. **1970** T. D. BROCK *Biol. Microorganisms* x. 260 Since the agar prevents the new virus particles from moving too far away, a localized area of lysis develops that contains no bacteria but many virus particles... This local area of lysis is called a plaque..and represents the end result of a chain of events initiated by one virus particle.

3. *Med.* A flat applicator designed to contain radium or one of its salts, formerly applied to the surface of the body over cancerous tissue for the curative effect of the radiation.

1919 [see *radium plaque*]. **1922** F. E. SIMPSON *Radium Therapy* xii. 110 The best type of metal applicator is made of silver, the radium salt being spread uniformly over a glazed surface which forms the face of the applicator. Lead free glass must be used. Plaques of this type are known as glazed radium applicators. **1931** G. E. BIRKETT *Radium Therapy* xi. 150 Superficial sclerosing type [of rodent ulcers].—In the early stages these may be successfully treated by the application of an unscreened radium plaque. **1950** WALTER & MILLER *Short Textbk. Radiotherapy* viii. 195 A beta-ray applicator is an example of a plaque. The one illustrated in Fig. 82 is made of brass, has an area of 4 sq. cm. and contains 5 mgm. of radium per square cm. It is covered by a filter of 0·1 mm. of monel metal.

plashily (plæ·ʃīli), *adv. rare.* [f. PLASHY *a.*[2] + -LY[2].] With a plashing noise.

1926 R. MACAULAY *Crewe Train* III. i. 240 Going away, going away, going away. The waves plashily said it over.

-plasia (-plē̆i·ziă), a word-forming element (f. Gr. πλάσις moulding, conformation (πλάσσειν to form, mould) + -IA[1]) used in medical and biological terms in the sense 'growth, development (of tissue)', as *DYSPLASIA, HETEROPLASIA. Occas. anglicized as **-plasy.**

plasma. Delete ‖ and add: **5*.** *Physics.* A gas in which there are positive ions and free negative electrons, usu. in approximately equal numbers throughout and therefore electrically neutral; *esp.* one exhibiting phenomena due to the collective interaction of the charges. Also, any analogous collection of charged particles in which one or both kinds are mobile, as the conduction electrons in a metal or the ions in a salt solution.

Electrical neutrality and collective phenomena are often made necessary characteristics of a plasma (e.g. quot. 1967[2]).

1928 I. LANGMUIR in *Proc. Nat. Acad. Sci.* XIV. 628 It seemed that these oscillations must be regarded as compressional electric waves somewhat analogous to sound waves. Except near the electrodes..the ionized gas contains ions and electrons in about equal numbers so that the resultant space charge is very small. We shall use the name *plasma* to describe this region containing balanced charges of ions and electrons. **1930** *Physical Rev.* XXXVII. 1467 The plasma used in this investigation was the positive column of a mercury arc. **1941** MILLMAN & SEELY *Electronics* x. 307 The largest portion of a glow discharge is the plasma. *Ibid.* 309 In addition to the electrons and ions that exist in equal concentrations, a plasma contains many gas molecules. **1958** *Engineering* 31 Jan. 134/2 The stable plasma reaches the high temperatures, of the order of 5 million deg. K., necessary for producing thermonuclear reactions. **1960** *Soviet Physics Doklady* V. 363 At a distance from the earth of 4 earth radii, a plasma with a temperature of not more than tens of thousands of degrees was detected. **1966** *McGraw-Hill Encycl. Sci. & Technol.* X. 386/1 If the over-all dimensions of a region containing a plasma are small compared to λ_D, only simple collisional or single-particle behavior is to be expected, the plasma will behave as an ordinary low-density gas, and collective processes will not be important. **1967** L. K. BRANSON *Introd. Electronics* ix. 315 The plasma consists of a mixture of positive, negative, and neutral particles and..in any given volume-element there are equal numbers of ions and electrons. Further, the plasma..fills the entire volume between anode and cathode except for a narrow region at the cathode called the sheath. **1967** CONDON & ODISHAW *Handbk. Physics* (ed. 2) IV. xi. 188/1 The phenomena that occur in a plasma and distinguish it from any arbitrary collection of charged particles are the near equality of positive and negative charges throughout the plasma volume and the ability of the charges to participate in plasma oscillations. **1969** STEELE & VURAL *Wave Interactions in Solid State Plasmas* i. 4 In a metal like copper, the free electrons comprising the plasma are electrically compensated by the positively ionized copper atoms. **1971** E. NASSER *Fundamentals of Gaseous Ionization & Plasma Electronics* xiv. 427 Liquid plasmas exist in salt solutions where the positive and negative ions move separately. **1974** R. C. DAVIDSON *Theory of Nonneutral Plasmas* p. xi, Nonneutral plasmas exhibit collective properties that are qualitatively similar to those of neutral plasmas. For example, in klystrons and traveling-wave tubes, the collective oscillations necessary for microwave generation and amplification are excited even under conditions in which the electron beams..are unneutralized. **1974** *Nature* 5 Apr. 494/2 In a cold plasma (which is a good approximation for most of the magnetosphere away from the equatorial region) there are two wave modes. **1976** T. BEER *Aerospace Environment* i. 16 The solar wind is a plasma of hydrogen ions (protons) and electrons travelling at speeds that range from 300 km s^{-1} to 1000 km s^{-1}, depending on solar activity.

5.** *Soil Sci.* (See quots.)

1958 I. W. CORNWALL *Soils for Archaeologist* xvii. 190 Intergranular spaces and conducting channels may be filled, or partly filled, with colloids and precipitates, conveyed and deposited..by percolating moisture. This is the soil-plasma, which constitutes in part the cement between adjacent grains and in part mere filling of available voids. **1976** COURTNEY & TRUDGILL *Soil* ii. 17/2 In thin sections under a microscope the soil plasma can be recognized... It is an amorphous combination of humus, clays and chemical compounds (e.g. iron oxide), and is produced by the secondary weathering processes..and by the incorporation of organic matter. *Ibid.*, The presence of mineral matter in the plasma distinguishes it from the overlying purely organic horizons.

6. *plasma cloud*; **plasma arc,** a very hot plasma jet produced by passing a noble gas through a nozzle that is one electrode of an electric arc, used in plasma torches; **plasma dynamics** (also as one word), the science of the dynamical properties and behaviour of gaseous plasmas; so **plasma-dynamic, -dynamical** adjs.; **plasma engine,** a form of jet engine that produces and ejects plasma; **plasma frequency,** the natural resonant frequency of a plasma oscillation, which is also the minimum frequency of electromagnetic waves that can travel through the plasma without attenuation and is approximately 8920 $\sqrt{n}$ Hz, where n is the number of free electrons per cc.; **plasma jet,** a high-speed stream of plasma (ionized gas) ejected from a plasma engine or plasma torch; **plasma membrane** *Biol.*, = *PLASMALEMMA; also, a similar membrane around an intracytoplasmic vacuole; **plasma oscillation,** a collective oscillation of the electrons in a plasma; **plasma physics,** the physics of plasmas such as ionized gases; hence **plasma physicist; plasma probe,** any device that is inserted or immersed in an ionized gas to investigate its physical properties; **plasma propulsion,** propulsion of a vehicle by means of a plasma engine; **plasma sheath,** a thin layer of space charge covering a surface in an ionized gas; **plasma torch,** a small device that produces a very hot plasma jet for use in cutting solids or coating them with refractory material.

1958 *Iron Age* 4 Dec. 136/1 Thanks to the development of the new plasma arc torch, a brand new method for fabricating shapes and applying ultra-high-temperature coatings is now a reality. **1963** H. R. CLAUSER *Encycl. Engin. Materials & Processes* 480/2 The cost of depositing the high-melting-point coatings with the plasma-arc process is comparable with that of the flame-sprayed coatings. **1973** *Materials & Technol.* VI. i. 60 A number of unconventional methods of cutting wood have been examined, largely with the object of reducing waste. These include the use of the plasma arc, which produces very high temperatures with a nozzle of very small diameter. **1960** *Aeroplane* XCVIII. 610/1 It appears that plasma clouds emitted from the Sun run up against the Earth's magnetic field, causing it to release previously trapped particles into the atmosphere. **1967** M. KENYON *Whole Hog* xviii. 181 Along comes space which everyone had thought was empty, a void, but it turns out it's not, it's filled with radiation belts and plasma clouds and solar winds. **1969** *Monthly Not. R. Astron. Soc.* CXLV. 328 In the Ryle-Longair model, plasma clouds formed by a strong explosion within a galaxy expand..until their dimensions exceed galactic dimensions, at which time they are ejected from the parent galaxy and henceforth evolve independently. **1964** E. STUHLINGER *Ion Propulsion* vi. 277 The heat energy absorbed by the coolant in the reactor may be used to heat..the plasma in a plasmadynamic converter, or the working fluid in a thermodynamic converter. **1959** *Astrophysical Jrnl.* CXXIX. 217 We are interested in the distance in which a stream of tenuous plasma, directed against another oppositely moving stream of tenuous plasma, is brought to rest as a consequence of plasma dynamical interaction. **1960** *Aeroplane* XCIX. 837/2 Such subjects as magneto-hydrodynamics, MHD mechanics, and plasmadynamics. **1970** *New Scientist* 5 Feb. 273/2 Plasma dynamics is a subject with applications in many branches of physics and technology..—space physics and the quest for thermonuclear fusion being but two examples. **1958** *S.A.E. Jrnl.* Apr. 93/2 Another phase of our investigations..is the development of a plasma engine, in which small amounts of plasma are ejected at extremely high velocities. **1967** *Electronics* 6 Mar. 8/2 He directed work on plasma engines and space suits. **1974** HAWKEY & BINGHAM *Wild Card* xv. 131 The propulsion pack was okay for the demonstration... We plan to replace it with a small plasma engine. [**1929** *Physical Rev.* XXXIII. 198 Thus the lower frequency limit for long waves coincides with the plasma-electron frequency.] **1949** *Ibid.* LXXV. 1852/1 For a typical density of 10^{12} electrons per cm^3, the plasma frequency is about 10^{10} c.p.s. **1964** D. B. NEWMAN *Space Vehicle Electronics* iv. 225 Above the plasma frequency the plasma has dielectric properties... Well below the plasma frequency, the plasma acts like a conductor. **1971** FERRY & FANNIN *Physical Electronics* vii. 96 For a metal, where the electron concentration is about 10^{28} m^{-3}, the plasma frequency is found to be about $5{\cdot}6 \times 10^{15}$ Hz, or in the ultraviolet region. **1957** G. M. GIANNINI *Plasma Jet & its Applications* (U.S.A.F. Office Scientific Res. Techn. Note 57-520) 22 The 'plasma jet' can be used for many of the purposes described... The jet is very hot, highly ionized, has a high velocity. **1960** *Aeroplane* XCVIII. 610/2 Because of the low thrust produced, the plasmajet cannot be employed for rocket-launching from Earth. It must be carried into orbit by a more powerful chemical rocket and started in the weightless environment of space. **1964** *Sci. News Let.* 12 Sept. 163 Plasma jets, the white-hot streams of gas used for such tasks as cutting and welding, may soon have yet another use, 'steering' satellites through space. **1972** D. G. SHEPHERD *Aerospace Propulsion* viii. 202 The arc jet or plasma jet utilizes the very high temperatures in arcs to heat the propellant. **1900** *Ann. Bot.* XIV. 352 The entire structure, antheridium, tube, and oogonium, have in reality become for a time a single cell bounded by a single continuous plasma-membrane. **1922** W. STILES in *New Phytologist* XXI. 141 The term plasma-membrane will be used to denote a surface layer of protoplasm which behaves as a membrane surrounding the bulk of the protoplasm, and which may exhibit different degrees of permeability to different substances... The membrane bounding the outside of the protoplast, and so in contact with the cell wall, will be called the external plasma-membrane, and that bounding the vacuole, the internal plasma-membrane. **1948** *New Biol.* V. 40 The plasma membrane is highly permeable to substances which are soluble in fats and in fat solvents. **1965** BELL & COOMBE tr. *Strasburger's Textbk. Bot.* 13 The inner plasma membrane surrounding the vacuole is known as the tonoplast, and that adjacent to the cell wall as the plasmalemma. **1968** R. RIEGER et al. *Gloss. Genetics & Cytogenetics* 341 In some cells (bacteria, plants), a cell wall..is universally recognized as a structure separate from the plasma membrane. **1970** AMBROSE & EASTY *Cell Biol.* viii. 258 The outer cell membrane, or plasma membrane (sometimes known as the cell membrane, or plasmalemma), has a unique role, since the cell interacts with its environment through it. **1928** I. LANGMUIR in *Proc. Nat. Acad. Sci.* XIV. 629 Plasma Oscillations.—If..we change the concentration of electrons by some transient external means, the resulting electric fields act..to equalize the concentration, but the potential energy of these fields is converted into kinetic energy of the electrons so that oscillations occur, and electric waves may result. **1970** W. A. HARRISON *Solid State Theory* iii. 288 Physically these plasma oscillations correspond to soundlike compression waves in the electron gas. **1972** AKASOFU & CHAPMAN *Solar–Terrestrial Physics* vii. 472 While the plasma cloud is streaming through the solar atmosphere it induces plasma oscillations there. These oscillations are observed at the earth as a Type II radio burst. **1968** *New Scientist* 24 Oct. 186 To bring about nuclear fusion..plasma with a density of 10^{14} nuclei per cu. cm. must be held together for about one second. To bring this about is the dream of plasma physicists. **1976** T. Beer *Aerospace Environment* i. 2 The plasma physicist can use the Earth's upper atmosphere as a gigantic laboratory to study the behaviour of a large-scale plasma being acted upon by the Earth's magnetic field. **1958** C. C. Adams *Space Flight* 345 Some scientists think that controlled fusion may be with us in 20 years or so, and if so we may completely bypass fission... Work in plasma physics will have to be carefully watched, and it is through research in this area that eventual success is expected. **1963** *Wall St. Jrnl.* 22 Jan., Kirtland researchers are delving into plasma physics—the study of partially ionized gases—to determine to what extent high-level nuclear blasts are likely to disrupt vital communications. **1970** G. K. Woodgate *Elem. Atomic Structure* i. 3 Quantitative calculations of the behaviour of free atoms are required for the less well-defined fields..of, for example, solid-state physics, plasma physics, and..astrophysics. **1961** *Flight* LXXIX. 462/2 Valuable information had been transmitted from the rubidium vapour magnetometer, two fluxgate magnetometers and the plasma probe. **1965** K. W. Gatland *Spacecraft & Boosters* II. 87/2 The radio-frequency plasma probe consisted of a pair of grid-like electrodes through which a radio-frequency electric field was applied to a small region near the satellite. **1977** *Sci. Amer.* Mar. 39/3 The first data available from the Ames Research Centre's plasma probe on *Pioneer 10* as it traversed interplanetary space were the hourly values of the speed of the [solar] wind. **1958** C. C. Adams *Space Flight* 54 A new Astronautics Research Laboratory with propulsion, astrophysical, and materials sections to study very high-energy fuels, including plasma propulsion systems. **1969** Boyd & Sanderson *Plasma Dynamics* v. 107 Space research has given a great impetus to the development of plasma propulsion since it has important potential advantages over conventional propellants, especially for long-range missions. **1961** *Aeroplane* C. 462/2 During hypersonic flight on the return from orbit, an ionized 'plasma sheath' will envelop the glider, impeding the reception and transmission of radio signals. **1969** M. A. Kasha *Ionosphere* iii. 48 A spacecraft is generally surrounded by some form of plasma sheath. This means that it is very difficult to measure..the electrical potential of the space plasma. **1959** *Welding Engineer* Feb. 50/2 Two 600-amp units power a 50-kw plasma torch, and voltage requirements are being set by the gas being used. **1961** *Jrnl. Appl. Physics* XXXII. 821/1 This article describes a plasma torch based on inductive coupling to an ionized gas... Conventional plasma torches require electrodes to carry energy to the gas. **1968** *Observer* 22 Dec. 4/5 The plasma torch, another torch device in industrial use, can virtually disintegrate material at a temperature of 36,000 degrees C.

b. Used *attrib.* to designate (the concentration of) substances in blood plasma.

1891 W. D. Halliburton *Text-bk. Chem. Physiol. & Path.* xv. 238 The globulin pre-existent in the blood plasma..may be termed plasma-globulin. **1927** [see *plasmapheresis]. **1941** *Amer. Jrnl. Path.* XVII. 360 The question whether increase of the plasma protein concentration would protect against heavy metal poisoning. **1956** *Nature* 4 Feb. 238/1 The amino-acid pattern of the urine from this cystinuric dog is.., apart from threonine, identical with that found in cases of human cystinuria, while the finding of a low plasma-cystine points to a similar etiology. **1961** *Lancet* 22 July 171/2 Because of the diurnal variation in plasma-cortisol (hydrocortisone) concentration, all blood samples were drawn between 9 A.M. and 10 A.M. **1969** E. Kelemen *Physiopath. & Therapy Human Blood Dis.* i. 105 About 80–90% of plasma proteins, i.e. fibrinogen, albumin, and certain globulins, including most of the plasma coagulation factors, are formed in the liver. **1975** J. W. Linman *Hematol.* v. 183/1 Plasma fibrinogen is increased in persons with valvular prostheses or homografts.

plasmablast, var. *plasmoblast b.

pla·sma cell. *Histology.* Also **plasma-cell.** [tr. G. *plasmazelle* (W. Waldeyer 1875, in *Arch. f. mikrosk. Anat.* XI. 189; also P. G. Unna 1891, in *Monatschr. f. prakt. Dermatol.* 1 Apr. 304): see Plasma and Cell *sb.*[1]] † **a.** (In Dict. s.v. Plasma 6.) *Obs.* **b.** A cell now recognized as the chief source of antibodies which is found in lymphoid tissue and at sites of chronic inflammation, and which has a strongly basophilic cytoplasm containing an extensive rough-surfaced endoplasmic reticulum and a usually eccentric nucleus. Cf. *plasmacyte.

1888 [in Dict. s.v. Plasma 6]. **1895** *Jrnl. R. Microsc. Soc.* 613 Waldeyer's plasma-cells correspond in staining reactions to Ehrlich's *Mastzellen*, but not to Unna's 'plasma-cells', and Waldeyer proposes to give up his use of the term as applied to normal elements of connective tissue. The cells he described as 'plasma-cells' are Ehrlich's *Mastzellen* and eosinophilous cells. **1906** *Jrnl. Amer. Med. Assoc.* 20 Oct. 1272/2 The intimæ of the smaller arteries are lifted or completely dissected off by an exudate composed chiefly of cells of the lymphocyte series, among which are examples of the typical plasma cell. **1929** *Amer. Jrnl. Ophthalm.* XII. 731/1 In 1875 Waldeyer applied the name of plasma cell to a poorly differentiated type of wandering cell which he found in chronically inflamed connective tissue. **1940** *Acta Med. Scand.* CIII. 569 When Waldeyer..used the term plasma cell in 1875, he elected to do so not because he thought these cells secreted part of the plasma, but because of the abundant protoplasm. He made use of the term plasma cell to designate a number of different cells rich in protoplasm, but in subsequent works, especially those of Unna..and Marschalko.., the use of the name was restricted to the cell that is now known as the plasma cell: a cell rich in protoplasm, with eccentrically placed nucleus, relatively small, round or oval, with five to eight bands of chromatine extending from the centre like the spokes of a wheel; around the nucleus is a lighter zone, whilst the abundant protoplasm is otherwise dark, basophile. **1960** *New Biol.* XXXI. 100 Absolute proof that plasma cells make antibody was furnished by an ingenious and elegant technique devised by Dr. A. H. Coons. **1968** Passmore & Robson *Compan. Med. Stud.* I. xvi. 5/1 Plasma cells are found where foreign proteins are likely to gain entrance to the body, e.g. beneath the epithelial membranes lining the respiratory and alimentary tracts. **1975** *Lancet* 3 May 1031/2 The plasma-cell is one of the effector cells of the B-lymphocyte system.

Hence **plasma-celled** *a.*, composed of plasma cells; **plasmace·llular** *a.* (also **plasmo-** and as two words), of or pertaining to plasma cells.

1929 *Jrnl. Path. & Bacteriol.* XXXII. 293 (*heading*) Two cases of myelomatosis: (1) Diffuse plasma-celled (2) with tumour-like nodules and visceral lesions. **1947** *Nature* 12 Apr. 499/1 (*heading*) Plasma cellular reaction and its relation to the formation of antibodies *in vitro*. **1948** R. A. Willis *Path. Tumours* l. 787 The multinucleated cells of plasma-celled and other myelomas are of no special histogenetic significance. **1957** *Jrnl. Amer. Med. Assoc.* 4 May 20/2 In the broadest sense of the term this plasmocellular barrier may be called an immune process. **1971** *Biol. Abstr.* LII. 6201/2 The ultrastructure of plasmacellular paracrystalline inclusions, detected in the sternal marrow of a patient with type k micromolecular myeloma was studied.

plasmacyte (plæ·zmăsəit). *Histology.* Also **plasmo-**. [f. Plasma + -cyte.] = *plasma cell b.

1941 I. N. Kugelmass *Blood Disorders in Children* xiii. 498 Accumulation of plasmacytes is probably due to migration from the blood stream and subsequent multiplication by fission. **1961** *Lancet* 16 Sept. 639/2 We must think of the effective (abnormal) cells in the spleen as plasmacytes which have taken on the character of low-grade tumour cells. **1976** *Ann. Rev. Microbiol.* XXX. 591, B cells were being transferred to the local granuloma as the site of antigen, and were then transferring to antibody-producing plasmacytes.

So **plasma-, plasmocytic** (-səi·tik, -si·tik) *a.*, of, pertaining to, or composed of plasmacytes; **plasma-, plasmocy·toid** *a.*, resembling (that of) a plasmacyte.

1932 *Jrnl. Path. & Bacteriol.* XXXV. 545 Histologically the tumour was plasmocytic and there were many cells of giant size. **1959** M. Burnet *Clonal Selection Theory of Acquired Immunity* iv. 61 Active proliferation to produce plasmacytoid cells and lymphocytes with active antibody-liberating capacity. **1970** R. T. Silver *Morphol. Blood & Marrow in Clin. Pract.* ix. 111 In other cases, the lymphocyte..may assume a plasmacytoid shape while still retaining the nuclear configuration of a lymphocyte. **1972** *Acta Med. Scand.* CXCII. 291/2 The morphological investigations have,..in most of the cases, revealed the presence of an infiltration in the lymphoreticular organs of plasmocytoid cells of varying maturity, probably responsible for the production of the monoclonal immunoglobulin. **1974** *Immunol.* XXVI. 486 The development of giant cells, epithelioid tubercles and plasmacytic infiltrates.

plasmacytoma (plæzmăsəitōu·mă). *Path.* Also **plasmo-**. Pl. **-omas, -omata.** [mod.L., ad. G. *plasmocytom* (H. Boit 1907, in *Frankfurter Zeitschr. f. Path.* I. 172): see prec. and *-oma.] A myeloma composed largely of plasma cells.

1907 *Index Medicus* V. Index 173/1 Plasmocytoma. **1931** *Amer. Jrnl. Med. Sci.* CLXXXI. 171 There appear in the literature cases of both extra-osseous and intra-osseous plasmocytomata. *Ibid.* 178 There is no very sharp line of demarcation between localized, benign plasmocytomata on the one hand, and the malignant, fatal multiple myelomata on the other. **1940** *Acta Med. Scand.* CIII. 569 There is no increase of globulin in a number of plasmocytoma cases. **1961** [see *myelomatosis]. **1972** [see *multiple *a.* and *sb.* A. 3d]. **1973** *Jrnl. Bone & Joint Surg.* A. LV. 1749 Most solitary plasmocytomas eventually become classic multiple myelomas if they are followed long enough. **1976** *Lancet* 6 Nov. 1003/2 Bone-resorbing factors have also been identified in cultured human tumour cells, including those from plasmacytomas.

plasmacytosis (plæzmăsəitōu·sis). *Path.* Also **plasmo-**. Pl. **-cytoses.** [f. as prec. + -osis.] The presence of more plasma cells than usual in a tissue.

1930 *Q. Cumulative Index Med.* VII. 1278/1 Uterine plasmocytoma and plasmocytosis. **1959** *New Engl. Jrnl. Med.* 5 Nov. 954/2 Multinucleate and atypical forms are observed in both the neoplastic and non-neoplastic plasmocytoses. **1963** *Lancet* 19 Jan. 125/2 In several cases there was bone-marrow plasmacytosis. **1968** *Amer. Jrnl. Clin. Path.* L. 304/1 Cytologically, all cases of plasmacytosis were characterized by proliferation of mature looking plasma cells in addition to more primitive lymphoreticular forms.

plasmagel (plæ·zmădʒel). Also **plasma gel.** *Biol.* [f. Plasma + *gel *sb.*] Gelatinous cytoplasm, such as surrounds the plasmasol in an amœboid cell. Cf. *plasmasol.

1923 S. O. Mast in *Proc. Nat. Acad. Sci.* IX. 258 By careful observations on Amœba proteus in motion the following structures can clearly be differentiated: (1) A central elongated fluid portion; (2) A solid layer surrounding the fluid portion; (3) A very thin elastic surface layer or membrane. The first I shall designate the *plasmasol*, the second the *plasmagel* and the third the *plasmalemma*. **1939** W. B. Yapp *Introd. Animal Physiol.* iv. 131 *Amœba* consists of three layers. On the outside there is a thin plasmalemma, which can be lifted off with needles and is of a dough-like consistency. Inside this is the plasmagel, which is solid, and includes the classical ectoplasm and some endoplasm; inside this again is liquid plasmasol. **1942** G. H. Bourne *Cytol. & Cell Physiol.* iii. 97 Streaming in [the aquatic plant] *Elodea* cells is associated with gelation, and..is abolished by application of sufficient hydrostatic pressure to liquefy the cortical plasmagel. **1970** Ambrose & Easty *Cell Biol.* xi. 360 The plasma gel is located near the plasma membrane; it is generally free from granules and other inclusions. **1973** N. Adresen in K. W. Jeon *Biol. of Amœba* iv. 102 The large pseudopodia ..may contain several parallel streams of plasmasol, each running in its own tube of plasmagel.

plasmagene (plæ·zmădʒīn). *Genetics.* [f. Plasma + *gene.] A supposed cytoplasmic entity having genetic properties.

1939 C. D. Darlington *Evol. Genetic Systems* xx. 121 The particles in the nucleus are genes; those in the plastids and cytoplasm may perhaps be treated more rigorously if we also think of them as genes—plastogenes and plasmagenes. **1952** [see *plasmid]. **1963** E. Mayr *Animal Species & Evolution* vii. 172 Like the chromosomal genes, plasmagenes seem to consist of nucleic acid molecules (including possibly RNA). **1965** Stern & Nanney *Biol. of Cells* xix. 529 The concept of the plasmagene, a gene-like cytoplasmic element capable of differential assortment and replication during development, provided a possible means of rationalization... Nevertheless, this interpretation was never generally accepted with enthusiasm. *Ibid.*, The episomes of bacteria..on occasion behave like plasmagenes. **1974** A. T. Soldo in W. J. Van Wagtendonk *Paramecium* 377 When kappa particles were first discovered they were generally regarded as cytoplasmic units of hereditary or 'plasmagenes'.

Hence **plasmage·nic** *a.*, of, pertaining to, or being a plasmagene.

1950 *Heredity* IV. 17 The existence of plasmagenic subunits in any one of the self-duplicating, cytoplasmic structures has not yet been established. **1968** J. A. Serra *Mod. Genetics* III. xix. 37 Probably infection is produced by a particle of the plasmagenic type.

plasmal (plæ·zmăl). *Biochem.* [a. G. *plasmal* (coined with *plasmalogen): see Plasm and *-al[2].] An aldehyde formed by the hydrolysis of a plasmalogen; chiefly used *attrib.* in **plasmal reaction**, a modification of the Feulgen reaction for detecting plasmalogens and aldehydes in tissue.

1925 *Chem. Abstr.* XIX. 1156 Plasmal is an aldehyde, a solid at ordinary temp. *Ibid.*, In accordance with its aldehyde nature, plasmal quickly takes a violet color when stained with H_2SO_3 fuchsin... This staining process is termed the plasmal reaction,..and the presence of plasmal, or its precursor plasmalogen, in frozen sections of tissues can be shown microscopically. **1949** *Stain Technol.* XXIV. 19 The plasmal reaction is here modified so that it is made specific for acetal lipids alone. **1966** *McGraw-Hill Encycl. Sci. & Technol.* X. 397/2 The Feulgen test for plasmalogens depends on the liberation of plasmal, principally palmitaldehyde and stearaldehyde, from these lipids by the action of mercuric chloride and acetic acid. **1969** *Acta Histochem.* XXXII. 425 (*heading*) Application of the plasmal reaction to suspensions.

plasmalemma (plæ·zmălemă). *Biol.* Pl. **-lemmas, -lemmæ.** [f. Plasma + Lemma[2].] The thin membrane immediately surrounding the cytoplasm of a cell, which restricts the passage of molecules into it.

1923 [see *plasmagel]. **1931** J. Q. Plowe in *Protoplasma* XII. 202 Mast (1924) has given us the term 'plasmalemma'... It has seemed permissible..to extend its use to the botanical world, and to employ it to denote a distinct, differentiated layer on the outer surface of the plant protoplast. **1965** K. Esau *Plant Anat.* (ed. 2) ii. 15 Surface membranes delimit the cytoplasm from the wall (plasma membrane, plasmalemma, or ectoplast) and from the vacuole (vacuolar membrane, or tonoplast). **1965**, **1970** [see *plasma membrane* s.v. *plasma 6 a]. **1970** *Austral. Jrnl. Bot.* XVIII. 285 Host plasmalemmae are invaginated by invading hyphae, and encapsulations are formed. **1971** *Proc. R. Soc.* B. CLXXVIII. 195 (*caption*) The plasmalemmas..of the two cells. **1976** *Nature* 9 Sept. 158/2 We report..a gradual increase in microviscosity of protoplast plasmalemma from petals of ageing rose flowers.

Hence **pla·smalemmal** *a.*, of, pertaining to, or formed from a plasmalemma.

1968 *Jrnl. Cell Biol.* XXXVII. 252 (*caption*) Grazing section of the endothelium in a blood capillary of rat myocardium showing the large accumulation of plasmalemmal vesicles..on the tissue front. **1975** *Ibid.* LXIV. 505/2 After fixation by immersion, the plasmalemmal membranes limiting the endothelial and epithelial cells were heavily delineated by the reaction product.

plasmale·mmasome. Also **-lemmo-**. *Cytology.* [f. prec. + *-some[4].] A plant or microbial cell organelle formed by invagination of the plasma membrane and composed of tissue derived from it.

1962 M. R. Edwards in *Abstr. 8th Internat. Congr. Microbiol.* 31/1 The ingrowths of the plasmalemma may branch repeatedly and anastomose to give rise to a complicated honeycomb-like organelle. This organelle may be termed a plasmalemmosome in view of its origin from the plasmalemma. **1968** *Ann. Bot.* XXXII. 468 In

higher plants, at least some of the plasmalemmasomes appear to have granular or fibrillar contents within well-defined vesicles. **1973** *Protoplasma* LXXVI. 235 The morphology of plasmalemmasomes in the species examined is variable and ranges from vesicles or tubules within the plasmalemma invagination to parallel arrays of membrane lamellae. Plasmalemmasomes thus appear to be primarily excess plasma membrane that has accumulated, perforce, endocellularly.

plasmalogen (plæzmæ·lodʒěn). *Biochem.* [a. G. *plasmalogen* (Feulgen & Voit 1924, in *Pflügers Arch. f. ges. Physiol.* CCVI. 399): see *plasmal and *-ogen.] Any of a class of phospholipids that yield an aldehyde on mild hydrolysis and are now regarded as having an unsaturated ether linkage in place of one of the fatty acid ester linkages.

1925 *Chem. Abstr.* XIX. 1155 Throughout the protoplasm of animal tissues there is to be found, very widely disseminated, a substance of lipoid character, insol. in water but sol. in org. solvents and extractable by alc., termed plasmalogen. When subjected to acids..this substance is split into undetd. components termed plasmal. **1964** A. White et al. *Princ. Biochem.* (ed. 3) v. 75 Plasmalogens without a nitrogenous base, *i.e.*, α, β-unsaturated ethers of phosphatidic acid, have also been reported in animal tissues, *e.g.*, liver. **1964** [see *cephalin[2]]. **1968** Passmore & Robson *Compan. Med. Stud.* I. x. 5/1 The plasmalogens have an unsaturated ether group, rather than an ester group, on the α position [of the glycerol molecule]. Those of ethanolamine and choline form a big proportion of the phospholipids of brain and heart.

Hence **pla:smaloge·nic** *a.*

1939 *Chem. Abstr.* XXXIII. 8635 This fraction contg. plasmalogenic acid (I) increases with the time of alk. hydrolysis. **1962** *Compar. Biochem. & Physiol.* V. 220 (*caption*) The arrow..points out the phosphatidic acid plasmalogen (plasmalogenic acid) characteristic of blood-fed leeches. **1970** R. W. McGilvery *Biochem.* xxiv. 600 The vinyl ether group [in plasmalogens] is believed to be formed by reduction of a diglyceride... The resultant plasmalogenic diglyceride reacts..to form phosphatidalcholine..in the same way that ordinary diglycerides react in the formation of the phosphatidylcholines.

plasmapause (plæ·zmăpǭz). [f. Plasma + Pause *sb.*] The outer limit of a plasmasphere, marked by a sudden change in the plasma density and lying wholly within the magnetosphere (in the case of the earth extending up to several earth radii from its centre at equatorial latitudes).

1966 [see *plasmasphere]. **1971** *New Scientist* 1 July 8/2 The time at which Io crosses the plasmapause [of Jupiter] and sets off a burst of radio emission therefore depends on the Sun's activity. **1976** T. Beer *Aerospace Environment* vi. 113 At around fifty-five degrees geographic latitude the plasmapause is in the F region and it is almost vertical. *Ibid.* vii. 137 We can expect a plasmapause on Mars somewhere above 400 km.

plasmapheresis (plæzmăfe·rīsis, -fĕrī·sis). *Med.* Also **-phoresis** (-fŏrī·sis), † **-pharesis.** [f. Plasma + Aphæresis; *plasmaphoresis* by alteration (cf. *-phoresis).] The removal of blood plasma from the body by the withdrawal of blood, its separation into plasma and cells in a centrifuge, and the reintroduction of the cells suspended in a harmless medium.

1920 G. H. Whipple et al. in *Amer. Jrnl. Physiol.* LII. 99 Bleeding a dog from a large artery and a simultaneous replacement of a red blood cell Locke's solution mixture may be called 'plasma depletion' or 'plasmapharesis'. **1927** M. Bodansky *Introd. Physiol. Chem.* vii. 168 Reduction of the plasma proteins by plasmapharesis..results in a condition of shock. **1935** H. Sobotka in Harrow & Sherwin *Textbk. Biochem.* iv. 144 Experimental anemia produced by repeated withdrawal of blood or plasma (plasmaphoresis). **1943** *Jrnl. Immunol.* XLIV. 112 Rabbits whose protein-reserves have been reduced by plasmapheresis and a low-protein diet (carrots) show a definitely lessened capacity to produce antibodies. **1971** *Nature* 27 Aug. 629/2 Before and during corticosteroid-induced labour large samples of plasma..were obtained by plasmaphoresis from chronically implanted catheters. **1974** Passmore & Robson *Compan. Med. Stud.* III. li. 16/2 The donation of blood by plasmapheresis is now a major part of blood transfusion practice. The main purpose is to procure large amounts of plasma rich in specific immunoglobulins..and blood grouping reagents.

plasma sheet (plæ·zmă ʃīt). Also **plasma-sheet.** [f. Plasma + Sheet *sb.*[1]] A layer of plasma in the magnetotail which lies in the equatorial plane of the earth some distance beyond the plasmapause and has two branches that diverge to reach the earth in polar latitudes.

1966 *Physical Rev. Lett.* XVI. 138 (*heading*) Electrons in the plasma sheet of the earth's magnetic tail. **1970** V. M. Vasyliunas in G. Skovli *Polar Ionosphere* ii. 27 Intense fluxes of electrons extend across the entire magnetotail, forming the plasma sheet, first detected by the Luna 2 space probe in 1959. **1974** *McGraw-Hill Yearbk. Sci. & Technol.* 276/2 The plasmasheet plays a key role in the development of the magnetospheric substorm.

plasmasol (plæ·zmăsǫl). Also **plasma sol.** *Biol.* [f. Plasma + *sol *sb.*[6]] Cytoplasm in the form of a sol, such as exists in the middle regions of amœboid cells. Cf. *plasmagel.

1923, etc. [see *plasmagel]. **1951** [see *gelate *v.*]. **1970** Ambrose & Easty *Cell Biol.* xi. 360 The relative viscosity of the plasma sol is quite low, being about two to ten times greater than that of water; on the other hand the gel is a moderately rigid structure which shows elasticity and breaks on application of a critical force.

plasmasphere (plæ·zmăsfiəɹ). [f. Plasma + Sphere *sb.*] The roughly toroidal region surrounding the earth at latitudes away from the poles in which there is a relatively dense plasma of low-energy electrons and protons that is thought to rotate with the earth; an analogous region around another planet. Cf. *plasmapause.

1966 D. L. Carpenter in *Jrnl. Geophysical Res.* LXXI. 695/1 With regard to nomenclature, the word 'plasmapause' will be used when the three-dimensionality of the knee phenomenon is emphasized... The word 'plasmasphere' will be used to indicate the dense region inside the plasmapause, and 'plasma trough' to indicate the tenuous region outside. **1971** *New Scientist* 1 July 8/1 The work..led to this detection of the Jovian plasmasphere from the now well-known modulation of Jupiter's radio emission caused by Io. **1974** *McGraw-Hill Yearbk. Sci. & Technol.* 345/2 The characteristic changes in size and shape of the plasmasphere which are observed by satellites and by ground-based vlf measurements have a direct effect on the dynamics of the F region of the ionosphere and on the location of regions of wave turbulence in the outer magnetosphere.

Hence **plasmasphe·ric** *a.*

1974 *Nature* 29 Mar. 401/1 He II λ303 Å emissions will be incident on the night-time upper atmosphere, arising from resonance scattering of the solar helium lines..by plasmaspheric He^+ ions. **1978** *Ibid.* 26 Jan 310/2 Inside the plasmasphere the characteristic emission is a featureless 'plasmaspheric hiss' which effectively fills the high density region but is believed to be generated by cyclotron resonant interactions with >30 keV electrons near the outer edge of the plasmasphere.

plasmid (plæ·zmid). *Biol.* [f. Plasm + *-id* (cf. Id, *chromatid).] Any genetic structure in a cell that can replicate independently of the chromosomes; *esp.* one in the cytoplasm of a bacterium.

1952 J. Lederberg in *Physiol. Rev.* XXXII. 403, I propose *plasmid* as a generic term for any extrachromosomal hereditary determinant. *Ibid.*, The taxonomic classification of plasmids as viruses, symbionts, or plasmagenes should not obscure careful descriptions of their function, hereditary or pathological, or both. *Ibid.* 414, κ, a plasmid in *Paramecium aurelia*. *Ibid.* 425 This review has contrasted the various forms of plasmid: the hereditary parasites as against the functionally co-ordinated plasmagenes, with the mutualistic endosymbionts somewhere between. **1964** *Daily Mirror* 24 July 8/2 The Rogue Bug's real name is RTF plasmid—R.T.F. stands for Resistance Transfer Factor. It is a tiny particle that appears to move from one bacterial cell to another, carrying with it a built-in resistance to new drugs. **1969** A. M. Campbell *Episomes* i. 13 Episomes are thus distinguished from chromosomal genes on the one hand and obligately cytoplasmic elements (plasmids) on the other. **1973** R. G. Krueger et al. *Introd. Microbiol.* xv. 421/1 There are..transmissible plasmids or sex factors like the F factor, which promote their own transfer to recipient bacteria, and there are nontransmissible plasmids which are incapable of transferring themselves to recipient cells. **1975** *Sci. Amer.* July 25 It has been called plasmid engineering, because it utilizes plasmids to introduce the foreign genes... Because of the method's potential for creating a wide variety of novel genetic combinations in microorganisms it is also known as genetic engineering. **1977** *Time* 18 Apr. 48/1 They possess much smaller closed loops of DNA, called plasmids—which consist of only a few genes.

plasmin. Add † *Obs.* to sense in Dict.

2. *Physiol.* A proteolytic enzyme which destroys blood clots by attacking fibrin.

1945 Christensen & Macleod in *Jrnl. Gen. Physiol.* XXVIII. 581 Under this scheme the activated enzyme may be termed 'plasmin' in comformity [*sic*] with common usage for proteases, where the prefix indicates the source of the enzyme, followed by -in... The inactive enzyme as it occurs in serum and plasma may be designated as 'plasminogen' to indicate its source, the plasma, and also to indicate that it is in an inactive, precursor state... The term 'plasmin' has been used in the past to designate a fraction of blood obtained by a special salting-out procedure. This usage, however, has become obsolete and the possibility of confusion with the proteolytic enzyme system is remote. **1962** [see *fibrinolysin* s.v. *fibrino-]. **1968** Passmore & Robson *Compan. Med. Stud.* I. xxvi. 16/1 The active enzyme in this system, plasmin, is proteolytic but possesses a preference for fibrin as substrate. It is formed from an inactive soluble blood protein precursor, plasminogen, by the action of plasminogen activator. **1976** *Nature* 22 Jan. 235/2 Plasminogen is the plasma proenzyme which, on conversion to its active form, plasmin, is considered responsible for lysis of fibrin deposits resulting from physiological or pathological activation of the coagulation cascade.

plasminogen (plæzmi·nŏdʒěn). *Physiol.* [f. prec. + *-ogen.] The inactive precursor, present in blood, of the enzyme plasmin.

1945 [see prec.]. **1962** *Lancet* 27 Jan. 191/1 The presence of plasminogen activator in tissue suggests that it may have a role in the removal of unwanted fibrin. **1968, 1976** [see prec.].

plasmoblast (plæ·zmoblɑst). *Histology.* [f. Plasmo- + -blast.] The precursor of a plasmacyte. † **a.** (See quot.; cf. *plasmocyte a.) *Obs.*

1897 G. Eisen in *Proc. Calif. Acad. Sci.* (*Zool.*) I. 16 The polar accumulations must, therefore, be considered as something entirely separate from the balance of the cytoplasm; they, in fact, give rise to the plasmocytes, and may, therefore, appropriately be called plasmocytoblasts, or for the sake of brevity, plasmoblasts.

b. Also **plasmablast.** [a. G. *plasmoblast* (S. Moeschlin 1940, in *Helv. Med. Acta* VII. 231).] An immature plasma cell. Cf. *plasmacyte.

1942 *Amer. Jrnl. Anat.* LXX. 485 Moeschlin believes that some of the youngest lymphoid reticular cells begin to assume plasma cell characters, and that beginning with this 'plasmoblast' there is a developmental sequence of stages leading to the mature plasma cell, and that this sequence is probably independent of the lymphocytic line. **1949** *Jrnl. Exper. Med.* XC. 165 Maturation of plasmoblasts to plasma cells was associated..with reduction in the size of the nucleus and disappearance of the PNA [*sc.* pentose nucleic acid] in the nucleolus. **1959** M. Burnet *Clonal Selection Theory of Acquired Immunity* vii. 113 In lymph nodes the first small foci of plasmoblasts appeared in a perivascular situation near the arterioles of the medullary cords. **1973** R. I. Weed tr. *M. Bessis's Living Blood Cells & their Ultrastructure* vii. 521/1 Plasmablast. On a smear, this cell measures 15 to 20 microns in diameter. Its principal characteristic is the profound basophilia of its cytoplasm.

Hence **plasmobla·stic** *a.*

1970 *Jrnl. Nuclear Med.* XI. 599/2 EH was a 54-year-old W.F. with plasmoblastic multiple myeloma with widespread skeletal involvement, [etc.]. **1975** *Biol. Abstr.* LIX. 6033/1 In lymphoblastic and plasmoblastic acute leukemias pathological cells were capable of synthesizing various Ig classes.

plasmocellular, var. *plasmacellular *a.*

Plasmochin (plæ·zmŏkin). *Pharm.* [f. Plasmodium + G. *chin*(*in* Quinine.] A proprietary name for *pamaquin. Cf. *Plasmoquine.

1926 *Trade Marks Jrnl.* 17 Mar. 631 Plasmochin... Chemical substances prepared for use in medicine and pharmacy. Bayer Products, Limited,..London,..merchants and manufacturers. **1926** *Lancet* 16 Oct. 825/2 The great merits of plasmochin are that it is cheaper than quinine, tastes better, and gives rise to less unpleasant secondary effects. **1926** *Official Gaz.* (U.S. Patent Office) 26 Oct. 715/1 I. G. Farbenindustrie Aktiengesellschaft..Germany. Filed Aug. 14, 1926. Plasmochin..Preparation for the Treatment of Malaria. Claims use since about December, 1925. **1948** J. H. Burn *Lect. Notes Pharmacol.* 90 Pamaquin was introduced by the Germans as plasmochin; it is a quinoline derivative. **1962** —— *Drugs, Med. & Man* xx. 197 Domagk was working in the Bayer fabrik in Elberfeld in Germany, where the advances had been made which led to the discovery of the antimalarial agents plasmochin and later atebrin.

plasmocyte (plæ·zmosəit). *Histology.* [f. Plasmo- + -cyte.] † **a.** (See quots.) *Obs.*

1897 G. Eisen in *Proc. Calif. Acad. Sci.* (*Zool.*) I. 4 A new corpuscle, which I have termed *plasmocyte*. *Ibid.* 13 (*heading*) Plasmocytes.—I apply this name to a hitherto undescribed element in the blood, first described by me in the blood of Batrachoseps. **1897** *Jrnl. R. Microsc. Soc.* 271 The plasmocyte may be defined as a corpuscle, generally without a cell-wall, always without a nucleus, consisting of the archosome and three spheres of cytoplasm. It shows power of growth, movement, phagocytosis, &c. **1900** E. B. Wilson *Cell* (ed. 2) i. 52 Eisen ('97) asserts that in the blood of a salamander..the attraction-sphere..containing the centrosomes may separate from the remainder of the cell (nucleated red corpuscles) to form an independent form of blood-corpuscle or 'plasmocyte', which leads an active life in the blood.

b. var. *plasmacyte.

plasmocytoma, -cytosis, varr. *plasmacytoma, -cytosis.

plasmodesma (plæzmode·zmă). *Bot.* Also anglicized as **pla·smodesm** (-dez'm). Pl. **plasmodesmata**; also **-desmæ**, ‖ **-desmen**, **-desms**; (erron.) **-desma.** [a. G. *plasmodesma*, pl. *plasmodesmen* (E. Strasburger 1901, in *Jahrb. f. wissensch. Bot.* XXXVI. 503 (sing. form on p. 607)): see Plasmo- and Desma.] A narrow thread of cytoplasm that passes through cell walls and affords communication between plant cells.

1905 *Amer. Naturalist* XXXIX. 220 A new point of view was introduced into the discussion by the very important paper of Strasburger, in 1901. He considered the protoplasmic connections as sufficiently clearly differentiated structures to rank as organs of the cell and proposed for them the name plasmodesmen. **1925** E. B. Wilson *Cell* (ed. 3) i. 103 It is probable that an important part in the coördination of the cell-activities is played by direct protoplasmic connections between cells ('cell-bridges', 'plasmodesms'). **1927** Fritsch & West *Treat. Brit. Freshwater Algae* (ed. 2) 40 The pit-membrane in such cases is probably traversed by plasmodesmae. **1931** *Stain Technol.* VI. 127 (*heading*) A technic for demonstrating plasmodesma. **1934** L. G. Livingstone in *Amer. Jrnl. Bot.* XXI. 707 The only way to arrive at a satisfactory understanding of the true nature of plasmodesmata in the living plant is to study the cell walls in an unaltered state. **1935** —— in *Ibid.* XXII. 75 The word plasmodesma is proposed for the singular, to designate an individual structure, and plasmodesmata is used in the plural sense. **1941** *Bot. Rev.* VII. 254 Plasmodesmata generally run straight from cell to cell, and only in the neighborhood of intercellular spaces or pit membranes of narrow pits are the outer plasmodesmata curved. **1951** F. Drouet in G. M. Smith *Man. Phycol.* viii. 163 Species of *Stigonema* have trichomes which in age become multiseriate throughout; the spherical or depressed-spherical cells are connected by strands of protoplasm usually considered to be plasmodesmen. **1966** *Protoplasma* LXI. 82 In longitudinal sections of the plasmodesms the real continuity of the cytoplasmic membranes can be observed. **1975** *Ann. Rev. Plant Physiol.* XXVI. 13 Despite other suggestions, the word plasmodesma has continued to be used to describe a protoplasmic connection. *Ibid.* 14 Plasmodesmata have been described in angiosperms, gymnosperms, pteridophytes, bryophytes, and many algae.

Hence **plasmode·smatal** *a.*, of, pertaining to, or being plasmodesmata.

1961 in Webster. **1964** J. Heslop-Harrison in H. F. Linskens *Pollen Physiol. & Fertilization* ii. 41 The peripheral archesporial cells show plasmodesmatal links with the tapetal cells, and these in turn with the cells of the inner wall layer of the anther. **1975** *Ann. Rev. Plant Physiol.* XXVI. 15 There are considerable technical difficulties involved in obtaining reliable estimates of plasmodesmatal frequency.

plasmogamy (plæzmǫ·gămi). *Biol.* [ad. G. *plasmogamie*: see Plasmo- and *-gamy.] The fusion of the cytoplasm of two or more cells.

1912 E. A. Minchin *Introd. Study Protozoa* viii. 128 In many cases, union of distinct individuals can be observed which have nothing to do with syngamy, since no fusion takes place of nuclei, but only of cytoplasm. Such unions are distinguished as plastogamy (or plasmogamy) from true syngamy. **1932** L. A. Borradaile et al. *Invertebrata* ii. 26 The union of nuclei is karyogamy: in most cases of syngamy it is accompanied by plasmogamy or the fusion of cytoplasm... Plastogamy..is plasmogamy without karyogamy. **1958** *Ibid.* (ed. 3) ii. 40 Here may be mentioned the union of individuals by fusion of their cytoplasm, the nuclei remaining distinct, which is practised by the Mycetozoa..and in some other cases. This process, which is not syngamy, is known as plasmogamy, and its product as a plasmodium. **1969** F. E. Round *Introd. Lower Plants* iv. 60 In most of the higher terrestrial fungi, plasmogamy, which is the fusion of small masses of cytoplasm containing the nuclei, is separated in time from karyogamy.

plasmoid (plæ·zmoid). *Physics* and *Astr.* [f. Plasm(a + -oid.] A coherent mass of plasma (*plasma 5*).

1956 W. H. Bostick in *Physical Rev.* CIV. 292/1 The plasma is emitted not as an amorphous blob, but in the form of a torus... We shall take the liberty of calling this toroidal structure a plasmoid, a word which means plasma-magnetic entity. The word plasmoid will be employed as a generic term for all plasma-magnetic entities. [*Note*] The term 'plasmon' (in line with the term 'geon' used by Wheeler) was originally proposed. However, David Pines (of Princeton University) has pointed out that the term 'plasmon' should be reserved for a quantum of plasma-oscillation energy. He kindly proposes the term 'plasmoid', which we adopt. **1962** Riley & Sailor *Space Systems Engin.* v. 129 Techniques have been developed..whereby doughnut-shaped 'blobs' of plasma or plasmoids can be projected by magnetic forces at speeds exceeding 10^7 cm/sec. **1971** S. Singer *Nature of Ball Lightning* viii. 119 Spherical plasmoids of relatively small dimensions, such as those reported for ball lightning or involved in experimental work, have presented greater difficulty for theoretical analysis. **1976** *Nature* 12 Feb. 451/1 As the galaxy ploughs through a dense intracluster gas at supersonic speed, it is envisaged to eject pairs of plasmoids in opposite directions, as commonly assumed for 'normal' double radio galaxies.

plasmolysis. Add: (Earlier example.) Hence also **plasmoly·tically** *adv.*, by means of plasmolysis; **pla·smolysed, -lysing** *ppl. adjs.*; **pla·smolysable** *a.*, capable of undergoing plasmolysis; **pla:smolysabi·lity.**

1883 *Q. Jrnl. Microsc. Sci.* XXIII. 151 (*heading*) On plasmolysis and its bearing upon the relations between cell wall and protoplasm. *Ibid.* 152 The protoplasmic body would appear to separate with a 'smooth surface' from the cell wall on treatment with the plasmolysing solution. *Ibid.* 153 Naegeli..described strings of protoplasm which connect the contracted protoplasmic body with the cell wall in plasmolysed cells. **1896** *Jrnl. Linnean Soc.: Bot.* XXXI. 370 A few of the younger leaves and leaf-cells are found to be living and plasmolysable, but show no assimilation. **1903** *Science* 1 May 706/2 A reduction of temperature gave rise to parthenogenetic spore formation.., as was also the case when water was plasmolytically withdrawn from the cells. **1955** P. J. Kramer in W. Ruhland *Encycl. Plant Physiol.* I. ii. 212 Errors in the plasmolytic method include adhesion of the cytoplasm to the wall, penetration of the plasmolyzing solute, and difficulty in measuring cell volume. **1960** L. Picken *Organization of Cells* iii. 77 The plasmolysability of the cells implies that the cell wall cannot contract below a certain area. **1964** J. Levitt in D. W. Newman *Instrumental Methods Exper. Biol.* xiii. 420 These very cells, with which the incipient plasmolysis method leads to difficulties, cannot have their osmotic potentials determined in any other way, and at least approximate values may be obtained plasmolytically. **1971** *Nature* 16 July 159/2 Attempts to fix and section plasmolysed sieve tubes have not so far yielded supporting evidence.

plasmolyte (plæ·zmoləit). *Physiol.* [f. Plasmo- + Gr. λυτ-ός loosed, soluble (f. λύειν to loosen), or f. G. *plasmolytikum* *plasmolyticum.] = next.

1927 *Biol. Abstr.* I. 238/2 The increase in volume after full plasmolysis is due to penetration of the plasmolyte. **1935** *Plant Physiol.* X. 119 That concentration of a harmless non-permeable plasmolyte which at osmotic equilibrium causes the protoplasm to recede ever so little from the cell wall. **1939** *Ibid.* XIV. 132 As a rule a minimal area form is not assumed during the process of plasmolysis but is reached either on standing in the plasmolyte, or at least after a slight degree of subsequent deplasmolysis. **1964** J. Levitt in D. W. Newman *Instrumental Methods Exper. Biol.* xiii. 419 Potassium nitrate was the plasmolyte used by many of the earlier investigators.

plasmolyticum (plæzmoli·tikɒm). *Physiol.* Pl. **-lytica.** [mod.L. (ad. G. *plasmolytikum*), f. Plasmo- + Gr. λυτικ-ός able to loosen (f. λύειν to loosen).] A substance used to produce plasmolysis.

1943 *Ann. Bot.* VII. 269 In experiments on rates of plasmolysis or deplasmolysis one would expect to find the wall-resistance exercising an effect when sucrose is used as the plasmolyticum. **1946** *New Phytologist* XLV. 7 The plasmolysis form taken up in any plasmolyticum, sucrose and KCl included, is much influenced by the distance of the cell from the cut edge of the strip of epidermis. **1973** Cocking & Evans in H. E. Street *Plant Tissue & Cell Culture* v. 100 Irreversible, damaging effects on the viability of isolated protoplasts readily result from the use of plasmolytica of either too high or too low an osmotic potential. **1974** A. J. Peel *Transport of Nutrients in Plants* x. 189 The sections were irrigated with plasmolytica composed of 0·5–2·0 M solutions of glucose or fructose.

plasmoma (plæzmōu·mă). *Path.* Pl. **plasmomata.** [f. Plasm(a + *-oma.] = *plasmacytoma.

1901 L. P. Hamburger in *Johns Hopkins Hosp. Bull.* XII. 43/1 Recently, Wright has described a myeloma in detail... The tumor elements, according to his research, really form a variety of plasma cells. A myeloma does not originate in the marrow cells as a whole, but in only one of its elements, the plasma cell. Following the results of this important contribution, the tumor may be classed as a plasmoma. **1931** *Amer. Jrnl. Med. Sci.* CLXXXI. 170 In our series of cases extra-osseous involvement occurred in..one of the plasmomata primary in a lymph node. **1961** *Arch. Path.* LXXI. 229 (*heading*) Extramedullary plasmacytoma of gastrointestinal tract with a case report of plasmoma of the rectum.

plasmon (plæ·zmǫn). [f. Plasm(a + *-on[1] (in sense 1 an arbitrary ending).] † **1.** Also **Plasmon.** A proprietary name of a soluble proteinaceous extract of milk; used *attrib.* to designate various foodstuffs made with this, as *Plasmon biscuit, chocolate, cocoa*. *Obs.*

1900 *Daily Express* 31 July 2/6 Plasmon is nothing more or less than milk dried after removing the cream and sugar. *Ibid.*, The writer has found Plasmon chocolate a most useful preparation in cycling. **1901** *Daily Tel.* 18 Mar. 11/6 (Advt.), Why Plasmon cocoa is a nourishing food. *Ibid.*, Plasmon is the albumen of pure fresh milk in the form of a dry, soluble, granulated white powder. **1906** *Trade Marks Jrnl.* 4 Apr. 477 *Plasmon*... Substances used as food or as ingredients in food. International Plasmon, Limited,..London,..food manufacturers. **1912** G. W. E. Russell *Afterthoughts* xvi. 157 Miss Larkins sups on strong coffee and a cigarette, and Meakin on hot milk with a plasmon biscuit. **1915** Galsworthy *Freelands* xxxvii. 335 Open your mouth and let me pop in one of these delicious little plasmon biscuits. They're perfect after travelling. **1921** R. Whymper *Cocoa & Chocolate* (ed. 2) xxiv. 359 Plasmon cocoa..a powder containing added matter in the form of dried milk solids (approximately sixty parts milk solids to forty parts cocoa powder). *Ibid.* 360 Plasmon, which was regarded as a soluble form of milk proteid, is rendered to some extent insoluble when added to cocoa. **1922** [see *hoosh *sb.*]. **1935** E. A. Knox *Reminisc. Octogenarian* x. 194 One of our side stole out during the debate, and fetched some 'Plasmon' biscuits. **1946** O. Sitwell *Scarlet Tree* iii. iii. 48 He had..altogether lost touch with his teeth, being confined by his doctor to a total diet of Plasmon biscuits, a health food.

2. *Genetics.* Also **-one.** [a. G. *plasmon* (F. von Wettstein 1927, in *Nachr. von der Ges. der Wissensch. zu Göttingen* (*Math.-Physik. Klasse*) 259).] The totality of cytoplasmic, or of extranuclear, genetic factors.

1932 *Proc. 6th Internat. Congr. Genetics* II. 281 The abnormal development of chlorophyll (pale color) may be caused by a non-Mendelian nuclear or cytoplasmic factor or by a certain state of the genome or plasmon. **1954, 1965** [see *plastome]. **1970** T. Dobzhansky *Genetics Evolutionary Process* x. 345 The plasmon of *E*[*pilobium*] *luteum* has retained its properties despite having carried an *E. hirsutum* genome for several generations. **1973** K. Mather *Genetical Structure of Populations* ii. 6 Together with the nuclear genotype, the plasmon can play its part in the process of adaptation especially..in the building up of barriers to crossing. **1976** Bell & Coombe tr. *Strasburger's Textbk. Bot.* 401 Plastome and plasmone mutations have been insufficiently studied..but their importance for evolution should not be underestimated.

3. *Astr.* and *Physics.* = *plasmoid.

1955 W. Bostick *Anal. of Plasmons* (U.S. Atomic Energy Comm., Rep. UCRL-4530) 2 Plasmons (plasma-magnetic entities) are toroidal packages of plasma wrapped up in their own magnetic fields. **1963** *Soviet Astron.—AJ* VI. 471/2 If the 'plasmons' ejected by the 'active' nuclei of radio galaxies experience no deceleration, the age of sources similar to Centaurus A and Fornax A will not exceed 10^8 years. **1971** R. C. Haymes *Introd. Space Sci.* xv. 447 The jet [of the galaxy M87] appears to be composed of a group of irregular concentrations of plasma, called 'plasmons' by some. **1974** *Nature* 23 Aug. 629/2 Christiansen has shown that a plasmon can travel a distance $D = M/(\pi \times 1{\cdot}67 \times 10^{-24}\, n_2\, r^2)$ before dispersing. Here M is the mass of the plasmon, r its radius and n_2 the number density of the gas surrounding the component.

4. *Physics.* The quantum or quasiparticle associated with a collective oscillation of charge density.

1956 D. Pines in *Rev. Mod. Physics* XXVIII. 184/1 The valence electrons in the solid..are capable of carrying out collective oscillations at a high frequency... The valence electron collective oscillations resemble closely the electronic plasma oscillations observed in gaseous discharges. We introduce the name 'plasmons' to describe the quantum of elementary excitation associated with this high-frequency collective motion. **1966** C. Kittel *Introd. Solid State Physics* (ed. 3) viii. 233 It is possible to excite a plasmon by passing an electron through a thin metallic film..or by reflecting an electron from the film. The reflected or transmitted electron will show an energy loss equal to integral multiples of the plasmon energy. **1972** F. Wooten *Optical Properties of Solids* ix. 220 The bulk plasmons we have considered so far are purely longitudinal. They cannot couple to transverse electromagnetic waves. However, at the surface of a solid an oscillation of surface charge density fluctuations is possible. These surface plasmons exist in a number of modes. **1976** J. Kleczek *Universe* ii. 71 A plasmon is a quantum of plasma waves, just as photons are quanta of electromagnetic radiation.

Plasmoquine (plæ·zmŏkwin, -kwīn). Also **-quin.** *Pharm.* [f. Plasmo(dium + Quin(in)e.] A proprietary name for *pamaquin. Cf. *plasmochin.

1926 *Trade Marks Jrnl.* 22 Dec. 2757/1 Plasmoquine... Chemical substances prepared for use in medicine and pharmacy. Bayer Products, Limited,..London,..merchants and manufacturers. **1927** *Proc. R. Soc. Med.* XX. 920 The detailed formula for plasmochin (plasmoquine), as it is now known, has not as yet been definitely stated. **1938** *Times* 22 Mar. 17/4 The striking success of acriflavine..is another case in point, and so also are..atebrin and plasmoquin used in malaria. **1945** *New Biol.* I. 102 Atebrin (mepacrine) and plasmoquine are the most useful of the synthetic anti-malarial drugs. **1963** F. Hawking in Schnitzer & Hawking *Exper. Chemotherapy* I. xix. 693 Experiments in rabbits have been carried out by Pols..who tried..chlortetracycline, babesin,..and plasmoquine without significant results.

plasmotomy (plæzmǫ·tŏmi). *Biol.* [ad. G. *plasmotomie* (F. Doflein 1898, in *Zool. Jahrb.* (*Abt. für Anat. und Ontogenie*) XI. 317): see Plasmo- and -tomy.] A mode of reproduction in certain protozoans, in which the organism divides into two or more multinucleate daughter cells.

1902 *Encycl. Brit.* XXXII. 817/2 Cohn and Döflein have discovered cases of plasmotomy, in which a kind of protoplasmic bud of ectosarc and endosarc containing some nuclei becomes detached. **1947** *Jrnl. Morphol.* LXXX. 96 Finally, the organism undergoes plasmotomy into from 2 to 6 individuals. **1973** M. A. Sleigh *Biol. Protozoa* iv. 72 Division of a cell is normally preceded by nuclear division, by either mitosis or meiosis, although in some multinucleate forms fission and nuclear division are not linked, so that for example new individuals may be formed by plasmotomy in which the body is simply separated into multinucleate masses—at any time some nuclei may be found in mitosis in such organisms.

Hence **plasmoto·mic** *a.*

1949 *Jrnl. Morphol.* LXXXV. 164 Plasmotomic division into two daughter individuals does not results [*sic*] in a 50 : 50 distribution of the nuclei.

plaster, *sb.* Add: **3. b.** *U.S.* Plaster of Paris, formerly used as a top-dressing for soils.

1787 G. Washington *Diary* 10 June (1925) III. 222 Where the Plaister had been spread the white and red clover was luxuriant. **1816** U. Brown *Jrnl.* 6 June in *Maryland Hist. Mag.* (1915) X. 264 A poor Hill Country well watered & adapted to Plaster. **1839** J. Buel *Farmer's Compan.* xxii. 213 Districts..in which clover and plaster..were first introduced..have unquestionably made the most rapid strides in agricultural improvement. **1880** *Harper's Mag.* June 67/2 Another glance detects the ..farmer sowing his load of plaster across the whitening field.

4. plaster saint, a virtuous person; freq. in ironical use, a person who makes a show of virtue; a hypocrite.

1890 Kipling *Barrack-Room Ballads* (1892) 8 Single men in barricks [*sic*] don't grow into plaster saints. **1898** G. B. Shaw *Philanderer* iv, in *Plays Unpleasant* 148 You fraud! You humbug! You miserable little plaster saint! **1934** —— *On Rocks* ii, in *Too True to be Good* 260 Theyd be sent back to Parliament by working class constituencies as if they were plaster saints. **1938** W. B. Yeats *New Poems* 13 My father upon the Abbey stage, before him a raging crowd: 'This Land of Saints,' and then as the applause died out, 'Of plaster Saints.' **1964** 'S. Woods' *This Little Measure* vi. 78 It's no good my setting up as a plaster saint..but I do think I'd draw the line at poison. **1965** *New Statesman* 30 Apr. 690/1 The total effect is that Tchaikovsky as plaster saint becomes a monster who couldn't have created anyone's music, let alone his own.

plaster, *v.* Add: **1. e.** (Further examples.)

1858 S. M. Schmucker *Public & Private Hist. Napoleon III* x. 154 In an hour every prominent place in the capital was plastered over with proclamations. **1907** G. B. Shaw *John Bull's Other Island* iii. 54 Ive seen them in that office, telling my father what a fine boy I was, and plastering him with compliments. **1924** A. Huxley *Let.* 29 Apr. (1969) 229 From Parma, which is a superb town, fairly plastered with Corregio's paintings, we went..to Mantua. **1953** *John o' London's Weekly* LXII. 3/4 They show two maps of middle England plastered with the names of remote villages and towns associated with the Lollard rising of 1413–14.

3. a. In other sports, to defeat utterly, to trounce.

1919 J. Masefield *Reynard the Fox* 30 He could plaster All those who boxed out Tencombe way. **1951** *Amer. Speech* XXVI. 230/2 Normal *plasters* Western. **1958** F. C. Avis *Boxing Ref. Dict.*, *Plaster*, to hit an opponent hard and often.

c. *trans.* To shell or bomb (a target) extensively or heavily.

1915 'I. Hay' *First Hundred Thousand* xviii. 262 The German front-line trenches had been 'plastered' from end to end. **1925** Fraser & Gibbons *Soldier & Sailor Words* 224 *Plaster*,..to shell heavily, *e.g.* 'The village was plastered badly last night'. **1941** *Hutchinson's Pict. Hist. War* 14 May–8 July 224 At night there is a concentrated attack on Bremen; the shipyards there and at Vegesack are plastered with bombs. **1942** E. Waugh *Put out More Flags* iii. 243 The bombers were not aiming at any particular target; they were plastering the ground in front of their cars. **1945** *Penguin New Writing* XXIV. 32 They've started firing... Here they come again. They're plastering the other side. **1957** 'N. Shute' *On Beach* vi. 185 You'd think with Boeing as the target all this area would have been well plastered. **1971** B. W. Aldiss *Soldier Erect* 249 Our gunners back at Zubza and Jotsoma kept plastering the heights from which the Japs plastered us.

5. c. Also, to treat (a crop) with plaster of Paris.

1814 J. Taylor *Arator* 155 [Bird-foot clover] among the plastered wheat will be three or four fold more luxuriant, than among the adjoining unplastered. **1852** *Trans. Mich. Agric. Soc.* III. 171 As soon as the corn came up, it was plastered on the hill.

plasterboard (plɑ·stəɹbōᵊɹd). Also **plaster board, plaster-board.** [f. Plaster *sb.* II + Board *sb.* I.] A light-weight building board made of gypsum plaster with a reinforcing or strengthening material, now usu. thick paper bonded to both sides of a plaster core. Also *attrib.* and *fig.*

1906 *Sci. Amer. Suppl.* 29 Sept. 25703/1 Thin plaster boards are nailed on the rafters. **1914** *Chem. Abstr.* VIII. 2620 Plastic composition for making 'plaster board', boxes, etc., formed of straw pulp. **1919** *Ibid.* XIII. 1916 A mixt. for the manuf. of plaster-board is formed of calcined gypsum.., ground tan bark.., a 'hastener'..and about 30 lbs. H_2O for each 100 lbs. of the other ingredients. **1929** W. C. Huntington *Building Constr.* xiv. 464 Plaster board consists of a gypsum plaster core and surfaces of fibrous felt sheets pressed together. **1936** *Archit. Rev.* LXXX. 192 On the inside this is sound-proofed with wood rock, plaster board and American rock-wool in blanket form. **1946** *Times* 9 Sept. 15/6 Plasterboard partitions in existing buildings would serve to provide married quarters. **1956** *Builders' & Decorators' Ref. Bk.* vi. 24 Plaster boards should be fixed with 1½-in. galvanized French wire nails No. 12 W.G. **1965** G. McInnes *Road to Gundagai* xi. 197 Almost as well known as the actors were the props:..the 'ruin' of doric plasterboard which was Ninny's Tomb, Cleopatra's palace and Henry V's tent. **1973** A. Ross *Dunfermline Affair* 72 Our separate rooms had once been one... The wall would probably be less solid than the others, perhaps even only plasterboard. **1974** *Times Lit. Suppl.* 22 Mar. 281/4 He dithers between the historical figure, who cannot fail at this stage to have some life in him, and the plasterboard Schuyler and his tedious sex life.

plaster-cast (plɑ·stəɹ,kɑ:st). Also **plaster cast, plastercast.** [f. Plaster *sb.* + Cast *sb.*] A reproduction in plaster made from a mould. Also *attrib.* and *fig.* and as *v. trans.* Hence **plaster-casting** *vbl. sb.* and *ppl. a.*

1825 [see Plaster *sb.* 4]. **1842** *Knickerbocker* XX. 468 The head of Miss Jewett is a portrait, taken from a plaster cast. **1856** Mrs. Stowe *Dred* I. 18 Bronzes and plaster-casts..gave evidence of artistic culture. **1859** [see Plaster *sb.* 4]. **1912** D. H. Lawrence *Phoenix II* (1968) 271 The method of translation, we are told, is the 'plaster-cast': that is, the outward form is strictly preserved. **1919** 'W. N. P. Barbellion' *Jrnl. Disappointed Man* 163 A plaster-cast mask of Voltaire when first hung up made him chuckle with indecent laughter. **1922** Joyce *Ulysses* 411 Plastercast reproductions..such as Venus and Apollo. **1959** P. & L. Murray *Dict. Art & Artists* 248 Plaster Casting is an intermediate stage in the production of a piece of sculpture which is often the last process actually to be carried out by the sculptor himself. **1970** *Oxf. Compan. Art* 879/2 Plaster casts of limbs..may also be made direct from the human body and then used as models by the stone-carver. *Ibid.* 880/1 The Vienna Academy..had a plaster-casting workshop of its own and devoted a whole floor of its palatial building to its collection of casts. **1977** 'M. Yorke' *Cost of Silence* xv. 115 Footprints..were very distinct, and a detective was making a plaster cast of one of the sharpest.

plastered, *ppl. a.* Add: **2.** *slang.* Highly inebriated; drunk.

1912 *Dialect Notes* III. 585 *Plastered*,..very drunk. **1924** Wodehouse *Bill the Conqueror* xv. 242 Freddy had got so plastered and tried to play the trap-drums. **1931** —— *Big Money* xiii. 309 You would have expected something better from a business man like J. B. Hoke, even if he had been getting steadily plastered all the afternoon. **1934** E. Waugh *Handful of Dust* iii. 110 The old boy's plastered. **1939** J. B. Priestly *Let People Sing* iii. 71 He's gone to cool off. He's very bottled, fairly plastered. **1942** E. Waugh *Put out More Flags* iii. 182 'If it had been anyone else but Angela, I should have thought she was tight.' 'Darling, she was plastered.' **1946** E. O'Neill *Iceman Cometh* ii. 118 Hanging around here getting plastered with you, Mac, is pleasant, I won't deny, but the old booze gets you in the end, if you keep lapping it up. **1953** L. Hobson *Celebrity* iii. 29 'My God,' he confided to the ceiling, 'I'm plastered.' **1958** [see *honkers *a.*]. **1964** N. Marsh *Dead Water* iii. 75 He's overdone it to-night. Flat out in the old bar parlour..he was plastered. **1966** J. Betjeman *High & Low* 66 You're barmy or plastered, I'll pass you, you bastard. **1979** G. Hammond *Dead Game* xiv. 180 'I'll probably get plastered.'..Keith carried his pint over to Constable Murchy.

plastering, *vbl. sb.* Add: **1. d.** The treatment of wines with gypsum.

1872 Thudichum & Dupré *Treat. Origin, Nature & Var. Wine* iv. 119 (*heading*) Plastering of wine and must. *Ibid.* 121 Diminution of yield is..not the only drawback connected with the plastering of wine. **1873** J. L. W. Thudichum *On Wines, their Production, Treatment & Use* ii. 16/1 Sherry..if properly treated, does not require either plastering, or the addition of..boiled must. **1895** S. P. Sadtler *Handbk. Industr. Org. Chem.* (ed. 2) 204 Of the methods of 'improving' wines, as it is termed, that known as 'plastering' is probably most largely practised. **1959** W. James *Word-bk. Wine* 145 It is said that plastering was given a bad name in England by whisky distillers jealous of sherry's growing popularity. **1967** A. Lichine *Encycl. Wines & Spirits* 405/1 *Plastering*, addition of plaster of Paris (gypsum or calcium sulphate) to low-acid musts to induce the necessary degree of acidity.

plasterless (plɑ·stəɹlės), *a.* [f. Plaster *sb.* + -less.] Of a building, structure, etc.: not provided with plaster; lacking plaster.

1866 'Mark Twain' *Lett. from Hawaii* (1967) 101 A huge gridiron of plasterless lathing droops from above. **1919** J. Masefield *Battle of Somme* 29 A few skeleton sheds of plasterless woodwork. **1926** *Public Opinion* 26 Mar. 310/1 A plasterless brick house would be preferable to a plasterless steel house.

plaster of Paris. Add: **b.** *U.S.* Formerly used as a top-dressing for soils.

1787 G. Washington *Diary* 10 June (1925) III. 222 We rid to the farm of one Jones, to see the effect of the plaister of Paris, which appeared obviously great. **1810** W. Thornton *Let.* 22 June in J. Steele *Papers* (1924) II. 627 Salt can be brought up the river in sufficient quantity, & plaister of paris if necessary to give a good coat of white clover on the soil.

pla·sterwork. Also **plaster-work, plaster work.** [f. Plaster *sb.* + Work *sb.*] The surface of a wall or other builders' work executed in plaster. Hence **pla·sterworker,** = Plasterer.

1600 [see Plaster *sb.* 4]. **1797** J. Woodforde *Diary* 3 Nov. (1931) V. 77 My Back-Kitchen..almost all taken down of the Stud and Plaister Work. **1845** *Gloss. Terms Archit.* (ed. 4) 272 In the market-place at Newark is a wooden house with small figures and canopies over them in plaster-work. **1897** W. Millar *Plastering Plain & Decorative* i. 34 From 1750 to 1780 A. Wilton did some fine plaster work at Cambridge. **1908** G. P. Bankart *Art of Plasterer* vii. 97 Exeter is rich in examples of seventeenth-century plasterwork. **1926** M. Jourdain *English Decorative Plasterwork of Renaissance* i. 6 The names of the plasterworkers that have so far come down to us are English. **1959** M. S. Briggs *Conc. Encycl. Archit.* 252 Apart from its normal function of providing a smooth external or internal surface for walls and ceilings, plasterwork has often reached a high level of artistic excellence. **1977** *New Yorker* 4 July 66/2 The plasterwork inside the hall itself blossoms with Manueline exuberance.

plastery, *a.* Add: **b.** Built with plaster, or in a manner suggestive of plaster.

1862 'G. Hamilton' *Country Living* 6 To move from this tumble-down old house..into a..plastery, shingly, stary, new one. **1907** *Daily Chron.* 18 Sept. 4/4 Plastery little red and white cottages and villas set at all angles among cabbage-plots.

plastery (plɑ·stəri), *sb.* [f. Plaster *sb.* + -y[3].] Plastered work; plastering.

1842 *Amer. Pioneer* I. 207 The stone work and plastery was done by major William Rutledge, a soldier of the revolutionary war.

plastic, *a.* Add: **I. 1. b.** (Earlier and later examples.) So *plastic surgeon*, one skilled in plastic surgery.

1839 *Brit. & Foreign Med. Rev.* VII. 388 We are, therefore, willing to confine ourselves, with our German author [*sc.* Zeis], to the use of the simple expression 'Plastic surgery', which is sufficient to imply generically all we mean. *Ibid.* 393 Syphilis, lupus, scrofula, &c. have made cases whereon to exercise the ingenuity of the plastic operator. **1853** J. Erichsen *Sci. & Art of Surg.* xlviii. 665 By plastic or reparative surgery is meant those processes by which mutilations are repaired, and loss of structure replaced. **1911** F. S. Kolle *Plastic & Cosmetic Surg.* i. 8 The successful plastic surgeon has become an imitator of nature's beauty to-day. **1935** W. de la Mare *Early One Morning* 271 Even if he could delete the scar of a wound of this kind which time has healed, how many of us would hasten to consult the plastic surgeon? **1941** *Ann. Surg.* CXIII. 642 The title of Eduard Zeis' (1807–1868) book, published in 1838, was Handbuch der Plastischen Chirurgie, and he says: 'As far as I know I was the first to use the words "plastic surgery".' **1972** *Daily Tel.* 18 July 3/3 After the accident she had plastic surgery, but found fashion jobs hard to get because of her scars. **1974** J. Grady *Six Days of Condor* 83 The plastic surgeons had done a marvelous job on his ear.

II. 4. b. Pertaining to, characterized by, or utilizing an ability to be permanently changed in shape, without fracture or rupture, by temporary pressure or tension; esp. in *plastic deformation* or *flow*.

1877 *Jrnl. Franklin Inst.* CIV. 228 (*heading*) Plastic flow. **1879** *Encycl. Brit.* IX. 240/2 More shapely bricks are thus produced than by plastic moulding. **1888** W. C. Unwin *Testing of Materials of Construction* i. 18 When a body is subjected to the action of external forces, it undergoes a deformation which is either a deformation which disappears if the load is removed (elastic deformation), or a deformation which remains after the load is removed (plastic deformation). **1923** Glazebrook *Dict. Appl. Physics* V. 395/1 Associated with plastic strain in many metals is the occurrence or formation of 'twinned' crystals. *Ibid.* 400/1 A viscous, under-cooled liquid, may..undergo deformation of a 'plastic' (*i.e.* non-elastic) nature. **1925** *Jrnl. Iron & Steel Inst.* CXII. 451 A consideration of the laws of plastic deformation of hot material during rolling and working. **1940** *New Statesman* 16 Mar. 360/1 In its simplest form we see plastic extrusion as combs or as cosmetic boxes, where the resin has been forced into a mould, with instantaneous cooling. **1951** *Gloss. Terms Plastics Industry* (*B.S.I.*) 24 *Plastic yield*, non-elastic deformation. **1963** E. S. Hills *Elements Structural Geol.* xii. 355 Plastic deformation of wall-rocks is exhibited around the Bald Rock batholith, California. **1967** M. Chandler *Ceramics in Mod. World* ii. 63 Shaping methods..include..plastic pressing, and extrusion. **1968** A. H. Cottrell *Introd. Metallurgy* xxi. 387 This leads to a plastic instability in which all subsequent deformation becomes concentrated in one short section..which stretches excessively and forms a narrow neck. **1968** R. W. Fairbridge *Encycl. Geomorphol.* 1191/2 Plastic flow will occur when an ice body attains a thickness of 100–150 feet. **1976** *Physics Bull.* Oct. 459/1 Most of part one is concerned with the elastic, plastic and fracture behaviour of minerals and rocks.

5. a. (Further examples.)

1881 *Engineering* 20 May 513/3 (*heading*) Richards' plastic metal. *Ibid.*, 'J. Richards' Plastic Metal' is being made by the J. Richards' Plastic Metal Company, of Charlotte-street, Birmingham. In general outward appearances it resembles..other varieties of white metal so largely used for lining bearings... Its special feature..is its great affinity for other metals..enabling it to be readily 'pasted on'. **1907** *Chem. Abstr.* I. 1077 Process of manufacturing a plastic material suitable for the production of fibers, pellicles, blocks, or plates, consisting in mixing together phenol, 17, casein, 40, pressing, heating.. and pressing again, then adding glycerol to give plasticity. **1908** L. Desvaux *Brit. Pat.* 9313 A plastic product for the manufacture of combs, molded objects of any kind and similar applications, composed of a mixture in variable proportions of nitrocellulose and camphor,..and of the food product extracted from maize by treating this substance with higher alcohols. **1921** *Chem. Abstr.* XV. 1770 A description of the classes of com. materials falling under the category of plastic masses. B. divides the important and most useful products into 6 classes, (1) glues, (2) papier maché, (3) wood products, (4) cellulose products, (5) egg white and casein, and (6) resins. The criterion for plastics is the condition of the raw material or of the final product... B. takes exception to the older classification of plastics based on some temporary condition during manuf.

b. (Earlier and later examples.) [tr. F. *argile plastique* (Cuvier & Brongniart 1808, in *Jrnl. des Mines* XXIII. 432).]

1812 T. Webster *Let.* 2 Aug. in H. C. Englefield *Description Isle of Wight* (1816) 210 The clay connected with this sand is frequently fit for the potter, and hence has been called the plastic clay. **1813** R. Jameson in R. Kerr tr. *Cuvier's Essay on Theory of Earth* 227 The fundamental rock or basis of the [Paris] district is chalk. This chalk is covered with plastic clay, and what is termed coarse marine limestone. **1929** P. G. H. Boswell in Evans & Stubblefield *Hand-bk. Geol. Gt. Brit.* 417 The lowest Eocene deposits are therefore the Reading Beds.. which..consist of mottled clays ('Plastic Clay') and sands with a glauconitic bed at the base. **1955** G. G. Woodford tr. *M. Gignoux's Stratigr. Geol.* ix. 477 Mammalian remains are present..at the base of the plastic clay. **1961** B. Kummel *Hist. Earth* 8/2 Of the strata above the Purbeck beds and below the Plastic clay, the most conspicuous unit is the Chalk.

c. *plastic explosive*, an explosive of putty-like consistency that can be shaped by hand and so placed in intimate contact with its target; so *plastic bomb*, one containing plastic explosive; *plastic-bomb* vb. trans., *-bombing* vbl. sb.

1906 C. E. Bichel *Brit. Pat.* 16,882 Add to the trinitrotoluol liquid resins..in such wise that..the crystalline trinitrotoluol with or without warming is worked in suitable mixing machines into a plastic explosive that detonates well. **1946** T. C. Ohart *Elements of Ammunition* ii. 37 A plastic explosive known as RDX-composition C is formed by mixing about 88% cyclonite with 12% plasticizer. **1955** G. Greene *Quiet American* iii. i. 185 That day all over Saigon innocent bicycle-pumps had proved to be plastic bombs and gone off at the stroke of eleven. **1961** *Times* 12 July 10/3 Bombs made with plastic explosive were discovered not far from the entrance to the Simplon. **1961** *Economist* 25 Nov. 777/1 One farmer has threatened to plastic-bomb the line. **1962** *Spectator* 23 Feb. 229/1 The imported disease of plastic-bombing the home of your adversary at the risk of maiming or killing his wife and children. **1962** *Daily Tel.* 14 June 14 Casual and indiscriminate plastic-bombing, in Paris as well as in Algeria, has been followed by attacks on a hospital..and on an oil-well. **1963** *Ann. Reg. 1962* 236 In France there were plastic bomb attacks, directed mainly against liberal politicians and journalists. **1976** N. Freeling *Lake Isle* xv. 113 A couple of cops stayed for a search. They got back..with an old army revolver, two-thirds of a kilo of plastic explosive, and a lot of gold coins.

d. *plastic bronze*, bronze containing a high proportion of lead, which is used for bearings on account of its softness.

1907 G. H. Clamer in *Chem. Engineer* Aug. 93 This alloy is largely sold under the name of 'plastic bronze'. **1939** [see *leaded *ppl. a.* e]. **1954** *Kempe's Engineer's Yearbk.* I. 633 'Plastic' bronze Cu 73 Sn 7 Pb 20.

e. *plastic wood*, a mouldable material that hardens to resemble wood and is used for filling knot holes, crevices, and the like.

1921 *Engineering* 9 Dec. 785 This material..is named by the firm 'Plastic Wood'. It is a collodion preparation made with very fine wood meal, and as supplied ready for use is of the consistency of soft putty. **1938** A. Durst *Wood Carving* 16 Knot-holes and blemishes of a like nature can be filled with plastic wood. This is a quickly-drying preparation of cellulose and wood pulp. **1974** J. Melville *Nun's Castle* ix. 205 The old wooden door frame had shrunk... The door around the lock had been built up with..plastic wood.

f. *plastic paint*, paint which is sufficiently thick and coarse when applied for it to retain a texture given to it with the aid of a brush, spatula, or the like.

1925 *Amer. Paint Trade Buyer's Guide* 208/2 Plastic Paint—*see* Plastic Relief Compositions. **1955** *Mod. Building Encycl.* 492/2 In addition to the excellent proprietary materials available, plastic paints may be prepared from equal parts of distemper and plaster-of-paris. **1974** E. McGirr *Murderous Journey* 28 A room painted with a dark shade of plastic paint.

g. *plastic crystal*, a soft substance in which the molecules occupy the points of a regular crystal lattice but have freedom of rotation about those points. (See also sense 5 in Dict.)

1961 *Physics & Chem. of Solids* XVIII. 8/2 In liquid crystals, by heating, the fluidity comes first, but in plastic crystals, the isotropy comes first. **1968** A. Bondi *Physical Properties of Molecular Crystals* vi. 140 The very small expansions of plastic crystals at their melting point generally result from the fact that a much larger expansion took place at a first-order transition of the crystal at some lower temperature. **1974** P. A. Winsor in Gray & Winsor *Liquid Crystals & Plastic Crystals* I. ii. 48 Plastic crystals separate in the crystal forms of the cubic system (rarely hexagonal) and to this extent resemble ordinary solid crystals. However, they show unusually low yield points. The most plastic..will flow under their own weight and although the majority are less soft, they may readily be cut with a knife or extruded through a small hole.

6. b. *Biol.* Pertaining to or (of an organism) exhibiting an adaptability to environmental changes.

1905 F. E. Clements *Research Methods in Ecol.* iii. 103 The amount of response to a stimulus is proportional to the intensity of the factor concerned. This does not mean that the same stimulus produces the same response in two distinct species, or..in two plants of one species. In these cases the rule holds only when the plants or species are equally plastic. *Ibid.* 146 Stable plants are less susceptible of evolution than plastic ones. **1930** *Jrnl. Ecol.* XVIII. 376 The broad-leaved plantain has proved, even within five months, exceedingly plastic. **1965** *Adv. Genetics* XIII. 133 The species that are plastic for leaf shape, that are able to produce both sorts of leaf, are all typically species of shallow water. *Ibid.* 137 Plastic response is able to provide adaptation to directional selection in some populations which in others is provided by genetic change.

IV. [Partly deriving from the sb. used attrib.] **9. a.** Made of plastic; of the nature of a plastic, or containing plastic as an essential ingredient.

1909 *Chem. Abstr.* III. 724 Artificial plastic materials industry... An interesting account..giving descriptions of the process for artificial rubber, leather, and substitutes; celluloid, viscoid, etc.; plastics obtained from cellulose and its compounds; and plastics from casein, maisin, albuminoids, and gelatins. **1911** E. C. Worden *Nitrocellulose Industry* II. xiv. 630 Formation of plastic rods and tubes was first successfully made by the patented process of I. and J. Hyatt. *Ibid.* 708 The manufacture of plastic cuffs and shirt bosoms. **1912** *Sci. Amer. Suppl.* 20 Apr. 246/1 The term 'plastic materials' is here employed in a restricted sense, including only such materials as celluloid and its numerous substitutes, which can easily be shaped by cutting and grinding, as well as by molding, and excluding artificial textile fibers and India rubber and its imitations. **1931** *Brit. Plastics Year Bk.* 17 We have pleasure in presenting to the Plastics Industry the first Year Book..dealing exclusively with Plastic Materials. **1940** *Economist* 29 June 1108/2 Plastic structural material has been introduced into the aircraft industry. **1943** *Times Weekly Ed.* 10 Feb. 17/1 Plastic bearings were going into the heaviest engineering applications. **1949** E. Coxhead *Wind in West* ii. 36 His wife, in a thin..plastic raincoat, looked perished with cold. *Ibid.* 41 Little Mrs. Turner, who had gone nearly as blue as her plastic mackintosh. **1951** A. Baron *Rosie Hogarth* 60 She..hung plastic curtains in his bedroom. **1957** *Daily Mail* 5 Sept. 11/5 Pre-cooked hamburgers..in their little frozen transparent plastic bags. **1958** *Engineering* 14 Mar. 349/1 Acid wastes are disposed of through plastic pipes. **1958** *Observer* 6 July 9/4 The light plastic mac, easily stuffed into pocket or bag, comes into its own during the British summer. **1961** *Ann. Reg. 1960* 510 Growers of flowers complained that imports of plastic flowers, mainly from Hong Kong, were having an adverse effect. **1966** 'G. Black' *You want to die, Johnny?* x. 193 There were..plastic tiles on the floor. **1969** W. R. R. Park *Plastics Film Technol.* vi. 147 With the current proliferation of..plastics films, the growing interest in and use of plastic laminates may seem somewhat surprising. **1972** *Guardian* 16 Oct. 9/4 Furtive gestures by elderly men in plastic macs. **1975** *New Yorker* 29 Sept. 43/1 The couch and the armchairs are protected by plastic covers. **1977** B. Pym *Quartet in Autumn* xiii. 109 A plastic bag lying on the kitchen table.

b. *fig.* Artificial; superficial, insincere.

1963 *Daily Tel.* 22 May 16 The plan's promoters must not take it amiss if, winking an eye, some of our elder oysters inquire whether plastic houses might not connote plastic people. **1967** *Harper's* Aug. 19 Now that so many of the young seem to wear their hearts on their sleeves, it is hard to tell which ones are real and which ones are plastic. **1970** *Observer* (Colour Suppl.) 15 Feb. 24/1 Sinister influences are at work to turn Fiji into another Hawaii, that plastic paradise further along the route. **1974** *Times Lit. Suppl.* 1 Mar. 219/5 The characters are by no means badly drawn and the girl in particular is notably less plastic than usual. **1977** *Daily Tel.* 16 Apr. 16 The flabby, chalky, doughy slabs of our unpalatable plastic muck which masquerades as bread.

B. *sb.*³ **I. 1.** A plastic material. **a.** A solid substance that can be readily moulded or shaped.

1905 E. H. Angle in E. C. Kirk *Amer. Text-bk. Oper. Dentistry* (ed. 3) xxiv. 720 Models sufficiently perfect cannot be made from impressions taken in modelling compound or other of the plastics. **1921** [see *A. 5]. **1923** *Blackw. Mag.* June 722/2 In the evenings Roupin constructed in plastic..a complete model of Haidar Pasha. **1933** L. F. Rahm *Plastic Molding* ii. 19 The molding properties of rubber are such as to make it one of the simplest plastics to handle. **1936** L. M. T. Bell *Making & Moulding of Plastics* i. 13 Dental uses of plastics... Stabalite [composed of china clay, rubber, sulphur, etc.] is used very largely for artificial palates. **1944** E. C. Jahn in L. E. Wise *Wood Chem.* xxiii. 820 In 1942..wood and lignin plastics are still largely in the developmental stage.

b. Any of a large and varied class of substances which are polymers of high molecular weight based on synthetic resins or modified natural polymers and may be obtained in a permanent or rigid form following moulding, extrusion, or similar treatment at a stage during manufacture or processing when they are mouldable or liquid; see also **laminated plastic*, **reinforced plastic*. Also used generically (without *a* and *pl.*): material of this kind.

In techn. usage the term is usu. held to exclude the synthetic rubbers (elastomers), and sometimes also any plastic in the form of fibres.

1909 L. H. Baekeland in *Jrnl. Industr. & Engin. Chem.* Mar. 156/2 As an insulator..it [*sc.* Bakelite] is far superior to hard rubber, casein, celluloid, shellac and in fact all plastics. *Ibid.* 157/1 It can be used for similar purposes like knobs, buttons, knife handles, for which plastics are generally used. **1911** E. C. Worden *Nitrocellulose Industry* II. xiv. 691 Pyroxylin plastic is extensively used for the bits of pipe stems, and consists of ordinary plastic containing..dyestuffs, picric acid, [etc.]. **1915** J. E. Crane in A. Rogers *Industr. Chem.* (ed. 2) xliv. 914 Pyroxylin plastics, variously called celluloid, xylonite,..viscoloid, and other names consist of a mixture or solid solution of cellulose nitrate and camphor. **1928** *Chem. Abstr.* XXII. 4209 Plastics are defined as materials that are horny and elastic at ordinary temp. but can be molded at higher temp. They include (1) cellulose plastics, (2) artificial resins and (3) protein plastics. **1935** *Economist* 7 Dec. 1140/1 The use of plastics in the motor accessory field will undoubtedly increase... Already the fitting of wireless sets as standard equipment on several cars has opened up a..new field for their application. **1941** *Electronic Engin.* XIV. 482 A large percentage of..plastics have good insulation properties, while at the same time the materials are available in a wide variety of forms..from lacquers..through rubber-like materials to the hard and rigid bakelite-type resins. **1945** *Daily Mirror* 27 Sept. 3/1 British-made women's shoes in 'patent leather' plastics may be on sale next summer. **1953** Kirk & Othmer *Encycl. Chem. Technol.* X. 798 When the resin itself is capable of being shaped into a finished article without a plasticizer.., as polystyrene, the terms resin and plastic are interchangeable for that material. **1955** *Observer* 13 Nov. 3/3 Nearly all plastics—except nylon stockings—crept into the house by the back door, disguised as 'cheap' substitutes for the real thing—china, glass, wood, metal, silk or wool. Now they have their own status, either as alternatives..or as new materials, to do a new job. **1963** H. R. Clauser *Encycl. Engin. Materials & Processes* 486/2 Silicones are unique among plastics, in that they are semiorganic, i.e., the molecular spine has alternating silicon and oxygen atoms with organic groups attached to the silicon. **1968** Kirk & Othmer *Encycl. Chem. Technol.* (ed. 2) XV. 790 Nylon and poly(ethylene terephthalate) are used both as fibers and plastics. **1973** *Materials & Technol.* VI. viii. 499 Twenty years later [*sc.* about 1890], casein plastics prepared by reacting together milk protein and formaldehyde were developed in Germany. **1973** *Sci. Amer.* Aug. 107/1 This container can be a baking pan made of sheet metal or plastic.

2. Plastic explosive.

1966 M. R. D. Foot *SOE in France* xi. 367 Though they had no plastic, they could get unlimited dynamite from the mines. **1968** D. Lampe *Last Ditch* vii. 75 Plastic is a form of cyclonite,..and is still today the standard military sabotage high explosive. **1973** D. Lees *Rape of Quiet Town* vii. 119 The bank manager type who'd been playing with the plastic had stayed behind. **1978** T. Allbeury *Lantern Network* ix. 112 Parker..showed them how to wire the plastic so that a whole length of track was taken out in a single explosion.

II. 3. *attrib.* in sense *1 b.

1911 E. C. Worden *Nitrocellulose Industry* II. xiv. 578 The general principles of plastic manufacture. *Ibid.* 660 The entire field of plastic molding. **1931** *Brit. Plastics Year Bk.* 69 The plastic trade consumes 1,200 tons of wood dust per annum for mouldings. **1956** A. H. Compton *Atomic Quest* 326 Paper, plastic, and textile plants. **1960** I. Wallach *Absence of Cello* 16 He was a trouble-shooter ..for a large plastic corporation. **1969** T. C. Thorstensen *Pract. Leather Technol.* xiv. 235 The increased 'plastic look' in leather may, in the long run, harm the marketing position of leather in its competition with synthetic materials.

4. Used *attrib.* in *pl.*, often to avoid possible confusion with branches I and II of the adj. **a.** Of, pertaining to, or concerned with plastics; = sense *3.

1925 *Plastics* Oct. 7/1 The plastics industry. **1935** *Economist* 4 Mar. 1042/2 Their interest in the plastics industry, through Mouldrite, Limited, continued to make progress. **1957** J. Braine *Room at Top* 61 He owned a plastics factory, a tannery, a bodywork builders.

b. = sense *9 of the adj.

1934 H. Read *Art & Industry* II. 90 The wireless cabinet is an example of the encroachment of new plastics materials, such as bakelite, on a province hitherto reserved for wood. **1958** *Engineering* 7 Mar. 320/2 Various tools with plastics handles. **1971** *Daily Tel.* 15 Feb. 4/8 Plastics windows to protect passengers from stone-throwing are being installed in trains in the New York area. **1974** *Brit. Standard 4998* (*title*) Moulded plastics dustbins.

5. *Comb.* Instrumental, as *plastic-coated*, *-covered*, *-lined*, *-tiled*, *-topped*, *-wrapped* (so *-wrap* vb.) adjs. Parasynthetic, as *plastic-macked*, *-mackintoshed*.

1960 *Farmer & Stockbreeder* 23 Feb. 69/1 The cab framework is constructed of precision steel tubing and the weatherproof roof of plastic-coated nylon fabric. **1977** D. MacKenzie *Raven & Kamikaze* iv. 56 The wire was plastic-coated and copper, the sort of thing used on a radio. **1961** *House & Garden* June 136/2 Even a neat, plastic-covered plunge is not exactly a joy to behold. **1973** R. Lewis *Blood Money* viii. 125 There's a plastic-covered card identifying the dead man. **1969** *Jane's Freight Containers 1968–69* 239/3 The walls are of plastic-lined plywood plates. **1979** *Tucson Mag.* Apr. 64/2 Another alternative is a plastic-lined pool. **1964** *Guardian* 9 Sept. 5/8 Plastic-macked parents and hordes of soggy children. **1973** J. Wainwright *Devil you Don't* 25 A plastic-mackintoshed young woman. **1962** *Listener* 10 May 831/3 Plastic-tiled and similar floors also need damp-washing. **1957** *Observer* 13 Oct. 1/2 The Queen and the Duke..walked to a plastic-topped limousine which drew away to drive to Government House. **1973** D. Francis *Slay-Ride* xiii. 159 We sat at a plastic topped table amid travellers with untidy hand luggage. **1968** *Economist* 11 May 69/3 The meat is cut, quick-frozen and plastic-wrapped under contract to supermarkets and to restaurant chains. **1975** M. Bradbury *Hist. Man* ix. 148 Contemporary, plastic-wrapped food. **1978** B. Norman *To nick Good Body* xvi. 132 A lump of plastic-wrapped, processed Cheddar.

plastic, *sb.*³: see *plastic *a.* B.

-plastic, *suffix.* [f. -plast(y or Gr. πλαστ-ός formed + -ic.] Forming adjectives that correspond to sbs. in *-plasty* (Gr. -πλαστία, f. πλαστός formed) or *-plasia* (Gr. πλασία, πλάσις formation).

plastically, *adv.* (Earlier and later examples.)

1835 *Southern Lit. Messenger* Dec. 43/2 Pictorially, or graphically, or as a German would say plastically. **1957** G. E. Hutchinson *Treat. Limnol.* I. vii. 532 A slow fall in temperature apparently permits the ice to flow plastically over the lake surface without cracking. **1966** C. R. Tottle *Sci. Engin. Materials* vii. 158 Many materials show only a very short elastic range and begin to deform plastically at comparatively low stresses. **1972** J. G. Dennis *Structural Geol.* vi. 109 'Brittle' rocks could yield plastically.

plasticate (plæ·stikĕit), *v.* [f. Plastic *a.* + -ate³, prob. after *masticate*.] **1.** *trans.* To change (particles of rubber or thermoplastic) into a homogeneous plastic (mouldable) mass by passing it through a suitable extruder and usu. simultaneously heating it.

1929 W. A. Gordon *Brit. Pat. 334,509* (*title*) Machines for plasticating materials. **1934** *Industr. & Engin. Chem.* Mar. 349/1 Rubber, plasticated in the Gordon machine, which gives the same y_s value..as a sample of mill-massed rubber, appears much stiffer in factory processing operations. **1968** *Encycl. Polymer Sci. & Technol.* IX. 8 Machines are either single-stage, in which plastication and injection are done by the same cylinder, or double-stage, in which the material is plasticated in one cylinder and fed to a second for injection into a mold. **1975** *Mod. Plastics Internat.* Nov. 25/2 In the twin-screw section, powder PVC or other material is plasticated at low shear with close control over temperature.

2. [ad. F. *plastiquer*.] *trans.* To blow up or destroy with a plastic bomb.

1962 *Guardian* 3 Jan. 7/4 Paris butchers are now using plastic bombs against colleagues... One butcher's shop.. was 'plasticated' early this morning. **1965** D. Francis *Odds Against* (1967) xix. 222, I couldn't get hold of Radnor on account of the office phones being plasticated.

Hence **pla·sticated** *ppl. a.*, **pla·sticating** *vbl. sb.* and *ppl. a.*

1934 *Industr. & Engin. Chem.* Mar. 349/1 There is need of a method for detecting the difference..between the plastic properties of plasticated and mill-massed rubber. **1953** *Ibid.* May 970/2 The combined operation of melting and extruding is called 'plasticating extrusion'. **1959** J. B. Paton in E. C. Bernhardt *Processing of Thermoplastic Materials* iv. 228 In the design of an extruder used for processing thermoplastic materials, the complex operations of plasticating, compacting, and conveying..must also be considered. **1970** Tadmor & Klein *Engin. Princ. Plasticating Extrusion* i. 8 Today's plasticating extruders operate mostly in the speed range of 20–200 rpm.., extruding up to 3500 lb/hr of polymer.

plastication (plæstikĕi·ʃən). [f. prec. + -tion.] The action of plasticating.

1939 *Ann. Rep. Progr. Rubber Technol.* II. 116 Knowledge of the mechanism of plastication has been carried a stage further by a study of mastication in an internal mixer at various temperatures. **1968** [see *plasticate *v.* 1].

plasticator (plæ·stikĕitəɹ). [f. as prec. + -or.] An extruder for plasticating rubber or thermoplastic particles, usu. by subjecting them simultaneously to pressure and heat.

1934 *Industr. & Engin. Chem.* Mar. 349/1 Two distinctly different types of masticating machines are employed in modern rubber plants, mills, and plasticators. **1968** *Encycl. Polymer Sci. & Technol.* IX. 48 The most widely used contemporary machines [for molding plastics] include: (a) the ram injection-molding machine..; (b) the plunger- or screw-type plasticator..; and (c) the reciprocating-screw injection machine. **1972** P. W. Allen *Natural Rubber & Synthetics* vii. 196 In place of the standard [masticating] machines shown..some factories use a 'plasticator' which fulfils essentially the same function.

plastician (plæsti·ʃăn). [f. Plastic *a.* + -ician.] An expert or specialist in plastic art, plastic surgery, etc.

1928 T. E. Lawrence *Let.* 16 Apr. (1938) 591 So many plasticians seem to admit to their notice the outside of machinery, and to exclude its purposefulness. **1933** *Archit. Rev.* LXXIII. 266/2 As a complement to the elaborate laboratory researches into the nature of thermo-plastic.., there is need of research and experiment directed to the proper development of design... It [*sc.* Lethaby's *Art and Workmanship*] should be in the hands of every 'plastician'. **1934** *Punch* 26 Dec. 718/2 And, by marvellous plasticians in mysterious robes arrayed, Faces are most wonderfully and most fearfully remade.

Plasticine. Now usu. pronounced (plæ·stisīn or plɑ·st-). Add further examples. Also *Comb.* and quasi-*adj.* (in *fig.* use).

1926 R. Macaulay *Crewe Train* II. vii. 152 She..idled about with toy soldiers or plasticine or meccano. **1935** H. G. Wells *Things to Come* xiii. 124 A nursery of children. Anno 2055. They play with plasticine, draw on sheets of paper.., build with bricks or run about after each other. **1958** *Spectator* 4 July 12/1 He was so pliant, so plasticine,..so insidiously seeing it all the other chap's way. **1967** H. Porter in *Coast to Coast 1965–6* 177 The sugar..infesting their plasticine-like texture tasted of garlic. **1976** *Times* 28 Jan. 1/3 The Russians..would respect us more if we were led by an iron lady rather than a Plasticine man.

plasticity. Add: *spec.* **a.** Capacity for being moulded or undergoing a permanent change in shape.

1782–3, etc. [in Dict.]. **1867** V. Ottolini et al. *Terra-cotta Archit. N. Italy* 5 Plasticity and homogeneity of ingredients are the two conditions essential to the composition of any ceramic paste. **1933** L. F. Rahm *Plastic Molding* ii. 24 In molding, shellac compounds are generally preheated to sufficient plasticity to require no further heat from the mold. **1935** G. E. Doan *Princ. Physical Metallurgy* v. 109 If deformation is continued until the temperature of the object is below the recrystallization temperature.., the plasticity of the metal may be insufficient and the object may crack in the operation. **1968** W. J. Patton *Materials in Industry* iii. 61 Most forming operations during manufacture require plasticity for their execution. **1971** B. Scharf *Engin. & its Lang.* i. 2 Ductility, malleability and plasticity are closely related properties.

b. *Biol.* Adaptability of an organism to changes in its environment.

1868 [in Dict.]. **1908** J. A. Thomson *Heredity* iii. 72 It is certain that many unicellular organisms are very plastic, and it seems reasonable to suppose that as differentiation increased, restrictions were placed on the primary plasticity, while a more specialised secondary plasticity was gained in many cases, where the organisms lived in environments liable to frequent vicissitudes. **1951** *Jrnl. Ecol.* XXXIX. 217 Ecological plasticity..may be defined as the potentialities of expression of physiological characters that determine what factors of the environment shall limit the distribution of a species or other taxon. **1976** Bell & Coombe tr. *Strasburger's Textbk. Bot.* 401 The wild strawberry..reproduces vegetatively by vigorous runners..and each clone can only develop the desirable genetic plasticity by means of stepwise somatic mutations (bud sparts). **1978** *Nature* 14 Sept. 140/2 The potential for plasticity of the developing mammalian visual system has been the subject of several investigations.

plasticization (plæstisəizĕi·ʃən). [f. Plastic *a.* + -ization.] The process of rendering (more) plastic or mouldable; *spec.* (*a*) addition of a plasticizer to a synthetic resin; (*b*) = *plastication.

1927 *Chem. Abstr.* XXI. 2535 The diff. methods of plasticization of a no. of natural and artificial plastics are discussed. **1937** A. Jones *Cellulose Lacquers* iv. 54 Plasticisation has long been studied in metallurgy in the production of soft alloys like solder and type metals. **1946** F. Marchionna *Butalastic Polymers* xiii. 414 Plasticization [of Buna-S] may also be effected by first masticating the sheet and then subjecting it to thermal softening. **1953** Kirk & Othmer *Encycl. Chem. Technol.* X. 773 The most important classes of thermoplastic resins requiring plasticization are the vinyls and the cellulosics. **1969** M. A. Wheelans in W. S. Penn *Injection Moulding of Elastomers* ix. 84 Plasticization by screw is quicker, more controllable and gives a more homogeneous distribution of heat and viscosity than that given by a simple plunger system. **1972** *Nature* 21 Apr. 405/1 A number of other insecticide chemicals also caused plasticization of the abdominal cuticle.

plasticize (plæ·stisəiz), *v.* [f. Plastic *a.* + -ize.] **1.** *trans.* To render plastic (mouldable); to produce or promote plasticity in (a substance), e.g. by addition of a plasticizer or by plastication.

1927 *Chem. Abstr.* XXI. 2535 Mech. and chem. methods for plasticizing such modern plastics as rubber, cellulose products, galilith [*sic*], phenol-formaldehyde resins, etc. **1931** *Engineering* 2 Jan. 26/1 The scrap [rubber] was then plasticised by treatment with steam in a horizontal heater. **1945** A. T. Birkby *Phenolic Plastics* viii. 92 Any attempt to plasticize a large mass of thermo-setting material at one time to a condition suitable for extrusion through a die in a continuous process would fail. **1957** H. R. Simonds *Conc. Guide to Plastics* iii. 111 Most resins are plasticized by heat, solvents, or plasticizers. **1969** M. A. Wheelans in W. S. Penn *Injection Moulding of Elastomers* ix. 84 The rubber is heated and plasticized as it progresses along a retractable screw. **1971** *Nature* 9 July 88/2 Textile fibres which are in bulk production are usually used. These are..usually copolymers containing a proportion of a second molecule to plasticize the fibre and to allow it to be spun and handled more easily. **1972** *Ibid.* 21 Apr. 405/1 The acetone-treated insects neither became paralysed nor was their abdominal cuticle plasticized.

2. *trans.* To treat or make with plastic.

1940 *New Statesman* 16 Mar. 361/1 An enterprising silk manufacturer..'plasticised' gold lace so that it became a permanent table-cloth when used as a table-surface. **1970** *Daily Tel.* 13 Jan. 15/6 Walls and ceiling are covered with a check fabric..and it is specially plasticised to resist condensation. **1977** *Austral. House & Garden* Jan. 49/1 (*caption*) Atel's modular kitchen units have an Italian Parawood teak veneer surface that has been plasticised for easy care.

Hence **pla·sticizing** *vbl. sb.* and *ppl. a.*

1925 *Paint, Oil & Chem. Rev.* 22 Jan. 10/2 The resin 'may or may not have been treated with..a plasticizing agent'. **1927** *Chem. Abstr.* XXI. 2535 (*heading*) The influence of plasticizing on the mechanical-elastic properties of natural and artificial plastics. **1948** Dalzell & Townsend *Masonry Simplified* I. i. 13 In order to yield a mortar, the lime or lime putty must exert a certain minimum plasticizing effect on the mortar. **1953** *Industr. & Engin. Chem.* May 989/1 A plasticizing extruder whose job is to transform the cold feed into a hot, formable melt. **1973** *Materials & Technol.* VI. viii. 606 For the plasticizing of the more polar cellulose nitrate the low-molecular phthalates are preferred.

plasticized (plæ·stisəizd), *ppl. a.* [f. prec. + -ed¹.] **1.** Rendered (more) plastic; treated with a plasticizer.

1943 *Industr. & Engin. Chem.* May 383/2 Paper and cloth coated with the plasticized esters may be crumpled without cracking the coating. **1971** *Nature* 3 Dec. 254/2 Plastic coatings may be applied to metals by dipping the preheated metal part into either a fluidized bed of polymer powder or into a liquid plastisol. The former process is applicable to polyethylene, nylon and unplasticized PVC, whereas the latter is applicable only to plasticized PVC compounds.

2. Treated or made with plastic.

1945 *Richmond* (Va.) *News-Leader* 26 July 27/2 Sheer originality is seen in the new plasticized beaver fur coat which will not mat or curl in the stormiest weather. **1972** *Daily Tel.* 7 Jan. 13 The cover-ups in plastic or plasticised cotton are good for sploshy activities like painting, attempts at cooking.

3. *fig.* = *plastic *a.* 9 b.

1974 *Globe & Mail* (Toronto) 12 Oct. 8/1 The CEGEPs are huge, modern, plasticized education factories of as many as 5,000 students with little discipline, relaxed standards and the best equipment money can buy. **1975** R. H. Rimmer *Premar Experiments* (1976) ii. 156 A simpler world of man experiencing man instead of machines and a plasticized environment?

plasticizer (plæ·stisəizəɹ). [f. Plasticize *v.* + -er¹.] Any substance which when added to another makes it (more) plastic or mouldable; *spec.* one (usu. a solvent) added to a synthetic resin to produce or promote plasticity and flexibility and to reduce brittleness.

1925 *Paint, Oil & Chem. Rev.* 29 Jan. 10/3 In order that the film may remain flexible..plasticizers—inert liquids of low vapor tension—are incorporated in the lacquer. **1943** H. R. Fleck *Plastics* vii. 160 Poly-vinyl chloride as formed is a hard brittle amorphous white powder which is useless for moulding purpose until plasticizer is added. **1947** J. C. Rich *Materials & Methods of Sculpture* xi. 342 Films of lacquer tend to harden and to become brittle as they age. For this reason plasticizers must be added to a lacquer formula. **1948** Dalzell & Townsend *Masonry Simplified* I. i. 16 In cement concrete, lime functions as a plasticizer. **1962** A. Nisbett *Technique Sound Studio* 248 Direct-cut disc... These are made of cellulose nitrate (which contains a castor oil plasticizer to soften it for easy cutting) but are often referred to as 'acetates'. **1967** M. Chandler *Ceramics in Mod. World* iv. 132 Most steatite electrical products..are most commonly shaped by dust-pressing, but because the body contains little clay the ceramist adds artificial plasticizers such as waxes or polyvinyl alcohol. **1971** *Materials & Technol.* II. v. 253 The so-called plasticisers which are added to mortars improve their working properties by causing air to be entrained into them at the time of gauging the mortar. **1972** *Ibid.* V. xxii. 844 Hairsprays nearly always include an important minor ingredient called a 'plasticizer' which softens the resin film and prevents it flaking on the hair.

plasticky (plæ·stiki), *a.* Also **plasticy.** [f. *plastic *sb.*³ + -y¹.] Suggestive of or resembling plastic.

1972 *Oxford Times* 12 May 20/4 His elder daughter Julie, 9, 'smelt plasticky fumes'. **1972** *Daily Tel.* 21 June 13/3 The interior [of the car] is rather 'plasticy' and the seats became hot in the sunshine. **1979** R. Rendell *Make Death love Me* xvi. 144 The gun..was a toy, as you could tell really by the plasticky look of it.

plasticware (plæ·stikwēəɹ). [f. Plastic *a.* + Ware *sb.*³] Articles made of plastic.

1972 *Science* 2 June 1039/2 We used plasticware or siliconized glassware..throughout the experiments. **1975** *New Yorker* 24 Nov. 102/2 D/R has imported a collection of opaque plasticware, in yellow, red, green, or white, from the firms of the celebrated English designer Terence Conran.

plasticy, var. *plasticky *a.*

plastidome (plæ·stidōᵘm). *Cytology.* [a. F. *plastidome* (P. A. Dangeard 1918, in *Compt. Rend.* CLXVI. 440), f. *plastide* Plastid *sb.* 2, after *chondriome*, *chromosome*, etc.] The plastids of a cell collectively.

1926 *Science* 18 June 620/2 Following..the non-committal terminology of the Dangeards, the components demonstrated are as follows: (1) Spindle fibers and cytoplasmic network... (2) Plastidome... (3) Spherome... (4) Vacuome. **1971** W. Stubbe in J. Reinert et al. *Origin & Continuity of Cell Organelles* 77 We assume that the division of the plastidome into a number of lentil-shaped chloroplasts and the consequent increase in surface may be the reason for the prevalence of this type among higher plants.

plastifier (plæ·stifəi,əɹ). [f. as next + -er¹.] = *plasticizer.

1919 H. Dreyfus *Brit. Pat. 160,225*, This invention has reference..to the manufacture of celluloid-like masses of any kind having a basis of cellulose ac[e]tate wherein high boiling solvents, called plastifiers, for the cellulose acetate, are incorporated with the mass in conjunction with one or more volatile liquids or diluents. **1971** *Materials & Technol.* II. ii. 111 Plastifiers and air-entraining agents. These materials improve the workability of concrete mixture.

plastify (plæ·stifəi), *v.* [f. Plastic *a.* + -ify.] = *plasticize *v.* 1, 2. Hence **pla·stifying** *vbl. sb.*

1919 H. Dreyfus *Brit. Pat. 160,225*, The solvent action on the cellulose acetate increases so that this is more and more dissolved and plastified until..only very little volatile diluent remains. **1963** H. R. Clauser et al. *Encycl. Engin. Materials & Processes* 348/1 A reciprocating plunger..forces the material into the plastifying cylinder. An equal quantity of fluid plastic is thus forced out of the front of the cylinder..into the mold. **1963** *Engineering* 13 Sept. 330/2 The concrete..is 'plastified' and remixed intensively by vibration. **1972** *Buenos Aires Herald* 4 Feb. 14/2 (Advt.), Floors polished, scraped, repaired, plastified.

plastigel (plæ·stidʒel). [f. Plasti(c *a.* + *gel *sb.*] A plastisol thickened to a putty-like consistency so that it retains its shape when heated.

1952 *Mod. Plastics* Jan. 99/1 Plastigels may be handled by many of the conventional methods of fabricating plastics but the pressures required are lower, leading to lower machine and mold costs. **1954** *Plastics Engin. Handbk.* x. 283 The control over the flow which is obtained with plastigels makes it possible to coat open-weave cloth or porous surfaces without excessive penetration. **1969** *Encycl. Polymer Sci. & Technol.* X. 246 With certain plasticizers or with modified paste resins the plastisol may become a gelatinous mass, or plastigel, under controlled conditions.

plastimeter, var. *plastometer.

plastique. Add: **2.** Statuesque poses or slow graceful movements in dancing; the art or technique of these.

1893 G. B. Shaw *Music in London 1890–94* (1932) III. 111 The unhappy students had been taught 'plastique' until they dared not call their arms and legs their own. The plastique professor..is almost as fatal a person as the harmony professor. **1897** —— *Our Theatres in Nineties* (1932) III. 146 Her interest in life and character will be supplanted by an interest in plastique and execution. **1947** *Ballet Ann.* I. 56 *Romeo and Juliet*..does contain both the passionate, lyrical dancing and strong, dramatic *plastique* demanded by the tragedy. **1960** *Times* 15 Mar. 6/1 Jive and an expressionistic development of *plastique* give more to these ballets than footwork. **1977** *New Yorker* 16 May 79/2 True to the 'Oriental' plastique of the period, the steps are all turned in.

3. Plastic explosive; a plastic bomb. Also *attrib.*

1968 L. W. Robinson *Assassin* (1969) xvi. 199 He planted another bomb... Bomb squad says it's made of *plastique*. **1969** E. Ambler *Intercom Conspiracy* (1970) vi. 136 They had no trouble..fixing the *plastique*, the bomb. **1974** *Publishers Weekly* 11 Feb. 64/2 To hold hostage a TV studio audience with a human plastique bomb.

‖ **plastiqueur** (plastikör). [Fr.] A person who plants or detonates a plastic bomb. Also *fig.*

1961 *Economist* 4 Nov. 435/1 Professor Palmer is the last and not the least of the *plastiqueurs*, as readers of the press learned not long ago, when he alleged that Evans had misstated some of his evidence in support of his pet theories. That explosion in the press was what gunners call a premature, and did Professor Palmer's discovery little good. **1962** *Times* 23 Apr. 9/2 The *plastiqueurs* were daily at work in metropolitan France. **1971** *Guardian* 17 July 9/4 The plastiqueurs were..soundly committed to private enterprise.

plastisol (plæ·stisǫl). [f. Plasti(c *a.* + *sol *sb.*⁶] A dispersion of particles of a synthetic resin in a non-volatile liquid consisting chiefly or entirely of plasticizer, which can be converted into a solid plastic simply by heating (cf. *organosol).

1946 *Mod. Packaging* Mar. 262/2 It is possible to prepare these dispersions without any volatile carrier. Such 100% solids dispersions are known as plastisols. **1954** [see *hot dipping* (*hot *a.* 12)]. **1963** H. R. Clauser *Encycl. Engin. Materials & Processes* 489/1 One of the most dramatic applications of plastisols is as a lining for kitchen dishwashers. **1971** [see *plasticized *ppl. a.* 1]. **1980** *Daily Tel.* 17 Jan. 11 (Advt.), A major manufacturer of PVC resins, compounds and plastisols.

plastochron (plæ·stǒkrǫn). *Bot.* Also **-chrone** (-krōᵘn). [a. G. *plastochron* (E. Askenasy 1878, in *Verhandl. des Natur-hist.-med. Vereins zu Heidelberg* II. ii. 76), f. Gr. πλαστό-ς formed, moulded (see -plast) + χρόν-ος time.] The interval of time between consecutive formations of leaf primordia (or of pairs of such primordia) in a growing shoot apex of a plant.

1929 *New Phytologist* XXVIII. 41 These differences may partly be accounted for by the fact that an interval elapses—called by some authors the 'plastochron'—between the initiation of successive primordia. **1938** Priestley & Scott *Introd. Bot.* xiii. 173 In the ⅖ system it will be noted..that each primordium is three plastochrones removed from its one neighbour, two from the other. **1957** *Amer. Jrnl. Bot.* XLIV. 298/1 A plastochron is conventionally defined as the time interval between initiation of two successive leaves. It might be more broadly defined as the interval between corresponding stages of development of successive leaves, and one might choose initiation, maturity, or any intermediate stage of development as the stage of reference. **1960** *Ibid.* XLVII. 707/1 This plastochron, averaging 3 wk. in length, delimits the 2 developmental stages of the heterophyllous shoot.

b. Special Combs.: **plastochron index**, an index of the developmental age of a shoot, being the number of leaves that are not less than some stated length, plus a fractional adjustment so calculated that the value of the index changes smoothly as the shoot grows; **plastochron ratio** (see quot. 1948).

1955 F. J. Michelini (*title of Ph.D. Dissertation, Univ. Pennsylvania*) The use of the plastochron index in studies of morphological and physiological development in *Xanthium italicum* Moretti. **1973** R. Maksymowych *Anal. Leaf Development* i. 5 The leaf plastochron index (LPI) can be used in developmental studies limited specifically to only one leaf. **1948** F. J. Richards in *Symp. Soc. Exper. Biol.* II. 226 The differences between the various orders of phyllotaxis have been referred to differences in the ratio of the distances of two successive primordia from the apical centre... The ratio will be referred to as the 'plastochrone ratio'. **1968** C. W. Wardlaw *Morphogenesis in Plants* ix. 231 As the several systems..in Fibonacci phyllotaxis have closely comparable divergence angles, i.e. about 137·5°, the essential differences between them are due to their plastochrone ratios. As the *P.R.* approaches unity, the phyllotaxis rises.

Hence **plastochro·nic** *a.*

1953 K. Esau *Plant Anat.* v. 104 The changes in the morphology of the shoot apex occurring during one plastochron may be referred to as plastochronic changes. **1957** *Amer. Jrnl. Bot.* XLIV. 302/1 Much would be gained by determining the precise plastochron age of each apex, rather than..judging the plastochronic phase subjectively.

plastocyanin (plæstosəi·ănin). *Biochem.* [f. *chloro)plast + -o + Cyanin.] A blue copper-containing protein (differing slightly from species to species) which is found in the chloroplasts of green plants and in certain bacteria, and is involved in photosynthesis.

1961 Katoh & Takamiya in *Nature* 25 Feb. 665/2 In view of its localization in the chloroplasts and its characteristic blue colour in the oxidized form, the name 'plastocyanin' is proposed for this copper protein. **1966** *Plant Physiol.* XLI. 1641/1 From its absorption spectrum, the plastocyanin of *C. reinhardi* resembles the plastocyanin of spinach. **1974** *Sci. Amer.* Dec. 74/1 (*caption*) The electron is passed through a series of carrier molecules including..plastoquinone..and cytochrome *f*... to plastocyanin. **1976** *Nature* 27 May 344/2 The 'small blue proteins'—azurins from bacteria, plastocyanins from photosynthetic cells.

plastome (plæ·stōᵘm). *Genetics.* Also **plastom.** [a. G. *plastom* (O. Renner 1929, in *Handb. d. Vererbungswissenschaft* IIa. 32), f. *plastid* Plastid *sb.* 2, after *genom* *genome.] The sum-total of the genetic factors or information in the plastids of a cell.

1954 P. Michaelis in *Adv. Genetics* VI. 290, I propose.. with Renner (1929) to include all extranuclear hereditary elements of the cell in the term plasmon, and to subdivide this into (1) the cytoplasmon, that is, the elements of the cytoplasm, and (2) the plastom, that is, the hereditary elements of the plastids, etc. **1965** Wettstein & Eriksson in S. J. Geerts *Genetics Today* xvi. 594 In higher plants two non-chromosomal genetic systems controlling chloroplast structure and function—the plastome and the plasmone—have long ago been recognized from the different modes of inheritance of certain chloroplast defects. **1967** Kirk & Tilney-Bassett *Plastids* ix. 277 There are as many as five genetically different plastoms within the subgenus *Euoenothera*. **1976** [see *plasmon 2].

plastometer (plæstǫ·mǐtəɹ). Also **plasti·meter.** [f. Plast(icity + -ometer.] An instrument for measuring the plasticity of a substance. Hence **plasto·metry.**

1919 Bingham & Green in *Proc. Amer. Soc. Testing Materials* XIX. ii. 645 As we wish, in a sense, to measure the 'plasticity' of a paint the apparatus for making the measurements has been called a 'plastometer'. **1922** E. C. Bingham *Fluidity & Plasticity* 319 (*heading*) Practical plastometry. **1933** *Physics* IV. 285/1 The theory of parallel-plate plastometry. **1940** *Brit. Standard Methods testing Latex* (*B.S.I.*) 26 Compress the pellet between thin sheets of paper under a load of 5 kg. in a parallel plate plastimeter. **1946** *Nature* 14 Sept. 371/1 The original type of plastometer, devised by Bingham, is still in use, with slight modifications, in many laboratories to-day. **1958** *New Scientist* 5 June 112/1 (*caption*) This machine..is a capillary plastometer..and is particularly suitable for studying the flow properties of semi-solid materials. **1971** J. A. C. Harwood in C. M. Blow *Rubber Technol. & Manuf.* iii. 59 The elastic recovery is measured by the height of the sample a fixed time after removal from the plastimeter.

plastoquinone (plæstokwi·nōᵘn). *Biochem.* [f. *chloro)plast + -o + Quinone.] Any of a homologous series of compounds which have a quinone nucleus with a terpenoid side-chain, and which occur in the chloroplasts of plants; *spec.* one having the formula $C_{48}H_{72}O_2$.

1958 F. L. Crane in *Plant Physiol.* XXXIV. 547/1 It is proposed that Q_{254} should be called plastoquinone... This name will serve to emphasize the localization in chloroplasts and possibly other plastid structures. **1968** *New Scientist* 25 Jan. 189/1 Terpenoids in chloroplasts of green leaves include..plastoquinone. **1973** *Biochim. & Biophys. Acta* CCCI. 36 Several homologues of plastoquinone..have been detected in algae and chloroplasts of higher plants... The detailed chemical structure of plastoquinones B and C is not known yet. **1975** D. Jarvis tr. *D. Hess's Plant Physiol.* 41 Another important redox system which is engaged in electron transport in photosynthesis is plastoquinone. **1978** *Sci. Amer.* Mar. 111/3 The two electrons that cross the membrane from P-680 are picked up at the outer surface by a hydrogen carrier similar in structure to ubiquinone but called plastoquinone, or PQ.

plastron. Add: **3. c.** *Ent.* [a. Fr. (F. Brocher 1912, in *Ann. Biol. Locustrine* V. 141).] In certain aquatic insects, a type of external gill formed by a patch of cuticle covered with hairs which retain a thin layer of air under water. Also *attrib.*

1947 Thorpe & Crisp in *Jrnl. Exper. Biol.* XXIV. 227 (*title*) Studies on plastron respiration. *Ibid.* 229 The volume of gas in the plastron is negligible. **1959** Southwood & Leston *Land & Water Bugs Brit. Is.* 367 The air film is self-renewing: oxygen continually diffuses in and out, the whole forming a plastron, or external gill. **1969** R. F. Chapman *Insects* xxiv. 481 The volume of the plastron is constant and usually small since it does not provide a source of air but acts as a gill. **1976** H. E. Hinton in H. R. Hepburn *Insect Integument* xxv. 482 In *Ptyopteryx*..the only diffraction lines are those formed by the plastron on the ventral surface of the abdomen.

plat, *sb.*² **I. 5.** (Earlier example.)

1788 J. Backus *Jrnl.* in W. W. Backus *Genealogical Mem. Backus Family* (1889) 20 A beautiful platt of a considerable extent.

plat, *sb.*³ **II. 2.** (Further examples.)

1954 *Ann. Assoc. Amer. Geographers* XLIV. 248 Areal boundaries on the congressional township plats do not always clearly differentiate the several areal units such as marshes, prairies, wet prairies, swamps, and timberlands. **1974** *Sumter* (S. Carolina) *Daily Item* 24 Apr. 15A/4 A plat showing where the land is located must also be submitted before any transfer of land can be made. **1977** *Sci. Amer.* Sept. 184/1 (*caption*) The computer-held data in turn can be fed to a plotter that will automatically convert days of field observations into a standard surveyor's plat.

plat, *sb.*⁶ Add: **b.** *plat du* (erron. *de*) *jour*: dish of the day; one of a restaurant's specialities on any particular occasion; also *fig.* and *ellipt.* as *plat*.

1906 W. J. Locke *Beloved Vagabond* (1907) vi. 71 The placarded list of each day's *plat du jour*. **1934** I. Stone *Lust for Life* v. xi. 374 The man scanned the menu, ordered a *plat du jour*, and within a moment was scooping up his soup with a large spoon. **1953** Wodehouse *Performing Flea* 213 We formed up in a queue, each man with a porcelain bowl for the *plat de jour* and a plate or a tin or a cigar-box for the potatoes. **1960** *Guardian* 3 Feb. 8/7 On Thursday the plat du jour will be paella. **1975** R. Rostand *D'Artagnan Signature* (1976) xxxvii. 210 Davis ordered the *plat du jour* and a full *pichet* of *vin rosé*. **1979** *Guardian* 26 Feb. 10/6 The Poetry Society seems to have achieved some success... The main plat du jour was the announcement of the winner of its new £1,000 prize. **1979** B. Peterson *Peripheral Spy* vi. 147 A menu..informed him that the *plat* today was *tendrons de veau*, a favourite of his.

platanna (plata·nā). *S. Afr.* Also **platana, plathander.** [Afrikaans, f. Plat *a.* + *-hander* Hander².] The clawed frog or toad, *Xenopus lævis*, belonging to the family Pipidæ and native to South Africa.

1898 *Empire* 24 Sept. (Pettman), It's a platana, one of them web-footed, flat-backed, smooth-skinned, yeller frogs, with a mouth that goes all round its neck. **1911** J. D. F. Gilchrist *S. Afr. Zool.* 224 The..Plathander (flat hand) or Clawed toad..occurs in most pools of water. **1949** *Cape Times* 21 Sept. 13/4 Two hundred female platanna frogs..left Stellenbosch yesterday on an adventurous journey to Canada. **1961** D. M. Cochran *Living Amphibians of World* 52/1 The three inner toes of the platanna's hind foot..bear the short black claws. **1971** C. A. du Toit in D. J. Potgieter et al. *Animal Life S. Afr.* 267/2 Just over thirty years ago physiologists in Cape Town, using the platanna as a test animal, discovered and perfected the first reliable test for pregnancy.

plate, *sb.* Add: **I. 1. d.** A number of animal skins sewn together, for making up into fur coats or for linings, trimmings, etc.

1910 *Encycl. Brit.* XI. 354/2 A very great feature of German and Russian work is the fur linings called rotondes, sacques or plates. **1957** M. B. Picken *Fashion Dict.* 256/1 *Plate*,..2. Skins sewn together, but not completely fitted or finished, for fur linings; also used to make garments or trimmings. **1972** *Guardian* 11 Aug. 7/8 The [import] ban did not include 'plates'—sections of fur coats ready to be made up. **1974** *Encycl. Brit. Macropædia* VII. 816/1 The less costly skin-on-skin method consists of sewing one full skin adjacent to another in a uniform alignment. This method is sometimes employed to sew the leftovers of full skins such as paws and flanks, into blanket-like 'plates' that are then fashioned into garments.

e. *Geol.* Each of the several nearly rigid pieces of lithosphere which are thought to make up the whole of the earth's surface and to be moving slowly relative to one another, the boundaries between adjacent ones being identified with well-defined belts of seismic, volcanic, and tectonic activity.

[**1904** H. B. C. Sollas tr. *Suess' Face of Earth* I. ii. xxii. 600 Towards the north [of North America], however, a very extensive 'plate' without folding appears, which stretches nearly to the Arctic archipelago. **1910** *Bull. Geol. Soc. Amer.* XXI. 191 As Suess says, we do not know the character of the platforms upon which lie the seas behind island arcs;..the platforms may be composed of ancient, crystalline rocks which moved as 'plates' without parallel foldings.] **1965** J. T. Wilson in *Nature* 24 July 343/1 Many geologists have maintained that movements of the Earth's crust are concentrated in mobile belts, which may take the form of mountains, mid-ocean ridges or major faults... This article suggests that these features are not isolated, that few come to dead ends, but that they are connected into a continuous network of mobile belts about the Earth which divide the surface into several large rigid plates. **1969** *Jrnl. Geophysical Res.* LXXIV. 4298/2 In the New Guinea mainland the zone of southerly dipping earthquakes can be associated with the northern edge of the Australian continent meeting the Pacific 'plate' and creating an overthrust zone of mountain building. **1972** *McGraw-Hill Yearbk. Sci. & Technol.* 305/1 Lithospheric plates are..segments of upper mantle and crust, varying in thickness from approximately 5 km at ridges to 150 km under central areas of continents, that are generated by growth of crust and mantle at oceanic ridges..and consumed in trenches. **1976** M. A. Khan *Global Geol.* viii. 147 The lower boundary of the plates is the base of the rigid lithosphere which moves over the plastic convecting asthenosphere. The plates are therefore often referred to as lithospheric plates.

4. f. Also, a similar portion of any orthodontic appliance; by extension, the whole denture or other appliance.

1932 E. Bowen *To North* v. 44 Her confirmation.., the fixing-in of a plate to correct prominent teeth,..had all been reported to him. **1973** M. Amis *Rachel Papers* 162, I had been coming down from Oxford about six times a year since I was ten so that he could put in and take out all the lousy braces and plates and other crap with which he tried to tame my mouth. **1977** B. Pym *Quartet in Autumn* v. 52 He had to visit the dentist, to adjust his new plate and to practise eating with it.

h. *Electr.* A metal plate that acts as a charge-storing electrode in a capacitor.

1782 *Phil. Trans. R. Soc.* LXXII. p. xxvii, An ample conductor, weakly electrified, imparts a considerable quantity of electricity to the metal plate of our condenser. **1801** *Encycl. Brit.* Suppl. I. 591/1 The mode of accumulating great quantities of fluid by means of parallel plates. **1887** P. Benjamin *Age of Electricity* xi. 259 The condenser will be charged with a quantity of electricity depending upon..the surface of the plates opposed to each other, and..the number of plates in the respective sets. **1923** E. W. Marchant *Radio Telegr. & Teleph.* ii. 15 A condenser can be charged by supplying positive electricity to one plate and negative electricity to the other. **1963** A. F. Abbott *Ordinary Level Physics* xxxv. 460 A parallel-plate capacitor is set up as shown.., one plate being earthed and the other..charged.

i. *Electr.* A metal electrode in a cell or battery, esp. one in the form of a plate or grid.

[**1801** H. Davy in *Phil. Trans. R. Soc.* XCI. 397, I have found that an accumulation of galvanic influence, exactly similar to the accumulation in the common pile, may be produced by the arrangement of single metallic plates, or arcs, with different strata of fluids.] **1807** —— in *Ibid.* XCVII. 15 The strong action of a battery of 150 pairs of plates of 4 inches square. **1828** F. Watkins *Pop. Sk. Electro-Magnetism* 15 Batteries of this construction usually consist of ten or twelve pairs of plates. **1923** Glazebrook *Dict. Appl. Physics* II. 71/1 The container [of a dry cell] is made of zinc and this is used as the zinc plate. **1963** A. F. Abbott *Ordinary Level Physics* xxxvi. 474 In modern commercial practice the plates [of a lead-acid cell] are made of grids of a lead-antimony alloy filled with paste... Red lead (Pb_3O_4) is used for the positive plates and litharge (PbO) for the negative plates. **1970** *AA Bk. Car* 82/4 When the surfaces of both plates have turned completely to lead sulphate, the battery is flat.

j. *Electronics.* The anode of a thermionic valve.

1905 J. A. Fleming in *Proc. R. Soc.* LXXIV. 477 It is preferable to use a metal plate carried on a platinum wire sealed into the glass bulb, the plate being bent into a cylinder which surrounds both the legs of the carbon loop. *Ibid.* 479 The resistance of these valves..may be anything from a few hundred ohms up to some megohms, depending on the state of incandescence of the filament.., as well as upon the size of the filament and the plate. **1915** *Electrician* 21 May 244/2 The plate of the pliotron oscillator is then connected to one of the terminals of the condenser. **1948** A. L. Albert *Radio Fundamentals* vi. 178 The plate usually surrounds the cathode in high-vacuum diodes. **1975** D. G. Fink *Electronics Engineers' Handbk.* vii. 21 The collector element for the electron flow is the anode, or plate.

5. d. The number plate of a motor vehicle.

1950 J. D. MacDonald *Brass Cupcake* (1955) iii. 23 She's got a grey Chevvy business coupé with Massachusetts plates. **1970** *Globe & Mail* (Toronto) 28 Sept. 27/4 (Advt.), 2 door sedan, 3 cylinder, 2 cycle. Not certified, no plates. **1973** R. Lewis *Blood Money* vi. 67 That car..ended in some garage with a bent mechanic stripping it, respraying it, changing the plates. **1975** *Drive* New Year 98/1 Secondhand plates are not expensive but they can be difficult to obtain.

9. (Earlier and later examples.)

1836 *Spirit of Times* 20 Feb. 6/2 Having the misfortune to break the plate on her left hind foot on one side,.. she was withdrawn after the first heat. **1937** E. Rickman *On & off Racecourse* vi. 130 If a horse is to be relieved of the considerable weight of these shoes, during a race they must be replaced by light plates made of aluminium or other suitable alloy. **1965** D. Francis *Odds Against* viii. 119 Horses race in thin light shoes called plates... Blacksmiths change them before and after, every time a horse runs.

12*. *Baseball* and *Softball.* A flat piece of metal or stone marking the home base; the home base itself. Also *fig.*

[**1857** *Spirit of Times* 28 Feb. 420/3 The home base and pitcher's point to be each marked by a flat circular iron plate, painted or enamelled white.] **1867** *Ball Players' Chron.* 5 Sept. 5/1 Thorne..pitched slow, 'drop' balls, many of which struck outside of 'the plate'. **1886** H. Chadwick *Art of Pitching & Fielding* 43 When the Umpire indicates the height of the ball required, the pitcher should send it in at once at the height required, but *not* over 'the plate'. **1902** *Encycl. Brit.* XXVI. 161/2 This corner is marked by a white plate a foot square, sunk level with the ground, and called the home base. **1917** C. Mathewson *Sec. Base Sloan* 172 Ellis walked to the plate and faced Chase grimly determined to get a hit. **1931** D. Runyon *Guys & Dolls* (1932) x. 224 Jo-jo squares away at the plate. **1936** *N.Y. Herald Tribune* 4 Oct. ii. 2/1 The Democrats have scored in their half of the last inning, but the Republicans still have a chance to bat. Alfred E. Smith has just come up to the plate for them. **1952** B. Malamud *Natural* (1963) 70 When Roy came up with Wonderboy, he hugged the plate too close to suit Fowler who was in there anyway only to help the batters find their timing. **1967** C. Potok *Chosen* i. 16, I went up to home plate for some batting practice. **1973** C. Sagan *Cosmic Connection* xv. 112 If he swings and misses—or, more likely, if the ball is wide of the plate—he can then go home for a two-hour nap, returning with his catcher's mitt to catch the ball. **1977** *Guernsey Weekly Press* 21 July 8/6 Rangers pushed five runs over the plate before going one down, but errors..were mainly responsible.

II. 14. c. Delete †*Obs.* and add later examples.

1880 L. Higgin *Handbk. Embroidery* i. 9 *Plate* consists of narrow plates of gold or silver stitched on to the embroidery by threads of silk. **1881** C. C. Harrison *Woman's Handiwork* I. 54 Bullion, passing, plate and spangles are employed in silk embroidery.

15. d. As a non-collective *sb.*: A thin coating of metal, esp. one applied electrolytically.

1915 *Chem. News* 10 Dec. 288/1 Plates on various stock pieces satisfactorily withstood the various bending, hammering, and burnishing tests. **1946** *Trans. Electrochem. Soc.* LXXXIX. 384 The nickel-cobalt plate is whiter, harder and more corrosion-resistant than nickel deposits. **1959** T. M. Rogers *Hand-bk. Pract. Electroplating* 14 The work is..given a thin plate of Rochelle copper. **1974** P. D. Groves *Electrochem.* xii. 92 A mixture of nickel (II) sulphate and nickel (II) chloride together with a boric acid buffer and a wetting agent..produces a good plate which resists wear and abrasion, even at high temperatures.

17. (Earlier and later examples.) Also *fig.*

1639 R. Verney *Let.* in F. P. Verney *Memoirs* (1892) I. viii. 185 'My Lord Carlile's white nagg,' says Ralph, 'hath beaten Dandy, and Sprat woone the cup, and Cricket the plate.' **1910** *Encycl. Brit.* XIII. 728/2 In 1739 an act was passed to prevent racing by ponies and weak horses,..which also prohibited prizes or plates of less value than £50. **1939** Joyce *Finnegans Wake* (1964) I. 39 The classic Encourage Hackney Plate was captured by two noses in a stablecloth finish, ek and nek. **1955** *Times* 5 Aug. 4/1 Magic Key may be favourite to open the card at Lewes in the selling race. He has been placed in his last three races, being unlucky enough to come up against horses above the average for selling plates. **1979** K. Bonfiglioli *After you with Pistol* v. 22 A diet of beefsteak, oysters and Guinness would soon lift me out of the selling-plate class.

III. 18. a. Also in colloq. phrases: *to hand* (something to someone) *on a plate* and varr.: to give (something to someone) without his asking or seeking or without requiring any effort or return from him; to present in ready-to-use form; *to have a lot* (*enough*, etc.) *on one's plate*: to have a lot (enough, etc.) to worry about or cope with.

1922 Joyce *Ulysses* 135 Gave it to them on a hot plate, Myles Crawford said, the whole bloody history. **1928** *Daily Express* 4 July 9/2 Can you tell me how many times in all she has forbidden you the house?—No, sir. Half a dozen times?—It might have been. I cannot say. I have a lot on my plate... Mr. Justice Horridge: A lot on your plate! What do you mean? Elton Pace: A lot of worry, my lord. **1935** Wodehouse *Right Ho, Jeeves* ii. 27 He can't get action when he's handed the thing on a plate. **1945** *Penguin New Writing* XXIV. 32 We haven't time to worry about D though, we shall probably have enough on our own plate any minute now. **1946** R. G. Collingwood *Idea of Hist.* 256 If anyone else..hands him on a plate a ready-made answer to his question, all he can do is to reject it. **1957** *Listener* 11 July 72 It is not often that radio is presented on a plate with such a fine natural script as these extracts made. **1959** 'R. Simons' *Houseboat Killings* xiv. 142 I'll leave you at it. I've got plenty on my plate at the moment. **1960** L. Cooper *Accomplices* I. vi. 67 That was an easy one—Steyne had handed it to us on a plate. **1963** T. Parker *Unknown Citizen* iii. 78 Duggie's got a lot on his plate just now, I didn't want to worry him. **1970** *Manch. Guardian Weekly* 11 July 4 If New Zealand has the EEC door slammed in her face..car factories of Japan..would have a new market handed to them on a plate. **1973** 'P. Malloch' *Kickback* xii. 78 You make that kind of mistake you're handing it on a plate to the cops. *Ibid.* xxiii. 145 The police..[have] got enough on their plate. We pull our job tomorrow while they're trying to tidy up their own mess.

b. Restrict † (*obs.*) to (*a*) and add later examples of (*b*).

1886 Kipling *Departmental Ditties* (ed. 2) 13 Who can raise a two-plate dinner off eight paltry 'dibs' a day? **1971** 'D. Shannon' *Ringer* (1972) viii. 138 'Oh—the low-calorie plate,' as the waiter came up. **1972** J. Wambaugh *Blue Knight* (1973) i. 19, I promised to come back Friday for the De luxe Businessman's Plate. **1974** D. E. Westlake *Help* (1975) xlii. 246 The man..recommended the roast beef plate..and the woman..said the turkey diet plate was first-rate.

d. *plates of meat*: feet; freq. ellipt. as *plates*; †*plate of meat*: a street (*obs.*). *Rhyming slang.*

1857 'Ducange Anglicus' *Vulgar Tongue* 15 *Plate of meat*,..street. **1887** *Referee* 6 Nov. 7/3 As she walked along the street With her little 'plates of meat'. **1889** J. S. Farmer *Americanisms* 425/2 *Plate of meat* (Cant),..in America does duty as the name, among thieves, for a street or highway. *a* **1896** A. R. Marshall in Farmer & Henley *Slang* (1902) V. 224/2 He is rocky on his plates, for he has forced them into 'sevens'. **1898** J. D. Brayshaw *Slum Silhouettes* 85 If a peeler heaves in sight..they'll.. take a rise out of 'im with a chorus of 'Boots!' alluding to his 'plates o' meat', as they calls 'em. **1917** W. Muir *Observations of Orderly* xiv. 222 To get your 'plates of meat' frostbitten wasn't such a 'cushy wound' as it was cracked up to be. **1948** C. Day Lewis *Otterbury Incident* ii. 17 'Your clodhopping feet.' 'Plates of meat,' murmured Dick Cozzens, who is an expert in slang. **1951** P. Branch *Lion in Cellar* ix. 105 He..took off his shoes. 'Heaven!' he sighed. 'My plates have been quite, quite killing me.' **1975** P. G. Winslow *Death of Angel* xii. 229 Gawd, I wore out my plates of meat.

e. *Biol.* and *Med.* A shallow vessel, usu. a Petri dish, used to contain a medium for the cultivation of micro-organisms; freq. used inclusively of this medium.

1886 E. M. Crookshank *Introd. Pract. Bacteriol.* v. 68 The glass plates are sterilised by filling the iron box..and placing it in the hot-air steriliser, at 150° C., from one to two hours. **1896** G. M. Sternberg *Text-Bk. Bacteriol.* viii. 72 By Koch's famous 'plate method' we obtain colonies of any particular microorganism which we desire to study. **1934** A. T. Henrici *Biol. Bacteria* xii. 203 The colonies which develop upon agar or gelatine plates exhibit specific characters by which one may often identify the organism of which they are composed. **1973** R. G. Krueger et al. *Introd. Microbiol.* xiv. 388/2 On mixed indicator plates phages that are wild type for the host range character (h⁺) form turbid plaques because they can only lyse one of the two kinds of bacteria in the mixture, whereas the h mutants, since they lyse both kinds of bacteria equally well, form clear plaques.

f. *U.S.* A place at a formal meal or banquet, for which one subscribes.

1925 L. S. Dunway in B. A. Botkin *Treas. S. Folklore* (1949) II. iii. 278 The committee on arrangements called on Jeff Davis at his office and wanted to know if the governor would like to have a plate at the banquet, the cost of which was $5. **1941** B. Schulberg *What makes Sammy Run?* xii. 288 They gave Sidney a testimonial dinner at the Ambassador at ten dollars a plate. **1964** Mrs. L. B. Johnson *White House Diary* 9 Apr. (1970) 105 The luncheon took place before an audience of twenty-eight hundred, who had paid $12·50 a plate. **1974** *News & Reporter* (Chester, S. Carolina) 24 Apr. 8-A/1 Tickets, priced at $3.50 per plate, are on sale with any members of the sponsoring Chester Girls Club and the Diversity Study Club.

g. *U.S. slang.* A gramophone record.

1935 *Vanity Fair* (U.S.) Nov. 38/1 None of these plates will be senders. **1937** *Amer. Speech* XII. 100 Behind the microphone they [*sc.* gramophone records] are referred to variously as *discs, E.T.'s, plates, platters, wax* and *cuts*. **1942** Berrey & Van den Bark *Amer. Thes. Slang.* §581/2 *Phonograph record*,..plate.

h. *N.Z.* and *Austral.* A plate of cakes, sandwiches, or the like contributed by a participant towards the catering at a social gathering.

1953 M. Scott *Breakfast at Six* viii. 70 Gents half-a-crown, ladies a plate... Larry explained what 'a plate' meant in the backblocks. **1962** S. Gore *Down Golden Mile* 110 We might start by having some sort of social. Nothing elaborate, you know. Just perhaps all the ladies could bring a plate. **1966** G. W. Turner *Eng. Lang. in Austral. & N.Z.* iii. 48 Newcomers to New Zealand country districts have been embarrassed by a failure to detect this semantic development [*sc.* metonymy] in the advertisements for country dances, used to ensure that the right amount of supper will be provided and the hire of the hall paid—'Gents 2/- —Ladies a plate'. Since New Zealand countrywomen are renowned for their cooking of fancy cakes, an empty plate shows up rather poorly.

IV. 19. a. (sense *1 e) *plate boundary*; (sense *4 j) *plate circuit, current, voltage*; (sense 12*) *plate umpire*.

1971 I. G. Gass et al. *Understanding Earth* xix. 263/1 We distinguish between a plate boundary, the surface trace of the zone of motion between two plates, and a plate margin, the marginal part of a particular plate. **1979** C. Kilian *Icequake* v. 70 It looks now as if the ice sheet put a strain..on a series of faults on the far side of the Queen Maud Range. Tim and I are pretty sure the faults mark the edge of a plate boundary. **1919** J. A. Fleming *Thermionic Valve* 224 In general the external E.M.F. required in the plate circuit of a very hard valve is 100 volts, or even more, to produce a plate current of 3 or 4 milliamperes with the grid at zero potential. **1974** *Encycl. Brit. Macropædia* VI. 688/1 Many of these secondary electrons are attracted to the screen grid and flow in its circuit, rather than in the plate circuit, where the output should flow for greatest circuit efficiency. **1915** *Electrician* 21 May 243/1 (*diagram*) Plate current. **1966** T. Korneff *Introd. Electronics* vi. 198 The plate current is a function of the screen grid voltage and does not depend too much on the plate voltage. **1967** *Boston Herald* 1 Apr. 17/4 The area around home plate was especially soft and with two sinker ball pitchers working, plate umpire Larry Nap had a terrible time. **1922** *Encycl. Brit.* XXXII. 1027/2 The plate voltage of the oscillating valve is not supplied by a high voltage battery but at most by a few cells. **1966** Plate voltage [see *plate current* above].

b. *plate-glazing, -making*; *plate-glazed* adj.

1915 J. Southward *Mod. Printing* (ed. 3) II. xxx. 258 Plate-glazed Paper is finished by being placed sheet by sheet between copper or zinc plates... The pile is pressed through powerful rollers. **1911** *Encycl. Brit.* XX. 734/2 The plate-glazing process is adopted mainly for the best grades of writing-papers, as it gives a smoother, higher and more permanent gloss than has yet been imitated by the roll-calender. **1962** F. T. Day *Introd. Paper* iv. 47 Plate glazing is carried out by passing the paper between zinc plates and pressing it to give the desired finish. **1939** R. R. Karch *Printing & Allied Trades* (ed. 2) xviii. 180 (*heading*) Offset Plate Making. **1967** E. Chambers *Photolitho-Offset* iii. 31 A good reproduction proof.. becomes copy and is photographed in the normal way for plate-making.

20. plate camera, a camera designed to take photographs on coated glass plates rather than film; **plate-clutch**, a form of clutch in which the engaging surfaces are flat metal plates; **plate count**, an estimate of cell density in milk, soil, etc., made by inoculating a plate (sense *18 e) with a suitably diluted sample and counting the number of colonies that appear; **plate cylinder**, in a rotary printing press, the cylinder to which printing plates are attached; **plate girder** (examples); **plate-line** = Plate-mark 2; **plate metal**, (*b*) = **plate pewter*; **plate mill** (examples); **plate pewter**, the hardest variety of pewter, used for plates and dishes; **plate pie** (see quot. 1946); **plate-powder** (earlier and later examples); **plate-printer**, a workman who prints from plates; **plate-roll**, a smooth roller for rolling metal plate or sheet; **plate-room**, (*a*) a room for keeping plate (sense 15); (*b*) = **plate-safe*; **plate-safe** (see quot.); **plate-shy**, *a.*, *Baseball* (see quots.); **plate tectonics** *Geol.*, a theory of the earth's surface based on the concepts of moving plates (*PLATE *sb.* 1 e) and sea-floor spreading, used to explain the distribution of earthquakes, mid-ocean ridges, deep-sea trenches, and orogenic belts; hence **plate-tectonic** *a.*, **plate tectonicist**.

1937 *Discovery* June 177/2 Really good second-hand plate-cameras can be bought quite cheaply. 1956 *Focal Encycl. Photogr.* 868/1 Plate cameras are generally larger than film cameras. 1977 *Times* 31 Aug. 10/4 A man peering through a large plate camera at them. 1906 Plate clutch [see *disc-clutch* s.v. *DISC *sb.* 8 f]. 1960 *Farmer & Stockbreeder* 5 Jan. 95/2 Power..is transmitted via an independent plate-clutch. 1901 *Jrnl. Hygiene* I. 301 The effect of the ice-packing upon the number of colonies appearing in the ordinary plate count has been already discussed. 1928 *Jrnl. Bacteriol.* XVI. 270 The manner of making plate counts which prevails in public-health and other laboratories where daily counts are made on a number of samples of milk. 1956 *Nature* 4 Feb. 221/1 The time of soaking and shaking the leaves before plating influenced the plate-count. 1972 *Ann. Rep. Freshwater Biol. Assoc.* XL. 41 Bacterial numbers as shown by plate counts. 1932 PLACE & CLUNES in W. Atkins *Art & Practice of Printing* II. xi. 192 The plate-Cylinder is made to carry the curved plates; plate-Cylinders have to be very accurately ground. 1973 J. MORAN *Printing Presses* xiii. 198 It [*sc.* the plate] was then placed in a bending box, where it received the appropriate curvature to enable it to lie on the plate cylinder. 1849 W. FAIRBURN *Acct. Construction Britannia & Conway Tubular Bridges* I. 176 Is there anything new in this application of wrought-iron plate girders? 1891 *Notes Building Construction* IV. viii. 154 The web of a plate girder being very thin can bear but a very small part of the direct stresses. 1950 *Engineering* 8 Dec. 465/3 Mr. Dean, in his report on metal under-bridges, concluded that plate girders are the most economical and satisfactory form of construction for the main girders. 1931 A. ESDAILE *Student's Man. Bibliogr.* v. 151 All intaglio engravings will show a 'plate-line'; the paper which is pressed by the plate is smooth and sunk, while beyond the edge of the plate it keeps its natural surface; the resulting line is called the plate-line. 1961 T. LANDAU *Encycl. Librarianship* (ed. 2) 281/1 *Plate line*, a characteristic mark in intaglio printing, especially of engravings, due to the great pressure exerted by the engraving press on the paper. 1668–9 in C. Welch *Hist. Worshipful Co. Pewterers* (1902) II. 140 It is..agreed..that..every person that taketh Hollow-ware of any workman & returneth not him for the same ½ plate mettle and ¼ London Trifles, shall pay unto such workman [etc.]. 1867 *Engineering* 4 Jan. 1/2 In the reversing plate mill at the London and North-Western Steel Works..the reversal of the motion of the rolls is effected by reversing the..engines. 1964 *Recent Progr. Metal Working* ii. 39 Automatic control of a plate mill entails..control of gauge in the last cross-rolling and last finishing pass. 1839 URE *Dict. Arts* 952 The plate pewter has a bright silvery lustre when polished. 1911 *Encycl. Brit.* XXI. 339/1 Plate pewter (100 parts of tin, 8 of antimony, 4 of copper and 4 of bismuth). 1946 F. M. MCNEILL *Recipes from Scotland* 13 A plate-pie, *i.e.*, with pastry above and below the filling. 1975 *Times* 19 July 11/4 Rhubarb plate pie. 1786 J. WOODFORDE *Diary* 24 Apr. (1926) II. 241 For some plate Powder at Chases pd 0. 1. 0. 1877 J. H. EWING in *Aunt Judy's Mag.* 146 The over-bearingness of the butler,..the inferior quality of the new plate-powder. 1976 *Century of Trade Marks* (Patent Office) 44 There is no longer any general need for such things as plate powder, polishing paste and blacking. 1889 *Cent. Dict.*, Plate-printer. 1902 *Encycl. Brit.* XXXIII. 414/2 Plate Printers' Union of United States, National Steel and Copper. 1921 *Dict. Occup. Terms* (1927) §529 *Printer, plate*, prints from copper or steel plates, on which design or lettering is sunk below surface of plate, instead of being raised as in letterpress work. 1861 W. FAIRBAIRN *Iron* 111 The cylindrical part B, for plate-rolls should be slightly concave. 1930 *Engineering* 7 Nov. 579/2 (*heading*) Plate-roll finishing machine. 1888 Plate room [see *plate-safe*]. 1931 *N. & Q.* 10 Oct. 262/2 The plate-room..is a strong steel and fireproof apartment. 1888 *Encycl. Brit.* XXIII. 710/1 The plate-safe or plate-room is the repository of the stereo and electro plates. 1912 C. MATHEWSON *Pitching* iv. 90 For a long time, 'Josh' Devore, the Giant's left-fielder was 'plate shy' with left-handers—that is, he stepped away. 1942 BERREY & VAN DEN BARK *Amer. Thes. Slang* §677/37 *Plate-shy*, afraid to stand close to the plate. 1972 *Sci. Amer.* May 59/1 According to the plate-tectonic view, continents and oceans are rafted along by the same crustal conveyor belt. 1976 *Nature* 9 Sept. 118/1 In the plate tectonic model, the Himalaya is considered to be the classic example of a continent–continent collision system. 1973 *Ibid.* 30 Nov. 263/1 They are in disagreement with the views of some plate tectonicists who wish to have the whole 500 km shift take place after mid-Miocene. [1966 *Bull. Geol. Soc. Amer.* LXXVII. 707 The folds and faults mapped at the surface [near the San Andreas fault] are attributed to raft tectonics whereby a passive surficial plate is deformed as it rides coupled to a moving undermass.] 1969 *Sci. Jrnl.* Aug. 40/2 Plate tectonics..has shown its ability to predict, amongst other things, the direction of the movement accompanying earthquakes. 1972 *Observer* (Colour Suppl.) 13 Feb. 12/1 During the past five years, there has been a revolution in the Earth Sciences, involving a theory called plate tectonics. 1976 *McGraw-Hill Yearbk. Sci. & Technol.* 314/1 The widespread acceptance by the earth sciences community of plate tectonics has had a revitalizing effect on many research fields, but none more so than paleobio-geography, the study of the factors controlling the distribution of fossil organisms.

plate, *v.* Add: **1. b.** *trans. Surg.* To treat (a fracture) by fixing a metal plate to both the fractured parts so as to hold them together; to attach a plate to (a bone).

1910 *Brit. Med. Jrnl.* 8 Oct. 1064/2 It..did the progress of surgery a disservice to suggest that to plate a fracture was a matter lightly to be undertaken. 1948 [see *FIXATION 3 c]. 1959 A. G. APLEY *Syst. Orthopaedics & Fractures* xi. 110 When a fracture has been plated, the technique of 'delayed splintage' is of great value. *Ibid.* xxi. 256 If closed reduction of a radius and ulna has failed..the bones should..be plated.

2. a. Add examples relating to the deposition of substances other than gold and silver.

1855 *Mechanics Mag.* 7 July 4/1 A patent has recently been obtained..for an improved process for plating or coating lead, iron, or other metals with tin, nickel, or alumina. 1919 *Rep. Progr. Appl. Chem.* IV. 255 A lead anode is employed when plating with lead. 1929 *Ibid.* XIV. 322 Automobile parts..to be plated with nickel or chromium. 1940 A. MORGAN *Things a Boy can do with Electrochem.* xii. 177 It is an easy matter to plate iron, steel, and brass articles with copper or nickel. 1966 *McGraw-Hill Encycl. Sci. & Technol.* IV. 531/2 Chromium plating is conducted from solutions containing chromic acid and sulfuric acid... For irregular shapes, auxiliary anodes must be used to plate the surface completely. 1968 *Rep. Progr. Appl. Chem.* LIII. 68 Smith and Lewis.. successfully plated toughened polystyrene.

b. Also, to deposit as a coating, esp. electrolytically. (Further examples.)

1919 *Rep. Progr. Appl. Chem.* IV. 255 Nickel can be plated directly on aluminium. 1947 *Electronic Engin.* XIX. 161/3 The chemical reduction method of plating nickel on steel is too expensive to replace electrodeposition. 1959 T. M. ROGERS *Hand-bk. Pract. Electroplating* 218 Nickel is normally plated from an acid solution. 1972 I. ASIMOV *Asimov's Guide to Science* (1975) I. v. 291 Attempting to plate a silicon layer on a platinum surface. ..The expected plating did not occur. 1979 *Sci. Amer.* May 71 (Advt.), Man has been plating chromium for over a century.

6. *Biol.* and *Med.* To inoculate (cells or infective material) *into* or *on to* a plate (*PLATE *sb.* 18 e), esp. with the object of purifying a particular strain or estimating viable cell numbers. Freq. with *out*.

1892 A. C. ABBOTT *Princ. Bacteriol.* xviii. 181 Again, 0·25 c.c. of this dilution is plated and we find 180 colonies on the plate. 1901 *Jrnl. Hygiene* I. 202 In order to isolate the organisms, one c.c. of each of the liquid stools was diluted 1–10,000 and 1–100,000 with distilled water, and 1/16 c.c., ½ c.c., and ½ c.c. of these dilutions were plated out in gelatine. 1905 *Ibid.* V. 342 The resulting cultures were plated and re-plated to ensure pure growths. 1930 [see *BROTH *sb.* 1 b]. 1971 *Nature* 10 Sept. 121/1 When these strains were plated on various drug agar plates..we obtained the growth pattern given in Table 1. 1972 *Ibid.* 18 Feb. 368/2 Rat lymph node cells are plated out onto mouse fibroblast monolayers *in vitro*.

7. To examine or test the distribution of shot from (a shot-gun) by firing at a pattern plate set at a suitable distance.

1904 *Kynoch Jrnl.* Oct.–Dec. 189 You can plate your gun with your favourite charge. 1932 G. BURRARD *Mod. Shotgun* III. 80 No record of such a thing has ever been noted on any pattern plate since the plating of guns first began.

8. To provide (a book) with a book-plate.

1906 [see *PLATING *vbl. sb.* 1 g]. 1930 *Publishers' Weekly* 1 Mar. 1095/2 After the latter book had been punched and plated, one of our catalogers discovered that..it was an exact duplicate of the former. 1941 *Amer. Speech* XVI. 311 Verbs are made from nouns, for instance *to plate*.., to furnish with bookplates.

9. *trans.* and *intr.* To practise fellatio or cunnilingus (on). *slang.*

1961 PARTRIDGE *Dict. Underworld* Add. 807/1 *Plate, v.* This and *french, go down, nosh*, are prostitutes' (esp. London) verbs, both transitive and, less commonly, intransitive, for 'to gamaruche' a man: C. 20. 1969 FABIAN & BYRNE *Groupie* i. 10, I wondered whether I should plate him. I hadn't done much of that, but I knew guys on the scene liked it because Nigel had told me so. 1969 B. PATTEN *Notes to Hurrying Man* 27 Guitarist from Mike's group Taught her how to plate correctly. 1971 J. MANDELKAU *Buttons* vii. 99 The various chapter prospects were showing everyone how well they could screw and plate her.

10. *trans.* To provide (a goods vehicle) with a plate recording particulars of weight, etc., according to government regulations.

1968 [see *PLATING *vbl. sb.* 1 j]. 1970 *Times* 29 Jan. 26/6 All trailers manufactured before January 1 last year should have been tested and plated by the Ministry within 12 months. 1972 [see *PLATING *vbl. sb.* 1 j]. 1976, etc. [see *PLATED *a.* 6]. 1977 'D. RUTHERFORD' *Return Load* ii. 32, I see it's plated at 43·3 tons. That's more than ten in excess of the UK limit.

11. *trans.* To put on a plate; to serve upon a plate.

1970, etc. [implied in *PLATED *a.* 5]. 1976 *Times* July 10/6 Dishes are plated in the kitchen, and mistiming is common. 1977 *Guernsey Weekly Press* 21 July 2/8 Mr Nugent said that when the policemen arrived the meals were ready and plated.

‖**platea** (plǣtĭ·ă). *Medieval Drama.* Also **placea**; pl. **plateæ**. [L., street; (late L.) courtyard, square; f. Gr. πλατεῖα.] An area before a raised stage, providing additional acting space as well as accommodation for the audience.

1831 J. P. COLLIER *Hist. Dramatic Poetry* II. 154 A castle and a ship were introduced [in the Digby Miracle-play of Mary Magdalen]. The 'place', termed *placea*, and a *mons* are also mentioned in the stage-directions. 1903 E. K. CHAMBERS *Mediaeval Stage* II. III. xx. 80 The *diabolus* thinks he is prevailing upon Adam. He joins the other demons and make [*sic*] sallies about the *plateae*. 1957 R. SOUTHERN *Medieval Theatre in Round* 235 The *platea* developed not into the stage of our modern theatre, but into the pit-and-stalls. 1978 *Amer. N. & Q.* Apr. 118/1 The Digby *Mary Magdalen* has the most complete stage directions in medieval drama, directions that provide several precise technical dramatic terms which distinguish exits from moves within and about the *platea*.

plateau. Add: Also with pronunc. (plæ·to). **1. b.** (Earlier and later examples.) More widely, a more or less level portion of a graph adjacent to a lower sloping portion; a condition or period that can be so represented, when there is neither an increase nor a decrease in something.

1894 W. EWART *Pulse-Sensations* IV. xiii. 277 In the cardiogram..this point occurs in the line of descent—or else in the 'plateau'. 1943 J. D. WHITE in H. L. Mencken *Amer. Lang.* Suppl. I. (1945) 416 Plateaus are the thing now. War production is on a plateau, meaning that it is way up and has been up long enough to establish a plateau in the curve of production figures. 1948 *Manch. Guardian Weekly* 30 Dec. 9 The men on the Stock Exchange modestly allow they will be content with 'a plateau'. 1959 *Listener* 1 Jan. 18/1 The Ionians..had already reached a high plateau of civilization. 1961 *Ann. Reg. 1960* 474 Although industrial production in aggregate was seemingly on a plateau, the fortunes of particular industries showed marked variations. 1969 H. PERKINS *Key Profession* v. 208 The post-1947 decline in the birth-rate would be succeeded by a further upturn, rising to a higher and more permanent 'plateau' from about 1960 onwards. 1976 *Nature* 8 July 83/2 The world total of annual military expenditure..has remained on a plateau of 210,000 million US dollars (at 1970 prices) for seven years now. *Ibid.* 146/1 Fig. 1*b* shows a large current with a plateau at around +0·8 V.

c. *Psychol.* A stage in learning when no apparent progress is made.

1897 BRYAN & HARTER in *Psychol. Rev.* IV. 52 All agree that just below the ability to understand what is spoken, there is a long discouraging plateau where many give up [learning telegraphy] in despair. 1936 *Brit. Jrnl. Psychol.* XXVI. 218 The occurrence of breathing-places and plateaux in the learning process has usually been attributed..to the gradual formation of low and high habits... A plateau means that the lower order habits are approaching their maximum development, but have not become sufficiently automatic to leave attention free to attack the higher order habits. 1964 P. M. FITTS in A. W. Melton *Categories of Human Learning* 265 The present writer knows of no evidence contrary to Keller's (1958) conclusion that a true plateau in skill learning has not been demonstrated, and that when such effects occasionally are reported they are artifacts. 1972 P. BACH-Y-RITA *Brain Mechanisms* iv. 77 Acquisition of a predominantly perceptual skill..and acquisition of a motor skill..are remarkably similar processes. Each is slow, and has several plateaus.

d. *Physics.* The range of applied voltage over which the counting rate of a Geiger counter remains approximately the same, for a given intensity of radiation.

1937 *Physical Rev.* LI. 1027/1 Reliable counting characteristics and long plateaus were obtainable only with counters whose cathode surfaces were completely cleaned ..previous to filling. 1953 H. H. STAUB in E. Segrè *Exper. Nucl. Physics* I. I. iv. 149 A good counter shows a plateau of 160 volts over which the counting rate should not increase by more than 3 percent. 1973 J. YARWOOD *Atomic & Nucl. Physics* xiv. 396 Well designed counter tubes have a plateau slope of about 2 per cent. increase in count rate for an operating voltage increase of 100 V. This flat plateau is valuable since it means that the operating voltage is not critical.

e. The second of four recognized stages of sexual intercourse (see quot. 1960), in which there is an intense sexual excitement lasting a variable but usually short time following a longer phase of increasing excitement and succeeded either by orgasm or by a longer period of decreasing excitement. Usu. *attrib.*, esp. in *plateau phase*.

1960 W. H. MASTERS in *Western Jrnl. Surg., Obstetr. & Gynecol.* LXVIII. 58 The four phases of the human female's sexual response cycle are in order of their development: (1) the excitement phase; (2) the plateau phase; (3) the orgasmic phase; and (4) the resolution phase. *Ibid.*, This plateau phase of sexual response is the base line from which the individual climbs with relative ease and rapidity to orgasm. 1966 MASTERS & JOHNSON *Human Sexual Response* i. 7 Some physiologic reactions.. may be confined to one particular phase of the cycle. Examples are the plateau-phase color changes of the minor labia in the female and the coronal engorgement of the penis in the male. 1972 *Encycl. Love & Sex* I. 8/2 One of the most common of all sexual problems occurs when the man's plateau phase is short,..so that he reaches his climax too soon to satisfy his partner. 1974 H. S. KAPLAN *New Sex Therapy* i. 9 During plateau, the local vasocongestive response of the primary sex organ is at its peak in both genders. *Ibid.* 21 The retarded ejaculator becomes excited, reaches the plateau, may experience the intense urge to proceed to orgasm which is characteristic at this time, but cannot ejaculate despite vigorous and effective stimulation. 1976 B. GOLDSTEIN *Human Sexuality* ix. 157 If sexual motivation is maintained by adequate erotic stimulation, the excitement phase accelerates to the plateau phase.

4. (sense 1 b) *plateau length, level, slope, value*; **plateau basalt**, basaltic lava extruded from fissures and forming sheets that cover

many square miles; in *Petrol.* freq. used with the implication of a tholeiitic or, formerly, an alkalic nature; **plateau gravel**, gravel occurring in a sheet on hilltops or a plateau, at a height that suggests it has been raised by earth movement since its deposition.

1888 *Proc. R. Soc. Edin.* XV. 347 In Antrim bosses of trachyte and pitchstone rise through the plateau-basalts. 1933 *Amer. Jrnl. Sci.* CCXXV. 241 Olivine-Basalt Magma-Type... Many of the Patagonian plateau basalts appear to be of this type and some, at least, of the Siberian Traps. *Ibid.* 242 The Deccan Traps and the majority of the plateau basalts which have been studied are of tholeiitic composition. 1944 A. HOLMES *Princ. Physical Geol.* xx. 458 Plateau basalts covering areas of 200,000 square miles or more occur in the Columbia and Snake River region of the north-western United States. 1972 G. A. MACDONALD *Volcanoes* xi. 255 The basaltic lava flows that built the plains have commonly been called plateau basalts. 1977 A. HALLAM *Planet Earth* 20/3 The materials of cratered plains [on the moon] resemble terrestrial plateau basalts. 1872 WOOD & HARMER in S. V. Wood *Suppl. Monogr. Crag Mollusca* p. xxvi. (*heading*) The Plateau gravel. 1881 *Proc. Geologists' Assoc.* VI. 33 On the top of Crawley (Portesbury Hill.. the plateau-gravel, with its overlying loam (loess) and its ferruginous layers, is well seen in the railway-cutting. 1970 R. J. SMALL *Study of Landforms* vii. 234 In the New Forest.., the Tertiary sands and clays are overlain at many points by thick plateau gravels of Quaternary age. 1965 *Wireless World* Aug. 382/2 This is particularly useful for Geiger tubes, as they tend to have individual working points and sometimes, if aged, a limited plateau length. 1957 G. E. HUTCHINSON *Treat. Limnol.* I. xii. 744 The absolute concentration..fell rapidly, reaching a plateau level which represents about 10 per cent of the original amount added. 1962 *Newnes Conc. Encycl. Nucl. Energy* 296/1 A good counter is one which has a low plateau slope, e.g. less than 5 per cent increase in count-rate for 100 V increase in applied voltage. 1964 L. WILETS *Theories Nucl. Fission* i. 3 As a function of excitation energy, the probability of fission frequently assumes the form of a barrier transmission curve.., rising to some plateau value.

5. *Med.* (Passing into adj.) Of the pulse: having a plateau (sense 1 b) in the rising portion of the sphygmomanometric tracing.

1923 W. D. REID *Heart in Mod. Practice* xviii. 231 If the pulse is anacrotic and plateau in type, the diagnosis [of aortic stenosis] obtains strong support. 1936 S. A. LEVINE *Clin. Heart Dis.* iv. 77 The plateau form of radial pulse is fairly characteristic of aortic stenosis. *Ibid.*, This will counteract the plateau character. 1972 PASSMORE & ROBSON *Compan. Med. Stud.* III. xvi. 6/1 The plateau pulse..is due to the slow ejection of blood from the left ventricle through the narrowed orifice.

Plateau[2] (plæ·to, ‖ plăt*ō*u·). *Math.* The name of J. A. F. *Plateau* (1801–83), Belgian physicist, used *attrib.*, in the possessive, and with *of* to designate the problem of finding the surface of smallest area bounded by any given closed curve.

1911 *Encycl. Brit.* XXVI. 123/2 The problem of finding a minimal surface to pass through a given curve in space, known as Plateau's problem, possesses an exceptional interest. 1927 *Bull. Amer. Math. Soc.* XXXIII. 259 This paper reduces the Plateau problem to a system of two integral equations. 1930 *Math. Zeitschr.* XXXII. 765 The problem of Plateau and the problem of least area have a common solution. 1976 *Sci. Amer.* July 82/3 In his honour an entire range of mathematical questions that deal with the geometry of soap-bubble-like and soap-film-like surfaces is referred to as Plateau's problem.

plateau (plæ·to, plăt*ō*u·), *v.* [f. the sb.] *intr.* To enter a period of stability or stagnation; to cease increasing or progressing, to level *out*. So **pla·teaued** (or **plateau·ed**) *ppl. a.*; **pla·teauing** (or *plateau·ing*) *vbl. sb.*

1952 *Proc. Soc. Exper. Biol. & Med.* LXXIX. 584/2 In each experiment 10 'plateaued' female rats of the Long-Evans strain were used. 1966 *Electronics* 3 Oct. 23 Many companies have diversified their activities or extended their product lines, after having been scared by the plateauing of business in 1962 and 1963. *Ibid.*, U.S. electronic companies have bitten deeply into local markets, causing the sales of European firms to plateau or slide on their home grounds. 1967 *Economist* 16 Sept. 1009/1 There is an ominous lull in military cargoes which the shipping companies believe is temporary but which the military expect to be permanent and which they describe as 'the pipeline plateauing out'. 1969 D. CLARK *Nobody's Perfect* ii. 60 The hope that some graph will plateau higher up the scale. 1973 *Maclean's Mag.* Aug. 63/1 Kids get into swimming and they have a lot of initial success and then they plateau—they stick at the same level for a long period, maybe six months. 1975 *Harvard Business Rev.* LIII. 30/3 We found a large number of managers who, in the judgment of their organization, have 'plateaued'. That is, there is little or no likelihood that they will be promoted. 1976 *New Scientist* 26 Aug. 439/2 The counts from radioactive carbon dioxide rose rapidly and then plateaued. 1978 *Jrnl. R. Soc. Arts* CXXVI. 483/2 Petroleum supply will peak or plateau at the end of the century.

plated, *a.* Add: **1. c.** *Surg.* Of a fracture: see *PLATE *v.* 1 b.

1916 E. W. H. GROVES *Mod. Methods Treating Fractures* vi. 198 In so many plated fractures removal of the plates and screws is required. 1959 A. G. APLEY *Syst. Orthopædics & Fractures* xxi. 256 If..a patient walks with a recently plated tibia unprotected by plaster, the plate will break.

2. (Further examples.) Also used with reference to metals other than gold and silver; *plated wire*, a wire of a non-magnetic metal such as copper having a thin coating of a magnetic alloy, used as an element in some computer memories. Cf. *chromium-plated* ppl. adj. s.v. *CHROMIUM 2.

1781 H. NEWDIGATE *Let.* 8 Oct. in A. E. Newdigate-Newdegate *Cheverels* (1898) iii. 42 All were well pleas'd to find we were only ten miles from Sheffield... We got there ..and saw the Plated Manufactory through all its branches. 1789 E. SHERIDAN *Jrnl.* (1960) 182 The plated candlesticks will also be very useful to us. 1899 J. W. URQUHART *Electro-Plating* (ed. 4) 222 The advantages of a thick deposit of nickel are that it can be manipulated by the polisher with confidence and the 'life' of the plated article is..greatly increased. 1931 E. F. BENSON *Mapp & Lucia* viii. 205 A magenta carpet and a nickel-plated mantelpiece. 1960 T. R. LONG in *Jrnl. Appl. Physics* XXXI. Suppl. 124S/2 After rinsing the plated wire, it is passed through a mercury ground contact and then through small testing coils. These coils..allow a continuous monitor of the nondestructive readout signal. 1967 *Times Rev. Industry* Feb. 104/2 (Advt.), UNIVAC 9000 Series computers utilize a plated-wire memory, basically a thin film electroplated on an extremely fine wire. 1967 *Engineering* 24 Feb. 306/3 Plated plastics parts can replace metal components in many domestic, industrial and automobile applications. 1976 M. WELLS *Computing Systems Hardware* ii. 49 Plated wire stores are substantially faster than ferrite core stores of the same capacity.

b. Also, produced by plating (sense *1 h).

c1890 *American Mail Order Fashions* (1961) 12 Ladies' Outsize Plated Silk Hose, black,..1.50. 1926 J. CHAMBERLAIN *Hosiery, Yarns & Fabrics* vi. 131 Innumerable colour designs are produced on a plated principle on plain and rib fabrics. 1963 A. J. HALL *Textile Sci.* iii. 152 Thus using cotton and wool yarns it can be arranged that the front of the fabric appears to be made of cotton with the back made of wool. Such fabrics are known as *plated* fabrics. 1968 J. IRONSIDE *Fashion Alphabet* 245 A knitted fabric which has a different kind of yarn as the face and back, is said to be plated.

4. Applied as a thin coating on another material.

1925 *Rep. Progr. Appl. Chem.* X. 301 Piersol finds plated chromium superior to any other metal for reflectors. 1966 *McGraw-Hill Encycl. Sci. & Technol.* IV. 533/1 The thickness of a plated coating is the most important factor in its protective value.

5. In which food is ready-served on a plate.

1970 *Drive* Spring 43/1 Plated service means that your meal arrives complete, on the plate. Semi-plated service is when the principal component of the meal arrives on the plate; vegetables are served from silver dishes. 1976 *Liverpool Echo* 23 Nov. 10/6 Some patients at Clatterbridge Hospital enjoy a plated meals service which gives them advance choice of menu and portion size. 1977 *Lancet* 16 July 130/1 Plated meals were packed in charcoal heated containers and dispatched at 11 A.M.

6. Of a goods vehicle: see *PLATE *v.* 10.

1976 *Liverpool Echo* 22 Nov. 11/4 (Advt.), Two Leyland Boxer Vans, plated,..M.O.T. until February 1977. 1977 *Horse & Hound* 14 Jan. 44/2 (Advt.), Bedford T.K. petrol, 2/3 horses, good box, very reliable, plated till January 1977. 1978 *Taxi* 16 Feb. 19/1 (Advt.), For Sale: 'J' reg. auto. Just rebored. Excellent condition, plated till July. 1980 *Daily Tel.* 6 Mar. 2 (Advt.), It's a big truck all right. But its plated weight is only 7·38 tons GVW.

plate-glass. Add: **b.** *spec.* (also with capital initial) used *attrib.* to denote any of the new British universities founded in the 1960s; also passing into adj., of or pertaining to such a university.

1968 M. BELOFF *Plateglass Universities* i. 20 The self-confident and colourful character of the Plateglass universities reflects the spirit of the high Macmillan age. 1968 —— in *Encounter* May 14/1 The New University explosion of the last decade has an element of illusion about it... Only seven..are new institutions... The difference between the Plateglass Universities and both their predecessors and upgraded successors was that in them alone was there the opportunity for pure experiment. 1968 *Economist* 1 June 47/1 Of the non-Oxbridge successful candidates, only four came from the new generation of plate-glass universities. 1971 C. DRIVER *Exploding University* I. iv. 187 Some time ago a Plate-glass professor suggested that a new university's potential for innovation fades after about three years. 1973 J. H. M. SCOTT *Dons & Students* ii. 17 Though the new universities have been dubbed 'plate-glass'..they have no monopoly of that material. 1979 *Times Higher Educ. Suppl.* 23 Nov. 27/5 Among universities, such labels as Oxbridge, Red Brick, Green Field and Plate Glass define origins and location rather than reputations.

Hence **plate-gla·sser**, a student or graduate of one of the new British universities.

1968 *Economist* 1 June 47/1 A man from Aberystwyth got into the Foreign Service, along with two plate-glassers, one red-brick man, one Dubliner, nine from Oxford and eleven from Cambridge.

platelet. Add: (A blood-platelet is usu. called a platelet simply.) Further examples. [(*blood-*)*platelet* tr. G. (*blut*)*plättchen* (G. Bizzozero in *Centralbl. für die med. Wissensch.* (1882) I. 17, 18), F. *petite plaque* (*du sang*) (*idem* in *Arch. ital. de Biol.* (1882) I. 1, 16).]

1910 H. W. ARMIT tr. *P. Ehrlich's Anæmia* iv. 203 He then determined the relative proportions of the platelets to the blood corpuscles. 1955 *Sci. News Let.* 14 May 320/3 Platelets, tiny cell fragments found in the blood, arrest bleeding by sticking to the edge of the wound and to each other until they pile up into a little cork that fills the hole. 1970 *Physics Bull.* July 322/1 The crystal..is then sliced into platelets 50 μm or so thick. 1978 *Detroit Free Press* 5 Mar. B4/1 Blood clots that block arteries are formed by platelets, the body's first line of defense against injury.

2. platelet count, the number of platelets in a stated volume of blood; a calculation of this.

1909 *Jrnl. Exper. Med.* XI. 542 In the experiments careful erythrocyte and leukocyte counts have been made and then the platelet count obtained by the indirect method, *i.e.*, by ascertaining their number relative to that of the red blood cells in fresh preparations and checking this result by the relative number in carefully made stained smears. *Ibid.* 544 In an animal which had received repeated doses of saponin intravenously..there was a high platelet count (1,400,000 per cubic millimeter). 1966 *Lancet* 24 Dec. 1384/1 The platelet-count gradually recovered after the drugs had been stopped.

plate-maker. 2. (Earlier examples.)

1772 J. WEDGWOOD *Let.* 22 July (1965) 127, I have ordered another sett of plate makers to work, and will if possible have a *sufficient stock* of those *every day* Articles. 1863 *1st Rep. Children's Employment Commission* 3 in *Parl. Papers* XVIII. 1 Flat pressers... It includes dish-makers, platemakers, saucermakers, and cup and bowl makers.

platen, platten, *sb.* Add: The form *platten* is now *Obs.* **1. b.** *Engin.* The movable table of a planing or milling machine.

1908 S. H. MOORE *Mech. Engin.* xiii. 298 Horizontal milling machines..resemble in a way the conventional planer with its deep bed and long platen or table. 1950 C. R. HINE *Machine Tools for Engineers* viii. 122 The worktable, or platen, moves back and forth on the bed ways and carries the work past the tool.

3. a. *spec.* A flat metal surface by means of which pressure is applied in a press. (Further examples.)

1927 KNIGHT & WULPI *Veneers & Plywood* xx. 210 The platens are alternately squeezed together to flatten the veneer, and opened up to allow the moisture to escape from the wood surfaces. 1936 H. W. ROWELL *Technol. of Plastics* xv. 97 The simple up-stroke hydraulic press with steam or electrically heated platens..is a cheap and economical machine. 1963 H. R. CLAUSER et al. *Encycl. Engin. Materials & Processes* 347/2 Two mold halves, which..combine to form one or more negative forms of the article to be molded, are tightly clamped between the platens of an injection-molding machine. 1964 B. LATHAM *Wood* xiii. 159 A modern hot press may have ten, twelve, or even twenty platens, so enabling up to twenty sheets of plywood to be manufactured at each pressing operation. 1975 A. D. DEUTSCHMAN et al. *Machine Design* iv. 170 Laminating... It is possible to produce shapes other than flat sheets by using shaped molding dies between the forming press platens.

b. In a typewriter, the surface against which the paper is held and the type strikes (a cylindrical roller in most machines). Freq. *attrib.*

1890 A. E. MORTON *Type-Writing & Type-Writers* 12 The paper-carriage is much smaller, and the platen or cylinder differs from that of other machines in that it presents a flat surface for the types to strike against. 1899 J. WARDLE *Universal Typewriter Man.* 45 *Platen knob*, for turning the Platen..either backward or forward. *Ibid.*, *Platen roller*, for the paper to rest upon when printing. 1907 F. H. BURNETT *Shuttle* xxiii. 227 The platen roller is easily removed without a long mechanical operation. 1909 G. C. MARES *Hist. Typewriter* i. 45 The platen cylinder is supported in a carriage that slides on a rod. 1928 M. CROCKS *Touch Typewriting for Teachers* xvii. 125 The student can be trained to get the right amount of 'flick' with the platen knob to enable the paper to enter the machine up to whatever writing point is desired. 1962 *Which?* Dec. 354/1 The bail bar, which should hold the paper against the platen, was thin and bent easily. 1976 J. FRASER *Who steals my Name?* x. 122 He pressed the activate key, and..paper began to spew from the platen, printed in neat lines at the rate of ten words a second.

4. platen press, a platen machine (not necessarily a printing-machine).

1888 *Encycl. Brit.* XXIII. 704/2 We may say that of platen presses there are the hand-press, the treadle platen press, and the steam or other power-driven press. 1927 KNIGHT & WULPI *Veneers & Plywood* xxvi. 294 Pressing plywood singly, even in the multiple platen presses,..is a slow and expensive process. 1967 V. STRAUSS *Printing Industry* vi. 278/1 Platen presses are the smallest and least complex of all widely used letterpress printing machines.

plate number. 1. *Philately.* (See quot. 1912.)

1912 *Gloss. Philatelic Terms* 19 *Plate Numbers*, numbers inserted in the margins of plates from which stamps are printed, indicating (in the case of British and British Colonial stamps) the order in which the plates for those particular values were made. 1934 *Neuphilologische Mitteilungen* XXXV. 130 Stamp-collecting..*plate number* 'number appearing in the margin of certain stamps'. 1971 D. POTTER *Brit. Eliz. Stamps* v. 63 In complete sheets all Bradbury Wilkinson printings bear a plate number, in the lower margin to the right; this consists of a number or a number and letter.

2. A serial number on the pages of some engraved music, which can indicate the chronological place of the score in the publisher's output.

1940 *Papers Amer. Musicological Soc. 1938* 114 [Hoffmeister] made arrangements in 1784 with the publisher.. to do the engraving for him... Work issued during this period either had no plate number engraved at the foot of

the plates, or else used as an identifying device the opus number of the composition. 1942 *Music Library Assoc. Notes* Dec. 1 Mathias Artaris..generally adds the initials 'M.A.' to his plate numbers. 1946 O. E. DEUTSCH *Music Publishers' Numbers* 7 This list..is..only a preliminary compilation of titles and plate-numbers known to the authors. 1965 NEIGHBOUR & TYSON (*title*) English music publishers' plate numbers in the first half of the nineteenth century. 1966 J. H. DAVIES *Musicalia* xv. 139 The dating of music by publishers' plate-numbers and paper-makers' watermarks begins to achieve some degree of accuracy.

3. The registration number of a motor vehicle, exhibited on a plate.

1973 'D. SHANNON' *No Holiday* (1974) iv. 64 Hackett.. got the plate number of the truck.

platform, *sb.* Add: **III. 6. c.** *spec.* in *Geol.* and *Physical Geogr.*: (i) A level or nearly level strip of land at the base of a cliff close to the water-level; occas., a similar terrace away from a body of water but thought to have been formed by the sea in such a situation.

1841 C. LYELL *Elem. of Geol.* (ed. 2) I. vi. 150 The sea is advancing upon the land, and removing annually small portions of undermined rock. By this agency a submarine platform is produced on which we may walk for some distance from the beach in shallow water, the increase of depth being very gradual, until we reach a point where the bottom plunges down suddenly. This platform is widened with more or less rapidity according to the hardness of the rocks, and when upraised it constitutes an inland terrace. 1901 *Bull. Geol. Soc. Amer.* XII. 212 A looped bar or ridge of gravel and sand formed on an old wave-cut platform. 1922 E. M. WARD *Eng. Coastal Evol.* ii. 34 There must come a time when further coast retreat would involve the total exhaustion of wave energy in crossing the shallow water of a wide wave-cut platform. 1944 A. HOLMES *Princ. Physical Geol.* xiv. 289 As the cliffs are worn back a wave-cut platform is left in front.., the upper part of which is visible as the rocky foreshore exposed at low tide. 1964 W. C. PUTNAM *Geol.* xiv. 387/2 Where the platform is mantled with sand, it is the beach. *Ibid.* 388/1 These coasts may be bordered by a whole flight of terraces, which are elevated wave-cut platforms. 1975 R. V. RUHE *Geomorphol.* ix. 178/2 There are five marine terraces in Santa Cruz, California; each platform was cut during rising sea level, and its cover of marine sediments was deposited during falling sea level.

(ii) *continental platform*: see *CONTINENTAL *a.* 1 d.

(iii) A former erosion surface or plateau represented by the common surface or summit level of neighbouring hills or other land forms.

1908 *Q. Jrnl. Geol. Soc.* LXIV. 384 Of the older topography..partly destroyed by post-Pliocene denudation, the most striking feature in the higher part of the area is presented by two well-marked high-level platforms, one at 750 feet above the sea, and the other may be called the 1000-foot platform, although it is really a little below this altitude. The latter was first recognized on Davidstow Moor,..but traces of it are to be seen on the surrounding high land in all directions. 1938 A. K. WELLS *Outl. Hist. Geol.* xvii. 226 In a few localities on the Chiltern dip slope remnants of the Lenham platform form a gently inclined shelf above the 400-feet platform. 1954 J. F. KIRKALDY *Gen. Princ. Geol.* ix. 96 Accordance of summit levels or the presence of platforms at lower levels can be inferred from the layout of the contours. 1966 J. I. CLARKE in G. H. Dury *Ess. Geomorphol.* 257 In Britain..there is a marked tendency to attribute platforms to marine erosion.., while in Australia..and elsewhere it is often held that uplift and rejuvenation of ancient surfaces are possible. *Ibid.* 270 Sparsely-distributed height-values rarely give..a good indication of erosion-platforms.

(iv) The part of a kratogen (craton) where the basement complex, elsewhere exposed as a shield, is overlain by a layer of more recent, relatively flat and undisturbed strata that are mainly sedimentary.

1908 tr. *Suess's Face of Earth* III. IV. ix. 376 The pre-Cambrian platform. In front of the Urals there extends the vast Russian plain. Its ancient foundation is not visible till we proceed a considerable distance to the west and south-west. 1923 L. D. STAMP *Introd. Stratigr.* iii. 36 The whole area of S.E. England consists of a blanket of Mesozoic Rocks resting on an eroded surface of Palæozoic rocks called the Palæozoic Platform. 1958 L. P. SMIRNOW in L. G. Weeks *Habitat of Oil* 1168 (*heading*) Oil-bearing basins on eastern edge of the Russian platform. 1968 C. R. TWIDALE *Geomorphol.* iii. 49 The Australian continent is built of a Shield, a Platform and an Orogen. 1972 B. B. BROCK *Global Approach to Geol.* iv. 35 The Russian platform, with a moderate thickness of flat-lying rocks covering the basement, brings the Baltic shield up to the normal size.

7. b. In a small boat or yacht: a light deck.

1950 R. MOORE *Candlemas Bay* I. 47 Otherwise she'd have come up and drained her platform through the scuppers, as soon as she floated. 1961 F. H. BURGESS *Dict. Sailing* 161 *Platform*, floor boards laid over the floors in small yachts to make a walking space.

8. b. (Further examples.)

1931 H. F. PRINGLE *Theodore Roosevelt* I. xv. 205 On the special train..a bugler appeared on the rear platform to sound the cavalry charge. 1932 *Atlantic Monthly* Apr. 437/2 As the train pulled out I saw from the back platform my two men. 1971 *Power Farming* Mar. 48/4 Illustrated recently in the trade press was a foot-propelled backward-travelling strawberry picking platform used in Israel. 1971 M. TAK *Truck Talk* 119 *Platform*, a flat bed trailer, a trailer that has a deck (or platform) on which cargo rests but no sides. 1978 R. L. HILL *Evil that Men Do* (1979) xx. 244 An early 'fifties flatbed farm truck sat beside the shack, its stacked platform serving as a temporary pen for two enormous sows. 1979 P. THEROUX *Old Patagonian Express* xviii. 286 We were supposed to have been in Mainara for three minutes... I sat on the steps of the platform and smoked my pipe.

d. A structure which is designed to stand on the bed of the sea (or a lake) and to provide a stable base above water level from which a number of offshore oil or gas wells can be drilled or regulated.

1938 *World Petroleum* May 76/3 The coast line..is exposed to strong winds and rough seas during six months of the year, so that a very substantial platform has to be provided when underwater drilling is to be done. 1955 *Rev. Petroleum Technol.* XIV. 24 Fixed-well platforms, unless capable of multiple-well work, are uneconomical 'at sea'. 1973 *Guardian* 23 May 13/1 In Scotland today the word platform means..a production unit weighing perhaps a quarter of a million tons, going up to 700 feet high..for use in the North Sea oilfields. 1974 *Esso Mag.* Summer 7 The next generation of platforms, now under construction, are concrete structures, in which a massive concrete cellular base, which doubles as oil storage, supports the towers which carry the production platform. 1975 *Sunday Times* 25 May 4/5 Britain's gas supplies are unlikely to be affected, even if some platforms have to stop production temporarily.

e. A gyroscopically stabilized mounting which is isolated from the angular motion of the craft carrying it and provides an inertial frame for the accelerometers of an inertial guidance system; the gyroscopes, accelerometers, and other instruments associated with this.

1946 WELLS & GLENNY in M. Davidson *Gyroscope* III. i. 174 The gyroscope and the magnetic compass assemblies are supported on platforms attached to a rectangular frame. 1954 *Aviation Age* Oct. 21/1 Gyros, by virtue of their ability to maintain a fixed direction in inertial space, provide a ready means for stabilizing the accelerometer platform against angular motion of the vehicle. 1964 C. F. O'DONNELL *Inertial Navig.* i. 20 To aid in platform stabilization, servos are used, with their input signals coming from pick-offs mounted on the precession or output axis of the gyroscopes. 1970 *Time* 27 Apr. 15 He charged up *Odyssey's* small re-entry batteries..and transferred the precise alignment of the command module's 'platform'—its complex of navigational gyroscopes and accelerometers—to a similar platform in the lunar lander. 1977 *Sci. Amer.* Feb. 21/2 A long-range cruise missile employs an inertial-guidance system consisting essentially of three or more accelerometers mounted on gyroscope-stabilized platforms, to guide it along a preassigned course.

f. A rigid diving-board fixed at any of a series of standard heights varying from 3 to 10 metres above the surface of the water; also, in a diving contest, the highboard event.

1971 L. KOPPETT *N.Y. Times Guide Spectator Sports* ix. 169 Off the platform, some dives are made from a handstand. 1973 *Tucson* (Arizona) *Daily Citizen* 21 Aug. 61/7 Finneran is a [*sic*] Olympic veteran, finishing fifth in the three-meters last year in Munich and ninth in the platform. 1974 *Encycl. Brit. Macropædia* XVII. 863/2 When diving first became part of the Olympic program in 1904, it was little more than plain high diving from five- and ten-metre fixed platforms. 1974 *Rules of Game* 202 Competitive diving... Highboard diving platforms and springboards are provided at the heights shown.

9. a. (Later *fig.* examples.)

1964 E. B. WHITE *Let.* 21 Feb. (1976) 517 A man is privileged to say anything he wants to about the magazine, but..he can't use one of my books as a platform. 1966 'W. HAGGARD' *Power House* vii. 72 *The Freeman* was important to him since it provided him with a platform. 1977 *It* May 29/4 Aims... To act as a platform for people with radical ideas and opinions.

b. (Earlier and later examples.) Also *transf.*

1803 *Massachusetts Spy* 27 Apr. (Th.), The platform of Federalism. 1837 *Liberator* 15 Dec. 203/3 We care not who is found upon this broad platform of our common nature. 1838 *Congress. Globe* 11 Jan. App. 73/1 We wanted no platform on which to stand, save the Constitution of our country. 1882 *Sydney Slang Dict.* 7/1 *Platform*, a standpoint, as 'Home rule's my platform'. Originally an Americanism. 1909 'O. HENRY' *Roads of Destiny* x. 166 He leaned on the desk and declared his platform to the clerk. He said he had come to Elmore to look for a location to go into business. 1924 H. G. WELLS *Dream* 142, I adopted Votes for Women as the first plank of my political platform. 1926 A. CONAN DOYLE *Hist. Spiritualism* I. ii. 25 The broad platform upon which his beliefs were constructed. 1937 [see *PLANK *sb.* 5]. 1964 GOULD & KOLB *Dict. Social Sci.* 484/2 The party platform is adopted before the candidates for President and Vice-President are nominated and..it can happen that the candidate and the platform disagree in important particulars. 1976 *Survey* Spring 87 The Communist Party of the United States of America..has held conventions to..discuss its strategy and approve a platform.

10. = **platform sole*. Also short for **platform shoe*.

1945 WEBSTER Add., *Platform*,..an outsole a half inch or more thick, made of wood, cork, etc., and usually covered with leather. 1946 *Sun* (Baltimore) 2 Nov. 3 (Advt.), Picture-Pretty Platforms... Two flattering styles to choose from..both mounted on black faille platforms. 1960 D. LESSING *In Pursuit of English* vii. 229, I could not keep my eyes off her shoes... The soles were platforms two inches deep. 1970 *New Yorker* 31 Oct. 125/1 A boot with a small platform in a contrasting color. 1973 *Times* 7 Nov. 18/3 An office manager wearing 4½ inch..platforms said 'they give you a masculine walk because you walk heavy'. *Ibid.* 18/4 He wore them to catch up with his girl friend's six inch platforms. 1977 C. MCFADDEN *Serial* (1978) v. 17/2 A woman in..eight-inch platforms that reminded him of the moon shot.

C. (sense 8 b) *platform body*; (sense *8 d) *platform leg, operator*; (sense *8 f) *platform diving*; (sense 9) *platform appeal, campaign, eloquence, engagement, -maker, manner(s), orator, plank, point, reply*; *platform-proud* adj.; (sense *10) *platform-wearer*; **platform-car** (earlier example); **platform machine** = *platform scale* (in Dict. and Suppl.); **platform paddle tennis** = **platform tennis*; **platform party**, the group of officials or distinguished persons who sit on the platform at a ceremony or a meeting; **platform rocker** orig. *U.S.*, a rocking chair constructed with a fixed stationary base; **platform sandal**, a sandal with a platform sole; **platform scale** (examples); **platform shoe**, a shoe with a platform sole; **platform sole**, a very thick outer shoe-sole; also *attrib.*; hence *platform-soled* adj.; **platform stage** *Theatr.*, (see quots. 1951) (cf. *apron stage* s.v. *APRON *sb.* 4 j); **platform tennis**, a form of paddle tennis (see *PADDLE *sb.*[1] 11) played on a platform, usu. of wood, enclosed by a wire fence; **platform ticket**, a ticket admitting a non-traveller to a railway station platform; **platform tree** *poet. nonce-use*, a tree with a wide-spreading, flat-topped crown; **platform truck**, a road transport vehicle having a platform body; **platform yard**, a yard where oil platforms are built.

1959 *Times* 18 June 13/2 Dame Christabel [Pankhurst] ..deliberately based her 'platform appeal' on charm rather than logic. 1973 *Amer. Speech 1969* XLIV. 207 *Platform body*, truck or trailer body with a floor but no sides or roof. 1977 *Horse & Hound* 14 Jan. 44/2 (Advt.), A 16 ft horse box frame, for Bedford T.K. platform body, rear ramp with springs. 1909 *Daily Chron.* 9 Feb. 1/7 The National Passive Resistance League is organising a platform campaign against the House of Lords. 1843 E. H. DERBY *Two Months Abroad* (1844) 20/1 By this, with the aid of a winch, diligences and private carriages are.. lifted, with their passengers and baggage, from the wheels and axles, and transferred to platform cars. 1971 L. KOPPETT *N.Y. Times Guide Spectator Sports* xx. 249 Platform and three-meter springboard diving. 1966 *Listener* 24 Nov. 783/3 Mr. Sandford's work was a ferocious contemporary indictment... *Cathy Come Home* may have done more in its hour and a quarter than the platform eloquence of half a year. 1907 G. ADE *Let.* 3 June (1973) 41, I have no hankering to undertake any platform engagements as long as I can get money doing something else. 1975 *Offshore* Aug. 51/2 Divers can be employed to hand risers or lead pipeline ends up into platform legs when lines are pulled. 1922 G. A. OWEN *Treat. Weighing Machines* x. 134 Platform machines and weighbridges..are used in the main for weighing above 1 cwt., and are distinguishable by a goods platform. 1969 T. J. METCALFE *Weighing Machines* I. x. 99 Platform machines may be portable or dormant. 1928 *Daily Tel.* 12 June 14/7 To-day the 'platform-makers' of both parties were trying to frame an election programme. 1969 B. TURNER *Circle of Squares* iii. 23 Anyone who was suspicious of Hirst's platform manner would be disarmed by that bulldozing statement. 1947 *Penguin Music Mag.* Sept. 34 As if Chopin was a puppet worked by a skilled ventriloquist of charming platform-manners. 1975 *BP Shield Internat.* May 1/3 A relatively new breed of oilman will be required. These are the platform operators, the men responsible for the day-to-day running of the platforms. 1866 J. C. PATTESON *Let.* in C. M. Yonge *Life J. C. Patteson* (1874) II. x. 207 Let no platform orator divulge the great secret. 1979 W. J. FISHMAN *Streets of E. London* 117/2 Eleanor Marx-Aveling['s]..remarkable qualities as teacher, platform orator and organiser. 1935 in F. S. Blanchard *Paddle Tennis* (1944) III. v. 56 The following rules..are the officially approved rules for Platform Paddle Tennis. 1959 —— (*title*) Platform paddle tennis: the official guide to platform tennis. 1967 O. H. DURRELL *Official Guide to Platform Tennis* i. 3 Originally called paddle tennis, it later became platform paddle tennis and was finally shortened to platform tennis although most old timers.. still refer to it as paddle. 1967 O. WYND *Walk Softly, Men Praying* xii. 186 You can come with us. As one of the platform party. 1976 C. BERMANT *Coming Home* II. iii. 143 The platform party enters, preceded by a mace bearer, and behind him..the Chancellor. 1931 H. F. PRINGLE *Theodore Roosevelt* I. xii. 161 He borrowed many a platform plank from the man he professed to hold in contempt. 1976 *National Observer* (U.S.) 28 Aug. 4/5 They swallowed platform planks calling for Constitutional amendments to bar abortions and school busing for racial balance. 1949 *Economist* 15 Oct. 825/2 His [*sc.* Lord Beaverbrook's] platform points are pure chauvinism. 1926 R. FROST *Let.* 11 Feb. (1964) 178 You should get so platform proud as to be undealable with. 1904 G. B. SHAW *Common Sense of Municipal Trading* x. 89 One of the keenest grievances of the commercial man who sees profitable branches of his own trade undertaken by the municipality is that it is competing against him 'with his own money', meaning that it forces him to pay rates, and then uses the rates to ruin him in his business. The effective platform reply to this is that the profitable municipal trades, far from costing the ratepayers anything, actually lighten their burden. 1969 J. GLOAG *Short. Dict. Furnit.* (rev. ed.) 564 A revolutionary design, invented in America about 1870, was the platform rocker. 1970 *Globe & Mail* (Toronto) 26 Sept. 45/7 (Advt.), Walnut platform rocker. 1958 *Times* 17 Oct. 17/1 Ivan is first shown wearing four-inch platform sandals. 1967 *Vogue* June 98 White patent platform sandals, 18 gns. 1834 *Mechanics' Mag.* 25 Oct. 248/2 E. & J. Fairbanks, a Concentrated Platform Scale—a diploma. 1851 C. CIST *Sk. Cincinnati in 1851* 227 Factories in which platform

scales are made. 1948 D. M. CONSIDINE *Industr. Weighing* iv. 67 The portable platform scale consists essentially of a rugged cast iron base mounted on four rubber tired wheels. 1969 T. J. METCALFE *Weighing Machines* I. x. 105 The compound lever machine shown..is properly described as a 'low pattern steelyard platform scale'. 1969 T. C. THORSTENSEN *Pract. Leather Technol.* xv. 248 Open-toed and platform shoes are more easily made by sliplasted procedures. 1977 D. WATKIN *Morality & Archit.* 12 An unhappy example of this [*sc.* public unconcern with what planners would deem practical] in costume would be the craze for 'platform shoes'. 1939 M. B. PICKEN *Lang. Fashion* 113/3 *Platform sole*, thick shoe sole, usually from ¼ inch to 3 inches in depth; often of cork or wood. 1941 *Amer. Speech* XVI. 98 The advertising writer reserves his best efforts for the finished products... Fine figures include..a platform sole about as thick as the wafer you get with your malted milk. 1960 R. P. JHABVALA *Householder* iii. 154 She would wear her platform-sole shoes and jasmine in her hair. 1977 *Monitor* (McAllen, Texas) 28 Mar. 7A/2 High platform soles..are being phased out although still available for young customers. 1973 *Woman's Own* 6 Jan. 61 Today's fashions, with their high-heeled, platform-soled shoes and long, straight trousers can easily make you look taller. 1974 'G. BLACK' *Golden Cockatrice* ix. 142 She was short even with platform-soled shoes. 1895 G. B. SHAW *Our Theatres in the Nineties* (1932) I. 189 The modern pictorial stage is not so favorable to Shakespearean acting and stage illusion as the platform stage. 1951 *Oxf. Compan. Theatre* 218/2 Before English actors had any settled homes they played chiefly in inn-yards.., and their first permanent buildings..were wooden structures, roughly circular, with a raised platform stage backing on to the wall and jutting out into the open space, still called a 'yard'. *Ibid.* 236/2 The success of Davenant's playhouse.. laid the foundations of the new style, and the Elizabethan platform stage was henceforth out of fashion. 1961 BOWMAN & BALL *Theatre Lang.* 261 *Platform stage*, a stage using an acting area which extends into the auditorium without a proscenium picture frame. 1955 *N.Y. Times* 14 Mar. 31/2 Platform tennis..is one of the fastest growing and most enjoyable of American sports. 1967 *Time* 3 Mar. 45 Platform tennis, more commonly called paddle tennis, is not only the newest addition to the family of tennis-type court games: it is unique in that it is played primarily in winter and always outdoors. 1972 *N.Y. Times* 27 Feb. v. 6 More than 250 players on 128 teams will gather at the 'home of platform tennis', the Fox Meadow Tennis Club in Scarsdale, N.Y., on Friday to compete in the 38th annual United States men's doubles championship. 1977 *Club Tennis* Mar. 13 (*title*) Platform tennis—the game of the 80's paddling its way to success. 1901 *Railway Engineer* XXII. 68/2 In Berlin, at all the railway stations, no one is allowed on the platform unless actually going by train or provided with a 'platform ticket'. 1929 *Station Accounts Instruction Bk.* (Gt. Western Railway) 4 Passengers travelling from Platform Ticket Stations without Railway Tickets must surrender their Platform Tickets, and excess fares be charged accordingly. 1935 C. WINCHESTER *Railway Wonders of World* I. 241/3 At the outset no charge was made for platform tickets by the English railways, but to-day a charge of one penny or thereabouts is usual. 1975 S. BRIGGS *Keep Smiling Through* 92/2 The Government had never intended the Tubes to be used permanently as shelters... However, there was nothing illegal in your buying a platform ticket for 1½d and not travelling. 1925 E. SITWELL *Troy Park* 67 All day in the limp helpless breeze Beneath the empty platform trees He sits with Brobdignagian asses. 1925 *Proc. Inst. Production Engineers* V. 144 If trucking is resorted to then use a platform truck. 1967 *Jane's Surface Skimmer Systems 1967–68* 7/1 *Accommodation.* It is available in three versions: a platform truck with a payload of 2,500 to 3,000 kg; or as a coach or bus with seats for twenty passengers and a driver. 1977 *Grimsby Even. Tel.* 14 May 8/5 (Advt.), 1969 Ford D800 platform truck (no test). 1973 A. PRICE *October Men* ix. 128 He's got a rig of his own... He's built a platform yard of his own at Hartlepool. 1977 *Offshore Engineer* Apr. 9/3 An end-of-contract bonus payment dispute flared up..at Highlands Fabricator's Nigg platform yard, where the steel platform for Chevron's Ninian field..is being completed.

platformate, platformer[2]: see next.

Platforming (plæ·tfǫɹmiŋ), *vbl. sb.*[2] Also **platforming**. [f. PLAT(INUM + RE)FORMING *vbl. sb.*] A proprietary name for a process for reforming petroleum using a platinum catalyst. Freq. *attrib.* Hence **pla·tformate** [after *distillate, filtrate*, etc.], the end product of the process; **pla·tformer[2]**, an installation for Platforming.

1949 E. F. NELSON in *Oil & Gas Jrnl.* 7 Apr. 95/1 Our [*sc.* Universal Oil Products'] research and development departments have become so used to referring to the process as 'platforming' that we have decided that at this time we would officially christen it. *Ibid.* 100/2 The platformate has an end point slightly above that of the charge. 1952 *Official Gaz.* (U.S. Patent Office) 29 Jan. 1178/2 Universal Oil Products Company, Chicago.. Platforming. For solid catalyst. Claims use since Aug. 29, 1947. 1954 *Ibid.* 19 Jan. 609/1 Platforming. For apparatus in the nature of a plant for the treatment of hydrocarbons. 1954 *Wall St. Jrnl.* 16 Aug. 5/3 Sunray Oil Corp., Tulsa, completed a new 5,000 barrel a day platforming unit at its Sunray Village refinery in Duncan, Okla. 1955 *Times* 8 June 9/4 The extension and modernization of the Suez refinery (including a 'platformer' to improve the quality of the motor benzine). 1957 *Trade Marks Jrnl.* 8 May 470/1 Platforming... Catalysts. Universal Oil Products Company.., Des Plaines, Illinois, United States of America; manufacturers. 1959 *Petroleum Handbk.* (ed. 4) 216 Platformate is used as a component in motor and aviation gasoline blends. 1973 S. A. BERRIDGE in Hobson & Pohl *Mod. Petroleum Technol.* (ed. 4) xi. 410 The Universal Oil Products Company has combined the Udex process with its Platforming process in an operation called Rexforming... The highly paraffinic raffinate from the Udex plant is recycled to the Platformer. 1978 *Trends in Oil & Gas Refining* (Shell Internat. Petroleum Co.) 5 In catalytic reforming..there has been 20 years of Shell experience in the design, development and operation of platformer units.

plathander, var. *PLATANNA.

plating, *vbl. sb.* Add: **1. a.** (Examples in *Surg.*: cf. *PLATE *v.* 1 b.)

1914 A. P. C. ASHHURST *Surgery* xii. 313 It is better not to plate a recent compound fracture.., but to postpone the plating until the soft parts have healed. 1971 W. J. W. SHARRARD *Paediatric Orthopaedics & Fractures* xx. 985 (*heading*) Sound union of both fractures 8 weeks after plating.

b. Also, the process of coating with a thin layer of any substance, *spec.* by means of electrolysis. (Further examples.)

1872 *Jrnl. Chem. Soc.* XXV. 1134 Plating with aluminium cannot be effected. 1946 *Trans. Electrochem. Soc.* LXXXIX. 384 In a recent cost analysis..on the plating of an electric flatiron, it cost $0·46 to apply nickel and chromium. 1966 *McGraw-Hill Encycl. Sci. & Technol.* IV. 531/1 It is difficult to apply zinc coatings thinner than about 0·002 in. by hot dipping... For many articles, thinner coatings are adequate and are applied by plating. 1968 R. W. BERRY *Thin Film Technol.* v. 266 In vapor plating, a volatile compound of the substance to be deposited is vaporized, then thermally decomposed at the substrate to yield the desired deposit. 1972 [see *PLATE *v.* 2 b].

d. (Later example.)

1951 E. RICKMAN *Come racing with Me* viii. 63 Principal items are veterinary charges, shoeing (or 'plating' as it is called).

f. *Biol.* and *Med.* The preparation of a culture on a plate (see *PLATE *v.* 6).

1898 *Public Health* (Papers & Rep. Amer. Public Health Assoc.) XXIII. 81 The technical difficulties in the way of successful 'plating' in agar are considerable. 1916 *Jrnl. Bacteriol.* I. 513 Eight different samples of raw and heated soil were selected for this examination, some samples being plated immediately, others being incubated at 37°C. for 48 hours before plating. 1934 A. T. HENRICI *Biol. Bacteria* xii. 201 The procedure most commonly used for obtaining pure cultures from mixtures of microbes is the process of plating introduced by Koch. *Ibid.* 203 Plating may be used also to measure or estimate the number of bacteria in a given substance. 1969 M. R. DROOP in Norris & Ribbons *Methods in Microbiol.* III B. xi. 276 The sample for isolation will usually require a considerable degree of dilution before plating.

g. The furnishing of a book with a book-plate.

1906 *Daily Chron.* 10 Aug. 3/2 'Plating'..would appear to be the process of affixing the book-plate to the inside of the first cover of the volumes. 1938 L. M. HARROD *Librarians' Gloss.* 116 *Plating*, the process of pasting labels in library books.

h. *Machine knitting.* (See quot. 1946.)

1946 A. J. HALL *Standard Handbk. Textiles* iii. 140 It is possible to run two threads at once into the knitting machine and arrange that one of these predominates in the back of the fabric whilst the other is mostly seen on the front... This method of knitting is known as plating. 1954 *Textile Terms & Defs.* (Textile Institute) 30 Plating usually involves the knitting of two yarns of different colour, different lustre, or different composition, so that only one of these yarns is visible on the face of the stitch.

i. = *FELLATIO, *CUNNILINGUS. *slang.*

1965 W. YOUNG *Eros Denied* xiv. 137 *Gamming*, from the French *gamahucher*, or *blowing*, or *plating*, or *noshing*. 1969 FABIAN & BYRNE *Groupie* ii. 16 Why do you think plating is perverted? Everyone I know does it.

j. With reference to a goods vehicle: see *PLATE *v.* 10.

1968 *Economist* 27 Jan. 61/2 All three firms have had a couple of prosperous years recently, benefiting from the rapid transition from rigid lorries to articulated vehicles and by the introduction of 'plating' and other new inspection requirements. 1972 *Police Rev.* 10 Nov. 1463/1 The Minister's approval certificate..will eventually obviate the necessity of attending the testing station for plating.

2. b. Also, any metal coating. (Further examples.)

1901 B. BLOUNT *Pract. Electro-Chem.* v. 268 Nickel plating is harder and more brittle than the metal in massive form. 1946 *Trans. Electrochem. Soc.* LXXXIX. 409 (*heading*) Plating deposited from nickel-cobalt chloride solution. 1962 *Engineering* 7 Sept. 321/3 (*heading*) Mirror surface on copper plating.

c. *Biol.* and *Med.* A culture on a plate (*PLATE *sb.* 18 e).

1901 *Jrnl. Hygiene* I. 298 Platings from this broth have then been made in litmus-lactose-agar. 1928 *Jrnl. Bacteriol.* XVI. 272 Platings were made of each sample of milk immediately after it was received.

3. plating certificate, a certificate stating that a goods vehicle has had a plating examination; **plating examination**, a legally-required inspection of a goods vehicle to establish weight, roadworthiness, etc.

1968 *Goods Vehicles (Plating & Testing) Regulations* (601) Reg. 2 'Ministry plate' means a plate issued by the Minister for a goods vehicle following the issue or amendment of a plating certificate. 1978 *Highway Code* 64 Before driving, make sure that..you have a current plating certificate for your goods vehicle. 1973 J. DUCKWORTH *Kitchin's Road Transport Law* (ed. 16) 94 The first examination consists of a plating examination in which the vehicle's axle and gross weights are assessed and recorded on a plate, followed by a test of roadworthiness.

platiniridium. Add: Also **platino-iridium**. (Examples.)

1848 J. D. DANA *Man. Mineral.* vi. 309 A similar platin-iridium has been obtained at Ava in the East Indies. 1888 *Encycl. Brit.* XXIV. 480/1 The new standard of the International Metric Commission is a line-standard of platino-iridium, 40 inches long. 1908 *Practitioner* Sept. 485 The best needle to use is one of platino-iridium, since it can be rendered absolutely sterile in the flame of a spirit lamp. 1965 G. J. WILLIAMS *Econ. Geol. N.Z.* x. 154/2 Farquharson (1910) quoted earlier reports on the occurrence of 'osmiridium' and 'platiniridium' in the Tertiary auriferous quartzose conglomerates..in north-western Nelson.

platinite. Add: **2.** *Metallurgy.* Also **Platinite**. An alloy of iron with 42 to 50 per cent nickel which has the same coefficient of expansion as platinum and has supplanted that metal in various electrical applications, esp. for metal-to-glass contacts in lamps.

1918 *Nature* 15 Aug. 471/1 The Germans..have found that for certain purposes an alloy of nickel and iron may replace platinum. The alloy—called 'platinite'—may be used in electric lamps. 1923 *Engineering* 23 Nov. 651/3 The iron-nickel alloy known as 'Platinite', containing 46 per cent. of nickel and about 0·15 per cent. of carbon, had practically the same coefficient of expansion as glass. 1929 [see *ELINVAR]. 1965 A. D. MERRIMAN *Conc. Encycl. Metallurgy* 731 Platinite..is used for lead-in wires in electric-lamp bulbs.

platino-. Add: **platinocy·anide**, any of a series of fluorescent salts which contain the anion $Pt(CN)_4^{2-}$; **platino-iridium**, var. PLATINIRIDIUM in Dict. and Suppl.

1845 W. GREGORY *Outl. Chem.* II. 306 The platinocyanides of barium, strontium, and calcium..crystallise readily in beautiful greenish yellow colour. 1926 *Sunday at Home* 677/2 The luminous paint..consists..of a mixture of some radium or thorium salt with some photo-sensitive substance like barium platinocyanide. 1974 *Sci. Amer.* Mar. 96/1 The screen would consist of a thin, translucent disk of mica, coated with a phosphor such as barium platinocyanide or zinc silicate.

platinum. Add: **1. b.** A greyish white colour like that of platinum.

1923 *Daily Mail* 1 Aug. 2 In the following colours: Black, White,..Suede, Platinum and Champagne. 1951 E. PAUL *Springtime in Paris* xvi. 313 Looking pensively down at the moving river surface, ebony and indigo. The moon, coming out thinly.., contributed platinum. 1976 *Milton Keynes Express* 11 June 38/2 (Advt.), 1974 'M' Vauxhall Victor 2300 Auto Saloon. Finished in platinum, fitted wing mirrors, radio.

c. = *platinum fox* s.v. sense 2 c below.

1948 A. L. RAND *Mammals Eastern Rockies* 105 Various other 'varieties' [of red fox] have been developed on fur farms, including the platinums and various white-spotted phases.

2. b. platinum-blue [tr. G. *platinblau* (Hofmann & Bugge 1908, in *Ber. d. Deut. Chem. Ges.* XLI. 312)], any of a class of dark blue polymeric complexes, a number of which have antitumour activity, which are formed by divalent platinum with amide ligands; orig. *spec.* one formed with acetamide; **platinum sponge**, a grey amorphous form of platinum which is obtained as spongy masses on heating ammonium chloroplatinate and is used as a catalyst.

1908 *Jrnl. Chem. Soc.* XCIV. I. 141 (*heading*) Platinum-Blue. 1964 *Ibid.* 2835 Platinum Blue is very soluble in water, methanol, and dimethylformamide from which it can be crystallised by the addition of dichloromethane. 1975 *Cancer Chemotherapy Rep.* I. LIX. 296/1 We tentatively conclude from these early results that the 'platinum blues' may have activity against a broad spectrum of tumors. 1976 *Cancer Res.* XXXVI. 3822/1 Platinum-uracil blue and platinum-thymine blue are prototype examples of platinum-blue complexes which have been shown to have a higher therapeutic index against ascites Sarcoma 180. 1826 W. HENRY *Elem. Exper. Chem.* (ed. 10) I. vii. 355 Into a mixture of carbonic oxide with a larger proportion of the explosive mixture, the platinum sponge cannot be introduced without causing detonation. 1894 G. S. NEWTH *Text-bk. Inorg. Chem.* III. xiv. 644 This action is more rapid in the case of platinum sponge, when a larger surface is brought into play, and a fragment of this material introduced into a detonating mixture of oxygen and hydrogen at once determines its explosion. 1968 A. A. BAKER *Unsaturation in Org. Chem.* ix. 125 In 1838 Frederic Kuhlmann produced ammonia by heating a mixture of nitric oxide and hydrogen in the presence of platinum sponge.

c. Platinum or platinum-blonde (see below) in colour, as *platinum hair, lace*; esp. of animals or their fur, as *platinum coney, fox, mink*; also *platinum-grey*; **platinum blond(e)** *a.*, (of the hair) silvery-blonde in colour; (of a person) having silvery-blonde hair; also as *sb.*, a person, esp. a woman, with platinum-blonde hair.

1931 *Daily Express* 15 Oct. 19/5 (*caption*) Miss Binnie Barnes, who appears as a platinum blonde in 'Cavalcade', is seen here as a brunette. Nature gave her auburn-red hair. **1934** R. Ferguson *Celebrated Sequels* 264 A costly platinum-blond young man from a famous night-club. **1934** F. Stark *Valleys of Assassins* ii. 187 It was a blue stream, as vivid in that thirsty solitude as a platinum blonde in a monastery. **1942** A. Christie *Body in Library* ii. 24 She had scarlet lips, blackened eyelashes, and a platinum-blonde head. **1966** J. S. Cox *Illustr. Dict. Hairdressing & Wigmaking* 118/1 *Platinum blonde*, a very fair, silvery hued colour popularized by Jean Harlow, the late curvaceous American film star. **1977** *Transatlantic Rev.* LX. 53 From ten to eleven, no one checked in except a commercial salesman with three suitcases of samples and a middle-aged gent with a platinum blonde. **1923** *Daily Mail* 14 Aug. 1 The wide collar and side panels..are made of the richest pulled Platinum Coney. **1946** A. Christie *Hollow* viii. 77 The platinum foxes that swathed her shoulders. **1908** *Westm. Gaz.* 25 Apr. 13/2 Such a suit is a very pleasant idea for the summer. I saw one the other day in a platinum-grey. **1951** Wodehouse *Old Reliable* xv. 171 This miserable creature, who has probably got platinum hair and a lisp. **1978** D. Francis *Trial Run* i. 11 Her fine-boned face and thick platinum hair. **1923** *Daily Mail* 20 June 8 In a platinum lace gown and cape. **1949** R. Chandler *Little Sister* xviii. 117 No big money,..no platinum mink, no name in neons. **1950** 'S. Ransome' *Deadly Miss Ashley* i. 12 She was wearing a fur piece...It was platinum mink.

platitudinal, *a.* (Earlier example.)
1870 O. Logan *Before Footlights* xxiii. 288 At the risk of uttering truisms and being altogether a platitudinal truist, I may mention that it requires a pretty strong organic construction to stand the ravages of an eight months' tour in the land of fast eaters.

platitudinary (plætitiū·dinări), *a.* [As if f. L. **platitudo, -din-* + -ary[1]; cf. Latitudinary *a.*] = Platitudinarian *a.*
1920 *Glasgow Herald* 2 Apr. 6 At a song-recital..the critic is again troubled by Elgar—this time by a 'tawdry catch-penny ballad'... At a Queen's Hall Concert the Elgar of the Second Symphony is 'platitudinary and tedious'. **1933** Dylan Thomas *Let.* Sept. (1966) 24 Wordsworth was..the humourless, the platitudinary reporter of Nature in her dullest moods.

platitu·dinist. *rare.* [f. Platitudinize *v.* + -ist.] A person who utters platitudes; a platitudinizer.
1905 W. J. Locke *Usurper* xx. 243 Jasper..was not sorry when the kind-hearted platitudinist had gone. **1905** —— *Morals of Marcus Ordeyne* ii. 22 If there is one platitudinist I dislike more than another, it is Marcus Aurelius.

Platonian. Restrict †*Obs.* to the sb. and add: **B.** *adj.* = Platonic *a.* 1. *rare.*
1942 B. Berenson *Jrnl.* 28 Feb. in *One Year's Reading for Fun* (1960) 30 Aristides speaks in the Platonian *Theages*.

Platonic, *a.* and *sb.* Add: **A.** *adj.* **1. c.** Appositive, as *Platonic-Christian* adj., both Platonic and Christian, of or pertaining to Christianity influenced by or fused with Platonism.
1933 A. N. Whitehead *Adventures of Ideas* iii. 40 In the hands of theologians..the Platonic-Christian tradition leant heavily towards its mystical religious side. **1948** L. Spitzer *Linguistics & Lit. Hist.* 55 To Dante, all dialects appeared as inferior..realizations of a Platonic-Christian ideal pattern of language. **1960** *Encounter* Feb. 49/1 This is at the root of the Platonic-Christian (or religious) tradition.
2. Now usu. with lower-case initial. (Earlier and further examples.) Also of affection for one of the same sex.
1631 Jonson *New Inne* iii. ii. sig. E5v, Most Socratick Lady! Or, if you will Ironick! gi' you ioy O' you Platonick loue here. c**1805** Jane Austen *Lady Susan* (1954) x. 258 We are advancing now towards some kind of confidence, and in short are likely to be engaged in a kind of platonic friendship. **1905** 'A. Cambridge' (*title*) A platonic friendship. *Ibid.* v. 67 What is known as a platonic friendship is generally nothing of the kind. **1919** G. B. Shaw *Heartbreak House* ii. 87 *Hector*... What do you get by it? Are you her lover? *Randall.* You must not misunderstand me. In a higher sense— *Hector.* Psha! Platonic sense! She makes you her servant; and when pay-day comes round, she bilks you: that is what you mean. **1924** 'W. Fabian' *Sailors' Wives* vii. 94 'You're taking a lot of notice of that dangerous young person, old bean,' remarked Dorr lightly. 'Platonic, purely. Couldn't well be anything else for a man with my prospects.' **1925** C. Connolly *Let.* 14 May in *Romantic Friendship* (1975) 78, I think I care more for Maurice than anyone else here—and the fact that such affection can be nothing but platonic enhances it, if anything. **1928** A. Huxley *Point Counter Point* xiii. 232 He had such a pure, childlike and platonic way of going to bed with women, that neither they nor he ever considered that the process really counted as going to bed. **1957** J. Braine *Room at Top* vii. 64 'Teddy wouldn't understand. Our relationship is strictly platonic.' 'Yes, I understand,' Teddy said, putting his arm round June's waist. 'I'm trying to take June on a platonic weekend. Of course, it'll be too bad if she has a platonic baby.' **1975** A. Price *Our Man in Camelot* v. 76 'Sharing a bedroom with a strange man in the line of duty. Kind of special relationship.' 'Special *platonic* relationship.'
3. a. Delete †*Obs.* and 'former', and add later examples. Now also called *Platonic solids*.
1745 E. Stone *Euclid's Elements* (ed. 2) II. p. xxiv, The thirteenth, fourteenth and fifteenth Books entertain us with curious and useful Speculations, relating to the five regular or platonick Bodies, in Regard to which, as Proclus tells us, Euclid compiled the whole Body of the Elements, the Platonicks having had them in wonderful Esteem. **1873** J. Booth *Treat. Some New Geom. Methods* I. p. xi, That the principle of Duality should not have been discovered by the great geometers of Ancient Greece is the more remarkable, as the five regular solids, the Platonic bodies as they were called, were with them a favourite subject of speculation. **1917** H. E. Dudeney *Amusements in Math.* 70/2 The icosahedron is another of the five regular, or Platonic, bodies having all their sides, angles, and planes similar and equal. **1952** Cundy & Rollett *Math. Models* iii. 70 The so-called Platonic solids..form the first and simplest group of polyhedra. **1952** G. Sarton *Hist. Sci.* I. xvii. 439 If the regular solids are restricted to five, those five bodies (later called the Platonic bodies) must each have some definite meaning. **1971** M. J. Wenninger *Polyhedron Models* i. 19 The dodecahedron is in some ways the most attractive of the five Platonic solids.
b. (Further example.)
1922 W. B. Yeats *Seven Poems* 23 Are not those who travel in the whirling dust also in the Platonic Year?
B. *sb.* **3.** (Later examples.)
1923 R. Macaulay *Told by Idiot* i. 11 To Vicky a young man *was* a young man, and no platonics about it. **1937** 'M. Innes' *Hamlet, Revenge!* ii. vi. 173 It is one of those affairs that are laced with long-term platonics.

platonically, *adv.* (Later examples.)
1901 Conrad & Hueffer *Inheritors* iii. 39 Gurnard I disliked platonically; perhaps because his face was a little enigmatic—a little repulsive. **1941** J. D. Carr *Case of Constant Suicides* iii. 34 There is nothing like spending the night with a girl, even platonically, to remove a sense of constraint. **1972** *Nature* 17 Mar. 92/1 It would be convenient if there were such a platonically ideal chemistry. **1973** *Listener* 6 Sept. 312/1 They stood in tasteful tableaux, their hands platonically resting on one another's nether regions.

Platonistic, *a.* Add: (Later examples.) Hence **Platoni·stically** *adv.*
1953 M. H. Abrams *Mirror & Lamp* i. 29 Shelley's Platonistic 'Defence of Poetry'. **1957** G. Ryle in C. Mace *Brit. Philos. in Mid-Cent.* 263 The difficulty is to steer between the Scylla of a Platonistic and the Charybdis of a lexicographical account of the business of philosophy and logic. **1959** P. F. Strawson *Individuals* viii. 234 No doubt some philosophers have deluded themselves with myths, have invested non-particulars with a character they do not really possess. There is Platonistic zeal as well as nominalistic zeal. But zeal of either kind is out of place. **1977** G. W. H. Lampe *God as Spirit* iv. 108 The Platonistically conceived Second Person of the Trinity in the classical formulations of the fourth and fifth centuries.

platoon, *sb.* Add: **1.** Revived in the British army for an organizational unit (usu. a quarter) of a company of infantry. Also used for comparable organizational units in other armies.
1913 *Army Order* No. 323. 16 Sept. §4 A company will be divided into four platoons, each commanded by a subaltern... Each platoon will be sub-divided under regulations to be issued later. **1915** D. O. Barnett *Let.* 18 Jan. 40 I've bought A. Coy. No. 4 Platoon. **1917** J. M. Barrie *Old Lady shows her Medals* 72 You have knitted enough things already to fit up my whole platoon. **1929** *Encycl. Brit.* XVIII. 64/2 In the U.S. cavalry a troop is divided into four platoons. **1938** 'I. Hay' *King's Service* xiv. 245 The number of platoons in a rifle company has been reduced from four to three. **1945** H. P. Samwell *Infantry Officer with Eighth Army* iv. 33 We had agreed that he should bring up Company H.Q. and the reserve platoon behind, while I led the forward platoons. **1948** N. Mailer *Naked & Dead* (1949) i. ii. 25 What a bunch of good old boys there were in the platoon, he told himself. **1964** H. D. Chaplin *Queen's Own Royal W. Kent Regiment 1951–1961* ii. 45 Four platoons under Major Crumplin spent Christmas and the New Year festival in the jungle. **1964** Clough & Cash tr. *Gorbatov's Years off my Life* x. 161, I told Kostevich to send a platoon from each battalion to man the line. **1965** I. Adamson *Forgotten Men* i. 15 Working with Animal Transport Platoons had been part of their training in the Bush Warfare School. **1971** E. Luttwak *Dict. Mod. War* 153/2 *Platoon*, an army formation subordinate to the battalion and comprising a number of squads or sections. Normally the smallest unit with an organizational identity, it varies in size from the 12 men of a Soviet army tank platoon to the 40 plus men of a U.S. army infantry platoon. **1972** J. Strawson *Battle for Ardennes* vi. 90 The critical feature was occupied..by a weak platoon numbering some eighteen men of the 394 US Infantry Regiment of 99th Division.
2. b. *Amer. Football.* A group of players trained to act together as a single unit of attack or defence and usu. sent into or withdrawn from the game as a body.
1941 *Charlottesville* (Va.) *Daily Progress* 14 Jan. 11 They [sc. football teams] can still send in as many players—platoons included—while the clock is stopped. **1948** *N.Y. Times* 28 Sept. 36/6 Eleven men may be sent in at a time now, even with the clock running..but Lou [Little] is opposed to this unlimited substitution rule in theory. Maj. Joel Stephens, of West Point,..said that Army has the 'two-platoon' system now. **1949** *Sun* (Baltimore) 2 Dec. 17/7 Schweder is one of the unusual football players of the 'platoon age' in that he plays on both offense and defense.
c. *Baseball.* (See quot.)
1976 *Webster's Sports Dict.* 318/2 *Platoon*..*Baseball*, two or more players who alternate at the same position. The players who make up a particular platoon are usually average players who are adequate fielders but who are not outstanding hitters.
3. *platoon commander, corporal, drill, leader, officer, sergeant, -training.*
1917 W. Owen *Let.* 23 Nov. (1967) 509 Interesting work but hardly 'lighter' than a Platoon Commander's. **1920** J. C. Chase *Soldiers All* 57 He received orders to proceed to Hill 182... He sent runners to notify his platoon commanders. **1974** G. Blaxland *Queen's Own Buffs* iii. 21 There was to be similar pairing of platoon commanders and platoon sergeants. **1941** A. Cotterell *What! No Morning Tea?* 29 After breakfast there was a joint lecture by our two platoon corporals on what to call the various parts of the rifle. **1935** I. Miller *School Tie* xv. 290, I could never get the hang of anything more evolutionary than platoon-drill. **1952** T. J. Mulvey *These are your Sons* iv. 85 He's got three rifle platoon leaders and one weapons platoon leader. **1923** Kipling *Irish Guards in Great War* I. 24 A newly appointed platoon-officer..admonished them unofficially. **1915** D. O. Barnett *Lett.* 41, I like the men awfully, especially my platoon sergeant. **1965** Brophy & Partridge *Long Trail* 224 The Staff..were known only by occasional glimpses. The platoon-sergeant, whatever his defects, was visible and human. **1974** Platoon sergeant [see *platoon commander]. **1923** Kipling *Irish Guards in Great War* II. 138 These Somme officers were accordingly told that most of their time should be given to platoon-training. **1942** E. Waugh *Put out More Flags* ii. 125 After the stand-easy they fell in for platoon training.

platoon, *v.* Restrict †*Obs.* to sense in Dict. and add: **2.** *trans.* To dispose in platoons.
1961 in Webster s.v., The advantages from platooning students in smaller schools.
3. *Baseball.* **a.** *trans.* To alternate (a player) with another in the same position. **b.** *intr.* To interchange with another player in the same position. Hence **platoo·ning** *vbl. sb.*
1967 Webster Add., *Platoon*.., to alternate (one player) with another player in the same position (as on a baseball team). **1969** *Time* 5 Sept. 52 Hodges decided to 'platoon' him by playing him only against lefthanded pitchers. **1971** L. Durocher in *Webster* Add. s.v., If I can't play him every day, I'll platoon him in left field. **1971** *Ibid.*, *Platoon*.., to alternate with another player in the same position. **1972** *N.Y. Times* 4 June v. 2/7 Buckner also shares first base with Wes Parker and Crawford platoons with Manny Mota in left field. *Ibid.* 3 Nov. 45/6 He might junk the platooning system used rigidly by Williams.

Platt: see next.

Plattdeutsch (plæt,doi·tʃ). Also **Platt-Deutsch.** [G., ad. Du. *Platduitsch* Low German, f. *plat* flat, low + *Duitsch* German.] The collective name of those dialects of Germany which are not High German (see German *sb.*[2] 2b). Also *attrib.* Also *ellipt.* as **Platt.**
1814 H. Weber *Illustr. North. Antiq.* 217 As the fragment [sc. *Hildebrandslied*] is evidently written in the dialect of the northern parts of Germany, now denominated Plat-t, or Low German, which was once nearly identical with the Anglo-Saxon, a great number of the words have been rendered into such as, with little variation, existed in the old English and Scottish. a**1834** Coleridge *Table Talk* (1835) I. 119 Originally..in the *Platt-Deutsch* of the north of Germany there were only two definite articles. **1867** J. Macgregor *Rob Roy on Baltic* xix. 239 What a linguist this critic must be before he attempts a voyage such as we have described! First he must learn Norwegian, then Swedish,..then Platt (on the Elbe), [etc.]. **1886** Strong & Meyer *Outl. Hist. German Lang.* 67 New-Low-German, or *Plattdeutsch*, so called from being spoken in the *platte land* or the low country. **1908** T. G. Tucker *Introd. Nat. Hist. Lang.* 220 The Saxonic dialects, under the name of *Nieder-Deutsch* or *Platt-Deutsch* are still in regular use among the populace of North Germany. **1939** L. H. Gray *Found. Lang.* 349 Lower Franconian is essentially Low Teutonic of the type represented by Flemish, Frisian, Old Saxon, and Plattdeutsch. **1942** *Amer. Pol. Sci. Rev.* XXXVI. 537 At present, the speakers of the Lower Saxon, Plattdeutsch vernacular of Gelderland, of Frisian in Friesland, and of genuine Lower Frankish, Dutch dialects are all united in using Dutch as the language of school and church and as the medium of their common national allegiance. **1953** *Trans. Philol. Soc.* 1952 135 The Plattdeutsch forms require discussion. **1970** L. Deighton *Bomber* xxiv. 349 [He was] trying to understand Voss's fast guttural *Plattdeutsch*, as much like Dutch as German. **1973** *Word* 1970 XXVI. 44 All informants felt that it was a loanword... A few said it was not good *Platt*, while others felt that it [sc. *hauptsächlich*] was thoroughly acceptable. **1977** *Trans. Philol. Soc.* 1975 187 At first sight North Frisian seems to have more in common with the Plattdeutsch spoken in Schleswig than with other forms of Frisian, and so the question arises as to what justification there is for not considering Frisian to be a variety of local Platt.

‖ **platteland** (pla·tələnt). *S. Afr.* Also with capital initial and hyphenated (**platte-land**). [Afrikaans, f. Du. *plat* flat + *land* country.] The rural areas of South Africa. Also *attrib.* Hence **plattelander** (pla·təla:ndəɹ), a native or inhabitant of a rural area.
1933 C. J. Uys *In Era of Shepstone* vii. 199 Like a leaven the discontent spread from town to town—leaving the *platteland* unaffected—while the Cape and Natal journals magnified the danger out of all proportion. **1934** *Sunday Times* (Johannesburg) 13 May, It is useless to take the platteland youngster and to teach him to enter a college or university. **1934** E. A. Walker *Gt. Trek* ii. 46 Generally speaking the plattelanders were healthy, as they had need to be if they were to survive. **1935** N. Giles *Dark Border* ii. i. 171 The cattle return to-morrow to the *platteland* and you will go with them. **1943** J. Y. T. Greig *Language at Work* 106 Political discussions..take place on the stoep or in the bar-parlour of a hotel in some dorpie of the platteland. **1954** W. K. Hancock *Country & Calling* vi. 173 Some idealists of the Dutch Reformed Church have recently deduced from *apartheid* the necessity of an immense economic investment in the Native Reserves. It is improbable that voters on the *platteland* will recognize the same necessity. **1955** J. H. Wellington *Southern Afr.* II. iii. xv. 215 The larger urban populations have no great sympathy with what they often regard as the reactionary and parochial attitudes of the plattelanders. **1958** A. Jackson *Trader on Veld* 20 The idea of leaving the comfort and relative sophistication of Port Elizabeth held no terrors for me, and when my Uncle mentioned that he might fix me up on the Platteland, I was eager to hear more. **1960** *Economist* 7 May 505/2 The prime minister's escape from assassination—an escape regarded by the platteland mentality as providential in a more than political sense. **1967** 'L. Black' *Two Ladies in Verona* ii. 28 She told him of her childhood in the platteland of South Africa. **1971** *Progress* (Cape Town) May 9/5 The Provincial Executive would..value any..contact which the plattelanders might make. **1971** *Rand Daily Mail* 4 Sept. 8/4 Platteland towns are being encouraged to join in the great property development game which has been largely confined until now to the bigger South African centres. **1977** *Time* 21 Nov. 10/2 'Man,' he shouted, 'this [*sc.* B. J. Vorster] is the man! This is the Churchill of the *platteland*!'

platter[1]. Restrict 'Now chiefly *arch.*' to primary sense in Dict. and add: Also in colloq. phr. *to hand* (something to someone) *on a (silver) platter* = *to hand on a plate* s.v. *Plate *sb.* 18a; *platters of meat* (Rhyming slang) = *plates of meat* s.v. *Plate *sb.* 18d; also ellipt. as *platters*.
1918 T. Wolfe *Let.* 18 Feb. (1958) 8 You don't get anything handed to you on a silver platter. **1923** J. Manchon *Le Slang* 227 *Plates of meat*,.. = *feet*. On dit aussi *platters of meat*. **1945** L. Shelly *Jive Talk Dict.* 16/1 *Platters*.., big feet. **1960** J. Franklyn *Dict. Rhyming Slang* 108/2 *Platters of meat*, feet. 20 C., a form of *plates*.. and not used in Cockney circles. **1960** G. Sanders *Mem. Professional Cad* i. vii. 55, I was otherwise engaged at the time building a telescope in my back garden and being, by vocation, a dilettante, this interested me far more than the golden future which Mr. Mayer was going to offer me on a silver platter. **1968** J. Updike *Couples* iii. 229 It's *you* who want to keep them down, to give them on a platter everything everybody else in this country has had to work for. **1970** C. Major *Dict. Afro-Amer. Slang* 92 *Platters*, feet. **1973** *New Yorker* 3 Feb. 49/1 Other things are handed to you on a platter.
b. platter-face (later example); **platter pull**, a type of ski-lift (see quots.).
1922 Joyce *Ulysses* 374 Some good matronly woman.. to mother him. Take him in tow, platter face and a large apron. **1951** *Amer. Ski Ann. & Skiing Jrnl.* 1952 30 (Advt.), Belleayre Mountain..Chair Lift Rope Tow Platter Pull Lifts. **1953** *Ibid.* 1954 117/2 The J- and T-bar are a form of seat. Later..came..the platter pull, a platter-shaped seat. These lifts are basically the same and tow the skier from an overhead cable. **1963** *Amer. Speech* XXXVIII. 206 *Platter pull*,..a kind of pull transporting skiers uphill. The skier places between the thighs a rubber or plastic disk which is suspended, usually on a bar, from a rope or a cable permanently fixed to the lift cable. **1970** M. Bennett *How to ski just a Little Bit* ii. 87 A platter pull, sometimes also called a poma, looks as if you took a T-bar, bent it a little, cut the crosspiece off, and attached instead a flat, slightly oval disc approximately six inches in diameter.
2. *slang.* A gramophone record. Cf. *Plate *sb.* 18g.
1931 H. Mutschmann *Gloss. Americanisms* 46/2 *Platter*, a gramophone record. **1935** *Vanity Fair* (N.Y.) Nov. 38/1 There ought to be a hot coupling on every platter. **1943** H. A. Smith *Putty Knife* 163, I bought a couple of Crosby platters. **1960** *Master Detective* July 83/2 Rock and Roll, that's what I'm good at. I got a terrific collection of platters. **1967** 'T. Wells' *Dead by Light of Moon* xviii. 184, I went into Fink Roth's pad and found treasures. Good old platters and stamps. I sold them. Got a good price for the records. The stamps were only so-so. **1977** *Sounds* 9 July 18/1 'Starz'—as the premier platter was called—was hardly the strongest product ever to find its way on to the record racks.
3. The metal disc of a turntable unit, on which the record is placed for playing.
1975 *Gramophone* Jan. 1297 (Advt.), With the heavy platter and extra thick turn-table mat, our final figures are impressive. **1975** *Hi-Fi Answers* Feb. 66/1 Trouble was experienced with ferrous platters inducing a high level of hum.

platy-. Add: **platycephaly** (examples); **platycra·nial** *a.*, broad-skulled; **platyhieric** *a.* (earlier example); **platypellic** *a.* (earlier and later examples).
1902 Platycephaly [see *brachycranial* adj. s.v. *Brachy-]. **1946** *Nature* 28 Sept. 428/1 Many of the skull characters are remarkably simian, including heavy projecting supra-orbital ridges, retreating forehead, marked platycephaly and massive jaws. **1902** *Biometrika* I. 462 Brachycephaly is associated with platycranial characters in both races. **1886** W. Turner in *Jrnl. Anat. & Path.* XX. 317 The following descriptive terms may conveniently express these differences in the relative length and breadth of the sacrum. As the Greek word ἱερόν is the equivalent of the Latin sacrum; the term *dolichohieric* would signify a sacrum in which the length exceeded the breadth, whilst *platyhieric* would signify a sacrum in which the breadth exceeded the length. **1885** —— in *Ibid.* XX. 128, I..shall make three divisions, two of which will represent extreme forms in opposite directions, whilst the third will be intermediate. I shall express these divisions in terms derived from the Greek, so that the nomenclature in pelvic classification may be as far as possible on the same lines as the well-known divisions of crania... By *dolichopellic* is to be understood a pelvis in which the conjugate diameter of the brim is either longer than the transverse or approaches closely to it; by *platypellic* a pelvis in which the transverse diameter of the brim greatly exceeds the conjugate; by *mesatipellic* a pelvis in which the transverse diameter is not so greatly in excess of the conjugate. **1966** Platypellic [see *Mesatipellic *a.*].

platycodon (plætikōu·dǫn). [mod.L. (A. de Candolle *Monographie des Campanulacées* (1830) 125), f. Platy- + Gr. κώδων bell.] A herbaceous perennial plant of the monotypic genus so called, belonging to the family Campanulaceæ, native to China and Japan, and bearing blue or white bell-shaped flowers; = *balloon-flower* s.v. *Balloon *sb.*[1] 10c.
1844 J. W. Loudon *Ladies' Flower-Garden Ornamental Perennials* II. 52 The Large-flowered platycodon. **1905** H. R. Ely *Another Hardy Garden Bk.* v. 141 By the third year the Platycodons become large, strong plants. **1937** Dunbar & Mahoney *Gardener's Choice* 139 The best way of increasing the Platycodon is from seed. **1959** *Times* 3 Jan. 9/5 The platycodon makes a fine pot plant for a cool greenhouse. **1977** J. Jeffreys *Perennials for Cutting* ii. 160 In a suitable site this platycodon is fully hardy and long-lived.

platykurtic (plætikɜ̄·ɹtik), *a.* *Statistics.* [f. Platy- + Gr. κυρτ-ός bulging + -ic.] Of a frequency distribution or its graphical representation: having less kurtosis than the normal distribution.
1905 [see *leptokurtic *a.*]. **1937** Yule & Kendall *Introd. Theory Statistics* (ed. 11) ix. 165 Platykurtic curves, like the platypus, are squat with short tails. Leptokurtic curves are high with long tails like the kangaroo—noted for 'lepping'! **1952** [see *kurtosis]. **1966** S. Beer *Decision & Control* xiii. 334 Moreover, the distributions may be either leptokurtic or platykurtic—that is, either too peaked or too flattened to be Gaussian. **1979** *Nature* 25 Jan. 297/1 Platykurtosis is not sufficient to demonstrate bimodality, but bimodal distributions are platykurtic.
Hence **pla:tyku·rtosis** [*kurtosis], the property of being platykurtic.
1939 A. E. Treloar *Elem. of Statistical Reasoning* ii. 34 Positive (or lepto-) kurtosis, mesokurtosis (that of the 'law of error' or normal curve), and negative (or platy-) kurtosis mean simply that the clustering at the center is respectively greater than, equal to, or less than that of the normal curve. **1949** [see *leptokurtosis]. **1979** [see above].

platymeria (plætimiə·riă). *Anat.* Also anglicized as **platymery** (plæ·timiəri). [ad. F. *platymèrie* (L. Manouvrier 1899, in *Compt. rend. Congr. internat. d'Anthropol. et d'Archéol. préhist.* (1891) 363), f. Gr. μηρία thigh bones: see Platy-, -ia[1], and -y[3].] The condition of a femur of which the antero-posterior diameter of its shaft is unusually small relative to the corresponding transverse diameter. Also **pla·tymerism**, in the same sense. Hence **platyme·ric** *a.*, of, pertaining to, or displaying platymery; esp. as *platymeric index*, the quotient of these diameters, multiplied by 100.
1895 *Proc. Soc. Antiquaries Scotland* V. 415 Dr Manouvrier of Paris..describes..femora from..neolithic burials..which showed the antero-posterior flattening in a very marked form, and to this condition he has given the name platymery (flat femur). *Ibid.* 416 The platymeria was very strongly marked. *Ibid.* 417 A platymeric femur is not necessarily associated with the squatting attitude. **1896** *Jrnl. Anat. & Pathol.* XXXI. 14 By far the most remarkable feature of this bone was its Platymeric index, which was almost as low as that of the lowest Maori indices. **1904** W. L. H. Duckworth *Morphol. & Anthropol.* xiii. 313 Platymeria implies flattening in two regions of the femoral shaft, viz., in an upper region, immediately below the level of the lesser trochanter, and in a lower region about 40 mm. above the highest level of the external portion of the anterior aspect of the condylar articular surface. **1934** J. Cameron *Skeleton of Brit. Neolithic Man* x. 159 The author wishes to mention..that the platymeric condition he is about to describe is that affecting the upper third of the femoral shaft. Platymeria involving the popliteal area at the lower end of the femur has also been described. *Ibid.* 165 A group of characteristic features that are more or less regularly present as concomitant phenomena of platymerism. **1971** *Nature* 6 Aug. 383/2 The flattening (platymeria) of the upper end [of the shaft of the femur] is exaggerated by the presence of a marked lateral expansion at the level of the gluteal tuberosity. **1972** J. T. Robinson *Early Hominid Posture & Locomotion* x. 143 If SK 97 belonged to a male, then it is not improbable that SK 82 belonged to a female since it has a smaller head, a smaller angle of the neck, a less robust shaft, and a lower platymeric index than has SK 97. **1976** *Nature* 17 June 575/1 The long slender platymeric shaft is unexpected and provides intriguing evidence for the possible variation of the femur in this genus.

platypussary (plæ·tipɒ̄sări). *Austral.* Also **platypusary, platypussery.** [f. Platypus + -ary[1].] An enclosure or building in which platypuses are kept.
1945 Baker *Austral. Lang.* xiv. 242 *Platypussary* has made its appearance only in recent years. **1960** —— *Drum* 135 Platypussery: A pen or specially prepared area in which platypuses are kept. Also, platypussary. **1966** G. Durrell *Two in Bush* v. 161 David's pair were housed in his specially designed Platypusary.

platyrrhine, *sb.* and *a.* In etym. substitute for *platyrrhinus* name of infraorder *Platyrrhini* (E. Geoffroy 1812, in *Ann. Mus. Hist. Nat.* XIX. 104). **1.** *Zool.* Substitute for def.: A monkey belonging to the infraorder Platyrrhini of the order Primates, distinguished by a flattened nose with widely separated nostrils facing outwards and including most of the New World monkeys. **b.** *adj.* Of or pertaining to this group of monkeys. (Later examples.)
1894 H. O. Forbes *Hand-bk. Primates* I. 127 The New World Monkeys..have the nose flat and the opening of their nostrils directed outwards, and the one nostril separated from the other by a broad cartilaginous septum, and they are therefore designated Platyrrhine Monkeys. **1920** *Proc. Zool. Soc.* 91 The observations recorded in this paper are based mainly upon the Platyrrhine Monkeys that have died in the Zoological Gardens during the past ten years. **1930** *Ann. & Mag. Nat. Hist.* VI. 387 In the more primitive Platyrrhines the brain is relatively small and has few sulci. **1934** W. E. Le Gros Clark *Early Forerunners of Man* vii. 177 In the Platyrrhine monkeys the nostrils are relatively wide apart. **1957** W. C. O. Hill *Primates* III. 91 The only platyrrhine monkey whose total musculature has been systematically studied is the Common Marmoset. **1957** I. T. Sanderson *Monkey Kingdom* viii. 80/1 All the living New World primates used to be called the Platyrrhines. **1967** J. R. & P. H. Napier *Handbk. Living Primates* i. 15 In the majority of Cebidæ the nostrils are wide apart (platyrrhine condition). **1978** *Nature* 11 May 173/3 The separation between the New World monkeys (platyrrhines) and the Old World monkeys and apes (catarrhines) is an ancient one.
2. (Earlier and later examples.)
1885 *Jrnl. R. Anthropol. Inst.* XIV. 71 Nasal Index... Leptorhine 47·0 and under. Mesorhine 47·1 to 51·0. Platyrhine 51·1 to 58·0. Hyperplatyrhine 58·1 and over. **1976** *Lancet* 25 Dec. 1394/2 The special features of the upper respiratory tract..in platyrrhine peoples.
platyr(r)hiny (further examples).
1902 [see *chamæconchy* s.v. *Chamæ-]. **1957** W. C. O. Hill *Primates* III. 85 The nasal opening in the skull gives little indication of the characteristic external platyrrhiny. **1970** F. Snowden *Blacks in Antiquity* i. 7 The platyrrhiny of the Ethiopian, like his color and his hair, was the norm for anthropological comparisons.

platytera (plætitḗ·ră). *Iconography.* [ad. Gr. πλατυτέρα she who is wider, compar. of πλατύς wide.] A type of icon of the Incarnation, also known as the Icon of the Sign, depicting the Mother of God, *orant*, and in front of her the Child, each usu. surrounded by a mandorla. Cf. Orant.
1911 O. M. Dalton *Byzantine Art & Archaeol.* xii. 674 The *orans* type holding the Christ-medallion over the breast..is sometimes described as Blacherniotissa, while that in which the Virgin stands and the medallion is unsupported is known as *Platytera*. **1943** C. de Tolnay *Michelangelo* I. 158 The idea of placing the Christ Child between the knees of His mother is exceptional. Michelangelo seems to approach in this, as well as in the severe vertical axis of the position, an artistic conception of the Middle Ages—the Platytera, or Virgin who carries the Child in her bosom. In the ordinary Platytera type the Child is surrounded by a mandorla... We may note the seated Platytera which occurs in the Etchmiadzin Gospels. **1944** *Burlington Mag.* July 176/1 Its form is derived from the 'platytera', showing the child as an abstract symbol affixed to but not inside the body of the mother... The 'platytera' proper appears in Byzantine art as early as the fifth or sixth century..and in many adapted forms spreads over Europe during the Middle Ages. **1963** von Herzfeld & Rice tr. *Onasch's Icons* 345/2 This older type shows the Virgin [*sc.* as *orans*] without a medallion before her, the later type represents her with it. This later type is called in Greek 'Platytera', the Ample One, with reference to a hymn: 'He made your womb more ample than the heavens'. In Russian icon painting it is called the 'Virgin of the Sign' (Znamenie).

plaur, var. *Plav.

plausible, *a.* (*sb.*) Add: **A.** *adj.* **3.** (Further examples without implication of mere appearance.)
1933 *Oxf. Eng. Dict.* Suppl. 403/2 One of the most plausible suggestions of etymology is F. *gâchette*. **1952** G. Sarton *Hist. Science* I. xi. 280 If Hippocrates actually wrote the first textbook of geometry, which is not only possible but plausible. **1969** *Daily Tel.* 17 Oct. 16/3 Something seriously missing here—the older actress, perhaps, needed to play this part, and certainly some more plausible explanation of the lady's behaviour. **1976** *Nature* 29 Apr. 813/3 Some ideas which have been suggested to explain this unexpected finding are plausible.
B. *absol.* or as *sb.* (Further example.)
1831 J. S. Mill in *Examiner* 6 Feb. 83/2, I mean the really profound and philosophic inquirers into history in France and Germany, not the Plausibles, who in our own land of shallowness and charlatanerie, babble about induction without having ever considered what it is.

‖ **plav** (plav). Also **plaur** (pla·ū̆ə̆ɹ). [a. Romanian *plav*, regional synonym of *pláur*.] One of the floating mats of reeds on the Danube, or the material of which it is composed.
1916 *Jrnl. Linnean Soc.* (*Bot.*) XLIII. 234 In Rumania, Plav is practically confined to the delta of the Danube. *Ibid.* 264 Reed of any size may in the delta of the Danube form an entire Plav, or portion of a Plav, from the base upwards. **1924** J. A. Hammerton *Countries of World* XVI. 1653/1 Many creatures perish, while such as do escape..eke out a desperate existence on floating islands of plaur or marsh weeds, so thickly matted and intertwined as to furnish a secure footing for both man and beast. **1961** *Times* 28 Oct. 9/7 We clambered out on to the *plaur*, the floating mass of decayed reeds.

play, *sb.* Add: **5. d.** Attention or patronage; a show of interest; publicity. *slang* (orig. *U.S.*).
1929 D. Runyon in *Hearst's International* July 57/1 Everybody goes to the Chicken Club now and then to give Tony Bertazzola, the owner, a friendly play. **1931** F. L. Allen *Only Yesterday* viii. 189 The insignificant Gray-Snyder murder trial got a bigger 'play' in the press than the sinking of the *Titanic*. **1935** J. O'Hara *Appointment in Samarra* 46 The Apollo [*sc.* a hotel] got a big play from salesmen who had their swindle sheets to think of. **1959** *Wall St. Jrnl.* 20 Nov. 17/2 Du Pont Co.'s nylon 501, brought out late last year, is getting a big play this fall from James Lees & Sons Co. and E. T. Barwick Mills, Inc., and other mills have nylon 501 carpets in production. **1970** *Washington Post* 30 Sept. B2/2 Asked her opinion on the 'youth revolt', she replied: 'I think it's such a minority—it gets far too much play.'
10. d. (Earlier and further examples.)
c**1788** *Laws of Cricket* §14, The Striker is out..if, in striking, or at any other time while the ball is in play, both his feet are over the popping-crease, and his wicket put down. **1816** W. Lambert *Instr. & Rules Cricket* 34 Always endeavour to hit the Ball on the same side on which it is bowled, and not draw it across the play. **1857** T. Hughes *Tom Brown's School Days* i. v. 109 As soon as the ball gets past them, it's in touch, and out of play. **1882** *Australians in England* 22 He got half way up the play, and just reached the ball with one hand. **1976** *South Notts Echo* 16 Dec. 7/3 The ball bounced across the goal line and into play.
g. An attempt to achieve or gain something; a move, manœuvre, or venture; *spec.* (*a*) *U.S. Sports*, an attacking move in a team game; an action that advances one's team's interest; (*b*) an attempt to attract or impress a person of the opposite sex; freq. in phr. *to make a play* (*for*). *slang* (orig. *U.S.*).
1868 H. Chadwick *Game of Base Ball* 46 A 'treble play' is made when three players are put out after the ball is hit, before it is pitched to the bat again. **1905** 'H. McHugh' *Get Next!* 75 His intentions are honorable and he wishes to prove them so by shooting his lady love if she renigs when he makes a play for her hand. **1906** H. Green *At Actor's Boarding House* 87 She had once made a play for the Swede, but he couldn't see her. **1912** C. Mathewson *Pitching* 174 Most clubs try to keep an umpire feeling hostile toward the team because, even if he means to see a play right, he is likely to call a close one against his enemies, not intending to be dishonest. **1930** *Amer. Mercury* Dec. 457/1 We make a play on their plant, but don't score. **1939** E. S. Gardner *D.A. draws Circle* ii. 26 Stall the thing along, make it casual, and be sure to back my play. **1943** D. Powell *Time to be Born* vi. 132 If you were twenty years younger I'd make a play for you, no fooling. **1961** P. Field *Rattlesnake Ridge* xiv. 170 It's the second time War Ax hands made a play for that money. **1961** *Dallas Morning News* 10 Oct. ii. 2 Gannon contributed saving plays on the Falcons' aerial thrusts in the late stages. **1966** Wodehouse *Plum Pie* i. 26 Grab the girl while the grabbing's good, because..your nephew Bertram is making a heavy play in her direction. **1969** *Official Baseball Rules* 13 A double play is a play by the defense in which two offensive players are put out as a result of continuous action. **1972** *Newsweek* 10 Jan. 30/2 In the U.S., a guard is supposed to handle the ball and set up plays. **1973** N. Moss *What's the Difference?* 45/1 *Play*, *n*, a team's action in American football, hence a strategic move towards a goal. **1973** E. Page *Fortnight by Sea* viii. 88 She'd been certain he would make a play for her the moment Lockwood took himself off. **1978** S. Brill *Teamsters* ii. 60 The attempt that finally worked was the play by Giacalone, to get Hoffa to a peace meeting.
14. a. Esp. in phr. *as good as a play*: very entertaining or amusing; *a play within a play*: a play acted as part of the action of another play (also with *the*, usu. with reference to Shakespeare's *Hamlet*).
[**1638** J. Taylor *Bull, Beare, & Horse* sig. C7, It was as good as a Comedy to him to see the trees fall.] **1672** Marvell *Rehearsal Transpros'd* 53 It was grown almost as good as a Play among us. **1827** T. Creevey *Let.* 22 Nov. in *Creevey Papers* (1963) xiii. 232 This morning after breakfast he has been as good as a play. **1871** [in Dict.]. **1875** [see Good *a.* 11 b]. **1925** A. Huxley *Those Barren Leaves* ii. iii. 111 He is the life and soul of Miss Carruthers's establishment... To see him with Fluffy—it's as good as a play. **1952** W. Plomer *Museum Pieces* xviii. 160 His eager account of the play was itself 'as good as a play'. **1975** D. M. Davin *Closing Times* vi. 129 'He's as good as a play,' my own parents would have said of him, had they known him.
1883 *Oxf. Mag.* 17 Oct. 308/1 He knew that the play within the play was meant for the conscience of the king. **1918** *Mod. Lang. Rev.* XIII. 151 The idea of having a play within the play is a famous one. **1935** *Ibid.* XXX. 433, I believe that in the circumstances surrounding the death of Francesco Maria I della Rovere, Duke of Urbino, we may well see the ultimate origin, not only of the play within the play, but of other elements in the plot of *Hamlet*. **1937** G. Rawson tr. *Schücking's Meaning of Hamlet* i. i. 3 The main action..reaches its apogee in the 'play within a play', a device that richly entertains both eye and ear. **1961** *Times* 24 Jan. 13/3 A play-within-a-play..convinces the audience that the actors are real addicts. **1973** *Listener* 26 Apr. 563/1 As a variation on the play-within-a-play we had the documentary-within-a-play.
16. b. The act of playing a gramophone record. *colloq.*
1961 in Webster. **1963** *Guardian* 15 June 3/7 The juke boxes each achieve 800 'plays' a week. **1967** *Melody Maker* 29 Apr. 10/4 It's nice party dance music..but the attention tends to wander after a few plays. **1974** *Listener* 3 Jan. 28/1 About eight records are played on each edition of *Top of the Pops*. That makes for four thousand 'plays' in ten years. **1978** *Oxford Times* (City ed.) 13 Jan. 15 A catchy tune with a sprightly arrangement that might make a hit if it gets enough plays on the radio.
17. *play-activity, -area, -centre, -clothes, -form, -garden, -hour* (earlier and later examples), *-impulse, -instinct, -lady, -language, -opera, -park, -poem, -producer, -reader* (later examples), *-reading, -sack, -shed, -song, -space, -spell* (earlier example), *-theory, -toy, -world, -wrecker, -yard*; *play-producing* adj.; **play-act** *v. intr.*, to act in a play; to be suitable for acting in a play; also *trans.*, to act (a scene, part, etc.); freq. *fig.*, to pretend, make-believe; to behave theatrically or insincerely; **play-acting**, also *fig.* and as *ppl. a.*; also (*Sc.*) **play-actoring** *ppl. a.*; **play-box**, a box in which a child, esp. at a boarding-school, keeps toys, books, and other personal possessions; also *transf.* and *fig.*; **playbroker** orig. *U.S.*, an agent who serves as an intermediary between playwrights and managers or actors; **playbus**, a bus adapted for children to play in; **play-by-play** *a.*, denoting a running commentary on a game; also *ellipt.* as *sb.*; **play-card**, (*a*) = Playbill; (*b*) repr. non-standard pronunc. of Placard *sb.* 2; **play-doctor**, a professional improver of other people's plays; **play-dough** orig. *U.S.*, a child's modelling clay; **play face**, an expression seen in apes or monkeys at play, in which the mouth is open but the teeth are hidden; **playfight**, a fight in play; hence **play-fighting** *sb.* and *a.*; **play-green** ?*Obs.*, a piece of land suitable for children to play on; **playland**, an area suitable for recreation; **play-leader**, an adult who leads or helps with children's play; the leader of, or a helper at, a play-group; so **play-leadership**; **play-lunch** *Austral.* and *N.Z.*, a snack taken by children to school for eating at playtime; **play-material**, (*a*) material used by children at play; (*b*) (see quot. 1969); **play-method** = **play-way*; **playmobile**, (*a*) (with capital initial) the proprietary name of a type of toy motor vehicle; (*b*) a vehicle containing facilities for a play-group; **play-night**, (*a*) a night on which a play is performed; (*b*) in Jamaica, a night of entertainment in connection with a funeral; **play-party**, (*a*) a party at which a play or plays are performed; (*b*) *U.S. dial.*, a party at which games are played, esp. dancing-games without music; also *attrib.*; **play-pen**, an enclosure in which a young child may play in safety; **play-pretty** *U.S. dial.*, a toy, plaything; **playscheme**, a local project offering play facilities for children, esp. during school holidays; **play school**, a nursery-school or kindergarten (orig. *U.S.*); **play-street**, a street closed to traffic so that children can play in it; **play suit**, (*a*) an actor's costume; *obs. rare*; (*b*) a light, casual outfit; **play therapy** *Psychol.*, therapy in which emotionally disturbed children are encouraged to act out and express their fantasies and feelings through play, aided by the therapist's interpretations; hence **play therapist**; **play-way**, an educational method which seeks to utilize play; **play-white** *S. Afr.* (see quot. 1956); **playwrite** *v. trans.*, to write in the form or style of a play; **playwriter** = Playwright; **playwriting** *vbl. sb.*
1896 G. B. Shaw *Let.* 6 Sept. (1965) I. 650, I always cut myself to the bone, reading the thing over and over until I have discovered the bits that can't be made to playact anyhow. **1901** N. Munro *Doom Castle* iv. 39 Very well pleased at the chance your coming gave him of play-acting the man of war. **1915** F. M. Hueffer *Good Soldier* iii. i. 140 She wished to appear like the heroine of a French comedy..she was always play-acting. **1938** S. V. Benét *Thirteen O'Clock* 321 They had to play-act whatever happened. **1962** I. Murdoch *Unofficial Rose* xv. 149 Or they might cold-bloodedly have play-acted the scene together, laughing about it afterwards. **1969** *Listener* 5 July 28/2 She wanted more dirty experience: could he not play-act a rapist? **1974** A. Price *Other Paths to Glory* ii. vi. 190 Most of us were play-acting—pretending to be soldiers. **1857** Trollope *Barchester T.* I. x. 138 Did you ever..hear anything so like play-acting as the way in which Mr. Harding sings the litany?.. There must be no more play-acting here. **1875** P. Ponder *Kirkumdoon* 142 Gettin' a vain play-actin' cretur to be oor minister. **1896** E. Terry *Let.* 4 Dec. in *Ellen Terry & Shaw* (1931) 133 Why don't you both come round after the play up to my room? Mayhap she doesnt like playacting folk? **1938** M. Allingham *Fashion in Shrouds* xi. 175 Georgia was doing no play-acting for Val. They were equals coming down to essentials. **1954** J. R. R. Tolkien *Fellowship of Ring* 183 You might be a play-acting spy, for all I can see, trying to get us to go with you. You might have done in the real Strider and took his clothes. **1972** J. Blackburn *For Fear of Little Men* iii. 45, I don't give a damn what you do, but I'm tired of play-acting. **1890** W. James *Princ. Psychol.* II. xxiv. 429 The immense extent of the play-activities in human life is too obvious to be more than mentioned. **1927** G. A. de Laguna *Speech, Its Function & Development* iv. 72 It was the adaptation of the play-activity to the needs of social coordination that was the essential agency in the process [of developing human speech]. **1836** J. M. Wilson *Tales of Borders* III. 29 Pittin sic daft-like notions intil a bairn's head as to read play-actorin books an' novels. **1968** *Punch* 13 Mar. 388/3 The day is not far off when all that will be left of unspoilt countryside will be designated 'Play Areas', with good parking and litter facilities and a free issue of blinkers to see what is left of the view. **1979** *Lore & Lang.* Jan. 1 Built before the turn of the century, the two tarmac play areas would nowadays be considered cramped. **1865** *Boy's Own Mag.* VI. 72/2, I had withdrawn from the school-room to 'the loft',..a long room above the school, where play-boxes were deposited. **1882** F. Anstey *Vice Versa* v. 103 Let every boarder go down into the box-room and fetch up his playbox, just as it is, and open it here before me. **1909** (*title*) The play-box, a picture reading book for little folks. **1923** Galsworthy *Captures* 56, I had taken them out of my playbox, together with the photographs of my parents and eldest sister. **1929** W. Deeping *Roper's Row* viii. 80 At her aunt's in Vane Street she had an attic which she called her studio, a young woman's play-box, and all that she knew she had taught herself by drawing things and yet more things. *Ibid.* xxi. 234 For, to Ruth Avery, No. 7 Roper's Row was a child's play-box, and much more than that—for it was the first playbox of her very own that she had possessed. **1949** I. Compton-Burnett *Two Worlds & their Ways* iv. 143 Here is the key of your playbox, Bacon. **1972** *Even. Telegram* (St. John's, Newfoundland) 23 June 3/1 When they tolerate a bunch of dandies sitting in that fancy playbox on..New Gower Street? **1910** *N.Y. Dramatic Mirror* 12 Mar. 9/4 Practically all of the new playwrights have been discovered by play-brokers. **1929** *Evening News* 9 Jan. 11/2 Major James Clare, a leading playbroker, who is also a dramatist. **1975** *Village* Winter 85 This interesting playbus experiment in Cumbria... The provision of holiday play facilities for village children. **1976** *Ann. Rep. Manpower Services Comm. 1975–76* iii. 23/2 On Merseyside, young people are working with craftsmen to convert buses into playbuses. **1927** *Amer. Speech* II. 241/2 The football extra, containing a 'play by play' story of all but the last few minutes, is locked on the press with a hole left in the plate. **1931** F. L. Allen *Only Yesterday* viii. 207 Thousands more sat in warm living-rooms to hear the play-by-play story over the radio. **1966** J. Ball *Cool Cottontail* (1967) x. 101 He turned on the radio and listened to a play-by-play of the California Angels. **1976** *Times Lit. Suppl.* 2 Jan. 13/3 The bulk of the book is given over to what looks like a play-by-play account of Hegel's thought. **1976** *Billings* (Montana) *Gaz.* 17 June 1-H/4 Announcers Lane Saunders and Bernie Lustig will provide the color and play-by-play, and a variety of guests will be lined up for half-time interviews. **1979** D. Anthony *Long Hard Cure* ix. 89, I began giving him a play by play of the events of Friday night. **1881** P. Fitzgerald *World behind Scenes* iv. 268 The Court Theatre, the Princesses, and the St. James have adopted square cards of a pale blue tint—an abnormal and inconvenient form. In the instance of the first-named house it is folded diagonally, it is a play card, and no longer a bill. **1934** T. S. Eliot *Rock* i. 40 On Christmas Day we can organize a Anti-God procession.. with playcards an' ex'ibitions exposin' all the dope o' Christianity. **1908** *Westm. Gaz.* 1 Feb. 7/3 The play-centres, far from tending to diminish the influence of home life, actually made the children appreciate it more. **1914** *Encycl. Relig. & Ethics* VII. 363/2 The 'play-centre', where, outside school hours, children who have no play-ground but the street, are taught organized games. **1936** G. M. Young *Victorian England* ix. 60 The Mechanics' Institutes..sank into play-centres for serious clerks. **1973** *Daily Tel.* 6 July 2/1 More than 80 London play centres are to be opened during the school summer holidays where children can play games, watch television, paint or read. **1919** *Ladies' Home Jrnl.* Mar. 62 Turn kids out in..Play Clothes—and let them play. **1959** *Times* 26 Jan. 11/1 Wit in styling, good fabrics and lovely colours are what the designers of 'play' clothes usually offer. **1971** *Times Lit. Suppl.* 20 Aug. 990/2 Japanese are not the only ones to feel irritation when they see elderly American tourists in gaudy 'play clothes' cavorting in Tokyo as if it were another Honolulu. **1922, 1938** Play doctor [see *doctor *sb.* 6 c]. **1967** P. McGerr *Murder is Absurd* ii. 32 A play doctor was brought in to rework Rex's unfinished script. **1978** I. B. Singer *Shosha* vii. 128 In America we have men who are called play doctors. They can't write a line themselves, but somehow they know how to rearrange a piece and make it right for the stage. **1959** J. Foster *Educ. in Kindergarten* (ed. 3) xi. 176 Clay, plasticine, play dough, sawdust and paste,..all afford the child the opportunity to make a three-dimensional impression of one sort or another. **1969** B. Ryan *Your Child & First Year of School* iii. 56 Play dough, if it is made from scratch from salt, flour, water, and perhaps a little alum powder as a preservative, gives an even broader experience of chemistry before little fingers begin to manipulate it. **1970** G. R. Taylor *Doomsday* vi. 126 Asbestos powder mixed with water is even given to children, in some schools, as play-dough. **1977** C. McFadden *Serial* xi. 29/2 She was standing at the sink digging the play-dough out of her demitasse cups. **1962** J. A. R. A. M. van Hooff in *Symp. Zool. Soc.* No. 8. 120 (*heading*) The Play Face. *Ibid.* 121 Suddenly one of the partners may..show the play face in the direction of the other who will immediately react by resuming the play. **1966** R. & D. Morris *Men & Apes* vi. 213 The play face..is performed by a number of species during vigorous bouts of playful wrestling and tumbling and is particularly obvious in young chimpanzees. **1971** J. van Lawick-Goodall *In Shadow of Man* ix. 99 He opened his mouth in the play-face or chimpanzee smile. **1973** *Observer* (Colour Suppl.) 16 Dec. 32/2 Certain facial expressions are also used [by monkeys to establish friendship with one another], such as the 'playface'—a smile with teeth covered. **1977** Savage & Rumbaugh in D. M. Rumbaugh *Language Learning by Chimpanzees* xvi. 300 A playface given with vocal laughter and head-covering can be used as a signal to continue tickling. **1922** Joyce *Ulysses* 442 An armless pair of them flop wrestling, growling, in maimed sodden playfight. **1932** S. Zuckerman *Social Life Monkeys & Apes* xvii. 277 The play-fighting activities and bodily examinations continued intermittently. **1953** *Psychological Rev.* LX. 293/1 Among the social interactions there were a few instances of serious aggression, many occurrences of bluffing or exhibitionistic behavior, a great deal of play-fighting, wrestling, [etc.]. **1911** J. A. Thomson *Biology of Seasons* ii. 224 For the endless task of finding out about the world has its play-form—which is obviously one of the roots of science. **1963** G. J. McCall in A. Dundes *Mother Wit* (1973) 425 As has happened with so many play-forms of games..it has become a 'multi-situated game', requiring a vast proliferation of goals, roles, and strategies. **1916** A. S. Neill *Dominie Dismissed* xiii. 153 The attraction of a play-garden school with its charms of social intercourse. c**1650** *Lillumwham* in Hales & Furnivall *Bishop Percy's Folio MS.* (1867) *Loose & Humorous Songs* 98 Other three on won play greene. **1800** M. Edgeworth *Parent's Assistant* (ed. 3) II. 178 All the children..were assembled in the play-green. **1812** —— *Absentee* in *Tales Fashionable Life* VI. ix. 131 He went to the village school—a pretty, cheerful house, with a neat garden and a play-green. **1741** Richardson *Pamela* IV. lxiv. 454 The Misses at their Books too, or their Needles; except at their Play-hours, when they were never rude, nor noisy. **1925** Blunden *English Poems* 19 And blushed for pride when other girls and boys Laughed at us sweethearts in the playhour's noise. **1890** W. James *Princ. Psychol.* II. xxiv. 427 The sexes differ somewhat in their play-impulses. **1927** W. E. Collinson *Contemp. Eng.* 127 The English play-impulse has certainly produced some remarkable forms. **1896** W. James *Will to Believe* (1897) 23 Mephistophelian scepticism..will satisfy the head's play-instincts much better than any vigorous idealism can. **1897** T. Ribot *Psychol. of Emotions* ii. 198 The play-instinct, if we use this word to designate the tendency to expend superfluous activity..is a stock which puts forth several branches. **1966** *New Statesman* 24 June 923/3 There is a 'play lady'..who spends her whole time seeing to the personal interests and difficulties of the children in the wards. **1976** *Amer. Speech* 1973 XLVIII. 208 At RT, he will meet *play ladies*, not to be confused with those in peds, but ones specially trained to teach physical activities. **1946** *Sun* (Baltimore) 26 Apr. 11/4 Baltimore is to be tied into schedules into the play-land of Michigan, bringing PCA flights out of the local airport to fourteen daily. **1974** *Sat. Rev. World* (U.S.) 2 Nov. 8/2 Cyprus..would be..a Mediterranean playland. **1977** *Time* 4 July 28/2 Remembering the rapacious play-lands of the past, where gambling, boozing and whoring were as rife as popcorn and pizza, most theme parks promote soft drinks and fast foods. **1934** O. Jespersen *Language* 149 Children..at first employ play language for its own sake. **1907** *Westm. Gaz.* 29 Aug. 3/1 A pressing need is for trained play-leaders who know how to play games and to organise the interests of children in ways that build the body and character as well. **1953** Play leader [see *adventure playground* s.v. *adventure *sb.* 10]. **1970** *Guardian* 14 May 11/4 Two mothers volunteered to take a play-leaders' course. **1975** *New Society* 18 Sept. 632/2 The Chells adventure playground, which is very large and has three full-time playleaders. *Ibid.* 13 Nov. 393/2 (Advt.), The successful applicant will be required to give general assistance within the Playleadership service. **1977** *Time Out* 28 Jan.–3 Feb. 53/1 (Advt.), Applicants, male or female, should be able to demonstrate experience in these areas, preferably within playleadership. **1962** *N.Z. Listener* 27 July 39/1 Children like to take the special little packets [of raisins and sultanas] to school for their playlunch. **1963** E. Spence *Green Laurel* ix. 109 She was not hungry enough to go back for her play-lunch, and to stay close to the class-room appeared to be the safest thing to do. **1974** *Age* (Melbourne) 12 Oct. 12/5 *Play lunch*, emergency rations for morning recess, usually being a piece of fruit, cake or some chocolate crackles. **1943** H. Read *Educ. through Art* v. 158 For example, in analysing the quantitative differences in the kinds of play-material used by children of age groups from 4 to 8, Dr. van Wylick found that..the use of human beings could not be related to the progressive age-groups. **1969** E. Ambler *Intercom Conspiracy* (1970) ii. 39 'Play material' was the jargon phrase used to describe the low-grade classified information fed back to the enemy through double agents. **1971** D. O'Connor *Eye of Eagle* xxii. 154 There'll be stretches on this tape with nothing on them but a lengthy silence. You could fill them in, if you wanted, with play material. **1914** H. C. Cook (*title*) First-fruits of the play method in prose. **1961** in *Amer. Speech* (1964) XXXIX. 79 Playmobile. **1963** *Official Gaz.* (U.S. Patent Off.) 21 May TM 148/1 DeLuxe Reading Corporation, Newark, N.J. *Playmobile* for toy miniature automobiles. First use Feb. 20, 1961. **1971** *Guardian* 16 Dec. 11/1 A children's playgroup in a converted double-decker bus?..The Playmobile will penetrate the drab streets of slumland. **1973** *Times* 25 July 13/2, I feel we should also provide the opportunity for the community to have far greater involvement in concern for others. Involvement such as taking playmobiles round to all our caravan sites. **1979** *Trade Marks Jrnl.* 4 July 1134/1 *Playmobil*... Toys having movable parts. Geobra Brandstätter GmbH & Co. K.G...Zirndorf..Germany; manufacturers and merchants.—17th March 1978. **1755** C. Charke *Life* 103 Those Assailants of Liberty..constantly attended every Play-Night there. **1786** J. Woodforde *Diary* 6 Apr. (1926) II. 238 It being Play Night we went to the Theatre. **1849** *Theatrical Programme* 2 July 43/1 His Majesty [*sc.* George II]..was pleased to order that the Guards should in future do duty every playnight, which custom has not yet been dispensed with. **1961** D. De Camp in R. B. Le Page *Creole Language Studies* II. iv. 72 Plie-nait. **1937** *Sun* (Baltimore) 12 Apr. 5/3 'The Second Hurricane' is neither grand nor light opera, and somebody had to think up a name for it. This turned out to be 'play opera'. **1954** *Grove's Dict. Music* (ed. 5) VIII. 125/2 The two other works of this period are the ballet-pantomime 'Schlagobers' .., and the autobiographical play-opera 'Intermezzo', for which Strauss wrote his own libretto. **1971** P. Young in J. Spencer *Eng. Lang. W. Afr.* 183 The dramatised version of *The Palm Wine Drinkard* was published in English originally, but performed as a play-opera in Yoruba at Ibadan. **1962** *Guardian* 31 Oct. 6/5 The playparks are a welcome addition to the other nursery facilities. **1964** *Ibid.* 30 Oct. 6/6 The number of play parks where children can 'let off steam' is to be increased after the Greater London Council takes over next year. **1977** *Cork Examiner* 6 June 1/8 In the past proceeds have gone to specific projects such as the building of a social room, gymnasium facilities, play park and equipment etc. **1879** L. Troubridge *Jrnl.* June in J. Hope-Nicholson *Life amongst Troubridges* (1966) 152 Met Amy and had quite a gay visit to Abbey Lodge, doing lots of plays. Uncle Hay failed us for a play party. **1902** *Dialect Notes* II. 241 *Play-party*, a party at which old-fashioned games are played. **1912** I. S. Cobb *Back Home* 44 Strict church members..wouldn't let their children..go to any parties except play parties. **1926** M. D. Lake in J. F. Dobie *Rainbow in Morning* (1965) 109 Parties of various kinds were indulged in at Christmas... Dominoes, candy pulls, corn poppings, play-parties, and dancing furnished additional amusement. **1937** B. A. Botkin *Amer. Play-Party Song* i. i. 16 The play-party..was a rural American social gathering for playing games, distinguished by the manner in which it was 'got up', by the age of its participants, and by the character of the games played. **1938** [see *caller *sb.* 1 e]. **1968** P. Oliver *Screening Blues* i. 31 It might be said.. that the blues singer rejoices in his folk-songs—his dance songs, play-party and game songs, ballads and stomps. **1973** Schafer & Riedel *Art of Ragtime* i. 13 This 'play-party' country ragtime style is of great age and hardiness. **1931** *Daily Express* 21 Sept. 7/5 (Advt.), Well built play-pens in best hard-wood. **1940** J. Betjeman *Old Lights for New Chancels* 47 White o'er the play pen the sheen of her dress. **1967** N. Freeling *Strike Out* 81 A child's playpen stood folded against the wall. **1972** J. Wilson *Hide & Seek* ii. 35 Jean picked Jamie out of the playpen and sat with him on her lap. **1976** W. H. Canaway *Willow-Pattern War* xiii. 136 A set of beads on wires, a bit similar to the set I'd had on my play-pen when I'd been smaller. **1907** *Daily Chron.* 17 May 3/7 A soulful little French play-poem, Coppée's 'Le Passant'. **1928** V. Woolf *Writer's Diary* 7 Nov. (1953) 137 Yes, but *The Moths?* That was to be an abstract mystical eyeless book: a playpoem. **1977** *Times Lit. Suppl.* 4 Feb. 123/1 Virginia Woolf described *The Waves* as a 'playpoem'. She was conscious of the hazards of poetic fiction, the dangers of uncontrolled fantasy, and wanted to minimize these by assimilating to prose fiction the structural tightness and compression of drama. **1905** *Dialect Notes* III. 90 The children want some play-pretties for Christmas. **1929** W. Faulkner *Sound & Fury* 36 Aint you shamed of yourself. Taking a baby's play pretty. **1935** R. Bass in A. Dundes *Mother Wit* (1973) 395 On the graves of little children are placed some play-pretties, little doll heads, small cups, or toy animals. **1942** J. Thomas *Blue Ridge Country* 160 The children's play-pretties—the poppet, a make-believe corn-shuck doll. **1976** *Publ. Amer. Dial. Soc.* 1973 lx. 17 One..called it [*sc.* a toy] *a pretty*, another *a play-pretty*. **1913** *Writer's Mag.* Dec. 253/1 For we are now in an era wherein the play-producer is on the alert for the young and virile writer. **1908** *Daily Chron.* 19 May 1/6 Several uncommercial play-producing societies..had done..good work recently. **1968** *Daily Tel.* 4 Nov. 9/5 And so we come back to the independent, unsubsidised play-producing companies. **1922** Play-reader [see *doctor *sb.* 6 c]. **1969** L. Hellman *Unfinished Woman* v. 53, I worked as a play reader for Anne Nichols,..who wanted to become a producer. **1913** F. H. Burnett *T. Tembarom* xix. 244 On still another evening they tried Shakespeare. He found play-reading difficult and Shakespearian language baffling. **1935** N. Mitchison *We have been Warned* ii. 154 The next pupil was late... 'Sorry,' he said, 'it was the play-reading society.' **1968** J. Haythorne *None of us cared for Kate* 61 We do try in our small way to keep the torch of culture flickering. Are you fond of play-reading? **1972** D. H. Laurence *Bernard Shaw: Coll. Lett. 1898–1910* 4 [Shaw]..joined Grant Allen and other neighbours in a play-reading society. **1970** *Daily Progress* (Charlottesville, Va.) 21 Mar. c2/5 If the age group is 4 to 10,..playsacks..are imaginative animal costumes that slip on easily. **1972** *Where* Apr. 104/2 There were in fact something over 400 playschemes in England and Wales last summer. **1975** *Village* Winter 83 You would have been looking at Cumbria's first mobile playscheme. **1935** *Sun* (Baltimore) 15 July 7/3 Three children's 'play schools' scheduled here were canceled. **1959** C. V. Good *Dict. Educ.* (ed. 2) 402/2 *Play school*,..an organized experience usually lasting for a short time to provide opportunities for high school and college students to observe and work with a small group of young children in a supervised situation. **1964** S. Bellow *Herzog* 267 I'm picking June up at noon tomorrow. She goes to a play school, half-days. **1973** *Times* 7 Mar. 10/3 Whoever coined the term 'play school' captured perfectly the ideal concept for the young child. **1977** D. MacKenzie *Raven & Kamikaze* iv. 49 The babble of children came from the open windows of the play school. **1906** *Macmillan's Mag.* Nov. 19 Rooms for the teachers and for the permanent staff, a covered play-shed, and all the outside accessories. **1932** *Times Educ. Suppl.* 20 Aug. 318/1 Playsheds can as a rule be omitted, but inexpensive bicycle sheds may be advisable and serve as shelters for the children against rainstorms. **1898** J. C. Harris *Tales of Home Folks in Peace & War* 19 The negroes made the night melodious with their play-songs. **1924** G. Parkes in M. W. Beckwith *Jamaica Anansi Stories* 110 Massah, me kyan' stop him singing, because it mus' of been his little play-song what he have singing. **1959** I. & P. Opie *Lore & Lang. Schoolch.* i. 3 The same continuity obtains in their games and play songs. **1958** *Times* 30 Aug. 7/4 The whole of the underside of the tall block is planned as a covered paved play space. **1959** [see *bed-sitting-room]. **1974** *Listener* 7 Mar. 296/3 Families..live around courtyards, half of which serve..as play-spaces for children. **1976** *Ilkeston Advertiser* 10 Dec. 10/3 The parents..told Broxtowe District Council about the lack of play space for children on Broad Oak Drive estate. **1845** S. Judd *Margaret* ii. i. 186 And her own play-spell comes, if, indeed, her whole life were not a play-spell. **1937** C. V. Godfrey *Roadsense for Children* viii. 64 Closing certain lesser thoroughfares to vehicular traffic and reserving them for the exclusive use of children ..is how Play Streets came into being. **1968** *Guardian* 25 Apr. 7/6 Some local authorities label streets as play streets when they have not enough money for playgrounds. **1977** *Wandsworth Borough News* 7 Oct. 14/5 It was a 'playstreet' with bollards to prevent all traffic, where children could safely play under the eyes of their parents. **1609** T. Dekker *Guls Horne-Booke* vi. 29 By sitting on the stage, you may..examine the play-suits lace. **1908** *Sears, Roebuck Catal.* 529/1 Play Suits. **1936** *New Yorker* 7 Mar. 64/1 Crisp, wearable and washable shirt-waist dresses, play suits, shorts, slacks, and skirts. **1942** *Capital* 3 Feb. 5/6 (*caption*) Ilyana Yankwich wearing a California-made playsuit; it's in vivid green and brown tones with a gay leaf motif. **1959** *New Statesman* 26 Sept. 384/2 Unguents, oils, special sun-bathing attire, bikinis and play-suits..have become big business. **1963** Playsuit [see *Garbo[1]]. **1925** I. A. Richards *Princ. Lit. Crit.* 233 The objection to the Play Theory..lies in its suggestion that the experiences of Art are in some way incomplete, that they are substitutes. **1960** C. Winick *Dict. Anthropol.* 535/2 *Play theory*, the theory that fine art is produced independently of the struggle for existence and that the imagination is exercised for the sake of the sense of freedom (Schiller), or power (Groos), or for conscious self-deception (Lange). **1942** *Brit. Jrnl. Psychol.* Jan. 262 A direct interpretation given to the child of the meaning of his play should be undertaken only..by the experienced play therapist. **1963** A. Heron *Towards Quaker View of Sex* 49 The child may have treatment with a play therapist. **1939** *Psychol. Abstr.* XIII. 111/1 The author believes that active play therapy offers rapid diagnostic and therapeutic assistance in the emotional problems of childhood. **1948** L. Kanner *Child Psychiatry* (ed. 2) xvii. 244 'Play therapy' thus becomes a form of participation, a means to an end, rather than an isolated technique. **1961** A. Huxley *Let.* 8 Jan. (1969) 903 One can imagine a genuinely realistic treatment of the mentally ill..by work and play therapy. **1978** M. T. Erickson *Child Psychopathol.* x. 225 Traditional clinicians have offered an array of treatments to children with learning disabilities: psychoanalysis,..play therapy, and group therapy. **1935** Z. N. Hurston *Mules & Men* 19 I'll put this play toy in his hand, and he will seize it and go away. Then I'll say my say and sing my song. **1914** H. C. Cook *First-Fruits of Play Method* 52 The boys do not object to learning anything, so long as they may do it in the Play way. **1920** T. P. Nunn *Education* viii. 92 Members of a rapidly growing company of pioneers..are all busily engaged in exploring the 'play-way' of teaching the several subjects of the curriculum. **1973** *Times* 13 Jan. 12/3 Nephew X, proud of himself for being tough with his daughter over the cello lessons, dismisses all this 'play-way' approach to education as a lot of soft nonsense. **1956** A. Sampson *Drum* xv. 205 Harry was only one of thousands of 'play-whites', as they call the light-skinned Coloureds who 'pass for white' and break away from the Coloured world. **1909** *Daily Chron.* 13 Dec. 3/4 Nature's kingdom is not all a reign of tooth and claw, but a play-world also. **1915** D. H. Lawrence *Rainbow* xi. 264 The religion, which had been another world for her, a glorious sort of play-world. **1962** W. Nowottny *Lang. Poets Use* iv. 89 This realization is articulated most clearly at the climax of the passage..and so too is the sense of the irruption into his play-world of intractable segments of reality. **1978** I. B. Singer *Shosha* vii. 135 There is no reason why hedonism, the cabala, polygamy, asceticism, even our friend Haiml's blend of eroticism and Hasidism could not exist in a play-city or play-world. **1901** *Chambers's Jrnl.* Aug. 545/2 Organised play-wreckers, who without uttering a word or an unseemly laugh have succeeded in destroying whatever chance of success a play may have had. **1949** *Sat. Rev. Lit.* (U.S.) 24 Dec. 24/3 One of the unique and beckoning characteristics of his plays was that they were written no less than playwritten. **1872** W. L. Collins *Aristophanes* iii. 41 To win the verdict of popular applause, which was the great aim of an Athenian play-writer, he must above all things hit the popular taste. **1903** W. B. Yeats *Let.* 27 July (1954) 408, I suppose every playwriter finds out the methods that suit him best. **1898** G. B. Shaw *Plays Unpleasant* Pref. p. v, I made a rough memorandum for my own guidance that unless I could produce at least half a dozen plays before I was forty, I had better let playwriting alone. **1935** *Discovery* May 130/2 Historical playwriting has had considerable vogue. **1959** *Times* 24 Oct. 9/2 *Pursue the Dry Stubble*, the winning entry in the playwriting competition organized by the Tower Theatre, was given the first of six performances there last night. **1976** *Radio Times* 27 Mar.–2 Apr. 37/1 Sir Terence Rattigan looks back over 40 years of playwriting from *French without Tears* to *Cause Célèbre*. **1960** J. J. Rowlands *Spindrift* iii. 186 One particularly important question was whether the play-yard in Heaven was equipped with an old dory with a mast and sail. **1973** *Jrnl. Genetic Psychol.* Sept. 160 Assertive behavior is a relatively stable characteristic of the preschool child, a characteristic which he brings to many situations and which can be seen even in 15 minutes in the play yard.

play, *v.* Add: **II. 10. d.** *to play around*: to amuse oneself; to behave in a playful or irresponsible manner; *spec.* to have a sexual relationship *with* (a person or persons of the opposite sex), esp. casually or extra-maritally. *colloq.* (orig. *U.S.*).

1929 D. Hammett *Red Harvest* xi. 109 Max was up there with a girl he used to play around with. 1932 G. Greene *Stamboul Train* iv. iv. 248 You mean you killed him.. just because he'd played around with your daughter? 1934 J. O'Hara *Appointment in Samarra* ii. 46 He played around a little, but Al knew Helene was the only one he really cared for, and Helene really cared for him. 1960 *Sunday Express* 14 Aug. 14/6, I went to all the parties; I played around. 1963 D. Gray *Murder in Mind* xv. 83 And if I found you were playing around, I'd give you a damned good hiding. 1973 S. Dobyns *Man of Little Evils* (1974) iii. 31 Ralph played around with other women but he liked one.

12. a. (Further examples.) *to play with fire*: see *FIRE *sb.* A. 3 g.

1945 *Tee Emm* (Air Ministry) V. 52 This will give you a little to play with and allow for a drop in barometric pressure. 1965 V. Canning *Whip Hand* iii. 33, I like a girl who doesn't play with her food or drink. 1976 Scott & Koski *Walk-In* (1977) xxxiii. 237 He was sweating now, all right. And is he playing with me? he wondered. Is the bastard playing with me? 1978 *Lancashire Life* Sept. 76/3 How could they possibly build docks when they had merely £60,000 to play with.

b. *to play with* (someone): to masturbate; usu. *refl.* *colloq.*

1922 Joyce *Ulysses* 552 You can apply your eye to the keyhole and play with yourself while I just go through her a few times. 1954 H. K. Fink *Long Journey* 14, I was going with girls..and I didn't feel the urge to play with myself. 1966 L. H. Farber *Ways of Will* iii. 58 This opening scene of a faceless woman silently playing with herself..sets the tone. 1967 A. Wilson *No Laughing Matter* II. 65 That kind of thinking can easily land you in the loony bin. It's worse than playing with yourself. 1969 H. Miller *Sexus* (1970) viii. 166 'Play with it a bit while I finish this.' 'You're filthy,' she said, but she did as I told her. 1971 'V. X. Scott' *Surrogate Wife* 54 He played with me. And little by little..I played with him. *Ibid.* 114 In bed, we played with each other. 1974 E. Tidyman *Dummy* xv. 199 I'd glance over at Donald and he'd be playing with himself in the courtroom.

III. 16. d. *to play politics*: to act on an issue for personal or political gain rather than from principle. orig. *U.S.*

1863 [in Dict., sense 16 c]. 1907 *Springfield* (Mass.) *Weekly Republican* 13 May 6 Mr. Balfour has seized the opportunity to play politics, and has apparently come out squarely in favor of trade preference. 1931 H. F. Pringle *Theodore Roosevelt* II. vii. 343 Roosevelt.. was playing politics in his own behalf. 1962 *Listener* 15 Nov. 798/1 It has been fashionable to claim that Mr Gaitskell..was deliberately playing politics with the Common Market issue. 1963 *Times* 11 Feb. 11/3 If it is too much to ask any Government to stop playing politics with the economy, the Opposition can at least be urged not to abet it in doing so. 1976 *Punch* 16 June 1070/2 There are a few people who find it disturbing that we are now the most heavily indebted nation in the industrial world—but as the Government would wish me to point out, they are just playing politics.

e. *to play the dozens*: to engage in a bout of verbal insults and ridicule with one or more other people: used of a ritualized form of dialogue customary among American Blacks.

1933 E. Caldwell *God's Little Acre* x. 142 If you want to play the dozens, you're at the right homestead. 1939 J. Dollard in *American Imago* Nov. 6 One asked the other, 'Do you want to play the Dozens?' The other boy said, 'Yes.' *Ibid.* 7 These reactions of concealment and shame convinced me that playing the Dozens is not an orgy of licentious expression for lower-class Negroes; all know that the themes treated are in general forbidden, some refuse to play the game and still others are very resentful and defensive at the mere thought of it. 1942 Z. N. Hurston in A. Dundes *Mother Wit* (1973) 24/1 The bookless may have difficulty in reading a paragraph in a newspaper, but when they get down to 'playing the dozens' they have no equal in America. 1962 R. D. Abrahams in *Ibid.* 298/1 'Playing the dozens' is one of the most interesting folkloristic phenomena found among contemporary Negroes. 1970 H. E. Roberts *Third Ear* 11/1 Playing the dozens, making derogatory..remarks about another's mother, parents. 1973 *Black World* Aug. 58/2 Could play the dozens for days, talk about your momma bad enough to make you cry. 1973 A. Dundes *Mother Wit* 141/2 A sample of some of the special techniques and forms of extended word play should convince even the most adamant sceptic that no black child who can signify or play the dozens can rightly be called lacking in verbal skills.

f. *to play pussy* (Aeronaut.): to fly under cover in order to avoid detection by another aircraft, etc. *slang.*

1942 *We speak from Air* 30, I wondered if he was playing pussy and intending to jink away. 1942 *Gen* 1 Sept. 14/1 Waiting in the air..he 'snakes about' or 'plays pussy' in the clouds. 1943 Hunt & Pringle *Service Slang* 52 *Play pussy*, to take advantage of cloud cover, jumping from cloud to cloud to shadow a potential victim or avoid recognition. 1948 Partridge *Dict. Forces' Slang* 143 *Play pussy*, to speed from one cloud to another in order to escape detection or to pounce upon a shadowed enemy aircraft.

17. f. *to play back, backward(s)*: in *Cricket*, said of the batsman: to move back before striking the ball; *play forward*: to move forward in making a stroke; *play through*: in *Golf*, to continue playing, passing other players who have agreed to suspend their game for this purpose.

1816 W. Lambert *Instr. & Rules Cricket* 27 [This] will direct him to play forward at the..bowling. *Ibid.* 29 If at these [short] kind of balls the Striker plays back about two feet behind the popping crease..it will afford him a little more time to judge how the Ball is coming. 1851 W. Clarke in W. Bolland *Cricket Notes* 135 It is the ball that catches him in two minds, so that he does not know whether to play forward or backward. 1899 W. G. Grace *Cricketing Reminiscences* x. 288 If a boy has once learned to play forward confidently he will soon adapt himself to playing backwards at balls that demand it. 1934 W. J. Lewis *Lang. Cricket* 198 *Play back*,..to step back with the right foot towards the wicket, playing the ball behind the popping crease. *Ibid.* 199 *Play forward*,..to reach forward, advancing the left foot and the bat, in making a stroke. 1967 M. Green *Art of Coarse Golf* x. 110 The general rule of etiquette in Coarse Golf seems to be that solo players have right of way over all matches. It is not normally necessary for them to ask permission to play through—they simply pound on round the course. 1970 H. Taylor *Golf Dict.* 159 If a ball is lost and cannot be found, the player with the consent of the other players signals to those following to 'play through'. 1973 A. MacVicar *Painted Doll Affair* viii. 89 The strangers came and putted... Duncan told them we were in no hurry and suggested they should play through. 1975 *Times* 29 Aug. 6/4 In breach of the game's etiquette, one fourball, finding itself behind the other at the second hole, attempted without invitation or request to 'play through' the slower group.

g. To co-operate, comply, agree; to do what is required of one; freq. in negative contexts. *colloq.*

1937 M. Allingham *Case of Late Pig* viii. 59 'Mr. Whippet,' she began breathlessly, 'he's gone! The body's gone!' What shall we do?'.. I was glad to see she wasn't playing, either. 'The body's gone,' she repeated. 1940 J. Reith *Diary* 17 Jan. (1975) v. 238 To see Attlee... Went over past troubles between his party and the Ministry of Information... I think I can make him play but of course he is weak. 1947 'N. Blake' *Minute for Murder* x. 223 Charles comes here to fetch Alice. He tells her Nita won't play. They decide to put their plan into operation. 1958 'A. Bridge' *Portuguese Escape* ix. 146 Tell me what's happened? Did the Duque play? 1961 E. Waugh *Unconditional Surrender* III. i. 218 The Air Force aren't playing until they know what's going on over there. 1967 'F. Clifford' *All Men are Lonely Now* I. ii. 30 'I've had another word with the Minister.' 'Will he play?' 'He's promised to do everything he can.' 1973 'M. Innes' *Appleby's Answer* III. xi. 105 Miss Pringle didn't want to play. She choked off her friend.

18. e. *to play favourites*: to show favouritism. *colloq.* (orig. *U.S.*).

1902 H. L. Wilson *Spenders* 201, I mustn't 'play favourites', as those slangy nephews of mine put it. 1905 R. E. Beach *Pardners* i. 31 Not wishing to play any favourites, I'd picked up a basket of tomatoes, a gunny-sack of pineapples, and a peck of green plums. 1973 *Black Panther* 7 July 7/2 The foreman plays favorites and only likes Blacks that act the way they want Blacks to act. 1974 'S. Woods' *Done to Death* 132, I decline to think that Lizzie—sorry, I'm not allowed to play favourites, am I?

f. *to play for safety, to play safe*: to act in such a way as to avoid risks (for orig. use in Billiards see SAFETY 1 g in Dict. and Suppl.); *to play for time*: to try to gain more time for oneself; to postpone an action or decision.

1906 Kipling *Puck of Pook's Hill* 212 The habit of playing for time sticks to a man! 1911 *Conc. Oxf. Dict.* 750/2 *Play for safety*, avoid risks in game or fig[uratively]. 1911 H. B. Wright *Winning of Barbara Worth* xxviii. 395 Greenfield is playing for time so that the strikers will make trouble. 1919 F. Hurst *Humoresque* 54 'Oh, anybody that plays as safe as you—' He raised his voice, shoving back his chair. 1919 R. W. Lardner *Real Dope* iv. 105 Its best to play safe..and see what comes off. 1930 *Engineering* 11 July 56/3 Consequently in 'playing for safety' in getting the casting through the machine shop the foundryman has tended towards using softer materials which give open and sometimes porous structures in the heavier sections. 1931 W. R. Inge *More Lay Thoughts* 85 A young man, we will suppose, is rather deficient in natural sympathy and has no expensive tastes. He is also of an anxious temperament and disposed to play for safety. 1942 M. B. Lowndes *Let.* 2 Apr. (1971) 229 He said he had come to the *Times* office last week hoping they would make a row, but of course they played for safety. 1944 'G. Graham' *Earth & High Heaven* (1945) 134 All she could do was to go on playing for time, trying to keep Marc from finding out what her family really thought of him, until, after a while, they thought a little better. *Ibid.* 268 The people who play safe don't change anything, they just sit tight and wait for someone else to change it. 1950 E. H. Gombrich *Story of Art* xxvii. 421 No artist can always 'play safe'. 1966 G. N. Leech *Eng. in Advertising* ix. 88 One of their functions is a non-communicative one—that of 'playing for time' under the pressure of extempore performance. 1975 S. Lauder *Killing Time on Corvo* iv. 41 'What are they doing out there?.. Playing for time?' I was inclined to say that they were playing for the gallery. 1976 *Southern Even. Echo* (Southampton) 11 Nov. 16/4 Mrs. Phillis Babey thought she was playing safe when she telephoned a hospital before leaving home to make sure there was a bed waiting for her.

g. Used with impersonal *it* as object, together with an adj., adv., or advb. phr., to denote a particular manner of behaviour; to deal with (something) in a specified way; esp. *to play it close to one's chest*: see *CHEST *sb.*[1] 9 c; *to play it cool*: to behave in a relaxed or unemotional manner (see *COOL *a.* 4 e); *to play it low (down)*: to behave meanly or despicably; *to play it safe* = *to play for safety* (see sense 18 f above).

1873 *Winfield* (Kansas) *Courier* 24 July 3/1 The horses attached to [the] hack which runs between this place and Wellington, one day last week concluded to 'play it alone'. 1882 [see sense 18 d in Dict.]. 1901 Conrad & Hueffer *Inheritors* i. 7 'Oh, come', I expostulated, 'this is playing it rather low down. You walk a convalescent out of breath and then propound riddles to him.' 1919 R. W. Lardner *Real Dope* iv. 117, I thought I would show them to Capt. Seeley and play it safe. *a* 1921 G. H. Gibson in *Penguin Bk. Austral. Ballads* (1964) 207 It's playin' it low on William, but perhaps he'll buckle-to. 1941 F. & R. Lockridge *Murder out of Turn* vi. 72 It's worth playing it that way until we find out different. 1951 *Manch. Guardian Weekly* 17 July 15 The Republicans are playing it safe. 1955 W. C. Gault *Ring around Rosa* vi. 77 Most gamblers I've met would play it cooler than that. 1960 L. Cooper *Accomplices* II. i. 73 Edwardes tried to play it just a bit too clever and that's what did him in. 1960 *Encounter* Nov. 30/2 My concern is that young people to-day, by 'playing it cool' and fearing to be thought 'squares', may create a style of life, not only in work but in every dimension of existence, which is less full, less committed, less complex, and less meaningful than mid-century opportunities allow. 1963 J. Prescot *Case for Hearing* x. 163 Let's wait until he's gone too far to draw back, and then we can produce our evidence and shoot him down in flames. That's how I'd like to play it. 1971 C. Bonington *Annapurna South Face* ix. 108 John Edwards dived for cover, but Jonathan Lane, the cameraman, played it cool, pausing to switch on the camera before getting out of the way. 1972 D. Craig *Double Take* xii. 149 Everyone knows we've got to play it your way, Mick. 1973 'D. Jordan' *Nile Green* i. 11, I let him play it his way. He was my boss. 1977 *Time* 10 Oct. 17/3 You have to follow your hunch. You can't play it safe.

20. b. (Earlier examples.)

1751 in H. T. Waghorn *Cricket Scores* (1899) 49 The Earl of Sandwich playes..eleven gentlemen of Eton College against any other eleven gentlemen in England which the Earl of March shall chuse. 1846 W. Denison *Cricket* 65 He has..long been played alone for his batting.

21. d. To bet or gamble at or on (races, cards, etc.); to take chances with. *colloq.* (orig. *U.S.*).

1858 D. C. Peters *Life of Kit Carson* 354 He'd bin playin' the papers (meaning gambling) and had lost everything. 1902 G. H. Lorimer *Lett. Self-Made Merchant* 115 When he chooses a father-in-law who plays the bucket shops, he needn't be surprised if his own son plays the races. 1925 E. Wallace *King by Night* vi. 21 We never say 'played the races' here; we say 'go racing'. 1932 Wodehouse *Hot Water* i. 25, I was a rich man myself at the time of our wedding. But unfortunately I played the Market. 1958 Blesh & Janis *They all played Ragtime* iii. 61 With ten to twenty a night in tips, a piano-player had more than he could spend so long as he didn't gamble or play the ponies. 1973 'R. MacLeod' *Burial in Portugal* iv. 73 He plays the stock market.

22. b. Also *transf.*, said of a 'hand', in reference to its effect upon the game.

1964 N. Squire *Bidding at Bridge* ii. 23 The hand may play better in either Spades or no-trumps. 1977 *Homes & Gardens* Feb. 17 Work out how the above hand would play opposite this typical Three Diamond opening.

c. (Earlier examples.)

1756 *Gentl. Mag.* XXVI. 489/1 From the Parthian steed, Not more unerring flew the barbed reed Than rolls the ball, with vary'd vigour play'd. 1816 W. Lambert *Instr. & Rules Cricket* 33 The Striker should move his right-foot back at the moment of hitting, playing the Ball between his left-leg and the wicket.

d. (Earlier and later examples.)

1858 *Bell's Life* 26 Sept. 7/4 Mr M'Dougall and Grundy caused a total of 20, when the latter 'played on'. 1963 *Times* 7 Feb. 3/3 Sir Donald Bradman hit one straight drive for four before playing on to Statham. 1977 *Times* 2 Dec. 10/5 Another [wicket] to the left arm spin bowler Iqbal Qasim when Willis played on, completed England's misery.

23. *to play oneself in* (further examples); also *fig.*

1900 W. J. Ford *Cricketer on Cricket* xii. 144 If he would only play himself in quietly..he would get 'lashings' of runs. 1928 A. Philips *Boy at Bank* I. i. 13 The cricket was slow to begin with; while the batsmen 'played themselves in' carefully. 1969 'J. Fraser' *Cock-pit of Roses* vii. 132 'Of course, the first day's always difficult.' 'Question of playing yourself in?' 1971 D. Ayerst *Guardian* xxx. 461 He was tied to the Manchester office and given little opportunity to play himself in as a public figure. 1974 A. Laski *Night Music* 122 'We'll start with the Mozart, play ourselves in.'.. He took the violin out of its case. 1975 *Times* 25 Aug. 9/3 He..went in in the second innings with no time to play himself in.

24. a. (Further examples.) Also, to fool, swindle; *to play* (someone) *for a sucker*: to treat (a person) as a dupe; to make a fool of; to cheat. Cf. sense 6 e in Dict.

1879 'Mark Twain' *Let.* 12 Nov. (1917) I. 369 You could have played him on a stranger for an effigy. 1886 *Lantern* (New Orleans) 20 Oct. 3/2 Some blokes can never see when they are being played for suckers. 1892 Kipling *Many Inventions* (1893) 168 We've played 'em for suckers so often that when it comes to the golden truth—I'd like to try this on a London paper. 1901 Conrad & Hueffer *Inheritors* vi. 95 It seemed to me that she was playing me with all this nonsense—as if she..were fooling me to the top of her bent. 1931 E. Linklater *Juan in America* II. xv. 167, I told him what would happen if he tried to play me for a sucker. 1938 *New Statesman* 8 Jan. 39/2 The 'steamer' (the victim) after being 'steered' (picked up) by one performer and 'played' (told the tale) by another, [etc.]. 1941 A. Christie *Evil under Sun* viii. 146 Crazy about the woman, idealising her, suddenly finding out he'd been played for a sucker. 1959 T. S. Eliot *Elder Statesman* I. 27 Stay out of politics, and play both parties: What you don't get from one you may get from the other. 1966 R. Stout *Death of Doxy* ii. 14 If the errand I had tackled for Orrie had been on the level, if he hadn't been playing me,..there would be fur flying soon.

1967 *New Yorker* 18 Mar. 50 Wise up. They're playing you for a bunch of saps! 1973 'D. Jordan' *Nile Green* xxxiv. 166 She's a fraud... She's working for the Russians... She's played me for a sucker.

b. For 'almost always' read 'frequently', and add further examples. Also in phr. *to play both ends against the middle*.

1938 E. Waugh *Scoop* II. iv. 211 The President kept his end up pretty well—played one company off against the other for months. 1950 O. Nash *Family Reunion* (1951) 46 The wise child handles father and mother By playing one against the other. 1965 *Listener* 10 June 852/1 Their deep African fear of a relapse into subordination makes them play off Eastern and Western contributors. 1972 T. P. McMahon *Issue of Bishop's Blood* v. 62 He would be the first member of the FBI who played both ends against the middle to get a personal belief buttressed. 1974 J. Stubbs *Painted Face* xiv. 192 Natalie..played one against the other for a few days, and reconciled them the following weekend. 1978 M. Puzo *Fools Die* xxv. 285 He was trying to play both ends against the middle, doing his friend the favor and yet trying to warn the reader off the book with an ambiguous quote.

e. *to play the field*: see *FIELD *sb.* 10 d.

V. 26. a. (Further examples without const.)

1816 Jane Austen *Emma* II. vi. 106 'Did you ever hear the young lady..play?'.. 'She plays charmingly.' 1907 G. B. Shaw *Major Barbara* I. 207 *Undershaft.* Do you play, Barbara? *Barbara.* Only the tambourine. 1920 D. H. Lawrence *Lost Girl* 17 She even taught heavy-handed but dauntless colliers, who were seized with passion to 'play'. *a* 1953 E. O'Neill *Long Day's Journey* (1956) III. 89, I couldn't play with such crippled fingers, even if I wanted to. *Ibid.* IV. 151, I play so badly now. I'm all out of practice. 1974 *Encycl. Brit. Macropædia* X. 1035/1 In March 1831 he [*sc.* Liszt] heard Paganini play for the first time.

27. (Further examples.)

1879 Grove *Dict. Mus.* I. 701/1 He [*sc.* Johann Michael Haydn] played the violin and organ. 1925 F. Scott Fitzgerald *Great Gatsby* v. 114 'Klipspringer plays the piano,' said Gatsby, cutting him off. 'Don't you, Ewing, old sport?' 1946 E. O'Neill *Iceman Cometh* II. 141 She was beautiful and she played the piano beautifully and she had a beautiful voice. 1959 'E. McBain' *'Til Death* xii. 162 My kid sister plays piano. 1973 *Publishers Weekly* 14 May 39/2 Cooke would be shown wandering around historical sites—playing piano in a former brothel in New Orleans, for instance. 1976 S. Brett *So much Blood* v. 69 He plays guitar too?

28. b. To cause (a gramophone record or a tape) to reproduce what is recorded on it; *to play back*, to play (a recording) after having made it; also *fig.*

1903 *Talking Machine News* Oct. 103/2 Each machine should play three records. 1907 [see *gramophone needle* s.v. *GRAMOPHONE 2]. 1932 *Times Educ. Suppl.* 1 Oct. 372/4 The record was 'played back' to him, and an expression of amazement dawned on his face. 1934 *B.B.C. Year-Bk.* 419 The ability to play-back a wax before processing is of great assistance in making records of running commentaries. 1939 *Electronics & Television* XII. 172/1 Automatic record changers..enable records to be played for almost three-quarters of an hour without attention. 1956 R. E. B. Hickman *Magn. Recording Handbk.* v. 124 A tape which has been in storage for some length of time should be re-spooled a short while before it is due to be played. 1957 *Technology* June 132/3 Magnetic tape is fed to a control unit associated with the machine tool. When played back the servomechanisms..carry out the demanded movements. 1958 *Listener* 4 Dec. 921/1 Having read what history books have to say about this person..he can play back as much of it as suits him as *The Confessions of*—for example—*Judas Maccabaeus.* 1962 A. Nisbett *Technique Sound Studio* i. 19 Even with a single microphone, it is still possible to make direct comparison tests by recording short snatches of the various music balances and playing them back. 1962 G. Lawton *John Wesley's English* 199 Many a time Wesley plays back to his readers..the common observations of familiar speech. 1973 'H. Howard' *Highway to Murder* xiii. 150 I've said no already. If you like I'll put it on tape and you can play it back to yourself. 1974 [see *PLAY *sb.* 16 b]. 1978 S. Brill *Teamsters* xii. 290 Barkett paused, as if to play back what he had just said.

c. *intr.* Of a gramophone record or a tape: to reproduce sound (esp. for a specified period).

1903 *Talking Machine News* Aug. 66/1 Most phonos finish the records almost as soon as one begins to enjoy them, but yours plays quite a long time. *Ibid.* Dec. 150/1 A record will play..without being screwed down. 1952 Godfrey & Amos *Sound Recording & Reproduction* vi. 163 When running at full speed, a reel of tape which plays for 21 minutes can be rewound in about 2 minutes. 1966 'R. Garioch' *Sel. Poems* 28 What a time a reel of tape can play!

d. *to play the piano* (see quot. 1933). *Austral.* and *N.Z. slang.*

1933 L. G. D. Acland in *Press* (Christchurch, N.Z.) 18 Nov. 15/7 *Play the piano*, to run the fingers over the sheeps' backs in order to find the softest and easiest to shear. 1966 Baker *Austral. Lang.* (ed. 2) iii. 55 An old hand at shearing can spot such a defect in a moment by what is known as *playing the piano*.

29. (Earlier and further examples.) Also, to pass (time) in playing.

1674 Head & Kirkman *English Rogue* III. xi. 136 Mine Host..causing them [*sc.* the 'fidlers'] to cease their playing..said..If you have played away my Guests, you shall pay their reckoning. 1823 *Spirit of Public Jrnls.* (1825) 354 Handel being once in a country church, asked the organist to permit him to play the people out. 1884 J. Hatton *Henry Irving's Impressions of America* I. III. 94 It is customary in American theatres for the orchestra to play the audience out as well as in. 1902 A. Machray *Night Side London* xiii. 196 When you go upstairs, you find more members up here playing the wee sma' 'oors away.

VI. 31. a. Also *intr.*, to be performed; to take a specified time to be performed.

1602, 1809 [in Dict.]. 1869 *Punch* 9 Jan. 10/2 Mr. Burnand's new Burlesque, now playing at the Haymarket, is called *The Frightful Hair.* 1929 *Radio Times* 8 Nov. 388/2 *Typhoon* plays for about an hour. 1935 E. Waugh *Edmund Campion* ii. 75 In 1577 a tragedy of his..was produced..before..the widow of Charles IX of France; it played for six hours. 1958 *Spectator* 31 Jan. 135/1 The new symphony plays for an hour. 1972 *New Yorker* 8 Apr. 32/2 Mr. Zeffirelli watched the action from the back of the auditorium, and he told us that, except for a few small details, the scene was playing well.

c. To perform a play or the like in (a specified town, theatre, etc.); to appear as a performer or entertainer at (a particular place). orig. *U.S.*

1896 *N.Y. Dramatic News* 29 Aug. 11/3 A troup of barnstormers..are playing the smaller towns in this vicinity. 1933 P. Godfrey *Back-Stage* xvi. 206 He writes for lodgings to the next town he is playing. 1936 N. Coward *'Red Peppers'* in *To-night at 8.30* I. 103 'I'll see you don't play this date any more.'.. 'I'd sooner play Ryde Pier in November.' 1959 *Manch. Guardian* 26 Feb. 8/7 Sir John Gielgud is back in London after a tour of Canada and the United States... He played sixty towns and gave 81 performances. 1965 *Listener* 18 Nov. 801/1 The trouble with Freud and his theory of economy of psychic endeavour is that Freud never played Glasgow Empire second house on a Friday night, and I have. 1973 *Times* 27 Jan. 11/8 It's my greatest dream, to play the Palladium. 1975 *New Yorker* 26 May 30/3 The Bolshoi Ballet, now visiting New York en masse for the first time in nine years, is a younger and considerably more experimental version of the company that played the old Met back in 1966.

32. b. (Further examples.) Also, to become worn out or extremely weak.

1872 *Rep. Vermont Board Agric.* I. 79 The old native fruit of our country is about playing out, as the saying is. 1924 R. J. Flaherty *My Eskimo Friends* III. ii. 93 The dogs almost played out before we reached the crest. 1964 Mrs. L. B. Johnson *White House Diary* 15 July (1970) 178 A little past one my enthusiasm played out and I put my head in the pillow.

36. b. (Further examples.)

1907 J. H. Elder-Duncan *House Beautiful & Useful* ii. 18 Many of our leading architects and decorative artists..'play up to', or subordinate everything to one feature in a room. 1927 Chesterton *Secret Fr. Brown* i. 40 There was something downright creepy about that little goblin with the yellow hair, that seemed to play up to the impression. 1929 Kipling *Limits & Renewals* (1932) 358 His mother did social small-talk without daring to stop, and Wilkie played up to her. 1972 J. Mosedale *Football* x. 142 While the pros proved that wasn't literally true, Trippi played up to the spirit of the comment.

c. *to play down to*: to lower one's standard, quality, etc., to suit the tastes or demands of (one's public); to bring oneself down to (a low standard, level, etc.).

1889 G. B. Shaw *London Music in 1888–9* (1937) 234 When a theatre has been playing down as nearly as possible to the music-hall level. 1906 Beerbohm *Around Theatres* (1924) II. 215 No dramatist, moreover, ever yet achieved popularity by deliberately 'playing down to' the public. 1930 *Cambridge Daily News* 24 Sept. 8/1 Let us avoid playing down to the public, lest it ask us for a better article than we can provide. *a* 1936 Kipling *Something of Myself* (1937) viii. 218 Never play down to your public.

d. *to play* (a person) *off the stage*: to act much better than (another actor); to dominate the stage at the expense of (another person). Also *fig.*

1895 G. B. Shaw *Let.* 9 Mar. (1965) I. 494 Our actor managers have a not unnatural reluctance to be played off their own stages by their leading ladies. 1905 Beerbohm *Around Theatres* (1924) II. 144 He played all the other people off the stage, figuratively. Literally, they remained there, I regret to say. 1920 G. B. Shaw *Let.* 22 Dec. in *B. Shaw & Mrs. Campbell* (1952) 216 You played Hackett off the stage, and made only a few blunders. 1979 P. Mason *Skinner* xi. 78 Perron was rather surly, a peasant..who is being played off the stage by a man with style.

e. *to play for laughs* (or *a laugh*): to try to arouse laughter in one's audience; also *trans.*, to depict or use (something) with the aim of arousing laughter.

In quot. 1905 used as *attrib.* phr. (without *to*).

1905 G. B. Shaw *Let.* 2 Oct. (1972) II. 565 The sooner we get John Bull off, the better... An abominable, coarse, play-for-laughs, third class suburban performance. 1906 Beerbohm *Around Theatres* (1924) II. 256 Mr. Shaw was not merely 'playing for a laugh'. He was trying to reproduce a thing that exists in life. 1963 *Listener* 14 Mar. 468/1 Joan Littlewood sensibly lets this plot look after itself. Her concern is to play for laughs. 1965 *New Statesman* 9 Apr. 580/3 Mr Donleavy..plays the genre for sad laughs.

VII. With adverbs.

37. play along. a. *intr.* = sense 17 g above; also, to pretend to agree or co-operate. Freq. const. *with.* *colloq.* (orig. *U.S.*).

1929 D. Hammett *Red Harvest* xi. 112 If the dick would play along, the hole in Tim's head from his own gun.. would smooth everything over pretty. 1935 S. Lewis *It can't happen Here* xix. 214 All we desire is for you to play along with us in your paper. 1947 J. Steinbeck *Wayward Bus* 45 There were only two things for Ernest to do—to laugh at her or play along. 1959 B. Kops *Hamlet of Stepney Green* II. ii. 50, I have no choice. I'll have to play along with him. 1965 *New Statesman* 23 Apr. 638/1 The Labour Party should stand no nonsense from the House of Lords. Although the Tory leadership there is still playing along, defeats of government business inflicted by gangs of Tory backwoodsmen could amount to a deliberate policy of obstruction. 1974 M. Birmingham *You can help Me* iv. 102 She seemed a little surprised at our enthusiasm for literature... But she was ready to play along with us.

b. *trans.* To deceive or tease (a person); to 'string *along*'. *colloq.*

1965 D. Francis *Odds Against* ii. 23, I smiled at him, and he guessed that I'd been playing him along. 1974 'J. le Carré' *Tinker, Tailor* vi. 51 'Wait till Percy sees that,' I tell her—playing her along, like.

38. play down. To minimize; to try to make (something) appear smaller or less important than it really is; to make little of.

1930 *New Statesman* 27 Dec. 351 They accused the Washington departments of being in league with the large employers to 'play down' the number of the unemployed and so encourage the too-ready optimism which continues to assert that prosperity is, once again, just round the corner. 1934 J. O'Hara *Appointment in Samarra* x. 295, I heard the boss tell you to play down the story. 1956 E. M. Forster *Marianne Thornton* 29 Personal immortality today may not be denied by orthodoxy but it is played down, it is felt to be self-centred and anti-social. 1958 *Listener* 18 Sept. 428/2 This impression is much diluted in the Arts Council's exhibition, a timid selection which tends to play down the more extreme and remarkable developments of Bomberg's art. 1973 'E. Ferrars' *Foot in Grave* x. 186 She might have..given the pair an exaggerated idea of their importance. She had been sure that Henry had been right to play the incident down. 1977 *Sunday Tel.* 4 Dec. 3/6 He accuses church leaders of playing down or disregarding these views and making those who hold them feel 'guilty and almost unchristian'.

39. play up. a. *intr.* To behave in a boisterous, unruly, or troublesome manner; to misbehave; *spec.* of a horse: to jump or frisk about. orig. *dial.*

1803 G. Colman *John Bull* II. iii. 23 (*Voices behind.*) *Bur.* They are playing up old Harry below; I'll run and see what's the matter. 1866 J. E. Brogden *Provincial Words in Lincolnshire* 151 He came home beery, and playing-up, broke the dolly. 1877 E. Peacock *Gloss. Words Manley & Corringham, Lincolnshire* 195/1 They're still enif when ther faather's at hoam, but they do play up when they're to their sens. 1886 R. E. G. Cole *Gloss. S.-W. Lincolnshire* 112 This pony does not play up at the trams as the other did. 1888 'R. Boldrewood' *Robbery under Arms* II. iii. 42 He could do more with a horse than any man I ever saw. They never seemed to play up with him. 1909 J. Swire *Anglo-French Horsemanship* 25 The secret of remaining on a horse when he 'plays up' is to drop the hands, press the heels down, [etc.]. 1931 L. A. G. Strong *Garden* 41 Paddy was always resentful of strangers, and played up with a redoubled vigour if he saw that they were afraid of him. 1968 [see *GET *v.* 62 m]. 1973 K. Giles *File on Death* iv. 108 Cucumber generally played up with the Chief Inspector. 1976 J. Snow *Cricket Rebel* 66 Back in England, before he had time to bid for a place against Australia, his left elbow started playing up and he was ordered to rest.

b. To behave manfully or heroically; to act in a helpful or co-operative manner. Cf. sense 36 b in Dict.

1897 H. Newbolt *Vitaï Lampada* in *Admirals All* 21 Play up! play up! and play the game! 1899 E. Wharton *Greater Inclination* viii. 249, I was in fact the only one of the three who did n't instantly 'play up'; but such virtuosity was inspiring, and by the time Vard had thrown off his coat and dropped into a senatorial pose, I was ready to pitch into my work. 1904 R. Fry *Let.* 9 Jan. (1972) I. 216 It is interesting to find that America is playing up so well and I can quite understand it if B.B. transfers his centre of gravity..to Boston. 1924 G. L. Mallory *Let.* 27 May in E. F. Norton *Fight for Everest: 1924* (1925) II. 236, I look back on tremendous effort and exhaustion... And yet there have been a good many things to set on the other side. The party has played up wonderfully. 1966 B. Kimenye *Kalasanda Revisited* 42 The other members played up nicely by expressing themselves as completely horrified. 1979 D. Gurr *Troika* vii. 42, I had to sound sensible. Adult... To hide the secret voice of the schoolboy yelling from the side lines to play up, play up.

c. *trans.* To make the most of; to emphasize; to exploit or trade upon, esp. in journalism and advertising. orig. *U.S.*

1909 R. Beach *Silver Horde* 106 It is a good newspaper story and I'll play it up. 1926 *Publishers' Weekly* 22 May 1687/1 Let us play up the habits, the appearance, the likes and dislikes, let us sell authors to our public. 1933 E. O'Neill *Ah, Wilderness!* (1934) I. 23 *Richard* (coming forward—seizing on the opportunity to play up his preoccupation..). 1945 [see *EASE *v.* 8 b]. 1961 *Los Angeles Times* 4 Aug. III. 4 The West Berlin crisis is being played up artificially because it is needed by the United States to justify its arms drive. 1973 'D. Jordan' *Nile Green* xi. 49 Guy always plays up the limey accent when he's in the States.

d. To tease, annoy, or irritate (someone); to make sport with; to give trouble to.

1924 Galsworthy *White Monkey* II. iv. 151 Did she choose that he should go away, thinking that she had 'played him up' just out of vanity? 1927 *Daily Express* 10 Dec. 1 The girls thought they had got hold of a soft-hearted fool, and they began to play me up. 1934 L. A. G. Strong *Corporal Tune* 138 His body was frightened of what it had undergone..; and now, having succeeded in making Ignatius aware of it, it played him up, throwing him into something approaching panic. 1964 A. Christie

Caribbean Mystery xxii. 223 That's the sort of thing you feel like when your husband's playing you up and you're terribly fond of him. 1974 J. Mitchell *Death & Bright Water* vi. 55 He wasn't in the mood for throwing, not with his back playing him up like it was. 1977 J. Aiken *Last Movement* xi. 230 They are trying to play me up. They believe that..I lose control. 1979 'M. Hebden' *Death set to Music* xv. 163 His stomach was playing him up again.

playa (pla·yă). orig. *U.S.* [a. Sp. *playa* shore, beach, coast, f. late L. *plagia*: cf. PLAGE[1].]

1. a. A flat silt- or sand-covered area, free of vegetation and usu. salty, that lies at the bottom of a desert basin and after rain becomes a temporary lake (*playa lake*). **b.** A playa lake.

1854 J. R. Bartlett *Pers. Narr. Explorations* I. 246 The playas..seemed to have an extent of twenty-five or thirty miles. 1856 in *Publ. S. Calif. Hist. Soc.* (1928) XIV. 124 We..stopped 1½ hours at the wagon & took breakfast & then pushed on to the playa & went ahead to hunt for water. 1885 I. C. Russell in *Monogr. U.S. Geol. Survey* XI. 10 Other lakes, which indicate still more pointedly the contrast between an arid and a humid climate, we may call playa lakes. These are sheets of shallow water, covering many square miles in winter season, but evaporating to dryness during the summer, their beds becoming hard, smooth mud-plains or playas. 1939 P. G. Worcester *Textbk. Geomorphol.* ix. 246 Ancient playas which have not been covered with water for many years are likely to have quite irregular surfaces. *Ibid.* 247 Two types of sloping plains usually connect the borders of desert basins with the flat central playa plains. These are bajadas and pediments. 1945 J. L. Marshall *Santa Fe* 188 In the *playas*—saucerlike depressions in the desert—were beds of glistening salt and gypsum. 1957 G. E. Hutchinson *Treat. Limnol.* I. i. 6 The lakes of southern Oregon and northern California already mentioned are shallow playas. 1969 Twidale & Foale *Landforms Illustrated* xvi. 110/1 Salts crystallise out and form a distinct layer covering the bed of the playa, which is called a saltpan... All the large playas of South and Western Australia are salty. 1975 McAlester & Hay *Physical Geol.* x. 330 In arid regions ..intermittent playa lakes are formed by infrequent downpours and quickly dry up, leaving an accumulation of evaporites.

2. A beach.

[1855 in *Publ. S. Calif. Hist. Soc.* (1934) XVI. 59 La Playa (the beach) is that part of the city nearest the mouth of the harbor.] 1856 'J. Phoenix' *Phoenixiana* 202 Three other small buildings,..a fence, and a grave-yard, constitute all the 'improvements' that have been made at the 'Playa'. 1857 in *Amer. Speech* (1941) XVI. 265 The following is a list of words and phrases frequently used in English conversation in California, and not unfrequently quoted in California newspapers:..*Playa*, beach. 1924 *Chambers's Jrnl.* Aug. 581/1 They turned eastward.. keeping to the *playas*, or little beaches. 1934 R. Macaulay *Going Abroad* i. 14, I think I shall go down to the playa and bathe. 1964 S. Blanc *Yellow Villa* (1965) 12 The South Seas setting that has made land along the northern *playa* so very expensive. 1966 M. Steen *Looking Glass* viii. 160 Little lights of fishing boats far away down on the *playa* [in Málaga].

3. (See quot. 1972.)

1898 R. T. Hill *Cuba & Porto Rico* v. 48 Occasionally a few acres of *playa*, or low alluvial land, may be found around the harbors, but the rivers are free from wide bottoms, and the land as a whole stands well above the sea. 1972 *Gloss. Geol.* (Amer. Geol. Inst.) 548/1 *Playa*, a flat, alluvial coastland, as distinguished from a beach.

playability. (Later example.)

1977 *Early Music* Apr. 151 It would be a useful piece of work if someone were to collect and analyse them [*sc.* wind instruments depicted in medieval manuscripts] according to types and theoretical playability.

playback (plēi·bæk). Also **play-back.** [f. vbl. phr. *to play back*: *PLAY *v.* 28 b.] **1. a.** The reproduction of a recording, esp. soon after it has been made. Also *fig.* and *attrib.*

1929 *Photoplay* Apr. 110/2 *Play-back*,..the immediate playing of the sound record after the taking of a scene in order that actors and directors may hear how it all sounds... The play-back is only possible from disc recording,..as in film recording the film must be developed before the sound can be heard. 1931 N. H. Slaughter in L. Cowan *Recording Sound for Motion Pictures* iv. 61 Disc recording permits play-backs, often a boon to the director and cast. 1934 *B.B.C. Year-Bk.* 419 For immediate play-backs the Blattner system requires a little time to rewind the tape before running off again. 1940 *N.Y. Times* 19 May IX. 10/2 As future generations listen to play-backs of the uses to which radio was put for propaganda purposes..they, too, may be amazed. 1941 B. Schulberg *What makes Sammy Run?* xi. 282 He felt he had to justify himself. He insisted upon giving me a playback of that historic interview. 1949 *Sun* (Baltimore) 5 Feb. 7 A courtroom playback of the broadcasts..brought out that Nazi propaganda was inserted between the GI's messages. 1949 *Electronic Engin.* XXI. 149/3 The recording and play-back heads are so arranged that two tracks of recording can be stored on the standard tape. 1957 C. MacInnes *City of Spades* II. v. 140 On and on she went, like a playback from a tape recorder. 1957 W. H. Whyte *Organization Man* 133 A young man's idle dream? It is a playback only mildly exaggerated of a vision of the future. 1959 *News Chron.* 25 Aug. 3/5 Visitors to the Radio Show ..will be able to see how TV shows are recorded ready for immediate playback. 1970 N. Armstrong et al. *First on Moon* ix. 203 We've been looking at your systems data on playback and everything is looking good. 1976 *Gramophone* Oct. 532 (Advt.), Pickering's exclusive new design development also makes it superior to other cartridges in the playback of stereo records. 1978 G. McDonald *Fletch's Fortune* x. 65 The playback volume was too high.

b. *Cinemat.* A technique of recording the voice of a singer for the soundtrack of a film as a substitute for that of an actor or actress when songs are called for. *playback singer*, a singer whose voice is so used; also *ellipt.*

1952 R. Spottiswoode *Film & its Techniques* 351 The technique of prescoring and playback with singer and orchestra. 1962 *Times* 26 Jan. (Survey of India) p. xiii/2 The well-loved singers of prewar days were displaced by a strange new tribe known as 'play-backs'. 1963 Barnouw & Krishnaswamy *Indian Film* 164 A Cine Writers Association and a Playback Singers Association had been launched but had ceased activity because of lack of support. 1966 R. P. Jhabvala *Star & Two Girls* in *Experience of India* (1971) 74 A playback singer rendered a love-song with..feeling. 1971 *Sunday Nation* (Nairobi) 11 Apr. 31/4 The best male playback singer trophy.

2. An apparatus for playing recordings.

1930 Pitkin & Marston *Art of Sound Pictures* 271 *Play-back*, a device which repeats the voices, recorded on a wax record. 1936 E. S. Besinger tr. *London's Film Music* IV. 114 The American 'play back' invention has the advantage of less danger, and accuracy more easily attained. 1945 L. A. G. Strong *Othello's Occupation* 44 It's a recording unit, with a play-back and ordinary controls for loud-speaker. 1960 *Aeroplane* 15 July 51 (Advt.), Permanent radar records with the AEI radar recorder and playback.

play-book. Add: **2.** A book of games and pastimes for children.

1694 (*title*) A play-book for children. 1761 A. Barclay (*title*) Tom Thumb's play book. 1886 (*title*) The golden playbook.

3. *Football.* A book containing various strategies and systems of play. *U.S.*

1967 *Time* 6 Jan. 64 On the field, Plimpton did the calisthenics and learned the playbook cold, but when the test came during an intra-squad scrimmage before a large crowd,..every play was botched. 1969 *Sunday Times* 28 Sept. 22 They spend most of their time watching films of their next opponents or studying the 'play-book' which sets out the dozens of moves they have to learn before the next match. 1972 J. Mosedale *Football* v. 57 He absorbed the fundamentals..out of a playbook more than 300 pages long.

playboy (plēi·boi). *colloq.* Also **play-boy.** [f. PLAY *sb.* + BOY *sb.*[1]] A man, esp. a wealthy man, who sets out to enjoy himself; a selfish pleasure-seeker (in quot. 1898 used of the devil). Also as *v. intr.* Hence **pla·yboyish** *a.*, **pla·yboyishness, pla·yboyism.**

Cf. also the obsolete sense s.v. PLAY *sb.* 17.

1829 G. Griffin *Collegians* viii. 161 The pretty Syl repeatedly told him that he was 'a funny gentleman' and 'a great play-boy'. 1898 J. MacManus *Bend of Road* 107 The divil sittin cheek be jowl with him in his own chimbley corner!..an' himself an' the playboy shoughed out o' the same pipe! 1907 J. M. Synge (*title*) The playboy of the western world. *Ibid.* II. 51 You're the walking playboy of the western world. 1926 C. Day Lewis in *Oxford Poetry* 20 Proud Playboy of his own complacency. 1926 *N.Y. Times* 11 Oct. 24/2 The playboy of baseball might have heard his name go ringing down the corridors of baseball as a man who won a series game with a home run. 1933 J. Cary *Amer. Visitor* 226 Jukes used to say all these officials were playboys moved by some impractical notion or other. 1936 M. de la Roche *Whiteoak Harvest* xi. 160 No matter how hard I worked I was looked on as a sort of playboy who couldn't do a man's job. 1939 Joyce *Finnegans Wake* (1964) I. 183 The house..was the worst, it is hoped, even in our western playboyish world for pure mousefarm filth. 1952 E. O'Neill *Moon for Misbegotten* I. 55 He is not the blatantly silly, playboy heir to millions whose antics make newspaper headlines. 1954 N. Coward *Future Indefinite* I. 4 Beneath a glittering veneer of..playboyishness, I had managed..to retain a few normal human instincts. 1959 M. Cumberland *Murmurs in Rue Morgue* xvii. 105 What's the matter with this age?.. Its bitterest insults are to call people playboys, pleasure-lovers, hedonists. 1960 Wodehouse *Jeeves in Offing* vi. 60 A New York playboy, accustomed from his earliest years to pursue blondes like a bloodhound. 1962 *Times* 28 Feb. 5/1 The elder boy, who has gone in for playboyism in a big way. 1963 V. Canning *Limbo Line* xvi. 218 Who wanted money? Amadeo to playboy around? 1973 'D. Jordan' *Nile Green* xx. 76 The Lebanese ladies..lusted after the trim, brown-tanned Beirut playboys. 1976 Botham & Donnelly *Valentino* xi. 86 Wearing tails as the Paris playboy, he was dancing his way into cinema history.

play-day. Add: **2.** *Theatr.* A day on which a play is performed.

1761 in G. C. D. Odell *Annals of N.Y. Stage* (1927) I. iv. 83 Those ladies who would have places kept in the boxes will please to send a sensible servant to the theatre at three o'clock on every play-day. 1888 G. O. Seilhamer *Hist. Amer. Theatre* I. xi. 111 Mr. Allyn's benefit took place on Saturday, instead of the regular play-day.

play-down, playdown (plēi·daun). orig. *Canad.* [f. PLAY *v.* + DOWN *adv.*] (See quot. 1939.)

1939 Webster Add., *Playdown*, one of a series of play-offs, as among the winning teams from different leagues or localities. *Canada.* 1970 *Globe & Mail* (Toronto) 28 Sept. 21/2 Senior hockey in Canada may abandon the Allan Cup playdowns in favor of an annual tournament with six to eight teams participating. 1973 *Courier & Advertiser* (Dundee) 26 Feb. 9/1 The north district play-down in the Scottish Curling championship was concluded at Perth Ice Rink yesterday.

played (plēid), *ppl. a.* [f. PLAY *v.* + -ED[1].] **1.** That has been played.

1833 J. Cairnie *Ess. Curling* 61 Every stone shall be reckoned as played, if the player part with the handle. 1877 *Encycl. Brit.* VI. 713/2 If a played stone rolls over, or stops, on its side or top, it shall be put off the ice. 1892 J. Brown *Man. Bowling* (ed. 2) 76 As soon as the last played bowl stops, the control of the rink is transferred to the other party. 1969 R. Welsh *Beginner's Guide to Curling* xvi. 102 If a curler touches a played stone belonging to his side, he himself will remove it from the ice.

2. a. *played-out*: see PLAY *v.* 32 c.

b. Exhausted, worn out; passé, finished. *U.S. colloq.*

1872 *Republican Rev.* 16 Mar. 2/4 The days of forked sticks for plows is about played. 1883 'Mark Twain' *Life on Mississippi* xliii. 439 That *used* to be, but that's all played now; that is, in this particular town. The Irish got to piling up hacks so, on their funerals, that a funeral left them ragged and hungry for two years afterward; so the priest pitched in and broke it all up. 1897 *Outing* XXIX. 421/2 He's about played.

3. *played-down*: see *PLAY *v.* 38.

1960 G. Charles in J. Pudney *Pick of Today's Short Stories* 45 He would have liked to have dropped a modest, played-down remark or two on his standing in his own community. 1973 J. Wainwright *Touch of Malice* 109 The deliberately played-down tone of his talk.

player[1]. Add: **I. 2. c.** (Earlier and later examples.) Also *transf.*

1806, etc. [see *GENTLEMAN 4 a]. 1976 W. Greatorex *Crossover* 31 Meade..seemed to like having a Player like Calder in among the Gentlemen of the section. 1978 B. Levin in K. Gregory *First Cuckoo* 12 We are all, gentlemen and players alike, engaged in the business..of expressing our views to thousands, or even millions, of people who have not invited us to do so. 1979 L. Meynell *Hooky & Villainous Chauffeur* viii. 106 That's how cricket was run in those days; it was gentlemen and players then.

II. 8*. A record-player.

1948 *Mod. Plastics* Mar. 84 (*heading*) Unique design of portable player, molded in phenolic, matches polystyrene record carrier. 1953 E. T. Canby *Home Music Systems* vi. 91 The greatest reason for using a manual player is to achieve better sound quality. 1963 J. Fowles *Collector* II. 167 G.P. jumped up and turned off the player. 1968 'E. Trevor' *Place for Wicked* i. 3 Alec said if music wasn't good enough to listen to without talking it wasn't worth putting on the player. 1976 J. Drummond *Funeral Urn* xxv. 127 Margot found him..listening to a recording by Led Zeppelin. He switched off the player and beckoned Margot in.

III. 9. player-coach, one who plays a game and also coaches his fellow-players; similarly, **player-manager, -trainer; player-piano,** a piano having a mechanical apparatus by which it can be played automatically.

1948 *Sporting Mirror* 21 May 11/1 Bobby Baxter, former Scottish international and player-coach to Leith Ath., has been appointed team manager to the Edinburgh speedway team. 1961 *Times* 12 May 4/7 The decision of the Toulston Club to engage an Argentine as player-coach. 1972 J. Mosedale *Football* ii. 21 Strictly functional equipment modeled by player-coach George Halas. 1905 Player-manager [in Dict.]. 1951 *Sport* 16–22 Mar. 3/1 He was..inundated with player-manager offers from non-League sides. 1977 *Western Morning News* 30 Aug. 12/4 On the eve of Exeter City's glamour second round Football League Cup-tie player-manager Bobby Saxton anxiously waits this morning for the latest fitness reports to know if he can field his strongest side to challenge the might of Cup holders Aston Villa. [1901 *Everybody's Mag.* Oct. 490/1 In the section devoted to musical instruments one can hear hourly concerts by mechanical piano-players.] 1907 *Strand* Nov. 103 It..is..the most remarkable achievement in player-piano construction. *Ibid.* 105 Melody Stops..distinguish the 'Autopiano' from all.. player-pianos. 1913 [see *music-roll* s.v. *MUSIC *sb.* 12 d]. 1922 S. Grew *Art of Player-Piano* 1 The player-piano, like the pianoforte and the organ, is a musical instrument. 1946 R. Blesh *Shining Trumpets* II. xi. 243 He made an unknown quantity of player-piano rolls. 1973 *Times* 25 Oct. 38/7 (Advt.), Steinway and Sons..are prepared to purchase or take part exchange pianos of their own or other makes except player pianos. 1956 *People* 13 May 13/6, I have had several offers to join clubs in Germany as player-coach or player-trainer.

Player[2] (plēi·ər). Also **Players, Player's.** The proprietary name of a cigarette made by the John Player Company. Also *Comb.*

[1885 *Trade Marks Jrnl.* 9 Dec. 1196 Player's Rough & Ready Mixture... John Player,..Nottingham; manufacturer. 1889 *Ibid.* 20 Mar. 285 Player... Manufactured tobacco, except snuff. The firm trading as John Player,.. Nottingham, tobacco manufacturers.] ?1932 Dylan Thomas *Sel. Lett.* (1966) 6 I've got a large Players, and my shoes are off. 1943 *R.A.F. Jrnl.* Aug. 29 'Why was that, Flight?' the Corporal asked, taking a packet of Players from his pocket and selecting one. The Flight Sergeant reached over and helped himself. 'In my days..corporals couldn't afford expensive fags like these.' 1945 Dylan Thomas *Let.* 28 Aug. (1966) 283, I raise one Player-coloured aspen hand to salute and supplicate. 1960 *Tobacco* June 37 A new Player pack... Player's Medium Navy Cut cigarettes, the big sellers, are now introduced in a new modernised pack. 1977 R. Barnard *Death on the High C's* xiii. 144 He..puffed his way through a third of his Player's. 1978 D. Bloodworth *Crosstalk* vii. 54 She.. lit a Player with a book match.

Playfair (plēi·fēər). The name of Lyon *Playfair*, 1st Lord Playfair (1818–98), British chemist and administrator, used *attrib.* and *absol.* to designate a cipher in which successive pairs of letters are replaced by pairs chosen in a prescribed manner from a matrix of 25 letters, usu. arranged in accordance with a key-word.

1922 J. C. H. Macbeth *Langie's Cryptography* iv. 166 M. Langie has omitted to give any reference to the 'Playfair' cipher, which has been extensively used for military purposes. This cipher is one of the substitution variety, and may be operated with one or more key-words. 1932 D. L. Sayers *Have his Carcase* xxvi. 344 Here's a cipher message. Probably Playfair. 1966 M. R. D. Foot *SOE in France* iv. 105 As simple as a Playfair code based on a single word. 1974 *Encycl. Brit. Micropædia* X. 643/3 He [*sc.* Sir Charles Wheatstone]..invented the Playfair cipher, which is based on substituting different pairs of letters for paired letters in the message. 1979 *Listener* 18 Jan. 131/1, I see that I have to encipher my answers to other [crossword] clues in Playfair without knowing the keyword... I have a friend who could solve two Playfair squares before breakfast.

play-game. Delete † *Obs.* and add later examples.

1972 *Evening Telegram* (St. John's, Newfoundland) 24 June 3/1, I won't be spending such a week as the one just past. All jokes aside, it has been no playgame. 1973 *Shooting Times & Country Mag.* 7 July 21/1 To my way of thinking, anything easy to acquire is never valued for long and, compared to what it was in my younger days, fishing is now almost a playgame.

playgirl (plēi·gɜːl). Also **play girl, play-girl.** *colloq.* [f. PLAY *sb.* + GIRL *sb.*] A woman who sets out to enjoy herself; a good-time girl. Cf. *PLAYBOY.

1934 *Sun* (Baltimore) 9 Oct. 7/3 Colletta..had better be good or she will be sent to a reformatory, her mother.. warned today as the Pittsburgh (Pa.) playgirl arrived in Manila. 1935 *Mademoiselle* May 15 Just trick yourself out in your wildest clothes and turn playgirl for one evening. 1935 A. J. Pollock *Underworld Speaks* 89/1 *Play girl*, a promiscuous female who goes out for a good time. 1939 F. Scott Fitzgerald *Let.* 4 Jan. (1964) 283 She has tendencies toward being a play-girl and has been put on probation. 1942 *Time* 16 Feb. 64/2 Victor Mature ..announced the breakup of his eight-month marriage to Martha Stephenson Kemp, complained that she was a 'playgirl'. 1963 I. Fleming *On H.M. Secret Service* v. 55 The worm of self-destruction..behind the wild, playgirl façade, was eating away..her soul. 1964 *Punch* 8 Jan. 66/3 A rich American playgirl. 1977 *Spare Rib* Jan. 41/3 And more and more they will try to con us into believing that, for playgirls as well as playboys, a taste for porn is a proof of liberation.

playground. Add: (Earlier example. Also, earlier and later examples in extended and fig. use.) Also *attrib.*

1780 A. Young *Tour in Ireland* 104 The school is a building of considerable extent..with..a spacious play-ground walled in. 1871 L. Stephen (*title*) The playground of Europe. 1910 *Westm. Gaz.* 31 Jan. 1/3 The 'playground' England, the England of great country houses, of game preserves, [etc.]. 1916 A. Huxley *Let.* 7 Sept. (1969) 111 They tell me..you are returning to.. England. Or else pacifically to what the Editor of *Truth* would call the *playground of Europe.* 1929 A. E. Ford *My Minnesota* iv. 194 Like hundreds of others we wish our vacation would never end, and we know now why Minnesota has become the 'Playground of the Nation'. 1963 W. Soyinka *Dance of Forests* 2 Ogun, they deify, for his playground is the battle field. 1968 *Listener* 19 Dec. 819/2, I gave midnight nudie bathing parties from my private yacht moored at the millionaires' playground of Grimsby. 1970 G. Greer *Female Eunuch* 271 The most telling playground for feelings of rejection about women is the joke department. 1971 'D. Halliday' *Dolly & Doctor Bird* vii. 89 Great Harbour Cay is an island.. undergoing transformation into a luxurious international playground for tropical sport. 1975 M. Simpson *Chrome Connection* ii. 10 A reminder about..playground duty. 1976 H. Nielsen *Brink of Murder* iii. 26 Simon sent off ten cables to ten European playgrounds where Jack Keith might be holidaying. 1977 *Wandsworth Borough News* 16 Sept. 15/1 Planning Proposals... St. Faith's School, Smardale-road, Wandsworth—erection of playground shelter.

play group (plēi·grūp). orig. *U.S.* [f. PLAY *sb.* 6 + GROUP *sb.* 3 a.] **a.** *Sociol.* A group formed naturally by young children in a neighbourhood for play and companionship.

1909 C. H. Cooley *Social Organization* (1913) iii. 24 The most important spheres of this intimate association and coöperation..are the family, the play-group of children, [etc.]. *Ibid.*, Nor can any one doubt the general prevalence of play-groups among children. 1939 F. J. Brown *Sociology of Childhood* 177 Miss S. Wisletzky was able to distinguish three factors in the formation of play groups. 1947 —— *Educ. Psychol.* x. 220 The first play group is small,..of a transitory character, formed only to carry on a specific activity of the moment. *Ibid.*, The activity itself is the basis for the organization of the play group. 1954 J. A. C. Brown *Social Psychol. of Industry* v. 120 Cooley gave as typical examples of primary groups the family, the play-groups of children, and the neighbourhood group of elders in the village community. 1964 Gould & Kolb *Dict. Social Sci.* 281/1 The dividing line between a clique or a play group and a gang is by no means clear and in practice such a group is whatever the researcher chooses to call it.

b. A group, freq. one organized informally by parents of pre-school children, formed with the object of providing supervised companionship for children while freeing their parents for other activities. Also *attrib.*

1942 C. Landreth *Educ. Young Child* I. i. 17 For this group probably the best immediate solution under 1942 conditions is parents' cooperative backyard play groups. 1962 *Guardian* 6 Nov. 18/3 A pre-school play group which a group of mothers are organising. 1968 D. Lawton *Social Class, Lang. & Educ.* iii. 24 Linguistic handicaps could be alleviated by separating the twins into different play-groups. 1969 *Times* 5 Nov. 13/4 Three years ago there were many Members of Parliament who thought playgroups had something to do with drama. 1973 B. Crowe *Playgroup Movement* v. 79 In some areas..the groups have not been allowed to start unless the playgroup leader is 'suitably qualified'. 1975 H. Jarecki *Playgroups* 10 Vivid pictures of children and their parents in the playgroup setting.

playhouse. Add: **1. b.** † **playhouse pay** (see quot. 1794). *Obs.*

1790 T. Wilkinson *Mem.* I. 146 The theatre being for the first month opened three nights in a week, my salary was only fifteen shillings as play-house pay, and when got to four nights, merely twenty shillings. 1794 C. Mathews *Let.* 19 June in A. Mathews *Mem. Charles Mathews* (1838) I. v. 90 Most of the salaries are what they call 'play-house pay'; that is, payment only each night they play; so that a man engaged at three pounds a-week, if he performs three times a-week only, has only half his salary. 1845 *Bentley's Misc.* June 600 In the year 1728 a first-rate singer, according to play-house pay, which means the actual nights of performance, could command no more than *forty-five pounds* annually.

2. a. (Freq. *play-house.*) A house, usu. small, in or with which children may play.

1792 A. Young *Trav. France* I. 108 These cases of models..have so much the air of children's playhouses, that I would not answer for my little girl..not crying for them. 1857 M. J. Holmes *Meadow-Brook* xxv. 317 At a short distance from the house was a tall cypress..where now was a play-house. 1908 G. Jekyll *Children & Gardens* ii. 11 A good play-house..is a little house somewhere in garden or shrubbery, consisting of a kitchen and a sitting-room. 1916 *Daily Colonist* (Victoria, B.C.) 13 July 8/3 She makes all sorts of toys; animals and dolls, rocking horses and play houses. 1965 *Guardian* 20 Nov. 3/8 (Advt.), Playhouse... 59/6... Takes 4/5 children. 1968 *Sunday Times* 16 June 61 The Peter Murray play house..is big enough to hold several children at once. 1978 *Detroit Free Press* 16 Apr. F11/1 (Advt.), 4 bdrm house,..att gar, play house for kids on 2 acres.

playing, *vbl. sb.* Add: **2.** *playing-life, -place* (later example), *-time.*

1957 *Records & Recording* Nov. 20/1 If it is already tracking properly, a sapphire should have a playing life of about 50 hours. 1852 W. J. Broderip *Leaves Notebk. of Naturalist* 152 On visiting the cedar-brushes of the Liverpool range, he discovered several of these bowers or playing-places. 1577 T. White *Sermon* (1578) sig. C viii *verso*, If it [*sc.* the Theatre] be not suppressed..it will make such a Tragedie, y[t] all London may well mourne..for it is no playing time..but time to pray rather. 1949 Frayne & Wolfe *Elements of Sound Recording* xxix. 600 Magnetic materials can be erased and reused..and..reasonable fidelity can be obtained with extremely long playing time. 1951 *Sport* 16–22 Mar. 2/1 Installation of large playing-time clocks on all major league grounds. 1961 *Jazz Monthly* Mar. 27/2 It doesn't amount to much qualitatively, or quantitatively in 36 minutes playing time. 1966 *Jrnl. Canad. Operational Res. Soc.* 117 *Ratio, playing time*, the ratio of playing time to combat time for an event or series of events.

playing, *ppl. a.* (Examples in *Cards.*)

1899 Somerville & 'Ross' *Some Experiences Irish R.M.* iv. 90 We were in the first game..and I was holding a very nice playing hand. 1959 Reese & Dormer *Bridge Player's Dict.* 166 Playing tricks are tricks that a hand may reasonably be expected to take when playing in its own best trump suit. 1964 *Official Encycl. Bridge* (Amer. Contract Bridge League) 430/2 *Playing trick*, an expected trick if the holder or his partner buys the contract. 1976 'Trevanian' *Main* ii. 24 He has a fair playing hand but no meld to speak of.

play-list (plēi·list). Also **playlist.** [f. PLAY *sb.* + LIST *sb.*[6]] **1.** A list of theatrical plays to be performed.

1962 *Times* 23 Jan. 13/3 The classical play-list consists of *Ruy Blas,..Amphitryon,* and *L'Homme à la Main de Fer.*

2. A shortlist of musical records that may be broadcast by a radio station in a given period. Also *attrib.*

1975 *Listener* 26 June 848/1 Radio 1 severely limits the number of records it plays... By and large you will hear only the few that are put on the mysterious 'play-list' each week. 1976 *New Musical Express* 17 Apr. 17/1 It makes most of the present Radio One playlist shrivel into insignificance. 1977 *Sounds* 1 Jan. 4/1 Stephen Bishop will triumph. Top of the Pops producers and BBC playlist folks be prepared.

play-maker. Restrict 'Now *rare*' to sense in Dict. and add: Also **playmaker. 1.** (Further examples.)

1812 *Dramatic Censor 1811* 182 This may be what our modern playmakers call *light* and *shade.* 1953 *Scrutiny* XIX. 204/3 The Restoration playmakers did not exercise their classical imitation by ordering *any* general issues into *any* pattern.

2. A player in a team game, esp. basketball, who leads an attack, or brings other players on his side into a position to score. Hence **playmaking** *a.* orig. *U.S.*

1942 Berrey & Van den Bark *Amer. Thes. Slang* § 662/2 Hockey... *Playmaker*, a player who does not attempt to score but puts the puck in place for the scorer. 1951 *Sun* (Baltimore) 12 Jan. 14/4 Bill Kelso, Patterson's playmaker at guard, is expected to be back in action tonight after missing the game against Southern. 1961 *Look* 15 Aug. 84 In high school..he was better known as a basketball playmaker than as a pitcher. 1967 *Boston Sunday Herald* 26 Mar. II. 1/5 Jones, the playmaker, turned gunner Saturday night... The Celtics went up 2-0 on New York in their best-of-five first-round Eastern Division playoffs. 1972 *N.Y. Times* 4 June 6/1 Ford, a playmaker on the varsity basketball team, is the son of Doug Ford. 1974 *Anderson* (S. Carolina) *Independent* 23 Apr. 7A/5 Little Ernie DiGregorio, the Buffalo Braves' playmaking guard, and big Ron Behagen, the Kansas City-Omaha Kings' brawny forward, were unanimous choices on the National Basketball Association's All-Rookie team for the 1973–74 season. 1975 *New Yorker* 7 Apr. 94/3 He was just the man the Bucks had wanted—an experienced ball handler and playmaker who would give the team the steadiness it needed in the backcourt. 1979 *Guardian* 26 Apr. 22/4 The Austrians could offer nothing in attack, where their elegant playmaker, Prohaska, continually dwelt on the ball.

pla·y-off, playoff. [f. PLAY *v.* + OFF *adv.*] **a.** An additional game or match played to decide a draw or tie; a replay. **b.** *N. Amer.* A series of games or matches played to decide a championship, etc. Also *transf.* and *fig.* Also as *v. intr.*

1895 *Outing* June 50/2 In the play-off for the championship of the city, the Sodality team won a bitterly contested game. 1906 *Liverpool Even. Express* 9 Mar., The play-off resulted in a win. 1915 *Literary Digest* (N.Y.) 21 Aug. 361/1 The race with the Cubs was a tie at the end of the season and a play-off game was necessary to decide the pennant. 1932 *Sun* (Baltimore) 6 Sept. 14/2 The play-off for the Middle Atlantic League baseball title will start Wednesday. 1939 *Beaver* June 25/2 He had a son with the St. Boniface Seals and..was very interested in the fact that they were in the Dominion Junior Hockey Championship play-offs. 1947 A. P. Gaskell *Big Game* 12 He spoke for a while about the traditions of the [Rugby] club and then about the honour of playing off for the championship. 1959 *Times* 29 May 5/2 P. Gill..won the..first prize.. after a six holes play-off. 1969 *John Edwards Mem. Foundation Quarterly* V. IV. 145 Guthrie Meade is presently studying fiddling contests... Many of the contests involved preliminary play-offs. 1970 G. F. Newman *Sir, You Bastard* vii. 190 He'd use the man in his play-off with Manso. 1973 *Courier & Advertiser* (Dundee) 1 Mar. 13/2 The six rinks who have qualified for the finals of the 1973 Scottish curling championships will play-off, on a league basis, for the right to represent Scotland at the world championship. 1978 *Morecambe Guardian* 14 Mar. 11/6 The third division could have to go to a play-off, depending upon the result of the..match, still to be played.

play-room (plēi·rūm). Also **play room, playroom.** [f. PLAY *sb.* + ROOM *sb.*[1]] A room used for children to play in; a nursery. Also *attrib.*

1819 M. Wilmot *Let.* 31 Oct. (1935) 25 The children's Nursery and play room are lovely. 1838 *Knickerbocker* XI. 12 One Saturday afternoon when seated with two or three other children in my little play-room. 1840 H. Cockton *Valentine Vox* xi. 82 He nevertheless contended within himself, that they were games which ought strictly to be confined to the play-room. 1847 J. H. Newman *Let.* 26 Jan. (1962) XII. 26 At Christmas..we find him in the playrooms (cameratas) with the little boys about him, they dressed up as the Magi. 1865 Mrs. Stowe *House & Home Papers* 45 Charlie and Jim..detesting the dingy lonely play-room, used to run the city streets. 1890 O. Wilde *Pict. Dorian Gray* viii, in *Lippincott's Monthly Mag.* July 61 He had used it first as a playroom when he was a child and then as a study. 1895 Kipling *Day's Work* (1898) 365 The two rooms..that had been his nursery and his play-room. 1905 F. H. Burnett *Little Princess* iii. 32 Is it true that you have a play-room all to yourself? 1909 *Westm. Gaz.* 22 Apr. 2/3 We will wander hand in hand, Like a boy and girl in a playroom land. 1927 *Ladies' Home Jrnl.* Dec. 12/1 Jessica's song and the children's noise, every sound in the play room, broke off short. 1932 *New Yorker* 23 July 22/2 My quarters would inevitably be under the play room, where my host's little ones would indulge in matutinal exercises. 1957 J. Masters *Far, Far the Mountain Peak* 105 Her brother..who..stands in a lordly manner in front of the play-room fire. 1966 *New Statesman* 13 May 691/3 There can be no denying that the most obvious method of bettering the circumstances of these working-class mothers is to improve their housing and to raise their standard of living, so that they have gardens or playrooms where the children may run loose. 1973 *Country Life* 14 June Suppl. 1 (Advt.), Georgian house..3 reception rooms..Playroom. Outside studio. 1979 G. St. Aubyn *Edward VII* ii. 114 Alix..doted on her son. All her instincts were to keep him in the playroom, while his wife tried to interest him in serious matters.

play-the-ball. *Rugby League.* Also **play the ball.** [PLAY *v.*] A move restarting play after a tackle, in which the tackled player kicks or heels the ball from the ground, with opponents in specified positions in front of him and a member of his team behind.

1959 *Observer* 13 Sept. 32/2 Some of their habits at the play-the-balls were suspiciously close to being offside. 1974 *Rules of Game* 158/5 The team in possession is allowed three successive play-the-balls (five in Britain and Australia). 1976 *Liverpool Echo* 6 Dec. 17/9 Midway through the period Les Gorley sneaked a smart try from a play-the-ball. **1978** *Times* 20 Nov. 9/2 In the first half Pepons nipped over from a play the ball, and Cronin kicked two penalties.

playtime. Add: **2.** (Earlier examples.) Also, the time during which a play is being performed.

1616 Jonson *Epicœne* iv. ii, in *Wks.* 572 Who will.. take vs to the cock-pit, and kisse our hands all the play-time? **1749** Smollett tr. *Le Sage's Gil Blas* IV. xii. i. 179, I waited impatiently for play-time, that I might go to the theatre.

playwrighting. (Further examples.)

1928 *Publishers' Weekly* 16 June 2445 Francis Brett Young, not content with writing distinguished novels and poems, has turned his attention to playwrighting. **1966** *Punch* 10 Aug. 230/1 Those grand old ample days of playwrighting when plots were clear and motives required no hard brainwork. **1973** E. Bullins *Theme is Blackness* 12 In the area of playwrighting, Ed Bullins, at this moment in time, is almost without peer in America.

plaza. Delete ‖ and add further examples. Also *U.S.*, a public square or open space; in extended uses (orig. and chiefly *N. Amer.*), a large paved area surrounded by or adjacent to buildings, esp. as a feature of a shopping complex. Also *attrib.*

Now usu. with pronunc. (plā·ză).

1844 J. Gregg *Commerce of Prairies* II. 77 Two or three miles above the plaza there is a dam of stone and brush. **1852** *Knickerbocker* XL. 197 The spirit-stirring fife and drum, and the roar of cannon on the plaza [at New Orleans], announce the hour for morning parade. **1856** G. H. Derby *Phoenixiana* 126 Every citizen..was aroused at 2 A.M. by the soul-stirring and tremendous report of the Plaza Artillery. **1884** Sweet & Knox *Through Texas* xxiii. 307 Old Gen. Ignacio Barterra 'cussed' a forty-foot steeple on the old church on the plaza. **1907** S. E. White *Arizona Nights* iii. 47 A freight outfit brought him to Tucson and dumped him down on the plaza. **1948** *Sun* (Baltimore) 20 Nov. 14 (*caption*) All tolls for travel across the Chesapeake Bay Bridge will be paid at booths on the 1000-foot toll plaza on the Western Shore approach to the bridge. **1957** *Times* 2 Dec. 13/1 Shopping..is simplified [in Canada] by the presence of suburban shopping plazas..an enormous parking lot encircled by branches of the down-town stores. **1959** *Ottawa Citizen* 21 July 21/7 Two plazas with parking facilities and offices for Customs and Immigration..are to be built. **1961** L. Mumford *City in Hist.* Note to plate 62, Ossip Zadkine's sculpture..placed on a plaza fronting the inner harbor of Rotterdam. **1966** T. Pynchon *Crying of Lot 49* v. 103 She came..into a plaza teeming with corduroy, denim, bare legs,..students in nose-to-nose dialogue. **1969** *Guardian* 15 July 10/6 There is no place nearer to hell than the all-under-one-roof shopping 'plaza'. **1969** *Parade* (N.Y.) 14 Dec. 18/3 It is not at all uncommon..to see hippies and Indians sitting together in the town's main plaza, sharing a canteen of water or a pack of cigarettes. **1974** *Sci. Amer.* Feb. 99/3 Bending and shear forces are maximum at the plaza (ground floor) level. **1975** *N.Y. Times* 16 Oct. 43/6 Scheduled for completion in one year, it will have 24 tennis courts, 14 tennis 'alleys' for practice, ..and a 13,000-square-foot covered plaza, including an arcade, between 56th and 57th Streets.

‖ **plaza de toros** (plā·þa, plā·za de to·ros). [Sp.: see Plaza.] In a Spanish-speaking country: a bull-ring. Also *fig.*

1846 R. Ford *Gatherings from Spain* xxi. 296 All the world crowds to the *Plaza de toros.* **1910** *Encycl. Brit.* IV. 789/2 Before the introduction of railways there were comparatively few bull-rings (*plazas de toros*) in Spain.... At the present day nearly every larger town and city in Spain has its *plaza de toros.* **1922** J. Hergesheimer *Bright Shawl* (1923) 70 At the Plaza de Toros..she was seated on an upper tier..over the entrance for the bulls. **1934** A. Huxley *Beyond Mexique Bay* 285 A temporary *Plaza de Toros* had been built..a circular fence with a precarious grandstand on the shadier side. **1965** C. D. Eby *Siege of Alcázar* iii. 58 Toledo was converted into a noisy *plaza de toros*, but the matador was never in danger of being gored by his victim. **1973** *Sat. Rev.* (U.S.) 25 Sept. 29/1, I remember clearly the first bullfight I ever saw, in Barcelona. .. The *plaza de toros* was packed.

plazolite (plæ·zoləit). *Min.* [f. Gr. πλάζ-ειν to perplex + -ite¹.] A calcium aluminosilicate that occurs as small, colourless to pale yellow, dodecahedral crystals with a vitreous lustre, and is probably a variety or impure form of hibschite.

1920 W. F. Foshag in *Amer. Mineralogist* V. 183 Included in some material collected by the writer near Riverside, California, were several specimens of vesuvianite associated with small colorless dodecahedrons. The latter proved to be distinct from any known species and the writer proposes the name plazolite from the greek *plazo*, to perplex, in allusion to the difficulty in interpreting its composition. **1941** *Amer. Mineralogist* XXVI. 451 Hibschite and plazolite are very much alike in their properties. Chemically, plazolite differs only by a small and rather variable content of CO_2, and geometrically by the form of its crystals. **1962** W. A. Deer et al. *Rock-Forming Minerals* I. 104 Plazolite originally was thought to contain essential CO_2, but this has been shown to be due to contamination and plazolite must be regarded as being similar to hibschite. **1968** *Mineral. Abstr.* XIX. 139/2 The so-called skarns accompanying some ore deposits in the western region of Honshu Island are revealed to be composed of many veinlets which contain hydrous silicate. The most remarkable..are scawtite, hillebrandite, plazolite, [etc.].

plea, *sb.* Add: **2. c.** Also, *plea-in-bar* (without *special*). (In quot. 1847 *fig.*)

1729 G. Jacob *New Law-Dict.* sig. L 1 *verso*, A Plea in Bar, not giving a full Answer to all the Matter contained in the Plaintiff's Declaration, is not good. *Ibid.*, If the Plea in Bar be to the Action it self, and the Plaintiff is barred by Judgment, &c. it is a Bar for ever in Personal Actions. **1847** Dickens *Dombey* (1848) xxiii. 229 A plea in bar that they would have valuable consideration for their kindness. **1963** *Times* 9 May 17/5 Connelly said yesterday that he wished to withdraw that plea of Not Guilty and to enter a plea-in-bar on the grounds of *autrefois acquit.*

d. *ellipt.* for 'a plea of guilty', *spec.* in *U.S.* slang phr. *to cop a plea*, to plead guilty, usu. as part of a bargain or agreement with the prosecution. Also *transf.* in Black English (see quot. 1970¹).

1927 *Amer. Speech* II. v. 281/1 *Cop a plea*, to tell the truth. **1929** Hostetter & Beesley *It's a Racket!* 222 *Cop a plea*, to plead guilty to a lesser crime than the one originally charged. **1941** J. Smiley *Hash House Lingo* 19 *Cop a plea*, acknowledge a complaint. **1959** Jowitt *Dict. Eng. Law* II. 1350/2 The word 'plea' is used colloquially to mean a plea of guilty. **1963** J. Prescot *Case for Hearing* viii. 123 As for the trial itself, I don't give it more than half a day. It's bound to end up in a plea. **1970** *Daily Tel.* (Colour Suppl.) 6 Mar. 19/2 The majority of accused pleaded guilty—the case then being known to practitioners as a 'plea' as distinct from a 'fight'—and for pleas there was no legal aid. **1970** C. Major *Dict. Afro-Amer. Slang* 41 *Cop a plea*, to be verbally evasive. **1970** J. Cole in A. Chapman *New Black Voices* (1972) 495 The street life style is the cool world... It is here we see the greatest development of stylized talking, sounding,..copping a plea and whupping game. **1972** J. L. Dillard *Black English* i. 4 Characteristic ghetto uses of language like.. 'coppin' a plea', and 'the dozens', are now fairly familiar. **1974** *Telegraph* (Brisbane) 4 June 14/7 Today he did what Americans call copped a plea, in return for pleading guilty to the least serious charge against him, all the other charges of involvement in Watergate and the burglary of Daniel Ellsberg's psychiatrist were dropped.

7. (in sense 2) *plea-roll*; **plea-bargaining** *vbl. sb.* (orig. *U.S.*), a practice whereby a defendant in criminal proceedings agrees to plead guilty to a charge in exchange for the prosecution's cooperation in securing a more lenient sentence or some other mitigation; hence (as a back-formation) **plea bargain** *sb.* and *v. intr.*

1969 *Northwestern Reporter* 2nd Ser. CLXV. 528/1 Court has proper role of discreet inquiry into propriety of settlement whereby defendant as a result of plea bargain agrees to plead guilty to lesser degree of offense than that with which he was charged. **1974** *Harper's Mag.* Jan. 8 The vast majority of criminal sentences in the United States.. are the result of 'plea bargains' in which the defendant 'waives' his constitutional right to trial in exchange for a 'good deal'. **1974** *Newsweek* 28 Jan. 14/2 Jaworski was plea-bargaining with a number of the principals—'dealing up' with reduced charges in return for their testimony against their betters. **1976** *National Observer* (U.S.) 22 May 3/3 That attitude, shared by some other judges here, is yet another reason for the sharp increase in jury trials. So is Connick's refusal to plea bargain with defendants in all but a relative handful of cases. **1978** *Globe & Mail* (Toronto) 2 Feb. 15/1 In a plea bargain, Polanski pleaded guilty Aug. 8 to one count of unlawful sexual intercourse with a minor. Five other counts were dismissed. **1964** *Univ. Pennsylvania Law Rev.* CXII. 865 Some prosecuting attorneys object to the use of the phrase 'plea bargaining'. One prosecutor indicated that 'by labelling the procedure "plea bargaining" you tend to make the procedure sound unethical and improper'. **1967** *Atlantic Reporter* 2nd Ser. CCXXIII. 703/1 Plea bargaining between the prosecution and the defense is a frequently resorted to technique. In exchange for a guilty plea, the prosecutor may agree to recommend a lighter sentence, to accept a plea to a lesser included offense, or to dismiss other pending charges. **1970** *Guardian* 25 Aug. 1/1 The Lord Chief Justice, Lord Parker, yesterday banned 'plea bargaining', where a court agrees to impose a lighter sentence if the accused pleads guilty. **1972** *N.Y. Times* 3 Nov. 18/3 Two-thirds of the addicts now in the program were admitted after being found guilty, usually in plea-bargaining situations in Criminal Court. **1973** *Reader's Digest* Nov. 169/1 They'll tell you that plea bargaining—in which a youngster pleads guilty to a lesser offense in return for a lighter sentence—is possible. **1975** P. Moyes *Black Widower* xii. 146 What about plea-bargaining?.. Suppose you told Martin that the police would only ask for a nominal fine on the streaking charge..if in return Martin would talk to me. **1873** A. C. Ewald *Our Public Records* 37 The Plea Rolls contain the general proceedings in causes, but are very defective, owing to the neglect of attorneys to bring the records in. **1886** [see *issue roll* s.v. *Issue *sb.* 15]. **1936** *Oxoniensia* I. 140 It is impossible to make intelligent use of a plea roll or a pipe roll without fully understanding the governmental machinery that produced it. **1959** Jowitt *Dict. Eng. Law* II. 1567/2 In the old common law practice the steps in every action were entered on a roll, which was called the plea roll, the issue roll, or the judgment roll, according to the stage which the action had reached. **1978** *Bodl. Libr. Rec.* X. 30 Chief clerks of the Court of Common Pleas were responsible for keeping the plea rolls.

pleach, *sb.* Delete *rare*⁻¹ and add later examples; *spec.* a flexible branch or stake or an intertwined arrangement of these, forming a hedge.

1823 E. Moor *Suffolk Words & Phrases* 283 *Pleach*, is described to be a branch of whitethorn brought down and laid horizontally in a fence to thicken a weak part. It is notched (or snotched) at the point of tact with the earth which is loosened to encourage the pleach to strike root. **1920** E. Pound *Umbra* 114 Come buds on bough and spalliard pleach. **1941** [see *Header 5 d]. **1968** J. Arnold *Shell Bk. Country Crafts* xx. 244 Stakes of cleft ash or chestnut..is [*sic*] driven into the ground, to form a rough 'weave' or pleach, depending on the flexibility of the stems. **1976** *Countryman* LXXXI. 1. 56 The hedges have been carefully layered and made stock-proof by the use of horizontal 'pleaches', the wide bottoms providing shelter for a variety of wild life.

plead, *v.* Add later examples of *pa. t.* and *pa. pple.* **pled** (no longer exclusively *Sc.* and *dial.*).

1929 R. S. Lynd *Middletown* ii. x. 122 It is not intended here to take the conventional forms under which divorce cases are pled as anything more than very roughly suggesting. **1932** E. Wilson *Devil take Hindmost* xii. 126 Irma went to his boss and pled with him. **1932** 'J. Aston' *They winter Abroad* xii. 192 The shades of Rugby and Caius, or wherever it was that he had been bred to be a Hawk, pled against him mutely. **1941** E. R. Eddison *Fish Dinner* vii. 103 Should a been unlorded long since,.. but the Vicar pled for him. **1943** S. Lewis *G. Planish* xxxi. 407 I've pled with them. **1955** W. Gaddis *Recognitions* i. v. 198 Is your name really Adeline? he pled. **1974** E. S. Gruen *Last Generation Roman Republic* viii. 327 Gabinius pled that his Egyptian adventure was in the interests of state.

7. c. (Examples with direct speech as object.)

1910 E. M. Forster *Howards End* xxvii. 235 'Don't you worry,' he pleaded. 'I can't bear that. We shall be all right if I get work. If I could only get work—something regular to do.' **1952** M. Laski *Village* xiii. 187 'I'm really sorry,' pleaded Margaret contritely. **1955** [see above]. **1976** B. Freemantle *November Man* viii. 108 'Stop it, Hannah,' he pleaded urgently.

d. Also *ellipt.* in sense 'to plead guilty'.

1959 Jowitt *Dict. Eng. Law* II. 1352/1 The word 'plead' is used colloquially to mean plead guilty. **1963** J. Prescot *Case for Hearing* vii. 109 The reek of spirits..met me like a wave... Dr. Depree depressed his right thumb. 'You'll have to plead to this one.' *Ibid.* viii. 119, I'll stake my pension on a conviction. In fact, if the lad's any sense he'll hold up his hand and plead. **1970** P. Laurie *Scotland Yard* vi. 137 Said he'd plead, then when he got in the box he gave you a grin and said, 'Not guilty'.

pleasant, *a.* (*adv.*) Add: **5.** *pleasant-looking*, *-spoken* (earlier and later examples).

1853 Mrs. Gaskell *Cranford* viii. 153 Lady Glenmire..who had been very pretty in the days of her youth, and who was even yet very pleasant-looking. *c* **1863** T. Taylor in M. R. Booth *Eng. Plays of 19th Cent.* (1969) II. 93 Ah, there's a pleasant looking party yonder. **1978** R. Ludlum *Holcroft Covenant* xiv. 162 He was in his early thirties, Noel guessed, and pleasant-looking. **1843** Dickens *Christmas Carol* 145 He is the pleasantest-spoken gentleman. **1873** 'Mark Twain' & Warner *Gilded Age* xx. 187 Senator Dilworthy was..a pleasant spoken man, a popular man with the people. **1959** T. S. Eliot *Elder Statesman* i. 22 A foreign person By the looks of him. But talks good English. A pleasant-spoken gentleman.

pleasantry. **2.** Restrict † *Obs.* to sense in Dict. and add: **b.** An instance of pleasantness or enjoyment; a pleasurable circumstance.

1790 [in Dict.]. **1925** T. Dreiser *Amer. Tragedy* (1926) I. i. iii. 17 A nerve plasm palpitation, that spoke loudly for all the seemingly material things of life, not for the thin pleasantries of heaven. **1959** *Kentucky Folklore Rec.* V. 118 The consumption of large quantities of watermelons during the day... With all these pleasantries, it is surprising that the afternoon [church] services are well attended.

please, *v.* Add: **3. b.** *please-it-you*: also in *arch.* use as *sb.*

1881 'Mark Twain' *Prince & Pauper* xii. 138 With never a by-your-leave or so-please-it-you, or anything of the sort.

4. b. *pleased to meet you*: a formula used in reply to an introduction. Cf. *Meet *v.* 4.

1914 C. Mackenzie *Sinister Street* II. iii. xv. 802 Doesn't it make you shiver? It's like the 'Pleased to meet you', of Americans and Tootingians. **1916** 'Taffrail' *Pincher Martin* vii. 102 Pleased ter meet yer, miss. **1934** A. Christie *Murder on Orient Express* ii. iv. 97 Mrs. Hubbard murmured: 'Pleased to meet you, I'm sure.' **1955** M. Allingham *Beckoning Lady* v. 68 Pleased to meet you... I thought I'd just step across..and touch my cap, so to speak. **1966** J. Cleary *High Commissioner* iii. 41 'Lady Porthleven, may I present Mr. Malone?' 'Pleased to meet you,' said Malone. 'Oh, really?' Lady Porthleven looked surprised: no one had ever actually *told* her he was pleased to meet her. **1974** A. Price *Other Paths to Glory* iii. vi. 185 'Mr.—Hayhoe?' Mitchell thrust out his hand... 'Pleased to meet you.' Hayhoe nodded easily... There was room neither for deference nor condescension in the greeting.

6. b. (Earlier and later examples of the sarcastic use.)

1816 Jane Austen *Emma* I. xii. 214 South End is prohibited, if you please. **1879** *Cornh. Mag.* XL. 558 He wants to pay his addresses, if you please, to Ursula! **1951** J. Cornish *Provincials* 57 In the winter the heating system was always going on the blink and then the headmistress would scurry round *opening* windows, if you please. **1973** *Math. Teacher* May 479/1 To a monotonous degree, then, each aspect of the operation of the Council is in the hands of the mathematics educators—the *teachers*, if you please. **1979** 'M. Yorke' *Death on Account* xi. 110 He's gone away for the weekend, if you please.

d. *as you please*, in comparative phrases.

1928 'Brent of Bin Bin' *Up Country* xvi. 273 The

native-born maids were as pretty and perky as you please. **1964** Mrs. L. B. Johnson *White House Diary* 23 Jan. (1970) 60 Lynda Bird got up..and said she had just come from the University of Texas where we had the Number One football team..to the house where she could listen to the Number One people of the nation.., just as poised as you please.

pleasure, *sb.* Add: **1. b.** (Earlier and further examples opp. *business.*)

1675 Wycherley *Country-Wife* ii. 32 Go, go, to your business, I say, pleasure, whilst I go to my pleasure, business. **1767** T. Hutchinson *Let.* 30 Sept. (1883) I. v. 243 Pleasure should always give way to business. **1804** M. Edgeworth *Pop. Tales* III. 30 Business was his aversion; pleasure was his business. **1837** C. G. F. Gore *Stokeshill Place* III. vi. 99 'Business before pleasure' is a golden rule which most of us regard as iron. **1853** R. S. Surtees *Handley Cross* xxii. 158 Business first, and pleasure afterwards. **1857** [see Business 13]. **1934** *Law Rep.* 27 Mar. 238 In my judgment, the word 'pleasure' is used in this policy in contradistinction to 'business'. **1941** F. Gruber *Hungry Dog* xv. 183 Pleasure before business. **1943** P. Cheyney *Farewell to Admiral* x. 238, I never believe in mixing business with pleasure. **1976** Hooker & Butterworth *M.A.S.H. goes to San Francisco* (1977) xiii. 170 Oh, how nice! And I think about *you*, too. But business before pleasure, as I always say.

e. *Psychol.* Used *attrib.* (esp. in *pleasure principle*) and as first element with *-pain* to denote the theory that the drives to achieve pleasure and to avoid pain are basic motivating forces in human and animal life; in psychoanalysis, the theory that the tension set up by unpleasure or the desire to achieve a pleasurable result forms the chief source of mental activity and is part of the life instinct, though frequently opposed by the reality-principle. Also in more general use.

1894 Creighton & Titchener tr. *Wundt's Human & Animal Psychol.* xiv. 211 The reference of feeling to a subjective condition of pleasure-pain. **1897** H. G. Wells *Under Knife* in *Plattner Story* 107 It occurred to me that the real meaning of this numbness might be a gradual slipping away from the pleasure-pain guidance of the animal man. **1912** *Amer. Jrnl. Psychol.* XXIII. 134 The sex impulses find no outlet before puberty. Until that time they remain under the control of the subconscious (pleasure principle). **1925** J. Riviere tr. *Freud's Papers on Metapsychology* in *Coll. Papers* IV. 14 It is called the pleasure-pain (*Lust-Unlust*) principle, or more shortly the pleasure-principle. **1951** S. F. Nadel *Found. Social Anthropol.* xi. 306 Emotions, sentiments, even the elementary pleasure-pain reactions, possess their dynamic properties..because they are the concomitants of instinctive tendencies. **1957** N. Frye *Anat. Criticism* ii. 75 In literature..the reality-principle is subordinate to the pleasure-principle. **1968** A. Laski *Keeper* xi. 133 Ralph's whole working life had been devoted to the pleasure principle. **1971** *Listener* 2 Sept. 299/1 Freud responded to the First World War by positing a death instinct beyond the pleasure principle. **1976** *Vogue* Jan. 5/2 The small son of the house witnesses Sigmund Freud looking askance at the pleasure principle embodied in Coney Island.

2. (*It is, was*, etc.,) *my pleasure*: a colloq. dismissal of thanks.

1950 L. Kaufman *Jubel's Children* xxi. 259 Think nothing of it. My pleasure. **1963** [see *Not at all]. **1975** R. Lewis *Double Take* i. 26 'I enjoyed the evening, Mr Hood.' 'It was my pleasure, Miss Stevens.'

5. g. *pleasures of the table*: see *Table *sb.* 6 c.

6. a. *pleasure-brake, -car* (examples), *-carriage, -cart* (earlier example), *-cottage, -craft, -cruise, -cruiser, -cruising, -day, -dome* (later examples), *-economy, -garden* (later examples), *-horse* (later example), *-land, -navy, -park, -party* (earlier example), *-path, -plane, -resort* (earlier example), *-ship, -sleigh, -steamer, -traveller, -trip* (examples), *-vessel, -visit, -voyage* (example), *-yacht.* **b.** *pleasure-hater, -hunter* (earlier example); *pleasure-crazed, -crowded, -mad, -minded* adjs.

1908 *Westm. Gaz.* 12 Aug. 8/3 She was cycling along the Bromley-road when a pleasure-brake..turned out of a side-street. **1833** *Amer. Railroad Jrnl.* II. 481/3 A pleasure car has been flying between this town and the river. **1960** C. Achebe *No longer at Ease* ii. 14 [In Lagos] if you don't want to walk you only have to wave your hand and a pleasure car stops for you. **1802** W. Priest *Trav. U.S.A.* 31 There are 806 two and four wheeled machines entered at the office, and pay duty, as *pleasure carriages.* **1844** Pleasure-carriage [see *Fiesta]. **1789** J. Woodforde *Diary* 27 Nov. (1927) III. 156 My Brother went in my little pleasure Cart with Briton. *a* **1828** D. Wordsworth *Jrnl.* (1941) II. 247 A charming spot for a pleasure-cottage. **1906** Conrad *Mirror of Sea* 38 Their striving for victory..has elevated the sailing of pleasure craft to the dignity of a fine art. **1943** J. S. Huxley *TVA* ix. 63 Nearby is a public lodge and a boat-house and dock for pleasure craft. **1932** *New Yorker* 23 July 12/2 It is not your idea..of a mad night on pleasure-crazed Broadway. **1906** B. von Hutten *What became of Pam* i. x. 73 The time that had seemed so long to her had quite naturally seemed to him, with his pleasure-crowded days, very short. **1926** *Scribner's Mag.* Aug. 12 (Advt.), Start planning now for pleasure-crowded days on cool, blue waters. **1909** *Daily Graphic* 26 July 2/1 (Advt.), P. & O. cheap return tickets pleasure cruises and round the world tours. **1976** H. MacInnes *Agent in Place* xxvi. 277 'Where are we going?' 'For a pleasure cruise.' **1926** *Daily Chron.* 13 May 3/6 (*heading*) Pleasure Cruisers on and off the rocks. **1945** Koestler *Yogi & Commissar* i. iii. 35 For he is a captain of a warship, not of a pleasure-cruiser. **1950** *Oxf. Jun. Encycl.* IX. 382/1 It was not.. until the early 1920's that modern pleasure cruising with its carefully planned itineraries really became established. *a* **1828** D. Wordsworth *Jrnl.* (1941) II. 292 The buoyancy of spirits felt in the earlier part of a pleasure-day's journey. **1957** *Observer* 3 Nov. 19/2 The triumphal renaissance, last Thursday night, of the New Shakespeare Theatre in Liverpool. This lambent pleasure-dome ..was built in the eighties. **1973** G. Beare *Snake on Grave* xvi. 93 It's a kind of floating pleasure dome, they have a restaurant, night-club, casino. **1977** *Time* 19 Dec. 41/1 Macy's.., the basement where women..battled with umbrellas for lingerie markdowns has become one of New York City's great gastronomic pleasure domes. **1910** W. James in *McClure's Mag.* Aug. 467 A permanently successful peace-economy cannot be a simple pleasure-economy. **1961** L. Mumford *City in History* xiii. 379 Such pleasure gardens were popular everywhere that court life was visibly on parade: the famous Tivoli Gardens in Copenhagen still bears witness to this. **1940** Auden *Another Time* 42 As a rule It was the pleasure-haters who became unjust. **1974** *Greenville* (S. Carolina) *News* 23 Apr. 11/3 The pleasure horse class will give spectators a chance to see the finest of the American saddlebred pleasure horses in competition. **1833** J. S. Mill in *Monthly Repos.* VII. 660 Few persons among the crowds of pleasure-hunters have diverged from the beaten track of the Rhine, Switzerland, and Italy. **1927** *Daily Tel.* 13 Sept. 12/2 Thirty years ago Piccadilly had still to establish its claim to be regarded as the centre of pleasure-land. **1925** *Scribner's Mag.* Oct. 373 It was exactly the kind of crowd which a dour philosopher might have described as typical of 'pleasure-mad America'. **1907** *Daily Chron.* 12 Oct. 4/7 Allah forfend, my pleasure-minded love, That aught shall harm thee in the Desert Lands. **1873** 'Vanderdecken' *Yachts & Yachting* xxix. 247 There are not a few sea-lawyers to be met with amongst the pleasure navy Jacks. **1904** R. J. Farrer *Garden of Asia* 70 Here we may fancy known beings resting in this pleasure-park of necessity. **1835** *Southern Lit. Messenger* IV. 303/1 Pleasure-parties to and from the Springs..were dashing along the well graded road. *c* **1806** D. Wordsworth *Jrnl.* (1941) I. 351 It is not easy to see the use of a pleasure-path leading to nothing. **1911** *Chambers's Jrnl.* Jan. 57/1 The aspect of the heavens will be wonderfully changed when the pleasure-plane of the air has arrived. **1883** 'Mark Twain' *Life on Miss.* xli. 427 Modern-style pleasure resorts. **1869** —— *Innoc. Abr.* lvii. 609 When I travel again, I wish to go in a pleasure ship. **1977** *New Scientist* 24 Mar. 707/2 The foundering of a pleasureship. **1774** 'J. H. St. John de Crevecœur' *Sk. 18th-Cent Amer.* (1925) 146 The pleasure-sleigh..can easily carry six persons. **1827** Pleasure-sleigh [see *Fiddle *sb.* 1 b]. **1872** B. Jerrold *London* iv. 43 The river..bright with the trifles of cockleboats and pleasure-steamers. **1948** *Brit. Birds* XLI. 314, I was on a pleasure steamer at the time. **1846** Dickens *Pictures from Italy* 150 Pleasure-travellers through life. **1936** *Discovery* Aug. 247/2 An area that is not well known to the general run of pleasure-travellers. **1833** *Chambers's Edin. Jrnl.* II. 285/2 The individuals who make pleasure-trips along the railway. **1926** *Daily Chron.* 13 May 3/6 The 20,000 ton R.M.S. steamer Otranto struck a rock on the way to the port of Athens... The Otranto is on a pleasure trip. **1906** Conrad *Mirror of Sea* 33 The writer praises that class of pleasure vessels, [*sc.* 52-foot linear raters] and I am willing to endorse his words. **1926** D. H. Lawrence *David* xiv. 103 My lord Jonathan comes too early for a pleasure visit. **1906** Conrad *Mirror of Sea* 39 For racing, a cutter; for a long pleasure voyage, a schooner; for cruising in home waters, the yawl. **1847** Dickens *Dombey* (1848) xxiii. 238 As many spars and bars and bolts..as you'd want an order for on Chatham-yard to build a pleasure-yacht with.

pleasure, *v.* Add: **1. a.** (Further examples.) *spec.* To gratify (someone) sexually; to have sexual intercourse with.

Revived in recent use.

c **1616** R. C. *Times' Whistle* (1871) 90 Silvius doth shew the citty dames brave sights, And they for that doe pleasure him a nightes. **1968** J. R. Ackerley *My Father & Myself* xii. 124 We entered together, quickly unbuttoned and pleasured each other. **1973** *Observer* 29 July 26/7 The rest of the treatment takes place in a hotel bed. Couples are first instructed to 'pleasure' each other by caressing, and not to attempt intercourse until they have learnt to recognise each other's body signals that express delight. **1975** *Times Lit. Suppl.* 9 May 503/1 Her first love, who took her to tea-dances..and pleasured her regularly at home on the brocade couch. **1977** *Observer* (Colour Suppl.) 27 Feb. 17/1 The brown Chippewa girl who was the first female he had ever pleasured.

b. Delete † *Obs.* and add later examples, *spec.* in prec. sense.

1908 C. W. Wallace *Children of Chapel at Blackfriars* ix. 112 Elizabeth intended the establishment of the Children of her Chapel as actors at Blackfriars..to pleasure herself and entertain the Court. **1938** M. K. Rawlings *Yearling* x. 89 'I'll bet we kin ketch us a cattywampus in one o' them ponds.' 'We kin sure pleasure ourselves tryin'.' **1947** *N.Y. Herald Tribune Weekly Bk. Rev.* 2 Mar. 10/3 Mordaunt Fitzmaurice Godolphin..has left Virginia because he pleasured himself with a married lady and then killed her husband in a duel. **1972** *Time* 17 Apr. 66/3 Pauline Tabor was smart enough to open up a house of her own. 'Pauline's' became a Kentucky institution—politicians went to pleasure themselves there.

c. In impersonal construction with *it* as subject (cf. Please *v.* 3).

1937 R. S. Morton *Woman Surgeon* xxxi. 346 A young carpenter said to me, 'It would not pleasure me if I could not see the cypress greening in the spring.' **1949** R. K. Marshall *Little Squire Junior* 249 Little Squire borrowed somethin of mine, and it pleasured me no end. **1951** L. Craig *Singing Hills* xiii. 124 It pleasures us a sight that you would come to see us. **1970** *New Yorker* 12 Sept. 109/3 It pleasured him to see the smoke.

pleasured, *ppl. a.* (Later examples). Also as *pa. pple.* (const. *up*).

1930 D. Runyon in *Collier's* 13 Sept. 8/2 They get all pleasured up over what he has to say. **1968** 'J. Welcome' *Hell is where you find It* iv. 61 He was wearing..the look of a pleasured tom-cat.

plea·suredrome. [f. Pleasure *sb.* + *-drome.] An amusement centre.

1959 *Spectator* 9 Oct. 467/1 The possibility of turning the Isle of Wight into some vast pleasuredrome (cf. Fr. *Baisodrome*). **1966** *Punch* 13 Mar. 392/2 He might bring his millions here and make West Hove the pleasuredrome of the Western world. **1973** *Daily Tel.* (Colour Suppl.) 18 Apr. 19/2 One corporation, which already operated 'pleasuredromes' in Dallas, Atlanta, and St Louis, is.. negotiating to open a park in New Jersey.

pleasure-house. (Earlier and later examples.)

1590 H. Wotton *Let.* in L. P. Smith *Life & Lett. H. Wotton* (1907) I. 247 The plot of his Majesty's pleasure-house shall in convenient time be provided. **1904** R. J. Farrer *Garden of Asia* 106 How few of the many Europeans who visit Japan, ever see the real pleasure-houses of the country! **1908** *Daily Chron.* 12 Dec. 4/6 The ballroom of a notorious eighteenth-century pleasure-house kept by a Mme. Cornelys. **1936** A. W. Clapham *Romanesque Archit.* iii. 55 The actual court of Palermo leaned heavily towards the Moslem element and the pleasure-houses and palaces of the Favara, Menani (Roger II), la Ziza (William I) and la Cuba (William II) were almost purely Moslem both in form and decoration.

pleasure-seeker. (Earlier examples.)

1825 [see Pleasure *sb.* 6]. **1846** *Swell's Night Guide* 14 At all hours of the night, the pleasure-seeker may gain admission.

pleasure-seeking *sb.* and *a.* (Earlier and later examples.)

1886 J. G. Matteson *Hist. Sk.* 63/1, I found the people in Norway far more religiously inclined than those in Denmark; they..are not so much given to pleasure-seeking. **1936** *Discovery* Oct. 301/1 Only a slight ripple on the normal pleasure-seeking surface of Blackpool was caused by the arrival and departure of the British Association.

pleater (plī·təɹ). [f. Pleat *v.* + -er¹.] (See quot. 1921.)

1921 *Dict. Occup. Terms* (1927) § 428 *Pleater*, pleats or folds material in pleats, by hand or by pleating machine. **1970** *Classification of Occupations* (Office of Population Censuses and Surveys) 66/2 Pleater.

pleb. Add: **A.** (Further examples.)

1911 H. G. Wells *New Machiavelli* i. iv. 104 They're Plebs and they know it. They haven't the Guts to get hold of things. **1922** *Dialect Notes* V. 189 At Annapolis, the natives are *crabs*, the freshmen *plebs*, the sophomores *youngsters.* **1928** A. Huxley *Point Counter Point* ix. 138 'A bit of a pleb, wasn't he?' put in the military friend. **1939** Joyce *Finnegans Wake* (1964) i. 175 The pleb was born a Quicklow and sank alowing till he stank out of sight. **1960** *Guardian* 29 Sept. 9/7 It all ends happily, with the squire and the pleb firm friends. **1973** *Nation Rev.* (Melbourne) 31 Aug. 1441/6 The YLA executive sat at a head table while the plebs and proles were strewn together en masse. **1977** J. Wainwright *Nest of Rats* i. xii. 92 You were..an aristocrat... And you turned pleb. Like grandfather.

B. *attrib.* or as *adj.* = Plebeian *a.* b. *colloq.*

1972 *Daily Tel.* 12 Feb. 11/2 Basically the situations, pursuit of boy friends, anxiety about the landlord, are identical, only here the background is pleb rather than deb. **1972** J. Symons *Players & Game* xxvi. 196 'What *was* his name? Barber?' 'No, some other pleb occupation. Taylor?' **1974** *New Statesman* 17 May 698 Orwell darkened the picture: Bowling is frankly pleb, Comstock descends from a line of sexless scrimpers.

Hence **ple·bbie, ple·bby** *a.*

1962 'J. le Carré' *Murder of Quality* i. 9 Mrs. Rode's quite decent..in a plebby sort of way: doyleys and china birds. **1977** J. McClure in *Winter's Crimes* 9 80 Portland Bill..all coach parties and orange peel... It does tend to be a bit plebbie.

plebbish, *a.* Add: Also **plebish.** (Examples.) **plebbishness** (further example).

1928 A. Huxley *Point Counter Point* xxi. 388 This is the sort of thing that really does make me feel rather pleb-ish. **1942** Berrey & Van den Bark *Amer. Thes. Slang* § 147/1 *Ungentility; plebeianism,..plebishness. Ibid.* § 147/5 *Plebeian; commonplace,..plebish.*

plebe. Add: **2.** (Earlier and later examples.) Also *attrib.*

1833 in *Mil. & Naval Mag.* (U.S.) (1834) Oct. 85 My drill master, a young stripling, told me I was not so '*gross*' as most other pleibs, the name of all new cadets. **1834** in *Ibid.* June 281, I was reckoned, already, as one of a class of cadets. To be sure, it was the 'plebe class'; but what of this? **1860** in *Amer. Hist. Rev.* (1928) XXXIII. 601 In most of our tents the cadets and plebes live together, 2 cadets, and 2 plebes to wait on them generally. **1947** *Newsweek* 6 Oct. 78/2 The 'plebe' system which gives upper classmen authority over newcomer midshipmen filled Smith with revulsion. **1948** Menjou & Musselman *It took 9 Tailors* 26 New arrivals are called plebes and a plebe is the dirt beneath an upperclassman's shoes; but to add insult to injury, a plebe has to clean and polish the shoes while he is being stepped on. **1970** N. Armstrong et al. *First on Moon* vii. 156 Buzz ranked number one in his class at the end of his plebe year. **1973** H. Gruppe *Truxton Cipher* iii. 31 Pozo was given to

salting his speech with naval maxims left over from his days as a plebe. **1977** *Time* 19 Sept. 39/3 That summer, it [*sc.* West Point] enrolled its first women plebes—and now has 177 female cadets.

plectonemic (plektonī·mik), *a. Biol.* [f. Gr. πλεκτός twisted + νῆμα thread + -ic.] Of, pertaining to, or designating two or more like helices coiled together side by side in such a way that they cannot be fully separated unless they are unwound. Opp. *Paranemic *a.*

1941 A. H. Sparrow et al. in *Canad. Jrnl. Res.* C. XIX. 325 Kuwada..describes two types of double-stranded spirals: (1) orthospirals, which are formed when the two threads being coiled have one end free so that internal twisting does not occur; (2) anorthospirals, which result when two strands with both ends fixed are coiled together and in consequence have a twist compensating for each gyre of the spiral... Orthospirals are interlocked and cannot be separated without untwisting; anorthospirals are independent and can readily be pulled apart or fitted into each other... The term 'paranemic' (*para* = beside) instead of anorthospiral will be used here since it is simpler and its implications are clear. Instead of orthospiral 'plectonemic' (*plektos* = twisted) will be used as this has the advantage of indicating the relationship of the strands. **1950** [see *Paranemic *a.*]. **1953** Watson & Crick in *Cold Spring Harbor Symp. Quant. Biol.* XVIII. 129/2 We therefore believe that if a helical structure is present [in DNA], the relationship between the helices will be plectonemic. **1971** *Nature* 22 Jan. 241/1 Tropocollagen..consists of three protein chains with similar, specific and very characteristic amino-acid compositions, twisted into a plectonemic triple helix of about 100 turns. **1974** [see *Paranemic *a.*].

Hence **plectone·mically** *adv.*

1953 *Cold Spring Harbor Symp. Quantitative Biol.* XVIII. 128/2 Apart from breaking the chains there are only two sorts of ways to separate two chains coiled plectonemically. **1966** Mahler & Cordes *Biol. Chem.* iv. 140 DNA in its B lattice configuration..consists of two right-handed helical polynucleotide chains of opposite polarity, plectonemically coiled around the same axis to form a double helix. **1979** *Nature* 26 Apr. 780/3 Sasisekharan *et al.*..have..proposed that the two strands of DNA do not coil plectonemically round one another.

plectrum. Add: **3.** *attrib.*, as *plectrum banjo, guitar, lute* (Lute *sb.*¹ 1 a).

1954 *Grove's Dict. Mus.* (ed. 5) I. 401/2 The banjo played with the fingers is referred to as the 'finger-style' banjo, to distinguish it from the instrument (called the 'plectrum banjo') which is played with a plectrum. **1961** A. Birch in A. Baines *Mus. Instrum. through Ages* 182 The standard instrument is the 'finger-style' banjo with five gut or nylon strings tuned *d'*, *b*, *g*, *c g'*, i.e. with the highest string on the left of the bass (and with its peg midway along the neck.) The 'plectrum banjo' omits this half-length string. **1956** I. Mairants in S. Traill *Play that Music* ix. 94 The plectrum guitar with wire strings for dance music and jazz. **1961** A. Birch in A. Baines *Mus. Instrum. through Ages* 182 In the early days of jazz, the banjo was replaced by the '*plectrum guitar*'... The plectrum guitar has metal strings fastened not usually to the bridge..but to a tailpiece screwed to the end of the instrument. **1970** P. Oliver *Savannah Syncopators* 109 *Kambrch*, plectrum lute played widely in the Savannah regions.

pledge, *sb.* Add: **5. b.** (Earlier and later examples.)

1833 *New Engl. Mag.* (Boston) Aug. 137 The Temperance Pledge. *Ibid.* 141 Has he signed the pledge? **1840** *Southern Lit. Messenger* VI. 325/1, I have signed the pledge, and since it is done I will make a virtue of necessity. **1843** in M. Miliband *Observer of 19th Cent.* (1966) 161 Father Mathew..called upon those who wished to take the 'pledge' to kneel down... About 3,000 persons took the pledge... From the appearance of many of them, we should say the total abstinence pledge was very necessary. **1914** G. B. Shaw *Fanny's First Play* iii. 214, I dont want any whisky and soda. I'll take the pledge if you like. **1922** Joyce *Ulysses* 348 Had her father only avoided the clutches of the demon drink, by taking the pledge or those powders the drink habit cured in Pearson's Weekly, she might now be rolling in her carriage, second to none. **1930** G. B. Shaw *Apple Cart* i. 43 Though none of us doubted that he would sign the pledge, we were not equally certain that the infirmities of his nature would allow him to keep it. **1970** J. H. Gray *Boy from Winnipeg* 126 It was only when bootleg beer became openly available in the downtown hotels after 1920 that he gradually slipped from the pledge.

c. *U.S. college slang.* A student who has promised to join a fraternity or sorority. Also *transf.*

1901 *Univ. of Chicago Weekly* 1 Aug. 1087/1 Still if the Kappas are as bad as you say—you say they lifted two pledges last year. **1930** *Randolph Enterprise* (Elkins, W. Virginia) 18 Dec. 1/1 [They]..have been announced as two of five pledges chosen by the University Dramatic club at Morgantown. **1945** W. Maxwell *Folded Leaf* 52 Shortly after seven o'clock the pledges appeared, one at a time, in the hotel lobby. **1949** *Reader's Digest* Aug. 71/1 The chapter might..keep Tom as a sort of permanent pledge. **1972** M. Mead *Blackberry Winter* viii. 98 For one thing, I had no dates; these were all arranged through commands to the freshman pledges of certain fraternities to date the freshman pledges of certain sororities.

7. *pledge-mania*; (sense *5 c) *pledge-master, pin, week*; **pledge card**, (*a*) in Dict. (further example); (*b*) *N. Amer.* a card on which one expresses willingness to contribute to a fund, sponsor a charity event, etc.

1958 *Times Lit. Suppl.* 10 Jan. 15/1 'Pledge' (temperance) cards and 'Decision' (conversion) cards interpolated their small crises. **1967** *Boston Sunday Herald Mag.* 26 Mar. 19/2 In 1960 Msgr. Leonard launched an $800,000 fund drive and more than 400 volunteers distributed pledge cards throughout the parish of about 3,000 families. **1970** *Toronto Daily Star* 24 Sept. 17/5 Pledge cards for the walk are available at any Dominion store while anyone wishing to enter a team in the skatathon can call 889-3967. **1832** Mill *Let.* 17 Sept. in *Wks.* (1963) XII. 121, I should say that the pledge-mania had been abated. **1949** *Sun* (Baltimore) 9 May 1/2 Santarelli..is pledge-master for Phi Theta Upsilon Fraternity at the Northern Illinois College of Optometry. **1944** *Chicago Daily News* 28 Oct. 1 After the incident, Soik turned in his pledge pin. **1949** *Time* 21 Mar. 47/2 As a finale to Brown's pledge week, fraternity men had made the rounds of chapter houses to 'congratulate' each other. **1964** *Amer. Speech* XXXIX. 193 The social affairs that are a major concern for most students, such as..pledge and rush weeks.

pledge, *v.* Add: **3. a.** Also *refl.*

1836 Dickens *Let.* ? 19 Nov. (1965) I. 198 He could not ..pledge himself whether it would appear this season, or whether they would begin with it, at the opening of the next.

b. *trans.* and *intr.* To enrol (a new student) in a college society. Of a student: to undertake to join a college society; to enrol in (a society). *U.S.*

1871 L. H. Bagg *Four Years at Yale* 62 They are very attentive to his wants and do not leave him until he is 'pledged'. **1887** *Lippincott's Mag.* Nov. 741 If as a result of several such interviews he is approved, he is asked to 'pledge', that is, to promise to join the society. **1901** *Munsey's Mag.* Feb. 734/2 The time and manner of pledging members to the fraternities vary with different colleges. **1949** *Reader's Digest* Aug. 69/1 The rushing season, during which freshmen are pledged to the various houses, was in full swing. **1977** *Rolling Stone* 19 May 67/2 Even though Hamilton went to the University of Georgia and pledged Phi Delta Theta, his exuberant intelligence wouldn't allow him to be satisfied with conformity.

4. c. To promise solemnly (*to* do something).

1928 *Sunday Dispatch* 2 Sept. 1/3 On my pledging not to disclose his name..he promptly handed over another cheque for £10,000.

pledging *vbl. sb.* (later examples).

1929 *Old Oregon* June 10 They went through rushing, pledging, moving, 'open house', freshman duties, in a cycle which at that time seemed to move ponderously over each event. **1959** *Ann. Reg. 1958* 152 At a special pledging conference in October 35 governments promised $27½ million for the Agency's work. **1964** *Amer. Speech* XXXIX. 194 The vocabulary of pledging, rushing.

pledgee. Add: **2.** One who takes a pledge, *spec.* = *Pledge *sb.* 5 c.

1937 (heard by Prof. A. L. Hench, Univ. of Virginia) 24 Nov., I needn't worry so long as I'm a pledgee. **1942** Berrey & Van den Bark *Amer. Thes. Slang* § 825/34 Pledge, pledgee,..a prospective fraternity member who has promised to join.

pleep (plīp). *slang* (? *obs.*). [? Echoic: see quot. 1948.] (See quots.)

1942 *Gen* 1 Sept. 14/2 A Heinkel pilot who shoots too soon or runs for home is definitely a 'poor type' or just a 'pleep'. **1943** Hunt & Pringle *Service Slang* 52 *Pleep*,..a Hun pilot who turns tail. **1948** Partridge *Dict. Forces' Slang* 144 *Pleep*, an enemy pilot that refuses aerial combat. Echoic of a timorous young bird.

-plegia, formative element, f. Gr. πληγ-ή blow, stroke (f. πλήσσειν to strike) + -ia¹, used with the sense 'paralysis', as in Hemiplegia, Paraplegia, *iridoplegia* s.v. *Irido-.

plein-air. Add: Usu. considered non-naturalized and written in italics. (Further examples.) Also used to designate work painted out of doors, or representing out-door scenes.

1930 *Observer* 6 Apr. 13 The giant Constable, the first of the plein air moderns. **1947** [see *Impressionistic *a.*]. **1970** *Daily Tel.* 9 July 12 Dame Laura never succeeded in liking town life. It was the wind and the sun and wild places that were her real loves and it was as a *plein air* painter of sunlight on landscapes and seascapes that she first succeeded. **1970** *Oxf. Compan. Art* 822/1 The expression '*plein air*'..implies a style of painting which emphasizes the impression of the open air and of spontaneity and naturalness. On the other hand it also indicates an actual technique of painting, which involved more than working in the open direct from nature instead of the older practice of composing a finished picture in the studio from rough sketches done on the spot. **1974** *Country Life* 6 June 1436/2 The impact of these *plein air* (in effect if not in fact) pictures.

plein-ai·rish *a.*, resembling or characteristic of the *plein-air* school of painting; **plein-airism**, ‖ **-isme** (-*ism*), the theories and practices of the plein-airists; **plein-airist** (earlier and later examples); ‖ **plein-airiste** (-*ist*) = *plein-airist.*

1891 *Academy* 6 June 544/3 'Impressionists', 'tâchistes', 'plein airistes', and 'pointillistes', to use the jargon of the day. **1897** *Daily Tel.* 10 Feb. 9/6 These pretty illustrations, from the designs of the well-known French *plein-airiste* and figure painter, Raphaël Collin, are delicate and graceful even to the verge of effeminacy. **1931** A. Huxley *Music at Night* 65 Bernini is, spiritually speaking, a *plein-airiste.* **1932** *New Statesman & Nation* 23 Jan. 93/1 Finally, with the emulation of his pleinairish friends, Manet loses not only his distinction of rhythm, but a great deal of his feeling for colour. *Ibid.*, It [*sc.* the Demoiselles au bord de la Loire]..marks the beginning of the long and triumphant development of pleinairism. **1946** *Penguin New Writing* XXVIII. 142 To imagine now an art of landscape into which plein-airisme had never intruded..it is impossible. **1959** *Listener* 9 Apr. 633/2 Go careful with the washing tub And do not spill the crystal slops Lest any of the escaping drops Should water down the *pleinairiste* Delights of Lady Tristram's feast! **1961** *Ibid.* 12 Oct. 571/2 It was necessary at the time..to talk about *plein airism* as though it was nothing else but a campaign against the time-worn clichés of the academic tradition. **1969** R. Mayer *Dict. Art Terms & Techniques* 299/2 The artists specifically called *pleinairistes* were a group of Impressionists of the 1880's and 1890's, notably Camille Pissaro.., Claude Monet.., Alfred Sisley.., and Pierre Auguste Renoir. **1972** D. Sutton *Lett. Roger Fry* I. 9 Fry was suspicious of Impressionism, by which he meant (I suspect) the watered-down *pleinairisme* of Bastien-Lepage, so popular in England, or the fragile pastiches of Whistler's followers. **1974** *Country Life* 28 Feb. 421/3 The historical emphasis..is on *plein air-ism* and the freshness and spontaneity of handling it produced. **1978** *Times* 17 Oct. 10/6 Some really wonderful paintings by the *plein-airists* of the 1880s on, whether it be McTaggart..or Guthrie.

‖ **plein jeu** (plen ʒö), *adv. phr.* and *sb. Mus.* [Fr. 'full play'.] **A.** *adv. phr.* As a direction: with full power; *spec.* in organ playing: without reeds.

1837 J. A. Hamilton *Dict. 2,000 Terms* (ed. 4) 54 *Plein jeu*.., full organ. **1938** *Oxf. Compan. Mus.* 670/2 *Plein jeu*, French for 'Full to Mixtures' (without reeds). *Ibid.* 739/1 *Plein jeu* (Fr.) 'full play', i.e. the whole power of the organ (or harmonium). **1954** *Grove's Dict. Mus.* (ed. 5) VII. 102/1 In the course of a piece it [*sc.* the term *organo pleno*] means the same as the French term *plein jeu.* **1960** *Times* 26 Apr. 16/2 The emotional stops are playing *plein jeu* in *Erwartung.* **1968** A. Niland *Introd. Organ* vi. 93 Bach writes at the beginning of several pieces, *pro organo pleno* (for full organ). This was undoubtedly the equivalent of the French *plein jeu.*

B. *sb.* A type of mixture stop on the organ; music written for the full organ. Also *fig.*

1855 E. J. Hopkins in Hopkins & Rimbault *Organ* ii. 328 The Madelaine [*sic*], at Paris... Clavier du Grand Orgue, 12 Stops... Plein-Jeu, X ranks. **1880** Grove *Dict. Mus.* II. 601/1 The scheme of Ducroquet's French organ stood as follows:—Great Organ. 10 stops... Bourdon.. Prestant..Plein jeu. **1898** J. Matthews *Handbk. Organ* ii. 29 Such compound stops are variously termed, Full, Grave, Acute or Sharp Mixture,..Fourniture, Plein Jeu, Cymbal, etc. **1919** G. A. Audsley *Organ of 20th Cent.* iii. 87 The largest Plein-Jeu known to us is that of ten ranks in the Grand division of the Organ in the Madeleine, Paris. **1944** W. Apel *Harvard Dict. Mus.* 588/1 *Plein-jeu*.., full organ. Also name for pieces written for the full organ. **1952** W. L. Sumner *Organ* x. 297 The plein-jeu is a large true mixture with sub-unison and unison ranks in the treble. **1958** *Times* 30 Sept. 3/5 He begins to draw out more stops in Siegfried..though reserving his *plein jeu*, so to speak, for a later date. **1963** Clutton & Niland *Brit. Organ* i. 38 Unlike the Germans, the French did not use their 'plein jeux' much for polyphonic music. **1975** *New Yorker* 28 Apr. 133/1 In addition, further pistons can bring on or silence the mixtures, the reeds, the *Plein Jeu*, the *Grand Jeu*, and 'Tutti'. **1978** *Gramophone* June 90/1 By the nature of the organs, many of the most effective pieces are dialogues between a group of reeds (the trompeteria) and the *plein jeu.*

pleione (plai,ōu·ni). Also **Pleione.** [mod.L. (D. Don *Prodromus Floræ Nepalensis* (1825) 36), f. Gr. Πλειόνη, the name of the mother of the Pleiades.] An orchid of the genus so called, belonging to the family Orchidaceæ, native to mountainous regions of northern India, Burma, and China, and bearing white, pink, or purple flowers, with plicate leaves which, in most species, fall before flowering begins.

1851 Lindley & Paxton *Paxton's Flower Garden* II. 5 The spotted Pleione has long been known to botanists as a species belonging to that Alpine group of so-called Cœlogynes. **1890** W. Watson *Orchids* li. 424 Pleiones are distinguished by their fleshy pseudo-bulbs, which are only of annual duration. **1930** T. W. Briscoe *Orchids for Amateurs* iv. 96 Pleiones are deciduous, and the leaves usually fall when the growths are matured. **1961** *New Statesman* 19 May 808/2 Cymbidiums..and pleiones..are cool-house plants. **1975** A. M. Coats *Treasury of Flowers* pl. 117 (*caption*) The Pleiones..are familiar to many gardeners, as they are both beautiful and easy to grow. **1979** *Country Life* 18 Jan. 158/4 Recently taxonomists have lumped together all the omnifarious pleione species under the aggregate specific epithet *P. bulbocodioides.*

pleiotropic (plai,otrǫ·pik, -trōu·pik), *a. Genetics.* [f. as next: see -ic.] Pertaining to, displaying, or being pleiotropy. Hence **pleiotro·pically** *adv.*

1938 *Proc. R. Soc.* B. CXXV. 138 The analysis of the pathological symptoms given above allows us to bring some order into the multitude of 'pleiotropic' effects produced by our lethal factor. **1956** C. Auerbach *Genetics in Atomic Age* 104 The gene for waltzing in the mouse acts pleiotropically on behaviour and hearing. Most or all genes have pleiotropic effects, often on such general characteristics as size, fertility, and longevity. **1964** *New Scientist* 17 Dec. 779/3 The streptomycin resistance is 'pleiotropic'—it brings about simultaneously one or more metabolically unrelated nutritional requirements. **1973** B. J. Williams *Evolution & Human Origins* ii. 30/1 A gene is said to have pleiotropic effects if it affects more

than one phenotypic trait. **1974** *Jrnl. Gen. Microbiol.* LXXXI. 165 Mutants in a gene..in *Aspergillus nidulans* pleiotropically affect the utilization of many nitrogen sources.

pleiotropism (plai,ǫ·tröpiz'm). *Genetics.* [f. as next: see -ism.] = *Pleiotropy.

1927 *Zeitschr. für Induktive Abstammungs- und Vererbungslehre* XLIII. 331 Plate..proposed the term 'pleiotrop'... Some of the highest authorities in the field of genetics have come to the conclusion that the pleiotropism of the genes is not an exception but rather the general rule. **1943** *Jrnl. Genetics* XLV. 6 A distinction has been made between genuine and spurious pleiotropism. **1970** T. Dobzhansky *Genetics of Evolutionary Process* vii. 210 Correlated responses due to pleiotropisms (physiological correlations) may make it impossible to endow a breed..with combinations of characteristics that would be desirable to man. **1973** B. J. Williams *Evolution & Human Origins* ii. 30/1 Pleiotropism does not interfere with a Mendelian analysis of inheritance. **1977** *Lancet* 29 Oct. 925/1 Gardner described such a family (no. 109) who also had extracolonic benign growths (fibromas, sebaceous and epidermal inclusion cysts, osteomas) which showed pleiotropism.

pleiotropy (plai,ǫ·tröpi). *Genetics.* [ad. G. *pleiotrop* (L. Plate 1910, in *Festschr. für R. Hertwig* II. 597), f. Gr. πλείων (see Pleio-) + τροπή turn, turning: see -y³.] The production by a single gene of two or more apparently unrelated phenotypic effects; an instance of this.

1939 C. H. Waddington *Introd. Mod. Genetics* vii. 162 (*heading*) Multiple effects of a factor or pleiotropy. **1957** —— *Strategy of Genes* 208 The more detailed the analysis, the more pleiotropy will be uncovered. **1973** *Nature* 21/28 Dec. 499/2 True pleiotropies of blood group genes are impossible to identify with the relatively crude characterisation of blood group specificities presently available. **1974** *Ibid.* 7 June 528/2 The suppression and temperature sensitivity characteristics were not separable, thus confirming that in each strain a single mutation was responsible for the pleiotropy. **1977** *Lancet* 9 Apr. 786/1 Genes always have more than one effect—a property known as pleiotropy.

pleiotypic (plai,oti·pik), *a. Biol.* [f. Pleio- + Typ(e *sb.*¹ + -ic.] Pertaining to the process whereby a single stimulus can elicit multiple unrelated responses from a living cell.

1971 *Nature New Biol.* 7 Apr. 162/1 Gordon Tomkins has proposed that mammalian cells may have a 'pleiotypic' control system to coordinate changes in the overall levels of synthesis and degradation of RNA and proteins. *Ibid.* 163/2 Dr. G. M. Tomkins (University of California, Medical Center, San Francisco) drew attention to a possible similarity between what he terms pleiotypic effects in mammalian cells and stringent control exhibited by bacteria deprived of nutrients. **1972** *Science* 5 May 486/1 The processes which have thus far been found to be under pleiotypic control are uridine uptake.., RNA synthesis.., polysome formation [etc.].

plenarium (plĭnēə·riŏm). Pl. **plenaria.** [ad. med.L. *plēnārium* in same sense, f. *plēnārius* complete: cf. Plenary *a.* (*sb.*).] A book or manuscript containing a complete set of sacred writings, e.g. all the gospels or all the epistles.

1908 W. G. Collingwood *Scandinavian Britain* 243 Bishop Patrick set forth to Iceland 'with wood for building a church, and a plenarium, and an iron bell'. **1911** F. Mershman in *Cath. Encycl.* XII. 164 Plenarium or Plenarius (*Liber*) is any book that contains completely all matters pertaining to one subject otherwise found scattered in several books... The entire mortuary office.. is called Plenarium. A complete copy of the four gospels was called an 'Evangelium plenarium'... Some Plenaria gave all the writings of the New Testament, others, those parts of the Sacred Scriptures that were commonly read in the Divine service and bore the name 'Lectionarium plenarium'. **1929** F. C. Thomas *Lay Folks' Hist. Liturgy* i. xvii. 87 In the form for Consecrating a Church we find.. a form for the consecration of a stole and of the Plenarium or Four Gospels as part of the Rite.

plenary, *a.* (*sb.*) Add: **B.** *ellipt.* as *sb.* **2.** *in plenary*: of an assembly, etc.: fully constituted or attended.

1969 D. Widgery in Cockburn & Blackburn *Student Power* 122 The infrequency of Council Sessions..and its unwieldy size in plenary means that the Executive alone takes the fundamental decisions about policy implementation and initiation.

3. Anglicized form of *Plenarium.

1909 *Encycl. Relig. & Ethics* II. 609/1 There were the select passages for Sunday in the so-called Plenaries, Postils, and Books of the Gospels and the Epistles. **1920** M. Deanesly *Lollard Bible* xii. 318 Of the three late fourteenth century English 'plenaries', or gospels and homilies, one is certainly Wycliffite.

‖ **plene administravit** (plī·ni ădministrē·vit). *Law.* [L., 'he has fully administered'.] (See quot. 1959.)

1729 G. Jacob *New Law-Dict.* s.v. *Executor*, If an Executor sued by several Creditors, pleads *Plene Administravit* to all at the same Time; and that he hath no Assets *præter* to pay one or two, he will make himself liable to all the debts. **1790** in Durnford & East *Rep. Cases King's Bench* III. 693 If an executor may plead *plene administravit* and neglect to do so, I see no difference between such a case and one where he does so plead and the plea is found against him. **1924** G. S. Bower *Res Judicata* i. vi. 108 Where a defendant, sued in the character of executor or administrator, omits to plead *plene administravit.* **1959** Jowitt *Dict. Eng. Law* II. 1354/2 *Plene administravit* (he has fully administered), a defence by an executor or administrator that he has fully administered all the assets which have come to his hands.

plenisphere (ple·nisfiəɹ). *rare*⁻¹. [f. L. *plēnus* full + Sphere *sb.*] A perfect sphere.

1912 E. Pound tr. *Calvacanti's Sonnets & Ballate* 99 Light I do see within my Lady's eyes And loving spirits in its plenisphere.

plentitude. (Further examples of this persistent (though erroneous) use.)

1939 Joyce *Finnegans Wake* (1964) ii. 241 A plentitude of house torts. **1944** Auden *For Time Being* (1945) 47 It was therefore only necessary for you to presuppose one genius, one unrivalled I to wish these wonders in all their endless plentitude and novelty. **1978** *Dædalus* Summer 197 This 'natural man' enjoys a plentitude of being.

plenty, *sb.* (*a.*, *adv.*). Add: **I.** *sb.* **2.** Also, a large amount, a great deal.

1939 R. Stout *Some buried Caesar* xiv. 164 The bill was $66.20, which was plenty. **1973** M. Yorke *Grave Matters* i. vi. 35 He must have paid plenty for the place, besides what they're going to lash out in alterations.

II. *adj.* or quasi-*adj.* **1. a.** (Further examples.)

1779 H. Cowley *Who's Dupe?* i. 1 When flowers are plenty, no body will buy 'em. **1794** N. Parry in *Reg. Kentucky Hist. Soc.* (1936) XXXIV. 390 Though much broken with Limestone, which is very plenty through these places. **1850** *New England Farmer* II. 123 The gopher..is very plenty on the west side of Mississippi. **1869** 'Mark Twain' *Innoc. Abr.* xxxiv. 368 Mosques are plenty, churches are plenty, graveyards are plenty, but morals and whisky are scarce. **1883** R. L. Stevenson *Silverado Squatters* 235 It is the same, they say, in the neighbourhood of all silver mines; the nature of that precious rock being stubborn with quartz and poisonous with cinnabar. Both were plenty in our Silverado.

b. (Later examples.)

1844 E. B. Browning *Poems* II. 181 What glory then for me In such a company?—Roses plenty, roses plenty, And one nightingale for twenty? **1922** Joyce *Ulysses* 609, I seen icebergs plenty, growlers. **1939** —— *Finnegans Wake* (1964) ii. 316 Besides proof plenty, over proof.

c. (Earlier and later examples.)

1857 E. Bandel *Diary* 28 May in R. P. Bieber *Frontier Life in Army* (1932) 138 A splendid country around us: plenty wood and water. **1899** 'S. Rudd' in Murdoch & Drake-Brockman *Austral. Short Stories* (1951) 111 The water they brought was a little thick..but Dad put plenty ashes in the cask to clear it. **1934** D. L. Sayers *Nine Tailors* 219 There's plenty farms now with the big brewing coppers still standing. **1939** Joyce *Finnegans Wake* (1964) iii. 443 Pretty knocks, I promise him with plenty burkes for his shins. **1942** 'M. Innes' *Daffodil Affair* ii. 47 I've known plenty men turn queer there. **1969** G. Greene *Travels with my Aunt* i. ix. 93 Leopard Society in Sierra Leone. They kill plenty people. **1973** *Sunday Express* (Trinidad) (Suppl.) 1 Apr. 12/3 When all dem fellas gambling and heap up plenty money, we..bawl out 'Police!'

3. Excellent. *slang.*

1933 *Fortune* Aug. 47/1 In sum, Mr. Brown plays *plenty trombone* or, as his friend suggested, a *gang o'horn.* **1941** R. P. Smith *So it doesn't Whistle* 53 When they want to say a man's good, they say he plays plenty sax or plenty drums. **1970** C. Major *Dict. Afro-Amer. Slang* 92 *Plenty*, good, excellent.

III. quasi-*adv.* (Further examples.)

1908 M. H. Morgan *How to dress Doll* (1973) xii. 85 Cut the hood..making it plenty large enough to slip on easily over Dolly's head. **1934** J. M. Cain *Postman always rings Twice* vi. 53, I was plenty blue around the gills. **1939** Joyce *Finnegans Wake* (1964) ii. 311 And plenty good enough, neighbour Norreys, every bit and grain. **1945** *Sun* (Baltimore) 13 June 8-O/7 Pavot just had the speed and the stamina, and turned it on plenty. **1956** B. Holiday *Lady sings Blues* (1973) iii. 35 Benny Goodman came around plenty, too, and eventually he asked me to make my first record with him. **1970** *New Yorker* 3 Oct. 32/1 You are wrong—but plenty. **1973** *Times* 27 July 8 It was not my business. I was plenty busy with other things. **1974** R. M. Pirsig *Zen & Art of Motorcycle Maintenance* iii. xxvi. 306 This notebook gets plenty grease-smeared and ugly.

plenum. Add: **1.** (Further examples.)

1827 J. Farey *Treat. Steam Engine* i. vi. 447 There is.. a plenum of steam in one compartment, and a vacuum, or exhaustion in the other. **1876** C. Slagg *Sanitary Wk.* ix. 102 The motion of the pan..disturbs the equilibrium between the air-pressure in the receiver and the outside, causing at times a partial vacuum and at other times a plenum of air in the receiver. **1956** E. H. Hutten *Lang. Mod. Physics* iii. 90 In Newton's theory it [*sc.* the aether] plays no rôle save to help visualisation: the victory of the Cartesian conception of space as a plenum as against the Aristotelian void. **1972** *Sci. Amer.* Apr. 115/2 It is not the ether as a medium that is denied by Einstein... We can deny only the Newtonian properties of the ether, in particular the linear addition of velocities. Quantum electrodynamics builds a real plenum in space.

b. (Further example.)

1949 E. Pound *Pisan Cantos* lxxvii. 61 Mind come to plenum when nothing more will go into it.

2. *spec.* a meeting of all the members of a communist party committee.

1948 J. Towster *Political Power in U.S.S.R.* x. 189 The plenums of the village and city soviets electing their own executive organs. **1950** D. W. Brogan *Era of Franklin D. Roosevelt* xv. 315 [The American Communist] regional party meetings had the strange title of 'plenums'. **1956** *Ann. Reg. 1955* 234 The Third Plenum of the Central Committee of the P.Z.P.R., held in Warsaw from 21 to 24 January. **1965** *New Statesman* 9 Apr. 566/2 At the recent Plenum, the first since Mr K was ousted, Ilyichev was booted upstairs as Deputy Foreign Minister. **1966** *Ibid.* 29 July 158/1 Ten years ago the man who failed Mao over the speed of collectivisation simply did not get promoted; today's failures suddenly find themselves dismissed and denounced, apparently without even the 'due process' of a central committee plenum. **1974** T. P. Whitney tr. *Solzhenitsyn's Gulag Archipelago* I. i. x. 417 Bukharin.. willingly assured the Plenum of his repentance, and immediately abandoned his hunger strike.

3. (Earlier and further examples.) Also applied to things connected with this method (which is used also for heating and air-conditioning).

1844 D. B. Reid *Illustr. Theory & Pract. Ventilation* ii. iii. 121 Plenum ventilation..can be sustained only by the constant use of machinery. **1894** J. Keith *Houses of Parl.: Rep. Heating & Ventilation* 3 The action of the Plenum fan..in blowing in the fresh air upwards through the grated floor of the Chamber. **1934** H. M. Vernon *Princ. Heating & Ventilation* ix. 177 A comparison of some thousands of observations made in factories ventilated by plenum air and by natural means showed very little defect of humidity in the plenum factories. **1948** T. Bedford *Basic Princ. Ventilation & Heating* xii. 198 In most of the plenum installations found in industry the air is untreated except that it is warmed in cold weather. **1967** W. P. Jones *Air Conditioning Engin.* xvi. 428 The primary air delivered by the nozzles escapes from the room through a plenum relief grille. **1968** *New Scientist* 7 Mar. 517/1 This [underwater] tunnel was driven by traditional methods, and the 'plenum' method (using compressed air behind the tunnelling shield) was not then known. **1978** Ld. Drogheda *Double Harness* xiii. 143 We had allowed for proper air conditioning... Instead we installed a horrible thing called plenum ventilation, which warmed the air in winter, filling the offices with smuts..but failing to cool it in summer.

b. (Not *attrib.*) = **plenum chamber* (*a*).

1940 W. H. Carrier et al. *Mod. Air-Conditioning* xviii. 425 The fan..is permitted to discharge into a plenum of section area at least ten times that of the fan discharge. **1970** *Toronto Daily Star* 24 Sept. 40/1 (Advt.), Humidifier... Coated pans fit sloped/vertical plenums. **1975** Croome-Gale & Roberts *Airconditioning & Ventilation of Buildings* vii. 272 There are three sources of pressure change within the plenum which must be kept under control to ensure uniform plenum pressure.

4. Special Combs.: **plenum chamber**, (*a*) in some plenum systems, an enclosed space into which the outside air is forced (after any conditioning) and from which ducts lead to the various outlets inside the building; (*b*) any analogous enclosure in which the pressure is maintained above that of the atmosphere by the forcing in of air, as in some air-cooled engines, a ram-jet, or a hovercraft; **plenum space** = **plenum chamber* (*a*).

1908 A. G. King *Pract. Steam & Hot Water Heating* xxi. 227 Separate ducts may be arranged to connect the main hot-air supply with the rising flues, or the heated air.. may be discharged under a slight pressure into a plenum chamber with which all supply pipes or warm-air ducts are connected. **1949** *Aircraft Engin.* Nov. 346/2 A disadvantage is that the pressure of the combustion chambers raises the temperature of the air in the plenum chamber, with a slight loss of power. **1959** *Motor* 30 Sept. 236/1 The entire engine, enclosed in the pressed-steel plenum chamber, is exposed to the cooling air, which escapes through vents below and behind. **1965** D. Herbert *How to design & install Warm Air Heating* iii. 47 A concrete or brick foundation should be laid to provide a flat and level base for the plenum chamber, which is that part of the heater to which the ducts are connected. **1967** *New Scientist* 31 Aug. 435/1 The cushion of air on which a hovercraft rides is created continuously by a flow of air delivered through ducts, which empty themselves into the plenum chamber by nozzles equally spaced around its edge. **1975** M. J. Nunney *Automotive Engine* v. 175 The cooling air flow entering the plenum chamber from the fan is directed downwards over the cylinders and cylinder heads. **1916** C. L. Hubbard *Ventilation Hand Bk.* ix. 154 The flues connecting the plenum space with the registers are..concealed in the leg of the pew. **1975** Croome-Gale & Roberts *Airconditioning & Ventilation of Buildings* vii. 278 The ventilated ceiling system is an all-air system which delivers supply air through..ductwork to a plenum space over a suspended ceiling, so that a relatively small plenum pressure may be used to evenly distribute the air through the ventilated ceiling to the room below.

pleochroic, *a.* Add: *pleochroic halo*, each of a series of concentric dark-coloured circles seen in sections of certain minerals and having a radioactive inclusion at their centre; usu. *pl.*

1894 [in Dict.]. **1909** F. P. Mennell *Introd. Petrol.* ix. 57 'Pleochroic halos'..often occur round crystals so small as otherwise to pass unnoticed. **1926** R. W. Lawson tr. *Hevesy & Paneth's Man. Radioactivity* xxvi. 216 (*heading*) Age determination from the intensity of coloration of pleochroic haloes. **1972** M. H. Battey *Mineral. for Students* iii. 93/2 Small crystals of zircon, containing thorium, embedded in biotite, may by their α-radiation produce an intensely pleochroic halo round the zircon grain.

pleocytosis (plī,osəitōu·sis). *Path.* [f. *pleo-*, PLEIO- + -CYT(E + -OSIS.] The presence of abnormally many cells, *spec.* of lymphocytes in the cerebro-spinal fluid.

1911 STEDMAN *Med. Dict.* 681/1 *Pleocytosis*, lymphocytosis in the cerebrospinal fluid in syphilitic and parasyphilitic diseases of the central nervous system. **1924** BROWNING & MACKENZIE *Recent Methods in Diagnosis & Treatment of Syphilis* (ed. 2) xvi. 304 Pleocytosis is found in practically all acute and chronic inflammatory processes affecting the meninges. **1976** *Lancet* 4 Dec. 1222/1 The cerebrospinal fluid in patient 3 showed pleocytosis (23 mononuclear cells/mm).

pleonastic, *a.* (Further examples.)

1898 H. SWEET *New Eng. Gram.* II. 54 The pleonastic genitive, as in *he is a friend of my brother's*, is generally partitive = 'one of the friends of my brother'. **1947** [see *INTERLINGUISTICS]. **1951** E. H. STURTEVANT *Compar. Gram. Hittite Lang.* (ed. 2) ii. 23 Pleonastic Vowels. In Akkadian, vowels are frequently written double (*U-UL* 'not', *BEE-EL* 'lord'). **1972** W. LABOV *Language in Inner City* iv. 146 The general nonstandard rule which operates here can be written as a simple pleonastic transformation.

pleophony (plī,ǫ·fōni). *Linguistics.* [f. *pleo-*, PLEIO- + *phony* after HOMOPHONY etc.] Vowel duplication; epenthesis of a vowel which harmonizes with that in the preceding syllable. Hence **pleopho·nic** *a.*

1949 *Archivum Linguisticum* I. 165 The East Slavonic languages..present the curious and striking phenomenon of pleophony or double vowelling. **1966** H. BIRNBAUM in Birnbaum & Puhvel *Anc. Indo-Europ. Dial.* 162 One such feature is 'pleophonic' /ToroT/ as a reflex of P[roto] Sl[avic] /TorT/. *Ibid.* 167 This change must have preceded in time the /TorT/>/ToroT/modification ('pleophony') mentioned above.

pleoptics (plī,ǫ·ptiks), *sb. pl.* (const. as *sing.* or *pl.*). *Ophthalm.* [ad. G. *pleoptik* (A. Bangerter 1953, in *Wiener klin. Wochenschr.* 20 Nov. 966/2), f. *pleo-*, PLEIO- + *optik* OPTICS.] A method of treatment for amblyopia and eccentric fixation employing the selective dazzling of parts of the retina in order to stimulate the use of the fovea and render it more sensitive.

1955 *Q. Cumulative Index Med.* LVIII. 1517/1 (*heading*) Pleoptics and orthoptics. **1962** J. W. HENDERSON in G. M. Haik *Strabismus* v. 111 For the eccentric fixator over 5 years of age, pleoptics rather than occlusion are thought to be the most acceptable method [of treating amblyopia]. **1964** A. SCHLOSSMAN in A. Sorby *Mod. Ophthalm.* III. 141 Pleoptics, which was initiated by Bangerter of St-Gallen, Switzerland, has its greatest value in the treatment of the large number of patients with eccentric fixation who cannot be managed by any other form of therapy. **1975** M. M. PARKS *Ocular Motility & Strabismus* xi. 98/1 An eccentrically fixating eye in some older children has been restored to central fixation using pleoptics after occlusion therapy has failed.

So **pleo·ptic** *a.*

1960 *Amer. Orthoptic Jrnl.* X. 7/1 The development of pleoptic methods and their therapeutic application preceded the discovery of some hitherto unknown characteristics of amblyopia. **1964** S. DUKE-ELDER *Parsons' Diseases of Eye* (ed. 14) XXX. 482 If eccentric fixation is well established, it is often well to occlude the affected eye for some weeks and then to stimulate the macula by special pleoptic methods (flashing devices, the production of after-images, etc.). **1967** J. L. C. MARTIN-DOYLE *Synopsis Ophthalm.* (ed. 3) 188 Pleoptic treatment is arduous and the instruments expensive. **1975** M. M. PARKS *Ocular Motility & Strabismus* xi. 98/1 The final result in the majority of amblyoptic patients is the same whether..simple occlusion therapy is used exclusively or if pleoptic exercises are used.

plereme (plī·rīm). *Linguistics.* [f. Gr. πλήρ-ης full: see *-EME.] **a.** = *full word. **b.** A unit of meaning. Hence **plerema·tic** *a.*, **plerema·tically** *adv.*, **plerema·tics** *sb. pl.*

1939 [see *GLOSSEMATIC *sb. pl.* and *a.*]. **1939**, etc. [see *KENEME]. **1949** C. E. BAZELL in E. P. Hamp et al. *Readings in Linguistics II* (1966) 209 No fusion in the expression of plerematic units is here involved. **1950** S. POTTER *Our Language* 87 Sememes (including *pleremes* and *kenemes*). **1957** — *Mod. Linguistics* vii. 143 Operators are sometimes called *kenemes* by those describing Chinese, as opposed to..*pleremes*. **1958** C. F. HOCKETT *Course in Modern Linguistics* lxiv. 575 Here the terms 'phonological' and 'grammatical' make too direct a reference to human language; it will be better to introduce two new terms for general applicability: *cenematic* and *plerematic*. The cenematic structure of language is phonology; the plerematic structure of language is grammar. Phonemes are linguistic *cenemes*; morphemes are linguistic *pleremes*. *Ibid.* 576 Productivity implies that some messages in the system..are *plerematically complex*: that they consist of an arrangement of two or more pleremes, instead of each consisting of a single indivisible plereme. If one starts with a system with no plerematic complexity, then there is only one variety of analogy by which a new message can be coined: *blending*. **1959** W. A. C. H. DOBSON *Late Archaic Chinese* i. 14 Undistributed, a plerematic word might be said to represent a notion undifferentiated by grammatical quality. **1961** F. W. HOUSEHOLDER in Saporta & Bastian *Psycholinguistics* 25/2 They [*sc.* distinctive features] may also be considered as 'cenemes' of which 'plerematic' phonemes are composed. **1967** C. L. WRENN *Word & Symbol* 4 The names of things, qualities or acts..are the words which linguists of the Danish school used to term *pleremes* (words of full or complete significance). **1969** *Word 1967* XXIII. 469 My observations apparently support the structuralist separation of cenetics and plerematics. *Ibid.* 471 In addition to their two cenetic and two plerematic systems, bilingual children naturally have to master two sets of form-to-meaning relationships. **1978** *Amer. Speech* LIII. 275 A communicative system has *duality of patterning*..if its meaningful signals (pleremes) are built out of some convenient stock of meaningless but differentiating pieces (cenemes). *Ibid.* 276 Some special sort of discourse in which there is a planned regularity of recurrence of cenematic features independent of their plerematic role.

plerocephalic (pliᵊrosĭfæ·lik), *a.* *Path.* [f. Gr. πλήρης, πληρο- full + κεφαλή head + -IC.] Of œdema: caused by increased intracranial pressure.

1927 H. M. TRAQUAIR *Introd. Clin. Perimetry* ix. 115 Plerocephalic œdema (œdema due to increased intracranial pressure)... As 'Papillœdema' and 'choked disc' refer to ophthalmoscopic appearances which may occur in local optic nerve disease this term has been chosen to indicate œdema of the disc due to increased intracranial pressure. **1976** *Proc. R. Soc. Med.* LXIX. 455/2 In the following 12 cases varying degrees of optic atrophy occurred in space-occupying lesions of the brain in children because of the development of plerocephalic œdema.

plerocercoid (pliᵊrosɜ̄·ɹkoid). *Zool.* [f. Gr. πλήρης, πληρο- full + κέρκος tail + -OID.] A larval form of certain tapeworms, in which the body is solid, lacking a bladder. Also *attrib.* or as *adj.*

1906 P. FALCKE tr. *Braun's Animal Parasites of Man* 219 Human beings, like other hosts, can only acquire the broad tapeworm by ingesting its plerocercoids. **1928** *Jrnl. Amer. Med. Assoc.* 30 June 2081/1, I found the plerocercoids in pike. **1961** SWELLENGREBEL & STERMAN *Animal Parasites in Man* xiii. 241 The plerocercoids have been recovered from various internal organs. **1962** J. D. SMYTH *Introd. Anim. Parasitol.* xix. 222 The lycophora.. develops into a procercoid and later a plerocercoid larva. **1973** T. C. CHENG *Gen. Parasitol.* xiv. 488/2 Procercoids develop into solid, wormlike plerocercoids, each with one invaginated scolex at one end.

plesiadapid (plĭzi,æ·dāpid), *sb.* and *a.* *Palæont.* [f. mod.L. family name *Plesiadapidæ*, f. the generic name *Plesiadapis* (P. Gervais 1877, in *Jrnl. Zool.* VI. 76), f. PLESIO- + *Adapis*, generic name of another fossil primate + -ID³.] **A.** *sb.* A primitive, extinct primate belonging to the family Plesiadapidæ, known from Palæocene fossil remains found in North America and Europe. **B.** *adj.* Of, pertaining to, or resembling an animal of this kind.

1945 A. S. ROMER *Vertebr. Paleontol.* (ed. 2) xviii. 343 The plesiadapids have been thought to be aberrant lemurs or tree shrews. **1949** W. E. LE GROS CLARK *Hist. Primates* 50 We know little as yet about the limb bones of the plesiadapids. **1963** E. L. SIMONS in J. Buettner-Janusch *Evolutionary & Genetic Biol. Primates* I. ii. 79 The anterior teeth of plesiadapids were enlarged and procumbent. **1970** [see *LORISID *sb.* and *a.*]. **1975** *Nature* 10 Jan. 111/2 Five plesiadapid lineages are known, at least two of which were common to both Europe and North America.

plesiadapoid (plĭzi,æ·dāpoid), *sb.* and *a.* *Palæont.* [f. mod.L. name of suborder *Plesiadapoidea*, f. the generic name *Plesiadapis* (see prec.) + -OID.] **A.** *sb.* A primitive, extinct primate belonging to the suborder Plesiadapoidea. **B.** *adj.* Of, pertaining to, or resembling an animal of this kind.

1966 A. S. ROMER *Vertebr. Paleontol.* (ed. 3) xviii. 217/2 (*heading*) Plesiadapoids. **1973** *Nature* 24 Aug. 518/1 The dental characteristics distinguishing primitive Early Eocene adapids and omomyids from other Eocene mammals are present in primitive plesiadapoids. **1974** B. J. STAHL *Vertebrate History* ix. 471 The plesiadapoids.. became extinct without issue before the close of the Eocene. *Ibid.* 472 A change in the climate is implicated in the failure of the plesiadapoid primates to survive.

Plesianthropus (plĭzi,æ·nþrǭpǭs). *Palæont.* [mod.L. (R. Broom 1938, in *Nature* 27 Aug. 377/1), f. PLESI(O- + Gr. ἄνθρωπος man.] An African fossil hominid of the genus formerly so called, now usually included in the species *Australopithecus africanus*.

1941 *Nature* 5 July 11/1 Plesianthropus..has a skull somewhat like that of the chimpanzee in size. **1948** A. L. KROEBER *Anthropol.* (rev. ed.) iii. 91 Australopithecus and Plesianthropus came to light in quarrying operations. **1960** E. WINICK *Dict. Anthropol.* 421/1 *Plesianthropus*, an African man-like ape fossil from the Pleistocene period. **1977** G. W. HEWES in D. M. Rumbaugh *Language Learning by Chimpanzee* i. 40 As for the Australopithecines (based on casts of the Sterkfontein individual formerly called 'Plesianthropus') no articulate language would have been possible for them.

plesiomorphous, *a.*, **-morphism.** (Earlier example of each.)

1833 *Rep. Brit. Assoc. 1831, 1832* 429 Plesiomorphism.—As the differences between the angles of the carbonates and sulphates above quoted cannot be accounted for by any accidental causes.., some crystallographers have been led to reject the term *isomorphous* as applied to such crystallized compounds, and to substitute in its place the term *plesiomorphous*.

-plet (plet), the ending of *triplet*, *multiplet*, etc., used with a prefixed numeral to denote a multiplet having the specified number of members.

1973 L. J. TASSIE *Physics Elementary Particles* xiii. 173 Taking the neutron and proton as members of the *SU*(6) 56-plet yields the ratio of their magnetic moments as.. −3/2 which is close to the experimental value of −1·46. **1975** *Physics Bull.* Apr. 180/3 The octet of mesons is now part of a 15plet which groups together one octet and one singlet with no charm, and two triplets with charm 1 and −1, respectively.

plethorous (ple·þǫ̈rəs), *a. rare.* [f. PLETHORA + -OUS.] = PLETHORIC *a.* 1.

1906 J. P. BARRY *At Gates of East* p. vii, But the book.. may do good in a practical way, if it weans the wearied, the plethorous and the valetudinarian from the Cult of the Spas.

plethy·smogram. *Physiol.* [f. as next + -GRAM.] The record produced by a plethysmograph.

1894 W. EWART *Heart Stud.* I. 181 The need for a separate study of shape, volume, and velocity of the pulse is rendered obvious by the difference between the three curves, which may be termed respectively the sphygmogram, the plethysmogram, and the tachogram, obtained by means of the three instruments bearing corresponding names. *Ibid.* 231 The plethysmogram (volumpuls) of the forearm..shows an increase in the size of the oscillations. **1929** *Proc. Soc. Exper. Biol. & Med.* XXVI. 711 (*heading*) A photographic method of recording plethysmograms. **1971** *Nature* 1 Oct. 340/2 Burton has ascribed the slow fluctuations..seen in human digital plethysmograms to the action of the thermoregulatory system.

plethysmograph. Add: (In mod. forms of the instrument other fluids, e.g. air, and other means of measuring its displacement are employed.) (Further examples.)

1957 *Clin. Sci.* XVI. 103 (*heading*) Venous collection in forearm and hand measured by the strain-gauge and volume plethysmograph. **1964** *Times* 5 Sept. 12/3 A small electronic instrument called a plethysmograph used for measuring changes of blood volume in the body's circulation. **1972** *English Studies* LIII. 76 Measurements were made..of the variations in volume of the air in the lungs. This..involved the use of a plethysmograph, a rigid airtight container enclosing the subject entirely except for head and neck.

Hence also **plethy:smogra·phically** *adv.*, by means of a plethysmograph; **plethysmography** (examples).

1897 *Jrnl. Exper. Med.* II. 334 The striking effect of such stimuli upon the volume of a limb when measured plethysmographically. **1898** *Jrnl. Physiol.* XXII. 380 A few experiments on intestinal plethysmography were made by Bayliss. **1930** *Amer. Jrnl. Physiol.* XCI. 717 That the abdominal venous reservoirs empty (into the chest) during systole can be demonstrated by abdominal plethysmography. **1970** *Nature* 18 July 276/2 The volume of the foot was measured plethysmographically before injection of carrageenin. **1977** *Lancet* 9 July 66/1 Strain-gauge plethysmography is probably an appropriate method for demonstrating arterial insufficiency in this condition.

pleurisy. 3. pleurisy-root (earlier and later examples).

1785 T. JEFFERSON *Notes on State of Virginia* 63 Pleurisy root, Asclepias decumbens. **1932** J. B. HARVEY *Wild Flowers Amer.* 55 Butterfly Weed or Pleurisy Root.. bears brilliant orange flowers, arranged in flat, terminal clusters.

pleuro. Add: (Earlier and later Austral. examples.) Also *attrib.*

1885 R. C. PRAED *Austral. Life* 244 Pleuro is very bad our way. I don't believe much in inoculation—do you? **1897** 'R. BOLDREWOOD' *My Run Home* xx. 176 'Do you ever have any pleuro among your cattle?' said I; 'I heard something about it in England.' **1917** A. B. PATERSON *Three Elephant Power* 42 Providence sends the pleuro, and big strong beasts slink away by themselves. **1944** W. E. HARNEY *Taboo* (ed. 3) 40 One day..a pleuro bullock chased him. **1965** *Bulletin* (Sydney) 13 Feb. 40/3 They used their weapons on the local pleuro bull which was the only large, live target they had a chance to assail. **1975** *Sunday Mail* (Brisbane) 20 July 10/5 Today he looks back with pride at the successful eradication of the killer disease 'pleuro' which threatened the cattle industry in 1970.

pleu:ropneumo·nia-like, *a.* *Biol.* Also with two or no hyphens. [f. PLEURO-PNEUMONIA + -LIKE.] *pleuropneumonia-like organism*: = *MYCOPLASMA. Abbrev. PPLO s.v. *P II.

1935 E. KLIENEBERGER in *Jrnl. Path. & Bacteriol.* XL. 93 (*heading*) The natural occurrence of pleuropneumonia-like organisms in apparent symbiosis with *Streptobacillus moniliformis* and other bacteria. **1951** *Jrnl. Bacteriol.* LXI. 395 A characteristic of the parasitic pleuropneumonialike organisms (PPLO) is the requirement of serum or ascitic fluid for growth *in vitro*. **1964** *New Scientist* 19 Nov. 497/1 The workers in Glasgow have grown pleuropneumonia-like organisms from cell cultures containing leukaemia 'virus'. **1973** *Nature* 9 Mar. 83/1 Mycoplasmas, which used to be known as pleuropneumonia-like organisms, are the smallest free-living organisms.

plexiform, *a.* Add: *plexiform layer* [tr. F. *couche plexiforme* (*externe, interne*) (S. Ramón y Cajal 1893, in *La Cellule* IX. 132)], either of two layers of the retina separated by the inner nuclear layer, the outer one of which contains synapses between the rods and cones and the neurones of the nuclear layer, whilst the inner one contains synapses between these neurones and ganglion cells; = *molecular layer* (*a*) s.v. *MOLECULAR *a.* 5.

1894 *Quain's Elem. Anat.* (ed. 10) III. III. 41 (*heading*) Inner molecular or inner plexiform layer, neurospongium. **1911** *Ophthalmoscope* IX. 437 The external plexiform layer remains very narrow. **1959** W. ANDREW *Text-bk. Compar. Histol.* xv. 604 The ten layers in order, from without in, i.e. toward the vitreous humor, are: (1) the pigmented epithelium, (2) the layer of rods and cones, ..(5) the outer plexiform layer, (6) the inner nuclear layer, (7) the inner plexiform layer,..and (10) the inner limiting membrane. **1972** THORPE & GLICKSTEIN in tr. *S. Ramón y Cajal's Structure of Retina* p. viii, The dendritic trees of the ganglion cells spread at different levels within the inner plexiform layer and form associations with the processes of the amacrine and bipolar cell processes, thereby creating the distinct laminar appearance of this layer so clearly described by Cajal.

Plexiglas (ple·ksiglas). Chiefly *U.S.* Also **Plexiglass**, and with lower-case initial. A proprietary name for the substance also sold under the names of *PERSPEX and *LUCITE. Freq. *attrib.*

1935 *Trade Marks Jrnl.* 10 Apr. 451 Plexiglass... Glass. Röhm & Haas, Aktiengesellschaft.., Darmstadt, Germany; manufacturers. *Ibid.* 31 July 968/2 Plexiglass... Glass substitutes made from artificial resins. Röhm & Haas Aktiengesellschaft.., Darmstadt, Germany; manufacturers. **1936** *Official Gaz.* (U.S. Patent Office) 30 June 999/1 Röhm and Haas Company, Philadelphia. *Plexiglas* for sheets of solid transparent resinous material to be used as a glass substitute. Claims use since June 5, 1935. **1936** *Jrnl. Aeronaut. Sci.* Nov. 13/1 For smaller aircraft, Plexiglas, Glyptal, and Vinylite are available in sheet form. **1941** *Product Engin.* May 267/3 Area of windows is 20 sq. ft., hence lighter Plexiglass instead of plate glass. **1943** J. STEINBECK in *N.Y. Herald Tribune* 31 Aug. 17/4 The sergeant had carved the handles of his gun from the plexiglass from the nose of a bomber. **1951** *Archit. Rev.* CX. 222/1 (*caption*) Roof, tar and gravel with plexiglass skylights. **1954** *Trade Marks Jrnl.* 15 Dec. 1264/2 Plexiglas... Chemical products used in industry, science and photography; fire extinguishing compositions, soldering preparations.., adhesives.., and synthetic resins. Röhm & Haas Gesellschaft mit beschränkter Haftung.., Darmstadt, Germany; manufacturers. **1955** *Ibid.* 13 April 387/1 Plexiglas... All goods..made of glass or of glass substitutes made from plastics. Röhm & Haas Gesellschaft. **1957** *Economist* 19 Oct. 229/2 The ceremonial drive in an open landau..or the neighbourhood tour in a grey Cadillac with plexiglass top. **1965** ZIGROSSER & GAEHDE *Guide Coll. Orig. Prints* vii. 115 Plexiglas is sometimes used for framing because it is unbreakable. The Plexiglas should be coated with an anti-static to eliminate its tendency to attract dust. **1970** HARARI & HAYWARD tr. *A. Amalrik's Involuntary Journey to Siberia* iv. 56 This door had a small panel of plexiglass. **1976** *So* (N.Y.) Feb. 30/2 The large full-feature backlighted digital clock has buzz signal alarm. All encased in a modern smoked plexiglas front panel. **1977** *Time* 31 Jan. 15/2 Cairo's flying squads of riot police with their Plexiglas face masks, shields and staves were in control.

plica. Add: **4.** *Medieval Mus.* (Also with pl. **plicas.**) A notational symbol, variously interpreted but now usu. considered to represent a type of ornament; the ornament indicated. Also *attrib.*

1782 C. BURNEY *Gen. Hist. Mus.* II. iii. 188 Few of the musical terms in the tract of Franco, are more difficult to comprehend or define than the word *Plica*, which he calls 'a note of *division* of the same sound, ascending or descending.' **1801** T. BUSBY *Dict. Mus.* s.v., *Plica*, the name formerly given to a kind of ligature used in the old music as a sign of hesitation, or pausing. **1881** GROVE *Dict. Mus.* III. 4/1 *Plica*..a character, mentioned by Franco of Cologne, Joannes de Muris, and other early writers, whose accounts of it are not always very easily reconciled to each other. Franco describes four kinds... Joannes de Muris describes the Plica as a sign of augmentation, similar in effect to the Point. Franco tells us that it may be added at will to the Long, or the Breve; but to the semibreve only when it appears in Ligature. Some other writers apply the term 'Plica' to the tail of a Large, or Long. The Descending Plica is sometimes identified with the Cephalicus. **1903** C. F. A. WILLIAMS *Story of Notation* vi. 101 In the sixth chapter Franco treats of the Plica. **1940** G. REESE *Music in Middle Ages* (1941) III. x. 283 Another type of ornament was notated by the *plica*. This was attached to either single notes..or ligatures... When it was applied to single notes, usually two parallel strokes of unequal length were added to the note-head. These strokes, enfolding the head, gave the *plica* its name. **1942** W. APEL *Notation of Polyphonic Music, 900–1600* (1944) III. iii. 234 The *plica* is a passing tone which is indicated ..by a downward or upward dash attached to the right of a note. In modal notation, the *plica* appears preferably in connection with ligatures. *Ibid.* 235 We shall carefully distinguish between *plica*-note and *plica*-tone. The former term refers to the written note to which the *plica*-dash is attached; the latter to the extra tone called for by the dash. **1954** *New Oxf. Hist. Music* II. 325 The *plica*.. is a short stroke which modifies the single square note, leading either upwards or downwards. *Ibid.*, The instructions of the medieval theorists most often quoted are those of the Anonymus of Paris, who in his *Quædam de arte discantandi* tells us that 'it should be formed in the throat with the epiglottis', and of Lambert, who wrote under the pen-name of Aristotle: 'The *plica* is made in the voice by compressing the epiglottis, combining it neatly with a repercussion of the voice.' *Ibid.*, The second, semi-vocal, note of the plica is not an ad libitum ornament..but has a time-value of its own, one-third or one-half of that which would belong to the parent note if it were not plicated. **1957** C. PARRISH *Notation of Medieval Music* (1958) v. 130 Plicas are used sparingly. **1979** *Early Music* Apr. 189/1 The *plica* is a note with stems on *both* sides of the note head and is meant to be sung as two pitches, the first one specified by the position of the note head, the second one unspecified, but lying above or below the first note, depending on the direction of the stems.

plicate, *v.* Add: **2.** *Medieval Mus.* To add a plica to. (Chiefly as ppl. adj.)

1903 C. F. A. WILLIAMS *Story of Notation* vii. 120 An imaginary dialogue dated 1326..shows a 'plicated' semibreve when three semibreves are used with one syllable. **1927** *Grove's Dict. Mus.* (ed. 3) IV. 210/2 Besides longs and breves, semibreves could be plicated when in ligature. **1954** [see *PLICA 4]. **1977** *Early Music* Apr. 199 The neum..is assumed to be the plicated equivalent of the *pressus*.

plicated, *ppl. a.* Add: **3.** (See *PLICATE *v.* 2.)

‖ **plié** (pli,e). *Ballet.* Also **plier.** [Fr., f. *plier* to bend.] A movement in which the dancer lowers the body, bending the knees outwards in line with the out-turned feet. Also as *v. intr.*, to execute such a movement.

1892 E. SCOTT *Dancing as Art & Pastime* vi. 76 The foot passes from the *fourth rearward position* to the *fourth in front* with a very decided *plié de genoux* as it comes into the *first position*. *Ibid.* 77 A movement consisting of two *pliés* in advancing and three forward steps. **1913** C. D'ALBERT *Dancing* 125 *Plié*,..flexion or bending of one or both knees in preparation for any step. **1920** *Dancing Times* July 793 The Russians take the grands battements at the end of the side practice instead of immediately after the 'plier'. **1922** BEAUMONT & IDZIKOWSKI *Man. Classical Theatr. Dancing* II. i. 37 The *plié*, or bend, may be small (*plié à quart*)..medium (*plié à demi* or *demi-plié*)..or large (*grand plié*). **1930** CRASKE & BEAUMONT *Theory & Pract. Allegro in Classical Ballet* 16 An *assemblé* is said to be *soutenu* when the knees are straightened and another *plié* is made before executing the next *pas*. **1949** A. CHUJOY *Dance Encycl.* 376/1 It is said that the entire technique of ballet consists in knowing when and how to do a plié. **1958** *Observer* 14 Sept. 14/7 Beautifully musical, he can turn a double *tour en l'air* into a deep *plié en arabesque* as trimly as if it were a phrase in a Mozart sonata. **1971** 'D. HALLIDAY' *Dolly & Doctor Bird* xi. 146 Krishtof would raise me..while I stood up. He would then plié round me. **1977** *New Yorker* 19 Sept. 43/1 The marble-floored reception area alone, on the ground floor, is vast enough to accommodate the entire corps de ballet of four or five major ballet companies, all doing their pliés and entrechats.

Pliensbachian (plĭnzbā·kiăn), *a.* *Geol.* [ad. G. *Pliensbachien* (A. Oppel 1858, in *Jahresh. des Vereins f. vaterländische Naturkunde in Württemberg* XIV. 249), f. *Pliensbach*, name of a locality near Boll, a village near Göppingen in Baden-Württemberg, W. Germany: see -IAN.] Of, pertaining to, or designating a stage of the Lower Jurassic in Europe comprising the Middle Lias and part of the Lower Lias. Freq. *absol.*

1903 *Q. Jrnl. Geol. Soc.* LIX. 455 In some districts—East Gloucestershire for instance—only a few feet of Toarcian are found separating the Inferior Oolite (Aalenian) from the Middle Lias (Pliensbachian). **1955** F. NEAVERSON *Stratigr. Palaeont.* (ed. 2) xii. 440 Pliensbachian brachiopods are most abundant in the limestone facies of Somerset. **1975** A. HALLAM *Jurassic Environments* ii. 13 No zonal subdivision for the Pliensbachian of southern Europe as a whole has yet been satisfactorily achieved. **1978** *Nature* 13 July 131/1 The initial Jurassic transgression here took place in the Pliensbachian.

plig (plig). *U.S. dial.* [Shortening of POLYGAMIST.] A polygamist, used esp. with reference to the practice of polygamy attributed to the Mormon Church.

1977 *Washington Post* 8 Aug. A14 Many citizens in Utah show an uncomfortable dualism toward polygamists, derisively calling them 'pligs'. **1978** *Observer* (Colour Suppl.) 12 Mar. 33/4 Polygamy..has made a surprising comeback of late. 'The Pligs are sprouting like weeds,' said a county sheriff.

plightage (pləi·tėdʒ). *rare*⁻¹. [f. PLIGHT *v.*¹ + -AGE.] The fact or state of being plighted or betrothed.

1908 HARDY *Dynasts* III. v. iv. 442 These vile tricks, to pluck you from Your nuptial plightage..Make me belch oaths!

plightful, *a.* Restrict †*Obs.* to sense in Dict. and add: **2.** Grievous; fraught with suffering. *rare.*

1906 HARDY *Dynasts* II. vi. v. 304 The tears that lie about this plightful scene Of heavy travail in a suffering soul.

Plimsoll. Add: Also *fig.* Also *Plimsoll's pancake* = *Plimsoll line*.

1894 [in Dict.]. **1896** *Nautical Mag.* Jan. 17 'Plimsoll's pancake' will ruin the colonial carrying trade. **1912** R. W. SERVICE *Rhymes of Rolling Stone* (1913) 78 Loaded to the Plimsoll mark With God's sunshine was that boy. **1920** P. L. WALDRON *Afloat & Ashore* 57 The ship was loaded down to 'Plimsoll's pancake'. **1957** R. CAMPBELL *Portugal* 29 It..will pierce a crocodile if shot on the 'Plimsoll-line' between back and belly. **1961** *John o' London's* 18 May 567/1 Wearing towel..and not wearing that a millimetre more than the censor demands either below or above the cheesecake plimsoll-line. **1964** C. WILLOCK *Enormous Zoo* iv. 59 Big bull elephants..stand up to their plimsoll lines in the lake. **1972** *Author* Winter 187/2 The secondary school performance was worse—only 9 'good' and 13 'reasonable', with 140 below the Plimsoll line. **1976** *National Observer* (U.S.) 31 July 15/2, I tried to teach my wife, Lila, to fish, but with only a mud turtle, a rusty bucket, and a two-inch bluegill to show for standing three hours up to her Plimsoll line in cold water, she became an incorrigible dropout. **1978** T. DE V. WHITE *My Name is Norval* i. 15 She had..[a] formula for..the restraining of male advances: a Plimsoll mark below which she was theoretically in danger.

2. (With lower-case initial.) A kind of rubber-soled canvas shoe. Usu. *pl.* Also (through association with SOLE *sb.*¹ 2) *plimsole*. Also *attrib.* Hence **pli·msol(l)ed** *a.*, wearing plimsolls.

1907 *Yesterday's Shopping* (1969) 326/1 The Plimsoll or Sand Shoes. **1922** *Times* 27 Dec. 7/7 When Seabrook appeared in court he was wearing white plimsolls. **1927** W. DEEPING *Kitty* xii. 148 These stealthy affairs..made him think of sneaking out in plimsolls and kidding some 'cop'. **1930** W. PETT RIDGE *Miss Collingwood* i. 12 She kicked off her plimsolls, and walked about in stockinged feet. **1936** G. POLLET *Song for Sixpence* xi. 89 For two pleasant days I have been holidaying it in plimsolls whilst my shoes have been upon the rack. **1939** JOYCE *Finnegans Wake* (1964) II. 397 Their blankets and materny mufflers and plimsoles. **1955** E. BLISHEN *Roaring Boys* iv. 183 Dancing on plimsolled toes like a boxer. **1959** F. BAINES *In Deep* 29 Moses and I put on plimsolls and slipped down the alley-way. **1963** 'R. EAST' *Pin Men* iii. 62 He lifted his plimsolled foot. **1965** D. FRANCIS *Odds Against* xii. 163 Chico had made a plimsole-shod inspection. **1968** J. IRONSIDE *Fashion Alphabet* 131 *Canvas shoes*..are usually 'plimsoll' or espadrille shape for sports, beach or leisure. **1973** 'D. JORDAN' *Nile Green* xxxvi. 181 She came in through the door like a commando in plimsolls. **1974** C. FREMLIN *By Horror Haunted* 157 The wary, plimsoled footstep.

pling (pliŋ), *v.* *U.S. slang.* [Origin unknown.] *intr.* and *trans.* To beg; to beg from (someone); *pling the stem* (see quot. 1927²). Hence **pli·nger**, a beggar; **pli·nging** *vbl. sb.*

1913 L. LIVINGSTON *Trail of Tramp* vii. 55 The other one will make a good assistant for me in plinging. *Ibid.* 56 It meant for James McDonald that he had become an apprentice for Kansas Shorty, the Plinger—a begging tramp. **1915** *N.Y. World* 9 May (Suppl.) 14/3 *Plinging*, to reach out for 'handouts'; to beg. **1927** *Amer. Speech* II. 390/1 A street is..a *stem*... Dinging the stem is known as *mooching, stemming* and *plinging*. *Plinging 'em right and left* is an arduous occupation, calling for gall, tact and sharp eyes. **1927** *Dialect Notes* V. 459 *Pling the stem*, to beg money on the street. **1931** G. IRWIN *Amer. Tramp & Underworld Slang* 148 *Pling*, to beg on the street, probably a corruption of 'pillinge'.

Plinian, *a.* and *sb.* Add: **A.** *adj.* **1.** (Later example.)

1962 D. HARDEN *Phoenicians* xi. 154 The Plinian tradition that glass was invented in Phoenicia.

2. Also **plinian.** Applied to (the stage of) a volcanic eruption in which a narrow blast of gas is ejected with great violence from a central vent to a height of several miles before it expands sideways. [In this sense ad. It. *Pliniano* (A. Stoppani *Corso di Geologia* (1871) I. II. v. 310); so called because the eruption of Vesuvius in A.D. 79, which killed Pliny the Elder and was described by his nephew the younger Pliny, was of this kind.]

Quots. 1884, 1897 refer specifically to the eruption of A.D. 79.

[**1884** H. J. JOHNSTON-LAVIS in *Q. Jrnl. Geol. Soc.* XL. 37 Some authors have supposed that the principal part of the Vesuvian cone was thrown up by the eruption which destroyed Pompeii... Let us imagine the condition of affairs towards the termination of the Plinian eruption. **1897** I. C. RUSSELL *Volcanoes N. Amer.* i. 16 Following the Plinian eruption Vesuvius became quiet once more.] **1903** A. GEIKIE *Text-bk. Geol.* (ed. 4) I. 278 Three phases of its [*sc.* Vesuvius'] energy are recognised... In the third and most vigorous phase, which has been termed Plinian,.. large volumes of steam, dust, ashes, scoriæ, bombs and blocks are expelled with great violence high into the air and fall around the crater, while occasionally streams of lava issue from rents in the cone. **1944** A. HOLMES *Princ. Physical Geol.* xx. 466 Four days after the paroxysm [of Vesuvius in 1906]—the Vesuvian phase—began, it culminated in a mighty uprush of gases—the Plinian phase..—which continued for the greater part of a day,.. tearing away the upper portions of the cone, and reaching a height of 8 miles. **1965** R. FURNEAUX *Krakatoa* iii. 37 Its occasional 'Plinian' outbursts bring Vesuvius within the same classification as Krakatoa. **1975** FIELDER & WILSON *Volcanoes of Earth, Moon & Mars* iv. 47/2 The pyroclastic blanket could easily be generated by a lunar plinian eruption. **1976** P. FRANCIS *Volcanoes* iii. 114 Probably the best modern example of a Plinian eruption was that of the Bezymianny volcano in Kamchatka.

plink (pliŋk), *v.* [Imit.] **a.** *intr.* To emit a short sharp metallic or ringing sound; to play a musical instrument in this manner.

1941 E. P. O'DONNELL *Great Big Doorstep* 194 A frog plinked, squirmed out, snapped open and away. **1945** B. MACDONALD *Egg & I* (1946) xii. 144 The berries..had begun to plink into the..buckets. **1976** *Gazette* (Montreal) 19 July 3/3 (*heading*) Pianist plinking for Canada. **1979** G. HAMMOND *Dead Game* v. 77 A bullet at full speed plinks like somebody leaned on the lid of a biscuit-tin.

b. *intr.* and *trans.* To shoot a gun *at* a target; to hit (a target) with a shot from a gun.

1966 R. THOMAS *Spy in Vodka* x. 94 It was an ugly gun, [not] designed..for plinking at rabbits. **1975** G. V. HIGGINS *City on Hill* vi. 160 The back was full of rats... I bought an air-pistol..and I'd plink at them. **1976** L. DEIGHTON *Twinkle, twinkle, Little Spy* iv. 41 Goddamned weather... I would have plinked him but for that damned patch of ice.

So **pli·nking** *ppl. a.* and *vbl. sb.*

1961 *Guns & Hunting* Dec. 11/2 You may therefore convert the gun into a small-game and plinking arm. **1965** *Listener* 30 Sept. 507/3 On summer Wednesdays we have supped full of horrors with heartbeats down the corridors, creaking boards, plinking music and plenty of the consequences of over-tidiness, over-mothering, and all that suppressed sex. **1977** J. CLEARY *High Road to China* iv. 139 The General practising his banjo..the plinking of the strings. **1977** R. E. HARRINGTON *Quintain* xiv. 162 Quintain heard a brittle plinking sound that went from a high tone to a low tone. **1978** *Detroit Free Press* 2 Apr. 8E/1 Say you owned a handgun and wanted to take it out for plinking or target practice.

plink (pliŋk), *sb.* [f. the vb.] The sound or action of plinking; a sharp metallic noise. Also quasi-*adv.* and as *int.*

1954 J. R. R. TOLKIEN *Two Towers* viii. 153 And plink! a silver drop falls. **1961** *Amer. Speech* XXXVI. 305 The smaller pebble goes *plink*, the larger, *plunk*. **1971** *Daily Tel.* (Colour Suppl.) 3 Dec. 7/1 No wonder..that the first plink of those conciliatory ping-pong balls produced rapturous applause. **1974** *Times* 7 Mar. 12/4, I cannot dance to electronic music, because I simply do not know how to hear those plinks and plonks and bumps.

plinth. Add: **1. c.** (Further example.) Also, a course of bricks or stones in a wall, above ground level, by which the part of the wall above is made to be set back in relation to the part below; = *plinth course* s.v. sense 3 below.

1968 W. G. NASH *Brickwork* II. v. 112 The golden rule to remember when setting out the bonding for plinth courses and the walling below is to set out the neat work immediately above the plinth course first and let the bonding below the plinth be bonded to suit the neat work. In other words, always bond downwards from the face-work and *not* upwards from the work below the plinth courses.

d. (Further examples.)

1904 R. J. FARRER *Garden of Asia* xvi. 150 That glorious cone towering up into the sky from its plinth of hills. **1935** 'E. QUEEN' *Spanish Cape Mystery* ix. 187 Ellery was sprawled on his side in an uncomfortable position, his eyes glued to a plinth of light.

2*. A shallow wooden cabinet in which a record deck is mounted.

1963 *Hi-Fi Year Bk.* 35 Both the GL58 and GL70 are now available on plinths. **1975** *Gramophone* Sept. 537/2 The model BDS80 is the first of the new BSR Belt Drive Series of transcription turntables; the unit under review was delivered already mounted in a plinth. **1976** *Southern Even. Echo* (Southampton) 2 Nov. 9/2 (Advt.), Pioneer SA5300 amplifier 2×12 watts RMS plus Garrard 125SB belt drive turntable, plinth, cover, fitted with Shure M75/6/SM magnetic cartridge.

3. *plinth-like* (example); **plinth block,** a block sited on the floor and forming part of the base of the moulding of a door or window; **plinth course** = sense 1 c above.

1893 J. P. ALLEN *Pract. Building Construction* xviii. 293 Plinth blocks or bases..are often put at the bottom of architraves..for the skirting to run up against, as well as for appearance. **1932** F. L. WRIGHT *Autobiogr.* II. 138 The whole exterior was bedeviled..with corner-boards, panel-boards, window-frames, corner-blocks, plinth blocks, rosettes, fantails, ingenious and jigger work in general. **1873** F. ROGERS *Specifications Pract. Archit.* II. xxv. 375 All plinth courses, jambs of doors and windows, and window-sills, strings, and chimney-cappings to be in red moulded bricks. **1968** Plinth course [see 1 c above]. **1905** *Harper's Mag.* July 195/1 Those short, stubbed girls and women..were of plinthlike bigness up and down.

Pliocene, *a.* (Earlier example.)

1831 [see *EOCENE *a.* 1].

Pliofilm (pləi·ofilm). Also **pliofilm.** A proprietary name for a type of transparent, waterproof membrane made of rubber hydrochloride and widely used for packaging, waterproofing, etc. Freq. *attrib.*

1934 *Official Gaz.* (U.S. Patent Office) 20 Nov. 537/2 Goodyear Tire & Rubber Company, Akron, Ohio. Filed Sept. 22, 1934. Pliofilm... Claim use since June 28, 1934. **1936** *Trade Marks Jrnl.* 19 Feb. 223/1 Pliofilm... Material in sheets, strips and films manufactured from a composition consisting principally of a derivative of india-rubber. The Goodyear Tire & Rubber Company.., Akron, Ohio, United States of America; manufacturers. **1938** *India Rubber World* XCVIII. III. 47/1 Pliofilm..is being used in direct contact with many food and pharmaceutical items. **1949** *Times* 26 Feb. 3/1 Aero engines and delicate instruments..were contained in Pliofilm envelopes during shipment. **1956** H. GOLD *Man who was not with It* xxxii. 310 Shoppers..wearing rubber or pliofilm galoshes on their shoes. **1961** L. MUMFORD *City in History* xvii. 546 A world in which he is insulated by glass, cellophane, pliofilm from the mortifications of living. **1969** J. H. STICKELMEYER in W. R. R. Park *Plastics Film Technol.* i. 7 'Pliofilm' is produced by adding hydrogen chloride gas to a special grade of natural crude rubber in benzene solution. **1974** H. MCCLOY *Sleepwalker* v. 73 All these furs were..in dry cleaner's pliofilm bags.

Plio-Pleistocene (pləi:opləi·stǒsīn), *a.* *Geol.* [f. PLIO(CENE *a.* (*sb.*) + PLEISTOCENE *a.* (*sb.*).] Of or pertaining to the end of the Pliocene and the beginning of the Pleistocene epochs, or the Pliocene and Pleistocene epochs together. Also *absol.*

1929 *Q. Jrnl. Geol. Soc.* LXXXV. 520, I am the more inclined to look on the remarkable conglomerate of Wadi Abu Nefukh..as of the late Pliocene or Plio-Pleistocene age. **1946** F. E. ZEUNER *Dating the Past* v. 135 Reasons have been given..for fixing the Plio-Pleistocene boundary just before the Early Glaciation. **1957** J. K. CHARLESWORTH *Quaternary Era* xl. 1087 The exchange of plants between Asia and North America and the spread of the horse and camel-llama from North America to Asia about the Plio-Pleistocene transition. **1975** *Nature* 4 Dec. 395/1 Plio-Pleistocene lacustrine and fluvial strata exposed along the east side of Lake Rudolf, Kenya, now known as 'East Rudolf', have been under study since 1969. **1977** *Offshore Engineer* May 37/3 Plio-Pleistocene sedimentation could not keep pace, and layers are very thin. **1978** *Nature* 17 Aug. 662/2 In the Plio-Pleistocene of East Africa, the various species of hippo are most common in river channel and lake margin sediments.

pliotron (pləi·ǒtrǫn). *Electronics.* [f. Gr. πλείω-ν more (see PLEIO-) + *-TRON.] A high-vacuum thermionic valve with one or more grids.

1915 I. LANGMUIR in *Electrician* 21 May 242/2 The term 'Pliotron' has been adopted to designate a Kenotron, in which a third electrode has been added for the purpose of controlling the current flowing between the anode and the cathode. **1918** *Wireless World* July 230 The characteristics of the pliotron depend upon the length of filament used. **1945** COOKE & MARKUS *Electronics Dict.* 278/2 *Pliotron*, a high vacuum thermionic tube in which one or more electrodes (grids) are employed... All amplifier tubes in radio sets are pliotrons, but the term has never gained extensive use. **1971** G. M. & R. D. CHUTE *Electronics in Industry* (ed. 4) x. 167 A third element, called the grid, is placed between the anode and cathode to form a high-vacuum type of triode (also called the pliotron).

plip-plop (plip,plǫp). [Imit.] A representation of a rhythmically regular sequence of sounds. Hence **pli·p-plopping** *a.*

1953 *John o' London's Weekly* 23 Jan. 75/3 Several percussion players produced delicate plip-plops, and three sopranos murmured the vocal line. **1961** *Countryman* Autumn 514 Plip-plop went the donkey's decorously pointing toes. *Ibid.* 516 Then a 'plip-plop'... The donkey..was tittupping round the dustbin. **1979** J. BARNETT *Backfire is Hostile!* xiv. 165 Rain began to fall slowly, in large separate plip-plopping drops.

‖ **plique à jour** (plīk a ʒur). [Fr.] A technique in enamelling in which small areas of translucent enamel are fused into the spaces of a wire framework to give an effect similar to stained glass.

1878 J. H. POLLEN *Anc. & Mod. Gold & Silver Smiths' Work* p. clxii, French writers give this kind of [transparent] enamel the name of 'plite' or 'plique à jour'. **1899** H. CUNYNGHAME *Art-Enamelling* iv. 95 By plique-à-jour we mean filigree-work executed in gold or silver, and filled up with transparent enamels. **1959** *Times* 9 Feb. 10/5 Her work in *cloisonné, plique-à-jour* and *champlevé* won medals and certificates in several countries. **1964** H. HODGES *Artifacts* iii. 63 To do this type of work, plique-à-jour, the areas for the enamel were fretted and given a temporary backing of sheet mica or some similar material to which the enamel would not adhere. **1973** *Country Life* 26 July 250/2 A gold filigree cup with a band of enamel in the exacting technique of *plique-à-jour*. **1975** *Times* 1 Aug. 17/6 A gold, *pliqué* [sic] *à jour* enamel and crystal kaleidoscope jewel made £750... It is an outstanding piece of French nineteenth-century craftsmanship.

plissé (plise), *sb.* and *a.* [Fr., pa. pple. of *plisser* to pleat.] **A.** *sb.* A piece of fabric shirred or gathered into narrow pleats; a gathering of pleats. Also in *Comb.*

1873 *Young Englishwoman* June 286/1 The front width of the skirt is trimmed with a blue plissé, headed with a deep flounce. **1880** [see *GOLD¹ 8 b]. **1920** *Glasgow Herald* 13 Nov. 4 Plissé is the last word in style, and the latest knitted frocks have wide borders done in ribbed work that spring out into plissés. **1954** *Sun* (Baltimore) 8 May 5 (Advt.), Glamour gown... Made of Strat-o-Sheer plisse. **1969** *Sears Catal.* Spring/Summer 11 Cool, comfortable Plissé-textured Pajamas of 80% cotton, 20% polyester.

B. *adj.* Formed into small pleats.

1875 [see *elbow-sleeve* s.v. *ELBOW *sb.* 5]. **1895** *Montgomery Ward Catal.* Spring & Summer 12/2 Printed plisse silk, 20 inches wide. This is the newest idea in gauffre or crinkled silks... It is a small, crinkly stripe pattern in white grounds over which is printed..floral designs. **1928** *Times* 9 May 10/6 A train of lime green and silver tissue, lined with plissé chiffon. **1962** J. T. MARSH *Self-Smoothing Fabrics* iv. 40 Combinations of various factors are capable of disguising creases, as in crepes, seersuckers and plissé fabrics, but other combinations may accentuate the crease. **1967** SINGHA & MASSEY *Indian Dances* xv. 130 Over these they wore plissé skirts made of stiff material in three tiers the longest of which reached several inches above the knee.

plock (plǫk), *sb.* [Imit.] A sharp click or report, as of one hard object striking another.

1936 C. DAY LEWIS *Friendly Tree* xii. 173 The cries of the boys, the curt shouts of the masters, the plock as bat met ball—all these sounds were somehow unsynchronized. **1969** H. R. F. KEATING *Inspector Ghote plays Joker* xvi. 210 From the lit table there came the soft double plock of one of the balls striking the other two. **1976** G. EWART *No Fool* II. 61 The plock of bat on ball penetrates outfields, calming to the mind.

plock (plǫk), *v. rare.* [Imit.: cf. prec.] *intr.* To make a sound as of taut fabric being pierced.

1931 V. SACKVILLE-WEST *All Passion Spent* II. 161 Sitting down by her, as her needle plocked in and out of her embroidery, he would gaze fondly at her bent head.

plod, *sb.*¹ Add: (Further examples.) Also with alliterative reduplication, as *plod-plod*.

1899 N. B. TARKINGTON *Gentleman from Indiana* xv. 266 What was there left but the weary plod, plod, and dust of years? **1926** *Blackw. Mag.* Apr. 519/2 The angles of the rungs become very painful under the slow plod-plod of the horse's movement. **1972** *Times Lit. Suppl.* 9 June 654/2 This plod-plod of approach and application isn't novel. **1975** M. BRADBURY *Hist. Man* ix. 148 The agenda has grown longer..a routine plod through matters of budgets..and examinations. **1979** *Guardian* 23 Oct. 15/8 Those who found the book a bit of a plod but hoped the screen might set it up.

plod (plǫd), *sb.*² *Austral.* [perh. PLOP *sb.* influenced by PLOT *sb.*: see PLOD *v.* 4.] **a.** A (particular) piece of ground worked by a miner. Also, a work sheet with information relevant to this.

1941 BAKER *Dict. Austral. Slang* 55 *Pitching the plod*, 'the exchange of words' between miners 'on the state of the ground when coming on or going off shifts'. **1948** K. S. PRICHARD *Golden Miles* 72 He had to go to the office for his plod—the card on which he filled in particulars of the work he was doing, its position in the mine, and the hours he was working.

b. A story or yarn; an excuse.

Plod is entered in the *Eng. Dial. Dict.* as a Cornish word meaning 'a short or dull story; a lying tale'.

1945 G. CASEY *Downhill is Easier* 136 'I suppose he told you the whole plod?' I sneered. **1954** T. A. G. HUNGERFORD *Sowers of Wind* 241 That's the plod he put up, anyway. **1975** X. HERBERT *Poor Fellow my Country* III. xxi. 1126 Put in a plod for me, mate.

-ploid (ploid), the ending of *HAPLOID *a.* (and *sb.*) and *DIPLOID *a.*, used to form analogous terms referring to the number of chromosome sets in a cell or organism (as *EUPLOID, *HEXAPLOID, *hyperploid* s.v. *HYPER- IV. *adjs.*); occas. used with prefixed arabic number as *16-ploid*. [*haploid, diploid* f. Gr. ἁπλόος single (also ἁπλοΐς, -ΐδος and ἁπλο-ειδής, f. εἶδος form), διπλόος double; the *-id* in these words Strasburger (*Progresses Rei Bot.* (1907) I. 137), their coiner, connected with *id, idant*, and *idioplasm* (f. Gr. ἰδιο-: see IDIO-).]

1928 *Hereditas* X. 245 Lately Blackburn.., investigating chromosome numbers in *Silene*, has described two races of *Silene ciliata* differing in chromosome number, the one being tetraploid (as compared with most other *Silene*-species), the other 16-ploid.

ploidy (ploi·di). *Biol.* [f. *HA)PLOIDY, *POLY)PLOIDY, etc.] The number of homologous sets of chromosomes in a cell, or in each cell of an organism.

1947 *Genetics* XXXII. 512 A state of indefinite 'ploidy'. **1953** *Jrnl. Gen. Microbiol.* VIII. 101 Another possibility is that there is a different degree of effective ploidy of the F+ and F− gametic cells, the F− gametic cell having a higher degree of ploidy (or, possibly, more nuclei) than the F+. **1961** *Lancet* 5 Aug. 318/1 All metaphases have been analysed. When an exact account was impossible the ploidy level has been estimated. **1970** *Watsonia* VIII. 140 As most of our counts are approximate the ploidy level rather than the chromosome number is given. **1976** *Nature* 29 Apr. 785/2 To confirm the ploidy, both strains were grown axenically and their chromosomes were stained.

plomb, var. *PLOMBE.

plombage (plǫmbā·ʒ). *Surg.* [a. F. *plombage* filling of teeth, f. *plomber* to fill, apply lead to (f. *plomb* lead: see PLUMB *sb.*) + *-age* -AGE.] The introduction of plombe into the cavity of the chest; treatment with plombe.

1933 *Tubercle* XV. 97 (*heading*) The operation of plombage in pulmonary tuberculosis. *Ibid.* 98 Before plombage could be considered it was necessary that extra-pleural thoracoplasty..should reach a stage of technical efficiency. **1957** P. DUFAULT *Diagnosis & Treatment of Pulmonary Tuberculosis* (ed. 2) xix. 352 Plombage for the purpose of collapsing a limited area in mid-lung has been replaced by wedge resection. **1975** *Amer. Rev. Respiratory Dis.* CXI. 270/1, 13 animals..underwent left pneumonectomy with wax plombage.

plombe (plǫm). *Surg.* Also **plomb.** [a. G. *plombe* seal, filling (of tooth), plombe, f. F. *plomb* lead, lead weight (see PLUMB *sb.*).] (A mass of) soft material inserted into a bone cavity or into the cavity of the chest around a collapsed lung.

1904 *St. Thomas's Hosp. Rep.* XXXII. 433 The material used for bone-plugging (iodoform-knochen plombe).. consists of a mixture of 60 parts of the finest pulverised iodoform and 40 parts each of spermaceti and oil of sesame. **1905** GOULD *Dict. New Med. Terms* 423/2 *Plomb*. ..*P.*, *iodoform* (of Mosetig-Moorhof), an antibacillary agent for filling bone-cavities after operations for tuberculosis or osteomyelitis. **1909** H. PRINZ *Dental Materia Medica & Therapeutics* III. 436 The material advocated by Mosetig..is known in general surgery as 'bone plombe'. *Ibid.*, The plombe must completely fill the cavity. **1931** *Surg., Gynecol. & Obstetr.* LII. 738/2 (*heading*) Sketches showing the effects of paraffin filling (plombe) on a cavity of the left lung. **1956** *Jrnl. Chronic Dis.* IV. 623 The material used in the plomb has varied over the years; currently the most popular material is methyl methacrylate (Lucite) spheres sealed in a polyethylene bag, or folded polyethylene sheets. The plomb is placed between the lung and periosteum centrally and the deperiostealized ribs peripherally. **1965** JENKINS & WOLINSKY in G. L. Baum *Textbk. Pulmonary Dis.* viii. 195/1 Stabilizing the subperiostally freed and collapsed chest wall with some type of plomb.

plombière (plǫmbi,ę̄r). [f. F. *Plombières-les-Bains*, name of a village in the Vosges Department of eastern France.] A kind of dessert made with ice cream and glacé fruits. Freq. *attrib.* or as *adj.*

1846 A. SOYER *Gastron. Regenerator* 573 Prepare..the.. plombière ice... Make a border of patisserie d'amande.. upon your dish, in the centre of which put a little of the plombière. **1849** THACKERAY *Pendennis* I. xxiv. 233 The ice was brought in—an ice of *plombière* and cherries. **1907** [see *CROÛTE]. **1958** W. BICKEL tr. *Hering's Dict. Classical & Mod. Cookery* 724 *Plombière*, vanilla ice cream mixed with salpicon of candied fruit macerated in kirsch, filled in parfait mould alternately with apricot jam. **1962** S. BECK et al. *Mastering Art French Cooking* x. 549 (*heading*) Plombières with Fresh Strawberries or Raspberries.

‖ **plongeur** (plǫñʒör). [Fr.] A boy who is employed as a menial in a restaurant or hotel.

1933 'G. ORWELL' *Down & Out in Paris & London* iv. 32 The whole staff, from the manager down to the *plongeurs*, was working twenty-one hours a day. **1954** J. ATKINS *George Orwell* iv. 94 Why does the *plongeur* exist if his work is largely useless? **1964** *Punch* 1 Apr. 494/1 He'd have to become a dish-washer, or perhaps even a *plongeur*. **1967** G. WOODCOCK *Crystal Spirit* II. v. 100 The world of common lodging-houses..as different from his [*sc.* Orwell's] life among the *plongeurs* as England is different from France. **1976** *Observer* 22 Feb. 22/3 It is over 40 years since George Orwell described a *plongeur's* life in 'Down and Out in Paris and London'. **1977** *Daily Tel.* 18 Mar. 18 Titles are nice but surely the Dorchester is going a little too far advertising for a 'Supervisor Plongeur' to head the washing up department.

plonk (plǫŋk), *v. dial.* and *colloq.* [Imit.: cf. PLUNK *v.* **1. a.** *trans.* To hit or strike with a plonk. Chiefly *dial.*

1874 in A. Easther *Gloss. Dial. Almondbury & Huddersfield* (1883) 103 There were three fighting when you plonked Wells in the face. **1883** A. EASTHER *Ibid.* 102 *Plonk*, to hit plump. Used especially of marbles, when the one shot strikes the other before touching the ground. **1891** *Leeds Mercury Weekly Suppl.* 3 Jan. 8/6 I'll plonk tha. **1896** *Ibid.* 21 Mar. 3/8 Plenk him one o' t'noase if he doesn't shut up. **1903** in *Eng. Dial. Dict.* IV. 549/2 I'll plonk tha, if I get hod on tha. **1925** FRASER & GIBBONS *Soldier & Sailor Words* 225 *Plonk*, to shell. Suggested by the sound of the impact and burst. **1941** *London Opinion* May 64/1, I plonked him good and hearty on the beak.

b. *intr.* To emit or cause something to emit an abrupt vibratory sound, *spec.* in playing a musical instrument. Freq. with *away*.

1927 *Melody Maker* May 489/2 Can you imagine..a saxophone section playing a nice ligato movement and the banjo plonking away for all he is worth..and killing the good work of the saxes. **1976** D. HEFFRON *Crusty Crossed* i. 8 By age three I was plonking away at the piano on my own. **1979** *Stand* XX. IV. 34/1 The band plonks away at sad, slow French and Italian numbers.

2. a. *trans.* To set or drop (something) in position with a heavy or clumsy gesture; to put *down* firmly. Also (with a person or object), to set (someone) abruptly in a particular place or set of circumstances; to seat (someone) hurriedly or unceremoniously. Cf. PLANK *v.* 2 a (in Dict. and Suppl.), *PLUNK *v.* 3 b.

1941 BAKER *Dict. Austral. Slang* 55 *Plonk down*, to put down. Also, 'plonk one's frame into a chair': to sit down. **1946** F. COOZE *Ten Bob Each Way* 22 So next time I plunged I forgot the tip And plonked my all on my own little pick. **1959** *Woman* 16 May 23/2 An officious nurse plonked down a gas and air mask on my face. **1959** *Sunday Times* 17 May 20/4 Jones has been plonked down in Gagland where the jokes are separate from the action and so has small chance of coming out strong as the comic actor we know him to be. **1967** *Spectator* 29 Sept. 358/2 A nasty-looking structure will be plonked down in front of King's Cross, thus ruining its two magnificent archways. **1972** G. DURRELL *Catch me Colobus* vii. 145 Then you'd lead her [*sc.* a piglet] carefully to the pan and she'd plonk both her stubby front feet into it, little hooves wide-spread, and dig her nose in and guzzle. **1976** P. CAVE *High Flying Birds* ii. 20 A litre bottle of red wine was plonked down on to the counter under my nose. Plonked plonk, in fact. **1977** *Sounds* 9 July 10/4 The 150 press-persons present were ushered into a darkened room, plonked on rows of chairs, told to put headphones on and left to listen.

b. *refl.* and *absol.* To sit (oneself) down heavily or unceremoniously.

1946 U. KRIGE *Way Out* v. 64 Handing them two cigarettes each, I plonked down beside them to tell them the whole story. **1976** M. SPARK *Takeover* xii. 178 Walter now plonked himself, tired from his walk, on the sofa. **1979** *Guardian* 9 Aug. 22/8 They would plonk themselves down, undress..to encourage support for local naturists.

Hence **plo·nking** *ppl. a.*, (*a*) large *dial.*; (*b*) that plonks; *spec.* (see quot. 1950); **plo·nking** *vbl. sb.*; **plo·nkingly** *adv.*

1896 *Leeds Mercury Weekly Suppl.* 21 Mar. 3/8 What a plonkin' hoile tha hes fer a bedrahm. Little Jimmy hes a plonkin' wife. **1903** in *Eng. Dial. Dict.* IV. 550/1 A gurt plonkin' cat. **1950** S. POTTER *Lifemanship* iii. 44 If you have nothing to say, or, rather, something extremely stupid and obvious, say it, but in a 'plonking' tone of voice—i.e. roundly, but hollowly and dogmatically. *Ibid.* 45 'Plonking' of a kind can be made by the right use of quotation or pretended quotation. **1957** *Economist* 5 Oct. 21/2 India, entangled in its own frontier troubles and engrossed with the problem of borrowing from abroad, has lately sounded a little less plonking in its pronouncements on international affairs. **1959** S. CLARK *Puma's Claw* xv. 181 Delivered with a gruff, passionate intensity (Potter would certainly call them plonking) those words always announced our arrival on a summit. **1965** *New Statesman* 19 Mar. 426/2 These reports, so far from being accurate, have been described by one member of the shadow cabinet as 'absolute poppycock', and by another even more plonkingly, as 'balls'. **1969** D. FRANCIS *Enquiry* ii. 31 'The bet was struck,' Gowery said plonkingly, pointing to the ledger. **1977** *Listener* 5 May 591/1 Presented with an argument that.. he intended to ignore, Lord Reith would say in a matter-of-fact way: 'I hear you.' It was an admirably plonking rhetorical device. **1977** *Chainsaw* Sept./Oct. 8/2 The singer is accompanied only by electric organ, regular drum beats, and plonking bass.

plonk (plǫŋk), *sb.*¹ [Imit.: cf. prec. and PLUNK *sb.*, *adv.*, *int.*] The sound of or as of one hard object hitting another; a heavy thud. Also as *adv.*, with a plonk, directly, and as *int.* So **plonk-plonk.**

1903 WODEHOUSE *Tales of St. Austin's* 9 There was a beautiful, musical *plonk*, and the ball soared to the very opposite quarter of the field. **1914** *Picture Fun* 26 Dec. 2 He unfortunately pinched it just as the waiter was passing with a tray of ices, and plonk came that kangaroo's hind paws,..bang agin the old chap's tummy! **1920** *Punch* 10 Mar. 199/2 A befogged Zeppelin laid a couple of bombs plonk into the homestead. **1928** *Manch. Guardian Weekly* 15 June 474/1 A patois that sounds like the plonk-plonk of ping-pong balls on a hard table. **1943** H. PEARSON *Conan Doyle* iii. 46 'Plonk' went the gun [*sc.* an airgun], and down went the medal. **1960** *Oxf. Mag.* 28 Apr. 248 (Advt.), The satisfying plonk of *The Observer* falling on the doormat. **1978** M. BIRMINGHAM *Sleep in Ditch* 118, I feel as if I'd thrown off an enormous weight. I hope it hasn't landed plonk on you.

plonk (plǫŋk), *sb.*² *colloq.* (orig. *Austral.*). [prob. a corruption of *blanc* in Fr. *vin blanc*.] Cheap wine, or wine of poor quality. Also *attrib.*

Various popular and humorous etymologies, such as that suggested in quot. 1967, are without foundation. Although it may be argued that the word denotes red wine more commonly than it does white wine, the etymology given above is attested by the earliest sources.

[**1919** W. H. DOWNING *Digger Dialects* 52 *Vin blank*, white wine. *Ibid.*, Von blink, a humorous corruption of vin blanc.] **1930** H. WILLIAMSON *Patriot's Progress* IV. 137 Nosey and Nobby shared a bottle of plinketty plonk, as *vin blanc* was called. **1933** *Bulletin* (Sydney) 11 Jan. 12 The man who drinks illicit brews or 'plonk' (otherwise known as 'madman's soup') by the quart does it in quiet spots or at home. **1940** A. L. HASKELL *Waltzing Matilda* 37 Fortified red wine of the kind that inebriates with speed and economy is 'pinky' or 'plonk'. **1941** K. TENNANT *Battlers* ix. 104 'Keep off the plonk,' Thirty-Bob said in an undertone to the Stray. 'They just spilt some on my boot and it burnt a hole.' **1946** D. STIVENS *Courtship of Uncle Henry* 72 Jessie's been on the plonk again... Goes round the wine bars at the Cross. **1949** *Here & Now* (N.Z.) Oct. 9/1 Rows of gaudily-labelled bottles of local 'plonk' stacked on shelves behind the bar. **1950** 'N. SHUTE' *Town like Alice* 322 He asked me if I would drink tea or beer or plonk. 'Plonk?' I asked. 'Red wine,' he said. **1953** A. UPFIELD *Murder must Wait* viii. 76 Mother gallivants about to plonk parties..plonk being Alice McGorr's designation of a sherry party. **1965** *New Statesman* 3 Dec. 873/3, I do not eat in restaurants, travel first-class, or buy fillet steak. But there are cheaper cuts of meat, and wine, though mostly poor plonk stuff in the South, is cheap enough. **1967** *Daily Tel.* 15 Nov. 21/8 Surely the word 'plonk' is onomatopoeic, being the noise made when a cork is withdrawn from the bottle? **1968** *Listener* 1 Aug. 134/3 Over the numerous bars were texts urging moderation and adverts pushing the cheapest and most potent plonk in Britain. **1970** *Times* 23 Mar. 25/5 Sales of his newly introduced Vin Plonque, or 'plonk' in the British vernacular, are soaring. **1973** E. MCGIRR *Bardel's Murder* ii. 29 A Miss Traylor, aged seventy, intelligent and given to plonk. **1976** *Scotsman* 24 Dec. (Weekend Suppl.) 3/6 The author is particularly scathing about Sainsbury's Spanish plonk, but does not mention the same chain's better-than-average range of Hocks and Moselles. **1977** *Time Out* 21 Jan. 3/3 Your review of 'party plonk'..misses out the largest 'chain' of off-licences in London, the independents who belong to no chain. **1979** *Globe & Mail* (Toronto) 25 Aug. 10/6 The only other customer was a construction worker who was buying a bottle of white plonk for about $1.40.

plonk (plǫŋk), *sb.*³ *R.A.F. slang.* [Origin uncertain.] An aircraftman second class.

1941 *New Statesman* 30 Aug. 218/3 *A.C. Plonk*—Lowest in the R.A.F., aircraftman 2nd class. **1943** C. H. WARD-JACKSON *Piece of Cake* 10 A/C Plonk, aircraftman 2nd class. In 1914–1918 'plonk' was Flanders slang for 'mud'. Hence, an A/C Plonk is an aircraftman literally in the mud or at the bottom—that is, lowest classification of the lowest rank in the R.A.F. **1946** *Slipstream* 62 Another synonym for an A.C.2 is A.C.Plonk. **1949** J. R. COLE *It was so Late* 61, I was only an A.C. plonk at the time.

plonked (plǫŋkt), *a. slang.* [f. *PLONK *sb.*² + -ED².] Intoxicated, drunk.

1943 *Life* 30 Aug. 70/2 A few badly plonked soldiers blearily unaware of just where they were. **1949** 'THE SARGE' *Excuse my Feet!* 42 George was difficult and Herbert was slightly *plonked*.

plonker (plǫ·ŋkəɹ). *dial.* and *slang.* [f. *PLONK *v.*] **1. a.** Something large or substantial of its kind. *dial.*

1861 C. C. ROBINSON *Dial. Leeds* 386 'A plonker' is an article having extraordinary substance. A piece of woven material unusually thick is 'a plonker'. **1885** *Pudsey Almanack & Hist. Reg.* Mar., Sitha Bill at that young woman's improver, isn't it a plonker? **1898** B. KIRKBY *Lakeland Words* 114 Noo that's a plonker. **1903** *Eng. Dial. Dict.* IV. 550/1 That turnip's a plonker.

b. A shell. *Austral.*

1941 BAKER *Dict. Austral. Slang* 55 Plonker, a shell. Diggers' slang. **1961** PARTRIDGE *Dict. Slang* Suppl. 1226/1 *Plonker*, a (cannon) shell: Australian soldiers': 1939+.

2. (See quot. 1970). *slang.*

1966 J. GASKELL *All Neat in Black Stockings* 72 If she'd been my daughter in fact I'd never have let her go out with an obvious plonker like myself. **1970** *Sunday Truth* (Brisbane) 13 Sept. 36/2 Do you know what a plonker is? —It's a chap who shares his ladyfriends with his mate.

plonko (plǫ·ŋko). *Austral. slang.* [f. *PLONK *sb.*² + *-O².] One who is addicted to 'plonk'.

1963 A. MARSHALL *In Mine Own Heart* 187 You end up a plonko with bells ringing in your head. **1965** W. DICK *Bunch of Ratbags* 69 We could go and see if there's any plonkos under Martin's Bridge and chuck rocks at 'em.

ploot. = *PLUTE.

1931 E. POUND *Let.* 6 Oct. (1971) 235 The only person in Amurikuh who *cd.* continue your periodical is Marianne. The necessary irreproachable respectability, the that against which no lousy ploot can object on the grounds of her not bein' a lady or bein' likely to pervert the growing school child, etc... Marianne has experience—quality dear to the cautious ploot. **1955** —— *Classic Anthol.* II. 106 Folk with no salary The heavens swat, while ploots can manage And the 'outs' cannot. **1959** —— *Thrones* 67 Ultramontaines Bitched France and then Austria, aristos are ignorant Plus illiteracy of the ploots.

plop, *sb.* and *adv.* Add: **A.** *sb.* (Later examples.) Also in *transf.* and extended uses.

1941 *Sun* (Baltimore) 4 Nov. 10/7 'Plops', the sound heard when 'p' or 'b' is spoken too loudly. **1965** *Wireless World* Aug. 379/2 Another effect is the possibility of a 10W 'plop' when switching on the power supply. **1967** P. A. PINCKNEY *Painting in Texas* v. 77 They might even hear the plop of acorns or pecans falling from trees along the creek. **1969** A. GLYN *Dragon Variation* vi. 171 The gas fire went out with a plop. **1974** HARVEY & BOHLMAN *Stereo F.M. Radio Handbk.* v. 104 The system is arranged so that the audio mute is activated before the i.f. mute and this staggering of the two muting levels reduces the edge-of-station 'plop'. **1979** W. NELSON *Minstrel Code* ix. 77 The automatic..was so silent that even the characteristic 'plop' of a silencer had been eliminated.

C. As first element in various alliterative combinations, as *plop-plop*, *plop-plump*, etc. Chiefly as *adv.* or *int.*

1893 'A. HOPE' in *Westm. Gaz.* 9 Dec. 2/1 Miss Phaeton flicked Rhino, and the groom behind went plop-plop on the seat. **1921** *Blackw. Mag.* Feb. 198/2 There is something peculiarly gratifying in the sound of the plop-plump of your naked feet in the round shallow pools of muddy water. **1922** JOYCE *Ulysses* 552 Whispering lovewords murmur liplapping loudly, poppysmic plopslop. **1928** J. M. BARRIE *Peter Pan* III. 75 There are many mermaids here, going plop-plop, and one might attempt to count the tails.

plop, *v.* Add: **1.** (Further examples in causative sense.) Also *refl.* and with *down*. Also *fig.*

1900 E. GLYN *Visits Eliz.* 66, I do hate to see a great hand..plopping a dish down..in the middle, so that one has to look at the next course all the time one is finishing the last one. **1960** V. NABOKOV *Invitation to Beheading* iii. 37 Emmie was gazing after them, while she lightly plopped the glossy red and blue ball in her hands. **1971** *Angling Times* 10 June 6/2, I plopped that lot under a bridge..and cheerfully expected to net every eel that passed downriver that night. **1975** A. BERGMAN *Hollywood & LeVine* vi. 71 She plopped herself comfortably onto the couch. **1975** *New Yorker* 28 July 8 (Advt.),

Something happens at The Biltmore that just doesn't happen in those plasti-glass, modular hotels that have plopped themselves down in every city in the country.

2. *intr.* To emit a sound or series of sounds suggestive of plopping.

1927 C. Connolly *Let.* 4 Jan. in *Romantic Friendship* (1975) 207, I got very depressed on Sunday evening and thought of..gas mantles plopping in evening chapel. **1972** R. Adams *Watership Down* iii. xxxviii. 316 All the surface of the river was winking and plopping in the rain.

plosh[1]. (Examples.)

1868 J. C. Atkinson *Gloss. Cleveland Dial.* 385 *Plosh* sb., puddle, liquid mire, like the sloppy mud on a road after much rain. **1895** J. Thomas *Randigal Rhymes* 22 Nor don't ee lag, or stag yourself By stanking through the plosh. **1930** H. Walpole *Rogue Herries* iii. 495 He found himself in the little dark wood,..his feet in plosh and mire.

plosh[2], var. Plash *sb.*[2]

1876 C. C. Robinson *Gloss. Dial. Mid-Yorks.* 103/2 *Plosh* is much more heard than 'plodge', and, as a substantive, bears relation to an object as well as an action. *Plosh* is anything of the nature and consistency of a puddle, into which, if a hasty foot be placed, or a stick let fall, there results a *plosh*. **1928** Blunden *Undertones of War* 138 The plosh of the whizzing fuse-top into the muck. **1935** S. Desmond *African Log* xliii. 208 'To listen to the silence of the forest', to hear the plosh in the dried herbage of the grasshoppers.

‖**ploshchadka** (plǫʃtʃa·dkā). *Archæol.* Pl. **ploshchadki.** [Russ., = ground, area, platform.] In Ukrainian sites of the Neolithic period, a raised area or platform, *spec.* one formed of burnt clay from the debris of collapsed buildings.

1913 E. H. Minns *Scythians & Greeks* vii. 134 The first finds were made about the village of Tripolje on the Dnĕpr forty miles below Kiev, whence this is called the Tripolje culture. The remains consist of so-called 'areas' (*ploshchádka*). **1923** *Nature* 26 May 726/1 The painted pottery comes either from large rectangular structures of wattle and daub called *ploshchadky* or from huts partly hollowed out in the earth. **1928** C. Dawson *Age of Gods* iii. 56 The clay figures..are found in Russia chiefly on the site of the curious buildings or platforms known as 'ploshchadki', which seem to have had a religious object. **1940** C. F. C. Hawkes *Prehist. Foundations of Europe* vi. 236 The Kiev region, where the peasants..made also.. rectangular structures ('ploshchadki') whose remains are always found burnt, without as yet any agreed explanation. **1957** V. G. Childe *Dawn Europ. Civilization* (ed. 6) viii. 137 The houses of later phases are represented by the celebrated *ploščadki*, areas of baked clay resulting from the burning and collapse of walls and floors.

plosion (plōu·ʒən). *Phonetics.* [f. Ex)plosion.] The eruption of breath involved in uttering a plosive. Hence **plo·sional** *a.*, of or pertaining to plosion.

1918 D. Jones *Outl. Eng. Phonetics* viii. 37 (*heading*) Faucal Plosion...Lateral Plosion. **1932** G. E. Fuhrken *Standard Eng. Speech* vi. 71 Nasal plosion is avoided if awkward combinations of consonants would result. **1935** B. Trnka *Phonol. Analysis Present-Day Stand. Eng.* 6 In the system of English consonantal phonemes there are two correlations, namely those of 1. voice and 2. plosion. **1946** [see *articulator 4]. **1961** L. F. Brosnahan *Sounds of Lang.* viii. 185 Consonants, produced primarily by plosional, frictional, or vibratory interference with the air stream. **1964** [see *alveolarity].

plosive (plōu·ziv), *sb.* and *a.* *Phonetics.* [f. Ex)plosive *a.* and *sb.*] **A.** *sb.* A consonantal sound in the formation of which the passage of air is completely obstructed and then suddenly released. **B.** *adj.* Of or pertaining to a plosive.

1899 W. Rippmann tr. *Vietor's Elem. Phonetics* 12 The passage may be completely closed. The breath is stopped for a moment..; but then it bursts through the obstacle with a little explosion. The result is a (voiced or voiceless) *stop* (or plosive or explosive). **1902** E. W. Scripture *Elem. Exper. Phonetics* xxix. 445 The *Association Phonétique Internationale* classifies and represents the consonants in the following way... Plosive [etc.]. **1909** D. Jones *Pronunc. of Eng.* 65 When we try to pronounce a breathed plosive, e.g. p, by itself, it is generally followed by a short breathed sound h. *Ibid.*, The explosion of a plosive consonant is formed by the air as it rushes out at the instant when contact is released. **1933** L. Bloomfield *Language* vi. 97 If we place the tongue or the lips (or the glottis) so as to leave no exit, and allow the breath to accumulate behind the closure, and then suddenly open the closure, the breath will come out with a slight pop or explosion; sounds formed in this way are *stops* (*plosives, explosives*), like our unvoiced [p, t, k] and our voiced [b, d, g]. **1961** *Amer. Speech* XXXVI. 220 Influence of preceding consonants negligible; that of following consonants, considerable. Vowels shorter before voiceless consonants than before voiced counterparts, before plosives than before fricatives. **1968** W. S. Allen *Vox Graeca* 2 If the speech-organs form a complete closure, during which air is prevented from passing until the closure is released, the resulting sound is termed a 'stop'. Stops are further subdivided into 'plosives' and 'affricates'. **1968** Chomsky & Halle *Sound Pattern Eng.* 321 Thus of the three nonnasal types with plosive efflux, one is aspirated and the other two are nonaspirated. **1968** J. Lyons *Introd. Theoretical Linguistics* iii. 104 If the obstruction in the air passage is complete, the resulting sounds are described as *stops* (or plosives). **1973** *Studies in Eng. Lit.: Eng. Number* (Tokyo) 20 The plosive sound in 'priest' in the third line is repeated in the succession of plosives, 'bound, brave, peril, doomed', in the fourth line. **1976** *Archivum Linguisticum* VII. 19 The Basque is not geminated, so if we wish to postulate that *sapo* had become a Castilian word before the intervocalic plosives had generally voiced, we might even be reduced to calling it 'culto'.

plot, *sb.* Add: **II. 3.** Delete '*Obs.* or *arch.* exc. in *U.S.*' and add further examples.

1899 Middleton & Chadwick *Treat. Surveying* I. iv. 146 It is often desirable to make a preliminary plot, as work progresses, to see how the work comes in. **1931** M. Hotine *Surv. from Air Photogr.* ix. 159 The minor control plot is the foundation of all subsequent detail plotting and will repay time and care spent on its construction. **1942** *R.A.F. Jrnl.* 13 June 11 Frequent advice from the plot was called for as to the safe course to steer. **1962** *Times* 22 Mar. 14/7, I had no idea where I was as there was no automatic plot in my ship and we had been too busy to keep an accurate reckoning. **1971** R. J. P. Wilson *Land Surveying* ii. 45 Once the pencil plot has been completed and checked the chain survey network of lines..is inked in. **1973** H. Gruppe *Truxton Cipher* xv. 154 Is it not standard procedure for the Combat Information Center to keep a plot of the ship's movements?

c. *Theatr.* A scheme or plan indicating the disposition and function of lighting and stage property in a particular production.

1883 D. Cook *On Stage* I. x. 219 The property-maker is duly furnished with a 'plot' or list of articles required of his department. **1949** T. Rattigan *Harlequinade* 56 The lighting for this scene has gone mad. This isn't our plot. There's far too much light. *Ibid.* 57 Check your plot, please. **1959** W. C. Lounsbury *Backstage from A to Z* 91 *Plot*, a floor plan or cue sheet or both, indicating location of lights, furniture, props, etc. Light plots, furniture plots, and prop plots should be made by the person responsible for each field, and notations of cues and changes should be clearly indicated.

d. A diagram showing the relation between two variable quantities each measured along one of a pair of axes usu. at right angles; = Graph *sb.*[1] 2 in Dict. and Suppl.

1912 *Jrnl. Amer. Chem. Soc.* XXXIV. 462 And reading from this plot by extrapolation to $C\Lambda = 0$, the value of $1/\Lambda_0$. **1947** E. E. Wahlstrom *Ign. Minerals & Rocks* viii. 240 (*caption*) Harker plot of analyses in Table 6. **1953** E. R. Peck *Electricity & Magnetism* ix. 279 Sometimes it is of interest not only to use the ratio B/H, but to plot in detail the B–H curve. Such plots are basic in the discussion of ferromagnetism. **1971** *Physics Bull.* Feb. 86/1 A ln σ against 1/T plot should, at the temperature of conversion, exhibit a change of slope.

e. *R.A.F. slang.* A group of enemy aircraft as represented on a radar screen.

1943 P. Brennan et al. *Spitfires over Malta* 44 We warned the new boys to be careful, as it was probably a big plot coming in and they would be certain to bomb Ta-kali. **1959** R. Collier *City that wouldn't Die* vii. 109 Every radar station reported a mass plot and the planes flew too high for visual checks.

IV. 8. (in sense 2) *plot-owner*; (in sense *3e) *plot-room*; (in sense 6) *plot-formula, -interest, -seller, -source, -spinning, -structure*; (in sense 7) *plot-maker*; **plot-line,** (*a*) the main features of a plot or story; a summary; (*b*) (see quot. 1961); **plot-ratio,** a ratio representing the density of building in a specified area of land (see quot. 1971).

1957 N. Frye *Anat. Criticism* 52 We may think of our romantic, high mimetic and low mimetic modes as a series of *displaced* myths, *mythoi* or plot-formulas. **1865** *Fortn. Rev.* 15 Dec. 354 The distinctive element in Fiction is that of plot-interest. The rest is vehicle. **1866** H. Sidgwick in A. & E. M. Sidgwick *Henry Sidgwick* (1906) 143 The plot-interest does not turn entirely on amativeness. **1961** M. McCarthy *On Contrary* (1962) 290 The chief plot interest in these books is to try to find out what happened before the book started. **1965** K. Graham *Eng. Criticism of Novel* iv. 99 By 1883..his [*sc.* Wilkie Collins's] brand of plot-interest was..out-moded. **1957** J. D. Salinger *Zooey* in *New Yorker* 4 May 33/3 The plot line [*sc.* of 'a sort of prose home movie']..is largely the result of a rather unholy collaborative effort. **1961** Bowman & Ball *Theatre Lang.* 267 *Plot line*, usually in the plural: Dialogue essential to the unfolding of the plot of a dramatic piece. **1962** *John o' London's* 12 Apr. 363/1 Its [*sc.* a film's] plot-line has the same grotesque implausibility. **1972** P. H. Kocher *Master of Middle-Earth* iv. 57 Like Greek drama or Miltonic epic which begin late along their plot lines, *The Lord of the Rings* begins just before the climax of Sauron's efforts to subdue the West. **1976** S. Hynes *Auden Generation* viii. 294 The main plot-line concerns an indigent painter..whose only completed picture..is seized by bailiffs. **1961** *Encounter* Apr. 71/1 Either Querry is right, or God is a plot-maker, working through his inferior priests,..ready to use any degree of absurdity..to get His own way. **1907** *Daily Chron.* 3 June 3/6 An association of..plot-owners has been formed for the purpose of improving their position. **1956** *Archit. Rev.* CXIX. 46 The main area of the site is developed according to the plot-ratio of 5:1 laid down by the City Corporation. **1958** *Listener* 23 Oct. 642/2 Plot ratios and daylight factors are now everyday tools at the disposal of the designer; only a decade ago such logical aids did not exist. **1971** P. Gresswell *Environment* 81 The density of business and commercial areas is usually expressed by total floor space of buildings divided by the area of the site, called 'floor space index' when local access roads are included, and *plot ratios* when they are not. **1973** *Geo. Abstr.* F. 65 The development of secondary dwelling areas should be subordinated to an overall objective for total land use intensity and the relation between gross and net plot ratios, together with the proportion of open space that is derived from this. **1978** *N.Y. Rev. Bks.* 23 Feb. 4/2 Leslie Martin, the most influential architectural teacher,..has for years pointed out that a far more effective use of a plot ratio can be devised than tower blocks. **1947** *Times* 3 Feb. 6/6 Below in the plot rooms radar engineers checked their screens for the position of the Home Fleet and exchanged bearings with the navigating officers. **1938** O. Sitwell *Trio* 60 The well-known plot-seller, Mr. X. Y. Z., is at present planning a new financial coup. **1885** J. O. Halliwell-Phillipps *Outl. Life Shakes.* (ed. 5) 566 The subtle devices..some of which..may be equally observed..in the original plot-sources of his dramas. **1962** *Times* 24 Apr. 14/7 Passages of plot-spinning conversation. **1957** N. Frye *Anat. Criticism* 207 The source of tragic effect must be sought.. in the tragic mythos or plot-structure. **1962** *Punch* 11 July 65/1 A good film..can hide even the most elaborate plot-structure by superficial casualness and naturalism. **1977** *Dædalus* Fall 107 Work on plot structure, the goal of which is a grammar of plots, has been carried out in many countries.

plot, *v.*[1] Add: **1. a.** *to plot* (one quantity) *against, versus,* etc. (another): to draw a graph in which ordinates (*y*-coordinates) represent values of the first quantity and abscissas (*x*-coordinates) represent those of the second.

1910 *Jrnl. Amer. Chem. Soc.* XXXII. 1015 The method of procedure employed has been to plot the values obtained by each observer for $\Delta t/N$ as ordinates against those of log N as abscissas. **1934** *Proc. R. Soc.* A. CXLV. 576 This would be shown clearly if the density were plotted against the logarithm of exposure, as is usually done for photographic plates. **1952** J. P. Casey *Pulp & Paper* II. xvi. 825 Stress-strain curves can be plotted as load scale reading versus angular deflection. **1976** *Sci. Amer.* Dec. 93/1 Data from spectrophotometry..enable one to plot the supernova's change in radius with respect to time.

b. Delete *rare* and add earlier and later examples.

1863 H. S. Merrett *Pract. Treat. Sci. Land & Engin. Surveying* ii. 124 For practical purposes, surveys should never be plotted to a less scale than three or four chains to the inch. **1906** Breed & Hosmer *Princ. & Pract. Surveying* I. xv. 397 The field maps of the U.S. Coast and Geodetic Survey are usually plotted on a scale of 1/10000. **1923** D. Clark *Plane & Geodetic Surveying* II. vi. 244 In plotting the map, the distances and elevations required must be obtained from the perspective dimensions on the photographs. **1963** W. K. Kilford *Elem. Air Survey* xi. 240 The radial-line plotter makes it possible to plot a map from the model formed in an ordinary mirror stereoscope. **1976** J. B. Garner et al. *Surveying* ii. 21 When the survey has been plotted it should, if possible, be checked visually on the ground.

2. a. (Further examples.) Also *fig.*

1915 W. Holt *Beacon for Blind* xiv. 140 When a proposed party was being plotted out he would say, 'Oh, don't ask the So-and-so's, they are such frumps'. **1928** *Oxford Poetry* 9 And, a week later, still, by plotting out The course of all the roadways round about 'In these some score of places he may be'. **1969** D. Widgery in Cockburn & Blackburn *Student Power* 130 The next period is worth some examination to plot out the Executive's responses to a fluid situation.

c. *Theatr.* (See *plot *sb.* 3 c.) To plan or devise (a stage production); to arrange lighting and stage property for (a production).

1933 P. Godfrey *Back-Stage* iv. 44 The amount of work [for the stage-manager] involved in organizing and plotting the complete stage arrangements for a simple play with a few changes of scene is considerable. **1974** *Times* 28 Dec. 7/3 At read-throughs and when you're plotting it [*sc.* a play], you stand there trembling..behind your script.

5. Delete † *Obs.* and add later examples. Also *absol.*

1943 *Writer* V. 99/1 Plotting problems are usually the chief difficulties fiction-writers have to face. **1951** F. Brown *Murder can be Fun* x. 138 There's a big difference in plotting soap operas and plotting magazine stories. **1962** *Listener* 30 Aug. 311/2 In *A Burnt Out Case* he [*sc.* Graham Greene] had kept the plotting simple, but it seemed somehow too strong all the same; 'perhaps it would have seemed less plotted if there had been more plot.' **1973** *Daily Tel.* 15 Mar. 8/5 The story is confusingly plotted. **1977** *Ibid.* 10 Sept. 7/2 Mrs Robins plots better but relies a bit much on coincidence.

plotch (plǫtʃ), *v. rare*[−1]. [Perh. f. Plotch *sb.*, or imit.] *trans.* To splash on to, to mark.

1922 Joyce *Ulysses* 745 All the mud plotching my boots.

Plotinian, *a.* (Earlier and later examples.)

1791 W. Enfield *Hist. Philos.* II. iii. ii. 69 We shall trace the progress of the Plotinian, or Eclectic, school. **1969** T. F. Torrance *Theol. Sci.* i. 18 This was the Augustinian doctrine of the sacramental universe, combining Plotinian and Ptolemaic notions in a 'Christian' cosmology.

Plotinist. (Earlier example.)

1871 J. S. Mill in *Fortn. Rev.* X. 524 A heap of useless and mostly unintelligible jargon, not of his own [*sc.* Berkeley's] but of the Plotinists.

plotless, *a* Add: **1.** (Further examples.)

1926 in C. Bailey *Mind of Rome* i. 167 Semi-dramatic productions, improved on the formal side but still plotless, became an established diversion. **1971** *Homes & Gardens* Sept. 134/1 Her story reads like a charming but plotless novel. **1979** A. Chisholm *Nancy Cunard* viii. 76 *Antic Hay*..is a deliberately plotless picture of..amusing, erratic, fundamentally desperate characters.

2. *Ecol.* Of a method of ecological sampling: not based on a defined unit of area.

1957 P. Greig-Smith *Quantitative Plant Ecol.* ii. 46 Considerable attention has recently been paid to a method of plotless sampling, particularly adapted to forest vegetation, where there are practical difficulties in delimiting the relatively large quadrats necessary for sampling trees. **1974** Mueller-Dombois & Ellenberg *Aims & Methods of Vegetation Ecol.* vii. 99 Plotless sampling means sampling without such a prescribed area unit. Plotless methods are available for all three commonly used quantitative parameters.

Plott (plǫt). The surname of Jonathan *Plott* (fl. 1750–80) and his descendants, used *attrib.* in **Plott hound** to designate a hunting dog belonging to a breed developed by this family from hounds brought from Germany to North Carolina in 1750, characterized by a smooth, dark brown coat, often with a black saddle or white feet, and large, drooping ears; it has long been used for bear and boar hunting in North Carolina and Tennessee, and recently the breed has become more widely known.

1945 C. L. B. Hubbard *Observer's Bk. Dogs* 202 Plott Hound. Although this breed is not yet given official status by the American Kennel Club, it is fairly popular among sporting men in the U.S.A... Mr. Geo. L. Gilkey of Wisconsin is piloting the modern Plott to popularity. **1949** H. P. Davis *Modern Dog Encycl.* 466/2 The present demand for Plott Hound stock is far in excess of the supply. **1969** E. H. Hart *Encycl. Dog Breeds* 257 The Plott Hound differs from all other hounds in colouring for he is a brindle dog with a black saddle. **1976** E. M. Schuler *Dog Lover's Answer Bk.* ii. 358 The Plott hound is a heavy-set, well-muscled dog with large ears.

plottable (plǫ·tăb'l), *a.* [f. Plot *v.*[1] + -able.] That may be plotted, in the various senses of the verb.

1968 A. L. Allan et al. *Pract. Field Surveying & Computations* ix. 456 The misclosure of a tacheometric traverse should always be within the plottable error of the plan and no adjustment is permitted. **1972** C. Mudie *Motor Boats & Boating* 116 Voyages out of sight of land only extend the distances between plottable positions and it is quite difficult nowadays to get out of range of the radio navigation beacons. **1976** S. R. Simpson *Land Law & Registration* viii. 145 The boundaries could only be replaced to the plottable accuracy of the map.

plottage (plǫ·tĕdʒ). [f. Plot *sb.* + -age.] **1.** Phr. *a mess of plottage*, used disparagingly of theatrical productions (by analogy with *a mess of pottage*: see Mess *sb.* 2 a).

1937 *Sun* (Baltimore) 7 Aug. 4/1 '*The Big Shot*', now showing at the Hippodrome Theater, is just another mess of plottage that some RKO Radio writers sold their birthrights for. **1958** *Spectator* 22 Aug. 249/3 The London Studio cast manage to make the whole mess of plottage taste mouldier even than it is.

2. The size or value of a specified piece of land, regarded in terms of the area accumulated from its constituent plots. Usu. in *plottage increment* (see quot. 1965).

1939 in Webster Add. **1952** 'Vigilans' *Chamber of Horrors* 100 *Plottage*, the area of a plot of land. (The deplorable result of an illicit union between *plot* and *acreage*.) **1961** R. U. Ratcliff *Real Estate Analysis* iii. 52 This value increment resulting from the assembly of a tract [of land] of sufficient size for a more intensive use is termed *plottage*. **1965** L. E. Davids *Dict. Insurance* 162 *Plottage increment*, the increase or appreciation of the unit value resulting from the joining together of smaller lots, parcels, or land units into one large single ownership, the resulting total value being larger than the individual unit values. **1973** *N.Y. Law Jrnl.* 2 Aug. 11/3 A plottage increment of 10 per cent is allowed for an appraised land area of 12,322 square feet.

plotted, *ppl. a.* Add: **3.** (Later examples.) Freq. with advs.

1970 *Daily Tel.* (Colour Suppl.) 11 Aug. 5/2 In our time it has become meaningless to produce a strongly plotted short story like Maupassant's *The Necklace* or a strongly plotted novel like *David Copperfield*. **1978** *Broadcast* 28 Aug. 8/1 Plays which were both tightly plotted and eminently actable. **1979** F. Kermode *Genesis of Secrecy* v. 113 We are more likely to remember a plotted narrative.

plotter. 1. Delete † *Obs. rare* and add to def.: one who plots points on a map. (Later examples.)

1908 *Geogr. Jrnl.* XXXI. 536 Some central organization at home would be required for plotting the results. This could be done..by a plotter and plotting machine attached to any existing photographic establishment. **1943** 'T. Dudley-Gordon' *Coastal Command at War* ii. 17 At a long table sits a W.A.A.F. officer, writing and making calculations. She is the 'plotter'... As the signals pour in..she translates the data into positions and..pin-points the chart, drawing pencil lines which 'lay off' courses of ships and aircraft. **1958** 'P. Bryant' *Two Hours to Doom* 48 Teams of plotters were at work, drawing in..the X points and target routes of the 843rd Wing. **1976** J. B. Garner et al. *Surveying* i. 7 The spacing of grid lines should be chosen so that it assists the plotter without becoming predominant on the completed sheet. **1978** *Daily Tel.* 19 Aug. 13 (*caption*) Mrs Vera Shaw, who was a Waaf plotter during the Battle of Britain.

4. An instrument or machine for making plots; *spec.* one for drawing maps or automatically plotting points on them.

[**1908** *Geogr. Jrnl.* XXXI. 544 A stereo-plotter..combines the offices of the stereo-comparator and plotting board.] **1926** R. M. Abraham *Surveying Instruments* x. 174 When the region to be mapped is rugged or mountainous the stereo method in conjunction with a suitable plotter is undoubtedly superior to all others. **1943** A. L. Higgins *Elem. Surveying* viii. 102 [With the plane table] angles are not observed in magnitude, as in the case of any goniometer,..such as the compass..and theodolite, but instead are constructed directly, so that the instrument is a goniograph, or angle plotter. **1948** *Rev. Sci. Instruments* XIX. 647/2 (*caption*) An automatic [electric] field plotter. **1959** *Engineering* 27 Feb. 262/2 The additional information derived from the true motion type of presentation is obtained from the movement of a target vessel across the PPI screen... This is not entirely satisfactory so a tube face reflection plotter..was developed. With its aid, true plots of own and several target vessels can be easily kept by the navigator. **1963** [see *plot *v.*[1] 1 b]. **1972** *Sci. Amer.* Mar. 2/2 (Advt.), Our 745 flatbed plotter will scribe lines equal to the tolerances and standards of the most skilled mapmaker's hand.

b. An instrument for automatically plotting a graph.

1956 *Proc. Eastern Joint Computer Conf.* 73/2 The Ballistic Research Laboratories have for some time been concerned with the development of a digital plotter capable of absorbing the output of..digital computers used in processing of missible ballistic data. **1970** O. Dopping *Computers & Data Processing* xi. 172 In some simple plotters, the paper is continuously fed in one direction, while a printing device, directly controlled by the computer, moves at right angle [*sic*] to the direction of paper motion. **1975** *Nature* 16 Oct. 559/2 Engineers, on the other hand, might well set up a mathematical model for a bridge—or an aircraft wing—and use the minicomputer to display on a visual display unit or graph plotter the consequences of certain input data (loading, sizes of beams, or whatever).

5. One who owns a plot of land, a plot-holder.

1927 *Smallholder* 20 Mar. 106/3 Every plotter should pull his weight, not only for his own sake but for the good of the national cause. **1976** D. G. Hessayon (*title*) Vegetable plotter.

plotting, *vbl. sb.*[1] Add: **b.** *plotting sheet*; **plotting board,** (*a*) a form of drawing board on which the positions or courses of objects may be plotted; (*b*) = *plotter 4 b; **plotting machine,** a machine for automatically plotting maps; **plotting rod,** a long rod made for moving the counters on a plotting table; **plotting table,** (*a*) a large table bearing a small-scale map of a region on which the positions of enemy aircraft may be represented by movable counters or the like; (*b*) = *plotter 4 b.

1903 *Jrnl. U.S. Artillery* Nov.–Dec. 253 A location of a fixed target is thus plotted, the corrections for atmospheric conditions and drift are determined by the ballistic board, the range and azimuth for the gun are determined by relocation on the plotting board, and transmitted to the gun by the gun telautograph. **1908** *Geogr. Jrnl.* XXXI. 544 The position in plan and height of any object can be plotted from readings of the three scales..and corresponding settings on a detached plotting-board. **1957** C. S. Forester *Naval War of 1812* 36 There had been exercises with a plotting-board in the Tripoli prison. **1961** S. Fifer *Analogue Computation* II. ix. 300 The Autograf.. is an example of a cylindrical plotting board. In this case the pen is controlled, as before, by one variable, but the moving carriage is replaced by a rotating drum which is driven by the second variable. **1908** Plotting machine [see *plotter 1]. **1963** W. K. Kilford *Elem. Air Survey* xi. 269 Some firms produce special plotting machines in which the relative orientation is already set for use with a particular pair of short-base cameras. **1977** J. M. Smith in P. G. J. van Sterkenburg et al. *Lexicologie* 243 Graphical output devices, based on the principles of plotting machines, can be used too. **1948** T. E. Winslow *Forewarned is Forearmed* i. 19 The size of a plotting table depends also on the 'stretch' of a plotter provided with a plotting rod. **1963** N. D. Smith *Royal Air Force* vi. 66 Plotters, using long-handled plotting rods, moved coloured magnetic counters on the large table map. **1926** *Blackw. Mag.* Dec. 830/2 By degrees there appeared on the plotting sheet a series of tiny needle-pricked marks. **1971** R. J. P. Wilson *Land Surveying* viii. 165 The plane table, essentially a drawing board,..carries the plotting sheet. **1943** P. F. M. Fellowes *Britain's Wonderful Air Force* xiii. 303 At each centre there is a centre room, with a plotting table on which is fitted a squared map. **1960** Rogers & Connolly *Analog Computation in Engin. Design* ii. 47 When it is desired to plot one variable against another, an *XY* plotting table is used. This unit employs a double servo system to drive a pen along an arm proportional to *Y*, and the arm along the plotting table proportional to *X*. **1968** *Listener* 15 Aug. 196/3 But it was all right to have mixed anti-aircraft batteries, with women tracking the target, and an ATS officer at the plotting table calculating the exact moment to shout the order 'Fire'. **1973** 'A. Hall' *Tango Briefing* xviii. 224 The whole of this area was on the plotting table at the Bureau. **1977** *Navy News* June 8 (*caption*) Prince Philip and Rear-Admiral D. W. Haslam (Hydrographer of the Navy) take a close look at the computer linked automatic plotting table being described by the commanding officer of H.M.S. Hecate.

plotty, *a.* Delete *nonce-wd.* and add earlier and later examples. Also, of a novel, play, or the like: having an elaborate or complicated plot.

1897 'S. Grand' *Beth Bk.* xl. 405, I would not write plotty-plotty books either. **1898** E. Pugh *Tony Drum* ix. 120 Novels of a common type, plotty and passionate, but gilt-edged with the proprieties. **1934** E. Bowen *Cat Jumps* 112 So plotty, so damned smart, so careful no one would see us who would remember, a different place every time. **1959** *Times Lit. Suppl.* 25 Sept. 546/5 His Phrygian Slave in the *Orestes*—admittedly a problem—talks a hair-raising amalgam of Wardour Street and hill-billy vernacular. 'God darn him dead For plotty sneaks,' this surprising menial observes at one point. **1973** L. Hellman *Pentimento* (1974) 197 What I thought was bite they [*sc.* theatre critics] thought sad, touching, or plotty and melodramatic. **1974** *Times* 18 Apr. 7/5 The basic plot is a bit plotty.

plo·twise, *adv.* [f. Plot *sb.*[1]: see Wise *sb.*[1] II.] As regards or in terms of a plot (in sense 6 of the sb.).

1955 *Times* 26 May 11/4 We gave what was, plotwise, a perfectly coherent rendering of *Romeo and Juliet* in 20-odd minutes.

plotz (plǫts), *v. U.S. slang.* [ad. Yiddish *platsen*, f. G. *platzen* to burst; in sense 1 influenced by G. *platz* place, seat.] **1.** *intr.* To sit down wearily, to flop; to slouch, loaf (*around*). Also *trans.* in causal sense.

1941 A. Kober *My Dear Bella* 199 At the end of a day I'm just like a wet rag... All I'm good for is to *plotz* in a chair. **1960** J. Kirkwood *There must be Pony* vi. 43 He just kind of plotzed around waiting to fall into some sort of a cushy job. **1976** *National Observer* (U.S.) 30 Oct. 20/1, I find there is still little to justify a parent plotzing his offspring in front of the tube for long hours on Saturday morning. **1978** R. Condon *Bandicoot* xv. 91 We are plotzing one night and I bring out a whole miniature roulette layout.

2. *intr.* To burst; usu. in *fig.* senses, esp. to 'explode' with frustration or annoyance, to demonstrate one's anger.

1967 P. Welles *Babyhip* iii. 46 'You're not smoking that filthy thing in here. I'll *plotz*,' Mrs Green said. **1968** L. Rosten *Joys of Yiddish* 292 So Pincus broke into a run, and he ran and he ran until he thought his heart would *plotz*. **1970** L. M. Feinsilver *Taste of Yiddish* 367 A recent ad for a Yiddish revue read: 'In Miami they're plotzing'—meaning, 'they're still howling over the preview'. **1978** J. Krantz *Scruples* vii. 214 She came back to pick them up today and *plotzed* for joy all over the studio.

plotzed (plǫtst), *a. slang.* [f. as prec. + -ed[2].] Intoxicated, drunk.

Not clearly associated with the senses of *plotz *v.*

1962 J. D. MacDonald *Key to Suite* (1968) ii. 16 If one of our boys gets plotzed, we run him off the team fast before any damage is done. **1974** 'M. Allen' *Super Tour* v. 175 Mimi got drunk that night..but something more than liquor knocked her off base... She was so loaded I had to put her to bed, and I know from my own experience that when I am plotzed I go out for the night.

plough, plow, *sb.*[1] Add: **5. h.** Any of various implements for deflecting (e.g. off a conveyor belt or a railway track) material against which they move, or which moves against them; in quots. 1860, 1975, a snow-plough.

1860 Clark & Colburn *Rec. Pract. Locomotive Engine* 68/2 In heavy snows, a plough of large size is fitted in front of the engine, to clear the line. *a* **1884** Knight *Dict. Mech.* Suppl. 173/2 Dowling's plow for unloading platform gravel-cars, is a V-shaped implement which has two flaring wings. **1901** M. M. Kirkman *Building & Repairing Railways* viii. 333 The Rodgers ballast car dumps the ballast in the center of the track, the last car in train of ballast cars having a plow for cleaning and flanging the track. **1922** F. V. Hetzel *Belt Conveyors & Belt Elevators* viii. 159 In some European boiler houses the bunker is served by a flat belt which runs through a movable carriage equipped with a V-point plow and a two-way chute. **1953** W. W. Hay *Railroad Engin.* xxii. 316 A spreader-type plow follows the unloading operation to spread the ballast where it is needed. **1971** B. Scharf *Engin. & its Language* xvi. 235 A plough (movable gate) may be provided across the belt so that the conveyor can be unloaded at that point or in order to deflect the material on to another conveyor. **1975** D. Pitts *Target Manhattan* (1976) xxviii. 117, I want your team of plows at Broadway and West 14th.

i. *Coal Mining.* A machine with cutting blades that remove a thin strip of coal when it is hauled along a coal face.

1950 *Trans. Inst. Mining Engineers* CIX. 273 The coal seams in this country are too hard to allow of the plough being successfully used. **1952** *Times* 16 Sept. 3/2 In his report for 1951, published to-day, Mr. G. Hoyle, North-Western Divisional Inspector of Mines, refers to revolutionary developments in mining technique... He instances the use of the plough and stripper for coal getting. **1964** A. Nelson *Dict. Mining* 335 Normally, on a wide face, and working 6 hr, a plough will produce 800 tons and more of coal in a 3 ft thick seam. **1971** *Daily Tel.* 20 Oct. 13/1 Machines, with strange names like Anderton shearers, trepanners, rapid ploughs or Huwood slicers, have replaced men underground.

7. a. *plough-collar, -culture, -horse, -mark, -ox* (later examples).

1908 *Sears, Roebuck Catal.* 137/4 A Southern Plow Collar. Made of heavy cotton duck with leather chafes on the side where the chain or trace attaches to the hame. **1942** W. Faulkner *Go down, Moses* 255 Plowlines and plow-collars and hames and trace-chains. **1937** R. H. Lowie *Hist. Ethnol. Theory* viii. 114 Thus was conceived the antithesis of primitive 'hoe-culture' and 'plough-

culture', the latter being the exclusive mark of higher civilizations. **1961** L. Mumford *City in Hist.* i. 27 Where hoe culture supported hamlets, plow culture could support whole cities and regions. **1539** Wyatt *Let.* in R. W. Bailey *Early Mod. English* (1978) 233/1 And I w[i]t[h] much ado apon plow horse In the diepe and fowle way gatt afore that nyght late to loshes. **1573** T. Tusser *Five Hundreth Points Good Husbandry* xviii, Sedge couers for plow horse, for lightnes of neck. **1744** W. Ellis *Mod. Husbandman* Jan. xxi. 56, I feed my Plough Horses with these green Thetches. **1817** Scott *Rob Roy* II. xiii. 280 There may be pasture aneugh for pleugh-horses, and owsen, and forty or fifty cows. **1880** *Harper's Mag.* Aug. 356/2 The next day the two girls, mounted on the plough horse and mare, followed an old Indian trail. **1911** R. D. Saunders *Col. Todhunter* ix. 118 A wall-eyed plow-horse with his tail full o'cuckle-burs. **1955** W. Moore *Bring Jubilee* i. 5 He would lay the reins on the plough-horse's back. **1930** W. Faulkner *As I lay Dying* 125 After a while she went on, stumbling a little on the plow-marks. **1963** *Field Archaeol.* (Ordnance Survey) (ed. 4) 56 A recent excavation has shown round huts..closely associated with plough-marks in underlying sand. **1973** *Nature* 23 Nov. 191/2 Further ambiguity was introduced by the presence of iceberg plough marks around the Rockall bank and evidence of ice-rafted deposition. **1906** Kipling *Puck of Pook's Hill* 237 Down would come the King's Officers, and take our plough-oxen to haul them [*sc.* guns] to the coast. **1946** E. Linklater *Private Angelo* xx. 257 Two pairs of matched plough-oxen arrived in Pontefiore.

8. plough-bullock, (*b*) (later example); also **plough-bullocker;** so **plough-bullocking** *vbl. sb.*; **plough grinding** *Cotton Spinning*, a way of grinding the wires of a cotton card (see quots.); so **plough-ground** *a.*; **plough-line,** (*b*) (earlier, later, and *fig.* examples); also (usu. *pl.*), the reins themselves; **plough pan** *Agric.* [Pan *sb.*[1] 8], a compacted layer in cultivated soil resulting from repeated ploughing; **plough-point** *U.S.*, the first (usu. detachable) share at the front of a plough; **plough-soil,** soil that has been thrown up by ploughing; **plough-stock,** the iron or metal frame of a plough.

1899 A. Nutt in H. Lowerison *Field & Folklore* 63 Certain players, distinguished by scarlet jackets, and known as plough-bullocks or boggins. **1905** *Eng. Dial. Dict.* IV. 552/1 *Plough..-bullockers,..-bullocking.* **1923** E. C. Pulbrook *Eng. Country Life* xiii. 194 At Whitby, the young men come in to celebrate the Plough Stots as of old, and the Plough Bullockers occasionally drive their decorated plough through the villages of Derbyshire, to the detriment of those who refuse largesse. **1838** W. Howitt *Rural Life of Eng.* II. iii. 144 Maying, guising, plough-bullocking, morris-dancing, were gone before.. Methodism appeared. **1892** J. Nasmith *Students' Cotton Spinning* iv. 135 The usual solution of the difficulty is found in the formation of a tooth with a chisel or knife edge, which is presented to the action of the cotton. This is usually obtained by what is called 'plough grinding'—that is, a method of passing between the teeth of the clothing a thin emery disc, which 'ploughs' deeply between them and grinds them on each side until they present a sharp edge to the cotton. **1923** T. Thornley *Adv. Cotton Spinning* (ed. 3) ii. 77 The plough grinding of the wire works is really side grinding carried to its most perfect degree, and producing a bevelled effect on each tooth from point to knee. **1965** W. G. Byerley et al. *Man. Cotton Spinning* III. vi. 108 In addition to surface- and side-grinding, reference must be made to 'plough-grinding'. This process was devised and patented by an English firm in 1880... The process..was superseded by the side-grinding process. **1896** W. S. Taggart *Cotton Spinning* I. vi. 176 A is the plough-ground wire, and is formed by grinding the sides away, almost to the bend, by special emery discs. **1923** T. Thornley *Adv. Cotton Spinning* (ed. 3) ii. 76 The plough ground tooth is obtained at the wire-making establishment by grinding away the sides of the teeth down to the knee or bend. **1777** *New Jersey Gaz.* 17 Dec. 3/3 Plough-lines, Bed-Lacings,..Sold by Edward Pole,..Burlington. **1886** F. T. Elworthy *W. Somerset Word-Bk.* 582 *Plough-lines*, or *plough-guides*,..the cords used as reins. **1935** Z. N. Hurston *Mules & Men* (1970) i. ii. 54 Y'all lady people ain't smarter *than* all men folks. You got plow lines on some of us, but some of us is too smart for you. **1940** W. Faulkner *Hamlet* i. 8 One afternoon he was in the store, cutting lengths of plow-line from a spool. **1969** G. E. Evans *Farm & Village* xi. 126 There they made rope for the plough-lines, the reins or *cords*, as the horsemen invariably called them. **1883** W. T. Lawrence *Princ. Agric.* i. 30 By repeatedly ploughing at about the same depth, their downward progress [*sc.* that of roots] is checked by the formation of a hard bottom called a plough-pan. **1924** Watson & More *Agric.* v. 86 Sub-soiling is absolutely necessary where a plough-pan has been formed. **1960** *Farmer & Stockbreeder* 19 Jan. 74/1 (*caption*) Years of cultivation at constant depth have resulted in some plough pan. Once this is broken up the crops will have a better chance of establishment. **1856** in G. N. Jones *Florida Plantation Rec.* (1927) 478 Paid Mr. Lem Jones 50 cts. on account of '54 maid by J. Evans for 2 Plow points. **1942** W. Faulkner *Go down, Moses* 168 The boy first remembered him as sitting in the door of the plantation black-smith shop, where he sharpened plow-points and mended tools. **1967** *Antiquaries Jrnl.* XLVII. 166 Today, the bank and the ditch..have been largely either ploughed flat or masked by comparatively modern accumulations of plough-soil. **1976** C. Thomas in P. H. Sawyer *Medieval Settlement* ii. xii. 149 Site XX is a field bounded by a ditch... Its date of use can be fixed by the dated broken pot sherds, part of the extensive domestic manure incorporated in the plough-soil. **1978** R. Bradley *Prehist. Settlement of Britain* 41/2 His argument was based on the foreign stones incorporated in the ploughsoil. **1786** G. Washington *Diary* 9 Jan. (1925) III. 5 [I] directed them to get me..scantling for Plow stocks. **1865** *Oregon State Jrnl.* 28 Oct. 4/2 Plow Stocks etc., made to order, on short notice. **1940** W. Faulkner *Hamlet* i. ii. 35 Ab..had snuck the wagon out the back way with the plow stocks and the sorghum mill in it. **1944** T. D. Clark *Pills, Petticoats & Plows* 276 Centre and rear passageways were blocked with piles of iron plows..plow stocks..and axes.

plough, plow, *v.* Add: **1.b.** Also *fig.*

1901 Ld. Rosebery in *Times* 20 July 15/5, I must proceed alone. I must plough my furrow alone. **1936** E. White *Wheel Spins* iii. 29 She always ploughed a straight furrow, right to its end. **1977** *Dædalus* Summer 149 In the United States, George Sarton had been plowing a lonely furrow at Harvard's Widener Library for about twenty-five years. **1978** *Lancashire Life* Nov. 39/2 No easy task, with everybody else ploughing the same furrow.

6. b. (Later examples with *through*.)

1952 C. Bardsley *Bishop's Move* xi. 119, I almost said 'plough through' the Bible. **1959** *Daily Tel.* 23 July 1/6 The Prime Minister..gave the House the impression that he was ploughing, with as much force and gaiety as he could muster, through an almost impenetrable bog. **1978** P. Boardman *Worlds of P. Geddes* xi. 408 One ploughs through the often complicated sentences of P.G.'s writings of 1926–30.

c. *intr.* Of a road vehicle, train, aeroplane, or the like: to move clumsily or laboriously, usu. at speed; to advance out of control *into* (or *through*, etc.) an obstacle.

1972 *Daily Tel.* 29 Dec. 2/5 A three-coach train was derailed..when it ploughed into a herd of cattle at 60 mph. **1973** *Times* 31 Dec. 5/5 The airliner..ploughed to a halt on the runway. **1976** *Southern Even. Echo* (Southampton) 11 Nov. 16/5 A Southampton lorry driver suffered only cuts and bruises last night when his lorry ploughed through a bridge and plunged 15 feet into a field at Wimborne. **1977** *Evening Gaz.* (Middlesbrough) 11 Jan. 1/9 Police in Cleveland are hunting the driver of a sports car which forced another car to plough into four people.

7. d. *Coal Mining.* To cut (coal) by means of a plough; to push (coal so obtained) away from the face by means of a plough.

1950 *Trans. Inst. Mining Engineers* CIX. 256 The first train of thought was to plough machine-cut coal on to a face conveyor. **1951** H. F. Banks in E. Mason *Pract. Coal Mining* (ed. 2) I. viii. 123/2 This device carries steel blades which shear or plane off the coal to a limited depth and ploughs it on to the face conveyor. **1964** A. Nelson *Dict. Mining* 335 Hard anthracite is being ploughed with only water infusion to soften the coal.

e. *trans.* To clear (an area) of snow using a snow-plough.

1961 'E. Lathen' *Banking on Death* (1962) xii. 99 'Don't know why they can't plow these streets,' he muttered as he pulled into the single lane left by the piles of snow. **1978** *Times* 23 Jan. 12/7 There was.. slush and compacted snow on roads the ploughs had not reached. It says much for the authorities in West Virginia..that they had ploughed all but about 40 miles of my route. **1979** J. van de Wetering *Maine Massacre* ii. 12 They may not have plowed the strip..last time... I had to circle while they pushed the old plow around.

9. f. *plough under*: to bury in the soil by ploughing.

1900 *Year-Bk. U.S. Dept. Agric.* 379 If crimson clover is grown, it should be plowed under rather early in the spring to get the best results. **1979** *Country Life* 6 Dec. 2141 The express way will bypass the old road..ploughing under landmarks that meant much to people.

g. *plough back*: to invest (income or profit) in the enterprise from which it emanates.

1930 *Economist* 24 May 1172/2 The extensive resort of American managements to the practice of 'ploughing back earnings into the business' further emphasises this tendency. **1945** *Richmond* (Va.) *Times-Dispatch* 25 Oct. 6/3 The proposed act would limit the annual dividends of such corporations to 6 per cent, requiring that all additional profits would have to be 'plowed back' into re-development. **1949** *Sun* (Baltimore) 26 Jan. 12/3 Profits are being plowed back into industry at unprecedented rates. **1955** *Times* 1 July 17/3 The profits that have accrued from this company have been largely ploughed back for further development and expansion. **1965** *Listener* 23 Dec. 1023/1 It was not long before we had functioning money-raising sweet shops, bargain stores—and the 'Green Dragon'. Profits were all ploughed back. **1970** *Physics Bull.* Mar. 99/2 For the services it renders the Centre charges small fees which are ploughed back into its operations. By this means it is planned to be selfsupporting within three years. **1974** N. Freeling *Dressing of Diamond* 96 We ploughed every penny back for ten years. **1976** *Milton Keynes Express* 30 July 11/4 He would not consider ploughing some of the £4 million back into the services and said he hoped the kitty would increase.

ploughed, plowed *ppl. a.*: also, (in sense *9g) **ploughed-back; ploughed-out,** obscured or destroyed by ploughing; (in sense 9e) **ploughed-up** (in quot. *fig.*). Also (in sense *9g) **ploughing-back** *vbl. sb.*

1920 J. Masefield *Enslaved* 120 From these ploughed-up souls the spirit brings Harvest at last. **1931** *Economist* 18 July 128/1 Reserves against remote contingencies, and those representing the 'ploughing back' of earnings into the business, should be set aside openly. **1944** J. S. Huxley *On living in Revol.* xii. 130 This compulsory ploughing-back of any excess profits is essential if the development of the area is to proceed at a reasonable rate. **1950** *Oxoniensia* XV. 7 Ploughed-out field-systems appear on air-photographs as a network of white lines which can often be recognized to some extent on the ground by bands of broken chalk and flints. **1957** *Times* 12 Dec. 18/1 Additional capital..has been injected into the business.. in the form of ploughed-back profits. **1958** *Spectator* 18 July 117/3 It is finding about 85 per cent. of its capital through ploughed-back profits. **1959** *Manch. Guardian* 7 Aug. 1/2 The suggestion that a reduction in selling prices might, in present circumstances, take precedence over the ploughing back of profits. **1974** C. Taylor *Fieldwork in Medieval Archaeol.* iv. 74 The discovery of Iron Age and Roman sherds scattered on a spur above the River Nene near Irthlingborough, Northamptonshire, led to the identification of ramparts round this spur as part of a ploughed out 'hill fort'. **1977** *Interim* IV. iv. 4 Then the light marks over ploughed-out walls or banks.. provide..an accurate deliniation of landscape elements.

plou·gh-back, plow-. *Econ.* Also as one word. [f. *to plough back*: see *plough *v.* 9g.] Investment of income or profit in the enterprise from which it emanates; the capital so invested.

1946 *Sun* (Baltimore) 28 Mar. 19/2 After payment of ..taxes and..dividends, the company showed a plow-back of $64.90 a share. **1961** *Times Lit. Suppl.* 6 Jan. 2/2 Less socialist-minded politicians..taxed such 'plough-backs' into companies' own coffers more heavily. **1966** *Economist* 26 Mar. 1264/2 Borax would have to justify itself by giving an assurance on the future flow of dollar dividends from the American ploughback. **1970** *Daily Tel.* 2 Jan. 14 Some new Government, if well advised, will be in a position to let the private sector have more of its head in..plough-backs, investment and profits. **1974** M. B. Brown *Economics of Imperialism* v. 123 The 7 per cent or so of Britain's national income, which was invested in the decades before the 1830s had..to be found almost entirely from the ploughback of profits.

plough-gear, plow-. Delete † *Obs.* and add later examples.

1644 *Archives Maryland* (1887) IV. 279 The ploughgeare sent of Engl[and]. *a* **1815** M. Lonsdale *Love in Cumberland* in *Westmoreland & Cumberland Dial.* (1839) 211 Thy plew-geer's aw liggin how-strow. **1885** 'C. E. Craddock' *Prophet Gt. Smoky Mts.* i. 15 The girl's hand trembled violently as she stepped swiftly to his horse and took off the plough-gear. **1940** W. Faulkner *Hamlet* iii. 64 He would give them credit for food and plow-gear when they needed it. **1952** *Oxf. Jun. Encycl.* VI. 243/1 Plough gear is generally simpler than chain gear; it has a transverse backband supporting the chains, and sometimes a loose bellyband.

ploughing, plowing, *vbl. sb.* Add: **2.** ploughing-match (later examples).

1845 *Ainsworth's Mag.* VII. 369 Unfortunately, my lord [*sc.* a servant] got a premium at a ploughing-match. **1882** [see Match *sb.*[1] 7]. **1949** E. Blunden *After Bombing* 41 Here was a set of anglers, there the ploughing-match. **1976** *Southern Even. Echo* (Southampton) 17 Nov. 6/5 The 1976 Hampshire County Ploughing Match will be at Upper Silkstead Farm, Poles Lane, Otterbourne, next Saturday.

plough-jogger, plow-. (Later U.S. examples.)

a **1852** F. M. Whitcher *Widow Bedott Papers* (1856) xx. 207, I wanted old Dawson's wife to see't I'd got a pardner ruther above a common plow-jogger, such as hern is. **1862** *Harper's Mag.* Nov. 782/2 City folks most generally fetch along a lot of traps and finery to show off afore us plowjoggers. **1865** *Trans. Illinois Agric. Soc.* V. 255 At least the old plow jogger will be mounted on his buggy seat.

ploughman, plow-. Add: **b. ploughman's lunch,** a cold snack, usu. including bread, cheese, and pickle, and freq. served in a public house at lunch-time. Also *ellipt.*, as *ploughman's*.

[**1837** J. G. Lockhart *Mem. Life W. Scott* IV. v. 161 The surprised poet swung forth to join them, with an extemporized sandwich, that looked like a ploughman's luncheon, in his hand.] **1970** R. Trehane in B. H. Axler *Cheese Handbk.* p. iii, English cheese and beer have for centuries formed a perfect combination enjoyed as the Ploughman's Lunch. **1971** *Oxford Mail* 27 Oct. 1/8 The loaf is to be cut into 200 pieces and eaten as part of 200 ploughman's lunches. **1973** P. Theroux *Saint Jack* v. 60 We had a ploughman's lunch in the village—beautiful old pub—and went back to London. **1975** *Times* 30 Aug. 10/2 The pubs specialize in lunchtime catering..and you can get a decent 'ploughman's' for between 20p and 30p. **1977** J. Thomson *Case Closed* v. 73 He treated himself to a ploughman's lunch of bread, cheese and pickle..at the Red Lion.

plough-tail, plow-. Add: Also *attrib.*

1912 *Chambers's Jrnl.* Sept. 564/1 No doubt the chie thought he cut a dash among the plough-tail lads.

plovery, *a.* Add: **b.** Of, characteristic of, or reminiscent of a plover.

1932 J. Joyce *Let.* 1 Jan. (1966) III. 239, I would be engaging you with my plovery soft accents.

ploy, *sb.*[3] **1.** For '*Sc.* and *north. Eng.*' read 'orig. *Sc.* and *north. Eng.*' and add later examples in general use.

1916 E. F. Benson *David Blaize* xi. 208, I think you're rather an ass, unless you prefer writing out the 'Æneid' to any other ploy. **1926** R. Macaulay *Crewe Train* ii. viii. 159 Whatever ploy she had on hand at the moment, such as lead casting, table tennis, or naval battles in the bath. **1930** *Times* 16 Apr. 15/4 Training in domestic ploys

such as household work. **1936** A. Christie *Cards on Table* xviii. 175 You'd gone off on your own ploys with the boy friend. **1936** 'J. Tey' *Shilling for Candles* xxiii. 257 Smuggling Edward Champneis might descend to, as a ploy, a mere bit of excitement. **1953** J. Trench *Docken Dead* ii. 27 He's obviously gone off on some ploy of his own. **1959** 'M. Neville' *Sweet Night for Murder* xi. 106 'Did you ever accompany her while she was shopping for clothes?'.. 'I wouldn't be much good at that sort of ploy... She didn't need anyone to help her choose what to wear.' **1979** *Country Life* 8 Nov. 1687/1 The search for the alternative life style is..a new rationalisation of a ruralising ploy.

2. A move or gambit suggested by particular circumstances and made in order to gain a calculated advantage, esp. self-advancement or the frustration of an opponent's intentions; a planned device or manœuvre.

1950 S. Potter *Lifemanship* 15 Each one of us can, by ploy or gambit, most naturally gain the advantage. *Ibid.* 90 P. Lewis, expert in Oxford Undergraduateship, has set it down as basic to this ploy that, where the Layman would concentrate on his subject, the Gamesman *concentrates on his tutor*. **1955** *Times* 24 June 10/3 Apart from claiming possible support in London and Glasgow and assuring the men that there would be 'important developments during the next 48 hours'—a ploy with which the strikers are becoming a little disillusioned—the speakers had nothing to offer. **1957** *Listener* 13 June 967/3 It is a common 'ploy' of reviewers to take the opportunity of a review to air their own views. **1957** *Economist* 5 Oct. 69/1 Conventional East–West political ploys that took the form of proposing, and rejecting rather more sharply than usual, the suggestion that non-members of the United Nations, such as Red China, should be admitted as observers. **1958** A. Wilson in *Times Lit. Suppl.* 15 Aug. p. viii/2 Whatever the ingenious and at times embarrassing ploys with which English novelists periodically assert their amateur, their unintellectual or their purely entertaining status..they are..concerned always to be serious. **1960** 'W. Haggard' *Closed Circuit* viii. 105 For suspects there is a standard ploy. You test them. **1966** *Listener* 13 Jan. 78/3 If West held the queen of spades, as he did, it would be East's duty to guard the suit. The only way to accomplish this end was by the unusual ploy of discarding a trump on the third round of clubs. **1970** G. F. Newman *Sir, You Bastard* 267 Perhaps she should cook it and leave it for him, but she recognized that thought as a ploy to delay herself, hoping that he'd return and prevent her departure.

pluck, *sb.*[1] Add: **I. 1. d.** *Naut.* A pull or tow.

1918 *Yachting Monthly* Jan. 155 A pluck out of dock, a fishing permit and a light breeze. **1934** 'Taffrail' *Seventy North* iii. 57 'D'ye want a rope's end, ole pal? We'll give ye a pluck home!' Sam's retort to this nautical insult was jocular but mostly unprintable. **1962** W. Granville *Dict. Sailors' Slang* 89/2 *Pluck*, a tow or tug. **1964** *Roving Commissions 1963* 176 A feeble little motor-boat gave us a half-hearted pluck and went away.

III. 7. c. Wine. *U.S. Black slang.*

1964 *N.Y. Times Mag.* 23 Aug. 64/2 *Pluck*, wine. **1967** *Trans-Action* Apr. 8/1 The dudes 'rap' and 'jive' (talk), gamble, and drink their 'pluck' (usually a cheap, sweet wine). **1969** H. R. Brown *Die Nigger Die!* ii. 24 We went and got some 'pluck' (wine) and I told him I was in college. **1973** *Black World* July 55/1 We want some pluck man, got any scratch? *Ibid.* 56/1 We was gittin away from the broke pluck bottles.

V. 9. Special Comb.: **pluck side** *Physical Geogr.*, the rough, 'downstream' side of a roche moutonnée from which rock has been plucked by a glacier.

1905 *Jrnl. Geol.* XIII. 6 We will no longer call the two sides of a *roche moutonnée* 'push side' and 'lee side', but we prefer the expressions 'scour side' and 'pluck side' introduced by Shaler. **1942** C. A. Cotton *Climatic Accidents Landscape-Making* xviii. 244 The lee side is termed also the 'pluck' side.

pluck, *v.* Add: **1. a.** Also *intr.* for *pass.*

1945 H. J. Massingham *Wisdom of Fields* viii. 163 It plucked dead ripe.

b. Substitute for def.: Of a glacier: to break loose (pieces of rock) by the mechanical action of ice which has formed around projections and in cavities in the rock; to erode (rock) by this process. Occas. also used of water (see quot. 1930). Freq. with adverbs, esp. *out*. (Examples.)

1893 *Bull. Mus. Compar. Zoöl. Harvard Coll.* XVI. 209 The pits which were left where masses of the rock were plucked out and borne away by the moving ice. **1915** L. V. Pirsson *Text-bk. Geol.* I. v. 124 The ice at the bottom of the névé fields being frozen into cracks and cavities and around projections in its stony bed, when motion begins, 'plucks' or quarries masses of rock and takes them forward with it. **1930** C. R. Longwell *Outl. Physical Geol.* iv. 43 In a stream flowing over horizontal layers of rock, corrasion along joints loosens large blocks, which are then torn or 'plucked' away by the current. In some situations this plucking action is much more important than wear by simple rasping. **1955** M. Hollander tr. *Kuenen's Realms Water* iv. 152 The glacier will from time to time pluck out large blocks from the lower part of the protuberance. **1971** I. G. Gass et al. *Understanding Earth* xv. 220/2 Rocks have been plucked into characteristic glacial shapes.

c. *Printing.* Of ink: to detach and remove the surface of paper during printing. Also *intr.* for *pass.*

1960 G. A. Glaister *Gloss. Bk.* 321/1 *Plucking*, a printing fault which is caused by the ink plucking the surface of the paper and leaving irregular white patches in printed areas. **1967** E. Chambers *Photolitho-Offset* i. 6 If 'washing-out' were omitted, the resin would cause trouble by causing the paper to 'pluck', owing to the resin sticking to the paper when printing.

2. a. (Further examples.)

1881 'Mark Twain' *Prince & Pauper* xv. 163 He is the stranger that plucked Giles Witt out of the Thames. **1975** *New Yorker* 19 May 120/2 After the war, the son got a job at the American Embassy, and one day in 1948 was plucked off the street and taken to prison.

3. c. *U.S. Mil. slang.* To cashier or retire (an officer).

1941 *Sun* (Baltimore) 5 Aug. 11/3 There are numerous retirements under way. However, it is highly unlikely that the army will make announcement of the officers who are being 'plucked' under the recent act permitting the Secretary of War to retire those whom a board has decreed to be 'unsuited for further active duty'. **1942** Berrey & Van den Bark *Amer. Thes. Slang* § 888/5 *Pluck*, to retire an officer.

5. b. To shape or thin (the eyebrows) by removing hairs.

c **1450** *Bk. of Knight of La Tour-Landry* (1868) 67 She hadde..plucked her browes, front, and forehed, to haue awey the here, to make her selff the fayrer to the plesinge of the worlde. **1926** F. Scott Fitzgerald *Great Gatsby* ii. 35 Her eyebrows had been plucked and then drawn on again at a more rakish angle. **1932** S. Gibbons *Cold Comfort Farm* xxii. 291 You shall not find me plucking my eyebrows, nor dieting. **1935** C. Isherwood *Mr. Norris changes Trains* ix. 147 He spent ten minutes.. thinning his eyebrows with a pair of pincers. ('Thinning, William: *not* plucking'). **1974** *Times* 22 Jan. 11/6 Whether you pluck your eyebrows depends on your type of looks.

7. Also *transf.*

1885 E. W. Hamilton *Diary* 10 June (1972) II. 880 Several Baronetcies are to be made. One or two claims were put aside; those who were not to the fore in the Division List on Monday were 'plucked', e.g. Mr. Palmer, which I regret because he has at other times rendered good service.

8. pluck up. c. Also, to get new courage, to take heart again.

1842 Dickens *Amer. Notes* I. ii. 20 Even those passengers who were most distrustful of themselves plucked up amazingly. **1890** W. C. Russell *Marriage at Sea* I. i. 10 But she had plucked up as she drew towards the close of her letter. **1901** G. B. Shaw *Caesar & Cleopatra* iii. 153 He eats another date, and plucks up a little.

9. Phr. *to pluck a rose*: of women, to visit the lavatory; to micturate or defecate. *slang.*

1613 Beaumont & Fletcher *Knight of Burning Pestle* ii, Then up and ride, Or if it will please you walke for your repose, Or sit, or if you will go plucke a rose. **1730** Swift *Panegyrick on the D—n* in *Miscellanies* V. (1735) 139 The bashful Maid, to hide her Blush; shall creep no more behind a Bush; Here unobserv'd, she boldly goes, As who should say to pluck a Rose. **1745** in J. R. Hetherington *Selina's Aunt* (1965) 21/1 Those ladies, who are so proud and lazy, that they will not be at the Pains of stepping into the Garden to pluck a Rose, but keep an odious Implement, sometimes in the Bed-chamber itself.. which they make Use of to ease their worst Necessities. **1768** Sterne *Sentimental Journey* I. 203 Grieve not, gentle traveller, to let Madame de Rambouliet p-ss on — And, ye fair mystic nymphs! Go each one *pluck your rose*. **1785** F. Grose *Classical Dict. Vulgar Tongue*, *To pluck a rose*, an expression said to be used by women, for going to the necessary house, which in the country usually stands in the garden. **1800** in *Proc. Amer. Antiquarian Soc.* (1897) XII. 248 Mrs. M. having occasion to pluck a rose as is usual with delicate women after a ride of 22 miles. **1937** Partridge *Dict. Slang* 641/1 *Pluck a rose*, to visit the privy.

plucked, *a.* **a.** (Earlier and further examples.)

1846 *Swell's Night Guide* 79 At a set to, he is a Dick Curtis the second; and an out and out plucked one. **1916** F. M. Ford *Let.* 23 Aug. (1965) 69 George V..really was in some danger... Still he gave the impression of a 'good plucked 'un'.

plucked, *ppl. a.* Add: **1. b.** *Textiles.* (See quot. 1940.)

1799 [in Dict., sense 1 a]. **1940** *Chambers's Techn. Dict.* 654/2 *Plucked*.., the term used to denote uneven thickness in a top, roving, or yarn, generally caused by excessive draft. **1974** H. McCloy *Sleepwalker* v. 72 A short coat of plucked nutria.

2. b. *plucked wool*: wool from a dead sheep.

1911 in Webster. **1932** E. Midgley *Technical Terms in Textile Trade* II. 156 *Plucked wool*, wool plucked from a sheep which has been dead a few days. Sometimes this term is applied to skin wool. **1957** M. B. Picken *Fashion Dict.* 258/1 *Plucked wool*, wool from dead sheep.

c. Of eyebrows: shaped or thinned by plucking out hairs.

1928 R. Hall *Well of Loneliness* xlviii. 449 A handsome young man with severely plucked eyebrows. **1935** R. Stout *League of Frightened Men* xvii. 237 Her face..with its broad flat nose and plucked eyebrows. **1938** L. MacNeice *I Crossed Minch* ii. 25 A few signpost details such as plucked eyebrows and lipstick. **1962** M. Barrett *Return of Cornish Sailor* ii. 15 The plucked eyebrow lifted. **1974** *Times* 26 Oct. 8/8 The corny peroxide blondes with their plucked eyebrows.

4. *Physical Geogr.* Eroded or broken off by plucking. (Cf. *pluck *v.* 1 b.)

1893 *Bull. Mus. Compar. Zoöl. Harvard Coll.* XVI. 210 The plucked out material carried away in the form of boulders amounts to as much as one fifth of that removed in the other forms of erosion. **1942** C. A. Cotton *Climatic Accidents Landscape-Making* xvi. 245 Shorn hills may.. present somewhat steep lee sides, perhaps plucked. **1957** J. K. Charlesworth *Quaternary Era* I. xi. The 249 boundary between abraded and plucked surfaces is sometimes that between different kinds of rock.

plucking, *vbl. sb.* **1.** (Further examples and later *attrib.* example.)

1893 *Bull. Mus. Compar. Zoöl. Harvard Coll.* XVI. 195 The southern end of Iron Hill is so much covered with glacial waste that it is not possible accurately to determine the relative amount of plucking which went on there. **1937** Wooldridge & Morgan *Physical Basis Geogr.* xxii. 366 Rock-masses on the floor or sides of a glacier may be frozen on to, or into, the ice and removed by 'plucking' in the onward motion of the glacier. **1952** A. Voet *Ink & Paper in Printing Process* xiv. 149 The separation of the paper-ink-plate union..occurs normally in the ink film... This phenomenon, characterized by partial or complete rupture of the paper stock, is known as 'picking' or 'plucking'. **1959** W. K. Richmond *Brit. Birds of Prey* xiii. 150 A cock Sparrow-hawk has several plucking-posts on his beat. **1959** *Washington Post* 31 July 1/6 Legislation passed this week providing for the 'plucking' of some 2500 senior officers was quietly amended..to abolish 'tombstone promotions'. **1960** *Gloss. Paper, Stationery Terms* (*B.S.I.*) 11 *Plucking*, the detachment of superficial zones of the sheet when the adhesion due to an applied external force is higher than the internal cohesion of the paper. **1968** *Gloss. Formwork Terms* (*B.S.I.*) 19 *Plucking*, spalling of the concrete face due to adhesion of concrete to the form.

plud-pludding, var. *plod-plodding* (s.v. Plodding *vbl. sb.*).

1912 W. Deeping *Sincerity* i. 1 The grinding of wheels and the 'plud-pludding' of drenched horses drifted along the high road.

plug, *sb.* Add: **1. b.** *spec.* One for temporarily stopping the waste pipe at the bottom of a sink, wash basin, or bath.

1860 T. Hamilton in T. L. Donaldson *Handbk. Specifications* i. 221 A neat wash-hand basin, with brass plug, socket, and chain. **1872** W. Eassie *Healthy Houses* iv. 37 Although not long introduced the ordinary troughs and basins with plugs at the bottom and with supply taps..are sufficiently familiar as not to need any description. **1901** G. L. Sutcliffe *Sanitary Fittings & Plumbing* vii. 48 The plug is of the type known as 'sunk', the stud and chain-ring being in the sunk portion of the plug, so as not to project above the bottom of the sink. **1965** A. L. Townsend *Plumbing Second Yr.* iii. 72 The bath will.. be fitted with a waste plug and chain. *Ibid.*, The 'pop-up' plug is operated by turning a handle incorporated in the overflow fitting.

c. A device designed to be inserted into a suitable socket to establish an electrical connection; *spec.* one for connecting the lead of an appliance to an electricity supply, consisting of an insulated casing with two or three pins (or, formerly, with one pin and a ring); also (chiefly *colloq.* or as *wall plug*), a socket fixed to or in a wall for receiving such a plug. Also *fig.* (For *to pull the plug* see next sense.)

1883 J. W. Urquhart *Electric Light* (ed. 2) ix. 286 When it is required to transmit the current to a particular lamp, a metal plug is inserted at the point where the bar connected with the lamp and the bar connected with the machine intersect. *Ibid.* 296 The 'safety fusible plugs' employed in the Edison and Swan systems usually consist of a short length of lead wire. Their function is to melt.. should an unduly strong current..be transmitted. **1888** D. Salomons *Management of Accumulators* (ed. 3) ii. ii. 97 Wall plugs are most useful about a house for attaching a portable lamp or small motor at will. *Ibid.* 98 The portable lamp has a reel of twin wire at its base, with the ends of wires going to the lamp-holder and a connector respectively. This connector fits the wall plug by pushing in the two pins it carries. **1890** *Ibid.* (ed. 5) ii. ii. 166 Mr. Taylor Smith's pattern of portable lamp has a reel of twin wire in its base, with the ends of the wires going to the lamp-holder and a connector plug respectively. The two pins of this plug are pushed into the wall connector..to obtain the light. **1891** F. C. Allsop *Telephones* vi. 97 When the plug is inserted between the two blocks.., the circuit is closed. **1892** —— *Pract. Electr.-Light Fitting* v. 72 When the plug..is inserted in the socket,..the lamp can be lighted. **1923** T. E. Herbert *Teleph.* xiii. 316 It is particularly important that during the insertion of the plug the two springs of the jack shall not be short circuited. **1929** E. A. Robertson *Three came Unarmed* vii. 111 Nonie was..stooping to fix into a wall-plug the flex of the standard lamp. **1945,** etc. [see *pin *sb.*[1] 1 m]. **1960** H. Pinter *Caretaker* ii. 48 There used to be a wall plug for this electrolux. **1972** *Village Voice* (N.Y.) 1 June 5/4 The whole point of the call, her thinking I was a plug into good connections. **1976** D. Phillips *Planning your Lighting* 13/1 A power point with an outcrop of plugs and flexes feeding a number of different items of equipment is still a common sight.

2. k. In some old types of water-closet, a stopper which kept the water in the pan and was pulled to let the contents fall into the soil pipe. Now *Obs.* exc. *Hist.* and in phr. *to pull the plug*, to flush the lavatory; also *fig.* and *transf.*, usu. referring to sudden release or (with allusion to sense *1 c) disconnection.

1859 F. Nightingale *Notes on Nursing* i. 13 As well might you have a sewer under the room, or think that in a water closet the plug need be pulled up but once a day. **1873** B. Latham *Sanitary Engin.* 331 When the handle H which lifts the plug is raised, everything in the basin is suddenly discharged into the trap below, and so into the drain. **1896** T. E. Coleman *Sanitary House Drainage* xiii. 97 Should a small piece of paper or other substance prevent the plug resting tightly upon its seat, the water above gradually escapes into the drain, and impure air is then free to enter the building. **1919** R. Fry *Let.* May (1972) II. 451 A real Victorian W.C. *with a pull up plug.* **1934** V. M. Yeates *Winged Victory* xix. 152 Showers of tracers..frightened him and made him pull the plug rather too soon, and..he..saw his bombs burst a long way from his target. **1935** A. J. Pollock *Underworld Speaks* 92/1 *Pull the plug*, to start negotiations; proceed; tell the narrative without delay. **1935** D. Sayers *Gaudy Night* ii. 43 It was not..an agreeable drawing... She took it..into the nearest lavatory, dropped it in and pulled the plug on it. **1939** R. Godden *Black Narcissus* xxv. 214 I've come to mend a loose joint in your pipe... The plug won't pull till I do. **1943** C. H. Ward-Jackson *It's a Piece of Cake* 50 *To pull the plug*, to release the bombs in one go, as distinct from playing the piano. **1943** G. Greene *Ministry of Fear* i. iii. 60 Pull the plug... Wait till the cistern refills, then pull the plug again. **1948** *Amer. Speech* XXIII. 38/2 [Submarines] *Pull the plug*, to dive or submerge. **1949** D. Smith *I capture Castle* (1950) iii. 31 It had a huge bath with a wide mahogany surround, and two mahogany-seated lavatories, side by side, with one lid to cover them both. The pottery parts showed views of Windsor Castle and when you pulled the plug the bottom of Windsor Castle fell out. **1961** C. Cockburn *View from West* vii. 81 The British statesman finds its [*sic*] nearly impossible to make a simple statement..which might not inadvertently pull the plug on himself and flush him..down the drain. **1964** C. Mackenzie *Life & Times* III. 75 'They must not hurt my seat.' He then pulled up the plug, and pushed it down again. 'Doulton you see.' **1965** *Ibid.* IV. 22 The plug in the water-closet seldom worked. **1972** *Guardian* 4 Sept. 11/1 Pauline Jones..was transferred from Holloway to open prison..and the plug was suddenly pulled out of the big public outcry over the sentence. **1973** *Houston Chron.* 21 Oct. 28 For the first time data are at hand on when to 'pull the plug' on an unconscious patient being sustained artificially. **1974** *Observer* 18 Aug. 11/4 Any prudent banker would have pulled the plug on Court Line long ago. **1977** *Spare Rib* Sept. 12/2 The older lady pulled the plug on her tormentors by prudently using the vibrator.

l. *Geol.* (i) A cylindrical mass of solidified lava occupying the vent of a volcano. Cf. Neck *sb.*[1] 11c.

1882 A. Geikie *Text-bk. Geol.* 256 If the tuff of a cone.. were swept away, we should find a central lava plug or core resembling the volcanic 'heads'..of Germany. **1900** *Q. Jrnl. Geol. Soc.* LVI. 221 Mount Kenya is an ancient much-eroded volcano; the highest peak is formed of the rocks of the central plug. **1944** A. Holmes *Princ. Physical Geol.* xx. 456 Later, the plug of the conduit was forced bodily upwards, through the dome, thus forming the celebrated 'spine' of Mont Pelée. **1976** P. Francis *Volcanoes* iii. 123 The vent may well become blocked with a slow-moving or stationary plug of lava.

(ii) A mass of rock, esp. salt, which has been forced upwards by tectonic pressures, lifting overlying strata into the form of a dome.

1906 *Prof. Papers U.S. Geol. Survey* No. 46. 67 Lee Hager..has suggested a hypothesis which explains the origin of these..domes..by the upthrust of an igneous plug. **1918** *Econ. Geol.* XIII. 452 The intrusion or formation of the salt plug has produced a sharp local doming or quaquaversal structure. **1944** *Nat. Geogr. Mag.* Jan. 16/2 Those 'salt domes' or 'plugs' that yield oil may also yield sulphur. **1970** W. G. Roberts *Quest for Oil* iii. 30 Salt.. flows relatively easily under the high pressures exerted by earth movements, and can be forced into a plug or dome.

m. A sparking plug.

1886 D. Clerk *Gas Engine* viii. 204 The igniting points ..consist of porcelain plugs. **1890** W. Robinson *Gas & Petroleum Engines* vii. 225 The igniter..consists of a brass tube..screwed into the end covers at the top of cylinder. This tube contains a plug of porcelain..to insulate the points of the platinum wires. **1902** J. E. Hutton in A. C. Harmsworth *Motors & Motor-Driving* viii. 151 An English firm has recently introduced a plug which contains no breakable insulators. **1922** J. Buchan *Bk. Escapes* viii. 151 They had flown all the way to Egypt without cleaning their plugs! **1948** A. Morgan *Boys' Bk. of Engines* ix. 110 When a spark jumps across the plug points it ignites the petrol-air mixture in the cylinder. **1973** F. Peterson *Hand-bk. Lawn Mower Repair* iii. 52 You can clean the old plug with a wire brush, then make sure the electrodes are set the proper distance apart.

6. a. (Earlier and later examples.)

1860 in A. H. Oldroyd *Lincoln's Campaign* (1896) 171 There's an old plow 'hoss' whose name is 'Dug'... He's short and thick and a regular 'plug'. **1930** V. Palmer *Men are Human* xxi. 195 There would be a moral rot and everyone would be looking for an old plug to ride. **1948** *Chicago Tribune* 12 Dec. (Grafic Mag.) 5/5 He was a hopeless plug and never ran in the money. **1972** *Dict. Contemp. & Colloq. Usage* (Eng.-Lang. Inst. Amer.) 22/3 *Plug*,..a worn-out old horse; a nag.

b. *transf.* An incompetent or undistinguished person. Also, a bloke, a fellow. Also *attrib.*

1848 *Ladies' Repository* VIII. 316/2 *Plug*,..a nickname for a homely man. **1863** in J. D. Billings *Hardtack & Coffee* (1887) 72 Next came General Meade, a slow old plug, For he let them away at Gettysburg. **1899** 'J. Flynt' *Tramping* ii. iv. 278 I'm always willing to be square to a square plug (fellow). *Ibid.* iv. 396 *Plug*, a fellow; synonymous with 'bloke' and 'stiff'. **1900** *Dialect Notes* II. 49 *Plug*... 2. A hard student... 3. A slow, disagreeable person. 4. A short, thick-set person. **1904** 'No. 1500' *Life in Sing Sing* 251/1 *Plug*, a fellow. **1920** S. Lewis *Main St.* 308 You figure I'm just a plug general practitioner. **1921** [see *gee *sb.*[4]]. **1935** N. Ersine *Underworld & Prison Slang* 58 *Plug*, a working stiff, an ox. **1948** *Redbook Mag.* (Chicago) Mar. 48/2 You—you broken reed! You doormat! Old steady, unimaginative, dumb *plug*!

c. A book which does not sell well, and becomes bad stock.

1889 *Cent. Dict.* (1890) VI. 4565/1 *Plug*,..a shelf-worn book. **1901** *Dialect Notes* II. 145 *Plug*, a book left on author's or publisher's hands. N.Y. City. **1928** *Publisher's Circular* 21 July 59/2 Out of the vast number of publications issued, some must, indeed, turn out to be plugs. **1930** *Publisher's Weekly* 15 Mar. 1546/1 The so-called plugs are weeded out..making room for new titles. **1948** H. L. Mencken *Amer. Lang.* Suppl. II. xi. 739 *Plug*, a good book that no one wants. **1970** R. K. Kent *Lang. Journalism* 104 *Plug*. 1. (a) a book sold at a reduced price by a publisher after sales have fallen off: in plural, also *remainders*.

d. A steady plodding course. (Cf. Plug *v.* 4.)

1903 *Eng. Dial. Dict.* IV. 557/2 *Plug*,..a long-continued pull. **1909** *Daily Chron.* 16 Sept. 3/4 The story is of the quiet plug of the prosaic Henry and the meteoric flight of the splendid Len. **1911** A. Cherry-Garrard *Jrnl.* 17 Dec. in *Worst Journey in World* (1922) II. x. 359 It was a hard plug up the waves.

7. (Earlier example.)

1848 *Ladies' Repository* VIII. 316/2 *Plug*, a hat.

8*. An advertisement; an instance of publicity; a method of drawing attention to (a product, an entertainment, etc.). *colloq.* (orig. *U.S.*).

1902 G. Ade *Girl Proposition* 50 They were friendly to the prosperous Bachelor and each one determined to put in a few quiet Plugs for Sis. **1929** *Variety* 10 July 1/5 Everything gets a Wrigley plug, for the benefit of his gum. **1937** [see *credit *sb.* 13 f]. **1946** *Sat. Rev. Lit.* 30 Nov. 5/1 Dale gets in a neat plug for the publisher's blurb on the dust jacket. **1953** *Recorder* 17 Nov. 4/5 Why do you give them [*sc.* Selfridges] a free plug? **1957** *Time* 2 Sept. 27/1 The policy emerged mostly as a clearly reasoned plug for the kind of development job private capital and U.S. aid have been doing in Latin America. **1958** *Spectator* 13 June 762/1 Nobody will be a penny the better off for the debate except Mr. Noel-Baker, whose new book on disarmament ..got a series of plugs that not even a film programme on BBC television could rival. **1965** N. Gulbenkian *Pantaraxia* xiv. 288 It was to give a 'plug' to Jack Barclay..and ..Panelcraft..that I agreed to appear in 'Tonight'. **1973** *Nation Rev.* (Melbourne) 31 Aug. 1442/5 The *Observer* more than compensated..in the same edition as the pious editorial and the color plug. **1978** *Jrnl. R. Soc. Arts* CXXVI. 418/1, I was interested in Sir Monty's plug for engineers to be involved as managers in top management.

8.** *Angling.* A lure with one or more hooks attached.

1932 *Kansas City Times* 13 May 22 There is some balm for the fellow who thinks he is paying too much for his plugs, flies and other equipment. **1944** 'N. Shute' *Pastoral* i. 5 To take his new rod and his new reel and his new plugs. **1960** M. Sharcott *Place of Many Winds* v. 100 These beaches have been known to yield many valuable articles, the best of which are probably the fishing lures, or plugs as fishermen call them. **1967** *Daily Tel.* 21 Oct. 14/6 Orange-coloured plugs are the most killing for late evening. **1976** *Norwich Mercury* 19 Nov. 9/5 Jeremy Epton and Anthony Raywood, 11, were having no success at all dipping a plug into the nearby River Wensum hoping for a pike.

9. (sense 8*) *plug number, schedule, song*; **plug-assist**, a heated plunger used in the vacuum moulding of plastics which forces the plastic partially into the mould cavity before the vacuum is applied; the technique of using this device; freq. *attrib.*; hence **plug-assisted** *a.*, using a plug-assist; **plug-bait** = *plug *sb.* 8**; **plug-board**, also, a similar piece of equipment used with data-processing apparatus, in which receptacles can be interconnected by lengths of wire with a plug at each end, and which is usually made to be removable to facilitate changes of program or function, the receptacles making electrical contact with fixed terminals on the machine when it is in place; (examples); **plug flow**, flow of a body of ice or other viscous fluid *en bloc*, with no shearing between adjacent layers; **plug fuse** *Electr.*, a fuse that is screwed into a socket; **plug gauge**, a gauge in the form of a plug which is used for measuring the diameter of a hole; **plug-hat** (earlier and later examples); (*b*) *Austral.*, a bowler-hat; **plug-hole**: also in *fig.* phrases (cf. *drain *sb.* 1 e); **plug horse** *N. Amer.* = Plug *sb.* 6; **plug nozzle**, in a rocket or jet engine, a nozzle containing a central plug that diverges towards the exit and then converges, so that the gas is expelled in a converging annular stream; also *attrib.*; **plug tobacco** (earlier example).

1958 *Brit. Plastics* XXXI. 20/2 Almost immediately the heated plug-assist is lowered into the bubble. **1958** *Times Rev. Industry* Aug. 57/2 This machine uses deep-draw, drape, Airslip, and drop-form plug assist techniques either individually or in combination. **1958** *Brit. Plastics* XXXI. 352/2 Drape and plug-assisted techniques are particularly suitable. **1965** L. A. H. Eastman tr. *Thiel's Princ. Vacuum Forming* iv. 41 For single mouldings plug-assisted forming seldom offers any advantage over air-slip forming. **1955** Plug-bait [see *ironmongery 1 d]. **1883** *4th Meeting U.S. Nat. Telephone Exchange Assoc., 1882* 38 Switchboards are generally classified as 'cord' or 'plug' boards. *Ibid.* 39 The plug boards were the favorites. **1946** *Ann. Computation Lab. Harvard Univ.* I. 21 The out-relay,..through which the product is read out of the multiply unit, connects to the buss through a plugboard provided to fix the decimal point relation between the product counter and the buss. **1957** D. D. McCracken *Digital Computer Programming* 231 A number of the small machines are controlled by a removable plugboard... There are ordinarily many of them with each machine, one for each recurring problem. **1970** O. Dopping *Computers & Data Processing* xv. 231 In older computers, the distribution of input card fields over different parts of memory was often effected with plugboards as in conventional punched card machines. **1977** *Sci. Amer.* Sept. 155/1 Hardware prototyping mechanisms commonly include wire-wrap breadboard models, plugboard setups or printed-circuit prototypes. **1951** *Proc. R. Soc.* A. CCVII. 560 There is then no relative movement except in the lowest layer and the block simply slides downhill as a rigid body... (The corresponding motion in a block between rough plates has been called 'plug flow'.) **1972** B. W. Sparks *Geomorphology* (ed. 2) xiii. 382 Temperate glaciers in summer are likely to be the most effective agents of erosion especially if they are sliding over their beds and undergoing plug flow. **1974** D. K. Smith in P. L. Moore et al. *Drilling Practices Manual* xvi. 437 This readily converts to a maximum pump rate in order to remain in plug flow. **1905** *Jrnl. Inst. Electr. Engineers* XXXV. 365 The earliest form of Edison plug fuse dates from 1880. *Ibid.* 405 Enclosed or cartridge fuses..have been developed from the Edison plug fuse. **1971** W. N. Alerich *Electr. Construction Wiring* xi. 283 (*heading*) Plug fuse designed to pass safely 15 amps. **1895** *Appleby's Illustr. Handbk. Machinery* iv. 129 The plug gauges above 2 inch are cored out for lightness. **1905** [see *lap *v.*[4]]. **1971** B. Scharf *Engin. & its Language* vii. 49 Plug gauges (plain, tapered or pin gauges) are used for checking holes, mainly in order to ensure that they are not too narrow at any point. **1863** in J. D. Burn *Three Yrs. Working Classes U.S.* (1965) 223 Fancy a ragged man.. with a gun, a knapsack, a butcher's knife and a plug hat. **1941** Baker *Dict. Austral. Slang* 55 *Plug hat*, a bowler hat. **1947** Plug-hat [see *cow-hide *sb.* 4]. **1977** *Time* 14 Mar. 30/1 He chuckled over the memory of seeing Tammany Democrats dressed in their long coats and plug hats but so broke they could not pay their hotel bills. **1968** *Listener* 22 Feb. 243/1 May I ask..whether the anonymity demanded by the ethics of the 'profession' is now to be regarded as having..gone down the plug-hole? **1973** *Guardian* 28 Mar. 10/6 Nothing escaped so completely as Warhol himself. A positive plug hole of a man around which the bath water swirled. **1973** *Times* 17 May 12/5 That [term] went down the plughole of progress. **1887** *Courier-Jrnl.* (Louisville, Ky.) 4 Feb. 3/5 Wanted—40 plug horses and mares at Lum Simon's Stables. **1969** N. W. Parsons *Sagebrush Harp* xviii. 97 Later, Papa bought another plug horse, giving a note due the following fall. **1960** *Astronautics* Apr. 100/2 There are many ways in which such combustors can be combined with a nozzle..to form attractive propulsion system configurations. One possible configuration is a segmented annular combustor with a 'plug nozzle'. **1970** A. V. Cleaver in N. H. Langton *Rocket Propulsion* II. iv. 136 There would be..appreciable savings in the length and weight of inter-stage vehicle structures if plug-nozzle engines were used in the upper stages; on the other hand,..the small performance gains would be associated more with the use of plug nozzles on bottom stages. **1933** P. Godfrey *Back-stage* xiv. 173 In pursuance of his theory of the value of reiteration de Courville instituted the feature of the 'plug number'. **1947** *Time* 24 Nov. 74/3 Music publishers and recording companies were getting together on 'plug schedules' to ration out the hits. **1939** *Melody Maker* 3 June 1/2 Each band..will be playing in any programme of ten items, no less than eight plug songs. **1952** B. Ulanov *Hist. Jazz Amer.* (1958) xiv. 159 The plug songs of the moment. **1814** in *Deb. Congress U.S.* (1855) 17th Congress 2 Sess., App. 1218 Plug tobacco manufactured at Columbia, one shilling and three pence per pound.

plug, *v.* Add: **1. c.** Also, to insert a fibre or plastic tube or cylinder for the same purpose.

d. (Further examples.)

1952 Dylan Thomas *Let.* 21 July (1966) 375 Now it's up to me & him to plug in lots more expenses. **1976** *National Observer* (U.S.) 17 Jan. 14/1 (Advt.), There's no need to rip out your old grass. Plug in Amazoy Zoysia Grass and let it spread into beautiful turf that never needs replacement.

e. Substitute for def.: (i) *trans.* To insert (a plug or the like) *into* a socket; to connect electrically (an appliance or apparatus) by inserting a plug *into* a socket; also *to plug in* (trans.), to connect electrically in this way. Also *absol.*, *transf.*, and *fig.*

1903 [in Dict.]. **1923** T. E. Herbert *Teleph.* xiii. 347 The operator plugs in with the service plug, restores the indicator, and ascertains the number required. **1925** P. J. Risdon *Crystal Receivers & Circuits* 15 A complete set of such coils will thus enable a big range of wave-lengths to be efficiently covered, by plugging in a coil most nearly corresponding to the wave-length required. **1925** *Scribner's Mag.* July 54/2 He wandered in to his radio, lighted the tubes and plugged in the ear phones. **1932** *Pictorial Weekly* 19 Mar. 201/1 Rescue vessels can 'plug in' to this buoy, and talk to the men below. **1934** *Archit. Rev.* LXXV. 108/3 (*caption*) A portable fire that can be 'plugged in' to a gas point in bedroom or bathroom. **1948** F. Thompson *Still glides Stream* i. 10 One here and there of her pupils had shown the sudden gleam of comprehension..which..she had referred to..as 'plugging in', or 'taking the bait'. **1960** *Farmer & Stockbreeder* 15 Mar. 143/1 A large range of implements can be 'plugged-in' for doing a number of other jobs. **1965** *Daily Tel.* 1 June 15/8 Extract this Danish machine and plug it into an AC power point. **1970** *Atlantic Monthly* July 88 Five children were plugged into a tape recorder, listening to a story and following it in the books in front of them. **1971** [see *Jack *sb.*[1] 15 d]. **1971** J. H. Smith *Digital Logic* ii. 18 A power supply..with facilities for plugging at least ten modules into it at any one time. **1972** *Sci. Amer.* Apr. 13/1 (Advt.), The most advanced automotive check-up in the world today... Your car will actually be plugged into a computer. **1972** *National Observer* (U.S.) 27 May

22/3 They tell us that to be entirely into the new literary life, one must in some way be plugged into films. **1972** *Sci. Amer.* July 105/1 All 10 digits are plugged into the expression ABC × DE = FGH × IJ. **1977** *Ibid.* Aug. 80/3 The placental embryo..is plugged into the maternal blood supply for nourishment.

(ii) *intr.* To be, or to be capable of being, plugged *in* or *into*. Also *fig.*

1956 T. E. Ivall *Electronic Computers* iv. 45 When the unit is put into the patch panel it also plugs into a d.c. amplifier at the rear. **1963** *Which?* Dec. 378/2 The *Timac* plugged directly into the mains socket. **1974** *Physics Bull.* Sept. 401/1 This assembly plugs into a choice of sockets dependent upon the position in which the instrument is to be used.

f. *trans.* To cut a cylindrical core from. Also *absol.* *U.S.*

1874 'Uncle Bob' *Lett. to Children* 19, I used to be a great hand to go into the patch, plug 'em before they were ripe, and then turn the cut side down. *a* **1910** 'Mark Twain' *Autobiogr.* (1924) I. 111, I know how to tell when it [*sc.* a watermelon] is ripe without 'plugging' it. **1948** *Chicago Tribune* 25 June ii. 3/3 The safest and best way to tell quality is to 'plug' the melon. **1969** *Times* 22 July (Moon Suppl.) p. i/1 It's a very soft surface, but here and there where I plug with the contingency sample collector, I run into a very hard surface.

g. *slang.* To copulate with.

1901 Farmer & Henley *Slang* V. 231/1 *Plug*,..to copulate. **1977** *Amer. Speech* 1975 L. 64 'I plugged her last night.' (male use).

h. *to plug off* or *back*: to seal off (an oil well or a rock formation) by inserting a plug. Also *absol.*

1919 *Summary of Operations Calif. Oil Fields* (Calif. State Mining Bureau) V. i. 9 *Plugged off.* Describes the condition existing when fluid encountered in a lower part of a well has been excluded from a higher part of the well by placing an effective plug between the two places. **1924** L. C. Uren *Textbk. Petroleum Production Engin.* ix. 276 It will be important..to estimate carefully the volume of that part of the well which it is desired to plug off. **1938** C. P. Parsons in A. E. Dunstan et al. *Sci. of Petroleum* I. ix. 472/2 The amount of cement left in the bottom of the tubing depends upon the amount of hole to be plugged back. **1938** Wilde & Moore in *Ibid.* xi. 573/2 Having located the water, it is important to plug off the water sand. **1976** L. St. Clair *Fortune in Death* i. 8 We've wasted enough time fishing drill pipe out of this hole. Let's plug back and slant-drill.

2. (Earlier and later examples.) Also, to fire (a bullet) *into* (example *fig.*).

1870 J. C. Duval *Adv. Big-Foot Wallace* xix. 99 Just at that instant Jeff plugged him with a half-ounce bullet. **1882** E. W. Hamilton *Diary* 27 Aug. (1972) I. 326 He has had a narrow escape of losing an eye, having been plugged in the face by Newport while grouse driving at Wharncliffe's. **1891** 'Mark Twain' in 'Twain' & Howells *Mark Twain—Howells Lett.* (1960) II. 635, I will plug into you at short range the first chapter of my new book. **1904** [see *lam *sb.*[2]]. **1924** [see *frame *sb.* 2]. **1936** G. Greene *Gun for Sale* i. 22 Don't say a word or I'll plug you... I don't care a damn if I plug one of you. **1969** C. Burke *God is Beautiful, Man* (1970) 28 They told their old man..if they didn't bring Ben back Simon would get plugged. **1973** W. M. Duncan *Big Timer* xxi. 137 That Carver packed a wallop, didn't he? I should have plugged him sooner.

3. Also, with a missile.

1971 Wodehouse *Much obliged, Jeeves* xvi. 169 Sidcup got a black eye. Somebody plugged him with a potato.

4. a. (Further examples.) Freq. const. with advbs.

1900 G. Ade *More Fables* 44 Any Husband could..get up every Morning ready to Plug for a Renaissance of their Early love. *Ibid.* 146 You take a Man who is Plugging along on a Salary. **1911** E. Ferber *Dawn O'Hara* vii. 99 Lots o' us are pluggin' an' savin' in the hopes that some day we'll have money enough to get back at some people we know. **1947** K. Tennant *Lost Haven* (1968) vi. 88 He was a mug to plug away at yet another new boat. **1953** Wodehouse *Performing Flea* 58, I am plugging along with *Hot Water* and have done 60,000 words. **1954** A. Huxley *Let.* 16 Sept. (1969) 711 Lacking the ability to write a text book, I have to plug on at these other, more precarious forms of literature. **1973** *Philadelphia Inquirer* 7 Oct. 19 Ronnie's not a quitter. He really plugs. **1977** *World of Cricket Monthly* June 11/1 Australia's bowlers plugged away, with Max Walker breaking through when Surrey were 2 wickets down for 147, and snaring 3 quick wickets for only 6 runs.

5. To prevent (a person) from carrying out a project by anticipating him or depriving him of his opportunity; to block (an action or design). *U.S.*

1880 *Scribner's Mag.* 492/2 One fisherman 'plugs' another when he puts out from shore and casts in ahead of him. **1896** G. Ade *Artie* xii. 110, I wouldn't like to start in and plug his game.

6. *intr.* (for *refl.*). To stick or jam; to become obstructed.

1902 S. E. White *Blazed Trail* xlviii. 338 Several times the jam started, but always 'plugged' before the motion had become irresistible. **1964** M. Gowing *Britain & Atomic Energy 1939–1945* viii. 222 The membranes must not 'plug', that is, get blocked.

7. a. *trans.* To popularize (a song) by having it played many times; to present (something) repeatedly; to publicize, emphasize, draw attention to. *colloq.* (orig. *U.S.*).

1906 H. Green *At Actors' Boarding House* 68, I ain't got any music, so you kin plug any publisher's stuff an' play what you wanter. **1927** *Daily Express* 9 Nov. 9/4, I.. thought it would encourage them to plug my songs. **1930** [see *cut *v.* 54 f]. **1940** *Brit. Jrnl. Psychol.* Oct. 118 The technique of 'plugging' trivial news about important personages. **1959** *Elizabethan* June 27/2, I gather from John that the other papers have plugged the crisis for all they're worth. **1967** *Wall St. Jrnl.* 12 Jan. 1/4 Mrs. Glick..now plugs Excedrin on television. **1970** G. F. Newman *Sir, You Bastard* i. 21 I'm obliged to listen to clients plugging their virility as relevant facts. **1975** C. James *Fate of Felicity Fark* v. 45 She found the concentration of rehearsal More challenging by far than plugging *Persil*.

b. *intr.* *to plug for*: to act in support of; to make favourable statements about. *U.S. colloq.*

1927 *Amer. Speech* II. 256/1 'Pluggers' or 'rooters', 'plug' or 'root' for their side or for their favorite players. **1929** D. Runyon in *Hearst's Internat.* July 58/1 Miss Missouri Martin keeps plugging for Dave the Dude with Miss Billy Perry. **1932** *Sun* (Baltimore) 27 Apr. 1/1 The secret subsidizing of newspaper financial writers to 'plug' for stocks in process of manipulation upward. **1943** *Amer. Speech* XVIII. 249 Judge James A. Chase, a Cashmere citizen, who had visited the Vale of Kashmir, plugged for the new name and won. **1974** *News & Courier* (Charleston, S. Carolina) 7 Apr. A-14/3 At present he is plugging for a written history of Dillon County.

pluggable (plɒ·găb'l), *a.* [f. Plug *v.* + -able.] Suitable for or capable of being plugged. **a.** Of a song or recording (see *plug *v.* 7 a).

1930 *Punch* 9 Apr. 414 One good rousing 'pluggable' air, 'The March of the Musketeers'. **1977** *Time* 19 Dec. 54/1 The first film..features a sound track overcrowded with highly pluggable Bee Gees songs.

b. Of an electrical device (see *plug *v.* 1 e).

1946 *Ann. Computation Lab. Harvard Univ.* I. 251 The L10 counter has a pluggable read-out from the counter into the buss. **1954** *Jrnl. Assoc. Computing Machinery* I. 13/2 Converting..consists principally of adding magnetic heads and pluggable units. **1977** *New Yorker* 12 Sept. 93/2 (Advt.), These remarkable, pluggable, portable 4-band radio-cassette recorders give you the wide sound of stereo FM.

plugged, *ppl. a.* Add: **2.** *U.S.* Of coins: having a portion removed and the space filled with base material.

1888 *Texas Siftings* 3 Nov., Ticket Agent—Can't sell you a ticket for that quarter; it's plugged. **1890** B. Hall *Turnover Club* 207 The first speaker..paid the price of his folly with a plugged quarter. **1909** 'O. Henry' *Options* 312 Mr. Hinkle told me..you'd never taken in a lead silver dollar or a plugged one. **1912** *Pearson's Mag.* (Amer. ed.) XXVII. 691/2 For a plugged peso I'd stay with you! **1923** C. E. Mulford *Black Buttes* 265 He says..he'll see us both in hell before he'll pay a plugged peso. **1936** C. Sandburg *People, Yes* 63 He seems to think he's the frog's tonsils but he looks to me like a plugged nickel. **1946** E. O'Neill *Iceman Cometh* (1947) iv. 210 Listen, you cockeyed old bum, for a plugged nickel I'd—. **1974** Hawkey & Bingham *Wild Card* viii. 81 If as much as a whisper gets out..none of our lives are going to be worth a plugged nickel.

3. plugged-in, electrically connected by means of a plug (see *plug *v.* 1 e); also *fig.*

1957 V. Nabokov *Pnin* i. 14 His devoutly plugged-in clock would make nonsense of his mornings after a storm in the middle of the night had paralyzed the local power station. **1970** E. McGirr *Death pays Wages* vi. 126 A box with a plugged-in pair of earphones. **1975** *Wentworth & Flexner's Dict. Amer. Slang* Suppl. 732/1 *Plugged-in* adj. = turned-on. **1977** M. Herr *Dispatches* (1978) 32 After a year I felt so plugged in to all the stories..that even the dead started telling me stories.

plugger. Add: **c.** One who extols or publicizes. Cf. *plug *v.* 7 b. orig. *U.S.*

1913 *Writer's Bulletin* Oct. 127/2 Publishers spend thousands..in order to attract the attention of out-of-town performers with whom, neither they nor their 'pluggers' ever come in contact. **1921** *Cleveland* (Ohio) *Enterprise* 4 June 1/3 Everybody out here is a booster and plugger for one common purpose. **1927** [see *plug *v.* 7 b]. **1958** [see A. and R. (*A III)]. **1972** P. Black *Biggest Aspidistra* i. iii. 29 The pluggers kept the initiative by inventing the request item. This was..almost impossible to identify as a proven plug.

d. *Angling.* One who fishes with a plug (sense 8**).

1967 *Daily Tel.* 21 Oct. 14/7 Many successful bass pluggers work on the principle that it is a fish with an easily aroused temper. So they use a 'teaser'.

plugging, *vbl. sb.* Add: **1.** (Further examples.)

1797 *Deb. Congress U.S.* 13 Dec. (1851) 718 They knew the silver coin circulated by tale, the gold by weight; the value of the latter had actually diminished by various means, such as sweating, plugging, clipping, &c. **1841** *Florist's Jrnl.* II. 266 The form is almost perfect, with great depth of petals, and an excellent rising centre, such as cannot be imitated by any of the usual plugging and trickery sometimes played off, or at least attempted, at exhibitions; and thus it is an excellent show Dahlia. **1908** *Animal Managem.* 211 Piece by piece the straw is wetted and forced into the body of the collar... This process is termed 'plugging'. **1908** K. McGaffey *Show-Girl* 109 Is it considered au fait for a bride-about-to-be to do a little plugging for wedding presents this early in the game? **1921** A. J. Empey *Madonna of Hills* xviii. 130 Plugging means to push the sale of songs by singing them in cabarets and places. **1926** Whiteman & McBride *Jazz* viii. 169 While plugging is important, the publishers contended recently that there can be too much of any good thing. **1934** *Evening News* 13 July 6/5, I wonder, by the way, whether Uncle Andre, in view of his hour tonight, has been asked to reduce the amount of 'plugging' in it. **1957** *Listener* 10 Oct. 583/1 This plugging of his plays, early and late, successful and otherwise. **1969** E. W. Hildick *Close Look at Advertising* 31 Some of the celebrities have been known to collect very large sums of money for 'letting slip' just such recommendations. It is known as plugging. **1974** G. S. Ormsby in P. L. Moore et al. *Drilling Practices Manual* vi. 167 Underflow plugging is usually caused either by a dry beach or by solids over-load.

3. (Examples in sense *7 a of verb.)

1929 *Melody Maker* Feb. 124/3 The listener is often served with all kinds of tripe songs, reiterated *ad nauseam* at the dictates of the wire pullers over an extensive plugging period. **1932** *Sun* (Baltimore) 27 Apr. 15/7 Checks..were identified..as payments for stock 'plugging' publicity. **1959** *Punch* 10 June 766/3, I suspect it may be a new plugging gimmick, a form of filibustering.

plu·g-in, *a.* and *sb.* [f. vbl. phr. *to plug in* (*plug *v.* 1 e).] **A.** *adj.* Designed to be plugged into a socket (esp. an electrical one); of or pertaining to such devices.

1922 *Wireless World* 8 July 458/1 One sees so many makeshift plug-in devices..in amateur sets. **1926** R. W. Hutchinson *Wireless* 187 The aerial inductance L_1 and the reaction coil L_2 are of the plug-in type. **1954** *Sun* (Baltimore) (B ed.) 19 June 15/1 Portable plug-in telephones with conveniently located outlets offer a modern convenience where a telephone is desired on a part-time basis in certain rooms. **1965** *Wireless World* Sept. 464/2 An encapsulated version..for plug-in applications is available. **1966** 'A. York' *Eliminator* iv. 57 A mahogany-boxed erinoid plug-in chess set lay on the coffee table. **1968** R. Petrie *MacLurg goes West* viii. 67 The hotel manager was knocking at MacLurg's door, followed by a porter with a plug-in telephone. **1968** B. Turner *Sex Trap* v. 28 His kitchen boasted one cup and a plug-in percolator. **1971** J. H. Smith *Digital Logic* ii. 18 A very easy method of construction is to use one of the commercially available rack construction systems with plug-in units. **1977** Warner & Bell in D. M. Rumbaugh *Language Learning by Chimpanzee* vii. 144 Purchasing the computer, its plug-in modules and peripherals proved to be a wise decision.

B. *sb.* A plug-in device or unit.

1950 *Sun* (Baltimore) 31 Aug. 8/7 The refrigerator units will operate either from car batteries or 110-volt or 220-volt plug-ins for use where electricity is available. **1955** *IRE Trans. Electronic Computers* IV. 3/1 One standard plug-in contains four flip-flops and eight cathode-followers for isolation. **1965** *Wireless World* July 54 (Advt.), Three new instruments..have all the in-built virtues of the original plug-ins. **1967** *Electronics* 6 Mar. 2 (Advt.), Sweep Oscillators with RF and marker plug-ins meet virtually all of your swept frequency testing requirements. **1971** C. Fick *Danziger Transcript* (1973) 171, I heated some coffee on my plug-in and we lit cigars. **1972** 'G. Black' *Bitter Tea* (1973) xiii. 212 The jacks for the phone plug-ins. **1975** *Sci. Amer.* Mar. 11/1 (Advt.), Priced at $800 for the main-frame and $110 to $250 for the plug-ins, the 5150A is easily the new price/performance leader in instrumentation recorders.

plugola (plɒgōᵘ·lă). orig. *U.S.* [f. *plug *sb.* 8* + *-ola.] Incidental or surreptitious promotion of a person or product, esp. on radio or television; a bribe for this. Also *attrib.* and *transf.*

1959 *Washington Post* 7 Dec. A12/4 The nasty charges of fees collected for plugola. **1963** R. I. McDavid *Mencken's Amer. Lang.* 213 *Plugola*, a subspecies [of payola]—payments to disk jockeys in return for frequently playing a record company's new recordings—was also disclosed during these scandals. **1972** *Times Lit. Suppl.* 29 Sept. 1138/2 If the garret doesn't get you, the plugola circuit will. **1972** *Guardian* 20 Dec. 4/5 Mr Whitehead spoke of 'ideological plugola' and went on: 'Station licensees have final responsibility for news balance.' **1977** *National Observer* (U.S.) 1 Jan. 2/2 The Federal Communications Commission said it will hold hearings next year into alleged 'payola' and 'plugola' by broadcasters. The FCC said it had received information that illegal payments had been made to disc jockeys for playing certain records and to broadcasters for mentioning products.

plug-ugly. For *U.S. slang* read *slang* (orig. and chiefly *U.S.*). More widely, a man of violence, one who adopts intimidatory methods. Also *attrib.* and as *adj.* (Earlier and further examples.)

1856 *Butte Rec.* (Oroville, Calif.) 29 Nov. 3/7 The.. Plug Uglies..went to Philadelphia on election day..to fight off and whip the democracy from the polls. **1857** *Lawrence* (Kansas) *Republican* 30 July 2 Only a pitiful minority of the actual voters of Kansas, cast their votes for delegates to the plug ugly convention. **1861** [see *blood-tub* (*blood *sb.* 19)]. **1909** *Dialect Notes* III. 358 *Plug-ugly*, an ugly person or thing, especially an ugly horse. **1915** *Ibid.* IV. 187 *Plug ugly*, adj. phr. Said of a very ugly man, not genteel. (Applied more commonly to an 'ugly customer'.—*Ed.*) **1923** Wodehouse *Adventures of Sally* xiv. 176 After he'd paid these two pluguglies their guarantees..he was just about cleaned out. **1935** *Punch* 13 Nov. 550/2 Readers who have led sheltered lives will think of plug-uglies, and I hope the cleaner kinds of plug-ugly will think of baths. **1953** [see *lunatic fringe* s.v. *lunatic *sb.* 1 c]. **1956** D. Karp *All Honorable Men* 18 The plug-uglies on the Right and the Left are the only people who don't give a damn about the opposing argument, and I don't count myself among them. **1972** K. Bonfiglioli *Don't point that Thing at Me* ii. 12 'Yes, Sir.' said Plug Ugly II. **1975** *New Yorker* 1 Dec. 136/2 When a director like Sam Peckinpah puts a group like his Wild Bunch on the screen, the men are so alive that the last thing that would ever come into your head is that some of them are plug-uglies. **1978** N. Marsh *Grave Mistake* ii. 62 Verity thought, I've..drunk their champagne so now I turn plug-ugly and refuse.

plum, *sb.* Add: **4. d.** (Further examples.) Also, the best part of a musical work.

1887 in G. Stimpson *Bk. about Amer. Politics* (1952) 258 The boys enjoying the plums will support anybody who is good for him or them. **1937** W. H. Saumarez Smith *Let.* 16 Oct. in *Young Man's Country* (1977) ii. 94 It [*sc.* the job] is definitely one of the three plums for the young civilian. **1967** *Boston Herald* 8 May 24/5 José has played his cards just right, and a rich little plum named Lucy falls into his outstretched arms. **1973** *Times* 20 Oct. 13/6 Its slow movement is its 'plum', a glorious, unbroken song. **1978** *Time* 3 July 42/2 Center directors receive only $11,000 a year, but Mendel offers them a plum: their kids can attend free.

5*. = *plum-colour*.

1878 *Trans. Illinois Dept. Agric.* XIV. 210 [Siamese Swine] varied in color from deep rich plum to dark slate and black. **1895** *Montgomery Ward Catal.* 3/1 Cashmere... All the fashionable colors..golden brown, medium plum, heliotrope. **1940** Graves & Hodge *Long Week-End* xvi. 278 Victorian colours—plum, maroon, and violet—were in favour. **1970** *Guardian* 5 June 9/2 She has featherweight car rugs in brushed wool in plum and purple at a modest three guineas each.

6. a. *plum brandy*, *-tart* (earlier example), *wine*. **b.** *plum-gathering*; *plum-dark*, *-purple* (earlier example), *-rich* adjs. **c.** *plum-coloured* (earlier example), *-stained* adjs. **d. plum-colour** (further examples); **plum-in-the-mouth** *a.* (*colloq.*), indistinctly articulated, esp. in a manner associated with the British upper classes (of speech, etc.); **plum pox** [tr. Bulgarian *sharka na slivite* (D. Atanasoff 1932, in *Godishnik na Sofiĭskiya Universitet Agronomski Fakultet* XI. 49)], a virus disease of plum trees characterized by yellow blotches on the leaves and pockets of dead tissue in the fruit; also known as *sharka*; **plum rains** [tr. Jap. *bai-u*] = *bai-u; applied also to the corresponding rains in southern China.

1950 *Chambers's Encycl.* IV. 326/1 Plums are grown as fresh fruit and for jam, though some are distilled into *slivovice* (plum brandy). **1958** A. L. Simon *Dict. Wines, Spirits & Liqueurs* 147/1 Slivovitz..is very similar to the Alsatian Plum Brandy called Quetsch. **1977** H. Fast *Immigrants* 11 The husband of the Polish woman.. hoarded plum brandy. **1898** G. B. Shaw *You never can Tell* iii. 274 The wall decoration of Lincrusta Walton in plum color and bronze lacquer. **1960** S. Plath *Colossus* (1967) 21, I squat..Counting the red stars and those of plum-colour. **1820** M. Edgeworth *Let.* 8 June (1979) 160 Fannys plum colored [*sic*] and Harrets lilac tabbinets. **1957** L. Durrell *Bitter Lemons* 103 Plum-dark mountain roses. **1958** —— *Balthazar* i. 13 A single plum-dark sail, moist, palpitant. **1963** *Glamour* Sept. 146 Plum-dark wool frames the soft blue silk collar of the coat, worn over a matching plum-dark flared skirt. **1928** C. Day Lewis *Country Comets* 25 At the time of plum-gathering When the hedge is plumy With Traveller's Joy. **1926** D. H. Lawrence *Plumed Serpent* vi. 118 She spoke rapidly, a rather plum-in-the-mouth Spanish. **1934** S. R. Nelson *All about Jazz* vii. 163 The lukewarm, plum-in-the-mouth style of some of the white vocalists. **1933** *Rev. Applied Mycol.* XII. 230 The disease, which the author [*sc.* D. Atanasoff] terms plum pox, was proved to be readily transmissible to healthy trees by grafting. **1943** *Bull. Min. Agric. & Fish.* CXXVI. 64 A single tree bearing foliage with symptoms corresponding to those of Plum Pox..was observed early in August 1934 at East Oakley, Hants. **1952** E. Ramsden tr. *Gram & Weber's Plant Diseases* ii. 204/2 A disease called plum pox, well known in Bulgaria, probably occurs as far north as Bohemia and Holland. **1976** *Nature* 12 Feb. 499/2 The Ministry of Agriculture reported that at least 150 acres of plums have some levels of infection with plum pox (sharka) virus, an aphid-borne virus. **1862** G. M. Hopkins *Vision of Mermaids* (1929), Plum-purple was the west. **1922, 1945** Plum rains [see *bai-u]. **1968** G. R. Rumney *Climatol. & World's Climates* xii. 235/1 The cloudiness, humidity, and generally oppressive conditions accompanying the start of the warm season's heavy rains in southern Japan combine to create a period of depressing, gloomy weather called here, as in south China, the plum rains (*Bai-u*). **1971** *Handbk. Aviation Meteorol.* (Meteorol. Office) xxiii. 378 In May, tropical air begins to advance northwards and is heralded by cyclonic activity and the widespread 'plum' rains of China and Japan. **1932** Auden *Orators* i. 3 The plum-rich red-earth valley of the Severn. **1922** Joyce *Ulysses* 564 Two trickies Frauenzimmer plumstained from pram falling bawling. **1770** J. Woodforde *Diary* 12 Oct. (1924) I. 102, I gave them for dinner a.. Plumb Tart and an Apple Tart. **1728** E. Smith *Compl. Housewife* (ed. 2) 208 To make Plum-wine. Take twenty pound of Malaga raisins..water..damson juice..at 4 or 5 months bottle it. **1976** 'M. Delving' *China Expert* i. 11 The guest list included..Chinese and Westerners, all.. eager to sample the stuffed, glazed chicken and fish in plum wine sauce.

e. passing into *adj.* = *plum-coloured* adj.

1922 Joyce *Ulysses* 551 In a flunkey's plum plush coat and kneebreeches, buff stockings and powdered wig. **1930** V. Sackville-West *Edwardians* v. 229 Buttoned into her plum velvet bodice, like the wife of any British tradesman. **1975** J. McClure *Snake* x. 133 There was a white Jaguar, a plum Datsun coupé and a..Land-Rover.

f. (sense 4 d) passing into *adj.* Choice, valuable, coveted.

1958 *Listener* 21 Aug. 277/1 Was the promotion of Chiappe to the plum governorship an easy method of shedding a dangerous man in a key position? **1959** *Economist* 2 May 455/2 While the aircraft industries of Britain and France are declining for lack of military orders, a plum military contract has been won by a company making its first serious venture into aircraft design since the war. **1966** *Listener* 26 May 746/2 After the Nationalists had come to power, they felt that they had to admit some Afrikaners to their boards and directorates. These were plum appointments and the Boers had been longing for them for years. **1970** *Financial Times* 13 Apr. 10/6 Europe (the present plum client in the German Railways advertising service). **1976** Botham & Donnelly *Valentino* xi. 85 The director..congratulated him on winning the plum role. **1977** *Time* 3 Jan. 50/2 The leader of L.D.P.'s largest faction, whose intellect had won him plum jobs in the Ministry of Finance before he turned to politics in 1952, has probably done exactly that.

plum, *v.* Add: **2. b.** To fill or stuff *up* (a person) with false information.

1921 *Chambers's Jrnl.* May 323/1 He ain't to know no different but what Jack's got prairie fever. Mind you plum him up stiff. **1927** *Observer* 20 Nov. 26/5 He has recently returned from Upper Silesia..and promptly puts into writing all that his clever German friends have been 'plumming' him up with.

plumasite (plū·măsəit). *Petrogr.* [f. the name of *Plumas* Co., California, where it was first found + -ite[1].] A coarse-grained, undersaturated, dike-rock consisting essentially of crystals of corundum in an oligoclase matrix.

1903 A. C. Lawson in *Bull. Dept. Geol. Univ. Calif.* III. 228 This particular type of rock magma does not appear to have been as yet recognized among the known occurrences of rocks, and it is, therefore, proposed to name it, for convenience in reference, *Plumasite*, from Plumas county in which it occurs. **1949** F. H. Hatch et al. *Petrol. Ign. Rocks* (ed. 10) iii. iii. 262 Dioritic rocks of unusual composition include plumasite. **1972** *Jrnl. Geol. Soc. India* XIII. 198 Investigations in the present case have proved that plumasite is a result of the local enrichment of alumina in the magma due to the reaction of the acid magma of the Peninsular gneissic period with the older magnesian country.

plumb, plum, *adv.* **2. c.** (Further examples.)

1846 S. F. Smith *Theatr. Apprenticeship* 213 Long before the time arrived..the house was plum, chock full—full to overflowing. *a* **1861** T. Winthrop *John Brent* xxviii. 296 When we got here, I paid their tickets plum through to York out of my own belt. **1901** F. Norris *Octopus* i. iii. 121 'I'll get plumb out of here,' he trumpeted. 'I won't stay here another minute.' **1926** 'R. Crompton' *William—the Conqueror* v. 89 Poor woman! She's sure plumb crazy! **1934** A. Christie *Murder on Orient Express* ii. ix. 136 'You are sure of that, M. Hardman?' 'I'm plumb certain.' **1967** G. F. Fiennes *I tried to run a Railway* vii. 76 In his presence I was tense, tongue-tied and often plumb stupid. **1973** E. Lemarchand *Let or Hindrance* xiv. 182 They must both be plumb crazy.

plumb, *v.* Add: **I. 1.** For † *Obs.* read *rare* and add later example.

1940 *Sat. Even. Post* 6 Apr. 17/3 [He] rolled down [from a house-top] & plumbed into the yard.

V. 9. *trans.* To connect (a domestic appliance or the like) permanently to the water supply and the drain. Usu. with *in*.

1963 *Which?* 6 Feb. 46/2 The Easiclene [dishwashing machine] would normally be plumbed in, but could be used with hoses from a tap and into a sink. **1976** I. Chanarin et al. *Blood & its Dis.* ix. 79 Automatic equipment using potentially toxic reagents should be plumbed-in. **1976** *Star* (Sheffield) 20 Nov. 10/7 (Advt.), Abbey Plumbing Emergency Service. Bursts and leaks, washing machines plumbed, gas fitting and alterations.

plumbane. Restrict † *Obs.* to sense in Dict. and add: **2.** [-ane 2 b.] **a.** Any of the alkyl compounds of lead or the hypothetical series of saturated lead hydrides (analogous to the alkanes) from which they are formally derived; freq. as a formative element in names of such compounds. **b.** *spec.* Lead tetrahydride, PbH_4, an extremely unstable gas.

1920 *Chem. Abstr.* XIV. 2182 The mol. wt. indicates that it has the double formula $[(C_6H_5)_3Pb]_2$ so that it might be considered as hexa-*p*-xylyldiplumbane. **1950** N. V. Sidgwick *Chem. Elements & Compounds* I. 554 Lead hydride, plumbane, presumably PbH_4..was prepared by Paneth in 1920. *Ibid.* 596 The diplumbane formed may change further into lead tetra-aryl and metallic lead. **1964** A. Rochow *Organometallic Chem.* i. 4 The system [of nomenclature] is so useful that it has been extended.. even to tin and lead (..$(C_6H_5)_2PbCl_2$, diphenyldichloroplumbane), although the older inorganic names like diphenyllead dichloride still persist to a considerable extent. **1967** P. L. Pauson *Organometallic Chem.* iii. 78 The rather unstable plumbanes have only been obtained by direct reduction of the halides with suitable hydrides. **1971** Wiberg & Amberger *Hydrides Elements of Main Groups I–IV* x. 757 Similar attempts to obtain plumbane, PbH_4,..were unsuccessful. *Ibid.* 760 Trimethylplumbane, Me_3PbH, and triethylplumbane, Et_3PbH, decompose according to kinetics that are first-order for $[R_3PbH]$. **1973** J. J. Lagowski *Mod. Inorg. Chem.* xi. 368 Plumbane has been prepared in only trace amounts from the reactions which yield the hydrides of the other elements in this family.

plumbate[2]. Substitute for def.: Any of various (salts of) oxyanions or hydroxyanions of quadrivalent lead, which are formed esp. by the action of alkalis on lead dioxide. Now also extended occas. to (salts of) any oxyanion of lead, the oxidation state being specified in brackets. (Earlier and later examples.) [First formed as F. *plombate* (E. Fremy 1843, in *Jrnl. de Pharm.* III. 32).]

1851 H. Watts tr. *Gmelin's Handbk. Chem.* V. 160 (*heading*) Plumbate of potash. **1889** I. Remsen *Inorg. Chem.* xxx. 645 Other salts derived from the acid $PbO(OH)_2$ are known, and are called plumbates. **1950** N. V. Sidgwick *Chem. Elements & Compounds* I. 603 Plumbic hydroxide, $Pb(OH)_4$. This..gives a series of salts, the plumbates, of various types, such as Ca_2PbO_4 and Na_2PbO_3. **1958** *Engineering* 21 Feb. 242/1 The manufacture, properties and applications of..rust-inhibiting calcium plumbate paint were outlined. **1959** *Nomencl. Inorg. Chem.* (I.U.P.A.C.) 32 By dissolving, for example, Sb_2O_3, SnO, or PbO in sodium hydroxide an antimonate(III), a stannate(II), a plumbate(II), *etc.*, is formed in the solution. **1962** Cotton & Wilkinson *Adv. Inorg. Chem.* xix. 364 Crystalline alkali metal stannates and plumbates can be obtained as trihydrates, for instance, $K_2SnO_3 \cdot 3H_2O$... Such materials..contain the anions $Sn(OH)_6^{2-}$ and $Pb(OH)_6^{2-}$. **1974** G. I. Brown *Introd. Inorg. Chem.* xviii. 210 Tin(II) and lead(II) oxides.. form salts with acids and stannates(II) and plumbates(II) with alkalis. *Ibid.* 211 Both oxides react with caustic alkalis to form stannates(IV) or plumbates(IV).

plumbate (plɒ·mbēit), *a.* [f. L. *plumb-um* lead + -ate[2].] Of or pertaining to a type of glazed and usu. lead-coloured pottery made in pre-Columbian central America. Also *absol.* as *sb.*[3]

1926 S. K. Lothrop *Pottery of Costa Rica & Nicaragua* I. i. iv. 116 The vitrified surface of the Plumbate Ware did not lend itself to painted decoration. **1936** —— *Zacualpa* 39 The typical vessel of pure plumbate clay is hard, well fired and has a glossy surface. **1940** J. E. S. Thompson in *Maya & their Neighbours* 129 The absence of plumbate and typical fine orange pottery, turquoise, and gold..hint that these sites are not contemporaneous with the Mexican Period in Yucatan. **1948** A. O. Shepard *Plumbate* 1/2 Plumbate effigies in particular show a fascinating versatility coupled with a tendency to follow fixed types. **1962** G. Kubler *Art & Archit. Anc. Amer.* ix. 205 A few 'glazed sherds' found during excavation: if these were plumbate, a Toltec Maya date would be in order. **1971** L. A. Boger *Dict. World Pottery & Porcelain* 345/2 Two outstanding types of pottery made in large quantities and traded wherever Toltec influence was evident are Fine Orange ware..and Plumbate. *Ibid.* 346/1 Plumbate ware dates from around 1000.

plumber. Add: **1.** In mod. use, a workman who installs and repairs piping and fittings to do with water supply, sanitation, and drainage.

2. *transf.* **a.** *Services' slang.* An armourer or engineering officer. **b.** *slang.* During the administration of United States President Richard M. Nixon (1969–74), a member of a White House special unit for investigating leaks of government secrets, which came to public notice after it was discovered that they had engaged in illegal practices, including the installation of concealed microphones; also *transf.*

a. 1941 D. Masters *So Few* xix. 224 The plumbers—the name by which the armourers are generally known in the service. **1943** C. H. Ward-Jackson *It's a Piece of Cake* 48 *Plumber*, the Engineering Officer. **1946** G. Hackforth-Jones *Sixteen Bells* i. vii. 106 To the average 'Plumber' the word Engine is synonymous [*sic*] with steam. **1962** *Flight* 11 Nov. 605/2, I am not an engineer (or 'plumber', as the Royal Air Force equivalent is unofficially called). **1979** *Navy News* Feb. 2/1 It would be of great help in this project if, among your readers, there were a few ex-Keyham 'plumbers' who would be prepared to turn out their photographs of those times for us to borrow.

b. 1972 *Time* 28 Aug. 24/2 The intelligence squad grew out of a team of so-called 'plumbers', originally recruited by the Administration to investigate leaks to the media. **1973** *Times* 15 May 7/1 President Nixon sent a personal letter to Mr J. Edgar Hoover, then Director of the Federal Bureau of Investigation (FBI), in June, 1971, telling him that Mr Egil Krogh had been put in charge of a special White House task force on national security and asking for FBI cooperation with it. The group later became known as 'the plumbers' because it tried to plug leaks of information. **1974** *Times* 9 Jan. 5/6 The investigating magistrate in the case of *Le Canard Enchaîné* questioned for two hours today a counter-espionage agent who has been allegedly identified as one of the 'plumbers' in the attempted bugging of the satirical weekly magazine. **1976** *Billings* (Montana) *Gaz.* 6 July 3A 1 Ehrlichman is now appealing a five-year sentence following his conviction for obstruction of justice for his part in the break-in of the office of Daniel Ellsberg's psychiatrist by the now-infamous 'plumbers'. **1977** *Time* 24 Jan. 52/1 He found a band of 'plumbers' busily installing listening devices.

plumbian (plɒ·mbiăn), *a.* *Min.* [f. L. *plumb-um* lead + *-ian 2.] Of a mineral: having a (small) proportion of a constituent element replaced by lead.

1930 W. T. Schaller in *Amer. Mineralogist* XV. 571 Lead—plumbian. **1944** C. Palache et al. *Dana's Syst. Min.* (ed. 7) I. 804 *Var.* Ordinary. Calcium and uranium. ..Plumbian... With Pb:U = 3:8 (anal. 1). Titanian. With Ti:Cb + Ta:Fe″ = 9:3:1 (anal. 6). **1968** I. Kostov *Mineralogy* ii. ii. 169 The following varieties are recognized: (*a*) of tetrahedrite—ordinary, zincian, ferroan,.. plumbian, nickelian, etc.

plumbic, *a.* **a.** For '(divalent)' read '(quadrivalent)'. Add: Formerly also applied by some writers to compounds of bivalent lead (cf. *plumbous *a.* 2). (Further examples.)

1854 R. D. Thomson *Cycl. Chem.* 328/1 Binoxide, Peroxide, Deutoxide, Brown or puce oxide, Plumbic acid.—PbO_2. **1868** W. A. Miller *Elem. Chem.* (ed. 4) ii. xviii. 717 Lead oxide, plumbic oxide, or protoxide of lead (PbO = 223)... This oxide is well known under the name of litharge. *Ibid.* 719 Plumbic dioxide, or peroxide of lead (PbO_2 = 239). **1900** W. A. Shenstone *Elem. Inorg. Chem.* xxix. 439 Plumbic oxide (PbO) is known as litharge or massicot. **1921** J. R. Partington *Text-bk. Inorg. Chem.* xliv. 925 A yellow liquid, which is lead tetrachloride, or plumbic chloride, $PbCl_4$, is deposited. **1935** *Chambers's Encycl.* VI. 564/1 Plumbic Oxide (monoxide of lead, massicot, litharge), PbO... Plumbic Peroxide..PbO_2... Plumbic Chloride (chloride of lead), $PbCl_2$. **1950** N. V. Sidgwick *Chem. Elements & Compounds* I. 587 The change from the plumbous alkyls to the plumbic is less easy with the heavier alkyls. **1962** P. J. & B. Durrant *Adv. Inorg. Chem.* xviii. 641 A list of certain alkyl and aryl compounds of lead in the plumbic condition. *Ibid.* 644 Lead tetra-acetate (plumbic acetate), $Pb(CH_3COO)_4$, is made by saturating hot glacial acetic acid with red lead.

plumbicon (plɒ·mbikɒn). *Television.* [f. L. *plumb-um* lead + *-icon*, after *vidicon.] A type of television camera tube similar to a vidicon but in which the photoconductive layer of the signal plate is of lead monoxide.

1962 E. F. de Haan in *Philips Technical Rev.* XXIV. 58/1 The photoconductive layer consists of vapour-deposited lead monoxide, PbO, and we have therefore called this tube the 'Plumbicon'. **1971** *Physics Bull.* Aug. 457/1 The other is the plumbicon colour TV tube now used in practically all colour TV cameras all over the world. **1976** *Wireless World* June 75/3 A camera based on three 2/3-in. Plumbicons with built-in image enhancement.

plumbing, *vbl. sb.* Add: **b.** Also *transf.* in various senses: (*a*) *colloq.*, a system of pipes, tubes, or ducts in an engine or other complicated apparatus or installation; (*b*) *Jazz slang*, a trumpet, trombone, or similar wind instrument; (*c*) *spec.* a lavatory; lavatory installations (*colloq.*); (*d*) *slang*, fillings in teeth; (*e*) *joc.*, the excretory tracts, the urinary system.

a. *a* **1929** in V. W. Pagé *Mod. Aviation Engines* (1929) II. xl. 1593 Courtney's flight, Franco's flight and the Polish flight were all three terminated by what is popularly known as 'plumbing' trouble... The engines stopped through failure of the feed lines. **1938** —— *Airplane Servicing Man.* xxii. 758 Aircraft space is limited, and.. complicated plumbing must be installed in very close quarters. **1950** J. L. Nayler *Mod. Aircraft Design* v. 78 Pumps and their connections have to be installed in places where they can be inspected and 'plumbing' is accordingly a big factor in their installation. **1955** *Sci. Amer.* Oct. 34/3 New materials—e.g., liquid sodium as the coolant, beryllium as the moderator, zirconium for the 'plumbing'..—may open the way for reactors..needing only a modest investment of fissile fuel. **1959** *New Scientist* 12 Feb. 359/2 The wave-guide, or 'plumbing' as the radar engineer calls it, is simply a metal tube down which radio waves will travel. **1971** H. E. Ennes *Telev. Broadcasting: Equipment, Syst. & Operating Fund.* xii. 548 Fig. 12–11 illustrates typical microwave plumbing.

b. 1935 *Vanity Fair* (N.Y.) Nov. 71/3 *Plumbing* or *piston* for trumpet. **1951** *Cosmopolitan* July 85 Hap said, 'You with the plumbin', what's your name?' **1955** *Vogue* 15 Sept. 124 Kai Winding and J. J. Johnson (above) pair their spruce, understated trombones ('just plumbing') against a backing of bass, drums and piano.

c. 1950 *Publ. Amer. Dial. Soc.* xiv. 50 *Outside plumbing*, euphemism for no plumbing, a privy. **1971** *Countryman* Winter 118 The house is progressing at last. We have bought a mountain of building supplies—pink plumbing and all. **1978** S. Allan *Inside Job* i. 18, I expect she wants to wash... Don't be obtuse... I want to show her the plumbing.

d. 1955 J. Thomas *No Banners* xiv. 117 The fillings in his teeth had been removed and replaced by unmistakable French 'plumbing'.

e. 1960 in Wentworth & Flexner *Dict. Amer. Slang* 398/1 This mild little medicine will fix your constipation, your stomach, your plumbing. **1963** M. McCarthy *Group* v. 103 Helena had known about sex from a very early age but treated it as a joke, like what she called your plumbing. **1972** G. Bell *Villains Galore* iii. 31 Excuse me, Albert, but I must see to the plumbing. Won't be a minute.

plumbite (plɒ·mbəit). *Chem.* [ad. F. *plombite* (E. Fremy 1843, in *Jrnl. de Pharm.* III. 31), f. *plomb* lead: see -ite[1].] Any of various (salts of) oxyanions or hydroxy anions of bivalent lead, which are formed esp. by the action of alkalis on lead(II) hydroxide and are freq. represented as PbO_2^{2-}; applied also to certain mixed oxides of bivalent lead and another metal.

1851 H. Watts tr. *Gmelin's Hand-bk. Chem.* V. 158 (*heading*) Plumbite of ammonia. **1903** H. C. Jones *Princ. Inorg. Chem.* xl. 480 Lead hydroxide..has..acid properties dissolving readily in strong bases. The salts have the composition M_2PbO_2 and are known as plumbites. **1927** J. W. Mellor *Comprehensive Treat. Inorg. & Theoret. Chem.* VII. xlvii. 665 The plumbites all appear to be salts of a monobasic acid analogous with the corresponding stannites and germanites. *Ibid.* 669 Grube obtained a mixture, $2MgO.PbO.3H_2O$, or magnesium plumbite, by the oxidation of magnesium hemiplumbide in the presence of moisture. **1950** N. V. Sidgwick *Chem. Elements & Compounds* I. 603 It is possible to make a plumbate by atmospheric oxidation of a plumbite. **1968** K. M. & R. A. Mackay *Introd. Mod. Inorg. Chem.* xv. 223/1 Addition of alkali to lead II solutions gives a precipitate of the hydrated oxide, which dissolves in excess alkali to give plumbites.

plumbless, *a.* (Further examples.)

1946 W. de la Mare *Traveller* 29 Into a plumbless deep of sleep he sank. **1960** C. Achebe *No Longer at Ease* iii. 26 For the first time since they had left Liverpool, the sea became really blue; a plumbless blue set off by the gleaming white tops of countless wavelets.

plumbly (plɒ·mli), *adv. rare.* [f. Plumb, Plum *a.* + -ly[2].] Vertically downwards.

1931 J. C. Gregory *Short Hist. Atomism* 7 The atoms that fell plumbly through the void were still restless.

plumbo-. Add: **plumboja·rosite** *Min.* [Jarosite], a basic sulphate of lead and ferric iron, $PbFe_6(SO_4)_4(OH)_{12}$, that occurs as brown hexagonal crystals, esp. as a secondary mineral in lead ores in arid regions.

1902 Plumbojarosite [see *natrojarosite* s.v. *natro-]. **1938** *Jrnl. R. Soc. W. Australia* XXIV. 112 This appears to be the first occasion on which either beudantite or plumbojarosite has been found in Australia. **1959** *Mineral Abstr.* XIV. 275/2 Plumbojarosite is found as compact and powdery aggregates of flaky crystals in the Akatuev deposit in eastern Transbaikal region.

plumbous, *a.* **2.** For 'has its lower valency', read 'is bivalent'. Add: Formerly also applied by some writers to substances in which lead was thought to be univalent (cf. *plumbic *a.* a). (Examples.)

1854 R. D. Thomson *Cycl. Chem.* 327/2 Protoxide of Lead. Yellow Oxide, Massicot, Litharge, Plumbous Acid. PbO. **1868** W. A. Miller *Elements of Chemistry* (ed. 4) ii. xviii. 721 When heated in closed vessels, part of the sulphur is expelled, and a subsulphide, or plumbous sulphide (Pb_2S) is left. **1883** C. L. Bloxam *Chem.* (ed. 5) 373 Oxide or protoxide of lead (plumbous oxide) is prepared on a large scale by heating lead in air. **1884** Frankland & Japp *Inorg. Chem.* xxxix. 699 Plumbous oxide, 'Pb'_2O', is best prepared by heating plumbic oxalate to 300° with exclusion of air... It is a black powder. **1892** Morley & Muir *Watts' Dict. Chem.* (rev. ed.) III. 128/2 Lead protoxide PbO. (Plumbous oxide. Litharge. Massicot.) **1900** W. A. Shenstone *Elem. Inorg. Chem.* xxix. 439 Plumbous oxide (Pb_2O). When oxalate of lead is heated to 300° out of contact with air, it leaves a grey residue which absorbs oxygen with avidity. This is the suboxide. **1926** P. C. L. Thorne tr. *Ephraim's Text-bk. Inorg. Chem.* xv. 359 When red lead, Pb_3O_4 (*i.e.* $2PbO.PbO_2$), is treated with nitric acid, the plumbous portion of this compound dissolves, leaving PbO_2 as a brown powder. **1950** [see *plumbic *a.* a]. **1962** P. J. & B. Durrant *Adv. Inorg. Chem.* xviii. 638 Plumbous nitrate, $(Pb(NO_3)_2$, is made by dissolving lead, lead oxide, or lead carbonate in dilute nitric acid... Concentrated sodium hydroxide solution dissolves plumbous chromate.

plume, *sb.* Add: **3. c.** Self-satisfaction, triumph. *rare.*

1910 W. De Morgan *Affair of Dishonour* iv. 66 He wanted..to choose his time, as a nobleman might then do..not only without shame or remorse, but even with some sense of plume or strut.

4. f. (i) A long streamer of smoke, vapour, or other fluid issuing from a localized source in the same or a different fluid and spreading out as it travels, esp. one with a degree of buoyancy in the ambient medium.

1878 [see sense 4 a in Dict.]. **1947** *Q. Jrnl. R. Meteorol. Soc.* LXXIII. 428 Smoke plumes from factory chimneys. **1955** *Trans. Amer. Soc. Mech. Engineers* LXXVII. 1/1 Under favorable weather conditions the plume from a smoke-stack will rise gradually as it flows downwind and the gases will be dispersed until only a negligible concentration prevails in the atmosphere. **1969** *Ann. Rev. Fluid Mech.* I. 30 Turbulent plumes (like jets) have a sharp boundary between the turbulent buoyant fluid and the surroundings; they increase their width through..entrainment of external fluid across this boundary by large eddies. **1970** *Nature* 7 Nov. 545/2 Ball lightning does not rise like hot air nor is it disrupted by convection into a thermal plume as are hot fireballs. **1975** *New Yorker* 12 May 70/3 At discharge, how fast should it come out, and at what level? Would fish die in the thermal plume? **1975** J. L. Pavoni et al. *Handbk. Solid Waste Disposal* iii. 127 Water vapor plumes are caused when the relative humidity of the effluent gas stream is significantly higher than the relative humidity of the ambient air around an incinerator chimney. **1977** *Daily Tel.* 25 Apr. 1/1 The awesome plumbing job..involves stuffing a 12 inch rubber plug into an 18-inch diameter well to staunch a plume of oil and gas sprouting 200 ft into the air.

(ii) *Geol.* A column of magma rising from the lower mantle and spreading sideways on reaching the base of the lithosphere, proposed as an explanation of the motion of lithospheric plates and of sites of volcanic activity away from plate margins.

1971 W. J. Morgan in *Nature* 5 Mar. 42/2, I now propose that these hotspots are manifestations of convection in the lower mantle which provides the motive force for continental drift. In my model there are about twenty deep mantle plumes bringing heat and relatively primordial material up to the asthenosphere and horizontal currents in the asthenosphere flow radially away from each of these plumes. **1975** *Sci. Amer.* Mar. 62/1 According to this argument, all upward movement of mantle material is confined to about 20 plumes, each plume a few hundred kilometers in diameter, rising from the core-mantle boundary. **1976** P. Francis *Volcanoes* i. 49 The plume effectively burns a hole through the overlying crustal plate..and a volcano results.

‖ **plumeau** (plümo). [Fr.] A duvet.

1892 C. A. M. Fennell *Stanford Dict.* 637/1 *Plumeau*,.. a thick quilt stuffed with feathers. **1931** E. Sackville-West *Simpson* iii. 250 Childeric was lying on his bed, half buried in the voluminous *plumeau*. **1967** T. La Chard *Sailor Hat* iii. 59, I subsided in sleep under my vast crimson *plumeau*, a towering feather bed gleaming through its white crochet covering.

plumed, *ppl. a.* Add: **3.** Special collocations: **plumed serpent**, a mythical creature depicted as part bird, part snake; *spec.* (freq. with capital initials) any of various deities in the religions of ancient Mesoamerica having this form, esp. Quetzlcoatl, the Aztec deity of vegetation and fertility; also *attrib.*

1915 *Amer. Anthropologist* XVII. 480 (*caption*) (*a*), Pelican design; (*b*), plumed serpent design; from La Bermuda. Collection [of pottery] of Señor Alberto Imery. *Ibid.* 481 Two bowls in the Imery collection..bear upon the outside an interesting representation of the plumed serpent. **1926** D. H. Lawrence (*title*) The Plumed Serpent. **1935** *Discovery* Sept. 270/1 A flint dagger, 8½ inches long, with a handle fashioned to represent the body and head of the Plumed Serpent. **1937** *Burlington Mag.* July 55/1 Another point of far-reaching significance is brought out in the discussion of the symbol of the plumed serpent. ..In China a bird is given the emphasis and the serpent appears on its wings..while among the Mayas the serpent dominates but has the wings and feet of a bird. **1959** C. S. Chinchilla in Kidder & Chinchilla *Art of Ancient Maya* 29 They do not worship Gukumatz (the plumed serpent) and Balau (the tiger) as animals, but as the deities that they embody under these forms. **1972** *Funk's Stand. Dict. Folklore* 915/1 *Quetzalcoatl.* The feathered or plumed serpent god: known over all Middle America, and surviving in mythology to this day. **1979** P. Theroux *Old Patagonian Express* iii. 47 The Indians liked him [*sc.* Maximilian] because he was blond, like Quetzalcoatl—Cortez enjoyed the same bizarre notoriety for his resemblance to the Plumed Serpent.

plumetty (plū·mĕti), *a.* and *sb.* *Her.* Also 5 plomte; plumeté. [ad. F. *plumeté*: see Plumetis.] (A heraldic device) with a motif of feathers (see quots.).

a **1500** in *Ancestor* (1903) Oct. 193 Gold and purpull plomte. **1780** J. Edmondson *Compl. Body Heraldry* II. Gloss., *Plumetty*. When the field is divided into fusils, filled with the ends of feathers, depicted in metal and colour alternately, such field is said to be *Plumetty*. **1892** Woodward & Burnett *Treat. Heraldry* I. iii. 71 Two curious forms of *Vair* occasionally met with in Italian or French coats are known as '*Plumeté*' and '*Papelonné*'. *Ibid.* 72 In *Plumeté* the field is apparently covered with feathers. **1929** *N. & Q.* 2 Nov. 317/2 'Papelonny'..and its analogue, 'plumetty',..are represented in English heraldry. Homologous in their tricking, the diapering of the one takes the form of fish scales; of the other, of the breasts of birds. **1969** Franklyn & Tanner *Encycl. Dict. Heraldry* 261/2 *Plumetty*, a field..of feathers.., a form that did not survive, and of which the few examples preserved from the feudal period contradict each other in appearance: one form is semé..of feathers; another consists of barwise rows of feathers conjoined laterally; yet a third displays a field totally covered with overlapping feathers.

Plummer-Vinson (plɒ·məɪ vi·nsən). *Path.* The names of H. S. *Plummer* (1874–1936), U.S. physician, and P. P. *Vinson* (1890–1959), U.S. surgeon, used *attrib.* to designate a syndrome characterized by dysphagia, glossitis, and iron-deficiency anæmia.

1926 A. F. Hurst in *Guy's Hospital Rep.* LXXVI. 426 (*heading*) The Plummer-Vinson syndrome (spasm of the pharyngo-œsophageal sphincter with anæmia and splenomegaly). **1958** Terracol & Sweet *Dis. Esophagus* xiii. 302 The Plummer-Vinson syndrome occurs predominantly in women in the proportion of approximately four to one. *Ibid.* 303 The Plummer-Vinson syndrome is a clinical complex, the cause of which though still not clearly understood has to do with avitaminosis, endocrine disturbances, and lack of available iron. *Ibid.* 306 There is strong clinical evidence to suggest that the mucosal changes characteristic of the Plummer-Vinson syndrome are of a precancerous nature. **1971** *Brit. Med. Bull.* XXVII. 34/1 The Paterson-Kelly or Plummer-Vinson syndrome was first described in 1919.

plummet, *sb.* Add: **7.** *plummet-deep*, *-measured* adjs.

1938 W. de la Mare *Memory* 38 Fleeter than Nereid, plummet-deep, Enticed by some long-sunken ship, She [*sc.* Memory]..laughs out to see The treasure she retrieves for me. **1939** W. B. Yeats *Last Poems* 16 And pressed at midnight in a public place Live lips upon a plummet-measured face.

plummet, *v.* Restrict *rare* to senses in Dict. and add: **4. a.** *trans.* To cause to drop rapidly, to hurl down. **b.** *intr.* To drop or fall rapidly, to plunge *down*. Also *fig.*

1933 *Sun* (Baltimore) 11 Apr. 1/1 The U.S.S. Akron had

reports that the weather was unfavorable for her purpose when she took off last Monday on the fatal flight that plummeted her into the sea from lightning-swept skies. **1939** Webster Add., *Plummet, v.i.*, to drop or plunge straight down. **1944** F. Leiber in D. Knight *100 Yrs. Sci. Fiction* (1969) 93 'I'm glad to see the last of that fellow,' he muttered,..as they plummeted toward the roof. **1953** A. Moorehead *Rum Jungle* vii. 96 They [*sc.* gulls] plummeted down with their beaks wide open. **1958** B. Nichols *Sweet & Twenties* x. 132 Even worse, waists plummeted nine inches, to remain suspended somewhere below the navel. **1959** *Daily Tel.* 21 Nov. 1/6 Capt. Kittinger plummeted towards the earth until his parachute opened automatically at 10,000 feet. **1961** *Time* (Atlantic ed.) 23 June 26 The price of new potatoes plummeted to 1 ¢ a pound. **1963** C. L. Cooper *Black* x. 153 Twice I stumbled over the garbage cans lining the walk, rising,..to plummet on, not daring to imagine the scene I might find in the apartment. **1972** G. Durrell *Catch me a Colobus* ix. 186 Great gouts of water plummeted down from the sky so that the road, which was an earth one, was immediately turned into a dangerous mire. **1978** D. Bloodworth *Crosstalk* xi. 91 Rumours that Mao's health is failing have sent the Hong Kong stock exchange plummeting. **1979** *Amer. N. & Q.* Apr. 127/1 The *Phillipe, Count of Darkness* stories..plummeted his [*sc.* F. Scott Fitzgerald's] story-asking price from $3000 to $300.

Hence **plu·mmeting** *vbl. sb.* and *ppl. a.*

1952 *Chambers's Jrnl.* May 262/1 The stone-like plummeting before the 'chute opens, the gentle floating in space, the enormous thud with which you hit the ground—these combine to give an experience unparalleled elsewhere. **1957** *Economist* 2 Nov. 380/2 The most calamitous thing..would be a world depression with plummeting prices. **1958** *Times Rev. Industry* Dec. 78/1 Spending..led to a rise of £80m. in the Union's imports, and the plummeting of the South African foreign exchange reserves.

plummy, *a.*² Add: **2. b.** Of the voice, then of sound gen.: thick-sounding, rich, 'fruity'; indistinct; with bass predominating.

1881 *Punch* 23 July 25/2 The same aged lover was bidding, with rather a 'plummy' voice, the More-than-Middle-Aged Heroine 'good bye for ever'. **1947** *Jrnl. Inst. Electrical Engin.* XCIV. IIIA 446/1 Such distortions can be tolerated..without serious loss of articulation, though the speech will usually sound rather 'plummy' and unnatural. **1951** K. Harris *Innocents from Abroad* 199 The rich, plummy voice of [actor] Edward Arnold. **1955** *Times* 3 May 14/4 A disc which sounds plummy and muffled in tone. **1965** G. McInnes *Road to Gundagai* xi. 197 His voice..was wonderfully plummy and Edwardian. **1970** *Daily Tel.* 1 Sept. 9/5 All India Radio—modelled..on the BBC, even down to the plummy accents of its announcers. **1975** *City Press* 1 May 16/5 Her duchess on the make is a finely pointed performance, the plummy vowels contrasting splendidly with consonants periodically marred by the lack of false teeth. **1977** *Early Mus.* Oct. 549/3 The plummy..tone [of Flemish virginals] is evidently more popular than the musically versatile but astringent Italian virginal. **1978** *Gramophone* Feb. 1439/1 His tone is mellow, but again, as in the Waltzes..the sound sometimes seems a bit plummy and close.

3. *Comb.*, as *plummy-voiced* adj.

1972 *Jazz & Blues* Oct. 8/3 His smooth, plummy-voiced style. **1978** *Times* 21 Jan. 14/7 The plummy-voiced announcers.

Hence **plu·mminess**²; **plu·mmily** *adv.*

1927 J. Masefield *Midnight Folk* 208 You haven't got such a thing as a seedless raisin about you?.. It's a real treat..to taste a bit of plumminess. **1953** *John o' London's Weekly* 13 Mar. 208/4, I was not persuaded at the première by his 'young Octavius', who had a curious plumminess. **1955** *Times* 31 Aug. 5/4 The creamy richness—free from all 'plumminess'—of her contralto voice is a constant delight. **1958** *Listener* 25 Dec. 1092/2 It was all splendidly true to type, and plummily theatrical. **1962** *Ibid.* 8 Feb. 268/3 The plumminess of the normal Wagnerian baritone. **1972** A. Ross *London Assignment* 17 'Go ahead, old boy,' he said plummily, 'He's all yours.'

plump, *a.*¹ Add: **II. 3. e.** *plump-bellied, -uddered* adjs.

1916 Joyce *Portrait of Artist* iii. 127 Thrust it out of men's sight into a long hole in the ground, into the grave, to rot, to feed the mass of its creeping worms and to be devoured by scuttling plump-bellied rats. **1922** —— *Ulysses* 537 A nannygoat passes, plumpuddered, buttytailed, dropping currants.

plump, *v.*¹ Add: **5.** Also *transf.* More widely, to opt *for*. Also, to decide or vote *against*.

1890 Barrère & Leland *Dict. Slang* II. 140/2 *Plump, to*.. (Racing), to lay one's money on one single horse. But I shall *plump* for Lord R. Ch.'s L'Abbesse de Jouarre, who has been well tried.—*Truth.* **1894** [see *votress¹]. **1929** C. Connolly *Let.* in *Romantic Friendship* (1975) 325, I have plumped against England. **1934** *Discovery* June 176/1 The more one knows..the less one is inclined to 'plump' for one particular solution rather than another. **1963** *Ann. Reg. 1962* 17 The largest union..had plumped two-to-one against a strike. **1966** *Observer* 3 Apr. 10/5 A large section of the electorate plumped for the Liberals. **1972** *Nature* 25 Feb. 424/3 The consultants recommend three road crossing schemes but plump for a Flint–Burton crossing with a coast road. **1976** *Which?* May 100/3 We don't think now is the best time to invest in equipment. Better to wait until one of the systems appears to be winning the battle and then plump for that.

plump, *v.*² Add: **1. a.** *spec.* of pillows, cushions, and other upholstery.

1848 [in Dict.]. **1848** Dickens *Dombey & Son* lvii. 571 Mrs. Miff resumes her dusting and plumps up her cushions. **1960** M. Spark *Ballad Peckham Rye* viii. 168 She turned and plumped out the cushion behind her. **1962** A. Sexton *All my Pretty Ones* 43 The houseboy, A quick-eyed Filipino..plumping up The down-upholstery. **1972** M. J. Bosse *Incident at Naha* i. 13 The bed, made with..carefully plumped-up pillows. **1975** L. Gillen *Return to Deepwater* iv. 68 The cushions on the settee freshly plumped.

plumpen, *v.* Add: Also *intr.*, to grow plump. Hence **plu·mpening** *vbl. sb.*, the action or process of making or becoming plump.

1926 *Spectator* 1 May 801/2 The plumpening of cherries on lichened wall. **1966** P. V. Price *France* 307 The chickens..wander about the roads in the Bresse plain, plumpening visibly.

plumper¹. Add: **c.** A preparation for causing hides to plump.

1903 L. A. Flemming *Pract. Tanning* xxii. 375 Quebracho is not a good plumper, and for this reason some material is necessary to plump the leather. *Ibid.* 376 Quebracho, being a sweet tan, is not of itself a plumper.

plumper². Add: **3. b.** *colloq.* An unusually large example of its type; a whopper. ? *Obs.*

1881 *Punch* 1 Oct. 155/1 Lovers of England..can hardly do better than help to fill that Purse, which Mr. Punch hopes will prove a 'plumper'.

plu·mping, *ppl. a.* *colloq.* [f. Plump *v.*² + -ing².] Very large, unusually big.

1903 'A. McNeill' *Egregious English* 91 You win by the skin of your teeth or with a plumping majority, as the case may be. **1978** *Guardian Weekly* 2 Apr. 20/3 The Master Builder is a sexless play in which a plumping girl with an alpenstock invades the home of an elderly architect.

plumpish, *a.* Delete *rare* and add later examples. Hence **plu·mpishness.**

1942 W. Faulkner *Go down, Moses* 226 The boy standing there looking down at the short plumpish grey-haired man. **1977** M. Hinxman *One-Way Cemetery* i. 7 She was nice, plumpish, friendly. **1979** 'G. Black' *Night Run from Java* iii. 27 Don't forget the plumpish redundancy payments they all want to collect. **1979** 'J. Ross' *Rattling of Old Bones* ii. 14 Plumpish..and a sensual plumpishness at that.

plumptitude (plʌ·mptitiūd). *joc.* ? *Obs.* Also **plumpitude.** [f. Plump *a.*¹ + *-titude* after *altitude, aptitude*, etc.: see -tude.] Plumpness.

1828 F. Kemble in G. Macpherson *Mem. Life Anna Jameson* (1878) ii. 44 To behold her sitting on a sofa in a very becoming state of blooming *plumpitude*. **1850** *Godey's Lady's Bk.* XLI. 180/2 At every pore I'm oozing—(I'm 'caving in' to-day)—My plumptitude I'm losing, And dripping fast away. **1854** G. Greenwood *Haps & Mishaps* ii. 43 The lord chancellor having formally announced that parliament stood prorogued until the 20th of August, Her Majesty rose as majestically as could be expected of one more remarkable for rosy plumptitude than regal altitude. **1890** *Daily News* 19 Aug. 3/1 Our own countrymen and countrywomen are more prone to a condition of 'blooming plumptitude', which too much sugar is likely to increase unduly.

plum pudding, plum-pudding. Add: **c.** **plum pudding mahogany**, mahogany with a mottled finish; **plum-pudding voyage**, delete 'including plum-duff'.

1924 G. O. Wheeler *Old English Furniture* (ed. 3) xii. 278 The plum mottle..signifies the spotted and mottled wood known as 'plum-pudding' mahogany. **1968** *Canad. Antiques Collector* Nov. 8/2 (Advt.), Sheraton style 'Plumb [*sic*] Pudding' Mahogany cabinet..circa 1790. **1976** *Country Life* 26 Feb. (Suppl.) 24c/1 An unusual Chippendale period supper table with two tier top. The table is of 'Plum pudding' mahogany, with carved detail.

d. (See quots.)

1851 H. Melville *Moby Dick* III. viii. 64 Plum-pudding is the term bestowed upon certain fragmentary parts of the whale's flesh. **1904** *Sci. Amer. Suppl.* 5 Mar. 23551/3 A muscular, fibrous substance known as 'plum pudding' permeates the blubber of the tongue of these two species of whales.

e. *Mil. slang.* A type of trench mortar shell.

1925 Fraser & Gibbons *Soldier & Sailor Words* 225 *Plum pudding*, the name for a type of trench mortar shell; suggested by its size and shape. **1928** Blunden *Undertones of War* 51 Now more serious and immediate omens of ordeal appeared in the mounds of trench mortar bombs—'plum puddings' or 'footballs', steely and shining.

plum-puddinger, also, a member of the crew of such a ship; (earlier and later examples).

1851 H. Melville *Moby Dick* I. xvii. 137 After listening to these plum-puddingers till nearly eleven o'clock, I went up stairs to go to bed. **1934** F. R. Dulles *Lowered Boats* iv. 45 Only the little 'plum-pud'ners' of Rhode Island remained wholly true to the Greenland whale.

plum-tree. Add: (Later examples.) Also *fig.*, the source of the spoils of political office; esp. in phr. *to shake the plum tree* (*U.S.*).

1905 D. G. Phillips *Plum Tree* ii. 24, I mentally called the roll—wealth, respectability, honor, all on their knees before Dominick, each with his eye upon the branch of the plum tree that bore the kind of fruit he fancied. **1906** U. Sinclair *Jungle* xxvii. 340 Those golden hours when he, too, had a place beneath the shadow of the plum tree. **1922** Joyce *Ulysses* 668 The name on the label is Plumtree. A plumtree in a meatpot, registered trade mark. **1933** *Sun* (Baltimore) 23 May 14/1 The thought uppermost in everybody's mind at the gathering at which James was the guest of honor was: When will you shake the plum tree? **1952** G. Stimpson *Bk. about Amer. Politics* 258 While he [*sc.* Matthew Stanley Quay] was State treasurer (1885–1887) he placed large sums of State money in the People's Bank of Philadelphia and used it to buy stocks on margin. On one occasion Quay and his associates piled up orders for the stock of the Metropolitan Railroad of New York until only about $10,000 was left in the bank. John S. Hopkins, the cashier, became frightened and protested to the political boss. Quay sent the cashier the message: 'Buy and carry a thousand Met for me and I will shake the plum tree.' **1959** *Chambers's Encycl.* X. 804/1 Pruning of standard plum trees simply means cutting out branches which intercross or thinning out overcrowded branches.

plunder, *sb.* Add: **3.** (Earlier and later examples.) Also in occas. wider use. Also *fig.*

1805 M. Lewis in *Lewis & Clark Exped.* (1904) II. 220, I dispatched Sergt. Ordway with 4 Canoes and 8 men to take up a load of baggage as far as Capt. Clark's camp and return for the remainder of our plunder. **1941** E. P. O'Donnell *Great Big Doorstep* 119 Your father puts on more every time he tells that there story. The plunder that man's got in his head! The plunder! **1948** E. N. Dick *Dixie Frontier* 113 Mules and a hardy tough breed of Indian and Spanish horse..were used to carry the money and plunder. **1962** W. Stegner *Wolf Willow* (1963) III. ii. 160 He gathers together his plunder and he hightails her off the mountain. **1972** O. Frederickson *Silence of North* viii. 64 We didn't have much plunder, and with only two grown-ups, a baby, and a pair of sled dogs on board, it rode high.

plunder, *v.*² Add: **1.** Also *fig.*

1896 A. Beardsley *Let.* 29 Oct. (1970) 193 How abominably she [*sc.* George Sand] has been plundered by everyone since. **1961** H. Adams in *Webster* s.v., Shakespeare and his fellow-dramatists plundered the Church legends. **1964** T. M. Andersson *Probl. Icelandic Saga Origins* v. 90 *Eiríks saga rauða* and *Grettis saga* combine to show that when *Landnáma* was used, it was plundered wholesale and not plucked for an occasional name.

plunderbund (plʌ·ndəɹbund). *U.S. colloq.* [f. Plunder *sb.* + G. *bund* alliance, league.] A corrupt alliance of political, commercial, and financial interests engaged in exploiting the public.

1914 *Voice of People* (New Orleans) 8 Jan. 1/1 The whole force of the Texan plunderbund..are howling at the heels of the dauntless army of workers. **1933** *Sun* (Baltimore) 27 Apr. 8/6 The..unemployed,..the losers in banks, stocks, homes,..and business people..are forever through with this systematic plunderbund. **1949** *Chicago Tribune* 2 Sept. 12/6 Hello, suckers who voted for the continuation of chaos and corruption and the plunderbund last November.

plunderous, *a.* Delete *rare* and add later examples.

1973 *Freedomways* XIII. 8 With the plunderous taxes..a large section of the people..are being returned to the bread lines. **1976** *Times* 30 Sept. 8/3, I would jot down all the best jokes from *Take It From Here* and write a new plot round them... So I can't really jibe when the writer of that Jimmy Edwards classic has the same kind of plunderous notion for making a book. **1977** *Evening News* 4 June 5/4 William Palmer..made a plunderous visit to the treasury.

plung (plʌŋ). *rare.* [Echoic.] A resonant noise as of a tennis racket striking a ball.

1952 J. Betjeman *First & Last Loves* ii. 13 So that the real Bournemouth is all pines and pines and pines and flowering shrubs, lawns, begonias, azaleas, bird-song, dance tunes, the plung of the racket and creak of the basket chair. **1954** —— *Few Late Chrysanthemums* 69 'Oh! Plung!' my tauten'd strings would call, 'Oh! Plung! my darling, break my strings.'

plunge, *sb.* Add: **I. 1. b.** = *plunge bed* (see sense 7 below).

1973 *Times* 22 Sept. 13/2 If we have..an Indian summer, water the 'plunge' regularly.

2. (Earlier and later fig. examples.) Esp. in phr. *to take* (less frequently *make*) *the plunge*, to take a decisive first step, to commit oneself irrevocably to a course of action.

1845 Dickens *Let.* 20 Oct. (1977) IV. 412 The venture is quite decided on; and I have made the Plunge. **1848** Thackeray *Pendennis* (1849) I. vi. 61 The poor boy had taken the plunge. Trembling with passionate emotion,..poor Pen had said those words which he could withhold no more. **1876** Trollope *Prime Minister* IV. x. 162 'You would not wish to live all your life in terror of seeing Arthur Fletcher?' 'Not all my life.' 'Take the plunge and it will be over.' *a* **1911** [see *limit *sb.* 2 h]. **1915** W. Owen *Let.* 15 Jan. (1967) 316 If I could devote myself to training in Music or Painting, I would take the plunge, were I never to read a book more. **1965** *New Statesman* 23 Apr. 630/3 At present only one local Co-op..is affiliated to Transport House, though London is thought to be contemplating taking the plunge. **1977** C. McCullough *Thorn Birds* xviii. 460, I think she's terrified of committing herself to..marriage... At least he's got the sense to wait until she's ready to take the plunge. **1979** G. Wagner *Barnardo* xi. 183 Samuel Smith..finally persuaded Barnardo to take the plunge and set up his own emigration scheme.

III. 6*. *Geol.* The angle a fold axis or linear feature makes with the horizontal, measured in a vertical plane. Cf. Pitch *sb.*² 24 b in Dict. and Suppl.

1913 W. Lindgren *Min. Deposits* xi. 142 The plunge¹..of an ore-body is the vertical angle between a horizontal plane and the line of maximum elongation of the body. In lenticular ore-bodies in metamorphic rocks which have undergone strong mechanical deformation, the plunge is an important factor, and often it is determined by the direction of the cleavage or schistosity. [*Note*] ¹Called 'pitch' or 'rake' by many authors. **1932** *Bull. Amer. Assoc. Petroleum Geologists* XVI. 209 The type of fold is an overturned anticline in which the plunge increases from a comparatively low degree until it becomes vertical and finally overturns. **1936** C. M. Nevin *Princ. Structural Geol.* (ed. 2) iii. 43 The angle of dip of the axial line [of a fold] is called the pitch or plunge. **1962** [see *pitch *sb.*² 24 b]. **1976** B. E. Hobbs et al. *Outl. Structural Geol.* iv. 177 The orientation of the hinge line or, since the fold is cylindrical, its axis, is expressed by its plunge and direction of plunge... The orientation of a line can also be expressed by means of its pitch... For pitch to be meaningful the orientation of the plane must be known.

IV. 7. **plunge basin** *Physical Geogr.*, a deep basin excavated at the foot of a waterfall by the action of the falling water; **plunge bed**, a flower-bed, often containing peat or other moisture-retaining material, in which plants in pots can be sunk; **plunge-board** *rare*, a diving board; **plunge cut** *Engin.*, a cut made by feeding a grinding wheel into the work-piece in the plane of rotation, without any traverse; usu. *attrib.* in *plunge-cut grinding* (= *plunge grinding*); so **plunge cutting; plunge grinding** *Engin.*, grinding by means of a wheel with no traverse of the work; **plunge neck, neckline** = *plunging neckline* s.v. *plunging *ppl. a.* e; hence **plunge-necked** *a.*; **plunge pool**, (*a*) *Physical Geogr.*, a plunge basin, or the water occupying one; freq. *attrib.*; (*b*) a cold-water pool, forming part of the equipment of a sauna bath.

1905 *Bull. Geol. Soc. Amer.* XVI. 24 A river can excavate a plunge-basin at the foot of a cataract. **1939** A. K. Lobeck *Geomorphol.* vi. 195 There are several other plunge basins along the course of this former stream. **1966** J. Wyckoff *Rock, Time, & Landforms* x. 231 Today there is likely to be a waterfall at the mouth of the hanging trough, and beneath the fall perhaps a plunge basin or an alluvial cone or fan. **1871** S. Hibberd *Amateur's Flower Garden* xiii. 242 The question now is about the formation of the plunge beds. **1935** A. G. L. Hellyer *Practical Gardening* xxx. 193 (*caption*) A useful plunge bed for pot plants. **1973** *Times* 22 Sept. 13/2 Pot them the same day if possible, and put them in their peat 'plunge bed'. **1908** *Daily Chron.* 15 Feb. 8/5 When a man wants to take a second plunge into the water he has to get out and remount the plunge-board. **1935** O. W. Boston *Engin. Shop Practice* II. v. 268 An automatic in-feed mechanism..is contained in the saddle... This is used for plunge cuts and is independent of the traverse in-feed mechanism. **1937** Colvin & Stanley *Grinding Practice* i. 7 The wheel in this instance is fed into within 0·001 to 0·0015 in. of size, using the plunge cut and making as many plunges as are necessary to cover each section ground. **1941** F. D. Jones *Engin. Encycl.* II. 971 *Plunge-cut grinding*, this term has been applied to grinding which is done by directly feeding into the work a wheel, the face of which is sufficiently wide to cover the entire surface being ground. **1964** S. Crawford *Basic Engin. Processes* vii. 190 The wheel is fed slowly into the work while the latter is oscillated a very slight amount to equalise the wheel wear. This method is known as plunge-cut grinding. **1972** E. N. Simons *Dict. Machining* 142 *Plunge cutting*, a method of grooving parts to close limits of size with an accurately-positioned brazed carbide tool. **1974** *Sci. Amer.* Jan. 35/3 A process known as plunge cutting is employed to cut holes in a metal casting. **1935** O. W. Boston *Engin. Shop Practice* II. v. 265 When the wheel face is as wide as the length of the surface being ground or when it is impracticable to traverse the work, the wheel may be fed in with no traverse of the wheel or work. This is called plunge grinding. **1958** *Times Rev. Industry* May 34/3 The workpiece is then released, so that it can be rotated by means of the control wheel while plunge-grinding is being carried out. **1967** *Industr. Diamond Rev.* XXVII. 437/2 In plunge grinding, the whole depth of profile is normally ground in one pass. **1951** *Sunday Times* 28 Oct. 11/3 Some [spencers], sleeveless and with a plunge neck, give never a sign of their comforting presence beneath cocktail dresses and chiffon blouses. **1959** *Woman's Own* 20 June 17/4 She was wearing a plunge-necked dress of some white material. **1949** *Sun* (Baltimore) 28 Jan. 4 (Advt.), Plunge neckline wool jersey glamour blouse. **1959** 'O. Mills' *Stairway to Murder* v. 50 Its plunge neckline was so low that Geoff was almost driven for safety to the handwoven wool and Fair Isle cardigan. **1977** M. Hinxman *One-Way Cemetery* xv. 112 Daphne delved into her deep plunge neckline and retrieved a gold chain. **1917** *Scientific Monthly* V. 559 Plunge pools are potholes, in general, of large size, occurring at the foot of a vertical or nearly vertical waterfall... In most plunge pools the water is much deeper than it is in the stream channel on their downstream side. **1932** *Jrnl. Geol.* XL. 333 By recession of the falls the plunge-pool hole is elongated upstream, forming the deeper part of the channel of a gorge, the upper part of which is produced by a caving in process as the plunge-pool undercuts the face of the fall. **1957** G. E. Hutchinson *Treat. Limnol.* I. i. 111 The most impressive of such plunge-pool lakes are on that part of the former course of the Columbia River that now forms the Grand Coulee in the state of Washington. **1970** Macdonald & Abbott *Volcanoes in Sea* xix. 367 (*caption*) Vertical valleys being cut by plunge-pool action of waterfalls. **1973** *Times* 21 Nov. 18/8 There is a plunge pool with two massage jets. **1976** 'D. Halliday' *Dolly & Nanny Bird* ix. 116 It was time for the plunge pool with the hydrojet massage.

plunge, *v.* Add: **4.** (Later examples.)

1935 A. G. L. Hellyer *Pract. Gardening* xxix. 183 It is an excellent plan to plunge the pots to their rims in a bed of ashes to reduce the necessity for frequent watering. **1965** *Sunday Mail Mag.* (Brisbane) 26 Sept. 15 You may read of pots being plunged. This refers to the burying of a pot up to its rim.

5. b. Also, to emerge or come *out* or *out of* (a place) impetuously or abruptly.

1891 C. Graves *Field of Tares* IV. vi. 241 The Norwich Express, plunging out of Liverpool Street Station. **1892** R. Buchanan *Come live with Me* xxiii. 256 Finally..he plunged out into the darkness and disappeared. **1896** C. M. Sheldon *His Brother's Keeper* viii. 226 They plunged right out of a great hole.

c. *spec.* In *Geol.*, of a fold: to have an axis that slopes or dips downwards, whether at a large or a small angle. Also said of the axis. (Further examples.)

1905 Chamberlin & Salisbury *Geol.* I. viii. 488 If the axis of a fold is not horizontal, that is, if it 'plunges'. **1932** *Bull. Amer. Assoc. Petroleum Geologists* XVI. 210 At this time all of the folds plunged consistently eastward at a low angle. **1942** M. P. Billings *Structural Geol.* iii. 46 In the southwest corner, the anticline plunges 15 degrees to the southwest. **1965** A. Holmes *Princ. Physical Geol.* (ed. 2) ix. 210 The axes of folds are not infrequently found to be tilted instead of horizontal; the folds are then said to pitch or plunge.

10. b. *trans.* To bet or speculate (a sum of money).

1922 Joyce *Ulysses* 320 Boylan plunged two quid on my tip *Sceptre* for himself and a lady friend.

11. *trans.* or *absol.* *Railways.* To release (signals or points, etc.) by depressing a plunger. Cf. *plunger 2 g.

[**1923** J. F. Gairns *Railways for All* xviii. 179 Sykes' 'Lock and Block' instruments..are operated by pressing a knob or plunger, hence the term 'plunging' used to describe their working by the signalmen.] **1926** C. J. Allen *Iron Road* xii. 180 Then the signalman in the next box cannot 'plunge' on his instrument until he has put his own starting signal lever back to danger in the lever frame behind the last preceding train. **1940** *Railway Signalling & Communications* iii. 93 All facing points unprotected by track circuit must be provided with a locking bar in addition to the plunger, and the points must be plunged before the signal reading over them can be cleared.

plungeon. Restrict †*Obs.* to sense in Dict. and add: **2.** *south-west. dial.* A ford across a rhine (Rhine¹).

c **1685** A. Paschall *Let.* in S. Heywood *Vindication of Mr. Fox's Hist.* (1811) p. xl, The horse, which the Lord Grey led marched towards the upper Plungeon. *Ibid.*, Whether Sir Francis were there..or came to the Plungeon afterwards..we do not know. **1894** Ld. Wolseley *Life Marlborough* I. xxxix. 314 They [*sc.* the Somerset 'rhines'] could only be crossed, even by single horsemen, at fords, called by the peasantry 'plungeons' or 'steanings'. **1933** W. S. Churchill *Marlborough* I. xii. 216 Grey's cavalry would branch off and..cross the Bussex Rhine at one of the plungeons to the east of the royal camp. **1955** *Hist. Today* V. 57/2 There were two easy crossings for cavalry, known as the upper..and lower 'plungeons'. **1969** C. C. Trench *Western Rising* x. 209 He [*sc.* Feversham] ordered the reserve cavalry in the village..to cross the Bussex Rhine by the upper plungeon. *Ibid.*, Similarly three troops of horse and one of dragoons crossed the lower plungeon.

plunger. Add: **I. 1. c.** *N. Amer.* A type of sailing boat (see quot. 1948).

1860 *North-West* (Port Townsend, Wash.) 12 July 3/1 The following craft were entered for the stakes:—Sloop H. L. Tibbals, Port Townsend;..and the plunger Star of the South. **1892** *Outing* Mar. 467/1 Yachting on the Pacific coast dates from about 1869, when the first club..was organised, though a few small plungers and sloops had long been owned on the bay. **1948** R. de Kerchove *Internat. Maritime Dict.* 541/2 *Plunger*, name given to various sailing craft employed in the Pacific coast oyster fisheries for transportation... Also called oyster sloop. Most of them are built with flush deck and a large central cockpit divided by a centerboard. The larger type is keel built... All are cat-rigged. **1969** *Islander* (Victoria, B.C.) 23 Mar. 4/2 The next day [24 Dec. 1860] a plunger brought a quantity of salvaged goods to Victoria, mostly in the form of cases of Old Tom gin.

2. g. *Railways.* Applied to various knobs or buttons used to operate signalling mechanisms and points; *esp.* (*a*) one with which a signalman operates an electric relay which releases locked signals or points, freq. in an adjacent block section; (*b*) a tapping key on a block instrument.

1881 [in Dict.]. **1897** W. E. Langdon *Applic. Electr. Railway Working* (new ed.) v. 103 The block instrument is fitted with a bell key or 'plunger'. **1899** J. Pigg *Railway Block Signalling* iv. 214 For the lines converging at the junction the plunger used to liberate the signal at the rear cabin of any one of the converging sections is arranged to be free for use only when the points have been set for a train coming from that direction. **1926** C. J. Allen *Iron Road* xii. 173 Below the instrument box is..a plunger or a 'tapper'..by the use of which the signalman exchanges the prescribed code of bell signals with his neighbour. *Ibid.* 180 Only the action of 'accepting' the train by the box next in advance—that is, of giving 'Line Clear' by means of the special plunger on the signalling instrument—will free the lock on the lever, and allow it to be pulled over. **1963** Kichenside & Williams *Brit. Railway Signalling* v. 44 If Mottingham can accept the train the signalman there acknowledges the 'Is line clear?' code and presses the plunger on his home signal instrument. This...unlocks Lee's starting signal and places the block indicator at Lee in the raised position. **1967** G. F. Fiennes *I tried to run Railway* iv. 46 As soon as he put his signals back and took the hook off his Sykes plunger Crowlands offered him the Mail and gave him On Line at once.

h. *Jazz slang.* A plunging device, resembling the type used by plumbers to clear blocked pipes, used as a mute for a trumpet or trombone. Freq. *attrib.* and *comb.*

1936 L. & E. Dowling *Panassié's Hot Jazz* i. 15 Some players..use a regular mute and place over it a rubber plunger—of the sort used by plumbers—which is manipulated by hand to produce 'wa-wa' effects. **1946** Mezzrow & Wolfe *Really Blues* 340 The trumpet got different tonal effects by using plungers and other home-made devices. **1949** L. Feather *Inside Be-Bop* III. 93 Took a couple of plunger solos on Decca. **1958** S. Dance in P. Gammond *Decca Bk. Jazz* xxiii. 293 Cootie at first found Miley's growl legacy altogether too strange... He had always previously played open horn, but since Duke had engaged him to fill Miley's chair, he felt bound to experiment with the plunger style. **1961** *John o' London's* 6 July 55/1 Booty Wood's melancholy plunger-muted trombone. **1966** *Crescendo* Jan. 6/1 A slow opening with Tricky Sam style plunger trombone (sounding here very much like a human voice).

III. 5. **plunger mute** = sense *2 h; **plunger-valve,** a valve having a plunging action.

1947 I. Lang *Jazz in Perspective* vii. 90 Miley was the pioneer and master of a new style of trumpet-playing—a powerful and strangely moving 'growl' obtained by the use of a rubber plunger mute. **1965** *New Yorker* 27 Feb. 123/1 Rudd, a former Dixieland trombonist,..uses the plunger mute effectively. **1977** *Ibid.* 20 June 95/1 Another muted passage (this time using a plunger mute), in which he mumbled funny gibberish through his instrument, all the while using his slide and seeming to blow out the words. **1908** *Westm. Gaz.* 2 Jan. 4/1 The crank-case, into the cover of which the cam-shaft and plunger-valves are built. **1931** *Engineering* 2 Oct. 5 (Advt.), Being a combination of a plunger valve with a mushroom valve, they possess a greater efficiency than is possible with either of these types alone.

plunging, *vbl. sb.* Add: **b.** *plunging system.*

1871 S. Hibberd *Amateur's Flower Garden* xiii. 238 The object of the plunging system is to keep up a rich display of flowers or leaves on the same spot the whole year round.

plunging, *ppl. a.* Add: **d.** *Geol.* Of a fold (see *plunge *v.* 5 c).

1905 Chamberlin & Salisbury *Geol.* I. viii. 483 Fig. 403 shows a doubly plunging anticline; that is, an anticline the axis of which dips down at either end. **1942** M. P. Billings *Structural Geol.* iii. 44 Although the larger plunging folds cannot be directly observed, they are easily recognized from their outcrop pattern. **1968** [see *pitching *ppl. a.* 1].

e. **plunging neckline,** a very deep-cut neckline on a woman's garment.

1949 *Sun* (Baltimore) 18 Nov. 14/2 Will you please express your opinion about the good taste of women wearing the so-called 'plunging necklines' to their business offices. **1959** *Sunday Express* 22 Nov. 6/6, I do wish Mr. Braine could tear his eyes away from girls' provocative, plunging necklines. **1973** 'D. Jordan' *Nile Green* xxxiv. 162 A temp with a plunging neckline and shaky shorthand.

plunk, *v.* Add: **I. 1.** Also, to play (a note) or pick *out* (a melody) on a stringed instrument.

1952 B. Ulanov *Hist. Jazz in Amer.* (1958) xx. 254 She reached her majority plunking two bass notes with her left hand. **1973** *Time* 25 June 94/3 Hungate got to tinkering at the piano one day and in 15 minutes plunked out a ditty he calls *Down at the Old Watergate.*

b. *intr.* To make a plunking sound.

1903 *Cosmopolitan* Sept. 484 Street pianos plunk away unweariedly. **1929** [see *gubble *v.*]. **1946** D. C. Peattie *Road of Naturalist* iii. 35 Into southern Utah they came trekking, with fine fat oxen, with meal sacks under the seats, and banjos plunking. **1978** G. Vidal *Kalki* vi. 153 Deafening was what H. V. W. would call the din from the rock stars' dressing rooms where electric guitars whined, drums rattled, sitars plunked.

II. 3. b. *trans.* To place or set *down* heavily. Also *refl.* to 'let oneself fall'. Cf. Plump *v.*¹ 2.

1899 G. W. Peck *Peck's Uncle Ike & Red Headed Boy* (1903) xxii. 194 The old man plunked down two dollars.., and they went and got seats on the bleachers. **1936** R. Lehmann *Weather in Streets* I. 222 The woman..plunked a great unappetising tray on my bed. **1943** A. Ransome *Picts & Martyrs* v. 46, I say, Dot, when you've plunked the roses in her room... Some on the dressing-table. **1963** *New Statesman* 11 Jan. 38/1 A pair of steel nuts that another worker inadvertently plunks on a plate. **1976** D. Heffron *Crusty Crossed* xi. 85, I went into the bedroom and plunked myself down on our bed. **1977** *Time* 29 Aug. 24/1 He [*sc.* Elvis Presley] had wandered into Sun Records with his guitar, two summers before, plunked down $4 to sing a couple of tunes to his mother.

c. *intr.* To opt *for*. Cf. *plump *v.*¹ 5.

1948 A. H. Vandenberg *Let.* 6 Dec. in Vandenberg & Morris *Private Papers of Senator Vandenberg* (1953) xxi. 414 We should smack down the Russkies more effectively in our speeches in the UN Council and Assembly... I plunk with you for 'curt decisiveness' mixed with 'derision'.

III. 4. b. *trans.* To hit, wound, shoot. *slang* (orig. *U.S.*).

1888 *Texas Siftings* 21 Apr. 12 (*caption*) He'd jest swallerd brother Bill afore I plunked him. **1891** *Outing* Nov. 138/2, I would plunk the big gobbler I could distinguish from where I lay. **1896** [see *out-of-sight *adj. phr.* 2]. **1916** C. J. Dennis *Songs of Sentimental Bloke* 42 Romeo..Plunks Tyball through the gizzard wiv 'is sword. **1916** H. L. Wilson *Somewhere in Red Gap* 120 Darned if he didn't up with this here air gun..and plunk me with a buckshot it carried. **1937** D. & H. Teilhet *Feather Cloak Murders* xvi. 286, I wish you'd killed Jeff instead of plunking him in the leg. **1978** L. Pryor *Viper* iii. 41 We..plunked about five hundred clay birds a day.

Hence **plu·nking** *vbl. sb.* and *ppl. a.*

1941 W. C. Handy *Father of Blues* xviii. 246, I had been everlastingly at the piano, forever picking out notes and chords for *Long Gone* but never playing anything consistently. A victim of all this plunking had been her parrot. **1952** B. Ulanov *Hist. Jazz in America* (1958) vi. 65 With this considerable rhythmic skill, however, went something less engaging, a plunking insistence on the beat. **1973** 'B. Mather' *Snowline* x. 123 The plunking of an out-of-tune guitar. **1979** G. Hammond *Dead Game* vi. 77 A bullet that's been deflected by passing through something..may make a plunking sort of sound.

plunk, *sb., adv., int.* Add: **A.** *sb.* **2. b.** (Earlier and later examples.)

1891 J. Maitland *Amer. Slang Dict.* 207 *Plunk* (Am.), a dollar. **1909** W. G. Davenport *Butte & Montana beneath X-Ray* 56 Make it 25 plunks and let it go at that. **1929** Wodehouse *Gentleman of Leisure* xiii. 107 Dere's a loidy here..dat's got a necklace of jools what's worth a hundred t'ousand plunks.

B. *adv.* and *int.* (Earlier and later examples.)

1892 'Mark Twain' *Amer. Claimant* xix. 196 Feel my pulse: plunk—plunk—plunk. **1936** A. Ransome *Pigeon Post* xxvii. 292 They heard the noise up in the top of the wood... Plunk, plunk, plunk, and the rhythmic scraw of a saw.

Also redupl. **plu·nkety-plu·nk**; also *attrib.*

1884 'Mark Twain' *Huck. Finn* viii. 63, I hear a *plunkety-plunk, plunkety-plunk*, and says to myself, horses coming. **1960** *Twentieth Cent.* Dec. 556 The plunkety-plunk banjo sound.

Plunket (plʌ·ŋkėt), *sb.*² *N.Z.* The name of Lady *Plunket*, wife of the Governor-General of New Zealand 1904–10 (see quot. 1938), used *attrib.* and *absol.* (with reference to the *Plunket Society*, a popular name for the Royal New Zealand Society for the Health of Women and Children), to designate a nurse trained in the methods of child feeding and care advocated by this society, a baby reared according to these methods, or a clinic following them.

1909 *Ann. Rep. Soc. for Promotion of Health of Women & Children* No. 1. 9 The doctor was pleased to have the assistance of the Plunket nurse, and at once consented to the children being fed on humanised milk. **1938** H. C. D. Somerset *Littledene* vii. 69 The Royal New Zealand Society for the Health of Women and Children..commonly called the Plunket Society from the name of its first president, Lady Plunket, was founded by Dr. Truby King in 1907. **1941** S. J. Baker *N.Z. Slang* vi. 58 No record..would be complete without reference to the famed organization, the Plunket Society. For the past twenty years or more it has been known as the Plunket **1944** F. L. W. Wood *Understanding N.Z.* iv. 48 The town will have..Plunket rooms (baby clinics). *Ibid.* ix. 130 Nearly three-quarters of the children born in New Zealand become 'Plunket babies'. **1945** R. M. Burdon *N.Z. Notables* II. ii. 41 By 1913..twenty-seven trained Plunket nurses were working from their appointed centres. **1958** *N.Z. News* 11 Mar. 3/1 In 1912, the Government gave him [*sc.* Truby King] six months' leave of absence to preach the movement throughout the country, and the number of 'Plunket' nurses rapidly multiplied. **1960** S. Ashton-Warner *Incense to Idols* 80 Organize societies for crippled children and the intellectually handicapped, Plunket for the babies, Heritage for the care of War Orphans. **1966** G. W. Turner *Eng. Lang. in Austral. & N.Z.* viii. 173 A New Zealander is likely to begin life as a 'Plunket baby'. His mother is visited periodically by the Plunket nurse from the Plunket Society, founded by Sir Truby King as an early experiment in socially supervised child care and named after the wife of the then Governor-General.

plup (plʌp). [Echoic: cf. Plop *sb.* and *adv.*] The sound of or as of a body falling on a hard surface, into liquid, etc. Also *fig.*

1911 R. Brooke in E. Marsh *Rupert Brooke* (1918) p. lxvii, The 'quaint' remarks fall all round one during meal times, with little soft plups like pats of butter. **1926** *Chambers's Jrnl.* Dec. 847/1 The surge of the water down below, and the *plup* of 'escape' above the roof, were but soft sounds. **1931** E. A. Robertson *Four Frightened People* v. 144 Gas bubbles rose with a soft 'plup-plup'-ing.

pluperfect, *a.* (*sb.*) Add: **2.** (Further examples.) Also (*slang*), used as an intensive.

1889 *Virginia University Mag.* Dec. 186 So take a drink, oh, Phæon, dear, we'll raise pluperfect Cain. **1917** 'Contact' *Airman's Outings* 204, I fully expect that we of the air service will lead the armies of pursuit and make ourselves a pluperfect nuisance to the armies of retreat. **1928** A. Philips *Boy at Bank* I. v. 49 Mrs. Fravalton, one of those mothers who believe their children pluperfect. **1933** E. Caldwell *God's Little Acre* x. 145 What in the pluperfect hell have you boys got to fight about so much, anyhow? **1977** *New Society* 5 May 238/2 They have done it with such bureaucratic precision... The erudition and workmanship are as impeccable, and absolutely deathly, as this kind of pluperfect reconstruction must always be.

plural, *a.* (*sb.*) Add: **2.** *plural community*, a community made up of culturally different ethnic groups; *plural democracy* (see quots.); *plural economy*, the economy of a plural society within which the different ethnic groups keep, to a great extent, their own economic systems; *plural marriage* (see Marriage 1 d); *plural society*, a society composed of different ethnic groups or cultural traditions; a society in which ethnic differences, etc., are reflected in the political structure (see quot. 1971); *plural voting* (further examples).

1869 *Utah Mag.* 18 Sept. 310/1 The Mormon proposition is not to make plural marriage obligatory on the world, but to declare its necessity and legitimacy under certain circumstances. **1906** I. Zangwill in *Times* 29 Oct. 10/5 In another leader on the very same page you defend plural voting on the ground of the necessity of 'the representation of local interests'. **1922** Joyce *Ulysses* 480 Dragging a lorry on which are..plaster figures..representing the new nine muses, Commerce, Operatic Music,..Plural Voting, Gastronomy. **1939** J. S. Furnivall *Netherlands India* xiii. 446 One finds a plural society also in independent states, such as Siam, where Natives, Chinese and Europeans have distinct economic functions, and live apart as separate social orders. *Ibid.*, Some Dutch writers use the term dual or plural economy..to connote the coexistence within the same political community of two or more distinct sets of different economic principles. **1947** *Sun* (Baltimore) 26 Nov. 5/2 Eighteen Utah residents..were charged with conspiring to advocate plural marriage. **1952** B. Davidson *Rep. Southern Afr.* I. vii. 79 They hold in their hands the saving—or the sinking—of a plural society in South Africa. *Ibid.* II. vii. 138 Clearly confronted with the fact of a plural community, they steadfastly refuse to recognize the plurality. **1963** W. N. Stephens *Family in Cross-Cultural Perspective* (1964) ii. 33 This chapter will deal with the last three forms: polygyny, polyandry, and group marriage, the forms of plural marriage. **1965** *New Statesman* 19 Nov. 796/2 The common theme..is the theory of the West Indies as 'plural societies', that is, societies lacking institutions, traditions and habits common to all members. **1966** Jacobs & Zink *Mod. Govt.* (ed. 3) I. vi. 70 Plural voting..resulted from the British belief that a person should be allowed to vote in every district where he was qualified... The Representation of the People Act of 1948 eliminated it completely. **1971** *Race* XII. 462 It appears theoretically useful, then, to partition the universe of culturally diverse societies into: (1) 'plural societies', in which politics tends (exclusively) to follow ethnic lines, and (2) 'pluralistic societies', in which politically relevant issues and actions do not always coincide with ethnic groups. **1977** *Time* 21 Nov. 12/1 What is at stake, ultimately, is whether the government will be able to carry on with the Afrikaners' grand scheme of apartheid—also known as 'separate development' and more recently as 'plural democracy'. **1978** *Guardian Weekly* 4 June 7/3 Apartheid—which has since [1948] undergone minor ideological mutations and several changes in name... Plural democracy, as apartheid is now known.

‖ **plurale tantum** (plūᵊrēⁱ·li tæ·ntŏm). *Gram.* Pl. **pluralia tantum.** [med.L. *plūrāle* the plural f. *plūrālis* Plural *a.* + L. *tantum* only.] A noun which, in any particular sense, is used only in plural form.

1930 T. Sasaki *Lang. R. Bridges' Poetry* II. iii. 57 The so-called 'pluralia tantum' (e.g. breeches, dregs). **1940** A. H. Gardiner *Theory of Proper Names* viii. 27 Here a *plurale tantum* has been resolved into its component individual members, each of whom is thus represented as a bearer of the proper name in question. **1957** R. W. Zandvoort *Handbk. Eng. Gram.* II. 95 Some nouns never occur without a plural suffix; they are known as *pluralia tantum* (sing. *plurale tantum*). Such are *riches, thanks, tongs. Ibid.* 340 Most of the *pluralia tantum* in Kruisinga's lists may also occur without *-s*, if usually in another sense: *compasses* 'instrument for describing circles', but *compass* 'instrument showing magnetic meridian'..*colours* 'flag', but *colour* 'hue'. **1969** *Language* XLV. 239 A collective is precisely to be expected in a word for 'clothes'; cf. the English plurale tantum as well as Gk. *heimata*, only as neuter plural in Homer. **1976** *Archivum Linguisticum* VII. 105 Because these so-called 'uncountable' nouns are in fact to be found in plural form, I dub them *pseudo-uncountables* to distinguish them from true uncountables like *thunder* and *heat*, and pluralia tantum like *scissors*.

pluralism. Add: **2.** (Earlier and later examples.) Also the theory that the knowable world is made up of a plurality of interacting things.

1882 W. James *Let.* 6 Dec. in R. B. Perry *Tht. & Char. W. James* (1935) I. 686 After all, pluralism and indeterminism seem to be but two ways of stating the same thing. **1884** T. H. Green tr. *Lotze's Metaphysic* I. vi. 125 The Pluralism with which our view of the world began has to give place to a Monism, through which the 'transeunt' operation, always unintelligible, passes into an 'immanent' operation. **1919** *Mind* XXVIII. 57 For pluralism, the living experience of the subject consists actually in his interaction with other subjects. **1955** R. Carnap in *Internat. Encycl. Unified Sci.* I. 49 It seems doubtful whether we can find any theoretical content in such philosophical questions as discussed by monism, dualism, and pluralism. **1969** N. Rescher *Many-Valued Logic* iii. 214 There is no balking the fact of *pluralism* in logic. *Ibid.*, Faced with the fact of pluralism, the step to relativism or conventionalism might seem short and easy. **1972** K. R. Popper *Objective Knowledge* iv. 153 Some philosophers have made a serious beginning towards a philosophical pluralism, but pointing out the existence of a *third world*.

3. a. *Pol. Sci.* A theory which opposes monolithic state power and advocates instead increased devolution and autonomy for the main organizations that represent man's involvement in society. Also, the belief that power should be shared among a number of political parties.

1919 H. J. Laski in *Philos. Rev.* XXVIII. 568 The monistic state is an hierarchical structure in which power is..collected at a single centre. The advocates of pluralism are convinced that this is both administratively incomplete and ethically inadequate. **1941** H. M. Magid *Eng. Political Pluralism* i. 8 If we deny the inevitability of totalitarianism, we may discover in pluralism..an attempt to analyze the problem of freedom in the light of modern world conditions. **1954** B. & R. North tr. *Duverger's Pol. Parties* II. i. 257 As a 'party system' the single party is obviously different from the multi-party system or 'pluralism'. **1969** M. Broady in *Architectural Assoc. Quarterly* I. 67 Political pluralism is concerned essentially with the relationship between the State and other kinds of social organization, and it draws attention to the limitations of State power in order to assert the importance in a democratic society of alternative and independent sources of power and foci of interest. **1977** M. Walker *National Front* i. 16 Britain is historically accustomed to one form or another of coalition government, which is the essence of pluralism.

b. The existence or toleration of diversity of ethnic or cultural groups within a society or state, of beliefs or attitudes within a body or institution, etc.

1933 *Sociol. & Social Research* XVIII. 103 (*title*) Social pluralism. **1956** H. M. Kallen *Cultural Pluralism* 46 'Cultural Pluralism' is a controversial expression. *Ibid.* 51 This pluralism is the kind always and everywhere characterizing men's undertakings. **1965** *New Statesman* 19 Nov. 796/2 The relevance of this 'pluralism' to Caribbean politics is plain. Analysis of social problems in simple economic terms..will not do. **1969** *Guardian* 20 Sept. 4/4 A good deal of discussion at the conference..has been about pluralism as opposed to assimilation. **1971** *Deb. House of Commons Canada* 8 Oct. 8580/2 Ethnic pluralism can help us overcome or prevent the homogenization and depersonalization of mass society. **1976** *Times* 7 Aug. 14/4 To be in favour of pluralism is to declare that one does not intend to hound or persecute theologians or catechists whose expression of faith differs from your own. **1977** F. Young in J. Hick *Myth of God Incarnate* ii. 42 The future seems to lie with pluralism in christology.

pluralist. Add: **1. b.** (Earlier example.)

1818 T. Brown *Brighton* II. 181 This long pole is but a walking stick to the magical wand of old Hurlothrumbo, the pluralist, who can fill so many places.

2. (Earlier and later examples.) Also *attrib.*

1885 W. James *Lit. Remains of H. James* 116 Their most serious enemy will be the *philosophic* pluralist. **1919** C. A. Richardson *Spiritual Pluralism* i. 12 The above is but a broad outline of the pluralist argument as applied to the inorganic world. **1919** *Mind* XXVIII. 58 The pluralist..recognizes that the fundamental fact from which the start must be made, is not a dualism of matter and mind, but the unity of the individual experience, which comprises a duality of subject and object. **1966** F. J. Copleston *Hist. Philos.* VIII. x. 250 The idea of God has the benefit of increasing the pluralist's confidence in the significance of finite existence.

3. *Pol.* and *Sociol.* One who upholds pluralism and is opposed to monolithic state power or to policies of cultural or ethnic assimilation. Also *attrib.* or as *adj.*, designating a system based on pluralism.

1920 *Amer. Pol. Sci. Rev.* XIV. 407 The all important fact, so consistently overlooked by the pluralist, that the truly federal state is a unitary state. *Ibid.*, The pluralist doctrine is timely in that it calls attention to the present bewildering development of groups within the body politic. **1957** M. P. Fogarty *Christian Democracy* I. iii. 29 A certain 'solidarist' conception of the individual's responsibility to and for the society around him, and, following from this, a 'federalist' or 'pluralist' ideal of the structure of society and the processes which go on within it. **1964** Gould & Kolb *Dict. Social Sci.* 244/2 Today attention is frequently paid to the difficulties of maintaining a pluralist society under the impact of mass society and culture. *Ibid.* 563/2 Pluralists define public opinion in terms of controversy. **1972** *Times Lit. Suppl.* 4 Feb. 115/2 The cheerful belief that the Negro was entering mainstream pluralist politics became incredible after Watts, Newark and Detroit. **1975** *N.Y. Times* 9 Oct. 3/3 The Market withheld similar aid in July for the Gonçalves Government on the ground that it did not represent a 'pluralist democracy'. **1977** M. Walker *National Front* i. 16 The great strength of the British pluralist system of channelling political allegiance into one of two wide coalitions is that many citizens who sympathize with totalitarian objectives are encouraged to stay within the political mainstream.

pluralistic, *a.* Add: (Earlier and later examples.) Also *spec.* in *Pol.* and *Sociol.* (see *pluralism 3); *pluralistic ignorance* (see quot. 1970).

1854 *Edin. Rev.* XCIX. 360 Even the 'pluralistic' marriage service has been published [by the Mormons]. The following is an extract from this novel rubric. **1909** W. James (*title*) Pluralistic universe. **1919** H. J. Laski in *Philos. Rev.* XXVIII. 562 (*title*) The pluralistic state. **1920** *Amer. Jrnl. Sociol.* XXV. 388 Pluralistic behavior, in distinction from individual behavior, has its own conditions, forms, and laws. **1933** *Sociol. & Social Research* XVIII. 107 The pluralistic conception of society carries

with it the principle of functionalism. **1950** T. M. Newcomb *Social Psychol.* xvi. 608 We have a condition for which F. H. Allport..suggested the term pluralistic ignorance. Everyone assumes that everyone except himself accepts the norms uncritically. **1963** F. Copleston *Hist. Philos.* VII. II. xii. 252 A pluralistic metaphysics which calls to mind the atoms of Democritus and the monads of Liebnitz. **1964** T. B. Bottomore *Elites & Society* vi. 119 Aron, when he urges the importance of the diffusion of power in the pluralistic democracies does not invoke only the principal elites. **1970** G. A. & A. G. Theodorson *Mod. Dict. Sociol.* 301 *Pluralistic ignorance*, a situation in which individual members of a group believe incorrectly that they are each alone..in believing or not believing in particular values. **1971** [see *plural society]. **1972** *Times Lit. Suppl.* 18 Feb. 195/2 We are now living in a morally pluralistic society;..attempts by one group to impose its morality on another are futile. **1975** *N.Y. Times* 10 Sept. 20/7 Mr. Kissinger added that the United States was 'working in closest harmony with our European allies' on the problem of encouraging 'the emergence of a pluralistic system' in Portugal. **1976** P. Donovan *Relig. Lang.* viii. 95 Apologists for the Baha'i World Faith appeal to the reasonableness of a syncretistic faith in today's pluralistic world. **1977** M. Walker *National Front* i. 15 The pluralistic kind of society has agreed to permit its citizens to differ about the society's objectives.

pluralization. (Later examples.)

1970 *English Studies* LI. 395 Participles used as adjectival modifiers were also disregarded, as were those used nominally (with actual or potential determiners, pluralisation, or government by a preposition). **1972** *Language* XLVIII. 356 All these authors, characteristically, regard the rules of Spanish pluralization as phonological ones. **1978** *Amer. Speech* LIII. 35 Personal singular *you* similarly takes what is formally a plural verb form (*are, were*), as do French *vous*, German *Sie*, and Russian *vy*; as with singular *they—themself*, this 'pluralization' of singular *you* can be said to follow reflexivization (whence a plural verb but a singular reflexive *yourself*).

pluralize, *v.* Add: **1. c.** *intr.* To express or form the plural.

1964 *Language* XL. 135 The way in which speakers of English pluralize.

pluralizer. Add: **2.** *Gram.* **a.** A pluralizing affix, inflexion, or word. **b.** A noun that may appear in plural form.

1933 *Times Lit. Suppl.* 16 Nov. 798/2 *Shang* is one of the six pluralizers. **1951** *Archivum Linguisticum* III. 66 Suffix: *-naka* ~ *-nak-* '(noun) pluralizer'. **1961** R. B. Long *Sentence & its Parts* ii. 39 The basic forms of pluralizer nouns are most often used as heads within nounal units, since they generally require determiners. *Ibid.*, Quantifiables such as *courage, fun,..machinery*, and *furniture* are not made plural, though it is true that some quantifiables have pluralizer status also. **1965** *Canad. Jrnl. Linguistics* Spring 172 The pluralizer *qə-* of Position 4. **1971** [see *intensifier].

pluranimity. *rare.* [f. L. *plus, plūr-* more, substituted for *un-* in Unanimity.] Diversity of opinions.

1907 W. De Morgan *Alice-For-Short* ix. 95 Whatever innate ideas on the subject of oil-painting he possessed, had been disorganised and carefully thrown out of gear by the want of unanimity, or presence of pluranimity, in his instructors.

plurative. **2.** (Earlier example.)

1849 W. Thomson *Laws of Thought* (ed. 2) II. 174 The judgment—'Most men are prejudiced' cannot..be considered as particular, for it implies not only that *some* men, but *more than the half* of mankind are prejudiced. These are termed *plurative* judgments.

pluri-. Add: **pluridi·sciplinary** *a.*, having or consisting of several disciplines or branches of learning; = *interdisciplinary *a.*; **plu·riform** *a.*, existing in many different forms; multiform; hence **plurifo·rmity**; **plurili·ngual** *a.* and *sb.* = *multilingual *a.* and *sb.*, Polyglot *a.* and *sb.*; hence **plurili·ngualism**; **plurilocular** (later examples); **plurimo·dal** *a.*, consisting of or involving more than one mode (in various senses); **plurisy·llable,** a word of two or more syllables; hence **plurisylla·bic** *a.*

1970 *Guardian Weekly* 14 May 12/1 The substitution of medium-sized pluri-disciplinary universities for the existing monstrous faculties. **1972** *Science* 12 May 621/3 They would also be 'pluridisciplinary' which meant..that the universities would 'associate wherever possible arts and letters with sciences and technics'. **1979** *Guardian* 12 June 8/7 This is a pluri-disciplinary show—which means that all the arts are being covered. **1973** *Times* 28 May 9/6 The remarks regarding churchmanship can only be described as naive; most Anglicans know their Church to be pluriform. **1974** *Times Lit. Suppl.* 22 Mar. 299/2 To say that all religions are equally true amounts to saying that all are equally false. But are not some more equal than others? Or is there a case for saying that truth can be pluriform? *Ibid.* 5 July 732/3 The New Testament writings are to be seen as pluriform. They were not composed..to produce a volume on a theme previously agreed. **1975** *Caribbean Contact* Feb. 4/3 One effective way of developing a pluriform approach in the context of the many needs of any single community is through the team ministry approach. **1947** *Theology* L. 419 The pluriformity of the churches is undoubtedly a sin of Christendom. **1975** *Church Times* 21 Feb. 14/2 No small part of his achievement has been the fundamental preservation of orthodoxy, despite the extensive adoption of pluriformity in forms of worship. **1976** *Times* 21 Feb. 15/6 We, like our Latin brothers, have come to accept pluriformity in belief as much as in cultural tradition. **1938** I. Goldberg *Wonder of Words* p. vii, It was the curiosity born of this pluri-lingual heritage that led me..to..a special interest in language. **1956** *Publ. Amer. Dial. Soc.* XXVI. 9 Strictly speaking, a bilingual..is one who knows two languages, but will here (as commonly) be used to include also the one who knows more than two, variously known as a *plurilingual*, a *multilingual*, or a *polyglot*. **1962** Y. Malkiel in Householder & Saporta *Probl. Lexicogr.* 11 Pluri-lingual dictionaries, which patently mark an increase in coverage. **1976** *Word 1971* XXVII. 407 The period of language 'acquisition' for plurilingual children is claimed to be longer than the period for monolingual children. **1971** *Incorporated Linguist* X. 42/1 Languages involved in societal systems of plurilingualism tend to stabilize their roles on a basis of complementary distribution—either spatially, functionally, or both. **1902** D. H. Campbell *Univ. Text-bk. Bot.* v. 129 In many of the Phæosporeæ..there are formed the plurilocular sporangia. **1961** R. W. Butcher *New Illustr. Brit. Flora* I. 19 Sometimes several carpels..are united along the flat sides, so forming a plurilocular ovary of 2—many cells. **1969** F. E. Round *Introd. Lower Plants* iii. 44 Plurilocular sporangia may also occur on the sporophyte. **1949** Wellek & Warren *Theory of Lit.* iii. 25 The alternative to these seems some bi-modal or pluri-modal truth. **1951** G. S. Carter *Animal Evolution* i. 26 If for the specimens collected at one horizon we plot a variability curve.., the curve..should have an apex for each of the mixed populations (pluri-modal) if the population consists of distinct but mixed elements. **1976** *Word 1971* XXVII. 195 Verbal expression, however, is only one facet of the plurimodal (multichannel) process of interpersonal communication. **1924** J. S. Kenyon *Amer. Pronunc.* 30 A Plurisyllable is a word of more than one syllable. **1934** Webster 1897/3 Plurisyllable..plurisyllabic. **1965** *Amer. Speech* XL. 12 There is also *tsetse*..and a fair number of other plurisyllables. *Ibid.* 13 *Sclaff*..is less well known than plurisyllabic *sclerosis*..and *sclerotic*.

pluriarc (plūᵊ·ri,ɑ̄ɹk). [a. Fr. (G. Montandon *La Généalogie des Instruments de Musique* (1919) 52), f. Pluri- + Arc.] A West African musical instrument made of a wooden resonator to which several curved rods holding taut strings are attached.

1961 A. Baines *Musical Instruments* i. 46 When played it [*sc.* the bow lute] is held like a harp, giving it the impression of being a harp in which each string has a separate neck... The French word *pluriarc*..expresses this well. **1970** *Western Folklore* XXIX. 237 It must be considered possible that the..American instruments of more than one string derive from pluriarcs and trough zithers.

pluripotency (plūᵊripōᵘ·tėnsi). *Biol.* [ad. G. *pluripotenz*: see Pluri- and Potency.] The property of being pluripotential; = *pluripotence, *pluripotentiality.

1927 *Biol. Abstr.* I. 217/1 The term pluripotency is used to describe the virtual ability possessed by every organism, under certain conditions, to strike out upon developmental paths deviating from the type. **1942** J. Needham *Biochem. & Morphogenesis* ii. 100 This condition of multiple potency of the parts of the early egg-cell has been termed pluripotency. **1974** *Cell* II. 166/1 To study the relationships between tumorigenicity, pluripotency, and growth control in vitro, eleven subclones of SIKR [*sc.* a cell line] were isolated in parallel.

pluripotent (plūᵊripōᵘ·tėnt), *a.* *Biol.* [f. Pluri- + Potent *a.*¹ and *sb.*²] = next. Hence **pluripo·tence** = *pluripotency.

1942 J. Needham *Biochem. & Morphogenesis* ii. 228 The retention of pluripotence through many cleavages, followed by separation of blastomeres. *Ibid.* iii. 435 They are at first pluripotent or relatively undetermined. **1963** J. W. Lash in M. Locke *Cytodifferentiation & Macromolecular Synthesis* 257 Even embryonic systems involving so-called 'pluripotent' reacting tissues are equational to the systems discussed in this paper. **1977** *Lancet* 26 Mar. 680/1 If we assume that the intrathymic stem-cells are really pluripotent and that their capacity to differentiate into various tissues is genetically controlled, this pattern of autoimmune pathogenesis need not be restricted to synaptic autoantigens and myasthenia.

pluripotential (plūᵊripote·nʃăl), *a.* *Biol.* [f. Pluri- + Potential *a.* and *sb.*] Of a cell, tissue, or organism: capable of developing in any of various directions; = *multipotential* s.v. *multi- 1 a.

1925 *Arch. Neurol. & Psychiatry* (Chicago) XIII. 468 Phylogenetically considered, this structure is a pluripotential organ, which in some of the vertebrates has become differentiated as an eye. **1939** *Nature* 27 May 903/1 The lymphocyte may be regarded as a pluripotential element capable of differentiation along several lines, the particular one followed being dependent upon the individual requirements. **1962** *Lancet* 27 Jan. 207/1 Many have regarded it [*sc.* the small lymphocyte] as a mature cell which could not become anything else, while others have regarded it as a temporarily resting stage of a primitive pluripotential cell capable of active development under suitable conditions. **1974** *Nature* 11 Oct. 518/2 We have recently proposed that the proliferative compartment of foetal epidermis is built up by 'pluripotential' stem cells which are replaced by 'committed' stem cells during adolescence.

Hence **plu:ripotentia·lity** = *pluripotency.

1956 R. W. Evans *Histol. Appearances of Tumours* xi. 165 The question of the pluripotentiality and transmutation of lymphocytes into other cells is considered by many authors to be concerned with the enigmatic disappearance of lymphocytes. **1971** *Jrnl. Immunol.* CVII. 1583 (*heading*) The pluripotentiality of mouse spleen lymphocytes.

plurisegmental (plūᵊ:risegme·ntăl), *a.* [f. Pluri- + Segmental *a.*] **1.** *Physiol.* That involves nerves from more than one segment of the spinal column.

1898 C. S. Sherrington in *Phil. Trans. R. Soc.* B. CXC. 151 It follows..that the reflex centrifugal discharge of the spinal cord is pluri-segmental. **1924** *Amer. Jrnl. Physiol.* LXIX. 649 The data indicate for the frog a plurisegmental innervation such as Agduhr..has shown to exist in cats and rabbits. **1969** J. P. Swazey *Reflexes & Motor Integration* v. 95 It is the pluri-segmental character of the motor nucleus, as expressed by reciprocal innervation, which gives the limb its functional solidarity.

2. *Linguistics.* = *suprasegmental *a.*

1964 R. H. Robins *Gen. Linguistics* 107 The most important features that must be treated as plurisegmental..are stress, pitch, and general voice quality. **1965** *Language* XLI. 309 The Firthian distinction of phonematic units (with 'unisegmental..relevance'..) and prosodies (with 'plurisegmental relevance') is probably more acceptable to some old-fashioned Americans than the Chomsky-Halle treatment of all features as equal.

Hence **plu:risegme·ntally** *adv.*

1930 *Jrnl. Physiol.* LXX. 210 Frogs which have been kept in the tank for several months and whose muscles are much atrophied, are likely to show a very low proportion of plurisegmentally innervated fibres. **1960** *Jrnl. Compar. Neurol.* CXV. 47/2 The neurons..occur plurosegmentally.

plurisign (plūᵊ·risəin). [f. Pluri- + Sign *sb.*] A 'sign' or word used with more than one meaning simultaneously: opp. *monosign. Hence **plurisigna·tion**, **plurisi·gnative** *a.*, **plurisi·gnatively** *adv.* Also **plu:risignifica·tion.**

1940 *Kenyon Rev.* II. 266 The atomic ingredient of poetic language tends to be the plurisign. *Ibid.* 268 We may distinguish three types of simultaneous plurisignation. *Ibid.* 282 There is also a third important dimension, which may be called the dimension of *plurisignative fullness*, or *poetic richness*. *Ibid.* 504 The realm of poetry-connotation-plurisignification. **1954** P. Wheelwright *Burning Fountain* iv. 61 The poetic symbol tends..to be plurisignative. *Ibid.* vi. 106 In plurisignation, a single verbal expression carries two or more meanings simultaneously. *Ibid.* vii. 149 A serious metaphoric pun, a poetic plurisign. *Ibid.* x. 217 The directions of mythic reference which..make a tragic action plurisignatively meaningful. **1964** J. Stone *Legal System & Lawyers' Reasonings* i. 34 (*heading*) 'Law properly so-called' and plurisignation of words. **1965** *Encycl. Poetry* 760/1 The depth symbol tends to be *plurisignative*; which is to say, its intended meanings are likely to be more or less multiple, yet so fused as to produce an integral meaning which radically transcends the sum of the ingredient meanings.

plurivalent (plūᵊrivēⁱ·lĕnt, plūᵊri·vălĕnt), *a.* [f. Pluri- + L. *valēnt-em*, pr. pple. of *valēre* to be worth.] = Multivalent *a.* in Dict. and Suppl., *spec.* sense *2.

1905 E. B. Wilson in *Jrnl. Exper. Zoöl.* II. 374 In the case of compound or plurivalent chromosomes..McClung's term 'chromatid' may conveniently be applied to each of their univalent constituents. **1952** *Archivum Linguisticum* IV. 169 The sorting out of plurivalent words like G. *Preis* between 'prize' and 'price'. **1976** *Nature* 1–8 Jan. 57/1 Aggregation of the minute somatic single chromosomes to large plurivalent chromosomes prevents mistakes in meiotic distribution.

pluri-valued (plūᵊrivæ·liūd), *a.* *Logic.* Also **plurivalued.** [f. Pluri- + Valued *ppl. a.*] Of a system of logic: using truth values in addition to those of true and false; many-valued.

1939 R. Carnap *Found. Logic & Math.* in *Internat. Encycl. Unified Sciences* I. III. § 12. 28 The systems of plurivalued logic as constructed by Lukasiewicz and Tarski, etc. **1949** A. Pap *Elements Anal. Philos.* v. 105 Is it possible to construct a logic which does not assume that every proposition is either true or false ('pluri-valued logic')? **1969** N. Rescher *Many-Valued Logic* i. 10 Two interesting procedures for creating new pluri-valued logics out of old ones—the methods of system *extension* and of forming the *product* of two systems—have been devised by Stanisław Jaśkowski. *Ibid.* ii. 83 The prospect remains that two-valued logic is somehow fundamental to the construction of all systems of pluri-valued logic in general.

plurry (plʌ·ri), *a.* and *adv.* *Austral.* and *N.Z.* *slang.* [Maori corruption of Bloody *a.* and *adv.*] = Bloody *a.* 10, *adv.* 2.

1900 H. Lawson *Verses, Pop. & Humorous* 227 And their language that day, I am sorry to say, Mostly consisted of 'plurry'. **1916** *Anzac Book* 137 By this time, being the twelfth month of the same year, it waxed 'plurry' cold, even unto a fall of snow. **1933** H. G. Wells *Bulpington of Blup* vi. 239 Gawd save us from any plurry mucking pushes. *Ibid.* 242 You're plurry well doing too much with your mouth... *Shut* it! **1938** R. D. Finlayson *Brown Man's Burden* 31 It's all right for Pakeha's to spout about Maori art but it won't help me to get manure for my plurry cow farm. **1959** *News Chron.* 6 July 4/2 'You watch your plurry step, man,' he said. 'We know your plurry kind.' **1966** G. W. Turner *Eng. Lang. in Austral. & N.Z.* x. 200 The literary convention of Maori

English interlarded with *plurry*..belongs to the language of journalists rather than the language of Maoris. **1977** *Sunday Times* 3 July 28/7 I'll bowl a plurry sight faster if they'd let me take my boots off.

plus. Add: **1.** (Earlier and later non-technical examples.)

1823 J. Keble *Let.* in G. Battiscombe *John Keble* (1963) iv. 71 A deaf handmaid, and a clerk-plus-gardener-plus-groom by no means dumb. **1922** Joyce *Ulysses* 642 A cup of Epp's cocoa and a shakedown for the night plus the use of a rug or two and overcoat doubled into a pillow. **1961** B. Fergusson *Watery Maze* xiv. 348 No. 1 Special Service Brigade under Lovat had one Marine and three Army Commandos, plus Lovat's personal piper. **1966** *Oxf. Univ. Gaz.* 23 Dec. 430/1 Just a reference from one meeting of Congregation to another slightly different meeting of Congregation, plus a few alumni. **1974** E. Ambler *Dr Frigo* III. 190 There..was a..plane waiting plus an army scout car and a guard of soldiers. **1975** N. Luard *Travelling Horseman* xiv. 247 That meant me plus Billy to drive me. It was a two-part job.

c. *ellipt.* Placed after a round number or a whole number to indicate a smaller or fractional amount more; with a positive amount added; or more, but not less. Also to indicate a slightly higher grade, as *beta plus* ($\beta+$). Also *colloq.* after other sbs. to indicate with extra qualities, better than usual.

[**1902**: see *alpha 4.] **1916** A. Huxley *Let.* 19 Mar. (1969) 94 Beta double plus is quite adequate for language. **1927** *Rep. Consultative Comm. Educ. Adolescent* 185 Raising the school age to 15 plus must lead either to the building of new schools or to the remodelling of existing schools. **1928** *Oxford Mag.* 25 Oct. 40 Till the University finds some benefactor willing to give '£100,000 plus' (to use modern phraseology). *a***1930** D. H. Lawrence *Collier's Friday Night* i. 20, I generally get an alpha plus. That's the highest, you know, mater. **1951** Wodehouse *Old Reliable* x. 120 You were right in mid-season form. It was battling plus. **1958** L. van der Post *Lost World Kalahari* vi. 104 'This,' I thought before sleeping, 'is Alpha Plus.' **1962** *Listener* 18 Jan. 112/2 To put forward a positive conception of coexistence—we might call it coexistence 'plus'—which includes as its principal ingredient the central idea..of *ideological* coexistence. **1962** *Guardian* 5 Dec. 6/4 Everything you could possibly want in make-up plus, for 1s 6d a time. **1966** *Listener* 30 June 951/2 Many of these can now be walled off, but not Garrincha of Brazil, who on his day is Stanley Matthews plus. **1972** J. Mosedale *Football* ii. 21 The $225-plus material worn by Kansas City tackle Buck Buchanan. **1973** P. Moyes *Curious Affair of Third Dog* i. 10 The Second World War, an unbelievable thirty-plus years ago. **1978** *Jrnl. R. Soc. Arts* CXXVI. 617/2 It will be a hundred years plus before we have a significant contribution.

d. quasi-*conj.* And in addition. *colloq.* (chiefly *N. Amer.*)

1968 N. Giovanni in W. King *Black Short Story Anthol.* (1972) 26 If, on the other hand, a tom tells you to get off the streets and you don't..plus if you can encourage him in a physical way to come on over to your side, then you've made a friend. **1972** C. L. Cooper in *Ibid.* 210 There was rain on the roof, and I was poor. Plus my mother was sick. **1973** *Black World* Jan. 63/2 All the ladies brought pies and cakes... Plus they had coffee and tea and punch. **1974** *Sci. Amer.* Jan. 1/1 (Advt.), Initial cost of our centerline design (WX 200) is $200K instead of the $600K of equivalent current models, plus it will deliver 100 hours MTBF (mean time between failures). **1974** *Maclean's Mag.* Dec. 16/2 But after all is said and done, take away the football game itself from the Grey Cup and what have you got? An early start home, for one thing. Plus you save maybe $30 on tickets. **1977** *Ripped & Torn* vi. 8/2, I was stereotyped as nothing but a drag act. Plus I was tired of doing it. **1978** *Daily Tel.* 9 Feb. 2 (Advt.), W. H. Smith have big names at big savings. Plus you get one year manufacturer's guarantee. **1978** *Detroit Free Press* 2 Apr. 6E/4 Plus they've added pitchers Rudy May and Ross Grimsley.

2. (Later example.)

1975 *Times* 15 Jan. 20/4 A 230-volt standard for electrical supplies throughout Europe was agreed yesterday... The new level has a tolerance of plus or minus 10 per cent.

3. *adj.* (Further examples.)

1928 *Publishers' Weekly* 30 June 2598 The material for the plus sale is always at hand in the book business. **1930** *Publishers' Circular* 2 Aug. 163/2 The whole book business should look on the reprint business as being *plus* business.

b. In various games, having an adverse handicap of a number of strokes or points.

1908 A. W. Myers *Compl. Lawn Tennis Player* 127 It will not take him long to discover the kind of decoy that will deliver the 'plus 15.3' men into his clutches. **1909** *Westm. Gaz.* 8 Feb. 12/4 Supposing a plus 3 man is partnered with a steady player whose handicap is 8, the two as a foursome side would be handicapped at 5. **1922** J. Cannan *Misty Valley* 203 Isn't it just like you to come up to the club-house..and to send a plus man in to fetch me out? **1927** R. J. B. Sellar *Play!* 32 The newly-joined member..asked if they might have a game together. 'Humph,' growled the plus player, 'perhaps. What's your handicap?'

c. Of superior quality; excellent of its kind.

1960 *Times* 10 Dec. 7/6 The country was surveyed to find 'plus trees' (superior examples of all useful species). **1968** *Globe & Mail* (Toronto) 17 Feb. 45 (Advt.), Living room, 15' dining room are plus features!

4. *sb.* **a.** *plus sign* (examples). Also *fig.* and in sense 4d below. **b.** (Later example.) **c.** (Earlier and later *fig.* examples.) **d.** An advantage.

1791 H. Walpole *Let.* 26 July (1905) XV. 25 The villain Paine..is engaged in a controversy with the Abbé Sieyès, about the *plus* or *minus* of the rebellion. **1902** W. James *Varieties Relig. Exper.* 166 Peace cannot be reached by the simple addition of pluses and elimination of minuses from life. **1943** *Sun* (Baltimore) 12 Feb. 17/1 American Telephone was up 1¾, and lesser plus signs were retained by Radio Corporation, Johns-Manville, Goodrich and United States Gypsum. **1959** *Washington Post* 20 July A 6/1 Radio city is one block from the Hotel Victoria. Other location plusses: Madison Square Garden is two blocks away, so are all subways. **1964** G. C. Kunzle *Parallel Bars* ix. 403 Its great plus is the fact that the superior difficulty is the dismount. **1971** *Guardian* 25 Nov. 15/4 Those whom Mr Blond has published are treated as human beings... This is a major plus in his favour. **1972** *Sci. Amer.* Jan. 83/2 The superscript plus sign denotes a positive ion. **1972** *Business Week* 18 Mar. 20/2 The ability to offer assurances of such long-range supplies..is a big plus for U.S. equipment builders. **1974** T. Allbeury *Snowball* xviii. 104 The friction between the governments—it's a plus but it's a side issue. **1975** *Daily Tel.* 4 Jan. 10 Fiction is not on the wane, but actually registered a plus in the number of books published in 1974. **1975** *Publishers Weekly* 31 Mar. 47/1 There are items here..that will interest only a handful of people west of Manhattan. But there are compensating plusses. **1975** *Studies in Eng. Lit.: Eng. Number* (Tokyo) 44 In Table 9 the frequency of occurrence is indicated by means of figures appended to plus signs (like + 1) for constructions which are themselves rare. **1976** *Evening Post* (Nottingham) 15 Dec. 24/7 But Glaxo 383p, Fisons 288p, Tubes 290p, and GEC 163p, still showed small plus signs. **1976** *Early Music* Oct. 382/2 Could..the same reader get something from an article on..the pluses and minuses of using gut strings? **1977** *Time* 7 Mar. 49/2 The scenery and the costumes..are a dazzling plus. **1978** C. Black *Asterisk Destiny* (1979) v. 122 All I am..is the plus sign in a necessary equation.

6. Special Combs.: **plus juncture** *Linguistics* = **open juncture* s.v. *Open *a.* 22c; **plus word**, a word which is characteristic of its user, or is used frequently.

1951 Trager & Smith *Outl. Eng. Struct.* 68 Within each phonemic phrase the constituents delimited by plus-junctures are noted. **1964** E. Bach *Introd. Transformational Gram.* vi. 129 One can save the situation by positing a phoneme of internal open juncture ('plus-juncture'). **1972** Hartmann & Stork *Dict. Lang. & Linguistics* 241/2 Open transition (open juncture),..often called plus juncture, which occurs at a word boundary. **1976** *Amer. Speech* 1974 XLIX. 15 Predictable pitch contours are not marked here and 'plus juncture' is indicated by conventional spacing. **1959** N. N. Holland *First Mod. Comedies* vi. 53 'Solid' became a 'plus' word because it suggested realness, the mass and volume the new physics could measure. **1962** A. Ellegård *Statistical Method for Determining Authorship* ii. 12 The typical, characteristic words of the writer—we shall..call his 'plus words'. **1963** *Times Lit. Suppl.* 25 Jan. 67/2 Words and expressions that were used particularly frequently (plus-words).

plusage (plɒ·sėdʒ). Also **plussage**. [f. Plus + -age.] **a.** (See quot. 1932). **b.** Something extra or added on; a bonus; a surcharge.

A word of restricted currency.—Ed.

1932 *N. & Q.* 30 Jan. 82/2, I have seen in print the word 'plus(s)age' used to denote a plurality of pluses in a collective sense, i.e., (15 per cent. plus 10 per cent. plus 15 per cent.) total plussage of 40 per cent. **1935** A. P. Herbert *What a Word!* iii. 83 From a well-known London shop: 'Should your order not exceed the sum of 17/6 an additional *plusage* of 25% will be charged.' **1959** *Times* 3 Nov. 13/4 When I took up a part-time job three weeks ago I was given a card with which to 'clock in' and 'out'. Among the list of payments was a heading 'Any further plusages'. **1962** *Engineering* 10 Aug. 181 Normal day rate together with a plusage for reaching a certain level of output. **1966** *New Statesman* 22 Apr. 563/1 A woman catcher's rate on a cigarette-making machine, for instance, is about £3 a week less than a man's, although the bonus or 'plussage' for both is equal.

‖ **plus ça change** (plü sa ʃɑ̃ʒ). In full, *plus ça change, plus c'est la même chose*. [Fr., 'the more it changes, the more it stays the same'.] A semi-proverbial phrase, expressing the fundamental immutability of human nature, institutions, etc. Hence **plus ça change-ness** *nonce*.

[**1849** A. Karr in *Les Guêpes* Sér. 6 (new ed.) 1859 Janv. 305 [After comment on recent political events in France] Après tant de bouleversements, de changements, il serait temps de s'apercevoir d'une chose, c'est que c'est comme au cabaret:—cachet vert, cachet rouge, etc.—On change quelquefois le prix, quelquefois le bouchon, mais c'est toujours la même piquette qu'on nous fait boire.—Plus ça change—plus c'est la même chose.] **1903** G. B. Shaw *Man & Superman* 182 The mere transfiguration of institutions, as from military and priestly dominance to commercial and scientific dominance, from commercial dominance to proletarian democracy,..are all but changes from Tweedledum to Tweedledee: *plus ça change, plus c'est la même chose*. **1955** *Times* 5 July 11/4 We have an uneasy suspicion that there is something to be said for the epic angle on history. After all, when boy met—or meets—girl, *plus ça change*. **1959** 'J. Byrom' *Take only as Directed* i. 8 'I don't like being out of date!' '*Plus ça change*,' I said. **1966** *New Statesman* 2 Sept. 306/1 Meanwhile it is a sort of lottery whether a murderer is hanged or not... Out of seven capital sentences passed at the Winter Assizes 1871/1872, *only one* was carried out. *Plus ça change*. *Ibid.* 312/3 Browsing in my current bedside book, *The Paston Letters*, I have been struck by the *plus ça change*-ness of the human animal. **1969** R. Blythe *Akenfield* 15 Where the strict village existence is concerned it is *Plus ça change, plus c'est la même chose*. **1970** W. Garner *Puppet-Masters* i. 14 *Plus ça change*—! The present French government's simply following tradition. **1978** *Language* LIV. 383 For both theories, then, plus ça change, plus c'est la même chose; and, interestingly, the même chose of each turns out to be the même chose for both.

plus-foured (plɒsfōᵊ·ɹd), *a.* [f. next + -ed².] Wearing or clad in plus fours.

1925 *N. & Q.* 28 Nov. 387/2 In *Truth* of Oct. 14 a writer ventured 'plus-foured'. **1926** *Daily Chron.* 13 May 2/2 'I'll be sorry to leave the old bus tonight,' said the plus foured pull-overed youth at the wheel of the 'General' yesterday afternoon. **1934** C. Lambert *Music Ho!* iv. 242 A begoggled, leather-coated and plus-foured figure. **1967** O. Lancaster *Eye to Future* iii. 64 The hearties, grey flannel-trousered or elaborately plus-foured. **1978** *Daily Tel.* 21 Jan. 9/1 On Saturday morning he stands tweedily plus-foured beside a gate, serving as honorary secretary of the beagles.

plus fours (plɒs fōᵊɹz). [f. Plus 3 + Four *a.* and *sb.*, since, to produce the overhang, four inches is normally added to the length required for ordinary knickerbockers.] A distinctive style of long, wide knickerbockers, or a suit having such knickerbockers, originally much worn by golfers and associated with outdoor pursuits. Also *transf.*, and *attrib.* (Also in form *plus-four*).

1920 *Isis* 25 Feb. 6/2 (*caption*) 'Plus Fours'. *Ibid.* 12 May 10/2 The desuetude of the traditional grey flannel 'bags' of the undergraduate... 'Plus fours' have succeeded them. **1921** *Isis* 1 June p. xii, Knicker (plus four) Suits from 8½ gns. **1922** J. Cannan *Misty Valley* 201 A tall man in plus fours and a yellow waistcoat. **1923** S. Herd *My Golfing Life* 151 The first time I saw a golfer wearing baggy 'plus 4's' I thought he looked like a lassie. **1928** *Sat. Evening Post* 10 Mar. 174/3 You can almost visualize the venerable Francis Joseph tweaking away at his plus-four whiskers. **1929** H. A. Vachell *Virgin* ix. 154 The Major got himself up 'to kill', wearing a new suit of 'plus fours'. **1934** G. B. Shaw *Village Wooing* ii. 119 He is in hiking costume..but wears well cut breeches (not plus fours) instead of shorts. **1939** Joyce *Finnegans Wake* (1964) i. 30 In topee, surcingle, solascarf and plaid, plus fours, puttees and bulldog boots. **1951** N. M. Gunn *Well at World's End* xix. 152 He wore a plusfour suiting of discreet checks and unpronounced bagginess. **1961** C. Willock *Death in Covert* i. 13 He wore knicker-bocker trousers that..were nearly plus-fours: say, plus-threes. **1972** *Country Life* 12 Oct. 926/2 Rust battle-jacket with plus-fours in Shetland tweed. By Christian Dior Monsieur. **1974** 'P. B. Yuill' *Bornless Keeper* ii. 19 Wood-pigeons strutted like fat old squires in plus fours. **1978** G. Sims *Rex Mundi* xxvi. 156 He was dressed in a brown plus-fours suit.

plush, *sb.* Add: **1. c.** Colloq. phr. *on* (or *in*) (*the*) *plush*: in comfortable circumstances.

1930 Wodehouse *Very Good, Jeeves* ix. 226 He was, to all appearances, absolutely on plush. He ate well, slept well, was happily married. **1945** *Richmond* (Va.) *Times-Dispatch* 23 Mar. 20/3 Morgenthau..made it plain that the drive would be aimed chiefly against 'the boys who live in the plush' rather than the small taxpayer.

3. a. Also, covered or upholstered in plush.

1895 *Montgomery Ward Catal.* 113/1 Silk plush work box with silvered center and corner ornaments. **1926** E. O'Neill *Great God Brown* i. iii. 89 At the left is a bald-spotted crimson plush chair. **1935** R. Macaulay *Personal Pleasures* 135 Pointing us to seats in the middle of an eagerly gazing row of persons, past whom we push, to subside into plush chairs and eagerly gaze too. **1935** *Punch* 4 Sept. 256/2 And, what is more, though you were seeing it in comfort from a plush seat, it was nearly as good as if you had been energetic and enthusiastic enough to go to Victoria. **1976–7** *Hamley's Catal.* 66/3 A fully jointed soft plush teddy.

b. *plush-bottomed, -bound, -capped, -clad, -fitted, -framed, -lined* adjs.; **plush horse**, used as a symbol of ostentation and over-elaborateness; also *attrib.*

1901 Conrad & Hueffer *Inheritors* xiii. 210, I sat on a plush-bottomed gilded chair. **1902** Conrad *Typhoon* xxiii. 185 She reclined in a plush-bottomed and gilt hammock-chair near a tiled fireplace. **1912** C. Mackenzie *Carnival* xxxiv. 347 These..were now propped dismally against the overmantel, individually obscured..by a plush-bound photograph of Mr. Lloyd George. **1876** G. M. Hopkins *Wreck of Deutschland* viii, in *Poems* (1967) 54 How a lush-kept plush-capped sloe Will, mouthed to flesh-burst, Gush! **1913** C. Mackenzie *Sinister Street* I. ii. vii. 259 He saw..plush-clad children who continually dropped Sunday-school books in the mud. **1927** A. Conan Doyle *Case Bk. Sherlock Holmes* i. 38 A butler..handed me over to a plush-clad footman, who ushered me into the Baron's presence. **1938** L. MacNeice *Zoo* 227 Plush-fitted theatres. **1937** E. Sitwell *I live under Black Sun* 98 A plush-framed photograph. **1922** S. Lewis *Babbitt* ii. 23 Just compare a real human like you and these neurotic birds like Lucile McKelvey—all high-brow talk and dressed up like a plush horse. **1936** J. Dos Passos *Big Money* 201 What I'd love more than anything in the world would be to get out and make my own living. I hate this plushhorse existence. **1946** P. Larkin *Jill* 184 The competition is very keen, because they're very, very plush-lined jobs.

B. *adj.* Luxurious, expensive, stylish. *colloq.*

1927 in Wentworth & Flexner *Dict. Amer. Slang* (1960) 398/2 'Plush' indicates..stylish. **1934** M. H. Weseen *Dict. Amer. Slang* 192 *Plush*, stylish. **1944** 'P. Quentin' *Puzzle for Puppets* xix. 141 We had the plushest hotel suite of any married couple in San Francisco. **1946** *Richmond* (Va.) *Times-Dispatch* 30 Oct. 1/2 The recently established plush eating place. **1952** W. R. Burnett *Vanity Row* (1953) v. 48 'Apartment address?' asked Roy. Joe gave it to him. 'H'm' said Roy. '..Pretty plush.' **1955** *Bull. Atomic Sci.* Feb. 52/2 These need not be plush or elaborate shelters. **1959** *Sunday Express* 1 Feb. 19/6 The word 'Set' in this context is an economic symbol—a plush version of what humbler people call 'The Gang'. **1959** H. Hobson *Mission House Murder* iii. 8 The sales-minded debutante who was earning pin-money in the plush establishment. **1977** *Guardian* 28 May 8/2 It was really plush, with its 3 swimming pools. **1978** *New York* 3 Apr. 40/1 Plush place with excellent service, a great chef, and strolling guitarist-singer.

plushery (plɒ·ʃəri). *U.S. Show-business slang.* [f. Plush *sb.* + -ery.] (See quots.)

1951 Green & Laurie *Show Biz* 579/2 *Plushery*, class joint (hotel, nitery, eatery). **1961** A. Berkman *Singers' Gloss. Show Business Jargon* 69 *Plushery*,..a classy restaurant or night club.

plushly (plɒ·ʃli), *adv.* [f. *Plush *a.* + -ly².] Richly, sumptuously, elegantly.

1964 W. Sansom *Face of Innocence* iii. 49 All the attitudes—stretching and crossing her long legs,..finding a green background to burn plushly behind her white profile—she was assured. **1969** *Observer* 10 Aug. 11/7 There it [sc. a countryside] all is, laid out so plushly for sale after sale. **1976** *Daily Tel.* (Colour Suppl.) 7 May 50/3 A lush settee and two armchairs, all plushly covered in fine green velvet.

plushy, *a.* Add: **c.** Luxurious, sumptuous, elegant. *colloq.*

1923 A. Huxley *Antic Hay* i. 14 Over the plushy floors of some vast and ignoble Ritz slowly he walked, at ease, with confidence. **1942** *Sat. Even. Post* (U.S.) 5 Dec. 12/3 Graff's plushy offices would lead the innocent visitor to conclude that successful publishing requires only some good titles and the organization to distribute them wholesale. **1946** R. Chandler *High Window* (1943) iv. 34 Runs a plushy night-club and gambling joint. **1947** *Yale Law Jrnl.* Dec. 188 Driving Fords rather than more plushy 'petit bourgeois' cars. **1953** W. Stevens *Let.* 4 June (1967) 779 The picture of her in her prime was plushy, but cold. **1961** 'Plushy, I must admit,' said Simon, as Paul entered his fantastically opulent-looking office. **1966** *Times* 19 Feb. 13/3 Much tongue wagging here [sc. Paris] from the plushier night haunts down to the homeliest bistro. **1979** 'D. Halliday' *Dolly & Cookie Bird* i. 7 The lawyer-trustee..took me to a nice plushy lunch at the Café Royal. **1977** *Sunday Tel.* 11 Sept. 19/7 In material quarters there are plenty of plushy berths to be filled.

Hence **plu·shily** *adv.*, **plu·shiness**.

1916 W. J. Locke *Wonderful Year* xiii. 183 The primly and plushily furnished salon. **1969** *Amateur Photographer* 21 May 53/1 Intimate exclusiveness, though not plushiness,..add to a feature of the club cinema.

plussage, var. *Plusage.

plus twos (plɒs tūz). [After *plus fours.] A narrower version of plus fours. Also (in form *plus two*) *attrib.*

1927 *Daily Tel.* 24 May 26/3 (*caption*) The high-waisted winter weather coat has a Russian influence, and the country outfit a single-breasted jacket and tartan plus-twos. **1970** P. Dickinson *Seals* vii. 136 Father in his plus-twos. **1977** *Field* 13 Jan. 68/4 (Advt.), Plus Two breeches.

plute (plūt). *slang* (chiefly *U.S.*). Abbrev. of Plutocrat. Cf. *ploot. Hence **plu·tish** *a.*, plutocratic.

1906 S. Ford *Side-Stepping with Shorty* i. 13 Then we bumps up against a real fly young plute, and gets a squint at his dinner check. **1910** Wodehouse *Gentleman of Leisure* i. He's got much more money than any man, except a professional plute, has any right to. **1922** S. Lewis *Babbitt* ii. 23 Our little bunch has a lot liver times than all those plutes. **1923** *Daily Mail* 29 May 8 'The plutes', as he [sc. Henry Ford] humorously nicknames the financial and industrial interests of the country, would never permit his nomination. **1926** 'Mixer' *Transport Workers' Song Bk.* 80 This plutish press condemns our class. **1932** *Randolph Enterprise* (Elkins, W. Va.) 7 Jan. 6/2 You'll hear the plutes cry out, 'The Yellow Peril' is Japan.

pluterperfect (plūːtəɹpəˑɹfekt), *a. nonce-wd.* [Poss. a corruption of Fr. *plus-que-parfait*: see Pluperfect *a.* (*sb.*).] = Pluperfect *a.* 2.

1922 Joyce *Ulysses* 347 The pluterperfect imperturbability of the department of agriculture. *Ibid.* 412 An omnivorous being which can..pass through the ordinary channel with pluterperfect imperturbability such multifarious aliments as cancerous females emaciated by parturition.

Pluto[1] (plū·to). [a. L. *Plūto*, Gr. *Πλούτων*, name of the god of the underworld, brother of Jupiter and Neptune.] **1.** A small planet of the solar system lying beyond the orbit of Neptune, discovered only in 1930 by C. W. Tombaugh.

1930 *N.Y. Times* 26 Mar. 10/3 'Pluto' is the provisional name that Italian astronomers have given to the new trans-Neptune planet discovered March 13 at Lowell Observatory in Flagstaff, Ariz... Hugh Rice..yesterday opposed naming the newly discovered planet 'Minerva', as has already been suggested, because an asteroid already bears that name. **1930** V. M. Slipher *Lowell Observatory Observation Circular* 1 May, Many names have been suggested and among them Minerva and Pluto have been very popular... Pluto seems very appropriate and we are proposing to the American Astronomical Society and to the Royal Astronomical Society, that this name be given to it. As far as we know Pluto was first suggested by Miss Venetia Burney, aged 11, of Oxford, England. [*Note*] Kindly cabled by Prof. H. H. Turner. **1930** H. H. Turner in *Times* 28 May 17/5 A post-card from the President of the Royal Astronomical Society offered this morning 'congratulations to the suggester of the name Pluto, now adopted'. The reference is to a telegram which I had the honour of sending to the Lowell Observatory on March 15..conveying the suggestion of Miss Venetia Burney, of Oxford, made at breakfast on that day to her grandfather, who sent it on to me... It was a brother of that same grandfather who suggested the names Deimos and Phobos for the satellites of Mars. **1933** D. L. Sayers *Hangman's Holiday* 120 Presently they had guessed among other things Miss Tomkins's mother's photograph,..the new planet Pluto,..and had failed to guess the Prime Minister's wireless speech. **1936** Skilling & Richardson *Astron.* i. 12 The length of seasons depends, of course, upon the length of time required for the planet to go around the sun—all the way from about 3 months for Mercury to 248 years for Pluto. **1953** A. F. O'D. Alexander in M. Davidson *Astron. for Everyman* iv. 203 Pluto's path ever crosses Neptune's..for its orbit-plane is inclined 17° to the ecliptic plane. **1972** P. L. Brown *Astron. in Colour* ii. 86 One theory accounts for the origin of Pluto by considering that it is an ex-satellite of Neptune which somehow broke free. **1976** J. Gribbin *Astron. for Amateur* v. 70 Because of its elliptical orbit Pluto sometimes passes within the orbit of Neptune, and just now (until the end of this century) Neptune is the furthest planet from the Sun.

2. The name of a cartoon dog that made its first appearance in Walt Disney's *Moose Hunt*, released in April 1931; a toy dog representing this character.

1932 W. Disney *Adventures Mickey Mouse* II. 14 Mickey watched, and Minnie too, and Pluto, who was very interested. **1936** —— *Mickey Mouse Fables* 3 'Now for my bone,' said Pluto, licking his lips. **1957** D. M. Miller *Walt Disney* viii. 137 Pluto got his name when Walt grew tired of calling him Rover. **1962** L. Deighton *Ipcress File* ii. 22 A cigarette-girl..offered me a pink felt pluto, too. **1972** *Guardian* 21 Aug. 11/5 Ten dogs took part, plus one yellow plastic Pluto on wheels. **1976** L. Maltin *Disney Films* iv. 267 As the 1930s wore on..Mickey played a progressively less important role in the proceedings, and supporting players took over by the character of Pluto.

Plu·to[2]. [Acronym f. initial letters of *Pipe Line under the Ocean*.] The code name for a system of pipe-lines laid in 1944 to carry petrol supplies from Britain to Allied forces in France. Also *attrib.*

1945 *News Chron.* 1 June 2/5 As in the Pluto pipeline, part of the basis of this secret weapon. Also Pluto the experts said at first that it could not be done. **1946** *Ann. Reg. 1945* 37 Foremost among these was the laying of a system of oil pipe-lines under the sea stretching across the Channel from the British coast to the Continent, and commonly known as 'Operation Pluto' (pipe line under the ocean). **1961** E. W. Gladstone *Army* vii. 103 Many clever inventions like the artificial 'Mulberry' harbours, and 'Pluto', the Pipe Line Under the Ocean to provide petrol, had been developed. **1978** A. Waugh *Best Wine Last* xv. 177 The Petroleum Warfare Department..invented 'PLUTO' (pipe line under the ocean) to carry oil across the Channel.

plutocratical (plūtokræ·tikăl), *a.* [f. Plutocratic *a.* + -al.] = Plutocratic *a.* So **plutocra·tically** *adv.*, in a plutocratical manner.

1893 J. S. Mill *Let.* 10 July in *Wks.* (1963) XII. 166 The anomaly of a democratic constitution in a plutocratical society. **1972** W. J. Locke *Stella Maris* xx. 238 Her grandmother..plutocratically owned vehicles. **1941** E. R. Eddison *Fish Dinner* xix. 334 An aristocratical plutocratical self-obtruding dilettante. **1978** *Country Life* 31 Aug. 589/4 In their heyday they certainly seem, plutocratically speaking, to have been a yesteryear of glory.

plutocratizing (plūtǫ·krătəiziŋ), *vbl. sb.* [f. Plutocrat + -ize + -ing¹.] The action or process of rendering plutocratic.

1896 *Westm. Gaz.* 9 Dec. 2/3 The 'plutocratising' of the Universities was a nineteenth century movement. **1929** *New Statesman* 28 Sept. 727/1 The Americanisation and plutocratizing of Europe.

pluto-democracy (plūːto,dimǫ·krăsi). Also **plutodemocracy**. [f. Gr. *πλοῦτος* wealth, riches + Democracy.] **1.** Plutocratic government which masquerades as democracy.

1895 in *Funk's Stand. Dict.* **1941** D. Wilson *Germany's 'New Order'* 24 German gibes at pluto-democracy. **1944** J. S. Huxley *On Living in Revol.* 41 Hitler has proclaimed his aims... They include..the destruction of what he is pleased to call 'pluto-democracy'. **1970** H. Arendt *On Violence* 72 A new system, which he [sc. Pareto] called 'Pluto-democracy'—a mixed form of government, plutocracy being the bourgeois regime and democracy the regime of the workers.

2. A country or state which purports to be a democracy but where power lies with the rich.

1902 *Nineteenth Century & After* July 119 If England be allowed to become a Plutodemocracy, then she has the tragic example of Venice to chasten and admonish her. **1940** *Common Sense* Aug. 6/2 The absolute power of a Stalin or a Hitler is seen to be even worse than the economic exploitation and political ineffectiveness of a 'pluto-democracy' with its checks and balances. **1942** L. B. Namier *Conflicts* 209 In the presence of social inequalities, parliamentary democracy without parties must inevitably result in a real 'pluto-democracy'. **1948** W. S. Churchill *Second World War* I. xx. 287 The Jew Litvinov was gone, and Hitler's dominant prejudice placated. From that moment the German Government ceased to define its foreign policy as anti-Bolshevism, and turned its abuse upon the 'pluto-democracies'.

Hence **plu:to-de·mocrat**, an adherent or advocate of pluto-democracy; **plu:to-democra·tic** *a.*

1940 *Economist* 24 Feb. 328/2 German propagandists are making steady use of..'Britain the pluto-democrat, living in luxury on the wealth extorted from 80 million native serfs'. *Ibid.* 29 June 1101/1 German propaganda among the French troops has followed the lines of 'plutodemocratic capitalist Britain'. **1946** R. Graves *Poems 1938–1945* 35 By every nastier manifestation Of plutodemocratic civilisation.

plutogogue (plū·tŏgǫg). [f. Gr. *πλοῦτος* wealth + *ἀγωγός* leading, leader, after *demagogue*.] A spokesman for the plutocrats; one who justifies or advocates the interests of the wealthy. Hence **plu·togogy**, the rule of plutogogues.

1894 *Westm. Gaz.* 17 Feb. 8/1 Mr. Williams, of Mississippi, said the opponents of the income tax are plutogogues and encouragers of plutogogy. *Ibid.*, A demagogue, he explained, appeals to the people; while the plutogogue appeals to the people who buy the mob. **1931** L. Steffens *Autobiogr.* II. iii. xvi. 474 Since the plutogogues could not fasten any crime on him they fell back on the all-sufficient charge that he was a demagogue. **1937** *Sun* (Baltimore) 30 Aug. 9/4 Smith called the demagogue, plutogogue and the theogogue, 'fearful trinity, constituting the very apotheosis of hypocrisy'. **1977** *New Yorker* 13 Sept. 29/5 Fascists and Communists did not strike him as any more dangerous than the plutogogue... The philosophy professor defined him as 'the voice of the wealthy when they can no longer speak for themselves, the successor of the plutocrat of other days'.

plutolater (plutǫ·lătəɹ). *rare.* [f. Plutolatry, after *idolater*.] One who worships wealth.

1938 S. Beckett *Murphy* ix. 168 All the self-made plutolaters who ever triumphed over empty pockets.

plutological (plūtǫlǫ·dʒikăl), *a. rare.* [f. Plutology + -ic + -al.] Of or pertaining to plutology.

1920 *Edin. Rev.* July 80 The whole plutological apparatus was developed—banking, investment, partnership, joint stock companies and even trusts. *Ibid.* 83 The economic difference between ancient, medieval and modern society is in scale, not in principle, and not in the plutological principle or the essential character of the process.

plutomanic (plūtomæ·nik), *a. rare.* [f. as Plutomania + *Manic *a.*] Characterized by the insane love or pursuit of wealth.

1938 S. Beckett *Murphy* v. 82 A colossal league of plutomanic caterers. **1963** B. S. Johnson *Travelling People* 159 His decision to enter the plutomanic world of commerce.

pluton (plū·tǫn). *Geol.* [a. G. *pluton* (H. Cloos 1928, in *Fennia* L. ii. 1), back-formation from *plutonisch* Plutonic *a.*] An intrusive body of igneous rock formed beneath the earth's surface, esp. a large one.

1936 R. A. Daly *Igneous Rocks & Depths of Earth* vi. 75 He [sc. H. Cloos] states that he does not assume floors for batholiths but on the next page defines his 'Plutone', which includes batholiths, as having visible or inferable floors. **1959** *Proc. Geol. Soc. Amer.* 1957 67 The granite appears to have come up along the edge of the mostly consolidated Pikes Peak mass and spread in a series of plutons and sills, the intrusion being largely governed by the rock invaded. **1964** M. P. Billings *Structural Geol.* xv. 296 The intrusion of large plutons may be associated with orogenic movements. **1962** F. H. Lahee *Field Geol.* (ed. 6) vi. 241 Concordant plutons include sills, laccoliths, lopoliths, and phacoliths. *Ibid.* 243 Discordant plutons include dikes, necks, chonoliths, batholiths, stocks, and bosses. **1965** [see *Greisen]. **1977** I. G. Gass et al. *Understanding Earth* x. 84/2 The observed linear relation between heat flow and heat production in granitic plutons may give direct evidence of temperatures as far down as the base of the crust. **1974** R. F. Garner *Origin of Landscapes* iii. 130/2 Metasomatic replacement is common along borders of larger intrusive masses (plutons).

Plutonism. Add: Also **plutonism**. **1.** (Examples.)

1910 *Encycl. Brit.* XI. 644/2 Wernerianism..rapidly declined in influence, while Plutonism came steadily to the front, where it has ever since remained. **1966** [see *Neptunism].

2. Geological activity associated with the formation of plutonic rocks.

1942 J. L. van Houten tr. *Umbgrove's Pulse of Earth* iv. 66 A local thickening of the sialic crust, resulting in the formation of a 'root' of sialic material... When this root's material begins to melt and invade the surface strata as batholithic intrusions, plutonism and volcanism cannot but be strongly acid ('pacific') in character. **1949** *Q. Jrnl. Geol. Soc.* CV. 104 Plutonism does not appear to be a steady process advancing at a uniform rate, but rather one of energetic pulses interrupted by operations

of diverse kinds, especially those of deformation. **1970** *Nature* 24 Jan. 315/1 After the late Mesozoic plutonism, the next phase of igneous activity in western America began in the Oligocene.

plutonium. Delete ‖ and add: **2.** *Chem.* † **a.** = Barium. *Obs.*

The word was introduced in a letter to the Royal Institution dated 1 Sept. 1816, but not published until 1817 (= quot. 1817 below). Its first occurrence in print was in *Ann. de Chim. et de Physique* (Sept. 1816) III. 61 in a Fr. transl. of it. Quot. 1816 is from a letter of 5 Oct. 1816 that refers to this (then unpublished) letter.

1816 E. D. Clarke in *Ann. Philos.* VIII. 358 The metal is in its purest state. It is the same metal for which I proposed the appellation of Plutonium; and I am glad to find the appellation generally approved. *Ibid.* 360 Plutonium, if fused in contact with platinum, always tarnishes the latter. **1817** —— in *Jrnl. Sci. & Arts* II. 120 The existence..of the metal of Barytes no longer admits of the smallest doubt... As any derivative from *Baρυς* would involve an error, if applied to a metal whose specific gravity is inferior to that of Manganese or Molybdenum, I have ventured to propose for it the appellation of *Plutonium*; because we owe it entirely to the dominion of fire.

b. A transuranic metallic element which is formed indirectly from uranium in nuclear reactors and occurs naturally in trace amounts, is chemically similar to uranium, and is very reactive; the longest-lived isotope (plutonium 244) has a half-life of 83 million years, and plutonium 239 is fissile and is produced for use in nuclear weapons and as fuel. Atomic number 94; symbol Pu. [So called from its being the next element after neptunium in the periodic table, as Pluto is the planet next beyond Neptune.]

1942 Seaborg & Wahl *Chem. Properties Elem. 94 & 93* (U.S. Office Sci. Res. & Devel. Report A-135) 17 Since such formulae are confusing when the symbols '93' and '94' are used, we have decided to use symbols of the conventional chemical type to designate these elements. Following McMillan, who has suggested the name neptunium..for element 93, we are using plutonium..for element 94. The corresponding chemical symbols would be Np and Pu. **1952** J. G. Feinberg *Atom Story* xx. 153 On 20th December 1943 the first batch of irradiated slugs were removed from the pile for plutonium extraction. In another month the pile was turning out about one-third ton of plutonium-enriched slugs a day. **1962** J. C. Wright *Metallurgy in Nucl. Power Technol.* iii. 60 The concentration of plutonium in minerals such as pitchblende is no more than one part in 10^{11}. **1964** R. L. Loftness *Nucl. Power Plants* ii. 63 The maximum permissible body burden for plutonium is 0·3 microgram, and such a low limit requires the use of very efficient enclosures, or glove-boxes, for all research, development, and production work. **1968** [see *neptunium]. **1969** *Times* 22 Apr. 6/6 Scientists..claimed to have synthesized element 104..by bombarding plutonium with neon. **1970** *Daily Tel.* 23 July 3/2 The nuclear-powered pacemaker—shaped like a small bullet and powered with plutonium 238—has a life of about 10 years. **1970** tr. *Vol'skii & Sterlin's Metallurgy of Plutonium* i. 19 Pure plutonium cannot be used as a nuclear fuel because of its low melting point, poor mechanical properties, high chemical activity, and the considerable volume changes caused by phase transformations... For reactors plutonium alloys and plutonium compounds are used. **1976** *Sci. Amer.* Dec. 30/2 In addition to uranium 235 the spent fuel contains between ·7 and 1 per cent of plutonium 239, synthesized from uranium 238 by the absorption of a neutron. **1979** *Ibid.* Apr. 33/2 A Precambrian mineral, a rare-earth fluocarbonate mined for cerium in California, has yielded the all but extinct isotope of plutonium with mass 244.

3. *attrib.* and *Comb.*, as (sense *2 b) *plutonium economy, enrichment*; **plutonium bomb**, an atomic bomb in which the fissile material is plutonium.

1948 C. Pincher *Into Atomic Age* 53 Three days after the attack on Hiroshima..the plutonium bomb was dropped on Nagasaki. *Ibid.* 54 The 'Mark II' plutonium bomb was a definite improvement. It would have devastated the flat, round target of Hiroshima more completely than did the now obsolete uranium 'Mark I'. **1973** *Guardian* 19 Apr. 14/3 To make a plutonium bomb, with which a criminal could hold a city to ransom..one needs just 16·2 kilos of the metal. **1976** *New Yorker* 9 Feb. 44/3 Such a 'plutonium economy', in which fissionable fuel is recycled between reactor and reprocessing plant, is essential to the operation of the breeder. **1977** *Jrnl. R. Soc. Arts* CXXVI. 15/2 The distinguished Royal Commission under Sir Brian Flowers has expressed disquiet about a 'plutonium economy', but that is not to condemn all nuclear development. **1977** *Time* 7 Mar. 7/1 West German officials are further incensed over Carter's public criticism of Bonn's deal to sell a plutonium-enrichment plant and eight nuclear reactors to Brazil.

pluto:nometamo·rphism. *Petrol.* [f. Pluton(ism + -o + Metamorphism.] Metamorphism that occurs at high temperatures and high pressures at great depths under the earth.

1889 A. Harker in *Geol. Mag.* Decade III. VI. 17 We may accordingly use the term plutono-metamorphism to describe the profound changes in rocks implied in the joint influence of very elevated temperature and enormous pressure. **1950** F. H. Hatch et al. *Petrol. Ign. Rocks* (ed. 10) vii. 336 The chief problem of the eclogites concerns their origin. They are believed to be products of 'plutono-metamorphism'—to lie on the borderland between the igneous and metamorphic.

plutonyl (plū·tonəil). *Chem.* [f. Pluton(ium + -yl.] The ion PuO_2^{2+}. Usu. *attrib.*

1947 B. G. Harvey et al. in *Jrnl. Chem. Soc.* 1013 On account of the similarity between the chemistry of uranium and of plutonium in their sexavalent states, the terms 'plutonate' and 'plutonyl' have been used in the present paper to describe the sexavalent compounds of plutonium. *Ibid.*, An orange solution, which..is believed to contain plutonyl nitrate, $PuO_2(NO_3)_2$. **1961** G. R. Choppin *Exper. Nuclear Chem.* xii. 192 The extent to which the internal irradiation by alpha particles causes reduction of plutonium ions to the (IV) state in an aqueous solution of PuO_2^{++} is necessary knowledge for the inorganic chemist studying plutonyl ion chemistry. **1976** *Sci. Amer.* Dec. 33/1 In the solvents that were used hexavalent uranyl ions, $(UO_2)^{++}$, and plutonyl ions, $(PuO_2)^{++}$, together with tetravalent plutonium ions, Pu^{4+}.., are soluble.

pluvial, *a.* Add: *spec.* designating periods of relatively high average rainfall in low and intermediate latitudes during the geological past (esp. the Pleistocene) which alternated with interpluvial periods in a cycle which may be correlated with or related to the better-known cycle of glacial and interglacial periods in higher latitudes. Cf. *interpluvial, *intrapluvial *adjs.* and *sbs.*

1868 A. Tylor in *Q. Jrnl. Geol. Soc.* XXIV. 105 Many of the Quaternary deposits in all countries..are of such great dimensions and elevation that they must have been formed under physical conditions very different from our own. They indicate a Pluvial period, just as clearly as the northern drift indicates a Glacial period. **1927** Peake & Fleure *Apes & Men* v. 75 He endeavoured to show that..in the valley of the Nile, there was evidence of four very wet periods, or pluvial periods. **1949** W. F. Albright *Archaeol. Palestine* 50 The cold phases are called 'glacial periods' in northern latitudes and 'pluvial periods' in the latitude of Palestine, where there was no glaciation, but instead a greatly increased rainfall. **1954** *New Biol.* XVII. 11 There is evidence of at least two great 'Pluvial' periods of heavy rainfall in the Pleistocene when these [East African] lakes reached their maximum size and depth. **1979** *Nature* 1 Mar. 80/1 The student interested in the Pleistocene will find many discredited ideas (for example, glacial = pluvial).

B. *sb.*[2] A pluvial period.

1929 *Nature* 6 July 9/2 A large mammalian fauna has been collected from the deposits of the various Pluvials. **1931** L. S. B. Leakey *Stone Age Cultures of Kenya Colony* ii. 13 The prehistoric tribes..moved down to the Rift Valley areas during the pluvials. **1959** *Bull. Geol. Soc. Amer.* LXX. 345/1 African pluvials are of great importance for climatic history, for Pleistocene correlation, and for meteorologic theory. **1970** Bray & Trump *Dict. Archaeol.* 184/1 Prolonged periods of high rainfall are called pluvials, and are marked by changes in lake levels and in flora and fauna.

pluvialine, *a.* (Example.)

1872 E. Coues *Key to N. Amer. Birds* 239 The pluvialine and scolopacine birds form the bulk of the division.

pluviometric, *a.* Add: (Further examples.) *pluviometric coefficient* (see quot. 1917).

1917 A. McAdie *Princ. Aërography* xv. 218 The term 'pluviometric coefficient' was introduced by Angot to indicate the ratio of the mean daily rainfall of a particular month to the mean daily rainfall of the whole year. **1944** S. Putnam tr. *E. da Cunha's Rebellion in Backlands* i. 46 Torrential rains..giving rise to higher pluviometric readings than in..fertile lands of plenty. **1959** R. E. Huschke *Gloss. Meteorol.* 428 Seen collectively, the twelve pluviometric coefficients describe the normal month-to-month distribution of the normal annual precipitation in terms of each month's 'share' of the annual amount.

Pluvius (plū·viɒs). Also with lower-case initial. [L. *pluvius* rainy, causing or bringing rain.] Used *attrib.* with ref. to the insurance of holidays, outdoor sports, entertainments, etc., against disruption by bad weather, as *Pluvius department, insurance, policy*.

1911 *Policy* 1 Apr. 206/2 The policies will be issued in respect of weather at the resorts..for any period from May 1st to September 30th. There will be four policies, known as the Pluvius Policies A, B, C, and D. **1920** *Post Mag. & Insurance Monitor* 4 Sept. 845/1 Now that Pluvius policies have come to stay, insurance men have of necessity to take an interest in the weather. **1949** *Policy Insurance Weekly* 2 June 373 In 1911 the brokers started Pluvius insurance. **1955** *Times* 13 May 18/2 Progress has been made in the Pluvius department. Despite extremely bad weather conditions over the British Isles and Europe, the experience was satisfactory. **1958** *Times* 3 July 15/4 The most direct effects of the recent rain will be on those firms who specialize in insuring against this sort of weather risk—so-called pluvius insurance. **1960** *News Chron.* 20 July 2/1 The Pluvius Department of the Eagle Star Insurance reports that business is 25% up on last year, and claims 50% up.

ply, *sb.* Add: Pl. **plies**, occas. **plys**. **I. 1.** Also, each of the layers that go to make up a multi-layer material such as plywood or laminated plastic; *single-, two-, three-ply*, etc., also, material (esp. plywood) composed of that number of layers. (Further examples.)

1910 *Timber Trades Jrnl.* 1 Jan. p. v (Advt.), Best Russian improved waterproof 3-ply. **1919** A. W. Judge *Handbk. Mod. Aeronaut.* iv. 235 Ordinary 3-ply (⅛ to ¼ in.) is used for the webs of aeroplane wing ribs. **1926** *Rep. & Memoranda Aeronaut. Res. Comm.* No. 1017. 7 The inner (gas) ply was separated for a few inches from a 2-in. strip of a three-ply rubbered balloon fabric and the single-ply gripped in the lower jaw. **1935** Dawson & Porritt *Rubber* 330/2 Sheets were calendered to produce the maximum grain in unvulcanised state and slabs built up..from plies with grain in same direction. **1936** H. W. Rowell *Technol. of Plastics* xxvii. 191 In order that all air may be expelled and the plys or laminations thoroughly consolidated, a slow curing resinoid with a long plastic stage is used. **1941** *Paper Trade Jrnl.* CXII. 33/2 In contrast to the tests in which the bonding between plies is measured normal to the sheet, there are tests that measure the force required to tear plies apart. **1944** *Chem. Abstr.* XXXVIII. 3875 (*heading*) Preparing and handling cord fabric plies in tire core manufacture. **1952** J. P. Casey *Pulp & Paper* II. xvi. 803 In laminating paper plys, it is sometimes desirable to place alternate layers of paper with their cross and machine directions at right angles. **1957** *Practical Wireless* XXXIII. 521/2 Five pieces of three-ply were used in the original. **1960** *Farmer & Stockbreeder* 15 Mar. 11/1, 1cwt 6-ply paper bags. **1970** [see *cross-ply* adj. s.v. *cross- B.]. **1975** [see *plywood]. **1977** *New Scientist* 6 Jan. 22/1 If the plies are set wholly radially..the tread squirms on the road far too much for acceptable roadholding. If..the plies are circumferential..the tyre..gives a harsh and unyielding ride.

b. = *plywood.

1929 A. Clarke *Pilgrimage* 31, I had a painted bedpost Of blue and yellow ply. **1957** *Practical Wireless* XXXIII. 542/1 The front panel should be of ⅛ in. ply or hardboard. **1974** T. R. Dennis in J. Burnett *Useful Toil* iii. 348 Drinking-water was kept in a bucket..with a piece of ply over to keep dust out. **1980** *Daily Tel.* 10 Mar. 18 Roughly half will come in the form of timber products: ..windows, flooring blocks, ply, hardboard and newsprint.

III. 5. Special Comb.: **ply rating**, a number indicative of the strength of a tyre casing (orig. the number of cord plies in it).

1952 E. C. Woods *Pneumatic Tyre Design* i. 5 Tyre markings now carry an indication of casing strength in the form of 'ply rating' (e.g., '8 ply rating') to serve the purpose of identifying a given tyre with its maximum recommended load. **1956** R. H. Spelman in McPherson & Klemin *Engin. Uses of Rubber* x. 299 Many standard over-the-road highway tires may be marked 10 ply rating and actually be made of 8 plies of rayon, or even of fewer plies of a stronger material, such as nylon. **1969** *Times* 12 May 16/1 Tractor tyre research is being carried out with a new mobile test rig designed to study the effects of size, ply rating, etc.

Plyglass (pləi·glɑs). Also **plyglass, ply-glass, ply glass**. A proprietary name for units consisting of two or more panes of glass enclosing one or more hermetically sealed spaces, which may contain dry air or be filled with a translucent material like glass fibre. Also *attrib.*

[**1949** *Trade Marks Jrnl.* 2 Mar. 191/1 *Plygloss* [sic]... Glassware.., plate glass, sheet glass, glass rods, glass tubes.., glass wool and spun glass. Plyglass Limited,.. Leyton, London,..manufacturers.] **1956** *Archit. Rev.* CXIX. 46 Normal windows are of galvanized steel; panels between windows, of cantilevered section, are plyglass. *Ibid.* CXX. 105 (*caption*) The panels between the first- and second-storey windows are of dark-green plyglass. **1958** *Engineering* 28 Feb. 276/1 Dark-blue ply glass panels are used beneath the windows on the north and south elevations. **1958** *Specification* (ed. 59) 1123/2 Uses of Plyglass diffusers are in roof lights, laylights and over-transom lights. **1960** *Trade Marks Jrnl.* 21 Sept. 1149/1 *Plyglass*... Units consisting of two or more panes of glass enclosing hermetically sealed cavities and being for use in the building industry. Plyglass Limited,.. Harlow, Essex; manufacturers. **1972** *Specification* (ed. 73) II. 441/2 Plyglass thermopane insulating glass (Plyglass Ltd.) consists of two or more panes of clear glass and most types of patterned glass enclosing a layer of dehydrated and filtered air which is sealed in with Bondermetic glass to metal seal.

Plym (plim), *sb.* [Shortened f. *Plymouth.] **1.** *humorous.* An inhabitant of Plymouth.

1913 W. Owen *Let.* 23 Apr. (1967) 185, I should enjoy treating at length of Plymouth and the Plyms, the charm of Devon.

2. *colloq.* A member of the Plymouth Brethren.

1953 'N. Blake' *Dreadful Hollow* ii. 20 Staying at a pub won't commend you to the Plyms. **1977** *Times Lit. Suppl.* 13 May 593/3 The solemn girl who turned him down, back in the 1890s, because of an apparent lack of religious conviction; she was what is known as a Plym.

plymetal (pləi·metăl). [f. Ply *sb.* or *ply(wood + Metal *sb.* (and *a.*).] A construction material consisting of plywood faced on both sides with aluminium.

1927 Knight & Wulpi *Veneers & Plywood* xv. 139 (*heading*) Plymetal. **1936** A. F. Collins *Motor Car Trailers* ii. 103 Where the duck joins the sheet steel bonnets of plymetal a 1" wide metal molding is used to cover the joint. **1940** Brimm & Boggess *Aircraft Maintenance* 17 Plymetal is used for floorboards, baggage compartment linings, etc. **1967** *Jane's Surface Skimmer Systems 1967–68* 63/1 The frame is of ply-metal panel, consisting of 1 in. (25·4 mm) exterior grade AC plywood faced both sides with 0·050 in. (1·3 mm) mill finished 3003-H12 aluminium.

Plymouth (pli·mɒþ). [Name of a city in Devon.] **1.** See Plymouth cloak.

2. See Plymouth Brethren.

3. a. Applied to the first hard-paste porcelain to be made in England, by a method patented in 1768 by W. Cookworthy of Plymouth (subsequently of Bristol: see *Bristol 2 b).

1816 W. Burt *Rev. Mercantile State Plymouth* xvi. 174 The substance serving as a base for the Plymouth porcelain was a granite..composed of a reddish felspar in pieces of a tolerable size, quartz in small grains, and black scaly mica. **1857** J. Marryat *Hist. Pottery & Porcelain* (ed. 2) xii. 289 The Plymouth china has become very scarce. **1869** C. Schreiber *Jrnl.* 11 Sept. (1911) I. 36 There C.S. discovered a coloured group of Venus and Cupid..valuable as being Plymouth. **1873** H. Owen *Two Centuries Ceramic Art in Bristol* vii. 193 The quantity of the Plymouth porcelain preserved to us, compared with the amount of Bristol manufacture in the possession of collectors, is relatively very small. **1946** F. S. Mackenna (*title*) Cookworthy's Plymouth and Bristol porcelain. **1974** *Country Life* 28 Feb. (Suppl.) 43 Set of four Plymouth figures of the Continents, 12½ in. high.

b. Applied to a coarse, brown and yellow earthenware manufactured at Plymouth in the 18th century.

1878 L. Jewitt *Ceramic Art Gt. Brit.* I. x. 338 (*heading*) Plymouth earthenware... The manufacture of china-ware having ceased in Plymouth in 1774 this useful and elegant art was lost to the town. Some years later rough common brown and yellow earthenware was made here. **1960** H. Hayward *Connoisseur's Handbk. Antique Collecting* 222/1 *Plymouth earthenware*, coarse earthenware (brown and yellow) was manufactured here in the 18th cent., but gave way about 1810 to the production of painted or printed cream colour.

4. Applied to a variety of gin orig. made in the west of England.

1864 C. Tovey *Brit. & Foreign Spirits* iii. 118 An imitation Hollands is made by some rectifiers, and meets with a sale in Cornwall and in the West of England. Plymouth Gin is somewhat of the character of Hollands. **1885** *Encycl. Brit.* XIX. 237/2 Plymouth has few manufactures, the principal being biscuits..and the celebrated Plymouth gin. **1920** G. Saintsbury *Notes on Cellar-Book* vii. 104 More recent conditions, when in England and Scotland, excellent brandy cost five shillings a bottle; ..and gin, whether 'squareface' or London or Plymouth, not much more than half a crown. **1951** R. Postgate *Plain Man's Guide to Wine* ix. 121 *Plymouth* gin..has rather more of the distinctive flat juniper taste. **1967** J. B. Priestley *It's Old Country* iv. 43 Large pink Plymouth for me, cobber. **1968** J. D. MacDonald *Girl in Plain Brown Wrapper* iv. 35, I took one of the big glasses and laid an impressive belt of Plymouth atop the cubes.

Plymouth-brethrenism. (Earlier example.)

1874 C. M. Yonge *Life J. C. Patteson* I. v. 161 Primitive Methodism and Plymouth Brethrenism supplied the void.

plywood (pləi·wud). Formerly also **ply-wood, ply wood.** [f. (*three-*, etc.) *ply wood*.] Board made of two or more thin layers of wood bonded together with the grain of adjacent layers crosswise to give increased strength and resistance to warping, there being either an odd number of layers including a thin central one or an even number sandwiching a thicker central core. Freq. *attrib.* or in *Comb.* Cf. *laminated wood* s.v. *laminated *ppl. a.*

1907 *Timber Trades Jrnl.* 13 Apr. 818/2 (Advt.), Agents for Swedish..wood goods..ply wood (oak, birch, alder, etc.). **1919** A. W. Judge *Handbk. Mod. Aeronautics* iv. 234 The ply-woods chiefly employed in aeroplane work may vary from ⅛ to ¼ in. in thickness, and are made of ash or birch with an intermediate layer of poplar..or white-wood. **1922** W. Schlich *Man. Forestry* (ed. 4) I. 228 These pines..are also used for butter boxes, plywood, etc. **1926** *Glasgow Herald* 1 Oct. 5 The prosperity of the plywood industry. **1940** H. R. Simonds *Industr. Plastics* x. 273 Formed or molded plywood is becoming increasingly important for the manufacture of trays, shallow dishes and bowls. **1957** *Times* 12 Nov. (Canada Suppl.) p. xvi/4 A new type of express cabin cruiser..costs less than $4,000, is mostly built of plywood, and can be easily hoisted onto a trailer. **1964** B. Latham *Wood* xiii. 163 In the large modern plywood mills..very little hand labour is employed. Another great plywood-producing area is the Baltic. **1969** *Jane's Freight Containers 1968–69* 523/1 The container is of welded steel frame and self-supporting plywood panel construction. **1975** W. P. K. Findlay *Timber* xii. 174 Thin plywood is generally made from three veneers. Multi-ply boards have an uneven number of plies so that the grain on the face and back always run in the same direction.

pneu (niū). *colloq.* (*a*) Abbrev. of Pneumonia. (*b*) Abbrev. of 'pneumatic cushion'. (*c*) Abbrev. of *pneumatique. Cf. *pneus.

1916 [see *gyppie, gyp(p)o, gyppy]. **1923** A. Huxley *Antic Hay* i. 7 An air cushion, a delicious pneu. **1928** J. Joyce *Let.* 19 Jan. (1966) III. 168, I got your pneu and shall follow your instructions. **1965** G. D. Painter *Proust* II. xvii. 362 Soon Reynaldo would come to write hurried *pneus* to Marcel's friends.

pneumatic, *a.* (*sb.*) Add: **1. a.** (Further examples.)

pneumatic dispatch (earlier example); *pneumatic drill*, a heavy drill for breaking stone by the rapidly repeated blows of a bit driven by compressed air.

1863 *Illustr. London News* 28 Feb. 217/3 (*heading*) Opening of the pneumatic despatch mail service. *Ibid.*, A company was registered in 1859 for the establishment in the metropolis of lines of pneumatic tube for the more speedy and convenient circulation of despatches and parcels. *a***1877** Knight *Dict. Mech.* II. 1753/1 By the pneumatic drill, the Mt. Cenis Tunnel, seven miles in length, was bored through the Alps. *Ibid.*, *Pneumatic hammer*, a hammer in which compressed air is the agent for lifting the helve or the head. **1894** W. Le Queux *Gt. War in Eng. in 1897* xxxi. 253 Small dynamite shells from Mackenzie's pneumatic gun had struck the car of the balloon. **1898** E. Howard *To-morrow* 48 Subways for sewerage and surface drainage,..pneumatic tubes for postal purposes, have come to be regarded as economic if not essential. **1902** *Encycl. Brit.* XXXI. 802/1 Sometimes, when only a small amount of work is to be done, pneumatic tools are brought to heavy pieces of material. *Ibid.* 803/1 The pneumatic jack..is placed below the piece to be lifted, and operates directly. **1911** *Ibid.* XXVII. 40/2 Pneumatic drills are usually worked by little motors having oscillating cylinders, by which the air and exhaust ports are covered and uncovered. **1927** *Jrnl. Amer. Med. Assoc.* 1 Oct. 1151/1 (*heading*) A pneumatic hammer for bone surgery. *Ibid.*, A light, compact, pneumatic hammer whose speed is controlled by a throttle..on the pistol-grip handle. **1930** *Daily Express* 9 Sept. 8/7 The noise of pneumatic drills has..been found to annoy the patients in a London Hospital. **1947** J. C. Rich *Materials & Methods of Sculpture* ix. 257 The vibration of a pneumatic tool, particularly after prolonged use, apparently affects the blood vessels of the hand and arm, frequently impairing the circulation and resulting in 'white fingers' or 'pneumatic hammer disease'. **1962** L. Zelikov tr. *G. Kamenshchikov's Forging Practice* viii. 192 Pneumatic hammers are mainly employed for hammer forging miscellaneous work and for forging in bolster dies. **1973** 'H. Carmichael' *Candles for Dead* xi. 134 Don't be quite so boisterous. Your voice has the effect of a pneumatic drill. **1978** R. Jansson *News Caper* viii. 76 His hands were shaking as if attached to a pneumatic drill.

e. Of a woman: having a well-rounded figure, esp. a large bosom; of or pertaining to a woman having such attributes.

1919 T. S. Eliot *Whispers of Immortality* in *Poems*, Grishkin is nice. . Uncorseted, her friendly bust Gives promise of pneumatic bliss. **1926** F. M. Ford *A Man could stand Up* i. i. 17 She didn't obviously offer—what was it the fellow called it?—promise of pneumatic bliss to the gentlemen. **1932** A. Huxley *Brave New World* vi. 108 'Every one says I'm awfully pneumatic,' said Lenina reflectively, patting her own legs... 'You don't think I'm *too* plump, do you?' **1951** J. C. Fennessy *Sonnet in Bottle* i. v. 25 A pneumatic pink and yellow bathing belle. **1961** A. Wilson *Old Men at Zoo* ii. 91 He looked at her..as though searching beneath her pneumatic form for the disguised contours of some familiar, leaner enemy. **1961** P. Ustinov *Loser* ii. 46 He became aware of the pneumatic warmth of that thigh. **1974** *Publishers Weekly* 21 Jan. 88/3 Sexologist Dr. Rhona Mitchell, she of the spectacularly pneumatic proportions. **1976** *Times Lit. Suppl.* 31 Dec. 1643/2 The pneumatic barmaid at their favourite wine-bar.

5. **pneumatic-drilled** *a.* (*nonce-wd.*), resembling a pneumatic drill or its action; † **pneumatic-shod** *a.*, fitted with pneumatic tyres; **pneumatic-tired** (-tyred) *a.* (further examples).

1947 Dylan Thomas *Let.* 14 July (1966) 316 The pick-axed and pneumatic-drilled mosquitoes in the guest's bedroom. **1909** *Westm. Gaz.* 3 June 4/2 Although they [sc. motor-cars] are pneumatic-shod, the tyres do not come into contact with the track. *Ibid.* 18 Nov. 4/2 The driving-wheels of this vehicle are fitted with Vilo wheels, in place of ordinary pneumatic-shod artillery wheels. **1956** *Nature* 4 Feb. 218/2 His research has covered..the dynamic instability of systems incorporating pneumatic-tyred wheels. **1967** *Gloss. Highway Engin. Terms* (*B.S.I.*) 34 *Pneumatic-tyred roller*, a roller in which the compacting weight is supported on wheels fitted with pneumatic tyres. The roller may be self-propelled or towed. **1979** *Guardian* 8 Nov. 22/2 A shunting locomotive with pneumatic-tyred wheels.

pneumatically, *adv.* Add: (Further examples.) *spec.* (in mod. use) by means of compressed air. Also *fig.*

1942 *R.A.F. Jrnl.* 3 Oct. 29 Four guns were installed, harmonized and fired pneumatically. **1958** *Times* 23 June 14/3 Equipping three Russian whalers with plant for cubing whale meat and transferring it pneumatically to storage vessels at sea. **1975** *Radiol.* CXV. 222/1 (*heading*) A pneumatically operated femoral artery compressor. **1975** D. O'Sullivan in D. Marcus *Best Irish Short Stories* (1977) II. 90 'It[sc. beer]'s blowing me out,' said Anne pneumatically.

‖ **pneumatique** (niumatī·k). [Fr.; see Pneumatic *a.* (*sb.*).] The pneumatic dispatch system in Paris (see Pneumatic *a.* (*sb.*) 1 a); a letter or message sent by this system.

1924 E. Hemingway *Let.* 9 Aug. in A. Mizener *Saddest Story* (1971) xxv. 342 Ford..had stayed up all night writing pneumatiques. **1938** F. Scott Fitzgerald *Let.* July (1964) 35 A *pneumatique* might reach Nanny at 23 rue Pascal-Lecointre. **1951** R. Senhouse tr. *Colette's Chéri* 76 *Pneumatiques* of four or five lines from sponging friends. **1964** *Sunday Mail* (Brisbane) 28 June 33/4 All the letters were sent by pneumatique—the giant system of suction pipes which links all the post offices in the Paris area and allows express letters to reach their destination in the city within 60 minutes. **1973** L. Snelling *Heresy* iii. ix. 191 I'll get off a *pneumatique* to his office.

pneumatization (niū:mătəizē·ʃən). *Med.* [f. next + -ation.] The development or presence of air-filled cavities in bone or other tissue.

1934 Lake & Marshall *Surg. Anat. & Physiol.* xv. 244 Pneumatisation of mastoid process varies between extreme limits. **1959** G. E. Shambaugh *Surg. of Ear* i. 22 It is not until air enters the middle ear at birth..that pneumatization accelerates, continuing throughout infancy and early childhood. **1976** *Nature* 19 Aug. 683/2 Another unusual feature in a dinosaur is the very strong pneumatisation of the skull bones, especially those of the roof and snout.

pneumatize, *v.* Restrict *rare* to sense 1 in Dict. and add: **2.** (Further examples.) Chiefly as **pneu·matized** *pa. pple.* Hence **pneu·matizing** *vbl. sb.*

1947 *Arch. Otolaryngol.* XLVI. 852 The complications of acute otitis media developed more often in temporal bones with extensive cell systems than in those that were poorly pneumatized. **1949** L. R. Boies et al. *Fund. Otolaryngol.* i. 11 The temporal bone is pneumatized by the extension of an epithelial sac from the middle ear into the bone marrow in the region of the mastoid process. *Ibid.* 12 Any interference with..development will cause a cessation of the pneumatizing process. **1971** *Sci. Amer.* Dec. 73/3 All birds have..lungs, air sacs and pneumatized bones. **1976** *Nature* 19 Aug. 683/2 [In a dinosaur] the nasals, in addition to being pneumatised as a whole, each bear a large air chamber connected with the external nare.

pneumatolysis (niūmătǫ·lisis). *Petrol.* [ad. G. *pneumatolyse* (R. Bunsen 1851, as next, p. 264): see *-lysis.] The chemical alteration of rock and formation of minerals by the action of hot magmatic gases and vapours.

1896 Phillips & Louis *Treat. Ore Deposits* (ed. 2) i. 129 He [sc. Vogt] uses the word *pneumatolysis* for this action. **1905** *Rep. Board of Regents Smithsonian Inst. 1904* 335 The importance of pneumatolysis in forming ore deposits was emphasized by the discovery..of a number of economically important deposits. **1934, 1962** [see *greisen]. **1966** C. A. Lamey *Metallic & Industrial Mineral Deposits* iii. 43 Various descriptions of cassiterite deposits and wolframite deposits indicate that their formation..was caused by pneumatolysis. **1970** R. J. Small *Study of Landforms* iv. 127 At many points the granite of the South-West Peninsula [of England] has been profoundly affected by the metamorphic process known as 'pneumatolysis'.

pneumatolytic (niū:mătoli·tik), *a.* *Petrol.* [ad. G. *pneumatolytisch* (R. Bunsen 1851, in *Ann. d. Physik* LXXXIII. 238): see Pneumato- and *-lytic.] Involving or formed by pneumatolysis.

1896 Phillips & Louis *Treat. Ore Deposits* (ed. 2) i. 173 There seems no urgent reason for adopting the theory of pneumatolytic, in preference to ordinary hydrothermal action. **1903** [see *fumarolic *a.*]. **1909** J. P. Iddings *Igneous Rocks* I. vii. 276 Those minerals that appear to depend largely upon the action of gases in magmas for their presence in igneous rocks have been called pneumatolytic. **1963** D. W. & E. E. Humphries tr. *Termier's Erosion & Sedimentation* vi. 134 The minerals of rocks can be altered without exposure at the surface by pneumatolytic or hydrothermal action. **1972** *Nature* 3 Mar. 13/1 They interpreted this variation as the result of pneumatolytic and gravitative differentiation taking place in the lava column.

Hence **pneumatoly·tically** *adv.*

1962 *Proc. Yorks. Geol. Soc.* XXXIII. 329 The systems of valleys which evolved would be to some extent guided by the prior existence of pneumatolytically altered zones. **1968** I. Kostov *Mineralogy* ii. v. 362 The lithium micas are typical products of pneumatolitically [*sic*] altered acid rocks of the 'greisen' type.

Pneumatomachian. b. *adj.* (Examples.)

1915 *Encycl. Relig. & Ethics* VIII. 225/2 The leading doctrine of the Macedonians is found in the thesis characterized by their opponents as 'Pneumatomachian', viz. that the Holy Ghost is not to be designated *Θεός*. **1957** *Oxf. Dict. Chr. Ch.* 1086/2 The historians Socrates and Sozomen..regard Macedonius as their founder, but..his name does not occur in contemporary anti-Pneumatomachian writings.

pneumectomy. (Examples.) Now usu. called *pneumonectomy* (see Pneumono- in Dict. and Suppl.).

1911 *Practitioner* Nov. 682 Incision and drainage of a pulmonary cavity was the first operation practised for pulmonary tuberculosis, and..it has been performed much more frequently than pneumectomy. **1932** [see *lobectomy].

pneumo-. Add: **pneumoconiosis** (or **-kon-**) is now the usual name, rather than *pneumonoconiosis*; (earlier and later examples); hence **pneu:moconio·tic** *a.*, affected with pneumoconiosis; *sb.*, a pneumoconiotic person.

1881 *Med. Times & Gaz.* 28 May 589/1 (*heading*) The pathology of pneumokonioses. **1908** T. Oliver *Dis. of Occupation* ix. 247 There are various forms of dust diseases of the lungs, or pneumokonioses, *e.g.*, anthracosis..; chalicosis and silicosis..; and byssinosis. **1953** *Jrnl. Path. & Bacteriol.* LXVI. 235 Simple pneumokoniosis ..is characterised by numerous small discrete aggregations of dust in which only a little fibrosis occurs. **1969** *Daily Tel.* 5 Feb. 18 (Advt.), The Industrial Injuries Advisory Council has been asked to consider whether any change should be made in the definition of pneumoconiosis in the National Insurance (Industrial Injuries) Act 1965. The disease is at present defined in the Act as fibrosis of the lungs due to silica dust, asbestos dust or other dust, and including the condition known as dust reticulation.

1976 *S. Wales Echo* 27 Nov. 6/9 The cause of death was given as broncho-pneumonia due to recumbency following a fracture of the left femur, and pneumoconiosis. **1944** *Rep. Advisory Cttee. Treatment & Rehab. Miners in Wales Region Suffering from Pneumokoniosis* 17 The pneumokoniotic lung, especially at the massive nodulation stage is in special degree liable to become tuberculous. **1948** *Hansard Commons* 15 Mar. 1846 Reference has been made to the..Grenfell factories in South Wales for silicotics and pneumoconiotics. **1963** K. M. A. Perry in Perry & Sellors *Chest Dis.* I. xxxi. 566 In 1,036 pneumoconiotic lungs which he examined, he found evidence of pulmonary tuberculosis in 43 per cent.

pneu:mo,ence·phalogram. *Med.* Also **pneumence·phalo-**. [f. as next: see -gram.] An X-ray taken by pneumoencephalography.

1935 *Bull. Neurol. Inst. N.Y.* IV. 261 In cases in which a septum pellucidum is present this structure forms in the anteroposterior and posteroanterior views of the pneumencephalograms a well demarcated vertical linear shadow. **1973** *Reader's Digest* Apr. 194/2 Additional tests had to be made—an angiogram and perhaps a pneumoencephalogram.

pneu:mo,encephalo·graphy. *Med.* Also **pneu:mencephalo-**. [f. Pneumo- + *encephalography* s.v. *encephalo-.] Radiography following the replacement of cerebrospinal fluid by air or oxygen.

1932 *Amer. Jrnl. Roentgenol.* XXVII. 657/1 Pneumoencephalography, following withdrawal of cerebrospinal fluid and injection of air by the method of spinal puncture, permits investigation of the cisterna magna and of the subarachnoid spaces, in addition to study of the ventricles. **1935** *Bull. Neurol. Inst. N.Y.* IV. 221 It seems essential..to establish definite criteria for the diagnosis of these lesions by pneumencephalography whether ventriculography, encephalography or both are utilized in a given case. **1974** Passmore & Robson *Compan. Med. Stud.* III. xxxiv. 23/1 Pneumoencephalography visualizes the ventricular system and CSF pathways of the brain. A lumbar puncture is performed and 15 ml of air is instilled slowly into the subarachnoid space with the patient sitting upright.

Hence **pneu:mo,encephalogra·phic** *a.*, **-encephalogra·phically** *adv.*

1935 *Bull. Neurol. Inst. N.Y.* IV. 261 From the above mentioned pneumencephalographic evidence one can now diagnose the presence of a perforated septum pellucidum. **1950** Davidoff & Epstein *Abnormal Pneumoencephalogram* xxix. 407 A hemiplegia with a pneumoencephalographically demonstrable porencephaly. **1961** *Lancet* 9 Sept. 569/2 Myeloencephalography..confirmed the pneumoencephalographic findings.

pneumogram. Add: **2.** An X-ray photograph made by pneumography (sense *2).

1921 *Johns Hopkins Hosp. Bull.* XXXII. 74/1 (*caption*) Cerebral pneumogram..of a case of communicating hydrocephalus; air has been injected into the spinal canal. **1964** Taveras & Wood *Diagnostic Neuroradiol.* 227 The straight anteroposterior pneumogram (C) reveals the forward portions of the lateral and the third ventricles outlined by gas.

pneumography. Add: **2.** *Med.* The radiography of tissues into which air or oxygen has been introduced.

1921 *Johns Hopkins Hosp. Bull.* XXXII. 70/1 The most frequent location for an obstruction in communicating hydrocephalus is in the cisternæ. This..will be seen in the results which are to follow in the patients who have been studied by cerebral pneumography. **1948** Lysholm & Bull in J. W. McLaren *Mod. Trends Diagnostic Radiol.* I. xxiv. 327 Subdural pneumography has proved to be valuable in the diagnosis of adhesions between brain and dura, and in subdural haematoma. **1970** *Sci. Jrnl.* Aug. 5/1 The cardiovascular system[s] of both cosmonauts were studied in detail by means of electrocardiography, seismocardiography, pneumography, [etc.].

pneumographic *a.* (further examples); **pneumogra·phically** *adv.*, by means of pneumography.

1921 *Jrnl. Amer. Med. Assoc.* 10 Dec. 1858/2 At times, it has been possible completely to remove such growths, when, except for the pneumographic record, there would have been absolutely no other way of knowing the situation of the growth. **1953** F. Lindgren in J. W. McLaren *Mod. Trends Diagnostic Radiol.* II. xix. 291 The pneumographic diagnosis of an expanding intracranial lesion is possible because the lesion, when it has attained a certain size, affects the central fluid-filled and gas-filled space. *Ibid.* 301 These tumours cannot be demonstrated pneumographically in any way other than by encephalography. **1964** Taveras & Wood *Diagnostic Neuroradiol.* 752/2 A pneumographic examination in any patient who presents a clinical syndrome that suggests a lesion in the posterior fossa. **1972** M. Gado in E. J. Potchen *Current Concepts Radiol.* xi. 261 Pneumographically differentiation between cerebellar and extra-axial masses behind the cerebellum is impossible.

pneumolysis (niūmǫ·lisis). *Surg.* [mod.L., f. Pneumo- + *-lysis.] The surgical separation of the parietal pleura either from the chest wall (*extrapleural pneumolysis*) or from the pulmonary pleura (*intrapleural pneumolysis*).

1913 *Index Med.* XI. 199/1 (Index), Pneumolysis. **1922** *Tubercle* III. 162 Since 1918, we have performed extrapleural pneumolysis in..cases of severe unilateral pulmonary tuberculosis. **1945** Keers & Rigden *Pulmonary Tuberculosis* xi. 193 Closed intrapleural pneumolysis..is the logical sequel to pneumothorax therapy where adhesion formation is preventing effective collapse of the lung. **1966** E. L. Farquharson *Textbk. Operative Surg.* (ed. 3) xiv. 447 Intra-pleural pneumolysis was employed mainly as an adjunct to artificial pneumothorax.

pneumonia. Add: **b. pneumonia blouse** *colloq.*, a woman's blouse made of thin or light material and having a low neck-line.

1902 in C. W. Cunnington *Eng. Women's Clothing* (1952) ii. 47 The questionable morality of the 'Pneumonia blouse'... A transparent blouse of muslin and lace with next to no collar. **1905** R. Broughton *Waif's Progress* xix. 209 'Catherine had a bad cold.' 'The result of a pneumonia blouse, I suppose!' **1958** *Listener* 2 Oct. 524/1 Clergy of all denominations denounced that shocking innovation of 1914, the open-necked 'pneumonia blouse'. **1961** *Guardian* 6 Mar. 10/6 Astrakhan and felt bootees shared a narrow bench with sandals and 'pneumonia' blouse.

pneumonitis. Add: (Further examples.) (See also quot. 1974.)

1918 *Surg., Gynecol. & Obstetr.* XXVI. 32/2 One of the most important predisposing causes of postoperative pneumonitis both in patients coming into the hospital with a recent cold or free from an active cold, is the exposure to which he is subjected during the first twenty-four to forty-eight hours of his stay in the hospital. **1947** *Radiology* XLIX. 284/1 There was no opacity of the eyes, no pneumonitis. **1962** A. Huxley *Let.* 17 Mar. (1969) 930 A cold on the chest which..turned out in the end to be what they called pneumonitis. **1971** *Sci. Amer.* Aug. 47/3 Inhalation of cadmium oxide fumes..can produce acute damage in the lungs in the form of pneumonitis or pulmonary edema. **1974** Passmore & Robson *Compan. Med. Stud.* III. xviii. 23/2 Pneumonia is a general term meaning inflammation of the lungs; it usually refers to a clinically acute condition caused by micro-organisms but it may result from damage by one or a number of chemical or physical agents. Pneumonitis is a similar term and is sometimes used to describe mild segmental pneumonia, but it is best avoided.

pneumono-. Add: **pneumonectomy** (examples); hence **pneumone·ctomized** *ppl. a.*, subjected to a pneumonectomy; **pneumonoconiosis**, add: [first formed as G. *pneumonokoniosis* (F. A. Zenker 1866, in *Deutsch. Arch. f. klin. Med.* II. 171)]; now generally superseded by *pneumoconiosis*; (later examples); hence **pneumo:noconio·tic** *a.* and *sb.*; **pneumo:no,ultramicrosco·picsi:licovolcanoconio·sis** (-konio·sis), a factitious word alleged to mean 'a lung disease caused by the inhalation of very fine silica dust' but occurring chiefly as an instance of a very long word.

1939 *Brit. Jrnl. Surg.* XXVII. 411 The possibility of maintaining a permanent pneumothorax on the pneumonectomized side is one that has not been fully investigated as yet. **1967** *Excerpta Medica: Radiol.* XXI. 138/2 (*heading*) Pre- and postoperative pulmonary angiography in pneumonectomized patients. **1890** *Ann. Universal Med. Sci.* III. B-27 The profession is urged by Zakharevitch to prosecute these investigations for the practice of pneumonectomy. **1938** *Brit. Jrnl. Surg.* XXVI. 190 (*caption*) Right lung, the site of bronchial carcinoma, successfully removed by dissection pneumonectomy. **1967** [see *lobectomy]. **1972** T. W. Shields *Gen. Thoracic Surg.* xx. 331/2 A pneumonectomy may be carried out by means of any one of the three standard thoracic positions: lateral, posterior, or anterior. **1977** *Lancet* 23 July 164/1 All the patients had a radical pneumonectomy or lobectomy. **1934** *Trans. Inst. Mining Engineers* LXXXVIII. 387 This pneumonoconiosis of coal-miners is a *progressive* condition, gradually passing on from its harmless early stages..to a condition of dyspnœa and breathlessness which makes work impossible. **1940** H. E. Collier *Outl. Industr. Med. Pract.* xxxix. 356 The general object of the tests has been to find a means of checking the interpretation of the X-ray appearances and thereby to provide an early method of recognizing the outset of any of the pneumoconioses before the onset of the disablement. **1933** E. M. Williams *Health of Old & Retired Coalminers in S. Wales* 80 Thirty-two men..were considered on clinical grounds to be definitely pneumonoconiotic. *Ibid.* 102 The proportion of probably tuberculous cases among these pneumonoconiotics is high. **1936** F. Scully *Bedside Manna* 87 Pneumonoultramicroscopicsilicovolcanakoniosis [*sic*], a disease caused by ultra-microscopic particles of sandy volcanic dust, might give even him laryngitis. **1966** *Word Study* Oct. 7/2 The resources of Greek have enriched the modern world as well as the ancient one. Perhaps this is most dramatically illustrated by the longest and most fantastic word now in an English dictionary (the Merriam-Webster's great Unabridged) which is forty-five letters in length: pneumonoultramicroscopicsilicovolcanoconiosis,..meaning 'a disease of the lungs caused by extremely small particles of ash and dust'. **1973** R. Megarry *Second Miscellany-at-Law* 160 It has been said that 'floccinaucinihilipilification' is the longest word in the English language... The word's proud title must yield to some technical terms, such as pneumonoultramicroscopicsilicovolcanokoniosis.

pneumonolysis (niūmǫnǫ·lisis). *Surg.* [f. Pneumono- + *-lysis.] = *pneumolysis.

1934 *Amer. Rev. Tuberculosis* XXIX. 270 We were opposed to..extrapleural pneumonolysis with paraffin filling because it is contrary to accepted surgical practice to place a foreign-body permanently in the tissues. **1975** W. L. Glenn et al. *Thoracic & Cardiovascular Surg.* (ed. 3) xi. 251/1 Pneumonolysis has..maintained a limited but definite field of usefulness.

pneu:moperitone·um. *Med.* Also with hyphen. [f. Pneumo- + Peritoneum.] The presence of air or gas in the peritoneal cavity, whether accidentally or artificially induced.

1896 *Trans. Amer. Assoc. Obstetr. & Gynecol.* VIII. 212 Sacculated pneumo-peritoneum is frequently found in conjunction with localized appendiceal abscesses. **1934** *Amer. Rev. Tuberculosis* XXIX. 625 Therapeutic pneumoperitoneum has proved to be a technically simple and practicable treatment for intestinal tuberculosis. **1974** Passmore & Robson *Compan. Med. Stud.* III. xxviii. 3/1 A pelvic pneumoperitoneum may be induced with CO_2 and the gas used as contrast medium to define the uterus and ovaries [for radiography], but this has been largely superseded by laparoscopy.

pneumotachograph (niūmotæ·kograf). [a. G. *pneumotachograph* (A. Fleisch 1925, in *Pflüger's Arch. für die ges. Physiol.* CCIX. 722), f. Gr. τάχος speed: see Pneumo- and -graph.] An apparatus for recording the rate of air flow during breathing. Also shortened as **pneu·motach**.

1926 *Index Med.* VI. 539/2 (*heading*) Pneumotachograph. **1929** *Arch. Internal Med.* XLIV. 295 The pneumotachograph..showed an increased respiratory rate. **1975** G. Ruppel *Man. Pulmonary Function Testing* viii. 80 Pneumotachs utilize various physical and chemical properties to produce an electrical output that can be integrated for measurement of volumes and flows. **1975** *Nature* 4 Sept. 51/1 Forced expiratory flow–volume curves were recorded by a Fleisch No. 3 pneumotachograph connected to a differential pressure transducer..and coupled to a carrier preamplifier.

So **pneumota·chogram** [A. Fleisch, *loc. cit.*, p. 718], a record produced by a pneumotachograph (e.g. a graph of the quantity of air expired or inspired as a function of time); **pneu:motachogra·phic** *a.*, of or pertaining to this apparatus or its use; **pneu:motacho·graphy**.

1926 *Q. Cumulative Index to Current Med. Lit.* XI. 447/2 (*heading*) Curve of speed of human respiratory air (pneumotachogram). **1928** *Q. Cumulative Index Med.* III. 854/1 (*heading*) Pneumotachographic picture in bronchial asthma. **1929** *Arch. Internal Med.* XLIV. 293 Patients with cardiovascular and pulmonary diseases..showed notchings of the pneumotachograms even during respiration standstill. **1930** *Jrnl. Amer. Med. Assoc.* 5 July 83/2 Pneumotachography in combination with electrocardiography is employed in the general examination of the circulation. **1955** *Jrnl. Physiol.* CXXX. 33P The method usually employed to record a pneumotachogram utilizes the pressure differences across an impedance placed in the air-flow stream. **1955** *Thorax* X. 258 (*heading*) Pneumotachographic measurement of breathing capacity. **1975** *Nature* 30 Oct. 787/2 The breathing pattern of the fish was recorded continuously by pneumotachography (Statham–Godart) and pressure measurements (Statham).

pneumothorax. Add: Pl. **-thoraces, -thoraxes.** (Earlier and later examples.) [a. F. *pneumothorax* (E. H. Itard *Dissertation sur le Pneumothorax ou les Congestions gazeuses qui se forment dans la Poitrine* (1803).)]

1821 J. Forbes tr. *Laennec's Treat. Dis. of Chest* I. ii. v. 204 The disease is named by M. Itard, Pneumo-thorax. *Ibid.*, The author..considers the pneumo-thorax as an affection always consequent to and depending on a latent phthisis. **1825** [see *pneumatothorax* s.v. Pneumato-]. **1974** R. J. McCormack in Smith & Williams *Surg. of Lung* x. 241 All but the smallest pneumothoraces should be treated initially with intercostal water-seal drainage.

b. Pneumothorax may be induced diagnostically, or therapeutically so as to cause relaxation of a lung; it is then usu. known as *artificial pneumothorax*.

1885 *Lancet* 16 May 894/2 Mr. Charters Symonds spoke of the induction of artificial pneumothorax as having been suggested by Dr. Mahomed in a case on which they had held a consultation. The aim of the operation was to be the arrest of phthisis. **1923** *Brit. Jrnl. Tuberculosis* XVII. 109 When inducing an artificial pneumothorax, the gas employed is usually oxygen, but after the first insufflation nitrogen or filtered air may be equally well used. **1927** H. T. Lowe-Porter tr. *T. Mann's Magic Mountain* I. iii. 66 That was Hermine Kleefeld, she whistles with her pneumothorax. **1937** A. J. Cronin *Citadel* iv. viii. 320 He was chary of using pneumothorax and his percentage of inductions was the lowest in the hospital. **1974** Passmore & Robson *Compan. Med. Stud.* III. xviii. 74/1 Once a mainstay of treatment for the control of tuberculous lesions, therapeutic artificial pneumothorax has now been abandoned. In modern medicine artificial pneumothorax is occasionally used as a diagnostic procedure.

pneu:moventriculo·graphy. *Med.* [f. Pneumo- + Ventricul(us + -o + -graphy.] The introduction of air or oxygen into the ventricles of the brain for radiographic purposes.

1918 W. E. Dandy in *Ann. Surg.* LXVIII. 6 From these and many other normal and pathological clinical demonstrations of the radiographic properties of air it is but a step to the injection of gas into the cerebral ventricles—pneumoventriculography. **1932** *Amer. Jrnl. Roentgenol.* XXVII. 657/1 Pneumoventriculography and encephalography are valuable aids in the diagnosis of cerebral lesions. **1974** Passmore & Robson *Compan. Med. Stud.* III. xxxiv. 24/1 Air may be introduced directly into the ventricle through a burr hole in the vault of the skull (pneumoventriculography).

pneus (niūz), *sb. pl.* ?*Obs.* [Short for Pneumatic *sb.* 3.] Pneumatic tyres.

1902 C. N. & A. M. Williamson *Lightning Conductor* 18 On roads like these of Dieppe it would be soothing to have 'pneus', as they call them. **1907** —— in *Strand Mag.* Nov. 500/1 [The motor-car], with heated pneus, topped a commanding hill. **1908** *Westm. Gaz.* 31 Oct. 13/1 Before it [*sc.* the motor-car] can cease from injuring,..you have to construct suitable roads, highways which it can no longer tear up with its ferocious pneus.

pnicogen (pni·kodʒěn). *Chem.* [f. Gr. πνίγ-ειν to choke, stifle (in allusion to nitrogen) + -o + -gen.] Any of a series of elements in group V of the periodic table, viz. nitrogen, phosphorus, arsenic, antimony, and bismuth.

1966 *Progress Sci. & Technol. Rare Earths* II. 76 The term 'pnicogen' seems to be gaining favor as a group name for the Va group of elements. **1971** *Nomencl. Inorg. Chem.* (I.U.P.A.C.) (ed. 2) 11 The use of the collective names 'pnicogen' (N, P, As, Sb and Bi) and 'pnictides' is not approved.

So **pni·ctide** [-ide], a binary compound of a pnicogen with a more electropositive element or radical.

1966 *Progress Sci. & Technol. Rare Earths* II. 35 (*heading*) Thermodynamic and magnetic properties of the rare earth chalcogenides and pnictides. **1973** *Physical Rev.* B. VIII. 5345 The fourth-order crystal-field parameters for the phosphides and for other rare-earth pnictides fall on a universal curve which is close to that predicted..for the light rare earths. **1977** *Sci. Amer.* May 41/1 Removing an electron from a trigonally bonded chalcogen atom leads to a low-energy state, because the atom then becomes structurally similar to a pnictide atom in its proper configuration.

p-n-p (pī:enpī·). *Electronics.* Also **pnp**, and in capitals. Designating a semiconductor device in which an *n*-type region is sandwiched between two *p*-type regions. Also *absol.*

1949 W. Shockley in *Bell Syst. Technical Jrnl.* XXVIII. 435 The principles and theory of a *p-n-p* transistor are described. *Ibid.* 474 The *p-n-p* transistor has the interesting feature of being calculable to a high degree. **1962** Simpson & Richards *Physical Princ. Junction Transistors* iii. 51 The diffused layer forms the base, and the original material the collector, of a high-frequency *p-n-p* transistor. **1967** [see *n-p-n]. **1975** D. G. Fink *Electronics Engineers' Handbk.* viii. 26 The low current gain of a lateral *pnp* can be improved by combining it with a monolithic *npn* transistor, to form a composite transistor structure.

po³ (pōᵘ). *colloq.* Pl. **poes, pos.** [ad. Fr. *pot* (*de chambre*).] A chamber-pot. Also *attrib.* and *Comb.*

1880 Longmuir & Donaldson *Jamieson's Etym. Dict. Sc. Lang.* (rev. ed.) III. 517/2 *Po, s*, a matula or urinal. **1911** A. Warrack *Scots Dial. Dict.* 420/1 *Po, n.*, a chamber-pot. **1937** Partridge *Dict. Slang* 643/1 *Po*, a chamber-pot... Ex the pronunciation of *pot* in Fr. *pot de chambre*. **1951** W. Sansom *Face of Innocence* iv. 56 'In bloody pos.' 'Pose?' 'Under-the-beds. The amusing ones, roses and all get out. Bloody jerries.' **1950** G. Wilson *Brave Company* (1951) xi. 201 There was a great white china po hanging on the wall near the roof. **1961** C. Willock *Death in Covert* i. 14 Do you make plastic poes? **1966** P. O'Donnell *Sabre-Tooth* iv. 68 On her head was a po-shaped cloche hat. **1970** *Daily Tel.* 17 Oct. 6/6 And the hats—those 'po' creations had to be seen to be believed! **1974** *Punch* 27 Mar. 510/1, I kneelin' by de bed..peein' in de smart Victorian po.

po' (pōᵘ), *a.* Repr. a U.S. dial. pronunc. of Poor *a.* (*sb.*).

1893 H. A. Shands *Some Peculiarities of Speech in Mississippi* 50 *Po'*.., Negro for *poor*. **1911** F. W. Rolt-Wheeler *Boy with U.S. Census* 47 The po' white..is goin' to be only a memory like the backwoodsman o' the time o' Dan'l Boone. **1926** *Opportunity* Mar. 84/1 Dey kilt my po' daddy. **1945** *Gumbo Ya Ya* (Writers' Program, Louisiana) ii. 27, I sell to the rich, I sell to the po'; I'm gonna sell the lady Standin' in that do'. **1968** *Globe Mag.* (Toronto) 17 Feb. 18/4 Grey Owl..was Archie Bellaney, born..to a ne'er-do-well Englishman and..a po' white girl from Florida.

‖**p'o** (po). Also **peh, po.** [Chinese *pò*.] Soul, spirit.

1850 *Chinese Repository* Nov. 611 The Chinese philosophers..regarded man as a microcosm, his constitution and nature formed their model. This model they conceived to consist of a body which had *hing*..form, and of a *hwan*,..animus, and a *peh*, anima. **1914** D. T. Suzuki *Brief Hist. Early Chinese Philos.* 44 It is one Animal Soul (*po*..) that becomes the drought in heaven, the metal on earth, and the animal soul in man. **1934** A. D. Waley *Way & its Power* 28 *P'o*, which originally meant the semen, becomes the female soul, which lodges in the tomb. **1973** J. Blofeld *Secret & Sublime* ii. 47 How they would rush about, seeking in vain some vehicle to save their *hun* and *p'o* (higher and lower souls) from gradual dissolution into nothingness! **1975** C.-Y. Chang *Tao* x. 32 According to Ho-shang Kung,..p'o is rendered as anima... Anima (p'o) is the dark yin-soul.

poa. Delete ‖. Insert in etym. after 'mod.L.' (Linnæus *Genera Plantarum* (1737) 20). Substitute for def.: An annual or perennial grass of the genus so called, which is widely distributed in temperate and cold regions; = *meadow-grass* s.v. Meadow *sb.* 4 c. (Later examples.)

1917 S. F. Armstrong *Brit. Grasses* vii. 125 *Poa maritima*, Huds. (Sea Poa.) A somewhat creeping perennial, frequent on the shores of the British Isles. **1952** A. R. Clapham et al. *Flora Brit. Isles* 1438 'Alpine Poa.' An erect, tufted perennial.

b. poa-grass (later examples).

1927 F. B. Young *Portrait of Clare* vii. 762 A stone bridge spanned a chalk-stream with emerald trailing poa-grass swimming in smooth flats beneath it. **1971** *Homes & Gardens* Sept. 128/1 There is a silver-striped poa grass that shows up brilliantly anywhere, but it must not be allowed to spread.

poach, *v.*¹ **1.** Add to def.: and simmering gently. Hence, to cook (fish, fruit, etc.) by simmering in water or another liquid. Also, to simmer or steam (an egg) in a poacher. Also *absol.*

a **1693** [in Dict.]. **1898** C. H. Senn *Senn's Culinary Encycl.* 74 *Poach (to)*.., to parboil or to boil slightly. Mode of cooking usually applied to eggs and quenelles of fish, meat or game. **1906** *Mrs. Beeton's Bk. Househ. Managem.* xxxviii. 1237 Put the tins in the oven, in a sauté-pan, surround them to half their depth with boiling water, and poach until the white is firm. **1907** *Yesterday's Shopping* (1969) 170/3 Steam Egg Poacher. A perfect way of poaching eggs. **1907** A. Escoffier *Guide Mod. Cookery* ii. xiv. 292 The poaching of fillets of sole must be effected without allowing the cooking-liquor to boil. **1940** A. L. Simon *Conc. Encycl. Gastron.* II. 18/1 Fresh Cod..is usually boiled, steamed or poached. **1959** *Listener* 26 Feb. 395/1 Peel and cut the pears in pieces and poach them very gently in the syrup. *Ibid.* 1 Oct. 551/2 Core the pears carefully. Poach until tender. **1963** Hume & Downes *Penguin Cordon Bleu Cookery* 432 Fresh peaches poached in a sugar syrup until tender. **1972** L. Davies *Easy Cooking* iii. 93 Very gently poach the sausages in the milk with the bay leaf, onion and seasoning, for 20 minutes.

poach, *v.*² Add: **III. 9. b.** In various ball games: to enter a partner's portion of the field or court and play a ball which he normally would have played.

1889 W. M. Brownlee *Lawn Tennis* 167 He need not be profusely apologetic when he poaches unsuccessfully. **1928** *Daily Express* 9 July 13/3 The pretty little Australian girl..would have won if her partner had not 'poached' and put himself out of position. **1960** *Times* 4 July 15/7 They must have noticed Osuna's swift ability to poach.

10. b. (Further examples.)

1955 *Times* 14 June 3/3 These are the men whom the N.A.S.D. are said to have 'poached' from the Transport and General Workers Union. **1958** *Listener* 11 Dec. 978/1 A girl doing it might later in life be tempted to poach service. **1979** *Internat. Jrnl. Sociol. of Law* VII. 176 Solicitors in the large criminal firms not only 'poach' clients, they also strive to poach each other.

poachable (pōᵘ·tʃăb'l), *a.* [f. Poach *v.*² + -able.] Of game or fish: that may be poached or carried off illegally; suitable for poaching.

1924 *Public Opinion* 22 Feb. 169/1 The open wood I seldom visited,—all that was poachable having been poached long before.

poached, *ppl. a.*¹ Add: **a.** (see *poach *v.*¹ 1). Also in extended use.

1940 A. L. Simon *Conc. Encycl. Gastron.* II. 18/1 Fresh Cod..may be served..with any of the sauces which are suitable for boiled, steamed or poached Turbot. **1978** *Chicago* June 225/1 For dessert, orange-sparked chocolate mousse, poached pear with brandy and whipped-cream sauce.

b. (*c*) *Cookery* (see quots.); (*d*) = *poached-egg flower* s.v. sense *d.

1951 *Good Housek. Home Encycl.* 592/1 'Poached eggs' (halved and glazed peaches on rounds of sponge cake, surrounded by a ring of whipped cream). **1959** *Listener* 24 Dec. 1135/1 'Poached eggs': rounds of sponge cake, covered with a halved tinned apricot with a ribbon of whipped cream piped round the edge. **1971** *Guardian* 17 Apr. 7/8 The low-growing annual *Limnanthes douglasii*, known to children as 'Poached Eggs'..will make a tapestry of lemon and white at the front of a sunny bed.

d. poached-egg flower, plant, a small annual herb, *Limnanthes douglasii*, belonging to the family Limnanthaceæ, native to California, and bearing fragrant white and yellow flowers.

1963 R. D. Meikle *Garden Flowers* 135 Poached-egg Flower... The cordate petals are yellow at the base and white at the apex, producing a parti-coloured effect which at once suggests (to the inartistic) the popular name. **1973** *Country Life* 30 Aug. 565/1 In Britain we call *Limnanthes douglasii* the poached-egg flower, but in America they use the much more appropriate name of meadow foam. **1977** M. Allan *Darwin & his Flowers* xv. 261 (*caption*) *Limnanthes douglasii*, the 'Poached Egg Plant', which Darwin found was self-pollinating. **1978** *Woman's Jrnl.* Dec. 15/4 Do consider..the poached-egg plant called *Limnanthes douglasii*, a dwarf annual with ravishing yellow and white flowers.

poached, *ppl. a.*² **b.** (Later example.)

1955 *Times* 6 June 4/3 The union should..attempt to reach a settlement..of the dispute over 'poached' members.

poacher¹. Add: **1. b.** *a poacher turned gamekeeper*: one who now preserves the interests he previously attacked; conversely, *a gamekeeper turned poacher*.

1945 *Times* 4 Aug. 5/2 Mr. Aneurin Bevan at the Ministry of Health..is conspicuously the poacher turned gamekeeper. **1977** *Times* 12 Feb. 12/6 Mr Camp has been working against the Railways Board in the interests of the unions... Is he then a poacher turned gamekeeper? **1978** *Broadcast* 27 Mar. 4/2 Stuart Wilson, former joint MD of Yorkshire Television..has been arguing with the ferocity of gamekeeper turned poacher against ITV's intended inroads into the fourth channel.

2. a. (Examples.)

1905 C. C. Townsend *Birds of Essex County, Mass.* x. 130 The Baldpate, being unable to dive, makes use of diving Ducks to obtain food in deep water, and has therefore received in some places the name of 'Poacher'. **1923** *Bull. U.S. Nat. Mus.* No. 126. 96 Such behavior has earned for the baldpate the local name of 'poacher'. **1973** *Nature West Coast* (Vancouver Nat. Hist. Soc.) 163 Some birds [*sc.* American widgeon] feed on sea-weeds, often snatching them from the bills of diving birds. Hence the popular name 'poacher'.

b. Substitute for def.: A small marine fish belonging to the family Agonidæ. (Later examples.) Cf. *sea-poacher* s.v. Sea *sb.* 23 d in Dict.

1961 E. S. Herald *Living Fishes of World* 252/2 The cold-water marine poachers and their relatives look much like some of the South American fresh-water armored catfishes. **1978** A. Wheeler *Key Fishes N. Europe* 224 Poachers or Pogges. A small family (Agonidae) of mainly Arctic marine fishes.

4. poacher('s) pocket, a large concealed pocket in a coat.

1925 [see *hare-pocket]. **1956** G. E. Evans *Ask Fellows who cut Hay* ii. 35 Inside the slop were two long hanging, or *poacher's*, pockets so that a shepherd could very easily conceal a couple of rabbits. **1974** *Country Life* 21 Mar. 688/2 The suit..is a three-piece garment with extremely wide lapels, poacher pockets, and baggy trousers with turn-ups. **1976** 'D. Halliday' *Dolly & Nanny Bird* xviii. 247 He took up..a manilla envelope which he zipped with care into a poacher's pocket on the inner side of his waterproof jacket.

poacher². Add to def.: usu. with shallow cup-like compartments in which an egg can be cooked over boiling water. (Earlier and later examples.) Also, a vessel or pan in which fish, etc. can be poached (see *poach *v.*¹ 1).

1861 Mrs. Beeton *Bk. Househ. Managem.* xxxiii. 827 For inexperienced cooks, a tin egg-poacher may be purchased. **1884** [see *egg-poacher* s.v. Egg *sb.* 6 b]. **1895** *Montgomery Ward Catal.* Spring & Summer 433/1 Buffalo steam egg poachers. **1951** *Good Housek. Home Encycl.* 452/1 Use a wide pan..or a specially constructed egg poacher. **1975** J. Beard et al. *Cooks' Catal.* 399 Poachers come in a range of sizes for everything from a miniature mackerel to a king-sized salmon... All poachers have a common feature—a rack that protects your fish from the direct heat source beneath and enables you to lift the whole fish from the pot intact. *Ibid.* 439 All of the so-called egg poachers on the market, however they operate, whatever they do, are not actually poaching at all. A true poached egg is cooked *in* water, not over it, and not in steam—an 'egg poacher', with its cuplike insert, actually steams an egg. **1976** *West Lancs. Evening Gaz.* 15 Dec. 1. 12/4 Two Poachers and steamer, 75p each.

poaching, *vbl. sb.*¹ (Later examples.)

1907 A. Escoffier *Guide Mod. Cookery* ii. xiv. 292 Where the exact amount of poaching-liquor is not given, allow one-quarter pint to every four fillets. **1913** J. M. Hill *Pract. Cooking* 69 The term poaching has come to be applied to the cooking of all articles containing eggs, either in the oven or on top of the range, in dishes that are surrounded with hot water. **1956** Spry & Hume *Constance Spry Cookery Bk.* iv. 70 *Poaching* is a term used to indicate a gentler cooking still [than simmering]. **1960** *Good Housek. Cookery Bk.* (rev. ed.) 80/2 Although this method of cooking fish is referred to as boiling, it is more accurately described as poaching, since the water should be kept only simmering, not actually boiling, while the fish is in it.

poaching, *vbl. sb.*² Add: **a.** Also *concr.*, a patch of mud. *poet.*

1929 J. Masefield *Poems* 703 Plastered with poachings, he rode on forsaken.

b. (Further examples.)

1899 E. H. Miles *Lessons in Lawn Tennis* xi. 69 The man should have no false modesty..about a certain amount of poaching. **1955** *Times* 27 May 11/6 He intimates that officers of the Transport and General Workers Union have been guilty of membership poaching offences. **1977** *World of Cricket Monthly* June 68/3 The approach had come initially from Imran, and there was no question of poaching on the part of Sussex.

Poale Zion (poăle· tsi,yo·n). [Heb., 'workers of Zion'.] Name of a predominantly left-wing Zionist labour movement which first emerged in Russia about 1899.

1919 N. Sokolow *Hist. Zionism* II. 367 The Poale Zion was established in 1901... The programme of the organization represents a synthesis of Zionism and Socialism. **1935** A. Revusky *Jews in Palestine* xii. 206 The left is represented by the Poale Zion and the Communists. **1973** *Jewish Chron.* 19 Jan. 42/3 Mr Burton was a member of Poale Zion from early youth.

pob² (pǫb). [Echoic.] An abrupt, heavy sound as when an inelastic body strikes a hard surface.

1911 J. Masefield *Jim Davis* xiii. 157 A heavy chewed slug would come 'pob' into the boat's side.

pobby (pǫ·bi), *a.* orig. *dial.* [cf. Pobs *sb. pl.*] Swollen, blown; also of food, pulpy, mushy.

1888 Kipling *Phantom 'Rickshaw* 30 There are, in this land, ghosts who take the form of fat, cold, pobby corpses. **1903** *Eng. Dial. Dict.* IV. 563/1 *Pobby, adj.* Yks..., swollen, *gen.* used of a soft swelling. **1937** V. K. Libby *How to care for Baby* vi. 49 If you want baby to have good teeth be careful not to give a lot of 'pobby' food. Never soak the rusks in milk, etc.

‖**poblacion** (pǫblasiǫ·n, -þiǫ·n). Also **población**. Pl. **poblaciones.** [Sp., = population; also, town, city, village.] **a.** In Spanish-speaking countries of South America: a community; a district of a town, etc. **b.** In the Republic of the Philippines: the principal community of a district; a town that is an administrative centre.

1926 J. Masefield *Odtaa* xiv. 236 The hut, like the other huts of the poblacion, was, at a guess, thirty feet long by fifteen broad. **1961** Webster, *Población*.., a center of a municipality in the Philippines that is usu. the barrio that gives the municipality its name and is the seat of government. **1964** A. Cutshall *Philippines* iii. 27 The postwar Philippine barrio and oftentimes the *poblacion*, the principal community in the municipality (county), is not appreciably different from the representative prewar village and town. **1967** M. C. Bello in M. D. Zamora *Stud. Philippine Anthropol.* 325 The site studied is the *poblacion* which refers to a group of contiguous villages found toward the middle of the region... The poblacion settlement is situated on the top of a mountain. **1978** *Listener* 23 Nov. 668/2 The *barrio* or *poblaciones*, the shanty-towns of the poor. **1979** E. Norman *Christianity & World Order* iv. 47 In 1978 I visited the working class *poblaciones* around Santiago, in Chile.

‖**poblador** (pǫblădǫ͡ə·ɹ). Pl. **pobladores.** [Sp.] In Spanish America, a settler, a colonist; *spec.* a country person who moves to settle or squat in a town.

1966 *Economist* 2 July 28/2 The *pobladores* (squatters) fled further up the hill. **1976** *New Yorker* 26 Apr. 122/1 Everyone seems to know precisely when his forefathers arrived—in the eighteen-fifties, as *pobladores*, or settlers, recruited..to colonize the million-acre Sangre de Cristo land grant.

pochade. Add: (Later example.) Also **pochade box** (see quot. 1961).

1959 M. Steen *Tower* II. iv. 206 My easel had been removed, and my pochade box. **1961** M. Levy *Studio Dict. Art Terms* 88 *Pochade box*, a small portable colour-box with panels fitted into the lid for quick sketching. **1972** D. Sutton in R. Fry *Lett.* I. 90 Fry..was usually more at ease with a *pochade* than with a finished work.

pochette (poʃe·t). [Fr.: see Pocket *sb.*] **1. a.** A small pocket.

1913 W. de la Mare *Peacock Pie* 80 A watch..He lifted from the hook where it was ticking And crammed in his Pochette. **1949** *Amer. Speech* XXIV. 38 The secret pockets which the conjurer is so often suspected of using are of two types, the *pochette* and the *profonde*. They differ in position and in size, the latter being the larger. **1964** *Punch* 1 Jan. 15/1 *Pochettes* in the [conjuror's] trousers.

b. A hand-bag. Also *pochette bag*.

1923 *Weekly Dispatch* 11 Mar. 15/5 When jewels are worn in the hair the vanity-bag becomes a satin or crêpe pochette, fastened with a buckle of jewels. **1930** *Daily Tel.* 9 Apr. 9/2 If you will make yourself pochettes to match your hats, you are adding..that extra touch of chic. **1972** *Times* 12 May 13/6 The little pochette..slides into a bigger bag. **1973** *Country Life* 1 Feb. 302/3 A shiny PVC coat and a great flat pochette, slung on a handy shoulder strap. **1976** *Evening Post* (Nottingham) 16 Dec. 10/4 You can even get a pochette bag with the words 'Gold-Rush' as a motif. **1977** *Times* 29 Oct. 12/2 More and more men use pochettes to avoid bulging..their trousers.

2. A small violin, supposedly once carried in the pocket by French dancing-masters; = Kit *sb.*²

1889 in *Cent. Dict.* **1976** D. Munrow *Instruments Middle Ages & Renaissance* 28/2 It was as an instrument for dancing that the small rebec survived as the *kit* or *pochette* into the eighteenth century.

‖**pochismo** (potʃī·zmo). [Mexican Sp., f. as next + *-ismo* -ism.] A form of slang used by speakers of Mexican Spanish and others along the border with the U.S., consisting of English words given a Spanish form or pronunciation; a word of this sort.

1944 *N.Y. Herald Tribune* 5 Aug. 10/3 The Mexican Academy has appointed a committee to eradicate 'pochismos'—that is English words and phrases used in speaking Spanish. **1946** *Mod. Lang. Jrnl.* Oct. 345 *Pochismo*, derived from *pocho*, an adjective which originally meant discolored, has now come to mean a type of popular slang in Mexico. **1966** N. S. Hayner *New Patterns in Old Mexico* xiv. 219 Students of language have a lively curiosity in the increasing use of 'pochismo', a type of popular slang, by even the most conservative of the Mexico City newspapers. **1976** F. A. & D. L. Latorre *Mexican Kickapoo Indians* ii. 29 They [*sc.* Kickapoos] have incorporated many Spanish words into their speech and many *pochismos*..such as *lonche* ('lunch').

‖**pocho** (po·tʃo). [Mexican Sp., Sp. *pocho* discoloured, faded, pale.] A citizen of the United States of Mexican origin; a culturally Americanized Mexican. Also *attrib.* or as *adj.*

1944 *Newsweek* 14 Aug. 76/3 A pocho in good standing will drag his fititoes (feet) up the estrita (street). *Ibid.* Slapstick actors like Tin Tan..who gets comic effects with pocho patter. **1960** *Time* 25 Jan. 92/2 Opium is..smuggled..from Mexico by special agents called 'pocho' women (generally Americans of Mexican descent) who cross the border into Mexico to shop. **1968** *Sunday Mail Mag.* (Brisbane) 25 Aug. 14/3 *Pocho*, derogatory term for American-Mexican (in U.S.A.). **1975** *Sat. Rev.* (U.S.) 8 Feb. 47/1, I was frequently labelled a *pocho*, a Mexican with gringo pretentions.

‖**pochoir** (pǫʃwār). [Fr., = stencil.] A process used in book illustration, especially for limited editions, in which a monochrome print is coloured by hand, using a series of stencils; a print made by this process. Also *attrib.*

1931 *Times Lit. Suppl.* 25 June (Salon International du Livre d'Art Suppl.) p. i/4 Emile Hazan is another publisher who has taken full advantage of the *pochoir* technique. *Ibid.* p. viii/3 Colette's 'Regarde'..with its drawings by Mathurin Méheut coloured by *pochoir*. **1932** J. Joyce *Let.* 10 Nov. (1966) III. 266 This *pochoir* reproduction can be done only in Paris and even here only by two or three firms. **1953** *Book Collector* II. 4 The unique coloured copy of Blake's *Jerusalem*..was recently reproduced in facsimile by the pochoir process. **1958** *Listener* 12 June 984/3 Lithographs and '*pochoirs*' by Dufy. **1965** Zigrosser & Gaehde *Guide Coll. Orig. Prints* iv. 56 In France around the 1920's,..an expert craftsman, J. Saudé, used the straight-stencil, or *pochoir*, process as a reproductive medium for illustrating books. **1973** *Times Lit. Suppl.* 26 Jan. 103/4 French books illustrated in pochoir..would be included. **1973** *Art Internat.* Mar. 22/2 He owned a pochoir by Gleizes and lithographs by Léger.

pock, *v.* Delete *rare* and add: Chiefly as pa. pple. or pa. ppl. adj. (Later examples.)

1938 *Proc. Prehist. Soc.* IV. 246 The chief tombs which exhibit incised & pocked designs on their walls & roofs are New Grange, Dowth, & those at Lochcrew. **1977** *Time* 31 Jan. 42/1 On went a coat of Viacryl, a synthetic polyurethane resin meant to protect the pocked and flawed surface of the 800-year-old glass.

Pockels (pǫ·kělz). Also erron. **Pockel.** The name of F. C. A. *Pockels* (1865–1913), German physicist, used, chiefly *attrib.*, with reference to an effect in certain crystals similar to the Kerr effect (sense *(*b*)) in liquids (see quot. 1975²); so *Pockels cell, constant, effect* (cf. *Kerr).

Described by Pockels in *Abhandl. der K. Ges. der Wissensch. zu Göttingen* (*Math.-phys. Klasse*) (1894) XXXI. 204.

1949 *Jrnl. Optical Soc. Amer.* XXXIX. 798/2 The constants t_{ij} are..related to Pockels' constants e_{ij}. **1957** D. E. Gray *Amer. Inst. Physics Handbk.* vi. 96 Twenty-one of the 32 crystal symmetry classes do not contain centers of symmetry, and of these, 20 may exhibit the linear Pockels effect. **1968** E. L. Steele *Optical Lasers in Electronics* v. 178 The electro-optical switches employ the photo-optical effect in a Kerr cell, Pockels cell, or Faraday rotator. **1975** D. G. Fink *Electronics Engineers' Handbk.* xi. 24 Linear Electrooptical Effect, or Pockel's Effect. Crystalline materials such as potassium dihydrogen phosphate (KDP) and lithium niobate are used... Metal electrodes are applied to rectangular blocks of these crystals, and the resultant structures are referred to as Pockel cells. *Ibid.* xiv. 60 In the Pockels effect the differential phase shift is linearly related to applied voltage; in the Kerr effect it is related to the voltage squared. **1976** *Nature* 22 Apr. 677/1 The emission intensities of the components parallel and perpendicular to the incident laser polarisation were determined using a Pockels cell to rotate the laser polarisation with respect to a fixed analyser.

pocket, *sb.* Add: **1.** (Further examples.)

1907 W. H. Koebel *Return of Joe* 239 A train of pack horses, heavily laden with the weighty 'pockets' of wool, toil meekly past. **1928** E. Wallace *Gunner* xxiv. 199 Bales of silk, chests of tea, pockets of rubber. **1940** E. C. Studholme *Te Waimate* 170 Over the hills [the wool] was carried by pack horses..each taking about 150 lbs. in what were called 'pockets'—loaf shaped packs slung one on each side of the saddle. **1953** *Word for Word* (Whitbread & Co.) 28/1 *Pocket*, a large sack made to contain roughly one and a half cwts. of dried hops.

3. b. *to be out of pocket*, also *transf.*, to be absent or out of reach. *U.S.*

1974 *Anderson* (S. Carolina) *Independent* 20 Apr. 1A/1 If you..have ever been sick and the only doctor is out of pocket for the weekend, then you know we need more doctors. **1978** *Internat. Herald Tribune* 24 July 14/4 Why does *out of pocket*..mean 'out of touch'?

d. esp. in phr. *to live in each other's pockets*: to live in excessively close proximity, to live in mutual dependence.

1959 [see *grandmother *sb.* 1 c]. **1965** F. Sargeson *Memoirs of Peon* vi. 158 We lived..in one another's pockets, so why should there be the privilege of privacy and seclusion for one and not for the other? **1978** *N.Y. Rev. Bks.* 23 Feb. 8/2 Architects and painters do not live in each other's pockets in England today.

e. *to put one's hand in one's pocket*: to (seek to) provide money from one's own resources.

1857 C. Kingsley *Two Yrs. Ago* I. p. xxii, There are other ways of being generous, besides putting your hand in your pocket. **1878** H. James *Europeans* I. iv. 147 Robert Acton would put his hand into his pocket every day in the week if that rattle-pated little sister of his should bid him. **1948** E. Waugh *Loved One* 28 We may have to put our hands in our pockets—I don't suppose old Frank has left much.

7. a. Also, an isolated body of opal or gum. *Austral.*

1873 J. E. Tinne *Wonderland of Antipodes* 54 If a man hits on a good 'pocket' of gum. **1910** *Lone Hand* (Sydney) Mar. 494 A dip in the seam, or some obstruction..temporarily dammed the stream, which thickened, solidified there, and formed a 'pocket'. **1971** J. S. Gunn *Opal Terminol.* 35 *Pocket*, small cluster of opal suddenly met in one place.

9. (Earlier example.)

a **1817** Jane Austen *Northanger Abbey* (1818) II. xiv. 285 With so much changing of chaises..I hope..you have not left anything behind you in any of the pockets.

11. b. *Amer.* and *Canad. Football.* A shielded area formed by blockers from which a player attempts to pass; the formation itself.

1963 *Time* 18 Oct. 94 Myers..is a drop-back 'pocket' passer. **1968** *Globe & Mail* (Toronto) 10 July 27/5 He is an accurate passer, either from the pocket or on the roll-out.

11*. = *air-pocket* s.v. *Air *sb.*¹ B. III. 1.

1911 G. C. Loening *Monoplanes & Biplanes* ii. 18 Everywhere in the atmosphere, and especially on windy days, there exist 'pockets' of high density and of low density. **1917** *Boy's Own Paper* Mar. 273/2 Evidently he had dropped into one of those air eddies which were so dangerous to flying men in the early days of aviation. They, in conjunction with 'pockets', accounted for the death of not a few pioneers in flight. **1919** C. P. Thompson *Cocktails* 46 The suddenly staggered Hun staggered and whirled in a treacherous 'pocket'. **1978** H. Kaplan *Damascus Cover* vii. 60 The plane rolled in a pocket of turbulent air.

11.** **a.** *Mil.* An area held by troops who are surrounded by opposing forces; an isolated concentration *of* resistance; also, the men themselves; esp. in phr. *pocket of resistance* (also *transf.*).

1918 *Observer* 29 Sept. 7/6 The Anglo-Belgian attack in the north..has reduced the enemy to the necessity of defending..a pocket such as those which brought him to disaster on the Marne and on the Avre. **1927** J. M. Keynes *Ess. in Biogr.* (1933) i. vi. 62 The strategic surrender, the deliberate withdrawal, the attempt to lure the enemy into a pocket where he could be taken in flank. **1941** *Manch. Guardian Weekly* 10 Jan. 20/2 The Australians engaged strong enemy defence pockets to the south-east of this line. **1943** *Times* (Weekly ed.) 24 Apr. 6 Here the Germans had a small pocket, based on the bridge which carried the highway to Gomel over the Dnieper. **1945** *Daily Express* 12 Apr. 4/8 Full aid to liberated Europe..must wait not only until the German army is beaten, but until pockets of resistance have been wiped out. **1959** *Listener* 12 Feb. 287/1 Except for provincial pockets of resistance it is now as successful as any architecture is ever likely to be. *Ibid.* 29 Oct. 740/2 An enemy 'pocket of resistance' was still occupying a wood about half-a-mile away. **1965** C. D. Eby *Siege of Alcázar* (1966) xi. 230 There was..a large pocket of militia barricaded in the seminary. **1966** T. Pynchon *Crying of Lot 49* iii. 61 A..battle of attrition in a minor pocket developed during the advance on Rome. **1975** G. St. George *Proteus Pact* (1976) i. 5 Kleist sat on top of an armored personnel carrier and watched the systematic elimination of pockets of resistance.

b. A small area contrasted with or differing from its surroundings in some respect; a local concentration *of* something. Cf. senses 7 a, b in Dict.

1926 *Scribner's Mag.* Aug. 163/1 The car swerved into the campus, that green, summer-deserted pocket of peace in the little, dusty, traffic-riddled village. **1932** *Times Educ. Suppl.* 2 Jan. p. iv/4 They walked into a pocket of gas and were asphyxiated. **1935** Huxley & Haddon *We Europeans* ii. 53 They do not form definite groups, but occur rather in local pockets where individuals still exhibit characters reminiscent of those remote times. **1937** *Brit. Jrnl. Psychol.* XXVII. 358 We may regard the adjusted group, if it has proceeded to the stage of new, reified institutions, as a small culture pocket or subculture within the larger culture. **1939** *British Birds* XXXIII. 102 The Black-tailed Godwit is increasingly occurring in the British Isles... The increase is greatest on the south coast of England with pockets elsewhere. **1945** *Daily Express* 22 May 2/4 What is to happen when, in the change from war to peace, there occur the inevitable pockets of unemployment? **1959** *Times* 4 Sept. 4/3 We must conclude that the existence of pockets of heat under the lunar surface is probable. **1976** W. H. Canaway *Willow-Pattern War* xi. 111 There were edelweiss in soil pockets on the rock outcrops. **1978** J. Blackburn *Dead Man's Handle* iii. 36 A pocket of upper-crust suburbia: detached, Regency-style residences with two-car garages.

12. a. (Further examples.)

1640 E. Verney *Let.* in F. P. Verney *Mem.* (1892) I. vii. 174, I pray be pleased to send mee a pocket prayer-book. **1726** Swift *Gulliver* II. iii. i. 7, I discovered by my pocket-glass several islands to the south-east. **1800** M. Edgeworth *Parent's Assistant* (ed. 3) IV. 145 Lady Augusta had just shown her a French pocket fan. **1827** J. F. Cooper *Prairie* I. vi. 179 *Quadruped*: seen..by the aid of a pocket-lamp, in the prairies. **1828** M. Wilmot *Let.* 23 Apr. (1935) 316 They all expect to be your pocket dictionarys and lionizers and walking sticks. **1832** *Chambers's Edin. Jrnl.* 14 Apr. 86/2, I scarcely recollect a single traveller without his pocket-comb. **1837** W. Irving *Capt. Bonneville* III. xi. 174 The captain now drew forth that grand lure in the eyes of the savage, a pocket mirror. **1848** in H. Howe *Hist. Coll. Ohio* 493 These little pocket editions of humanity are well cared for by kind dames. **1866** 'Mark Twain' *Lett. from Hawaii* (1967) 45 We..ran by a pocket compass in the hands of Captain Fish. **1874** *Eng. Mech.*

1 May 165/2 (*title*) How to make a pocket camera. **1885** *Sam Scaramouch* 12 Dec. 247/2 If you must have a drink, gentlemen, carry your pocket flasks. **1903** C. E. Wolff *Mod. Locomotive Practice* p. ii (Advt.), The 'Mechanical Engineer' Pocket Calculator. **1906** M. Corelli *Treasure of Heaven* 43 Mrs. Sorrel..drew out a black pocket-fan and fanned herself vigorously. **1913** *Punch* 17 Sept. 252 Portrait of gentleman using pocket-clipper to trim beard at back of neck. **1916** *Daily Colonist* (Victoria, B.C.) 9 July 7/1 (Advt.), Hudson's Bay Company Fine Old Irish Whisky Per Pocket Flasks..50¢. **1917** G. B. McCutcheon *Green Fancy* 56 Barnes found his electric pocket torch and dressed hurriedly. **1921** *Daily Colonist* (Victoria, B.C.) 25 Oct. 8/1 (Advt.), We are now showing an open face pocket watch, in strong nickel case. **1923** Conrad *Rover* iv. 61 The lieutenant,..with a pocket glass glued to his eye, growled angrily: 'You can see her now, can't you?' **1926** *Daily Colonist* (Victoria, B.C.) 16 July 10/7 (Advt.), Matches are dangerous. When you are camping in or around the forests, use a pocket lighter. **1927** S. Ertz *Now East, now West* xi. 173 Again that keen glance, over her pocket mirror. **1933** M. Arlen *Man's Mortality* 21 Taking out his pocket-transmitter, he held it near the light. **1939** T. S. Eliot *Family Reunion* i. i. 21 Reflecting a pocket-torch of observation Upon each other's opacity. **1955** C. Smith *Speaking Eye* iii. 33 He took out a pocket comb and ran it through his dark wavy hair. **1957** *Practical Wireless* XXXIII. 534/2 Messrs. Cossar announce a neat printed circuit transistor pocket radio. **1972** *Gloss. Electrotechnical, Power Terms* (B.S.I.) iv. iii. 21 *Pocket lamp*, portable luminaire embodying a miniature lamp fed by a dry battery or accumulator. **1972** D. Bloodworth *Any Number can Play* xi. 92 Ivansong..kept the pocket radio close to his ear. **1977** *Jrnl. R. Soc. Arts* CXXV. 71/1 Aldus used it [*sc.* italic type] for the pocket editions of classical authors which he printed in Venice around the year 1500. **1977** J. Hedgecoe *Photographer's Handbk.* 20 Most pocket cameras use a fixed lens with a focal length of about 25 mm. **1978** *Daily Tel.* 19 Sept. 8/6 Semiconductors..are an essential part of microprocessor technology and are used in pocket calculators, electronic watches, mini-computers, and many other modern devices.

b. (Further examples.)

1820 M. Edgeworth *Let.* 14 May (1979) 128 My dear little pocket Prince de Beauvoau for me!—worth all the Russian bears and giants put together. **1936** *Sun* (Baltimore) 18 May 8/3 No decisive victory in the Austrian pocket dictators' duel is possible without profound repercussions. **1951** N. M. Gunn *Well at World's End* xxi. 174 The garden of a gentleman farmer, a pocket laird. **1972** *Listener* 28 Dec. 898/1 The producer is not altogether the little pocket-dictator... He is assisted by a colleague, usually an expert on the topic to be discussed. **1977** *Time* 8 Aug. 30/1 Charles [Prince, of Luxembourg] was long active in promoting business and industry in his pocket principality.

e. Of warships: armoured and equipped like a ship of the class named, but smaller.

1930 *Economist* 1 Feb. 227/1 The technical progress represented by Germany's 'pocket battleship' of identical tonnage but heavier armament. **1932** *Sun* (Baltimore) 17 Sept. 1/1 (*heading*) New pocket cruiser to be started October 1. **1941** *Hutchinson's Pictorial Hist. War* 14 May–8 July 89 Some of the latest and fastest motor launches..are known as 'pocket destroyers'. **1942** H. Richmond *War at Sea Today* 27 The 'pocket battleships' are protected by armour against which 6-in.-gun fire could not be expected to be fully effective at long ranges. **1951** *Chambers's Jrnl.* Oct. 632/1 Aided by George (now Sir George) Binney, three 'pocket freighters' were rapidly built in British shipyards. **1974** G. Jenkins *Bridge of Magpies* ii. 40 We heard the sound of heavy guns: a raider or a pocket battleship, we thought.

13. † **pocket allowance** = Pocket-money; **pocket beach** *Physical Geogr.*, a small, narrow beach between two headlands or in a similar sheltered position; **pocket billiards,** (*a*) a North American type of pool (Pool *sb.*³ 3); (*b*) *slang* (orig. *Schoolboys'*), manipulation of the male genitals (cf. *ball *sb.*¹ 15 b) by the pocketed hands; also phr. *to play pocket billiards*; **pocket-expenses** (later example); **pocket-gopher** = Gopher *sb.*¹ 1; **pocket-hunter** (see quot. 1906); **pocket-miner** *U.S.* = *pocket-hunter*; so **pocket-mining** *vbl. sb.*; **pocket pager** = *pager *sb.*³; so **pocket-paging** *vbl. sb.*; **pocket passer** *Amer.* and *Canad. Football*, one who passes from the pocket (*pocket *sb.* 11 b); **pocket rot,** a fungus infection causing localized decay in the trunks or roots of trees; also with prefixed defining word (e.g. *brown pocket rot*); **pocket-size** *a.*, of a size suitable for carrying in the pocket; hence *fig.*, petty, small-scale; also *pocket-sized* adj.; **pocket stay** (see quot.); **pocket valley** *Physical Geogr.*, a steep-sided, usu. flat-floored valley at the head of which a stream emerges at the base of a steep slope; **pocket Venus,** a small and beautiful woman; also *transf.*; **pocket veto** (earlier and later examples).

1813 Jane Austen *Pride & Prejudice* III. viii. 143 Her board and pocket allowance, and the continual presents in money, which passed to her, through her mother's hands. **1893** N. S. Shaler in *Ann. Rep. U.S. Geol. Survey* XIII. 141 Where..there are islands or shoals lying on either side of a considerable reentrant, a curious action arises, which leads to the formation of what we may term 'pocket beaches'. **1932** W. H. Emmons et al. *Geol.* ix. 254 On rocky coasts beaches of boulders and cobblestones commonly form at the heads of indentations, although occasionally sand may occupy such positions. These pocket beaches..are found along the coast of California at Carmel, at La Jolla, and at many other places. **1976** A. N. Strahler *Princ. Earth Sci.* xvii. 250 Shingle beaches form in the most sheltered locations—in bays between rocky promontories—and are called pocket beaches... They are typically crescent-shaped and are concave toward the sea. **1913** J. T. Stoddard *Science of Billiards* vii. 152 Pocket Billiards. The more common pool games are played on tables with pockets, and with balls 2¼ inches in diameter,—slightly smaller than the ball used in billiards. **1917** *Billiards Mag.* Dec. 13/1 Greenleaf..began to play pocket billiards 6 years ago... His ambition is to win the pocket billiard championship. **1940** S. Spender *Backward Son* ii. 94 He paused, feeling in his trouser pockets with his hands, with a familiar gesture of the class room which the boys knew as 'pocket billiards'. **1949** F. Sargeson *I saw in my Dream* i. 40, I don't remember nothing about when school was in except him playing pocket billiards. **1963** *Landfall* Mar. 14 A pillar of our..community,..addicted to long volleys of handball, I mean pocket billiards. **1971** A. Burgess *MF* ii. 31 Saint Face, as if wishing to play pocket-billiards with my balls, thrust his hands in from the rear. **1905** G. B. Shaw in *Grand Mag.* Feb. 116 If you keep the pocket expenses down to twelve and six. **1873** E. Eggleston *Myst. Metropolisville* iv. 37 She would..explain how the pocket-gophers built their mounds. **1932** S. Zuckerman *Social Life Monkeys & Apes* iv. 59 The animals of the first sub-group are those that spend the anœstrus in solitude... Examples are the jaguar of Central America and the pocket gopher of the United States. **1977** R. B. Cowles *Desert Jrnl.* xx. 209 A pocket gopher or..a beaver burrowed through the bank. **1906** *Chambers's Jrnl.* Feb. 159/1 They [*sc.* prospectors] include the 'pocket-hunter' who disdains to search for gold except in the form of pockets. **1947** *Field & Stream* June 30/3 Now and then a 'pocket-hunter'..will find a place in the hills containing perhaps a few hundred dollars in gold. **1902** J. London *Daughter of Snows* 207 The pocket-miner's eyes sparkled. **1909** 'Mark Twain' *Is Shakes. Dead?* vii. 75, I have been a 'pocket' miner—a sort of gold mining not findable in any but one little spot in the world. **1872** —— *Roughing It* 436 In that one little corner of California is found a species of mining..called 'pocket-mining'. **1975** *Kingston* (Ontario) *Whig-Standard* 6 Dec. 25/1 Off in a side office, MacLean Hunter manager Jack French talks to a client about ordering 'pocket-pagers'. **1977** Pocket pager [see *pager *sb.*³]. **1973** *Times* 11 Jan. 15/2 The control room is the centre of a vast and flexible intercom system which complements and extends the telephone network..and the pocket-paging system. **1963** Pocket passer [see *pocket *sb.* 11 b]. **1977** *Globe & Mail* (Toronto) 26 Nov. 53/6 If Wade, a pocket passer, is harassed by the Eskimos' blitzing linebackers, he could throw a number of interceptions. **1926** *Jrnl. Agric. Res.* XXXIII. 687 As the brown cubical rot produced [by *Trametes subrosea*]..is more or less restricted to definite pockets in the wood, it has been called brown-pocket rot. **1938** J. S. Boyce *Forest Pathol.* xvii. 451 Pocket dry rot..is a brown pocket rot of the heartwood of living incense cedar caused by *Polyporus amarus*. **1956** F. W. Jane *Structure of Wood* ix. 208 In the pocket rots the areas of decay are confined to pockets. **1972** *Mycologia* LXIV. 1258 [*Phellinus torulosus*] occurs..mostly on south-western white pine..in which it causes a white pocket rot of the roots. **1909** *Daily Chron.* 13 Nov. 3/2 They form a serviceable little group of pocket-size manuals. **1973** P. Evans *Bodyguard Man* xxx. 183 You know what you are... Just a latter-day Judas, pocket size. **1907** *Daily Chron.* 14 Aug. 3/2 A new pocket-sized edition of Mr. Edward Hulme's 'Wild Fruits of the Countryside' is being published. **1954** Koestler *Invisible Writing* xxiii. 253 Our Jan..was a pocket-sized Stalin. **1964** *McCall's Sewing* ii. 31/2 *Pocket stay*, a strip of interfacing sewn to the wrong side of a pocket opening for reinforcement. **1942** O. D. von Engeln *Geomorphol.* xxii. 569 Seepage at the bases of the outer scarps [of uplifted coral reefs] promotes the formation of gullies by headward erosion through solution-sapping. Pocket valleys, with vertical head and side walls and a nearly flat floor, are..opened up by this process. **1966** J. Wyckoff *Rock, Time, & Landforms* xii. 281 Whereas the blind valley ends at a blank wall, the so-called pocket valley begins at a blank wall. A pocket valley forms where water emerges near the foot of the slope, dissolving out the rock around and below the point of emergence. **1971** J. N. Jennings *Karst* vi. 112 A distinction is sometimes made between steepheads incised to an impervious basement and pocket valleys of the same general nature but within the karst rock outcrop. **1869** S. R. Hole *Bk. about Roses* viii. 125 The lovely little Banksian Rose..this pocket, or rather button-hole, Venus. **1921** W. de la Mare *Memoirs of Midget* xxxiii. 229 Aunt Alice calls you her 'pocket Venus', and she means it, too, in her own sly way. **1969** H. K. Fleming *Day they kidnapped Queen Victoria* vi. 106 Four years had gone by, since, as the 'Pocket Venus', she had been the rage and toast of society. **1979** 'P. O'Connor' *Into Strong City* ii. xxvii. 98 Nancy was dark and petite, perfectly formed—the proverbial pocket venus. **1842** *Ohio Statesman* 19 Dec. 3/1 (*heading*) The pocket vetoes. **1973** *Time* 25 June 20/3 As presidential counsel, he worked out the legal basis for Nixon's impoundment of funds, broad use of pocket vetoes and Executive privilege.

pocketability (pǫ:kětăbi·lĭti). [f. Pocketable *a.*: see -ity.] The capacity to be put or carried in the pocket.

1970 *Amateur Photographer* 22 Apr. 21/1 No modern range-finder camera with a comparable lens..and a built-in selenium meter has its priceless asset of pocketability. **1977** *New Scientist* 10 Feb. p. ii (Advt.), If real pocketability matters, the Sinclair Cambridge Scientific offers a comparable range of functions in a uniquely compact format.

pocketa-pocketa (pǫ·kětă,pǫ·kětă). Also **pockety.** [Echoic.] An imitation of the regular sound made by a smoothly-running internal combustion engine; also *transf.* Also *attrib.* and *adv.*

1939 J. Thurber in *New Yorker* 18 Mar. 19/1 The pounding of the cylinders increased: ta-pocketa-pocketa-pocketa-*pocketa-pocketa*. **1968** L. Deighton *Only when I Larf* i. 11 A helicopter was warming up... Pockety, pockety, pockety. **1971** *Flying* Apr. 93/1, I brought the power back from the standard 1,700 run-up rpm to a pocketa-pocketa idle. **1977** J. Aiken *Last Movement* ix. 174, I could hear his heart going pocketa-pocketa.

pocket-book. Add: **1.** In recent U.S. use, also a cheap edition, esp. paper-bound. Also *attrib.*

1953 *Amer. Scholar* XXIII. 10 The *Galaxy* serial 'Gravy Planet' (recently republished in a pocket book as *The Space Merchants*). **1959** N. Mailer *Advts. for Myself* (1961) 280 Anyone looking for a fairly close portrait of that outfit is invited to read *The Day the Century Ended*—in pocket-book called *Between Heaven and Hell*. **1962** A. Buchwald *How much is that in Dollars?* 167 Always try to get a large advance on pocket-books. **1979** *Maledicta* III. 15, I was paid $50,000 by New American Library in 1968 for the U.S. pocketbook rights of my last novel, *The Man Who Loved Women*.

2. (Further examples.) Also, a woman's hand-bag or other container for bank-notes or coins. Also *fig.* Now chiefly *U.S.*

1816 *Niles' Reg.* X. 216/1 Two methodist preachers were lately robbed of their pocket-books, containing very considerable sums in bank notes. **1862** O. L. Jackson in *Colonel's Diary* (1922) v. 67, I..thought it best to take my pocket-book out of my pocket and put it under my head. **1897** *Sears, Roebuck Catal.* 223/1 At 50c. Our price poultice for tired..pocket books. **1907** *St. Nicholas* Sept. 1007/2 In her pretty pocket-book..she had found a crisp one-dollar bill. **1936** C. Sandburg *People, Yes* 111 So dumb he spent his last dollar buying a pocketbook to put it in. **1960** A. Sexton *To Bedlam & Part Way Back* 13 You guided past groups of robbers..clutching your pocketbook. **1976** *National Observer* (U.S.) 13 Mar. 18/4 Why should women whose pocketbooks are flatter have to jeopardize their lives..by patronizing sleazy, unqualified nonprofessionals.

pocketed, *ppl. a.* (In Dict. s.v. Pocket *v.*) (Earlier and later examples.)

1879 Trollope *John Caldigate* II. xxi. 288 They who were less privileged had fed themselves with pocketed sandwiches. **1934** T. S. Eliot *Rock* i. 10 With pocketed hands..We stand about in open places. **1974** C. Ryan *Bridge too Far* i. iii. 29 Though pocketed—the sea lay behind them to the north and west, and Canadians and British were pressing in from the south and east—they nevertheless controlled most of the southern bank of the estuary.

pocket-handkerchief. Add: **1.** (Earlier examples.)

1645 in *Essex* (Mass.) *Inst. Hist. Coll.* (1914) L. 326, 3 pocket handkerchiefs. **1680** Rochester *Poems* 16 Where Critick-like, he sits and squints, Steals Pocket-Handkerchiefs, and hints. **1760** Washington *Diary* 15 Feb. (1925) I. 126 Pockethandkerchiefs servd the purposes of Table Cloths and Napkins.

2. *transf.* and *fig.* **a.** A very small area (*of* land, etc.).

1866 [see sense 3 below]. **1891** 'Ganconagh' *John Sherman & Dhoya* 70 This pocket-handkerchief of a garden. **1949** T. Rattigan *Browning Version* 27 It's only a pocket handkerchief, I'm afraid, but it's very useful to Andrew. He often works out there. **1961** L. Mumford *City in History* xv. 465 The more respectable quarters..with a soiled pocket-handkerchief of grass before their houses. **1973** *Daily Tel.* 8 Dec. 12/3 There is a pocket handkerchief of a dance floor.

b. A light sail.

1936 B. Adams *Ships & Women* xi. 239 Martin told me to 'Go get the pocket handkerchiefs off her'. So I called out the watch and hauled down the jib topsail. **1941** *Penguin New Writing* II. 17, I could ride out bad weather with two of the smallest pocket-handkerchiefs.

3. *attrib.* and *Comb.*

1866 Geo. Eliot *Felix Holt* II. xx. 87 A mere pocket-handkerchief..Pod's End. **1935** N. L. McClung *Clearing in West* iv. 32 This kind of pocket handkerchief-farming makes people mean. **1953** Wodehouse *Performing Flea* 53 Bungalows..each with a little lawn in front and a pocket-handkerchief garden at the back. **1961** *Daily Tel.* 21 Oct. 6/2 The knowledge..helped the Russians to turn the screw..on economic and technical aid to this pocket-handkerchief State [*sc.* Albania]. **1973** N. Graham *Murder in Dark Room* v. 33 A row of little old-fashioned houses with pocket handkerchief lawns.

pocketing, *vbl. sb.* (s.v. Pocket *v.*) Add: **a.** (Later examples in various senses of the vb.)

1885 [see Pocket *v.* 3 b]. **1960** E. Ennion *House on Shore* ix. 108 They would have the greatest difficulty in taking off again: pocketing in snow or sand might easily prevent it. **1963** J. Osborne *Dental Mechanics for Students* (ed. 5) ix. 165 The extent of periodontal pocketing will also be shown.

b. Material for pockets.

1933 J. E. Liberty *Practical Tailoring* v. 64 The pocketing should be about 15½ in. to 16 in. long. **1968** J. Ironside *Fashion Alphabet* 97 *Pocketing*, strong cotton used for men's pockets.

pocket-money. (Further examples.)

1892 A. W. Pinero *Magistrate* iii. i. 114 All my pocket money is in my overcoat at the Hôtel des Princes. **1926** T. E. Lawrence *Seven Pillars* (1935) xii. 89 They made pocket-money during their service, if they were ingenious. **1973** *Times* 13 Dec. 21/7 Baroness Marie-Anne..bought from her pocket money Van Gogh's L'arlésienne'. **1979** *Jrnl. R. Soc. Arts* CXXVII. 135/2 We already live in a pocket-money economy where essentials are provided and we work for the extras.

Hence **po·cket-mo:neyless** *a.*
1925 A. S. M. Hutchinson *One Increasing Purpose* I. xv. 90 The kind of children, well-bred, entirely pocket-moneyless, that retired Anglo-Indians often have.

pocket-picking, *ppl. a.* That picks pockets.
1868 Geo. Eliot in *Blackw. Mag.* Jan. 3/2 A poor pocket-picking scoundrel, who will steal your loose pence while you are listening round the platform.

po·cketwards, *adv.* [-wards.] In the direction of one's pocket.
1909 H. G. Wells *Tono-Bungay* III. i. 280 He made a motion pocketwards, that gave us an invincible persuasion that he had a sample upon him.

pockety, *a.* Add: Also **pocketty. 1.** Also *fig.*
1920 Galsworthy *In Chancery* II. x. 204 The atmosphere of his house was strange and pocketty when Jolyon came in and told them of the dog Balthasar's death. The news had a unifying effect.
2. Also, characterized by secluded hollows.
1929 J. Buchan *Courts of Morning* II. 257 Days with a bobbery pack of hounds in difficult pockety country. **1932** —— *Gap in Curtain* iv. 190 The river valley was pockety and swampy.

pock-mark (pǫ·kmāɹk), *sb.* Also **pock mark, pockmark.** [f. Pock *sb.* + Mark *sb.*[1]] A scar, mark, or 'pit' left by a pustule, esp. of small-pox. Also *fig.*
1673, 1851 [in Dict. s.v. Pock *sb.* 4]. **1952** G. Wilson *Julien Ware* i. 5 In the yard outside the bail a second cow..stumbled uncertainly..over the sun-dried pock-marks and ridges carved by her own hoofs..when the mud of winter lay there. **1954** G. Durrell *Bafut Beagles* ix. 156 A steady downpour..turned the red earth of the great courtyard into a shimmering sea of blood-red clay freckled with pockmarks of the falling rain. **1966** L. Cohen *Beautiful Losers* I. 23 Can I yearn after pimples and pock marks? **1976** M. Green *Children of Sun* iv. 135 Orwell, while..at Eton, had no powerful defences against the stimulus to dandyism, and..a few pock-marks remained all his life in testimony of his inoculation. **1979** R. Blythe *View in Winter* 85 The pock-marks of the shots are still to be seen today on the crinkle-crankle wall.
So **po·ck-marked** *a.* (see Pock *sb.* 4); also *fig.*; **po·ck-mark** *v. trans.*, to mark or disfigure with pock-marks; also *fig.*
1756, 1899 Pock-marked [in Dict. s.v. Pock *sb.* 4]. **1908** *Flag* (Union Jack Club) 39 The floors lower down were pock-marked with splashes of the liquid. **1928** *Daily Express* 17 Apr. 10/2 Petrol pumps that pockmark the English countryside. **1952** V. Canning *House of Seven Flies* xi. 155 The oars pock-marking the dark current with white eddies. **1957** L. Durrell *Justine* I. 68 The silence pock-marked by the sound of our horses' hooves. **1963** V. Nabokov *Gift* v. 306 A Georgian socialist with a pockmarked face. **1964** A. Wykes *Gambling* i. 10 He risked his own somewhat pockmarked career. **1973** *Country Life* 14 June 1751/1 Hollyhocks can be sprayed..to suppress rust-disease, whose orange pustules otherwise would soon pock-mark the foliage. **1977** H. Innes *Big Footprints* I. i. 9 Walls pock-marked with bullets. **1979** V. Canning *Satan Sampler* ix. 183 A fierce spring shower was pock-marking the surface of the lake.

‖ **poco** (pǫ·ko), *adv. Mus.* [It.] A little, rather: used in musical directions.
1724 *Explication of Foreign Words in Musick Bks.* 56 *Poco*, a little less, and is just contrary to the forgoing Word *Piu*. **1760** L. Sterne *Tristram Shandy* II. vi. 48 How does the *Poco piu* and the *Poco meno* of the *Italian* artists;—the insensible, more or less, determine the precise line of beauty. *a* **1817** [see *allegretto]. **1884** F. Niecks *Conc. Dict. Mus. Terms* 194 Poco (It.), a little—*Poco a poco*, little by little; *poco allegro*, somewhat quick. **1963** *Times* 16 Jan. 11/1 Perhaps his tone in *mezzo* or *poco forte* should have been more incisive. **1969** *Listener* 4 Sept. 320/2 In the melodic lines of the *Poco Adagio*..the pervading interval not only shapes the..themes but provides the movement of the harmonic bass lines.

Pocomania (pokǒmēɪ·niă). Also with lower-case initial. [Origin unknown: prob. Hispanicized form of native name, with second element interpreted as Mania.] A Jamaican religious rite combining revivalism with ancestor-worship and spirit possession; the cult in which this rite is practised. Also *attrib.* Hence **pocoma·niac**, an adherent of this cult; **pocomani·acal** *a.*
[**1929** M. W. Beckwith *Black Roadways* 176 Revivalist and Obeah Man unite in the particular religious cult known as the Pukkumerian... The Pukkumerians hold their meetings near a grave-yard, and it is to the ghosts of their own membership that they appeal when spirits are summoned to a meeting... 'They jump and dance and sing and talk in a secret language because the spirits do not talk our language.'] **1938** Z. Hurston *Voodoo Gods* (1939) ii. 10, I went to the various 'tables' set in Pocomania, which boils down to a mixture of African obeah and Christianity enlivened by very beautiful singing. **1957** F. Henriques *Jamaica* x. 184 Of the specifically native cult groups Pocomania is the most powerful and active. *Ibid.* 185 Native churches..incorporated a degree of physical enthusiasm, evinced in dancing, violent singing, and clapping... It is from this type of activity that such cult groups as Pocomania arose. **1957** *Times Lit. Suppl.* 11 Oct. 612/4 The second part of the book is a study of contemporary Jamaican life... He deals with..the significance of cult-groups such as the Pocomaniacs. **1959** A. Salkey *Quality of Violence* ii. 36 He told them that the Jamaican celebration of Pocomania closely resembled Haitian Voodoo. *Ibid.* iv. 54 There were about twenty-five people gyrating and uttering Pocomaniacal prayers. **1971** J. Brunner *Honky in Woodpile* iv. 31 They broke up religious ceremonies, in particular *pocomania*, rites of possession using the *yoma-xi* drug. **1974** L. Watson *Romeo Error* I. iii. 80 In Zambia, traditional healers cast out evil spirits by holding a patient's head under a blanket and over a smoking brazier, where he is forced to hyperventilate by breathing very rapidly and shallowly... The pocomania ceremonies in Jamaica are built round 'tromping', which is a rhythm of foot-stamping and peculiar breathing sounds. **1976** Boot & Thomas *Jamaica* 82/2 When..the Pocomaniacs feel the spirit quickening in them, they jump and shout and testify till they get so drunk with righteous heaven-sent electricity they froth at the mouth.

pocosin, poquosin. Add: Also 7 **pocosen.** (Earlier examples.)
1634 in *Amer. Speech* (1940) XV. 296/2 From that runn along the side of the Pocoson or great Otter pond soe called. **1681** *Rec. Court of New Castle on Delaware* (1904) 504, 74 perches to a Corner marked spannish oake standing neare a pocosen.

poculum (pǫ·kiuləm). *Rom. Antiq.* Pl. **pocula.** [L.] A cup or drinking-vessel.
1863 W. Chaffers *Marks & Monograms Pott. & Porc.* 15 Figure 16 is a poculum of the Castor Ware of white paste. **1884** A. Rich *Dict. Rom. & Gr. Antiq.* 514/1 *Poculum*, a general term for any description of vessel employed as a drinking-cup, and thus including all the special ones. **1965** *Amer. N. & Q.* Mar. 106/1 Visitors can admire a series of cases, bright vases..and decorated 'pocula'.

pod, *sb.*[2] Add: **1. b.** Colloq. phr. *in pod*: pregnant; also *fig.*
1890 Barrère & Leland *Dict. Slang* II. 141/2 *Pod, in*, in the family way. **1922** Joyce *Ulysses* 385 Costello..would sing a bawdy catch..about a wench that was put in pod of a jolly swashbuckler. **1935** L. Durrell *Spirit of Place* (1969) 33, I am in pod again and am pupping a novel. **1958** P. Mortimer *Daddy's gone a-Hunting* xi. 60, I married you because you were in pod. **1968** M. Bragg *Without City Wall* xxvii. 245 Your working-class lad is still a bit worried if he gets his girl in pod. **1972** 'R. Gordon' *Doctor on Brain* xxvi. 190 But why didn't Josephine just tell you she suspected she was in pod? **1976** J. McNeish *Glass Zoo* II. xvi. 179 It wasn't Leonard who got Marsh's sister in pod.
c. *slang.* Marijuana. Cf. *pot *sb.*[6]
1952 [see *jive *sb.* 4]. **1952** [see *joint *sb.* 14 c]. **1955** *Amer. Speech* XXX. 304, 'I got no eyes for turning on with pod' (I have absolutely no use for smoking marijuana). **1959** W. Burroughs *Naked Lunch* 8 A square wants to come on hip... Talks about 'pod', and smokes it now and then. **1979** *High Times* Mar. 19/1 Pod. Say it aloud. It's so much better than 'pot'. A marijuana mantra. *Ibid.*, Pod suggests seeds, buds, pollen, odors, all the multi-dimensional sensual life of the fine plant, while pot ought to remain a word for a thing you plant pod in.
5*. *Geol.* A body of ore or rock whose length greatly exceeds its other dimensions.
1942 T. P. Thayer in *Bull. U.S. Geol. Survey* No. 935. 23 Chromite deposits of the sack-form variety are notable for their variation in size and shape... The majority..might be termed lenses or pods, as their length greatly exceeds their width. **1969** Bennison & Wright *Geol. Hist. Brit. Isles* iii. 42 They occur as isolated pods distributed very widely through the Scourian granulites and their Inverian derivatives. **1977** *Bulletin* (Sydney) 22 Jan. 69/3 The Darling Range deposits grade only between 28 and 36 percent aluminium oxide. Also, they occur in pods, instead of more or less continuous orebodies.
5.** An elongated, streamlined compartment attached to an aircraft and containing an engine, fuel tanks, or the like; a detachable compartment in a spacecraft; also, any protruding or detachable casing on or in a craft or vehicle.
1950 J. V. Casamassa *Jet Aircraft Power Systems* 318 Jet pods are mounted beneath the wings. **1951** *Engineering* 6 July 8/2 One of the reasons for mounting jet engines on 'pods'..was because it was impossible to bury the engines in thin wings. **1955** Pohl & Kornbluth *Space Merchants* ii. 23 The cargo pod of the 'copter hit the concrete a yard from where we stood. **1963** *New Scientist* 9 May 320/3 Rides are being 'hitch-hiked' on Atlas rockets for pods of space instruments. **1965** *Guardian* 25 Aug. 9/2 They [*sc.* two astronauts] had ejected the radar evaluation pod, a 76 lb. satellite. **1967** *Jane's Surface Skimmer Systems 1967–68* 91/2 Power is transmitted through a mechanical right-angle drive transmission to a propeller at the aft end of a strut and pod assembly. **1973** *Sci. Amer.* Aug. 13/1 A rotating radome, or radar pod, is mounted on two struts above the rear section of the fuselage. **1976** *Good Motoring* May 21/2 A particularly quiet car, allowing most of the sound from speakers front and rear to be appreciated. Three pairs fitted at the rear offer a choice from large flush, small flush or pods. **1977** W. Marshall *Thin Air* xi. 148 The fuel from the main tanks ran out and was replaced a moment later by the Kerosene from the secondary pods.
6. (sense *1 c) *pod smoke, -smoker*; **pod corn,** a variety of maize, *Zea mays* var. *tunicata*, in which each kernel, as well as the whole ear, is enclosed in a husk; **pod maize** = prec.
1893 *Science* 17 Nov. 268/1 The kernel of the pod corn does not present structural differences markedly unlike that of the flint corn..but this type differs from all others in that each kernel has a husk of its own..; hence the name pod corn. **1923** Wallace & Bressmann *Corn & Corn Growing* xxvi. 147 Pod corn—each kernel enclosed by a husk as well as the entire ear. **1957** E. Hyams *Speaking Garden* viii. 106 There occurs as a kind of aberration among maize a plant called pod-corn. **1976** R. W. Jugenheimer *Corn* iii. 41 Podcorn is not being grown commercially. **1904** T. F. Hunt *Cereals in Amer.* x. 164 Pod maize is rarely grown. **1914** J. Burtt-Davy *Maize* iv. 103 In 'pod maize'..the glumes are large, completely enclosing the ovary and persisting around the ripe grain. **1979** *High Times* Mar. 19/1 Early jazz-musician pod smokers. *Ibid.*, The culture that made it possible for jazz musicians to turn sweet pod smoke into sweet soul sounds.

podargus (pǫdā·ɪgəs). *Ornith.* [mod.L. (L. J. P. Vieillot in *Nouveau Dict. Hist. Nat.* (ed. 2, 1818) XXVII. 151), f. Gr. πόδαργος swift-footed.] A nocturnal, greyish-brown bird of the genus so called, belonging to the family Podargidæ and found in Australia, New Guinea, and the Solomon Islands; esp. the Australian tawny frogmouth, *Podargus strigoides*. Cf. Frog-mouth, frog's mouth 2.
1837 *Proc. Zool. Soc.* V. 67 The sclerotic ring of the great *Podargus* does not present the slightest appearance of distinct plates. **1901** A. J. Campbell *Nests & Eggs Austral. Birds* II. 539 Under the heading of the Tawny-shouldered Podargus..is included *P. cuvieri*, *P. gouldi*, and the ever doubtful *P. megacephalus*. **1933** *Bulletin* (Sydney) 19 Apr. 21/4 My choice for the quietest bush bird is the tawny frogmouth, or podargus. **1961** *Coast to Coast 1959–60* 66 'You know the Podargus.' 'It's a bird. The tawny-shouldered frogmouth.'

podded, *a.* Add: **3.** *Aeronaut.* Mounted in a pod or pods.
1959 *Times* 26 Feb. 10/6 It has not been British practice to build aircraft with podded engines. **1960** *New Scientist* 5 May 1115/3 Unlike the United States, which has remained faithful to podded engines slung beneath the wing, Russia has resorted to both podded and buried engines in placing the power units of the *Bounder*.

poddle (pǫ·d'l), dial. var. Paddle *v.*[1] 4.
1827 J. Clare *Shepherd's Calendar* 69 The ruddy child, nursed in the lap of care,..Beside its mother poddles o'er the land. **1842** C. Ridley *Let.* 5 Mar. in *Cecilia* (1958) vii. 89, I..spend a great deal of time in poddling about the garden. **1869** R. D. Blackmore *Lorna Doone* I. x. 109 Now I am uncommonly fond of ducks..and it is a fine sight to behold them walk, poddling one after other. **1976** Scollins & Titford *Ey up, mi Duck!* I. 59 *Poddlin'*, walking; implies a comical gait. Usually describes a small child, or a little old man, etc. As in: 'Eh wer *poddlin'* along wi'aht a care int wold!'

poddy, *a.* Add: **A.** *adj.* **2.** (See sense B. 1 b below.)
B. *sb.* (Austral.) **1. a.** An unbranded calf.
1893 K. Mackay *Out Back* (ed. 2) I. v. 75, I did occasionally put my brand by mistake on one of Massey's 'poddies'. **1907** G. B. Lancaster *Tracks we Tread* iii. 52 [The wild cattle] were a mixed haul: two-year-olds, poddies and pikers. **1950** [see *clean-skins* s.v. *clean- 2].
b. In full *poddy calf* (*foal*, etc.). A calf (less commonly a lamb or foal) fed by hand.
1898 *Bulletin* (Sydney) 8 Jan. (Red Page), Prof. Morris [in *Austral English*] defines 'Poddy' as 'a Vic. name for sand-mullet', but leaves out its meaning of motherless calf or foal (common in the bush). A poddy calf or a poddy foal is heard all over Australia. **1901** M. Franklin *My Brilliant Career* v. 24 It was my duty to 'rare the poddies'. **1908** *Bulletin* (Sydney) 30 Jan. 14, I saw a boy..driving back to pasture his flock of sixty or seventy newly shorn 'poddies', and it reminded me that the ewe is about the most indifferent mother in the bush. **1911** E. M. Clowes *On Wallaby* iii. 66 He drives off with the separated milk—due from the day before for his poddy-calves. **1927** B. Cronin *Red Dawson* xliii. 194 His whole outfit was five old cows and a coupler poddies. **1930** H. S. Palmer *Men are Human* xxv. 235 He's tame as a poddy calf. **1963** A. Lubbock *Austral. Roundabout* 5 The kitchen range..had saucepans of milk, and babies' bottles and teats, boiling on the top to feed the 'poddies', as hand-fed calves and lambs are called.
2. *attrib.* and *Comb.*, as *poddy swill*; *poddy-rearing* vbl. sb.; **poddy-dodger,** one who steals unbranded calves; a cattle rustler; also *poddy-dodging* vbl. sb.
1934 *Bulletin* (Sydney) 1 Aug. 46/3 Nine poddy-dodgers out of ten gets caught the same way. **1953** A. Moorehead *Rum Jungle* ii. 30 The cattle rustlers—known as 'poddy dodgers'—followed close behind. **1970** *Sunday Mail Mag.* (Brisbane) 30 Aug. 3/5 His practice, as a 'poddy-dodger', was to steal branded cows and cleanskin calves from neighbours, then remove the calves from their mothers before they were ready. **1945** T. Ronan *Strangers on Ophir* 9 He'll be a doctor or a lawyer or a banker with no need to go poddy-dodging for a living like his old jail-bird of a Dad. **1950** 'N. Shute' *Town like Alice* ix. 263 They'll come on to your station and round up the poddys and drive them off on to their own land, and then there's nothing to say they're yours. That's poddy-dodging. **1957** R. S. Porteous *Brigalow* 61 Mick did a bit of poddy-dodging when things were slack... He might lift a few head of cleanskins now and then. **1901** M. Franklin *My Brilliant Career* iii. 17 They do all the milkin', and pig-feedin', and poddy-rarin'. **1941** *Coast to Coast* 108 Tug Treloar carrying buckets of poddy swill.

poddy (pǫ·di), *v. Austral. colloq.* [f. *poddy *sb.*] *trans.* To feed (a young animal) by hand.
1896 H. Lawson *While Billy Boils* (1897) 61 Then he 'poddies'—hand-feeds—the calves which have been weaned too early. **1901** M. Franklin *My Brilliant Career* iv. 20 He procured fifty milch-cows, the calves of which had to be 'poddied'. **1908** *Bulletin* (Sydney) 30 Jan. 14 The squatter knows that a deserted lamb will die, also he has no time to 'poddy' it. **1931** V. Palmer *Separate Lives* 176 When her [*sc.* the filly's] mother died, that old drover said I could have her... And I did rear her, bought condensed milk to poddy her for months.

‖ **podere** (podẹ̄·re). Pl. **poderi** (-*i*). [It.] A farm or estate.
1884 S. & J. Horner *Walks in Florence & its Environs* II. xxi. 299 Villas and their surrounding vineyards, olive gardens, and farms (*Poderi*) are enclosed within these walls. **1891** L. Scott *Vincigliata & Maiano* 13 The Podere of the Torre..with its courts, trees, house, huts and vineyards. *Ibid.* 15 Niccolò, as heir to his father, had a larger claim..on several lots, especially the Castle of Vincigliata and its *poderi*. **1904** H. James *Golden Bowl* I. iii. 57 The dear contadini of the *podere*, the little girls and the other peasants of the next podere. **1926** D. H. Lawrence *Sun* v. 18 Yet she missed him when he did not come to work on the podere. **1961** W. Vaughan-Thomas *Anzio* vii. 137 The peasant farmers had remained in the *poderi* of the countryside. **1970** I. Origo *Images & Shadows* v. 116 The steep terraces of the *podere*, partly cultivated with plots of wheat.

podge, *sb.* Add: Freq. a plump child. (Later examples.) Also, excess weight, fat.
1967 M. Summerton *Memory of Darkness* i. 17 She used to be a horrid little podge, always whining. **1967** A. Wilson *No Laughing Matter* i. 16 Don't tease the poor Podge. Time enough when she loses all that puppy fat. **1976** *Leicester Chron.* 26 Nov., The average man is putting on too much weight, and..needs clothes designed to help him hide the podge.

podger[2] (pǫ·dʒəɪ). [f. Podge *sb.* + -er[1].] Any of various tools having the form of a short bar (see quots.).
1888 J. G. Horner *Dict. Terms Mech. Engin.* 377 *Tommy*, a pointed round iron bar or lever used for insertion in the holes drilled in the circular back nuts of lathes and other machines, for the purpose of tightening them up. Also a metal rod kept for insertion in the eyes of the tightening screws of hand-rest sockets, for tightening the T rest. Sometimes called a podger. **1893** —— *Princ. of Fitting* ii. 22 Fig. 18 shows a podger, employed for two purposes. The tapered end A, of round section, is used for pulling drilled or punched holes into line, so that their bolts or rivets can be inserted. The flattened end B is used like the end of a crowbar for lifting up and slipping along a casting or forging into position, or for prising open..plates that are in close contact. **1894** W. J. Lineham *Textbk. Mech. Engin.* vii. 286 Before riveting a seam, the plates, if punched or drilled separately, are brought into alignment by the podger and bolted. **1971** B. Scharf *Engin. & its Language* ix. 66 A special type of single ended spanner is the podger or prong-ended spanner. This has a long, tapering handle which can be used..to align holes which do not entirely coincide (e.g. rivet holes in plates).

podiatry (pǒdəi·ătri). orig. and chiefly *U.S.* [f. Gr. ποδ-, πούς foot + ἰατρεία healing: see -y[3].] The diagnosis and treatment of disorders of the foot; chiropody.
1914 F. von Oefele in M. J. Lewi *Text Bk. Chiropody* i. 3 The practice of foot lesions may hereafter be styled 'Helotomy' or, 'Heliatry', or more generally 'Podiatry'. **1947** P. Lewin *Foot & Ankle* (ed. 3) xxxii. 717 Podiatry is the science of treatment of certain disorders of the feet. **1958** *Technology* Feb. 425/4 The National Association of Chiropodists has announced that it has changed its title to The American Podiatry Association and that its members will henceforth be known by the 'more dignified' style of podiatrists. **1968** F. Weinstein *Princ. & Pract. Podiatry* i. 6/1 More and more hospitals [in the U.S.A.] are working toward the establishment of a regular department or division of podiatry. **1979** *Arizona Daily Star* 22 July B1/6 Podiatry problems can trigger one of many vicious circles affecting the health of the elderly.
So **podia·tric** *a.*; **podi·atrist,** one who practises podiatry; a chiropodist.
1914 F. von Oefele in M. J. Lewi *Text Bk. Chiropody* i. 50 We should prevent the possibility of such a ridiculous misunderstanding by substituting the word 'podiatrist' (physician of the foot) for the unscientific term 'chiropodist'. **1922** *Jrnl. Nat. Assoc. Chiropodists* XII. 35 (Advt.), Bachelet podiatric triplex generator. **1929** E. W. Springs *Above Bright Blue Sky* 97 I've got to hobble along and see my podiatrist. **1950** K. Newton *Geriatric Nursing* xv. 287 The aged patient will avoid serious difficulties with corns and calluses by having his feet cared for regularly by a podiatrist. **1968** T. R. Amberry in F. Weinstein *Princ. & Pract. Podiatry* viii. 138 All too often these factors are either ignored or overlooked by some surgeons, both medical and podiatric. **1974** *Telegraph* (Brisbane) 21 Mar. 16/7 The Australian Chiropody Council wants chiropodists to be called podiatrists. **1976** *Jrnl. Amer. Podiatry Assoc.* LXVI. 15 A wide variety of medications may be necessary during the hospitalization of podiatric surgical patients. **1978** *Detroit Free Press* 16 Apr. 1c/5 Podiatrists and athletes debate endlessly about proper footwear.

podium. Delete ‖ and add: Also with pl. **podiums. 1. b.** (Earlier example.)
1743 W. Stukeley *Abury* vii. 28 This was as the *podium* of an amphitheater, for the lower tire of spectators.
d. A raised platform or dais at the front of a hall or stage; *spec.* that occupied by the conductor of an orchestra.
1947 A. Einstein *Music in Romantic Era* xv. 215 The longer Chopin continued in his career, the more he avoided the concert podium and the expectant masses. **1955** R. Crane *Hero's Walk* v. 94 There was no applause when Dr. Werner took his place at the podium. **1972** *N.Y. Times* 3 Nov. 30/2 Mr. Steinberg stands there, all but motionless,..and manages to get more from an orchestra than a squadron of podium monkeys jumping up and down. **1973** W. H. Hallahan *Ross Forgery* (1977) iv. 65 The auctioneer stepped off the podium. **1977** *Kuwait Times* 23 Nov. 5/3, I address the following appeal from this podium to the people of Israel. **1978** *Gramophone* Oct. 658/3 He has a genre of courage and determination that enabled him to fight back after a particularly bad car accident and reappear on the podium long before the time appointed by his doctor.
e. A projecting lower structure around the base of a tower block.
1962 *Times* 19 Mar. 13/7 The podium-and-tower pattern is not only a product of daylighting codes. **1970** *Daily Tel.* 21 May 7 At the base of the 220 ft-high tower will be a podium of two and three-storey buildings.

podo (pōu·do). [f. generic name *Podo(carpus* (see *podocarp 2).] An evergreen tree belonging to one of several East African species of the genus *Podocarpus*, or the softwood timber obtained from it. Also *attrib.*
1922 W. Schlich *Man. Forestry* (ed. 4) I. 289 Podo, a medium-sized tree, has a wide distribution. **1940** W. H. Eggeling *Indigenous Trees of Uganda Protectorate* 179 Recently the tendency in Uganda has been to regard musizi..as a better timber than Podo. **1947** *E. African Ann. 1946–7* 41/1 The north-west point of the Game Reserve..is an impressive mountain with cedar, olive and podo on its summit. **1957** *Handbk. of Softwoods* (Forest Prod. Res. Lab.) 48 Podo is widely used for joinery and interior fittings. **1961** *New Scientist* 24 Aug. 451/3 The cedar, olive and podo forest that covers most of the Mau is being systematically destroyed.

podocarp (pōu·dokāɹp). *Bot.* [f. Podo- + Gr. καρπ-ός fruit.] **1.** (In Dict. s.v. Podo-.)
2. [f. mod.L. *Podocarpus* (C. L. L'Héritier de Brutelle in C. H. Persoon *Synopsis Plantarum* (1807) II. 580).] A coniferous tree or shrub belonging to the genus *Podocarpus* or the family Podocarpaceæ, including several valuable timber trees native to parts of the Southern Hemisphere. So **podocarpa·ceous, -ca·rpous** *adjs.*, belonging to the family Podocarpaceæ. Cf. *podo.
1858 G. Gordon *Pinetum* 268 (*heading*) *Eupodocarpus*, Endlicher, or the true podocarps. **1882** Podocarpous [in Dict. s.v. Podo-]. **1918** A. D. Webster *Coniferous Trees* i. 158 The Alpine Podocarp... This distinct and very interesting Tasmanian conifer may be seen in excellent condition in several gardens in the neighbourhood of London. **1923** A. Rehder in L. H. Bailey *Cultivated Evergreens* II. v. 179 (*heading*) Longleaf Podocarp. **1939** M. Hornibrook *Dwarf & Slow-Growing Conifers* (ed. 2) 216 Of the many podocarps which inhabit Australia and South America, not many of them are fully hardy in north European climates. **1959** A. H. McLintock *Descr. Atlas N.Z.* p. xv, The forest is of silver and mountain beech, with traces of red beech and podocarps at low altitude. **1975** *Nature* 27 Nov. 305/2 Among conifers, the mixture of podocarpaceous and araucarian elements is typical of the Southern Hemisphere early Tertiary. **1977** *N.Z. Listener* 15 Jan. 21/2 Cut down the last of those giant podocarps and you cut down about 800 years of growth. **1978** *Sci. Amer.* July 102/3 At the lower elevations and latitudes the forests were a mixture of broadleaf trees and podocarp conifers, a group of trees peculiar to the southwest Pacific and Chile.

Podolian (pǒdōu·liăn), *a.* [f. *Podolia* (see below) + -an.] Of or pertaining to Podolia, a region in south-west Ukraine (see quot. 1974).
c **1850** *Bee Keeper* 18 (*heading*) The construction of the Podolian hive. **1881** *Encycl. Brit.* XIII. 451/2 The breed of cattle most widely distributed throughout Italy is that known as the Podolian. **1911** *Ibid.* XXI. 875/1 The Dniester is an important channel for trade, corn, spirits and timber being exported from Podolian river ports. **1920** *Glasgow Herald* 7/2 Another great Bolshevik offensive in the Southern Polish and Podolian sector has failed absolutely. **1936** C. Roth *Short Hist. Jewish People* xxvi. 332 The revivalist movement in Poland permeated all sections of society, until it touched a simple Podolian lime-digger, Israel ben Eliezer (1700–1760). **1974** *Encycl. Brit. Micropædia* VIII. 60/2 *Podolia*..was under Polish rule until 1772, when the part west of the Zbruch River became Austrian; the rest became the Podolian *guberniya* (province) in Russia in 1793. After World War I the region continued to be divided at the Zbruch, between Poland and the U.S.S.R.; after World War II, it was entirely incorporated into the U.S.S.R.

Podsnap (pǫ·dsnæp). The name of a character in Dickens's 'Our Mutual Friend' (1864–5) used allusively for a person embodying insular complacency and self-satisfaction and refusal to' face up to unpleasant facts. So **Podsna·ppery, Podsna·p(p)ian** *a.*
1864 Dickens *Mut. Fr.* (1865) I. I. xi. 97 Mr. Podsnap..stood very high in Mr. Podsnap's opinion... Mr. Podsnap settled that whatever he put behind him he put out of existence... Mr. Podsnap's world was not a very large world..he considered other countries..a mistake... Mr. Podsnap was sensible of its being required of him to take Providence under his protection...What Providence meant was invariably what Mr. Podsnap meant.] *Ibid.* 98 These may be said to have been the articles of a faith and school which the present chapter takes the liberty of calling after its representative man, Podsnappery. [*Ibid.*, A certain institution in Mr. Podsnap's mind which he called 'the young person'..was an inconvenient and exacting institution... The question about everything was, would it bring a blush into the cheek of the young person.] **1880** *Daily Tel.* 1 Dec. 5/2 It is all very well to sneer at 'Chauvinism' and 'Podsnappery' but the claim of British children to supremacy among their kind must be resolutely upheld. **1901** *Daily News* 16 Feb. 3/3 The Podsnappian findings of a Commission which..reported ..that a certain hospital had been infested with bugs, but not in such a degree as to incommode the patients. **1902** *Westm. Gaz.* 21 Nov. 9/3 Podsnap General, Podsnap Financier, Podsnap Statesman we know, but it is my joy to have perceived our old friend in his last incarnation... Podsnap then, Podsnap in crimson tie of The School, 'The Red Badge of Culture', but Podsnap still, pervading the Press. **1905** G. K. Chesterton *Heretics* 29 No one will accuse the author of the 'Inferno' of..a Podsnapian optimism. **1929** A. Huxley *Do what you Will* 31 Those detestable Puritans to whom we owe..Grundyism and Podsnappery. **1944** A. Thirkell *Headmistress* vi. 149 *As You Like It* might be a suitable play..for the Young Person in a Podsnappian sense. **1960** *Times* 12 July 13/4 There are too many Podsnaps of officialdom who treat athletes and *aficionados* alike as noisy nuisances. **1972** G. S. Fraser in Cox & Dyson *20th-Cent. Mind* II. xi. 401 The solid Podsnappian complacencies and unbudgingesses of the male. **1977** P. N. Furbank *E. M. Forster* viii. 138 They had a jokey and facetious relationship, based on the fiction of Thompson's extreme Podsnappery and Philistinism.

podsol, var. *podzol.

‖ **podu** (pǫ·dū). [Telugu.] = *kumri.
[**1855** H. H. Wilson *Gloss. Indian Terms* 420 *Podu*, Tel., land or lands recently cleared from thickets and prepared for cultivation.] **1938** [see *kumri]. **1954** O. H. K. Spate *India & Pakistan* viii. 203 Shifting agriculture—the *jhum* of Assam, the *kumri* or *podu* of the Peninsula—conforms to the standard pattern so widespread in tropical regions.

Podunk (pōu·dəŋk). [Algonquian place-name.] **1.** Also 7 **Potunck.** A small tribe of Indians formerly inhabiting an area around the Podunk river in Hartford County, Connecticut (chiefly *pl.* or *collect.*) Also *attrib.* or *adj.*
1656 in *Public Records of Colony of Connecticut* (1850) I. 305 The Court wearied wth their speeches pressed the Potunck Indians to deliver up the murtherer. **1761** E. Stiles *Extracts from Itineraries* (1916) 136 Podunk Tribe at the dividing Line between Windsor & Hartford East side; between 2 & 300 Men in Philips War; went off & never returned. **1842** W. L. Stone *Uncas & Miantonomoh* 31 The Podunks resided upon the lands now comprised in the town of East Hartford. **1859** *N.Y. Weekly Tribune* 8 Oct. 3/1 The Numkatunks, Quinnipiacs, Podunks, and Quinnebogs, were present from New Haven and vicinity. **1910** F. W. Hodge et al. *Handbk. Amer. Indians* II. 271/1 Podunk. A band or small tribe on Podunk r., in Hartford co., Conn., closely related to the Poquonnoc. **1935** *Colony of Connecticut* (Connecticut Board of Educ.) (Senate Doc. 53, 74th Congr., 1st Sess.) 1 A few years previously in 1631 the chief of one of the Indian tribes in the Connecticut Valley, the Podunks, had journeyed to the Massachusetts and Plymouth Colonies to invite them to see the fertile Connecticut Valley. **1937** *Bull. Archaeol. Soc. Connecticut* Apr. 2 The Podunk tribe had two permanent villages during early colonial times... The largest Podunk village was located in South Windsor.
2. *U.S. colloq.* Name for a fictive insignificant, out-of-the-way town; a typical small town.
1846 *Daily National Pilot* (Buffalo, N.Y.) 13 Jan. 2/2 Messrs. Editors: I hear you ask, 'Where in the world is Podunk?' It is in the world, sir; and more than that, it is a little world of itself. It stands 'high up the big Pigeon', a bright and shining light amid the surrounding darkness. I look back, sir, with pride upon the day when I located in the then unincorporated burgh of Podunk. *Ibid.* 20 Jan. 2a, The distinguished festival of Podunk is the *candy bee*. *Ibid.* 6 Mar. 2b, Podunk is a huge town, not distinguished exactly as the geographies have it, for its 'fertile soil, salubrious and healthy climate', but for some of the characters that here do congregate. **1865** O. W. Norton *Army Lett.* (1903) 277, I presume that just about this time of day you are sitting in one of the slips in the 'Podunk' or 'Chachunk' (what do you call it?) 'meeting house'. **1901** *Harper's Weekly* 7 Sept. 903/2 He might just as well have been John Smith, of Podunk Centre. **1933** *Review of Reviews* Feb. 33 It [*sc.* the 18th amendment] required one rule for Podunk, Kansas, and one for New York City. **1947** [see *Main Street, Main streeter]. **1960** *Times Lit. Suppl.* 15 Jan. 27/3 A diploma from Harvard is much more marketable than a diploma from Podunk College. **1976** *National Observer* (U.S.) 9 Oct. 15/5 It won't be just the Podunk liberal-arts colleges that have to hustle. Some of the big state schools and some well-known large universities, will be out there too—to find new publics.

podzol (pǫ·dzǫl). *Soil Sci.* Also **podsol,** and formerly also with capital initial. [a. Russ. *podzól*, f. *pod-* under- + *zolá* ash.] An acidic, generally infertile soil which is characterized by a well-marked white or grey ash-like subsurface layer from which minerals have been leached into a lower dark-coloured layer, and which occurs esp. under coniferous trees or heath vegetation in moist, usu. temperate climates (typically in parts of N. Russia). Orig. applied only to the ash-like layer itself.
[**1906** E. W. Hilgard *Soils* x. 186 Woodlands of northern countries bearing beech and oak are especially apt to be benefited by the action of lime on the 'raw', acid humous

soil and underlying hardpan, which is commonly underlaid by a leaden-blue sandy subsoil ('Bleisand' of the Germans, 'Podzol' of the Russians) colored brown by earth humates.] **1908** *Jrnl. Agric. Sci.* III. 83 The most characteristic feature of the Podzol..is the dissolution and removal of soluble parts of silicates..and an increase in the percentage of insoluble silica. **1912** H. B. Woodward *Geol. Soils & Substrata* vii. 82 Of mixed soils, the *Podzol* of Russia, as described by Professor Glinka, consists of sands, loams, and clays, locally calcareous, but generally poor in mineral plant food. **1927** C. F. Marbut tr. *Glinka's Great Soil Groups of World* 71 All these profiles of Russian soils belong to the Podsols. This term is used to designate soils which have a pronounced and well developed whitish A₂ horizon. If this horizon is not well developed, and the corresponding horizon contains whitish specks and stringers the soil is said to be Podsolic. **1928** *Ecology* IX. 177 Originally, the term podsol applied more specifically to the gray-colored zone, though now it is commonly used to describe the entire profile. **1934** *Forestry* VIII. 25 Podzol soils owe their name to the presence of an ashy-grey layer which underlies the surface layer of dead vegetation and plant roots. **1936** *Nature* 17 Oct. 692/2 This utilization of the physical character and colour of the soil..is a novelty to glacialists from the leached podsol areas of the north-west. **1946** F. E. Zeuner *Dating Past* v. 124 Brownearth and podsol soils are characteristic of the humid-temperate countries. **1972** J. G. Cruickshank *Soil Geogr.* ii. 63 A coarse sandy deposit in Sherwood Forest..also showed signs of podsol features only 25–30 years after replanting with pine. **1973** *Sci. Amer.* Dec. 64/2 The tropical podzols are useless even for shifting agriculture; the Dayak peoples of Borneo call them *kerangas*: 'land on which one cannot grow rice'.
Hence **podzo·lic** (or **-ds-**) *a.*, of the nature of or resembling a podzol in possessing a layer from which some leaching of bases has occurred.
1927 C. F. Marbut tr. *Glinka's Great Soil Groups of World* 44 We can see again the change from Tschernosem to gray forest soils and the latter into Podsolic soils in the vicinity of Borshom. **1932** G. W. Robinson *Soils* xvi. 315 The soils of Great Britain belong mainly to the podsolic group. **1952** P. W. Richards *Tropical Rain Forest* ix. 209 If the American view is accepted, a lateritic soil can also be podzolic. **1973** P. A. Colinvaux *Introd. Ecol.* iii. 46 Some heath lands of northern Europe, with acid litter and leached soils, reveal podzolic profiles.

podzolize (pǫ·dzǒləiz), *v. Soil Sci.* Also **-sol-.** [f. prec. + -ize.] *trans.* and *intr.* To render or become podzolic. Chiefly as **po·dzolized** *ppl. a.*, **po·dzolizing** *vbl. sb.*
1923 *Soil Sci.* XVI. 97 It is the presence of the acid layer in a definite position in the profile and the strongly marked gray or podsolized horizon which chiefly distinguishes the typical northern from the typical southern profile. **1927** C. F. Marbut tr. *Glinka's Great Soil Groups of World* 100 Traces of the Podsolizing processes are shown in shallow depressions... The Podsolizing and leaching has reached a more advanced stage of the development. **1932** *Technical Communications Imperial Bureau Soil Sci.* No. 24. 11 Under conditions of poor drainage..resilication may give rise to a kaolinite red earth which..may be podsolised into a quartzose, bleached surface soil. **1938** *Geogr. Jrnl.* XCI. 163 The soils of the lower Trent Vale are podsolized sands. **1957** *Soil Sci.* LXXXIII. 215 The parent materials of Swartswood sand loam podzolize with comparative ease. **1976** *Sci. Amer.* Apr. 56/3 The dunes are partly podzolized, and they sustain pines, dwarfed redwoods and shrubs. **1976** *Nature* 15 Apr. 602/2 It would seem to result from percolating solutions removing surface coverings of ferric oxides from quartz sand grains and depositing them again within a short distance, that is, in a typical podzolising process.
Hence **po:dzoliza·tion,** the leaching of bases out of the upper parts of a soil and their deposition lower down; the formation of a podzolic soil.
1923 *Soil Sci.* XVI. 103 It is evident..that podsolization may take place in acid or alkaline [s]oils. **1934** *Discovery* July 200/2 The leaching and subsequent deposition of iron (and aluminium) oxides is a characteristic of all podsols, and is..generally referred to as 'podsolization'. **1952** P. W. Richards *Tropical Rain Forest* ix. 209 The removal of sesquioxides of iron and aluminium with the accumulation of silica is called podzolization. **1970** E. M. Bridges *World Soils* iii. 21/1 The process of podzolization is prevalent in the soils of the cool humid parts of the world and produces soils of the podzolic group and the podzols in particular.

pœcilitic, *a.* Add: † **2.** *Petrogr.* = *poikilitic *a.* 2. *Obs.*
1887 G. H. Williams in *Amer. Jrnl. Sci.* CXXXIII. 139 Here..we have another example of the structure which the writer has distinguished as pœcilitic* in describing the hornblende of the Cortlandt peridotites. [*Note*] *The word is here changed to the accepted form. **1909** F. P. Mennell *Introd. Petrol.* xii. 87 (*caption*) Poecilitic structure, in picrite, Belingwe, Rhodesia. (Olivine enclosed in enstatite).

poee-poee, var. *poipoi.

Poeesque (pōu·e·sk), *a.* Also **Poesque, Poe-esque.** [f. the name of Edgar Allan *Poe* (1809–49), American author, + -esque.] Of, pertaining to, or resembling E. A. Poe or his work. Also as quasi-*sb.*
1919 *Times Lit. Suppl.* 19 June 335/1 Mr. Harvey's 'The Beast with Five Fingers' is tinged with the Poesque. **1934** R. Campbell *Broken Record* 38 To a stranger they [*sc.* crows] give the English countryside a terrible and sinister, Poeesque atmosphere. **1959** R. Fuller *Ruined Boys* II. xi. 154 He was at a loss to imagine precisely what would indicate any impending disaster, short of some obvious and Poesque symptom like a great fissure in the school walls. **1977** *Times Lit. Suppl.* 20 May 611/3 [Edmund] Wilson's future biographers may want to make something of the Poe-esque inclination.
Similarly **Poeana** (pōuă·nă), objects associated with E. A. Poe; publications by or about Poe; **Poe·ish** *a.*; **Poe·ishly** *adv.*; **Poe·ist,** a student or devotee of Poe's works; **Poe--like** *a.*
1908 *Westm. Gaz.* 12 Feb. 4/2 There appeared in Paris, in 1836, Gautier's Poe-like story 'La Morte Amoureuse'. **1925** H. Acton in *Oxf. Poetry* 7 And oh, the Poe-like harmonies of bells! **1929** Wyndham Lewis *Let.* 20 Feb. (1963) 187 My reply to this Poeist..is that Poe would not have found my writing 'difficult'. **1955** *N. & Q.* May 223/1 All Poeana has been enthusiastically collected for seventy years. **1976** *National Observer* (U.S.) 9 Oct. 24/3, I like the plantation owner brooding Poeishly over the corpse of his too-well-beloved sister. **1977** *Amer. N. & Q.* XV. 70/1 While several contemporary reviewers saw 'Poeish' similarities in Melville's writings, modern critics ..have found little reason to suspect that Melville was influenced by Poe.

‖ **poêlée** (pwẹle). Also *erron.* **poêle.** [Fr., a panful; cf. *poêler* to cook in a pan.] A broth or stock (see quots.).
1830 R. Dolby *Cook's Dict.* 414 *Poelee.* Take two pounds of veal, two pounds of bacon, two large carrots and three onions; cut all these into dice and put them into a stewpan with a pound of butter, the juice of three or four lemons, four cloves, two bay-leaves, bruised, a little thyme, salt, and pepper. **1845** E. Acton *Mod. Cookery* vii. 185 *Poêlée.* Cut into large dice..lean veal..fat bacon..carrots..onions;..add..butter..; pour..boiling broth..; strain the *poêlée* through a fine sieve... Use instead of water for boiling. **1861** Mrs. Beeton *Bk. Househ. Managem.* 46 *Poêlée*, stock used instead of water for boiling turkeys, sweetbreads, fowls, and vegetables, to render them less insipid. **1877** E. S. Dallas *Kettner's Bk. of Table* 182 The following receipt is nearly identical with what the French cooks call Poêle... Take two carrots..onions, two cloves..sweet-herbs; mince all finely with half a pound of beef fat, and melt it. **1889** J. Whitehead *Steward's Handbk.* IV. 405/1 Poêle is white or colorless broth of bacon and ham with vegetables used to boil chickens, sweetbreads, etc., in instead of water.

poem. Add: **3.** *poem-book* (later examples).
1887 W. B. Yeats *Let.* 25 June in *Lett. to K. Tynan* (1953) 31 Sparling knows and much admires your 'Flight of the Wild Geese', from which I conclude it will figure in his poem-book. **1949** E. E. Cummings *Let.* 23 Aug. (1969) 193 Poems are nonsellable enough..without calling the poembook by some foreign word.

poemscape (pōu·ėmskēɪp). [f. Poem + Scape *sb.*[3]] The imaginary world envisaged in a poem.
1958 K. Patchen (*title*) Poemscapes. **1960** H. Kenner *Invisible Poet* v. 225 The Vrotting field-mouse is a figure in a poemscape, not like the rat in *The Hollow Men* the synecdoche of some omnipresent world.

‖ **pœna** (pī·nă). Also **pena.** In phrases freq. in L. inflected forms. [a. L. *pœna* penalty.] **a.** Chiefly *Law* and *Theology.* A punishment.
1632 in *Decisions Court of Session* (1805) XXIV. 10036 They..were content to pay the L. 100 *loco penae*. **1678** G. Mackenzie *Laws & Customs Scotl.* II. xxx. 560 Skeen..observes, that *pœna extraordinaria*, may be sometimes extended to death. **1757** in *Decisions Court of Session* (1805) XXIV. 10049 Expenses of plea..are in no case due or exigible, unless the Court finds that a party has been litigious, and specially subjects him to the costs of his opponent, *in pœnam* of his offence. **1859** T. C. Sandars *Inst. Justinian* (ed. 2) 492 *Pœna* is a punishment imposed by some general law, affecting possibly the *caput* and *existimatio* of the person punished. **1863** *Scottish Jurist* XXXV. 588/2 It is plain that here *pœna*—penalty—is used to mean a sum stipulated to be paid in the event of breach of contract. **1916** Joyce *Portrait of Artist* (1969) 128 This is the greatest torment which the created soul is capable of bearing, *pœna damni*, the pain of loss. **1953** F. de Zulueta *Inst. Gaius* II. III. 207 In these cases of negative interest there is no question of dividing the action or the *poena* recovered. **1959** Jowitt *Dict. Eng. Law* II. 1357/1 *Pœna*, a penalty, the punishment of an offence; generally inflicted for delicts.
b. *School slang.* = Imposition 5 c.
1842 *Eton Bureau* 251 Then the luxury, when told to write out and translate my lesson, to know that beneath the ill-formed letters that disfigured my 'pœna', might safely lurk an intimation to posterity that the reader—charitably supposed to be the inflictor of the punishment—'was a fool'. **1865** *Etoniana* xiv. 201 To such a boy, of course, the usual '*pœna*' of lines from a Greek or Latin poet to learn by heart could be no kind of punishment at all. **1870** 'Etonian' *Recoll. Eton* I. viii. 87 He got a pœna for coming in late for morning school one day. **1877** G. N. Bankes *Day of My Life* xiii. 136 It's wretched bad practice for handwriting, this *pœna* writing. **1911** R. Nevill *Floreat Etona* x. 301 Their usual practice being either to set some tremendous 'poena', which they afterwards revoked, or settle upon the wrong boy. **1941** L. A. G. Strong *Bay* 70 If you were in disgrace he..helped you with your poena and shooed you out of the empty classroom.

poet. Add: **1. c.** *poet's poet*, a poet whose poetry is generally considered to appeal chiefly to other poets.
1844 L. Hunt *Imagination & Fancy* 75 Spenser..has always been felt by his countrymen to be what Charles Lamb called him, the 'Poet's Poet'. He has had more idolatry and imitation from his brethren than all the rest put together. *Ibid.* 107 Spenser emulated the Raphaels and Titians in a profusion of pictures... They give the Poet's Poet a claim to a new title,—that of Poet of the Painters. **1867** O. W. Holmes *Guardian Angel* I. xviii. 280 Master Gridley lifted his eyebrows very slightly, remembering that some had called Spenser the poet's poet. **1930** *Times Lit. Suppl.* 27 Feb. 149/2 Assuredly, in Lamb's day Spenser was the poet's poet. **1932** J. Buchan *Sir W. Scott* iv. 79 Dryden was not a poet's poet, any more than his editor. **1958** *Reporter* 10 July 38/2 (*heading*) A poet's poet looks at his art.
e. *poet-in-residence*, a poet working in or associated with a university or college or a community (see *residence *sb.*[1] 2 b).
1972 *Guardian* 8 Feb. 24/5 W. H. Auden..returns to Christ Church, Oxford... Mr Auden..will be what the Americans like to call 'poet in residence'. **1973** *Black World* Jan. 28/2 Buford..is now poet-in-residence at Cleveland State University. **1977** *Canad. N. & Q.* Dec. 15/1 In January 1921 Robert Frost..was invited to visit Queen's and become the first poet-in-residence to occupy such an office in any Canadian University.
2. a. *poet-bishop, -composer, -critic, -king, -musician* (examples), *-novelist* (examples), *-painter* (examples), *-prophet, -singer* (later example), *-warrior* (example).
1909 *Westm. Gaz.* 2 June 5/1 The oldest existing wine club, the 'Phœnix', of which the poet-bishop, Heber, was once a luminary. **1947** A. Einstein *Music in Romantic Era* xvi. 256 Lortzing..in more modest proportions was comparable to Wagner as a poet-composer. **1968** *Jrnl. Mus. Acad. Madras* XXXIX. 102 The Tirupati poet-composer. **1977** *Early Music* Oct. 469/2 Machaut..maintained a dual role: one of..poet-composer, detached from the story. **1956** *Essays in Criticism* VI. 212 Poet-critics as dissimilar as Arthur Symons and Mr. Eliot. **1964** *English Studies* XLV. 290 Of course a poet-critic may be allowed to speak in images. **1859** W. Bagehot *Coll. Works* (1965) II. 114 The poet-king of Israel..David. **1903** A. W. Patterson *Schumann* 140 May not the shadow of the gloom that already brooded over him..already have been overclouding the mental vision of the poet-musician? **1947** A. Einstein *Music in Romantic Era* iii. 28 Wagner, all his life, thought of himself not merely as a poet-musician. **1957** N. Frye *Sound & Poetry* I. 5 The poet-musician of the Renaissance disappeared, and with few exceptions the major poets of the period gave little thought to the possibilities of musical setting. **1931** R. L. Mégroz *Joseph Conrad's Mind & Method* 154 Three modern poet-novelists..might perhaps be bracketed with Wells among the competitors. **1948** F. R. Leavis *Great Tradition* iii. 128 It was the profundity of the pondering that I had in mind when I referred to him [*sc.* Henry James] as a 'poet-novelist'. **1881** O. Wilde *Grave of Keats* in *Poems* 145 O poet-painter of our English Land! **1943** F. Thompson *Candleford Green* v. 75 Dante Gabriel Rossetti..that poet-painter. **1963** M. H. Abrams in N. Frye *Romanticism Reconsidered* 41 This voice is that of the poet-prophets of the Old and New Testaments, now descending on Blake from..John Milton. **1903** L. F. Anderson *Anglo-Saxon Scop* 27 To have seen many lands, to have had a wide and varied experience was considered a qualification for the poet-singer's calling. **1934** *Publ. Mod. Lang. Assoc.* XLIX. 365 A poet-warrior sings, adding the name of Grendel's conqueror to the role of Germanic heroes.
e. **Poets' Corner,** (*b*) (earlier example); **poet's daffodil** = *poets' narcissus*; **poets'** (or **poet's**) **narcissus** (examples); also = Pheasant's eye 2.
1781 W. Cowper *Let.* 27 Feb. (1908) 60 If you please you may send it [*sc.* a poem] to the Poet's Corner. [**1772** R. Weston *Universal Botanist* III. 504 Poetic or Common pale Daffodil, or Narcissus.] **1870** W. Robinson *Wild Garden* II. 112 Poet's Daffodil. *Narcissus poeticus.* Southern Europe. **1841** Poet's narcissus [see *hoop-petticoat 2]. **1883** W. Robinson *Eng. Flower Garden* 192/2 The finer types of the Poet's Narcissus should be grown for cutting. **1936** L. B. Wilder *Adventures with Hardy Bulbs* 19 The Poets Narcissus is perfect for dampish locations. **1965** H. Ramsbotham tr. *Schauenberg's Bulb Bk.* III. 229 This is the Poet's narcissus, one of the most widely distributed European species. **1977** R. Genders *Scented Flora of World* 322/2 (*heading*) The Poet's Narcissus and Hybrids.

poetast (pōu·ėtæst). Also **poetaste.** [Shortened form of Poetaster.] = Poetaster. Also as *v. trans.* and *intr.*, to write in the manner of a poetaster; hence **po·etasting** *ppl. a.*
1892 Beerbohm *Let.* June (1964) 22 She [*sc.* Mrs Grundy] demands that 'they bring unto her by and by the head of Oscar the Poëtast on a charger'. **1908** *Daily Chron.* 10 Apr. 3/2 In the spring the poetaster poetastes as sure as fate. **1909** G. B. Shaw *Admirable Bashville* Pref. 290, I poetasted The Admirable Bashville in the rigmarole style. **1969** *Daily Tel.* (Colour Suppl.) 14 Nov. 36/1 George Herbert is not there, although..Herbert's poetasting brother, Lord Herbert of Cherbury, is.

poetaz (pōu·ėtæz). [f. *poet(icus* + *taz(etta*, the specific epithets of two species of *Narcissus*.] In full, *poetaz daffodil* or *narcissus*. A narcissus (*Narcissus* × *medioluteus*) belonging to a group of hybrids produced by crossing *Narcissus poeticus* and *N. tazetta*, bearing fragrant white or yellow flowers in clusters. Also *attrib.*
1906 *Gardeners' Chron.* 17 Mar. 169/3 Narcissus 'Poetaz'.—This name applies to a race derived by crossing *N. poeticus* and *N. Tazetta*. **1910** *Ibid.* 3 Dec. 406/1 Tazetta and Tazetta Hybrids. To include N. Tridymus, Poetaz varieties, [etc.]. **1913** *Daffodil Year-bk.* 51 These Poetaz Daffodils are now well to the front, and are really good decorative plants. **1934** E. A. Bowles *Handbk. Narcissus* xvi. 176 The present-day race classed as Poetaz varieties owes its origin to the Dutch firm of Messrs. R. Van der Schoot. In 1885 large stocks of Tazetta varieties were growing in their Nursery alongside beds of N. poeticus ornatus, and the experiment of crossing them was decided upon. **1959** S. Gibbons *Pink Front Door* xx. 234 The little sitting-room..smelt of Poetaz narcissus. **1971** *Daffodil & Tulip Year Bk.* XXXVI. 44 There are..not more than a couple of dozen poetaz in commerce today. **1977** R. Genders *Scented Flora of World* 322/2 Geranium. A poetaz hybrid..unsurpassed amongst all flowers for its perfume.

‖ **poète maudit** (poet mōdi). [Fr., = cursed poet.] A poet who is insufficiently appreciated by his contemporaries. Also *transf.*
[**1884** P. Verlaine (*title*) Les poètes maudits.] **1930** L. P. Shanks *Baudelaire* p. vii, Certainly Baudelaire was a *poète maudit*, pursued by the disaster which pursued his fellow-poet Edgar Poe. **1949** M. Turnell tr. *J.-P. Sartre's Baudelaire* 155 The proud free criminal, the Don Juan of hell, the rebel was also at the same time the *poète maudit*, the Devil's marionette. **1958** *Listener* 17 July 98/2 A *poète maudit* is doubly accursed when he exhibits himself and his world with the dry precision of classical prosody. **1963** *Times* 8 Feb. 14/1 It is a loosely assembled collection of episodes,..following the self-destructive career of a late romantic hero—drunkard, sexual athlete and *poète maudit*. **1977** *Time* 26 Dec. 52/1 Once the ignored art, photography now stands robed in puffery, and armored with analysis; like painting, it has acquired its cast of heroes and *poètes maudits*.

poetese (pōuėtī·z). [f. Poet + -ese.] The mannered style of language supposed to be characteristic of poets.
1948 I. Brown *No Idle Words* 53 Has Empery, 'the power, status, or dominion of an Emperor', become entirely 'poetese'? In a leading article it would look a trifle odd. **1958** *Listener* 18 Dec. 1049/1 Mr. Bridson, choosing rhymed verse, never let it aspire to the high-poetic style or fall to 'poetese'.

poetette (pōuėte·t). *rare.* [f. Poet + -ette.] A young or minor poetess.
1913 E. Pound *Let.* Nov. (1971) 26, I seem to spend most of my time attending to other peoples' affairs, weaning young poetettes from obscurity into the glowing pages of divers rotten publications, etc.

poetic, *a.* and *sb.* Add: **A.** *adj.* **1.** *poetic diction*, diction used in or considered to be proper to poetry (see Diction 4).
1800 Wordsworth *Lyrical Ballads* (ed. 2) I. Pref. p. xxi, There will..be found in these volumes little of what is usually called poetic diction. *Ibid.* p. xxvii, The distinction of rhyme and metre is regular and uniform, and not, like that which is produced by what is usually called poetic diction, arbitrary and subject to infinite caprices. **1815** G. F. Nott *Works of Henry Howard & Sir T. Wyatt* I. p. clxxxviii, Chaucer did much towards refining our poetic diction, but he left it..open to subsequent innovation and experiment. **1886** *Encycl. Brit.* XX. 859/2 As a mere question of methods, a reaction against the poetic diction of Pope and his followers was inevitable. **1928** O. Barfield *Poetic Diction: Study in Meaning* 177 The stale Miltonics, which lay at the bottom of so much eighteenth-century 'poetic diction'. **1938** A. Campbell *Battle of Brunanburh* 41 Despite the wealth of poetic diction at his command, he can be, at times, astonishingly simple and direct. **1951** C. Day Lewis *Poet's Task* 5, I hope to devote a number of..lectures to what is called, somewhat uninvitingly, 'poetic diction'. **1970** M. Swanton *Dream of Rood* 59 The highly formalised nature of Old English poetic diction.
4. Also, fond of poetry, able to appreciate poetry.
1817 Jane Austen *Sanditon* (1925) 91, I have read several of Burn's Poems with great delight..but I am not poetic enough to separate a Man's poetry entirely from his Character.
B. *sb.* **2.** Also in extended senses.
1917 T. S. Eliot *Prufrock & Other Observations* 38 With your air indifferent and imperious At a stroke our mad poetics to confute. **1973** *Word 1970* XXVI. 66 Jakobson avoids the term *stylistics*, preferring instead *poetics*. **1976** *Times Lit. Suppl.* 2 Jan. 11/2 To subscribe to this poetic was to doubt the validity of art and the veracity of dreams. **1976** *Daily Tel.* 5 July 10/3 So autonomous are the poetics of Krzysztof Penderecki's compositional techniques, I found it hardly possible to reconcile words and music in his 'Canticum Canticorum Salomonis'. **1976** *Times Lit. Suppl.* 12 Nov. 1411/2 It is developed theoretically into an alternative poetic, for literature that classical and Coleridgean poetics are unable to treat with justice: a poetic of architectural as against organic form. **1977** A. Sheridan tr. *J. Lacan's Écrits* iii. 102 This notion must be approached through its resonances in what I shall call the poetics of the Freudian corpus.

poeticality. Delete † and add later examples.
1923 J. M. Murry *Pencillings* 200 The novel with which he was so enchanted was full of vague poeticalities. **1950** *Scrutiny* XVII. 188 He avoids the opposite danger of bolstering the commonplace by an extraneous poeticality. **1976** *Observer* 17 Oct. 32/2 What he was in fact up to a good deal of the time was spouting frigid poeticalities.

poeticism. Delete *nonce-wd.* and add: **1.** (Further example.)
1972 *Daily Tel.* 22 June 7/1 The book harks back to the luminous poeticism of much German romantic writing.
2. A poetical expression; an example of poetic diction.
1926 Fowler *Mod. Eng. Usage* 442/1 *Poeticisms.* Simple reference of words to this article warns the reader that to use them in ordinary prose contexts is dangerous. **1956** *Essays in Criticism* VI. 156 His language avoids conventional poeticisms.

poeticize, *v.* Add: Hence **poe·ticized** *ppl. a.*; **poe·ticizing** *vbl. sb.* Also **poe·ticizable** *a.*; **poeticiza·tion, poe·ticizer.**
1923 J. M. Murry *Pencillings* 189 What he is really lamenting is the absence of poeticisation, of what is called 'imaginative writing'. **1926** Fowler *Mod. Eng. Usage* 442/2 *Poeticize* makes *-zable*. **1961** *Encounter* XVII. 68 Rossetti appears to me a poeticiser. **1965** *New Statesman* 22 Oct. 617/3 In the studio, he [*sc.* Corot] dropped into tame poeticising, due no doubt to what Mr. Gould calls his 'vein of emotional immaturity'. **1973** *Human World* Feb. 7 The anxiety Mr Maddox senses in the environmentalists comes to no more than a yearning for a universal poeticized prosperity. **1975** *Listener* 11 Sept. 348/1 Francis Ponge..is obsessed with the external world as a series of poetic or poeticisable objects. *Ibid.* 4 Dec. 772/2, I found the sub-Dylan Thomas-ish poeticising here off-putting.

poetico-. Add: *poetico-commercial, -metaphysical* adjs.
1837 J. S. Mill in *Westm. Rev.* XXVII. 50 Much genuine philosophy, disguised though it often be in a poetico-metaphysical vesture of a most questionable kind. **1930** Blunden *Leigh Hunt* viii. 99 It [*sc. Rimini*] almost became fashionable..but..by no means challenged the poetico-commercial achievements of Rogers.

poetism (pōu·ėtiz'm). [f. Poet + -ism.] **a.** = Poeticism 1. **b.** = *poeticism 2.
1848 A. H. Clough *Let.* 16 July in T. Arnold *N.Z. Lett.* (1966) 111 Matt[hew Arnold]..has become sadly cynical again of late. However I think the poetism goes on favourably. **1867** G. Meredith *Vittoria* I. xii. 202 Agostino smiles and chuckles, and talks his poetisms. **1977** P. B. & J. S. Medawar *Life Sci.* ii. 23 The myth that geese might be born of such organisms as the attractive barnacle-like crustacean the goose barnacle, *Lepas anatifera*. Such notions belong to 'poetism'.

poetize, *v.* Add: **1. a.** (Later examples.)
1895 *Wales* May 240/2 A shoemaker from Llandwrog was with me..the person who poetized to Mr. Williams in the Bangor paper. **1917** J. B. Cabell *Cream of Jest* I. ii. 13 So Horvendile descended, still poetizing: '*Pus ab mi dons no m pot valer.*' **1939** L. M. Montgomery *Anne of Ingleside* xxii. 147 He doesn't look clever but he can poetize.

poetizing, *vbl. sb.* (Further examples.)
1888 F. H. Williams *Atman* (1891) 243 He is probably bilious, but that is no excuse for his threadbare poetizings. **1936** F. Clune *Roaming round Darling* viii. 74 A freezing night in August and eating cold corned beef! That's what poetizing does to a man. **1948** F. R. Leavis *Great Tradition* iii. 129 *The Waves* and *The Years*—works that offer something like the equivalent of Georgian poetizing. **1961** *PMLA* LXXVI. I. 309/1 None of these glimpses of poetizing without writing is intended to incorporate a signature into the epic matter.

poetolatry (pōuėtǫ·lătri). [f. Poet + -olatry.] The worship of poets; immoderate veneration for poets. So **poeto·later,** one who practises poetolatry.
1936 C. S. Lewis in *Essays & Studies* XXI. 165 There is yet another way in which Personal Heresy offends against personality;..I am referring to the growth of what may be called Poetolatry. *Ibid.* 167 Most poetolaters hold that a dead man has no consciousness. **1939** Lewis & Tillyard *Personal Heresy* v. 104 Naturalism..wants poets to be a separate race of great souls or *mahatmas*. Poetolatry is the natural result, for if there were such a race..those who know no higher deity would do well to worship them.

poetry. Add: **5.** (Earlier and further examples.) Phr. *poetry of the foot* or *of motion*: dancing.
1664 Dryden *Rival Ladies* III. 32 The Poetry of the foot takes most of late. **1813** Lady Morgan *Wild Irish Girl* (ed. 5) II. xix. 156, 'I seldom dance,' said I—'Ill health has for some time coincided with my inclination, which seldom led me to try my skill at the *Poetry of Motion*.' **1818** Byron *Childe Harold* IV. lviii. 32 That music in itself, whose sounds are song, The poetry of speech. *c* **1863** T. Taylor *Ticket-of-leave Man* in M. R. Booth *Eng. Plays of 19th Cent.* (1969) II. 101 Come along, Emily, if you're at liberty to give your Montague a lesson in the poetry of motion. **1874** Hardy *Far from Madding Crowd* I. ii. 13 The poetry of motion is a phrase much in use. **1946** D. C. Peattie *Road of Naturalist* iv. 42 There is left only the poetry of speed and wind. **1959** E. H. Clements *High Tension* i. 10 He had never been back there. He had not..seen poetry in the small exploit. **1975** *Times* 6 Mar. 13/5 There is a moment of poetry in a sequence where the dancers simply walk about carrying umbrellas. **1975** *Listener* 4 Dec. 747/2 Raffles..compares the poetry of cricket with the poetry of burglary. **1977** *Zigzag* Apr. 30/3 This song captures what Television are all about: a kind of poetry in motion with a scorching musical backdrop.
6. (Earlier and further examples.)
1629, 1773 [see *grammar *sb.* 5 c]. **1838** C. Waterton *Essays on Nat. Hist.* p. xxiv, One day, when I was in the class of poetry..about two years before I left the college.., he called me up to his room. **1887** *Stonyhurst Mag.* Nov. 34/1 Poetry..were granted a most unexpected but none the less welcome holiday on Thursday October 20th. **1946** D. Gwynn *Bishop Challoner* iii. 39 By the summer of 1708 he had passed through the two higher classes of Poetry and Rhetoric.
7. *poetry professorship, reader, school, workshop; poetry-loving* adj.; **poetry-book,** a book containing a collection of poems, esp. one used in schools; **poetry reading,** the reading of poetry, esp. to an audience; a poetry recital; **poetry recital,** a public performance of poetry; **poetry-voice,** a pompous or mannered style of writing poetry or reading it aloud.
1847 Thackeray *Van. Fair* (1848) xii. 103 She wrote whole pages out of poetry-books without the least pity. **1877** A. B. Edwards (*title*) A poetry-book of elder poets. **1881** R. L. Stevenson *Virginibus Puerisque* 176 Whether we regard life as a lane leading to a dead wall..or puling little atheistic poetry-books about its vanity and brevity [etc.]. **1903** 'A. McNeill' *Egregious English* 102 The demand for poetry-books by new writers has practically ceased to exist. **1935** E. Farjeon *Nursery in Nineties* iv. 271 The poem was 'good enough' for the Poetry-Book. **1980** G. Nelson *Charity's Child* vi. 86 The poetry book, sir. **1941** Blunden *T. Hardy* ii. 34 The change was natural to the period and the poetry-loving author. **1979** E. Koch *Good Night Little Spy* iv. 24 A poetry-loving moon faced charmer. **1793** W. B. Stevens *Jrnl.* 11 May (1965) I. 72 Received a College letter, requesting me to support the pretensions of Mr. Hurdies to the Poetry Professorship which will be vacant in Michaelmas Term next. **1940** R. S. Lambert *Ariel & all his Quality* v. 121 Few poetry-readers win its [*sc.* the audience's] general approbation. **1975** 'G. Black' *Big Wind for Summer* ii. 32 The voice of the British Broadcasting Corporation's top poetry reader. **1917** A. Huxley *Let.* 11 Dec. (1969) 140 After that to Eliot, whom I found as haggard..as usual; we held a council of war about a poetry reading which both of us are supposed to be performing. **1945** 'G. Orwell' in *New Saxon Pamphlets* III. 35 That grisly thing, a 'poetry reading'. **1975** O. Sela *Bengali Inheritance* xvi. 139 She..organised poetry readings and prescribed reading books. **1966** J. Betjeman *High & Low* 25 A poetry recital we are giving to the troops. **1978** J. Symons *Blackheath Poisonings* I. 34 The Rink Hall in the village, where the poetry recital was to take place. **1970** Poetry school [see *poetry workshop* below]. **1971** *Guardian* 28 Dec. 13/5, I hate the poetry-voice; the poetry should speak for itself. **1972** *Country Life* 1 June 1418/3 Stevie Smith..was not one for the Poetry Voice. She mixed nonsense and its opposite. **1976** *Times* 1 Mar. 3/1 Mr. Lovibond and his supporters..operate a poetry school and workshop. **1977** *Time Out* 28 Jan.–3 Feb. 40/5 Audio-visual poetry workshop last Fri of month.

po-faced (pōu·fēɪst), *a.* [Perh. f. Poh *int.* or *po[3] + Faced *ppl. a.*[2], infl. by *poker-faced adj.] Having or assuming an expressionless or impassive face, poker-faced; priggish, narrow-minded, or smug. So **po--face.**
1934 C. Lambert *Music Ho!* III. 191, I do not wish, when faced with exoticism, to adopt an attitude which can best be described by the admirable expression 'po-faced'. We cannot live perpetually in the rarefied atmosphere of the austerer classics. **1937** T. Rattigan *French without Tears* I. 11 What's he like, though, really? Po-faced, I suppose? **1951** N. Balchin *Way through Wood* xvii. 239 You sounded po-faced on the telephone. **1958** *Economist* 11 Oct. 133/3 That 'middle class' which Low used to symbolise by a po-faced woman in a big fur coat. **1959** [see *galumph *v.*]. **1967** *Listener* 24 Aug. 232/2, I glimpsed [him]..being lightly grilled by a deliciously po-faced Joan Bakewell. **1973** K. Giles *File on Death* iii. 75 Her po-face. **1973** *Times* 11 Dec. 5/3 We do not want to appear po-faced or ministry Scrooges, but on the other hand we should not like to see the country flooded with blazes of extravagant light. **1975** *Globe & Mail* (Toronto) 1 Sept. 7/1 Ted Stuebing..had one of his minions come on the tube all po-faced Saturday night to say that CFTO couldn't let the press in because it would have cluttered up the production studio.

‖ **poffertje** (pǫ·fəɪtʸə). Also (*S. Afr.*) **poffartje, poffertjie** (-kʸə). [a. Du. *poffertje*, Afrikaans *poffertjie*, f. Fr. *pouffer* to blow up.] A small light doughnut or fritter dusted with sugar. Also *attrib.*
1872 *Cape Monthly Mag.* V. 230 Shall we take offence if an English host..set before our craving appetite..the dainty though untranslatable pumpkin 'poffartjes'? **1890** *Cent. Mag.* Nov. 49 Beside her..sat Jan Wiskerke watching a buxom darky in a scarlet turban frying poffertjes. *Ibid.*, The kitchen walls were covered with waffle-irons, poffertje pans. **1905** *Speaker* 11 Mar. 565/1 Poffertjes..are little round pancaky blobs, twisted and covered with butter and sugar. **1942** H. W. Van Loon *Van Loon's Lives* 850 There is no use trying to describe them to outsiders, for *poffertjes* are the one dish that was made only in the Low Countries. They are a cross between a very small kind of pancake and a fritter and are the main delicacy of..village fairs. **1949** [see *melktert]. **1970** D. Winsor *Death Convention* viii. 68, I took a poffertje. It was a rather basic lump of dough fried in butter and dusted in icing sugar.

poge, var. *pogue.

pogey (pōu·gi). *N. Amer. slang.* Also **pogie, pogy.** [Origin unknown.] **a.** A hostel for the needy or disabled; a poorhouse; a local relief centre or welfare office. Now chiefly *Hist.*
1891 *Contemporary Rev.* Aug. II. 255 Begging is called 'battering for chewing'; railway brakemen, 'brakies'; poorhouses, 'pogies'. **1927** *Amer. Speech* June 387/2 A *pogey* is a poorhouse. Government homes for disabled veterans are also known as pogies. **1936** K. Mackenzie *Living Rough* 269, I was in the Pogey a couple of nights. It stinks. **1953** D. M. LeBourdais *Nation of North* 211 Thousands of self-respecting workmen..sat at home while

their wives made the dreary pilgrimage to the 'pogie'. **1959** *Maclean's Mag.* (Toronto) 15 Aug. 21/1 Lean and hungry alley-cat men swung down from the freights and headed for a fifteen-cent mission meal or the innumerable pogies and scratch houses for a ten-cent cot. **1974** P. Gzowski *Bk. about This Country* 18 We had lived in teacherages, pogies and scratch houses and boarding houses. **1976** *Whig-Standard* (Kingston, Ont.) 9 Jan. 1/1 Unemployment insurance has indeed come a long way since the days of the pogey houses of turn-of-the-century England—dank, gloomy places where the indigent could do menial tasks in exchange for food and lodging.

b. Relief given to the needy from national or local funds; unemployment benefit.

1960 *Maclean's Mag.* (Toronto) 2 Apr. 54/2 Today unemployment-insurance payments are often referred to as pogey. But pogey in the depths of the thirties meant something as different from present-day unemployment insurance as panhandling is from drawing money from your bank account. **1961** *Time* 31 Mar. 9/2 Said a jobless Hamilton steelworker, father of six children: 'Why should I sweat for $40 a week? I'm getting more than that from the pogey, the welfare and the baby bonus.' **1964** H. T. Barker *Ice Road* 49 During the winter we lived on turnips, potatoes, canned clams and the pogy, and Mother and I would hook rugs for the tourist trade. **1976** *Whig-Standard* (Kingston, Ont.) 6 Jan. 1/6 The Kingston area's fourth largest and fastest growing industry is unemployment insurance—pogey or, if you wish, the dole.

pogey bait (pōᵘ·gi bēɪt). *U.S. slang.* Also **poggy, pogie, poguey bait.** [perh. f. Pogy + Bait *sb.*[1]] Candy, sweets. (See also quot. 1970.)

1918 L. E. Ruggles *Navy Explained* 88 While going through the war zone, the pockets are used for ciggies and poggy bait. **1929** *Papers Mich. Acad. Sci., Arts & Lett.* X. 315 *Poggy bait*, the sailor's designation for sweets. **1935** A. J. Pollock *Underworld Speaks* 90/1 *Pogey bait*, candy. **1953** M. Dibner *Deep Six* xv. 154 A candy bar's called poguey-bait. **1970** *Esquire* Nov. 116 Pogie bait is any snack that is not prepared..in a government mess-hall.

Poggendorff (pǫ·gĕndǫɹf). [Name of J. C. *Poggendorff* (1796–1877), German physicist and chemist, who pointed out the illusion (see *Ann. der Physik und Chem.* (1860) CX. 502).] *Poggendorff illusion*: an optical illusion in which the two ends of a straight line whose central portion is obscured by a rectangular strip crossing at an angle seem not to be in line.

1898 *Psychol. Rev.* V. 540 The Poggendorff illusion... This illusion is due to the fact that the two ends of the transversal, sundered by the vertical strip, are made to belong to two different systems, each suggesting its own perspective interpretation and being under no necessity.. of appearing as parts of a continuous line. **1976** *Nature* 1 Apr. 397/2 (*caption*) The Poggendorff illusion in geological contexts; the illusory perceptual data are more likely to mislead when they favour an investigator's geological hypothesis than when they do not.

poggle (pǫ·g'l), *sb.* and *a. slang* (orig. Anglo-Ind. *colloq.*). Also **puggle, puggly, pugley.** [ad. Hind. *pāgal, paglā* (fem. *paglī*) madman, idiot.] **A.** *sb.* A crazy or foolish person, an idiot. ?*Obs.*

1829 J. Shipp *Memoirs* II. viii. 233 It's true, the people call me, I know not why, the 'pugley'. **1863** 'N. Broughton' *Dawk Bungalow* II. 37, I was foolish enough to pay these hurrumzarders beforehand, and they have thrown me over. I must have been a poggle to do it! **1886** Yule & Burnell *Hobson-Jobson* 542 *Poggle, puggly*, etc... Properly Hind. *pāgal*, a madman, an idiot; often used colloquially by Anglo-Indians.

B. *adj.* Mentally unbalanced, crazy; also, drunk, mad-drunk.

1925 Fraser & Gibbons *Soldier & Sailor Words* 225 *Poggle* (also *puggle*), mad. An idiot. An old Army term. **1936** H. Graham *Private Life of Gregory Gorm* 229, I can't quite make up my mind whether he is a genius, as some people seem to think, or only slightly poggle. **1971** B. W. Aldiss *Soldier Erect* 80 A woman in this bloody dump? You're going *puggle*, Page, that's your trouble! Too much tropical sun.

poggled (pǫ·g'ld), *a. Army slang.* [f. prec. + -ed[2].] = *poggle *a.* Cf. *puggled *a.*

1933 Partridge *Words, Words, Words!* III. 202 *Poggle(d), puggled*, 'rattled' as well as eccentric and mad-drunk, is a pre-war Regular-Army word.

poggy, var. Pogy in Dict. and Suppl.

poggy bait, var. *pogey bait.

pogie, var. *pogey.

pogo (pōᵘ·go). Also **Pogo.** [Orig. unknown.] **1.** A stilt-like pole (also called a *pogo stick*) on which one jumps about (see quot. 1921); the pastime of jumping on or as on such a pole. Also *attrib.* Also applied to dancing or a dance with movements suggestive of jumping on a pogo stick. So as *v. trans.* to traverse on a pogo stick; *intr.* to jump on or as on a pogo; to perform a pogo dance. Hence **po·goer; po·going** *vbl. sb.* and *ppl. a.* Formerly registered in the U.S. as a proprietary term.

1921 *Glasgow Herald* 30 Aug. 7 What is a Pogo? It is a four-foot pole, hollowed at the foot for the insertion of a strong spring, with a rubber cushion at the end of it. About half a foot above the spring are two steps. To Pogo you place one foot on each step, clutch the top of the pole firmly in both hands, and hop. *Ibid.*, On the Continent there are Pogo clubs, which conduct Pogo carnivals where the principal items are the high and the long jumps, and there are halls where you Pogo under much the same conditions as obtained on the roller skating rinks. **1921** *Punch* 21 Sept. 225/1 Three men..expressed their intention of 'Pogoing' the Channel. *Ibid.*, Charlie Chaplin intends to give a 'Pogo' to each of the children who are now attending his old schools. *Ibid.*, A small girl has 'pogo-ed' five hundred miles. **1921** *Oxford Times* 11 Nov. 16/4 On Thursday afternoon two undergraduates were seen racing along Cornmarket Street on Pogo sticks. **1924** *Punch* 24 Sept. 338 A dozen well-mounted pogo-players. **1942** *Official Gaz.* (U.S. Patent Office) 1 Sept. 14/2 Philip de Journo, Forest Hills, Long Island, N.Y. *Pogo* for jumping sticks. Claims use since Feb. 28, 1941. **1958** *Daily Mail* 8 July 6/4 On stilts and pogo sticks (their latest craze). **1968** *New Scientist* 19 Dec. 653/2 The first stage of the three-stage rocket went into 'pogo-stick' oscillations. **1973** *Nature* 30 Nov. 313/1 The hopping of kangaroos is reminiscent of a bouncing ball or the action of a pogo stick. **1977** *Zigzag* Mar. 8/1 I've never subscribed to the theory that up to the age of thirty it was all pogo dancing down the Roxy. *Ibid.* June 6/4 'I want him thrown out,' he snapped to a roadie, pointing to the pogoing culprit. **1977** *Oxford Times* (City ed.) 30 Sept. 16 The dancers at the front jumping up and down on one leg (they call it pogoing, m'dear). *Ibid.*, Despite the vigorous pogoers, many members of the audience stood on the sidelines. **1977** *New Wave Mag.* No. 7. 3 They just arrived and pogoed through the front door. **1977** *Chicago Tribune* 2 Oct. vi. 14/4 The..punks' hopping 'pogo' dance. **1978** *New Society* 19 Jan. 115/2 The wildly exuberant punk dance—the pogo—..derives from the celebratory 'knees-up' of the football terraces. Basically, the participants leap up and down, two-footed, some of them alone, more in two or threes, with their arms linked... It's hot work, pogoing.

2. Used *attrib.* with reference to low-frequency longitudinal oscillations of a space rocket. (Cf. quot. 1968 above.)

1971 *Nature* 10 Dec. 316/2 The Diamant B [booster rocket] has been used successfully on three previous occasions, but a strong vibrational 'pogo effect' was noticed. **1976** Sutton & Ross *Rocket Propulsion Elements* (ed. 4) viii. 259 Techniques for damping Pogo instability tendency include..properly designed engine, interstage, and payload support structures.

pogonion (pŏgōᵘ·niǫn). *Anat.* [f. Gr. πώγων beard + *-ion[2].] The foremost point on the midline of the chin.

1897 in *Lippincott's Med. Dict.* **1920** H. H. Wilder *Lab. Man. Anthropometry* I. i. 47 *Pogonion* (pg), the most projecting median point of the chin, on the anterior surface (mental process). **1977** *Proc. R. Soc. Med.* LXX. 433/2 Mandibular length is of particular interest following mandibular osteotomy and it was measured from articulare to pogonion on cephalometric radiographs, using calipers.

pogonophore (pŏgōᵘ·nofǫɹ). [a. mod.L. *Pogonophora* (K. E. Johannson 1937, in *Zool. Bidrag från Uppsala* XVIII. 253), f. Gk. πώγων beard + -φόρος bearing: see -phore.] A worm-like marine invertebrate belonging to the phylum Pogonophora. So **pogono·phoran, pogono·phorous** *adjs.*, of or pertaining to an animal of this kind or the phylum as a whole.

1955 *Systematic Zool.* III. 184/1 The large size of pogonophorous animals is interesting. **1958** *New Scientist* 20 Nov. 1303/2 All pogonophores look like bits of string or trawl twine. **1959** L. H. Hyman *Invertebrates* V. xviii. 210 Pogonophores are exclusively marine and are mostly limited to abyssal depths. **1964** *New Scientist* 24 Dec. 842/3 Pogonophores are to be found in all seas, even in those places where the marine fauna had been previously studied with care. **1973** *Nature* 21 Dec. 452/1 Pogonophoran Phylogeny... The first international symposium on Pogonophora..took place in the University of Copenhagen on November 1–3. *Ibid.* 452/3 Gupta finally expressed a doubt that pogonophores are coelomate animals at all. **1976** *Ibid.* 18 Mar. 218/2 The author devotes most space to the few species of animals, such as the ostracod *Gigantocypris* and the pogonophoran worm *Siboglinum*, that have been used by physiologists.

pogonotomy. Add to def.: shaving. (Later examples.)

1942 Berrey & Van den Bark *Amer. Thes. Slang* §125/3 *Pogonotomy*, shaving. **1960** *Times* 28 Sept. (Advertising Suppl.) p. iii/2 This is the age, in fact, of pogonotomy. **1966** J. S. Cox *Dict. Hairdressing & Wigmaking* 119/1 *Pogonotomy*, beard-cutting or shaving.

pogonotrophy. (Further example.)

1966 *Daily Mail* 29 Oct. 1/1 This week's picture of Beatle George Harrison wearing a moustache—and a particularly sad, droopy looking one at that—caught students of pogonotrophy the world over in two minds.

pogrom, *sb.* Delete ‖ and add: Also with pronunc. (pǫ·grǫm). **a.** (Further examples.)

1919 N. Sokolow *Hist. Zionism* II. p. li, Not even the dark ages extracted so heavy a toll of Jewish blood: something like 1400 pogroms took place all over the Ghetto. **1968** *New Left Rev.* Jan.–Feb. 65 Then came the years of galloping inflation, of the pogroms, of acute social, political and intellectual ferment. **1979** O. Sela *Petrograd Consignment* 142 Wasn't he eager to go back to Russia..to read the Protocols of the Elders of Zion again; wasn't another pogrom all he lived for.

b. In general use: an organized, officially tolerated, attack on any community or group. Also *fig.*

1906 *Tribune* 16 June 7/2 This was the immediate signal for a *pogrom*, or organized riot. **1920** H. J. C. Grierson in *Proc. Brit. Acad. 1919–1920* 433 Only Henley refused to take part in the 'pogrom'; and he alas! died before completing his work as champion, critic, and editor of Byron. **1928** 'S. S. Van Dine' *Greene Murder Case* i. 13, I note that our upliftin' Press bedecked its front pages this morning with headlines about a pogrom at the old Greene mansion last night. **1936** H. A. L. Fisher *Hist. Europe* I. xviii. 232 The Greek Empire..had disgraced itself by a *pogrom* against the French and Italian colony in Constantinople. **1964** *New Statesman* 13 Mar. 405/1 On 20 March 1914 58 British cavalry officers, stationed in Ireland, announced that they would not obey the orders of their lawful superiors... The cry of 'mutiny' was answered by the charge that there had been a plot—a 'pogrom' in the contemporary phrase—to crush Ulster's resistance to Home Rule by force of arms. **1967** T. Gunn *Touch* 27 Am I Your mother or The nearest human being to Hold on to in a Dreamed pogrom. **1971** *Sunday Times* 13 June 12/4 The army units, after clearing out the rebels, pursued the pogrom in the towns and villages. **1975** R. Browning *Emperor Julian* iii. 51 Hannibalianus had been killed in 337 in the pogrom of his relations engineered by Constantius.

c. *attrib.* and *Comb.*

1931 *Times Lit. Suppl.* 5 Nov. 855/2 Refugees to England from pogrom-haunted Russia. **1941** Koestler *Scum of Earth* 85 The French Government discovered a welcome diversion from the general discontent by exploiting the people's natural hostility to foreigners, and appealing to their pogrom instincts. **1949** —— *Promise & Fulfilment* I. vii. 69 Many of these young men had been members of the Jewish self-defence organizations in the pogrom-threatened small towns of Russia. **1978** D. Murphy *Place Apart* viii. 167 Few of us would wish to see our army crossing the [Irish] border to fight Loyalist paramilitaries... If another 'pogrom' situation did arise ..it would make more sense to welcome..refugees into the Republic.

Hence **pogro·mist** (also stressed *po·gromist*), an organizer of or participant in a pogrom.

1907 *Athenæum* 26 Jan. 99 Small wonder that the 'pogromists' laugh at Europe, and now pursue their work without intermission or disguise. **1960** S. Becker tr. *A. Schwarz-Bart's Last of Just* (1961) II. 87 The pogromists were White Guards. **1962** *Guardian* 13 Oct. 6/3 Hate-mongers and pogromists. **1963** *Times* 24 Jan. 8/7 However, he criticized the 'complete lack of publicity in the Soviet press' and said that neither the pogromists nor the local police and prosecutors who abetted them had been punished or reprimanded. **1978** I. B. Singer *Shosha* xiv. 254 People sacrificed themselves for Stalin, for Petlura, for Machno, for every pogromist.

pogrom (pogrǫ·m, pǫ·grǫm), *v.* [f. the sb.] *trans.* To massacre (persons) in a pogrom; to destroy (a place) in a pogrom. Hence **pogromed, pogrommed** *ppl. a.*, that has experienced a pogrom.

1915 *Boston Jrnl.* 2 Feb. 3/2 [The Jews in Galicia] are being..pogromed. **1918** G. Frankau *One of Them* II. xv. 107 Its East End drab bits..Where toiled..The pogromed horde, and multiplied like rabbits. **1919** *Daily Chron.* 10 Oct. 1/1 The total number of places pogrommed was 353, and the Jews killed 20,500. **1946** Koestler *Thieves in Night* 217 They [sc. the Jews] are the most admirable salesmen in the world, regardless of whether they sell carpets, Marxism,..or their own pogromed infants. **1979** *Country Life* 8 Nov. 1688 Such a gathering is, to Mr Pirates, vile enough to justify his decision to pogrom the lot.

pogue (pōug). *slang.* Also **poge.** [Perh. related to Pough *sb.*] A bag, purse, wallet or container. Also by metonymy, money, takings. Also *attrib.*, as **pogue-hunter,** a thief who steals purses, a pickpocket.

1812 J. H. Vaux *Vocab. Flash Lang.* in *Mem.* (1964) 259 *Pogue*, a bag, (probably a corruption of poke.) **1879** J. W. Horsley in *Macm. Mag.* XL. 504/2, I went out the next day to Maidenhead, and touched for some wedge and a poge (purse), with over five quid in it. **1896** A. Morrison *Child of Jago* xxiii. 229 The pogue-hunter, emptying the pogues in his pocket by sense of touch. **1906** E. Pugh *Spoilers* vi. 66 When the tiggies made a raid for a 'ot poge-hunter or snidesman. **1942** Berrey & Van den Bark *Amer. Thes. Slang* §88/15 Purse, dummy, hide,..poge, poke. **1975** M. Crichton *Great Train Robbery* v. 29 What's your pogue up there, anyway? *Ibid.* vii. 39 It was the stickman's job to take the pogue once Teddy had snaffled it, thus leaving Teddy clean, should..a constable stop him.

poguey bait, var. *pogey bait.

pogy. Add: Also **poggy.** (Earlier and later examples.) Also *attrib.* and *Comb.*, as *pogy-fisherman, steamer*; **pogy chum** (see quot. 1858); **pogy-press,** a press for extracting oil from fish; **pogy-seiner,** a boat used in seining for pogy (see Seine *v.*).

1858, 1859 [see *chum *sb.*[3]]. **1864** *Rep. Maine Board Agric.* 42 Rock weed, muscle bed and pogy chum will make grass grow. **1880** *Harper's Mag.* Aug. 341/1 A cast-off 'pogy'-press..had been piled upon an old wharf. *Ibid.* 347/1 The 'pogy' business was the catching of porgies and menhaden for their oil. **1913** *Oysterman & Fisherman* 10 July 31/1 The new pogy steamer E. B. Thomas the largest

ever built for the menhaden fishing industry made her official trip off Portland Monday. **1949** *Sun* (Baltimore) 29 July 2/1 Owners of the Virginia Pogy-Seiner Pluck today planned to advertise for a crew of licensed Maine fishermen. *Ibid.*, Newspaper ads would be tried tomorrow to see whether enough trained pogy-fishermen are available in this area. *Ibid.*, Pogies, or menhaden, are good only for their oil and fish-meal content.

pogy, var. *pogey.

pohutukawa (pohu:tukā·wā). [Maori.] A New Zealand evergreen tree, *Metrosideros excelsus*, belonging to the family Myrtaceæ and bearing clusters of red flowers with projecting stamens; also called the Christmas tree, as it flowers in December and January. Also *attrib.*

1832 G. Bennett in *London Med. Gaz.* 7 Jan. 508/1 There is an unpublished species of Metrosideros..named Pohu-takawa by the natives of New Zealand. **1851** V. Lush *Jrnl.* 1 Sept. (1971) 84 Passing now at the foot of lofty perpendicular cliffs, in the crevices and the tops of which were the beautiful Pohutukawa trees. **1867** [see *Christmas-tree b]. **1886** J. A. Froude *Oceana* xviii. 308 Low down on the shores the graceful native Pohutukawa was left undisturbed. **1935** J. Guthrie *Little Country* xxi. 316 Along the edge of the coast road are the pohutukawa trees, flowering like bonfires. **1944** [see *Christmas-tree b]. **1959** M. Shadbolt *New Zealanders* 22 We..picnicked in the shade of a crimson-blooming pohutukawa grove. **1965** F. Sargeson *Memoirs of Peon* iv. 62 A travelling Frenchman..had not long since visited the Auckland beaches and painted the pohutukawas in flower. **1977** *N.Z. Herald* 5 Jan. 1-6/10 The cliffs rise behind, dotted with tiny pohutukawas, some little bigger than their first blooms.

poi[1]. (Earlier and further examples.)

1823 C. S. Stewart *Jrnl.* 18 May (1828) vi. 133 This immense bulk of person is supposed to arise..from the abundance and nutritious quality of their food, especially that of poe, a kind of paste made from the taro, an esculent root, a principal article of diet. **1826** W. Ellis *Tour through Hawaii* xi. 293 The house..was soon furnished. A sleeping mat spread on the ground,..a few calabashes for water and poë. **1829** [see *poipoi]. **1833** A. Smith in M. D. Frear *Lowell & Abigail* (1934) 72 Their [sc. the Hawaiians'] 'staff of life' is poi, which is made by baking taro in the ground and pounding to the consistency of thin flour paste. **1862** M. Hopkins *Hawaii* iii. 34 This succulent root [sc. taro] was sometimes cooked, but was more generally pounded into a semi-fluid mess, and allowed partially to ferment, when it was called *poi*. **1905** [see *luau]. **1924** J. M. Brown *Riddle of Pacific* xvii. 182 The low coralline islands had not breadfruit, and yet made the fermented paste called sometimes *poi* and sometimes *mai*; and even volcanic islands like the Marquesas, which had breadfruit, preferred to make their *poi* of *taro*. **1951** *Amer. Speech* XXVI. 23 *Poi* dog (a nondescript cur; formerly the native breed of dog was fattened on 'poi' and served at feasts). **1954** [see *laulau]. **1964** *Asia Mag.* 16 Aug. 20/1 *Poi*... Made from starchy taro roots, it is a pinkish-gray goo of one, two, or three-finger thickness depending upon the consistency of the paste. **1968** O. Wyndette *Islands of Destiny* II. 85 All five of his wives were enjoying with him a typical Hawaiian meal: baked pork, poi, and sweet potatoes.

poi[2] (poi). *N.Z.* [Maori.] A ball made of leaves and fibre attached to a string; a dance to the accompaniment of traditional songs, performed by Maori women and girls using such a ball. Also *attrib.*

1843 E. Dieffenbach *Travels in N.Z.* II. iv. 57 Another game is with one ball (poi) suspended from a string. **1859** A. S. Thomson *Story of N.Z.* I. I. x. 196 Poi is a game played with variegated balls, about the size of large oranges, to which strings are attached. The string is held in one hand and the ball is struck with the other. **1905** W. Baucke *Where White Man Treads* 87 When the feasting and gaiety had subsided..we all lazily awaited the lining up of the poi maidens. **1935** J. Guthrie *Little Country* vii. 147 The tiny pois danced against arms and shoulders and breasts. **1938** [see *haka]. **1943** N. Marsh *Colour Scheme* ii. 36 He's going round your younger lot talking about teams of *poi* girls. **1945** R. Park in *Coast to Coast 1944* 42 There'll be girls dancing the poi. **1949** P. Buck *Coming of Maori* (1950) II. ix. 243 The women's *poi* dance..used an accessory in the form of the *poi* ball... The *poi* balls in common use in modern times are made of dry bullrush leaves (*raupo*), about the size of an orange but slightly elongated, and with a short string... The movements with the long *poi* were slower than with the modern short *poi*. The string of the *poi* was held in the right hand and the ball was twirled and beaten back with the left hand while various movements were made over the shoulder, to the sides, the thighs, the knees, the head, the *poi* balls being kept twirling in perfect time to the songs sung by the leaders. **1950** *N.Z. Jrnl. Agric.* Jan. 89/1 One of the pois..is interwoven with raw doghide, a custom which, because it pertained only to very early times, makes this poi a rarity. **1958** S. Ashton-Warner *Spinster* 47 The Maori love-songs..and the poi tunes and the melodies they use in canoes. **1977** *N.Z. Woman's Weekly* 10 Jan. 35/2 A calendar which features Maori poi dancing at Rotorua.

poi, var. *pwe.

poiesis (poi,ī·sis). [f. Gr. ποίησις a making, creation; cf. Poesis, Poesy *sb.*] Creative production, esp. of a work of art.

1934 in Webster. **1962** *Listener* 24 May 901/2 The tutelary figure of all that belongs to poiesis. **1971** G. Steiner *In Bluebeard's Castle* III. 72 The equivocations between *poiesis*—the artist's, the thinker's creation—and death. **1973** Matias & Willemen tr. Cegarra & Metz in *Screen* Spring/Summer 152 Metz uses the term realism to characterise both types [of filmic modernity]: in the case of Godard, 'a copiously disorganised *realism*, a brilliant and euphoric avatar of poiesis'.

-poiesis (poi,ī·sis), comb. form of Gr. ποίησις (see prec.), used to form terms in *Med.*, as *hæmopoiesis* s.v. *hæmo-, hemo-, *lymphopoiesis* s.v. *lympho-.

poignance. Add: Now also with pronunc. (poi·nʸăns). (Earlier and further examples.)

1769 O. Ruffhead *Life A. Pope* 114 The solemn air.. greatly heightens the poignance of the ridicule. *Ibid.* 119 The poet's closing the climax with the highest disaster of all, gives additional poignance to the ridicule. **1812** E. Wynne *Let.* 6 Apr. in *Wynne Diaries* (1952) xxx. 514 All the poignance of his sufferings.

poignancy. Add: Now also with pronunc. (poi·nʸănsi). **3.** (Further example.)

1934 M. Bodkin *Archetypal Patterns in Poetry* 310 In each poem the lovely image gains poignancy from its imagined background of frustration and pain.

poignant, *a.* Add: Now also with pronunc. (poi·nʸănt).

poikilitic, *a.* Mark sense in Dict. † **1.** *Obs.* and add: **2.** *Petrogr.* Applied to the structure or texture of a rock (now only an igneous rock), or to the rock itself, in which small crystals of one mineral are enclosed within crystals of another. Cf. *pœcilitic *a.* 2.

1886 G. H. Williams in *Amer. Jrnl. Sci.* CXXXI. 30 This structure is so common in many massive rocks, especially in the more basic kinds, that I would venture to suggest the use of the term '*poicilitic*' (derived from the Greek ποικίλος, mottled) for it. **1893** —— in *Jrnl. Geol.* I. 176 (*heading*) On the use of the terms poikilitic and micropoikilitic in petrography. *Ibid.* 177 This term was at first incorrectly spelled *poicilitic* and subsequently corrected by Prof. Dana to its Latin form, *poecilitic*... Its preferable orthography is, however, that given above [*sc. poikilitic*]. At the time it was proposed the writer was not familiar.. with the designations..*poecilitic* and *poikilitic*, given successively..to the 'New Red' sandstone. **1920** [see *poikiloblastic* adj. s.v. *poikilo-]. **1923** *Jrnl. Geol.* XXXI. 177 The feldspar grains of the arkose are set, in a sort of poikilitic fashion, in a matrix of much more coarsely crystalline quartz. **1954** H. Williams et al. *Petrography* ii. 19 (*caption*) Poikilitic texture in hornblende peridotite, Odenwald, Germany. **1970** *Nature* 25 July 366/1 The poikilitic chromitite harzburgite..could well be formed from dunite by permeation by silica-bearing solutions.

Hence **poikili·tically** *adv.*

1912 *Meddelelser om Grønland* XXXVIII. 150 Each anhedron encloses a large number of sodalite-crystals poikilitically. **1932** F. F. Grout *Petrogr. & Petrol.* vi. 371 Less commonly some rounded granules in a hornfels may be enclosed in a coarser mineral developed poikilitically. **1963** W. A. Deer et al. *Rock-Forming Min.* IV. 263 Kalsilite..occurs as an interstitial mineral poikilitically enclosing the ferromagnesian constituents.

poikilo-. Add: **poikiloblast,** (*b*) (also **pœcilo-**) *Petrol.*, each of the inclusions in a poikiloblastic rock; **poi:kilobla·stic** *a.* (also **pœcilo-**) *Petrol.* [ad. G. *poikiloblastisch* (F. Becke 1903, in *Compt. Rend. IX Sess. Congr. Géol. Internat.* (1904) II. 570): see -blast], applied to the structure or texture of a metamorphic rock, or to the rock itself, in which small crystals of an original mineral occur within crystals of its metamorphic product (cf. *poikilitic *a.* 2).

1944 *Trans. R. Soc. Edin.* LXI. 225 Accessories are apatite and sphene,..the latter in rounded red-brown pleochroic poeciloblasts in the same minerals. **1969** A. Spry *Metamorphic Textures* 169 (*heading*) Poikiloblasts. **1920** A. Holmes *Nomencl. Petrol.* 187 *Poikiloblastic*, a metamorphic texture due to the development, during re-crystallisation, of a new mineral around numerous relics of the original minerals, thus simulating the poikilitic texture of igneous rocks. **1932** A. Harker *Metamorphism* xiii. 191 (*caption*) A large porphyroblast of green hornblende showing typical pœciloblastic or sieve-structure. **1954** H. Williams et al. *Petrography* ix. 168 (*caption*) Poikiloblastic (sieve) texture in skarn, Doubtful Sound, New Zealand. **1969** A. Spry *Metamorphic Textures* vi. 170 Inclusions increase the total free energy and thus a poikiloblastic crystal is not in its most-stable..condition.

poikiloderma (poi:kilŏdə·ɪmă). *Path.* [mod.L., ad G. *poikilodermia* (E. Jacobi 1907, in *Verhandl. der deutsch. dermatol. Gesellschaft: IX. Kongr.* 1906 322), f. Gr. ποικίλο-ς variegated + δέρμα skin.] An atrophic condition of the skin characterized by reticular pigmentation and associated with telangiectasia. Hence **poi:kilode·rmatous** *a.*

1907 *Index Medicus* 173/2 (Index), Poikiloderma. **1936** *Arch. Dermatol. & Syphilol.* XXXIII. 289 The word poikiloderma..is used indiscriminately, to indicate variegated lesions, instead of being limited to eruptions presenting a definite picture of poikiloderma atrophicans vasculare. *Ibid.* 290 Poikilodermatous changes may be part of the picture of many chronic inflammatory dermatoses. **1967** H. Montgomery *Dermatopath.* xi. 246/2 Various types of poikiloderma, including poikiloderma of Civatte, do not reveal the typical changes in the blood vessels or in the collagenous fibers although telangiectasia is present clinically and histologically. **1968** A. Rook in A. Rook et al. *Textbk. Dermatol.* II. xlii. 1271/2 Some inflammatory dermatoses, such as lichen planus may also give rise to poikilodermatous changes.

poikilosmotic (poi:kilǫzmǫ·tik), *a. Physiol.* Also **poikilo-osmotic** (poi:kilo,ǫzmǫ·tik). [ad. G. *poikilosmotisch* (R. Höber *Physikal. Chem. der Zelle und der Gewebe* (1902) ii. 26): see Poikilo- and Osmotic *a.*] Of an animal: that allows the concentration of solute in its body fluids to vary with fluctuations in that in the surrounding medium. Opp. *homœo-osmotic* s.v. *homœo-.

1905 *Biol. Bull.* VIII. 262 All of the marine invertebrates which we have worked with are truly 'poikilosmotic'. Two factors may be at work in producing the variations in the osmotic pressure, viz., the interchange of water, and of salts. **1931** *Biol. Rev.* VI. 473 Characteristically marine forms such as *Hyas, Cancer* and the polychaete *Nereis cultrifera* only respond to lowered external salinity by passive swelling till their blood is isotonic with the medium. Such 'poikilosmotic' organisms show no change in oxygen consumption during this process. **1953** E. Palmer tr. *Ekman's Zoogeogr. Sea* vi. 118 The marine invertebrates were until recently considered to be poikilosmotic. **1971** *Biol. Abstr.* LII. 5434/1 Poikilosmotic animals are actually slightly hyperosmotic as against their external medium in the normal circumstances of the animal's existence.

Hence **poi:kilosmo·sis, poi:kilosmoti·city,** the state or property of being poikilosmotic.

1935 *Proc. Linnean Soc. N.S.W.* LX. 244 A 'law of poikilosmoticity' is by no means applicable to all marine invertebrates when the surrounding water is of lower salinity than ordinary sea-water. **1939** A. Krogh *Osmotic Regulation in Aquatic Animals* 242 (Index), Poikilosmosis. **1955** Elkinton & Danowski *Body Fluids* ii. 50 In many invertebrates the poikilosmoticity is only relative and.. steady states are maintained with respect to both total osmolarity and individual ionic levels. **1971** *Biol. Abstr.* LII. 5434/1 (*heading*) On poikilosmosis and iso-osmosis.

poikilothermal, -thermic: substitute for entry s.v. Poikilo-:

poikilothermic (poi:kilopɔ̄·ɪmik), *a. Physiol.* [ad. G. *pōkilotherm* (now *poik-*) (C. Bergmann 1847, in *Göttinger Studien* I. 613): see Poikilo- and Thermic *a.*] Characterized by a body temperature that varies with the temperature of the environment; cold-blooded. Also **poi:kilothe·rmal, -the·rmous** *adjs.*, in the same sense. Opp. *homœothermic *a.*

1884 Poikilothermic [in Dict. s.v. Poikilo-]. **1885** W. Stirling tr. *Landois's Text-bk. Human Physiol.* I. vi. 426 The so-called cold-blooded animals are called poikilothermal. **1928** Pearse & Hall *Homoiothermism* i. 2 The change from the poikilothermic to the homoiothermic condition. *Ibid.* iv. 27 The extreme minimum temperature which poikilothermal animals can tolerate. **1933** R. H. Wolcott *Animal Biol.* lxiv. 456 Animals which can maintain a constant temperature are termed homoiothermous..; animals unable to do so are termed poikilothermous, or cold-blooded. **1956** *Sci. News* XL. 71 Under ordinary circumstances most poikilothermous animals have body temperatures that fall when the environment becomes cooler, and rises when it becomes warmer. **1963** *Lancet* 5 Jan. 29/1 Newborn infants are poikilothermic, and their temperatures may be rapidly reduced. **1973** *Marine Biol.* XXI. 262/2 Lower poikilothermal animals (protozoans, sponges, some coelenterates). **1973** P. A. Colinvaux *Introd. Ecol.* xx. 288 Poikilothermous lizards are not excluded from the Arctic because they are poikilothermous, but because, on the average, they cannot balance their heat budgets if the ambient temperature is low.

Hence **poi·kilotherm,** a poikilothermic animal.

1934 in Webster. **1950** C. L. Prosser in C. L. Prosser et al. *Compar. Animal Physiol.* x. 349 Aquatic poikilotherms follow changes in environmental temperature rapidly and precisely. **1965** B. E. Freeman tr. *Vandel's Biospeleol.* xix. 326 Resistance to starvation is a general property of poikilotherms. **1968** [see *homœotherm].

poikilothermia (poi:kilopɔ̄·ɪmiă). *Physiol.* Also anglicized as **poi·kilothermy.** [f. prec.: see -ia[1], -y[3].] The state or property of being poikilothermic. Also **poi:kilothe·rmism** (*rare*).

1903 *Jrnl. Physiol.* XXIX. 369 (*heading*) Poikilothermism in rabies. **1921** *Physiol. Rev.* I. 304 Increased surface blood flow promotes poikilothermia because it facilitates the conduction of heat either to or from the body. **1939** *Nature* 22 Apr. 684/1 Torpidity [in humming-birds] appears to be a temporary poikilothermy, rather similar to that already described for bats. **1968** D. W. Wood *Princ. Animal Physiol.* viii. 122 Social methods of overcoming the disadvantages of poikilothermy are also found. Larvae of the butterfly *Vanessa* cluster together in the cold. **1974** *Nature* 22 Feb. 568/1 In most newborn animals there is a physiological hypothermia and poikilothermia.

poile. Restrict † *Obs. rare* to sense in Dict. and add later example in spelling *poil*.

1806 M. Lewis *Jrnl.* 15 May in *Orig. Jrnls. Lewis & Clark Expedition* (1905) V. xxviii. 38 The poil of these [grizzly] bear were infinitely longer finer and thicker than the black bear.

b. poil(e) de chèvre (see quot. 1960).

1873 F. B. Palliser *Descr. Catal. Lace S. Kensington Mus.* (ed. 2) 9 Lace is made of gold, silver, silk, cotton, and flax, to which may be added poil-de-chèvre. **1927** E. Sitwell *Rustic Elegies* 86 And splashed the red and white striped poil de chèvre short gown. **1960** C. W. Cunnington et al. *Dict. Eng. Costume* 269/2 *Poile de Chèvre*,..a textile of goat's hair (weft) and silk (warp) in plain weave; having a shiny satin-like lace.

Poilite (poi·ləit). Also **poilite.** A proprietary name for a building material made of asbestos and cement, used in the form of tiles, sheets, etc.

1903 *Trade Marks Jrnl.* 11 Feb. 154 Poilite... A fire-resisting lining and roofing material.., manufactured of a mixture of asbestos, lime, and Portland cement, for use in building. Bell's Asbestos Company, Limited,..London. **1922** *Daily Mail* 7 Nov. 3 The amalgamation of the company's poilite (asbestos-cement) section with the British Everite and Asbestilite Works. **1925** E. G. Blake *Roof Coverings* vii. 157 The cost of Poilite slates at the present time is about twenty per cent. less than that of the ordinary Welsh slates. **1936** *Discovery* Apr. 117/1 Wainscotings, skirtings, panelling, poilite sheets, etc., are loosened to allow entry of the gas. **1968** *Laxton's Building Price Bk.* (ed. 141) 726/1 'Poilite' Asbestos-Cement Flat Sheets &c., Turners Asbestos-Cement Co. Ltd.

poilu (pwalü). *colloq.* [Fr., hairy, virile.] A soldier in the French army, *esp.* one who fought in the war of 1914–18. Also *attrib.* and *Comb.*

1914 in *Further Lett. from Man of No Importance* (1932) 15, I hear dear old Madame Waddington is busy as a bee with comforts for *poilus*. **1915** G. Adam *Behind Scenes at Front* 183 France has every reason to be proud of her infantry, the 'poilus' as they have been called in this war. **1916** 'Taffrail' *Pincher Martin* xvi. 303 *Poilus* in their slate-blue uniforms. **1918** E. M. Roberts *Flying Fighter* 54 We were away north of the French lines, but that made no difference to the poilus, who also were to attend the ceremony. **1923** *Daily Mail* 28 Feb. 1 (Advt.), In shades of Coral,..Poilu-Blue, Lemon, Brique, Mole. **1926** *Glasgow Herald* 30 Nov. 8 He depicts..the adventures of two poilus who miss the train that should have brought them back to barracks in time for roll-call. **1930** Kipling *Limits & Renewals* (1932) 324 He pointed downward to the little cast-iron *poilu*, which seemed to be standard pattern for War memorials in that region. **1940** 'Gun Buster' *Return via Dunkirk* I. x. 79 The room was packed with poilus, singing songs. **1946** G. Millar *Horned Pigeon* xvii. 263 There were fierce yellowish photographs of men posed in *poilu* uniform all round the walls of her room. **1966** J. Dos Passos *Best Times* (1968) ii. 43 The Boche..scattered a few salvoes of artillery..just to keep the poilus on their toes. **1975** P. Fussell *Gt. War & Mod. Memory* (1977) vii. 240 During the 1917 mutinies in the French army, the *poilus* being marched up to the line frequently made loud *baa*-ing noises.

poinciana (poinsĭ,ā·nă). [mod.L. (J. P. de Tournefort *Institutiones Rei Herbariæ* (1700) I. 619), f. the name of M. de *Poinci*, a 17th-century governor of the Antilles + -*ana*.] An evergreen tree or a deciduous prickly shrub of the genus formerly so called, now divided between *Delonix* and *Cæsalpinia*, belonging to the family Leguminosæ, native to the West Indies, Madagascar, or northern Africa, and bearing racemes of scarlet or yellow flowers.

1731 P. Miller *Gardeners Dict.* s.v. Poinciana; Barbadoes Flower-Fence, or Spanish Carnations. **1807** *Curtis's Bot. Mag.* XXV. 995 The name of Poinciana was given to this splendid shrub by Tournefort. **1824** H. E. Lloyd tr. *Spix & Martius's Travels in Brazil* I. II. i. 174 The beautiful bushes of the poinciana are planted. **1859** J. Froebel *Seven Years' Travel in Central Amer.* iii. 34 The way down from the city passes through thickets of shrubbery covered with the most splendid flowers, amongst which the Poinciana, with scarlet panicles, is most prominent. **1908** *Daily Chron.* 11 Sept. 7/2 The Royal Poinciana crowns itself with cardinal magnificence. **1927** *Transition* Apr. 101 Under the poinciana, of a noon or afternoon Let fiery blossoms clot the light. **1954** *Coast to Coast 1953–54* 86 The house was..enclosed by gardens and shaded by tamarind- and poinciana-trees. **1969** T. H. Everett *Living Trees of World* 196/2 A native of Madagascar, but freely planted as an ornamental in most warm climates, the royal poinciana is one of the showiest of flowering trees. **1978** 'A. York' *Tallant for Disaster* xi. 167 A larger than average wave..smashing at the base of the poinciana trees.

poinsettia. Substitute for etym.: [mod.L. (R. Graham 1836, in *Edinb. New Philos. Jrnl.* XX. 412), f. the name of J. R. *Poinsett* (1779–1851), American minister to Mexico + -ia[1].] (Earlier and later examples.) Also *attrib.*

1836 *Curtis's Bot. Mag.* LXIII. 3493 (*heading*) Showy Poinsettia. **1906** P. Pennington *Jrnl.* 9 Mar. in *Woman Rice Planter* (1913) ix. 302 Mr. Poinsett..brought many rare plants from Mexico, among others the Flor de la Noche Buena, which has borne in this country the name Poinsettia in his honor. **1947** K. Tennant *Lost Haven* vii. 99 The town on the slope was aflame with red poinsettias. **1964** 'R. Macdonald' in H. Waugh *Merchants of Menace* (1969) 78 Plaster painted adobe color, poinsettia-red curtains. **1968** J. C. Holmes *Nothing More to Declare* 43 A few luxuriant poinsettias bloomed among the crumbling buildings and the blistered streets. **1977** Ward & Wellsted *Indoor Plants* 25 (*caption*) These [developments] culminated in the dwarf poinsettias now so common in florists' displays. **1978** *Detroit Free Press* 16 Apr. 10/2 It is an eclectic and nurturing environment where poinsettias thrive long after Christmas.

point, *sb.*[1] Add: **A. I. 3. a.** (Earlier example of *full point*.)

1587 F. Clement *Petie Schole* 25 The perfect pause, or full point is set down in the line immediatly after the last word.

II. 5. a. Also used with preceding numeral to form an *attrib. phr.* designating a statement or document that has the number of items specified by the numeral.

1945 *Richmond* (Va.) *Times-Dispatch* 26 Oct. 5/1 He proposed, in its stead, a six-point program as a foundation for world peace. **1961** *Chicago Daily Tribune* 25 Oct. I. 16/3 If the parties failed to sign an eight point protocol agreeing on Gen. Gursel as president. **1975** *New Yorker* 21 Apr. 134/2 The basic policy of the Communists, set forth in a ten-point statement.., is constantly rebroadcast. **1977** *Time* 10 Jan. 20/2 The CPI supported Mrs. Gandhi's 20-point program for social reform, but pointedly witheld support from Sanjay's five-point youth program.

III. 13. e. *on points* (Boxing): according to or as a result of the points scored in a number of rounds, esp. in phr. *to beat* (or *defeat*) *on points*: to beat (an opponent) in a contest by winning more points and not by achieving a knockout. Also *to lose* (or *win*) *on points*, etc. Also *fig.*

1904 *C. B. Fry's Mag.* June 301/1 Aeneas called 'time', and gave a decision..'on points'. **1929** *Daily Express* 7 Nov. 13/5 Young Stribling, the American boxer, defeated Maurice Griselle, France, on points in a ten-round.. contest. **1929** *Evening News* 18 Nov. 16/4 Rolland..beat Wilhelm Bech on points. **1930** *Cambridge Daily News* 25 Sept. 7/4 Campolo..will probably retire for good..if Sharkey gives him the full count, or if he loses on points. **1948** J. B. Priestley *Linden Tree* I. 38 'Daddy had a blazing row with the man at the bookshop. Didn't you, Daddy?'.. 'Yes, but I thought he won on points.' **1955** *Times* 12 May 4/3 He landed a fair number of swings and a few straight lefts... It was not enough and Eddington won on points. **1957** E. Gowers *H. W. Fowler* 11 The draftsman was attacked for using this construction... If I had been the referee in that contest I should have awarded Jespersen a win on points. **1968** *Listener* 18 July 90/3 Mrs Vlachou was as icily contemptuous of the colonels' intentions as she used once to be of British intentions at the time of the Cyprus troubles; Mr Sparrow was eloquent in their defence. On points I should give the victory to Mrs Vlachou. **1975** *Oxf. Compan. Sports & Games* 116/1 This championship, for which Clay and Frazier, the winner on points over 15 rounds, each received $2,500,000, must have grossed more than $20 million.

14. b. A unit of credit towards an award or benefit, *spec.* (*a*) an academic qualification (*U.S.*); (*b*) allocation of local authority housing; (*c*) discharge from the armed forces or return from overseas service.

1903 *N.Y. Times* 29 Aug. 3/4 For university credit, each 30 hours' course counts one point, and laboratory work, at the rate of 60 hours, to one point. **1950** B. Wootton *Testament for Social Science* ii. 41 The local authority's decision to give x points for size of family, y points for service in the armed forces, plus z points for being bombed out may be regarded as entirely subjective value-judgments. **1953** *Manch. Guardian Weekly* 5 Feb. 7 Into San Francisco come the wounded, come the soldiers who have accumulated enough rotation 'points' to be sent home. **1959** N. Mailer *Advts. for Myself* (1961) 120 The regiment was disbanded, and those men who did not have enough points to go home were sent to other outfits. **1963** [see *credit *sb.* 13d]. **1974** M. Birmingham *You can help Me* i. 10 The Bengalis..have not enough points yet to live in tower blocks so they pay exorbitant rents to private landlords. **1974** *Times* 14 Nov. 17/7 Housing lists are so long..I think..they [*sc.* agricultural workers] should be given points for the number of years they have lived in a tied house. **1977** E. Ambler *Send no more Roses* v. 89 It's to be first in, first out, with a points bonus for every month of overseas service. **1977** *Jrnl. R. Soc. Arts* CXXV. 550/2 Usually points were allocated according to the inadequacy of the existing accommodation, the degree of overcrowding, the health of the occupier, the length of time that these conditions had prevailed, and the extent to which the dwelling was unsuited to the needs of the occupier. When a household had acquired enough points to reach the top of the housing list, then their own preference as to location and dwelling type would be taken into account.

c. *Bridge.* A unit by which a hand is evaluated.

1959 *Listener* 5 Mar. 434/2 A balanced hand with less than 25 points is considered insufficient to make a mandatory force to game. *Ibid.* 434/3 A ten-point hand. **1964** *Official Encycl. Bridge* 431/1 This [*sc.* the high-card valuation] gives a total of 40 points in the pack, and makes an average hand worth ten points. **1977** *Times* 10 Dec. 13/4 The text-books advise you to pass, because you have four points only and six points are needed for a positive response.

15. a. (Later examples.) Also used in quoting variations in interest rates (one point representing one per cent) and exchange rates (one point being one-hundredth of the smallest monetary unit).

1906 L. C. Cornford *Defenceless Islands* 98 Prices have dropped six points. A point is the hundredth part of a penny. **1930** M. Clark *Home Trade* 163 Prices in the case of spot transactions are not stated in terms of pence per lb., but as so many 'points on' or 'points off' the price of cotton.., a 'point' being one hundredth of a penny. **1971** *Daily Tel.* 5 Apr. 7/2 Bank Rate is now 2½ points higher than at the beginning of 1956, and mortgage rates are also 2½ points higher over the same period. *Ibid.*, This would save the building societies about ¼ of a point of interest. **1974** *Ibid.* 23 Feb. 19/4 Metal Box at 206p managed an 11-point rise while Pilkington jumped 13 to 302p. **1980** *Times* 12 Feb. 19/3 Sterling..continued to maintain a firm position closing 60 points ahead at 2·3045.

b. A unit of value and exchange in rationing; *on points*, (rationed) on the basis of such units.

1940 *Economist* 31 Aug. 280/2 Textiles are sharply rationed [in Holland]... On August 12th, the German system of a clothing card of 100 points was introduced. **1942** *Business Week* 9 May 15/1 The rationing method—according to current OPA thinking—will be to establish a secondary currency of points alongside the dollar currency. Essential articles..will be listed and given a price in points. Then every person in the country will be given a book of stamps representing a certain number of points. **1944** M. Laski *Love on Supertax* i. 13 You always seem to forget that breakfast cereals cost points. **1944** *Times* 23 Feb. 2/3 From April 2 imported tinned marmalade will be available on points, and will not be, as hitherto, part of the preserve ration. **1947** *Ann. Reg. 1946* 55 Surplus [Bread] Units would be exchangeable for Points entitling to other foodstuffs. **1948** J. Bell *Wonderful Mrs Marriott* ii. 36 Mrs Dale's worries with points and coupons. **1950** 'P. Woodruff' *Island of Chamba* vii. 109 The Sultana sounds like something you get from the grocer if you have enough points. **1965** N. Freeling *Criminal Conversation* II. xv. 169 England during the reign of Sir Stafford Cripps..with points and coupons and austerity. **1975** S. Briggs *Keep Smiling Through* (1976) 155 People could distribute their 16 monthly points as they liked, sometimes spending the lot on a delicious tin of salmon, sometimes cautiously stocking up on sensible spam and pilchards.

17*. A measure of weight used for diamonds and other precious stones, equal to one hundredth of a carat.

1931 Kraus & Holden *Gems* (ed. 2) vii. 99 The weight of a diamond is often expressed in points. Thus, a stone weighing 65 points actually weighs 0·65 carats. **1974** *Encycl. Brit. Micropædia* II. 546/3 The metric carat, equal to 0·200 grams, and the point, equal to 0·01 carat, was adopted by the U.S. in 1913 and, subsequently, by most other countries. **1979** *Guardian* 3 Nov. 17/7 (Advt.), A dazzling 1½ point diamond, handset in a brilliant star-burst of gold 'vermeil'.

IV. 19. b. (Further examples.) Also, a rallying point or rendezvous for police, military personnel, etc.

1898 J. D. Brayshaw *Slum Silhouettes* 201 'Here, John,' he shouts to the potman, 'fetch the man from the point.'.. In a few minutes up comes the potman with a sergeant an' p'liceman. **1963** N. Marsh *Dead Water* (1964) vii. 191 Shall I return to my point, sir? **1967** 'S. Woods' *And shame Devil* ii. 36, I made my point with t'sergeant... Corner of Badger's Way, that was. **1968** P. N. Walker *Carnaby & Gaol-breakers* viii. 80 P.C. Williams... His last point was in Romanby village. **1972** J. Rossiter *Rope for General Dietz* ix. 128 They're waiting until half-past ten. That's when the *Guardia* make their point near the *Bar El Toro Blanco* and wait for thirty minutes.

c. (Later examples.)

1920 *Blackw. Mag.* Jan. 108/1 These marshy channels.. are the invariable point of any hunted boar. **1939** *Country Life* 11 Feb. p. xxxii/1 After running in all for an hour and forty minutes and making a six and a half miles point. **1977** *Field* 13 Jan. 52/1 Our fox crossed the valley and made his point to Moorhill.

d. *pl.* Localities or places considered in some special connection, esp. as being in a particular direction from a specified place. (Influenced by sense B. 9.)

1885 U. S. Grant *Personal Mem.* I. xxx. 422 From there [*sc.* Vicksburg] a railroad runs east, connecting with other roads leading to all points of the Southern States. **1895** *Montgomery Ward Catal.* Spring & Summer 589/1 Freight paid by us to all points east of the west line of Dakota, Kansas, Nebraska and Louisiana. To points farther west we apply $1.00 on the payment of freight. **1903** *N.Y. Even. Post* 19 Aug. 7/6 The number here is now estimated at 21,000 persons from Eastern points, with fully 35,000 persons in addition from California. **1926** *Publishers' Weekly* 22 May 1684/2 Some of us here get supplies from other points that they know nothing about. *Ibid.* 18 Dec. 2256 The business must be going to distant points—New York, Chicago, etc. **1933** *Fortune* Aug. 94 Loring F. ('Red') Nichols, of Cleveland and points Mid-West, is a crack director and trumpeter. **1969** R. Tashkent *Ambiguous Man* i. 14 She and a friend had started off for Athens and points East. **1969** D. Barron *Man who was There* i. 11 I've been in Pakistan and points East for six weeks. **1973** A. Ross *Dunfermline Affair* 38 We..took the road through Kilmany and Auchtermuchty and all points south west. **1978** *Jrnl. R. Soc. Arts* CXXVI. 712/1, I will be testing that later this month in Washington and other points West.

e. (See quot. 1926); *spec.* a socket fixed in a wall or the like which is connected to an electricity supply and designed to receive the plug of an electrical appliance; = *outlet *sb.* 1 d. Cf. also *power point* s.v. *power *sb.*[1] 18 f.

1904 H. Walter *Electric Lighting for Inexperienced* viii. 82 The cost per point depends..on the materials used for protecting the wires, and whether the wires are run on or under the surface of walls. **1913** D. S. Munro *Practice of Electr. Wiring* xiv. 126 The stamped boxes are not so useful for variety of outlet point. **1921** J. H. Havelock *Electr. Installation Work* xiii. 155 The lamps may be varied at one of the switch points. **1925** J. C. Connan *Electr. Estimating* ix. 167 If the area in square feet to be illuminated is divided by the total number of points (including ceiling roses, wall brackets, and wall plugs), the average area illuminated per point is obtained. **1926**

Gloss. Terms Electr. Engin. (Brit. Engin. Stand. Assoc.) 111 *Point*, in wiring. The termination of the wiring intended for attachment to a fitting for one or more lamps or other consuming devices. **1940** G. D. H. & M. Cole *Murder at Munition Works* 138 Presumably there was a lead to that reading lamp point over there. **1967** *Listener* 21 Dec. 831/2 There is no electric point in her room, so she uses the ceiling light festooned with wires to plug in her iron. **1972** M. Babson *Murder on Show* ix. 107 Helena Keswick plugged an electric kettle into a point underneath the table. **1976** *Cumberland & Westmorland Herald* 4 Dec. 16/6 (Advt.), Kitchen/breakfast room,.., floor units and matching wall cupboards and electric cooker point.

20. Restrict *Her.* to senses in Dict. and add: **d.** *Sculpture.* Any one of a series of holes drilled in a piece of stone or marble or on the model to be copied to the depth to which the material has to be cut away. Also, the position of such a hole.

1841 *Penny Cycl.* XXI. 142/1 This process is repeated till the numerous points at fixed depths, corresponding throughout with the surface of the model, are attained, and a rough copy of the sculptor's original work is thus mechanically made. **1911** A. Toft *Modelling & Sculpture* 254 A good pointer will keep all his 'points' a little 'full', by never allowing the needle to go quite home. **1947** J. C. Rich *Materials & Methods of Sculpture* ix. 261 The indirect method of stone carving involves the use of previously prepared three-dimensional models, built up in most cases of plastic clay and then cast in a more durable substance, such as plaster of Paris. The casts are then utilized as master models from which to take points or otherwise copy. **1970** *Oxf. Compan. Art* 884/2 Sometimes hundreds and even thousands of points will be taken to ensure a meticulously exact copy. **1974** *Encycl. Brit. Micropædia* VIII. 68/1 The final points on the stone are usually left about $\frac{1}{16}$ inch (about one millimetre) higher than those on the model to enable the sculptor to put the finishing touches on the stone.

22. c. Phr. *up to a* (*certain*) *point*: to a certain extent, but by no means absolutely.

1823 Byron *Don Juan* xiii. lxxxi. 95 For good society Is no less famed for tolerance than piety: That is, up to a certain point; which point Forms the most difficult in punctuation. **1916** G. B. Shaw *Androcles & Lion* p. xv, As they [*sc.* Savonarolas and Knipperdollings] know, very sensibly, that a little religion is good for children and serves morality, keeping the poor in good-humor or in awe by promising rewards in heaven or threatening torments in hell, they encourage the religious people up to a certain point. **1936** —— *Simpleton* II. 53 Well, it has worked, up to a point. **1951** E. Paul *Springtime in Paris* xvi. 296 He had been an understanding husband up to a certain point. **1961** *Chicago Daily Tribune* 10 Feb. III. 9/7 But he had in Walter Hendl a willing conductor able only up to a point. **1978** P. McCutchan *Blackmail North* vi. 69 'There's been a threat, Shard!' 'Being taken seriously?' 'Up to a point.'

V. 26. b. *pl.* Of persons and things: good features, advantages; usu. in phr. *to have one's* (or *its*) *points*.

1897 A. Beardsley *Let.* 16 Sept. (1970) 369 It was a sad moment when I tore myself from Dieppe... Paris however has points and I am forgetting my sorrow. **1931** *Times Lit. Suppl.* 26 Feb. 157/1 A simple story, but it has its points. **1934** E. O'Neill *Days without End* I. 24 What the devil's got into Walter lately, anyway? Getting drunk as a pastime may have its points, but as an exclusive occupation—. **1946** —— *Iceman Cometh* I. 24 *Parritt.* (*With a disparaging glance around*) Must be hard up for a place to hang out. *Larry.* It has its points for him. He never runs into anyone he knows in his business here. **1953** B. Gordon-Cumming *Gentle Rain* 38 She had her points, certainly... In her occasional dreamy moods she was lovely. **1961** F. & R. Lockridge (*title*) Murder has its points.

c. An area of contrasting colour in the fur of certain cats, usually on the face, paws and tail. Cf. also *seal point.

1903 F. Simpson *Bk. of Cat* xxiii. 259/2 The [Siamese] kittens are born absolutely white..and gradually all the points come. **1935** E. B. Simmons *Cats* xxix. 149 Blue points are rare, a sort of 'sport'. **1955** R. Tenent *Pedigree Cats* vi. 53 The points—marking the mask, ears, legs, feet, and tail—are all a dense and clearly defined seal-brown. **1972** Ing & Pond *Champion Cats of World* II. 87/2 The colouring and points [of Birmans] are as for the Siamese.

28. a. (Further examples.) Also, (the expression of) an important fact or truth; a noteworthy comment. Phr. *to have a point*: to have made a convincing or significant remark; to be correct (in a particular matter).

c **1939** A. D. Lindsay *Let.* in D. Scott *A. D. Lindsay* (1971) xv. 258, I have now read the article with interest and appreciation but it doesn't meet my point at all. **1962** *Listener* 22 Feb. 342/2 Is it possible that the Doctor had a point? **1963** *Ibid.* 21 Feb. 350/2 What most convinced me they had a point was the line taken by the interviewer. **1978** L. Thomas *Ormerod's Landing* ii. 20 'Right,' he agreed sportingly. 'You've got a point, Ormerod.'

c. *to take* (*someone's*) *point* (and variants): to understand the import or significance of what is being said; to concede the truth or value of a particular contention.

1898 G. B. Shaw *You never can Tell* II. 254 Do I take your point rightly, Mr McComas? **1901** —— *Capt. Brassbound's Conversion* III. 276 Rankin (cannily). I take your point, Leddy Ceecily. It alters the case. **1916** Joyce *Portrait of Artist* (1969) v. 187, I see. I quite see your point. *Ibid.* 188 Yes, yes: I see, said the dean quickly, I quite catch the point. **1943** N. Marsh *Colour Scheme* x. 187 'The point is quite well taken,' he said at last. **1961** C. Willock *Death in Covert* viii. 168 'You have absolutely nothing to go on except your sixth sense.' 'Point taken.' **1964** R. Braddon *Year Angry Rabbit* xii. 104 'But this is dusk,' Fitzgerald objected. 'Yes, dear,' Karen got the point quickly, 'which for nocturnals is what dawn is to us.' **1966** *Listener* 3 Nov. 658/2, I take his point about Laszlo Rajk. **1969** V. Gielgud *Necessary End* xxiii. 205 'A First Officer who doesn't play along to an extent with female passengers would probably be considered to be neglecting his job.' 'Point taken.' **1974** E. Lemarchand *Buried in Past* x. 168 'The affair'll have to be shelved.' 'I take your point, sir,' Pollard replied. **1976** J. Wainwright *Bastard* ii. 35 He nods and says, 'Okay. Point taken.'

d. Sense, purpose, or advantage (*in* or *of* a course of action, state of affairs, etc.). Chiefly in negative contexts, esp. in phr. *there is no point in*, it has no purpose, it is pointless.

In some cases there are connotations of sense B. 10.

1903 G. B. Shaw *Man & Superman* II. 60 Look here, Ann: if theres no harm in it theres no point in doing it. **1923** W. S. Maugham *Our Betters* II. 85 Thornton has plenty of money. Do you think there is any point in his spending his life making more? **1934** J. B. Priestley *Eden End* I. 6 What's the point of reading if it makes you feel uncomfortable? **1947** —— *Inspector Calls* II. 49 *Inspector.* And if her story is true—that he was stealing money—. *Mrs. Birling.*.There's no point in assuming that. **1953** K. Amis *Lucky Jim* xix. 208 'Give me your address, Christine.' She looked at him scornfully... 'That'll do no good at all. What on earth would be the point?' **1957** J. Osborne *Look Back in Anger* III. ii. 90 *Helena.* There doesn't seem much point in trying to explain everything, does there? **1966** M. Frayn *Russian Interpreter* xxxvii. 207 Was there any possible point for Manning in trying to deny all knowledge of those activities? **1968** C. Churchill in *New Eng. Dramatists* XII. 96 *Tim.* Was it nice in the aeroplane? *Grandfather.* I didn't really notice. *Tim.* What's the point of being in it then? **1971** P. Mortimer *Home* viii. 80 'Will you..get married?'.. 'There doesn't seem any point.' **1973** G. Greene *Hon. Consul* v. ii. 262 Tell Pablo to come in. If they have spotted us, there is no point in leaving him outside to be picked off alone. **1977** *Times* 1 June 17/5 If..it were true that many Anglican or Roman Catholic Christians could not accept this fundamental affirmation.. we should have to ask whether there is any point in continuing to search for unity.

e. *debating point*: see *debating *vbl. sb.* b.

VI. 32. A marking on a Hudson's Bay or Mackinaw blanket indicating weight.

1780 in *Beaver* (1935) June 47 [They] had misunderstood him about the price of the pointed plankets as the points were known to every Indian to be the price to be paid for each as 2½ points, 2½ beaver, 3 points, 3 beaver, etc. **1818** T. L. McKenney *Mem.* (1846) I. 309 Northwest Company blankets—so called—three points, to measure six feet six inches long. **1921** *Outing* Nov. 82/1 Hudson Bay blankets run as follows: Three point, 60 × 72 inches, double, weight, 8¼ lbs.; 3½ point, 63 × 81 inches, double, weight, 10 lbs.; 4 point, 72 × 90 inches, double, weight, 12 lbs. 'Points' refer to the markings on the blankets and indicate their size. **1935** *Beaver* June 47 The 'point' on the blanket in its present standardized form is comparatively modern, being introduced in 1850. Prior to that date blankets of the Hudson's Bay Company were made with the bar only by individuals in their own homes, each maker putting a distinctive mark, a 'point' on his product to show the size and weight. These 'points' were usually in coloured wools and usually about one inch long. **1954** E. E. Rich *Moose Fort Jrnl., 1783–85* 371 Originally the points and staves of the blankets were blue, but the colour was changed to red in 1786.

B. I. 1. a. Also of a pencil.

1826 M. R. Mitford *Our Village* II. 248 A pencil without a point. **1886** C. F. Woolson *East Angels* vii. 129 He sharpened all the pencils industriously, taking pains to give each one a very fine point. **1963** C. Fremlin *Trouble Makers* xviii. 140 Hundreds of coloured pencils without points went back into their dozens of boxes without lids. **1979** 'J. Le Carré' *Smiley's People* (1980) xvi. 189 Herr Kretzschmar owned a fine gold pencil... He popped out the point and..drew a pure circle.

d. (Further examples.)

1911 H. S. Harrison *Queed* iv. 45 The *Post*, not to put too fine a point upon it, had for a time run fast to seed. **1926** F. W. Crofts *Inspector French & Cheyne Mystery* iii. 40 Not to put too fine a point on it the situation is this: You are there, and you can't get out, and you can't attract attention to your predicament. **1935** C. Isherwood *Mr. Norris changes Trains* x. 155 He seems to have suggested, not to put too fine a point upon it, that you were an accomplice in my nefarious crimes. **1952** A. J. Cronin *Adventures in Two Worlds* I. vi. 52 This outbreak of scarlet fever... It's spreading, you know, and I find that in all my cases.., well, not to put too fine a point on it—the milk has come from Shawhead. **1971** 'E. Candy' *Words for Murder Perhaps* iv. 44 One of the doctor's most cherished personal finds, now happily in the City Museum, was, not to put too fine a point on it, a fake. **1977** *It* May 10/3 Not to put too fine a point on it, one could say that the real picture consists of nothing but exceptions to the rule.

e. Phr. *to a* (*fine*) *point*: to a precise form; completely.

1888 in Farmer & Henley *Slang* (1902) V. 241/2 Boiled down to a fine point bondsmen are in demand. **1902** G. H. Lorimer *Lett. Self-Made Merchant* xvii. 253 When she was through I knew that I'd been licked—polished right off to a point.

2. b. Also *transf.* (chiefly *U.S.*), such a feature on a river; the tapering extremity of any woodland reaching down into a prairie or other treeless area; any tapering extremity *of* land, or *of* rocks, woods, etc., constituting a special feature of this.

1637 in *Amer. Speech* (1940) XV. 297/2 Easterly butting out with a point of wood. **1660** *Early Rec. Warwick, Rhode Island* (1926) 256 His point of Meddowe on the south side of Occupessuatuxet Cove. **1682** *Early Rec. Providence, Rhode Island* (1899) XIV. 101 A black Oake tree standing upon a point of Rocks. **1772** D. Taitt in N. D. Mereness *Trav. Amer. Col.* (1916) 501, I..viewed this Town which Stands upon a point of Land on the North west side of the River. **1826** T. Flint *Recollections* 15 You hear of..sawyers, and points, and bends, and shoots. *Ibid.* 258 The entire uniformity of the meanders of the rivers [in Arkansas] called, in the phrase of the country, 'points and bends'. **1837** W. Irving *Capt. Bonneville* II. vii. 108 The whole band soon disappeared behind a point of woods. **1857** P. Cartwright *Autobiogr.* xxi. 328 We rode two miles, and the point of timber was plain in view. **1859** 'Mark Twain' in *New Orleans Daily Crescent* 17 May, The point at Cairo, which has not even been moistened by the river since 1813, is now entirely under water. **1883** —— *Life on Miss.* iii. 61 The big raft was away out of sight around the point. **1964** W. C. Putnam *Geol.* xi. 275/2 The common name for a broadly curving part of a river is a bend. The convex bank in such a curve is a point.

f. *the Point*: the United States Military Academy at West Point, N.Y. *U.S. colloq.*

1828 J. F. Cooper *Notions of Americans* I. 274 To these relics of a former age, must be added the actual and flourishing establishment at the 'Point', which comprises a village of academic buildings, barracks, and other adjuncts. **1922** *Frontier* (Missoula, Montana) Nov. 14 Ada's father had been C.O. when we were in the Point, and nearly every member of the class had been at one time or another in love with her. **1968** *Michelin: New York City* 139 West Point... Among the war heroes who graduated from the Point are Generals MacArthur (1903), Patton (1909) and Eisenhower (1915). **1971** C. Fick *Danziger Transcript* (1973) 167 Sam had worked his way into Dartmouth..and then transferred to the Point. **1973** E. Pace *Any War* (1974) III. 239 'I've been learning karate... We had the fundamentals at the Point.' 'You went to West Point?'

g. The tip of the lower jaw; the spot on which a knock-out blow is dealt.

1898 [see *out *adv.* 19 e]. **1901** R. Fitzsimmons *Phys. Cult. & Self-Defense* 159, I saw Fitzsimmons' right hand reach the point of Corbett's jaw. **1915** E. Corri *30 Yrs. Boxing Referee* 229 There is no sleeping-draught like a punch on the point. **1923** *Daily Mail* 16 Feb. 8 He once caught Lewis with a hard right near the point. **1924** *Truth* (Sydney) 27 Apr. 6 *Point*, sensitive portion of the jaw. **1942** Berrey & Van den Bark *Amer. Thes. Slang* §121/18 Point,..the vulnerable point of the chin.

h. Either of the extensions at the front end of a saddle-tree.

1908 *Animal Managem.* 166 The front arch extends below the side bars; the extension is known as the 'points', and these are intended to help the girths and prevent the saddle from heeling over.

i. *Ballet.* The tips of the toes. Usu. with *on* and *pl.* = *pointe.

1912 *Dancing Times* Aug. 449/2 *Points*.—Exposition of Principles. **1928** A. L. Haskell *Some Studies in Ballet* 153 A solo on the points. **1936** L. Sokolova in 'C. Brahms' *Footnotes to Ballet* v. 227, I held myself poised on my points, before literally collapsing exhausted in the middle of the stage. **1936** N. Streatfeild *Ballet Shoes* iv. 52 The children were most impressed by the way the children in the photographs stood on their points. **1949** A. Christie *Crooked House* x. 71 'He stopped me learning to be a ballet dancer.'.. She..kicked off her shoes and endeavoured to get on to what are called technically..her points. **1967** 'La Meri' *Spanish Dancing* (ed. 2) vii. 89 Very rarely it [*sc.* the Bolero] is danced on point. **1975** *New Yorker* 26 May 31/1 Mr. Griforovich, who was watching a tiny blond ballerina rise on point with the single-minded intensity of an adult star, put a hand across his face to conceal a smile. **1977** *Time* 24 Jan. 36/3 In a pas de deux with Ted Kivitt, she stepped majestically on point..as if there were magnets concealed in her toe shoes.

j. *U.S.* The position at the front of a herd of cattle, etc.; the position at the head of a column or wedge of troops (cf. sense B. 2 d in Dict.); also quasi-*adv.* in phrases *to ride* or *walk point*. Also *fig.*

1916 'B. M. Bower' *Phantom Herd* xiv. 245 You see a herd drifting before a storm maybe—a blizzard like yesterday, with your pal riding point. **1927** *Scribner's Mag.* Feb. 178/1 Consider the passing herd, anyone of the many that went up the Long Trail. At the 'point' ride two men, at the 'drag' two more, while other horsemen loiter on either flank. **1959** C. Ogburn *Marauders* (1960) ii. 64 Major Osborne looked around and his eye alighted on me. 'You take the point,' said he... He thought the way to use a communications officer was to have him lead the battalion column into action. **1962** J. Onslow *Bowler-Hatted Cowboy* xxi. 205, I was riding 'point', not as a leader to the cattle, but to warn oncoming traffic of the herd. **1964** F. O'Rourke *Mule for Marquesa* 118 Fardan trotted past the mules and took the point. **1969** I. Kemp *Brit. G.I. in Vietnam* v. 102 Goad walked point and I..took the tail, with the rest of squad well spaced out between us. **1970** *Times* 9 May 8/8 Daniel Lepointe, aged 21, a sergeant.. was walking point—the front position of his platoon— when..a helicopter observer told him about two huts. **1975** W. Safire *Before Fall* I. iv. 45 He said no: 'Let Romney take the point.' (In military tactics, the soldier 'on the point' of a wedge is the most likely to be shot.) **1977** 'J. Le Carré' *Hon. Schoolboy* iv. 100 Let me send you an advocate. Somebody who can ride point for you, draft your submission, carry it to the barricades.

k. *N.Z.* The hocks of a sheep; the wool that grows on them.

1922 W. Perry *Sheepfarming in N.Z.* iv. 44 The wool should be..well spread on the back, belly, and points. a **1948** L. G. D. Acland *Early Canterbury Runs* (1951) 379 *Fribby*,..the yolky locks round the points taken off by the

POINT

roller from a decently skirted fleece. **1956** G. Bowen *Wool Away!* (ed. 2) xii. 132 The Corriedale..grows more wool than hair on the hocks, thereby making it imperative that they be shorn trimmed to the feet. These points or socks do not have a tendency to lift or rise off the skin.

3. e. (Earlier and further examples.) *spec.* (*a*) the tapering extremity of a lightning conductor; (*b*) in an internal-combustion engine, either of the metal pieces on a sparking plug between which the spark jumps, or either of the metal surfaces of a contact-breaker which touch to complete the circuit; usu. *pl.*

1766 in *Essex Inst. Hist. Coll.* (1916) LII. 275 A new Meeting-House building..was struck with Lightning; it had Points and a Conductor as far as the Bellfree. **1775** in *Ibid.* (1877) XIII. 208 They have a handsome clock, points to the house, a fine walk on the top, [etc.]. **1870** 'Mark Twain' in *Galaxy* Sept. 424/1 He said it would be necessary to know exactly how many 'points' I wanted put up, what parts of the house I wanted them on, and what quality of rod I preferred. **1902** J. E. Hutton in A. C. Harmsworth *Motors & Motor-Driving* viii. 150 The points may be sooty and require cleaning. **1927** R. T. Nicholson *Austin Seven Bk.* xiii. 116 The rapid separation of the points of the contact-breaker. **1961** J. Mills *Car Repair & Maintenance* iv. 41 If this happened, there would not be a suitably fat spark at the plug points and the contact breaker points would become burned. **1968** K. Weatherly *Roo Shooter* 111, I must have dried the flamin' plugs and points twenty times.

o. *Archæol.* (See quot. 1959.)

1912 *Archaeologia* LXIII. 129 (*caption*) A symmetrical 'point' of laurel-leaf form.., but the surface flaking only partial. **1932** *Antiquity* VI. 190 The true Mousterian industries are characterized by flake-tools such as side-scrapers, points and Levallois flakes. **1943** J. & C. Hawkes *Prehist. Britain* i. 19 The Levalloisian, an outstanding culture during the last interglacial, when it was responsible for beautifully finished points and scrapers which..were partially trimmed before the flake was struck from the parent core. **1949** W. F. Albright *Archaeol. Palestine* iii. 59 The Natufian was a thorough-going microlithic culture, consisting largely of blades and points. **1959** J. D. Clark *Prehist. S. Africa* ii. 42 *Point*, a pointed flake or blade, often with careful secondary retouch, or a pointed tool of bone. Believed to have sometimes formed the heads of spears and arrows. **1963** J. Hawkes in Hawkes & Woolley *Prehist. & Beginnings of Civilization* I. iii. 71 Implements found..in the Solo valley..include..points and picks made from bone and antler. **1971** J. Bordaz *Tools of Old & New Stone Age* iv. 31 In regions where good stone is comparatively abundant,..it is not unusual to find Levallois points up to six inches long.

IV. 10. a. (Earlier example.)

1694 J. Savage tr. C. de St. Evremont in T. Brown *Misc. Ess. M. de St. Evremont* II. II. 96 Points, Antithesis's and Paradoxes.

b. (Further example.)

1901 H. James *Sacred Fount* 17 Having a reputation for 'point' to keep up, she was always under arms.

c. *Theatr.* A gesture, vocal inflection, or some other piece of theatrical technique used to underline a climactic moment in a speech, rôle, or situation; a moment so underlined. Usu. used with the implication that the integrity of the performance as a whole is being subordinated to the desire for immediate applause. Also *fig.*

1822 C. Mathews *Let.* 4 Oct. in A. Mathews *Mem. Charles Mathews* (1839) III. 314, I don't know an instance of a point failing which I considered to be really good myself. **1864** [see *gravy 2 c]. **1870** O. Logan *Before Footlights & Behind Scenes* 135, I began to practice the effects, the stage walks, the managing of the voice, the general bearing of the person, the making of 'points', the attaining of 'climax'. **1897** G. B. Shaw *Our Theatres in Nineties* (1932) III. 124 It lends itself to people talking at each other rhetorically from opposite sides of the stage, taking long sweeping walks up to their 'points'. *Ibid.* 132 He succumbed to the temptation to utter the two or three most fatuously conceited of Helmer's utterances as 'points'. **1900** T. E. Pemberton *Kendals* ix. 259 So natural is she at all times that she never seems to be 'making points' after the crude fashion of inferior actresses. **1916** J. R. Towse *Sixty Years of Theater* 29 Woe to the unfortunate actor who was not on his appointed spot and instant in his speech when he was a factor in one of Macready's laboriously calculated 'points'. **1952** Granville *Dict. Theatrical Terms* 139 A player who is not capable of 'making his *points*' (i.e. stressing his lines at the right time) will never get over.

V. 11. a. (Earlier and later examples.)

1816 W. Lambert *Cricketer's Guide* (ed. 6) 41 The Fieldsmen... The Point. The person who stands at the Point should place himself in a line with the popping crease, about seven yards from the striker. **1904**, etc. [see *backward a. 1]. **1916** *Anzac Book* 128 Was it a boundary hit or a catch at point? **1951** *People* 3 June 8/7 When he was out—to a brilliant catch by Ken Graveney at point—the bowling had been tamed.

b. (Earlier and later examples.)

1868 *Chambers's Encycl.* X. 597/1 In the arrangement of the men on each side, the *goal-keeper* defends the goal; *point* is the first man out from the goal; *cover-point* is a little in advance of point. **1935** *Encycl. Sports* 379/1 'Point' takes his position immediately in front of goal. **1967** *Globe & Mail* (Toronto) 16 May 39/9 Actually, the goaltender led a charmed life. Most of the danger was involved with the fellow who played between point and cover-point. **1975** *Oxf. Compan. Sports & Games* 588/2 The ten-a-side team consists of a goalkeeper, three defences (known historically as point, cover point, and third man), a left wing defence, [etc.].

D. 1. g. *at this* (or *that*) *point in time*: at this (or that) particular moment (cf. *moment sb. 1 c).

1974 R. B. Parker *Godwulf Manuscript* viii. 68 You don't understand the situation..at this point in time. **1975** *Atlantic Monthly* Jan. 32/2 The phrase 'at that point in time'..quickly became an early trademark of the whole Watergate affair. **1975** G. V. Higgins *City on Hill* iii. 89 'At that point in time I came away with the impression' that she was the best thing in the world. **1977** *Irish Times* 8 June 12/2 At this point in time the private rented sector of the housing market was shrinking.

11. point-to-point, *a.* **a.** (Further examples.) Also, from one point to another in turn (not necessarily in a direct line). Hence **point-to-pointer, point-to-pointing.**

1920 *Isis* 10 Mar. 2/2 No one..would go to a Point-to-point without a hat. **1930** *Telegr. & Teleph. Jrnl.* XVI. 110/2 The State wireless services only undertake 'Ship-to-shore'..and 'Point-to-point' internal traffic. **1934** *Sun* (Baltimore) 30 Mar. 3/2 The running of the Maryland Hunt Cup Point-to-Point. **1945** *Salt* 10 Sept. 10 The point-to-point voyages of air transport. **1952** F. A. Brown *Sport from Within* v. 163 Why not do the job properly and buy a racehorse instead of a point-to-pointer? *Ibid.* 177 It might be as well to confine hunters to Point-to-Pointing for the future, instead of encouraging racehorses to masquerade as hunters at race meetings. **1957** *Practical Wireless* XXXIII. 520/1 The arrangement and layout of main components is illustrated in Figs. 2 and 3, which also show partial point-to-point wiring. **1960** *Times* 12 Mar. 9/7 One of the great criticisms of point-to-pointing is the 'readying' of some of the horses. **1962** *Listener* 12 July 62/3 This instrument [*sc.* the electron microprobe] can provide information about the composition of, say, a copper-zinc alloy by point-to-point exploration of tiny areas one-thousandth of a millimetre square. **1964** A. Wykes *Gambling* viii. 187 Hunting's most direct descendant is a form of horse racing called 'point-to-point'. **1967** M. Chandler *Ceramics in Mod. World* iv. 117 These pores..make it possible for charged ions to carry a current over a comparatively easy point-to-point route through it [*sc.* the insulator]. **1970** M. Williams *Continuing Story of Point-to-Point Racing* xxiii. 146 The champion point-to-pointer of the 1964 season was Mr. W. J. A. Shepherd's Straight Lady. **1975** *Oxf. Compan. Sports & Games* 496/2 The major event of the point-to-point season is the Player's Gold Leaf Championship. **1976** *Horse & Hound* 3 Dec. 55/2 She got much pleasure and fun out of owning some useful point-to-pointers, in particular her good horse Brough. **1979** *Guardian* 31 Aug. 10/1 BMW..decided to sponsor point-to-points. *Ibid.*, The point-to-pointer interviewed thought it all very subtle.

c. In every particular or respect.

1934 C. Lambert *Music Ho!* v. 326 A book that..has purposely avoided a point-to-point analysis of individual works. **1949** R. K. Merton *Social Theory & Social Structure* xiv. 330 Discussions of the why and wherefore of science bore a point-to-point correlation with the Puritan teachings. **1958** *Listener* 27 Nov. 885/3 Changes of shape in the retinal image will necessarily be reproduced in the occipital lobe of the brain since there is a point-to-point correspondence between them.

12. point of view. (Earlier and later *fig.* examples, *esp.* in Literary Criticism.) Also *attrib.*

1760 Sterne *Sermons* II. xi. 112 Look at a man in one light...behold him in another point of view. **1793** Burke *Remarks on Policy of Allies* in *Three Memorials on French Affairs* (1797) 193 It is not the point of view in which we are in the habit of viewing guilt. **1905** S. L. Whitcomb *Study of a Novel* iii. 66 The narrator takes some general point of view for the entire action, and specific points of view for every part of it, in reference to time, place,..etc. The unity of a passage or a plot depends largely on the clearness and stability of this position. **1909** H. James *Wings of Dove* I. p. xvi, There is no economy of treatment without an adopted, a related point of view, and though I understand, under certain degrees of pressure, a represented community of vision between several parties to the action when it makes for concentration, I understand no breaking-up of the register, no sacrifice of the recording consistency, that doesn't rather scatter and weaken. **1921** P. Lubbock *Craft of Fiction* xvii. 251 The whole intricate question of method, in the craft of fiction, I take to be governed by the question of the point of view—the question of the relation in which the narrator stands to the story. **1927** E. M. Forster *Aspects of Novel* iv. 109 The problem of a point of view..is peculiar to the novel. **1948** M. Schorer in *Hudson Review* I. 69 Let it [*sc.* technique in fiction] be thought of in two respects particularly: the uses to which language, as language, is put to express the quality of the experience in question; and the uses of point of view not only as a mode of dramatic delimitation, but more particularly, of thematic definition. **1958** *N. & Q.* CCIII. 85/2 The experimentation with dramatic forms in *The Blithedale Romance* is clearly a prefiguration of the point-of-view technique. Eschewing the novelist's omniscience, Hawthorne had his narrator cloud in vague terms the nature of Moodie's early crime. **1961** W. J. Harvey *Art of George Eliot* i. 14 His [*sc.* Henry James's] insistence on dramatic representation, point of view, elimination of the author..has undergone a subtle critical change into something like dogma. **1973** R. Fowler *Dict. Mod. Critical Terms* 149 Some contemporary experimental novelists..transcend the issue altogether by abrupt and unsignposted shifts from one point of view to another. **1976** *Amer. Speech* 1974 XLIX. 232 *Stephen Hero* is written from a heavily omniscient point of view.

12*. point of order. In a debate, meeting, etc., an objection or query respecting procedure.

a **1751**, **1781** [see *order sb. 18]. **1885** *Encycl. Brit.* XVIII. 312/1 A member may speak once only to any question, except to explain, or upon a point of order, [etc.]. **1903** G. B. Shaw *Man & Superman* III. 75 The anarchist (*rising*) A point of order, Mendoza—. Mendoza (*forcibly*) No, by thunder: your last point of order took half an hour. **1952** *Oxf. Jun. Encycl.* X. 194/1 If a member wishes to raise a 'point of order', that is, to suggest that a rule of debate is being broken, he must remain seated and put on a hat to call the attention of the Speaker. **1974** *Encycl. Brit. Micropædia* VIII. 68/1 If the point of order is overruled by the presiding officer, the speaker resumes the floor.

12. point of departure.** *fig.* The starting point of a thought or action; the initial assumption, procedure, etc., which is developed. Also (with hyphens) *attrib.*

1857 Dickens *Dorrit* II. xiii. 438 In the relief of having this companion, and of feeling that he could trust him, he passed on to both [subjects], and both brought him round again..to his point of departure. **1876** [see Departure 5]. **1927** R. H. Wilenski *Mod. Movement in Art* 31 The French romantics of the early nineteenth century made the romantic elements in his art their point of departure. **1959** J. Kirkup tr. *S. de Beauvoir's Memoirs of Dutiful Daughter* (1963) III. 218 He looked upon marriage as a solution and not as a point of departure. **1961** *Listener* 17 Aug. 257/1 Schönberg began as an heir to Wagner. His point of departure was the intense chromaticism of *Tristan* and *Parsifal*. **1962** *Amer. Speech* XXXVII. 216 Every grammarian analyzes, using his own language as a point of departure. **1965** *Language* XLI. 189 The original point-of-departure vocabulary. **1976** *Early Music* Oct. 469/1 Seebass's point of departure is the tonary contained in the manuscript Paris, Bibliothèque nationale, fonds lat. 1118 from the mid-11th century.

12*. point of no return.** (See quot. 1941.) Freq. *transf.* and *fig.*

1941 *Jrnl. R. Aeronaut. Soc.* XLV. 306 This three-engined operation data is used to determine our so-called 'Point of No Return'. Laymen are inevitably intrigued by this fatalistic expression. As a matter of fact it is merely a designation of that limit-point, before which any engine failure requires an immediate turn around and return to the point of departure, and beyond which such return is no longer practical. **1946** E. Hodgins *Mr Blandings* x. 141 It would be delightful..to..die of old age in a rented apartment... But..he had reached and passed the crucial mark known, in the poetic language of the air navigator, as the Point of No Return. **1948** 'N. Shute' *No Highway* iv. 98 They passed the point of no return, and as a routine matter the navigator reported to him. **1953** R. Lehmann *Echoing Grove* 281 You were admitting to one another, weren't you, your secret knowledge that you—hadn't reached, together, the—the point of no return? **1956** *Jrnl. Educ.* July 312 But now he stopped [stealing] long enough to write a book (this is usually the point of no return) about his approved school, Borstal and prison experiences. **1958** *Times* 9 Jan. 10/3 When we were on the way to the Pole we received a request from Dr. Fuchs to establish a further depôt, but we were 240 miles from the Pole, and beyond the point of no return. **1960** J. Lehmann *I am my Brother* vii. 314, I finally decided that Leonard and I had reached a point of no return: if our partnership remained the same..not only would the Hogarth Press come to a standstill, but my own career would finally be frustrated. **1966** D. Varaday *Gara-Yaka's Domain* ix. 105 To my consternation I realised that I had let the matter run to the 'point of no return'. For me to refuse to go on now would..hurt his feelings. **1970** *Times* 7 May 12/7 Forbes..says all the A.B.P. films have a financial 'point of no return'. **1977** *Oxford Diocesan Mag.* Oct. 20/3 Scholars may well 'have passed the point of no return' in this matter.

12**. point-of-lay,** *a.* The stage of a hen's life-cycle at which it is able to begin laying eggs. Chiefly *attrib.*

1950 *Starting Poultry Keeping* (Poultry World) (ed. 8) 108 (Advt.), One of the largest suppliers of laying and point-of-lay pullets..in England. **1953** L. Robinson *Mod. Poultry Husbandry* (ed. 3) xxi. 612 During both the autumn and spring the early hatches produced birds of heavier weights at point of lay. **1960** *Farmer & Stockbreeder* 16 Feb. 147/3 Losses to point-of-lay have averaged only 2½ per cent. *Ibid.*, This system of rearing..enables him to offer a point-of-lay pullet at a really economical price. **1964** J. Portsmouth *Practical Poultry Keeping* (ed. 6) iii. 44 As a pullet approaches the point of lay stage its body undergoes great changes. **1975** A. C. Stewart *Dark Dove* vi. 41 Rhode Island Red pullets, six, point-of-lay—thirty-five shillings. **1977** D. Kay *Poultry Keeping for Beginners* v. 70 At the age of 20 weeks..the pullet becomes a point of lay bird.

12***. point-of-sale.** The place at which retail transactions are made. Chiefly *attrib.*

1953 D. Riesman *Individualism Reconsidered* (1955) 222 Using the retail store as the point-of-sale as in the Supermarket. **1959** *Design* Sept. 49/1 The coin operated food vending machine is required to carry its own point-of-sale appeal. **1960** *Times* 28 Sept. (Advt. Suppl.) p. ii/5 Posters and point-of-sale displays. **1962** H. E. Beecheno *Introd. Bus. Stud.* x. 88 Point-of-sale advertising consists of using special display material in shop windows. **1974** *Encycl. Brit. Micropædia* VIII. 68/1 Early point-of-sale displays consisted of stock designs (that is, the brand name of the product or a picture of the product or the factory) on strawboard placed in shop windows and on stands and counters. **1978** *Bookseller* 17 June 3197/3 The time when point-of-sale data was run against a local stock-control system.

13. *point-current, -hole* (examples), *-making* (earlier and later examples), *-size, -strap, -system* (further examples); *points system*; *point-free, -leafed, -tipped* adjs. In Phonetics, *point consonant* (earlier and later examples), *-element, -open, -stop, -teeth* (examples), *-trill*; *point-lingual* adj.

1867 Point consonant [see *back a. 1 c]. **1902** Point consonant [see *fan consonant* s.v. *fan sb.¹ 11]. **1857**

Dickens *Perils Eng. Prisoners* iii, in *Househ. Words* Extra Christmas No., 7 Dec. 30/2 The off-settings and point-currents of the stream. **1933** O. Jespersen *Essentials Eng. Gram.* iv. 39 Sometimes the point-element [of *r*] remains though without any trill. **1947** *People* 22 June 1/3 Small quantities of biscuits originally intended for the Services are point-free—if you can get them. **1770** P. Luckombe *Conc. Hist. Printing* 500 Point-holes, holes made by the Points in a worked off sheet of paper. **1940** *Chambers's Techn. Dict.* 658/1 *Point holes* (*Print.*), punctures made in the printed sheet by the spurs of the register points. **1932** R. Lehmann *Invitation to Waltz* I. 5 And there growing up the side of the house..is that kind of thick, bristling, woody, point-leafed shrub. **1931** G. O. Russell *Speech & Voice* xiv. 133 Point-lingual fricative consonants. **1889** G. B. Shaw *London Music in 1888–89* (1937) 129 Signor Novara, who played the part with unexpected success..acting without any senseless posturing and point-making. **1975** *New Yorker* 28 Apr. 124/2 The cut between the two scenes is not a piece of easy point-making. **1877** H. Sweet *Handbk. Phonetics* II. 37, *rh, r* (point-open). **1927** J. J. Hogan *Eng. Lang. in Ireland* 29 Point-open and stops: *thedynge* 'tiding,' *onther* 'under,' *tanked* 'thanked'. **1934** —— *Outl. Eng. Philol.* I. i. 8 English has two Point-Opens, þ as in *think*, ð as in *then*. **1931** A. Esdaile *Student's Man. Bibliogr.* 143 Easily available to English printers,..are the following types, which can, of course, be had in different normal point-sizes and with their own italic. **1973** S. Jennett *Making of Books* (ed. 5) ii. 41 These two lines appear to be set in two quite different sizes of type, yet they are both in the same point size and a dozen lines of either would occupy exactly the same depth of space. **1944** *Ann. Reg. 1943* 40 The points system was securing an equitable distribution of non-perishable foodstuffs. **1964** *Listener* 1 Oct. 505/2 The 'points system'..determines the amount of money which can be spent for special allowances for teaching staff and other purposes. **1974** *Guardian* 23 Jan. 9/3 Council houses are allocated on a points system. **1899** W. Rippmann tr. *Viëtor's Elements of Phonetics* 142/1 Point stops, etc. = dental stops, etc. **1934** J. J. Hogan *Outl. Eng. Philol.* I. i. 6 The English Point Stops, *t, d, n.* **1963** E. H. Edwards *Saddlery* xv. 111 An additional girth known as a 'point strap' and fixed under the point itself, which allows the girth to be fastened to this strap and..will place the girth much farther forward and help to keep it and the saddle in place. **1888** Point system [see Pointsman 2]. **1931** A. Esdaile *Student's Man. Bibliogr.* iv. 131 Simon Pierre Fournier,..best known by..being the first author of the Continental point-system of measuring types. **1941** *New Statesman* 26 Apr. 430/1 The 'point system' is based on the allocation of so many points per head and if you squander them on caviare instead of on corned beef, it is just too bad for you. **1953** R. J. C. Atkinson *Field Archaeol.* ii. 50 *The point system*, a fairly close grid of pegs is laid out to divide the site into squares. On one side of each peg a small pit is dug. **1959** *Chambers's Encycl.* XI. 533/1 Under the point system [of rationing], there is obviously greater freedom of choice than under specific rationing. **1888** H. Sweet *Hist. Eng. Sounds* (rev. ed.) 5 þ (as in *thin*) [is] a point-teeth consonant. **1906** H. C. Wyld *Hist. Study Mother Tongue* ii. 32 *th* (þ) in 'think', made between the *Point* of the tongue and the *Teeth* (Point-Teeth-Open). **1952** C. L. B. Hubbard *Pembrokeshire Corgi Handbk.* 2 Erect and point-tipped ears. **1877** H. Sweet *Handbk. Phonetics* II. 49, *rhr* (point-trill). **1927** J. J. Hogan *Eng. Lang. in Ireland* 75, *r*. This consonant is everywhere retained [in Irish] as in M.E... A strong point-trill is heard in the South. **1933** O. Jespersen *Essentials Eng. Gram.* iv. 39 Originally *r* was a full point-trill everywhere.

14. point-action *Gram.*, applied to an aspect which is not durative; **point bar,** (*b*) *Physical Geogr.*, an alluvial deposit that forms by accretion inside the loop of a river as the loop expands outwards, usu. consisting of low, curved, parallel ridges; **point blanket,** a Hudson's Bay or Mackinaw blanket with points (sense *A. 32) to indicate weight; **point block,** a high building with flats, offices, etc., built around a central lift or staircase; **point break** *Surfing*, a type of wave characteristic of a coast with a headland; **point charge** *Electr.*, a charge regarded as concentrated in a mathematical point, without spatial extent; **point contact,** the state of touching at a point only; *spec.* in *Electronics*, the contact of a metal point with the surface of a semiconductor so as to form a rectifying junction; freq. *attrib.*; **point-count** *Bridge*, the value of a hand in points; also, any system of allocating points to a hand; hence *point-counting* vbl. sb.; **point-counter** *Physics*, an early version of the Geiger counter in which discharges occur between positively-charged chamber walls and a central, earthed, metal point; **point defect** *Cryst.*, any defect in a crystal structure which involves only one lattice site; **point discharge,** an electrical discharge in which current flows between an earthed pointed object and the surrounding gas; also *attrib.*; so **point discharger,** such an object; **point-event,** something conceived of as having a definite position in space and time but no extent or duration; **point focus** *Physics*, a focus (of a beam of light or particles, etc.) which is small enough to be considered as a point; **Point Four** (or **IV, 4**) *U.S. Pol.*, the fourth point of President Harry S. Truman's Fair Deal programme (*Fair Deal) which made provision for technical aid to underdeveloped countries; freq. *attrib.*; **point ground,** in lace-making, a type of *réseau* ground; also *attrib.*; **point-instant,** the minimal unit of space-time; a mere position in space-time; **point load** *Engin.*, a load that acts at a single point; **point mass** *Physics*, a mass regarded as concentrated in a mathematical point, without spatial extent; also *attrib.*; **point mutation** *Genetics*, a mutation not distinguishable by recombinational analysis from a point change within a gene; **point net** (further examples); **point number,** in a musical, a song which is integral to the action; **point paper** (examples); **point-policeman** (further examples); **point rationing,** a system of rationing whereby goods are priced in terms of points (sense *A. 15 b) and a certain number of points are assigned to each consumer; so *point-rationed* adj.; **point resistance** *Engin.*, the upward force exerted by soil on the base of a pile; **points food,** rationed food available on points only; **point shoes,** shoes with pointed toes, *spec.* of a type used by ballet dancers; **point-shooting,** shooting game from a fixed point; **point source,** a source (as of light or sound) of negligible dimensions; **points rationing** = **point rationing*; **points value** = **point value*; **points victory,** a victory won on points; **points win** = **points victory*; **point value,** value in terms of rationing points; **point-work** *Ballet*, dancing on the points. Also *point group.

1925 G. O. Curme *College English Gram.* II. 56 The point-action aspect calls attention, not to an act as a whole, but to only one point, either the beginning or the final point. **1932** *Jrnl. Eng. & Gmc. Philol.* XXXI. 251 Most scholars recognize in English a durative aspect and a point-action aspect. **1970** *Language* XLVI. 300 It is particularly noteworthy in the latter case that the headline to this news item read 'Man may not have smelt killer gas', the *-t* form co-occurring with an effective or 'point-action' aspect. **1977** *Ibid.* LIII. 437 The ungrammaticality of 3d would be accounted for by the fact that *arrive*, like all (non-iterable) point-action predicates, does not co-occur with duration adverbials like *until*. **1945** H. N. Fisk in *Geol. Investigation Alluvial Valley of Lower Mississippi River* (U.S. Mississippi River Commission) 20 The point bar, the composite accretion within a bend, consists of an alternation of sand bar ridges, capped with thin top-stratum, and swales underlain by clay plugs. **1963** D. W. & E. E. Humphries tr. *Termier's Erosion & Sedimentation* v. 110 Alluvial plains appear to be composed chiefly (80–90%) of 'point bars'. **1974** C. H. Crickmay *Work of River* ii. 25 The land round which a meander winds is termed the tongue, and the tip of the tongue is the point, whence the term point bar. **1976** K. W. Butzer *Geomorphol. from Earth* viii. 157 Point bars, developed as a series of low levees on the inside meander bends, form arcuate or parallel ridges and swales. **1783** in E. E. Rich *Moose Fort Jrnl.* (1954) 152, I have enclosed instructions for your Guidance and the Standard of the point Blankets I now send you. **1797** in *Georgia Hist. Soc. Collections* (1916) IX. 347, 2 2½ point blankets. **1855** J. H. Chambers in *Contrib. Montana Hist. Soc.* (1940) X. 116 We have..30 prs. 3 pt. blkts 20 Pr W 20 1 Blkt 10 blue blkt 18 Scan & 25 Hudson Bay blkts. **1926** *Beaver* Dec. 22 The earliest reference found in a search of the minutes of the..Hudson's Bay [Co.] to Hudson's Bay 'Point' Blankets is one dated 22nd December 1779, where a notation is made of an order for one hundred pairs of each of five sorts of pointed blankets. **1962** W. Stegner *Wolf Willow* II. v. 67 The somewhat obscure source of the red point blankets that I slept under. **1954** *Ann. Reg. 1953* 371 The London County Council's large sites..were of special interest on account of their carefully landscaped mixture of terrace houses, maisonettes, and 11-storey 'point-blocks'. **1958** *Listener* 6 Nov. 727/1 Eight-storey point-blocks are contrasted with low, gable-ended, four-storey blocks. **1970** *New Scientist* 5 Mar. 460/1 A series of disasters, or near disasters,..involving a wide range of modern structures from cooling-tower shells to point blocks of flats. **1975** *Times Lit. Suppl.* 5 Sept. 988/4 The London skyline today, peppered with point blocks, tower-blocks, and slabs. **1966** *Surfabout* III. vi. 8/1 The original concept was to have three separate events. A point break, reef break, and a beach break contest. **1968** W. Warwick *Surfriding in N.Z.* 21/1 There are four main types of breaks in New Zealand. They are the beach break, the reef break, point break and the river bar break. *Ibid.* 21/2 The point break wave..is formed when swells move almost at right angles along a peninsula or headland. **1970** *Studies in English* (Univ. of Cape Town) I. 26 A headland, point or pier, bends the wave into a *point break*, which gives a consistent ride in one direction, often in a perfect tubing shape, so that the surfer slides along the face of the wave with a tube of water continually breaking behind him. **1903** S. J. Barrett *Electro-Magn. Theory* 66 The law of inverse squares..is due to the continuity of the electric displacement.., the flux from a point-charge being distributed equally in all directions. **1975** D. G. Fink *Electronics Engineers' Handbk.* I. 11 A combination of two point charges of equal magnitude and opposite polarity separated by a distance small compared to that at which the field of the dipole is to be determined is called an electric dipole. **1914** *Proc. Physical Soc.* XXVII. 70 The present paper relates..to the conductivity of 'point contacts' when a steady, or slowly varying, E.M.F. is applied. *Ibid.*, The behaviour of a typical point contact (zincite and tellurium). **1945** R. K. Allan *Rolling Bearings* vi. 143 In roller bearings..we..have 'line contact' as distinct from the 'point contact' of ball bearings. **1947** C. F. Edwards in *Proc. IRE* XXXV. 1181/2 In view of the fact that the crystalline state of the silicon is more nearly like that of iron and copper, which are not ordinarily regarded as crystals,..it seems desirable to eliminate the terms 'crystal' and 'crystal detector' and designate these devices by the term 'point-contact rectifier'. **1948** *Physical Rev.* LXXIV. 230/2 When the two point contacts are placed close together on the surface and d.c. bias potentials are applied, there is a mutual influence which makes it possible to use the device to amplify a.c. signals. **1959** [see *junction transistor* s.v. *junction sb. 4]. **1962** Simpson & Richards *Physical Princ. Junction Transistors* i. 1 The event that opened this era of 'solid-state electronics' was the invention of the point-contact transistor by Bardeen and Brattain in 1948. **1970** H. J. Watson *Mod. Gear Production* ii. 24 Crossed helical gears theoretically make point contact only which, under load sufficient to cause deflection of the contacting surfaces, becomes a line. **1975** D. G. Fink *Electronic Engineers' Handbk.* ix. 61 The point contact is fabricated by a metal whisker forming a rectifying junction in contact with the semiconductor. **1959** T. Reese *Bridge Player's Dict.* 149 The Milton Work point-count remains the most popular..because of its simplicity and convenience. **1963** *Times* 9 Jan. 12/7 The Souths who were influenced more by their point-count than their distribution. **1979** *Guardian* 5 Sept. 2/6 The Bridge Challenger..run by two micro-processors..bids the point-count system. **1925** *Proc. Cambr. Philos. Soc.* XXII. 676 The point counter has also been studied as a unit of an electrical circuit containing capacity and resistance, and an analogy established between its discharges, and the 'flashing' of a neon lamp. **1938** R. W. Lawson tr. *Hevesy & Paneth's Man. Radioactivity* (ed. 2) i. 18 When we are concerned with the measurement of a very small flux of radiation,..the sensitivity of the point-counter is often inadequate. **1964** J. B. Birks *Theory & Practice Scintillation Counting* x. 394 The observation of the time relationship between two or more ionizing events, as represented by pulses from radiation detectors, forms the basis of the coincidence method... It was originally introduced..for use with point-counters and Geiger counters. **1963** *Times* 6 Mar. 13/1 The conventional openings, combined with point-counting, brought greater refinements into bidding. **1973** *Times* 6 Jan. 9/3, I have written on many occasions that no expert relies upon point-counting alone to value his hand. **1960** H. G. Van Bueren *Imperfections in Crystals* ii. 41 Point defects can be introduced in large numbers into solids by plastic deformation. **1974** D. M. Adams *Inorg. Solids* ix. 292 For many years it seemed that non-stoichiometric systems of wide compositional range could be understood in terms of a parent lattice with a high concentration of randomly distributed point defects. **1886** R. Wormell *Electricity in Service of Man* 49 (*heading*) Point discharge. **1927** *Proc. R. Soc.* A. CXV. 443 The important part played by the point-discharge currents in the total exchange of electricity between the earth and the atmosphere. **1938** R. W. Lawson tr. *Hevesy & Paneth's Man. Radioactivity* (ed. 2) i. 17 Each time an ionizing particle..passes the neighbourhood of the point [of the needle], a point discharge takes place by virtue of ionization by collision. **1973** R. H. Golde *Lightning Protection* ii. 8 Point-discharge currents and the resulting space charges play an important part in the development of the lighting discharge and in the action of a lightning conductor. **1928** *Proc. R. Soc.* A. CXVIII. 255 Wormell..used a single point-discharger at a height of 8 metres, which is stated to be likely to produce similar effects to those from a small tree. **1965** S. C. Coroniti *Problems of Atmospheric & Space Electricity* iii. 174 (*heading*) The behavior of trees as point dischargers. **1920** A. S. Eddington *Space, Time & Gravit.* xii. 186 In the relativity theory of nature, the most elementary concept is the *point-event*. In ordinary language a point-event is an instant of time at a point in space. *Ibid.*, The aggregate of all the point-events is called the *world*. **1928** C. E. M. Joad *Future of Life* 36 Faced with a universe consisting of ephemeral point-events, the mind selects from it certain characteristics which have a particular interest for it. **1948** *Mind* LVII. 298 The idea of a mile or a day is an everyday idea. So are those of above and below, east and west, or before and after. Those of a point, an instant, and still more of a point-event, are relatively sophisticated, and much less used. **1965** P. Caws *Philos. of Sci.* xli. 316 This calls for a solution of the problem of rendering point events (particles interacting and so on) in terms of experiences. **1908** L. Laurance *Gen. & Pract. Optics* xii. 329 Thus, rays in the pencil do not have a point focus, since there are two focal lines. **1923** Glazebrook *Dict. Appl. Physics* IV. 213/1 Whilst the geometrical theory claims a perfect point-focus, the undulatory theory merely demonstrates that there is a maximum of intensity at that point. **1966** D. G. Brandon *Mod. Techniques Metallogr.* 135 The incident intensity can be increased by using a point-focus X-ray source with the sample close to the source. **1949** *Manch. Guardian Weekly* 27 Oct. 15 The much admired 'point four' also fell by the wayside. **1953** A. Huxley *Let.* 21 June (1969) 675 It [*sc.* an M.A. in Public Health] would surely be helpful in all manner of fields—e.g. the UN if you wanted to enter it later, or any other of the international agencies, or Point Four. **1955** *Bull. Atomic Sci.* Mar. 98/2 The Point IV program of technical assistance stirred the imagination throughout the undeveloped areas of the world. **1961** L. D. Stamp *Gloss. Geogr. Terms* 466/1 An official definition used to determine countries qualifying for American Point IV aid was national income per capita. **1965** H. S. Truman *Memoirs* II. xvi. 269 The Point Four program was a practical expression of our attitude toward the countries threatened by Communist domination. **1972** *Times* 27 Dec. 5/1 The Point 4 programme inaugurated technical aid to underdeveloped countries. **1832** J. R. McCulloch *Dict. Commerce* 697 About 1777, or 1778, quite a *new* ground was attempted by the inhabitants of Buckingham and its neighbourhood, which quickly superseded all the others; this was the *point ground*, which had (as is supposed) been imported from the Netherlands. **1865** F. B. Palliser *Hist. Lace* xxx. 361 In 1778, according to M'Culloch, was introduced the 'point' ground, as it is locally termed, from which period dates the staple pillow lace trade of these counties. **1968** J.

Arnold *Shell Book of Country Crafts* 305 'Buckingham Point' or 'Point-ground' lace was that lace containing Lille or Mechlin elements. **1969** E. H. Pinto *Treen* 311 *A* is a 'Trolly'..used for the gimp thread—the thick, soft thread which outlines the design in point ground lace. **1920** S. Alexander *Space, Time, & Deity* I. 58 It is assumed that at each point-instant (the name is due to Mr. Lorentz, *Ortzeit*) there exists some perceptible 'substance'. **1959** *Listener* 8 Jan. 57/2 The suggestions he [*sc.* Samuel Alexander] made about these 'pervasive features'—for instance, that everything in the world is ultimately composed of complexes of 'space-time point-instants'—are certainly not such as could be either reached or confirmed by any kind of direct observation. **1941** *Jrnl. Inst. Civil Engin.* XVI. 524 A single load was connected with a much higher ultimate moment, owing to a better stress-distribution and a rapid decrease of the moment, than was obtained with two point-loads. **1976** Attewell & Farmer *Princ. Engin. Geol.* iv. 193 It is possible to relate the point load strength index to the uniaxial compressive strength. **1955** W. Heisenberg in W. Pauli et al. *Niels Bohr* 17 For Bohm, the particles are 'objectively real' structures, like the point masses of classical mechanics. **1968** M. S. Livingston *Particle Physics* iii. 47 This hypothesis had a built-in 'irrationality'..in attempting to correlate the divergent concepts of point-mass particles and of wave motion. **1968** R. A. Lyttleton *Myst. Solar Syst.* i. 20 The action of the stars once the collision is over can safely be regarded as pure attraction by point-masses. **1925** *Genetics* X. 117 If one thinks of mutations as being simply inherited changes, it becomes necessary to distinguish changes that involve whole chromosomes.., changes that involve several adjacent genes.., and what have been called 'point-mutations' or 'gene-mutations'. **1928** *Jrnl. Genetics* XIX. 223 If such parallel forms were due simply to the special action of a single factor, 'point-mutation' might afford a reasonable basis of explanation. **1974** Goodenough & Levine *Genetics* v. 195 A mutation that affects only one or a few nucleotides in a chromosome is known as a point mutation. **1977** *Sci. Amer.* Dec. 94/3 Such mutants of influenza virus are considered to be 'point mutations' that might affect only one of the nucleotide building blocks of the RNA. **1865** F. B. Palliser *Hist. Lace* xxxvi. 418 In 1777, Else and Harvey introduced at Nottingham the 'pin' or point net machine, so named because made on sharp pins or points. **1953** M. Powys *Lace & Lace-Making* xi. 196 In working *point net ground* the pin is not enclosed, but after it is placed, two extra twists are given to the pairs which have formed the half stitch. **1937** N. Coward *Present Indicative* III. 105 There was a sentimental ballad..and a bright 'Point' number. **1958** B. Nichols *Sweet & Twenties* 16, I heard a cabaret artist use the word Hiroshima as a comic gag in a point number. **1960** A. Kimmins *Lugs O'Leary* iii. 35 It was a 'point' number, and she was singing it like a 'pop'. **1960** *News Chron.* 31 Mar. 4/5 'Point numbers'..arise directly out of the play's action—often carrying the plot..forward while they are being sung. **1967** *Stage* 2 Mar. 6/2 (Advt.), Glam. Singer: Pop, Ballad, Point Nos. **1899** J. W. Mackail *Life W. Morris* II. xiii. 44 'Point-paper'—paper, that is, divided into minute spaces, each representing a single knot of the carpet. **1940** *Chambers's Techn. Dict.* 658/1 *Point paper*,..ruled paper upon which the interlacing of the threads in a fabric is shown. **1928** *Evening News* 18 Aug. 6/3 The point policemen at spacious centres like the Marble Arch and Hyde Park Corner seem to be waving the wide world around and the exhilaration of moderate speed can be enjoyed. **1939** H. Hodge *Cab, Sir?* 236 People don't realise how dangerous it is for a point-policeman. **1944** *Times* 15 Feb. 2/2 The increase has been due to higher expenditure on unrationed or 'point' rationed foods. **1944** A. M. Taylor *Lang. World War II* 56 *Point Rationing*, a rationing method announced by the Office of Price Administration (OPA), establishing a secondary currency of points alongside the dollar currency. **1959** *Chambers's Encycl.* XI. 533/1 Under point rationing, the group may be extended to cover a combination of several different kinds of article. **1943** K. Terzaghi *Theoret. Soil Mech.* viii. 136 One part Q_f of the total load on the pile is carried by the skin friction. The balance Q_p is transferred onto the soil through the base or the point of the pile and is called the point resistance. **1972** L. Zeevaert *Foundation Engin. for Difficult Subsoil Conditions* v. 278 The lower part of the piles will work under ultimate point resistance and positive friction. **1941** *New Statesman* 6 Dec. 475 When the retailer sells 'points' food, he can replenish his supplies only by handing over the 'points' that he has collected. **1948** *Hansard Commons* 8 Mar. 795 [He] has been given a licence to sell points foods. **1957** G. B. L. Wilson *Dict. Ballet* 218 *Point shoes*, the silk or satin shoes, tying up the ankle with ribbon, used by dancers when dancing sur les pointes. *Ibid.*, Point shoes cost about 17s a pair. **1970** R. Lowell *Notebook* 150 My coat limp chestnut-colored suede Cut to match my point shoes that hurt my toes. **1977** *Times* 20 Jan. 8/6 American mothers just want to see their children in point shoes as early as possible. **1874** J. W. Long *Wild-Fowl Shooting* 71 For point-shooting, shooting from a blind on shore, or in the edge of the willows from a boat, a few hints may be welcome. **1876** *Fur, Fin & Feather* Sept. 90 We prepared to move out into the clear water onto a log, and there get some point shooting. **1903** *Nature* 1 Jan. 202/2 for a point source of strength Q, the V it produces is [etc.]. **1949** [see *angular a. 2 c]. **1971** I. G. Gass et al. *Understanding Earth* xiii. 178 Contrast this with a deep ocean floor, lacking in relief, and far from any strong point-sources of debris. **1941** *New Statesman* 6 Dec. 475 'Points' rationing of canned meat, canned fish etc., from December 1st onwards makes disposal of these goods to 'black market' merchants singularly unattractive. **1950** *Times* 20 May 4/1 The ending..of the points rationing system, which has been in operation for more than eight years, was announced by the Minister of Food yesterday. **1947** *People* 22 June 1/3 More 'ups' and 'downs' in the points values of food come into force today. **1929** *Daily Express* 7 Nov. 13/2 Jackson's points victory was about the most easily gained of the night. **1976** *New Musical Express* 12 Feb. 30/3 Theoretically this bout should have provided at least a points-win decision in favour of Harold Melvin's The Blue Notes. **1976** *Rhyl Jrnl. & Advertiser* 9 Dec. 31/1 One of Rhyl Star Boys Club's most promising boxers, Philip Siddall, gained a classy points win over Peter Williams, of Llay ABC, at Llay last week. **1946** *Mod. Lang. Notes* LXI. 443 Prestige value, cash value, point value. **1959** *Chambers's Encycl.* XI. 533/1 In the case of foodstuffs, the point value may be related to calorie value. **1936** S. J. Simon tr. A. Benois in 'C. Brahms' *Footnotes to Ballet* iv. 206 The 'classical ballet' in the steel support does endow point-work with a special brilliance. **1957** G. B. L. Wilson *Dict. Ballet* 217 Dancing on the point or 'point work' is said to have been practised by the cossacks since time immemorial. *Ibid.* 218 A danced stage performance need not necessarily include 'point work' to be termed ballet. **1975** *New Yorker* 26 May 89/2 Its a question whether 'The Sleeping Beauty' can be danced within the limitations of Bolshoi technique, which stresses big jumps at the expense of brilliance in turns and pointwork.

point, *sb.*³ Add: **point d'appui,** also *fig.*, esp. *Mil.*, a strategic point; *point d'attache*, point of connection; *point de départ* = *point of departure* s.v. *point sb.¹ D. 12**; *point de repère* (further examples); *point d'orgue* (see quots. 1876 and 1883).

1823 Byron *Don Juan* xiv. lxxxiv. 7 This I could prove beyond a single doubt, Were there a jot of sense among mankind; But till that point *d'appui* is found, alas! Like Archimedes, I leave earth as 'twas. **1833** *Edin. Rev.* LVI. 383 She [*sc.* the Bank of England] is then, as it were, the *point d'appui* of the whole moneyed and commercial interests. **1836** H. Greville *Diary* 20 Feb. (1883) I. 88 England being now in the hands of *Democrats*, she is no longer useful as a *point d'appui* to France. **1876** C. M. Yonge *Three Brides* II. ii. 35 Raymond used to arm himself with the newspapers as the safest *point d'appui*. **1895** E. Malet *Dispatch* 7 Sept. in *F.O.* 64/1351 No. 201 (Public Record Office), The acquisition of a 'point d'appui' in East Asia. **1915** 'I. Hay' *First Hundred Thousand* II. xviii. 264 The wood itself is a *point d'appui*, or fortified post. **1920** S. Alexander *Space, Time, & Deity* II. 136 Some *point d'appui* is needed for our thinking. **1934** C. Lambert *Music Ho!* II. 127 His [*sc.* Satie's] progressions have..no trace of the *point d'appui* that we usually associate with the word progression. **1945** R. Hargreaves *Enemy at Gate* 103 Gibraltar..furnished a perfect example of the value of sea power, combined with that ability to retain physical possession of a *point d'appui*. **1967** G. F. Fiennes *I tried to run Railway* vii. 87 The point d'appui was the results in general. **1973** D. Aaron *Unwritten War* VII. xix. 294 This defense of the past was..a *point d'appui* for the Southern traditionalist. **1939** A. Toynbee *Study of Hist.* V. 624 For this linguistic legacy of the Napoleonic Empire—a legacy which is the *point d'attache* of the present Annex to the main thread of this Study of History—see V. C.. above. **1923** G. Arthur *Further Lett. from Man of no Importance* (1932) 146 King Edward was, however, equally determined that his *point de départ* for the Vatican should be the British Embassy and not the English College which Cardinal Rampollá (I think) urged. **1933** *Psychoanalytic Q.* II. 156 However this may be, Freud's formulation serves as the *point de départ* for the author's subsequent discussion of the subject. **1956** *N. & Q.* CCI. 254/2 The Restoration of King Charles II..provided an invaluable *point de départ* for all those who had economic schemes to present to government and people. **1977** *N.Y. Rev. Bks.* 13 Oct. 27/3 History, in this instance, will largely be the reflection of the historian's *point de départ*. **1904** H. James *Golden Bowl* I. xvi. 288 You give me a *point de repère* outside myself—which is where I like it. Now I can work round you. **1933** *Times Lit. Suppl.* 3 Aug. 527/1 It is this very uncertainty that causes him to regret at the outset the absence of *points de repère* in modern literature, which would enable him to fit his authors into schools and categories. **1937** *Burlington Mag.* Apr. 192/2 This picture..will now take its place as the *point de repère* for the identification of this painter's individual style. **1958** *Listener* 20 Nov. 846/1 There are, of course, a few *points de repère*, significant works that represent a moment of apocalyptic vision. **1967** V. Nabokov *Speak, Memory* (rev. ed.) iv. 85 Where a crack or a shadow afforded a *point de repère* for the eye. **1876** Stainer & Barrett *Dict. Mus. Terms* 363/2 *Point d'orgue*, ..a pedal-point. **1883** Grove *Dict. Mus.* III. 6/2 *Point d'orgue*, organ point, appears..to be used (1) for an organ point or pedal, that is, a succession of harmonies carried over a holding note..; and (2),..for the cadenza in a concerto. **1893** G. B. Shaw *Music in London 1890–94* (1932) III. 12 It gives him [*sc.* Brahms] no trouble to pile up *points d'orgue*, as in the Requiem. **1902** —— *Perfect Wagnerite* (ed. 2) 61 A specifically contrapuntal theme, *points d'orgue*, and a high C for the soprano. **1977** *Listener* 15 Dec. 797/3 Fluctuations of pitch..must either be confined to a single line of melody, or the accompaniment must simply be a drone—as is recognised in our classical system with the pedal-point or *point d'orgue*.

b. *point d'Angleterre* (earlier and later examples), *point d'Argentan, point de France, point de Paris*; *point d'esprit* (earlier and later examples); *point coupé*, cut work; *point russe*; *point à l'aiguille* (earlier examples), *point de neige, de reprise* (earlier examples).

1842 Point d'Argentan [see *bertha, berthe]. **1865** Point coupé [see *lacis]. **1865** F. B. Palliser *Hist. Lace* iii. 28 Point also means a particular kind of stitch, as point de Paris, point de neige, [etc.]. *Ibid.* vii. 102 They [*sc.* English lace merchants] bought up the choicest laces of the Brussels market, and then smuggling them over to England, sold them under the name of Point d'Angleterre, or 'English Point'. *Ibid.* 106 There are two kinds of flowers: those made with the needle are called 'point à l'aiguille'. *Ibid.* ix. 143 The point de France supplanted that of Venice; but its price confined its use to the rich. *Ibid.* xvii. 216 Point de Paris, mignonette, bisette, and other narrow cheap laces were made. *Ibid.* xxxvi. 424 This was followed by the 'spot', or 'point d'esprit', and various other fancy nets. **1872** *Young Englishwoman* Oct. 555/1 Stars worked in point russe. *Ibid.* Nov. 611/2 Fill up the grey rows..with scarlet wool in point de reprise. **1879** *Sylvia's Embroidery Bk.* 242 The chain stitch and point russe embroidery is worked with red silk. **1881** C. C. Harrison *Woman's Handiwork* I. 43 *Point russe* is best known by small block patterns worked in fine back stitch. *Ibid.* 59 Surplices of fine lace resembling point d'esprit. *Ibid.* 89 Modern point coupé..is made on a shut linen foundation, of which some of the threads are cut away and the remainder worked over with buttonhole stitch, making regular square spaces. **1882** *Encycl. Brit.* XIV. 186/2 'Point d'Argentan' has been thought to be especially distinguished on account of its ground of hexagonally arranged *brides*. **1882** Point de reprise [see *darning-stitch* (*b*) s.v. *darning vbl. sb. 3]. **1895** *Montgomery Ward Catal.* Spring & Summer 78/1 Point de Paris Ivory Lace. **1919** Point d'esprit [see *leadwork 2]. **1953** M. Powys *Lace & Lace-Making* iv. 14 *Point d'Argentan*... This lace is generally known from the Point d'Alençon by the ground, the Brides Bouclées, a hexagonal ground with buttonhole stitches on each of the six sides. *Ibid.* vi. 47 Point de France was used by distinguished prelates in the 18th century. **1958** *Times* 18 Nov. 12/3 All wore Empire dresses of white *point d'esprit* net over pale blue trimmed with blue satin ribbons. **1959** *Chambers's Encycl.* VIII. 293/2 *Point de neige*—a variety of *rosalino*. **1960** H. Hayward *Antique Coll.* 223/1 *Point d'Angleterre lace*: this is not an English lace but is a pillow lace made in Brussels. **1971** *Country Life* 4 Nov. 1197/3 At first the same as the Venetian product, this lace had its own evolution into distinctive *Point de France*. **1974** *Encycl. Brit. Micropædia* VIII. 66/3 In modern usage, *point de Paris* has come to mean any bobbin-made lace..with a six-pointed star mesh that is twisted, as opposed to that of Chantilly, which is plaited. **1975** *Oxf. Compan. Decorative Arts* 522/1 It was not until Colbert under Louis XIV set up his state factory at Lonray near Alençon and Argentan that French Needlepoint lace under the general name of *point de France* established its reputation.

point, *v.*¹ Add: **5. c.** *to point up*: to emphasize, draw attention to. orig. *U.S.*

1934 in Webster. **1940** *Sun* (Baltimore) 3 Feb. 8/1 The warnings which Finnish spokesmen have recently given to the world that this resistance cannot continue indefinitely have been sharply pointed up by the renewal of very heavy Russian assaults along all sections of the front. **1941** L. Trilling in D. Lodge *20th Cent. Lit. Crit.* (1972) 286 An analysis of this sort is not momentous and not exclusive of other meanings; perhaps it does no more than point up and formulate what we all have already seen. **1951** *Jrnl. Aeronaut. Sci.* XVIII. 622/1 The introduction of the automatic control system, particularly in pilotless aircraft, points up another basic change in our philosophy of aircraft control. **1958** *Listener* 20 Nov. 814/2 The Federal Government..should point up the dangers of prolonging the *status quo*. **1967** *Ibid.* 22 June 826/3 Another instance of this is the imagery at the beginning of *Pincher Martin*, which the authors ingeniously point up by comparison with a passage from *Robinson Crusoe*. **1969** A. Cockburn in Cockburn & Blackburn *Student Power* 20 In a key concluding section he points up the lessons to be drawn from this record. **1972** *Daily Tel.* 17 Nov. 19/7 The [census] returns point up the discrepancies which occur in unemployment figures. **1978** *Verbatim* May 5/2 He points up the ambiguity in the application of the epithet *unabridged*.

7. b. *Sculpture.* To mark at a series of points on (a block of stone or marble) the depth to which the initial working or roughing-out is to be done.

1841 *Penny Cycl.* XXI. 142/1 The statue being rudely blocked out or pointed, the marble is in this state put into the hands of a superior workman called a carver. **1877** A. B. Edwards *Thousand Miles up Nile* 423 A recent writer..is of opinion that the Egyptian sculptors did not even 'point' their work beforehand. **1911** A. Toft *Modelling & Sculpture* 254 The appearance of a work when pointed is not pleasing, covered all over with innumerable holes, and little mounds of marble projecting between these holes. **1947** J. C. Rich *Materials & Methods of Sculpture* ix. 261 Occasionally an indirect sculptor may personally point a work, or have his studio assistants or students do this for him.

9. c. *trans.* To indicate or state.

1928 *Publishers' Weekly* 12 May 1957 The effect on books by established authors like Galsworthy's 'Silver Spoon' and Ferber's 'Show Boat' could not be as clearly pointed. **1975** *Language for Life* (Dept. Educ. & Sci.) xxi. 303 There should be an appraisal of the kind of support schools can be given, and this points the need for close consultation between the education authority's advisers and the schools themselves, and the library staff.

12. a. (Further examples with a person as direct object.)

1611 Shakes. *Wint. T.* IV. iv. 539 On mine honor, Ile point you where you shall have such receiuing As shall become your Highnesse. **1842** Tennyson *Love & Duty* in *Poems* II. 87 Should it [*sc.* my shadow] cross thy dreams, So might it..point thee forward to a distant light. **1850** E. B. Browning *Prometheus Bound* in *Poems* I. 175 Point me not to a good, To leave me straight bereaved. **1922** Joyce *Ulysses* 399 Yes, Pious had told him of that land and Chaste had pointed him to the way. **1972** J. Philips *Vanishing Senator* (1973) I. ii. 16 They're simply using Lloyd to point us in the wrong direction. **1976** S. George *Fatal Shadows* 106 Someone..had pointed me in her direction, had wanted me to see what she was doing.

b. Fig. phr. *to point the finger* (*of scorn*) (*at a person*).

1829 P. Egan *Boxiana* 2nd Ser. II. 499 It was a shame..that pure and honourable men should be suspected of such doings..for even at him the finger of scorn had been pointed. *a* **1862** [see sense 12 a in Dict.]. **1939** G. B. Shaw *Geneva* III. 81 You can point the finger of the whole world at the slayer of my husband and say 'You are guilty of murder.' *a* **1966** 'M. na Gopaleen' *Best of Myles* (1968) 93 The finger of scorn is pointed at you. **1978** 'J. Horbury' *Diplomatic Affair* x. 126 We [*sc.* the

British] naturally hesitate to point the finger..but the fact is that all these..leaks have been date-lined Washington.

c. *Anthrop.* Esp. in phr. *to point the bone.* Amongst Australian Aborigines, to will or to bring about the death of a person by a ritual involving special bones or sticks and incantations. Also *transf.* and *fig.* Cf. *pointing-bone* s.v. *pointing vbl. sb.¹ 11.

1897 W. E. Roth *Ethnol. Stud. N.-W.-Central Queensland Aborigines* xi. 156 It is most important to remember that in all cases while the death bone is being 'pointed', the blood of the victim passes invisibly across the intervening space to the 'pointer', and so along the connecting string, into the receptacle..: at the same time one of the doctor's gew-gaws, or..bone, pebble, &c., passes invisibly from the 'pointer' to be inserted in the body of the victim. **1904** Spencer & Gillen *Northern Tribes of Central Australia* xiv. 458 If it were known that any one had 'pointed the bone', that man would at once be killed. **1913** J. G. Frazer *Golden Bough: Balder the Beautiful* (ed. 3) I. 14 The magical bone, which the native sorcerer points at his victim as a means of killing him, is never by any chance allowed to touch the earth. *Ibid.*, The custom of killing a man by pointing a bone or stick at him, while the sorcerer utters appropriate curses. **1934** [see *bone sb. 1 e]. **1939** Joyce *Finnegans Wake* I. 193 He points the deathbone and the quick are still. **1953** A. Upfield *Murder must Wait* xvii. 148 Buttons and ends came from me, are a part of me, are necessary objects required for the practice of pointing the bone. **1967** B. Jefferis *One Black Summer* x. 184 You're asking me to point the bone at someone on no real evidence at all. **1974** J. Cleary *Peter's Pence* vii. 197 He was..convinced that the bloody God-botherers had pointed the bone at him just like the blacks did back home in Aussie.

13. c. *U.S.* In sport, to make special preparations *for* a particular opponent or game.

1933 *Sun* (Baltimore) 15 Nov. 12/7, I suppose there is a good deal of hooey in the talk of Army and Navy 'pointing' for each other: they do not like to be licked at any time. **1937** *Ibid.* 1 Sept. 18/2 We are not pointing for any team in particular, but are trying to develop for our major games without being knocked off. **1944** *Ibid.* 19 Oct. 21/2 The Jackets also are pointing for another bowl bid and defeats are anathema to gridsters with January 1 on their minds.

† 17. *intr. Cricket.* To field at point. *Obs.*

1862 *Baily's Mag.* Aug. 85 The Surrey people..selecting..a Lyttelton to bowl; a John Walker to keep; an F. Burbidge to point. **1863** *Ibid.* Sept. 44 The bowling of Tarrant and Grundy, the wicket-keeping of Lockyer, the pointing of Carpenter,..was all cricket in perfection.

18. *Naut.* Of a sailing vessel: to lay a course close to the wind. Freq. with *up* or in phr. *to point high.*

1899 in *Cent. Dict.* **1941** H. I. Chapelle *Boatbuilding* i. 37 The sailing qualities of the V-bottom hull are somewhat like those of the flat-bottom types, but with improved windward qualities if well designed. They will rarely point as high as a well-designed round-bottom boat. **1947** A. Ransome *Great Northern?* xiii. 162 'The Gael's castle is behind the ridge beyond it,' said Dorothea. 'She won't point up for our inlet,' said John. **1954** G. Bradford *Gloss. Sea Terms* 146/2 A vessel *points well* if she sails close to the wind. **1950** E. C. Hiscock *Cruising under Sail* I. iv. 73 It [*sc.* the ketch] cannot point so high as a sloop, cutter, or yawl.

19. *trans. U.S.* To turn, guide, or deflect (cattle) in a particular direction.

1903 A. Adams *Log of Cowboy* iv. 42 Priest sent Officer to the left and myself to the right to point in the leaders. **1916** 'B. M. Bower' *Phantom Herd* xiv. 244 You're trying to point the herd then. **1947** C. Price *Trails I Rode* 184 One time we were pointing a herd, Bill on one side and I on the other.

20. To insert white hairs (into a self-coloured fur). Usu. as pa. pple.

1911 in Webster. **1913** [see *pointed ppl. a.¹ 5 c]. **1916** *Fur Trade Directory* (N.Y.) 95 We point either skins or made-up goods. **1922** W. E. Austin *Princ. & Pract. Fur Dressing* i. 10 They are dyed black in imitation of the black fox, or these when pointed with badger or other white hair to imitate the silver fox. **1936** F. Grover *Practical Fur Cutting & Furriery* xv. 22 A black fox is often pointed with silver badger hairs. **1957** M. B. Picken *Fashion Dict.* 260/1 Fox fur pointed to imitate silver fox.

pointage (poi·ntĕdʒ). [f. Point sb.¹ + -age.] Points collectively, *spec.* the number of ration points needed to make a particular purchase.

1934 in Webster. **1944** *Times* 23 Feb. 2/3 From April 2 imported tinned marmalade will be available on points... The pointage will be announced later. **1946** *Evening News* 15 Aug. 1/7 Mr. Strachey said..that he had not contemplated increasing pointage to compensate for the estimated demand for points.

poi·nt angle. [f. Point sb.¹ + Angle sb.²] The angle at a vertex of a solid body; *spec.* (*a*) the angle between two diametrically opposite edges or surfaces at the tip of a tool; (*b*) the re-entrant solid angle at a vertex of an artificial cavity in a tooth.

1869 [see Point sb.¹ sense D. 13]. **1908** G. V. Black *Operative Dentistry* II. 9 All point angles are formed by the junction of three walls at a point, and are named by joining the names of the walls forming the angle. They are, therefore, named in three terms. *Ibid.*, The point angles are formed where the line angles of one set meet the other set at the angle of the cavity. **1919** G. W. Burley *Machine & Fitting Shop Practice* II. v. 142 The standard point-angle of lathe centres is 60° for small and medium-sized work-pieces. **1950** P. Gates in A. W. Judge *Machine Tools & Operations* I. iv. 149 When drilling soft or ductile materials, drills ground with a smaller point angle tend to throw up considerable burrs on the back of the work. **1973** W. W. Howard *Rev. Operative Dentistry* i. 9 Point angles are named for the line angles that form them.

point-blank, *a.*, *sb.*, and *adv.* Add: **C.** *adv.* **3. a.** (Earlier and later examples.)

1598 J. Florio *Worlde of Wordes* 104/2 *A dirittura*, foorthright, point blanke. *a* **1914** Joyce *Stephen Hero* (1944) 45 McCann always represented a member of the Opposition and he spoke point-blank. Then a member would protest. **1938** E. Waugh *Scoop* II. v. 233, I read the newspapers with lively interest. It is seldom that they are absolutely, point blank wrong. **1968** *Punch* 6 Nov. 658/2 However, I took the oranges home just the same, with marmalade in mind, but my wife refused pointblank.

point-blanker. [f. Point-blank a. + -er¹.] A point-blank shot.

1830 J. F. Cooper *Water Witch* II. vii. 202 'Run in the quoin, and..give her a point-blanker!' said the gruff old seaman... 'None of your geometry calculations for me!'

point-device, *a.* (Further examples.)

1885 R. L. Stevenson *Prince Otto* II. i. 64 Otto looked so gay, and walked so airily, he was so well dressed and brushed and frizzled, so point-de-vice, and of such a sovereign elegance. **1900** C. M. Yonge *Modern Broods* xvii. 161 He is a little too point device, too obviously got up for the occasion! **1928** D. L. Sayers *Ld. Peter views Body* xii. 283 His double-breasted suit of navy-blue and his socks, tie, and handkerchief, all scrupulously matched, were a trifle more point-device than the best taste approves. **1958** *Observer* 28 Dec. 7/7 Odd how neat and point-device and sane they all looked.

point-duty. Add: Also *attrib.*

1908 *Daily Chron.* 30 June 1/3 A point-duty constable..was knocked down..by a taxi-cab. **1967** N. Lucas *C.I.D.* vi. 72 Ignoring a point-duty policeman's signal to stop. **1974** *Drive* Autumn 62/2 These days, alas, only small boys want to be point-duty policemen!

‖**pointe** (pwænt). *Dancing.* [Fr.] The tip of the toes. Also, a dance-movement executed on the tip of the toes. Also *attrib.*

1830 R. Barton tr. *Blasis's Code of Terpsichore* VI. 505 To waltz properly, all the beats, or *tems*, should be clearly marked, being attentive not to turn upon *les pointes*, or toes, in the same beats. **1846** *Musical World* 21 Nov. 587/2 The *pirouette* and the *pointe*, which she achieves with great facility. **1912** J. C. Flitch *Mod. Dancing & Dancers* iv. 58 The *pointes*..came to be regarded as the highest form of accomplishment. **1912** *Dancing Times* Aug. 449/2 Various Pointe Steps and..Enchaînements of Steps on the Pointes. **1920** *Ibid.* Feb. 413/1 Second and fourth couples run on the pointe under arch round first and third couples. *Ibid.* Dec. 182/2 The damage that can be done by too early pointe work is incalculable. **1949** *Ballet Ann.* III. 93 The advent of the Italian ballerina and her steel-like *pointes* and brilliant *fouettés*. **1959** T. Mara *On your Toes!* (1961) 18/1 In standing correctly on the pointes the dancer stands on the pads of the toes. **1961** S. Lester in *Ballet here & Now* i. 29 The swift exploitation of the use of the *pointe* for the female dancer. **1974** P. L. Moldon *Your Bk. of Ballet* ii. 27 Dancing on the tips of the toes (on *pointe*)..enhanced the ballerina's lightness.

pointed, *ppl. a.*¹ Add: **5.** (Further examples.)

1934 Priebsch & Collinson *German Lang.* 357 In German documents it [*sc.* Anglo-Saxon type] occurs usually in the *pointed* (Lowe: *miniscule*) form. **1947** *Hansard Commons* 15 Dec. 1451 Mr. Lipson asked..if smaller tins of pointed foodstuffs can be provided. **1957** *Encycl. Brit.* XIX. 664/2 The pointed runes were generally known and used in the whole of the Scandinavian North throughout the Middle Ages as the writing of cultured laymen.

b. *pointed blanket* = **point blanket*.

1779 in *Beaver* (1935) June 47 Sends samples of five different sorts of Pointed blankets with their respective prices per pair. **1780** in E. E. Rich *Moose Fort Jrnl.*, *1783–85* 356 We now send..pointed Blankets of different sizes..to be delivered to him. **1926** [see *point blanket* s.v. *point sb.¹ D. 14]. **1956** *Beaver* Summer 50 It can be no coincidence that while Maugenest was in London enquiries were put in hand for *Pointed* blankets.

c. *pointed fox*: (see quots.). Cf. *point v.¹ 20.

1911 *Directory Fur Trade* (N.Y.) 11 Fine Kamchatka and American Foxes in Sitka, Pointed, Black and Baum-marten. **1913** J. W. Jones *Fur-Farming in Canada* iv. 100 Latterly, the Germans have developed a large trade in 'pointed fox', which is an ordinary cheap fox dyed black, and afterwards 'pointed' by sewing in white hairs. **1939** M. Byers *Designing Women* III. xix. 250 Pointed Fox is red fox dyed black with white hairs stuck into it artificially. **1952** Lapick & Gelle *Scientific Fur Servicing* 7 In pointed fox the inserted hairs are generally all white. **1969** R. T. Wilcox *Dict. Costume* (1970) 142/1 *Fox, pointed*, the common red fox dyed black and pointed with silvery badger hairs to simulate silver fox.

6. *pointed-butted, -toed* adjs.

1928 Peake & Fleure *Steppe & Sown* iii. 46 Pointed-butted axes of jadeite and other hard stone came into use as well as flint. **1891** 'L. Malet' *Wages of Sin* III. v. vi. 33 Presenting his cousin with a fine view of a pointed-toed shoe sole. **1962** *Times* 30 Jan. 12/5 Pointed-toed backless shoes.

pointer. 3. f. For *U.S.* read 'orig. *U.S.*' and add further examples.

1957 V. Brittain *Testament of Experience* I. iii. 104, I liked the sound of the Bill not better than she; it was another 'pointer' towards the impending shadow of war and its threat to human liberty. **1961** *Lancet* 29 July 245/1 Experimenta evidence provides some pointers. **1977** J. F. Fixx *Compl. Bk. Running* vii. 82 An experienced runner may occasionally offer some pointers to a friend, but most of the time we coach ourselves. **1977** *Oxford Diocesan Mag.* Nov. 8/3 Other significant pointers to the future.

h. (See quot. 1872.)

1866 in *Nebraska Hist. Mag.* (1932) XIII. 149 After awhile I get my last pointer yoked, drive the whole team around and hitch it on the wagon tongue. **1872** C. H. Eden *My Wife & I in Queensland* 36 Twelve bullocks is the usual number in a team, the two polers and the leaders being steady old stagers; the pair next to the pole are called the 'pointers'. **1941** Baker *Dict. Austral. Slang* 55 *Pointers*, two of the bullocks in a team, placed next to the 'polers'.

4. Also *attrib.*

1797 J. Woodforde *Diary* 15 Jan. (1931) V. 5 One of his Pointer Dogs..was gone mad. **1811** Jane Austen *Sense & S.* III. viii. 181 He reminded me of an old promise about a pointer puppy. **1822** J. Woods *Two Years' Residence Eng. Prairie* 287, I lately saw a young pointer-dog,..that had three days before been bitten by a rattle-snake. **1849** J. W. Audubon *Western Jrnl.* (1906) 46 Some of the men had stolen a valuable pointer dog. **1979** *Times* 24 Nov. 26/6 (Advt.), *For sale*: Pointer puppies.

9. d. A person who does the ornamental work on the backs of gloves.

1903 *Sci. Amer.* Suppl. 24 Jan. 22629/3 Some make the gloves,..others, called 'pointers', work the ornamental lines on the back. **1921** *Dict. Occup. Terms* (1927) § 411.

e. The person who lays or points a gun. *U.S.*

1904 *Sci. Amer.* 18 June 475 The turrets are trained by one man, the trainer; and each gun is pointed by another man, the pointer, who fires the gun.

f. The person who 'points' a block of marble, etc.

1911 [see *point sb.¹ A. 20 d].

g. A person who points furs.

1929 *Fur Trade Directory* (N.Y.) 73 James Feuerlicht, Inc. 'The Old Reliable Fur Pointer'. **1930** M. Bachrach *Fur* xxiii. 431 The other use of the Badger is by a separate branch of the Industry known as the Pointers. These people furnish new hairs for a peltry by a method known as pointing. **1936** F. Grover *Practical Fur Cutting & Furriery* xv. 23 Badger hairs are also used for pointing of other furs; the pointer taking the skins for pointing ascertains the sort of pointing necessary. **1953** *Fur Trade Directory* (N.Y.) 66 Artistic Fur Pointers. Lydia Silver Fox Pointing Co.

11. *U.S.* A herdsman riding at the head of a herd of cattle on the march to keep it going in the desired direction.

1869 *Overland Monthly* III. 126 On the march the mighty herd sometimes strings out miles in length, and then it has 'pointers', who ride abreast of the head of the column, and 'siders', who keep the stragglers out of the chaparral. **1908** *Pacific Monthly* Mar. 324/2 The pointer is the herdsman who rides at the head of a straggling herd of cattle on the march, a sort of Cowboy John the Baptist. **1943** L. V. Hamner *Short Grass* 50 Two men, his best, were put near the front of the line... These were the pointers.

12. Either of two sharks belonging to the family Isuridæ, the blue pointer, *Isurus glaucus*, or the white pointer, *Carcharodon carcharias*.

1882 J. E. Tenison-Woods *Fishes & Fisheries N.S.W.* iv. 95 On the appearance of a 'blue pointer' among boats fishing for schnapper..the general cry is raised. **1896** F. G. Aflalo *Sketches Nat. Hist. Austral.* III. 222 The Blue-Pointer..is the favourite shark of that particular form of fiction which passes for popular natural history. *Ibid.* 223 Such are the chief sharks, the others being the huge White-Pointer [etc.]. **1963** H. W. McCormick et al. *Shadows in Sea* iv. 101 The Mako shark of the western Atlantic is a very close relative of the Blue Pointer (*Isurus glaucus*) of the Indian and Pacific Oceans. *Ibid.*, The Blue Pointer, in turn, is a name given by some South African Fishermen to the shark elsewhere known as the Great White, called in Australia the White Death or White Pointer. **1973** *Parade* (Melbourne) Sept. 63/1 The shark was a white pointer, of the maneater variety. **1975** *Daily Colonist* (Victoria, B.C.) 20 Dec. 7/6 Terry Page also offered odds in favour of Australian diver Wally Gibbons killing the white pointer shark before it kills him.

13. *Canad.* A rowing boat, pointed at both ends and having a shallow draught, used by loggers.

1901 'R. Connor' *Man from Glengarry* 13 Swiftly the pointer shot down the current, the swaying bodies and swinging oars in perfect rhythm with the song that rose and fell with melancholy but musical cadence. **1947** J. J. Rowlands *Cache Lake Country* 96 And all the while men in pointers..big boats with high pointed ends, move about picking up drivers, working logs off rocks. **1950** J. Hambleton *Abitibi Adventure* 140 They would use 'pointers', twenty-foot heavily built craft with steeply sloping sides, which seemed very tippy but actually were the most practical craft ever devised for Canadian logging. **1961** Price & Kennedy *Notes on Hist. Renfrew County* 156 At the suggestion of J. R. Booth, who wanted a useful river craft, John Cockburn in 1883 designed the 'pointer', a sharp-pointed boat that is still widely used in the lumbering business. **1964** *Canad. Geogr. Jrnl.* Feb. 67/3 If you ask Emmett Chartrand about the driving boat or 'pointer' he has a faraway look when he answers, '—used them, wore them out, broke them.' **1970** *Canad. Geogr. Mag.* Feb. 51/1 But now that Mr. John A. Cockburn has retired and his family's boatbuilding workshop has been demolished, the red Pointer will slowly disappear from the Canadian scene.

14. *Comb.* **pointer reading,** the reading of an instrument as shown by a pointer; also *fig.*

1933 *Mind* XLII. 108 The reduction of the objects of

physics to pointer-readings is not for him..a mark of the finality of physics. **1961** *Listener* 17 Aug. 245/2 There are no public pointer-readings in religion. **1967** *Philos.* XLII. 285 The doctrine that science is all about pointer-readings.

pointful (poi·ntful), *a.* [f. POINT *sb.*[1] + -FUL.] Full of point; apposite, pertinent. So **poi·ntfulness.**

1897 *Daily Tel.* 4 Jan. 5/4 Similarly, and with greater pointfulness, it was remarked that the Select Committee.. never consulted any person who was not in full work. **1931** *Cath. Gaz.* Feb. 71/2 The story—old, even apocryphal, it may be, but certainly typical and pointful—of Queen Victoria. **1960** *Guardian* 15 July 6/5 The guileless burbling is often very entertaining, and sometimes even pointful. **1970** *Daily Tel.* 27 June 7 Boulez will agree that for such randomness to be pointful the difference between one realisation and another needs to be rendered perceptible. **1976** P. DONOVAN *Relig. Lang.* vii. 81 A surrounding describable in words sets the act in the appropriate light, making it profound and religiously pointful. **1977** *Gramophone* Dec. 1164/1 Another pointful contrast came when we compared this latest RCA version of *La forza del destino* with that company's 1965 version also recorded with Leontyne Price. **1978** *Hi-Fi News* Dec. 159/3 The finale is particularly fine, imbued with an urgency and pointfulness that has rarely been equalled.

poi·nt group. *Math.* and *Cryst.* [f. POINT *sb.*[1] + GROUP *sb.*] † **a.** [tr. G. *punktgruppe* (Brill & Noether 1873, in *Nachrichten von der K. Ges. der Wiss. zu Göttingen* 117).] A set of points. *Obs.*

1895 *Proc. London Math. Soc.* XXVI. 495 The following paper deals with the properties of point-groups in relation to algebraic curves drawn through them. **1900** *Rep. Brit. Assoc. Adv. Sci.* 1900 121 (*heading*) Report on the present state of the theory of point-groups. **1900** K. FINK *Hist. Math.* 240 This theorem..introduced point groups, or systems of points of intersection of two curves, into geometry.

b. Any subgroup of the symmetry group of a sphere; *esp.* any of the thirty-two sets of symmetry operations which are used to classify crystal types.

[**1896** *Amer. Jrnl. Math.* XVIII. 172 We mark on the faces of the regular bodies a general point group.] **1903** H. HILTON *Math. Crystallogr.* iv. 48 Finite groups whose operations all leave one point unmoved are called point-groups. **1924** R. W. G. WYCKOFF *Structure of Crystals* i. 21 The point groups, or classes of crystal symmetry, thus formed may be uniquely defined either by stating their symmetry properties or, more analytically, by giving the coördinates of the equivalent points which arise from any arbitrarily chosen point by the operation of each of their elements of symmetry. **1950** W. J. MOORE *Physical Chem.* xiii. 364 The possible combinations of these symmetry elements that can occur in crystals have been shown to number exactly 32. These define the 32 crystallographic point groups, which determine the 32 crystal classes. **1973** T. JANSSEN *Crystallographic Groups* iii. 72 Each subgroup of *O*(3) is called a point group, because it leaves the origin invariant. Point groups occur as symmetry groups of molecules and of atoms in a crystal.

pointillé, *a.* (Further examples.) Also as *sb.* (example). Also *fig.*

1931 A. ESDAILE *Student's Man. Bibliogr.* vi. 207 The sprays are lighter, not only in themselves but because impressed by pointillé tools, i.e. tools with a dotted surface. **1933** T. E. LAWRENCE *Let.* 1 Aug. (1938) 773 His style disintegrates and not integrates, this time. Too pointillé. **1940** *Proc. Prehistoric Soc.* VI. 173 Sherds ornamented with three bands decorated in pointillé. **1964** H. HODGES *Artifacts* iv. 78 Strictly speaking, a surface that has been backgrounded by stabbing all over with a small pointed punch is referred to as *pointillé*, although some archaeologists are in the habit of using the term to describe any surface with a fine dot pattern, be the dots bosses or pits. **1974** *Times Lit. Suppl.* 23 Aug. 910/2 The English binders showed a liveliness and originality... The new designs included..the all-over pattern, introduced in the 1670s and generally composed of *pointillé* flowers and drawer-handle tools.

pointillism, -isme. (Further examples.)

1901 [see *NEO-IMPRESSIONISM]. **1947** BERGSTRÖM & TAYLOR tr. *Bergström's Dutch Still-Life Painting 17th Cent.* vi. 232 His broad and free handling is set off by a spirited *pointillisme* in some passages. **1976** *New Yorker* 15 Mar. 28/2 The murals are a triumph of Japanese pointillism.

2. *transf.* spec. *Mus.*, the breaking up of the musical texture into thematic, rhythmic, and tonal fragments.

1934 C. LAMBERT *Music Ho!* I. 32 In the pointillism of its scoring, *La Mer* represents the apex of Debussy's impressionist manner. **1959** *Times* 4 Apr. 10/2 Hamilton's music has been moving towards serialism for some time, and in this sonata he finds himself up to the elbows in neo-Webernist *pointillisme*. **1972** S. HYNES *Edwardian Occasions* 166 Mrs [Beatrice] Webb..created another character..by a large number of small strokes—a kind of literary pointillism. **1973** *Daily Tel.* 19 Mar. 14/2 A happy synthesis of recent techniques and time-hallowed devices, of chords and clusters, speech-slides and sung intervals, pointillism and canon. **1976** *Visible Language* X. 59 A piece of the once-voguish art of typewriter pointillisme.

pointillist, *sb.* Add: Also ‖**-iste.** (Earlier and later examples.)

1891 *Academy* 6 June 544/3 'Impressionists', 'tâchistes', 'plein airistes', and 'pointillistes', to use the jargon of the day. **1892** *Mag. of Art* p. xxxv/1 Among the painters who devote themselves to the application of colour in minute subdivision—called *pointillistes*, which may be freely rendered stipplers—the most conspicuous were M. Signac and M. Van Rysselberghe. **1916** A. BENNETT *Lion's Share* xii. 93 Also she was acquainted with the names and styles of all known modern painters from pointillistes to cubistes. **1929** E. LINKLATER *Poet's Pub* ii. 34 We ought to learn from the pointillistes. Put your colour on pure, in spots, and you get a luminous spectrum in verse as well as in painting. **1954** W. LEWIS *Demon of Progress in Arts* I. v. 23 The theory was that the colours would mix in the eye of the spectator—they must never be allowed to mix anywhere else... The Pointillistes adhered most rigidly to this rule. **1961** *Encycl. World Art* V. 184 By employing the color theories of Michel Chevreul..and refining the principle of optical mixture from brush strokes to tiny dots, the pointillists arrived at..'the logical consequence of impressionism'. **1963** F. GETTINGS *Golden Pleasure Book of Art* 59 Because of the way Seurat and his friends used small points of colour, they were called *Pointillistes*.

B. *attrib.* passing into *adj.*

1902 *Encycl. Brit.* XXIX. 414/1 There are several fallacies however, theoretical and practical, in this 'spectral palette' and pointillist method. **1905** [in Dict.]. **1934** C. LAMBERT *Music Ho!* I. 25 The methods of the pointillist painters have something in common with the use of the orchestra as displayed in Debussy's works. **1940** O. SITWELL *Left Hand, Right Hand!* II. 132 Those figures, full of latent movement, seen in a pointillist picture. **1971** [see *DIVISIONISM]. **1975** D. THOMAS *Impressionists* 87 He [*sc.* Van Gogh] met Pissaro, who was then moving towards the pointillist technique of Seurat.

b. *transf.*

1921 E. SAPIR *Language* 243 We cannot assimilate the luxurious periods of Latin nor the pointilliste style of the Chinese classics. **1934** C. LAMBERT *Music Ho!* I. 46 This pointillist orchestration gives to many of Schönberg's works an impressionist effect in performance that an inspection of the score with the eye alone would hardly lead one to expect. **1958** *Observer* 27 Apr. 16/4 Thackeray's.. effective 'pointillist' method. **1959** *Ibid.* 14 June 24/3 Although its style was largely *pointilliste*, the music did not strike me as derivative. **1970** *Daily Tel.* 30 Apr. 16/5 A certain lack of eloquence became most clearly discernible during a few melodious lines given to the oboe in this largely pointillist piece. **1976** *Times Lit. Suppl.* 1 Oct. 1229/4 Neither the satire nor the pointillist plotting adds up to much that is coherent and compelling.

Hence **pointilli·stic** *a.*

1922 H. CRANE *Let.* 19 Apr. (1965) 84 All this talk of Matty's is..metallic and pointillistic. **1954** *Grove's Dict. Mus.* (ed. 5) IX. 226/1 His [*sc.* Webern's] next compositions..are still more highly concentrated, still more transparent and pointillistic in texture. **1972** *Village Voice* (N.Y.) 1 June 44/3 Philip Corner adds sporadic pointillistic piano gestures. **1977** *Times* 22 Oct. 10/6 A new and exclusive pointillistic dyeing method... You can shade your carpet from dark to pale.

pointing, *vbl. sb.*[1] Add: **1. e.** (Earlier and later examples.) Also **pointing-up.**

1845 *Encycl. Metrop.* V. 465/2 The construction of these instruments for pointing is not always the same, but the principle on which they act is exactly similar. **1947** J. C. RICH *Materials & Methods of Sculpture* ix. 272 The process of pointing consists of marking with pencil all projections and recessions or 'points' on the model, which is generally a plaster cast, and making position measurements with the adjustable rods of the pointing machine, which are capable of measuring at intervals a fraction of an inch apart. **1969** L. R. ROGERS *Sculpture* vi. 199 The pointing-up in marble of clay models in a mechanical fashion by craftsmen assistants. **1974** *Encycl. Brit. Micropædia* VIII. 68/1 Although pointing has limitations as a technique of creative sculpture, it has been used widely, especially in the 19th century.

8. b. The disposition and colouring of the points in a cat's fur. Cf. *POINT *sb.*[1] A. 26 c.

1978 R. HILL *Pinch of Snuff* v. 50 The room was filled with cats..of various ages and pointings.

10. (Later example.)

1902 *Act 2 Edw. VII.* c. 29 § 2 Fishing for trout..by what is known as double rod fishing, or cross line fishing, or set lines,..or by striking the fish..or by pointing.

10*. *Furs.* The insertion of hairs into a pelt, usu. to repair damage or to simulate another fur. Cf. *POINT *v.*[1] 20.

1900 *Fur Trade Rev.* 1 May 213/2 Pointing will be one of the popular methods of enhancing the attractiveness of some of the most desirable furs. **1910** *Encycl. Brit.* XI. 354/1 The process of inserting white hairs is called in the trade 'pointing', and is either done by stitching them in with a needle or by adhesive caoutchouc. **1922** W. E. AUSTIN *Princ. & Pract. Fur Dressing* i. 7 Badger hair is very extensively used for 'pointing'. **1930** M. BACHRACH *Fur* xix. 275 Rubbed spots on the rumps are judged according to the damaged area, which, in the Silver Fox, can be somewhat restored to its original appearance by the process of pointing. **1936** F. GROVER *Practical Fur Cutting & Furriery* xv. 22 Pointing of furs is often resorted to, to make an imitation of another fur. **1950** *N.Z. Jrnl. Agric.* June 597/1 Pointing refers to the practice of glueing in white hairs to improve inferior silver fox furs. **1952** LAPICK & GELLE *Scientific Fur Servicing* 4 Pointing may be detected by the fact that the badger hairs used are glued in groups of two or three to the natural fox hairs and the leather.

11. pointing-bone *Anthrop.*, a bone, or apparatus consisting of bones, and usually a string made of woven hair, used by Australian Aborigines in a secret ritual to bring about the death or illness of the person at whom it is pointed; also *attrib.*; **pointing-machine** (*a*) (in Dict.: see quot. 1875); (*b*) an apparatus used in sculpture for taking points; **pointing-stick** *Anthrop.*, a stick used for the same purpose as a pointing-bone.

1904 SPENCER & GILLEN *Northern Tribes of Central Austral.* xiv. 459 The pointing apparatus..consists of a long strand of human hair-string, to one end of which five small pointing-bones are affixed. **1959** S. H. COURTIER *Death in Dream Time* v. 50 He..saw a piece of bone, sharp at one end, blunt at the other, and bound round the middle with strands of black fur or hair..a native pointing bone or death bone. **1965** *Austral. Encycl.* I. 67/1 The 'pointing bone' sorcery (widely distributed throughout southern and central northern Australia). **1886** *Encycl. Brit.* XXI. 571/2 Partly by eye and partly with the constant help of the pointing machine, which is used to give any required measurements, the workman almost completes the marble statue, leaving only the finishing touches to be done by the sculptor. **1947** J. C. RICH *Materials & Methods of Sculpture* ix. 271 There is evidence that during the Late Hellenistic period the Romans used a primitive version of the pointing machine, and this mechanical device may partly account for the large number of Roman copies of Greek sculpture. **1959** P. & L. MURRAY *Dict. Art & Artists* 249 A pointing machine measures the depth from a given vertical, of, for example, the receding planes of the nose of a plaster cast and transfers the measurement to a hole drilled in the marble block. **1904** SPENCER & GILLEN *Northern Tribes of Central Austral.* xiii. 433 A Thakomara man of the Karinji totem.. had special pointing sticks. **1959** *Chambers's Encycl.* XIV. 629/2 He [*sc.* the person who has caused illness or death] can be put out of the way by means of the 'pointing stick' or 'pointing bone' made efficacious by the spells of witch or sorcerer (usually a man).

pointless, *a.* Add: **2.** (Later examples.) Also (in the sense of *POINT *sb.*[1] A. 28 d), without purpose or advantage; having no good effect.

1934 E. WAUGH *Handful of Dust* ii. 33 It sometimes seems to me rather pointless keeping up a house this size if we don't now and then ask some other people to stay in it. **1954** I. MURDOCH *Under Net* xiii. 179 It seems pointless to conceal our identity. Sammy would guess it anyway as soon as we announced our terms. **1966** H. ROSEVEARE *Give me this Mountain* i. 18 Without Him, it was a weary, stupid, empty, pointless, useless life. **1970** G. F. NEWMAN *Sir, You Bastard* v. 147 Flowers didn't occur to Sneed until he had arrived at the hospital, and there the gesture was pointless. **1978** M. BIRMINGHAM *Sleep in Ditch* 193 'He died..soon after six.'.. 'S-so all last night's h-horror was quite pointless.'

pointman. [f. POINT *sb.*[1] or *v.*[1] + MAN *sb.*[1]]

1. = *POINTER II.

1903 A. ADAMS *Log of Cowboy* iii. 28 Two riders, known as point men rode out and well back from the lead cattle. **1942** BERREY & VAN DEN BARK *Amer. Thes. Slang* § 913/10 Point men or riders, cowboys riding near the head of a herd of cattle. **1977** M. HERR *Dispatches* iii. 35 Classic essential American types; point men, *isolatos* and outriders.

2. = POINTSMAN 2.

1927 *Observer* 20 Nov. 11/3 It..bore a number of legends; on the footboard, 'Step on the gas'; on the bonnet, 'Don't look inside', and 'Pointman, let us pass'.

3. (Usu. as two words.) The soldier who walks at the head of a patrol. Also *transf.*

1944 *Yank* 4 Feb. 9/1 The Jap point man was on the scene before any camouflaging could be done. **1969** I. KEMP *Brit. G.I. in Vietnam* v. 96 Next to the commander..the point man was the most important member of the squad on jungle patrols. It was his duty to walk at the head of the formation..and to act as scout for the rest. **1970** *Time* 5 Jan. 14 Cisneros survived 42 patrols in Viet Nam, mostly as the exposed point man, and saw his unit chewed up behind him several times. **1977** *Time* 21 Mar. 56/1 In Young's view, his role is to be a kind of 'point man' for the Administration in foreign policy, charged with getting out in front with new ideas and possibilities. **1978** *Guardian Weekly* 25 June 18/2 Although Carter started his term as no friend of Kennedy.. Kennedy has, ironically, become Carter's point man in the senate.

4. = POINTSMAN I.

1945 G. MILLAR *Maquis* ix. 180 This pointman would not presumably be antagonistic to us, since he would be an ordinary 'cheminot'.

Pointolite (poi·ntŏlait). Also **pointolite.** A proprietary name for a type of lamp containing a small but bright source of light, produced by the incandescence of a small knob of tungsten heated by an arc struck between it and a cathode.

1916 *Illuminating Engineer* IX. 32/1 A very interesting new form of lamp for projection purposes..has just been brought out by the Edison and Swan United Electric Light Co., Ltd. The 'Pointolite' lamp is in reality an arc within a sealed bulb. **1923** *Trade Marks Jrnl.* 25 July 1558 Pointolite... Electric lamps... The Edison Swan Electric Company, Limited,..London..; manufacturers. **1926** W. E. WOODWARD *Metallogr. Steel & Cast Iron* i. 42 For taking photomicrographs a 500 c.p. 'pointolite' gives excellent results. **1949** A. PEREIRA *Man. Sub-Standard Cinematogr.* vi. 234 A form of arc light which has found wide application for comparatively low power projection is the enclosed tungsten arc known as Pointolite. **1964** M. HYNES *Med. Bacteriol.* (ed. 8) v. 57 A good illumination of sufficient intensity is essential. This is given by an arc-light or a 'pointolite' lamp.

pointswoman. (In Dict. s.v. POINTSMAN.) (Earlier example.)

1865 A. MUNBY *Diary* 17 Mar. in M. Hiley *Victorian Working Women* (1979) II. i. 86/2 In the cutting I found a girl..the pointswoman, whose duty it is to mind the rails whenever a train of coalwaggons goes by.

pointwise, *adv.* Add: **2.** *Math.* With regard to individual points; esp. in *pointwise convergent*, that converges for each individual point in a space, but not necessarily for the space as a whole (so *pointwise convergence*).

1932 *Colloquium Publ. Amer. Math. Soc.* XV. i. 26 The term 'convergence in the mean' was introduced to distinguish convergence in the space $\mathfrak{L}_2$..from ordinary point-wise convergence. **1955** HALL & SPENCER *Elem. Topology* iv. 133 This sequence is said to converge pointwise to a mapping $f: S \to T$, if and only if, for each point x of S, the sequence $\{f_n(x)\}$ of points of T converges to the point $f(x)$ of T. **1966** A. FEINSTEIN tr. *G. Choquet's Topology* i. 92 Every sequence (f_n) which converges uniformly to f converges pointwise to f, but it is important to note that the converse is incorrect. **1970** *Manifold* VI. 17 A student produced three new specimens... They came from adding the blackboard examples pointwise. **1971** POWELL & HIGMAN *Finite Simple Groups* iii. 138 And r leaves a hyperplane pointwise fixed. **1974** G. J. O. JAMESON *Topology & Normed Spaces* ix. 90 A decreasing sequence of continuous functions on a compact space can certainly be pointwise convergent to a discontinuous function.

pointy, *a.* Add: **1.** (Later examples.) Also *fig.*

1906 *Daily Chron.* 2 Oct. 3/4 Puck of the pointy-ears. **1927** J. MASEFIELD *Midnight Folk* 113 She has..pointy, black shiny shoes. **1953** A. MILLER *Crucible* I. 20 Let either of you breathe a word,..and I will bring a pointy reckoning that will shudder you. **1969** M. PUZO *Godfather* IV. xix. 263 Conny stood with hands on hips, her face pointy and white with rage. **1975** J. MCCLURE *Snake* ix. 116 Her nails..were long and pointy. **1977** *Rolling Stone* 19 May 69/3 Zbigniew Brzezinski, the president's national security adviser and a man whose every feature is pointy, enters the office.

4. Special collocations, as **pointy-head** *U.S. colloq.*, a supposed expert or intellectual (used derogatorily); also *attrib.*; hence *pointy-headed* adj.

1972 *Guardian* 21 Feb. 2/4 George Wallace..attacked Muskie, Humphrey, and his other 'pointy-head' opponents. **1972** *Times* 5 May 6/3 Mr Wallace..dismissed it quickly at the end of his address as 'the most callous, asinine, stupid thing that was ever conceived by some pointy-head in Washington DC'. **1972** *National Observer* (N.Y.) 27 May 12/1, I am not ashamed of being a 'pointy headed intellectual snob' who notices the correlation between our 'leadership' and the many distressing phenomena 'coinciding' with those sages' descriptions. **1973** J. DI MONA *Last Man at Arlington* (1974) 178 Idiots... Pointy heads that didn't know how to operate. **1975** *N.Y. Times* 12 Nov. 42/2 Let the dust gather on the pointy-headed bureaucrats and all the other props from yesteryear.

b. *Comb.*, as *pointy-eared*, *-toed* adjs.: see also *pointy-headed* adj. above.

1906 KIPLING *Puck of Pook's Hill* 7 They saw a small, brown, broad-shouldered, pointy-eared person with a snub nose. **1968** A. YOUNG in A. Chapman *New Black Voices* (1972) 147 Sport shirt, creased pants, shiny black pointy-toed stetsons [*sc.* shoes]. **1971** E. BULLINS in W. King *Black Short Story Anthol.* (1972) 61 His snake-skinned, pointy-toed shoes. **1978** *N.Y. Times* 18 Jan. 1/3 A scene normally dominated by men with Western twangs in ten-gallon hats and pointy-toed boots.

‖**poipoi** (poi-poi). Also **poee-poee, poi-poi, popoi.** [Polynesian word.] A Polynesian dish, usu. made from fermented bread-fruit (see quot. 1829). Also *attrib.* Cf. POI.

1829 W. ELLIS *Polynesian Researches* I. xiii. 377 The most general dish in the Southern Islands is what they call *popoi*, nearly resembling the *poe* of the Sandwich Islands. It is made with the ripe mountain plantain, either raw or baked, beaten up into a paste or jelly, and diluted with cocoa-nut milk. Another kind of *popoi* is made with bread-fruit, or *opio*, beaten up and diluted with cocoa-nut or plain water. **1846** H. MELVILLE *Narr. Residence Marquesas Islands* xv. 128 The great staple articles of food into which the breadfruit is converted by these natives are known respectively by the names of Amar and Poee-Poee. **1888** M. I. STEVENSON *Let.* 25 Aug. in M. C. Balfour *From Saranac to Marquesas* (1903) 121 Madame Stanislas gave Fanny a very finely carved *poi-poi* bowl of *mio* wood. **1910** F. W. CHRISTIAN *Eastern Pacific Lands* xii. 160 To a European palate *Popoi* tastes like an acid custard. **1919** *Century Mag.* Aug. 450/1 The players halted briefly to eat a bowl of *poipoi*. **1970** I. GOLDMAN *Ancient Polynesian Society* vii. 125 The fermented breadfruit mash (*poipoi*) stored well in leaf-lined pits. **1974** T. HEYERDAHL *Fatu-Hiva* iii. 122 *Poipoi*, the staple diet in most of Polynesia. Nowhere else was *poipoi* made as strong as in the Marquesas group.

poire (pwār). [abbrev. of Canad. Fr. *poire sauvage*, lit. 'wild pear'.] A name formerly used in Canada for a tree or shrub of the genus *Amelanchier*, belonging to the family Rosaceæ, or its fruit, a blue-black berry; = SASKATOON, *shad-bush* s.v. SHAD *sb.* 4 b.

1789 A. MACKENZIE *Jrnl.* 14 Aug. in *Voyages from Montreal* (1801) vii. 107 There were plenty of berries, which my people called *poires*; they are of a purple hue, somewhat bigger than a pea, and of a luscious taste. **1807** G. KEITH *Let.* 7 Jan. in L. F. R. Masson *Les Bourgeois de la Compagnie du Nord-Ouest* (1890) II. 66 There are *poire*, gooseberry and raspberry bushes. **1837** *Trans. Lit. & Hist. Soc. Quebec* III. 126 In the country parts this small fruit is dignified with the name of poire, more from its fine flavor, it is presumed, than from any resemblance to pears. **1865** MILTON & CHEADLE *N.W. Passage by Land* xiv. 266 The Indians brought in a plentiful supply of the poire, wild pear, or service berry. **1951** W. O'MEARA *Grand Portage* xviii. 102 He found..a large purple berry called the poire that was the most delicious of all.

poise (pwāz, poiz), *sb.*[2] *Physics.* Pl. **poises, poise.** [f. *POISE(UILLE.] The unit of (dynamic) viscosity in the C.G.S. system, equal to one gramme per centimetre-second; in the International System of Units replaced by the pascal second (equal to 10 poises). Symbol P.

1913 DEELEY & PARR in *Phil. Mag.* XXVI. 87 It would be a distinct advantage to have a name for the unit of viscosity expressed in C.G.S. units, and we would suggest that the word Poise be used for this; for it is to Poiseuille that we owe the experimental demonstration that when a liquid flows through a capillary tube..at constant temperature, the viscosity is constant at all rates of shear, provided that the flow is not turbulent. *Ibid.* 89 R. M. Deeley obtained $\eta = 6 \times 10^{11}$ poises for the viscosity at 0° C. of fine crystalline ice. **1939** *Nature* 6 May (Suppl.) p. i (Advt.), The Goodeve Thixoviscometer... Range 1 centipoise to 1 megapoise (0·01 to 10^6 poises). **1957** G. E. HUTCHINSON *Treat. Limnol.* I. iii. 209 The viscosity of water, 0·0114 poise at 15°C., is also very high for a liquid of low molecular weight. **1964** *Times Sci. Rev.* Spring 4/2 As a result of its relatively low viscosity—probably about 10^2 to 10^3 poises—..the lava flows freely. **1971** *Nature* 10 Sept. 101/1 Liquefaction implies a reduction in the viscosity of the sediment from that of a plastic solid (say, 10^6 poise) to that of a thick soupy liquid (say, 10 poise).

poise, *v.* Add: **6. d.** *pass.* To be ready *for* (or *to do*) something; to be about (to do something).

1932 W. FAULKNER *Light in August* (1933) xvii. 381 She looked exactly like a rock poised to plunge over a precipice. **1956** A. H. COMPTON *Atomic Quest* 284 The boys who had been poised for the invasion of Japan. **1961** *National Rev.* 30 Dec. 462/3 The Free Chinese know that the situation on the Mainland is in flux, and are poised to strike. **1977** A. THWAITE *Portion for Foxes* 38 A scornful phrase Poised to put down the parasite or bore. **1979** *Daily Tel.* 3 Feb. 1/1 British Petroleum was poised last night to make further reductions in oil deliveries to customers around the world.

7. Also, to hover or be poised in readiness *for* (something).

1898 C. M. SHELDON *His Brother's Keeper* iii. 64 The gravity of events that were evidently poising for a crisis left little room for anything but sober feeling.

poised, *ppl. a.* Add: Also, of persons, their behaviour, etc.: composed, self-assured.

1928 E. O'NEILL *Strange Interlude* I. 12 His manner is cool and poised. He speaks with a careful ease. **1961** J. MERCIER *Whatever you Do* ii. 28 Somehow managing to get out a cool, poised, 'Won't you hold on a second, please', I covered up the mouthpiece, [etc]. **1974** E. FERRERS *Hanged Man's House* v. 45 She was very poised, and had a terrific social manner.

Poiseuille (pwāzöy). *Physics.* The name of J. L. M. *Poiseuille* (1799–1869), French physiologist, used *attrib.* and in the possessive to designate concepts and phenomena related to his work on fluid flow, as **Poiseuille('s) equation, expression, formula,** or **law,** the relation between the volume V of fluid flowing per second through a long cylinder of length l and radius r under conditions of Poiseuille flow, viz. $V = \pi r^4 p/8\eta l$, where p is the pressure drop along the cylinder and η is the viscosity of the fluid (given by Poiseuille in *Compt. Rend.* (1840) XI. 1047, (1841) XII. 114); **Poiseuille flow,** laminar flow of a viscous fluid of negligible compressibility, esp. through a long, narrow cylinder.

1883 *Phil. Trans. R. Soc.* CLXXIV. 946 The discharge from the pipes agreed exactly with those given by Poiseuille's formula for capillary tubes. *Ibid.* 981 With all the smaller tubes Poiseuille's law held throughout his experiments. **1931** G. BARR *Monogr. of Viscometry* ii. 24 Let the viscosity..calculated from an experiment by means of the uncorrected Poiseuille equation be denoted by η'. **1946** *Q. of Applied Math.* III. 119 The Poiseuille flow in a circular pipe was studied..with a conclusion of stability. **1958** CONDON & ODISHAW *Handbk. Physics* v. vi. 78/1 Viscous, laminar flow of gases through cylindrical tubes is governed by Poiseuille's law. **1967** MARGERISON & EAST *Introd. Polymer Chem.* ii. 103 In these viscometers the time taken for the level of solvent or solution to pass between two fixed marks is determined and this is related to the viscosity by Poiseuille's equation. **1975** *Jrnl. Surg. Res.* XIX. 26/1 The resistance of varying degrees of obstruction placed within the circumflex coronary artery was determined by employing Poiseuille's Law. **1975** *Microvascular Res.* X. 153 *In vivo* RBC [*sc.* red blood cell] velocity profiles for mammalian arterioles and venules.. are time variant and more blunted than would be anticipated for Poiseuille flow.

poison, *sb.* (*a.*) Add: **2.** (Earlier and later examples of *to hate like poison.*)

1530 J. PALSGRAVE *Lesclarcissement de la Langue Francoyse* fol. cclviii[v], He gyueth me fayre wordes and yet he hateth me lyke poyson. **1864** A. TROLLOPE in *Good Words* Dec. 931/1 Everybody liked Barty,—excepting only Mally Trenglos, and she hated him like poison. **1905** H. A. VACHELL *Hill* i. 20 'He hates me like poison,' said Duff. **1974** 'M. INNES' *Appleby's Other Story* xii. 97 Enormous sums vanishing in bitter law-suits, which is a thought the wealthy hate like poison.

b. *colloq.* (orig. *U.S.*). Alcoholic liquor; an alcoholic drink; esp. in phr. *to name one's poison*, to say what drink one would like; also *transf.*

1805 'RED JACKET' in *Freemason's Mag.* (Philadelphia) II. 388 We gave them corn and meat; they gave us poison in return. [*Note*] Alluding it is supposed to ardent spirits. **1866** 'MARK TWAIN' *Lett. from Hawaii* (1967) 85 In Washoe, when you are..invited to take 'your regular pison', etiquette admonishes you to touch glasses. **1876** *Carson Valley News* (Genoa, Nev.) 2 June 2/2 Nominate your poison, gents: it's my treat. **1876** J. MILLER *First Fam'lies of Sierras* 128 A true Californian of Sierras.. heads straight up to the bar,..hoists his Poison, throws back his head, and then falls back wiping his mouth. **1914** JOYCE *Dubliners* 113 Just as they were naming their poisons who should come in but Higgins! **1951** T. STERLING *House without Door* ii. 12 Name your poison, lady. Chocolate, vanilla, pistachio, maple cream. **1965** E. BROWN *Big Man* xvii. 157 'What's your poison tonight, miss?' 'Make it a gin and bitter lemon.' **1973** J. ASHFORD *Double Run* v. 37 Come right in and name your poison.

c. *Chem.* A substance which destroys or reduces the activity of a catalyst.

1913 in C. ELLIS *Hydrogenation Oils* (1914) 316 Sulphur is a 'poison' to the catalyst. **1938** *Thorpe's Dict. Appl. Chem.* (ed. 4) II. 426/2 Nickel is in general very effective, but..is sensitive to 'poisons', particularly sulphur compounds and carbon monoxide. **1966** *McGraw-Hill Encycl. Sci. & Technol.* II. 548/1 Catalysts gradually lose catalytic activity. Traces of impurity in the feed, called poisons, may be strongly adsorbed and exclude the reactants from the surface.

d. *Nuclear Sci.* A fission product or an impurity in a nuclear reactor which interacts with neutrons and thus slows the intended reaction; also, an element with this property which is added to the fuel in order to facilitate control of the reaction.

1952 S. GLASSTONE *Elements Nuclear Reactor Theory* xi. 315 Some of these [*sc.* fission products] may have large cross sections for the absorption of neutrons, and so they can act as poisons. **1961** J. F. HILL *Textbk. Reactor Physics* vii. 201 It is sometimes of advantage deliberately to introduce a high neutron absorbing material into a reactor to increase the intervals between recharging the reactor with fuel. A material used in this way is called a 'burnable poison'. **1963** J. F. HOGERTON *Atomic Energy Deskbk.* 406/1 It should be noted that some poisons are classified as undesirable whereas others are deliberately introduced into the system... The major fission product poisons are xenon-135 and samarium-149. **1966** *McGraw-Hill Encycl. Sci. & Technol.* IX. 185/1 Fission-product poisons (neutron absorbers) can be lowered by frequent processing of fuel. **1978** *Nature* 26 Jan. 306/3 The higher natural abundance of ^{235}U..prevailing at that time.., and effective absence of neutron-absorbing 'poisons', constituted favourable conditions for the occurrence of a self-propagating fission reaction.

3. b. Applied to a person who exerts a baneful influence or who is detested.

1910 W. M. RAINE *Bucky O'Connor* 28 They say he's part Spanish and part Indian, but all pisen. **1964** L. DEIGHTON *Funeral in Berlin* xlii. 262 You are poison to Gehlen... There isn't a place left in the whole world where you would get a sniff of a job. **1974** A. WILLIAMS *Gentleman Traitor* xiii. 194 Philby's poison, whichever side he's on. **1977** R. BARNARD *Death on High C's* xvi. 164 One knows the type... Simply eaten up with egotism... They're complete poison, wherever they go.

4. a. *poison bottle, pill.*

1854 DICKENS *Hard T.* xiii. 104 It were the Poison-bottle on table. **1978** P. LOVESEY *Waxwork* 64 That struck me as peculiar..that a man committing suicide would put the poison bottle back in the cabinet. **1946** P. BOTTOME *Lifeline* xxxv. 269 With his hands tied securely behind him Mark could not reach the poison pill he had been given for such emergencies. **1975** *Times* 29 Aug. 6/8 There are many organizations working against Mrs Gandhi... Ours is serious... We all carry poison pills in our pockets.

5. poison book = *poison register*; **poison gas,** any chemical that is released into the atmosphere as a gas or vapour to harm those who inhale it or absorb it through their skin; also *attrib.*; hence as *vb. trans.*; **poison green,** a bright, sharp shade of green; **poison oracle,** a form of divination in which a Zande witch-doctor administers poison to a fowl and draws inferences from its effect on the bird; **poison pen,** one who writes anonymous letters with malicious, libellous, or scurrilous intent; also *attrib.*, of or pertaining to such a person or letter; **poison** (also **poisons, poisons'**) **register,** a register of the names of those to whom a poison or poisons have been made available.

1930 D. L. SAYERS *Strong Poison* i. 12 She signed the poison-book in the name of Mary Slater, and the handwriting has been identified as that of the prisoner. **1943** G. GREENE *Ministry of Fear* I. iii. 33 We'll look into the poison books. **1947** A. CHRISTIE *Labours of Hercules* ii. 62, I never said anything about the missing arsenic. I even cooked the poison book! **1950** 'A. GILBERT' *Is she Dead Too?* iii. 46 He had brought a prescription that required..a drug only to be obtained by signing the Poison Book. **1978** J. SYMONS *Blackheath Poisonings* III. 152 His poison book's all in order, and there's this entry in it for arsenic. **1915** H. W. WILSON *Great War* IV. 336/2 After the great chemical experiment with poison gas in April, the Germans had been able to advance to the manor-house. *Ibid.*, The Duke of Würtemberg..had apparently become convinced, after his poison-gas victory in April, that chemical methods of making war were the most successful. **1922** D. H. LAWRENCE *Fantasia of Unconscious* xi. 207 The problem of the future is a question of the strongest poison-gas. **1924** T. HARDY *Winter Words* (1928) 171 After two thousand years of mass We've got as far as poison-gas. **1970** R. STETLER *Battle of Bogside* 179 President Johnson..called a press conference to deny the poison gas charge. **1970** G. JACKSON *Let.* 4 Apr. in *Soledad Brother* (1971) 211 An enemy that would starve his body, ..chain his body,..and poison-gas it. **1975** tr. *Melchior's Sleeper Agent* (1976) II. 39 Stacks of incendiary bombs and poison gas projectiles. **1926** S. LEWIS *Mantrap* x. 117 The poison-green tufted velvet couch. **1937** [see *candy-pink* s.v. *CANDY *sb.*[1] 2]. **1975** P. G. WINSLOW *Death of Angel* x. 212 He drives a poison-green two-seater. **1937** E. E. EVANS-PRITCHARD *Witchcraft, Oracles & Magic among Azande* 10 The principal Zande oracles are: (*a*) *benge*, poison oracle, which operates through the administration of strychnine to fowls, and formerly to human beings also. **1955** M. GLUCKMAN *Custom & Conflict in Africa* iv. 88 Each question is framed to allow of a 'yes' or 'no' answer to the problem, thus: 'if X is the witch who is making my son ill, poison-oracle, kill the chicken; if X is not the witch, poison-oracle, spare the chicken'. **1972** M. D. MCLEOD in Singer & Street *Zande Themes* 167 The Zande clearly considered the rubbing-board oracle less accurate than both the termite oracle and the poison oracle. **1914** *N.Y. World* 11 Mar. 5/1 Women..crowded the Union County Court room..hoping to hear some plausible elucidation of the 'poison pen' mystery. **1929** M. LIEF *Hangover* 302 The King of the Tabloids sat in his counting-house counting up the two and a half million circulation gained through the blood and scandal shed by ..poison-pen letters. **1935** D. L. SAYERS *Gaudy Night* v. 100 Isn't our poison-pen rather silly to get all her spelling right? **1956** A. WILSON *Anglo-Saxon Attitudes* II. iii. 388 To all the other clergymen she was busy addressing poison-pen letters. **1973** J. THOMSON *Death Cap* vii. 93 She had seemed..a perfect front runner in the poison-pen stakes, the classic example of the embittered spinster. **1975** D. LODGE *Changing Places* iii. 124 I've had what I believe is called a poison-pen letter from Euphoria, an anonymous letter. **1936** COOK & LAWALL *Remington's Pract. Pharm.* (ed. 8) lxxxiv. 1357 The poison register must be always open for inspection by the proper authorities. **1978** J. SYMONS *Blackheath Poisonings* III. 150 He sent..Sergeant Miles to look at the poison registers. **1907** *Yesterday's Shopping* (1969) 499 In the case of Poisons being required it is absolutely necessary..that the Poisons Register be signed at the time of purchase. **1957** *Encycl. Brit.* XVII. 693/1 These poisons in their uncompounded form may only be supplied to persons known to the pharmacist and their sale must be recorded in the poisons' register. **1958** H. G. MOSS *Retail Pharmacist's Handbk.* xxiii. 360 Certain professional and trade users may obtain First Schedule poisons on a signed order instead of attending and signing the Poisons Register. **1971** GILBERT & SHARP *Pharmaceuticals* xi. 140 First Schedule poisons may be sold without any prescription, but only if the purchaser is known to the pharmacist and signs the Poisons Register.

b. poison-bulb, substitute for def.: one of several South African bulbous plants belonging to the family Amaryllidaceæ, esp. *Boophane disticha*; (earlier and later examples); **poison-bush,** (*c*) *Austral.*, one of several plants bearing leaves harmful to cattle, esp. a species of *Gastrolobium*; **poison-dogwood** (examples); **poison-elder** (earlier and later examples); **poison-ivy,** substitute for def.: one of several trailing or climbing North American shrubs belonging to the genus *Rhus* (or *Toxicodendron*), esp. *R. toxicodendron* (or *T. toxicaria*), bearing leaves resembling ivy, and greenish flowers followed by white berries, and producing inflammation of the skin and other reactions when touched; (earlier and later examples); also *fig.*, an unpleasant person; **poison-oak** (earlier and later examples); **poison-plant,** also, a name used for various plants harmful to man or livestock; (later examples); **poison-sumac,** for *Rhus venenata* substitute *Rhus vernix* or *Toxicodendron vernis*; (earlier and later examples) **poison vine,** (*b*) (examples); **poison-weed,** delete † and ? and add later example.

1822 W. BURCHELL *Trav. Interior S. Africa* I. xxi. 539 Plants of *Amaryllis toxicaria* were..very abundant... This plant is well known to the Bushmen, on account of the virulent poison contained in its bulb. It is also known to the Colonists and Hottentots, by the name of *Gift-bol* (Poison-bulb). **1966** E. PALMER *Plains of Camdeboo* v. 82 The Poison Bulb, with its innocent blue-green fan of leaves, that they [*sc.* Bushmen] pounded for its deadly juice. **1889** J. H. MAIDEN *Useful Native Plants Austral.* 129 *Gastrolobium* spp... These plants are dangerous to stock and are hence called 'Poison Bushes'. Large numbers of cattle are lost annually in Western Australia through eating them. **1927** M. M. BENNETT *Christison of Lammermoor* xx. 185 There were quicksands and the dreaded poison-bush, *Gastrolobium grandiflorum*. **1965** *Austral. Encycl.* VII. 157/2 Many species of *Gastrolobium* have, and nearly all species deserve, the name poison-bush or poison-plant. **1814** J. BIGELOW *Florula Bostoniensis* 72 *Rhus vernix*. Poison dogwood. Swamp Sumach... Grows in bunches in wet swamps. **1958** Poison-dogwood [see *poison-elder*]. **1822** A. EATON *Man. Bot.* (ed. 3) 428 *Rhus vernix*, poison sumach, poison elder... Berries green, at length whitish. **1958** G. A. PETRIDES *Field Guide to Trees & Shrubs* 84 Names in common use, such as Poison-elder or Poison-dogwood, usually refer to Poison Sumac. **1784** *Mem. Amer. Acad.* I. 422 Poison Ivy..produces the same kind of inflammations and eruptions..as the poison wood tree. **1832** W. D. WILLIAMSON *Hist. State Maine* I. 130 Poison Ivy..is a dangerous medicine. **1891** M. E. FREEMAN *N. England Nun* 191 [She] saw Joseph Tenney's face through branches of pink dog-bane and over masses of poison-ivy. **1935** M. DE LA ROCHE *Young Renny* xxvi. 265 Bright-coloured tendrils of poison ivy stretched toward their path. **1939** 'B. GRAY' *Miss Dynamite* xvi. 179 So this is the charming little prairie flower that Norman's fallen in love with!.. Primrose, my foot! Her name's Poison Ivy! **1963** W. BLUNT *Of Flowers & Village* 29 We mayn't have these growing wild in England, nor the American poison ivy. **1971** *Rhodora* LXXIII. 76 More than 350,000 cases of poison-ivy dermatitis are estimated for the United States per year. **1976** F. GREENLAND *Misericordia Drop* II. viii. 138 Those amiable characters, my personal poison ivy, who so conscientiously compile our Code of Procedure. **1743** J. CLAYTON *Flora Virginica* 33 *Rhus*... Poison-Oak. **1905** G. E. COLE *Early Oregon* 29 Having been poisoned with poison oak so that I was completely blind, the others advised me to return. **1958** G. A. PETRIDES *Field Guide to Trees & Shrubs* 81 Some authorities believe differences between the several forms of Poison-oak and Poison-ivy are inconsequential. **1971** *Rhodora* LXXIII. 523 As the finer particles become less prevalent, the soil becomes more conducive to the growth of poison-oak. **1881** F. OATES *Metabele Land & Victoria Falls* xi. 243 The 'poison plant', growing low, and bearing a yellow plum-like fruit, was gathered on one occasion near the waggon-track. **1927** J. MASEFIELD *Sard Harker* III. 121 Dangling from the boughs, there were strings of withered poison-ivy... He dodged the poison-plant. **1965** Poison-plant [see *poison-bush*]. **1817** A. EATON *Man. Bot.* 34 *Rhus... vernix*, (poison sumach) glabrous panicle few-flowered. **1820** J. C. GILLELAND *Ohio & Mississippi Pilot* 261 Sumach... Most common in bottoms that are rich or at least moderately so... *R*[*hus*] *pumilum* (poison sumach). **1832** W. D. WILLIAMSON *Hist. State Maine* I. 118 The poison Sumach occurs in the western, but very seldom, if ever, in the eastern part of the State. **1901** C. T. MOHR *Plant Life Alabama* 600 Poison Sumach, Poison Elder... Alleghenian, Carolinian, and Louisianian areas. **1978** *Washington Post* 4 Aug. (Weekend Suppl.) 27/2 Poison ivy, poison sumac, and some species of baneberry have white fruits and are poisonous. **1709** J. LAWSON *New Voyage to Carolina* 101 The Poison Vine is so called, because it colours the Hands of those who handle it. **1803** A. ELLICOTT *Jrnl.* viii. 212 My journey up the river was disagreeable and painful, by being blistered by the rhus radicans (poison vine) from head to feet. **1891** M. E. RYAN *Told in Hills* II. i. 24 Here and there a poison-vine flashed back defiance under its crimson banners. **1935** *Yale Review* Sept. 174, I hear them [*sc.* horses] snortin' up the land where the pizen-vines grow around the sycamore stumps. **1856** L. J. F. JAEGER *Jrnl.* 20 Sept. in *Publ. Hist. Soc. S. Calif.* (1928) XIV. 128, 2 of the mules died at the Tinajas Altas—I think they ate some of the poison weed also.

B. *adj.* **2.** Wicked, dangerous; hateful, objectionable. *U.S. dial.*

1839 C. F. BRIGGS *Adventures of Harry Franco* I. 18 'I presume there's no occasion for hurrying,' said the driver. 'Yes there is though, you pisen critter,' said a passenger. **1850** 'M. TENSAS' *Odd Leaves from Life of Louisiana 'Swamp Doctor'* 152 Lizey Johnson's middle darter, Prinsanna,..left her husband in the state of Georgy, and kum to Luzaanny an' got marred to a nother man, the pisen varmint, to do sich as that and her own laful husband. **1880** 'MARK TWAIN' *Tramp Abroad* 225 B'long to a *church*! Why boss he's ben the pizenest kind of a Free-will Babtis' for forty year. They ain't no pizener ones 'n' what *he* is.

C. *adv.* Intensely, extremely. Chiefly *U.S. dial.*

1840 C. F. HOFFMAN *Greyslaer* I. 61 The night was pison cold, I tell ye. **1884** 'MARK TWAIN' *Huck. Finn* xxvii. 275 The funeral sermon was very good, but pison long and tiresome. **1892** R. L. STEVENSON *Let.* 31 Jan. in *Wks.* (1923) XXXIII. 23 This is a poison bad world for the romancer, this Anglo Saxon world. **1894** 'MARK TWAIN' *Pudd'nhead Wilson* xiv. 194 You's got to be pison good, en let him see it. **1926** in H. Wentworth *Amer. Dial. Dict.* (1944) 464/2 Pizen-neat.

poison, *v.* Add: **4. b.** *Chem.* Of a substance: to reduce or destroy the activity of (a catalyst, or occas. an electrode). Cf. *POISON *sb.* 2 c.

1913 in C. ELLIS *Hydrogenation Oils* (1914) 311 The use of chlorine would 'poison' the catalyst. **1921** G. G. HENDERSON *Catalysis in Industr. Chem.* iv. 72 Infinitesimal quantities of chlorine, bromine or iodine absolutely poison the metal, the presence of even a minute trace of bromine in phenol, for instance, preventing the latter being changed into cyclohexanol. **1937** *Jrnl. Inst. Electr. Engineers* LXXX. 198/2 An antimony electrode..lends itself particularly well to the recording of hydrogen-ion concentration, since this electrode is 'poisoned' by very few substances. **1965** H. H. WILLARD et al. *Instrumental Methods Chem. Analysis* (ed. 4) xxii. 588 The quinhydrone electrode is quickly prepared, develops its potential rapidly, and is not readily poisoned. **1972** *Times* 27 Sept. 20/3 Lead contaminants in fuel tend to 'poison' catalytic elements that help burn exhaust more completely in a converter mounted in the exhaust pipe. **1974** BANDTOCK & HANSON *Success in Chem.* xiv. 318 Vanadium (V) oxide is a reasonably efficient catalyst for the oxidation of sulphur dioxide and is not readily poisoned.

c. *Nuclear Sci.* To act as a poison in (a nuclear reactor or fuel). Also occas., to add a poison to (a reactor). Cf. *POISON *sb.* 2 d.

1945 H. D. SMYTH *Gen. Acct. Devel. Atomic Energy Mil. Purposes* viii. 80 Other fission products are being produced also. These consist typically of unstable and relatively unfamiliar nuclei so that it was originally impossible to predict how great an undesirable effect they would have on the multiplication constant. Such deleterious effects are called poisoning. **1948** C. PINCHER *Into Atomic Age* 38 Fragments from the split uranium 235 atoms collect in the slugs and..are said to 'poison' the uranium. **1960** WEHR & RICHARDS *Physics of Atom* xi. 328 This radioactivity is due principally to the fission products which poisoned the fuel element. **1968** F. KERTESZ *Lang. Nucl. Sci.* (Oak Ridge Nat. Lab. TM 2367) 23 Nuclear jargon is filled with gloomy, funereal terms: fuel elements are transported in coffins and reactors are poisoned to control them.

poisoner. Add: **b.** A cook, esp. for large numbers. *joc.* (*Austral.* and *N.Z.*).

1905 E. C. BULEY *Austral. Life in Town & Country* 23 The shearers' cook is always a competent man and supplies his clients with the best fare obtainable, utterly 'belying' the name of 'poisoner', usually bestowed upon him. **1936** A. RUSSELL *Gone Nomad* 14, I had to take my turn..as 'slushy' to 'Doughboy' Terry, the cook—'camp poisoner', as we affectionately called him. **1969** L. HADOW *Full Cycle* 208 'I'm not much good at cooking but I'll try.' 'Never you mind about that. Up north we've got the best poisoners in the country.'

poisoning, *vbl. sb.* Add: **b. blood-poisoning:** see *BLOOD *sb.* 19.

poisonous, *a.* Add: **2.** Also in trivial use, unpleasant, nasty.

1906 [in Dict.]. **1912** E. PUGH *Harry the Cockney* xi. 121 Foolish habit to think at any time, Weaver. But to think on an empty stomach—it's poisonous. Poisonous! **1916** 'TAFFRAIL' *Pincher Martin* xii. 225, I thought the weather was absolutely poisonous. **1929** P. GIBBS *Hidden City* xxxix. 189 It's something to do with that poisonous little beast Benito..the boy she dances with.

poisonwood. Add: Also *attrib.* (Later examples.)

1930 R. MACAULAY *Staying with Relations* ix. 127 There ..was..the nettle that one chews when one has inadvertently been spattered by the milky juice of the poison-wood tree. **1965** F. KNEBEL *Night of Camp David* xv. 248 A crescent moon perched above the thicket of gumbo limbo and poisonwood trees.

Poisson (pwasoṅ). *Math.* [The name of S. D. *Poisson* (1781–1840), French mathematician and physicist.] **1.** Used, chiefly *attrib.*, with reference to a discrete frequency distribution defined by $e^{-m}m^x/x!$, which gives the probability of x events occurring, m being its only parameter; it has mean m and variance m^2, and is appropriate if the events occur independently and there is no upper limit to their number, so that the distribution is the limit of the binomial distribution as the number of trials increases, the probability of success at each decreases, and the average number of successes tends to m; so *Poisson('s) approximation, distribution, form, law*, etc.; *Poisson-distributed* adj. Also passing into adj. (= *POISSONIAN) and used predicatively. [Described in Poisson's *Recherches sur la Probabilité des Jugements* (1837).]

1911 *Ark. för Matem., Astr. och Fysik* VII. xvii. 8, I shall generally understand with Poisson's theorem the expressions..giving the probability for obtaining in s trials with variable chances in all m white and n black balls (where $m + n = s$). **1914** *Biometrika* X. 36 (*heading*) On the Poisson law of small numbers. **1919** G. U. YULE *Theory Statistics* (ed. 5) 372 [This] may be termed Poisson's limit to the binomial. **1922** *Ann. Appl. Biol.* IX. 331 When the statistical examination of these data was commenced it was not anticipated that any clear relationship with the Poisson distribution would be obtained. *Ibid.* 334 The curves strongly suggest that the departures in these data from the Poisson samples were not..systematic. **1928** [see *NORMAL *a.* 2 e]. **1931** L. H. C. TIPPETT *Meth. Statistics* ii. 34 This is known as Poisson's Limit to the Binomial, the Poisson Series, or as the Law of Small Numbers. **1939** H. JEFFREYS *Theory of Probability* ii. 75 Put $x = r/n$ and let n tend to infinity; then the law tends to the Poisson form. **1948** H. E. FREEMAN et al. *Sampling Inspection* xvii. 185 For such small values of n the Poisson approximation is not adequate. **1950** W. FELLER *Introd. Probability Theory* I. vi. 119 A radioactive substance emits α-particles, and the number of particles reaching a given portion of space during time t is the best-known example of random events obeying the Poisson law. *Ibid.* xvii. 367 In the Poisson process the probability of a change during $(t, t + h)$ is independent of the number of changes during $(0, t)$. **1954** [see *ERLANG 1]. **1958** CONDON & ODISHAW *Handbk. Physics* I. xii. 155/2 The discrete distribution.. termed the Poisson exponential distribution, is (with the normal, and binomial distributions) one of the three principal distributions of probability theory. **1966** *McGraw-Hill Encycl. Sci. & Technol.* X. 631/2 If p is so small that the mean np is of the order of unity in any given application, Bernoulli's distribution is then approximated by Poisson's law. **1968** P. A. P. MORAN *Introd. Probability Theory* iii. 162 We suppose that customers arrive at a servicing counter in a Poisson process with mean λ, i.e. the number arriving in any interval of length T has a Poisson distribution with mean λT, and the numbers arriving in different intervals are independent. **1971** J. B. CARROLL et al. *Word Freq. Bk.* p. xxxvi, The remaining entries show that..9,436 [words] would be expected not to appear at all in the AHI Corpus, but that 3,826 would appear once, 776 would appear twice, 105 would appear 3 times, 10 would appear 4 times, and 1 would appear 5 times. (These numbers are predicted by the Poisson distribution.) **1976** E. J. DUDEWICZ *Introd. Statistics & Probability* iii. 56 Suppose that X is binomial with parameters n and p... Then X is approximately Poisson with $\lambda = np$. **1979** *Nature* 15 Feb. 533/1, n_i, the number of aberrations in the i-th culture, is Poisson-distributed.

b. *ellipt.* for *Poisson distribution.*

1962 S. R. CALABRO *Reliability Princ. & Pract.* vi. 65 If the expected number of failures..is substituted in the Poisson, then it is possible to calculate the probability of 0, 1, 2, 3, etc., failures. **1975** R. M. BETHEA et al. *Statistical Methods for Scientists & Engineers* iii. 57 We can estimate the probability of getting less than two adverse reactions using the Poisson as follows.

2. Special Comb.: **Poisson bracket,** a function $[u, v]$ of two dynamical variables $u(p_1, p_2, \ldots p_n, q_1, q_2, \ldots q_n)$ and $v(p_1, p_2, \ldots p_n, q_1, q_2, \ldots q_n)$ equal to $\sum_{i}\left(\frac{\partial u}{\partial q_i}\frac{\partial v}{\partial p_i} - \frac{\partial u}{\partial p_i}\frac{\partial v}{\partial q_i}\right)$; **Poisson's equation** [discussed by Poisson in *Nouveau Bull. des Sci. par la Soc. philomath. de Paris* (1813) III. 390], the generalization of Laplace's equation produced by replacing the zero of the right hand side by a constant, or, more generally, by a specified function of position; **Poisson's ratio** [discussed by Poisson in *Ann. de Chim. et de Physique* (1827) XXXVI. 385], the ratio of the proportional decrease in a lateral measurement to the proportional increase in length in a sample of material that is elastically stretched.

1904 E. T. WHITTAKER *Treat. Analytical Dynamics* xi. 309 If ϕ and ψ are two integrals of the system, the Poisson-bracket (ϕ, ψ) is constant throughout the motion. **1960** DICKE & WITTKE *Introd. Quantum Mech.* v. 86 The Poisson bracket provides a powerful tool in formulating quantum theory. **1976** MATHEWS & VENKATESAN *Textbk. Quantum Mech.* 351 Canonically conjugate coordinate-momentum pairs are..characterized by a unit value for the Poisson bracket. [**1872** *Trans. R. Soc. Edin.* XXVI. 71 If P be the potential at ρ, and if r be the density of the attracting matter, &c., at ρ, $\nabla\sigma = \nabla^2 P = 4\pi r$ by Poisson's extension of Laplace's equation.] **1873** J. C. MAXWELL *Treat. Electr. & Magn.* I. I. ii. 80 This equation, in the case in which the density is zero, is called Laplace's Equation. In its more general form it was first given by Poisson... We may express Poisson's equation in words by saying that the electric density multiplied by 4π is the concentration of the potential. **1916** F. B. PIDDUCK *Treat. Electr.* iii. 61 This becomes..$\Delta V = -4\pi\rho$, which is known as Poisson's equation. **1971** C. R. CHESTER *Techniques in Partial Differential Equations* iii. 87 The nonhomogeneous potential equation, $\nabla^2 u = F(x, y, z)$ is called Poisson's equation. **1886** J. D. EVERETT *Units & Physical Constants* (ed. 2) v. 62 The following values of Poisson's ratio have been found. **1930** *Engineering* 11 Apr. 465/1 The modern theory of the elasticity of isotropic materials makes use of a number of physical constants, all of which are definitely related to Young's Modulus E and Poisson's ratio $\eta = 1/m$ the latter of which is sometimes known as the 'stretch-squeeze' ratio. **1966** C. R. TOTTLE *Sci. Engin. Materials* vii. 153 An orthorhombic crystal can thus be defined by nine independent constants, three elastic moduli, three moduli of rigidity, and three values of Poisson's ratio.

Poissonian (pwasōu·niăn), *a.* [f. prec. + -IAN.] Of, pertaining to, or being the Poisson distribution.

1914 *Ark. för Matem., Astr. och Fysik* IX. xxv. 16 Repeating the experiment r times we obtain respectively $m_1, m_2, m_3, \ldots, m_r$ white balls in these r sets. These numbers form what is called a Poissonian series. **1951** *Jrnl. R. Statistical Soc.* B. XIII. 168 If they call for service in a random manner the input-process (while q customers are waiting or being served) will be Poissonian, but will depend on the value of q. **1972** *Science* 2 June 1034/1 The Poissonian probability times the efficiency of plating gave an actual distribution of ≤0·3 cell per well. **1976** *Physics Bull.* Mar. 111/1 If T was short compared with the inverse spectral linewidth of a random source, a geometric distribution was obtained while if T was long the photon counting distribution became Poissonian due to averaging of the photon arrivals.

Poitevin (pwa·tėvin, ‖ pwatəvaṅ), *sb.* and *a.* Also 7 **Poictevin.** [Fr.] **A.** *sb.* **a.** A native or inhabitant of Poitou, an ancient province of west central France roughly corresponding to the modern *départements* of Vienne, Deux-Sèvres, and Vendée, or of its capital Poitiers, now capital of Vienne. **b.** The French dialect of Poitou or Poitiers. **B.** *adj.* Of, pertaining to, or characteristic of Poitou or Poitiers or the dialect spoken there.

1642 J. HOWELL *Instructions for Forreine Travell* x. 127 The law Norman useth to contract many words..and the Poictevin will mince the word. **1653** [see *BRETON *sb.* and *a.*]. **1866** C. M. YONGE *Prince & Page* i. 7 A Poitevin, a falconer at Kenilworth. **1880** —— *One Will & Three Ways* iii, in *Bye-words* 65 Stories..of Poitevin castles won by escalade. **1934** M. K. POPE *From Latin to Mod. French* ii. 18 This sound was not, however, diphthongised, cf. Poitevin *amar*. *Ibid.* xvii. 211 The southern border dialects (e.g. Poitevin). **1957** A. DUGGAN *Devil's Brood* vi. 68 Geoffrey had been born..in..1158... He grew up to be another fair handsome northerner, obviously more Norman than Poitevin. **1968** F. WHITE *Ways of Aquitaine* iv. 52 These delicate Poitevin carvings must be regarded primarily as adornment of the churches. **1974** P. RICKARD *Hist. French Lang.* ii. 33 Poitevin, which shows close affinity with Occitan up to and including the tenth century, can by the twelfth century be considered essentially a dialect of the *langue d'oïl*. *Ibid.* vii. 127 Non-standard constructions, too, may be used; for instance.. the Poitevin *les enfants sont après jouer* (= *en train de jouer*). **1976** *Sat. Rev.* (U.S.) 30 Oct. 25 The population [of St. Barthélemy] is 90 per cent Caucasian, descendants of Normans, Bretons, Poitevins, and the Swedes owners of St. Barts for a century.

poitrel. (Further examples.)

1805 C. JAMES *Mil. Dict.* (ed. 2) *Poitrel*, armour for the breast of a horse. It is generally written *poitrail*. **1824** S. R. MEYRICK *Critical Inquiry into Antient Armour* III. Gloss. s.v. Pectorale, The poitrail, a steel plate for the protection of a horse's chest. **1830** J. SKELTON *Engraved Illustrations Antient Arms & Armour* Pl. LIV (fig. 2), The fleur-de-lis is seen on the horse furniture, twice on the croupière-base, and as an armorial bearing on that of the poitral. **1918** E. S. FARROW *Dict. Mil. Terms* 459 *Poitrail*, that portion of the horse armor which covers the breast, fitted either with hinges or like a flounce. **1920** G. F. LAKING *Rec. European Armour & Arms* III. xxii. 178 For the head there was..the chanfron, for the neck the crinet, for the chest the poitrel.

poitrinaire. (Earlier and later examples.)

1856 *Sat. Rev.* 25 Oct. 578/2 As a *poitrinaire*, he was the most devoted, extravagant, sentimental of lovers and husbands. **1969** *Sunday Tel.* 12 Jan. 15/8 Although she cannot help suggesting a *poitrinaire* in the most comfortable health, deft make-up and still more a skilful modification of the voice suggested the growing physical weariness.

poitrinal (poi·trinăl). Now *Hist.* or *arch.* [ad. Fr. *poitrinal*: see PETRONEL.] **1.** = POITREL.

1786 [see *PECTORAL *sb.* 1 c]. **1869** J. R. PLANCHÉ *Catal. Meyrick Coll. exhibited at S. Kensington Museum* p. x, The barding of the horse is also very fine, consisting of chamfron, crevette, and poitrinal, all fluted and in fine condition. **1918** E. S. FARROW *Dict. Mil. Terms* 459 *Poitrinal*, in ancient armor, the horse's breastplate, formed of metal plates, riveted together, as a covering for the breast and shoulders.

2. = PETRONEL.

1824 [see PETRONEL]. **1829** *Archæologia* XXII. 86 The President Fauchet..introduces to our notice a piece called a Petronel or Poitrinal, because it was rested on the poitrine or chest..and thence fired.

‖**pokal** (pokā·l). [G., ad. L. *poculum*, a drinking-cup.] A large German glass tankard, often with a lid. Also *transf.*

1868 C. G. LELAND *Hans Breitmann's Party* 11 How stately rode der Breitmann oop!—how lordly he kit down? How glorious from de great *pokal* he drink de bier so prown! **1869** G. A. SALA in *More Yankee Drolleries* 25 'Pokal', drinking-cup: in this case the large glass pint beer-mug used in America. **1950** *Chambers's Encycl.* VI. 421/1 The great tankards and pokals were intended for display; their size was so great that they were too heavy to drink from. **1969** *Canadian Antiques Collector* Oct. 5/1 Two ceremonial cups or pokals, late 17th century, silver gilt. **1974** *Encycl. Brit. Micropædia* VI. 891/2 (*caption*) Milk glass pokal, German, 17th century.

pok-a-tok: var. *POK-TA-POK.

poke, *sb.*[1] Add: **1. a.** (Further examples.)

1910 R. SERVICE *Trail of '98* 347 The girl will pry him loose from his poke. **1922** G. PRINGLE *Tillicums of Trail* 250 It wasn't safe to come out by way of Skagway with your gold,..you were likely to be relieved of your 'poke' by desperados. **1935** A. J. CRONIN *Stars look Down* I. ix. 68 He had pie, too, in his poke. **1948** C. W. HOLLIDAY *Valley of Youth* 144 A miner might come into a store for provisions with no money, but a little poke of gold dust. **1966** *Indians, Eskimos, & Aleuts of Alaska* (U.S. Bureau of Indian Affairs) 12 (*caption*) These villagers are carrying a sealskin 'poke' filled with seal oil. **1976** *Islander* (Victoria, B.C.) 15 Aug. 12/2 Nearby lay an empty buckskin 'poke' such as early miners favored for their gold dust. **1978** *Guardian* 14 Aug. 7/2 You may also find yourself at a temporary disadvantage if, after buying several items from a shop, the young lady assistant asks if you would 'like a poke'.

d. A purse or wallet; a pocketbook. *N. Amer. slang.*

1859 G. W. MATSELL *Vocabulum* 68 *Poke*, a pocket; a purse. **1883** *Echo* 25 Jan. 2/3 The poke, which a pick-pocket glories in having appropriated, is the Saxon bag or purse. **1908** J. M. SULLIVAN *Criminal Slang* 18 *Poke of leather*, a pocketbook. **1931** 'D. STIFF' *Milk & Honey Route* 211 *Poke*, a leather wallet. **1939** J. O'HARA *Pal Joey* 50 There I was with only about $85 in my poke. **1953** 'W. BURROUGHS' *Junkie* iv. 48 He took a crumpled mass of bills from his pocket and counted out eight dollars... 'Had it in his pants pocket. I couldn't find a poke.' **1976** 'TREVANIAN' *Main* (1977) vi. 123, I notice his wallet's half out of his pocket...it comes to me that I might as well lift his poke... So I reach over and pull it out.

e. A roll of bank-notes; money. *slang.*

1926 J. BLACK *You can't Win* xiv. 190 My hand was on the big fat 'poke'. **1933** E. SEAGO *Circus Company* 295 *Poke*, money. **1965** L. J. CUNLIFFE *Having it Away* iv. 38 It's a very satisfying feeling knowing you can put your finger on a bit of poke. (Which is more slang for money: get it, poke, loot, poppy—any of them will do!) **1974** *Evening News* (Edinburgh) 8 Oct. 3/2 Colgan asked him: 'Have you got your poke?' obviously referring to the money.

poke, *sb.*[3] Add: **1.** Delete *obs.* from '*obs. slang*, A blow with the fist' and add 'esp. in phr. *to take a poke at*'. (Later examples.) Also colloq. phr. *better than a poke in the eye* (and variants), used of something minimally desirable. Also *fig.*

1852 GEO. ELIOT *Let.* 4 Dec. (1954) II. 71 'Then,' he said..'Here are those "Letters from Ireland" which I hope will be something better than a *poke in the eye*.' **1936** J. STEINBECK *In Dubious Battle* viii. 120 'They got those cops here quick,' said Burke. 'I'd like to take a poke at a few of 'em.' **1941** BAKER *Dict. Austral. Slang* 55 *Poke*, to hit a person. Also, 'a poke': a blow with the fist. **1944** J. CARY *Horse's Mouth* xvi. 81 Anarchists who love God always fall for Spinoza because he tells them that God doesn't love them. This is just what they need. A poke in the eye. To a real anarchist, a poke in the eye is better than a bunch of flowers. It makes him see stars. **1956** B. HOLIDAY *Lady sings Blues* (1973) ii. 24 She tried to get at me. I took a poke at her, and down the stairs she went. **1969** *Listener* 10 Apr. 478/1 What sort of salute, I wondered, amounted to a poke in the eye? On this occasion compromise was reached: it was agreed that the occasion should be marked by a second broadcast of my *Salute to Stalingrad*. **1974** *Bulletin* (Sydney) 6 July 44 An Australian way of expressing ecstasy is to say: 'It's better than a poke in the eye with a burnt stick.' **1976** *N.Y. Times Mag.* 10 Oct. 111/4 Better than a poke in the eye with a sharp stick.

b. poke-out, (*b*) *slang*, a bag of food handed to a beggar; a lunch; (*c*) *slang* (see quot. 1960).

1894 'J. FLYNT' in *Century Mag.* Mar. 713/2 He returned with a 'poke-out' (food given at the door). **1907** J. LONDON *Road* 12, I could 'throw my feet' with the next one when it came to 'slamming a gate' for a 'poke-out' or a 'set-down'. **1918** H. A. VACHELL *Some Happenings* i. 4 [He] finished what was left of a 'poke out' (cold food) handed to him by a good Samaritan. **1936** *New Republic* 15 July 289/1 Sympathetic women will often cook a meal for tramps,..and 'lumps' or 'poke-outs' are possible at any time during the day. **1960** WENTWORTH & FLEXNER *Dict. Amer. Slang* 399/2 *Poke-out*,..2 An outdoor dinner cooked over wood or charcoal; a gathering for the purpose of preparing and eating such a meal; any long hike or camping trip which includes such meals. **1964** J. L. KORNBLUH *Rebel Voices* 407/2 *Pokeout*, handout.

c. *Cricket.* A batting-stroke made by jabbing at the ball.

1853 F. GALE *Public School Matches* 54 Sticker gets his runs by quiet little pokes one at a time. **1896** W. J. FORD in *Badminton Mag.* Sept. 278 Besides..'the draw', there was another weapon forged for the armoury of him for whom leg-hitting was not—viz. the 'Cambridge Poke', so called, I believe, in contemptuous irony. **1960** J. FINGLETON *Four Chukkas to Australia* xvi. 133 He was confusing the cut with the 'poke', a disastrous nibbling by so many Englishmen.

d. An act of sexual intercourse; also, a woman with whom one has sexual intercourse, a 'lay'; = *FUCK *sb.* 1. *slang.*

1902 FARMER & HENLEY *Slang* V. 242 *Poke*,..an act of coition. **1958** N. LEVINE *Canada made Me* ii. 82 When I met her I only want a poke. Then she tell me a baby made. **1968** H. C. RAE *Few Small Bones* II. i. 77 'Caroline', said Derek..'wouldn't make a good poke for a blind hunchback.' **1970** L. MEYNELL *Curious Crime* xii. 160 Landladies can nearly always be paid in kind. Services in lieu of rent. A poke a night. **1977** *Listener* 11 Aug. 184/4 Turning a series of squalid pokes into a series of honourable combats.

e. *fig.* Power, horsepower. *slang.*

1965 R. T. BICKERS *Hellions* vi. 69 With all that extra poke under the bonnet. **1977** *Drive* Mar.–Apr. 54/2 The Scirocco gives a worst figure of 28 mpg, using all the poke its free-revving engine will deliver. **1979** *Sunday Mail Mag.* (Brisbane) 1 July 3/5, I expect you'd prefer something with a bit more poke. A Ferrari say, or an Aston Martin.

2. (Earlier and later examples.)

1809 E. A. KENDALL *Travels through Northern Parts of U.S.* II. 198 A hog..by some mischance had turned his poke, so that his throat was squeezed into one of the acuter angles. **1949** R. J. SIM *Pages from Past* 105 Such a rig is known as a 'poke'. It is put on the neck of a critter with fence-jumping inclinations. **1956** W. R. BIRD *Off-Trail in Nova Scotia* viii. 220 And here were some sheep, too, with pokes. **1969** K. M. WELLS *Owl Pen Reader* I. 67 A poke is supposed..to make it impossible for any living thing to get through even an ordinary wire fence.

b. *transf.* A collar. *slang.*

1908 'O. HENRY' *Man Higher Up* in *Gentle Grafter* 147 With only feetwear and a dozen 15½ English pokes in his shopping bag. **1924** *Truth* (Sydney) 27 Apr. 6 *Poke*, a collar.

4. *attrib.* and *Comb.*, as **poke-check** *Ice hockey*, a defensive play made by poking the puck off an opposing player's stick; hence as *v. intr.*; **poke-checking** *vbl. sb.*

1945 R. FONTAINE *Happy Time* 45 Frank Nighbor, Canada's immortal poke-check genius. **1964** *Maclean's Mag.* 2 May 46/1 To me some of the most fascinating moves in hockey..are poke-checks, or well timed interceptions, or expeditions of forechecking. **1966** *Hockey News* (Montreal) 1 Jan. 13/2 He poke-checks and sweep-checks like the oldtimers. **1963** A. O'BRIEN *Headline Hockey* 60 At that point the defenceman will likely resort to poke-checking.

poke, *sb.*[4] Add: **2. a.** For *P. decandra* substitute *P. americana*. (Later examples.)

1945 *Chicago Tribune* 13 May VII. 1/3 Opal had found the first tightly curled leaves of poke, the best known of all Ozark greens. **1977** LEWIS & ELVIN-LEWIS *Med. Bot.* iv. 90/1 Poke..has long been a favorite spring potherb in the southern United States.

3. poke-berry, substitute for def.: the dark

purple berry of *Phytolacca americana*, or the plant itself; also *attrib.*; (earlier and later examples); **poke-greens**, the young leaves of poke-weed used as a vegetable; **poke-salad, -salat, -sallet**, the young leaves of poke-weed used as a salad; **poke-weed** (earlier and later examples).

1774 P. V. FITHIAN *Jrnl.* 15 Oct. (1900) 269 To Day Harry boil'd up a Compound of Poke-Berries, Vinegar, Sugar &c to make a red Ink or Liquid. **1834** W. A. CARRUTHERS *Kentuckian in N.Y.* II. 215 His face looks like it was boiled in poke-berry juice and indigo. **1911** G. S. PORTER *Harvester* xiii. 252 Pokeberry!.. Roots bring five cents a pound. Good blood purifier. *Ibid.* xv. 334 A few pokeberry plants for the colour. **1974** A. DILLARD *Pilgrim at Tinker Creek* xiv. 249 A skin-colored sandstone ledge beside me was stained with pokeberry juice. **1848** *Knickerbocker* XXXI. 222 The southern negro will dance after eating his poke-greens and bacon. **1938** M. K. RAWLINGS *Yearling* i. 12 There were poke-greens with bits of white bacon buried in them. **1881** J. C. HARRIS *Uncle Remus* 197, I got mustard, en poke salid, en lam's quarter in dat baskit. **1892** Poke salad [see *CALLALOO]. **1913** H. KEPHART *Our Southern Highlanders* xiii. 282 This poke salat eats good. **1947** *Sun* (Baltimore) 20 May 18/3, I was introduced to poke, as poke-salat, by a Southern Maryland family. **1751** *Gentl. Mag.* July 306/2 Tho' the Phytolacca be known to almost every one in America, by the name of pokeweed,.. yet I think it proper..to give a description of it. **1886** M. ARNOLD *Let.* 29 July (1895) II. 341 The pokeweed (*Phytolacca*) is, I think, American too. **1945** *New England Homestead* 13 Oct. 6/4 Pokeweed, huckle and blueberries, wild roses, bittersweet and hazelnut bushes are also appreciated. **1976** *National Observer* (U.S.) 3 July 15/2 A beautiful black woman..supplied some delectable recipes (several ways to serve pokeweed, for instance).

poke, *v.*[1] Add: **1. e.** *Cricket.* To hit (the ball) with a jabbing stroke.

1836 *New Sporting Mag.* Oct. 360 He was very successful in poking leg stump balls for one run. **1862** J. PYCROFT *Cricket Tutor* 45 See, he is longing to poke the ball to the on-side. **1872** *Baily's Mag.* Aug. 166 The Eton men hit with freedom..the same bowling that the day before.. they only poked or played with tameness and hesitation.

f. To have sexual intercourse with (a woman). *slang.*

1868 *Index Expurgatorius of Martial* 27 Saufeia,.. though she was willing to be poked, would not enter the bath with the poet for decency's sake. *c* **1888–94** [see *GET *v.* 38 h]. **1962** J. BRAINE *Life at Top* ix. 129, I wanted to poke Lucy so I poked her. **1967** L. MEYNELL *Mauve Front Door* ii. 24 Your uncle was..as randy as a goat... He poked them everywhere. **1971** R. FALKIRK *Chill Factor* xiv. 149 Are you out of your mind poking an Icelandic girl while you're on this sort of mission? **1975** N. LUARD *Robespierre Serial* xvi. 144 They're far from sure she's the one this GI poked.

g. To hit, strike (someone). *colloq.*

1906 *Dialect Notes* III. 122, I *poked* him on the nose. **1959** S. J. BAKER *Drum* 136 *Poke*, v., to hit a person with the fist.

h. *Baseball.* To hit.

1908 *Atlantic Monthly* Aug. 229/1 Sharky poked a bingle. **1951** in Wentworth & Flexner *Dict. Amer. Slang* (1960) 399/2 Jackie Robinson poked a pitch out of the park.

4. b. *Cricket.* To make pokes at the ball (see *POKE *sb.*[3] 1 c). Also const. *about.*

1851 J. PYCROFT *Cricket Field* vii. 114 Mere stopping balls and poking about in the blockhole is not cricket. **1899** E. V. LUCAS *Open Road* 146 (The Cricket Ball Sings) Perish the muff and the little tin Shrewsbury, Meanly contented to potter and poke. **1906** A. E. KNIGHT *Compl. Cricketer* viii. 268 His drive is a clean honest lift straight from the shoulders; he never pokes, 'puddling about his crease'. **1927** M. A. NOBLE *Those 'Ashes'* 193 His usual aggression was missing and he poked about, mistiming and apparently being unable to make a clean stroke.

c. Of a man: to have sexual intercourse with a woman. *slang.*

1973 *Nation Rev.* (Melbourne) 24–30 Aug. 1417/2 Working class morality where the male never 'pokes' after marriage but lusts away in obscenity and dirty jokes.

poker, *sb.*[1] Add: **6. b.** = *FUCKER. *slang.*

1879–80 *Pearl* (1970) 214 I've been told by jokers That the ladies they do all agree that he's the prince of pokers.

c. *Cricket.* A batsman who 'pokes' (*POKE *v.*[1] 4 b).

1888 A. G. STEEL in Steel & Lyttelton *Cricket* iii. 143 But to the poker, the man who refuses to do anything but stick his bat in front of the wicket..the high-dropping full-pitch is an excellent ball.

7*. = POKE *sb.*[3] 2. *rare.*

1805 T. B. HAZARD *Nailer Tom's Diary* (1930) 260/2 Put Poker on one of my oxen.

8. *poker-stiff, -straight* adjs.; **poker back**, (*a*) a perfectly straight back; (*b*) *Path.* (see quot. 1973); **poker spine** *Path.* = *poker back* (*b*) above.

1931 M. ALLINGHAM *Look to Lady* xxvi. 276 A single slim aristocratic figure, with the unmistakable poker back of the old regime. **1960** H. EDWARDS *Spirit Healing* x. 87 The healing of certain troubles, as with poker-back spines. **1973** TAYLOR & COTTON *Short Textbk. Surg.* (ed. 2) xl. 539 [In ankylosing spondylitis] the normal spine curvatures become replaced by a single kyphosis, occasionally so acutely angled that the patient's back becomes horizontal (poker back). **1917** *Brit. Med. Jrnl.* 30 June 860/1 Dr. John Drummond (Liverpool) asks for suggestions as to treatment in a case of poker spine in a man 30 years of age... The back is now immobile. **1960** S. PLATH *Colossus* (1967) 27 Rigged poker-stiff on her back. **1962** 'K. ORVIS' *Damned & Destroyed* xi. 77 Frankie's back was poker-stiff. **1966** J. S. COX *Illustr. Dict. Hairdressing* 119/1 *Poker straight*, without a vestige of curl. **1979** N. FREELING *Widow* ii. 3 The hair was poker-straight.

poker, *sb.*[4] Add: (Earlier examples.) Also *fig.*

1836 J. HILDRETH *Campaigns Rocky Mts.* I. xv. 128 The M— lost some cool hundreds last night at poker. **1842** *Knickerbocker* XX. 305 Squeezing a great deal of boisterous amusement out of a game of 'poker'. **1978** *Time* 3 July 42/3 The Justice Department was in no mood to be bluffed, even by troubled steelmakers, and talks dragged on and on in a months-long game of high-stakes political poker.

b. *poker-deck, -game* (examples), *hand, player* (examples), *table*; **poker chip**, a chip [CHIP *sb.*[1] 2 d] used as a stake in poker; **poker dice**, (*a*) dice with the representation of a playing card on at least two of their faces; (*b*) a dice game, played with either poker or regular dice, in which the thrower aims for combinations which would constitute a winning hand in poker; **poker face**, an inscrutable face appropriate to a poker-player; a face in which a person's thoughts or feelings are not revealed; also, a person with such a face; hence as *v. trans.* (*rare*[−1]) to regard with a poker-face; **poker-faced** *a.* (cf. *PO-FACED *a.*); **poker machine** *Austral.*, a type of 'one-armed bandit' bearing card symbols; **poker patience**, a form of competitive patience the object of which is to form winning poker combinations in each row and column; **poker school**, a group of people meeting to play poker.

1879 *News & Press* (Cimarron, New Mexico) 20 Nov. 4/3 The toughest thing we have heard about any candidate in this section is that he got his poker chips cashed after he 'experienced religion'. **1929** WODEHOUSE *Mr. Mulliner Speaking* iv. 122 At the end of five minutes, Osbert was mildly surprised to find himself in possession of a smoking-cap, three boxes of poker-chips, some polo sticks, [etc.]. **1973** E. PACE *Any War* (1974) III. 189 He heard..no laughter, no rattle of poker chips. **1844** J. COWELL *Thirty Years among Players* 94 He was, apparently, quietly shuffling and cutting the poker-deck for his own amusement. **1874** Poker dice [see *DIE *sb.*[1] 1 a]. **1901** *Game of Poker Dice* 1 The only Implements required are Sets of The Poker Dice and Cups, according to the number of players. **1926** E. HEMINGWAY *Sun also Rises* I. vi. 43 Harvey had won two hundred francs from me shaking poker dice. **1975** D. BLOODWORTH *Clients of Omega* x. 87 Poker dice, of course, man..*Strip* poker. **1885** *Encycl. Brit.* XIX. 283/2 A good *poker face* is essential; the countenance should not betray the nature of the hand. **1919** G. A. MILLER *Prowling about Panama* xiii. 198 (*caption*) San Blas Indians have 'poker faces'. **1926** H. C. WITWER *Roughly Speaking* 243 His teeth clicked and he gave me a long, thoughtful look, but I poker-faced him and went on plugging my [switch-]board. **1934** E. O'NEILL *Days without End* I. 20 His features automatically assume the meaninglessly affable expression which is the American business man's welcoming poker face. **1950** G. B. SHAW *Buoyant Billions* III. 28 Sunday clothes and poker faces. No peace, no joy. **1974** 'J. MELVILLE' *Nun's Castle* i. 21, I..kept a poker-face. Inside, however, I was deeply distressed. **1976** P. DICKINSON *King & Joker* vii. 104, I hardly need say it to you, because you're such an old poker-face anyway, but..you have to..behave as though you are the only person who knows. **1923** *Nation* (N.Y.) 18 July 61 The picture of that poker-faced gentleman placidly smoking a Pittsburg stogie. **1949** *Time* 12 Sept. 20/1 The poker-faced fellow was putting up a terrific fight. **1973** D. WESTHEIMER *Going Public* ix. 134 'We'd send them a letter, see,' said Margo poker-faced. 'Telling them how to commit suicide.' **1932** T. S. ELIOT *Sweeney Agonistes* 18 What about that poker game? eh what Sam? What about that poker game in Bordeaux? **1957** *Times Lit. Suppl.* 13 Dec. 753/2 Ward politics, big poker-games, prostitution and murder. **1977** H. FAST *Immigrants* II. 97 He remembered such faces from poker games. **1935** *Encycl. Sports* 467/1 The object of the game [*sc.* poker patience] is so to place the cards as they are played that finally each row and each column will form a poker hand. **1963** G. F. HERVEY *Handbk. Card Games* 231 There are nine possible poker hands at which to aim. **1974** *Encycl. Brit. Macropædia* XIV. 623/1 A Poker hand consists of five cards. **1964** A. WYKES *Gambling* 330 Gamblers also managed to spend about $1,500,000 on 'poker machines' (a kind of slot machine that bears card symbols). **1973** *Sunday Mail* (Brisbane) 29 July 5/3 Canberra soon may be the first city in Australia to have poker machines in its hotels. **1976** *Daily Mirror* (Sydney) 14 Oct., Young poker-machine players should be given a warning about how much they could lose. **1912** 'SAKI' *Stampeding of Lady Bastable* in *Chronicles of Clovis* 55 He particularly wanted to teach the MacGregor boys.. poker-patience. **1932** R. FRASER *Marriage in Heaven* III. ii. 292 The whole party joined in a game of poker-patience. **1972** A. CHRISTIE *Elephants can Remember* v. 78 They played picquet, and poker patience with each other. **1844** J. COWELL *Thirty Years passed among Players in England & Amer.* 94 The cabin was entirely cleared.. with the exception of one of the poker players. **1912** M. NICHOLSON *Hoosier Chron.* 137 He had the reputation of being a poor poker player, but 'a good loser'. **1963** G. F. HERVEY *Handbk. Card Games* 237 Experienced poker players often say that what counts is not so much what they win as what the other players lose. **1949** J. R. COLE *It was so Late* 91 Men from the camp..make up a poker school. **1968** E. MCGIRR *Lead-Lined Coffin* iii. 129 Pope joined one of the large poker schools. **1977** A. C. H. SMITH *Jericho Gun* xi. 139 'Sorry to bust up the poker school.'..'I don't mind... I was two pounds ahead.' *a* **1861** T. WINTHROP *John Brent* (1862) 295 He sot his white head down to the poker-table, and stuck thar. **1930** J. DOS PASSOS *42nd Parallel* VII. 94 Pokertables piled with new silver dollars. **1977** *Times* 29 Aug. 6/2 Batesy was an old hand at marshalling her clients, bringing together seven to make up a full poker table.

poker, *v.* Add: **3.** *trans.* Of a verger, etc: to escort (a church dignitary) ceremoniously. Cf. POKER *sb.*[1] 3 b.

1924 C. LANG *Let.* in R. C. D. Jasper *G. Bell, Bishop of Chichester* (1967) iii. 36, I shall feel more free to laugh when I see you clothed in apron and gaiters and being pokered at Canterbury. **1975** *Theology* LXXV. 260 Hamling was also verger, and did all the old establishment things like pokering the preacher to the pulpit, and generally gave the services tone.

pokerish, *a.*[2] (Earlier examples.)

1827 *Massachusetts Spy* 21 Nov. (Th.), A patriarchal ram, who would fight anything but a pokerish looking ducking gun. **1833** H. BARNARD in *Maryland Hist. Mag.* (1918) XIII. 352, I feel quite pokerish in this region.

poker-work. Add: (Further examples.) Also *fig.* Hence **poker-worked** *a.* (in example applied to a design resembling poker-work).

1914 [see *GEORGIANISM]. **1929** E. BOWEN *Last September* ix. 104 Cushions..with poker-worked kittens. **1942** C. BARRETT *On Wallaby* x. 193 He..does poker-work with red-hot wire, finishing off with tiny colour sketches. **1958** L. DURRELL *Balthazar* v. 104 He has had an immense and vivid firescreen made for the flat..in poker-work. **1966** *Listener* 29 Dec. 959/2 The play relapses into what it looked as if it was going to satirize. By the end we are left with a stuffy, inhibiting piece of Victorian poker-work as a message. **1973** *Times* 31 July 14/8 She remembered well the famous Breton school of artists at Pont-Aven which included Gauguin. He made her a pair of 'sabots' in poker work. **1977** *Listener* 11 Aug. 171/3 An element of family creativity here, not too dissimilar to the Victorian samplers and poker-work.

pokey[1] (pōᵘ·ki). *slang* (chiefly *U.S.*). Also **poky**. [Alteration of *POGEY, prob. infl. by POKY *a.*[1]] Prison, gaol.

1919 C. H. DARLING *Jargon Book* 26 *Pokey*, a jail. **1929** D. RUNYON in *Hearst's International* Aug. 73/2 He hears riding rum is illegal and may land a guy in the pokey. **1947** *Daily Progress* (Charlottesville, Virginia) 24 June 6/2 They gave the police a list of the phone numbers to call if it became necessary to turn any of the old grads in the direction of the pokey. **1955** 'S. RANSOME' *Deadly Bedfellows* viii. 70 Instead of thanking him, you've threatened to throw us both into the pokey. **1957** M. MILLAR *Soft Talkers* 151 This isn't the Royal York Hotel, but it's better than a cell in the local pokey. **1965** 'D. SHANNON' *Death-Bringers* (1966) iv. 50, I find that our star sleuth..has..carted him off to the pokey. **1974** *Maclean's* (Toronto) Dec. 30/2 A number of revered figures sat out the Depression in the poky, because they fiddled with other people's money. **1976** *National Observer* (U.S.) 29 May 18/4 Were it possible to prosecute an actor for stealing scenes, *The Missouri Breaks* (United Artists) would land Marlon Brando in the pokey for life.

pokey[2], **pokie.** *Austral.* [Familiar corruption of POKER *sb.*[4]] = *poker machine* s.v. *POKER *sb.*[4] b.

1967 D. HORNE *Southern Exposure* 44 In the clubs of Sydney the poker machines ('the pokies') stand up in dozens and more beer flows than in a hotel. **1968** *TV Times* (Brisbane) 27 Nov. 6/2 In his unmarried days Henderson was surefire meat for bandits (the one-armed type). 'Never play the pokies now,' he says. **1969** *Telegraph* (Brisbane) 4 Jan. 6/2 He bought a beer and walked over to the nearest 'pokey' with the change from a £5 ($10) note. He put this through the machine and tripled his money. **1969** *Australian* 24 May 40/3 He painted a glowing picture of Melbournites banking their money or investing it in homes while the degenerate New South Welshmen frittered away their cash on the pokeys. **1976** *Sydney Morning Herald* 23 Sept. (Advt.), Entertainment... There are pokies, casino, disco, movies.

poking, *vbl. sb.* Add: **1. b.** Sexual intercourse. *slang.*

1968 J. SANGSTER *Foreign Exchange* i. 16 'There's no law against poking,' I said. 'There is when you poke a fifteen-year-old,' he said. **1978** J. I. M. STEWART *Full Term* xvi. 181 He was petting her future mother in the heather long before she was born. And later..another young hopeful went from petting to poking. **1979** L. MEYNELL *Hooky & Villainous Chauffeur* xiv. 187 Loverboy is going to be busy (pounds before poking any day).

poking, *ppl. a.* Add: **2. b.** *Cricket.* With a batting style characterized by 'pokes' (*POKE *sb.*[3] 1 c).

1836 *New Sporting Mag.* Oct. 360 A remarkably bad poking back player, with no hit in him at all. **1898** J. A. GIBBS *Cotswold Village* xi. 241 If only something could be done to..rid us of that awful nuisance the poking, time-wasting batsman, there would be little improvement possible.

‖**pok-ta-pok** (pǫ·k,ta,pǫ·k). Also **pok-a-tok.** [Maya.] The Maya name of the sacred ball game of Middle America, called *TLACHTLI by the Aztecs, which was played on a court as a religious ritual. The object of the game was to knock a rubber ball through a stone ring, using only the hips, knees, and elbows.

1959 *Times* 27 Apr. (Rubber Industry Suppl.) p. v/1

The Mayan game of Pok ta Pok. *Ibid.* 8 June (Latin America Suppl.) p. iv/5 It is recorded that the early Spaniards found the Indians of Guatemala playing a curious game with a ball called *pok-ta-pok*, but in which the players used not their feet, head, or hands, but their posteriors. **1962** V. W. VON HAGEN *Ancient Sun Kingdoms of Americas* ix. 161 The passion of the Maya, and one that they shared with most Central American Indians, was the game of *pok-a-tok*, it was not unlike the modern basket-ball. **1963** C. GALLENKAMP *Maya* ii. 32 Every city had a ball court consisting of a playing-field enclosed by viewing platforms where spectators could watch a game called *tlaxtli* or *pok-ta-pok*. **1973** *Times* 27 Oct. 14/4 Maya life is illustrated by dozens of pottery figurines showing the killing of a deer, musicians, animals and the helmeted men of Lubaantun who are thought to be participants in the sacred ball-game *pok-ta-pok*.

poky, *a.*[1] Add: Also **pokey**. **1. a.** (Earlier examples.)

1853 LADY LYTTELTON *Let.* 20 Aug. in *Corr. Sarah Spencer* (1912) xvi. 413 All *I* want is to love *more*, and to smile *more*, and to be *more* amused and *more* merry, and less poky and morose and dry and grave! **1854** E. TWISLETON *Let.* 29 June (1928) xi. 213 A dreadfully stiff and pokey set of people.

b. (Further examples.)

1930 J. B. PRIESTLEY *Angel Pavement* v. 209 All this for less than it would cost to live in some dingy and dismal boarding-house or the pokiest of pokey flats. **1971** *Daily Tel.* 28 Sept. 2/5 A pokey, little, highly rented flat.

2. (Earlier example.)

1888 A. G. STEEL in Steel & Lyttelton *Cricket* iii. 142 To the pokey, nervous style of batsman it [*sc.* the high-dropping full-pitch] is fraught with considerable uneasiness.

pol (pǫl). *N. Amer.* colloq. abbrev. of POLITICIAN.

1942 BERREY & VAN DEN BARK *Amer. Thes. Slang* § 854/1 *Politician*,..pol, polly, poly. **1965** F. KNEBEL *Night of Camp David* ii. 54 The clutter of pols and stale whisky glasses in the hotel suites. **1966** *Economist* 18 June 1315/2 Gossip has it that the 'pols', as the state's professional politicians (particularly Democrats) are called, felt guilty about the shabby treatment delivered to Mr Peabody when he was Governor [of Massachusetts]. **1972** *Time* 17 July 15/3 The young pols beat them at their own game. **1976** *Toronto Star* 14 Feb. B1/2 Can a bunch of battle-scarred old pols—including a couple of Liberal party retreads—gang up to stop a brash young lawyer named Brian Mulroney? **1978** J. CARROLL *Mortal Friends* II. ii. 139 What had he become? A two-bit pol, flashing about other people's corridors, waiting for his break?

POL: see *P II.

Polab (pōᵘ·lāb). Also **Polabe.** [Slav., cf. Pol. *po* on, *Labe* Elbe.] **a.** A member of a Slavonic people once inhabiting the region around the lower Elbe. **b.** The West Slavonic language of this people, now extinct. Also *attrib.*

1882 *Encycl. Brit.* XIV. 347/2 The earlier inhabitants of Lauenburg were a Slavic tribe known by the name of Polabes. **1895** *Funk's Stand. Dict.*, *Polabian*,..one of an ancient Slavic people dwelling on the lower Elbe, now wholly Germanized... *Polab.* **1911** *Encycl. Brit.* XXI. 902/1 *Polabs*,..the Slavs..who dwelt upon the Elbe and eastwards to the Oder. **1911** [see *LECH, LEKH *sb.*[4] and *a.*]. **1934** [see *LECHITIC *sb.* and *a.*]. **1974** *Encycl. Brit. Micropædia* VIII. 72/3 By the early 9th century the Polabs were organized into two confederations or principalities.

Polabian (polēⁱ·biăn), *sb.* and *a.* [f. as prec. + -IAN.] **A.** *sb.* = prec. **B.** *adj.* Of or pertaining to the Polabs or their language.

1866 *Chambers's Encycl.* VIII. 767/1 The Polabians never attained any distinct political footing. **1888** [see *LECHISH *sb.* and *a.*]. **1891** M. MÜLLER *Sci. Lang.* I. vii. 270 The Polabian dialect became gradually extinct at the beginning of the last century. **1911** *Encycl. Brit.* XXI. 902/1 Polabian agrees mostly with Polish and Kašube with its nasalized vowels and highly palatalized consonants. **1925** P. RADIN tr. *Vendryès's Language* 287 Polabian has been absorbed into German, as Cornish into English. **1929** [see *LECH, LEKH *sb.*[4] and *a.*]. **1934** O. JESPERSEN *Language* vi. 117 The now extinct Polabian language. **1939–40, 1950** [see *LECHITIC *sb.* and *a.*]. **1955** R. JAKOBSON *Slavic Lang.* (ed. 2) 2 The last remnant of Polabian on the left bank of the lower Elbe..died out toward the middle of the eighteenth century, but is known through a few vocabulary lists and short texts recorded about 1700. **1972** W. B. LOCKWOOD *Panorama Indo-European Lang.* 157 A diminutive islet of Elbe Slavonic or Polabian (*po* 'on', *Laba* 'Elbe') lingered on into the first decades of the eighteenth century. It was situated just west of the Elbe in the Lüneburg Wendland north of Salzwedel.

Polabish (polā·biʃ), *sb.* Also **Polabisch.** [ad. G. *polabisch.*] = *POLAB b.

1877 A. H. KEANE tr. *Hovelacque's Sci. of Lang.* 280 We may conclude this notice by mentioning the old dialects of the Elbe Slavonians, known by the name of *Polabish*, idioms now extinct, and whose scanty records, greatly affected by German influence, date from the seventeenth and beginning of the eighteenth century. **1890** W. R. MORFILL *Ess. on Importance of Study of Slavonic Lang.* 15 The extinct Polabish, a language once spoken on the Elbe.., was restored from some fragments by Schleicher. **1908** T. G. TUCKER *Introd. Nat. Hist. Lang.* 226 *Polabish*, once spoken by Slavs on the lower Elbe, is now extinct. **1955** *Trans. Philol. Soc. 1954* 87 Cornish and polabisch exist in a modern period but, for the present purpose, they naturally cannot rank as 'modern' since they are no longer spoken.

polacca[1]. Add: (Earlier and later examples.) Also applied more widely to other music of a (supposed) Polish character. Also *attrib.* and in phr. *alla polacca.*

1806 T. BUSBY *Dict. Mus.* (ed. 2) *Polacca*, a Polish movement of three crochets in a bar, chiefly characterised by its emphasis being laid on the fifth quaver of the bar. *c* **1807** W. CROTCH *Specimens Various Styles Music* I. 10 Some modern composers have given the title Polacca to movements which would sound very foreign to the ear of a Polander. **1812** J. M. WILLIAMS *Dramatic Censor* 41 Master Byrne and Miss Smith executed a *pas de deux* (a polacca) in the second act. **1954** *Grove's Dict. Mus.* (ed. 5) VI. 836/1 Polaccas may be defined as polonaises treated in a denationalized manner, but still retaining much of the rhythm characteristic of their Polish origin. *Ibid.* 836/2 Instrumental movements with the tempo indication *alla polacca* also occur. **1970** W. APEL *Harvard Dict. Mus.* 683/2 The 'Polacca' in Bach's Brandenburg Concerto no. 1 shows hardly any affinity to the polonaise. **1975** *Gramophone* July 174/2 In the finale with its polacca rhythms, and particularly in the obviously Slavonic episodes.., the Broadwood does increasingly suggest a Hungarian cymbalom.

Polack, *sb.* (*a.*) Delete †*Obs.* and add: Also with pronunc. (pōᵘ·læk, pōᵘ·lāk). Also 6–7 **Polake**, 9 **Pollack, Pullack** and with lower-case initial. **A.** *sb.* **1.** (Earlier and later examples.)

1574 SIDNEY *Let.* 27 Nov. in *Wks.* (1968) III. 99 The Polakes hartily repente their so fur fetcht election. **1895** *Funk's Stand. Dict.*, *Polack*, same as *Pole*. **1922** M. F. LIDDELL in *Contemp. Rev.* Dec. 770 Danzig fears and hates the 'Polacks' and still more the French. **1933** S. K. PADOVER *Let Day Perish* 140 You cowardly little sneak! It's craven pups like you that make the Polacks trample on us! If we Jews would learn to..kill..like they do, the —Polacks would grovel at our feet—!

2. A Jew from Poland.

1834 *Manch. Old Hebrew Congregation Acct. Bk.* in B. Williams *Making of Manchester Jewry* (1976) iii. 71 Given him the Polack for leaving Town 8/6. **1892** [see *LITVAK]. **1909** *Cent. Dict.* (Suppl.) *Polack*, a name given to the Jews of the Polish provinces, by their Lithuanian co-religionists. **1971** M. A. SHULVASS *From East to West* i. 23 It is hard to arrive at any accurate estimate of the numbers of Jews who emigrated westward in this period... The strongest indication that they came in considerable numbers is the fact that the nickname Pollack was current both in the Germanies and in the Hapsburg monarchy. *Ibid.* iv. 111 Following the great influx of *Betteljuden* from Poland to the West, the nickname Pollack assumed a more derogatory connotation than ever before.

3. *N. Amer.* A (usu. disparaging) term for a Polish immigrant or person of Polish descent.

1898 F. P. DUNNE *Mr. Dooley in Peace & War* 234 'Well,' said Mr. Dooley, 'ye'er thoughts on this subject is inthrestin', but not conclusive, as Dorsey said to th' Pollack, that thought he cud lick him.' **1900** *Congress. Rec.* 7 Feb. 1625/2, I have some Polacks in my district, and..the blood of Pulaski, the brave Pole who fell at Savannah in the defense of American liberty, has never been avenged. **1905** [see *COLD *a.* 1 c]. **1922** E. E. CUMMINGS *Enormous Room* iv. 61 Get out of the way you damn Polak! **1935** W. SAROYAN *Daring Young Man* 108 All that mattered was this moment, Wolinsky in love, alive, walking down Ventura Avenue, in America, Wolinsky of the universe, the crazy Polak with the broken nose. **1944** *Sun* (Baltimore) 2 Aug. 2/3 'You know, I sure did hate to shoot him,' said the sergeant, 'Because he might have been a Polack, but he wouldn't stop.' **1952** F. L. ALLEN *Big Change* iii. 53 They were scornfully known as Dagoes, Polacks, Hunkies, Kikes. **1965** P. DE VRIES *Let me count Ways* vii. 101, I now recognized him as a blond Polak I had seen around town. **1971** [see *HUNK *sb.*[4]]. **1976** *National Observer* (U.S.) 26 June 1/3 The Crusher's a clean-living Polack from Milwaukee who don't truck with no drugs or bad women.

B. *adj.* (Earlier and later examples.) Also, of Polish origin or descent.

1602 SHAKES. *Hamlet* v. ii. 388 You from the Polake warres, and you from England Are heere arriued. **1928** [see *FLY *v.*[1] 5 a]. **1930** [see *BOHUNK]. **1966** E. V. CUNNINGHAM *Helen* iv. 45 You're some cheap Polack hooker that was tossed out of a parochial school fer diddling little boys. **1974** L. DEIGHTON *Spy Story* xix. 199 Any sign of that goddamn Polack sub?

Poland[2]. The name of a town in Maine, U.S., used *attrib.* and *absol.* to designate the variety of mineral water obtained from springs there.

1881 J. G. BLAINE *Let.* 6 Sept. in H. Ricker *Poland Mineral Spring Water* (1883) 37 Send two more cases Poland to the President [*sc.* Garfield]... The President will drink no other water. **1883** H. RICKER *Poland Spring, Maine* 18 The well-known effects following the use of Poland water were not discovered in a day. **1893** G. H. HAYNES *State of Maine* 41 All parties are requested to examine the fine display of Poland Water..at this Columbian Exposition. **1917** H. RICKER *Poland Spring* Poland water can be obtained in dining-cars, transatlantic steamships and in the leading cities throughout Europe. **1937** *Maine: Guide 'Down East'* III. 362 Poland Water is one of the few bottled waters that has continued to maintain a popularity. **1967** H. JOHNSON in C. Ray *Compleat Imbiber* IX. 148 Poland water is America's Perrier. It is not so fizzy but just as smart. **1968** J. LEASOR *Passport for Pilgrim* x. 181 He had a bottle of Poland water and a glass... He never drank anything but Poland water. He had a weak stomach, and it was comforting to him. **1977** *Times Lit. Suppl.* 4 Feb. 120/1 The American billionaire, Howard Hughes... All he drank was Poland water (whatever that may be).

polar, *a.* (*sb.*) Add: **1. a.** Also, of or pertaining to the poles of another heavenly body. (Further examples.)

1894 [see *polar cap* in 1 b below]. **1922** H. S. JONES *Gen. Astron.* v. 134 The structure of the corona is very complex; it has no definite boundary and is usually symmetrical with respect neither to the centre of the Sun nor to the Sun's polar axis. **1973** [see *polar wandering* in 1 b below].

b. *polar bear*, substitute for def.: the white bear of Arctic regions, *Ursus* (or *Thalarctos*) *maritimus*, or its fur; also *attrib.*, *comb.*, and *fig.*; (examples); *polar cap*, a large region of ice or other frozen matter surrounding a pole of a planet; *polar flattening*, the extent to which the polar diameter of a planet is shorter than the mean equatorial diameter; *polar front* (Meteorol.), a front between polar and equatorial air masses; *polar hare*, also called the Arctic hare; (examples); *polar orbit*, an orbit that passes over polar regions; *spec.* one whose plane contains the polar axis; so *polar-orbiting* adj.; *polar wandering*, the slow, erratic movement of the earth's poles relative to the continents which is thought to have occurred throughout geological time and is ascribed largely to continental drift; also extended to corresponding movement on other planets.

1781 T. PENNANT *Hist. Quadrupeds* II. 290 The Polar bear might have been one [*sc.* an animal natural to a rigorous climate]. **1829** [see SEA-BEAR 3]. **1834** DICKENS *Sk. Boz* (1836) 1st Ser. I. 210 In their shaggy white coats they look just like Polar bears. **1847** T. ARNOLD *Let.* 23 Oct. in *N.Z. Lett.* (1966) 10 In Prince Edward's Island, the winter..is enough to deter anyone but a polar bear. **1910** E. T. SETON *Life-Histories Northern Animals* II. 1034 It [*sc.* the grizzly bear] is easily distinguished.. from the Polar-bear by the latter's white colour. **1917** R. FRY *Let.* 2 Mar. (1972) II. 404 Lady Scott, the widow of the Antarctic man came in yesterday with Peter Scott, the most wonderful little monster of a polar bear cub. **1959** G. D. PAINTER *Proust* I. ix. 126 Montesquiou..had a room decorated as a snow-scene, with a polar-bear rug. **1968** A. DIMENT *Bang Bang Birds* iii. 37 The living room, with its nylon polar bearskin rug. **1974** P. DICKINSON *Poison Oracle* i. 17 The polar bear was swimming, huge in its tiny pool. **1976** H. L. GUNDERSON *Mammalogy* xvi. 375 The female and sometimes the male polar bears..become dormant throughout the winter. **1894** *Astron. & Astrophysics* XIII. 542 So much for the terrestrial conditions under which the observations were made. The Martian ones were such as to make the polar cap and its accompanying phenomena the centre of interest upon the planet. **1932** [see *fast ice* s.v. *FAST *a.* 11]. **1967** K. LASSEN in B. M. McCormac *Aurora & Airglow* v. 453 We define the Polar Cap as the area with corrected geomagnetic latitude..greater than some 70°. **1968** S. GLASSTONE *Bk. of Mars* vi. 107 Even if the polar caps are largely carbon dioxide, it does not mean that they do not also contain some solidified water. **1895** *Astrophysical Jrnl.* II. 136 Micrometric measures of the diameters of Mars..give as the most probable value for the equatorial diameter of the planet at distance unity: 9″.40 ± ·007; for the polar one: 9″.35 ± ·005; and for the polar flattening 1/215 of the equatorial diameter. **1899** G. W. MYERS tr. *Lommel's Exper. Physics* i. 85 From the values of the acceleration resulting from pendulum experiments and the magnitude of the centrifugal force, the polar flattening may be computed. **1966** *McGraw-Hill Encycl. Sci. & Technol.* XIII. 516/2 Clairaut's formula for polar flattening... $\alpha = \frac{5}{2}m - \beta$ in which *m* is the ratio of centrifugal force to gravity at the equator..and *β* is the coefficient of the principal latitude term. **1920** V. BJERKNES in *Nature* 24 June 524/1 This line shows how far the cold air has succeeded in penetrating; it is a kind of polar front line. *Ibid.* 524/2 All meteorological events of the temperate zone..are derived from the general atmospheric circulation..as we know it from the polar front. **1935** C. F. BROOKS et al. *Why the Weather?* (ed. 2) v. 50 The polar front is the forward edge of a moving mass of cold dry air, usually coming more or less directly from polar or subpolar regions. **1973** R. G. & A. H. PERRY *Synoptic Climatol.* iii. 184 The classical view of tropical and polar air, separated by the polar front, does not accord well with modern knowledge of the general circulation. **1823** J. FRANKLIN *Narr. Journey to Shores of Polar Sea* 664 The Polar hare appears to vary much in size, and consequently in weight. **1866** W. R. KING *Sportsman & Naturalist in Canada* 26 In this respect it differs from the Polar-hare, the finer and softer fur of which is in winter pure white to the roots. **1895** [see *blue hare* (*BLUE *a.* 12 a)]. **1911** E. T. SETON *Arctic Prairies* 231 It was only a Polar Hare, the second we had seen. **1961** *Times Rev. Industry* Feb. 26/3 There are a number of possible satellite systems using polar, inclined and equatorial orbits. **1966** *McGraw-Hill Yearbk. Sci. & Technol.* 171/1 NASA is presently planning six or seven observatories to be launched alternately into the highly eccentric equatorial orbits..and the low polar orbits. **1978** *Times* 28 July 16/1 Europe's first lunar mission..would put a satellite in polar orbit round the moon. **1964** *Yearbk. Astron. 1965* 141 Transmissions from the United States were being sent by conventional means to Jodrell Bank for reflection to Gorky via the polar-orbiting balloon-satellite. **1968** *New Scientist* 24 Oct. 175/3 It should now be possible, by means of a polar-orbiting satellite, to reap..data on..cloud heights. **1924** J. G. A. SKERL tr. *A. Wegener's Origin Continents & Oceans* viii. 123 Extensive, even if slow, polar wanderings are then able to take place. **1969** *Times* 23 Apr. 7/4 Some of the more strange implications of the early studies of magnetism in ancient rocks—polar wandering, continental drift and the like. **1973** *Science* 9 Mar. 997 Polar wandering during the past 10^8 years may be recorded by unique quasi-circular structures in the polar regions of Mars.

3. b. *Chem.* Applied variously in cases where

bonding electrons are unequally shared between atoms in a molecule, so that there is some separation of electric charge: (i) applied *spec.* to electrovalent or ionic bonds, and to substances (usu. solids) in which bonding of this type predominates; (ii) applied to covalent bonds in which electrons are unequally shared between the atoms, to molecules or groups which contain such bonds, esp. those which possess a resulting electric dipole moment, and to substances (usu. liquids) which consist of such molecules.

1913 *Jrnl. Amer. Chem. Soc.* XXXV. 1443 In the preceding paragraphs we have suggested that there are two distinct types of union between atoms: polar, in which an electron has passed from one atom to the other, and non-polar, in which there is no motion of an electron. **1924** O. MAASS in H. S. Taylor *Treat. Physical Chem.* I. iv. 130 A polar molecule, one in which the molecular force of attraction is more concentrated in one particular part, so that if these molecules [of propionic acid] are oblong in shape, the field of force around one end, the —COOH end, will be more pronounced. **1927** N. V. SIDGWICK *Electronic Theory Valency* iv. 52 Polar or ionizable linkages between the oppositely charged ions of a salt. **1950** S. GLASSTONE *Elem. Physical Chem.* iii. 69 Compounds containing the groups —OH, —CN, —COOH and $—NO_2$, which are examples of polar groups, are generally highly polar in character, unless they happen to be completely symmetrical. *Ibid.*, Polar liquids have relatively high boiling points. **1950** W. J. MOORE *Physical Chem.* xi. 289 The polar compounds, of which NaCl was a prime example, could be adequately explained as being composed of positive and negative ions held together by coulombic attraction. **1951** I. L. FINAR *Org. Chem.* ii. 14 A symmetrical molecule is non-polar, although it may contain polar bonds. **1962** CORSON & LORRAIN *Introd. Electromagn. Fields* iii. 82 A water molecule..possesses just such a permanent dipole moment and is thus called a polar molecule. **1966** GUCKER & SEIFERT *Physical Chem.* (1967) xi. 276 Both the Trouton and Hildebrand constants are abnormally high for many liquids like water and ammonia, which are known to be polar. **1970** S. W. BENSON *Atoms, Molecules, & Chem. Reactions* iv. 109 An extreme example of polar bonds occurs in the case of the alkali metal halides. **1974** J. S. BLAKEMORE *Solid State Physics* (ed. 2) iv. 338 Optical phonon scattering..is especially important for a solid with a polar (partially or completely ionic) lattice. **1975** HUGHES & POOLEY *Real Solids & Radiation* ii. 15 Crystalline solids which are held together by electrostatic forces between oppositely charged ions are known as ionic or polar crystals.

5. *polar body*, one of the small cells which bud off from an oocyte at the two meiotic divisions and do not develop into ova; = *POLOCYTE. Cf. *OOCYTE, *OOTID.

1888 [in Dict.]. **1908** [see *OOTID]. **1927** HALDANE & HUXLEY *Animal Biol.* ii. 60 In order to retain the large size of the egg, three of every four gametes produced are minute and non-functional, and are called polar bodies, while only one becomes a functional ovum. **1945, 1946** [see *OOCYTE]. **1964** [see *OOTID]. **1974** *Sci. Amer.* Sept. 54/3 The remaining 23 [human chromosomes] replicate once more, and it is only after a sperm makes contact with the surface of the egg that a second polar body is expelled.

6. *polar diagram*, a diagram in which the length of the radius joining a fixed point to any point of a curve represents the magnitude of something (as the sensitivity of an aerial or the brightness of a lamp) measured in the direction of the radius; *polar vector* (see *VECTOR).

1895 *Electrician* 10 May 43/1 (*heading*) Representation of periodic currents by polar diagrams. **1923** GLAZEBROOK *Dict. Applied Physics* IV. 429/2 These diagrams are what are generally termed 'polar diagrams of light distribution'. In these curves the length of the radius vector at any angle gives the candle-power at that angle. **1943** *Electronic Engin.* XVI. 241/1 The field strength relations and polar diagrams of several aerials..were discussed. **1962** A. NISBETT *Technique Sound Studio* i. 20 (*caption*) In a polar diagram such as this the curve indicates the output of the microphone for a given sound arriving from any angle.

7. b. (Later examples.)

1953 T. PARSONS et al. *Working Papers in Theory of Action* 208 The instrumental and the system-integrative norms, which very closely characterize what..have been thought of as polar types of institutional structure. **1959** MCKINNEY & LOOMIS in J. S. Roucek *Contemp. Sociol.* 557 The polar extremes in point are clearly ideal or constructed types. *Ibid.* 558 The polar type formulations.. have firmly established the point that the *continuum* is a vital notion in the comparative analysis of social phenomena. **1964** E. A. NIDA *Toward Sci. Transl.* ii. 24 The differences between literal and free translating are, however, no mere positive-negative dichotomy, but rather a polar distinction with many grades between them. **1965** *Language* XLI. 275 Only the conjunction of 'polar' adjectives in contexts of this kind seems odd. **1972** *Sci. Amer.* Jan. 35/1 Although sex-role ideologies form a continuum, we grouped the respondents into two polar categories, which we labeled 'traditional' and 'contemporary'. **1975** *Language* LI. 1 Polar interrogative sentences ('yes/no questions') are different from the corresponding declarative sentences not only pragmatically ..but also semantically.

polari-. Add: **polari-(bi)locular**, of a lichen spore: (see quots. 1921[1] and 1967).

1871 W. LEIGHTON *Lichen Flora Gt. Brit.* 175 Spores 8, colourless, ellipsoid, polari-bilocular. **1921** A. L. SMITH *Handbk. Brit. Lichens* 9 In the family, Physciaceæ, the cross wall of the septate spore is so thickened that the lumen of each cell is reduced to a small area at the ends; hence the term polari-bilocular. **1921** —— *Lichens* 422 In such a phylum as the Physciaceæ (with colourless polarilocular spores) there is a clear example of a closely connected series. **1967** M. E. HALE *Biol. Lichens* ii. 37 Polarilocular spores: two-celled spores with a thick median wall and a thin isthmus, or conversely a single-celled spore with a median constriction.

polarimetry. Add: (Examples referring to electromagnetic radiation other than visible light.) Hence **polarime·trically** *adv.*

1930 [see *FOUCAULT]. **1937** *Jrnl. Org. Chem.* II. 431 The mutarotation of glucose was measured polarimetrically, in 200 mm. tubes. **1973** *Jrnl. Biol. Chem.* CCXLVIII. 4165/1 Each solution was then diluted to 12 ml..and the further progress of the reaction followed polarimetrically. **1975** *Nature* 14 Aug. 537/1 In the past, the best-known argument for X-ray polarimetry has been that it is often symptomatic of non-thermal synchrotron-tye emission.

Polaris (polā·ris). [a. med.L. *polāris* polar.] The name of a type of guided missile developed for the U.S. Navy, having a nuclear warhead and designed to be carried by submarines and launched under water. Freq. *attrib.* and *Comb.*, as *Polaris missile, submarine; Polaris-carrying* adj.

1957 *N.Y. Times* 1 Jan. 1/3 The Navy is developing a ballistic missile to be fired from submerged submarines at targets hundreds of miles away, it was disclosed today. The missile is named the Polaris. **1957** *Life* 21 Jan. 121/2 Although Polaris can be launched from surface ships it will find its greatest strategic value with fast new nuclear-powered submarines. **1958** *Observer* 10 Aug. 3/4 Submarines of the Nautilus type equipped with 'Polaris' guided missiles could clearly use the Arctic Ocean as a base from which to threaten, with virtual impunity, the northern coasts of Russia. **1958** *New Statesman* 16 Aug. 181/1 They will be equipped with the deadly Polaris missile. **1960** *Daily Tel.* 22 Apr. 1/3 Britain has sounded the United States Defence Department on the possibility of having her own Polaris-carrying submarines, it was learned to-day. **1965** *New Statesman* 14 May 752/1 Last October, *The Times* felt able to predict with confidence that all Labour would do with Polaris would be to assign it irrevocably to Nato. **1965** H. KAHN *On Escalation* ii. 48 One can conceive of a slow-motion counterforce war lasting for weeks or months during which Polaris submarines are hunted down. **1973** D. KYLE *Raft of Swords* (1974) iii. 18 The Americans developed their Polaris programme very quickly... The Polaris rocket went from design to deployment in one and a half years.

polariscope. (Earlier example.)

1829 *Amer. Jrnl. Sci.* XV. 369 (*heading*) Description of the polariscope, an instrument for observing some of the most interesting phenomena of polarised light.

polariton (polæ·ritǫn). *Physics.* [f. POLAR(IZATION and related words + *-iton*, prob. after *EXCITON.] A quasiparticle in an ionic crystal consisting of a photon strongly coupled to a quasiparticle such as a phonon or exciton.

1958 J. J. HOPFIELD in *Physical Rev.* CXII. 1558/2 The polarization field 'particles' analogous to photons will be called 'polaritons'. (Excitons will be shown to be one kind of polariton... Optical phonons are another example of polaritons.) **1963** R. S. KNOX *Theory of Excitons* iii. 133 An external photon excites a crystal to a state described by a polariton packet... If this new particle of excitation fails to interact with any energy sinks, it re-excites an external photon on the other side of the crystal. However, by virtue of its exciton component, it can decay into states which do not couple as easily to external photons (i.e., 'nonradiative' states). **1971** J. I. PANKOVE *Optical Processes in Semiconductors* i. 16 A polariton is the complex resulting from the polarizing interaction between an electromagnetic wave and an oscillator resonant at the same frequency... Although polaritons initially have designated the interaction between excitons and photons, they can also represent the interaction between photons and optical phonons and between photons and plasmons. *Ibid.* 17 The polariton is not to be confused with the polaron of ionic crystals, which results from an interaction between the electron and the lattice. **1972** *Physics Bull.* Aug. 490/1 Two chapters..cover.. light scattering by polaritons.

polarity. Add: **2. e.** *Biol.* The tendency of living matter to assume a specific form; the property observed in animals from which parts have been severed, and in severed parts of animals and plants, of regenerating the missing parts.

1864 G. J. ALLMAN in *Rep. Brit. Assoc. Adv. Sci. 1863* 392 The lower segment, on the other hand, instead of pushing forth from the cut extremity a simple continuation of the coenosarc, developes from this extremity a polypite. There is thus manifested in the formative force of the Tubularia-stem a well-marked polarity, which is rendered very apparent if a segment be cut out from the centre of the stem. **1864** H. SPENCER *Princ. Biol.* I. II. iv. 181 The vitalized molecules composing the tissues, show their proclivity towards a particular arrangement... For this property there is no fit term. If we accept the word polarity, as a name for the force by which inorganic units are aggregated into a form peculiar to them; we may apply this word to the analogous force displayed by organic units. **1895** *Jrnl. Morphol.* X. 322 If we assume the polarity of the egg to be pre-determined from the beginning, we must admit that the polarity determines the position of the segmentation-nucleus. **1924** E. G. CONKLIN in E. V. Cowdry *Gen. Cytol.* ix. 558 The polarity of the egg is the earliest recognizable and most fundamental differentiation of morphogenesis. **1926** J. S. HUXLEY *Ess. Pop. Sci.* xviii. 251 When small pieces of a planarian regenerate, they exhibit what we may call polarity; for (with a few special exceptions) the new head is formed from that region of the piece which was nearest to the old head, the new tail from that region which was nearest to the old tail. **1975** R. L. PETERSON in Torrey & Clarkson *Devel. & Function of Roots* vii. 146 The inherent polarity in most root segments, which manifests itself by the regeneration of buds at the proximal end and roots at the distal end, may be due to the polar distribution of more than one endogenous hormone.

5. b. (Further examples.)

1934 C. HARTSHORNE *Philos. & Psychol. of Sensation* iv. 134 Feeling involves an 'opposition' of positive and negative, liking and disliking. Does sensation exhibit a similar polarity?.. It is, as we have seen, precisely in terms of polarities that the facts of sensation are to be described. **1945** *Downside Rev.* 131 Be it remarked in passing that the relation between God and the world is not here conceived as one of polarity. **1950** D. RIESMAN *Lonely Crowd* (1952) i. 20 To what extent, in establishing America's polarity from Europe, he [*sc.* de Tocqueville] tendentiously noticed those things that were different rather than those that were the same. **1957** *New Statesman* 2 Nov. 555/3 We must ask ourselves what we can do to break this polarity [between the U.S. and Russia]. **1963** *Times Lit. Suppl.* 26 Apr. 306/5 Polarity, or the use of contrast as an artistic means. **1970** B. M. H. STRANG *Hist. Eng.* 134 The polarity *mental/physical*. *Ibid.* 237 It [*sc.* Northern English] thus heightened the polarity of tense-contrast, which in strong verbs was far less clear-cut in the south. **1972** *Encycl. Psychol.* III. 19/2 *Polarity*, a relationship between features or traits which are antithetical pairs. **1972** R. PLANT in Cox & Dyson *20th-Cent. Mind* III. iii. 69 The wholly necessary or the wholly contingent, the two polarities of empiricist epistemology.

polarizability. Add: *spec.* the degree to which an atom or molecule can be polarized, expressed in terms of the electric dipole moment induced by unit electric field. (Further examples.)

1930 PAULING & GOUDSMIT *Structure Line Spectra* iii. 45 The electric moment of the induced dipole is *α*F, in which *α* is called the 'polarizability' of the atom or ion. **1947** SLATER & FRANK *Electromagnetism* iv. 44 We thus have means for finding the dielectric constant of a material, if we know the polarizability of its molecules. **1964** PHILLIPS & WILLIAMS *Inorg. Chem.* I. iv. 132 The simplest theoretical treatment of the London effect leads to an energy between identical atoms or molecules given by $-\frac{3}{4}\frac{h\nu\alpha^2}{r^6}$. *α*, the polarizability of the atom or molecule, may be determined from its molecular refraction. **1974** *Nature* 23 Aug. 686/3 The fourth chapter..is an excellent account of the determination of dipole moments and polarisabilities of molecules in excited electronic states.

polarization. Add: **I. 1.** (Further examples.) Also used of other kinds of wave.

1923 H. L. BROSE tr. *Sommerfeld's Atomic Structure & Spectral Lines* i. 23 Polarisation signifies that a ray favours a certain plane passing through it more than the one perpendicular to this plane. In the case of longitudinal vibrations..there is a symmetry about the ray and no such preference can be imagined. Longitudinal radiation must therefore be unpolarised. In the case of transversal vibrations..a favoured plane..is determined by the direction of vibration and the direction of the ray. **1929** *Physical Rev.* XXXIII. 760 (*heading*) A test for polarization of electron waves by reflection. **1936** *Wireless World* 16 Oct. 396/3 Although one would expect a slight departure from vertical polarisation of the received waves in this location, a vertical aerial was found to be 6 to 10 db. better than a horizontal. **1976** *IEEE Trans. Antennas & Propagation* XXIV. 5/1 In microwave communication links above 10 GHz the employment of frequency reuse in orthagonal polarizations is limited by cross polarization.

b. = *optical activity* s.v. *OPTICAL *a.* 6. Now used chiefly with reference to sugar solutions.

1845 *Mem. & Proc. Chem. Soc.* II. 29, I shall explain.. what is meant by the deviation or rotation of the rays of polarized light when transmitted through fluids said to possess circular polarization. **1862** *Jrnl. Chem. Soc.* XV. 308 An experiment made..while examining the circular polarization of camphoric acid. **1912** C. A. BROWNE *Handbk. Sugar Anal.* ix. 236 The reading multiplied by 1·3 gives the polarization (degrees Ventzke) of the sugar cane. **1935** *Economist* 20 Apr. 906/2 Raw sugar of 97° polarisation..pays a duty of 8s. 4 3/10 d. per cwt. **1963** D. BECKER in P. Honig *Princ. Sugar Technol.* III. ix. 455 In contrast to raw cane sugars, raw beet sugars for refining cannot be definitely characterized and graded by stating the polarization alone.

c. Measurement of the optical activity of a sugar solution. Cf. *POLARIZE *v.* 1 c.

1905 G. W. ROLFE *Polariscope* 96 Polarizations made at average room temperature by the standard commercial methods give with requisite accuracy the per cent of sucrose in the sample. **1945** A. L. & K. B. WINTON *Anal. of Foods* 640/1 Addition of solid sodium carbonate to slight alkaline reaction after the immediate polarization.. destroys the mutarotation without changing the dilution. **1973** SNELL & ETTRE *Encycl. Industr. Chem. Anal.* XVIII. 345 Direct polarization of the sugar solution or polarization before and after hydrolysis are commonly used assay methods.

1*. *Physics.* A partial or complete alignment of the spin axes of particles; the degree to which this exists.

1928 *Proc. R. Soc.* A. CXVIII. 675 It will..suffice to treat of only one type of polarisation and we shall take that corresponding to *z*. **1929** *Ibid.* CXXIV. 427 This polarisation could be detected by letting the scattered beam fall on a second target. **1956** *Rev. Mod. Physics* XXVIII. 279/1 We shall speak of transverse polarization of an electron beam if the direction of the spin is perpendicular to the momentum, of longitudinal polarization if the spin is parallel or antiparallel to the momentum. **1963** K. NISHIJIMA *Fundamental Particles* vii. 380 Hyperon polarization is transverse rather than longitudinal unless parity is violated in production. **1979** *Sci. Amer.* May 64/3 The polarization is defined as the difference between the number of spin-up particles and the number of spin-down ones, divided by the total number of protons.

II. 2. (Further examples.) *spec.* the partial separation of positive and negative electric charge produced in a dielectric by an electric field, and expressed by a vector quantity equal to the electric dipole moment per unit volume of the dielectric; also, a similar state in an individual atom or molecule.

1916 F. B. PIDDUCK *Treat. Electricity* iii. 93 The total electric moment of an element of volume $d\tau$ of a dielectric near the point (x, y, z) has components $P_x d\tau$, $P_y d\tau$, $P_z d\tau$, where the vector $P(P_x, P_y, P_z)$ is called the polarization at the point. **1933** N. V. SIDGWICK *Physical Properties of Covalent Link* v. 129 The polarization of the molecules in an electric field can take place in three ways: (1) The arrangement of the electrons will be displaced with respect to the nuclei... (2)..The nuclei themselves will be to some extent displaced with respect to one another... (3)..If the molecule has a permanent dipole moment of its own—if it is polar—the field will tend to orient it along the direction of the lines of force. **1935** J. DOUGALL tr. *Born's Atomic Physics* viii. 230 The polarization **P** is connected with Maxwell's displacement vector **D** by the relation $\mathbf{D} = \mathbf{E} + 4\pi\mathbf{P}$; on the other hand, by definition, $\mathbf{D} = \epsilon\mathbf{E}$, where *ε* is the dielectric constant. **1966** *McGraw-Hill Encycl. Sci. & Technol.* IV. 111/2 The dielectric constant of a material depends on its polarization in an applied field or, microscopically, on the relative displacements, in the field direction, of the electrons and nuclei comprising the molecules of the dielectric. **1973** P. C. CLEMMOW *Introd. Electromagn. Theory* vi. 232 **P** and **M**..represent, respectively, electric and magnetic dipole moment densities. **P** is called the (electric) polarization, and **M** the magnetization or magnetic polarization.

4. b. The accentuation of a difference between two things or groups; the process of division into two groups representing the extremes of opinion, wealth, or the like.

1945 KOESTLER *Yogi & Commissar* II. v. 117 False polarizations and national splits which merely reflect latent conflicts. **1947** *Sun* (Baltimore) 15 Aug. 12/6 The same polarization of thought which is going on in the rest of the world is seen in Korea in extreme form. **1951** Y. MALKIEL in *Language* XXVII. 485 Lexical polarization is used in this paper as a convenient label for the influence exerted by one word on its semantic opposite. **1957** *Atlantic Monthly* Aug. 8/1 Any outside disturbance of the evolving polarization between moderate and extreme Arab nationalists may well bring disaster to the whole region. **1960** *Daily Tel.* 8 Nov. 12/2 But this does not alter the fact that every tendency towards polarisation of trade in separate camps in Europe, necessary and inevitable while the division stands, is in the long run a disaster for all. **1964** T. BOTTOMORE *Elites & Society* ii. 19 The development of capitalism brings about a more radical polarization of classes than has existed in any other type of society. **1970** *Daily Tel.* 7 Oct. 5/2 There were already signs in the central districts [of London] of 'social polarisation', areas peopled only by the richest and the poorest. **1972** M. L. SAMUELS *Ling. Evolution* iii. 39 Further phonetic divergence ('polarisation' of the existing difference) to /k/ and /tʃ/. **1975** *Chinese Econ. Studies* VIII. IV. 90 (*heading*) The polarization between the rich and the poor is a general law of capitalist accumulation.

III. 5. polarization charge, the charge that appears on the surface of a dielectric when it is polarized in a direction not parallel to the surface.

1947 SLATER & FRANK *Electromagnetism* iv. 45 We have the result that the normal component of *P*, pointing out of the dielectric, equals the surface polarization charge that appears on the surface as a result of the polarization. **1975** GRANT & PHILLIPS *Electromagnetism* ii. 53 Polarization charges induced on the surface of a dielectric material make a contribution to the macroscopic electric field inside the material...The sign of the induced charge always ensures that the field just inside the dielectric surface is less than the field just outside.

polarize, *v.* Add: **I. 1. a.** (Further examples.)

1939 *Wireless World* 26 Jan. 83/1 Considerable attention has been paid to the question of whether the waves of the Alexandra Palace television transmitter should be vertically or horizontally polarised. **1966** *McGraw-Hill Encycl. Sci. & Technol.* X. 448/1 Electromagnetic radiation is difficult to polarize in certain spectral regions.

b. Also, to rotate the plane of polarization of plane-polarized light by (a specified amount) under standard conditions.

1900 *Bull. Div. of Chem., U.S. Dept. Agric.* No. 59. 52 The wine is fermented... [If] it polarizes −3° after fermentation. It contains only levorotatory sugar. **1945** A. L. & K. B. WINTON *Anal. of Foods* 616/1 Wash by decantation until the washings polarize zero in a 200-mm. tube.

c. *trans.* To measure the optical activity of (a solution) in order to determine the concentration of sugar in it. Also *ellipt.*

1905 G. W. ROLFE *Polariscope* 87 The sample which the chemist polarizes must be strictly representative of the total lot of sugar. **1945** A. L. & K. B. WINTON *Anal. of Foods* 615/1 Invert Polarization... Cool to about 20°, fill exactly to the mark, mix, and polarize at 20° in a 200-mm. tube. **1963** TRIEBOLD & AURAND *Food Composition & Qual.* iii. 60 According to their standard, a solution containing 26·000 grams of sucrose (normal weight) in 100 ml of solution at 20°C., and polarized in a 200-mm tube at 20°C., should give a reading of 100° on the saccharimeter. **1973** SNELL & ETTRE *Encycl. Industr. Chem. Anal.* XVIII. 348 Dilute the solution to 100 ml at 20°C. Mix well and polarize in a 200-mm tube.

1*. *Physics.* To produce a partial or complete alignment of the spins of (particles).

1932 *Proc. R. Soc.* A. CXXXV. 431 The theoretical existence of these methods provides some evidence that electron beams can be polarised. **1953** *Progr. Nuclear Physics* III. 72 In order to polarize elementary particles, there must be some sort of coupling of the particle spin with a fixed spatial direction. **1974** FRAUENFELDER & HENLEY *Subatomic Physics* ix. 206 In a radioactive source at room temperature, the nuclear spins are randomly oriented. It is necessary to polarize the nuclei so that all spins J point in the same direction. **1979** *Sci. Amer.* May 64/2 The next stage polarizes the protons and depolarizes the electrons.

II. 2. (Further examples.) *spec.* to induce an electric dipole moment in (a substance, or an atom or molecule).

1887 W. LARDEN *Electr.* x. 161 Lines along which the molecules of the dielectrics are 'polarised' by a separation of + and − charges in them. **1945** A. F. WELLS *Structural Inorg. Chem.* ii. 52 An atom is polarized when placed in an electric field. **1962** CORSON & LORRAIN *Introd. Electromagn. Fields* iii. 126 Consider a large block of dielectric polarized uniformly with a dipole moment per unit volume **P**. **1977** *Sci. Amer.* Feb. 91/3 The separation is accomplished by passing the atomic beam through an inhomogeneous electric or magnetic field, which deflects those atoms that are more readily polarized or have a larger magnetic moment.

b. Also *intr.* (of a cell), to exhibit polarization and consequently suffer a decrease in e.m.f.

1903 *Electr. World & Engin.* 24 Jan. 150/2 There is yet not a battery of this class known that will not polarize in a short time, which means that it has temporarily exhausted itself and must rest for awhile before it is as efficient as it was. **1969** J. J. DEFRANCE *Electr. Fundamentals* xi. 124 The prevention is not complete, and with age a cell does polarize.

III. 3. c. *trans.* To accentuate a division within (a group, system, etc.); to separate into two (or occas. several) opposing groups, extremes of opinion, or the like. Also *intr.*, to undergo or exhibit such a process.

1949 KOESTLER *Promise & Fulfilment* I. xi. 125 The controversy about Zionism would have become as 'polarized' between pro- and anti-Russians as, say, the controversy about Poland. **1957** R. N. CAREW HUNT *Guide to Communist Jargon* xxvii. 96 There could be no place for neutrality, and indeed such an attitude is explicitly excluded by the Marxist-Leninist dialectic which polarizes every issue and denies the possibility of intermediate solutions. *Ibid.* xxxvi. 122 As a result of the Industrial Revolution, society had polarized into two hostile classes, the bourgeoisie and the proletariat. **1957** *New Statesman* 2 Nov. 555/2 If we walked out of the nuclear arms race then the world would be 'polarised' between America and the Soviet Union. **1969** *N.Y. Rev. Books* 2 Jan. 41/1 New York was racially polarized as never before; hundreds of thousands of public school children were not being taught. **1972** *Nature* 3 Mar. 39/2 The problems are polarized into three areas... Only the first of these is really meteorological in character. **1972** M. L. SAMUELS *Ling. Evolution* vi. 132 Changes gather momentum when the upper classes polarise differences to maintain their value as prestige-markers. **1973** *Sci. Amer.* Apr. 86/3 Yet Henri labored..to avoid the utter disintegration of a France polarized and torn by civil war. **1977** *Time* 19 Dec. 8/2 From the first, Soares had insisted on governing without political alliances. Any compromise, he feared, would further polarize the country's politics.

polarized, *ppl. a.* Add: **1. a.** Also used of other kinds of wave. (Further examples.)

1923 H. L. BROSE tr. *Sommerfeld's Atomic Structure & Spectral Lines* i. 23 Barkla discovered that primary X-rays are partially polarised, secondary X-rays are wholly polarised in certain directions. **1946** *Wireless World* Aug. 251/1 If the receiving aerial is horizontal it will receive nothing from a vertically polarized wave. **1966** *McGraw-Hill Encycl. Sci. & Technol.* VIII. 418/1 A circular wave guide is particularly useful in transforming plane-polarized electric intensity into circularly polarized electric intensity.

b. *Physics.* Of a particle or beam of particles: exhibiting an alignment of the spins.

1929 *Proc. R. Soc.* A. CXXIV. 426 Our only hope of observing the moment of a free electron is to obtain a 'polarised' beam, in which all the spin axes are pointing in the same direction, or at any rate more in one direction than another. **1953** *Progr. Nuclear Physics* III. 75 It may be possible to produce polarized protons by using polarized thermal neutrons..as projectiles in an (*n*, *p*) reaction. **1975** *Nature* 5 June 514/1 These treatments will be of increasing value as polarised beams of deuterons and other ions become available.

c. Of a substance or device: causing the polarization of light passing through it; = POLARIZING *ppl. a.* 1.

1936 *Discovery* Oct. 302/1 Polarised glass has for some time been used..in certain scientific instruments. **1955** *Ann. Reg. 1954* 420 More than one critic referred to 1954 as the year of the decline of 3D on account of the general discomfort of polarized spectacles. **1977** *Sci. Amer.* Dec. 172/2 If a motorist wants to eliminate the glare from a road surface, he can wear polarized sunglasses.

4. b. Characterized by division into opposing groups or principles. Cf. *POLARIZE *v.* 3 c.

1957 *New Statesman* 2 Nov. 555/3 This gives the world something quite different from the polarised powers. **1962** E. CLEAVER in A. Dundes *Mother Wit* (1973) 19/1 What we term *The Polarized Western Mind* derives from the symbolism attached to the two colors, black and white, in the mind of Western man. *Ibid.* 20/1 An obvious and striking example of polarized thinking.

polarogram (polā·ro-, pōᵘ·lārogræm). *Chem.* [f. as next + -GRAM.] A graphical record of current against voltage produced by a polarograph (see next).

1925 HEYROVSKÝ & SHIKATA in *Recueil Travaux Chimiques Pays-Bas* XLIV. 496 The authors have set up a photographic auto-registering machine, which..records the polarisation curves, giving such 'polarograms' in less than 20 minutes. **1946** *Nature* 20 July 96/1 If several elements can be isolated simultaneously, there is the possibility of estimating them in a single polarogram. **1968** PECSOK & SHIELDS *Mod. Methods Chem. Analysis* xiv. 339 The steps in the polarogram are called 'polarographic waves'. **1974** S. E. ALLEN *Chem. Analysis Ecological Materials* 398 If poor quality polarograms are produced even after passing nitrogen further bubbling may improve the display.

polarograph (polā·ro-, pōᵘ·lārograf). *Chem.* [f. POLAR(IZATION + -O + -GRAPH.] An apparatus for automatic chemical analysis in which a sample solution is electrolysed using a steadily increasing voltage, and a graph, known as a polarogram, of current against voltage is produced; this usu. shows a series of steps each of which occurs at a voltage characteristic of a particular component and has a height proportional to the concentration of that component.

Polarograph is a proprietary term in the U.S.A.

1925 HEYROVSKÝ & SHIKATA in *Recueil Travaux Chimiques Pays-Bas* XLIV. 496 (*heading*) The polarograph. **1933** *Official Gaz.* (U.S. Patent Office) 15 Aug. 527/1 Polarograph Laboratories of America, Berkeley, Calif. Filed June 5, 1933. Polarograph. For Electrical Analyzing Apparatus... Claims use since Feb. 1, 1927. **1937** *Jrnl. Iron & Steel Inst.* CXXXVI. 109A From the current-potential curve the deposition potential for a given level on the curve can be determined, so that the polarograph can be used as a qualitative test. **1946** *Nature* 13 July 59/1 Iron can be determined satisfactorily on the polarograph if the solution contains triethanolamine. **1953** *Electronic Engin.* XXV. 314/1 One such instrument is the polarograph, which is used to determine the concentration of reducible ions in a solution, by measurement of the current that flows when the ions are electrolytically reduced. **1974** S. E. ALLEN *Chem. Analysis Ecol. Materials* 400 Oxygen is the only dissolved gas of interest to biologists which can be readily determined on the polarograph.

Hence **polarogra·phic** *a.*, of, pertaining to, or used in a polarograph or polarography; **polarogra·phically** *adv.*; **polaro·graphy**, the technique of using the polarograph.

1926 *Brit. Chem. Abstr.* A. 1184/1 (*heading*) Polarographic methods in biology. **1930** *Chem. News* 19 Dec. 388/2 Current-voltage curves obtained polarographically.. reveal the presence of proteins in solutions. **1936** *Nature* 30 May 889/2 The last section, on polarography by Prof. J. Heyrovský, gives a comprehensive survey of the researches with the dropping mercury cathode. **1952** KOLTHOFF & LINGANE *Polarography* (ed. 2) xviii. 397 This reaction may be utilized to remove dissolved oxygen from polarographic solutions. **1959** *Times* 30 Jan. 3/7 This involves supervision of staff engaged on a variety of analytical work, including..infra-red spectroscopy and polarography. **1966** *McGraw-Hill Encycl. Sci. & Technol.* X. 455/1 The most widely used polarographic indicator electrode is the dropping-mercury electrode. **1970** C. N. GRAYMORE *Biochem. Eye* x. 651 Noell..measured oxygen tension polarographically. **1973** *Nature* 10 Aug. 370/2 Characteristic polarographic half-wave potential.. values for proteins were attributed..to exposed, reactive sulphydryl..groups. **1975** M. R. JENKINS in Williams & Wilson *Biologist's Guide to Princ. & Techniques Pract. Biochem.* vii. 200 The main use of polarography in biochemistry is not, however, the measurement of oxidation-reduction potentials, but rather the qualitative and quantitative analysis of various compounds.

Polaroid (pōᵘ·lāroid). Also **polaroid.** [Proprietary name.] **A.** *sb.* **1. a.** A material which in the form of thin sheets produces a high degree of plane polarization in light passing through it.

1936 *Official Gaz.* (U.S. Patent Office) 26 May 760/1 Sheet Polarizer Company, Inc., Union City, N.J... *Polaroid* for composite material comprising suspensions of crystalline particles in a light-transmitting medium adapted to be used in connection with optical devices such as microscope eye-pieces, glare eliminators,..gem testers, and the like. Claims use since Nov. 19, 1935. **1936** *Nature* 22 Aug. 312/2 Another [firm] manufactures ophthalmic instruments employing Polaroid. **1937** *Ann. Reg. 1936* II. 65 Land (*Nature*, Aug. 22) developed a new device for producing plane polarised light. Available commercially under the name Polaroid it consisted of a layer of ultramicroscopic needle-shaped crystals of herapathite, oriented with their axes parallel on a cellulose film. **1940** *Trade Marks Jrnl.* 5 June 362/1 *Polaroid*...

Materials specially prepared for use in the polarization of light. Polaroid Corporation.., Dover, State of Delaware, United States of America; manufacturers. **1942** *Chem. Abstr.* XXXVI. 2183 By use of film polarizers, e.g., polaroid, the expense is small. **1946** F. Schneider *Qualitative Organic Microanalysis* iv. 119 The sections of Polaroid are cut so that their planes of polarization include an angle of approximately 5° when the segments are mounted in place with a slight overlap. **1949** H. C. Weston *Sight, Light & Efficiency* iv. 133 If they are placed between two thin plates of polaroid, or between crossed Nicol prisms, so that the light passing through them is polarised, the presence of strain is shown by the formation of a coloured pattern due to double refraction by the strained glass. **1956** *Nature* 3 Mar. 434/1 A cylindrical unit..is formed by cementing a calcite crystal between two pieces of 'Polaroid'. **1976** *Nature* 19 Aug. 709/2 All stimuli were plane polarised by a 60-cm diameter rotatable disk of polaroid positioned between the screen and the eye.

b. A piece of 'Polaroid'.

1955 *Physical Rev.* XCIX. 1694/1 The action on polarized light of the rotating λ/2 plate followed by a polaroid is to produce an interruption of the light at four times the rotation frequency. **1967** H. von Klüber in J. N. Xanthakis *Solar Physics* ix. 261 For nearly all analysers used in the detection of such inverse Zeeman effects—such as polaroids, double-splitting crystals, quarter- or half-wave plates, etc.—the result..is just the same as in the emission case. **1976** *Nature* 11 Mar. 155/1 The relative intensities of the red and green components could then be varied by rotating a Polaroid interposed in the common beam.

2. *pl.* Sunglasses containing 'Polaroid'.

[**1942** *Official Gaz.* (U.S. Patent Office) 20 Oct. 463/1 Polaroid Corporation, Cambridge, Mass... *Polaroid*..for viewing devices—namely, filters, lenses, eyeglasses, and goggles. Claims use..on eyeglasses since July, 1936; and on goggles since December, 1937.] **1959** C. MacInnes *Absolute Beginners* 163 The water sparkling so that I had on my Polaroids. **1959** *New Statesman* 19 Sept. 354/2 The light..beat back from the limestone with a spectral, otherworldly intensity that polaroids only served to accentuate. **1967** H. Hunter *Case for Punishment* i. 17 Cummings..was sporting a pair of heavy, black-rimmed spectacles. Inspector Shade had acquired a pair of green-tinted Polaroids. **1972** *Country Life* 23 Mar. 697/3 That afternoon my wife, wearing Polaroids, stationed herself opposite their lie and watched to see their reactions.

3. a. A kind of camera which develops the negative and produces a positive print within a short time of the exposure's being made.

1961 A. Gordon *Cipher* (1962) iv. 58 How about a nice picture, sir?.. I use a Polaroid... I'll have a print for you in a minute. **1966** H. B. Taylor *Triumvirate* xxxi. 168, I took a couple of pictures with my Polaroid, chatted a little. **1968** [see *Long Tom 5]. **1977** *N.Y. Rev. Bks.* 23 June 25/4 From the first Kodak, when it took weeks before a developed roll of film was returned to the amateur photographer, to the Polaroid, which ejects the image in a few seconds.

b. A photograph produced by such a camera.

1972 D. Marlowe *Do You remember England?* iv. 57 All I got out of it were two Polaroids of her. **1975** *New Yorker* 19 May 12/3 (Advt.), Large toned prints of an abstract character. Manipulated color Polaroids of bizarre activities. **1977** *Rolling Stone* 13 Jan. 22/2 Grace snapped a couple of Polaroids for the wedding album.

B. *attrib.* **1.** Applied to the polarizing material (see *A. 1) and articles in which it is employed.

1936 *Nature* 22 Aug. 312/2 It is by some such means described by Land that the Polaroid Corporation of Boston, U.S.A., has succeeded in manufacturing the 'Polaroid' sheets of polarizing material now on the market. *Ibid.*, One firm supplies Polaroid analysers and polarizers for the microscope. *Ibid.* 313/2 The details of reflecting objects can be seen more clearly through Polaroid films. **1951** *Electronic Engin.* XXIII. 10/2 The images.., plane polarized in mutually perpendicular directions by polaroid film, are superposed by a semi-reflecting mirror and are viewed through polaroid spectacles. **1954** *Ann. Reg. 1953* 365 Stereoscopic films viewed through polaroid spectacles were no novelty. **1958** *Woman* 9 Aug. 14/2 Polaroid sun-shield. **1961** E. N. Cameron *Ore Microsc.* ii. 20 The polarizer is a polaroid plate or a calcite prism. **1962** L. S. Sasieni *Optical Dispensing* xiii. 326 And Polaroid lenses are used as plano protective lenses (sunglasses), or as sighted lenses to prescription. **1965** *Wireless World* July 340/2 Many of the digital instruments.. incorporate polaroid filters to reduce reflected light.

2. Applied to a type of camera and photographs taken with it (see *A. 3).

1963 L. Deighton *Horse under Water* xxxv. 136 Two armed policemen..photographed me with a Polaroid camera and filed the photo. **1965** H. C. Shands in J. H. Masserman *Sci. & Psychoanalysis* VIII. 135 The final 'discovery' in science is more like a cooperative collage than it is like a two-second Polaroid snapshot. **1973** C. Sagan *Cosmic Connection* (1975) xv. 109 Page 100 shows a Polaroid photo of the video-monitor image of Phobos. **1976** *Early Music* IV. 451/1 Polaroid cameras produce a result very fast, but only the most sophisticated models provide a negative so that duplicate prints and enlargements are usually unobtainable. **1976** A. Grey *Bulgarian Exclusive* iv. 30 Two polaroid colour photographs of a dark-haired man.

polaron (pǝʊ·lărǫn). *Physics.* [f. Polar(ization and related words + *-on¹; orig. formed as Russ. *polyarón* (S. Pekar 1946, in *Zh. eksper. i teoret. Fiziki* XVI. 344).] A quasi-particle consisting of a free electron in an ionic crystal and the associated distortion of the crystal lattice.

1946 S. Pekar in *Jrnl. Physics* (Moscow) X. 343/2 Such local self-consistent quantum states of the electron in a crystal we shall briefly call polarons. **1955** *Physical Rev.* XCVII. 660/1 An electron moving with its accompanying distortion of the lattice has sometimes been called a polaron. **1969** *Nature* 15 Nov. 641/2 The interaction between an electron and its strain field in a polar solid gives rise to the excitation known as a polaron, the presence of which can be detected by irregularities in the shape of the conduction band. **1971** Mott & Davis *Electronic Processes in Non-Crystalline Materials* iv. 115 At low temperatures a polaron, whether large or small,..will behave exactly like a heavy particle, being scattered by impurities or lattice vibrations; moreover a high density of polarons can form a degenerate gas.

Hence **polaro·nic** *a.*

1978 *Nature* 16 Feb. 647/1 There will..be some stored energy involved in the system due to the polaronic nature of the moving charge carrier.

pole, *sb.*¹ Add: **1. b.** Colloq. phr. *up the* (or *a*) *pole*: in trouble or difficulty; in confusion, in error; drunk; mad, crazy; pregnant but unmarried.

1886–96 A. R. Marshall in Farmer & Henley *Slang* (1902) V. ii. 245/1 But, one cruel day, behind two slops he chanced to take a stroll, And..he heard himself alluded to as being up the pole. **1896** *Daily News* 1 Apr. 7/6 She remonstrated with the latter, and told him he was 'up a pole'—i.e. in the wrong. **1897** *Daily Tel.* 11 Dec. 10/4 Plaintiff:..but your little girl was frequently saying that you were 'up the poll [*sic*]'... The Judge: Up the what?.. The High Bailiff explained that the term was a slang one for being intoxicated. **1899** *Daily Mail* 29 Mar. 5/1 When there are nineteen Frenchmen to four Englishmen they were slightly up the pole. **1904** *Westm. Gaz.* 19 Mar. 7/2 Plaintiff's definition of the phrase 'up the pole' differed from that of her cousin..who said it meant being drunk. Mrs. Frasier said that it..meant being crazy. **1905** *Daily Chron.* 14 Dec. 6/4 Alec went to football smoker. Came home up the pole at one a.m... 'Up the pole,' Mrs. Norman said, was one of her husband's slang terms for a person under the influence of drink. **1906** E. Dyson *Fact'ry 'Ands* xiv. 188 Then, as a bright afterthought, she added, 'Yer fair up the pole!' **1915** C. J. Dennis *Songs of Sentimental Bloke* 49 The dreams I dreamed, the dilly thoughts I thunk Is up the pole, an' joy 'as done a bunk. **1916** E. V. Lucas *Vermillion Box* 165 It must require an awful lot of pluck... Either pluck or so much panic that one was practically up the pole with it. **1917** [see *blotto *a.*]. **1922** Joyce *Ulysses* 23 That red Carlisle girl, Lily... Spooning with him last night on the pier. The father is rotto with money.—Is she up the pole?—Better ask Seymour that. **1922** *Daily Mail* 20 Dec. 3 Keith came to her, saying he was 'up the pole and in a frightful mess'. **1932** D. L. Sayers *Have his Carcase* xxii. 205, I think we may take it for granted that our friend Weldon is a bit up the pole financially. **1950** J. Cleary *Just let me Be* xi. 108 If I go and see 'em now, tell 'em what I done and why I done it, I'd be well and truly up the pole. **1961** 'F. O'Brien' *Hard Life* v. 37 To say nothing of a lot of crooked Popes with their armies and their papal states, putting duchesses and nuns up the pole, and having all Italy littered with their bastards. **1965** W. Dick *Bunch of Ratbags* vi. 92 'Right,' said Curly, agreeing with Ronnie's logic for once. He generally thought Ronnie was all up the pole when giving advice to someone. **1970** R. Beilby *No Medals for Aphrodite* vi. 244 We'd 'a' been up the pole without him, that's why we didn't send him on his way. **1974** G. Moffat *Corpse Road* x. 142 'Do you really suspect that Pilgrim—*Pilgrim!*—killed the girl?..' 'You're up the pole,' Mrs Kent said to Page.

c. colloq. phr. *I wouldn't touch him* (*it*) *with a forty-foot* (or *ten-foot*) *pole* (and varr.): I refuse to have anything to do with him (it). Cf. *barge-pole.

1903 'T. Collins' *Such is Life* 22 The young feller he used to come sometimes an' just shake hands with her, but otherways he wouldn't touch her with a forty-foot pole. **1909** *Dialect Notes* III. 383, I wouldn't touch it with a ten-foot pole. **1941** *Coast to Coast* 167 'Me take the harness off him!' my mother said, surprised. 'Why, I wouldn't touch that mad thing with a forty-foot pole.' **1958** E. O. Schlunke *Village Hampden* 26 Attracting a lot of business of the more or less shady sort that our reputable men wouldn't touch with a forty-foot pole. **1974** P. Erdman *Silver Bears* i. 11 No respectable bank.. would touch our business with a ten-foot pole.

2. e. (*a*) *Racing.* The inside fence surrounding a racecourse; the starting position closest to the inside fence; (*b*) *Motor racing.* esp. in phr. *pole position*, the grid position which is on the front row and on the inside of the first bend; also *fig.*, any advantageous position.

1851 *Fraser's Mag.* XLIII. 657/1 The distance round is calculated at a mile,..For a saddle horse that has the pole, it comes practically to a little less... A horse 'has the pole', means that he has drawn the place nearest the inside boundary-fence of the track. *Ibid.* 659/1 'What a beauty she is!' says Harry. 'And she has the pole too.' **1868** H. Woodruff *Trotting Horse* xxiv. 206, I had the pole with Kemble Jackson, and soon took the lead. **1902** A. D. McFaul *Ike Glidden* xxii. 198 This stroke apparently gave the friends of the colt more confidence in the result, as drawing the pole was a position in favor of the colt. **1953** *Motor* 22 July 857/3 The newest B.R.M... had to be worked on all night and brought back to the course only just in time to come to the pole position on the starting line. **1963** *Times* 27 Apr. 3/3 J. Clark set a new unofficial lap record in his fuel injected Lotus 25..which gives him pole position on the starting grid. **1966** *Telegraph* (Brisbane) 14 May 17/3 Australian Jack Brabham, ..former world champion racing driver, snatched pole position for tomorrow's international formula one race. **1968** *Globe & Mail* (Toronto) 3 Feb. 36/2 Two Ford GT40s broke a course record and grabbed the pole positions for this weekend's 24 Hours of Daytona at Daytona Beach, Fla. **1969** *Australian* 24 May 34/4 Won Mobile 12¼f event here four starts back. Place claims from the 'pole'. **1971** *Sunday Times* 12 Sept. 50/3 The German company retained a pole position in hormone research which led to the Pill. **1976** *Milton Keynes Express* 11 June 42/7 Colin Hawker's Cosworth Grand Prix-engined VW was on pole. **1976** *Listener* 8 July 7/3 Brazil's foreign investment needs..would double its foreign borrowing, and take Brazil to pole position in the big league of world debtors. **1977** *News of World* 17 Apr. 24/7 Ipswich relinquished their hold on the pole position to champions Liverpool.

5. a. *pole barn, bridge* (earlier and later examples), *corral, fence* (earlier and later examples), *plantation*. **b.** *pole jump* (earlier and later examples), *-jumper* (examples), *-jumping, leap, -leaping* (examples).

1950 Pole barn [see *pole corral*]. **1960** *Farmer & Stock-breeder* (Suppl.) 19 Jan. 41/1 In winter they are in pens of 45 in pole-barns. **1969** *Times* 24 Feb. 12/1 They [*sc.* bullocks]..were present by the score,..making their systematic way to adjacent pole-barn silos. **1972** *N.Y. Law Jrnl.* 22 Aug. 13/8 (Advt.), Five bedrooms, 14 room modern home..on 12½ acres,..pole barn suitable for horses, riding trails nearby, would make ideal ski lodge. **1793** J. Lindley in *Michigan Pioneer & Hist. Coll.* (1892) XVII. 574 Pole bridges..made the tour very disagreeable. **1850** *Congress. Globe* 29 Jan. 240/1 Now, his colleague must start by the most direct route the 23rd of November, making provision for..contingencies of travel over corduroy roads, pole bridges, mud turnpikes, etc. **1974** D. Sears *Lark in Clear Air* ii. 34 We crossed a pole bridge over the Perch. **1950** *Amer. Speech* XXV. 85 No longer heard is *old land*, though *pole barn* and *pole corral* still describe forms of construction. **1962** J. Onslow *Bowler-Hatted Cowboy* v. 51 Two saddle horses which we turned into a small pole-corral. **1973** *Whig-Standard* (Kingston, Ont.) 11 Aug. 7/4 His..sheep..didn't want to bunch in the pole corral for the night. **1788** G. Washington *Diaries* (1925) III. 346 All hands were..finishing the Pole fence round the Barley and Pease in field No. 1. **1959** W. R. Bird *These are Maritimes* x. 278 We saw..many ancient pole fences crossing the fields. **1973** L. Russell *Everyday Life Colonial Canada* ii. 32 Probably the pole fence was the kind most frequently constructed at this stage [of land clearing]. In this, the trunks of moderate-sized trees were used, cut in lengths of 12 to 15 feet. **1886** *Year's Sport* 21 The other winners..were..pole jump, T. Ray; and wide jump, J. Purcell. **1912, 1920** Pole jump [see *decathlon]. **1887** M. Shearman *Athletics & Football* v. 163 Ulverston..has produced many fine pole-jumpers. **1908** *Westm. Gaz.* 1 July 8/4 Quite recently, Szathmary, the pole-jumper, broke the Hungarian record. **1868** H. F. Wilkinson *Mod. Athletics* viii. 88 Pole jumping. The leaping pole should be made of fir or ash. **1912** [see *field events* s.v. *field *sb.* 21]. **1931** E. Linklater *Juan in America* i. vi. 59 He could not help being tickled by the thought of an American university; pole-jumping and cheer-leaders. **1868** *Kendal Times* 5 Sept. 2/5 (*heading*) Grasmere annual sports... High pole leap for £1. **1869** *Ibid.* 18 Sept. 4/3 The high pole leaping..was gone through in capital style. **1885** F. Gale *Mod. Eng. Sports* vi. 67 Running, jumping, and pole-leaping were often the outcome of a very old-fashioned sport, 'Follow my leader'. **1888** T. Bright *Pole Plantations & Underwoods* i. 1 A pole plantation is an assemblage of young trees, the produce of plants that have been inserted in the soil at regular distances, or of the stems formed from such plants after their having been cut for poles.

c. pole-bean (examples); **pole-board,** a board or placard carried on poles like a banner; **pole-boat** now chiefly *Hist.*, a river-boat propelled by means of a pole or poles; so **pole-boating** *vbl. sb.*; **pole-bullock,** a bullock that is harnessed alongside the pole of a wagon; **pole-cure** *v.*, to cure (tobacco) by hanging it on poles; **pole-dray,** a dray furnished with or drawn by means of a pole; **pole-horse** (example); **pole-lathe** (further examples); **pole-masted** (further examples); **pole-mule,** a mule harnessed alongside the pole of a wagon; **pole-rose** = *pillar rose* s.v. Pillar *sb.* 12 in Dict. and Suppl.; **pole-screen** (further examples); **pole-trailer** (see quot. 1971); **pole-trap,** a bird-catching device which consists of a trap fixed to the top of a pole; **pole vault,** a jump over a horizontal bar which is achieved by means of a pole; = *pole-jump*; also *attrib.*; hence as *v. trans.* and *intr.*; so **pole-vaulter**; also *fig.*; **pole-vaulting** *vbl. sb.*; **pole-wagon,** a wagon furnished with or drawn by means of a pole; **pole-wound** *rare*, a wound that has been inflicted with a pole.

c **1770** 'J. H. St. J. de Crèvecœur' *Sk. 18th-Cent. Amer.* (1925) 120, I had once some hops and pole-beans, about twenty feet high. **1857** *Trans. Ill. Agric. Soc.* (1859) III. 503 There are many varieties of pole beans. **1865** [see *Lima b]. **1871** Mrs. Stowe *Oldtown Fireside Stories* 246 There was thick pole-beans quite up to the buttery-door. **1941** J. Stuart *Men of Mountains* 192 We drove down past Shelton's polebean patch. **1976** *Washington Post* 19 Apr. A10/7 (Advt.), Fresh Florida pole beans. 4 lbs. $1. **1909** *Westm. Gaz.* 29 Dec. 6/4 Others, again, carrying pole-boards setting forth all deceased's honours and titles. **1827** A. Sherwood *Gaz. Georgia* (1939) 22/1 Cargoes.. are thrown into pole boats. **1835** W. G. Simms *Partisan* II. ii. 12 At this point the river ceased to be navigable even for the common poleboats of the country. **1841** —— *Kinsmen* I. xiv. 163 Wherever a pole-boat had made its way, there had the name of Jack Bannister found repeated echoes. **1968** R. F. Adams *Western Words* 232/1 *Pole boat*, a river boat; so called because of the means by

which it was propelled upstream. **1837** A. Sherwood *Gaz. Georgia* (ed. 3) 193 A revolution in the mode and manner of transshipping goods must take place. The slow, tedious and expensive process of pole-boating will be exploded. **1930** L. G. D. Acland *Early Canterbury Runs* 1st Ser. vii. 150 In 1868 Strawberry, one of the pole-bullocks, died after working ten years and seven months on the station. **1933** *Press* (Christchurch, N.Z.) 18 Nov. 15/7 *Pole-bullocks*, the two bullocks which work on the p[ole] of a dray or waggon. **1899** *Rep. U.S. Dept. Agric.* No. 62. 30 The present method of manipulating these tobaccos after they are pole-cured is quite different from what it was years ago. **1848** H. W. Haygarth *Bush Life Australia* v. 49 In some districts..shaft-drays are used; but pole-drays are found to be more suitable to the nature of the country. **1889** *Harper's Mag.* June 160/2 The leaders sprang upward and onward.., the pole-horses simultaneously crashing backward and downward. **1859** Pole-lathe [see *back-rest 1]. **1932** G. M. Boumphrey *Story of Wheel* 38 When the lathe came here, it was altered into what is called the pole-lathe. **1968** J. Arnold *Shell Book of Country Crafts* xv. 200 Whether a turner uses a pole-lathe or the treadle type seems a matter for individual inclination... The extremely primitive pole-lathe came firstly with a primitive society, but its retention appears largely due to its simplicity and easy portability. **1915** *Yachting Monthly* XIX. 366/1 This necessitated a change of rig, and a pole-masted ketch was decided on. **1970** *Mariner's Mirror* LXVI. 165 Evidence for the pole-masted brigantine rig was found at the Dubrovnik museum. **1862** O. W. Norton *Army Lett.* (1903) 106 A driver riding the near pole mule and guiding his team with one line. **1848** W. Paul *Rose Garden* 67 Pillar or Pole Roses. **1937** *Burlington Mag.* Apr. p. xxiv/1 A mahogany Polescreen on a beautifully carved tripod stand. **1960** B. Snook *English Historical Embroidery* 98 Pole screens were not unconnected with vanity. Created in the 18th century to protect the make-up worn by elegant women from the heat of large open fires, they were very practical little pieces of furniture. **1976** *Northumberland Gaz.* 26 Nov. 18/4 (Advt.), Pair Regency mahogany pole screens. **1969** Pole trailer [see *expandable *a.*]. **1971** M. Tak *Truck Talk* 120 *Pole trailer*, a trailer composed of a single telescopic pole, a tandem rear-wheel unit and a coupling device used to join the trailer to a tractor. **1909** *Westm. Gaz.* 17 Feb. 5/1 The catching of birds with hooks in Cornwall has been stopped by an Act passed last year, and efforts have been made to abolish entirely the illegal pole-trap. **1972** *Guardian* 3 May 7/2 The illegal and viciously cruel pole trap..a spring trap fixed by a short chain to the top of a pole which kills the captured bird in an extremely slow and painful way—was made illegal in 1904... It appears to be still in common use. **1975** *Country Life* 2 Oct. 849/2 Even in Britain..prosecutions are still being brought for the illegal use of pole-traps. **1893** Pole vault [in Dict., sense 5 b]. **1935** *Encycl. Sports* 467/1 In front of the pole-vault bar is a slideway, in which the base of the pole is placed when jumping. *Ibid.* 467/2 The world's record pole vault is 14 ft. 4½ in., made by W. Graber..in 1932. **1951** *Time* 29 Jan. 85 Did, or did not, the Rev. Robert Richards become the second man in history to pole-vault 15 feet? **1973** *Times* 2 May 11/2 She..will train regularly with Michael Bull, who holds the British pole vault record and who is preparing for the decathlon in the Christchurch Commonwealth Games next January. **1977** *Western Morning News* 30 Aug. 12/7 Brian Hooper edged a little closer to the magical 18-foot pole vault as another landslide of wins carried Britain to overwhelming victory against West Germany. **1978** *Times* 18 July 2/8 Ape pole-vaults over park moat... A chimpanzee..escaped from an island by pole-vaulting across a moat using an 8 ft. pole. **1888** A. Randall-Diehl *2000 Words* 165 *Pole Vaulter*, one who practices leaping by aid of a pole. **1891** W. M'Combie Smith *Athletes & Athletic Sports of Scotland* viii. 88 From the moment he takes hold of the pole as he commences his run till he lets it go as he crosses the bar the pole-vaulter never shifts his hands. **1893** [in Dict., sense 5 b]. **1956** *Sun* (Baltimore) 13 Oct. (B ed.) 12/2 The United States Air Force unit responsible for making these jaunts across the roof of the world a daily routing is the 58th Weather Reconnaissance Squadron—more colloquially known as 'the pole vaulters'. **1976** *Milton Keynes Express* 30 July 43/2 Saturday's performance was helped by the presence of two new pole vaulters. **1877** *N.Y. Times* 5 Apr. 8/4 The following programme was adopted: one-mile walk, two-mile walk, one-mile run,..pole-vaulting. **1891** W. M'Combie Smith *Athletes & Athletic Sports of Scotland* viii. 88 Pole-vaulting is comparatively a recent introduction at Scottish sports. **1932** Webster & Heys *Exercises for Athletes* viii. 161 Now let us consider which are the muscles principally involved in the feat of pole-vaulting. **1968** M. Watman *Hist. Brit. Athletics* x. 168 Pole vaulting for height, as distinct from distance, was pioneered in the mid-nineteenth century by members of the Ulverston (Lancashire) Cricket and Football Club. **1908** S. Ford *Side-Stepping with Shorty* vi. 90 The pole waggon brings up the rear. **1910** G. B. McCutcheon *Rose in Ring* i. i. 6 Here and there, in the gloomy background, stood the canvas and pole wagons. **1968** J. Arnold *Shell Book of Country Crafts* v. 76 The pole-wagon, which was the fore-runner of the present-day motor-hauled trolley, had iron stanchions at its four corners of two transverse beams, each above its respective axle. **1908** Hardy *Dynasts* III. iv. vi. 417 Who knows but that we should have been kings too, but for my crooked legs and your running pole-wound?

pole, *sb.*² Add: **1.** (Later *fig.* example.)

1916 K. J. Saunders *Adventures Christian Soul* 68 When God's will is thy heart's pole, Then is Christ thy very soul.

2. d. colloq. phr. *poles apart* (or, less commonly, *removed*), completely opposite to or different from (someone or something).

1917 A. Huxley *Let.* 30 Sept. (1969) 134 They are deeply engaged in something very far removed from the sordid present, poles apart from any clap trap I may be talking about English literature. **1922** Joyce *Ulysses* 618 On this knotty point, however, the views of the pair, poles apart as they were, both in schooling and everything else, ..clashed. **1935** *Discovery* Feb. 52/2 Mr Dunne is poles apart from the dry-as-dusts who care not whether or no what they write is read. **1957** J. S. Huxley *Relig. without Revelation* iv. 95 Bringing together whole realms of fact.., which had hitherto seemed poles apart. **1966** *Listener* 13 Jan. 75/2 The world which his symphony inhabits is poles apart from that of Bruckner's symphonic 'confession'. **1971** *Scope* (S. Afr.) 19 Mar. 32/3 He is poles removed from men like Carel de Wet on the one side and Douglas Mitchell on the other.

9. (Further examples.)

1898 G. B. Shaw *Plays Pleasant* p. xi, These are the opposite poles of our system, represented in practice by our first rate managements at the one end, and the syndicates which exploit pornographic musical farces at the other. **1935** B. Malinowski in M. Black *Importance of Lang.* (1962) 77 These two poles of linguistic effectiveness, the magical and the pragmatic. **1965** *New Statesman* 30 Apr. 690/2 At the opposite pole to Tchaikovsky's introversion stands Verdi.

9*. *Math.* A point c in whose neighbourhood the magnitude of a function $f(z)$ becomes infinite, but in such a way that, were the function multiplied by an appropriate power of $(z-c)$, it would remain finite.

1879 *Encycl. Brit.* IX. 819/2 A rational (non-integral) function has a certain number of infinities, or poles, each of them of a given multiplicity. **1893** Harkness & Morley *Treat. Theory of Functions* iii. 112 If $f(z)$ be infinite at c, while $(z-c)^m f(z)$ is regular at c, m being a finite positive integer, c is said to be a pole of the function, and m is said to be the order of multiplicity, or simply the order, of the pole. **1935** E. T. Copson *Introd. Theory of Functions of Complex Variable* iv. 79 If $f(z)$ has a pole at a, $|f(z)|$ tends to infinity as z tends to a in any manner. Moreover, if $f(z)$ has a pole of order m at a, $1/f(z)$ is regular and has a zero of order m there. **1968** P. A. P. Moran *Introd. Probability Theory* vii. 299 $\phi_1(z)$ is therefore an analytic function with no zeros or poles. **1973** L. J. Tassie *Physics Elem. Particles* xii. 162 $F(l, k)$ is an analytic function of l except for poles above or on the real axis. These poles in the complex angular momentum plane are called Regge poles. The positions of the poles are analytic functions of the energy.

10. pole-cell, delete '(sense 6)' and add: any of the cells which move to the posterior end of the embryo in certain invertebrate species, and subsequently give rise to the germ line; (further examples); [tr. G. *polzelle* (A. Weismann 1863, in *Zeitschr. f. wissensch. Zool.* XIII. 111)]; **pole-changer** (earlier example); **pole figure** *Metallurgy* [tr. G. *polfigur* (F. Wever 1924, in *Zeitschr. f. Physik* XXVIII. 72)], a circular diagram that is a stereographic projection of a sphere showing the positions of the poles of one or more lattice planes of a crystal or crystalline substance, the intensity of any spot in the diagram being proportional to the number of planes having the corresponding orientation; **pole-finding paper,** impregnated paper which can be used to identify the sign of an electric terminal or the like by the change of colour it undergoes when brought into contact with it; **pole-hunting** *vbl. sb.*, the act of going on an expedition to either the North or South Poles; **pole-piece** (later example); **pole-shoe** *Electr.*, a detachable extension of a pole-piece.

1941 Johannsen & Butt *Embryol. Insects & Myriapods* ii. 10 This mass of cells are [*sic*] the germ cells.., also called 'polar globules' or 'pole cells' by earlier writers. **1969** R. F. Chapman *Insects* xviii. 365 In the Nematocera all the pole cells migrate in to form the germ cells in the gonads. **1839** Pole changer [see *alternating *ppl. a.* d]. [**1938** *Mem. Geol. Soc. Amer.* VI. 101 These oriented textures have been..recorded in the pole-diagram of metallography.] **1943** *Proc. Amer. Soc. Testing Materials* XLIII. 785/1 The only feasible method for studying two-dimensional preferment such as this is to make a pole figure for the specimen. Such a pole figure is a summarization by stereographic projection of data obtained for one set of crystal planes from a series of X-ray photographs taken at different angles to the sheet. **1962** R. E. Smallman *Mod. Physical Metallurgy* vi. 205 The scatter about the ideal orientation can only be represented by means of a pole-figure which shows the spread of orientation about the ideal for a particular set of (*hkl*) poles. **1902** J. E. Hutton in A. C. Harmsworth *Motors* viii. 145 'Pole-finding' paper may also be used for this purpose. **1963** G. M. B. Dobson *Explor. Atmos.* v. 84 These instruments recorded on 'pole-finding' paper the sign of the electric current flowing through a long wire hanging from the balloon. **1907** *Daily Chron.* 30 July 4/6 The Nimrod..sails from the East India Dock today to pick up Lieutenant Shackleton..and convey him towards the South Pole... But the point is not merely that the Nimrod is to go Pole-hunting. **1920** *Glasgow Herald* Aug. 4/2 Such an expedition [to the Antarctic], undertaken not for Pole-hunting but for observation and collection in all possible branches of science, accumulates abundant material. **1962** M. G. Say *Conc. Encycl. Electr. Engin.* 337/1 The magnets..consist of a circular yoke of cast iron to which inwardly projecting laminated main pole pieces are bolted. **1892** S. P. Thompson *Dynamo-Electr. Machinery* (ed. 4) xxiii. 657 Field-magnet cores, 8¼ inches long, 4½ inches diameter; pole-shoes, 8 inches by 3½ inches. **1901** Sheldon & Mason *Dynamo Electr. Machinery* iv. 75 Pole shoes are put on the ends of the pole pieces to distribute this flux over a wider area where it has to pass through the air. **1957** *Encycl. Brit.* VIII. 148/2 The enlarged portions of the poles near the armature are the pole shoes.

pole, *sb.*³ **1.** Substitute for def.: A marine flatfish, *Glyptocephalus cynoglossus*, of the family Pleuronectidæ, found in north-west European waters and on the North American side of the Atlantic; = Witch *sb.*⁴ (Later examples.)

1836 W. Yarrell *Hist. Brit. Fishes* II. 227 (*heading*) The Pole, or Craig Fluke. **1864** J. Couch *Hist. Fishes Brit. Islands* III. 190 The Pole is a fish of the Arctic Sea.

2. *Comb.* **pole-dab, -flounder** = prec. sense.

1838 *Mem. Wernerian Soc.* VII. 370 The Pole Dab is distinguished from the plaise [*sic*] in having no tubercles on the head. **1896** J. T. Cunningham *Marketable Marine Fishes* 233 The witch..has been called the pole dab, pole flounder, and long flounder by English naturalists. **1925** J. T. Jenkins *Fishes Brit. Isles* 184 The Witch, or Pole Dab, may be recognised by the fact that the eyes are on the right side of the head. **1969** A. Wheeler *Fishes Brit. Isles & N.-W. Europe* 542 (*heading*) Witch (Pole Dab). **1888, 1890** Pole Flounder [in Dict.]. **1896** [see *pole-dab* above].

pole, *v.*¹ Add: **4. c.** *Baseball.* To hit (the ball, a shot) hard. Also with *out*.

1905 C. Dryden *Champion Athletics* 40 At a tight spot in the game Hoffman poled out a vicious liner. **1943** *Amer. Speech* XVIII. 104 A batter who hits a line drive..is said ..to *pole one out*. **1976** *Webster's Sports Dict.* 326/1 *Pole*, to hit the ball hard.

8. To take advantage of someone; to impose or sponge *on*. *Austral. colloq.*

1906 E. Dyson *Fact'ry 'Ands* vi. 66 'What rot, girls; why don't yer get er shift on?' cried Feathers virtuously... ''Taint ther mealy pertater, polin' on the firm like this.' **1919** W. H. Downing *Digger Dialects* 38 *Poll*, to take advantage of another's good nature. **1938** X. Herbert *Capricornia* (1939) xxxii. 486 Call me a wastrel, would ya? You—why, you're poling on Jesus Christ! **1945** Baker *Austral. Lang.* v. 107 The N.S.W. Libraries Advisory Committee (1939)..said that inter-library loans had been summed up as 'poling instead of pooling'. **1946** K. Tennant *Lost Haven* (1947) xiv. 214 Only his own obstinacy kept him working, but Launce was as independent as any other man in Lost Haven. He wasn't going to pole on Alec. **1947** V. Palmer *Hail Tomorrow* i. 10, I asked him why he should come up north and pole on men who were trying to win decent conditions for themselves, but he said he wanted a holiday. **1953** 'Caddie' *Sydney Barmaid* xxxviii. 220 'And while there's anything in the Sutton cupboard, Caddie,' he assured me when I said I couldn't stay and pole on them, 'it's yours.' **1957** 'N. Culotta' *They're Weird Mob* (1958) xiii. 203 It [*sc.* bludger] means that you..'pole on yer mates'.

pole (pǝʊl), *v.*² *Physics.* [f. Pole *sb.*², or a back-formation from *poling *vbl. sb.*²] *trans.* To render (a ferroelectric material) electrically polar by the temporary application of a strong electric field.

1961 *Proc. IRE* XLIX. 1162/1 Certain polycrystalline ferroelectric substances..can be given lasting polar properties, including pyroelectric and piezoelectric effects, by treatment with high electric fields for a short time. The term 'to pole' is recommended for this treatment. **1963** *IEEE Trans. Ultrasonic Engin.* X. 38/2 The shell is poled in the radial direction. **1976** *Ibid.* XXIII. 394/1 The bar is poled perpendicular to the length direction.

So **po·ling** *ppl. a.*

1956 *IRE Trans. Ultrasonic Engin.* IV. 55 The stress is applied along the axis of the poling field. **1961** *Proc. IRE* XLIX. 1166/2 The shear deformation..occurs in the plane containing both the poling and signal fields.

Pol. Econ., colloq. abbrev. of *Political Economy*. See Economy 3.

1893 W. K. Post *Harvard Stories* 12, I have not been tutoring you in Pol. Econ. **1900** *Dialect Notes* II. 16 Some tendencies in student English..are worth a passing remark. Most significant..is the tendency to use abbreviations... The subjects he studies..are all clipped... Thus he studies..*pol-econ* (political economy).

poled (pǝʊld), *a.*¹ [f. Pole *sb.*¹ or *v.*] **1.** Provided with or supported by a pole or poles.

1864 E. A. Parkes *Man. Pract. Hygiene* i. ix. 287 This is a two-poled tent, with a connecting ridge-pole. *Ibid.* 288 The first is a single poled pyramidal tent, with a second pole to sustain the entrance flap. **1894** *Westm. Gaz.* 1 Jan. 2/1 But it is pointed out that the proportion of 'poled' steamers, and of sailing ships with masts that can be readily lowered, is always increasing.

2. = Poleaxed *ppl. a.*

1920 *Outward Bound* Nov. 20/2 It caught him fairly above the ear so that he fell like a poled ox.

poled (pǝʊld), *a.*² *Physics.* [f. Pole *sb.*², *pole *v.*² + -ed¹, -ed².] Of a ferroelectric material: rendered electrically polar (see *pole *v.*²).

1961 *Proc. IRE* XLIX. 1162/1 Poled ferroelectric ceramics have become an important component in electromechanical devices. **1975** D. G. Fink *Electronics Engineers' Handbk.* vii. 58 Poled ferroelectric devices are capable of doing electric work when driven mechanically, or mechanical work when driven electrically.

polemicize (pole·misǝiz), *v.* [f. Polemic *a.* and *sb.* + -ize.] *intr.* = Polemize *v.* Hence **pole·micizing** *vbl. sb.*

1953 *Encounter* Nov. 49/2 But Vigolo's edition arrived just in time to save Roman literary honour, and..to polemicise against her American view of the poet. **1968** *Listener* 6 June 728/3 You might want to polemicise and say: 'Hang on a minute! What about the swallowing up of small farms? What about the fact that British agriculture as a whole is subsidised by the taxpayer to the tune of something like three million pounds a year?' The trouble is that there isn't anyone to polemicise with. **1969** A. Walicki in Ionescu & Gellner *Populism* 69 He sharply polemicized with Chernyshevsky and Dubroliubov, defending the spiritual heritage of the 'superfluous men' from the gentry. **1970** R. J. Hollingdale tr. *Schopenhauer's Essays* 215 The polemicizing against the assumption of a life force which is now becoming fashionable.

polemology (pǫlɪmǫ·lǒdʒi). [f. Gr. πόλεμο- combining form of πόλεμος war + Gr. -λογία, -logy.] The science or study of war. Hence **polemological** *a.*; **polemo·logist.**

1938 *Nature* 23 Apr. 728/1 Last of all, perhaps, has been a lively call to solve the problems of polemology (*sit venia verbo*). **1968** *Sunday Times* 4 Feb. 53/3 Students of polemology will not find anything startlingly original in what he has to say. **1968** *N.Y. Times* 26 Aug. 35 There is a French Institute of Polemology in Paris and a Polemological Institute at Groningen in the Netherlands. **1970** H. Arendt *On Violence* 59 A brand-new science, called 'polemology', has emerged. *Ibid.* 64 This..is precisely what the psychiatrists and polemologists concerned with human aggressiveness recommend. **1970** *Harper's* Dec. 28 Professor Gaston Bouthoul of the *Ecole des Hautes Etudes Sociales* in Paris..the founder of Polemology (a word he coined), or the study of war. **1976** *Times Lit. Suppl.* 2 Apr. 363/2 *Urban Guerrilla* contains the papers read at a conference in 1974 organized by the Polemological Centre of the Free University of Brussels.

polemonium (pǫlǐmǝʊ·ni,ǒm). [mod.L. (J. P. de Tournefort *Institutiones Rei Herbariæ* (1700) I. 146), a. Gk. πολεμώνιον: see Polemoniaceous *a.*] An annual or perennial herb of the genus so called, belonging to the family Polemoniaceæ, native to America, Asia, or Europe, and bearing single or clustered bell-shaped flowers. Cf. Jacob's ladder 1.

1900 J. M. Abbott in W. D. Drury *Bk. Gardening* viii. 279 Polemoniums are free-flowering plants. **1931** *Times Educ. Suppl.* 25 July (Home & Classroom Sect.) p. iv/2 We fairly waded through flowers, including pale blue borage, mauve polemonium,..and blue primula. **1957** *Dict. Gardening* (R. Hort. Soc.) III. 1620/1 The Polemoniums are hardy and several good for the rock-garden, the larger ones for the border. **1968** J. Berrisford *Very Small Garden* viii. 95 Polemoniums, with their 'Jacob's ladder' leaves, and the taller geranium species are good.

Polenske (pōle·nskǝ). [Name of E. *Polenske* (fl. 1904), German public health chemist.] *Polenske number* or *value*: a number expressing the proportion of volatile, water-insoluble fatty acids in a fat (see quot. 1973).

1906 *Analyst* XXXI. 259 When a low Reichert-Wollny figure was associated with fluid insoluble volatile acids, and especially when the Polenske number exceeded 2, the proof of the presence of cocoanut oil [in the butter] was fairly definite. *Ibid.* 260 He knew nothing about the Polenske value under these conditions. **1928** [see *Kirschner]. **1936** *Analyst* LXVI. 407 (*heading*) Polenske (or insoluble volatile acid) value. **1973** D. Pearson *Lab. Techniques Food Anal.* vi. 153 The proportion of butter fat present is frequently assessed from the Reichert and Polenske values... The Reichert value is the volume of 0·1 N alkali in millilitres required to neutralise the water-soluble volatile fatty acids distilled from 5 g of fat under specified conditions. From the same distillation, the Polenske value is the volume of 0·1 N alkali in millilitres required to neutralise the water-insoluble volatile fatty acids.

poler. Add: **3.** (Earlier and later N.Z. and Austral. examples.)

1863 S. Butler *First Year in Canterbury Settlement* vii. 105 The leaders..slewed sharply round, and tied themselves into an inextricable knot with the polars. **1878** E. S. Elwell *Boy Colonists* 234 The polers, that is, the bullocks attached to the pole of the dray, and accustomed to bear the weight of the dray-load on their necks, are always the quietest. **1929** W. S. Smyth *Bonzer Jones* 97 The 'polers' pulled back. **1936** I. L. Idriess *Cattle King* viii. 68 They call the two polers the pin bullocks, because they swing the turntable of the wagon! **1941** —— *Great Boomerang* vi. 48 The leaders had reared right back on the polers, the team in terrified confusion. **1959** H. P. Tritton *Time means Tucker* ii. 36 A bullock-team is made up in four parts: polers, pin, body and leaders... The polers..have the job of steadying the dray or wagon while the pin-bullocks take the pull. **1972** *Sunday Mail Mag.* (Brisbane) 21 May 2/2 The team was made up as follows: Polers on the wagon pole were usually low set and chunky.

4. b. A boat propelled with a pole or poles.

1925 *Chambers's Jrnl.* Apr. 253/1, I was travelling by 'poler' because no steamer was available.

5. *Austral. slang.* A cadger, sponger; one who shirks work. Cf. *pole *v.*¹ 8.

1938 X. Herbert *Capricornia* (1939) xxxii. 486 'You long-jawed poler,' Norman roared. 'Living on the fat of the land, while your poor damn flock feeds on soup and coconuts and what they can root out of the bush.' **1945** [see *hummer *sb.*² 2]. **1947** I. Douglas *Opportunity in Australia* 89 *Poler*, one who does not pull his weight. **1952** A. G. Mitchell in *Chambers's Shorter Eng. Dict.* Suppl., *Poler*, one who sponges on another, or avoids his fair share of work... (The polers in a bullock team are yoked to the pole and often leave most of the pulling to the leading bullocks.)

pole-star. Add: **1.** (Later examples.)

The star α Ursæ Minoris is now (1982) within 1° of the north celestial pole, and will remain so until the 23rd c. A.D.

1946 Dylan Thomas *Deaths & Entrances* 24 Your pole-star neighbour, sun of another street, Will dive up to his tears. **1955** *Sci. News Let.* 27 Aug. 138/1 This group [*sc.* Ursa Minor] contains the little dipper, with Polaris, the pole star, at the end of the handle. **1979** R. Laidlaw *Lion is Rampant* xiii. 108 As long as I could see The Plough I could get a fix on the Pole Star and check my direction.

polewards, *adv.* (Examples referring to another sense of Pole *sb.*²).

1895 *Anat. Anzeiger* XI. 74 The divergent limbs which travel polewards along the spindle are in reality the original halves of a chromosome, which has become doubled, or bent, upon itself. **1925** *Jrnl. Genetics* XV. 252 The constituent chromosomes of the bivalents are pulled polewards into loops.

poley, polley, *a.* and *sb.* Add: Also *N.Z.* and *U.S.* Also **polly. A.** *adj.* (Further examples.) Also *fig.*

1901 M. Franklin *My Brilliant Career* xxviii. 233 A couple of dirty knives and forks, a pair of cracked plates, two poley cups and chipped saucers. *Ibid.* 234 A cup was broken, and another, also a poley, was put in its stead. **1922** Joyce *Ulysses* 289 Angus heifers and polly bullocks of immaculate pedigree. **1930** L. G. D. Acland *Early Canterbury Runs* 1st Ser. vii. 173 A new-chum had a theory that if you turned your back on a beast and stooped down, and looked at him through your legs he wouldn't charge, and how a noted poley bullock completely exploded his theory. **1954** *Coast to Coast 1953–54* 14 Home made cartridges? There they go again! No. Poley chokes on their guns. That American idea for greater range.

B. *sb.* A type of saddle (see quots. 1958, 1966).

1958 *Amer. Speech* XXXIII. 167 *Poley*, a saddle without a pommel. **1966** Baker *Austral. Lang.* (ed. 2) iii. 67 Another type of saddle is the *poley*. This is a saddle without kneepads—'like a poley bull without horns, but generally considered to be derived from polo saddle'. **1975** *Sunday Mail Mag.* (Brisbane) 26 Jan. 15/1 My own poley had had its day... Good second-hand saddles were not easy to come by.

polianite (pǝʊ·liănǝit). *Min.* [ad. G. *polianit* (A. Breithaupt 1844, in *Ann. der Physik* LXI. 191), f. Gr. πολιά greyness (in allusion to its colour): see -ite¹.] A variety of pyrolusite that occurs as large well-formed crystals.

1849 J. Nicol *Man. Mineral.* 420 Polianite..acts like pure hyperoxide of manganese. **1937** *Mineral. Mag.* XXIV. 521 (*heading*) X-ray studies on pyrolusite (including polianite) and psilomelane. *Ibid.*, The present study had for its main objective an attempt to settle..the uncertainty existing as to the relationship between polianite and pyrolusite. **1969** *Mineral. Abstr.* XX. 289/1 The Mn oxides..have a microrhythmic structure..; in the veins they include polianite.

police, *sb.* Add: Also in reduced forms **polie** (*Sc.*), **p'leece,** etc. See also *polis². **3. c.** (Earlier example.)

1834 J. Kemper in *Wisconsin Hist. Coll.* (1898) XIV. 412 The towels, basins &c here are not what they ought to be. The police of the boat is bad.

5. a. (Further examples of *pl.* const.)

1922 Joyce *Ulysses* 162 Squads of police marching out, back. **1970** *Daily Tel.* 27 June 1/4 One hundred police and 200 civilians yesterday searched lonely country around Stephen's home. **1973** *Ibid.* 21 Mar. 10 The factory-gate clash between 400 pickets and 232 police.. foreshadows a new style of strike demonstration. **1976** *Daily Record* (Glasgow) 4 Dec. 32/2 The police then gave evidence, after being told they need not answer questions.

c. (As a count noun.) A policeman. Chiefly *Sc.* and *U.S. colloq.*

1839 *Chicago American* 5 Sept., There is a police in attendance..in the theatre. **1856** 'Mark Twain' *Adv. Thomas Jefferson Snodgrass* (1928) 8 He was a police. **1890** J. Kerr *Reminisc.* I. 98 Then for a while the loon to jail Was taken by a polie, O. **1904,** etc. [see *polis¹]. **1951** M. McLuhan *Mechanical Bride* 107/2 Joyce's famous remark that, 'though he might have been more humble, there's no police like Holmes,' contains a world of insight. **1960** *Huntly Express* 19 Aug. 7 It was all over the market that 'the unco man wis a p'leece wi' plain claes'. **1964** J. H. Clarke *Harlem* 277 He crawled out of th' door hollerin' for a police to save him. **1975** *Caribbean Contact* Feb. 14/1 His father was a police and his mother, familiarly known as 'Sister Lu', a laundry worker.

6. *police agent, ball, boat, cadet, car, cell, charge-sheet, chief, college, commissioner, cordon, courtroom, department, doctor, force* (earlier and later examples), *headquarters, horse, house* (examples), *jeep, laboratory, launch, medal, patrol, photograph, photographer, power, procedure, protection* (examples), *radio, raid, regulation, report, spy* (earlier and later examples), *spying, surgeon, system, van, woman* (further examples), *work*; also *police-aided, -controlled, -protected* adjs.; **police action,** (*a*) the deeds or activity of the police; (*b*) military intervention without a formal declaration of war when a nation or group within a nation is considered to be violating international law and peace; **police bail** (see quot. 1976); **police blotter:** see *blotter 4; **police box,** (*a*) a box or kiosk containing a telephone specially for the use of police or of members of the public wishing to contact the police; (*b*) a reinforced shelter on London streets during the 1939–45 war for the protection of policemen on duty during an air raid; **police boy,** in European colonies or former European territories, a 'native' police assistant or security officer; **police captain,** (examples); also *Ireland*; **police cruiser** *N. Amer.*, a police patrol car; **police dispatcher** *U.S.*, a member of the staff of a police station who receives information about crimes and transmits it to police patrols; **police dog,** (*a*) a dog employed by the police to track and capture criminals, to find lost persons, etc.; (*b*) = *Alsatian *sb.* B.2; **police grip** *rare*, a grip or hold used by policemen; **police informer,** a criminal who gives information about crime to the police; **police judge,** also *U.S.*; **police lock,** (see quot. 1975); **police matron,** a policewoman who takes charge of women or juveniles at a police station or in court; **police message** (see quot. 1941); **police novel,** a detective novel in which police procedures in detecting crime form the central interest; **police officer** (later example); **police orphanage,** a home for the orphans of policemen; **police positive,** a type of Colt's pistol; **police procedural** *a.*, of or pertaining to police procedure, applied *spec.* to a type of crime detection story; also as *sb.*, = *police novel*; **police record,** a dossier kept by the police on all persons convicted of crime; hence, a past which includes some conviction for crime; **police reporter,** a newspaper reporter who concentrates on stories concerning crime and police activity; **police science,** the science dealing with the investigation of crime; so **police scientist; police siren,** the siren on a police vehicle; **police special,** a type of revolver; **police trap,** an arrangement made by police for detecting motorists who exceed the speed limit, or for apprehending criminals or other wanted persons; also *fig.*; so **police-trapped** *a.*; **police whistle,** a special type of loud whistle used by the police; **police-witness,** a witness whose testimony supports a police prosecution.

1885 *Encycl. Brit.* XIX. 334/1 Police action in relation to the serious matters constituting crime is familiar knowledge. **1933** *Week-end Rev.* 1 July 17/1 Blurring the distinction between war the duel and 'police action'. **1959** *Chambers's Encycl.* VII. 512/2 The Dutch started the first 'police action' and occupied large parts of the republic. After United Nations intervention a truce agreement was signed. **1968** *Listener* 19 Dec. 821/3 What's the difference..between police action and war, if the soldiers of the two sides are killing each other? **1978** G. Vidal *Kalki* iii. 51 He would have been able to avoid the Vietnam war—or 'police action', to properly designate that valiant attempt to save Southeast Asia for the free world. **1852** E. E. Hale *If, Yes, & Perhaps* (1868) 44, I had told the police agent he might send it to the St. Nicholas. **1930** G. B. Shaw *Apple Cart* p. xvi, Proletariats are never revolutionary, and..their direct action, when it is controlled at all, is usually controlled by police agents. **1910** *Times* 21 Mar. 13/1 Mr. Winston Churchill will visit Brighton on April 2 to make himself acquainted with the Brighton police aided scheme for clothing poor children, in which the King recently showed an interest. **1922** Joyce *Ulysses* 670 Embroidery, darning or knitting for the policeaided clothing society. **1976** *Daily Tel.* 6 Feb. 2/8 Where the police are unable to complete their inquiries immediately, they also have power under the Magistrates' Courts Act 1952 to grant what is known as 'police bail'. The person in custody is required to enter into a recognizance, with or without sureties, to appear again at a police station at a certain time for further questioning. **1977** *Gay News* 24 Mar. 1/3 All the customers were released on police bail. **1969** C. Watson *Bump in Night* ii. 26 Don't forget it's the police ball on the 14th. **1973** *Guardian* 12 Mar. 1/8 Saturday night's killings occurred while most of Bermuda's unarmed police force were at the semi-annual police ball. **1873** G. Lening *Dark Side of N.Y. Life* 155 The six police-boats cannot be usefully employed on the rivers in the interest of commerce. **1890** A. Conan Doyle *Sign of Four* ix. 182, I shall want a fast police-boat—a steam-launch—to be at the Westminster Stairs at seven o'clock. **1943** *Sun* (Baltimore) 13 Dec. 5/1 The children and guarding against sabotage now present the greatest number of police-boat problems. **1975** 'A. Hall' *Mandarin Cypher* vii. 105 A police boat had put to sea at full speed. **1932** *Daily Tel.* 23 May 12/2 A policeman was cut off in a police-box by 10ft of water and had to be rescued. **1940** R. Morrish *Police & Crime-Detection* iii. 36 Many police districts now possess their police-box system, by means of which officers performing duty in outlying districts are able to

communicate with their stations. **1941** *Newsweek* 13 Oct. 29 One of many air-raid precautions taken in the British capital for the expected winter Luftwaffe attacks is the building of 'police boxes' at street intersections. The reinforced brick shelters will protect London Bobbies on duty during Nazi air raids. **1971** R. Amberley *Ordinary Accident* xiii. 116 Someone, evidently ringing from the police box on the Banbury road. **1946** C. B. Jeppe *Gold Mining on Witwatersrand* II. xvi. 1758 On all the mines of the Witwatersrand (1943) there were 142 Europeans and 1,887 native 'police boys' in the police organisation (underground and surface). **1961** *Quest* Oct./Dec. 33/2 Motu prevailed over the candidates firstly because it was based on the speech of the police-boys who formed the larger part of the total mobilized section of the population, and secondly because these police-boys were much more accessible to the ordinary people than the English-speaking rulers. **1971** *Sunday Times* (Johannesburg) 28 Mar. 16/5 This clause shall not apply to..labourers and watch men or police boys. **1959** M. Gilbert *Blood & Judgement* xii. 131 A police cadet motor-cyclist was propping his machine up. **1976** *Southern Even. Echo* (Southampton) 12 Nov. 15/6 He wanted to become a police cadet—and had even been to the police station to find out details of what was involved in following this career. **1834** M. Edgeworth *Let.* in *Tour in Connemara & Martins of Ballinahinch* (1950) 93 The cottage at the end of the walk to Swinnerton, in which I believe a police-captain Henderson lived in your time. **1902** *Chambers's Jrnl.* Oct. 674/1 The next grade above is that of sergeant. Above this comes the police captain. **1976** H. Nielsen *Brink of Murder* xxi. 187 He's smart. He knew the odds against nailing a police captain. **1924** A. Christie *Poirot Investigates* viii. 221 A large police car was waiting for us, with some plain-clothes men. **1931** E. S. Gardner *Candy Kid* in *Case of Crying Swallow* (1972) 92 The two men in the police car glimpsed boxes of candy in the rear of the sedan. **1971** B. Patten *Irrelevant Song* 62 The sirens wailing on police-cars. **1898** *McClure's Mag.* X. 547/2 He was still in the infirmary attached to the police-cells. **1965** D. Francis *For Kicks* xix. 240 Four nights and three days in a police cell. **1977** J. Wainwright *Do Nothin'* xiii. 220 They are already on their way to some all-mod-con police cell. **1922** Joyce *Ulysses* 180 Police chargesheets crammed with cases get their percentage manufacturing crime. **1961** Copp & Peck *Betrayal at UN* iv. xxi. 208 Colonel Frank Begley, former Police Chief of Farmington, Conn., and now head of U.N. special police. **1974** E. Ambler *Dr Frigo* ii. 104 How does he reward him? Make him police chief or head of intelligence? **1977** H. Fast *Immigrants* i. 65 He could have Mayor McCarthy and Police Chief Martin as guests in his house. **1936** N. Marsh *Death in Ecstasy* ix. 109 Have you been through the Police College? **1958** S. Hyland *Who goes Hang?* i. v. 24 The senior lecturer in forensic studies at any of your admirable police colleges. **1976** *Daily Mail* (Hull) 16 Dec., He was chosen as the first member of that rank to go to the Police College at Ryton on Dunsmore. **1869** *Harper's Mag.* Oct. 754/1 The system of relief that has now become the perplexity of the Police Commissioners. **1911** *Daily Colonist* (Victoria, B.C.) 11 Apr. 7/1 A meeting of the police commissioners will be held this afternoon..when the department's estimates for the year will be considered. **1977** *Hongkong Standard* 12 Apr. 8/3 The courageous, determined and almost unique efforts of former Police Commissioner C.S. to root out internal graft backfired. **1939** H. Hodge *Cab, Sir?* 236 A police-controlled cross-roads. **1961** *Times* 6 Dec. 15/5 Let him [*sc.* the Minister of Transport] give us more police-controlled and light-controlled crossings. **1970** 'D. Halliday' *Dolly & Cookie Bird* xii. 193 I'd put a police cordon round the house with orders not to let anyone through. **1979** J. Gardner *Nostradamus Traitor* xvi. 59 He carried ID which would get him through any police cordon. **1866** 'Mark Twain' *Lett. from Hawaii* (1967) 28 The old police courtroom in San Francisco. **1937** C. Himes *Black on Black* (1973) 143 He broke away..before the police cruiser got there. **1958** 'Castle' & 'Hailey' *Flight into Danger* viii. 107 At the turn-off from the main highway..a police cruiser stood..its roof-light blinking a constant warning. **1974** *Globe & Mail* (Toronto) 16 Jan. 8/5 With the red light flashing, the police cruiser tails the motorist down the street and flags him over to the curb. **1976** *Billings* (Montana) *Gaz.* 2 July 8-B/4 In Louisville, police began a slowdown and 41 police cruisers notified headquarters they had been disabled by flat tires. **1810** *Rec. Early Hist. Boston* (1904) XXXIII. 495 In the estimate of expences for the..year, a sum was named for the police department. **1931** E. S. Gardner *Vanishing Corpse* in *Case of Crying Swallow* (1972) 107 [He] was on terms of intimacy with most of the police department heads. **1963** 'E. McBain' *Ten Plus One* (1964) 24 The police department is a vast organization, and a detective is only an organization man. **1977** H. Fast *Immigrants* ii. 110 In the annals of the San Francisco Police Department, a dead Chinese was statistically different from a dead Caucasian. **1973** *Tucson* (Arizona) *Daily Citizen* 22 Aug. 11/1 Police dispatcher Bill Pyles called an aid car, then began telling Mrs. Sweet how to keep little Jeremy from dying. **1976** *Billings* (Montana) *Gaz.* 2 July 1-A/2 Heagerty was about to be fined when a police dispatcher intervened and told the judge of Heagerty's plans. **1934** M. Allingham *Death of Ghost* vii. 86 The altruistic murderer is rare, and of course I couldn't say what the chances of your being one were until we have the evidence of the police doctor. **1972** K. Royce *Miniatures Frame* viii. 103 If I knew anything of police doctors the fuzz would ring around for hours until one reluctantly agreed to come. **1908** *Daily Chron.* 28 Aug. 7/2 Most of the principal German towns possess police dogs. **1911** *Chambers's Jrnl.* Feb. 136/1 What is needed as an ideal police-dog is an animal that can not only track well, but that can attack the criminal. **1925** F. Scott Fitzgerald *Great Gatsby* ii. 32 I'd like to get one of those police dogs; I don't suppose you got that kind? **1974** *Encycl. Brit. Macropædia* XIV. 672/1 Police dogs. The training of dogs for police work was originally developed in Ghent, Belgium, about 1900 and was soon copied elsewhere. **1838** Dickens in *Bentley's Misc.* Sept. 222 Professor Nogo wished to be informed what amount of automaton police force it was proposed to raise. **1956** H. Nicolson *Diary* 2 Nov. (1968) 314 There is a suggestion..that UNO should police the area, but it will take a long time before the police-force can be assembled. **1968** *Listener* 21 Nov. 667/1 As I saw it, the UN must move quickly to set up some kind of international police force. **1976** *Daily Record* (Glasgow) 4 Dec. 2/5 More than 1000 officers from every prison, police force, and state mental institution in the country turned up for the funeral service at Carnwath village in Lanarkshire. **1910** Police grip [see *ju-jitsu *sb.*]. **1875** M. D. Landon *E. Perkins* 237 The first thing you must do after the child is lost is to go to the Police Headquarters. **1952** Auden *Nones* 37 Between the burnt-out Law Courts and Police Headquarters. **1971** W. J. Burley *Guilt Edged* i. 9 The new police headquarters on the outskirts of the city. **1935** N. Mitchison *We have been Warned* iv. 453 She was knocked down.. almost under the nose of a police horse. **1973** R. Busby *Pattern of Violence* x. 157 A troop of police horses from the mounted branch held the crowd back. **1969** J. Wainwright *Big Tickle* 52 To live in his own house and not a 'police house' gave Cohen a freedom denied most police officers. **1974** L. Lamb *Man in Mist* viii. 51 Its [*sic*] P.C. Marchant, sir, who has the police house at Buntingbury. **1965** N. Freeling *Criminal Conversation* ii. xx. 191 [He] looked like a gentleman... Janus had never imagined he might be a police informer. **1969** G. Greene *Travels with my Aunt* i. vii. 60 He suspected me to be some unsavoury police informer. **1977** M. Kenyon *Rapist* vi. 73 Is it police informer I'm to be now, Sergeant? **1972** E. Hargreaves *Fair Green Weed* vii. 89 A sergeant, a corporal and a young police doctor arrived in the police jeep. **1956** B. Holiday *Lady sings Blues* (1973) ii. 23 It was Magistrate Jean Hortense Norris, the first woman police judge in New York, a tough hard-faced old dame. **1976** *Pioneer* (Big Timber, Montana) 30 June 12/2 He served in the Armed Forces during World War II, was a Montana Highway Patrolman and a police judge for the City of Billings. **1937** D. & H. Teilhet *Feather Cloak Murders* v. 85 The Chinese gentleman in the police laboratory removed the wicked little steel dart from under the Zeiss microscope. **1974** 'R. Tate' *Birds of Blooded Feather* iii. 53 He's a policeman... Something to do with the police laboratories. **1940** R. Morrish *Police & Crime-Detection* iii. 39 Some of the police-launches on the river Thames are also fitted with radio. **1972** *Police Rev.* 1 Dec. 1565/2 Section Officer Marshall swept the pale blue and white Police launch across the lake surface. **1975** J. Aiken *Voices in Empty House* 15 The apartment was further defended by a police lock, a rod which fastened into place on the back of the door and hooked into the floor, so that the door would not open beyond a certain point. **1976** *New Yorker* 1 Mar. 33/2 She slammed the door and I heard the police lock snap into place, then silence. **1934** Webster, Police matron. **1942** A. Christie *Body in Library* xiv. 133 In the corner of Superintendent Harper's office sat an elderly lady... She was certainly no police matron. **1972** R. Bloch *Night-World* (1974) xvii. 113 You'd have a police matron assigned to you, but they wouldn't put you in a cell. **1955** M. Allingham *Beckoning Lady* ii. 72 Divisional Detective Chief Inspector Charles Luke..had emerged from hospital with..a recommendation for the coveted Police Medal. **1977** 'A. York' *Tallant for Trouble* ii. 13 He had proved..a good policeman..his courage during one of Guyana's mini-revolutions had earned him a Queen's Police Medal. **1938** N. Marsh *Artists in Crime* xv. 219 The B.B.C. had instructions to send out a police message. **1941** *B.B.C. Gloss. Broadcasting Terms* 23 *Police Message*, message broadcast at the request of the police authorities. **1968** T. Stoppard *Real Inspector Hound* 13 We interrupt our programme for a special police message. **1896** G. B. Shaw *Our Theatres in Nineties* (1932) II. 223 As a novel, I can pass my idle hour with it, just as Bismarck used to pass his with the police novels of Du Boisgobey. **1908** G. K. Chesterton *All Things Considered* 116 The police novel..permits privacy only to explode and smash privacy. **1972** J. Symons *Bloody Murder* xiv. 197 The police novel, or the police-procedural as it has recently been called, concentrates upon the detailed investigation of a crime from the point of view of the police. **1976** M. Underwood *Menaces* xii. 114 Having been a police officer, she was much better equipped than most to be the wife of one. **1938** M. Allingham *Fashion in Shrouds* xx. 360 [He] made the suggestion as if he were announcing a rich gift to the Police Orphanage. **1972** M. Gilbert *Body of Girl* xxiii. 199 Any unclaimed money goes to the Police Orphanage. **1899** Somerville & 'Ross' *Some Experiences Irish R.M.* iii. 63 Flurry espied the police patrol on the road. **1936** 'N. Blake' *Thou Shell of Death* xiv. 258 On the main road he'd have to go straight for a bit, and the police patrols would be out. **1947** *Sun* (Baltimore) 7 May 7/3 At first, the jail visitors balked at riding to the prison in a police patrol wagon. **1969** M. Pugh *Last Place Left* xxix. 209, I suppose you'll recognize a police patrol car? **1974** *Times* 4 Oct. 3/3 She was given the task of observing the strength of police patrols. **1943** G. Greene *Ministry of Fear* iii. i. 163 A police photograph is like a passport photograph... We protest: This isn't me. **1965** E. S. Gardner *Case of Beautiful Beggar* (1972) 7 He took us into the police laboratory and museum, and started explaining the cases on which he had worked..all..illustrated by police photographs. **1978** 'A. Garve' *Counterstroke* i. 38 There were two police photographs, full face and profile. **1931** M. Allingham *Police at Funeral* xv. 206 Mr. Bowditch and a police photographer had completed their work on the footprint. **1970** W. J. Burley *To kill a Cat* i. 21 The police photographers..had already taken pictures of the undisturbed room. **1907** *Yesterday's Shopping* (1969) 642/1 Colt's new police positive revolver. **1931** 'G. Trevor' *Murder at School* xii. 244 This is what is called a Colt Point 22 Police Positive. Not a nice thing to be plugged with. **1975** J. Gores *Hammett* (1976) xxxii. 221 He took out the long-barreled police positive... He thumbed back the hammer. **1827** *U.S. Supreme Court Rep.* XXV. 442 The power to direct the removal of gun-powder is a branch of the police power which unquestionably remains..with the states. **1932** N. M. Butler *Looking Forward* xi. 168 'Police power'—which in American law means the principle that the public interest often requires the extension of government authority in repression..of individual activity or habit. **1964** Gould & Kolb *Dict. Social Sciences* 508/2 *Police power* may be defined as the broad and elastic power of government especially of one of the states of the United States, to restrict, control, regulate, and restrain individuals and groups in the use of their liberty and property in order to protect and promote the health, safety, morals, convenience, peace, order, and general welfare of other individuals and the public generally. **1967** *Punch* 16 Aug. 256/1 Police-procedural thrillers must be original to compete, yet, of their nature, ordinarily plausible. **1972** 'L. Egan' *Paper Chase* (1973) iii. 40 She'd always read mysteries, but mostly..what they call the police-procedural ones. **1972** [see *police novel* above]. **1977** *Time* 27 June 56/2 *Laidlaw* is also the first police procedural by Scottish Author William McIlvanney. **1885** *Encycl. Brit.* XIX. 337/1 If they [*sc.* orders having the approval of the secretary of state for the government of the metropolitan police] are carefully considered and prepared, their issue must produce a uniform code of police procedure for the force. **1974** *Publishers Weekly* 5 Aug. 51/2 European police procedure buffs might enjoy this, but the author makes it hard for the reader to grapple with the evidence. **1976** 'J. Charlton' *Remington Set* iii. 18, I wish I had a book somewhere I could look up police procedure in. **1901** *Sketch* 17 July 518/2 Herr Kubelik..will have to be police-protected against the patrons of Señor Sarasate. **1908** *London Mag.* Oct. 240, I would demand police protection. **1942** E. Paul *Narrow St.* xvi. 125 Mariette.. footed the bills of the establishment,..arranged police protection and gave the place its personality and reputation for fair play. **1972** C. Drummond *Death at Bar* i. 40 Hundreds of people in this wicked city would relish the idea of police protection. **1958** A. Budrys in *Venture Sci. Fict.* Mar. 14/2 He'd used a police radio often enough. **1975** *Listener* 2 Jan. 26/1 The loudspeakers.. crackled with voices from a police radio in the streets. **1919** G. B. Shaw *Heartbreak House* p. xx, The ordinary law was superseded by Acts under which newspapers were seized and their printing machinery destroyed by simple police raids *à la Russe*. **1933** J. Buchan *Prince of Captivity* iii. i. 277 That cheerful party broke up in confusion... Yes, a police raid. **1975** *Country Life* 16 Jan. 155/1 A recent police raid led to the arrest of a well-organized gang. *a* **1911** D. G. Phillips *Susan Lenox* (1917) II. xi. 277 Once in the pariah class, once with a 'police record'. **1950** T. Walsh *Nightmare in Manhattan* iv. 106 Calhoun was examining his police record, which was long and bad. **1972** J. Wainwright *Night is Time to Die* 92 He has no police record. No previous convictions. **1802** C. Wilmot *Diary* in *Irish Peer on Continent* (1920) 61 It is a good display of the Police regulations, for such order, method, and tranquillity I cou'd not have imagined. **1853** E. Twisleton *Let.* 23 May (1928) v. 85 Owing to the recent revolts, all the police-regulations were doubled in stringency. **1938** R. G. Collingwood *Princ. Art* xi. 255 Philosophical controversies are not to be settled by a kind of police-regulation governing people's choice of words. **1978** N. Freeling *Night Lords* i. 7 It was forbidden to park on the bridge. A lot Castang cared for municipal police regulations—. **1837** H. Martineau *Society in America* I. i. iii. 77 The disgustingly jocose tone of their [*sc.* British] police reports, where crimes are treated as entertainments. **1882** C. M. Yonge *Three Brides* xix. 265 Such disgraces to England as I see in your police reports—brutal mechanics beating their wives. **1915** F. M. Hueffer *Good Soldier* i. iii. 31, I used..to inspect the little police reports that each guest was expected to sign upon taking a room. **1976** J. Lee *Ninth Man* 257 Most of our leads so far have come from police reports. **1834** *Sun* (N.Y.) 23 July 2/3 Your police reporter be one dam liar. **1959** [see *corn bell* s.v. *corn *sb.*¹ 11]. **1977** L. Meyer *Capitol Crime* i. 9 The story..was Sid Jacobson's, our police reporter. **1961** Webster, Police science. **1971** *Publishers' Weekly* 6 Dec. 22/3 The Glencoe Press..is doing very well with curricula in such new areas as fire science and police science. **1979** *Internat. Jrnl. Sociol. of Law* Feb. 112 Solomon's Soviet criminologists seem to be very similar to American 'police science' experts and to government scientific officers anywhere in the West. **1957** *Encycl. Brit.* XII. 562/2 A police scientist does not have to be an expert in every branch of chemical analysis to use these instruments effectively. **1976** *Billings* (Montana) *Gaz.* 16 June 3-B/2 Ticks enclosed in an envelope received in the mail..were being studied by police scientists Tuesday to see if they were..infected with numerous diseases. **1937** 'M. Innes' *Hamlet, Revenge!* ii. i. 98 Billups would not have thought to requisition a fire-engine, the Prime Minister had. Its bell, he explained, gained more respect than did a police siren. **1956** B. Holiday *Lady sings Blues* (1973) viii. 77 When the bus pulled out, you could hear the old sheriff's police siren coming after us. **1974** M. Birmingham *You can help Me* v. 132 The police sirens wailed up the street. **1959** I. Jefferies *Thirteen Days* viii. 100 The Legion officers had thirty-eight police specials, the peak of revolver achievement. **1970** 'J. Morris' *Candywine Development* xxiv. 263 That guy; he's holding something heavy. Police Special, maybe, or a '44 magnum. **1849** J. S. Mill in *Westm. Rev.* LI. 4 Chenu..is now admitted to have been..a police spy. *c* **1874** D. Boucicault in M. R. Booth *Eng. Plays of 19th Cent.* (1969) II. 193 The police spy—Harvey Duff—the man that denounced me. **1922** J. Hergesheimer *Bright Shawl* (1923) 66 Probably we are all ruined... The police spies will be waiting for us at home. **1973** G. Sims *Hunters Point* xviii. 165 You police spies don't seem to be a very efficient bunch, letting an old man be drowned while you are supposed to be keeping a watch on him. *Ibid.* 166 It's just more police spying activity. **1868** *N.Y. Herald* 2 July 8/3 Dr. Waterman, police surgeon, was called and dressed the wounds. **1928** D. L. Sayers *Unpleasantness at Bellona Club* xxi. 274 'Nervous shock with well-marked delusions,' said the police surgeon. **1978** N. Freeling *Night Lords* v. 23 I'll have the Parquet and the police surgeon standing by. Autopsy and lab report. **1885** *Encycl. Brit.* XIX. 336/2 The police system of necessity involves the existence in a district of police stations or lock-ups, for the temporary detention of prisoners. *a* **1963** C. S. Lewis *Discarded Image* (1964) v. 94 On the imaginative and emotional level it makes a great difference whether, with the medievals, we project upon the universe our strivings and desires, or with the moderns, our police-system and our traffic regulations. **1903** *World's Work* July 123/2 To set police traps for a man going thirteen miles an hour on an open road is sheer idiocy. **1914** Beerbohm *James Pethel* in *English Rev.* Dec. 18 In France he always rather missed the British police-traps. **1966** M. R. D. Foot *SOE in France* vii. 173 The others fell successively into a Vichy police trap..at the Villa des Bois. **1974** J. Wainwright *Evidence* iv. 23 Motorways..were..police traps: box canyons into which a man on the run could be flushed and cornered. **1902** *Pall Mall Mag.* XXVIII. 410/2 Every police-constable on the much-police-trapped Ripley Road. **1859** G. A. Sala *Twice round Clock* 201 Then..troublesome bodies..are securely shackled and straight-waistcoated up, and carted away in police-vans. **1976** Deakin & Willis *Johnny go Home* ii. 49 She was in a police van on her way to Holloway. **1884** Sweet & Knox *Through Texas* iv. 50 He began blowing a police-whistle. **1922** Joyce *Ulysses* 160 Police whistle in my ears still. **1967** O. Wynd *Walk Softly* x. 163, I..blew a long blast on a police whistle. **1979** J. Scott *Clutch of Vipers* x. 175 A whistle blew. A good old-fashioned police whistle such as is rarely heard nowadays. **1932** 'Solicitor' *Eng. Justice* iii. 94 On his version, supported by witnesses, he was clearly innocent. He seemed surprised when I said he was to plead 'Not guilty', and said, 'But there's a police witness.' ..He got off. **1979** *Internat. Jrnl. Sociol. of Law* Feb. 71 The average Nigerian..believes that to be a police witness is almost as bad as being the criminal offender. *a* **1930** D. H. Lawrence *Apocalypse* (1931) xvi. 233 Ah woman, you have known many bitter experiences. But never, never before have you been condemned by the old dragon to be a policewoman. **1955** W. Gaddis *Recognitions* ii. vi. 560 A policewoman handed that nomadic laundress over to the stronger arm of the law. **1976** *Birmingham Post* 16 Dec. 3/5 Policewoman Susan Oliver, who was the first police officer on the scene, said lights on the side of the lorry's trailer were illuminated at the time of the accident. **1937** 'M. Innes' *Hamlet, Revenge!* ii. ii. 112 In plain police-work you could usually go straight for the truth. **1960** 'E. McBain' *Give Boys Great Big Hand* (1962) iii. 20 Police work is not for you..if you..believe that corpses 'look just like they're sleeping'. **1977** L. Meynell *Hooky gets Wooden Spoon* xiii. 152 The basic rule of police work—get it down on paper.

policeless *a.* (earlier example).

1882 E. W. Hamilton *Diary* 2 Sept. (1972) I. 331 It was determined yesterday to dismiss summarily those insubordinate men; and the act of their dismissal was followed by the resignation of the force *en masse*. Accordingly Dublin is police-less.

police, *v.* Add: **2. b.** (Earlier and later examples.) Also const. *up,* as of an area: cf. *clean *v.* 6 c.

1851 *Colburn's United Service Mag.* LXVII. 57 All hands were then distributed in separate parties..to 'police' or clean round the garrison. **1930** *Amer. Speech* V. 380 Nearly all our old army expressions were taken to France with the A.E.F. *Bunk fatigue,.. police up,* and *salvage* are some of these. **1956** *Ibid.* XXXI. 108 *Police up* (clean up an area). **1968** *Listener* 9 May 594/2 'Last night we *policed up* two sampans, killing six enemy,' said an Airborne major in modest triumph. **1977** M. Herr *Dispatches* i. 26 Some troops..were pissing on the ground... The men finished..and walked away laughing, leaving the captain alone shouting orders to police up the filth, thousands of empty and half-eaten ration cans, soggy clots of *Stars and Stripes,* and M-16 that someone had just left lying there and, worse, evidence of a carelessness unimaginable to the captain, it stank even in the cold rain, but it would police itself in an hour or two if the rain kept up.

3. c. (Earlier and later examples.)

1885 F. W. Maitland *Justice & Police* x. 112 The Cornish St. Ives, without a commission of the peace, polices, or lately policed, itself. **1928** R. Stranger *Wireless* xii. 150 Being policed by the aerial potentials the filament electrons will pass through in varying numbers. **1943** *Sun* (Baltimore) 13 Dec. 18/1 Lucien E. D. Gaudreau, area rent director, said yesterday that the agency definitely will not 'police' rent regulations. **1946** *R.A.F. Jrnl.* May 176 The future work of B.A.F.O. is to police the Reich in the sky. **1946** *Rep. Internat. Control Atomic Energy* (U.S. Dept. of State) ii. iii. 15 While suppression is not possible where we are dealing with the quest for knowledge, this thirst to know (that cannot be 'policed' out of existence) *can* be used. **1970** T. Lupton *Managem. & Social Sci.* (ed. 2) ii. 52 The organization cannot effectively police every individual item of behaviour. **1974** *Socialist Worker* 26 Oct. 14/5 It seems now that the social contract is to be policed by trade union leaders. **1977** *Time* 15 Aug. 37/2 He believes that fund members will approve some new articles that will enable him to police currency exchange rates.

policeable (pǫlī·sǎb'l), *a.* [f. Police *v.* + -able.] That can be policed.

1926 H. W. Fowler *Mod. Eng. Usage* 443/1 *Police,* vb., makes *-ceable.* **1976** *Listener* 23 Dec. 802/2 When those frontiers sucked forces too far inland for their logistical support, disasters like Kabul or Majuba compelled London to fall back on more policeable limits. **1979** *Guardian* 15 Jan. 12/2 There was no question at the moment of any action against secondary picketing... 'One has to find out if it is illegal. And if you make it illegal, is it policeable?'

police court. (Further examples.)

1930 D. H. Lawrence *Nettles* 19 And Mr. Mead..said: 'Gross! coarse! hideous!'—and I, like a silly Thought he meant the faces of the police-court officials. **1964** [see *form *sb.* 16 c]. **1965** [see *magistrate *sb.* 3]. **1979** S. Weintraub *London Yankees* iv. 113 English readers discovered Frederic in the newspaper transcripts from Croydon Police Court in 1898.

policedom. (Earlier example.)

1866 *Chambers's Jrnl.* 22 Sept. 608/1 Of Antoine the imperturbable, when he returned home,.. policedom could make nothing.

policeman. Add: **1. a.** (Earlier examples.)

1801 R. Musgrave *Mem. Different Rebellions in Ireland* I. 227 They boasted that they had killed the policemen at Dunboyne. **1824** J. S. Mill in *Westm. Rev.* II. 22 Thus they went on..till November 1822, when,..in rushed the sheriff with a number of police men.

c. A police informer. *slang.*

[**1874** Hotten *Slang Dict.* 257 *Policeman,*..among the dangerous classes, a man who is unworthy of confidence, a sneak or mean fellow.] **1923** E. Wallace *Missing Million* xv. 128 Being an amateur, he left his finger-prints, and it's no job of mine to let 'em stay. 'Live and let live' is one of my mottoes, and 'Thou shalt not be a policeman' is another. *Ibid.* xvi. 134 Casey is a born 'policeman', and would sell his own mother if you paid him the right kind of money. **1924** [see *copper *v.*² 2].

d. *Naut.* (See quots.)

1933 P. A. Eaddy *Hull Down* 179 One of the boys, who was acting as 'policeman' (one of the watch who is told off to rouse the rest, should the Mate suddenly want them), came banging on the lamp-locker door. **1962** A. G. Course *Dict. Nautical Terms* 151 *Policeman,* a name used in sailing ships for the member of the watch on deck who kept awake to hear the officer's orders and call his shipmates.

e. *fig.* A person or object regarded as a deterrent or obstacle. In phr. *sleeping policeman*: a ramp in the road intended to jolt a moving motor vehicle, thereby encouraging motorists to reduce their speed.

1951 E. Coxhead *One Green Bottle* iii. 79 The climb above the crux was even more delightful; a knife-edge..then a stone policeman barring the way. **1969** G. E. Evans *Farm & Village* viii. 80 After the farm-workers had cleared a field of the crop they left one sheaf standing on it. This last sheaf was called the *Policeman,* and it was understood that no gleaner could enter the field while the Policeman was on guard. **1969** L. Thompson in R. Blythe *Akenfield* i. 35 The policeman was the name given to the last trave or stook which the farmers would leave standing in the middle of the field so they could have time to rake-up all the loose corn they could before the gleaners arrived. There was one farmer who made a habit of keeping the gleaners waiting and one night a young man stole the 'policeman'. **1973** *Guardian* 27 Feb. 1/4 Labour intends to..build 'sleeping policemen' (ridges) into residential and shopping streets to slow traffic down. **1974** *Times* 24 July 4/7 The government would proceed with experiments in the use of 'sleeping policemen'—road humps to slow motorists.

2. *Chem.* A glass rod or tube with a short piece of rubber tubing on one end (see also quot. 1963).

1916 F. A. Gooch *Representative Procedures in Quantitative Chem. Analysis* iii. 64 The precipitate, transferred from the container to the filter..with the aid of a rubber-tipped glass rod (the 'policeman'). **1930** W. T. Hall *Textbk. Quantitative Analysis* xi. 149 This so-called policeman may be made by sticking together the end of a piece of rubber tubing that fits the rod tightly. **1963** N. L. Parr *Lab. Handbk.* vii. 103 Other means of separation [of solids] are sometimes adopted. Among them are dialysis, and the 'policeman' for dealing with solids in microchemistry, which is a small snipe feather mounted in a short length of glass tubing. **1974** *Nature* 6 Dec. 498/2 Attempts to passage stage 2 cells..by scraping with a rubber policeman, resulted in a rapid degeneration of the cells in a new flask.

3. *attrib.* and *Comb.*, as **policeman fly** *Austral.*, a small wasp belonging to the family Nyssonidæ, Arpactidæ, or Stizidæ, which preys on flies; **policeman's helmet,** the Himalayan balsam, *Impatiens glandulifera,* a purple-flowered annual plant belonging to the family Balsaminaceæ.

1907 W. W. Froggatt *Austral. Insects* 108 Several of these [small wasps] are known in the bush as policemen flies. **1926** R. J. Tillyard *Insects Australia & N.Z.* xxii. 299 These wasps [*sc.* Arpactidæ] and the members of the following family [*sc.* Nyssonidæ] are known as 'Policeman Flies' in Australia. **1933** *Bulletin* (Sydney) 26 July 21/3 The robber fly, known [erroneously] in the back country as the policeman fly. **1969** *Courier-Mail* (Brisbane) 11 Jan. 6/8 Mr. Barrand found a fly which chases away other flies. He hopes to install tape-recorded sounds of the policeman fly in a transistorized unit. **1950** J. Hutchinson *Uncommon Wild Flowers* 226 Jumping Jack, Policeman's Helmet... A tall, sometimes very strong growing herb up to 6ft. or more. **1958** P. Lewis *Brit. Wild Flowers* 125 Policeman's Helmet, a Himalayan plant grown in cottage gardens, has become naturalized along river-banks and in waste places. **1961** E. Salisbury *Weeds & Aliens* iii. 65 The Policeman's Helmet is now to be met with in semi-wild stations in nearly half the British counties. **1971** *Country Life* 18 Feb. 384/2 An incongruous coloniser from the Himalayas, that Impatiens so appropriately known as policeman's helmet, attracting bumble-bees until the frosts come.

Hence **poli·cemanish** *a.*, suggestive of a policeman; **poli·cemanism,** the methods or conduct of policemen; **poli·cemanly** *a.*, appropriate to or characteristic of a policeman.

1891 *Star* 31 Oct. 4/3 Instances of policemanism crop up daily. **1908** *Daily Chron.* 30 Mar. 3/3 (*heading*) Policemanism. The Prince..went on to denounce a Government influenced by 'men with the education of a policeman, and with the convictions of a pogromshchik (official of massacre)'. **1916** A. Bennett *Lion's Share* iii. 27 The heavy policemanish step of Mr. Cowl was heard on the landing. **1936** 'M. Innes' *Death at President's Lodging* ix. 171 The policemanly demand at last. **1973** —— *Appleby's Answer* xx. 172 The day had passed when she judged it amusing to join her husband in policemanly scampers. **1975** W. Marshall *Yellowthread Street* 64 He had been..firm and policeman-ish. **1977** 'E. Crispin' *Glimpses of Moon* x. 194 Policemanly instincts..reasserted themselves. Single-handed, he would make an arrest

Police Motu (pǫlī·s mōᵘ·tū). [f. Police *sb.* + *Motu *sb.*] A Papuan pidgin, based on Motu.

1950 P. Chatterton (*title*) A primer of Police Motu. **1962** R. Brett et al. (*title*) A dictionary of Police Motu. **1965** B. A. Hooley in *Language* XLI. 168 Police Motu, almost as important as a lingua franca in Papua as Neomelanesian is in..New Guinea. *Ibid.* 169 Police Motu is derived from Motu, a member of the Malayo-Polynesian family. **1974** L. Todd *Pidgins & Creoles* i. 7 In the Pacific, in PNG [*sc.* Papua New Guinea], Police Motu arose from the contact between speakers of Motu and other Papuan vernaculars. It has recently expanded its vocabulary by adopting words from Neo-Melanesian. **1977** C. F. & F. M. Voegelin *Classification & Index of World's Languages* 47 Police Motu, a pidgin widely used as a lingua franca in Papua, is based on Motu.

police state. A state regulated by means of a national police force having secret supervision and control of the activities of citizens. Also *attrib.*

1865 *Times* 6 Sept. 10/3 Austria was long known on the Continent as the 'police State', and..M. von Weiss will again obtain for her that unenviable title. **1896** B. Russell *German Social Democracy* iv. 114 This infamous Law, the crowning endeavour of the enlightened police state. **1938** *New Statesman* 15 Jan. 74/1 Meanwhile, the atmosphere of the 'police state' is already with us. **1939** *War Illustr.* 28 Oct. 217/1 Spies are everywhere; indeed, Germany is the modern exemplification of the 'police state' in action. **1947** *Life* 7 June 37/1 They have failed in France and Italy because the peoples have had a chance to show their preference for Western democracy over a police state. **1950** G. B. Shaw *Farfetched Fables* 79 In the imagination of our amateur politician England is a Utopia in which everything and everybody is 'free' and all other countries 'police States'. I, being Irish, know better. **1959** *Listener* 8 Oct. 573/2 The Devlin Commission reported that Nyasaland has been turned into a 'police state'. **1967** Coulthard & Smith in Wills & Yearsley *Handbk. Management Technol.* 197 Short of introducing a police-state, in which people are directed to jobs and change them only with State approval, we must accept a degree of labour turnover as necessary and desirable. **1973** *Times* 12 Apr. 19/4 That Rhodesia is a police state, where the rights of individuals, the rule of law and the right to speak or report the truth count for nothing, is a fact for which there has long been ample evidence. **1975** *Times* 11 Jan. 12/6 A campaign to spot the terrorists before they act..requires measures which smack of the police state.

police station. Add: (Earlier example.) Also *attrib.*

1846 Dickens *Pictures from Italy* 53 The hall..is as dirty as a police-station in London. *a* **1930** D. H. Lawrence *Last Poems* (1932) 175 He has..stepped down To the police-station cell.

‖ **policier** (pǫlisi,e). [a. F. *policier* detective novel.] A film based on a police novel. Cf. *roman policier* s.v. *roman *sb.*⁴

1975 *Listener* 21 Aug. 249/3 A *policier* of rare complexity... The plot threads of the film must..be followed down some labyrinthine byways. **1977** *Time* 18 Apr. 23/1 Not so in *Man on the Roof,* the Swedish-made *policier* based on one of the Martin Beck novels.

policy, *sb.*¹ Add: **III. 8.** *attrib.* and *Comb.*, as (sense 5) *policy decision, document, -maker, -making, statement*; *policy-making* adj.; **policy science** (see quot. 1951); hence **policy scientist.**

1960 I. Jefferies *Dignity & Purity* iv. 66 Their purpose is the application of scientific method to policy decisions. **1964** Gould & Kolb *Dict. Soc. Sci.* 510/1 Current interest centres on such questions as the nature of policy decisions... Policy decisions are contrasted, for instance, with judicial decisions by reference to the relatively greater freedom of choice in the former. **1974** S. Gulliver *Vulcan Bulletins* 11 A policy decision had meant more careful buying. **1976** *Burnham-on-Sea Gaz.* 20 Apr., Mr Shore..can hardly have had time to read the policy documents before he was expected to stand up and defend them in the House. **1948** J. Towster *Political Power in U.S.S.R.* iii. xiii. 314 High-income executants, not policy-makers. **1975** *Times* 19 Feb. 14/6 American energy policy-makers find their own outlooks discouraging. **1978** *Dædalus* Fall 50 As citizens and policymakers, we can make explicit the potential personal and societal consequences of legislation. **1943** J. S. Huxley *TVA* xix. 137 The Board was always a policy-making body. **1946** *Nature* 9 Nov. 646/1 Authoritative information which those..at the policy-making or executive level might be expected to need. **1950** *N.Y. Times* 20 Apr. 1/3 The cataloging of persons eligible for policy-making positions would be..done without regard to their party affiliations. **1968** E. A. Powdrill *Vocab. Land Planning* ii. 5 Policy-making and technique are a symbiosis, but it must be supported by wise and sound administration. **1976** *Times* 21 May 4/1 Mr. Len Murray..told the policy-making conference of the Society of Graphical and Allied Trades that the T.U.C. would expect the Government to take action. **1951** H. D. Lasswell in Lerner & Lasswell *Policy Sciences* i. 4/1 We may use the term 'policy sciences' for the purpose of designating the content of the policy orientation during any given period. The policy sciences includes (1) the methods by which the policy process is investigated, (2) the results of the study of policy, and (3) the findings of the disciplines making the most important contributions to the intelligence needs of the time. **1964** I. L. Horowitz *New Sociology* 30 Sociology cannot be a 'policy science' until and unless there is a

sociology of ethics. **1977** *Dædalus* Summer 59 It should move away from the contemporary, toward the past; ..away from the impossible quest for stability; from the glide into policy science. **1970** *Nature* 19 Sept. 1189/2 There will have to be changes in the ways in which 'prime television time' is allocated so that the policy scientists can have their say..when people are most likely to be glued to their television sets. **1979** *Bull. Amer. Acad. Arts & Sci.* Mar. 28 International consultants and policy scientists serve as the conveyors and preservers of these untested staff papers until their ideas, approaches, and methodologies develop a life of their own. **1960** *Times* 1 Feb. 11/2 Mr. Macleod's speech..will be the most important policy statement that has been made on East Africa for many a long year. **1966** N. Nicolson *Diaries & Lett. H. Nicolson 1930–1939* 258 He wrote the main policy-statement of the National Labour Party.

policy, *sb.*² Add: **1. c.** (Examples.) Freq. in phr. *to write policy.*

1830 [see **policy certificate*]. **1879** *Rep. N.Y. State Court of Appeals* LXXIV. 64 He testified that he paid to the defendant, at different times, sums amounting to $3,601.08 for tickets in a Kentucky lottery and in 'playing policy', as it is called. **1890** J. A. Riis *How Other Half Lives* (1891) xiii. 155 The game of policy is a kind of unlawful penny lottery. **1944** *Crisis* June 189/2 He even tried writing policy, but the players didn't like him. He couldn't shop the proper degree of sympathy when someone played 341 and 342 came out. **1949** *Amer. Speech* XXIV. 190 The oldest of the games, and very likely the parent of most contemporary forms, is policy, which is believed to have been in existence in England as early as the first half of the eighteenth century. **1968** P. Oliver *Screening Blues* 133 During the Depression..many impoverished Negroes wrote policy in the hopes of winning sufficient to feed their families. **1972** *Times* 23 Nov. 9/8 Its most spectacular proposal is that gambling, prostitution and 'policy'—an illegal betting game—should all be legalized.

3. (sense 1 a) *policy business, -holder* (earlier and later examples); (sense 1 c) *policy certificate, game, player, -playing, racket*; *policy play* vb. trans.; **policy blues** *U.S.*, a blues melody concerning the game of policy; **policy king** *U.S.*, one who has profited greatly from running policy games; **policy office** *U.S.*, = *policy-shop*; **policy-shop** (earlier example); **policy-slip** (examples); **policy wheel** *U.S.*, a revolving drum used in the selection of winning numbers at policy; **policy writer** *U.S.*, one who collects bets from those playing policy.

1928 J. Jackson (*song-title*) Policy Blues. **1968** P. Oliver *Screening Blues* iv. 134 Reflecting the popularity of the numbers game were innumerable policy blues. **1844** G. Wilkes *Mysteries of Tombs* 52/2 He is an old offender in the policy business. **1883** 'Mark Twain' *Life on Miss.* xliii. 437 Dull policy-business till next fire. **1830** *Baltimore Amer.* 26 Aug. 3/2 To Adventurers and the Public, Policy Certificates, in the greatest variety. **1885** *Rep. Massachusetts Supreme Judicial Court* CXXXVII. 250 The defendant has been convicted of setting up and promoting a certain lottery for money; and the only question raised by his exceptions is whether the jury were warranted in finding that a game popularly known as the policy or envelope game is a lottery within the Pub. Sts. c. 209, § 1. **1934** *Sun* (Baltimore) 30 Apr. 6/5 Skilled investigators have revealed that the slot machines and the policy games take $2,000,000 out of Richmond each year. **1964** A. Wykes *Gambling* 344 In 1957, nearly 12,000 people were convicted on policy-game charges. **1851** C. Cist *Cincinnati* 98 Penn Mutual Life Insurance Co... All the profits divided among the policy holders every year. **1915** [see *chinchy *a.*] **1970** J. Hansen *Fadeout* i. 4 My company—every insurance company—sends out investigators in cases like this... Where the policyholder's body can't be found. **1949** *Collier's* 15 Jan. 21/1 Thousands of other suckers..are not only making millionaires out of a few dozen policy kings, they also pay for the corruption of many police officials. **1968** P. Oliver *Screening Blues* iv. 133 Policy kings made occasional magnanimous gestures with large donations to charities or churches and a mystique developed in which they were viewed as benefactors rather than as parasites. **1843** J. H. Greene *Exposure of Arts & Miseries of Gambling* 283 These swindling shops are numerous, and are sometimes called *policy offices.* **1926** C. Jackson in P. Oliver *Screening Blues* (1968) iv. 130, I looked in my purse t'see if I had a little dough, So I could policy play 4-11-44. **1847** C. White (*title*) The policy players. An Ethiopian sketch. **1901** E. Harrigan *Mulligans* 65 A policy player's chances are a hundred to one against him. **1972** *Sci. Amer.* Oct. 112/3 Thousands of Manhattan policy players bet on 932 and won. **1887** *Gen. Statutes Connecticut* cliv. 563 The court of common council of any city..shall have power to make, alter, and repeal ordinances or by-laws to suppress and punish all kinds of gambling and gaming, pool selling, policy playing, [etc.]. **1949** *Amer. Speech* XXIV. 190 In the early days policy playing was associated with the regular number lotteries, being a device whereby people unable to afford a regular lottery ticket could wager small amounts on the outcome of the drawing. **1938** *Sun* (Baltimore) 1 Sept. 1/4 Davis, the broken mouthpiece of the once-powerful Dutch Schultz policy racket, swore..that James J. Hines..was paid thousands of dollars by the mob. **1968** P. Oliver *Screening Blues* iv. 133 The policy racket had a folk-lore of its own. **1858** 'Q. K. P. Doesticks' *Witches of N.Y.* 54 The propinquity of the 'lottery agency' and the 'policy-shop', just round the corner. **1934** *Sun* (Baltimore) 30 Apr. 6/5 The second fact can be done away with by making it a criminal offense to sell policy slips to minors. **1972** 'T. Coe' *Don't lie to Me* xi. 103, I knew..he could make it stick. Find heroin in my car. Shake me down and find policy slips. **1906** *Southwestern Reporter* XCI. 785/1 Evidence that accused was seen..in the house where people were betting at a lottery, and that he at one time turned the policy wheel. **1968** *Sunday Tel.* 1 Sept. 13/6 An excursion into the cabalistic number symbolism employed by bettors on the 'policy wheels'—those intricate gambling devices surreptitiously played by millions of Americans. **1949** *Collier's* 15 Jan. 21/2 In Detroit, one auto-plant policy writer explained to me, 'I been in this racket for twenny years.' **1968** P. Oliver *Screening Blues* iv. 142 Within the lower class the policy writer was considered a parasite who lived off his fellows.

polie: see Police *sb.* in Dict. and Suppl.

poliencephalitis. Add: **polioencephalitis** is now the more usual form. (Earlier and later examples.) [First formed in Ger. (C. Wernicke *Lehrb. d. Gehirnkrankheiten* (1881) II. ii. 229).]

1885 *London Med. Rec.* 16 Feb. 44/1 (*heading*) Strümpell on acute infantile polio-encephalitis (cerebral paralysis of children). **1885** J. Ross *Handbk. Dis. of Nervous Syst.* 621 (*heading*) Poliencephalitis acuta infantum. **1911** A. Bruce tr. *H. Oppenheimer's Text-bk. Nervous Dis.* II. 837 The diagnosis of poliencephalitis..should be made only with great reserve. **1923** J. H. Parsons *Diseases of Eye* (ed. 4) xxix. 554 Acute polioencephalitis accounts for not infrequent cases of paralytic squint following a febrile attack in young children. **1934** *Times Educ. Suppl.* 22 Sept. 341/1 Of polioencephalitis there were 88 notifications. **1971** *Jrnl. Neurol. Sci.* XII. 414 This disorder is a polioencephalitis of viral origin.

poling, *vbl. sb.*¹ Add: **3. poling boat** *N. Amer.*, = *pole boat* s.v. *pole *sb.*¹ 5 c.

1900 J. London *Son of Wolf* 163 Madelaine shook the dust of the Lower River from her moccasins, and with her husband, in a poling-boat, went to live on the Upper River. **1973** D. Andersen *Ways Harsh & Wild* ii. 70 We whipsawed lumber from birch logs and put together two large poling boats.

poling (pōᵘ·liŋ), *vbl. sb.*² *Physics.* [f. Pole *sb.*² + -ing¹.] The process of polarizing a ferroelectric material (see *pole *v.*²).

1954 *Jrnl. Appl. Physics* XXV. 1166/2 The crystals ..were grown in our laboratory and the *c* axis was made to be perpendicular to the face of the crystals by means of a high-field poling technique. **1965** *IEEE Trans. Sonics & Ultrasonics* XII. 7/1 During the initial poling both 109°/71° and 180° domain switching occurs. **1976** *Ibid.* XXIII. 393/2 The common practice..is to analyse the immittance behavior of a resonator with a certain geometry and poling axis.

polio (pōᵘ·lio). *colloq.* [Abbrev. of Poliomyelitis.] **1.** Poliomyelitis, esp. the paralytic form. Freq. *attrib.*

1931 *Survey* 15 Oct. 93/1 (*heading*) Panic and polio. **1934** *Ladies' Home Jrnl.* Feb. 10/1 How did the polio fighter..come to catch it? **1940** *Time* 2 Sept. 37/2 (*heading*) Polio scare. **1949** *New Harmony* (Indiana) *Times* 5 Aug. 1/5 New Harmony was doused..by a spraying plane..as a precautionary measure against polio. **1955** G. Greene *Quiet American* iv. ii. 241 My son's got polio. He's bad. **1955** *Sci. News Let.* 23 July 51/1 Children are the principal carriers of polio, and if enough children are immunized, it would probably not be necessary to vaccinate the adults in order to stamp out the disease. **1962** *Observer* 11 Mar. 8/4 Last year I had a play..out on tour... There was a polio scare in Hull, and the star didn't want to play. **1977** *Daily Tel.* 9 Mar. 8/4 Doctors have cleared two girl polio suspects in Greater Manchester.

2. A person who has, or has had, polio. *rare.*

1934 *Ladies' Home Jrnl.* Feb. 107/1 Health departments of cities and states poured out money to buy serum from recovered polios to try to cure already sick babies. **1962** *Guardian* 26 Sept. 8/5 'Polio',.. a person who has been paralysed by polio.

polioencephalitis: see Poliencephalitis in Dict. and Suppl.

poliomyelitis. Add pronunc. (pōᵘ·lio-) and substitute for def.: A disease caused by a neurotropic virus (infection with which usu. produces no symptoms) which may give rise to a temporary meningitis, with fever and delirium, or, esp. in older patients, a permanent and sometimes fatal localized paralysis as a result of the infection and death of groups of nerve cells in the spinal cord or brain stem. Also *attrib.* (Earlier and later examples.)

1878 *Amer. Jrnl. Med. Sci.* LXXV. 411 The case was..one of acute polio-myelitis. **1934** R. W. Fairbrother *Handbk. Filterable Dis.* vi. 95 The poliomyelitis virus only produces spontaneous disease in man. **1955** [see *infantile paralysis* s.v. *infantile *a.* 3]. **1966** Wright & Symmers *Systemic Path.* II. xxxiv. 1192/2 The non-paralytic form of poliomyelitis—the main symptoms of which are those of catarrhal inflammation of the upper parts of the respiratory and alimentary tracts—is the commonest manifestation of the infection during epidemics. **1972** B. A. Curtis et al. *Introd. Neurosci.* viii. 186/1 The majority of patients with poliomyelitis virus infection make a complete or significant recovery. *Ibid.* xxii. 590/1 With the development of effective vaccines..poliomyelitis has become a preventable disease.

Hence **po:liomyeli·tic** *a.*, of or affected with poliomyelitis.

1911 *Jrnl. Exper. Med.* XIV. 117 Flexner and Clark showed that the poliomyelitic virus would survive for a period of days in the subcutaneous tissues of the rabbit. **1940** *Ann. Reg. 1939* 376 Evidence suggested that poliomyelitic virus may occur in urban sewage. **1958** *Jrnl. Bone & Joint Surg.* XL. a. 513 In a few convalescent poliomyelitic patients..the Milwaukee brace has aided in reducing the curve and holding it in check until muscle balance has been re-established. **1971** *Biol. Abstr.* LII. 10390/1 (*heading*) Study of the circulation of poliomyelitic and other enteric viruses in waste water.

poliorcetic, *a.* Delete *rare* and add further examples.

1936 H. A. L. Fisher *Hist. Europe* i. xv. 187 To the marine skill of the Scandinavians they [*sc.* the Normans] added all that was then known of cavalry warfare and the poliorcetic art. **1975** C. J. Bishko in K. M. Setton *Hist. Crusades* III. xii. 410 Its [*sc.* Lisbon's] stubborn defense ..against the combined Portuguese and crusader resources, including the northerners' heavy siege machines and poliorcetic skills.

poliorcetics, *sb. pl.* (Further examples.)

1936 H. A. L. Fisher *Hist. Europe* ii. xvii. 628 The sieges of Hertogenbosch, of Maestricht, and of Breda showed that in the art of poliorcetics the Dutch had lost none of their ancient cunning. **1958** M. Roberts *Gustavus Adolphus* II. iii. 185 Maurice [of Orange] was a great innovator in siege warfare. His massive barrages against a strictly limited objective enabled him greatly to quicken the pace of siege operations, and marked an epoch in poliorcetics.

poliosis (pǫliōᵘ·sis). *Med.* [mod.L., f. Gr. πολι-ός grey + -osis.] Partial or general greyness or whiteness of the hair, esp. if premature.

1813 T. Young *Introd. Med. Lit.* lxiv. 370 *Spilosis,* discolorations of the skin, or of the cuticular substances, without constitutional disease... *S. poliósis,* grey hairs. **1817** J. M. Good *Physiol. Syst. Nosology* 501 *Poliósis,* hairs prematurely grey or hoary. *Poliosis,* auct. var. **1868** *Jrnl. Cutaneous Med.* I. 277 Poliosis, or canities, was present in 9 cases in the two thousand... The white hairs were pretty uniformly dispersed through the rest of the hair, giving rise to incipient greyness. **1940** Becker & Obermayer *Mod. Dermatol. & Syphilol.* xxx. 531/1 Poliosis is definitely hereditary, and..is due to complete lack of pigment function. **1968** Ebling & Rook in A. Rook et al. *Textbk. Dermatol.* II. xlvi. 1414/2 Poliosis results from absence or deficiency of melanin in a group of neighbouring hair follicles. Clinically it presents as a strand or mesh of white hairs.

poliovirus (pōᵘ·liovəiᵊrǒs). *Med.* Also **polio virus.** [f. *polio + Virus.] Any of a group of enteroviruses that includes those that cause the various forms of poliomyelitis. Also *attrib.*

[**1954** *Nature* 3 Apr. 621/1 Proposed 'non-Linnæan' binomials for some animal viruses... *Poliovirus hominis* (human poliomyelitis). *Poliovirus muris* (mouse encephalomyelitis, *TO* type).] **1955** *Virology* I. 186 (*heading*) The poliovirus group. **1958** *Economist* 26 July 283 To achieve that we must apparently wait for the successful development of vaccines made from live attenuated polioviruses. **1961, 1965** [see *echo virus]. **1968** Rhodes & van Rooyen *Textbk. Virol.* (ed. 5) v. iii. 541 The feature distinguishing polioviruses from other members of the enterovirus group is their capacity to produce poliomyelitis in man and other primates. **1970** Passmore & Robson *Compan. Med. Stud.* II. xviii. 113/2 At the beginning of the century, infection with poliovirus was widespread among children under 5 years of age, but only an occasional patient developed the typical infantile paralysis. **1973** R. G. Krueger et al. *Introd. Microbiol.* xix. 528/1 Polio virus will adsorb to certain cells of certain organs of primates but not to nonprimate cells.

Hence **po·lioviral** *a.*

1977 *Lancet* 19 Feb. 434/1 Polioviral antigen has been detected in 3 patients.

poliphant: see Polyphone in Dict. and Suppl.

‖ **polis** (pǫ·lis), *sb.*¹ *Hist.* [Gr.] A Greek city-state; *spec.* such a state considered in its ideal form. Also *transf.*

1894 A. Holm *Hist. Greece* I. xx. 252 In Greece state and city are one, and both are designated by the word 'Polis'. **1929** N. Mallinson tr. *Glotz's Greek City* 2 From the association of many villages, the complete State was created, the perfect community, the *polis.* **1941** Auden *New Year Letter* iii. 51 We can at least serve other ends, Can love the polis of our friends. **1941** H. G. Wells *You can't be too Careful* iv. i. 222 Homo Tewler does not behave as a political animal should do, participating with the utmost fullness in the collective life of his *polis.* *Ibid.* 223 The *polis* of Aristotle. **1958** *N. & Q.* CCIII. 507/2 She appears an irreverent seductress pursuing Merlin in order to destroy the Arthurian polis. **1959** *Listener* 12 Feb. 292/1 The fully developed, classical *polis* was a new and original institution. **1968** *Ibid.* 5 Sept. 313/1 The epic and tragic were becoming increasingly hard to realise; an art that had emerged from the social organisation of the Greek *polis* no longer applied to the mysterious forces of capitalism. **1972** A. P. Hinchliffe in Cox & Dyson *20th-Cent. Mind* III. xiv. 416 The idea that no man can prosper unless his *polis* prospers. **1978** A. Sanders *Victorian Historical Novel* viii. 172 The Florence of the novel [*sc. Romola*] is the Renaissance *polis.*

polis (pǫ·lis, pō·lis), *sb.*² Chiefly *Ir.* and *Sc.* Also **poliss, pollis.** [Repr. regional pronunc. of Police *sb.* (see etym. note).] = Police *sb.* 5 a; a member or (construed as *pl.*) members of a police force. Also **po·lisman.**

c **1874** D. Boucicault *Shaughraun* (1884) i. iii. 7/2 The polis were in my cabin today about ye. **1892** Kipling *Barrack-Room Ballads* 43 They sent the Polis there, The English were too drunk to know, the Irish didn't care. **1900** E. H. Strain *Elmslie's Drag-Net* 115 The poliss is no far awa. **1904** 'H. Foulis' *Erchie* 124 Her niece Sarah, and Macrae the nicht polis. **1907** J. M. Synge *Playboy* i. 10 Is it yourself is fearing the polis? You're wanting, maybe? **1919** J. Buchan *Mr Standfast* iv. 84 We'll get a good turn-out at your meeting..but they're sayin' that the polis will interfere. **1922** Joyce *Ulysses* 434 Don't be all night before the polis in plain clothes sees us. **1928** J. Buchan *Runagates Club* ii. 73 The pollisman..says they're looking for a man that personated an inmate. *a* **1930** N. Munro *Para Handy Tales* (1955) 109 'There's no a polisman in the island of Barra,' said Para Handy. 'If there wass any need for polismen they would have to send to Lochmaddy.' **1931** D. L. Sayers *Five Red Herrings* ii. 20 Juist tummled intae the burn..an' drooned himself... The pollis'll be up there now. **1967** H. Calvin *Nice Friendly Town* viii. 104 'But I'll have to get on to the police,' I protested, and Jumbo..pointed to Eddie Bone and said: 'He's a polis. Get on to him.' **1973** J. Patrick *Glasgow Gang Observed* iii. 34 Tim bragged: 'We've chibbed aboot seven polis.' **1979** *Listener* 31 May 749/4 The odd *poliss* and such like..could be confusing to English ears.

Polisario (pǫlisā·rio). [f. the initial letters of Sp. *Frente Popular para la Liberacion de Sagnia el-Hamra y Rio de Oro* Popular Front for the Liberation of Sagnia el-Hamra and Río de Oro.] An independence movement in Western Sahara, formed in May, 1973. Also *attrib.*

1975 *Guardian* 17 Oct. 4/6 The Front for Liberation and Unity..has vowed to break up separatist movements such as the Algerian-backed Polisario Front, which is campaigning for independence for the territory. **1977** *Time* 3 Jan. 49/1 Polisario is fighting to gain independence for a new 'Saharan Arab Democratic Republic' and the 100,000 people, most Reguibet tribesmen, it would represent. **1977** *Kuwait Times* 23 Oct. 4/7 Spain ceded sovereignty of the Western Sahara to Morocco and Mauritania nearly two years ago but Algeria disputes the agreement and supports guerrillas of the Polisario Front fighting for independence for the territory.

polish, *sb.* Add: **3. b.** Short for *nail polish* s.v. *nail *sb.* 13 a.

1917 *Harrods Gen. Catal.* 410 (*heading*) Manicure preparations and sundries... Majestic Polish..1/6. **1924** M. A. Burbridge *Road to Beauty* 115 Finish by using a bit more of the tinted polish and rub up with the buffer. **1937** H. Rubenstein *This Way to Beauty* viii. 119 A very white hand is flattered by a very dark polish. **1957** N. Williams *Powder & Paint* v. 144 Until 1930..more 'colourless' polish was sold than of the three shades of pink that made up the manicurist's palette. **1976** 'M. Albrand' *Taste of Terror* x. 61 The salesgirls never look at a customer. They're always staring at their nails to see if the polish is chipped.

4. polish remover, a preparation used for removing nail varnish.

1935 D. Cocks *Help Yourself to Beauty* xii. 249 Moisten a small pad of absorbent cotton with polish remover. **1973** M. Mackintosh *King & Two Queens* iv. 58 He played absently with the small bottle of polish remover..running the tiny brush over a thumb-nail.

Polish, *a.* Add: **a.** (Earlier and further examples.)

1674 R. South *Acct. of Travels into Poland in 1674* in *Posthumous Wks.* (1717) 26 The Queen is now about 33 Years of Age..and speaks the Polish Language full as well as her own natural Tongue. **1795** S. Jones *Hist. Poland* i. xix. 24 The Pater-noster in the Polish language is of the following tenor. **1842** *Penny Cycl.* XXII. 115/1 The Polish language is considered to be more flexible and euphonic than the other Slavonic dialects. **1884** W. R. Morfill *Simplified Gram. Polish Lang.* i. 3 The sounds of the Polish language may be grouped as *hard* and *soft.* **1944** S. Grabski *Poland* iii. xiv. 107 The magnificent song of the Knights..to the glory of the Virgin Mary, the first literary monument in the Polish language. **1975** *Language* LI. 407 These languages..differ from the Polish dialects discussed above in not permitting alternatives with singular and plural.

b. Polish draughts (later examples); **Polish wheat** (further examples).

1960 R. C. Bell *Board & Table Games* ii. 75 Modifications in the rules have been made and as now played, Polish draughts must rank as one of the great board games of the world. **1971** *Country Life* 25 Nov. 1429/1 The board on this games table is arranged for Polish draughts..popular in England throughout the 18th century, the boards having 100 squares and each player 20 men. **1843** Polish wheat [see *hard wheat* s.v. *hard *a.* (*sb.*) 22 a]. **1908** Polish wheat [see *durum].

d. Applied to logical theories, methods, or systems developed esp. in Lwow, Breslau, and Warsaw before the war of 1939–45, and to the related symbolism, as **Polish notation,** a bracketless and unpunctuated system of formula notation now freq. used in computing science to represent the order in which arithmetical operations are performed in many computers and calculators, often with operators following their operands (*reverse Polish*) instead of preceding them.

1940 *Jrnl. Symbolic Logic* V. 77 The second section is devoted to various present currents of thought: Hilbert formalism, German exact thought.., Polish logic.., the unity of science movement. **1954** I. M. Copi *Symbolic Logic* 254 The Polish notation has the obvious advantage of dispensing with all special punctuation marks. **1965** N. Kretzmann *Elements of Formal Logic* p. vi, We use Łukasiewicz's so-called Polish notation for..the logic of statements. **1966** Y. Bar-Hillel in *Automatic Transl. of Lang.* (NATO Summer School, Venice 1962) 15 In Polish notation calculi you cannot introduce syntactic ambiguity..by omitting symbols, since there are no special scoping symbols to omit. **1968** *Jrnl. Assoc. Computing Machinery* XV. 466 The following is a simple Polish transduction grammar. **1975** D. G. Fink *Electronics Engineers' Handbk.* xxiii. 85 One possible Polish string would be AB+C+E=. In this string, the system would find A and B and, as determined by the plus sign *following* the two operands, add them. The result is then combined with C under addition called for by the second plus sign. The E= symbols indicate that the result is to be stored in E. **1975** *Physics Bull.* May 227/1 The Oxford 300 scientific calculator with either floating point or scientific notation and algebraic logic (as opposed to the reverse Polish logic used on the more familiar Sinclair Scientific).

B. *absol.* or as *sb.*² **1.** (See Polish *a.* c.)

2. The Polish language.

1784 W. Coxe *Travels into Poland, Russia, Sweden, & Denmark* I. ii. iii. 176 The king informed me, that they had no good history of their country in Polish. **1807** G. Burnett *Present State of Poland* xvi. 277 Within thirty or forty, or perhaps fifty miles of Dantzic, I found that the people knew scarcely a word of Polish **1861** Max Müller *Lect. Sci. of Lang.* 1st Ser. v. 187 The oldest specimen of Polish belongs to the fourteenth century: the Psalter of Margarite. **1925** P. de Soissons *Polish Self-Taught* 1 This work is issued as a practical introduction to Polish, a language spoken by about thirty millions of people. **1943** S. Segal *Nazi Rule in Poland* iv. 70 Only a few shows in Polish are at present permitted in Warsaw and Krakow. **1968** B. Newman *New Poland* ii. 31 If a Lithuanian youth wanted an education, he had to learn Polish to get it. **1975** *Language* LI. 363 The passive construction is the same for both aspects in Polish and Bulgarian.

Hence **Po·lishness,** the quality or state of being Polish or of displaying Polish characteristics.

1958 *Listener* 2 Oct. 499/2 They [*sc.* the Poles] need to reaffirm their past, their Polishness. **1964** E. Huxley *Back Street New Worlds* iii. 35 Few of their children attend Saturday schools or sustain their Polishness.

polish, *v.* Add: **1. a.** Also *absol.* or *intr.*

1803 T. Sheraton *Cabinet Dict.* 289 At other times they polish with soft wax. **1828** Webster, *Polisher, n.,* the person or instrument that polishes. **1902** D. C. Peel *How to Keep House* xii. 205 Dip a small piece of flannel lightly in colza oil and then into the bath-brick, and rub all the bright part well, wipe off with a soft cloth and polish with a leather. **1919** L. R. Balderston *Housewifery* vi. 133 Rottenstone is a fine gray powder... Like any gritty substance, it works best with a lubricator like oil. In this way it cleans and polishes. **1957** M. Dodd *America's Homemaking Bk.* xxiii. 189 There are commercial scratch removers that can be applied like a polish ..or mix with varnish an oil paint to match the wood... Let it dry and then polish. **1961** *Modern Maturity* IV. vi. 19/2 Some women get a real thrill out of housework. They love to dust, scrub, polish, wax floors, move the furniture around from place to place, [etc.]. **1962** *Home Managem.* (Homecraft Series) 34/1 It can be sprayed straight on to a dusty surface and enables you to dust and polish in one operation.

c. To wipe or scrape up and eat every morsel of food on (one's plate, bowl, etc.).

1908 A. J. Dawson *Finn* xix. 289 Finn polish the tin dish clean and bright. **1962** M. Duffy *That's how it Was* x. 85 The little Reeses polished their plates after every meal. **1972** M. Babson *Murder on Show* vii. 79 Pandora ate her dinner..and began polishing her bowl.

2. a. (Later examples.) Also *absol.*

1946 K. Tennant *Lost Haven* (1947) xv. 235 Mrs. Ayre was thrown back for employment on a series of maids, and was always to be found hard at work 'polishing' some raw girl. **1961** *Vogue* (N.Y.) July 100/3 They discovered any number of ways in which they wanted to polish their own interview techniques. **1961** E. Streeter *Chairman of Bored* xxiii. 221 Did men become perfectionists as they grew older, polishing, polishing, reluctant to let go? **1961** *PMLA* LXXVI. i. 310 The poet in a written tradition who generally never blots a line may once in a while pause and polish without incurring blame.

Polish-American, *sb.* and *a.* [f. Polish *a.* + American *a.* and *sb.*] **A.** *sb.* An American of Polish origin.

1898 F. Baringer in *Memorial of Polish-American Organizations in U.S.* 29 The Polish-Americans in this city are very industrious. **1938** W. Seabrook *Those Foreigners* v. 258 The late Congressman Zionchek who went beserk on the White House lawn was a Polish American. **1943** *Who's Who in Polish America* (ed. 3) 9 This..is the outgrowth of a project begun modestly in 1939, with the publication of a brochure containing brief biographies of 100 leading Polish-Americans. **1945** *Polonia* 83 The New Poland that came out of the War proved disappointing to many Polish Americans. **1974** D. MacKenzie *Zaleski's Percentage* ii. 40 A handful of Polish-Americans made the pilgrimage. **1977** B. Garfield *Recoil* i. 18 'She's married to a Polack.' 'Polish-American, Charlie.'

B. *adj.* Of or pertaining to Americans of Polish origin.

1911 *Cath. Encycl.* XII. 207/2 The most typical of the Polish American laymen to achieve distinction was Peter Kiolbassa. **1936** Mencken *Amer. Lang.* (ed. 4) 673 The Polish-American journalists are rather more careful than most. **1949** *Dziennik Związkowy* (Chicago, Ill.) 19 Nov. 6/1 A banquet in honour of the Polish American public official. **1961** 'E. Lathen' *Banking on Death* x. 80 The Buffalo Polish-American Democratic Club. **1976** *National Observer* (U.S.) 1 May B6/4 Television cameras invaded a Polish-American working-class home on the south side of Milwaukee.

polished, *ppl. a.* Add: **1. c.** Of rice: having the outer layers of the grain removed.

1922 W. G. R. Francillon *Good Cookery* (ed. 2) xii. 236 Unpolished rice is cheaper and more nutritious than polished rice. **1948** *Good Housek. Cookery Bk.* ii. 400 The natural rice has rather more flavour and food value than polished rice. **1979** P. B. Medawar *Advice to Young Scientist* vi. 32 Scientists..demonstrated that..unpolished rice is much better for us than polished white rice.

polishing, *vbl. sb.* Add: **1. b.** Also, the outer layers of rice grain usually removed during the milling process.

1912 *Chambers's Jrnl.* Apr. 237/2 If the birds were fed on the milled rice mixed with the outer husks or 'polishings' which had been removed, the disease did not manifest itself. **1937** *Discovery* Nov. 348/2 Rice, or rather that part of it known as 'polishings', is the source of a vitamin product... Rice polishings are the external layers of the rice grain, usually removed by milling or 'polishing' in preparing the cereal for the market.

2. c. The filtration of the last traces of suspended solids from a liquid at the final stage in a process, *spec.* in the brewing of beer, and in the purification of effluent.

1938 *Jrnl. Inst. Brewing* XLIV. 466/2 One type of sheet filter composed of 160 sheets, 80 on each side, one side for 'roughing' and the other for 'polishing'. **1956** L. B. Escritt *Sewerage & Sewage Disposal* xviii. 361 The waterworks process of sand filtration has also been found satisfactory for effluent polishing. **1957** K. Barton-Wright tr. *J. de Clerck's Textbk. Brewing* I. xxii. 465 These sheet filters hold back yeast much better, and polishing is frequently omitted, whereas the metal mesh grids invariably let yeast pass through and a polishing filtration must always be carried out. **1958** *New Biol.* XXV. 88 Similar oxidation ponds are used in this country, e.g. in Bradford, for the final 'polishing' of a treated effluent, where the low flow in the receiving water demands this. **1971** R. L. & G. L. Culp *Adv. Wastewater Treatment* iii. 49 Effluent chlorination can provide more efficient disinfection of the effluent, which negates the only remaining virtue of the polishing pond for larger plants.

3. *polishing-paste* (further examples), *-powder* (earlier and later examples).

1916 'Taffrail' *Pincher Martin* vi. 86 A convenient receptacle for dirty cotton-waste, polishing-paste, bath-brick, and emery-paper. **1969** *Gloss. Terms Dentistry (B.S.I.)* 64 *Polishing paste,* a blend of fine abrasive particles with bonding agents and flavouring. It is used for cleaning and polishing surfaces of teeth, restorations and appliances. **1849** C. Brontë *Shirley* III. xiv. 299 The cup and platter he burnished up with the best polishing-powder. **1895** [see *blueing, bluing *vbl. sb.* 2]. **1969** *Gloss. Terms Dentistry (B.S.I.)* 64 *Polishing powder,* a material containing fine abrasive particles for polishing teeth.

‖ **polisson** (polisoṅ). [Fr.] An urchin or scamp; an ill-bred and uncouth person. Also *attrib.*

1866 G. J. Whyte-Melville *Cerise* II. x. 148 He was discovered as a coxcomb, an intruder and a *polisson.* **1897** G. du Maurier *Martian* 15 The polisson picked up his pocket-handkerchief and went. **1905** W. James *Let.* 8 Feb. in R. B. Perry *Tht. & Char. W. James* (1935) II. 398 When will either you or I, to whom Locke's mind was that of a street *polisson* in point of subtlety and 'truth', have statuettes? **1915** M. F. Sandars *Life & Times of Queen Adelaide* vii. 111 Instead of the polisson manner for which he used to be celebrated, he is now quiet and well-behaved, like anybody else.

Politbureau, -buro (pǫ·litbiūᵊ:rōᵘ). Also **politbureau, -buro.** [a. Russ. *politbyuró,* f. *polit*(*icheskoe* political + *byuró* bureau.] The highest policy-making committee of the U.S.S.R., or of some other Communist country or party (in quot. 1930, a district committee). Also *attrib., transf.,* and *fig.*

[**1926** *Encycl. Brit.* III. 428/1 This 'plenum' elects the Political Bureau of nine members with five deputies.] **1927** *Daily Express* 19 July 3/4 Stalin has packed the Politburo, which..is practically the Cabinet, with his friends. **1930** *Morning Post* 13 Aug. 13 The factory Soviet asked the local politbureau for lecturers to explain the causes of the shortage. **1937** E. Snow *Red Star over China* iv. v. 165, I [*sc.* Mao Tse-tung] was dismissed from the Politbureau, and also from the Party Front Committee. **1949** *Ann. Reg. 1948* 285 The names of the *Politburo* of the [Yugoslav Communist] party were..made public. **1949** G. B. Shaw *Sixteen Self Sketches* xi. 67 For some years the leaders in the Politbureau or Thinking Cabinet of the Fabian policy were Webb, Olivier, Wallas, Shaw. **1952** [see *Bolshevik *a.*]. **1953** *Encounter* Oct. 67/1 Malenkov..had..a role more prominent than all other living Politburo members. **1962** *Listener* 6 Sept. 354/2 Mr Ben Bella's Politburo..ordered nationalist troops to advance on Algiers. **1967** I. Grey *First Fifty Years* xxviii. 479 The congress [of 1966] revived the name 'Politburo', which had been changed to 'presidium' by the sixteenth congress in 1952. **1974** J. White tr. *Poulantzas's Fascism & Dictatorship* iv. ii. 171 Insurrection was then decided on by the Comintern and the majority of the Russian politburo for October 1923. **1977** 'S. Leys' *Chinese Shadows* (1978) v. 119 The keenness and energy with which..old men cling to their seats on the Politburo. **1978** *Whitaker's Almanack 1979* 959/1 The real power of the Party is vested, however, in the *Politbureau,* the *Secretariat* and the permanent Departments of the Central Committee.

polite, *a.* Add: **3.** *absol.* or as *sb.* In colloq. phr. *to do the polite*: to perform a polite action (freq. with *thing* understood); to behave politely.

1856 [see Do *v.* 11 j]. **1933** D. L. Sayers *Murder must Advertise* vi. 95, I saw you doing the polite to Miss Rossiter. **1935** G. Greene *England made Me* iv. 199 They are leaving at the end of the week. I've got to do the polite. **1939** 'M. Innes' *Stop Press* i. vi. 136 Some chaps over there. Must do the polite.

politic, *a.* and *sb.* Add: **B.** *sb.* **3.** *pl.* politics. **c.** Delete †*Obs.* and add later examples in unfavourable sense. Phr. *to play politics*: see *play *v.* 16 d.

1930 N. W. Stephenson *Nelson W. Aldrich* xx. 327 Northern enemies were quick to draw a conclusion; the expulsion of the Brownsville soldiers was mere politics, a play to the gallery to make sure the hold of the administration on the Southern Republican machine. **1952** *Manch. Guardian Weekly* 11 Dec. 13 The 'politics' involved..in key posts are not the private appetites of machine politicians or rarely that.

g. Used as a singular noun.

1906 *Daily Chron.* 7 Dec. 6/4 She [*sc.* Australia] has a politics of her own, and Europe is all the poorer for being out of touch with it. **1931** M. de la Bedoyère *Drift of Democracy* ii. 16 This politics is the vaguest of disciplines. **1970** I. L. Horowitz *Masses in Lat. Amer.* i. 23 If the United States model is to succeed in Latin America..a pluralistic politics of competitive, numerous, but autonomous groups must emerge.

h. *Comb.*, as *politics-conscious, -free, -infested, -mad, -ridden* adjs.

1957 H. Read *Tenth Muse* xxxv. 303 Futurism was more conscious of its environment, machine-conscious, politics-conscious. **1977** *New Yorker* 4 July 85/1 Considering the period it covers, the 'Memoir' is politics-free to an amazing degree. **1949** Koestler *Promise & Fulfilment* v. 270 The good, clean academic atmosphere acted like a disinfectant on our politics-infested minds. **1937** W. B. Yeats *Let.* 8 Feb. (1954) 880 He says that in England the educated classes are politics-mad. **1946** R. Blesh *Shining Trumpets* (1949) xiii. 299 The politics-ridden, local relief of Depression days.

politic (pǫ·litik), *v.* Also **politick.** [f. Politic *sb.* or (esp. in later use) a back-formation f. *politicking *vbl. sb.*] *intr.* To engage in political activity, esp. in order to strike political bargains or to seek votes (for oneself or another). Also *trans.* (rare).

1917 O. Douglas *Setons* xiv. 225 He has been politic-ing down in Ayrshire. **1967** [see *drop *v.* 28 a and b]. **1974** *Observer* (Colour Suppl.) 3 Nov. 31/2 Within the same square mile..Richard of Gloucester politicked. **1977** R. L. Duncan *Temple Dogs* (1978) ii. i. 164 He was having to politic the old man to keep him from swerving away from his beliefs.

political, *a.* (*sb.*) Add: **1. a.** (Further examples.)

1825 J. S. Mill in *Westm. Rev.* Apr. 291 The subjects on which it is the interest of rulers that the people *should* be misled; the political religion of the country, its political institutions, [etc.]. **1846** —— in *Edin. Rev.* LXXXIV. 344 They [*sc.* the Greeks] were..the originators of political freedom. **1882** G. A. Sala *Amer. Revisited* i. xi. 161 For some mysterious reason..the Consumers' Ice Company figured as a political organisation in this astounding Parade. **1907** L. H. Morgan *Anc. Society* ii. xiii. 335 City wards and country townships..would have become the basis of the new political system. **1934** T. S. Eliot *Rock* i. 15 Political religion is like invalid port: you calls it a medicine but it's soon just a 'abit. *a* **1942** B. Malinowski *Scientific Theory of Culture* (1944) v. 50 The Chinese civilization differs from ours..in the economic and political organization of the country. **1949** M. Mead in M. Fortes *Social Structure* 20 We have blocked out conceptually a large number of such areas as: the relationship between the representations of family structure and political structure in the psychology of the individual. **1957** P. Worsley *Trumpet shall Sound* 227 They occur..among people living in..societies..which lack centralized political institutions. **1962** P. Diesing *Reason in Society* v. 170 The political structure of a group is the organization of forces which determines how its decisions are made. **1976** *Times* 21 May 2/5 The United Unionist M.P.s who provide most of Northern Ireland's political representation at Westminster.

4. (Earlier and further examples.) Also (freq. in derogatory use), serving the ends of (party) politics; having regard or consideration for the interests of politics rather than questions of principle.

1749 [see Machine *sb.* 4 *fig.*] **1861** J. S. Mill *Repr. Govt.* i. 4 Political machinery does not act of itself. **1909** C. F. G. Masterman *Condition of England* iii. 83 They appear as the mainstay of the political machine in suburban districts. **1912** *Out West* (Los Angeles) June 401 Are you interested..in the way people are defrauded, and bunkoed, and swindled, and played with by the political bosses? **1934** L. Mumford in W. Frank et al. *Amer. & Alfred Stieglitz* i. ii. 34 The political boss and his underlings. **1974** G. Woodbridge in H. van Thal *Prime Ministers* I. 349 He [*sc.* Lord Grey] concluded with a highly political and very clever speech, suggesting that all that was at stake was the acceptance of the principle that there should be some reform, leaving the exact details..to be settled in the committee stage. **1977** *Times* 27 Jan. 5/1 If the Government chose the path of ill-considered and largely political legislation aimed at achieving a union takeover of private industry, it would have dealt a damaging blow. **1978** *Times* 3 Jan. 1/3 The union's vice-president..told a rally of 1000 striking firemen that the employers were playing 'a political game'.

6. *political animal* [tr. Gr. πολιτικὸν ζῷον (Aristotle *Politics* i. ii. §9) an animal intended to live in a city, a social animal] man, as acting in concert with others; a person who is interested in, or who participates in, politics; *political anthropology,* the study of the origins, forms, and exercise of community authority as it has evolved in aboriginal or isolated societies; *political asylum,* the condition of being, or permission to remain in a country as, a political refugee; *political commissar,* in China, a representative appointed to a military unit to be responsible for political education and organization; *political football,* a subject of contentious political debate; *political hostess,* a hostess at a party or gathering attended by politicians; *political morality,* public ethics; *political novel,* a fictitious political narrative, a novel about imaginary politicians; *political philosophy,* that department of philosophy which treats of politics or public ethics; hence *political philosopher*; *political police,* a police force concerned with offences against the state; *political prisoner* (examples); *political refugee,* a refugee from an oppressive government; *political science,* the study of the factors involved in politics (see Politic *sb.* 3) or the scientific analysis of political activity and behaviour; hence *political scientist*; *political sociology* (see quot. 1968); hence *political sociologist*; *political theory,* theory that is concerned with philosophical ideas of political power and with the history, forms, and activity of the state; hence *political theorist*; *political trial,* a trial of a defendant charged with a political offence, or a trial conducted for political reasons; *political warfare,* propaganda against another state, calculated to weaken it.

1776 W. Ellis tr. *Aristotle's Treat. Govt.* i. ii. 6 Hence it is evident, that a city is a natural production, and that man is naturally a political animal. **1892** I. Zangwill *Childr. Ghetto* II. 113 The East End Jew is only slowly becoming a political animal. **1960** *Victorian Studies* June 348 Characters who necessarily act and feel as isolated individuals as well as political animals. **1968** L. Durrell *Tunc* ii. 89 If we get in again it will be to try and prove..that the key to the political animal is magnanimity. **1975** *Times* 17 July 19/1 Lady Young is..very much a political animal. **1970** A. M. S. Smith tr. *Balandier's Polit. Anthrop.* p. vii, This book is also intended to show how political anthropology is contributing to a clearer definition and a better knowledge of the political field. **1975** in Beattie & Lienhardt *Stud. in Soc. Anthrop.* xiv. 336 What Evans-Pritchard treated as no more than one sub-system among others..now seems..to have become a specialization on its own, a 'political anthropology'. **1954** *Times* 14 Dec. 9/3 A steady consensus of judicial interpretation sustains the tradition of political asylum. **1962** *Weekly Law Reports* III. 1016 *Per* Viscount Radcliffe. In my opinion the idea that lies behind the phrase 'offence of a political character' is..the analogy of..'political refugee', 'political asylum' or 'political prisoner'. It does indicate..that the requesting State is after him for reasons other than the enforcement of criminal law. **1973** *Ann. Reg. 1972* i. iii. 25 Two officers of the Royal Moroccan Air Force..arrived in Gibraltar and sought political asylum. **1937** E. Snow *Red Star over China* viii. vi. 299 The discussion continued for over an hour. Occasionally the commander or political commissar interrupted to summarize what had been said. **1956** F. F. Liu *Milit. Hist. Mod. China* ii. 19 The Kuomintang..adopted a plan of gathering armed forces..and then inserting trained political commissars at the various levels of the newly absorbed groups. **1965** J. Ch'en *Mao & Chinese Revolution* (1967) i. viii. 175 At Iyang there was the 10th Red Army under Fang Chih-min with Shao Shih-p'ing as the political commissar. **1978** H. McLeave *Borderline Case* (1979) xiii. 134 Yao was a political commissar and was running a military setup. **1971** *Financial Mail* (Johannesburg) 26 Feb. 673/3 The whole question of new negotiations seems to have become a political football and little more. **1975** *Australasian Express* 27 Mar. 7/1 In a strong attack, Mr Perkins claimed Aborigines were a 'political football' in Australia. **1977** *New Scientist* 17 Mar. 641/2 Recombinant DNA research is rapidly assuming the shape of a political football. **1883** E. W. Hamilton *Diary* 18 Mar. (1972) II. 411 Lady Hayter has played the part of political hostess this year excellently well. **1944** C. Dilke in *Wine & Food* Spring 24 The house of a well-known political hostess. **1977** J. Aiken *Last Movement* v. 92 Her..husband had suddenly ditched her in favour of a fat blonde political hostess. **1827** J. S. Mill in *Arch. f. Sozialwissensch.* (1929) 450 There always ought to be..a certain difference of opinion in every ministry. Let any one consider what the effect would be if the contrary maxim were received as a rule of political morality. **1861** —— *Repr. Govt.* x. 193 Undoubtedly neither this nor any other maxim of political morality is absolutely inviolable. **1866** *Times* 26 June 6/3 *Felix Holt, the Radical,* is not..a political novel, though it necessarily touches on politics. **1976** *Hiroshima Stud. Eng. Lang. & Lit.* XXI. 69 It [*sc.* Nostromo] is regarded as a 'political novel', a 'historical novel' or a 'philosophical novel'. **1833** Lytton *England & English* II. 338 It is necessary..to distinguish between Mr. Bentham's practical conclusions..and his systematic views as a political philosopher. **1924** V. L. O. Chittick *T. C. Haliburton* xiii. 326 It was unquestionably not Sam Slick the political philosopher..that gave to his 'Sayings and Doings' the surprising vogue they formerly enjoyed. **1961** Kaplan & Katzenbach *Pol. Foundation of Internat. Law* i. iii. 64 Political philosophers will recognize its origins in the rejected doctrines of Hobbes. **1785** W. Paley (*title*) The principles of moral and political philosophy. **1825** J. S. Mill in *Westm. Rev.* Apr. 286 The general question, to what extent restraints upon the freedom of the press can be considered as warranted by sound principles of political philosophy. **1958** A. R. Radcliffe-Brown *Method in Social Anthropology* ii. i. 139 There is an abundant literature on the subjects of social philosophy, political philosophy,..and the philosophy of art. **1976** *Listener* 3 June 705/1 There has been a real revival in political philosophy..particularly..in the U.S.A. **1910** R. Anderson *Lighter Side of my Official Life* xv. 246 Before coming to England as Ambassador, Count Schouvaloff was head of the Political Police at St. Petersburg. **1953** *Encounter* Oct. 67/1 Never during the Stalin era was the political police charged with illegal extortion of evidence. **1974** J. White tr. *Poulantzas's Fascism & Dictatorship* vii. v. 341 None of this reorganization of the State apparatus can be understood without taking into account the growing and dominant role of the political police. **1860** Dickens in *All Yr. Round* 13 Oct. 14/2 All the town knew about the Englishman and his political prisoner. **1927** B. Russell *Outl. Philos.* iii. 44 Perhaps in time the State will perform these experiments with the children of political prisoners. **1972** *Guardian* 8 May 1/2 The loyalists have already sent a letter to Mr Whitelaw demanding political prisoner status. **1941** Koestler *Scum of Earth* 137 Paragraph 19 of the Armistice Treaty, providing for the extradition of political refugees. **1974** N. Freeling *Dressing of Diamond* 48 Marrying a political refugee..hadn't done him any good. **1779** Hume *Dial. Nat. Relig.* i. 16 This is their practice in all natural, mathematical, moral, and political science. **1794** A. Ferguson *Princ. Moral & Polit. Sci.* II. ii. vi. 411 *Salus populi, suprema lex esto,* is the fundamental principle of political science. **1836** J. S. Mill *Ess. Pol. Econ.* (1844) v. 146 This we can seldom do in ethical, and scarcely ever in political science. We cannot try forces of government and systems of national policy on a diminutive scale in our laboratories. **1898** G. B. Shaw *Let.* 18 Oct. (1972) II. 66 You and I have been confronted often enough with the follies of current political science. **1958** A. R. Radcliffe-Brown *Method in Social Anthropology* i. iv. 102 Political systems, economic systems, and systems of law are studied in social anthropology and also in economics, political science and jurisprudence. **1974** A. Barbrook *Patterns Polit. Behav.* i. 3 We have been left with a considerable range of behavioural theory, some of which has been completely accepted into the methodology of political science. **1902** *Amer. Jrnl. Sociol.* Jan. 564 There is not a political evolution through steps a, b, c, d, etc., which the political scientist can account for. **1974** *Listener* 24 Oct. 578/1 Professor Richard Rose..is the most productive political scientist in Britain. **1972** Dowse & Hughes *Political Sociol.* i. 7 As a matter of fact political sociologists do tend to concentrate on seeing in what ways society affects the state. **1957** *Current Sociol.* VI. 79 Political sociology is one of these recent additions. The label is perhaps more novel than the field. **1968** *Encycl. Soc. Sci.* XII. 298/2 Broadly conceived, political sociology is concerned with the social basis of power in all institutional sectors of society. In this tradition, political sociology deals with patterns of social stratification and their consequences in organized politics... By contrast, in narrower terms, political sociology focuses on the organizational analysis of political groups and..leadership. **1977** W. J. Goode *Princ. Sociol.* xiii. 399/2 Political sociology and political science as special fields have overlapped comfortably for several decades. **1951** D. Easton in Gould & Thursby *Contemp. Polit. Thought* (1969) xvii. 309 Speculations of the best political theorists have always been founded on acute observations of the contemporary political scene and a knowledge of human history. **1896** J. N. Figgis *Divine Right of Kings* x. 256 It is a far cry from the conception expressed in the Holy Roman Empire, that theology is the source of political theory. **1974** H. M. Drucker *Pol. Uses of Ideology* iv. 37 When a political theory points out that one kind of political system has such an advantage over another, we prefer the former. **1973** *Listener* 14 June 793/1 Every political trial has a long hidden history of what went on behind the scenes. **1974** J. Banning *How I fooled World* xi. 52 There were the political trials in Czechoslovakia. **1950** *Chambers's Encycl.* XI. 254/2 In 1941..Eden,..Bracken, the minister of information, and Dalton, who supervised propaganda to the enemy, constituted a Political Warfare Executive. **1977** E. Ambler *Send no more Roses* iv. 70 The only higher-ups who took any real interest in our findings were the political warfare people.

7. In *Comb.*, prefixed to an adj. to denote: **a.** 'politically, as applied to politics', as *political-ethical, -moral, -strategic*; **b.** 'political and...', as *political-bureaucratic, -cultural, -economic, -juridical, -military, -religious, -social.*

1970 C. Furtado in I. L. Horowitz *Masses in Lat. Amer.* ii. 31 The political-bureaucratic structure exterted [*sic*] a strong influence within the society. **1959** C. W. Mills in —— *New Sociology* (1964) 85 [The] joint political-cultural struggle must be waged in intellectual and moral ways. **1937** *Science & Society* I. 153 There are abundant instances of the relation between 'standard' speech and the political-economic character of the ruling class. **1970** J. Cotler in I. L. Horowitz *Masses in Lat. Amer.* xii. 437 Due to the political-economic limitations in Mancha India, social opportunities for the Cholo are somewhat limited. **1936** L. Wirth in K. Mannheim *Ideology & Utopia* p. xviii, Political-ethical norms not only cannot be derived from the direct contemplation of the facts, but themselves exert a moulding influence upon the very modes of perceiving the facts. **1971** R. apRoberts *Trollope* iv. 78 The political-ethical dilemmas of that novel [sc. *Ralph the Heir*]. **1919** J. T. Garvin *Econ. Foundations of Peace* x. 231 The means..will not be provided by the political-juridical part of the coming Constitution of the League. **1965** H. Kahn *On Escalation* vi. 125 Its possible worth in fulfilling European political-military objectives. **1970** H. Trevelyan *Middle East in Revolution* p. x, The withdrawal from Aden was a political-military operation conducted jointly by Headquarters, Middle East and the High Commission. **1953** S. Spender *Creative Element* 9 In the 1930's..I wrote of a 'political-moral' theme in modern literature. **1970** R. Stavenhagen in I. L. Horowitz *Masses in Lat. Amer.* vii. 259 Individual economic pre-eminence..arises, individually, through positions held in the political-religious structure. **1965** *English Studies* XLVI. 395 Melville is working in cosmic-religious, rather than political-social, terms. **1965** H. Kahn *On Escalation* vi. 122 It is not an improbable international political-strategic order for the future.

B. *sb.* **1. a.** (Later examples.)

1926 [see *fiddly *a.*]. **1939** *Times* 1 Aug. 13/3 Sir John Maffey..was an Indian political. **1958** L. Durrell *Mountolive* iv. 91 Pursewarden as political feels that the Embassy has also in a way inherited Maskelyne's department. **1979** C. Allen *Tales from Dark Continent* viii. 110 Most administrators—other than the Sudan politicals—regarded themselves as badly paid.

c. (Later examples.)

1938 *New Statesman* 19 Feb. 273/2 There are only 15 'politicals' still in gaol in the United Provinces and only 26 in Bihar. **1968** *Guardian* 22 Nov. 9/4 We started off being D Group prisoners, the lowest grade which only applies to politicals.

politicalization. (Further examples.)

1935 *Sun* (Baltimore) 19 Dec. 12/1 The current strong tendency toward politicalization of the intellectuals. **1947** *Partisan Rev.* XIV. 485 The ever-growing politicalization of intellectual life makes more and more difficult a disinterested theoretical approach. **1959** *Times Lit. Suppl.* 5 June 330/4 This is what Professor Marcuse calls the 'externalization' or the 'politicalization' of ethics. **1974** *Nature* 1 Mar. 1/1 A move toward what NIH scientists refer to as 'politicalisation of research'.

politicalized, *ppl. a.* [f. Politicalize *v.*] Made political in character.

1926 *Public Opinion* 13 Aug. 147/3 We are to have a politicalised Civil Service in this country. **1949** *Sun* (Baltimore) 19 Nov. 6/1 Does Congress wish to encourage big credit-seekers to turn..to a kind of politicalized or socialized banking setup?

politically, *adv.* Add: **3.** *Comb.*, as *politically-active, -inclined, -minded, -motivated* adjs.

1974 *Disturbances Univ. of Essex: Rep. Annan Enquiry* 17 Students protested that politically active ringleaders were singled out. **1969** J. Mander *Static Soc.* iii. 95 Enough for the politically-inclined tourist. **1907** *Westm. Gaz.* 11 Dec. 2/1 The politically-minded stay-at-home citizen. **1973** W. J. Burley *Death in Salubrious Place* iv. 73 The suave, politically-minded Bellings. **1972** *Listener* 21 Dec. 854/2 A rising level of politically-motivated violence. **1975** *Ibid.* 21 Aug. 233/2 We are not dealing with..an irresponsible, politically-motivated organisation in trade unions.

politicalness. (Earlier and later examples.)

1678 Cudworth *Intell. Syst.* i. v. 890 Not so much as any the least *seeds*, either of Politicalness, or Ethicalness at all in it. **1935** *Discovery* May 128/2 Notwithstanding all his politicalness and his zest for the letters and society ..it is in the campaigns and battles that Mr Trevelyan is happiest.

politicization (poli:tisəizēi·ʃən). [f. Politicize *v.*] The action or process of rendering political or of establishing upon a political basis; the fact of being politicized.

1934 *Times Lit. Suppl.* 25 Oct. 724/2 The attempted politicization of the German Protestant Church. **1938** *Downside Rev.* LVI. 384 The totalitarian supremacy of the State is the outcome of that 'total politicization' of man and his activities which was of the essence of Marxist theory. **1962** *Times Lit. Suppl.* 4 May 308/1 The politicization of private life. **1968** *Internat. Encycl. Soc. Sci.* X. 284 The major political process of recent years [in 'Middle America'] has been referred to as 'politicization', the recognition of the State as the ultimate authority and the recognition of legitimacy of certain governmental processes. **1976** F. Zweig *New Acquisitive Society* i. v. 57 The rapidly growing 'pressure group' movement..definitely leads to politicization of economic life. **1978** *Listener* 2 Nov. 564/2 Politicisation does not mean mere organised political activity... Politicisation of religion means the internal transformation of the faith itself, so that..it becomes..concerned with social morality rather than with the ethereal qualities of immortality.

politicize, *v.* Add: **2.** (Further examples.)

1962 S. E. Finer *Man on Horseback* xii. 236 Such parties..seek to dominate and politicize all politically important voluntary bodies outside their ranks. **1970** *Guardian* 5 June 11/7 The skinheads who Mr Powell has managed to politicise. **1978** *Time* 3 July 38/1 We're not out to politicize the White House..but we've got to use the political resources we have better than before.

Hence **poli·ticized** *ppl. a.*, interested or involved in politics, politically motivated; **poli·ticizing** *vbl. sb.*, the action or process of rendering political.

1887 [see Politicize *v.* 2]. **1971** *Daily Tel.* 17 Aug. 10 Any danger in this arises from the majority of viewers being..unaware of the bias as they watch it because most people are manifestly less politicised. **1975** R. Butt in Cox & Boyson *Black Paper 1975* 43/1 In this single statement, the politicizing of education in an entirely new sense—namely that it is now the vehicle used by those who, in varying degrees, wish to change the cultural basis of society—is explicit. **1977** *N.Y. Rev. Bks.* 12 May 50/4 (Advt.), I am a politicized, socialist, would-be scholar deeply isolated from academic radicalism.

politicking (pǫ-litikiŋ), *vbl. sb.* [f. *politic *v.* or Politic *a.* and *sb.* + -ing[1].] The action or fact of engaging in (esp. partisan) political activity.

1928 M. H. Weseen *Crowell's Dict. Eng. Gram.* 481 *Politicking*, a coined word that has no recognized standing. **1934** *Sun* (Baltimore) 10 July 1/3 Mr. Farley..confided to 'the boys' that he expects to do considerable 'politicking' along the way. **1943** *Sat. Even. Post* 30 Jan. 90/2 The politicking had started the minute his back was turned. **1957** *Economist* 5 Oct. 15/2 To dangle before the tenants..the idea that a Labour government will 'promptly' redress their grievances..might politely be described as politicking. **1975** F. Heer *Charlemagne & his World* x. 149 This is the elevated ideal that lay behind all the politicking and manoeuvering for position that took place in Rome in 800.

politico. Delete 'now *rare*' and add further examples. Also, a political agent; a person holding strong political views. Pl. **politicoes, politicos.**

1929 *Times* 14 Aug. 11/4 The old 'politicos' and the members of the more progressive parties, may be recommended to remember the object-lesson of the fate of the Aventine Opposition in Italy. **1941** E. Wilson *Wound & Bow* v. 215 Going back over Hemingway's books to-day, we can see clearly what an error of the politicos it was to accuse him of an indifference to society. **1951** I. Asimov *Foundation* v. xiii. 218 You're an old dog of a politico. **1955** J. Cary *Not Honour More* 42 It's a lot deeper than politics. And I wish you'd get all these politicos out of my house. **1956** W. Lewis *Red Priest* xxvii. 229 'The Mahatma was a very unusual man.' Father Makepeace replied. 'I referred to ordinary politicos.' 'There are no ordinary politicos in India,' said the elderly interrupter. **1958** *Listener* 18 Sept. 436/1 On Channel One..it was colour-bar week, with the Jamaican politico, Norman Manley, given a starring role. **1960** *Guardian* 8 Mar. 1/5 The press is here..and surprisingly important politicoes in ineffective disguises. **1967** *Observer* 26 Mar. 9 Hippies are sympathetic, but they are not politicos. **1973** *Black Panther* 11 Aug. 8/2 While publicly apologizing for Nixon, Colson managed to privately leak that the politicos Haldeman and Ehrlichman were guilty along with Mitchell and Dean. **1975** D. Lodge *Changing Places* iv. 145 Politicoes, frat rats, sallys and jocks and mommas for peace are..touching each other's hearts.

politico-. Add: **a.** *politico-aesthetic, -artistic, -moral* (further examples), *mythic, -Philistine, -racial*; **b.** *politico-bureaucratic, -criminal, -diplomatic, -legal, -literary, -military* (further examples), *-philosophical, -physiographical, -religious, -social* (further examples).

1974 *Times* 8 Jan. 14/6 Almost immediately he started giving long and fiery politico-aesthetic lectures to adoring left-wing students. **1961** *Times* 31 Oct. 14/3 *Accatone* [sc. a film]..has become the subject of one of those politico-artistic controversies. **1964** P. Worsley in I. C. Horowitz *New Sociology* 377 Politico-bureaucratic machines are in the saddle from the beginning, and there is no 'heroic' period of Cuban-type mass participation in government. **1955** D. W. Maurer in *Publ. Amer. Dial. Soc.* xxiv. 26 Very good connections with the politico-criminal liaison,..which extends like a network across the country. **1856** Dickens *Little Dorrit* (1857) i. x. 84 He fully understood the Department to be a politico diplomatico hocus pocus piece of machinery. **1973** P. A. Allum *Politics & Society in Post-War Naples* vii. 219 Their electoral and personal politico-diplomatic activities cannot be justified in straight *Gesellschaft* normative language. **1926** Galsworthy *Silver Spoon* i. ii. 9 Lady Alison's politico-legal coterie no longer counted. **1970** B. Brewster tr. *Althusser & Balibar's Reading Capital* ii. v. 133 In fact, this is to attribute to the concept 'superstructure' a breadth Marx never allowed, for he only ranged within it: (1) the politico-legal superstructure, and (2) the ideological superstructure. **1924** Galsworthy *White Monkey* 73 She..picked out the biggest 'bug', politico-literary, and waited to pin him. **1978** D. Daiches *Edinburgh* x. 182 There was some pretty savage politico-literary fighting, especially in the Tory *Blackwood's*. **1931** *Times Lit. Suppl.* 29 Oct. 827/1 One of the objects of the Prussian campaign, a masterpiece of politico-military strategy, was to separate Prussia and Saxony. **1975** *Times* 1 Dec. 10/5 The then politico-military concepts of revolutionary war. **1920** B. Russell *Pract. & Theory Bolshevism* i. ii. 30 The sincere Communists..are not unlike the Puritan soldiers in their stern politico-moral purpose. **1965** *Mod. Law Rev.* XXVIII. v. 536 There would seem to be much more force in Professor H. L. A. Hart's argument that the judges do not know or at least cannot evaluate what the 'politico-moral' principles behind their decisions are. **1974** *Publishers Weekly* 4 Nov. 62/1 This politico-mythic tale..expresses wishes rather than facts. **1960** *Times* 3 Oct. 13/3 They [sc. universities] may save themselves from the politico-Philistine interference they fear. **1936** *New Yorker* 14 Mar. 24/1 Proletarianism.. superimposed on some remarkably interesting politico-philosophical formulae. **1973** *Nation Rev.* (Melbourne) 31 Aug. 1455/2 These uncertainties..are the ruination of most politico-philosophical series. **1935** *Univ. Mich. Publ. Lang. & Lit.* XIII. 43 South of this point it coincides with the strong politico-physiographical frontier already indicated. **1959** *Times Lit. Suppl.* 22 May 303/3 Lynchings and murders and other forms of politico-racial violence have significantly diminished since the war. **1908** *Daily Chron.* 19 June 4/6 The Great Powers.. should recall some of their agents and those politico-religious missionaries, who, instead of putting out the fire, secretly throw fuel into the flames. **1937, 1953** [see *Caodaism]. **1978** D. Murphy *Place Apart* xi. 235 In the midst of all the politico-religious dissension [in Ulster], one tends to overlook the general social problems. **1958** Politico-social [see *Miltonizing *ppl. a.* and *vbl. sb.*]. **1950** M. Crosland tr. *Rovan's Germany* 99 The D.G.B... remains the richest and most powerful union confederation in Europe, constituting a genuine politico-social power.

c. prefixed to a noun, forming a quasi-*adj.*, as *politico-crime, politico-travel*.

1952 Turkus & Feder *Murder, Inc.* vi. 100 In November of 1950, an investigation of politico-crime tie-ups was launched in New York. *Ibid.* xiii. 291 The investigation shocked the public into a new consciousness of the politico-crime danger. **1970** *Daily Tel.* 14 May 6/2 A politico-travel diary, the book is useful background to the present crisis and conflict.

poli·tico-econo·mic, *a.* [Politico-.] = Politico-economical *a.*

1839 Carlyle *Chartism* x. 97 Paralytic Radicalism.. which..sounds with Philosophic Politico-Economic plummet the deep dark sea of troubles. **a 1854** Mill *Draft Autobiogr.* (1961) 82 The Benthamic & politico-economic form of Liberalism. **1910** J. W. Welsford (*title*) The strength of England: a politico-economic history of England from Saxon times to the reign of Charles I. **1933** E. E. Evans-Pritchard in *Africa* VI. 372 Exchange of blood in such situations sacralizes and endows with sanctions a politico-economic transaction. **1955** *Bull. Atomic Sci.* June 205/2 A complete understanding of political and sociologic conditions will have to be used to decide what politico-economic incentives and assurances may be brought to bear. **1974** P. Gore-Booth *With Great Truth & Respect* 380 Such are the problems of politico-economic diplomacy.

politico-economical, *a.* Add: (Further examples.)

1837 *Democratic Review* I. 113 In spite of the plain principles of politico-economical truth. **1858** [see *economical *a.* 3 b].

Hence **politico-economist,** a student of, or writer on, political economics (cf. Economist 4 b).

1885 W. Harris *Hist. Radical Party* vii. 141 It is worth noticing that Ricardo, the politico-economist, was in the minority.

politique. Add: **1. a.** (Later examples in extended senses.)

1958 *Times Lit. Suppl.* 11 July 390/4 It certainly attracted a number of outstanding political leaders to it. Dr. Zeldin introduces us to many of these latter-day politiques. **1959** *Encounter* July 45/2 Some presenting him [sc. Odysseus] as an enlightened statesman, others as a machiavellian *politique*. **1972** K. B. McFarlane *Lancastrian Kings & Lollard Knights* i. ii. 24, I would maintain that in politics Henry [IV] was not a man of constitutional principle at all but an opportunist and a *politique*.

2. A political concept or doctrine; an expression of political ideas.

1958 A. Dru tr. *Péguy's Temporal & Eternal* 27 A country, a *régime* does not need you, it does not need mystics, a *mystique*, or its *mystique*... It needs a sound *politique*, which means a good government policy. **1958** *Times Lit. Suppl.* 27 June 366/2 Péguy used the witness of the few independent supporters of Dreyfus to illustrate that radical distinction between *mystique* and *politique* which is the essential clue to all his thought... For him the Revolution and the Christian religion were in origin and in essence both *mystiques* that were profoundly true... But they had become *politiques*: 'It is one and the same movement which makes people no longer believe in the Republic and no longer believe in God.' *Ibid.* 366/3 The withering criticism of the clergy is always in terms of their compromising with the *mystique* which alone makes Christianity valid: the Church has devoted its energies instead to propagating a *politique*. **1959** *Listener* 4 June 999/1 His analysis of Communist politique had a tension which lent itself to radio dramatization. **1977** A. Ecclestone *Staircase for Silence* v. 88 *Politique*..set in as a process of dislocation,..a choosing to go it alone. It became an end in itself, ever seeking an aggrandisement of its own power, suspicious of and ready to suppress whatever challenged its own authority.

polje (pǭu·lyĕ). *Physical Geogr.* Also **polye.** Pl. **poljes,** ‖**polja,** (after Ger.) **poljen.** [Serbo-Croat *polje* field.] An enclosed plain in a karstic region, esp. Yugoslavia, that is larger than a uvala and usu. has steep enclosing walls and a covering of alluvium.

1894 *Geogr. Jrnl.* III. 323 The *poljen* occur at low levels, and therefore receive an enormous supply of 'ground water', especially at the times of the autumn or winter rains, which the underground outlet cannot carry it [*sic*] off fast enough. **1902** *Ibid.* XX. 428 In spring..the floors of the polyes are flooded. **1918** D. W. Johnson *Topogr. & Strategy in War* xiv. 169 In the rear of the Serbian armies ..runs the straight subsidiary trench formed by the Lepenatz valley, Kosovo Polye, and the Ibar valley. **1926** *Geogr. Jrnl.* LXVII. 197 Lakes in the high calcareous Alps of Switzerland occupy dolines or polyes, the bottom of which have been more or less filled by deposits of impervious material derived from the ground moraines of the ancient glaciers. **1934** *Discovery* Sept. 247/1 Some of these *polja* are periodic lakes. **1954** W. D. Thornbury *Princ. Geomorphol.* xiii. 324 The largest polje in the Western Balkans, the Livno polje, is 40 miles long and 3 to 7 miles wide. **1958** *Geogr. Jrnl.* CXXIV. 41 Some of the largest polja are found among the Dinaric Alps in the hinterland of Split. **1960** B. W. Sparks *Geomorphol.* vii. 155 The largest depressions of Yugoslavia, the poljes, are probably not solution forms at all but tectonic depressions modified by solution of the limestone preserved in them. **1972** *Science* 12 May 664/3 The perennial flooding of the farmlands in the poljes of Yugoslavia.

polk, *v.* Add: Hence **po·lking** *vbl. sb.*

1848 E. Gray *Let.* in M. Lutyens *Ruskins & Grays* (1972) x. 95, I got introduced to good partners and got some good polking. **1853** J. R. Planché *Mr. Buckstone's Ascent of Mount Parnassus* 30 Each night to some Casino.. Where I am pulled by fast young men about And with eternal polking quite worn out.

polka, *sb.*[1] Add: **1.** *polka-mazurka* (examples); *polka-time*.

1861 H. Rhys *Theatrical Trip for Wager!* xiii. 120 They advanced in line, in polka time, then right-about-turned. **1928** A. M. M. Douton *Bk. with Seven Seals* i. 112 The course of calisthenics..terminated with lessons in the ..polka-mazurka. **1957** G. B. L. Wilson *Dict. Ballet* 219 *Polka-mazurka*, dance derived from the polka, from which it differs in that it is in 3/4 time, and from the mazurka, by which it is distinguished by having an accent on the 3rd instead of the 2nd beat. **1967** Chujoy & Manchester *Dance Encycl.* (rev. ed.) 738/2 *Polka-Mazurka*, a Polish variation of the polka, in 3/4 time, danced as a ballroom dance in countries of Eastern Europe.

3. polka-dot (further examples); also *fig.*, *attrib.* or as *adj.*, and as *v. trans.*; hence *polka-dotted* adj.

1895 *Montgomery Ward Catal.* Spring & Summer 9/3 Polka Dot Chambray, linen finish. **1906** 'O. Henry' *Four Million* 136 The next day a person with red hands and a blue polka-dot necktie..called. **1908** W. G. Davenport *Butte & Montana beneath X-Ray* 9 Miss P—— received her guests in a lovely polka dotted frock. **1924** W. M. Raine *Troubled Waters* xxi. 224 He took off his big white hat and rubbed a polka-dot handkerchief over his bald head. **1928** F. N. Hart *Bellamy Trial* i. 3 He wore a shabby tweed suit, a polka-dotted tie. **1956** *Daily Mail* 19 July 6/1 Camping sites are scattered like polka dots all over the Riviera. **1957** V. Nabokov *Pnin* vi. 138 Amber-brown Monarch butterflies flapped.., their incompletely retracted black legs hanging rather low beneath their polka-dotted bodies. **1964** D. Varaday *Gara-Yaka* ii. 23 Polka dots were becoming clearly defined on her tawny, golden hide. **1966** Mrs. L. B. Johnson *White House Diary* 3 Apr. (1970) 382 A young newspaper-woman in a black-and-white polka-dot bikini, with a figure to suit it. **1969** *Better Homes & Gardens* (U.S.) Apr. 83 Tiny bright red fruits that polka-dot the long branches. **1970** *New Yorker* 16 May 18/1 (Advt.), Men often are remembered for odd reasons. Like our 11th President. Little did he dream when he designed the polka dot that one day it would become the epitome of fashion. **1972** *Daily Tel.* 16 Aug. 1/1 A blonde girl wearing a green polka dot bikini was found dead in the sea..yesterday. **1972** *Suttons Seeds* 16 Polka Dot mixed. These bushy Cornflowers stand up to weather well, with a good range of colours. **1978** J. Wainwright *Jury People* xxxvi. 111 Kids are out. Because the world ain't yet ready for polka-dot kids. Black is only beautiful in places. **1979** C. Wood *James Bond & Moonraker* xi. 112 She had big puffed sleeves..and a petticoat effect of over-lapping polka-dotted skirts.

polka *v.* (earlier example).

1846 Dickens *Let.* 5 July (1977) IV. 580 The common people waltzed and polka'd without cessation, to the music of a band.

poll, *sb.*[1] Add: **1. β.** (Further examples of *Sc.* and *north. dial.* form *pow.*)

1876 F. K. Robinson *Gloss. Whitby* 146/2 *Pow*, the poll; the human head. **1901** G. B. Shaw *Devil's Disciple* iii. 79 The Devil's Disciple here will start presently as the Reverend Richard Dudgeon, and wag his pow in my old pulpit. **1919** *Kelso Chron.* 4 Apr. 3 My blood's not chill, though near the night, And grey-haired is my pow. **1940** E. Pound *Cantos* lxxi. 186 His daughter told me he had burnt all his papers In melancholia May be from that swat on the pow. **1947** E. A. McCourt *Flaming Hour* ix. 54 'Weren't ye lyin' locked up in yon shed no later than yesterday mornin' with a bump on your pow that should have kept ye still for a fortnight?' he demanded ferociously. **1963** G. Thomson *Crocus & Meadowlark Country* xi. 75 Ethel added a drawing she made of Jim one cold day when he made his way there with Chaddy's red tam-o'-shanter pulled down on his red pow. **1965** *Buchan Observer* 12 Jan. 2 Ye'll hum an' hae an' claw your pow. **1973** *People's Jrnl.* (Inverness & Northern Counties ed.) 1 Dec. 4/5 I'd got no further than filling my pow with rollers and covering my head with a woollen head square.

7. d. A poll taken to estimate public opinion on a specified issue by questioning a sample intended to be representative of the whole people (see *Gallup); *spec.*, (*a*) = *popularity poll* s.v. *popularity 8; (*b*) a poll intended to forecast the result of a presidential, parliamentary, or other election (see **opinion poll* s.v. *opinion *sb.* 1 b). Also *attrib.*

1902 F. Clarke tr. *Ostrogorski's Democracy and Organization of Political Parties* II. v. 306 The poll taken in each locality is of general import for the whole Union, as well as of special significance for each political subdivision in the States. **1940** Gallup & Rae *Pulse of Democracy* iii. 35 In this poll [of July 1824] Andrew Jackson received 335 votes; John Quincy Adams, 169; Henry Clay, 19, and William H. Crawford, 9. **1944** *Times* 9 June 5/5 The recent British Medical Association poll of members' opinions with regard to medical interests caused considerable controversy. **1950** *Times* 8 Feb. 7/6 Public opinion polls and investigations carried out by the Japanese Press show that the Yoshida Cabinet no longer enjoys as much support as it did at the time of the General Election in January, 1949. **1964** Gould & Kolb *Dict. Social Sci.* 517/1 Poll denotes..the canvas of opinions, prior to an election, by simple or complex interviewing. **1973** *Melody Maker* 31 Mar. 18 It's time to vote in the Melody Maker Jazz Poll. This..is your opportunity to register your appreciation of the musicians, bands, and singers whom you think have made the finest contributions to jazz over the last year. **1974** *Times* 11 Feb. 15 The poll firms showed their anxiety by conducting their own inquest into their failure... Poll findings are being presented as *predictions* of the result. **1977** *New Yorker* 9 May 136/2 Polls conducted in April show that Carter currently has the approval of about two-thirds of the public. *Ibid.* 24 Oct. 42/3 Fetching coffee in paper cups for the poll watchers.

10. (sense 7 b) **poll-card,** an official notification informing voters of the place and date of voting; (sense *7 d) **poll-rating,** the popularity of a person (usu. a political leader) as indicated by a poll; **poll-taker,** usu. in *pl.*, a newspaper or other organization which conducts an opinion poll, = *pollster; hence **poll-taking** *vbl. sb.*; **pollwinner,** a successful candidate in a poll; so **poll-winning** *ppl. a.*

1908 *Westm. Gaz.* 4 May 2/2 Mr. Amery's final appeal.. s going with the poll-card to every elector. **1975** *Times* 27 Feb. 4/3 Poll cards..remind the elector of his right to vote and they tell him his voting number and where his polling station is. **1967** *Guardian* 16 Oct. 6/1 Mr Heath's poll-ratings were unsatisfactory. **1959** *Spectator* 4 Sept. 288/2 When poll-takers put the question directly to the citizenry, it seems that vast numbers think it good. **1964** *Economist* 12 Sept. 1021/1 A victory for Mr Goldwater would be the greatest upset for the poll-takers since the *Literary Digest* predicted a landslide for Mr Landon, the Republican who was crushed by President Roosevelt in 1936. **1976** *National Observer* (U.S.) 11 Dec. 5/3 Poll-taker Teeter told the governors that 'our Presidential elections have become nonpartisan media events'. **1964** I. L. Horowitz *New Sociology* 31 Problems of this kind can be multiplied a hundredfold—in every sphere of sociology, from poll-taking to theory-making. **1966** *Melody Maker* 7 May 4 Pollwinner Tubby Hayes heads a five-man British contingent which will join Austrian composer Friedrich Gulda's all-star international band on its tour of the continent this summer. **1958** P. Gammond *Decca Bk. of Jazz* xviii. 218 Jackson in particular was the equivalent of the poll-winning trumpet men of today. **1962** *Melody Maker* 7 July 8 Pollwinning bandleader Chris Barber has firm views on the subject.

Pollack, var. Polack *sb.* (*a.*) in Dict. and Suppl.

pollakanthic (pǫlăkæ-nþik), *a.* *Bot.* [f. Gr. πολλάκ-ις many times + ἄνθ-ος flower + -ic 1.] = *polycarpic *a.* a. Also **pollaka·nthous** *a.* in the same sense.

1909 Groom & Balfour tr. *Warming's Oecol. Plants* ii. 6 In recent times these Candollean terms have been suppressed often in favour of..Kjellman's 'pollakanthic'. **1965** Bell & Coombe tr. *Strasburger's Textbk. Bot.* 355 The production of vegetative members by the growing points changes over to that of flowers, fruits and seeds,.. periodically, as in pollakanthous plants. **1973** McLean & Ivimey-Cook *Textbk. Theoret. Bot.* IV. xlii. 3353 Polycarpic (pollakanthic) plants..flower and fruit repeatedly.

polled, *ppl. a.* **2.** (Further examples.)

1842 [see *Angus]. **1867** W. McCombie *Cattle* iv. 138 Mr Lyell..has a very good herd of polled Angus cattle. **1891** R. Wallace *Rural Econ. Austral. & N.Z.* xxxii. 415 Polled Angus cattle have come very rapidly into favour during the last few years. **1909** J. Wilson *Evolution of Brit. Cattle* v. 57 The Sutherland polled cattle are long extinct. **1940** J. Hammond *Farm Animals* viii. 149 If we mate a polled red Aberdeen Angus bull to a Shorthorn cow we shall obtain polled calves. **1976** *Cumberland & Westmorland Herald* 4 Dec. 16/2 (Advt.), Fortnightly sale of Friesian, Hereford cross, Charolais' cross and polled bullocks and heifers of all ages.

pollee (pǭuli·). [f. Poll *sb.*[1] + -ee[1].] One who is questioned in a poll (sense *7 d).

1940 *Propaganda Analysis* IV. 4/2 The question, 'For whom will you vote for President of the United States?' illustrates Objectivism. It gives to the pollee full liberty to name any candidate. **1941** *Amer. Speech* XVI. 306 *Pollee*, one polled by a public-opinion 'institute'. **1945** *Richmond* (Va.) *Times-Dispatch* 25 Oct. 10/6 Mr. Gallup.. should consult his little band of pollees and come up with their prophecy as to which of our service teams will go home with the goal posts. **1962** *Guardian* 3 Aug. 14/6 The 54 pollees who had voted last time had come within one percentage of splitting their 1960 votes.

pollen, *sb.* Add: **3.** *pollen-content(s), -zone; pollen-bearing* (example), *-dated, -free* adjs.; **pollen analysis** = *palynology; hence **pollen analyst,** a scientist who uses the techniques of pollen analysis; **pollen-analytic(al)** *adjs.*; **pollen-analytically** *adv.*; **pollen count,** an index of the quantity of pollen in the air, obtained by counting the grains collected on a given area of a coated glass plate exposed for twenty-four hours, and published as a warning to those allergic to it; also, an indication of the frequency of pollen in an archæological site; also *fig.*; **pollen diagram,** a sequence of pollen spectra from one site, showing changes in the frequencies of various types of pollen; **pollen graph** = **pollen diagram*; **pollen index** = **pollen count*; **pollen mother cell,** a cell in a seed plant which undergoes meiosis to yield four pollen grains; **pollen parent,** the plant from which pollen is taken to fertilize another plant in an attempt to produce a hybrid; **pollen profile** = **pollen diagram*; **pollen spectrum,** the relative frequencies of the various types of pollen in a single sample.

1924 G. Erdtman in *Jrnl. Linn. Soc.: Bot.* XLVI. 450 The study of micro-fossils (and especially of fossil pollen-grains), upon which von Post's method of pollen-analysis is based. **1935** *Discovery* Apr. 100/1 The methods of pollen-analysis..enable one to know of the afforestation and in consequence of the climate of the period during which any particular stratum in a peat deposit was laid down. **1944, 1958** [see *palynology]. **1973** *Microscopy* XXXII. 321 Erdtman prepared and presented the first doctoral thesis ever to be based on pollen analysis. **1943** G. Erdtman *Introd. Pollen Analysis* i. 1 Still more remarkable appear the performances accomplished by the pollen analyst today. **1973** *Microscopy* XXXII. 320 Gustav Lagerheim (1860–1926) was one of the earliest pollen analysts. **1949** *Bull. Geol. Soc. Amer.* LX. 1359/2 Events ..may have been contemporaneous in the astronomic as well as the pollen-analytic sense. **1946** F. E. Zeuner *Dating Past* iii. 59 Further important remarks on the principles and system of pollen-analytical datings are included in a great many papers. **1976** *Nature* 24 June 628/1 Some new information..has been gleaned..as a result of a pollen analytical study of upper Miocene lignites. **1936** *Proc. Prehistoric Soc.* II. 146 One may quote an instance from the Baltic..pollen-analytically dated to Boreal times. **1946** F. E. Zeuner *Dating Past* iv. 78 A good many localities have been studied pollen-analytically. *Ibid.* 80 Nilsson has studied the connexion of pollen-bearing deposits with raised beaches. *Ibid.* iii. 57 The pollen-contents of a peat are more or less characteristic of the tree-associations that grew in the neighbourhood of the spot under investigation. **1954** S. Piggott *Neolithic Cultures* i. 3 The evidence for the natural conditions of vegetation in Atlantic and Sub-Boreal times is based most reliably on the pollen-content of stratified peats. [**1873** C. J. Blackley *Experimental Researches on Catarrhus Æstivus* iv. 122 After being exposed for twenty-four hours, each slip was placed under the microscope, and any deposit it contained was carefully examined, and the number of pollen grains counted.] **1926** Koessler & Durham in *Jrnl. Amer. Med. Assoc.* LXXXVI. 1205/1 For the air study..differential pollen counts were made daily. **1944** Urbach & Gottlieb *Allergy* xix. 607 In order to become acquainted with the local flora, the physician should undertake pollen counts himself. **1965** *Punch* 15 Sept. 375/2 Throughout the summer, New York newspapers forecast daily..a 'pollen count' for hay-fever victims. **1975** G. W. Dimbleby in R. Bruce-Mitford *Sutton Hoo Ship-Burial* I. i. 68 (*heading*) Soil under ship-barrow pollen counts. **1978** *Times* 7 July 2/8 The pollen count issued in London yesterday by the Asthma Research Council was one, very low. **1936** *Proc. Prehistoric Soc.* II. 239 The forest culture of S.E. Britain, best known from the pollen-dated site of Lower Halstow, had diverged very markedly. **1924** G. Erdtman in *Jrnl. Linn. Soc.: Bot.* XLVI. 453 By means of the percentage numbers a pollen-diagram is constructed. **1954** S. Piggott *Neolithic Cultures* i. 4 The pollen diagrams constructed from stratified peats in many parts of the British Isles show a consistent evolution of forest assemblages. **1973** Proctor & Yeo *Pollination of Flowers* viii. 276 The pollen diagrams from different sites show striking correspondences in the course of events. **1963** *New Yorker* 15 June 117 Heated swimming pool... Pollen-free air. **1975** G. W. Dimbleby in R. Bruce-Mitford *Sutton Hoo Ship-Burial* I. i. 55 This pollen profile appears to have developed in a pollen-free sand deposit. This would accord with the suggestion that at some time during or since the Anglo-Saxon period the soil was truncated down to the pollen-free subsoil. **1959** J. D. Clark *Prehist. Southern Afr.* vi. 160 The grasslands of our Central Plateau Region which, the pollen-graphs from Florisbad tell us, must have still been open country even at the height of the pluvial. **1973** 'D. Jordan' *Nile Green* ii. 12 Most of us have hay fever and the pollen index was high. **1884** *Jrnl. R. Microsc. Soc.* 714 We should have a case similar to that seen in the pollen-mother-cells of *Fritillaria persica*. **1889** *Bot. Gaz.* XIV. 109 If any person has experienced difficulty in obtaining pollen mother-cells in excellent condition for study, their attention is called to the young anthers of *Negundo aceroides* Moench. **1926** *Ibid.* LXXXI. 154 In a given loculus of an anther the pollen mother cells of the apple show little variation in stage of development. **1976** Bell & Coombe tr. *Strasburger's Textbk. Bot.* 39 (*caption*) Prophase of the first meiotic divisions of a pollen mother cell. **1910** *Nat. Rose Society's Rose Ann.* 50 Place the top of the finger upon the anthers of the variety it is proposed to use as the pollen parent. **1933** *Lily Year-bk.* 173 The plant showed no trace of the pollen parent and was discarded. **1976** *Lilies* 36 The plant with narrow leaves had a long chromosome..of a type found in the pollen parent. **1967** M. J. Coe *Ecology Alpine Zone Mt. Kenya* 52 If the pollen profile is correlated with a time scale..it has been possible..to demonstrate an interesting sequence of vegetation zone depression. **1972** *Computers & Humanities* VII. 40 Pollen analysis is now being aided by the computer, either to compile and print the pollen profiles..or to perform statistical analysis of pollen data. **1924** G. Erdtman in *Jrnl. Linn. Soc.: Bot.* XLVI. 453 The relative frequency-numbers..constitute the pollen-spectrum of the sample. **1946** F. E. Zeuner *Dating Past* iii. 59 A circle with sectors giving the frequency of the most important species in the pollen-spectrum can be inserted on a map. **1977** *New Phytologist* LXXVIII. 711 The sequence of four pollen spectra from the Coralline Crag shows no great variation from one level to the next. **1946** F. E. Zeuner *Dating Past* 389, 23 finds..dated according to pollen zones. **1973** P. A. Colinvaux *Introd. Ecol.* vii. 100 A nine or ten pollen-zone sequence..named in Roman numerals.

pollenin. Delete † *Obs.* and add: Cf. *sporopollenin. (Further examples.)

Quot. 1931 does not represent a new sense.

1931 *Chem. Abstr.* XXV. 2455 Qual[itative] tests on many pollens showed that their hulls or membranes are similar chemically to lycopodium-sporenin [*sic*], and the name *pollenin* is proposed for such compds. **1964** [see *sporonin]. **1971** Chaloner & Orbell in J. Brooks et al. *Sporopollenin* 274 In terms of modern usage, both John and Braconnot were using the word pollenin not only for the exine, but also for the underlying intine of cellulose—in fact the whole of the sporoderm or spore coat. **1974** Stanley & Linskens *Pollen* ix. 138 Early studies..reported pollen to contain about 40% pollenin.

pollenizer (pǫ·lėnəizəɹ). [f. Pollenize *v.* + -er[1].] = *pollinator.

1897 *Bull. Central Exper. Farm Dept. Agric.* (Canada) No. 27. 18 This [variety of strawberry] is valuable as a pollenizer.

poll-evil. Add: (Later examples.) Also † *fig.*, an obsession with elections.

1794 J. Byng *Torrington Diaries* (1938) IV. 76 Sir G. at present is plagued by the pollevil—reverting to the past election with all the possibilities of a future one. **1873** J. H. Beadle *Undevel. West* xxvi. 565, I..reined up my horse suddenly and again butted him in the back of the head, at the imminent risk of giving us both the poll-evil. **1970** Miller & West *Black's Vet. Dict.* (ed. 9) 730/2 'Poll evil' is an old, colloquial name sometimes incorrectly applied to any swelling in the poll region.

pollex. Add: **1.** (Later examples.)

1897 Parker & Haswell *Text-bk. Zool.* II. xiii. 77 The first digit of the fore-limb is distinguished as the pollex or thumb. **1909** W. Bateson *Mendel's Princ. Heredity* xii. 213 The case is more probably to be regarded as a homoeotic variation of the digits into the likeness of the hallux and pollex. **1959** [see *alula 1]. **1971** A. Burgess *MF* xv. 169 She clutched her bag between index and pollex. **1975** *Nature* 17 Jan. 192/1 *Proteles* differs from *Hyaena* principally in having a dentition much reduced in size, and in retaining the pollex (a digit lost in both *Hyaena* and *Crocuta*).

2. *Zool.* The movable part of the forceps in some crustaceans.

1895 F. H. Herrick *Amer. Lobster* ix. 147 The pollux [*sic*] is depressed, so that when the claw is closed it falls almost exactly midway between the normal and first superadded digit. **1904** *Biol. Bull.* VI. 75 The added structure [of an aberrant limb of a crayfish] is..a movable piece with two immobile prongs that otherwise resemble the index and pollex of a forceps.

pollinate, *v.* Add: (Later examples.) Also *absol.*

1919 J. N. Martin *Bot. for Agric. Students* iv. 49 These plants are not successfully pollinated when they are wet. **1935** C. Zirkle *Beginnings of Plant Hybridization* i. 6 It is not stated specifically in the description of this building [sc. Solomon's Temple] that the cherubim were engaged in pollinating the flowers of the palms. **1942** Hayes & Immer *Methods of Plant Breeding* iv. 63 With wheat it is equally satisfactory to pollinate during the day. **1973** Proctor & Yeo *Pollination of Flowers* i. 19 The role of insects in pollinating flowers is a commonplace. **1974** A. Huxley *Plant & Planet* xiii. 132 There is just a little nectar in each flower, so that the insects..pollinate more actively.

pollinating (pǫ-linēitiŋ), *ppl. a.* [f. Pollinate *v.* + -ing[2].] That pollinates or facilitates pollination.

1911 F. O. Bower *Plant-Life on Land* 69 The very genesis of the forms of flowers, their tints, and scents is in strict accordance with their efficiency as pollinating mechanisms.

pollination. Add: (Further examples.) Also *attrib.* and *fig.*

1924 Holman & Robbins *Textbk. Gen. Bot.* vii. 283 The transfer of pollen from the anther to the stigma is called pollination. **1941** D. C. Peattie *Road of Naturalist* i. 19 The paper-bag bush, too, had gone to pod, just a few of its purple mint flowers left, where I had seen the humming-birds at pollination. **1971** E. Mavor *Ladies of Llangollen* v. 90 Romantic pollinations..were taking place between an ever growing number of cultivated women. **1974** A. Huxley *Plant & Planet* xiii. 131 The second classic pollination symbiosis is that of the American Yuccas with certain moths.

pollinator (pǫ-linēitəɹ). [f. Pollinate *v.* + -or.] Any insect or other agent that pollinates plants.

1903 *Amer. Naturalist* XXXVII. 368 The small concealed flowers of Gaultheria..do not want for pollinators. **1924** *Chambers's Jrnl.* Aug. 501/2 The value of bees as pollinators is appreciated by progressive fruit growers. **1955** *Sci. Amer.* Aug. 52/1 It would be more appropriate to think of them [sc. bees] first of all as the great pollinators. **1977** M. Allan *Darwin & his Flowers* xi. 202 The pollinator is not known for any *Cryptophoranthus*.

pollinctor. Restrict †*Obs.* to sense in Dict. and add: **b.** *U.S. Blacks.* One employed by a funeral director to prepare bodies for cremation or burial; an undertaker.

1969 *Liberator* Dec. 13/2, I left the funeral home along with the body and the pollinctor.

polling, *vbl. sb.* Add: **II. 5. b.** The action or process of conducting an opinion poll.

1937 *Public Opinion Q.* Jan. 38 Scientific polling on individual issues fills a great gap in the democratic form of government. **1939** G. Gallup *Public Opinion in Democracy* 13 The research man who..devises an accurate and efficient method of polling 'ballot cattle' is obviously not contributing much to better government. **1944** —— *Guide to Public Opinion Polls* 12 Although the layman doesn't recognize it as polling, he is himself daily conducting his own private poll of public opinion. **1951** 'A. Garve' *Murder in Moscow* vii. 83 You go into the streets and do a little polling on the subject... You'll get a hundred per cent 'Yes'—not a single 'Don't know'. **1968** W. Safire *New Lang. of Politics* 108/2 The power of polling, both on nose counts and in depth, was never more vividly demonstrated than in the 1968 New Hampshire primary campaign. **1974** *Encycl. Brit. Macropædia* XV. 214/1 Polling can..reveal something about the intensity with which opinions are held. *Ibid.*, Polling..is unlikely to provide very much information about the elites who may have played an important part in developing the opinion.

III. 6. b. *polling-day* (examples), *-district* (examples), *-station* (further example).

1865 K. Amberley *Diary* 13 July in B. & P. Russell *Amberley Papers* (1937) I. viii. 399 Polling Day—cloudy & windy. **1974** *Times* 6 Sept. 1/4 He hinted heavily several times that polling day was only a matter of weeks away. **1895** C. Porritt in *Elections* (Yorks. Union of Conservative Assocs.) 12 An agent..should keep a list of gentlemen willing to lend conveyances to the various polling district committees. **1976** *Western Mail* (Cardiff) 27 Nov. (Advt.), Before you can trace your name in the lists you must know in which Polling District you live. **1977** *Grimsby Even. Tel.* 5 May 1/2 Unfortunately not enough publicity was given by the parties to the fact that the polling stations close at 9 instead of 10.

polliwog, pollywog. Add: Also **pollywiggle, pollywoggle.**

1823 E. Moor *Suffolk Words & Phrases* 288 *Polly-wiggle*, the tad-pole—in Norfolk called *potladle*. **1881** S. Evans *Evans's Leicestershire Words* (new ed.) 216 *Polly-wig*, or *pollywiggle*,..a tadpole. 'Poddywig' is, I think, the commoner form. **1933** H. G. Wells *Bulpington of Blup* ii. 45 These things you call pollywiggles and pollywoggles. **1965** *East Anglian* May 242/1 Tadpoles were.. *pollywiggles*.

polloi (pǫloi·). *slang.* [ad. Gr. πολλοί many: see *hoi polloi.] *pl.* (const. as *sing.*). A crowd, a rabble (see also quot. 1940).

1940 M. Marples *Public School Slang* 90 *Polloi* (Cheltenham).., the lowest football set. **1948** E. E. Cummings *Let.* 27 Aug. (1969) 185 A very refreshingly authentic polloi, & some good lively winds.

pollone: see *polone.

pollster (pǭu·lstəɹ). orig. *U.S.* [f. *poll *sb.*[1] 7 d + -ster.] One who conducts an opinion poll; an analyst of such polls, or of voting patterns generally.

1939 *Time* 9 Oct. 11/3 Gallup pollsters reported that 43% of the voters want Mr. Roosevelt to run again. **1941** *Time* 6 Jan. 11 According to Pollster Gallup's figures 60 per cent of U.S. voters now want to aid Britain even at the risk of war. **1951** M. McLuhan *Mech. Bride* (1967) 46/2 On November 3, 1948..the pollsters were proved wrong in forecasting a Dewey victory over Truman. **1959** *Times Lit. Suppl.* 2 Oct. 556/3 This is not a swing towards Communism (as the pollsters would say), but an impulsive reaction to success. **1968** [see *motivation 3]. **1972** M. Williams *Inside Number 10* xiii. 343 Possibly there was a reaction against the pollsters, a desire to prove the computers wrong. **1977** *Time* 21 Nov. 63/2 The pollsters asked people to make judgments on a series of actions, deciding whether such actions were morally wrong or not a moral issue.

pollutant. Delete *rare* and add further examples.

1936 *Nature* 29 Feb. 353/2 The presence of pollutants in the atmosphere is proof of wastage of fuel. **1958** *Times* 13 Nov. 9/1 The even more harmful gaseous pollutants present a problem which will tax the resources of all engaged in the present campaign. **1966** *Listener* 3 Nov. 656/2 Fluoride..can be used for the improvement of human welfare. Out of place it is a more serious pollutant than chlorinated hydrocarbons, in that it is indestructible. **1970** *Daily Tel.* 30 Dec. 3/1 Mercury is now the most dangerous environmental pollutant. **1977** *Guernsey Weekly Press* 21 July 3/1 Petrol could become a pollutant of the past if the device fitted to a local car is marketed worldwide.

pollute, *v.* Add: **1. a.** *spec.* To contaminate (the environment, atmosphere, etc.) with harmful or objectionable substances. Also *absol.* (Further examples.)

1954 *Thorpe's Dict. Appl. Chem.* (ed. 4) XI. 885/1 Poisons like arsenic, etc., are rarely present in natural waters unless polluted by trade wastes or agricultural washes. **1966** *Petroleum Handbk.* (Shell Internat. Petroleum Co.) (ed. 5) 144/1 The absence of sulphur ensures that the products of combustion are non-corrosive..and do not pollute the atmosphere. **1973** *New Earth Catal.* 59/3 Plant—don't pollute.

polluted, *ppl. a.* Add: **b.** *slang* (orig. *U.S.*). Intoxicated, drunk; under the influence of drugs, 'high'.

1912 *Dialect Notes* III. 585 *Polluted*,..very drunk. The same as *pickled* and *plastered*. **1914** 'High Jinks, Jr.' *Choice Slang* 16 *Polluted*, intoxicated, drunk, 'pifflicated', 'soused'. **1927** *New Republic* 9 Mar. 71/2 The following is a partial list of words denoting drunkenness now in common use in the United States..polluted. **1938** *Amer. Speech* XIII. 185/2 High..usually infers that the addict is noticeably under the influence of drugs... The same is true of the following equivalents:..lit, polluted, shot up. **1974** Wodehouse *Aunts aren't Gentlemen* iii. 20, I was helping a pal to celebrate the happy conclusion of love's young dream, and it may be that I became a mite polluted.

polluter. Add: **a.** *spec.* a person or organization that causes pollution of the environment.

1970 *New Society* 5 Feb. 209/2 The polluters often have a strong commercial lobby on their side, while the anti-polluters must rely on voluntary effort. **1970** *Toronto Daily Star* 24 Sept. 22/2 The federal government is in a much stronger position to deal with big industrial polluters than most provincial administrations are likely to be. **1974** *Times* 15 Mar. 6 (*heading*) Making Europe's polluters foot the bill. **1977** 'E. Crispin' *Glimpses of Moon* xi. 215 'Polluters,' said the hunt saboteuse.

b. A pollutant.

1975 *Physics Bull.* Mar. 100/3 Noise is now acknowledged as a major polluter of the environment.

pollution. Add: **1. a.** *spec.* The presence in the environment, or the introduction into it, of products of human activity which have harmful or objectionable effects. (Further examples.)

1877 Roscoe & Schorlemmer *Treat. Chem.* I. 255 The running water seldom reaches the sea in its natural or pure state, but is largely contaminated with the sewage of towns, or the refuse from manufactures or mines. So serious..is this state of things becoming that some steps are about to be taken to prevent the further pollution of the rivers. **1934, 1947** [see *atmospheric *a.* 2]. **1955** *Sci. Amer.* May 63/3 As our economy uses more and more organic chemicals, air pollution by volatile organic compounds becomes more and more of a problem. **1969** *Financial Times* 9 Jan. 4/6 The danger of 'thermal pollution' is greatest where electric and other power plants return to rivers and streams water that has been heated by between six and 16 degrees Centigrade. This often proves deadly to fish. **1970** *New Society* 5 Feb. 209/3 At American universities, pollution has been a student rallying cry for some months now. **1970**, etc. [see *noise pollution* s.v. *noise *sb.* 7]. **1975** *Physics Bull.* June 256/1 Noise pollution from aircraft and motorways and the design of speech and music reinforcement of St. Paul's cathedral are two of the varied aspects of noise and sound which have occupied Mr Allen.

b. (Earlier example.)

1605 Bacon *Adv. Learn.* ii. fol. 9 The Sunne..passeth through pollutions, and it selfe remaines as pure as before.

4. *attrib.* and *Comb.*, as *pollution control*; *pollution-free* adj.

1961 *San Francisco Chron.* 27 Mar. 32 Stronger water pollution control programs. **1969** *New Scientist* 9 Oct. 90/1 Pollution-control measures are only an extra charge on the expenses of a company and have little direct return. **1978** *N.Y. Times* 30 Mar. d14/4 A $150 million pollution-control program in that state. **1974** *Spartanburg* (S. Carolina) *Herald* 25 Apr. c2/3 Engineers for Japan's Honda and Mazda car makers have developed some of the world's most pollution-free auto engines which are getting wide attention in the United States.

pollutional (pǫliū-ʃənăl), *a.* [f. Pollution + -al.] Causing or constituting pollution.

1921 *Bull. Nat. Hist. Survey Illinois* XIV. 40 Usually tolerant or pollutional species that have come into Peoria Lake since 1915 seem to be several in number. **1941** *Sewage Works Jrnl.* XIII. 270 (*heading*) The natural purification of river muds and pollutional sediments.

pollutive (pǫliū-tiv), *a.* [f. L. *pollūt-* (see Pollute *v.*) + -ive.] Causing environmental pollution.

1970 *New Scientist* 2 July 12/2 The diesel engine..is a naturally less pollutive engine than the spark ignition one. **1972** Ward & Dubos *Only One Earth* xi. 247 Developing countries are hard put to it to raise the capital even for existing cheaper, though more pollutive, technologies and energy systems.

Polly[2] (pǫ·li). *slang.* Also **'polly.** Abbrev. of *Apollinaris; a bottle or glass of Apollinaris water. Also *attrib.*

1852 Dickens *Bleak Ho.* (1853) xx. 200 Four small rums is eight and three, and three Pollys is eight and six. **1878** [see *Apollinaris]. **1899** *Westm. Gaz.* 10 June 6/1 The dividend on Polly shares was announced too late to be public property in business hours yesterday. **1905** *Daily Chron.* 6 Sept. 4/7 'Johnny and Polly' is a common order in Piccadilly. **1907** *Westm. Gaz.* 17 Aug. 3/2 'Wagner and 'polly', or 'Tchaikowsky and soda'..would be both less stimulating and less harmful than some of the more popular intoxicants. **1954** J. Betjeman *Few Late Chrysanthemums* 43 And a host of little spiders Ran a race across the ciders To a box of baby 'pollies by the beer. **1969** S. Hyland *Top Bloody Secret* ii. 165 Do not call it Apollinaris [*sic*] water... Nobody else..calls it that. They all say 'Polly water'. **1973** D. Kyle *Raft of Swords* (1974) viii. 77 'Soda? Apollinaris?' 'Whisky and Polly... I haven't had one in years.' 'It becomes increasingly difficult to come by.'

polly[3] (pǫ·li). *U.S.* and *Austral. slang.* Abbrev. of Politician.

1942 Berrey & Van den Bark *Amer. Thes. Slang* §854/1 *Politician*,..polly. **1955** *Publ. Amer. Dial. Soc.* xxiv. 151 Perhaps some *polly* (politician) can be interested sufficiently to intervene with the judge. **1969** *Courier-Mail* (Brisbane) 18 Oct. 13/7 The pollies..have been using television as an electronic soap box. **1974** *Bulletin* (Sydney) 12 Oct. 12 Pollies peel off the tax perks. **1978** *Sunday Sun* (Brisbane) 4 June 5/1 The eight pollies are members of an all-Party Parliamentary delegation led by Industry Minister Norm Lee.

Pollyanna (pǫ:li,æ·nă). Also **Polyanna** and with lower-case initial. The name of the heroine of stories written by Eleanor Hodgman Porter (1868–1920), American children's author, used with allusion to her skill at the 'glad game' of finding cause for happiness in the most disastrous situations; one who is unduly optimistic or achieves happiness through self-delusion. Also *attrib.*, *Comb.*, and as *adj.*

[**1913** E. H. Porter *Pollyanna* xv. 148 'Her name is Pollyanna Whittier.'.. 'And what are the special ingredients of this wonder-working—tonic of hers?'.. 'As near as I can find out it is an overwhelming, unquenchable gladness... Her quaint speeches are constantly being repeated to me, and, as near as I can make out, 'just being glad' is the tenor of most of them.'] **1921** *Collier's* 11 June 11/1, I should not like to hold stock in a company with Pollyanna as president. **1925** Wodehouse *Carry On, Jeeves* ix. 214 Uncle Thomas, when his gastric juices have been giving him the elbow, can make Schopenhauer look like Pollyanna. **1926** B. Barton *Bk. Nobody Knows* ii. 42 Job's crops are destroyed, his barns burned, his children taken sick, and he himself breaks out all over with horrid boils. In this condition he is visited by a group of three friends—professional moralists and Pollyannas. **1931** E. Pound *Let.* 27 Dec. (1971) 238 With Possum Eliot apptd. to Hawvud, he won't bring the glad polyanna yawp, but the ignorance of the Stork-Auslander-Mabie-Canby period can't continue. **1937** M. Hillis *Orchids on your Budget* i. 14 The Pollyanna-like theory that you can have just as much fun with very little money as you can with a lot of it. **1939** Wodehouse *Uncle Fred* xv. 218 The Ovens home-brewed is a liquid Pollyanna, for ever pointing out the bright side and indicating silver linings. **1940** O. Nash *Face is Familiar* 203 Some people are just naturally Pollyanna, While others call for sugar and cream and strawberries on their manna. **1944** T. Rattigan *While Sun Shines* i. 32 You're not fooling anyone, Babe, but yourself with this Pollyanna stuff. I want another drink, you want another drink, so we both have another little drink. **1953** *Amer. Scholar* XXIII. i. 22 As intellectuals, we the 'Pollyannas' inevitably..ask ourselves if we can be right, when the country..is with us. **1959** A. Huxley *Let.* 13 Feb. (1969) 866 A short section..devoted to what may be called negative quotations—utterances of pure nonsense, pollyanna uplift, anti-intelligence and anti-liberty. **1962** A. H. Maslow *Summer Notes on Social Psychology* 3 Talk, talk, talk, not very effective, more optimistic and Pollyanna rather than realistic. **1963** 'E. Lathen' *Place for Murder* ii. 20 Thatcher decided that a Pollyanna tone was necessary... 'I am looking forward to tomorrow night,' he said firmly. **1964** B. Hardy *Appropriate Form* iii. 57 Robinson Crusoe is a prototype Pollyanna, comparing what is bad with what might have been much worse. **1971** 'D. Shannon' *Ringer* ix. 155 You're not exactly a Pollyanna today, Luis. **1977** *Time* 12 Sept. 60/2 Authors who try generally find themselves accused of going soft, of frivolously aping the Pollyanna fadeouts of popular schlock.

Hence **Pollya·nn(a)ish** *a.*, naïvely cheerful or optimistic; **Pollya·nnaism,** a statement characteristic of (a) Pollyanna.

1922 E. E. Cummings *Let.* 26 Feb. (1969) 83 Three Soldiers having, in his absence, been rendered Polyannish..by the highly moral Doran. **1923** *Grey Towers* 277, I wrote a paper for English 198 and the reader put on the outside, 'All right but Polly-Annaish.' **1946** G. Stimpson *Bk. about Thousand Things* 33 We hear now not only of pollyannas and pollyannaisms, but of pollyanna statements, pollyanna propaganda. **1948** *Time* 6 Dec. 90/2 Mildly Saroyanesque throughout and a trifle Pollyannaish at the end, its best scenes *The Silver Whistle* is genuinely funny. **1967** R. A. Epstein *Theory of Gambling & Statistical Logic* xi. 411 Efforts to develop a complete and rigorous axiomatic treatment for psychological probability theory are sanguine, if not pollyannish. **1975** *Times* 18 June 7/5 The special brand of United Nations Pollyannaism. **1976** H. H. Humphrey *Educ. Public Man* ii. 25, I sound Pollyanna-ish when I speak of my childhood, my family, and Doland, but I think it is real and not simply nostalgia.

polo[1]. Add: **2.** (Earlier and later examples of hockey played on skates.)

1883 *Boston Daily Globe* 18 Nov. 6/2 (*heading*) The Winslow Rules Governing Polo on Skates. *Ibid.* 6/3 The American Roller..[is] to be published in the interests of roller skating, polo, and other popular sports. **1885** *Providence* (Rhode Island) *Jrnl.* 29 Oct. 3/3 The game of polo at the Skating Academy last evening..was an active game. **1906** H. P. Burchell *Official Roller Skating Guide* 108 (Advt.), The Spalding 'Rink Polo' stick is made of the best..material.

2*. Short for **polo hat*, **polo-neck*, etc.

1905 *Daily Chron.* 30 Jan. 8/1 Among the Victorian revivals..are various items of dress from the sixties... The small round hat that the French milliners call the 'polo' and we in this country term the pork-pie, is a detail in point. **1967** *Harper's Bazaar* Sept. 45 The neck..high everywhere—emphasized by polos, wrapped, petal or stand-up collars. **1975** *Country Life* 29 May 1424/1 For summer, we are now diving into silky polos. **1976** *Woman's Weekly* 6 Nov. 36/2 (Advt.), Worn here over classic Polo in black, red, antrim, green, or cassis.

3. *polo boot, club* (earlier and later examples), *-ground*; also designating garments with a polo-neck, as *polo jersey, -jumper, sweater* (hence *-sweatered* adj.); **polo cloth,** a soft, loosely-woven camel's-hair cloth; **polo coat,** a type of camel's-hair coat; **polo collar,** (*a*) (see quot. 1960); (*b*) = **polo neck* (a); hence *polo-collared* adj.; **polo hat,** a small round hat worn esp. in the latter part of the nineteenth century; **polo-neck,** (*a*) a high, close-fitting roll collar; also *attrib.*; (*b*) a jersey with such a collar; hence *polo-necked* adj; **polo shirt,** (*a*) a shirt of the kind worn by polo-players; (*b*) a shirt having a polo-neck.

1894 *Country Gentlemen's Catal.* 154 Hunting top-boots & butcher boots. Polo boots. **1963** E. H. Edwards *Saddlery* xx. 149 Should the animal require greater protection..any of the heavier polo boots..may be more suitable. **1910** *Dry Goods Reporter* 22 Oct. 21 (Advt.), The polo coat, 55 inches long, made in a complete line of mixtures..as well as the regular polo cloth. **1919** *Official Gaz.* (U.S. Patent Office) 23 Dec. 689/2 Polo cloth. Worumbo Mfg. Co., Bath, Me. Claims use since Jan. 10, 1910. Woolen goods in the piece. **1926** *Daily Colonist* (Victoria, B.C.) 13 Jan. 7/6 (Advt.), Plush and Polo Cloth Hats for Boys. Smart little shapes in black plush and light lovat polo cloth. **1879** *Scribner's Monthly* June 309/2 Three summers ago, some young men in New York formed the Polo Club, and built a sumptuous house at Fordham, with grounds especially laid out for the game. **1885** *Providence* (Rhode Island) *Jrnl.* 1 Nov. 8/4 (*heading*) The Roller Skating Season... The members of the Chelsea Polo Club..have been engaged to form a Providence team. **1935** *Encycl. Sports* 468/1 The first polo club in London was at Lillie Bridge. **1975** *Oxf. Compan. Sports & Games* 788/2 British officers..began to establish polo clubs in India. **1910** Polo coat [see **polo cloth*]. **1953** M. McCarthy *Groves of Academe* iii. 45 She jumped up.. and seized her polo coat from the coat-rack. **1913** C. Mackenzie *Sinister Street* I. ii. v. 209 In his blue serge suit, wearing what the shops called a Polo or Shakespeare collar, Michael felt more at ease. **1937** A. Thirkell *Summer Half* ix. 253 They take off their detachable polo collars and look just like us, only nastier. **1960** C. W. Cunnington et al. *Dict. English Costume* 169/1 Polo collar. C. 1899. A starched white stand-fall collar, the fronts sloping apart. **1955** J. Cannan *Long Shadows* vi. 102 I've corduroys..and a polo-collared jersey. **1968** *Guardian* 22 Aug. 7/4 The current fashion for men to wear polo-collared shirts even with formal wear. **1895** Kipling in *Cosmopolitan* July 303 The hard, dusty Umballa polo-ground was lined with thousands of soldiers. **1897** *Daily Tel.* 8 Oct. 7/1 Lieutenant Rattray galloped in from the Khar polo-ground to take command of the post. **1971** *Shankar's Weekly* (Delhi) 18 Apr. 19/3 A vaguely green building on the other side of the polo-ground. **1905** *Daily Chron.* 30 Jan. 8/4 (Advt.), A three-quarters redingote.. and the revived polo hat. **1929** D. L. Moore *Pandora's Letter Box* v. 92 Polo jerseys were abandoned by normal men..because they were seen on many who were known to be unmanly. **1960** *Woman* 23 Apr. 73/4 What gave him most joy was a yellow polo jersey that Frances knitted for him. **1937** A. Christie *Death on Nile* vi. 77 He was wearing..a high-necked polo jumper. **1949** 'M. Innes' *Journeying Boy* viii. 94 Two undergraduates in demodé polo-jumpers. **1944** 'N. Shute' *Pastoral* i. 5 A grey jumper with a polo neck. **1951** 'A. Garve' *Murder in Moscow* ix. 96, I changed it for a polo-neck sweater. **1968** J. Ironside *Fashion Alphabet* 48 One [woman] may look very chic in a polo neck and another may look as if she were a kennel-maid manqué. **1971** *Vogue* 15 Oct. 73 Black poloneck, sleeves and back of thin black stripes. **1973** M. Amis *Rachel Papers* 128 He was wearing a fashionable black polo-neck jersey (fashionable, that is, among the weasly middle-aged). **1973** 'D. Halliday' *Dolly & Starry Bird* iv. 55 Wife and bambinos in suede jackets and knitted jackets and..enough polo necks to outfit the entire British Raj. **1955** M. Hastings *Cork & Serpent* vi. 79 She had changed into a pair of black ski trousers and a polo-necked sweater. **1974** A. Price *Other Paths to Glory* ii. i. 111 An equally pink polo-necked sweater. **1920** F. Scott Fitzgerald *This Side of Paradise* (1921) i. ii. 44 The faces indistinct above the polo shirts. **1938** 'E. Queen' *Four of Hearts* (1939) i. ii. 11 You left out this wine-coloured polo shirt. **1974** R. B. Parker *God save Child* (1975) ii. 16 Six polo shirts of different colors with the sleeves neatly folded under. **1979** R. Gillespie *Crossword Mystery* iv. 100 He threw on a polo shirt and slacks. **1950** A. Koestler in *God that Failed* i. 58 A member of the intelligentsia could never become a real proletarian, but his duty was to become as nearly one as he could. Some tried to achieve this by forsaking neckties, by wearing polo sweaters and black fingernails. **1955** N. Fitzgerald *House is Falling* ix. 156, I always wear.. flannel bags, a polo sweater. **1963** Polo-sweater [see *Chelsea 3]. **1950** A. Wilson *Such Darling Dodos* 136 The polo-sweatered organist.

polo[2] (pǭu·lo). [Sp.] An Andalusian folk-dance, or the music which accompanies this dance.

1883 Grove *Dict. Mus.* III. 9/2 *Polo*, a Spanish dance accompanied by singing, which took its origin in Andalusia. **1902** *Encycl. Brit.* XXVII. 374/2 Other provincial dances now in existence are..the *Palotéa*, the *Polo*, the *Gallegada*, [etc.]. **1926** D. C. Parker *Georges Bizet* iv. 233 Gaudier..asserts that Bizet has..made use of a *polo*, sung by a serenading student. **1934** W. Starkie *Spanish Raggle-Taggle* vii. 76 She began to play a medley of Spanish airs—*polos, boleros, tangos, malagueñas*. **1967** 'La Meri' *Spanish Dancing* (ed. 2) vi. 81 The Polo dates back to the ancient and unaccompanied Andalucian songs.

polocrosse (pǭu·lokrǫ-s). Also **polo crosse.** [Blend of Polo and Lacrosse.] A team game played on horseback with a rubber ball and a stick with a head like that of a lacrosse-stick.

1952 J. B. Pick *Phoenix Dict. Games* 105 The ground required by polo crosse is smaller than a polo field. **1965** *Newsweek* 19 Apr. 94 For a battering, breath-taking roughhouse on horseback, have a go at polocrosse in New South Wales. **1974** *Courier-Mail* (Brisbane) 16 Aug. 19 Polocrosse was far from being a poor cousin of the better known millionaire sport of polo. **1975** *Oxf. Compan. Sports & Games* 792/1 There were at least ten polo crosse clubs in Britain soon after the second world war, most of them in the West Country, but there is little evidence of the game in Britain now.

polocyte (pǭu·losəit). *Biol.* [f. Pol(e *sb.*[2] + -o + -cyte.] = *polar body* s.v. *polar *a.* 5.

1915 C. W. Prentiss *Lab. Manual & Text-bk. Embryol.* i. 28 The cleavage is unequal and, instead of four cells of equal size resulting, there are formed one large ripe ovum or oöcyte and three rudimentary or abortive ova known as polar bodies or polocytes. **1927** [see *ootid]. **1931** J. E. Frazer *Man. Embryol.* i. 6 Among the lower animals it seems to be usual for maturation not to be completed (*i.e.*, the second polocyte not to be cast off) before conjugation with the male cell is effected. **1969** F. D. Allan *Essentials Human Embryol.* (ed. 2) i. 14 Polocytes consist of a nuclear mass and a delicate cytoplasmic sheath and have no known function in addition to the passive role they play in meiosis, i.e., reduction or elimination of excess chromosomal material.

poloidal (pǭloi·dăl), *a.* [f. Pol(ar *a.* (*sb.*) + Tor)oidal *a.*] Being or representing a magnetic field of the form associated with a circular current loop, in which each line of force is confined to a radial or meridian plane.

1946 W. M. Elsasser in *Physical Rev.* LXIX. 108/1 We shall now introduce names for these three types of vectors. They will be designated as scaloidal (U), toroidal (T), and poloidal (S) vector fields. The electric field and vector potential pertaining to a poloidal magnetic field are toroidal, and *vice versa*. **1967** *McGraw-Hill Yearbk. Sci. & Technol.* 225 The magnetic field at the Earth's surface has a radial component and is thus of the poloidal type. **1972** *Sci. Amer.* July 67/3 A toroidal electric field maintains a toroidal current that flows inside the plasma, and this current in turn generates a magnetic-field component that is poloidal... The combination of the poloidal field with the toroidal field produces helical magnetic-field lines that lie on closed magnetic surfaces (nested toroids of circular cross section). **1973** *Nature* 20 Apr. 516/2 There are on the Moon weak, local, fairly randomly oriented magnetic fields but no overall poloidal field.

pology (pǭulǫ-lŏdʒi). *Nuclear Physics.* [f. Pol(e *sb.*[2] + -ology.] The theory of Regge poles.

1961 *Nuovo Cimento* XXII. 214 (*heading*) Pology and ND^{-1} solutions in the multiple-channel problems. **1964** S. DeBenedetti *Nucl. Interactions* vii. 471 This simple argument forms the basis of much of the present-day thinking about strong forces: perturbation methods may not be justified in general, but they are expected to give correct results near the predicted 'resonances' or 'poles'. The theory developed along these lines has received the informal name of 'pology'. **1966** D. Park *Introd. Strong Interactions* vii. 134 (*heading*) Pology and the determination of g.

Polonaise, *sb.* Add: **1. d.** A cloth of a silk and cotton mixture. Cf. Polonese *sb.* 1.

1894 J. E. Davis *Elem. Mod. Dressmaking* v. 93 Polonaise, a mixture of silk and cotton, which has the appearance of a soft dull silk with a distinct serge-like twill, is very much used as a skirt-lining for rich materials. **1923** *Daily Mail* 13 Aug. 4 (Advt.), The lining of the coat is silk Polonaise. **1932** G. Heyer *Devil's Cub* x. 152 Lady Fanny..in a négligée of Irish polonaise, with a gauze apron.

3. *Cookery.* Applied *absol.*, *attrib.*, or as *adj.* to dishes supposedly cooked in a Polish style. Also *à la Polonaise*.

1889 J. Whitehead *Steward's Handbk.* iv. 405/2 Polonaise (*a la*), in Polish style. **1950** E. Brunet tr. *Saulnier's Répertoire de la Cuisine* 199 (*heading*) *Asparagus ..Polonaise.*—Dished in rows, sprinkle with hard boiled eggs and parsley chopped, pour over some bread crumbs tossed in butter nicely browned. *Ibid.* 203 (*heading*) *Cauliflowers..Polonaise.*—Dressed cauliflower on buttered dish, sprinkle with chopped parsley and hard boiled eggs, and bread crumbs, [etc.]. **1965** *House & Garden* Dec. 84/2 *Polonaise*, lightly fried breadcrumbs used in among other ingredients and not just as a topping or garnish. **1969** G. Payton *Proper Names* 355/1 *Polonaise* (Cooking), with beetroot and sour cream.

4. Used *attrib.* or as *adj.* to designate a type of rug woven in Iran during the sixteenth and seventeenth centuries using silver and gold warp threads. Also *absol.* as *sb.*

1911 G. G. Lewis *Pract. Bk. Oriental Rugs* xxi. 322 According to Dr. Valentiner the so-called Polanaise [*sic*] and Ispahan rugs belong to the 17th century. **1913** W. A. Hawley *Oriental Rugs* vii. 88 It was doubtless after.. Shah Abbas I had begun to embellish his capital at Ispahan, that were made the famous 'Polish' silk or 'Polonaise' carpets about which there has been so much controversy. **1931** A. U. Dilley *Oriental Rugs & Carpets* ii. 44 Many of the king's rugs—the ones containing gold and silver, now called Polonaise—were woven at Kashan. **1962** C. W. Jacobsen *Oriental Rugs* 272 All Polonaise rugs belong to the 17th century, particularly the first half. The name Polonaise dates from the Paris exposition in 1878, when several rugs of this type were exhibited by Prince Czartorski of Warsaw. **1973** M. S. Dimand *Oriental Rugs in Metropolitan Museum of Art* v. 65/1 Because some of the rugs bear Polish coats of arms, they, too, have been called 'Polish' or 'Polonaise'. **1976** *Times* 15 Apr. 16/3 Christie's sale of Eastern rugs. A sixteenth-century Polonaise was unsold at £16,500.

POLONE

polone, polony[3] (pǫ̆lōu·ni). Also **pollone.** Varr. *PALONE.

1937 PARTRIDGE *Dict. Slang* 646/2 *Polone*,..a girl or woman. 1938 H. W. WICKS *Prisoner Speaks* vi. 95 The jogars and griddlers..had a language of their own, such as: omie, meaning man, pollone—woman, feelia—boy,.. and so on. 1938 G. GREENE *Brighton Rock* I. ii. 30 'What about that polony he was with?' 'She doesn't matter,' the Boy said. 'She's just a buer.' *Ibid.* II. ii. 88 'Napoleon the Third used to have this room,' Mr. Colleoni said, 'and Eugenie.' 'Who was she?' 'Oh,' Mr. Colleoni said vaguely, 'one of those foreign polonies.' 1979 R. RENDELL *Make Death love Me* iv. 37 'When the polone comes to lock up, Groombridge'll be due to split.'..'Call her a girl, can't you? You're not a poove.'

Polong (polǫ·ŋ). [Mal.] A Malayan spirit or imp (see quots.).

1839 T. J. NEWBOLD *Straits of Malacca* II. xii. 191 The Polong is a small sprite which can be domesticated. 1900 W. W. SKEAT *Malay Magic* vi. 320 The Polong..is described as a diminutive but malicious species of bottle-imp. 1972 *Daily Tel.* (Colour Suppl.) 12 May 58/3 Finally there is the Polong, a vampiristic imp who, in association with the demoniac house-cricket, the Pelasit, enters the victim's body at will.

Polonial (polōu·niăl), *a. rare.* [f. POLONIA + -AL.] = POLISH *a.*

1922 *Blackw. Mag.* June 801/2 A very intelligent-looking secretary to the Polonial Embassy to the Vatican assured me [etc.].

polonium. Add: (The name is now recognized as that of an element, of atomic number 84.)

Polovtsy (polǫ·vtsi), *collect. pl.* Also **Polovtsi, Polovtzi, Polovzi.** [Russ.] A union of the nomad tribes belonging to the Kipchak Turks, which inhabited the steppes between the Danube and the Volga in the 11th–13th centuries. So **Polove·tsian, Polo·vtsian** *a.*, of or pertaining to these people or their language; also as *sb.*

1799 W. TOOKE *View Russ. Empire* III. XI. ii. 581 The trade of the Krimea was heretofore uncommonly gainful and extensive; for, in the eleventh century, when a part of this peninsula fell under the dominion of the Polovtzi, better known from the byzantine history under the appellation of the Romanians, they granted the Genoese ..the permission to erect warehouses. 1803 H. CARD *Hist. Revolutions of Russia* 64 The Polovtsi took the imprudent resolution of observing a strict neutrality. *Ibid.* 66 One of the Polovtsian princes..demanded and received the sacrament of baptism. 1878 *Encycl. Brit.* VIII. Pl. XII opp. p. 715, Cumanians or Polovzians. 1885 *Ibid.* XIX. 286/1 He entered Kieff with the Polovtzi as his auxiliaries. *Ibid.* 410/2 As early as 988 the Russians erected several towns on the Sula and Trubezh for their protection against the Petchenegs and Polovtsy, who held the south-eastern steppes. 1938 *Oxf. Compan. Mus.* 750/2 Igor may go free if he will not make war on the Polovtsy again. 1954 *Grove's Dict. Mus.* (ed. 5) I. 821/1 Toward the end of 1874 Borodin's interest in 'Igor' was revived; the 'Polovtsian March' was composed, and in the following summer the famous dances. 1965 N. POPPE *Introd. Altaic Linguistics* 72 Kuman (Polovetsian, called so after the Russian name for Kumans) is also a Middle Turkic language. It was spoken in the XII–XVI centuries by Turkic nomads in Southern Russia, including the Crimea, and parts of Central Asia, and also by turkicized Armenians in the XV–XVIII centuries... There are no speakers of Kuman at the present time. 1968 M. GUYBON tr. *Solzhenitsyn's First Circle* l. 308 Half Polovtsian by blood, Prince Igor was for years an ally of the Polovtsians. 1973 R. C. HOWES *Tale of Campaign of Igor* 2 The Polovetsians (called Polovtsy, Cumans, or Kipchak Turks) had been moving into the steppes north of the Black Sea since the 1050s. *Ibid.* 3 The organization of two Polovetsian political units in the steppes..also boded ill for Kiev. 1974 T. SZAMUELY *Russian Tradition* I. ii. 13 The Kievan state had been engaged in perpetual warfare since its foundation. Khazars, Pechenegs, Polovtsy —one wave followed the other.

Pol Roger (pǫlˌrǫʒe). The proprietary name of a champagne produced in Épernay.

1889 *Trade Marks Jrnl.* 25 Sept. 916/2 *Pol Roger*... Champagne. Pol Roger, Albert Roger, and Veuve Julie Roger, trading as Pol Roger & Co., Epernay, France..; Champagne shippers. 1891 [see *KRUG[2]]. 1912 C. MACKENZIE *Carnival* xx. 209 Maurice..in a quandary of taste between Pol Roger and Perrier Jouet. 1925 H. ACTON in *Oxf. Poetry* 8 For after belching Pol-Roger the bile Will wreak revenge. 1957 L. DURRELL *Justine* III. 185, I was of course drunk by this time and exhausted—drunk as much on Justine as upon the thin-paper-bodied *Pol Roget* [sic]. 1968 E. HYAMS *Mischief Makers* x. 184 A steak and chips and a pint of Pol Roger will meet this case. 1979 P. ALEXANDER *Show me Hero* xv. 166 A maid had entered with a refreshment trolley. On it were two bottles of Pol Roger in ice-buckets.

polrumptious (pǫlrʊ·mpʃəs), *a. dial.* and *slang.* ? *Obs.* Also **pollrumptious.** [Perh. f. POLL *sb.*[1] + RUMPTI(ON + -OUS.] (See quots. 1787 and 1902.)

1787 F. GROSE *Provincial Gloss* s.v. *Pollrumptious*, restive, unruly. 1818 SCOTT *Heart Midl.* in *Tales my Landlord* 2nd Ser. III. vii. 170, I think thou doest not look so polrumptious as thy play-fellow yonder. 1888 'Q.' *Troy Town* xvii. 213 I'll get the loan o' the Dearloves' blunderbust in case they gets polrumptious. 1902 FARMER & HENLEY *Slang* V. II. 247/1 *Pollrumptious*,..restive: unruly; foolishly confident.

‖ **pol sambol** (pǫl sæ·mbǫl). [Sinhalese.] A spicy Indonesian dish: see *SAMBAL.

1962 *Housewife* (Ceylon) Feb. 28 Learn to make a pol sambol. 1971 *Ceylon Daily Mirror* (Colombo) 4 Oct. 8/4 Miss Munasinghe had consumed a meal consisting of..Pol sambol with maldive fish.

polska (pǫ·lskă). [Sw., f. *Polsk* Polish.] A Swedish folk dance of Polish origin in ¾ time; the music which accompanies such a dance.

1883 GROVE *Dict. Mus.* III. 11/2 Polskas are usually written in minor keys. 1910 C. WAERN *Mediæval Sicily* p. xiv, In Old Sweden the gaily decked Maypole is, or was, set up on Midsummer Eve, and people dance round it.. the brisk national 'polska' or the characteristic 'ring dance'. 1925 [see *HAMBO]. 1947 A. EINSTEIN *Mus. in Romantic Era* xvii. 319 Here too the Romantic movement began with the collecting and publishing of the old national treasures of folk song—full of feeling and dancelike, the folk songs of sentiment being often modal, the dancelike ones (polskas) being related to the mazurka.

Poltalloch (pǫltæ·lǫ̆χ). The name of an estate in Argyll, Scotland, used *absol.* or *attrib.* to designate a small, stocky, rough-coated, white terrier belonging to a breed developed there by the Malcolm family, esp. Colonel E. D. Malcolm (1837–1930), and now usually called the West Highland White terrier.

1887 D. J. THOMSON GRAY *Dogs of Scotland* (1891) iii. 51 A white variety of the Scottish terrier existed at one time (and stray specimens may still exist) under the cognomen of Poltalloch terriers. 1920 R. MACAULAY *Potterism* II. i. 63 'We have two Pekingese, a King Charles, and a pug...' I answered with some inanity about my mother's Poltalloch. 1922 [see *HIGHLAND *a.* 2 a]. 1932 E. WEEKLEY *Words & Names* ix. 129 We have the new class-names which come into existence, as new breeds of terriers are evolved to suit changing fashions, e.g...the poltalloch, first bred by Colonel Malcolm, of Poltalloch (Argyll). 1950 A. C. SMITH *Dogs since 1900* xi. 187 West Highland White Terriers..were at first known as Poltalloch Terriers. 1968 C. G. E. WIMHURST *Bk. Terriers* xxvii. 189 These [sc. the Duke of Argyll's terriers] were not related to the Poltalloch terriers.

poltergeist. Delete ‖ and add earlier and later examples. Also *attrib.* and *fig.*

1848 C. CROWE *Night Side of Nature* II. vi. 238 (*heading*) The poltergeist of the Germans, and possession. *Ibid.* 239 The annoyances appear rather like the tricks of a mischievous imp. I refer to what the Germans call the *Poltergeist*, or racketing spectre. 1863 *Q. Rev.* July 193 It seems a suspicious circumstance that the old-fashioned visible ghost has in these modern *séances* been almost entirely superseded by the *Poltergeist* or noise-making spirit. 1927 J. S. HUXLEY *Relig. without Revelation* vi. 187 Exorcism is magic: the Rumanian poltergeist medium, Eleonore Zugun, whose case was recently investigated in London, was the subject of exorcist rites by Rumanian priests. 1940 [see *DOPPELGANGER]. 1979 M. BABSON *So soon done For* i. 8 A disturbed adolescent in an area of intense poltergeist activity.

Hence **po·ltergeistic** *a.*, of, pertaining to, or suggestive of a poltergeist; **po·ltergeistism**, the manifestation or activity of poltergeists.

1952 R. F. JONES in 'E. Crispin' *Best SF Five* (1963) 44 Reports on poltergeistism at Leander Castle near London. 1973 L. M. BOSTON *Memory in House* v. 61 The house.. from time to time sent feelers out from its darkest corners, such as slight poltergeistic displacements, footsteps up the wooden stairs,..etc.

polthogue (pǫltōu·g, -ōu·χ). *Ir. colloq.* Also **palthogue.** [f. Ir. *palltóg*.] A blow with the fist; a thump or punch.

1830 W. CARLETON *Traits & Stories Irish Peasantry* II. 59 John Grimes hot him a *palthoge* on the sconce wid the but-end of a gun. *c* 1874 D. BOUCICAULT in M. R. Booth *Eng. Plays of 19th Cent.* (1969) II. 181 Be jabers, I'd have liked to see in your face when you got that polthogue in the gob. 1898 J. D. BRAYSHAW *Slum Silhouettes* 25 Faix, there's a polthogue will knock sinse into yez. 1899 S. MACMANUS *In Chimney Corners* 164 He draws the flail one polthogue at the lad in the door and just barely missed him.

po'ly, poly (pō·li), *a.* Repr. a U.S. dial. pronunc. POORLY *adv.* and *a.*

1890 *Dialect Notes* I. 69 'How d'you do?' 'I am po'ly to-day.' 1893 H. A. SHANDS *Some Peculiarities of Speech in Mississippi* 50 A negro, when asked about his health, if not well, will answer, 'Poly, poly, bress God!' 1929 W. FAULKNER *Sartoris* I. 26 'How is you?' 'Po'ly, ladies; po'ly.' 1935 Z. N. HURSTON *Mules & Men* (1970) I. vi. 140 My wife is po'ly.

poly[2] (pǫ·li), colloq. abbrev. of POLYTECHNIC *sb.* 2 in Dict. and Suppl.

1858 M. TUCKETT *Diary* 30 Sept. (*c* 1975) 9 We came into Falmouth..and then went with Aunt to the Poly. 1882 Q. HOGG *Let.* Sept. in E. M. Hogg *Quintin Hogg* (1904) vi. 145 Our Poly. boys came to speak their word in the matter. 1892 *Scribner's Mag.* Feb. 170/1 The Young Men's Polytechnic Institute, which is universally known among young people in London as the 'Poly'. 1932 E. M. WOOD *Polytechnic & its Founder Quintin Hogg* (rev. ed.) 16, I have just tried to keep out of the light and let the Poly tell its own history. 1967 *Times Rev. Industry* May 115/2 The new polys will come into being largely as regroupings of existing colleges. 1970 *Guardian* 19 Aug. 9/2 Fowler..is taking the poly job and saying a regretful 'no thanks' to LSE. 1978 I. MURDOCH *Sea* 226 When he left school he went into the poly, you know, the polytechnic... He had a student grant. 1979 *Jrnl. R. Soc. Arts* July 489/1 So what we should like is for those institutions such as polys throughout the Commonwealth that are already running business management courses to devote part of them to the skills of publishing.

poly[3] (pǫ·li), colloq. abbrev. of *POLYTHENE and *POLYETHYLENE b. Chiefly *attrib.* and *Comb.*, as *poly bag* (*polybag*), *poly-wrapped* adj.

1965 *Supermarket & Self-Service* (Johannesburg) June/July 13/3 Quick-frozen polywrapped broilers. 1968 *Punch* 12 June 858/1 Little poly-wrapped trays in the frozen food counter. 1971 C. BONINGTON *Annapurna South Face* ix. 103 Tom and I finished packing our rucksacks, leaving a proportion of our gear in polybags outside the boxes to reduce our loads. 1976 *Canad. Forces Sentinel* (Ottawa) XII. III. 6/3 One three-pound polybag of the stuff served as a week-long protein supplement for..six. 1976 *Southern Even. Echo* (Southampton) 11 Nov. (Advt. Suppl.) 7/1 Fresh meat from the farm for your freezer: ..All poly wrapped and cut to your requirements. 1977 *Undercurrents* June–July 34/3 Find an old boiler suit,..a small shovel, and a big thick poly bag or a cut away poly jerry can. 1978 *Detroit Free Press* 5 Mar. (Spring Fashion Suppl.) 14/1 (Advt.), Sleek soft sandals..latest styles, Spring colors, and the newest poly bottoms.

poly-. Add: **1. polyabolo** (pǫliˌæ·bǒlo) [f. *DI)ABOLO by deliberately false analogy (see quot.); cf. *PENTOMINO], any planar shape formed by joining a number of identical right-angled isosceles triangles by their edges; **polya·ctine**, a sponge spicule having numerous rays; **po:lyallomo·rphic** *a. Philol.*, having several allomorphs (*ALLOMORPH[2]); **po:lyalphabe·tic** *a. Cryptography*, employing more than one alphabet, so that each letter of the alphabet may be represented in a code by any of two or more letters or other characters; **polya·nion** *Chem.*, a negatively charged polyion (see below); hence **po:lyanio·nic** *a.*; **po:lyarteri·tis** *Path.* [mod.L., coined in Ger. as *polyarteritis acuta nodosa* (E. Ferrari 1903, in *Beiträge zur path. Anat. und zur allgem. Path.* XXXIV. 383) to replace *periarteritis*] = *periarteritis* s.v. PERI- I c in Dict. and Suppl.; **polyaxon** *sb.* (example); **polyblast**, (*b*) *Histology* [a. G. *polyblast* (A. Maximow 1902, in *Beiträge zur path. Anat.: Suppl.* V. 43)], a wandering macrophage; **polyca·tion** *Chem.*, a positively charged polyion (see below); hence **po:lycatio·nic** *a.*; **polycho·ral** *a. Mus.*, in which the choral ensemble is divided into groups who sing alternately (and, properly, jointly also); **polycistro·nic** *a. Genetics*, comprising or derived from more than one cistron and so containing the information for more than one gene product; hence **polycistro·nically** *adv.*; **polyclo·nal** *a. Biol.* and *Med.*, (of a population of organisms) comprising many clones; (of a population of cells) comprising several cell lines of separate origins; of or pertaining to the products of such cell lines; hence **polyclona·lity**; **polyclo·nally** *adv.*; **po·lyclone** *Biol.* and *Med.* [*CLONE *sb.*], a group of cells all descended from one or other of an initial small group of cells; **poly·cratism** = *Polycracy*; **po·lycross** *Bot.* and *Agric.*, a cross made by planting two or more mutually fertile varieties together and allowing free natural pollination; freq. *attrib.*; **polycythæmia** (earlier example); [ad. G. *polycythaemie* (J. Vogel 1854, in R. Virchow *Handb. der speciellen Path. und Therapie* I. iv. 377)]; hence **polycythæ·mic** *a.* of, involving, or suffering from polycythæmia; **polyde·ntate** *a. Chem.* [L. *dentātus*: see DENTATE *a.*], (of a ligand) forming two or more separate bonds (usu. but not necessarily with the same central atom); (of a molecule or complex) formed by such a ligand; **polyelectro·nic** *a. Chem.*, containing or consisting of more than one electron; **polye·ndocrine** *Path.*, characterized by the involvement of several endocrine glands; **po:lyendocrino·pathy** *Path.*, a polyendocrine disorder; **polye·nergid** *a. Biol.* [*ENERGID], having many complete sets of chromosomes; multinucleate or polyploid; **polye·thism** *Ent.* [Gr. ἦθος character], the display of different patterns of behaviour by particular individuals within a social group; **polyfu·nctional** *a. Chem.*, having two or more different functional groups in the molecule; orig. also applied to reactions involving two or

more such compounds; hence **po:lyfunctiona·lity**; **polygla·cial** *a.*, involving (a belief in) more than one ice age; hence **polygla·cialism**, the polyglacial theory; **polygla·cialist** *a.* and *sb.*, (of or pertaining to) a supporter of this theory; **polyha·ploid** *Bot.* [*HAPLOID *a.* (and *sb.*)], a plant descended from polyploids that has half of the set of chromosomes that would normally be expected from its ancestry; **polyhedroid** (-hī·droid, -he·droid) *Math.* [POLYHEDR(ON + -OID] = *polytope* s.v. POLY- I in Dict. and Suppl.; **polyhex** (pǫ·liheks) [f. HEX(AGON], any planar shape formed by joining a number of identical regular hexagons by their edges; **polyiamond** (pǫliˌəi·ămənd) [f. D)IAMOND by deliberately false analogy; cf. *PENTOMINO], any planar shape formed by joining a number of identical equilateral triangles by their edges; **polyion** (pǫ·liˌəi:ǫ̆n) *Chem.*, an ion which consists of or contains a number of atoms of its parent element, or a large ion derived from a polyelectrolyte; so **polyio·nic** *a.*; **polyka·ryocyte** *Biol.* [KARYO- + -CYTE], an osteoclast, esp. a large osteoclast with many nuclei; hence **po:lykaryocy·tic** *a.*; **polyle·ctal** *a. Linguistics* [*-LECT], having or recognizing many regional or social varieties (within a language); **polyli·ngual** *a.* = *MULTILINGUAL *a.*; **polyli·ngualism** = *MULTILINGUALISM; **polylithic** *a.*, add: also, containing several kinds of stone or rock; also *fig.* (cf. *MONOLITHIC *a.* 4); (further examples); **polyli·thionite** *Min.* [ad. G. *polylithionit* (J. Lorenzen 1884, in *Zeitschr. f. Krist. und Min.* IX. 253), f. *lithion* lithia], a variety of lepidolite; **polymeta·llic** *a.*, containing (ores of) several metals; **polymicrian** *a.* (delete *nonce-wd.*; earlier examples); **po·lymineral** *a. Petrol.*, composed of or containing more than one mineral; **polymorphe·mic** *a. Linguistics*, consisting of two or more morphemes; **polymya·lgia rheuma·tica** *Path.* [MYALGIA] (see quot. 1957); **polymyositis** (earlier example); **polyneuri·tic** *a.*, of, pertaining to, or suffering from polyneuritis; **polyneuro·pathy** *Path.*, a general degeneration of peripheral nerves that starts distally and spreads proximally; **polyœstrous** *a.* (further examples); also, in mod. use, ovulating more than once each year; **polyomino** (pǫliˌǫ·mino) [f. D)OMINO by deliberately false analogy; cf. *PENTOMINO], any planar shape formed by joining a number of identical squares by their edges; **polypneu·stic** [Gr. πνευστ-ιάω to pant] *a. Ent.*, bearing many respiratory spiracles; **polypnœa** (further examples); also **polypnea**; hence **polypnœ·ic** *a.*; **po·lyprotein** *Biochem.*, a protein which is composed of a number of smaller proteins; **polyri·bosome** *Biol.* = *POLYSOME; hence **po:lyriboso·mal** *a.*; **po:lysapro·bic** *a. Ecol.* (ad. G. *polysaprob* (Kolkwitz & Marsson 1902, in *Mittheilungen aus der K. Prüfungsanstalt f. Wasserversorgung und Abwässerbeseitigung* I. 46): see *SAPROBE], of, being, or inhabiting an aquatic environment having in solution much reducing decayed organic matter and little or no oxygen; **polysemantic** *a.* (further examples); = *POLYSEMIC, *POLYSEMOUS *adjs.*; **po:lysemanti·city, polysema·ntism** = *POLYSEMY; **po:lyseros·itis** *Path.*, inflammation of serous membranes; **po·lysoap**, a detergent whose molecules are polymeric chains to which soap molecules are attached; **po·lyspike** [SPIKE *sb.*[2]] (see quot. 1950); **po:lysympto·ma·tic** *a. Med.*, involving or exhibiting many symptoms; **polythe·lia, polythe·lism, po·lythely** *Med.* [ad. F. *polythélie*, f. Gr. θηλή nipple], the condition of having one or more supernumerary nipples; **polytope** (examples); [ad. G. *polytop* (R. Hoppe 1882, in *Arch. der Math. und Physik* LXVIII. 30)]; hence **polyto·pal** *a.*; **polyto·pic** *a. Biol.* [Gr. τόπος place], of or pertaining to (the independent origin of a species in) several places; **polyto·pical** *a.*, dealing with many subjects; **polyuria, -uric** *a.* (earlier examples); **polyxe·nic** *a. Biol.* [Gr. ξένος stranger], applied to a culture, or the cultivation, of an organism in the presence of more than one other species.

1967 *Sci. Amer.* June 129/1 The pieces had been suggested to him by S. J. Collins of Bristol, England, who gave the name 'tetraboloes' to the order-4 set because the Diabolo, a juggling toy, has two isosceles right triangles in its cross section. This implies the generic name 'polyaboloes'. 1902 *Encycl. Brit.* XXXII. 813/1 Fig. 5A, typical polyactine. 1949 E. A. NIDA *Morphol.* (ed. 2) iv. 98 The suffix *-al* is polyallomorphic /əl/ and /æl/. 1972 *Archivum Linguisticum* III. 40 In order to simplify their analysis one of the allomorphs of poly-allomorphic morphemes is designated as the basic one, and the changes are described on that basis. 1927 *Daily Express* 24 Nov. 13 The 'polyalphabetic' codes..are much more difficult to decipher, as a letter is often represented in a cryptogram by a dozen different signs, letters or numerals. 1939 H. F. GAINES *Elem. Cryptanalysis* (1940) viii. 68 Multiple-alphabet substitution (also called double-key substitution, polyalphabetic substitution, etc.) makes use of several different cipher alphabets. 1962 MOORE & WALLER *Cloak & Cipher* xv. 138 Edgar Allan Poe..seems to have had a blind faith in polyalphabetic ciphers. 1931 *Chem. Abstr.* XXV. 3261 The submicrons detected by the ultramicroscope are negatively charged and consist of aggregates of the polyanions $[Pb_9]^{4-}$ or $[Sn_9]^{4-}$. 1948 *Jrnl. Polymer Sci.* III. 261 In the presence of excess electrolyte, the polyanion would be completely associated and behave approximately like an uncharged macromolecule. 1965 [see *polycation* below]. 1972 COTTON & WILKINSON *Adv. Inorg. Chem.* (ed. 3) xxv. 823 The decavanadate ion is only one example of the type of polyanion generally called isopolyanions. 1930 *Chem. Abstr.* XXIV. 2077 [Such] elements..will combine in liquid NH_3 with Na to give polysulfide-like compds.,..to which the name 'polyanionic' salts is given. 1974 *Amer. Jrnl. Anat.* CXXXIX. 404/1 Staining with ruthenium red..reveals that a polyanionic surface coat, probably mucopolysaccharide in nature, covers all the microvilli..on all cell types found in the nasal cavities. 1907 *Jrnl. Path. & Bacteriol.* XII. 54 Polyarteritis acuta nodosa.—Characterised by the formation upon the smaller and medium-sized arteries of small localised nodules. 1951 E. N. CHAMBERLAIN *Text-bk. Med.* vi. 448 Sometimes known as polyarteritis nodosa, this is a rare disease generally affecting persons before mid-life. 1974 PASSMORE & ROBSON *Compan. Med. Stud.* III. xxv. 29/1 The term polyarteritis includes a number of uncommon disorders in which the changes are focal, segmental inflammation and necrosis of arteries, arterioles or capillaries... Besides the classical form, also known as polyarteritis nodosa.., there are five recognized variants. 1940 L. H. HYMAN *Invertebrates* I. vi. 299 Polyaxons..are spicules in which several equal rays radiate from a central point. 1904 Polyblast [in Dict.]. 1959 F. M. BURNET *Clonal Selection Theory* vii. 115 Macrophages..include fixed macrophages, wandering tissue macrophages or polyblasts and the blood monocytes. 1967 *Biol. Abstr.* XLVIII. 9803/2 By the 6th day, a barrier of connective tissue, formed mainly from polyblasts and hist[i]ocytes, was beginning to be formed. 1948 *Jrnl. Polymer Sci.* III. 259 Due to the high concentration of charge in the polycation, which is itself quite large, electrostatic forces can be transferred over much greater distances than in solutions of ordinary electrolytes. 1965 PHILLIPS & WILLIAMS *Inorg. Chem.* I. xii. 465 The formation of polyanions is quite common among the heavier non-metals (e.g. polysulphides and selenides), although the formation of polycations from uncomplexed metals appears to be limited to Hg_2^{2+}. 1949 *Science* 25 Nov. 553/1 When excess acrylate is added, more polyanions attach themselves to the polycationic exterior of the precipitate particles. 1970 R. W. MCGILVERY *Biochem.* xix. 451 They [sc. spermine and spermidine] appear to occur in association with nucleic acids, as might be expected from their polycationic character. 1944 W. APEL *Harvard Dict. Mus.* 593/1 Early adumbrations of polychoral treatment occur in the works of Josquin des Près who frequently interrupts the full-voiced writing in four parts..by 'antiphonal' passages in which two half-choruses..perform a short phrase twice, in an echo-like manner. 1963 *Times* 9 May 16/5 'Jauchzet dem Herren',..a big polychoral 'concerto' from the *Psalmen Davids* of 1619, was a casualty, its elaborate antiphonies of voices and instruments blurred by the cathedral's hopelessly over-resonant sound. 1975 *Gramophone* Nov. 869/3 In 1628, Salzburg Cathedral was re-opened... Its inaugural Mass was a grand polychoral affair. 1978 *Early Music* Apr. 170 Venice and the grand manner of the polychoral motet seem so inseparable that it is hard to consider them apart. 1963 OHTAKA & SPIEGELMAN in *Science* 25 Oct. 493/2 An RNA molecule which can be translated into two or more proteins may be referred to as a 'polycistronic' message. 1968 H. HARRIS *Nucleus & Cytoplasm* ii. 23 The idea of a 'polycistronic' template, that is, one which can specify the amino acid sequences of a group of related proteins, now enjoys considerable popularity. 1974 *Nature* 1 Nov. 75/2 Kennel *et al.* concluded that each cistron in these polycistronic RNAs has a unique site that is vulnerable to attack. *Ibid.*, There is..an indication that some tRNAs are made polycistronically. 1978 *Nature* 25 May 304/2 The polyclonality of B cell responses to LPS has excluded the participation of immunoglobulin combining sites in the process of triggering. 1977 *Lancet* 5 Nov. 958/2 We suggest that immunosuppression in this syndrome is the result of the collective immunosuppressive effects of trypanosome-derived immune-modulating free fatty acids, polyclonally stimulating B-cell mitogen, and complement-activating factors. 1975 CRICK & LAWRENCE in *Science* 1 Aug. 341/3 The progeny of a cell marked at about the time of the drawing of boundary lines never fills a compartment completely, but often occupies an appreciable proportion of it. A compartment is thus made by the descendants of a small group of cells. We propose to call the cells in the compartment a polyclone. Just as a clone is a group of cells which are all, without exception, the descendants of a single cell, so a polyclone is a group of cells that are descended from a certain (small) group of cells—the founder cells—which were present in the embryo at an earlier time. 1979 *Sci. Amer.* July 93/1 Each compartment is made by a set of complete clones, which we call a polyclone, that develops from a few founder cells. 1914 W. E. AGAR in *Phil. Trans. R. Soc.* B. CCV. 422 When a population..is composed of a number of clones each descended from an original ancestor not asexually connected with the original ancestors of the other clones, the population may be called polyclonal. 1961 *Harvey Lect.* 1960–61 LVI. 221 He had a broad-banded, polyclonal γ-globulin with a rich serological picture. 1973 *Sci. Amer.* Aug. 44/2 If plaques were a simple response to an injury of some kind, as has been proposed, their cells should be polyclonal, the Benditts point out. 1921 *19th Cent.* July 148 The maximalists, of course, are for polycratism, provincial rule, insubordination and importation of foreign ideas. 1948 TYSDAL & CRANDALL in *Jrnl. Amer. Soc. Agronomy* XL. 294 The present paper deals with methods for determining the combining ability of the components of a hybrid or variety. For convenience, the method is referred to as the 'polycross method'. *Ibid.*, The single crosses and polycrosses exhibited even greater superiority over the checks. *Ibid.* 297 Polycross seed is the seed produced on selected clones interpollinated at random in isolation. 1977 *Crop Sci.* XVII. 909/2 Twenty-one clones whose polycross progenies ranked high for rate of seedling emergence under field conditions or had high forage yield..were selected for this study. Polycross seed from these clones was produced in isolated blocks. 1857 DUNGLISON *Dict. Med. Sci.* (rev. ed.) 741/2 *Polycythæmia*, a condition of the blood in which there is an increase of the red corpuscles. 1906 *Lancet* 7 July 20/2 The following case is published as a contribution to the study of the polycythæmic condition. 1935 [see *hypovolæmia* s.v. *HYPO- II]. 1962 *Lancet* 26 May 1098/2 The patient was polycythæmic until 1958, when a leukæmic picture first appeared. 1937 *Chem. Rev.* XXI. 39 The simple variation of acidic and coördinating groups in the polydentate molecules has escaped investigation. 1961 G. R. CHOPPIN *Exper. Nucl. Chem.* ix. 147 Complexes with a high degree of covalent character are formed by the interaction of metal ions with polydentate organic ions. 1972 *Nature* 21 Jan. 181/1 Recently his expertise in coordination chemistry was extended to the complexes formed by the alkali metals and alkaline earth elements with a variety of polydentate ligands. 1909 *Cent. Dict. Suppl.*, Polyelectronic. 1939 L. PAULING *Nature Chem. Bond* i. 29 The electron distribution function for a poly-electronic atom or ion shows the presence of electron shells as regions of maximum electron density. 1947 *Amer. Scientist* XXXV. 185 Just as little tested in the laboratory is the conclusion that positrons, like protons..can form short lived polyelectronic entities of the type e^+e^-, $e^+(e^-)_2$, [etc.]. 1964 J. W. LINNETT *Electronic Struct. Molecules* i. 9 The most important factor governing the electronic structures of the ground states of polyelectronic atoms is the effect summarized in the Pauli Principle. 1965 *Amer. Surgeon* XXXI. 695 (*heading*) Polyendocrine adenomatosis with Zollinger-Ellison syndrome. 1967 S. L. ROBBINS *Path.* (ed. 3) xxix. 1243/2 (*heading*) Multiple endocrine adenomatosis (polyendocrine adenomas). 1976 *Lancet* 11 Dec. 1273/1 Antibodies reacting with normal human pancreatic islet cells have been described in patients with diabetes associated with autoimmune polyendocrine disease. 1964 *Medicine* XLIII. 176/1 It is suggested that Schmidt's syndrome with diabetes mellitus may be a polyendocrinopathy. 1973 *Acta Endocrinol.* LXXII. 411 As for the theoretical implications of poly-endocrinopathies, the possibility of common aetiological factors lies near at hand. 1920 W. E. AGAR *Cytol.* vii. 209 Examples of such polyenergid nuclei (Hartmann, 1909) are afforded by the great nuclei of the Radiolaria. 1939 *Nature* 14 Jan. 47/2 Schussnig..reaffirms..his view that the Conjugales are derived from a polyenergid ancestry, and a similar origin is suggested for the Red Algæ. 1961 MACKINNON & HAWES *Introd. Study Protozoa* 66 The gigantic nucleus [of *Aulacantha*] is remarkable for the number of its chromosomes, of which there are some 1,500... This remarkable structure, according to Grell, is really polyenergid. 1976 BELL & COOMBE tr. *Strasburger's Textbk. Bot.* (rev. ed.) 44 Free nuclear divisions, that is, divisions not accompanied by cell division, occur in those Thallophyta showing the polyenergid condition. 1967 J. H. SUDD *Introd. Behaviour Ants* viii. 154 Animals which show these variations in behaviour from one to another can be said to show polyethism—a word formed by analogy with polymorphism. 1973 J. P. SPRADBURY *Wasps* vi. 155 There may occur several forms of polyethism, namely those based on age, physiological condition, and size. 1929 W. H. CAROTHERS in *Jrnl. Amer. Chem. Soc.* LI. 2550 (*heading*) Polyfunctional compounds. *Ibid.*, All these may be classed together as polyfunctional reactions. 1962 J. T. MARSH *Self-Smoothing Fabrics* v. 45 The finishing process..consists of impregnating the fabrics with the appropriate polyfunctional compound and a catalyst, drying, heating and washing. 1963 J. OSBORNE *Dental Mech.* (ed. 5) i. 23 The basic ingredient is a polyfunctional mercaptan with..the average formula $HS(R—S—S)_{23}—R—SH$. 1964 N. G. CLARK *Mod. Org. Chem.* xv. 301 By employing polyfunctional halides in place of alkyl halides depicted above, more complex ketones are obtained. 1936 *Trans. Faraday Soc.* XXXII. 39 (*heading*) Polymers & polyfunctionality. 1961 SORENSON & CAMPBELL *Prep. Methods Polymer Chem.* iii. 59 The qualitative aspects of condensation polymerization, including..effect of polyfunctionality on branching and gelation, have been thoroughly treated. 1927 PEAKE & FLEURE *Apes & Men* 69 This polyglacial, or preferably multiglacial, view was not well received, and considerable opposition was offered to it. 1937 *Geogr. Jrnl.* XC. 180 Formerly James Geikie, almost alone, insistently voiced the case for the polyglacial view and perhaps he strained it by over-statement. 1972 SPARKS & WEST *Ice Age in Brit.* v. 123 The limits of the successive glaciations of this polyglacial sequence, imperfectly known at present, are shown. *Ibid.*, The evidence for polyglacialism lay not so much in evidence for different end-moraines of successive ice advances..but in the finding of non-glacial sediments between glacial deposits. 1946, 1968 Polyglacialist [see *monoglacialist* sb. and adj. s.v. *MONO- 1]. 1972 SPARKS & WEST *Ice Age in Brit.* v. 123 The supporters of the monoglacial theory..were eclipsed by polyglacialists, though some survived till a few years ago. 1935 Y. KATAYAMA in *Jap. Jrnl. Bot.* VII. 374 The writer has classified (though provisionally) the haploid plants as follows... If the haploid has occurred from allopolyploids, it is classified under the name of polyhaploid. 1955 *Nature* 12 Mar. 469/1 This plant had the chromosome number 2*n* = 24, suggesting that it might be a polyhaploid of *S*[*olanum*] *polytrichon*, having arisen by haploid partheno-

genesis. 1975 *Ibid.* 17 Apr. 596/1 In wheat and oats, the polyhaploids show very little chiasmate pairing because genetic control is effective in the hemizygous state. 1880 W. I. STRINGHAM in *Amer. Jrnl. Math.* III. 2 It will be convenient to designate as an *n*-fold polyhedroid the *n*-dimensional figure which is bounded by (*n*−1)-fold flat (not curved) figures. 1914 H. P. MANNING *Geom. Four Dimensions* viii. 289 A regular polyhedroid..consists of equal regular polyhedrons together with their interiors, the polyhedrons being joined by their faces so as to enclose a portion of hyperspace, and the hyperplane angles formed at the faces by the half-hyperplanes of adjacent polyhedrons being all equal to one another. 1972 C. S. OGILVY *Tomorrow's Math* (ed. 2) iv. 79 A polytope is an *n*-dimensional polyhedroid. 1967 *Sci. Amer.* June 124/3 Other names have been proposed, but it seems to me that the best is 'polyhexes', the name adopted by David Klarmer, who was one of the first to investigate them. 1975 *Ibid.* July 114/3 Combinatorial geometers have given special attention in recent years to tiling with polyominoes and their cousins the polyiamonds and polyhexes. 1967 *Ibid.* June 124/2 By joining equilateral triangles along their edges one obtains another well-explored family of shapes known as polyiamonds. 1975 [see *polyhex* above]. 1938 A. I. OPARIN *Origin of Life* vi. 138 Regarding every living cell as a 'single chemical particle or, more correctly, as a colossal poly-ion'. 1947 *Jrnl. Polymer Sci.* II. 12 Both negative and positive polyions may be made; the former as polycarboxylic or sulfonic acids and their salts and the latter, for example, as onium salts of polymers such as vinylpyridine. 1959 *Acta Crystallogr.* XII. 165/2 The crystal structure of inyoite contains isolated polyions, $[B_3O_3(OH)_5]^{-2}$. 1963 *New Scientist* 11 Apr. 103/3 The chemistry of polyions, such as proteins, mucopolysaccharides, [etc.]..is very relevant to understanding the behaviour of the cell surface. 1972 COTTON & WILKINSON *Adv. Inorg. Chem.* (ed. 3) xvi. 486 There is slight evidence in the bromine system for Br_3^-, but the series of polyions I_3^-, I_5^- and I_7^- is well-established for iodine. 1907 *Publ. Carnegie Inst.* No. 63. XII. 352 A fuller experimental investigation of the properties of dissolved salts, especially of those of polyionic types. 1970 FOX & FRIED tr. *Staudinger's From Org. Chem. to Macromolecules* B. vii. 134 These anomalous phenomena in solutions of polyelectrolytes were termed 'polyionic viscosity phenomena'. 1890 W. H. HOWELL in *Jrnl. Morphol.* IV. 118 The first class might be named polykaryocytes, or multinucleated giant cells. 1946 *Blood* I. 29 Morone sharply differentiated the polykaryocytes from osteoclasts, but in this he was disputed by Lambin and Lamers. 1968 E. KELEMEN *Physiopath. & Therapy Human Blood Dis.* (1969) i. 36 Even larger cells, resembling megakaryocytes, are the osteoclasts or polykaryocytes. These two cells have to do with bone formation and are more often seen in a trephine biopsy. 1947 *Jrnl. Lab. & Clin. Med.* XXXII. 664 The concept of a polykaryocytic origin of the megakaryocyte has not received general acceptance. 1964 *Biol. Abstr.* XLV. 3106/2 Injection of confluent polykaryocytic cultures into chicks resulted in the appearance of sarcomas which contained transformed cells and polykaryocytes. 1972 B. BICKERTON in *Georgetown Univ. Ser. Lang. & Linguistics* (1973) xxv. 34 The demonstration of similarities between Black English and Guyanese Creole..was simply a by-product of the attempt to write a polylectal grammar of the latter. 1972 C.-J. N. BAILEY in Stockwell & Macaulay *Linguistic Change & Generative Theory* 24 Rule changes of the sort being described could never occur in a homogeneous grammar... Without the retention of the older forms in a different style or in a different class lect known to a speaker long enough for a rule change to be generalized, such a generalization could not occur. Only a polylectal grammar is adequate for historical linguistics. 1977 *Word* 1972 XXVIII. 166 The subject matter of dialectology may be viewed..I. As a complex of shared and differentiated items which function within a single diasystem (a pan-dialectal or polylectal system). 1978 *Archivum Linguisticum* IX. 37 That implicational (polylectal) patterning obtains in Table 1(b) is clear: some speakers are invariable users of the feature S, others are invariable users of the feature P, while a third group of speakers alternate S and P. 1933 'E. CAMBRIDGE' *Hostages to Fortune* III. vi. 184 Foreign students, slow south Germans, French boys from Rennes, polylingual Swedes. 1958 *Times* 5 Dec. 16/3 Polytextual, polylingual wrestlings with canti fermi. 1978 *Amer. N. & Q.* XVI. 146/2 A few other bilingual and polylingual glossaries. 1956 J. WHATMOUGH *Language* 241 Correlation methods may be used to show how much..the polylingualism of an interlingua may safely draw from different languages. 1977 *Word* 1972 XXVIII. 193 Borrowing, especially when related to bilingualism or polylingualism, increases the number of opportunities for metanalytical processes to take place, both at the time of borrowing and subsequently. 1908 *Sci. Amer. Suppl.* 25 Jan. 61/1 These crevices and fissures are filled with a polylithic mass of brown and white 'calcic spar'. 1961 *Economist* 11 Nov. 538/2 Somewhere in the essentially 'polylithic' variety of the sisterhood there must be an answer. 1886 *Jrnl. Chem. Soc.* L. 677 Minerals from Kangerdluarsuk, in Greenland... Polylithionite (lithium mica). 1927 *Amer. Mineralogist* XII. 275 Polylithionite is of doubtful stability, but mica of approximately this composition has been described from Greenland. 1962 *Geochemistry* XI. 1197 In lithium micas (polylithionites) the geochemical similarity of lithium and magnesium does not play an important part and lithium occupies an independent position in the structure of the mica lattice. 1892 *Dental Rec.* XII. 488 Amalgams consist of the combination of either one or several metals with mercury,..the bulk of a polymetallic amalgam usually consisting of Tin and Silver. 1956 *Mineral. Abstr.* XIII. 38 Nests and lenses of plumbojarosite are found in the oxidized zone of polymetallic ore deposits. 1968 BETHELL & BURG tr. *Solzhenitsyn's Cancer Ward* I. xv. 236 My theory is that you can discover deposits of polymetallic ore by looking for radioactive water. 1974 *Nature* 16 Aug. 545/1 The polymetallic province is particularly enriched in silver north of boundary 1. 1829 W. GREENFIELD (*title*) Polymicrian lexicon to the New Testament. 1838 *Bagster's Catal.* 22 Polymicrian series of New Testaments, Concordances, Lexicons, and Psalters, Small Pocket Volumes. 1938 *Mem. Geol. Soc. Amer.* VI. 134 Except for a few monomineral fabrics, such as those of pure quartzite.., most rock fabrics are polymineral. 1975 *Nature* 25 Dec. 690/1 This suite of rather unusual minerals has received wide attention because the minerals have been identified in light-coloured, millimetre-sized polymineral inclusions present in carbonaceous chondrites. 1949 E. A. NIDA *Morphol.* (ed. 2) iv. 97 Simple structures consist of a single morpheme, free or bound. Complex structures consist of more than one morpheme. Simple structures may be called 'monomorphemic' and complex structures 'polymorphemic'. 1962 H. C. CONKLIN in J. A. Fishman *Readings Sociol. of Lang.* (1968) 416 Single morphemes are necessarily lexemes, but for polymorphemic constructions the decision depends on meaning and use. 1964 R. H. ROBINS *Gen. Linguistics* 206 Polymorphemic words may consist wholly of free morphemes. 1957 H. S. BARBER in *Ann. Rheumatic Dis.* XVI. 237/2, (1) A condition characterized by widespread muscular pains without arthritis but accompanied by a high erythrocyte sedimentation rate and occasional pyrexia is described. (2) The relationship to rheumatoid disease is discussed and it is concluded that this is probably a clinical entity within the rheumatic group of diseases. (3) It is proposed to term the syndrome 'polymyalgia rheumatica'. 1971 BOYLE & BUCHANAN *Clin. Rheumatol.* xvi. 434/2 Polymyalgia rheumatica.. affects subjects in the later years of life, the average age of onset being the late sixties. 1878 D. F. LINCOLN tr. A. Eulenburg in *Ziemssen's Cycl. Pract. Med.* XIV. 133 According to these, the disease consists in an essentially inflammatory process, a 'polymyositis chronica progressiva'. 1895 *Jrnl. Nervous & Mental Dis.* XXII. 316 (*heading*) Polyneuritic psychoses. 1932 *Times Lit. Suppl.* 7 Jan. 14/3 His results..show that brain tissue from polyneuritic pigeons..has *in vitro* a lower power of oxygen uptake. 1968 M. PYKE *Food & Society* ii. 17 The remarkable effects of a few milligrams of thiamine on a polyneuritic pigeon. 1938 I. S. WECHSLER in *Jrnl. Amer. Med. Assoc.* 4 June 1913/2 It is suggested that the term multiple neuropathy, polyneuropathy or peripheral neuropathy be substituted for multiple neuritis in those cases in which both the cause and the pathologic changes point to a degenerative process. 1954 *Jrnl. Neuropath. & Exper. Neurol.* XIII. 168 Severe polyneuropathy with massive involvement of the large nerve trunks may not only appear in association with the more chronic forms of diffuse connective tissue disease,..but may even dominate the clinical picture so as to obscure the diagnosis of the underlying disease. 1974 PASSMORE & ROBSON *Compan. Med. Stud.* III. xxxiv. 36/2 Polyneuropathy arises from dietary deficiencies, chemical poisoning and may be a manifestation of numerous diseases. 1919 *Amer. Jrnl. Anat.* XXVI. 131 The females of the wild swine of Europe are monoestrous, according to Kaeppeli ('08), having but one period of heat in the year; but under domestication the sow becomes polyoestrous, coming in heat at intervals of two to four weeks. 1975 *Sci. Amer.* July 77/1 The particular response of each species to light seems to depend on whether the species is monestrous or polyestrous, that is, on whether it normally ovulates once a year (in the spring or fall) or at regular intervals throughout the year. 1954 S. W. GOLOMB in *Amer. Math. Monthly* LXI. 675 We shall generalize the 'domino' to the 'polyomino'... We *define* an *n*-omino as a simply-connected set of *n* squares of the checker-board which are 'rook-wise connected'; that is, a rook placed at any square of the *n*-omino must be able to get to any other square, in a finite number of moves. 1965 —— *Polyominoes* 13 Ever since I 'invented' polyominoes in 1953 in a talk to the Harvard Mathematics Club, I have found myself irrevocably committed to their care and feeding. 1972 W. F. LUNNON in R. C. Read *Graph Theory & Computing* 108 Free polyominoes whose symmetry groups contain no improper elements..are enantiomorphic. 1974 *Sci. Amer.* Feb. 106 Solomon W. Golomb's polyomino-placing game..has finally reached the marketplace. 1918 R. NEWSTEAD in *Ann. Trop. Med. & Parasitol.* XII. 93 The main pair of stigmata..lie in the deep cup-shaped cavity or pit between the polypneustic lobes. *Ibid.* 95 The low-convex anal lobes or callosities were distinctly polypneustic in character. 1925 A. D. IMMS *Gen. Textbk. Entomol.* 110 In *Glossina* there are about 500 of these pores to a side which form the sculpturing on a pair of polypneustic lobes. 1962 GORDON & LAVOIPIERRE *Entomol. for Students of Med.* xxix. 182 The larva contracts considerably to form a barrel-shaped object varying between 5 and 8 mm. in length, with the prominent polypneustic lobes of the larva still clearly visible. 1921 *Physiol. Rev.* I. 296 Marsupials are the lowest mammals capable of 'heat polypnea'. 1966 *Amer. Jrnl. Physiol.* CCX. 1270/1 Free-breathing cats demonstrated a thermal polypnea similar to that reported for dogs, cattle, and monkeys. 1975 J. J. GROEN in L. Levi *Society, Stress & Dis.* II. xxxiv. 350/2 The child may substitute in situations of frustration another form of respiratory behaviour such as apnoea, polypnoea, or a peculiar kind of pressing with the abdominal muscles during expiration which..produces an expiratory wheeze. 1909 *Cent. Dict. Suppl.*, Polypnœic, polypneic. 1934 *Amer. Jrnl. Physiol.* CIX. 528 Any type of panting (the polypneic or the hyperpneic) that occurs after decortication is dependent on a rise in blood temperature. 1975 *Biol. Abstr.* LIX. 1795/1 (*heading*) Induction of polypneic threshold by heating during cat sleep cycle. 1974 *Jrnl. Virol.* XIV. 261 (*heading*) Cleavage of mengovirus polyproteins in vivo. 1975 *Sci. Amer.* May 27/3 This huge protein, really a polyprotein, is then systematically cleaved by proteolytic enzymes. 1962 *Science* 28 Dec. 1401/2 An intrinsic property of the polyribosomal unit. 1970 *Sci. Jrnl.* Apr. 36 In the unaltered cytoplasm surrounding these areas ribosomes were gathered into polyribosomal aggregates, indicating very active synthesis of protein. 1962 Polyribosome [see *POLYSOME]. 1973 *Sci. Amer.* Apr. 41/1 As the synthesis of the messenger RNA proceeds, giving rise to a longer strand of RNA, more ribosomes attach themselves to the strand. They form a string called a polyribosome, which continues the translation of the elongating messenger RNA. 1925 *Bull. Illinois Nat. Hist. Survey* XV. 440 The septic or grossly polluted portions of a stream... The organisms of this zone are those which have been termed by Kolkwitz and Marsson..polysaprobic and by Forbes and Richardson.. septic or saprobic. 1932 *Trans. Brit. Mycol. Soc.* XVII. 112 The association of polysaprobic organisms occurs in waters rich in decaying organic matter. 1933 [see *oligosaprobic* adj. s.v. *OLIGO-]. 1946 *Jrnl. Ecol.* XXXIII. 274 Judging from data obtained from other rivers the amounts of algal growth appear to fall into four groups,..in the polysaprobic waters (e.g. the river Tame) the numbers.. are low. 1950 *Folia Limnologica Scandinavica* V. 76 The polysaprobic zone is defined in a chemical respect as the zone in which reduction of the polluting substances takes place. 1973 M. A. SLEIGH *Biol. Protozoa* xi. 265 The largest numbers of protozoan organisms occur in polysaprobic conditions. 1939 L. H. GRAY *Foundations of Lang.* 255 Words are very frequently polysemantic. 1960 E. DELAVENAY *Introd. Machine Transl.* vi. 81 In recent concise dictionaries the vocabulary of the English language comprises some 60,000 word entries: this number may run four times as high if each meaning of each polysemantic word is entered separately. 1961 *Amer. Speech* XXXVI. 5 (*heading*) Polysemantic extensions of 'dog' and allied terms. 1966 S. CECCATO in *Automatic Transl. of Lang.* (NATO Summer School, Venice, 1962) 75 First of all there is the problem of the polysemanticity of the individual words. 1939 L. H. GRAY *Foundations of Lang.* ix. 258 The principle of analogy or metaphor in polysemantism..appears when the name of a well known historical..figure is extended to persons supposed to resemble that character. 1946 *Word* II. 124 *Synchronic* semasiology..deals with..problems like homonyms, homophones, synonyms, polysemantism. 1900 *Brit. Med. Jrnl.* 15 Dec. 1693/2 Italian physicians..have given a name or names to this multiple inflammation of the serous cavities... The names are polyserositis and polyorromenitis. 1915 *Amer. Jrnl. Med. Sci.* CL. 518 (*heading*) Chronic lead-poisoning in guinea-pigs: with special reference to nephritis, cirrhosis, and polyserositis. 1966 WRIGHT & SYMMERS *Systemic Path.* I. i. 5/1 Occasionally, as a result of the compression of the inferior vena cava by the coarse fibrous tissue, a syndrome known as 'polyserositis' or Concato's disease develops in which fluid gradually collects in the pleural and peritoneal cavities. 1951 STRAUSS & JACKSON in *Jrnl. Polymer Sci.* VI. Polysoaps are defined as polymers to whose chain soap molecules are attached. 1976 *Nature* 5 Aug. 519/2 Some enzymes may be converted into surface-active amphipathic conjugates by covalent coupling to certain types of polymeric detergents (polysoaps). 1950 H. GASTAUT in *Electroencephalogr. & Clin. Neurophysiol.* II. 250/1 In the EEG is a burst of very large amplitude rhythmic spikes of a frequency equal to that of the flashes; these are bilateral and synchronized, and appear predominantly in the precentral and frontal regions where they can in fact be localized... These spikes are sometimes quite pure and thus constitute the complex for which we have proposed the name 'polyspike'. 1975 S. ARIETI *Amer. Handbk. Psychiatry* (ed. 2) IV. xiii. 320/2 The state is associated with prolonged EEG discharge of the 3-Hz. spike-wave type as well as..slower and faster components with polyspikes. 1952 *New Biol.* XII. 28 The only method at present available for diagnosing monozygosity involves comparison of as many morphological and physiological characters as possible—the so called 'polysymptomatic similarity' method which is that normally used for determining the zygotic nature of twins in man. 1962 A. BOURNE *Doctor's Creed* vi. 117 The unhappy woman is the victim of constant physical troubles which present a polysymptomatic picture which is quite incurable by the ordinary methods of clinical medicine. 1977 *Lancet* 24/31 Dec. 1340/1 What do you do with a polysymptomatic patient in whom the only positive finding is an enlarged liver? 1894 W. R. WILLIAMS *Dis. Breast* iv. 56 In other cases one or more supernumerary nipples, each with its own areola, have been met with, in various positions, on a single breast (intramammary polythelia). 1928 [see *POLYMASTIA]. 1970 H. P. LEIS *Diagnosis & Treatm. Breast Lesions* iv. 60 Polythelia or accessory nipples may occur along the 'milk line' from the axilla to the symphysis pubis or anywhere over a given breast. 1886 Polythelism [see *POLYMASTISM]. 1928 F. Z. SNOOP *From Monotremes to Madonna* 23 Polythely. This last form is commoner in men than women. 1974 *Tetrahedron* XXX. 1596/1 The concept of shape is presented in terms of a dihedral angle relationship between adjacent polytopal faces. This procedure was first employed..to map out structural form in the relatively complicated 8-atom family. 1908 *Proc. Sect. Sci. Koninkl. Akad. van Wetensch. Amsterdam* X. 689 This leads us gradually to the question, whether it is not possible to point out one or more polytopes—if not quite regular ones—which with C_5 fill the fourdimensional space. 1929 D. M. Y. SOMMERVILLE *Introd. Geom. N Dimensions* x. 190 In a plane there are an unlimited number of regular polygons and 3 regular networks, in space of three dimensions there are 5 regular polyhedra and one regular honeycomb, in S_4 there are 6 regular polytopes and three regular honeycombs, in space of more than four dimensions there are just three regular polytopes and one regular honeycomb. 1974 H. S. M. COXETER *Regular Complex Polytopes* xiii. 141 The regular polytopes and honeycombs so far considered are the only ones that can exist in unitary spaces. 1904 *Science* 10 June 885/1 The idea that a species may originate in more than one place.. did not originate with Briquet, but he resuscitated it and christened it the polytopic theory. 1939 *Geogr. Jrnl.* XCIII. 271 We are forced to fall back on the theory of polyphyletic and polytopic evolution. 1970 *Watsonia* VIII. 143 The distribution of the hexaploids in Britain does not show any obvious pattern.., and this agrees with Rousi's suggestion that hexaploids have had a polytopic origin from the tetraploids. 1876 C. A. CUTTER *Rules for Printed Dict. Catal.* 14 It will be well to have both words,—polygraphic denoting (as now) collections of several works by one or many authors, polytopical denoting works on many subjects. 1961 T. LANDAU *Encycl. Librarianship* (ed. 2) 282/1 Polytopical. Descriptive of a book treating of several subjects. 1842 DUNGLISON *Dict. Med. Sci.* (ed. 3) 562/1 *Polyuria*, diabetes. 1870 J. R. CORMACK tr. *Trousseau's Lect. Clin. Med.* III. lxv. 533 Polyuria, saccharine diabetes, and also sometimes albuminuria, may, in succession, attack the same individual. *Ibid.* 536, I have..had the pain to see nearly all the polyuric patients whom I had to treat, waste away rapidly. 1953 *Parasitology* XLII. 260 (*table*) Polyxenic. [Number of associated organisms] Several. [Source of

term] New. 1976 *Ann. Rev. Microbiol.* XXX. 128 Laboratory stocks of many protozoa are maintained on mono- to polyxenic substrates. In an attempt to establish axenic cultures of *Entamoeba histolytica*, spontaneous and sporadic lysis in these amoeba developed.

2. a, b. Add: Now used esp. to form the names of polymers and other types of compound which have a number of identical groups in their structure. The second element is in some cases a suffix. Usage of the prefix is restricted by some authors to cases where the number of constituent groups is large (in contrast to OLIGO-), but there is no uniformity in this respect. The prefix is now very abundant in *Chem.*, and only the more widely occurring formations are included below.

poly-A, poly A, etc. = *polyadenylic acid* below; **polya·cetal**, any of a class of polymers containing the repeating group —O·CH(R)—, which are prepared by addition polymerization of aldehydes and are typically strong thermoplastics used as moulding materials; **polyace·tylene**, any organic compound containing two or more carbon–carbon triple bonds; hence **po:lyacetyle·nic** *a.*; **polyacry·lamide** (or *po:lyacryla·mide*) [ACRYL + AMIDE], any of the polymers of acrylamide, CH_2═$CH·CONH_2$, or its substituted derivatives, which are water-soluble polymers widely used to form or stabilize gels and as thickening, suspending, or clarifying agents, etc.; **polyacry·lic** *a.* (and *sb.*), designating compounds which are polymers of acrylic acid or its esters, or thermoplastic materials consisting of or made from such polymers; hence **polya·crylate**, an ester or salt of polyacrylic acid; a polymer of an acrylic ester or, *loosely*, of acrylic acid; **po:lyacryloni·trile**, any of the polymers of acrylonitrile, many of which are used commercially, esp. as man-made fibres; **po:lyadeny·lic acid** *Biochem.* [*ADENYL], a polynucleotide formed from adenosine monophosphate by the action of polynucleotide polymerase, and isolated as fibres; abbrev. **poly-A*; **polya·lcohol**, a polyhydric alcohol; **polya·llomer** [f. ALLOMERISM after *polymer*], any of a class of crystalline thermoplastics which are copolymers of two or more different alkenes, esp. ethylene and propylene; **polya·mine**, any organic compound which contains two or more amine groups; **polyanhy·dride**, any of a class of polymers in which the units are linked through the anhydride group, —CO·O·CO—, and which includes many resins used commercially, esp. as fibres; **polyanion, anionic** *a.* (see sense 1 above); **polybro·minated** *a.*, applied to compounds in which two or more hydrogen atoms have been replaced by bromine atoms; **po:lybutadi·ene**, any of the polymers of 1,3-butadiene or its derivatives; also, any of the class of synthetic rubbers consisting of or made from such polymers; **po:lycarboxy·lic** *a.*, having more than one carboxyl group in the molecule; **polycation, -cationic** *a.* (see sense 1 above); **polychlo·roprene**, any of the polymers of chloroprene; also, any of the class of synthetic rubbers (esp. neoprene) consisting of or made from such polymers; **polydi·ene** *Chem.*, any polymer of a conjugated diene, esp. any of those forming a number of types of synthetic rubber; **polyene** (pǫ·liˌīn) [-ENE], any organic compound containing two or more carbon–carbon double bonds; hence **polye·nic, polyeno·ic** *adjs.*; **polye·thenoid** *a.* = **polyenic* adj.; also as *sb.*; **polyfo·rmal, -forma·ldehyde** = *polyoxymethylene* below; **polygluta·mic acid** *Biochem.*, a synthetic polypeptide consisting of glutamic acid residues; **polygly·cine** *Biochem.*, any oligopeptide or polypeptide composed of glycine residues; *spec.* a synthetic crystalline long-chain polypeptide having this structure; **polygly·col** = *polyethylene glycol* s.v. *POLYETHYLENE a; **polyglyco·lic acid**, a polyester fibre which is made by polymerizing glycolic acid, CH_2OH·$COOH$, and is used in surgery for ligatures, as it is slowly and harmlessly absorbed by the body; **polyhydroxy(-)**, *prefix* used to designate compounds or groups containing more than one hydroxyl group; also (without hyphen, as an independent word) as quasi-*adj.*; **polyhydro·xyl** = prec. (quasi-*adj.* use); **po:lyisobu·tylene**, any polymer of isobutylene; also, any of the large class of synthetic rubbers consisting of or made from such polymers; **polyly·sine** *Biochem.*, a synthetic polypeptide consisting of lysine residues; **po:lymethacry·lic acid**, any polymer of methacrylic acid; hence **polymetha·crylate**, a salt or ester of polymethacrylic acid; also, any of the synthetic resins made by polymerizing esters of methacrylic acid; **polyol** (pǫ·liˌǫl) (*a*) [*OL], an ol complex which contains more than one bridging hydroxyl ligand; now *rare*; (*b*) [-OL], a polyhydric alcohol; **polyo·lefin(e**, any polymer of an olefin, esp. any of the commercially important synthetic resins of this type; also = *polyene* above; **po:lyoxye·thylene**, designating, or used in the names of, compounds containing the polymeric group $—(CH_2·CH_2·O)_n—$; **po:lyoxyme·thylene**, any of a number of white, crystalline polymers which are prepared from formaldehyde and in which the repeating unit is $—CH_2·O—$; *esp.* any of the tough, strong thermoplastics of this type which are used as moulding materials; **polyphe·nol**, any compound which contains more than one phenolic hydroxyl group; hence **polyphe·nolic** *a.*; **polyphenol oxidase, po:lyphenolo·xidase** *Biochem.* [ad. G. *polyphenoloxydase* (Battelli & Stern 1912, in *Ergebnisse d. Physiol.* XII. 96): see *OXIDASE], any of the phenolases which oxidize polyphenols; *loosely*, any phenolase; **polyphe·nylene**, any polymer in which the repeating unit is or contains the *para*-phenylene group; *polyphenylene oxide*, a thermoplastic having the structure $—(p\text{-}C_6H_4O)_n—$, which is used as a moulding material; **po:lyphospho·ric acid**, any oxyacid of pentavalent phosphorus which contains two or more phosphorus atoms, spec. a mixture of polymers of orthophosphoric acid which is used in organic chemistry esp. as a mild dehydrating agent; hence **polypho·sphate**, a salt, anion, or ester of any such acid; **polysilo·xane** [*SILOXANE], any polymer which is based upon a chain of alternating silicon and oxygen atoms; *esp.* a silicone; **polysty·rol** = *POLYSTYRENE; **polysu·lphide**, (*a*) a salt or other compound which contains two or more sulphur atoms bonded together, as an anion or group; also, such an anion or group; (*b*) any polymer in which the units are linked through polysulphide groups, esp. any of a class of synthetic rubbers with this structure; freq. *attrib.*; **polysu·lphone**, any polymer in which the units contain the sulphone linkage, $—SO_2—$, esp. a type of thermosetting synthetic resin which has this structure and is used as a moulding material, esp. in electrical and electronic applications; **polyte·rpene** [ad. G. *polyterpen* (O. Wallach 1885, in *Ann. d. Chem.* CCXXVII. 302)], any of the higher members of the terpene series, $(C_5H_8)_n$; a polymer of a terpene; **polyte·rpenoid** = prec.; **polyu·ronide** [*URONIC *a.* + -IDE], any polysaccharide which consists of uronic acid residues, usu. in combination with simple monosaccharides.

1957 *Jrnl. Amer. Chem. Soc.* LXXIX. 2023/2 Experiments were carried out with poly-A and poly-U prepared.. with polyribonucleotide phosphorylase from *E. coli.* 1968 W. MÜLLER in E. Harbers et al. *Introd. Nucleic Acids* iii. 51 Below pH 6, polyA yields fibers with high negative birefringence. 1975 Poly(A) [see *polyadenylic acid* below]. 1931 *Chem. Rev.* VIII. 371 The reaction between glycols and acetaldehyde (or acetylene) presents the possibility of forming cyclic acetals..or polyacetals. 1967 *Times Rev. Industry* June 72/3 There are in hand expansion programmes covering polyvinyl chloride, polyethylene, polystyrene, butadiene, polyacetals, and polyesters. 1973 *Materials & Technol.* VI. viii. 578 Polyacetal is largely crystalline and not transparent. It resists weathering well and..shows little cold flow. 1885 *Jrnl. Chem. Soc.* XLVIII. 759 (*heading*) Polyacetylene compounds. 1952 *Jrnl. Amer. Chem. Soc.* LXXIV. 1588/2 From these spectra, Dr. Sörensen identified our compounds as polyacetylenes. 1967 *New Scientist* 13 Apr. 95/2 In recent years the widespread occurrence in fungi and plants..of straight-chain 'polyacetylenes'..has been recognized. 1978 *Sci. Amer.* Dec. 66/1 It is conceivable that a polyacetylene film could replace ordinary metal conductors in some special circumstances, such as where weight or resistance to corrosion is important. 1952 *Jrnl. Amer. Chem. Soc.* LXXIV. 1588/2 The isolation of polyacetylenic compounds from several genera of *Compositae*. 1961 *Chem. Nat. Products* (I.U.P.A.C.) I. 570 The *cis*-lachnophyllum ester..is also most unusual in being a polyacetylenic compound which has been used in industry. 1944 *Jrnl. Org. Chem.* IX. 501 Another possible source of polyvinylamine would be the hypobromite degradation of polyacrylamide. 1962 H. BLOEMENDAL et al. in A. Pirie *Lens Metabolism Rel. Cataract* 300 The large size of α-crystallin is responsible for its electrophoretic behaviour..in polyacrylamide gel. 1976 *Nature* 18 Nov. 264/1 Polypeptides were identified by polyacrylamide slab gel electrophoresis and autoradiography. 1932 *Chem. Abstr.* XXVI. 1249 With the polyacrylate salts,..the process is approx. reversible. 1946 [see *polymethacrylate* below]. 1974 P. L. MOORE et al. *Drilling Practices Manual* v. 117 Polymers of the colloidal type..do not aggregate solids, as do the polyacrylates. 1930 *Chem. Abstr.* XXIV. 1563 Colloid mols. may be homeopolar (polystyrols, rubber), or heteropolar (polyacrylic acid salts, albuminoids). 1939 *Jrnl. R. Aeronaut. Soc.* XLIII. 241 The article deals with the mechanical and physical properties of four representative transparent plastic resins: Cellulose nitrate, cellulose acetate, polymer mixtures and polyacrylic esters. 1943 *Ibid.* XLVII. 140 The polyacrylic resins are thermoplastic, and articles can be made of them by moulding or extrusion. 1959 *Times Rev. Industry* Sept. 4/1 The last 20 years have seen the development of non-cellulosic fibres such as..polyacrylics (Courtelle, Orlon, Acrilan). 1973 *Materials & Technol.* VI. viii. 560 The chemical composition of polyacrylic acid, as the basic polymer of the whole class of acrylics, permits the production of many derivatives. 1935 C. ELLIS *Chem. Synthetic Resins* II. 1072 Hydrolysis of polyacrylonitrile in the presence of water also gives an aqueous solution of the polymerized acid. 1963 A. J. HALL *Textile Sci.* ii. 89 Great difficulty has attended the devising of a satisfactory process for spinning Orlon from polyacrylonitrile. 1969 *Nature* 25 Jan. 357/2 Cellulose and polyacrylonitrile.. have been found to produce carbon fibre of good strength and modulus. 1956 *Jrnl. Amer. Chem. Soc.* LXXVIII. 3548/2 While studying the X-ray diffraction patterns of synthetic nucleotide polymers, we mixed together the sodium salts of polyadenylic acid and polyuridylic acid. 1961 *Jrnl. Molecular Biol.* III. 78 The helical molecule of polyadenylic acid consists of two polynucleotide chains organized about a twofold rotation axis. 1975 *Nature* 27 Nov. 357/1 Polyadenylic acid (poly(A)) is present at the 3' terminus of most classes of cytoplasmic messenger RNAs.. in all eukaryotic organisms reported so far. 1900 E. F. SMITH tr. *V. von Richter's Org. Chem.* (ed. 3) II. 247 Of the aromatic polyalcohols, having the hydroxyl groups attached to different carbon atoms of the same side-chain, it is only the glycols and their oxidation products which have been studied in any sense completely. 1974 *Nature* 19 Apr. 668/1 The donor and acceptor groups were aliphatic hydroxyl groups in polyalcohols, saccharides and related compounds. 1962 *New Scientist* 22 Mar. 697/2 They are described as stereoregular crystalline plastics and have been given the name polyallomers because their highly crystalline structure differs in chemical composition from other crystalline plastics. 1962 H. J. HAGEMEYER in *Mod. Plastics* June 157/2 The term polyallomer was coined [by the writer] to identify this new class of polymers and to distinguish them from previous known homopolymers and copolymers... These new polymers are examples of allomerism in polymer chemistry. 1975 C. A. HARPER *Handbk. Plastics & Elastomers* I. 91 Polyallomers are superior to polyethylene in flow characteristics, moldability, softening point, hardness, stress-crack resistance, and mold shrinkage. 1975 *Nature* 18 Dec. 638/2 CsCl powder (5·4g) was added to the solution in a siliconised polyallomer tube. 1861 *Proc. R. Soc.* XI. 281 (*heading*) Monacid polyamines. 1875 *Chem. News* 2 July 1/1 (*heading*) The mono character of ethylen and other polyamines. 1910 N. V. SIDGWICK *Org. Chem. Nitrogen* iii. 72 This looseness of attachment of the nitrogen is characteristic of these poly-amines. 1965 *New Scientist* 25 Mar. 795/2 Polyamines such as spermine and putrescine have been found in vegetable embryos and the seeds of various plants. 1931 *Chem. Rev.* VIII. 371 (*heading*) Polyanhydrides. 1932 *Jrnl. Amer. Chem. Soc.* LIV. 1584 Polyanhydrides derived from dibasic acids of the series $HOOC(CH_2)_x$·$COOH$ are especially easy to obtain in the superpolymeric state. 1972 *Encycl. Polymer Sci. & Technol.* X. 649 The best fiber-forming properties are found..in the series of polyanhydrides prepared from di(*p*-carboxyphenoxy)-α,ω-alkanes. 1940 *U.S. Patent* 2,199,397 4/2 A process for producing new surface active products which comprises reacting trimethylamine with a polybrominated palmitic acid. 1977 *Time* 4 Apr. 56/3 Michigan farmers..last year lost thousands of cattle to poisoning when a fire retardant called polybrominated biphenyl (PBB) was accidentally mixed with feed. 1977 *Lancet* 9 Apr. 790/2 When, in the U.S.A., flame resistance in children's sleeping clothes became mandatory, industry responded promptly, using those polychlorinated or polybrominated compounds which were to hand. 1935 *Chem. Abstr.* XXIX. 3976 Polyethylene sulfone..and polypropylene sulfone.. decompose..above 300°, polybutadiene sulfone, polyisoprene sulfone,..at 200–20°. 1946 F. MARCHIONNA *Butalastic Polymers* vii. 209 In this method there are obtained 1-ethenyl-3-cyclohexene and low molecular weight polybutadiene. 1960 *Times* 28 Sept. 21/6 When the Shell Chemical Company announced that they would be making polybutadiene and poly-isoprene in the United Kingdom it marked another important step in the production of synthetic rubbers in Britain. 1975 *Sci. Amer.* Dec. 101/1 Polymers that exhibit rubbery behaviour at room temperature include polyisoprene (natural rubber) and polybutadiene (a synthetic rubber). 1898 *Proc. Chem. Soc.* XIV. 179 The preparation of a number of salts of polycarboxylic acids..is described. 1947 [see *polyion* s.v. *POLY- 1]. 1970 *Jrnl. Polymer Sci.* A. VIII. 1483 The polycarboxylic resins act in a manner similar to some monomeric polycarboxylic acids. 1931 W. H. CAROTHERS et al. in *Jrnl. Amer. Chem. Soc.* LIII. 4206 We will call this product μ-polychloroprene to distinguish it from other chloroprene polymers that will be described later in this paper. 1951 *Engineering* 7 Sept. 289/3 The insulation employed includes..vulcanised rubber with sheaths of lead alloy or polychloroprene compound. 1970 *Cabinet Maker & Retail Furnisher* 30 Oct. 208/3 Adhesives based on neoprene or more generally polychloroprene have been used for many years for bonding decorative laminates to various core materials. 1946 *Nature* 17 Aug. 224/1 Quite a number of vinyl polymers, poly-esters, polyamides, and polydienes give well-defined patterns indicative of a high degree of internal order, provided they are stretched as in rubber or are drawn into fibres as in polyamides. 1960 *Times Rev. Industry* May 53/3 Also at the end of March came the announcement from Shell Chemicals that it is to build a plant..for the manufacture of polydiene rubbers, polybutadiene and polyisoprene. 1960 *Economist* 28 May 896/2 The main increase will be in butyl rubber..and in the new polydienes—polyisoprene and polybutadiene rubber—which the makers hope will prove suitable for heavy-duty tyres. 1928 *Chem. Abstr.* XXII. 1768 (*heading*) Addition of hydrogen and bromine to the poly-enes. 1934 *Science* 25 May 489/1 The names which have been given to almost all the known polyene pigments have had a taxonomic origin in either botany or zoology. 1970 *New Scientist* 5 Nov. 260/1 The polyene macrolides are an important group of antibiotics. 1961 WEBSTER, Polyenic. 1972 *Nature* 21 Jan. 132/1 A quite different kind of molecule..is the polyenic visual pigment constituent, retinal. 1976 *Chem. Physics Lett.* XLIII. 270 The Raman spectra show..large shifts of vibrational frequencies relative to other polyenic polymers and oligomers. 1949 *Arch. Biochem.* XX. 333 Table II shows the composition of the various tissue fatty acids with respect to dienoic, trienoic, and polyenoic fatty acids. 1964 *Oceanogr. & Marine Biol.* II. 181 Gas-liquid chromatography revealed so little polyenoic C_{18} acids that the mean unsaturation could generally only be expressed as monethenoid. 1935 *Biochem. Jrnl.* XXIX. 1553 (*heading*) Polyethenoid acids. 1951 H. J. DEUEL *Lipids* I. ii. 20 From a quantitative standpoint, linoleic acid is the most important of the polyethenoid acids found in vegetable oils. 1957 *Lancet* 13 Apr. 787/1 The poly-ethenoids in fish oils are so different from those in the other food fats. 1964 *Oceanogr. & Marine Biol.* II. 177 The polyethenoid alcohols never seem to amount to more than traces [in the depot lipids of fish]. *Ibid.* 179 There are some suggestive findings to indicate that polyethenoids and long-chain homologues may..be greatly reduced in amount [in the castor-oil fish, *Ruvettus pretiosus*]. 1935 HILL & CAROTHERS in *Jrnl. Amer. Chem. Soc.* LVII. 925/2 Compared with the polyesters derived from carbonic acid, the rate of distillation was quite slow; in this respect, the polyformals resembled the..polyesters derived from the higher dibasic acids such as sebacic. 1962 J. T. MARSH *Self-Smoothing Fabrics* vii. 91 When fabrics are treated with the polyformal they have a softer handle and better resistance to abrasion than when treated with formaldehyde alone. 1959 *Trans. Faraday Soc.* LV. 1484 The polarized spectra of oriented films of Delrin, a commercial polyformaldehyde resin have been investigated. 1965 HASLAM & WILLIS *Identif. & Anal. Plastics* viii. 248 Polyformaldehyde is inherently an unstable polymer and if unmodified decomposes rapidly on heating. 1973 E. H. IMMERGUT tr. *Vollmert's Polymer Chem.* ii. 253 Polyformaldehyde..can be transformed to transparent and hard plastics with high mechanical strength. 1945 H. FRAENKEL-CONRAT et al. in *Jrnl. Amer. Chem. Soc.* LXVII. 317/1 Polyglutamic acid was prepared..from a bacterial culture medium. 1970 A. L. LEHNINGER *Biochem.* vi. 113 Polyglutamic acid is a random coil at pH 7·0 because its R groups at that pH are all negatively charged. However, at pH 2·0, its R groups have no charge, and it readily forms an α-helix. 1906 *Jrnl. Chem. Soc.* XC. I. 403 (*heading*) Action of nitrous acid on polyglycine esters. 1956 *Nature* 18 Feb. 326/1 The two crystallographic forms of polyglycine..have recently been reinvestigated. 1968 E. J. DUPRAW *Cell & Molecular Biol.* xii. 290 (*caption*) A molecule of polyglycine, the simplest possible polypeptide chain. 1889 G. M'GOWAN tr. *Bernthsen's Text-bk. Org. Chem.* 193 Ethylene glycol combines with glycol to form the so-called Polyglycols, e.g. Di-ethylene glycol, $C_2H_4(OH)—O—C_2H_4(OH)$. 1959 *Times* 3 Mar. 7/6 Shell chemicals are already extensively used..in hydraulic brake fluids (glycols, glycol ethers and polyglycols). 1961 H. R. SIMONDS *Source Bk. New Plastics* II. iv. 49 The polyglycols are receiving increased attention as intermediates in plastics compounds. 1956 *Chem. Abstr.* L. 11349 $HOCH_2CO_2H$..(7·6 g.) standing with 100 cc. dioxane-Et_2O (1:3) satd. with HCl at room temp. gave 0·8 g. of a polyglycolic acid $H(OCH_2CO)_nOH$.., m. 126–8°. 1969 *Brit. Med. Jrnl.* 3 May 308/1 A synthetic absorbable suture material made of polyglycolic acid (P.G.A.) has recently been developed. 1977 *Lancet* 28 May 1128/1 The results of using interrupted nylon skin sutures or subcuticular polyglycolic acid (P.G.A.) sutures after appendicectomy were compared in a prospective controlled trial in 127 patients. 1895 THOMSON & BLOXAM *Bloxam's Chem.* (ed. 8) 587 (*heading*) Polyhydroxy-monobasic acids. 1913 *Jrnl. Chem. Soc.* CIV. I. 1147 (*heading*) The spatial arrangement of the hydroxyl groups of polyhydroxy-compounds. 1945 [see *POLYISOCYANATE]. 1965 PHILLIPS & WILLIAMS *Inorg. Chem.* I. xiv. 545 A number of elements and compounds have the ability to form glasses. Examples include..most polymeric materials such as polystyrene, and many poly-hydroxy compounds, e.g. water, glycol and glycerol. 1967 *New Scientist* 4 May 270/3 The darkening in afrormosia, another African hardwood, is due to certain polyhydroxystilbenes. 1951 L. H. LONG tr. *Hückel's Struct. Chem. Inorg. Compounds* II. xi. 916 A further example is the intensification of the acidity of the very weak boric acid by complex-formation with organic polyhydroxyl compounds. 1957 B. A. DOMBROW *Polyurethanes* i. 2 If we take the urethane group, and, instead of..a simple alcohol,..utilize a polyhydroxyl material like glycol, etc., a point of growth is produced. 1957 G. E. HUTCHINSON *Treat. Limnol.* I. xi. 710 Various organic substances, either colloids, such as gum arabic, or crystalloids, such as ascorbic acid and other polyhydroxyl compounds.., stabilize ferric hydroxide sols. 1931 *Jrnl. Physical Chem.* XXXV. 1893 (*table*) Polyisobutylene. 1935 C. ELLIS *Chem. Synthetic Resins* I. ix. 166 Staudinger and Brunner have examined isobutylene polymerized in the presence of floridin... They separated the resulting mixture..into tri-isobutylene, pentaisobutylene, and a polyisobutylene. 1942 *Industr. & Engin. Chem.* Oct. 1192/1 The high degree of chemical stability and excellent dielectric properties of polyisobutylene have led to its widespread commercial use. 1966 *Economist* 1 Oct. 84/3 Later the two companies may co-operate in making polyisobutylene, plastic foam and other products. 1947 E. KATCHALSKI et al. in *Jrnl. Amer. Chem. Soc.* LXIX. 2564/2 On extending experiments concerning polymerization of amino acids to basic amino acids, we succeeded in preparing poly-lysine. 1964 G. H. HAGGIS et al. *Introd. Molecular Biol.* iii. 55 In acid solution,..polylysine forms a flexible chain, the repulsion between the side chains preventing helix formation. 1973 *Nature* 6 Apr. 361/1 He [*sc.* E. Katchalsky] was the first to synthesize polylysine, a molecule that is much used in immunological research. 1935 C. ELLIS *Chem. Synthetic Resins* II. liii. 1078 The metallic..polymethacrylates are said to possess useful thermoplastic and film-forming properties. 1946 *Nature* 17 Aug. 224/1 The most notable exceptions are polyacrylates and polymethacrylates, and polyvinyl acetate. The X-ray diffraction patterns produced by these amorphous polymers yield practically no information regarding their constitution. 1973 *Sci. Amer.* Aug. 111/3 Lenses, mirrors and fiber optics of plastics, usually polymethacrylate and polystyrene, are often made by glass-working techniques. 1897 *Jrnl. Chem. Soc.* LXXII. I. 399 (*heading*) Polymethacrylic acid. 1935 C. ELLIS *Chem. Synthetic Resins* II. liii. 1080 Polymethacrylic acid begins to decompose at 200°C. 1973 *Materials & Technol.* VI. viii. 559 In polyacrylic acid, polymethacrylic acid and the amide derivatives..the polymers decompose on heating as the softening point is reached. 1931 *Jrnl. Physical Chem.* XXXIV. 44 To account for lack of mobility in the shift of equilibrium, we can picture the aluminum oxychloride sol as resembling the polyol basic chromic salts reported by Bjerrum. 1948 W. PIGMAN *Chem. Carbohydrates* vi. 232 The designation 'polyols' introduced here is synonymous with the longer, customary term, polyhydric alcohols. 1962 R. VAN HEYNINGEN in A. Pirie *Lens Metabolism Rel. Cataract* 396 A comprehensive review on the biochemistry of acyclic polyols has just been published. 1975 *Nature* 17 Jan. 194/1 Since glycerol causes cell fusion, other polyols have been investigated for fusogenic properties. 1930 *Industr. & Engin. Chem.* June 591/1 Polymerization (A polymers). Examples: Olefins and poly-olefins, unsaturated hydrocarbons, azo-compounds. 1936 *Trans. Faraday Soc.* XXXII. 5 Dimerisation is observed with the olefines and it is possible that the polyolefines may be built up in this way. 1959 *Economist* 7 Mar. 895/2 By 1961, with these new plants, British capacity in 'polyolefin' plastics will be over 150,000 tons a year. 1962 *B.S.I. News* Dec. 19/2 The viscosity number of polyolefines. 1969 L. S. MOUNTS in W. R. R. Park *Plastics Film Technol.* v. 122 Low density polyethylene films..constitute the largest segment of the polyolefin film market. 1939 *Jrnl. Amer. Chem. Soc.* LXI. 1905/2 This method..was used in the present work to synthesize the 6-, 18- and 42-membered polyoxyethylene glycols. 1952 *Martindale's Extra Pharmacopœia* (ed. 23) I. 574 The polyoxyethylene derivatives are mostly soluble or dispersible in water. 1960 A. E. BENDER *Dict. Nutrition* 100/1 Monoglycerides are soluble in fat, but by reacting with ethylene oxide the resulting polyoxyethylene derivatives become water-soluble to whatever degree is required. 1972 *Materials & Technol.* V. ix. 265 Polyoxyethylene dioleate and polyoxyethylene lauryl ether are viscous liquids and act as non-ionic emulsifiers. *Ibid.* x. 309 The polyoxyethylene alcohol surfactants range in solubility from completely oil-soluble to completely water-soluble, depending on the number of moles of ethylene oxide added. 1908 *Jrnl. Chem. Soc.* XCIV. I. 131 When heated in the open, these three polyoxymethylenes volatilise without first melting. 1930 *Chem. News* 24 Oct. 264/2 The ends of the long chains might be saturated by groups such as hydroxyl, methoxyl, or acid residues, as in the case with poly-oxymethylenes. 1952 *New Biol.* XII. 109 The clouds in the atmosphere of Venus..are said by some to be dust and by others to be polyoxymethylene. 1959 *Jrnl. Appl. Physics* XXX. 1516/1 (*caption*) Single crystals of an acetal resin, polyoxymethylene. 1975 *Sci. Amer.* Dec. 104/2 Examples of drawable semicrystalline polymers are polyethylene, polypropylene, polyoxymethylene, and nylon. 1894 *Jrnl. Chem. Soc.* LXVI. I. 415 It is noteworthy that all polyphenols derived from pyrogallol yield blue compounds. 1947 *Sci. News* V. 90 Many substances to be found in soil will reduce manganese dioxide, for example, polyphenols and sulphydryl compounds. 1973 *Sci. Amer.* Dec. 62/1 The abundance in the leaves of plants of distasteful and toxic compounds such as alkaloids and polyphenols. 1928 *Chem. Abstr.* XXII. 411 (*heading*) Complexes of uranyl with polyphenolic acids. 1958 *Times* 22 Dec. 1/5 (Advt.), Synthesis and testing of natural polyphenolic compounds as anti-oxidants. 1913 *Chem. Abstr.* VII. 796 They [*sc.* Battelli and Stern]..defend the use of the term polyphenoloxidase used to designate the enzyme which oxidizes chiefly the polyphenols and polyamines. 1956 *New Biol.* XX. 96 Browning [of tomatoes] may be associated with phenolic substances..and their subsequent polymerization to melanins by the action of polyphenoloxidases present in the host cells. 1973 F. B. ABELES *Ethylene in Plant Biol.* viii. 205 Polyphenol oxidase is a copper-containing enzyme which catalyzes the oxidation of phenols such as tyrosine and is responsible for the blackening of cut raw potatoes on exposure to air. 1974 R. G. S. BIDWELL *Plant Physiol.* vi. 119 Several enzymes that oxidize phenols to quinones are known. Two of the most important are monophenol oxidase (tyrosinase) and polyphenol oxidase (catechol oxidase). 1931 *Chem. Rev.* VIII. 375 In a similar way the oxidation of phenols may lead to the formation of polyphenylene ethers. 1965 *Jrnl. Appl. Polymer Sci.* IX. 513 Polyphenylenes tend to be brittle and intractable. 1965 *Mod. Plastics Encycl.* 1966 303/1 Polyphenylene oxide (PPO) is a new high performance engineering thermoplastic, with a unique combination of properties. 1971 *New Scientist* 24 June 761/1 Printed circuitry utilises..polyphenylene oxide parts. 1975 J. A. BRYDSON *Plastics Materials* (ed. 3) xxi. 470 Several substituted linear polyphenylenes have also been prepared but none appear to have the resistance to thermal decomposition shown by the simple poly-*p*-phenylene. 1908 *Jrnl. Chem. Soc.* XCIV. II. 838 The various supposed polyphosphates can be considered theoretically as formed by the union of pyrophosphate and metaphosphate in various proportions. 1960 A. E. BENDER *Dict. Nutrition* 100/2 *Polyphosphates*, complex phosphates added to foods, in particular to meat products... Include pyrophosphate ($Na_4P_2O_7$), tripolyphosphate ($Na_5P_3O_{10}$), longer phosphate chains of 100 phosphate units, polyphosphate glasses prepared by rapid quenching of $Na_2O–P_2O_5$ melts. 1962 COTTON & WILKINSON *Adv. Inorg. Chem.* xx. 397 Linear polyphosphates..are salts of anions of general formula $[P_nO_{3n+1}]^{(n+2)-}$. *Ibid.* 398 Cyclic polyphosphates..are salts of anions of general formula $[P_nO_{3n}]^{n-}$. 1895 *Jrnl. Chem. Soc.* LXVIII. II. 445 (*heading*) New polyphosphoric acid, $H_6P_4O_{13}$, and its salts. 1950 *Jrnl. Amer. Chem. Soc.* LXXII. 2962/2 In order to test this hypothesis the reaction was carried out in polyphosphoric acid, a commercially available mixture of the 'strong phosphoric acids'. 1967 I. L. FINAR *Org. Chem.* (ed. 5) I. ix. 229 Snyder *et al.* (1954) have shown that the hydrolysis of cyanides with polyphosphoric acid gives very good yields of amide. 1944 *Mod. Plastics* Nov. 124/1 (*caption*) The formation of silicones... Condensation to siloxanes... A polysiloxane. 1946 *Industr. & Engin. Chem.* Nov. 1117/1 Industrial attention has been directed to the liquid polysiloxanes since the announcement in 1944 that silicones were in commercial production. 1955 BROWN & DEY *India's Mineral Wealth* (ed. 3) 391 The resultant organosilicon chlorides are hydrolysed to silanols which condense into the polysiloxanes or silicones. 1959 B. S. GARVEY in M. Morton *Introd. Rubber Technol.* i. 33 The silicone rubbers are polysiloxanes. 1932 *Nature* 19 Nov. 756/1 He [*sc.* H. Staudinger] has prepared a polystyrol $(C_8H_8)_{5000}$, with about 100,000 atoms in the molecule and a molecular weight of 600,000. 1940 'PLASTES' *Plastics in Industry* vi. 73 Polystyrol..is mechanically somewhat weaker than cellulose acetate. 1966 *Economist* 16 July 263/1 (Advt.), Technical synthesis of styrol for polystyrol and Buna synthesis. 1849 H. WATTS tr. *Gmelin's Hand-bk. Chem.* III. ii. 98 The aqueous solution of the polysulphide of sodium is yellow. 1871, 1882 [in Dict. s.v. POLY- 2 a]. 1935 C. ELLIS *Chem. Synthetic Resins* II. lviii. 1170 As some of the commercial sulphur resins are polysulphides, Thomas and Riding's work on the alkyl polysulphides should be considered. 1959 B. S. GARVEY in M. MORTON *Introd. Rubber Technol.* i. 33 The Thiokols are polysulfides of organic dihalides. 1959 J. S. JORCZAK in *Ibid.* xv. 363 Polysulfide polymers were first introduced in 1930. 1963 C. R. COWELL et al. *Inlays, Crowns, & Bridges* v. 50 One brand of polysulphide rubber is supplied in two viscosities, a more fluid grade for use with a syringe and injection into the cavity, and a more viscous grade for use in an impression tray. 1965 PHILLIPS & WILLIAMS *Inorg. Chem.* I. xvi. 578 The sulphides redissolving in excess sulphide mostly give rise to polysulphide anions, e.g. $[SnS_3]^{2-}$. 1934 *Jrnl. Amer. Chem. Soc.* LVI. 1815/2 Seyer and King have suggested a polysulfone structure..for the addition product of cyclohexene and sulfur dioxide. 1967 *Times Rev. Industry* June 68/1 In the main these are specialist materials offering advances in thermal, mechanical or electrical properties, and include the phenoxy polymers, the polysulphones, methyl pentene polymers, to mention a few. 1971 *New Scientist* 24 June 761/1 Polysulphone moulded components perform satisfactorily in various aircraft parts. 1885 *Jrnl. Chem. Soc.* XLVIII. I. 551 The author [*sc.* O. Wallach] proposes to classify the terpenes as follows:..C *Polyterpenes*. 1. Tripentenes, $C_{15}H_{24}$.. 2. Tetrapentenes, $C_{20}H_{32}$.. 3. Polyterpenes, $(C_{10}H_{16})_x$, such as caoutchouc, &c. 1956 I. L. FINAR *Org. Chem.* II. viii. 250 Rubber is the most important polyterpene. 1970 *Encycl. Polymer Sci. & Technol.* XIII. 577 Serious attempts are being made by the resin manufacturers to develop a substitute for polyterpenes from petroleum distillates. *Ibid.* 591 Polyterpene and terpene–urethan resins are used as additives in the preparation of hot-melt coating mixtures. 1936 L. F. FIESER *Chem. Nat. Products related to Phenanthrene* 358 Previously polyterpenoid compounds had been known to occur only in plants. 1964 *New Scientist* 22 Oct. 220/1 A feature of polyterpenoids which had already been noted by chemists was that the carbon skeletons of their molecules could usually be dissected into five-carbon units with branched chains. 1971 G. P. MOSS in K. H. Overton *Terpenoids & Steroids* I. v. 198 Although the best-known polyterpenoid is rubber, recent work has demonstrated a range of polyprenols and related compounds such as vitamins E and K. 1930 *Jrnl. Amer. Chem. Soc.* LII. 2474 Conjugated uronic acids, the so-called polyuronides found in pectins, gums, alginic acids, the specific polysaccharide substances of certain micro-organisms, and other plant materials also yield carbon dioxide when heated with 12·0% hydrochloric acid. 1957 G. E. HUTCHINSON *Treat. Limnol.* I. xvii. 889 The main part of the humus of acid peat is not a lignin derivative but apparently consists of hemicelluloses or polyuronides. 1975 *Nature* 11 Dec. 483/2 The natural soil 'cements' include polysaccharides and polyuronides.

3. Words in which *poly-* represents or is derived from another Eng. word beginning with the element, as *poly-cotton* [f. *POLY-(ESTER]; **po·lyreaction** *Chem.*, any reaction that yields a polymer; **po·lyrod** *Radio* [see quot. 1950], an antenna consisting of a rod of dielectric material (usu. tapered) projecting from a waveguide. Cf. *POLY[3].

1978 *Country Life* 28 Dec. 2237/1 The basic Burberry trenchcoat..now costs around £99, made in poly-cotton. 1979 *Times* 1 Dec. 10/1 (Advt.), Dreamy Nightwear... The skirts of both garments are polycotton; 65% polyester and 35% cotton. 1941 MARK & RAFF *High Polymeric Reactions* 3 To avoid the lengthy expressions, polymerization reactions and polycondensation reactions, ..we shall designate both as polyreactions. 1959 *New Scientist* 2 July 34/2 The remaining lectures were divided into two groups with the general titles 'Physics and Physical Chemistry of Macromolecules' and 'Polyreactions'. 1945 *Electronic Industries* Sept. 222 *Polyrod antenna*, an antenna in which the radiating element is a rod of polystyrene. 1947 *Bell Syst. Technical Jrnl.* XXVI. 844 The principal defect of the uniform polyrod is the strong minor lobes. 1950 H. P. WILLIAMS *Antenna Theory & Design* II. iv. 190 This form of dielectric antenna is commonly called a 'polyrod' antenna, since the dielectric material is often polystyrol. 1967 E. L. GRUENBERG *Handbk. Telemetry & Remote Control* iv. 130 Plastics, foams, and ceramics can be used for the radiating element. The most common type is called a polyrod and the usual construction contains a dielectric that is linearly tapered over slightly more than half its length.

poly-A: see *POLY- 2. **polyabolo:** *POLY- 1. **polyacetal, -acetylene, -acetylenic:** *POLY- 2.

polyacid (pǫli,æ·sid), *a.* and *sb. Chem.* Also **poly-acid.** [f. POLY- + ACID.] † **A.** *adj.* Applied to a base which requires more than one equivalent of acid for neutralization, and to a salt of such a base; of an alcohol, polyhydric. *Obs.*

1858 *Proc. R. Soc.* IX. 152 It became extremely probable that the action of ammonia upon poly-acid alcohols would give rise to poly-ammonium bases. 1880 [in Dict. s.v. POLY- 2 b]. 1904 *Jrnl. Chem. Soc.* LXXXVI. I. 698 (*heading*) The polyacid salts of rosaniline. 1912 E. FEILMANN tr. *Molinari's Treat. Inorg. Chem.* 423 When the salt is formed by the action of an acid on a base which has more than one hydroxyl (OH) group, that is, in the case of polyacid bases, such as $Bi(OH)_3$, $Pb(OH)_2$,..various types of salts may be formed. 1920 [see B]. 1926 N. H. FURMAN *Kolthoff's Indicators* iv. 117 When all of the dissociation constants of polybasic acids, or polyacid bases are large, they behave like strong monobasic or acid compounds upon neutralization.

B. *sb.* A compound which has more than one acidic group; *esp.* an acid containing polymeric anions. Occas. also *attrib.* or as *adj.*

1911 *Jrnl. Chem. Soc.* C. I. 265 The extension of Werner's co-ordination theory to poly-acids..facilitates the correct formulation of these acids. 1920 T. H. POPE tr. *Molinari's Treat. Inorg. Chem.* (ed. 2) 274 Several molecules of a polybasic acid are able to condense and form polyacids, such as pyroantimonic, pyrophosphoric, and pyrosilicic acid, etc., and in the same way it is known that the polyacid bases, also called polyhydric bases, condense to form polybases or polyhydroxides. 1939 L. PAULING *Nature Chem. Bond* vii. 226 The pyro, meta, and other polyacids of the second-row atoms contain MO_4 tetrahedra condensed by sharing oxygen atoms. 1950 N. V. SIDGWICK *Chem. Elements* II. 999 The molybdates and tungstates go much further, forming the highly condensed polyacids and hetero-poly-acids. 1964 *Biophysical Jrnl.* IV. I. Suppl. 11 The polyelectrolyte nature of some polyacid and polybase polypeptides endows them with inhibitory activity in enzymatic reactions. 1974 D. M. ADAMS *Inorg. Solids* vii. 239 The main features of polyacid chemistry have been recognized for a long time, but not understood.

polyacrylic and derivatives: see *POLY- 2. **polyactine:** *POLY- 1.

polyad. Add: **2.** *Philos.* A relative containing more than two elements.

a 1914 C. S. PEIRCE *Coll. Papers* (1931) I. 146 A thorough study of the logic of relatives..shows that logical terms are either monads, dyads, or polyads, and that these last do not introduce any radically different elements from..triads. *Ibid.* (1933) IV. 301 If the number of blanks exceeds two, I term it a Polyad, or Plural Relative.

polyaddition (pǫli,ădi·ʃən). *Chem.* [a. G. *polyaddition* (O. Bayer 1947, in *Angewandte Chem.* A. LIX. 263/2), f. *poly-* POLY- + *addition* ADDITION.] An addition reaction between two compounds which yields a polymeric product.

1948 *Chem. Abstr.* XLII. 6160 (*heading*) The diisocyanate polyaddition process. 1970 E. L. MCCAFFERY *Lab. Prep. for Macromolecular Chem.* 58 Condensation polymerization..may take place either by a polycondensation reaction, whereby low molecular-weight by-product is formed along with the polymer,..or by a polyaddition reaction in which the total reactants are incorporated in the polymer chain, as is typified by polyurethane formation. 1973 K. J. SAUNDERS *Org. Polymer Chem.* i. 23 Some authors..apply the term rearrangement polymerization (or, sometimes, polyaddition) to polymerization which proceeds through the interaction of functional groups without elimination of a small molecule.

polyadenylic acid: see *POLY- 2.

polyadic (pǫli,æ·dik), *a.* [f. POLYAD + -IC.] Involving many (usu., three or more) quantities or elements. Hence **polya·dically** *adv.*; **polyadi·city**, the state or quality of being polyadic.

1906 C. S. PEIRCE in *Monist* XVI. 512 A Predicate is either non-relative, or a monad..as is 'black'; or it is a dyadic relative, or dyad, such as 'kills', or it is a polyadic relative, such as 'gives'. 1919 A. N. WHITEHEAD *Enquiry Princ. Nat. Knowl.* vii. 84 A percipient event in the polyadic relation of a sense-object to nature is the percipient event of an awareness which includes this recognition of that sense-object. 1933 C. D. BROAD *Exam. McTaggart's Philos.* I. IV. xv. 282 We must distinguish extent of application, which belongs to *all* characteristics, from 'polyadicity', which belongs only to relations. 1950 W. V. QUINE *Methods of Logic* (1952) III. 135 We *also* have occasion to speak of monadic and polyadic schemata, referring thereby rather to the absence or presence of polyadically occurring predicate letters. 1964 E. A. NIDA *Toward Sci. Transl.* v. 111 For the most part these differences are binary or dyadic, but they may be singular or multiple (or polyadic). 1972 W. V. QUINE *Methods of Logic* (ed. 3) III. xxv. 141 The polyadic ingredients—'*Hxy*' and its suite—are what are new.

polyalcohol, -allomer: see *POLY- 2. **polyallomorphic, -alphabetic:** *POLY- 1.

polyamide (pǫli,ǣi·məid). [f. POLY- + AMIDE.] Any of a large class of polymers in which the units are linked by an amide group, —CO·NH—, and which includes many synthetic resins used commercially, notably fibres of the nylon group. Also *attrib.*

1929 *Jrnl. Amer. Chem. Soc.* LI. 2550 Polyintermolecular condensation requires as starting materials compounds in which at least two functional groups are present in the same molecule (*e.g.*, hydroxy acids..might lead to polyesters,..amino-acids, to poly-amides..). 1942 *Endeavour* Apr. 72/2 *Nylon*, the new and truly synthetic fibre (a polyamide) that is now rivalling natural silk. 1963 H. R. CLAUSER *Encycl. Engin. Materials* 501 The presence of the polyamide resin brings about the gelation and provides thixotropy to the paint. 1969 L. S. MOUNTS in W. R. R. Park *Plastics Film Technol.* v. 136 Polyamide films are produced from two types of nylon polymer. 1971 *Daily Tel.* (Colour Suppl.) 30 Apr. 8/3 The fabric..is an industrial nylon polyamide, tightly woven with a glistening surface and faintly translucent. 1972 *Sci. Amer.* Dec. 48/3 Some 10 different basic types of nylon are now produced, and the worldwide consumption of these polyamide fibres exceeds four billion pounds per year.

Hence **po:lyamida·tion**, a reaction or process which yields a polyamide.

1946 *Chem. Rev.* XXXIX. 146 The polyamidation reaction parallels closely the rate, temperature coefficient, and reaction order of monoamidations. 1961 SORENSON & CAMPBELL *Prep. Methods Polymer Chem.* iii. 63 An alternative method for carrying out polyamidations from a nylon salt is to use a pipe autoclave.

polyamine, -anhydride: see *POLY- 2. **polyanion(ic:** *POLY- 1.

Polyanna, var. *POLLYANNA.

polyantha (pǫli,æ·nþă). [f. POLY- + Gr. ἄνθος flower.] A small shrub rose or a climbing rose belonging to a group of hybrids of *Rosa chinensis* and *R. multiflora* and bearing flowers in clusters. Freq. *attrib.*

1889 *Jrnl. R. Hort. Soc.* XI. p. cv, The pretty and interesting Polyantha Roses comprised such sorts as Mignonette, Golden Fairy, [etc.]. 1894 A. FOSTER-MELLIAR *Bk. of Rose* ii. 26 Turner's Crimson Rambler.. is a very strong-climbing perpetual Polyantha from Japan. 1931 E. S. ROHDE *Scented Garden* v. 134 The double flowered white and pink multiflora or polyantha roses were introduced as cultivated plants from Chinese gardens more than a hundred years ago. 1934 *Times Educ. Suppl.* 24 Nov. (Home & Classroom Sect.) p. iv/1 A new group of hybrid *polyantha* varieties has been created. 1945 G. GRAVES *Trees, Shrubs, & Vines for Northeastern U.S.* 175 The polyanthas have the ruggedness and the ability to thrive under conditions that would discourage the more tender hybrid teas. 1955 G. S. THOMAS *Old Shrub Roses* ix. 92 'Floribunda' is now accepted as a group name for these large-flowered 'polyanthas'. 1962 R. PAGE *Educ. Gardener* iv. 129 Roses, particularly the polyantha or floribunda kinds, are good plants for these small formal gardens. 1976 *Hortus Third* (L. H. Bailey Hortorium) 980/2 The Floribundas..owe their origin to crosses between Polyanthas, Hybrid Teas, and other classes.

polyarch, *a.* (Later examples.)

1914 [see *hexarch* adj. s.v. *HEXA-]. 1964 H. J. DITTMER *Phylogeny & Form in Plant Kingdom* xxiii. 563 The xylem may be termed diarch when only two points are present, triarch with three, tetrarch with four, or polyarch with many.

polyarchic (pǫli,ā·ɹkik), *a.* [f. POLYARCH(Y + -IC.] Of, pertaining to, or having the characteristics of a polyarchy.

1892 G. GISSING *Born in Exile* II. IV. i. 164 He could admit that such men as Runcorn and Kenyon—the one with his polycharchic commercialism, the other with his demagogic violence—had possibly a useful part to play. 1970 R. A. H. ROBINSON *Origins of Franco's Spain* 22 Society was 'polyarchic', being composed of a series of 'intermediate republics' (guild, municipality, region) between the 'monarchies' of the State and family. 1974 *Govt. & Opposition* IX. 29 During the 1950s, the polyarchic formula, which took for granted the unconditional adherence of individuals and groups to majority rule insofar as it assumed that there was no permanent majority or minority on any particular question, seemed to be about to take root in a great many liberal democracies.

polyarchism (pǫ·li,āɹkiz'm). [f. POLYARCH(Y + -ISM.] The principles or practice of polyarchy.

1915 E. BARKER in *Political Q.* Feb. 120 This may seem anarchism. Really it is polyarchism. And as for the problem of polyarchism—why..it is likely to be settled by the needs of mere ordered life. 1917 H. J. LASKI *Let.* 20 Nov. in *Holmes–Laski Lett.* (1953) I. 110, I shall preach anarchism or rather polyarchism in the guise of political theory.

polyarteritis: see *POLY- 1.

polyatomic, *a.* Add: In mod. use (composed of molecules) containing many atoms (usu., three or more). Also as *sb.*, such a molecule. (Further examples.)

1937 L. S. STEBBING *Philos. & Physicists* ix. 215 Human beings are polyatomic. 1958 *Oxf. Univ. Gaz.* 23 Apr. 880 Vibration-rotational bands of polyatomic molecules. 1961 G. R. CHOPPIN *Exper. Nucl. Chem.* iii. 37 Counting gases are usually one of the noble gases mixed with a small amount of a polyatomic gas. 1971 *Nature* 3 Dec. 277/1 There are chapters on diatomic radicals, linear polyatomic ones, non-linear polyatomics and a discussion of dissociation.

polyaxon: see *POLY- 1. **polybag:** *POLY[3].

polybase (pǫ·libēis). *Chem.* [f. POLY- 2 + BASE *sb.*[1]] A compound which contains more than one basic group. Cf. POLYBASIC *a.* in Dict. and Suppl.

1920 [see *POLYACID *sb.*]. 1956 *Nature* 24 Mar. 586/1 This polybase, $—[—CH_2—CH_2—NH_2—]_n—$, has the advantage of being stained by naphthalene black 12*B* as well as by sudan black. 1966 G. M. FLECK *Equilibria in Solution* v. 83 Ion-exchange resins are familiar examples of polyacids and polybases.

polybasic, *a.* Substitute for def.: Requiring more than one equivalent of base for neutralization; containing two or more atoms of hydrogen capable of replacement by a base. Formerly also applied to salts derived from polybasic acids by replacement of more than one hydrogen atom. (Further examples.)

1851 H. WATTS tr. *Gmelin's Hand-bk. Chem.* V. 225 Phosphate of ferric oxide, or ferric phosphate... Polybasic. 1889 G. M'GOWAN tr. *Bernthsen's Text-bk. Org. Chem.* 206 The tri- and polybasic alcohols are capable of yielding the most various products upon oxidation. 1926 H. G. RULE tr. *J. Schmidt's Textbk. Org. Chem.* 59 These differences are so considerable that they may be used to distinguish between mono- and polybasic acids. 1972 NORMAN & WADDINGTON *Mod. Org. Chem.* xxi. 321 Polyester fibres are formed by condensation of polyhydric alcohols and polybasic acids.

polybasicity (examples).

1912 E. FEILMANN tr. *Molinari's Treat. Inorg. Chem.* 46 In 1835 Graham demonstrated the polybasicity of phosphoric acid. 1931 *Jrnl. Physical Chem.* XXXV. 2226 (*heading*) The polybasicity of several common sugars.

polyblast: see *POLY- 1. **polybrominated, -butadiene:** *POLY- 2.

polycarbonate (pǫlikā·ɹbǫnēit). *Chem.* [f. POLY- + CARBONATE *sb.*] † **1.** A carbonate containing several equivalents of the acid radical. *Obs.*

1886 T. S. HUNT *Mineral Physiol. & Physiogr.* viii. 289 It was further declared that the carbon-spars must be represented as polycarbonates, having not less than from 'twelve to eighteen equivalents of base replaceable so as to give rise to a great number of species'. 1891 *Chem. News* 23 Oct. 212/2 (*heading*) Contamination of alkaline polycarbonates by heavy metals.

2. Any of a class of polymers in which the units are linked by the carbonate group, —O·CO·O—, many of which are thermoplastic resins widely used esp. as moulding materials and films. Also *attrib.*

1930 *Jrnl. Amer. Chem. Soc.* LII. 315 It seems quite certain, therefore, that these polycarbonates are also open chains. 1959 *Mod. Plastics* XXXVI. 39 Farben fabriken-Bayer..has been manufacturing polycarbonates on a commercial scale in Germany for nearly a year. 1965 *Wireless World* Aug. 409/2 In the manufacture of an optical filter gold has been sputtered onto a polycarbonate substrate. 1969 L. S. MOUNTS in W. R. R. Park *Plastics Film Technol.* v. 138 Due to good forming characteristics, heat resistance, and toughness, polycarbonate film and sheet are used in skin and blister packaging. 1976 *Sci. Amer.* Feb. 19/2 (Advt.), The instrument is housed in a tough, high-impact polycarbonate case. 1978 *Skatcat's Quiz Bk.* (R. Soc. Prevention of Accidents) 2/1 Also O.K. are polycarbonate laminates of glass or carbon-fibre reinforced resins with plywood cores.

polycarboxylic: see *POLY- 2.

polycarpic, *a.* Delete *rare* and substitute for def.: **a.** = POLYCARPOUS *a.* a. (Further examples.) † **b.** = POLYCARPOUS *a.* b. *Obs. rare.*

1909 GROOM & BALFOUR tr. *Warming's Oecol. Plants* ii. 6 These plants are divided into monocarpic and polycarpic: the former produce flower and fruit (or spores) once, and then die; the latter may produce fruit repeatedly before death claims them. 1951 MCLEAN & IVIMEY-COOK *Textbk. Theoret. Bot.* I. xxi. 834 The common factor in all perennials is that they are polycarpic, that is to say, they flower and fruit repeatedly.

polycation(ic: see *POLY- 1.

polycentric, *a.* Delete entry in Dict. (s.v. POLY- 1) and substitute:

polycentric (pǫlise·ntrik), *a.* [f. POLY- 1 + *-CENTRIC.] = *multicentric* adj. s.v. *MULTI- 1 a; in *Politics*, characterized by polycentrism. Also as *sb.*, a polycentric chromosome or chromatid.

1887 [in Dict. s.v. POLY- 1]. 1909 *Q. Rev.* Apr. 499 The Messina earthquake belongs to the class known in Italy as that of 'polycentric' earthquakes. 1945 M. J. D. WHITE *Animal Cytol. & Evol.* iv. 57 The example of *Ascaris megalocephala* shows us that in some other organisms polycentric chromosomes can exist in nature. 1953 *Heredity* VI. Suppl. 73 The occurrence of high polycentrics such as C_3, C_4, C_5 and rarely even C_6..with a high frequency after treatment at meiosis shows..that reunion is not low. 1956 tr. *Togliatti's Probl. Devel. Socialist Democracy* 12/2 The advance towards socialism is an objective for which there is a concentration of forces from

different movements... The whole system is becoming polycentric, and even in the Communist movement we cannot speak of a single guide. 1965 *N.Y. Times* 18 Jan. 34 After the success of Tito's revolt against Moscow Togliatti conceived of 'polycentric' Communism. This means adjusting party methods in each country to national traditions and requirements. 1965 *Economist* 3 Apr. 43/2 The reality of an increasingly 'polycentric' world, marked by the growing self-assertion of the middle and smaller states. 1967 G. STEINER *Lang. & Silence* 80 African English, Australian English..represent a complicated, polycentric field of linguistic force. 1968 R. RIEGER et al. *Gloss. Genetics & Cytogenetics* 351 Polycentrics are in most cases the result of chromosome mutations. 1971 *Physics Bull.* Jan. 16/2 There are several nuclei so that orbitals are not monocentric but polycentric. 1976 T. EAGLETON *Crit. & Ideology* v. 164 Literature is multiple and polycentric, saturating the very textures of our social life.

Hence **po:lycentri·city**, the fact of being polycentric.

1959 [see *HOLOGENESIS]. 1973 *Times Lit. Suppl.* 3 Aug. 897/2 The year 1963 was also one in which polycentricity became a conspicuous feature of international relations.

polycentrism (pǫlise·ntriz'm). [f. POLY- + CENTR(E + -ISM.] In Communist political theory, the idea first promulgated by P. Togliatti (1892–1964) in 1956 that each separate Communist party has the right of full national autonomy and that the Soviet model need not be binding for all Communist parties. Also in extended use. Hence **polyce·ntrist** *a.* and *sb.*

1961 *Economist* 2 Dec. 918 He [*sc.* Gomulka] rejected the Italian idea of 'polycentrism'. 1963 *Ann. Reg. 1962* 199 'Polycentrist' tendencies within the world Communist movement. 1963 *Spectator* 28 June 829 Pelung is in effect returning to the position of support for East European polycentrism against Moscow. 1966 *Times* 7 July 7 Many more responsible non-communists were becoming polycentrists, including General de Gaulle. 1968 *Economist* 24 Feb. 36/1 The monolithic structure of the international communist movement has collapsed under the combined pressure of the Sino-Soviet conflict, and the bid for greater freedom of the ex-satellites in Eastern Europe. The result is the rise of 'polycentrism'. 1969 P. ALLUM in Henig & Pinder *European Political Parties* 206 Khrushchev's denunciation of Stalinist errors..permitted Togliatti to formulate explicitly the doctrines of 'polycentrism' and the 'Italian way to Socialism'. 1971 W. LAQUEUR *Dict. Politics* 568 The final break with China as well as Albania came at the second Moscow Conference of Communist parties in Nov. 1960... After the 1960 Conference, polycentrism..became a fact. 1974 *Times* 19 Feb. (European Defence Suppl.) p. i/1 Long before the Middle East war it had become commonplace to contrast the bipolarity of the past decade with the new pattern of polycentrism as developing forces in Europe, China and Japan impinged upon the distribution of world power.

polychlorinated (pǫliklō³·rinēited), *a. Chem.* [f. POLY- 2 + CHLORINATED *ppl. a.*] Applied to compounds in which two or more hydrogen atoms have been replaced by chlorine atoms; esp. in *polychlorinated biphenyl* [*biphenyl* = DIPHENYL], any of a class of such compounds derived from biphenyl, $(C_6H_5)_2$, or its derivatives, which have a wide variety of industrial applications and are persistent environmental pollutants; abbrev. *PCB* (s.v. *P II d).

1935 C. ELLIS *Chem. Synthetic Resins* II. lviii. 1178 Amino and hydroxy compounds are formed when polychlorinated paraffins are heated with solutions of ammonia or caustic alkali, respectively. 1951 H. H. SHEPHARD *Chem. & Action Insecticides* xi. 271 The higher polychlorinated naphthalenes are utilized as noninflammable waxes in electric insulation. 1962 *Chem. Abstr.* LVII. 8498 (*heading*) Polychlorinated biphenyl derivatives. 1968 *Times* 17 Dec. 10/5 The chemicals, called polychlorinated biphenyls, are in some ways similar to chlorine containing pesticides such as DDT and..can poison the liver even in minute amounts. 1971 P. GRESSWELL *Environment* 201 Polychlorinated biphenyls were found in high concentration in birds which died in 1969 in the Irish Sea.

polychloroprene, -choral: see *POLY- 2, 1.

polychoric (pǫlikō³·rik), *a. Statistics.* [f. POLY- + Gr. χώρ-ισις separation (f. χωρίζειν to separate) + -IC; cf. *TETRACHORIC *a.*] Used to describe a table in which data are divided into three or more classes by each of two criteria; of or pertaining to such a table; applied *esp.* to an estimate of the product-moment coefficient derived from such a table, and to concepts used in obtaining such an estimate.

1918 *Biometrika* XII. 93 (*heading*) The correlation coefficient of a polychoric table. 1922 *Ibid.* XIV. 149 What we actually desire is to compare the observations and the regression lines as given by the present polychoric method with those obtained by product-moment methods. 1964 *Psychometrika* XXIX. 386 In what follows, the tetrachoric series method is generalized to produce a new polychoric estimate of ρ.

polychromasia (pǫ:likromēⁱ·ziă). *Med.* [mod.L., back-formation from POLYCHROMATIC *a.* (see -IA[1]).] = *POLYCHROMATOPHILIA.

1909 R. J. M. BUCHANAN *Blood in Health & Dis.* xi. 196 Polychromasia is common; with stains containing methyl blue and eosin such cells may be a light violet or even a distinct blue, with methylene blue and iodine the erythrocytes in this disease exhibit a green colour not usually met with in other forms of anæmia. 1935 WHITBY & BRITTON *Disorders of Blood* iii. 64 For many years polychromasia was considered to be a degeneration until Hawes (1909) showed that the number of polychromatic cells was always approximately parallel to the number of reticulocytes. 1956 [see *anisocytosis* s.v. *ANISO-]. 1973 WOODLIFF & HERRMANN *Conc. Haematol.* i. 18 In many cells polychromasia of the cytoplasm remains after loss of the nucleus. 1977 *Proc. R. Soc. Med.* LXX. 284/2 Film was leukoerythroblastic and showed polychromasia, anisopoikilocytosis, occasional erythroblasts..and tear-drop cells.

Hence **po:lychroma·sic** *a.* = *POLYCHROMATOPHIL *a.*; cf. *POLYCHROMATIC *a.* 2.

1911 *Jrnl. Path. & Bacteriol.* XV. 9 Degenerate forms with vacuolated or irregularly stained and polychromasic cytoplasm are often seen. 1933 [see *POLYCHROMATOPHILIA]. 1942 M. M. WINTROBE *Clin. Hematol.* ii. 56 A close parallelism between the numbers of polychromasic and reticulated cells in various samples of blood has been found, although the proportion of reticulocytes is always higher.

polychromatic, *a.* Add: **1. b.** Of radiation: containing a number of wavelengths, not monochromatic.

1935 H. HARRIS *Metallic Arc Welding* iii. 13 Asterism is caused by the diffraction of a polychromatic X-ray beam by a deformed crystal. 1976 *Nature* 12 Aug. 541/2 A parallel beam of white (polychromatic) radiation falls on an oriented sample.

2. *Med.* = *POLYCHROMATOPHIL *a.*; *esp.* as *polychromatic normoblast*, an immature erythrocyte. Cf. *POLYCHROMASIC *a.*

1899 *Jrnl. R. Microsc. Soc.* 379 Polychromatic normoblasts which become violet in eosin and methylen-blue and red in triacid. 1935 [see *POLYCHROMASIA]. 1938 W. MAGNER *Textbk. Hematol.* i. 4 Normoblasts showing this mixture of red and blue in their cytoplasm are known as polychromatic normoblasts. 1958 G. C. DE GRUCHY *Clin. Haematol.* ii. 43 Polychromatic cells are young red cells which have not yet completely lost their ribose nucleic acid; they are normally present in only small numbers in the peripheral blood (0·2–2·0 per cent). 1973 B. A. BROWN *Hematol.* ii. 28/1 The production of heme and globin takes place independently of each other, beginning in the polychromatic normoblast, and ending in the reticulocyte stage.

polychromatism (pǫlikrōᵘ·mătiz'm). [f. as POLYCHROMATIC *a.*: see -ISM.] The property of having or responding to many colours.

1950 *Sci. News* XV. 26 There may be, as it were, polychromatism of the retinal sense organs but trichromatism of the brain, and therefore of the organ of vision as a whole. 1965 B. E. FREEMAN tr. *Vandel's Biospeleol.* xxv. 408 Populations of *Asellus aquaticus cavernicolus*.. show extreme polychromatism.

polychromatophil (pǫlikrōᵘ·mătofil), *a. Med.* Also **-phile** (-fəil). [a. G. *polychromatophil* (G. Gabritschewsky 1890, in *Arch. f. exper. Path. und Pharmakol.* XXVIII. 86), f. Gr. πολυχρώματ-ος many-coloured: see -PHIL, -PHILE.] Of an erythrocyte: having an affinity for basic as well as for acidic stains, and so recognizable by its appearance when a mixed stain is used. Of or pertaining to such erythrocytes. Also **po:lychromatophi·lic** *a.*, in the same sense. So **po:lychromatophi·lia**, polychromatophil condition.

1897 R. C. CABOT *Guide Clin. Exam. Blood* 77 The typical megaloblast is an abnormally large cell.., frequently showing marks of degeneration (polychromatophilia) in its protoplasm, which is therefore brownish or purplish with the Ehrlich-Biondi stain. *Ibid.* 124 More common than in any other form of anæmia are the polychromatophilic red corpuscles (see Plate IV.). 1898 A. C. COLES *Blood* Pl. I (*caption*) Polychromatophile corpuscles. 1908 *Practitioner* Aug. 324 In polychromatophil degeneration, the stained cell may vary in colour from lilac to quite deep blue (when Jenner's stain has been used). 1933 A. PINEY tr. *Morawitz's Blood Dis. in Clin. Pract.* ii. 12 Among the non-nucleated red corpuscles Jenner's stain often shows elements which are not purely red, but violet: this is polychromatophilia... More rarely erythrocytes with blue stippling are found: these have the same significance as the polychromasic ones. 1947 *Jrnl. Lab. & Clin. Med.* XXXII. 765 The polychromatophile normoblasts likewise were identified by their nuclei..and not by the cytoplasm which was basophilic rather than polychromatophilic. 1956 E. PONDER tr. *Bessis's Cytol. Blood* ix. 246 In the normal state, the circulating blood contains an extremely small number of polychromatophil red cells... In pathological conditions..polychromatophil red cells can appear in the blood in large numbers and their basophil material can take the form of little structures. 1973 B. A. BROWN *Hematol.* ii. 34/2 Polychromatophilia..indicates red blood cells containing RNA. They will stain a pinkish-gray to pinkish-blue color. 1974 *Exper. Parasitol.* XXXVI. 6 Early in the course of the developing anemia, many polychromatophilic erythrocytes and occasional normoblasts were found in the blood.

polychrome, *a.* and *sb.* Add: **A.** *adj.* **1.** Also *fig.*

1959 [see *DEBUSSYAN *a.* and *sb.*]. 1962 I. MURDOCH *Unofficial Rose* vii. 75 Her polychrome being fell into an authoritative pattern which proclaimed her free.

2. *Biol.* Of a stain or dye: containing quantities of a number of derivatives which differ in colours from the parent compound; esp. in *polychrome methylene blue*.

1895 G. W. CALE tr. P. G. Unna in *St. Louis Med. & Surg. Jrnl.* July 30 All that is necessary is to decolorize with a concentrated tannin solution the sections overstained with my polychrome-methyl-blue solution. 1896 *Ibid.* Feb. 83 This secondary effect of the polychrome methylene blue solution proves its value because it made the differential diagnosis of mast-cells (red) and plasma cells (blue) a very easy matter. 1909 *Boston Med. & Surg. Jrnl.* 7 Oct. 494/1 Widal, Abrami and Brulé recommend the use of polychrome blue of Unna 1:10. 1925 H. J. CONN *Biol. Stains* v. 47 The polychrome properties just mentioned are likely to develop in a methylene blue solution upon standing. Anyone who has had much experience with the stain is familiar with the occasional green tones from methylene green, the reddish shades of methylene azure..and methylene violet. Such a solution is known as 'polychrome methylene blue'. 1960 E. GURR *Encycl. Microsc. Stains* 415 Wright's stain..is prepared by neutralising polychrome methylene blue with eosin. 1960 JACOBS & GERSTEIN *Handbk. Microbiol.* 248/2 *Jenner's Stain*, a polychrome stain prepared by mixing 0·5 per cent eosin in methyl alcohol with 0·5 per cent methylene blue in methyl alcohol in the ratio 5:4.

B. *sb.* **1.** (Later examples.)

1959 [see *BICHROME *a.* and *sb.*[2]]. 1972 *Trans. Oriental Ceramics Soc.* XXXVIII. 41 It was difficult, apart from the fine Ch'êng-hua *tou-ts'ai*, to discover really good pieces of the more usual polychromes.

2. (Earlier example.)

1870 C. SCHREIBER *Jrnl.* (1911) I. 101 He had procured for us..a dish with house in blue,..and a larger one with pastoral subjects in polychrome.

4. *Med.* A polychromatophil erythrocyte or normoblast.

1909 *Boston Med. & Surg. Jrnl.* 7 Oct. 495/1 The blood showed 6% of reticulated forms, with ·4% of polychromes. 1933 *Lancet* 3 June 1173/1 Hawes..hazarded the suggestion that the stippled cell was merely a variant of the polychrome.

polychrome (pǫ·likrōᵘm), *v. Biol.* [f. prec.] To convert (a stain or dye) to a polychrome form (sense *A. 2). Hence **po·lychroming** *vbl. sb.*

1925 H. J. CONN et al. *Biol. Stains* v. 48 Methylene blue should be partly polychromed in order to have its best staining powers. 1958 J. R. BAKER *Princ. Biol. Microtechnique* xiv. 268 It was his [*sc.* G. Giemsa's] purpose to avoid methylene blue that had been polychromed at random, and to use instead known quantities of known dyes. *Ibid.* 271 No dye that does not arise spontaneously in the polychroming of methylene blue has any special virtue in Romanowsky dyeing. 1963 M. J. LYNCH et al. *Med. Lab. Technol.* xiii. 256/1 The methylene blue is polychromed by heating with sodium bicarbonate. *Ibid.* xxxv. 630/2 The polychroming involves the oxidation of the methylene blue so that methyl groups are lost and formaldehyde gas is given off.

polychromed (pǫ·likrōᵘmd), *ppl. a.* [f. prec. + -ED[1].] Rendered polychrome: chiefly *Biol.*, of stains and dyes.

1924 *Jrnl. Bacteriol.* IX. 405 The method involves pipetting about one cc. of a well polychromed, Loeffler's methylene blue over the agar slant culture and tilting the tube so that the stain comes in contact with the bacterial film. *Ibid.* 407 A polychrome methylene blue divised [*sic*] by Novy has given results surpassing in clearness those obtained by the ordinary polychromed methylene blue. 1958 J. R. BAKER *Princ. Biol. Microtechnique* xiv. 268 It is necessary to know what substances besides methylene blue itself are present in the polychromed dye. 1978 K. BONFIGLIOLI *All Tea in China* iv. 44 An incomparable saucer, polychromed, yet from the very earliest part of the Ming Dynasty.

polychromed (pǫ·likrōᵘmd), *a.* [f. POLYCHROME *a.* and *sb.* + -ED[2].] = POLYCHROME *a.* 1.

1922 *19th Cent.* May 804 The polychromed wooden statue of St. Paul recalls a work in stone from the hand of Vecchietta. 1936 *Burlington Mag.* July 48/2 The low relief polychromed stucco ornament. 1947 J. C. RICH *Materials & Methods Sculpture* x. 308 The Chinese and Japanese produced a large amount of polychromed wood sculpture. 1972 K. BONFIGLIOLI *Don't point that Thing at Me* xiii. 101 A great polychromed Mexican carving of an agonized Madonna.

polychromism (pǫlikrōᵘ·miz'm). [f. as POLYCHROMIC *a.*: see -ISM.] = *POLYCHROMATISM.

1903 *Amer. Naturalist* XXXVII. 295 (*heading*) Albinism, partial albinism and polychromism in hag-fishes. 1933 THORPE & LINSTEAD *Synthetic Dyestuffs* (ed. 7) viii. 72 Polychromism..is the name assigned to the phenomenon of colour variation in salts.

polycistronic(ally: see *POLY- 1.

polyclad, *a.* and *sb.* Substitute for etym.: [ad. mod.L. name of suborder *Polycladidea* (A. Lang 1884, in *Fauna und Flora des Golfes von Neapel* XI. 1), f. POLY- + Gr. κλάδος branch.] (Later examples.)

1896 F. W. GAMBLE in *Cambr. Nat. Hist.* II. i. 16 The Polyclads were so called by Lang on account of the numerous primary branches of their intestine. They are free-living, purely marine Platyhelminthes. 1918 *Proc. Nat.*

Acad. Sci. IV. 381 Muscular creeping operations are probably general among polyclads. 1941 J. STEINBECK *Sea of Cortez* xi. 111 Beautiful purple polyclad worms crawled over lawns of purple tunicates. 1975 *Nature* 28 Aug. 737/1 Polyclads have limited powers of regeneration. *Ibid.* 11 Dec. 518/1 We..found them [*sc.* reddish-brown streaks in the sea] to consist of high densities of the dinoflagellate *Noctiluca* and of a polyclad turbellarian.

polyclimax (pǫliklai·mæks). *Ecol.* [f. POLY- + CLIMAX *sb.*] The presence of several distinct stable communities of plants within a given region. Usu. *attrib.*

1934 *Empire Forestry Jrnl.* XIII. 21 An appreciation of some, at least, of these truths, has led many ecologists to adopt a poly-climax theory as a working hypothesis. 1953 *Ecol. Monogr.* XXIII. 43/1 A significant difference in climax interpretation has..been described in the monoclimax and polyclimax theories. 1960 N. POLUNIN *Introd. Plant Geogr.* xi. 331 It seems best to admit the likelihood, in any one region, of several different climax communities as representing what may then be termed a 'polyclimax'. 1973 P. A. COLINVAUX *Introd. Ecol.* xl. 551 Whether you believed in the monoclimax required by dogmatic interpretations of Clements' writings, or whether you took the pragmatic view that there could be many local climaxes in any clima[c]tic region (so-called polyclimax theory) you tacitly assumed that succession was a process of social organization; that it led to an organized entity which you called a society.

polyclinic. Add: **1.** In mod. use, a clinic not attached to a hospital where specialists in various branches of medicine are available to outpatients. (Further examples.)

1963 *Spectator* 11 Oct. 446 Other nations whose medical services are based primarily on the hospital or the polyclinic are beginning to see the virtues of a personal doctor. 1967 *Guardian* 15 July 10/3 These doctors [in the U.S.S.R.] work from *polyclinics* which are a combination of general practice, as we know it, and hospital outpatient departments. 1973 *Times Lit. Suppl.* 3 Aug. 896/2 The Soviet Union..has based its system on the use of the polyclinic, with its specialist staff. Here there is no 'physician of primary contact' or general practitioner service as we know it. 1975 J. DE BRES tr. *Mandel's Late Capitalism* xii. 385 [In the age of late capitalism] the independent general medical practitioner is replaced by a polyclinic with affiliated specialists. 1976 *Times Lit. Suppl.* 7 May 555/2 Cuba is perhaps nearest to ordinary Western experience... Much is made of the rate of increase of district hospitals, polyclinics and local health centres.

polyclonal(ity, -clonally: see *POLY- 1.

po:lycondensa·tion. *Chem.* Also **poly-condensation.** [ad. G. *polykondensation* (H. Staudinger *Hochmolekularen Verbindungen* (1932) II. 255): see POLY- and CONDENSATION.] A condensation reaction between molecules each having at least two functional groups which yields a polymer, or a process based on such a reaction. Freq. *attrib.*

1936 *Trans. Faraday Soc.* XXXII. 52 The formation of high polymeric products can also take place by a polycondensation process in which numerous reactive molecules condense. 1949 *Jrnl. Textile Inst.* XL. A307 Polycondensation products, such as phenolic plastics, phenol-acetylene resins, urea-formaldehyde resins, linear and branched polyesters, and super polyamides. 1962 *Times* 26 Feb. (Canada Suppl.) p. vii/2 (Advt.), Complete chemical plant and equipment including plant for thermosetting and thermoplastic polymers and synthetic fibres from polymers and polycondensation products. 1963 A. J. HALL *Textile Sci.* ii. 79 It [*sc.* nylon] is made by the melt-spinning of polyhexamethylene adipamide which in its turn is produced by the polycondensation of equimolecular proportions of adipic acid..and hexamethylene diamine. 1971 *Jrnl. Oil & Colour Chemists' Assoc.* LIV. 888 The volatile components (solvents, monomers and low molecular weight poly condensation products) evaporate almost completely from the liquid paint during the process of film formation. 1978 *Nature* 29 June 738/2 Polymers of related structures have long been known, however, prepared by polycondensation of alkali metal polysulphides and α, ω-dihaloalkanes.

Hence (as back-formations) **polyconde·nse** *v. trans.*, to cause to undergo polycondensation; **polyconde·nsed** *ppl. a.* Also **poly-co·ndensate** [after *filtrate*, *precipitate*, etc.], a product or preparation resulting from polycondensation.

1942 *Jrnl. Amer. Chem. Soc.* LXIV. 2269/2 The results ..show that the poly-condensates are built up quantitatively from alanine units linked by —CONH— bonds. 1967 *Immunochem.* IV. 77 (*heading*) Sensitization of erythrocytes by polycondensed proteins of immune serum and their use for determining antigen content. 1968 J. CHAMBION in H. F. Mark et al. *Chem. Man-Made Fibers* II. 448 For higher homologs of polyamide 6, the corresponding monomer amino acid can be polycondensed. 1971 B. BUCK tr. *Ludewig's Polyester Fibres* iv. 90 The whole process from dissolution to extrusion can be carried out in one single apparatus in which the dimethyl terephthalate is dissolved in glycol, transesterified, and polycondensed. *Ibid.* 125 Moisture promotes degradation of the polyester melt in the further processing of the polycondensate. 1976 *Nature* 6 May 76/3 Structures..much more easily oxidisable than polycondensed aromatics.

poly-cotton: see *POLY- 3. **polycratism, -cross:** *POLY- 1.

polycrystal (pǫ·likristăl). [f. POLY- + CRYSTAL *sb.* and *a.*] A polycrystalline body. Also *attrib.*, = next.

1925 *Physical Rev.* XXV. 248 The crystal was then hammered and swaged to change it into polycrystal copper. 1932 *Proc. R. Soc.* A. CXXXVIII. 358 The complete change [in density] from single crystal to polycrystal ..is..of the order of 1 in 3000 for, *e.g.*, iron, nickel and aluminium. 1966 C. R. TOTTLE *Sci. Engin. Materials* iv. 93 More than one nucleus inevitably gives rise to a polycrystal, where several points of growth lead to crystallites of different crystallographic orientation. *Ibid.* vi. 136 A polycrystal of iron is not always magnetized.

polycrystalline (pǫlikri·stăləin), *a.* [f. POLY- + CRYSTALLINE *a.* and *sb.*] Composed of many crystals or crystallites; having a crystalline structure in which there is a random variation in the orientation of different parts.

1925 *Proc. R. Soc.* A. CIX. 144 The figures for single crystals are higher than for the polycrystalline test piece. 1932 *Ibid.* CXXXVIII. 364 A polycrystalline wire can be considered as an assemblage of crystallites in which the axes of crystal symmetry are oriented at random. 1950 *Sci. News* XV. 61 Distorted lattice planes will not slide over one another easily and hence polycrystalline metals are often hard. 1959 *Times Rev. Industry* Mar. 32/3 Fine grinding stone..consisting of a sintered polycrystalline ruby material. 1970 *Sci. Jrnl.* Feb. 46/1 Bulk metals are polycrystalline and the grains are randomly oriented.

Hence **po:lycrystalli·nity**, polycrystalline condition or structure.

1955 *Rep. Progress Physics* XVIII. 233 Research on these materials has been impeded by the anomalous effects of polycrystallinity. 1978 *Nature* 19 Oct. 634/1 There was evidence of polycrystallinity with definite, but usually very weak, X-ray lines assignable to diamond.

polyculture (pǫ·likɒltiŭɹ). [irreg. f. POLY- + CULTURE *sb.* Cf. *MONOCULTURE.] **a.** The simultaneous cultivation or exploitation of several crops or animals. **b.** An area in which this is practised. Opp. *MONOCULTURE. Hence **polycu·ltural** *a.*

1915 C. R. ENOCK *Tropics* xl. 439 It is not to 'monoculture' but to 'polyculture'—that is, to varied production as contrasted with single products—that any community must look for its economic and social security. The peculiar conditions of the tropics must always of necessity call for a considerable exercise of monoculture.. but it must be balanced by equitable regard for the native producer and the exercise to the utmost possible extent of polycultural principles, whereby a supply of all products and all articles necessary for life are producible locally. 1967 *Geo. Abstr.* D. 15 Much of the area has a polyculture of nuts, fruits and vines. 1973 *Country Life* 21 June 1818/3 The polycultural tradition in Italy may be seen from the fact that Frescobaldi [*sc.* a wine producer] have considerable tree plantations and 70,000 olive trees. 1974 *Oryx* XII. 358 A polyculture of wild ungulates can be likened to a sophisticated system of crop rotation at a secondary level.

polycyclic (pǫlisai·klik), *a.* [f. POLY- + CYCLIC *a.*] † **1.** *Math.* (See quots.) *Obs.*

1869 W. THOMSON in *Trans. R. Soc. Edin.* XXV. 253 When the function is cyclic with reference to several different mutually irreconcilable circuits, it is called polycyclic. *Ibid.*, Irrotational motion may be either acyclic or cyclic. If cyclic it is monocyclic if there is only one distinct circuit, or polycyclic if there are several distinct circuits, in which there is circulation. 1888 A. B. BASSET *Treat. Hydrodynamics* I. iv. 74 If φ be a polycyclic velocity potential, the circulation round any closed curve, which does not cut any of the barriers is..zero. *Ibid.* 75 Every polycyclic function may be expressed as the sum of..monocyclic functions.

2. (In Dict. s.v. POLY- 1.)

3. *Chem.* Having more than one ring of atoms in the molecule. Also as *sb.*, a polycyclic compound.

1903 *Chem. News* 13 Mar. 130/2 (*heading*) A method for the synthesis of polycyclic hydrocarbides. 1909 C. A. KEANE *Mod. Org. Chem.* xv. 414 The most important of these polycyclic substances are built up of benzene rings. 1943 *Endeavour* Jan. 32/2 Carcinogenic potency is by no means confined to polycyclic compounds. 1968 I. L. FINAR *Org. Chem.* (ed. 4) II. xi. 462 This is only possible if rings B and C are fused together in a *trans* manner (*cf.* polycyclics, §11 d. IV). 1971 *Daily Tel.* (Colour Suppl.) 28 May 27/2 Other potential dangers in exhaust fumes are polycyclic hydrocarbons—which can cause cancer in laboratory animals. 1976 *Nature* 9 Sept. 93/1 Binding of carcinogenic polycyclics to DNA..has been well established.

4. *Electr. Engin.* Involving the simultaneous transmission along a conductor of currents of different frequencies and voltages.

1903 *Jrnl. Inst. Electr. Engin.* XXXII. 751 The advantages of low frequency for power work and of high frequency for lighting are combined in this polycyclic system. 1906 A. RUSSELL *Treat. Theory Alternating Currents* II. xvii. 478 (*heading*) Three phase polycyclic system of distribution. 1913 S. P. SMITH tr. *La Cour & Bragstad's Theory & Calculation Electr. Currents* xvi. 289 The object of the polycyclic system..is to simultaneously transmit electrical energy by means of currents at different pressures and frequencies through one and the same conductor, and to distribute the same without their affecting one another.

5. *Geol.* [ad. F. *polycyclique* (E. Argand 1922, in *Compt. Rend. XIII Session, Congrès Géol. Internat.* (1924) I. 365).] Produced by or having undergone many cycles, esp. of erosion and deposition. Cf. *multicyclic* adj. s.v. *MULTI- 1 a.

1935 H. BAULIG *Changing Sea Level* ii. 13 (*caption*) Block-diagram of a polycyclic valley. 1947 *Trans. R. Geol. Soc. Cornwall* XVII. 341 Polycyclic forms in the valleys still witness the later Pliocene stillstands despite intense peri-glacial solifluction. 1958 [see *MORPHOMETRY]. 1968 R. W. FAIRBRIDGE *Encycl. Geomorphol.* 554/2 A lateritic crust..reflects a polycyclic regime, usually as a result of repeated alternation of hot, humid conditions..with dry, evaporating conditions. 1975 *Nature* 13 Feb. 502/3 Such sediments may incorporate.. duplicated or polycyclic layers.

6. *Biol.* Producing several generations during a year by sexual reproduction.

1957 *New Biol.* XXIII. 58 These polycyclic populations are found in complex environments like ponds where there are successive outbursts of different planktonic algae which cause fluctuations in the amount of food available. 1965 J. JOHNSTON tr. *Danilevskii's Photoperiodism & Seasonal Devel. Insects* vii. 194 They are all potentially polycyclic, and in southern provinces may have several generations a year. 1970 *Amer. Naturalist* CIV. 398 In most polycyclic insects of the temperate zone, long days cause continuous development, while short days cause diapause.

polycythæmia, -ic, polydentate: see *POLY- 1. **polyde(s)oxyribo-:** see s.v. *POLYRIBO-. **polydiene:** see *POLY- 2.

polydisperse (pǫlidispə̄·ɹs), *a.* [ad. G. *polydispers* (W. O. Ostwald *Grundriss d. Kolloidchem.* (ed. 2, 1911) i. 39), f. *poly-* POLY- + *dispers* disperse (cf. *DISPERSE *v.* 9).] Existing in the form of or containing colloidal particles (which may be macromolecules) having a range of sizes; applied esp. to macromolecular substances in which there is a simple distribution of particle size (or occas. some other specified physical property) with one peak; also applied to such a property or its distribution. Also **polydispe·rsed** *a.* Cf. *POLYMOLECULAR *a.* d.

1915 M. H. FISCHER tr. *Ostwald's Handbk. Colloid-Chem.* i. 35 These systems in which the disperse phase is composed of particles having different degrees of dispersion may be called polydisperse systems. 1938 *Nature* 4 June 1001/2 The specific polysaccharide of type I pneumococcus is polydisperse with an average weight of approximately 225,000. 1941 *Jrnl. Franklin Inst.* CCXXXI. 1 Fine powders and colloidally dispersed material are also typically polydispersed. 1948 *Chem. Abstr.* XLII. 1787 (*heading*) Birefringence of flow in polydispersed mixtures. 1962 H. BLOEMENDAL et al. in A. Pirie *Lens Metabolism Rel. Cataract* 303 Prolonged dialysis of 7 M urea-treated α-crystallin against 1·25 M urea results in the appearance of a polydisperse peak with sedimentation coefficient of 10 S. 1963 *Nature* 16 Nov. 665/2 A preparation may be monodisperse with respect to one parameter, polydisperse with respect to a second and heterogeneous with respect to a third; the foregoing terms are only meaningful if the parameter measured is..specified. 1968 H. HARRIS *Nucleus & Cytoplasm* iii. 56 Preparations of the total RNA in the cell cytoplasm..did not show any fraction which had either the kinetic characteristics or the polydisperse sedimentation of the rapidly labelled nuclear RNA. 1973 *Sci. Amer.* June 112/1 With a calibrated grid in the viewing field of the microscope one can measure the relative sizes of the particles in a polydisperse suspension (a suspension containing particles of widely varying size).

Hence **polydispe·rsity**, the condition or property of being polydisperse.

1927 H. S. VAN KLOOSTER tr. *Kruyt's Colloids* x. 154 Viewed in the ultramicroscope, this sol presents a very colorful picture which reveals the poly-dispersity very plainly. 1936 *Trans. Faraday Soc.* XXXII. 49 It appears that determinations of polydispersity by measurement of sedimentation velocity can only be effected in extremely dilute solutions. 1963 *Nature* 16 Nov. 665/2 Some degree of random variability in secondary or tertiary structure may confer polydispersity with respect to frictional coefficient on some proteins. 1967 MARGERISON & EAST *Introd. Polymer Chem.* ii. 47 The value of the parameter $\overline{M}_w/\overline{M}_n$ in a practical case thus gives a measure of the range of molecular weights in the sample;..a value of two or greater indicates considerable polydispersity. 1976 [see *POLYMOLECULARITY].

polyelectrolyte (pǫli,ĭle·ktrǒləit). *Chem.* [f. POLY- + ELECTROLYTE.] A substance which consists of large, usu. polymeric molecules containing several ionizable groups.

1947 *Jrnl. Polymer Sci.* II. 12 By using salts of quaternary nitrogen compounds, polyelectrolytes can be made which are soluble in organic solvents. 1959 *Times* 3 Dec. 9/4 There are water softeners using poly-electrolyte membranes with an output of 50 cubic feet an hour. 1964 *Biophysical Jrnl.* IV. I. Suppl. 10 Many important plant products such as the pectic acids of fruit jellies and alginic acid of seaweed are typical polyelectrolytes. 1968 *New Scientist* 26 Dec. 703/3 Pure poly-electrolytes also promise to be valuable additives in many foodstuffs and beverages. 1973 *Nature* 2 Nov. 34/1 Polyelectrolytes are increasingly being used as flocculants for electrostatically stabilised dispersions of colloidal particles.

Hence **po:lyelectroly·tic** *a.*

1948 *Science* 19 Nov. 548/2 We are thus led to the assumption that the high viscosity of polyelectrolytic solutions and its concentration dependence are due to the presence of the high charge density at the polyions. **1958** *New Scientist* 29 May 75/1 Some of the defence mechanisms of the living body are polyelectrolytic in nature.

polyembryony. Add: (Further examples.) Also *Zool.*, the development of more than one embryo from a single egg; **polyembryonic** *a.* (examples).

1906 *Science* 21 Dec. 813/2 Polyembryony [in *Encyrtus* eggs] reaches its greatest intensity..when the young larvæ of the *Hyponomeuta* leave their winter shelter. **1912** *Anatomischer Anzeiger* XLI. 369 (*heading*) The demonstration of polyembryonic development in the armadillo. **1925** A. D. IMMS *Gen. Textbk. Entomol.* I. 154 Polyembryony is met with among insects in certain parasitic families of Hymenoptera. **1947** *Arch. Path.* XLIV. 501 There has been much speculation concerning the cause of polyembryony (monozygotic multiple birth) as it occurs regularly in the armadillo. **1956** *Nature* 28 Jan. 191/1 An examination of the seeds of different species of *Cassia* ..has shown the occurrence of polyembryony to be geographically widespread. **1969** R. F. CHAPMAN *Insects* xix. 377 The effect of polyembryony is to increase the reproductive potential of the insect, but the net effect is not always much greater than in related monoembryonic species because polyembryonic forms tend to lay fewer eggs.

polyendocrine, -endocrinopathy: see *POLY- 1. **polyene:** *POLY- 2. **polyenergid:** *POLY- 1. **polyenic, -enoic:** *POLY- 2.

Polyergus (pǫli,ǝ·ɹgɒs). *Ent.* Also **polyergus.** [mod.L. (P. A. Latreille 1804, in *Nouveau Dict. Hist. Nat.* XXIV. 179), f. Gr. πολύεργος hard-working.] A slave-making ant of the genus so called, found in Europe and North America; = AMAZON-ANT.

1882 J. LUBBOCK *Ants, Bees, & Wasps* vii. 180, I presented a slave of *Polyergus* with a dead fly pinned down. **1908** *Westm. Gaz.* 21 Jan. 2/1 The polyergus seems to lose even the faculty of making a nest. **1924** J. A. THOMSON *Sci. Old & New* xiii. 72 Slave-keeping is much more marked among the Amazon Ants, of which the European Polyergus is a good representative. **1945** C. P. HASKINS *Of Ants & Men* ix. 175 Not only is *Polyergus* little interested in its nesting site, but it is compelled to abandon it and to move about at frequent intervals. **1954** BORROR & DELONG *Introd. Study Insects* 729 The *Polyergus* ants are often called amazons.

polyester (pǫli,e·stǝɹ). [f. POLY- + *ESTER.] Any polymer in which the units are joined by the ester linkage, —CO·O—; also (*a*) (more fully *polyester fibre*), a man-made fibre consisting of a polyester; (*b*) (more fully *polyester resin*), any of numerous synthetic resins or plastics consisting of or made from a polyester, different kinds of which are used as fibres or films, in paint, and as moulding materials or reinforced plastics. Freq. *attrib.*

1929 [see *POLYAMIDE]. **1935** C. ELLIS *Chem. Synthetic Resins* II. l. 1002 Strong, oriented fibers were obtained from polyesters of molecular weights greater than 9330. **1952** *Jrnl. R. Aeronaut. Soc.* LVI. 707 Working with glass fibres, the Americans have used the polyester or 'contact resins' almost exclusively. **1955** *Times* 4 May 21/3 Production items based on polyester resins and glass fibres..include panels and bodies for the transport and aircraft industries, mouldings for boat hulls, [etc.]. **1958** *Times Rev. Industry* Feb. 77/1 The advent of resins of the polyester and epoxide types..meant that laminates could be produced without the application of high pressures. **1958** *Engineering* 7 Mar. 311/1 Cellulose-acetate and, more recently, polyester are used at present for data-recording tape bases. **1958** *Manch. Guardian* 25 Sept. 1/3 Polyester fibre was discovered in 1941 by Mr. J. R. Whinfield and Dr J. T. Dickson in the laboratories of the Calico Printers' Association. **1964** *Which?* Aug. 253/3 Fabrics made from polyester tend to attract dirt but can be easily washed. **1968** J. IRONSIDE *Fashion Alphabet* 210 Polyester fibres have very similar properties to the polyamides but they are perhaps even more durable. **1969** L. S. MOUNTS in W. R. R. Park *Plastics Film Technol.* v. 135 Polyester films have great strength and good aging characteristics. **1973** *Materials & Technol.* VI. viii. 568 The polyesters are usually marketed dissolved in the cross-linking monomer in the form of a syrup, to which inhibitors are added. **1975** *Guardian* 27 Jan. 16/3 Today the accent is on..blends of a polyester fibre with cotton or polynosic rayon. **1977** *R.A.F. News* 22 June–5 July 2 (Advt.), 'Tootal' Polyester/Cotton Wedgwood Blue Shirts.

Hence **po:lyestera·mide,** any polymer which contains both ester and amide linkages, esp. any of various rubbery materials of this type which are usu. made by mixed condensation reactions and can be drawn into fibres; **po:lyesterifica·tion,** a reaction or process which yields a polyester.

1932 *Jrnl. Amer. Chem. Soc.* LIV. 1560 The polyesterification thus consists in a series of intermolecular couplings resulting in the formation of progressively longer chains. **1943** H. FOSTER *U.S. Pat.* 2,333,922 1/1 The above objects are accomplished through the production of electrically insulated conductors in which the insulation is a polyester-amide which has been treated with polyisocyanate. **1950** *Thorpe's Dict. Appl. Chem.* (ed. 4) X. 45/2 Among the second class of amide interpolymers are the polyesteramides. These are made by condensing together di-reactive amide-forming and ester-forming components, or more specifically by condensing di-reactive components containing carboxy groups in proportions equivalent to the sum of the amine and hydroxyl groups. **1958** *Technology* Mar. 12/3 Not long after Carothers' pioneering work, J. R. Whinfield took up the study of similar polyesterification processes, the main idea being to use starting materials which might give some sense of rigidity to the high-polymer substances made by these reactions. **1974** K. F. HEINISCH *Dict. Rubber* 508/2 The production of the polyesteramide is performed without a deficiency of dicarboxylic acid, as for Vulcolan. **1976** *Nature* 24 June 658/2 This demonstrated that ester interchange reactions were slow compared with polyesterification in the conditions used to make commercial alkyd resins.

polyethenoid: see *POLY- 2.

polyether (pǫli,ī·þǝɹ). [f. POLY- + ETHER.] Any of a variety of polymers in which the repeating unit contains an ether linkage, C—O—C, many of which are used commercially, esp. as plastic foams and epoxy resins. So *polyether foam*.

1922 *Chem. Abstr.* XVI. 1741 (*heading*) Polyethers of trimethyleneglycol. **1962** *Listener* 8 Mar. 451/1 The less expensive polyether foam is edging its way in. **1966** [see *EPOXY]. **1968** *Times* 28 Mar. 9 All three versions are available in either Burma teak, or prime beech and are fitted with specially composed polyether foam cushions. **1972** *Physics Bull.* Oct. 582/3 A fluorocarbon polyether, recently used as a rotary pump lubricant.., has allowed pump operation at above 100 °C. **1975** P. BROWNE *Bodywork Maintenance* vi. 80/1 Polyether is mainly used as stuffing in cheap upholstery.

polyethism: see *POLY- 1.

polyethylene (pǫli,e·þilīn). [Back-formation from *POLYETHYLENIC *a.* Cf. F. *polyéthylène* (Berthelot 1867, in *Jrnl. de Pharm. et de Chim.* VI. 28).] **a.** *Chem.* Used *attrib.* in the names of polymeric substances prepared from derivatives of ethylene, as † **polyethylene alcohol** = next; **polyethylene glycol,** any polymer of ethylene glycol; *esp.* any of a series of water-soluble oligomers and polymers which have the structure $H—(OCH_2CH_2)_n—OH$, of which the lower members are used as solvents and the higher esp. as waxes; **polyethylene oxide,** any polymer having the structure $—(OCH_2CH_2)_n—$; *esp.* any of the thermoplastics of high molecular weight made from ethylene oxide or ethylene glycol (usu. by copolymerization of both) and used esp. as water-soluble films; **polyethylene terephthalate,** a thermoplastic condensation polymer of ethylene glycol and terephthalic acid which is widely used to make polyester fibres.

1862 Polyethylene-alcohol [in Dict. s.v. POLY- 2 a]. **1884** ROSCOE & SCHORLEMMER *Treat. Chem.* III. II. 37 Ethylene glycol,..in its turn, combines with the free ethylene oxide to form polyethylene alcohols. **1886** E. F. SMITH tr. *V. von Richter's Org. Chem.* 258 (*heading*) Polyethylene glycols or alcohols. **1947** R. L. WAKEMAN *Chem. Commercial Plastics* xxvi. 779 Polyethylene glycols are..soluble in both water and aromatic hydrocarbons. **1971** P. TOOLEY *High Polymers* ii. 43 Polyethylene glycols..vary from oily liquids to waxy solids according to their degree of polymerization. **1930** *Chem. Abstr.* XXIV. 3003 On cooling the fused polyethylene oxides they crystallize like paraffin. **1935** C. ELLIS *Chem. Synthetic Resins* II. l. 991 All the polyethylene oxides were shown by X-ray diagrams to be crystalline and viscosity measurements indicated a meandering structure for the molecule. **1969** L. S. MOUNTS in W. R. R. Park *Plastics Film Technol.* v. 140 The solubility of PVA and polyethylene oxide films increases with increasing water temperature. **1946** ASTBURY & BROWN in *Nature* 14 Dec. 871 Polyethylene terephthalate (terylene) gives a well-oriented x-ray fibre diagram. **1967** E. CHAMBERS *Photolitho-Offset* ix. 118 The more recent introduction of the thermoplastic material polyethylene terephthalate.. as a support for photographic emulsions has increased considerably the dimensional stability of films. **1967** *Times Rev. Industry* June 70/3 Polysulphones, polyphenylene oxide and polyethylene terephthalate..are claimed to have improved creep resistance and temperature stability.

b. Chiefly *N. Amer.* = *POLYTHENE. Also *attrib.* and *Comb.*

1939 *Plastics* III. 289/1 Compounding polyethylenes with polyisobutylenes has yielded some very useful compounds. **1942** *Electronic Engin.* Mar. 668/2 Polyethylene and polyisobutylene..are finding extensive use as flexible coatings for wires intended for low-loss work. **1956** *Nature* 25 Feb. 393/1 The reaction was also measured in a polyethylene flask. *Ibid.* 3 Mar. 440/1 A polyethylene-covered magnetic stirring bar. **1966** T. PYNCHON *Crying of Lot 49* iii. 56 Suddenly, a dozen boats away, a form, covered with a blue polyethylene tarp, rose up. **1967** *Economist* 30 Sept. 1225/1 Heavy over-capacity seems to be on the way in Europe in the production of high-density polyethylene (polythene). **1968** J. UPDIKE in *Transatlantic Rev.* XXVIII. 6 The hydra's swollen coelenteron..veiled the preceding meals like polyethylene film protecting a rack of dry-cleaned suits. **1969** L. S. MOUNTS in W. R. R. Park *Plastics Film Technol.* v. 122 Polyethylene films also are used in many nonpackaging applications. **1969** *Daily Colonist* (Victoria, B.C.) 7 Dec. 26/5 A new television antenna designed for boats... An entire signal receiving system anchored in polyethylene foam. **1976** *Country Life* 22 Jan. 211/1 A fabric of the future... Spun, bonded polyethylene, a synthetic material made from bonded, random fibres.

polyethylenic (pǫ:li,eþilī·nik), *a. Chem.* [ad. F. *polyéthylénique* (A. Wurtz 1860, in *Compt. Rend.* L. 1195), f. *poly-* POLY- + *éthylène* ETHYLENE + *-ique* -IC.] †**1.** Used to designate polymeric compounds prepared from derivatives of ethylene. Cf. *POLYETHYLENE a. *Obs.*

1860 A. WURTZ in *Chem. News* 25 Aug. 121/1 Glycolic, lactic, and oxalic acids..are acids of very simple constitution; but such is not the case with those obtained by oxidising the complex glycols, which I have called polyethylenic alcohols. **1878** C. M. TIDY *Handbk. Mod. Chem.* xxiii. 574 By heating glycol with ethylenic oxide in sealed tubes, a series of compounds called polyethylenic glycols (or alcohols) are formed. **1895** *Jrnl. Chem. Soc.* LXVIII. I. 320 Ethylenic oxide and sodium ethoxide.. give a product which boils at 110–120°..; this..must therefore be regarded as a complex ethylic polyethylenic oxide.

2. Polyunsaturated; of the nature of polyunsaturation.

1928 *Proc. R. Soc.* A. CXXII. 564 Two polyethylenic acids.., namely, linoleic acid of soya bean or cotton-seed oil..linolenic acid of linseed oil. **1958** *Jrnl. Sci. of Food & Agric.* IX. 777 The polyethylenic C_{20} and C_{22} acids, so characteristic of the animal glycerophosphatides. **1964** *Oceanogr. & Marine Biol.* II. 176 The pattern of polyethylenic unsaturation in the fatty acids of marine invertebrates is probably the same as that in fish lipids. **1965** E. F. JANSEN in Bonner & Varner *Plant Biochem.* xxv. 655 The hexahydromatricaria ester is included to illustrate the occurrence of homologous polyethylenic compounds.

polyformal(dehyde: see *POLY- 2.

Polyfoto (pǫ·li,fōʊto). Also **-photo.** [f. POLY- + *foto*, alteration of PHOTO.] A proprietary name for a kiosk in which a person can sit and have a number of photographs taken of himself in quick succession (now usu. automatically). Also *transf.* and as *vb. trans.*

1938 *Trade Marks Jrnl.* 2 Feb. 131/1 Polyfoto... Photographs. Polyfoto (International) Limited,..London,..merchants. **1945** 'A. GILBERT' *Don't open Door* xiii. 105 'Have you got a photograph of the girl?' 'She was polyfotoed the other day to please me.' **1962** —— *No Dust in Attic* v. 63 They ran against Miss Malpas coming out of the Polyfoto kiosk... Out of forty-eight positions there must be one that would reproduce satisfactorily. **1964** G. SIMS *Terrible Door* xxv. 132 My husband was one of the first photographers to make strips of little films, rather like Polyphotos nowadays. **1965** E. J. HOWARD *After Julius* xviii. 277 He got himself polyphotoed..for his passport. **1976** 'A. HALL' *Kobra Manifesto* iv. 44 Worn Polyphoto of current girl friend.

polyfunctional(ity: see *POLY- 1.

polygala (pǫli·gălă). [mod.L., f. Gr. πολύ much + γάλα milk.] An annual or perennial herb or a shrub belonging to the large genus so called, which is a member of the family Polygalaceæ and is widely distributed in most regions of the world; = MILKWORT 1 (in quot. *a* 1661 = SAINFOIN).

1578 [see MILKWORT 1]. *a* **1661** FULLER *Worthies* (1662) Kent 57 Saint-Foine or Holy-hay..otherwise called *Polygala*, which I may English *much Milk*, as causing the cattle to give abundance thereof. **1671** [see MILKWORT 1]. **1823** *Curtis's Bot. Mag.* L. 2437 The common Polygala is so very variable a plant. **1840** [see *IVY-BERRY b]. **1870** W. ROBINSON *Wild Garden* I. 9 The small sapphire buds of the alpine Polygala. **1936** *Discovery* Feb. 48/2 Large anchusas were beginning to flower.., huge polygala, and I don't know what else. **1977** S. DEALLER *Wild Flowers for Garden* iv. 67, I think of the Polygalas as generally trouble free, non-aggressive, slightly vulnerable plants.

polygamical, *a.* Delete (? *obs.*) and add later example.

1914 CHESTERTON *Flying Inn* 69 Why should you shrink then, ladies, from this great polygamical experiment?

polygene (pǫ·lidʒīn). *Genetics.* [Back-formation from *POLYGENIC *a.* 3; cf. *GENE.] A gene whose individual effect on the phenotype of a single organism is too small to be observed, but which can act together with other, non-allelic polygenes to produce observable phenotypic variation in a quantitative character.

1941 *Jrnl. Genetics* XLI. 163 The inbreeding this line had undergone had in fact made it homozygous for the hair-number polygenes. **1949** [see *POLYGENIC *a.* 3]. **1961** *Lancet* 9 Sept. 601/2 One might envisage a system of pleiotropic masculinising polygenes located more or less randomly along the whole of the Y chromosome. **1973** *Nature* 21/28 Dec. 498/2 Techniques for the location of polygenes, developed in *Drosophila* but also very successful in mice and wheat ,have not been applied to human material.

polygenesis. Add: **b.** *Linguistics.* The theory that there is a plurality of independent sources for languages. Opp. *MONOGENESIS 3.

1936, 1949 [see *MONOGENESIS 3]. **1979** *Amer. Speech* 1978 LIII. 247 My gut feeling is that radical polygenesis is impossible. It would require either a biologically unrealistic degree of parallelism, or else that languages be much more varied in their groundplans than in fact they are.

c. *gen.*

1962 G. KUBLER *Art & Archit. Anc. Amer.* i. 11 (*heading*) Diffusion or polygenesis?

polygenetic, *a.* Add: **2.** (Further examples.)

1889 *Amer. Jrnl. Sci.* XXXVII. 431 The present topography..is an uncompleted advance in a second cycle of development, with recent complications by glacial action and slight changes of level. Like mountains of repeated growth, this topography may be called 'polygenetic'. **1943** *Ibid.* CCXLI. 486 Polygenetic complex soils and Late Quaternary alluvial bodies may preserve duplicate records of climatic changes for a given region. **1969** BENNISON & WRIGHT *Geol. Hist. Brit. Isles* ix. 201 In the thrust block of Roman Fell, the Roman Fell Beds..rest on the Polygenetic Conglomerate of unknown age but with pebbles including andesites probably derived from the Old Red Sandstone. **1975** *Nature* 27 Mar. 369/2 Coasts and mountains are treated separately as polygenetic landscapes, the former being the zone of contact between two distinct environments and the latter the areas within which internal uplift conflicts markedly with external erosion.

b. (See quot. 1959.) [After G. *polygen* in this sense (A. Stübel *Die Vulkanberge von Ecuador* (1897) III. 352).]

[**1903** A. GEIKIE *Text-bk. Geol.* (ed. 4) I. 322 A volcano formed in this way he terms monogene; while where it has been built up by the gradual accumulations of successive eruptions he calls it polygene.] **1959** A. A. G. SCHIEFERDECKER *Geol. Nomencl.* 239 *Polygenetic volcano*, formed by several volcanic outbursts. **1962** E. A. VINCENT tr. *Rittmann's Volcanoes* iii. 114 In the case of the great polygenetic volcanoes, the old topography is completely buried. **1976** *New Scientist* 9 Sept. 527/1 Those which erupt many times are called polygenetic volcanoes.

c. *Linguistics.* Deriving historically from a number of sources; having more than one antecedent.

1952 *Archivum Linguisticum* IV. 93 The outstanding trait in the history of -(*i*)*ego*..is its polygenetic character.

polygenic, *a.* Add: **3.** *Genetics.* [ad. G. *polygen* (L. Plate *Vererbungslehre mit besonderer Berücksichtigung des Menschen* (1913) ii. 75).] Of, pertaining to, or determined by polygenes.

[**1927** *Zeitschr. für Induktive Abstammungs- und Vererbungslehre* XLIII. 331 The discovery of polygenous characters, especially the phenomenon of polymeria discovered by Nilsson Ehle, made it necessary to recognize, that at least a great many characters are determined not by one but by several and even many genes.] **1941** *Jrnl. Genetics* XLI. 160 Qualitative variation is usually monogenic or digenic in inheritance. Cases of trigenic and tetragenic inheritance are known, but are relatively rare. In contrast with these, quantitative variation may be said to be polygenic, and this term will be adopted. **1949** DARLINGTON & MATHER *Elem. Genetics* iii. 66 These genes are inherited in the mendelian way, but their differences have effects which are small in relation to those of non-heritable agencies (or at least in relation to the total variation), similar to one another and supplementary to one another... Such a set of genes constitutes a polygenic system, and its individual members may be conveniently termed polygenes. **1954** *Antiquity* XXVIII. 197 A genetic analysis of such polygenic characters as bone size and shape. **1961** *Brit. Med. Bull.* XVII. 241 All polygenic inheritance is polymeric, but not all polymeric inheritance is polygenic. **1971** *Heythrop Jrnl.* Apr. 171 The analysis of complex (polygenic) hereditary determination is most commonly conducted by means of twin studies. **1978** *Dædalus* Spring 215 Much more complex choices will confront us when we think we know enough to tackle polygenic traits which may have important expression in emotional behavior or intellectual capacity.

Hence **polyge·nically** *adv.*, by means of or with regard to polygenes.

1943 *Biol. Rev.* XVIII. 61 Polygenically controlled differences are quantitative rather than qualitative and do not lead to the sharp segregation shown by the more familiar genetical differences. **1957** *Heredity* XI. 392 Such an approximately even distribution..might even be anticipated given that any sizeable block of heterochromatin near the centromere is, mitotic length for length, about as polygenically active as the euchromatin of more distal regions. **1976** *Nature* 23 Sept. 317/1 A common feature of many selection experiments, when polygenically determined traits are involved, is a reduction in the reproductive fitness of the selected strains.

polyglacial(ism, -ist: see *POLY- 1.

polyglot, *a.* and *sb.* Add: **A.** *adj.* **2. b.** Characterized by the use of a plurality of languages, or of elements derived from a plurality of languages.

1952 W. D. JACOBS *William Barnes* i. 12 Dorset gave Barnes the material by which to judge the polyglot English of our day. **1957** *Thought* XXXII. 240 In his early polyglot days..he [*sc.* T. S. Eliot] wrote often with odd patches and tags of French, German—and even Sanskrit —as well as Latin and Greek. **1965** *Economist* 3 Apr. 44/2 The use in, say, Cyprus of a polyglot (or mixed-manned) force. **1972** W. B. LOCKWOOD *Panorama Indo-European Lang.* 214 Here some 900,000 of them compete with the Gujaratis for the position of second largest linguistic group in that remarkably polyglot city [*sc.* Bombay].

So **po·lyglotter** (*nonce-wd.*), a polyglot person; **polyglo·ttery** = POLYGLOTTISM; **polyglo·ttically** *adv.* = POLYGLOTTALLY *adv.*

1910 W. J. LOCKE *Simon* vi. 71 Mr. Papadopoulos polyglottically acknowledged the honour I had conferred upon him. **1912** 'R. DEHAN' *Between Two Thieves* 616 That white haired Polyglotter in the shabby togs..is a queer kind of chap. **1915** *Singapore Free Press* 14 Jan., If its polyglottery were all that was wrong with it [*sc.* Austria-Hungary], it still might be possible to jog along in a sort of mutual unintelligibility. **1931** *Time & Tide* 8 Aug. 943 (*heading*) Polyglottery [*review of W. Gerhardi's* Memoirs of a Polyglot]. **1962** *Times Lit. Suppl.* 14 Sept. 685/3 Rebuses abound, as do polyglottery (classical and modern), Finnegans wakefulness, and enormous catalogues *à la* Rabelais.

polyglutamic, -glycine, -glycol(ic: see *POLY- 2.

polygon, *sb.* and *a.* Add: **A.** *sb.* **2. b.** *Physical Geogr.* One of the approximately polygonal figures characteristic of patterned ground (cf. *PATTERNED *ppl. a.* b).

1913 *Amer. Jrnl. Sci.* CLXXXVI. 459 The shale surfaces are flat, the polygons not being concave upward as is commonly the case in the Mauch Chunk and Newark shales. **1914** T. THORODDSEN in Rosenvinge & Warming *Bot. Iceland* I. I. 258 The surface is divided into squares or less regularly formed polygons, by bands of small stones or gravel, while the clay of the interior of the squares or polygons is destitute of stones. **1921** *Geogr. Jrnl.* LVIII. 308 The snow falls, and when it melts, the spare gravel that remains on the clay substratum has taken upon itself the same system of polygons..which we see in Polar lands. **1950** [see *PATTERNED *ppl. a.* b]. **1960** B. W. SPARKS *Geomorphol.* xiv. 318 Boulders weighing..over one hundredweight..have been observed in some Greenland polygons. **1974** *Environmental Conservation* I. 58/1 He found no serious thermokarst conditions along the trails, except where ice-wedges of polygons were crossed.

polygonal, *a.* (*sb.*) Add: **1. b.** Containing or forming polygonal features.

1924 *Geogr. Jrnl.* LXIII. 213 One is apt to mistake this polygonal system of ice-wedges for a continuous sheet of ground-ice. **1930** *Ibid.* LXXVI. 417 Polygonal shrinkage fissures in clay are widespread. **1960** B. W. SPARKS *Geomorphol.* iii. 33 Basalt..often displays a very well defined, polygonal, vertical joint pattern. **1974** T. L. PÉWÉ in Smiley & Zumberge *Polar Deserts & Mod. Man* iii. 42/1 One of the most widespread geomorphic features associated with permafrost in the polar deserts is the microrelief pattern of the ground generally called polygonal ground or tundra polygons.

polygonboden (pǫ·ligǫ̆nbōʊ:dǝn). *Physical Geogr.* Const. as *pl.* [a. G. *polygonboden*, lit. 'polygon ground', tr. Sw. *rutmark* chequered ground (F. R. Kjellman 1879, in *Öfversigt af K. Vetenskaps-Akad. Förhandl.* XXXVI. IX. 11).] Polygons (sense *A. 2b).

1924 *Geogr. Jrnl.* LXIII. 226 Schimper..saw well-developed 'polygonboden' and arrangement of vegetation with respect to the polygons, in the high plateaux of the Pamirs. **1933** *Bull. Geol. Soc. Amer.* XLIV. 949 Numerous minor forms, including polygonboden, increase the contrasts between alpine surfaces and those of lower elevations. **1947** *Geogr. Rev.* XXXVII. 640 Plainly visible..are the polygenboden, a phenomenon of frozen ground. **1970** MACDONALD & ABBOTT *Volcanoes in Sea* viii. 150/1 The fragments..are arranged in stripes of varying coarseness, or in polygonal patterns ('Polygonboden'), that..are the result of frost action.

polygonization (pǫ:ligǫ̆naizē¹·ʃǝn). *Metallurgy.* [f. POLYGON + -IZATION.] The formation of smaller grains within the grains of a metal as a result of the migration of dislocations following deformation and annealing.

1948 R. W. CAHN in *Rep. Conf. Strength of Solids, 1947* 137 The x-ray and microscopical results together lead to an interpretation as follows: the deformation leads to elastic bending of glide lamellae; when the crystal is annealed, the lamellae undergo 'polygonization'. **1949** —— in *Jrnl. Inst. Metals* LXXVI. 136 It is suggested that the process be named polygonization, to distinguish it from other forms of recrystallization. The name derives from the fact that the crystallographic direction which coincides with the wire axis, while bent along an arc before annealing, becomes part of a polygon afterwards. **1950** *Engineering* 5 May 499/3 Polygonisation could be obtained in hexagonal metals only after plastic bending and not after deformation in tension. **1957** D. MCLEAN *Grain Boundaries in Metals* vii. 204 The word 'polygonization' has lost its original geometrical significance and now refers to the process of dislocations gathering into stable arrays. **1976** COTTERILL & MOULD *Recrystallization & Grain Growth in Metals* iv. 74 The basic differences between the formation of a substructure by polygonization, during annealing, and the formation of subgrains, directly, during cold-work have been summarized by Byrne.

polygonize (pǫ·ligǫ̆naiz), *v. Metallurgy.* [Backformation from prec.] *intr.* To undergo or give rise to polygonization.

1949 *Jrnl. Inst. Metals* LXXVI. 138 Zinc crystals bent at −180° C. would not polygonize on subsequent annealing. **1967** A. H. COTTRELL *Introd. Metall.* xxi. 402 Dislocations of opposite signs come together and annihilate each other. Those of the same sign polygonize into well-defined cell walls.

Hence **po·lygonized** *ppl. a.*

1950 *Progress Metal Physics* II. 188 The grains..were in a polygonized condition. **1954** A. R. BAILEY *Text-bk. Metall.* v. 110 (*caption*) Polygonised structure in super-purity aluminium, developed by creep at 200° C. **1976** COTTERILL & MOULD *Recrystallization & Grain Growth in Metals* iv. 74 Although the energy per dislocation is reduced, the total energy of the boundary is increased, and that of the regions away from the boundary is decreased, by the conversion from the deformed structure to the polygonized one.

polygram. Add: **2.** A recording made with a polygraph (sense 3 in Dict. and Suppl.).

1923 W. D. REID *Heart in Mod. Pract.* iii. 60 It is frequently not possible to determine from the polygram whether the tachycardia originates in the auricle or in the ventricle unless the tracing shows the onset or cessation of the paroxysm. **1966** R. J. FERGUSON *Polygraph & Private Industry* vii. 299 Extreme caution must be used in interpreting polygrams.

polygraph, *sb.* Add: **I. 3.** Also used to obtain tracings of other physiological characteristics (such as rates of pulse and respiration, or the electrical conductivity of the skin), and made to serve as a lie-detector. (Earlier and later examples.)

1871 *Lancet* 25 Nov. 739/1 The most direct method for recording the heart's motion is that which we owe to Chauveau and Marey. These ingenious experimenters have supplied us with many instruments adapted for the registration of movements, but none more generally useful than the following, which has been well named the 'polygraph'. **1923** J. A. LARSON in *Jrnl. Exper. Psychol.* VI. 424 A deception test based upon the correlation between the physiological and emotional activities... The technique consists of securing a continuous blood pressure curve (secured by an Erlanger sphygmomanometer or more preferably by a modification of the McKenzie or the Jacquet polygraphs) taken synchronously with a respiratory and a timing curve. **1942** F. E. INBAU *Lie Detection* i. 5 Until 1939 the Keeler Polygraph consisted only of the blood pressure-pulse-respiration units; since then it has been obtainable either with or without a galvanometer unit for recording electrodermal responses. **1959** M. DOLINSKY *There is no Silence* iii. 47 Anxiety causes subtle and involuntary increases in the heartbeat respiration,..and blood pressure which the polygraph records. **1971** *Nature* 9 July 124/2 Instantaneous blood flows to the two hindlimbs..were displayed on a 'Grass P7' polygraph together with the instantaneous and mean (integrated) blood pressure. **1973** *N.Y. Law Jrnl.* 20 Mar., The court then explained that there is no scientific proof that lying heightens anxiety, and that a subject's emotional responses as measured by the polygraph are no proof either of truthfulness or lying. **1976** *Time* 27 Dec. 41/2 Patients spend the night hooked up to a polygraph, a lie-detector-like machine that monitors sleep-related physiological functions (breathing, muscle twitching, rapid eye movement). **1977** *New Yorker* 30 May 29/1 A lot of firms..now routinely give standard polygraph tests to prospective employees.

II. 4. (Earlier example.)

1854 A. G. HENDERSON tr. *Cousin's Philos. of Kant* i. 8 Leibnitz..was led away by a passion for universal knowledge... Wolf endeavoured to bring all the scattered views of the great polygraph to a common centre.

III. 6. *Cryptography.* A group of two or more letters; also, in *Phonetics*, a group of three or more letters expressing a simple sound of speech.

1943 L. D. SMITH *Cryptogr.* iv. 82 A method that represents a distinct departure from Vignère's..is found in polygraph substitution—that is, the substitution of cipher digraphs or trigraphs for the plain-text digraphs or trigraphs. **1959** *Brno Studies in English* I. 18 The following stage..replaced the cumbersome digraphs (and polygraphs) by simple but diacriticized graphemes. **1974** *Encycl. Brit. Macropædia* V. 332/1 In substitution ciphers, the characteristic relative frequencies of single letters, digraphs, and longer polygraphs serve as a basis for the assignment of plaintext equivalents to cipher values.

polygraph (pǫ·ligrɑf), *v.* [f. prec. sb.] **a.** *intr.* To perform (satisfactorily, etc.) when examined with a polygraph. **b.** *trans.* To examine with a polygraph, esp. for truthfulness. Hence **po·lygraphing** *vbl. sb.*

1969 H. H. COOPER *Cave with Two Exits* I. 68 The checks on him were being completed. He polygraphed okay, for what that's worth. **1978** 'W. WINGATE' *Bloodbath* ii. 15 He was..a mystery, and this despite all his debriefings..and polygraphings. *Ibid.* 16 In the two years since his defection, Yazov had been repeatedly polygraphed. **1979** P. FRIEDMAN *Termination Order* (1980) i. 14 Your superiors..will want to polygraph me. *Ibid.* iv. 57 All this nonsense—calling people back for polygraphing, or debriefing or whatever.

polygrapher. Restrict †*Obs.* to sense 2 in Dict. and add later example of sense 1 (still *rare*).

1871 L. B. PHILLIPS *Dict. Biogr. Ref.* 298 Cunaeus, Peter (*Van der Kun*), Dutch savant and polygrapher; 1586–1638.

3. A user of a polygraph (sense 3).

1954 *Reporter* (N.Y.) 8 June 14/3 (*heading*) Eminent polygraphers. **1962** *Harvard Business Rev.* Nov.–Dec. 128/2 The National Board of Polygraph Examiners, an organization of polygraphers..set forth minimum standards for polygraphs. **1973** *Biomed. Engin.* VIII. 155/3 Regarding the opposing views on the skin-resistance channel; many practising polygraphers consider it to be too sensitive an indicator of emotional responses. **1978** 'W. WINGATE' *Bloodbath* ii. 25 Kruger watched the polygrapher flop into the easy chair.

polygraphic, *a.* Add: **5.** Of, pertaining to, or involving a polygraph (sense 3 in Dict. and Suppl.).

1871 *Lancet* 25 Nov. 741/2 When the capsule is pressed firmly upon the skin, the interior of the instrument is converted into a closed chamber, and every shock of the heart expels air out of the chamber into the polygraphic tambour. **1927** *Welfare Mag.* May 667/1 The exact value of this deception technique can only be determined by.. the study of all possible types of pathology associated with polygraphic reactions. **1954** *Reporter* (N.Y.) 8 June 13/2 The fear of being found out and/or conscious efforts to deceive are the main causes of significant reactions in polygraphic tests of deception. **1972** *Science* 29 Sept. 1205/3 Polygraphic sleep data were scored according to standardized criteria.

6. Applied to a cipher in which the letters of the plain text are enciphered or deciphered two or more at a time.

1929 L. S. HILL in *Amer. Math. Monthly* XXXVI. 312 If polygraphic ciphers based upon normal transformations (linear ciphers) prove to be of real interest, we shall indicate a surprising way in which these ciphers may be manipulated easily and quickly. **1967** D. KAHN *Codebreakers* xiii. 406 Such a system is genuinely polygraphic, and its cryptographic security is substantial. *Ibid.* 407 A polygraphic encipherment of this magnitude is possible only with a Hill transformation. **1968** A. SINKOV *Elem. Cryptanal.* iv. 113 A system of cryptography in which a group of *n* plain text letters is replaced as a unit by a group of *n* cipher letters is called a polygraphic system.

So **polygra·phically** *adv.*, by means of a polygraph.

1911 T. LEWIS *Mechanism of Heart Beat* v. 45 (*caption*) An irregularity due to premature contractions arising in the ventricle (the actual events were determined polygraphically). **1970** *Psychophysiol.* VII. 323/1 Plasma samples were obtained from 4 infants during four behavioral states..monitored polygraphically.

polygraphist. [f. POLYGRAPH + -IST.] = *POLYGRAPHER 3.

1954 *Reporter* (N.Y.) 8 June 13/2 Only thirty-six percent of the psychologists (as against seventy-five percent of the polygraphists) agreed with this statement.

polygraphy. Add: **3. c.** The use of a polygraph (sense 3).

1923 W. D. REID *Heart in Mod. Pract.* iii. 63 Polygraph tracings are also taken from the apex of the heart, the carotid artery, and over the liver, but their importance does not warrant description in this limited presentation of polygraphy. **1954** *Reporter* (N.Y.) 22 June 22/2 The most acute current problem in polygraphy..is how to set and maintain professional standards. **1973** *Biomed. Engin.* VIII. 155/3 To readers outside of the field of polygraphy, it will perhaps seem remarkable that electronic recording is not used for all three channels and that other physiological events are not recorded routinely.

polyhaploid: see *POLY- 1.

polyhedral, *a.* Add: **4.** *polyhedral disease* = *POLYHEDROSIS.

1913 *Jrnl. Econ. Entomol.* VI. 482 In Europe there is a tendency to group all the caterpillar diseases which are characterized by the formation of polyhedral bodies under the name of 'polyederkrankheit' or polyhedral diseases. **1971** *Indian Jrnl. Entomol.* XXXIII. 111 (*heading*) Investigations on the nuclear polyhedrosis of *Prodenia litura* Fabricius: I. Nature of the polyhedral disease.

polyhedroid: see *POLY- 1.

polyhedrosis (pǫlihīdrōʊ·sis, -hedrōʊ·sis). *Ent.* Pl. **-oses** (-ōʊ·sīz). [f. POLYHEDR(AL *a.* + -OSIS.] A fatal disease of caterpillars, characterized by the presence of polyhedral virus particles.

1947 *Science* 3 Oct. 323/2 The fragility of the integument and the marked internal liquefaction of tissues, so characteristic of polyhedroses, is absent. **1967** K. M. SMITH *Insect Virology* ii. 8 The polyhedroses are subdivided into nuclear and cytoplasmic diseases. **1973** *Nature* 2 Nov. 5/3 The outbreak tends to end abruptly in the spring of the third year, because of the explosive spread of a nuclear polyhedrosis virus which kills off some of the caterpillars at the end of their first moult (instar).

polyhex: see *POLY- 1.

polyhybrid (pǫlihai·brid), *sb.* and *a. Biol.* [f. POLY- + HYBRID *sb.* and *a.*] **A.** *sb.* A hybrid that is heterozygous at several genetic loci. **B.** *adj.* Of, pertaining to, or characteristic of such a hybrid, or a cross resulting in one.

1911 FARMER & DARBISHIRE tr. *H. De Vries's Mutation Theory* II. 586 The di-polyhybrids are mongrels whose parents differ from one another in respect of two or more elementary characters. *Ibid.* 681/1 (Index), Polyhybrids, 586. **1922** *Hereditas* III. 233 The segregation is evidently polyhybrid showing transgression in one direction. **1965** BELL & COOMBE tr. *Strasburger's Textbk. Bot.* 334 The genetic consequences of crossing both dihybrids and polyhybrids are directly dependent upon the manner in which the chromosomes are distributed at the reductional division.

polyhydramnios (pǫ:lihaidræ·mniǫ̆s). *Obstetrics.* [f. POLY- + HYDR(O- + AMNIOS.] = *HYDRAMNIOS.

1889 R. R. RENTOUL *Causes & Treatm. Abortion* iv. 114 Delore holds that polyhydramnios exists when the quantity exceeds four and two-tenths of a pint. *Ibid.* 115 In a few rare cases of twins, polyhydramnios was found with one child, while oligohydramnios accompanied the other. **1923** E. P. DAVIS *Complications of Pregnancy* xxi. 256 At the stage of fetal development, very often polyhydramnios may be mistaken for ascites as well as ovarian cyst. **1976** *Lancet* 30 Oct. 960/2 She was admitted to hospital with significant polyhydramnios.

polyhydric (pǫlihai·drik), *a. Chem.* [f. POLY- + HYDRIC *a.*] Containing more than one hydroxyl group in the molecule.

1879 *Jrnl. Chem. Soc.* XXXVI. 1033 The action of sulphuric monochloride on monhydric alcohols gives rise to ethereal sulphates.., and a similar reaction takes place with polyhydric alcohols. **1937** *Discovery* Sept. 284/1 On this type of polyhydric phenol resin..a further..discovery has been made. **1946** *Nature* 3 Aug. 155/2 Manganese dioxide is reduced with great ease to form divalent manganese ion by..polyhydric phenols such as quinol, catechol, gallic acid, etc. **1951** I. L. FINAR *Org. Chem.* I. xi. 208 The polyhydric alcohols chemically resemble glycerol in many ways. **1972** *Materials & Technol.* V. ix. 266 The epoxy resins differ from most of the other polyhydric alcohols in that they are polymeric.

polyhydroxy(l, -iamond: see *POLY- 2, 1.

polyimide (pǫli,i·maid). [f. POLY- + IMIDE.] Any polymer in which the units contain imide groups, usu. in the form

$$—CO—\overset{|}{N}—CO—;$$

esp. any of a class of thermosetting resins widely used for heat-resistant films and coatings.

1945 *Brit. Pat.* 570,858, The products are referred to as polyamide-imides. When a tetracarboxylic acid of the above mentioned kind is used each pair of carboxyl groups reacts with an amino group of the diprimary diamine through imide-formation so that the polymer is a polyimide rather than a polyamide-imide. **1955** *U.S. Pat.* 2,710,853 1 This invention relates to a novel group of linear polymeric polyimides,..said polyimides differing from previously known polyimides in properties which are important in injection molding applications. **1965** *New Scientist* 19 Aug. 443/1 Du Pont's polyimide known as H-film, or Kapton, a plastic film that finds its chief use as a high-temperature electrical insulation. **1975** WEN-HSUAN CHANG et al. in M. Lewin et al. *Flame-Retardant Polymeric Materials* x. 442 Decorative coatings for cookware are possible with polyimide coatings because of their ability to withstand direct contact with flame.

polyion(ic, -isobutylene: see *POLY- 1, 2.

polyisocyanate (pǫ:li,aisosai·ăně¹t). [f. POLY- + *isocyanate* s.v. *ISO- b.] Any organic compound containing two or more isocyanate groups; also applied to polymers prepared from such compounds, esp. polyurethanes.

1943 [see *POLYESTERAMIDE]. **1945** *Chem. & Engin. News* 25 Sept. 1615/2 The polyurethanes..are produced by the reaction of polyisocyanates with polyhydroxy compounds such as polyesters. **1958** *Times Rev. Industry* June 37/3 An I.C.I. special product known as Suprasec K is a 50 per cent. solution in ethyl acetate of a newly developed polyisocyanate designed for use in high-quality lacquers for wood, rubber, and other materials. **1959** *Times* 27 Apr. (Rubber Industry Suppl.) p. vi/7 Natural rubber, neoprene and polyisocyanates play the same part in ships as in hotels—for luxury upholstery, non-slipping table covering, fire-resisting foam bedding, floor covering, &c. **1968** E. R. WELLS in C. R. Martens *Technol. Paints, Varnishes & Lacquers* xiii. 205 Urethane is now the accepted description for a group of polymers that are sometimes called polyurethanes, isocyanates or polyisocyanates.

polyisoprene (pǫli,ai·soprīn). [f. POLY- + *ISOPRENE.] Any of the polymers of isoprene, which include the major constituent of natural rubber and some synthetic rubbers very similar to it.

1935 [see *polybutadiene* s.v. *POLY- 2]. **1946** F. MARCHIONNA *Butalastic Polymers* xvi. 483 The sodium butadiene rubber can be admixed with 50 per cent of a polyisoprene or purified natural rubber. **1959** *Economist* 21 Mar. 1101/1 Shell Chemical Corporation has announced that it will shortly begin commercial production of polyisoprene. **1963** H. R. CLAUSER *Encycl. Engin. Materials* 580/1 New 'synthetic natural rubber', cis-polyisoprene is said to have performance characteristics virtually identical with those of natural rubber. **1975** *Sci. Amer.* Dec. 101/1 Polymers that exhibit rubbery behaviour at room temperature include polyisoprene (natural rubber) and polybutadiene (a synthetic rubber).

polykaryocyte(-cytic): see *POLY- 1. **polylectal, -lingual(ism, -lithic, -lithionite:** see *POLY- 1.

polylogue (pǫ·lilǫg). [f. POLY- + -LOGUE.] A discussion between more than two persons.

1941 *Horizon* 15 Mar. 207 English people..tend to think of America..as a polylogue of cigars. **1961** *Lancet* 29 July 251/1 The polylogue here reported is not word-perfect, but the stenographer has tried to catch the to-and-fro of animated discussion. **1964** *Amer. Speech* XXXIX. 206 It is important as another voice in what has..become a 'polylogue' on a matter having wide interest. **1969** P. L. BERGER *Rumor of Angels* iv. 100 The ecumenical movement..has tried to bring Eastern Christianity more and more into the 'polylogue' (if the term will be permitted). **1977** *N.Y. Rev. Bks.* 23 June 30/2 For us there are dispersed, interchangeable 'points of view'; photography is a polylogue.

polylysine: see *POLY- 2.

polymastia (pǫlimæ·stiă). *Med.* Also anglicized as **-masty.** [mod.L., ad. G. *polymastie*, f. Gr. πολυ- POLY- + μαστός breast.] The condition of having more than two breasts (the supernumerary ones being generally very small). Cf. *POLYMASTISM.

1878 *Glasgow Med. Jrnl.* X. 70 Polymastia is not extremely rare in the female sex, but is so in the male. **1904** G. S. HALL *Adolescence* (1908) I. vi. 421 Polymasty or supernumerary breasts occurs about once in five hundred persons. **1928** M. SUMMERS *Discovery of Witches* 40 A large number of cases may be explained by polymastia and polythelia, anatomical divagations which are far commoner than is generally supposed. **1970** H. P. LEIS *Diagnosis & Treatm. Breast Lesions* i. 15 Polymastia, presenting as more than one breast on one or both sides, is due to the persistence of part of the milk ridge and these supernumerary breasts can occur anywhere along the milk line.

Hence **polyma·stic** *a.* and *sb.*, (a person or animal) having more than two breasts; **polyma·stoid** *a.* (*rare*).

1879 *Jrnl. Anat. & Physiol.* XIII. 434 In the dog, which is the only polymastoid animal readily available, the position of the nipples seems to be irregular and unsymetrical [*sic*]. **1891** *Ibid.* XXV. 228 The additional mammary structures do not develop just anywhere; but they appear only in certain definite positions, which almost invariably correspond with those occupied normally by the glands of polymastic animals. **1918** DEAVER & MCFARLAND *Breast* iii. 54 Robert..points out that the mother of his famous case of supernumerary mammae on the outer side of the thigh, was a polymastic. **1934** *Jrnl. R. Anthrop. Inst.* LXIV. 93 The best known examples of the first form are the paleolithic figures at Laussel and the polymastic Diana of the Ephesians. **1943** C. F. GESCHICKTER *Dis. Breast* i. 12 Most authors have found the majority of the accessory mammae below the normally situated pair... Iwai, however, in 511 Japanese polymastics, found 88 per cent of the supernumerary breasts above the normal ones.

polyma·stism. *Med.* [f. as prec. + -ISM.] = *POLYMASTIA.

1886 W. N. PARKER tr. *Wiedersheim's Elem. Compar. Anat. Vertebr.* A. 28 The occasional existence in men of supernumerary teats, and in women of supernumerary mammæ and teats (polymastism and polythelism) is very remarkable. **1903** *Lancet* 28 Feb. 613/2 It may be concluded..that human polymastism is a reversion to a primitive condition in which many glands were developed and many young were brought forth at a birth. **1936** NEAL & RAND *Compar. Anat.* IV. 181 (*caption*) The presence of supernumerary teats (polymastism) in man supports the theory of the animal origin of the human body.

polymath, *sb.* (*a.*) **b.** *attrib.* or as *adj.* (Further examples.)

1919 T. S. ELIOT *Poems*, The masters of the subtle schools are controversial, polymath. **1959** *New Scientist* 15 Oct. 712/2 Polymath historians, who command the language and literature and science of the period. **1976** *Nature* 16 Sept. 261/3 The book is polymath, up-to-date and non-mathematical.

polymathic *a.* (further examples).

1976 *Publishers Weekly* 16 Feb. 82/2 Crichton's polymathic talent has settled on another exotic site for a fanciful story. **1976** S. HYNES *Auden Generation* viii. 259 A synthesizing, diagnostic, polymathic mind, given to schematizing knowledge and history in elaborate diagrams.

polymenorrhœa (pǫ:limenorī·ă). *Path.* Also **-rrhea.** [f. POLY- + MENORRHŒA.] Excessively frequent or unduly profuse menstrual bleeding.

1931 *Endocrinol.* XV. 180 The problem of polymenorrhea, too profuse and too frequent menses,..is a very vital one to the gynecologist. **1956** C. F. FLUHMANN *Managem. Menstrual Disorders* xvii. 196 Polymenorrhea may supervene as the result of an endocrine disorder such as hypothyroidism. **1964** L. MARTIN *Clin. Endocrinol.* (ed. 4) viii. 247 (*heading*) Polymenorrhoea and postponement of menstruation. **1968** VORYS & NERI in J. J. Gold *Textbk. Gynecol. Endocrinol.* xi. 253 Polymenorrhea may also be seen frequently as a premenopausal symptom.

Hence **polymenorrhœ·al, -rrhœ·ic** *adjs.*

1963 *Jrnl. Jap. Obstetr. & Gynecol. Soc.* X. 40/2 Frequency of corpus luteum insufficiency..differed according to difference in the length of the cycle, being higher in polymenorrheal cycle of 15–24 days and in oligomenorrheal cycle of 39–43 days. **1968** VORYS & NERI in J. J. Gold *Textbk. Gynecol. Endocrinol.* xi. 261 Polymenorrheic cycles of less than 20 days.

polymer. Add: In mod. use, any substance which has a molecular structure built up largely or completely from a number (freq. very large) of similar polyatomic units bonded together. (Further examples.)

The repeating units are often molecules of a single compound, but polymers may also result from combination of two or more different constituents; in either case linking of units may be accompanied by elimination of small molecules, and the formula and molecular weight of the polymer are not necessarily exact multiples of those of any component monomer. Also, esp. in contexts of Applied Chemistry, the use of the term is sometimes restricted to cases where the number of units in the structure is large.

1929 W. H. CAROTHERS in *Jrnl. Amer. Chem. Soc.* LI. 2548 Whatever the term polymer may mean now, it does not mean precisely what Berzelius intended, and the conditions which he set up are not sufficient to define it. *Ibid.* 2549 Two types of polymers may be distinguished... (1) Addition or A polymers. The molecular formula of the monomer is identical with that of the structural unit... (2) Condensation or C polymers: the molecular formula of the monomer differs from that of the structural unit. **1935** C. ELLIS *Chem. Synthetic Resins* I. iv. 53 In a condensation reaction, the polymer is no longer a multiple of the monomer as in the case of the addition polymer. **1958** *Times* 21 Nov. 5/6 All of them are formed by the joining together of simpler molecules; they are therefore polymers, as are plastics and synthetic fibres. **1967** MARGERISON & EAST *Introd. Polymer Chem.* iii. 131 As the conversion of the monomers into polymer proceeds, the number of possible ways in which the monomers may add on to the growing polymer increases rapidly. **1969** *Times* 3 July 7/7 The substance is a polymer of ordinary water molecules linked together by an unusual kind of chemical bond. **1973** *Nature* 6 Apr. 420/1 Natural rubber is still the preferred polymer for many high performance applications. **1974** *Sci. Amer.* Mar. 66/3 Glass is an inorganic polymer made up of rings and chains of repeating silicate units. **1974** D. M. ADAMS *Inorg. Solids* vii. 239 The most stable polymers for other metals are $M_6O_{19}^{8-}$ (M = Nb, Ta), $Mo_7O_{24}^{6-}$, $Mo_8O_{26}^{4-}$ and $HW_6O_{21}^{5-}$. **1978** *Prospects for Polymers* (Shell Internat. Petroleum Co.) 1 Not all polymers are man-made: wool, rubber, cotton and silk are examples of natural polymers.

2. *attrib.* and *Comb.*, as *polymer chemistry*, that branch of chemistry concerned with the preparation and properties of polymers; so *polymer chemist*.

1929 *Industr. & Engin. Chem.* Feb. 131/1 Beyond certain degrees of polymerization x-ray methods do not longer enable estimations of polymer length. **1933** *Ibid.* Feb. 132/1 Development of the best binder—be it resin or another member of the polymer family—is the next step. **1948** *Science* 19 Nov. 545/2 In recent years polymer chemists have shown how it is possible to synthesize compounds of high molecular weight..from known compounds (monomers) of low molecular weight. **1950** *Nature* 22 Apr. 634/1 (*heading*) Polymer chemistry as applied to plastics. **1953** *Endeavour* Apr. 92/1 In terms of polymer-chemistry concepts, the molecular shape is changed by the electrostatic field. **1966** C. R. TOTTLE *Sci. Engin. Materials* p. vi, An enormous quantity of synthetic polymer materials has replaced much of the metal in the domestic kitchen, leather, wool, and cotton in furnishings, and natural rubber, wood, and metal in vehicles, prime movers. **1967** MARGERISON & EAST *Introd. Polymer Chem.* i. 16 However, in the case of a real polymer chain, the finite size of the backbone carbon atoms and the substituents cannot be neglected. *Ibid.* 22 Polymer molecules in the pure liquid state can easily be pictured in the terms used to describe concentrated solutions. **1975** *Nature* 31 July 443/3 For those wishing to get a feel for what is going on in a selection of other areas of polymer science this volume will provide much of interest.

polymerase (pǫ·limĕrēiz, pǫli·mĕrēiz). *Biochem.* [f. POLYMER + *-ASE.] Any enzyme which catalyses the formation of a polymer, esp. a polynucleotide.

1958 I. R. LEHMAN et al. in *Jrnl. Biol. Chem.* CCXXXIII. 163/2 In order to facilitate reference in this report, the enzyme responsible for deoxyribonucleotide incorporation is designated as 'polymerase'. **1964** *New Scientist* 23 Jan. 211/1 DNA has been bio-synthesised from its four basic building blocks put together with the enzyme polymerase and some natural DNA to act as a template. **1973** B. J. WILLIAMS *Evolution & Human Origins* vi. 89/1 RNA polymerase is active when the cell is carrying on its normal metabolic functions.

polymeria, var. *POLYMERY.

polymeric, *a.* Add: In mod. use, of the nature of or characteristic of a polymer; consisting of a polymer or polymers. Of a reaction: giving rise to a polymer. Cf. *POLYMER. (Earlier and later examples.)

1833 *Rep. Brit. Assoc. Adv. Sci. 1831–32* 435 To designate compounds approaching in atomic constitution very nearly to those properly called Isomeric, Berzelius has proposed the introduction of two new terms, polymeric and metameric. **1845** W. GREGORY *Outl. Chem.* II. 394 Hydrated lactic acid is a syrupy liquid... It is..polymeric with dry grape sugar and with gum. **1935** C. ELLIS *Chem. Synthetic Resins* I. ix. 169 Natural rubber is supposed to consist of a mixture of polymeric hydrocarbons. **1941** MARK & RAFF (*title*) High polymeric reactions. **1951** UVAROV & CHAPMAN *Dict. Sci.* (ed. 2) 171 Many important products, such as plastics and textile fibres, consist of polymeric substances, either natural (e.g. cellulose..) or synthetic (e.g. nylon). **1967** MARGERISON & EAST *Introd. Polymer Chem.* iii. 137 At the early stages of the reaction, the mixture consists mainly of low molecular weight species, dimers, trimers, etc., with only a few molecules of polymeric size. **1974** *Environmental Conservation* I. 63/1 The term resin is usually applied to the long chains of repeating units in the polymeric material.

2. *Genetics.* Of, pertaining to, or displaying polymery.

1949 DARLINGTON & MATHER *Elem. Genetics* iii. 68 Genes such as those with which Nilsson-Ehle was concerned have two of the properties of polygenes. They are of similar and supplementary action. But they cannot be described as polygenes because their effects are so large as to cause a sharp discontinuity in the variation, a discontinuity which permits the analysis of the system by the mendelian method. Genes like these are often termed polymeric genes. All polygenes are therefore polymeric, but not all polymeric genes are polygenes. **1961** [see *POLYGENIC *a.* 3]. **1970** *Genetika* VI. x. 124 It is established that complementary maltose factors MA_{1-p} and MA_{2-g} are recessive alleles of polymeric genes, MA_1 and MA_2, respectively.

polymeride. Add: (Later examples.) Now †*Obs.*

1906 *Nature* 21 June 190/2 A semi-solid brown substance..formed when acetylene is subjected to discharges... It is apparently a polymeride of acetylene. **1943** R. HILL in R. S. Morrell et al. *Synthetic Resins* (ed. 2) v. 194 A defect of polystyrene is a tendency for mouldings and castings to develop fine cracks... It is said that this is due to the evaporation of low molecular weight polymerides.

polymerism. Add: Now †*Obs.* **1.** (Earlier example.)

1833 *Rep. Brit. Assoc. Adv. Sci. 1831–32* 435 (*heading*) Polymerism and metamerism.

polymerizable (pǫ·limĕraizăb'l), *a. Chem.* [f. POLYMERIZ(E *v.* + -ABLE.] Capable of being polymerized. Hence **po:lymerizabi·lity.**

1884 *Jrnl. Chem. Soc.* XLV. 419 Just as the pentine is polymerisable by heat or by the action of sulphuric acid into a dipentine, so the heptine may be converted into a diheptine. **1928** *Proc. & Trans. R. Soc. Canada* XXII. III. 39 The influence of methyl substitution on the polymerizability of butadiene. *Ibid.*, Both the α- and the β-methyl butadienes are polymerizable to caoutchoucs. **1939** *Brit. Plastics* XI. 320/1 Methacrylic acid amide copolymerized with other polymerizable compounds. **1948** C. E. H. BAWN *Chem. High Polymers* i. 15 The polymerizability of diolefins is determined by the relative positions of the double bonds. **1963** A. J. HALL *Textile Sci.* ii. 88 It [*sc.* acrylonitrile] very readily undergoes polymerisation by itself (homopolymerisation) and with other polymerisable compounds (copolymerisation). **1978** (*title*) Solventless polymerizable resinous compounds used for electrical insulation. (Brit. Standards Inst.)

polymerizate (pǫ·limĕraizē¹t). [f. next + *-ate*, after *filtrate*, *precipitate*, etc.] A product or mixture of products obtained from a polymerization reaction or process.

1931 *Brit. Chem. Abstr.* B. 264/1 By polymerising a mixture of an aliphatic conjugated diolefine such as butadiene..with styrene..an intimately mixed polymerisate is obtained. **1959** *Times Rev. Industry* Feb. 57/1 The later Ziegler developments in the higher olefin field relate only to a process which yields polymerisates of practically no commercial interest. **1973** *Materials & Technol.* VI. viii. 558 The polymerizates of acrylic acid,.. methacrylic acid,..and their derivatives such as acrylonitrile,..are known collectively as acrylics.

polymerize, *v.* Add: **1.** Cf. *POLYMER. (Further examples.)

1867 *Proc. R. Soc.* XVI. 158 If we remember the facility with which the aldehydes are polimerized [*sic*], the question presents itself, whether the aldehyde formed by the slow combustion of methylic alcohol is represented by the formula CH_2O, or a multiple thereof. **1893** [see *AFTERGLOW b]. **1910** *Encycl. Brit.* X. 667/2 It is not possible to obtain the aldehyde in a pure condition, since it readily polymerizes. **1930** H. F. LEWIS *Fund. Org. Chem.* viii. 95 Three molecules of acetaldehyde polymerize under the influence of a small amount of concentrated sulfuric acid. **1936** H. W. ROWELL *Technol. Plastics* ix. 57 The primary ester may be polymerized by heating in the presence of a peroxide catalyst. **1957** *Times* 10 Sept. 11/1 The effect was to produce reactive groups which could be used to polymerize a different material with the result that mixed polymers of very varied properties could be produced. **1971** *Brit. Printer* Jan. 64/2 The use of ultra-violet radiation for polymerising polyester wood lacquers has been known for some time. **1972** *Nature* 18 Feb. 404/3 This leads into a subsidiary argument: which were the first to polymerize, amino-acids or polynucleotides?

polymerization (earlier and later examples); also, the state of being polymeric; hence also **po·lymerized** *ppl. a.*, **po·lymerizing** *vbl. sb.* and *ppl. a.* Also **po·lymerizer,** an apparatus or installation in which polymerization occurs.

1867 BLOXAM *Chem.* 665/2 (Index), Polymerising by sulphuric acid. **1872** *Jrnl. Chem. Soc.* XXV. 433 Besides the olefines which are derived from compounds having the general formula $C_nH_{2n+1}R$ (R representing a monad radical), by the abstraction of HR, there exists another group, formed by the polymerisation of the members of the first group. **1879** *Ibid.* XXXV. 743 We purposely avoided distilling the polymerised product until entirely freed from substances volatile in a current of steam. **1923** B. D. W. LUFF *Chem. of Rubber* vii. 74 In the meantime it had been observed that not only isoprene but many other unsaturated hydrocarbons were capable of polymerising. **1929** *Jrnl. Amer. Chem. Soc.* LI. 2549 Polymerization then is the chemical union of many similar molecules either..without or..with the elimination of simpler molecules. **1930** *Chem. Abstr.* XXIV. 612 $SnCl_4$ was employed as the polymerizing agent. **1933** *Industr. & Engin. Chem.* Feb. 126/2 The length and shape of the chain must be considered when comparing two different resins or the same resin in different degrees of polymerization. **1940** A. N. SACHANEN *Conversion of Petroleum* i. 64 The absorption of olefins by sulphuric acid and the polymerization take place in the reactor. No special polymerizer is necessary. **1945** A. T. BIRKBY *Phenolic Plastics* ii. 16 The larger or polymerised molecule possesses different physical properties from the constituent smaller molecules. **1946** J. GRANT in *Mod. Petroleum Technol.* (Inst. Petroleum) 165 In the separator the acid phase settles to the bottom and is withdrawn to the polymerizer. **1950** *Thorpe's Dict. Appl. Chem.* (ed. 4) X. 21/1 A polymerising unit of such polyfunctionality that gelation will ensue through the cross-linking which accompanies the oxidation. **1956** *Atlantic Monthly* Sept. 24/3 It is a polymerized ethylene glycol that differs from what you pour into your radiator only in being solid at room temperature. **1965** P. W. MORGAN *Condensation Polymers* iii. 107 Many continuous polymerizers are comprised of a cascade system of reactors or a reacting chamber provided with high-speed stirring followed by a larger holding reservoir with lower speed agitation. **1967** MARGERISON & EAST *Introd. Polymer Chem.* iii. 146 When 50% of the original carboxylic acid groups have reacted, the average degree of polymerization has only doubled. **1971** *Brit. Printer* Jan. 65/1 These highly reactive molecular units can then react with unsaturation in resin molecules to cause polymerisation and hence drying. **1974** E. AMBLER *Dr. Frigo* II. 112 Polymerisation is..a change of state, a molecular rearrangement. Raw rubber becomes vulcanized rubber, say... The second is a polymer of the first. **1977** *Lancet* 19 Nov. 1070/2 How much of a toxic gas such as vinyl chloride will escape from the polymeriser pots where it is turned into P.V.C.?

polymery (pǫli·mĕri). *Genetics.* Also as mod. L. **polyme·ria.** [ad. G. *polymerie* (A. Lang 1911, in *Zeitschr. für induktive Abstammungs- und Vererbungslehre* V. 113), ad. Gk. πολυμέρεια a consisting of many parts: see -IA[1], -Y[3].] The phenomenon whereby a number of non-allelic genes can act together to produce a single effect.

1914 *Zeitschr. für induktive Abstammungs- und Vererbungslehre* XII. 118 The phenomenon of plurality of genes having a similar function, i.e., independently producing the same character, is called by Lang (1911) 'polymery'... Johannsen (1913) suggests that..[this term] be retained..for the phenomenon in general. **1927** [see *POLYGENIC *a.* 3]. **1929** [see *ISOLATION 3].

polymetallic: see *POLY- 1. **polymeter** var. *POLYMETER. **polymethacrylate, -acrylic:** see *POLY- 2.

polymethyl (pǫlime·þil, -mī·þail). [f. POLY- + METHYL.] **a.** *polymethyl acrylate* (also as one word): a resinous material obtained by polymerizing the methyl ester of acrylic acid.

1936 *Industr. & Engin. Chem.* Oct. 269/2 Polymethyl acrylate is a colorless, transparent substance. **1950** *Thorpe's Dict. Appl. Chem.* (ed. 4) X. 107/1 Polymethyl acrylate is a tough, transparent, and colourless material, highly extensible, and to a limited degree rubber-like. **1973** *Materials & Technol.* VI. viii. 559 The softening point of polymethylmethacrylate lies accordingly about 90 degrees higher than that of polymethylacrylate.

b. *polymethyl methacrylate* (also as one word): = *methyl methacrylate* (*b*) s.v. *METHYL.

1936 *Industr. & Engin. Chem.* Oct. 270/1 Polymethyl methacrylate is a very hard, tough mass which can be sawed, carved, or worked on a lathe with ease. **1942** *Endeavour* I. 111/2 Cellulose acetate and polymethylmethacrylate are fabricated into cockpit covers..for aircraft by this process. **1960** *Times* 2 Sept. 14/1 They could be eating sandwiches from polyethylene packs with polymethylmethacrylate dentures. **1965** ZIGROSSER & GAEHDE *Guide to Collecting Orig. Prints* vii. 111 The most durable are the acrylic sheets, particularly Plexiglas (polymethyl methacrylate), which is produced in thicknesses from 1·5 to 25·0 mm. **1973** [see sense a above]. **1975** *Sci. Amer.* Dec. 96/2 Familiar thermoplastic polymers include polyethylene,..polymethyl methacrylate,.. and nylon.

polymethylene (pǫlime·þilīn). *Chem.* [f. POLY- + METHYLENE.] A compound, group, or polymeric structure which consists of or contains a chain of methylene groups, $—(CH_2)_n—$; orig. *spec.* any of the series of saturated cyclic hydrocarbons of formula $(CH_2)_n$. Freq. *attrib.*

1892 *Jrnl. Chem. Soc.* LXII. 1310 In this paper the author discusses the question of the identity of the naphthenes from Caucasus petroleum and the polymethylenes. *Ibid.*, The naphthenes in general may also include derivatives of other polymethylene rings. **1910** N. V. SIDGWICK *Org. Chem. Nitrogen* ii. 21 In the polymethylene derivatives the ring is often affected. **1930** *Jrnl. Amer. Chem. Soc.* LII. 317 Those poly-esters in which the structural units contain polymethylene chains $(CH_2)_x$, in which *x* is greater than 3–5, show great solubility in benzene. **1953** R. J. W. REYNOLDS in R. Hill *Fibres from Synthetic Polymers* v. 98 Other addition reactions which can give rise to linear polymers are those involving the polymerisation of..diazomethane or diazoalkanes to give polymethylenes or their alkyl substituted products. **1967** MARGERISON & EAST *Introd. Polymer*

Chem. i. 17 Taking a portion of the polymethylene chain as the simplest case, the configuration..in which the CH_2 groups on adjacent carbon atoms are staggered relative to one another is more frequently assumed..than any other.

polymetre (pǫ·limītəɪ). *Mus.* Also (*U.S.*) -meter. [f. POLY- + METRE *sb.*¹] **a.** The succession of different metrical patterns in sixteenth-century vocal music. **b.** Music using two or more different time-signatures simultaneously. So **polyme·tric, -me·trical** *adjs.*; also **po·lymetered** *a.*

1922 S. GREW in *Contemp. Rev.* Aug. 226 The first voice of the above has a 'metre' of four pulses, the second has a 'metre' of three; on the authority of the musical terms *polyphonic* and *polytonic*, I have ventured to coin and use the word *polymetrical*. Students of polymetre do not appear to have sufficiently considered the fact that in certain words Elizabethan accent was different from ours. 1944 W. APEL *Harvard Dict. Mus.* 594/1 Twice in the history of music have polymetric designs played a prominent role: around 1400, and in present-day music. 1946 R. BLESH *Shining Trumpets* (1949) 344 The second type of rhythmic peculiarity, technically known as a polymetric, is the cross-rhythm or overrhythm. 1947 W. RUSSELL in R. de Toledano *Frontiers of Jazz* 60 The ability of Lux to create great swing and rhythmic effect from apparently so simple a polymetrical device. 1966 C. KEIL in T. Kochman *Rappin' & Stylin' Out* (1972) 90 It is a subjective pulse that Richard Waterman is speaking of when he uses the concept 'metronome sense' as the ordering principle in the polymetered rhythms of West African ensembles. In jazz groups polymeter or even a sense of polymeter may or may not exist, but the subjective pulse or metronomic sense remains. 1970 P. OLIVER *Savannah Syncopators* 15 These [sc. characteristics of African music] included: dominance of percussion; polymeter; off-beat phrasing of melodic accents [etc.].

polymicrian: see *POLY- 1.

polymict (pǫ·limikt), *a. Petrol.* [f. POLY- + Gr. μικτ-ός mixed, perh. after G. *polymikt* (H. Rosenbusch *Elemente der Gesteinlehre* (1898) 17).] = *POLYMICTIC *a.* 1.

[1931 A. JOHANNSEN *Descr. Petrogr. Igneous Rocks* I. i. 7 Rocks may be composed of a single mineral only..or they may be composed of aggregates of several minerals. The former are called monomineral rocks by Vogt and monomikt by Rosenbusch, and the latter polymikt by Rosenbusch.] 1952 W. WAHL in *Geochim. et Cosmochim. Acta* II. 91 It is proposed to call..breccias in which the enclosed fragments are of a foreign material as compared with the surrounding principal mass of stone..'polymict breccias'. 1958 *Proc. Geologists' Assoc.* LXIX. 85 Two kinds of conglomerate are worth distinguishing, first the oligomict..and, second, the polymict, with a variety of pebbles of unstable rocks undergoing decay and most likely formed by rapid deposition of material worn quickly from high mountains. 1973 *Phil. Trans. R. Soc.* A. CCLXXIII. 392 The conglomerates are characterized by the extreme polymict nature of the boulders, cobbles and pebbles. 1975 TINDALL & THORNHILL *Rock & Mineral Guide* iii. 167 A rock with grains of various materials is polymict.

polymictic (pǫlimi·ktik), *a.* [f. as prec. + -IC.] **1.** *Petrol.* [ad. Russ. *polimiktovȳĭ* (M. S. Shvetsov *Petrografiya Osadochnȳkh Porod* (1934) viii. 155).] (See quot. 1935.)

1935, 1949 [see *OLIGOMICTIC *a.* 1]. 1959 W. W. MOORHOUSE *Study of Rocks in Thin Section* xix. 337 The polymictic conglomerate..comprises a great variety of pebbles, including granite, schist, sediments such as shale, slate, sandstone, and even limestone. 1969 S. H. HAUGHTON *Geol. Hist. Southern Afr.* iv. 89 At various horizons above the Intermediate Reefs bands of polymictic conglomerates occur.

2. *Limnology.* Applied to a lake that has no stable thermal stratification but exhibits perennial circulation.

1956 [see *OLIGOMICTIC *a.* 2]. 1966 *McGraw-Hill Encycl. Sci. & Technol.* V. 523/2 In addition there are..low-altitude tropical oligomictic lakes with irregular circulation, and high-altitude tropical polymictic lakes with continuous circulation.

polymineral: see *POLY- 1.

polymitosis (pǫlimaitōu·sis). *Biol.* [f. POLY- + MITOSIS.] The occurrence of multiple mitotic cell divisions, esp. following meiosis in microsporogenesis; one of these divisions. So **polymito·tic** *a.*, pertaining to, affected by, or being such cell divisions.

1931 G. W. BEADLE *Mem. Cornell Agric. Exper. Station* No. 135. 3 This genetic factor..has been given the name *polymitotic* for which the genetic symbol *po* is used. *Ibid.*, On first thought, the term *polymitotic* may seem to be inappropriate on the grounds that normal maize plants have many mitoses. It may be pointed out that the polymitotic character is expressed only in the gametophytic generation in which there are normally but two mitotic divisions in maize. 1932 C. D. DARLINGTON *Recent Adv. Cytol.* xiii. 367 This property of having 'polymitosis' is inherited as a mendelian recessive character. 1937 *Ibid.* (ed. 2) ix. 399 The spermatids are originally tetraploid through double non-reduction and by three polymitotic divisions become 32*x*. 1948 *Nature* 5 June 874/1 Plants heterozygous for this gene must produce two genetically different kinds of pollen, one of which will be like the lethal polymitotic pollen borne on the homozygous plant. But none of this pollen shows polymitotic behaviour. Thus polymitosis is like incompatibility in heterostyled plants. 1973 *Cytologia* XXXVIII. 515 A gene responsible for supernumerary cell division following meiosis which Beadle..termed the polymitotic division gene (*po*). *Ibid.* 518 Postmeiotic polymitoses were observed in reciprocal interspecific hybrids between two diploid *Clarkia* species... The genetic system responsible for these polymitoses is most likely not a simple recessive gene as is the case in maize but rather genic disharmony between the two parental sets of chromosomes.

polymixin, var. *POLYMYXIN.

polymodal (pǫlimōu·dăl), *a.* [f. POLY- + MOD(E *sb.* + -AL.] **1.** *Mus.* Of, pertaining to, or designating music using two or more modes.

1929 W. W. COBBETT *Cycl. Survey Chamber Music* I. 44/1 Since we are polymodal as well as polytonal, each of these combinations may take the following four different forms according as one or the other triad is major or minor. 1938 *Scrutiny* VII. 174 It becomes quite obvious that his [sc. Roussel's] polymodal melodic thinking must condition his scheme of harmony. 1957 W. MELLERS *Romanticism & 20th Cent.* ii. v. 157 Roussel in his later work has more self-consciously to recover a tradition: which may explain the more strenuous quality of his (often polymodal) melodic power.

2. = *MULTIMODAL *a.* a.

1934 *Jrnl. Sedimentary Petrol.* IV. 73/1 Curve C represents a glacial till, in which more than one mode occurs, and is an example of a polymodal frequency curve. 1975 *Nature* 28 Aug. 723/1 In general it seems that cells of different sizes, whether they follow a normal or a polymodal distribution, are randomly distributed throughout the tissue.

Hence **polymoda·lity** = *MULTIMODALITY; **polymo·dally** *adv.*

1929 W. W. COBBETT *Cycl. Survey Chamber Music* I. 38/2 It is at this point that polymodality commences to impinge upon polytonality. *Ibid.* 45/2 A total of five major and five minor keys, representing polymodally six tonalities. 1934 C. LAMBERT *Music Ho!* v. 289 The slow destruction of the key system that we find in Milhaud's polytonality or Vaughan Williams' polymodality. 1952 R. STEVENSON *Music in Mexico* i. 7 For those whose ears have become conditioned by long familiarity with the European diatonic system, the 'polymodality' of indigenous music inevitably sounds as if it were 'polytonality'. 1962 *Lancet* 26 May 1092/2 Hoobler's remarks, together with the recent discussions about polymodality in frequency distributions of blood-pressure, prompted me to examine some experimental data on 'pressor responses'. 1975 *Nature* 28 Aug. 724/1 Techniques based on sorting cells after digestion of the tissue would certainly reveal polymodality.

polymolecular (pǫlimŏle·kiŭlăɪ), *a. Chem.* [ad. G. *polymolekular* (Van 't Hoff & Cohen *Studien zu chem. Dynamik* (1896) 4): see MOLECULAR *a.*] **a.** In chemical kinetics: having or pertaining to an order or a molecularity of more than one.

1896 T. EWAN tr. *Van 't Hoff & Cohen's Stud. Chem. Dynamics* 4 We will call a change in which the interaction of several molecules is required, polymolecular. 1896 *Jrnl. Chem. Soc.* LXX. ii. 158 (*heading*) The velocity law of polymolecular reactions. 1937 *Jrnl. Amer. Chem. Soc.* LIX. 2530/2 The polymolecular nature of the solvolysis is strongly evidenced by the fact that it occurs only with a high concentration of hydroxylic molecules.

b. Consisting of or built up from more than one molecule.

1930 *Jrnl. Amer. Chem. Soc.* LII. 4110 The direct preparation of this anhydride by the removal of water from the acid should give a polymolecular product. 1936 *Trans. Faraday Soc.* XXXII. 116 Oleic acid is bimolecular in apolar solvents, and forms polymolecular micelles in aqueous alkaline solution.

c. With reference to a film or layer: being more than one molecule in thickness. Of adsorption: characterized by the formation of such a layer.

1931 *Jrnl. Physical Chem.* XXXV. 869 There was no evidence that the layer of adsorbed molecules was of polymolecular thickness. 1931 J. W. MCBAIN *Sorption of Gases & Vapours by Solids* x. 325 Such built-up polymolecular layers differ in principle from the third conception,..that all the molecules coming within a certain range of the solid are directly attracted. 1972 M. M. DUBININ in F. Ricca *Adsorption–Desorption Phenomena* 4 On the surface of intermediate pores there occurs monomolecular and polymolecular adsorption of vapours.

d. Consisting of macromolecules which have similar polymeric structures but differing molecular weights. Cf. *POLYDISPERSE *a.*

1940 *Chem. Abstr.* XXXIV. 1228 Such compds. of high mol. wt. are also polymolecular and yield polydisperse solns. Since compds. whose mols. are identical with respect to structure and mol. wt...may form in soln. colloidal particles of varying size, polydispersivity is a condition or state whereas polymer homogeneity and polymolecularity are properties of mols. 1943 H. M. SPURLIN in E. Ott *Cellulose* V. ix. 930 The expression 'polymolecular' is preferable to 'polydisperse' as a term to describe systems composed of molecules all having substantially the same chemical composition and mode of linkage but differing in chain length. 1970 FOCK & FRIED tr. *Staudinger's From Org. Chem. to Macromolecules* 116 In order to characterize a polymolecular macromolecular material with accuracy, it is necessary to know its distribution and to determine how many low and high molecular parts it contains. 1976 H.-G. ELIAS in K. Solc *Order in Polymer Solutions* 218 Most synthetic polymers are polymolecular, *i.e.*, the unimers possess a distribution of degrees of polymerization.

Hence **polymolecula·rity**, the condition or property of being polymolecular (esp. in sense *d); as a back-formation **polymo·lecule**, a polymeric molecule.

1938 *Chem. Abstr.* XXXII. 2810 A distinction is made between the terms polydispersion and polymolecularity. 1940 [see sense d above]. 1943 H. M. SPURLIN in E. Ott *Cellulose* V. ix. 930 The quantitative relationships between such physical properties as viscosity in solution, physical strength, and flexibility of films may differ as the degree of polymolecularity is changed. 1951 *Jrnl. Polymer Sci.* VII. 400 Let us take a momentary picture of an assembly of equal polymolecules of polymerization degree *P* each carrying ν charged groups. 1964 *Biophysical Jrnl.* IV. i. Suppl. 11 As a first demonstration of the joint operation of polymeric and electrical properties within polyelectrolyte molecules, we shall consider the shape dynamics of charged polymolecules. 1976 H.-G. ELIAS in K. Solc *Order in Polymer Solutions* 218 The problem on [*sic*] how the polymolecularity of the unimers influences the polydispersity of the multimers has been solved recently for end-to-end and segment-to-segment associations.

polymorph. Add: **1.** (Later examples.) Also *attrib.*

1950 *Evolution* IV. 298/1 There is an interesting correlation between habitat differences and shifts in polymorph frequencies. 1958 *Proc. Zool. Soc.* CXXXI. 87 Comparisons are often made between different polymorphs in the same population. 1975 *Zool. Jrnl.* CLXXVII. 334 This species [sc. *Charaxes zoolina*] is a simple dual polymorph with respect to wing shape.

2. (Examples.) In mod. use, each of the different forms of such a substance.

1902 *Jrnl. Chem. Soc.* LXXXII. ii. 448 With polymorphs..a quite different behaviour is observed. The melting point of the stable form is scarcely altered by the addition of the unstable modification. 1944 [see *KRISOBITE]. 1957 G. E. HUTCHINSON *Treat. Limnol.* I. x. 660 $CaCO_3$ exists under ordinary conditions in nature in two crystalline polymorphs, calcite and aragonite. 1973 *Nature* 23 Mar. 241/1 Anhydrite, $CaSO_4$, is the higher temperature polymorph of calcium sulphate.

3. (Further examples.)

1970 *Nature* 5 Sept. 1052/1 Red cells, monocytes and the majority of polymorphs had grain densities within the range of control preparations. 1977 *Lancet* 29 Jan. 225/1 Evidence indicates that steroids prevent the accumulation of macrophages and polymorphs in inflammatory areas.

polymorphemic: see *POLY- 1.

polymorphic, *a.* Add: **2.** (Later examples.)

1925 A. D. IMMS *Gen. Textbk. Entomol.* iii. 257 Termites live together in large communities composed of polymorphic individuals. 1940 E. B. FORD in J. S. Huxley *New Systematics* 503 Though polymorphic forms are to be distinguished from geographical variation, they may be a function of it. 1976 R. A. GOLDSBY *Basic Biol.* xx. 330/1 (*caption*) Polymorphic variation in one species of snail, *Helicella virgata*: different banded and unbanded forms on one plant.

3. *Chem.* and *Min.* = POLYMORPHOUS *a.* 3 (in Dict. and Suppl.).

Now the more usual adj. in this sense.

1895 C. S. PALMER tr. *Nernst's Theoret. Chem.* i. iii. 86 The different polymorphic modifications may exist together..if they are not easily convertible into each other, as diamond and graphite. 1924 A. E. HILL in H. S. Taylor *Treat. Physical Chem.* I. ix. 380 At high pressures there exist several polymorphic forms of ice, differing from the common variety in density, heat of formation, crystalline structure and other physical properties. 1974 K. FRYE *Mod. Mineral.* ii. 83 Quartz, tridymite, and cristobalite have high- and low-temperature polymorphs... The high-low polymorphic inversion is rapid and nonquenchable, since it involves no rupture of Si–O bonds.

polymorphism. Add: **2.** (Later examples.)

1913 *Phil. Trans. R. Soc.* B. CCIV. 227 The following paper gives an account of a series of breeding experiments..in the course of one and a half years' research on insect polymorphism. 1940 E. B. FORD in J. S. Huxley *New Systematics* 505 It is important to distinguish true polymorphism..from the existence of multiple phases attained at different stages of development. 1976 R. A. GOLDSBY *Biol.* xxiii. 552/2 Recent analysis of enzymes and other proteins has..revealed a previously unsuspected number of genetic polymorphisms in protein chains.

3. (Earlier and later examples.)

1848 *Mem. & Proc. Chem. Soc.* III. 93 (*heading*) Dimorphism and polymorphism. 1966 PHILLIPS & WILLIAMS *Inorg. Chem.* II. xix. 7 Polymorphism is commonly found among metals, and for one metal there is in general very little difference in energy between one structure and another. 1971 I. G. GASS et al. *Understanding Earth* iii. 60/2 This property, polymorphism, occurs in the minerals of the mantle.

polymorpho-. Add: **polymorpho-nuclear** *a.* (usu. written as one word), also used esp. to designate a class of leucocyte (see quot. 1968); also *ellipt.* as *sb.*; (earlier and later examples).

1897 R. C. CABOT *Clin. Exam. Blood* i. v. 49 Next in age come the cells usually known as 'polynuclear' but more properly called *polymorphonuclear neutrophiles*.

These cells constitute the vast majority of those found in ordinary pus. 1903 *Amer. Jrnl. Med. Sci.* CXXVI. 190 A differential count of the leucocytes shows a slight increase in the polymorphonuclears and a diminution in the small mononuclears since the previous record. 1950 *Brain* LXXIII. 144 An early puncture gave 10 polymorphonuclears and 3 lymphocytes per c.mm. 1961 R. D. BAKER *Essent. Path.* ii. 12 The inflammatory lesion is rich first in polymorphonuclear cells and later in macrophages. 1968 PASSMORE & ROBSON *Compan. Med. Stud.* I. xxvi. 2/1 There are three varieties of white cell, the polymorphonuclear leucocyte or .polymorph, the lymphocyte and the monocyte.

polymorphous, *a.* Add: **2.** (Later example.)

1928 C. K. OGDEN tr. *Forel's Social World of Ants* II. v. 337 The formicary is a society of females and their polymorphous derivative forms.

3. (Earlier and later examples.) Also, of or pertaining to polymorphism (sense 3).

1848 *Mem. & Proc. Chem. Soc.* III. 57 (*heading*) On the relation in volumes between simple bodies, their oxides and sulphurets, and on the differences exhibited by polymorphous and allotropic substances. 1895 C. S. PALMER tr. *Nernst's Theoret. Chem.* i. iii. 86 The different kinds of crystals of a polymorphous substance, are to be regarded as different modifications analogous to the different states of aggregation. 1906 J. P. IDDINGS *Rock Minerals* i. i. 19 Silica (SiO_2) is certainly dimorphous and possibly polymorphous. 1964 J. SINKANKAS *Mineral. for Amateurs* vi. 170 Another polymorphous pair also shows marked though less striking differences in hardness: calcite (H 2½–3) and aragonite (H 3½–4).

5. *Psychol.* Phr. *polymorphous-perverse, polymorphously perverse* (see next); so *polymorphous perversity*.

1909 A. A. BRILL tr. *Freud's Sel. Papers on Hysteria* ix. 191 The constitutional sexual predisposition of the child is more irregularly multifarious than one would expect, that it deserves to be called 'polymorphous-perverse', and that from this predisposition the so-called normal behavior of the sexual functions results through a repression of certain components. 1910 —— tr. *Freud's Three Contrib. to Sexual Theory* ii. 49 Under the influence of seduction the child may become polymorphous-perverse. *Ibid.*, The child does not behave differently from the average uncivilized woman in whom the same polymorphous-perverse disposition persists. 1954 W. MAYER-GROSS et al. *Clin. Psychiatry* iv. 179 The active male and the passive female [homosexual]..adopt their homosexual behaviour as a *pis aller*, or, as frequently occurs, out of an abundance of sexual urge and interest and as part of a polymorphous perversity. 1954 D. RIESMAN *Individualism Reconsidered* (1955) vi. xxii. 355 He [sc. Freud] makes the famous charge that children are 'polymorphous-perverse'—that is, that their sexual life is not confined to the genital zone. 1963 AUDEN *Dyer's Hand* 411 Three kinds of erotic life are possible... The polymorphous-perverse promiscuous sexuality of childhood, courting couples whose relation is potential,..and the chastity of natural celibates who are without desire. 1974 *Encycl. Brit. Macropædia* XVI. 610/1 Most people with a polymorphous-perverse personality are either nearly or wholly psychotic persons or nonpsychotic persons who, in sexual and nonsexual areas of living, are unable to develop lasting, affectionate relations with others.

polymorphously (pǫlimǭ·ɪfəsli), *adv.* [f. prec. + -LY².] *polymorphously perverse* (Psychol.): characterized by a diffuse sexuality that can be excited and gratified in many ways and is normal in young children but regarded as perverted in adults. Also *transf.*

1949 J. STRACHEY tr. *Freud's Three Ess. Theory of Sexuality* ii. 69 Under the influence of seduction children can become polymorphously perverse, and can be led into all possible kinds of sexual irregularities... In this respect children behave in the same kind of way as an average uncultivated woman in whom the same polymorphously perverse disposition persists. 1957 *Ann. N.Y. Acad. Sci.* LXVI. 429 Is not a child infinitely potential rather than polymorphously perverse? 1980 *Church Times* 22 Feb. 11/3 Traditional Christianity..offers us encouragement about living with shame and dependence, which, in the polymorphously perverse landscape of chaos, need some careful (and painful) thought.

polymorphy. (Later example.)

1902 D. H. CAMPBELL *Univ. Text-bk. Bot.* vi. 176 Rusts are characterized by the production of several quite different forms [of spores]. This polymorphy is complicated in some species by heterœcism.

polymyalgia, -myositis: see *POLY- 1.

polymyxin (pǫlimi·ksin). *Pharm.* Also **polymixin.** [f. mod.L. *polymyxa*, specific epithet (f. POLY- + Gr. μύξα mucus, slime) + -IN¹.] Any of a class of antibiotics (*polymyxin A, B*, etc.) which are polypeptides obtained from strains of the soil bacterium *Bacillus polymyxa* and are used against Gram-negative bacteria in infections of the urinary tract and the skin.

1947 P. G. STANSLY et al. in *Bull. Johns Hopkins Hosp.* LXXXI. 43 The antibiotic-producing organism has been identified as *Bacillus polymyxa* and the antibiotic substance accordingly designated 'Polymyxin'. 1950 *Lancet* 17 June 1139/2 Polymyxin D in doses of 40 mg. per kg. for ten days produced definite injury in canine kidneys. 1956 *New Biol.* XXI. 18 Various methods for purifying pitching yeast have been in vogue for some years... The latest approach to this problem is the suggested use of antibiotics such as polymixin. 1974 M. C. GERALD *Pharmacol.* xxvii. 473 Polymyxin attaches to the cell membrane of bacteria, disrupting its function and causing the loss of essential intracellular materials.

Polynesian, *a.* and *sb.* Add: **a.** *adj.* (Earlier and further examples.)

1812 W. MARSDEN *Gram. Malayan Lang.* p. xxii, The Polynesian or general East-insular language..does not include those spoken by the description of people termed *Papúa* and *Samang*. 1820 J. CRAWFURD *Hist. Indian Archipelago* II. v. v. 93 The Sanskrit language exists indeed embodied in writing, while the Polynesian language can be traced only as it is scattered over a thousand living dialects. 1863 J. C. PATTESON *Let.* 8 Aug. in C. M. Yonge *Life J. C. Patteson* (1874) II. ix. 69 One might almost get together all the *disjecta membra*, and reconstruct the original Polynesian tongue. 1874 TROLLOPE *Harry Heathcote* iv. 89 A gang of Polynesian labourers..from the South Sea Islands. 1901 *Chambers's Jrnl.* May 343/1 With me was a young Polynesian half-caste named Alan, about twenty-two years of age. 1931 R. CAMPBELL *Georgiad* iii. 52 The huge jaws of Polynesian clams. 1960 T. & L. DAVIS *Makutu* i. i. 13 It had to be a Polynesian island. 1978 B. PRIESTLEY *Island Emperor* iii. 26 'My great-grandfather used to eat men..' He seemed rather ashamed..talking about the darker side of the Polynesian past.

b. *sb.* (Earlier and later examples.) Also, the language of Polynesia.

1812 W. MARSDEN *Gram. Malayan Lang.* p. xviii, This language..may be conveniently termed the Polynesian, and distinguished..into the *Hither* (frequently termed the *East insular* language) and the *Further* Polynesian. 1820 J. CRAWFURD *Hist. Indian Archipelago* II. v. v. 84 All agree in borrowing from the same source—from the great Polynesian. 1874 TROLLOPE *Harry Heathcote* iv. 91 Jacky was one of the Polynesians, who at once started on his errand. 1901 *Chambers's Jrnl.* Dec. 799/2 His eyesight, like that of all Polynesians, was better than that of any white man. 1923 A. L. KROEBER *Anthropol.* v. 121 Articles..recur in Semitic, in Polynesian, and in several groups of American languages. 1960 T. & L. DAVIS *Makutu* i. iii. 43 Family relationships are important among the Polynesians and families are large. 1962 [see *MORIORI]. 1976 'M. DELVING' *China Expert* i. 16 He had several times been mistaken for a Japanese or a Polynesian.

polyneuritic, -neuropathy: see *POLY- 1.

polynia. Add: Now usu. written **polynya.** Pl. **polynyas** (rarely ‖ **polynyi**). (Further examples.)

1957 *Sat. Even. Post* 8 Dec. 7 From the deck of the A-sub, here surfaced in a 'polynya'—a hole in the ice-pack—crewmen took a close look at the perpetually frozen Arctic Ocean. 1963 G. L. PICKARD *Descriptive Physical Oceanogr.* vii. 154 Some of this cap ice melts in the summer... Open water spaces, 'polynyas', may form. 1963 *Sunday Tel.* 22 Sept. 15 Learning to find holes, or *polynias*, was one of the primary tasks of the two British submarines Porpoise and Grampus. 1971 *Nature* 1 Jan. 37/2 The present study was undertaken to measure the actual distribution of CO_2 between the atmosphere and the sea over open leads and polynyi in the ice-covered Bering Sea. 1974 L. DEIGHTON *Spy Story* xix. 206 We found a suitably large polynya—which is the proper name for a lagoon in the ice—and..the Captain began surfacing procedures.

polynomial, *a.* and *sb.* Add: **A.** *adj.* **2.** (Later example.)

1964 *Huntia* I. 34 He makes the essential distinction between the old Aristotelian polynomial phrase-names and the new trivial names.

B. *sb.* **1.** The terms are usually taken to be multiples of powers, finitely many in number. (Later examples.)

1941, 1966 [see *FACTOR *v.* 2].

2. (Later examples.)

1951 G. H. M. LAWRENCE *Taxon. Vascular Plants* ix. 194 Before the middle of the eighteenth century the names of plants commonly were polynomials. 1971 W. T. STEARN in W. Blunt *Compleat Naturalist* 248/1 Such a polynomial determines the application of the binomial.

Polynosic (pǫlinōu·zik), *a.* and *sb.* Also **polynosic.** [ad. F. *polynosique*, contraction of *polymère d'un glucose* + *-ique* -IC (see N. Drisch 1959, in *Reyon, Zellwolle u. andere Chemiefasern* IX. 436).] **A.** *adj.* A proprietary term applied to fibres of a type made from regenerated cellulose and resembling cotton in such properties as a high wet modulus, alkali-resistance, and a crystalline multi-fibrillar structure. **B.** *sb.* A fibre of this type.

A number of erroneous accounts of the etymology occur in the literature.

1959 *Chem. & Engin. News* 24 Aug. 23/1 Hartford Fibres, a division of Bigelow-Sanford Carpet, takes the wraps off a new 'polynosic' fiber. 1959 *Skinner's Silk & Rayon Rec.* Nov. 1084/1 Recent reports from the Continent about a new class of cellulosic fibres, the polynosics,..have aroused interest in the British trade. 1963 *Trade Marks Jrnl.* 16 Jan. 82/1 *Polynosic*... Threads made of synthetic or natural textile materials. Association Internationale Polynosic.., Geneva, Switzerland; merchants. 1964 *Financial Times* 3 Mar. 15/4 Courtauld's polynosic fibre is Vincel, which has already passed from the pilot plant stage to commercial production. 1964 *Economist* 7 Mar. 913/3 The two companies will collaborate on the research, development and production of viscose rayon; particularly important will be their joint work on the 'polynosics', modified viscose fibres with properties similar to cotton. 1967 *Encycl. Polymer Sci. & Technol.* VI. 547 Within the last few years, regenerated cellulosic fibers having higher moduli values than the standard rayon fibers have been developed... These fibers have been given the generic name of polynosics. 1970 *Which?* Oct. 311/2 Vincel and Zaryl are polynosic rayons, as strong as cotton and with many of its wet strength properties.

polynuclear, *a.* (in Dict. s.v. POLY- 1). Add: (Further examples.) Also, applied *spec.* to polymorphonuclear leucocytes. Also as *sb.*

1891 [see *MYELOCYTE 2]. 1894 *Jrnl. Physiol.* XVII. 85 The other term which was applied to the finely granular oxyphile cells by Metschnikoff and others, namely, 'polynuclear leucocyte' is not satisfactory, seeing that..the different nuclear masses..are, in point of fact, joined by threads or bars of nuclear substance so that the cell is really mononuclear with a very much branched nucleus. 1897 [see *POLYMORPHO-NUCLEAR *a.*]. 1907 *Med. Rev.* X. 364/1 Centrifugalisation showed 68 per cent. of polynuclears, 14 per cent. of large mononuclears, and 18 per cent. of lymphocytes. 1935 *Trans. R. Soc. Trop. Med. & Hygiene* XXVIII. 477 The polynuclear count can also be used in the study of populations in which some pathological element exists. 1936 *Ibid.* XXX. 273 Infection provides a strong stimulus in modifying the percentage of neutrophiles with a segmented nucleus (polynuclear cells). 1967 *Jrnl. Reticuloendothelial Soc.* IV. 168 (*heading*) Influence of serum on intracellular digestion of *Staphylococcus aureus* by polynuclear neutrophils from the guinea pig.

b. *Chem.* Of a complex: containing more than one metal atom. Of a compound: having more than one nucleus (*NUCLEUS *sb.* 8).

1908 *Chem. Abstr.* II. 1101 (*heading*) Polynuclear metalammonias. 1924 W. THOMAS *Complex Salts* v. 54 This latter compound is an example of what may be termed a polynuclear complex, since the complex contains more than one central atom. 1933 *Jrnl. Soc. Chem. Industry* 8 Dec. 422T/2 Any polynuclear compound from, say, phenol possesses free positions in internal rings at which further condensation with formaldehyde..is possible. 1951 I. L. FINAR *Org. Chem.* I. xxix. 571 Polynuclear hydrocarbons may be divided into two groups, those in which the rings are isolated,..and those in which two or more rings are fused together in the *o*-positions. 1971 *Nature* 20 Aug. 539/1 Studies of a synthetic and structural nature of polynuclear metal complexes containing hydrogen and organic groups as ligands may eventually allow a better understanding of heterogeneous reactions.

polynucleated, *a.* (in Dict. s.v. POLY- 1). Add: **b.** Designating an urban area planned in the form of a number of smaller, self-contained communities.

1938 [see *MONONUCLEATED *a.*]. 1965 *Listener* 27 May 774/2 The Clyde Valley plan proposals were for a polynucleated urban system designed on a tight pattern and set in a green background.

polynucleotide (pǫliniū·kliŏtəid). *Biochem.* [ad. G. *polynucleotid* (Levene & Mandel 1908, in *Ber. d. Deut. Chem. Ges.* XLI. 1906): see POLY- and *NUCLEOTIDE.] A polymeric compound whose molecules are composed of a number (usu. large) of nucleotides.

1911 *Jrnl. Biol. Chem.* IX. 394 Yeast nucleic acid is a polynucleotide. 1916 [see *HEXOSE]. 1953 S. E. LURIA *Gen. Virol.* v. 100 Watson and Crick..have proposed for DNA a structure consisting of two helical polynucleotide chains coiled around the same axis and held together by bonds between the purine and pyrimidine bases. 1964 G. H. HAGGIS et al. *Introd. Molecular Biol.* ix. 225 Nucleic acids formed by *in vivo* synthesis with restricted base composition are termed polynucleotides. *Ibid.* 226 Synthetic polynucleotides form a variety of double and triple helices. 1970 R. W. MCGILVERY *Biochem.* iii. 23 A particular arrangement of three nucleotide units in the DNA polynucleotide specifies a particular amino acid. 1971 *Sci. Amer.* July 28/2 Antibody responses in animals were enhanced by certain synthetic polynucleotides. These are analogues of the nucleic acids DNA or RNA that are made in the laboratory by combining nucleotides (the subunits of nucleic acids) in arbitrary ways.

polynya, var. POLYNIA in Dict. and Suppl.

polyœstrous-: see *POLY- 1. **polyol, -olefin(e**: *POLY- 2.

polyoma (pǫli,ōu·mă). *Microbiol.* [f. POLY- + *-OMA.] In full *polyoma virus*. A papovavirus that is endemic in mice without producing tumours but which can produce many kinds of tumour in young rodents.

1958 B. E. EDDY et al. in *Proc. Soc. Exper. Biol. & Med.* XCVIII. 848/1 A virus, which we shall refer to as SE polyoma virus, was recovered from tissue cultures inoculated with tumor material from mice and was shown to induce multiple tumors in mice..and hamsters. 1962 [see *PAPOVAVIRUS]. 1962 *Times* 27 July 21/6 In many fully formed growths induced by the polyoma virus, no sign of the virus can be found under the electron microscope. 1967 AMBROSE & EASTY in E. J. Ambrose et al. *Cancer Cell in Vitro* v. 41 Several viruses will produce malignant transformations *in vitro* when grown on cells of animal origin, for example, polyoma virus, Rous sarcoma virus, and SV 40 virus. The polyoma transformation is the one most extensively studied. 1969 A. M. CAMPBELL *Episomes* xiv. 169 Mammalian DNA viruses such as polyoma and SV40 cause the formation of tumors. 1973 R. G. KRUEGER et al. *Introd. Microbiol.* xxviii. 699/1 All of the tumors induced by polyoma virus in various different

mouse strains and hamsters have a common antigen. 1975 MELIEF & SCHWARTZ in F. F. Becker *Cancer* I. v. 123 The polyoma virus, which commonly infects both wild and laboratory mice.., does not produce tumors under natural conditions even though it is potentially very oncogenic.

polyomino: see *POLY- 1.

polyonymosity (pǫ:li,ǫnimǫ·siti). *rare*⁻¹. [f. as POLYONYMOUS *a.* + -ITY.] The availability of different names for the same person or thing.

1923 W. DE LA MARE in *Times Lit. Suppl.* 3 May 293/4 But how happy is the country polyonymosity that hails it [*sc. Oxalis acetosella*] also as sheep-sorrel, cuckoo-spice, hallelujah, [etc.].

Polyox (pǫ·li,ǫks). A proprietary name for polyethylene oxide resin.

1957 *Official Gaz.* (U.S. Patent Office) 6 Aug. TM4/2 Union Carbide Corporation, New York..Polyox. For water soluble resins. First use on or about Mar. 4, 1957. 1958 *Industr. & Engin. Chem.* Jan. 9 (*caption*) A little Polyox resin goes a long way in thickening water. 1973 *Trade Marks Jrnl.* 6 June 1072/2 Polyox... Synthetic water-soluble resins. Union Carbide Corporation.., New York,..United States of America; manufacturers and merchants. 1976 *Nature* 1 July 47/1 We have reported that polyethyleneoxide (Polyox) in solution reduced the damping of free oscillations in a semicircular manometer.

polyoxyethylene, -methylene: see *POLY- 2.

polyp, polype. 2. Delete *rare* and add further examples.

1955 *Sci. News Let.* 1 Oct. 217/1 Polyps are small growths which may be non-cancerous but which are believed capable of developing into cancers. 1961 [see *POLYPUS 2]. 1966 *Economist* 12 Nov. 654/3 Power can corrupt—the Far Eastern tour apparently made both Mr Johnson's incisional hernia and the polyp in his throat worse. 1974 PASSMORE & ROBSON *Compan. Med. Stud.* III. xxviii. 43/1 Endometrial polyps are frequently asymptomatic and discovered in the course of a curettage... Recurring polyps associated with adenomatous hyperplasia in the postmenopausal patient should be regarded as premalignant and treated by hysterectomy.

3. polyp-tree = *polyp-stem*.

1915 E. R. LANKESTER *Diversions of Naturalist* xi. 97 The little jelly-fish are the ripe individuals of the polyps, and produce eggs and sperm which grow to be polyp-trees.

polypectomy (pǫlipe·ktŏmi). *Surg.* [f. POLYP + *-ECTOMY.] Excision of a polyp.

1950 *Surg. Clinics N. Amer.* XXX. 661 For nasal polypectomy..anesthesia is accomplished by painting the polyp and its area of attachment or pedicle with a saturated solution of cocaine. 1974 PASSMORE & ROBSON *Compan. Med. Stud.* III. xxviii. 43/1 The diagnosis is established by exploration of the uterus with polypectomy forceps during diagnostic curettage.

polypeptide (pǫlipe·ptəid). *Biochem.* [ad. G. *polypeptid* (E. Fischer 1903, in *Sitzungsber. d. k. preuss. Akad. d. Wissensch.* 389, after *di-, tripeptid*, etc. (Fischer 1902: see *PEPTIDE)).] Any peptide in which the number of amino-acid residues that go to make up the molecule is not small (cf. *oligopeptide* s.v. *OLIGO-), but is not so large that it can be regarded as a protein; **polypeptide chain** = *peptide chain* s.v. *PEPTIDE 2.

1903 *Jrnl. Chem. Soc.* LXXXIV. i. 466 These acid chlorides combine easily with glycylglycine esters and similar compounds to form chains of amino-acids joined together by an anhydride linking. Such are termed polypeptides. 1935 R. H. A. PLIMMER in Harrow & Sherwin *Textbk. Biochem.* v. 177 Fischer held the view that the polypeptide chain was not long enough for attack by pepsin. 1949 H. W. FLOREY et al. *Antibiotics* I. i. 38 The active preparation made from this organism..contains two antibiotics, gramicidin and tyrocidine, both crystalline polypeptides. 1951 *New Biol.* XI. 99 Larger molecules that show biological activity when applied to living organisms are exemplified by polypeptides, such as A.C.T.H. (the pituitary hormone which controls the secretion of cortisone). 1959 *Times* 2 Jan. 11/3 Further evidence suggested that this particular polypeptide was a precursor of the cell wall of the staphylococcus. 1961 *Ann. Reg.* 1960 401 The polypeptide chain, the backbone of the protein molecule, was found to be coiled in a helix-like spiral spring with only a space inside. 1978 *Sci. Amer.* Dec. 68/2 A hemoglobin molecule is made up of four polypeptide chains, two alpha chains of 141 amino acid residues each and two beta chains of 146 residues each.

Hence **polype·ptidase** [*-ASE], any enzyme which hydrolyses polypeptides.

1922 *Chem. Abstr.* XVI. 3491 (*heading*) Influence of materials obtained from yeast cells and organs on the rate of hydrolysis of substrates by polypeptidases, carbohydratases and esterases. 1929 R. P. WALTON tr. *Waldschmidt-Leitz's Enzyme Actions* 159 The yeast polypeptidases are totally inactive against dipeptides. 1940, 1961 [see *EREPSIN].

polyphage. Delete †*Obs. rare*⁻⁰ and add examples.

1924 *Scribner's Mag.* Aug. 156/2 The flimsy telegraph copy of a presidential message fluttered out of the window and was lost... 'Oh, say that the office cat ate it.'.. The animal immediately became popular as a polyphage in hundreds of other newspaper-offices. 1965 B. E. FREEMAN tr. *Vandel's Biospeleol.* xxx. 472 Only the polyphages..have a chance of subsisting underground.

polyphagia. For 'Rarely' read 'Also' and add: **1.** (Later example.)

1946 *Nature* 28 Sept. 454/1 Such animals [*sc.* diabetic rabbits]..exhibited classical symptoms of diabetes mellitus—hyperglycæmia, glycosuria, polyuria, polyphagia.

2. (Examples.)

1907 W. R. FISHER *Schlich's Man. Forestry* (ed. 2) IV. iv. 158 Observations are not yet complete regarding the monophagy, or polyphagy of certain insects. 1950 *New Biol.* VIII. 64 Predaceous insects, spiders, birds and so on, exhibiting various degrees of polyphagy (i.e. eating more than one kind of food) usually come into the picture. 1965 B. E. FREEMAN tr. *Vandel's Biospeleol.* xix. 337 The categories which have been recognised must remain fluid because of the marked tendency of cavernicoles towards polyphagia. 1970 K. R. NORRIS in *Insects of Australia* (Commonwealth Sci. & Industr. Res. Org., Australia) v. 114/2 An example of polyphagy is afforded by the scale insect *Ceroplastes rubens* feeding on hundreds of different host plants.

polyphant: see *POLYPHONE.

polypharmaceutical (pǫ:lifā:ɪmăsiū·tikăl), *sb.* and *a. Med.* [f. POLY- + PHARMACEUTICAL.] **A.** *sb.* A medicinal preparation containing several drugs. **B.** *adj.* Of or pertaining to polypharmacy.

Usu. disparaging; cf. next.

1961 *Lancet* 16 Sept. 658/2 The [pharmaceuticals] industry, say some doctors, makes excessive profits;..indulges in excessive and irrelevant promotion of its products;..goes in for dubious polypharmaceuticals. 1974 M. C. GERALD *Pharmacol.* i. 6 A very simple preparation was Paracelsus' laudanum, which contained opium, gold, and pearls. (Note that even Paracelsus retained vestiges of polypharmaceutical formulation.)

polypharmacy. Add to def.: Freq. with the suggestion of indiscriminate, unscientific, or excessive prescription. (Further examples.)

1906 H. SAINSBURY *Principia Therapeutica* vi. 109 The purist..whilst limiting himself scrupulously to the use of one drug at a time, will seldom hesitate to prescribe the crude drugs,—opium, digitalis, bark, [etc.]..entirely oblivious of the fact that in so doing he is guilty of the most flagrant polypharmacy. 1928 SOLIS-COHEN & GITHENS *Pharmacotherapeutics* v. 379 There is a tendency at the present time to decry the association of remedies as 'polypharmacy', and to advocate the use of 'single medicines'. 1953 J. L. SIMONSEN *Plant Products & Utilisation* (Univ. Nottingham: Sir Jesse Boot Found. Lect.) 4 There is less polypharmacy now than formerly, but I am satisfied that there is less good prescribing now than in my student days. 1977 *Lancet* 26 Mar. 685/2 Therapeutic misadventures..are more likely in the elderly because of inappropriate dosage,..erratic pill-taking, and polypharmacy for multiple diseases.

polypharmacal *a.*, delete †*Obs.* and add examples.

1927 C. H. LA WALL *4,000 Yrs. Pharmacy* iii. 93 The Arabians perpetuated the polypharmacal combinations which had come down from the Egyptians. 1966 G. WATSON *Theriac & Mithridatium* iii. 114 Texts of ancient medical writers with polypharmacal formulae had become available.

polyphase, *a.* (*sb.*). Add: **b.** Consisting of or occurring in a number of separate stages.

1936 *Proc. Prehist. Soc.* II. 155 In 1932..I attempted an analysis of the evidence for a polyphase Ice Age. 1938 *Mem. Geol. Soc. Amer.* VI. 84 Heteroaxial symmetry means, therefore, a sideward drag in the course of tectonic flow, or a polyphase deformation. 1958 R. S. WOODWORTH *Dynamics of Behavior* ii. 39 The child's developing purposiveness spreads in the opposite direction. It is visible first in the little two-phase and polyphase acts, their time span being only a few seconds. 1969 BENNISON & WRIGHT *Geol. Hist. Brit. Isles* iv. 85 The Manx Slates have been affected by polyphase folding and low-grade metamorphism.

c. Consisting of or involving a number of different phases of matter.

1940 *Jrnl. R. Aeronaut. Soc.* XLIV. 538 Precipitation hardening leads generally to the formation of polyphase systems, and a solution hardened metal shows distinct advantages. 1950 *Proc. Amer. Acad. Arts & Sci.* LXXVIII. 167 Poly-phase, poly-component chemical systems. 1975 *Physics Bull.* May 225/1 The last chapter deals with microstructural and polyphase effects.

polyphasic (pǫlifēi·zik), *a. Physiol.* [f. POLY- + PHAS(E + -IC.] Having several successive peaks.

1922 *Amer. Jrnl. Physiol.* LIX. 278 The fall of temperature..ought to occupy about the same place in the diphasic as it does in the polyphasic thermocardiograms. 1936 *Brit. Jrnl. Psychol.* XXVII. 71 We observed three types of voluntary movements: The 'motor impulse effect', the polyphasic movement and the amorphous movement. 1968 *Brit. Med. Bull.* XXIV. 257/2 An example..is the so-called 'polyphasic' potentials occurring following partial denervation, in which a normal spike with a duration of several msec. is replaced by a repetitive series of much shorter spikes.

polyphenol(ic, -phenylene: see *POLY- 2.

po:lyphiloproge·nitive, *a.* [f. POLY- + PHILOPROGENITIVE *a.*] Very prolific, *spec.* of a person's talent, imagination, inventive powers, etc.

Quot. 1919 is perhaps influenced also by PHILOPROGENITIVE *a.* 2.

1919 T. S. ELIOT *Poems*, Polyphiloprogenitive The sapient sutlers of the Lord Drift across the window-panes. 1947 [see *NATTER *sb.*]. 1953 G. WILLIAMSON *Reader's Guide to Eliot* iv. 93 The first line, 'Polyphiloprogenitive', is not merely a *tour de force*, but a learned word which derides the quality that unites the modern Church functionary with the caterpillar world. 1963 *Punch* 4 Sept. 358/3 There remains his polyphiloprogenitive invention, which occasionally pushes a story across the starting line if you can bear the writing. 1966 *Ibid.* 31 Aug. 339/2, I find Heinlein too sentimental and full of cracker-barrel drollery for my taste; but he certainly has a polyphiloprogenitive talent.

polyphloisbic (pǫlifloi·zbik), *a. rare*⁻¹. [f. as POLYPHLOISBOIAN *a.* + -IC.] = POLYPHLOISBOIAN *a.*

1915 R. BROOKE *Lett.* (1968) 662 Will the sea be polyphloisbic and wine dark and unvintageable (you, of course, know if it is)?

polyphonal (pǫli·fŏnăl), *a. Mus.* [f. as POLYPHONE + -AL, after ANTIPHONAL *a.* and *sb.*] = POLYPHONIC *a.* 1. Hence **poly·phonally** *adv.*

1946 R. BLESH *Shining Trumpets* (1949) i. 8 The ultra-modern polyphonal and dissonantal school of today. *Ibid.* iii. 68 A woman's chorus..that sings, part antiphonally, part polyphonally, in undulating lines of chain-fourths.

polyphone. For '7–9 -phon' read '7– -phon'. In sense 1 a the forms **poliphant, polyphant** are usual. **1. a.** For *Obs.* read '*Obs.* except *Hist.*' and add later examples.

1954 *Grove's Dict. Mus.* (ed. 5) VI. 838/2 Queen Elizabeth was particularly partial to the poliphant. 1968 *New Oxf. Hist. Music* IV. xiii. 727 The drawing of the polyphant in Randle Holmes's *Academy of Armory* suggests a flat bandora-body surmounted by a harp-like frame. 1977 D. GILL *Wire-Strung Plucked Instruments contemp. with Lute* 19 Two other contemporary wire instruments have to be mentioned. One is the 'polyphant' or 'polyphone'. *Ibid.* 20 The 1671 inventory of Belvoir Castle does not list a polyphant.

c. (Further examples.)

1954 *Grove's Dict. Mus.* (ed. 5) VI. 848/1 In the 1880s the Polyphon was invented, in which projections punched up on a steel disc were used..to pluck the teeth of the comb. 1973 A. W. J. G. ORD-HUME *Clockwork Music* 108 Probably the best known of the musical box dealers and wholesalers was Henry Klein... His main business was in Polyphons and amusement machines. 1975 *Country Life* 11 Dec. 1715/2 (Advt.), Antique clocks, musical boxes, polyphones.

2. (Further example.)

1937 *Antiquity* XI. 273 Many of the Sumerian word-signs were polyphons.

polyphonic, *a.* Add: **2.** (*fig.* example.)

1920 H. CRANE *Let.* 15 Jan. (1965) 31 Your aristocrat is much more vital and admirable than the polyphonic God, chosen to symbolize the artist.

b. Of prose: written to sound pleasant and melodious.

1916 J. G. FLETCHER in *Poetry* Apr. 35 It seems fitting that a new name should be given to these poems of hers [*sc.* Amy Lowell's], which, printed as prose, or as prose and verse interspersed, display all the colors of the chromatic palette. The title that fits them best is that of Polyphonic Prose. 1917 A. LOWELL in *N. Amer. Rev.* Jan. 115 Metre, cadence, and rhyme are some of the many 'voices' employed in 'polyphonic prose'. Others are assonance, alliteration, and return. 1920 H. CRANE *Let.* 18 Aug. (1965) 41 Conrad's *Nigger of the Narcissus* seems to me all polyphonic prose. 1925 I. A. RICHARDS *Princ. Lit. Crit.* 135 Even the most highly organised lyrical or 'polyphonic' prose raises as it advances only a very ambiguous expectation. 1940 C. STRATTON *Handbk. Eng.* 249/2 *Polyphonic prose*, prose very carefully written to make the sounds pleasant and harmonious... The sound is obtained by attention to combinations and sequences of letters and syllables. 1977 *Amer. N. & Q.* XVI. 39/2 It is not improbable that the master of polyphonic prose was conscious of some metempsychosis which had taken place.

polyphonist. Restrict *rare* to sense 1 and add examples of sense 2.

1944 *Scrutiny* XII. 205 They [*sc.* Gesualdo's phrases] are related not so much to the great polyphonists as to the new technique of Monteverdi. 1954 *Grove's Dict. Mus.* (ed. 5) VI. 863/2 His [*sc.* Tallis's] music..has that expressive power characteristic of the later English polyphonists in its feeling for tonality and harmonic progression. 1968 *New Oxf. Hist. Mus.* IV. vii. 375 The Spanish polyphonists of this century [*sc.* the 16th] express a religious devotion and mystic fervour parallel to that of the painters and religious poets and prose-writers of the same epoch.

polyphonous, *a.* **3.** (Further example.)

1956 *Jrnl. Theol. Stud.* VII. 87 Transliterations would have been a great help to a Babylonian in enabling him to read ideograms and to determine the value of polyphonous signs.

polyphony. Add: **2.** (Further examples.) Also *transf.* and *fig.*

1965 *New Statesman* 10 Dec. 939/1 A polyphony of death, art and 'incorporeal love' was perhaps the most

exalted solution for a girl who would have disliked fulfilment. 1968 COATES & ABRAHAM in *New Oxf. Hist. Mus.* IV. vi. 329 The movement gradually expands into a suave polyphony..and reaches a fine climax at 'Gloria in excelsis'. 1973 C. D. GARRATT *Masterpieces in Steam* 133 The Austerity created a steady, even blast whilst the ancient lady in front wheezed and rasped away in a totally different rhythmic pattern, so creating a marvellous polyphony of sound. 1977 A. SHERIDAN tr. *Lacan's Écrits* ii. 17 This formal fixation..is the very condition that extends indefinitely his world and his power, by giving his objects their instrumental polyvalence and symbolic polyphony.

polyphonical *a.* (further example); also **polypho·nically** *adv.*, as regards polyphony, in a polyphonic manner.

1936 *Jrnl. Theol. Stud.* XXXVII. 168 This is exactly the point needed to explain the presence of a set of polyphonical Sequelae in our MS. 1936 *Scrutiny* V. 268 The increasing tendency in Beethoven's music to think of harmony..vertically and dramatically instead of horizontally and polyphonically. 1942 Polyphonically [see *MONOPHONICALLY *adv.* a]. 1946 R. BLESH *Shining Trumpets* (1949) iv. 87 The response lines begin to lose their strictly harmonic division into set chords and separate into independent melodic lines woven together polyphonically. 1959 *Listener* 8 Jan. 80/1 The polyphonically derived harmony intensifies the seventeenth-century partiality for modal variety and false relation.

polyphosphate, -phosphoric: see *POLY- 2.

Polyphoto: see *POLYFOTO.

polyphylesis (pǫ:lifəilī·sis). *Biol.* [Back-formation f. POLYPHYLETIC *a.*, after GENESIS.] The polyphyletic development of a species or other taxon. Also **polyphy·letism.**

1897 *Amer. Naturalist* XXXI. 281 Reinke..labors constantly under the delusion that those who contend for the distribution of the lichens, deny their polyphylesis. 1905 F. E. CLEMENTS *Res. Methods Ecol.* iv. 232 All have ignored the fact that the polyphylesis of genera carries with it the admission of such origin for species. 1926 *Jrnl. Bot.* LXIV. 119 The difficulties which arise from polyphylesis complicate the problem. 1951 G. H. M. LAWRENCE *Taxon. Vascular Plants* vii. 164 Polyphylesis is the situation represented by a polyphyletic origin. 1969 *Biol. Rev.* XLIV. 576 The appearance of bifid and trifid lobed forms [of ammonoids]..with the shell sculpture were seen as evidence of polyphyletism. 1978 *BioSystems* X. 82/1 This paradox may be due to polyphyletism within the chytrids.

polyphyly (pǫ·lifəi:li). *Biol.* [f. POLY- + Gr. φυλή tribe.] = prec.

1927 *Q. Jrnl. Geol. Soc.* LXXXII. p. ci, The whole of our System..is riddled through and through with polyphyly and convergence. 1961 G. G. SIMPSON *Princ. Animal Taxon.* iv. 124 The level of polyphyly is specified by the category of the highest ranking taxa two or more of which were immediately ancestral to the taxon in question. 1963 DAVIS & HEYWOOD *Princ. Angiosperm Taxon.* ii. 47 Proved or suspected polyphyly may lead us to see if there is a way of reclassifying the group into monophyletic units. 1978 *BioSystems* X. 110/1 Later modifications..considerably reduce this polyphyly, but at the expense of making the Protista even more of a rag bag.

polypier. Add *fig.* example.

1904 A. L. TEIXEIRA DE MATTOS tr. *Maeterlinck's Double Garden* 85 All nations have the natural right to pass through this phase of the political evolution of the human polypier.

polyplacophoran, *sb.* and *a.* (Later examples.)

1962 D. NICHOLS *Echinoderms* xii. 163 It is only fair to mention that other animal groups, such as the polyplacophoran molluscs,..have claimed them. 1973 P. TASCH *Paleobiol. Invertebr.* viii. 329/1 The polyplacophoran shell is composed of aragonite. 1976 *Nature* 2 Sept. 50/1 (*caption*) A polyplacophoran (*Chiton*) has removed sediment cover.

polyploid (pǫ·liploid), *a.* (*sb.*) *Biol.* [a. G. *polyploid* (H. Winkler 1916, in *Zeitschr. f. Bot.* VIII. 422): see POLY- and *-PLOID.] Having more than two homologous sets of chromosomes (in each cell nucleus). Also as *sb.*, a polyploid organism.

1920 W. E. AGAR *Cytol.* vii. 209 In the Protista..it appears that the nucleus may be polyploid, containing, not one or two, but a great number of series of elements. 1924 *Hereditas* V. 168 The chromosome number of a polyploid species must necessarily contain a certain number of complete haploid chromosome sets and it must have arisen through addition of such sets. 1928 [see *ALLOPOLYPLOIDY]. 1936 *Discovery* May 162/1 An account of the breeding behaviour of polyploid plants would be of almost universal interest since this class includes important crops such as wheat, oats, and tobacco. 1949 A. G. SANDERS in H. W. Florey et al. *Antibiotics* II. xvi. 683 In the field of agriculture polyploid forms of plants produced by chemical means have been of great importance as they are larger and often more vigorous than the normal plant. 1956 [see *EUPLOID *a.*]. 1963 E. MAYR *Animal Species & Evolution* xv. 439 Two types of polyploids are distinguished that have a rather different significance in evolution: autopolyploids and allopolyploids. 1975 J. B. JENKINS *Genetics* iv. 135 Although some animal tissues are commonly polyploid (the liver, for example), polyploid animals are rare. Most animal polyploids reproduce either hermaphroditically..or parthenogenetically.

Hence **po·lyploidy** [cf. G. *polyploidie* (E. Strasburger 1910, in *Flora* C. 406)], the condition of being polyploid.

1922 *Genetics* VII. 545 The value of polyploidy may have certain limitations. 1942, 1943 [see *endomitosis* s.v. *ENDO-]. 1973 *Nature* 11 May 87/2 Polyploidy may increase the flexibility of a species.

polyploidize (pǫ·liploidəiz), *v. Biol.* [f. prec. + -IZE.] *trans.* To render polyploid. Chiefly as **po·lyploidizing** *ppl. a.* Hence **po:lyploidiza·tion.**

1941 *Amer. Naturalist* LXXV. 128 It is often assumed that treatment with colchicine or other polyploidizing agents, if effective, will induce an exact doubling of each chromosome so that a balanced 4*n* condition, for example, will result throughout the part of the plant affected. This is very far from what actually happens. 1945 *Bot. Rev.* XI. 162 Schmuck found that wheat and barley seeds were polyploidized by acenaphthene, acenaphthylene, [etc.]. 1968 G. B. WILSON *Elem. Cytogenetics* v. 58 Many species of plants have been polyploidized by man through the use of conditions which disrupt mitosis and meiosis by preventing anaphase separation. 1974 *Oncology* XXIX. 520 Besides the loss and acquisition of chromosomes and polyploidization, these tumours were mainly characterized by the occurrence of multiple more or less complex translocations. 1975 *Nature* 31 Jan. 361/2 Cells treated with any one of the polyploidising agents gave rise to colonies, 13 of which were isolated and developed into established lines.

polyploidogenic (pǫ:liploidŏdʒe·nik), *a. Cytology.* [f. as prec. + -O + -GENIC.] Tending to produce polyploidy.

1944 [see *DIBENZANTHRACENE]. 1974 *Biol. Abstr.* LVII. 3982/2 In the sprouts there appear morphological changes (tumor-like thickenings in some parts) similar to those evoked by polyploidogenic substances.

polypneustic, -pnœa, -pnœic: see *POLY- 1.

polypod, *a.* Add: (*e*) [tr. It. *polipodo* (A. Berlese 1913, in *Redia* IX. 127)], esp., of a phase in the development of certain insect larvæ, having a segmented abdomen with rudimentary or functional appendages.

1925 A. D. IMMS *Gen. Textbk. Entomol.* 179 In the polypod phase the abdomen has acquired its complete segmentation and full number of appendages. 1969 R. F. CHAPMAN *Insects* xx. 400 A second basic form is the polypod larva... The larvae of Lepidoptera, Mecoptera and Tenthredinidae are of the polypod type.

polypore (pǫ·lipōəɪ). [f. mod.L. *Polyporus* (P. A. Micheli *Nova Plantarum Genera* (1729) 129), f. POLY- + Gr. πόρος pore.] A bracket-fungus belonging to the genus *Polyporus* or the family Polyporaceæ.

1902 *Science* 12 Dec. 954/1 A correspondent..sent me a fine specimen of a polypore which he found on the trunk of a tall tree. 1923 F. DICKSON in L. H. Bailey *Cultivated Evergreens* iv. 150 Most of the fungi causing these wood-rots are of one general type commonly known as 'bracket-fungi' or 'polypores'. 1946 *Nature* 7 Sept. 325/1 The paper contains an extensive key to the principal white re-supinate polypores in culture. 1971 P. H. B. TALBOT *Princ. Fungal Taxon.* xii. 201 The polypores and hydnums are typically woody, corky or membranous in texture. 1976 G. C. AINSWORTH *Introd. Hist. Mycol.* iii. 35 Greek and Roman writers certainly distinguished between agarics, polypores, and truffles.

polyporic (pǫli·pŏrik), *a. Biochem.* [tr. G. *polyporsäure* polyporic acid (C. Stahlschmidt 1877, in *Ann. d. Chem. u. Pharm.* CLXXXVII. 180), f. mod.L. *polypor-us* (see prec.) + -IC.] *polyporic acid*: a bronze-coloured crystalline solid, 3,6-dihydroxy-2,5-diphenyl-*p*-benzoquinone, $C_{18}H_{12}O_4$, which is a colouring matter found in certain fungi and lichens, and was first isolated from a fungus of the genus *Polyporus*.

1877 *Jrnl. Chem. Soc.* XXXII. ii. 620 The fungus..seems to be closely allied to Polyporus purpurascens... When brought into contact with dilute ammonia, its colour changes to a fine deep violet, and it yields a solution from which hydrochloric acid throws down a yellow precipitate of the new acid—polyporic acid. 1931 *Jrnl. Amer. Chem. Soc.* LIII. 2373 Two of the important coloring matters found in fungi are polyporic acid..obtained from *Polyporous* [sic] *nidulans* and atromentin..from *Paxillus atrotomentosus*. 1957 R. H. THOMSON *Naturally occurring Quinones* ii. 26 Polyporic acid also occurs in the fungus *Peniophora filamentosa*..and..has been found in two lichens. 1967 M. E. HALE *Biol. Lichens* viii. 106 Polyporic acid is known from *Sticta coronata* and *Polyporus nidulans*.

polyporus (pǫli·pŏrŏs). [mod.L.: see *POLYPORE.] **a.** = *POLYPORE.

1887 [see POLYPOROID *a.*]. 1907 T. R. SIM *Forests & Forest Flora Cape Good Hope* iii. 31 The attack of a Polyporus..softens and disintegrates the living heartwood.

b. A slice of a dried fungus of the genus *Polyporus*, esp. *P. betulinus*, used as a mount for particularly delicate insects. Also *attrib.*

1900 *Instructions for Collecting Insects* (Brit. Mus. (Nat. Hist.)) 6 A very useful material for staging is a fungus called *Polyporus*, which is cut into strips sold for the purpose by dealers. 1940 J. SMART *Instructions for Collectors, No. 4A: Insects* (Brit. Mus. (Nat. Hist.)) iv. 153 The smaller forms should be pinned on fine stainless steel points which are stuck into small strips of polyporus. 1951 COLYER & HAMMOND *Flies Brit. Isles* 334 If it is desired to pin medium-sized or small flies with short pins, they can be subsequently 'staged' on polyporus or slips of celluloid. 1962 GORDON & LAVOIPIERRE *Entomol. for Students of Med.* l. 303 Other forms of very useful..specialised equipment are entomological forceps and polyporus strips.

polyposis (pǫlip,ōu·sis). *Path.* [f. POLYP + -OSIS.] A condition characterized by the presence of numerous internal polyps, *esp.* a hereditary disease in which the large intestine is so affected and which becomes malignant if untreated.

1914 *Surg., Gynecol. & Obstetr.* XIX. 31/2 The symptoms of intestinal polyposis vary within wide limits. 1952 *Ann. Eugenics* XVII. 1 Polyposis is caused by an excessive proliferation of the glandular epithelium in the mucous membrane of the colon and rectum. 1961 [see *POLYPUS 2]. 1974 PASSMORE & ROBSON *Compan. Med. Stud.* III. xix. 112/2 Isolated cases of colonic polyposis without a family history are attributed to gene mutation.

polypropylene (pǫliprōu·pilīn). [f. POLY- + PROPYLENE. Cf. F. *polypropylène* (Berthelot 1867, in *Jrnl. de Pharm. et de Chim.* VI. 31).] Any of the polymers of propylene, which include a number of thermoplastic materials widely used as films, fibres, or moulding materials.

1935 [see *polybutadiene* s.v. *POLY- 2]. 1957 *Times* 17 July 7/3 As dense polythene advances in commercial use another type of plastic—the polypropylenes—appears to be established at the laboratory stage. 1964 *Which?* Aug. 253/3 Fabrics made from polypropylene can be easily washed and boiled. 1973 *Materials & Technol.* VI. viii. 528 Polypropylene is an eminently suitable material from which to prepare thin films. 1976 E. SCARROW *N.Z. Vegetable Gardening Guide* 14 Many modern spades have steel handles, with a polypropylene 'D' grip fitted to the handle.

polyprotein: see *POLY- 1.

polyprotic (pǫliprōu·tik), *a. Chem.* [f. POLY- + *PROTON 2 + -IC.] Of an acid: capable of donating more than one proton to a base; polybasic; occas. also used of bases which can accept more than one proton.

1944 J. A. TIMM *Gen. Chem.* xxxii. 343 Acids whose molecules may donate more than one proton are called polyprotic acids. 1968 J. G. MORRIS *Biologist's Physical Chem.* v. 114 In biological media we frequently encounter polyprotic, weak acids (e.g. carbonic, phosphoric, citric acids), whose complete neutralization requires the addition of two or more equivalents of sodium hydroxide. 1969 H. T. EVANS tr. *Hägg's Gen. & Inorg. Chem.* xii. 316 Acids and bases that can give up or take up more than one proton are called polyvalent..or polyprotic.

polypus. Add: **2.** (Further examples.) Cf. POLYP, POLYPE 2 in Dict. and Suppl.

1961 R. D. BAKER *Essent. Path.* xvi. 390 Adenomatous polyps and gastric polyposis are quite like their counterparts in the large bowel. The polypi are pedunculated or sessile and are composed of mucosa like that of the gastric wall. 1974 PASSMORE & ROBSON *Compan. Med. Stud.* III. xix. 52/2 A subtotal gastrectomy is recommended if the polypi occur in the mid and lower stomach.

polyreaction: see *POLY- 3.

polyrhythm (pǫ·liriðm). *Mus.* [f. POLY- + RHYTHM *sb.*] The use of two or more different rhythms simultaneously; music using such rhythms.

1929 P. ROSENFELD *Hour with Amer. Mus.* i. 12 Its alternation of bars of three and four and five units, the so-called jazz polyrhythm, is sheer willful contrast and change. 1942 *Scrutiny* XI. 12 The thirteenth century composer may teach the composer of the twentieth century how polytonalities and polyrhythms..may be reconciled with..the natural resources of the art of sound. 1949 *Funk's Stand. Dict. Folklore* I. 151/1 Musically, the blues are distinguished by..syncopation and polyrhythm characteristic of Negro music. 1956 W. MELLERS in A. Pryce-Jones *New Outl. Mod. Knowl.* iii. 363 Messiaen has created some fascinating noises out of the complex scales and polyrhythms of Indian music. 1973 *Black World* Sept. 37 The polyrhythms of 'bop' jazz are..in the contrasted regular rhythms of the first stanza and the irregularly punctuated rhythms of the second stanza. 1979 *Daily Tel.* 1 Nov. 15/6 In its simultaneous use of conflicting time signatures, of extremely intricate syncopations and polyrhythms, the late 14th century produced music of a complexity that has hardly been equalled until our own time.

polyrhythmic (pǫliri·ðmik), *a.* Chiefly *Mus.* [f. POLY- + RHYTHMIC *a.* and *sb.*] Involving or using two or more different rhythms, esp. at the same time. Also **polyrhy·thmical** *a.*

1893 J. S. SHEDLOCK tr. *Riemann's Dict. Mus.* 609/2 *Polyrhythmical*, i.e. containing a mixture of various rhythms. 1917 E. C. FARNSWORTH *Ideals & Tendencies Mod. Art* 69 That ultra phase of poetry *vers libre*, or, as some prefer, 'unrhymed cadence' or, what is more impressive, 'polyrhythmical poetry'. 1932 L. Saminsky *Music of our Day* i. 40 Jazz has shown that synthetic rhythm embraces not only straight polyrhythmic structures. 1942 *Scrutiny* XI. 15 A tonal structure based, however..polyrhythmic the music may grow, on the absolute and perfect consonances rather than on the diatonic triad. 1944 W. Apel *Harvard Dict. Mus.* 593/2 Properly speaking, all truly contrapuntal or polyphonic music is polyrhythmic, since rhythmic variety in simultaneous parts more than anything else contributes to giving the voice-parts that quality of individuality which is essential to polyphonic style. 1958 P. Gammond *Decca Bk. Jazz* xv. 178 He maintains an incredibly difficult polyrhythmical contrivance, playing in 3/4 time in the bass against the normal 4/4 in the treble. 1958 *Times* 9 Sept. 5/4 The relevant departments of the London Philharmonic proved inadequate to Bartok's polyrhythmic counterpoint. 1970 P. Oliver *Savannah Syncopators* 6 It was possible to agree on generalities concerning..the polyrhythmic texture of piano, guitar, bass and drums.

Hence **polyrhy·thmically** *adv.*

1946 R. Blesh *Shining Trumpets* (1949) xiii. 314 The ragtime left hand..is beyond the ability of the classically trained executant, let alone its combination polyrhythmically with the right. 1963 *Listener* 14 Mar. 457/1 African drumming relies on the interweaving of different strands of rhythm each of fixed beats, conflicting polyrhythmically with each other.

polyribo- (pǫliraibo). *Biochem.* [f. Poly- + *ribo-.] Formative element used in the names of polymers of ribonucleotides, as *polyri:boadeny·lic, -cytidy·lic, -inosi·nic*, etc., *acid*; also *po:lyribonu·cleotide*. Cf. *polyribosome* s.v. *poly- i.

1956 *Nature* 11 Feb. 271/1 Some 10–20 per cent of the total polyribonucleotide content of the bacteria was extracted by this procedure. 1959 *Times* 10 Nov. (Guinness Suppl.) p. ii/6 An important series of papers on the synthesis of polyribonucleotides which have an important function in cellular metabolism. 1961 Steiner & Beers *Polynucleotides* i. 6 The equimolar complexes formed by polyriboadenylic acid with polyribouridylic acid and with polyriboinosinic acid appear to have doubly stranded helical structures. *Ibid.* viii. 263 Even less is known of the detailed fine structure of polyribocytidylic acid... All that can be said is that some helical structure is present. 1964 G. H. Haggis et al. *Introd. Molecular Biol.* ix. 228 (*caption*) T = polyribothymidylic acid (a polyribonucleotide containing only the base thymine found in natural DNA but not in natural RNA). 1970 *New Scientist* 15 Jan. 96/2 Poly I:C—a combination of polyriboinosinic and polyribocytidylic acids—would stimulate interferon production both in cell cultures and in animals. 1976 *Nature* 15 Jan. 141/2 Antibodies to native DNA, double-stranded RNA, and various synthetic polyribonucleotides occur with great frequency in patients with systemic lupus erythematosus (SLE).

So also **polyde(s)oxyribo-** (see *deoxy-), in names of polymers of deoxyribonucleotides, as *po:lyde(s)oxyribonu·cleotide*.

1956 *Federation Proc.* XV. 291/2 To define the chemical events in the development of a bacterial virus, we have explored the pathways of polydesoxyribonucleotide synthesis in normal and infected cells. 1961 Steiner & Beers *Polynucleotides* i. 5 The primary structures of polydeoxyribonucleotides and polyribonucleotides are identical except for the absence of the hydroxyl group on C_2' of deoxyribose. 1976 W. Guschlbauer *Nucleic Acid Struct.* vi. 86 Single-stranded polydeoxyribonucleotides are, as a rule, less stacked and structured than their ribo counterparts.

polysaccharide (pǫlisæ·kārəid). *Chem.* Formerly also **-id.** [ad. G. *polysaccharid* (B. Tollens *Kurzes Handbuch d. Kohlenhydrate* (1888) 16), f. Poly- + *saccharid* Saccharide (in Dict. and Suppl.).] Any carbohydrate whose molecules consist of a number of monosaccharide residues (or their simple derivatives) bonded together, usu. in a chain structure, and esp. one of high molecular weight; also applied to such a structure which forms part of a larger molecule.

1892 E. F. Smith tr. *V. von Richter's Org. Chem.* (ed. 2) 512 It is very probable that the polysaccharides having the empirical formula $C_6H_{10}O_5$, really possess a much higher molecular weight, $(C_6H_{10}O_5)_n$. 1895 *Jrnl. Chem. Soc.* LXVIII. ii. 322 It appears probable that the fermentation of the polysaccharides by saccharomycetes is preceded by their conversion into monosaccharides through the agency of enzymes. 1902 *Encycl. Brit.* XXXI. 723/2 By further polymerization and loss of water the group of polysaccharides..is produced. 1947 *Endeavour* VI. 89/2 Other substances of high molecular weight, such as the polysaccharides, consist mostly of molecules of continuously varying size. 1951 *Sci. News* XXI. 72 Any carbohydrate, such as glucose, cellulose, cane-sugar, or starch, can be represented by the formula $C_n(H_2O)_m$ where *m* is equal, or very nearly equal, to *n*... For polysaccharides such as starch or cellulose, *n* and *m* may run up to hundreds. 1960 *New Biol.* XXXI. 72 The virulent and avirulent types [of pneumococcus] can be quickly and easily distinguished because the virulent cells are enclosed in a polysaccharide capsule that can be seen under the microscope. 1968 A. White et al. *Princ. Biochem.* (ed. 4) xli. 910 (*caption*) Two units from neighboring polysaccharide chains can be bridged by a peptide. 1969 *New Scientist* 7 Aug. 270/1 Some natural polysaccharides show startlingly similar conformational behaviour to proteins and nucleic acids. 1973 R. G. Krueger et al. *Introd. Microbiol.* xxiv. 590/2 The major antigenic components of bacteria and their products are polysaccharides of one sort or another.

† **polysa·ccharose.** *Chem. Obs.* [f. Poly- + Saccharose.] = prec.

1894 Perkin & Kipping *Org. Chem.* I. xv. 275 The polysaccharoses do not ferment with yeast, and do not reduce Fehling's solution. 1931 E. C. Miller *Plant Physiol.* viii. 411 The polysaccharoses have the general formula $(C_6H_{10}O_5)_n$ or $(C_5H_8O_4)_n$ depending on whether they yield hexoses or pentoses on hydrolysis.

polysaprobic, -semantic(ity, -ism: see *poly- i.

polysemy (pǫlisīmi). *Linguistics.* Also in mod.L. form **polyse·mia.** [ad. F. *polysémie* (M. Bréal *Essai de Sémantique* (1897) xiv. 155), f. med.L. *polysēmus* (see *polysemous* adj. s.v. Poly- i and below): see -y³, -ia¹.] The fact of having several meanings; the possession of multiple meanings.

1900 N. Cust tr. *Bréal's Semantics* xiv. 140 The new meaning of a word, whatever it may be, does not make an end of the old. They exist alongside of one another... In proportion as a new signification is given to a word, it appears to multiply and produce fresh examples, similar in form, but differing in value. We shall call this phenomenon of multiplication *Polysemia*. All the languages of civilised nations have their part in it. 1928 O. Jespersen *Monosyllabism in Eng.* 26 We now see the reason why polysemy is found so often in small words to an extent which would not be tolerable in longer words. 1931 G. Stern *Meaning & Change of Meaning* iv. 74 *I wish you luck*... This polysemy is quite different from the polysemy..of *Hund*, signifying either 'dog', or 'kind of cart used in mines'. 1937 J. Orr tr. *Iordan's Introd. Romance Linguistics* 192 In Gilliéron's view, sound-change, with all its consequences, homonymy, polysemia, and the like, are causes of disease in words. 1950 S. Potter *Our Language* 110 If we assume that the central meaning of *place* is still 'square' and that these other diverse uses *radiate* from that centre, we might equally well put it into our third semantic category: radiation, polysemia, or multiplication. 1951 S. Ullmann *Princ. Semantics* ii. 115 Should one describe 'a straight line' and 'shipping line, air line' as radical shifts in application or as mild cases of polysemy? *Ibid.* 117 Polysemy is the pivot of semantic analysis. 1960 W. F. Twaddell *Eng. Verb Auxiliaries* 3 We can acknowledge the existence of meaningful lexical verbs in our syntax, and gracefully recognize a linguistically reasonable polysemia of our grammatical signals within different lexical contexts. 1972 M. L. Samuels *Linguistic Evol.* v. 75 The effect of polysemy is in principle the same as that of homonymy—the representation of two or more meanings by a single form. 1975 *Times Lit. Suppl.* 16 May 531/1 Matters are complicated by the polysemy of the noun *linguist*, both 'polyglot' and 'scientific student of language'. 1977 *Dædalus* Summer 77 Thus symbol is distinguished from sign both by the multiplicity (multivocality, polysemy) of its signifieds, and by the nature of its signification.

So **po·lyseme**, a word having several or multiple meanings; **polyse·mic, polyse·mous** *adjs.* [see s.v. Poly- i], of or pertaining to polysemy; having several meanings, exhibiting polysemy.

1884 Polysemous [in Dict. s.v. Poly- i]. 1930 *S.P.E. Tract* xxxiv. 463 Even the names of concrete things are nearly always polysemic, though this may not be perceptible until we compare them with corresponding words in other languages. The word *leg*, for instance, may be applied to the supports of a table or chair, and the legs of an insect in English, but not in French. 1931 G. Stern *Meaning & Change of Meaning* iii. 32 Different meanings can be expressed by the same word, as instanced by *crown* or any other polysemous word. 1953 *Trans. Philol. Soc.* 60 Identifications of this type are..most convincing when parallel translation-pairs are found as homonyms within a single language, which one would then wish to consider as polysemes. 1954 *Eng. Stud.* XXXV. 170 The cropping up of new senses may lead to polysemic conflicts, causing older senses to disappear. 1957 N. Frye *Anat. Crit.* 72 The principle of manifold or 'polysemous' meaning, as Dante calls it, is not a theory any more..but an established fact. 1969 *Times Lit. Suppl.* 18 Dec. 1445 The earnest digging goes on, and we are..grateful for the unearthing of new polysemes [in *Finnegans Wake*]. 1973 *Ibid.* 8 June 640/2 The book begins with an essay..which makes the point that a work of thought is polysemous, that its meaning is not a message but rather the effect which it has on its readers. 1974 *Amer. Speech* 1971 XLVI. 125 Polysemes, or terms that exhibit more than one denotation each, even though their connotations are synonymous in their negativism. 1976 G. Steiner in D. Villiers *Next Year in Jerusalem* 67 The elaborate investigations of the Kabbalists into the polysemic nature of the written word.

polyserositis, -siloxane, -soap: see *poly- i, 2, i.

polysomatic (pǫlisomæ·tik), *a.* [f. Gr. πολυσώματ-ος with many bodies: see -ic.]

1. *Petrol.* [ad. G. *polysomatisch* (G. Tschermak *Die mikrosk. Beschaffenheit der Meteoriten* (1885) i. 12).] Consisting of more than one grain or more than one mineral.

1888 *Amer. Geologist* I. 201 The boundaries between the different members of these 'polysomatic' masses of augite are traceable only with difficulty. 1910 *Mineral. Mag.* XV. 356 The commonest are polysomatic olivine chondrules with either granular or porphyritic structure. 1920 *Proc. Nat. Acad. Sci.* VI. 455 The porphyritic forms [of chondrule]..pass gradually into those which are almost or quite holocrystalline and polysomatic. 1973 G. J. H. McCall *Meteorites & their Origins* xv. 191 Polysomatic chondrules may consist of numerous grains of a single mineral..or of more than one mineral species.

2. *Biol.* [ad. G. *polysomatisch* (O. F. I. Langlet 1927, in *Svensk Bot. Tidskr.* XXI. 3).] Of, pertaining to, or exhibiting polysomaty.

1937 *Cytologia* VIII. 270 Although the possibility exists that polysomatic cells may arise in a number of different ways, there is evidence, if we can consider as significant the paired condition of polysomatic metaphase chromosomes, that the process involving two successive cleavages of the chromosomes is more widespread in plants than heretofore supposed. 1948 *Nature* 17 Jan. 80/2 These first studies..indicated that a very large proportion of differentiated cells behind the meristematic region are polysomatic. 1969 Brown & Bertke *Textbk. Cytol.* xxiii. 538/1 Examples of the second category, replication, are differentiating cells that produce polytene and/or polysomatic nuclei..[etc.].

polysomaty (pǫlisǝu·māti). *Biol.* [f. as prec.: see -y³.] The occurrence of polyploid cells together with diploid cells in the same somatic tissue.

1937 *Cytologia* VIII. 247 In *Kochia scoparia* L...the large periblem cells of the root were found to exhibit the same phenomenon of polysomaty as those of *Spinacia*. 1962 *Lancet* 12 May 1005/1 Very possibly, spindle formation is upset in the divisions preceding the formation of orthochromatic erythroblasts, thus giving rise to polysomaty. 1969 Brown & Bertke *Textbk. Cytol.* xix. 424/2 In numerous species such as spinach, *Cannabis*, potato, onion, beet, etc., polysomaty occurs typically in root tips and in many species in shoot tips and leaves.

polysome (pǫ·lisǝum). *Biol.* [f. Poly- + *ribo)some.] A cluster of ribosomes, held together by a strand of messenger RNA which each is translating; = *polyribosome* s.v. *poly- i.

1962 J. R. Warner et al. in *Science* 28 Dec. 1399/2 We have been able to show that the site of hemoglobin synthesis in vivo is not the single ribosome but rather a cluster of ribosomal particles, which we have called a 'polyribosome' or simply a polysome. 1970 *New Scientist* 15 Oct. 113/1 This reticulum is a series of intracellular membranes which carry on them the polysomes (messenger RNA and ribosomes) thought to be responsible for the biosynthesis of proteins destined for export from the cell. 1971, 1972 [see *monosome 2]. 1973 R. G. Krueger et al. *Introd. Microbiol.* xi. 336/2 Each ribosome in the polysome has a growing polypeptide attached to it, and the ribosomes near the 3' terminus of the mRNA molecule will have the most nearly complete polypeptide chains attached to them.

Hence **polyso·mal** *a.*, of or pertaining to a polysome.

1962 *Science* 28 Dec. 1402/1 Dialysis followed by spraying causes considerable degradation of the polysomal structure. 1972 *Nature* 31 Mar. 237/2 The ratio of cytoplasmic to chloroplast polysomal RNA is increased two to three times 24h after inoculation.

polysomic (pǫlisǝu·mik), *sb.* and *a.* *Cytology.* [f. Poly- + *-some⁴ + -ic.] **A.** *adj.* Having one or a few normal chromosomes in excess of the usual diploid or polyploid complement; being such a chromosome. **B.** *sb.* A polysomic organism.

1932 C. D. Darlington *Rec. Adv. Cytol.* iii. 66 Polysomic forms arise in a diploid through two daughter chromosomes passing to the same pole at mitosis, or at meiosis. 1937 T. Dobzhansky *Genetics & Origin of Species* iv. 80 Spontaneous polysomics, monosomics, polyploids, haploids, and translocations were observed in *Datura stramonium* by Blakeslee (1922). 1939 *Jrnl. Genetics* XXXVIII. 409 All five asynaptic polysomics possessed three extra chromosomes. 1949 R. A. Fisher *Theory of Inbreeding* iv. 78 Polysomic organisms differ from disomic in having more than two chromosomes mutually homologous and capable of pairing, and of interchange of segments. 1949 K. Mather *Biometrical Genetics* x. 305 A general treatment of polysomic inheritance has not been attempted because of its inherent complexity. 1966 J. A. Serra *Mod. Genetics* II. xii. 35 Cases in which one chromosome of the set becomes polysomic (above the normal number) are known in which the chromosome remains euchromatic and other cases in which it is heterochromatinized. Genetic imbalance and diminished viability or more or less marked lethality accompany, as a rule, the cases of polysomy when the chromosome remains euchromatic, while heterochromatic supernumeraries do not, in general, produce such effects and, probably, are of use to the cell.

polysomy (pǫ·lisǝumi). *Cytology.* [f. as prec. + -y³.] The state of being polysomic.

1932 C. D. Darlington *Rec. Adv. Cytol.* iii. 59 Reduplication of some of the chromosomes of a set beyond the normal diploid number is called polysomy. 1946 *Nature* 17 Aug. 239/2 It is not impossible that polysomy and failure of pairing may be jointly responsible for the abnormal numbers [of chromosomes] observed. 1966 [see prec.]. 1973 *Cytogenetics & Cell Genetics* XII. 87 The first recorded observation of XYY polysomy in man was made by Sandberg et al. in 1961.

polyspermic (pǫlispɜ̄·ɹmik), *a.* *Physiol.* [f. Polysperm(y + -ic.] Involving or exhibiting polyspermy.

1890 Billings *Med. Dict.* II. 368/2 *Polyspermic*, requiring more than one spermatozoon to fructify the egg. 1894 *Anatomischer Anzeiger* IX. 146 Stained preparations of these eggs show them to be, previous to division, polyspermic. 1953 *Austral. Jrnl. Biol. Sci.* VI. 674 In the

polyspermic rat egg, the chromosome complements from the female pronucleus and the male pronuclei all take part in the formation of the first cleavage spindle. 1975 *Nature* 8 May 112/1 Since silkworms are polyspermic the opportunity then arises for two different male pronuclei to fuse and form a diploid zygote nucleus.

polyspike: see *poly- i.

polyspory (pǫ·lispǭəri). *Bot.* [f. Poly- + Spor(e + -y³.] The production of unusually many spores.

1929 *Genetics* XIV. 213 Giant spores, dyads, quartets and polyspory with and without extra, small nuclei or chromatin masses were repeatedly observed. 1959 *Canad. Jrnl. Plant Sci.* XXXIX. 272 A case of polyspory in a *Triticum-Agropyron* hybrid line..proved to be due to spindle misfunction. 1970 *Indian Jrnl. Exper. Biol.* VIII. 128/2 The chromosome fragmentation and lack of active polar movement result in laggards which persist as micronuclei and consequently lead to polyspory.

polystelic (pǫlistī·lik), *a.* *Bot.* [a. F. *polystélique* (P. van Tieghem & H. Douliot 1886, in *Ann. Sci. Nat. Bot.* 7 ser. III. 276), f. Poly- + Stele 2 + -ic.] Of a stem or root: having more than one internal vascular cylinder or stele. So **po·lystele** (see quot. 1965); **po·lystely,** polystelic condition.

1891 *Ann. Bot.* V. 515 In the Cryptogams above cited..the original cylinder branches and the stem becomes polystelic. *Ibid.* 516 The two Dicotyledonous genera, in which alone, so far as we know, polystely prevails, belong to families remote from each other. 1896 Polystelic [see Poly- i]. 1902 *Phil. Trans. R. Soc.* B. CXCV. 128 In *Anemia phyllitidis* the adult stem is characterised by the presence of so-called polystelic structure. 1902 *Encycl. Brit.* XXV. 413/2 This is the condition of astely, entirely parallel with polystely except that the separate strands are usually all or mostly leaf-traces. 1908 Boodle & Fritsch tr. *Solereder's Systematic Anat. Dicotyledons* II. vi. 1156 When the axis shows several steles in a transverse section, it is said to be polystelic. 1925 Eames & MacDaniels *Introd. Plant Anat.* v. 133 In contrast with the monostele was the 'polystele', a type of stele in which the vascular tissues are in the form of strands. 1938 *Current Sci.* VI. 383 Polystely occurs in the cortical region of the stolons. *Ibid.* 384 The polystelic condition observed in these plants might be regarded as the anatomical expression of the renewed adoption of the terrestrial habit. 1965 Bell & Coombe tr. *Strasburger's Textbk. Bot.* 136 We come ..to the polystele, consisting of a system of individual vascular bundles, distributed over the whole of the transverse section. 1969 F. E. Round *Introd. Lower Plants* x. 125 (*caption*) Stellar arrangements in the vascular cryptogams... N_1 polystele.

polystyrene (pǫlistəi·rīn). [f. Poly- + *styrene.] Any polymer of styrene, esp. a hard, colourless thermoplastic resin; also, any of various plastics made from or containing this, which are widely used as moulding materials, films, and rigid foams. Also *attrib.*

1927 G. S. Whitby in *India-Rubber Jrnl.* 16 Apr. 16/2 Poly-styrene becomes markedly elastic when warmed or swollen. *Ibid.*, Unstretched poly-styrene gave only an 'amorphous ring' when subjected to X-ray examination. 1939 *Nature* 13 May 787/2 The vinyl and polystyrene resins have valuable properties some of which are not possessed by the other resinoids. 1945 *Electronic Engin.* XVII. 698/1 Polystyrene..has long been considered to be the lightest in weight of all plastics. 1959 *Economist* 7 Mar. 895/2 The present output of polystyrene is about 28,000 tons a year, divided between Shell, Distillers and Monsanto. 1961 *Wall St. Jrnl.* 9 June 20/3 Koppers Co. is displaying a complete 12-foot sailboat made of polystyrene foam. 1970 N. Saunders *Alternative London* 18 Sack chairs..consist of a chair-shaped bag three-quarters full of expanded polystyrene granules. 1973 *Materials & Technol.* VI. viii. 530 Polystyrene is readily polymerized exothermally to the straight polystyrene (homopolymer). *Ibid.* 533 Polystyrenes are manufactured in two stages. In the first stage the monomer styrene, possibly together with other monomers, is converted into the polymer... In the second stage the polymer is granulated, other substances frequently being added at the same time; rubbers,..oxidizers, pigments and the like, are such additions.

polystyrol, -sulphide, -sulphone: see *poly- 2. **polysymptomatic:** see *poly- i.

polysynthetic, *a.* Add: **1.** Also applied to twinning of this kind. (Further example.)

1944 *Amer. Mineralogist* XXIX. 199 Under the microscope, hydrotungstite is seen to occur as tiny green plate-like crystals which show polysynthetic twinning.

2. (Earlier and later examples.)

1816 P. S. Duponceau *Let.* 30 Aug. in *Trans. Hist. & Lit. Comm. Amer. Philos. Soc.* (1819) I. 430 Crantz and Egede prove in the most incontrovertible manner that the language of Greenland is formed on the same *syntactic* or *polysynthetic* model. 1977 *Language* LIII. 10 Particularly interesting would be a polysynthetic language with many layers of morphology built into a single word.

polysynthetically *adv.* (examples in *Cryst.*).

1903 *Amer. Geologist* XXXII. 67 Within it are small, triclinic, polysynthetically twinned feldspars which are rather vaguely crystallized. 1968 I. Kostov *Mineral.* 304 The crystals are almost invariably polysynthetically twinned.

polysystemic (pǫlisistī·mic), *a.* *Linguistics.* [f. Poly- + Systemic *a.*] Composed of, characterized by, or recognizing many systems; used esp. with ref. to prosodic analysis.

1949 J. R. Firth in *Trans. Philol. Soc. 1948* 151 The monosystemic analysis based on a paradigmatic technique of oppositions and phonemes with allophones has reached, even overstepped, its limits! The time has come to try fresh hypotheses of a polysystemic character... The phonological structure of the sentence and the words which comprise it are to be expressed as a plurality of systems of interrelated phonematic and prosodic categories. 1957 *Year's Work Eng. Stud. 1955* 28/2 It remains a matter for regret that Firth's polysystemic approach to language has not yet been fully and explicitly formulated, for it is clear..that it has much to offer us. 1964 *Language* XL. 315 Firth..indicated that it is a property of language itself to be irreducibly multistructural and polysystemic. 1970 B. M. H. Strang *Hist. English* ii Language..is not mono- but polysystemic. 1973 *Archivum Linguisticum* IV. 21 Such a three-choice system is inherent in all polysystemic approaches to linguistic structure.

So **polysyste·mically** *adv.*, in a polysystemic manner; **po:lysystemi·city, polysy·stemy,** the fact or condition of being polysystemic.

1964 *Archivum Linguisticum* XVI. 72 Martinet's recognition of the polysystemicity of language (instanced by the different vowel contrast systems in French final open and closed syllables..) puts him in some agreement on this point..with the Firthian position in linguistic analysis. 1964 *Language* XL. 315 There is at least one instance..in which the appeal to linguistic polysystemy has the consequence of preventing the analyst from making an adequate descriptive statement. 1966 T. Hill in C. E. Bazell *In Memory of J. R. Firth* 218 The examples will also illustrate the principle of polysystemicity as applied to different grammatical categories. 1973 *Archivum Linguisticum* IV. 21 The choice of tone at each place in the structure can be considered polysystemically.

Polytec, -tech (pǫlite·k), colloq. abbrev. of Polytechnic *sb.* 2 (in Dict. and Suppl.); = *poly².

1911 O. Onions *Widdershins* x. 192, I don't think I shall go to the Polytec. to-night. 1974 'E. Lathen' *Sweet & Low* xix. 182 Why shouldn't I speak English well? I went to Rensselaer Polytech. 1977 *N.Z. Woman's Weekly* 10 Jan. 16/3 The weaving course at the Polytec has generated a great interest in looms. 1977 *Belfast Tel.* 14 Feb. 22/3 Now they have given the Polytech's first big romance a fairy tale ending by getting engaged.

polytechnic, *a.* and *sb.* Add: **A.** *adj.* (Further examples.)

1837 *Penny Mechanic* II. 92/2 A Sample School, to be called the Polytechnic University, No. 1 for 2000 students. 1921 Beerbohm *Lett. to R. Turner* (1964) 258 The incredible job [*sc.* H. G. Wells's *History of the World*], done so neatly..in a very awful cheap sciolistic polytechnic way. 1965 *Economist* 11 Sept. 1000/1 In the eyes of authority, naval history has remained a soft option in a polytechnic world.

B. *sb.* **2.** (Earlier and later examples.) In mod. use, a kind of institution of higher education offering courses mainly in technical and vocational subjects (see quot. 1973).

1836 C. Fox *Jrnl.* 31 Aug. (1972) 31 Dr. Buckland..came on to the Polytechnic and stayed with us. 1841 M. Edgeworth *Let.* 25 May (1971) 593 Lestock..took Honora and Captain Beaufort and me to the Polytechnic and we all had our likenesses taken. 1850 W. Howitt *Yr.-Bk. Country* iv. 111 Such places as Saint Paul's and Westminster Abbey should stand wide open; the Colosseum and the Polytechnic be accessible at the smallest price. 1857 C. Kingsley *Two Yrs. Ago* I. vii. 171 He would thrust his head into lectures at the Polytechnic and the British Institution. 1934 G. B. Shaw *On Rocks* ii. 237 Jafna's grandsons will go to Eton. Mine will go to a Polytechnic. 1967 *Listener* 6 July 5/1 Mr. Crosland and his advisers envisage rather an eternal separation between the universities and an entirely new race of animals they have created called the polytechnics. 1973 *Times* 4 Oct. 4/4 Polytechnics differ from universities in that they are not centrally financed, teach courses for degrees of the Council for National Academic Awards (CNAA), have a substantial proportion of students on courses below degree level, do much less research, and have, in theory, a greater commitment to the vocational aspect of higher education. 1975 *Physics Bull.* Jan. 6/2 It is natural to think of the polytechnics as being primarily concerned with science and technology but this is not so. Only one third of the work of the average polytechnic lies in these fields. 1975 *Guardian* 27 Jan. 5/1 The higher education building programme..will contain a bias in favour of the polytechnics.

4. *attrib.*

1839 C. Fox *Jrnl.* 8 Oct. (1972) 58 The Bucklands dined with us, after a Polytechnic morning. 1911 O. Onions *Widdershins* 184 It was of Polytechnic classes that he spoke. *Ibid.* 189 The young Polytechnic student. 1972 R. K. Kelsall et al. *Graduates* i. 53 Parents of..university students..find university education more acceptable than do parents of..Polytechnic students. 1972 *Accountant* 19 Oct. 483/2 Polytechnic lecturers on a secondment period of six weeks for updating in auditing techniques. 1973 *Times Higher Educ. Suppl.* 20 July 12/1 With the formal inauguration of the Association of Polytechnic Teachers..yet another teachers' organization has emerged. 1979 V. S. Naipaul *Bend in River* x. 170 The polytechnic term was over.

Hence **po:lytechniza·tion,** the action or process of making (some activity) polytechnic; *spec.* in Communist countries, the process of educating children in technical and industrial subjects considered essential for the proper running of the State.

1932 *Times Lit. Suppl.* 17 Mar. 204/2 It would have been of advantage to provide a fuller and more detailed account..of the system of 'polytechnization' in the primary schools [of the U.S.S.R.]. 1933 *Times Educ. Suppl.* 25 Feb. 57/4 Polytechnization 'aims at producing a nation of socialistically thinking technical specialists'. 1949 K. Davis *Human Society* viii. 229 Economically, through 'polytechnization', the school is geared with productive life. 1974 *Encycl. Brit. Macropædia* VI. 375/2 From the 1950s onward, much attention has been paid [in Communist education] to the ideal of 'polytechnization'.

polytene (pǫ·litīn), *a.* *Cytology.* [Poly- + *-tene.] Applied to giant chromosomes found in certain interphase nuclei, esp. in dipterous insects, and composed of many parallel copies of the genetic material, in which the active regions may be identifiable microscopically.

1935 P. C. Koller in *Proc. R. Soc.* B. CXVIII. 372 We can regard these chromosomes as corresponding with paired pachytene chromosomes at meiosis in which the intercalary parts between chromosomes have been stretched and separated into smaller units, and in which, instead of two threads lying side by side, we have 16 or even more. Hence they are 'polytene' rather than pachytene; I do not, however, propose to use this term; I shall refer to them as 'multiple threads'. 1959 C. M. M. Begg *Introd. Genetics* iii. 27 The specificity of this pairing [of homologous chromosomes] is shown very strikingly in the case of the exceptional polytene threads of the much enlarged salivary gland nuclei of Diptera. 1971 *Nature* 21 May 184/2 Individuals of the species which breeds in freshwater can only be identified with certainty by the banding patterns of their polytene chromosomes. 1976 Bell & Coombe tr. *Strasburger's Textbk. Bot.* (rev. ed.) 31 (*caption*) Giant chromosome, polytene and consisting of about 2048 strands, from the suspensor of the embryo of *Phaseolus vulgaris.*

Hence **polytenic** (-tī·nik) *a.*, in the same sense; **po·lyteny,** the state of being polytene; also **po:lyteniza·tion,** the production of polyteny; **polyte·nized** *ppl. a.* (of a chromosomal constituent) reduplicated owing to the polyteny of the chromosome.

1942 *Proc. R. Soc. Edin.* B. LXI. 318 The degree of polyteny found in the tissues of a dipteran larva varies with the species. 1953 W. Hovanitz *Text-bk. Genetics* vii. 100 The terms salivary chromosomes and polytenic chromosomes have been applied to them [*sc.* giant chromosomes]. 1958 Polyteny [see *mixoploidy]. 1966 *Proc. R. Soc.* B. CLXIV. 280 After many steps of polytenization (12 or 14 consecutive replications) such a labelled chromosome thread may still extend as a single linear unit from the one end of the giant chromosome to the other end. 1968 H. Harris *Nucleus & Cytoplasm* iv. 77 The chromosomes in these [salivary] glands are extraordinary for two reasons. The first is that they are grossly polytenic. Each chromosome contains thousands of identical parallel copies of the basic diploid genetic structure. The extent of polyteny may vary in different species, but for *Chironomus tentans* it has been estimated that each chromosome contains about 16 000 times the normal diploid amount of DNA. 1974 *Nature* 29 Mar. 446/1 Mutational data also rule out simple polyteny and polyploidy, where every single gene on every chromosome is polytenised or where every chromosome is present in two or more copies. 1976 *Ibid.* 17 June 614/2 Ciliata and Diptera..show a high tendency for endopolyploidisation and polytenisation. 1976 Bell & Coombe tr. *Strasburger's Textbk. Bot.* (rev. ed.) 31 In rare cases repeated multiplication of chromonemata leads to multistranded (polytenic), cable-like, giant chromosomes.

polyterpene, -terpenoid: see *poly- 2.

po:lytetraflu:oroe·thylene. *Chem.* Also **-fluorethylene.** [f. Poly- + *tetrafluoroethylene.] A highly crystalline resinous polymer of tetrafluoroethylene having the structure $—(CF_2CF_2)_n—$, which is tough, resistant to chemicals, stable over a wide range of temperature, and widely used as a moulding material, esp. under the trade name *Teflon*; abbr. *P.T.F.E.* (s.v. *P II. d).

1946 *Industr. & Engin. Chem.* Sept. 877/1 Although pure polytetrafluoroethylene is white in color, frequently commercially fabricated articles are somewhat gray and speckled, apparently as a result of minute traces of contamination. 1951 *Electronic Engin.* XXIII. 370 In order to obtain a smooth bearing, use has been made of polytetrafluoroethylene for the bearing surface. 1959 *Times Rev. Industry* June 22/1 Gaskets of polytetrafluorethylene or Fluon have proved invaluable in difficult services, since this material is resistant to all 'searching' chemical reagents except molten sodium and gaseous fluorine. 1961 *Lancet* 19 Aug. 410/1 Polytetrafluoroethylene (P.T.F.E., 'Fluon', 'Teflon') is at present the most suitable material for permanent cannulation. 1962 *Which?* Aug. 255/1 There are two kinds of non-stick frying pans—those with a silicone finish and those with a plastic called polytetrafluoroethylene, or PTFE. 1975 *Sci. Amer.* July 63/1 Non-metals such as..polytetrafluoroethylene and carbon-graphites are successful bearing materials because of their excellent resistance to scoring and corrosion. 1978 J. Sherwood *Limericks of Lachasse* i. 8 The plant..made polytetrafluorethylene, a polymer with three times the density of polypropylene and enormously strong.

polytheistically, *adv.* (Example.)

1909 W. James *Pluralistic Universe* viii. 310 It [*sc.* the superhuman consciousness] may be polytheistically or it may be monotheistically conceived of.

polythelia, -ism, -y: see *poly- i.

polythene (pǫ·liþīn). [Contraction of *polyethylene.] A tough, light, translucent thermoplastic made by polymerizing ethylene and used esp. for moulded and extruded articles, as film for packaging, and as a coating. Freq. *attrib.* and in *Comb.*

1939 *Plastics* III. 231/2 Polythene, the new polymerized ethylenic resin. *Ibid.* 289/1 Polymerized ethylenes manufactured..under the trade name of Polythene by I.C.I. have excellent electrical properties... These new compounds are now being carefully considered by well-known cable companies. 1943 *Brit. Plastics* XV. 417 Polythene is a general term for a range of solid polymers of ethylene, first discovered and prepared in I.C.I. Research Laboratories by subjecting ethylene to extremely high pressures under carefully controlled conditions. These products are sold under the registered name Alkathene'. 1945 *Times* 25 May 8/6 An outstanding I.C.I. achievement in the field of plastics was the discovery and development of polythene or polymerised ethylene. Polythene was seen to be a most valuable insulating material for high-frequency radio and television. 1951 *Catal. of Exhibits, South Bank Exhib., Festival of Britain* 109/1 Polythene ice cube mould. 1957 E. Bone *Seven Years' Solitary* xiv. 211, I had never heard of polythene bags. 1958 *Economist* 29 Mar. 1138/1 Sealed polythene bags offering protection and convenience in handling, have become commonplace for many small consumer products which used to be sold unpackaged. 1959 *Spectator* 25 Sept. 409/1 Polythene-wrapped food may be staler than it looks. 1960 *Farmer & Stockbreeder* 8 Mar. 105/2 There is a water trough supplied by the polythene pipe laid up one grass verge of the road. 1973 R. Fiennes *Headless Valley* vii. 129 We camped above the river beneath a thin sheet of polythene and slept. 1973 *Materials & Technol.* VI. viii. 524 Polythene has the simplest chemical structure of all plastics, consisting only of carbon and hydrogen in a straight chain. *Ibid.* 525 There are now many manufacturers of polythene, which is marketed under trade names such as Alkathene (UK), Bakelite, Alathon (USA), Lupolen (Germany) and so on. 1973 J. Rossiter *Manipulators* xxii. 213 She put on a glistening white polythene raincoat.

polythetic (pǫlipe·tik), *a.* [f. Poly- + Gr. θετ-ός placed, arranged + -ic i.] Sharing a number of common characteristics, without any one of these being essential for membership of the group or class in question. So **polythe·tically** *adv.*

1962 P. H. A. Sneath in Ainsworth & Sneath *Microbial Classification* 291 Such groups, in which several sets of characters occur, are called 'polytypic' by Beckner, but are better called polythetic. Phenetic taxa are always in theory polythetic. 1963 Sokal & Sneath *Princ. Numerical Taxon.* ii. 15 Polythetic groups can..be arranged polythetically to give higher polythetic groups, as is done in building a hierarchy in the natural system. 1969 E. Mayr *Princ. Systematic Zool.* iv. 83 No single feature is essential for membership in a polythetically defined taxon. 1972 S. Themerson *Special Branch* 75 Let it take a polythetic way of classifying: let it consider the largest possible number of characteristics for each fact or object. 1972 C. Renfrew *Emergence of Civilisation* ii It is now possible to make a statement about civilisations which does not seek to define them in terms of a single principal culture trait, or even polythetically, in terms of, for example, two out of three traits. 1976 *Brit. Jrnl. Psychol.* LXVII. 379 Most human concepts are polythetic..which means that no particular attribute or combination of attributes need necessarily be present (or absent) for an object to belong to a given category.

polytocous, *a.* Add: Also **-tokous. a.** (Examples.)

1932 *Proc. 6th Internat. Congr. Genetics* I. 188 In polytokous mammals we may expect that genes favoring rapid embryonic growth will spread, as slower growing embryos are at a disadvantage. 1936 *Nature* 12 Sept. 451/2 This..process [*sc.* fœtalization] has been essential for the evolution of man:..it could not have occurred in a polytocous form, where slowing of early growth would be prevented by intra-uterine selection. 1956 *Ibid.* 11 Feb. 288/2 Many observers have stressed the importance of a change in the nutrition of polytocous ewes during late pregnancy in the etiology of the disease. 1971 *Ibid.* 8 Oct. 379/1 In the polytocous species such as the small mammals and even the pig, it is not normally necessary to induce superovulation in order to provide the requisite number of eggs.

polytomy. 2. (Earlier example.)

*a*1856 W. Hamilton *Lect. Metaphysics & Logic* (1860) IV. xxv. 23 If a division has only two members, it is called a dichotomy..; if three, a trichotomy..; if four, a tetrachotomy; if many, a polytomy, &c.

polytonality (pǫlitonæ·lĭti). *Mus.* [f. Poly- + Tonality.] The simultaneous use of two or more keys in a musical composition.

1923 [see *atonal *a.*]. 1934 S. R. Nelson *All about Jazz* i. 12 Polytonality, dissonance, and the more lurid forms of Expressionism are the shibboleths of Schönberg, Stravinsky, Bartók and the moderns. 1946 G. Abraham in A. L. Bacharach *Brit. Music* iii. 61 The experiment is in polytonality—the flute plays in A, the oboe in A flat, and the viola in C. 1955 L. Feather *Encycl. Jazz* 32 Many young jazzmen were experimenting with atonality and polytonality. 1969 *Daily Tel.* 8 Nov. 9/3 Cold admiration seems to have been the response to his consciously manufactured rhythms, systematic polytonality and pattern building. 1977 J. Crosby *Company of Friends* xxxiv. 220 'Roger found out, did he?' asked Sascha idly, playing simultaneous D-minor and C-sharp triads. Polytonality—while naked. Very sexual.

So **polyto·nal** *a.*, containing or pertaining to polytonality; **polyto·nalist,** one who writes or advocates polytonal music.

1924 P. A. Scholes *Crotchets* 164 The device of 'canon' sometimes pointed to a polytonal future for music. 1934 C. Lambert *Music Ho!* ii. 68 The polytonal choral writing of Milhaud. 1938 *Oxf. Compan. Mus.* 406/1 The polytonalists appear to claim that the value of their work lies in the significance of the horizontal lines. 1949 *Penguin Music Mag.* July 24 The polytonal writing, in which different parts in the polyphony may have, not only their individual harmonization, but also each their own tonality, is chiefly a tentative realization of a stage that music is steadily approaching. 1952 B. Ulanov *Hist. Jazz in Amer.* (1958) xxiii. 332 All of the trio and octet scorings and performances partake..of the controlled but not stifling disciplines of a music which is polytonal, polyrhythmic at times, and spontaneous too. *Ibid.*, Dave..is another polytonalist. 1976 *New Yorker* 15 Nov. 194/2 Rich in feeling, sometimes polytonal (or, at any rate, of indefinite tonality), the piece is not without interest.

polytonic (pǫlitǫ·nik), *a.* [f. Poly- + Tonic *a.* and *sb.*] Using or having several (musical or vocal) tones.

1948 D. Diringer *Alphabet* i. vi. 99 So characteristic are the tones in the Tibeto-Chinese languages, that some scholars have suggested to term them 'polytonic'. 1955 *Sci. News Let.* 13 Aug. 102/1 A new signaling system for telephone dialing..has been developed... The device, called a polytonic coder, sends out digits just about as fast as they can theoretically be packed into a line. *Ibid.* 102/2 The tests showed the polytonic signal could be used on all except a few telephone connections in this country. 1961 J. Blades in A. Baines *Mus. Instruments* 342 The ingenious steel drums recently introduced in Trinidad are polytonic gongs... Each sector is tuned to a different note.

polytopal, -tope, -topic(al: see *poly- i.

polytrope (pǫ·litrǝup). *Physics* and *Astr.* [Back-formation from Polytropic *a.*, or a. G. *polytrope* in the same sense (R. Emden *Gaskugeln* (1907) i. i. 13).] A polytropic body of gas (see *polytropic *a.* 4).

1926 A. S. Eddington *Internal Constitution of Stars* iv. 86 The polytrope $n = 3$, which is believed to correspond nearly to the actual conditions of the stars. 1939 S. Chandrasekhar *Introd. Study Stellar Struct.* iv. 170 So far we have considered only complete polytropes. We shall now proceed to a consideration of composite polytropes, i.e., configurations which consist of different zones each characterized by a different value of the index *n*. 1975 *Nature* 27 Mar. 295/1 The energy required to break up a star of mass *m* and radius *R* is..$\frac{3}{5}(Gm^2/R)$ for a polytrope of index $n = 3\frac{1}{2}$.

polytropic, *a.* Add: **3.** [ad. G. *polytrope* (E. Loew 1884, in *Jahrbuch K. Bot. Gartens Berlin*).] Of a bee: collecting nectar from many kinds of flower.

1899 C. Robertson in *Bot. Gaz.* XXVIII. 29 If a bee has a long flight it must be regarded as polytropic. 1919 J. H. Lovell *Flower & Bee* 120 There are also on the wing at the same time 6 species which are polytropic.

4. *Physics* and *Astr.* [ad. G. *polytropisch* (G. Zeuner *Technische Thermodynamik* (ed. 3, 1887) I. xxix. 143).] Pertaining to or designating a body of gas or a process in which pressure and volume change in such a way that a specific heat remains constant. Also as *sb.*, a graph showing such a variation of pressure and volume.

1907 J. F. Klein tr. *Zeuner's Technical Thermodynamics* I. xxix. 152 If the initial condition is given by p_1 and v_1, we accordingly have $pv^n = p_1v_1^n$..as the equation of the sought pressure curve, which we will hereafter call the polytropic curve. 1926 A. S. Eddington *Internal Constitution of Stars* iv. 94 A class of problems arises in which the polytropic condition..applies only to part of the star. 1933 *Monthly Notices R. Astron. Soc.* XCIII. 390 Emden's well-known researches on the equilibrium of polytropic gas spheres has been of fundamental importance in its repercussions on the modern theories of stellar structure. 1939 S. Chandrasekhar *Introd. Study Stellar Struct.* ii. 40 An adiabatic..is a polytropic of zero specific heat, and an isothermal a polytropic of infinite heat capacity. 1952 W. M. Deans tr. *Prandtl's Essent. Fluid Dynamics* v. 391 If we assume a polytropic law for the variation of density with height, $p/p_0 = (\rho/\rho_0)^n$, where the suffix *g* refers to the ground. 1968 Cox & Giuli *Princ. Stellar Struct.* II. xxiii. 703 A polytropic star is one which obeys an equation of the form $P = K\rho^{(n+1)/n}$, where *n*, the polytropic index..and *K* are constants throughout the star. 1976 *Nature* 17 June 561/1 The moment of inertia of a spherical star can be written $I = km_pNR^2$ where for stable polytropic stars $0{\cdot}04 \leq k \leq 0{\cdot}132$.

polytype. Add: **2.** *Cryst.* A polytypic form of a substance.

In quot. 1916 rendering G. *polytypie* polytypism.

1916 *Chem. Abstr.* X. 872 (*heading*) The various modifications of carborundum and the occurrence of polytypes. 1922 *Mineral. Abstr.* I. 318 Three modifications or polytypes of carborundum crystals are distinguished as types I, II, and III. 1951 *Phil. Mag.* XLII. 1019 Unstable structures will not grow if the supersaturation is small: but probably most of the observed polytypes of carborundum differ in thermodynamic potential by an amount negligible compared with the supersaturation actually occurring. 1972 *Physics Bull.* Dec. 712/1 The difficult areas of x ray crystallography, such as in the study of martensitically transformed materials, mixed polytypes of a mineral or certain nonstoichiometric oxides, can be opened up by an electron microscope technique. 1974 *Nature* 22 Feb. 537/2 The basic structural unit of tin sulphide polytypes consists of two layers of hexagonal close-packed sulphide ions with smaller tin ions nested between them.

polytypic, *a.* **1.** Substitute for def.: Having several variant forms; esp., of a species, including several subspecies or other lower taxa. (Later examples.)

1945 *Bull. Amer. Mus. Nat. Hist.* LXXXV. 16/2 This morphological scope may be almost entirely filled or exploited by known species if the genus has many (is polytypic). 1953 [see *allopatric *a.*]. 1959 *New Biol.* XXVIII. 81 This process [*sc.* the variation of isolated populations] might produce simply what the systematist calls a polytypic species, consisting of a sequence of subspecies each occupying a distinct geographical region and each somewhat different in ecology. 1970 *Nature* 5 Sept. 1065/1 It is now recognized that *Papio* is a single polytypic species with morphologically different subspecies interbreeding wherever they meet. 1975 *Trans. R. Entomol. Soc.* CXXVI. 613 We will discuss the data for this polytypic species under the four form names.

2. *Cryst.* Exhibiting polytypism (sense *1); of the nature of a polytype.

1944 [see *polytypism 1]. 1974 Verma & Trigundayat in C.N.R. Rao *Solid State Chem.* ii. 52 The list of polytypic substances includes minerals, layer silicates, chalcogenides, and several other inorganic and organic compounds.

3. = *polythetic *a.*

1959 M. Beckner *Biol. Way of Thought* ii. 25 Polytypic concepts are found in many branches of biological theory, but the clearest instances are afforded by taxonomy. 1961 G. G. Simpson *Princ. Animal Taxon.* ii. 43 The defining attributes do not appear in all individuals... The principle (which he calls 'polytypic') has been elucidated at greater length and in ultramodern terms by Beckner.

polytypism (pǫlitəi·piz'm). [f. Polytyp(ic *a.* + -ism.] **1.** [After G. *polytypie* (H. Baumhauer 1915, in *Zeitschr. f. Krist. u. Min.* LV. 252).] A kind of polymorphism in which a substance occurs in a number of crystalline modifications (polytypes) which differ only in one of the dimensions of the unit cell.

1944 N. W. Thibault in *Amer. Mineralogist* XXIX. 266 Following Baumhauer, the phenomenon is called 'polytypism' the adjective being 'polytypic' and each one of the different modifications a 'type'. 1951 *Phil. Mag.* XLII. 1016 It has long been recognized that the relationship between the various forms of carborundum is so close as to deserve a special name 'polytypism'. 1973 L. Suchow in P. F. Weller *Solid State Chem. & Physics* I. ii. 130 All compounds exhibiting polytypism are..of the layer-structure type. 1974 Verma & Trigundayat in C. N. R. Rao *Solid State Chem.* ii. 52 The modifications, called polytypes or simply types, differ only in the arrangement of packing of the close-packed planes of the solids... In this sense the polytypism may be termed one-dimensional polymorphism, although its origin seems to be distinctly different from the latter.

2. *Biol.* The occurrence of several variant forms within a single species. Also **po·lytypy** in the same sense.

1949 W. F. Albright *Archaeol. of Palestine* iii. 57 The polytypism now known to be characteristic of early man is equally characteristic of dogs, whose classification as a single species is not disputed by zoologists. 1961 G. G. Simpson *Princ. Animal Taxon.* iv. 135 In cases of extreme polytypy, first consideration should be given to making the taxon less unwieldy by use of intermediate lower taxa or subgroups.

polyunsaturated (pǫli,ɒnsæ·tiŭrēitĕd), *a.* *Chem.* Also **poly-unsaturated.** [f. Poly- + Unsaturated *ppl. a.*] Containing more than one multiple bond between carbon atoms at which addition can normally occur; applied esp. to fatty acids in which the hydrocarbon chain has more than one multiple bond, which occur esp. in some vegetable oils.

1932 *Biochem. Jrnl.* XXVI. 1978 The marine flora contains poly-unsaturated fatty acids of high molecular weight similar to those found in the fat of fishes, although in very small amounts. 1958 *Jrnl. Sci. Food & Agric.* IX. 779 In animal tissues the glycerophosphatides are outstanding in their high content of long-chain polyunsaturated acids. 1960 *Farmer & Stockbreeder* 22 Mar. 133/1 Polyunsaturated fats for peak condition, healthy skin and coat, more efficient metabolism. 1961 *Sunday Times* 7 May 36/3 American housewives,..concerned about their husbands' health, are shopping for 'polyunsaturated fats'. The rush is on because medical research has established a connection between arterial diseases and the level of cholesterol in the blood—and cholesterol seems to be formed chiefly from the intake of the 'saturated fats', particularly animal and dairy fats. 1970 *New Scientist* 19 Feb. 356/2 It might also be possible to increase the proportion of poly-unsaturated acids in ruminant tissue fats. 1973 *Which?* Feb. 54 Butter generally has 2 per cent polyunsaturated fatty acids. 1977 B. Pym *Quartet in Autumn* xiii. 115 A tub of polyunsaturated margarine. 1979 *Times* 7 Dec. 10/3 'Polyunsaturated' has become a totem word... The idea that heart attacks can be avoided by replacing saturated fats in the diet with polyunsaturated fats has been promoted with great skill and large budgets.

So **polyunsa·turate** *sb.*, a polyunsaturated fatty acid; also as *adj.*

1950 *Arch. Biochem.* XXV. 6 Polyunsaturates were also concentrated in high amounts in the liver fatty acids. 1962 *Seattle Sunday Times* (Sunday Pictorial) 25 Mar. 18/2 (Advt.), It's a face cream that contains essential polyunsaturates. That's right—polyunsaturates, the natural elements you've been reading so much about that are so important to your health. 1962 *Chicago Sun-Times* 20 July 28/3 Maybe, so current thinking goes, we would be better off in the long run to reduce our costly animal fats and substitute the vegetable (polyunsaturate) fats and oils. 1974 *Observer* (Colour Suppl.) 17 Mar. 15/1 There are doubts..about the side-effects of the various polyunsaturate oils used. *Ibid.*, Only a few vegetable oils are high in polyunsaturates.

polyurethane (pǫli,yū·rĕþēin). Formerly also **-an.** [f. Poly- + Urethane.] Any of a large class of synthetic resins and plastics consisting of or made from polymers with the units linked by the group —NH·CO·O—, which are made esp. by the reaction of polyisocyanates with polyhydroxy compounds and are important commercially as plastic foams, as fibres, and in paints, adhesives, synthetic rubbers, films, etc. Also *attrib.* and *Comb.*, esp. *polyurethane foam.*

1944 *Chem. Abstr.* XXXVIII. 381 (*heading*) Polyurethans and polyureas. 1945 *Chem. & Engin. News* 25 Sept. 1615/1 Among the new or less well-known polymers which they [*sc.* the Germans] have been manufacturing are polyvinyl ethers..and polyurethanes. 1945 *Mod. Plastics* Oct. 152F/2 Polyurethane resin was used on a small scale as an adhesive in aircraft construction. 1957 B. A. Dombrow *Polyurethanes* i. 8 In the latter part of 1947..Lockheed Aircraft Corporation began independently to develop polyurethane foam systems for use in radome construction and in filling aircraft components. 1958 *Engineering* 7 Feb. 167/2 A lightweight portable shelter suitable for aircraft, vehicles or stored goods, consisting of a tubular light-alloy framework covered with polyurethane-proofed nylon. 1964 *Which?* Aug. 254/1 Polyurethane has become known in the form of plastic foam for sponges among other things, but is now finding its way into the textile field. 1968 W. Warwick *Surfriding in N.Z.* 16/3 At present surfboards are being constructed from the material polyurethane. 1969 *Sears Catal.* Spring/Summer 9 Exclusive 3-inch thick cushions..softly padded with shredded Serofoam polyurethane. 1971 *New Scientist* 10 June 630/2 Polyurethane-coated fabrics are still at an early stage of development. 1974 *Nature* 16 Aug. 526/3 Polyurethane foam is sprayed on to the inside of a lightweight aluminium mould.

Hence as *v. trans.*, to coat or protect with polyurethane.

1977 B. Randall *Fan* 22 Don't go in the dining room, I polyurethaned the floor. 1978 J. Updike *Coup* (1979) iii. 109 I've seen blueprints, concrete wings supported by polyurethaned nylon cables. 1979 M. Page *Pilate Plot* ix. 137 Washed Chinese carpets on the sanded and polyurethaned floorboards.

polyuria, -uric: see *poly- i. **polyuronide:** *poly- 2.

polyvalent. Add: **1.** Now usu. with pronunc. (pǫli·vĕilĕnt). (Further examples.)

1950 N. V. Sidgwick *Chem. Elements* I. 177 The polyvalent state of gold..is much more stable than the monovalent. 1964 N. G. Clark *Mod. Org. Chem.* xxiv. 510 With all other polyvalent species of atom accounted for, the side chain..must consist of three linked carbon atoms. 1975 *Nature* 6 Nov. 19/1 The formation of a ring structure by a protein factor or possibly polyvalent ions, the use of the ring as a nucleating centre, [etc.]..were discussed.

b. *Med.* = *multivalent *a.* b.

1912 *Amer. Jrnl. Med. Sci.* CXLIV. 815 The complement fixation test (using a polyvalent antigen) for gonococcus antibodies should prove an addition to our methods of diagnosticating between..gonococcus infection and..other causes. 1951 Whitby & Hynes *Med. Bacteriol.* (ed. 5) vi. 72 When an antigen in colloidal solution is mixed with antibody under suitable conditions a precipitate of antigen-antibody complex appears. Most of the precipitate consists of antibody, since antibody molecules in general are larger than those of antigens, whilst antigens are polyvalent and each molecule combines with two or more molecules of antibody. 1975 *Nature* 11 Sept. 103/1 Interaction of divalent or polyvalent ligands (for example, antibodies, lectins) with cell surface structures on lymphocytes and many other cell types induces redistribution of the largely diffuse binding sites into clusters and into highly polarised single aggregates or 'caps'.

2. Delete ? and add to def.: or affording immunity against various species of micro-organism. (Later examples.)

1929 [see *botulin]. 1965 C. Andrewes *Common Cold* xix. 168 'Polyvalent' vaccines—those effective against many bacteria or viruses—are very fine in theory. 1966 *Jrnl. Immunol.* XCVII. 517/2 A third antigen preparation consisted of a single polyvalent antigen prepared by combining equal amounts of the 12 flexner stock antigens and a stock *Shigella sonnei* antigen.

3. = *multivalent *a.* 3.

1957 S. Ullmann *Style in French Novel* 20 Stylistic elements, it will be remembered, are 'polyvalent'; the same device may produce several effects, and conversely, the same effect may be obtained from several devices. 1960 E. Delavenay *Introd. Machine Transl.* iii. 36 The grammatical value of polyvalent words. 1966 *New Statesman* 25 Feb. 266/2 The colour-symbolism in [Tarsis'] *Red and Black* is pervasive and bafflingly polyvalent. 1975 J. De Bres tr. *Mandel's Late Capitalism* viii. 268 Full automation reduces the number of semi-skilled workers and gives rise to a new and highly skilled polyvalent work force. 1977 *Word 1972* XXVIII. 294 The analyst will be tempted to ignore the systems based on very general concepts yielding the simplest description for all languages in favor of a very specific system adapted to the specific language of the texts to be recorded: the kind of simplicity he will look for is of the monovalent rather than of the polyvalent type.

polyvalence (further example); also stressed *polyva·lence*.

1971 *Archivum Linguisticum* II. 104 Now he writes..about the polyvalence of the injunctive—including *s*, *d* and *ē* stems. In spite of this polyvalence he thinks that there is a common core in the subjunctive and preterite use of these stems.

polyvinyl (pǫlivəi·nĭl, -vəi·nəil). [f. Poly- + Vinyl.] **a.** Used *attrib.* in the names of polymeric substances derived from vinyl compounds, as **polyvinyl acetal,** any of a class of synthetic resins prepared by condensing polyvinyl alcohol with an aldehyde (sometimes *spec.* acetaldehyde), and mainly used, esp. polyvinyl butyral (see below), in safety glass and in lacquers and paints; **polyvinyl acetate,** a fairly soft plastic having the structure $—[CH_2—CH(O{\cdot}CO{\cdot}CH_3)]_n—$, which is made by polymerizing vinyl acetate and is used chiefly in paints and adhesives; abbrev. *PVA* (s.v. *P II. d); **polyvinyl alcohol,** any of a series of polymers consisting wholly or largely of the repeating unit $—[CH_2—CHOH]—$ which are prepared by hydrolysis of polyvinyl esters and have a wide range of uses, e.g. as emulsifiers, adhesives, coatings, films, and fibres; abbrev. *PVA* (s.v. *P II. d); **polyvinyl butyral,** the most widely used of the polyvinyl acetals (see above), which is prepared from butyraldehyde; **polyvinyl chloride,** any of various thermoplastics consisting of or made from a polymer having the structure $—[CH_2—CHCl]_n—$ and made by polymerizing vinyl chloride, which are produced in a wide variety of rigid and plasticized forms and are characterized esp. by their toughness, chemical inertness, and electrical resistivity; abbrev. *PVC* (s.v. *P II. d); **polyvinyl pyrrolidone,** a water-soluble polymer of vinyl pyrrolidone which is physiologically harmless and has a great variety of applications, esp. in solution, e.g. as a synthetic blood plasma substitute, as a thickening, suspending, or binding agent in the cosmetic, drug, and food processing industries, and in fibres and films; abbrev. *PVP* (s.v. *P II. d).

All these often occur as single words.

1933 *Industr. & Engin. Chem.* Feb. 129/1 The structure of polyvinyl acetal. 1944 [see *phenolic *a.* b]. 1973 *Materials & Technol.* VI. viii. 552 The use of polyvinyl butyral for making safety glass was patented in 1938 and this has since become one of the most important applications, certainly in America. The next important consumer of polyvinyl acetals is the lacquer and paint industry (for wash-primers). 1927 *Brit. Chem. Abstr.* A. 1051/2 Hydrolysis of polyvinyl acetate yields polyvinyl alcohol which resembles starch in its insolubility in organic media. 1958 W. M. Smith *Vinyl Resins* i. 19 Polyvinyl acetate emulsions as a replacement for conventional grain starch are stated to withstand as many as 15 launderings. 1967 W. Gaunt *Compan. Painting* ii. 85 Some modern artists have..found congenial qualities in such synthetic paints as ripolin and polyvinyl acetate. 1972 *Homes & Gardens* Mar. 106/2 In their infancy, emulsion paints were regarded with some suspicion, partly because early ones, based on a polyvinylacetate (pva) medium, tended to become brittle and flake off. 1927 Polyvinyl alcohol [see *polyvinyl acetate* above]. 1962 J. T. Marsh *Self-Smoothing Fabrics* x. 133 There are occasions where a slightly stiff finish is required and polyvinyl alcohol is a popular choice of additive in this connection. 1971 D. Potter *Brit. Eliz. Stamps* vi. 68 A synthetic adhesive was invented. Completely non-toxic, colourless, odourless and tasteless, it is known as Polyvinyl Alcohol, and familiarly named pva by philatelists. 1973 *Materials & Technol.* VI. viii. 548 The usual method of preparing polyvinyl alcohol is by the hydrolysis of a polyvinyl ester, generally polyvinyl acetate. 1943 *Industr. & Engin. Chem.* Feb. 175/2 In polyvinyl butyral resin..certain fillers appear to have reinforcing action. 1955 *Sci. News Let.* 16 Apr. 247/1 Adhesives for bonding two pieces of glass together are best made of polyvinyl-butyral resin mixed with modified phenol. 1973 Polyvinyl butyral [see *polyvinyl acetal* above]. 1933 *Industr. & Engin. Chem.* Feb. 129/1 Co-polymerization of vinyl chloride and vinyl acetate gives a better resin than the mere admixture of polyvinyl chloride and polyvinyl acetate. 1937 *Chem. Abstr.* XXXI. 6434 A mixt. of water with polyvinylchloride..is used for coating glass plates and is dried upon them, and then the coated plates are united by the action of heat and pressure. 1945 *Electronic Engin.* XVI. 499/3 Plastic sleeving of the polyvinylchloride type is the real answer to all these problems. 1958 *Engineering* 7 Mar. 294/2 The folding top is in leather-cloth, coated with polyvinyl chloride. 1976 *Nature* 3 June 409/1 Each trough held 50 oysters on a false bottom consisting of polyethylene grids resting on polyvinylchloride pipe. 1945 *Chem. Abstr.* XXXIX. 5408 Such solns. [i.e. blood plasma substitutes] are prepared by adding polyvinylpyrrolidone..to a physiol. salt soln. 1957 *Financial Times Ann. Rev. Brit. Industry* 86/3 Experiments have shown that polyvinyl pyrrolidone..used as a blood thickener can be readily broken down..to..the most useful molecular size for medical use. 1974 M. C. Gerald *Pharmacol.* ix. 164 The plasma substitutes, dextran and polyvinylpyrrolidone (PVP), are also histamine releasers.

b. Any of the plastics or synthetic resins made by polymerizing a compound containing the vinyl group, $CH_2{=}CH—$; applied *esp.* to those derived from a compound usu. designated as vinyl (see prec. sense), but sometimes extended to include polyvinylidenes, polystyrene, etc. Freq. *attrib.*

1933 *Industr. & Engin. Chem.* Feb. 129/1 To render the normal polyvinyl resin more water-resistant, it is converted partly or wholly to the acetal by reaction with an aldehyde, usually formaldehyde or acetaldehyde. 1937 *Brit. Plastics* IX. 172/1 Vinyl esters of both inorganic and organic acids are well known..raw materials for conversion to polyvinyls. 1940 'Plastes' *Plastics in Industry* ii. 15 More recently polyvinyl plastics have been employed as floorcloth, long-playing phonograph records, [etc.]. *Ibid.*, An interesting series of thermo-plastics which resemble the polyvinyls closely. 1946 *Nature* 16 Nov. 689/2 The use of methacrylate esters and the various polyvinyls is increasing in these fields [*sc.* arts and crafts]. 1963 [see *polyvinylidene b]. 1966 *Punch* 26 Jan. 116/3 Palpitations on finding sleek strangers at the front door in answer to my inquiry about polyvinyl tiling. 1969 R. Mayer *Dict. Art Terms & Techniques* 422/2 Few of the vinyl products are practicable for use in artists' materials... One exception is a polyvinyl isolating varnish containing isopropyl alcohol that thins with alcohol.

polyvinylidene (pǫ:livəini·lidīn, -vəinəi·lidīn). *Chem.* [f. Poly- + *vinylidene.] **a.** Used *attrib.* in the names of substances which are polymers of vinylidene compounds, esp. *polyvinylidene chloride*, any of a class of resinous polymers of vinylidene chloride which have the structure $—[CH_2—CCl_2]_n—$, and have a wide range of applications, esp. as impact- and chemical-resistant films and fibres.

1940 *Chem. Rev.* XXVI. 163 Natta and Rigamonti..found..polyvinylidene chloride to be quite highly crystalline. 1942 *Industr. & Engin. Chem.* Mar. 327/2 These polyvinylidene chloride plastics are known by the trade name, 'Saran'. 1964 *Discovery* Oct. 17/1 The lamination consists of an aluminum foil and three plastic films—one of polythene, another of polyvinylidene chloride and a third of polyester. 1966 *McGraw-Hill Encycl. Sci. & Technol.* X. 496/2 Films of polyvinylidene chloride, and especially the copolymer containing about 15% of vinyl chloride, are resistant to moisture and gases. 1973 *Materials & Technol.* VI. viii. 547 Because of its resistance to chemicals, polyvinylidene fluoride is used for tubes, valves, and pumps in the chemical industry. 1975 P. Browne *Bodywork Maintenance* vi. 79/2 Cloths such as Tygan, woven from polyvinylidene chloride filaments, need an interlayer..to prevent a chemical reaction between the Tygan and the rubber.

b. Any of the synthetic resins prepared from vinylidene compounds. Freq. *attrib.*

1941 *Plastics* V. 249/2 (*caption*) Flow sheet showing production of polyvinylidene polymers from petroleum and brine. 1960 *Times Rev. Industry* Sept. 65/3 The tendency seems to be for polythene yarns to replace some of the earlier types such as polyvinylidene. 1963 A. J. Hall *Textile Sci.* ii. 18 (*table*) Polyvinyls and polyvinylidenes. 1972 *Homes & Gardens* Mar. 107 One firm has now introduced a full gloss emulsion paint based on a polyvinyledene [*sic*] medium.

polywater (pǫ·liwǭtəɹ). [f. Poly- + Water *sb.*] A supposed polymeric form of water having properties markedly different from those of ordinary water and reported to have been found in fine capillary tubes.

1969 E. R. Lippincott et al. in *Science* 27 June 1482/3 The properties, therefore, are no longer anomalous but rather, those of a newly found substance—polymeric water or polywater. 1969 *Daily Tel.* 13 Sept. 22/3 American scientists have confirmed the existence of a new kind of water which freezes at 40deg C and boils at 500deg C... The new water, known as 'polywater' has a density of about 40 per cent. greater than ordinary water. 1972 *Britannica Yearbk. Sci. & Future 1971* 216 Anomalous water (or polywater), which has attracted much attention in recent months, was shown to be not an unusual form of water but ordinary water containing ionic impurities. 1972 F. Franks *Water* I. i. 13 During 1969 the 'polywater' bandwagon began to roll. *Ibid.*, A climax was provided by a warning to scientists against experimenting with polywater as this might affect the oceans, turning them solid. 1975 J. Clifford in *Ibid.* V. ii. 119 Defenders of the water polymer idea have suggested that polywater could be a weakly bonded complex that disintegrates under electron impact and does not appear in the mass spectrum.

polyxenic: see *poly- i.

polyzoal, *a.* (Example.)

1856 G. J. Allman *Monogr. Fresh-Water Polyzoa* 3 The investigations of Trembley and Baker..clearly demonstrated..all the essential characters of polyzoal structure.

polyzoan, *a.* and *sb.* Add: = Bryozoan *a.* and *sb.* (Earlier and later examples.)

1856 P. H. Gosse *Man. Marine Zool. Brit. Isles* II. 5

A tiny Annelid or other animal, caught by the bird's-head of a Polyzoan and tightly held, would presently die. **1880** T. Hincks *Hist. Brit. Marine Polyzoa* I. p. xvii, The tentacular sheath is an important element..of the ordinary Polyzoan type. **1924** S. J. Hickson *Introd. Study Recent Corals* viii. 159 Zooecia..build up the various kinds of branching, net-like, or encrusting structures of the Polyzoan corals. **1959** *Jrnl. Exper. Biol.* XXXVI. 613 (*heading*) Experiments on the selection of algal substrates by polyzoan larvae.

polyzoism. (Earlier example.)
1890 W. James *Princ. Psychol.* I. vi. 179 It may be called the theory of polyzoism or multiple monadism.

Pom[1] (pǫm). Also **pom.** Colloq. abbrev. of Pomeranian *sb.*
1904 *Outing* Feb. 484/2 Collies and 'poms' in America have hardly maintained their status because of this coat trouble. **1910** *Bazaar, Exchange & Mart* 10 June 1523/2 (*heading*) Coming shows... Dogs... Manchester (Poms). **1911** F. T. Barton *My Bk. Little Dogs* iii. 33 The Pekinese and the Pom are the most popular toy dogs at the present time. **1923** R. Macaulay *Told by Idiot* ii. xxi. 138 Rome..drove elegantly in hansoms, often with an enormous wolf-hound or a couple of poms. **1939** T. S. Eliot *Old Possum's Pract. Cats* 27 And the Pugs and the Poms.. will now and again join in to the fray. **1956** E. Berckman *Beckoning Dream* vi. 46 Lydia..bred miniature Poms. In the vast living-room..twenty little dogs disported themselves. **1973** R. Hill *Ruling Passion* ii. i. 85 'Not much of a guard-dog,' he said. 'It's a pom,' Pascoe said patiently.

Pom[2] (pǫm). *Austral.* and *N.Z. colloq.* Also **pom.** Abbrev. of *Pommy *sb.* (*a.*)
1919 W. H. Downing *Digger Dial.* 38 *Pommy*, an English soldier. *Pom*, see *Pommy*. **1941** Baker *Dict. Austral. Slang* 56 *Pom*, an Englishman. **1946** F. Sargeson *That Summer* 92 He was a big matelot, though not a Pom, it was easy to tell he was a Pig Islander. **1957** *Economist* 9 Nov. 510/2 New British migrants are more readily assimilated than continental Europeans. Australians do not consider the 'Poms' as foreigners. **1963** A. Lubbock *Austral. Roundabout* 83 'Be seein' yer soon in England... Good on yer, Pom.' **1975** D. Bloodworth *Clients of Omega* ix. 84 You a Pom or something, sister?.. You've got a swine of a Pom accent. **1977** *Bulletin* (Sydney) 22 Jan. 20/1 And there's New Zealand to come; and then the Centenary Test against the Poms.

Pom[3] (pǫm). Also **pom.** The proprietary name of a brand of dried and powdered cooked potato.
1947 *Trade Marks Jrnl.* 23 Apr. 234/1 Pom... Cooked potatoes and potato preparations..M.P.P. (Products) Limited,..Norwich; Manufacturers. **1955** G. Band *Road to Rakaposhi* xii. 136 The menu was..complicated.. stewed steak and pom, fruit cake and tea. **1968** *Economist* 15 June 64/3 Oddly enough, the greatest potential market for dried food is those old wartime standbys—dried milk and dried potatoes. Dehydrated potato, indistinguishable from the 'pom' of the 1940s,..now has a market worth £2 million a year. **1970** *Times* 26 Nov. 17/4 The late André Simon compiled his Dictionary of Gastronomy during the War years, presumably to cheer himself up among the snoek and pom.

Poma[2] (pōu·mă). *N. Amer.* Also **poma.** [f. the name of J. *Pomagalski*, its inventor.] The proprietary name of a type of ski-lift having detachable hangers; so *Poma lift*, *Pomalift*.
1954 *Amer. Ski Ann. & Skiing Jrnl.* Jan. 55/1 The poma lift at Arapahoe Basin, once the wooden towers were in place, was erected in sixteen days in the worst possible weather. **1955** *Ski Mag.* Oct. 51/2 (Advt.), Big news at Big Bromley..a new 2,190 ft. Poma lift has been installed. **1957** T. Kesting *Outdoor Encycl.* 403/2 On the Pomalift—a kind of platter-pull—the hangers are stored at the bottom and clamped onto the cable as needed. **1963** *Amer. Speech* XXXVIII. 206 Skiers usually do not differentiate between *platter pull* and *poma*; the latter seems to have become the universally accepted term for both types working with disks. **1968** *Globe & Mail* (Toronto) 13 Jan. 35/3 Other Muskoka resorts offering modern accommodation as well as T-bar or pomalift include Muskoka Sands, [etc.]. **1970** *Official Gaz.* (U.S. Patent Office) 14 Apr. TM 80/2 Jean Pomagalski S.A., Fontaine-Grenoble, France... *Poma* for cable transport or towing apparatus and installations—namely, cable cars, gondola-lifts, chair-lifts, ski-lifts. **1973** P. A. Whitney *Snowfire* vi. 105 A chair lift, with T-bar, J-bar and poma lift off to our right. **1977** *Forest Sci.* XXIII. 168 Area 1..is only affected by skiing activity (ski runs, poma lifts).

pomace. Add: **4. pomace-fly** = *Drosophila.
1897 J. H. Comstock *Insect Life* 185 As these insects are often abundant about pomace in cider-mills and wineries, they have been termed pomace-flies. **1924** J. A. Thomson *Sci. Old & New* xxvii. 152 When the pomace-fly, Drosophila, is feeding on fermenting fruit, it must have yeasts to help it. **1946** C. T. Brues *Insect Dietary* v. 194 The pomace fly, *Drosophila*, so successfully used by geneticists to elucidate the processes of inheritance, has likewise served..to demonstrate some of the food relations of microphagous insects.

Pomak (pōu·mæk). [Bulg.] A Muslim Bulgarian.
1887 *Encycl. Brit.* XXII. 149/2 Those Bulgarians who have embraced Islam are called Pomaks,—a word of which no satisfactory derivation has been given. **1897** E. A. Bartlett *Battlefields Thessaly* iii. 49 The local militia were mostly Pomaks, or Mussulman Bulgarians. **1900** 'Odysseus' *Turkey in Europe* viii. 363 The country between Seres and Philippopoli is inhabited by people called Pomaks, who are commonly described as Mohammedan Bulgarians. **1921** *Contemp. Rev.* May 587 It is not unusual to find that in any computation made by the Greeks,..the Pomaks—*i.e.*, Bulgarians who have embraced the Mohammedan faith, are reckoned with the Turks. **1972** D. Dakin *Unification of Greece* 269 The Slav minority, which included 16,000 Pomaks, was about 80,000.

pomander. Restrict 'Now *Hist.*' to senses in Dict. and add: **1. a.** (Later examples.) Also, a piece of fruit, esp. an orange, stuck with cloves and usu. tied with ribbon, which is hung or placed in a wardrobe.
1931 E. S. Rohde *Scented Garden* viii. 219 *Pomanders, Etc.*.. The orange..will scent a drawer deliciously for well over a year. **1946** J. De Both *Mod. Househ. Encycl.* 237/1 Pomanders may be made from apples, oranges, or lemons.—*to make*, select firm fruit and stick whole cloves into entire surface; hang in clothes closet or place in dresser drawers. **1963** *Good Housek. Home Encycl.* (rev. ed.) 367/2 The pomander...looks prettier if tied round with ribbon or tinsel, with a loop for hanging it up. **1974** Westland & Critchley *Art of Dried & Pressed Flowers* ix. 80 Hang the pomander in a wardrobe, on a coat hanger or over your dressing table.
2. a. (Later examples.) Also, a small perforated ceramic container filled with potpourri or other aromatic substances, for hanging in a wardrobe, placing on a dressing-table, etc.
1973 *Woman's Jrnl.* Dec. 108 (Advt.), Colognes, bath essences, soaps, pot-pourri and pomanders from..J. Floris Ltd. **1975** *Lady* 6 Nov. p. vii/2 (Advt.), Bone china pomander, traditional long-lasting perfume. **1976** *S. Wales Echo* 25 Nov. 8/2 (Advt.), There are pots of French herbs.., jams laced with whisky, silk scarves and pomanders.
b. (Later example.)
1895 E. Nesbit (*title*) A pomander of verse.

Pomard, Pommard. Add: Also with small initial. (Further examples.)
1875 [see *Beaune]. **1889** [see *Corton]. **1920** G. Saintsbury *Notes on Cellar-Bk.* iv. 56 Less distinguished representatives of the Slope of Gow [*sc.* Côte d'Or].. Pommard, Santenay, Chenas and others. **1962** R. Jeffries *Exhibit No. 13* v. 48 He..picked up a bottle of Pommard. **1979** I. S. Black *Journey to Safe Place* ix. 95 We'll have another bottle of this Pommard.

pomato (pomā·to, -ēi·to). [f. P(otato + T)omato.] A name used by Luther Burbank (1849–1926), American horticulturist, for the fruit of a hybrid potato, which resembled a tomato; later used to designate the result of attempts to hybridize the potato and the tomato, by grafting or other methods. Also *attrib.*
1905 *Century Mag.* Mar. 668/1 The 'pomato', one of the most wonderful creations, now under way. This may be called a tomato growing upon a potato. It produces in abundance a white, fragrant, succulent, delicious fruit upon potato tops. **1914** L. Burbank in J. Whitson et al. *L. Burbank: his Methods & Discoveries* II. ix. 283 As the fruit grew on a hybrid potato vine, and in itself had much the appearance of a tomato, it was christened the 'Pomato'. The name..led to the unauthorized assumption that the fruit was really a cross between the tomato and the potato. In point of fact, I have never been able to cross these two plants. *Ibid.* 284 The pomato plant produced fruit abundantly, but very few tubers. **1971** *New Scientist* 27 Apr. 263 (*caption*) Protoplasts of potato and tomato are fused... These cells are then cultured, and from them are grown embryoids and eventually whole 'pomatoes'. At the moment, research has got as far as the protoplast-fusion stage. **1976** *Nat. Geographic* Sept. 388/1 'Pomatoes'—hypothetical vegetables with the fruit-bearing foliage of a tomato and the tuberous roots of a potato. To some botanists, it is just a matter of time until almost any type of plant can be hybridized with another. **1980** *Garden* CV. 46/2 Am I the only person who doesn't grow pomatoes?

pombe. Add: (Later examples.) Also *attrib.* and *Comb.*
1935 H. Thurnwald in R. C. Thurnwald *Black & White in E. Afr.* iv. 158 The brewing of beer (pombe) is not done by women everywhere. It is a complicated process requiring time and performed especially for feasts. **1952** *Chambers's Jrnl.* May 278/1 The local fermented beverage will be flowing freely—be it coconut pombe, or barley talla, or honey tej, depending on where in Africa the ceremony is taking place. **1966** C. Sweeney *Scurrying Bush* v. 67 The local beer, called *pombe*, and made, in this case, from sorghum grain. **1969** *Tanzania Notes & Rec.* July 26 In 1947 women pombe sellers..brought their case before Kimalando... He saw no justification for mounting a patrol in the African pombe markets. **1977** D. Beaty *Excellency* iv. 54 A big black policeman with breath that smelled strongly of banana pombe. **1978** J. Updike *Coup* (1979) v. 213 An entire American boom town, with.. *pombe*-dispensing saloons.

pome (pōum), *sb.*[2] A jocular alteration of 'poem'.
1861 G. Meredith *Let.* 16 Aug. (1970) I. 100 Did you get the Pome I sent? **1897** A. R. Marshall (*title*) Pomes from the Pink 'un. **1927** Joyce (*title*) Pomes Penyeach. **1959** *News Chron.* 12 Aug. 4/5 My next pome,..is dedicated to a very fine poet. **1975** A. Coren *Further Bull. Pres. Idi Amin* 7 Come on out, John Milton!..wot about dis year's jumbo pome you lazy bum?

pomegranate. Add: Now freq. with pronunc. (pǫ·megræ:nėt). **1. d.** A colour resembling that of the pomegranate. Usu. *attrib.* or as *adj.*
a **1855** C. Brontë *Emma* in *Cornh. Mag.* (1860) I. 495 Miss Wilcox..in her blue merino dress and pomegranate ribbon. **1881** C. C. Harrison *Woman's Handiwork* i. 47 Pomegranate, Bokhara red, Damascus blue..are some of the colours to be had in plushes. **1906** W. J. Locke *Beloved Vagabond* vii. 83 A beautiful gipsy, holding fascinating allurements in lustrous eyes and pomegranate lips. **1927** [see *Ash *sb.*[1] 1 e]. **1955** E. Bowen *World of Love* v. 94 Mamie's pomegranate toenails. **1958** J. Cannan *And be a Villain* iv. 100 A high-waisted pomegranate satin with gold lace sleeves. **1972** *Guardian* 17 Oct. 13/4 The walls are pomegranate with pomegranate velvet lighting in the recesses.

pomellated (pǫ·mĕlēi:tėd), *a. rare*[−1]. [Presumably ad. Fr. *pommelé* 'dappled'.] ? Dappled, mottled.
1922 Joyce *Ulysses* 289 Thither the extremely large wains bring foison of the fields, flaskets of cauliflowers.. and red green yellow brown russet sweet big bitter ripe pomellated apples.

pomelo. a. For *Citrus decumana* substitute *Citrus grandis*. **b.** Also = *Grapefruit. Also *attrib.* (Later examples.)
1908 R. W. Chambers *Firing Line* v. 54 Is that the pomelo grove? **1943** Webber & Batchelor *Citrus Industry* I. v. 568 The grapefruits and the pummelos, or shaddocks, are very closely related. **1968** J. W. Purseglove *Trop. Crops: Dicotyledons* II. 502 *C*[*itrus*] *grandis* is highly esteemed in the East as a dessert fruit, where it is known as the Pomelo. **1971** 'A. Burgess' *MF* vi. 68 A papier-mâché cornucopia spilling bananas, pomelos..and jackfruit.

Pomeranchon (pǫmĕræ·ntʃǫn). *Nuclear Physics.* [f. next + *-on[1].] = *Pomeron.
1967 *Physical Rev. Lett.* XIX. 1061/1 The observed nonshrinkage of diffraction peaks has indicated that the Pomeranchon has an anomalously small slope, so that at present it is the only trajectory generally accepted by Regge phenomenologists which has no particles assigned to it. **1973** *Physical Rev.* D. VII. 1496 Pomeranchon exchange is assumed to be mediated by an isoscalar scalar σ meson.

Pomeranchuk (pǫmĕræ·ntʃuk). *Physics.* [Name of Isaak Yakovlevich *Pomeranchuk* (1913–66), Russian physicist.] **a.** Used *attrib.* with reference to the cooling that a mixture of liquid and solid helium 3 undergoes when it is solidified by compression. [Described by Pomeranchuk in *Zh. éksper. i teoret. Fiziki* (1950) XX. 919.]
1958 *Chem. Abstr.* LII. 17841 (*heading*) Theory of the Pomeranchuk effect in helium-3. **1971** *McGraw-Hill Yearbk. Sci. & Technol.* 86/2 First proposed in 1950, the Pomeranchuk method is based on the unusual thermodynamic properties of a solid-liquid mixture of He[3] at low temperatures. **1974** *Nature* 6 Dec. 441/3 Compressional solidification of [3]He, known as Pomeranchuk cooling,..restricts experiments to the solidification pressure of 34 atmospheres. **1976** *Ibid.* 23 Sept. 276/1 A pair of Pomeranchuk cells was used both for cooling the [3]He into the superfluid A-phase and also for inducing a flow of liquid through the narrow tube which connected them together.
b. Used, chiefly *attrib.*, to designate certain concepts relating to the scattering of sub-atomic particles at high energies, as **Pomeranchuk pole**, a special Regge pole with $\alpha(0) = 1$ and even signature, and with zero isospin, charge, hypercharge, and baryon number (α being the trajectory function); **Pomeranchuk('s) theorem**, a theorem according to which the reaction cross-sections for a particle and for its anti-particle incident on the same target particle should approach the same constant value as the energy of the incident particle is increased; (proposed by Pomeranchuk in *Zh. éksper. i teoret. Fiziki* (1958) XXXIV. 725); **Pomeranchuk trajectory**, the trajectory traced by a Pomeranchuk pole as α increases.
1961 *Physical Rev.* CXXIV. 2049/2 We shall present.. a rigorous generalization of Pomeranchuk's theorem. **1962** *Nuovo Cimento* XXV. 735 We are assuming *either* that the coupling constant of the ρ Regge pole is much smaller than that of the Pomeranchuk (vacuum) pole, *or* that their trajectories lie close together. **1963** *Physical Rev.* CXXIX. 1456/1 Further insight into the behavior of two Pomeranchuk trajectories can be achieved by evaluating the two leading terms in the high-energy behavior of the total elastic cross section. **1973** *Physics Bull.* Mar. 183/2 There is some doubt about the validity of the Pomeranchuk theorem on the constancy of cross sections for K⁺, K⁻ production. **1973** M. Leon *Particle Physics* xii. 244 Elastic scattering..is supposed to be dominated by the Pomeranchuk pole. **1976** N. W. Dean *Introd. Strong Interactions* xvi. 303 If all total cross sections are to become asymptotically constant, the Pomeranchuk trajectory must be present in all elastic scattering amplitudes. *Ibid.*, The f(1270) and the f′(1514) mesons have the correct quantum numbers, but there is considerable question whether either of them actually belongs to the Pomeranchuk trajectory.

Pomeranchukon (pǫmĕræ·ntʃukǫn). *Nuclear Physics.* [f. prec. + *-on[1].] = *Pomeron.
1968 *Physical Rev. Lett.* XX. 236/2 Assuming the continuum contribution to the FESR for the amplitude *B* to be confined to the Pomeranchukon, one could subtract the Pomeranchuk contribution on the right-hand side and the continuum contribution on the left-hand side of the FESR. **1973** *Physical Rev.* D. VIII. 3050 The *S* matrix for the Pomeranchukon enters here because we must include the possibility of Regge-pole exchange without any diffractive interaction between the incoming or outgoing particles.

Pomeranian, *a.* (*sb.*) Add: Also **Pomoranian.**
The form *Pomoranian*, reflecting Pol. *Pomorze* 'Pomerania' (f. *po* 'on', *morze* 'sea') is used chiefly in linguistic writings.
A. *adj.* (Further examples.)
1880 E. W. Hamilton *Diary* 23 June (1972) I. 21 It seems that Bismarck, though unwilling to 'sacrifice a single Pomeranian soldier' in the cause of Greece, will give Germany's moral support, at any rate, to a demonstration. **1919** G. B. Shaw *Augustus does his Bit* in *Heartbreak House* 235 The Colonel of the Pomeranian regiment which captured me. **1934** Priebsch & Collinson *German Lang.* i. i. 11 The Western group [of the Slavonic languages]..includes..the Cassubian and almost extinct Slovinzian (brought by Lorentz under the collective name of Pomoranian) along the Baltic coast of Pomerania. **1935** F. Lorentz in F. Lorentz et al. *Cassubian Civilization* 6 The whole Pomeranian language is divided into seventy-six dialects, which are, in many cases, very different from one another. **1955** R. Jakobson *Slavic Lang.* (ed. 2) 2 These are the remnants of the Pomoranian group. **1957** [see *Kashube]. **1965** G. Y. Shevelov *Prehist. of Slavic* 1 Pomoranian or Baltic Sl[avic] dialects of such Sl[avic] tribes as Vilci-Veletians, Obodrites, etc. **1972** [see *Kashube]. **1974** *Encycl. Brit. Macropædia* XVI. 867/1 Kashubian dialects (including Slovincian) are considered to be the remnants of a Pomeranian subgroup that belonged to the Lekhitic group.
B. *sb.* **b.** A native or inhabitant of Pomerania.
1870 W. B. Ullathorne in C. Butler *Vatican Council* (1930) I. xii. 237 Then there was a Pomeranian, who gave ..an interesting and pathetic account of the difficulties of religion in his country. **1919** [see *Kashube]. **1939** [see *Lech, Lekh *sb.*[2] and *a.*].
c. The West Slavonic dialect of Pomerania, a subgroup of Lechitic, now represented only by Kashube (cf. *Lechitic *sb.* and *a.*, *Kashube).
1934 [see Kashube]. **1935** F. Lorentz in F. Lorentz et al. *Cassubian Civilization* 5 Popular speech..is nowhere uniform... A classic instance of this is furnished by Cassubian, or, as it is more scientifically termed, Pomeranian. This language is divided into Northern Pomeranian and Southern Pomeranian. **1935** T. Lehr-Spławinski in *Ibid.* iii. i. 347 The dialects spoken in the Middle Ages..by the ancestors of the modern Cassubians constituted an intermediate belt between the dialects of Pomeranian properly speaking and those of Polish.

Pomerol (pǫ·mərǫl). The name of a commune in the department of Gironde in SW. France, used *attrib.* or *absol.* to designate the red wine produced there.
[**1833** C. Redding *Hist. Mod. Wines* v. 141 With this quality of wines also may be ranked those grown on the level grounds where the soil is sand and gravel. The most in repute are those of Pommerol and of the environs of Libourne.] **1951** R. Postgate *Plain Man's Guide to Wine* iv. 75 As there is no classification of Pomerols, I am reduced to making a personal list. **1959** W. James *Word-bk. Wine* 146 All are agreed on one point—Pomerol wines taste of truffles. **1969** J. Wainwright *Take-Over Men* vi. 92 He swirled the wine in his glass,..moistened his lips with it,..then said: 'Graves, I think.'..'No. I rather think Pomerol.' **1971** P. Purser *Holy Father's Navy* viii. 43 The meal was indifferent but Father Freeloader's choice of wine impeccable, even if they had brought him the '64 Pomerol and not the '62. **1974** *Times* 29 Oct. 1/5 The alleged fraud consisted in transforming ordinary table wines of the Languedoc into nobler Pomerols or Medocs.

pomeron (pǫ·mĕrǫn). *Nuclear Physics.* Also **Pomeron.** [f. *Pomer(anchuk + *-on[1].] The Pomeranchuk pole or trajectory, or a virtual particle regarded as exchanged in the type of scattering they represent.
1967 R. J. Eden *High Energy Collisions* ix. 234 If total cross-sections are asymptotically constant, there must be a Regge pole of *even* signature..having, $l = \alpha_1{}^+(0) = 1$... The corresponding trajectory is called the Pomeranchuk trajectory. The object exchanged at $t = 0$ is called the 'Pomeron'. It is not a physical particle. **1971** *Physics Bull.* Sept. 517/2 With a linear trajectory this required that the pomeron has a slope of 0·64 GeV⁻². Several factors however have led to the abandoning of the association of the pomeron with the f⁰. **1973** B. H. Bransden et al. *Fund. Particles* ix. 181 At present it does not seem that the Pomeron corresponds to a physical particle and it probably should be thought of as a device to bring diffraction scattering into the same exchange framework as other reactions. **1974** M. L. Perl *High Energy Hadron Physics* xvii. 408 It also became common to speak of the Pomeranchuk trajectory as representing the exchange of a virtual particle called the pomeron.

‖ **pomme** (pǫm). *Gastron.* [Shortened form of next.] A potato. Chiefly *pl.*, esp. in *pommes allumettes*, 'potato matchsticks', i.e. matchstick-thin chips; *pommes frites*, 'fried potatoes', i.e. chips.
1910 A. Mario *Easy French Cookery* 140 *French Fried Potatoes* (Pommes frites). Cut some potatoes in strips.., and cook as for Straw Potatoes. **1931** A. de Croze *What to eat & drink in France* xxix. 260 Pommes frites à la Gasconne (diced potatoes fried with goose dripping or lard). **1952** G. Maurois *Cooking with French Touch* vii. 118 There are forty different ways of preparing potatoes known to the French cuisine. *Pommes frites* is probably the best known of all. **1962** L. Deighton *Ipcress File* xix. 127 The steak was..served with asparagus tips and *pommes allumettes*. **1966** *Vogue* Nov. 148/2 *Pommes frites* (they are thinner than chips half fried). **1975** P. Organ *House on Cheyne Walk* xv. 126 The smell of the beach and *pommes allumettes*.

‖ **pomme de terre** (pǫm də tęr). *Gastron.* Pl. **pommes de terre.** [Fr.] A potato.
1823 [see *Ground-pea]. **1846** A. Soyer *Gastronomic Regenerator* 470 (*heading*) Pommes de terre sautées au beurre. **1877** [see *Lyonnais B *adj.* b]. **1963** R. Carrier *Great Dishes of World* 211/2 *Pommes de terre Anna* sets overlapping layers of sliced raw potatoes in a small buttered baking dish or round mould, each layer dotted with butter and the whole then baked. **1977** *Zigzag* Mar. 8/1 We ate mashed pommes de terre.

pommel, *sb.* Add: **5. b.** Either of a pair of removable curved handgrips fitted to a vaulting horse.
1887 A. Alexander *Mod. Gymnastic Exercises* 137 The Vaulting Horse..contains a set of pommels, which are removable if required. **1895** W. Maclaren *A. Maclaren's Physical Educ.* (new ed.) 103 For vaulting with one hand, circling, feint exercises &c..., it is customary to have pommels fitted on the horse. **1908** *Man. Physical Training* (H.M.S.O.) viii. 184 Bend the knees and spring quickly from the ground up to the 'First position', with the hands gripping the pommels. **1920** Naylor & Temple *Mod. Physical Educ.* 125 The starting position is taken by grasping the pommels with 'inward-grip'. **1932** T. McDowell *Vaulting* vii. 28 This vault may also be performed with one hand on a pommel and the other on the horse proper. **1971** L. Koppett *N.Y. Times Guide Spectator Sports* 242 The gymnast balances himself on the pommels..and performs various maneuvers with leg movements, handstands, and so forth. **1972** B. Taylor et al. *Olympic Gymnastics for Men & Women* viii. 181/2 The left arm pushes off the left pommel enabling the gymnast to gain the necessary height.
8. *pommel vault*; **pommel horse**, a vaulting horse having pommels; also **pommelled horse.**
1908 *Man. Physical Training* (H.M.S.O.) viii. 185 Progression should be obtained by gradually raising the height of the pommel horse till it is somewhat higher than the average troop horse. **1932** T. McDowell *Vaulting* p. v, Where the teacher has a 'box horse' and not a 'pommelled horse', it will be found that many of the vaults are adjustable to the apparatus available. *Ibid.* p. vi, Then comes the 'pommel horse' with pad. *Ibid.* vii. 30 *Pommel Vault.* Take off from both feet as the hands grasp the pommels. **1957** *Encycl. Brit.* XI. 20/2 The 'Olympic six' for men comprise floor exercises, work on the horizontal bar, parallel bars and rings, pommelled horse and vaulting. **1962** [see *Horse 6c]. **1964** D. M. Kunzle in G. C. Kunzle *Parallel Bars* 13 All sorts of climbing instruments—ropes, bars, beams and the pommelled horse. **1971** *Sportsweek* (Bombay) 21 Feb. 9/1 Jim Prestidge recalls..the basic pommel horse exercises developed by German Ludwig Jahn, the father of the sport, in the last century. **1972** B. Taylor et al. *Olympic Gymnastics for Men & Women* viii. 180/1 Place the pommel horse under the parallel bars.

pommer (pǫ·məɹ). *Mus.* [G., altered form of Bombard *sb.*] A type of shawm; = Bombard *sb.* 4.
1878 [see Bombard *sb.* 4]. **1884** *Encycl. Brit.* XVII. 706/2 The little schalmey and tenor pommer seem to have disappeared in the 17th century. **1911** E. F. Cook *Dict. Mus. Terms* 202 *Pommer*, an ancient wooden wind reed instrument of various sizes. **1950** *Oxf. Jun. Encycl.* IX. 487/1 The lower ones were called 'pommers' and 'bombards'. In the 17th and 18th centuries hautboys.. developed from shawms, and 18th-century bassoons from the pommers. **1967** *Daily Tel.* 10 Feb. 19/4 The strength of the group lies..in the diversity and precision of its wind instruments, which include such rareties as the shawm, pommer, crumhorn and cornet. **1976** D. Munrow *Instruments Middle Ages & Renaissance* vi. 40 Praetorius says that shawms are designated by the term *bombarde* or *Pommer* irrespective of their size. **1977** *Early Music* July 347/1 A collection of different pirouettes may be useful. I have five (made from corks) for my soprano pommer.

Pommery (pǫ·mĕri). Also **Pommery and Greno.** The proprietary name of a brand of champagne produced by the firm of Pommery & Greno, founded in Rheims in 1836.
1882 *Official Gaz.* (U.S. Patent Office) 7 Nov. 1540/2 Champagne-Wine.—Veuve Pommery & Fils, Reims, France... The designation 'Pommery & Greno'. **1887** *Trade Marks Jrnl.* 27 July 867 A. Pommery. Wines... Pommery & Cie. Reims. **1891** [see *Perrier Jouët]. **1892** A. W. Pinero *Magistrate* ii. 55 Cis:..You'll look better after a glass or two of Pommery, Guv. *Mr. Posket*: No, no, Cis—now, no champagne. **1896** *Westm. Gaz.* 8 Dec. 2/1 'It is so awful to think of the suffering among the masses and to feel that one can do so little to alleviate it,' he said as he sipped his Dry Pommery. **1907** [see *Mumm]. **1918** G. Frankau *One of Them* xix. 144 Crimson, the orchids flaunted; gold, the chalice Bubbled with Pommery's unstinted measure. **1920** [see *Heidsieck]. **1951** R. Senhouse tr. *Colette's Chéri* 86 'What d'you drink, now you're married?'..'Pommery,' Chéri said. **1966** M. Brewer *Man against Fear* vii. 70 And a bottle of the Pommery. Properly iced, please. **1975** *Woman's Jrnl.* Sept. 73/2 They drank Pommery & Greno 1889.

Pommy (pǫ·mi), *sb.* (*a.*) *Austral.* and *N.Z. colloq.* Also **Pommie** and with lower-case initial. [Origin obscure.] **A.** *sb.* A derogatory term for an immigrant from the United Kingdom; an Englishman or Englishwoman, a Briton. **B.** *attrib.* or as *adj.* Of or pertaining to a Pommy; British, English, *spec.* (often as a term of affectionate abuse) in *Pommy bastard*. Cf. *Pom[2].
The most widely held derivation of this term, for which, however, there is no firm evidence, is that which connects it with *pomegranate* (see quots. 1923, 1963). A discussion of this and of other theories may be found in W. S. Ramson *Australian English* (1966) 63.
1915 in B. Gammage *Broken Years* (1974) 86 We call the Regulars—Indians and Australians—'British'—but Pommies are nondescript. **1916** in *Ibid.* 240 They're only a b—— lot of Pommie Jackeroos and just as hopeless. **1916** *Anzac Bk.* 31 A Pommy can't go wrong out there if he isn't too lazy to work. **1920** D. O'Reilly in Murdoch & Drake-Brockman *Austral. Short Stories* (1951) 144 The 'Pommy' parson made good, as a good man always will. **1923** D. H. Lawrence *Kangaroo* vii. 162 Pommy is supposed to be short for pomegranate. Pomegranate, pronounced invariably pommygranate, is a near enough rhyme to immigrant, in a naturally rhyming country. Furthermore, immigrants are known in their first months, before their blood 'thins down', by their round and ruddy cheeks. So we are told. *Ibid.* 164 In this way Mr Somers had to take himself to task, for his Pommy stupidity. **1926** Galsworthy *Silver Spoon* ii. iv. 137 They call us Pommies and treat us as if we'd took a liberty in coming to their blooming country. **1933** 'P. Cadey' *Broken Pattern* xii. 130 'You should have heard the English accent!' 'Pommy gab, eh?' commented his mate. **1938** N. Marsh *Artists in Crime* ix. 128 She was always shooting off her mouth about the way the Aussies don't know a good thing when they see it. These pommies! She gave me the jitters. **1946** B. James in *Coast to Coast 1945* 63 He was an Englishman, not a 'pommy', mind you. It seemed he hadn't even reached to that dignity. **1947** B. Mason in D. M. Davin *N.Z. Short Stories* (1953) 333 What time we had left was spent on fruitless errands for the Pommie matelots. **1949** F. Sargeson *I saw in my Dream* ii. xiii. 118 Look at Wally's ma—she got over her Pommy ways. **1951** D. Stivens *Jimmy Brockett* 214 Like most of these pommy bastards, he had funny ways but he wasn't a bad old bloke at heart. **1957** *New Scientist* 23 May 13/3 There is..an elusive background of strangeness, imbued with an element of timelessness, which comes home to the sensitive 'new chum', or 'pommy', only after he has lived for a while in this new-old southern continent. **1962** J. Frame *Edge of Alphabet* vi. 47 Look at the foreigners flooding the country on every immigrant ship, la-di-da Pommies and all. **1963** X. Herbert *Disturbing Element* vi. 91 He still wore the heavy clumsy British type of clothing of the day [before 1914]. When we kids saw people on the street dressed like that we would yell at them: 'Jimmygrants, Pommygranates, Pommies!' **1966** R. D. Eagleson in *Southerly* XXVI. 200 Lest British readers should be misled, *pommy* is frequently pejorative. **1974** P. McCutchan *Call for Simon Shard* iv. 36 I'm Australian born and bred, not a pommie immigrant... Now, grand-dad, 'e *was* a pommie bastard! **1975** *Times* 27 Aug 10/8 Colin Shaw..has just sent Ernest Whitehouse an explanation of how God came to be described in the television programme *Beneath the News* as a 'Pommy bastard'... Shaw adds that 'Pommy bastard' is an 'affectionate colloquialism' in Australia. **1979** *Guardian* 31 Oct. 3/2 British Leyland reacted angrily ..to antipodean 'pommy-bashing' about the quality of buses.
Hence **Po·mmyland**, Britain, England.
1957 R. Stow *Bystander* 21 I'm a Pommy. And going back to Pommy-land, after twenty-four years. **1967** F. Hardy *Billy Borker yarns Again* 61 Sir Robert himself wanted to be a whiskey-taster at the Melbourne show, but ended up as some kind of wharfie over in Pommy Land. **1973** *Times* 12 Oct. 15/7 An adaptation of Barry Humphries's cult strip cartoon about the life of darkest Pommie-land seen through the eyes of an antipodean innocent. **1979** M. Kaufman *Container* iii. 31, I suppose you'll head off back to Pommyland now?

Pomo (pōu·mo), *sb.* and *a.* [See quot. 1978.] **A.** *sb.* **a.** An Indian people of Northern California; a member of this people. **b.** Any of the languages of this people. **B.** *adj.* Of, pertaining to, or designating this people or their languages. Hence **Pomo·an**, the group of Pomo languages. Also as *adj.*
[**1852** G. Gibbs *Jrnl.* 2 Feb. in H.R. Schoolcraft *Hist. & Stat. Information Indian Tribes* (1853) III. 112 Four bands consented to enter into a treaty, viz., the Sah-nel, Yukai, Pomo, and Masu-ta-kaya; numbering in all, as was supposed, 1042 souls.] **1872** *Overland Monthly* Apr. 328/1 The great family of Pomos on Russian River..have many dialects, and a name for each—as Ballo Ki Pomos Cahto Pomos, etc. **1875** H. H. Bancroft *Native Races Pacific States* I. 362 The *Pomos*, which name signifies 'people', and is the collective appellation of a number of tribes living in Potter Valley... Each tribe of the nation takes a distinguishing prefix to the name of Pomo, as, the *Castel Pomos* and *Ki Pomos*. **1881** [see *Klamath]. **1910** F. W. Hodge *Handbk. Amer. Indians* II. 277/1 The Pomo were the most southerly stock on the coast not brought under the mission influence of the Franciscans. **1913** [see *Hokan]. **1933** M. R. Harrington *Gypsum Cave, Nevada* 87 The Pomo 'tee-weave' is somewhat similar. **1936** G. A. Reichard *Navajo Shepherd & Weaver* 149 To demonstrate

his skill..a Pomo Indian basket-maker fashions a basket so small it must be kept in a tiny bottle. **1959** E. Tunis *Indians* 113/2 One tribe, the Pomos, made baskets that were possibly the finest ever made in the world. **1965** *Language* XLI. 304 Well-known families such as Pomoan, Chumashan, and Yuman. *Ibid.* 305 Pomo..shows no initial vowel in any of the languages. **1973** A. H. Whiteford *N. Amer. Indian Arts* 39 One-rod coiling was done by the Pomo and Paiute. **1977** *Language* LIII. 260/2 This work..is the first descriptive account ever published of the phonology and grammar of Southeastern Pomo, one of seven distinct languages comprising the Pomoan family within the Hokan stock. **1978** *Handbk. N. Amer. Indians* VIII. 277/1 The word Pomo originated in two Northern Pomo forms that are quite distinct in the native language but that became confused in early writings. The earliest known recordings..give Pomo as the name of an Indian group on the east fork of the Russian River. For a village in southern Potter Valley, on the east fork of that river, Vihman..provides the full phonemic form: *phó·mo·* 'at red earth hole'... A second source..is based on Northern Pomo *phóʔmaʔ*..., which is added to place-names to designate those that live at that place.

pomology. Add: **pomological** *a.* (later example); **pomologist** (earlier and later examples); also **pomolo·gically** *adv.*
1833 *Chambers's Edin. Jrnl.* II. 96/1 It is..the chief object of the modern pomologist to obtain..new varieties. **1920** R. Frost *Let.* 14 May (1964) 105 Their report was that pomologically it was all right, but poetically not. **1976** *Jrnl. R. Soc. Arts* CXXIV. 577/1 Pomologists are now busy 'taking the fruit tree back to the drawing board', seeking better ways of intercepting light. **1976** *Nature* 12 Aug. 574/1 Pomological literature contains two reports of the influence of grafted scions on the size, colour and ripening season of apples borne on the stock portions of topworked trees.

pomonal (pomōu·năl), *a. rare.* [f. Pomona + -al.] Of or pertaining to fruit-trees; pomonic.
1859 *Trans. Illinois Agric. Soc.* III. 354 We may proudly claim this land..as the favorite seat of horticultural and pomonal progress.

Pomoranian, var. *Pomeranian *a.* (*sb.*).

pomp, *v.*[1] Delete † *Obs.* and add later *poet.* examples. Hence **pomped** (pǫmpt) *a.*, honoured with pomp, celebrated: **pomping** *a.*, restrict † to sense in Dict. and add: (*b*) *dial.*, involved in acting.
1919 W. de la Mare *Flora*, Mount to the porch the pomped grandees In lonely state, by twos, and threes. **1922** Hardy *Late Lyrics* 48 And once or twice she has cast me As she pomped along the street Court-clad,..A glance from her chariot-seat. **1937** G. Frankau *More of Us* xiv. 153 And all that day, despising fun and frolic With Janes or Joans, he pomped about the ship. **1969** G. Macbeth *War Quartet* 26 So few yards Beyond this dust-whirl, those pomped victors. **1976** *Birmingham Post* 16 Dec. 2/4 Rover, one of the pomping folk, thinks principally in Shakespearean tags. **1976** J. C. Trewin in D. V. Baker *Cornish Short Stories* 134 It [*sc.* the rain] hammered against the side of the tent where the pomping folk had prepared hopelessly for the evening.

Pompadour, *sb.* Add: **1.** (Further examples.)
1885 A. Edwardes *Girton Girl* I. xii. 238 It was not Louis Seize furniture, or Pompadour cabinets..that Marjorie missed. **1909** *Daily Chron.* 17 Sept. 5/3 Charged ..with stealing..two silver pompadour boxes. **1925** O. Sitwell in E. Sitwell et al. *Poor Young People* 35 Through her pompadour-peruke. **1971** S. Jepson *Let. to Dead Girl* xii. 143 She opened the skirt of a pompadour doll on a side table, pulled out a white telephone.
2. (Later examples.)
1968 *N.Y. City* (Michelin Tire Corp.) 57 The most precious types of Sèvres porcelain, in the pink known as 'Pompadour'. **1976** N. Roberts *Face of France* x. 107 Balloons, scarlet, orange, blue and..Pompadour pink.
5. a. (Earlier and later examples.)
1887 *Evening Sun* (N.Y.) 15 Apr. 3/4 A tall, slender young man, with a full blonde beard and pompadour hair. **1920** S. Lewis *Main St.* 76 The meek ambitiousness.. clouded like an aura his pale face, flap ears, and sandy pompadour. **1955** W. Gaddis *Recognitions* i. vi. 208 His hair, a shiny black pompadour which he wore like a hat. **1976** *New Yorker* 24 May 107/1 Reagan looks good at the rostrum: a tall figure with ruddy cheeks, his reddish-brown hair swept back in a slight pompadour.
b. (Later example.)
1975 J. Drummond *Slowly the Poison* i. 97 Her hair.. was not worn in the current high pompadour style, but cut short.

po·mpadour, *v.* [f. Pompadour *sb.* 5.] *trans.* To dress (hair) in the pompadour style; to arrange (hair) in a pompadour. Chiefly as *pa. pple.* or *ppl. adj.*
1908 *London Opinion* 22 Aug. 362/2 She was large, plumply built, with grey hair artfully pompadoured and undulated. **1909** L. M. Montgomery *Anne of Avonlea* xiv. 153 Gertie Pye swept in, pompadoured and frilled. *a* **1913** F. Rolfe *Desire & Pursuit of Whole* (1934) xvii. 178 Some pretentious pompadoured image trailing satin. **1957** V. J. Kehoe *Technique Film & Television Make-Up* ix. 111 Other tribes pompadoured the front of the hair and had two side partings. **1971** *Daily Tel.* (Colour Suppl.) 3 Sept. 42/4 They wore their hair very long, very high, pompadoured, swept up..and back. **1973** R. Rosenblum *Mushroom Cave* (1974) 18 The squat bully with greasy pompadoured black hair.

pompano. 1. Substitute for def.: A North American or West Indian marine fish belonging to the genera *Trachinotus*, *Parona*, or *Zalocys*, of the family Carangidæ, esp. *Trachinotus carolinus*, the common pompano, found near south-eastern coasts of North America. (Earlier and further examples.)
1778 tr. *J. Chappe d'Auteroche's Voy. California* 24 The pampano is very plenty in the southern part of the gulph of Mexico. **1840** *Picayune* (New Orleans) 1 Sept. 2/1 Pompanos were plentiful, and sparkling hock flew about. **1851** A. O. Hall *Manhattaner in New Orleans* 161 We forgot our military sighings in the discussion..of the momentous question whether it was orthodox to eat rum-omelette with 'pompano'-fish. **1883** 'Mark Twain' *Life on Mississippi* xliv. 445 We had dinner [in New Orleans].—the chief dish the renowned fish called pompano, delicious as the less criminal forms of sin. **1892** Stevenson & Osbourne *Wrecker* xix. 289 There we sat.. eating pompino and drinking iced champagne. **1965** A. J. McClane *Standard Fishing Encycl.* 693/2 Pompano cookery is an art in Florida and Louisiana. **1973** *New Yorker* 17 Feb. 30/2 Hundreds of red snappers, Carolina mullets,..pompano, Palm Beach mackerels, and others, ..lie in neat rows on a bed of shaved ice. **1977** *Time* 19 Dec. 42/2 Newly appreciated..are such home-grown marvels as Long Island duckling,..Chesapeake oysters, Gulf shrimp and pompano, [etc.].

Pompeian, *a.*[1] Add: Now usu. with pronunc. (pǫmpē·ăn). (Further examples.) Also, characteristic or imitative of the architecture or painting of Pompeii, esp. frescoes.
1869 D. G. Rossetti *Let.* 21 Aug. (1965) II. 716 She built..a Pompeian house for the schoolmaster. **1879** A. Holt *Fancy Dresses* 60 *Pompeian lady.* White llama skirt, with Grecian border worked in purple. **1881** C. C. Harrison *Woman's Handiwork* i. 20 Pompeian red velvet, for portières. **1939** A. Thirkell *Before Lunch* iv. 106 The ceiling..painted in what were called Pompeian colours. **1962** *Listener* 18 Oct. 632/2 The story of this horrifying episode was told in 'Hurricane!'.. The presentation..had a subtly Pompeian quality about it. *a* **1967** A. Ransome *Autobiogr.* (1976) xxvii. 230 A well-designed preparation for life in this Pompeian society, Pompeian in the sense that all these people were living as it were on the slopes of a volcano. **1972** *Sci. Amer.* Sept. 86/1 The Pompeiian mosaic in the museum in Naples.
B. *sb.*[1] A native or inhabitant of Pompeii.
1823 Lady Blessington *Jrnl.* 12 Aug. in E. Clay *Lady Blessington at Naples* (1979) 62 The repairs speak little for the taste of the Pompeians. **1840** *Penny Cycl.* XVIII. 380/2 The emperor Nero..adjudged that the Pompeians should be deprived of all theatrical amusements for ten years. **1869** 'Mark Twain' *Innoc. Abr.* xxxi. 330 Those Pompeiians were very luxurious in their tastes and habits. **1974** *Encycl. Brit. Macropædia* XIV. 789/2 Pompeians and colonists seem to have adjusted with a minimum of friction. **1976** *Times* 23 Nov. 11/3 The faces of the Pompeians as they recorded themselves are reminiscent of those one can still see in Campania.

Pompeian (pǫmpē·ăn, pǫmpī·ăn), *sb.*[2] and *a.*[2] [ad. L. *Pompēiān-us*, f. the name *Pompēius* Pompey.] **A.** *sb.* A follower of Gnaeus Pompeius Magnus (106–48 B.C.), Roman consul, or of his son. **B.** *adj.* Of or pertaining to Pompey or his party.
1845 J. H. Newman in *Encycl. Metrop.* X. 281/2 Bent on retiring to the Pompeians in Sicily. *Ibid.* 282/1 The remains of the Pompeian party. **1908** W. W. Fowler *Social Life at Rome in Age of Cicero* iii. 87 Some £17,500, all of which, while in deposit at Ephesus, was seized by the Pompeians in the Civil War. **1913** D. Hannay *Navy & Sea Power* ii. 31 When Julius Caesar followed Pompey into Thrace he crossed the Adriatic under the very nose of a superior Pompeian fleet. **1949** L. R. Taylor *Party Politics in Age of Caesar* iii. 68 In the year 61, when Pompey was pouring out money to elect his man to the consulship, Cato charged in the senate that the money was being distributed from the house of the Pompeian consul in office. *Ibid.* viii. 171 Caesar also contributed by the policy he followed when..he came home from his victory over the Pompeians in Spain. **1974** E. S. Gruen *Last Generation of Roman Republic* ii. 56 Bibulus and Domitius could be counted on by Cato for steady cooperation in reducing the influence of Cæsarians and Pompeians in Roman politics. *Ibid.* 62 It was a standard Pompeian practice to express his own ambitions in terms of the needs and desires of his soldiers.

Pompey (pǫ·mpi). [Origin unknown.] A nickname for: **a.** The town and dockyard of Portsmouth, in Hampshire. **b.** Portsmouth Football Club. Also *attrib.*
1899 *Evening News* 9 Dec. 3/6 Wilkie, amid tremendous cheering from the Pompey lads, won the toss, and played with the wind in their favour. **1916** 'Taffrail' *Pincher Martin* iii. 40 The *Belligerent* was a 'Pompey' ship. **1930** *Daily Express* 6 Oct. 16/7 Despite their undeniable superiority Portsmouth could not penetrate the Derby defence... A brilliant Pompey could do everything except score. **1943** C. S. Forester *Ship* xviii. 109 The grim wife he had in Pompey. **1944** Williams & Savage *Second Penguin Problems Bk.* 160 Shouting 'Good old Pompey,' Portsmouth supporters went home. **1959** *Observer* 22 Mar. 17/5 An old roadman..: 'There's been a lot of unemployment in Pompey and Southampton.' **1966** (*title*) Pompey chimes: the journal of the Portsmouth West Conservative Association. **1972** E. Grierson *Confessions of Country Magistrate* xv. 149 That Plymouth should possess a second-tier court and Portsmouth only a third-tier one will infuriate, and with reason, the good citizens of Pompey. **1976** *Oxf. Compan. Ships & Sea* 659/1 '*Pompey*', the sailors' slang name for Portsmouth... It is not known how or when the name came into being, one theory being that it owes its origin to the fact that the local fire brigade, known as the Pompiers, used to exercise on Southsea Common, adjacent to the town of Portsmouth. **1977** *Navy News* June 6/3 Is there any chance of recreating the Bluejacket Band at Pompey?
Hence **Po·mpeyite**, a sailor from Portsmouth.
1916 'Taffrail' *Pincher Martin* iv. 63 Down wi' the Pompeyites!

pompier. Add: **1.** (Further examples.)
1871 E. G. E. Ward *Jrnl.* 24 May in D. P. Carew *Many Years, Many Girls* (1967) i. 51 All the 'pompiers' from St. Germain are gone to Paris, but the fire is enormous. **1958** L. Durrell *Balthazar* x. 203 The hall was full of fancy-dress figures of *pompiers* with hatchets and buckets.
2. *transf.* An artist who paints in an academic, imitative, vulgarly neo-classical style. Also *attrib.*
1924 A. Huxley *Let.* 9 Aug. (1969) 231 It may be mere folie de grandeur and pompier prejudice on my part. **1950** Wyndham Lewis *Let.* 22 Mar. (1963) 520 The greatest news of all is that you have taken to the brush and palette! Are you a *Douanier*, or a *pompier*? **1974** *Times* 23 Nov. 14/1 The names of the so-called Pompier artists—late nineteenth-century French academic painters—are on every lip... The term 'Pompier'..is thought to derive from the helmets worn by the Greek gods and heroes depicted in the canvases of the late Classical painters and their close similarity to that of the Paris firemen, or *pompiers*. *Ibid.* 14/4 The Pompiers are exotic and flamboyant. **1977** *Times* 30 Mar. 12/4 The art of the *Pompiers* as they have been contemptuously called... The exhibition of *French Nineteenth-century Paintings* at the Alpine Club Gallery..is..welcome.

pompilid (pǫ·mpilid), *sb.* and *a. Zool.* Also **Pompilid.** [ad. mod.L. family name *Pompiliidæ*, f. the generic name *Pompilius* (J. C. Fabricius *Supplementum Entomologiæ Systematicæ* (1798) 212), f. Gr. πομπίλ-ος a fish which followed ships + -id[3].] **A.** *sb.* A predatory fossorial wasp belonging to the family Pompiliidæ. **B.** *adj.* Of or pertaining to this group of insects.
1909 *Cent. Dict.* Suppl., *Pompilid* n. & a. **1913** *Oxf. Univ. Gaz.* 4 June 952/1 The specimens show the resemblance of the black red-banded Hemipteron to a common pattern of the Pompilid group of Fossorial Hymenoptera. **1924** J. A. Thomson *Sci. Old & New* xxxv. 200 There are predatory wasps called Pompilids that hunt spiders. **1941** W. S. Bristowe *Comity of Spiders* II. vii. 354 Spiders are the sole prey of Pompilids. **1962** *Oxf. Univ. Gaz.* 19 Mar. 852/1 A Pompilid wasp and its spider prey from Trinidad. **1973** J. P. Spradbery *Wasps* xiv. 307 Among some pompilids at least, considerable powers of orientation have been evolved.

pom-pom. Add: **1.** In later use: any of various heavier guns, esp. if multi-barrelled or one of a group.
1916 'Boyd Cable' *Action Front* 131 The muzzles of the two pounder pom-poms moved slowly after their target. **1940** 'N. Shute' *Landfall* vii. 175, I should think the multiple pom-poms would have got the machine. **1944** *R.A.F. Jrnl.* Aug. 272 (*caption*) Battleship..; carries forty pom-poms in multiple mountings. **1973** J. Quick *Dict. Weapons* 353/1 *Pom-pom.* 1. A rack of antiaircraft cannons, usually mounted in fours, as on the deck of a ship. 2. An automatic cannon.
2. A representation of a repetitive sound, e.g. the beat of a popular tune or poem. Also *pom-pom-pom*, etc.
1909 Beerbohm *Lett. to R. Turner* (1964) 181 They have been re-printing *Yet Again*. Second impression ready within a few days. Pom-pom-pom. **1916** A. Huxley *Let.* 31 Mar. (1969) 95 Time percolates with a distressing rapidity through the coffee-machines of life... So you see, pom, pom, where we are?..as the old song says. **1945** W. Stevens *Let.* 26 Jan. (1967) 485 Many lines exist because I enjoy their clickety-clack in contrast with the more decorous pom-pom-pom that people expect. **1978** M. Kenyon *Deep Pocket* ix. 105 *Pom* pom pom *pom* pom, something sweet Willies.
Hence **po·m-pom** *v. intr.*, to fire a pom-pom; **pom-po·mming** *vbl. sb.*
1901 'Linesman' *Words by Eyewitness* vii. 147 Continuous sniping, pom-pomming, and occasional shelling. *Ibid.* ix. 191 Boers can fire shrapnel, Britons can pom-pom with the best.

pompon. Add: The spelling with final -m is now common. **1.** (Later and *attrib.* examples.)
1904 P. N. Hasluck *Upholstery* 19 Pom-poms are used for ornamenting upholsterers' work. To make a pom-pom lap a wool or cardboard washer with three or four thicknesses of fibres. **1927** [see *Evzone, evzone]. **1924** D. H. Lawrence *Pansies* 148 My! the bloomin' pom-poms! Even as trimmings they're stale. **1952** Granville *Dict. Theatr. Terms* 139 *Pom-pom dress*, the conventional Pierrot costume of white pantaloons, white jacket, decked with black pom-poms, or any other good combinations of colour. **1968** *Listener* 3 Oct. 445/3 The making of Bunny Girls' pom-poms for export. **1975** *Times* 9 Dec. 9/8 Mink jacket..has..mink pom-pom ties at the neck. **1977** *Time* 4 Apr. 42/2 He has rigged it with 100,000 steel darts, which, if detonated at just the right moment, can wipe out everybody in the stadium, down to the last pompon girl.
2. (Later examples.) Also, one of a group

of dwarf varieties of *Rosa centifolia* with small double flowers. Also with capital initial.
1843 *Florist's Jrnl.* IV. 106 *R*[*osa*] *centifolia* (the Provence or Cabbage rose), with its varieties, including the 'mossy' and 'pompone' roses. **1869** [see *Miniature *a.* a]. **1894** A. Foster-Melliar *Bk. of Rose* ii. 15 A sub-variety of the Provence is the Pompon Rose. **1908** E. J. Banfield *Confessions of Beachcomber* i. ix. 130 Stalkless mushrooms [of coral], gills uppermost,..blossom as pom-pom chrysanthemums. **1922** Mrs. C. H. Stout *Amateur's Bk. Dahlia* i. 7 At that time [*sc.* 1870] appeared a tiny ball-shaped blossom, originating probably with Hartweg of Karlsruhe, which he called 'pompon'. **1943** F. Thompson *Candleford Green* i. 20 Old-fashioned pompom dahlias in autumn. **1952** W. E. Shewell-Cooper *Chrysanthemum Growing* xv. 161 These chrysanthemums are said to have been quite popular in the Victorian era... They are called 'Pompoms' in some parts of the country, 'Pompons' in other districts, and even 'Pompones'. **1955** C. C. Hurst in G. S. Thomas *Old Shrub Roses* ix. 91 Both varieties were dwarf Pompons a few inches high. **1958** [see *Collarette c]. **1961** [see **Korean chrysanthemum*]. **1974** J. Berrisford *Window Box & Container Gardening* v. 48 The smaller-growing of the Pompoms [*sc.* chrysanthemums] are also useful.

‖ **pomposo** (pǫmpōu·zo), *adv.* (*adj.*) and *sb.* [It.] **A.** *adv. Mus.* In a stately manner. Also as *adj.*, affected, pompous.
1801 Busby *Dict. Mus.*, *Pomposo*,..a word implying that the movement to which it is prefixed is to be performed in a grand and dignified style. **1847** Webster 845/2 *Pomposo*,..grand and dignified. **1876** Stainer & Barrett *Dict. Mus. Terms* 364/1 *Pomposamente*, *pomposo* ..pompously. **1959** *Collins Mus. Encycl.* 508/2 *Pomposo*, in a pompous manner. **1960** 'A. Bridge' *Numbered Account* 220 Don't be so pomposo, Colin—really you bore me.
B. *sb.* **1.** An affected, self-important person.
1930 *New Statesman* 15 Nov. 176/1 Their satyrs, pomposos, and ninnies..became the delight of future generations. **1938** C. S. Forester *Ship of Line* ix. 116 'And now,' said Bolton, 'we must await in idleness the arrival of Sir Mucho Pomposo, Rear Admiral of the Red.'
2. *Mus.* A movement, musical passage, or the like to be played in a stately manner; a piece of music marked *pomposo*.
1966 E. R. Reilly tr. *Quantz's On Playing Flute* xvii. 231 A Maestoso, Pomposo, Affetuoso, or Adagio spiritoso must be played seriously, and with a rather heavy and sharp stroke.

'pon, pon (pǫn), aphetic form of Upon *prep.*
1557 in *15th Rep. R. Comm. Hist. Manuscripts* App. iii. 39 in *Parl. Papers 1897* (C. 8364) XLVIII. 71 Suche impositions as the lorde deputie for the tyme beinge shall taxe and set pon them. *c* **1560** [see Matter *sb.*[1] 25 c]. **1796** F. Burney *Camilla* iv. 119 Much obliged to him, 'pon honour! **1821** M. Edgeworth *Let.* 5 Dec. (1971) 287 Fanny quite well p'on honor. **1850** F. E. Smedley *Frank Fairleigh* v. 47, I didn't think you had it in you; 'pon my word, I didn't. **1901** M. Franklin *My Brilliant Career* xiii. 78 'Pon my honour, Miss Melvyn, I had no idea it was you. **1914** 'Bartimeus' *Naval Occasions* xxiv. 244 Have you any rich aunts, Guns? 'Pon my word, I might get off this afternoon. **1924** D. Moore *Fen's First Term* x. 108 'Pon my word, I can't say. **1973** 'M. Innes' *Appleby's Answer* ii. 18 A delightful-looking creature, madam, 'pon my soul.

ponask (pōu·nɒsk, pū·nɒsk), *v. Canad.* Also **poonask.** [Algonquian.] *trans.* To cook (game or fish) by splitting it and roasting it on a spit or stick over an open fire. Hence **po·nasked** *ppl. a.*; **po·nasking** *vbl. sb.*
1922 *Beaver* Mar. 39/2 As we had no kettle..we were forced to 'ponask' the fish on a pointed stick before a bright fire. **1934** P. H. Godsell *Arctic Trader* 46 She had, therefore, taken the heart, impaled it on a stick, and ponasked it as one would roast a duck. **1944** C. Clay *Phantom Fur Thieves* 31 Thus were the two pieces of duck held up to the blaze and heat. 'That's called 'ponasking', Dave,' said the old trapper. **1961** J. W. Anderson *Fur Trader's Story* viii. 66 With the addition of salt, the ponasked fish was a delightful repast. **1963** G. S. McTavish *Behind Palisades* 90 While the kettles would be boiling their meat, they [*sc.* Indians] would be 'Poonasking' strips of meat and delicacies like leg-bones in front of the fire. *Ibid.*, 'Poonasking' is a method of cooking before a campfire by splitting meat or game, impaling on a pointed stick, where it is quickly roasted from the intense heat.

ponce, *sb.* Substitute for def.: **a.** One who lives off a prostitute's earnings; a prostitute's protector; a pimp. (Later examples.)
1914 C. Mackenzie *Sinister St.* II. iv. ii. 868 You're nothing more than a dirty ponce. I've gone five years without keeping a fellow yet. **1916** W. S. Maugham *Writer's Notebk.* (1949) 98 A raid was made, and fourteen ponces were arrested. **1957** C. MacInnes *City of Spades* ii. iv. 127 These whores are always masters of their ponces. One word to the Law, and the lucky boy's inside. **1965** *New Statesman* 23 Apr. 642/2 If a girl has to get 10 clients for a male ponce, she will need 20 to satisfy the monetary demands of a woman ponce. From my observations at least one ponce in four in London is a lesbian. **1970** G. Greer *Female Eunuch* 131 The role of the ponce ..is too established for us to suppose that prostitutes have found a self-regulating lifestyle. **1975** J. Symons *Three Pipe Problem* xviii. 182 What do you think I am, a tart trying to find a ponce?
b. A male homosexual; a lazy or effeminate man. Also as a vague term of abuse.
1932 Auden *Orators* iii. 98 Dyers and bakers And boiler-tube makers, Poofs and ponces, All of them dunces. **1953** K. Amis *Lucky Jim* xi. 119 As if I'd have said a word in front of that little ponce. **1969** N. Cohn *AWopBopaLooBop* (1970) xix. 185 Mods thought that Rockers were yobs, Rockers thought that Mods were ponces. **1974** P. Wright *Lang. Brit. Industry* xi. 95 An infuriated spectator may shout at a plump, sleek referee, 'You nasty little ponce!' **1978** in P. Marsh et al. *Rules of Disorder* ii. 46 Anybody that works in a lesson.. that you know you're going to doss about in,..you get called 'ponce' and everything.

ponce (pǫns), *v. slang.* [f. the sb.] *intr.* To act as, or behave like, a ponce; to live *on* the earnings of a prostitute; *fig.* to sponge (on), take advantage (of). Usu. const. *on* or *off*. Also, *to ponce about*, to act in an effeminate or languid manner; to fool or mess about; *to ponce up*, to tart up, to make effeminate or effete. Hence **po·ncing** *vbl. sb.* and *ppl. a.*; **ponced up** *ppl. a.*
1932 G. S. Moncrieff *Café Bar* iv. 35 Lou left her periodically, usually to live with some other tart, poncing. *Ibid.*, Now he was unemployed and they were saying to her that he was poncing on her. **1936** J. Curtis *Gilt Kid* ii. 23, I didn't say no one was poncing on her. **1936** [see *Kite *sb.* 4 c]. **1937** J. Curtis *You're in Racket* i. 13 Why the hell don't you buy some for yourself instead of poncing on other people? **1938** G. Kersh *Night & City* iii. 42, I don't ponce it orf 'em. **1953** P. Scott *Alien Sky* i. ii. 18 Urdu's a man's language... Don't ponce it up with that bastard higher standard muck. **1954** Ponce about [see **brothel-creeper* (*shoe*)]. **1955** 'C. H. Rolph' *Women of Streets* x. 114 He was arrested a third time for poncing on the girl and sent to prison. *Ibid.* 115 The man was sentenced to two years for his third poncing offence. **1957** C. MacInnes *City of Spades* ii. ii. 120 Best of all..is poncing on some woman, but I haven't got the beauty enough for that. **1966** J. Wainwright *Evil Intent* 46 'Why the hell can't they stick to plain facts?' he snarled. 'Why must they ponce everything up to suit their own ends?' **1969** N. Cohn *AWopBopaLooBop* (1970) vi. 57 No poncing about, no dressing up or one-shot gimmicking. **1970** G. Lord *Marshmallow Pie* v. 49 What do you think you look like, all ponced up like that? Fucking queer! **1971** *Guardian* 24 June 13/2 Let's face it, New Zealand has been poncing on us for years. **1972** D. Lees *Zodiac* 132 If my own mother had been murdered I wouldn't ponce about like you're doing. **1973** K. Giles *File on Death* vi. 150 Part-time poncing as well, but he thought book-making was more honest. **1974** J. Gardner *Corner Men* vii. 54 Their Rolls is in the Dean Street car park and Chung Yin's sitting in it ponced up like the sweet and sour faggot he is. **1977** *Zigzag* Aug. 14/1, I mean the one before that was just like a stopgap when we was poncin' around, so much to do and nothing was being done for us. **1977** M. Kenyon *Rapist* v. 52 Poncing rapist English.. thinking they owned the place.

poncey (pǫ·nsi), *a. slang.* Also **poncy.** [f. Ponce + -y[1].] Of, pertaining to, or resembling a ponce (sense b); effete, homosexual.
1964 J. Hale *Grudge Fight* xi. 179 'Come on, sissy boy,' says Brooks, 'come on you poncy bastard.' **1970** 'D. Craig' *Young Men may Die* xii. 118 Stephen read.. from his notes in that poncy briefing voice he could put on. **1970** L. Henderson *Sitting Target* xii. 106 This smells like a poncey brothel. **1973** M. Amis *Rachel Papers* 174 You haven't half got poncy mates. **1973** J. Wainwright *Pride of Pigs* 160 Wot yer bring this poncey gear for, anyway? **1977** *Listener* 6 Jan. 13/1 If you are an intellectual sort of chap, not well-versed in small-talk and tittle-tattle, you cannot make much headway with the dumb blonde. You will scare her off by being too poncey or too high-hat.

poncho. Add: **a.** (Earlier and further examples.) Now in common use as a fashion garment.
1717 tr. *Frezier's Voy. to South-Sea* ii. 71 The Spaniards have taken up the Use of the *Chony*, or *Poncho*..to ride in, because the Poncho keeps out the Rain. **1885** C. M. Yonge *Two Sides of Shield* I. iv. 40 'Here are some overshoes and Poncho.'..Poncho..turned out to be a sort of cape. **1907** *Yesterday's Shopping* (1969) 320 c The Mercedes motor cycling poncho. In fawn cashmere, fitted with rubber neck band and wrists, when worn with overalls, rendering the wearer absolutely proof against rain, dust, or wind. **1929** F. A. Pottle *Stretchers* (1930) 40 We had now been issued ponchos... A poncho is simply a rectangular sheet of water proofed material, with a hole in the center to put one's head through. **1952** W. R. Burnett *Vanity Row* (1953) i. 9 The little Italian newsboy, wearing a black rubber poncho and cursing the weather, was trying to make up his mind to go home and the hell with it! **1956** G. Durrell *Drunken Forest* i. 18, I found him clad in pyjamas and a *poncho*, that useful Argentine garment that resembles a blanket with a hole in the middle through which you stick your head. **1967** *Observer* 26 Mar. 9 Girls..wear what men wear, though Inca ponchos have a strong appeal. **1969** I. Kemp *Brit. G.I. in Vietnam* iv. 84, I sat out in the open, wrapped miserably in my poncho. **1974** *Times* 4 Oct. 7/3 There will be no central heating in the Elysée until October 15... The staff has been allowed to sport..polo-necked pullovers and South American ponchos. **1976** T. Sharpe *Wilt* ii. 13, I was thinking of trying Felicity Fashions for a shantung poncho.
b. poncho dress, (*a*) a costume including a poncho; (*b*) a dress made like a poncho; **poncho liner,** a lining garment worn under a poncho.
1811 W. Walton *Hist. & Descr. Acct. Peruvian Sheep* ii. 52 The Indian driver in this plate is also represented in his proper *poncho* dress. **1968** *Vogue* 15 Apr. 28 Poncho dress..25 gns. **1969** I. Kemp *Brit. G.I. in Vietnam* iv. 76 Nights at this altitude were beautifully cool and we often slept under our padded poncho liners.

‖ **poncif** (ponsīf). [Fr.] Stereotyped or conventional literary ideas, plot, character, etc.
1923 J. M. Murry *Pencillings* 136 The modern spirit, with its almost fanatical desire to get rid of the *poncif*, might make a fine thing of classical translation. **1940** *Scrutiny* IX. 258 He [*sc.* Verlaine] revived some of the oldest and loveliest verse-forms; and he managed, in his best work, to escape from the Romantic *poncif* and to go back to something more human.

poncy, var. *Poncey *a.*

pond, *sb.* Add: **1. b.** (Further N. Amer. examples.)
1693 H. Kelsey *Kelsey Papers* (1929) 3 This wood is poplo ridges with small ponds of water. There is beavour in abundance but no Otter. **1794** A. Thomas *Newfoundland Jrnl.* (1968) 27 In this Island is a fresh water pond a full mile in length, and in it are large Eels and other Fish. **1801** J. Quincy in *Proc. Mass. Hist. Soc.* (1889) 2nd Ser. IV. 132 Nantucket whale-fishers pursuing perch in a pond half a mile in circumference are objects ludicrous enough. **1831** J. J. Audubon *Ornith. Biogr.* I. 479 It searches for food..by the margins of such inland lakes as, on account of their small size, are called by us ponds. **1948** *Canad. Geogr. Jrnl.* Mar. 49/1 Everyone knows what a lake is and there are lakes of all sizes from coast to coast, but if you happen to reside in the Eastern Townships of Quebec you may find your *lake* is called a *pond*. **1969** H. Horwood *Newfoundland* 220 In Newfoundland almost all lakes, no matter how large, are called 'ponds'. **1974** *Maclean's Mag.* Dec. 83/2 The Syncrude pond will cover nine square miles.
d. = *Lagoon[1] 2**.
1956 K. Imhoff et al. *Disposal of Sewage* xii. 205 The area of pond required for waste purification may be computed by means of the oxygen balance. **1961** Bolton & Klein *Sewage Treatm.* vi. 88. If properly operated, the ponds are reasonably free from bad smells, due possibly to the deodorizing effect of the chlorophyll in the algae. **1973** T. H. Y. Tebbutt *Water Sci. & Technol.* ix. 138 In warm climates biological treatment is sometimes achieved in oxidation ponds. **1978** *Coal Option* (Shell Internat. Petroleum Co.) 8 Substantial research is also going into agglomeration processes to recover coal from potential waste material, such as the effluent streams from existing colliery washeries, and coal from existing slurry ponds and tips.
4. *pond-keeper*; **pond-barrow** *Archæol.* (further examples); **pond-culture**, the keeping of fish in ponds; hence **pond-cultured** *a.*; **pond-skater**, an aquatic insect belonging to the family Gerridæ, found on the surface of fresh or salt water; **pond-snail**, esp. one belonging to the genus *Limnæa*; (earlier and later examples).
1941 *Proc. Prehist. Soc.* VII. 89 The so-called pond-barrow consists of a slight depression,..the material from which has been placed round the circumference to form an embanked rim. **1963** *Field Archaeol.* (Ordnance Survey) (ed. 4) 47 The pond barrow appears as a regular circular shallow depression,..surrounded by a small bank. **1885** *Encycl. Brit.* XIX. 127/2 Pond-culture..has been practised for many centuries. **1977** *Undercurrents* June–July 30/3 The mirror carp is by far the best fish for pond culture in Britain. **1972** *Country Life* 7 Dec. 1565/1 There are no public-health worries about eating pond-cultured fish. **1779** G. White *Let.* 7 May in *Selborne* (1789) i. 259 Five of those most rare birds..were shot upon the verge of Frinsham-pond... The pond keeper says there were three brace in the flock. **1909** *Westm. Gaz.* 12 Jan. 5/2 The pondkeeper was unavoidably absent from his post. **1895** L. C. Miall *Nat. Hist. Aquatic Insects* xiii. 382 The Pond-skaters stand or run upon the surface of the water, which they dimple but do not break. **1923** E. A. Butler *Biol. Brit. Hemiptera-Heteroptera* 244 Popularly known as pond-skaters or water-measurers, they attract the attention of even the least observant by the free and easy way in which they dart along over the surface of the water. **1952** J. Clegg *Freshwater Life* xiii. 198 The Pond Skaters feed largely on dead or dying insects that fall on the water. **1973** *Nature* 9 Mar. 132/1 The family Gerridae..includes the common pond-skaters or water-striders. **1855** C. Kingsley *Glaucus* 159 A few of the delicate pond-snails (unless they devour your *Vallisneria* too rapidly). **1952** J. Clegg *Freshwater Life* xvi. 261 The Pond Snails proper..belong mainly to the genus *Limnaea*.

pond, *v.* Add: **1. b.** To produce (a lake) by forming or acting as a dam.
1949 *Bull. Geol. Soc. Amer.* LX. 1383/2 In southern Ohio it is claimed that certain coastal-plain plants..are still avoiding the deposits of the proglacial lakes that are supposed to have been ponded by the advancing Nebraskan ice in the upper drainage of the Teays River. **1971** *Nature* 8 Oct. 391/1 Potassium-argon determinations on trachyte lavas which possibly ponded the former Chemoigut lake gave results of 1·1 and 1·2 m.y.

ponded, *ppl. a.* (s.v. Pond *v.*). Add: **b.** Of a sewage filter: blocked; under a depth of liquid.
1940 Imhoff & Fair *Sewage Treatm.* vi. 106 *Psychode* prefers an open bed and *Achorutes* a ponded surface. **1967** *Jrnl. Inst. Public Health Engineers* LXVI. 170 This large accumulation of film did not affect the performance of the filters very markedly, even though at times the surfaces of both were quite badly ponded. **1971** T. H. Y. Tebbutt *Princ. Water Quality Control* xii. 122 A ponded

filter can be brought back into use by applying the partially stabilized effluent from another filter.

ponder, *sb.*[2] (Later examples.)

1970 [see *COME *v.* 69 m]. 1976 T. McClure *Rogue Eagle* iv. 66 The obese sunbather.. went away to think about it. Buchanan had a bit of a ponder himself.

po·nderate, *a. rare.* [f. L. *ponderāt-*, ppl. stem of *ponderāre* to weigh, consider.] Careful; deliberate.

1922 *Times* 7 Oct. 11/2 It is a time for calm and ponderate consideration of the issues involved. 1970 P. O'Brian *Master & Commander* x. 257 The mature, the ponderate mind does not embark itself upon a man-of-war—is not to be found wandering about the face of the ocean in quest of violence.

ponderomotive, *a.* Add to def.: usu. applied *spec.* to such forces exerted upon bodies by electric or magnetic fields. (Earlier and further examples.)

1881 *Phil. Mag.* XII. 17 The force with which one quantity of electricity acts upon another quantity of electricity may be regarded either as an electromotive force or as a true ponderomotive force; for it tends to move electricity, and also to move matter, if matter be associated with the electricity. 1934 I. M. Freeman tr. *Joos's Theoret. Physics* xv. 296 The current in the segment, the magnetic field, and the ponderomotive force must form a right-handed orthogonal system in this order (Fleming's left-hand rule). 1964 R. R. Birss *Electr. & Magnetic Forces* i. 1 Ponderomotive forces are also exerted on dielectric bodies in electric fields. 1964 S. K. Runcorn in A. E. M. Nairn *Probl. Palaeoclimatol.* 192 The varying fields generated in the core.. will also generate induced currents in the mantle, and the ponderomotive forces resulting from these will cause angular acceleration and deceleration of the mantle on time-scales of 100 years. 1978 *Nature* 23 Mar. 316/2 In this case, ponderomotive forces were used to overcome the power barrier.

ponderosa (pɒndǝrǝ̄u·ză, -să). [a. the specific epithet of *Pinus ponderosa* (P. & C. Lawson *Agriculturist's Manual* (1836) 354), f. L. *ponderōsus* heavy.] In full, *ponderosa pine.* A large conifer, *Pinus ponderosa* or western yellow pine, native to western North America and widely cultivated elsewhere; also, the timber of this tree. Also *attrib.*

1878 R. J. Hinton *Hand-bk. Arizona* 292 *Ponderosa* reaches a height of 70 feet; some firs are higher. 1937 *Range Plant Handbk.* (U.S. Dept. Agric. Forest Service) B-44 Deerbrush is most commonly found in the ponderosa pine and mixed conifer belts. 1949 *Democrat* 2 June 3/1 Ponderosa wood is light in color, varying from creamy white to straw. 1951 V. Nabokov *Speak, Memory* vi. 96 Mariposa lilies bloomed under Ponderosa pines. 1957 *Handbk. Softwoods* (Forest Prod. Res. Lab.) 41 Canadian-grown ponderosa pine is about 20 per cent more resistant to splitting along the rings than Baltic redwood. 1966 Mrs. L. B. Johnson *White House Diary* 2 Apr. (1970) 379 He described the 'relic forest' of maple.. and ponderosa pine with huge trunks. 1971 *New Scientist* 10 June 628/3 The study considered growing douglas fir and ponderosa pine, but concluded that softwoods would not be economically feasible. 1976 *Billings* (Montana) *Gaz.* 20 June 9-A/1 (Advt.), Enjoy living in a beautiful natural setting, abundant with ponderosas, junipers, chokecherries, and wild roses. 1978 *Times* 11 Mar. 3/3 A finely balanced relationship has evolved between scale insects and ponderosa pine trees in the north-western United States.

ponding, *vbl. sb.* (s.v. Pond *v.*). Add: **b.** Blockage *of* a sewage filter; the accumulation of liquid above a filter.

1939 L. B. Escritt *Sewerage Engin.* vii. 133 Excessive flows.. are often the cause of 'ponding' of the filters. 1953 E. W. Steel *Water Supply & Sewerage* (ed. 3) xxiv. 488 Media of small size will furnish more surface, but unloading will be less complete and ponding on the surface [of the filters] is more likely. 1971 T. H. Y. Tebbutt *Princ. Water Quality Control* xii. 120 Film growths may result in blockage of the voids causing ponding of the filter and anaerobic conditions. *Ibid.* 122 When the first filter shows signs of ponding the direction of flow.. is reversed.

pond-lily. *orig. U.S.* Also **pond lily.** [f. Pond *sb.* + (Water-)lily.] A water-lily, esp. the common yellow spatterdock, *Nuphar advena.*

1748 J. Eliot *Ess. Field-Husbandry New Eng.* (1760) i. 5 A natural Pond.. over grown with Pond Lillies. 1778 [see Pond *sb.* 4]. 1827 *Western Monthly Rev.* I. 251 The flowers are large, of a pure white, nearest resembling the northern pond-lily. 1846 *Knickerbocker* XXVII. 52 A little mill-pond.. is covered all over with pond-lilies and rank grasses. 1873 T. B. Aldrich *Marjorie Daw* 14 All this splendor goes into that hammock, and sways there like a pond-lily. 1911 G. Stratton-Porter *Harvester* vii. 121 The pond lilies are just beginning to open. 1938 F. Perry *Water Gardening* vii. 82 The Yellow Pond Lily will flourish in shady positions. 1947 E. Paul *Linden on Saugus Branch* 367 On the Linden ponds, frogs' eggs, turtles, pond lilies with flat leaves, not shaped like plates. 1974 H. W. Rickett *Wild Flowers U.S.* VI. i. 130 The conspicuous part of the yellow pond-lilies is the calyx.

Pondo (pɒ·ndo). [Nguni.] **a.** A member of a Xhosa-speaking Nguni people in the eastern part of the Cape Province in South Africa. Cf. *Amapondo. **b.** The language spoken by the Pondos, a dialect of Xhosa. Also *attrib.*

[1835 A. Steedman *Wanderings Interior S. Afr.* (1966) I. p. iv, The third division are the Amapondo tribes,.. whose territories extend from the Bashee to the River Umsikalia. 1876 *Encycl. Brit.* V. 42/2 The Amapondo country of Kaffraria. 1884 K. Johnston *Africa* (ed. 3) xxiv. 399 The remaining portions of Kafraria, including.. that of the Amapondo extending across the St. John's River between the Umtata and the Umtamfuna, the boundary river of Natal.] 1919 H. H. Johnston *Compar. Study Bantu & Semi-Bantu Lang.* I. iii. 298 The [Kafir] dialects include Feṅgu, Baẓa and Pondω words. *Ibid.* v. 797 The divergent dialects of ʔōsa, such as Isi-pondω, Isi-baẓa, Feṅgu, &c. 1950 *Cape Argus* 18 Mar. (Mag. Section) 7/7 The area from Umtata down to the sea, and northwards to the borders of Natal, belongs to the Pondos—a tribe that would have been exterminated long ago by raiding bands of Zulus if the British Government had not intervened. 1973 *Times* 19 Feb. 12/1 The hillsides are dotted with clusters of neat.. round huts.., the homes of the local Africans who are a mixture of Zulu and Pondo tribes. 1979 A. McCoy *Insurrectionist* vi. 56 The grinning one is a Pondo. There is bad blood already between the Swazi and the Pondo.

pondok (pɒ·ndɒk). *S. Afr.* Also **pondokkie, pandokkie** (-ɒ·ki), **pandok, pondhock, pondhok, bond-hoek.** [Afrikaans, f. Malay.] A hut or shack made of oddments of wood, corrugated iron, etc.; a mean house or hovel, esp. one inhabited by non-whites. Also *attrib.*

1815 A. Plumptre tr. *Lichtenstein's Trav. S. Afr.* II. xli. 185 Near it stand six or eight *pandokken*, as they are called, a kind of huts made of reeds woven into a wooden frame, which are inhabited by the principal Bastard-Hottentots. 1818 C. I. Latrobe *Jrnl. Visit S. Afr.* 218 The present dwelling.. is a hovel, not much better than a Hottentot's bondhoek. 1832 A. Smith in P. R. Kirby *Andrew Smith & Natal* (1955) 35 In the afternoon to the Komga River to the 'pondok' (straw hut) of a trader. 1843 J. C. Chase *Cape Good Hope* III. 235 The Hottentots.. planted themselves at the outskirts of the country villages in small pondhoks, or huts, partly covered with old rags, decayed hides, sugar bags, and occasionally a little thatch. 1899 *Eastern Province Herald* (S. Afr.) 4 Nov. (Pettman), The poor burghers are living in pandokkies. 1911 *State* (Cape Town) Dec. 612 In the morning we found that a dozen or more Hottentots had pitched their pondhocks close to the wagon. 1944 *Cape Argus* 23 June, The people who are living in overcrowded shanties and pondokkies. 1948 L. G. Green *To River's End* xiv. 161, I built myself a pondok in a lonely kloof and became tame. 1952 E. H. Burrows *Overberg Outspan* iv. 103 They were the original herdsmen. Each inhabited his own *pondok*, and each was master of his own field. 1960 D. Lessing *In Pursuit of English* 29 He painted.. pondokkies. In other words, African huts, slums, broken-down villages. 1971 *Post* (S. Afr., Cape ed.) 9 May 4/6 Here we are.. in our well ventilated pondok in the bundu. 1974 *Cape Times* 2 Aug. 3/6 He said that he had read reports about alleged 'pondok farming' published in the Cape Times recently.

pond pine. *U.S.* [f. Pond *sb.* + Pine *sb.*[2]] A conifer belonging to the species *Pinus serotina*, growing on wet or marshy ground in south-eastern parts of the United States.

1810 F. A. Michaux *Hist. Arbres Forestiers de l'Amérique Septentrionale* I. 17 Pond pine (Pin des mares). 1832 D. J. Browne *Sylva Amer.* 240 The Pond Pine frequently recurs in the maritime parts of the Southern States. 1858 J. A. Warder *Hedges & Evergreens* II. 249 *Pinus serotina*, or Pond Pine, is thirty five or forty feet high. 1860 M. A. Curtis *Woody Plants N. Carolina* 21 Pond Pine.. has considerable resemblance to the Pitch Pine. 1940 *Amer. Forests* Oct. 462/2 Pond pine bears such local names as marsh pine, bay pine, and pocoson pine. 1967 N. T. Mirov *Genus Pinus* iii. 187 *Pinus serotina*, or pond pine, grows in the Coastal Plain from south-eastern Virginia south to central and southeastern Alabama.

pondy, *a.* Add: **2.** Belonging to or suggestive of a pond.

1922 *Chambers's Jrnl.* July 440/1 The peculiar 'pondy' smell of the bird [*sc.* moorhen] does not suggest that it would prove a great delicacy.

pone[3]. Add: **c.** *attrib.* as *pone bread.*

c 1785 in *Maryland Hist. Mag.* (1907) II. 258, I procured some milk and excellent pone bread from a hut. 1833 J. Neal *Down-Easters* I. 47, I should like to know.. what upon irth he means by.. hoe-cakes an pone bread. 1879 *Scribner's Monthly* June 223/1 Now that the wagons were up and 'pone' bread and beef stews had re-appeared in the menu, the Foot Cavalry, feeling its keep, waxed fat and kicked. 1935 Z. N. Hurston *Mules & Men* (1970) i. viii. 175 Nobody.. don't take de fork and turn over every fish in de dish in order to pick de best one. You does dat wid yo' eye whilst youse choosin' yo' pone bread. 1936 J. C. Chenault *Old Cane Springs* xviii. 78 Then pone bread, light bread, and biscuits were brought out.

‖**ponente** (pone·nte). [It.: see Ponent *a.* (*sb.*).] (See quot. 1959.) Also *ponente wind.*

1906 W. Marriott *Hints to Meteorol. Observers* (ed. 6) 67/2 *Ponente*, a Westerly wind in the Mediterranean. 1959 R. E. Huschke *Gloss. Meteorol.* 433 *Ponente*, a west wind on the Côte d'Azur.., the northern Roussillon region, and Corsica. On the Côte d'Azur it is a weakened mistral and brings clear skies. In northern Roussillon it is the land breeze of early morning, changing to southeast during the day. 1974 *Country Life* 25 Apr. 996/3 On some days the drying mistral blows, on others the damp *ponente* wind. 1978 S. Sheldon *Bloodline* vii. 84 The winds blew... The mistral and the *ponente*, the tramontana and the *grecale* and the levanter.

ponerid, *sb.* (Example.) Cf. next.

1895 J. H. & A. Comstock *Man. Study Insects* xxii. 642 The Ponerids... The ants of this family resemble [the Formicidæ].. in that the peduncle of the abdomen consists of a single segment.

ponerine (pɒ·nǝrǝin, -in), *a.* and *sb. Ent.* Also **Ponerine.** [ad. mod.L. name of subfamily *Ponerinæ*, f. generic name *Ponera* (P. A. Latreille 1804, in *Nouveau Dict. Hist. Nat.* XXIV. 179), f. Gr. πονηρά, fem. of πονηρός wicked + -ine[1].] **A.** *adj.* Of, or pertaining to, or designating ants of the subfamily Ponerinæ, which includes mainly tropical species. **B.** *sb.* A ponerine ant.

1910 W. M. Wheeler *Ants* ii. 26 The base of the abdomen is more primitive and more like that of certain Ponerine ants. 1933 *Discovery* Sept. 286/1 Professor W. M. Wheeler.. paid special attention to the colony-foundation among the primitive Ponerine ants. 1945 C. P. Haskins *Of Ants & Men* xi. 207 Australia is pre-eminently the home of the Ponerines of today. 1966 C. Sweeney *Scurrying Bush* vii. 101 *Platythrea cribinodis*, a common, large, dull-black ponerine ant. *Ibid.* 103 As with the giant solitary black ponerine,.. the black stink ants were not attacked because of their strong odour. 1977 M. V. Brian *Ants* iii. 39 All ponerines have a constriction between the first and second segments of the gaster. *Ibid.*, The Myrmicinæ are thought to have evolved from ponerine ants.

pong, *sb.*[1] Add: **b.** Abbrev. Ping-pong *sb.* Also, an electronic game resembling ping-pong, played on a pinball machine or a television screen. Also *attrib.*

1968–70 *Current Slang* (Univ. S. Dakota) III–IV. 94 *Pong, n.* ping-pong. 1976 *Washington Post* 19 Apr. A15/2 (Advt.), Now the whole family can play the new & exciting pong on your home TV. 1976 *Billings* (Montana) *Gaz.* 16 June 11-c/6 (Advt.), The new amusement machine, Pachinco, is sweeping the country like the pong games did five years ago. 1978 *Chicago* June 36/1 Game room with pong and pinball machines used by neighborhood folks.

pong (pɒŋ), *sb.*[2] *colloq.* Also (*rare*) **ponk.** [Etym. obscure.] An unpleasant smell; a stink.

1919 W. H. Downing *Digger Dial.* 38 *Pong*,.. stink. 1925 Fraser & Gibbons *Soldier & Sailor Words* 226 *Pong*, a stink. 1936 F. Clune *Roaming round Darling* xxiv. 257 Avoid the smell of camel. They were complete with permanent, pyramid, and perfume, commonly called pong. 1941 Baker *Dict. Austral. Slang* 56 *Ponk*, a stink. As verb, to stink. 1957 J. Braine *Room at Top* iv. 44 'What a pong,' he said. 'Don't know how you stand it.' 1960 H. Pinter *Dumb Waiter* 130 What, you mean it might be my pong? (He sniffs sheets.) Yes... It could be my pong I suppose. 1973 G. Moffat *Deviant Death* i. v. 64 She's burning the feathers... She only does it when the wind takes the smell away from us... The pong's not bothering us. 1974 J. Gardner *Return of Moriarty* 292 There ain't half a pong down here.

Pong (pɒŋ), *sb.*[3] Chiefly *Austral. slang.* [Origin uncertain.] A derogatory name for a Chinese.

1931 V. Palmer *Separate Lives* 221 Blow into one of those Chow joints.. and call for a dollar's worth of duck and fowl. Enough for two those pongs always give you. 1938 X. Herbert *Capricornia* 339 Your grandmother was a lubra and your grandfather was a Pong. 1941 Baker *Dict. Austral. Slang* 56 *Pong*, a Chinese. 1957 D. Stivens *Scholarly Mouse* 65 He was too tall to be a Pong or an Eyetoe. 1962 J. Franklyn *Dict. Nicknames* 86/1 *Pong* is a nickname given to a Chinaman in Australia—punning the *ong* sound in some Chinese words, and *pong*, a bad smell. 1970 'B. Mather' *Break in Line* i. 11 I'm the only Pong I know who wouldn't say Charling Closs.

pong (pɒŋ), *v.*[2] *colloq.* Also (*rare*) **ponk.** [f. *Pong *sb.*[2]] *intr.* To stink. Also *fig.*

1927 [see *Hum *v.*[2]]. 1939 R. Campbell *Flowering Rifle* 21 What matters most to them is—'Does it Pong?' 1941 [see *Pong *sb.*[2]] 1944 'N. Shute' *Pastoral* ii. 17, I think it looks ugly as sin, and it's starting to ponk a bit. 1948 M. Allingham *More Work for Undertaker* xiii. 164 The old boy never bought a sausage that didn't pong. 1950 A. Baron *There's no Home* i. 16 'It don't 'alf pong,' he observed. 1960 [see *Drain *sb.* 1 e] 1972 P. Cleife *Slick & Dead* ii. 22 This loving thing could be a cover story for any old racket?.. Pongs a bit, don't you think? 1979 R. Rendell *Make Death love Me* ix. 85 The place.. just pongs of dirty clothes.

ponga (pɒ·ŋā). Also **bunger, bungy, punga.** [Maori.] An evergreen New Zealand tree-fern, *Cyathea dealbata*, belonging to the family Cyatheaceæ; also *attrib.*

1832 G. Bennett in *London Med. Gaz.* 22 Sept. 793/2 This fern.. is named Ponga by the natives, who use the trunks as posts in the erection of their houses. 1855 R. Taylor *Te Ika a Maui* viii. 115 Some of the trees themselves.. held down their heads, and have never been able to hold them up since; amongst these, were the *ponga* (a fern tree) and the *kareao* (supple jack), whose tender shoots are now always bent. 1874 J. White *Te Rou* xi. 179 Round two sides and one end [of the hangis, or ovens] a ponga fence is put. 1892 E. S. Brookes

Frontier Life xv. 139 The Survey department graded a zigzag track up the side to the top, fixing in punga steps. 1898 Morris *Austral Eng.* 65/1 Bunga or Bungy,.. a New Zealand settlers' corruption of the Maori word *punga.* 1905 W. B. *Where White Man Treads* 232 It irks to go back to ponga whare and earthen floors. 1926 *Trans. N.Z. Inst.* LVI. 670 In some instances the Maori name has been adopted but corrupted:.. 'bunger' (now fortunately seldom heard) for 'ponga'. 1933 *Press* (Christchurch, N.Z.) 23 Sept. 13/7 Bungy.—Tree fern. I [*sc.* L. G. D. Acland] have only heard this word used on the West Coast; it is just as often pronounced pungy. 1935 'J. Guthrie' *Little Country* iii. 58 Tall punga ferns spread their proud fronds. 1949 F. Sargeson *I saw in my Dream* ii. xiv. 172 On the ridge it was too dry for pungas. 1959 *Times* 10 June 12/6 We strained our eyes.. to see the punga fern under which the first white child was born in New Zealand. 1963 B. Pearson *Coal Flat* xxii. 376 Peter was urinating urgently against a ponga. 1966 *Encycl. N.Z.* I. 650/1 *C[yathea] dealbata* or ponga has distinctive whitish undersurfaces to the leaves. 1966 G. W. Turner *Eng. Lang. Austral. & N.Z.* viii. 168 Other Maori words in changed form are cockabully from kokopu,.. bunger (Maori ponga), a common name for the treefern [etc.]. 1968 *N.Z. Listener* 11 Apr. 10/1 He built himself a ten-foot-high punga fence and was snug. 1977 *N.Z. Herald* 5 Jan. 2-16/8 (Advt.), Genuine bush with pongas, totara and kauri surrounding this most impressive 4-brm contemporary home.

ponga, var. *Panga[1].

pongal (pɒ·ŋgăl). Also **pongol, pongul, ponkal.** [ad. Tamil *poṅkal*, 'boiling'.] The Tamil New Year festival at which new rice is cooked; hence, a dish of cooked rice.

1788 F. Magnus tr. *Sonnerat's Voy. East-Indies & China* I. v. 142 The second day the festival is called Maddou-Pongol, or Pongol of cows:—they paint the horns of these animals, cover them with flowers, make them run in the streets, and lastly make the Pongol at home for them. 1809 *Asiatic Ann. Reg. 1807* Misc. 141 (*heading*) An interesting account of the great Hindu festival Pongal, by Teroovercadoo Mootiah. *Ibid.* 144/1 The Hindoos visit and compliment each other, wishing a happy Pongal, or many returns of that Pongal, for the preservation of each other. 1855 H. H. Wilson *Gloss. Indian Terms* 421/1 *Pongal*.., a boiling or bubbling up, the boiling of rice, whence it becomes the name of the popular festival held by the Hindus in the Madras provinces. 1877 M. Monier-Williams *Hinduism* xii. 182 In the South of India this festival is commonly called 'Pongal', and is the commencement of the Tamil year. 1897 H. K. Beauchamp tr. *Dubois's Hindu Manners* II. iii. iii. 580 The pongul, or Maha-sankranti, always takes place during the winter solstice, the period when the sun, having finished its course towards the southern hemisphere, turns to the north again and comes back to visit the people of India. 1906 W. Crooke *Things Indian* 211 The opening of the agricultural year is marked.. by the *Pongol* of South India. 1913 *Encycl. Relig. & Ethics* VI. 44/1 The central rite of the great Pongol festival of S. India consists in cooking new rice, some of which is offered to Gaṇeśa, the remainder being eaten by the family. 1961 [see *Idli]. 1968 P. Lal *Indian Recipes* 58 Many delicacies are prepared for the Pongal feast... The six months following Pongal are considered auspicious for marriage. 1974 F. W. Clothey tr. L. S. Ramamirtham in *New Writing in India* 97 No matter how poor a man is, it's only at Poṅkal that his rice boils in a new pot.

pongelo (pɒ·ŋgǝlo). *slang.* Also **pongelow.** [Etym. obscure.] Beer. Also *attrib.*

1864 Hotten *Slang Dict.* 204 *Pongelow*, beer, half-and-half. 1880 M. E. Braddon *Just as I Am* I. ix. 130 'He stood sam for a pot o' pongelo,' continued Mr. Scaffers, 'and narchurly we got talkin'.' 1898 A. M. Binstead *Pink 'Un & Pelican* viii. 185 Some well-known publican has given twenty thousand pounds for the local pongelo palace, with the plate-glass saloon bars. 1899 *Westm. Gaz.* 9 Jan. 5/2, I hope **** is quite well and keeping himself allright [*sic*] and not soaking to [*sic*] much pongelow. 1899 H. Wyndham *Soldiers of Queen* 256 One night I had a drop too much 'pongelow', and there was a bit of a row. 1905 *Daily Chron.* 2 Mar. 4/5 You said, 'What're you goin' to 'ave?' an' the pongelos flow'd free as advice. 1909 J. R. Ware *Passing Eng.* 199/1 *Pongelo*, (Anglo-Indian Army), pale ale—but relatively any beer.

pongid (pɒ·ŋgid), *sb.* and *a. Zool.* [f. mod.L. family name *Pongidæ*, f. generic name *Pongo* (B. G. E. de La V. Lacépède *Tableau des Mammifères* (1799) 4: see etym. of Pongo) + -id[3].] **a.** *sb.* An anthropoid ape belonging to the family Pongidæ, which includes the gorilla, the chimpanzee, and the orang-utan. **b.** *adj.* Of or pertaining to this group of apes.

1955 W. E. Le Gros Clark *Fossil Evidence Human Evolution* iv. 141 The differential characters of the dentition.., on the basis of the comparative study of large numbers of hominids and pongids.., have been established. 1957 *Antiquity* XXXI. 191 Another tarsioid line produced the primitive apes (the pongid line). 1963 *New Scientist* 27 June 737/1 Dr Leakey revealed that there were true pongids (apes) living in East Africa during the Miocene period. 1968 *Nature* 9 Nov. 548/1 To split the African apes from the Pongidae and place them in the Hominidae would ignore the extraordinary change in the hominid line since it split off from the pongid line. 1973 B. J. Williams *Evolution & Human Origins* ix. 127/2 The genus *Dryopithecus* is ancestral to the present-day African pongids, the gorilla and chimpanzee. *Ibid.* 129/1 *Ramapithecus* is smaller than most species of the fossil pongid genus *Dryopithecus*. 1977 A. Hallam *Planet Earth* 284 The hominids are a family containing Man and his close relatives, distinct from the family of apes, the pongids.

pongo. Add: **1. b.** Substitute for def.: Later transferred to the orang-utan, an anthropoid ape native to Borneo and Sumatra, and adopted as its generic name by Lacépède in 1799; see *Pongid *sb.* and *a.* (Earlier and later examples.)

1798 *Phil. Mag.* I. 238 The orang outang of the large kind, or the pongo of Buffon, is not common even in its native country Borneo. 1913 D. G. Elliott *Rev. Primates* I. [Errata] The premier genus of the Great Apes is Pongo. 1972 D. Bloodworth *Any Number can Play* xii. 104 An intelligent pongo digging a twig into a wasp's nest.

2. a. *Naut. slang.* A marine, a soldier. Also *attrib.*

1917 'B. Copplestone' *Lost Naval Papers* vi. 85 You could pass as a naval officer more easily than you could as a Pongo. 1919 W. H. Downing *Digger Dial.* 38 *Pongo*, a soldier; one of the rank and file. 1923 F. H. Vizetelly *Desk-Bk. Idioms* 328 *Pongo*, a marine: in playful British usage, from a native African name for an anthropoid ape. 1943 D. Glover in *Penguin New Writing* XVI. 15 What about that bloody pongo what's been loafing round since I come ashore last? 1946 J. Irving *Royal Navalese* 137 *Pongo*, the matelot's name for a soldier. In the First World War it was his name for a Royal Marine. 1955 'N. Shute' *Requiem for Wren* iii. 92 Each service at that time had its own slang; to her the army were all Pongoes. 1961 B. Fergusson *Watery Maze* i. 40 Captain (afterwards Admiral Sir Frederick) Dalrymple-Hamilton having an elder brother in the Army, and consequently, as he said, 'a soft spot for Pongoes'. 1964 J. Hale *Grudge Fight* ii. 33 Just before the admiral's car got to the pier the pongos blew a great hole in it. 1975 *Canad. Forces Sentinel* (Ottawa) XI. vi. 7/2 Cpl. Don Lyons won praise from naval techs who described him as the only 'pongo stoker' with the maintenance team. 1977 *Daily Mail* 17 Nov. 15/1 Fourteen youths.. went out looking for soldiers to beat up... Favourite expressions of the gang were 'squaddy bashing' and 'pongo bashing'.

b. *Austral.* and *N.Z. slang.* An Englishman. Also *attrib.*

1942 *2 N.Z.E.F. Times* 7 Sept. 5 A big bronzed Pongo came in. 1945 J. Henderson *Gunner Inglorious* 148 The successful applicant [for the position of Batman], an elderly, quiet-spoken Pongo, was a dinkum butler. 1947 D. M. Davin *For Rest of Our Lives* xxxii. 165 That pongo cobber of yours, the homo. 1947 — *Gorse blooms Pale* 208 The poor old pongos are probably still indenting in triplicate for mine-detectors. 1949 *Here & Now* (N.Z.) Oct. 11/2 He long ago began featuring himself as a 'New Zealander born and bred', a sop to the vague public feeling against Pommies, Pongoes, Homies. 1964 *Courier-Mail* (Brisbane) 19 Nov. 12 Mr. Arthur Bryan.. dislikes what he calls the 'pongo' Englishman. 'This is the bloke who is so bound by tradition and the old establishment that he can think of nothing else,' he says. 1969 *Private Eye* 19 Dec. 5/2 The *pongos* are shooting through like streaks of *weasle piss!* 1972 G. W. Turner *Eng. Lang. Austral. & N.Z.* (ed. 2) 109 Like Australians, New Zealanders call the English *pommies*, but also have a variant *pongo* which seems rather less tolerant in its tone.

c. *Mil. slang.* An army officer.

1943 B. J. Hurren *Eastern Med.* xii. 139 In the slang of the desert the word 'pongo' is used for all Army officers. 1943 C. H. Ward-Jackson *Piece of Cake* 48 *Pongo*, an army officer. 1949 F. Maclean *Eastern Approaches* ii. vii. 269 'Operation pongo'.. was the code-name I had chosen for the abduction of the General. 1965 O. Manning *Friends & Heroes* xxii. 237 What were you doing walking about holding on to that bloody little pongo?

d. *slang.* A Negro. Also *attrib.*

1968 L. Deighton *Only when I Larf* vii. 99 You wouldn't want no breech block blowing back and crippling some poor pongo, no matter what country he's in. 1972 M. Woodhouse *Mama Doll* viii. 89 Our Pongo brothers in darkest Africa.

pongo, var. *Panga[1].

pongol, var. *Pongal.

Pongola (pɒŋgǝ̄u·lā). Also **Pangola,** and with lower-case initial. In full, *Pongola* or *Pangola (finger-)grass.* The name of a South African river used *absol.* and *attrib.* to designate a variety of the perennial grass *Digitaria decumbens*, originally native to regions near the river, but now widely cultivated in tropical areas.

1947 D. B. D. Meredith *Effect of Fertilisers on Grasses in S. Afr.* v. 115 The area selected had been planted.. to alternate rows of the Pongola Finger grass and *Digitaria Smutsii.* 1952 M. A. Flores in *Proc. 6th Internat. Grassland Congr.* II. 1435 Pangola grass, introduced within recent years from Africa through Florida, U.S.A., is finding favor with livestock producers because of its resistance to drought. 1959 *Agronomy Jrnl.* LI. 111/1 Napier or elephant grass.. guineagrass.. and pangolagrass.. are among the most important of the forage grasses of the Tropics. 1966 A. M. M. Rees in Davies & Skidmore *Trop. Pastures* xi. 171 Recorded gross outputs of over £50 per acre were obtained.. from milk and beef produced from Pangola pastures. 1972 E. Hargreaves *Fair Green Weed* vii. 93 We're putting in pangola to make pastures of the old banana plantations. 1972 *Stand. Encycl. S. Afr.* V. 320/2 The quick-grasses.., Pongola finger-grass.. and Swaziland finger-grass.. are amongst those grasses that are commonly planted as lawns or on sports grounds. 1977 A. V. Bogdan *Trop. Pasture & Fodder Plants* 113 The name Pangola grass has been derived from the Pongola River in the Piet Retief district of eastern Transvaal.. and some authors suggested the name Pongola grass, possibly with a view of avoiding confusion with any other forms of *Digitaria* which may come from the Pangola River area of western Transvaal. It has also been suggested that the name Pangola grass should be applied only to those clones which.. were originally brought to USA in 1935 and which are now widely grown in a number of countries.

pongul, var. *Pongal.

pongy (pɒ·ŋi), *a. colloq.* [f. *Pong *sb.*[2] + -y[1].] Malodorous; smelly.

1936 'Taffrail' *Mystery at Milford Haven* xi. 153 'Kippers!' she groaned. 'They are a bit pongy sometimes,' Victor had to confess. 1960 *Times Lit. Suppl.* 4 Nov. 714/2 A cheap forty-eight-hour excursion in contemporary Gauguin-land complete with Papeete night-spots, pongy with frangipani. 1965 G. McInnes *Road to Gundagai* xii. 215 Dad.. kept turning up.. with loot from the Prahran market: strings of saveloys and frankfurters, pongy cheeses,.. and huge Portuguese sardines. 1975 *Islander* (Victoria, B.C.) 3 Aug. 2/2 After lunch the pongy wharf became too much for us.

pongyi, var. Poonghie, Phoongyee in Dict. and Suppl.

ponhaus (pɒ·nhǭs). *U.S. dial.* Also **pawnhaus, ponhaws, ponhoss.** [ad. Ger. *panhas*, f. *pfanne* frying pan + *hase* rabbit. Said to be used in a similar sense in Ger. dial.: see M. B. Lambert *Dict. Non-English Words of Pennsylvania-German Dial.* (1924) 117.] = Scrapple *sb.*[2] Also *attrib.*

1869 *Atlantic Monthly* Oct. 483/1 Some make *pawnhaus* from the liquor in which the pudding was boiled; adding thereto corn-meal. 1882 P. H. Gibbons *Pennsylvania Dutch* (ed. 3) 423 Mr. W. liked the fried *pawn-haus* although he found it rather rich. 1923 *Dialect Notes* V. 236 *Ponhaws*, scrapple. 1931 *Sun* (Baltimore) 11 Mar. 8/7 He's goin' to have one more grand ponhaus celebration this season. 1943 *Chicago Daily News* 8 Sept. 25 Originally, Ponhaws or scrapple was made from the head of the freshly killed porker, but good, fresh, lean pork of any cut may be used. 1944 *Sun* (Baltimore) 4 Dec. 8-0/3 The Pennsylvania Germans.. called it 'Pfännhaas', which.. was corrupted into.. 'Ponhaus',.. which simply means.. 'pan rabbit'. This is in line with the use of 'Welsh rabbit'. 1953 *Amer. Speech* XXVIII. 244 Certainly *ponhaus* and *smearcase*, standard food items from colonial days, are old words in the area [*sc.* Bedford, Pennsylvania].

ponk, var. *Pong *sb.*[2] or *v.*[2]

ponkal, var. *Pongal.

ponor (pǭu·nǫɹ). *Physical Geogr.* [Serbo-Croat.] A steep natural shaft leading from the surface of the ground in a karstic region.

[1921 *Geogr. Rev.* XI. 593 The article of Professor Cvijić marks a step forward in the science of physiography; but it is far from easy reading for the average geographer since many unfamiliar terms, such as 'bogaz' and 'ponor', are used without either definition or explanation by synonyms. *Ibid.* 600 The shaftlike aperture Cvijić called 'ponor'.] 1922 *Geol. Mag.* XIX. 406 The funnel-shaped hollows which are so frequently met with on the surface of the karst are termed ponors. 1937 Wooldridge & Morgan *Physical Basis Geogr.* xix. 290 They [*sc.* limestone caves] commonly form part of a complex system of channels, widening locally into chambers, and fall broadly into two sets, viz. roughly horizontal galleries, and vertical or steeply inclined shafts, of which the higher members are the 'ponors' communicating with the surface. 1971 J. N. Jennings *Karst* vi. 139 In some poljes certain ponors change function for a period in the wet season and spew out water. 1976 S. T. Trudgill in E. Derbyshire *Geomorphol. & Climate* iii. 92 They observed in the Kuh-E-Parau limestone area of Iran how the overall form on a large scale is solutional in origin, with rounded hills, dolines and ponors.

Ponsonby rule (pɒ·nsǝ̆nbi rūl). [Named after Arthur A. W. H. *Ponsonby* (1871–1946), 1st Lord Ponsonby, English politician.] A rule by which the Government may authorize an agreement without Parliamentary approval (see quot. 1976).

1957 *Erskine May's Law of Parl.* (ed. 16) xiii. 275 This practice, which is known as the 'Ponsonby rule', had its origin in a departmental minute dated 1 February 1924 and signed by Mr. Arthur Ponsonby, then Under-Secretary of State for Foreign Affairs. 1967 P. G. Richards *Parliament & Foreign Affairs* iii. 43 These arrangements constituted the so-called Ponsonby Rule. 1976 H. Wilson *Governance of Britain* x. 185 Under the so-called Ponsonby rules specific parliamentary ratification is not required. The Government assumes authority in respect of any treaty or agreement it has negotiated if Parliament has not reacted within twenty-one days.

Pontet-Canet (pǫ̃nte kane). The name of a château in the Pauillac commune of the Médoc, applied to a claret produced there.

1883 H. James in *Atlantic Monthly* Oct. 458/1 There is a touch of French reason, French completeness, in a glass of Pontet-Canet. 1891 in C. Ray *Compleat Imbiber* (1967) IX. 122 *Claret*.. St. Estèphe.. St. Emilion.. Pontet Canet. 1912 'Saki' *Chron. Clovis* 184 Waiter, a bottle of *Pontet Canet.* 1920 G. Saintsbury *Notes on Cellar-Bk.* iv. 64 Pontet Canet.. became.. a 'literary' wine a good

many years ago. 1966 H. Yoxall *Fashion of Life* xxv. 240 A Soho restaurant where.. they served a pre-war Ch. Pontet-Canet at 2*s.* 9*d.* a half-bottle. 1967 A. Lichine *Encycl. Wines* 409/1 Classified a Fifth Growth.. in 1855, and traditionally at the head of the Fifths, Pontet-Canet actually sells with the Seconds and Thirds. 1975 'D. Jordan' *Black Account* xiii. 65 The House had.. lashed out on a Pontet-Canet.. and a port which set out to impress the Minister.

Pontian (pɒ·ntiăn), *a. Geol.* [ad. Russ. *Ponticheskiĭ* (N. Barbot de Marny *Geol. ocherk' Khersonskoi Gubernii* (1869) xiv. 106), f. as Pontic *a.*[1]: see -ian.] Of, pertaining to, or designating the uppermost stage of the Miocene series in Europe (sometimes regarded as the lowest of the Pliocene series). Also *absol.*

1893 P. Lake tr. *Kayser's Text Bk. Compar. Geol.* iv. 361 Congeria or Pontian series. 1895 J. D. Dana *Man. Geol.* (ed. 4) iv. iv. 927 Above the Tortonian, the stages Sarmatian and Pontian are recognized in Dauphiné, Austria and Italy. 1903 A. Geikie *Text-bk. Geol.* (ed. 4) II. 1291 The top of the Miocene series (Pontian stage). 1940 A. W. Grabau *Rhythm of Ages* xxxviii. 466 The Pontian Hipparion clays, which were formerly considered in part Miocene, are here placed in the base of the Pliocene. 1971 *Nature* 30 Apr. 562/1 In 1929, Matthew.. pointing to the occurrence of a relatively primitive *Hipparion* fauna in the lower part of the stratotype Pontian at Sebastopol,.. argued that the first appearance of *Hipparion* could thus be used to define the base of the Pliocene in continental mammalian successions. In this way, 'early Pliocene' and 'Pontian' became equivalent terms in vertebrate biostratigraphy. 1973 *Ibid.* 15 June 391/1 Estimates by most vertebrate palaeontologists have ranged between 10–12 m.y. because of the supposed initial appearance of the three-toed *Hipparion* in the lower part of the stratotype Pontian of the eastern Mediterranean.

‖**pontianak**[1] (pɒntiā·næk). [a. Malay *pontianak*, f. *pati-anak* child-killer.] A type of vampire (see quots.). Cf. *Langsuir, *Penanggalan.

1839 T. J. Newbold *Pol. & Statistical Acct. Straits of Malacca* II. xii. 191 Spirits.. supposed to exert a baneful influence over them [*sc.* Malays] in this sublunary world. First, the Plissit and the Pontianak. 1900 W. W. Skeat *Malay Magic* vi. 320 The Pontianak or Mati-anak.. is also a night-owl, and is supposed to be a child of the Langsuir. 1965 C. Shuttleworth *Malayan Safari* vi. 86 Perhaps the most fearsome of all superstitions is that of the *pontianak* or vampire, widely prevalent throughout Malaya. 1966 D. Forbes *Heart of Malaya* xiii. 185 She had turned into what they call a *pontianak*. 1972 *Daily Tel.* (Colour Suppl.) 12 May 58/3 The Malayan vampire family includes.. the Pontianak (the stillborn child of the Langsuir) which adopts the shape of a night owl.

pontianak[2] (pɒntiā·næk). Also **Pont-, -ac.** [The name of a city and formerly of a sultanate on the island of Borneo.] = *gutta-jelutong* s.v. *Gutta[2] 2 (orig. that from Borneo).

1911 *India Rubber World* XLIII. 130/2 Different qualities of jelutong are known in the trade, according to the districts from which they are derived, as Palembang (Sumatra), Pontianak (South Borneo), Sarawak, and so on. 1923 D. W. Luff *Chem. of Rubber* iii. 35 An inferior rubber which in the days of high rubber prices became of importance industrially is that known variously as Jelutong, Gutta Jelutong, Pontianac, Bresk, or Dead Borneo, which is obtained from the *Dyera costulata*, a large tree growing in Borneo, Sumatra and Malaya. 1927 [see *gutta-jelutong]. 1947 H. Barron *Mod. Rubber Chem.* (ed. 2) iii. 27 Jelutong (or pontianak) comes chiefly from Sumatra.

Pontic, *a.*[1] Add: **1. a.** (Further examples.)

1726 Swift *Let.* 15 Oct. in Pope *Corr.* (1956) II. 407 They must have been pontic mice, which as Olavs Magnus assures us always devours whatever is green. 1887 A. T. de Vere *Legends & Rec. Church & Empire* 208 Thou Pontic Paradise! 1895 W. Robinson *Eng. Flower Garden* (ed. 4) i. ix. 126 We too often see the common pontic kind [of rhododendron]. 1906 Kipling *Puck of Pook's Hill* 167 I've tramped Britain and I've tramped Gaul, And the Pontic shore where the snow-flakes fall. 1935 *Discovery* July 199/1 There is, e.g., another orthopter which lives in the Pontic and west Mediterranean areas, and in some places in Central Europe as a 'Pontic relic'. 1956 R. Macaulay *Towers of Trebizond* xv. 178, I lay in a swoon, pretending to be dead, because the barbarous Pontic natives, the Mossynoici, were all about.

c. Of or pertaining to the ancient kingdom of Pontus, its kings (see Mithridatic *a.*), its people, or the dialect of Greek attributed to them. Also as *sb.*

1665 D. Lloyd tr. *Plutarch's Worthies* 372 According to the Pontick Kings dream of floating on the waters. 1816 Byron *Dream* viii, in *Prisoner of Chillon* 44 Like to the Pontic monarch of old days, He fed on poisons, and they had no power. 1939 [see *Median *sb.* 2]. 1972 W. B. Lockwood *Panorama Indo-European Lang.* 267 It would not be surprising if the language known to the Greeks as Pontic were a descendant of Kaskian. 1974 *Encycl. Brit. Micropædia* VIII. 115/2 An independent Pontic kingdom with its capital at Amaseia was established at the end of the 4th century BC in the wake of Alexander's conquests. *Ibid. Macropædia* VIII. 396/2 The Asia Minor dialects [of Greek] also display archaic features (*e.g.*, Pontic *e* for ancient *ē* in certain word elements).

d. *Anthropol.* Designating a type of peoples identified in the Balkans and southern Russia (see quots.). Also as *sb.*

[1932 V. Bunak in *Zeitschr. für Morphologie u. Anthropologie* XXX. 471 Die zwei südlichen analogen Kombinationen—nordkaukasische und ostbalkanische—sind untereinander ähnlicher und stehen von den nordpontischen Varietät weiter ab. Sie bilden eine andere Rasse des östlichen mediterranen Zweigs, den ich vorläufig als pontische Rasse.. bezeichnen werde.] 1939 C. S. Coon *Races Europe* xii. 617 The Mediterranean racial divison which the Russian anthropologists call Pontic.. is with little doubt of Neolithic date in southern Russia, Rumania, Bulgaria, and the Hellespont region, and probably in Greece and the Aegean. *Ibid.* 679 Pontic. A variety of Mediterranean or Atlanto-Mediterranean,.. is concentrated in Bulgaria and in the Rumanian lowlands; it also is found in the Caucasus and Ukraine and westward sporadically as far as Germany, Poland, and Lithuania.

pontic, *a.*[2] Add: **B.** *sb. Dentistry.* An artificial tooth that forms part of a dental bridge, being held in place by attachment to its neighbouring teeth, and not fixed directly to the jaw.

1916 J. H. Prothero *Prosthetic Dentistry* (ed. 2) xxix. 785 The term 'pontic' has been suggested as a substitute for 'dummy' in describing a bridge tooth replacement. The term seems scarcely appropriate, since practically all fixed bridges are of the rigid truss type. 1932 F. R. Felcher *Art of Porcelain in Dentistry* xi. 133 Pontics should not be so built that they extend too far into the sockets, as a recession will usually result if this is done. 1956 J. N. Anderson *Appl. Dental Materials* xiii. 128 When making a bridge, the 'pontic' or bridging part is joined to the supports or 'retainers'. 1974 D. H. Roberts in Harty & Roberts *Restorative Procedures Practising Dentist* xxii. 327 Where metal-ceramic full crowns are used as retainers then the same material is normally employed for the pontic.

ponticello (pɒntitʃe·lo). *Mus.* [a. It. *ponticello* little bridge.] **a.** The bridge of a stringed instrument.

1740 J. Grassineau *Mus. Dict.* 182 *Ponticella* [*sic*], a small bridge. 1849 *Hamilton's Celebrated Dict.* 92 *Ponticello*.., the bridge, in speaking of the violin, guitar, etc. 1961 A. Baines *Mus. Instruments* 356 *Ponticello*,.. the bridge of a violin, etc.

b. Phr. *sul ponticello*: a direction in a musical score that bowing should be close to the bridge. Also *ellipt.* as *ponticello.* Also applied *attrib.* to the sound produced by such bowing.

1849 *Hamilton's Celebrated Dict.* 112 *Sul ponticello*, on or near the bridge. 1883 Grove *Dict. Mus.* III. 15/2 *Ponticello*.. or *sul ponticello*, a term indicating that a passage on the violin, tenor, or violoncello, is to be played by crossing the strings with the bow close to the bridge. 1931 G. Jacob *Orchestral Technique* ii. 6 The ponticello tremolo in which a most eerie effect is produced by bowing the strings nearer to the bridge than the normal position. 1959 *Collins Mus. Encycl.* 509/1 *Sul ponticello* (or *ponticello* alone), on the bridge, *i.e.* play near the bridge, thus producing a glassy, brittle tone. 1967 *Listener* 8 June 769/2 The famous passage in the finale where the first violin skips aloft, high over a sinister progression of rapid *ponticello* chords on the lower strings. 1977 'E. Crispin' *Glimpses of Moon* iii. 41 'And then *erk, skerk*,' he added, possibly attempting to convey ponticello strings.

ponticum (pɒ·ntikǒm). Also **Ponticum** [a. mod.L. specific epithet of *Rhododendron ponticum* (Linnæus *Species Plantarum* (ed. 2, 1762) II. 562): see Pontic *a.*[1]] A mauve-flowered, evergreen shrub, *Rhododendron ponticum*, of the family Ericaceæ, native to Spain, Portugal, and Asia Minor and naturalized in many other temperate regions. Also *attrib.*

1875 H. Fraser *Handy Bk. Ornamental Conifers* 185 The best and most commonly-used stocks for grafting are free-grown seedlings of the robust form of the common Ponticum. 1917 J. G. Millais *Rhododendrons* I. viii. 227/2 Enormous numbers of young ponticums are annually used as stocks for the finer varieties. 1962 R. Page *Educ. Gardener* 27 We uprooted the ponticums and burned them. 1972 'I. Drummond' *Frog in Moonflower* 5 A heavy clump of Ponticum rhododendrons grew unexpectedly on top of the little cliff. 1976 J. Lees-Milne *William Beckford* iv. 59 Today the lake, swathed in rampaging ponticum.., lies dark and almost unapproachable. 1977 *Evening Standard* 22 Apr. 19/2 Those ponticums.. are not the ordinary purple ones, but very rare.

Pontifex. Add: **1.** (Further examples.) Also *transf.*

1777 P. Thicknesse *Year's Journey* II. xiii. 83 The consecration of the Roman Pontifex Maximus. 1934 *New Statesman* 3 Nov. 614/2 Stalin has exiled Trotsky and become the Pontifex Maximus of the new Russo-Catholic Church of Communism. 1957 *Oxf. Dict. Chr. Ch.* 1089/2 *Pontifex Maximus*,.. originally a pagan title of the chief priest at Rome, Tertullian used it satirically.. of the Pope, and from the 5th cent. onwards it was a regular title of honour for the Popes, and occasionally used also of other bishops.

pontifical. Add: **B.** *sb.* **7.** Short for *pontifical mass.*

1923 R. Seton *Memories Many Yrs.* 291 The most interesting of my pontificals was in San Nicola *in carcere.*

pontificate, *v.* Add: **1. a.** (Later example.)

1928 G. B. Shaw *Intelligent Woman's Guide Socialism* 439 The Russian archbishop.. is now presumably pontificating much more freely than the Archbishop of Canterbury.

2. a. (Later examples.)

1909 *Englishwoman* Apr. 296 The need of such a group as that which pontificates from Villa Wahnfried is past. 1921 R. Hichens *Spirit of Time* v. 76 Why should I allow this young woman to pontificate about human nature. 1952 *Times Lit. Suppl.* 4 Jan. 1/4 Success made him pontificate more than ever. 1979 *Kansas City Times* 22 May 6A/1 They [*sc.* senators] must think they are pontificating on the moon or Mars or somewhere remote from Jefferson City.

b. *trans.* To say or utter (something) in a pontifical manner.

1922 A. S. M. Hutchinson *This Freedom* iv. i. 252 All modern teaching, if this new stuff that they pontificate may be called teaching, offers us [etc.]. 1973 *N.Y. Law Jrnl.* 24 July 4/5 The court pontificated, 'One cannot look at a rainbow with mud on his shoes'. 1976 *Verbatim* Dec. 15/1 He also pontificated, 'The Reds are favored to win, and, as we all know, everybody hates a favorite.'

So **ponti·ficating** *vbl. sb.* and *ppl. a.*; **ponti·ficator.**

1825 Pontificating [see Pontificate *v.* 2]. 1926 W. J. Locke *Stories Near & Far* 156 Pontifex—Pontifex something.. a playful title given him by her mother, for his possible pontificating aims as a young man. 1930 *Radio Times* 17 Jan. 127/2 Nine out of ten people are fond of pontificating. 1934 B. Dobrée *Mod. Prose Style* iv. i. 221 If we examine the writings of the pontificators, people skilled in 'a way of saying things', we invariably find that their style is bad. 1972 *Daily Tel.* (Colour Suppl.) 10 Nov. 7/1 Highbrows—the pontificators about Television—are apt not merely to condone but to applaud: the gratuitous nastiness of allegedly 'serious' plays and aggressive documentaries. *Ibid.*, The pontificators make it so clear that they never watch television for pleasure and don't intend that other people should.

pontification. Restrict † *Obs. rare*[−1] to sense in Dict. and add: **2.** The act or an instance of pontificating (Pontificate *v.* 2).

1925 C. D. Broad *Mind & its Place* viii. 389 It is a pity to create prejudice.. by ignorant pontifications about 'the New Psychology'. 1959 *Spectator* 4 Sept. 307/1 They will resent his careless pontification ('Marxian materialism and Freudian psychology are excuses for laziness').

pontificial. B. *sb.* **2.** (Later example.)

1920 *Trans. Scottish Ecclesiol. Soc.* VI. 79 We are enabled to do this, as the pontificial or book of offices used by him has been printed.

pontil. Add: **1.** (Later examples.)

1918 [see *Gadget 1 d]. 1961 E. M. Elville *Collector's Dict. Glass* 190/1 In the eighteenth century.., the foot of a three-piece glass was attached to a 'pontil', for the finishing operation in the chair. The pontil is a solid rod of iron about the same length but not quite so thick as the blow-iron. 1977 *Lancashire Life* Feb. 35/1 (*caption*) At Cumbria Crystal,.. the base of a stem having been formed, it is transferred to a 'pontil' for finishing.

2. *attrib.* and *Comb.*, as *pontil mark*; **pontil rod,** a pontil.

1937 *Burlington Mag.* Nov. 221/1 Several of these have a round base with a mere trace of pontil mark. 1968 *Canad. Antiques Collector* Oct. 27/2 The long window of crown glass in one of the farm buildings where the pontil marks are evident in each pane. 1975 A. A. C. Hedges *Bottles* 23/2 There is little chance of them being mistaken for free-blown bottles with all their 'blemishes',.. pontil marks and kick-ups. 1955 G. Stevens *In Canad. Attic* 60 The basic tools necessary to a glass blower are a blowpipe, a pontil rod. 1970 *Awake!* 8 Jan. 23/2 During the final forming the glass is attached to a long, solid 'pontil' rod that leaves a mark in the base.

Pontine (pɒ·ntain, pɒ·ntĭn), *a.*[2] [f. L. *Pontus*, Gr. Πόντος, the Black Sea + -ine[1].] = Pontic *a.*[1]

1920 *Q. Rev.* Jan. 244 It would be necessary to guarantee a local autonomy to the Greeks of the Pontine littoral. 1968 S. Johnson *Turkish Panorama* x. 96, I was now in the very foothills of the Pontine Mountains.

‖ **Pont l'Évêque** (poṅ levɛk). The name of a town in Normandy, northern France, used *attrib.* and *absol.* to designate a type of sweet, soft cheese made there.

[1881 J. P. Sheldon *Dairy Farming* xxxv. 515/2 *Fromage de Pont l'Évêque*, this cheese was made as long ago as the thirteenth century.] 1896 Long & Benson *Cheese* v. 57 Pont l'Évêque cheese is a variety with a great local reputation in the north.. of France. 1910 [see *Gorgonzola]. 1932 R. Fraser *Marriage in Heaven* i. xiv. 85 Taste this Pont L'Eveque... A friend brought it to me yesterday from Normandy. 1967 T. A. Layton *Wine & Food Soc. Guide Cheese & Cheese Cookery* 86 Pont l'Evêque requires from 15–24 days to reach maturity. 1979 N. & I. Lyons *Champagne Blues* 75 Is this a Pont-l'Évêque I see before me?

Pontocaine (pɒ·ntokēin). *Pharm.* Also **pontocaine.** Formerly also **-cain.** A proprietary name in the U.S. for *pantocain.

1935 *Surg. Clinics N. Amer.* Dec. 1501 Anesthetists have recently turned to the longer-acting drugs pontocain and nupercaine. 1935 *Official Gaz.* (U.S. Patent Office) 10 Dec. 259/2 Winthrop Chemical Company, Inc., New York,.. *Pontocaine* for anesthetic. Claims use since Aug. 26, 1935. 1938 *New Eng. Jrnl. Med.* 27 Jan. 170/1 For the past four years at the Faulkner Hospital, pontocaine and novocain have been combined for spinal anesthesia. 1946 *Anesthesiology* VII. 500 Since 1943 we have employed

ephedrine in combination with pontocaine hydrochloride for spinal anesthesia in over 2,500 cases. 1975 *Nature* 24 Apr. 710/2 Twelve Dorset and Western ewes at days 67–147 of gestation were starved for 48 h and then placed under pentobarbitol sedation (5 mg kg^{-1}) and spinal anaesthesia (6 mg pontocaine in hyperbaric glucose).

pontoon, *sb.*[1] Add: **1.** More widely, any structure designed to provide buoyancy in the water. (Further examples.)

1941 *Sun* (Baltimore) 15 Sept. 13/1 Just before the regatta ended, he was driving Onwego, a hydroplane, out of the pits and ran his pontoon well over the side of one of the Coast Guard picket boats. 1975 *North Sea Background Notes* (Brit. Petroleum Co.) 11 The hull platform rests on a number of legs which have at their bases pontoons. During moves from one location to another, the entire vessel floats on the sea surface, but on reaching the new location the pontoons are then ballasted with water so that they sink. 1976 *Offshore Platforms & Pipelining* 121/1 Pipe leaves the barge via the curved ramp and a straight or curved pontoon and progresses to the sea floor.

pontoon (pɒntū·n), *sb.*[2] [Appar. corrupted from Vingt-et-un, vingt-un.] A popular name for the card game Vingt-et-un, vingt-un.

1917 A. G. Empey *Over Top* 304 *Pontoon*, a card game, in America known as 'Black Jack' or 'Twenty One'. The bank is the only winner. 1927 *Daily Express* 26 July 9/5 A ghostly platoon wouldn't frighten me!.. perhaps they'd be playing pontoon. 1961 A. Wykes *Gambling* vii. 177 The three modern banking games—baccarat (or chemin-de-fer), *blackjack*, (or vingt-et-un or *pontoon*), and *seven-and-a-half*—are all complicated versions of European games of the 15th and 16th centuries. 1973 J. Wood *North Beat* x. 134 The locker-room table.. used for pontoon and brag sessions. 1976 J. Bingham *God's Defector* iii. 28 He was playing pontoon and drinking with four friends in a back room... Rob Flint had just laid two cards face up on the table, an ace and a king.

b. A prison sentence or term of twenty one months (occas. twenty one years). *slang* (chiefly *Criminals'*).

1950 C. Franklin *She'll love you Dead* vii. 90 'They'll get me a pontoon for assault when Mr. Garfield tells 'is story,' said Al miserably. 'A pontoon?' 'Twenty-one months,' explained Garfield. 1958 F. Norman *Bang to Rights* 177 This geezer was doing a pontoon. 1962 *John o' London's* 25 Jan. 82/3, 21 months [imprisonment], *pontoon.* 1977 'E. Crispin' *Glimpses of Moon* xii. 235 He had been put away three times.. the third for a pontoon.

Pontypool (pɒntipū·l). Also 8–9 **Ponty-pool, Pont-y-Pool, Pont y Pool.** The name of a town in Gwent, Wales, used *attrib.* and *absol.* to designate a type of Japanned metal ware originally produced there or items made from this Japanned metal.

[1734 C. H. Williams *Let.* Dec. in L. T. Davies *Men of Monmouthshire* (1933) I. 72 Tom Allgood has found a new way of japanning which I think so beautiful that I'll send you a couple of pieces of it. 1763 *Gloucester Jrnl.* 4 July 2/1 (Advt.), By Allgood, Davies, and Edwards, all Sorts of the real and most durable Japan Ware is continued to be made and sold at the Manufactory at Pont-y-Pool.] 1764 in W. D. John *Pontypool & Usk Japanned Wares* (1953) iv. 36 (Advt.), Great Variety of Ponty-pool Goods Sold by Henry Johns, At his Ponty-pool Warehouse... Great variety of Snuff-Boxes, Japan'd Waiters, Bread Baskets, Tea Kitchens, Tea Kettles, and Lamps, Coffee Pots. 1781 J. Byng *Diary* 16 June in E. Burton *Georgians at Home* (1967) iv. 183, I bought a Pontypool snuff box, a beautiful and dear ware. 1801 W. Coxe *Hist. Tour Monmouthshire* II. xxv. 234 The town.. is likewise remarkable for the japan manufacture, known by the name of Pont y Pool ware. 1872 *Art Jrnl.* XI. 24 (*caption*) The premises in which the Allgoods last manufactured Pontypool ware. *Ibid.* 24/1 'Now,' said Old Billy in the highest glee, 'now you shall see what real Pontypool Japan is!' 1928 *Daily Express* 6 Oct. 11/7 The secret formula for the production of the artistic lacquer work known as 'Pontypool japan', which had been missing since 1864, has been found. 1953 *Ann. Sci.* IX. 218 Several firms [in the Midlands] that paid special attention to the production of wares of high artistic merit were calling themselves 'Pontypool makers' in the early years of the nineteenth century, when the trade was in a flourishing condition. 1960 *House & Garden* May 56/4 Red and gold decorated Pontypool tray, £65. 1969 *Canad. Antiques Collector* Jan. 8/1 This japanned tinware was generally known by the name Pontypool (even when it was later manufactured at Wolverhampton and Birmingham). 1971 H. Huth *Lacquer of West* viii. 112 A color effect typical of Pontypool but later much imitated was a tortoiseshell ground made by placing irregular pieces of silver foil under brown lacquer, thus giving it the appearance of gold.

pony, *sb.* Add: **1. b.** A race-horse. Usu. *pl.* *slang* (chiefly *U.S.*).

1907 J. London in *Cosmopolitan* May 17/2, I had been out to the race-track watching the ponies run. 1942 Berrey & Van den Bark *Amer. Thes. Slang* § 731/1 *Race horse*.., pony. a 1953 E. O'Neill *Long Day's Journey into Night* (1956) i. 21 If it takes my snoring to make you remember Shakespeare instead of the dope sheet on the ponies, I hope I'll keep on with it. 1958 [see *Play *v.* 21 d]. 1961 *Dallas Morning News* 17 Feb. 1. 5 Rep. Berry, an ex-gambler from San Antonio, got elected on his advocacy of betting on the ponies.

2. (Further examples.)

1892 *Pall Mall Gaz.* 23 Mar. 6/3 Mr. Kisch said the bets were two ponies. The Master of the Rolls: What? Two what? Mr. Kisch said a pony was £25. 1928 D. Byrne *Destiny Bay* vii. 318 It would have to be done very carefully,.. in ponies and fifties and hundreds. 1958 *Times* 18 Feb. 5/1 Heath said that for a 'pony' (£25) he would see what could be done. 1966 B. Naughton *Alfie* xxix. 188 'A pony is neither here nor there to me,' I said. 'It's just that I want to give somebody something.' 1976 J. O'Connor *Eleventh Commandment* xiv. 178 'Bet you the next three guys that come by do that,' he said. 'Make it a pony (£25),' said Charlie.

3. (Earlier and later examples.) Also *transf.* (see quot. 1977).

1827 *Harvard Reg.* Sept. 194 I'll tell you what I mean to do. Leave off my lazy habits.. and stick to the law, Tom, without a *Poney*. 1931 W. Faulkner *Sanctuary* xviii. 182 She kept the dates written down in her Latin 'pony'. 1952 G. Sarton *Hist. Sci.* I. iii. 89 The tablets were used not so much for study as for recapitulation and remembrance, like cribs or ponies. 1972 *Catholic Biblical Q.* Jan. 93 The Hebrew syntax is 'Akkadianized'... The result is that the book provides an excellent 'pony' for the student who is weak in Akkadian. 1977 *Sounds* 9 July 22/2 After leaving college his vaguely literary ambitions found him earning a living by turning out 'ponies', the Stateside word for those little revision booklets English (or US, in this case) Lit. students buy when they haven't read, say, 'Bleak House' and there's an exam tomorrow morning.

b. *U.S.* Used *attrib.* to designate an abridged news report or the service whereby such reports are supplied to particular news agencies. The service has appar. been discontinued.

1877 *Harper's Mag.* Dec. 57/1 Condensed abstracts, known as 'pony' reports, are made and forwarded to smaller towns. 1909 *Census Bull.* (U.S.) No. 216. 67 Besides the full reports delivered to large papers are the 'pony' reports—condensations of the full reports, sold at a cheaper rate. 1915 G. M. Hyde *Newspaper Editing* vi. 199 Certain members, too small to be full members [of the Associated Press], receive a daily 'pony' service—a condensed version of the world's news to the extent of a few hundred or few thousand words—and pay proportionately. 1923 M. V. Atwood *Country Newspaper* 133 The writer wonders if it may not be that the country daily.. furnishing a reasonably adequate service of telegraph news through a 'pony' service.. may not become of increasing importance. 1931 C. E. Rogers *Journalistic Vocations* iii. 57 The United Press developed the use of the telephone for delivering abbreviated news reports—P.N.T. (public news transmission) or pony service. *Ibid.* 61 There are shorter leased wire reports, too, and pony reports. 1942 Radder & Stempel *Newspaper Editing, Make-up & Headlines* (ed. 2) vii. 125 Such reports, known as pony calls, usually amount to only 15 minutes of service (1,000 to 1,500 words) once or twice a day. The pony service still is used by a number of smaller newspapers, and some rely on a bulletin service or pony service by telegraph.

4. (Earlier and later examples.)

1849 G. G. Foster *N.Y. in Slices* 81 The game is kept up, mollified now and then by a choice swig at the 'poney'. 1943 *Harper's Mag.* Dec. 44/2 Dr. Stuker rapidly downed two ponies of brandy. 1959 G. Hamilton *Summer Glare* 155 Os pulled a beer each for me and Tommy, and a pony for himself. He always drank small beers. 1966 [see *Pot *sb.*[1] 3].

b. A small chorus girl or dancer. *slang.*

1908 K. McGaffey *Sorrows of Show Girl* 118, I went into the pony ballet of a LaSalle Theatre show—can you see me as a pony? 1920 [see *Drag *sb.* 7 f]. 1930 *Daily Express* 23 May 10/5 We have what are known in stage parlance as 'ponies'—a troupe of girls, ages ranging from sixteen to twenty-three or four. 1948 *Sat. Even. Post* 3 July 63/2 In the chorus of ponies—the smallest sized dancers—there was a pert redhead named Gracie Barrett. 1950 Blesh & Janis *They all played Ragtime* (1958) ix. 180 The music that follows is a 'rush-on' of the period, so called because it was the cue for the high-stepping, brown-skinned 'ponies' to get out on the stage.

5*. A dance originating in the U.S. and popular in the early 1960s.

1963 *N.Y. Times Mag.* 27 Oct. 104/2 That brings us to our own young and the Twist, the Pony, the Slop, the Mashed Potato. 1968 M. & J. Stearns *Jazz Dance* i. 5 The Pony employed bits of the Slow Drag. 1969 N. Cohn *A WopBopaLooBop* (1970) ix. 85 Dance crazes bossed pop right up until the Beatles broke. There was the Hully Gully, the Madison, the Fly, the Pony [etc.].

6. a. *pony-boy, -carriage* (examples), *-chair* (earlier example), *-chaise* (earlier example), *-man, -phaeton* (earlier example), *race, ride; pony-penning, -racing, -rider, -riding*; see also sense *3 b. **b. pony club,** a club founded in 1929 and now run by the British Horse Society for young people with ponies; hence as *vb. trans.* (chiefly *pass.*) to enter (a pony) for a pony club competition; **pony clubber,** a member of a pony club; **pony clubbing,** participating in pony club activities; the pony club movement; **pony express** (earlier and later examples); also *attrib.*; **pony post** = *pony express*; **pony-skin,** the (dressed) hide of a pony; also *attrib.*; **pony-tail,** a hair-style in which the hair is gathered back through a band or other fastening to resemble the shape of the tail of a horse or pony (cf. *Horse-tail 1 c); also *attrib.*; hence *pony-tailed* adj.; **pony-trekking,** pony-riding for long distances across country, esp. as undertaken as a group holiday activity; hence *pony-trekker.*

1909 *Daily Chron.* 16 Feb. 5/1 Murten.. is employed as a pony-boy in the Woolley Colliery, Barnsley. 1946 B. Naughton (*title*) Pony boy. 1831 M. Edgeworth *Let.* 11 Apr. (1971) 520 Dr. Fitton in the pony carriage behind me was giving.. another derivation to Fanny from the German. 1870 Geo. Eliot *Jrnl.* 26 May in *Geo. Eliot Lett.* (1956) V. 100 Mrs. Pattison took me a drive in her little pony carriage. 1905 'P. Pennington' *Woman Rice Planter* (1913) iv. 150, I sent Chloe to Gregory in the pony carriage, and she brought back the money. 1827 T. Hamilton *Youth & Manhood C. Thornton* I. xvi. 282 Mr Pynsent with some difficulty did so, pleading an engagement to drive Lady Amersham in her pony-chair. 1831 Disraeli *Young Duke* I. ii. x. 239 A pony-chaise was Lady Faulconcourt's delight. 1929 *Horse* I. 60 The Pony Club.. has been inaugurated for the purpose of interesting young people in riding and sport. 1936 A. Thirkell *August Folly* ii. 50 Pony Clubs! No pony clubs when I was young. You got on and you fell off, and there you were. 1941 M. Treadgold *We couldn't leave Dinah* iii. 50 Pony club members and their guests were sedately walking their ponies round and round the lawn. 1972 J. McClure *Caterpillar Cop* i. 12 'You're strong,' he murmured. 'Riding,' she said, 'I'm in the pony club.' 1976 *Horse & Hound* 3 Dec. 63/1 (Advt.), Moonmaster... Very pretty strawberry roan gelding. 10 yrs. Leading rein, gymkhanaed, Pony Clubbed. Good in traffic. 1977 *Ibid.* 14 Jan. 40/2 (Advt.), Gelding... He jumps well, has hunted, and been Pony Clubbed etc. 1970 J. Campbell *World of Ponies* 127 (*caption*) Australian Pony Clubbers are always sure of good weather for their outdoor activities. 1977 *Horse & Hound* 14 Jan. 33/2 Always a keen 'pony clubber', Alison has competed in numerous inter-branch competitions. 1970 J. Campbell *World of Ponies* 125 It has been the ponies of all breeds, cross-breeds, shapes and sizes, that have made Pony Clubbing. 1977 *Horse & Hound* 14 Jan. 46/3 (Advt.), Four children 3–9, 4 ponies, 1 horse. Pony Clubbing, showing etc. 1847 *N.Y. Weekly Tribune* 18 Dec. 4/5 By our Pony Express from the South, we have intelligence from New Orleans to the afternoon of the 2d. 1860 *San Francisco National* 19 Mar. 2/3 The Central Overland Pony Express Co. will start their Letter Express from San Francisco to New York and intermediate points, on Tuesday, the 3rd day of April next. 1948 *Chicago Daily News* 26 Aug. 4/1 St. Joseph, Mo... The original Pony Express stable was put up for sale for \$442.32 but no one bid on it. 1976 *Times* 23 July 11/6 Buffalo Bill Cody.. had been in turn horse wrangler, pony express rider, unlucky prospector, [etc.]. 1900 *Geogr. Jrnl.* XV. 563 Group of Astor pony-men at Lob jungle. 1968 *Economist* 10 Aug. 45/1 Four ponies need four ponymen which adds another £10 a day to the bill. 1971 *Daily Tel.* 10 June 9/3 The future livelihood of ghillies, stalkers, gamekeepers and ponymen is threatened by the 'punitive' proposed rating reassessments of estates with sporting facilities, the Scottish Landowners' Federation claims. 1932 *Sun* (Baltimore) 27 July 4/3 Pony-penning has attracted thousands to the island. 1958 *Washington Post* 30 July A 24/1 The pony penning dates back to 1835 but the firemen took it over in 1924 as their fund raising project. 1799 Malthus *Diary* 1 July (1966) 109 Mr A had insisted on our taking his small poney phaeton. 1893 M. H. Cushing *Story of our Post Office* 420 Before railroads led to every part of the country the only communication was by pony post. 1901 W. Churchill *Crisis* ii. vii. 178 Three-weeks letters from San Francisco, come by the pony post to Lexington. 1765 J. Woodforde *Diary* 27 May (1924) I. 47 After dinner Jack went to Wincanton to a Pony Race. 1824 J. Decastro *Mem.* 155 The pony races were brought out, and they had a more than usual run for a whole season. 1949 *Sun* (Baltimore) 29 July 20/6 The spectators stayed on for the day-long program of pony sales.. pony races.. and band concerts. 1827 W. Clarke *Every Night Bk.* 174 Furnishes a neat stud for poney racing. 1943 *Sporting Life* 12 July 5 Pony racing will have an eager Turf Authority, a second-to-none race course at its disposal, and cash in the bank. 1969 *Pony* Sept. 57/2 Two centuries ago.. Charles O'Neill.. established a pony-racing event on Broughshane race-course, near Ballymena. 1819 M. Edgeworth *Let.* 17 Apr. (1971) 200 She has just come in from her poney ride. 1840 C. Fox *Jrnl.* 16 Feb. (1972) 67 They joined Mamma and Anna Maria in a pony ride. 1871 'Mark Twain' *Lett. to Publishers* (1967) 62 Stretching our necks and watching for the pony-rider. 1975 *Country Life* 13 Feb. 393/1 Pony riders will have to keep on to the country lanes. 1949 R. Colville (*title*) Pony riding. 1908 *Daily Chron.* 26 Dec. 3/4 Among the novelties are the pony-skin suits. 1960 *Times* 26 Sept. 17/2 In Mongolian ponyskin.. it [*sc.* a coat] is very hard-wearing. 1971 'A. Burgess' *MF* viii. 94 The upholstery was black-and-white ponyskin. 1976 'J. Ross' *I know what it's like to Die* xvi. 102 Your lady-friend with the ponyskin coat. 1872 Trollope *Eustace Diamonds* (1873) II. xxxiv. 100 'How a man can like to kiss a face with a dirty horse's tail all whizzling about it...' 'I haven't even a pony's tail,' said Lucy. 1952 *Sun* (Baltimore) 23 Feb. 2 The panel of high-school boys and girls discusses the latest teen-age fashions, including.. the pony tail. 1954 J. Trench *Dishonoured Bones* iv. 150 She pulled her own hair back and fastened it into a pony-tail. 1957 *New Yorker* 16 Nov. 104/2 The young lady.. was wearing a ponytail hairdo. 1971 M. Spark *Not to Disturb* iii. 88 She loosens her hair which has been pulled back, pony-tail style. 1975 *New Yorker* 28 Apr. 31/2 She, too, had her hair in a ponytail, held by a rubber band. 1977 *Time* 30 May 40/3 The counter-culture ponytail is gone, sacrificed to the heat of arena lights and the sizzling sweat of the fast-break pace. 1956 *Time* 26 Mar. 72/2 Pony-tailed Carol stood aside. 1958 S. Ellin *Eighth Circle* ii. i. 18 She bore the sallow-complexioned, nail-bitten, pony-tailed earmarks of adolescence. 1974 *Times* 14 Nov. 16/6 A bearded, pony-tailed, 42-year-old.. is not everyone's idea of a nanny-raised child. 1972 *Guardian* 3 July 7/3 Pathways were being worn down by pony-trekkers and others. 1959 *Sunday Times* 8 Mar. 20/1 Golf, Tennis, Fishing, Ballroom, Pony-trekking, Dinghy sailing. *Ibid.* 29 Mar. 18/1 Guided pony-trekking is a fine way to explore wild country in the company of other adventure-seekers. 1962 *Times* 21 Apr. 11/3 Over the past few years.. great has been the demand for pony-trekking holidays. 1971 *New Yorker* 27 Feb. 21/1 (Advt.), Come to Ireland. Go pony-trekking through Macgillycuddy's Reeks.

pony, *v.* Add: **2.** Also *intr.* **b.** *trans.* To give extra tuition to. *rare.*

1847 in W. G. Hammond *Remembrance of Amherst* (1946) 153 The others are pooying most unmercifully. *c*1853 in Root & Lombard *Songs of Yale* 23 If you poney he will see. 1865 *Harper's Mag.* July 213/2 A classmate, whom..I had ponied through term after term, in Latin, Greek, and mathematics. 1908 W. G. Davenport *Butte & Montana* 134 It were a hundred times better to teach the average boy how to build a fence..than to..'pony' his way through three or four years of Latin.

Ponzi scheme (pǫ·nzi). *U.S.* [f. the name of Charles *Ponzi*, who perpetrated such a fraud 1919–20.] A form of fraud in which belief in the success of a fictive enterprise is fostered by payment of quick returns to first investors from money invested by others.

[1957 *Encycl. Brit.* IX. 708/1 The Ponzi Scheme... Beginning in Dec. 1919 Ponzi..produced a scheme involving the purchase of International Postal Reply coupons in countries where the exchange was low, trading them in for postage stamps at their face value in a country where the rate was high, and then selling the stamps at a great profit... The slogan of the swindle was 50% in 90 days.. Actually..Ponzi made no purchases whatever of International Postal Reply coupons.] 1973 *Guardian* 14 Apr. 2/4 The indictments..allege that Mackell's staff invested in what is called a Ponzi scheme, a confidence game named after a famous Italian. 1976 *Billings* (Montana) *Gaz.* 27 June 9-G/1 The Home-Stake scandal is a form of the 'Ponzi' scheme, named for a self-educated, slight but dapper Italian immigrant named Charles Ponzi whose intricate schemes in the 1920s were front-page stuff. 1978 *National Observer* (U.S.) 10 July 8/3 'We are operating a Ponzi scheme,' says Michael Mustokoff, chief of the unit. The first few investors were paid 'dividends' out of the money provided by people who came in later, and word spread that the club was raking in the bucks.

Ponzo (pǫ·nzo). [The name of Mario *Ponzo* (b. 1882), Italian psychologist.] *Ponzo illusion*: an optical illusion in which two parallel straight lines of equal length appear to be of unequal length when seen side by side against a triangular background (such as a set of straight lines radiating from a single point and passing through the two parallel lines).

1942 *Jrnl. Exper. Psychol.* XXX. 84 (*heading*) Experimental evidence for the electrical character of visual fields derived from a quantitative analysis of the Ponzo illusion. 1968 *Science* 22 Mar. 1375/1 The Ponzo illusion increases in magnitude between childhood and adulthood. 1976 *Sci. Amer.* Apr. 50/1 In the Ponzo illusion, although both vertical lines are the same length, the effect of the subjective triangle is to make the line at the left appear to be longer.

pooay, var. *PWE.

pooch (pūtʃ), *sb.* and *a. colloq.* (orig. *U.S.*) [Etym. obscure.] **A.** *sb.* A dog, esp. a mongrel. **B.** *adj.* Mongrel. *rare.*

1924 B. Hecht *Cutie* vi. 46 All you do is sink your teeth in my shoulder and make noises like a basket full of hungry pooches. 1927 *Collier's* 3 Dec. 32/4 Therefore, at home, the trick pooch got all the attention, eating at the table with the family. 1941 Baker *N.Z. Slang* vi. 60 Hundreds of Australian terms are unused here... pooch, for instance, a greyhound. 1951 C. Armstrong *Black-Eyed Stranger* (1952) iii. 27 It wasn't even my dog... But..I'd more or less met the pooch. 1962 *Country Life* 15 Apr. 895/1 The training of dogs, whether pedigree or pooch, has assumed considerable importance during the last 80 years. 1963 D. Ireland *Animal Life & Love* i. 15 There is one very old claim of an ancient pooch of 34 years. 1973 *Sunday Australian* 8 Aug. 30/4 You've got some useful ammunition to aim at that noisy cherished little pooch next door. 1977 *Cornish Times* 19 Aug. 15/1, I tend to fury when children cannot play games on fields intended for their use without falling on some pooch's revolting mess. 1977 J. Wambaugh *Black Marble* (1978) v. 68, I got more invested in that pooch than you *made* in the last five years.

poodle, *sb.* Add: **1. b.** *fig.* A lackey or cat's-paw.

1907 Lloyd George in *Hansard Commons* 26 June 1429 The House of Lords consented. This is the defender of property! This is the leal and trusty mastiff which is to watch over our interests... A mastiff? It is the right hon. Gentleman's poodle. It fetches and carries for him. It barks for him. It bites anybody that he sets it on to. 1944 J. Jones *Man David* vii. 144 There were certain barriers to progress, the greatest being 'that Tory poodle', the House of Lords. 1954 R. Jenkins (*title*) Mr. Balfour's poodle. An account of the struggle between the House of Lords and the government of Mr. Asquith. 1967 *Daily Tel.* 10 Feb. 30/2 Labour MPs did not appear to find the speech objectionable. One comment was that Prince Philip had shown himself to be 'nobody's poodle'. 1968 *Guardian* 9 Aug. 10/1 Mr Curran..vigorously denied suggestions that he would be Lord Hill's 'poodle'. 1969 G. Black' *Cold Jungle* viii. 114 Bill would have been more likely to have a heart attack being chased by a thug down on the Riviera. 1974 Ld. Aldington *Advising BBC* 13 The suspicious will say that such a close link between the advisers and the advised..ensures that at least the Chairman of GAC, if not all its members, become the poodle of the BBC. 1976 *Times* 12 Nov. 14/4 Mr Foot is happy to act as Mr Jones's poodle in introducing the Bill.

3. *poodle-dog* (earlier examples), *-fashion*; **poodle cloth,** a woolly sort of cloth; also *attrib.*; cf. sense 2 in Dict.; **poodle-cut,** a hair-style in which the hair is cut short and curled all over.

1957 M. B. Picken *Fashion Dict.* 261/2 *Poodle cloth*, a coating of knotted yarn or loopy bouclé. Woven in all fibers and also knitted. Originally was made only in wool. 1959 *Observer* 13 Dec. 14/6 For country-house wear a poodlecloth wool is in allied tones, such as violet and amethyst. 1977 *New Society* 30 June 665/3, I managed to purchase a square-cut, early sixties coat in a fabric we used to call poodle cloth. 1953 *Sun* (Baltimore) 25 Mar. 3/2 (*caption*) In this recent picture, Mrs. Truman wears a little painted face and her yellow poodle cut. 1975 R. L. Simon *Wild Turkey* (1976) x. 8 A well-dressed woman with a poodle cut. 1820 in *Amer. Speech* (1965) XL. 131 Called e Hog, a Poodle dog' all the sailors joking me. 1821 M. Edgeworth *Let.* 27 Jan. (1971) 336 A new poodle dog..milk white silken curls all over except the poor shorn half that is sacrificed to poodle-fashion.

poodle, *v.* Add: **1.** Also, to overdress, to dress up.

1962 N. Streatfeild *Apple Bough* iii. 39 Why do they want to poodle the poor kid up?

2. *intr.* (Usu. with advbs.) To move or travel in a leisurely manner. *colloq.*

1938 F. D. Sharpe *Sharpe of Flying Squad* i. 10 The long, lazy summer days, when the streets. *Ibid.* ii. 30 For the last half dozen years or so, the prosperous poodle about in their sleek, unobtrusive-looking cars poodle about the Metropolis well under 30 m.p.h. 1960 M. Cecil *Something in Common* xii. 131 She tells the patrons which gangway and then they poodle off the opposite way. 1972 *Police Rev.* 8 Dec. 1598/1 What will happen to the chap who wants to quietly poodle along at 50 m.p.h. even when there is no fog? 1973 *Radio Times* 22 Mar. 36/2 (Advt.), A sports jacket should protect you from dawn till dusk on a Scottish moor. Not merely while you're poodling down to the local on Sunday morning. 1975 *New Society* 2 Oct. 26/3 One member of each two-man [bicycle racing] team doing the racing while the other poodles round high on the banking until it's his turn to take over the attack. 1976 J. O'Connor *Eleventh Commandment* i. 24, I went indoors, messed around, poodled about for quite a while.

Hence **poo·dler** *slang*, a small motor vehicle. 1961 *Road Services Mag.* Dec. 94/2 *Poodler*, small vehicle. 1968 *Drive* Spring 113/1 A poodler [is] a small vehicle, like a cyclecar, a small light wagon, and Billy Bunter is a shunter.

poo·dle-faker. *slang* (chiefly *Services*). [f. Poodle *sb.* 1 + Faker.] A man who cultivates female society, esp. for the purpose of professional advancement; a ladies' man; a socialite; also, a young, newly commissioned officer. So **poo·dle-faking** *vbl. sb.* and *ppl. a.*

1902 *T.C.D.* 22 Nov., The 'poodle-faker' is just as much a social necessity as tea-cakes. 1914 'I. Hay' *Knight on Wheels* (ed. 2) iii. xiii. 228 And now my lad, you are going to put on your best duds and come poodle-faking with me! 1915 'Bartimeus' *Tall Ship* iv. 75 Don't tell me the lad is going poodle-faking! 1918 A. H. Chute *Real Front* xiv. 240 He [sc. a big Australian private] encountered a pink-faced English youth, who had just got his commission, one of the Percival or Cuthbert type, whom we here refer to in the army as 'poodle-fakers'. 1918 'Taffrail' *Watch Below* 57 The ladies' men or 'poodle-fakers', as we called them, had their tea-parties, dinner-parties, and dances more often than was good for them. 1925 F. C. Bowen *King's Navy* 39 The tea party to which the other sex is invited from 'the beach' is a 'tea-fight' or 'bun-worry', while paying calls ashore is 'poodle-faking'. 1930 A. B. C. Cator in *Hudson's Rev.* Jan. 53 Man is primaevally a killer; by the word man I mean a real man, not one long haired poodle faking, over dressed idiot, all too common at home in these post-war days. 1938 L. C. Morgan *Flashing Stream* i. 59 Brissing, you're the poodle-faker in this mess. 1939 A. Powell *What's become of Waring?* vii. 200 But what did you live on?.. A bit of journalism here and there, a good deal of poodle-faking. 1949 H. Pakington *Young W. Washbourne* 38 John had got quite frankly that one didn't always want to be saddled with one's cousins, however charming, and William had retorted that he'd be damned if he'd go round poodle-faking all the time, and what was a flag-lieutenant for except to amuse the Admiral's guests. 1965 Marsh *Mask Tapestry* 21 Women are not admitted to the main club premises except once a year at the ball. Traditionally, married members are confined to the club between 7.30 and 8.30 pm while poodle-fakers daily with their wives in their victorias outside on the lan. 1969 L. Marsh *Dead Water* (1964) vi. 134, I tell my regiment. Took on this damned poodlefaking instead. 1967 D. Bush *Craft of Diplomacy* vii. 184 In Britain it is still widely assumed, perhaps largely because of ignorant or malevolent press comment, that diplomacy is poodle-faking..private means and anyhow is only poodlefaking. 1977 J. Potter *Who the Heck* xi. 97 There's some blooming Parisian couturier coming to see her... To hear her talk you'd think a bunch of corn slicers and foreign poodle-fakers was more important than solving the crime of the century. 1978 M. M. Kaye *Far Pavilions* xi. 175 He could go and shoot in Kashmir..which would do him a lot more good than poodle-faking at tea-parties.

poof (puf, pūf), *sb.*[1] *slang.* Also **pouff, pouf,** etc. [Prob. a corruption of *puff* (see *Puff *sb.* 8 d).] An effeminate man; a male homosexual; a man who acts or speaks in an affected manner. Also *attrib.* Similarly **poove** (pūv) *sb.*; also as *v. intr.*, to act like a poof, to speak or behave in an effeminate or affected manner; **pooved-up** *ppl. a.*

Often considered offensive.

*c*1850–60 in G. R. Taylor *Angel-Makers* (1958) iv. 80 These monsters in the shape of men, commonly designated Margeries, Pooffs, &c. 1932 Auden *Orators* iii. 98 Poofs and ponces, All of them dunces. 1951 I. Shaw *Troubled Air* xvi. 272 Don't be a traitorous old poof. 1952 A. Wilson *Hemlock & After* ii. 37 'Bloody little pouff,' said Ron aloud. 1955 'C. H. Rolph' *Women of Streets* x. 131 Although I never met any who lived with male homosexuals, several..girls..referred to other prostitutes living with 'pouffs' who reciprocally kept them by their earnings when necessary. 1955 G. Greene *Quiet American* iv. ii. 241 He made a feeble attempt to mock my accent. 'You all talk like poufs. You're so damned superior.' 1959 C. MacInnes *Absolute Beginners* 51 The Hoplite has been in business with some of the city's top poof raves. 1962 *Private Eye* 30 Nov. 15/3, I may be a poove but I'm a terrific engineer. 1964 *New Statesman* 6 Mar. 374/1 We have a pooved-up tenor introducing a parade of Variety Girls. 1967 J. Rathbone *Diamonds Bid* xiii. 116 'Do you remember meeting Stephen Hamilton-Rose..?' 'A fat poove?' I asked. 1968 A. Diment *Gt. Spy Race* iii. xi. 206 The woolly-headed pooves in the widely various Ministry of Defence networks are all completely mad with jealousy. 1968 *Listener* 19 Sept. 372/3 On the first occasion the loved object..was an able-bodied seaman,..who never manifested the slightest interest in girls but who nevertheless was totally remote from the world of pansies, pouffs and queans. 1971 F. Forsyth *Day of Jackal* xx. 336 You bloody pooves make me sick. 1971 *Melody Maker* 9 Oct. 11/1 He reckoned they pooved around a little, but commented that..their music wasn't all that rough after all. 1974 J. Betjeman *Nip in Air* 45 Touching the little children, better pooves Or murderers, they said. 1975 J. Symons *Three Pipe Problem* ix. 60 It's some poove who's been done, named Sonny Halliwell. 1976 A. Richards *Former Miss Merthyr Tydfil* 14 A young man..had been heard in the showers to refer to Elgar as 'a bit of a pouf'. 1977 W. McIlvanney *Laidlaw* xxviii. 128 Harry Rayburn's a poof... Whit's a poof doin' wi' a lassie? 1978 R. Rendell *Sleeping Life* xiii. 109 All you can do is get your picture in the papers like some poove of a film actor.

Hence **poo·fdom** *nonce*, the state or condition of being a homosexual.

1972 F. Raphael *April, June & Nov.* 466 He's a late convert to the joys of poofdom.

poof, *int.* Add: Also **pouff. A.** (Earlier and later examples.)

1824 J. Morier *Adventures Hajji Baba* II. i. 39 Putting up her five fingers to his face, she said, 'Poof! I spit on such a face.' 1829 G. Griffin *Collegians* I. viii. 159 Gi' me the hat, sir, an' I'll hang it up—poof, it's full of dust. 1905 E. Glyn *Viciss. Evangeline* 62 'Pouff!' I said, and I pointed at him. 1921 H. Williamson *Beautiful Yrs.* 80 'Pouff, what a lot of rot,' scoffed Willie. 1935 M. de la Roche *Young Renny* x. 82 Pouf! You don't know anything. 1949 P. Hastings *Cases in Court* v. 281 He said the three shots were fired in rapid succession or, as he put it somewhat dramatically, 'pouf, pouf, pouf'. 1951 M. Kennedy *Lucy Carmichael* ii. i. 88, I get quite interested ..for about 5 minutes and then—poof! I go flat like a burst balloon. 1968 C. M. Vines *Little Nut-Brown Man* xiii. 237 'Pouf!' he said when he had recovered from choking, 'supposing I had been dead before the water got here?' 1974 S. Coulter *Chateau* ii. xii. 359 Oh, pouff! Bravado, Madame. Sheer bravado. 1979 J. Rathbone *Euro-Killers* v. 55 Poof! Rubbish! This is some ruse.

B. as *sb.*[2] An utterance of 'poof'; a short sharp puff.

1908 *Westm. Gaz.* 25 May 5/2, I was riding on the back of the balloon..when suddenly I heard a 'pouff', as if someone had blown a blast from a bellows. 1915 D. H. Lawrence *Rainbow* vii. 158 She burst into a 'Poof' of ridiculing laughter. 1951 Koestler *Age of Longing* ii. iv. 142 Father Millet..gave a scornful poof. 1972 B. Malamud *Tenants* 68 [He] left so cleanly,..it seemed to Lesser as though he had willed his disappearance in a prestidigitated poof. 1973 J. McKenzie *Man against Tutte* iii. 196 Refinements on the use of a dash of poison now go well beyond a squirt of spray or a poof of dust to the habitat of the fly.

poof (puf), *v. colloq.* Also **pouff.** [f. Poof *int.*] **intr.** **A.** To blow up, to peter out. Also *refl.*

1915 *N.Y. World Mag.* 9 May 14 *Poof*, to blow up. 1923 Wodehouse *Inimitable Jeeves* xviii. 242 The fact is, I suppose, I'd seen so many of young Bingo's love affairs start off with a whoop and a rattle and poof themselves out that I couldn't imagine this one lasting long. 1934 —— *Right Ho, Jeeves* xxi. 231 Then the dialogue sort of poofed out once more, and we stood eating cheese straws and cold eggs respectively in silence.

b. To utter a 'poof'. *rare.*

1915 D. H. Lawrence *Rainbow* vi. 148 Again she poufed with mockery.

poofter (pu·ftəɹ, pū·ftəɹ). *slang* (chiefly *Austral.*). Also **pooftah, poufter, pufter.** [Fanciful extension of *Poof *sb.*[1]] A homosexual; an effeminate man. Also used as a general term of abuse to a man. Also *attrib.* So **poo·fteroo-** [see *-eroo].

Often considered offensive.

1900 O'Brien & Stephens *Materials for Dict. Australasian Slang* 1900-10 (typescript), *Pouf* or *poufter*, a sodomite or effeminate man. 1941 Baker *Dict. Austral. Slang* 56 *Poofter*, a homosexual. 1945 —— *Austral. Lang.* 123 A procurer for homosexuals is known as a *poofter rorter*. 1952 *Here & Now* (N.Z.) Jan. 19/1 The butcher's assistant was as likely to have a star bird dog or rabbit dog as the richest pooftah. 1953 T. A. G. Hungerford *Riverslake* iii. 49 He hawked disgustingly and spat on the floor between his feet. 'They want men in the unions, not poofters!' 1955 D. Niland *Shiralee* 207 They'd play around like poofters, with the kid gloves and the soft soap. 1961

P. White *Riders in Chariot* iv. xi. 392 'They will tell you,' she said, 'that Norman is a pufter... Norm could not impress a woman even if he tried.' *Ibid.* 401 'You are a proper pufter rotter, Hannah!' Reen had to remark, because..'You pommy poofter.'.. Bond said mildly, 'What's a poofter?' 'What you'd call a pansy.' 1966 *Punch* 9 Mar. 352/1 It is illegal to ingratiate oneself with the Examiner.' Ingratiate oneself! Good heavens! It was slowly beginning to dawn on me that this man was a poofteroo. 1966 P. White *Solid Mandala* 18 You ought to move in with the poofters. 1968 *Daily Mirror* 25 Jan. 17 We once had a poofter here. 1973 A. Broinowski *Take One Ambassador* v. 53 The feller's a bloody poofter. Queen's a four pound note. 1974 R. Gadney *Something Worth Fighting For* xvi. 139 Looking at all them poofters. I was thinking what some people do for money. 1976 *Telegraph* (Brisbane) 21 Sept. 50/1 'Poofter-bashing' is a strong and blunt phrase, but the only way to describe the latest wave of discrimination against homosexual men and women in Brisbane. 1977 *Listener* (N.Z.) 15 Jan. 61 The white-legged Pommy poofter is an integral part of one view of the English. 1978 J. Barnett *Head of Force* ix. 82 He was..a poofter... He was having it off with some body he called Soldier.

poofy (pū·fi, pu·fi), *a. slang.* Also **poovey, poovy, pouffy, poufy.** [f. *Poof *sb.*[1] + -y[1].] Of, pertaining to, or characteristic of a poof; effeminate or homosexual. Also *Comb.*

1970 J. Hale *Grudge Fight* v. 74 Before lunch the soap is scented, pink stuff sent to him by Momma in the last parcel. 'Very poofy,' says Yang, the gas man. 1974 I. Jan. 24/7 Wailed a poovey young author named Ned, 'I'd rather be dead than unread.' 1968 A. Diment *Gt. Spy Race* ii. vii. 99 He was just what I wanted. Stupid, poor, poovey. 1969 J. Gardner *Founder Member* x. 160 'Get into those poufy drawers and..hurry.' Boysie pulled on the nylon briefs. 1970 *Guardian* 8 July 8/4 The material..makes fun of..old-fashioned 'pouffy' homosexuals. 1970 M. Tripp *Man without Friends* vi. 61 In our cockeyed civilisation it's regarded as poovey for a man to enjoy housework. 1972 D. Lees *Zodiac* 73 One hand raised..like a poofy traffic cop. 1976 J. O'Connor *Eleventh Commandment* i. 17 After being given the once over by a poofy-looking scout who gave me a pat on my arse, I was sworn in.

pooh, *sb.* Add: **2.** *slang.* Excrement, faeces. Also *transf.* and *fig.*; *in the pooh* = *in the shit* s.v. *shit *sb.* 1 d.

1960 Wentworth & Flexner *Dict. Amer. Slang* 401/1 *Poo*..., feces. 1961 'J. Danvers' *Living come First* x. 177 'You're rather in the pooh with the Adelaide police.' 'How much do I stink with them?' 1967 Partridge *Dict. Slang* Suppl. 1303/2 *Pooh*, anything smelly or disgusting, esp. *faeces*: Australian juvenile. 1970 R. Beilby *No Medals for Aphrodite* 229 If they catch you with her, then you're really in the pooh. 1975 X. Herbert *Poor Fellow my Country* 873 She'll put you in the poo if she writes anything 'bout you. 1976 J. McClure *Rogue Eagle* ii. 33 'But what..if someone..gave him the money and support he needed?' 'We might be right in the poo.'

Pooh Bah (pū bā). Also with small initials. [Name of a character in W. S. Gilbert's *Mikado*.] A person who holds a large number of offices at the same time. Also in extended use, a person or body with much influence or many functions; a self-important person. Also *attrib.* Hence **Poo-Ba·hism.**

1888 L. D. Powles *Land of Pink Pearl* 77 To the first of these 'vacancies' the Governor appointed an English jeweller, named Brown, to the second one of the local 'Pooh Bahs!' named Crawford. 1923 *Westm. Gaz.* 4 May 1 6 (*heading*) Pooh-Bah Role for Local Bodies. 1927 M. Terry *Through Land of Promise* 44 Although principally protectors of aboriginals, stock inspection, mining wardens' responsibilities and a host of other offices make them a collection of veritable 'Poo-Bahs'. 1949 F. Swinnerton *Doctor's Wife comes to Stay* 163, I do a bit of painting, myself: enough to take the boys in art. You see, we're so short-handed—' 'Good God! You must be the Pooh Bah of this school.' 1956 *Newsweek* 7 May 59/2 TV Pooh-Bahs expect this year's giveaway shows to reach a total hand-out of $8 million. 1962 S. E. Finer *Man on Horseback* xi. 157 In his Pooh-bah capacities as Prime Minister, Minister of War and Marine, Commander-in-Chief and Military Governor of Egypt, Neguib's power now rested on two pillars, the military and the civilian. 1964 C. Duffy *Wild Goose & Eagle* xii. 175 In a magnificent display of Pooh-Bahism Francis Stephen, as Emperor, supported strictness and vigilance, but in his capacity as Grand Duke of Tuscany protested against interference with vessels on their peaceful way from Tuscan ports to Genoa. 1972 W. A. Pantin *Oxford Life* v. 65 In the late eighteenth century a serious moral case..would now be dealt with by the Hebdomadal Board, that all-purpose, constitutional Pooh-Bah of the period. 1972 *Publishers' Weekly* 13 Nov. 16/1 His first book..was a tough, illuminating picture of what happens in the surgical wards of a metropolitan hospital, and it did not exactly endear him to the pooh-bahs of medicine.

pooh-pooh, *v.* Add: (Further examples.)

1926 [see *à la *phr.* c]. 1957 *Observer* 29 Sept. 13/4 It is one thing to pooh-pooh the final scene as melodrama.. and quite another thing to remain detached as Salome lies, on your own fireside, intoxicated with passion. 1962 *Hovering Craft & Hydrofoil* Nov. 20/2 A few years ago most of us would have pooh-poohed the idea of a modern version of Jules Verne's atom-powered 'Nautilus', which has since become a reality several times over. 1971 *Petticoat* 24 July 39/1 If he refuses, or pooh-poohs your concern, go to a new family doctor, and try again. 1977 *New Yorker* 27 June 30/3 My companion pooh-poohed the mishap and bade me choose an apéritif.

Hence **pooh-poohing** *vbl. sb.* (further examples), **pooh-poohe·r** (earlier and later examples); **pooh-poohingly** *adv.*, in a dismissive or contemptuous manner. **pooh-poohy** *a.* (*nonce*), inclined to pooh-pooh.

1841 Dickens *Let.* 2 Apr. (1969) II. 249 The pooh-poohers and Lord Burleighs have it hollow, all the world over. 1858 H. Derby *Diary* 21 Oct. in C. L. Graves *Hubert Parry* (1926) I. 169 Before the performance I met Otto Goldschmidt, and he was rather pooh-poohy about it. 1898 W. James *Coll. Ess. & Rev.* (1920) 423 Concerning this question, at any rate, the positivists and pooh-poohers of metaphysics are in the wrong. 1911 G. B. Shaw *Doctor's Dilemma* p. lvii, The moment his practice is touched down to its source in human passion there is a great and quite sincere poohpoohing..from the mass of the public. 1939 Joyce *Finnegans Wake* iii. 498 The poohpoohor old booslooster, with his arthurious claycoves... bunted to the world at large. 1956 Poohpoohingly [see *Borrovian *sb.* and *a.*]. 1959 *Economist* 7 Feb. 490/1 For all his pooh-poohing of 'sentimentalism', he admitted that he never heard 'God Save the Queen'..without feeling tears in his eyes.

Pooh-sticks, pooh-sticks (pū·stiks). [f. the name of Winnie-the-*Pooh*, a character created by A. A. Milne + Stick *sb.*[1]] A game in which sticks are thrown over one side of a bridge into a stream and the first to emerge on the other side wins.

1928 A. A. Milne *House at Pooh Corner* vi. 94 And that was the beginning of the game called Poohsticks, which Pooh invented, and which he and his friends used to play on the edge of the Forest. But they played with sticks instead of fir-cones, because they were easier to mark. 1972 *Times* 16 May 10/4 He spends most of his time playing pooh-sticks. 1974 C. Milne *Enchanted Places* i. 15 The bridge where they had played Pooh-sticks was a real bridge, looking just like the drawing in the book. *Ibid.* viii. 58 We used to stand on Pooh-sticks Bridge throwing sticks into the water and watching them float away.. until they re-emerged on the other side. 1979 *Guardian* 16 May 1/1 The horde of visitors who..chuck[ed] a twig or two into the sun-dappled stream in honour of the bear of very little brain who invented Poohsticks.

poojah, puja. Add: Also **pujah**; 7 **poujah, pudgiah.** (Earlier and later examples.) Also **puja pantai** (pantəi·) [Malay *pantai*, = beach, seashore] (see quot. 1965).

1681 R. Knox *Ceylon* iii. iv. 80 In this Poujah or Sacrifice the King seems to take delight. *Ibid.* v. 85 They reckon the chief poynts of goodness to consist in giving to the Priests, in making Pudgiahs, sacrifices to their Gods, in forbearing shedding the blood of any creature. 1800 S. Turner *Embassy Tibet* iv. 243 It was only the Gylongs at their *pooja*, or religious exercises. 1893 Kipling *Day's Work* (1898) 10 In London I did poojah to the big temple by the river for the sake of the God within. 1909 M. Diver *Candles in Wind* i. i. 14 Is it permitted that we kill a goat and make *poojah*? 1913 E. M. Forster *Let.* 6 Mar. in *Hill of Devi* (1953) 28 The Rajah is doing Pujah after his bath. 1936 J. Nehru *Autobiogr.* 8 The women of the family indulged in various ceremonies and *pujas*. 1951 *Chambers's Jrnl.* Oct. 611/2 The main religious ceremonies or 'poojas'..take place in the temple sanctum, and..eventually terminate with the bath and the fire-walk. 1951 *Jrnl. Malayan Branch R. Asiatic Soc.* XXIV. iii. 33 The *pelas negri* ceremony of Perak was, in Malaya, known by other names too:..in Kelantan, and in Province Wellesley, for fishermen, the ceremony is known as *puja pantai*. 1965 C. Shuttleworth *Malayan Safari* vi. 77 A rare and colourful ceremony known as the *Puja Pantai*, it is performed by Malay fishermen when they believe that the sea gods are angry with them and need to be propitiated. 1968 *Jrnl. Music Acad. Madras* XXXIX. 59 Every home has its family idol and the pujas are performed according to the..16 types of ceremonies. 1969 *Cultural News from India* Nov. 20 The heavy and intermittent rains which, a few days ago, threatened to dampen the Puja spirit are now over. *Ibid.*, Calcutta, or for that matter, entire Bengal is in the grip of the Puja celebrations. 1971 *Northern India Patrika* 1 Feb. 1/1 Sub-inspector Makhan Lal Dutta..was attacked..near a Saraswat puja pandal. 1971 *Illustr. Weekly India* 18 Apr. 29/3 (*caption*) Gujarati women sway and swoon at a *puja*. 1972 M. Sheppard *Taman Indera* 197 The general name which incorporated all the ceremonies and offerings on such an occasion [sc. a Malay folk festival] was *Puja Pantai*. 1977 W. H. S. Smith *Young Man's Country* ii. 36 The Puja holiday was the biggest Hindu festival in the Bengali year.

poojari, var. *pujari.

pool, *sb.*[1] Add: **1. f.** = *oil pool* s.v. *oil *sb.*[1] 6 e.

1902 *Bull. U.S. Geol. Survey* No. 198. 23 North and northeast of the Snyder pool six wells have been sunk. 1976 M. Machlin *Pipeline* ii. 32 It tests over two thousand barrels a day—and God knows how much gas. I don't know how big the pool is.

g. A swimming pool.

1921 A. Huxley *Crome Yellow* iv. 33 That part of the garden that sloped down from the foot of the terrace to the pool. 1941 B. Schulberg *What makes Sammy Run?* iv. 61 Collier was urging him to come early and try the pool. 1961 J. S. Salak *Dict. Amer. Sports* 337 It is recommended that pools for championship meets should be at least 75 feet in length and 42 feet in width. 1974 R. Thomas *Porkchoppers* v. 35 He..lived..in a house with a pool, two Russian wolfhounds, and his wife.

2. *the Pool*, (*a*) (later examples); (*b*) Liverpool.

1966 D. Francis *Flying Finish* (1968) ii. 11 A small wharf down in the Pool. 1978 K. Bonfiglioli *All Tea in China* ix. 88 We could drop down-river to the Pool.. without feeling a pilot. 1963 *Austral. T.V. Times* 18 Apr. 10/4 The pool, the port of Liverpool. 1969 R. Busby *Robbery Blue* xxii. 151 I'd reckon on Liverpool... I'd head for the 'Pool, get myself swallowed up in a big city. 1978 F. Driscoll *Wilby Conspiracy* (1973) ii. 29 His origins: a street of back-to-backs..off the Scotland Road, the toughest part of the Pool to grow up in. 1975 *Times Lit. Suppl.* 31 Jan. 100/1 A football team. 1975 Aug. i 1 (Advt.), At Alf we are darrevry Scouse Big'ead's broad special for 'Pool, like.

3. *pool-side* (later examples); *pool-clear* adj.; **pool cathode** *Electronics*, a cathode consisting of a pool of mercury used in certain types of discharge tube; **pool house, poolhouse** chiefly *U.S.*, (*a*) a house by a swimming pool, for the use of bathers; (*b*) a building with a swimming pool in it; **pool party,** a party at which the guests bathe; **pool room,** a room with a pool in it.

1934 *Electrical Engineering* (N.Y.) Jan. 75/2 If a rectifier having a single anode and mercury pool cathode within a separate small tank were to be built, no continuous back current could flow to the anode which is a ... 1966 *McGraw-Hill Encycl. Sci. & Technol.* VIII. 236/2 Tubes with a pool cathode have a higher current capacity and longer life than the hot-cathode tubes because of the indestructible nature of the mercury pool cathode... Pool-cathode mercury-arc tubes are widely used for medium and high-power applications in welding and rectifier service. 1924 E. Sitwell *Sleeping Beauty* xiv. 51 Pierced through the pool-clear heart. 1957 P. Quentin *Suspicious Circumstances* ii. 24 He was way off, down by the pool... They'd eaten at the poolhouse. 1975 A. Bergman *Hollywood & the Vine* (1976) ix. 123 The poolhouse had showers, marked 'Fillies' and 'Stallions'. 1978 R. Moore *Big Paddle* (1979) xxix. 269 'Seems to me I remember a pool house around here somewhere.' 'You're planning on taking a swim, sir?' 1973 *Ottawa Jrnl.* 16 July 27/5 Arch, if Veronica's having a pool party, why don't I wear a bathing suit? 1925 F. Scott Fitzgerald *Great Gatsby* (1926) v. 110 Through dressing-rooms and poolrooms, and bathrooms with sunken baths. 1968 J. Sangster *Touchfeather* xiv. 151 Where's the pool room? 1921 W. de la Mare *Veil* 6 Wan glow-worms greened the pool-side grass. 1963 *New Yorker* 22 June 109 Poolside buffet luncheon daily. 1968 *Globe & Mail* (Toronto) 17 Feb. 31 Sitting by the poolside at the Trinidad Hilton. 1970 P. Zelver *Honey Bunch* (1971) vii. 34 Trays of hot hors d'œuvres..which she brought to the poolside. 1973 H. Nielsen *Severed Key* iv. 39 Zachariah O'Hara..sat at a poolside table.

pool, *sb.*[3] Add: **3.** *shoot pool*: see *shoot *v.*

b. Colloq. phr. *to play* etc. *dirty pool*: to use unfair tactics; to be dishonest. *N. Amer.*

1951 H. Wouk *Caine Mutiny* xxxvii. 445, I played pretty dirty pool, you know, in court. 1973 *Maclean's Mag.* Feb. 85/2 'You use as much dirty pool as possible,' says an alumnus cheerfully. 1976 *Times Lit. Suppl.* 19 Mar. 326/4 It is, of course, a combination of petty carping and dirty pool to demand of the author of a scholarly work information that does not fall within the limits he has meticulously drawn up for his work.

4. b. (Earlier and later examples.) Also, *auction pool*, the total sum realized when the names of horses in a race, or likely winners in other contests, are sold by auction to those who wish to hold them; *to scoop the pool*: see *scoop *v.*[1] 5 a.

1868 *N.Y. Herald* 3 July 10/1 Let us take a glance at the pool stand before the races begin. 1874 'Mark Twain' *Sk. New & Old* 310 No pools permitted on the run of the comet—no gambling of any kind. 1913 A. Bennett *Regent* ii. x. 311 The *Lithuania* was lagging... Every day, in the auction-pool on the ship's run, it was the holder of the lower field that pocketed the money. 1928 *Daily Mail* 7 Aug. 12/5 Stewards are stationed at different points to give weary travellers a welcome lift and prevent them from getting off the beaten track and missing the auction pool. 1949 *Radio Times* 15 July 6/1 Wilfred acts as auctioneer on board the 'Queen Mary' during the pool on the ship's daily run. 1955 *Times* 30 Aug. 5/1 Under the Act, the balance-sheets, to be deposited with the local authority, must show the aggregate total stakes in all pools,.. or, at the option of the promoter, the percentage of the total stakes. 1973 *Irish Times* 2 Mar. 2/6 The unexpected victory of Game Sauce in the second division scuttled jackpot hunters and the pool of £1,027 goes forward to Naas tomorrow.

c. = *football pool. Usu. *pl.*

1938 *Mass-Observation: First Year's Work 1937–38* iv. 39 The Pools provide an outlet for personal frustration, ambition and faith. 1947 [see *banker[2] 5]. 1948 M. Allingham *More Work for Undertaker* xiv. 174, I wouldn't have had this happen, not for a thirty-thousand win in the pools. 1957 *London Mag.* May 48 They see themselves being eaten alive by this ignorant creature, with his telly and his pools, swallowing up all culture. 1958 *Listener* 28 Aug. 308/1 He is telling us about the important things about the working class, how they feel about pools and pubs as well as about socialism, trade unionism, and religion. 1966 A. E. Lindop *I start Counting* xxi. 266 I'm saving up to buy her a big book on birds... We had a nuthatch last Friday, and you'd think she'd won the pools. 1974 A. Fowles *Pastime* iv. 39 He sat at the main desk doing his pools... He'd never won a sausage. 1979 *Times* 20 Dec. 15/6 Nearly 60 per cent of

people in the country as a whole replied to the question, 'How often do you do the pools?' with 'Never'.

5. b. A common reservoir of commodities, resources, etc. Cf. *gene pool* s.v. *gene 2.

1917 'Contact' *Airman's Outings* 127 Before they join a squadron pilots fresh from their instruction in England gain experience on service machines belonging to the 'pool' at Saint Gregoire. 1940 *Times* (Weekly ed.) 7 June 15 A rice pool has been formed from August 1 through which all rice imported into Singapore must pass. The pool will be used as a means for turning over the emergency stocks of rice which the Government have now acquired. 1943 J. S. Huxley *Evolutionary Ethics* vi. 46 The more individuals there exist whose desirable potentialities are fully developed, the more health, vigour, knowledge, wisdom, happiness, beauty and the rest can go into the common pool, and the better that common pool will work. 1946 S. Spender *European Witness* i. 11 People like myself could draw on an alleged 'pool' of cars to take them on journeys. 1958 *Observer* 15 June 13/7 Each animal and plant population draws on a large pool of genes. 1963 *Higher Educ.: Rep. Comm. under Ld. Robbins 1961–3* 53 in *Parl. Papers 1962–3* (Cmnd. 2154) XI. 639 The increase has been almost as great among the children of professional parents, where the pool of ability might have been thought more nearly exhausted. 1967 E. Short *Embroidery & Fabric Collage* i. 28 A useful 'pool' of ideas will soon be built up to be referred to when needed.

c. A group of persons any one of whose abilities or services may be drawn upon or who share duties; spec. *typing, typists'*, etc., *pool*: a number of typists in an organization, department, etc., among whom work is distributed. Hence also, the office or building where such typists work.

1928 I. Curtis in *Schools of England* xvii. 333 The staff as a whole is organized as a 'pool' for various miscellaneous duties, such as examination work, the preparation of the tutorial courses. *Ibid.* 334 All the examination work which falls to the Education Board,..is carried out by the pool as part of the regular work. 1937 V. Bartlett *This is my Life* x. 150 The typists [of the League of Nations] were..relegated to a 'pool' at the top of the building. 1942 N. Balchin *Darkness falls from Air* vi. 104 How many people are there in your typing pool? 1944 'N. Shute' *Pastoral* iv. 75 Chap with a face like a burglar—came in with the last lot from the pool. 1949 *Manch. Guardian Weekly* 20 Jan. 3 A secretarial 'pool'. 1958 G. Greene *Our Man in Havana* 58 The secretaries' pool should have been informed. *Ibid.* v. vi. 260 I'll try to stay in the typists' pool. 1959 *Times* 3 Sept. 14/1 A married woman, aged 23, who has settled down in the audio-typing pool after only a few years as a secretary, confessed that she hated it at first but soon changed her mind. 1960 *Guardian* 11 July 3/3 The Derby force opens a new typing pool where the reports..will be typed. 1960 M. Spark *Ballad of Peckham Rye* iii. 42 That Miss Coverdale in the pool..is working Dixie to death. 1970 G. Greer *Female Eunuch* 126 She can command better money and have time off as well if she would only walk out of her typing pool. 1972 K. Benton *Spy in Chancery* iv. 32 Diana's a competent girl, and she doesn't chatter about her work to the typists' pool.

d. = *pool petrol.

1940 M. Nicholson *How Britain's Resources are Mobilized* 4 In war it [petrol] all goes round in grey tankers and is all called 'Pool'. 1944 *Amer. Speech* XIX. 294 The word 'pool', printed on a strip of paper pasted on a gasoline pump, announces that the pump contains an unidentified brand of gasoline from the nation's pooled supplies, instead of the brand advertised on the pump. 1952 *Economist* 6 Sept. 581 Early in the war all petrol was of uniform quality, the Pool.

e. In a sporting tournament, a group the members of which play against each other to decide which of the competitors or teams qualify for the next round; a minor league.

1955 *Times* 26 May 12/5 His many successes in the matches and 'pools' inaugurated at that time. 1972 *Sunday Tel.* 30 Apr. 34/7 The team flies to Groningen tomorrow, drawn in a tough pool with Poland, Spain and Hungary.

f. A register of free-lance dockworkers seeking employment.

1958 *Engineering* 14 Mar. 329/2 Members of the National Dock Labour Board's 'pool' at the..group of docks decided to ban overtime in order to spread the work available among the registered workers. 1964 O. E. Middleton in C. K. Stead *N.Z. Short Stories* (1966) 194 The usual round and the same stale answers on every ship: 'All hands are hired through the Pool. Are you established members of the Pool?' 1972 *Guardian* 17 Aug. 1/4, 2,000 [dockers] are going to be put out in Liverpool, and there are 1,200 in the Royal Group on the 'pool' (the temporary, unattached register which is to be abolished under the new proposals).

g. *Biochem.* A quantity of one or more metabolites in some definite part or tissue of the body which is continually being diminished and replenished by cellular activity.

1961 in Webster. 1962 *Bacteriol. Rev.* XXVI. 292/1 Bacteria maintain internally synthesized small molecules at high internal concentrations and in addition have the capacity to concentrate many compounds from the environment. Since the majority of these compounds are intermediates in synthesis, they are collectively termed the pool of metabolic intermediates or, simply, the 'pool'. 1968 *Brit. Med. Bull.* XXIV. 249/2 Fowle, Matthews and Campbell (1964)..put forward a model which separated the body CO_2 into intracellular and extracellular pools. 1971 *Nature* 30 July 329/1 Thus shorter chain acids.. would have been produced by partial degradation of palmitic acid, with concomitant dilution of label by the cellular pool of palmitic acid.

h. A small number of reporters who have access to news sources and pass information to other journalists.

1967 R. J. Serling *President's Plane is Missing* (1968) vi. 102 Call a press conference and lay it on the line. They can choose themselves between a permanent pool or a one-shot visit. 1973 *Washington Post* 13 Jan. c 2/3 Instead of limiting coverage to a selected 'pool' of a few reporters, all accredited reporters and photographers—more than 100—were allowed to attend.

8. a. (sense 3) *pool hall, joint, parlour, -shooter, -shooting*; (sense 4 b) *pool box, -seller* (earlier examples), *-selling* (earlier examples); (sense *4 c) *pool-betting, coupon, promoter*; (sense *5 b) *pool currency, driver, product, service, transport*; (sense *5 c) *pool typist*; (sense *5 f) *pool office*; (sense *5 h) *pool group, policy, reporter, representative.*

1955 T. H. Pear *Eng. Social Differences* xiii. 293 It is easy to understand why the popularity of all-in wrestling and speedway racing has not spread 'upwards', while that of pool-betting has. 1957 *Encycl. Brit.* IX. 998/2 In Great Britain the Pool Betting act of 1954 set certain requirements for the conduct of pool betting, including the registration of operators and inspection and publication of financial details. 1878 M. Long *Life Mason Long* vi. 102 The field won, and after the race I drew six hundred and twenty-five dollars from the pool-box. 1902 A. D. McFaul *Ike Glidden* 171 The vehement cheers of those about the pool box seemed more deafening as the race progressed. 1951 'M. Innes' *Operation Pax* ii. iv. 63 A few brought pool coupons from their pockets and studied them. 1955 *Times* 15 Aug. 6/4 The first Brazilian auction of 'pool' currency will be tendered for on Thursday, the statement continues, and the currency thereby made available,.. will be available for the three participating countries. 1973 A. Mann *Tiara* x. 91 Four or five are in the motor pool for the use of Vatican people on official journeys, but they would have pool drivers. 1970 *Globe & Mail* (Toronto) 26 Sept. 10/5 The patients spoke mostly in Arabic to an Arab journalist in a pool group of reporters who were taken to the hospital by the Jordanian army. 1928 *Collier's* 29 Dec. 43/2 He entered a pool-hall speak-easy. 1944 J. S. Pennell *Hist. Rome Hanks* 63, I heard a young pool-hall lounger standing idly in Pawnee street refer to him as Old Man Beckham. 1951 [see *joke book]. 1973 *Black World* May 77/1 Their penchant for open debate in the chop bars..reappears in our barber shops and pool halls. 1975 *Listener* 18 Dec. 834/3 An endless succession of dreary bars, pool-halls and discos. 1930 H. Zink *City Bosses in U.S.* 137 Money paid by saloons, gambling and pool joints, and houses of the underworld. 1964 O. E. Middleton in C. K. Stead *N.Z. Short Stories* (1966) 194 Why do they think we are tramping fifteen miles of dockside, when the other way, you have only to show your papers at the Pool office, and wait for a ship? 1912 Pool parlour [see *cabaret[1] 2 b]. 1932 J. Dos Passos *1919* 413 Izzy had gotten to loafing in poolparlours. 1973 *Washington Post* 13 Jan. c 2/3 The new 'pool' policies under which Washington Post reporter Dorothy McCardle was excluded from five previous social events involving President and Mrs. Nixon. 1939 *New Statesman* 18 Nov. 740 The only mystery about the Petroleum Board (the wartime name for the powerful trade committee which fixes distributors' quotas and agrees prices) is that it is unable to get permission from the Government to charge what prices it likes for the imports of 'pool' products. 1940 Harrison & Madge *War begins at Home* x. 279 Despite all the efforts of Pool-promoters, postal facilities were denied, and the vast Pool vested interests were closed down. 1967 Mrs. L. B. Johnson *White House Diary* 14 Mar. (1970) 496 There were John Gardner and Liz and I and a pool reporter..in the tiny room. 1972 *Guardian* 9 June 13/8 The pool reporter..is the duty man who attends the senator in his intimate and most insignificant moments. Clearly five hundred pressmen..cannot accompany the senator into a small shirt shop: the pool reporter is their eyes and ears. 1974 *Times* 17 Apr. 14/7 The briefing was open to all Israel newspapers..but the Foreign Press Association was told it must choose one pool representative. 1887 *Advance* (Chicago) 13 Oct. 6/1 No less than 15 poolsellers were in the grand stand. 1888 *Outing* May 118/1 John Hatfield is a bookmaker and pool-seller in St. Louis. 1869 J. H. Browne *Great Metropolis* 573 Pool selling is managed in this way. 1872 *Alabamian & Times* (Tuscumbia, Alabama) 26 Sept. 4/2 Pool selling is lively and fine sport is anticipated. 1964 *B.E.A. Advance Timetable* (Summer), Pool Services from London: Flights to Denmark, Norway and Sweden are in co-operation with S.A.S...to Prague with C.S.A. and to Warsaw with LOT. 1961 *John o' London's* 2 Nov. 495/1 A professional pool-shooter. 1974 *Greenville* (S. Carolina) *News-Piedmont* 20 Apr. 3/2 Rudolph Wanderone, known internationally as Minnesota Fats, king of the pool-shooting hustlers, was reported in serious condition Friday after undergoing emergency surgery. 1868 Pool stand [see sense 4 b above]. 1945 *News Chron.* 7 June 4/3 The bicycles which are being used in these country trips are needed to provide pool transport for mine workers and farm labourers. 1942 N. Balchin *Darkness falls from Air* ix. 121 Pearce rang up with a long tale of woe about being very short of pool typists. 1979 G. Hammond *Dead Game* x. 140 McLure got one of the pool typists put onto it.

b. Pl. (sense *4 c), as *pools coupon, entry, investor, panel, win, winner.*

1951 A. Baron *Rosie Hogarth* v. iv. 337 Children will be educated.., not just to be office boys and fill in the pools coupon. 1978 J. Galway *Autobiogr.* ii. 20 The set.. became the whole focal point of his life every Saturday afternoon when it was time to check the football results for his pools coupon. 1972 A. Draper *Death Penalty* i. 5 The copy coupon of his pools entry. 1958 Pools investor [see *banker[2] 5]. 1976 *Daily Record* (Glasgow) 4 Dec. 30/2 And with the arctic weather spreading south, the Pools Panel have been put on active stand-by. 1963 *Times* 14 Feb. 15/3 An innocent from academe, made footloose by a pools win, incautiously agrees to become guest dramatic critic to the *Evening Gazette*. 1977 J. Wainwright *Pool of Tears* 203 You know about the pools win. 1960 I. Jefferies *Dignity & Purity* xii. 187, I suddenly felt almost like a pools winner. 1973 *Guardian* 1 Mar. 28/5 The biggest pools winner in history..came to London yesterday to receive his cheque for £542,252—won from a £1 stake.

c. Special Combs., as **pool butter,** the butter of uniform quality available in Britain during the war of 1939–45; **pool car,** (*a*) a freight wagon shared by a number of hirers; (*b*) a car available for the use of a number of drivers; **pool-drive** *v.*, to share vehicles and driving duties on a regular journey; **pool petrol,** the unbranded petrol which was the only grade available in Britain during and just after the 1939–45 war; **pool room,** (*a*) room with billiard tables where pool can be played, usu. for a fee; (*b*) a room where a betting pool is held; a betting shop; **pool shark** *U.S. colloq.*, an expert pool player; one who makes money by winning at pool; **pool train** *Canada*, a train run jointly by more than one railway company; also **pool passenger train.**

1940 Pool butter [see *pool petrol* below]. 1926 *Daily Colonist* (Victoria, B.C.) 18 July 3/5 Victoria Baggage Company. Furniture Moved, Crated and Shipped. Pool Cars for Prairies and All Points East. 1967 *Lebende Sprachen* XII. 186/1 Pool car..= *Sammelwagen*. 1973 A. Mann *Tiara* xiii. 118 There are five pool cars. 1974 *Globe & Mail* (Toronto) 18 Mar. s6/1 Most go back and forth from Oshawa—working at General Motors is a way of life. They pool-drive. 1959 *Kingston* (Ontario) *Whig-Standard* 28 Sept. 1/5 His death was the fifth caused by the collision in which a CNR freight ripped open the side of a dining car on a pool passenger train being shunted in the yards. 1940 *Weekly Chron.* (Newcastle) 23 Mar., The war has given us special meanings of a number of words already in use—*pool* (for *pool-butter*, *-petrol* etc.). 1949 *Punch* 28 Sept. p. iv, When the days of 'Pool petrol only' are over. 1968 B. Foster *Changing Eng. Lang.* iii. 123 Wartime conditions frequently give a new twist to an existing term, e.g. 'austerity, pool (as in 'pool petrol', the unbranded petrol of wartime days), [etc.].' 1861 T. Hughes *Tom Brown at Oxf.* II. iii. 51 He could go and smoke a cigar in the pool room. 1887, 1892 [in Dict., sense 8]. 1931 D. Runyon in *Collier's* 26 Sept. 57/2 We get the race results by phone off a pool room downtown as fast as they come off. 1944 Auden *For Time Being* (1945) 50 Back to the upland mill town.. with its grope-movie and its pool-room lit by gas. 1959 N. Mailer *Advts. for Myself* (1961) 74 The poolroom itself was down in the cellar. 1973 *Black World* June 79/1 This style..emanates from the urban street world of pimps, prostitutes, poolroom sharks. 1978 *Detroit Free Press* 5 Mar. B 1/1 My best friend Antoine..and I were hanging around O'Quinn's poolroom. 1908 *Busy Man's Mag.* Mar. 128 The Pool Shark. Bide Dudley... Blue Book. 1944 W. Russell in *Needle* July 21/2 On the Gulf Coast they'd call him a pool shark and gambler. 1971 J. H. Gray *Red Lights on Prairies* i. 13 Sometimes they [sc. prostitutes] travelled from town to town with pool sharks. 1965 *Globe & Mail* (Toronto) 10 Nov. 19/8 At one time the pool trains ran over the CNR to Brockville and then by CPR tracks to Ottawa... As at the end of October, the pool trains ceased to exist and the two railways went on their separate ways.

pool, *v.*[1] Add: **1.** (Examples of revived use.) Also *transf.*

1973 D. Andersen *Ways Harsh & Wild* iv. 107 Behind our cabin there was a meadow that gradually pooled with water. 1977 *Rolling Stone* 13 Jan. 38/3 An afternoon sun pools warmly on the hardwood floor in the rambling frame house. 1978 C. Tomlinson *Shaft* 20 The brook.. entered the garden, pooling.

3. *intr.* Of blood: to accumulate in parts of the venous system, e.g. as a result of the forces produced by continuous acceleration.

1933 *Jrnl. R. Aeronaut. Soc.* XXXVII. 398 Fig. 6 shows a sketch of a device which I suggest might serve to counteract the tendency of centrifugal force to make the blood leave the head and pool in the thin-walled abdominal vessels. 1945 [see *anti-gravity]. 1962 [see *G-suit, g-suit]. 1962 F. I. Ordway et al. *Basic Astronautics* xii. 463 The blood..increases in weight to the point where the heart can no longer pump it, and it pools in the extremities of the body. 1973 Towler & Butler-Manuel *Mod. Obstetr.* xvii. 469 Here the blood 'pools' in the large muscles and is effectively lost to the circulation.

pool, *v.*[2] Add: (Further examples.) Also *transf.*

1921 G. B. Shaw *Back to Methuselah* p. lxxvii, What we should do, then, is to pool our legends and make a delightful stock of religious folk-lore on an honest basis for all mankind. 1926 *Amer. Mercury* Dec. 462/2 Sime countered by pooling his stories with the other fellow. 1927 E. Thompson *These Men, thy Friends* 245 Hart and Kenrick pooled friends. 1940 *Hutchinson's Pict. Hist. War* 14 Feb.–4 Sept. 68 Petrol was pooled, buildings were seized, children were sent to safety. 1955 *Times* 16 Aug. 9/3 The sterling, Deutschmarks, and guilders which the Brazilian exchange control authorities secure from the proceeds of Brazilian exports will be 'pooled', and when in future they are 'auctioned' importers who buy them will be free to use them for imports from any one of the three countries. 1978 S. Brill *Teamsters* vi. 220 In a rare display of cooperation,..the IRS and FBI agreed to pool their efforts. 1979 *Canad. Jrnl. Linguistics* XXIV. i. 35 Unfortunately, Brown pooled the data, so that if any

developmental trends were present, they were obscured in the analysis.

2. *Austral. slang.* To implicate; involve a person against his will; inform on.

1919 W. H. Downing *Digger Dial.* 39 *Pool*, to involve; cast blame or a burden on. 1928 A. Wright *Good Recovery* 117 Leave the sheilas alone; they're sure to pool a man sooner or later. 1932 W. Hatfield *Ginger Murdoch* 282 To rig that evidence against him—pool him. 1942 L. Mann *Go-Getter* 313, 'I got pooled into it,' he explained. 1967 K. Tennant *Tell Morning This* 85 A man thought he'd do the decent thing and tide a girl over a patch of trouble, and she pools him every time. You can't prove it isn't your kid.

pooled *ppl. a.* (further examples); **pooling** *vbl. sb.* (further examples).

1928 *Manch. Guardian Weekly* 10 Aug. 104/2 Washington, too, is shown by the dispatches to suspect something like an Anglo-French alliance or merger or pooling behind the text of the naval compromise. 1936 *Discovery* Aug. 232/1 It was to be hoped that such pooling of knowledge might become world-wide. 1943 J. S. Huxley *Evolutionary Ethics* 39 Part of the blind struggle for existence between separate individuals or groups is transposed into conflict in consciousness..within the tradition which is the vehicle of pooled social consciousness. *Ibid.* 45 The pooling of experience and co-operative action in a cumulative tradition. 1946 J. W. Day *Harvest Adventure* xvi. 278 As to the waste of 'pooled' machinery, this is a sore point in many counties. 1951 *N.Y. Times* 21 Aug. 1/6 A pooled dispatch said the services had 'an almost carnival atmosphere'. 1955 *Bull. Atomic Sci.* Apr. 146/1 In science, only a pooling of all new ideas can lead to progress. 1967 *Oceanogr. & Marine Biol.* V. 162 When animals are small, biochemical study of a pooled sample from many specimens is imperative. 1971 *Jrnl. Gen. Psychol.* LXXXV. 111 The baseline performance of the pooled controls. 1975 *Language for Life* (Dept. Educ. & Sci.) iii. 40 It is an essential principle of the item pooling system that assessment can reflect changes in the use of language and stylistic differences over the years.

Poole (pūl). The name of a town in Dorset, used *attrib.* to designate a type of clay suitable for pottery found near there, or pottery manufactured there.

1878 *Encycl. Brit.* VIII. 229/2 Of importance next to it [sc. kaolin], as potter's material, is the 'Poole clay' of Dorsetshire. 1924 *Design in Mod. Industry: Year-Bk. of Design & Industries Assoc. 1923–24* 51 (*caption*) The painted dish in the upper illustration is a piece of Poole pottery, and shows how essentially decorative simple colour groupings may appear on surfaces uncomplicated by needless corrugations. 1938 *Decorative Art* p. x, Poole pottery... If there is a mantleshelf..it is a narrow one, but there are Poole Pots to meet the difficulty. 1957 [see *Dorset]. 1971 R. Rendell *No More Dying Then* i. 11 There were flowers..in the Poole pottery vases.

pooled (pūld), *a. rare.* [f. Pool *sb.*[1] + -ed[2].] Furnished with or placed in a pool. Also *fig.*

1947 Dylan Thomas *Let.* 20 May (1966) 307 The pooled ponded..garden. 1967 F. Warner *Madrigals* 10 You stand Pooled in discarded clothes.

pooly (pū·li), *sb. rare.* [f. Pool *sb.*[1] + -y[6].] A small pool, *spec.* of urine.

1922 Joyce *Ulysses* 529 It's as limp as a boy of six's doing his pooly behind a cart.

poon (pūn), *sb.*[2] *slang* (chiefly *Austral.*). [Origin obscure.] **a.** A simple or foolish person. **b.** A person living alone in the outback.

1940 M. Marples *Public School Slang* 60 Another considerable group of words in recent use has a definite transatlantic flavour, as, for example: boob..goof..mutt.. poon (Dulwich, 1930+). 1941 Baker *Dict. Austral. Slang* 56 *Poon*, a lonely, somewhat crazy dweller in the Outer Beyond... A simpleton or fool. 1945 —— *Austral. Lang.* v. 97 Another outback term for a person who lives alone is *poon*. 1972 G. Morley *Jockey rides Honest Race* 73 They don't look for the guts of a lecture; just the mistakes. Then they can get up and shoot their mouths off and everybody else nods wisely and tries to pick up the mistakes of the poon that's just said his piece. 1974 D. Williamson *Jugglers Three* in *Three Plays* 69 What possessed Keren to shack up with a poon like you?

poon, *sb.*[3] *slang.* Abbrev. of *poontang *sb.*

1969 J. Leasor *They don't make them like that any More* vi. 192 It's against my principles to pay for poon: if I can't get it for what is laughingly called love, then I'll do without. 1972 J. Wambaugh *Blue Knight* (1973) i. 16 Watching all that young poon.

poon (pūn), *v. Austral. slang.* [Origin obscure.] To dress *up*; esp. to dress flashily. Also in pa. pple. *pooned up.*

1943 Baker *Dict. Austral. Slang* (ed. 3) 61 *Poon up*, to dress up, especially in flashy fashion. 1945 —— *Austral. Lang.* 206 School slang... *poon up*, to dress up, especially with considerable care. 1951 D. Stivens *Jimmy Brockett* 48 Some of 'em were young lairs, all pooned up to kill. 1972 A. Chipper *Aussie Swearer's Guide* 48 *Pooned up*, dressed to impress, often with sexual success in view.

poonac. Add: (Example.)

1927 *Trop. Agriculturalist* LXVIII. 279 As regards Phosphorus, there are many food-stuffs, available in Ceylon, rich in this element. For example, the various poonacs and pollard.

Poonah. Add: Also **Poona** (now the usual form). Also as *adj.*, with allusion to the attitudes, way of life, etc., held to be characteristic of the Army officers stationed there during British rule.

1939 G. Treast in *Best One-Act Plays 1938* 63 Major Manners is discovered playing patience... He is an elderly, heavily jovial man, who, in spite of twenty years' Isolation, remains essentially pukka and Poona. 1944 'N. Shute' *Pastoral* vi. 138 They're county people, all frightfully toffee-nosed and Poona. 1973 'B. Mather' *Snowline* iv. 49 Blimey! Class consciousness rearing its ugly head... How Poonah-Poonah can you get? 1975 A. Christie *Curtain* i. 7 One of your so British old Colonels—very 'old school tie' and 'Poona'.

poonask, var. *ponask *v.*

poonghie, phoongyee, phungyi. Add: Also **pongyi.** (Further examples.)

1929 F. T. Jesse *Lacquer Lady* i. v. 36 The King had allowed it [sc. the house] to be built with a triple roof, a thing usually only permitted to Princes and poongyis. *Ibid.* ii. i. 118 After the litter came sixty-five poongyis..; they walked with downcast eyes, holding their fans up so that they should not by any chance catch sight of a woman. 1930 *Aberdeen Press & Jrnl.* 29 May 7/5 Last night a pongyi (Burmese priest) attempted to stab a military policeman. 1951 'N. Shute' *Round Bend* 122 He has been a Buddhist monk, a pongyi we call them, for over thirty years. 1966 D. Forbes *Heart of Malaya* xi. 129 Miss Khan..took me across the lane to the Burmese *wat* and kowtowed to the *pongyi* in the temple hall.

poontang (pū·ntæŋ), *sb. U.S. slang.* Also *poon tang* and with capital initial(s). [Prob. ad. Fr. *putain* prostitute.] Sexual intercourse, sex; women collectively, or a woman, regarded as a means of sexual gratification. Also *attrib.* Hence as *v. intr.*, to copulate.

1929 T. Wolfe *Look Homeward, Angel* 343 A fellow's got to have a little Poon Tang. 1947 C. Willingham *End as Man* ii. vii. 78 Poley looked out the window and saw a pretty Negro girl on the sidewalk... 'Eye that poon tang there,' he said. 1959 R. Condon *Manchurian Candidate* ii. 21 Every now and then I think about you coming all the way to Korea from New Jersey to get your first piece of poontang. 1966 C. Hines *Heat's On* xv. 122 That ain't our racket. We just sells poontang here. 1968 E. J. Gaines *Bloodline* 144 Yesterday this time I was poontanging like a dog. 1970 D. Dodge *Hatchetman* x. 127 'Is it true what they say about gook women?'.. 'I heah it changes youah luck, though. Like black poontang.' 1972 'T. Coe' *Don't lie to Me* iv. 44 May be you're some kind of poontang sex maniac. 1972 *Listener* 22 June 845/2 Massa gonna smack yo black ass, nigger. You can't go chasing white poontang all night long. 1976 *Honolulu Star-Bull.* 21 Dec. e-10/6 (Advt.), The other girls majored in home ec... but Debby majored in Poon-tang.

poop, *sb.*[1] Add: **3. poop-break,** the front of the poop of a ship; **poop-ornament** *Naut. slang*, a ship's apprentice.

1912 J. Masefield in *Eng. Rev.* Oct. 353 Under the poop break, sheltering from the rain. 1902 *Athenæum* 8 Feb. 177/1 He [sc. the apprentice in the merchant service] was and is emphatically the ship's loblolly-boy..miscalled 'a blarsted poop ornament', the drudge even of ordinary seamen. 1934 J. Masefield *Taking of Gry* 75, I looked at these fellows, and concluded that the lieutenant was a young poop-ornament and that the men were slacking.

poop, *sb.*[2] Delete †*Obs.* and add: Also (*S. Afr.*) ‖ **poep. 1.** (Later examples.) Also, the report of a gun.

1908 K. Grahame *Wind in Willows* vi. 128, I faithfully promise that the very first motor-car I see, poop-poop! off I go in it! 1919 W. Deeping *Second Youth* xxviii. 240 The faint 'poop-poop' of distant anti-aircraft guns..brought Laverack sharply back to the immediate present.

2. *slang* (orig. children's). An act of breaking wind or of defecation; faeces. Also *fig.*

Quot. *c*1744 is an *interj.*

*c*1744 [see *hole *sb.* 8]. 1937 Partridge *Dict. Slang* 648/2 *Poop*,..a breaking of wind. 1948 *Amer. Speech* XXIII. 264 We used the words *poop* and *poot* with their onomatopoeic significance in regard to bodily discharges. 1974 *Amer. Speech 1971* XLVI. 82 *Fart*,..poop. 1976 *Telegraph* (Brisbane) 7 June 9/2 A young woman claims a 'bird poop treatment' has cured her of a chronic dandruff... She's been free of dandruff since a mynah bird relieved himself on her head during lunch one day. 1977 J. McClure *Sunday Hangman* x. 105 Ja, we scared the poop out of him that time. 1977 *Listener* 22–29 Dec. 842/3 That's just a set-up... He just says all that kind of poop about you in the beginning.

3. *attrib.* and *Comb.*, as **poop-butt** (see quot.); **poophead** *U.S.* (see quot.); **poop-hole** (in quot. *S. Afr.*), anus; **poop-pusher** (see quot.); **poop-scared** *a.* (in quot. *S. Afr.*), extremely frightened, = *shit-scared adj.; **poop scoop** *joc.*, an implement used for clearing up faeces; so **pooper scooper; poop-stick,** a fool, ineffectual person (cf. *poop *sb.*[4]).

1973 C. & R. Milner *Black Players* ii. 42 A *poop-butt* is a lazy person. 1976 *Maclean's Mag.* 12 Jan. 34/1 A man on a motor-cycle with a pooper-scooper would be dispatched to clean up. 1977 J. Wambaugh *Black Marble* (1978) viii. 112 Bring your pooper-scoopers, boys. The dogs are covering the red carpet in a sea of shit. 1978 *Daily Tel.* 17 Aug. 3/3 The 'pooper-scooper' law in New York requires dog-owners to pick up anything their dogs drop. 1977 *Amer. Speech 1975* L. 64 *Poophead*, person regarded as dull or stupid. 1969 A. Fugard *Boesman & Lena* i. 26 That tickey deposit heart of his is tight like his *poephol* and his fist. 1966 'L. Lane' *ABZ of Scouse* 83 *Poop-pusher*, a laxative, especially one of satisfactory violence. 1976 in J. Branford *Dict. S. Afr. Eng.* (1978) 188/1 OK so they give up at last, but..never been so poep-scared in my whole entire life before or since. 1978 *Daily Tel.* 24 July 13/6 It may soon be common to see people holding a dog lead in one hand and a 'poop scoop' in the other. 1930 'Hay' & Wodehouse *Baa, Baa, Black Sheep* ii. 55, I believe she really does care for that poop-stick... Fancy loving a man called Osbert Bassington-Bassington! 1932 P. MacDonald *Rope to Spare* viii. 100 'You make me sick!' he said. 'Let a little poop-stick like that walk all over you!'

poop (pūp), *sb.*[4] *colloq.* [Perh. abbrev. of Nincompoop.] A stupid or ineffectual person; a fool, a bore.

1915 V. Woolf *Voyage Out* iii. 51 They talk about art, and think us such poops for dressing in the evening. 1924 'Sapper' *Third Round* vii. 189 The genuine Professor Scheidstrun appeared to be a harmless old poop, who was more sinned against than sinning. 1936 C. S. Forester *General* xv. 155 Every little poop of a temporary major-general. 1942 Wodehouse *Money in Bank* (1946) xx. 169 For God's sake, don't put your trust in that poop. 1952 S. Kauffmann *Philanderer* (1953) iii. 42 Yes, yell for your father. I'd like to get a look at the poop. I'd like to pick you up and throw you at him. 1966 H. Marriott *Cariboo Cowboy* ix. 86 It seemed to me that a real snotty-nosed poop like him was a poor character to have in that sort of job. 1971 R. Dentry *Encounter at Kharmel* xii. 211 Those stupid bloody Yankee poops blew the panic whistle and the whole shebang went sky-high. 1972 D. Delman *Sudden Death* iv. 95 'Honey, I do declare you've turned into nothing but an old poop,' she said.

poop (pūp), *sb.*[5] *slang* (orig. and chiefly *U.S.*). [Origin obscure.] Up-to-date or 'inside' information, 'low-down'. Freq. *attrib.* in **poop-sheet,** a written notice, bulletin, or report.

1941 *Amer. Speech* XVI. 167/2 *Poop sheet*, drill schedule or any written announcement. 1947 *Ibid.* XXII. 216 The word *poop*, which indicated the latest information, whether official or unofficial, was also incorporated into *poop sheet*, denoting the latest bulletin or directive. 1950 'D. Divine' *King of Fassarai* ii. 14 Have you picked up any poop about where the ship's going..? *Ibid.* xxiv. 205 'Have you seen the poop sheet anywhere?..' 'Aw, hell... There ain't any news anyway.' 1961 B. Malamud *New Life* (1962) 300, I sent out a poop sheet this afternoon.. with the names of the nominees. 1963 H. Slesar *Bridge of Lions* (1964) v. 84 I've asked Sandy to send me a poop sheet on him, but maybe you can add something. 1973 R. Hayes *Hungarian Game* xxi. 130 How did you get the poop on Kovács? 1974 'M. Allen' *Super Tour* vii. 264 He sends in a report—straight facts, no frills, and a minimum use of adjectives. What he says is included in the mimeographed poop sheet the organization sends out every month. 1976 *Sounds* 11 Dec. 33 (*heading*) Hot poop on old punks.

poop, *v.*[1] Add: **1. b.** (Later examples.)

1937 Partridge *Dict. Slang* 648/2 *Poop*,..to defecate.. mostly of and by children. 1972 L. Hancock *There's a Seal in my Sleeping Bag* vii. 165 The joys of motherhood were considerably lessened when a baby murre pooped with regularity between my shirt and bare warm skin. 1974 *Cape Times* 1 Aug. 11/5 Five-year-old eyes grow round with wonder at the memory of the elephant 'pooping' on the carpet.

2. a. *trans.* To fire (a bullet, shell, or other missile); to discharge (a gun); to shoot (a person or animal). Freq. with *off*. Also *transf.* and *fig.*

1917 W. Owen *Let.* 6 Dec. (1967) 514, I shall continue to poop off heavy stuff at you, till you get my range at Scarborough, and so silence me. 1929 [see *heat *sb.* 12 b]. 1937 P. B. Hawk *Off Racket* ii. 99 At any rate he pooped the ball into the [tennis] net. 1940 *Manch. Guardian Weekly* 5 Apr. 277 An old wildfowler..earns a precarious living by stealing down the estuaries with a home-made blunderbuss or duck-gun and pooping it off at..sea-fowl. 1940 'N. Shute' *Landfall* 142 'Can't we fly it over a known ship and poop it off?' he said. 'Poop off half a dozen of them.' 1960 *Guardian* 9 Nov. 7/7 The emergency code word has just been flashed through... Is it all right by London if they poop off their Polarises? 1974 A. Price *Other Paths to Glory* ii. vi. 187 There was this Jerry prisoner..and this Aussie comes up..and he poops him... Kills him—shoots him.

b. *intr.* Of a person: to fire a gun, to shoot; also in weakened sense: to go. Of a gun or similar device: to go off, to fire. Usu. with adv., as *away, off*, etc.

1919 *Chambers's Jrnl.* Jan. 43/1 As soon as the artillery opens up, poop off for all you're worth. Let 'em have a hurricane. 1928 Blunden *Undertones of War* ii. 22 A field battery glaring brutally out would 'poop off'. 1930 R. Pertwee *Pursuit* 59, I arrived about eight last night and the guns were pooping away like mad. 1931 N. Coward *Post Mortem* i. 13 If it only stays quiet the way it has the last three nights, and that machine-gun from the sunken road doesn't start pooping at us—we'll get through it in a few hours. 1942 T. Rattigan *Flare Path* i. 112 What in hell was the idea of pooping off the Station like that? They told you this morning something still might come through. 1945 *Tee Emm* (Air Ministry) V. 40 It will not..poop off automatically. 1956 'Taffrail' *Arctic Convoy* xxiii. 237 See to it you don't open up on anything unless you've a damn good chance of hitting. No wild pooping off at targets out of range—understand? 1961 *Encounter* June 21/1 Take getting up in the morning. 'I arise,' he says..'and poop along..to the end of the passage..to get in the milk and the papers.'

poop, *v.*³ Add: **a.** (Later examples.) Also *transf.*

1916 'TAFFRAIL' *Pincher Martin* xi. 210 There is a grave risk of the craft being pooped by a heavy sea. **1955** *Times* 3 May 5/2 The worst seas they encountered in the whole great voyage, however, were between the Start and Portland Bill. They were pooped. **1972** *Daily Tel.* 22 Jan. 11/4 Returning home via a less dizzy gradient (it is only 1-in-4), we [*sc.* the writer and his dog] faced the same sort of spate that had tried to poop us while we were descending.

poop (pūp), *v.*⁴ *colloq.* (orig. *U.S.*). [Origin unknown.] **a.** *intr.* To break down, 'conk' out. **b.** *trans.* To tire, to exhaust. So **pooped** *ppl. a.*, exhausted, worn out. Freq. with adv., esp. *out.*

1931 *Technol. Rev.* Nov. 65/2 If his engine *poops* or *konks*, he will be *forced down*. **1932** *Amer. Speech* VII. 335 *Pooped; all pooped*, tired out; exhausted. **1934** J. T. FARRELL *Young Manhood* xii. 187 He was tired and pooped. *Ibid.* xxii. 377 Studs took a large rocker, and carried it slowly downstairs... When he set it down in the alley, he was breathless, and all pooped out. **1938** 'E. QUEEN' *Four of Hearts* (1939) iv. 57 He ain't had a drink in five days. That would poop up any guy. **1944** E. B. WHITE *Let.* 15 May (1976) 253 This would be a very bad time to pull our exhaustion on our readers, a lot of whom are pretty well pooped out themselves for one reason or another. **1949** R. CHANDLER *Little Sister* xxx. 222 'Tired?' he asked. 'Pooped.' **1955** M. DICKENS *Winds of Heaven* iv. 93 He'd better be..or he'll find his mother-in-law in the hospital with him. You've really pooped yourself, mother. **1957** D. KARP *Leave me Alone* xviii. 274, I don't think he understood me. The poor old guy is pooped out. **1959** N. MAILER *Advts. for Myself* (1961) 45 He remembered the old man sitting on the porch..all pooped out after work. **1960** *Sunday Express* 24 July 4/2 Bringing up eight kids..really has me pooped. **1966** *New Scientist* 22 Sept. 658/1 Lt Cdr Richard Gordon's space walk was cut short because..'he was blinded by sweat and felt pooped'. **1967** *Time* 2 June 33 Paley Park offers pooped passers-by a respite at little white tables and chairs in a setting of geraniums, honey locust trees, and a 20-ft. waterfall. **1971** B. MALAMUD *Tenants* 7 If it [*sc.* the heating system] pooped out, and it pooped often—the furnace had celebrated its fiftieth birthday—you called the complaint number of Rent and Housing Maintenance. *Ibid.* 183 His electric heater has pooped out and is being repaired. **1977** *Time* 18 Apr. 64/3 Pheidippides..was so pooped by his performance that he staggered into Athens.

poor, *a.* (*sb.*) Add: **I. e.** Phr. *poor but honest.*

1748 SMOLLETT *R. Random* I. xviii. 150, I am a poor, but honest cobler's son. **1824** KNAPP & BALDWIN *Newgate Calendar* I. 149/1 John Hawkins was born of poor but honest parents. **1869** 'MARK TWAIN' *Innoc. Abr.* xxi. 211 'He was the son of——' 'Poor but honest parents—that is all right—never mind the particulars—go on with the legend.' **1922** W. J. LOCKE *Tale of Triona* viii. 90, I was born—I shan't tell you the year—of poor but honest parents. **1939** A. THIRKELL *Before Lunch* v. 126 'Do you mean to say you ride one of those things.' Daphne said she was poor but honest, and why not. **1972** C. WESTON *Poor, Poor Ophelia* (1973) iii. 18 'So he's a slave, too,' she commented. 'Right on for Poor-But-Honest headed for the top.'

3. a. (U.S. examples.)

1778 *Maryland Jrnl.* 10 Feb. 4/2 [The sheep] are very poor, and appear to have been out all winter. **1878** J. H. BEADLE *Western Wilds* xvii. 276 They get poor as snakes on such food; but it does keep body and soul together for awhile.

5. a. (Examples in phr. *poor-quality* used *attrib.*)

1892 [see *IGNORANT *a.* 1 b]. **1948** C. L. B. HUBBARD *Dogs in Brit.* 234 The English Setter appears to be in danger of deteriorating into a very pretty but poor-quality worker. **1960** *Farmer & Stockbreeder* 9 Feb. 57/3 When I was a boy we used to chaff poor-quality hay and mix it with molasses. **1966** G. GREENE *Comedians* I. i. 23 Cynicism is cheap..it's built into all poor-quality goods.

e. Used with *little* in depreciatory (and freq. ironical) senses, esp. in the phrases *poor little guy*, the ordinary individual, the 'man in the street'; *poor little me* (see *LITTLE *a.* 13); *poor little rich boy, girl*, used (sometimes ironically) of a person whose wealth has not brought happiness.

1925 N. COWARD *Poor Little Rich Girl* (song) 3 Poor little rich girl, You're a bewitched girl, Better beware! **1934** 'G. ORWELL' *Burmese Days* v. 91 Unmanly whinings; poor-little-rich-girl stuff. **1940** GRAVES & HODGE *Long Week-End* xvii. 300 Spender wrote poor-little-rich-boy poems, full of genuine pity for the exploited poor, and for himself. **1958** [see *MASS *sb.*² 4 a]. **1961** *Guardian* 28 Apr. 30/6 A disturbing flavour of the poor little rich boy, of the attitude which claims privilege for me just because I am me. **1967** *Boston Sunday Herald* 26 Mar. I. 9/7 Only the poor little guy is subject to the zoning code. **1973** *Times* 23 Mar. 17/6 A comedy-weepie about a poor little rich girl. **1974** J. MANN *Sticking Place* ii. 37 There was still something pathetic about her..a poor little rich girl. **1977** *Daily Tel.* 4 Mar. 3/5 A Conservative M.P.'s daughter on a heroin charge was 'really just a poor little rich girl..who has had an unhappy life' a magistrate said yesterday.

f. *to take a poor view*, to have a low opinion (*of* something); to regard unfavourably.

1943 HUNT & PRINGLE *Service Slang* 52 If you do not agree with a statement or with your C.O.'s ruling..or, in fact, with the world in general, you take a poor view. **1944** 'N. SHUTE' *Pastoral* i. 4 The Wing Commander had taken a poor view of that. **1946** E. LINKLATER *Private Angelo* x. 115 The Germans are about to do something that we take a poor view of, and I'm going to see if I can put a stop to it. *a* **1966** 'M. NA GOPALEEN' *Best of Myles* (1968) 41 The brother took a very poor view and said she'd be a sorry woman.

7. a. (Further examples.) Freq. with a preceding epithet, as *the aged poor, the deserving poor, the good poor, the respectable poor, the sick poor, the undeserving poor.* Also, *the very poor.*

c **1658** in F. J. Furnivall *Harrison's Descr. Eng.* (1908) IV. 207 Cures Colledge..with maintenance for 16. aged poore of the parish. **1823** E. WEETON *Let.* 16 Apr. in *Jrnl. of Governess* (1969) II. 217 Going about..to visit the sick poor as a Member of the Benevolent Society. **1845** E. SMITH *Jrnl.* 28 Oct. (1980) 81 Sick poor, destitute poor, idle, prejudiced poor, oppress me. **1852** DICKENS *Bleak Ho.* (1853) vi. 48 It is said that the children of the very poor are not brought up, but dragged up. **1907** G. B. SHAW *Major Barbara* Pref. 154 'The respectable poor', and such phrases are as intolerable and as immoral as 'drunken but amiable' [etc.]. **1909** W. J. LOCKE *Septimus* i. 6 Cousin Jane held distinct views on the cut of under-clothes for the deserving poor. **1910** E. M. FORSTER *Howards End* vi. 53 We are not concerned with the very poor. They are unthinkable and only to be approached by the statistician or the poet. **1928** A. M. M. DOUTON *Bk. with Seven Seals* I. 14 In those days pews were only for those who could pay for them, and free benches..were occupied by the respectable poor. **1937** The sick poor [see *HOSPITALIZATION]. **1972** *Listener* 9 Mar. 317/1 The Tolpuddle Martyrs..were..the good poor—and little enough they got by it. **1972** J. MANN *Mrs. Knox's Profession* xiii. 102 The man was obviously one of the old-fashioned 'good poor'. **1973** *Guardian* 21 Mar. 10/4, I have no sympathy at all with the kind of people who..do not believe in private property. There is a difference between the undeserving poor and the deserving. *Ibid.* 18 May 16/2 Many of those who want to be owner-occupiers could properly be described as the deserving poor. **1974** *Times* 11 Apr. 20 (*heading*) How can we decide who are the deserving poor? **1977** P. LASLETT *Family Life Earlier Generations* iv. 171 During the next 2 years of life, ages 4 and 5, infants remain very dependent, and even amongst the very poor in pre-industrial society were extremely unlikely to be sent out of the home. **1979** P. THEROUX *Old Patagonian Express* xv. 220 On the higher harder-to-reach slopes..were the huts of the very poor.

8. poor boy (sandwich) *Southern U.S.*, a large sandwich containing a wide variety of simple but substantial ingredients; **poor do** *U.S.*, a dish made up of scraps of food; a hash; **poor mouth** *v. trans.* and *intr.*, (*a*) to claim to be poor; to make demands (on someone) alleging poverty; (*b*) to deprecate, make little of (something); so **poor-mouthing** *vbl. sb.*; *to make* (*put on*, etc.) *a poor mouth*; see MOUTH *sb.* 3 m in Dict. and Suppl.; *to talk poor-mouth* (U.S.), to plead poverty; **poor relation** (earlier and later examples; also *fig.* and *attrib.*).

1952 *New Orleans Item* 28 Feb. 17/5 'Way back yonder when a poorboy sandwich was just that—namely, a five-cent filling of bread, meat and mixed pickles for a poor boy. **1954** *Newsweek* 15 Mar. 99 In the South, a poor boy is a frankfurter in a long bun. **1962** E. WASON *Cooks, Gluttons & Gourmets* 271 Eventually they became known ..as 'poor boy' sandwiches, as they are called in New Orleans to this day. **1968** *Amer. Speech 1967* XLII. 286 A term [for sandwich] which is used primarily in the South is *poor boy*... The usual interpretation of this term is that those who eat the sandwich are in the lower social and economic classes. **1976** *National Observer* (U.S.) 6 Nov., Exploring Greenwich Village, I found Poor Boys, salami and cheese and chili peppers on great hunks of Italian bread. **1909** *Pioneer Days in Southwest 1850–1879* 253 When we had hogmeat we would fry a few pieces, take the grease and crumble corn bread in it, putting in water and salt, and we had a pot of soup called 'poor doo'. **1913** H. KEPHART *Our Southern Highlanders* 292 The old Germans taught their Scotch and English neighbors the merits of scrapple, but here it is known as poor-do. **1965** *Lebende Sprachen* X. 37/1, I am intrigued..by the abundance of 'poor' edibles which we have in the States...*poor-do* is scrapple, etc. **1967** WEBSTER Add., Poor mouth v. **1968** *New Yorker* 21 Sept. 169 [Eugene] McCarthy's advertising campaign, despite the McCarthy camp's constant poor-mouthing on the subject, wasn't exactly modest. **1970** *Globe & Mail* (Toronto) 26 Sept. 9/1 Six months ago, it was sacrilegious to poormouth the fight against air pollution. It was akin to being against motherhood. **1972** T. ARDIES *This Suitcase* xvi. 176 What prompted the Professor to start poor mouthing me? **1972** *Time* 17 Apr. 25/1 Some democrats..are already poor-mouthing his victory. **1976** *Times Lit. Suppl.* 26 Mar. 339/2 The latter [book] came to be poormouthed by its author as 'a literary exercise'. **1941** in H. Wentworth *Amer. Dial. Dict.* (1944) 469/1 College professors are supposed to talk po' mouth. **1961** *Newsweek* 14 Aug. 15/1 Because they are politicians, they like to talk as poor-mouth as the lowliest voter. **1965** *N.Y. Times* 29 July 16 It is hard to talk poor mouth just after the papers have written of your daughter's coming-out party for 2,000 guests. **1978** R. THOMAS *Chinaman's Chance* xv. 168 My teeth hurt whenever you start talking poormouth. **1968** *Guardian* 28 Jan. 16/8 Poor-mouthing at home feeds the doubts of foreigners abroad..and gloomy prognostications from abroad in turn further depress the spirits of the British. **1969** 'R. MACDONALD' *Goodbye Look* xxviii. 163 She became a bit of a miser... Her poor-mouthing actually had me convinced. But of course she'd been quite wealthy all along. **1720** DEFOE *Capt. Singleton* 338 Seeing..he had some poor Relations in England..he would write to know..what Condition they were in. **1748** SMOLLETT *R. Random* I. i. 1 My father..fell in love with a poor relation..whom he privately espoused. **1906** J. M. SYNGE *Lett. to Molly* (1971) 13, I dont like hanging about their house as a poor relation. **1962** *Rep. Comm. Broadcasting 1960* 26 in *Parl. Papers 1961–2* (Cmnd. 1753) IX. 259 The suggestion or fear that sound radio was becoming the 'poor relation' of broadcasting. **1970** J. EARL *Tuners & Amplifiers* ii. 36 Most attention..is usually focused on to the f.m. department where the quality potential truly exists, the a.m. department of a composite model then being very much a 'poor relation'. **1972** *Guardian* 14 Jan. 11/1 In many cases servicing and spare parts were the poor relations subsidised by sales. **1977** D. FRANCIS *Risk* viii. 100 Novice hurdles were customarily first or last..the poor-relation races for the mediocre majority. **1978** J. SYMONS *Blackheath Poisonings* I. 28 He resented..the mother who had inconsiderately died and left him a poor relation.

9. c. *poor-ass* (*U.S.*), *poor-looking* (examples).

1957 J. KEROUAC *On Road* (1958) 113 Find out just what he's poor-ass pondering about this year's turnip greens. **1970** R. D. ABRAHAMS *Positively Black* iii. 72 Colored man went to the store and bought him one of them poor-ass damned roosters. **1973** E. BULLINS *Theme is Blackness* 163 I'm only a weak little old pore/ass black woman. **1799** MALTHUS *Diary* 11 July (1966) 139 Saw some poor-looking houses. **1847** E. SMITH *Jrnl.* 13 Jan. (1980) 116 Poor looking house with three rooms.

10. poor-basket, a basket containing material from which clothes for the poor could be made; **poor-farm** (examples); **poor-work,** work done to provide clothes etc. for the poor.

1814 JANE AUSTEN *Mansf. Park* I. vii. 147 If you have no work of your own, I can supply you from the poor-basket. **1852** J. W. GUNNISON *Hist. Mormons* 145 A Poor Farm of forty acres is in the centre, controlled by the bishops. **1895** A. BROWN *Meadow-Grass* 168 The latter had actually taken to her bed..announcing that 'she'd rather go to the poor-farm and done with it than resk her life there another night'. **1949** *Chicago Tribune* 2 Dec. 20/7 It used to be a disgrace to go to the 'poor farm' and be cared for by the rest of society. **1961** N. LOFTS *House at Old Vine* VI. vi. 380 I've thought about what I owed you. .. You'd have fared better at the Poor Farm! **1854** C. M. YONGE *Castle Builders* v. 69 Each good lady had a great basket full of poor-work. **1876** *Monthly Packet* Feb. App. 5 The Sisters at Kilburn are glad to have 'poor work' done for them, and..will provide the material.

poor-box. (Earlier example of the form *poor-box*.)

1738 POPE *First Epistle of First Bk. of Horace Imitated* 15 The rest, some farm the Poor-box, some the Pews.

poor-cod (pūə·ɹˌkɒd). [f. POOR *a.* 5 + COD *sb.*³] A small marine fish, *Trisopterus minutus*, belonging to the cod family Gadidæ, and found in coastal waters of north-western Europe; = POWER *sb.*²

[**1828** J. FLEMING *Hist. Brit. Animals* 191 Morhua. Cod... *M. minuta.* Poor.—Nine punctures on each side of the jaws and gill-covers.] **1836** [see POWER *sb.*³]. **1925** J. T. JENKINS *Fishes Brit. Isles* 149 The Poor Cod ranges from Trondhjem to the Mediterranean. **1959** A. HARDY *Open Sea* II. xi. 227 The poor-cod..is the smallest of all our gadoid fish, rarely exceeding 8 inches in length. **1969** A. WHEELER *Fishes Brit. Isles & N.-W. Europe* 272/1 The poor-cod is caught mainly in trawls.

poor man. Add: **1.** Example in *attrib.* use.

1831 J. BANIM *Smuggler* I. xi. 127 What have you to do with..my poor-man sneers at a viscount?

3*. Also **poorman.** = *poorman('s orange* (see sense 4 a below).

1912 *Jrnl. Dept. Agric. N.Z.* IV. 141 He has several varieties all doing well, amongst them Paramatta, Poor Man, Navel. **1956** F. T. BOWMAN *Citrus-Growing in Austral.* ii. 20 Poorman was mentioned in Shepherd's catalogue (1851) as having been recently introduced from Shanghai by a Captain Simpson.

4. a. poor man's diggings *U.S.*, *Austral.*, and *N.Z.* (see quot. 1941); **poorman('s orange,** a variety of grapefruit, *Citrus paradisi*, once cultivated in New Zealand; **poor man's orchid,** an annual or biennial plant belonging to the genus *Schizanthus* of the family Solanaceæ, native to Chile and bearing flowers thought to resemble orchids; **poor man's sauce** (further example); **poor man's torment** *U.S.* (see quot.).

1875 *Chicago Tribune* 14 Oct. 7/3 If it did pay, it would be what is called poor man's diggings, for it was no place where capital could be successfully employed. **1876** R. I. DODGE *Black Hills* 109 It has passed into a proverb that 'placer' mining is the poor man's diggings, while 'quartz' mining is only for the rich. **1941** BAKER *Dict. Austral. Slang* 56 *Poor man's diggings*, alluvial gold deposits, i.e., gold which a poor man can work, contrasting with reef-gold which requires capital to develop. **1884** G. E. ALDERTON *Treat. & Handbk. Orange-Culture in Auckland* 66 The Poor Man's Orange is only good for preserving. **1929** *Jrnl. N.Z. Inst. Hort.* I. 65 The Poorman Orange is really a Pomelo. **1949** R. PARK (*title*) Poor man's orange. **1966** *Encycl. N.Z.* I. 758/2 The main kinds of citrus grown commercially in New Zealand include..so-called New Zealand grapefruit ('Poorman' orange, selected strains). **1959** *Listener* 20 Aug. 298/3 Now is the time to sow schizanthus—the 'poor man's orchid'—for next May. **1976** *Hortus Third* (L. H. Bailey Hortorium) 1018/1 Schizanthus... Butterfly Flower, Poorman's orchid. **1723** J. NOTT *Cook's & Confectioner's Dict.* sig. Mm3 Poor Man's Sauce, *i.e.* a Shalot cut small, white Pepper, Vinegar, and Oil. **1899** W. STEVENS *Jrnl.* 17 July in *Lett.* (1967) 28 Snap-dragon, or as it is vulgarly known: the weed—'poor man's torment' is a close-knit, yellow, tumbled sort of thing.

b. Now commonly used *fig.* in Combs. to denote a cheaper, usu. simpler or inferior version or imitation of something, or a less satisfactory substitute for something or someone.

1854 H. MELVILLE in *Harper's Mag.* June 95/2 A cup of cold rain water..is called by housewives a 'Poor Man's Egg'. *Ibid.* 97/1 'It is only rice, milk, and salt boiled together.' 'Ah, what they call "Poor Man's Pudding", I suppose you mean.' **1891** *Tit-Bits* 8 Aug. 277/2 There are thousands of costers who earn a livelihood by the sale of ..mussels, which are regarded as the poor man's oyster. **1906** *Dialect Notes* III. 151 *Poor man's pudding*,..cottage pudding. **1924** R. LARDNER in *Cosmopolitan* July 60/2 Another nickname for the town [*sc.* St. Petersburg, Florida, U.S.A.] is the Poor Man's Palm Beach. **1949** *Amer. Speech* XXIV. 94 The cheapness and abundance of rabbit pelts..have made them the 'poor man's mink'. **1951** M. MCLUHAN *Mech. Bride* (1967) 63/1 Huck Finn, the poor man's Thoreau, is to be read there, too. **1959** *Observer* 9 Aug. 11/2 I.T.V.'s *This Week* has recently become not much better than a very poor man's 'Panorama'. **1962** A. HUXLEY *Island* v. 46 Chemical and biological weapons—Colonel Dipa calls them the poor man's H-bombs. **1963** *Guardian* 8 Feb. 9/3 The long, many-scened story..is superficially like a poor man's 'Peer Gynt'. **1963** 'R. ERSKINE' *Passion Flowers in Italy* iv. 42 The porter was heavy-set, with burning Latin eyes: a kind of poor man's Marlon Brando. **1971** *Jrnl. Chem. Documentation* X. 249/1 A general-purpose text-editing system can be a valuable 'poor man's' information-handling tool. **1973** *Times* 21 Apr. 12/1 'Good King Henry' or 'Poor Man's Spinage' must have been tried out for centuries before being used traditionally and regularly, in spring 'messes'.

poorshouse (pūə·ɹzhaus). Also **poors house, poor's house.** Mainly *Sc.* var. of POORHOUSE.

1745 *Sessions Papers* (Donaldson v. Home) 3 July 2 Samuel Neilson, late Deacon of the Masons, Undertaker for building of the Poors House. **1756** *Bristol* (Virginia) *Vestry Bk.* (1898) 164 Ordered that Stephen Dewey..agree in settleing the Terms of the Poors House. **1820** J. FLINT *Lett. from Amer.* (1822) 192 Some paupers in a poor's house at Cincinnati refused to carry water for their own use. **1870** J. NICHOLSON *The Puir's-Hoose Laddie* in *Idylls* 45, I was glad to become a wee Puir's-hoose laddie. **1899** E. F. HEDDLE *Marget* ii. 10 She..is to gang to the sale; but she's no' to gang to the puirs-hoose. **1907** [see POOR *a.* 7 d]. **1923** T. JOHNSTON *Hist. Working Classes in Scotl.* 35 There were 30 hospitals or poorshouses from Turriff to the Lowlands.

poort. Add to def.: *esp.* one cut by a stream or river. (Earlier and later examples.)

1796 tr. *F. Le Vaillant's New Trav. Afr.* II. 194 We issued from the mountains through a sort of passage, or defile, which is called the *Poort.* **1801** J. BARROW *Acct. Trav. S. Afr.* I. ii. 109 The Poort may be considered as the entrance into Camdeboo. **1932** C. FULLER *Louis Trigardt's Trek* vi. 68 Once through the poort, the junction of the spruit with the river is but a few hundred yards off. **1949** L. G. GREEN *In Land of Afternoon* i. 21 A poort is different from a pass, for it is a passage through the mountains along the bed of a stream.

pooter (pū·təɹ), *v.* [Etym. unknown.] *intr.* To depart in a hurry; to hasten away. Also with *off.*

1907 *Dialect Notes* III. 196 *Pooter*,..to depart speedily. 'I told him to git, and he just *pooter*, I can tell you.' **1966** *Punch* 6 July 32/3 The ex-bookseller, his fortune depleted, is left on the last page pootering off to his ex-girl-friend.

Pooterish (pū·təriʃ), *a.* [f. the name *Pooter* (see below) + -ISH¹.] Resembling, characteristic of, or associated with Charles Pooter, an assistant in a mercantile firm, whose mundane domestic, social, and business troubles are the subject of the fictional *Diary of a Nobody* by George and Weedon Grossmith (1892).

1966 *New Statesman* 11 Mar. 349/3 Take a Pooterish Little Man with sexual and cultural ambitions outside his class, [etc.]. **1976** *Times Lit. Suppl.* 31 Dec. 1626/2 So many square miles of vapid and banal and Pooterish suburb. **1977** *Times* 14 May 10/4 George VI's deadpan account of Pooterish bishops blundering through his coronation. **1978** *Times Lit. Suppl.* 24 Feb. 229/3 The Pooterish touch in 'inexpensive' betrays a lack of awareness.

pooty (pū·ti), *a.* (*sb.*) Affected or childish var. PRETTY *a.* (*sb.*)

1825 [see *CROSS LOTS *advb. phr.*]. **1848** J. R. LOWELL *The Courtin'* in *Biglow Papers* 1st Ser. 10 The wannut logs shot sparkles out Towards the pootiest, bless her! **1849** C. BRONTË *Shirley* II. iv. 105 Purchase in his stead some sweetly pooty pug or poodle. **1850** THACKERAY *Pendennis* II. xv. 147 She's a little money too..a pooty little bit of money. *a* **1854** [see *country jake* s.v. *COUNTRY 16]. **1906** GALSWORTHY *Man of Property* II. iii. 149 'You'll have room here,' he said, 'for six or seven hundred dozen—a very pooty little cellar!' **1932** AUDEN *Orators* III. 104 That piss-proud prophet, that pooty redeemer. **1961** PARTRIDGE *Dict. Slang* Suppl. 1229/1 Pooty is a favourite mid-Victorian adjective meaning 'pretty'—of which, via *purty*, it is a perversion. **1980** G. NELSON *Charity's Child* iii. 45 Do 'e remember..the pooty shells I collected?

poove: see *POOF *sb.*¹

poovey, poovy, varr. *POOFY *a.*

pop, *sb.*¹ Add: **1. c.** In Baseball: a ball hit high into the air but close to the batter, thus providing an easy catch. Usu. *attrib.*, as *pop fly* [FLY *sb.*² 2 b], etc. *N. Amer.*

1935 J. T. FARRELL *Judgment Day* viii. 185 A line single was driven to left, the pitcher picked a pop out of the air. **1945** *Sun* (Baltimore) 12 Mar. 10-0/5 A pamphlet which knocked the Doubleday legend higher than one of Babe Ruth's pop fouls. **1961** *Rocky Mountain News* (Denver, Colorado) 2 May 50 The White Sox had taken a 5–4 lead in the top of the sixth on a pair of pop fly hits. **1969** *Sci. Amer.* Jan. 49/1 The outfielder is watching..a pop fly to the infield. **1972** *N.Y. Times* 4 June v. 2/5 Gentry retired the great man on a pop foul to Mays. **1975** *New Yorker* 14 Apr. 98/2 Jay Kleven, a young non-roster catcher, hit two pop flies to center. **1978** *Verbatim* Feb. 2/2 One of my favorites is the phrase for a towering pop fly that shoots straight up to the sky and comes down in the same area, usually caught by the catcher. The announcer says, 'He could have hit that ball in a silo.'

d. An injection of a narcotic drug. *slang.*

1935 A. J. POLLOCK *Underworld Speaks* 118/1 *Take a pop*, to take an injection of morphine. **1953** W. BURROUGHS *Junkie* Gloss. 14 *Pop, bang, shot, fix*... Injection of junk. **1956** R. THORP *Viper* vi. 92 'Care for a pop now and again?' This was a kick I hadn't made, I told him. **1970** N. MARSH *When in Rome* v. 126 I'm not hooked. Just the odd pop. Only a fun thing.

2. b. *on the pop of*, about to, on the point of. *rare.*

1922 JOYCE *Ulysses* 66, I was on the pop of writing Blazes Boylan's.

c. A turn (at doing something); an attempt; a 'go'.

1868 'MARK TWAIN' *Let.* 20 Nov. (1917) I. ix. 156, I am simply lecturing for societies, at $100 a pop. **1904** W. N. HARBEN *Georgians* 2 Ef I don't whack it to you this pop, old hoss, I'll eat my hat. **1916** 'TAFFRAIL' *Pincher Martin* xv. 271 'Why doesn't we 'ave a pop at 'er?' ''Ave a pop at 'er! She's twenty mile orf, if she's a hinch, an' yer knows as well as I does that none o' our ships 'ere 'as got hanti-haircraft guns wot'll 'it 'er at that range.' **1928** WODEHOUSE *Money for Nothing* ii. 35 He decided to have a pop at it. **1946** F. SARGESON *That Summer* 66, I thought no, the going's good, I'll give it one more pop. **1954** WODEHOUSE *Jeeves & Feudal Spirit* i. 12 But why didn't Florence tell Percy to go and have a pop at Stilton Cheesewright? **1971** *Southerly* XXXI. 136 But I couldn't keep that game up for too long; at five cents a pop you can't afford to waste too many. **1976** R. BARNARD *Little Local Murder* x. 133, I don't suppose he makes much more than seventy-five pee a pop for them.

d. The rapid opening of a pop-valve.

1901 M. M. KIRKMAN *Locomotive Appliances* 122 Should the valve close with too much drop of boiler pressure, move the screw-ring (C) to the left..until sufficient change has been accomplished. To increase the pop, move ring (C) to the right. **1905** C. S. LAKE *World's Locomotives* vi. 112 The screw-down valve is set so that the limit of 'pop' action is 2 lbs. per sq.in. above the nominal boiler pressure, and the valves close when that pressure has been reduced to 2 lbs. per sq.in. less than the nominal boiler pressure. **1951** E. A. STEEL *Greenly's Model Steam Locomotives* (rev. ed.) xiii. 234 A 'pop' action (an accelerated discharge of the valve) can be obtained by making the head of the valve nearly fit a cylindrical recess in the seating.

5. (Later examples.)

1926 *Scribner's Mag.* Aug. 116/2 Senior officers may for dignity's sake get off with light treatment and a fine of cigars or pop (it was beer in the good old days). **1931** W. S. MAUGHAM in *Hearst's International* Oct. 51/2 A bottle of pop tonight, my pet, and a slap-up dinner. **1969** L. KENNEDY *Very Lovely People* ii. 106 The waiter said, 'All I got is bottled pop. Take your choice.' **1976** A. HILL *Summer's End* i. 18 We sat in the stern drinking the pop, trying to count the bubbles as they rose behind our noses.

pop, *sb.*⁴ (Later examples.)

1898 *Westm. Gaz.* 19 Dec. 10/2 A Dohnanyi 'Pop'. In every respect Mr Ernest von Dohnanyi was the hero of Saturday's Popular Concert at St James's Hall. **1934** M. H. WESEEN *Dict. Amer. Slang* 381 Pop..a popular concert.

pop (pɒp), *sb.*⁶ *colloq.* (chiefly *U.S.*). **a.** Abbreviation of POPPA.

1838 in *Southwestern Hist. Q.* (1926) XXX. 147 Sent my packet..to pop in the post office at N Orleans. **1840** *Knickerbocker* XVI. 207 'Pop!' screamed a white-headed urchin from the house, 'Mam says supper's ready.' **1904** H. R. MARTIN *Tillie* iii. 33 Are you feelin' too mean to go help pop? **1911** [see *MOM]. **1948** *Denison* (Texas) *Herald* 1 July 1/3 Butch..was vacationing with his pop at the popular National Park Service Lake Texoma resort. **1958** H. E. BATES *Darling Buds of May* i. 11 'Larkin, that's me,' Pop said... 'Larkin by name, Larkin by nature.' *Ibid.* ii. 39 'We're in the library,' Ma said. 'Pop, look at the library.' **1962**, etc. [see *MOM]. **1973** P. DICKINSON *Gift* v. 77 'Oh yes, Pop, *please*,' said Sonia. **1979** R. RENDELL *Make Death love Me* i. 12 His father-in-law came in... Alan and Pam called him Pop, and Christopher and Jillian called him Grandpop.

b. Hence in extended use, an elderly man.

1844 in *Amer. Speech* (1965) XL. 131 And I'll go down to ole birginy, And marry pop Miller's sister. **1889** *Sporting Life* (Philadelphia) 29 May 2/6 'Pop' Chadwick is among those who are opposed to the wire. **1943** K. TENNANT *Ride on Stranger* (1968) vii. 78 You've just told us, pop,..that if the cops catch up on you, you'll be lining a cell. **1947** *Daily Oklahoman* (Oklahoma City) 28 Dec. 5/8 'Pop', as he is known in this area, will use the 'fancy' cane to help guide his sightless way during his strolls along Shamrock streets. **1980** P. GOSLING *Zero Trap* iii. 29 'Can somebody give me a hand with Pop, here?' He still wants to stay sleepies for a while.

pop (pɒp), *sb.*⁷ Abbrev. of POPPYCOCK.

1890 KIPLING *Barrack-Room Ballads* (1892) 11 All we ever got from such as they Was pop to what the Fuzzy made us swaller. **1924** GALSWORTHY *White Monkey* II. iv. 151 Nobody pitied her; why, then, should she pity them? Besides, pity was 'pop', as Amabel would say.

pop (pɒp), *a.* (*sb.*⁸) *colloq.* [Abbrev. of POPULAR *a.* (*sb.*): cf. POP *sb.*⁴] **I. a.** Designating music (esp. song) having or regarded as having a wide popular appeal (see *POPULAR *a.* (*sb.*) 6 b). Freq. *absol.* as *sb.*, a popular song or piece of music; popular music collectively.

Quot. 1862 is an isolated nonce-use influenced by and alluding to POP *sb.*⁴

1862 GEO. ELIOT *Let.* 26 Nov. (1956) IV. 67 There is too much 'Pop' for the thorough enjoyment of the chamber music. **1926** *Amer. Mercury* Dec. 465/1 She coos a pop song. **1935** *Hot News* Aug. 19/1 Turn the record over and you have another winner—'Add a Little Wiggle'—a masterpiece made out of a song-and-dance 'pop'. **1945** S. HUGHES in C. Madge *Pilot Papers* 78 Cole Porter's 'Begin the Beguine'..has twice the regulation number of bars that a good 'pop' should have. **1947** A. J. MCCARTHY *Jazzbook 1947* 119 Jelly would play one of his new 'pop' songs, watching..for its effect. **1954** *Unicorn Bk. 1953* 320/1 A magazine..each December publishes a list of the year's top pop music and musicians. *Ibid.* (*heading*) Top pop tunes. **1954** *Billboard* 13 Nov. 38 It is interesting to note that the preponderance of local over national sponsorship of deejay programs varies according to the program category, with the weight of local sponsorship most evident in rhythm & blues then country & western and finally pop. **1957** D. HAGUE in S. Traill *Concerning Jazz* 129 The veteran Lizzie Miles from New Orleans has evoked nostalgia with her selections of blues and early 'pops'. **1959** J. BRAINE *Vodi* iv. 63 At this time there'd be some pop tunes... They could sometimes induce a vapid cheerfulness. **1959** D. COOKE *Lang. Mus.* ii. 62 The Irving Berlin tune..is a rare example of minor 'pop' music. **1962** D. LESSING *Golden Notebk.* I. 102, I remember the sharp feeling of dislocation it gave me to hear the pop-song in London, after Willi's sad nostalgic humming of what he told us was 'A song we used to sing when I was a child'. **1963** *Daily Tel.* 7 Dec. 9/7 A 'pop music' dispute between song writers and concert promoters is to go before the Performing Right Tribunal in London on Monday next. **1967** *Crescendo* Feb. 23/2 A pop that will only last a couple of weeks. **1970** *Observer* 20 Sept. 26/1 In the world of pop, the death of Jimi Hendrix on Friday from a suspected overdose of drugs will seem as if Tchaikovsky or Mozart had also been struck down at only 24. **1973** *Country Life* 13 Dec. 2015/1 Pop-song writers masquerading as composers in the grand manner. **1974** J. COOPER *Women & Super Women* 9 During the holidays they..play pop music too loudly for their parents' liking. **1975** *Gramophone* Jan. 1357/2 Incidentally the 'pop' purchaser may well be disconcerted that the battery and carillon at the end of '1812' are relatively restrained. **1976** H. NIELSEN *Brink of Murder* i. 12 An aged spinster..not only refused to sell to Pucci but insulted the dignity of his project by leasing the premises to a group of pop musicians. **1977** *Rolling Stone* 21 Apr. 91/1 He..makes a misguided stab at pop blues in 'Bluesman'.

b. Phr. *top of the pops*, applied to the most popular or the best-selling gramophone record over a given period; also *transf.* and *fig.*, highly successful or popular.

1958 *Punch* 8 Oct. 483/1 'Wagon Train' stays top of the pops in ITV features on every channel. **1964** [see *DOLLY *a.* c]. **1965** [see *CHART *sb.* 3 c]. **1970** J. PORTER *Rather Common Sort of Crime* vi. 64 Your little friend Rodney was a dodo, a brontosaurus, last week's top of the pops..but dead, finished, a stale bun. **1978** G. GREENE *Human Factor* II. iii. 84 The top of the pops for any given year came as readily to Davis's memory as a Derby winner.

c. In various special collocations: attributive, as *pop album, ballad, band, concert, disc, fan, festival, group, lyric, number, opera, record, single, star, world*; objective, as *pop-singer, -singing* adj.; similative, as *pop-style(d)* adj.

Quot. 1880 for *pop-concert* is properly in the sense of POP *sb.*⁴

1949 *Billboard* 8 Oct. 26/2 (*heading*) Pop albums. **1955** L. FEATHER *Encycl. Jazz* 100 Basie also used a girl singer, usually for the pop ballads. **1964** *Punch* 28 Oct. 658/2 Those sentimental pop-ballads of the 'thirties. **1958** *Amer. Speech* XXXIII. 225 A *mickey* or *Mickey Mouse* band is..the kind of pop band that sounds as if it is playing background for an animated cartoon. **1967** *Listener* 16 Feb. 229/1 Some acoustical engineers in the United States believe that the sound produced by teenage pop bands is actually damaging to human ears. **1880** GEO. ELIOT *Jrnl.* in *Lett.* (1956) VII. 342 Went to our first Pop-Concert and heard Norman Neruda, Piatti, etc. **1963** 'D. SHANNON' *Death of Busybody* iv. 51, I went to the Hollywood Bowl... It was a pop-concert night, Gershwin. **1973** R. PARKES *Guardians* ii. 49 He imagined continental pop-concerts had something to do with the youth counter-culture. **1957** *Times* 19 Dec. 5/1 The most recent phenomenon in the world of the 'pop disc' has been the astonishing rise of the 'teen-age' singer. **1973** R. PARKES *Guardians* iii. 64 His income as a disc-jockey; his profits from the few pop-discs he cut; and..his chairman's salary. **1960** *Guardian* 13 Apr. 3/3 An out and out 'pop' fan. **1966** *B.B.C. Handbk.* 44 They have to cater for..the 'pop' fan. **1979** S. SMITH *Survivor* xvii. 176, I was obviously beyond the age group of the average hippy or pop fan. **1970** *Guardian* 31 July 9/5 Pop festivals..are big business. **1975** *Times* 8 Aug. 1/1 The Government has ordered an urgent review of public policies on pop festivals. **1965** M. BRADBURY *Stepping Westward* i. 65 A pop group, called the Haters, were tunelessly celebrating dim proletarian adolescent oestrus. **1967** *Listener* 18 May 644/1 Two of the Rolling Stones 'pop group' are sent for trial on drugs charges. **1977** 'E. CRISPIN' *Glimpses of Moon* viii. 147 Ten minutes alone inside the tent, with Miss Bale to keep intruders away, and that pop group to cover up any noise. **1960** *Guardian* 22 July 10/2 The committee have found that pop lyrics are drivel and often debasing. **1966** *Vogue* Oct. 177/1 Almost the only simple, open-hearted verse we now have are pop-lyrics. **1945** S. HUGHES in C. Madge *Pilot Papers* 76 The term Dance Music is used here to denote..

the playing and singing of 'pop' numbers as opposed to the cult of 'Jazz'. **1958** P. GAMMOND *Decca Bk. Jazz* ii. 38 The popularity of jangle-piano, and of pop numbers performed in cool ragtime style. **1960** *News Chron.* 31 Mar. 4/4 Pop numbers..can be sung and understood outside the story's context. **1969** N. COHN *A WopBopaLooBop* (1970) xviii. 172 Townshend has finally written a full-scale pop opera. **1976** *Cumberland News* 3 Dec., 56 11-year-olds..were practising an ambitious production of 'Smike', a pop opera based on the Dickens novel, 'Nicholas Nickleby'. **1950** *Billboard* 7 Oct. II. 27/2 (*heading*) Top pop records of the year. **1961** H. E. BATES *Day of Tortoise* 60 She played pop records such as *What Do You Want If You Don't Want Money?* **1973** L. COOPER *Tea on Sunday* ii. 27 The..strident noise of pop records. **1948** *Billboard* 25 Dec. 38 At press time it was learned that Apollo had re-signed pop singer Mary Small for another year with options. **1955** L. FEATHER *Encycl. Jazz* 79 The Decca company began to record him..in duets with pop singers. **1958** J. TOWNSEND *Young Devils* 8 The sickly exhortations of 'pop' singers. **1973** J. WAINWRIGHT *Pride of Pigs* 30 The pop singer finished his protest song and there was a thin ripple of applause. **1955** L. FEATHER *Encycl. Jazz* 160 Grotesque distortions..valid more as entertainment than as jazz or pop singing. **1962** *Times* 28 Feb. 5/4 An atmosphere more suggestive of pop-singing..than great artistry. **1949** *Billboard* 8 Oct. 26/1 (*heading*) Best-selling pop singles. **1962** A. NISBETT *Technique Sound Studio* 252 Pop singles contain the same amount as a 10-inch 78, whereas e.p. records contain perhaps double. **1978** *Sunday Times* 29 Jan. 43/1 A record by two Jamaican girls is currently No. 2 in the BBC's top twenty pop singles. **1967** *Listener* 23 Feb. 271/2 We were taken, step by step, through the process of manufacturing a pop star. **1972** J. MCCLURE *Caterpillar Cop* v. 71 She was behaving as if Boetie had become a pop star, rather than a corpse. **1955** L. FEATHER *Encycl. Jazz* 159/2 Eddie [Heywood], Jr...formed own sextet late '43, made name through pop-style arr. of *Begin the Beguine*. **1963** *Times* 24 May 15/7 The pop-style hymn-settings of John Gardner. **1974** *Publishers Weekly* 26 Aug. 302/2 It's a pop-styled run-through of the big moments, great plays and subway series heroics. **1959** 'F. NEWTON' *Jazz Scene* i. 22 Jazz has made much of its way as part of the pop world. **1967** M. DRABBLE *Jerusalem the Golden* vii. 170 She had as resolutely and as puritanically scorned the pop world..as her mother had done before her. **1973** *Melody Maker* 25 Aug. 27 In the pop world, the rule is that musicians are a special breed.

2. *pop art*, art that uses themes drawn from popular culture, *spec.* an art form characterized by the depiction of commonplace subjects using strong colour and imagery, sharp features, and a photographic technique of representation (see also quot. 1967). Also *ellipt.*, as *pop*. Hence *pop artist*, *-painter*; *pop-painting* vbl. sb.

1957 *Listener* 26 Sept. 464/1 A sophisticated apologia for subtopia is to call it 'pop art' which the middle-aged are perverse to frustrate. *Ibid.* 470/1 Some people even defend Subtopia as a type of vigorous folk art—or 'pop art'—to be fostered. **1958** *Archit. Rev.* CXXIII. 208/1 Four chairs..would not have been known to the designer of this room had they not been published in the popular magazine *Look*, which gave the *chaise-longue* version the full pop-art treatment. **1962** *Listener* 9 Aug. 217/3 All three of the painters are adherents of the new school of 'pop'. *Ibid.* 30 Aug. 324/1 Certain of the 'pop' painters can apparently be paired off with artists on the other side of the Atlantic. *Ibid.* 324/2 The tendency of 'pop' paintings, Hockney's for instance, to resort to the use of words in order to help out the images is in itself significant. *Ibid.* 27 Dec. 1087/1 The third wave of pop artists use their imagery to differentiate themselves from the regular audience for art. **1964**, etc. [see *OP¹]. **1966** 'H. MACDIARMID' *Company I've Kept* iii. 78 The pop artist does not address any audience, does not represent any point of view; he has staked everything on nothingness. **1967** L. ALLOWAY in L. R. Lippard *Pop Art* 27 The term 'Pop Art' is credited to me, but I don't know precisely when it was first used. (One writer has stated that 'Lawrence Alloway first coined the phrase "Pop Art" in 1954'; this is too early.) Furthermore, what I meant by it then is not what it means now. I used the term, and also 'Pop Culture', to refer to the products of the mass media, not to works of art that draw upon popular culture. In any case, sometime between the winter of 1954–55 and 1957 the phrase acquired currency in conversation, in connection with the shared work and discussion among members of the Independent group. **1968** *New Yorker* 24 Feb. 100 There were about a hundred and fifty paintings on view in the huge Main Hall..and they ranged from Op and Pop to Picasso. **1971** 'A. BURGESS' *MF* xiii. 144 There was a big pop-art poster whose crude yellows and blues were an obscenity. **1972** E. LUCIE-SMITH in Cox & Dyson *20th-Cent. Mind* III. xvi. 470 The first example of Pop is now generally conceded to have been a small collage made by the English painter Richard Hamilton..in 1956. **1976** *New Yorker* 22 Mar. 107/1 Among the Pop artists shown, Claes Oldenburg is by far the most gifted as a draftsman. **1977** *Jrnl. R. Soc. Arts* CXXVI. 47/2 Out of Léger came aspects of Pop: in particular that aspect known as Roy Lichtenstein.

3. Appealing to or expected to appeal to popular taste generally (chiefly in the senses of *POPULAR *a.* (*sb.*) 4a). Also *absol.* as *sb.* Spec. *pop culture*, culture based on popular taste and disseminated widely and usu. on a commercialized basis; hence *pop-cultural* adj. Also, of a technical subject, etc.: popularized, presented in a popular form, as *pop psychology* (hence *pop-psycher*, *-psychologist*).

1958 *Spectator* 14 Feb. 197/2 The promoters of 'pop' fiction must ruthlessly wipe out any tragedy that remains unique and personal. **1958** *Observer* 23 Mar. 14/3 As a sop to pop, the gallants on the benches at the sides of the stage could be TV personalities. **1958** *Ibid.* 25 May 14/2 His admirable pop science *New Horizon* series. **1959** C. MACINNES *Absolute Beginners* 73 It's my aim..to bring quality culture material to the pop culture masses. **1962** *Observer* 20 May 12/7 Pop archaeology books sell like hot cakes. **1962** *Punch* 12 Sept. 390/2 A highly competent performer on these pop-science occasions. **1963** *Ibid.* 3 July 30/2 Pop religion is the dreariest mixture imaginable. **1963** *Dædalus* Winter 22 Available critiques of pop-cultural depravities (from *Playboy* to the *National Geographic*) and compilations of economic facts about massification..are, to be sure, of some help. **1964** *Punch* 5 Feb. 211/1 An almost naïvely sensational bit of pop-psychology sex. **1966** D. JENKINS *Educated Society* ii. 58 That commercialized 'pop culture' which is a form of anti-culture. **1967** *New Scientist* 25 May 473/1 Expo is dominated by technology, but it is a gay, often pop technology that you meet, technology that is confident enough to laugh at itself. **1968** *Punch* 27 Nov. 753/1 Pop-psychologists are saying that certain trigger phrases used by Enoch Powell expose a subconscious racial prejudice. **1969** *Listener* 17 July 92/2 If Pop means mass media and consumer goods, ads and comics, Coke bottles and plastic, what then can it have to do with Art? **1970** G. GREER *Female Eunuch* 171 The pop revolution..has replaced sentiment with lust. **1970** K. MILLETT *Sexual Politics* (1971) II. iv. 186 In such cases Freud and his school after him will do all in their power to convince her of the errors of her ways:..by the actual mental policing of 'pop psych'. **1971** *Time* 14 June 16/2 The fact that Ed [Cox] proposed so quickly after Tricia [Nixon] began her new life at the White House might suggest to pop-psychers that he was afraid of losing her. **1972** *Nature* 25 Aug. 471/2 The author has shrugged off..practically everything that animal ethologists have tried to contribute to our understanding..—dismissing..[Desmond] Morris's books as 'pop ethology'. **1973** J. WAINWRIGHT *High-Class Kill* 41 Pop culture: garbage done up in poster-colours and caterwauling to badly played guitars. **1975** *Imperial Oil Rev.* IV. 30/2 How to make work more satisfying or, to use the word of pop sociology, how to 'humanize' it. **1975** *New Yorker* 21 Apr. 111/1 No one has yet piously complained of too much violence in the Ngorongoro Crater or tried to shroud a beehive in pop psychology. **1977** P. JOHNSON *Enemies of Society* xi. 160 At various levels, too, psychology and sociology have become the pop-science, or folk-science, of the western urban masses. **1977** *Time* 14 Mar. 43/1 Most of the Morgan message is standard to all the pop self-help books that publishers have been churning out ever since Dale Carnegie and Norman Vincent Peale reaped their first millions. **1978** *Encounter* July 96/1 On the debit side..are the evils of 'development' and the pap of pop culture.

pop, *v.*¹ Add: **2. b.** Delete † and add later examples.

1931 E. O'NEILL *Mourning becomes Electra* (1932) 217 Small comes tearing out and down the portico steps, his face chalky white and his eyes popping. **1940** W. FAULKNER *Hamlet* I. ii. 37 They looked exactly like two fellows that had done hung themselves in one of these here suicide pacts, with their heads snubbed up together and pointing straight up..and their eyes popping. **1951** M. KENNEDY *Lucy Carmichael* II. i. 73 Pray Bess, what was he like? Oh, says Bess, her eyes popping, he's *terrific!* **1961** N. MANERO *Cook-Out Barbecue Bk.* 16 You'll have the neighbor's eyes popping as well as their mouths watering! **1979** G. HAMMOND *Dead Game* xi. 143 He sold the Dickson Round Action [gun] there for a price that made Molly's eyes pop.

3. *to pop corn* (earlier and further examples).

1850 *Quincy* (Illinois) *Whig* 12 Nov. 4/1 One barrel of rice corn will make 32 barrels after popping. **1853** *Harper's Mag.* May 853/1 A little boy sat by the kitchen-fire, A-popping corn in the ashes. **1873** 'S. COOLIDGE' *What Katy Did* x. 201 'I popped the corn!' cried Philly. **1907** *St. Nicholas* May 614/1 Grandma lives on a farm and we used to have great fun popping corn whenever we went to see her. **1949** *Sat. Even. Post* 21 May 36/1 Last year American farmers grew some 300,000,000 pounds of popcorn. This, when popped, is enough to fill 2,400,000,000 ten-cent bags. **1979** *Sunset* Apr. 129/2 (Advt.), The Popaire hot air popper pops 4 quarts of light, fluffy popcorn in 5 minutes!

5. (Further examples.)

1858 *Punch* 20 Nov. 206 If you will pop on your hats..I'll take you and your friend out for a drive. **1891** B. POTTER *Let.* in J. Mackenzie *Victorian Courtship* (1979) ix. 125, I popped on an old skirt and a mackintosh and trudged through the rain. **1977** B. PYM *Quartet in Autumn* ii. 22 'I should put the bacon in a cooler place if I were you,' said Letty. 'Yes, I'll pop it in one of the filing cabinets.' **1977** K. O'HARA *Ghost of T. Penry* viii. 67 Sit you down and I'll pop the kettle on.

6. a. (Later examples.)

1900 M. SHARP *Something Light* vii. 64, I haven't actually..popped, yet. **1972** *N.Y. Times* 3 Nov. 7/1 (Advt.), Now's the time to pop the question! 20% off diamond engagement rings. **1976** *Daily Mirror* 16 Mar. 9/3 The thought of popping the question to Princess Marie Therese de Bourbon Parma 'has never entered my head,' he added.

b. *intr. to pop off*, to speak hastily, angrily, or wildly; to state one's opinions vociferously; to complain loudly. *U.S. colloq.*

1933 PARTRIDGE *Slang To-day & Yesterday* 455 *Pop off, to*, talk wildly, threateningly, argumentatively. C20. **1934** M. H. WESEEN *Dict. Amer. Slang* 381 *Pop off*, to lose one's temper; to give vent to anger. **1943** *Sun* (Baltimore) 20 Sept. 16/8 The dealer 'popped off without knowing what he was talking about'. **1951** R. S. PRATHER *Bodies in Bedlam* vi. 47, I popped off to Brane last night, but I didn't kill him. **1970** *Daily Tel.* 7 Feb. 16/2 Company chairmen have been popping-off about the iniquities of selective employment tax for four years. **1977** *Time* 7 Mar. 40/3 Most Plains residents dismissed Billy's charge. 'He was just poppin' off,' said one woman. **1977** J. WAMBAUGH *Black Marble* (1978) x. 241 He remembered what happened to him today in a phone booth when he popped off, so he bit his lip and kept quiet.

8. (Further examples with *in*, *out*, *up*). Also const. *off*, *over*. Also (*Austral. colloq.*) phr. *how are you popping (up)?*: how are you getting on?

1860 F. NIGHTINGALE *Notes on Nursing* iv. 29 Many of the accidents which happen from feeble patients tumbling down stairs..happen..from the nurse popping out of a door. **1904** E. NESBIT *Phoenix & Carpet* xii. 224 If you'll excuse me, I'll just pop out and see what I can do. **1913** C. MACKENZIE *Sinister St.* I. I. iv. 54 Nurse..had acquired a habit..of popping out of the back-door on secret errands. **1919** WODEHOUSE *Damsel in Distress* xv. 186 'And now you get along,' said the man. 'You pop off.' **1934** C. LAMBERT *Music Ho!* III. 156 He [*sc.* Glinka] was more than a gifted amateur who happened to pop up at the right time. **1942** *Tee Emm* (Air Ministry) II. 88 A cunning safety switch..pops up when there is a short. **1960** *Guardian* 26 Feb. 5/4 Mrs Harris popped out to do some shopping. **1960** I. JEFFERIES *Dignity & Purity* ii. 28 Let's pop off for a drive. **1965** *Listener* 2 Dec. 934/2 'Afternoon Theatre', sometimes infinitely trivial, popped up with a winner in *The Aquarium on Platform Two*, by Peter Preston. **1968** 'N. BLAKE' *Private Wound* v. 79 Maire'll look after you till I get back. I just have to pop out and see a fella for a minute or two. **1977** B. PYM *Quartet in Autumn* vii. 63 'Goodbye, then,' she said. 'I'll pop in again some time.' **1978** J. THOMSON *Question of Identity* xiv. 146 Will you pop over to my tent and bring me my little box?

1894 H. LAWSON *Short Stories in Prose & Verse* 89 'How are yer?' 'Oh! I'm alright!' he says. 'How are ye poppin' up!' **1907** N. SPIELVOGEL *Cocky Farmer* 16 Whatto, Joe. How are you popping up? **1933** N. LINDSAY *Saturdee* 10 What-oh, Stinker, how you poppin' up? **1942** S. CAMPION *Bonanza* 207 Howya poppin', cobber?

c. (Further examples.) Also ellipt. *to pop*. Also *trans.*, to kill, destroy.

1824 J. HOGG *Private Mem. Justified Sinner* 253 Might we not..pop him off in private and quietness? **1922** E. WALLACE *Flying Fifty-Five* x. 58 'If he'd only popped off in the war, Jacques.'..'You might have been 'popped off' yourself if you'd only got within range of a bullet.' **1928** D. L. SAYERS *Unpleasantness at Bellona Club* ix. 110 Perhaps it's just as well he popped off when he did. He might have cut me off with a shilling. **1940** G. S. GORDON *Let.* 24 May (1943) 221, I have joined the Defence Volunteers, and hope to pop a parachutist before the business ends. **1945** J. B. PRIESTLEY *Three Men in New Suits* v. 65 He fancies he might pop off at any time. **1952** W. R. BURNETT *Vanity Row* (1953) v. 45 She'd be worrying how to knock me off. Or trying to get me het up..so's I'd pop. **1975** J. GOULET *Oh's Profit* vi. 36 Oh popped a carpenter ant, chewed. **1975** *New Yorker* 26 May 32/2, I agreed not to say 'death', 'dying',..'go home feet first', 'pop off the hooks'. **1977** *Navy News* Sept. 21/5 It is possible for a Seacat or Seaslug missile to get close enough to topple the target off course and 'pop' the parachute recovery system.

d. *to pop in and out*: to visit or come and go frequently or casually.

1858 MRS. GASKELL *Let.* 19 Oct. (1966) 517 We have more people popping in & out than we expected. **1926** WODEHOUSE *Heart of Goof* iv. 126 He drew a picture of their little home, with Crispin for ever popping in and out. **1971** N. FREELING *Over High Side* I. 40 Martinez was not altogether unknown... He had often 'popped in and out'. **1974** 'S. WOODS' *Done to Death* 14 He can't keep popping in and out... But if she had a companion—. **1979** 'M. HEBDEN' *Death set to Music* iii. 26 He spent most of his time off duty popping in and out of bed with any pretty woman he could find.

e. *Cricket.* Of the ball: to rise sharply off the pitch when bowled; to get up (GET *v.* 72 h). Also *to pop up*.

1871 'THOMSONBY' *Cricketers in Council* 39 'Spin' is not twist, it is that which gives the ball a tendency to twist, break back, shoot, pop up, or, in fact, do something eccentric. **1888** STEEL & LYTTELTON *Cricket* iii. 153 The ball will twist a great deal on this class of wicket [hard and crumbled]... It is also inclined both to 'pop' and keep low. **1906** A. E. KNIGHT *Compl. Cricketer* iii. 119 The ball, too, will rear up quickly, kick or 'pop up'. **1921** P. F. WARNER *My Cricketing Life* vi. 126 On a sticky wicket he was capable of sending down a difficult off break, and of making the ball pop. **1926** J. B. HOBBS *Test Match Surprise* xxi. 211 Then the ball commenced to 'pop', in cricket parlance—to 'stop and look at them'—and Grimmell..had the two brilliant batsmen in difficulties. **1959** *Times* 29 May 4/2 Nicholls skied a catch..aiming across the line at one that popped.

f. To pay (*for*). *slang.*

1959 R. BLOCH *Big Kick in Blood runs Cold* (1961) 213 He popped for three jugs tonight, just to get in. Likes to make the scene. *Ibid.* 218 He didn't pop... I said we were leaving..and all he did was smile. **1968** L. J. BRAUN *Cat who turned on & Off* (1969) xxi. 182 Hell. I didn't buy you anything, but I'll pop for lunch.

9. *trans.* In Baseball: to hit (a ball) high into the air but close to the batter, thus providing an easy catch. Also *intr.*, to get put out by hitting a high ball that is caught by an opponent. *N. Amer.*

1867 *Ball Players' Chron.* 6 June 2/3 On Hunniwell popping one up which fell into Sumner's hands, Smith had to retire, a double play putting both out. **1886** [see *FAN *v.* 8 b]. **1912** C. MATHEWSON *Pitching in a Pinch* 204 Then Doyle popped up a weak foul behind the catcher. **1931** *Kansas City* (Missouri) *Times* 19 Oct., Hallahan replaced Grimes on the mound for the Cardinals and then Bishop popped out to 'Pepper' Martin. **1947** *Los Angeles Times* 3 Oct. II. 1/7 Johnson swung and popped up to end the inning. **1948** *Chicago Tribune* 7 Mar. II. 1/4 Lupien popped to Johnson. **1974** *Los Angeles Times* 13 Oct. III. 10/2 Bando struck out. Messersmith grounded to the pitcher. Rudi popped to short.

10. *trans.* To inject (a narcotic drug). Also, to take (a narcotic drug). Also *intr. slang.*

1956 R. THORP *Viper* vi. 92 Nearly everyone there seemed to be popping. There were so many needles working you might have thought it was a tailors shop. **1959** W. BURROUGHS *Naked Lunch* 29 Ever pop coke in the mainline? **1962** 'K. ORVIS' *Damned & Destroyed* xii. 79 Was there ever a junkie..that was too pooped to pop? **1968** *N.Y. Times* 2 Aug. 46 Executives of finance and insurance companies are popping pills these days to tranquilize their nerves. **1968** M. WOODHOUSE *Rock Baby* ii. 109 For him the day..started when he swallowed the first pill or popped the first vein. **1972** *Sunday Sun* (Brisbane) 2 July 14/3 The addict..now bangs, pops, shoots and jabs his veins with the hypodermic needle. **1976** R. ROSENBLUM *Sweetheart Deal* iv. 46 The half-million ghetto kids who'll start popping junk this year. **1977** *Amer. Speech 1975* L. 64 *Pop*,..take an amphetamine in order to stay awake (to study). 'I popped for my history final.'

pop, *int.*, *adv.* Add: *spec.* in phr. *to go off pop* (N.Z. colloq.), to break into angry speech.

1933 'P. CADEY' *Broken Pattern* xii. 126 There's no need to go off pop like that. **1940** F. SARGESON *Man & Wife* (1944) 65 He'd do things wrong too, and every chance he got he'd pick on me and go off pop. And of course I'd tell him off back.

2. In repeated form, with (the action or sounds of) a series of pops. Also as *adj.* and *sb.*, the sound of such a series.

1928 V. WOOLF *Writer's Diary* 22 Mar. (1953) 124 A rabbit that passes across a shooting gallery, and one's friends go pop-pop. **1951** J. FRAME *Lagoon* 10 The pop-pop boats we used to whizz round in the bath on Christmas morning. **1957** —— *Owls do Cry* 29 Sometimes the coal makes a pop-pop.

pop-. Add: **pop-beer**, ginger-beer or some other aerated drink; **pop-bottle**, a bottle for an aerated drink; **pop-call** *U.S.*, a sudden or unexpected visit; **popcorked** *a.*, provided with a cork which pops when drawn; **pop-eye** (earlier examples); **pop-hole**, a hole in a hedge, fence, etc., through which animals can pass; **pop-off**, (*b*) used *attrib.* to designate a safety valve which operates with a pop; **pop-rivet**, a kind of tubular rivet used for fastenings where only one side of the work is accessible, and which is inserted into the hole and then clinched by the action of withdrawing a central mandrel; hence as *v. trans.*; **pop-riveting** *vbl. sb.*; **pop safety valve** = *pop-valve* in Dict. and Suppl.; **pop-top** *U.S.* = *ring-opener* (*RING *sb.*¹ 18); also *attrib.*; hence **pop-topping** *vbl. sb.*; **pop-valve**, delete '= PUPPET-VALVE' and substitute: a spring-loaded safety valve designed to open or close very rapidly at a predetermined pressure; (further examples).

1887 C. D. WARNER *Their Pilgrimage* (1888) ii. 40 Shooting-galleries, pop-beer and cigar shops, restaurants, [etc.]. **1900** ADE *Fables in Slang* 28 More than once he had let drive with a pop bottle at the umpire. **1921** A. G. EMPEY *Madonna of Hills* 2 The occasional noise of an empty 'pop' bottle kicked over under the seats, indicated the audience was restless. **1946** C. HIMES *Black on Black* (1973) 259, I looked around and saw cases of pop bottles stacked against the wall. **1971** *Country Life* 23 Dec. 1788/1 That cherished book, its olive green covers carefully clothed in brown paper as a safeguard against pop-bottle rings. **1941** W. C. HANDY *Father of Blues* (1957) iii. 27 The pop-calls of policemen dropping in to catch vagrants. **1974** *News & Reporter* (Chester, S. Carolina) 24 Apr. 4-B/8 Mr. and Mrs. Mark Winchester from Charlotte made 'pop calls' at the Fergusons, W. C. Gladdens, and Countermans. **1922** JOYCE *Ulysses* 260 Pat paid for diner's popcorked bottle: and over tumbler tray and popcorked bottle ere he went he whispered. **1828** A. ROYALL *Black Bk.* II. 377 But the lawyer..is a shrimp in size, a sallow complexion, small face, and little blue pop eyes. **1885** 'C. E. CRADDOCK' *Prophet Gt. Smoky Mts.* ii. 45 He had wide pop-eyes, and long ears, and a rabbit-like aspect. **1944** *Living off Land* ii. 30 Kangaroos, often hard to shoot, are fairly easy to snare in country where there are netting fences, as often their pads run alongside the fence to a spot where they either jump it or crawl through a 'pop-hole'. **1945** 'G. ORWELL' *Animal Farm* i. 9 Mr Jones, of the Manor Farm, had locked the hen-houses for the night, but was too drunk to remember to shut the pop-holes. **1949** D. M. DAVIN *Roads from Home* 159 'One thing I will say for a pop-hole,' Paddy said, 'once you've found it you're pretty well set. They always make for it.' **1963** *Times* 14 Jan. 13/2 They [*sc.* pigs] get two-thirds of their food outside, having free access through a pop-hole, the rest they forage for in the litter. **1944** H. KERWIN *Arc & Acetylene Welding* ii. 6 Keep all pop-off valves on the [carbide] generator in good working order. **1977** *Brit. Jrnl. Anaesthesia* XLIX. 71/1 There is a cooling coil and a safety pop-off valve, by which extra pressure is released and extra moisture in the circuit is expelled intermittently. **1932** *Air Ann. Brit. Empire 1932–33* 396 The most interesting type of rivet which has been specially developed is a tubular rivet known as the pop-rivet. **1953** *Flight* 18 Sept. 410/2 A corrugated core is sandwiched between two skins, the outer skin being spot-welded and the inner skin pop-riveted to this core. **1967** J. MILLS *Low-Cost Car Repairs* xii. 243 Pop-rivets are easily and quickly drilled out in a matter of seconds. **1973** P. REVERE *Do Your Own Car Body Repairs* vii. 36 Any area to be pop riveted which is on a surface liable to to be seen, will have to be countersunk first. **1978** *Daily Tel.* 18 Nov. 14/6 His 1924 version was the first all-steel aircraft produced in Britain and necessitated the now universally-used pop-rivet. **1934** M. LANGLEY *Metal Aircraft Construction* (ed. 2) ix. 309 Some very ingenious methods of tubular or 'pop' riveting have been devised by the A.T.S. Co., Ltd., which..combines the patented processes for metal construction of the Armstrong-Whitworth, Boulton & Paul, and Gloster firms. **1973** P. REVERE *Do Your Own Car Body Repairs* vii. 37 The cost of a pop riveting gun is about a third the price of a new tyre. **1908** G. F. GEBHARDT *Steam Power Plant Engin.* xv. 598 (*heading*) Consolidated pop safety valve. **1961** F. A. S. BROWN *N. Gresley* xvii. 123 One of the new 'B12s', No. 8579, was the subject of extensive rebuilding, and the opportunity was taken to mount a larger boiler..and a round-topped fire-box with pop safety valves. **1970** *Time* 21 Sept. 60/3 Insert finger, tug and quaff: in those few seconds, the aluminium ring atop a pop-top can of beer or soda fulfills its function. **1972** *Washington Post* (Potomac Suppl.) 29 Oct. 15/3 Acres and acres of sweeping lawn without a styrofoam cup, a pop-top, [etc.]. **1977** C. MCFADDEN *Serial* (1978) xxxiii. 72/2 Joan snapped the pop top on the last can of beer. **1975** *Publishers Weekly* 27 Jan. 42/3 (Advt.), Pop-topping... New craft with pull-tabs from beverage cans. [**1881** *Engineering News* 10 Sept. 362/2 The great peculiarity of the 'Pop' nickel-seated safety valve..is that by the use of a stricture the recoil action of the steam is made available to overcome the increased pressure of the spring on the valve-head as it rises.] **1908** V. PENDRED *Railway Locomotive* xxiv. 185 On some lines 'Pop' valves have been tried. They are so called, because instead of rising gradually as the pressure increases after they have begun to blow off, they lift suddenly with a 'pop' and blow off hard for a minute or so. **1927** E. L. AHRONS *Brit. Steam Railway Locomotive 1825–1925* xxiii. 364/2 Ramsbottom safety valves, though still used, are rapidly giving place to pop valves of the Ross pattern. **1968** J. H. WHITE *Amer. Locomotives* viii. 148 The Richardson valve was said to open more than twice as far as an ordinary safety valve. Because of its quick opening it became popularly known as the 'pop' valve.

popadam. Add: Also **papadam, papadum, papodam, papodum, poppadom, puppadum, puppodum.** (Further examples.)

1906 *Mrs. Beeton's Bk. Househ. Managem.* lviii. 1601 Thin wafer-like cakes called Papodums. **1928** *Daily Express* 19 July 5/2 There are Bombay ducks and papodams. **1928** *Sunday Express* 12 Feb. 10 And then add the curry... The hot chutney of Madras is the best accompaniment, and with it you may take, if you will, poppadums, Bombay ducks, and a little powdered mint. **1931** *Punch* 13 May 506/3 A puppodum is a thin wafer-like cake made of lentil-flour or something like that. **1932** *Times Lit. Suppl.* 22 Dec. 978/1 *Papadums* are essential to a curry; but to make a *papadum* requires the accumulated experience of a few generations. **1936** *Times Lit. Suppl.* 3 Oct. 792/2 A reputable firm will supply..curry powder, puppadums, [etc.]. **1951** *Good Housek. Home Encycl.* 432/2 Chupatties and papadum are types of Indian breads often served with curries. **1959** *Good Food Guide* 316 Curried chicken Madras with poppadoms and Bombay duck. **1962** *Housewife* (Ceylon) Feb. 33 A meal devoid of appetizers, in the form of a salad, papadams or a blob of cream on the fruit salad, is, I feel, a miserable failure. **1969** *Guardian* 21 Mar. 11/3 A new play opens at the Arts Laboratory..'Chicken Curry and Poppadoms' by Richard Huggett. **1974** 'J. LE CARRÉ' *Tinker, Tailor* xxviii. 243 Jerry Westerby with his enormous hands shattered a popadam on to the hottest curry on the menu.

pop-corn. Add: (Further examples.) Also *transf.* and *fig.*

1823 W. FAUX *Memorable Days Amer.* 302, I crossed the Big Wabash..at La Valette's ferry, where is beautiful land, fine young orchards, and two lonely families of naked-legged French settlers, from whom I received two curious ears of poss [*sic*] corn. **1850** *Quincy* (Illinois) *Whig* 12 Nov. 4/1 Pop corn is dependent for its peculiar powers..upon the quantity of oil which its whole contains. **1855** 'Q. K. P. DOESTICKS' *Doesticks, what he Says* xix. 257 [He] had just pawned his coat and a spare shirt to get money to set himself up in business again, as a pop-corn merchant. **1875** *Chicago Tribune* 21 Nov. 2/6 Each one had grown tired of jaw-breakers and popcorn balls. **1922** 'R. CROMPTON' *Just—William* xi. 220 He purchased a large bag of popcorn. **1947** *Downtown Shopping News* (Chicago) 2 Jan. 16/3 Pop corn balls are the delight of every child. **1949** F. PETO *Amer. Quilts & Coverlets* ii. 23 The loom-made tufted type used candlewicking for the warp of the foundation and..the characteristic 'popcorn' decoration. **1953** [see *HUSTLER 1]. **1965** MRS. L. B. JOHNSON *White House Diary* 20 Dec. (1970) 340 There were garlands of popcorn. **1968** *Guardian* 15 Nov. 1/6 Miss Plummer won through on dazzling good looks. As far as a popcorn poll went, they appeared to be a popular choice. **1972** *Country Life* 9 Mar. 591/1 Popcorn is an extremely hard form of maize whose corns expand and 'pop' on heating; it is used only in confectionery. **1973** C. & R. MILNER *Black Players* ii. 41 *Popcorn* is a humorous insult which may be translated as 'light-weight'.

pope, *sb.*¹ Add: **I. 1. a.** (Earlier and later examples of *Black Pope*.)

1873 *Times* 30 May 8/1 The only practical result has been an almost unanimous vote by which the General of the Jesuits, Father Becks—the 'Black Pope' as he is called—will be instantly..turned out of the apartments. **1911** *Encycl. Brit.* XV. 339/2 It is said that the general of the Jesuits is independent of the pope; and his popular name, 'the black pope', has gone to confirm this idea. **1976** P. VAN RJNDT *Tetramachus Collection* (1977) i. 15 Political details gleaned by the ranks of the 'black pope', ..head of the Society of Jesus.

3. a. Delete † *Obs.* and add later examples.

1902 *Encycl. Brit.* XXVII. 237/2 'The most holy Pope and patriarch of the great city of Alexandria and of all the land of Egypt, of Jerusalem the holy city, of Nubia, Abyssinia, and Pentapolis, and all the preaching of St. Mark', as he is still called. **1925** [see *BEATITUDE 1 b]. **1976** *Daily Tel.* 24 Aug. 4/5 Pope Shenouda the Third, Patriarch of the Egyptian Coptic Church, said..that the deposition of Abuna Theophilos was 'illegal and inhuman'.

6*. A hot spiced drink of mull based on any of various wines. Cf. BISHOP *sb.* 8, CARDINAL *sb.* 5.

1920 G. SAINTSBURY *Notes on Cellar-bk.* xi. 162 'Pope', *i.e.* mulled burgundy, is Antichristian, from no mere Protestant point of view. **1965** O. A. MENDELSOHN *Dict. Drink* 264 *Pope*, a spiced drink made from tokay.., ginger, honey and roasted orange. **1976** *Times* 15 Jan. 12/8 Many of these hot drinks have clerical names—Bishop being a type of mulled port, Cardinal using claret, and Pope Champagne. **1977** *Centuryan* (Office Cleaning Services) Christmas 8/2 A mull..using Tokay, the famed Hungarian dessert wine, was known as 'The Pope'.

7. b. *pope's hat*, applied to the head-dress of the Grenadier Guards. *Literary.*

1886 R. L. STEVENSON *Kidnapped* ii. 5/1 An old red-faced general on a grey horse at the one end, and at the other the company of Grenadiers with their Pope's-hats.

pope, *v.* Delete † *Obs.* and add: **1.** (Later example.)

1966 *Duckett's Reg.* Feb. 14/2 He [*sc.* Pope John XXIII] would pope it in his own way, God guiding him.

2. b. To be converted to Roman Catholicism; to become a Roman Catholic.

c **1916** in E. Waugh *Life R. Knox* (1959) II. i. 142 I'm *not* going to 'Pope' until after the war (if I'm alive). **1954** R. MACAULAY *Last Lett. to Friend* (1962) 163, I was..very sorry that your friend..has 'poped', as we call it here. **1961** *Spectator* 19 May 709 In another generation the Upper Chamber may be riddled with families who have poped. **1966** J. BETJEMAN *High & Low* 37 Kensit threatens and has Sam Gurney poped? **1977** *Observer* 27 Nov. 28/6 Wilfred [Knox], an Anglo-Catholic priest who never showed the least inclination to follow his younger brother into the Roman Church—or 'to Pope' as it was facetiously called among the undergraduates of Ronnie's generation.

Popean, Popeian varr. POPIAN *a.* (in Dict. and Suppl.).

pop-eyed, *a.* orig. *U.S.* [POP-.] Having bulging or prominent eyes; wide-eyed (with amazement, etc.).

1830 A. ROYALL *Lett. from Alabama* 176 The first countenance I caught, was Senator Foot of Connecticut—a handsome middle-sized black pop-eyed Yankee. **1860** [in Dict. s.v. POP-]. **1906** *Atlantic Monthly* Oct. 573 The class was open-mouthed, and the professor pop-eyed with wonder. **1923** R. D. PAINE *Comrades of Rolling Ocean* ix. 152 They are simply pop-eyed to hear all about the speedy apprentice. **1937** *John o' London's Weekly* 22 Jan. 682/1 The king [*sc.* George III] fumbled for a few years with makeshift ministries,..and then committed his affairs to Lord North, an amiable pop-eyed creature, rather like himself in appearance. **1947** V. NABOKOV *Bend Sinister* II. 12 Paying perhaps terrific fines, but stopping the train. Say, why did you do it? the popeyed conductor might ask. **1952** WODEHOUSE *Pigs have Wings* iv. 74 The landlord of the Emsworth Arms..and the half dozen Shropshire lads who were propping up the establishment's outer wall had stamped her with the seal of their popeyed approval. **1973** 'E. MCBAIN' *Let's hear It* viii. 124 Her audience..watched her every move in pop-eyed fascination. **1978** A. PRICE *'44 Vintage* x. 131 He stared pop-eyed at Butler..then started to wave madly.

pop-gun, popgun. Add: **1.** (Later example.)

1967 D. ABERCROMBIE *Elem. Gen. Phonetics* ii. 24 An air-stream mechanism can be compared to a fruit-spray, a Flit gun, a syringe, or a child's pop-gun.

2. (Later example.) Also *transf.* (contextually an aeroplane).

1919 M. BEER *Hist. Brit. Socialism* I. II. vii. 240 To equip men with pop-guns for a hunting expedition in the jungle. **1929** W. FAULKNER *Sartoris* I. 43, I tried to keep him from going up there on that goddam little popgun... I tried to keep him from going up there on that damn Camel.

3. (Later examples.)

1844 *Knickerbocker* XXIII. 182 To the United States in reference to the pop-gun shots of foreign tourists, might be addressed the warning which Peter Plymley thundered against Bonaparte. **1874** E. EGGLESTON *Circuit Rider* ix. 87 He had been flogged in boyhood for shooting pop-gun wads into the face of a portrait of the reigning monarch. **1895** *Montreal Med. Jrnl.* XXIII. 565 The physician without physiology and chemistry flounders along in an aimless fashion, never able to gain any accurate conception of disease, practising a sort of pop-gun pharmacy, hitting now the malady and again the patient, he himself not knowing which. **1963** *Times* 5 Jan. 12/3 A complicated regimen of treatment which Dr. Walton described as 'pop-gun polypharmacy'. **1968** *Wall St. Jrnl.* 29 Mar. 1 (*headings*) Centers in Slums Offer Legal, Employment Assistance; 'Outreachers' Make Rounds. A Popgun Effort, Critics Say.

Popian, *a.* Add: (Later examples.) Also as *sb.*, an imitator of the poet Alexander Pope.

a **1849** H. COLERIDGE *Ess. & Marginalia* (1851) II. 121 Neither Rogers nor Campbell are Popeans. They belong to another school—the sentimental. **1895** W. D. HOWELLS *My Literary Passions* 55, I..hammered away at my blessed Popean heroics till nine, when I went regularly to bed, to rise again at five. **1914** J. A. ROY *Cowper & his Poetry* 54 He [*sc.* Johnson] failed to remark the absence of the Popeian inversions in the seemingly orthodox verse. **1953** R. FULLER *Second Curtain* v. 74 What nourishment could he give his Popean young poet? **1975** *Times Lit. Suppl.* 14 Mar. 275/1 There is no philistine repudiation of the achievements of Popeian scholarship.

popinac (pǫ·pinæk). *U.S.* Also **popinack.** [f. OPOPANAX.] A tropical or subtropical leguminous shrub, *Acacia farnesiana*, whose fragrant yellow flowers yield an essential oil used in perfume; also called the opopanax tree.

1900 L. H. BAILEY *Cycl. Amer. Hort.* I. 8/2 Acacia.. Farnesiana... Popinac, Opopanax, Cassie... Grown in S. France for perfumery. **1945** R. P. WODEHOUSE *Hayfever Plants* iii. 113 One of the best known [acacias] is the opopanax or huisache,..also called popinack and cassie. It is a small tree, 20 to 30 feet high, with spreading spiny branches bearing bright yellow flowers closely compacted in small globular heads. **1952** [see *CASSIE*].

Popish (pōu·pif), *a.*² Also **Popeish.** [f. *Pope*, proper name + -ISH¹.] = POPIAN *a.*

1825 *Gentl. Mag.* XCV. I. 334 In this *Popish* controversy, though Mr. Bowles may affix the term 'final' to his Appeal, we have some doubt whether he will be permitted to have the last word. **1882** MRS. OLIPHANT *Lit. Hist. Eng.* I. 76 The very words of the Popish era still lingered on Cowper's tongue. *c* **1885** —— in A. L. Coghill *Autobiogr. & Lett. Mrs. Oliphant* (1899) 13 She..was fond of quoting Pope, so that we used to call her Popish in afterdays when I knew what Popish in this sense meant. **1944** *New Yorker* 9 Dec. 97/2 The 'Times' vigorous and somewhat Popeish effort included this stern couplet.

poplar. Add: **3. poplar-borer** *U.S.*, substitute for def.: the larva of a beetle, *Saperda calcarata*, which attacks the trunk and branches of poplar and certain other trees; (examples).

1884 *Rep. Comm. Agric.* (U.S. Dept. Agric.) 383 The Poplar-Borer..has been destructive to poplar trees on the shore of Casco Bay. **1942** S. W. FROST *Gen. Entomol.* xix. 381 The poplar borer..and the carpenter worm..keep at least a portion of their burrows free from frass and other waste material. **1972** SWAN & PAPP *Common Insects N. Amer.* 455 Poplar Borer... The larvae work in the trunk and large limbs of felled and weakened poplar.

Poplarism (pǫ·plăriz'm). [f. *Poplar*, the name of a district, formerly a borough in the East End of London + -ISM.] The policy of giving out-relief on a generous or extravagant scale, practised by the Board of Guardians of Poplar about 1919 and later; any similar policy which lays a heavy burden on ratepayers. Hence **Po·plarist,** one who practises or advocates Poplarism; also *attrib.*; **Po:plariza·tion,** the adoption or implementation of Poplarism; **Po·plarize** *v. trans.*, to make like Poplar; to subject to Poplarism.

1922 *Glasgow Herald* 3 Nov. 8 The hard-headed workers of Yorkshire..have learned the lesson of Poplarism. **1923** *Daily Mail* 31 July 5/3 'Poplarism' was a portent of the changing of the modern state. **1923** R. MACAULAY *Told by Idiot* I. 44 So Poplarised..did she become that she took to speaking of her parental home in Bloomsbury as being in the West End. **1923** *Glasgow Herald* 1 Oct. 8/6 A decision in the opposite sense would simply mean an indefinite continuance of the Highland variety of 'Poplarisation' in the Lewis. **1924** *Ibid.* 7 Apr. 12/4 Mr. Wheatley ..had been accused of desiring to 'Poplarise' the British people. **1925** *Ibid.* 6 Mar. 9 Even the cautious prophets.. foretell the announcement of a rebuff to the Poplarists tomorrow. **1928** *Daily Tel.* 6 Nov. 12/6 Those..will demand increased subsidies, allowances, and 'Poplarised' social services, to be paid for out of the proceeds of very high taxation. **1931** *Times* 20 Feb. 7/6 The chief issue of the election is whether or not the policy of 'Poplarism' advocated by the Labour-Socialist Party is to be applied to London government. **1962** B. KEITH-LUCAS in *Public Law* Spring 67 The central government was finding that the weapons available to it were inadequate for the purpose of defeating Poplarism. The Poplar Order was being defied week by week. **1973** G. W. JONES in *Ibid.* Spring 28 Morrison had been opposed to the three elements of 'Poplarism', the payment of excessive wages and excessive relief and the refusal to honour legal obligations. **1976** H. WILSON *Governance of Britain* ii. 29 It was all done by agreement, Snowden recording that, as George Lansbury had to be found a job, despite his 'Poplarist' reputation, it was he who suggested Lansbury for the Office of Works.

popocracy¹ (popǫ·krăsi). *U.S. Obs.* exc. *Hist.* Also with capital initial. [f. POP(ULIST + DEM)OCRACY 4.] The rule or policy of the Populists or People's Party in the United States. So **po·pocrat**¹ [DEM]OCRAT 2], a member or supporter of the People's Party; a Populist; **popocra·tic** *a.*, of or pertaining to the Popocrats. Also earlier **poplocracy** (poplǫ·krăsi); so **poplocra·tic** *a.*

1895 T. R. SMITH in *Voice* (N.Y.) 18 July 5/3 Our fight will be for poplocracy, popular rule... I think no more significant name could be found than the Poplocratic Party. **1896** *Chicago Tribune* 4 Aug. 1/1 The first returns are always in favor of the Popocrats. *Ibid.*, Incomplete returns..indicate Popocratic gains. **1896** *Boston Jrnl.* 24 Oct. 7/3 (*heading*) He is ready to support Popocracy. *Ibid.* 31 Oct. 4/3 (*heading*) Popocratic claims about Iowa. **1896** *North Amer. Rev.* CLXIII. 744 The threats..of the Popocrats to change..our financial system. **1904** *Omaha Bee* 16 Aug. 4 If it is so important that the people of Nebraska move cautiously in the selection of their chief executive this year, why did not the popocratic conventions discover the fact before? **1972** *Time* 17 Apr. 32/1 Populists who joined the Democrats were known as 'popocrats'.

po·pocrat². *Temporary.* [f. *POP *a.* (*sb.*⁸) 2 + ARIST)OCRAT.] A leading figure in fashionable pop culture or society. So **popo·cracy**².

1970 *Tailor & Cutter* 4 Dec. 1180/2 Mick Jagger has moved far away from his former scruffy image—in fact, he is one of our more elegant popocrats. **1973** *Observer* 7 Oct. 39/1 He is a Nobel prizewinner, dresses with the Byronic panache of the popocracy.

popoi, var. *POIPOI.

‖ **po po po** (pǫ pǫ pǫ), *int.* An exclamation used in Greece expressing surprise, astonishment, commiseration, etc.

1936 L. DURRELL *Spirit of Place* (1969) 45 I'm almost tempted to come back and join you. Only the English weather po po po as they say here, shaking their rueful heads. **1972** J. AIKEN *Butterfly Picnic* i. 16, I..said, 'Po, po, po', which is the Greek equivalent of the French 'oh la la', or the English 'tut tut'. **1973** 'M. YORKE' *Grave Matters* i. i. 12 'The *Kiria* had no sons.' '*Po, po, po.*' The Greek tossed his head in sympathy. **1974** —— *Mortal Remains* III. vi. 83 Their talk was punctuated with cries of '*po, po, po*'.

‖ **popote** (popot). [Fr.] A French military kitchen or canteen.

1928 *Observer* 11 Mar. 17/7 By 1870 the young cook had become a chef, and he had charge of the kitchen of the Headquarters of the Army of the Rhine and later of MacMahon's Headquarters Mess, or 'popote', at Versailles during the Commune. **1934** A. WOOLLCOTT *While Rome Burns* 74 This, then, is the story..as they tell it..in the smoky *popotes* of the French army. **1966** J. DOS PASSOS *Best Times* (1968) ii. 43 The French cooks were getting dinner at the popote behind me.

pop-out (pǫ·paut). [POP-.] **1.** (In Dict. s.v. POP-.)

2. In surfing: a mass-produced surfboard.

1963 *Surfing Yearbk.* 42/2 *Popouts*, mass produced surfboards. **1965** *N.Z. Listener* 17 Dec. 5/1 Surfers will advise you to buy a custom-built board..rather than the mass-produced 'pop-out'. **1969** *Observer* 3 Aug. 31/2 Mass-produced fibreglass surfboards, known as 'pop outs', were soon being produced in this country as well as special custom-built boards to meet more individual requirements. Pop outs today sell from about £27 and custom-built boards from £37. **1970** *Studies in English* (Univ. Cape Town) I. 28 The air of individuality surrounding a surfer extends to his board, and so *pop-out*, meaning a mass-produced surfboard 'popped out' of a mould, is usually a derogatory term.

3. *attrib.* or as *adj.* Designating that which pops out (in various senses); *spec.* of the windscreen of a motor vehicle: designed to 'pop out' on impact.

1963 *Lebende Sprachen* VIII. 167/1 Pop-out windshield. **1967** *Autocar* 5 Oct. 45 (Advt.), The same people who care about all-round disc brakes, 'pop-out' safety windscreen, rally-proved suspension and seats designed by an orthopaedic surgeon. **1968** *Listener* 21 Mar. 391/3 The new high-performance laminated windscreens which since 1966 have cut injuries substantially in the States, or the 'pop-out' screens favoured by Mercedes-Benz and Volvo. **1968** A. DIMENT *Bang-Bang Birds* v. 72, I rejected..pop-out piano wire..and poison gas spray.

pop-over. Add: **1.** Also **popover, pop over.** (Earlier and later examples.) Also *attrib.*

1876 M. N. HENDERSON *Pract. Cooking* 71 Breakfast Puffs, or Pop-overs... May be baked in roll-pans. **1906** *Mrs. Beeton's Bk. Househ. Managem.* lix. 1626 *Pop overs*..white flour..milk..egg..salt... Pop-over tins are similar to sheets of patty pans. **1932** J. DOS PASSOS *1919* 251 Then she set down the popovers and went out. **1935** M. DE LA ROCHE *Young Renny* xiv. 122 Two golden popovers wrapped in a snow-white napkin. **1945** B. MACDONALD *Egg & I* (1946) 66 Recipes for pop-overs, cup cakes and other hot oven delicacies. **1958** L. WHISHAW *As Far as You'll take Me* ii. 15 They gave me a meal of chicken and popovers. **1973** E. TAYLOR *Serpent under It* (1974) v. 73 Comforted by two cups of coffee, six popovers, and the promise of veal paprika for dinner. **1976** *Woman's Day* (N.Y.) Nov. 156/1 Serve pop-overs at once with lemon-honey butter.

2. A loose casual garment put on by slipping it over the head.

1945 *Sun* (Baltimore) 11 Jan. 7-0/3 Miss McCardell, author of the bareback sun dress, the wraparound 'pop-over' housedress, [etc.]. **1963** *Harper's Bazaar* Oct. 44/1 A well-shaped pop-over—superbly smart in warm wool. **1968** [see *boat neck (-line)*]. **1973** *New Yorker* 3 Dec. 68/1 (Advt.), Sunbound popover, to be paired with pants,..loosely string-tied, and loosely kimono-sleeved. **1974** *Index-Jrnl.* (Greenwood, S. Carolina) 23 Apr. 3/2 (Advt.), Baby popover & pantie. **1976** *Times* 8 Apr. 7/4 Waterproof pop-overs fit over his blanket coats.

poppa. Add: (Earlier and later examples.) Also *transf.*

1897 G. B. SHAW *Our Theatres in Nineties* (1932) III. 20 We are permitted to take to our bosoms an American girl, because, to gratify her Poppa's love of a title without forfeiting her own self-respect, she has heroically refused a silly young Duke and married a venal old Earl. **1910** *Punch* 28 Sept. 227 Say, if that's poppa's notion of 'literary calm' I wish he'd never come home. **1956** [see *Father's Day* (*FATHER *sb.* 12)]. **1959** *Listener* 10 Sept. 375/1 It is a strange feeling, unfamiliar I imagine to most other countries, which..divide the rituals of the state and its political leadership between at least two dignitaries, so that Poppa is always at home. **1962** *Amer. Speech* XXXVII. 35 Another song which became popular sometime early in the century was entitled 'I Wonder Where My Sweet, Sweet Poppa's Gone'. The 'poppa' of this song is again a male lover. **1970** G. GREER *Female Eunuch* 158 Baby-talk, even to the extent of calling the husband..'poppa'. **1977** *N.Z. Herald* 5 Jan. 2-20/3 [Deaths] Lucich, Stipan Bartul... dear poppa of 19 grandchildren; in his 81st year.

b. *poppa stoppa* (U.S. Blacks' slang): see quots.

1944 D. BURLEY in A. Dundes *Mother Wit* (1973) 210 Elderly man—*Poppa Stoppa.* **1945** L. SHELLY *Jive Talk Dict.* 31/1 *Poppa stoppa*, smart old man. **1970** C. MAJOR *Dict. Afro-Amer. Slang* 92 *Poppa-stoppa*,..any old man who is effective at what he does.

popped (pǫpt), *ppl. a.* [f. POP *v.*¹ + -ED¹.]

a. Of eyes: bulging; protruding.

1927 *Scribner's Mag.* Apr. 383/2 Prentice's slightly popped blue eyes wandered to the colored folders.

b. *U.S. slang.* Arrested; apprehended by the police.

1960 R. G. REISNER *Jazz Titans* 163 *Popped*, caught (with drugs in one's possession). Example: I got popped. **1968–70** *Current Slang* (Univ. S. Dakota) III–IV. 94 *Popped*, arrested by the police.

popper, *sb.* Add: **3.** (Later examples.) See also *corn popper* s.v. CORN *sb.*¹ 11 in Dict. and Suppl.

1893 W. D. HOWELLS *Coast of Bohemia* 207 She bought a popper and three ears of corn. **1911** S. E. WHITE *Bobby Orde* (1916) xviii. 201 The pan..was replenished with popcorn, Bobby unhooked the long-handled wire popper from its nail..and set to work over the open fire. **1949** *Sat. Even. Post* 21 May 36/2 It operates popcorn machines on a concession basis..and turns out home poppers for the kitchen trade. **1957** HESELTINE & DOW *New Basic Cook Bk.* (rev. ed.) 592 Success in popping corn depends upon the quality of the popcorn and the equipment used as well as the skill of the person doing the popping. For popping over coals, a wire popper may be used; for cooking on a gas or electric range, a pressure saucepan or a heavy frying pan with a tightly fitting lid is more satisfactory. Electric corn poppers are convenient. **1972** F. VAN W. MASON *Roads to Liberty* 179 When she bent over to pop a popper of corn [etc.].

5. In Cricket, a ball that pops (*POP *v.*¹ 8 e) when bowled.

1857 *Bell's Life in London* 19 July 7/5 Mortlock defended his wicket well against the 'breakers' and 'poppers', which had by that time commenced their work. **1870** *Baily's Monthly Mag.* July 295 Mr. Grace was caught at point off a 'popper' of Emmett's. **1921** G. R. C. HARRIS *Few Short Runs* ii. 38 In my first Eton *v.* Harrow Match I calculated the batsman had to stop something like three shooters every eight balls, and at the same time one had to look out for poppers.

6. (The snapper on) a whip-lash. *U.S.*

1870 *Great Trans-Continental Tourist's Guide* (rev. ed.) 27/1 How often the sharp ring of the 'popper' aroused the timid hare or graceful antelope? **1877** H. RUEDE *Sod-House Days* (1937) 80 The lash is about 1½ inches thick at the handle, and tapers to the popper, and a good hand will make them crack like a pistol. **1934** *Amer. Ballads & Folk Songs* 375 And the stage-driver loves the popper of his whip. **1935** [see *bull-whip* (*BULL *sb.*¹ 11)].

7. A press-fastener.

1959 *Woman* 9 May 46/4 Sandwich a length of plastic foam between two layers of canvas held together with poppers. **1970** *Guardian* 17 June 13/6 He would help me dress him in his night things, laboriously doing up every popper and zip. **1973** *Times* 15 May 20/2 (*caption*) Beach bloomers in striped lawn have big sleeves to save shoulders from burning and practical poppers between the legs. **1974** N. FREELING *Dressing of Diamond* 33 Bernard, stop it, you're bursting my poppers.

Popperian (pǫpīə·riăn), *sb.* and *a.* [The name of Sir Karl Raimund *Popper* (b. 1902), philosopher + -IAN.] **A.** *sb.* A person who advocates the theories or methods of Popper. **B.** *adj.* Of or pertaining to Popper's theories or methods, esp. the theory that scientific laws are justified only by their resistance to falsification, and his criticism of the philosophical basis of Marxism and other ideologies which entail limitations on freedom. So **Po·pperism,** the theory or practice of Popper's philosophical ideas.

1962 T. S. KUHN in *Internat. Encycl. Unified Sci.* II. II. xii. 146 If only severe failure to fit justifies theory rejection, then the Popperians will require some criterion of 'improbability' or of 'degree of falsification'. **1963** R. M. HARE *Freedom & Reason* vi. 91 We must..notice an analogy between it and the Popperian theory of scientific method. **1966** *Sci. & Society* XXX. I. 1 (*heading*) Popperism: the scarcity of reason. **1971** *Nature* 28 May 269/2 Feyerabend goes on to criticize the Popperians for not making good their claim to exhibit science as a rational enterprise. **1972** *Ibid.* 10 Nov. 110/1 The author then delineates in turn three theories of the logical structure of science—inductivism, Popperian falsificationism and positivism. **1973** B. MAGEE *Popper* iii. 41 [T. S.] Kuhn's theory..about the working activity of scientists..is not irreconcilable with Popperism. *Ibid.* vi. 85 The Popperian approach has this consequence..: instead of encouraging one to think about building Utopia it makes one seek out ..the specific social evils under which human beings are suffering. **1977** A. GIDDENS *Stud. in Social & Polit. Theory* i. 72 Lakatos's studies, although nominally directed at supporting main elements of the Popperian standpoint, show how wide the discrepancy is.

poppet, *sb.* **1.** Delete 'Now *dial.*' and add later examples.

1849 DICKENS *Dav. Copp.* (1850) iv. 45 Davy, dear. If I ain't been azackly as intimate with you. Lately, as I used to be. It ain't because I don't love you. Just as well and more, my pretty poppet. **1937** M. ALLINGHAM *Dancers in Mourning* ii. 26 'That how you see it, poppet?' he said. **1964** A. SETON *Katherine* xii. 201 'Whist, poppet!' Hawise stroked the girl's arm. **1965** E. H. CLEMENTS *High Tension* vii. 65 'Cheer up, poppet, it's going to be all right.' **1976** 'M. UNDERWOOD' *Reward for Defector* i. 11 She cast a doting glance at his wife. 'Well, poppet, it's time we were off.' **1978** D. DEVINE *Sunk without Trace* iii. 33 'No, you don't eat the spoon, poppet.' She hoisted the child out of his chair and put him in the play-pen.

poppet-valve: see PUPPET-VALVE in Dict. and Suppl.

poppied, *a.* Add: **1.** (Later examples.)

1935 E. R. EDDISON *Mistress of Mistresses* viii. 143 The poppied frieze, the walls, the very floor of marble, seemed to waver. **1943** C. DAY LEWIS *Word over All* 15 Over the corn, over the poppied plains.

2. (Later example.)

1881 O. WILDE *Poems* 75 O for Medea with her poppied spell!

popping, *vbl. sb.*¹ (Further examples.)

1929 J. MASEFIELD *Hawbucks* 80 'What do you think about popping?' 'You mean proposing?' 'Yes, popping.' **1957** C. MACINNES *City of Spades* I. v. 65 'Charging is different from popping.'..'Popping?' 'With needles. White stuff—man, that's danger!' [illegible]

poppit (pǫ·pit). Also **poppet.** [f. POP *v.*¹ + -IT.] A kind of bead (see quot. 1968).

1958 N. MARSH *Singing in the Shrouds* (1959) v. 85 He started in on her rope of big beads, which, being poppets, [illegible]. **1968** J. IRONSIDE *Fashion Alphabet* 178 Poppits were beads with a small bar which pushed into a hole in the next bead so that one could make necklaces or bracelets. **1969** *Observer* 23 Feb. 31/1 There was a time when everyone wore diamonds or poppit-beads. **1974** *Sci. Amer.* Oct. 48/3 [illegible] Poppit beads in a row to form protofilaments.

popple, *sb.*¹ **b.** (Later examples.)

1931 S. WHITE *Rules of Game* xii. 66 The remains of the forest, overgrown with scrub oak and popple thickets, pushed down to the right-of-way.

poppy, *sb.* Add: **1. b.** *tall poppy*: in Australia, an especially well-paid, privileged, or distinguished person; also *transf.*

1902 *New South Wales Parl. Debates* 30 July 1840 The Premier cannot truthfully say that a measure which deals with a certain section of the community which he refers to as the privileged class and as 'tall poppies' is in accord with the Melbourne agreement. **1965** *Times* 22 Mar. (Australia Suppl.) p. xviii/2 [illegible] **1967** J. YEOMANS *Scarce Australians* viii. 85 If there is one place where the genuine eccentric is crushed, the tall poppy lopped and the penetrating discussion stifled, it is Australia. **1969** *Listener* 13 Nov. [illegible] **1975** *Sydney Morning Herald* 8 Apr. 6 Labor is obsessed with the 'tall poppies', and seems determined to pull them down.

c. = *Flanders poppy* s.v. *FLANDERS 2 b.

1921 *Times* 3 Nov. 9/3 Lord Haig..visited yesterday the headquarters of the British Legion, where the work of distributing poppies throughout the country for Poppy Day is being carried out. [illegible] **1976** *Wymondham & Attleborough Express* 10 Dec. 5/2 The sale of poppies in Oct... raised £25.

d. Money; *slang.*

1943 *Police Jrnl.* XVI. 69 Poppy, money. **1959** A. WESKER *Roots* I. 29 How's poppy?.. Tight as ever. **1960** [see *CABBAGE *sb.*¹ 1 e]. **1963** *Autocar* 6 Sept. 427/1 A good many British families which run their own cars must spend at least 13 per cent of the family poppy on that. **1972** L. HENDERSON *Cage until Tame* xvii. 148, I don't know why he's around without the gelt, because Tolly's not the boy to be parted from the poppy.

e. = Revived in *slang* use in the sense 'opium'.

Cf. quots. 1604, 1621–3 in Dict.

1935 A. J. POLLOCK *Underworld Speaks* 90/2 Poppy, opium. **1967** H. E. GOLDIN *Dict. Amer. Underworld Lingo* 162/2 Poppy, opium. **1977** H. OSBORNE *White Poppy* xv. 114 The village people would see nothing wrong in what the smugglers were doing—most of the other men still smoked the poppy.

b. A perfume derived from the poppy.

1905 *Smart Set* Sept. 113 1 Wistaria, oil of cloves,.. poppy and crab-apple. **1923** W. A. POUCHER *Perfumes & Cosmetics* I. 96 Oakmoss resin..is a liquid of characteristic odour... It is useful in oriental bouquets, particularly those of the 'poppy' type. **1954** A. J. KRAJKEMAN tr. *Jellinek's Pract. Mod. Perfumery* i. 72 (*heading*) Tables of Perfume Complexes... Peau D'Espagne... Poppy.

6. (Later examples.)

1949 *Dict. Colours Interior Decoration* (Brit. Colour Council) III. 22/1 Poppy, a colour standardised by B.C.C. in 1934, matched to the flower. Similar to Gules. **1971** *Vogue* 15 Sept. 130/2 Dress sizes: 36–42 in.; colours: green, cactus, poppy.

8. *poppy-land, -life, -syrup* (later example); (sense *1 c) *poppy appeal, cross, organizer, seller*; instrumental, as *poppy-crowned* (earlier and later examples), *-hung, -laden* adjs.; similative, as *poppy-drowsy, -glossy, -shallow, -sleepy*; **poppy-coloured** *a.* (earlier example); **Poppy Day,** a day (= *Remembrance Day) on which those killed in the world wars of 1914–18 and 1939–45 are commemorated by the wearing of a Flanders poppy (see sense 1 c above and *FLANDERS 2 b); also *attrib.*; **poppy mallow** (examples); **poppy oil** (earlier and later examples); **poppy straw,** poppy plants, or a plant, from which the seeds have been removed.

1977 *Belfast Tel.* 14 Feb. 4/6 Area chairman Mr. G. A. R. Finlay thanked the people of Northern Ireland for their support to the Legion's Poppy Appeal. [illegible]

poppy, *a.* Restrict *rare* to sense in Dict. and add: **b.** Of eyes: protuberant.

1907 *Westm. Gaz.* 11 Dec. 12/1 An American exclaiming before a family picture: 'My, what poppy eyes these Churchills have got!' **1915** *Pearson's Mag.* Jan. 106/1 Hair dark and curly; eyes poppy; lips, full. **1968** J. R. ACKERLEY *My Father & Myself* 29 A red foreign nobleman with rather poppy eyes.

poppycock. For *U.S. slang* read *slang* (orig. *U.S.*) and add earlier and later examples.

1865 C. F. BROWNE in 'A. Ward' *His Travels* [illegible] **1914** [illegible] 3 Sept. 416/2 The Headmaster of Eton became aware that he was talking poppycock for boys. **1924** M. KENNEDY *Constant Nymph* iii. 54 Sometimes, you know, you talk..poppycock. **1935** *Punch* 9 Jan. 30/1, I am not going to..ruin the perfect cadences of my English prose by pointing out to you in courteous and dignified language that your objections are all poppycock and my eye. **1955** *Times* 24 June 4/5 The peculiar capacity for pumping generals into jobs for which they were never suited continued the poppycock started by the Labour Government. **1973** *Nation Rev.* (Melbourne) 31 Aug. 1443/6 He was..a 'dangerous, ravine, psychotic, stupid, vicious, scheming writer of poppycock'. **1977** *Punch* 31 Aug.–6 Sept. 335/3 If you still think that harmonisation is so much Brussels poppycock..then draw comfort from this statistic.

poppy-seed. Add: **1.** Esp. the seed of cultivated varieties of *Papaver somniferum*, the opium poppy, used as a flavouring, filling, or garnish for cakes, bread, etc. (Later examples.)

1932 L. GOLDING *Magnolia St.* I. viii. 127 The special bread sprinkled with poppy-seed for the Sabbath repast. **1947** F. GREENBERG *Coshery Bk.* 384 Brush over with beaten egg, sprinkle with poppy seeds, and bake..about 30 minutes. **1978** *Observer* (Colour Suppl.) 3 Dec. 81/2 For poppy seeds, you have to go to the delicatessen, and be prepared to grind them at home in an electric coffee mill.

3. Also *attrib.*: *poppy-seed cake, cookie, roll*; **poppy-seed oil** = *poppy oil* (POPPY *sb.* 8 in Dict. and Suppl.).

1943 A. L. SIMON *Conc. Encycl. Gastron.* iv. 100/1 (*heading*) Poppy-seed cake. **1976** 'B. SHELBY' *Great Pebble Affair* 134 The religious group..supported themselves by..baking poppy seed cakes. **1965** *Tamarack Rev.* Summer 7 Hey, poppyseed cookies! Real stuffed fish! **1893** G. TERRY *Pigments, Paint, & Painting* xi. 305 Poppy-seed Oil.—Oil is yielded by the seeds of three kinds of poppy—the opium-poppy (*Papaver somniferum*), the spiny-poppy (*Argemone mexicana*), and the yellow-horn poppy (*Glaucium luteum*). **1942** GETTENS & STOUT *Painting Materials* 50 The cold-drawn oil is pale in colour and is the 'white poppy-seed oil' of commerce. **1974** G. USHER *Dict. Plants used by Man* 438/1 [illegible] **1973** *Listener* 20 Sept. 377/2 Tea was served by Auntie Golda..smoked salmon on black bread, poppy-seed rolls coated with cream cheese.

poppy-show. *dial.* and *Sc.* Also **puppie-show.** [f. *poppy, puppy*, dial. var. of PUPPET *sb.*: cf. PUPPY *sb.* 1 d.] A puppet-show; a peep-show. Also *transf.* and *fig.*

1798 D. CRAWFORD *Poems* 88 You'd mak a noble poppy-show. **1828** D. M. MOIR *Mansie Wauch* vii. 62 They..let me in with a grudge for twopence..to see a punch and puppie-show business. [illegible] **1887** T. DARLINGTON *Folk-Speech S. Cheshire* 298 A pin to see a poppy-show. [illegible] **1937** PARTRIDGE *Dict. Slang* 650/1 *Poppy-show*, a display, esp. if accidental, of underclothes. [illegible]

poppysmic (pǫpi·zmik), *a.* rare⁻¹. [f. L. *poppysma, -ysmus* (see POPPISM) + -IC.] Produced with smacking of the lips.

1922 JOYCE *Ulysses* 552 Florry whispers to her. Whispering lovewords murmur liplapping loudly, poppysmic plopslop.

pops², var. *POP *sb.*⁶ [see *-S².] Also used in Jazz slang as a form of address to a man.

1928 BARRIE *Half an Hour* in *Plays* 616, I never heard how much you paid Pops for me? **1932** E. M. DELAFIELD *Gay Life* ii. 32 My Pops says I'm ever such a lucky girl to have such heaps of friends. **1944** C. CALLOWAY *Hepsters Dict.*, *Pops*, salutation for all males. **1944** [see *FATSO]. **1948** [see *DADDY-O]. [illegible] **1976** S. HARVESTER *Siberian Road* ii. 30 Me a defenceless girl..without my Mom and Pops.

pop-shop. (Later examples.)

1919 G. B. SHAW *O'Flaherty V.C.* in *Heartbreak House* 183 She hadnt half the jewelry of Mrs Sullivan that keeps the popshop in Drumpogue. **1942** WODEHOUSE *Money in Bank* (1946) iv. 36 This makes me feel like a pawnbroker. [illegible] a necessity.

popsicle (pǫ·psik'l). orig. and chiefly *U.S.* Also **popsickle** and with capital initial. [Fanciful name.] An ice-lolly. Also *fig.* and *attrib.*

Popsicle is a proprietary name in the U.S.

1923 *Official Gaz.* (U.S. Patent Office) 25 Sept. 600/1 Trade-mark 'Popsicle'. Particular description of goods.—Lollypops. Claims use since May 28, 1923. **1932** H. H. SOMMER *Theory & Pract. Ice Cream Making* xxi. 302 Frozen Suckers, Popsicles, etc.,..patterned after candy suckers, or lollypops, are made by freezing ice mixes in suitable forms with wooden sticks inserted to serve as a handle. **1941** S. V. BENÉT in *Life* 7 July 90/1 The usual crowd. Kidding the local cop and eating popsicles. **1945** A. KOBER *Parm Me* 177 Leon Blatt..had just consumed a Popsicle and was now busy scraping its stick with his tongue. **1952** E. B. WHITE *Charlotte's Web* xvi. 123 In the hard-packed dirt of the midway..you will find a veritable treasure of..salted almonds, popsicles, partially gnawed ice cream cones, and the wooden sticks of lolly-pops. **1961** A. SEXTON in *Poetry* (Chicago) Dec. 161 The maid As thin as a popsicle stick Holds dinner as usual. **1962** J. LUDWIG in R. Weaver *Canad. Short Stories* (1968) 2nd Ser. 250 'I look like a big popsicle, eh Josef?' Mrs Goffman said to her chauffeur. **1963** E. B. WHITE *Let.* 22 July (1976) 504, I am always astonished to discover that my haphazard literary popsicles have found takers in places like Helsinki. **1967** M. MCCARTHY in *Observer* 30 Apr. 11/7 All the children who had gathered round to buy popsicles..from the popsicle man. **1972** *Last Whole Earth Catalog* (Portola Inst.) 291/3 Every bookstore has a line of ocean popsicles for your refreshment now. **1973** *Nation Rev.* (Melbourne) 31 Aug. 1460/3 Perhaps you would prefer one of these nice laudanum popsicles? **1977** *Time* 2 May 42/3 All they have to show an employer is a Social Security card, which is about as hard to acquire as a Popsicle. *Ibid.* 18 July 47/2 The delectable nymphet Lolita has a cruel, popsicle heart.

po·pskull. *N. Amer. slang.* [f. POP *v.*¹ + SKULL¹.] A powerful or unwholesome (esp. home-made) liquor; inferior whisky. Also *attrib.*

1867 G. W. HARRIS *Sut Lovingood* 222 Well, Maje cum blowin mad intu the doggery, an seein nobody, he jis' grabbed a bottel, an' tuck hisself a buckload ove popskull. **1917** *Dialect Notes* IV. 415 *Pop-skull*, bad whiskey. **1932** V. RANDOLPH in B. A. Botkin *Folk-Say* IV. 237 He died from drinking too much popskull whisky. **1946** *Amer. Speech* XXI. 195 Distillers never refer to a still coil as a 'worm', as did the bootleggers who manufactured popskull and rotgut during Prohibition. **1950** *Amer. Legion Mag.* Apr. 19/1 Although much of the 'panther sweat' or 'popskull scotch' brewed today is distributed to customers who know it's moonshine and don't care, thousands of gallons are sold..to unsuspecting customers. **1956** *Wall St. Jrnl.* 21 Sept. 1/1 About one out of every four gallons of hard liquor..was moonshine—also known by devotees as corn squeezins, white lightning, popskull, bumblebee stew and mountain dew. **1973** B. BROADFOOT *Ten Lost Years* xxiv. 286 Old pop skull, the kind we made by freezing a milk pail of cider and when the..alcohol had collected in the middle of the ice block, then we drained that off.

popster (pǫ·pstəɹ). [f. *POP *a.* (*sb.*⁸) + -STER.] A pop musician or artist; an enthusiast for pop music, pop art, or pop culture in general.

1963 *Meet the Beatles* 37 12.45 p.m. Popsters posing in the park. **1965** [see *OPSTER]. **1967** [see *HIPPIE, HIPPY *sb.* and *a.*]. **1967** *Listener* 1 June 727/1 There were quizzes,..teenage popsters (though in skirts down to the knee), professionally sincere teeth-gleamers—the lot. **1968** *Economist* 14 Dec. p. vii/2 The school-teachers of Monmouthshire, banning transistor radios from the playground for fear of the linguistic pollution of Anglo-Saxon popsters.

popsy. Add: Also **poppsie, popsie.** Also *gen.*, a woman or girl; a casual female acquaintance, girl-friend. Also *attrib.*

1931 C. LITHGOW *Simple Sailor* xv. 194 Chase me, you fast women; ginger yourselves up, you slow 'uns!.. Lord, but I *like* a good popsy! **1943** J. HILLIER in *Penguin New Writing* XVI. 26 He ached too much for her to be satisfied in regarding her as a short-term poppsie, and yet ops. really permitted only a poppsie outlook. **1944** M. LASKI *Love on Supertax* viii. 82 American colonels with their popsies. **1953** *Chambers's Jrnl.* June 325/1, I usually line up a local popsy. **1959** 'J. WELCOME' *Stop at Nothing* ix. 135 The blonde popsie. **1968** *Listener* 10 Oct. 475/2 The Christine of *Lucky Jim* is a somewhat conventionally pretty high-class popsy. **1973** WODEHOUSE *Bachelors Anonymous* xii. 155 The door was opened by a rather personable popsy, who proved to be a girl who lives with the Fitch. **1978** J. KRANTZ *Scruples* iii. 75 Meanwhile, he had his popsies and he had his friend, Valentine, whose cozy, crazy stage set of a Paris attic had become a special refuge for him.

popular, *a.* (*sb.*) Add: **4. a.** (Further examples.) *spec.* in *popular (news)paper, press, romance*, etc., designating literature and ephemeral publications intended for a general readership.

1817 COLERIDGE *Biog. Lit.* I. xii. 253 To an Esquimaux or New Zealander our most popular philosophy would be wholly unintelligible. **1835** J. S. MILL in *London Rev.* II. 273 Not only has it no leaders in Parliament, but it has none in the popular press. **1841** T. WRIGHT (*title*) Popular treatises on science. *Ibid.* p. vii, They [*sc.* the treatises] are important documents of the history of popular science. **1865** J. S. MILL *Auguste Comte* 48 The truths which popular philosophy calls by the misleading name of Contingent. **1876** W. JAMES *Let.* 21 Sept. (1920) I. 190 The free-thinking tendency which the 'Popular Science Monthly'..represents. **1890** —— *Princ. Psychol.* I. iii. 81 The popular-science notions of cells and fibres are almost wholly wide of the truth. **1901** CHESTERTON *Defendant* 16 The coarse and thin texture of mere current popular romance. **1901** G. B. SHAW *Three Plays for Puritans* p. vii, They read a good deal, and are at home in the fool's paradise of popular romance. **1907** *Boston Med. & Surg. Jrnl.* 26 Dec. 847/2 A second means of dissemination of knowledge of the psychotherapeutic movement is through the medium of the popular press. **1937** W. S. CHURCHILL *Let.* 20 Sept. in *Second World War* (1948) I. I. xiv. 193, I was very glad to see that Neville [Chamberlain] has been backing you up, and not, as represented by the Popular Press, holding you back by the coat-tails. **1937** *Discovery* Sept. 292/2 A contribution to popular science. **1952** H. HERD *March of Journalism* xvii. 326 Many popular newspapers..aim to interest everyman without indulging in sensationalism... Most of the 'populars' come within this classification. **1957** R. HOGGART *Uses of Literacy* vi. 144 The leader-writers of the popular Press make great play with horizons, new dawns, broad highways, forward movements..and forward-lookers. *Ibid.* 149 The popular papers, always identifying themselves with 'the people', conduct polls on this matter and questionnaires on that matter among their readers, and so elevate the counting of heads into a substitute for judgement. **1960** K. AMIS *New Maps of Hell* (1961) ii. 53 A popular-science article on atomic physics. **1964** HALL & WHANNEL *Popular Arts* vii. 165 Popular romance..is full of variants on the Romeo-and-Juliet or Cinderella themes. **1972** S. HYNES *Edwardian Occasions* 178 Hewlett..had begun by working entirely within the established conventions of the popular romance. **1976** *Conservation News* Sept./Oct. 9/1 The wartime slogan 'Digging for Victory' has reappeared in the popular press. **1977** *New Yorker* 19 Sept. 133/1 Why is it that there is not more good popular-science writing?

b. (Further examples of phr. *popular price.*) Also *attrib.*

1885 C. E. PASCOE *London of To-day* iv. 67 The multitude which invades the 'Zoo' on Monday, which is the 'popular-price' day, when a sixpence opens the gate to the neediest. **1890** *Lady's Pictorial* 15 Mar. 347/3 The book is to be produced at the popular price of one shilling. **1911** G. B. SHAW *Doctor's Dilemma* p. xxxvii, Yet people expect to find vaccines..retailed at 'popular prices' in private enterprise shops just as they expect to find ounces of tobacco and papers of pins. **1916** *Variety* 27 Oct. 12/1 Sid Grauman's 'Night at the World's Fair' is drawing well at the Majestic, attendance doubtless being encouraged by popular prices. **1971** L. LAMB *Worse than Death* ii. 23 'Teas at popular prices?' 'Oh, the teas were cheap enough.'

6. b. Designating (aspects of) art and culture whose forms appeal to or are favoured by people generally; esp. in *popular art, music, song*, etc.

Also influenced by sense 4.

1841 S. BAMFORD *Passages in Life of Radical* (ed. 2) I. xxxiii. 200 A hundred or two of our handsomest girls.. danced to the music, or sung snatches of popular songs. **1855** W. CHAPPELL (*title*) Popular music of the olden time. **1866** C. ENGEL *Introd. Study National Mus.* v. 168 The peculiar character of the popular music of a nation appears to be in great measure determined by the climate of the country, by the occupation and habits of the people, and even by the food upon which they principally subsist. **1898** G. B. SHAW *Plays Pleasant & Unpleasant* I. p. v, I had no taste for what is called popular art, no respect for popular morality, [etc.]. **1911** H. G. HEWLETT *Chorley's National Mus. of World* (ed. 3) 201 The large share,..which popular, if not Church, music has taken and takes in mourning for the dead in Ireland, is a characteristic not to be overlooked. **1927** R. H. WILENSKI *Mod. Movement in Art* 28 The nineteenth century produced original and popular art of the romantic and descriptive kinds. **1934** A. HUXLEY *Beyond Mexique Bay* 267 Where popular art is vulgar, there the life of the people is also essentially vulgar in its emotional quality. **1935** *Vanity Fair* (N.Y.) Nov. 38/1 Our jazzmen have had no attention except for the exploitation of a few Tin-Pan Alley terms concerning the popular song industry. **1941** *Musical Q.* XXVII. 48 The prevalent false dichotomy of 'classical' and 'popular' based on a belief in the inferiority of the latter as music. **1947** *Sat. Rev. Lit.* (U.S.) 10 May 9/2 By popular art we mean creative work that measures success by the size of its audience and the profit it brings to its makers. **1956** B. NETTL *Mus. in Primitive Culture* ix. 121 American Negro material..has had its effect on folk music, on popular music in the form of jazz, and on a good deal of cultivated music. **1957** R. HOGGART *Uses of Literacy* I. v. 129 The finest period in English urban popular song seems to have been between 1880 and 1910, when each great music-hall star had errand boys and earls singing his or her characteristic songs. **1959** *News Chron.* 10 July 3/2 'Search your attics, turn out your cupboards,' exhorted the B.B.C., 'and join in a television treasure hunt.'.. This is a first-class idea for popular culture. **1962** A. NISBETT *Technique Sound Studio* iii. 65 This method is employed a great deal in the recording of popular music—but very rarely indeed for serious music. **1964** HALL & WHANNEL *Popular Arts* iii. 66 In the previous chapter we tried to show the continuity between folk art and popular art. Then, by following the line of continuity into the early cinema and Chaplin, we indicated the way in which this popular art emerged within the new media. **1966** D. JENKINS *Educated Society* ii. 58 Popular culture, which..is to be sharply distinguished from..commercialized 'pop culture'..is the style of life of the majority of the members of a community. **1978** J. PASCALL *Illustr. Hist. Rock Music* 12 Popular music has never existed to be analysed. It has existed purely to give pleasure. Rock & roll, more than any other popular music, defies intellectual examination. **1979** *Jrnl. R. Soc. Arts* July 511/2 It is a catalogue of one of the largest collections of Indian popular painting outside India itself.

7. b. *popular etymology* [tr. G. *Volksetymologie*] = *folk etymology* (FOLK 6).

1880 A. H. SAYCE *Introd. Sci. of Lang.* II. ix. 246 Such myths are created by those popular etymologies—that *Volksetymologie* as the Germans call it—which play so large a part in local names. **1901** H. OERTEL *Lect. Study of Lang.* iii. 187 In all cases of so-called popular etymology it is necessary that the meaning of one of the two words should be unknown, that of the other familiar. **1926** FOWLER *Mod. Eng. Usage* 227/1 It is true..that *-yard* [in *halyard*] is no better than a popular-etymology corruption. **1933** L. BLOOMFIELD *Language* xxiii. 423 So-called popular etymologies are largely adaptive and contaminative. An irregular or semantically obscure form is replaced by a new form of more normal structure and some semantic content—though the latter is often far-fetched. **1934** S. ROBERTSON *Devel. Mod. Eng.* (1936) xi. 456 Words altered by popular etymology have often.. displaced the original forms and become thoroughly accepted in standard speech. **1958** A. S. C. ROSS *Etym.* i. 68 There is..another subject to which the General Public applies the name *etymology*, a subject which philologists often call *Popular Etymology*. This subject is one quite without value but..it is one of the great breeders of popular fallacies.

9. *popular-minded, -priced* (later examples).

1837 J. S. MILL in *Westm. Rev.* XXXVII. 8 This want is most felt..by the most popular-minded public men. **1916** *Variety* 27 Oct. 12/1 'The Little Girl that God Forgot', the popular priced attraction at the Crescent opened Sunday. **1958** *Newnes Compl. Amat. Photogr.* xvi. 158 The more popular-priced cameras fitted with lenses of f/3·5 or f/4·5, will be fast enough for instantaneous exposures in artificial light.

B. *absol.* or as *sb.* **c.** (Earlier and later examples.)

1865 *Punch* 4 Mar. 92/1 Pity poor Lucy! Obliged to go to the Monday Popular with Cousin Bess (from the country). **1894** [see *BRAHMSIAN *a.* and *sb.*].

d. Short for *popular newspaper*: see sense 4 a above.

1952 [see sense 4 a above]. **1961** *Times* 11 Feb. 5/5 (Advt.), Choosing one's Sunday newspaper seems to have become a shade less straightforward in 1961. For forty years the alternative has had such a bonny simplicity about it: on the one hand the 'posh Sundays', on the other the populars. **1964** 'W. HAGGARD' *Antagonists* xviii. 169 The Press was besieging Nikola Mitrovic... He was hinting at women and money... The populars would run it hard. **1968** *Economist* 7 Sept. 67/2 If the decline of the populars continues, Fleet Street stands to lose a third of its present newspapers. **1976** T. HEALD *Let Sleeping Dogs Die* vi. 110 Bognor..picked up the paper. It was one of the populars.

Popular Front. Also with lower-case initials. [tr. Sp. *frente popular*, F. *front populaire* in the same sense: see *FRONT *sb.* 5 g.] An international political alliance of Communist, radical, and Socialist elements formed in 1935 and gaining power in France (1936–38), Spain (1936), and Chile (1938–42), although in Europe it was largely ineffective after 1938. Also *transf.*, of other radical or popular movements.

1936, etc. [see *FRONT *sb.* 5 g]. **1936** *Age* (Melbourne) 5 May 11/7 The crowds last week were predominantly hostile to the Popular Front. **1937** 'G. ORWELL' in *New English Weekly* 29 July 308/1 The worker and the bourgeois,..are fighting side by side. This uneasy alliance is known as the Popular Front. **1940** H. G. WELLS *Babes in Darkling Wood* IV. iii. 380 The idea seems to be to make it the working credo of one world-wide popular front. **1942** E. PAUL *Narrow St.* xxxi. 280 The Popular Front election, April 16 to 21, 1936, was the last held in France, as the Popular Front election in February of that same year was the last held in Spain. **1958** *Spectator* 6 June 721/1 A Popular Front drifting into Communism. **1960** C. DAY LEWIS *Buried Day* x. 219 The Spanish war began: the Popular Front was formed 'against fascism and war'. In Cheltenham the Party group sought to engage the local Labour Party in popular-front activities. **1963** *Listener* 14 Mar. 450/1 He [*sc.* Khrushchev] looks forward to the appearance of strong Popular Front movements, similar to those in the late nineteen-thirties. **1966** K. MARTIN *Father Figures* x. 205 Was it true that a Popular Front was the only hope of salvation? **1970** *Guardian* 25 Feb. 1/2 The 'hardline' guerrilla groups are led by the Popular Front for the Liberation of Palestine. **1971** I. DEUTSCHER *Marxism in our Time* (1972) 291 The Popular Front was Stalinism's reaction against its own ultraleft follies through which it had smoothed Hitler's road to power. **1976** S. HYNES *Auden Generation* vii. 210 At the Albert Hall in February 1937..Harry Pollitt.. spoke in support of the Spanish Loyalists and the Popular Front.

Hence **Popular Fronter,** one who supports the Popular Front; **Popular Fronting,** activity associated with the Popular Front; **Popular Frontism,** the principles or policies maintained by the Popular Front.

1938 *Nation* (N.Y.) 14 May 555/1 He [*sc.* Philip F. La Follette] wants no more to be tied to trade unionism, as the British progressives are, than he wants to be tied to popular frontism, as the French progressives are. **1940** *Economist* 13 July 36/1 The Russians were liquidating Popular Frontism at home. **1941** 'G. ORWELL' in *Partisan Rev.* Mar.–Apr. 110 It was extremely amusing to watch the behaviour of orthodox Popular Front-ers, who were exclaiming dolefully 'It's going to be another Munich.' **1957** *New Republican* 7 Jan. 14/2 What kind of liberals will you find in the average large university? A handful of ADA people, most of them far from firebrands; *perhaps* one liberal of the sort who feels a nostalgic attachment to Popular Frontism. **1969** M. STEED in Henig & Pinder *European Pol. Parties* 157 There has been no suggestion that any formal links should be created between the FGDS and the Communist Party... Whilst the popular fronting of 1966–68 was happily accepted by both sides.., it broke up in May 1969, when rival presidential candidates were chosen.

popularism. Add: **b.** = POPULISM a in Dict. and Suppl.

1961 *Sewanee Rev.* Autumn 519 French Egalitarianism had had only nominal influence in this country before the days of Popularism. **1962** *Listener* 5 Apr. 584/1 For a clearer understanding of what is happening we should be ready to play down these uneasy notions of right and left, and, instead, study British political attitudes in terms of republicanism and popularism.

popularist. Add: **a.** (Later *attrib.* examples.) **c.** *attrib.* or as *adj.*, concerning or appealing to the people generally, popular.

1922 *New Witness* 25 Aug. 120/2 The P.P.I. must now be reckoned with as the most powerful political force in Italy... When one comes finally to examine the main points in the Popularist programme, one meets old friends often discussed in these pages. **1962** *Listener* 5 Apr. 585/2 The popularist fear of going into Europe is the fear of being swamped in another Holy Roman Empire. **1970** *New Yorker* 10 Oct. 150 What emerged was a popularist outlook (Hightower called it 'human') and the determination to dissolve painting and sculpture into broader aesthetic streams. **1970** *Guardian Weekly* 14 Nov. 3/4 There is no doubt that, after two years of passive and effacingly popularist administration, the President of All the People [*sc.* Nixon] tried to become President of Most of the People; a President who saw an opening to the..right.

popularity. Add: **8.** *popularity-hunting* (later example); **popularity contest,** a competition in which the popularity of the contenders is judged; freq. *transf.* and in allusive uses with reference (chiefly in negative contexts) to one's supposed popularity; **popularity poll** [*POLL *sb.*¹ 7 d], a poll taken from a section of a population in order to assess the popularity of a particular person or proposition in terms of the population as a whole; **popularity rating,** an assessment of popularity based on the findings of a *popularity poll.

1941 B. SCHULBERG *What makes Sammy Run?* i. 15 I'm not running a popularity contest; I'm running a business office. **1952** *Manch. Guardian Weekly* 24 Apr. 8/4 This is a 'popularity contest' only; the results are not binding on the delegates of either party. **1959** R. CONDON *Manchurian Candidate* (1960) xvii. 211 Life isn't a popularity contest... I didn't ask them to like me. **1964** E. AMBLER *Kind of Anger* i. 16 You wouldn't win a popularity contest where he's concerned, and..he can be a vindictive old bastard. **1973** 'E. MCBAIN' *Hail to Chief* i. 15 The decisions I make ain't always popular, but..I'm not running no popularity contest. **1976** *National Observer* (U.S.) 1 May 4/5 Unless he can really blast a hole through this interest screen in the Democratic Party by winning handsomely in the primaries, not just in the popularity contests but in the delegate races as well, then I suspect he could be squeezed out at the convention. **1946** W. S. CHURCHILL *Victory* 162 Sir, I trust there will be no popularity-hunting at the public expense. **1938** 'E. QUEEN' *Four of Hearts* ix. 127 Three outstanding stars (selected..on the basis of the latest popularity poll conducted by Paula Paris for the newspaper syndicate). **1958** *Punch* 1 Jan. 50/1 Continuing to top the popularity poll for the masses was Princess Margaret. **1962** *Sunday Times* (Colour Suppl.) 10 June 3 A fact recognised in this year's *Melody Maker* popularity poll. **1972** *Country Life* 10 Feb. 347/3 Let us pray that they [*sc.* poodles] will never find themselves leading the popularity poll. **1979** J. WAINWRIGHT *Duty Elsewhere* lxiii. 176 Calling an assistant chief constable a flaming liar doesn't win popularity polls. *a* **1974** R. CROSSMAN *Diaries* (1977) III. 327 *The Times* has a poll showing a 20 per cent lead for the Tories, with a new popularity rating specially designed so that Ted Heath can be ahead of Harold Wilson. **1974** *Listener* 17 Jan. 70/1 The Japanese Prime Minister..enjoys an even lower popularity-rating..than President Nixon.

popularization. Add: (Later examples.) *spec.* (in sense 2 c of the vb.), the adapting of ideas or theories to the level of an educated but non-specialist public; freq. with derogatory connotations, the over-simplification of a subject to suit popular taste (cf. *POPULAR *a.* (*sb.*) 4 a). Also, the result or product of this process.

1926 WYNDHAM LEWIS *Art of Being Ruled* XIII. iv. 423 It is plainly the popularization of science that is responsible for the fever and instability apparent on all sides. **1951** M. MCLUHAN *Mech. Bride* (1967) 28/1 The lethal psychological and social effects..arise not from science but from its popularization. **1962** *Listener* 18 Jan. 119/1 Popularization is, then, a word with a moral identity... It implies..that the cases it describes are those where the subject is misrepresented by this treatment; where the truth of the matter has been diluted, if not falsified. **1973** *Nature* 23 Mar. 280/1 This book is more than an extraordinarily successful popularization. **1974** B. PEARCE tr. *Amin's Accumulation on World Scale* II. 591 Mandel places alongside a popularization of *Capital* a diatribe against the Soviet bureaucracy. **1977** M. COHEN *Sensible Words* 157 John W. Yolton..discusses Gildon's *Deist's Manual* (1705) as an important imitation and popularization of Locke's epistemology.

popularize, *v.* Add: **2. c.** (Earlier and later examples.) Also *absol.*

1833 J. S. MILL in *Monthly Repos.* VII. 266 The peculiar 'mission' of this age..is to popularize among the many, the more immediately practical results of the thought and experience of the few. **1916** G. B. SHAW *Pygmalion* 100 His [*sc.* Henry Sweet's] great ability as a phonetician..would have entitled him to high official recognition, and perhaps enabled him to popularize his subject, but for his Satanic contempt for all academic dignitaries and persons in general who thought more of Greek than of phonetics. **1923** *Times Lit. Suppl.* 4 Jan. 10/3 True-blue musicians; they knew their facts and.. looked at them steadily in order to check their theories, and they did not popularize.

d. To make (a word) generally known, *spec.* to use or encourage the use of (technical vocabulary) in general contexts or in everyday language.

1921 G. B. SHAW *Back to Methuselah* p. xliii, Why did not Erasmus Darwin popularize the word Evolution as effectively as Charles? **1926** [implied in *POPULARIZED *ppl. a.*]. **1965** E. GOWERS *Fowler's Mod. Eng. Usage* (ed. 2) 461/1 Our interest in our bodies has always made us prone to popularize medical terms.

popularized *ppl. a.* (later examples); **popularizer** (later examples).

1919 M. BEER *Hist. Brit. Socialism* I. I. vi. 80 Adam Smith and Abraham Tucker's populariser, Paley, either use and interpret natural law in a conservative sense or draw its social-revolutionary teeth. **1926** FOWLER *Mod. Eng. Usage* 444/2 A few examples of these popularized technicalities may be gathered together. **1934** H. G. WELLS *Exper. Autobiogr.* II. viii. 546 He..began writing books for the general reader and essays in natural history. He was a successful popularizer. **1951** M. MCLUHAN *Mech. Bride* (1967) 48/1 Professor Kinsey's book is a *carte blanche* for maximal genital activity. As popularized science, that is its entire drift. **1974** *Nature* 30 Aug. 754/1 The populariser of science, as he functions today, cannot disseminate the subtle ideas of science. **1978** *Ibid.* 31 Aug. 930/3 Popularisers have a duty to be scrupulous in matters of fact and clarity of presentation.

population². Add: **2. c.** (Further examples.) Also of other entities.

1956 A. H. COMPTON *Atomic Quest* ii. 89 If the neutron population is increasing generation after generation, the reaction grows in intensity. **1968** *Brit. Med. Bull.* XXIV. 244/1 Much study has therefore gone into the effect of radiation on proliferating cell populations in various environments.

d. *Statistics.* A totality of objects or individuals under consideration, of which the statistical attributes may be estimated by the study of a sample or samples drawn from it.

1877 F. GALTON in *Nature* 12 Apr. 513/2 The number of pellets in each compartment represents the relative number in a *population* of seeds, whose weight deviates from the average, within the limits expressed by the distances of the sides of that compartment from the middle point. **1903** *Biometrika* II. 273 If the whole of a population were taken we should have certain values for its statistical constants, but in actual practice we are only able to take a sample, which should if possible be a random sample. **1922** *Phil. Trans. R. Soc.* A. CCXXII. 329 It is unfortunate that in this memoir no sufficient distinction is drawn between the *population* and the *sample*. **1939** D. D. PATERSON *Statistical Technique in Agric. Res.* i. 1 The sum total of all the units of any one kind is called, in statistical terminology, the population. **1970** *Jrnl. Gen. Psychol.* LXXXIII. 14 The experimenter notes that Ss used in this study had experienced only a brief period of hospitalization... Consequently, he cautions against the generalization of his findings to other neurotic and schizophrenic populations. **1977** *Accountants Weekly* 29 July 17/1 The tests built into the program ensured that sample data extracted for audit checks came from a complete population, a fundamental requirement in auditing.

e. *Genetics.* A breeding group of animals, plants, or humans; **population biology, genetics,** the branches of biology and genetics which treat such groups statistically. So *population biologist, geneticist.*

1889 F. GALTON *Nat. Inheritance* iv. 35 The science of heredity is concerned with Fraternities and large Populations rather than with individuals, and must treat them as units. **1949** C. STERN *Princ. Human Genetics* x. 168 Both fields, population genetics and pedigree genetics, are significant and both rest on the Mendelian analysis of inheritance. *Ibid.* 594 The genic constitution of the later populations will obviously depend on the genotypes of their second-generation ancestors. **1960** *Biol. Abstr.* XXXV. 1729/1 (*heading*) Data on ecology and population biology of mosquitoes. **1966** MACARTHUR & CONNELL *Biol. of Populations* p. x, Experiments performed by population biologists indicate another difference between population biology and other branches of science. **1966** R. ARDREY *Territorial Imperative* iv. 138 A population, in biology, is a reproductive community. More sharply stated, it is any group of individuals who have a modest probability, within any generation, of meeting and mating. **1968** R. C. LEWONTIN *Population Biol. & Evolution* 2 It is a problem of population biology to discover under what circumstances a one-to-one sex ratio is evolved. **1972** *Sci. Amer.* Jan. 100/3 He maintained four populations of *Drosophila* in the laboratory over a 48-month period. **1973** *Listener* 28 June 850/2 He [*sc.* A. Jensen] now knows a good deal more about population genetics than he did..in 1968. **1977** *Nature* 6 Jan. 26/2 The basic interests of the population geneticist of course lie in the realm of population dynamics, and the knowledge of allele frequencies is an essential prerequisite to further work. **1979** *Ibid.* 9 Aug. 455 (*heading*) Population biology of infectious diseases.

f. *Physics.* The (number of) atoms or subatomic particles that occupy any particular energy state.

1931 *Physical Rev.* XXXVII. 143 (*caption*) Illustrating the relation between spectral intensity distribution in the Compton line (left) and population of electron speed states (right). **1938** R. W. LAWSON tr. *Hevesy & Paneth's Man. Radioactivity* (ed. 2) viii. 88 The resulting gap in the atom, due to the incomplete population of the *K*-shell, may now be filled by the transition of an electron from the *L*-level into the *K*-level. **1961** [see *INVERSION 2 l]. **1971** *Sci. Amer.* June 22/1 By elevating more atoms to an upper energy level than exist at a lower level the absorption of excitation radiation produces an 'inverted' atomic population in the laser.

g. *Astr.* Either of the two groups into which stars can be approximately divided: those of *population I* are formed from the debris of other stars, those of *population II* are coeval with their galaxy.

[**1944** W. BAADE in *Astrophysical Jrnl.* C. 137 The stellar populations of the galaxies fall into two distinct groups, one represented by the..stars in our solar neighborhood (the slow-moving stars), the other by that of the globular clusters. Characteristic of the first group (type I) are highly luminous O- and B-type stars and open clusters; of the second (type II), short-period Cepheids and globular clusters.] **1951** *Astrophysical Jrnl.* CXIII. 413 Highly luminous stars of population I. **1952** C. PAYNE-GAPOSCHKIN *Stars in Making* iv. 74 The names 'Populations I and II' were originally given by Baade to the two groups of stars. As we shall see, they probably represent extremes rather than an absolute distinction. **1974** F. W. COLE *Fund. Astron.* xiii. 344/1 Two very different populations of stars can be recognized: population I consisting of the normal, metal-rich stars found in spiral arms of galaxies; and population II, the metal-poor stars found in globular clusters, as isolated stars in the galactic halo, and in the central galactic nucleus.

h. The general body of inmates in a prison, rehabilitation centre, etc. (see quot. 1950). Freq. in phr. *in population.*

1950 H. E. GOLDIN *Dict. Amer. Underworld Lingo* 162/2 *Population*,..the general body of inmates as differentiated from convicts in trusty jobs, hospital patients, and occupants of punishment or psychopathic observation wings. **1953** W. BURROUGHS *Junkie* viii. 79 After eight days, you get a sendoff shot and go over in 'population'... You are allowed seven days to rest in population after medication stops. **1956** J. RESKO *Reprieve* (1959) III. xvi. 135 Our friends out in population took care of us. **1971** *Black Scholar* June 54/1 The officials told me that I would never be returned to the general population (Auburn) and advised me to put in for transfer to another institution. **1973** *Philadelphia Inquirer* (Today Suppl.) 7 Oct. 50/3 Sprague tried to block Soleni's move into general population. **1977** *New Yorker* 24 Oct. 68/2 Collectively, the inmates are referred to as 'population'.

4. *population census, control, cycle, distribution, drift, growth, increase, planning, policy, pressure, question, survey, theory, trend*; **population biology**: see sense 2 e above; **population curve,** a graph showing the variation of population with time; **population explosion,** a rapid or sudden marked increase in the size of a population; hence *population-explosive* adj.; **population genetics**: see sense 2 e above; **population inversion**: see *INVERSION 2 l; **population pyramid,** a roughly triangular figure on a level base, the width of which at any height is proportional to the numbers having an age proportional to that height.

1968 *Internat. Encycl. Social Sci.* XII. 369/1 In most advanced countries illiteracy has been almost eliminated, and therefore questions on literacy are no longer included in population censuses. **1931** J. S. HUXLEY *What dare I Think?* v. 166 Might it not have been better to have left the death side of nature's population-control to itself until we have some future policy for dealing simultaneously with birth? **1959** *New Statesman* 21 Mar. 401/1 Countries like India and Japan have made population-control a central feature in national policy, because they know that without it they are headed for disaster. **1973** J. M. WHITE *Garden Game* 117 War is becoming a necessary instrument of population-control. **1889** F. GALTON *Nat. Inheritance* 245 The population curve will..be a straight line. **1935** *Proc. Prehist. Soc.* I. 11 Advances of critical importance to humanity should be followed by such a multiplication as to be conspicuously reflected in the population curve. **1968** *N.Y. City* (Michelin Tire Corp.) 139 Long Island... The population curve is constantly rising. **1969** N. W. PIRIE *Food Resources* ii. 70 A few plant and animal species..go through fairly regular population cycles. **1968** R. A. LYTTLETON *Mysteries Solar Syst.* vi. 185 The density of finds does show some relation to the population-distribution. **1974** *Encycl. Brit. Macropædia* XVII. 67/2 The pattern of population distribution and density closely follows that of the rainfall pattern. **1964** *Ann. Reg. 1963* 22 The Minister of Housing and Local Government..was wrestling with the problems of population drift, regional planning, and slum clearance. **1977** *Modern Railways* Dec. 480/1 Population drift from the London area to central and north-east Essex necessitated a fundamental restructure of Great Eastern line services. **1953** *Population Bull.* Oct. 65 (*heading*) Latin America: area of population explosion. **1953** [see *EXPLOSION 4 b]. **1964** 'J. MELVILLE' *Murderers' Houses* vii. 116 Emily knew all about the Bomb and the Pill and could advocate one remedy for the declining Middle Classes and another for the Population Explosion. **1970** *Oxf. Univ. Gaz.* C. Suppl. vi. 5 A kind of documentary population-explosion, in fact. **1970** *Daily Tel.* 18 July 11/7 Population explosions in summer may result in up to 10 million aphides taking to the air each day from an acre. **1974** *Times* 21 Jan. 6 (*heading*) Leading the fight to head off the population explosion. **1967** *Punch* 3 May 637/3 This two-way flow, once started, would never stop; no government, much less any shipping company, could ever stop such a flow, population-explosive, travel-agency-prodded and democratic. **1927** J. S. HUXLEY *Relig. without Revelation* ix. 325 The study of heredity and population-growth. **1970** G. GERMANI in I. L. Horowitz *Masses in Lat. Amer.* viii. 295 With regard to the other Latin American states, it is clear that immigration made a crucial contribution to population growth. **1978** R. MITCHISON *Life in Scotland* vi. 108 Industrialization..was..the 'answer' to that population growth. **1931** J. S. HUXLEY *What dare I Think?* iv. 135 Population-increase cannot go on indefinitely. **1959** *New Statesman* 21 Mar. 401/1 In under-developed countries, excessive population-increase reduces the possibility of an economic break-through. **1974** *Times* 21 Jan. 6/2 The idea of population planning antagonized many countries. 1944 J. S. Huxley *On living in Revolution* xii. 131 It is very important that there should be a well-thought-out population policy for backward areas. 1974 *Times* 21 Jan. 6/5 Opposition to..population policies is led by countries with..large natural resources. 1931 J. S. Huxley *What dare I Think?* v. 165 Causing more babies to live and so creating greater population-pressure. 1969 *Times* 26 June 14/7 The migrations..could be a means of relieving the population pressure in a particular area. 1950 *Chambers's Encycl.* XI. 93/2 The age distribution in populations of plaice and other fish has been examined; and attempts have been made to study honey-bee colonies and wireworm populations from this point of view. All show, under natural conditions, the expected 'population pyramid', formed by large numbers of young individuals and gradually decreasing numbers of individuals of the higher age-groups. 1976 *Nature* 1 July 19/1 As a result of their high crude birth rate..the population pyramid has a relatively broad base; 51·8% are less than 20 yr old and 28·7% are less than 10 yr old. 1885 *Encycl. Brit.* XIX. 517/2 We cannot here deal with what is known as the population question'... The 'population question' is a question of conduct. 1911 G. B. Shaw *Getting Married* 116 St. Paul's reluctant sanction of marriage;.. his contemptuous 'better to marry than to burn' is only out of date in respect of his belief that the end of the world was at hand and that there was therefore no longer any population question. 1953 Population survey [see *attractant]. 1966 *Economist* 17 Dec. 1253/3 Except in the matter of population theory, where he anticipated and influenced Malthus, Steuart had little or no effect on later economic thought in Britain. 1933 Thompson & Whelpton (*title*) Population trends in the United States. 1950 Theimer & Campbell *Encycl. World Politics* 347/1 In the U.S.S.R., Eastern Europe and Latin America population trends are like those of the West in the later nineteenth century. 1976 J. S. Moore *Goods & Chattels of our Forefathers* 9 The main population-trends can already be seen from a study of the data contained in the numerous ecclesiastical and private censuses taken in Gloucestershire in the early modern period.

populationist. Delete *nonce-wd.* and add later examples. Also, one who considers the population to be a significant element in a state's power.

1949 K. Davis *Human Society* 552 If the populationist stopped here, however, his work would have little to do with social science. 1968 *Internat. Encycl. Social Sci.* XII. 350/2 In Germany, Hermann Conring also attributed the power of states mainly to population. In England.. another confirmed populationist was William Petty, who founded the science of 'political arithmetic', or demography.

populism. Add: (Further examples.) Also *transf.*

1960 *Encounter* July 13 Russian Populism is the name.. of a widespread radical movement in Russia in the middle of the 19th century. 1969 [see *populistic a.]. 1972 *Time* 17 Apr. 31/1 Populism is a label that covers disparate policies and passions: among many others, New Deal reforms, consumer rage against business, ethnic belligerence. Often it is merely a catch phrase. Yet it describes something real: the politics of the little guy against the big guy—the classic struggle of the haves against the have-nots or the have-not-enoughs. 1973 *Black Panther* 21 July 3/2 The result was the defeat of western Populism and the further entrenchment of racism. 1976 T. Eagleton *Crit. & Ideology* v. 166 Populism and theoreticism, in aesthetics as in politics, are familiar deformations of Marxist-Leninism. 1977 *Time* 1 Jan. 71 His creed combines traditionally antithetical elements of help-the-deprived populism and deny-thyself fiscal conservatism.

b. The theories and practices of the populist movement in French literature.

[1929 L. Lemonnier in *Revue Mondiale* 1 Oct. 282 (*title*) Du naturalisme au populisme.] 1930 — in *This Quarter* Mar. 440 (*title*) Populism. *Ibid.* 443 At last, we hit upon the word 'populism'. It clearly expressed the fact that we meant to depict the people; it was not altogether a new word in French, inasmuch as it had been used to translate the name of the German political party *Volkspartei*, but it had never as yet been applied to any artistic, political or literary movement specifically French. Having then dubbed ourselves populists, we decided to write a manifesto. 1932 *French Rev.* IV. 473 Since the opening of the twentieth century, only three schools have counted [in French literature], unanimism, between 1908 and 1911, surrealism, about 1924, and populism in 1929. 1932 *Ibid.* V. 389 Populism is the antonym of 'snobbisme'. 1934 F. Walter in *Lemonnier's Populisme* in *PMLA* XLIX. 356 Populism is a reaction founded on the realistic tradition and directed against the literature of analysis. 1934 *N. & Q.* 26 May 361/2 The beginnings of Populism, admittedly somewhat obscurely in 1924, came out into shape in 1929 under the initiative of M. André Thérive and M. Léon Lemonnier. *Ibid.* 361/2 In Thérive's 'Le Baiser de Satan,' Populism has attempted what Naturalism shied from the historical novel.

populist. Add: **3.** A member of a group of French novelists in the late 1920s and early 1930s who placed emphasis upon observation of and sympathy with ordinary people.

[1929 L. Lemonnier in *Revue Mondiale* 1 Oct. 285 Tous ces romanciers viennent de se grouper et de se donner un nom: ils veulent être les *romanciers populistes*. Ils entendent le mot dans un sens très large.] 1930 [see *populism b]. 1934 *PMLA* XLIX. 361 A sort of Tolstoyan sympathy is a cardinal virtue in the eyes of the Populists. 1934 *N. & Q.* 26 May 361/2 Eugène Dabit (a genuine proletarian), though a populist, begins somewhat to abandon the political and social neutrality.

4. One who seeks to represent the views of the mass of ordinary people.

1961 *Listener* 30 Nov. 897/2 They are not Populists or Poujadists. 1972 *New Society* 20 Jan. 131/1 LBJ was a true populist, as he recognised himself, remarking tartly (and justly enough) that it is 'the term some liberals reserve for Progressives who come from the southern and western parts of the nation'. 1977 *Time* 7 Mar. 7/1 Brogan questions whether Carter is a bona fide populist at all.

5. *attrib.* or as *adj.*

1893 [in Dict., sense 1]. 1898 *Nation* (N.Y.) 7 July 6/2 The Populist Governor abused his power by appointing as commissioners only men of his own party. 1924 *Glasgow Herald* 4 July 7 He [*sc.* the Russian intellectual] has lost much of his former 'populist' idealism, of his old worship of the people. 1928 [see *centrist b]. 1931 *French Rev.* IV. 473 This paper will give an account of the rise and origins of the populist school. 1934 R. Michaud *Mod. Thought & Lit. in France* xi. 228 A so-called 'populist movement' was launched in 1929 by Léon Lemonnier and André Thérive, as a protest against the précieux and individualistic novel and as a return to the great naturalistic traditions of Zola. 1954 E. A. Shils in Christie & Jahoda *Stud. Scope & Method of 'The Authoritarian Personality'* 45 A vein of xenophobia, populist, anti-urban and anti-plutocratic sentiment. 1955 H. Peyre *Contemp. French Novel* ii. 47 Duhamel never made a speciality of the study of misery, as did the proletarian novelists and later a short-lived group of 'populist' novelists (Henri Poulaille, André Thérive, Eugène Dabit). 1961 *Listener* 7 Dec. 997/1 There is the ascending conception [of law and government], according to which the law-creating power may be ascribed to the community or people—the populist theory. 1968 W. Safire *New Lang. Politics* 346/2 When the Populist candidate, General James B. Weaver, won 22 Electoral College and 1,029,846 popular votes in the 1892 election, many people..were fearful of impending revolution. 1969 R. Blackburn in Cockburn & Blackburn *Student Power* 190 A wholesale revision of classic liberal democratic theory to eliminate its dangerously populist tendencies and to accommodate the elitist features of contemporary capitalist society. 1974 M. B. Brown *Econ. of Imperialism* xi. 275 Populist forms of government in the ex-colonial underdeveloped lands were overthrown mainly because they failed to develop their economies. 1976 *Survey* Summer–Autumn 15 There is little doubt that US policy..will be dominated by self-consciously populist politicians. 1977 *Time* 21 Mar. 53/3 Ironically, the very success of Carter's populist appeal may cause him special backlash problems.

populistic *a.* (further examples); **populi·stically** *adv.*

1969 D. MacRae in Ionescu & Gellner *Populism* 162 That Ireland did not produce a full-fledged populism—as distinct from populistic themes that continue through De Valera to the present—is a paradox of European history. 1971 S. Cavell *World Viewed* 54 A film like *Mr. Smith Goes to Washington*..suggests, populistically, ..that they are curable by the individual or mass goodness of the little people. 1976 *Times Lit. Suppl.* 16 Apr. 457/2 A man with his own blend of simple, populistic dignity and even honesty.

pop-up (pǫ·pʌp), *sb.* and *a.* [f. Pop *v.*[1] + Up *adv.*[1]] **A.** *sb.* **1.** *Baseball.* A ball which is hit softly up into the air and is easily caught.

1906 *Spalding's Offic. Base Ball Guide* 126 A trapped ball play was made when runners were on bases, and a 'pop-up' fly ball was expected to be caught. 1926 *Amer. Speech* I. 369/1 A 'pop-up', for an infantile attempt to hit, is beautifully characteristic of the contempt prompting the name. [illegible] 1978 *Time* 3 July 60/1 One of their catchers, Frank Mancuso, was a former lieutenant who had injured his back during parachute training; he could neither remain in the Army nor look skyward for a pop-up.

2. A pop-up toaster, trailer, etc.

1970 *New Yorker* 14 Nov. 49/1 Browned off like a piece of toast in a broken pop-up. 1976 *Times Lit. Suppl.* 6 Aug. 991 (Advt.), We manufacture jigsaw puzzles, pop-ups and other educational toys. 1978 *Sunday Sun-Times* (Chicago) 1 Jan. 121/1 Another popular RV type is the camping trailer, or 'pop-up'. 1978 *Detroit Free Press* 2 Apr. 17F/1, 75 Apache pop-up, slps 8, refrig, stove, heater, awning, like new.

B. *adj.* **a.** Designed to pop up or having a component that pops up. **b.** With a mechanism which causes something to pop up; esp. *pop-up toaster*: an electric toaster in which the sliced bread pops up when it is ready.

1934 in Webster. 1959 C. Oburn *Marauders* (1960) ii. 65 We shot off quantities of ammunition, mostly at informal or pop-up targets. 1959 'E. McBain' *Pusher* v. 42 The precinct house..did not boast climatic curtains or pop-up toasters. 1960 S. Kauffmann *If it be Love* i. vi. 90 An electric dishwasher and deep freeze and pop-up toaster. 1962 N. Freeling *Love in Amsterdam* i. 29 The pretty girl..flipped a pop-up file like those for telephone numbers. 1962 *Which? Car Suppl.* Apr. 55/1 The Austin A60 was the only car which did not have pop-up knobs..which show easily whether the doors are locked or not. 1963 S. Marshall *Exper. in Educ.* iv. 153 Every illustration is conceived and executed as a 'pop-up' scene. 1966 *Listener* 24 Nov. 755/1 This is a toaster of standard performance—the pop-up kind. 1972 *Times* 6 Dec. 22/6 A birthday card, with a pop-up centre. 1973 *Country Life* 19 July 151/3 For large lawns the most advanced system is the pop-up sprinkler..small spray nozzles which only appear above turf level when the water is turned on. 1975 *Evening News* 26 Apr. 4/1 London motorists may soon face..pop-up metal barriers ..to keep them off bus-only lanes. 1978 *Dumfries Courier* 20 Oct. 27/3 (Advt.), Hoover pop-up toaster in very good order, £8.50. 1978 *Times Lit. Suppl.* 1 Dec. 1400/3 Pop-up books are a different roaming ground for the imagination.

poral, *a.* (Further examples.)

1926 *Jrnl. Bot.* LXIV. 144 The poral outline is much like that of *Pseudonavicella*. 1961 D. M. Pillsbury et al. *Man. Cutaneous Med.* ii. 49 In acne the sebaceous gland may atrophy as a result of concomitant poral occlusion.

porcelain. Add: **1.** (Examples of its use in *Dentistry*.)

1845 C. A. Harris *Princ. & Pract. Dental Surg.* (ed. 2) vi. i. 541 A want of resemblance to the other teeth, in colour, transparency, and animation, was the great objection, that was urged against the porcelain. 1863 *Trans. Odontol. Soc.* III. 228 The universal use of porcelain as a material of which to construct artificial teeth. 1911 G. H. Wilson *Man. Dental Prosthetics* viii. 311 The materials entering into dental porcelain are feldspar, silica, kaolin or clay, alkalies, and pigments. 1956 J. N. Anderson *Appl. Dental Materials* xxiv. 324 Today, porcelain is finding a resurgence of life, particularly in its application to crowns. 1965 *Brit. Dental Jrnl.* CXIX. 251/1 One of the main criticisms levelled at dental porcelain is the liability to fracture under low impact stress.

3. b. A variety of pigeon, having dark brown and cream plumage.

1855 [see *hyacinth 3b]. 1876 in R. Fulton *Illustr. Bk. Pigeons* xxv. 348 Porcelains..are closely allied to Suabians... These birds are of a nice rich brown.., the under parts being of an ashen tint.

5. *porcelain-blue* (earlier example); *porcelain-like* adj. (later examples); **porcelain enamel,** †(*a*) = Glaze *sb.* 1; (*b*) = Enamel *sb.* 1 a; so **porcelain-enamelled** *a.*; hence **porcelain-enamel** *v. trans.*, **porcelain enamelling** *vbl. sb.*; **porcelain-kiln** = *porcelain oven*; **porcelain-paper** (examples); **porcelain tooth,** a false tooth made of porcelain.

1703 tr. *H. van Oosten's Dutch Gardener* ii. xxxviii. 91 The Hyacinth that is handsom, must have a clear Porcelin, or China Blew, or near white Colour. 1883 *Jrnl. Chem. Soc.* XLIV. 397 (*heading*) Composition of porcelain enamels. 1921 *Chem. & Metall. Engin.* XXIV. 486/1 In 1860 the enameling of sheet iron was begun, and upon the advent of the drawing press and clay muffle in 1870 the process of porcelain enameling steel as we know it today developed into the enameling industry. 1924 E. G. Blake *Plumbing* II. iii. 40 Porcelain enamel is practically everlasting, does not chip with reasonable treatment, and can be kept perfectly clean with very little trouble. 1946 Simonds & Bregman *Finishing Metal Products* (ed. 2) xxxiii. 332 The compound word 'porcelain enamel' was made necessary by the wrong use of the word 'enamel' by manufacturers of paints. 1951 *Good Housek. Home Encycl.* 224/2 Porcelain enamel, being essentially a glass fused to a metal, has the properties of glass. 1955 Insley & Fréchette *Microsc. Ceramics & Cements* xi. 211 Porcelain enamel may be defined as an inorganic, glassy coating on a metal base, prepared by covering the preformed metal with a wet or dry powder coating of suitable composition and firing it for a few minutes to melt and smooth the surface. 1968 *Engineering* 26 July 171/3 Electron microbe examinations were made on a number of the magnesium-bearing alloys which had been porcelain enamelled without prior [illegible] short period of time, porcelain teeth must be used if the succeeding dentures are to be serviceable for an adequate period.

porcelainic (pǭəɹsēlǣ·nik), *a.* [f. Porcelain + -ic.] = Porcellanic *a.*

1839 H. T. de la Beche *Rep. Geol. Cornwall, Devon & W. Somerset* ix. 267 The fragments of them included in the greenstone of Kellan Head have a porcelainic appearance. 1971 *Materials & Technol.* II. v. 314 Ceramic ware of the porcelainic type was made in China from very early times.

porcellanic, *a.* Add: **b.** Characteristic or suggestive of porcelain.

1939 J. Cannan *No Walls of Jasper* 29 His tooth brush dropped into its stand with the accustomed porcellanic chink.

porch. Add: **1. a.** (Later examples.)

1898 G. B. Shaw *Candida* i. 80 The parsonage is semi-detached, with a front garden and a porch. 1916 Joyce *Portrait of Artist* (1969) 162 He pushed open the latchless door of the porch and passed through the naked hallway into the kitchen. 1929 G. B. Shaw *O'Flaherty V.C.* in *Heartbreak House* 167 The porch, painted white, projects into the drive. 1980 R. McCrum *In Secret State* vii. 69 He returned to the porch, unlocked the front door and stepped inside.

c. A small platform outside the hatch of a spacecraft.

1969 *Daily Tel.* 14 July 16/5 Wearing their bulky suits and life-support packs, they will open the narrow hatch. Armstrong will squeeze himself out on to a small platform called the 'porch'. 1970 N. Armstrong et al. *First on Moon* xi. 266 Armstrong: 'Yes. Got it... Okay, Houston, I'm on the porch.'

3. b. (Earlier and later examples.) Also *Canad.*

1832 J. P. Kennedy *Swallow Barn* II. 42 Hafen Blok was regaling his circle of auditors in the porch at Swallow Barn. 1901 S. E. White *Westerners* 251 Then there was the gambler, the faro man, who sat on the hotel 'porch'. 1916 H. L. Wilson *Somewhere in Red Gap* v. 195 Wilbur Todd had once endeavoured to hold her hand on the porch after a country-club dance. 1925 F. Scott Fitzgerald *Great Gatsby* (1926) i. 14 The two young women preceded me out onto a rosy-colored porch, open toward the sunset, where four candles flickered on the table in the diminished wind. 1932 *Atlantic Monthly* Feb. 193/2 Broad porches ran the length of the house on both sides. 1948 *Manch. Guardian Weekly* 30 Dec. 13/1 President Truman has left the non-Roman slabs of Washington to go home to Independence in Missouri, where he can feel more comfortable sitting on the back porch of an old frame house. 1968 *Globe & Mail* (Toronto) 13 Jan. 13/1 Raymond Souster would be the amiable fellow on the porch reminiscing with compleasant nostalgia for lost times. 1978 C. MacLeod *Rest you Merry* ix. 60 The student had dumped the suitcases on the short walk in front of the brick house, and was studying the porch.

c. A small utility room attached to the back of a house. *N. Amer. dial.*

1916 *Dialect Notes* IV. 335 [Nantucket] *Porch*, an ell kitchen. 1929 *Amer. Speech* V. 124 'Piazzer' was the only term applied to a veranda [*sc.* in the dialect of Maine]. The 'porch' was a sort of extra shed-kitchen used as a laundry. 1969 in Halpert & Story *Christmas Mumming in Newfoundland* 211 The 'porch' is a small room at the rear of the house used for storing wood, hanging coats, cooking utensils, and so on. A door, which is always kept closed, leads from the porch into the kitchen.

6*. *Television.* In a video signal, either of the two periods of line blanking immediately before and after the line-synchronizing pulse; known respectively as the *front porch* and *back porch*.

1941 *Proc. IRE* XXIX. 30/1 The difference between 0·008H and 0·07H, namely 2 per cent of H, is the 'front porch' of the pedestal. 1963 Amos & Birkinshaw *Television Engin.* I. ii. 32 The period of blanking level immediately following the line-sync signal is termed the back porch. *Ibid.*, There is a brief period of blanking level occurring immediately before each line-sync signal. This is known as the front porch. 1965 *Wireless World* Aug. 389/1 The phasing of the oscillator is determined by the duration of the front porch of the composite video waveform, the flyback time of the line circuit and the tightness of lock. 1966 [see *line blanking* s.v. *line sb.[2] 33].

7. *porch chair, rail, roof*; **porch-climber** *N. Amer. slang*, a burglar; hence **porch-climbing** *ppl. adj.*

1908 Sears, Roebuck *Catal.* 753/2 Folding porch chair, made of wood frame with denim body. 1911 *Daily Colonist* (Victoria, B.C.) 25 Apr. 6/7 (Advt.), Porch and Verandah Chairs. There are no chairs more suitable for the porch or verandah than Sea Grass or Rattan. 1948 *Democrat* 21 Apr. 1/7 Porch and Lawn Chairs, Swings, Gliders and Metal Tables. 1902 Ade *More Fables* 218 He had a Chinaman for a Servant, because he was the Author, but supposed him to be a Retired Porch-Climber. 1901 'J. Flynt' *World of Graft* 37 The remaining third of Chicago's professional thieves are good, bad, and indifferent 'sneaks', 'porch-climbers', [etc.]. 1916 *Daily Colonist* (Victoria, B.C.) 27 July 6/3 Some well-intentioned citizens see a potential second-story man or porch-climber in everyone who is not within doors after the stroke of midnight. 1927 *Scribner's Mag.* Feb. 180/1 The depredations of porch-climbers, safe-blowers, and common thieves were a source of alarm. 1932 *Collier's* 28 Dec. 15/3 Beware of the beautiful ladies who fear porch-climbing, safe-blowing pals. 1929 *Oxf. Poetry* 33 Brown meadow grass and cat-tails My banisters and porch rails—All these belonged to me. 1948 E. Pound *Pisan Cantos* (1949) lxxvi. 43 As the cat walked the porch rail at Gardone. 1869 'Mark Twain' *Innoc. Abr.* xliii. 448 The porch-roof is composed of tremendous slabs of stone.

porcupine, *sb.* Add: **1. δ.** (Later example in humorous use.)

1936 T. S. Eliot *Coll. Poems 1909–1935* 147 How unpleasant to meet Mr. Eliot! With a bobtail cur In a coat of fur And a porpentine cat.

5. a. Substitute for def.: A small Australian monotreme, the echidna or spiny ant-eater, *Tachyglossus aculeatus*; (earlier and later examples).

1832 J. Bischoff *Sk. Hist. Van Diemen's Land* ii. 29 The native porcupine or echidna is not very common. 1843 J. Backhouse *Narr. Visit Austral. Colonies* vii. 89 The Porcupine of this land..is a squat species of ant-eater, with short quills among its hair. 1924 *Truth* (Sydney) 27 Apr. 6 Porcupine, a rather large rodent with spiked quills, and which feeds chiefly upon bark, leaves and ants. 1944 *Living off Land* ii. 29 The spiny ant-eater (sometimes called porcupine) is a good food. 1970 W. D. L. Ride *Guide Native Mammals Austral.* xii. 191 The most widespread Australian monotreme is the Echidna or Spiny-anteater which is called 'Porcupine' by many people.

6. *porcupine ant-eater*, substitute for def.: sense *5 a; (earlier examples).

1847 [see Echidna]. 1860 G. Bennett *Gatherings of Naturalist in Australasia* 147 The Porcupine Ant-eater of Australia..and the Ornithorhynchus..form the only two genera of the order Monotremata.

Hence **po·rcupinal** *a.*, suggestive of a porcupine, prickly (in quot. *fig.*).

1846 R. Ford *Gatherings from Spain* xii. 139 The very tightness into the catgut of an overstrung fiddle, getting attuned to the porcupinal irritability of the tension of the mind.

pore, *sb.*[1] Add: **4.** *pore-size, -space*; **pore pressure,** the pressure of pore water; **pore water,** water contained in pores in soil or rock.

1947 D. P. Krynine *Soil Mech.* (ed. 2) iv. 112 (*caption*) Measuring apparatus (*a*) pore pressure in clay and (*b*) hydraulic uplift in a dam. 1969 C. R. Scott *Introd. Soil Mech. & Foundations* ix. 201 The stability of a retaining wall is adversely affected by large pore pressures in the soil behind it. 1977 A. Hallam *Planet Earth* 64/3 Deformation of a rock by pore pressure..the resistance of the [illegible] *Jrnl. Soil Sci.* II. 175 If a series of filters of known pore-size is used, the size of particles which can pass through them..can be obtained. 1915 L. V. Pirsson *Text-bk. Geol.* I. xvi. 388 The rocks are penetrated by cracks, crevices and jointing planes, and on a minute scale they have the pore spaces between grains. 1975 *Nature* 17 Apr. 585/2 Usually the pore-space in a porous medium is visualised as a more or less complicated assemblage of isolated or interconnected capillaries, and such pore-space models are used in describing transport phenomena in porous media. 1936 *Proc. 1st Internat. Conf. Soil Mech.* III. 13 The ultimate deformation and the conditions of failure of cohesive soils are governed mainly by the principal effective stresses, defined as the differences between total principal stresses and the hydrostatic pressure of the pore water. 1943 K. von Terzaghi *Theoret. Soil Mech.* When dealing with [illegible] compute the pressure [illegible] the point of failure [illegible] 1971 L. Zeevaert *Foundation Engin. for Difficult Subsoil Conditions* Upon dissipation of the excess pore water pressures, the soil tends to resume its initial elevation. 1977 *Offshore Engineer* June 35/2 The engineers are looking for tell-tale signs of soil fluidisation by large increases in pore water pressure caused by water trapped within the steel skirts and forced out beneath the skirt.

pored, *a.* (Later examples.)

1930 *Engineering* 23 May 670/1 A wall built with Fletton bricks and very finely pored mortar. 1963 *Agra Univ. Jrnl. Res.* XII. 53 (*heading*) A note on the double pored epidermal cage worms of some of the Indian carnivores. 1973 B. Shelmire *Art of Looking Younger* (1974) iv. 45 This cellular buildup causes the whole layer to become thicker and, in turn, makes the skin look coarse, leathery, and large-pored.

porge, *v.* Substitute for etym.: [ad. Judæo-Spanish *porgar*, Sp. *purgar* to cleanse, f. L. *purgāre* Purge *v.*] (Later examples.) Also *absol.* **porger** (earlier example).

1773 J. H. Bomira *Kehilath Jaakob* 110/2 Porger of Meat.—Purgador. 1871 [see *kosher a.]. 1932 C. Roth *Hist. Marranos* vii. 179 The children of Israel did not eat the sinew of the thigh; and it was customary..to 'porge' the leg before preparing it for food. 1973 *Jewish Chron.* 18 May 32/4 (Advt.), Shomer. Able to porge and knowledge of kashrus requirements for butchers and poulterers.

porgy. Add: **2.** *attrib.*, as *porgy boat, fleet, steamer*; *porgy-hunting* vbl. sb. (see quot. 1904).

1904 *N.Y. Even. Post* 28 Mar. (Saturday Suppl.) 1/2 The 'porgy' boats, dirty, snub-nosed..are far..removed in standing from their fellows. 1906 J. J. Rowlands *Spindrift* 211 The porgy boats are built and operated for just one purpose—fishing. 1914 W. D. Steele *Storm* 191 ..the porgie fleet moved in around Long Point. 1904 *Scribner's Mag.* May 548 When we cruise about, looking out for any fish we can catch, and at any price we can get for it, that's porgy hunting. 1880 Porgy steamer [see *packetting vbl. sb.]. 1914 *Oysterman & Fisherman* Jan. 7/3 The porgie steamer Long Island which..has been undergoing repairs.

porina (poraɪ·nă). *N.Z.* [mod.L. (F. Walker *List Specimens Lepidopterous Insects in Brit. Mus.* (1856) VII. 1572).] The larva of a moth formerly belonging to the genus *Porina*, now usually included in the genus *Oxycanus*, which damages grassland. Freq. *attrib.*

1929 W. Martin *N.Z. Nature Bk.* I. xvi. 153 The Porinas, Swifts, or Bull Moths... The subterranean larvae thus attacked [by fungi] have long been known as Vegetable Caterpillars. 1940 *N.Z. Jrnl. Agric.* Apr. 245/1 The Porina moth makes its first appearance in early October... It is the caterpillar..which causes the damage during the winter. *Ibid.* 246/1 Porina larvae feed almost entirely on surface leafage. 1950 *Ibid.* Apr. 357/1 Porina..and grass-grub are a serious hindrance. 1966 *Encycl. N.Z.* I. 319/1 The larval stages of the moths commonly called porina moths..are important pasture pests in New Zealand. They live in tunnels in the soil during the day and feed at night at the surface of the soil on grass foliage. 1969 [see *grass grub* s.v. *grass sb.[1] 13]. 1973 A. D. Lowe in G. R. Williams *Nat. Hist. N.Z.* viii. 198/2 The practice of closing paddocks for hay and seed crops frequently appears to have accentuated the porina problem.

porion (pǭə·riǫn). *Anat.* Pl. **poria.** [f. Gr. πόρ-ος way through, passageway + *-ion[2].] (See quot. 1937.)

1909 in *Cent. Dict.* Suppl. 1920 H. H. Wilder *Lab. Man. Anthropometry* 47 *Porion*, the uppermost point in the margin of the auditory meatus. 1933 *Jrnl. R. Anthrop. Inst.* LXIII. 417 The poria [are] widely separated. 1937 *Amer. Jrnl. Physical Anthrop.* XXII. 485 *Porion*, a point on the upper margin of the external auditory meatus which according to Martin is vertically over the middle of the meatus, and according to Wilder is simply the uppermost point. 1970 *Monogr. Soc. Res. Child Devel.* XXXV. iii. 48 As a rule, the two poria are instrumentally determined by contact of the upper margin of the rounded ear rods of the headspanner as they are inserted into the *S*'s external auditory meatus.

pork[1]. Add: **2. b.** *U.S. slang.* Federal funds obtained for particular areas or individuals on the basis of political patronage. Cf. *pork barrel.

[1862 in D. W. Mitchell *Ten Yrs. in U.S.* xv. 271 To put myself in a position in which every wretch entitled to a vote would feel himself privileged to hold me under special obligations, would be giving rather too much pork for a shilling.] 1879 *Congress Rec.* 28 Feb. 2131/1 St. Louis is going to have some of the 'pork' indirectly; but it will not do any good. 1916 *N.Y. Even. Post* 12 May 8/2 'Pork' has hitherto stood for just one process, the parcelling out of Federal moneys for court houses, post offices, and waterways, not by States, but by Congressional districts. 1949 *Marshfield* (Wisconsin) *News-Herald* 19 July 4/3 That difference of more than $54,000,000 includes a lot of pork for individual senators. 1962 *Economist* 20 Oct. 252/1 Pork is the generic name for the tasty morsels of federal spending..which a member of congress likes to bring back to his constituents. 1964 D. M. Berman *In Congress Assembled* xii. 323 One of the first facts of congressional life is that it does not pay to antagonize the committee to which one will someday have to appeal for funds to support a local project. Such projects are commonly referred to as 'pork'.

c. Phr. *pork and beans* (Mil. slang), a name given to Portuguese soldiers serving in the war of 1914–18.

1919 *Athenæum* 8 Aug. 727/2 He [*sc.* the soldier] gave nicknames to the Overseas troops, as..'Chinks' for Chinese labourers..and 'Pork and Beans' for Portuguese. 1919 W. H. Downing *Digger Dial.* 39 *Pork-and-beans*, Portuguese soldiers. 1925 Fraser & Gibbons *Soldier & Sailor Words* 228 *Pork and beans*, a nickname for the Portuguese troops serving on the Western Front.

3. *pork steak, trade*; *pork-packer* (earlier and later examples), *-packing* (earlier examples), *raiser, raising*; **porkburger,** a kind of hamburger made from pork; **pork-butcher,** (*a*) (earlier and later examples); (*b*) a shop-keeper who specializes in pork; so **pork-butchering** *vbl. sb.*, **-butchery**; **pork-eater** *Canad.*, a canoeman engaged on the run between Montreal and Grand Portage; also, by extension, any canoeman, esp. a new recruit; *obs.* exc. *hist.*; **pork house,** a business house trading in pork; **pork king,** a magnate in the pork trade; **pork-knocker, porknocker,** in Guyana (formerly British Guiana), an independent or casual prospector for gold or diamonds; hence **pork-knocking** *vbl. sb.*, the activity of a pork-knocker.

1939 *Amer. Speech* XIV. 154/2 *Porkburger*, ground pork, in other words, sausage! 1969 R. & D. De Sola *Dict. Cooking* 180/1 *Porkburger*, pork patty fried and eaten like a hamburger. 1807 Southey *Lett. from Eng.* III. lxiii. 182 The pork-butchers are commonly Jews. 1922 Joyce *Ulysses* 59 The porkbutcher snapped two sheets from the pile, wrapped up her prime sausages and made a red grimace. 1925 W. de la Mare *Two Tales* 40 Yet to judge from some poets' faces, you might be easily justified in supposing they would have flourished better in the pork-butchering line. 1935 *Times Lit. Suppl.* 24 Oct. 673/1 [They] set a rabbinical winkle-seller on the road to fortune which leads to pork-butchery. 1793 J. MacDonell *Diary* 5 July in C. M. Gates *Five Fur Traders* (1933) 94 Between two and three hundred yards to the East of the N.W. Fort beyond the Pork eaters camp is the spot Mess[rs] David and Peter Grant have selected to build upon. 1801 A. Mackenzie *Voy. from Montreal* p. xxvii, Of these, five clerks, eighteen guides, and three hundred and fifty canoe men, were employed for the summer season in going from Montreal to the Grande Portage, in canoes, part of whom proceeded from thence to Rainy Lake,.. and are called Pork-eaters, or Goers and Comers. 1823 J. Franklin *Narr. Journey Shores Polar Sea* vii. 281 There is a pride amongst 'Old Voyagers', which makes them consider the state of being frost-bitten as effeminate, and only excusable in a 'Pork-eater', or one newly come into the country. 1829 J. McLoughlin *Let.* 8 Dec. (1948) 69 By this opportunity I send you all you requested..and four Pork eaters. 1859 P. Kane *Wanderings of Artist* 34 The men who usually work this brigade of [Hudson Bay Company] canoes are hired at Lachine, and are called by the uncouth names of mangeurs du lard, or pork-eaters. 1953 *Beaver* Dec. 50 The provisions for the Crew were Pork & Biscuits; from which circumstance the young recruits were called 'Pork Eaters' to distinguish them

from the old Winterers, who feed chiefly on 'Pemican'. 1969 E. W. Morse *Fur Trade Canoe Routes* i. ii. 23 The voyageurs plying the run between Montreal and Lake Superior were known derisively among the tougher breed wintering in the North West as 'pork eaters', *mangeurs de lard*. 1837 W. Jenkins *Ohio Gazetteer* 171 Eaton contains ..four pork houses. 1848 *Rep. Comm. Patents 1847* (U.S.) 527 The hogs are taken into the pork house from the wagons and piled up in rows. 1890 W. D. Howells *Boy's Town* 36 Cooper-shops where the barrels were made, alternated with the pork-houses. 1893 M. Elliott *Honor* 155 Gwendoline O'Shaunessey, the daughter of old O'Shaunessey the Western pork-king. 1930 R. Macaulay *Staying with Relations* xv. 222, I should like to go off with a president,..or a film or pork king. 1910 M. B. & C. W. Beebe *Our Search for Wilderness* vi. 187 The universal Guianian name for this type of independent miner is 'pork-knocker', the explanation being that by knocking the rocks to pieces, they find just enough gold to procure the pork upon which they live. 1930 *Times* 14 Mar. 5/3 The pork-knockers make a night of it before they go up into the bush. 1949 P. Hastings *Cases in Court* iii. 130 These gentlemen employed a number of natives who enjoyed the somewhat peculiar title of 'Pork knockers'. 1957 [see *inboard a.]. 1972 *Guardian* 1 Dec. 14/1 The famous gold and diamond prospectors of the interior (the so-called 'porknockers' whose name derives from their salt pork rations). 1974 H. MacInnes *Climb to Lost World* iii. 42 A couple of prospectors, or 'porknockers', were staying in the Park Hotel... They are called porknockers because salted pork was their staple diet; when they had a run of bad luck, they used to borrow, or 'knock' pork from their more prosperous friends. 1965 'Lauchmonen' *Old Thom's Harvest* vii. 95 Winston, man, you better had go back to your pork-knocking. 1974 H. MacInnes *Climb to Lost World* xii. 221 We had some Brazilian natives with us... They..had come over for the diamond prospecting but, since the water had been too high for porknocking, they had agreed to work for us instead. 1838 *N.Y. Advertiser & Express* 7 Feb. 3/3 It is due to that enterprising class of citizens, the pork packers, that the error should be corrected. 1949 *Boston Sunday Globe* 26 June (Fiction Mag.) 2/1 These corporations were principally distillers, manufacturers of tobacco, and, especially, beef and pork packers. 1851 C. Cist *Sk. Cincinnati in 1851* 228 Pork and Beef Packing. 1870 *Trans. Illinois Agric. Soc. 1867–68* VII. 475 The only reliable statement of the pork-packing of the West we have any knowledge of. 1839 *Jrnl. Indiana Ho. Representatives* 8 Jan. 231 The scarcity..is likely to prove so mischievous to the interests of our pork raisers and dealers. 1872 *Trans. Illinois Dept. Agric. 1871* IX. 390 Dark, cold, damp Piggeries are a nuisance to any farmer or pork raiser. *Ibid.* 354 He had said that pork raising stood preeminent as a branch of stock raising in our State. 1880 G. T. Ingham *Digging Gold* 203 Is this the *honor* of Western pork-raisers? 1783 J. Woodforde *Diary* 8 Apr. (1926) II. 68 We had for Dinner..Mutton Stakes, Pork Stakes, Peas Soup. 1922 Joyce *Ulysses* 230 Master Patrick Aloysius Dignam came out of Mangan's..carrying a pound and a half of porksteaks. *Ibid.* 251 Master Patrick Aloysius Dignam..raised also his new black cap with fingers greased by porksteak paper. 1957 *Encycl. Brit.* XVIII. 245/2 Pork chops or steaks are usually browned in a hot skillet. 1851 A. O. Hall *Manhattaner* 13 Here, too, is modest beauty from Ohio (papa in the pork trade).

pork barrel. orig. and chiefly *U.S.* [Pork[1] 3.] **1.** A barrel in which pork is kept. Also *fig.*, a supply of money; the source of one's livelihood.

1801 *Farmer's Almanack 1802* (Boston) sig. C2 *Better spare at the brim, than at the bottom*, is an old proverb, and should teach us to mind our pork and cider barrels. 1842 *Joliet* (Illinois) *Courier* 2 Feb. 3/1 Farmers can be accommodated with a very good Pork Barrel in exchange for Oats Butter or Wheat. 1861 *Harper's Mag.* Oct. 643/1, I came very near tumbling over a pork-barrel, and made a remark concerning obstructions in the street which was more forcible than chaste. 1863 *Frank Leslie's Illustr. Newspaper* 24 Jan. 277/3 We find that those who work honestly, and only seek a man's fair average of life, or a woman's, get that average... And thus we find that when an extraordinary contingency arises in life..we have only to go to our pork-barrel and the fish rises to our hook or spear. 1909 *Chambers's Jrnl.* Mar. 178/2 We had bought ten empty pork-barrels. 1946 S. Newton *Paul Bunyan* xxvii. 158 It was as big around as a pork barrel. 1978 M. Puzo *Fools Die* xi. 106 The Army Reserve of the United States was a great pork barrel. By just coming to a meeting for two hours a week you got a full day's pay.

2. *fig.* The state's financial resources regarded as a source of distribution to meet regional expenditure; the provision of funds (in U.S., Federal funds) for a particular area achieved through political representation or influence. Freq. *attrib.* or as *adj.*

1909 *Westm. Gaz.* 1 June 2/1 The Democratic Party.. has periodically inveighed against the extravagance of the administration, but its representatives in the Legislature have exercised no critical surveillance over the appropriations. They have preferred to take for their own constituencies whatever could be got out of the Congressional 'pork barrel'. 1913 R. M. La Follette *Autobiogr.* 60 It was on the so-called 'pork-barrel' bill for river and harbor appropriations. 1916 *N.Y. Even. Post* 12 May 8/2 The River and Harbor bill is the pork barrel par excellence, and the rivers and harbors are manipulated by Federal machinery and not by State machinery. 1926 R. Luce *Congress* 82 Undoubtedly there was once a 'pork-barrel', a metaphorical barrel from which legislators pulled out 'pork' to satisfy the ravenous appetites of greedy constituents. 1950 *Reader's Digest* Jan. 96/2 The Army Civil Functions appropriation bill—once known as the Rivers and Harbors bill and still called the 'pork barrel' bill—this year provided for 275 projects. 1950 *Sun* (Baltimore) 24 Aug. 4/3 The section of the bill is sometimes called the 'pork-barrel', and as it contains funds for projects in virtually every state, it is one that is the hardest to cut item by item. 1953 *Manch. Guardian Weekly* 11 June 10/4 We are shown..the way in which members of Congress..deal with the problem of public works (coarsely called the pork barrel). 1960 *Economist* 15 Oct. 266/2 It [*sc.* the Macmillan Government] has treated some nationalised industries almost as if they were its positive enemies, while a quite considerable pork-barrel has been opened up for a growing number of private firms. 1961 D. L. Munby *God & Rich Society* vi. 122 'Pork-barrel' politics, by which governments step in to help particular economic groups in the community. 1973 *Sat. Rev. Sci.* (U.S.) Mar. 29/3 Present fire allocations..have reached pork-barrel dimensions. 1976 H. Wilson *Governance of Britain* x. 172 In Westminster, the Government has complete control over expenditure... Thus, in Britain, 'pork barrel' expenditure is ruled out.

Hence **pork-barrelling** *vbl. sb.*, the process of providing regional funds by these means.

1967 *Economist* 14 Oct. 133/3 The one piece of regional pork-barrelling at last week's Labour party conference was the Prime Minister's promise that two aluminium smelters, using subsidised electricity, would be built, probably in development areas. 1974 *Camden* (S. Carolina) *Chron.* 22 Apr. 2/1 The commission needs stronger authority than it has in that field if legislative pork-barreling on higher education is to be kept to a minimum. 1976 *Guardian Weekly* 12 Sept. 7/5 Pork barrelling—the way Congressmen with political influence can obtain Federal funds for local projects. 1977 *N.Y. Rev. Bks.* 26 May 36/3 Such reallocation would sorely affect agricultural usage of the land and stimulate the construction of various water aqueducts, dams, and other projects currently denounced as mere congressional pork-barreling.

pork chop. [Pork[1].] **1.** (See Pork[1] 3 in Dict.)

2. An American black who accepts an inferior position in relation to whites. Chiefly *attrib.* *U.S. slang.*

1970 *Rep. 20th Ann. Round Table Meeting Lang. & Ling. Stud., Georgetown Univ.* 9 *CR*: Who can make magic? *Greg*: The son of po'—... I'm sayin' the po'k chop God! He only a po'k chop God! *Ibid.* 36 A *pork chop* is a Negro who has not lost traditional subservient ideology of the South..and the *pork chop God* would be the traditional God of the Southern Baptists. 1977 *N.Y. Rev. Bks.* 4 Aug. 35/1 This is the year of the Bionic Black, and porkchop nationalists have lost prestige.

Hence **po·rk-cho·pper** *U.S. slang*, a full-time union official (see quots).

1946 *N.Y. Times* 11 Aug. iv. 7/7 In the UAW, the rank and file call those who live by income derived from the union 'pork-choppers'. 1953 Berrey & Van den Bark *Amer. Thes. Slang* (1954) §527/2 *Pork-chopper*, an official who is in the union for self-interested reasons. 1960 Wentworth & Flexner *Dict. Amer. Slang* 403/1 *Pork-chopper*, a political appointee, union official, or relative or friend of a politician, union officer, or the like, who receives payment for little or no work; one who is put on a payroll as a favor or as a return for past services. 1968 *Economist* 2 Nov. 30/2 They feel mostly contempt for the 'pork-chopper'—the former factory workers who have become full-time members of the union staff. 1977 *Time* 17 Jan. 32/3 Rank-and-filers have never considered him a 'pork-chopper', their term for a high-hat leader.

porker. Add: **1. b.** *fig.* A fat or porcine person.

1892 [see *Balzacian a.]. 1959 I. & P. Opie *Lore & Lang. Schoolch.* ix. 168 The unfortunate fat boy.., is known as: back end of a bus..porker, [etc.]. 1959 *Good Food Guide* 42 So many restaurants in the Thames Valley have been ruined by the expense-account porkers, who neither care what they pay nor know what they eat.

pork-pie. Add: **2.** (Further examples.) Now usu. applied to a kind of hat worn by men. Also *pork-pie cap*.

1888 C. M. Yonge *Beechcroft at Rockstone* II. xv. 41 There certainly was a figure in somewhat close proximity, the ulster and pork-pie hat being such as to make the gender doubtful. 1910 *Blackw. Mag.* Jan. 113/1 In the dreadful mustard-coloured uniform and pork-pie cap which the Government has ordained for these unusually fat servants. 1937 *Evening News* 12 Feb. 8/3, I seem to remember that porters at the entrances of big hotels once wore greatcoats reaching almost to their ankles and that each had a pork-pie cap with a peak and a little round button on top. 1940 Graves & Hodge *Long Week-End* xxi. 376 Low-crowned pork-pie hats were in fashion again. 1943 R. Chandler *Lady in Lake* xxi. 118 His blue pork-pie hat was set very square on his head. 1948 M. Allingham *More Work for Undertaker* xvii. 199 The inevitable green demob pork-pie sat a little too far back from his lined forehead. 1955 *Times* 16 Aug. 8/7 The men had white handkerchiefs on their faces, and wore raincoats and 'pork-pie' hats. 1968 R. Clapperton *No News on Monday* vi. 60, I..limped down the hot pavement with a straw pork pie tipped forward over my eyes. 1977 *New Yorker* 25 July 61/2 Lawford, the arrant individualist, sometimes showed up on the court wearing a pork-pie hat, a striped jersey, tight knickers, and long stockings.

porky (pǭə·ɹki), *sb.* Also **porcy.** Abbrev. of Porcupine *sb.*

1902 W. D. Hulbert *Forest Neighbors* 146 We found the Porky asleep in the sunshine. 1921 *Chambers's Jrnl.* May 290/2 An encounter in which 'Porcy' had the best of it. 1936 D. McCowan *Animals Canad. Rockies* xxvii. 235 A pair of shoes left carelessly outside a tent forms a tasty meal for a prowling Porky. 1956 W. R. Bird *Off-Trail in Nova Scotia* x. 267 They heard the familiar complainings of a 'porky' on the move.

Porlock (pǭ·ɹlǫk). The name of a town in Somerset, used allusively (see quot. 1816) in phr. *a person* etc. *from Porlock*, a person who interrupts at an inconvenient moment.

[1816 Coleridge *Kubla Khan* in *Christabel* 53 At this moment he [*sc.* Coleridge] was unfortunately called out by a person on business from Porlock, and detained by him above an hour.] 1959 *Listener* 1 Jan. 37/3 All the incidental distractions—the telephone-bell, the Christmas carollers at the door, the gentleman or lady 'from Porlock' —to which one is subject.

porn (pǭɹn), **porno** (pǭ·ɹno), *a.* and *sb. colloq.*

A. *adj.* Abbrevs. of Pornographic *a.*

1952 N. Mailer *Man who studied Yoga* in *Advts. for Myself* (1961) 175, It is dirty, downright porno dirty, it is a lewd slop-brush slapped through the middle of domestic exasperations and breakfast eggs. 1963 'D. Cory' *Hammerhead* ii. 21 Judging from the script..it might be just the *tiniest* bit porno. 1964 N. Freeling *Double-Barrel* v. i. 148 Looking at Miss Burger through binoculars was porno..anything porno is so hatefully sad. 1970 *Time* 16 Nov. 92 So busy are the makers of porn films in San Francisco that they have depressed the market for imported sex movies. 1972 J. Brown *Chancer* xv. 208 First, take the evidence. The joints, the photos, the porno magazines. 1972 *Screw* 12 June 19/1 This week I am reviewing the very best porn film ever made, so superior to others that it defies comparison. 1972 *New Yorker* 2 Dec. 164/2 They use their porno fantasies as part of the case they make for the slaughter of the whites. 1973 'D. Halliday' *Dolly & Starry Bird* iii. 36 Jacko had put away his porn pictures. 1974 *Publishers Weekly* 18 Nov. 45/2 A teenage girl who had been trying to make the big time by performing at Beverly Hills orgies and doing porno films. 1976 *Daily Tel.* 13 Feb. 13/5 If only he would feed her, or take her to a porno movie—or even for a walk! 1978 S. Sheldon *Bloodline* xxii. 249 It's a porno film.

B. *sb.* **1.** Abbrevs. of Pornographer. *rare.*

1958 L. Durrell *Balthazar* vi. 144 'The old Porn himself!' (He had coined this nickname from the word 'pornographer'.) 1969 *Time* 5 Sept. 27 The right-wing National Democratic Party derides [Gunther] Grass as a 'porno', because his works are peppered with four-letter words.

2. Abbrevs. of Pornography. Also *attrib.* and qualified by *hard* or *soft* (cf. *pornography 2a).

1962 *John o' London's* 10 May 456/2 The central character and narrator, the Captain, is a seedy but not at all unsympathetic individual who makes a precarious living by writing 'porn'. 1964 *New Society* 13 Feb. 5/3 The stuff men pass round in barrack rooms as 'a nice bit of porn'. *Ibid.* 6/2 'There's nothing odd about our customers,' the porn shop assistant said. 1967 L. Deighton *London Dossier* 192 Books..divided into two main sections, straight 'porn' and sadism. 1968 *Punch* 31 July 144/1 Gavin, the Old Etonian ex-monk who is now one of London's most around publishers ('Not porno, love, the word's *engagé*, ha-ha-ha!'). 1969 R. Airth *Snatch!* ii. 21 Cognac, cigarettes, porn—small-time, Harry, small-time. 1970 *Guardian* 17 Jan. 10/1 However thin and tippling the Lennon lithographs..they were not pure tosh or porn. *Ibid.* 26 Nov. 4/4 The porno and sex wave crashes relentlessly on. 1972 *Daily Mirror* 12 Oct. 1 An all-out attack is to be mounted against the porn-pushers in Britain's High Streets. 1973 L. Meynell *Thirteen Trumpeters* i. 8 A large and flourishing porn shop. 1974 *Publishers Weekly* 4 Mar. 67/3 The condition of the cinema today (lousy stories, no real stars, too much porno). 1975 D. Lodge *Changing Places* ii. 96 Standing amid the alien porn of Soho. 1976 T. Heald *Let Sleeping Dogs Die* viii. 154 An elderly woman..looked up from the nudie magazine... More soft porn lay around her. 1977 E. J. Trimmer et al. *Visual Dict. Sex* (1978) xxiv. 274 Perhaps the institution of an Oscar for the best hard porn movie of the year might give the producers an incentive to quit conning their public. 1978 M. Puzo *Fools Die* xii. 130, I could write the soft-porn love stories for the top-of-the-line magazine. 1980 G. Greene *Dr. Fischer* xv. 108, I..sat for an hour before a soft porn film.

pornie (pǭ·ɹni). *slang.* [f. *porn *a.* and *sb.* + -ie.] A pornographic film.

1966 R. H. Rimmer *Harrad Experiment* (1967) 157 'What the devil are pornies?' Beth demanded. Jack grinned. 'Flickers a little on the pornographic side.' 1967 I. Hamilton *Man with Brown Paper Face* ii. 27 We're planning a pornie for the blue movie racket. 1975 *Publishers Weekly* 17 Feb. 80/3 A nice California kid until she was conned into filming pornies to pay off her lover's addict brother's connection.

porno-, comb. form of Pornography or Pornographic *a.*, prefixed to sbs. and adjs., as *porno-biography, -film, -movie, -photographer*; *porno-chic, -gothic* adjs. (also as sb.); **po·rnogram,** a pornographic short poem; **pornoma·nia,** a mania for pornography; **po·rnophile** a lover of pornography; **pornopho·bic** *a.*, having a horror of pornography (in quot. *absol.*); **Po·rnosec,** a name given to a department of the Ministry of Truth by George Orwell in his novel 'Nineteen Eighty-Four'; **pornoto·pia,** an ideal setting for the activities described in pornographic literature.

1970 *Sat. Rev.* (U.S.) 17 Oct. 32 The mildly lascivious may be grateful that he gives the longest plot summary of *Glenarvon* I know of, and prints the entire text of *Don Leon*, a not very titillating piece of pornobiography. 1971 *New Yorker* 11 Dec. 24 The Music Lovers—Ken Russell seems to have invented a new genre of pornobiography. In this film, Tchaikovsky is the chief victim of Russell's baroque vulgarity. 1973 *Times* 17 Feb. 12/8, I went to see *Deep Throat*, which is the big porno-chic movie of all time. 1969 *Truth* (Melbourne) 12 July 13/2 In these two

countries [*sc.* America and England], there are something like 350 theatres and cinema clubs which feature and openly advertise movies under the name of pornofilms. 1968 *Tribune* 16 Feb. 11/1 So one must, I think acquit James Saunders of all but the technical responsibilities of adapting 'The Italian Girl', a pornogothick novel by Iris Murdoch, for the stage. 1975 *New Yorker* 19 May 23/1 *The Night Porter*—A porno gothic, set in Vienna in 1957 and veneered with redeeming social values. 1936 C. S. Lewis *Allegory of Love* vi. 251 He [*sc.* Dunbar] practises every form from satiric pornogram to devotional lyric. 1979 K. Bonfiglioni *After you with Pistol* xvi. 119 Death..lays on others the chore of hiding the pornograms, the illegal firearms, the incriminating letters. 1969 *Daily Tel.* 25 Apr. 20/3 Haven't we got a steadily increasing amount of violence, bigotry and gang warfare (to say nothing of..nihilism and pornomania)? 1969 *Truth* (Melbourne) 12 July 13/5 Pornomovies which are being shown in theatres in America, England and Western Europe known as sex houses. 1976 *Publishers Weekly* 16 Feb. 80/2 He comes in contact with porno movies and a girl who makes them. 1977 M. Drabble *Ice Age* i. 61 He'd been had up for offering bribes to council employees: the whole story had been ridiculous, tales of..call girls and twenty-pound notes, of tax evasion and porno-movies. 1960 *20th Cent.* May 434 The great social surveys ..are no diet for the pornophile. *a* 1966 E. Waugh in D. Pryce-Jones *Evelyn Waugh* (1973) xiv. 226 Will not this bit in the *Sunday Times* excite your pornophiles to fancy prices? 1973 *Austral. Humanist* xxvi. 4/2 Mary Whitehouse could provide fresh impetus for the pornophobic. 1965 *Listener* 3 June 837/1 A sequence of Gaby..untrussing for Pierre Louys in a room full of Oriental knick-knacks, then being somewhat primly caressed by the great pornophotographer. 1949 'G. Orwell' *Nineteen Eighty-Four* i. 46 There was even a whole sub-section [*sc.* of the Ministry of Truth]—*Pornosec*, it was called in Newspeak—engaged in producing the lowest kind of pornography. *Ibid.* ii. 132 Pornosec, the sub-section of the Fiction Department which turned out cheap pornography for distribution among the proles. 1966 S. Marcus *Other Victorians* v. 216 The results..are to turn the novel in the direction of pornotopia—that vision which regards all of human experience as a series of exclusively sexual events or conveniences.

pornograph, *sb.* (*a.*) **2.** (Later examples.)

1955 M. Allingham *Beckoning Lady* vii. 109 Through the window, a line of bladders, now a trifle flabby, were.. visible. 'Listen.' Tonker seized the mouthgrip... Mr. Campion pulled himself together. 'It's horrible,' he said. 'A pornograph, Tonker.' 1967 *Spectator* 1 Dec. 683/1 A pornograph can be either verbal or visual, but the visual stimulus is generally more intense than the verbal one.

pornographic, *a.* Add: Hence **pornogra·phica,** pornographic literature or art; **pornogra·phical** *a.* (in quot. *fig.*); **pornogra·phically** *adv.*; **pornographico-** *comb. form*, as *pornogra·phico-devo·tional* adj.; **porno·graphize** *v. trans.*, to make pornographic in character.

1917 J. B. Cabell *Cream of Jest* ii. v. 65 The latest masterpiece of a pornographically gifted genius. 1921 *Times Lit. Suppl.* 10 Feb. 90/4 She introduced him to a coward, an alienist who was himself mad, a pornographically minded professor. 1939 A. Toynbee *Study of Hist.* V. 531 Apuleius's pornographico-devotional romance [*sc.* *Metamorphoses*]. 1966 *New Statesman* 16 Dec. 912/3 Within half a mile of Great Marlborough Street, shops with red neon signs offering 'Books' go on selling printed pornographica at eight quid a time. 1968 *Economist* 7 Dec. 79/3 Facing the exchange rate problem, he maintained, does not mean that 'pornographical financial thoughts are being harboured'. 1971 S. Cavell *World Viewed* xiii. 95 Godard perceives..that our tastes and convictions in love have become pornographized. 1973 J. Money in Zubin & Money *Contemp. Sexual Behavior* 414 *True Confessions* and *True Love* magazine stories are, in fact, the genuine pornography of women. So women have been allowed to enjoy themselves pornographically as much as they desire. 1978 D. Murray *Place Apart* ix. 199, I stood in the middle of the road being pornographically insulted.

pornography. Add: **2. a.** (Later examples.) Also qualified by *hard* or *soft*, with reference to *hard core* (b) s.v. *hard *a.* 22 b, *soft core* s.v. *soft *a.* 27, to denote pornography of a more, or less, obscene kind. Also *transf.*

1930 W. S. Maugham *Gent. in Parlour* xii. 64 Pornography rather than brevity is the soul of wit. 1968 *Sat. Rev.* (U.S.) 19 Oct. 23 In recent years the movies and television have developed a pornography of violence far more demoralizing than the pornography of sex, which still seizes the primary attention of the guardians of civic virtue. 1972 *Times Lit. Suppl.* 7 Jan. 12/2 Of course pornography should never be treated as if only its sexual aspects mattered—that is, as if no other kind of stimulus offered by the written word could be as socially or ethically significant. 1976 *Time* (Canada ed.) 5 Apr. 36/1 What pornography is can endlessly be debated. One rough definition: explicit books, films and other materials (including, by extension, performances) designed chiefly for sexual arousal. 1977 *Broadcast* 30 May 3/3 [Italian] 'pirate' TV stations which flourish on..'soft pornography'. 1977 *Lancet* 11 June 1241/2 A distinction could be drawn between erotic art (or soft pornography)..and hard pornography, which by connecting sex with violence, hatred, pain, and humiliation, stimulated gratification of sexual desire in deviant ways.

b. In *transf.* and extended uses.

1968 [see above]. 1977 *Listener* 17 Nov. 655/4 Turgid moralising..is the real English vice, the pornography of our day.

porny (pǭ·ɹni), *a. slang.* [f. *porn *sb* + -y[1].] Of, pertaining to, or characteristic of pornography; pornographic.

1961 S. Price *Just for Record* viii. 65 He had a real porny article... Not just dirty, mind you, but Art. 1967 [see *kinky a. B]. 1973 J. Wilson *Truth or Dare* i. 12 You make it sound like one of those porny books—'His hand caressed her silken knee' and all that rubbish. 1974 *Daily Tel.* 18 Oct. 16 A reduction in repeats, inane quizzes and cheap porny [television] programmes could do nothing but good. 1977 L. Meynell *Hooky gets Wooden Spoon* xiii. 157 Once he starts looking at those porny photos of his he can't think of anything else.

Poro (pǫ·ro). Also **poro, Porro, purra, Purrow.** [W. Afr.] The name of a secret tribal cult for men, based on circumcision and a school of initiation, which is widespread amongst tribes in Sierra Leone and Liberia, and is socially powerful; the head of such a tribal group; also *attrib.* Cf. *Sande.

1788 J. Newton *Thoughts upon Afr. Slave Trade* 15 The Purrow has both the legislative and executive authority. *Ibid.*, Everything belonging to the Purrow is mysterious and severe. 1803 T. Winterbottom *Acct. Native Africans Sierra Leone* I. viii. 135 They [*sc.* the Bulloms] have a superior, or head purra man, assisted by a grand council. 1925 T. N. Goddard *Handbk. Sierra Leone* iii. 57 Initiation into the Porro society takes place in youth. While boys remain in the Porro bush they are taught the arts and crafts of their tribe. 1930 R. P. Strong *Afr. Republ. Liberia* I. v. 83 The *poro* or presiding official of the society also uses signs to indicate the entrance to the bush school. 1954 E. Warner *Trial by Sasswood* (1955) ix. 161 The Loma have the most vigorous of Poro cults and are credited with having introduced the Poro into Liberia. 1962 C. Fyfe *Hist. Sierra Leone* xx. 571 Those who planned the rising had used the Poro as a cover to ensure secrecy. 1968 Harris & Sawyerr *Springs of Mende Belief* i. 1 The principal tribal cults, *poro* and *sande*... Poro..is made manifest by a series of rites and ceremonies depicting death to the early stages of life, and re-birth through resurrection to adulthood.

porocyte (pǫ·rosəit). *Zool.* [f. Gr. πόρο-ς Pore *sb.*[1] + -cyte.] In sponges, a cell containing a pore.

1898 E. A. Minchin in *Q. Jrnl. Microsc. Sci.* XL. 485 The especial functions and consequent peculiar form of the pore-cells, or porocytes, as they may be termed generally, ..enable us to regard the pore-cells as constituting a distinct class of cell-elements. 1932 Borradaile & Potts *Invertebrata* iii. 111 Other cells, known as porocytes, of a conical shape, extend through the jelly, having their base in the covering layer. 1940 G. S. Carter *Gen. Zool. Invertebrates* xix. 384 These pores pass through the substance of cells which surround them—the porocytes. 1966 *McGraw-Hill Encycl. Sci. & Technol.* II. 393/1 Cells called porocytes..pierce the walls [of calcareous sponges] at intervals and allow water to enter the central cavity.

porogamous (pǫrǫ·găməs), *a.* *Bot.* [f. as Porogamic *a.* + -ous.] = Porogamic *a.* So **poro·gamy,** fertilization of this kind.

1902 *Encycl. Brit.* XXV. 436/1 The pollen-tube normally reaches the apex of the embryo-sac through the micropyle (acrogamy or porogamy). 1905 I. B. Balfour tr. *C. E. von Goebel's Organogr. Plants* II. 615 The micropyle in all porogamous plants evidently conducts the pollen-tube. 1950 P. Maheshwari *Introd. Embryol. Angiosperms* vi. 183 The tube may enter the ovule either through the micropyle or by some other route. The former is the usual condition and is known as porogamy. *Ibid.* 184 Even in plants classed as porogamous, there are several modifications. 1965 Bell & Coombe tr. *Strasburger's Textbk. Bot.* 631 The pollen tube may reach the micropyle and hence the embryo sac by traversing the cavity of the ovule at a point where it is commonly filled with mucilage (porogamy, Fig. 757).

porokeratosis (pǫ:ro-, pǭə:rokerătōu·sis). *Path.* [mod.L., ad. It. *porocheratosi* (V. Mibelli 1893, in *Giorn. Ital. delle Malattie Veneree* XXVIII. 340), f. Gr. πόρο-ς Pore *sb.*[1]: see *keratosis* s.v. *kerato-.] A skin disease in which the lesions are annular horny ridges enclosing an atrophic area and (in the rare classic form of the disease) usu. occur early in life on the hands and feet.

1893 tr. V. Mibelli in P. G. Unna et al. *Internationaler Atlas Seltener Hautkrankheiten* xxvii. (1) 10/2 As regards the microscopic examination the case presents itself as a pathological fact *sui generis*, inasmuch as the principal alteration consists in hyperkeratosis of the sudoriferous ducts; and it is on this account that it deserves to be considered apart under the designation of 'Porokeratosis' which I propose provisionally as it has the merit of indicating clearly the anatomo-pathological significance of the alteration itself. 1899 *Allbutt's Syst. Med.* VIII. 478 In connection with these ringed nodules of the hands the peculiar disease known as Porokeratosis should be mentioned. 1943 *Arch. Dermatol. & Syphilol.* XLVII. 2 The extension of a lesion of porokeratosis is usually centrifugal, slow and insidious in onset, leaving a somewhat atrophic area in the center free of hair and glandular structures. 1960 J. Marshall *Dis. Skin* xxix. 777 Porokeratosis is a misnomer as the lesions are not, as Mibelli believed, related to sweat pores. 1967 *Arch. Dermatol.* XCVI. 611/1 This study of 31 patients presents disseminated superficial actinic porokeratosis (DSAP) as a distinctive and recognizable entity characterized by many uniformly small, minimal, annular, anhidrotic, keratotic lesions developing during the third or fourth decade of life on sun-exposed areas of skin. 1973 *Internat. Jrnl. Dermatol.* XII. 152/1 Our recent studies indicate that disseminated superficial actinic porokeratosis (DSAP) is a distinct entity which differs in many aspects from classic porokeratosis described by Mibelli and by Respighi.

Hence **po:rokerato·tic** *a.*

1943 *Arch. Dermatol. & Syphilol.* XLVII. 14 This consideration excludes the necessity of calling the lesion a nevus, in spite of the strong tendency of porokeratotic lesions to arrange themselves in linear and systematized configuration. 1972 *Brit. Jrnl. Plastic Surg.* XXV. 325 The tumours in both our cases were multifocal, and appeared to have arisen in the atrophic zone of the Porokeratotic lesion.

poromeric (pǭərome·rik), *a.* and *sb.* [f. Poro(us *a.* + Poly)meric *a.*] **A.** *adj.* Applied to synthetic leather-like materials that are permeable to water vapour. **B.** *sb.* A poromeric material.

1963 *Boot & Shoe Recorder* 1 Oct. 117/3 Du Pont has coined a generic term for it [*sc.* 'Corfam']: 'poromeric', which means a 'microporous and permeable coriaceous sheet material'. 1964 *Times Rev. Industry* Mar. 23/2 Another new Du Pont development is Corfam, a 'poromeric' material which has brought a breakthrough in the manufacture of shoe uppers, since it 'breathes' like leather. 1966 *Chem. & Engin. News* 19 Dec. 13/1 Farbwerke Hoechst's subsidiary, Kalle, A.G., has a poromeric at the development stage. 1967 *Economist* 4 Nov. 544/3 Both Du Pont's and ICI's products are porous, or poromeric, as the trade chooses to say,..but Courtaulds' Quox is not, and is therefore only suitable for sandals. 1971 *New Scientist* 12 Aug. 370/2 The collagenous poromerics currently under development..are mostly in the form of a fibrous mat composed of intermingled collagen and synthetic fibres agglomerated by a polymeric bonding agent. 1978 *Times* 26 Apr. 12/2 The high quality poromeric synthetic materials..launched in the 1960s never fulfilled..expectations. *Ibid.* 12/4 The breakthrough..came with the poromerics..which, the makers claimed breathe like leather.

porometer (porǫ·mītəɹ). [f. Gr. πόρο-ς Pore *sb.*[1] + -meter.] An instrument for measuring the degree of porosity; *spec.* one for estimating the sizes of the stomata of leaves by measuring the rate at which air can be passed through them.

1911 Darwin & Pertz in *Proc. R. Soc.* B. LXXXIV. 137 We believe that a much more intimate knowledge of the *living* stoma and its movements would be necessary to prove his contention..that 'the regulatory function is almost *nil*'. With a view to testing the question we have designed an instrument which we propose to call a porometer. The idea is to estimate changes in the stomata by recording the change in the velocity of a current of air drawn through them in the living leaf. 1939 *Geogr. Jrnl.* XCIV. 124 The effect of the continuous illumination of the arctic summer in the stomatal mechanism of leaves was studied by porometers. 1970 *Nature* 25 July 377/2 The stomatal activity of fully hydrated..broad bean plants.. was investigated using a recording porometer attached to the second mature leaves from the apex. 1975 G. Anderson *Coring* iv. 77 The basic instruments [for core analysis] are the porometer for porosity measurements, the permeameter for permeability measurements, and the saturation retorts for saturation measurements.

poroporo (pǫ·ropǫro). Substitute for def.: A shrub, *Solanum aviculare*, belonging to the family Solanaceæ, native to Australia and New Zealand, and bearing violet-blue flowers followed by large orange berries; also called bullibulli or kangaroo apple. (Earlier and later examples.)

1853 J. D. Hooker *Bot. Antarctic Voy.: Flora Novæ-Zelandiæ* I. 182 Nat[ive] name [of *Solanum aviculare*], 'Poroporo' in the northern, and 'Kohoho' in the southern parts of the Islands. 1882 W. D. Hay *Brighter Britain!* II. 152 Among indigenous vegetable productions came.. the berries of poroporo..and other trees. 1921 H. Guthrie-Smith *Tutira* xxii. 216 At other times of the year kowhai and hill-flax..will provide nectar,..poroporo.. and other native plants, seeds and berries. 1965 [see *gunyang]. 1970 M. E. Fisher et al. *Gardening with N.Z. Plants* 112 The poroporo is a leafy shrub.

porosity. Add: Also, the degree to which a substance is porous (see quots.).

1939 *U.S. Dept. Agric. Yearbk. 1938* 1174 *Porosity, soil*, degree to which the soil mass is permeated with pores or cavities. 1971 *Gloss. Soil Sci. Terms* (Soil Sci. Soc. Amer.) 13/2 *Porosity*, the volume percentage of the total bulk not occupied by solid particles. 1975 G. Anderson *Coring* i. 2 Porosity is a measure of the space in a rock not occupied by the solid structure or framework of the rock. It is defined as the fraction of the total bulk volume of the rock not occupied by solids. *Ibid.*, A commercial oil-bearing sandstone can have varying porosities... The formation should contain at least 8–10% porosity before it can be considered commercially interesting.

porphin (pǭ·ɹfin). *Chem.* Also **-ine.** [a. G. *porphin* (Fischer & Halbig 1926, in *Ann. der Chem.* CDXLVIII. 194), f. *porph(yr)in* *porphyrin.] A synthetic, purple, crystalline solid, $C_{20}H_{14}N_4$, which has a macrocyclic aromatic molecule consisting of four pyrrole residues linked by $-CH=$ groups, and from which the porphyrins are formally derived; now also = *porphyrin.

1926 *Brit. Chem. Abstr.* A. 963/1 The name 'porphin' is suggested for the parent substance of (II), containing no β-substituents. 1939 *Nature* 10 June 967/1 The chlorophyll molecule may be likened to a signet, or rubber stamp, the disk representing a so-called porphine ring. 1939 *Thorpe's Dict. Appl. Chem.* (ed. 4) III. 82/2 Porphin and the porphyrins..possess a characteristic type of absorption spectrum, and the power to form stable derivatives with metals, in which the two imino-hydrogen atoms are substituted, e.g. by a divalent metal atom. 1954 A. White et al. *Princ. Biochem.* x. 201 Porphin was first synthesized by Hans Fischer and Gleim in 1935; the compound is not known to occur in nature. 1966 *Q. Rev. Chem. Soc.* XX. 211 When each of the four pyrrole rings of a porphin bears two different substituents A and B in the β-positions, then four isomers exist. 1968 I. L. Finar *Org. Chem.* (ed. 4) II. xx. 797 Fischer et al. (1965) have also examined porphin by X-ray analysis, and found the molecule is nearly planar, the observed small deviations from planarity not being large enough to be significant. *Ibid.*, Substituted porphins are known as porphyrins. 1973 *Jrnl. Amer. Chem. Soc.* XCV. 8506/2 The porphine used in this study was prepared by the acid-catalyzed condensation of pyrrole-2-carbinol in 10% acetic acid–xylene solution.

porphobilin (pǭɹfobai·lin). *Biochem.* [a. G. *porphobilin* (Waldenström & Vahlquist 1939, in *Zeitschr. f. physiol. Chem.* CCLX. 191), f. *porph-yrie* *PORPHYRIA + -o + L. *bil-is* BILE: see -IN¹.] Any of a group of red-brown pigments derived from porphobilinogen.

1939 *Chem. Abstr.* XXXIII. 8777/5 Boiling in acid destroys I by conversion to a red porphobilin (II) with a mol. wt. of 750 to 800. 1952 *Science* 4 May 496/1 Solutions of porphobilinogen obtained by the above method, after having been acidified to pH 4·0 and boiled for 30 min, gave rise only to porphobilin, a dark-brown pigment without characteristic absorption spectrum, and any porphyrin characteristics. 1964 A. White et al. *Princ. Biochem.* (ed. 3) xiii. 793 The urine darkens markedly on standing, because of the conversion of abnormal quantities of porphyrinogens and other heme precursors into porphyrins, porphobilins, and other unidentified pigments.

porphobilinogen (pǭɹfobaili·nǒdʒĕn). *Biochem.* [a. G. *porphobilinogen* (Waldenström & Vahlquist 1939, in *Zeitschr. f. physiol. Chem.* CCLX. 191): see prec. and *-OGEN.] A colourless, crystalline, substituted pyrrole, $C_{10}H_{14}N_2O_4$ (see quot. 1972), which in animals is a precursor of porphyrins and is excreted in the commoner forms of porphyria.

1939 *Chem. Abstr.* XXXIII. 8777/1 In acute porphyria the urine contains a colorless substance (porphobilinogen), which is a pyrrole deriv. and is converted to uroporphyrin on standing. 1961 *Lancet* 22 July 175/2 The clinical signs disappeared quickly, but the patient continued to excrete rather a lot of porphobilinogen and uroporphyrin in the urine. 1964 A. White et al. *Princ. Biochem.* (ed. 3) xiii. 790 Enzymic preparations utilizing porphobilinogen for porphyrin synthesis have been isolated from plant, bacterial, and animal sources. 1970 R. W. McGilvery *Biochem.* xxi. 496 The parent indole ring is made in the next step by the condensation of two molecules of δ-aminolevulinate to form porphobilinogen. 1972 *Stedman's Med. Dict.* (ed. 22) 1004/2 *Porphobilinogen*, 2-aminomethyl-4-(2'-carboxyethyl)-3-carboxymethylpyrrole.

Hence po:rphobili·nogenu·ria, the presence of porphobilinogen in the urine.

1961 *Ann. Rev. Med.* XII. 265 The probable relationship between this syndrome [sc. toxic porphyria] and the consumption of seed wheat treated with hexachlorobenzene-containing fungicides is supported by the production of massive uroporphyrinuria and porphobilinogenuria in rats by mixing hexachlorobenzene with the food. 1970 Passmore & Robson *Compan. Med. Stud.* II. xxx. 4/2 In hepatic porphyria of Swedish type, new methods of measuring aminolævulinic acid and porphobilinogen enable subclinical cases to be detected. Such cases are said to have a low degree of expressivity by comparison with their relatives who have obvious clinical manifestations and readily detectable porphobilinogenuria.

porphyr-, porphyro-. Add: **po:rphyro·blastic** *a. Petrol.* [ad. G. *porphyroblastisch* (F. Becke 1903, in *Compt. Rend. IX Sess. Congr. Géol. Internat.* (1904) II. 570): see *-BLAST], applied to (the texture of) rock (usu. metamorphic) in which larger grains formed by recrystallization occur in a finer groundmass; so **po·rphyroblast**, one of these larger crystals; **po:rphyrocla·stic** *a. Petrol.* [ad. G. *porphyroklastisch* (F. Becke 1903, in *Compt. Rend. IX Sess Congr. Géol. Internat.* (1904) II. 570): cf. CLASTIC *a.*], applied to (the texture of) rock which has undergone dynamic metamorphism and in which larger grains remain in a finer groundmass; hence **po·rphyroclast**, one of these larger crystals; **po:rphyro·psin** *Biochem.* [*RHODOPSIN], any of a class of light-sensitive pigments found in the retinas of freshwater vertebrates, differing from rhodopsin in containing the aldehyde of vitamin A_2 rather than that of A_1 and in having a maximum absorption at a slightly longer wavelength.

1920 A. Holmes *Nomencl. Petrol.* 188 *Porphyroblast*,.. a term given to the pseudo-porphyritic crystals of rocks produced by thermodynamic metamorphism. The corresponding texture is called *porphyroblastic*. 1926 G. W. Tyrrell *Princ. Petrol.* xvi. 270 When idioblasts form large crystals embedded in a fine-grained groundmass, like the phenocrysts of a porphyritic igneous rock, the term *porphyroblastic* is used to describe the texture. *Ibid.* 272 Maculose structure is that in which porphyroblasts of strong minerals such as andalusite, cordierite, etc., are well developed. 1966 *McGraw-Hill Encycl. Sci. & Technol.* VII. 11/2 The large crystals of some plutonic rocks are probably more properly classed as porphyroblasts. They may have formed essentially in solid rock by recrystallization aided by residual fluids from the solidifying magma. 1975 G. Andersson *Geolog.* ii. 31 Mottled dolomite is the result of incomplete dolomitization and exhibits itself as porphyroblasts of dolomite in a matrix or as scattered patches of dolomite. 1920 A. Holmes *Nomencl. Petrol.* 188 Porphyroclastic structure..results from the crushing of the solid constituents of a rock..resistant minerals..or fragments..may be crushed, and may stand out in a pseudo-porphyritic manner from the finer material produced by the crushing of the softer constituents. This structure is called porphyroclastic. 1954 R. L. Parker tr. *Niggli's Rocks & Min. Deposits* xii. 239 Porphyroclastic, with porphyroclasts..lying in a finer groundmass. 1975 *Nature* 20 Feb. 598/2 Rocks of the amphibolite-sole association..are separated from an amphibolite facies gneiss complex..by sheared mylonite. 1933 G. Wald in *Nature* 13 June 1017/1 The visual purple of freshwater fishes possesses different spectral properties. I shall refer to it as porphyropsin. 1962 K. F. Lagler et al. *Ichthyol.* xi. 376 The retinas of fishes yield two kinds of light-sensitive pigments, rhodopsin and porphyropsin. 1975 *Compar. Biochem. & Physiol.* A. LI. 720/2 While terrestrially..appears to act as a selective force for the predominance of rhodopsin, the function of the porphyropsin and mixed porphyropsin-rhodopsin systems remains obscure.

porphyria (pǭɹfi·riă). *Path.* [mod.L., f. *PORPHYRIN + -IA¹.] Any of various metabolic disorders characterized by the excretion of abnormally large quantities of porphyrins.

1923 A. E. Garrod *Inborn Errors of Metabolism* (ed. 2) viii. 144 Pigmentation of the enamel is a rare phenomenon.. Even in congenital porphyria it is evidently an exceptional sign. 1944 *Jrnl. Biol. Chem.* CLVII. 330 A complex mixture of pigments occurs in porphyria. 1969 Macalpine & Hunter *George III & Mad-Business* iii. 176 The galaxy of pigments demonstrated in these pages is due to our method of selection and must not create the impression that porphyria is common among the great, than the not so great. 1970 Passmore & Robson *Compan. Med. Stud.* II. xxxi. 10/2 The Swedish type of hepatic porphyria..may appear clinically only after administration of certain drugs, particularly barbiturates. 1975 *Victoria* (B.C.) *Times* 29 Apr. 14/2 Hematologists..had been treating the man who lived in a soldiers' home for..porphyria.

Hence **porphy·ric** *a.*, of, pertaining to, or affected with porphyria; also as *sb.*, a person so affected.

1954 *Acta Med. Scand.* LXXXIII. 286 Hans Fischer..has found a violet substance in the gallstones of the famous porphyric Petry. 1944 *Ibid.* CXVII. 8 A latent porphyric may suffer from..acute abdominal malady than a porphyric colic. 1974 *Nature* 9 Aug. 564/1 We noticed that skin lesions were provoked in porphyric rats, whose coats were shaved.

porphyrin (pǭɹfirin). *Chem.* [a. G. *porphyrin* (Willstätter & Fritzsche 1909, in *Ann. d. Chem.* CCCLXXI. 33), f. *haematoporphyrin* hæmatoporphyrin (s.v. *HÆMATO-). f. Gr. πόρφυρ-α purple + -IN¹.] Any of a large class of deeply-coloured red or purple fluorescent crystalline pigments that are substituted derivatives of porphin, many of which occur widely in nature, both in the free state and as complexes with metals (as in the hæms).

1910 *Jrnl. Chem. Soc.* XCVIII. i. 127 The phyllins are converted by acids into the corresponding porphyrins, compounds which do not contain magnesium. 1939 [see *PORPHIN]. 1940 *Endeavour* VIII. 83/1 In both [sc. adult and foetal haemoglobin] the prosthetic group (to which the oxygen becomes attached) is the same iron porphyrin compound. 1950 *Sci. News* XV. 95 Porphyrin is the coloured part of haemoglobin minus the iron. 1955 It is especially related because it shows an intense red fluorescence when held before an ultra-violet lamp. 1961 *Lancet* 22 July 175/1 A deficiency in purine synthesis could result, during porphyria, from the increased formation of porphyrins. 1970 Ambrose & Easty *Cell Biol.* vii. 228 The biologically important compounds chlorophyll, haemoglobin, and the cytochromes all contain a common cyclic structure, called a porphyrin, which consists of four pyrrole rings linked by methine bridges (—CH=). 1974 *Sci. Amer.* Dec. 73/1 In chlorophyll the central cavity of the porphyrin is occupied by a magnesium atom... Porphyrin rings in the blood protein hemoglobin and in the cytochromes contain an atom of iron instead.

Hence **po:rphyrino·pathy** (see quot. 1950).

1950 *Thorpe's Dict. Appl. Chem.* (ed. 4) X. 132/2 Porphyrinopathies, or diseases accompanied by abnormal porphyrin production or excretion. 1961 *Ann. Rev. Med.* XII. 258 In this review, the main porphyrinopathies will be listed using Waldenström's classification.

porphyrinogen (pǭɹfiri·nǒdʒĕn). *Biochem.* [a. G. *porphyrinogen* (Fischer & Bartholomäus 1913, in *Ber. d. Deut. Chem. Ges.* XLVI. 512), f. *porphyrin* *PORPHYRIN: see *-OGEN.] Any of the colourless, reduced derivatives of porphyrins in which the four pyrrole nuclei are linked by methylene groups, —CH_2—.

In quot. 1913 applied *spec.* to the first such compound to be prepared.

1913 *Jrnl. Chem. Soc.* CIV. i. 409 A colourless, crystalline reduction product, $C_{34}H_{44}O_4N_4$, of high molecular weight is obtained. This is termed porphyrinogen as it is very readily converted into a red product having the spectroscopic properties of porphyrin. 1938 *Chem. Abstr.* XXXII. 384 Until this work was carried out, only three crystalline leuco-compounds of porphyrins, or porphyrinogens, were known, and none of these was derived from the closely related precursors, possibly the more highly reduced porphyrinogens. 1968 Passmore & Robson *Compan. Med. Stud.* I. xxvi. 7/2 Porphyrinogens are similar [to porphyrins] but in these compounds the pyrrole rings are linked by methylene (—CH_2—) bridges.

porphyrinuria (pǭːɹfirinĭū·riă). *Path.* [mod. L., f. *PORPHYRIN + -URIA.] The presence of excessive or abnormal porphyrins in the urine.

1926 *Chem. Abstr.* X. 1671 The term porphyrinuria is used in place of hematoporphyrinuria because the substance isolated from the urine is not identical with hematoporphyrin. 1944 *Jrnl. Amer. Med. Assoc.* 29 Jan. 287/1 The term porphyrinuria should be reserved to denote a disturbance of porphyrin... They occur naturally in the urine are present in amounts above the normal range, a condition which exists in a great variety of diseases. 1967 *Amer. Jrnl. Med.* XLII. 476/1 Marked porphyrinuria was an additional feature in this case, and the urine also consistently contained another brown pigment.

porphyrism (pǭɹfiriz'm). *Path.* [ad. G. *porphyrismus* (H. Günther 1920, in *Deutsch. Arch. f. klin. Med.* CXXXIV. 257): see -ISM.] = *PORPHYRIA.

1923 A. E. Garrod *Inborn Errors of Metabolism* (ed. 2) viii. 139 In a more recent paper the same author puts forward the view that a constitutional anomaly, which he styles 'porphyrism', underlies acute and chronic cases alike. 1934 *Acta Med. Scand.* LXXXIII. 298 No signs of porphyrism were found.

porphyry. Add: **5.** *porphyry-red, -smooth* adjs.; *porphyry-born a.*, born in the purple (see *PURPLE *sb.* 2d).

1964 Auden in *Listener* 1 Oct. 525/1 Neither of our Dads, like Horace's, Wiped his nose on his forearm. Neither was porphyry-born. 1963 A. Lubbock *Austral. Roundabout* 194 The range thrust its porphyry-red buttresses into the forested valley. 1930 E. Pound *XXX Cantos* vii. 78 And the cave salt-white, and glare-purple, cool, porphyry smooth.

porpoise, *sb.* Add: **1. b.** *porpoise beef* (earlier Canad. example), *-dive* v. intr.

1833 W. F. Tolmie *Jrnl.* 21 Jan. (1963) 97 Ate some porpoise beef at breakfast. 1973 'C. Canning' *Flight of Grey Goose* vii. 138 He..took a deep breath, and porpoise-dived down, swimming strongly.

2. (See quots.)

1929 F. C. Bowen *Sea Slang* 105 *Porpoise, doing a*, said when a submarine dives down nose first at a sharp angle. 1931 *Times* 21 Aug. 7/1 It [sc. a seaplane] dropped back on the water... The aircraft then reappeared again, doubled to the height of her first bounce. 1958 F. H. Burgess *Dict. Sailing* 74 *Diving a porpoise*, said of a submarine taking a sharp dive. 1963 *Amer. Speech* XXXVIII. 119 *Porpoise*, an undesired landing in which the airplane bobs up and down like a porpoise playing in the waves, caused by landing on the nose gear first.

porpoise (pǭ·ɹpəs), *v.* [f. the *sb.*] *intr.* To move like a porpoise; *spec.* **a.** Of an aircraft, esp. a seaplane: to touch the water or the ground and rise again. **b.** To move through the water like a porpoise, alternately rising above it and submerging.

1909 H. G. Wells *Tono-Bungay* i. iii. 110 'Just as though an old Porpoise like him would ever make money,' she said... 'He'll just porpoise about.' 1917 *Adv. Comm. Aeronaut. No.* 437. 6 The author has seen a machine..porpoise very badly in waves of only one to two feet high. 1930 P. White *How to fly Aeroplane* xiv. 216 Sometimes, students fail to level off at all. This is an error which is bound to result either in a crash, or in a 'wheel' landing from which the plane will bounce or 'porpoise' quite high. 1931 [see *PORPOISE *sb.* 2]. 1934 H. G. Jones *Nelson's Renowned* 46 The Cocswain and the Second Coxswain had their work cut out to prevent the vessel [sc. a submarine] from 'porpoising'. 1944 *Richmond* (Virginia) *Times-Dispatch* 21 Sept. 6/6 The Liberator [bomber]..touched the water at a speed of approximately 100 miles per hour, porpoised [bounced] once, and struck again tail-first. 1968 F. W. Holiday *Great Orm of Loch Ness* v. 39 These objects were moving south-west... They were 'porpoising'—rolling under the surface and then reappearing. 1969 D. Bagley *Spoilers* vii. 206 I've set her [sc. a torpedo] to run at twelve feet. Any less than that an' she's likely to porpoise—jump in an' out o' the water. 1976 *Province* (Vancouver) 8 Mar. 1/3 From 500 feet there appeared an island full of seals and an ocean full of an enormous whale porpoising through his domain before diving out of sight. 1977 *Modern Boating* (Aus-

tral.) Jan. 34/2 With the trim..right out the boat begins to porpoise badly.

So **po·rpoising** *vbl. sb.*

1915 G. C. Loening *Military Aeroplanes* xi. 134 The latter condition, causing sudden changes in the angle of the bottom and its planing pressure,..gives rise to the disagreeable effect of 'porpoising'—a fore and aft rocking and jumping. 1920 L. Bairstow *Appl. Aerodynamics* ii. 55 Such phenomena as the depression of the bow due to switching on the engine and 'porpoising' are reproduced in the model with sufficient accuracy for the phenomena to be kept under control in the design stages of a flying boat. 1933 [see *HYDROFOIL 1]. 1974 P. Lovesey *Invitation to Dynamite Party* xii. 152 We have to steer by coming to the surface at intervals. It's a process known as 'porpoising'.

porridge, *sb.* Add: **3. a.** (Later examples.)

1972 *Listener* 18 May 662/2 Sometimes the programme has been a radio porridge, sometimes a shapely..necklace of sound, but never anything really remarkable. 1976 *Brit. Jrnl. Sociol.* XXVII. 36 On the other side of the great divide are the empiricists who..correlate vaguely-worded, interchangeable scales with each other and call the subsequent statistical porridge, alienation.

b. (Later examples.)

1870 *Scribner's Monthly* I. 154 While the engineers were floundering in the porridge at the west end, they wisely resolved to..sink a shaft to grade. 1966 H. Sheppard *Dict. Railway Slang* (ed. 2) 9 *Porridge*, sludge removed from drains.

c. A prison sentence; a term of imprisonment. *slang.*

1954 *Britannica Bk. of Year* 637/1 Several examples of underworld slang, probably of a date earlier than 1953, appeared in the newspapers. Thus, the reader learned that *porridge* meant a term of imprisonment. 1955 D. Webb *Deadline for Crime* i. 16 He did his porridge quietly, peacefully, earned full remission and came out. 1958 F. Norman *Bang to Rights* iii. 171 Week excuses that's all you get, when you go away to do a bit of porridge. 1968 J. Wainwright *Edge of Extinction* 92 D'you think I'd forget the frigging jack 'ot sent me down for two years' porridge? 1972 J. Brown *Chancer* xiii. 136 You've served your time..You've done your porridge. 'E. Crispin' *Glimpses of Moon* xii. 236 His emotions at the prospect..of yet another dose of porridge were such that he was..incapable of thinking clearly.

4. *to keep* (etc.) *one's breath to cool one's* (*own*) *porridge* (later examples); *to make a porridge*, to blunder, to make a mess of something.

1883 R. Cleland *Inchbracken* xii. 92 If our young Captain has wance ta'en the notion, thy may save their breath to cool their parritch, that would gainsay him. 1924 G. B. Shaw *Saint Joan* iii. 39 If you are going to say 'Son of St Louis: gird on the sword of your ancestors, and lead us to victory' you may spare your breath to cool your porridge. 1930 E. Pound *XXX Cantos* xxii. 99 He said he would save his breath to cool his own porridge. 1969 D. Clark *Nobody's Perfect* iii. 79 'Three months from retirement does serious notice.' Hunt said soberly. 'For a man who has made a porridge, perhaps.' 1971 'H. Calvin' *Poison Chasers* ii. 28 These buffins have made a porridge of the thing. 1976 A. White *Long Silence* xi. 101, I knew I would make a porridge of explaining it.

5. *porridge bowl* (examples), *saucepan*; *-coloured* adj.

1900 Farmer & Henley *Slang* V. 258/2 *Porridge-bowl*.. the stomach. 1925 *Heal & Son Catal.: Table Wares*, Porridge Bowl..1/-. 1936 J. Buchan *Island of Sheep* iii. 35 Archie Roylance..looked up sympathetically from his porridge bowl. 1974 P. Lovesey *Invitation to Dynamite Party* iii. 110 A tin porridge-bowl..and a mug. 1977 'J. Fraser' *Hearts Ease* xv. 166 You can go to that orphanage..stand in line and hold out your porridge bowl at breakfast time. 1909 E. Coxhead *Wind in West* i. 11 Grimy porridge-coloured upholstery. 1979 *Homes & Gardens* June 37/3 The village grocer, with a porridge-coloured haired granddaughter..and a blacksmith in belted greys. 1958–7 *Army & Navy Stores Catal.* 193/1 Double milk or porridge saucepan..4 pt. 9/-. 1975 C. Franklin *Long Shadow* iii. 24 A porridge saucepan soaking in the sink.

porridge, *v.* Add: **c.** *trans.*, to send to prison (*slang*). Cf. *PORRIDGE *sb.* 3c.

1966 B. Knox *Taste of Proof* i. 27 Jean reckoned you blokes had porridged the wrong fella when you pulled in Frank for the Glen Ault job.

Porro, var. *PORO.

‖ **porron** (porɔ·n). Pl. **porrones, porrons.** [Sp.] In Spain, a wine-flask with a long spout from which the contents are drunk directly.

1845 R. Ford *Hand-bk. for Travellers Spain* I. vi. 479/1 They are..drinking out of *Porrones*. 1936 *Burlington Mag.* June 301/1 An interesting fourteenth or fifteenth-century ancestor of the Spanish *porron* which is still used in Soho. 1968 K. Bird *Smash Glass Image* iv. 50 Francisco tilted the teapot-shaped *porrón* and ejected a stream of wine from the spout. It curved through the air..into his open mouth. 1974 *Times* [Dublin] 'D. Craig' *Double Take* x. 133 Really, it should be a skin, but this is a porron. 1976 *Scotsman* 24 Dec. (Weekend Suppl.) 2/5 We..drink our modest bottle of vino corriente while the crowd below swill porrons and pigskins of the stuff which makes them very vocal indeed.

Porson (pǭ·ɹsən). The name of Richard *Porson* (1759–1808), English classical scholar, used (*a*) in the possessive to designate a metrical law formulated by him and governing Greek tragic trimeters (see quots.); (*b*) *attrib.* and *absol.* to designate certain founts of Greek type influenced by his handwriting.

(*a*) 1894 W. W. Goodwin *Greek Gram.* (ed. 2) v. 358 When the *tragic* trimeter ends in a word forming a cretic (—⏑—), this is regularly preceded by a short syllable or by a monosyllable... This rule is known as 'Porson's rule'. 1949 *Osf. Classical Dict.* 565/2 The comic trimeter..is not bound by the 'Porson's Law'. 1958 H. Lloyd-Jones tr. *Maas's Greek Metre* §5 The rule..is known in this connexion as 'Maas's Law', as it is known as 'Porson's Law' with respect to the trimeter, and 'Havet's Law' with respect to the tetrameter. 1973 A. H. Sommerstein *Sound Pattern Anc. Greek* 136 The iambic trimeter..is governed in tragedy by, among other constraints, a rule known as Porson's Law. 1976 R. Pfeiffer *Hist. Classical Scholarship 1300–1850* xii. 160 Bentley had led the way in the scholarly treatment of Greek and Latin metre, and Porson was the first to make a further substantial advance; we may say that his claim to immortality is based above all on the rule called 'Porson's Law': that no word may end after a long *anceps* in the last iambus of the tragic iambic trimeter.

(*b*) 1894 *Amer. Dict. Printing & Bookmaking* 440/2 Greek type is cast by one foundry in New York... Some kinds now in use in the United States have been imported from England and from Germany... A heavy face, generally known as a Porson, is in use for headings and emphasis. 1927 D. B. Updike *Printing Types* II. 184 A..Porson Greek. 1960 Monotype Porson Greek. 1960 *Penrose Ann.* LIV. 36 Porson..used for the Porson greek type in London.

Hence **Porso·nian** (-ō^ən-) *a.*, pertaining to or characteristic of Porson, his work, or the fount of Greek type named after him. Also *absol.* as *sb.*

1840 *Penny Cycl.* XVIII. 420/2 Porson's great reputation during his lifetime served in a certain degree to fix the Porsonian school of critics', as they have been termed, 'three many impediments in the way of sound and comprehensive scholarship. 1896 *Dict. Nat. Biogr.* XLVI. 163/1 There was a reaction..against the Porsonian school. 1929 *N. & Q.* 13 Apr. 267/1 A stronger and more dignified type than the..commonly current Porsonian. 1976 R. Pfeiffer *Hist. Classical Scholarship 1300–1850* xv. 179 In his editions of Greek tragedies he [sc. Hermann] was..to rival Porson and the Porsonians or even surpass them.

port, *sb.*¹ Add: **1. c.** Phr. *port in a* (*or the*) *storm*, a refuge in difficulties or troubled circumstances (also in weakened senses); *spec.* in proverbial phr. *any port in a storm*, any refuge or escape (is welcomed) in adverse circumstances.

1749 J. Cleland *Mem. Woman of Pleasure* II. 133, I feeling pretty sensibly that it was going by the right door, and knocking desperately at the wrong one, I told him of it: 'Pooh,' says he my dear, any port in a storm.' 1787 J. Cobb *First Floor* II. ii. 51 Here is a door open, i' faith—any port in a storm, they say. 1821 Scott *Pirate* I. iv. 60 As this Scotsman's bow'l lies right under your lee, why, take any port in a storm. 1897 R. L. Stevenson *St. Ives* xxv. 188 'Any port in a storm' was the principle on which I had acted. 1936 R. Adams *Shardik* x. 229 'How do you like Maggie Cuddeford?' she asked. I replied, 'Any port in a storm, I suppose. It's better'. 1965 J. Porter *Dover Three* ii. 19 It was not quite the sort of company with which Dover would mix from choice but, as the jolly sailors say, any port in a storm. 1970 *Guardian* 7 Apr. 18/1 Midnight cries of scorn and indignation rang round the Commons, which were intensified as its return to work with a row over the Ports Bill... Well, any port in a storm these days. 1977 A. Morice *Murder in Mimicry* v. 111. 67 Henry and I moved on to our next port in the storm, which was a bar round the corner.

2. c. *port of entry*: a port by which people and goods may enter a country. Cf. *PORT *sb.*³ 4 b.

1840 *Niles' Reg.* 23 May 188/1 Mr. King..reported a bill for the establishment of ports of entry in the states of Missouri and Arkansas. 1936 *Phytopathology* XXVI. 476 The suggested system of field inspection and certification will..relieve the port-of-entry inspection services of the sole responsibility of passing materials offered for import. 1977 *Arab Times* 13 Nov. 4/7 Ercan is not an internationally recognised port of entry into Cyprus. Only the Turkish state airline THY and the tiny Turkish Cypriot airline, KTHY use it.

d. *port of call*: a port visited by a vessel in the course of a voyage. Also *transf.*

1810 *Sea Calb.* 5 a. 1910 W. T. Grenfell *Labrador Doctor* (1920) viii. 168 We..put down our helm..to avoid the wash... The last port of call was Henley, or Château, where formerly the British had placed a fort. 1978 J. B. Hilton *Anathema Stone* ix. 93 Waiting for me to be out of earshot..so that I would not know what was to be her next port of call.

6. a. *port-log, -officer.*

1891 Kipling *Barrack-Room Ballads* (1892) 206 O the mutter overside, when the port-fog holds us tied. 1923 Moored and helpless, a mile from the pier. 1901 *Chambers's Jrnl.* Aug. 531/2 The port officer..and the two Eurasian residents, came to the office:..to interview us.

port, *sb.*³ Add: **2. e.** *U.S.* An aperture in the body of an aircraft (see quots.).

1946 *Aeroplane Spotter* 21 Sept. 226/1 (*caption*) This photograph shows well the fabric covering the three machine gun ports in each wing. 1954 D. M. Desoutter *All about Aircraft* 415/2 Details of the armament and interior arrangements are sparse, but two large gun ports are visible in pictures. 1958 *N.Y. Times Mag.* 6 Apr. 68/4 The bombardier tightens the canvas over his ports. 1959 F. D. Adams *Aeronaut. Dict.* 128/1 *Port*, a circular window in the side of an aircraft fuselage, hull, or cabin, or a side aperture for a gun, a camera, etc.

3. b. (Earlier and later examples.)

1789 D. Davidson *Thoughts on Seasons* 169 They closed fast on every side—A port could scarce be found. 1811 J. Ramsay *Acct. Game of Curling* 10 Whether they will have to draw, strike, wick, or *enter a port*, they will seldom deviate an inch from their aim. [Note] To *enter a port*, is to make a stone pass through an opening made by two others lying opposite to one another. 1937 T. Henderson *Lockerbie* iii. 63 If ye mak' yersel' sma' ye can squeeze through the port. Here's the tee; noo canny. 1975 *Scotsman* 17 Mar. (Curling Suppl.) v. 1/7 The whys and wherefores of shots made and lost, the backing take-outs, in-wicks, out-wicks and draws through narrow ports were all double Dutch.

4. Also, an aperture by which the mixture enters the cylinder or combustion chamber of an internal-combustion engine, or by which the exhaust gases leave it. (Further examples.)

1886 D. Clerk *Gas Engine* vii. 168 An exhaust valve, leading into the space by a port, is also actuated at suitable times from the secondary shaft. 1913 *Autocar Handbk.* (ed. 5) ii. 33 During the compression and firing strokes all four ports are out of line, so that the cylinder is completely closed. 1956 F. Frenton *Pract. Car-Owner* i. 19/1 The upward stroke not only drives out burnt gas through an exhaust port in the cylinder wall but also draws in fresh mixture..through an inlet port. 1966 B. D. Power *High Vacuum Pumping Equipment* ii. 387 Conditions remote from the pumping port are being considered. 1967 L. Holmes *Odhams New Motor Man.* i. 34/1 Valves and a camshaft are not needed, as there are ports in the cylinder walls which are uncovered by the moving pistons to let fuel mixture in and let the exhaust gases out of the cylinders. 1978 L. Pryor *Flyer* (1979) ii. 25 Around the perimeter there were 72 ports. The instant the flames in one port, exploded between ports, then is expelled through the other port.

b. *Med.* = *PORTAL *sb.*¹ 1 f. Also *port of entry* (cf. *PORT *sb.*¹ 2 c).

1908 [see *CRYPTOGENETIC a.]. 1928 B. J. Leggett *Theory & Pract. Radiol.* II. viii. 220 Risk [of injury to surrounding tissues] becomes smaller the greater the number of fields or ports of entry. 1928 *Amer. Jrnl. Roentgenol.* XX. 135/2 It is not really necessary to have two separate ports for the useful radiation. 1936 B. J. M. Harrison *Textbk. Roentgenol.* iii. 50 Considering the physical conditions of the technique adopted, the milliamperage, the kilovoltage..and the size of the area treated (port of entry). 1962 Ross & Moore in *Surg. Pract. Lahey Clinic* (ed. 3) 169 If necessary biopsies are desired, the biopsy port is reopened by strong negative pressure applied on 'B' syringe for 5 seconds. 1977 *Radiologia Clinica* XLVI. 225 In order to obtain greater homogeneity of biological effects within the treatment volume, all prescribed ports should be used at each treatment session.

c. An aperture in any kind of container or vessel for the entry or egress of fluid.

1944 *Plastics* Jan. 18/2 In transfer moulding the material is placed in a heated pot from which it is forced through a narrow port into the actual mould. 1962 V. Grissom in *Into Orbit* 152 In the rush to get out before I sank I had not closed the air inlet port in the belly of my suit, where the oxygen tube fits inside the capsule. 1971 *Sci. Amer.* Sept. 222/3 A filter should be inserted between the inlet port of the compressor and the gas outlet of the laser.

d. An aperture in a loudspeaker enclosure.

1949 Frayne & Wolfe *Elem. Sound Recording* xxx. 627 Ports are provided at the front of the enclosure in order to utilize some of the backradiated energy to reinforce the energy from the horn at the lower frequencies. 1975 G. J. King *Audio Handbk.* vi. 143 The box has two main apertures, one to accommodate the driver unit and the other, called the vent or port, which allows air to move in and out of the enclosure in sympathy with the air pressure changes inside.

e. (i) *Electr.* A pair of terminals where a signal enters or leaves a network or device, the current flowing into one terminal at any instant being equal to that flowing out of the other. Freq. *ellipt.* with preceding numeral adj.

1953 Wheeler & Dettinger in *Wheeler Monogr.* ix. 7 After considering many alternatives, the writer has adopted the term 'portal' or simply 'port' as the general designation of the terminal pair of a network. A self-impedance becomes a 'one-port'. The usual transducer becomes a 'two-port' with one 'in-port' and one 'out-port'. The general network is designated a 'multi-port'. 1958 N. Balabanian *Network Synthesis* i. 9 The simplest network is the one-terminal pair, or one-port. 1966 L. A. Manning *Electr. Circuits* xii. 256 A two-port network may be driven by either a voltage or a current source of input, and either voltage or current may be measured at the output. 1973 *Nature* 3 Aug. 264/1 The switching element was a four-port ferrite switch driven at 1 kHz. 1975 D. G. Fink *Electronics Engineers' Handbk.* iii. 43 A transistor is a two-port network, although it has three terminals. Connecting an extra wire to one of the terminals provides the extra terminal without violating any network laws.

(ii) A place where signals enter or leave a data-transmission system or a device in such a system.

1970 C. S. Carr et al. in *Proc. AFIPS Conf.* XXXVI. 592/2 We assume here that a process has several input-output paths which we will call ports. Each port may be connected to a sequential I/O device, and while connected, transmits information in only one direction. 1972 *Proc.*

IEEE LX. 1409/1 The combiner may have a fixed number of input ports to which the terminals are either always connected, or to which they may be connected, if not already occupied. *Ibid.* 1412/1 Each remote TYMSAT is capable of accommodating up to 31 simultaneous users... In addition, each CPU has 60 input ports, each corresponding to a different user. 1976 *U.S. Agric. Outlook 1977* (Nat. Agric. Outlook Conf., U.S.) 366 The University user can lease either a 10-character per second or a 30-character per second port. The monthly rate varies.., depending on the speed. 1976 *Rep. Computer Board of Managem.*, 1975–76 (University Coll., London, Computer Centre) I. 4 For several years we have had a single dial-up line, operating at only 1200 bands. This single port was heavily used.

6. port-light (see quot. 1927); **port-mouth,** delete † and add later examples; also *attrib.*

1926 *Chambers's Jrnl.* July 478/2 Portlights as fitted to deck cabins have some drawbacks. 1927 G. Bradford *Gloss. Sea Terms* 132/2 The usual round openings closed with glass for light and air are called ports. The glass is set in a hinged brass frame called the port light. 1908 *Animal Managem.* 140 Swimming mounted, requires a capable horseman, who should be a good swimmer himself. Before riding in, it is well to remove the portmouth bit if one is worn. 1965 C. E. G. Hope *Riding* v. 62 The best-known variety of the Pelham must be the British military bit, the *port-mouth universal, reversible.*

port, *sb.*⁴ Add: **9.** esp. in phr. *at the high port*; also *transf.* and *fig.*

[1918 E. S. Farrow *Dict. Mil. Terms* 294 *High port*, a position in bayonet training.] 1937 Partridge *Dict. Slang* 19/2 *At the high port*, at once; vigorously; unhesitatingly; very much: military; from ca. 1925. I.e. in fine style. 1956 D. M. Davin *Sullen Bell* ii. vi. 148 'You seem very much at the high port,' Hugh said. 'I haven't seen you so bright since the evening you flung the smoke bomb into the Yank mess at Caserta.' 1970 *Daily Tel.* 28 Apr. 2/5 He began to climb the stairs with the gun at the 'high port' position. 1971 S. Mays *No More Soldiering for Me* xv. 153 He spun round with fists at the high port.

port, *sb.*⁶ (*a.*) Add: **2. a.** Also used with reference to aircraft.

1917 R. B. Matthews *Aviation Pocket-bk.* vi. 164 The leading edges of the port and starboard top wing should be in a straight line. 1939 [see *flight engineer* s.v. *FLIGHT *sb.*¹ 15]. 1948 [see *ASSEMBLY 1 c]. 1971 R. Dentry *Encounter at Kharm* ·l ii. 25 He had landed at Peshawar.. because the port motor was running too roughly to warrant continuing the flight. 1976 J. McClure *Rogue Eagle* iv. 69 The landscape sliding away beneath the port wing. 1977 *R.A.F. News* 27 Apr.–10 May 8/2 Then the port engine burst into flames.

b. *port-watch*: see quot. 1883.

1867 [see Larboard *sb.* (*a.*) B]. 1883 *Man. Seamanship for Boys' Training Ships R. Navy* (Admiralty) (1886) 5 The starboard watch work the starboard side of the deck, and the port watch the port side of the deck. 1953 C. S. Forester *Hornblower & Atropos* xvi. 228 Port watch wins!.. Starboard watch provides the entertainment tomorrow night!

port, *sb.*⁷ Add: **a.** (Later examples.) Also, a drink of port; a glass used for port.

1907 *Yesterday's Shopping* (1969) 937 Table glass services... 12 Sherries... 12 Ports... 12 Clarets [etc.]. 1925 [see *LIQUEUR *sb.* 2]. 1938 G. Greene *Brighton Rock* III. i. 98 Give me another port. 1974 *Times* 5 Apr. 12/3 The goblet is £4.25... A claret is £3.15 and a sherry/port £2.95.

port (pōᵊɹt), *sb.*¹⁰ *Austral.* colloq. abbrev. of PORTMANTEAU *sb.*

1908 E. G. Murphy *Jarrahland Jingles* 82 Silently they packed their 'ports' and flitted to the West. 1915 J. P. Bourke *Off Bluebush* 122 They see a young chap with a 'port' on his back. 1928 J. Devanny *Dawn Beloved* ix. 107 'Get my working togs out of my port, will you?'.. Dawn..opened his old portmanteau and took out the things. 1934 T. Wood *Cobbers* xviii. 236 A dignitary festooned in silver lace opened the door and asked me if I had any more ports. in the brake. 1946 D. Stivens *Courtship Uncle Henry* 53 You take your port up and come back to the car. 1954 G. Dutton in *Coast to Coast 1953–54* 149 Well grab your ports, and I'll take you out to the huts. 1967 *Sunday Truth* (Brisbane) 17 Sept. 16/3 She went back to her hut and happily unpacked her ports. 1972 R. Magoffin *Chops & Gravy* 46 Roly grabbed his port..charged towards the bus.

port, *v.*¹ **1.** Delete † *Obs.* and add later examples.

1973 W. H. Hallahan *Ross Forgery* iv. 53 The skids.. had been ported into the press rooms. 1979 'E. Peters' *One Corpse too Many* vii. 113 The boat..was of the light, withy-and-hide type that could be ported easily overland.

2. *port arms* (examples as *sb.*; also *attrib.*).

1918 E. S. Farrow *Dict. Mil. Terms* 462 *Port arms*, a position in the Manual of Arms. 1973 D. Barnes *See the Woman* (1974) I. 38 The..white-helmeted officers..stood with batons in a port-arms position, facing the crowd. *Ibid.* 81 Johnson held the shotgun at port-arms. 1974 D. E. Westlake *Help* (1975) xix. 128 The sentry..was still at port arms as though frozen in that position.

‖ **port-a-beul** (porʃt a biəl). *Sc.* Pl. (also used erron. as sing.) *puirt-a-beul.* [Gael., lit. 'music from mouth'.] 'A quick tune, gen. a reel-tune or the like, of Lowland Sc. orig. to which Gael. words of a quick repetitive nature have been added to make it easier to sing, now occasionally used as an accompaniment to dancing in the absence of instrumental music' (*Sc. Nat. Dict.*). Also *transf.* and *attrib.*

1901 K. N. MacDonald *Puirt-a-beul* 3 Puirt-a-beul, 'mouth-tunes', or 'tunes for dancing'. 1938 [see *mouth music* s.v. *MOUTH *sb.* 21]. 1945 B. Fergusson *Lowland Soldier* 12 The burn's making ever its own port-a-beul. 1952 N. Mitchison *Lobsters on Agenda* viii. 94 The three from the Glen, Janet, Sheila and young Mrs Macrae, were trying over a port a beul, very lightly, a living breath of humming. 1957 *Scottish Stud.* I. 133 The *Puirt-a-beul* are popularly supposed to have originated as a result of the religious opposition to musical instruments such as the bagpipes and the fiddle, which was at its strongest in the middle of the nineteenth century. 1964 *Listener* 15 Oct. 595/2, I shall never forget.., in a Roman street, a childish jet of water dancing, like MacDiarmid's duck, to its own *port a beul.* c 1970 A. MacPhee *Story of Highland Bagpipe* (An Comunn Gaidhealach) 8 All Gaelic music and pipe tunes are not sad and plaintive. Merely listen to a good 'puirt-a-beul' (mouth-music) singer. 1974 *People's Jrnl.* (Inverness & Northern Counties ed.) 5 Jan. 13/2 Other lively numbers are 'Ta-Ra-Ra Bhoom Di-Ay', written in Gaelic when that rhythm was the fashion and 'Tha na Cailean Meallda', a swinging puirt-a-beul.

portability. Add: (Later examples.) Also *transf.* and *fig.* in the senses of *PORTABLE *a.*

1955 *Sci. News Let.* 5 Mar. 160/1 Portable rink for outdoor ice skating makes this winter sport possible in some areas from April to November. Installed or dismantled in six days, the portability is achieved by more than eight and one-half miles of plastic piping. 1969 *Courier-Mail* (Brisbane) 19 May 11/4 Advocates of portability of superannuation benefits overlook the fact that private schemes have always been voluntary. 1970 A. Cameron et al. *Computers & Old Eng. Concordances* 24 Commenting on portability, I'm as great a sinner as anyone. 1972 *Sci. Amer.* Nov. 100/2 Nine-ounce portability and advanced computational capability. 1975 *Times* 14 Oct. 19/3 To make computer applications more independent of the manufacturer by developing 'software portability' (the ability to carry programmes, in effect, from one maker's machine to another). 1977 *Daily News* (Perth, Austral.) 19 Jan. 2/3 Seek long service leave after 10 years' continuous service, with 'portability' in the building industry and in the government.

portable, *a.* Add: **A. 1. a.** Also used to distinguish mechanical devices or electrical apparatus manufactured in forms smaller and lighter than normal, to enable them to be easily carried about.

1913 *Wireless World* Apr. p. xxxiv/2 The hon. secretary showed some model Marconi apparatus and a portable set. 1926 *Scribner's Mag.* Aug. 76/2 (Advt.), Portable receiving sets..now make it possible to carry this fine radio entertainment to summer camps and cottages. 1929 *Radio Times* 8 Nov. 444/1 What fun you can have with a portable gramophone. 1932 A. Christie *Peril at End House* ii. 31 There was a gramophone and..a portable wireless. 1937 E. Wharton *Ghosts* 22 In the middle of the carefully scoured table stood a portable wireless. 1951 *Catal. of Exhibits, South Bank Exhib., Festival of Britain* 128/2 Portable radio..mains or battery. 1961 *Lebende Sprachen* VI. 70/1 Portable typewriter. 1976 H. Nielsen *Brink of Murder* i. 11 The sportscast on the portable TV was in progress. 1977 *Wandsworth Borough News* 16 Sept. 17/4 To permit portable radios and accept the inevitability of dog mess, whilst stridently prohibiting children from cycling seems to reflect odd standards.

d. Of a building or the like: not of a permanent construction; capable of being dismantled and re-erected elsewhere.

1860 *Players* I. v. 39 (Advt.), Portable Theatres with scenery, Gas Fittings, &c. fitted up in town or country. 1955 [see *PORTABILITY]. 1968 *Globe & Mail* (Toronto) 3 Feb. 50/1 All have temporary lodging, two in portable halls and one in Woodbine Junior High School. 1972 [see *MODULAR *a.* 1 b].

e. *fig.* Of rights, privileges, information, etc.: capable of being transferred or adapted in changed circumstances.

1965 *Economist* 13 Feb. 671 Pension rights..should not be lost when an employee is sacked or moves to another firm; ideally they should be 'portable'. 1967 *Wall St. Jrnl.* 5 Jan. 2/3 (*heading*) Hoffa to seek portable pensions for teamsters moving into new jobs. 1970 A. Cameron et al. *Computers & Old Eng. Concordances* 22 The proposal never says..what care he's going to take to make sure that his work is portable, that his work really can be used by people at other institutions.

B. *sb.* That which is portable; *spec.* a piece of machinery that is portable (in sense 1 a above); usu. ellipt. for *portable camera, gramophone, radio, typewriter*, etc.

1883 J. Hay in *Century Mag.* Dec. 281/2, I don't doubt ..but what we could pay ourselves well for the job,—spoil the 'Gyptians, you know,—forage on the enemy. Plenty of portables in them houses, eh! 1918 C. Stone *Let.* 3 Apr. in C. Mackenzie *My Life & Times* (1966) V. 132 The gramophone (a Decca portable) is going again this evening after a fortnight's silence. 1926 *Wireless World* 26 May 16/1 (Advt.), For sale... Portables. 1930 T. E. Lawrence *Let.* 8 Jan. (1938) 677 An Italian bad-hat dashed me one of those electric gramophones... All the same that portable was good at Miranshah. 1931 B. Brown *Talking Pictures* v. 132 The Western Electric portable for sound-on-film. 1933 *Hearst's International* Mar. 101/2, I doubt if Mrs. Norris could type out a really good chapter on her rickety portable. 1952 M. Laski *Village* vi. 103 We've got an old portable we used to use for the farm correspondence... But she can't do touch-typing. 1957 *Practical Wireless* XXXIII. 530/2 Next comes the portable. It is probably here that it [sc. the transistor] finds its greatest application. 1960 *Life* 5 Dec. 8 (Advt.), This is the new Motorola portable—forerunner of all TV to come and gift idea of a lifetime. 1966 Auden *About House* 18 The Olivetti portable, The dictionaries (the very Best money can buy). 1973 *Times* 30 Oct. 32/9 (Advt.), Colour T.V. portables. 1976 J. Lee *Ninth Man* ii. 201 He would steal some of these carbons. Sarah had a little portable at his apartment. He would retype them.

portacabin, var. *PORTAKABIN.

portacaval (pōᵊɹtăkē^i·văl), *a. Surg.* Also **portocaval,** and with hyphen. [f. Portal *a.* (+ -o) + Cav(a + -al.] Applied to an anastomosis between the portal vein and one of the venæ cavæ, esp. an artificial one made so that blood in the former bypasses the liver.

1945 *Ann. Surg.* CXXII. 488 Every one of the ten cases of portacaval shunts..went through a successful postoperative convalescence. 1958 A. H. Hunt *Contrib. Study Portal Hypertension* xii. 99 The history of the modern operative treatment of portal hypertension begins with Ecle who suggested in 1877 that his portacaval anastomosis applied to human beings would provide the correct treatment for portal stasis. 1961 *Lancet* 19 Aug. 389/2 Portocaval anastomosis sometimes offers a way out of this quandary. 1968 Passmore & Robson *Compan. Med. Stud.* I. xxx. 47/1 The tributaries of the portal vein anastomose in certain sites with adjacent caval veins which return blood to the heart through either the superior or the inferior vena cava. Little blood passes through these porto-caval anastomoses in health. 1974 Marin & Ostrow in F. P. Brooks *Gastrointestinal Pathophysiol.* vi. 205 Surgeons have expended much time and effort in attempts to reduce portal hypertension by bypassing the hepatic sinusoidal bed, especially with portocaval shunts.

Portagee, var. *PORTUGUEE.

Portakabin (pōᵊ·ɹtăkæ:bin). Also **portacabin** [f. Porta(ble *a.* + *kabin* altered f. Cabin *sb.*] The proprietary name of a make of portable building. Cf. *PORTABLE *a.* 1 d.

1963 *Trade Marks Jrnl.* 4 Dec. 1742/1 *Portakabin* B 851, 268. Buildings (not being fixed metal structures) and parts thereof included in Class 19. Portasilo Limited, Blue Bridge Lane, York..; Manufacturers. 1975 *Times* 23 Sept. 6/1 The Portakabin which served as his drawing office. 1977 'R. Rostand' *Killing in Rome* xiii. 69 A large Portakabin, set on concrete blocks at the edge of the apron..the head office of Essex Air Ltd. 1979 *Jrnl. R. Soc. Arts* July 500/1 We created an artists' village—a collection of portacabins used in arctic oil exploration.

portal, *sb.*¹ Add: **1. d.** *Engin.* A rigid structural frame consisting essentially of two uprights connected at the top by a third member; orig. such a frame forming the end of a truss bridge.

1876 *Trans. Amer. Soc. Civil Engineers* V. 178 This bill of materials is calculated: chords, latticing, joint and reinforcing plates..85,912 pounds... Struts and portals.. 6,000 [pounds]. 1882 *Min. Proc. Inst. Civil Engineers* LXIX. 101 The Author [C. B. Bender]..believes it to be conducive to greater stiffness to put the material needed for knees and gussets into two effective end-portals and into the lateral top and bottom bracings. 1908 A. Tolhausen tr. *Böttcher's Cranes* vi. 245 The portal is to cover double railway lines of normal gauge. 1937 *Sunday Express* 14 Feb. 11/1 A series of vast concrete underground bridges, 'portals' they were called technically, were built. There were sixty in all, and each one bridged a railway tunnel under the earth and made a platform on which the building could be built. 1950 *Engineering* 31 Mar. 366/1 A simple portal structure built from broad-flanged beams may be used for spans up to about 40 ft. 1971 *Timber Trades Jrnl.* 21 Aug. 23/3 After the gale had been blowing for a whole week, the temporary bracing finally gave way and two portals were destroyed.

e. (The structural frame forming) the entrance to a tunnel.

1881 *Engineering* 25 Mar. 296/3 The geologist of the St. Gothard Tunnel..has been giving careful attention to the variations in the air currents between the two portals at Goeschenen and Airolo. 1909 J. W. Orrock *Railroad Struct.* iv. 85 The end portals for the tunnel consist of 12″ × 12″ posts..for a distance of 8 feet from the ends, with 12″ × 12″ timbers built over and across the end posts, to form retaining wall on top. 1941 Richardson & Mayo *Pract. Tunnel Driving* xxi. 364 Some kind of parapet over the portal is necessary to catch loose rocks rolling into the cut. 1971 K. G. Messenger *Flora of Rutland* 109/1 At the southern portal, the cutting is wider and deeper and it is obvious that it offered serious drainage problems to the engineers.

f. *Med.* Usu. *portal of entry* or *entrance* (or *entrance portal*), *portal of exit* (or *exit portal*). (*a*) The place where a micro-organism or drug enters or leaves the system. (*b*) The area of the body where a beam of radiation enters or leaves it. = *PORT *sb.*³ 4 b.

1910 *Jrnl. Amer. Med. Assoc.* 24 Sept. 1109/2 We have been led..to view the nasopharynx as the location in the body to be regarded with special suspicion as being the portal of entry of the virus. 1919 *Jrnl. Exper. Med.* XXIX. 380 Other portals of experimental infection were ..disclosed, such as the large nerves, subcutis, subarachnoid space, nasal mucosa, eye, and..the general blood. 1930 J. S. Friedenwald *Path. of Eye* xi. 227 An occasional individual contracting the infection in spite of having taken every conceivable precaution to protect against infection through all other known portals. 1931 *Jrnl. Amer. Med. Assoc.* 23 May 1756/1 If the portal is of limited area, the lateral scatter [of X-rays] is small. 1960

A. L. Smith *Carter's Microbiol. & Path.* (ed. 7) xii. 135/1 Pathogenic agents..have rather definite routes of discharge from the body, known as portals of exit, which, to a great extent, depend on the part of the body that is the site of disease. 1963 S. E. Wedberg *Paramedical Microbiol.* xix. 410 Portal requirements limit the number of opportunities provided for pathogens to cause damage to tissues. 1966 A. A. de Lorimier in *Radiol. in World War II* (U.S. Army) iv. 86 Except for the portal of exit of the primary beam, the roentgen tube housing is impregnated with material possessing a protective equivalence of no less than 1·5 mm. of lead. 1973 Fletcher & Tapley in G. H. Fletcher *Textbk. Radiotherapy* (ed. 2) i. 65/1 With a 22 Mev beam and a single homolateral portal, the skin reaction is minimal on the entrance side and is moderate on the opposite side. 1977 *Lancet* 8 Jan. 78/1 Disposable devices are now becoming available for prolonged controlled delivery of appropriate drugs at other portals of entry such as the eye and uterus.

g. *U.S. Theatr.* (See quots. 1947, 1959.)

1947 *Gloss. Technical Theatr. Terms* (Strand Electr. & Engin. Co.) 23 *Portal*, German and American terms for pros[cenium] opening. 1959 W. C. Lounsbury *Backstage from A to Z* 94 *Portal*, a gate, door, or entrance, usually downstage on either side of the stage. Portals may be scenery constructed for the play, or they may be a permanent part of the proscenium. In many theatres of newer design, portals are built to accommodate spotlights for sidelighting. 1978 *English Jrnl.* Dec. 44/1 'Gel the lights in the upstage right portal' are heard shouted across the auditorium.

4. portal bracing = sense 1 d above; also, the technique of using such a frame; **portal crane,** a crane mounted on a portal frame, so as to allow the passage of vehicles underneath; **portal frame** = sense 1 d above; **portal strut,** a horizontal member rigidly joining the tops of two uprights, esp. in a portal frame.

1881 *Trans. Amer. Soc. Civil Engineers* X. 164 Strong top lateral and portal bracing would greatly increase the strength and durability of the bridge. 1908 M. S. Ketchum *Design of Highway Bridges* vii. 112 Portal bracing is placed at the ends of through bridges in the planes of the end-posts to transfer the wind loads from the upper lateral system to the abutments. 1928 W. A. Mitchell *Civil Engineering* xvi. 479 The sway bracing at the entrance of the span..is called portal bracing. 1974 *Sci. Amer.* Feb. 95/1 The Crystal Palace was..the first [building] in which a light frame was made rigid against wind loads by the technique that came to be known as portal bracing. 1908 A. Tolhausen tr. *Böttcher's Cranes* vi. 245 (*heading*) Hydraulic portal crane. 1958 *Times Rev. Industry* Oct. 20/3 No. 21 Quay at Alexandra Dock has been opened this year after being re-equipped with five 6/3 ton electric portal cranes. 1908 A. Tolhausen tr. *Böttcher's Cranes* vi. 245 The portal or gantry frame.. shall be of built-up plates, and shall carry the platform on its top side. 1949 *Archit. Rev.* CVI. 287 The mullions in front of the portal frames are bright ultramarine. 1971 *Timber Trades Jrnl.* 21 Aug. 23/3 Each portal frame is constructed using 50mm (2in) nominal timber throughout. 1894 W. H. Warren *Engineering Construction* xix. 294 The perpendicular distance between the end strut of the top lateral system and the intermediate portal strut. 1938 C. T. Bishop *Structural Design* x. 194 Portal struts are used at the ends of through bridges to transmit top-lateral stresses to the abutments through the end posts acting as girders.

b. portal-to-portal *attrib.*, *spec.* of workers' pay: pertaining to the time spent on the premises of one's place of work, for example in travelling to and from the entrance, changing, or washing, as distinguished from the time spent working. *U.S.*

1943 *Time* 25 Oct. 21/3 He emerged with proposed Contract No. 3: an intricate formula which cagily skirts any mention of increased hourly wages or 'portal-to-portal' pay. 1944 *Birmingham* (Alabama) *News* 27 Mar. 1/5 The Supreme Court ruled Monday that underground iron ore miners are entitled to 'portal-to-portal' pay for the time spent traveling between the mouth of the mine and the place where the ore is actually mined. 1948 *Ann. Reg. 1947* 231 Long-pending retrospective claims for portal-to-portal pay (*i.e.* pay for time spent inside factory gates but not actually on the job) were unequivocally disallowed. 1965 *McGraw-Hill Dict. Mod. Econ.* 385 Proponents of portal-to-portal pay insist that the worker should be paid for the time involved in necessary activities before or after actual on-the-job time on the ground that otherwise the work could not be done.

‖ **portal** (pǭɹtā·l), *sb.*² Also **portale, portales.** [ad. Sp. *portal*, pl. *portales*, porch, portico, piazza.] In S. America and the southwestern U.S., a veranda, portico, or arcade.

1844 J. Gregg *Commerce of Prairies* I. 144 The only attempt at anything like architectural compactness and precision, consists in..buildings, whose fronts are shaded with a fringe of *portales* or *corredores*. 1892 C. F. Lummis *Tramp across Continent* 153 Outside, in the long *portal*, was enough blue, and red, and white corn to feed an army of horses. 1910 [see *MAJOR-DOMO c]. 1927 W. Cather *Death comes for Archbishop* ii. i. 51 Under this *portale* the adobe wall was hung with bridles, saddles [etc.]. 1927 *South Amer.* Nov.–Dec. 181/1 Our hall not being large enough, the *portales*—a large corridor with arches running down one side—was swept and tastefully decorated. 1948 *Southwest Rev.* Summer 245/2 What are now empty mule stalls then used to be the *portales* of a convent. 1973 D. Hamilton *Intriguers* ix. 59, 'I..crawled to where I could watch the long porch outside the living room.'.. I said, 'Around these parts [*sc.* Arizona], that porch is known as a portal, ma'am. Accent on the last syllable.'

portal, *a.* Add: **2.** *portal system*, also, any other system of blood vessels which runs directly from one system of smaller vessels to another. (Examples.)

1930 *Jrnl. Anat.* LXV. 88 These vessels of the portal system lose their heavy neuroglial wrapping and open out into a network of very fine channels. 1974 M. Hildebrand *Analysis Vertebr. Struct.* xii. 262 In several places in the body (digestive organs, kidneys, hypophysis) blood that has passed a capillary bed elsewhere enters a second capillary bed before reaching the heart. The veins between two capillary networks constitute a portal system. 1974 D. & M. Webster *Compar. Vertebr. Morphol.* xvi. 416 This is a renal portal system, filtering blood from the tail through a kidney capillary system before sending it to the [piscine] heart.

Portal (pōᵊ·ɹtăl). Name of Lord *Portal* (1885–1949), Minister of Works and Planning and First Commissioner of Works and Public Buildings 1942–4: **Portal house,** a steel-framed type of prefabricated house proposed in 1944. Also *ellipt.*

1944 *Archit. Rev.* Sept. p. lii/1 It is not clear whether the 250,000 [houses] will all be of the Portal ('Churchill') design which..is far from perfect of its kind. 1945 *Ann. Reg. 1944* 66 The Government had chosen the so-called Portal house of steel... The model Portal house..had been seen by about 30,000 persons. 1945 *Punch* 16 May 425/1 But my aunt was good for me when I was a child, and will possibly offer me accommodation if I cannot secure a Portal when I go home, so I suppose I must oblige her. 1948 A. M. Taylor *Lang. World War II* (rev. ed.) 158 *Portal houses*, proposed prefabricated houses for England, so called after Lord Portal, whose ministry was in charge of housing. Punch played up the name in jokes about 'crossing one's portal'.

‖ **portamento.** (Earlier and later examples.)

1771 C. Burney *Present State of Mus. France & Italy* 18 The French voice never comes further than from the throat; there is no *voce di petto*, no true *portamento* or direction of the voice, on any of the stages. 1926 *Amer. Speech* I. 500/2 Two Italian words, *glissando* and *portamento* are similar in meaning to the word 'smear', the principal difference being that the last-named is used [in trombone-playing] for a comic effect while the others are used for carrying the voice or sliding the fingers on the violin from one stop to the next. 1931 *Times Lit. Suppl.* 11 June 461/1 The notation of the tunes [in Bartok's *Hungarian Folk Music*] includes a number of signs to signify graces, quarter-tones, *portamento* and other of the folk-singer's idiosyncracies. 1961 C. Bunting in A. Baines *Mus. Instruments* vi. iii. 142 The cellists were trying to..achieve the eloquence and directness of the human voice... There was felt a need to 'carry' the music (*portamento*) through the intervals without a break. 1971 *Guardian* 26 Aug. 10/1 Imrat..is a master of fluid portamento effects and of creating an illusion of sustained legato—obtained by pulling a sitar-string sideways across the frets. 1975 *New Yorker* 20 Jan. 86/3 She oscillated in portamento between the A on the staff and the C sharp above it, swinging pendulum-true. 1976 *Gramophone* Aug. 337/2 The sheer certainty of the vocal delivery is remarkable, with the singer's famous portamento much in evidence.

‖ **portantina** (pǭɹtantī·nă). [It.] = Sedan chair.

1758 M. W. Montagu *Let.* May (1967) III. 149 He hopes you took nothing ill, tho' you refused the Portantina. 1937 *Tablet* 11 Dec. 800/1 Carried in the *portantina*—the Sedan chair which he now invariably uses in the palace—the Pope passed through the *Salone Sistino* into the Sacred Museum.

portapak (pōᵊ·ɹtăpæk). [f. Porta(ble *a.* + Pa(c)k *sb.*¹] A portable system comprising a small television camera and a video tape recorder.

1974 *Cablevision* (Rediffusion) 10 Since October 1973 Jessica Stanley Clarke..has spent most of her time in Knowle West with one of the station's portapaks. *Ibid.* 12 The Youth Department had purchased a Sony portapak. 1974 *New Society* 26 Dec. 805/1, I showed them how to set up and operate the portapak. 1975 *Listener* 9 Jan. 40/1 The miniature and monochrome amateurism of the 'portapak' suitcase studio.

portasystemic, var. *PORTOSYSTEMIC *a.*

‖ **port de bras** (por də brā). *Ballet.* Pl. **ports de bras.** [Fr., lit. 'carriage of the arms'.] The act or manner of moving and poising the arms; also, one of a series of exercises designed to develop the graceful movement and poising of the arms.

1912 *Dancing Times* Aug. 449/2 Arms... The Port de Bras, and exercises thereon. 1920 *Ibid.* Dec. 181 In operatic dancing..certain exercises have been evolved..known as the side and centre practice, the ports de bras and the adage. 1922 Beaumont & Idzikowski *Man. Classical Theatr. Dancing* ii. ii. 56 *Port de Bras* deals with the positions and movements of the arms. 1940 C. W. Beaumont *Diaghilev Ballet in London* ix. 203 The attention paid to such details..makes the dance a work of art and not a mere combination of steps and *port de bras* rendered by a woman in ballet costume. 1948 *Ballet Ann.* II. 68 The eight *ports de bras*, or exercises to develop the graceful movement and co-ordination of the arms. 1952 [see *ÉPAULEMENT]. 1958 *Times* 22 Aug. 6/3 So gracious and erect was her head held on her shoulders, so beautifully poised her *ports de bras*. 1963 *Times* 9 May 16/7 Her footwork proved exemplary... Her *ports de bras* were a little inelegant, sometimes giving a slightly gauche appearance to her line. 1975 *New Yorker* 18 Aug. 77/3 The grandeur of her carriage makes even a simple rhythmic port de bras, such as the one she does standing in a cluster of Wilis, an exciting dance experience.

‖ **port de voix** (por də vwa). *Mus.* [Fr., lit. 'carrying of the voice'.] A kind of appoggiatura (see quot. 1944).

1740 J. Grassineau *Mus. Dict.* 182 *Port de voix*, a French term, which signifies the faculty and habitude of making shakes, passages, and diminutions, wherein the beauty of a song or piece of music greatly consists. 1876 Stainer & Barrett *Dict. Mus. Terms* 364/2 *Port de voix* (Fr.), a kind of appoggiatura combined with the Pincé. 1944 W. Apel *Harvard Dict. Mus.* 595/1 *Port de voix*,.. one of the most important French *agréments* of the 17th and 18th centuries. Essentially it is an upward-resolved suspension or appoggiatura... Usually..both appoggiatura and resolution are repeated, so that the ornament consists of four notes, the last three forming a *pincé*. 1978 *Early Music* Oct. 518/2 The *port de voix*, being firstly a connecting note and secondly a fluctuation of pitch, usually upwards, but also downwards, can be viewed as the most rudimentary form of decoration. 1979 *Ibid.* Jan. 22/1 When the *Port de voix* was shown as a rising grace note it could often be followed by a *pincé* without the need for a separate sign.

Port du Salut, var. *Port Salut* s.v. *Port Salut 2.

porte-. Add: *porte-bouquet* (earlier example); **porte-parole** (-parol) [F. *parole* word], a spokesman, a mouthpiece.

1839 C. Schreiber *Jrnl.* (1950) 93 The Ex-Chancellor.. took the nosegay in his hand, extricated it from the porte-bouquet..and instead of giving the flowers into Mr. Pamther's expectant hands, he smelt them himself. 1946 J. Flanner in *New Yorker* 23 Mar. 74/2 Milch gave the appearance of being Göring's *porte-parole*. 1966 T. Reese *Story of Accusation* xi. 158 He is a lawyer. He is only a *porte-parole*.

ported, *a.*¹ Restrict *rare* to sense in Dict. and add: **2.** Having one or more ports or apertures; freq. in comb. with preceding numeral or adj.

1850 J. Bourne *Catechism of Steam Engine* 67 Of the slide valve there are many varieties; but the kinds most in use are the D valve..and the three ported valve. 1884 *Engineering* 19 Dec. 566/2 The face on the cylinder is double ported. 1897 C. Hurst *Valves* ii. i. 98 The exhaust valves may assume the double-ported form. This type permits a considerable reduction of travel compared with the single-ported valve. 1952 H. F. Olson *Musical Engin.* ix. 319 (*heading*) Phase inverter or ported cabinet. 1975 G. J. King *Audio Handbk.* vi. 143 Because an aperture is an important part of the enclosure, the terms 'vented-box', 'ported' and 'tunnelled' are sometimes used to describe the system.

po·rted, *a.*² [f. Port *sb.*⁷ + -ed².] Supplied with port-wine.

1929 J. Masefield *Hawbucks* 27 We're all dined and ported, thanks.

portée (pōᵊ·ɹtē^i, ‖ porte). Also **portee.** [a. Fr. *portée* (in various senses), f. *porter* to bear, carry.] **1.** The importance or weight (of a theory, an argument, etc.); the (far-reaching) consequences (of an action or an event).

1894 A. Lang *Cock Lane* 9 It is with this majority, if they choose to find time, and can muster inclination for the task of prolonged and patient experiment, that the ultimate decision as to the *portée* and significance of the facts must rest. 1899 W. James *Let.* 28 Jan. in R. B. Perry *Tht. & Char. W. James* (1935) II. 136 You seem to take my intention in the lecture to have had a wider *portée* than I ever thought of. 1904 H. James *Golden Bowl* II. xxv. 10 She called it [*sc.* her action] names, the invidious, the grotesque attitude, holding it up to her own ridicule, reducing so far as she could the *portée* of what had followed it.

2. In hand-loom weaving, a specified number of threads grouped together to form the warp. Also *attrib.*, as *portee cross.*

1910 L. Hooper *Hand-Loom Weaving* i. ii. 37 It only remains to take the group of eight threads *below* and *over* peg A in order to finish the first *portee*, as such a collection of threads warped in one round is called. *Ibid.* iv. 57 An excellent way of keeping account of the portees as they are warped is shown at fig. 27. *Ibid.* The fifty threads, taken all together, will pass above the first peg W, below the last one, then round it, and under the first, which completes the portee cross. 1954 H. J. Brown *Hand-Weaving* v. 76 This second leash is called the portee cross. 1954 M. E. Pritchard *Short Dict. Weaving* 67 One portee makes two warp 'ends'. The word..dates back to the days of the Huguenot weavers in England, deriving from the French *portee*, meaning 'carried'. 1958 A. Hindson *Designer's Drawloom* vi. 55 The portee, or grouped, crosses must not be split when they are spaced in the raddle. 1965 J. Tovey *Technique of Weaving* ii. 30/1 For convenience in keeping count of the number of threads warped, the ends of a strong yarn of a contrasting colour are crossed between groups of portées.

3. *Mil.* A self-propelled vehicle on which an anti-tank gun can be mounted. Also *attrib.* and in phr. ‖ *en portée* (aṅ) [Fr. *en* in].

1942 *Times* 22 Apr. 4/6 Finally, only two guns remained in action... Immediately afterwards one of these was destroyed and the portee of another was set on fire. 1944 *Return to Attack* (Army Board, N.Z.) 22 (*caption*) Anti-tank gun on portee at night. *Ibid.* 28/2 Portee anti-tank guns. **1948** R. Farran *Winged Dagger* viii. 141 It was a large command for a subaltern—ten armoured cars of mixed varieties, eight Bofors guns, a two-pounder portee and a tank. **1948** E. H. Smith *Guns against Tanks* (N.Z. Dept. Internal Affairs) 3 The two-pounders were carried on the decks of specially constructed lorries, termed *portées*, which were fitted with ramps and winches to enable the guns to be quickly hoisted into place. Special fittings..enabled the trail and spade to be clamped firmly to the deck so that the gun, pointing over the rear of the *portée*, was ready for immediate action. *Ibid.*, Great attention should be paid to training the gun crews in fighting the two-pounders from the decks of the *portées*: that is, *en portée*. **1952** *Times* 30 Aug. 5/6 Through it the light vehicles of 7th Armoured Division, portees, carriers, and light tanks, began to pass across our front as they skilfully drew on the enemy. **1958** M. K. Joseph *I'll soldier no More* viii. 151 It was eight o'clock before they piled into the portee-waggon.

‖ **Porteño** (pɔrte·nʸo). Also **porteño**. Fem. **Porteña**. [Sp.] A native or inhabitant of Buenos Aires, the capital of Argentina and (until 1884) of the province of Buenos Aires. Also *attrib.* or as *adj.*

1884 R. G. Watson *Spanish & Portuguese S. Amer.* II. xviii. 274 The great majority of the people of Buenos Ayres were not..passive spectators... The colonial society opened its salons to the English officers, and the *Porteña* beauties were not displeased to number them among their admirers. **1904** C. E. Akers *Hist. S. Amer.* ii. 37 The provincial representatives, whilst entertaining most vindictive feelings towards Rosas, had no real sympathy with the Porteños... *Ibid.* 39 He was to check *Porteño* influence that the majority of the provinces joined hands against Mitre. **1910** *Encycl. Brit.* II. 474/1 The national army..assaulted the *porteños* posted before Buenos Aires. **1959** *Chambers's Encycl.* I. 580/2 The relatively wealthy and cultivated oligarchy of merchants and professional men, the *porteños*, who dominated Buenos Aires (which then meant both the present city and the present province of that name). *Ibid.* 581/2 Rosas was by birth a member of the *porteño* oligarchy. *Ibid.* 582/1 The *porteño* army under Bartolomé Mitre..was defeated and Buenos Aires was incorporated in the Confederation in 1859. **1973** G. M. D. Howat *Dict. World Hist.* 1212/1 *Porteño* municipal loyalty was clearly manifested in 1806–7, when..local leaders organized the defeat of the British expeditions. *Ibid.*, *Porteños* demanded free trade and free movement of European capital, technology, immigrants, and ideas. **1978** *Times* 4 Feb. 5/3 Rats..can be spotted worrying refuse bags in the streets. For the *Porteños*..have waited to the last moment to start installing the [refuse] compactors now required in large buildings.

portentious (pǫəte·nʃəs), *a.* [Corruption of Portentous *a.*, infl. by Pretentious *a.*] Pretentious, pompous; portentous. Hence **porte·ntiously** *adv.*

1863 K. Stone *Jrnl.* 16 July in *Brokenburn* (1955) 227 The earth, the air, the sky, all are a dull dead grey..The sun seems to emit neither heat nor light, gleaming with a red glare like a blood-red moon... Some think it portentious, a sign of great victories or defeats. **1937** E. Partridge *Dict. Slang.* **1949** *Sun* (Baltimore) 21 Jan. 3/1 A portentious feature of this first telecast of a presidential inauguration was the use made of it in schools. **1956** *Ibid.* 20 Feb. (B ed.) 8/3 If you were lucky enough to have seen them, you witnessed a portentious enlargement of mankind's field of knowledge. **1958** *Times* 29 Oct. 3/1 A poem..was hammered home as a portentious statement containing the whole truth about the meaning, or meaninglessness, of life. **1962** *John o' London's* 17 Feb. 163/4 An Italian send-up of the portentious I.Q. flummery. **1975** *Publishers Weekly* 3 Mar. 64/2 Loving Ty whom she sees as somehow portentiously East Coast.

porter, *sb.*¹ Add: **3.** porter's chair (see quot. 1969).

1939 A. Christie *Ten Little Niggers* viii. 109 On the main terrace, Mr. Justice Wargrave sat huddled in a porter's chair. **1953** J. Cary *Except the Lord* xxv. 152, I was able to assist her own efforts enough to get her into her usual armchair, one of those great leather chairs with a high domed back, which were called porter's chairs and were valued in farmhouses for their power of keeping out draughts. **1969** J. Gloag *Short Dict. Furnit.* (rev. ed.) 532 *Porter's chair*, a high-backed armchair with wings raised to an arched hood, upholstered in leather, and placed in the hall of a town or country house, so that the porter or page boy on door duty could sit protected from draughts. Such chairs were introduced in the 16th century, and during the Georgian period were found in the entrance hall to every well-furnished house. **1972** N. Marsh *Tied up in Tinsel* viii. 195 He..sat down in one of two great porter's chairs that flanked the fireplace.

porter, *sb.*² Add: **1. d.** In full, *hospital porter*: a person employed by a hospital to convey patients and to carry out other general duties.

1950 G. B. Shaw *Farfetched Fables* iii. 109 They offered me a job as hospital porter because I'm physically strong. **1964** D. Francis *Nerve* vi. 64 The nurse came back with a stretcher trolley and two khaki-overalled porters... We waited outside in the hall, and saw them trundle Pip off towards the open lift. **1975** *Oxford Times* 7 Nov. 4/8 An Oxford hospital porter with a grudge against his employers used cheques he had stolen and forged..to obtain over £200 from banks in the city.

porter, *sb.*³ (Later examples.)

1903 Somerville & 'Ross' *All on Irish Shore* 73 Mrs Brennan added another spoonful of the brown sugar to the porter that she was mulling in a saucepan on the range. **1919** G. B. Shaw *O'Flaherty V.C.* in *Heartbreak House* 179 And look at your fine new uniform stained already with..the porter youve been drinking. **1922** E. O'Neill *Anna Christie* (1923) i. 6 Johnny draws the lager and porter and sets the big, foaming schooners before them. **1939** Joyce *Finnegans Wake* (1964) iii. 511 I've a big suggestion it was about the pint of porter. **1973** [see *porter-house 2].

b. *porter bar, -bottle, -pot* (earlier example).

1953 Dylan Thomas *Let.* July (1966) 350 One day a week I shall walk the miles to Glendormatie where there is a shop and porter bar. **1922** Joyce *Ulysses* 42 A porterbottle stood up, stogged to its waist, in the cakey sand dough. **1807** Southey *Lett. from Eng.* I. viii. 99 A transparency..which represented a loaf of bread saying to a pot of porter, I am coming down; to which the porter-pot made answer, So am I.

porter, *v.*² Add: Also, (of any person) to carry from one place to another; = Portage *v.*; **portering** *vbl. sb.* (later examples); also as *ppl. a.*

1927 *Glasgow Herald* 13 Aug. 8 This impressionable passenger thinks that it must be a rule of the railway portering brotherhood that each should 'portal' 'tell' to its uttermost. **1966** O. Varaday *Gara-Yaka's Domain* xiii. 142 Tan decided to camp there and..strengthen Cwgki and save himself the unwelcome bother of portering his riches. **1967** D. Cooper *Psychiatry & Anti-Psychiatry* v. 99 A member of the portering staff who witnessed the incident called a nurse who took her back to her ward. **1971** C. Bonington *Annapurna South Face* iv. 46 The Sherpas, who hold a monopoly in high-altitude portering, come from Sola Khumbu just opposite Everest. **1974** *Daily Tel.* 13 Dec. 14 We portered all the rapids. We did this because it would have been foolhardy for two relatively inexperienced canoeists to attempt them. **1977** J. I. M. Stewart *Madonna of Astrolabe* iii. 60 I've just been talking to a couple of young men who have been doing some portering for you. **1978** *Morecambe Guardian* 4 Apr. 29/2 (Advt.), Applicants should be fit and active as the work entails some portering and assisting the physiotherapists with the walking of elderly patients.

porterage, *sb.*¹ Add: **1.** Also, the manpower available for hire as porters.

1924 E. F. Norton *Fight for Everest: 1924* 122 He went down with Bruce and Irvine that same day to Camp III, intent on investigating afresh with Bruce's aid the question of available porterage.

3. *porterage work* (later example).

1957 M. Barrow *W. Afr. City* ii. 31 Members of this tribe had started coming to Freetown to pick up a living by doing porterage work in the streets.

porterage, *sb.*² Delete *rare* and add: Also, the availability of the services of a porter or caretaker.

1975 *Country Life* 22 May (Suppl.) 48/1 Luxury flats..24 hour porterage. **1980** *Sunday Times* 20 Jan. 49/1 Purpose-built studio flat..CH, CHW, porterage.

porter-house. (Earlier and later examples.)

c1758 S. Foote *Diversions of Morning* in T. Wilkinson *Wandering Patentee* (1795) IV. 239, I heard a goodish-looking well-dress'd man, that sat in the next box at the porter-house..affirm, that to his knowledge, if you proceeded to exhibit, you and your pupils would be all sent to Bridewell. **1780**, **1875** [see Ordinary *sb.* 18]. **1939** Joyce *Finnegans Wake* 405 He was immense, topping swell for he was after having a great time of it..in a porterhouse. If you want to know, Saint Lawzenge of Toole's, the Wheel of Fortune.

b. porter-house steak (earlier and later examples).

1842 C. Mathews *Career of Puffer Hopkins* xiii. 90 But I guess I'll take a small porter-house steak, without the bone. **1883** *Harper's Mag.* Aug. 464/2 A porter-house steak learned to expect him on the moon of every day. **1904** *Nation* (N.Y.) 17 Mar. 245 The picture was drawn of the country stepping up to the butchers' block and demanding porterhouse and sirloin steaks in abundance. **1955** [see *Chateaubriand]. **1970** W. H. Canaway *Solid Gold Buddha* xxiii. 148 Two porterhouse steaks..washed down with three bottles of Rhône wine.

2. *ellipt.* **a.** porter-house steak. Also *attrib.*

1884 *Harper's Mag.* Jan. 269/1 Will you have it very well-done? Shall it be a porter-house? **1908** G. H. Lorimer *Jack Spurlock* iv. 63 That [dream] in which the waiter is just taking the covers off a double porter-house, medium, with fresh mushrooms on top. **1911** [see *rare *a* and *sb.*³ 5]. **1931** F. P. Jones *Servant's Prob.* 17 This is a juicy porterhouse neighborhood. **1973** G. Beare *Snake on Grass* vii. 98 He would never pass up a pint of porter for a pound of Porterhouse.

po·rterless, *a.* [f. Porter *sb.*³ + -less.] Lacking a porter or porters.

1885 [see *gare *sb.*²]. **1973** *Daily Tel.* 30 June 12/6 Here is my timetable: Arrive Weymouth Town Station shortly before midnight..; bump baggage up long, dark, porterless platform and hurl it bodily into waiting pantechnicon.

‖ **porteur** (portǫ̈r). *Ballet.* [Fr., lit. 'one who carries'.] (See quot. 1957.)

1862 A. Haskell in C. Brahms' *Footnotes to Ballet* 5 Taglioni (1822) was frail and ethereal, and..the shoes were designed to conceal the fact that the ballerina..was lifted..the nose was soon degraded to the rôle of *porteur* from which Nijinsky finally rescued it. **1949** *Ballet Ann.* III. 51 With the advent of the Italian ballerina..ballet became even more a vehicle for the virtuosity of the female dancer backed by rows of pretty girls in the *corps de ballet*. The male dancer was..relegated to the position of *porteur*. **1957** G. B. L. Wilson *Dict. Ballet* 220 *Porteur*, Fr. lit. one who carries, a porter. A male dancer whose sole function is to lift the ballerina in supported leaps and similar movements. **1961** *Ibid.* (rev. ed.) 235 The male dancer in Western Europe was relegated to the rôle of a porteur during the half century preceding the coming of the Diaghileff Ballet and male dancing rôles were taken by the danseuse travestie.

portfolio. Add: **1. b.** The collection of securities held by an investing institution or individual; also, a list of such securities.

1930 *Economist* 4 Jan. 8/1 This fall is partly due to the banks' failure to secure any of last week's Treasury Bills, which is forcing them to replenish their portfolios in the market. *Ibid.* 11 Jan. 83/2 The more notable changes in the portfolio comprise an increase in the number of Chartered shares. **1948** *Ibid.* 3 Jan. 29/1 The banks last year took a hard knock on their gilt-edged portfolios. *Ibid.* 30/1 These examples demonstrate how important it is for the discount houses to keep their portfolios short. **1955** *Times* 3 June 10/3 The investment policy of the Company is under the constant review of the board and the portfolio periodically examined by a special committee of directors. **1967** [see *investment trust* s.v. *Investment* 5 b]. **1967** *Listener* 5 Oct. 423 (Advt.), M & G General Trust Fund is a very big Fund (over £26m.) with a portfolio largely of British 'blue chips', plus a number of overseas stocks. **1969** J. Argenti *Managem. Techniques* 199 Just as an investor tries to select a portfolio of shares to give him a good return without too much risk, so the Board of a company tries to do the same thing when choosing a 'portfolio' of subsidiary companies. **1972** *Accountant* 17 Aug. 206/1 A valuation of the entire portfolio, the result to be incorporated in the current year's accounts. **1973** C. Willans *Man on Leash* (1974) 83 He was..buying more stock all the time. His portfolio was worth a million or a little over. **1978** *N.Y. Times* 30 Mar. D 1/2 Analysts said that institutions stepped up their stock purchases yesterday to dress up their portfolios before the end of the first quarter.

2. Also in phr. *without portfolio*, (of a government minister), not being in charge of a specific department of state. Cf. *Minister without Portfolio* s.v. *Minister sb.* 5.

1891 W. Fraser *Disraeli & his Day* 370 At the time when Lord John Russell held office under Lord Aberdeen 'without portfolio', Disraeli made a very happy hit.

3. *portfolio capital, manager, official, policy, security, selection*; **portfolio investment**, the purchase of stocks and shares in a variety of companies.

1964 *Economist* 30 May 1015/1 Flows of portfolio capital. **1958** *Times* 29 July 5/2 It was a very complicated question because United States investment might be direct or it might be portfolio investment. **1965** *Guardian* 8 Jan. 1 (Advt.), Provides for portfolio investment in Australia. **1972** *Accountant* 27 May 587/2 Last month Kingside negotiated a $4 million, seven-year, multi-currency loan facility with Lloyds and Bolsa International Bank primarily for the purpose of portfolio investment in the United States and Europe. **1969** *Times* 5 May (Wall St. Suppl.) p. vi/3 Mutual funds are being bought as equity equivalents, and..I'm frankly very worried about it,' said a portfolio manager for one of the large trusts. **1973** *N.Y. Law Jrnl.* 2 Aug. 3/3 Another bull market similar to the one that occurred following Thanksgiving 1971, says Kenneth Herlihy, portfolio manager of Philadelphia-based Decatur Income Fund. **1969** H. G. Wells *Babes in Darkling Wood* iv. i. 322 They [sc. the Bolsheviks] only half did the job, they cleaned up the site and then they jerry-built some unattractive sheds... A sort of mushroom growth of nasty little portfolio officials may have sprung up on the stumps. **1969** *Times* 5 May (Wall St. Suppl.) p. iii/3 Credit demanders and institutional investors may not yet be convinced that there are downside risks in the aggressive implementation of economic plans and portfolio policies. **1959** *Listener* 12 Feb. 272/1 American portfolio securities. **1966** H. I. Ansoff *Corporate Strategy* (1968) ii. 30 While CIT [sc. capital investment theory] deals with selection of physical assets for the firm, portfolio selection concerns selecting securities, either for an individual investor or for an investment firm. **1969** J. Argenti *Managem. Techniques* 199 Portfolio Selection methods are very advanced mathematically. They are similar to Linear Programming..but include a means of balancing risks and returns.

port-hole. Add: Also **porthole**. **1. b.** A small glazed window, often round, in the side of an aircraft or spacecraft.

1936 W. A. Heflin *U.S. Air Force Dict.* 393/1 *Porthole*, a naval term sometimes applied to a circular window in an aircraft. **1962** W. Schirra in *Into Orbit* 33 They pointed out that they had already stuck on a periscope and a couple of small port-holes, but we all felt strongly that a pilot ought to have a clear, visual reference to his surroundings. **1968** *Listener* 27 June 827/1 Departure by air could involve hazards quite separate from the taking-off fears..of being sucked, à la James Bond, out of a porthole. **1970** T. Hughes *Crow* 13 It was cosy in the rocket, he could not see much But he peered out through the portholes at Creation. **1970** W. Smith *God Must Wait* 79 The Boeing began to roll forward. Manfred twisted his head against the neck rest and peered through the Perspex porthole.

2. c. *Austral.* and *N.Z.* An aperture in the wall of a shearing shed through which each shearer passes the sheep when shorn into his individual counting-out pen.

1898 Armstrong & Campbell *Austral. Sheep Husbandry* xv. 175 Upon the opposite side of the shearing board, 'port-holes', or small doorways, are made (one for each shearer), through which the sheep are turned when shorn. **1933** L. G. D. Acland in *Press* (Christchurch, N.Z.) 30 Sept. 15/7 *Counting out pens*. Each shearer has his own and passes his sheep through a *porthole* into his, so that each man's tally may be counted. **1956** G. Bowen *Wool Away!* (ed. 2) iii. 43 A lot of time and effort can be wasted in switching off and kicking sheep out the porthole. **1965** [see *chute *sb.* 5 b].

d. *Archæol.* A hole in a slab or two adjacent slabs of stone, large enough to allow the passage of a body into a chambered tomb.

1940 *Proc. Prehist. Soc.* VI. 133 Problems associated with the nature and origin of port-holes in megalithic tombs in Europe. **1954** S. Piggott *Neolithic Cultures Brit. Isles* v. 136 The chamber entrance was formed by a 'port-hole' made by hollowing the edges of two adjacent slabs to form an oval hole through which it is just possible to gain access to the burial chamber. **1958** G. Daniel *Megalith Builders W. Europe* ii. 44 Port-holes occur in southern Iberia and in a small number of tombs in France and Britain, as well as in the Gallery Graves of southern Sweden. **1963** *Field Archæol.* (Ordnance Survey) ed. 4) 39 Space does not permit a description of all the variations of entrance true and false, forecourts, passages, portholes, different forms of chamber, horn features, revetments, etc.

4. (sense *2 d) *porthole cist, slab, stone.*

1939 V. G. Childe *Dawn European Civilisation* (ed. 3) ix. 168 Forssander seems inclined to explain Pontic elements in Central Europe by a migration from the Caucasus of the makers of Globular Amphoræ who would also have brought the idea of the porthole cist and the pit-cave tombs. *Ibid.* xii. 206 A porthole stone often enhances the resemblance of a built tomb's doorway to the entry into a natural or artificial cave. The desire to emphasize the similarity has in fact been suggested as an explanation for the porthole stone's origin. **1970** Bray & Trump *Dict. Archæol.* 185 *Port-hole slab*, a stone slab with a hole cut through it, forming an entrance to a burial chamber. **1970** Bray & Trump *Dict. Archæol.* forming the entrance to a chamber tomb. Sometimes the hole is square, or the entrance is made from two slabs set side by side with notches cut from their adjoining edges.

Hence **po·rt-holed** *a.*, provided with a porthole or port-holes.

1938 *Antiquity* XII. 302 Some of these (e.g. Züschen, Fritzlar) have a porthole septal slab. **1940** *Proc. Prehist. Soc.* VI. 135 These figures should give the lie to the frequent assertion..that Britain is peculiarly rich in portholed tombs. **1969** *Daily Tel.* (Colour Suppl.) 11 Apr. 40/2 Their port-holed superstructure of the National Maritime Union.

portia¹ (pǭ·ɹʃiă). [f. Tamil *puarassu* flower-king.] In full, *portia tree*. An Indian name for *Thespesia populnea*, a tropical evergreen tree belonging to the family Malvaceæ and bearing yellow flowers.

1861 H. Cleghorn *Forests & Gardens S. India* 197 It is usual to plant large branches of the portia..and banyan trees in such a slovenly manner, that there is little probability of the trees thriving or living ornamental. **1883** J. S. Gamble *Man. Indian Timbers* 43 *T[hespesia] populnea Corr.*..the Portia or Tulip Tree... A moderate-sized evergreen tree. **1921** R. S. Troup *Silvicult. Indian Trees* I. 150 Tulip tree, portia tree... A small or moderate-sized evergreen tree of the coast forests of India and Burma. **1969** T. H. Everett *Living Trees of World* 237/1 The most important of the genus [sc. *Thespesia*]..is the Portia tree...Common along the coasts of tropical Asia and the Pacific Islands, this evergreen attains a height of 60 feet.

Portia² (pǭ·ɹʃiă). The name of the heroine of Shakespeare's *Merchant of Venice* used as the type of a female advocate or barrister. Hence **Po·rtian** *a.*, pertaining or resembling Portia.

1901 *Westm. Gaz.* 22 Jan. 12/2 The Paris Portia's First Success... Mdlle. Chauvan, the young lady barrister, made her first appearance yesterday in the Paris courts. **1909** *Ibid.* 17 Aug. 2/1 China..then took refuge in an interpretation of the letter of the Treaty which was quite Portian in its meticulousness. **1923** *Brewer's Dict. Phr. & Fable* (new ed.) 868/1 *Portia*, a rich heiress and 'lady barrister' in Shakespeare's *Merchant of Venice*... Her name is often used for a lady barrister or advocate. **1952** E. Weekley *Words & Names* 33 *Portia*, which is good journalese for a lady barrister. **1977** J. Mitchell *Half-Life* ii. 39 Good Christ, they've brought back the people's Portia.

portico. Add: **3.** *attrib.* as *portico area*; **portico thief** = *cat-burglar* (s.v. *cat *sb.*¹ 16 and 19).

1977 *N.Z. Herald* 5 Jan. 2-16/8 (Advt.), Portico area for outdoor entertainment. **1934** P. Savage *Savage of Scotland Yard* xxiv. 260 Suspicion fell on half a dozen or more cat burglars—or portico thieves as they were officially called. **1938** F. D. Sharpe *Sharpe of Flying Squad* xxv. 254 Cat-burglars have existed since there have been houses to climb, but until comparatively recently they were always known by the more prosaic title of Portico Thieves.

porticus. Add: Pl. **porticus**, **porticuses**. **1.** Delete *Obs.* and add later example.

1850 J. H. Parker *Gloss. Terms Grecian, Roman, Italian, & Gothic Archit.* (ed. 5) I. 371 In the middle ages the word porticus was used for the entrance porch of a church, and for the *apses*... The structure over a tomb was termed *porticus* and *porticus*. But *porticus* also retained its original sense of a long ambulatory... This porticus [by Cuthbert Tunstall at Durham] is a long gallery in the south side.

2. *spec.* in *Anglo-Saxon Archit.*, an aisle or transept on the north or south side of a church, containing a chapel.

1888 C. C. Hodges *Abbey of St. Andrew, Hexham* iii. 16 We may assume the word *porticus* to mean side chapels at the east and west ends of the aisles, as at Brixworth, on transepts, as at Norton, Stow, Sompting, and the church in Dover Castle. **1911** A. H. Thompson *Ground Plan Eng. Parish Ch.* ii. 13 A feature of the early cathedral and of St Pancras at Canterbury, was the provision of *porticus*, porches or side chapels, from the nave. These were entered by archways pierced in the centre of the lateral walls. **1925** A. W. Clapham *Romanesque Archit. Eng.* 8 Early Anglo-Saxon buildings... The southern group of churches..are distinguished by a simple aisle-less plan with an apsidal chancel and a series of annexes called 'porticus' adjoining or surrounding the nave. **1950** H. M. Taylor in P. Clemoes *Anglo-Saxons* 141 From the earliest days the Saxon builders showed a fondness for separate chapels, or *porticus*, opening from the naves or chancels of their churches through comparatively small doorways. **1961** E. W. Parker *Great Ch. of St. Mary, Stow* 20 Hence 11 What, we may ask, is the reason for the Saxon doorways into the transept? Did it lead to a *porticus* or chapel? **1971** D. M. Wilson *Anglo-Saxons* (rev. ed.) iii. 49 Various ancillary elements were added to this basic pattern—porches, *porticuses*, crypts, towers, western aisles and even, in the latest period, transepts. **1978** *Archaeologia Aeliana* III. 123 The church [of St. Peter, Bywell, Northumberland] formerly had *porticus* overlapping the junction of nave and chancel on the north and the south. The roof-raggle of the north *porticus* is still visible. **1977** E. Morris in Barley & Hanson *Christ. in Britain 300–700* 161 The Anglo-Saxon porticus and its origins. **1979** *Country Life* 19 Nov. 1859/2 Before that there was a church to combat rising damp may destroy the remains of an Anglo-Saxon *porticus*.

portière. Add: (Earlier and later examples.) Also *fig.*

1843 *Ainsworth's Mag.* IV. 111 Her ladyship's cosy house refurnished with *portières* to all the doors. **1905** *Spectator* 7 Jan. 11/2 The waters go chasing down the cliffs in deep descending channels hung with curtains and *portières* of moss. **1909** *Chambers's Jrnl.* Oct. 665/2 The Cashmere shops are specially fine with portières and tapestries for walls. **1927** F. B. Young *Portrait of Clare* v. 534 A chair went over with a crash, the portière was ripped from its hanging. **1944** S. Bellow *Dangling Man* 184, I sat down at a desk in a corner, next to one of the portieres. **1975** *New Yorker* 11 Aug. 71/1 At the age..that the men have got a black backcloth that opened into portieres.

Hence **portiè·red** *a.*, furnished with a *portière*.

1923 F. L. Packard *Four Stragglers* v. 184 She turned her head a little, facing the portiered window beside the fireplace of the living-room in which they stood.

portiforium (pǭɹtifǭə·riəm). Pl. **portiforia**. [a. med.L. *portiforium* portable breviary.] = Portas 1, Portuary.

1880, **1884** [see Portuary]. **1916** H. M. Bannister in C. H. Turner *Early Worcester MSS.* p. lv, The traditional title *Portiforium S. Oswaldi*..is clearly wrong, as that Saint's name appears by the first hand in its kalendar; it might justly be called the *Portiforium S. Wulfstani*. **1920** *Jrnl. Theol. Stud.* XXI. 174 In the library of Corpus Christi College..is preserved a book..known since the middle of the seventeenth century as the *Portiforium Wulstani*; ..is the *Portiforium* [or *Breviarium*] *Oswaldi*. *Ibid.* 175 The book..is a stout vellum volume... It is in fact almost a Breviary, or *Portiforium*. **1931** A. Ebdale *Student's Man. Bibliogr.* v. 100 The small breviaries, intended to be carried about and hence named 'portiforia', the chemise was at the head. **1956** *In Great Tradition* (Benedictines of Stanbrook) ix. 176 She [sc. Dame Laurentia McLachlan] had contributed..an article on the so-called *Portiforium Oswaldi*, an eleventh-century volume. **1974** *Bodl. Libr. Rec.* Dec. Dr. Emden's Register provides the information that John Neele, master of Holy Trinity College, Arundel, from 1484 until his death in 1498, bequeathed all his books at Arundel, with the exception of his bible and portiforium, to Magdalen College.

porting (pǭə·ɹtiŋ), *vbl. sb.* [f. Port *sb.*³ + -ing¹.] The arrangement, size, etc., of the ports in an internal-combustion engine.

1960 C. F. Taylor *Internal-Combustion Engine* I. vii. 236 (*caption*) Two-stroke engine porting. **1972** *Proc. Inst. Mech. Engin.* CLXXXVI. 746/2 The only cycle for which the porting requirements have compromised the rotor shape is the compressor-expander.

portion, *sb.* Add: **5. b.** *Judaism.* The section of the Pentateuch or of the Prophets appointed to be read on a particular Sabbath or Festival.

1892 I. Zangwill *Childr. Ghetto* II. 87 Reb Shemuel was already poring over a Pentateuch in his Friday night duty of reading the portion twice in Hebrew and once in Chaldaic. **1901** M. Gaster *Bk. of Prayer* I. 37 The Minister then reads the first section of the Parashá (Portion of the Law for the ensuing Sabbath). **1922** *Glasgow Magazine* Ct. 1. ii. 17 She should have kept Sam in..to say the week's portion out of the Pentateuch, to do his soul good. **1938** H. Kemelman *Thursday the Rabbi walked Out* x. 57 After the portion is read, you say another blessing. **1970** *Sat Rev.* 11 May 46 His sister and bar Mitzvah boy chants the portion from the Prophets. **1980** *Jewish Chron.* 15 Feb. 22/1 In next week's portion of the law, 'Terumah' (Exodus 25). *Ibid.* 39/4 Saturday, February 16 (Shevat 30)..Portion of the Law (Torah)..*Parashat Shekalim*..Portion of Prophets (Haftara) [etc.].

portionable, *a.* Restrict † *Obs. rare* to sense in Dict. and add: **2.** Designating a woman endowed with a marriage portion or dowry.

1875 Paley & Sandys *Select Private Orations of Demosthenes* II. 162 'Orphan-sons or heiresses', meaning by the latter 'orphan-daughters' 'portionable-sisters'. **1910** *Blackw. Mag.* Aug. 200/1 Priti had carried off a portionable lady of the island on whom Sidd also had cast his eye.

Port Jackson (pǭɹt dʒæ·ksən). The name of the harbour of Sydney, Australia, used *attrib.* and *absol.* in the names of plants and animals native to the region, esp. **Port Jackson fig**, a small tree, *Ficus rubiginosa*, of the family Moraceæ; **Port Jackson (shark)**, a small bull-head shark belonging to the genus *Heterodontus*, esp. *H. portusjacksoni*, which is light brown with black markings; **Port Jackson (willow)** *S. Afr.*, a large shrub, *Acacia cyanophylla*, of the family Leguminosæ, which was introduced to South Africa from Australia and has become naturalized there.

1889 J. H. Maiden *Useful Native Plants Austral.* iv. 225 *Ficus rubiginosa*..'Port Jackson Fig'..This fig, like other figs, evolves a juice which might be worth attention. **1904** [see *Illawarra]. **1954** *Coast to Coast 1953–54* 133 Today Ellen planted the Port Jackson Fig. **1965** *Port Jackson Fig* [see *Illawarra]. **1880** A. C. L. G. Günther *Introd. Study of Fishes* 716 (Index), Port Jackson Shark. **1932** *Nat. Geogr. Mag.* Sept. 369/2 There were sharks there in abundance—all sorts and sizes—'gummies', 'angels', and 'Port Jacksons'. **1974** D. & M. Webster *Compar. Vertebr. Morphol.* iv. 63 (*caption*) The jaws of a chondrichthyean, the Port Jackson shark, and of a teleost, the sheepshead. **1902** *Trans. S. Afr. Philos. Soc.* XI. 62 The value of Port Jacksons bark on trees still standing..is worth 6s. per acre. **1950** *Cape Times* 12 Dec. 9/7 Great masses of rooikrantz and Port Jackson willow grow to within a few feet of the houses. **1959** *Ibid.* 27 Mar. 1/7 A man was shot dead at Durbanville last night after a 40-yard police chase at dusk through thick Port Jackson bush. **1973** Palmer & Pitman *Trees S. Afr.* II. 731 Some Australian species, such as the aggressive and fast-spreading Port Jackson willow, *Acacia cyanophylla* Lindl., are cultivated in South Africa.

Portland¹. Add: **Portland stone** (earlier and later examples); also *ellipt.*; also, applied to the colour of Portland stone.

1673 J. Ray *Observations Journey Low-Countries* 120 These figured Bodies were of very different Substances as to hardness..some soft Stone, others..as hard as Portland Stone. **a 1706** Evelyn *Diary* an. 1666 (1955) III. 459 All the Cornices..Columnes, freeses, Capitels & projectures of massie Portland stone flew off. **1711** J. Thornhill *Jrnl.* 22 May in *Proc. Suffolk Inst. Archæol. & Nat. Hist.* (1907) XIII. 3 St. Martin having cap't his peers with Keitan stone & made pedestals of Portland..his face was Portland freestone stone. **1860** *Bradshaw's Railway Manual* XXI. 2. xxiv. (Advt.), Anticorrosion paint, White, Light Stone, Bath do., Cream Colour, Light Portland Stone, Drab or Portland do. **1963** *Times* 17 May 15/7 Viewed with a shaft of sunlight on it, the Portland stone is startlingly white.

portlandite (pǭə·ɹtlǎndait). *Min.* [f. Portland¹ (see quot. 1933) + -ite¹.] A form of calcium hydroxide that occurs naturally as minute hexagonal colourless plates and is a common product of the hydration of Portland cement.

1933 C. E. Tilley in *Mineral. Mag.* XXIII. 420 The occurrence of $Ca(OH)_2$ as a well-defined species in the contact-zone of Scawt Hill (Co. Antrim) merits a new mineral name, and in view of its occurrence as a common product of hydration of Portland cement, and furthermore as crystals from this source have provided the most reliable physical and optical data on crystalline $Ca(OH)_2$, it is fitting that the mineral name should bear record of these facts. The name Portlandite is accordingly proposed. **1968** I. Kostov *Mineral.* 221 Periclase group... The calcium members are rare minerals; portlandite occurs in metasomatic deposits in limestones. **1977** *Sci. Amer.* Aug. 84 In a hardened cement [the gel]..represents about 50 percent by volume of the hydrated material... The major part of the other crystalline products consists of calcium hydroxide (portlandite).

Portland Place (pǭɹtlænd plēis). The name of a street in London, applied allusively to the B.B.C., whose headquarters are there. Also *attrib.*

1937 D. L. Sayers *Busman's Honeymoon* xvi. 324 Now, my little ministrels of Portland Place! Strike, you myrtle-crowned boys! **1939** 'N. Blake' *Smiler with Knife* ii. 29 Bursting out..into Somerset folk-songs in perfect harmony and Portland Place accents. **1941** C. King *Diary* 17 July in *With Malice toward None* (1970) 135 Walker had been spending the morning with the B.B.C.... These birds from Portland Place..took his criticisms with a very bad grace. **1974** *Times* 18 Oct. 5/5 The most popular disc jockey..might almost have been born in Portland Place. **1974** C. Hill *Behind Screen* xxxi. 266 If it is intercontinentally 'we' have a more powerful..Board of Governors, then as interventionist chairman I was at Portland Place. **1977** T. Jackson in *Rep. Comm. Future of Broadcasting* iii. xiv. 227 Without the local stations Portland Place ought well be a London-based organization.

portmanteau, *sb.* Add: **4. a.** *portmanteau manufacturer, -trunk* (later example).

1879 M. Eyckworth *Let.* 1 June (1971) 210 The boxes and small portmanteau trunk..have not arrived. **1885** *List of Subscribers, Classified* (United Telephone Co.) 6 213 (*heading*) Trunk and Portmanteau Manufacturers.

b. (Further examples.)

1896 [see *brunch]. **1929** *Amer. Speech* 1968 XLIII. 201 He was particularly concerned with..portmanteau forms of the sort illustrated by *motel* for 'motor hotel'. **1973** *Sci. Amer.* Dec. 116/2 A more interesting blend, called a portmanteau word by Lewis Carroll, combines two words with similar meanings into one: 'instantaneous' and 'momentary' into 'momentaneous', 'splutter' and 'blisters' into 'splisters', 'shifting' and 'switching' into 'swifting' and 'edited' and 'annotated' into 'editated'. **1978** *Dædalus* Fall 93 But such names are more often nothing more than the portmanteau terms describing a group of contemporaries who come together for a few years and then go their own ways.

c. Applied *attrib.* to a general description or category, or to a word or expression which has a general or generalized meaning.

1909 *Daily Chron.* 21 Feb. 4/7 You may notice the same 'portmanteau' descriptions of persons wanted by the police. They would fit a dozen men in every hundred yards of London. **1949** [see *tabloid *a.* (sb.)]. **1955** *Times* 12 May 7/3 The phrase *Britannicarum omnium*, which had appeared on the coinage since 1953, was discontinued from the beginning of 1954. The words were a portmanteau expression designed as a free rendering in Latin of 'The British Dominions beyond the seas'. **1957** *Listener* 18 July 86/2 The Act of 1897..is one of those portmanteau measures under which a person can be charged with any action. **1960** [see *classroom 1]. **1960** *Times* 14 Oct. 15/6 Such portmanteau terms as 'sprays', 'seed dressings', and 'insecticides' have been used untidily..**1973** *Listener* 19 Apr. 674/1 The concept, or rather the concepts, of 'culture'. This portmanteau word has been indispensable to intellectuals for too years at least.

d. portmanteau morph *Linguistics*, a morph which represents two morphemes simultaneously; also ellipt. *portmanteau*.

1947 [see *morphemic a.]. **1950** *Language* XXVI. 84 If we find it more convenient to regard these forms as single morphs, we must at least take them to be portmanteaus, and not completely arbitrary ones. **1955** C. E. Bazell *Linguistic Form* 54 Furthermore this terminology renders superfluous the family of subtractive morphs, zero-morphs, and portmanteau-morphs. **1972** Hartmann & Stork *Dict. Lang. & Linguistics* 180/1 *Portmanteau morph*, a single morph which stands for two morphemes. The best known example is French *au* [o] 'to the' which represents *à* + *le*.

portmanteau (pǫɹtmæ·nto), *v.* [f. Portmanteau *sb.* 4 b.] *trans.* To combine. Also *intr.* for *pass.*

1902 *Westm. Gaz.* 28 May 2/2 We are amused at the attempt to portmanteau into one (as Lewis Carroll would say) the Education Bill and the Bread Tax. **1906** *Daily Chron.* 22 Mar. 6/7 Hotten's Slang Dictionary..has only two [words] for threepence—'thrums' and 'thrups'—neither of which will portmanteau with 'telegram' comfortably. **1967** G. F. Fiennes *I tried to run a Railway* v. 58 Chingford, Enfield, Hertford and Bishops Stortford [portmanteaued as the 'Chenford'].

portmanto·logism. [f. Portmanteau *sb.* 4 b + -olog(y + -ism.] = Portmanteau *sb.* 4 b. Hence **portmanto·logist**, one who utters or studies portmanteau words.

1887 *Spectator* 9 Apr. 492/2 An allusion to the 'Terrible Zone' which is one of the most beautiful of portmantologisms. **1890** T. Nicklin *Sounds of Standl. Eng.* 85 Sometimes we may surmise that these constructions are what may be called 'portmantologisms'. **1954** *Times* 18 Mar. 15/5, I wonder how many 'portmantologists' realize that the Russian language as spoken in the U.S.S.R. is resorting more and more to the invention of portmantologisms.

portmote. Add: Also **portmoot.**

1951 D. M. Stenton *Eng. Society Early Middle Ages* 177 The ancient borough court, the portmoot, was presided over by the reeve.

‖ **porto** (pǭ·ɹto). [a. Pg. *pôrto* port wine.] **a.** = Port *sb.*⁵ **b.** In full *porto français*: an aperitif made from port; also *attrib.*

1847 Disraeli *Tancred* I. i. 9 A capon in every platter, with some fountains of ale and good Porto. **1857** Dickens *Dorrit* II. xxviii. 564 Bring Port wine! I'll drink nothing but Porto-Porto. **1926** R. Firbank *Concerning Eccentricities Cardinal Pirelli* vi. 63 All this Porto and stuff to keep awake make a woman liverish. **1935** Schoonmaker & Marvel *Compl. Wine Bk.* iii. 88 The dark red Priorato of Catalonia, which was formerly sold as 'Tarragona Port', is a wonderfully good wine in its own right. **1958** R. Postgate *Plain Man's Guide to Wine* iii. 62 If he is with a Frenchman he can offer him 'un porto' which is like a cold tub of port, or 'un malaga français'..But he should not drink them himself. **1961** N. Freud et al. tr. *Montagné's Larousse Gastronomique* 761/1 *Port*, *Porto*, Portuguese wine... Sweet and soft, and in France it is sometimes drunk as an aperitif. It is also used in the preparation of certain dishes, particularly to flavour sauces. **1970** N. Freeling *Kitchen Bk.* ii. 16 Porto, the invariable bourgeois aperitif. **1972** A. L. Simon *Gazetteer of Wines* 114/1 Port, Porto or Vinho do Porto can only be fortified wines made in the area [sc. Douro]. **1974** S. Coulter *Château* I. xxii. 172 It was foie gras with porto jelly.

portocaval, var. *portacaval *a.*

‖ **portolano.** Add: Also **portolan** (pǭɹ·tolān), which is now the usual form. Also *attrib.* and *Comb.*

1897 F. A. Bather tr. *Nordenskiöld's Periplus* 18 The portolan-manufacturer or draughtsman used by preference goatskin..and bright colours. **1898** *Geogr. Jrnl.* XII. 384 We then have..a Series of World-maps and Mediterranean portolans. **1935** *Ibid.* LXXXV. 105, 430 portolans dating from the fourteenth to the sixteenth centuries. **1961** *Antiquity* XV. 186 The portolan charts were intended for the use of mariners. **1972** *Daily Tel.* 22 Dec. 14/6 Dolphin Book Company, of Oxford, paid £7,000 for a complete Mediterranean portolan atlas signed Joannes Oliva and dated Messina 1582. **1978** *Nature* 1 June 409/1 For many the intriguing question remains: whether the keenly observant craftsmen-sailors of Northern Europe did not have a recorded lore of their own—effective but, like the portolan charts, tardily acknowledged by the churchmen?

Porto Rican: see *Puerto Rican *sb.* and *a.*

portosystemic (pǭɹtosistę·mik, -ī·mik), *a.* *Surg.* Also porta-, and with hyphen. [f. as Porto-pyæmic *a.* + Systemic *a.*] Applied to an anastomosis between the portal vein and a systemic vein.

1961 *Lancet* 25 Dec. 1289/2 These studies have shown abnormalities of serum-L.L.A..in patients with portal hypertension and its associated portosystemic bypass. **1974** Passmore & Robson *Compan. Med. Stud.* III. xx. 27/4 Another site for porta-systemic anastomosis is in the territory of the junction of the inferior mesenteric and the inferior rectal veins. **1976** *Lancet* 11 Dec. 1268/2 An explanation was attempted..but was largely abandoned because of massive blood-loss due to extensive portosystemic shunts and bleeding tendency.

portrait, *sb.* Add: **3. c.** *Typogr.* A format in which the height of an illustration or page is greater than the width; cf. Upright *a.* 5 c. Often used as quasi-*adj.* or quasi-*adv.*

1930 [see *landscape *sb.* 1 c]. **1960** H. Williamson *Methods Bk. Design* iii. 16 The book is in fact taller than it is wide, and by analogy with the painter's method these proportions are sometimes called 'portrait'. **1975** J. Butcher *Copy-Editing* 304 *Portrait*, (of) the shape of a book or illustration, is referred to as 'portrait' when its height is greater than its width; (2) if a table is 'set portrait' it is set upright on the page and not turned to read up the page.

4. *portrait-group* (examples); **portrait-painter** (earlier and later examples); **portrait-painting** (earlier example); *fig.*; hence as *ppl. a.*

1911 *Encycl. Brit.* XXII. 129/1 The magnificent portrait groups at Haarlem by Hals..must also be mentioned. **1937** *Burlington Mag.* Jan. 47/2 Concerning Cotes's portrait-group (represented by a colour plate), one may suspend judgment. **1970** *Oxf. Compan. Art* 452/1 [sc. Cranbrook] Taylor also painted some small portrait groups in landscape settings. **1758** *N.Y. Gaz.* 21 Aug. 3/1 ..Thomas Milworth, Portrait Painter, Has removed to the House of Mr. Samuel Deall in Broad-Street. **1980** J. Wedgwood *Let.* 21 Oct. (1965) 260 Methinks I would not be a portrait painter upon any consideration whatever. **1969** *Observer* 29 Mar. 7/2 You have no idea what portrait painters suffer from the vanity of their sitters. **1765** T. H. Croker et al. *Compl. Dict. Arts & Sci.* II. s.v. *Portrait*, We see the term portrait-painting in contradistinction to history painting. **1821** H. C. Robinson *Diary* 27 June (1967) 71, I have finished *Waverley*... Its merit lies in portrait and scene painting. **1841** Dickens *Let.* 2 Apr. (1974) II. 253 The portrait-painting..

portrait, *v.* **1.** (Later example.)

1908 *Daily Chron.* 3 Apr. 4/4 We are not puffed and paragraphed and portraited in the papers.

portraitist. Add: (Later examples.) Also *fig.*

1976 *Amer. N. & Q.* XIV. 151/2 JEB [sc. James E. Butterworth] was..an authentic ship portraitist, in response to demands from the owners of the shipping lines. **1977** V. S. Pritchett *Gentle Barbarian* iii. 119 Turgenev..is a portraitist who gives the surface of people.

‖ **portrait parlé** (portrę pārle). Pl. **portraits parlés.** [Fr., = spoken portrait.] A detailed description of a person's physical characteristics in mainly anthropometric terms, esp. one of a type used in the identification of criminals and developed by Alphonse Bertillon (see *Bertillonage). Also *transf.*

1913 A. B. Reeve *Poisoned Pen* v. 141 Neither the 'portrait parlé' nor the ordinary photography nor any other system will suffice alone against the arch-criminal. **1940** N. Marsh *Surfeit of Lampreys* (1941) xiv. 205 Is this the B-b-Bertillon [sic] system?..P-portrait parlé? **1956** H. T. F. Rhodes *Alphonse Bertillon* xiv. 105 The portrait parlé is a derivative of the anthropometric system. **1963** T. Tullett *Inside Interpol* iii. 35 One serious shortcoming of this system was that the Portrait Parlé method of facial identification and the famous Bertillon system of measurement of certain key parts of the human frame. **1973** R. Cobb *Reactions to French Revolution* iii. 67 *Portraits parlés* tell us perhaps more about the police, their habits of eye..more than these visible passports were so painstakingly fixed. **1973** *Listener* 3 May 583/1 They left an enormous volume of papers. Martin Gilbert has made of them not so much a biography as a *portrait parlé*. **1975** *Encycl. Brit. Macropædia* XIV. 671/1 Long before the birth of Christ, Egyptians used detailed word descriptions of individuals, a concept known today as 'portrait parlé'.

portrayist (pǫɹtrē·ist). *rare.* [f. Portray *v.* + -ist.] = Portrayer.

1924 *Glasgow Herald* 25 Sept. 4 His considerable skill as portrayist and his great humour.

Port Salut. Restrict † *Obs. rare* to sense in Dict. and add: **2. Port Salut** (por salü). [f. the name of a Trappist monastery, *Port du Salut* (also used), in Mayenne, N.W. France, where it was first produced.] A kind of soft pressed cheese.

1881 J. P. Sheldon *Dairy Farming* xxx. 512 Cooked, heated, and pressed cheese... Gruyère, Parmesan, Port du Salut, Kangiport. **1896** Loch & Benson *Cheese* ii. 15 Camembert, Brie, Bondon, Neufchâtel, and Port du Salut, all..hail from France. **1902** *Encycl. Brit.* XXVII. 355/2 In France the pressed varieties of cheese with hard rinds include Gruyère..and Port Salut. **1932** [see *bel paese]. **1950** E. Waugh *Helena* xi. 208 In the store of cheese was forthcoming (a Port Salut that increased my respect for the fact of who did Port Salut daily marketing). **1951** [see *Oka*² 91 c (b)]. **1965** *Sunday Times* 21 Nov. 38/1 A half moon, pallid cheese like Port Salut. **1966** P. V. Price *France* 297 There are..two sorts of cheese involved; Port du Salut is that made by the Trappist monks at Entrammes, near Laval... Other Trappists, however, and other cheese makers..are able to make and market a cheese under the name of Port Salut, which, unlike Port du Salut, is not a trade mark. **1972** *Sat. Rev.* (U.S.) 24 June 77/3 French Trappist monks in northern Touraine have been making Port Salut cheese for 150 years.

port-si·der. *N. Amer. colloq.* [f. Port *sb.*⁶ (*a.*) + Sider².] A left-handed person.

1936 *Amer. Speech* I. 369/2 They [sc. baseball players] are 'south-paws' or 'port-siders' or 'side-wheelers' when they are left-handed. **1945** *Record* (Philadelphia) 28 Oct. 8/2 We doubt that portsiders will ever get their rights. **1946** *Sun* (Baltimore) 13 June 20/2 The firm had been printing check-books for Baltimore portsiders for many years. **1975** *Daily Colonist* (Victoria, B.C.) 29 July 21/1 [He] was a portsider of my acquaintance. They're called southpaws, wily-siders and port-siders.

‖ **Portugaise** (portüġę·z). [Fr., = Portuguese.] **1.** *Cookery.* Chiefly in phr. *à la Portugaise*, designating food prepared in a Portuguese style. Also *attrib.* or as *adj.*

1845 E. Acton *Mod. Cookery* xx. 348 *Arcot Dock* (or *Sweet Rice. A la Portugaise*)... This is quite the best sweet preparation of rice that we have ever eaten, and it is a very favourite dish in Portugal, whence the receipt was derived. **1880** A. B. Marshall *Cookery Bk.* x. 54 *Calves' Head à la Portugaise*. **1907** Cabanes French plums and leeks as below, then stir and add to them sufficient clear stock, boil up all together, and serve the soup very hot. **1907** G. A. Escoffier *Guide Mod. Cookery* xvi. 489 *Poularde à la Portugaise*. Stuff the pullet with three-quarters lb. of rice, combined with five oz. of peeled and concassed tomatoes, cooked in butter. *Poële* the pullet. Dish it; coat it with a Portugaise sauce..and surround it with a garnish of..tomatoes, stuffed with rice 'à la Portugaise'. **1929** G. Vinton *French Cooking for Me* 53 *Sole à la Portugaise*..Butter a baking-dish, lay the fish in it and cover with a little chopped onion and cover with diced tomatoes. **1931** A. de Croze *What to eat & drink in France* ii. 20 In French cookery 'Portugaise' means that chopped meat or forcemeat is used as a bed or cover with eggs or vegetables, shell-fish, or fish. **1953** J. Conil *Haute Cuisine* x. 171 Rabbit..Tomato sauce or Portugaise. **1964** E. Bowen *Little Girls* i. iv. 60 An eight-egg omelette, Portugaise, had been contrived in the kitchen. **1970** Simon & Howe *Dict. Gastron.* 307/2 *Portugaise, à la*, (culinary term) a Portuguese dish featuring tomatoes above all else, but usually also garlic and onions.

2. = *Portuguese oyster.

1942 E. Paul *Narrow St.* xxv. 221 The price tags on the turquoise green Portugaises and the fat grey Marennes of incomparable flavour would have caused a riot fifty years or so before. **1961** L. Deighton *Ipcress File* xxxi. 200 Seallops and fat oysters and portugaises that looked like pieces of rock. **1966** P. V. Price *France: Food & Wine Guide* 40 There is the Portugaise, with its bumpy, longish and deep shell. **1970** *Sunday Times* 18 Oct. 21/2 The big, fat portugais [sic] in their craggy misshapen armour. **1978** *Observer* (*Colour Suppl.*) 23 July 34/4 You can eat Pacific oysters, small as the 'portugaises' in Paris.

Portugal. Add: **6. b.** **Portugal laurel**, substitute for def.: an evergreen shrub, *Prunus lusitanica*, native to Spain and Portugal; (later examples); **Portugal onion**, a variety of onion, esp. young seedlings of this variety used as spring onions; **Portugal oyster** = *Portuguese oyster*; **Portugal peach**, quince (earlier examples).

1914 W. J. Bean *Trees & Shrubs Hardy in Brit. Isles* II. 241 In all but the coldest parts of Great Britain the Portugal laurel is one of the handsomest and most effective of evergreens. **1972** F. Perry *Flowers of World* 262/1 The Portugal Laurel..makes a tree of 3–6 m. (10–20 ft.) with long-elliptic evergreen leaves and large racemes of dull white, heavily scented flowers. **1847** W. Lilly *Christian Astrol.* ii. iv. 398, I..heavily complained to the woman for seven Portugall Onyons which I lost; [she] not knowing of my loss at that time. **1841** C. Ramsay *Rose of Dimma* 136 [The] Portugall Onions, and [they] have the Strong of the heat ripening. **1967** M. & I. L. Harrington *the Portugal Onions* as in it dos 2 ft. *See* **1853** *Chambers's Edin. Jrnl.* 30 Apr. 283/2 A root of rye, with a glass of red wine. **1783** J. Woodforde *Diary* 21 Nov. (1926) II. 107 Mr. Priest..made me a present of..fine Portugal Peach. **1845** E. Acton *Mod. Cookery* xx. 97 The meat may then be stewed..with a Portugal onion. **1885** W. Miller tr. *Vilmorin-Andrieux's Veget. Garden* 363 White Portugal Onion... Bulb of a dull-white colour. **1890** J. R. Philpots *Oysters* I. xxv. 570 The Portugal oyster has appeared for the last two or three years in our markets. **1664** Evelyn *Kalendarium Hortense* in *Sylva* 72 August..Fruits in Prime..Portugal Peach, Crown Peach, Bourdeaux Peach. **1629** J. Parkinson *Parad.* xv. 574 The Portingall Apple Quince is a great yellow Quince... The Portugal Quince hath more..but is not fit to be eaten raw like the former.

Portuguee (pǭɹtiŭgī·). *U.S.* Also **Portagee**, **Portugee**. Repr. a spurious 'singular' form of Portuguese *a.* and *sb.*, this being regarded as a plural.

1830 J. F. Cooper *Water Witch* II. vi. 197 I'll being altogether unreasonable to suppose that a Portuguee should do what an Englishman had not yet thought of doing. **1860** *Atlantic Monthly* Dec. 735/1 Somehow I can't help misthrustin' them Portugee-lookin' fellahs. **1878** [in Dict. s.v. Portuguese 1]. **1880** *Harper's Mag.* Sept. 595/1 At one place was a 'Portugee' of the Western Islands. **1913** C. C. Martindale *In God's Army* i. II. 143 China is a Portugees was a forbidden land. **1975** J. Gores *Hammett* iii. 28 'Keep your eye on the Portugee,' said Hammett. But Dancing Frankie..put the Portuguee body on the canvas for a six count with a roundhouse right that wasn't fully Portuguese.

Portuguese, *a.* and *sb.* Add: **A.** *adj.* **a.** (Later examples.) Also *spec.*, of or pertaining to Sephardic Jews whose ancestors came from Portugal.

1851 [see *Sephardi]. **1866** Geo. Eliot *Let.* 10 Aug. (1956) IV. 298 We looked about for the Portuguese Synagogue where Spinoza was nearly assassinated... There are..three Portuguese Synagogues now [in Amsterdam]. **1904** *Ibid.* etc. [see *Spanish *a.* 1 d]. **1937**, **1960** [see *Marrano].

b. **Portuguese oyster**, a type of oyster, *Crassostrea angulata*, which has a bumpy, greenish shell and is native to Portugal although it is cultivated elsewhere, esp. in France; also *ellipt.*; **Portuguese parliament** *Naut. slang*, a discussion in which many speak simultaneously; hubbub.

1890 J. R. Philpots *Oysters* I. xxv. 590 The mollusc known under the name of the Portuguese oyster does not belong to the same genus as our indigenous oyster. **1928** Russell & Yonge *Seas* xiv. 301 Of recent years it [sc. the French oyster] has been almost completely ousted in the more southern beds by the Portuguese oyster. **1960** C. M. Yonge *Oysters* iv. 166 Some Portuguese oysters are ready for sale when only two years old. **1964** E. Clark *Oysters of Locmariaquer* i. 4 The Portuguese oyster..has been gradually moving farther north. **1976** N. Roberts *Less of Fourier Lett.* 124 The Portuguese oysters of Arcachon... turn up its nose at the most modest meal, whether..the delicately flavoured 'flats' or the plumper but more common-place Portuguese. **1897** F. B. Williams *On Many Seas* xvi. 162 Here they always had their meeting and every one..had more to say than the other, and all talked louder than another that I ever heard, that was the worst. It was like what sailors call a Portuguese Parliament. **1898** H. E. A. Coate *Realities of Sea Life* 133 They [sc. wild monkeys] could only be compared with the 'members of a Portuguese parliament' where, according to Jack's idea, they are 'all talkers and no listeners'. **1962** Granville *Dict. Sailors' Slang* 91 *Portuguese parliament*, rowdy discussion in which everybody talks and nobody listens.

B. *sb.* **2.** (Earlier and later examples.)

1615 T. Roe *Jrnl.* 21 July in *Embassy to Court of Gt. Mogul* (1899) I. 19 The Enterprizers were certaine Magadoreaumens..and Portugueses which were sent to them. **1888** W. W. Skeat *Etym. Dict. Eng. Lang.* p. xxiii, The other Romance languages..are Italian, Spanish, Portuguese, Provençal, Romansch, and Wallachian. **1933** L. Bloomfield *Language* xxv. 474 The descendants of..a runaway slaves who settled on the island of San Thomé off the coast of West Africa, spoke a creolized Portuguese. **1950** J. H. Steward *Handbk. S. Amer. Indians* VI. 168 It has been estimated that 15 percent of the vocabulary of Brazilian Portuguese is of Tupi origin. **1974** *Encycl. Brit. Macropædia* XV. 1031/2 Portuguese owes its importance largely to its position as the language of Brazil. **1980** G. Greene *Dr. Fischer* viii. 47 Two letters in Portuguese were sent me to translate, although I knew no Portuguese.

portulaca. Add: (Later examples.) Also (erron.) portulacca.

1947 M. M. Bennett *Christison* xi. 116 They ate portulacca to keep off scurvy. **1939** R. Godden *Black Narcissus* xix. 173 Lupins, delphiniums..and the portulaca she had grown to love in the plains. **1942** *E. Afr. Ann. 1941–2* 44/1 The indigenous portulaca found all round Nairobi with its masses of purple flowers is a lovely ground cover. **1959** *Arena* (Wellington, N.Z.) xxxv. 4 'And your lovely portulaca,' Hester said gently, 'I had forgotten you always grew it.' **1962** [see *cleome]. **1974** J. Berrisford *Window Box & Container Gardening* xix. 143 The annual portulacas..are suitable for seasonal use. **1978** *Detroit Free Press* 16 Apr. (Gardening Guide) 10/1 If you want to enjoy your garden in the evening, don't plant flowers such as day lilies, morning glories and portulacca that are closed at night.

portulace, -lack. Delete † *Obs.* and add later example of spelling *portulac*.

1911 C. E. M. Bean *'Dreadnought' of Darling* xxviii. 234 They helped their provisions by eating a good deal of a shrub known as portulac.

port-wine. Add: **a.** (Later examples.)

1930 H. Craddock *Savoy Cocktail Bk.* 192 Port Wine Sangaree... Port Wine negus. **1976** *Daily Tel.* (Colour Suppl.) 13 Dec. 41/4 Portwine is being poured over chunks of ice.

b. *port-wine negus* (example); **port-wine magnolia** *Austral.*, an evergreen shrub, *Michelia figo* of the family Magnoliaceæ, native to temperate or tropical Asia and bearing scented brownish-yellow or purple flowers.

1858 Trollope *Three Clerks* III. ii. 38 Mrs. Davis was mixing port-wine negus as fast as her hands could make it. **1903** M. Tennant *Ride on Stranger* viii. 86 At Lindfield there were port-wine magnolias. **1977** *Austral. House & Garden* Jan. 17/1 Check these out, along with your port-wine magnolia.

port-winy *a.* (later examples); hence **port-wi·ni·ly** *adv.*; **port-wi·ner**, an habitual drinker of port.

1904 *Daily Chron.* 20 Oct. 5/5 A magnificently Georgian figure in his court-robes, with a red wig and generally 'port-winey' make-up. **1909** *Ibid.* 16 Apr. 3/4 The port-winiest of the..port-winers. **1930** S. Beresford *Faced of Cheechako* (1930) 54, I smoked and sat as I marvelled at the sky's 'port-winey' glow. **1931** G. E. R. Buckle *Life Disraeli* V. iv. 96 B. Beauchamp..warned Disraeli that the High Church party other than 'the old port-winers' were holding aloof from the political contest. **1942** A. Huxley *Crome Yellow* ii. 18 He was a large, comfortable, stolidly, port-winily English, with its great sloping-rafter face. **1963** [see *cigary a.]. **1938** S. Beckett *Murphy* 266 'A proper port-winer,' said the Coroner. 'The afterglow is unmistakable.'

pos. (Examples.)

1711 *Spectator* 4 Aug. 2/1 It is perhaps..speaking no more than we needs must, which has so miserably curtailed some of our Words, that..they sometimes lose all but their first Syllables, as in *Mob. rep. pos. incog.* and the like. **1716** [see Pos]. **1922** Joyce *Ulysses* 418 Got a pectoral trauma, eh, Dix? Pos. fact. **1940** [see *Mammy *a.* 1]. **1951** [see *rebolt *sb.*]. **1969** 'R. Crawford' *Cockleburr* iii. vi. 126 'Are you sure you weren't mistaken?' 'Pos.' **1977** *Hot Car* Oct. 99/4 (Advt.), Smiths electronic tacho (pos earth). **1979** *SLR Camera* Mar. 34/1 Having said that, there are of course the negpos experts who will tell you how to produce a 10 × 16 print from a quarter inch portion of a 35mm negative.

pos², var. *poss *a.*

posada. Add: **1.** (Later examples.)

1924 C. Guedalla *Palm* xi. 236 An obliging innkeeper rode twenty miles to tell him that Claus was safely lodged in two posadas. **1961** H. Grenfell *Face of Spain* iv. 95, I had not reached the posada till after midnight. **1962** *Times* 14 July 12/7 Dying in poverty in 1556, Luis..lived in the posada in Valladolid. **1965** *New Statesman* 7 May 734/1, I suppose it must have been Balzac who, when Cervantes' posadas and Fielding's coaching-inns were beginning to pall, discovered the potentialities of the boarding-house. **1966** *Listener* 4 Aug. 164/3 I was recently on my way to..the host in Mexico..to the ancient site of Teotihuacan..and afterwards to lunch at a posada in the neighbouring village. **1974** A. Lawson *Matthew's Hand* xxxiii. 161 He went a posada in the Spanish city.

2. In Mexico, each of a series of visits traditionally paid to different friends during the days before Christmas, representing Mary and Joseph's search for a lodging in Bethlehem.

1930 R. Redfield *Tepoztlan* vi. 128 There is time to..now, and..so the Christmas *fiestas*... In many houses..the nine posadas are celebrated as they are celebrated in all over Mexico... The matrons of the neighborhood have arranged to make a little chant to each of the nine fiestas. **1932** Chase & Tyler *Mexico* ii. 202 December 16. Christmas fiesta. Nine days. Posadas. Processions, dances, singing. **1959** *Listener* 24 Dec. 1105/1 Those and other decorations will be used in the posadas that are held..a posada is enacted, each of its seven.. in the posadas which go on night after night for the twelve days before Christmas. The *posadas* are supposed to represent the vain search of Mary and Joseph for a lodging in Bethlehem. **1978** *Tucson Mag.* Dec. 105/1 You were..the Los Posadas celebration, December 12 beginning at 7:30 p.m. The candle-lighted procession of Catholic Mexican families..began at the school [sc. Carrillo School] and continue through one of the city's oldest neighborhoods, in emulation of the journey of Mary and Joseph seeking shelter. The 'inn' or posada for this evening's festivities will be the school itself, where refreshments will be served and a piñata will be broken.

posadaship (posā·dăʃip). [f. Posada + -ship.] The position of the keeper of a posada.

1923 *Blackw. Mag.* Nov. 700/2 The details of how from posadaship she had fallen to this minute eating-house were unknown.

‖ **posadero** (posadē·ro). [Sp.] In Spanish-speaking countries: an innkeeper.

1904 Conrad *Nostromo* x. viii. 116 The posadero in Rincon swore that on calm nights..he could catch the sound in his doorway. **1912** W. G. Lawrence in T. E. Lawrence *Home Lett.* (1954) 423 After lunch..Juan the Posadero came out to talk with us.

pose, *sb.*² Add: **4.** *N. Amer.* A resting place on a portage; the distance between two such rests. *Obs.* except *Hist.*

1793 I. Macdonell in C. M. Gates *Five Fur Traders* (1933) 96 The portage is full of hills is divided by the voyageurs into sixteen Poses or resting places. **c 1840** D. Thompson *Narr. Explorations W. Amer. 1784–1812* (1916) xx. 218, A Rest, or Pose, is the distance the cargo of a canoe is carried from place to place and then rest. **1848** *Porter's Spirit of Times* 30 Jan. 338/1 In crossing a long portage, they do not go through the whole distance with one load, but divide it into 'poses' or rests; and carry in succession each load to the first 'pose', and then carry them all to the second one, and so on, so that they can rest in walking back for the loads. **1933** C. M. Gates *Five Fur Traders* 97 Inasmuch as the same places were used as poses by all who passed, it came to be the common thing to measure the length of a portage by the number of poses along the trail. **1941** J. F. McDermott *Gloss. Mississippi Valley French 1673–1850* 120 The average length of a *pose* was about one-half of a mile. **1969** E. W. Morse *Fur Trade Canoe Routes* i. 4 It was not unusual for more than half a mile [a ten-minute carry], the voyageur, in order to break the long carry into manageable periods, dropped his packs at what was known as a *pose*, and went back for the next load. *Poses* were about half a mile apart.

posé, *a.* Add: **2.** Composed, poised, self-possessed.

1958 Queen Victoria *Let.* 28 Apr. in R. Fulford *Dearest Child* (1964) 98 She..improves so much—is become so quiet and posée. **1966** Crown Princess of Prussia *Let.* 8 Apr. in R. Fulford *Dearest Mama* (1968) 50 Valerie..could hardly believe she was so young—she is so posée, quiet and well-bred.

3. *Ballet.* Of a position, 'held', prolonged.

1968 A. Gregory *Dance Encycl.* 1963 *Posé*, a step to any direction or position, made by stepping, with the knee straight, on the toe, to the point, or half-pointe; a ballet step 'held', such as an arabesque posé, from one's position.

4. Adopted as a pose; = *posed *ppl. a.*¹ c. *rare.*

1958 L. Durrell *Balthazar* iii. 63 A world..which could afford to cultivate emotions purely as poses.

B. *sb.* **1.** *Ballet.* (See quots. 1949 and 1957.)

1927 [see *jeté]. **1930** Craske & Beaumont *Theory & Pract. Allegro in Classical Ballet* 42 (*heading*) Posé en avant... A *posé* is a step in which one leg is stepped [with] the leg fully stretched on to the half-pointe, or pointe... **1949** Kersley & Sinclair *Dict. Ballet Terms* 93 *Posé*, Posé en avant, a slow movement employed in *adagio*, whereas a *posé* is a quick movement employed in *allegro*, executed *pied à pointe* or *pied à quarte pointe*..[see 3 above]. **1957** G. B. L. Wilson *Dict. Ballet* 220 *Posé* (*en avant*, *en arrière*), the poising of the body (forwards, sidewards, or backwards) by stepping with a straight leg on to the toe or half pointe. **1973** J. Colton *Acrobatic Enchainements* (Assoc. Amer. Dancing) (ed. 2) 18 Present your Posé Down as near as possible to a Star. Do not confuse it with Posé.

2. = *pose *sb.*⁵ 4.

1931 G. L. Nute *Voyageur* 46 The length of a portage was computed by voyageurs in a characteristic way. The canoe and goods were carried about a third of a mile and put down, or *posé*; two or more trips often being required to transport all the load to this spot. Then, without resting, the men shouldered their burdens and went on to the next *posé*, and so on till all the *posés* had been passed. **1961** *Canad. Geogr. Jrnl.* July 5/2 He puts down his load at a set place known as a 'posé'. **1968** *Beaver* Autumn 43/2 We..made the traditional poses or rests after each half mile.

posed *ppl. a.*¹ Add: **c.** Assumed as a pose; deliberately adopted or put on.

1909 M. B. Saunders *Litany Lane* i. iv. 43 There was also a nun-like acquiescence in her bearing, prim for her thirty-three years, and possibly a trifle posed.

‖ **pose plastique** (pǫz plasti·k). Also **posé plastique**. [Fr., lit. 'flexible pose'.] A type of *tableau vivant*, freq. featuring near-naked women. Also *transf.*

c 1846 J. R. Planché *Invisible Prince* (1852) v. 25 The figure..was alive and did both sing and speak... Ah! then it must have been a pose plastique!—A pose plastique! what's that? **1851** Dickens & Collins *Lazy Tour* v, in *Househ. Words* 31 Oct. 410/1 Poses Plastiques in the ecclesiastical 'Theatre Royal'..they have seen from the music-halls tableaux vivants the lads wirt this stage evening. **1859** G. A. Sala *Twice round Clock* 77 The vast shop-fronts..offer a reverse of pose plastique..of nude, the remaining..there is..a series of Poses Plastiques in the shape of young ladies in morning costume, and young gentlemen..who are..'dressing' the shop window. **1862** H. Mayhew *London Labour* Extra vol. p. xxxi, The demoralizing influence of low theatres, music-halls, singing saloons, dancing rooms, and *tableaux vivants*, and Poses Plastiques, might surely afford scope for rigorous investigation by the Society's operations. **1865** *M. & Q.* 9 Jan. 3/1 In 1846 Nicholson was again back at the Garrick's Head, where he added to his usual attractions 'poses plastiques and tableaux vivants'. **1903** *Rep. R. Comm. Alien Immigration* III. 107/2 The *poses plastiques*..are there girls half naked. **1963** P. Fryer *Mrs. Grundy* xxi. 242 The men of the Victorian age were enthusiastic patrons of the *poses plastiques* and *tableaux vivants* (the precise distinction between these is far from clear). **1969** W. G. Kiernan *Lords of Human Kind* iv. 136 There was a vogue of *poses plastiques*, resembling such as 'the Sultan's favourite returning from her bath' were popular. **1974** H. R. F. Keating *Underside* iii. 17 The poses plastiques inside, sir, you're very welcome to step in.

poseur. Add: (Earlier and later examples.) **poseuse** (earlier and later examples).

1872 B. Jerrold *London* xx. 174 'As one cannot go to bed in the middle of the afternoon—11.30 p.m.—it is necessary to go somewhere after the opera,' is the declaration of a well-known *poseur*. **1888** Mrs. H. Ward *R. Elsmere* III. xiii. 260 *Poseuses*, schemer, woman of the world that she was. **1926** G. S. Stern *Thunderstorm* ii. 32 If you let yourself be beaten by that trickster and *poseur*, some of your Socialist fads..you'd never..turn a tide, be stuff for a play. **1977** *Times* 28 Feb. 11/3 Eleanor Bron's disastrous Elena converts that passive enchantress into an assertive Bloomsbury *poseuse*. **1977** *Time* 8 Aug. 46/3 But his is the self-styled genius and unrepentant poseur continues to tantalize.

† **posh** (pǫʃ), *sb.*³ *Obs.* [App. thieves' slang (cf. Romany *posh* half.)] **1.** *slang.* Money; *spec.* a halfpenny; a coin of small value.

Eng. Dial. Dict. s.v. *posh sb.*³ provides evidence of *dial.* currency.

1830 *Sessions Papers Old Bailey, 1824–33* VI. 590/1 He had not got the *posh* (which means money) yet. 1839 H. Brandon *Poverty, Mendicity & Crime* 164/2 *Posh*—a stiver, lowest price of money. *Ibid.* 168 The paper makers get the tats and never tip the motts a posh. [*Translation*] Thieves who pretend to belong to paper mills get the rags and never pay the women a farthing. 1846 *Swell's Night Guide* 68 As I used to doss there sometimes, her nibs got sweet on me, and in course we did our reg'lars and the dossing mongary, lush and posh. 1859 Hotten *Dict. Slang* 76 *Posh*, a half-penny, or trifling coin. 1886 W. Newton *Secrets Tramp Life Revealed* 8 Posh..money of all kind. 1888 in Farmer & Henley *Slang* (1902) V. 261/1 They used such funny terms: 'brads', and 'dibbs', and 'mopusses', and 'posh'..at last it was borne in upon me that they were talking about money. 1892 M. Williams *Round London* 67 That sort of patter..is the thing to get the posh. 1905 *Daily Chron.* 2 Mar. 4/5 'But if I'd a brighfull o' posh,' she said, 'I wouldn't parker no wedge to you.'

2. *slang.* A dandy.

Perh. a different word.

[1867 E. FitzGerald *Let.* 5 Jan. in T. Wright *Life E. FitzGerald* (1904) II. 81, I believe I have smoked my pipe every evening but one with Posh [*sc.* the nickname of FitzGerald's fisherman, Joseph Fletcher] at his house.] 1890 Barrère & Leland *Dict. Slang* II. 146/2 *Posh*,..a dandy. [1892 G. & W. Grossmith *Diary of Nobody* 197 Frank..said..he had a friend waiting outside for him, named Murray Posh, adding he was quite a swell.] 1902 in Farmer & Henley *Slang* V. 261/1.

posh (pɒʃ), *sb.*[3] [Etym. unknown.] Balderdash, rubbish, bosh.

1924 Galsworthy *White Monkey* II. xii. 214 Was he a fool? Could he not let well alone? Pity was posh! And yet! 1953 A. Miller *Crucible* (1956) I. 20 Oh, posh!.. She took fright, is all. 1957 J. Kerouac *On Road* (1958) I. viii. 49 You're talking absolute bullshit and Wolfean romantic posh!

posh (pɒʃ), *a. slang.* [Of obscure origin, but cf. *posh *sb.*[2] The suggestion that this word is derived from the initials of 'port outward, starboard home', referring to the more expensive side for accommodation on ships formerly travelling between England and India, is often put forward but lacks foundation. The main objections to this derivation are listed by G. Chowdharay-Best in *Mariner's Mirror* (1971) Jan. 91–2.] Smart, 'swell', 'classy'; fine, splendid, stylish; first-rate. Also *absol.* as *sb.*

The pronunciation (pəʊʃ), a supposedly 'posh' or facetious way of saying the word, is occasionally heard. Quot. 1903 may exemplify a different word.

[1903 Wodehouse *Tales of St. Austin's* 37 That waistcoat..being quite the most push thing of the sort in Cambridge.] 1918 *Punch* 25 Sept. 204 Oh, yes, Mater, we had a posh time of it down there. 1923 Wodehouse *Inimitable Jeeves* vii. 72 Practically every posh family in the country has called him in at one time or another. 1925 W. Deeping *Sorrell & Son* ii. 22 Tips. Don't forget the tips. If a man's obliging—... It's a posh job. 1927 —— *Doomsday* xiv. 153 'You like it.' 'It's the poshest thing I've ever seen, old chap.' 1927 —— *Kitty* xxix. 372, I say—that's a posh show. 1927 *Daily Tel.* 24 May 9 It was a club in Ham-yard—not a very posh club. 1929 J. B. Priestley *Good Companions* II. v. 391 I'd like to have..a very cosy car, small but frightfully posh. 1930 Auden *Poems* 63 To be posh, we gather, One should have no father. 1935 C. Day Lewis *Time to Dance* 32 For no silver posh Plane was their pigeon..But a craft of obsolete design. 1941 Baker *Dict. Austral. Slang* 56 *Posh, do the*, to spend lavishly, do something 'in style'. 1954 [see *ground *sb.* 11 b]. 1957 J. Osborne *Look back in Anger* I. 13 *Jimmy*: Haven't you read the other posh paper yet? *Cliff*: Which? *Jimmy*: Well, there are only two posh papers on a Sunday. 1958 K. Amis *I like it Here* viii. 98 This railway..though posher and faster, had often reminded him of the tram-like train. 1959 *20th Cent.* Nov. 377 At the time..none of the posh papers carried television reviews. 1970 [see *chez]. 1977 *Lancashire Life* Nov. 153/1 The poshest Granada Ghias..have electric windows. 1979 E. Cave *Blood Bond* I. iii. 35 Charles's penchant for the grand—the 'posh' as Tom would have called it.

posh (pɒʃ), *v.* [f. the adj.] *trans.* To smarten *up.* Freq. *refl.* and *pass.* Hence **po·shed-up** *a.*

1919 *War Slang* in *Athenæum* 22 Aug. 791/2 To 'posh up' was to make oneself look as smart as possible. 1932 D. L. Sayers *Have his Carcase* xv. 197, I don't get time to posh myself up of a morning. 1955 *Times* 29 June 12/5 Beautifully 'poshed-up' for this she is, though certainly she has never been allowed to look shabby overlong. 1959 *Oxf. Mail* 1 June 3/6, I knew most of that stuff has been stolen; I wanted to posh up my old car. 1967 R. Mackay *House & Day* 30 He was poshed up. That meant..a trip to the town. 1967 P. Purser *Twentymen* xvi. 104 We..had dined at a rotten, poshed-up Thames pub. 1968 T. Parker *People of Streets* 35 He'd..turn up at a big state ball all poshed up. 1975 J. McClure *Snake* iii. 45 Who goes to the trouble to posh himself up for the postman?

posh (pɒʃ), *adv.* [f. the adj.] In a 'posh' manner.

1957 *Listener* 7 Nov. 739/1 A device to enable you to 'write posh', as it were. You look up the 'vulgar' words, and against them you find the more elegant. 1961 P. Strevens *Papers in Lang.* (1965) xi. 130 The recent popularity of *My Fair Lady* led to another legend, that of phonetics as a system of teaching people to 'speak posh'. 1972 'M. Innes' *Open House* ii. 20 A crowd of electricians might well take him for a cunning crook talking posh. 1979 R. Barnard *Posthumous Papers* iii. 33 Why do you sometimes talk Lancashire and sometimes talk posh?.. Not everyone can do the posh as well as you.

‖**po shan lu** (bo ʃan lū). Also bo shan lu. [Chinese.] A type of Chinese bronze censer, made during the Han dynasty and having the form of a mythical mountain of immortality.

1915 R. L. Hobson *Chinese Pott. & Porc.* I. ii. 12 Fig. 3 is an incense burner... It is a variation of the more usual 'hill censer' (*po shan lu*). 1954 S. H. Hansford *Gloss. Chinese Art & Archæol.* 13 *Po shan lu*, hill censer; censer shaped as a stem cup with cover in the form of a mountain peak. 1973 *Genius of China* 104/2 Bronze censer in the shape of a cosmic mountain *po-shan-lu*, held up by a man seated on a monster. 1978 *Nagel's Encycl.-Guide: China* 216 The *bo shan lu* or incense-burner with a conical lid suggesting the rocks on the island where the Taoist Immortals lived.

po·shness. *slang.* [f. *posh *a.* + -ness.] The state or quality of being 'posh'.

1931 C. Lithgow *Simple Sailor* xvii. 218 In his pet dark blue morning coat, with a gardenia in its buttonhole.., he was rather a splendid figure. No doubt whatever was there of his 'poshness'. 1960 C. S. Lewis *Studies in Words* vii. 179 'A general simplicity in our dress, our discourse, and our behaviour.' Sincerity (not affectation)? Plainness (not 'poshness')? The easy (not the hard)? 1965 M. Morse *Unattached* iv. 123 The real focus of dislike was the teachers' 'poshness', which the girls felt was often assumed in order to demonstrate that a social gap existed between teacher and pupil. 1978 R. Hill *Pinch of Snuff* xiii. 132 An extra element of 'poshness' was inherent in the 'Private Road' sign.

‖**posho** (pɒ·ʃo). *East Africa.* [Swahili, = daily rations.] **a.** Daily rations given to porters on safari. **b.** A kind of maize flour; a porridge made from this. Also *attrib.*

1892 M. French-Sheldon *Sultan to Sultan* v. 118 The method of dealing out rice, which is carried for *posho* or rations..is somewhat peculiar. *Ibid.*, Porters, carrying the heavy loads, are paid the least and receive the smallest *posho.* 1927 *Chambers's Jrnl.* XVII. 761/1 The safari boys were squatting round the cooking-pots, stirring their 'posho'. 1946 G. Hanley *Monsoon Victory* x. 106 In Africa they [*sc.* askaris] ate *posho*, which is maize flour cooked into a porridge. 1964 C. Willock *Enormous Zoo* ix. 157 He searched two fully occupied huts in the labour lines looking for *posho*. 1971 *E. Afr. Standard* (Nairobi) 13 Apr. 9/3 An Assistant Minister for Agriculture..was officially opening a posho mill. 1974 *Observer* (Colour Suppl.) 28 Apr. 45/3 Posho (ground corn on the cob)... It is cooked with water to make a kind of porridge. 1975 *Times* 17 Nov. 13/7 The [Ugandan] guards were already sharing their beans and posho and lumps of lean meat with me. 1977 H. Innes *Big Footprints* II. i. 122 Sacks of maize flour they called posho.

posigrade (pɒ·zigreɪd), *a. Astronautics.* [f. Posi(tive *a.* and *sb.* + *-grade*, after Retrograde *a.* and *sb.*] Of, pertaining to, or designating a small rocket that can be fired briefly to give forward thrust to a spacecraft.

1961 *Aeroplane* C. 152/1 Next, the three small 'posigrade' rockets on the blunt face of the capsule are fired to achieve separation from the booster. 1962 S. Carpenter in *Into Orbit* 53 This sends out an electrical impulse which explodes the bolts and fires the posigrade rockets that shove the capsule and booster apart. 1965 *Time* 24 Dec. 35 To change his elliptical orbit into a circle that reached closer to Gemini 7, he made several 'posigrade' burns. 1965 *World Bk. Year Bk.* 48/2 It was also necessary to take into account..another figure for the thrust of the small, posigrade rockets that give the spacecraft a gentle separation push away from the booster.

posish (pozi·ʃ). orig. *U.S.* Also **pozish.** Colloq. abbrev. Position *sb.*

1862 O. W. Norton *Army Lett.* (1903) 113 Snorting their impatience to 'get into posish', came the Monitor, the Galena and others. 1865 J. Pike *Scout & Ranger* i. 25 Some of the managers and their wives, feeling the importance of their 'posish', put on considerable 'style'. 1914 *Dialect Notes* IV. 130 What did I tell ya about standing—Oh what a pozish! 1923 Wodehouse *Inimitable Jeeves* xviii. 251, I wanted to find out the general posish of affairs. 1936 —— *Laughing Gas* xii. 139 Well, you've gone and landed yourself in a nice posish. A dashed nice posish, I don't think. 1960 J. Grant *Come again, Nurse* x. 55 What's the coffee posish?

posit (pɒ·zit), *sb. Philos.* [f. the vb.] A statement which is made on the assumption that it will prove valid (see quots. 1949).

1949 Hutten & Reichenbach tr. *H. Reichenbach's Theory of Probability* ix. 373 A posit is a statement with which we deal as true, although the truth value is unknown. *Ibid.*, We do not say *B* will occur, but we posit *B*... The word 'posit' is used here in the same sense as the word 'wager' or 'bet'... We do not want to say..that it is true that the horse will win, but we behave as though it were true by staking money on it. 1953 W. V. Quine *From Logical Point of View* ii. 45 Physical objects, small and large, are not the only posits... The abstract entities which are the substance of mathematics are another posit in the same spirit. 1976 *Sci. Amer.* Mar. 119/3 He proposed a set of five posits about the structure of the world that he believed were sufficient to justify induction.

positing, *vbl. sb.* (Earlier and later examples.)

1854 Geo. Eliot tr. *Feuerbach's Essence Christianity* xxii. 213 This negativing of limits by the imagination is the positing of omniscience as a divine power and reality. 1967 *Listener* 5 Oct. 430/1 If subject does not respond to direct approach try seemingly more casual positings of kindred questions at the macro-level.

position, *sb.* Add: **I. 5.** *spec.* (*a*) the disposition of the limbs in a dance step (see also *first position* (b) s.v. *first C, *fourth position* s.v. *fourth C); (*b*) the posture adopted during sexual intercourse.

(*a*) 1778 *English Mag.* Feb. 59/2 A woman who was ignorant that her first curtsey should be in the third position. 1819 M. Edgeworth *Let.* 17 Apr. (1971) 199 She seems evermore as if she had the fear of the five positions before her eyes. 1884 D. Anderson *Compl. Ball-Room Guide* 10 *Second position*, put out right foot in a straight line with left heel, right heel about four inches from left heel. 1922 [see *à terre *adv.* and *adj. phr.*]. 1930 Craske & Beaumont *Theory & Pract. Allegro in Classical Ballet* 15 Lower the arms to the *fifth* position *en bas.* 1971 'D. Halliday' *Dolly & Doctor Bird* xi. 143 Krishtof Bey rose to his feet..and struck the fifth position, brown arms outflung. 1979 A. Morice *Murder in Outline* iii. 26 Carefully placing his feet, right heel to left instep, in the number two position.

(*b*) 1883 tr. *Kama Sutra of Vatsyayana* II. vi. 65 When the woman forcibly holds in her yoni the lingam after it is in, it is called the 'mare's position'. This is learnt by practice only. 1933 E. A. Robertson *Ordinary Families* vi. 112, I show de shentleman de twenty-seex poseetions of lof? 1969, 1971 [see *missionary *a.* 1 b]. 1974 W. Garner *Big enough Wreath* v. 60 My pa..always warned me about the Chinese position. 1976 *Sounds* 11 Dec. 4/1 Mickey Gallagher was being particularly adventurous one bedtime with his lady friend, and collapsed from Position Number 368 (look it up yerselves, you cheeky...) onto the floor and broke his wrist. 1977 *Times* 26 Mar. 12/5 There was actually—this was, maybe, 1938—a chapter on positions. Wow!

II. 7. b. *line of position* or *position line*: a line on which the observer is computed to be after having taken a bearing.

1863 W. Chauvenet *Man. Spherical & Pract. Astron.* I. viii. 428 Let the first observation give the position line *AA′* (Fig. 35), and let *Aa* represent, in direction and length, the ship's course and distance sailed between the observations. 1865 J. H. C. Coffin *Navigation & Nautical Astron.* (ed. 2) ix. 224 The nearer the body is to the prime vertical, the more nearly the line of position coincides with a meridian. 1919 G. C. Comstock *Summer Line* p. iii, The line of position, or Summer line, is generally recognized as the best method for fixing the ship's place by observation of the sun or stars. 1920 J. E. Dumbleton *Princ. & Pract. Aerial Navigation* i. 13 Owing to small errors three position lines will rarely intersect at a point, but a small triangle is formed known as a 'cocked hat'. 1962 *Flight Handbk.* (ed. 6) xiii. 292 One bearing gives a position-line; two or more are needed for a fix. 1974 K. Wilkes *Pract. Yacht Navigator* ix. 115/1 A position line from observation of a single identifiable object can be established by taking a compass bearing of it.

9. b. *spec.* in *social position.*

c 1832 J. S. Mill in F. A. von Hayek *J. S. Mill & Harriet Taylor* (1951) iii. 61 For a long time the indissolubility of marriage acted powerfully to elevate the social position of women. 1853 C. Brontë *Villette* II. xxviii. 293 Pedigree, social position, and recondite intellectual acquisition, occupied about the same space and place in my interests and thoughts. 1949 M. Mead in M. Fortes *Social Struct.* 18, I found it impossible to give an adequate sociological statement which did not include the specification of each actor in terms both of his social position and of his personality. 1971 P. J. Keating *Working Classes in Victorian Fiction* iii. 73 The physical and spiritual struggles inherent in their social position. 1976 G. Butler *Vesey Inheritance* vi. 176, I am a young woman of education and social position.

III. 10. position change *Genetics*, any change in the order of the genes along a chromosome; **position effect** *Genetics*, an effect on the phenotypic expression of a gene produced by a difference in its chromosomal position, esp. by its proximity to a mutant gene or to heterochromatin; **position-finding,** the process of ascertaining one's position or that of a distant object, esp. automatically by radio or similar means; usu. *attrib.*; **position line** (see *7 b); **position mark,** a mark made on a stone or other component part of a structure to indicate the position it is designed to occupy; **position paper** orig. *U.S.*, a written statement of attitude or intentions; **position play** *Chess* (see quot. 1960); **position player** (*a*) *Chess*, one who adopts position play; (*b*) *Austral. Football* (see quot. 1969); **position vector** *Math.*, a vector which defines the position of a point.

1937 *Nature* 30 Oct. 761/2 The primary structural change of inversion gives rise to secondary changes such as reduplication and deficiency. These are changes of 'balance', and rank with intra-genic changes and position changes as one of the three effective means of variation. 1952 C. P. Blacker *Eugenics* x. 245 These alterations of chromosome structure resulting from one or more breakages and recombinations have been called..position changes. 1930 *Jrnl. Genetics* XXII. 315 Since the addition of different deletions results in much the same effects, regardless of exactly where the breakage occurred, these are not 'position effects' caused by displacement of certain genes from others previously adjacent to them. 1952 Srb & Owen *Gen. Genetics* x. 201 In many ways, the numerous position effects described in the literature of genetics appear as a bewildering array of vaguely related phenomena. 1974 *Genetic Res.* XXIII. 291 Position effect variegation is now regarded as a general phenomenon but it is in *Drosophila* that by far the largest number of cases

have been described. 1918 E. S. Farrow *Dict. Mil. Terms* 463 *Position finding system*, the term applied to the system used in determining the range and direction to any target from a battery or station. 1947 Crowther & Whiddington *Sci. at War* 57 Another position-finding system in which, however, the aircraft 'interrogates' by sending out pulses. 1959 A. Hardy *Fish & Fisheries* vii. 160 The deep-water trawlers have been getting larger, more powerful and more efficiently equipped with echo-sounding and position-finding apparatus. 1928 G. G. Coulton *Art & Reformation* viii. 145 An inspection..will convince us that the rare marks found otherwise than on the surface are not banker-marks, but position-marks. 1965 *Guardian* 4 Sept. 9/5 Republican leavers..got out an eleven-page policy statement, called a 'position paper'. 1972 Ld. Gladwyn *Mem.* xiii. 226 The idea was that all the political Under-Secretaries..should meet every so often and discuss what I suppose would now be called 'position papers'. 1977 *Time* 7 Mar. 13/1 Carter is just beginning to receive position papers from his advisers on what his policies should be. 1932 E. Lasker *Man. Chess* iv. 166/1 Whereas by combination values are transformed, they are proved and confirmed by 'position play'. Thus, position play is antagonistic to combination, as becomes evident when a 'combinative player' meets with his counterpart, the 'position player'. 1960 Horowitz & Mott-Smith *Point Count Chess* (1973) 356 *Position play* is a strategic move or plan as distinguished from a tactical (combination). 1969 Eagleson & McKie *Terminol. Austral. Nat. Football* III. 4 *Position player*, a variant for *placed man* [*sc.* a player who is allocated a fixed position on the field], recorded by four informants. 1961 C. C. T. Baker *Dict. Math.* 242 If *P* is the position of a point at any time, and *O* is a fixed point, the line *OP*, having length and direction, is a vector, and is denoted by $\overline{OP}$, or **OP**. If *P* is the point (*x, y, z*), the position vector is **R** = *x***i** + *y***j** + *z***k**. 1969 Wade & Taylor *Contemp. Analytic Geom.* vii. 243 When we speak of a position vector it is to be understood that the vector has only one representative and that the initial point of this representative is at the origin.

position, *v.* **1. a.** (Later examples.)

1955 *Sun* (Baltimore) 12 Jan. 12/4 The straw is baled, elevated, and positioned on an accompanying truck. 1955 *Sci. Amer.* May 124/1 You first position the hairline of the slider over the caret between the first four balls (1, 2, 3, 4) and the second four (5, 6, 7, 8) in the bottom tier of this rule. 1959 *Listener* 5 Mar. 432/3 *Il Tabarro* was directed.. by Charles Rogers and produced (*i.e.* positioned, rehearsed dramatically, and so on) by Colin Graham. 1960 *Practical Wireless* XXXVI. 429/1 Beginners would be well advised to position the amplifier so that the underneath parts can be inspected while the power is on. 1967 *Times Rev. Industry* Feb. 90/3 Three two-jet engines positioned in much the same way as the British Trident. 1979 J. Wainwright *Tension* 62 Uniformed constables had been positioned to re-direct traffic.

positional, *a.* Add: (Further examples.) *spec.* (*a*) *Linguistics* = *isolating *ppl. a.* 1; (*b*) *Chess*, characterized by position play.

1908 T. G. Tucker *Introd. Nat. Hist. Lang.* 92 Languages which express grammar and modification of sense by position, without external or internal modification of the 'roots'..may be called *Inorganic* or *Positional. Ibid.* 93 The *Positional* languages include Chinese, Burmese, Anamese, and their group. *Ibid.* 117 The line of linguistic ease naturally taken by a language which finds the purely positional structure inadequate to it [*sic*] needs. 1929 *Times* 2 Nov. 6/7 Newman increased a run of 312.. to 349, and then missed a positional red winner. 1937 M. Euwe *Strategy & Tactics in Chess* 18 We call games such as the preceding one, in which strategy plays such an important part, *positional* games, in contrast to *combinative* games, in which the strategy is of minor importance. 1937 J. R. Firth *Tongues of Men* vii. 88 Three types of language structure: (i) Meanings indicated by words: relations by position... These were called *Isolating* and *Positional* Languages. *E.g.*, Chinese, English. 1937 *Language* XIII. 3 Not uncommonly, the same language has long consonants of different phonemic types: for example, English has long consonants both as positional variants and as geminate clusters. 1938 *Times Lit. Suppl.* 5 Mar. 157/2 It secured the dominance of the positional over the combinational school. 1945 *Diamond Track* (Army Board, N.Z.) 37/2 The Allied success..made the position of large forces of the enemy's positional infantry on our front most precarious. 1946 *Sunday Dispatch* 8 Sept. 6/2 There was no weakness in United, who played clever positional football. 1952 A. Cohen *Phonemes of Eng.* 90 'Positional diphthongs' are characterised by the preservation of the individual character of the component parts. 1960 C. Barnett *Desert Generals* III. ii. 103 The mentality of the army's senior and rising officers could not be similarly converted by decree to suit mobile armoured warfare. The stiff, positional war of the Western Front between 1914 and 1918 was the only..influence. 1964 R. B. Lees *Gram. Eng. Nominalizations* p. xli, Subcategorization is obviously a different kind of constraint on syntactic constituents from positional constraints in a tree or domination by a certain kind of node. 1970 *Nature* 12 Dec. 1121/1 In fact, the positional accuracy achieved is typically a small fraction of the vehicle stability. 1971 *Physics Bull.* July 397/2 Three instruments incorporating laser interferometers for calibration have been built at NPL. The first of these, for measuring the positional errors of the lines on precision scales and gratings, has been in use for several years. 1972 G. Green *Great Moments in Sport: Soccer* iv. 58 Cohen and Ray Wilson, the full backs, overlapping down the flanks ..as the whole side bamboozled Spain with mobile positional play. 1979 E. H. Gombrich *Sense of Order* ix. 243 Repetition devalues elements while isolation in the centre will emphasize them. In the 'field of force' we can observe the effects of 'positional enhancement'.

b. Special collocations: *positional goods* (Econ.): see quots.; *positional player* (Chess) = **position player* (a).

1976 F. Hirsch *Social Limits to Growth* iii. 27 The positional economy..relates to all aspects of goods, services, work positions, and other social relationships that are either (1) scarce in some absolute or socially imposed sense or (2) subject to congestion or crowding through more extensive use... If..positional goods remain in fixed supply while material goods become more plentiful, the price of positional goods will rise, as consumers' relative intensity of demand for them increases in terms of material goods. 1976 *Economist* 11 Dec. 129/2 Many of the things which are valued in our society are hierarchical. They are what Professor Hirsch calls 'positional goods'. 1977 *N.Y. Times Bk. Rev.* 13 Feb. 10/3 'Positional goods'—a house at the shore, tenure on the Harvard faculty, a Picasso on the wall. 1977 *Econ. Jrnl.* Sept. 574 Positional goods are defined as those to which access is a function of an individual's income relative to other people's. 1933 M. A. Schwendemann tr. *Reti's Masters of Chess Board* 4 Typical positional players like Steinitz and Rubinstein are of the opinion that this variation of the King's Gambit is in favour of White.

Hence **posi·tionally** *adv.*

1923 C. D. Broad *Sci. Thought* xi. 408 Some of these strands may be positionally uniform. 1961 *Times* 13 Feb. 4/1 The new open side wing forward, Rogers, was positionally at sea. 1964 *Amer. Speech* XXXIX. 35 The main adjective class is positionally any word that goes either as a prepositive or as a complement of a copulative. 1971 D. Crystal *Linguistics* 190 The..is positionally fixed, preceding the noun it modifies. 1975 *Nature* 31 Jan. 310/2 The molecules are positionally ordered but orientationally disordered and mobile.

positioner (pozi·ʃənəɹ). [f. Position *v.* + -er[1].] One who or that which positions; *spec.* a device or machine for mechanically moving an object into position and keeping it there.

1934 in Webster. 1957 E. B. Jones *Instrument Technol.* III. II. 133 The positioner also corrects for movements of the valve stem owing to unbalanced forces on the valve plug. 1969 *Engineering* 29 Aug. 216/1 A universal positioner,..with vertical table adjustment under full load, has been added to the range of manipulative equipment... Capacity is from 1 to 10 tons. 1972 *Physics Bull.* June 363/1 A micrometer stage is used to traverse the resistor under an RF probe which is held by a vertical positioner.

positive, *a.* and *sb.* Add: **A.** *adj.* **II. 5. c.** Functioning for the special purpose required; having or being a well-defined and effective action.

1903 *Sci. Amer.* 21 Feb. 134/1 Instead of depending on splash lubrication alone for oiling every part of the engine, positive oil feeds are led to each of the crankshaft bearings. 1938 L. V. W. Clark in A. E. Dunstan et al. *Sci. of Petroleum* I. IX. 434/1 Blow-out preventers of the first group have been..quite satisfactory for drilling in areas of normal pressure, but where higher pressures are encountered it becomes necessary for a positive control to be available. 1958 *Times* 1 July 6/6 The steering, which used to be somewhat indefinite, is now light and pleasantly positive in action. 1972 *Physics Bull.* Apr. 230/2 Minor but important details have been considered—the cable is five feet long and very flexible, the grip on the bench (with rubber ball feet) is positive. 1977 *Offshore Engineer* Apr. 27/2 The port has a step-down diameter which provides a positive stop during pipe pull-in. 1977 *Sci. Amer.* Aug. 106/1 The unassisted drum brakes are balanced and positive in action but require heavy foot pressure.

IV. 8. a. (Further examples.) Also, consisting in or characterized by constructive action or attitudes; see also *positive thinking* below.

1930 H. Crane *Let.* 22 May (1965) 351 The poem [*sc. The Bridge*]..is, I think, an affirmation of experience, and to that extent is 'positive' rather than 'negative' in the sense that *The Waste Land* is negative. 1961 *Oregonian* (Portland, Oregon) 24 Oct. 8 The Portland school board was asked..to take a positive stand towards developing ..more plans for the city's schools in event of attack. 1971 *Times* 15 Feb. 9/3 Ireland were the more positive side throughout and the same XV has been chosen for the game against Scotland. *Ibid.* 9/4 All the positive rugby after the interval came from Ireland. 1973 *Howard Jrnl.* XIII. 310 The opportunity to have a positive experience of learning may be very significant. 1976 M. Millar *Ask for me Tomorrow* (1977) xvi. 132 Please try to take a more positive attitude.

c. Designating a copy or likeness of an object with the same relief as that of the original, as opposed to the reverse relief of a mould.

1911 [see *negative *a.* 11 a]. 1931 [see *negative *sb.* 8 c]. 1940 [see *negative *a.* 11 a]. 1973 [see *negative *sb.* 8 b].

d. Psychol. *positive thinking*: the practice or result of concentrating one's mind affirmatively on what is constructive and good, thereby eliminating from it negative or destructive thoughts and emotions; also *attrib.*; *positive transfer*: the transfer of effects from the learning of one skill that facilitate the subsequent learning of another skill; *positive transference*: transference in which the feelings involved are of a positive or affectionate nature.

1916 C. E. Long tr. *Jung's Coll. Papers Analytical Psychol.* ix. 270 As long as it is a question of the so-called 'positive' transference, the infantile-erotic character can usually be recognised without difficulty. 1921 F. N. Freeman *Exper. Educ.* ii. 47 There is..positive transfer again from Set 3 to Set 4. 1924, etc. [see *negative transference* s.v. *negative *a.* 8 c]. 1933 R. W. Bruce in *Jrnl. Exper. Psychol.* XVI. 351 There is a marked positive transfer in learning to make an old response to a new stimulus. 1953 N. V. Peale (*title*) The power of positive thinking. *Ibid.* i. 2, I listened to your speech tonight in which you talked about the power of positive thinking. 1959 N. V. Peale *Amazing Results of Positive Thinking* p. vii, By the application of positive thinking principles to their own life situations, they have mastered fear, healed personal relationships,..and gained strong new confidence. 1970 B. C. Mathis et al. *Psychol. Found. Educ.* iii. 83 If the experimental group is superior to the control group on B..positive transfer of training is said to have occurred. 1970 A. Janov *Primal Scream* xiv. 246 When the therapist is helpful and warm and offers a bit of advice, he is encouraging the 'positive' transference. 1974 *Country Life* 21 Nov. 1616/2, I believe that positive thinking can help one overcome many difficulties in life.

e. Pol. *positive neutralism* or *neutrality*: a policy adopted by some of the poorer and less developed countries of maintaining relations with each of the major powers while remaining neutral in regard to their rivalry (see quot. 1968). So *positive neutralist.*

[1957 *Political Sci. Q.* LXXII. 266 A communiqué condemning the development of Power blocs in international affairs and urging neutralism as a positive way towards the establishment of international peace.] 1960 *Sunday Times* 28 Aug. 5/1 'We Africans are positive neutralists.'.. These phrases..have cropped up constantly in the speeches so far: 'positive neutralism' [etc.]. 1961 *NATO or Neutrality* (Fabian Soc.) iv. 20 A policy of positive neutrality would be of immense assistance in helping Britain establish good post-imperial relations with the new countries. 1968 P. Calvocoressi *World Politics since 1945* IV. xiii. 256 Neutralism and non-alignment, therefore, as distinct from neutrality, were the expression of an attitude towards a particular and present conflict: they entailed, first, equivalent relations with both sides and, secondly—in the phase called positive neutralism—attempts to mediate and abate the dangerous quarrels of the great.

9. c. *positive logic*: (*a*) [tr. G. *positive logik* (Hilbert & Bernays *Grundlagen der Math.* (1934) I. iii. 68)] (see quots. 1943, 1947); (*b*) circuit logic in which the larger or most positive signal is taken as representing 1 and the smaller signal 0.

1943 *Mind* LII. 49 Of other 'rudimentary systems' I mention only the so-called positive logic, which does not operate with negations. 1947 *Mind* LVI. 215 Rules 4.1 to 4.5 constitute the positive logic of compound statements, that is to say, they suffice for that part of propositional logic which is independent of negation. 1955 A. N. Prior *Formal Logic* III. ii. 258 This segment of Heyting's calculus forming what Hilbert and Bernays have called 'positive logic'. 1958 *Proc. IRE* XLVI. 1249 In the following discussion, positive logic is used. A binary '1' is defined as the most positive signal potential, and a binary '0' as the most negative. 1962, 1968 [see *logic *sb.* 3*]. 1974 D. A. Calahan et al. *Introd. Mod. Circuit Anal.* iii. 47/2 In theory, we must be told which type of logic, positive or negative, is assumed before we can determine the function of a logical device; in practice, positive logic is assumed unless otherwise specified. 1976 Belove & Drossman *Systems & Circuits for Electr. Engin. Technol.* xiii. 325 We shall use the positive logic convention whereby a relatively low voltage represents logical 0 and a relatively high voltage represents logical 1.

VI. 15. Other collocations: *positive discrimination*, the making of distinctions in favour of groups considered disadvantaged or underprivileged, esp. in the allocation of resources and opportunities; *positive electron*, a particle analogous to the ordinary negative electron but having a positive charge: orig. applied to the proton, now to the positron; *positive eugenics*, an attempt to encourage the birth of children to parents having qualities considered desirable to the community; *positive feedback* (see *feedback, feed-back *sb.* a); *positive pressure* (Med.), pressure greater than that of the atmosphere, used to force air or oxygen into the lungs intermittently to supplement or replace natural inspiration; freq. *attrib.*; *positive ray*, a stream of positively-charged ions which are produced in a gas discharge tube and move towards the cathode; *pl.* except when *attrib.*

1967 *Children & their Primary Schools* (Central Advisory Council for Educ.) I. v. 57 We ask for 'positive discrimination' in favour of such schools [in deprived areas] and the children in them, going well beyond an attempt to equalise resources. 1974 *Observer* 21 Apr. 14/6 Israeli educationists regard this as a challenge, and positive discrimination—in theory at least—has become an article of faith. 1977 *Film & Television Technician* Mar. 8/4 Ms Betty Lockwood, Chairman of the Equal Opportunities Commission, told a WEA seminar of trade union officials that positive discrimination inside unions and in training should be encouraged. That means discrimination in favour of either women or men, though in practice it is likely to mean women. 1978 *Daily Tel.* 28 Jan. 16 Camden Council's announcement that henceforth members of immigrant minorities, even when less qualified, will be preferred for employment to indigenous citizens will bring to the boil the simmering debate on what is euphemistically known as positive discrimination or affirmative action. 1900 Ld. Kelvin in *Phil. Mag.* L. 306 For atoms of electricity, which, following Larmor, I at present call electrons, it inevitably occurs to suggest a special class of atoms... A positive electron would be an atom which by attraction condenses ether into the space occupied by its volume; and a negative electron would be an atom which, by repulsion, rarefies the ether remaining

in the space occupied by its volume. 1902 [see *electron[2]]. 1903 O. Lodge *Mod. Views on Matter* 12 The chief defect in the electrical theory of matter at present is that the *positive* electron, if it exists, has never yet been isolated from the rest of an atom. 1921 *Phil. Mag.* XLII. 307 So far as the writer can learn, the word proton was suggested by Rutherford for use in designating the hydrogen nucleus. This will also be designated as the positive electron in the present paper. 1932 *Science* 9 Sept. 238/1 Up to the present a positive electron has always been found with an associated mass 1,850 times that associated with the negative electron. 1964 M. Gowing *Britain & Atomic Energy, 1939–1945* 18 This positive electron, or positron as it is often called, was first found in..cosmic rays. [1907 C. W. Saleeby in *Sociol. Papers* III. 31 We must.. preserve the two-fold aspect of eugenics, the one positive—the encouragement of the better; the other negative—the discouragement of the worse.] 1909 —— *Parenthood & Race Culture* xi. 172 We must clearly divide our proposals, as the present writer did some years ago, with Mr. Galton's approval, into two classes: positive eugenics and negative eugenics. *Ibid.*, In regard to positive eugenics I ..cannot believe in the propriety of attempting to bribe into parenthood people who have no love of children. 1914 [see *negative eugenics* s.v. *negative *a.* 8 c]. 1952 C. P. Blacker *Eugenics* v. 111 The statement made [by Galton] in 1901 says that positive eugenics is more *important* than negative and that made in 1908 declares that negative eugenics is more *pressing* than positive. 1970 *Sci. Amer.* Mar. 107 Conceivably a changed social climate and increased knowledge will make it possible for positive eugenics to be practiced on man. 1972 P. B. Medawar *Hope of Progress* 71 The case for 'positive eugenics', that is for constructive rather than merely remedial eugenics, is based on the model of stockbreeding. 1885 I. B. Yeo tr. *Oertel's Respiratory Therapeutics* II. 607 (*heading*) Action of positive pressure on the surface of the thorax. 1909 *Arch. Internal Med.* III. 369 During positive pressure respiration, the so-called artificial respiration of laboratory procedure, the blood pressure falls during inspiration and rises during expiration. 1948 *Anesthesiol.* IX. 29 Positive pressure respiration decreases the venous return, and is in effect a way of applying tourniquets not only to all four extremities but also to the head and abdomen. 1970 *Jrnl. Pediatrics* LXXVI. 183 The indications for intermittent positive pressure ventilation were asphyxia on admission, a single asphyxial attack, [etc.]. 1903 J. J. Thomson *Conduction Electr. through Gases* xvii. 522 On the view of the discharge given in Chap. xvi. there is a stream of positively charged molecules moving towards the cathode, causing this to emit cathode rays; if the cathode is perforated, part of this stream may pass through the holes, producing in the gas behind the cathode luminosity, forming in fact the Canalstrahlen, or positive rays as we may call them, if we think this view of their constitution sufficiently established. 1920 [see *mass spectrograph* s.v. *mass *sb.*[2] 10 d]. 1922 Glazebrook *Dict. Appl. Physics* II. 602/1 Positive rays were discovered by Goldstein in 1886 in electrical discharge at low pressure. 1955 C. G. Darwin in W. Pauli *Niels Bohr* 9 Many elements in a positive-ray tube form temporary hydrides. 1968 M. S. Livingston *Particle Physics* ii. 20 Thomson's studies of the positive rays from ionized hydrogen gas were the first experiments in which the proton was isolated and identified as a particle.

16. Comb.: **positive definite** *adj. phr. Math.*, positive (formerly, positive or zero) in all cases; (of a matrix) having all its eigenvalues positive; hence **positive-definiteness; positive-going** *a.*, increasing in magnitude in the direction of positive polarity; becoming less negative or more positive; **positive-negative** *a.*, exhibiting both positive and negative characteristics.

1907 M. Bôcher *Introd. Higher Algebra* xi. 150 A positive definite form is positive or zero for all real values of the variables. 1948 W. V. Houston *Princ. Math. Physics* (ed. 2) vii. 120 The potential energy will be a quadratic expression in the coordinates that, if the equilibrium is stable, will be a positive definite expression. 1957 L. Fox *Numerical Solution Two-Point Boundary Probl.* vii. 179 If all the λ_r are positive, which is the case in many physical problems, and corresponds to some structure of the differential system corresponding to a positive-definite matrix *A*.., we can also assert [etc.]. 1970 G. Sposito *Introd. Quantum Physics* iii. 53 We further restrict this scalar product by stipulating that it be positive-definite: $(f, f) \geq 0$. 1968 Fox & Mayers *Computing Methods for Scientists & Engineers* i. 6 The associated matrix may be 'general', or it may have special properties such as symmetry, with or without positive-definiteness. 1957 *Wireless World* Jan. 10/2 The area under the positive-going excursion is nearly equal to that under the negative-going excursion. 1979 *Sci. Amer.* Mar. 104/1 (*caption*) A positive-going (but not negative-going) shift in membrane voltage causes a brief outward gating current that coincides with the opening of the sodium channels. 1946 C. Morris *Signs, Lang. & Behavior* 82 Appraisors signify along a positive-negative continuum. 1964 E. A. Nida *Toward Sci. Transl.* ii. 24 The differences between literal and free translating are, however, no mere positive-negative dichotomy, but rather a polar distinction with many grades between them.

positively, *adv.* Add: **3.** (Later examples.)

1961 C. E. Vincent *Unmarried Mothers* x. 254 Her inability to resolve these various identity crises positively. 1972 *Jrnl. Social Psychol.* LXXXVII. 34 It has generally been assumed that level of fear is related positively to response in a potential panic situation.

5. Also *ellipt.*; *colloq.* used as an emphatic affirmative: yes, indeed.

1922 H. S. Walpole *Cathedral* II. ii. 188 She hasn't an idea in her head. I don't believe that she knows it's Jubilee Year. Positively! 1942 Partridge *Usage & Abusage* 4/2 *Absolutely* and *positively*... In slang their meaning is *yes* (popularized by a famous vaudeville duet between 'Mr. Gallagher and Mr. Sheehan').

positivism. Add: **1.** (Earlier and further examples.) Also, the name given generally nowadays to the view, held by Bacon and Hume amongst others (including Comte), that every rationally justifiable assertion can be scientifically verified or is capable of logical or mathematical proof; that philosophy can do no more than attest to the logical and exact use of language through which such observation or verification can be expressed. Also *ellipt.* for *logical positivism* (see *logical *a.* (and *sb.*) 7).

1847 J. D. Morell *Hist. View Philos.* (ed. 2) I. I. i. 88 Let those who claim Bacon as the apostle of *positivism*, give us an interpretation of this whole division of his system. *a* 1866 J. Grote *Exam. Utilitarian Philos.* (1870) 2 A way of thinking about morals, which may be roughly called by the name *Positivism*; by which I mean the line of thought which endeavours to construct a system of morals..from observation and experience of fact alone. 1934 W. M. Malisoff tr. R. Carnap in *Jrnl. Philos. of Sci.* I. 16 In the following example we deal with the conflict of two theses..which correspond more or less to positivism and to realism. 1945 K. R. Popper *Open Society* I. v. 59 Ethical positivism..maintains that..what is, is good. (Might is right.) 1961 M. Čapek *Philos. Impact Contemp. Physics* xvi. 297 The positivism prevailing amongst contemporary physicists, who insist on a consistent elimination of *all* unobservable factors. 1964 Fodor & Katz in R. Klibansky *Contemp. Philos.* (1969) III. 303 We shall therefore examine the two dominant schools of thought in recent philosophy of language, ordinary-language philosophy and positivism. 1967 *Encycl. Philos.* VI. 415 Both share the general idea of progress, but whereas social positivism deduces progress from a consideration of society and history, evolutionary positivism deduces it from the fields of physics and biology. 1974 H. Wang *From Math. to Philos.* p. ix, The much publicized juxtaposition of logic with positivism (or empiricism or 'analytic' philosophy) has burdened logic with a guilt by association.

3. *Law.* A term derived from positive law (cf. Positive *a.* 1) and applied to theories concerned with the enactment of law, the reaching of legal decisions, the binding nature of legal rules and the study of existing law; which postulate that legal rules are valid because they are enacted by the 'sovereign' or derive logically from existing decisions, and deny that ideal or moral considerations (such as those of natural law, or that a rule is unjust) should in any way limit the operation or scope of the law.

1927 M. R. Cohen in *Proc. 6th Internat. Congr. Philos., 1926* 469 (*title*) Positivism and the limits of idealism in the law. 1944 W. Friedmann *Legal Theory* xv. 135 Positivism in jurisprudence comprises legal movements, poles apart in every respect. 1945 H. Kelsen *Gen. Theory Law* I. iii. 52 No sanction without a legal norm providing this sanction, no delict without a legal norm determining that delict. These principles are the expression of legal positivism in the field of criminal law. 1959 Jowitt *Dict. Eng. Law* II. 1366/2 *Positivism*, in international law, this means the method which attempts to present law as actually applied in State practice. 1961 H. L. A. Hart *Concept of Law* i. 7 Some contemporary legal theory which is critical of the legal 'positivism' inherited from Austin. 1967 *Encycl. Philos.* IV. 419/1 The definition of law as the command of the 'sovereign' is no doubt the most prominent example of a form of positivism. *Ibid.*, Sometimes 'legal positivism' is used to refer to the view that correct legal decisions are uniquely determined by pre-existing legal rules. 1969 M. Moritz in R. Klibansky *Contemp. Philos.* IV. 140 The author intends to give an empirical account of what is a legal order. He regards the distinction between natural law theory and legal positivism as being of secondary importance. 1971 *Mod. Law Rev.* XXXIV. vi. 632 Positivism regards law as a system of comprehensive and closely defined rules.

positivist. Add: (Further examples.) See also *logical positivist* s.v. *logical *a.* (and *sb.*) 7.

1892 [see *criticist]. 1936 A. J. Ayer *Lang., Truth & Logic* i. 23 Some positivists have adopted the heroic course of saying that these general propositions are indeed pieces of nonsense. 1958 G. J. Warnock *Eng. Philos. since 1900* 58 The Positivists were also engaged in linguistic analysis, officially without metaphysical ambitions; theirs was supposed to be the two-sided task, on the one hand of exposing the muddles of metaphysicians, and on the other hand of humbly clarifying the vocabularies of the scientist and the mathematician. 1971 J. H. Haddox *Antonio Caso* 86 To the cowardly positivists frightened by the idea of 'mental anarchy' I say, no.

b. (Further examples.)

1934 *Philos. of Sci.* I. 16 In using the formal mode of expression the pseudo-problem 'What is a thing?' disappears, and therewith the opposition between the positivist and the realist answer disappears. 1943 W. G. Hardy *Some Semantic Theories* in *Cornell Univ. Abstr. of Theses* 56 Bridgman's operational theory of meaning amounts to a positivist demand that meanings be assigned according to the operations performed. 1960 J. O. Urmson *Conc. Encycl. Western Philos.* 324/1 In the twenties of the twentieth century Hume's positivist arguments were revived and strengthened. 1969 F. Halliday in Cockburn & Blackburn *Student Power* 298 A bid to introduce IQ tests was made, but this positivist attack was repelled when students occupied the main building of the campus for two months. 1974 *Nature* 16 Aug. 609/1 Most philosophers of science, at least within the dominant positivist schools, take the Comtean view, of physics as the paradigmatic science. 1977 *New Yorker* 9 May 145/1, I suspect that we remain, in our hearts, medieval people: our assumptions are Aristotelian, not positivist or existentialist.

2. An adherent or supporter of legal positivism (see *positivism 3).

1927 M. R. Cohen in *Proc. 6th Internat. Congr. Philos., 1926* 469 It is therefore easy..to show that other positivists are full of hidden or unavowed natural law. 1971 *Mod. Law Rev.* XXXIV. vi. 631 Most positivists, and certainly Hart, would argue that legal rules can never be spelled out in terms of all the situations to which they might be relevant. 1973 I. M. Sinclair *Vienna Convention on Law of Treaties* v. 112 They [*sc.* the school of jurists led by Bynkershoek, Moser and Martens] did not wholly deny the role of natural law in filling gaps, but their emphasis on the constituent elements of positive international law gave them the title 'positivists'.

b. *attrib.* or as *adj.*

1923 R. Pound *Interpretations of Legal Hist.* iv. 78 The positivist ethnological interpretation [of legal history].. was given a comparative basis. 1944 W. Friedmann *Legal Theory* xv. 135 The number and variety of positivist legal theories is as great as that of the sciences. 1963 S. I. Shuman *Legal Positivism* i. 11 Friedmann who speaks.. of 'Austin's positivist system'. 1976 *Howard Jrnl.* XV. I. 51 The change from a classical to a positivist approach to criminology..took the form of a belief in the biological and social causation of crime and the necessity for early prevention.

positivistic, *a.* **1.** (Further examples.)

1927 M. R. Cohen in *Proc. 6th Internat. Congr. Philos., 1926* 471 The same system of legal rights and duties may be expressible in positivistic or in idealistic language. 1935 [see *anti-metaphysical* s.v. *anti-[1] 3 b and c]. 1938 B. F. Skinner *Behavior of Organisms* ii. 44 So far as scientific method is concerned, the system set up..is positivistic. It confines itself to description rather than explanation. 1956 J. O. Urmson *Philos. Analysis* viii. 119 It might be thought that such a characterization of positivistic analysis embodied an illegitimate nostalgia. 1961 M. Čapek *Philos. Impact Contemp. Physics* xvi. 297 A similar positivistic motive is conspicuously present in the minds of physicists dealing with the problems of determinism in quantum mechanics. 1975 *Sci. Amer.* Feb. 101/1 Thinkers of a mystical turn of mind..consider it the ..most fundamental of all metaphysical questions... Those of a positivistic, pragmatic turn of mind consider it trivial.

positivi·stically, *adv.* [f. Positivistic *a.* + -ly[2].] In a positivistic manner.

1890 W. James *Princ. Psychol.* I. vi. 177 Nevertheless, this formula which is so unobjectionable if taken vaguely, positivistically, or scientifically, as a mere empirical law of concomitance..tumbles to pieces entirely if we assume to represent anything more intimate or ultimate by it. 1935 *Mind* XLIV. 128 The difference between the subjective and the objective may be interpreted positivistically by the category of degree which rules the newer physics. 1946 R. G. Collingwood *Idea of Hist.* IV. ii. 182 The facts are positivistically conceived as isolated from each other. 1976 *Times Lit. Suppl.* 17 Dec. 1590/5 It is just such ordinariness that has so often been lost from psychology in its efforts to deal positivistically with isolated variables.

positon (pɒ·zitɒn). *Physics.* [f. Posit(ive *a.* and *sb.* + *-on[1].] † **a.** Proposed as an alternative name for the proton. *rare*[—1]. *Obs.* **b.** [a. G. *positon* (P. Gruner 1935, in *Helvetica Physica Acta* VIII. 326).] = next.

1928 [see *negaton]. 1937 *Chem. Abstr.* XXXI. 2917/1 The following names are proposed for corpuscles of mass approx. 10^{-27} g. (present names in parentheses): nulliton (neutrino), positon (positron), negaton (electron). 1938, etc. [see *negaton]. 1952 B. Rossi *High-Energy Particles* i. 2 Positive electrons (or positons) are identical to negatons, except for their opposite sign of charge. 1956 *Nuclear Physics* I. 72 The word 'electron' applies to particles with an elementary charge of either sign. When it is desirable to emphasize the sign of the charge, the words 'positon' and 'negaton' are used. Thus the illogical phrase 'electron-positron pair' is replaced by 'electron pair', or..'positon-negaton pair'. 1960 *Ibid.* XVI. 683 (*heading*) Branching ratios of K capture to positon emission in non-unique first forbidden $2^- \to 2^+$ beta transitions. 1974 *Ibid.* A. CXXXII. 230 (*heading*) Electron capture to positon decay ratios and second-class currents.

positron (pɒ·zitrɒn). *Physics.* [f. Posi(tive *a.* and *sb.* + *elec)tron[2].] The anti-particle of the ordinary (negative) electron, having the same mass and a numerically equal but positive charge.

1933 C. D. Anderson in *Science* 5 May 432/2 Experiments have been carried out which gave conclusive evidence that positrons are ejected from lead by the γ-radiation of ThC″. [*Note*] The contraction positron is here used to denote the free positive electron. 1933 *Times* 9 Dec. 9/3 The unit of heavy hydrogen, the deuteron, or deuton, as it has been called, bids fair to rival in interest its recently found cousins, the neutron and positron. 1934 *Discovery* May 123/2 It appears from investigations that the earth is being bombarded by streams of positrons and electrons of very high energy. 1946 *Ann. Reg. 1945* 356 The Bethe carbon cycle (in which carbon acts as a catalyst for the conversion of hydrogen into helium and positrons). 1958 *Spectator* 13 June 778/3 When the electron and the positron meet they annihilate each other. 1960 Chalmers & Quarrell *Physical Examination of Metals* (ed. 2) xvi. 765 Positrons, sometimes called 'positive electrons', result from proton–neutron transitions. 1968 C. G. Kuper *Introd. Theory Superconductivity* xii. 193 The conduction electrons and holes of semiconductor theory are closely analogous to the electrons and positrons of the Dirac theory. 1974 G. Reece tr. *Hund's Hist. Quantum*

Theory xv. 203 Not until the experimental observation of the positron..was the Dirac 'hole' theory generally believed.

Hence **positro·nic** *a.*

1948 I. Asimov in *Astounding Sci. Fiction* Feb. 44/2 If the field were a trifle stronger, the robot would never reach the technician concerned, since its positronic brain would collapse under gamma radiations—and then we would be out one expensive and hard-to-replace robot. 1957 —— *Naked Sun* (1958) ii. 32 He knew a positronic brain..nestled in the hollow of the skull. He knew that Daniel's 'thoughts' were only short-lived positronic currents. 1968 *Punch* 23 Oct. 592/2 Barbarella (Jane Fonda), a respected astronaut in 40,000 AD, is being briefed for her mission to find an important scientist from Earth who has disappeared among the planets with his great invention the positronic ray. 1974 I. Asimov in George & Humphries *Robots are Coming* 5 My positronic robot stories.

positronium (pɒzitrəʊ·niəm). *Nuclear Physics.* [f. prec. + -ium.] A short-lived neutral system, analogous to an atom, consisting of a positron and a negative electron bound together.

1945 A. E. Ruark in *Physical Rev.* LXVIII. 278/1 In 1937 I conceived the idea that an unstable atom composed of a positron and a negative electron may exist in quantities sufficient for spectroscopic detection. The name positronium is suggested. The spectrum of positronium would have lines at wave-lengths twice as great as the hydrogen lines. 1957 *New Scientist* 26 Dec. 28/3 When positronium is in free (i.e., near-vacuum) conditions, spectroscopic measurement shows that it has a lifetime of about one and a half ten-millionths of a second. 1966 *McGraw-Hill Encycl. Sci. & Technol.* X. 524/1 Positronium is of particular interest because it is the two-body system to which quantum electrodynamics is applicable, and its study has served as an important confirmation of the theory of quantum electrodynamics. 1970 [see *hydrogenic *a.* b]. 1976 *Science* 19 Nov. 826/3 If ψ were a bound system of two spin ½ particles, there could be excited levels with a spectrum similar to that of positronium (the system of electron and positron, bound to each other by the electromagnetic force).

posnjakite (pɒ·znyākəit). *Min.* [ad. Russ. *poznyakit* (Komkov & Nefedov 1967, in *Zap. Vsesoyuz. Min. Obshch.* XCVI. 58): see quot. 1967 and -ite[1].] A hydrated basic copper sulphate, $Cu_4(SO_4)(OH)_6 \cdot H_2O$, that occurs as dark blue crystals similar to langite.

1967 *Mineral Abstr.* XVIII. 285/2 A new mineral..is named posnjakite in honour of E. W. Posnjak (1888–1949), well known for work on copper sulphates. 1970 *Mineral. Mag.* XXXVII. 740 (*heading*) Posnjakite from Cornwall. *Ibid.*, While examining and surveying the disused workings of the Drakewalls mine, Gunnislake, we have found specimens of the recently described species posnjakite in old stopes..above the deep adit level.

poss (pɒs), *a.* Also **pos.** Colloq. abbrev. Possible *a.* Chiefly in phrases *if poss, as soon as poss.*

1886–96 in Farmer & Henley *Slang* (1902) V. 260/1 While the public morals-shaper Thinks of writing to the paper To upset the show, if pos. 1909 *Punch* 3 Mar. 160/1 People tell me I ought to have all the amusement *poss* to prevent me from brooding, so I'm making an effort. 1916 A. Huxley *Let.* 30 June (1969) 104 Think over this and let us have it as soon as poss. 1959 P. Bull *I know Face* i. 27, I only came back with four ambitions in mind: to learn a little about acting (if poss.), [etc.]. 1972 D. Francis *Smokescreen* ii. 30 'Why the rush?' 'Well, I don't know, darling. She just said, could we come as soon as poss.'

possess, *v.* Add: **I. 3. b.** *spec.* To have sexual intercourse with (a woman). Also *absol.*

This sense, suggested in private correspondence in 1969 by Professor W. Empson, may not have been intended by the writers themselves in some of the examples that follow.—Ed.

1592 Kyd *Sp. Trag.* I. i. 10 By duteous seruice and deseruing love, In secret I possest a worthy dame. 1600 Shakes *A.Y.L.* IV. i. 144 Now tell me how long you would haue her, after you haue possest her? 1680 Rochester *Poems* 87 Mad to possess himself he threw, On the defenceless lovely Maid! *c* 1707 T. D'Urfey *Wit & Mirth* (1719) IV. 332 And tho' I let Loobies Oft finger my Bubbies: Who think when they kiss me, That they shall possess me. 1749 Smollett tr. *Le Sage's Gil Blas* II. v. ii. 197 The four banditti expressed an equal desire of possessing the lady who had fallen into their hands, and talked of casting lots for her. 1876 *Romance of Lust* IV. 39 Her delight and surprise at finding the dear Egerton had equally desired to possess her. 1922 Joyce *Ulysses* 72 Possess her once take the starch out of her. *Ibid.* 491 All the male brutes that have possessed her. 1961 *Partisan Rev.* XXVIII. 648 A conflict having to do with father-murder and the wish to possess the father's woman.

possessed, *ppl. a.* Add: **2. d.** *like all possessed*: with great force, vehemence, energy or spirit. *U.S.*

1833 S. Smith *Life & Writings J. Downing* 209 [He] struck his fists together like all possessed. 1916 E. Porter *Just David* 280 He danced and laughed and clapped his hands,..an' carried on like all possessed.

posse·ssingly, *adv.* [f. Possessing *ppl. a.* + -ly[2].] So as to possess or captivate (one); fascinatingly.

1927 *Observer* 11 Dec. 10/4 Miss Jenkins's diary..is nothing worth in itself, but how possessingly dramatised by the identity of her correspondent!

possession, *sb.* Add: **1. c.** *to take possession of*: see also Take *v.* 71. Conversely, *to give possession.*

1849 *Illustr. London News* 22 Dec. 406/1 Dan Sheedey and five or six men come to tumble my house; they wanted me to give possession. I said that I would not.

f. *U.S. colloq.* ellipt. for 'possession of narcotic drugs'.

1970 *N.Y. Times Mag.* 15 Feb. 19/1 John E. Ingersoll ..suggested that the penalty for simple possession for personal use be reduced to that of a misdemeanor. 1973 R. L. Simon *Big Fix* x. 71 What's a few years in the cooler for possession. 1977 D. E. Westlake *Enough!* i. 21 Her freak got busted on possession and went away for an extended rest.

8. possession order, an order made by a court of law directing that possession of a property be given to the owner.

1971 *Times* 24 June 3/8 Southwark council was granted possession orders against nine families squatting in eight properties in the borough in a High Court action. 1973 *Times* 11 Dec. 8/4 He granted a possession order to the GLC. The women took over the house..three weeks ago. 1977 F. Branston *Up & Coming Man* ii. 17 Their large Edwardian semi-detached (three mortgages, possession orders pending on two of them).

possessionless, *a.* Delete *rare* and add further examples.

1898 A. P. Atterbury tr. *Sombart's Socialism & Social Movement in 19th Cent.* i. 9 Troops of possessionless workers..herded in great undertakings. 1938 Dylan Thomas *Let.* Dec. (1966) 211 We are..completely possessionless. 1944 I. Origo *Diary* 24 June in *War in Val d'Orcia* (1947) 224 It is a very odd feeling to be entirely possessionless. 1969 G. Leff *Hist. & Social Theory* ix. 174 As defined by Marxism a class is a group which stands in a certain relation to the means of production, either as possessors or possessionless or as independent of either state.

possessive, *a.* Add: **2. a.** Also, showing a desire to possess or to retain what one possesses.

1924 E. O'Neill *Desire under Elms* I. iv, in *Compl. Wks.* II. 164 Eben..stares around him with glowing, possessive eyes... It's purty! It's damned purty! It's mine! 1931 —— *Hunted* I, in *Mourning becomes Electra* (1932) 121 You know how possessive Vinnie is with Orin. She's always been jealous of you. I warn you she'll do everything she can to keep him from marrying you. 1958 P. Gibbs *Curtains of Yesterday* xx. 170 One of those possessive women who wants to grab everything within reach. 1977 C. Storr *Tales from Psychiatrist's Couch* x. 104 A classical case of the possessive Jewish Mum... She didn't like the boy to go out of the house without telling her.

possibilism (pɒ·sibiliz'm). [ad. F. *possibilisme*: see -ism.] A possibilist doctrine or view; **a.** in *Politics*; **b.** in *Geogr.*

a. 1915 G. B. Shaw in *New Statesman* 23 Jan. 386/2 Having noticed that modern Secularism, Materialism, Rationalism: in short, Possibilism, have brought the minds of Mr Blatchford and Mr McCabe to a dead stop.. they [*sc.* the Chestertons]..have frankly embraced Impossibilism. 1954 G. D. H. Cole *Hist. Socialist Thought* II. xv. 248 Paul Brousse's Possibilism, which stressed the importance of reform within capitalism, was definitely unorthodox doctrine. 1974 tr. *Wertheim's Evolution & Revolution* iii. 376, I could not endorse possibilism if it claims that all solutions have equal chances of materializing.

b. 1925 H. Berr in Mountford & Paxton tr. *Febvre's Geogr. Introd. Hist.* p. xi, He [*sc.* Febvre] has found striking formulae in which to state the question precisely. Against the geographical determinism of Ratzel he sets the possibilism of Vidal de la Blache. 1951 G. Tatham in T. G. Taylor *Geogr. in 20th Cent.* vi. 151 The development of Possibilism is closely linked with the writings of Vidal de la Blache and Brunnes in France. 1965 H. & M. Sprout *Ecol. Perspective Human Affairs* v. 83 The doctrine called environmental possibilism, or simply possibilism, represents an historic reversal of perspective towards man-milieu relationships. 1974 Kolars & Nystuen *Geogr.* xx. 375 The message of possibilism is that the environment offers not one, but many, paths for human activities and development.

possibilist. Add: **1.** (Further examples.)

1936 *Sat. Rev. Lit.* (U.S.) 15 Feb. 11/2 He [*sc.* Mazzini] was never what is called in modern phrase a 'possibilist'. He was..that most inspiring and most dangerous product of mankind, an 'idealist'. 1940, 1966 [see *Guesdist]. 1973 *Times* 26 Nov. 15/2 The Labour Party would be irreparably split between its moderate possibilists and its left-wing extremists.

2. *Geogr.* One who emphasizes man's freedom of action in cultural development and minimizes the effects and restrictions of the environment. Also *attrib.* or as *adj.*

1925 Mountford & Paxton tr. *Febvre's Geogr. Introd. Hist.* 20 We will not ask whether there are not really any cracks in the geographical edifice, and whether it is possible to follow at the same time..the 'determinists' after the manner of Ratzel, and what we may perhaps call the 'possibilists' after the pattern of Vidal. 1951 G. Tatham in T. G. Taylor *Geogr. in 20th Cent.* vi. 155 Possibilists do not, nor have they ever claimed, that man can free himself from all environmental influences. *Ibid.*, Possibilist statements published during the last fifty years, make quite clear the contention that Nature does not drive man along one particular road. 1964 *Welsh Hist. Rev.* II. 275 He begins by disavowing any intention of arguing for geographical determinism and affirms his allegiance to the 'possibilist' school of geographers.

possibilistic (pɒsibili·stik), *a.* [f. prec. + -ic.] Of or pertaining to possibilism.

1965 H. & M. Sprout *Ecol. Perspective Human Affairs* v. 83 A possibilistic analysis directs attention to those factors of the milieu that may affect the operational result. 1974 *Encycl. Brit. Micropædia* III. 912/3 Contemporary environmentalists recognize that physical surroundings are only part of a total environment that includes social and economic factors... Their approach is probabilistic, rather than deterministic or possibilistic.

possibility. Add: **4.** Special Comb.: **possibility theorem** = **impossibility theorem.*

1950 [see **impossibility theorem*]. 1961 J. Rothenberg *Measurement of Social Welfare* ii. 24 We shall give a sketch of Arrow's proof of the General Possibility Theorem. 1964 C. E. Ferguson *Macroecon. Theory of Workable Competition* i. 10 (*heading*) The possibility theorem and rigorous proof of the competitive optimum.

possible, *a.* (*sb.*, *adv.*) Add: **2. c.** *Philos.* Logically conceivable; that which, whether or not it actually exists, is not excluded from existence by being logically contradictory or against reason. Freq. in phr. *possible world*; also *attrib.* Also in gen. use, orig. with allusion to Voltaire's *Candide* (see quot. 1759).

1738 tr. *Bayle's Gen. Dict.* VI. 674/1 That cause must also be intelligent; for this world, which actually exists, being contingent, and an infinite number of other worlds being equally possible; the cause of the world must have considered all these possible worlds to pitch upon one. *Ibid.* 674/2 It will be true still..that there is an infinity of possible worlds. 1759 W. Rider tr. *Voltaire's Candidus* i. 3 Pangloss read Lectures in Metaphisico-theologo-cosmolonigology. He demonstrated that there can be no Effect without a Cause, that in this best of *possible* Worlds, the Baron's Castle was the finest, and my Lady the best of all possible Baronesses. 1878 S. H. Hodgson *Philos. of Reflection* I. i. 79 There is then, beside our determinate world, a world indeterminate to us, but *possible* if there should be other modes of consciousness than ours, that is possible to our thought since we imagine its condition, and *actual* to those other modes, if they are actually existing. 1900 Russell *Crit. Expos. Philos. Leibniz* v. 68 It may be well, for the sake of clearness, to enumerate the principal respects in which all possible worlds agree, and the respects in which other possible worlds might differ from the actual world. 1911 G. B. Shaw *Blanco Posnet* 299 The administrative departments were consuming miles of red tape in the correctest forms of activity, and..everything was for the best in the best of all possible worlds. 1914 —— *Misalliance* p. xlii, A life's work is like a day's work: it can begin early and leave off early or begin late and leave off late, or, as with us, begin too early and never leave off at all, obviously the worst of all possible plans. 1922 tr. *Wittgenstein's Tractatus* 127 Everything which is possible in logic is also permitted. 1924 A. Huxley *Little Mexican* 166 Next to the intimate and trusted friend, the perfect stranger is the best of all possible confidants. 1926 J. B. Cabell *Silver Stallion* xxvi. 112 The optimist proclaims that we live in the best of all possible worlds; and the pessimist fears this is true. 1928 R. Lynd *Green Man* xviii. 147 It was impossible not to believe that this was the best of all possible worlds, for a world in which young men enjoy playing bad cricket is clearly a far happier place than a world in which young men would enjoy playing only good cricket. 1949 A. Pap *Elem. Analytic Philos.* ix. 177 Suppose there existed just one individual, *a*, that might be characterized by either one of the properties A, B and C. Then we can imagine the following 'possible worlds': (1) Aa. Ba. Ca [etc.]. 1966 R. F. Anderson *Hume's First Princ.* i. 3 (*heading*) Whatever is conceivable is possible. 1968 Hughes & Cresswell *Introd. Modal Logic* iv. 77 This notion of one possible world's being accessible to another has at first sight a certain air of fantasy or science fiction about it. 1973 J. J. Zeman *Modal Logic* xv. 276 The system B,..whose possible world semantics involve an accessibility which is reflexive and symmetrical but not transitive. 1977 *Canad. Jrnl. Linguistics 1976* XXI. II. 136 Now, we should be aware of the fact that the specific reading of (12) doesn't imply that the fish *a* which belongs to the set I (μ_1) of individuals in the possible world μ_1 belongs to the set I (μ_0) of individuals in the world μ_0 too.

3. (Later example.)

a 1817 Jane Austen *Northanger Abbey* (1818) II. xiv. 273 The only offence against him of which she could accuse herself, had been such as was scarcely possible to reach his knowledge.

5. Delete *rare* and add further examples.

1929 A. Huxley *Let.* 26 Aug. (1969) 317 One is at 3500 feet in a rather primitive but quite possible little hotel. 1934 H. G. Wells *Exper. Autobiogr.* I. vi. 313, I went the round of the scholastic agents,..and I answered many impossible and some possible advertisements. 1968 A. Munro in R. Weaver *Canad. Short Stories* (1968) 2nd Ser. 300 Leaving Miss Marsalles and her no longer possible parties behind, quite certainly forever.

B. *absol.* or as *sb.* **1. a.** Phr. *the art of the possible.*

The equivalent G. phr. *Die Politik ist die Lehre von Möglichen* is attributed to Bismarck (1867).

1969 D. C. Hague *Managerial Econ.* i. 12 Management, like politics, is the art of the possible. 1979 *Oxf. Dict. Quotations* (ed. 3) 84/2 Politics is the art of the possible. 1979 *Guardian* 31 Oct. 4/4 Britain's strong suit is jurisprudence. France's is the art of the possible.

b. (Later examples.)

1970 *Morning Star* 5 Mar. 2 Michael Parkinson looks at award-winning film possibles in 'Cinema' on Granada at

10.30 tonight. **1974** A. Morice *Killing with Kindness* ii. 14 We were going to have a look at some boats... He'd marked one or two possibles in the local paper.

c. (Earlier and later examples.)

1792 H. More *Lett.* (1925) 175, I thought to have sent a line to Mr. T. but I have done my *possible* in writing for to-day. **1922** E. E. Cummings *Let.* 26 Feb. (1969) 83 Dos's first words to me were a grim assurance that..his possible would be done to save The Chambre énorme from any similar fate.

d. A person who possibly may have done or may do something or attain some position; a possible candidate, member of a team, sexual partner, suspect, etc.

1915 J. Buchan *39 Steps* vii. 180 You're in no danger from the law of this land... they have dropped you from the list of possibles. **1923** *Daily Mail* 3 Mar. 13 C. L. Spackman..and H. J. Still as reserve backs are possibles. **1948** *Sporting Mirror* 21 May 13/3 Olympic 'possibles', especially those recognised to be in the first flight, are going to be in great demand everywhere. **1948** 'J. Tey' *Franchise Affair* xi. 117 He hadn't even thought of her when he sat down... She just wouldn't occur to any man as a possible. **1959** *Times Lit. Suppl.* 6 Nov. p. xx/4 Most of the presidential possibles in this year are college graduates. **1973** D. Westheimer *Going Public* ix. 136 Some files they rejected..others they read through. A considerable stack of possibles began to mount. **1975** T. Allbeury *Special Collection* i. 4 They'd spent almost a month..checking..for suitable candidates. There had been three 'possibles'.

possie, pozzy[1] (pǫ·si, pǫ·zi). *slang* (orig. and chiefly *Austral.* and *N.Z.*). Also **possy, pozzie.** [f. Pos(ition *sb.* + -y[6], -ie.] A position (orig. in sense 7c); the space a person occupies; a location; a place of residence; an appointment; an occupation. Hence as *v. trans.*

1915 T. Skeyhill *Soldier Songs from Anzac* (1916) 15 'E climbs up stunted pine-trees, An' snipes away as us. But 'e never shows 'is pozzy. **1916** *Anzac Bk.* 10 The new sniper's pozzy down at the creek. *Ibid.*, *Pozzy* or *Possie*, Australian warrior's short for 'position' or 'lair'. *Ibid.* 102 His mates used to take a mean advantage of his good nature, and would shunt all the work, such as sweeping out the 'possie', or trenches, on to him. **1918** *Chrons. N.Z.E.F.* 25 Oct. 149/2 We were 'possied' some distance in rear of the front line. **1919** *Ibid.* 10 Jan. 284/2 In the small hours we reached our next 'possie'—a shell-torn gully near Pusieux. **1919** W. H. Downing *Digger Dial.* 39 *Possy*, position; place; dugout; home. **1925** A. Wright *Boy from Bullarah* 99 Quick, get a pozzy with the machine. **1926** 'J. Doone' *Timely Tips for New Australians* Gloss., *Possie*, a slang contraction of the word position which denotes a place. A job. *c* **1926** 'Mixer' *Transport Workers' Song Bk.* 70 You'll hear him in his bar-room 'possie', As a bloke comes in for a 'wet'. **1929** A. W. Wheen tr. *Remarque's All Quiet on W. Front* v. 101 Then we change our possy and lie down again to play cards. **1934** *Bulletin* (Sydney) 3 Jan. 21/3 A bad set in a good possie will bring nothing save an occasional adventurous doe. **1937** N. Marsh *Vintage Murder* xxi. 238 Messing about on the scene of the crime... She's going to find herself in a very, very uncomfortable..pozzy, is Miss Caroline. **1941** *Coast to Coast* 204 S'pose you'll be after some easy stuff next time. A possie where there's a bit of fresh air. **1941** K. Tennant *Battlers* xxvii. 301 Going to get a possie in the cannery? **1956** D. M. Davin *Sullen Bell* 47 'I've brought a picnic,' he said. 'So you watch out for a nice little pozzy while a good husband keeps his eyes on the road.' **1962** E. Salter *Voice of Peacock* vii. 77 He couldn't have fallen into a possy like that by himself. **1970** P. White *Vivisector* 620 Should have got here early—got us a good pozzy. Never be in the picture now. **1971** *N.Z. Listener* 19 Apr. 56/5 They found a possie in a bit of a trog and boiled-up.

possie, var. *pozzy[2].

possum, *sb.*[1] Add: **1. a.** (Further examples.)

1880 'Mark Twain' *Tramp Abroad* iii. 38 He cocked his head to one side, shut one eye and put the other one to the hole, like a 'possum looking down a pig. **1905** N. Davis *Northerner* 156 Falls ate his possum with the appetite which the ride in the cold air had given him. **1932** W. Faulkner *Light in August* (1933) i. 19, I reckon it ain't any human in this country is going to dispute them hens with you, lessen it's the possums and the snakes. **1948** *Sat. Rev.* (U.S.) 29 May 4/2 The possum broke out of his cage. **1971** *Black World* Oct. 64/2, I suggested castor oil and fricassee possum in the milkshakes. **1975** E. Wigginton *Foxfire 3* 25, I wish I could get a good possum.

b. For (U.S. colloq.) read *colloq.* (orig. *U.S.*). (Earlier and later examples.)

1822 W. H. Simmons *Notices E. Florida* 40 After being severely wounded, they have been known to lie for several hours as if dead... Hence, the expression of 'playing possum' is common among the inhabitants, being applied to those who act with cunning and duplicity. **1924** [see *death-feigning* (*death *sb.* 19)]. **1949** *Time* 5 Sept. 13/1 By last week, in the Senate investigation of Washington five-percenters, it became plain that John had been playing possum the whole time. **1961** G. H. Coxe *Error of Judgment* iii. 26, I knew the only way I could beat you was to play possum, but it was a good try, kid.

c. (Further examples.)

1864 R. Henning *Let.* 27 Nov. (1966) 184 Several [aborigines] were carrying possums which they had caught on their way. **1878** *Punch* 10 Aug. 59/2 Australian beef, and kangaroos—and 'possums, wombats, and ornithorhyncusses. **1901** M. Franklin *My Brilliant Career* xxx. 258 There would be more life in trapping 'possums out on Timlinbilly. **1916** J. B. Cooper *Coo-oo-ee* xii. 174 'Tim' was pushing his cold nose into Jack's hand to coax him to open the back door to let him out to race round the hut to bark at the 'possums. **1928** 'Brent of Bin Bin' *Up Country* i. 6 'Possums when excited by the bark of dogs at the foot of trees where they have refuge will run up and down every branch in turn. **1941** I. L. Idriess *Great Boomerang* xiii. 93 His arm caught in a hollow limb reaching for possum or cockatoo's nest. **1959** *Post-Primary School Bulletin* (Wellington, N.Z.) XII. iii. 22 A 'possum is born only sixteen days after the egg starts developing within the mother's body. **1966** J. K. Baxter *Pig Island Lett.* 4 He sets his trap for possums And whistles to his dog. **1966** *New Scientist* 29 Sept. 713/3 Although still popularly known as possums to distinguish them from the opossums of the New World, the Australasian family has been given the apt name of Phalangeridae. **1970** *Southerly* XXX. 303 One night a possum broke into the kitchen, through a window left just slightly ajar. **1977** D. P. Gilmore in Stonehouse & Gilmore *Biol. Marsupials* x. 171 Today the possum is probably the most numerous mammal in New Zealand.

c*. *fig.* In various slang uses (see quots.).

1833 *N.Y. Mirror* 7 Sept. 80 *A 'possum*, the western phrase for a paltry fellow—a coward. **1900** *Dialect Notes* II. 51 *Possum*, a negro, or negress. **1943** Baker *Dict. Austral. Slang* (ed. 3) 61 *Possum*, a 'ring-in'. **1945** —— *Austral. Lang.* vi. 130 Fools of one kind and another.. *flathead, possum, gammy*, [etc.]. *Ibid.* vii. 138 Thieves are described variously as..*dwelling dancers, stoops* and *possums*. *Ibid.* 142 *Jay* and *possum*, a trickster's victim.

c.** *like a possum up a gum-tree*: contented; (see also quot. 1898); *to stir* (or *rouse*) *the possum*: to stir up controversy, to liven things up. *Austral. colloq.*

See also Gum-tree 2 in Dict. and Suppl.

1898 *Bulletin* (Sydney) 17 Dec. (Red Page), 'Like a possum up a gum tree' is not bad to express quickness or cleverness in doing anything. **1907** C. MacAlister *Old Pioneering Days in Sunny South* 51 Sometimes..an ambitious carrier or drover would 'rouse the 'possum' by giving some long-winded ditty of the time. **1908** E. S. Sorenson *Squatter's Ward* 144, I mean to stir the 'possum in Sultan Susman from this out. **1941** Baker *Dict. Austral. Slang* 56 *Like a possum up a gumtree*, completely happy, in the best of spirits and contentment. **1958** 'W. Henry' *Seven Men at Mimbres Springs* (1960) viii. 94 They made 'more racket about it than six pickaninnies with a possum up a gum tree'. **1972** *Sydney Morning Herald* 31 Oct. 3 'I could be sitting in Parliament now without any great cost provided I forgot this idea of stirring the possum,' he said. **1976** *Courier-Mail* (Brisbane) 27 Mar. 18/2 Mr. Bob Hawke sees an opening to stir the possum and to step up union wage demands.

d. Also *possum beard, hunt* (also as *vb. intr.*), *hunting, rug* (earlier and later examples), *scalper, skin, snare, token*; **possum belly** *U.S. slang* (see quots.).

1928 'Brent of Bin Bin' *Up Country* xvi. 273 His 'possum beard and his slouch hat, out of which his mild blue eyes looked with incontestable simplicity. **1926** Maines & Grant *Wise-Crack Dict.* 12/1 *Possum belly*, tent stake box carried under circus railroad cars. **1939** P. A. Rollins *Gone Haywire* 66 There was a sufficient supply of firewood in the 'cooney' or 'possum belly' (a baggy, dried cowhide fastened horizontally beneath the wagon box and used for carrying a reserve of fuel). **1973** *Amer. Speech* 1969 XLIV. 207 *Possum belly*, livestock trailer with a drop frame to haul small animals underneath heavy cattle. **1841** *Spirit of Times* 17 July 235/1 A 'possum Hunt. **1900** *Congress. Rec.* 11 Jan. 784/1, I used to 'possum hunt. **1949** *Natural Hist.* May 223/1 According to song and story, most 'possum hunts end at the foot of a 'simmon tree. **1976** C. S. Brown *Gloss. Faulkner's South* 154 An axe is a regular part of a possum hunt. **1840** *Southern Lit. Messenger* VI. 784/1 He is fond of possum, rabbit, and coon-hunting. **1873** J. H. H. St.John *Pakeha Rambles through Maori Lands* vii. 128 With a blanket, or better still, a 'possum rug,..the traveller may jog along very comfortably. **1942** C. Barrett *On Wallaby* iii. 36 'Fifty quid,' he said reflectively. 'More than a possum rug's worth.' **1946** F. D. Davison *Dusty* 164 He was a possum scalper, now. **1911** C. E. W. Bean *'Dreadnought' of Darling* xxix. 249 They did..sometimes sew themselves sort of elementary clothes out of possum skins threaded together with kangaroo tendons. **1966** 'J. Hackston' *Father clears Out* 13, I remembered the honey and the fruit, the rabbits, the possum-skins, the money—all the help my little world had given me. *Ibid.* 8 Before the dawn I went round possum snares. **1961** B. Crump *Hang on a Minute* 145 Well, a man could sell deerskins and possum tokens.

Possum (pǫ·sǝm), *sb.*[2] Also **possum.** [f. the initial letters of *p*atient *o*perated *s*elector *m*echanism, after Possum *sb.*[1]] A proprietary name for any of various electronic devices, operated in different ways, which enable disabled persons to operate or control domestic fittings, machines, or other equipment.

1961 *New Scientist* 8 June 561/1 The basis of the device, which is known as 'Possum', is that a small steady suck or blow can operate a sensitive switch which, in turn, operates a rotary switch connected to a grid. The grid sets out letters, numbers and punctuation symbols. **1961** *Trade Marks Jrnl.* 12 July 918/1 *Possum*... All goods included in class 10 [*i.e.* surgical, medical, dental and veterinary instruments and apparatus, including artificial limbs, eyes, and teeth]. Reginald George Maling,..Aylesbury, Buckinghamshire; merchant. **1972** *Daily Tel.* 31 Oct. 18 Electronic devices with which severely disabled people can operate typewriters and other machines by mouth, finger or toe, possums..enable the handicapped to lead useful lives. **1974** *Listener* 25 Apr. 525/3 The Possum machine with which, by sucking and blowing into a tube, she can type, telephone, open the front door..or buzz for help. **1976** *Responaut* Autumn 13/2 My window on the world, apart from my eyes, is 15 inches by 7, the size of the tilting mirror attached to my respirator. Reflected through this I can see my two Possum indicators revealing the 21 electrical devices I am able to control.

possum, *v.* Add: Also *Austral.* **1.** (Further examples.)

1846 R. Levinge *Echoes from Backwoods* II. 32 'Possuming is become an idiom; a term signifying any one who is humbugging or deceiving. **1862** *Harper's Mag.* Dec. 99/2 So you see you must endure it to the end—fur thar's no possumin' thar. **1912** W. H. Thomas in J. F. Dobie *Rainbow in Morning* (1965) ii. 4 Now when that nigger comes to, if she's been possumin', she sho' will be hungry.

b. *trans.* To feign, to simulate.

1853 J. G. Baldwin *Flush Times Alabama* 150 All this time I was possuming sleep..as innocent as a lamb.

2. (Further examples.)

1933 *Bulletin* (Sydney) 4 Jan. 11 A short-sighted young man went out 'possuming. **1942** C. Barrett *On Wallaby* iii. 36 He may have done a bit of cyaniding and out of season possuming.

possy, var. *possie, pozzy[1].

post, *sb.*[1] Add: **I. 4. c.** (Further *fig.* examples.) Phr. *first past the post*, used *attrib.* and *absol.* to designate the electoral system whereby the candidate with the largest number of votes, or the party with the largest number of seats, wins an election; *to pip on* (or *at*) *the post*: see *pip *v.*[3] 1 c.

1818 C. Grenville *Let.* 19 Dec. (1920) 228 The 2nd Miss Morgan *expects* to marry Lord Rodney, if he does not again *jib at the Post.* **1921** E. O'Neill *Emperor Jones* i. 161 Den de revolution is at de post. **1935** 'N. Blake' *Question of Proof* x. 197 'After all,' he continued, 'the Business-As-Usual slogan gets the British middle-class where they live—it has just the right combination of backs-to-the-wall bulldog courage and commercial *savoir-faire*. In this case it will leave its only rival—the respect-for-the-dead ballyhoo—at the post.' **1949** J. D. Carr *Below Suspicion* xvii. 208 'One of the things I like about you,' commented Butler,..'is the pellucid clarity of your style. Addison is nowhere. Macaulay is left at the post.' **1952** L. Overacker *Austral. Party System* viii. 221 At that time the 'first past the post' system of election was in use. **1958** C. P. Snow *Conscience of Rich* v. 37 In strength of character we were about the same. In everything but natural gifts, he had so much start that I was left at the post. **1965** *Austral. Encycl.* III. 367/1 In 1892, Queensland became the Australian pioneer of one of the chief improvements on first-past-the-post, namely the alternative or contingent vote. **1966** *Encycl. N.Z.* I. 864/2 The..first-past-the-post electoral system. Under this system, which has been in operation for most of this century, minor parties are crushed. **1976** *Times* 20 Aug. 13/1 The existing electoral system, based on the 'first past the post' principle which has shown itself to be so anomalous at Westminster.

e. A goal-post.

1867 [see *poster[2]]. **1878** *Chambers's Encycl.* IV. 414/1 He will touch it down as near as he can to the goal, if possible between the posts. **1880** *Times* 15 Mar. 6/5 For some little time after this the English kept play in close proximity to their rivals' posts, causing the goal-keeper some anxiety. **1900** A. E. T. Watson *Young Sportsman* 284 *Poster*,..a place kick which..would have hit the posts produced upward and rebounded into the field of play. **1972** G. Green *Great Moments in Sport: Soccer* xviii. 156 It ended with Nordahl turning Puis's chip to the near post against Wilson's upright, with the goalkeeper helpless. **1978** *Rugby World* Apr. 7/3 The Scots would..have been awarded a penalty try.., with the conversion being taken from in front of the posts, instead of from the more difficult position farther out.

f. A leg of a chair. *U.S.*

1902 W. N. Harben *Abner Daniel* 202 Something like a groan escaped Bishop's lips as he lowered the front posts of his chair to the floor.

IV. 8. b. *post and pan* (further examples); *post and panel* (further example). **j.** *post and beam*: applied to a mode of construction in which the framework consists of upright and horizontal beams.

b. 1954 S. Piggott *Neolithic Cultures Brit. Isles* vi. 163 with the façade formed by orthostats ascending in height to the portals and originally linked by dry-stone walling in a 'post and panel' technique. **1975** *Country Life* 6 Feb. 319/3 Black and white timber and plaster work of the post-and-pan variety. **j. 1958** *Listener* 25 Sept. 459/1 The other structural method is the application of the simple post-and-beam technique to form a framed structure similar to that obtained by steel or reinforced concrete. **1978** *N.Y. Times* 30 Mar. c 8/2 The designer, Donald Davidson, assembled cedar slats post-and-beam style to make an armchair.

V. 9. *post-maker*; **post-mill** (later examples); **post-retained crown** *Dentistry* = *post crown; **post-sitter** *Austral.* = *post-boy 3; **post time** *N. Amer.*, the starting time for a horse-race; **post-windmill** = *post-mill*.

1845 Thoreau *Jrnl.* 14 July in *Writings* (1906) VII. 365 A woodchopper, a post-maker. **1934** *Archit. Rev.* LXXVI. 165/3 The Post mill is the earliest known form of mill. The structure is box-like in shape and carries the machinery and the sails. Supporting this structure is a single upright post on which the mill revolves. **1968** J. Arnold *Shell Bk. Country Crafts* 169 The oldest mills, those in existence at the time of the Domesday book, were all post-mills. **1974** C. Taylor *Fieldwork in Medieval Archaeol.* 12 The circular mound on a hilltop may be a Bronze Age barrow, or it may be the base of a medieval post-mill, or it may be both. **1963** C. R. Cowell et al. *Inlays, Crowns, & Bridges* viii. 84 A post-retained crown is commonly indicated for a root-filled anterior tooth the natural crown of which has become discoloured. **1974** C. L. Sturridge in Harty & Roberts *Restorative Procedures Practising Dentist* ix. 141 In the front of the mouth

a post-retained crown will be the treatment of choice if the tooth is non-vital. **1901** A. J. Campbell *Nests & Eggs Austral. Birds* I. 106. The Brown Flycatcher or 'Post Sitter'..begins to breed [in] September or October. **1911**, etc. Post-sitter [see *post-boy 3]. **1941** *Sun* (Baltimore) 30 Aug. 13/1 Everything is in readiness for the opening of business about an hour before post time tomorrow. **1968** *Globe & Mail* (Toronto) 17 Feb. 42/1 (Advt.), Post time 7:45. **1931** *Times Educ. Suppl.* 19 Dec. (Home & Classroom Suppl.) p.iv/2 A Cambridgeshire post-windmill.. revolves in an artificial breeze to show wind-power. **1974** C. Taylor *Fieldwork in Medieval Archaeol.* vi. 119 A circular mound, discovered on the ground or from air photographs, can be proved to have been the site of a post-windmill if an old estate map depicts a windmill there.

post, *sb.*[2] Add: **I. 5. b.** (Further examples.)

1785 J. Woodforde *Diary* 11 Nov. (1926) II. 214, I.. put it [*sc.* a letter] into the Post myself. **1835** Dickens *Let.* 4 May (1965) I. 60, I am in great haste having scarce time to get this letter in the Post. **1887** W. B. Yeats *Lett.* (1954) 54, I must finish to catch the post. **1921** G. B. Shaw *Back to Methuselah* ii. 40 Excuse me, sir; but the letters must go to catch the post.

III. 8. c. (Earlier and later examples without the definite article.)

1789 J. Woodforde *Diary* 3 July (1927) III. 118 Received a Letter..desiring an answer by return of Post. **1792** F. Burney *Jrnl.* Aug. (1972) I. 225, I wrote her my good wishes, which she answered by return of post. **1980** Alexander & Anand *Queen Victoria's Maharajah* x. 183 The Maharajah replied by return of post.

V. 12. b. *post-carriage* (later example), *-girl*, *-packet*. **d.** *post-time* (earlier examples).

1872 *Argosy* XIV. 208 There was no railroad then. The ladies and the girls crammed themselves into a post-carriage from the Star. **1850** C. M. Yonge *Henrietta's Wish* v. 55 The post girl could take the jelly. **1944** *Coast to Coast 1943* 112 Living only for the next time the postgirl's whistle sent its shrill stab through her nerves. **1977** 'J. Gash' *Judas Pair* ix. 113 There were a couple of letters..on the doormat, so the post girl had called. **1819** Keats *Let.* 12 Mar. (1958) II. 71 The sail of the Post-Packet to new york or Philadelphia. **1772** J. Wedgwood *Let.* 1 Sept. (1965) 133, I have been so long in these and other particulars this morning that post time is at hand. **1836** F. Witts *Diary* 14 July (1978) 117 B., as usual, not appearing till nearly the luncheon hour and at post time, when I received a joint letter from my wife and Edward.

13. post-box (later examples); (*b*) a box to which post-office mail, newspapers, etc., are delivered; (*c*) any box where papers, etc., are left for collection; **post-bus**, a post-office vehicle which also carries passengers; also *attrib.*; **post-lady** = *post-woman*; **post-paid** *a.* (earlier and later examples); also *fig.*; **post-rider** (earlier example); **post-village** (earlier examples).

1954 J. Collin-Smith *Scorpion on Stone* ii. 42, I say! I've been to the post-box. My papers have come. **1955** *Times* 3 May 5/4 Posting the letters at different post-boxes on the way. **1960** G. Martelli *Agent Extraordinary* v. 85 The *réseau* was entirely self-sufficient... There were no parachute drops, no wireless transmitters, no system of internal couriers or 'post-boxes'. **1963** *Times* 18 May 8/5 There were four other explosions in Westmount and one in the suburb of Pointe aux Trembles during the night, all in post boxes. **1964** L. Linton *Of Days & Driftwood* iv. 27 The car was ready and able to take us beyond the post box for the first time in many days. **1968** 'S. Jay' *Sleepers can Kill* xv. 149 It's a post-box. The agents can deliver reports there, someone picks them up and sends them on. **1978** G. Greene *Human Factor* III. iii. 126 Muller was on his own in a strange town, in a foreign land, where the post boxes bore the initials of a sovereign E II. **1960** *Guardian* 22 Feb. 6/4 Trains connect with post-bus services. **1968** A. Marin *Clash of Distant Thunder* (1969) x. 80 'How did you get to Geneva?'.. 'By post bus from Bourg,' I said. **1972** *Times* 24 Oct. 5/4 Crundale, Kent... At 6 am the new Royal Mail Post bus began its morning run. **1975** *Scottish Field* Apr. 88/3 The inauguration of the 50th Scottish postbus is a milestone in an inspired scheme by the postal service to help those who live out in the wilds. **1975** *Oxford Times* 25 July 18/5 (*heading*) Postlady is dog's best friend!.. Mrs Kathy Hilsdon,..a postwoman for nearly 17 years. **1979** *Guardian* 30 Mar. 2/7 Elstead's postlady Mrs Pam Moss is confronted with one of the giant house numbers. **1653** T. Bateman *Let.* 13 Dec. in H. Ellis *Orig. Lett. Illustr. Eng. Hist.* (1827) 2nd Ser. III. 373 Post payd. **1689** in *12th Rep. R. Comm. Hist. Manuscripts* App. vii. 265 in *Parl. Papers 1889* (C. 5889) XLV. 533 Cannon bullets flew as fast as you could count them, and as soon as we took up their bulletts we sent them back again post paid. **1708** *Boston News-Let.* 11 Oct. 4/2 Whereas several persons do write upon their Letters Post paid ..without ever paying the Postage of the said Letters. **1762** Goldsmith *Life R. Nash* 117 This description.. must be sent in a letter post-paid. **1814** *Niles' Reg.* V. 369/1 Letters to the editor must be post-paid. **1926** *Scribner's Mag.* Sept. 24/1 (Advt.), Italian tooled cigarette-cases are smart and light... Colors: brown, dark red and dark green. Postpaid. **1973** *Sci. Amer.* Oct. 118/3 The Math Shop..will supply postpaid (on prepaid orders) 100 plastic cubes. **1976** *Physics Bull.* Feb. 69/1 (Advt.), Price £6.50 postpaid. **1705** *Boston News-Let.* 19 Nov. 2/2 Strayed..a sorrel Mare... Whoever can give any true intelligence of her to..the Post-rider..shall be sufficiently Rewarded. **1827** A. Sherwood *Gazetteer Georgia* p. v, Post Village. **1847** H. Howe *Hist. Coll. Ohio* 264 Allensville, Middleton, Oak Hill and Charleston are small post villages.

post, *sb.*[3] **2. d.** *post-trader* (earlier examples); **post exchange** *U.S.*, a shop at a military post where goods and services are available to military personnel and authorized civilians.

1871 *Republican Rev.* (Albuquerque, New Mexico) 1 Apr. 2/1 Indians stole Levinsky's buggy horses from the Post trader's corral. **1873** J. H. Beadle *Undevel. West* xxv. 525 Mr. Lionel Ayres fills the position of Post Trader. **1892** *Ann. Rep. Secretary of War* (U.S.) I. 57 In February last, upon the ground that the term 'canteen' possibly conveyed to the public mind a meaning which, though foreign to the main purpose of the institution, has been for years associated in other armies with a place of conviviality and dissipation, the Secretary of War decided to change the name of such establishments to that of 'post exchange'. **1919** *Lit. Digest* 22 Nov. 70/2 The Y.W.C.A. hostess house has been turned into a post exchange. **1973** H. Gruppe *Truxton Cipher* iii. 25 He added that he intended to remain in the Naval Reserve so as to retain his..post-exchange privileges.

post, *sb.*[5] **1.** Delete ?*Obs.* and add further examples.

1906 R. W. Sindall *Paper Technol.* 21 The 'coucher', who transfers the wet sheet from mould to felt and builds up the pile or 'post' of alternate wet sheets and felts. **1965** Zigrosser & Gaehde *Guide to Collecting Orig. Prints* iv. 64 When 144 sheets [of paper] have been formed, they and their protective pads (the stack being known as the *post*) are conveyed to a press to squeeze out more water.

post (pōust), *sb.*[11] U.S. slang abbrev. of Post-graduate *sb.*

1900 *Dialect Notes* II. 51 *Post, n.*,.. 2. A post-graduate student. **1914** *Ibid.* IV. 134 It must be nice to be a *post*,—they have so many privileges. **1930** *Amer. Speech* V. 242 *Post*, post graduate.

post, *sb.*[12] *colloq.* (chiefly *U.S.*). [Abbrev. of Post mortem, post-mortem *sb.*] An autopsy, post-mortem. Hence as *v. trans.*, to perform an autopsy on (someone).

1942 Berrey & Van den Bark *Amer. Thes. Slang* § 534/2 *Post*,..post-mortem examination. **1961** *Amer. Speech* XXXVI. 145 The patient died last night and will be posted this morning. **1968** J. Hudson *Case of Need* i. v. 41 The post hadn't been started. **1969** 'F. Richards' *Risky Way to Kill* (1970) xii. 147 She died last night. Overdose, probably. They're doing a post. **1979** R. Cook *Sphinx* 177 They had no internal organs. Just a shell of a body. When a post is done the shell is only cursorily examined.

post, *v.*[2] Add: **4.** (Further example.)

1975 *Publishers Weekly* 10 Feb. 45/1 This poster was mailed to ABA members in the hope that they will post it.

5. e. To achieve, 'notch up'. *N. Amer.*

1949 *Richmond* (Virginia) *Times-Dispatch* 10 Oct. 13/5 William and Mary, which Saturday posted a 54-6 decision over the Keydets to tie North Carolina for the conference lead (each has a 2-0 record), has one remaining State battle. **1968** *Globe & Mail* (Toronto) 5 Feb. 18/9 John Armstrong of Oshawa posted the longest jump of 110 feet on his way to second place. **1972** *Time* 13 Mar. 48/3 In 1944 he [*sc.* a basketball pitcher]..posted the lowest earned run average in the major leagues. **1973** *Internat. Herald Tribune* 15 June 15/3 Wise, posting his eighth victory against three defeats, struck out four, walked three and retired 14 consecutive batters at one stretch. **1975** *New Yorker* 23 June 43/1 He won nineteen games for the Pirates and lost only eight, posting an earned-run average of 2.48.

f. To announce, publish. *N. Amer.*

1961 *Los Angeles Times* 21 June iv. 6/6 Gains of 2¾ were posted for Teleprompter and Republic Foil. **1962** *Economist* 19 May 697/2 Producer governments certainly cannot get better prices for their crude oil than those 'posted' by the companies controlled by the international groups there. **1973** *Time* 25 June 23/4 Companies that posted big price increases during Phase III will be audited. **1976** *Billings* (Montana) *Gaz.* 17 June 7-E/6 The stock market shook off Tuesday's spell of profit taking Wednesday and posted a modest gain in moderately active trading.

6. (Later U.S. examples.)

1967 *Boston Sunday Herald* 26 Mar. ii. 9/1 Highway arteries have been posted, warning us that stiff fines will be imposed if we toss our leavings out of the car windows. **1976** *Billings* (Montana) *Gaz.* 30 June 8-A/1 We have posted all the bars and put up signs that make it clear that no drinking is allowed in any public place.

post, *v.*[5] Add: Also, *spec.* of bail money.

1974 *Observer* 7 Apr. 4/8 Immediately after posting five million francs..bail money..he took a private plane home from Geneva. **1974** *Progress* (Easley, S. Carolina) 24 Apr. 10/1 Arrested and charged with illegal possession and sale of piranha, the dealer posted bond and awaits trial which should come this week. **1978** S. Brill *Teamsters* vi. 223 The other defendants flew in from around the country to plead innocent and post bail.

post, *Latin preposition.* Add: **post bellum** (earlier and later examples). Also *fig.*

1874 *Southern Mag.* XIV. 37 It [*sc.* Atlanta] looks so little like a post-bellum town. **1920** *Czecho-Slovak Trade Jrnl.* Apr. 5 The post-bellum difficulties..do not yet permit her to throb with full vigour and strength. **1940** Beerbohm *Mainly on Air* (1946) 97 The future, the post-bellum period, is to be perfectly splendid. **1974** 'M. Innes' *Appleby's Other Story* xvi. 127 A post-bellum relationship. .. That was it. Mr Charles Carter..was no longer Miss Kenterell's quarry.

post coitum, after sexual intercourse; also as *sb.*; *spec.* with allusion to the proverb *post coitum omne animal triste est* (and variants) 'after sexual intercourse every animal is sad'. Also *transf.*

The phrase as such does not occur in classical Latin, but cf. [Aristotle] *Problems* 877 b 9 διὰ τί οἱ νέοι, ὅταν πρῶτον ἀφροδισιάζειν ἄρχωνται, οἷς ἂν ὁμιλήσωσι, μετὰ τὴν πρᾶξιν μισοῦσιν; 'Why do young men, on first having sexual intercourse, afterwards hate those with whom they have just been associated?'; Pliny *Nat. Hist.* x. lxxxiii. homini tantum primi coitus paenitentia 'man alone experiences regret after first having intercourse'.

1762 Sterne *Tr. Shandy* V. xxxvi. 126 The oily and balsamous parts are of a lively heat and spirit, which accounts for the observation of Aristotle, '*Quod omne animal post coitum est triste.*' [**1920** A. Huxley *Leda* 34 Some..Mount up on wings as frail and misty As passion's all-too-transient kiss (Though afterwards—oh, *omne animal triste*!)] **1928** *Jrnl. Morphol.* XLVI. 171 Embryos with six to ten somites are to be expected in the opossum about 8½ days post coitum. **1933** M. Lowry *Ultramarine* iii. 158 Whether life was worth living or not was a matter for an embryo rather than a man. 'Post coitum omne animalia triste est. Omne? Supinus pertundo tunicam. **1959** *Listener* 15 Oct. 651/2 A remorseful attack of *post coitum triste.* **1966** C. M. Bowra *Memories* xi. 246 Once Clark was visiting a college farm, and the party witnessed a bull servicing a cow. Clark..said, 'Blakeway, *omne animal post coitum triste.* There was, Blakeway, a firm of solicitors in London called Mann, Rogers, and Greaves.' **1967** *Listener* 12 Jan. 71/2 Elizabeth Sellars.. had to..register such stock emotions as there was time for—fear (lover dead *post coitum*?) and gratification (*post coitum* and dialogue apparently working). **1975** M. Bradbury *History Man* i. 6 He is in that flat state of literary post coitum that affects those who spend too much time with their own lonely structures and plots. **1979** *Guardian* 13 June 9/5 If you think there's no post coitum triste in angling, try chatting up a tench fisherman after he's spent a night by the lake.

post eventum = Post factum.

1846 Geo. Eliot tr. *Strauss's Life of Jesus* III. 166 Thus renouncing what is narrated..as composed *post eventum.* **1920** *Glasgow Herald* 13 July 6 Mr Asquith's post-eventum reproofs..leave us cold when we recall that ..he and his colleagues might have done much to make the preaching of their historic dogma on retrenchment an effective feature. **1958** *Listener* 25 Dec. 1092/2, I should have liked to put my oar, *post eventum*, into the argument. **1961** J. B. Wilson *Reason & Morals* i. 8 What seems to happen is that the philosopher arrives on the scene several centuries too late, and explains the advance of knowledge *post eventum.*

post festum = prec. (In quot. 1966 lit. 'after the festival'.)

1887 Moore & Aveling tr. *Marx's Capital* I. i. 47 He begins, post festum, with the results of the process of development ready to hand before him. **1935** H. Straumann *Newspaper Headlines* ii. 55 It must be kept in mind that it is an interpretation *post festum.* **1958** W. Stark *Sociol. of Knowl.* 193 They [*sc.* derivations] are merely the accompaniments of action..the *post festum* rationalizations of human conduct which, in itself, is anything but rational. **1966** *New Statesman* 11 Nov. 718/3 The reference is to a mate in 17 which the problem's author very kindly meant to mark the 17th anniversary of this column some months ago... So let's 'celebrate' it, even though a bit *post festum.* **1970** B. Brewster tr. *Althusser & Balibar's Reading Capital* (1975) ii. v. 122 It begins, *post festum*, with already established givens.

post partum (earlier and later examples); *post-partum depression* = *postnatal depression* s.v. *postnatal *a.* Also *fig.*

1844 *Medico-Chirurg. Rev.* XLI. 267 On using the catheter, 36 hours post partum, Dr. Crosse found the uterus inverted in the vagina. **1846** *Northern Jrnl. Med.* IV. 1 (*heading*) Some suggestions regarding the anatomical source and pathological nature of post-partum hemorrhage. **1911** R. Jardine *Delayed & Complicated Labour* xiv. 185 If the third stage of labour is properly conducted, post-partum hæmorrhage can be largely prevented. **1929** *Amer. Jrnl. Psychiatry* VIII. 767, I didn't study the infanticidal impulses of many women, because these impulses are more prominent in post-partum depressions. **1957** A. Guttmacher *Pregnancy & Birth* xvi. 250 'Post partum blues', beginning a day or two after delivery and lasting several days, are frequently encountered... Your doctor..can reassure you he has met postpartum blues..in people whose depression cleared up. **1959** N. Mailer *Advts. for Myself* (1961) 244, I was beginning to feel the empty winds of a post-partum gloom. **1974** *Publishers Weekly* 18 Mar. 44/3 The scary-funny business of bearing a baby and coping postpartum. **1976** *Dissertation Abstr. Internat.* B. XXXVI. 5793/1 Hostile attitudes toward mother figures predispose some women toward experiences of postpartum depression. **1977** *Lancet* 28 May 1126/1 A 31-year-old woman, 6 months post-partum, complained of fatiguability.

6. With English words and phrases. [Cf. *post- B. 1 d.]

Usu. found in contexts where *after* would be equally appropriate and more agreeable.—Ed.

1965 *Listener* 16 Sept. 432/3 *Der Ferne Klang* is post-Wagnerian, and post just about everything else that was happening at the turn of the century. **1973** *Nature* 26 Jan. 273/1 Medium was replaced two days post plating and the number of foci determined on the third day. **1974** *Daily Tel.* 7 Jan. 13/3 Now, post the increase [in the price of oil],..future gold price prospects far outweigh individual share fundamentals. **1979** *Ibid.* 19 July 21/4 Post the Geneva meeting of Opec the OECD reckons that its 24 member countries..can expect average economic growth of only two p.c. over the next 12 months.

post-, *prefix.* Add: **A. 1. a.** (*a*) with verbs or pa. pples. *post-process* vb.; *post-stressed* ppl. a.; also, in nonce-wds. formed after verbs or pa. pples. in *ante-*: as *posticipated* (opp. to *anticipated*). **post-acce·lerate** *v. trans. Electronics*, to accelerate (an electron or electron beam) after it is deflected in a cathode-ray

tube; so **post-acce·lerating** *ppl. a.*; **postmultiply** *v.* (examples); **post-o·smicate** *v. trans. Biol.*, to postfix with a solution of osmium tetroxide; **post-preci·pitate** *v. trans.*, to deposit by post-precipitation (see b); *intr.*, to be so deposited; **post-produ·ce** *v. trans.*, to subject (film) to post-production (see b).

b. In *quasi*-adjectival relation to a sb.: *post-processing, -stressing.* In nonce-words formed after nouns in *pre-*: *postface* (later example), *postference, postmonition.* Also **post-accelera·tion** *Electronics*, in a cathode-ray tube, acceleration of the electron beam after its deflection; also **post-acce·lerator**; **post-ge·nitive** *Gram.*, a possessive noun following the noun it qualifies; cf. *post-possessive* below; **post-heat·ing** *vbl. sb.*, the heating of metal after welding, in order to relieve stresses; **postmultiplica·tion** *Math.*, multiplication by a postfactor; **postosmica·tion** *Biol.*, postfixation with a solution of osmium tetroxide; **post-posse·ssive** *Gram.*, a possessive pronoun following the noun it qualifies; cf. *post-genitive* above; **post-precipitation** *Chem.*, precipitation of a compound spontaneously following that of another for the same solution; **post-pro·cessor** *Computers* (see quot. 1977); **post-produ·ction**, film-production effected after the completion of shooting.

1946 *Proc. IRE* XXXIV. 433/2 The intensifier electrode or electrodes, sometimes called post-accelerating electrodes, provide acceleration after deflection. **1971** Klemperer & Barnett *Electron Optics* (ed. 3) x. 397 Electrons may be deflected while they are of very small velocity, and they may be 'post-accelerated' later on to whatever speed is required. *Ibid.*, The spot size should not be increased by the post-accelerating field. **1940** *Philips Technical Rev.* V. 245 (*heading*) A cathode ray tube with post-acceleration. *Ibid.* 249/1 The strength of the lens increases with increasing post-acceleration voltage. **1971** Klemperer & Barnett *Electron Optics* (ed. 3) x. 399 The best post-acceleration is effected by means of fine meshes. **1956** *Proc. IRE* XLIV. 665/2 Conventional post-accelerators, also called intensifiers, operate with conductive coatings of band or spiral shape, plated on the inside of the bulb. **1959** Rider & Uslan *Encycl. Cathode-Ray Oscilloscopes* (ed. 2) i. 20/1 The beam..traverses the gap between the accelerator and post-accelerator areas. **1974** *Nature* 20 Sept. 262/1 In the postface (this must be one of the few nearly Latin words invented in the twentieth century!), Eugene Skolnikoff outlines the increasing involvement..of MIT in the field of science and public policy. **1877** Swinburne *Let.* 21 Apr. (1960) III. 326 To compare either with Shelley or Hugo for preference or postference, is purely absurd. **1922** E. Kruisinga *Handbk. Present-Day Eng.* (ed. 3) ii. i. 361 Nouns preceded by a definite article are not seldom used with a post-genitive. **1957** R. W. Zandvoort *Handbk. Eng. Gram.* ii. ii. 105 The only case where an English genitive may be said to follow its headword is when it is the principal part of an *of*-adjunct to a preceding noun... The construction is known as the *post-genitive.* I gave him an old raincoat of my brother's. **1938** *Times* 4 Feb. 11/2 The post-heating process greatly improves an already satisfactory weld by allowing the metal at the weld to recover its original structure after the severe treatment given by the welding operation. **1966** C. R. Tottle *Sci. Engin. Materials* x. 224 Temperature gradients are reduced ..in welding by preheating the parent metal or post-heating immediately after welding. **1922** Joyce *Ulysses* 663 The anticipated diamond jubilee of Queen Victoria ..and the posticipated opening of the new municipal fish market. **1938** S. Beckett *Murphy* ix. 176 In the morning nothing remained of the dream but a post-monition of calamity. **1862** *Phil. Trans. R. Soc.* CLI. 312 Every matrix of the type *n*×*n* is equivalent (by post-multiplication) to one, and only one, of the reduced matrices included in the formula (62.). **1968** Fox & Mayers *Computing Methods for Scientists & Engineers* v. 100 For the general eigenvalue problem we can perform virtually the same operation, though with post-multiplication included to preserve similarity. **1862** *Phil. Trans. R. Soc.* CLI. 312 These numbers will remain unchanged, when the given matrix is premultiplied by any unit-matrix, and post-multiplied by any matrix whatsoever. **1939** *Brit. Jrnl. Psychol.* XXIX. 302 After post-multiplying..by *ℓ*′ we get..[etc.]. **1978** *Nature* 20 Apr. 740/1 We are told, as though it were surprising, that pre- and post-multiplying a matrix by the matrix of its eigenvectors produces a diagonal matrix. **1965** *Jrnl. R. Microsc. Soc.* LXXXIV. 129 Sucrose was first used with formaldehyde in fixation for electron microscopy by Holt and Hicks.., who postosmicated the tissues that had been treated this way. **1971** *Nature* 2 Apr. 334/2 Random pieces of grossly normal thyroid tissue..were diced.., fixed in 1·5% glutaraldehyde.., post-osmicated, dehydrated and embedded in 'Araldite 502'. **1963** *Jrnl. Cell Biol.* XVII. 54/2 The final dense product could be easily referable to fine structure in sections of material embedded without postosmication. **1943** *Eng. Stud.* XXV. 103 The construction called by.. Curme 'double genitive', by Kruisinga 'post genitive', and, in the case of a pronoun (*a friend of mine*), 'post-possessive'. **1957** R. W. Zandvoort *Handbk. Eng. Gram.* iii. ii. 140 The construction may be denoted as the post-possessive. a. I gave him an old raincoat of mine. He hated that pride of hers. b. It was no fault of theirs. **1936** Kolthoff & Sandell *Textbk. Quantitative Inorg. Analysis* viii. 105 This second phase is therefore not coprecipitated but postprecipitated. **1939** A. I. Vogel *Text-bk. Quantitative Inorg. Analysis* i. 148 Zinc sulphide is slowly post-precipitated. **1960** Belcher & Nutten *Quantitative Inorg. Analysis* (ed. 2) x. 61 Examples of precipitates which tend to post-precipitate are zinc sulphide on mercury sulphide, and magnesium oxalate on calcium oxalate. **1932** I. M. Kolthoff in *Jrnl. Physical Chem.* XXXVI. 861 First of all, calcium oxalate precipitates and then on standing magnesium oxalate crystallizes out slowly. Therefore, we are not dealing here with a case of coprecipitation, but of post-precipitation, the crystals of calcium oxalate being not at all or only slightly contaminated by magnesium. **1963** G. Svehla tr. *Erdey's Gravimetric Analysis* i. iii. 183 Tin(IV) sulphide tends to carry down nickel, cobalt and iron ions from a hydrochloric acid containing solution by post-precipitation. **1977** *Sci. Amer.* June 9/2 (Advt.), With the HP 5420A, you can post-process measurement results using the four basic arithmetic functions. **1966** C. J. Sippl *Computer Dict. & Handbk.* 278 The real-time system relieves the larger system of time consuming input and output functions as well as performing pre-processing and postprocessing functions, such as validity editing and formatting for print. **1967** *Economist* 12 Aug. 588/2 Molins, Ferranti and IBM have told the Ministry of Technology that they can jointly produce a post-processor for the Molins machine within twelve months flat. **1977** *Gloss. Terms Data Processing* (*B.S.I.*) vii. 2/1 *Postprocessor*, a computer program that effects some final computation or organization. **1976** *Broadcast* 12 Jan. 1/1 Bring in your film and we'll post-produce it on tape. *Ibid.* 1/2 Spend a day on post-production using time code computer editing. **1953** K. Reisz *Technique Film Editing* i. 49 A post-production break-down of the finished sequence. **1953** *Archit. Rev.* CXIII. 377/2 (*caption*) Auditorium spanned by post-stressed beams. **1965** *Language* XLI. 473 As many as five syllables occur prestressed, but only one syllable occurs post-stressed, except rarely when a postclitic follows a suffix. **1941** *Concrete & Constructional Engin.* XXXVI. 93/1 Dr. Abeles suggests stretching the hard steel wires after hardening and setting, which he calls 'post-stressing'. **1976** *Offshore Engineer* Mar. 26/3 The Kishorn structure departs from its smaller brothers by having a two-stage wall poststressing. The first stage will be stressed at the 15m high level, the remainder when the walls reach full height.

B. 1. a. With substantives, forming adjectives. *post-attack, -bop, -Christmas, -coition, -college, -contact, -crash, -creole, -election* (examples), *-experience, -flu, -game, -harvest, -holiday, -independence, -injury, -language, -lunch, -luncheon, -menopause* (also as *sb.*), *-midnight, -operation* (also as *adv.*), *-orgasm, -ovulation, -publication, -Reformation* (earlier and later examples), *-Renaissance, -revolution* (examples), *-school, -seizure, -Sputnik* (also as *adv.*), *-surrealist, -symbolist, -Watergate, -World War II.*

b. With adjs., or formed from *post* + a L. or Gr. sb. with an adjectival ending. *post-Aristotelian, -Cartesian* (further examples), *Chomskyan, -Darwinian* (examples), *-Hegelian, -Homeric* (earlier and later examples), *-Humian, -Jamesian, -Kantian* (earlier and later examples), *-Keynesian, -Marxist, -Nietzschean, -Saussurean, -Wagnerian*; *post-adolescent, -anæsthetic, -analytic(al), -atomic, -capitalist, -climacteric, -cognitive, -coital, -collegiate, -colonial, -conciliar, -conditional, -encephalitic, -epileptic* (earlier and later examples), *-eruptive, -experimental, -feudal, -industrial, -junctural, -marital, -menopausal, -orgasmic, -paroxysmal* (examples), *-pausal, -revolutionary* (later examples), *-romantic, -teenage*; adjs. formed as in senses 1 a and b above are occas. used *ellipt.* as *sbs.*; such adjs. may also have adverbial forms, as *post-coitally, -maritally.*

Also **postabo·rtal** *Med.*, occurring or performed after an abortion; **postabo·rtion** (also as *adv.*) *Med.*, (occurring or performed) after an abortion; **post-abo·rtum** (also as *adv.*) *Med.* [L. *abortus* abortion] = prec.; **post-abso·rptive** *Med.*, occurring after food has been absorbed into the body; **post-cenal** (further example); **post-co·nquest**, applied to periods after a conquest, *spec.* (with capital), after the Norman Conquest; **post-conque·stual, -Conquestual**, after the Norman Conquest; **post-consona·ntal, -consona·ntic**, after a consonant; hence **post-consona·ntally** *adv.*; **post-convu·lsive** *Path.*, subsequent to a convulsion; **postcy·clic, -ical**, occurring or operating after the termination of a cycle or cycles (esp. in *Transformational Gram.*); hence **postcy·clically** *adv.*; **post-defle·ction** *Electronics*, pertaining to or being acceleration of an electron beam after its deflection in a cathode-ray tube; **post-depositional** *Geol.*, occurring after the deposition of sediment; **post-eme·rgence**, occurring, performed, or applied after the emergence of seedlings from the soil; also *absol.*; **post-i·ctal** *Med.*, subsequent to a stroke or fit, esp. an epileptic fit; hence **post-i·ctally** *adv.*; **post-infe·ctious**, subsequent to an infection; *esp.* caused by an infection but arising after it has ceased; **post-lapsarian**, delete † *Obs.* and add: also *gen.*, after the Fall of Man; **postli·teral**, following a letter of the alphabet; **post-meio·tic** *Cytology*, occurring subsequent to meiosis; **post-mena·rchal, -mena·rcheal** *Med.*, of, pertaining to, or designating a girl who has menstruated; **post-metamo·rphic** *Geol.*, occurring or existing after metamorphism; **post-mi·neral**, occurring after the formation of a mineral deposit; **postneona·tal** *Med.*, pertaining to or designating the period between the end of the neonatal period, four weeks after birth, and the end of the first year of life; **post-no·minal**, following a substantive or a proper name; also *ellipt.* as *sb.*; **post-o·vulative, -ovula·tory** *Med.*, subsequent to ovulation; **post-pai·nterly** *Art* [cf. *painterly *a.*], of or characterized by a style of abstract painting that employs traditional qualities of colour, form, and texture; **post-pu·beral** = next; **post-pu·bertal**, subsequent to puberty; **post-pu·berty** *a.* = prec.; **postra·dical** *Philol.* following a root or root-word; also as *sb.*, a postradical element or word; **post-reprodu·ctive**, occurring after the period of life when a female can bear offspring; **post-tecto·nic** *Geol.*, occurring or existing after tectonic activity; **post-tonic** (further examples); **post-ve·rbal**, following a verb; hence **postve·rbally** *adv.*; **postvoca·lic** *Philol.*, following a vowel; hence **postvoca·lically** *adv.*

c. With sbs., forming sbs. *postcreation*; **post-a·rticle** *Gram.*, one of a set of words that can follow an article in a noun phrase; **post-climax** *Ecol.* [*climax *sb.* 4 b], the point in a plant succession at which development has continued beyond the balanced state of climax.

d. Adjs. of the type in sense B. 1 a above are sometimes used adverbially (cf. *post *Latin prep.* 6), as *postabortion, -abortum, -operation, -Sputnik* above, *postflight *adv.*

1910 *Surg., Gynecol. & Obstetr.* July 55/1 Each case of post-partum or post-abortal infection must be studied individually. **1973** I. M. Cushner et al. in H. J. & J. D. Osofsky *Abortion Experience* vi. 147 This newer approach toward postabortal laparascopic sterilization..may very well lead to further reduction in the need for major surgical procedures in abortion. **1963** *Amer. Jrnl. Psychiatry* CXIX. 982/2 No known attempts to document statistically the incidence of post-abortion psychiatric illness have been found in American medical literature. **1973** E. C. Payne et al. in H. J. & J. D. Osofsky *Abortion Experience* xii. 272 In the first reported prospective study..a single postabortion interview took place during a period of 3 to 6 months after the procedure. In even more recent studies, women have been interviewed at a relatively specific time postabortion. **1910** F. J. Taussig *Prevention & Treatm. Abortion* vi. 44 There was no fever or odor to the discharge, so that the diagnosis was clearly endometritis post-abortum due to decidual remnants. **1950** *Proc. Soc. Study Fertility* I. 26 (*heading*) The re-establishment of ovulation, post-partum and post-abortum. **1972** *Biol. Abstr.* LIII. 5727/2 The results of treatment of 2 groups of patients with post abortum acute insufficiency of the kidneys are discussed. **1919** *Proc. R. Soc.* B. XCI. 45 The energy expenditure during sleep may be assumed..to be only slightly smaller than that during complete muscular rest in the post-absorptive condition, *i.e.*, 12–14 hours after the last meal. **1972** *New England Jrnl. Med.* 5 Oct. 678/1 Recordings were made in the postabsorptive state between 9:30 a.m. and noon. **1936** *Jrnl. Pediatrics* VIII. 52 The change from the straight, boyish figure of the preadolescent girl to the more rounded, mature figure of the postadolescent girl. **1977** *New Yorker* 29 Aug. 19/2 The new false post-adolescent authority that needs to be blown away by somebody. **1910** *Practitioner* Feb. 253 The post-anaesthetic condition. *Ibid.* Mar. 361 Ether..rarely causes post-anaesthetic vomiting. **1965** J. Pollitt *Depression & its Treatm.* vi. 80 Patients should be carefully questioned about this to avoid the risk of post-anaesthetic vomiting. **1934** *Mind* XLIII. 136 [A sense of 'theory of knowledge'] is essentially a 'post-analytic' *evaluation*, not a description, of knowledge. **1927** J. Loewenberg in *Jrnl. Philos.* XXIV. 5 Designating by 'pre-analytical', whatever is given *for* analysis, and by 'post-analytical', whatever is given *through* analysis, we wish to ascertain whether there is any possibility of assimilating to one another these two classes of data. **1934** Webster, Post-Aristotelian. **1936** J. R. Kantor *Objective Psychol. Gram.* viii. 100 The post-Aristotelian subjectivists divided the individual into soul and body. **1957** C. Vereker *Devel. Polit. Theory* i. 40 This doctrine of equality, the emergence of which is sometimes held to be the distinguishing mark of post-Aristotelian social thought. **1965** N. Chomsky *Aspects of Theory of Syntax* ii. 107 Det→(pre-Article—*of*) Article (post-Article). **1971** Post-article [see *non-lexical* s.v. *non- 3]. **1948** *Britannica Bk. of Year* (U.S.) 805/2 *Post-atom(ic)*, subsequent to the dropping of the atomic bomb. **1954** A. Huxley in *Encounter* Feb. 5/2 This..[temple] will be standing in the Western desert, an object, to the neo-Neolithic savages, of post-atomic times, of uncomprehending reverence and superstitious alarm. **1956** Auden & Kallman *Magic Flute* (1957) 58 The form of suite For piano in a Post-

Atomic Age. **1961** *Washington Post* 1 June 24 The speaker suggested that the desolation of a post-attack world would be too awful to face. **1964** Dentler & Cutright in I. L. Horowitz *New Sociol.* 424 Our discussion of the effects of a nuclear attack on the population makes it clear that the composition of the postattack population would be so different from that of today's population that this factor alone would make for differences in the postattack society. **1976** *Sci. Amer.* Nov. 33/3 In the immediate postattack period the fallout levels could vary greatly from one place to another. **1955** Keepnews & Grauer *Pict. Hist. Jazz* xix. 250 A young star who came along in the post-bop 1950s. **1977** *Rolling Stone* 5 May 24/2 Today his style is the antithesis of classic post-bop horn players like Coltrane. **1964** I. L. Horowitz *New Sociol.* 81 His critique of America could only be relevant and fruitful if it was made..with the projected hopes and ideals of a post-capitalist era. **1976** N. O'Sullivan *Conservatism* v. 122 The contemporary discussion about the nature of 'post-capitalist' (or 'post-industrial') society. **1931** *Times Lit. Suppl.* 10 Sept. 674/2 A particularly satisfying classification of the great seminal post-Cartesian theories of knowledge. **1963** *Ibid.* 1 Mar. 150/1 The general context of post-Cartesian thought. **1922** Joyce *Ulysses* 720 A temporary concussion caused by a falsely calculated movement in the course of a postcenal gymnastic display. **1970** *Jrnl. Linguistics* VI. 130 With a side glance at some post-Chomskyan developments. **1975** *Amer. Speech 1973* XLVIII. 154 According to the post-Chomskyan revisionist Charles J. Fillmore, however, 'there are reasons for questioning the deep-structure validity of the traditional division between subject and predicate'. **1959** *Encounter* Feb. 74/2 All these books are written with a light touch: just the thing for post-Christmas hang-overs. **1961** *Wall St. Jrnl.* 23 Jan. 2/2 Steel's post-Christmas production recovery is running out of steam. **1977** *Times* 21 Dec. 17/8 The post-Christmas clearance sale. **1897** A. D. L. Napier *Menopause* iv. 100 Of 500 post-climacteric cases, 36·5 per cent. had a return of hæmorrhage after the menopause had been established a year or more. **1973** E. A. Moscovic tr. *J. Botella-Llusiá's Endocrinol. of Woman* xvii. 364/2 The classic concept held that in postmenopausal woman estrogenic activity had been wiped out completely by the time woman entered the postclimacteric phase. **1916** F. E. Clements *Plant Succession* vi. 110 If a change of climate results in increased water-content..the sere..continues the development by replacing the climax and it may be termed the post-climax. **1928** *Jrnl. Ecol.* XVI. 26 Any well shaded ravine is occupied by *Picea albertiana* Stew. Br. which may be looked upon as a post-climax. **1964** V. J. Chapman *Coastal Vegetation* ix. 215 One must regard the Plantaginatum maritimi as either a post-climax or more properly as a.. deflected climax. **1949** *Mind* LVIII. 220 A tendency to 'spatio-temporal scatter' is characteristic of paranormal cognition, the displacement manifesting itself as 'post-cognitive' or 'precognitive' telepathy. **1974** *Listener* 3 Jan. 22/3 If I make a 'correct' guess before the target has appeared..that's a pre-cognitive hit; if I get it right but several shots in arrears, that's post-cognitive. **1922** C. G. Child *Sterility & Conception* viii. 68 (*heading*) Postcoital tests for sterility. **1947** E. Hyams *William Medium* 34 Smoking a post-coital cigarette. **1975** B. Garfield *Hopscotch* ii. 26 'I'm distressed to see you so lackluster, old friend.' 'It's only post-coital *tristesse*.' **1968** M. R. Cohen in J. J. Gold *Textbk. Gynecol. Endocrinol.* xxv. 546 Ovulation timing..disclosed good mucorrhea at midcycle with excellent longevity of spermatozoa postcoitally. **1977** S. Schoenbaum *W. Shakespeare* (rev. ed.) vii. 85 A shotgun wedding boomed for the post-coitally chastened Will. **1953** N. Tinbergen *Herring Gull's World* xiii. 109 There is no post-coition display. **1883** 'Mark Twain' in *Century Mag.* Dec. 235/1 He..had finished a post-college course in an Eastern law school. **1973** *Jrnl. Genetic Psychol.* June 183 The relevance of undergraduate courses for meeting the demands of postcollege life. **1960** *Encounter* Nov. 26 A face-to-face group—the post-collegiate fraternity of the small suburbs. **1934** Webster, Post-colonial. **1959** *Daily Tel.* 12 Dec. 6/2 It was probably inevitable that India, in the full flush of post-colonial sensitivity, should fear that association with the America of that period might involve her unnecessarily in troubles which were little to do with Asia. **1969** *Times* (Uganda Suppl.) 15 Sept. p. i/5 Behind the imposing physical presence is a mind that has been described as one of the shrewdest in post-colonial Africa. **1974** 'G. Black' *Golden Cockatrice* iii. 57 If there's one thing worse than..rampant colonialism..it's post-colonial dictatorship. **1968** *Times* 24 Feb. 9/5 The post-conciliar attitude seems to draw attention..to the value of personal and social relationships. **1976** *Times* 9 Aug. 11/1 The post-conciliar church polishes characteristically pre-conciliar weapons. **1939** Joyce *Finnegans Wake* (1964) ii. 270 All them fine clauses in Lindley's and Murrey's never braught the participle of a present to a desponent hortatrixy,..from her postconditional future. **1922** E. Ekwall *Place-Names Lancs.* 1 The county of Lancaster developed out of the post-Conquest honour of Roger of Poitou. **1940** *Burlington Mag.* Aug. 56 Peruvian post-conquest tapestry work. **1959** E. A. Fisher *Introd. Anglo-Saxon Archit. & Sculpture* 94 Such post-Conquest Anglo-Saxon influence on art is outside the scope of this book. **1976** *Jrnl. Medieval Hist.* II. 1/1 The balance between French and English in post-Conquest England is still being discussed. **1920** *Contemp. Rev.* Dec. 898 The manuscript is post-conquestual. **1924** *Ibid.* Nov. 673 Welsh life in post-Conquestual days. **1934** Priebsch & Collinson *German Lang.* ii. i. 93 The position in the word (initial, post-consonantal, inter-vocalic). **1964** Y. Malkiel in *Archivum Linguisticum* XVI. 27 World-medial, postconsonantal *cl-*. **1953** K. Jackson *Lang. & Hist. Early Brit.* 470 (*heading*) IE. *g*... This became *b* in CC. initially except before *u*..; intervocally and preconsonantally, *g*; post-consonantally, *b*. **1969** *Word* XXV. 25 *Tana-dg-usim had-a-n* 'towards the village', *had-a-n* 'towards it or him,' the *-a-* being a postconsonantic variant of the post-vocalic *-·*. **1934** Webster, Post-contact adj. **1946** *Nature* 30 Nov. 769/2 Each article presents chronologically..the data available from earliest times onwards through four hundred years of contact with White civilization... post-contact change and the absorption of the tribes into European civilization are revealed and traced in as much detail as possible. **1907** W. A. Turner *Epilepsy* vi. 129 A third epileptic, whose post-convulsive symptoms were mainly of the nature of cataleptic rigidity and dementia. **1974** E. Niedermeyer *Compendium of Epilepsies* xi. 187 A post-convulsive sleep may ensue for a few hours. **1966** *Economist* 24 Dec. 1329/3 An effort to prevent post-crash injuries by fire. **1977** *Hongkong Standard* 12 Apr. 9/3 The prevention and control of both in-flight and postcrash fuel system fires and explosions. **1922** Joyce *Ulysses* 385 In woman's womb word is made flesh but in the spirit of the maker all flesh that passes becomes the word that shall not pass away. This is the postcreation. **1968** D. Decamp in *Lat. Amer. Research Rev.* III. 38 If any term is needed to distinguish the situation in Jamaica from that in Surinam and Haiti, then I suggest that we call Jamaica a *post-creole* community. **1977** *Language* LIII. 330 Speakers in a post-creole community are triply pressured: to avoid the basilect, to acquire the acrolect, and to vary the mesolect. **1967** J. Ross in *To Honor Roman Jakobson* III. 1672 If it cannot apply before the cycle, it must either apply in the cycle or after all cyclic rules have been applied—rules of this last type are called *post-cyclic*. *Ibid.* 1677 It is only if *pronominalization* is formulated as a post-cyclic rule that some constraint on forward pronominalization becomes necessary. **1971** J. W. Bresnan in *Language* XLVII. 276 One might think of ordering the NSR after the entire transformational cycle but before the postcyclic transformations. **1976** *Archivum Linguisticum* VII. 138 This obviously, would call for..an explicit formulation of the concept of 'transformation' one uses, that is whether one believes in prelexical, cyclic, precyclic, and postcyclic transformations. **1972** *Language* XLVIII. 310 Primary stress assignment must *precede* postcyclic transformations. **1972** *Ibid.* 301 It is preferable for a rule to apply postcyclically rather than cyclically. **1899** T. Veblen *Theory of Leisure Class* xi. 288 The ostensibly post-Darwinian concept of a meliorative trend in the process of evolution. **1939** *Mind* XLVIII. 528 In continuity with his previous books, Dewey is anxious to be a post-Darwinian Mill developing 'the science of evidence' in close connection with all the sciences. **1972** S. Hynes *Edwardian Occasions* 7 Much important Edwardian literature implies..a post-Darwinian, post-Hegelian way of looking at the linear shape of time. **1943** F. E. Terman *Radio Engineers' Handbk.* iv. 342 In the post-deflection arrangement the beam is deflected at low velocity. **1950** *Electronic Engin.* XXII. 461/1 Two main advantages..in the use of commercially available post deflexion accelerator (P.D.A.) cathode-ray tubes..are: (i) The attainment of higher screen brightness..and (ii) The problem of insulation in the glass pinch and the base are eased. **1959** Rider & Uslan *Encycl. Cathode-Ray Oscilloscopes* (ed. 2) i. 18/1 These post-deflection accelerating anodes are rings or wide bands of electrically conductive material painted on the inside surface of the envelope. **1971** Klemperer & Barnett *Electron Optics* (ed. 3) x. 398 Post-deflexion acceleration..appears to offer the means to increase the sensitivity. **1949** F. J. Pettijohn *Sedimentary Rocks* i. 7 Diastrophism plays the dominant role in controlling the production and deposition of a sediment—and to some extent its post-depositional history, also. **1965** G. J. Williams *Econ. Geol. N.Z.* xiii. 206/2 Post-depositional alteration is indicated by large chlorite plates that are discordant in relation to the foliation in schist pebbles. **1962** 'K. Orvis' *Damned & Destroyed* i. 9 His opinion would be the first post-election one I would hear. **1976** *New Yorker* 15 Nov. 204/2 Post-election surveys show that almost ninety per cent of the Republicans who voted backed the ticket. **1940** *Phytopathology* XXX. 334 Experience over a period of years with various chemical methods of damping-off control has emphasized the need for some soil disinfectant that will control both pre- and post-emergence damping-off. **1955** *Sci. News Let.* 2 July 10/3 There are three prescribed methods for fighting the weed war. They are known as pre-planting, pre-emergence and post-emergence. Pre-planting treatment is made on the soil before any seed is planted in the ground. Pre-emergence control is done after seeds have been sown, but before a desired plant pushes up through the ground. Post-emergence is designed to kill undesirable plants that exist in areas where plants are already growing. **1962** *Times* 12 Nov. 17/4 Another avenue of research which is leading to useful results is that involving pre-emergence spraying of cereal crops instead of the traditional post-emergence method. **1977** *Protecting World's Crops* (Shell Internat. Petroleum Co.) 8 Crop herbicides are often divided into pre-plant, pre-emergence and post-emergence products. **1928** E. F. Buzzard in H. French *Index Differential Diagnosis Main Symptoms* (ed. 4) 880 The tremor in post-encephalitic Parkinsonism closely resembles that of paralysis agitans. **1932** W. Boyd *Textbk. Path.* xxx. 812 Other postencephalitic conditions ..are narcolepsy and oculogyric crises. **1961** *Lancet* 23 Sept. 683/2 Since then, she has had postencephalitic epilepsy. **1971** G. W. Voeller in G. Birdwood et al. *Parkinson's Dis.* iii. 50 Fifteen years ago the post-encephalitic forms [of Parkinsonism] were predominant.., today, the idiopathic or arteriosclerotic forms predominate. **1875** J. H. Jackson in *W. Riding Lunatic Asylum Med. Rep.* V. 111 The automatism in these cases is not, I think, ever epileptic, but always post-epileptic. **1954** L. Fairfield *Epilepsy* i. 13 In some cases the post-epileptic deep sleep or 'coma', lasts an hour or more. **1936** *Discovery* May 148 (*heading*) Post-eruptive movements of the earth's crust. **1946** *Nature* 19 Oct. 560/1 Some powerful beneficial influence had been at work during the post-eruptive as well as the developmental period. **1964** *Economist* 31 Oct. 502/1 The needs of the 'post-experience' students and..immediate postgraduates. **1977** P. Strevens *New Orientations Teaching Eng.* viii. 90 In all types of occupational ESP a distinction emerges between *pre-experience* and *post-experience* courses. **1970** *Jrnl. Gen. Psychol.* Apr. 253 All Ss were asked in a post-experimental questionnaire to indicate the two words they thought earned them points. **1949** Koestler *Promise & Fulfilment* i. iii. 29 We are always apt to forget that nationalism is a product of a relatively recent, post-feudal European development. **1970** R. Stavenhagen in I. L. Horowitz *Masses in Lat. Amer.* vii. 267 The Spanish Conquest was..part of the political and economic expansion in post-feudal and mercantalistic Europe. **1918** A. Huxley *Let.* 25 Nov. (1969) 171 The whiskey bottle seems the only refuge from that post-flu depression. **1971** P. D. James *Shroud for Nightingale* v. 163 She had post-flu depression and she felt she couldn't cope with the baby. **1966** *Jrnl. Canad. Operational Res. Soc.* 114 *Analysis, post game*, use of data generated during a *game* or series of games to derive conclusions about the problems to which play was directed. **1976** *Springfield* (Mass.) *Daily News* 22 Apr. 39/1 When his club gets beaten 7–1 he is a post game press conference no-show. **1962** *Times* 20 Mar. 3/2 The study of post-harvest physiological changes in pasture plants. **1976** *National Observer* (U.S.) 25 Dec. 1/4 Last week the FAO council called for 'prompt action..on reducing post-harvest losses'. **1909** W. James *Pluralistic Universe* v. 184 Royce makes by far the manliest of the post-hegelian attempts to read some empirically apprehensible content into the notion of our relation to the absolute mind. **1964** P. Meadows in I. L. Horowitz *New Sociol.* 446 The Hegelians and the post-Hegelians (of either idealistic or materialistic breed). **1972** Post-Hegelian [see *post-Darwinian]. **1960** *Farmer & Stockbreeder* 5 Jan. 16/1 A post-holiday lull prevailed at most markets. **1976** H. Ferguson *Confessions Long Distance Acid Head* 52 Gordon..had brought back the usual amount of cigarettes and liquor, and another rare gift—the post-holiday feeling of exuberance. **1810** C. Lamb *Lett.* (1935) II. 97, I should suspect these personifications are the Translator's. They sound *post*-Homeric. **1959** *Encounter* July 46/1 This is the element of allegory or symbolism which so many post-homeric writers have found in the Lotus Eaters. **1909** W. James *Pluralistic Universe* v. 210 This being our post-humian and post-kantian state of mind, I will ask your permission to leave the soul wholly out of the present discussion. **1961** *Encounter* Jan. 16 The Humian and post-Humian side. **1941** Penfield & Erickson *Epilepsy & Cerebral Localization* (1942) ii. 19 The state in which an individual is deprived of, or released from conscious control, is called automatism. If it occurs following a seizure it may be called post-ictal automatism. **1961** *Lancet* 29 July 242/2, 36 hours later.. the E.E.G. showed no seizure activity, but there was general post-ictal disorganisation. **1972** M. Crichton *Terminal Man* iv. xi. 186 He's in a post-seizure state—post-ictal, we call it. **1959** *Brain* LXXXII. 152 Post-ictally the activity of the focus was very much reduced. **1975** *Electroencephalogr. & Clin. Neurophysiol.* XXXVIII. 601/2 Motor activity was observed to increase postictally concomitantly with the increase in EEG frequency and amplitude. **1961** *Middle East Jrnl.* Winter 3 The people were still experiencing post-independence let-down and suffering the after effects of poor harvests in 1957. **1976** *Times Lit. Suppl.* 13 Feb. 164/2 The post-independence state is even more centralized than the colonial one. **1947** *Partisan Rev.* XIV. 230 Industrial organization and the postindustrial state are here to stay. **1977** *Times* 21 Feb. 11/4 We are already laying the foundation for the post-industrial future. **1928** *Trans. Chicago Path. Soc.* XIII. 15 (*heading*) The pathology of post-infectious acute toxic encephalitis in children. **1946** A. B. Christie *Infectious Dis.* xiii. 108 Encephalitis of the so-called post-infectious type occasionally occurs after chicken-pox. **1974** S. L. Robbins *Path. Basis Dis.* xxxii. 1532/1 Injection of brain tissue and adjuvants..can set up a reaction in the central nervous system of experimental animals, which has a marked resemblance to post-infectious encephalomyelitis. **1951** Postinjury [see *Kümmell's disease]. **1976** *National Observer* (U.S.) 3 July 12/6 Whenever you injure a joint..there is a possibility of a permanent residual stiffness or postinjury arthritis. **1960** J. Bayley *Characters of Love* iv. 258 Both D. H. Lawrence and E. M. Forster use them [*sc.* symbolic patterns] in a discernibly post-Jamesian manner. **1959** E. P. Hamp in *Studia Linguistica* XIII. 34 Those post-junctural syllabics which did not themselves bear a primary. **1964** *Eng. Stud.* XLV. 385 Postjunctural pre-vocalic /ß/ began to be distributionally a spirant. **1843** Mill *Logic* I. i. iii. 79 His philosophical views are generally those of the post-Kantian movement, represented by Schelling and Hegel. **1946** *Nature* 7 Sept. 322/2 James Marsh..did much to acclimatize Kantian and post-Kantian philosophy in the United States. **1977** L. Houlden in J. Hick *Myth of God Incarnate* vi. 128 There is a strong element of post-Kantian consciousness in distinguishing the two approaches at all. **1960** *New Left Rev.* May–June 5/1 The more sophisticated elaboration of post-Keynesian evolutionary theory. **1975** *Times Lit. Suppl.* 18 July 811/3 Post-Keynesians who instinctively treat saving and investment as different activities. **1977** *Dædalus* Fall 61 Post-Keynesian and econometric studies in economics. **1946** C. Morris *Signs, Lang. & Behavior* ii. 47 The latter [*sc.* proprioceptive stimuli] are..not themselves language signs; since they are substitute signs synonymous with language signs they are properly called post-language symbols. **1950** *Eng. Stud.* XXXI. 63 What would the corrupt post-lapsarian variety of this attitude be? **1972** *Times Lit. Suppl.* 29 Dec. 1587/2 A prelude..to the deplorable history of postlapsarian man. **1953** *Archivum Linguisticum* V. 68 He [*sc.* Marr] indicates..labialization with a postliteral circle. **1958** J. Berry in J. A. Fishman *Readings Sociol. of Lang.* (1968) 742 Diacritically modified letters (i.e. 'simple' letters with e.g...postliteral circle or apostrophe). **1974** E. Ambler *Dr. Frigo* iii. 186 Saw patient while he was taking his post-lunch bed rest. **1959** N. Marsh *False Scent* (1960) ii. 59 A long gloomy post-luncheon talk. **1903** R. Whiteing *Let.* 9 Feb. in D. L. Moore *E. Nesbit* (1933) xi. 177 The love scenes, for such they are..though they are post marital. **1957** V. W. Turner *Schism & Continuity in Afr. Society* p. xviii, 'Uxorilocal' refers to the post-marital residence of a man in his wife's village. **1975** R. H. Rimmer *Premar Experiments* i. 111 They aren't fully aware of the impact that freer and open sexuality, premaritally and postmaritally, will have on human goals and values. **1949** G. B. Shaw *Buoyant Billions* i. 17 In your time the young were post-Marxists and their fathers pre-Marxists. **1963** M. H. Abrams in N. Frye *Romanticism Reconsidered* 29 It may be useful, then, to have a new look at the obvious as it appeared, not to post-Marxist historians, but to intelligent observers at the time. **1978** *Bull. Amer. Acad. Arts & Sci.* Jan. 35 He considers himself post-Risorgimento, post-Marxist. **1905** *Q. Jrnl. Microsc. Sci.* XLVIII. 490 In animals there are (normally) no post-maiotic divisions, whereas in plants there may be, and often are, a large number. **1905** [see *pre-meiotic* s.v. *PRE- B. 1]. **1934** L. W. Sharp *Introd. Cytol.* (ed. 3) xvi. 265 In certain cases evidence has been brought forward to show that the chromonema in each chromatid at the close of the second meiotic mitosis is already split 'in preparation for' the first postmeiotic division. **1973** *Genetical Res.* XXII. 285 A characteristic feature of recombination in *Sordaria brevicollis* is the relatively high proportion of recombination events which exhibit post-meiotic segregation. **1968** N. Vorys et al. in J. J. Gold *Gynecol. Endocrinol.* xi. 253 Postmenarchal ovarian hypersensitivity may cause short proliferative phases. **1937** Post-menarcheal [see *premenarcheal* s.v. *PRE- B. 1]. **1977** *Yearbk. Obstetr. & Gynecol.* 332 Their mean chronological age was 13·02 years and 11 were postmenarcheal. **1928** E. Novak in *Gynecol. & Obstetr. Monogr.: Cumul. Suppl. & Composite Index* 32 (*heading*) Postmenopausal bleeding with ovarian cancer. **1949** M. Mead *Male & Female* viii. 180 In Bali..the post-menopausal woman and the virgin girl work together at ceremonies from which women of child-bearing age are debarred. **1975** *Lancet* 5 July 7/2, 11 women were postmenopausal (by at least three years); their ages ranged from forty-eight to sixty-five years of age. **1925** *Practitioner* July 43 Post-menopause vulvitis is thus set up. **1975** *Acta Endocrinologica* LXXX. 262 Levels of plasma PRL rose with puberty and decreased during post-menopause and in elderly men. **1956** *Q. Jrnl. Geol. Soc.* CXII. 115 The gneiss..is transected by the Boyne Line.., and all the rocks appear to have suffered a certain amount of post-metamorphic shearing. **1965** G. J. Williams *Econ. Geol. N.Z.* vi. 66/1 Grindley (1963) noted that post-metamorphic folding in southern Westland was accompanied by axial-plane cleavage on mesoscopic shear folds. **1943** L. B. Lyon *Evening in Stepney* 18 Post-midnight hours, be born of costlier reverence. **1970** I. Petite *Meander to Alaska* ii. xi. 108 A ride..ending with a postmidnight ride back on a bicycle. **1907** *Technical Lit.* Sept. 189/1 It is important to look for evidence of recent or post-mineral faulting that may be connected with the secondary enrichment of the deposits. **1965** G. J. Williams *Econ. Geol. N.Z.* iii. 26/1 From the descriptions which are available the pattern of pre- and post-mineral faulting cannot be resolved in detail. **1958** *Jrnl. Amer. Med. Assoc.* 21 June 937/2 Infant deaths..may be divided into two groups—deaths of infants in the neonatal period (under 28 days) and those in the postneonatal period (28 days to one year). **1965** S. Peller in Glass & Eversley *Population in Hist.* V. 93 The difference between the rates for the ruling families and for the general population..was larger during the post-neonatal period of infancy than in the neo- or peri-natal period. **1973** Puffer & Serrano *Patterns of Mortality in Childhood* v. 76 The infant period is divided into neonatal (0–27 days of age) and postneonatal (28 days through 11 months). **1928** A. Huxley in *Vogue* 28 Nov. 122/3 A form which the critical intelligence of post-Nietzschean youth can respect. **1977** *N.Y. Rev. Bks.* 15 Sept. 41/1 To the post-Nietzschean Bloom, these are the essential qualities of human life under the shadow of belatedness. **1935** H. Straumann *Newspaper Headlines* ii. 51 As the word preceding it is a nominal, this position may more exactly be called *postnominal position*. **1952** *R.A.F. Rev.* Jan. 9/2 The use of the post-nominal letters 'T.D.' is peculiar to the Territorial Army. **1961** *Amer. Speech* XXXVI. 159 Finally certain postnominal modifiers, such as relative clauses. **1975** *Daily Tel.* (Colour Suppl.) 3 Jan. 11/3 In the higher ranks of the Forces and the Civil Service the appropriate title, or post-nominal letters, almost automatically follow appointment to a senior post. **1978** E. St. Johnston *One Policeman's Story* xii. 287 The third problem affecting Honours..was the right of any officer holding the Queen's Police Medal to put the initials, QPM, after his name. This is technically known as wearing 'post-nominals'. **1969** E. H. Pinto *Treen* 17 Operation Pegs... They are still used for plugging post-operation tubes. **1979** *Nature* 25 Jan. 327/3 The nodules were found in all rats by 6 months post-operation. **1973** S. Fisher *Female Orgasm* vii. 190 The judgments typically portrayed the orgasm and post-orgasm experience as having favorable..connotations. **1953** A. C. Kinsey et al. *Sexual Behav. Human Female* xv. 638 Sometimes, especially in youth, the post-orgasmic relaxation is hardly more than momentary. **1973** S. Fisher *Female Orgasm* ix. 290 How relaxed she feels during the postorgasmic state. **1923** *Amer. Jrnl. Physiol.* LXVI. 325 When a bird is killed for such post-ovulation stages care must be taken that a new ovulation stage is not initiated in the meantime. **1951** *Jrnl. Clin. Endocrinol.* XI. 937 Administration of chorionic gonadotropin was begun early in the postovulation period. **1933** *Amer. Jrnl. Anat.* LII. 610 The existence of a double uterine cycle explains the occasional appearance of intermenstrual post-ovulative bleedings. **1968** M. R. Cohen in J. J. Gold *Textbk. Gynecol. Endocrinol.* xxv. 543 During the postovulative phase. **1922** H. M. Evans in L. F. Barker et al. *Endocrinol. & Metabolism* II. vi. 581 In the postovulatory period it is the corpora lutea which are responsible for further marked changes which now regularly occur in the genitalia of some other mammals. **1975** *Biol. Reproduction* XII. 573 The cervical epithelium of postovulatory rabbits consists of ciliated cells and nonciliated cells with bulbous apical processes. **1965** *N.Y. Times Mag.* 21 Feb. 12/2 Post-painterly or 'hard-edge' abstraction cleaned up the gooey mess and substituted neatly defined geometrical shapes in chaste combinations. **1969** *New Yorker* 6 Dec. 184 At the Metropolitan, the largest displays are by so-called object-makers—'post-painterly' canvases; that is to say, smooth-surfaced, cool, and tending to blend with their setting. **1972** E. Lucie-Smith in Cox & Dyson *20th-Cent. Mind* III. xvi. 473 The man who formed a link between Abstract Expressionism and what came to be called 'post-painterly abstraction' was Morris Louis (1912–62). **1977** P. Johnson *Enemies of Society* xvi. 218 Morris Louis, by his own description a 'post-painterly abstractionist', simply painted streaks on the edge of the canvas, leaving the rest untouched. **1876** J. H. Jackson in *W. Riding Lunatic Asylum Med. Rep.* VI. 266 The temporary state of the patient immediately after the paroxysm—which will be called the post-paroxysmal condition. **1967** *Biol. Abstr.* XLVIII. 7986/2 (*heading*) Character of the clinical progress of adolescent rheumatism in post-paroxysmal period. **1966** W. S. Allen in C. E. Bazell *In Memory of J. R. Firth* 11 This difficulty could be met only by assuming that, except in post-pausal position, a high tone required a lower pitch to precede it. **1968** A. F. Fraser *Reproductive Behaviour in Ungulates* v. 75 A group of bulls which were in their first post-puberal year. **1886** *Buck's Handbk. Med. Sci.* III. 396/2 The post-pubertal falling off in growth is more rapid in girls than in boys. **1955** *New Biol.* XVIII. 32 It is possible that the post-pubertal mammal behaves like an insect imago.., and that the fundamental change which leads to eventual senescence has already taken place at puberty. **1978** *Homes and Gardens* Oct. 117/2 There are some effects of the pill which we know... Some girls suffer from post-pubertal amenorrhœa and do not conceive easily afterwards. **1943** Koestler in *Horizon* VII. 230 But that was still in his early period, a hang-over from adolescence, the nihilistic post-puberty pose. **1964** E. A. Nida *Toward Sci. Transl.* xi. 251 Incorporating postpublication corrections into subsequent printings. **1946** L. Bloomfield in H. Hoijer et al. *Ling. Struct. Native Amer.* 121 Some roots appear in *postradical* extensions... Most postradicals..have no clear meaning. **1958** *Archivum Linguisticum* X. 170 In some words Postradical..elements are recognized. **1850** *Dublin Rev.* Mar. 145 The elucidation of the post-reformation history of Ireland. **1964** P. F. Anson *Bishops at Large* i. 31 The post-Reformation *Ecclesia Anglicana*. **1978** R. Strong *And when did you last see your Father?* 153/2 The rise of the gentry and the establishment of a new post-Reformation aristocracy. **1941** *Listener* 19 June 882/2 The literature of post-Renaissance Europe. **1964** R. H. Robins *Gen. Linguistics* viii. 315 The post-Renaissance process of creating learned vocabulary from classical sources. **1978** D. Daiches *Edinburgh* ii. 43 The College..was from the beginning a post-Renaissance, post-Reformation university. **1900** *Q. Jrnl. Microsc. Sci.* XLIV. 3 Reproductive Period.—I have used this expression to denote the whole of that period in the life of a mammal.. during which its generative organs are capable of the reproductive function; and in contrast to the Pre-reproductive and Post-reproductive periods which severally precede and follow it, during which the generative organs are either not fully developed or are degenerative. **1963** *Lancet* 5 Jan. 2/1 As we reach the postreproductive years, a man's chance of dying is much greater than a woman's..so that at 60 a man has almost twice the likelihood of dying before 61 than a woman has. **1928** *Manch. Guardian Weekly* 7 Sept. 73/4 The Duce's personal implication in numerous ugly stories of the post-Revolution period. **1957** *Times Lit. Suppl.* 25 Oct. 685/3 Apart from a single passing reference to Zinoviev and Kamenev, Lenin, Trotsky, Stalin, Beria and Kruschev seem to be the only Bolsheviks named in the post-revolution chapters. **1938** *Burlington Mag.* June 270/1 Ducreux' post-Revolutionary studio. **1966** F. Schurmann *Ideol. & Organization in Communist China* p. xxxiii, Political centralization is one of the forms that post-revolutionary organization has taken. **1970** S. L. Barraclough in I. L. Horowitz *Masses in Lat. Amer.* iv. 155 Where land reform has been rapid..as was the case in post-revolutionary Mexico and Bolivia—some lines of production temporarily decreased. **1943** Y. Winters *Anatomy of Nonsense* 19 The method of the Post-Romantics, whether French Symbolists or American Experimentalists. **1947** A. Einstein *Mus. Romantic Era* xvii. 330 The post-Romantic period, when these musicians can be classified roughly according to their training in Germany or in Paris. **1965** *Times Lit. Suppl.* 25 Nov. 1063/2 Most readers..in this post-romantic age. **1949** Post-Saussurean [see *POST-BLOOMFIELDIAN *a.* and *sb.*]. **1977** *Language* LIII. 394 Each of these topics is divided in turn into three parts..the last one to post-Saussurean developments. **1934** Webster, Postschool. **1939** H. M. Miner *St. Denis* ix. 193 The young school child wants to have cards, ice skates, and a bicycle, because these amusements are those of postschool children. **1968** *Economist* 17 Aug. 17/2 The trendy thing for the past ten years has been to work on post-school education. **1975** *Language for Life* (Dept. Educ. & Sci.) xix. 278 Some schools do successfully guide their pupils to post-school opportunities. **1959** *Brain* LXXXII. 181 During the stage of post-seizure cortical hypoxia with flattening of the EEG loud sounds failed to produce spike responses in the strychninized animal. **1957** *N.Y. Herald Tribune* 25 Nov. 19/1 There is no doubt that post-Sputnik Washington is a different city and a different atmosphere. *Ibid.* 3 Dec. 24/1 He [*sc.* R. M. Nixon] has defeated the pinch-penny economizers, who, even post-Sputnik, are still in the..mood of providing too-little, too-late. **1977** *Dædalus* Fall 80 Many of the post-Sputnik educational programs were in fact based on this conclusion. **1938** Post-surrealist [see *muck-pot* s.v. *MUCK *sb.*¹ 5]. **1951** N. Rorem *Paris Diary* (1967) i. 7 She encouraged gaudy and exhibitionistic comportment..to give herself an identity with the post-surrealist gang she hung out with. **1952** Koestler *Arrow in Blue* xxiv. 224 But none of the existentialists, post-surrealists..had the guts to speak his opinion. **1953** S. Spender *Creative Element* i. 22 Writers on the Symbolists and Post-Symbolists.. note the tendency of a creative impulse which begins with a religious intensity. **1955** D. Davie in C. Tomlinson *Necklace* 1 Charles Tomlinson has taken note of the experiments and achievements of French symbolism. This does not mean that he belongs to the post-symbolist 'school' or the post-symbolist 'movement', if there are such things. **1958** J. Press *Chequer'd Shade* viii. 183 Whether the systematic employment of post-symbolist technique has weakened or strengthened poetry is likely to remain a matter of dispute. **1977** *Radio Times* 29 Oct.–4 Nov. 13/4 Nerval's sonnet sequences, *Les Chimères*..is now recognised as the source of all post-Symbolist and Surrealist poetry. **1938** *Mem. Geol. Soc. Amer.* No. 6. 108 The crystallization in such fabrics can be termed: pre-tectonic, posttectonic, or paratectonic with reference to the particular mineral that is under consideration. **1956** *Q. Jrnl. Geol. Soc.* CXII. 125 The well-known post-tectonic growth of andalusite in the eastern zone may have been accompanied by recrystallization which destroyed any incipient fabric. **1971** I. G. Gass et al. *Understanding Earth* xx. 296/1 These granites..are both syntectonic and post-tectonic. **1973** M. Amis *Rachel Papers* 22 Firstly, I assume I'm right in saying that teenage sex is quite different from post-teenage sex? **1976** *Sounds* 11 Dec. 40/2 Personally, I was in the throes of extreme post-teenage depression at the total unadventurousness of yer average British audience. **1953** K. Jackson *Lang. & Hist. Early Brit.* ii. 268 Syncope of the post-tonic penultimate syllable. **1973** A. H. Sommerstein *Sound Pattern Anc. Greek* v. 123 Post-tonic vowels (not just svarita vowels, but all vowels at any distance to the right of the accent in a word). **1934** Webster, Post-verbal. **1948** [see *COMPLEMENTATION]. **1965** N. Chomsky *Aspects of Theory of Syntax* 228 A somewhat different analysis of post-Verbal Adjectives in English. **1978** *Language* LIV. 85 The most unmarked order of adverbs in English is generally considered to be manner, place, and time in post-verbal position. **1971** *Ibid.* XLVII. 532 The principle involved is that the experiencer in certain types of sentences cannot be extracted if the complement sentence ends up postverbally. **1892** J. Wright *Primer Gothic Lang.* i. 10 Final postvocalic *g* and *g* in the final combination *gs* was probably a voiceless spirant. **1956** *Archivum Linguisticum* VII. 94 Icelandic words of the type *epli* being equivalent to those of the form *lappa* in having the quantitative peak on the post-vocalic consonant. **1964** R. H. Robins *Gen. Linguistics* iii. 101 In Scots English, /r/ occurs both prevocalically and postvocalically (*cart*, standard /kaːt/, Scots /kart/). **1895** G. B. Shaw in *Liberty* (N.Y.) 27 July 3/1 A post-Wagnerian reaction. **1965** *New Statesman* 7 May 736/1 The technical mêlée of.. early Renaissance polyphony and sensuous post-Wagnerian harmony. **1977** *Time* 14 Mar. 26/1 The post-Watergate Congress is in trouble with its constituents. **1978** *N.Y. Times* 30 Mar. B 5/1 Governor Carey's executive order and the city law were issued at a time of post-Watergate morality. **1957** K. Rexroth in *New World Writing* XI. 32 Many of the post-World War II abstract expressionists..look alike, and do look like accidents. **1970** I. L. Horowitz *Masses in Lat. Amer.* i. 5 This definition..strangely enough became the chief ideological tool of post-World War Two 'neo-Marxism'. **1979** *Dædalus* Winter 119 The successive discard, in the post-World War II world, of the totalitarian scheme.

2. In adjs. (rarely sbs.), chiefly *Anat.* and *Zool.* **post-alveo·lar,** behind the teeth-ridge; in *Phonetics* applied to a consonant articulated with the tongue against the back part of the alveolar ridge; **post-auri·cular,** behind the ear; **post-central** (further examples); hence **post-ce·ntrally** *adv.*; **postcra·nial,** situated posterior to the cranium; also as *sb. pl.*, the postcranial remains of an animal; hence **postcra·nially** *adv.*; **post-cri·coid,** posterior to the cricoid cartilage; **post-dental** (earlier and later examples); also as *sb.*; **post-humeral** (examples); **postpalatal,** also in *Phonetics*, applied to a consonant articulated with the tip or middle of the tongue against the hard palate; also *ellipt.* as *sb.*; **post-ve·lar,** behind or at the back of the velum; in *Phonetics*, applied to a consonant articulated with the tongue against the rear half of the velum or soft palate; also *ellipt.* as *sb.*

1932 D. Jones *Outl. Eng. Phonetics* (ed. 3) ix. 44 Post-alveolar: articulated by the tip of the tongue against the back part of the teeth-ridge. **1964** I. Dahl in D. Abercrombie et al. *Daniel Jones* 314 [c, ɟ] are post-alveolar. **1973** *Amer. Speech 1969* XLIV. 265 Heavy retroflexion is understood to be an *r* produced by passing the breath between the underside of the apex of the tongue and the postalveolar or prepalatal region. **1903** *Ann. & Mag. Nat. Hist.* XII. 342 *Mus hypoxanthus bacchante*... Fine hairs of ears rufous; no postauricular patch. **1934** F. Stark *Valleys of Assassins* ii. 193 The post-auricular length..is about one-third of the total length. **1977** *Proc. R. Soc. Med.* LXX. 399/1 Smaller defects in the centre of the face are therefore sometimes repaired with free grafts of full-thickness postauricular skin. **1959** Schuell & Jenkins in Saporta & Bastian *Psycholinguistics* (1961) 428/2 *Verbal aphasia*, resulting from lesions of the pre- and post-central convolutions. **1967** G. M. Wyburn et al. *Conc. Anat.* vii. 193 It..separates the precentral gyrus from the postcentral gyrus. **1968** *Brit. Med. Bull.* XXIV. 202/1 A normally responsive alpha rhythm at 7–8 cyc./sec. is present postcentrally, with slightly higher voltage on the right side. **1913** *Bull. Amer. Museum Nat. Hist.* XXXII. 563 The post-cranial skeleton. **1956** *Biologia* II. 231 The skull of the common lizard..has been described in detail previously... The present paper deals with its postcranial skeleton, i.e., the vertebral column, ribs, sternum, girdles and limbs. **1971** *Nature* 5 Feb. 407/2 It has been thought for some time that Miocene hominoids differed post-cranially from living hominoids, resembling instead the cercopithecoid monkeys. But careful examination of the post-cranials of *Limnopithecus*..and the European fossil *Pliopithecus* indicates that such non-hominoid structural features as they have are ceboid-like rather than specifically cercopithecoid-like. **1978** *Sci. Amer.* July 105/1 Anatomical studies..relating respectively to the birds' postcranial skeleton and their skull. **1971** Cappell & Anderson *Muir's Textbk. Path.* (ed. 9) xviii. 485/1 There is also the relationship between the iron-deficiency anaemia..and post-cricoid carcinoma in women. **1971** *Brit. Med. Bull.* XXVII. 34/1 Difficulty in swallowing is associated with a post-cricoid web of mucous membrane. **1899** W. Rippmann tr. *Viëtor's Elements of Phonetics* 77 As a rule the English sounds [in *thou, thin* etc.] are *postdental*, the narrowing being between the tongue point (with apical articulation) and the back of the front upper teeth. **1933** [see *DOMAL *a.* 3]. **1933** Bloomfield *Language* vi. 102 French speaks its [n] in postdental position. **1961** L. F. Brosnahan *Sounds of Language* vi. 138 The order of appearance of consonants is generally from back to front, in the order: glottal, velar, post-dental, palatal, labial, and labio-dental. **1906** J. B. Smith *Explanation Terms Entomol.* 106 Post-humeral bristles: in *Diptera*, are usually two. **1961** Post-humeral [see *ACROSTICHAL *a.* 2]. **1899** W. Rippmann tr. *Viëtor's Elements of Phonetics* 142/1 Postpalatal stops, etc. = back stops, etc. *Ibid.* 67 The *gutturals* (more strictly: postpalatals or velars). **1902** [see *medio-palatal* adj. s.v. *MEDIO- 2]. **1925** [see *pre-palatal* s.v. *PRE- B. 3]. **1942** Post-palatal [see *medio-palatal* adj. s.v. *MEDIO- 2]. **1934** Webster, Post-velar *a.* **1934** J. J. Hogan *Outl. Eng. Philol.* i. 7 If the back-stops are made farther back than the normal series, they are called Post-Velars. **1942** Bloch & Trager *Outl. Linguistic Analysis* ii. 16 Different points of articulation are designated by the terms *pre-velar*, *mediovelar*, and *postvelar* (or *uvular*). **1964** E. Palmer tr. *Martinet's Elem. Gen. Linguistics* ii. 50 A dorsal may also be..post-velar or uvular as in the initial sound of *rouge* in the Parisian pronunciation. **1966** M. Pei *Gloss. Linguistic Terminol.* 215 *Post-velar*, a consonant produced with the tongue farther back than the velar position, and the articulation against the rear half of the velum, or soft palate (Arabic *q*).

b. Rarely in *quasi*-adjectival relation to a sb. forming the second element: = occurring behind or posteriorly, as *POSTFIXATION *sb.* 1.

postable, *a.* Delete *rare*—0 and add example.
1926 *Glasgow Herald* 23 Mar. 9/1 The £40 limitation is wholly inadequate for jewellers and others, whose goods, though of 'postable' dimensions, are of considerable value.

postabortal, -abortion, -abortum, -absorptive: see *POST- B. 1 b. **post-accelerate:** *POST- A. 1 a. **post-acceleration, -accelerator:** *POST- A. 1 b.

postage¹. Add: **6.** **postage meter** *N. Amer.* = *FRANKING MACHINE; hence *postage-metered* adj.
1927, 1961 [see *FRANKING MACHINE]. **1972** *Times* 18 May 27/1 Pitney Bowes..claims to be the world's largest producer and marketer of postage meters and mailing machines. *Ibid.*, It was 50 years ago that Walter H. Wheeler sold the British Post Office on the idea of postage metered mail. **1974** P. Gzowski *Bk. about This Country* 212 I'm sort of in the same position as the stamp-licker watching a postage meter being brought in the door.

postage stamp. Add: **a*.** Also *transf.* and *fig.*
1908 [see *DOD-]. **1930** R. Graves *Ten Poems More* 4 It is a large patch,..The postage-stamp of its departure,..closing in now To a plain countryside of less and less. **1971** M. Tak *Truck Talk* 121 *Postage stamp*, state permits in the form of a small decals that must be placed on a tractor. **1978** J. Gores *Gone, no Forwarding* vii. 42 Kearny began pacing the postage stamp of space behind his desk.
b. *postage-stamp-sized* adj.
1968 A. Diment *Gt. Spy Race* vi. 78 She wore nothing but the white, postage stamp-sized panties. **1968** 'R. Raine' *Night of Hawk* viii. 38 A postage-stamp-sized moustache.
B. as *adj.*, used to denote something very small.
1962 *Housewife* (Ceylon) Feb. 33, I am certainly a veteran at providing the family with reasonably good food on a postage stamp budget. **1965** *New Society* 14 Oct. 5/3 Postage-stamp photos of smiling typists. **1968** J. Wainwright *Web of Silence* 137 A postage-stamp dance floor. **1971** R. Busby *Deadlock* iii. 32 Chrysanthemums were blooming in the postage stamp garden. **1973** 'R. MacLeod' *Nest of Vultures* ii. 40 Spotlights were trained on a postage-stamp stage.
Hence **postage-stamped** *ppl. a.*, supplied with a postage stamp.
1942 Partridge *Usage & Abusage* 241/2 In the British Empire, a postcard may be already postage-stamped or it may require a postage stamp; in the U.S.A., it requires one, an already postage-stamped card being in the States a *postal card*.

postal, *a.* (*sb.*) Add: **b.** **postal ballot,** a method of voting by post; also *attrib.*; **postal card** (U.K. example); **postal code** = *POSTCODE; hence as *v. trans.*, to write a postcode on (a letter, etc.); **postal draft,** †(*a*) in 1914 the form used at Post Offices for the payment of Navy and Army Separation Allowances, later called 'allowance form'; (*b*) a draft or cheque drawn on the Postmaster General, introduced in Jan. 1925 for the payment of National Health Insurance benefits, and later extended to certain Government Departments; **postal note,** (*b*) *Austral.* and *N.Z.*, an order issued by a post office for any required sum and payable at any other post office; **postal order** (further examples); **postal trade,** trade in which orders are received and goods dispatched by post; **postal tube,** trade name for a cardboard tube designed to protect documents, plans, etc., during transmission by post; **postal vote,** a vote in an election, on a resolution, etc., submitted on a special form by post; so *postal voting* vbl. sb.
1945 *Times* 25 May 4/1 An elaborate procedure has been devised for checking all service votes so as to eliminate any proxy vote cast on behalf of a service voter from whom a postal ballot paper is received. **1973** A. Broinowski *Take One Ambassador* i. 15 [The] returning officer for the elections..may be able to put his hand on a postal ballot paper for you. **1974** *Times* 12 Feb. 4/8 Union members should have the opportunity of electing their leaders by postal ballot. **1876** C. M. Yonge *Womankind* xix. 151 Do not come down to slap-dash notes and postal-cards. **1968** *Internat. List P.O.* (Universal Postal Union) (Eng. ed.) ii. 7 In recent years several countries have worked out postal codes designed to facilitate the sorting, routeing and delivery of mail. **1978** S. Naipaul *North of South* i. ii. 63 West 11..If we use only the postal code..they mightn't make the connection. **1969** P. West *Words for Deaf Daughter* v. 139 A two-page spread..sent postage-due, incorrectly postal-coded from some college. **1973** *Times* 17 July 11/5 The Post Office has published a booklet with the postal codes in it. **1929** *Post Office Guide* July 144 Remittances are made by certain Government Departments, etc., by means of Postal Drafts. **1885** *Victorian Year-Bk. for 1884–5* 481 Postal notes were first issued on the 1st January, 1885. **1926** *Austral. Encycl.* II. 318/1 In 1893 [in New South Wales] an inland and intercolonial parcels post was established and the postal-note system introduced. **1962** J. R. Bernard in *Southerly* XXII. ii. 98 A number of words compounded from standard words and attracting to themselves specific meanings are not treated at all in the dictionary. Among them are *bin-boy, bushfire, postal note*, [etc.]. **1973** *Bulletin* (Sydney) 25 Aug. 3 Enclosed please find my cheque/postal note. **1916** A. Huxley *Let.* c 12 July (1969) 106 Business first...this postal order, is *not* for you. ..au contraire, for me. **1974** *Encycl. Brit. Macropædia* XIV. 887/1 Postal orders were introduced in 1881. **1902** *Encycl. Brit.* XXV. 99/2 What is called in England 'postal trade', and in America 'mail order business', is growing very rapidly. **1894** *Country Gentlemen's Catal.* 166 Postal Pockets..Direction Labels, Postal Tubes. **1945** *Times* 25 May 4/1 There will be some inevitable duplication of postal and proxy votes. **1955** *Times* 11 May 14/4 Ministers in Argyll, who have been called to attend the General Assembly of the Church of Scotland in Edinburgh during the time they should vote in the general election, have been refused postal votes. **1971** *Oxf. Univ. Gaz.* 18 Feb. 671/2 More than fifty members of Congregation have required a postal vote on the resolution. **1974** *Times* 15 Jan. 2/6 An individual can claim a postal vote so long as he has moved from one local authority area to another. **1945** *Times* 25 May 4/1 Before these arrangements were made for postal voting by members of the forces a high proportion of them had appointed proxies. **1974** *Times* 15 Jan. 2/6 If an election is called..all claims for postal voting must be filed..a fortnight before election day.

postalize (pōu·stăləiz), *v.* [f. Postal *a.* (*sb.*) + -ize.] To make like the postal system in respect of its fixed prices for delivery, regardless of distance. So **postaliza·tion.**
1893 *Review of Reviews* Oct. 394 Why not postalize Railway Traffic, and go as far as you like for 2½d? **1939** *Sun* (Baltimore) 19 Jan. 14/7 (*heading*) I.C.C. seeks order for investigating postalized rail fares. *Ibid.*, Chairman Marion M. Caskie..suggested today that the commission be given a mandate from Congress before investigating the Hastings plan for postalization of the railroads. **1950** *Economist* 16 Sept. 472/2 There is also some pressure for extending the system of 'postalisation' whereby United Kingdom consumers as a whole bear the transport charges. **1953** *Ibid.* 27 June 894/1 It takes a good deal of economic sophistication nowadays to see anything wrong with the principle of 'postalisation'—the principle of the postal service, that the consumers in thickly populated areas who can be served cheaply should subsidise the high-cost consumers elsewhere. **1966** *Ibid.* 17 Sept. 1166/1 This American gas expert, incidentally, would suggest selling to all boards at the same price, which might certainly interest the boards remote from Britain's east coast; but this idea of 'postalising gas' seems an ancillary frill on this argument.

postally (pōu·stăli), *adv.* [f. Postal *a.* + -ly².] For postal purposes; in the post; as far as postal matters are concerned.
1896 *Rep. Exhib. Sheffield Philatelic Soc.*, There were two letters postally used in 1768 and 1772. **1930** *Observer* 20 Apr. 15/5 It might..be better to show him our very latest additions to Whitehall—although they occur, postally at least, in that part of it called Charing Cross. **1970** *Daily Tel.* 17 Oct. 11/7 Those used for stamp duty, bills, receipts, licences and other fiscal purposes have *some* value, but it is very small indeed compared with the value of postally-used copies. **1972** *Police Rev.* 1 Dec. 1558/3 Neither was it acceptable to display mint (unused) stamps on the same page as those which had been postally used.

post-alveolar: see *POST- B. 2.

post and rail. (In Dict. s.v. Post *sb.*¹ 8 c.) Also hyphenated and in pl. **1.** A post and rail fence (see quot. 1823 in Dict.). Also, materials for post and rail fencing.
[**1641** *Rec. Colony & Plantation New Haven* (1857) 54 Fencing with..strong and substantiall posts and rales.. nott above 18d.] **1778** 'J. H. St. John de Crèvecœur' *Sk. 18th-Cent. Amer.* (1925) 81 Our present modes of making fences are very bad... I have often observed whole lengths of posts and rails raised from the ground in the spring, and the labours of weeks thus destroyed. **1797** H. Newdigate *Let.* 13 July in A. E. Newdigate-Newdegate *Cheverels* (1898) xiii. 184 As far as yᵉ Road is near the Cliff..there is a strong post & Rail all yᵉ Way. **1823** Byron *Don Juan* viii. lv. 138 So was his blood stirred..As is the hunter's at the five-bar gate, Or double post and rail. **1865** [see Post *sb.*¹ 8 c]. **1936** 'J. Tey' *Shilling for Candles* vii. 85 Flight fell with me at a post-and-rails last winter. **1959** J. Verney *Friday's Tunnel* xxxi. 292 The lane ended at the wood's edge in a broken post and rails.
2. *Austral. slang.* **a.** A wooden match.
a **1890** D. B. W. Sladen in Barrère & Leland *Dict. Slang* (1890) II. 147/1 'Alf,' said a great friend of mine to a companion who was engaged with us on a shooting expedition down in Bulu-Bulu, one of the eastern provinces of Victoria, 'Have you got a match?' 'Only a post-and-rails,' was the deprecating reply, responded to with a patronising 'Never mind.' **1941** Baker *Dict. Austral. Slang* 56 *Post-and-rails*,... (2) Wooden matches.
b. = **post-and-rail tea*.
1899 W. T. Goodge *Hits! Skits! & Jingles!* 75 There is 'post and rails' and 'brownie' For yer breakfast now, yer know. **1904** T. Petrie *Reminisc. Early Queensland* ii. iv. 241 The tea then was all green tea, and very coarse, like bits of stick—indeed it was christened 'post and rails'. **1934** *Bulletin* (Sydney) 12 Sept. 9/2 A 'bark hut or log cabin will be erected' so that royalty may not get sunstroke while sipping his 'post-and-rail'. **1966** Baker *Austral. Lang.* (ed. 2) iv. 85 Three old expressions are *ration tea*,..*post-and-rails* (also called *post-and-rail tea*) and *jack the painter*; the second is derived from the pieces of stalk and leaf floating on top.
c. [Rhyming slang for 'fairy tale'.] A lie.
1945 Baker *Austral. Lang.* xv. 271 *Post-and-rail*, a lie (by rhyme on 'fairytale').
3. *attrib.* (sense *1), as *post-and-rail fence, fencing, paddock*.
1684 *Public Rec. Colony of Connecticut* (1859) III. 512 Great parte of my post and rayle fences being feched and burnt by the sowders. **1765** G. Washington *Diary* 6 Nov. (1925) I. 216 Sowing..19 Bushls. in ye large cut within the Post and Rail fence. **1850** H. C. Watson *Camp-Fires of Revolution* 43 A party of our men.. pulled up a post-and-rail fence. **1914** Conrad *Chance* i. ii. 40 She had taken the trouble to climb over two post-and-rail fences only for the fun of being reckless. **1944** M. Morris in *Coast to Coast 1943* 85 Bare paddocks tufted with winter-whitened grass and endless post-and-rail fences regular as printed staves of music. **1973** E. Egleton *Seven Days to Killing* xx. 210 The post-and-rail fence at the bottom of the yard. **1786** G. Washington *Diary* 18 Mar. (1925) III. 30 Post and rail fencing lately erected as yards for my Stud horses. **1944** E. Dithmack in *Coast to Coast 1943* 26 Inside the post-and-rail paddock the leaves of the box-trees glittered. **1976** *Horse & Hound* 3 Dec. 70/4 (Advt.), Good stabling, grazing in post-and-rail paddocks in exclusive parkland setting.
b. Special Combs., as **post-and-rail tea** *Austral.*, strong, roughly made tea with stalks, etc., floating on the top.
1851 [see Post *sb.*¹ 8 c]. **1887** *All Year Round* 30 July 66 The tea so made [in a billy can] is naturally of rather a rough and ready description, and when the stalks and coarse particles of the fragrant leaf float thickly thereon, it is sometimes graphically styled 'post-and-rails' tea. **1936** A. Russell *Gone Nomad* iv. 24 Flour, 'post and rail' tea (the cheapest kind), black sugar, salt and meat, were the only rations provided. **1959** H. P. Tritton *Time means Tucker* i. 10/2 We got the jobs,..at a pound a week and tucker,..the tucker being mutton and damper, post-and-rail tea and brown sugar.

post-A·rmistice, *a.* [Post- B. 1 a.] Of or pertaining to the period immediately following the Armistice of 11 Nov. 1918.
1929 *Times* 22 July 15/2 The reckless capital flotation during the short-lived post-Armistice 'boom'. **1965** G. McInnes *Road to Gundagai* iii. 48 When Terence sat at the keyboard to 'tickle the ivories', rip-roaring post-Armistice America with all its heady intoxication came vividly alive.

post-article, -auricular: see *POST- B. 1 c, 2.

po·stbase. *Linguistics.* [f. Post- B. 1 c + Base *sb.*¹] A derivational suffix.
1958 A. A. Hill *Introd. Linguistic Struct.* viii. 121 The other class of morphemes which can follow a base, we shall call postbases... Postbases, like suffixes, are non-initial. Since more than one postbase can be added to a single base, as in *boy-ish-ness*, a postbase must be described as a morpheme which can follow a base or another postbase. **1962** *Canad. Jrnl. Linguistics* VII. ii. 94 This study shows that the morphophonemics of English words having post-bases ('derivational suffixes') of learned origin can be systematized by a comparatively small number of statements of wide application. **1967** *Ibid.* XIII. i. 22 OE adverbs also fall into subsets marked by comparative and superlative postbases.

post-Bloomfie·ldian, *a.* and *sb.* *Linguistics.* [f. Post- B. 1 b + the name *Bloomfield* (see below) + -ian.] **A.** *adj.* Subsequent to the work of the American linguist Leonard Bloomfield (1887–1949), freq. applied *spec.* to American structural linguistics in the 1950s. **B.** *sb.* An American structuralist.
[**1949** J. R. Firth in *Archivum Linguisticum* I. 110 Slav linguistics..are certainly post-Saussurean and showing signs even in America of becoming post-Bloomfield.] **1961** F. W. Householder in Saporta & Bastian *Psycholinguistics* 16/1 It would seem to a naive observer that the question 'what is the grammar for?' is an obvious one. Nevertheless, until recent years, no serious attempt to answer it seems to have been made by post-Bloomfieldian linguists. **1963** F. G. Lounsbury in J. A. Fishman *Readings Sociol. of Lang.* (1968) 63 The post-Bloomfieldian dogma in American linguistics has long upheld the desideratum of exclusion of semantic considerations from linguistics. **1965** *Foundations of Lang.* I. 92 It remains quite clear that post-Bloomfieldian linguistics was preoccupied with accounting for the corpus of speech-utterances. **1966** A. A. Hill *Promises & Limitations Newest Type Gram. Anal.* 11 Post-Bloomfieldian structuralists produced rather remarkably few grammars. **1970** *Jrnl. Linguistics* VI. ii. 287 Structuralist phonemics of the so-called Post-Bloomfieldians, for instance, is only a set of practical procedures. **1970** G. C. Lepschy *Survey Structural Linguistics* 11 'Structural linguistics' in a more limited sense (the post-Bloomfieldians). **1971** D. Crystal *Linguistics* iv. 208 The term 'structuralism' was often used, in fact, in a narrow sense to refer to the kind of emphases which characterized post-Bloomfieldian linguistics (in a broad sense, of course, *all* linguistics is structural).

post-boarding, *vbl. sb.* [f. Post- A. 1 a + Boarding *vbl. sb.*] The shaping of a garment by heating it on a form after it is dyed, rather than before. So **post-board** *v. trans.*
1952 *Dyeing of Nylon Textiles* (I.C.I.) xii. 135 When high-temperature post-boarding techniques are employed the scarlet and blue components..are not sufficiently fast to sublimation. **1963** Meitner & Kertess tr. *Schmidlin's Preparation & Dyeing Synthetic Fibres* ii. 39 After..setting..the stockings are dyed and subjected to post-boarding to impart the right type of handle. **1963, 1970** [see *PREBOARD *v.*].

post-boy. Add: **3.** *Austral.* = *Jacky Winter* s.v. *Jacky 3.
1911 J. A. Leach *Austral. Bird Bk.* 121 Australian Brown Flycatcher, Jacky Winter, Post-boy, Post-sitter. **1931** N. W. Cayley *What Bird is That?* 64 Brown Flycatcher... Also called Peter-Peter, Post-boy, Post-sitter. **1969** [see *Jacky Winter* s.v. *Jacky 3].

postcard. Add: **2.** *attrib.*, as *postcard album, flower, -monger, -photograph, poll, portrait, stand, survey, system*; *postcard-size* adj.
1899 Postcard album [in Dict.]. **1907** *Yesterday's Shopping* (1969) 436c Post Card Albums. **1929** R. Graves *Poems* 21 Post-card flower of Kodak mud. **1938** *New Statesman* 13 Aug. 241/2 He can make his way..to..the Museums, outside which eager postcard-mongers will sell him views of Westminster Abbey and the Tower. **1920** T. P. Nunn *Education* x. 126 A postcard-photograph of a yacht. **1909** *Daily Chron.* 19 Mar. 1/6 There had been strong opposition..to the Sunday concerts, and a postcard poll was taken. **1907** *Yesterday's Shopping* (1969) 436c These Albums have been designed to hold the Post Card Portraits now so popular. **1926** *Paper Terminol.* (Spalding & Hodge, Ltd.) 21 Post Card size... Generally applied to a board measuring 22½ × 28 in., out of which 32 official post cards may be cut. **1973** R. Busby *Pattern of Violence* iv. 60 A handful of postcard-size prints. **1907** *Yesterday's Shopping* (1969) 428 Post card stand. **1948** *Shell Aviation News* No. 120. 5/3 The study is based on a post-card survey of about 17,000 aircraft owners, and shows that an estimated 9,800,000 hours were flown by non-scheduled aircraft in 1946. **1897** Post-card system [in Dict.].
b. Designating something picturesque, as *postcard land, sky, view*.
1958 *Spectator* 14 Feb. 204/1 A postcard land of blossom, and bridges humped over gurgling streams. **1959** *Woman's Own* 16 May 13/1 The shining sea. The postcard sky. **1959** *Listener* 12 Mar. 459/1 He [*sc.* Utrillo] was undoubtedly clever at enlarging, squaring-up, colouring and feeling his way into postcard views of paintable *motifs*. **1979** M. A. Sharp *Sunflower* v. 45 They were crossing the Triboro Bridge, with its postcard view of the city.
c. Special Combs., as **postcard beauty,** a fashionable beauty whose picture appeared on postcards which were collected by admirers.
1924 G. B. Shaw *Let.* in *To a Young Actress* (1960) 66, I think you will gravitate towards literature after a reign as a postcard beauty. **1958** *Sunday Times* 28 Sept. 4/7 She was included among the postcard beauties of her musical-comedy days.

po·stcard, *v.* [f. the sb.] **a.** *trans.* To communicate with or inform by postcard. **b.** *intr.* To send a postcard.
1910 *Westm. Gaz.* 2 Feb. 5/3 (Advt.), Patterns ready for sending by return post. Postcard us to-day. **1947** *Ki-grams* (Washington, D.C. Kiwanis Club) 6 Feb., Zeddie Blackistone post-cards about the flowers, the sunshine and golf at Palm Beach.

post-cart. (Further examples.)
1926 O. Schreiner *From Man to Man* ix. 339 At half past four the postcart driver had inspanned his horses. **1949** L. G. Green *In Land of Afternoon* ii. 33 The old post-cart drivers used to halt between Ashton and Montagu.

post-centrally: see *POST- B. 2.

post-Chauce·rian, *a.* and *sb.* [Post- B. 1 b.] **A.** *adj.* After the lifetime of Chaucer; *spec.*, of a poet: writing after, and influenced by, Chaucer. **B.** *sb.* A post-Chaucerian poet.
1933 R. Tuve *Seasons & Months* ii. 70 One may see this Lydgatean influence in most of the poets of the 'post-Chaucerian' school. *Ibid.* iii. 71 English poetry from the time of Chaucer and the post-Chaucerians to that of the eighteenth-century pastoral. **1966** *Eng. Stud.* XLVII. 172 The foregoing are the only instances of post-Chaucerian *spiced conscience* that I have been able to discover. **1967** P. J. Bawcutt *Shorter Poems of Gavin Douglas* p. xxx, Douglas must have been well read, not only in Chaucer and the post-Chaucerian poets of the fifteenth century, but in medieval Latin poetry.

post-Chri·stian, *a.* and *sb.* [Post- B. 1b.]
A. *adj.* **a.** Subsequent to the lifetime of Christ, or to the rise of Christianity. **b.** Subsequent to the decline or rejection of Christianity.
1864 [see Post- B. 1]. **1888** T. K. Cheyne et al. *Bible* (Variorum Teacher's ed.) (1893) Pref., The vowel-points merely represent a valuable, but still post-Christian, exegetical tradition. **1929** H. Keyserling *America Set Free* ii. 583 This is not a pre-Christian, but a post-Christian state. **1945** *Downside Rev.* LXIII. 201 He exemplifies that paralysis from which post-Christian philosophy is suffering. **1956** K. Clark *Nude* iii. 109 The yearning for another world had entered the post-Christian spirit. **1974** *Listener* 24 Jan. 121/3 Heliogabalus..reduced Rome to a kind of post-Christian Sodom and Gomorrah.
B. *sb.* A person in a nominally Christian society who has no professed religion.
1946 *Downside Rev.* LXIV. 117 Or we may have wondered for whom the book was written, for Mgr Knox addresses sometimes the post-Christian, sometimes the 'Sunday Mass' Catholic and sometimes his fellow priests. **1958** *Spectator* 4 July 10/3 A generation of men and women who have grown up completely outside institutional Christianity. They are not lapsed Christians... But neither are they formal agnostics. They are best defined as post-Christians. **1961** *Christian Century* 18 Jan. 80/2 Australian Catholics sincerely believe that their countrymen would be better off as Catholics than as post-Christians.

post-classic, *a.* Add: **b.** *spec.* Usu. with initial capital. Applied to that period (*c* 900 to 1520) of Meso-American civilization succeeding the Classic.
[**1956** J. E. S. Thompson *Rise & Fall of Maya Civilization* ii. 96 Burials have also been found in collapsed rooms at other sites, notably a burial with pottery of post-classical types at Copan.] **1965** R. F. Spencer et al. *Native Americans* xi. 443/1 The Post-Classic period of Mesoamerican culture history essentially amounts to reorientation in Mesoamerican cultural foci... Many of the characteristics which are typical of the Post-Classic were present in the Classic Period. **1973** *Times* 26 July 18/4 Several pottery vessels..and the architecture of the house suggested connexions with the northern part of Yucatan in the Early and the Middle Postclassic periods. **1975** *Sci. Amer.* Oct. 73/2 The most neutral of scholars' terms for this period in Middle America is Last Post-classic; many simply call it 'the decadent period'.

post-classical, *a.* (Examples in *Mus.*)
1947 A. Einstein *Mus. Romantic Era* xii. 167 But Bruckner did not write 'Post Classical' or Romantic church music in the style of Schubert. **1969** *Listener* 4 Sept. 320/1 In one week, between 4 and 10 September, the Proms give a most interesting perspective of post-classical symphonic development.

post-climax: see *POST- B. 1 c.

postclitic (pōustˌkli·tik). *Linguistics.* [f. Post- A. 1 b + *CLITIC.] An unemphatic word stressed as part of the preceding word.
1963 Bloomfield & Newmark *Linguistic Introd. Hist. Eng.* iv. 147 *The* is a proclitic in NE..and *one*, as in *a bad one* or *the green one*, is a postclitic. **1965** *Language* XLI. 473 As many as five syllables occur prestressed, but only one syllable occurs poststressed, except rarely when a postclitic follows a suffix.

post-cly·peus. *Ent.* [Post- A. 2 b.] In certain insects, the upper section of a divided clypeus.
1888 J. H. Comstock *Introd. Entomol.* 10 When these [*sc.* two sclerites of the clypeus] are distinct they are designated as the ante-clypeus and post-clypeus respectively. **1926** R. J. Tillyard *Insects Australia & N.Z.* i. 12 In others [*sc.* insects] the clypeus may be divided into two parts, an anteclypeus and a postclypeus. **1937** C. Longfield *Dragonflies Brit. Isles* 13 In the *Anisoptera* the frons is the most prominent portion of the two, in the *Zygoptera* it is the postclypeus. **1957** Richards & Davies *Imms's Textbk. Entomol.* (ed. 9) i. 19 In some insects the clypeus is partially or completely divided by a transverse suture into two sclerites—the post-clypeus and the ante-clypeus.

postcode (pōu·stˌkōud). Also with capital initial. [f. Post *sb.*² + Code *sb.*¹] A series of letters or numbers, or both, allocated to postal areas to facilitate the automatic sorting and speedy delivery of mail. Also *attrib.* Hence **po·stcoding** *vbl. sb.*
1967 *Telegraph* (Brisbane) 18 May 3/1 The Post Office will allocate every city, town, suburb and small centre in Australia a four-figure postal location number called Postcode. Announcing this today, the Postmaster-General..said the Postcode system would enable the Post Office to handle the growing volume of mail more quickly, by taking full advantage of electronic mail coding equipment. **1968** *Times* 24 July 9/7 On the back of it is stamped the instruction 'Remember to use the postcode'. **1969** *Daily Tel.* 5 June 21/6 Postcoding is to be extended to cover all of Britain's 20,000,000 addresses instead of only 75 main centres. **1969** *Guardian* 9 Aug. 8/3 The first two letters of the Postcode route the letter to the distant mail-handling centre. **1971** *Sunday Times* (Colour Suppl.) 20 June 17 A postman at the letter-coding desk has each envelope put in front of him and types out its postcode on a keyboard at his fingertips. **1973** *Guardian* 7 Feb. 1/1 Postcodes are to be put on road signs. The experiment will start in the new town of Milton Keynes, Bucks.

post-common. (Later Hist. examples.)
1879 T. F. Simmons *Lay Folks Mass Bk.* 307 They were to kneel again during the post-common, as was the English name for the prayer after the communion. **1882** G. H. Forbes *Missale Drummondiense* 26 This Postcommon is found without any variation in Gerbert p. 294 b.

post-conquest(ual, -consonantal, etc., **-convulsive:** see *POST- B. 1. **postcranial(ly, -cricoid:** *POST- B. 2.

post crown. *Dentistry.* Also with hyphen and as one word. [f. Post *sb.*¹ + Crown *sb.*] A prosthetic dental crown held in place by a post or wire sunk into the root of the tooth.
1905 G. Evans *Pract. Treat. Artific. Crown-, Bridge-, & Porcelain Work* ii. 58 Each of these conditions causes fracture of roots carrying post or dowel crowns. **1936** J. R. Schwartz *Cavity Prep. & Abutment Construction in Bridgework* xvii. 183 The earliest attempts to incorporate esthetics in the mechanical procedures of tooth restoration was the banded dowel, or postcrown, with a porcelain facing, presented by Dr. C. M. Richmond, many years ago. **1963** C. R. Cowell et al. *Inlays, Crowns, & Bridges* viii. 90 An effective temporary post-crown consists of acrylic resin on a preformed post or wire. **1974** L. J. Leggett in Harty & Roberts *Restorative Procedures for Practising Dentist* xvi. 216 A post crown is indicated, in a root-filled tooth, where there is insufficient bulk or strength in the remaining crown to support a jacket crown.

post-Cu·bist, *a.* and *sb.* *Art.* [Post- B. 1b.] **A.** *adj.* Subsequent to Cubism. **B.** *sb.* A post-Cubist painter.
1927 R. H. Wilenski *Mod. Movement in Art* 147 Imitations of..Post-Cubist art are not confined to actual works of painting. **1937** *Burlington Mag.* Dec. p. xx/1 Cubists and Post-Cubists. **1959** H. Read *Conc. Hist. Mod. Painting* v. 147 His post-Cubist development cannot be dissociated from the typical manifestations of Surrealism.

po·stcure, *sb.* (and *v.*). [f. Post- A. 1 b + Cure *sb.*¹] The action or process of further curing plastic (or an article of it) after fabrication, so as to complete the cure.
1957 B. A. Dombrow *Polyurethanes* iii. 35 In a matter of minutes, the entire foam is produced in place and ready for post-cure, if such is necessary. **1958** D. J. Duffin *Laminated Plastics* iv. 75 It has been found advisable to place the part in a second oven at lower temperatures..to effect a postcure. **1967** *Electronics* 6 Mar. 324 The finished coating requires no post-cure, has very high cut-through temperature and superior resistance to high temperatures. **1968** *Encycl. Polymer Sci. & Technol.* IX. 18 A typical postcure..might involve 2 hr at 275°F, followed by 2 hr at 150°F.
So as *v. trans.*; **po·stcuring** *vbl. sb.*
1956 *Brit. Plastics* XXIX. 453/1 Post curing, for example, three hours at 80°C, will eliminate all residual peroxide and bring many laminates to a stable condition. **1964** Saunders & Frisch *Polyurethanes* II. vii. 115 Postcuring of the sponge was carried out for 1 hr. at 100°C. **1969** Monroe & Chitwood in G. Lubin *Handbk. Fiberglass & Advanced Plastics Composites* xiv. 325 If warpage is a problem, then the part must be post-cured on a post-cure fixture which holds the part to contour. **1975** A. T. Radcliffe in Whelan & Brydson *Devel. with Thermosetting Plastics* v. 61 The laminate should be initially cured at 50°C and subsequently postcured at 80°C.

postcyclic(al, -cyclically: see *POST- B. 1.

post-da·m, *v.* *Dentistry.* [f. Post(erior *a.* and *sb.* + Dam *sb.*¹] *trans.* and *absol.* To construct a ridge along the upper posterior border of a palatal denture, which will press against the soft palate and form a seal. So **post-da·mming** *vbl. sb.*, the construction of such a ridge.
1910 J. W. Greene *Clin. Course Dental Prosthesis* 47 We must have a way to post-dam, in such cases, without the patient's help. *Ibid.* 50 After you post-dammed its [*sc.* the impression's] rear end, it drops. *Ibid.*, You overstrained the palate in post-damming. **1945** F. W. Craddock *Prosthetic Dentistry* iii. 56. Finally, the upper impression is post-dammed with a further addition of softened wax or compound across the posterior palatal edge. **1953** H. R. B. Fenn et al. *Clin. Dental Prosthetics* v. 115 Post-damming is a means of increasing pressure over an area in order, either to control an impression material, or to improve peripheral seal.
Hence **post-dam** (also **post dam**) *sb.*, a ridge of this kind.
1932 E. H. Mauk in Turner & Anthony *Amer. Textbk. Prosthetic Dentistry* (ed. 6) iii. 133 Dr. J. Ewell Neal.. makes special reference to the notch or groove in the bone posterior to the tuberosity on each side where the quantity and character of the soft tissue are such as to tolerate, without injury, more 'post-dam' than was usually employed. **1940** M. G. Swenson *Compl. Dentures* xiii. 167 The anterior line of the post dam area is determined by palpating in the mouth, in order to keep the post dam pressure on soft tissue. **1945** F. W. Craddock *Prosthetic Dentistry* iii. 65 The post-dam extends from tuberosity to tuberosity and will generally be about ⅛ inch wide and $\frac{1}{32}$ to $\frac{1}{16}$ inch deep. **1953** H. R. B. Fenn et al. *Clin. Dental Prosthetics* x. 277 Indirectly the post-dam seal influences phonation, for if it is inadequate the denture may become unseated during the formation of those sounds having an explosive effect. **1976** R. M. Basker et al. *Prosthetic Treatm. Edentulous Patient* ix. 110 This projection, the post-dam, compresses the palatal mucosa once the denture is placed in the mouth and thus creates a border seal.

post-date, *v.* Add: **2.** *trans.* To belong to a later date than (something).
1909 in Webster. **1955** [see *FERRICRETE]. **1971** *Daily Tel.* 5 Aug. 8/1 The mineral miracle, which post-dates his first visit to the Antipodes in 1965, has now made it clear that Australia has more to offer..than sun and surf. **1971** I. G. Gass et al. *Understanding Earth* i. 37 (*caption*) The granites post-date metamorphic rocks. **1971** *World Archaeol.* III. 170 The hearth must have post-dated the last use of the pit.
Hence **post-da·ting** *vbl. sb.*
1963 *N. & Q.* Jan. 16/2 (*heading*) A post-dating. **1968** *Times Lit. Suppl.* 11 Jan. 37/4 There are post-datings, that is, evidence for the later existence of words that OED supposes obsolete.

post-deflection, -depositional: see *POST- B. 1.

postdiction (pōu·stdikʃən). [f. Post- A. 1 b + L. *dictiōn-em* a saying, speaking, after Prediction *sb.*] (The making of) an assertion or deduction about something in the past. So (as a back-formation) **postdi·ct** *v. trans.*, to assert or imply something about (something in the past or the present).
1940 J. Laird *Theism & Cosmology* v. 169 If, however, the future be indeterminate before it occurs, it cannot be fixed before it occurs. For there is nothing determinate to fix. Hence inferential prediction has quite a different status from inferential post-diction. **1952** *Mind* LXI. 40 Inductive sentences may also postdict a past event, *e.g.*..'It is probable that Caesar crossed the Rubicon.' **1960** *Commentary* June 487/1 The Gluecks' five-factor scale in effect *post*-dicted delinquency in the Boston boys from whom it was developed. *Ibid.*, Post-diction is easier than prediction. **1966** *Amer. Speech* XLI. 208 The auxiliary of prediction *will* combines with inflectional ending of postdiction *-ed* to produce a form, *would*. **1971** *Computers & Humanities* V. 192 Discriminant analysis of 30 content variables and 192 editorials..postdicted correctly the masthead of most editorials.

postdoc (pōᵘstˌdǫ·k), *sb.* and *a.* Colloq. abbrev. of *POST-DOCTORAL *a.* and *sb.*
1970 *New Scientist* 21 May 368/3 In 1970 55 per cent of them are 'postdocs' as compared with 25 per cent in 1967. *Ibid.* 17 Dec. 519/3 Today the 'post-doc' appointment is referred to as a 'holding pattern'. **1971** *Ibid.* 27 Apr. 264/1 Meanwhile, the team has been enlarged by three new post-docs. **1974** *Nature* 8 Nov. p. xxix/2 (Advt.), Wanted. A postdoc in population biology interested in genetic structure of populations and species. **1978** *Ibid.* 18 May 182/1 [In the EMBL central laboratory] there is also a considerable number of 'fellows'—mainly post-docs who bring their own funds—and visiting scientists.

post-do·ctoral, *a.* [Post- B. 1 b.] Of, pertaining to, or for the purpose of advanced research by persons already holding a doctor's degree. Also as *sb.*, one who is undertaking or has undertaken a period of study or research subsequent to obtaining his or her doctor's degree.
1939 Webster Add., Postdoctoral *adj.* **1956** *Nature* 10 Mar. 456/1 As in Britain, there is considerable competition for the junior academic posts and post-doctoral fellowships. **1958** *Times* 8 May 2/3 (Advt.), Applications are invited for post-doctoral grants for research under the general direction of Professor R. N. Hazeldine. **1961** *Times* 26 Sept. 1/4 Postgraduate or post-doctoral work. **1966** *Rep. Comm. Inquiry Univ. Oxf.* I. 127 They [*sc.* junior research fellowships] are awarded to encourage individuals at the D.Phil. or immediate post-doctoral stage. **1971** *Nature* 10 Sept. 87/2 Physicists are the worst hit single group of scientists, with 2·9 per cent of the PhDs and 3·3 per cent of the postdoctorals out of a job. **1977** *Time* 14 Mar. 40/2 During his two-year stint as a post-doctoral fellow at the Max Planck Institute for Biochemistry near Munich, R.G., 27, impressed his boss as 'probably the most diligent man I've ever known'.

post-docto·rate, *a.* [f. Post- B. 1 a + Doctorate *sb.*¹] = *POST-DOCTORAL *a.*
1939 in Webster Add. **1957** [see *BOOSTER 2 d]. **1959** *Times* 17 Mar. 2/5 Applicants should be graduates (preferably post-doctorate).

post-e·cho. [Post- A. 1 b.] A faint repetition of a loud sound occurring in a recording soon after the original as a result of the accidental transfer of signals in a recording medium. Cf. *PRE-ECHO.
1956 G. Slot *Hi-Fi from Microphone to Ear* xi. 151 The pre-echoes and post-echoes sometimes heard on gramophone records are usually due to this effect. **1962** *Times Lit. Suppl.* 19 Oct. 810/5 The vocal perspective is ingeniously varied to suggest the changing scene, but the fortissimo passages are occasionally marred by post-echo. **1977** *Gramophone* Mar. 1476/1 Each layer of tape tends to magnetize its recorded signal on to the adjacent layers during storage—in the process known as 'print-through'—to produce faint pre-echoes and post-echoes.

post-e·ditor. [Post- A. 1 b.] Someone who edits text that has been produced or processed by a machine.
1953 A. D. & K. H. V. Booth *Automatic Digital Calculators* xvii. 206 Reiffler suggests the use of a pre-editor who need not know the (*T.L.*) at all but who, reading his own language (our *F.L.*), eliminates from the text all such alternatives... In the same way a post-editor might be used to turn the translation into acceptable *T.L.* **1960** E. Delavenay *Introd. Machine Transl.* 121 For each Russian polysemantic word the machine has given alternative translations; the one selected by the post-editor has been given first. **1966** D. G. Hays in *Automatic Transl. of Lang.* (NATO Summer School, Venice, 1962) 146 Before text reaches the posteditor, it has been put through automatic dictionary lookup and automatic sentence-structure determination.
So **post-edit** *v. trans.*, to edit in such circumstances; **post-e·diting** *vbl. sb.*
1957 *Year's Work Eng. Stud. 1955* 41/2 His suggestions for mechanized translation involve a much heavier reliance upon 'post-editing' than experts at present feel they need to assume. **1960** E. Delavenay *Introd. Machine Transl.* iii. 29 He postulated the necessity both for pre-editing texts before translation and for post-editing them when translated. **1962** *Times Lit. Suppl.* 20 Apr. 268/3 The machine output will have to be 'post-edited', before submission of the finished product to the readers. **1971** *Computers & Humanities* VI. 45 A printer's computer tape would probably require some post-editing.

posteen, postin. (Further examples of forms poshteen, -tin.)
1895 Kipling *Day's Work* (1898) 212 William, wrapped in a *poshteen*—silk-embroidered sheepskin jacket trimmed with rough astrakhan—looked out with moist eyes and nostrils that dilated joyously. **1910** *Encycl. Brit.* I. 314/1 Poshtins (sheepskin clothing) and the many varieties of camel and goat's hair cloth..are still the chief local products of that part of Afghanistan. **1961** P. Fleming *Bayonets to Lhasa* xi. 151 Saved from death by his thick poshteen. **1973** 'W. Haggard' *Old Masters* viii. 99 Bentinck was wearing moleskin trousers and..a sort of poshteen which would keep out the cold.

post-Einstei·nian, *a.* [Post- B. 1b.] Subsequent to the work of Albert Einstein (see *Einstein); involving or pertaining to concepts developed later than the theories of relativity.
1938 S. Chase *Tyranny of Words* viii. 89 Bridgman develops various concepts for 'length' in post-Einsteinian terms. **1953** *Authentic Sci. Fiction Monthly* Apr. 137 With such rare exceptions as Bradbury, writers seem..to accept a relativity of moral standards which, while it may display a misleading harmony with the post-Einsteinian scientific outlook, is in fact simply a well of poison. **1973** B. Magee *Popper* vii. 97 Post-Einsteinian physics,.. post-Freudian psychology,..post-Keynesian economics, ..post-Frege logic.

post-emergence: see *POST- B. 1.

post-e·ntry, *a.* [Post- B. 1 a.] Occurring after entry; *spec.* applied to a closed shop in which new employees are required to join a union after appointment. Cf. *PRE-ENTRY *a.*
1964 [see *PRE-ENTRY *a.*]. **1972** H. Williamson *Trade Unions* (ed. 2) ii. 22 In a post-entry closed shop membership of a particular union is only insisted upon after the worker has been appointed. **1976** *Milton Keynes Express* 16 July 13/2 (Advt.), Clerical Assistant (Income Section). .. National conditions; superannuation; five day week; post-entry training facilities. **1978** R. Taylor *Fifth Estate* i. 18 Employers..have written post-entry closed-shop clauses into their collective agreements with unions.

poster¹. **3.** (Earlier example.)
1797 H. Newdigate *Let.* 9 July in A. E. Newdigate-Newdegate *Cheverels* (1898) xiii. 181 A good dinner..was ready to come upon Table when yᵉ Posters arrived.

poster². Add: **3.** *poster art, design, designer* (examples), *-designing, panel*; **poster colour** = next; **poster paint,** an opaque paint with a water-soluble binder such as is used on posters; **poster paper,** paper used in bills or posters; **poster session,** a meeting of scientists at which their work is represented by displayed pictures (see quot. 1977).
1920 Hardie & Sabin *War Posters* ii. 7 Poster art and pictorial art have essentially different aims. **1974** *Listener* 28 Feb. 283/3 Shostakovich's Twelfth Symphony. 'Poster art' it may be; but if so, some more Russian posters, please. **1925** *Studio* (Art & Publicity No.) Autumn (verso front cover), The 'Kingsway' Poster Colours..supplied in Tubes and Glass Pots. **1973** Poster colour [see *POP *a.* (*sb.*⁴) 3]. **1920** Hardie & Sabin *War Posters* ii. 14 The number of poster designs from his hand is at least fifty. **1970** *Oxf. Compan. Art* 900/1 English poster design in the 20th c. varies between the illustrative..and the work of great merit done for Shell-Mex. **1901** W. S. Rogers *Bk. of Poster* xix. 133 The standard of art-education amongst the poster designers is higher than in this country. **1935** [see *copy-writer*]. **1954** T. Eckersley *Poster Design* 7 The illustrations in this book are by leading poster designers. **1901** W. S. Rogers *Bk. of Poster* xix. 129 The scope of this work does not embrace the idea of treating the subject of poster designing at any great length. **1933** L. Richmond *Technique of Poster* p. v, Poster designing is a serious but most interesting art. **1939** W. Clemence *Man. Postercraft* xii. 92 Good quality flat poster paints, thinned with a little turpentine..blend well, are smooth of application, and dry in excellent time and condition. **1974** I. Murdoch *Sacred & Profane Love Machine* 103 She.. laid out his coloured pencils, his poster paints, brushes, water. **1929** *Encycl. Brit.* XVIII. 319/2 On the poster panels of to-day, in the United States, may be seen the work of Harrison Fisher. **1901** J. Beveridge *Paper-makers' Pocket Bk.* i. 26 German classification and sizes of papers..Affichenpapiere (thin poster paper). **1959** R. Hostettler et al. *Technical Terms Printing Industry* (ed. 3) 130/2 *Poster paper*,..papier pour affiches. **1976** *Physics Bull.* Aug. 335/3 Although it is clear that the technique derives from the scientific exhibition, the exact origin of poster sessions is unknown. **1977** *Sci. Amer.* Apr. 51/2 (Advt.), All contributed 'papers' are to be given in poster sessions, where from a visual display one decides whether to stop and exchange thoughts person to person.

poster³ (pōu·stəɹ). *Rugby Football.* [f. Post *sb.*¹ + -er¹.] A ball that passes directly over the top of one of the goal-posts.
1867 *Routledge's Handbk. Football* 35 If the ball..rises directly over the end of one of the posts, it is called a *poster*, and is no goal. **1930** *Times* 14 Mar. 7/1 He played in three International matches, dropped a goal in two, and scoring [*sic*] a 'poster' in the other.

postered (pōu·stəɹd), *a.* [f. Poster² + -ed².] Depicted or described on posters; adorned or disfigured with posters.
1916 S. Kaye-Smith *Sussex Gorse* iii. 173 Rye electors were confronted with the postered virtues and vices of Captain Mackinnon (Radical) and Colonel MacDonald (Conservative). **1927** *Scots Observer* 7 May 9/2 The real blemishes of Glasgow are raw and postered gable-ends, untended waste grounds, [etc.].

posterior, *a.* and *sb.* (*adv.*) Add: **A.** *adj.* **1. b.** *Statistics.* Applied to the result of a calculation made subsequent to, and in consideration of, some observation(s); *posterior probability*, the probability that a hypothesis is true, calculated in the light of relevant observations. Opp. *PRIOR *a.* (*adv.*) A. c.
1921 *Phil. Mag.* XLII. 387 Even if the prior probabilities of two laws with different domains are notably different, the effect of several verifications of each is able to make the posterior probabilities of the two laws practically equal to each other and to unity. **1931** H. Jeffreys *Sci. Inference* ii. 18 The posterior probability of *p* is the prior probability of *p* divided by the prior probability of the consequence. **1943** M. G. Kendall *Adv. Theory Statistics* I. vii. 179 A further difficulty arises if *θ* can lie in an infinite range, for then Bayes' postulate apparently leads to the conclusion that prior probabilities in any finite range are zero and hence so are posterior probabilities. **1972** A. W. F. Edwards *Likelihood* iv. 46 The posterior odds of two hypotheses on some data is equal to the product of the prior odds and the likelihood ratio. *Ibid.* iv. 48 In practice it is determined by the fact that both the prior and the posterior distributions of *θ* integrate to unity. **1977** *Lancet* 13 Aug. 339/1 If the sister of a hæmophiliac initially has two unaffected boys, the 'posterior' probability of her being a carrier falls from 1/2 to 1/5.
B. *sb.* **2. a.** Now usu. *sing.* The rump or backside (of a person). *colloq.*
1936 G. B. Shaw *Simpleton* Prologue ii. 27 He shoots his foot against the E.O.'s posterior and sends him over the cliff. **1976–7** *Sea Spray* (N.Z.) Dec./Jan. 90/1 (Advt.), It is soft so that a crewman winding the spinnaker sheet winch down aft can rest his posterior on it.

posterish (pōu·stəriʃ), *a.* [f. Poster² + -ish¹.] Characteristic or suggestive of a poster or posters. So **po·sterishness.**
1930 *Aberdeen Press & Jrnl.* 25 Apr. 6 Norah Neilson Gray still seems to us to be straying too much towards a sweet but too pretty posterishness in her work. **1931** *Times Lit. Suppl.* 25 June (Suppl.) p. vi/4 Several cover-designs achieve posterish attractiveness. **1947** *Sat. Rev. Lit.* (U.S.) 1 Mar. 30 Figures in a flat, cylindrical, posterish style. **1958** *Times* 15 Dec. 3/5 There is a painting in Sickert's posterish vein. **1967** *Listener* 16 Mar. 355/2 The formalized, abstracted figures have become posterish and realistic like those in a Soviet hero painting.

posterist (pōu·stərist). [f. Poster² + -ist.] A designer of posters; a poster-artist.
1901 W. S. Rogers *Bk. of Poster* p. v, If I may seem too enthusiastic on the side of Continental work, it is from no lack of sympathy with British posterists. **1968** *Times* 21 Dec. 20/4 The lack of sensitivity apparent in his academic studies..was almost an asset to a posterist. **1972** *Guardian* 19 June 11/4 The best known posterist of the period was John Hassall... His most famous design, 'Skegness is *so* Bracing', was..designed in 1909.

posterize (pōu·stəraiz), *v.* *Photogr.* [f. Poster² + -ize.] *trans.* To print (a photograph) so that there is only a small number of different tones. Also *absol.* Hence **po·sterized** *ppl. a.*, **po·sterizing** *vbl. sb.*
1943 C. I. Jacobson *Enlarging* (ed. 4) 278 Posterising Photographs... One method published by R. W. Wade, achieves something more than simple tone-separation. It does produce in addition, a 'posterised' picture, which is.. a picture in which all the elements are built up out of one or other of four tones only... The range of subjects suitable for posterising is practically unlimited. **1948** F. H. Smith *Photographs & Printer* 135 The photo-engraver can

POSTERO-

readily reproduce a photograph which has already been posterised. 1977 R. Hattersley *Photographic Printing* xviii. 132 Once you know how to make high-contrast negatives with Kodalith..it is relatively easy to posterize. *Ibid.* 133 With four or more tones a posterized print may not even look posterized. Instead it may look like an ordinary print with something just a wee bit odd about the tones.

Also **posteriza·tion**, the process of posterizing.

1950 O. R. Croy *Compl. Art of Printing & Enlarging* ii. 72 (*caption*) Posterization. The print consists only of three tones. 1977 J. Hedgecoe *Photographer's Handbk.* 245 Posterization or tone separation means turning a normal, continuous tone photograph into an image consisting of clearly distinguished areas of flat tone.

postero-. Add: (*a*) *postero-medial*; (*b*) also forming advs., as **posterolaterally, -ventrally.**

1959 *Bull. Mus. Compar. Zool. Harvard* CXX. 185 The segment anterior to Jacobson's organ no longer appears as a transverse slit in frontal section, but now runs antero-medially to posterolaterally. 1967 G. M. Wyburn et al. *Conc. Anat.* iv. 112/2 The great wing of the sphenoid terminating posterolaterally in the spine. 1901 *Proc. Zool. Soc.* I. 263 The characteristic features of this cavity.. are:—..(3) the characteristic position of its postero-medial wall, as seen from behind. 1967 G. M. Wyburn et al. *Conc. Anat.* i. 28/2 The duodenal papilla—on the postero-medial aspect is the situation of the opening of the bile duct. 1902 *Proc. Zool. Soc.* I. 89 The blue sides are margined posteroventrally with a black line. 1961 J. E. Collin *Empididae* I. 31 Middle femora with a comb-like row of tiny black bristles posteroventrally. 1974 D. & M. Webster *Compar. Vertebr. Morphol.* vii. 133 The external oblique muscle..runs anterodorsally to posteroventrally.

post eventum: see *post *Latin prep.*

post-existent, *a.* (Later example.)

1977 G. W. H. Lampe *God as Spirit* iii. 71 Luke was, in fact, unable to make a simple identification of the glorified, 'post-existent' Jesus with the Spirit in the Church.

post factum. Add: *adv. phr.* (Earlier and later examples.) Also as *adj.*

1692 Locke *Some Considerations Money* 12 Unless you intend to break in only upon Mortgages and Contracts already made, and (which is not to be supposed) by a Law, *post factum*, void Bargains lawfully made. 1927 *Mod. Philol.* Nov. 227 New locutions..constantly replace old ones, which, viewed *post factum*, would have been unintelligible, had they remained in use. 1949 Koestler *Insight & Outlook* p. vii, But textbooks are post-factum rearrangements of long and devious processes of inquiry. 1964 V. Nabokov *Defence* viii. 125 I'm very glad *post actum*..but..for the moment it somehow disturbs me. 1972 *New Yorker* 16 Jan. 28/2 Not wanting to use the police emergency number—911—in a post-factum situation.

post festum: see *post *Latin prep.*

po·st-final. *Linguistics.* [Post- B. 1 c.] In a word-final consonant cluster, a consonant following a main final consonant. Cf. *pre-final *sb.*

1933, 1965 [see *pre-final *sb.*].

postfix, *v.* Add: **2.** *Biol.* To fix again after a previous fixation; to treat with a second fixative.

1960 D. C. Pease *Histol. Techniques Electron Microscopy* iii. 49 The small blocks are washed..in the buffer, and then postfixed in buffered osmium tetroxide. 1969 *Jrnl. Cell Sci.* IV. 439 After perfusion blocks were cut from the cerebral cortex which were post-fixed by placing them without washing in a 2% solution of OsO_4. 1976 *Nature* 5 Aug. 494/2 The tissue was immediately fixed in collidine-buffered glutaraldehyde and after a short buffer rinse, postfixed in osmium tetroxide.

postfixa·tion, *a.* and *sb.* **A.** *adj. Biol.* [Post- B. 1 a.] Carried out or used after the fixation of tissue.

1958 J. R. Baker *Princ. Biol. Microtechnique* i. 30 The fixative..may be unsuitable for indefinite preservation... In such cases a post-fixation preservative may be used. 1961 *Jrnl. Biophysical & Biochem. Cytol.* XI. 492 (*heading*) Method for obtaining increased contrast in Araldite sections by using postfixation staining of tissues with potassium permanganate. 1966 R. Mahoney *Lab. Techniques Zool.* v. 285 (*heading*) Post-fixation preservatives.

B. *sb.* **1.** *Anat.* [*post- B. 2 b.] The state of a nerve of being postfixed (sense *2).

1953 G. A. G. Mitchell *Anat. Autonomic Nervous Syst.* xiv. 202 The sympathetic preganglionic fibres issuing from the cord are found chiefly in all the thoracic and upper two lumbar ventral nerve roots, although occasionally prefixation or postfixation may occur in association with coincident shifts of somatic nerves. 1957 *Ann. R. Coll. Surgeons England* XXI. 367 Table II shows that.. there is a much higher incidence of 'postfixation'.

2. *Biol.* [Post- A. 1 b.] Fixation of tissue that has already been treated with a fixative.

1963 *Jrnl. Cell Biol.* XVII. 28/2 For adequate post-fixation in osmium tetroxide of tissues fixed in the aldehydes ..the blocks were washed in buffered sucrose solution. 1967 *Jrnl. Cell. Sci.* II. 379 Post-fixation in osmium tetroxide was carried out for not less than 1 h. 1975 *Nature* 18 Dec. 613/2 Double fixation for blood cells with glutaraldehyde and osmium, followed by uranyl acetate postfixation, was also used.

po·stfixed, *ppl. a.* **1.** [f. Postfix *v.* + -ed[1].] Affixed at the end of or after a word, root, or stem.

1874 [in Dict. s.v. Postfix *v.*]. 1975 *Amer. Speech 1969* XLIV. 107 The endings of the English verbs have evolved in a continuing process from earliest Indo-European times, or perhaps more remotely from earliest Proto-Indo-European-Uralic-Altaic (at which time, one may surmise, they represented nothing more than postfixed pronominal forms).

2. *Anat.* [f. Post- A. 2 a + Fixed *ppl. a.*] Of a nerve: connected to the spinal cord relatively caudally. Cf. *prefixed, prefixt *ppl. a.* 3.

1892 C. S. Sherrington in *Jrnl. Physiol.* XIII. 635 A plexus and its trunks and branches will..be referred to as prefixed if containing spinal root-filaments attached to the cord further forward (headward) than are the root-filaments entering the corresponding trunks and branches of a converse class of plexus which will be referred to conversely as postfixed. 1931 *Proc. R. Soc.* B. CVII. 511 In animals with a 'post-fixed' sacral plexus this has been done by dividing extradurally the 6th and 7th post-thoracic dorsal roots... In animals with a normal or 'pre-fixed' sacral plexus only the 6th post-thoracic dorsal root has been divided.

3. *Biol.* [f. *postfix *v.* 2 + -ed[1].] Treated with a second fixative.

1968 *Jrnl. Cell Sci.* III. 579 The tissue elements in the glutaraldehyde-perfused and OsO_4 post-fixed cortex are separated by narrow extracellular spaces.

postflight, *a.* and *adv.* [Flight *sb.*[1]] **A.** *adj.* [Post- B. 1 a.] Existing or occurring after (a) flight. **B.** *adv.* [*post- B. 1 d.] After flight.

1970 N. Armstrong et al. *First on Moon* iv. 84 There was some suspicion, lingering in the postflight shock of the first Sputnik, that this was the road the Soviet Union had chosen. 1971 *New Scientist* 29 July 249/1 Postflight there was an important development... The cosmonauts.. suffered from severe orthostatic hypotension.

postfo·rming, *vbl. sb.* [f. Post- A. 1 b + Forming *vbl. sb.*] Shaping of thermosetting laminated plastic carried out upon reheating before setting is complete. So **post-fo·rmed** *ppl. a.*, **post-fo·rm** *v. trans.*

1945 *Metals & Alloys* XXI. 392/1 (*heading*) Post-formed laminated phenolic plastics. *Ibid.*, 392/2 The term of post-forming is generally used to designate the procedure by which this material is formed after the laminations have been heat-set. 1945 *Plastics* IX. 266/2 The first important reference to laminated plastics for hot post-forming was probably the announcement of 'Micarta 444' by the Westinghouse Electric and Manufacturing Co. of America. 1965 *Encycl. Polymer Sci. & Technol.* II. 53 To facilitate this postforming ability, a melamine-formaldehyde resin with an added plasticizer is used for impregnation. 1967 P. B. Schubert *Die Methods* II. ii. 53 (*heading*) Designing dies for postforming thermosetting laminates. 1970 *Interior Design* Dec. 753/1 The 78 feet long, chevron-shaped post office counter is entirely clad in postforming grade Formica laminate. 1971 P. Tooley *High Polymers* ii. 60 Polymethyl methacrylate..softens at about 120°C and becomes quite pliable at 160°C. This enables the sheet polymer to be 'post-formed' in the fabrication of smooth streamlined shapes such as windows. 1977 *Cleethorpes News* 27 May 6/7 (Advt.), Worth a second look are..the wide range of worktops in ceramics square and postformed laminates.

post-free, *a.* Add: (Examples.) Also as *adv.*, without postal charge.

1723 *Boston News-Let.* 7 Mar. 2/2 The Publisher... Desires them to send their Accounts Post-Free. 1743 Pope *Dunciad* i. 65 [*Note*] It was a practise so to give the Daily Gazetteer and ministerial pamphlets..and to send them *Post-free* to all the Towns in the kingdom. 1873 *Young Englishwoman* Apr. 202/2 Patterns sent post-free. 1929 *Times* 1 Nov. 16/6 The post-free price for copies ordered direct from the publisher is 5s. 9d. 1980 *Radio Times* 12–18 Jan. 10/1 (Advt.), Send for Littlewoods new catalogue... Post-free. No stamp needed.

post-Freu·dian, *sb.* and *a.* [Post- B. 1 b.] **A.** *sb.* Someone whose psychotherapeutic ideas or practice have developed and diverged from strictly Freudian doctrine; someone whose views have been influenced as a result of Freudian theory. **B.** *adj.* Subsequent to the impact and influence of Freudian ideas. Cf. *Freudian *a.* (*sb.*).

1938 *Essays & Stud. 1937* XXIII. 82 He [*sc.* Balzac] had little to learn of normal psychology from the post-Freudians. 1961 J. A. C. Brown (*title*) Freud and the post-Freudians. 1964 E. Becker in I. L. Horowitz *New Sociol.* 114 Mills has here failed to push on to a fully post-Freudian social psychology. 1966 *Listener* 13 Jan. 64/3 We should have no post-Freudian difficulties in reconciling patience, honour, and kindness with destiny, treachery, and retribution. 1973 *Times* 29 Dec. 12/6 Heavenly rewards hereafter make us post-Freudians suspicious. 1977 R. L. Wolff *Gains & Losses* i. 46 The emotional language frequent in nineteenth-century friendships between persons of the same sex..arouses thoughts in post-Freudian minds that sometimes were not present..in pre-Freudian minds.

postganglio·nic, *a. Anat.* [Post- B. 2.] Of a nerve of the autonomic nervous system: running from a ganglion to the organ which it innervates.

1897 [see *preganglionic *a.*]. 1908 *Jrnl. Physiol.* XXXVII. 139 Prof. Langley remarks that according to him all reflexes in the sympathetic system isolated from the spinal cord are such axon-reflexes either in pre- or postganglionic fibres. 1946 *Nature* 19 Oct. 556/1 Giant fibres which..are distributed, one in each postganglionic stellar nerve, to the mantle musculature. 1968 Passmore & Robson *Compan. Med. Stud.* I. xxiv. 88/2 Postganglionic fibres are unmyelinated. 1974 [see *preganglionic *a.*].

Hence **postganglio·nically** *adv.*

1967 [see *preganglionically *adv.*].

post-genitive: see *post- A. 1 b.

post-glacial, *a.* Add: Also as *sb.*, a post-glacial deposit or period; **post-gla·cially** *adv.*

1928 *Funk's Stand. Dict.*, *Post-glacial* n., a sedimentary deposit resulting from the retreat of a continental glacier. 1937 A. L. Du Toit *Our Wandering Continents* iv. 77 Over extensive areas the ice-front discharged into the ocean or else the sea lay not far away, as indicated by the marine post-glacials. 1949 *Bull. Geol. Soc. Amer.* LX. 1369/2 Some overlapping of ranges has taken place post-glacially, wherever reproductive isolation has gone far enough to permit it. 1957 J. K. Charlesworth *Quaternary Era* II. xliv. 1231 The Dogger Bank..was post-glacially dry land or a vast freshwater fen. 1975 J. G. Evans *Environment Early Man Brit. Isles* i. 8 We at present are in a period optimistically termed the Post-glacial, but which may be an interglacial.

post-grad (pōustgræ·d), *sb.* and *a.* Colloq. abbrev. of Post-graduate *a.* (*sb.*).

1950 [see *grad[1]]. 1973 'D. Shannon' *No Holiday for Crime* (1974) vi. 93 Ron set up another one, one of the postgrad students. 1976 *Eastern Daily Press* (Norwich) 19 Nov. 9/6 (Advt.), Four Cambridge post-grads seek accommodation, Wymondham area. 1979 *Guardian* 26 Oct. 10/2 Best four years of your life..your postgrad days.

post-graduate, *a.* (*sb.*) Add: **A.** *adj.* (Further examples.) Also, *spec.* with reference to a second or further degree. Also *transf.* and *fig.*

1900 *Dialect Notes* II. 48 *P.G.* i.e. *post-graduate*, or *pretty girl*, n. 1. A post-graduate student. 2. A pretty girl. 1931 J. van Dernoot (*title*) Postgraduate contract bridge: advanced points for advanced players. 1952 A. Huxley *Let.* 12 Oct. (1969) 657 Let it be a post-graduate school. 1955 *Publ. Amer. Dial. Soc.* xxiv. 30 Only after they [*sc.* amateur adult criminals] have been exposed to the postgraduate curriculum of prison life do a few of them make possible recruits for the ranks of the professionals. 1965 *Nursing Times* 5 Feb. p. lxii (Advt.), *Post-graduate nurses*... Applications invited for this course from S.R.N.s..and Enrolled Nurses. 1974 *Times* 31 Jan. 3/2 Strong criticism of proposals to give loans to postgraduate students to supplement state grants. 1975 *Language for Life* (Dept. Educ. & Sci.) xxiii. 338 A form of teacher training which has shown a considerable expansion in recent years is the one-year course for graduates, leading to a post-graduate certificate of education. 1978 *Time* 3 July 48/1 Beatty is also a health-food enthusiast and, as Nichols notes, 'a postgraduate hypochondriac'.

B. *sb.* (Examples.) Also *transf.*, and loosely, one who has received a higher degree.

1900 *Congress. Rec.* 19 Feb. 1917/1 Now, the Senator is a senior, a post-graduate of great distinction of the academy of which he is now a member. 1904 M. E. Waller *Wood-Carver of 'Lympus* 178 Marking out the work for the post-graduates..has filled my time. 1932 *Daily Express* 20 Sept. 7/2 Able young post-graduates in America..have a love of knowledge. 1959 *New Statesman* 23 May 730/2 There is the 'grand tour' of the post-graduate, working off his money at the end of a year at Harvard or Princeton. 1972 *Daily Tel.* 22 Sept. 3/7 It..was the brainchild of John Fauvel, 24, a postgraduate in mathematics. 1975 *Physics Bull.* Mar. 129/3 By issuing in both cased and paperback editions, Macmillan have ensured that post-graduates should be able to afford it.

Hence **post-gradua·tion.**

1920 H. Crane *Let.* 14 Apr. (1965) 37 His last letter told of splenetic days following his post-graduation from Columbia.

post-heating: see *post- A. 1 b.

post-histo·ric, *a.* [f. Post- B. 1 b, after *prehistoric.*] Of, belonging to, or pertaining to, an imagined period beyond the close of recorded history or otherwise subsequent to the present historical period; also *transf.* Hence **po·st-history.**

1918 S. Graham *Quest of Face* i. 50 There never leaves his eyes the gleam of something beyond this time, the post-historic. 1953 S. Spender *Creative Element* 133 We live in an epoch without historic precedent, which may indeed be a kind of post-history. 1957 *Times Lit. Suppl.* 15 Nov. 689/3 When the scholars of post-historic China or India study English literature in the twentieth century. 1961 L. Mumford *City in Hist.* i. 4 His dehumanized alter ego, 'Post-historic Man'. 1977 *N.Y. Rev. Bks.* 27 Oct. 31/1 Benjamin's study of what might be called the post-history or the afterlife of works of literature has spurred the recent interest in the 'history of reception' among younger critics in Germany.

postholder. Add: **2.** [Reconstituted from native elements.] One who occupies a post or office; an official.

1976 *Cumberland News* 3 Dec. 32/7 (Advt.), Our Regional Officer..requires a person to look after the Carlisle office for a minimum of two months while the current postholder is away.

post-hole. Add: **1.** Also spec. *Archæol.*, a hole orig. dug to receive a wooden post and usu. packed with stones or clay to support the post. Also *attrib.* other than of implements, as *post-hole evidence, pattern.*

1932 *Times Lit. Suppl.* 28 July 542/3 The misprint 'portholes' for 'postholes'..should not betray the unwary reader into attributing too nautical a flavour to the Iron Age fort of Salmonsbury. 1936 *Discovery* Apr. 99/2 New house-sites were brought to light: some were of wattle-and-daub, and another with a compacted floor and post-holes was similar in plan to one discovered in 1934. 1962 H. R. Loyn *Anglo-Saxon Eng.* i. 43 Photography..and analysis of post-hole evidence have disclosed a series of royal halls. 1963 *Field Archaeol.* (Ordnance Survey) (ed. 4) 37 There are no stone settings or any indications of post-holes. 1971 *World Archaeol.* III. 120 Posthole patterns will be confused. 1977 *Antiquaries Jrnl.* LVII. 261 'Stakehole' is used here to mean the void made by the decay of timber post driven into the ground; 'posthole' means a larger hole excavated in order to insert a post.

2. A hole or well drilled to not very great depth.

1932 *Amer. Speech* VII. 269 [Language of California oil fields.] *Post-hole*, a shallow hole. *Post-hole territory*, shallow productive territory—i.e., in which the oil sands lie at depths up to 2500 feet, 'just under the grass roots'. 1965 G. J. Williams *Econ. Geol. N.Z.* ix. 133/1 The only published information directly relating to the ironsands represents the pioneering work of Nicholson, Fyfe (1958) who put down post-hole bores to depths of not more than 25 ft.

post-I·bsen, *a.* [f. Post- B. 1 a + the name of Henrik *Ibsen* (see *Ibsenism).] Subsequent to the life or works of Ibsen; influenced by the style or views (esp. concerning social reform) of Ibsen. Hence **post-I·bsenist** *a.*, pertaining to or characteristic of drama subsequent to, and influenced by, Ibsen; also as *sb.*, a developer of the methods or views of Ibsen.

1913 G. B. Shaw *Quintessence of Ibsenism* (Completed ed.) 199 The post-Ibsen playwrights. *Ibid.* 206 What I have called post-Ibsenist plays. *Ibid.* 209 The post-Ibsenists. 1919 —— *Heartbreak House* p. xxxvii, The most advanced post-Ibsen plays in the most artistic settings. 1930 E. Pound *XXX Cantos* xxviii. 131 So Loica went out and died there After her time in the post-Ibsen movement.

postical, *a.* For *Obs.* read *rare* and add later examples.

1940 *Chambers's Techn. Dict.* 666/1 *Postical*, relating to or belonging to the back or lower part of a leaf or stem. 1965 F. E. Round *Introd. Lower Plants* viii. 105 Branching may occur at the apex or it may be intercalary. In the former, new three-sided apical cells are formed by cleavage in the leaf of initial cells—usually in the ventral half, thus eliminating the ventral (i.e. postical) lobe of the leaf.

postiche, *a.* and *sb.* Add: **B.** *sb.* **c.** A piece of false hair worn as an adornment. Also *attrib.*

1886 C. E. Pascoe *London of To-day* (ed. 3) xl. 345 False tresses have been imported by cart-loads..and postiches and other mysteries of the toilette have been brought to that perfection to which competition so greatly conduces. 1908 *Westm. Gaz.* 21 Nov. 15/2 The postiches in use must be carefully manipulated to afford the exact size demanded. 1928 *Sunday Dispatch* 9 Dec. 8 The permanent wave has already given place to the permanent curl. And the little *postiches* (buns) which have made their appearance in Paris are being eagerly adopted. 1966 J. S. Cox *Illustr. Dict. Hairdressing* 120/2 *Postiche clip*, a small, flat spring clip to which are attached small postiches; used to secure them to the growing hair of a head. 1970 J. G. Vermandel *Dine with Devil* i. 10 Try that big hat; it might be good for a waiflike effect, very ingenue. And the postiche with..what's that long white thing?

post-ictal(ly: see *post- B. 1.

postie (pōu·sti). *slang.* Also **posty.** [f. Post *sb.*[2] + -ie, -y[6].] A familiar name for a postman.

1871 S. S. Jones *Northumberland* 84/1 Tom Buglehorn, the postie,..when he cam wi' the letters. 1886 H. Baumann *Londinismen* 143/2 Posty. 1892 G. Stewart *Shetland Fireside Tales* (ed. 2) 227, I mind when I saw 'Posty' come, My heart began ta beat. 1898 [see *nuff 1 b]. 1916 'Taffrail' *Pincher Martin* ix. 163 The marine postman..was delayed... ''Ere, posty!' shouted some one, 'got my *Dispatch*?' 1939 F. Thompson *Lark Rise* vi. 106 There was one postal delivery a day, and towards ten o'clock, the heads of the women..would be turned.. to watch for 'Old Postie'. 1953 E. Simon *Past Masters* ii. 76 If Postie did not offer one a lift on the road—that could not be imagination. 1962 *Coast to Coast 1961–62* 73 They had a long run on outdoor workers at one stage, and it was often a cop or a postie. 1975 D. Clark *Premedicated Murder* ii. 29 The postman would read the name rather than the number. So how did the posty..give Harte a card addressed..to Rencory? 1977 *S. Wales Guardian* 27 Oct. 6/2 He was missed by the upper valley residents on his transfer down to Ammanford, where he has been a 'postie' for the past 13 years.

postilion, postillion. Add: **6.** *postilion back* (earlier example).

1872 *Young Englishwoman* Dec. 651/2 A dress of olive-brown..had a basque bodice with a postilion back.

posti·l(l)ion, *v. slang.* [f. the sb.] *trans.* To insert and manipulate a finger in the anus of (a sexual partner) as a means of sexual excitement. Hence **posti·l(l)ioning** *vbl. sb.*

1888 tr. *Tableaux Vivants* xi. 95 The fair houri was postillioning me. 1969 G. Legman *Oragenitalism* i. 90 Postillioning can best be done by the middle finger.

post-impre·ssionism. Freq. with capital initials. [Post- B. 1 c.] The theory or practice of the post-impressionist school in art; *spec.* a style of painting favoured in the early years of the twentieth century in which the artist sought to reveal the structural form of his subject without strict fidelity to its natural appearance; a movement or group of aims in art which constitutes a development away from impressionism.

1910 C. J. Holmes *Notes on Post-Impressionist Painters* 11 The tradition of Post-Impressionism,..if any principles so youthful can be called a tradition, is the expression of personal vision. 1910 *Connoisseur* Dec. 315/2 The committee..wisely diluted the post-impressionism of the pictures in the entrance room by the inclusion of a dozen or more examples by Manet. 1932 *John o' London's Weekly* 30 Jan. 678/3 'Harbour of Gravelines' shows him [*sc.* Seurat] trying..to impose upon impressionism its missing sense of form. But that triumph goes to the two masters of post-impressionism, to Cezanne [*sic*] and Gauguin. 1948 R. O. Dunlop *Understanding Pictures* iii. 20 This painter's name was, of course, Paul Cézanne, the unconscious founder of Post-Impressionism. 1957 *Observer* 3 Nov. 14/3 The deficiencies of Impressionism.. were clear to many in the 1880s, and the term Post-Impressionism covers the often very different reactions of artists. 1972 S. Hynes *Edwardian Occasions* 10 Bloomsbury..supported Roger Fry's efforts to publicize Post-Impressionism. 1978 *Antiques & Art Monitor* 28 Oct. 11/3 A major exhibition entitled 'Post-Impressionism and Europe' is scheduled for the Royal Academy, London, next year.

post-impre·ssionist. Freq. with capital initials. [Post- B. 1 b.] An artist whose work exhibits one or more of the facets of post-impressionism; also *transf.* and *attrib.* or as *adj.* Hence **post-impressioni·stic** *a.*, characteristic of the post-impressionists.

1910 *Poster* in *Lett. R. Fry* (1972) I. Pl. 47 Grafton Gallery. Manet and the Post-Impressionists. 1910 C. J. Holmes *Notes on Post-Impressionists* 10 In the first Post-Impressionist painters we have a reaction from the materialism which limited the original Impressionists to the rendering of natural effects of light and colour with the greatest attainable scientific truth. 1911 *Athenæum* 28 Jan. 104/3 Post-Impressionist Sculpture. 1913 A. E. Housman *Let.* 8 Mar. (1971) 129 An exhibition of post-impressionist undergraduate art, which is calculated to frighten the Germans. 1913 *Punch* 16 July 70/1 They grumble at the ladies' skirts, The Post-Impressionistic settings. 1914 H. Holley (*title*) Creation: Post-Impressionist poems. 1922 C. Bell *Since Cézanne* 81, I can't think why you don't like it: its Post-Impressionist isn't it? 1928 [see *Cubism]. 1934 C. Lambert *Music Ho!* iii. 22 The post-impressionist harmonic experiments, the austerities and asperities of Stravinsky and Bartók. 1945 D. MacCarthy *Memories* (1953) 181 'What was the exhibition to be called?'.. At last Roger [Fry], losing patience, said: 'Oh, let's just call them post-impressionists; at any rate, they come after the impressionists.' 1957 Manvell & Huntley *Technique Film Music* iii. 70 In each sequence décor is derived from the style of one or other of the post-Impressionist painters. 1974 *Impressionism* (R. Acad.) 53/1 Increasingly affected by Post-Impressionists 1909+, with rest of Camden Town Group. 1978 *Jrnl. R. Soc. Arts* CXXVI. 701/2 Mrs. Potter Palmer had been buying Impressionist and post-Impressionist paintings with great discernment almost a century ago.

post-infectious: see *post- B. 1.

posting, *vbl. sb.*[1] Add: **I. 3.** Also, the (amount of) mail posted during a given period.

1909 *Daily Chron.* 30 Dec. 3/6 During the Christmas week of last year the postings in London alone totalled upwards of 70,000,000. 1971 D. Potter *Brit. Eliz. Stamps* i. 14 Beginning with the Shakespeare set in 1964, special envelopes were..used. Stratford-on-Avon must have had one of their heaviest postings ever.

posting, *vbl. sb.*[2] Add: **2. b.** *posting-board.*

1838 *Actors by Daylight* I. 182 A pair or two of wooden posting-boards.

post-la·rval, *a. Zool.* [Post- B. 1 b.] Belonging or pertaining to those stages in the development of certain animals in which some larval characteristics may be retained, before the adult form is reached. So **post-la·rva,** an animal, esp. a fish, during this period of its development.

1898 *Proc. Zool. Soc.* 204 (*title*) On the early post-larval stages of the Common Crab. 1929 *Amer. Naturalist* LXIII. 160 Some..interesting discoveries..have resulted from the writer's investigations into the habits of young postlarval lobsters. 1942 *Copeia* 126 (*heading*) The occurrence of flounder post larvae in fish stomachs. 1962 K. F. Lagler et al. *Ichthyol.* x. 313 Differences between postlarvae and adults in some fishes are so trenchant that they have resulted from time to time in taxonomic confusion. 1967 *Oceanogr. & Marine Biol.* V. 239 The post-larval development of Copepods similarly aroused Mazza's interest. 1972 *Aquaculture* I. 179 (*caption*) Third postlarval stage with 18 mm of total length. 1973 *Ibid.* I. 363 Postlarvae and juvenile siganids are attracted by artificial light at night.

postliteral: see *post- B. 1.

postlude. Restrict *Mus.* to sense in Dict. and add: **a.** (Further examples.)

1947 A. Einstein *Mus. Romantic Era* xiv. 187 The task..of supplying a commentary in the prelude and, particularly, the postlude. 1955 G. Abraham in H. Van Thal *Fanfare for E. Newman* 26 The second version of the piano postlude to 'Schmerzen' was sent to Frau Wesendonk on a separate piece of paper. 1976 *Gramophone* Jan. 1229/1 Why the ties in the penultimate bar of the prelude and postlude of 'Pause'?

b. A written or spoken epilogue; an afterword, conclusion; an envoy.

1928 M. Williams *Catholicism & Mod. Mind* 339 (*heading*) Postlude: Easter in Gethsemani. 1934 *Punch* 2 May 503/2 Miss Bowen's 'postlude', a carefully reasoned essay in historical criticism, is a drastic and devastating but on many points convincing analysis of romantic legend. 1939 Joyce *Finnegans Wake* (1964) 426 As the wisest postlude course he could playact, collaspsed [*sic*] in ensemble and rolled buoyantly backwards. 1959 *Times* 11 Sept. 16/6 Mr. Gerald Moore..whose little summarizing postludes to many of the songs..were miracles of concentrated wisdom. 1974 *Times* 15 Apr. 5/6 The rumpus was mostly caused by a very offensive postlude, spoken by one of the actors.

Hence **po·stlude** *v. intr.*, to supply a postlude; **postlu·dial** *a.*

1960 'A. Burgess' *Doctor is Sick* xxi. 174 The psalmist ended, postluded. 1961 *Times* 10 Nov. 18/6 One of the poems is set twice: another has a preludial and postludial movement.

postman[1]. Add: **1. d.** In the possessive, as **postman's knock**: (*a*) a sharp knock or rap upon a door, typically made by a postman; (*b*) *transf.*, esp. a parlour game in which the participants in turn take the role of postman and deliver letters which are paid for by kisses.

1835 [see sense 1 b in Dict.]. 1837 Dickens *Pickw.* xxix. 312 Sam Weller..was displaying that beautiful feat of fancy sliding which is currently denominated 'knocking at the cobbler's door', and which is achieved by skimming over the ice on one foot, and occasionally giving a two-penny postman's knock upon it, with the other. 1847 *Sporting Life* 11 Dec. 204/1 The *postman's knock*—a quick rat-tap, which only a postman can execute. 1873 C. M. Yonge *Pillars of House* III. xxxii. 199 A postman's knock made her start. 1927 W. E. Collinson *Contemp. Eng.* 12, I was interested to see the kissing-forfeit game of postman's knock under the guise of 'American post'. 1928 *Daily Express* 30 July 13/6 Rose's left glove was seldom out of his opponent's face, and he often brought off a punch which in the old days was called 'The Postman's Knock'. 1954 M. Sharp *Gipsy in Parlour* xii. 125 Postman's Knock found me..maladroit: to one pimply youth who called me out I presented such a face of scorn that he never kissed me.

postmaster[1]. Add: **3.** *Canad.* The master of a fur-trading post.

1832 in R. H. Fleming *Minutes Council Northern Dept. Rupert Land* (Hudson Bay Co.) (1940) 236 Post Masters Are a Class which ranks in the Service between Interpreters & Clerks. They are generally persons who while filling the office of common Labourers..were..raised from the 'ranks' and placed in charge of small Posts at Salys. from 35 to £45 p. Annm. 1953 A. R. M. Lower *Unconventional Voy.* 33 The postmaster at Attawapiskat ..was minus the toes on his right foot.

postmaster general. Add: On 1 Oct. 1969 the office of postmaster general was abolished in Great Britain and responsibility for executing its functions was transferred to the newly constituted Post Office Board.

postmaster-generalship (earlier example).

1882 E. W. Hamilton *Diary* 26 Nov. (1972) I. 365 In this list he had included..the Postmaster Generalship.

postmatu·re, *a.* [f. Post- B. 1 b; cf. Premature *a.* in Dict. and Suppl.] **1.** *Obstetrics.* = *post-term *a.*

1895 F. A. Stahl in *Amer. Jrnl. Obstetr. & Dis. Women* XXXI. 843, I take the liberty of suggesting the adoption of the term postmature labor to apply to labor which takes place after term. 1937 *Amer. Jrnl. Obstetr. & Gynecol.* XXXIV. 37 When pregnancy is prolonged by new corpora lutea or by progesterone, large postmature fetuses are produced. 1972 S. B. Korones *High-Risk Newborn Infants* iv. 89 Approximately 75% to 85% of all deaths among postmature babies occur during labor.

2. Pertaining to or designating a person who is over the age of maturity.

1897 A. D. L. Napier *Menopause* iv. 79 In the post-mature woman it is partly in the early senile changes that we must look for a solution of the climacteric. 1941 J. S. Huxley *Uniqueness of Man* i. 19 A large proportion of the leaders of the community are always post-mature. 1971 *Biochim. & Biophys. Acta* CCXXXVI. 458 (*heading*) Evidence for progressive, age-related structural changes in post-mature human collagen.

Hence **postmatu·rely** *adv.*; **postmatu·rity,** the state of being postmature.

1902 *Jrnl. Obstetr. & Gynæcol.* II. 524 Neither the dimensions of the fœtus, nor the degree of development of his tissues and organs, nor the history of the pregnancy, can be taken as certain proof of postmaturity. 1933 *Jrnl. Pediatrics* II. 677 Similar factors of safety and stability seem to operate when the infant is postmaturely born, for in his behavior equipment he is advanced even though birth is postponed. 1937 *Amer. Jrnl. Obstetr. & Gynecol.*

XXXIV. 36 Kaern found that there was a definite relationship between postmaturity, excessive size of the fetus, and intrauterine death. 1941 J. S. Huxley *Uniqueness of Man* i. 18 Another point in which man is biologically unique is the length and relative importance of his period of what we may call 'post-maturity.' 1972 E. D. Morris in C. J. Dewhurst *Integrated Obstetr. & Gynaecol.* xxii. 383/1 Postmaturity can, in general, be said to be present when the pregnancy has continued so long beyond term that an extra risk to the foetus exists. 1977 *Lancet* 3 Dec. 1169/2 Labour was induced, often before term, for obstetric reasons..and also for post-maturity—which was defined as 40 weeks plus 7 days.

postmedial, *a.* Add: Also in *Linguistics* and *absol.* as *sb.*

1946 L. Bloomfield in H. Hoijer et al. *Ling. Struct. Native Amer.* 104 Suffixes appear in divergent forms, so that we set off accretive elements: premedials, postmedials, prefinals. 1958 *Archivum Linguisticum* X. 170 In some words Postradical, Medial, Postmedial, and Prefinal elements are recognized.

post-meiotic, -menarch(e)al: see *post- B. 1.

postme·nstrual, *a. Med.* [Post- B. 1 b.] Occurring after menstruation.

1885 *Brit. Med. Jrnl.* 14 Feb. 342/2 Taking the 'menstrual period' as lasting about four days, he marked off on his temperature-charts four days before the 'show' as the 'premenstrual period', and four days after as the 'post-menstrual period'. 1901 A. E. Giles *Menstruation* ii. 17 The metabolic changes may thus be summarised:..in the post-menstrual week, gradual return to the normal condition. 1922 Joyce *Ulysses* 411 The postmenstrual period. 1948 *Amer. Jrnl. Obstetr. & Gynecol.* LV. 38 In other regions the endometrium was of the usual postmenstrual type, and was covered with surface epithelium.

postme·nstruum. *Med.* [mod. L.: see Post- B. 1 c and Menstruum.] The stage of the menstrual cycle which follows menstruation.

1910 *Trans. N.Y. Obstetr. Soc. 1909–11* 229 They [*sc.* Hitchmann and Adler] divide the monthly cycle into four phases. The first phase, the postmenstruum, corresponds to the picture of the normal endometrium of our text-books. The glands are small and regular, round in cross-section. 1933 *Amer. Jrnl. Anat.* LII. 564 Jaeger.. and Blumenfeld..found conception most frequent in postmenstruum (63·6 per cent up to the fourteenth day). 1964 K. Dalton *Premenstrual Syndrome* iv. 23 Symptoms limited to the postmenstruum are extremely rare.

post-metamorphic, -mineral: see *post- B. 1.

postmito·tic, *a. Cytology.* [Post- B. 1 b.] After mitosis; *spec.* (of a cell) having ceased (reversibly or irreversibly) to display cell division. Also *absol.*, a cell which is unlikely or unable to divide again.

1942 E. V. Cowdry *Probl. Ageing* (ed. 2) xxiv. 628 The highly specialized cells, formed from the last differentiating intermitotics at the end of the series,..are 'end cells'. Their individual lives begin..after the last mitosis... In a word, they are postmitotics. *Ibid.* 631 The length of life of the majority of postmitotic nerve cells can be taken to be that of the body less one year. 1962 *Lancet* 27 Jan. 211/1 The granulocytes are post-mitotic and can be ruled out, and so we are restricted to the 'mononuclears'. 1968 H. Harris *Nucleus & Cytoplasm* v. 92 When all the nuclei in the heterokaryon enter mitosis together, post-mitotic reconstitution may produce a single large nucleus at one step. 1971 J. Z. Young *Introd. Study Man* xxii. 288 Errors in the hereditary instructions and their transcription may be of especial importance in post-mitotic cells.

post-mo·dern, *a.* Also **post-Modern.** [Post- B. 1 b.] Subsequent to, or later than, what is 'modern'; *spec.* in the arts, esp. *Archit.*, applied to a movement in reaction against that designated 'modern' (cf. *modern *a.* 2 h). Hence **post-mo·dernism, post-mo·dernist** *a.* and *sb.*

1949 J. Hudnut *Archit. & Spirit of Man* ix. 108 (*heading*) Post-modern house. *Ibid.* 119 He shall be a modern owner, a post-modern owner, if such a thing is conceivable. Free from all sentimentality or fantasy or caprice. 1956 A. Toynbee *Historian's Approach to Relig.* ii. xi. 146 Our post-Modern Age of Western history. 1959 C. W. Mills *Sociol. Imagination* ix. 166 Just as Antiquity was followed by several centuries of Oriental ascendancy ..so now the Modern Age is being succeeded by a post-modern period. Perhaps we may call it: The Fourth Epoch. 1965 L. A. Fiedler in *Partisan Rev.* XXXII. 508, I am not now interested in analyzing..the diction and imagery which have passed from Science Fiction into post-Modernist literature. 1966 F. Kermode in *Encounter* Apr. 73/1 Pop fiction demonstrates 'a growing sense of the irrelevance of the past' and Top [*sic*] writers ('post-Modernists') are catching on. 1966 N. Pevsner in *Listener* 29 Dec. 955/2 The fact that my enthusiasms cannot be roused by..Churchill College.., does not blind me to the existence today of a new style, successor to my International Modern of the nineteen-thirties, a post-modern style, I would be tempted to call it, but the legitimate style of the nineteen-fifties and nineteen-sixties. 1977 *N.Y. Rev. Bks.* 28 Apr. 30/3 A process that culminates, by a curious but inexorable logic, in the post-modernist demand for the abolition of art and its assimilation to 'reality'. 1979 *Jrnl. R. Soc. Arts* Nov. 743/1 Many Post-Modern architects use motifs..in questionable taste. *Ibid.* 751/1 Post-Modernists have substituted the body metaphor for the machine metaphor, because so much research has shown that we unconsciously project bodily states into architecture. 1979 *Time* 8 Jan. 53/1 The nearest man Post-Modernism has to a senior partner is, in fact, the leading American architect of his generation: Philip Cortelyou Johnson. 1980 *Times Higher Educ. Suppl.* 7 Mar. 16/1 Postmodernism, structuralism, and neo-dada (formerly known as 'concrete poetry') all represent a reaction against modernism.

po:st-modifica·tion. *Linguistics.* [Post- A. 1 b.] The qualification or limitation of the sense of one word or phrase by another coming after. Also *attrib.* Hence **post-mo·difier; post-mo·difying** *vbl. sb.*

1962 R. Quirk *Use of English* xi. 181 In such units as *bravery of all kinds* or *bravery in the struggle against barbarism*, we meet..postmodification. 1965 *Language* XLI. 205 Deictic + head + post-modifier. 1970 *Eng. Stud.* LI. 404 The category noun modification is a large one (105 instances), almost half of them with postmodifying infinitives. 1975 *Amer. Speech 1971* XLVI. 224 A postmodification rule is necessary to transform relative clauses containing an infinitive, a prepositional phrase, or an adverb.

post mortem, post-mortem, *adv. phr.*, *a.*, and *sb.* Add: **B.** *adj.* (Further examples.) Also *transf.*

1909 *Westm. Gaz.* 8 June 5/1 M. Chauchard,..who is to sleep his last sleep..in a tomb which has cost nearly £4,000, has had many predecessors in post-mortem luxury. 1922 Joyce *Ulysses* 609 Her brandnew arrival is on her knee, *post mortem* child. 1962 *Gloss. Terms Automatic Data Processing* (*B.S.I.*) 45 *Post-mortem routine*, a diagnostic routine which may be used to indicate the contents of selected locations after a program has stopped. 1965 in Bessinger & Creed *Medieval & Linguistic Stud.* 54 What conception of the human spirit and its *post mortem* future is implied in our inscriptions? 1969 P. B. Jordain *Condensed Computer Encycl.* 389 Usually, when an unexpected or inexplicable difficulty is encountered, a postmortem dump is taken to record all available information about the failed state of a program: then a post-mortem analysis is made to discover the cause of the difficulty. 1979 R. Rendell *Means of Evil* 68 Doreen Betts's denial had..been..a post-mortem white-washing of her mother's character.

C. *sb.* **2.** *transf.* A searching (and freq. recriminatory) analysis or discussion of a past event, as an examination or card-game. Cf. *inquest *sb.* 3 c.

In quot. 1844, a re-sit examination at Cambridge University.

1844 in Farmer & Henley *Slang* (1902) V. 264/1 I've passed the Post-mortem at last. 1907 R. Dunn *Shameless Diary of Explorer* ix. 111 Here in camp, we've been holding a post-mortem of the day. 1922 A. E. M. Foster *Light Side Auction Bridge* xxxvii. 155 The post-mortem fiend simply will not be denied. 1930 A. P. Herbert *Water Gipsies* viii. 82 'I knew,' he said at the family *post mortem* in the evening. 'I knew the colt had the legs of the field, if he only had the luck.' 1943 K. Tennant *Ride on Stranger* viii. 82 They drew in to the table and..began a family post mortem of the party. 1960 V. Jenkins *Lions Down Under* 114 The post-mortem at a team-talk in Timaru was a searching one. 1972 R. Markus *Aces & Places* 105 The post-mortem centred the blame on East for not ducking the jack of diamonds at trick two. 1974 L. Deighton *Spy Story* xii. 119 It's all right for you... You won't have to do the post-mortem with these guys.

Hence **post-morte·mity,** the state of death (*nonce-wd.*); **post-mo·rtemizing** *vbl. sb.*

1851 H. Melville *Moby Dick* III. ix. 67 At a certain juncture of this post-mortemizing of the whale. 1922 Joyce *Ulysses* 387 In the nights of prenativity and post-mortemity.

post-mo·rtem, *v.* [f. Post-Mortem *a.*] *trans.* To subject to a post-mortem examination (in the senses of Dict. and Suppl.). Hence **post-mo·rteming** *vbl. sb.*

1871 M. Clarke *His Natural Life* (1874) iii. xv. 291 'Strange that he should drop like that.'..'Yes, unless he had any internal disease... I'll *post-mortem* him and see.' 1900 *Jrnl. Compar. Path. & Therapeutics* XIII. 2 Hundreds of horses dead of horse-sickness had been post-mortemed by farriers and shoeing-smiths. 1910 H. G. Wells *Hist. Mr. Polly* iv. 105 You didn't, I suppose, Mr. Polly, think to 'ave your poor dear father *post-mortemed.* 1934 R. A. Knox *Still Dead* v. 67 The corpse..was taken up to the house and post-mortem'd and buried. 1971 D. E. Westlake *I gave at the Office* 123 If you people in the legal department manage to distill Truth from your post-morteming you'll be better than Solomon. 1977 N. Freeling *Gadget* i. 3 Who looks twice at a couple post-morteming a traffic scrape.

postmultiplication: see *post- A. 1 b.

postnatal, *a.* Add: *postnatal depression*, depression in a woman caused by a recent confinement, characterized by fatigue, irritability, and fits of crying. Hence **post-na·tally** *adv.*, after birth.

1927 *Jrnl. Anat.* LXI. 321 The two parts of which the foetal suprarenal is composed—the true cortex which persists post-natally and the foetal cortex which atrophies after birth—are generally considered to have a common origin. 1934 *Times Lit. Suppl.* 29 Mar. 222/2 The child.. contracts the actual disease only when post-natally brought into contact with the specific germ. 1966 *Ann. Rev. Med.* XVII. 221 The inhibition seems to disappear as the animal matures postnatally. 1973 *Guardian* 22 Feb. 13/2 Some 88,000 women a year were recently deemed in need of treatment for postnatal depression. 1978 J. Mann *Sting of Death* viii. 69 Anna Buxton was suffering severely from post-natal depression..a textbook case: Emmy was four months old.

postneonatal: see *post- B. 1.

post-Newto·nian, *a.* [f. Post- B. 1 b + Newtonian *a.* and *sb.*] Subsequent to the life or work of Sir Isaac Newton (1642–1727); *spec.* in *Physics* (see quots. 1964, 1973).

1865 Mill *Exam. Hamilton's Philos.* xxvii. 542 Applied mathematics in its post-Newtonian development does nothing to strengthen..these errors. 1963 N. Frye *Romanticism Reconsidered* 5 A post-Newtonian poet has to think of gravitation. 1964 *Astrophysical Jrnl.* CXL. 428 A post-Newtonian approximation, in which the effects of general relativity are treated as first-order corrections. 1973 *Nature* 31 Aug. 537/1 The only precise tests of the applicability of the theory of general relativity to the actual physical world are those long ago proposed by Einstein, and certain other tests proposed more recently that are closely related to Einstein's... They test the theory only to the so-called post-Newtonian approximation, roughly speaking to terms in $1/c^2$, where c is the speed of light. 1977 *Astrophysical Jrnl.* CCXVI. 914 The tidal radiation can be comparable to the radiation from post-Newtonian effects.

post-nominal: see *post- B. 1.

postno·tum. *Ent.* [f. Post- A. 2 b + Notum.] = Postscutellum.

1926 R. J. Tillyard *Insects Austral. & N.Z.* i. 18 In some cases..a short posterior sclerite is also developed, called the postnotum or postscutellum. 1964 R. M. & J. W. Fox *Introd. Compar. Entomol.* ii. 52 The notum continues forward to cover the posterior part of the next segment where it is termed the postnotum.

post-nu·clear, *a.* (*sb.*) [Post- B. 1 b.] **1.** *Phonetics.* Situated after a nucleus. Also *absol.* as *sb.*

1961 [see *internuclear *a.* 3]. 1961 Y. Olsson *On Syntax Eng. Verb* vii. 189 The variation between the post-nuclears *er* and *ee*..corresponds to a distinction between (2) and (1). 1968 *Language* XLIV. 80 Allotone 2..of the glide occurs in postnuclear syllables of the word. 1976 *Archivum Linguisticum* VII. 38 One feature both sub-classes share is their insistence that any non-nuclear occurrence be in prenuclear position rather than in tail (postnuclear) position.

2. Subsequent to the development or use of nuclear weaponry.

1963 *Economist* 7 Sept. 813/2 China was dreaming of a post-nuclear heaven. 1965 *Punch* 17 Feb. 259/3 Another post-nuclear society, this time dominated by farmers.

postnuptial, *a.* Add: Also, subsequent to mating (of animals).

1956 *Nature* 21 Jan. 143/1 Our experiment was begun on July 23..at about the estimated end of the post-nuptial refractory period and before the next season's spermatogenesis began.

post-oak. For *Quercus obtusiloba* substitute *Quercus stellata.* Also *attrib.* (Earlier and further examples.)

1775 B. Romans *Conc. Nat. Hist. E. & W. Florida* 18 The principal however are the following:..Virginian white oak... Dwarf white oak, or post oak. *a* 1816 B. Hawkins *Sk. Creek Country* (1848) 19 The trees are post oak, white and black oak, pine [etc.]. *Ibid.* 20 Between these rivers, there is some good post and black oak land. 1836 D. B. Edwards *Hist. Texas* 46 They are protected.. by..post-oak ridges. 1892 J. C. Duval *Young Explorers* in *Early Times in Texas* ii. ii. 14 About noon we came to a small stream bordered by a strip of post oak woods. 1906 'O. Henry' *Four Million* 58 Joe Larrabee came out of the post-oak flats of the Middle West. 1945 B. A. Botkin *Lay my Burden Down* 263 They found the body of a white man hanging to a post oak. 1969 T. H. Everett *Living Trees of World* 118/2 The post oak..is widely distributed in dryish uplands from Massachusetts to Nebraska. 1975 *New Yorker* 5 May 101/1 All but six of the thirty-six holes are set off by themselves, framed by borders of post oak—a pretty tree that loses its large leaves in winter but retains its attractiveness because of the pleasing contortions of its branches.

Post Office, Post-Office. Add: Also with small initials. **2. b.** *transf.* A person who receives information and either transmits it or holds it for collection, esp. in espionage; also = *drop *sb.* 17 d. *slang.*

1885 E. W. Hamilton *Diary* 12 Apr. (1972) II. 835 M. Lessar suggests that Brett should be asked to be the post office of the Russian Embassy. Accordingly, Lessar goes to Brett, and hands him a Memorandum... Brett forwards the Memorandum here. 1919 J. Buchan *Mr. Standfast* vii. 145, I had got precisely what Blenkiron wanted, a post office for the enemy... I could see the juiciest lies passing that way to the *Grosses Hauptquartier.* 1935 A. J. Pollock *Underworld Speaks* 90/2 *Post office*, a person who receives or delivers letters to crooks. 1945 *Tee Emm* (Air Ministry) V. 55 Beware of becoming a 'post office', simply passing on everything that comes in, happy in the knowledge that there is a higher authority behind you. 1965 D. Williams *Not in Public Interest* vii. 133 It became evident in 1911 that the hairdresser's shop of Karl Gustav Ernst was being used as a 'post office' or clearing-house for German espionage agents in this country. 1974 T. Allbeury *Snowball* iv. 20 Just a low-grade courier, a dead letterbox and a post office.

2*. *U.S.* A parlour game in which the participants in turn act as postmaster or postmistress and pretend to deliver letters which are paid for by kisses. Cf. *postman's knock* (*postman[1] 1 d (*b*)).

The sense in quot. 1851 is uncertain.

1851 J. H. Green *Twelve Days in Tombs* 157 How often have the professors of Christianity violated all moral principles in the..game of Post-office, where we find stationed some beautiful sister as post mistress, whose duty it is to write the names of those from whom she thinks she can secure the postage. 1855 *Quincy* (California) *Prospector* 31 Mar. 2/1 We are astonished to see men and women who are looked upon as samples for the rising generation, join in such childish plays as.. 'Post office', &c. 1899 *Amer. Physical Educ. Rev.* 361 Those who select love games are at the dawn of adolescence. 'Drop-the-handkerchief' and 'post-office' are the two favorites of this group. 1904 C. S. Darrow *Farmington* 163 We had to keep still, and couldn't go outdoors, and had to play 'needle's eye' and 'post-office'. 1914 B. Tarkington in *Cosmopolitan* Mar. 489/2 'We'd have been playing "Quaker meeting", "clap in, clap out", or "going to Jerusalem", I suppose.' 'Yes, or "post-office" and "drop the handkerchief",' said Mrs. Schofield. 1949 *Sat. Even. Post* 12 Mar. 60/3 After a time this palled and they played Post Office.

3. *post-office directory*; also in the names of colours associated with various Post Office services, as *post-office green, red, yellow*; **post-office box** (*b*) = *post-office bridge*; **post-office bridge,** a portable self-contained form of Wheatstone bridge containing a large number of resistors which are selected by means of plugs; **post-office order** (earlier example); **post-office packet** *Hist.*, a packet-boat carrying mail for the Post Office; **post-office savings-bank,** add to def.: since 1 Oct. 1969, known as the National Savings Bank; hence *post-office savings(-bank)-book.*

1894 W. A. Price *Measurement Electr. Resistance* 81 Bridge ratios of 1000 and ·001 are available in addition to those in the Post Office box. 1914 *Phil. Mag.* XXVIII. 470 The resistances of the films of low resistance were measured by a post-office box in the ordinary way. 1965 G. A. G. Bennet *Electr. & Mod. Physics* viii. 129/2 A Wheatstone type network is connected up as shown ..with the galvanometer as one of the four resistances; a Post Office box would be suitable for this purpose. 1891 H. L. Webb *Testing of Insulated Wires & Cables* vi. 38 A small clamp for holding the battery key down permanently would be a useful addition to a Post-office bridge. 1931 W. L. Upson *Electr. Lab. Stud.* iii. 45 There is a second form of Wheatstone bridge known as the 'post-office bridge'. In this type there is no wire giving wide variability to the ratio of *B* to *A*, but there are fixed resistance coils so that the ratio may be made 1 to 1, 1 to 10, [etc.]. 1803 *Post-Office Annual Directory: London* 3 The Editors of the Post-Office Directory..present their most sincere and grateful Acknowledgements. 1852 Dickens *Bleak Ho.* (1853) viii. 70 It appeared to us that some of them must pass their whole lives in dealing out subscription-cards to the whole Post-office Directory. 1963 *Ophthalmic Optician* 20 Apr. 408/1 If the hole exposes telephone services and cables, the perimeter will be lined with machinery painted in Post Office green. 1843 Dickens *Let.* 30 Dec. (1974) III. 617 For a post office order there is no time. 1780 A. Young *Tour in Ireland* i. 342 It is much to be wished, that there were some means of being secure of packets sailing regularly..; with the post-office packets there is this satisfaction. 1855 Dickens *Holly Tree Inn* in *Househ. Words* Extra Christmas No., 15 Dec. 1/2 The Post-office packet for the United States was to depart from Liverpool. 1930 *Times Educ. Suppl.* 18 Jan. (Suppl.) p. iv/2 Red offers a fertile field for flights of descriptive fancy... Post Office red, and sealing wax red may be hackneyed words, but their tone is not in doubt. 1978 *Lancashire Life* Sept. 89/2 Newspapers, medicines, grocery orders—all are piled aboard the valley's post-office-red lifeline for delivery *en route.* 1861 *Act* 24 & 25 *Vict.* c. 14 Post Office Savings Banks..An Act to grant additional Facilities for depositing small Savings at Interest, with the Security of the Government for due Repayment thereof [17th May 1861]. Whereas it is expedient..to make the General Post Office available for that Purpose. 1885 *Encycl. Brit.* XIX. 572/2 The establishment of post-office savings banks was practically suggested in the year 1860 by Mr Charles William Sykes of Huddersfield, whose suggestion was cordially received by Mr Gladstone... Half a century earlier (1807) it had been proposed to utilize the then existing..money-order branch of the post-office for the collection and transmission of savings..to a central savings bank to be established in London. 1936 M. Allingham *Flowers for Judge* xi. 172 Mr. Campion turned over the battered cardboard-backed book... 'Post Office Savings Bank?.. Whose is it?' *Ibid.* xviii. 260 One day you find her Post Office Savings Bank-book. 1966 B. Kimenye *Kalasanda Revisited* 21 His Post Office Savings book boasted the grand total of 600s. 1973 P. Moyes *Curious Affair of Third Dog* viii. 104 A Post Office savings book showing a balance of some ten pounds. 1976 *Scotsman* 15 Dec. 14/3 (Advt.), The Telecommunications showroom..is decorated internally in Post Office yellow, with relief panels throughout in silver.

post-offi·cial, *a.* and *sb.* [f. Post office, Post-office + -ial, with play upon Official *a.* and *sb.*] **A.** *adj.* Of or pertaining to a post office or its staff. **B.** *sb.* A post-office employee.

1938 Dylan Thomas *Let.* 31 Dec. (1966) 219, I don't know why your letter returned 'unknown', unless it was a post-official hint at the subsidence of my..reputation. *a* 1939 E. G. Murphy in *Penguin Bk. Austral. Ballads* (1964) 209 For the Smiths rolled up..To drive the post-officials mad.

post-op (pōu·st,ǫ·p), colloq. abbrev. of *postoperative *a.* (*sb.*).

1971 E. Candy *Words for Murder Perhaps* vii. 84, I can take temperatures and..make lovely beds for post-ops. 1974 *Country Life* 3–10 Jan. 58/3 Nursing Home...medical and post-op. patients. 1977 D. Bennett *Jigsaw Man* 14 He felt he had been sawn in half. The post-op drugs took over.

posto·perative, *a.* (*sb.*) [Post- B. 1 b.] Occurring in or pertaining to the period following a surgical operation; having recently undergone an operation. Also as *sb.*, a person who has recently had an operation.

1889 in *Cent. Dict.* 1898, 1900 [in Dict. s.v. Post- B. 1 b]. 1925 E. Hemingway *In our Time* (1926) 24 'I'm terribly sorry I brought you along, Nickie,' said his father, all his post-operative exhilaration gone. 1951 *Science* 19 Oct. 416/1 Prothrombin levels remained normal throughout the postoperative period in each animal. 1956 K. Hulme *Nun's Story* x. 160 Sister Luke examined some of his post-operatives. 1962 *Lancet* 5 May 936/1 During the postoperative period he had a chest infection. 1973 R. Hayes *Hungarian Game* xiv. 332 They didn't want post-operatives to smoke. 1977 *New Yorker* 12 Sept. 114/2 Postoperative patients are always complaining about the quality of hospital food.

Hence **posto·peratively** *adv.*, after an operation.

1908 *Practitioner* Sept. 435 The nephrectomies shown to have a normal freezing point before operation invariably demonstrated post-operatively that the kidney, which was left, was functionally adequate. 1931 *Arch. Surg.* XXII. 552 The basal metabolic rate was not determined preoperatively or postoperatively. 1961 *Lancet* 26 Aug. 491/1 Why not give sedatives postoperatively by mouth instead of by injection. 1975 *Nature* 8 May 152/2 The occurrence of ptosis and miosis postoperatively was accepted as evidence that surgery had been successful.

post-osmicate, -osmication: see *post- A. 1 a, b. **post-ovulative, -ovulatory, -painterly:** *post- B. 1.

postponable, *a.* Delete *rare*—[0] and add further examples.

1963 *Punch* 6 Feb. 206/1 So much of its expenditure is postponable. 1965 P. Wylie *They both were Naked* i. iv. 189, I am in the middle of something. A novelette. Postponable, though. 1971 *Human World* Aug. 9 The accepted right and power of the country to decide by majority vote in free elections, not indefinitely postponable, that it has..had enough of the present government.

postpone, *v.* **2.** Delete † *Obs.* and add later example.

1874 H. J. Roby *Gram. Latin Lang.* II. 351 Most prepositions are prefixed to the substantive; a few are always postponed; others are occasionally but rarely postponed in prose.

postponedly (poustpōu·nėdli), *adv. rare.* [f. Postponed *ppl. a.* + -ly[2].] At a late time; belatedly.

1851 H. Melville *Moby Dick* III. xxvi. 171 He was an old man who..had postponedly encountered that thing in sorrow's technicals, called ruin.

postpo·neless, *a. rare*—[1]. [f. Postpone *v.* + -less.] That may not be postponed or averted.

c 1862 E. Dickinson *Poems* (1955) I. 307 It's Coming—the postponeless Creature—It gains the Block—and now—it gains the Door.

postpose, *v.* Restrict † *Obs.* to sense b in Dict. and add: **a.** Now usu. *Gram.*

1930 T. Sasaki *On Lang. R. Bridges' Poetry* i. i. 7 Not a single adj. in '-able' or '-ible' is postposed in Bridges. 1962 R. Quirk *Use of English* xiv. 241 To postpone an adjective as in 'the young man carbuncular'. 1978 *Studies in Eng. Lit.: Eng. Number* (Tokyo) 106 It is obvious that there are similarities..between the rule of extraposition which postposes the string *into the heavens* in (46b) and the one which postposes *into the clouds* in (42b).

Hence **postpo·sed** *ppl. a.*; **postpo·sing** *vbl. sb.*

1927 O. Jespersen *Mod. Eng. Gram.* III. i. 19 Postposed adjectives are not in general accord with colloquial English. 1972 W. Labov *Language in Inner City* iv. 143 There are some postposed expressions which seem quite straightforward, even unmarked. 1975 *Language* LI. 815 Passivization..may involve not one but two transformational operations—subject postposing and object preposing. 1978 *Ibid.* LIV. 76 On the other hand, response-stance verbs, and verbs that are not stance verbs at all, do not allow such postposings.

postposition. Add: **2.** (Further examples.)

1928 H. Poutsma *Gram. Late Mod. Eng.* (ed. 2) I. i. viii. 488 The influence of Latin Grammars makes itself felt in the post-position of the adjective. 1930 T. Sasaki *On Lang. R. Bridges' Poetry* i. i. 5 The postposition of two attributes joined together by means of 'and'. 1975 *Amer. Speech 1971* XLVI. 226 Simple modifiers may appear in postposition in both English and German.

3. (Further examples.)

1925 Grattan & Gurrey *Our Living Lang.* i. xiii. 83 Look at the word *at* in the following sentences:—..(*d*) These are the remarks they laughed at... We shall therefore avoid confusion of thought if we call it [*sc.* 'at'] a Postposition. 1976 J. S. Gruber *Lexical Struct. Syntax & Semantics* ii. iii. 343 In Japanese, there are some pieces of evidence..that postpositions, quantifiers, and other things which manifest left-branching..actually form one word.

Hence **postposi·tioning** *vbl. sb.*; **postpositional** *a.* (further examples); also *absol.* as *sb.*

1968 *Canad. Jrnl. Linguistics* XIV. i. 50 In this example we see that the head NP is identical to the object of the postpositional phrase. 1972 *Language* XLVIII. 390, I am suggesting that the morphemes which appear as postpositionals are really functionally like verbs. 1974 L. Todd *Pidgins & Creoles* ii. 15 In the Atlantic varieties plurality can be overtly marked by the postpositioning of *dem* immediately after the noun. 1975 *Language* LI. 797 The putative specified subject can never surface, either as a subject or as a postpositional object. 1976 J. S. Gruber *Lexical Struct. Syntax & Semantics* i. iii. 67 It appears that we always have the form with *from* to the left, but for *in* and *on* post-positioning is possible or obligatory.

postpositive, *a.* (*sb.*) Add: **A.** *adj.* (Further examples.)

1930 T. Sasaki *On Lang. R. Bridges' Poetry* i. iv. 17 All these adjj. have been abundantly used through all the stages of English poetry... This fact appears partly to account for their frequent occurrence as postpositive attributes. 1936 *Amer. Speech* XI. 363/2 A discussion of the postpositive use of adjectives in such groups as *law ecclesiastical.* 1951 *Archivum Linguisticum* III. 24 Post-positive adjectives are in fact common in ON. 1963 [see *enclisis]. 1978 *Language* LIV. 281 A particular clause-modifying particle has a fixed position, either clause-initial or 'post-positive'—i.e. placed (along with indefinites and enclitic personal pronouns) immediately after the initial item in the clause.

Hence **postpo·sitively** *adv.*

1961 R. B. Long *Sentence & its Parts* xi. 256 Superlatives in *most* are now felt as compounds in which a modifying auxiliary pronoun has been united, postpositively, with a basic-form adjective head. 1964 *Amer. Speech* XXXIX. 36 Verbs contrast with adjectives by their ability to go postpositively.

post-possessive: see *post- A. 1 b. **post-precipitate:** *post- A. 1 a. **post-precipitation:** *post- A. 1 b.

post-pri·mary, *a.* [Post- B. 1 b.] Of education or schools: subsequent to that which is primary. Of a pupil: receiving such education.

1919 *App. Jrnls. House of Representatives N.Z.* E. vi. 15 Suffice to say the main aim we have had in view is the bringing of the post-primary work into closer touch with the vocational needs of the pupils. 1926 *Rep. Consult. Comm. Educ. Adolescent* ii. 36 Is it possible..to ensure that all normal children may pursue some kind of post-primary course for a period of not less than three, and preferably four, years from the age of 11+. 1929 *N.Z. Educ. Gaz.* 1 Mar. 36/1 The teaching of French pronunciation in our post-primary schools. 1930 *Times Educ. Suppl.* 18 Jan. 21/3 A post-primary course. 1961 J. K. Hunn *Rep. Dept. Maori Affairs* (N.Z.) 23 Maori indifference to post-primary and university education. 1975 *Language for Life* (Dept. Educ. & Sci.) xiv. 213 The Hadow Report urged that there should be no sharp division between infant, junior, and post-primary stages. 1977 *Daily Times* (Lagos) 27 Aug. 4/5 The education committee would help to propagate the importance of ideal family life by organising lectures for post-primary school pupils.

post-processor: see *post- A. 1 b. **post-produce:** *post- A. 1 a. **post-production:** *post- A. 1 b. **post-puberal, -pubertal, -puberty, -radical:** see *post- B. 1.

postredu·ction. *Genetics.* [a. G. *postreduction* (Korschelt & Heider *Lehrb. der Vergleichenden Entwicklungsgeschichte der Wirbellosen Thiere* (Allgemeiner Theil) (1903) II. vi. 595): see Post- A. 1 b.] **a.** Reduction of chromosome number at the second of the two meiotic cell divisions, rather than at the first. Opp. *prereduction a.

1905 *Proc. Acad. Nat. Sci. Philadelphia* LVII. 188 To the idea of postreduction we can apply the criticism 'not proven'. 1915 *Jrnl. Morphol.* XXVI. 122 In regard to the question of pre- and post-reduction, I have evidence here that the first maturation division is the reduction division. 1921 L. W. Sharp *Introd. Cytol.* xi. 245 Such tetrads separate reductionally at the second mitosis (postreduction) rather than at the first (prereduction) in certain orthopterans.

b. Separation of homologous chromatids or genes at the second of the two meiotic cell divisions, rather than at the first. Opp. *prereduction b.

1934 L. W. Sharp *Introd. Cytol.* (ed. 3) xvi. 254 Disjunction in [meiosis] I is called 'prereduction'; that in II, 'postreduction'. 1950 *Adv. Genetics* III. 221 The prereductional or postreductional division of the heterozygous allele pair depends on its situation in the bivalent. Let us consider only the most common bivalent type, a bivalent with prereduction in respect to centromere, and provided with one chiasma. (In contradistinction to this, the Lecanium species have postreduction.) 1967 U. Mittwoch *Sex Chromosomes* ix. 152 The sex chromosomes of *Apodemus* frequently exhibit postreduction, in contrast to most other mammals in which the X- and Y-chromosome show prereduction.

Hence **postredu·ctional** *a.*, involving or pertaining to postreduction; **postredu·ctionally** *adv.*

1905 *Proc. Acad. Nat. Sci. Philadelphia* LVII. 187 He is the solitary worker on the spermatogenesis of the Hemiptera who has taken the postreductional view. 1950 [see *postreduction b]. 1950 *Adv. Genetics* III. 221 Since the number and situation of chiasmata, as a rule, varies even in the eggs of a single female, certain parts of a bivalent and consequently the gene pairs included in them may separate either prereductionally or postreductionally depending on the number and position of chiasmata.

‖ **post rem.** *Philos.* [med.L. (Albertus Magnus) lit. 'after the thing'.] Used, often post-positively, of universals considered as concepts intuited from individual instances, as opposed to their having real existence either prior to the individual instance (see *ante rem) or only as experienced in individual instances (see *in re* (*c*) s.v. *in *Latin prep.*).
1902 W. James *Var. Relig. Exper.* 523 For religion generally..the word 'judgment' here means no such bare academic verdict or platonic appreciation as it means in Vedantic or modern absolutist systems; it carries, on the contrary, *execution* with it, is *in rebus* as well as *post rem*, and operates 'causally' as partial factor in the total fact. 1927 [see *ante rem]. 1931 S. Beckett *Proust* 60 The Baudelarian [*sic*] unity is a unity 'post rem', a unity abstracted from plurality. 1941 E. C. Thomas *Hist. Schoolmen* vi. 123 There remains a third aspect of Universals, generally described as Nominalist, which posits the Universal '*post rem*'. 1952 R. I. Aaron *Theory of Universals* xii. 218 Nevertheless it [*sc.* the quality 'human'] is *post rem* and, to a certain degree, 'the workmanship of the mind'.

postreproductive: see *post- B. 1.

postscript, *sb.* Add: **c.** Delete *rare* and add: Also, an additional or conclusory remark or action, an afterthought, a sequel.
1926 C. Hamilton in *Hutchinson's Best Story Mag.* Nov. 16/1 'We are to keep each other company until my son returns,' she added. And as a postscript, 'It is his wish.' 1932 E. V. Lucas *Reading, Writing & Remembering* ix. 153 *The Gentlest Art* led to an amusing postscript. A firm of drapers..sent me..a specimen of epistolary gentleness of the highest order. 1949 M. Mead *Male & Female* xvi. 340 Some couples attempt a last child, for which there are..slang phrases—'little postscript', 'little frost blossom'. 1963 A. Ross *Australia 63* iii. 76 Benaud, who fancies these kind of brief postscripts against weary bowlers, drove Statham and hooked Coldwell. 1965 *Listener* 23 Sept. 463/3 Would he have expanded during his sixty-odd extra years, or remained as much a postscript from the 'nineties as Max?
d. A short talk broadcast after a B.B.C. radio news bulletin.
They began in 1940 and were discontinued in 1944.
1940 *Radio Times* 18 Oct. 3/1 Priestley fans in this country..hear him only once a week, when he gives his Sunday-night postscript. 1943 *New Statesman* 20 Nov. 328 A sensible postscript by Barbara Ward. 1961 E. Waugh *Unconditional Surrender* I. iii. 48 The BBC don't want to renew 'The Voice of Trimmer' Sunday evening postscripts. 1972 P. Black *Biggest Aspidistra* II. i. 95 The Postscripts began in March [1940], following the Nine O'clock News on the Home Service as a counter-attraction to Haw-Haw.

postscript, *v.* For *rare*⁻¹ read *rare* and add: Also, to furnish as a postscript.
1970 D. Marlowe *Echoes of Celandine* vii. 127 He remembered writing a letter to her... *Suddenly one realizes*, he had postscripted, *that there is a sadness.*

‖ **post scriptum.** Also postscriptum, pl. -ta. [L. *post scriptum* (see Postscript *sb.*).] = Postscript *sb.* a and b. Also *attrib.* and as *adv. phr.*
1523, 1535, 1586 [see Postscript *sb.* a]. 1853 Mrs. Gaskell *Cranford* v. 91 In a post-scriptum note in his handwriting, it was stated that the Ode had appeared in the 'Gentleman's Magazine', December, 1782. 1899 C. J. C. Hyne *Further Adventures Capt. Kettle* x. 191 The letter ..ended in a post-scriptum tag. 1946 R. Capell *Simiomata* I. 11 Postscriptum.—While the April mutinies were Communist-inspired..they were not unprecedented. 1977 *Listener* 7 Apr. 447/3 A *postscriptum* has been pasted into the book registering Tambimuttu's disapproval.

post-sele·ction, *sb.* and *a.* **A.** *sb.* [Post- A. 1 b.] Selection, *spec.* natural selection, occurring subsequently. **B.** *adj.* [Post- B. 1 a.] Occurring after selection; of or pertaining to a time after selection.
1928 *Funk's Stand. Dict.*, *Post-selection*, natural selection carried on after the character of the animal has appeared. 1941 J. S. Huxley *Uniqueness of Man* ii. 56 Once the immigrants were established in the country, selection continued. This post-selection..must on the whole have encouraged and discouraged the same qualities favoured by pre-selection. 1946 *Nature* 7 Sept. 320/2 The main business of post-selection training for the Colonial Service will be carried out for the present in the Universities of Oxford, Cambridge and London. 1977 A. W. F. Edwards *Found. Math. Genetics* iii. 23 Starting at the point representing zygotic proportions before selection.., the population will move to the corresponding post-selection point.

post-Sta·lin, *a.* [Post- B. 1 a.] Of Russia or of Communism subsequent to the time of Stalin (died 1953); following the death of Stalin. Hence **post-Sta·linist** *a.*
1955 H. Hodgkinson *Doubletalk* 86 Mr. Molotov's colleagues in the post-Stalin leadership. 1958 *Times Lit. Suppl.* 17 Jan. 27/1 It is not clear how much of the post-Stalinist course in the Soviet Union would fall under the same condemnation. 1964 T. B. Bottomore in I. L. Horowitz *New Sociol.* 365 The post-Stalinist Communist societies. 1965 *Language* XLI. 125 The Soviet discussions on structuralism and other linguistic debates of the post-Stalin era. 1974 tr. *Wertheim's Evolution & Revolution* 332 What characterizes the post-Stalin period is that what should have been a temporary compromise is increasingly being accepted as a lasting characteristic of Soviet society.

post-stre·tching, *vbl. sb.* *Building.* [Post- A. 1 b.] = *post-tensioning *vbl. sb.* So **post-stre·tch** *v. trans.*; **post-stre·tched** *ppl. a.*
1941 *Concrete & Constructional Engin.* XXXVI. 78/2 Test results with structures post-stretched according to Mr. Hewett's method. 1946 *Ibid.* XLI. 147 Post-stretching is carried out against the hardened concrete, no immediate loss owing to elastic deformation of the concrete occurring, and that owing to shrinkage being reduced. *Ibid.* 195 The post-stretched mild steel reinforcement. 1949 P. W. Abeles *Princ. & Pract. Prestressed Concrete* xii. 92/1 Two kinds of transfer of prestress can be distinguished:—(a) instantaneous release of prestress by severing wires, previously tensioned independently of the members to be prestressed, a process which is called 'pre-tensioning' or 'pre-stretching', and (b) gradual release of the prestress by tensioning the wires directly against the hardened body of the member to be prestressed, a process which is called 'post-tensioning' or 'post-stretching'.

† **postsyna·psis.** *Cytology.* *Obs.* [Post- B. 1 c.] (See quots.)
1898 T. H. Montgomery in *Zool. Jahrbücher: Anat. u. Ontogenie* XI. 20 The anaphase itself may be subdivided into 3 well marked periods: the early anaphase, the synapsis, and the postsynapsis. *Ibid.* 28 The postsynapsis, a term introduced here for the first time, is a well marked stage of the anaphase, distinguishable alike from the preceding synapsis as from the following telophase. 1911 *Jrnl. Morphol.* XXII. 753 My postsynapsis stage, equivalent to the diplotene.

postsyna·ptic, *a.* **1.** *Cytology.* [Post- B. 1 b.] Subsequent to meiotic synapsis.
1909 [see *presynaptic *a.* 1]. 1912 *Jrnl. Exper. Zool.* XIII. 377 (*heading*) The post-synaptic spireme. Pachytene and diplotene. 1921 *Ann. Bot.* XXXV. 366 During the period when the mother-cells are in synapsis and the postsynaptic spireme stages, the tapetal cells vary greatly in appearance. 1931 *Cytologia* II. 353 Secondary association..is a post-synaptic phenomenon.
2. *Physiol.* [Post- B. 2.] Of, pertaining to, or designating a neurone that receives a nerve impulse at a synapse. Opp. *presynaptic *a.* 2.
1937, 1965 [see *presynaptic *a.* 2]. 1974 M. C. Gerald *Pharmacol.* v. 100 This interaction may produce one of two types of changes in the permeability of the postsynaptic membrane.
Hence **postsyna·ptically** *adv.*
1952 *Jrnl. Physiol.* CXVII. 115 The fact that a moderate dose of curarine abolishes 'external' as well as 'internal' miniature potentials, strongly indicates that they both arise 'post-synaptically', in the end-plate. 1973 *Nature* 8 June 355/1 Lead can only block postsynaptically in conditions in which the density of ACh-receptor complexes is much lower than it is during an end-plate potential. 1978 *Ibid.* 22 June 674/1 These neurones lie postsynaptically to high densities of nerve terminals shown by immunohistochemical techniques to contain somatostatin-like material.

post-sync(h (-siŋk), abbrev. of *post-synchronization, *post-synchronize *v.* Hence **post-sy·nc(h)ed** *ppl. a.*; **post-sy·nc(h)ing** *vbl. sb.*
1960 O. Skilbeck *ABC of Film & TV* 98 *Post Synch.*, to make apparently Synchronous sound for existing Mute film with the aid of a Guide Track, Wild track or otherwise. 1962 *Movie* June 26/1 He wasn't helped there by the post-sync sound, which was often unrelated to what was on the screen. 1963 *Movie* Jan. 21/3 The post-synching of the Italian and French versions..was done without any reference to me. 1968 J. Bingham *I love, I Kill* xii. 160 Doing bits and pieces of reshooting and recording the dialogue, post syncing, as they call it. 1968 *Punch* 31 Jan. 153/2 'Post-synch' may be necessary for various reasons, perhaps because the sound track of an outdoor scene has been ruined by aircraft noise, perhaps because a foreign actor's speech is so thick as to call for a substitute voice. 1972 I. Hamilton *Thrill Machine* xiv. 58 We could always post-sync the questions or re-cut the answers. 1978 *New York* 3 Apr. 68/2 He..plays the piano with self-dramatizing virtuosity that isn't helped by the poorly matched post-synced studio sound recording. 1979 D. Lowden *Budapesti* 3 iii. 22 'So tell me what you want?.. A post-synch and dubbing theatre that'll take at least eight tracks?'.. He got them [*sc.* the letters] back, neatly typed..with the words 'dubbing tracks' and 'post-synch loops' intruding strangely in the text.

post-synchroniza·tion. *Cinemat.* and *Television.* [Post- A. 1 b.] The addition of a sound recording to the corresponding images of a film or video recording after the latter has been made. So **post-sy·nchronize** *v. trans.*; **post-sy·nchronizing** *vbl. sb.*
1933 A. Brunel *Filmcraft* 163 *Post-synchronise*, to add sounds, music or dialogue, after the mute film (pictorial image) has been shot. 1936 —— *Film Production* 177 For this purpose the artist, whose voice is to be used for post-synchronisation, is placed before a microphone in the studio and on to a screen is projected the scene to be post-synchronised. 1953 K. Reisz *Technique Film Editing* 275 This process involves getting the actors to speak the lines in synchronisation with the picture and is known as post-synchronising. Post-synchronisation is not used only to cover up faulty recording. *Ibid.* 276 In some Hollywood studios it has become common to post-synchronise most of the spoken lines. 1959 P. Bull *I know Face* v. 93, I..said the two words. Two days later I was asked to post-synchronise them, as they were totally incoherent. 1960 D. Wilson *Television Playwright* 16 The advantage the director gains by retaking doubtful scenes..by the post-synchronizing of music and sound effects. 1965 *Wireless World* July 34 (Advt.), The PRO 70 for multi-channel recording and for dubbing, post-synchronising and transfer. 1968 *Punch* 31 Jan. 153/2 Dubbing is a specialised form of post-synchronisation.

post-tectonic: see *post- B. 1.

post-te·nsioning, *vbl. sb.* *Building.* [Post- A. 1 b.] Strengthening of reinforced concrete by application of tension to the reinforcing rods after the concrete has set. So **post-te·nsion** *v. trans.*; **post-te·nsioned** *ppl. a.*
1948 *Concrete & Constructional Engin.* XLIII. 155 For post-tensioning there is no limitation of the size of the bars, but for pre-tensioning plain wires of 0·2 in. diameter are the largest that have been successfully employed. 1950 *Engineering* 13 Jan. 54/2 Adequate reasons exist to justify the claim that prestressed post-tensioned beams can successfully sustain the effects of fire. 1954 *Archit. Rev.* CXVI. 110 The gymnasium has a post-tensioned prestressed concrete frame and the changing rooms have load-bearing walls and precast concrete roof units. 1958 *Times Rev. Industry* July 25/1 Post-tensioning methods require relatively little capital expenditure. 1965 [see *pre-tensioned *ppl. a.*]. 1974 Lin & Zia in B. Bresler *Reinforced Concrete Engin.* I. vi. 302 The amount of concrete posttensioned at the job site in the U.S.A...is estimated to have been about 700,000 cu yd in 1972. 1976 G. S. Ramaswamy *Mod. Prestressed Concrete Design* vii. 86 The member is post-tensioned after the concrete has sufficiently hardened.

postterm (pōᵘst,tǝ̄ɹm), *a.* *Obstetrics.* [Post- B. 1 a.] Born or occurring after a pregnancy that lasted significantly longer than normal.
1933 *Jrnl. Pediatrics* II. 677 There is a stable substrate of maturation..which tends to keep the underweight, the preterm and the postterm infant close to his normal maturity levels in the field of behavior. 1971 Pierog & Ferrara *Approach to Med. Care of Sick Newborn* i. 7 The preterm infant is one born less than 38 weeks by gestational age from the first day of the last menstrual period; the postterm infant is one born 42 weeks or more from the first day of the last menstrual period. 1976 L. O. Lubchenco *High Risk Infant* vii. 163 There are relatively more infants who are large for gestational age in postterm deliveries than in term deliveries.

posttest (pōᵘ·st,test), *a.* and *sb.* *Psychol.* [f. Post- + Test *sb.*³] **A.** *adj.* [Post- B. 1 a.] That comes after or is subsequent to a test. **B.** *sb.* [Post- A. 1 b.] A subsequent test designed to measure the effects of or changes since the initial test. Hence **postte·sting** *vbl. sb.*
1966 *Jrnl. Personality & Social Psychol.* IV. 175, 60 male and female college Ss were..tested..in a social pressure plus reinforcement session and again in a posttest reinforcement session 2 wk. later. 1966 J. S. Bruner *Beyond Information Given* (1974) xviii. 321 This was followed by some training trials and then there was a posttest. 1972 *Jrnl. Social Psychol.* LXXXVI. 12 Endler ..observed posttest behavior consistent with his reinforcement schedules. 1973 *Jrnl. Genetic Psychol.* CXXIII. 90 The Ss were reassembled for the posttesting session, and the TAT and Test Anxiety Questionnaire were given. 1974 *Florida FL Reporter* XIII. 76/1 The recognition test of singular and plural SE forms was administered after the three-week training session and after the posttest.

Post Toasties. orig. *U.S.* Also with lower-case initials and sing. **Post Toasty, -ie.** The proprietary name of a breakfast cereal first marketed by Charles William *Post* (1854–1914), American manufacturer (cf. *Postum). Also *transf.* and *fig.*
[1907 *Trade Marks Jrnl.* 2 Oct. 1941 *Toasties*... [Substances used as food or as ingredients in food]. Charles William Post,..London, England, and..Michigan, United States of America; manufacturer.] 1908 *Official Gaz.* (U.S. Patent Office) 31 Mar. 1177/2 *Postum Cereal Co. Limited*, Battle Creek, Mich. Filed 28, 1908. *Post Toasties*... Cereal Breakfast Foods. 1914 H. W. Wiley *1001 Tests of Foods* vi. 72 Many are the letters received in regard to the cereal breakfast foods, especially for children's use. One mother writes me: 'Two small youngsters are anxiously awaiting your opinion in regard to their favorite shredded wheat, grape nuts, and post toasties.' 1926–7 *Army & Navy Stores Catal.* 2/1 *Breakfast cereals*, Post Toasties—pkt. -/9. 1943 G. Greene *Ministry of Fear* III. iii. 202 There was milk and post-toasties and bread and marmalade. 1945 L. Shelly *Jive Talk Dict.* 31 *Post toasty*, corny character. 1950 A. Wilson *Such Darling Dodos* 134 Little pyramids of chocolate powder and post toasties..called 'Coconut Kisses'. 1969 *Listener* 22 May 732/1 They range from chilling glimpses of the world in the next century to golden, Post-Toastie pieces of life in which..the ancient decencies are reinforced. 1979 P. Niesewand *Member of Club* iv. 30 The economy-size box of Post Toasties on his breakfast table.

post-town. Add: **1.** Also, a town with its own postcode.
1973 *Guardian* 7 Feb. 1/1 The Post Office will recognise Milton Keynes as a separate post town on Monday.

po·st-treatment, *sb.* and *a.* **A.** *sb.* [Post- A. 1 b.] Treatment carried out subsequently. **B.** *adj.* [Post- B. 1 a.] Existing or occurring after treatment.
1946 *Genetics* XXXI. 375 Studies of hatchability of eggs deposited by females that had been inseminated by males treated with..2000 r of X-rays..suggest that the effect of posttreatment in reducing the frequency of chromosomal rearrangements is attributable to increased restitution. 1962 *Radiation Bot.* I. 132/2 Immediate ultraviolet post-treatments at 2967 Å..completely overcame 1 kr X-ray depression of embryo viability. 1967 E. Chambers *Photolitho-Offset* xiii. 198 It is used mainly as a 'post treatment' after developing and before etching. 1972 *Gloss. Terms Timber (B.S.I.)* 25 *Post-treatment*, preservative treatment of ordinary plywood after it has been made. 1972 *Jrnl. Social Psychol.* LXXXVI. 84 Any differences in posttreatment attitude can therefore be attributed to the effects of the experimental treatment.

postulational (pǫstiŭlēi·ʃənăl), *a.* [f. Postulation + -al.] Of or pertaining to postulation; based on or involving deduction from a set of postulates.
1926 L. Bloomfield in *Language* II. 153 Nevertheless, the postulational method can further the study of language, because it forces us to state explicitly whatever we assume. 1932 M. H. Stone *Linear Transformations in Hilbert Space* i. 2 The postulational treatment was carried through only recently by J. v. Neumann. 1933 L. S. Stebbing *Mod. Introd. Logic* (ed. 2) 506 A set of primitive propositions may be called 'a set of postulates', and the system thus constructed may be called 'a postulational system'. 1956 E. H. Hutten *Lang. Mod. Physics* iii. 75 The three laws of motion, together with the definitions, are given [by Newton] as a postulational system. 1961 E. Nagel *Struct. of Sci.* v. 91 A familiar example of an abstract calculus is demonstrative Euclidean geometry developed in a postulational manner.
Hence **postula·tionally** *adv.*
1936 *Mind* XLV. 174 In general, if *x* is defined postulationally, any proposition containing *x* will *implicate* the postulational definition of *x*. 1966 H. V. Guenther *Tibetan Buddhism* vi. 89 The distinction between a 'co-emergent belief in things' and a 'postulationally defined belief in things'..is concisely stated.

postulator. (Earlier and later examples.)
1863 tr. *Luquet's Life Anna Maria Taigi* 2 The undersigned Bishop of Hesebon, Postulator of the suit of the servant of God, Anna Maria Taigi. 1973 *Times* 2 Nov. 5/5 The required proof of two miracles is said to have brought to the postulator of the cause 'an embarrassment of choices'.

Postum (pōᵘ·stǝm). orig. *U.S.* Also **postum.** [Pseudo-Lat. formation f. the name of Charles William *Post* (see *Post Toasties) + L. *-um.*] The proprietary name of a coffee substitute. Also *attrib.*
1895 *Official Gaz.* (U.S. Patent Office) 3 Dec. 1549/1, 27,402 *Food Drinks.* Postum Cereal Co., Limited, Battle Creek, Mich. Filed Oct. 7, 1895 *Postum Cereal.* Used since February 1895. 1912 [see *instant *a.* 4 c]. 1922 *Hotel World* 15 Apr. 15/2 Choice of Coffee, Tea, Milk, Postum, Cocoa. 1962 *Flight Internat.* LXXXI. 871/1 Carpenter drank normal coffee instead of postum. 1967 *Listener* 21 Sept. 370/1 His frugal breakfast—perhaps a cup of Postum coffee and some biscuits. 1977 *New Yorker* 20 June 67/1, I will have a drink of Postum or coffee. 1978 J. Carroll *Mortal Friends* v. vi. 572 None of that cappuccino stuff, Jack... See if they have Postum, will you?..Well, lace it with a lot of milk, will you? Make it like American.

postural, *a.* Add: **1. c.** *postural integration*, 'some form of "body work", a term that applies to various specialized kinds of massage..that claim to bring about mental well-being through physical manipulation and body realignment' (private communication, Cyra McFadden, 15 Aug. 1979).
1977 C. McFadden *Serial* xviii. 42/1 She had like *mutated* over the years through Gurdjieff..Human Life Styling, postural integration, [etc.]. 1977 *Rolling Stone* 19 May 32/3 There was postural integration and the Alexander Technique and Aural Ecology and the Chiropractic Information Bureau and Rolfing and even one discipline called 'Prosperity Training'. 1979 *San Francisco Sunday Examiner & Chron.* (Sunday mag.) 26 Aug. 29, I used to do Gestalt therapy. I did body work, postural integration.

posturant (pǫ·stiŭrănt). *nonce-wd.* [f. Posture *v.* + -ant¹.] One who adopts an intellectual or æsthetic posture; a poser.
1934 E. Sitwell *Aspects Mod. Poetry* i. 19 A school of American-Greek posturants..began to exude a thin stream of carefully chosen watery words.

posture, *sb.* Add: **1. b.** Among animals, a particular pose which is a signal of a specific pattern of behaviour.
1940 H. F. Witherby et al. *Handbk. Brit. Birds* II. 329 In terrifying posture feathers [of long-eared owls] are ruffled up. 1953 N. Tinbergen *Herring Gull's World* ii. 7 Some of these movements and postures are not difficult even for the human observer to appreciate. 1962 *Symp. Zool. Soc.* No. 8. 71 A rat on the defensive in the upright posture does not look at its opponent. 1964 A. L. Thomson *New Dict. Birds* 282/1 Behaviour by which the plumage is exposed to the sun by special postures seems to be widespread in birds. 1974 I. C. J. Galbraith tr. *Dorst's Life of Birds* I. ix. 175 The sexes may be recognized by certain postures characteristic of each mate.
3. Also, *spec.* a military or political attitude; a condition of armed readiness.
1962 *Listener* 29 Mar. 545/2 What is important about the Soviet posture is that the Soviet armed forces aim at flexibility. 1964 *Ann. Reg. 1963* 140 The public denunciations and rigidities of the cold war appeared to have been set aside in favour of more traditional diplomatic methods and more relaxed diplomatic postures. 1974 *Times* 5 Mar. 7/1 Renewed anxiety over Nato's conventional defences was expressed today..in the annual Pentagon 'posture' report to Congress. 1976 *National Observer* (U.S.) 1 May 16/3 DEA's mission is..to have foreign governments in a responsive posture.

Posturepedic (pǫstiŭǝpī·dik). *U.S.* Also **posturepedic.** [f. Posture *sb.* + Ortho)pedic *a.*] The proprietary name of a Sealy mattress designed to give proper support to the relaxed body.
1952 *Official Gaz.* (U.S. Patent Office) 14 Oct. 289/2 Sealy, Incorporated, Chicago, Ill. Filed July 24, 1951. *Posturepedic.* The representation of the human figure is fanciful... For mattresses and box springs. Claims use since June 1, 1950. 1967 *N.Y. Times* 9 Apr. 63 Choose your Posturepedic at one of these fine sleep centers. 1976 *Laurel* (Montana) *Outlook* 23 June 19/5 (Advt.), Refinished walnut dining table, wooden high chair, posturepedic crib mattress. 1977 C. McFadden *Serial* xxxii. 71/1 He staggered down the hall to his Posturepedic.

posturing, *vbl. sb.* Add: **2.** Among birds, the use of particular poses as signals of specific patterns of behaviour.
1929 H. E. Howard *Introd. Study Bird Behaviour* i. 17 Posturing..is regarded as a manifestation of the affective aspect of the functioning of the sexual response. 1940 H. F. Witherby et al. *Handbk. Brit. Birds* I. 8 Display and Posturing.—Aerial evolutions..are a feature of courtship [of the raven]. 1953 N. Tinbergen *Herring Gull's World* ii. 7 In general, the movements playing a part in posturing are based on flattening or raising of the plumage as a whole, on eye movements and on the attitude of head, neck and wings. 1973 R. A. Hinde in D. S. Farmer et al. *Avian Biol.* III. viii. 493 Ambivalent Posturing. Sometimes the bird adopts a posture that simultaneously expresses both tendencies.

post-velar: see *post- B. 2. **post-verbal(ly:** *post- B. 1.

post-Victorian, *a.* and *sb.* [f. Post- B. 1 b.] **A.** *adj.* Subsequent to the reign of Queen Victoria (died 1901) or the Victorian era (esp. in style or manners). **B.** *sb.* One who lives in the post-Victorian era.
1938 H. Palmer *Post-Victorian Poetry* p. ix, Because of the amount of space that has been taken up by the verse dating from 1900..the subject-matter seems to justify the selection of the title, *Post-Victorian Poetry.* 1960 *Times Lit. Suppl.* 24 June 400/2 Hulme was able to dress a post-Victorian change of mood in conceptual clothing which seemed..to fit. 1974 *Nature* 1 Mar. 86/2 Bowlby..offers a..less punitive approach to the child, than is generally taken..by the post-Victorians or the Freudians.

postvocalic(ally: see *post- B. 1.

postvo·calized, *ppl. a.* *Philol.* [f. Post- A. 1 a + Vocalized *ppl. a.*] Followed (as a consonant) by a vowel.
1876 [see Prevocalized *ppl. a.*].

po·st-war, *a.* (*sb.*) Also **postwar.** [f. Post- B. 1 a + War *sb.*¹] **A.** *adj.* Of, pertaining to, or characteristic of, the period after a war (esp. those of 1914–18 or 1939–45); *post-war credit*, a system of additional personal taxation introduced by the British Government in 1941 to supplement wartime expenditure and repayable during the post-war period; a sum of money or promissory note associated with this scheme.
1908 *Daily Chron.* 16 July 5/1 There has been a reduction of some £2,000,000 since 1904–5, the first post-war year. 1915 *Political Q.* May 118 The plans outlined..admit most easily of deviation into whatever may approve itself as the desired norm of post-war politics and economics. 1919 J. M. Keynes *Econ. Consequences Peace* 84 Our hypothetical calculations leave us with post-war human requirements, on the basis of a pre-war efficiency of railways and industry. 1938 *Encycl. Brit. Bk. of Year* 22/1 The sheep population reached a post-war maximum in 1932. 1942 *Times* 9 July 2/3 (*heading*) Post-War Credit Certificates. *Ibid.*, The issue of over 9,000,000 certificates for post-war credit in respect of income-tax payments for the year 1941–42 has begun. 1959 *Daily Tel.* 9 Apr. 1/7 The Trades Union Congress yesterday welcomed the Chancellor's cuts in purchase tax and his proposals to speed up post-war credit payments. 1964 F. Bowers *Bibliogr. & Textual Crit.* I. v. 34 Some resistance to accepting the validity of post-war pioneering studies. 1972 *Times* 9 Aug. 6/6, I propose..that repayment of post-war credits to people who have not been able to produce a certificate should begin on January 1, 1973. I envisage that this final stage of the repayment should be spread over the 12 months ending December 31, 1973. 1973 D. Aaron *Unwritten War* III. vi. 103 Parkman did not attempt to predict the post-War fate of the Brahmin class.
B. *ellipt.* as *sb.* *U.S.* The period following a war.
1944 *Sun* (Baltimore) 18 July 8/7 They will glibly write ..of projects to be undertaken 'at post war'. 1945 *Time* 24 Sept. 15/2 Management-labor unity..must be continued in the postwar. 1947 *Partisan Rev.* XIV. 454 The tank driver of the War becomes the tractor driver of the Postwar.

posty, var. *postie.

pot, *sb.*¹ Add: **1. e.** (Earlier and later examples.) Also, the pan of a close stool; a lavatory pan.
1598 Florio *Worlde of Wordes* 280/1 *Pitale*,..the pan or pot of a close stoole. 1915 *Dialect Notes* IV. 228 *Pot*,..very common for *chamber*. 1954 A. S. C. Ross in *Neuphilol. Mitt.* LV. 42 U-speakers use ['dɛəri]..or *pot*. 1956 *New Statesman* 8 Dec. 740/2 The old woman on the next floor fighting over who choked the cawsy by pitching cinders down the pot. 1958 [see *article 14 c].
h. A protuberant stomach, a paunch; = Pot-belly 1.
1928 'Brent of Bin Bin' *Up Country* ii. 40 Mazere..was happy that he could turn to manual work again himself, and felt the better for it. 'It's taking a little of the pot off me,' he would exclaim. 1929 Kipling in *London Mag.* Dec. 631/1 Keede patted his round little pot. 1942 D. Powell *Time to be Born* (1943) i. 22 At forty-eight..he had no pot at all due to his Yogi exercises. 1952 [see *gussy *v.*]. 1959 *Encounter* Dec. 31/1 There was a time when I had a little professorial pot; later I was 'stout'..; then I became, I suppose, definitely pot-bellied. 1965 G. McInnes *Road to Gundagai* xiii. 222 The door opened carefully and revealed a tall man with a florid face, a large Roman nose,..and a big pot. 1973 'D. Jordan' *Nile Green* xxix. 129 His pot was hanging obscenely over the lip of a pair of scarlet bathing trunks.
i. (See quots.) *slang.*
1941 *Amer. Speech* XVI. 240 *Pot*, carburetor. 1945 Baker *Austral. Lang.* viii. 160 Here is a brief list of indigenous Air Force language:..*pot*, a cylinder. 1961 Partridge *Dict. Slang* Suppl. 1230/1 *Pot*, n., a cylinder, esp. in one of the old rotary engines: R.F.C.–R.A.F.: 1914–18. They tended to split or to fly off. Hence, any aeroplane-engine cylinder. 1966 'L. Lane' *ABZ of Scouse* 83 A pot is also a carburettor. *To tickle ther pot*: To prime a carburettor.
2. a. (Later examples.) Also allusively in *fig.* phr. *pot of gold* (see quot. 1895).
1895 *Brewer's Dict. Phr. & Fable* (new ed.) 1036/2 *Rainbow chasers*, problematical politicians and reformers, who chase rainbows, which cannot possibly be caught, to 'find the pot of gold at the foot thereof'. This alludes to an old joke, that a pot of gold can be dug up where the rainbow touches the earth. 1966 A. E. Lindop *I start Counting* xxi. 266 Matron makes a pot of tea quite late at night and lets me go and have a cup with her. 1971 *Cape Herald* 15 May 14/1 Francis Lee, the Manchester City and England striker has hit the 'pot-of-gold' in more ways than one. 1974 'E. Lathen' *Sweet & Low* i. 11 The Japanese stock market..had been the legendary pot of gold at the end of some local rainbow. 1978 *Observer* 26 Mar. 11/7 The tendency has been to look at the North Sea in terms of its immediate isolated wealth, as a pot of gold. 1979 W. H. Canaway *Solid Gold Buddha* iii. 27 Miller..made a pot of tea.
b. Also, a pot of tea.
1831 E. J. Trelawny *Adventures Younger Son* III. xxiii. 152 My wife always turned in three spoonsful,—one for I, one for her, and t'other for the pot. 1973 'E. Ferrars' *Foot in Grave* ix. 166 The tea had got cold, so Christine made a fresh pot.
3. (Later *Austral.* examples.)
1943 [see *handle *sb.*¹ 2 c]. 1966 G. W. Turner *Eng. Lang. Austral. & N.Z.* viii. 163 In addition Sydney has a *pony* of five ounces.., Melbourne a *pot* of ten ounces (but a pot is eleven ounces in Brisbane), Adelaide a *butcher* (six ounces) and Perth a *big pot* (fifteen ounces, which would be a *schooner* in Sydney or a *pint* in Adelaide). 1973 *Parade* (Austral.) Oct. 35/1 'Oh, yes,' said the barman. 'We like their money, but we don't have middies, we sell pots.'
9. a. (Earlier example.)
1856 *Knickerbocker* XLVIII. 619 They had hauled down a big pot and intended henceforth to live as jolly as clams.
d. (Earlier and later examples.)
1880 Hardy *Trumpet-Major* I. viii. 135 When Festus put on the big pot, as it is classically called, he was quite blinded *ipso facto* to the diverting effect of that mood and manner upon others. 1885 *Punch* 12 Sept. 131/2 Oh, Yorkshire and Lancashire both are big pots. But Cricket's top honours again go to Notts. 1909 J. R. Ware *Passing Eng.* 28 Big pot (*Music-hall* 1878–82)... This phrase is probably one of the few that filter down in the world from Oxford, where, in the 50's it was the abbreviation of potentate. It referred to a college don, or a social magnate. 1947 'A. P. Gaskell' *Big Game* 24, I don't feel at home with these big pots. 1979 R. Rendell *Make Death love Me* iii. 29 Some general at the head of it. Some big pot who means business.
e. (Examples of sense in Dict.); also, orig. U.S., the betting pool in poker and other gambling games. Also *fig.* Cf. sense 9 b in Dict., *jack-pot* s.v. Jack *sb.*¹ 33 a in Dict. and Suppl.
1847 J. H. Greene *Gambling Unmasked* (rev. ed.) 196 He won the first twenty '*pots*', that is to say, the stake [in poker]. 1856 G. W. Bagby *Old Virginia Gentleman* (1910) 228 He has no great faith in 'cases', but believes in betting on three cards at a time, and has a special hankering for 'the pot' [in faro]. 1868 [see *keno b]. 1878 [see *bullet *sb.*¹ 6]. 1889 'Mark Twain' *Speeches* (1923) 147 What is still more irregular, the man that loses a game gets the pot. 1890 J. P. Quinn *Fools of Fortune* II. ii. 194 In the [faro] 'lay-out'..the six, seven and eight are called the pot.

1892 [see *chip *v.*¹ 8 b]. 1895 *Funk's Stand. Dict.*, *Pot*,..5. *Card-playing*, (1) The amount of stakes played for; the pool. (2) In faro, the six, seven, and eight of the layout, collectively. 1935 *Encycl. Sports* 466/2 If no player opens there is a fresh deal, each player once more contributing to the pot, and so on until the pot is opened. 1951 *Amer. Speech* XXVI. 100/1 *Open the pot*, to make the first bet after the ante [in poker]. *Ibid.* 100/2 *Pot*, the total accumulation of all bets. It rests in the centre of the table, equi-distant from each player. 1963 *Richmond* (Virginia) *Times-Dispatch* 16 Dec. 19/2 When a poker player has absolute confidence in his hand he shoves into the pot every chip he has. 1971 *Black World* June 73/1 The sergeant put up the house, got the men into the game and took half the pot. 1977 I. Shaw *Beggarman, Thief* II. i. 118 'And if you succeed, then what?' he said. 'Russia takes the whole pot.'
f. *old pot*: see *old *a.* 1 c.
10. (Later examples of spelling *pott*, which is now usual.)
1890 in Webster. 1911 *Encycl. Brit.* XX. 735/1 Writing papers. Pott..12½ × 15. 1926 *Paper Terminol.* (Spalding & Hodge, Ltd.) 21 *Pott*, a standard size of printing paper measuring 15½ × 12½ in. (with slight variations). The term is derived from an ancient 'pot' watermark, which represented the Sangraal. 1962 F. T. Day *Introd. to Paper* vii. 69 Sizes of paper in the United Kingdom centre round fifteen designations: Foolscap, Demy, Medium,..Pott, Elephant,..Eagle and Columbier.
11. (In *hopscotch.*) (Examples.) Also, the game itself or a part of the game. Also *local U.S.*
1866 W. Gregor *Dial. Banffshire* 132 *Pot*,..the last division in the game of *hippin'-beds* [*sc.* hopscotch]. 1893–4 R. O. Heslop *Northumb. Words* II. 549 *Pot*, the heading written at the top of the game called 'beds', or, locally, 'hitchey dabber'... To achieve it is to get 'pot'. 1895 *Funk's Stand. Dict.*, *Pot*..*pl.* (Local, U.S.) (1) The game of hop-scotch. 1920 Webster, *Pot*.., a piece of pottery or earthenware, as a marble or piece for playing hop scotch. 1936 *Glasgow Herald* 10 Nov., 'Hopscotch', however is an English name. We Scots called it 'peever',..or 'pot', or 'the beds'... In some parts of Scotland beds 7 and 8 were called 'the kail pats', and this may be one reason why the game is sometimes called 'pot'. Another explanation is that a piece of broken pot or earthenware was often used as a peever.
13. f. *to go to pot.* (Further examples.) For *vulgar* read *colloq.* Also, to deteriorate, to go to pieces.
1846 *Swell's Night Guide* 120/2 *Gone to pot*, become poor in circumstances, gone to the dogs. 1889 *Cornh. Mag.* July 46 For the potato is really going to pot... Constitutional disease and the Colorado beetle have preyed too long upon its delicate organism. 1910 E. M. Forster *Howards End* xxv. 205 Evie heard of her father's engagement when she was in for a tennis tournament, and her play went simply to pot. 1923 S. Kaye-Smith *End of House of Alard* iv. 327 If we hung on now, still further crippled by death-duties, the land would simply go to pot. 1942 E. Paul *Narrow St.* iv. 37 The Comédie Française.. went to pot artistically and remained a travesty of its former self until reorganized about 1938. 1953 E. Simon *Past Masters* II. 81 Discipline's gone all to pot at the camp. 1956 W. Graham *Sleeping Partner* 57 She could go in and clean up once in a while... The house wouldn't go to pot in a week or so. 1968 *Globe & Mail Mag.* (Toronto) 13 Jan. 11/3 Only a quarter of the Sames now depend on reindeer herding... Some of them, like some of Canada's native people, go to pot. 1979 *Truck & Bus Transportation* (Austral.) Apr. 65/2 It's [*sc.* the brake is] there to do its job, but it can throw a spanner in the works if the adjustment setting goes to pot.
k. *Austral.* and *N.Z. slang.* *to put* (a person's) *pot on*: to inform against, to tell tales; to destroy the prospects of. Cf. *pot *v.*¹ 6 b.
1913 A. J. Rees *Merry Marauders* xi. 206 You ought to put the Liquor Party's pot on. 1919 W. H. Downing *Digger Dial.* 40 *Put his pot on*—Report him. 1928 W. S. Smyth *Jean of Tussock Country* vi. 59 Dalton has put your pot on. 1935 Davison & Nicholls *Blue Coast Caravan* 178 He saw some blacks..standing on the platform under guard of a policeman. 'Hullo, what's up?' One of them replied, 'Aw, somebody's been putting our pot on.' 1948 *Landfall* II. 110 'Got a ten bob rise last week,' Duggan said. 'Funny that, you know,' Larry said. 'I been there about the same time as you, Tom, and I haven't had a rise yet. Wonder if Myers put my pot on.' 1957 V. Palmer *Seedtime* 119 There's an election coming on, and there's a chance I'll be dumped... This afternoon's work has probably put my pot on.
l. *to get off the pot* (and extensions): see quot. 1972. *coarse slang* (chiefly *N. Amer.*).
1961 Partridge *Dict. Slang* Suppl. 1269/2 *Shit or get off the pot!* A Canadian Army c.p. (1939–45), directed at a dice-player unable to 'crap out'. 1966 'A. Hall' *9th Directive* x. 90 Get some definite information for me. Tell the Ambassador to get off the bloody pot. 1972 *Dict. Contemp. & Colloq. Usage* (Eng.-Lang. Inst. Amer.) 26/2 *Shit or get off the pot*, vulgar. A command that someone either complete an action in process or abandon the attempt and give someone else the opportunity to try. 1973 W. McCarthy *Detail* ii. 112 You've got forty-eight hours, that's all. Either get it, or get off the pot. 1974 *Farm & Country* 26 Mar. 4/2 To put it bluntly: the Ottawa politicians had better perform at once or get off the pot. 1977 'J. Le Carré' *Hon. Schoolboy* xii. 275 You better tell some of those limousine liberals back in Langley Virginia it's time for them to shit or get off the pot.
14. *pot-grown* adj.; **pot annealing** *vbl. sb.* = *box annealing; also *attrib.*; hence **pot-anneal** *v. trans.*; **pot-bunker** *Golf*, an artificially constructed pot-shaped bunker; **pot-burial**, a prehistoric form of burial found in Crete (see quot.); **pot clay** (later examples); also (freq. with capital initials), a bed of this kind of clay near the base of the English coal measures; **pot courage** = *Dutch courage* s.v. Courage *sb.* 4 d; **pot cupboard**, a bedside cupboard designed to hold a chamber pot; **pot-drum** (see quot.); **pot-furnace**, (further examples); also, any furnace in which crucibles are heated; **pot-green** = Pot-herb; **pot-gut(s)**, (*a*) = Pot-belly 2; also, a pot-bellied animal; (*b*) = Pot-belly 1; **pot-gutted** *a.* (later examples); **pot-lace** (examples); **pot-layering**, a method of propagating trees or shrubs in which a ball of soil, sometimes held within a split pot, is attached to a cut on a branch until enough roots have grown for the branch to be planted independently; **pot-licker, potlicker** *N. Amer.*, a mongrel dog; also *attrib.*; hence **pot-licking, potlicking**, toadying; **pot life**, the length of time that a glue, resin, or the like remains usable after preparation; **pot-line**, a line of retorts used for the electrolytic production of aluminium; **pot marigold**, the common marigold, *Calendula officinalis*, whose petals may be used to colour or flavour food; **pot marjoram**, substitute for def.: a small shrub, *Origanum onites*, whose leaves are used to flavour food; (earlier and later examples); **pot-mess** *Naut. slang*, a stew concocted from various scraps; *transf.*, a state of confusion or complete disorder; **pot rassler, rastler,** U.S. var. **pot-wrestler**; so **pot rassling, rastling** *vbl. sb.* (see quot.); **pot-sick** *a.*, (*b*) *nonce-use*, pot-bound; cramped or starved of nutrient; **pot stand**, a stand designed to hold pots or potted plants; **pot still** (earlier and later examples); so **pot-stilled** *adj.*; **pot-training** *vbl. sb.*, the training of a small child to use a chamber pot; hence (as a back-formation) **pot-train** *v.*, **pot-trained** *ppl. a.*; **pot-washings** *sb. pl.*, food removed from pots by washing; **pot-woman**, (*a*) a woman who sells pots; (*b*) *obs.*, a barmaid; (*c*) a woman who works at pottery; **pot-work** (earlier and later examples); **pot wrestler** (examples); (*c*) a chef; so **pot wrestling** *vbl. sb.*
1928 H. M. Boylston *Introd. Metall. Iron & Steel* xv. 519 Tool steels are sometimes annealed in open-type furnaces of fairly small size, but in many cases are pot annealed. 1938 C. G. Johnson *Forging Pract.* 111 The steel is cooled very slowly either with the furnace..or cooled in a pot surrounded by heat insulating material (pot annealed). 1925 *Jrnl. Iron & Steel Inst.* CXII. 453 A newly designed installation for the drying of wire bundles..is..described, in which the chambers are heated with the waste gases from pot annealing furnaces. 1934 *Ibid.* CXXIX. 519 (*heading*) The heat conditions for the pot-annealing of steel hoops. 1909 *Westm. Gaz.* 30 Apr. 4/2 Had its original whins been forest-trees we should not now be digging pot-bunkers. 1963 *Times* 5 June 5/2 He..found a pot bunker and took seven which enabled Pirie..to halve. 1921 *Discovery* Feb. 33/1 A simpler form of burial, known as the 'pot-burial', was effected by trussing up the body, placing it under an inverted jar, and then burying it in the earth. 1860 E. Hull *Geol. Leicestershire Coalfield* (Mem. Geol. Survey) vi. 35 (*caption*) A. Loose breccia... B. Purple marl forming base of New Red Sandstone. C. Sandy shale. D. Coal 3 feet thick. E. Pot-clay, with rootlets stretching from the coal. 1913 *Geol. Derbyshire Coalfield* (Mem. Geol. Survey) viii. 124 Pot-clay suitable for the manufacture of stoneware occurs below a thin coal above the Alton seam. *Ibid.* 125 A bastard gannister, unsuited for use either as a pot-clay or as a gannister. 1939 tr. *E. N. Marais's My Friends the Baboons* vii. 81 Some were busy digging pot-clay from the hole with their hands while others were fashioning oxen and other animals from the clay. 1951 *Concealed Coalfield of Yorkshire & Nottinghamshire* (ed. 3) (Mem. Geol. Survey) iii. 13 The Coal Measures rest conformably on the Millstone Grit Series, the dividing line being placed, by international agreement (Jongmans 1928, p. xliv), at the Pot Clay or *Gastrioceras Subcrenatum* Marine Band. 1968 M. A. Calver in Murchison & Westoll *Coal* viii. 173 The Pot Clay fauna lacks the typical benthonic assemblage exhibited by the other kinds of marine band. 1806 C. Wilmot *Let.* 14 Oct. in *Russ. Jrnls.* (1934) II. 231 In a fit of *Pot Courage* no doubt he stroked his paunch & felt himself a Hero! 1867 E. Cust *Lives Warriors* II. I. 190 One of the best officers..became so drunk, that in his pot-courage he wrapped the napkin about his head, and.. went out in this guise into the trenches to attack the foe. 1794 T. Sheraton *Cabinet-Maker & Upholsterer's Drawing Bk.* III. 364 This left-hand drawer is..sometimes made very short, to give place to a pot-cupboard behind, which opens by a door at the end of the sideboard. 1973 *Country Life* 26 Apr. (Suppl.) 59/4 Late 18th century mahogany pot-cupboard. 1912 *Encycl. Relig. & Ethics* V. 90/1 The pot-drum is an earthenware vessel headed with a membrane. 1905 W. Macfarlane *Lab. Notes Pract. Metall.* 11 (*heading*) Exercises in a crucible or 'pot' furnace. *Ibid.* 13 Gas coke is often good enough for a pot furnace. 1930 *Engineering* 18 Apr. 525/3 Sillimanite sieges in pot furnaces were more common. 1971 *Materials & Technol.* II. vi. 366 *Pot furnaces.* These are used for melting optical glass and other special glasses in quantities of up to half a ton at a time. 1742 W. Ellis *Mod. Husbandman* July iii. 36 They proved..sweet Eating, when no other Pot Greens could be hardly got. 1972 *Sci. Amer.* Nov. 130/3 She studied our Southern cookbooks, too,..coming to the conclusion that 'distinctive features of Southern cooking are African in origin': gumbos and burgoo, hush puppies and pot greens, to begin with. 1915 W. L. Howard (*title*) Rest period studies with pot-grown woody plants (Missouri Agric. Exper. Station Res. Bull. No. 16). 1946 G. A. R. Phillips *Rock Garden* iv. 64 Pot grown plants may be planted with comparative safety at any time of the year. 1960 *Farmer & Stock-breeder* 9 Feb. 119/2 Start with small trees. In the case of *macrocarpa*, which are usually pot-grown, you could plant trees of 1½–2ft. 1977 C. Lloyd *Clematis* vii. 114 Pot-grown clematis have the advantage of needing very little root disturbance in the process of planting. 1909 *Dialect Notes* III. 359 *Pot-gut*, *n.*, a pot-bellied person. *c* 1926 'Mixer' *Transport Workers' Song Bk.* 72 She's seated in a motor With some 'pot-gut' by her side. 1926 G. Frankau *My Unsentimental Journey* xiv. 182 Those little squirrels they call 'pot-guts' scuttle fatly across the well-made road. 1942 Berrey & Van den Bark *Amer. Thes. Slang* §429/2 *Fat person*,..pot-gut(s). 1951 R. Campbell *Light on Dark Horse* 75 Then his old pot-guts would shake like a jelly. 1845 *Spirit of Times* 2 Aug. 267/1 *Ar* you a goin to tumtum all nite on that pot-gutted old pine box of a fiddle, *say*? 1909 *Dialect Notes* III. 359 *Pot-gutted, adj.*, pot-bellied. 1912 *Ibid.* 586 Look at that pot-gutted beer fly, will you. 1941 Baker *Dict. Austral. Slang* 56 *Pot-gutted*, fat, paunchy. 1865 Pot lace [see *Antwerp]. 1960 H. Hayward *Antique Coll.* 15/2 The style of pattern has given this [*sc.* Antwerp lace] the name of pot lace..from the substantial two-handled vase usually prominent in it. 1912 A. F. Broun *Sylviculture in Tropics* II. iv. 146 'Pot-layering' is employed for branches which are either too high up a tree or too brittle to be bent into the ground. 1934 [see *air-layering* (*air *sb.*¹ II)]. 1961 *Amat. Gardening* 30 Sept. (Suppl.) 2/3 Air-layering. Also known as pot-layering... A means of rooting branches or shoots. 1932 V. Randolph *Ozark Mountain Folks* 223 Jethro was splitting wood as I rode into his little clearing, heralded by a great number of pot-licker dogs. 1947 *Clarke County Democrat* (Grove Hill, Alabama) 30 Oct. 4/3 A hound is a hound, regardless of whether he is July, Red Bone, Walker, potlicker or just plain hush-puppy. 1948 W. Faulkner *Intruder in Dust* (1949) i. 5 A true rabbit dog, some hound, a good deal of hound, maybe mostly hound, redbone and black-and-tan with maybe a little pointer somewhere once, a potlicker, a nigger dog. 1971 W. Hillen *Blackwater River* ii. 10 One man was walking through the village with his three pot-lickers when they met a housecat. 1929 Potlicking [see *hole *v.*¹ 7]. 1968–70 *Current Slang* (Univ. S. Dakota) III–IV. 95 *Pot-licking*, *n.* Oversolicitous behavior... He made his way to the top only by *pot-licking*. 1945 H. Barron *Mod. Plastics* viii. 199 When the hardener is mixed into the resin then the mixture has a very limited pot life. 1969 T. C. Thorstensen *Pract. Leather Technol.* xiv. 226 In this kind of finish the reactive components are usually mixed shortly before application, due to the limited pot life of the components. 1976 G. S. Ramaswamy *Mod. Prestressed Concrete Design* vii. 86 The pot-life of the mixture was 20 to 25 minutes at an ambient temperature of 30°C. [1936 *Industr. & Engin. Chem.* Feb. 148/1 Shutting down a 'line' of aluminum cells or 'pots'..is not a difficult or lengthy operation if properly performed.] 1951 *Economist* 29 Sept. 748/1 The drought had forced the huge hydro-electric installations on the Columbia River to reduce..power to the plants that provide aluminium. Already three 'potlines' have been closed down. 1957 *Times* 12 Nov. (Canada Suppl.) p. vi/5 The power is brought down on the other side of the mountain, 10 miles away, to feed the potlines located by the deep water estuary of the Douglas channel. 1965 *Wall St. Jrnl.* 25 Feb. 32/2 Aluminum Co. of America said it will install a third 33,000-ton-a-year potline for producing primary aluminum at its Warrick County, Ind. plant. 1814 J. Green *Address on Bot. U.S.* 41/2 *Calendula officinalis.* Pot Marygold, Common. 1883 *Encycl. Brit.* XV. 544/1 The pot-marigold..is the familiar garden plant with large orange-coloured blossoms. 1910 *Daily Chron.* 19 Feb. 9/6 Among the best annuals for town gardens are the..French and African marigolds..and the calendula or pot marigold. 1936 E. S. Rohde *Herbs & Herb Gardening* ix. 128 Marigolds, including the old Pot Marigold.., have for some years been coming into favour again. 1966 G. B. Foster *Herbs for Every Garden* iii. 77 The petals of pot marigold were used to color butter, cheese, custards and sauces. 1597 Pot marjoram [see Marjoram a γ]. 1629 J. Parkinson *Parad.* II. cxxvi. 447 This kind of Marierome belongeth to that sort is called..In English Winter Marierome, or pot Marierome. 1936 E. S. Rohde *Herbs & Herb Gardening* vii. 73 Pot Marjoram..is a larger and more branching plant than Sweet Marjoram. 1974 Page & Stearn *Culinary Herbs* 24 Pot Marjoram..is a dwarf shrub with erect densely hairy stems. 1914 'Bartimeus' *Naval Occasions* xxiv. 238 What an awful pot-mess my cabin is in. 1916 'Taffrail' *Carry On!* 64 ''Strewth!', he murmurs under his breath, gazing at the littered floor in dismay.. ''ere's a fine pot mess.' 1926 *Blackw. Mag.* Dec. 835/2 The resulting pot-mess vanished all too soon. 1961 F. H. Burgess *Dict. Sailing* 163 *Potmess*, a mixture of any- and everything; a big heap; confusion. 1962 Granville *Dict. Sailors' Slang* 90/2 *Pot mess*, kind of stew very popular on the mess decks. 1942 Berrey & Van den Bark *Amer. Thes. Slang* §460/15 Pot rassler or rastler, a *dishwasher*. 1968 R. F. Adams *Western Words* (rev. ed.) 234/2 *Pot rastler*, a logger's name for a dishwasher. 1942 Berrey & Van den Bark *Amer. Thes. Slang* §819/1 Pot rassling, -rastling *or* wrestling,..*dishwashing or cooking.* 1872 Hardy *Under Greenw. Tree* I. II. iii. 157 Every morning I see her eyes mooning out through the panes of glass like a pot-sick winder-flower. 1907 *Yesterday's Shopping* (1969) 148/2 Pot Stands, triangular wood..each 0/3. 1947 [see *back-drop* (*back- B)]. 1971 *Cambr. Anc. Hist.* (ed. 3) I. II. xviii. 398 A biconical potstand has remote parallels at Büyük Güllücek, and in the Khirbet-Karak wares of the 'Amūq and Palestine. [1799 *Rep. Comm. Distilleries Scotl.* in *Parl. Papers* 1803 XI. 727/2 Private families distilled Whiskey for their own use; and the Still they used was a large pot, globular.] *Ibid.* 730/2 Suppose then that Fig. 5 represents an old fashioned Pot-Still. 1939 Joyce *Finnegans Wake* 246 Ansighosa pokes in her potstill to souse at the sop be sodden enow and to hear to all the bubbles besaying. 1958 P. Kemp *No Colours or Crest* viii. 150 A very small, wizened old man crouched over a crude pot-still..; on the ground beside him stood a number of small flasks filled with the clear spirit. 1965 Pot-stilled [see *Mao tai]. 1972 J. Gathorne-Hardy *Rise & Fall Brit. Nanny* viii. 265 It is actually physiologically impossible to pot train a child before the age of about six months. 1975 H. Jolly *Bk. Child Care* I. xi. 179 You cannot truly pot train a baby before he is physically mature enough to exercise some control over his bladder and bowels. 1961 *Spectator* 17 Feb. 218 One-year-olds are pot-trained. 1966 'K. Nicholson' *Hook, Line & Sinker* 11 Three sisters, the youngest only two and a half, and not fully pot-trained. 1960 L. Durrell *Clea* II. ii. 126 Have you managed to annul your early pot-training? 1975 H. Jolly *Bk. Child Care* I. xv. 217 If as a mother you never respond to your toddler's signals when he is about to wet his pants, he may give up trying to take the initiative himself and will probably be less co-operative with your efforts at pot training. 1912 C. N. Moody *Saints of Formosa* ix. 195 They threatened to..feed her on the pot-washings with which the pigs are nourished. 1802 D. Wordsworth *Jrnl.* (1941) I. 182 We then went to the Pot-woman's and bought 2 jugs and a dish. 1918 *Pall Mall Gaz.* 29 June 5/4 A 'potwoman' at a public house applied for a summons for wages in lieu of notice. 1979 *Listener* 20 & 27 Dec. 854/4 The Thistle..had a three-cornered taproom. I once saw a pot-woman dance an impromptu fertility dance there. 1765 J. Wedgwood *Let.* 2 Mar. (1965) 29 This trade to our Colonies we are apprehensive of losing in a few years as they have set on foot some Potworks there already. 1902 A. Bennett *Anna of Five Towns* xii. 328 Behind it..was a small, disused potworks. 1965 *Punch* 17 Feb. 244/1 The potworks expose their raw materials on their open marlbanks. 1860 Bartlett *Dict. Amer.* (ed. 3) 335 *Pot-wrestler*, a scullion. Pennsylvania. 1873 *Kansas Mag.* Aug. 139/1 'Bullwhackers', .., 'pot-wrestlers' and 'ink-slingers' are but a few of the common pet English names a politician must use in addressing his audience, in order to show them he is sufficiently familiar with their language. 1889 Farmer *Americanisms* 434/2 *Pot-wrestler*, a Pennsylvanian equivalent of the English 'pot-walloper'; a scullion. 1902 [see Potwalloper 2]. 1941 J. Smiley *Hash House Lingo* 44 *Pot wrestler*, dish washer. 1947 *N.Y. Jrnl. American* 18 Mar. 17/4 The off-center meatball has been endorsed by the chefs of the old world. No less a pot-wrassler than the King's own glorified the chuckwagon croquette as the ambrosia of the parked gulp. 1942 Pot wrestling [see *pot-rassling].

pot, *sb.*⁴ Add: **b.** *Rugby Football.* A dropped goal (see *dropped *ppl. a.* 1 a). *N.Z.*
1959 *N.Z. Listener* 24 July 6/4 Five potted goals—that was when a pot was worth four points.

pot (pǫt), *sb.*⁵ *slang* (orig. *U.S.*). [prob. f. Mexican Sp. *potiguaya* marijuana leaves; perhaps influenced by Pot *sb.*¹; cf. *pod *sb.*² 1 c.] **1.** = *marijuana 1 a.
1938 [see *jagged *a.*² b]. 1951 *N.Y. Times* 13 June 24/4 Progression from sneaky pete to pot to horse to banging. 1952 *Amer. Speech* XXVII. 28 Pot,..marijuana cigarettes... (Maurer, *potiguaya*, marijuana leaves after pods removed.) 1959 *Oxford Mail* 9 Nov. 1/6 The detectives invited their Greenwich Village pals to a big party..with poetry readings, bongo drums and plenty of 'pot' (marijuana) to smoke. 1961 N. Mailer *Advts. for Myself* 209 In Mexico, however, down in my depression with a bad liver, pot gave me a sense of something new. 1964 S. Bellow *Herzog* (1965) i. 87 You have to do more than take a little gas, or slash the wrists. Pot? Zero! Daisy chains? Nothing! Debauchery? A museum word. 1965 *Punch* 22 Dec. 930/3 'Like a spot of pot, Jean?' he said. 'I never touch it,' I said archly. 1966 T. Pynchon *Crying of Lot 49* iii. 63 'But we don't repeat what we hear,' said another girl. 'None of us smoke Beaconsfields anyway. We're all on pot.' 1968 *New Scientist* 22 Aug. 371/2, I am not defending the behaviour of these students and of course I don't agree with them that 'pot' is harmless. 1972 J. L. Dillard *Black English* vii. 289 The teacher probably should not let illusions of knowledge about the ghetto tempt him into using sentences like *John and Mary are smoking pot.* 1973 [see *LSD¹]. 1975 D. Lodge *Changing Places* ii. 63 He took an extreme radical position on all such issues as pot, sex, race. 1977 *Jrnl. R. Soc. Arts* CXXV. 464/2 His son occasionally smokes pot.
2. *attrib.* and *Comb.*, as *pot liquor, smoke, -smoker, -smoking* vbl. sb. and ppl. a.; **pot-head, pothead** (see *head *sb.* 7 e); **pot party**, a party held for the smoking of marijuana.
1959, etc. [see *head *sb.* 7 e]. 1974 *Times Lit. Suppl.* 1 Mar. 219/4 A girl..herself something of a pothead, who introduces Evans to the joys of the weed. 1970 C. Major *Dict. Afro-Amer. Slang* 92 *Pot liquor*..: brew from marijuana seeds and stalks. 1967 *Boston Sunday Herald* 26 Mar. I. 40/8 Authorities said that as many as 25 teenagers attended drug sessions held twice a month at different Randolph homes by one group, and that others also held 'pot' parties. 1971 D. E. Westlake *I gave at the Office* (1972) 219 The media have some leniency..as for instance when magazines stage pot parties. 1976 B. Lecomber *Dead Weight* v. 68, I couldn't quite imagine even an Antiguan Customs officer suspecting Herr Ruchter of holding riotous pot-parties. 1978 J. Krantz *Scruples* iii. 66 An occasional pot party was as anti-establishment a gathering as he attended. 1966 T. Pynchon *Crying of Lot 49* iii. 64 Their rising coils and clouds of pot smoke. 1967 *Guardian* 8 July 6/4 The bored pot-smoker looking for some better kick. 1971 *New Scientist* 11 Mar. 533/3 Pot smokers have a lowered sexual drive. 1976 H. Nielsen *Brink of Murder* xi. 99 You must be calling on a pot smoker. Take me along? 1964 *Punch* 18 Mar. 413/1 'Pot-smoking' parties, gaming sessions. 1967 *Guardian* 3 Feb. 4/4 They said, there is no evidence of a direct link between pot-smoking and addiction to hard narcotics. 1970 G. F. Newman *Sir, You Bastard* ii. 39 He didn't have the stereotype-copper look..but appeared more as one expected a pot-smoking nymphomaniac to look. 1977 *Rolling Stone* 13 Jan. 37/2 Kerr-McGee officials later quoted from the letter when they described Silkwood's pot-smoking proclivities to reporters.

pot (pǫt), *sb.*⁶ Colloq. abbrev. of Potentiometer.
1943 C. L. Boltz *Basic Radio* iii. 56 Radio workers always refer to the 'pot'. 1948 *Electronics* July 120/2 The output voltage *x* of the multiplying pot is proportional to its angular rotation. 1953 R. Bretz *Techniques Television Production* xix. 401 The first thing the student operator learns about the audio console is that the fader dials, or 'pots', in the middle of the panel each controls [*sic*] a separate source of sound. 1967 *Electronics* 6 Mar. 311/2 (Advt.), No matter what your pot requirements, they take a turn for the better when you contact Gamewell Division.

pot, *v.*¹ Add: **II. 2. a.** (Further *fig.* example.) Also, to summarize, put into 'potted' form (see *potted *ppl. a.* 3 a).
1927 *Year's Work Eng. Stud.* 1925 42 After preliminaries, the matter is divided into: the effect of Function upon Sound..; of Emotion upon Sound..; and of *Gliederung*... The statistics and argument can hardly be 'potted' here.
c. To encapsulate (an electrical component or circuit) in a liquid insulating material, usu. a synthetic resin, which sets solid.
1950 W. W. Stifler *High-Speed Computing Devices* xvi. 426 Tests..showed that it was possible to pot printed circuits in a special casting resin in such a fashion as to permit the plugging in of the complete subassemblies. 1962 F. I. Ordway et al. *Basic Astronautics* vi. 279 The plastics in which electronic components are potted usually contain bacteria. 1971 J. H. Smith *Digital Logic* i. 5 The whole assembly is usually 'potted' into a block to form a module.
3. b. (Further examples.) Also, *pot off*, to transplant seedlings into individual pots; *pot on*, to move a plant from one pot into a larger one; *pot out*, to plunge potted plants into a bed in the open garden; *pot up*, to move seedlings or larger plants into pots.
1870 W. Robinson *Alpine Flowers for Eng. Gardener* I. 63 This is a better way than sowing in pots,..from which they require to be 'potted off'. 1916 M. Hampden *Flower Culture* ii. 39 Now [*sc.* March] is the time for.. potting up clumps of hardy plants from the garden. 1926 E. T. Brown *Year in my Flower Garden* 58 Lift and pot roots of Solomon's Seal. 1950 *N.Z. Jrnl. Agric.* Aug. 167/1 When the leaves of the small plants touch each other in the boxes it is time to pot them up. 1950 O. Sitwell *Noble Essences* IX. vi. 139 The gentle gold of the industrial haze lay lightly on the rich beds of tulips, carnations or begonias, so neatly potted out. 1952 C. E. L. Phillips *Small Garden* ix. 77 Nurselings in small pots are 'potted-on' into bigger ones when their roots have filled up the first one. 1958 *Listener* 12 June 982/2 They [*sc.* amaryllis plants] will throw off bulbils from the main bulb. These can be taken off and potted up. 1978 R. Gorer *Growing Plants from Seed* v. 69 The seedlings are potted up separately in very small pots and progressively potted on.
5. c. (Earlier example.) Also, in *Rugby Football*, to score (a dropped goal). *N.Z.*
1862 W. B. Cheadle *Jrnl. Trip across Canada* (1931) 48 Worked at harness-making & potted much money during term. 1959 [see *pot *sb.*⁴ b].
III. 6. b. *Austral. slang.* To hand (someone) over for trial, to inform on. Cf. *pot *sb.*¹ 13 k.
1911 A. Wright *Gambler's Gold* 138 Why should I pot the bloke? He done me a good turn, an' th' police is no good to me. 1916 J. B. Cooper *Coo-oo-ee* ix. 108 'Yer see,' he explained, 'they've got to try to hang some cove or else they'd lose their job. The more men they pot the better they're fixed in their jobs. See?' 1945 Baker *Austral. Lang.* II. xi. 207 A few general expressions concerned with school life:..*to pot someone* or *to put someone's pot on*, to inform on. 1953 'Caddie' *Sydney Barmaid* xl. 230 What dirty swine has potted me?
IV. 8. (Examples.) Also freq. *intr.*
1914 R. Fry *Lett.* (1972) II. 377 Vanessa and I have been potting all day... We went when the potter wasn't there and got the man to turn the wheel. 1967 B. Jefferis *One Black Summer* (1968) i. 1 The grounds and buildings would be full of summer school students; doctors who longed to pot; dressmakers who yearned to try their hands at sculpture. 1968 *Canad. Antiques Collector* June 12/1 The Rockingham China Works..began to produce china (porcelain) in 1826. The factory ceased to manufacture towards the end of 1841. Many fine porcelain wares..were potted in this relatively brief period. 1971 J. White *Left for Dead* 68 All I've got to do is to teach myself to pot... I've always been interested in making pottery. 1976 J. G. Hurst in D. M. Wilson *Archaeol. Anglo-Saxon Eng.* vii. 323 Stamford ware is finely potted on a fast wheel and fired in a developed single-flue kiln.
V. 9. *colloq.* To cause (a baby or young child) to use a chamber-pot. Also *absol.*
1943 A. Medley *Your First Baby* xviii. 180 With children who hardly wake up, perform with their eyes closed and drop off again, it is well worth while to pot them and to know that they will sleep dry and comfortable. 1948 B. Goolden *Jig-Saw* ii. 9, I prefer them [*sc.* babies] house-trained... One feeds and pots automatically. 1957 H. Croome *Forgotten Place* iv. 51 I'm not going to pot him or anything. I think early habit training is such a mistake. 1973 *Daily Tel.* (Colour Suppl.) 26 Oct. 7/2 She has poured the last coffee and sat back for the first time since potting the baby at 7.30 that morning.

‖ **potager** (potaʒe). [Fr.: see Potagere.] = Potagere.
c 1786 T. Blaikie *Diary Scotch Gardener* (1931) 199 This garden..goes with a narrow stripe of potager to the point du jour where there is another potager. 1792 A. Young *Trav. France* I. 99 There is a town, and a great *potager* to remove before it would be consonant with English ideas. 1885 H. James *Little Tour in France* iii. 20 Your eye wanders over the neighbouring *potagers*. 1926 *Spectator* 9 Oct. 581/2 The herb garden..lost its supremacy to the *potager* or vegetable garden less than two centuries ago. 1958 L. Durrell *Spirit of Place* (1969) 146, I calculate that with ten chickens and the excellent *potager* out there I shall just squeeze by. 1966 A. Christie *Third Girl* iv. 35 In England..you do not love your *potager* as much as you love your flowers. 1978 J. Lees-Milne *Round the Clock* 55 The north garden.. would be called the *potager*, were it in France. Vegetables, espaliers, roses and lawns were neatly compartmented.

potamoplankton (pǫ:tămoplæ·ŋktǫn). [a. G. (C. Zimmer 1898, in *Biol. Centralbl.* XVIII. 522), f. Gr. ποταμό-ς river + Plankton.] Plankton found in rivers or streams.
1902 *Ann. Bot.* XVI. 584 These backwaters thus form a kind of transition from the typical Potamoplankton of the flowing river to the Heleoplankton of the ponds of the Thames valley. 1909 E. Warming *Oecol. Plants* xxxviii. 161 Other subdivisions [of limnoplankton] may be recognized, such as potamoplankton, heloplankton, and probably several more. 1923 *Jrnl. Ecol.* XI. 209 Potamoplankton is the plankton of rivers. It is usually benthoplanktonic in character, but at times may show limnoplanktonic features. 1969 G. W. Prescott *Algae* i. 17 Plankton of rivers is called potamoplankton.

pot and pan. [Rhyming slang for 'old man'.] One's father or husband.
1906 E. Dyson *Fact'ry 'Ands* xii. 153 How's yer ole pot-'n'-pan, Tutsie? 1935 A. J. Pollock *Underworld Speaks* 90/2 *Pot and pan*, the old man (otpay and anpay). 1938 L. Ortzen *Down Donkey Row* 12 *Pot and Pan*, old man; meaning husband or father. 1945 Baker *Austral. Lang.* 270 *Pot and pan*, old man. 1971 *National Times* (Austral.) 13 Dec. 20 From this handy guide you could soon establish the meaning of..'her pot and pan is in the bucket and pail'... Translated, it means '..her old man is in jail.'

potarite (pǫ̆tā·rəit). *Min.* [f. the name of the *Potaro* River, Guyana, where first found: see -ite¹.] Palladium amalgam, occurring as brittle, silvery grains and nuggets and reported to consist of two distinct phases.
1925 *W. India Committee Circular* 22 Oct. 429/2 The second report deals with the occurrence of Palladium-Mercuride, a mineral new to science in British Guiana. Up to the present it has only been found in certain diamond-bearing gravels on the Potaro river; and so it was named 'Potarite'. 1928 *Mineral. Mag.* XXI. 397 He [*sc.* Sir John Harrison] proposed the name potarite (letter of April 3, 1925), but he did not himself record this name in print. 1960 *Amer. Mineralogist* XLV. 1094 Samples of PdHg which are identical with the mineral potarite may be prepared either by displacement of palladium from $Pd(NO_3)_2$ solution by mercury or by direct synthesis from powdered palladium and mercury in sealed evacuated tubes. 1968 I. Kostov *Mineral.* II. i. 95 Mercury group... Gold amalgam..and potarite (PdHg, tetragonal) with about 35% Pd, are rare representatives of the group.

potashery. (Earlier and later examples.)
1799 *Canada Constellation* (Niagara, Ontario) 8 Nov. 4/1 For field ashes, 9d. at the potashery, and 6d. if he goes for them. 1882 W. M. Thayer *From Log-Cabin to White House* 150 A pot-ashery was an establishment containing vats for leeching ashes, and large kettles for boiling the lye. 1930 *Canad. Hist. Rev.* XI. 39 The only buildings were..two saw-mills, a carding shop, a pot-ashery, [etc.]. 1979 *Jrnl. R. Soc. Arts* Dec. 60/2 A syndicate of Liverpool-Manchester merchants..opened potasheries in New York and Philadelphia.

potassic, *a.* Add: **b.** *Geol.* Of a mineral or rock: containing an appreciable or a greater-than-average quantity of potassium, often as compared with sodium. Also applied to a metamorphic process in which such minerals are formed.
1903 W. Cross et al. *Quantitative Classification Igneous Rocks* 227 The orthoclase may be considered as wholly potassic and reckoned as pure orthoclase. 1932 A. Johannsen *Descr. Petrogr. Igneous Rocks* II. 63 The term orthorhyolite was originally suggested in 1919 for rhyolites whose only feldspar is potassic. 1967 *Amer. Mineralogist* LII. 828 The early part of the alteration sequence featured chiefly potassic metasomatism (biotite, potash feldspar), whereas the latter part was characterized by sodic metasomatism (aegirine, crocidolite). 1971 I. G. Gass et al. *Understanding Earth* i. 18/1 Two compositionally different feldspars, one more sodic and the other more potassic. 1974 P. G. Harris in H. Sørensen *Alkaline Rocks* vi. i. 434/1 The anomalous $^{87}Sr/^{86}Sr$ ratios of many suites of potassic rocks suggests..a multistage origin.

potassium. Add: **c.** Special comb.: **potassium-argon**, used *attrib.* to designate a method of isotopic dating, or results obtained from it, based upon measurement of the relative amounts in rock of potassium 40 and its decay (electron capture) product, argon 40.
1953 *Bull. Geol. Soc. Amer.* LXIV. 1473 (*heading*)

Potassium argon studies at the University of Toronto. **1955** *Ibid.* LXVI. 1711 (*heading*) Potassium-argon ages of metamorphic and igneous rocks from the Southern Appalachians. **1968** *Times* 3 Oct. 13/5 The duration of the various magnetic reversals is known from potassium-argon dating of land rocks. **1969** Bennison & Wright *Geol. Hist. Brit. Isles* xvi. 362 Earlier pleistocene deposits can only be dated, as yet, by the use of the potassium-argon method..which, because of the long half-life of potassium, is not suited to dating such relatively recent events. **1977** *Time* 7 Nov. 52/2 The age of a fossil can often be determined by analyzing the layer of rock or soil in which it was found and determining, often by the so-called potassium-argon method, just how old the layer is.

potato, *sb.* Add: **2. d.** δ. (Further examples in spellings *pratie, praty.*)

In quot. 1966 the use is *fig.*

1781 W. Dyott *Diary* 8 Sept. (1907) I. 5 In short, I think them [*sc.* the Irish recruits] calculated merely to eat potatoes, or 'pratys', as they call them. **1826** 'N. Nondescript' *The* — 15 Apr. 56 'I was just thinking,' said he in a whimpering tone, 'what we poor Irish would do, if we hadn't paraties.' **1830** *Constellation* II. 1/1 She took my advice, and doubling herself up in the blanket, was asleep before your Honour'd say praties. **1927** in C. Sandburg *Amer. Songbag* 463 O, I met her in the mornin' And I'll have yez all to know That I met her in the garden Where the praties grow. **1949** C. Graves *Ireland Revisited* vii. 82 Nobody uses the word 'begorrah,' and a potato is a 'spud' not a 'praty'. **1966** *Listener* 12 May 687/1 A sentimental domestic melodrama—what Irish audiences..call a 'pratie', or potato. **1972** *Islander* (Victoria, B.C.) 12 Mar. 8/1 We call them 'spuds'. The Irish affectionately call them 'praties' and they sometimes call mashed potatoes 'poundies'. **1973** *Times* 29 Aug. 7/6 Do you fancy some German sausage in the garden where the praties grow?

5. a. *hot potato*; (to drop something) *like a hot potato*: see *hot *a.* 12 a; *small potatoes* (further examples); also in sing. and in phr. *small potatoes and few in a* (or *the*) *hill.*

1836 D. Crockett *Exploits & Adventures Texas* ii. 25 This is what I call small potatoes and few of a hill. **1839** *Boston* (Mass.) *Morning Post* 23 July 1/1 The Conservatives in Maine have held a convention and nominated F. O. J. Smith, for Governor. S.P. (small potatos). **1880** [see *ham-fatter* s.v. *ham *sb.*[1] 3]. **1886** *Galaxy* 1 Oct. 272 Insignificant people are 'small potatoes, and few in a hill'. **1914** 'Bartimeus' *Naval Occasions* xiii. 101 In the beginning he was an Assistant Clerk—which is a very small potato indeed. **1923** Conrad *Rover* x. 160 Then indeed that corvette, the big factor of everyday life on that stretch of coast, would become very small potatoes indeed. **1926** M. J. Atkinson in J. F. Dobie *Rainbow in Morning* (1965) 80 He's a mighty small potato in my estimation. He's mighty small potatoes and few in a hill. **1927** H. T. Lowe-Porter tr. *Mann's Magic Mountain* II. vii. 787 If some first-class excitement doesn't come along every day, you pull a face as though you were saying: 'H'm, small potatoes *and* few in the hill!'. **1962** A. Buchwald *How much is that in Dollars?* 122 Mary Soo, by tradition but not contract, has the garbage concession of all United States Navy ships entering Hong Kong, which out here is no small potatoes. **1968** *Globe & Mail* (Toronto) 3 Feb. 7/6 When the conference gets around to 'other matters' on Wednesday, the current dispute over medicare should seem like small potatoes. **1973** H. Nielsen *Severed Key* iv. 49 'Morry Sacks is going to tell the law we found a million dollars on the beach?' 'No! Morry is too small potatoes.' **1974** *Publishers Weekly* 26 Aug. 300/2 A milieu where the crime is petty and municipal corruption small potatoes. **1976** *Gramophone* July 153/3 Serenus is small potatoes by CBS or RCA standards but its albums are tastefully produced and carefully annotated.

c. Also, (*the*) *clean potato*: a person or thing whose character or excellence is beyond reproach. Hence the phr. *not* (*quite*) *the clean potato* and vars., not completely sound or reliable; not (quite) the right or real thing.

1880 [in Dict.]. **1881** G. H. Gibson in *Bulletin* (Sydney) 16 Mar. 8 You weren't quite the cleanly potato, Sam Holt. **1890** 'R. Boldrewood' *Colonial Reformer* III. xxvii. 104 'Well,' said Mr. Cottonbush,..'it ain't quite the clean potato, of course [*sc.* to steal a neighbour's grass]; but if your sheep's dying at home, what can you do?' **1913** Galsworthy *Dark Flower* II. vii. 137 A suspicion he had always entertained, that Cramier was not by breeding 'quite the clean potato'. **1921** K. S. Prichard *Black Opal* xvi. 148, I ain't always been what you might call the clean potato. **1929** J. Masefield *Hawbucks* 165 We'll shake hands, clean potato, and be good friends. **1931** M. Franklin *Back to Bool Bool* 233 She was only the great-granddaughter of old Larry Healey of Little River, none so clean a potato, if rumour was correct. **1933** G. Heyer *Why shoot a Butler?* vi. 86 Not strictly the clean potato, is it?.. Guest in the man's house, you know. The Public School Spirit, and Playing for the Side, and all that wash. **1939** —— *No Wind of Blame* iv. 80 It isn't at all the clean potato. In fact, it's very dishonourable. **1941** Baker *Dict. Austral. Slang* 56 *A clean potato*, a free or unconvicted person, one with unblemished character. **1962** T. Ronan *Deep of Sky* 42 Some of the grand old pioneers and land-takers of history were not quite the clean potato.

d. A large or conspicuous hole in a sock or stocking through which the flesh shows.

1885 *Eng. Illustr. Mag.* June 616/1 The gladiators wore pasteboard helmets..and fleshings for legs and arms, with—what are vulgarly termed 'potatoes', that is, holes in the fleshings perceptible in many places. **1886** H. Baumann *Londinismen* 144/1 *Potatoes*, grosse Löcher in den Strümpfen. **1902** Farmer & Henley *Slang* V. 270/1 *Potato*,..used esp. for a heel through an undarned sock or stocking. **1949** D. M. Davin *Roads from Home* III. iv. 241 It was a mystery the way that Paddy went through his stockings... A great big potato staring out over the heel. **1973** *Country Gentlemen's Estate Mag.* Mar. 156/1 Gumboots..will hole a 'potato' like a cannon-ball in the heels of a new pair of socks in an afternoon.

e. In pl. *U.S. slang.* Money. Occas. in more specific use, dollars ('pounds' in quot. 1939).

1931 D. Runyon in *Collier's* 4 Nov. 8/2 'Listen, Sam,' I say, 'you have seven duckets, and we are only six, and here is a little doll who is stood up by her guy, and has no ducket, and no potatoes to buy one with, so what about taking her with us?' **1932** —— in *Collier's* 26 Mar. 7/4 Many citizens are figuring that maybe he suddenly discovers all his potatoes are counterfeit, because nobody can think of anything that will worry Sorrowful except money. **1933** *Sun* (Baltimore) 15 Sept. 1/2 Nobody gives fifteen thousand 'potatoes' to a party committee without wanting something. **1935** A. J. Pollock *Underworld Speaks* 90/2 *Potatoes*, money. **1939** Wodehouse *Uncle Fred in Springtime* i. 9 Was it conceivable..that any man, even to oblige a future brother-in-law, would cough up the colossal sum of two hundred potatoes? **1976** *National Observer* (U.S.) 8 May 14/2 Usually he [*sc.* a horse] runs with a price tag of about $3,500. With those kind of potatoes, it can be hard to get respect.

f. *Austral. slang.* Also in spelling *potater.* [Shortening of *potato peeler*, rhyming slang for *sheila.] A girl, a woman.

1957 D. Niland *Call me when Cross turns Over* ii. 69 Snow told him not to be a mug, the sheila had him in her sights because she thought he was a bit of all right. That would be the day, Locky retorted, when some bloody potater..had him stringing along with her. **1970** G. Greer *Female Eunuch* 266 Terms..often extended to the female herself. Who likes to be called..a *potato*? **1970** *Private Eye* 2 Jan. 12 He's been endeavouring to commit intimacy with your *potato*. **1971** *Ibid.* 2 July 16 As for this potato I must guide her footsteps back into the paths of righteousness.

6. a. *potato-croquette, -flour* (later examples), *-fritter, -harvest, -house, -land, -sack* (examples), *-soup* (earlier and later examples). **b.** *potato-chipper, -digger* (earlier and later examples), *-peeler* (examples), *-peeling* (further examples), *-planting.*

1895 *Montgomery Ward Catal.* 436/1 Potato Chipper, can be used as a..chopper for potatoes. **1951** *Catal. of Exhibits, South Bank Exhib., Festival of Britain* 52/1 Potato chipper; Thos. A. Nutbrown Ltd., Walker Street, Blackpool, Lancs. **1977** *Western Mail* (Cardiff) 5 Mar. 12/2 (Advt.), Hobart Potato Chipper, 40lb. per minute. Reconditioned. **1876** M. N. Henderson *Pract. Cooking* 194 Potato Croquettes. Add to four or five mashed potatoes..the beaten yolk of one egg. **1942** C. Spry *Come into Garden, Cook* iv. 38 The best croquettes in the world, *Potato Croquettes.* **1845** *Quincy* (Illinois) *Whig* 18 Dec. 2/5 A new potatoe digger was recently exhibited in operation at Salem, West Jersey. **1858** J. Brown *Let.* 13 May (1912) 160 And is the delightfullest of potato-diggers already digging? **1945** *Hardin* (Montana) *Tribune-Herald* 15 Feb. 2/4, I will sell at public auction..1 potato digger. **1906** U. Sinclair *Jungle* xi. 139 Potato-flour is the waste of potato after the starch and alcohol have been extracted; it has no more food value than so much wood. **1911** *Daily Colonist* (Victoria, B.C.) 4 Apr. 3/1 (Advt.), Potato Flour, Health Brand, packet 20c. **1845** E. Acton *Mod. Cookery* xix. 417 Potato fritters. (Entremets.) See directions for potato puddings. The same mixture dropped in fritters into boiling butter, and fried until firm on both sides will be found very good. **1966** P. V. Price *France* 308 Small sweetened potato fritters. **1808** E. Weeton *Let.* 1 Apr. (1969) I. 78 An uncommonly plentiful potatoe harvest. **1979** *Country Life* 2 May 1375 Mr Hurd works on the early potato harvest. **1791** W. Bartram *Trav. N. & S. Carolina* 192 The lowest or ground part is a potatoe house. **1861** C. M. Yonge *Stokesley Secret* vi. 89 There was a bonfire by the potato-house. **1921** *Proc. 3rd National Country Life Conf. 1920* (U.S.) 155 Potato houses..are isolated and located with special reference to the good of the products involved. **1970** S. Trueman *Intimate Hist. New Brunswick* iii. 56 The 'potato houses' looking like dwellings that have sunk so deep into the ground that they now consist mainly of high-peaked roofs. **1780** A. Young *Tour in Ireland* I. 344 Plough the potatoe land once or twice for barley. **1855** *Trans. Amer. Inst. City of N.Y. 1854* 168 Salt..will kill grubs, and may be used to advantage on potato land. **1965** K. H. Connell in Glass & Eversley *Population in Hist.* xvii. 428 The practice of letting farms by auction in a country [*sc.* Ireland] where land was almost the only resource encouraged tenants to outbid one another in the tribute they offered to acquire the right to potato-land. **1895** *Montgomery Ward Catal.* 436/1 The Peerless Potato Peeler. This is an entirely new and novel article for peeling and slicing potatoes. **1951** *Good Housek. Home Encycl.* 610/1 Potatoes should be peeled..with either a special potato peeler or a sharp, short-bladed knife. **1961** *Which?* Mar. 61 (*heading*) Potato peelers. *Ibid.* 61/1 In this report, CA discusses five potato peeling devices on the market when our tests began. **1975** L. Gillen *Return to Deepwater* x. 178 She impatiently brushed them [*sc.* tears] away with the back of one hand before resuming her potato-peeling. **1885** A. Edwardes *Girton Girl* III. xiii. 221 The Seigneur..taking part in his potato-planting and his vraic harvest. **1951** R. Firth *Elem. Social Organiz.* iv. 142 To revert to the Irish peasantry..—there is a form of non-monetary co-operation..in tasks such as..potato-planting. **1859** G. A. Sala *Twice round Clock* 40 There are tall potato-sacks, propped up in dark corners. **1939** F. Thompson *Lark Rise* i. 4 A superannuated potato-sack thrown down by way of hearthrug. **1979** *Guardian* 28 Feb. 13/3 Ladies..manage to knit passable potato sacks to cover their nether limbs. **1845** E. Acton *Mod. Cookery* i. 20 Potato soup. Mash to a smooth paste three pounds of good mealy potatoes..; mix with them..two quarts of boiling broth,.. add pepper and salt. **1861** Mrs. Beeton *Bk. Househ. Managem.* 76 Potato soup... When the potatoes are boiled, mash them smoothly..and gradually put them to the boiling stock [etc.]. **1960** *Good Housek. Cookery Bk.* (rev. ed.) 72/2 *Potato soup*... Peel and slice the potatoes and chop the onion.

7. potato-ball, (*a*) (examples); (*b*) also in *sing.* and *attrib.* (examples); **potato-beetle**², (*a*) (examples); substitute for def.: the Colorado beetle, *Leptinotarsa decemlineata*, a brown beetle with black spots and stripes which attacks the leaves of the potato and related plants; **potato bread** (earlier and later examples); **potato-bug,** (*a*) = *potato-fly*; (*b*) = *potato-beetle*²; (earlier and later examples); **potato-cake** (earlier examples); **potato chip,** (*a*) = *chip *sb.*[1] 2 b; (*b*) *U.S.* = **potato crisp* (see also quot. 1975); also (usu. with hyphen) *attrib.*; **potato clay,** a variety of clay used by the Hopi Indians in making pigments; **potato creeper** = *potato vine* (*b*); **potato crisp** (see *crisp *sb.* 7); **potato dumpling,** a dumpling whose ingredients include sieved cooked potatoes; **potato-eater,** a derogatory nickname usu. applied to an Irishman (see also quot. 1871); **potato failure** = **potato famine*; **potato famine,** a dearth of potatoes caused by crop failure; *spec.* (usu. with capital initials) that which occurred in Ireland in 1846–7; **potato flake** (usu. in pl.) (see quot. 1955); **potato-fly,** for *Lytta* substitute *Epicauta* (examples); **potato hook** (examples); **potato latke** [*latke], a pancake made with grated potato; **potato masher,** a device consisting of a set of wires or a perforated flat plate (formerly, a solid wooden cylinder) attached to a handle, for mashing potatoes; also *transf.*, (*a*) in full **potato-masher grenade,** a type of hand grenade whose shape resembles that of a potato masher; (*b*) (see quot. 1945); **potato moth** = **potato tuber moth*; **potato-mouth** *v. trans.*, to mutter; also **potato-mouthed** *a.* = Mealy-mouthed *a.*; **potato onion,** substitute for def.: a variety of the common onion, *Allium cepa*, in which new bulbs are produced at the base; (examples); **potato pancake,** a pancake in which sieved mashed potato is the basic ingredient; also, = **potato latke*; **potato patch,** a plot of ground on which potatoes are grown; **potato peelings,** strips of the peeled skin of potatoes; **potato pie,** (*a*) (further examples); (*b*) (later example); **potato puff,** a kind of potato crisp in the form of a puff (see Puff *sb.* 5); **potato race** (examples); **potato rot** (earlier examples); **potato salad,** pieces of cold cooked potato mixed with salad dressing and other ingredients; **potato scone,** a scone made with sieved cooked potatoes; **potato set** = Set *sb.* 23 b; **potato stick,** a small crisp potato chip; **potato straw,** a very thin stick of potato, fried until crisp; **potato tuber moth,** the moth whose larva is the potato tuberworm; **potato tuberworm** *U.S.*, the pinkish-white caterpillar of the moth *Gnorimoschema operculella*; **potato-vine,** substitute for def.: (*a*) a potato plant, *Solanum tuberosum*; (*b*) one of several South and Central American climbing plants, esp. *Solanum jasminoides* or *S. wendlandii*, bearing blue or white flowers; (examples); **potato worm,** substitute for def.: = **potato tuberworm*; (examples).

1823 T. B. Hazard [illegible] a Potatoe Ball. **1824** [illegible] 120 Potato Balls. [illegible] of an egg, roll them [illegible] *Mod. Cooking* (ed. 2) [illegible] *Jewish Manual, or* [illegible] *Cookery* v. 91 Potatoes [illegible] into balls glazed with the yolk of egg, and browned with a salamander. **1850** *Rep. Comm. Patents: Agric. 1849* (U.S.) 198 In 1847, he planted a single potato-ball or apple; only one seed grew. **1877** *Rep. Vermont Board Agric.* IV. 33 Nature can make potato balls, but she couldn't make the Early Rose. **1912** M. B. Brown *Just Use-it-Up* vi. 140 Potato balls. **1948** *Good Housek. Cookery Bk.* II. 289 Scoop out the potato balls with a Parisian potato cutter. **1963** R. Carrier *Great Dishes of World* 213/2 Roll potato balls in flour and then in beaten egg. **1969** E. H. Pinto *Treen* 141 The hardwood handle with, at each end, a steel bowl with a hole in the base..is a potato ball maker, probably 18th-century. **1866** *Pract. Entomologist* I. 105/2 One day last week a gentleman left at our office a stalk from a potato hill, which was literally covered with the larva of the new potato beetle. **1868** *Amer. Entomologist* I. 44/1 It might perhaps be desirable..to get people to call it [*sc.* the Colorado potato-bug] a 'potato-beetle'. **1876** *Times* 29 Aug. 6/5 The fact of its surviving in a letter posted at Listowel, Ontario, and delivered at Stranraer, Wigtonshire, N.B., shows that the potato beetle possesses great powers of endurance. **1906**

J. W. Folsom *Entomol.* xii. 382 From Colorado the well-known potato beetle..has worked eastward since 1840. **1932** Z. P. Metcalf *Fest-bk. Econ. Zool.* viii. 259 The Colorado potato beetle..often completely destroys whole fields of unsprayed potatoes. **1974** L. E. Chadwick tr. *Lindenmeyer's Insects of World* 165/2 The Colorado potato beetle..was imported accidentally into Europe from America. **1743** W. Ellis *Mod. Husbandman* Sept. xxv. 119 Potatoe Bread. This Root has often been employed, like the Turnep, towards making Loaves of Bread in the scarce Times of Corn. **1915** *Chambers's Jrnl.* Oct. 643/2 There is a rather large group of words that have come at us with a rush, and cannot be classified,..that fine phrase of Mr. Lloyd George's, 'the potato-bread spirit'. **1950** *Mrs. Beeton's Bk. Househ. Managem.* xxx. 695 *Potato bread.*—The adhesive tendency of the flour of the potato prevents it being baked or kneaded without being mixed with wheaten flour or meal. **1799** E. Drinker *Jrnl.* (1889) 347 They call them [*sc.* a species of Cantharides] here..the Potato-Bug, being numerous on the potato tops. **1838** *Hesperian* (Columbus, Ohio) I. 145 This company..formed for the praiseworthy purpose of encouraging the growth of potatoe-bugs, and manufacturing potato-bug oil. **1864** *Trans. N.Y. State Agric. Soc. 1863* 798 Some have been discouraged from planting potatoes, the ravages of this potato-bug have been so great. **1865** *Pract. Entomologist* I. 3/1 The new Potato Bug is not what naturalists call a bug, but a true Beetle. **1907** L. H. Bailey *Cycl. Amer. Agric.* II. 524/1 The old-fashioned potato bug or blister-beetle..is combated in the same way as the Colorado potato-beetle. It is now rarely seen. **1908** *Springfield* (Mass.) *Republ.* 2 Sept. 14/6 Potato bugs on the rails..stalled eight trolly cars. **1974** *N.Y. Times Bk. Rev.* 2 June 14/2 It was settled that I should receive 1 cent per hundred for picking potato bugs. **1979** R. Thomas *Eighth Dwarf* xxi. 210 The U.S. Constabulary..were swarming over the Opel plant ..like so many potato bugs. **1747** H. Glasse *Art of Cookery* ix. 98 Potatoe-Cakes. Take Potatoes boil them ..mix them with Yolks of Eggs [etc.]. **1824** E. Weeton *Jrnl.* (1969) II. 33 Often I have..been obliged to live on potatoes..for weeks. **1848** Mrs. Gaskell *Mary Barton* II. xviii. 186 The potatoe-cakes [illegible]. **1878** *Amer. Home Cook Bk.* 67 Put around potato chips prepared as follows. **1886** [see *chip *sb.*[1] 2 b]. **1934** Webster, *Potato chips,* thin slices of raw potato fried crisp in deep fat. **1955** *Sci. News Let.* 5 Mar. 153/1 Scientists at the Eastern Laboratory are also responsible for the potato-chip bar developed primarily as a 'high-calorie, high-density military ration with taste appeal'. The potato-chip bar takes up only one twentieth the space needed for an equivalent amount of ordinary potato chips. **1972** C. Weston *Poor, Poor Ophelia* (1973) vi. 29 Two barefoot hippies were sharing a bag of potato chips. **1975** *N.Y. Times* 30 Nov. III. 1/2 The F.D.A. gave Procter & Gamble permission to go ahead with..the use of the words 'potato chips' on its product..its potatoes are dehydrated, then turned into a mush and pressed and fried. ..a long way toward solving a basic potato-chip problem.. The natural chips are easily broken. **1898** *Internat. Folk-Lore Congr. World's Columbian Exposition 1893* I. 264 The corn having boiled about three-quarters of an hour, the pot is taken from the fire and its content poured upon the sieve, through which the purple-stained boiling water is strained upon the sumac berries. Some of the talc-like substance called potato-clay is then produced, and the operator puts a piece about the size of a walnut in his mouth. **1925** H. F. Macmillan *Trop. Gardening & Planting* (ed. 3) 129 (*caption*) *Solanum wendlandii.* Giant Potato-creeper. **1928** K. Gough *Garden Bk. for Malaya* xii. 205 Several attractive flowering creepers, sometimes called 'Potato Creepers' belong to this large genus. **1929**, etc. Potato crisp [see *crisp *sb.* 7]. **1940** Graves & Hodge *Long Week-End* xiv. 231 Potato crisps were a popular new food. **1970** *New Yorker* 20 Sept. 123/1 His mum..leaves him a florin..for some ginger pop and potato crisps. **1973** J. Thomson *Death Cap* v. 73 A stiff little breeze blew the empty potato crisp packets across the paving-stones. **1976** W. Trevor *Children of Dynmouth* v. 114 Having eaten two packets of bacon-flavoured potato crisps, he had purchased another tube of Rowntree's Fruit Gums. **1912** M. B. Brown *Just Use-it-Up* vi. 139 Potato dumplings. This is a very well-known dish in North Germany. **1948** *Good Housek. Cookery Bk.* ii. 284 Potato dumplings (to serve with Meat or Vegetable Casserole). **1972** *Sat. Rev.* (U.S.) 25 Mar. 36/1 Roast goose with potato dumplings. **1974** 'D. Craig' *Dead Liberty* xvii. 100 Dravier asked for roast goose and potato dumplings. **1822** W. Cobbett in *Weekly Reg.* 9 Aug. 356 Never, in this country, will the people be base enough to be potato-eaters, starvation under the operation of the extreme unction! Nothing but a potato-eater will ever do that. **1871** J. Mackenzie *Ten Years North of Orange River* I. vi. I have heard 'potato-eater' employed by them [*sc.* Dutch farmers] as a contemptuous epithet. **1978** *Maledicta* II. 160 *Potato-eater*, anyone from Ireland, or of Irish descent, after the Irish dietary staple. **1845** E. S. Cayley *Lett. to Ld. John Russell* (1846) i. 9 The deficiency caused by the potato failure will be in some measure compensated by the unusually large crops of oats, barley, and beans. **1846** *Times* 7 Feb. 5/1 The extreme variety in the extent of the potato failure, and the..insulated subdivisions of land in which it prevails, lead us to..doubt whether any adjustment of public works can be made to meet the need wherever it may occur. **1846** *Illustr. London News* 12 Sept. 170/3 The hon. and learned gentlemen adverted in the first place to the potato failure. **1978** R. Mitchison *Life in Scotland* vi. 112 If we compare the state of crofting families—sustained, in squalid poverty, through the potato failure of 1846 by their landowners—with the hardships of the..urban poor..in the early 1840's [etc.]. **1875** J. O'Rourke *Hist. Great Irish Famine* vii. 196 To have met the Potato Famine with anything like complete success, would have been a Herculean task for any government. **1881** J. A. Froude *English in Ireland* (new ed.) III. p. xiii, The potato famine, and responsibility of England. **1970** R. Lowell *Notebk.* 106 We're burnt, black chips knocked from the blackest stock: Potato-famine Irish-Puritan, and Puritan—gold made them smile like pigs once. **1974** P. Lovesey *Invitation to Dynamite Party* ii. 27 He was an Irishman whose family emigrated at the time of the potato famine. **1955** *Sci. News Let.* 5 Feb. 89/3 To potato chips, French fries and home fries can now be added 'potato flakes', a new kind of dehydrated mashed potato... The flakes are made by drying cooked mashed potatoes on the rolls of a steam-heated double-drum drier. **1961** *Coast to Coast 1959–60* 165 A whole family of people..had spread a rug beside the path and were drinking coloured drinks from bottles and eating potato-flakes from bags. **1806** in R. B. Thomas *Farmer's Almanack for 1807*, The potatoe fly, or bug, appears about the first of July. **1832** W. D. Williamson *Hist. State Maine* I. 172 Potato Fly (looks like a Spanish Fly). **1854** E. Emmons *Agric. N.Y.* V. 96 *Cantharidæ*...are at times abundant upon potato vines, whence they have acquired the name of potato fly. **1856** *Trans. Mich. Agric. Soc.* VII. 53 D. O. & W. S. Penfield..[exhibited] six Partridge's potatoe hooks. **1874** *Ann. Rep. Vermont Board Agric.* II. 551 Then with axes, potato hooks, and bog hoes, the turf was all peeled off. **1927**, **1974** Potato latke [see *latke]. **1855** *Chicago Times* 16 Jan. 4/1 Butter moulds and stamps, ladles, rolling pins, potato mashers..at Hollister's Bazaar. **1895** *Montgomery Ward Catal.* 439/3 Tinned wire Potato Masher, wood handle. **1906** *Daily Colonist* (Victoria, B.C.) 26 Jan. 4/6 (Advt.), Kitchen utilities..Potato Masher, wood. Potato Masher, metal. **1915** J. Webster *Dear Enemy* (1916) 238, I casually picked up the potato masher this morning while I was commenting upon last night's over-salty soup. **1919** H. G. Proctor *Iron Division* xii. 188 The German trench bombs were known..as 'potato mashers', because they are about the size of a can of sweet corn, fastened on the end of a short stick. **1925** Fraser & Gibbons *Soldier & Sailor Words* 229 *Potato-masher grenade*, the name given a species of German hand-grenade, resembling in form a domestic potato-masher. **1929** F. A. Pottle *Stretchers* (1930) x. 266 We saw bushels of potato-masher grenades, minenwerfer shells, and a machine gun belt of cartridges all of twenty feet long. **1929** W. T. Scanlon *God have Mercy on Us!* xxvi. 160 We had instructions on the use of every kind of grenade, including the German potato-mashers. **1945** L. Shelly *Jive Talk Dict.* 31/1 *Potato masher*, drumstick. **1967** N. Freeling *Strike Out* 65 An old enamel saucepan,..and an oval metal affair with zigzag holes punched in it..he recognised it as a potato-masher. **1969** I. Kemp *Brit. G.I. in Vietnam* vi. 131 One..had begun to lob in grenades at us; these were of the 'potato masher' type, which sometimes failed to explode. **1969** E. H. Pinto *Treen* 141 Potato mashers were always turned from a single block,..like this one, they were often made *en suite* with a rolling pin..because in olden times, the pair was considered a lucky wedding gift. **1891** *Agric. Gaz. New South Wales* II. 158 Mr. A. Bragg..and Mr. T. B. Linley..have forwarded potatoes infested with the larvae or grubs of the potato moth. **1926** R. J. Tillyard *Insects Austral. & N.Z.* xxviii. 426 The Potato Moth..is an introduced pest of potatoes, tomatoes and tobacco in both countries. **1965** *Austral. Encycl.* V. 88/2 The larvae of the potato-moth..tunnel in the leaf tissue. **1937** *Daily Express* 17 Mar. 6/4 Lewis, square, heavy-browed, stentorian, also potato-mouthed some words, seemed bothered by having to stick to text. **1930** J. Dos Passos *42nd Parallel* IV. 313 You know what we'd do if we had a man in the WhiteHouse instead of a yellow-bellied potatomouthed reformer...? We'd..clean this place up. **1845** J. C. Loudon *Suburban Horticulturist* III. v. 661 The potato-onion may be planted in February. **1855**, **1866** [see Onion *sb.* 2 a]. **1890** E. Watts *Mod. Pract. Gardening* I. xiii. 68 The underground, or potato onion,..is so called from its habit of increasing at the bulb. **1955** W. E. Shewell-Cooper *Complete Veg. Grower* x. 135 The Potato Onion..is more difficult to get hold of today. **1935** L. Zara *Blessed is Man* I. iii. 103 She made him a heaping plateful of the fried potato pancakes so closely associated with this holiday [*sc.* Chanukah]. **1941** L. Hellman *Watch on Rhine* II. 76, I love a good potato pancake. **1960** *Good Housek. Cookery Bk.* (rev. ed.) 535/2 *Potato pancakes*... Put the mashed potato through a sieve if at all lumpy, add seasoning, and work the flour into it to make a smooth dough. **1962** S. V. Thompson *Let.* 30 Jan. in G. Marx *Groucho Lett.* (1967) 237 Mother thinks that with this you serve potato pancakes and onions with peas. True? **1978** *Detroit Free Press* 16 Apr. (Detroit Suppl.) 28/2 With that, they bring potato pancakes which are fresh and moist on the inside, with a good crusty exterior. **1794** E. Drinker *Jrnl.* 25 June (1889) 229 John brought in a Mole he found in a potato patch he was laying out. **1807** *Salmagundi* 15 Oct. 331 Some..enjoy the varied and romantick scenery of..potatoe patches and log huts. **1863** A. D. Whitney *Faith Gartney's Girlhood* xxii. 207 A hollow, beyond which were the cornfields and potato-patches. **1913** J. London *Valley of Moon* 404 Hall put Billy to work on the potato patch—a matter of three acres which the poet farmed erratically. **1919** G. B. Shaw *O'Flaherty V.C.* in *Heartbreak House* 165, I..gave him for his mother a Volumnia of the potato patch rather than an affectionate parent from whom he could not so easily have torn himself away. **1972** R. Adams *Watership Down* xxxiv. 263 The cottager..shot him [*sc.* a rabbit] as he came through the potato-patch at dawn. **1875** Trollope *Way we live Now* II. c. 314 If her future husband would consent to live on potatoes, she would be quite satisfied with the potato-peelings. **1959** I. & P. Opie *Lore & Lang. Schoolch.* xiii. 295 Tales are told of..forcing a new boy into a box or dustbin half-filled with fish-heads, potato peelings..and making him stay there for an hour. **1975** *Sunday Times* 16 Nov. 44/3 My husband was having fun posting old tomatoes and potato peelings down the waste disposal. **1609** Dekker *Guls Horne-Booke* i. 7 Potato-pies and Custards, stood like the sinfull suburbs of Cookery. **1728** E. Smith *Compl. Housewife* (ed. 2) sig. A7[v] (*heading*) A Bill of Fare..For June... Second Course... Potato-Pye. **1842** *Ainsworth's Mag.* I. 2 A large remnant of a potato-pie in a brown earthenware dish. **1965** in P. Jennings *Living Village* (1968) 61 Weather permitting, potato pies are opened and the potatoes sold to the merchants. **1972** K. Bonfiglioli *Don't point that Thing at Me* xviii. 152 We took a cheap night flight to Blackpool... I had potato pie for supper. **1883** *Girls' Own Paper* 14 July 654/1 *Potato puffs.*—Chop..some cold meat or fish. Mash some potatoes and make them into a paste with an egg... Fold... Fry. **1972** *V.A.T.: Scope & Coverage* (H.M. Customs) 22 Any of the following when packaged for human consumption without further preparation, namely,..potato sticks, potato puffs and similar products made from the potato. **1882** G. B. Bartlett *New Games for Parlor & Lawn* 212 The Potato Race... This amusing out-of-door game requires a swift runner, with his feet well under control. **1946** *R.A.F. Jrnl.* May 167 The..Inter-Command Sports Meeting will include the following events..:—potato race,..relay race. **1978** 'F. Parrish' *Sting of Honeybee* i. 8 The potato race starts in three minutes. All entries to the collecting ring. *Ibid.* 9 The strange grey pony was not in the potato race. **1848** *Rep. Comm. Patents 1847* (U.S.) I. 136 The potato rot seems likewise to have been felt to a considerable extent among the common potato. **1854** B. P. Shillaber *Life & Sayings Mrs. Partington* 43 A more disastrous havoc of potato rot has never since transpired than assailed her crops. **1861** Mrs. Beeton *Bk. Househ. Managem.* 593 *Potato Salad*... Cut the potatoes into slices about ½ inch in thickness; put these into a salad-bowl with oil and vinegar [etc.]. **1877** E. S. Dallas *Kettner's Bk. of Table* 399, I should weary the reader if I went on to sound the praises of the mustard and cress salad,..the potato salad, and the *salade de légumes.* **1952** B. Malamud *Natural* (1963) 185 Turkey, potato salad, cheese, and pickles. **1967** M. Gilbert *Dust & Heat* III. 246 He stopped..at a delicatessen store in Soho where he bought..a carton of cold potato salad. **1978** J. Symons *Blackheath Poisonings* I. 39 There was a tongue..a potato salad, a Russian salad and a green salad. **1885** A. Edwardes *Girton Girl* I. vi. 136 Boiled mullet, hot potato scones, with other indigenous Guernsey dishes. **1931** D. L. Sayers *Five Red Herrings* Foreword, We shall come back next summer to eat some more potato scones. **1973** *Perthshire Advertiser* 8 Aug. 17/6 Baking... Potato scones. **1972** Potato stick [see *potato puff* above]. **1895** M. Ronald *Century Cook Bk.* I. facing p. 82 (*caption*) Fluted knife for cutting potato straws. *c*1950 *Mrs. Beeton's Bk. Househ. Managem.* xxx. 692 *Potato straws*... Slice the potatoes thinly, cut them into strips about 1½ inches long... Fry the straws..until crisp. **1892** *Insect Life* IV. 239 (*heading*) The Potato-Tuber Moth (*Lita solanella*). **1928** Metcalf & Flint *Destructive & Useful Insects* xvi. 481 Potato tuber moth..is very destructive to potatoes in warm dry regions. **1932** *Jrnl. Econ. Entomol.* XXV. 625 During the past nine years the potato tuber moth..has caused economic losses to potato growers. **1950** *N.Z. Jrnl. Agric.* Sept. 221/3 The district is not suited to the growing of late potatoes because from January onward the potato tuber moth is very active. **1939** Metcalf & Flint *Destructive & Useful Insects* (ed. 2) xvi. 516 Potato tuberworm..is very destructive to potatoes in warm dry regions. **1960** *Jrnl. Econ. Entomol.* LIII. 868/1 The potato tuberworm, *Gnorimoschema operculella*.., has long been a pest of potatoes in California. **1774** P. V. Fithian *Jrnl.* 19 Sept. (1900) 257, I took a Walk thro' the Pumpkin & Potatoe Vines. **1870** *Amer. Naturalist* III. 92 The early frosts..nearly killed the potato vines. **1902** L. H. Bailey *Cycl. Amer. Hort.* IV. 1680/2 *Solanum*..*jasminoides*, Paxt. Potato Vine (from the fl[ower]s.). Fine greenhouse climbing shrub. **1939** L. & J. Bush-Brown *America's Garden Bk.* xxiv. 807 Potato Vine. Solanum jasminoides. Annual... Star-shaped, white flowers. **1947** *Southern Folklore Q.* XI. 264 Proverbial remarks disparaging to a person's character [include]..this pronunciamento from central Mississippi: 'He ought to have been hung when a potato vine would hang him.' **1963** Robertson & Gooding *Bot. for Caribbean* xxiii. 197 Climbers... Chalice Vine... Potato Vine (*Solanum wendlandii*). **1971** B. Clark in E. L. Wardman *Bermuda Jubilee Garden* i. 6 There is a very wide selection to be made from..potato-vines (*Solanum seaforthianum* & *S. wendlandii*), and chalice-vines. **1977** P. Moyes *To kill a Coconut* xv. 207 Henry was eating breakfast under an arbour of potato-vines and goat's foot. **1842** T. W. Harris *Treat. Insects Injurious to Vegetation* 226 Every farmer's boy knows the potato-worm. **1879** *Scribner's Monthly* Dec. 242/1 This white grub, which the farmers often call the 'potato worm' is..the strawberry's most formidable foe.

potato-ring. (Further examples.)

1901 *Chambers's Jrnl.* Feb. 103/2 Old Irish potato-rings are also much sought after by collectors; at recent sales they have sold for nearly £5 an ounce. **1907** *Daily Chron.* 1 June 4/5 Upon the tables were immense silver baskets and old silver potato rings filled with pink and red carnations. **1932** *Times Lit. Suppl.* 12 May 350/4 Certain characteristic features of Irish silver, notably the so-called 'potato' rings made in large numbers in Dublin from about 1760 to 1820. **1960** H. Hayward *Antique Coll.* 98/1 'Potato ring'... Circular silver dish or bowl stand with straight or incurved sides usually pierced and chased with pastoral or classical motifs. **1968** *Canad. Antiques Collector* Sept. 22/1 Delicate piercing [of silverware] might be suited to cake baskets, the so-called 'potato' rings, bottle coasters or the gallery of a tray. **1973** *Country Life* 8 Mar. 608/2 *Irish Silver in the Rococo Period* will..get out of his [*sc.* the average English-man's] head that so-called potato rings have anything to do with potatoes... The fact is that such things are dish rings.

‖ **pot-au-feu** (potofö). [Fr., = pot on the fire.] A large cooking pot of a kind common in France; the soup or broth cooked in it, *spec.* the traditional French recipe associated with this. Also *attrib.* and *fig.*

1792 C. Smith *Desmond* III. xxiii. 278 The pot *au feu* was brought forward to receive a supply of leeks. **1841** C'tess Blessington *Idler in France* I. ii. 32 Our good hostess..served up a plentiful dinner, consisting of an excellent *pot au feu*, followed by fish, fowl, and flesh. **1868** C. M. Yonge *Chaplet of Pearls* I. xxiv. 311 To eat of the savoury mess in the great *pot-au-feu.* **1868** M. Jewry *Warne's Model Cookery* 55/2 *Pot-au-feu*, the stock pot. **1909** H. James *Novels & Tales* XV. Pref. p. xvii, We can surely account for nothing in the novelist's work that..hasn't, in that perpetually simmering cauldron his intellectual *pot-au-feu*, been reduced to savoury fusion. **1934** H. Hiler *Notes Technique Painting* iii. 183 A large glazed earthenware pot (*pot-au-feu*) which has not been used for any other purpose. **1948** 'J. Tey' *Franchise Affair* xvi. 188 The dish was a pot-au-feu chicken with all its vegetables round it. **1960** *News Chron.* 30 Mar. 6/5, I get out a vast blackened saucepan and make the old-fashioned, traditional French pot-au-feu. **1962** *Economist* 3 Nov. 485/2 This is an intimate family story, relating.. Saturday evening pot-au-feu dinners. **1975** [see *New England boiled dinner* s.v. *New England b].

Potawatomi (pɒtɑ̄wǫ-tǭmi), *sb.* Also **Pattawatami, Pottiwatomie, Poutouatami,** etc. [Native name.] **a.** A member of an Algonquian Indian tribe located in the Great Lakes region of the northern U.S.A., principally in Michigan and Wisconsin. **b.** The language of this people. Also *attrib.* or as *adj.*

1698 tr. *Hennepin's New Discovery* I. xxiii. 74 We sent afterwards three Men to buy Provisions in the Village with the *Calumet* or Pipe of Peace, which the Poutouatami's of the Island had given us. **1722** D. Coxe *Descr. Carolana* 48 The Nations who dwell on this River, are.. the Poutouatomis beforemention'd. **1789** *Deb. Congr. U.S.* 25 May (1834) 41 The treaties of Fort Harmar.. with the Sachems and warriors of the Wyandot,..Pattiwatima, and Sac nations. **1805** J. Wilkinson *Let.* 12 July in *Deb. Congr. U.S.* (1852) 10th Congr. 1 Sess. 575 All hopes of the speedy recovery of their prisoners from the hands of the Pattawatamies, being at an end. **1808** Governor Harrison *Let.* 14 Apr. in *Deb. Congr. U.S.* (1853) 12th Congr. 1 Sess. 1857 A young man from the Delaware towns came to inform me that a Pottawatomie Indian had arrived at the towns. **1835** [see *Kickapoo]. **1838** [see *long knife 1]. **1868** *N.Y. Herald* 31 July 5/4 The Senate..ratified treaties with the Potawatamies. **1877** L. H. Morgan *Anc. Society* II. iv. 105 The Potawattamies have eight gentes of the same name with eight among the Ojibwas. **1927** L. Bloomfield in *Amer. Speech* II. 437/1 She knows only a few words of English, but speaks Ojibwa and Potawatomi fluently, and, I believe, a little Winnebago. **1933** —— *Language* 72 The Algonquian family covers the northeastern part of the continent and includes the languages of..the Great Lakes region (Ojibwa, Potawatomi..and so on). **1946** E. A. Nida *Morphol.* p.v, Linguists will readily recognize many of the problems as being drawn from Greek, Latin, French,..Potawatomi, Cherokee, and Navaho. **1965** *Language* XLI. 75 Pike and Erickson's work on Potawatomi..studies the field structures of certain lexical oppositions within given orders of Potawatomi verb affixes. **1972** *Ibid.* XLVIII. 847, 5 c [in a table of five-vowel systems] is given by Hockett for Potawatomi. **1978** *Maledicta* II. 233, I am told that the name [of Waukesha in Wisconsin] is from the Potawatomi word for 'fox'. **1979** *Tucson* (Arizona) *Citizen* 20 Sept. (Old Pueblo Suppl.) 3/1 The woman saw the 6-foot-4, 260-pound Potawatomi Indian walking down a Phoenix street.

pot-bellied, *a.* Add: (Further examples.) Also, = *pot-belly 2 b.

1899 [see *merchant *sb.* 4]. **1959** [see *pot *sb.*[1] 1 h]. **1964** Mrs. L. B. Johnson *White House Diary* 21 May (1970) 143 Not only was the schoolhouse the same as the one at Fern, but there was a big pot-bellied stove inside. **1973** *Nation Rev.* (Melbourne) 31 Aug. 1455/4 He revealed an ancient potbellied stove. **1979** O. Sela *Petrograd Consignment* 5 A burly, pot-bellied Swiss.

pot-belly. Add: **2. b.** Used *attrib.* to designate a kind of domestic stove made in the shape of a barrel.

1973 L. Russell *Everyday Life Colonial Canada* vii. 76 About the middle of the nineteenth century a new kind of heating stove appeared, inelegantly known as the pot-belly stove. It was barrel-shaped, with a flat top on which a pot or kettle could be heated. **1976** *Columbus* (Montana) *News* 1 July 7/6 (Advt.), One..cast iron pot belly stove. **1976** *Morecambe Guardian* 7 Dec. 33/2 (Advt.), Wood burning or coke burning stove (pot belly type) preferably with chimney.

pot-boil, *v.* (Earlier example.)

1867 D. G. Rossetti *Let.* 22 Mar. (1965) II. 618, I have been pot-boiling to an extent lately that does not hold out much hope of estate buying.

pot-boiler. Add: **2. a.** (Earlier and later examples.) Also applied to musical compositions, plays, and films.

1864 D. G. Rossetti *Let.* 25 June (1965) II. 509 Small things and water-colours I never should have done at all, except for the long continuance of a necessity for 'pot-boilers'. **1915** W. S. Maugham *Of Human Bondage* 256 You hear of men painting pot-boilers to keep an aged mother. **1934** C. Lambert *Music Ho!* v. 306 A certain number of works that were neither potboilers nor works of individual genius. **1973** *Times* 14 Mar. 18/7 In the next three years he directed five pot-boilers and did some screen writing. **1975** *Listener* 31 July 152/3 Ayckbourn's name could become associated with middle-brow, comedy potboilers. **1977** *Time* 10 Oct. 61/1 Condon works on his potboilers seven hours a day, seven days a week for ten weeks at a stretch.

pot-bound, *a.* (Later examples.)

1913 Mrs. G. De H. Vaizey *College Girl* v. 66 The red-brown earth..was too tempting to be resisted when she thought of her poor pot-bound plants at home. **1919** [see *aerotropism* (*aero-)]. **1925** W. Deeping *Sorrell & Son* vi. 56 You can get many a good hint from a man who dislikes you if you are not too pot-bound to soak it up. **1966** Rochford & Gorer *Rochford Bk. Flowering Pot Plants* i. 17 More frequent waterings will be required when the plant is pot-bound.

potch (pɒtʃ), *sb.* Also † **potsh.** [Origin unknown.] In full *potch opal.* Opal that has no play of colour and is of no value; also, a flat colour characteristic of this; *potch and* or *with colour* (see quot. 1971).

1897 *Jrnl. & Proc. R. Soc. New South Wales* XXX. 256 The dull, milky, and opaque stones are called 'potsh' by the miners. **1900** in J. S. Gunn *Opal Terminol.* (1971) 35 Demand for potch with color being active. **1902** *Chambers's Jrnl.* V. 494/2 'It's only potch, an' not worth a drink, the whole durned lot.'.. Occasionally I cut through seams of opal matrix carrying stones of beautiful red tints and sometimes of a peculiar blend of almost every colour. For all these Dan had but one contemptuous name, 'potch', which is the miner's term for inferior opal. **1912** *Empire Mag.* Nov. 282/1 A pocketful of 'potch-and-colour'—that is, 'potch' with a slight 'colour' of opal. **1921** [see *knobby *sb.*]. **1936** A. Russell *Gone Nomad* vii. 58 The value of a pocket varied according to the size, quantity and quality of the opal stones it contained... 'Potch' or immature opal could be found by the ton. **1940** [see *noodle *v.*[2]]. **1958** M. D. Berrington *Stones of Fire* 24 'What's potch?'.. 'Opal crystal without fire. Looks something like pieces of crockery.' **1962** R. Webster *Gems* I. x. 189 The colourful precious opal is found in irregular patches in the thin veins of potch—the miners' term for opal which may be colourful but not showing the play of colour, or as they say 'not alive'—which fills the joints and bedding planes of the sandstone. **1971** J. S. Gunn *Opal Terminol.* 35 *Potch-and-colour, potch-with-colour*, opal potch with a slight colour of opal showing through. **1976** *Sci. Amer.* Apr. 94/3 The small amounts of precious opal are accompanied by an enormous quantity of valueless 'potch' opal, which looks like opal but shows no colour and is generally discarded by the miners... Electron microscopy shows that it usually consists of rounded particles of silica that are not well enough shaped, sized or ordered to form light-diffracting arrays.

potch (pɒtʃ), *v.* [ad. Yiddish *patshn*, ad. G. *patschen* to slap.] *trans.* To slap or smack; hence **potch** *sb.*, **po·tching** *ppl. a.*

1892 [see Poach *v.*[2] 1 d]. **1966** R. H. Rimmer *Harrad Experiment* (1967) 150, I told you, Saul, Harry's not too old for a *potch* before he becomes a *paskudnick.* **1968** L. Rosten *Joys of Yiddish* 293 Don't be fresh or I'll potch you. **1969** K. Vonnegut *Slaughterhouse-Five* iv. 73 Her palm on his little jelly belly made potching sounds.

pot cheese. orig. and chiefly *U.S.* [f. Pot *sb.*[1] 14.] A type of cottage cheese.

1812 'H. Bull-Us' *Diverting Hist. John Bull & Bro. Jonathan* xiv. 111 Tell me, thou heart of cork,..and brain of pot-cheese. **1859** Bartlett *Dict. Amer.* (ed. 2) 420 *Smear case*,..a preparation of milk..; otherwise called Cottage-Cheese. In New York it is called Pot-cheese. **1878** *Rep. Indian Affairs* (U.S.) 19 They learn to milk and make butter and pot-cheese, which they relish highly. **1935** *Colony of Connecticut* (Connecticut Board of Educ.) (Senate Doc. 53, 74th Congr., 1st Sess.) 12 Pot cheese, Dutch cheese, and bonnyclabber were the same. These were generally sweetened with maple sugar. **1964** W. Markfield *To Early Grave* xii. 245 She has eggs for me. *And* pot cheese. **1965** T. Fitzgibbon *Art Brit. Cooking* 133 Pot cheese is a type of cottage cheese made with sour milk and butter milk..cream and salt is worked in. It is rolled into small balls.

‖ **pot de chambre** (po də ʃɑ̃br). [Fr.] = Chamber-pot. Also *attrib.*

1777 P. Thicknesse *Year's Journey* I. i. 6 The priest put the present under the side of the bed..it was only a *pot de chambre.* **1891** A. T. Ritchie *Lett.* (1924) x. 220 The Dame du Palais..escorting the King to bed, carrying a sword, a lamp and a pot de chambre. **1894** G. N. Curzon *Probl. Far East* v. 140 The national implement of Korea, a circular brass pot, with a lid, but no handle, which..serves alternately as pillow, candlestick, ash-plate, spittoon, and *pot de chambre.* **1931** E. Waugh *Remote People* 65 A pedlar..with a bundle of bootlaces in one hand and an enamelled *pot de chambre* in the other. **1934** C. Lambert *Music Ho!* II. 74 Those two symbols of an alien civilization, the top hat and the *pot de chambre.* **1938** L. Bemelmans *Life Class* 20 The *pot de chambre* humor of Regensburg. **1969** N. Freeling *Tsing-Boum* xxi. 146 It [*sc.* tea] arrived, in massive pot-de-chambre porcelain with little roses on it. **1972** 'I. Drummond' *Frog in Moonflower* ii. 47 Coffee in cups like *pots-de-chambre.* **1976** [see *routier[2] 2].

Potemkin (pote·mkin). The name of Grigory Aleksandrovich *Potemkin* (1739–91), favourite of Empress Catherine II of Russia, used *attrib.* to designate the sham villages reputed to have been erected, on his orders, for Catherine's tour of the Crimea in 1787. Also *transf.* and *fig.*

1938 G. Soloveytchik *Potemkin* xiv. 283 Potemkin's detractors have asserted that he built whole sham villages, with cardboard houses and paste palaces..in order to create a false picture of progress and prosperity. ..The originator of these stories..was the Saxon diplomat Helbig, and the legend of 'Potemkin Villages'..as a synonym of sham owes its inception to him. **1954** Koestler *Invis. Writing* xi. 132 This was not a Potemkin village. It was something more curious. **1965** B. Pearce tr. *Preobrazhensky's New Econ.* 39 To lull the vigilance of..the working class, to keep it in the dark about the dangers which threaten it, and to weaken its will with Potemkin villages of childish optimism when it needs to continue to wage the heroic struggle of October. **1967** I. Marder *Paris Bit* i. 27 Paris is above all a city of façades, an enormous Potemkin village. **1973** J. Shub *Moscow by Nightmare* vii. 77 The new Kalinin Prospekt—the latest Potemkin Village of soaring glass and aluminium. **1974** *Guardian* 21 Mar. 3/8 It is good diplomacy ..to pretend that the EEC is a political entity... But don't expect serious decisions from a political Potemkin village.

potence[2]. **5.** (Later example.)

1978 *Erddig* (National Trust) 7 The building, shown on an estate plan of 1739, is complete with its potence (the revolving arm supporting the ladder needed to collect eggs and squabs) and several pairs of nesting fantails.

potency. Add: **1. c.** *Homœopathy.* The degree of dilution of a drug, taken as a measure of its efficacy (a high dilution being regarded as more efficacious).

[**1833** J. B. Gilchrist *Pract. Appeal* 54 Homœopathic medicines, extreme, attenuated, and minimissimised, acquire a potency in the inverse ratio of their attenuation and diminution.] **1846** C. J. Hempel *Homœopathic Domestic Physician* p. vii, Homœopathic drugs have now been potentialized up to the 200th and many of them up to 300th, 400th, 500th. etc., until the 2000th potency. **1906** *Homeopathic World* XLI. 109 From this tinctures of ever-augmenting potency (where disease is concerned) can be prepared. **1938** D. Shepherd *Magic of Minimum Dose* 4 *Nux vomica* has been a stand-by and valuable help in other cases of acute sinus trouble.., both in low potencies (1x) and high potencies. **1975** C. H. Sharma *Man. Homoeopathy* I. iv. 29 Whether the remedy has been chosen for you or you have chosen it yourself, always keep a careful record of what it is, the potency, frequency and quantity, and the symptoms which called for it.

d. Ability to have orgasm in sexual intercourse. Opp. Impotence 2 b.

[**1900** *Yale Med. Jrnl.* VI. 126 There are two forms of potency—the *potentia coeundi* and the *potentia generandi.*] **1901** F. R. Sturgis *Sexual Debility in Man* x. 293 My patient was..an ardent admirer of women, in whose company he indulged himself freely with perfect potency. **1929** G. R. Scott *Sex & its Mysteries* xii. 108 Anything which causes a lowering of the vitality is sufficient to induce impotence, hence the recommendation of meat, eggs and oysters for the generation of sexual potency. *Ibid.* 110 Nor does general disease affect woman's potency. **1939** G. V. Hamilton in E. V. Cowdry *Probl. Ageing* xvi. 469 The decline in sexual potency experienced by men during the ageing period. **1966** *Listener* 10 Mar. 352/1 It may..be true to say that young men use the car or motor-cycle as a potency symbol. **1977** E. J. Trimmer et al. *Visual Dict. Sex* (1978) xxii. 262 Sterility is inevitable if both testes are removed... Desire and potency may be lost, but this is not inevitable.

e. *Genetics.* The extent of the contribution of an allele towards the production of some phenotypic character. Also *attrib.*

1905 *Publ. Carnegie Inst.* No. 23. 59 From these cases it seems clear that the production of partial-rough young was due to some unusual potency of the gametes bearing the smooth character. **1916** *Proc. Nat. Acad. Sci.* II. 53 The appearance of gynandromorphism in certain crosses found its right explanation in the hypothesis of a quantitatively different behavior or a different potency of the male sex-factors in the different races. **1944** *Genetics* XXIX. 528 Using the morphological guide of bristle length..we might assign to gene b^c a potency of about 34, b^d a potency of 50,..and b^f of 54. *Ibid.*, A potency series was set by Stern (1929b) to represent the additive effects of bobbed alleles in *D*[*rosophila*] *melanogaster.* **1955** R. B. Goldschmidt *Theoret. Genetics* III. v. 368 The potencies thus discovered..turned out to be of the orderly type, that is, acting like dosage and thus acting in different combinations in an orderly and parallel way.

f. *Pharm.* The strength of a drug, as measured by the amount needed to produce a certain response.

1933 *Med. Res. Council Special Rep. Ser.* No. 183. 25 An approximate estimate of the potency of a preparation can be obtained by administering a series of doses, each to a single animal. **1968** A. Goldstein et al. *Princ. Drug Action* v. 351 The essential attribute we seek in a drug is not potency, but efficacy at a safe dose. **1978** F. F. Cowan *Pharmacol. for Dental Hygienist* ii. 18 The position of graded dose-response curve along the dose axis is a measure of the drug's potency, i.e., how much of the drug it takes to produce a certain intensity of response... The maximum effect, or efficacy, of a drug is of greater clinical interest.

3. b. *Embryol.* A capacity in embryonic tissue for developing into a particular kind of specialized tissue or organ.

1908 F. R. Lillie *Devel. of Chick* 9 A very important property of primordia in many animals is their capacity for subdivision, each part retaining the potencies of the whole. **1926** J. S. Huxley *Essays Pop. Sci.* 263 The potency of forming limbs is confined to a definite area of the flank. **1958** B. M. Patten *Found. Embryol.* v. 108 If an area where a particular potency has been located is exposed in more detail, it is found that there is a certain central part of it from which practically all the explants exhibit the potency in question. **1968** C. W. Bodemer *Mod. Embryol.* ix. 139 During gastrulation the entire nervous system becomes limited in its potencies and can no longer develop into other structures.

5. *Math.* (See quot. 1959.)

1906 [see *factor *sb.* 6]. **1959** G. & R. C. James *Math. Dict.* 41/1 The cardinal number of a set is also called the potency of the set and the power of the set.

potent, *a.*[1]: **3.** Substitute for def.: Capable of orgasm in sexual intercourse: applied chiefly to men. Opp. Impotent *a.* 2 b. (Further examples.)

1893 E. Martin *Impotence & Sexual Weakness* 74 He..took to himself a wife, and showed by subsequent events, that he was both potent and fertile. **1929** G. R. Scott *Sex & its Mysteries* xii. 110 So long as there is no disease or malformation of the genital organs a woman is potent until practically her dying days. **1975** L. B. Hobson *Examination of Patient* ix. 360 Sexual arousal, erection, and even ejaculation..are emotional as well as hormonal, and a man castrated in later life is still able to have sexual intercourse; that is, he remains potent. He is, however, sterile, since he produces no sperm.

potentia. (Further examples of phr. *in potentia.*)

1612 Jonson *Alchemist* II. iii. sig. E1 The Egg's.. a Chicken, in Potentia. **1674** N. Fairfax *Treat. Bulk & Selvedge* vi. 170 Gods bare Essence must be forthwith or actu, but his everlasting Essence..must be forth-coming or in potentia. **1797** *Encycl. Brit.* XV. 441/2 Potential, in the schools, is used to denote and distinguish a kind of qualities, which are supposed to exist in the body *in potentia* only. **1948** *Mind* LVII. 486 When I claim to *know* I simply make emphatic that I am logically committed, *in potentia*, in certain determinate ways. **1957** L. Durrell *Justine* I. 75 Life, the new material, is only lived *in potentia* until the artist deploys it in his work.

potential, *a.* and *sb.* Add: **A.** *adj.* **4. a.** (Further examples).

1945 *Language* XXI. 2 In English what may be called the potential mode of the verb is an overt category marked by the morpheme *can* or *could.* **1946** [see *handle *v.*[1] 1 b]. **1964** P. Healey in F. W. Householder *Syntactic Theory I* (1972) III. xv. 216 It will be noted that the Potential tense may occur in the Quote of the Saying sub-type.

5. a. Delete 'Now usually called simply *potential*' and add further examples of *potential function.*

1931 *Rev. Mod. Physics* III. 288 He..used several forms for the potential function; one in which the potential depended only upon central force fields..and others in which the potential was also a function of the apex angle. **1939** L. Bairstow *Appl. Aerodynamics* (ed. 2) vii. 320 The boundary conditions to be satisfied are easily deduced; a solid boundary must be a stream-line and along it ψ must be constant or $d\phi/ds$ zero. In the case of the potential function the condition takes the form that the normal velocity, i.e. $d\phi/dh$ must be zero. **1960** Houghton & Brock *Aerodynamics* xi. 267 The velocity potential and the stream function are combined in a new function called the complex potential function. **1967** N. M. Queen *Vector Anal.* vii. 66 An auxiliary function from which a given vector field **V** can be derived by a suitable differential operation is sometimes called a potential function for **V**; in particular, ϕ and **A** in equation (7.2) are known as the scalar and vector potentials for **V**.

c. *potential temperature* [tr. G. *potentielle temperatur* (W. von Bezold 1888, in *Sitzungsb. der K. Preuss. Akad. der Wissensch. zu Berlin* 1190)]: the temperature that a given body of gas or liquid would have if it were brought adiabatically to a standard pressure of 1 bar or 1 atmosphere.

1891 C. Abbe tr. W. von Bezold in *Mechanics of Earth's Atmosphere* xvi. 243 Von Helmholtz recognized the objection..and proposed that the word 'wärmegehalt' should be replaced by the evidently much more proper expression 'potential temperature'. When without gain or loss of heat it is adiabatically or pseudo-adiabatically reduced to the normal pressure. **1937** N. A. V. Piercy *Aerodynamics* i. 21 An atmosphere is stable when the potential temperature is greater, the greater the altitude. **1962** W. S. von Arx *Introd. Physical Oceanogr.* v. 128 In the very deepest part of the ocean such as in the trenches flanking island arcs, it can be shown that the potential temperature of the water is virtually uniform from the depth of the surrounding ocean floor to the bottom of the trench, even though the actual temperature increases somewhat with depth. **1967** P. Groen *Waters of Sea* vii. 290 At 3500 meters, where the temperature is 1·6°C, the potential temperature is only 0·3°C lower. Whereas strictly speaking the temperature is not invariable even if there is no heat exchange, the potential temperature is, if the water proceeds to other depths and is therefore subjected to a different pressure.

B. *sb.* **2.** (Further examples.) Also, resources that can be used or developed; freq. preceded by a defining word.

1941 *Sun* (Baltimore) 24 June 10/2 The vast armored power, mobile tactics and industrial potential of the Nazi armies have been exhibited and proved in the Low Countries, France and the Balkans. **1943** *Times* 10 Dec. 5/3 The whole war potential of the German Reich. **1958** *Spectator* 14 Feb. 196/1 Industrial potential has multiplied six times since currency reform. **1958** *Listener* 27 Nov. 897/1 Mr. Cooper managed to slip the theme of indifference and its potential into this play without breaking its back. **1959** *Ibid.* 23 July 122/2 There is thought to be enormous oil potential. **1965** *Ibid.* 1 July 22/3 All collections have a built-in boredom potential. **1969** H. MacInnes *Salzburg Connection* xx. 281 His record..has been excellent. His potential was high—very high. **1970** *Nature* 26 Dec. 1248/2 Although oceans and seas cover some 360 thousand km²..., exploitation of their vast potential is only just beginning. **1978** *Observer* (Colour Suppl.) 12 Nov. 58/4 Children without these experiences..are likely to be handicapped in terms of the development of their full potential. **1979** *Country Life* 21 June 2047/3 This ever-perceived dark potential is surely part of the reason why the cat is the intellectuals' favourite beast.

4. a. (Further examples.) More generally, any function from which a vector field **F** can be derived by differentiation, esp. the scalar potential ϕ, where $\mathbf{F} = -\text{grad}\ \phi$, and the vector potential **A**, where $\mathbf{F} = \text{curl}\ \mathbf{A}$.

1873 J. C. Maxwell *Treat. Electr. & Magn.* II. III. ii. 27 (*heading*) The Vector-Potential of Magnetic induction. **1909** J. G. Coffin *Vector Anal.* vi. 173 We obtain $\mathbf{H} = ..\nabla \times \mathbf{Q}$... Q is called the potential due to the current distribution q, or the vector-potential belonging to the magnetic force H... The force vector H is obtained from the vector Q in a manner analogous to the way the force vector F is obtained from the scalar V, where..$\mathbf{F} = \nabla V$. **1933** H. B. Phillips *Vector Anal.* v. 102 When the potential is known the velocity can be obtained by differentiation. If a potential exists it is simpler to describe the motion by means of it rather than the velocity. **1966** *McGraw-Hill Encycl. Sci. & Technol.* X. 540/2 If the acceleration a satisfies a relation such as..$\mathbf{a} = -\text{grad}\ \Phi$, the Φ is called an acceleration potential. **1971** W. Hauser *Introd. Princ. Electromagnetism* iii. 77 Vector function $\mathbf{F}_1$ is a curl-less vector function... It is therefore expressible as the negative gradient of a scalar function of position. We thus set $\mathbf{F}_1 = -\nabla\psi(\mathbf{r})$, where the function $\psi(x, y, z)$ is referred to as the scalar potential of $\mathbf{F}_1$. *Ibid.*, The divergenceless vector function $\mathbf{F}_2$ is..expressible as the curl of a vector function $\mathbf{A}, \mathbf{F}_2 = \nabla \times \mathbf{A}$, where **A** is referred to as the vector potential of $\mathbf{F}_2$.

b. Any of a group of thermodynamic functions mathematically analogous to electric and gravitational potentials, viz. the Gibbs free energy G (or ζ), the Helmholtz free energy A (or F or ψ), the enthalpy H (or χ), the internal energy U (or E or ϵ), and the chemical potential μ.

The Gibbs and the Helmholtz functions are given respectively by $G = U + PV - TS$ and $A = U - TS$, where U is the internal energy, P the pressure, V the volume, T the temperature, and S the entropy of the system. The chemical potential μ_i of a component i in a given phase is equal to $(\partial G/\partial m_i)_{P,T}$, where m_i is the quantity of the component present in the phase and the quantities of all other components remain constant.

1878 J. W. Gibbs in *Trans. Connecticut Acad. Arts & Sci.* III. 119 If we call a quantity μ_x, as defined by such an equation as (12), the potential for the substance S_x in the homogeneous mass considered, these conditions may be expressed as follows:— The potential for each component substance must be constant throughout the whole mass. *Ibid.* 149 If to any homogeneous mass we suppose an infinitesimal quantity of any substance to be added, the mass remaining homogeneous and its entropy and volume remaining unchanged, the increase of the energy of the mass divided by the quantity of the substance added is the potential of that substance in the mass considered... In the above definition we may evidently substitute for entropy, volume, and energy, respectively, either temperature, volume, and the function ψ; or entropy, pressure, and the function χ; or temperature, pressure, and the function ζ. **1917** G. E. Gibson tr. *Sackur's Textbk. Thermo-Chem. & Thermodynamics* vi. 178 The component I can go spontaneously from B to A if its chemical potential in A is less than its chemical potential in B. *Ibid.* 179 In chemistry it is usual to take the mol of each component as the unit of mass, and we may then define the chemical potential..of the component I in the solution A as the change in the thermodynamic potential of a very large mass of A when 1 mol of the component I is added to it without changing the temperature, the pressure or the masses of the other components. **1924** H. S. Taylor *Treat. Physical Chem.* I. ii. 67 The thermodynamic potential of all spontaneously occurring processes decreases. **1937** M. W. Zemansky *Heat & Thermodynamics* xvii. 321 If a substance is not present in a phase, it does not follow that its chemical potential is zero. The chemical potential is a measure of the effect on the Gibbs function when a substance is introduced. **1950** E. O. Hercus *Elem. Thermodynamics & Statistical Mech.* iv. 24 Two new thermodynamical quantities dependent only on the state of a system can be defined from the entropy. These are: Free Energy, $F = U - TS$. Thermodynamic Potential, $G = H - TS = U + pv - TS$. **1960** Hall & Ibele *Engin. Thermodynamics* x. 183 The stability of thermodynamic systems can be examined with reference to a set of quantities known as thermodynamic potentials. **1973** D. C. Kelly *Thermodynamics & Statistical Physics* viii. 147 The four thermodynamic potentials were invented to make thermodynamics 'easy'. Each potential is the natural energy variable for certain classes of physical processes.

c. Special Comb.: **potential barrier,** a region in a field of force in which the potential is significantly higher than at points either side of it, so that a particle requires energy to pass through it; *spec.* that surrounding the potential well of an atomic nucleus; **potential flow,** flow which is irrotational and for which there therefore exists a velocity potential; **potential gradient,** (the rate of) change of (electrical) potential with distance; **potential scattering** *Nuclear Physics*, elastic scattering of a particle by an atomic nucleus in which the scattering cross-section varies smoothly with the energy of the incident particle (cf. **resonance scattering*); **potential wall,** a region in a field of force in which the potential increases sharply; **potential well,** a region in a field of force in which the potential is significantly lower than at points immediately outside it, so that a particle in it is likely to remain there unless it gains a relatively large amount of energy; *spec.* that in which an atomic nucleus is situated.

1929 *Physical Rev.* XXXIII. 134 The particle in the internal region received energy sufficient to raise it over the potential barrier. **1931** G. Gamow *Constitution of Atomic Nuclei* ii. 37 How can such an α-particle get out from the nucleus if it has to cross on its way a potential barrier which is certainly higher than the total energy of the α-particle itself? **1966** H. J. Reich et al. *Theory & Applications Active Devices* iii. 64 An electron encounters a potential barrier in moving from left to right across the junction. **1973** V. Acosta et al. *Essent. Mod. Physics* xvi. 223 (*heading*) A beam of particles of kinetic energy E incident on a potential barrier $V > E$ with a width $OA = t$. **1937** N. A. V. Piercy *Aerodynamics* vi. 140 Irrotational flow is often called potential flow. **1962** Walshaw & Jobson *Mech. of Fluids* viii. 211 'Potential' flows..neglect viscous actions and merely provide a framework of reference against which the behaviour of a real fluid may be compared. **1975** *Sci. Amer.* Nov. 85/1 An airfoil or a wing in steady motion through the air is a device by means of which circulation is created and maintained in the form of a vortex bound to the wing. The bound vortex is then superposed on the flow pattern that the wing profile would produce in an ideal fluid. The pattern is termed the potential flow. **1895** A. Daniell *Text-bk. Princ. Physics* (ed. 3) xvi. 585 This Electric Force on Unit Quantity, $\phi = (V_1 - V_2) \div d$, is the Potential-Slope or Potential-Gradient. **1931** *Discovery* July 212/1 Measurements of potential gradient have been made in balloons up to a height of nine kilometres. The gradient falls off rapidly, most of the positive charge being in the lower strata. **1963** E. V. Vernon in Zepler & Punnett *Electron Devices & Networks* i. 25 The potential gradient along the material is due to its ohmic resistance. **1973** R. Brown *Electricity & Atomic Physics* xii. 270 There is a potential gradient in the depletion layer, positive on the *n* side and negative on the *p* side, and this represents a potential barrier. **1937** Bethe & Placzek in *Physical Rev.* LI. 460/2 It amounts..to a scattering cross section $\sigma_1 = 4\pi R^2$... This part may be called potential scattering in the narrower sense. *Ibid.* 462/1 The total potential scattering is $\sigma_{pot} = (\sigma_1 \frac{1}{2} - \sigma_2 \frac{1}{2})$. **1955** A. E. S. Green *Nuclear Physics* xiii. 433 Within the category of elastic scattering we may distinguish two types of processes, namely, potential scattering and resonant scattering. **1971** B. L. Cohen *Concepts of Nuclear Physics* xii. 331 Between the resonances in Fig. 13-7 we see the effects of potential scattering only. **1931** *Proc. R. Soc.* A. CXXXIII. 238 If this theory of the resonance levels is correct, it is difficult to reconcile the results of Pose, who finds quite sharp resonance levels in Al, with the results of experiments on α-particles of sufficient energy to pass over the top of the potential wall. **1973** V. Acosta et al. *Essent. Mod. Physics* xvi. 222 This situation may be treated in a simplified manner by using a thin potential wall—a potential barrier. [**1931** G. Gamow *Constitution of Atomic Nuclei* i. 18 The potential..must be more or less constant inside the nucleus and increase sharply at the boundary, the distribution forming a 'potential hole' of the shape shown.] **1935** *Physical Rev.* XLVII. 852/1 The positive valued parameters A and α are to be determined to fit the binding energies of the deuteron and the alpha-particle. Evidently A and $1/\alpha$ are directly proportional to the depth and breadth, respectively, of the potential well. **1952** Blatt & Weisskopf *Theoret. Nuclear Physics* ii. 49 Even very refined experiments at low energies do not suffice to determine more than an 'effective range' and 'depth' of the potential well, leaving the detailed shape completely indeterminate. **1972** DePuy & Chapman *Molec. Reactions & Photochem.* i. 1 Within this potential well the molecule can occupy any of a number of discrete vibrational energy levels.

potentialness. (Later example.)

1930 G. Greene *Two Witnesses* 135 The turning of potentialness into creative life.

potentiate, *v.* Add: **3.** *trans.* To increase the effect of (a drug or its action); to act synergistically with; also, to promote or enhance (a physiological or biochemical phenomenon). Also *fig.*

1917 T. Sollmann *Man. Pharmacol.* 96 Mansfield and Hamburger..believe..that ether potentiates chloral or morphine, by favoring the distribution of these agents in the nervous system. **1941** *Cancer Res.* I. 107/2 Numerous experiments have shown that various exogenous influences, each in itself carcinogenic, can 'potentiate' each other's influence by simultaneous action. **1962** *Lancet* 13 Jan. 112/1 He showed that cocaine potentiated the pressor action of adrenaline. **1969** D. Clark *Nobody's Perfect* iii. 109 Though the phenobarbitone had caused death, it had been potentiated by the alcohol. Without the alcohol he might have lived. **1973** R. G. Krueger et al. *Introd. Microbiol.* xix. 524/2 The membrane may serve to potentiate cell-virus association and thereby enhance infection. **1974** Hawkey & Bingham *Wild Card* xi. 107 What do you intend doing to potentiate the virus's cytolytic properties? **1977** *Observer* 21 Aug. 21/5 New York potentiates music the same way that alcohol potentiates certain pills.

potentiated *ppl. a.* (further examples); also **pote·ntiating** *ppl. a.* and *vbl. sb.*; **potentiation,** (*b*) the phenomenon whereby the simultaneous effect of two drugs or other agents may exceed the sum of their individual effects.

1914 J. T. Halsey tr. *Meyer & Gottlieb's Pharmacol.* xviii. 576 Honigmann's experiments with mixed narcosis with ether and chloroform or ether and alcohol apparently indicate a potentiation. **1917** T. Sollmann *Man. Pharmacol.* 96 When several drugs are administered together, each may act independently, as if it were present alone... In many cases, however, the combined action is greater or smaller than would be calculated (potentiated or deficient summation). **1941** *Cancer Res.* I. 107/2 The 'potentiation' seen in the simultaneous action of tumor-producing virus and other carcinogenic influence. *Ibid.*, The above-mentioned 'potentiating' influence of two different exogenous influences. **1958** *Bull. Math. Bio-*

physics XX. 1 Two drugs may act together synergistically or they may act antagonistically. The former action we shall refer to as positive potentiation; the latter as negative potentiation. **1970** J. Murdoch *Fairly Honourable Defeat* x. 165 Her body seemed to be weighted and pinned to the sloping bank by a potentiated force of gravity. **1971** P. Hudson *Devil's Picturebk.* ii. 64 Frequent use of a complex mnemotechnic system was believed to result in the enlivening or 'potentiating' of the imagination. **1974** *Aquaculture* IV. 410 Other workers..have also found the in vivo efficacy of a potentiated sulphonamide..to be far superior to sulphonamide, used alone, in the treatment of furunculosis. **1974** M. C. Gerald *Pharmacol.* iii. 63 Potentiation may result from two agents acting by different mechanisms or one drug facilitating the action of the other. **1975** *European Jrnl. Pharmacol.* XXXIV. 169 (*heading*) Potentiating effect of lithium chloride on aggressive behaviour induced in mice. **1977** *Amer. N. & Q.* XV. 83/1 These properties include curative power,..potentiation,..and the prevention of decay or corruption.

potentiator (pote·nʃiˌeɪtəɹ). *Pharm.* [f. prec. + -or.] An agent that increases the effect of a drug.

1955 *Sci. News Let.* 9 July 19/3 Animals that had just recovered from a barbiturate almost immediately went back into deep hypnosis when given chlorpromazine. This shows that it is a true potentiator, and not merely a prolonger of the action of the barbiturate. **1975** *Acta Endocrinol.* LXXIX. 311 (*heading*) Normal recognition of glucose as a potentiator in subjects with low insulin response and in mild diabetes.

potentiometer. a. Substitute for def.: A device for measuring potential difference or an e.m.f. by balancing it against a variable potential difference of known value produced by passing a known (usu. fixed) current through a known (usu. variable) resistance. (Further examples.)

1922 Glazebrook *Dict. Appl. Physics* II. 611/2 A Weston cell having an E.M.F. of 1·0183 volts would be balanced across 10 coils and 0·183 of the total length of the slide wire, the pressure drop across each coil of the potentiometer being then 0·1 volt. **1935** Turner & Banner *Electr. Measurements* xi. 127 The Crompton type is one of the most widely used potentiometers; in this, instead of a continuous slide wire covering the whole voltage range, resistance coils are used with a selector switch. **1975** D. G. Fink *Electronics Engineers' Handbk.* xvii. 6 The constant-resistance potentiometer..uses a variable current through a fixed resistance to generate a voltage for obtaining a null with the unknown emf.

b. A voltage divider which is regulated by varying a resistance; also, *loosely*, a rheostat.

1910 G. W. Pierce *Princ. Wireless Telegr.* xxvii. 324 The accurate adjustment of the local voltage is achieved by the use of a potentiometer. **1914** R. Stanley *Text-bk. Wireless Telegr.* xvi. 221 Such an arrangement of battery and wire is called a potentiometer; by means of it we can obtain any voltage, to apply to our apparatus, from zero up to the full voltage of the battery. **1955** *Sci. Amer.* Aug. 96/2 Connect the power supply to the solution [in the ice-box dishes] through the carbon electrodes and adjust the potentiometer or tapped resistor to the prescribed potential of 200 volts. **1962** F. I. Ordway et al. *Basic Astronautics* ix. 375 The voltage in a current varies with the displacement of the mass by means of a potentiometer. **1966** *McGraw-Hill Encycl. Sci. & Technol.* X. 542/1 By using only the movable and one fixed connection, a potentiometer may be used as a rheostat. **1968** A. Marcus *Electricity for Technicians* iv. 59 Such variable resistors are called rheostats or potentiometers.

2. *attrib.* and *Comb.*

1881 [in Dict.]. **1916** C. C. Garrard *Electric Switch & Controlling Gear* iv. 261 Potentiometer-type regulators are used when it is desired to reduce the voltage applied to the terminals of the field coil to zero, as is the case, for example, with boosters. **1920** *Whittaker's Electr. Engineer's Pocket-bk.* (ed. 4) 434 These curves can be used for any potentiometer regulator which has a resistance 3½ times the field coil. **1922** Glazebrook *Dict. Appl. Physics* II. 615/2 The total resistance of the potentiometer circuit remains unchanged whatever the setting of the dials. **1961** G. V. Sadler in G. F. Tagg *Pract. Electr. Engin.* III. 215 Special precautions must also be taken when potentiometer control is used on a crane whose d.c. supply is obtained from a static rectifier. **1962** G. A. T. Burdett *Automatic Control Handbk.* ix. 4 Potentiometer pressure transducers are made in a number of forms. **1966** *McGraw Hill Encycl. Sci. & Technol.* X. 543/2 Potentiometer measurement of current is accomplished by passing current through a standardized resistor of appropriate value and measuring the potential difference across this resistor. **1979** *Sci. Digest* July 35/1 When potentiometer controls are used with microprocessor games the analog voltage due to the potentiometer must be converted into a digital quantity.

Hence **potentiome·tric** *a.*, of or pertaining to a potentiometer; employing, or obtained by means of, a potentiometer; *potentiometric titration*, a titration which is followed by measuring the change in potential of an electrode immersed in the sample solution; **potentiome·trically** *adv.*; **potentio·metry**, the technique of measurement with potentiometers, esp. in chemical analysis.

1915 *Jrnl. Chem. Soc.* CVIII. ii. 307 (*heading*) Potentiometric arrangement for electrochemical investigations. **1926** Kolthoff & Furman *Potentiometric Titrations* viii. 151 Everyone who has had experience with the performance of potentiometric titrations knows that, near the equivalence-point especially,..the potential does not become constant immediately after the addition of the reagent. *Ibid.* xi. 247 Manganous salts may be titrated potentiometrically according to the Volhard-Wolff method. **1931** I. M. Kolthoff *Colorimetric & Potentiometric Determination of pH* vii. 129 (*heading*) Problems in potentiometry. **1946** L. Michaelis in A. Weissberger *Physical Methods Org. Chem.* II. xxii. 1052 Potentiometry consists of the measurement of the electromotive force of a galvanic cell composed of two half-cells one of which is a reference half-cell of known composition; and the other an electrode immersed in the solution to be investigated. **1966** Elvidge & Sammes *Course Mod. Techniques Org. Chem.* (ed. 2) xxxiv. 284 Potentiometry makes possible the titration of very weak acids and bases. **1967** *Times Rev. Industry* Oct. 49/3 The length of the cable is then determined by a potentiometric method and may be indicated at any remote control position. **1972** Grunwald & Kirschenbaum *Introd. Quantitative Chem. Anal.* xviii. 337 A silver–silver halide electrode is used for the potentiometric titration of a mixture of halide salts with silver nitrate. **1975** D. G. Fink *Electronics Engineers' Handbk.* x. 13 Potentiometric displacement transducers are widely used because of their relative simplicity of construction and their ability to provide a high-level output. **1978** *Nature* 17 Aug. p. ix/1 Finally the controller starts the titration to titrate the free iodine potentiometrically using sodium thiosulphate.

potentiostat (pŏte·nʃiostæt). [f. Potenti(al *a.* and *sb.* + -o + *-stat.] A device used to regulate automatically the potential difference between electrodes in electrolysis.

1942 A. Hickling in *Trans. Faraday Soc.* XXXVIII. 27 The electrical circuit of the device, which will subsequently be referred to as a 'potentiostat', is shown in Fig. 1. **1949** *Analytical Chem.* XXI. 497/1 The instrument...functions as a potentiostat in automatically maintaining the potential of a working electrode constant during an electrolysis. **1965** *New Scientist* 8 Apr. 99/1 These potentiostats..have a very rapid response to voltage fluctuations and now make the continuous control of commercial processes look very attractive. **1979** *Nature* 15 Mar. 239/1 The cinnabar electrode is connected to a platinum counter electrode and a potential of 0 V maintained between the two by means of a potentiostat.

Hence **potentiosta·tic** *a.*, under the control of or employing a potentiostat; with the potential difference between electrodes held constant; **potentiosta·tically** *adv.*

1955 *Chem. Abstr.* XLIX. 5166 The potentiostatic method was compared with other methods. **1961** *Trans. Symposium Electrode Processes* (U.S. Electrochem. Soc.) 164 An attempt was made to develop apparatus in which potentiostatic conditions could be achieved in a time short enough to study the growth and decay of metal monolayers at constant potential. **1967** tr. *K. J. Vetter's Electrochem. Kinetics* (Theoret. Aspects) ii. 364 The potentiostatically applied overvoltage must..be pure charge-transfer overvoltage at time $t = 0$. **1975** *Nature* 27 Mar. 322/2 Specimens held under potentiostatic control in the electrolyte were subjected to square wave strains in the range ±0·02 to ±0·05 at strain rates of ~0·30 s⁻¹. **1976** *Ibid.* 19 Aug. 681/1 Anodisation could be accomplished both galvanostatically and potentiostatically.

potentize, *v.* Add: Hence **po·tentized** *ppl. a.* Also **po:tentiza·tion**, dilution of a drug in order to increase its power or efficacy.

1850 C. J. Hempel *New Homœopathic Pharmacopœia* i. 10 By the former, these successive developments of the original substance are called dynamizations, or potentizations; by the latter, attenuations. **1864** *Trans. Homœopathic Med. Soc. N.Y.* 56 The administration of potentized remedies. **1938** D. Shepherd *Magic of Minimum Dose* 7 The first law is the law of simillimum, which is followed by (2) the principle of the minimum dose, (3) the principle of potentization. **1972** D. V. Tansley *Radionics* viii. 80 To augment the violet one may add an ampoule of potentized onyx to the treatment. **1974** *Homoeopathy* June/July 87 Symptoms were noted when the drug was given in material and potentised doses. *Ibid.* 88 As the substance becomes more dilute with succeeding potentisations it becomes soluble.

potestas. Add: **4.** *Philol.* The phonetic or phonemic value of a letter in an alphabet.

1949 J. R. Firth in *Trans. Philol. Soc. 1948* 135 Each Arabic letter has..syllabic value, the value or *potestas* in the most general terms being consonant plus vowel, including vowel zero, or zero vowel. **1963** *English Studies* XLIV. 4 Thus if there is a contrast..between the North and elsewhere, then..it is best treated as a contrast in graphemes irrespective of their phonemic 'value', or, to speak in more mediaeval terms, as a contrast in *figurae* irrespective of the *potestas* of each. **1967** R. H. Robins *Short Hist. Linguistics* iii. 56 Priscian worked systematically through his subject, the description of the language of classical Latin literature. Pronunciation and syllable structure are covered by a description of the letters (*litterae*), defined as the smallest parts of articulate speech, of which the properties are *nōmen*, the name of the letter, *figūra*, its written shape, and *potestās*, its phonetic value. **1972** H. Benediktsson *First Grammatical Treat.* iii. 90 He is willing to establish new letters with a value (*potestas*) of a combination of two consecutive letters.

‖ **pot-et-fleur** (pot,e,flör). [Fr., lit. 'pot and flower'.] A style of floral decoration using pot-plants together with cut flowers. Also *attrib.*

[**1960** V. Stevenson in *Daily Tel.* 11 Feb. 11/2 There's a new mood afoot in flower arrangement—to combine cut flowers with indoor pot plants. I have been using this style of arrangement for ages, without finding a name for it. Can readers make any suggestions?] **1963** —— *Decorating with Flowers & Plants* vii. 71 When I wanted to describe this style of decoration about which I am writing, I invited the readers of the Woman's Page in the *Daily Telegraph* to coin a name... Our new style became pot-et-fleur, a term already recognized by the National Association of Flower Arrangement Societies of Great Britain, who have included a class in their national competitions. *Ibid.* 73 You will have made your first pot-et-fleur arrangement. **1968** *Flower Arranger* Sept. 125 There were strong line arrangements in the tall slit windows behind the choir-stalls; pot-et-fleurs were mostly in the [illegible]. **1973** *Jrnl. R. Hort. Soc.* XCVII. 373 Bonsai and pot-et-fleur are two main classes. **1978** T. Taylor *Flower Arranging* 66/1 A pot-et-fleur is a decoration that combines plants and cut flowers. The plants can be grown permanently and the flowers can be added from time to time.

pot fisherman. Add: (Examples of sense in Dict.); also, = *fish-potter* s.v. Fish *sb.*¹ 7.

1895 *Funk's Stand. Dict.* 1392/1 *Pot-fisherman*, one who fishes while floating buoyed up by an earthen pot, as on Asiatic rivers. **1970** *Times* 1 Sept. 10/5 A much brighter prospect has suddenly opened up for Guernsey's traditional pot fishermen, who looked as if they were doomed to go gradually out of business.

pot-head. Add: **2.** *Canada.* In full, *pothead whale.* = Ca'ing-whale, *caa'ing whale, *pilot-whale* s.v. Pilot *sb.* 6 in Dict. and Suppl.

1863 *Islander* (Charlottetown, Prince Edward Island) 14 Aug. 2/3 Large numbers of Potheads are in the Bay, which probably accounts for the squid panic. **1964** *Canad. Geogr. Jrnl.* Mar. 92/3 Entire herds of 40-foot potheads whales have been known to run aground on the beaches. **1975** *Sunday Monday* (McAllen, Texas) 16 July 14/1 More than 170 pothead whales beached themselves on Sunday.

3. *Electr.* An insulated connector used for making a sealed joint between conductors, esp. between insulated and uninsulated lines.

1901 *Electr. World* 19 Jan. 138/1 The pothead (in figure) Pot head splice. **1903** *Phil. Mag.* V. 327 The cables were brought out to pot-heads..and each were terminated in a screw-cap. **1930** H. P. Seelye *Electr. Distribution Engin.* xvi. 497 The pothead usually consists of a cast-iron tank or pot into which the cable is inserted... The conductors pass through the pot and are either brought out through porcelain bushings, or are attached to conducting terminals fixed in the bushings. The pot is then filled with insulating compound. **1963** K. Neville in D. Knight 100 *Yrs. Sci. Fiction* (1969) 73 Ramon Lopez, one of the truck crew, was killed today hosing down a high-voltage pothead.

4. (See *pot *sb.*¹ 2.)

pother, *sb.* Add: **2.** (Later example of the form *pudder*.)

1956 Auden & Kallman *Magic Flute* i. iv. 32 (From within the temple comes the sound of singing)..What's that? What's that pudder? I fancy I heard..

pothery, *a.* Add: **1.** Also puthery.

1855 Mrs. Gaskell *Let.* Feb. (1966) 332 It is so *puthery* here, I can hardly walk.

pot-hole. Add: Also pothole, pot hole. **1.** (Earlier example.)

1826 T. L. McKenney *Sk. Tour to Lakes* (1827) 54 The waters were once, in many places, some fifty feet above their present level; for their action upon the rocks is plainly seen in the pot holes, as the excavations are called, which are made by the action of pebbles upon the rocks.

3. *N. Amer.* A pond formed by a natural hollow in the ground in which water has collected. Cf. Slew *sb.*¹ 1, Slough *sb.*¹ 4. Also *attrib.*

1902 *Saskatchewan Hist.* (1956) IX. 31 In natural depressions of the soil..water had gathered and formed so-called 'pot-holes'—circular basins of a swampy nature, generally rather shallow. The 'potholes' usually have a heavy growth of wild hay. **1923** S. C. Ells *Northland Trails* 85 It is a land of mysterious pot-hole lakes and ponds with neither inlets nor outlets. **1946** *Sun* (Baltimore) 5 July 11/3 Experiment planting of fish with airplanes gives promise for the 'back in there' pothole lakes and spring holes. **1955** E. P. Hobson *Nothing too Good for Cowboy* (1956) vi. 76 The pothole lakes were still some five miles away from the proposed line. **1962** *Daily Progress* (Charlottesville, Va.) 17 Aug. 4/2 Minnesota and the Dakotas..are dotted with innumerable small ponds called 'potholes'. **1963** *Globe & Mail* (Toronto) 13 Sept. 35/1 When the time comes you'll see the shivers when half-a-dozen greenhead mallards start side-slipping into the pothole in front of him.

4. A depression or hollow part forming a defect in the surface of a road or track. Also *pothole.*

1909 *Westm. Gaz.* 30 Aug. 1/3 We are also beginning to see how much our urban and suburban macadam roads suffer from artificial watering and constant scavenging, for all road engineers are agreed that the uneven surfaces and pot-holes..are practically confined to the districts where the water-cart reigns supreme. **1930** *Motor Cycle* 30 Sept. 384/2 On the outward journey the pot-holes between Edinburgh and Stirling seemed appalling. **1955** *Times* 5 July 3/7 On the other hand, the going was rather heavy going of potholes like those caused by recessed manhole covers. **1972** 'J. E. Borrett' *No Time to Kill* v. 32 The grey car veered to the left and slowed to a crawl as it met the potholes. **1978** *New York* 3 Apr. 100/4 The locals of D.C. 37 represent social-service workers, botanical garden employees, city engineers, pothole repairmen, hospital employees, [etc.].

5. *Austral.* A shallow hole dug in the ground in prospecting for opal dirt (see *opal 4). Also *attrib.*

1940 I. L. Idriess *Lightning Ridge* 90 For a time I sank pot-holes alone then went mates with little Archie Campbell. **1967** —— *Opals & Sapphires* 112 Keep the find quiet until you have sunk more potholes to prove that it is worth while pegging out.

po·t-hole, *v.* Also **pothole.** [f. the sb.] **1.** *trans.* To produce pot-holes in.

1909 *Cent. Dict. Suppl.*, *Pot-hole*, to produce in (a solid rock mass) a hole by the action of stones and silt whirled round in an eddy of water. **1975** *Wales on Mar.* 189/2 It was channelled and potholed by running water.

2. *a. trans.* To explore (pot-holes or the like). **b.** *intr.* To engage in pot-holing.

1970 *Observer* (Colour Suppl.) 25 Jan. 30/4 Its underworld labyrinth of rivers and trolls consists of tunnels to be potholed as a boy. **1970** *Times* 26 Oct. 4 They went with to others potholing on the Pennines at Casterton, near Kirkby Lonsdale.

po·t-holed, *a.* [f. as prec. + -ed².] Having pot-holes.

1933 Auden *Poems* (ed. 2) 43 By pot-holed becks. **1939** Joyce *Finnegans Wake* 31 [He] had been meaning to inquire what, in effect, had caused you causeway to be thus potholed. **1952** C. Barnes *Britain's Bishop's Move* x. 114 It took the car of civilization away from the main road on to a humpy, pot-holed track. **1960** I. Cross *Backward Sex* i. 33 The lines of the tram tracks were pot-holed and rutted. **1976** S. Wales Echo 27 Nov. 6/3 People will have to go on putting up with a badly pot-holed road.

po·t-holer. Also **potholer.** [f. as prec. + -er¹.] Someone who goes pot-holing.

1900 *Jrnl. Yorkshire Ramblers' Club* I. 134 The temptation to remain above ground certainly was very strong, and some of the pot-holers gazed rather wistfully after the men who, at the parting of the ways, left them for Ingleborough. **1968** *Pearson's Mag.* Mar. 282/1 Then a party of pot-holers on the warpath you would think, from the mass of spars and cordage, that they had got some idea. **1969** *Listener* 27 July 114 At Derby-day cave explorers and pot-holers from several parts of England will meet to found a British Speleological Association. **1957** *Times* 10 Dec. 10/5 (*heading*) Six potholers brought out. Rescuers still all out. **1969** *Community singing in cave.* **1973** D. Orgill *Jasius Pursuit* xi. 107 This kind of cat-run, in the potholers' experience, often led to a larger chamber.

po·t-holey, *a.* [f. as prec. + -y¹.] Having many pot-holes.

1921 *Blackw. Mag.* Nov. 641/2 We lurched along over a very pot-holey road. **1925** *Motor* 15 Dec. 1001/3 Some rough stretches of wavy and pot-holey surface. **1936** F. Clune *Roaming round Darling* x. 89 Road, neither pot-holey nor unholey for thirty-five miles.

po·t-holing, *vbl. sb.* Also **potholing.** [f. as prec. + -ing¹.] **1.** The exploration of underground pot-holes.

1899 *Jrnl. Yorkshire Ramblers' Club* I. 65 The sport of cave-hunting and pot-holing, apart from its scientific interest, has fascinations and charms peculiar to itself. **1908** *Pearson's Mag.* Mar. 282/1 Subterranean mountaineering, which is, in the vernacular, pot-holing, is indisputably possesses a fascination of its own. **1925** *Jrnl. Fell & Rock Climbing Club* VII. 69 An afternoon's 'pot-holing'..was the only mountaineering done. **1935** *Times* 27 July 13/4 The pot-holing clubs of the Yorkshire and Lancashire have already joined to form a rescue organization. **1963** *Potholing & Caving* (Know the Game Ser.) 11/1 The words caving and potholing are used loosely to mean the same thing... Technically a cave is an underground system which may or may not be large enough to admit a human being... A pothole is a natural hole in the ground or a cave system, containing shafts or pitches. **1975** *Times* 9 Sept. 6/2 Mr Tony Harrison, aged 19, a member of Lancaster University's potholing team, has been killed in a 75 ft fall.

2. The formation of pot-holes.

1903 *Geogr. Jrnl.* XXI. 672 Of this [erosion] Mr. Ball thinks that at least two-thirds is accounted for by the pot-holing action. **1941** *Jrnl. Geomorphol.* IV. 71 (*heading*) Pot-holing of limestone by development of solution cups.

pot-hook, *sb.* Add: **2.** (Further examples.) Also *pl.* = Shorthand, *colloq.*

1868 *Sewell's Night Guide* 128/1 Potlooks and hangers, short hand characters. **1937** Partridge *Dict. Slang* 653/1 *Pot-hooks and hangers*, shorthand. **1939** Joyce *Finnegans Wake* 181 Instead of clothering those model household plain wholesome pothooks (a thing he never possessed of his Nigerian own) what do you think Vulgariano did but copy all their various styles of signature. **1962** *Spectator* 13 July 58/3 Dear Sir, Dashing, if this is a formal meeting, I guess you had better make pothooks. **1962** G. Lawton *John Wesley's English* 230 His [sc. Wesley's] works are full of proverbs, potboiling, allusions, idioms, colloquialisms, pothooks, and slang. **1963** C. Mackenzie *My Life & Times* I. 156 From her I learnt to write pothooks and hangers and very soon to pass from pothooks and hangers to real letters.

b. *attrib.* Delete †*Obs.* and add later example (with reference to shorthand).

1950 W. Owen *Let.* 2 May (1967) 236 Thanks for the Catalogue of Pothook-books [Pitman's catalogue].

pot-housey, *a.* [f. Pot-house + -y¹.] Suggestive of or appropriate to a pot-house.

1875 Hardy *Under Greenw. Tree* I. i. viii. 105 If I strip by myself and not necessary, 'tis rather pot-housey, I own.

pot-hunt, *v.* [Back-formation f. Pot-hunter.] *intr.* To hunt 'for the pot' (see Pot *sb.*¹ 1 *b*); to be a pot-hunter.

1926 *Chambers's Jrnl.* July 418/1 You..prefer to pot-hunt—luckily for us, with six hefty Gurkhas and the servants to feed, as well as ourselves! **1936** Auden & Isherwood *Ascent of F6* 46 Where we were boys at school, I saw him..win the prizes... We are older now...But James..has still the last gap in his great collection And pot-hunt for his brother.

pot-hunter. Add: **2.** (Further examples.) Also *fig.*

1856 *Porter's Spirit of Times* 25 Oct. 126/1 It is disgusting to every lover of fair play to witness the ravages committed by the pot-hunter, who coolly murders the deer by torch-light from a dug-out or canoe. **1878** C. Hallock *Amer. Club List & Sportsman's Gloss.* p. 18/1 *Pot-hunter*, one who hunts or fishes for profit, regardless of close seasons, the waste of game, or the pleasure to be derived from the pursuit. **1905** 'O. Henry' in *N.Y. World Mag.* 22 Oct. 8/1 He was an old man, with a slow and limping gait, so a pot-hunter of a newly-licensed chauffeur ran him down one day when livelier game was scarce. **1922** Joyce *Ulysses* 158 Lady Mountcashel.. rode out with the Ward Union staghounds... Unstable fox. Pothunters too. **1946** *Sun* (Baltimore) 22 Feb. 12/5 Hundreds of ducks flying northward have been slaughtered. The chief offenders..are the so-called 'pot-hunters', who hunt for profit and rent duck blinds to sportsmen.

3. (Later examples.)

1891 A. J. Bennett *Meador* 21 He used to be what they called a pot-hunter', a racing bicyclist. **1948** Berrey & Van den Bark *Amer. Thes. Slang* (1953) 636/21 Contestant for a prize, *mug hunter* or chaser, pot hunter.

4. One who finds or obtains objects of archaeological interest or value, esp. by unscientific or illicit methods, and for the purpose of private collection or profit.

1958 *N.Y. Times Mag.* 16 Feb. 48/3 This satisfaction.. is the big distinction between the new species of digger and the acquisitive 'pot-hunter', that bane of archaeological scientists. Souvenir-hunting amateurs are still with us, of course. **1965** Rounds & Irvine *Amat. Archaeologist's Handbk.* v. 83 This chapter might carry the subtitle: '[Potfinders, beware!' A pot-hunter is a person who visits a site in order to find and hoard as many 'relics' as possible. **1967** L. Deuel *Conquistadors without Swords* iv. 52 Museum collections all over the world owe most of their treasures to disreputable pothunters. **1973** *New Yorker* 24 Mar. 100/2 Stewart L. Peckham, chief archeologist for the Museum of New Mexico,..was distressed to find that almost all the major Mimbres sites..had been ravaged by pot-hunters.

pot-hunting. Add: (Further examples.) Also in sense corresponding to *pot-hunter 4.

1866 *Harper's Mag.* Jan. 157/1 This is regard not only entitled to the name of sport, and only to be resorted to for the purpose of securing the meat needed for food. **1958** L. P. Hartley *Skin Illusions* ii. 47 I'm doing a bit of pot-hunting and have to attend classes and pow-wows. **1966** A. R. Kenn in *Amer. Antiquity* Apr. 423/1 Scientific archaeology carries more satisfaction than 'pot-hunting' and 'treasure troving'. **1973** *New Yorker* 24 Mar. 100/3 Poverty and desperation are extenuating factors in Bangladesh, but not in the United States, where pot-hunting is a weekend hobby of the more affluent. The rise in prices for pre-Columbian-Indian art..has had disastrous consequences for scholarship.

‖ **potiche** (poti·ʃ). [Fr., in the same sense.] A large porcelain vase, usu. rounded in shape with bulging shoulders and a wide mouth, freq. having a lid, originally produced in China during the Ming dynasty.

1891 in *Funk's Stand. Dict.* **1933** *Burlington Mag.* Nov. 202 (*caption*) Semi-Celadon Potiche of Chien-Swankakik Ware. **1939** *Ibid.* June p. xxi/1 A royal potiche ornamented with a band of wave scrolls on the neck. **1960** H. Hayward *Antique Coll.* 226/2 *Potiche*, large broad-mouthed Chinese porcelain jar of 'baluster' shape, with cover; favoured from Ming times. **1967** P. Wirth in *Coast to Coast* 1965–66 231 Gunnerina back of her mistress rush at the Sibelids, dust the *potiches* her mind, grubbed? Gunnera... **1973** *Trans. Oriental Ceramics Soc.* XXXVIII. 126 The jar of *potiche* form with broad base, short wide neck and low domed cover.

potichomanist (potiʃō·mănist). *rare.* [f. Potichomania + -ist.] A person who practises potichomania.

1884 *Decorator's Assistant* 122 Potichomanists have found the art capable of greater results than the mere imitation of porcelain vases, by the introduction of glass panels [etc.].

‖ **potin**¹ (potæ̃n). *rare.* [Fr.] A piece of gossip; a rumpus, a row.

1922 M. Arlen *Piracy* ii. vi. 111 He would hear of great dinners and dances and *potins*. **1938** G. Arthur *Not Worth Reading* vi. 86 No shred of evidence could ever be adduced to reinforce the *potin* that Fred Archer was the natural son of a peer. **1948** E. Waugh *Brideshead Revisited* I. vi. 136 'What's going on?' 'Oh, just another boring little *potin*. Sebastian got tight again.'

potlatch. Add: **b.** (Earlier and later examples.) Also, an extravagant giving away or throwing away of possessions to enhance one's prestige or establish one's position. Also *attrib.*, *transf.*, and *fig.*

1865 C. C. Leighton *Jrnl.* 30 Aug. in *Life at Puget Sound* (1884) 25 There was going to be a great *potlach* at the coal-mines, where a large quantity of *iktas* would be given away,—tin pans, guns, blankets, canoes, and money... It seems that anyone who aspires to be a chief must first give a *potlach* to his tribe. **1902** H. L. Wilson *Spenders* xxx. 357 This life of idleness you been leadin'—one continual potlatch the whole time—it wa'n't doin' you a bit of good. **1916** [see *give-away 1]. **1934** R. Benedict *Patterns of Culture* (1935) vi. 202 A variant of this type of potlatch was that which was given upon the adolescence of the woman of highest rank. **1957** *Times* 12 Nov. (Canada Suppl.) p. v/3 Another change was the removal of the prohibition on potlatches and on some traditional religious ceremonies, an ironic measure, as they have now practically disappeared. **1965** H. Kahn *On Escalation* xiii. 270 'Potlatch' wars. Competitions in conspicuous consumption of resources or spectacular successes in such areas as space, economic growth, and 'showy' military systems are employed to gain prestige and influence events. **1969** *Times* 22 Sept. 14/3 Potlatch was an obligation to anyone caught out in a misdemeanour, or who had suffered loss of face through some mishap. Only by a parade of wealth or wild generosity or conspicuous waste could such a man regain his shattered image. **1970** *Globe & Mail* (Toronto) 26 Sept. 29/3 (Advt.), The game has become secondary to a potlatch ceremony called tailgate picknicking... This..requires that the host participants outdo their neighbours in the quality and variety of food and drink and the elegance of serving accessories. **1976** *New Yorker* 22 Mar. 44/3 But in September the potlatch ends and there is no one left but old men and women caring for babies whose parents cannot afford to keep them abroad.

po·tlatch, *v.* [f. the sb.] To give; *spec.* to establish one's name or position by the extravagant giving or throwing away of goods or by holding a feast, which entails some form of reciprocity or return.

1901 *Daily Colonist* (Victoria, B.C.) 6 Oct. 2/2 He had $120 coming to him, and he went to Mr. Landsberg and asked if he would potlatch one half of his wages to him and give the other half to the Duke. **1911** *Ibid.* 30 Apr. 8/3 The forty-one family heads of the Songhees band of Indians..had the pleasure of collecting the last of the moneys potlached to them by the provincial government. **1934** R. Benedict *Patterns of Culture* (1935) vi. 203 Potlatching for an heir on the North West coast. **1943** W. H. Chase *Sourdough Pot* xxiii. 171 The deal was closed, the butter potlatched to her father. **1958** A. R. Radcliffe-Brown *Method in Social Anthropol.* I. v. 123 Amongst the Tlingit..it is members of one moiety who potlatch against members of the other moiety. **1964** Gould & Kolb *Dict. Social Sci.* 523/1 If..a person were humiliated by an accident which made him appear ridiculous, or if he were taken in war and made a slave, he or his relatives must *potlatch* in his name in order to reinstate him in public esteem.

potlatching. (Later examples.)

1964 Gould & Kolb *Dict. Social Sci.* 523/1 Potlatching demanded reciprocity. **1975** H. White *Raincoast Chron.* (1976) 182/1 There was incessant potlatching.

pot-lid. 1. (Later *Pottery* examples.)

1924 Clarke & Wrench *Colour Pictures on Pot Lids* i. 4 The collecting of pot lids and other Staffordshire pottery adorned with these pictures has gone on in a quiet way for some considerable time. **1957** Mankowitz & Haggar *Conc. Encycl. Eng. Pott. & Porc.* 181/1 Polychrome colour-printing on pottery was developed..from 1848..and was used extensively for the decoration of pot-lids. **1972** *Times* 7 July 14/7 These Staffordshire pot lids were a nineteenth-century packaging gimmick for products such as fish paste or cold cream.

po·tlikker, colloq. and dial. (chiefly *U.S.*) var. Pot-liquor; also *fig.*

1930 *Sun* (Baltimore) 17 May 1/3 Pot likker is the best way of preventing nearly all the diseases that we are heir to. **1932** *Ibid.* 8 Apr. 1/4 The Louisianian contended that the corn pone should be 'dunked' into the pot likker. **1950** *Amer. Speech* XXV. 230 *Pot likker*, the juice in which black-eyed peas are cooked. **1963** *New Statesman* 12 July 37/3 During the years when McCarthy's briefcase was a national badge of shame, Americans looked to Britain for an example of sanity and regard for civil liberties. The present anxiety may be unfounded, but some honourable Americans succumbed to the tempting potlikker of the loyalty issue with all its hazy innuendoes. **1974** *Black World* Jan. 53 The boiler at school threatening to blow/flannel gowns and slips Pot likker and cornbread/bundles of kindling for the stove.

pot-liquor. (Earlier and later examples.)

1744 W. Ellis *Mod. Husbandman* July xvii. 101 Mix fine Pollard with fresh Pot-liquor. **1909** *Dialect Notes* III. 359 *Pot-liquor*,..liquor from boiled greens or field peas and fat meat. **1931** *Sun* (Baltimore) 4 Mar. 12/7 Maryland pot liquor is the liquor left over in the pot after a boiling of ham and cabbage. **1949** C. Himes *Black on Black* (1973) 277 They sat in the hot kitchen and ate greens and side meat and rice and..drank the pot-liquor with the corn bread.

po·t-ma:king, *vbl. sb.* The making of pots or pottery.

1767 J. Wedgwood *Let.* 26 July (1965) 58, I..am sorry to find the heat..was much more intense than can be made use of for Pottmaking. **1927** Peake & Fleure *Peasants & Potters* 141 We think that they had picked up the art of agriculture, weaving and pot-making, etc. **1951** E. E. Evans-Pritchard *Social Anthropol.* v. 101 A man who..fails in some enterprise, such as pot-making, through lack of skill is responsible for the penalties or failures his actions incur. **1957** P. Worsley *Trumpet shall Sound* vi. 115 Malasang, the centre of the pot-making industry, ceased production.

potman. Add: **2.** (Further examples.)

1898 G. B. Shaw *You never can Tell* ii. 255 He's at the Bar... A potman, eh?... No, sir: the other bar. **1925** D. Garnett *Sailor's Return* 25 The next morning the potman came to the inn. **1938** Mencken *Amer. Lang.* (ed. 4) 243 Barmaids do the work, with maybe a *barman*, *potman* or *cellarman* to help. **1972** *Classification of Occupations* (Dept. Employment) II. 399/2 Bar potmen (wash glasses in licensed bar).

4. In various manufacturing processes: a man who attends to the filling, emptying, firing, etc., of pots (see quots.).

1874 J. A. Phillips *Elem. Metallurgy* 581 In order to desilverize by the aid of this arrangement, the potman sinks the ladle sideways to the bottom of the kettle. **1921** *Dict. Occup. Terms* (1927) §143 *Pot man* (alkali);..charges shallow iron pans with salt, adds sulphuric acid, attends to firing, and supervises process until product is ready for salt cake furnace. [illegible] *Pot man* (lead); puts pig and scrap lead into pot and tends and feeds fire beneath it which melts lead [etc.]. **1932** *Amer. Speech* VII. 269 *Pot man*, the man who tends the boilers in which steam is generated for drilling.

potoroo. Substitute for def.: A small nocturnal marsupial belonging to the genus *Potorous*, esp. *P. tridactylus*, found in areas of dense vegetation in Australia; = Kangaroo-rat 1, *rat-kangaroo* s.v. Rat *sb.*¹ 7 e. (Later examples.)

1907 P. Fountain *Rambles of Austral. Naturalist* iv. 38 There are several local varieties of the potoroo. **1924** F. W. Jones *Mammals S. Austral.* 218 The remaining *potoroos* should be carefully protected. **1943** C. Barrett *Austral. Animal Bk.* 111 100 The dark rat-kangaroo, known to the natives of Port Jackson as the potoroo, is a rare little animal. **1965** *Courier-Mail* (Brisbane) 24 Aug. 13 The potoroo is one of the rarest native animals. **1970** W. D. L. Ride *Guide Native Mammals Austral.* 64 Potoroos apparently require the dense natural vegetation of their habitat for protection. **1975** *Nature* 8 May 141/2 The bones we exhibited included forms similar to those of living potoroos and phalangers.

pot-oven. Add: **2.** A kiln in which pottery is fired.

1878 L. Jewitt *Ceramic Art Gt. Brit.* xiii. 528 In his will, dated 30th of January, 1728..this Caleb Glover 'of Lane-end, potmaker', leaves all his 'working tools belonging to the trade of a potmaker, and the 'pot oven'. **1971** F. C. D. Breaks *Eng. Country Pott.* 225 In 1798, John Morton & Sons assessed for their pot-ovens at Sinderhills.

pot-pie. (Later examples.)

1902 *Amer. Illustr. Mag.* Feb. 465/1, I was out huntin' for squirrels to make a potpie out o', for squirrel potpie is just lickin' good. **1933** M. de la Roche *Master of Jalna* ii. 16 From there came the appetizing smell of chicken pot-pie. **1935** A. J. Cronin *Stars look Down* iii. ix. 564 I know you've had no breakfast. Are you stove eating a pit-pot-pie? **1940** Mencken *Happy Days* iv. 59 The Bennett [Hotel] also offered an oyster pot-pie that had its points. **1975** *New Yorker* 20 Oct. 31/3 'He just bein' chicken', says his father. 'Some of them non-needy chirrun might sneak in there and grab some of that chicken potpie.'

pot-plant. 1. (Earlier and later examples.)

1816 'A. Singleton' *Let.* (1824) 63 The young Virginian ladies take pleasure in nurturing..and fostering their parlour pot-plants. **1870** W. Robinson *Alpine Flowers* i. 56 Our pot-plants are far too often the worst of their sort. **1871** S. Hibberd *Amateur's Flower Garden* xiii. 240 Well-grown pot-plants..have a much higher effect, a much more artistic and finished appearance, than plants of the same kind equally well grown in open ground. **1914** W. F. Rowles *Garden under Glass* xii. 221 The loam..will be more suitable for the growth of the majority of pot plants. **1966** Rochford & Gorer *Redward Bk. Flowering Pot Plants* i. 11 Flowering pot plants are usually regarded as expendable.

pot-pourri. Add: **2. a.** (Further examples.) Also *fig.*

1855 S. Whiting *Heliondé* 71 A couch, which..was made of pot-pourri, or some mixture of dried flowers peculiar to the Sun. **1921** D. H. Lawrence *White Peacock* ii. iii. 244 The squire's lady has written in a little vase some of its potpourri. [illegible] **1938** W. S. Maugham *Cakes & Ale* iv. 51 On little Chippendale tables stood large Oriental bowls filled with potpourri. **1960** E. Hemphill *Fragrance & Flavour* 79 If the pot-pourri becomes too dry add more salt, and if too moist add more orris root powder.

b. *transf.* A container designed to hold pot-pourri.

1770 J. Wedgwood *Let.* 20 Aug. (1965) 94 Or have an Auction of Statues,..Lamps, Potpourris [etc.]. **1960** R. G. Haggar *Conc. Encycl. Continental Pott. & Porc.* 383/1 Faïence in the Swedish-Dutch style was made, including 'fruit-platters',..and ornamental pieces (pot-pourris, tureens, etc.). **1971** L. A. Boger *Dict. World Pott. & Porc.* 268 *Potpourri*..In ceramics the name is given to a container designed with a delicate perforated cover used to hold potpourri, a mixture of dried petals. **1975** *Times* 19 Dec. 16/3 Among Minton's copies of Sèvres porcelain, a pair of *bleu-de-roi* ground pot-pourris and covers made £1,375.

3. a. (Earlier and further examples.) **b.** (Further examples.) **c.** Any diverse collection or assortment (of people or things).

1855 *Illustr. London News* 29 Dec. 755/1 The overture to the pantomime consisted of a pot-pourri of those popular tunes which the professors of the street-organ have been insisting on during the past year. **1866** G. H. Lewes *Let.* 1 July in *Geo. Eliot Lett.* (1956) IV. 282 There drink the sparkling water and lounge in the sun listening to the tolerable band performing overtures..pot-pourris and waltzes. **1874** G. B. Shaw in *Fortnightly Rev.* Feb. 256 My first Richard III..turned out to be a wild potpourri of [illegible]. **1921** H. Crane *Let.* 25 Dec. (1965) 73 We have a houseful of indiscriminate relatives and it has been hard to collect myself for even this postcard. **1940** *N.Y. Fed. Writers' Proj.* [illegible] printed it is as a *pot-pourri* of comment and opinion on various life and manners. **1973** *World Archaeol.* III. 193 All the work, except firing, takes place..amid the characteristic pot-pourri of tools, vessels, mats and clutter typical in such household work areas. **1977** *Time* 28 Feb. 7/1 The election of December 1973 split the Folketing among a potpourri of ten parties. **1978** P. Mclean *Take Nine Spies* ii. 49 A *pot-pourri* from Franz Lehar's latest opera.

4. *pot-pourri jar* (earlier example), *vase.*

1889 *Daily Tel.* 8 Mar. 3/3 There were bowls, so-called, the 'Two Roses' come as fresh as a bright June morning,..but some there are who,..will find instead merely a 'potpourri' jar full of dried rose leaves. **1957** E. Meteyard *Wedgwood Handbk.* 108 Pot-Pourri Vases. Decorated vessels for containing rose leaves and other scents were formerly much used. **1957** Mankowitz & Haggar *Encycl. Eng. Pott. & Porc.* 180/2 *Pot-pourri vases*, vessels with lids and pierced shoulders to allow the perfume of the mixture of dried rose-petals,..to escape. **1974** Savage & Newman *Illustr. Dict. Ceramics* 230/2 *Pot-pourri vase*, a vase, sometimes richly mounted in gilt-bronze, which is characterized by pierced decoration on the shoulder or cover, or both.

potrero (potrē·ro). [Sp., f. Potro.] **1.** In the S.W. United States and South America, a paddock or pasture for horses or cattle.

1848 J. A. Sutter *New Helvetia Diary* (MS.) 7 Mar., 16 men have been sent from the Race, on usual, having no tools, employed them to get the small potrero repaired. **1864** *Notes & Queries* [illegible]..where the cattle..are let out, they come to some little *potrero* where pine-clad hills inclose a soft green meadow, [etc.]. **1892** *Daniel Nolan* I. 299 *Potrero*, a pasture, generally for colts and young horses. Also a piece of land easily fenced in, situated in the bend of a stream or in a valley with a narrow pass for entrance. **1943** C. F. Saunders *Southern Sierras Calif.* iii. 105 A broad, open field or pasture known as a *potrero*. **1970** T. Lilley *Hist. & Gen.* [illegible] *Potrero*—a verdant, wild-flowery bowl rimmed around with mountains. **1931** *Times Lit. Suppl.* 19 Mar. 214/2 He was found to be capable of milking exceptionally large numbers of untamed cows, which were driven straight from a 'potrero' into a 'corral'. **1953** A. F. Tschiffely *Tschiffely's Ride* (1934) 5 Large herds of cattle grazing in the 'potreros' (paddocks). **1950** H. Backhouse *Among Gauchos* xiii. 117 It is customary to herd with the cattle in the *cuarrera potrero* or paddock reserved for cattle that are to be slaughtered. **1965** G. J. Butland *Lat. Amer.* xx. 287 This is the Uruguayan pastoral region *par excellence*; a land of cattle and sheep *estancias* with *potreros* or paddocks fenced and managed to secure the best seasonal use of the grass.

2. In the S.W. United States, a narrow steep-sided plateau or mesa.

1879 Lt. G. Bourke *Diary* (MS.) 3 Dec., Hills break away in potreros. **1886** A. F. Bandelier *Southwestern Jrnls.* (1966) ii. 135 There are three mesas..one, to the Potrero de las Vacas, about one day's journey northwest. **1946** —— in *Papers Archaeol. Inst. Amer.* IV. 158 These cliffs appear like pillars or gigantic posts; hence their Spanish name 'Potrero'. The one forming the southern wall of the Cuesta Colorado gulch of the San Julian is called Potrero Chato. **1940** E. Ferguson *Our Southwest* xiii. 264 A potrero is a narrow ridge between canyons, and a saddle is a sag between peaks. **1953** B. P. Dutton *Hewett's Pajarito Plateau* (ed. 2) iii. 104 The potreros are like narrow potrero bounding the canyon on the north is carried out for one a distance of nearly a mile, thus throwing into one separate open park the width of two small canyons and the formerly intervening mesa. **1966** in A. F. Bandelier *Southwestern Jrnls.* Gloss. 397 *Potrero*, pastureland; in New Mexico, a tongue of high ground.

pot roast. orig. *U.S.* [f. Pot *sb.*¹ + Roast *sb.*] A piece of beef or other meat cooked slowly in a closed container. Also *attrib.*

1881 F. E. Owens *Cook Bk.* 59 Pot Roast of Beef. Get a solid piece from the round [etc.]. **1897** *Librarian Cook Bk.* 48 (*heading*) Mutton Pot Roast. **1920** [see *hamburger 1]. **1929** E. Wilson *I thought of Daisy* iv. 232 My dinner was an admirable pot-roast, with onions, potatoes, and carrots. **1936** L. C. Douglas *White Banners* ii. 59 'We've had breast of lamb three times in the past two weeks.' 'Sorry,' said Hannah. 'I'll have a pot roast to-morrow.' **1955** Paradine & Hawkes *Journey down Rainbow* 23 They are taking your order for Navy Bean soup and pot roast. **1966** *House & Garden* Nov. 100/3 Normandy pot roast of beef, bacon... 1969 pot roast chicken [illegible]. **1973** G. Sims *Hunters Point* vi. 49 Can you stay for lunch?..Only a pot-roast but..I make it fairly tasty. **1976** *National Observer* (U.S.) 9 Oct. 14/3 The meat of pot roast coming to a climax would drive me wild. *Ibid.*, I'm starving for a pot roast sandwich.

Hence **pot-roa·st** *v. trans.*, to cook (meat) in this way; **pot-roasted** *ppl. a.*; **pot-roasting** *vbl. sb.*

1917 M. Green *Better Meals* ii. 15 *Pot roasting* is cooking in an iron kettle or earthen pot in a small amount of water or steam in a closed vessel... Meat..cooks slowly until very tender, with or without vegetables. **1945** *ABC of Cookery* (Ministry of Food) ix. 35 Pot roasting is..cooking meat in a little fat in a saucepan with its lid on. **1951** N. M. Gunn *Well at World's End* xii. 124 A cold roast of pot roasted beef. **1964** *Good Housek. Cookery Book* (rev. ed.) 134/2 Pot-roasted ham, we then pot-roast, boil, stew, braise, or pot-roast the meat or bird. *Ibid.*, Use the rack when pot-roasting or boiling. **1964** *House & Garden* Nov. 100/1 If he intended to pot roast, he left the lid empty. **1972** *Guardian* 9 Aug. 9/6 Pot roasting, as everyone knows, keeps juices and flavour in the meat..reducing the risk of oven sides. **1979** *Country Life* 13 Sept. 806/4 Chickens and pheasants were..pot-roasted.

po·t-shoot, *v.* *U.S.* [On analogy with Pot-shot *sb.*¹] *trans.* and *intr.* To take a pot-shot at (someone); to take pot-shots. Also *fig.* So **pot-shooter**; **pot-shooting** *vbl. sb.*

1907 J. London *Road* 73 [illegible]. **1913** *Outing* [illegible] pot-shooting..the deer..the fox..and the bears like ducks.. **1921** *Outing* [illegible]. **1926** *Outing* [illegible] hundred paces make fine pot-shooting. **1928** *Outing* [illegible]. **1931** *Newsweek* [illegible]. **1932** *Coming of Cassidy* [illegible] 'I've got to come back there is this place with a man who would pot-shoot. **1925** *Sun* (Baltimore) [illegible] ..Indian pot-shooters, who are..among the tribes. **1938** *Sun* (Baltimore) 2 July 4/1 The pot-shooters..and those..political pot-shooters. **1945** *Ibid.* 20 Nov. 1/5 What the campaign will pot-shoot at [illegible]. **1966** *English Jrnl.* Dec. 1309 The young men who have enjoyed pot-shooting policemen for the past few summers.

pot-shot, *sb.*¹ Add: **1.** (Further *transf.* examples.) Also, a random blow or punch.

1907 C. E. Mulford *Bar-20* iii. 31 A pot shot at Hopalong sent that gentleman's rifle barking. **1938** *Sun* (Baltimore) 28 Jan. 1/5 He pleaded guilty to firing two 'pot shots' into the bed of a loaded school bus. **1950** J. Dempsey *Championship Fighting* ii. 23 Gunning for a knockout. **1973** A. C. Kay *Siege of Alcatraz* (1966) iii. 70 A drunken *muchacho*..took a pot shot at him. **1979** A. Blaisdell *No Villain need be* iii. 50 Some nut..taking potshots at couples in cars.

b. *fig.* **a.** *U.S.* A piece of criticism or verbal attack, freq. one which is random or opportunistic. **b.** A random attempt.

1926 *Forum* (N.Y.) Nov. 757 But I don't think much of the pot-shot method of refutation. **1927** *Christian Science Monitor* 7 July 8/8 Let him take lusty potshots, though, at some of the progressive ghost of bygone years, and he is hailed as brilliant, erudite, and—extraordinary—daring! **1940** Berrey & Van den Bark *Amer. Thes. Slang* §179/6 Potshot, shot in the dark, a *wild guess*. **1953** *Commonweal* 30 Apr. 98 It is not just that Mr. Wilkie is taking pot-shots at the British Empire—and please let's not swell up with anti-British self-righteousness. **1960** *Keesom* [illegible] XII. 158/1 At a time when the intellect was at the tips of the senses' and so on. Admittedly these are literary phrases, therefore a kind of pot shot at the real and genuine. **1965** *New Statesman* 7 Mar. 71 The tobacco industry is leaving no radioisotope unturned to counter medical pot-shots at its product and to bolster sales. **1968** *Times* 20 June 3/4 Thus certain canvases are less pot-shots at *profundis* and the pot-shots at pictures. **1971** A. & A. Silverman *Cats against having Children* vi. 189 All we could know there is to take potshots at the Myth of the Working Mother.

Hence **pot-shot** *v.* *trans.* and *intr.*, to take a pot-shot at (a target), to take pot-shots; **pot-shooter**; **pot-shotting** *ppl. a.* and *vbl. sb.*

1904 P. Fountain *Gt. North-west* iv. 27 The breech-loader is the weapon of the dandy pot-shooter. **1918** E. Pound *Let.* 16 Dec. (1971) 143 And what the deuce of your punctuation!—You must deliberate, and therefore..to be taken (by me) with studious meticulousness?? I cant go on potshooting at it..you must pot-shot. **1925** Kipling *Irish Guards in Gt. War* II. 60 Snipers were pot-shotting from the other side. **1927** *Amer. Speech* III. 19 *Right hand pot-shooters*. **1938** Berrey & Van den Bark *Amer. Thes. Slang* §394/3 *Careless person*, ..potshooter. **1945** J. Wolfert *Battle for Solomons* v. 65 However, it is likely that the battle which started as an old-fashioned 'better 'ole' kind of war may have even more old-fashioned..and become a ruthless, tracking, potshotting, Indian kind of war. **1947** Harrow & Black *Pract. Public Relations* 179 Competing companies will make it their business to 'pot-shot' both purchasing and supplying organizations. **1950** *Atlantic Monthly* Apr. 21 What is really disturbing is the constant potshotting of the administrative departments from Capitol Hill. **1954** W. Cowley *Lit. Situation* x. 178 At last he is likely to decide that the expenses are beyond his powers of computation; he'll simply pot-shot at them, hoping that his guess won't be too far wrong. **1958** *Times* 18 June 36/3 After six years of marriage and nearly five of potshotting between their armed camps, they braced for the showdown. **1966** H. Waugh *Pure Poison* vi. 41 She goes for the lunatic theory—the lunatic pot-shotting her. **1970** *National Rev.* (U.S.) 30 June 685/2 Like Tocqueville, Toynbee potshots, and hits most, the trivial and mundane, filling space with more profound insights.

potstick. Add: (Later examples.) Also, a stick used for moving washing about in a pot.

1894 S. R. Crockett *Lilac Sunbonnet* xxvii. 225 She turns roon' wi' the pat-stick i' her hand. **1903** Somerville & Ross *All on Irish Shore* 9 He..had, in addition, filled the meal for the hounds with a knowledge of proportion and an untiring devotion to the use of the potstick which produced 'stirabout' of a smoothness and excellence that Miss Barnet herself might have been proud of. **1922** Joyce *Ulysses* 223 Maggy at the range rammed down a greyish mass beneath bubbling suds twice with her potstick and wiped her brow. **1961** F. G. Cassidy *Jamaica Talk* x. 186 To stir cooking food one uses a pot-stick.

potsy (pǫ·tsi). [Etym. obscure, but cf. *pot *sb.*¹ 11.] **1. a.** The object thrown in the game of the same name. **b.** The name of a children's game similar to hopscotch.

1931 *Recreation* Mar. 672/2 *Potsy* is an adaptation of Hop Scotch, which now rivals its progenitor in popularity in the city. The 'potsy' is a piece of tin, a rock or a puck. **1932** *Sun* (N.Y.) 26 Mar. 18/3 As any New Yorker will recognize, the potsy refers to the piece of tin can, doubled and redoubled and stamped flat with the heel, which is kicked from flagstone to flagstone of the sidewalk by the hopping, juvenile player of the game potsy. **1943** B. Smith *Tree grows in Brooklyn* xiii. 100 Potsy was a game that the boys started and the girls finished. A couple of boys would put a tin can on the car track and let a car run over it. **1948** *New Yorker* [illegible] ..the potsy..the potsy..with boxes marked off on the sidewalk and the game was turned over to the girls who hopped on one foot pushing the potsy from square to square. **1955** F. M. Evans *Hopscotch* 5 Then you feel around and pick up your potsie without stepping on your potsie. **1965** E. Fay *Ghosts Lit. Paul* 61 [illegible] a kid..potsy..in the playground..to watch the kids play potsy and skipping. **1967** T. Pynchon *V.* x. 177 'What are you guys doing,' Profane said, 'playing potsy?'

2. The badge worn by a policeman or fireman.

1929 *Sun* (N.Y.) 26 Mar. 18/3, I recently bought a Sun which gave a vocabulary of firemen's slang, in which 'potsy' was the word for 'badge'. I have also found policemen and detectives who referred to their badges in this manner. **1936** *Baltimore News-Post* 18 Apr. 18/1 Most fire-fighters in town never polish their gold badges (yellow potsys). **1940** *Amer. Speech* [illegible] 'potsy'. **1953** *Times* [illegible]. **1955** *New Yorker* [illegible]. **1962** *Ibid.* 24 Jan. 23/1 This boullace has been wearing his potsy as house dick for years. **1970** L. Sanders *Anderson Tapes* xiii. 39 Ernie goes in the lobby and flashes his potsy.

Pott (pot). *Med.* The name of Sir Percivall Pott (1713–88), English physician, used in the possessive to designate phenomena described by him, as **Pott's disease**, a form of paraplegia caused by tuberculous disease of the spine (described in *Med. & Philos. Commentaries* (1779) VI. 316–24); **Pott's fracture**, a fracture of the fibula close to the ankle, of a type described by Pott (in *Remarks on Fractures & Dislocations* (1769) 57–64) and due to eversion of the foot; *loosely*, any fracture of the lower fibula.

1835 *Lancet* 12 Sept. 775/1 Patients attacked by Pott's disease frequently..are attacked..of other affections before the former has made much progress. **1897** T. L. Stedman *Twentieth Cent. Pract. Med.* X. 513 Pott's disease of the spine is also of tuberculous nature..affects the bodies of the vertebræ, to the affected nerve roots of the spinal cord and to the spinal cord itself. **1976** C. B. T. Adams in R. Ross *Hist. et al. Surgery* xiv. 186 Tuberculosis (Pott's disease..of the spine) or a pyogenic extradural abscess may also cause cord compression. **1849** *Proc. Path. Soc.* Dublin II. 13 Mr Robert W. Smith exhibited a remarkable specimen of Pott's fracture, taken from the body of a man, aet. 70, who had lately died in the hospital of the South Union Poor House. **1884** W. Pye *Surg. Handicraft* xiv. 209 It is generally produced by..a 'Pott's fracture' of the fibula, attended by a subsequently produced 'partial dislocation' of the ankle joint internally. **1924** *Arch. Surg.* IV. 56 Pott's fracture, as generally understood, is a fracture of the fibula, attended by a subsequently produced 'partial dislocation' of the ankle joint internally. **1933** J. G. Bonnin *Injuries to Ankle* viii. 181 I believe that the term 'Pott's fracture' should be applied as a generic term to any and all fractures of the ankle due to external rotation violence. [illegible]

potted, *ppl. a.* Add: **1.** (Further examples.)

1922 Joyce *Ulysses* 368 Potted herrings gone stale. **1922** J. Buchan *Huntingtower* iv. 69 There was new milk..and most of the dainties which had appeared at tea, supplemented by a noble dish of shimmering 'potted-head'. **1954** *Special Sch. Rep.: Fisheries* (U.S. Dept. Interior: Fish & Wildlife Service) No. 104. 31 Potted tuna compares favorably with..cooked with potatoes and carrots. **1960** E. David *French Provincial Cooking* 284 As an alternative to a home-made pâté, *rillettes*, which might be described as a kind of potted pork, are quite easy to make at home. **1977** *Observer* 12 June 19/4 'I'm so excited,' she said; her sentences came in potted phrases.

2. (Later examples.)

1939 A. H. Wood *Grow them Indoors* p. xi, In Carpaccio's..painting of St. Ursula's vision, two potted plants appear in the window of her room. **1976** *Times* 12 Apr. 11/4 The gorgeous store..collapsed amid a welter of potted palms and recriminatory statements.

3. *fig.* **a.** Of a piece of information, work of literature, or historical or descriptive account: put into a short and easily assimilable form; condensed, summarized, abridged. Also *transf.*

1883 [in Dict., sense 1]. **1901** F. Gardner *Pure Folly* I. 1 Pélissier..in April, 1907, produced his first 'potted play'... It had been described as 'Raffles' a Prize Pan-tomime.' Needless to say, the skit was a blend of the two plays 'Raffles' and 'Peter Pan'. **1931** *Times* 25 Feb. 8/1 Some of the [illegible] which he described as 'potted plays'. *Cf.* **1959** J. Braine [illegible] 'potted' versions of [illegible]. **1939** *Cheese Wrapped* [illegible] the potted..on potted in 'a potted history'. **1939** *Morning Post* 2 Oct. 10/7 Previously such questions had merely served as an excuse for potted lectures on the iniquity of the British Empire. **1937** 'A. Bridge' *Enchanter's Nightshade* 32 Those little potted abstracts for the general reader. **1946** *R.A.F. Jrnl.* May 146 A potted history of the *Journal* from its infancy up to this final issue. **1957** *Listener* 24 Oct. 641/1 Even potted biographies are seen more in children's books. **1968** *Ibid.* 23 June 321/3 Photographs of all the county teams, a list of records, potted careers of most of the current players, [etc.]. **1975** *Physics Bull.* May 197/3 The first chapter..attempts to provide a very potted treatment of energy physics.

4. = *canned *ppl. a.* b.

1928 *Melody Maker* 23 [illegible] The delightful art of piano-playing..is in immediate danger of being usurped by the 'potted' music of wireless and gramophone. **1928** T. E. Lawrence *Let.* 23 Apr. (1938) 593 Only gramophone music, but the potted stuff is very well, for people away abroad. **1935** P. MacLean *Eastern Approaches* I. ii. 29 It was then that I grasped that the cheering was potted, synthetic cheering, issuing from loudspeakers..and conveniently obviating the need for unhygienic, insecure spectators.

5. Of pottery or porcelain, with defining adv.: (well, beautifully, etc.) fashioned or manufactured.

1902 *Encycl. Brit.* XXXI. 874/2 The ware is thin, light, beautifully potted, and of the utmost durability. **1969** *Canad. Antiques Collector* Mar. 9/3 The fifteen wares of the Koryo period are thinly potted and covered with the well known celadon glaze. **1972** *Country Life* 30 Mar. 788/1 The flat, elaborately pierced, thinly potted bowl..only 3 in. high, has been..simplified.

5. a. *N. Amer. slang.* Drunk, intoxicated.

1924 P. Marks *Plastic Age* xiv. 149 I'd a [been] potted about half the time. [*Ibid.* xxvi. 302, I don't get potted regularly. **1930** *Collier's Humor* Aug. 125/2 Did I ever tell you..to go to getting potted like you were last night?] **1947** *Sun* (Baltimore) 14 Aug. 6/4 Awful calamity at the Park bird bath..when somebody discovered the birds were being mad to some members of the Mint Julep Association having emptied their julep glasses in the fountain. **1949** *Amer. Speech* XXXIV. 195 Gators never merely drink; instead, they sop. They may later be.. potted. **1974** J. Dowell *Look-off Bear* 90 He was potted, plastered, stinko.

b. *U.S. slang.* Under the influence of marijuana (cf. *pot *sb.*⁸).

1960 Wentworth & Flexner *Dict. Amer. Slang* 404/1 *Potted*, under the influence of a marijuana cigarette. **1968** Busby & Holtham *Main Line Kill* v. 59 The Jamaicans..didn't appear to be potted. **1968–70** *Current Slang* (Univ. S. Dakota) III/IV. 95 *Potted*, v. High on marijuana... I was potted out of my mind yesterday. **1973** *Publ. Amer. Dial. Soc.* xli. 54 Potted.

6. Of an electrical component or circuit: encapsulated in an insulating material (cf. *pot *v.*¹ 2 e).

1947 *Plastics* July 71/1 Several practical applications of resin-potted circuits at the Bureau have given information comparable to that of conventionally constructed devices. **1953** Wh. Ir. Styles *High-Speed Computing Devices* xvi. 427 Mass production of potted plug-in units. **1955** *Brit. Plastics* XXVIII. 418/1 Extreme care must be exercised in the removal of even traces of atmospheric moisture from the surface of the potted component. **1967** *Electronics* 6 Mar. 193/1 (Advt.), Special Assemblies.. Rectifier stacks, potted bridges, [etc.].

potter, *sb.*¹ Add: **2.** Also, in northern England, a vagrant, a kind of tramp or gypsy.

1903 *Cf.* Rev. CXXII. 378 The 'potters'..a kind of indigenous gipsies, often curiously bearing the names of the great Northern families. **1889** *Specimens Westmoreland Dial.* 36 A com at a potter tent in' great camp. **1953** *Sat. Cumberland Times* 28 Jan. 5/2 He was described as a 'potter'. He had been potters camping on the main road. You mean tramps or gipsies?—Yes, something of that kind. **1972** *Times Lit. Suppl.* 3 Mar. 245/3 The travellers and vagrants—gipsies, 'potters', pedlars, beggars, Irish labourers looking for work—who tramped the roads of Lakeland.

3. *potter's* (or *potters'*) *rot*, silicosis or other lung disease caused by the continued inhalation of dust in a pottery.

1902 T. Oliver *Dis. of Occupation* x. 307 As far back as the days of Ramazzini (1690) the lung troubles of the potters had been recognised and described. With the terms 'potters' rot' and 'potters' asthma' the public are quite familiar. **1966** Wright & Symmers *Systemic Path.* II. 1. 405/1 Soon after exposure to the dust..in the manufacture of porcelain, at about the middle of the 18th century, 'potter's rot' made its appearance and as silicosis in the potteries is..the most sinister of the hazards. **1975** *Evening Sentinel* [illegible] *Stoke-on-Trent* [illegible] ..silicosis, known in Stoke as 'potter's rot', was a scourge in Dudley.

potter, *sb.*³ (in Dict. s.v. Potter *v.*). Add: (Further examples.) Also, a gentle stroll or saunter. Also *fig.*

1901 'L. Malet' *Hist. R. Calmady* III. v. 210 But Camp, who missing Richard, had followed his mistress out of the house for a leisurely morning potter, came back slowly. **1949** E. Bowen *Heat of Day* xiv. 248 A potter through the boundary woods. **1954** M. Allingham *Beckoning Lady* v. 84 The prospect of a glorious potter about was alive in her. **1966** C. Noyes *School of Liars* v. 113 'You and Joe will have to have a little potter'. **1968** *Sunday Express* [illegible]. **1972** *Ibid.* [illegible]. **1976** D. Bell *Virginia Woolf* II. ii. 33 Leslie's favourite exercise..[illegible] potter.

potter, *v.* Add: **5. a.** (Further examples.)

1860 G. H. Lewes *Jrnl.* 26 Sept. in *Geo. Eliot Lett.* (1954) III. 349 To-day..I wrote some letters and pottered. **1922** Joyce *Ulysses* 737 He prefers pottering about the house. **1947** W. S. Maugham *Creatures of Circumstance* 94 Then he would turn his business over to his son and retire with his wife to a little house in the country where he could potter about till death claimed him at a ripe old age. **1979** *Times* 3 Oct. 13/3 Randall..pottered about for 26 in 1½ hours before giving a catch to..

b. (Further examples.)

1867 T. Hughes *Tom Brown's School Days* I. ii. 32 Past the old church and down the footpath, pottered the old man and the child hand-in-hand. **1903** G. B. Shaw *Man & Superman* iv. 161 Mrs Whitefield, who has been pottering round the Granada shops, and has a net full of little parcels in her hand, comes in through the gate and sees him. **1928** Galsworthy *Five Tales* 271 He..pottered in and out of his dressing-room. **1938** E. Waugh *Black Mischief* vii. 257 The Envoy Extraordinary finished his second cup of coffee, filled and lit his pipe, and avoiding the social life of the lawn, pottered round by the back way to the Chancery.

pottery. Add: **4.** *pottery clay* = *pot clay* s.v. Pot *sb.*¹ 14 in Dict. and Suppl., *potter's clay* s.v. Potter *sb.*¹ 3.

1909 *Geol. Surv. Mem. S. Staffs. Coalfields* (Mem. Geol. Survey) [illegible] 'Potter's clay'. **1954** W. Harris [illegible] **1961** *S.W. Coalfield* (Mem. Geol. Survey) XIII. iii. 160 The isolated Bovey Tracey Beds of the western part..of the county..of pottery clays..are supposed to have been laid down in a small, local lake-basin. **1964** W. Stegner *Wolf Willow* iv. ii. 250 Floods of settlers..all figured in the dreams, and so did potteries with the deposits of pottery clay.

pottiness (pǫ·tinės). [f. *potty *a.* + -ness.] The state or condition of being potty.

1933 Wodehouse *Heavy Weather* iii. 47 It was not primarily his pottiness that led him to steal the Empress. **1935** D. L. Sayers *Gaudy Night* v. 98 We shall all feel perfectly splendid if you really think the pottiness of our conversation doesn't sound a little potty. It's the pottiness, you know, that's so awful. **1957** E. Hyams *Into Dreams* [illegible] We..got an impression of pottiness from the old politician's remarks. **1974** *Times* 19 Jan. 7/8 The dementia of children who enter..[illegible] **1976** *Times* 17 [illegible] ..pottiness..sees..any flaws in oneself is a branch of pottiness.

potting, *vbl. sb.*¹ Add: **2.** (Further examples.)

1979 [see *Ming b]. **1976** P. Flower *Crisscross* i. 6 How Sibyl loved it all: visiting nurseries, potting and [illegible].

3. c. *Woollen Manufacture.* (See quots.)

1904 J. M. Matthews *Application of Dyestuffs* i. 66 An operation very similar to that of decatizing is known as potting, which consists in treating the cloth with steam and hot water for the purpose of producing a particular character of finish. **1937** Horsfall & Lawrie *Dyeing of Textile Fibres* iv. 175 Potting. This process is likened to the old 'crabbing' or 'roll boiling'... The cloth is wound..on to a perforated roll which..is placed upright in a cistern in cold water. **1951** *Rev. Textile Progress* III. 205 This is..known as the potting..process..or potting, *i.e.*, winding the cloth on a roller and heating in water or dye liquor, is still used at temperatures of up to four days. **1961** Blackshaw & Brightman *Dict. Dyeing* 137 *Potting*, a finishing process for wool cloths in which a roll of fabric is treated in water at 60°C for several hours, then allowed to cool slowly and finally immersed in cold water to set the fabric.

d. The act or process of abridging, condensing, or summarizing.

1909 *Daily Chron.* 20 Oct. 1/6 Drury Lane Dignity is down on Apollo Impudence for using the title of 'The Whip' in the 'potting' department of 'the Follies', and the seriousness of the whole business is expressed in Mr. Pelissier's letter to the British public: 'Hurry up! We have had a letter from Messrs. Hamilton and Raleigh's solicitor objecting to our burlesque of "The Whip", so we may be locked up at any moment.' **1966** *Punch* 5 Oct. 523/3 The enormous subject is covered by rapid potting.. The potting is efficient: the course of the 1914–18 War on the Western Front..is explained without oversimplification.

5. b. *dial.* The catching of lobsters in pots.

1971 *Nat. Geographic* Apr. 556/1 We were out potting. Potting? Lobstering, bringing in the pots.

6. b. *Billiards.* The act or process of knocking a ball into one of the pockets.

1909 in *Cent. Dict. Suppl.* **1934** *Encycl. Sports* 571/1 The practical application of this knowledge will make potting easy for many who find it exceedingly difficult at present. **1950** *Hoyle's Games Modernized* (ed. 20) iii. 340 The more successful [pool] player is not necessarily he who can 'pot' with the deadliest accuracy, but he who can bring positions with effectively playing for position. **1958** B. [illegible] Second Sport *to* in Sn. Gilroy put an exquisite edge on his potting, though he did have a tendency to overcut in spot on the..

6*. Encapsulation of a circuit or component in an insulating material (cf. *pot *v.*¹ 2 c). Freq. *attrib.*

1947 *Plastics* July 57/1 During the war, exacting mechanical and electrical stability requirements led to the 'potting' of circuit components. **1959** *Rep. Progress Appl. Chem.* XL. 213 A typical potting resin comprises a mixture of 2:4-tolylene diisocyanate (20–50%), monomeric styrene (up to 25%), a polyester and..a trace of polymerization catalyst. **1962** A. C. Volk et al. *Electr. Encapsulation* i. 5 Potting is the simplest every-day process. Procedure consists of positioning the component in its container ('pot'), adding encapsulant to fill the pot, then curing to polymerise or harden the encapsulant.

6**. The action or an instance of causing a baby or young child to sit on a chamber-pot.
1948 *Practitioner* Dec. 505 It is as well that the infant should suffer the experience of 'potting' quite early, and should come to appreciate that the environment reacts less encouragingly to random disposal of excreta than to its opposite. 1953 R. S. Illingworth *Normal Child* xxvii. 283 Provided that there is never a fight to keep the child on the chamber and the child does not resist, 'potting' is a harmless procedure. 1958 *Observer* 20 Apr. 8/4 Much else that controverts the views of Sir Truby King, and his 'Mothercraft' disciples, on early potting, bodily guilt, schedule feeding. 1960 *Guardian* 1 Apr. 7/1, I attended to the telephone, 'potting' of small child, and comforting child after fall. 1967 *Ibid.* 1 May 4/4 The job extends to getting the children up... No doubt the next 10 years will see minor changes—no more potting, for example. 1972 J. Gathorne-Hardy *Rise & Fall of Brit. Nanny* viii. 265 Nearly incessant potting..produces a high..quota of wins.
7. *potting bench, business, compost, -shed* (examples), *soil*; **potting-pot** (earlier example).
1874 *Gardeners' Chron.* 17 Jan. 95/3 A movable wooden tray, shaped like the top of a potting bench..will answer the purpose. 1935 A. G. L. Hellyer *Pract. Gardening* xxx. 185 (*caption*) The principal ingredients may well be stored in bins under the potting beach. 1766 J. Wedgwood *Let.* 15 Sept. (1965) 42 As our connections are to become extensive in the Potting business, it is absolutely necessary you should visit the Manufacture. 1916 M. Hampden *Flower Culture* ii. 38 Silver sand and old manure, chopped fine, may make up one part of the ordinary potting compost. 1971 P. D. James *Shroud for Nightingale* vi. 183 There were balls of green twine,..packets of seed,..a small plastic bag of potting compost and one of fertilizer. 1976 J. Berrisford *Backyards & Tiny Gardens* xv. 107 Bedding plants do well also in John Innes potting compost No. 2. 1739 E. Smith *Compl. Housewife* (ed. 9) 52 *Machrel to pot*..place them close in your potting-pots, and pour clarified butter on the top. 1874 *Gardeners' Chron.* 17 Jan. 95/3 The manure and compost yard should include a potting shed. 1902 W. B. Yeats *Where there is Nothing* (1903) I. 14 Come over here to the potting shed. 1907 E. Gosse *Father & Son* vii. 181 My Father would..bolt..round the garden into the potting-shed. 1940 J. Betjeman *Old Lights for New Chancels* 48 Back down the Avenue, back to the pottingshed. 1976 *Derbyshire Times* (Peak ed.) 3 Sept. 18/1 (Advt.), Detached garage, well-established gardens to front and rear, potting shed. 1908 *Daily Chron.* 29 Feb. 9/1 This material [*sc.* manure from mushroom growing] is in excellent condition for mixing with potting soil as a fertiliser. 1936 *Forestry* X. 12 The potting soil used was a standard mixture.

potty (pǫ·ti), *a. colloq.* [f. Pot *sb.*¹ + -y¹.]
1. a. As a general term of disparagement: indifferent, feeble; petty, insignificant, esp. in *potty little*.
1860 Hotten *Dict. Slang* (ed. 2) 193 *Potty*, indifferent, bad looking. [*ed. 3* (*1864*), indifferent, bad looking,—said of a rotten or unsound scheme.] 1870 *Times* 12 Aug. 10/3 Then came a single, and then a catch from a 'potty' hit. 1899 E. Phillpotts *Human Boy* 72 It is such a potty little place, hardly worth calling a wood. 1904 Kipling *Traffics & Discov.* 270 'Think they'll take you an' your potty quick-firers? 1907 Galsworthy *Country House* III. iv. 246 We stand on our petty rights here, And our potty dignity there. 1927 Chesterton *Secret of Father Brown* v. 178 Who would, or could, have killed him up in that potty little place? 1930 —— *Four Faultless Felons* 236 It was within reasonable distance of revolution; not a potty little palace revolution. 1939 G. B. Shaw *Geneva* III. 57 What I think of the mob of bagmen from fifty potty little foreign states that calls itself a League of Nations. 1980 'M. Innes' *Going it Alone* xx. 178 They've..smashed their way in. Just the idea I had with that potty little motor-mower.
b. Easy to manage, accomplish, or deal with; easy, simple.
1899 E. Phillpotts *Human Boy* 127 Ferrars..got regularly muddled over a potty question about Jacob. 1916 E. F. Benson *David Blaize* iv. 70 It was quite certain that Helmsworth would have won had not that ass Blazes..dropped the 'pottiest' catch ever seen. 1922 *Blackw. Mag.* July 55/2 It's potty on this scaffolding.., no end of cross-pieces to hold on to.
2. a. Crazy, mad; out of one's senses; eccentric, 'dotty'.
1920 *Cornh. Mag.* Jan. 7 Next day, at tea time, the producer of the comedy solemnly thanked Jess for saving a situation past praying for... In your potty part, you put 'em all to bed. How did you do it? 1921 *Chambers's Jrnl.* July 511/2 Pull yourself together. You'll be going potty if you don't get a move on. 1925 *Punch* 7 Jan. 7 Hear about Mary, Mums? Gone potty! Broke off her engagement 'cos her people disapproved. 1929 W. Faulkner *Sartoris* III. 170 Aunt Sally, a potty little woman. 1930 'R. Crompton' *William the Bad* iv. 113 'I don't know what you're talking about,' she said. 'You seem to me quite potty today.' 1942 *Sun* (Baltimore) 25 July 8/1 Confronted, in the final scene, with the prospect of winning and wedding this incurably potty creature, he lets out an anguished yawp. 1952 [see *Baconian *a.* and *sb.* 2]. 1960 *Times* 21 Sept. 3/7, I realized Floss had put the boot in. I have been going potty about this. When I 'jumped' him we didn't mean to kill or hurt him. 1976 K. Bonfiglioli *Something Nasty in Woodshed* i. 19 Violet's mother..is, I suppose, either potty or an alcoholic. 1977 *Daily Mirror* 12 Apr. 17/1 He played the joyously potty day-dreamer.
b. Madly in love; mad *about*, gone *on* (someone or something).
1923 E. V. Lucas *Advisory Ben* xxxix. 206 I'm potty about her. 1926 *Punch* 8 Dec. 622/3, I suppose you're potty about the poor fish? 1928 'R. Crompton' *William—the Good* ix. 228 'Hector's potty on her too,' said Ginger. 1930 A. P. Herbert *Water Gipsies* xx. 292 Jane..confessed to herself that she was mad about Mr. Bryan—just potty. 1975 *Reveille* 20 June 11/7 Women are potty about pans—they can't resist buying them.
Hence **po·ttily** *adv.*
1977 *Times Lit. Suppl.* 3 June 686/3 Contributors to anthologies can pursue minor themes more selectively, if occasionally pottily.

potty (pǫ·ti), *sb.* Also **pottie.** [f. Pot *sb.*¹ + -y⁶.] A nursery word for a chamber-pot; occas. applied also to a lavatory.
1942 Berrey & Van den Bark *Amer. Thes. Slang* §84/14 *Chamber pot*,..potty. 1952 C. Hayes in S. Rogers *Children & Language* (1975) iv. 283 Don't flush the soap down the potty, dear. 1959 *Observer* 22 Mar. 21/5 Pram bedding, waterproof sheets, potties. 1960 S. Foot *Emergency Exit* ii. 20 Awful little bedside tables which..have place for potty. 1960 *News Chron.* 20 Aug. 6/8 Granny soaking her corns and Baby..shaking his pottie. 1966 Auden *About House* 26 Lifted off the potty, Infants from their mothers Hear their first impartial Words of worldly praise. 1976 *Milton Keynes Express* 4 June 22/3 (Advt.), White baby bath and potty, £1.25.
2. *attrib.* and *Comb.*, as **potty-chair**, a child's commode; **potty-mouth** *U.S.* (see quot. 1968–70); **potty-seat** = *potty-chair*; **potty-training** *vbl. sb.*, the training of a child to use a chamber-pot; also *fig.*; hence (as a back-formation) **potty-train** *v.*; **potty-trained** *ppl. a.*
1961 Webster, Potty-chair. 1965 M. Shadbolt *Among Cinders* ii. 11 The only thing I really saw was a little kid's potty-chair. 1977 *Time* 19 Sept. 41/3 The junk in his basement, including 'one used potty-chair, a tricycle with no handle bars, one broken ski, an old doorknob and six bags of leaves'. 1968–70 *Current Slang* (Univ. S. Dakota) III–IV. 95 *Potty mouth*, *n.* A person who repeatedly uses foul language. 1976 *Time* 10 May 79/2 Potentially even more annoying is the widespread abuse of the channels—especially by so-called potty mouths using obscenities. 1961 L. Mumford *City in Hist.* Note to plate 10, Full of objects that bring the daily life near, from a clay sausage-griddle to a ceramic potty-seat with holes for the child's legs. 1974 E. Tidyman *Dummy* xiv. 195 He did go to school for a while, but..he wasn't potty trained at the time. 1978 F. Weldon *Praxis* xxii. 211 Justin was not, as they said in the nursery world, 'potty-trained'. 1958 R. Martin *Before the Baby—and After* xiii. 274 The number of modern young mothers who have erroneous views on 'potty' training is amazing. 1969 *New Scientist* 13 Mar. 31/1 There is no necessary link between outstanding cleanliness and tidiness and outstanding research work. Strict potty-training is not the key to the Royal Society. 1978 P. O'Donnell *Dragon's Claw* xii. 262, I expect you suffered from poor potty training as a baby.

potzer (pǫ·tsəɹ). *U.S. slang.* = *patzer.
1948 *Chess Rev.* Apr. 5/2 Immediately, spectators inquired, 'Didn't you see that win?' 'Yes,' was the impudent reply. 'But, with such a potzer, I draw when I will, not when he wills.' 1962 *Chess* 12 Mar. 190/1 When I meet these Russian potzers I'll put them in their place. 1966 *New Yorker* 12 Nov. 70/1 He was at work on what in the language of the park is called a 'potzer'—a relatively weak player with an inflated ego. 1970 J. Hansen *Fadeout* (1972) v. 41 'Do you..play chess, Mr...Brand..stetter?' 'I'm what's called a potzer.'

pou, var. Pouw (in Dict. and Suppl.).

pouch, *sb.* Add: **1. c.** (Earlier example.)
1879 *Post Master General's Rep.* in *Parl. Papers 1878–9* (C. 2405) XXI. 197 The..number of pouches exchanged with these Travelling Post Offices..in 24 hours is now 1090.
f. = *diplomatic bag* s.v. *diplomatic *a.* and *sb.* A. 3.
1958 L. Durrell *Mountolive* vi. 140 When the Syrians want to be clever, they don't use a diplomatic courier; they confide their pouch to a lady, the vice-consul's niece. 1967 M. Childs *Taint of Innocence* iv. 240 He put the locked briefcase on Wyant's desk. 'Watrous said he thought you wouldn't mind sending it back to him by pouch.' 1968 D. Torr *Treason Line* 163 I've been down here for the past hour checking the airgrams for the Washington pouch. 1974 *Lebende Sprachen* XIX. 39/2 US *pouch*—BE/US *diplomatic bag*. Kuriersack, -tasche.
3. c. = *bag *sb.* 12 c.
1928 J. Buchan *Runagates Club* iv. 134 There were dark pouches under his eyes. *a* 1953 E. O'Neill *Hughie* (1959) 8 His blue eyes have drooping lids and puffy pouches under them. 1980 P. Harcourt *Tomorrow's Treason* I. ii. 37 Pouches under his eyes as if he hadn't slept.

pouch, *v.* Add: **1. c.** *Cricket.* To catch (the ball); also with the batsman as object.
1910 A. A. Milne *Day's Play* 114, I heard Slip call 'Mine' and he pouched the ball. 1963 *Times* 13 June 13/3 A series of pulls which ended with a catch at the wicket would appear in this form: 'After several cow-shots into the Great Beyond, Basher was neatly pouched by the timber-watcher.' 1970 *Times* 12 Jan. 7/7 Neither catch that Fletcher dropped was at all easy, but..Bobby Simpson or Philip Sharpe might well have pouched them both.

pouched, *a.* Add: **4.** *pouched mouse*, a small Australian marsupial resembling a mouse, belonging to the family Dasyuridæ, esp. the genera *Antechinus* and *Sminthopsis*; also called kangaroo-mouse.
1888 O. Thomas *Catal. Marsupialia in Brit. Mus.* 287 Little Pouched Mouse. Size rather small, general form murine. 1896 F. G. Aflalo *Sk. Nat. Hist. Austral.* II. 28 In some of the pouched mice, the tribal badge is either replaced by a mere fold of bristles or else entirely wanting. 1907 P. Fountain *Rambles Austral. Naturalist* vi. 60 The pouched-mouse varies its habits with its locality. 1942 C. Barrett *On Wallaby* iii. 40 There are perhaps a dozen kinds of pouched-mice, all Australian natives. 1970 W. D. L. Ride *Guide Native Mammals Austral.* viii. 112 Kowari, Byrne's Pouched Mouse, *Dasyuroides byrnei*. Central Australia..; desert associations and grasslands.

poudre (pudr'). *Canad.* [Fr. *poudre* powder.] Light, powdery snow, = *powder snow. *poudre day*, a day on which light, fine snow falls.
1791 E. P. Simcoe *Diary* 31 Dec. (1911) 71 There is little wind here, except with a snowstorm of fine snow. The French call it poudre or powdered snow, and to travel with that blowing in one's face is very disagreeable. 1873 W. F. Butler *Wild North Land* xiv. 152 The sun, which on one of these 'poudre' days in the North seems to exert as much influence upon the war of cold and storm as some good bishop in the Middle Ages was wont to exercise over the belligerents at Cressy or Poictiers..muffled himself up in the nearest cloud and went fast asleep until the fight was over. 1901 G. Parker *Right of Way* 83 It was a goodly scene..the flowery tracery of frost hanging like cobwebs everywhere; the poudre sparkle in the air. 1951 W. O'Meara *Grand Portage* xxxiii. 214 It was a real 'poudre day'. Only a few inches of dry, fine snow lay on the prairie.

‖**poudré** (pudre), *a.* Fem. **poudrée.** [Fr.] Powdered, *spec.* of the hair or a wig.
1827 Disraeli *Viv. Grey* III. v. vi. 104 A little old, odd-looking man, with a very *poudré* head, and dressed in a costume in which the glories of the vieille cour seemed to retire with reluctance. 1852 E. Ruskin *Let.* 24 Feb. in M. Lutyens *Effie in Venice* (1965) II. 278 Marmont poudré welcomed me. 1906 Beerbohm *Around Theatres* (1924) II. 243 Poudrée, she sings of Brittany; and in a crinoline she warbles of Parthenay. 1909 A. Lancastre Saldanha *Recoll.* iv. 34 My hair was exquisitely *poudré*.

‖**poudreuse** (pudrǭz). [Fr.] A lady's dressing table of a kind made in France in the time of Louis XV (see quot. 1966).
1929 V. Woolf *Let.* 5 May (1978) IV. 54 There are two pieces of furniture I must buy, so do mark them down—one a poudreuse, not necessarily in good condition. 1929 G. G. Gould *Period Furnit. Handbk.* vi. 82 *Toilette-coiffeuse*, especially for powdering, new in *Régence*; later popular as *poudreuse*. 1936 *Burlington Mag.* Aug. 88/2 *Berceuses* or *poudreuses* show the delicate outlines of the curved ornament of the French cabinet-makers. 1947 *Antique Dealer* Jan. 34 (*caption*) Louis XV Poudreuse by E. J. Gibrekyer. 1959 *House & Garden* Dec.–Jan. 27/1 The so-called *poudreuse* (dreadful word) with twin adjustable mirrors and pivoting cosmetic trays. 1960 *Times* 21 June 22/7 A Louis XV marquetry poudreuse. 1966 M. M. Pegler *Dict. Interior Design* (1967) 351 *Poudreuse*, a lady's powder or toilet table, often equipped with a mirrored lid in the center which lifts up. A Louis XV period innovation. 1973 *Country Life* 13 Dec. 32f (Advt.), An early 19th century Austrian mahogany poudreuse.

pouf¹. Add: **2.** (Later examples.) Also *attrib.*
1906 *Queen* 28 Apr. p. viii/3 A quaint pouf sleeve. 1976 *State Jrnl.* (Lansing, Mich.) 11 July D 1/6 Designed with deep *pouf* sleeves. 1977 *New Yorker* 11 July 84/3 Noelle demonstrates her virginal contempt by choosing a white strapless evening dress with a pouf of fabric.
3. Now usu. a low stuffed or padded seat or cushion. (Further examples.)
1919 'C. Dane' *Legend* 32 Mrs Howe was in the chair... The Baxter girl crouched on the pouf. 1925 C. S. Taylor *Pract. Upholstery* xi. 96 The pouffe, or floor cushion, is much in favour. 1949 A. Christie *Crooked House* xiv. 106 Roger was astride a big pouffé [*sic*] by the fireplace. 1956 M. Shulman in *Good Housekeeping* July 51/2 You rented an apartment in Greenwich village and sat on a pouf. 1962 A. Sampson *Anat. Brit.* I. i. 22 Between the two sides is something which looks like a huge red pouff, with a back-rest in the middle of it: this is 'the Woolsack'. 1968 C. M. Vines *Little Nut-Brown Man* iii. 58 He held out the cutting towards me, down to where I was sitting on the pouffe, his thumbnail indicating a line of it. 1974 S. Coulter *Château* II. vi. 281 This little room..stuffed with Algerian poufs and cushions and ottomans.

pouf², var. *poof *sb.*¹

pouf(f, poufter, varr. Poof *int.*, *sb.*² and *v.* in Dict. and Suppl., *poofter.

poughite (pǭᵘ·ait). *Min.* [f. the name of Frederick H. *Pough* (b. 1906), U.S. mineralogist + -ite¹.] An iron tellurite, $Fe_2(TeO_3)_2(SO_4)\cdot 3H_2O$, found as yellow orthorhombic crystals.
1968 R. V. Gaines in *Amer. Mineralogist* LIII. 1075 In 1944..Frondel and Pough mentioned a further new iron tellurite from Honduras... This mineral had first been observed by Dana and Wells in 1890 and thought to be tellurite... Its macroscopic definition [etc.]..enabled the author to establish its identity with material found at the Moctezuma mine in Sonora, Mexico. The name poughite is proposed for this new mineral. 1968 I. Kostov *Mineral.* 518 Poughite is orthorhombic..with perfect 010 cleavage.

poui (pu·i). Also **pui.** [Local name in Trinidad.] A tree belonging to the genus *Tabebuia*, of the family Bignoniaceæ, esp. *T. rosea*, the pink poui, and *T. serratifolia*, the yellow poui, native to the West Indies and Central America, and bearing terminal clusters of trumpet-shaped flowers; also, the wood of this tree.
1864 A. H. R. Grisebach *Flora Brit. W. Indian Islands* 787 *Tecoma serratifolia* Pony [*sic*]. 1924 Record & Mell *Timbers Trop. Amer.* II. 541 The 'pui' or 'poui'..is one of the best known timbers of Trinidad. 1939 R. C. Marshall *Silviculture of Trees of Trinidad & Tobago* 182 One frequently comes across patches of young poui in the forest. 1952 S. Selvon *Brighter Sun* i. 5 In April..pouis blossomed and keskidees sang for rain. 1962 *Times* 31 Aug. (Trinidad Suppl.) p. iv/4 The visitor will see some of the most lovely flowering trees and shrubs in the world—the bougainvillaea, the clear clean beauty of the yellow and pink poui. 1968 E. Lovelace *Schoolmaster* I. i. 7 The poui is dropping rich yellow flowers. 1973 E. G. B. Gooding *Wayside Trees & Shrubs of Barbados* 94 The Yellow Poui was introduced into Barbados early this century.

Pouilly (pūiyi). Also with small initial. Any of various dry white wines produced in central France and named after villages called *Pouilly*, *spec.* (*a*) in full *Pouilly Fumé* (fūme), that produced near Pouilly-sur-Loire in the Nièvre department; (*b*) in full *Pouilly-Fuissé* (fwīse), a white Burgundy produced near Pouilly-sous-Charlieu in the Mâcon region of the Loire department.
[1814 M. Birkbeck *Notes Journey through France* I. 29 Pouilly, [renowned] for its Vin Blanc.] 1833 C. Redding *Hist. Mod. Wines* v. 160 Cosne is best known for its white wines called Pouilly, in considerable repute in Paris. 1924 D. H. Lawrence *Let.* 29 Oct. (1962) II. 816 Dinner at Coyoacán, and drank absinthe, gin, pouilly, chablis, beaune, port and whisky from beginning to end of an evening, and was not comforted. 1927 [see *Moulin-à-Vent]. 1935 M. Morphy *Recipes of All Nations* 104 The Pouilly comes from Saône-et-Loire: it is usually sold as Pouilly-Fuissé to differentiate it from the white wines of Pouilly-sur-Loire, sold as Pouilly-Fumé. 1942 E. Paul *Narrow St.* i. 4 Noël pointed out to me once, over a bottle of Pouilly, that men and women, like gods, choose pets in their own image. 1952 W. Plomer *Museum Pieces* 117 A cut of salmon, some wisps of chicken, a bottle of Pouilly. 1959 E. Linklater *Merry Muse* iv. 81 The fish and the Pouilly fumé were succeeded by a bevy of partridges and a tenderly decanted Léoville Poyferré of 1947. 1976 *Times* 20 Mar. 3/3 (Advt.), Mâcon Blanc, a dry, very fruity wine, typified by Pouilly-Fuissé and St. Véran.

Poujadism (pū·ʒādiz'm). Also with lower-case initial. [ad. Fr. *Poujadisme* (also used): see below.] The mainly reactionary and conservative political philosophy and methods advocated by Pierre *Poujade* (b. 1920), French publisher and bookseller, who in 1953 founded a movement for the protection of artisans and small shopkeepers (Union de Défense des Commerçants et Artisans), protesting chiefly against the French tax system then in force. Also in extended and allusive applications of similar movements acting in the interests of small-scale commercial enterprise.
1955 *Life* 18 Apr. 63/1 The mushrooming political strength of Poujadism last month forced Premier Edgar Faure's government to promise sweeping exemptions to the small shopkeepers. 1957 *Times Lit. Suppl.* 11 Oct. 607/2 The work of Chancellor Maupeou was undone..and what was no better than an aristocratic form of Poujadism grew in power of mischief. 1961 *Listener* 7 Sept. 337/1 France presents the contrast between a strongly proletarian industrial society and a peasantry that finds its sporadic expression in Poujadism or the recent Breton uprisings. 1964 *Economist* 1 Feb. 429/1 The ensuing process of concentration in commerce and industry..contributed to the political spread of *poujadisme* in the next few years. 1970 *Rev. Politics* XXXII. 174 Once organized Poujadism began its spread throughout France, it shared its founder's preoccupation with class defense. 1974 tr. *Wertheim's Evolution & Revolution* 149 Poujadism in France and the Farmers' Party in the Netherlands have..been openly conservative or even reactionary. 1976 *Times* 9 Feb. 13/4 The small business sector..is surely the very class which in other countries has..turned to 'poujadism'.
Hence **Pou·jadist**, an advocate or supporter of Poujadism; also *attrib.* or as *adj.*; **Poujadi·stic** *a.*
1957 *Observer* 27 Oct. 13/2 Miss S. is not irresponsible, not a Poujadist... But the spirit of Poujadism is awake... 'The Liberal vote is a "Poujadist", disgruntlement vote.' 1958 *Economist* 6 Dec. 867/1 Electing poujadists under the name of conservatives does not mean a change. 1960 *20th Cent.* Apr. 302 In France there is a strong Poujadist element; small shopkeepers, small factory owners, the depressed bourgeoisie. 1962 *Daily Tel.* 25 June 10/2 Unless Britain is going to become Poujadist in the full sense—that is, inimical to all politicians as such—then the electorate will have to accept the sense expressed by the Prime Minister. 1967 M. Dogan in Lipset & Rokkan *Party Syst. & Voter Alignments* iv. 181 The Poujadistic social epidemic spread particularly in the small sleepy towns. 1972 *New Statesman* 28 Jan. 100/2 Another mass movement of angry shopkeepers, the Poujadist wave of the 1950s. 1977 *Guardian Weekly* 27 Feb. 8/3 The Poujadist Progress party, led by Mr Mogens Glistrup, the controversial Copenhagen tax-lawyer, is once again the second largest party with 26 seats.

‖**poule²** (pul). [Fr., = 'hen'.] **1.** *poule au pot*, in Gastronomy, a boiled chicken casserole. Also *fig.*
1884 N. Lake *Menus made Easy* iv. 99 A fowl boiled and served with Bourgeoise sauce is called *Poule au pot*. 1930 M. T. & L. Bonney *French Cooking for Eng. Kitchens* vii. 150 The Poule au Pot has as dignified a place on a menu as the Poussin. 1962 *Harper's Bazaar* Dec. 97/1 Chicken served as *poule au pot*. 1971 A. Mizener *Saddest Story* xxxii. 433 An absolute monarchy committed to all Ford's favorite Tory doctrines—the *poule au pot* of the French peasant instead of the 'boiled fowls out of a pot that is no pot but a can' of American mass production.
2. *slang.* A girl or young woman, esp. one regarded as being promiscuous. Also *poule-de-luxe*, a prostitute.
1926 E. Hemingway *Sun also Rises* iii. 14 The *poules* going by, singly and in pairs. 1937 [see *chichi *sb.*² and *a.*]. 1946 'S. Russell' *To Bed with Grand Music* vi. 87 All I know of the tricks used by poules-de-luxe. 1949 J. B. Priestley *Home is Tomorrow* 43 He is probably amusing himself somewhere with that little brown poule of his. 1955 D. Barton *Glorious Life* i. 12 'If I had thought she would have understood,' said Swindlehurst, 'I would have called her to her face "the typing *poule*".' 1958 L. Durrell *Mountolive* v. 102, I answer him in the voice of a *poule* from the Midi. 1976 *Times Lit. Suppl.* 24 Sept. 1213/3 Returns to France to find that his wife has remarried and that his daughter is in business as a *poule de luxe* and doing very well. 1979 *Ibid.* 23 Nov. 14/5 The archimandrite who regularly used the express between Sofia and Belgrade to entertain his *poules*.

poulet. Add: **2.** A chicken or a chicken dish, esp. in French Gastronomy.
The spelling in quot. 1840 represents a jocular anglicization.
1840 Thackeray *Barber Cox* in *Comic Almanack* 213 Eating, for my share,..a pully bashymall, and other French dishes. 1861 Mrs. Beeton *Bk. Househ. Managem.* xxi. 474 *Poulet aux cressons*... A fowl, a large bunch of water-cresses, 3 tablespoonfuls of vinegar, ½ pint of gravy. 1894 G. du Maurier *Trilby* II. v. 111 A delightful little Franco-Italian pothouse near Leicester Square, where they had..a *poulet rôti*, which is *such* a different affair from a roast fowl! 1900 C. M. Yonge *Mod. Broods* xiii. 120 Here are eggs, and some milk..four *poulets*, such as they are. 1933 D. C. Peel *Life's Enchanted Cup* viii. 87 She made a delicious *Poulet Marengo* with a suspicion of garlic in it. 1957 P. Wildeblood *Main Chance* 53 The pressure cooker burst... It suddenly went off bang and threw bits of *poulet en cocotte* all over the place like confetti. 1978 *Chicago* June 218/2 Poulet Provençal ($6.50) comes with pearl onions, fresh mushrooms, olives, tomato purée.

‖**poulette** (pulet). [Fr. *poulette* young hen.] In full, *poulette sauce*. In French Gastronomy, a sauce made with butter, cream and egg-yolks.
1813 L. E. Ude *French Cook* 389 The *poulette* is made with a little *sauce tournée*, which you reduce, and next thicken with the yolks of two eggs, to which you add a little parsley chopped very fine. 1877 E. S. Dallas *Kettner's Bk. of Table* 376 *Poulette Relish*..consists of shalots and mushrooms passed in butter, and served with chopped parsley in a Poulette sauce. 1907 A. Escoffier *Guide Mod. Cookery* iii. 42 *Poulette Sauce*. Boil for a few minutes one pint of Sauce Allemande, and add six tablespoonfuls of mushroom liquor. Finish..with two oz. of butter, a few drops of lemon-juice, and one teaspoonful of chopped parsley. 1957 M. McCarthy *Memories Catholic Girlhood* viii. 204 Sweetbreads, with patty shells, and with a poulette sauce. 1977 Rossant & Davis tr. *Bocuse's New Cuisine* 330 *Poulette Sauce*..butter..flour..bouillon..mushroom.

Poulsen arc (pǭu·lsĕn ā̦ɹk). [Named after its inventor, Valdemar *Poulsen* (1869–1942), Danish electrical engineer.] An arc discharge between a carbon electrode and a water-cooled copper electrode operated in an atmosphere of hydrogen or hydrocarbon vapour and a strong magnetic field in order to make possible the emission of waves short enough to be used for radiotelegraphy.
1906 *Electrician* 21 Dec. 375/1 The physical..circumstances of the Poulsen arc are rather forbidding. 1916 *Ibid.* 14 Jan. 535/2 In wireless telegraphy there has been a notable advance in the use of the Poulsen arc. 1931 [see *keying *vbl. sb.* 1]. 1949 W. R. Maclaurin *Invention & Innovation in Radio Industry* iv. 60 The patents on the Poulsen arc were held by enterprises that were competing with Marconi.

poult (pult), *sb.*² [Shortening of Poult-de-soie.] = Poult-de-soie. Also, a similar fabric manufactured from man-made fibres. Also *attrib.*
1938 *Times* 7 July 22/1 A gown..with a bodice of white poult. 1951 *Good Housek. Home Encycl.* 199/1 Nylon Fabric is now available in many different finishes, such as tricot, net, poult. 1959 *Times* 21 Sept. 12/4 A short dress of coral-coloured poult. 1960 *Woman's Own* 9 Apr. 7/1 At a dance recently I wore my poult dress. 1970 *Trafford Spring & Summer Catal.* 276 See through lace sleeves and a 40 denier Nylon poult skirt.

poulter. **1. b.** Delete † *Obs.* and add later examples.
1900 W. Raleigh in W. E. Henley *Castiglione's Bk. of Courtier* p. xlv, The one-legged poulter's measure is not responsible for all the horrors of this. 1962 G. K. Hunter *John Lyly* iv. 244 The case with 'fourteeners' or 'poulter's measure' is even more obvious. 1972 *Times Lit. Suppl.* 10 Nov. 1363/1 Poulter's measure (the form of verse with alternating lines of twelve and fourteen syllables).

poulterer. Add: **b.** *poulterer's measure*, = *poulter's measure* s.v. Poulter 1 b in Dict. and Suppl. Chiefly *Hist.*
1841 R. G. Latham *Eng. Lang.* v. 382 *Poulterer's Measure*,—Alexandrines and Service Measures alternately. Found in the poetry of Henry the Eighth's time. 1957 N. Frye *Anat. Crit.* 263 There were some comparative failures, such as poulterer's measure.

poultice, *sb.* Add: **2.** *Austral. slang.* A mortgage.
1932 K. S. Prichard *Dark Horse of Darran* in *Kiss on Lips* 184 Mick Mallane..sayin' if the bank wanted his farm, poultice or no poultice, it'd have to go out and take it from him, and he'd be waitin' for 'm with his gun loaded. 1934 T. Wood *Cobbers* xi. 134 Men talked about their blister, or their poultice, which means a mortgage, with complacency. 1958 *Coast to Coast 1957–1958* 137 When the farm was free of its 'poultice', her father had promised to hand over to Sam.
b. A (large) sum of money; a bribe.
1951 E. Lambert *Twenty Thousand Thieves* III. xii. 235 It's only two days to pay day and I've got a poultice in that pay-book of mine. 1957 'N. Culotta' *They're a Weird Mob* (1958) v. 73 'Reckon 'e pulled 'im?' 'That's wot I reckon.'..'Yer can't prove ut.' 'Somebody slipped him a poultice, I bet.' 'They're all crooked.' 1979 *Sun-Herald* 24 June 143 A bloke who made a poultice in recent times was when he sold Rupert a quarter of a million Channel Ten shares.

poultry. Add: **4.** *poultry breeder*, *-fancier* (earlier example), *-house* (later examples), *-keeper* (examples), *meat*, *-raising* (examples), *-run*, *-show* (examples), *-yard* (earlier examples); **poultry-man** (later examples); also **poultry-woman.**
c 1882 W. Cook (*title*) Practical poultry breeder and feeder. 1976 *Evening Post* (Nottingham) 14 Dec. 6/6 Derbyshire police advises poultry breeders to have better lighting as a deterrent. 1865 *Atlantic Monthly* June 661/2 My experiments with chickens have been attended with..a success so brilliant that unfortunate poultry-fanciers have appealed to me for assistance. 1819 *Amer. Farmer* (Baltimore) I. 46 Respecting the cleansing of poultry-houses from vermin, or chicken-lice. 1921 *Proc. 3rd Nat. Country Life Conf. 1920* (U.S.) 155 No argument is necessary to justify poultry houses. 1942 C. Milburn *Diary* 11 July (1979) 145 Mr Willoughby came later with a copy of the *Smallholder* and we looked at an advertisement for a 'Utility Poultry House'. 1867 L. Wright (*title*) The practical poultry-keeper. 1935 *Discovery* Oct. 302/2 The poultry keeper and game preserver. 1960 *Farmer & Stockbreeder* 9 Feb. 123/3 Some poultry-keepers prefer to use no preventive drugs and to rely upon curative drugs, if and when an outbreak of coccidiosis occurs. 1976 *Newmarket Jrnl.* 16 Dec., He served on the committee of the allotments association and was an active gardener and poultry-keeper. 1889 *St. Landry Democrat* (Opelousas, Louisiana) 7 Dec. 2/5 Mr. Felix..the well-known poultryman, has a Scotch collie dog. 1960 *Farmer & Stockbreeder* 9 Feb. 125/2 Mr. Thornber remembered that the poultryman slept for 21 days with the first mammoth they installed. *Ibid.* 16 Feb. 79/1 The amount of poultry meat marketed continued to shoot upwards. 1978 *Daily Tel.* 19 Sept. 19/7 Meat consumption in Britain is now at its lowest level since 1955—replaced largely by poultry meat. 1850 in D. J. Browne *Amer. Poultry Yard* 304 There is profit attending poultry raising when undertaken on a moderate scale. 1870 'Mark Twain' in *Buffalo Express* 4 June 2/3 From early youth I have taken an especial interest in the subject of poultry-raising. 1896 *Rep. Vermont Board Agric.* XV. 28 Poultry raising has of late years attracted the attention of the farmer. 1916 *Daily Colonist* (Victoria, B.C.) 23 July 8/5 There is room near every industrial centre in the province for many market gardens, poultry-runs, orchards and aviaries. 1930 R. Campbell *Adamastor* 64 Heaven in which our senses swim, Aviary of aviators And poultry-run of seraphim! 1868 M. H. Smith *Sunshine & Shadow in N.Y.* lxxiv. 597 He has gotten up baby-shows, poultry-shows, and dog-shows. 1873 F. Kilvert *Diary* 6 Aug. (1944) 223, I went..to Garth to attend the Garth Flower Show, Dog Show, Poultry Show, Bazaar and Athletic Sports, all in one. 1851 *Harper's Mag.* Apr. 662/1 The poultry-woman must be changed. 1748 Richardson *Clarissa* I. ix. 52 You must remember the Green Lane..that runs by the side of the wood-house and poultry-yard. 1811 Jane Austen *Sense & Sens.* III. vi. 114 The rest of the morning was..whiled away..in visiting her poultry-yard.

‖**pounamu** (po,unamu). *N.Z.* Also 9 punamu. [Maori.] Nephrite.
[1773 J. Hawkesworth *Acct. Voy.* II. II. vi. 400 This land..consisted of two Whennuas or islands..which he [*sc.* Cook] called Tovy Poenammoo; the literal translation of this word is, 'the water of green talc'.] 1835 W. Yate *Acct. N.Z.* (ed. 2) v. 271, I have put on board the Buffalo a mere pounamu and two garments. 1867 E. Sauter tr. *F. von Hochstetter's New Zealand* xvii. 362 A magnificent Mere punamu, a battle axe..cut out of the most beautiful, transparent nephrite. 1905 W.B. *Where White Man Treads* 39 'Pounamu' (greenstone) was only to be found in the creeks of the inland ranges of the Middle Island. 1965 G. J. Williams *Econ. Geol. N.Z.* v. 45/2 Pounamu Ultramafics occur within the rocks of this area. 1973 *Times Lit. Suppl.* 9 Feb. 141/3 The repeated word of the title, *pounamu*, however translucent to New Zealanders, will be opaque to the English reader. It means 'greenstone', a New Zealand form of jade, evocative of a past when it was the prized material of axe and ornament.

pounce, *sb.*¹ Add: **3.** *pounce-bag* (examples), *-box* (later example), *pattern*, *pot*.
1855 W. Williams *Transparency Painting on Linen* 20 Pounce patterns..are formed of outlines perforated through the paper on which they are drawn, by a succession of small needle holes. *Ibid.* 27 The pounce-bag is made by tying a little fine, dry, black powder, in two or three squares of the muslin..used for painting on... The perforated pattern being placed on the cloth, the pounce-bag is lightly tapped on the surface, so as to force the powder..through all the perforations of the pattern. 1866 W. Davidge *Footlight Flashes* xv. 146 A pack of cards, a piece of rosin, a flute case, a pounce, or sand box, and a newspaper printed in the German language. 1939 *Newsweek* 13 Mar. 37/3 And (using a 'pouncebag' filled with dry pigment) [the painter] rules the outline through the perforations onto the preliminary coat of mortar. 1957 Mankowitz & Haggar *Conc. Encycl. Eng. Pott. & Porc.* 181/2 *Pounce pot*, a small box with a perforated lid for perfumes; a powder box or jar; a pierced-topped jar for sprinkling powdered pumice used to dry ink (before the popularity of blotting paper). 1971 *Country Life* 29 July 263/1 The top [of an inkstand] is inset with an ormolu tray flanked by inkpot and pounce pot.

pounce, *v.*³ **2.** (Further examples.)
1855 W. Williams *Transparency Painting on Linen* 28 If an accident..occur, it is only necessary to dust the powder off the muslin, to re-adjust the pattern, and again pounce in the design. 1960 B. Snook *Eng. Hist. Embroidery* 51 The design was either pricked and pounced and drawn with a clear black line, or it was printed from an engraving direct upon the linen.

pounced, *ppl. a.*³ **1.** (Later example.)
1855 W. Williams *Transparency Painting on Linen* 28 The pattern being removed, the pounced design is secured by being traced with a soft black-lead pencil, and drawn in with a reed pen.

pouncer² (pau·nsəɹ). [f. Pounce *v.*³ + -er¹.] A pouncing-tool; a pounce-bag.
1881 *Sylvia's Bk. Artistic Knicknacks* 371 Place the design on the canvas and pin it down; then take your pouncer, which is filled with fine charcoal or powdered colour, and dab..all over your perforated outline. The pouncer is..made by half-filling a thick muslin bag with charcoal or soot, and tying it tightly round. 1960 G. Lewis *Handbk. Crafts* 16 Using a pouncer (a roll of felt) dipped into black pounce (powdered charcoal) for light materials..gently rub the powder through the perforations, then lift away the tracing.

pound, *sb.*¹ **I. 1. a.** For 'weight' read 'weight and mass'. (Further examples.)
1959 *Nature* 10 Jan. 80 To secure identical values for each of these units in precise measurements for science and technology, it has been agreed [by standards laboratories in many countries] to adopt an international yard and an international pound..: the international pound equals 0·453 592 37 kgm... The yard and pound units to be used in trade are the imperial units laid down in the Weights and Measures Act, 1878. *Ibid.* 81 With regard to the pound, the values currently in use..are: 1 imperial standard pound = 0·453 592 338 kgm.; 1 Canadian pound = 0·453 592 43 kgm.; 1 United States pound = 0·453 592 4277 kgm. There is evidence that the imperial standard pound has diminished by about 7 parts in 10 millions since 1846. 1961 [see *lb]. 1963 Jerrard & McNeill *Dict. Sci. Units* 109 In 1963 the pound was defined as being equal to '0·45359237 Kilogramme exactly' by the Weights and Measures Act... This pound is identical with the International pound adopted in 1959 by Standards Laboratories.
f. *pound of flesh*: also (with hyphens) as *attrib. phr.*
1958 *Listener* 13 Nov. 775/2 He is entirely consistent.. In the application of this pound-of-flesh attitude. 1963 Auden *Dyer's Hand* 228 *Pecorone* or other versions of the pound-of-flesh story.
g. *pound and pint* (Naut. slang), a sailor's ration as determined by the Board of Trade's Scale of Provisions. So *pound and pinter*, a ship on which rations were provided on this scale; *pound-and-pint idler* (see quot. *a* 1865). *Obs.* exc. *Hist.*
a 1865 Smyth *Sailor's Word-bk.* (1867) 540 *Pound-and-pint-idler*, a sobriquet applied to the purser. 1902 W. Runciman *Windjammers & Sea Tramps* vii. 90 Their 'whack', or to be strictly accurate, the phrase commonly used was 'your pound and pint'. 1910 D. W. Bone *Brassbounder* 168 A pound and pint ruddy limejuicer. 1938 W. E. Dexter *Rope-Yarns* v. 31 It seemed my lot to mostly sail in what we called 'hungry-gutted ships', 'pound and pinters'. 1952 'Sinbad' *Sargasso Sam* xxviii. 211 Wot about tucker? We never come aboard this old wagon to eat deepwater muck. Looks like we're gettin' pound an' pint and no more.
h. [from *pound of lead*, rhyming slang for 'head'.] The human head.
1933 F. Richards *Old Soldiers never Die* xiv. 180 We old hands often used to remark that when we did get hit it would either be a bullet through the pound or stop a five-point-nine all on our own.
II. 3. Since 1971, of the value of 100 new pence.
d. Also applied to units of currency originally valued at par with the pound sterling.
1949 *Britannica Bk. of Year* 364/2 On Aug. 17, 1948, a new Israeli pound..displaced the Palestine pound. 1955 *Ibid.* 141/1 Cyprus... Monetary unit: Cyprus pound (= £1 sterling). 1958 *Spectator* 15 Aug. 216/3 No one wanted to lend any money in terms of the Israeli Pound. 1975 *Times* 25 Nov. 7/1 The Israeli pound is officially fixed at seven to the dollar.
f. Five dollars; a five-dollar note. *U.S. slang.*
1935 J. Hargan *Gloss. Prison Lang.* 6 *Pound*, a five dollar bill. 1950 *New Yorker* 25 Feb. 76 A pound off of thirty-four-fifty would still leave twenty-nine-fifty. 1970 H. E. Roberts *Third Ear* 11/1 Pounds, money; five dollars.
4. **pound brush** (earlier example); **pound-force** (pl. *pounds-force*), a unit of force equal to the weight of a mass of 1 pound avoirdupois, esp. under standard gravity; **pound note**, delete '(such as are issued in Scotland and Ireland)'; **pound-noteish** *a.* (*slang*), affected, pompous; **pound-weight** (pl. *pounds-weight*) = *pound-force* above.
1830 G. Colman *Random Rec.* I. ii. 35 My pictures are only sketches, and dabs of the pound-brush. 1896 T. W. Wright *Elem. Mech.* ii. 62 The word pound has a..variety of meanings. We speak of a pound weight, a pound force, and of a certain body itself as 'a pound'. 1909 J. M. Jameson *Elem. Pract. Mech.* ix. 149 These two units of force, the pound force and the gram force are sometimes called Gravitational Units of Force. 1949 W. Ernst *Oil Hydraulic Power* i. 2 The pound force imparts 32·174 feet per sec² to the pound mass. 1961 [see *lb]. 1972 *Physics Bull.* May 285/1 The subsequent addition of small weights permits forces to be obtained directly in both tons-force and pounds-force. 1977 *Daily Tel.* 16 Dec. 2/3 As Britain moves towards complete metrication motorists will have to get used to checking their car tyre pressures in atmospheric bars instead of pounds force per square inch. 1936 J. Curtis *Gilt Kid* vi. 63 Her pound-noteish voice both annoyed and amused the Gilt Kid. 1966 Auden *About House* 28 When we get pound-noteish..send us some deflating Image. [1871 J. C. Maxwell *Theory of Heat* iv. 83 In all countries the first measurements of forces were made in this way, and a force was described as a force of so many pounds' weight or grammes' weight. 1877 W. H. Besant *Treat. Hydromech.* i. 9 The unit of force is 750 lbs. weight.] 1891 J. G. Easton *First Bk. Mech.* iv. 59 It is sometimes convenient..to speak of a force as of so many pounds weight. 1907 Franklin & MacNutt *Elem. Mech.* viii. 174 (*heading*) Values of the stretch modulus of various substances. (In pounds-weight per square inch.) 1936 A. W. Hirst *Electr. & Magn.* i. 4 A force of one pound-weight = 32·2 poundals. 1960 F. Land *Lang. Math.* vi. 71 When I buy a pound of apples, the weight of the apples is 1 pound-weight..and its mass is 1 pound mass. 1976 *Daily Tel.* 4 Mar. 2/6 The Metrication Board..warned the Government that unless it introduced a sense of urgency into replacing feet for [*sic*] metres..and pounds weight for kilogrammes, then the target of 1980 for completion of the programme could never be met.

pound, *sb.*² Add: **1. d.** An enclosure in which vehicles impounded by the police are kept.
1970 P. Laurie *Scotland Yard* iii. 75 Civilian cars that have been stolen or in accidents..stand in a pound nearby. 1970 *Globe & Mail* (Toronto) 25 Sept. 39/2 (Advt.), Permanent part time dispatcher for police auto pound, Saturday, Sunday and Monday nights. 1972 *Daily Tel.* 16 Mar. 17/6 The Vauxhall Viva had been towed to a pound because it was found parked on an urban clearway. 1974 *Times* 18 Feb. 17 I'm going to sell my car... No more police towing [it]..to a car pound.
6. **pound-boat** (examples); **pound-master** (earlier example); **pound net** (earlier and later examples).
1884 *Bull. U.S. Nat. Museum* No. 27. 700 Lake Erie pound boat... Their peculiar construction enables them to carry large quantities of fish in shallow water and to lift the bowl of the pounds without upsetting. 1891 *Rep. U.S. Comm. Fisheries 1887* 27 The pound-boat has two tall, tapering masts. 1792 *Southampton* (N.Y.) *Records* (1878) III. 335 John Cooper Samuel Cooper Henry Corwithe Poundmasters. 1865 *Michigan Gen. Statutes* (1882) I. 577 The penalties of this section shall not apply or work injury to persons who are the present owners of pound or trap nets. 1973 *Fisheries Fact Sheet* (Environment Canada Fisheries & Marine Service) No. 1. 4/3 Gill-nets and pound-nets are the chief gear.

pound, *v.*¹ Add: **2. a.** (Later examples.)
1908 *Smart Set* June 21/2 She stopped at the door of the house and pounded the knocker vigorously. 1951 *Amer. Speech* XXVI. 230/2 St. Joseph *pounds* Mansfield. 1960 M. Spark *Bachelors* xii. 224 The typist in the corner listlessly pounded her silent machine. 1967 *Boston Sunday Globe* 23 Apr. 17/4 Air Force and Navy jets pounded North Vietnam in 118 missions Friday. 1968 *Globe & Mail* (Toronto) 5 Feb. 17/4 Detroit..pounded Minnesota North Stars 8–1. 1972 'E. Ferrars' *Breath of Suspicion* vii. 101 I'll be working..pounding my typewriter.
d. phr. *to pound one's ear*: to sleep. *slang* (orig. *U.S.*).
1899 'J. Flynt' *Tramping with Tramps* iv. 396 *Pound the ear*, to sleep. 1900 *Dialect Notes* II. 51 [College slang] '*Pound* one's ear, or one's pillow,' to sleep. 1907 [see *flop *v.* 2 c]. 1926 M. Walsh *Key above Door* xii. 128 'Only just awakened,' I admitted..'and how are my comrades in misfortune?'..'Still pounding their ears, no doubt.' 1927 C. Samolar in *Amer. Speech* II. 290/2 To sleep is to *pound the ear*. I think this phrase originated with railroaders. Sleeping in a caboose on a fast-moving train actually consists of pounding one's ear. 1947 J. Steinbeck *Wayward Bus* xx. 300 Listen to the old bastard snore. He's pounding his ear.
e. To produce or turn *out* by 'pounding' a typewriter or the like.
1904 F. Lynde *Grafters* v. 38 He sat down at the typewriter to pound out a letter to the general counsel, resigning his sinecure. 1941 B. Schulberg *What makes Sammy Run?* ix. 162, I was back in the old groove, pounding it out for the *Record* again. 1973 W. McCarthy *Detail* i. 48 He had just enough time to pound out two or three short paragraphs.
f. To walk upon; to cover (a distance or area) on foot; *spec.* of a policeman: to patrol (a beat). *colloq.* (orig. *U.S.*).
In quot. 1959 the use is *fig.* in punning allusion to the poetry of Ezra Pound (see *Poundian *a.*).
1906 A. H. Lewis *Confessions of Detective* iv. 44 It's worth while to pound a beat, when one has such kindly and appreciative superiors. 1909 'O. Henry' *Options* (1916) 30 I'm pounding the asphalt for another job. 1923 L. J. Vance *Baroque* vi. 33, I won't get sent back to pound sidewalks for what I'm pulling off tonight. 1935 A. J. Pollock *Underworld Speaks* 91/1 *Pounding the pavement*, a prostitute soliciting men on the street. 1946 [see *kriegie]. 1959 *Times Lit. Suppl.* 25 Sept. 546/5 An awful warning to any future translator tempted to indulge in the pleasures of what, metrically speaking, might be described as Pounding the beat. 1974 S. Marcus *Minding Store* (1975) ii. 26 He personally pounded the pavements calling on fellow businessmen. 1978 J. Gardner *Dancing Dodo* xxxiv. 270, I shall personally arrange for you to be back pounding the beat, in uniform.

poundage¹. Add: **6. b.** A person's weight, esp. that which is regarded as excess.
1930 Wodehouse *Very Good, Jeeves!* iv. 93 Women who have anything to do with opera..always appear to run to surplus poundage. 1971 *Time* 5 Apr. 44/3 With his hair transplant and added poundage.

poundage². Add: **2.** The keeping of cattle in a pound or enclosure; an enclosure in which cattle are kept.
1867 C. Tomlinson *Cycl. Useful Arts* I. 3/2 [The slaughterman] only paying for the poundage of his beasts according to the requirements of his business. 1902 *Encycl. Brit.* XXXII. 644/1 The bye-laws usually provide..for the poundage to have floor-space sufficient for each animal.

pound-cake. (Earlier and later examples.)
1747 H. Glasse *Art of Cookery* xv. 138 Pound Cake. Take a Pound of Butter..twelve Eggs..a Pound of Flour..a Pound of Sugar [etc.]. 1807 M. E. Rundell *New Syst. Domestic Cookery* 217 (*heading*) A good pound cake. 1942 B. Robertson *Red Hills & Cotton* iii. 69 We liked..cornbread with chitterlings, ambrosia, stuffed eggs, pound cake. 1951 T. Capote *Grass Harp* (1952) i. 12 Dolly, who lived off sweet foods, was always baking a pound cake. 1977 *Time* 24 Jan. 5/2 Pound cake will remain just that, no matter how many grams the ingredients weigh.

pounded, *ppl. a.*¹ Add: *pounded meat* (earlier and later examples); *spec.* (*U.S.* and *Canad.*) the flesh of buffalo or other game cut up, dried, and pulverized into powder to form the basic ingredient of pemmican.
1775 S. Hearne *Jrnl.* 5 Sept. (1934) 177 One Cannoe came with some Dry'd & Pownded Meate. 1805 *Deb. Congress U.S.* 9th Congress 2 Sess. App. 1066 Buffalo robes, tallow, dried and pounded meat and grease. 1898 F. Russell *Explor. Far North* 163, I saw large quantities of pounded meat, grease, and tongues eaten. 1922 *Beaver* Nov. 51/1 Pounded meat was made from the dried meat by beating it with flails until it became as small as desired and then stored away in bags. 1956 H. S. M. Kemp *Northern Trader* 93 The meat so acquired would either be dried, converted into pounded meat, or mixed with fat and cranberries and made up as pemmican.

pounder, *sb.*¹ Add: **2. b.** A policeman. *U.S. slang.*
1938 *New Yorker* 12 Mar. 38/2 Letting the sickly-sweet odor of burning marijuana into the street for the first passing pounder, or patrolman, to smell. 1970 C. Major *Dict. Afro-Amer. Slang* 93 *Pounder*, a policeman or detective.
c. *Surfing slang.* (See quot. 1967.)
1967 J. Severson *Great Surfing* Gloss., *Pounder*, an unusually hard-breaking wave. 1970 [see *greenie].

pounder, *sb.*⁴ **2.** (Earlier and further examples.) Cf. Six-pounder, Ten-pounder.
1684 [see Four C. 2]. 1747 [see Nine *a.* 5 b]. 1771 [see Two IV. 1]. 1845 [see *one-pounder* s.v. *one B. 33]. 1862 *Rambler* Mar. 414 A large number of 100-pounder Armstrong guns. 1896 [see Seven C. 3]. 1915 A. D. Gillespie *Let.* 14 June in *Lett. from Flanders* (1916) 196 They started with 33-pounder bombs, like a big turnip with a long handle, and we watched them sailing through the air. 1915 C. Mackenzie *Guy & Pauline* 264 'I know a man..who caught a four pounder with a bumble-bee.' 'I caught a six pounder at Oxford with a mouse's head myself.' 1977 F. Parrish *Fire in Barley* ii. 25 He had sometimes seen very big trout here, three and four pounders. 1978 K. Bonfiglioli *All Tea in China* x. 127 The gunner ambled towards the long brass Armstrong 68-pounder.

Poundian (pau·ndiăn), *a.* [f. the name *Pound* (see below) + -ian.] Of, pertaining to, or characteristic of the American writer and poet Ezra Pound (1885–1972) or his work; resembling or influenced by the style of Pound. Also *absol.* as *sb.*
1939 E. H. W. Meyerstein *Let.* 4 Apr. (1959) 221, I never thought I should come round to Eliot as a poet. Here he has dropped his Poundian Babel-tongues. 1958 *N. & Q.* June 265/2 Sappho—a well-known source of Poundian inspiration. 1960 N. Stock in *Agenda* June 1 The usual Poundian emphasis On medieval money and wages. 1965 *Times Lit. Suppl.* 25 Nov. 1070/4 Mr. Charles Olson..can often be peculiarly irritating with his Poundian mannerisms. 1971 *Guardian* 27 May 9/6 His Poundian hankerings after aristocracy. 1975 P. Fussell *Gt. War & Mod. Memory* ix. 313 The reader in search of innovation will find it in..Jones's Eliotic and Poundian juxtapositions. 1976 *Times Lit. Suppl.* 23 July 908/1 In this new short book he is more a Poundian than a critic.

poundiferous (paundi·fĕrəs), *a. rare.* [f. Pound *sb.*³: see -ferous.] Accompanied by pounding.
1871 'Mark Twain' *Let.* 28 Jan. (1917) I. 183 A long, vociferous, poundiferous and vitreous jingling of applause announces the conclusion.

pounding, *vbl. sb.*¹ **4.** *pounding-mill* (earlier and later examples).
1785 T. Jefferson *Notes Virginia* vi. 43 A good situation on a creek for a pounding mill. 1849 C. Lanman *Lett. from Alleghany Mts.* i. 17 The vein gold is brought to light by means of what is called a pounding mill. 1913 'P. Pennington' *Woman Rice Planter* (1913) 142 The cows and pigs are fed on the flour, a gray substance that comes from the grain as the chaff is removed in the pounding mill.

Poupart (pū·pāɹ). *Anat.* [The name of François *Poupart* (1661–1708), French surgeon.] *Poupart's ligament*: the inguinal ligament which extends from the anterior superior spine of the ilium to the pubic tubercle.
1756 P. Pott *Treat. Ruptures* I. 6 What is called Poupart's or Fallopius's ligament, is nothing more than the lower border of this tendon stretched from the anterior part of the os ileum or haunch bone, to the pubis. 1804 A. Cooper *Anat. & Surg. Treatm. Inguinal & Congenital Hernia* xiv. 48 On examination, a fulness could be perceived above Poupart's ligament. 1844 A. Colles in S. Mac Coy *Lect. Theory & Pract. Surg.* I. xx. 30 In the operation for femoral hernia you should always begin your first incision at least an inch above Poupart's ligament. 1910 *Practitioner* June 848 The opening in the peritoneum was closed and the operation was completed by suturing Poupart's ligament..to the thickened edge of the pectineal fascia. 1970 I. L. Lichtenstein *Hernia Repair* vi. 72 Poupart's ligament will move when the patient strains, allowing a necessary resiliency to the body's exertion.

poupeton. Add: See also *pupton.

pour, *v.* Add: **1.** (Further examples.)
1881 Mrs. J. H. Riddell *Senior Partner* II. x. 203 An old, old pug..took no notice of Mr. McCullagh or of anyone else, till Janey poured him out a glass of milk. 1961 E. Banks *Mystery F. Farrington* 54 Pour me some tea, Nancy, and tell me about your play.
c. *spec.* with ellipsis of the name of the thing poured.
1906 W. S. Maugham *Bishop's Apron* ix. 61/2 Mrs. Railing stirred the tea, put milk in each cup, and poured out. 1919 'C. Dane' *Legend* 5, I used to pour out when interesting people came to tea. 1925 E. H. Young *William* viii. 81 Lydia immediately got into a hansom... 'I can eat here, but..I can't pour out. Dora can do it.' 1930 A. Bennett *Imperial Palace* lvi. 420 'Will you pour?' she asked... He poured out the tea. 1961 R. Fuller *Image of Society* ix. 226 'Shall I pour?' she said. 1962 *Woman's Own* 31 Mar. 89/2, I think Alison should pour today. 1965 R. Petrie *Running Deep* x. 115 'Mrs. Fairfield poured,' Ian reminded her. 1973 'D. Halliday' *Dolly & Starry Bird* iii. 43 'I shall pour,' Johnson said, and pushed a cup under my nose. 'Black coffee and Saxon pride and good intentions.'
3. c. *to pour* (*on*) *the coal* (Aeronaut. slang), to cause an aircraft to accelerate; to pilot an aircraft at high speed. Also *transf.*
1937 E. C. Parsons *Gt. Adventure* xix. 233, I poured coal into the old Hispano and lit out like a scared jackrabbit. 1944 T. H. Wisdom *Triumph over Tunisia* xxii. 183 The bombs gone away, Jimmy put the nose down and poured on the coal to escape. 1961 J. M. Foster *Hell in Heavens* 58 He poured the coal to his plane and banked to avoid passing too close. 1971 M. Tak *Truck Talk* 122 *Pour on the coals*, to drive a truck at high speed.

pourable (pǭᵊ·rāb'l), *a.* [f. Pour *v.* + -able.] That may be poured; that flows easily.
1946 *Nature* 2 Nov. 636/1 The experimental preparation used was..a dilution in water prepared..in proportions to give an easily pourable, miscible oil. 1957 V. J. Kehoe *Technique Film & Television Make-Up* xii. 183 All Vinamold grades melt down to easily pourable liquids. 1959 *New Scientist* 22 Oct. 750/2 A thick syrup at room temperature, SAIB becomes pourable as the temperature is raised. 1970 *Sci. Jrnl.* May 19/1 Even with temperatures of −40°C the Astrolites did not freeze, remained pourable and were easily and safely handled.

pourboire. (Earlier and later examples.)
1817 H. C. B. Campbell *Journey to Florence* (1951) 81 The expence of the *pour boire* at each of these places is very great... The people of the custom houses..regularly ask for it. 1963 [see *mancia]. 1964 *Economist* 5 Dec. 1133/1 A *pourboire* in the shape of high rebates. 1979 *Times* 5 Dec. 14/3 Commissioners may acquire the much coveted Cabinet boxes..if they pay for them. Yet [1978]..the boxes were a kind of pourboire.

‖**pour encourager les autres** (pur ãnkuraʒe lez ōtr'), *phr.* [Fr., 'in order to encourage the others', with reference to the execution of Admiral John Byng in 1757: cf. Voltaire *Candide* xxiii. 212 Il est bon de tuer de tems en tems un Amiral pour encourager les autres.] 'In order to encourage (or deter) the others', said of an exemplary punishment or sacrifice. Also *transf.* and in weakened senses.
1804 Wellington *Disp.* (1844) II. 1032 The destruction of the band is complete, but I wished to hang some of their chiefs, *pour encourager les autres*. 1825 H. Wilson *Mem.* (ed. 2) II. 194 We talked a great deal. Hertford's subject was death, pour encourager les autres. 1896 G. B. Shaw *Let.* 29 Jan. (1965) I. 589, I am not alluding in any way to the 'Sign of the Cross' question, or to you. I have simply flicked the insect off the window pane in passing, *pour encourager les autres*. 1915 A. D. Gillespie *Lett. from Flanders* (1916) 118 We had another concert last night, and since some officer had to be the victim 'pour encourager les autres', I had to..sing them the Skye boat song. 1917 T. E. Lawrence in *Lett.* (1938) 235 This did not fall in with our ideas, since (pour encourager les autres) we wanted the news to get about that the Arabs accepted prisoners. 1963 *New Society* 21 Nov. 20/2 So *pour encourager les autres* the institute..has just published its conclusions. 1969 Y. Carter *Mr. Campion's Farthing* xviii. 178 A cruel gesture, made, I suppose, *pour encourager les autres*. He committed suicide? 1973 'M. Underwood' *Reward for Defector* viii. 71, I must be punished '*pour encourager les autres*'. 1977 *Rolling Stone* 30 June 69/3 He sent the old man away to prison all the same, *pour encourager*, as Voltaire bitterly put it in another context, *les autres*.

pouring, *vbl. sb.* Add: **b.** **pouring cream**, cream that flows readily; single cream.
1966 P. V. Price *France: Food & Wine Guide* 245 *Fontainebleau* is a very light cream cheese, often eaten with sugar and/or pouring cream. 1971 *Islander* (Victoria, B.C.) 13 June 8/1 Strawberries and cream go hand in hand..strawberries and pouring cream; strawberries and ice cream [etc.].

‖**pour le sport** (pur lə spor), *phr.* [Fr., 'for sport'.] For fun, amusement, or sport.
1924 M. Arlen *Green Hat* i. 6 A green hat, of a sort of felt..being, no doubt, one of those that women who have many hats affect *pour le sport*. 1936 A. Christie *ABC Murders* xiii. 97 He kills, it would seem from his letters, *pour le sport*—to amuse himself. 1939 *Time* 22 May 10/2 He has never gained sufficient maturity to be anything more than a conscientious individualist 'pour le sport'. 1955 M. Allingham *Beckoning Lady* ii. 20 Mr. Magersfontein Lugg himself, garbed tastefully *pour le sport*. 1973 L. Meynell *Thirteen Trumpeters* ii. 22 Lucian had automatically made tentative exploratory investigations as to her readiness *pour le sport* and bedworthiness.

‖**pour passer le temps** (pur pase lə tã), *phr.* [Fr.] To pass the time; to amuse oneself.
1681 Otway *Souldiers Fortune* I. i. 3 Some little inconsiderable questions *pour posser* [sic] *le temps*. 1777 P. Thicknesse *Year's Journey* II. lii. 153 The gay part of the French women love none, but receive all, *pour passer le tems*. 1823 Scott *Quentin Durward* I. p. xxiv, Although he admitted he read them *pour passer le temps*, yet..it was not without execrating the tendency. 1846 Geo. Eliot *Let.* 5 Nov. (1954) I. 225, I wanted some kind of worship pour passer le temps. 1889 E. Dowson *Let.* 1 Apr. (1967) 59, I still scribble occasionally—pour passer le temps, and devote most of my leisure to the French Renaissance. 1895 C. M. Yonge *Long Vacation* xiii. 128 'Yet you are helping on this concern.' 'True, but partly *pour passer le temps*.' 1976 M. Birmingham *Heat of Sun* ix. 158 Cynthia asked if she could go too: '*Pour passer le temps*,' she said... 'I love riding in cars at night.'

pour-point (pǭᵊ·ɹpoint), *sb.*² Also **pour point.** [f. Pour *v.* + Point *sb.*¹] The temperature below which an oil is too viscous to be poured.
1922 T. G. Delbridge in D. T. Day *Handbk. Petroleum Industry* I. 660 Pour point of a petroleum oil is the lowest temperature at which this oil will pour or flow when it is drilled without disturbance under certain definite specified conditions. 1936 W. L. Nelson *Petroleum Refinery Engin.* v. 48 Since 1929..the average specified pour-point for lubricating oils has been lowered from 30 to about 10°F. 1972 *Daily Colonist* (Victoria, B.C.) 3 Feb. 12/8 The reservoir has..a low sulphur content and a pour point of minus 55 degrees Fahrenheit. 1972 L. M. Harris *Introd. Deepwater Floating Drilling Operations* xvi. 171 Devices for heating and treating the emulsions are helpful, particularly, with high pour-point crudes.

‖**pour rire** (pur rīr), *a.* (*adv.*) *phr.* [Fr., lit. 'in order to laugh'.] Of a kind or in a manner that causes amusement or derision, or suggests light-hearted and jocular pretence; not serious or in a serious manner.
1872 B. Jerrold *London* x. 92 The laugh is general..over Smug who swept his own office once—and is no Liberal *pour rire*. 1884 *Sat. Rev.* 3 May 562/2 The author of a motion admits that it is only a motion *pour rire*. 1909 Mrs. H. Ward *Daphne* III. ix. 219 'Then there was some local scandal?'..'Possibly. Scandal *pour rire*! Not a soul believed that there was anything..in it.' 1926 A. Huxley *Jesting Pilate* I. 115 Hereditary aristocracies still exist in the West—exist, but *pour rire*. 1931 *Times Lit. Suppl.* 5 Feb. 96/4 These and other essays in the book are full of those analogies *pour rire* and elegant verbal jokes that so delight us nowadays. 1942 J. Lees-Milne in *Ancestral Voices* (1975) 65 The King washed up after dinner, or rather carried some glasses into the pantry and made a gesture of washing up, '*pour rire*'. 1946 A. L. Rowse *Use of Hist.* 173 They all came over in a coach... The lady who descended from it..was got up to look like one of the *louche* ladies of the Restoration Court of Charles II. It was all *pour rire*. 1959 *Punch* 25 Mar. 411/2 A flight of facetious fancy, a suggestion *pour rire*. 1980 *Country Life* 28 Feb. 609/2 The jokes about such jargon as 'a fried-chicken taste that's lip-lickin' good' have long ceased to be *pour rire*.

‖**pourriture** (puritūr). [Fr.: see Pouriture.] **a.** Rottenness, a person in a 'rotten' or unwell condition.
1890 E. Dowson *Let.* 11 Feb. (1967) 136, I have a bad cold, leaden spirits, and in fine feel all round a pretty fair 'pourriture'.
b. *pourriture noble* [lit. 'noble rot': see *noble *a.* 7 f]: a common grey mould, *Botrytis cinerea*, affecting grapes, which is deliberately cultivated to perfect certain French and German wines, and Tokay; the condition of being affected by this mould; also *ellipt.* as *pourriture*.
1911 *Encycl. Brit.* XXVIII. 722/2 The peculiar character of the Sauternes, for during the latter period of ripening a specific micro-organism termed *Botrytis cinerea* develops on the grape, causing a peculiar condition termed *pourriture noble* (German *Edelfäule*). 1924, 1935 [see *noble *a.* 7 f]. 1951 R. Postgate *Plain Man's Guide to Wine* vi. 98 A full and scented wine, the product of the *pourriture noble*. 1963 *Times* 8 Feb. 12/5 Wines made from..over-ripe single grapes which are, in effect, sun-dried raisins, in a state of *pourriture noble*, or princely decay. 1966 P. V. Price *France: Food & Wine Guide* 175 Several great properties have begun to make dry wines again..by picking the grapes before the pourriture has started to form. 1972 *Country Life* 14 Dec. 1658/3 The *vin jaune* grapes are not harvested until November, by which time they have been withered by the *pourriture noble*..that affects Sauternes and the great sweet wines of Germany.

‖**pour-soi** (pūr,swa). *Philos.* [F., lit. 'for itself', 'for oneself'.] A phr. used by J.-P. Sartre in *L'Être et le Néant* (1943) to designate conscious, free being; being-for-itself; contrasted with *en-soi*, being-itself.
1947 *Jrnl. Philos.* XLIV. 720 The latter [*sc.* Heidegger], seeing the given always turning again into the abstract, concludes that human existence alone is *pour soi*. 1950 *Mind* LIX. 270 Sartre..in *L'Être et le Néant*..discusses the historical character of the *pour-soi*. 1962 *Listener* 30 Aug. 317/1 Sartre sees existence in the world as a polarity, a struggle for power. To survive, the *pour-soi* or individual consciousness has to reject and deny the *en-soi* of mere identity, of being what I have to be. 1964 C. Smith *Contemp. French Philos.* ii. 33, I doubt if Sartre would agree that there can be any possibility of 'lucidly knowing' the *pour-soi*. 1966 A. Manser *Sartre* iii. 45 Even though Sartre distinguishes between his approach and metaphysics..his division of being into two kinds, *l'en-soi* and *le pour-soi*, together with the question he asks in his introduction..seem to indicate that what follows can only be muddled and obscure. 1977 Warren & Ponse in Douglas & Johnson *Existential Sociol.* 304 This tension between the *en-soi* and the *pour-soi* is common in gay liberation writings and experience.

‖**pousada** (pōusā·dā). [Pg. *pousada* resting-place, f. *pousar* to rest.] An inn or hotel in Portugal, esp. one of a chain of hotels administered by the State. Also *attrib.*
1934 J. & C. Gordon *Portuguese Somersault* iv. 35 'Is there no such thing as a good *posada* [sic]?' asked Jo. 'Hotel?' said the cord porter. 1949 Bridge & Lowndes *Selective Traveller in Portugal* vii. 115 Just outside the main gate of Elvas, with a fine view of the aqueduct, is one of the nicest of the Government *pousadas* or inns. 1954 G. Hogg *Portuguese Journey* 44 As we began to descend the hill into S. Tiago we passed the government pousada. 1967 *Guardian* 16 Dec. 9/2 Nights in pousadas will have to be rationed. But I have stayed in little *pensoes* and inns. 1972 *Times* 17 June (Portugal Suppl.) p. ii/4 The *pousada* restaurants cater for motorists. 1976 *Good Motoring* Jan.–Feb. 16/2 Hotel prices [in Portugal] vary enormously but the State-owned pousadas are good value.

pousse-café. Add: (Earlier and later examples.) Also *fig.*
1880 *Harper's Mag.* June 25/2 There is no easier way of solving a social problem than through the medium of a mild and fragrant cigar and a *pousse-café*. 1897 *N.Y. Dramatic Mirror* 27 Nov. 22/3 (Advt.), Weber and Fields' Music Hall..Next Burlesque, commencing Nov. 29, Pousse Cafe; or, the Worst Born. 1948 *Sun* (Baltimore) 1 Jan. 15/1 The sophisticate who goes to the tourist traps makes the barkeep unhappy by asking for fancy things like a pousse cafe (six different liqueurs which are poured gently to avoid mixing the colors). 1959 W. Burroughs *Naked Lunch* 72 Men and women in evening dress sip pousse-cafés. 1962 E. Lanham *No Hiding Place* ii. 20 The sophisticated liqueurs of a pousse-café. 1965 *New Statesman* 24 Dec. 1011/2 The other glamour-hunter's night out was the Australian Ballet's *Raymonda*... It's tempting to be funny about..the interior of Rank's New Victoria, a storm in a pousse-café, sinking knee-deep in ice-cream kups and sweet-wrappings. 1973 *Anglican Theol. Rev.* (Evanston, Illinois) Sept. 35 In the position of Episcopalians in the sociological *pousse-café* of American life, such ecclesiastical Anglophilia was acceptable, even pleasant. 1977 *Time* 12 Sept. 46/3 Some striped jobs look like *pousse-café* or rugby sweaters gone south.

‖**poussin** (pusæ̃). [Fr., a newly-born chicken.] In *Gastronomy*, a baby chicken. See also *petit poussin* s.v. *petit *a.* (*sb.*) 5.
c 1938 *Fortnum & Mason Price List* 72/1 Poussin and Green Peas—per head 5/6. 1957 'P. Quentin' *Suspicious Circumstances* vi. 64 Lunch that day was so gourmet that I thought I would choke on it... Mother took little bird pecks at her Poussin Marie Louise or whatever it was. 1964 M. Kelly *March to Gallows* vii. 72, I looked at the menu card..saw the words *half poussin*. 1969 *Daily Tel.*

11 Jan. 14/2 In Dominica try 'mountain chicken'... Well, yes, I know it's a frog, but a special frog, tasting like *poussin* when grilled. 1974 *Times* 7 Mar. 13/5 Fresh *poussins*—small chickens four to eight weeks old are available.

Poussinesque (pusane·sk), *a.* [f. the name *Poussin* (see below) + -ESQUE.] Pertaining to or characteristic of the French landscape painter Nicolas Poussin (1594–1665) or his work; resembling or influenced by the style of Poussin.

1919 R. FRY *Let.* 9 Feb. (1972) II. 446 *La Jeune Parque* seems to me to have a quality of pure beauty—in the Miltonic Poussinesque direction that hardly any modern work possesses. 1934 *Burlington Mag.* Dec. 297/2 We hear nothing of Pierre Lamaire..whose variant of the Poussinesque landscape formula is yet quite a notable and personal one. 1944 *Ibid.* Aug. 186/2 It is no longer possible ..to point to individual figures as Poussinesque. 1955 *Times* 12 May 5/5 There is a brilliant little canvas by him of women bathing in a Poussinesque landscape. 1964 *Listener* 9 Jan. 57/1 The combination of rectilinear and diagonal movements in plane and inferred space..attains a complexity and variety of such a high order that we recognize it as 'Poussinesque'. 1979 *Basildon Park* (National Trust) 18 The Poussin-esque landscape by Francisque Millet (1642–79) on the west wall.

pou sto. (Later examples.)

1890 E. E. C. JONES *Elem. Logic* xi. 90 The interpretation of our proposition is liable to involve us in an infinite regress; we have no ποῦ στῶ. 1911 J. WARD *Realm of Ends* iii. 67 This is the physical basis which is supposed to furnish teleology with its indispensable ποῦ στῶ. 1963 W. SELLARS *Sci., Perception & Reality* iii. 89 It is here, rather than at the level of sense contents, that we find a *pou sto* for the apparatus of hypothetico-deductive explanation. 1967 *Philos. Rev.* LXXVI. 181 The self-stultifying form appears to be shorn of its ποῦ στῶ.

pout-net. (Later examples.)

1859 *Act* 22 & 23 *Vict.* c. 70 § 14 To kill Salmon in or from the River by means of any Pout Net, Rake Hook, or similar Engine. 1911 A. WARRACK *Scots Dial. Dict.* 425/2 *Pout-net*, a round net fastened to two poles, thrust under the banks of a river to force out fish.

pouty (pau·ti), *a.* *U.S.* [f. POUT *sb.*^{2} or *v.*^{1} + -Y^1.] Inclined to pout (said of a person or of the mouth); hence (as a personal attribute), sullen or petulant.

1863 'G. HAMILTON' *Gala-Days* 221 They never were tired when anything was to be done, or..peevish, or pouty, or 'offish'. 1897 R. M. STUART *In Simpkinsville* 23 This stove's ez dull-eyed and pouty ez any other woman ef she's neglected. 1912 R. A. WASON *Friar Tuck* 77 With a pouty look on his face, Tank sez: 'It's time we fixed up an' moved out into the dark.' 1971 C. FICK *Danziger Transcript* 44 Stella as in starlight was direct brown eyes and a pouty lower lip. 1976 'O. BLEECK' *No Questions Asked* xiii. 147 She had a full red mouth that..might have looked kissable to some, but it looked only pouty to me.

pouw. Add: Also **pou, pow.** Cf. *GOMPAAUW, PAAUW, *PAUW. (Earlier and later examples.)

1798 A. BARNARD *Jrnl.* 13 May in A. W. C. Lindsay *Lives of Lindsays* (1849) III. 439, I..tied the *pow*, or wild peacock, to the waggon—a very fine bird indeed, of grave colours, but rich brown. 1858 [see *DIKKOP 1]. 1959 *Cape Argus Mag.* 3 Jan. 8/2 Even the beginner in bird lore would not look for..a pou in the Knysna forest. 1966 [see *NARTJIE].

poverty. Add: **III. 8. poverty-grass,** (*a*) substitute for def.: one of several North American grasses that grow on poor soil, esp. *Aristida dichotoma*; (earlier and later examples); **poverty level** = *poverty line*; **poverty line,** the estimated minimum income sufficient for obtaining the necessities of life; **poverty programme** *U.S.*, a programme or policy designed to alleviate poverty; **poverty shop** (see quots.); **poverty trap,** a situation in which an earned increase to a low income is offset by the consequent loss of means-tested state benefits.

1832 *Boston Even. Transcript* 30 Apr. 2/3 Fields..long given up to barrenness and poverty-grass, are now broken up in readiness to receive the grain. 1906 J. C. LINCOLN *Mr. Pratt* vi. 95 He owned the sheds and barn..and the beach grass and the poverty grass. 1939 H. H. BENNETT *Soil Conservation* xvii. 418 Scores of plants, known as weeds, enter into this far-reaching cover of volunteer vegetation: goldenrod, ragweed, poverty grass, [etc.]. 1973 H. McCLOY *Change of Heart* iii. 23 The soil is still so sandy that only scrub pine, beach plum, bayberry bushes, and poverty grass grow wild there. 1976 *Billings* (Montana) *Gaz.* 27 June 12-G/1 Statistics for last year show that over five million persons past age 65, or one out of every six, live on a poverty level—defined as an income of under $46 a week for a single aged person or $57 for an aged couple. 1977 *Rolling Stone* 30 June 59/1 The poverty level varies according to family status. 1901 W. S. CHURCHILL in R. S. Churchill *Winston Churchill* (1969) II. Compan. I. ii. 108 Families who cannot provide this necessary sum, or who, providing it do not select their food with like discrimination are *underfed* and come below the 'poverty' line. 1901 B. S. ROWNTREE *Poverty* iv. 114 The recipients of charity are the poor, i.e. those who from causes 'primary' or 'secondary' are below the poverty line. 1904 [in Dict.]. 1932 *Discovery* June 181/2, 21 shillings in Charles Booth's time..[was] the income for an 'ordinary family' at or about the poverty line. 1941 *Economist* 19 Apr. 522/1 The most fortunate group were the old age pensioners of Group C, who might..have an income of 32s., but even this, though well above the poverty line, is nowhere near the human needs level. 1968 E. BRILL *Old Cotswold* vi. 89 He had little of the sharp business acumen that goes with the making of money on a big scale, but this is not evidence that he or his family were ever on the poverty line. 1973 *Observer* (Colour Suppl.) 19 Aug. 26/2 In 1960 about 11 per cent of the population..were living below the poverty line, defined as basic National Assistance rates plus 40 per cent. 1967 *Freedomways* VII. 104 A few years ago..it seemed as if there was a real promise of hope for the poor—both black and white—through the Poverty Program. 1970 *New Yorker* 29 Aug. 57/1 Some of the people..being in a position to welcome any kind of investment, even poverty-program investment..have decided Miller is entitled to his ideas. 1971 *Black Scholar* Apr.–May 10 Who subsist on such foreign aid programs as 'welfare' and 'poverty programs'. 1948 R. GLASS *Social Background of Plan* IV. ii. 161 Newport Road is famous for its 'poverty shops': fried fish shops, pawnbrokers and junkshops. 1956 J. M. MOGEY *Family & Neighbourhood* i. 10 It has become customary to call certain types of shops 'poverty shops': these are fried-fish shops, pawnbrokers, and junkshops. 1961 Poverty shop [see *fish and chips* s.v. *FISH *sb.*^{1} 7]. 1972 *Daily Tel.* 21 Nov. 1/8 The idea was to prevent families falling into the 'poverty trap'—the situation in which a pay rise can mean the poor are worse off because they lose a disproportionate number of State benefits. 1973 *Guardian* 30 Mar. 6 The abolition of the 'poverty trap' called for the reintroduction of reduced tax rates for those near the threshold of tax. 1977 *Times* 23 Mar. 16/6 The 'poverty trap'..arises from the fact that the level of supplementary benefits judged to be the minimum.. acceptable can..total more than the personal allowance against tax.

poverty-stricken, *a.* (Earlier and later examples.)

1803 M. WILMOT *Let.* 1 Oct. in *Russ. Jrnls.* (1934) I. 55 Amidst such a multitude of titles a count or countess is often the merest poverty stricken low bread [*sic*] animal that ever was known. 1956 *Railway Mag.* Nov. 739/2 The original promoters of the Port & Pier Railway could hardly have visualised their poverty-stricken child playing such an important role as it was destined to do!

poverty struck, *a.* (Earlier example.)

1801 D. WORDSWORTH *Jrnl.* 29 Dec. (1941) I. 98 A miserable, poverty-struck looking woman.

pow (pau), *int.* orig. *U.S.* [Echoic.] The sound of a punch, blow, shot, or the like. Also *fig.*, used to denote the impact of an emotion or an idea. Also as *sb.*

1881 J. C. HARRIS in *Scribner's Monthly* June 244/1 He step en hit de hoss a rap—*pow!* De hoss 'uz dat s'prise at dat kinder doin's dat he make one jump, en lan' on his footses. 1914 B. TARKINGTON *Penrod* xxii. 207 Herman rubbed his smitten cheek. 'Pow!' he exclaimed. 'Pow-ee! You cert'ny did lan' me good one *nat* time.' 1931 *Technol. Rev.* Nov. 66/1 That class of comic-strip words like *zowie* and *pow*. 1955 H. KURNITZ *Invasion of Privacy* (1956) xxvi. 163 'A man hit him.' She went through the motion with both little fists. 'Pow!' 1961 J. HELLER *Catch-22* (1962) v. 42 He called me a wise guy and punched me in the nose..knocked me flat on my ass. Pow! Just like that. 1964 'E. McBAIN' *Ax* iv. 72 He got himself an axe someplace... Pow, good-bye janitor. 1968 L. DEIGHTON *Only when I Larf* i. 16 Imagine beating that typewriter.. for a hundred a week and all the pencils you can take home. Pow. Not me. 1970 G. GREER *Female Eunuch* 173 Perhaps they will not fall in love all at once but feel a tenderness growing until one day *pow!* that amazing kiss. 1970 [see *KER-]. 1976 *Leicester Chron.* 26 Nov. 16/3 In some cases that does not mean films which are more sanguinary, but poorly made action stuff with entire reliance on the pows and kerplunks.

powder, *sb.*^{1} Add: **1. e.** = *POWDER SNOW.

1948 *Sun* (N.Y.) 30 Dec. 16 North Conway. 3 inch new powder. Skiing fair. 1973 P. A. WHITNEY *Snowfire* xii. 235 The average skier..didn't care for loose powder. But there was still powder on the steepest slope. 1973 R. HAYES *Hungarian Game* xliv. 257, I came here to ski and I'd hate to miss all this nice fresh powder. 1974 G. MOFFAT *Corpse Road* ii. 34 We had a light fall at Christmas ..just powder on frozen grass.

2. d. *in powder*, wearing hair-powder; also *out of powder*.

1792 LADY TEMPLETOWN *Let.* 11 June in A. E. Newdigate-Newdegate *Cheverels* (1898) vii. 103 Mr Romney.. has acquitted himself well in respect to Lady Newdigate... The hair is of an agreable *duskiness* that is neither in nor out of powder. 1849 THACKERAY *Pendennis* I. xxiii. 219 Two superior officers in black..now in livery with their hair in powder. 1863 CROWN PRINCESS OF PRUSSIA *Let.* 11 May in R. Fulford *Dearest Mama* (1968) 211 We give a ball in powder tomorrow as this old Palais will have been up 100 years. 1874 L. TROUBRIDGE *Life amongst Troubridges* (1966) ix. 78 The day began hatefully by Grobee telling us that we were not to go to the Cresswells' dance in powder..what *possible* difference could it make to him whether we went with black hair or white. 1924 M. IRWIN *Still she wished for Company* xix. 233 Slovenly Lady Catherine Grey drove over, out of powder at four in the afternoon. 1954 H. ASHTON (*title*) Footman in powder.

g. Denoting other preparations in the form of a powder, chiefly in cookery, hygiene, perfumery, etc., and usu. as the second element of a Comb., as *baby powder, baking powder, curry powder, flea powder, insect powder, milk powder, soap powder, talcum powder, tooth powder, washing powder*, etc.: see under the headwords.

h. Slang phr. *to take a powder*: to depart, absent oneself; to abscond. See also *RUN-OUT 4. orig. and chiefly *U.S.*

1934 J. PROSKAUER *Suckers All* xxiv. 279 The smartest guy in the office took a walk out powder this morning. 1940 J. O'HARA *Pal Joey* 72 And take a powder out of here that day. 1941 R. CHANDLER in *Street & Smith's Detective Story Mag.* Sept. 25 Why are you taking a powder? 1954 'N. BLAKE' *Whisper in Gloom* xvi. 220 'Where's the Yank?'..'Gone. He took a powder.' 1961 J. MACLAREN-ROSS *Doomsday Bk.* v. 65 Phoned four times—no reply. Seems as if..Passman's taken a temporary powder. 1972 'H. HOWARD' *Nice Day for Funeral* iii. 39 If he'd dumped it [*sc.* a corpse] in the river..everybody was bound to think Frankie had taken a powder to dodge the grand jury. 1979 P. ABLEMAN *Shoestring* i. 14 The very minute that I first looked into her..eyes.. Philip Marlowe took a powder and Shoestring, the womanless, took over.

3. b. Fig. phr. *to keep one's powder dry*, with allusion to the advice said to have been given by Oliver Cromwell to his troops: to adopt a practical or realistic policy; to act prudentially or cautiously, be on the alert.

[1834 COL. BLACKER *Oliver's Advice* in E. Hayes *Ballads of Ireland* (1856) I. 192 The Pow'r that led great William, Boyne's reddening torrent thro',—In his protecting aid confide, and every foe defy—Then put your trust in God, my boys, and keep your powder dry. 1856 E. HAYES *Ballads of Ireland* (ed. 2) I. 191 There is a well-authenticated anecdote of Cromwell. On a certain occasion, when his troops were about crossing a river to attack the enemy, he concluded an address, couched in the usual fanatic terms in use among them, with these words —'put your trust in God; but mind to keep your powder dry'.] 1908 *Times Lit. Suppl.* 5 Nov. 383/1 In thus keeping his powder dry the bishop acted most wisely, though he himself ascribes the happy result entirely to observance of the other half of Cromwell's maxim. 1931 F. L. ALLEN *Only Yesterday* ii. 40 An inheritor of Theodore Roosevelt's creed of fearing God and keeping your powder dry. 1948 A. TOYNBEE *Civilization on Trial* x. 193 A 'Zealotism' tempered by a belief in keeping his powder dry. 1954 C. P. SNOW *New Men* ii. 17 It doesn't sound like business for this time. Still it won't do any harm to watch out and keep our powder dry. 1955 *Times* 6 Aug. 5/1 It is clear that M. Faure, to judge by what he did not say today, is keeping his powder dry. 1968 *Listener* 27 June 833/3, I seem to have been resigned most of my poetic life to the virtues of keeping one's powder dry rather than trying to fire the big guns.

5. a. *powder mark, scales, -smoke* (example); *powder-dry, -light* adjs.; (sense 2 d) *powder bowl*; *powder-dusted* adj.

1919 in G. Howell *In Vogue* (1975) 34/1 Porcelain powder bowls, for dusting powder. 1930 'R. CROMPTON' *William's Happy Days* ix. 218 Powder bowls and dolls and cushions. 1972 *Daily Tel.* 10 Oct. 13 Today's young people hardly know what a rose-bowl is, and few possess a cut-glass powder-bowl for loose powder and feathery puff. 1934 T. WOOD *Cobbers* xvi. 215 They worked themselves powder-dry. 1942 W. FAULKNER *Go down, Moses* 100 The pale, powder-light, powder-dry dust of August. 1917 V. WOOLF *Mark on Wall* in *Two Stories* 20 The miniature of a lady with white powdered curls, powder-dusted cheeks, and lips like red carnations. 1937 D. & H. TEILHET *Feather Cloak Murders* i. 14 The revolver bullet left a clean hole when shot close, always with powder marks. 1975 G. LYALL *Judas Country* ix. 68 If I'd been faking a suicide, I'd've put the gun in a more obvious place. Anyway, can't you test his hand for powder marks? 1976 *Shooting Times & Country Mag.* 16–22 Dec. 30/1 Whereas almost every large, local gun-dealer stocks reloading machines, very few stock powder scales. 1905 T. COLLINS in Murdoch & Drake-Brockman *Austral. Short Stories* (1951) 16 The explosion came off, nearly smothering me with powder-smoke.

b. powder base = *foundation cream* s.v. *FOUNDATION 7 d; also (with hyphen) *attrib.*; **powder-burn,** a burn made by the hot gases emitted by a firearm; so **powder-burn** *v. trans.*; **powder cake,** a block of compressed face-powder; **powder chamber,** (*a*) (earlier example); **powder closet** *obs.* exc. *hist.*, a small room formerly used for powdering hair or wigs; **powder colour,** (*a*) an opaque water-colour in powder form; (*b*) (see quot. 1966); **powder compact** = *COMPACT *sb.*^{2} e; **powder-house** (earlier and later examples); also *fig.*; so **powder-house-keeper**; **powder keg,** a small barrel or container for holding gun powder or blasting powder; also *fig.*; **powder magazine** (further examples); also *fig.*; **powder metallurgy,** the branch of metallurgy which is concerned with the production of metals as fine powders and their subsequent pressing and sintering into compact forms; hence **powder-metallurgical** *a.*; **powder metallurgist**; **powder paint** = *powder colour* (*a*); **powder pattern,** (*a*) *Cryst.* (see 5 c below); (*b*) a pattern indicative of the domain structure of a magnetized solid, formed when a colloidal magnetic powder is allowed to settle on it; **powder-post** (later example); **powder-post beetle,** a small brown beetle belonging to the family Lyctidæ, the larva of which bores tunnels in seasoned timber, reducing it to powder; **powder rag,** a piece of cloth used for applying

face-powder; **powder slope,** a slope covered in powder snow.

1927–8 *T. Eaton & Co. Catal.* Fall & Winter 367 A greaseless Vanishing Cream, for use as a protective or powder base. 1932 [see *BASE *sb.*^{1} 11 b]. 1955 M. ALLINGHAM *Beckoning Lady* iv. 66 A good foundation of that powder-base stuff. 1972 *Vogue* 1 Mar. 52/1 An ideal powder-base—inimitable beneath modern make-up to ensure a flawless, perfectly matt finish. 1846 J. W. WEBB *Altowan* I. iv. 125 He might powder-burn the bear by the nearness of the shot. 1847 in H. Howe *Hist. Coll. Ohio* 99 In this struggle, Lytle..had..his face powder burnt. 1927 *Scribner's Mag.* Feb. 176/2 In the pursuit, the Rangers literally carried out their leader's orders to 'powder-burn' them. 1969 G. MACBETH *War Quartet* 72 With his gun One braised his leg three times. The doctor saw Powder-burns there, and left him. 1975 M. BABSON *There must be Some Mistake* xx. 182 You have gloves... Put them on now. The only powder burns must be found on those two that the police may re-enact the scene. 1961 'A. A. FAIR' *Stop at Red Light* (1962) v. 82 Parts of a powder cake were on the floor, and bits of glass from the broken mirror. 1803 *Jrnl. Natural Philos.* IV. 251 As soon as the lever has arrived at the position N, the powder chamber P is exactly opposite the ball, and ready to be discharged against it. 1905 *Pall Mall Mag.* Dec. 746/1 Violante..lay dozing in the powder closet which opened out of Donna Carlotta's bedroom. 1927 *Daily Express* 12 Dec. 4 Methley Park..has one or two unusual features, however; and among these are some queer old powder closets. 1929 S. ERTZ *Galaxy* i. 11 The house contained a powder-closet for the dressing of ladies' hair in earlier days. 1980 *Country Life* 28 Feb. 609/1 The emergence of indecencies from the powder closet to the respectable page. 1862 *Illustr. Catal. Internat. Exhib., Industr. Dept., Brit. Div.* II. No. 5512 Colours prepared for missal-painting, and illumination in soluble powder-colour. 1913 R. FRY *Let.* 28 Dec. (1972) II. 376 Grind up your powder colours in water very stiffly and use with the yolks of eggs. 1963 S. MARSHALL *Exper. in Educ.* iii. 88 Powder colour has proved its worth. 1966 J. S. COX *Illustr. Dict. Hairdressing* 121/1 *Powder-colour*, colour rinses in powder form. 1927 [see *COMPACT *sb.*^{2} e]. 1937 D. L. SAYERS *Busman's Honeymoon* xv. 259 He laid a powder-compact aside on the what-not. 1978 *Cornish Guardian* 27 Apr. 32/4 The new president, Mrs. Tippitt, presented Mrs. M. Williams, the retiring president, with a powder compact. 1720 in *Mass. House of Representatives Jrnl.* (1921) II. 288 Daniel Powning, keeper of the Powder-House. 1848 *Knickerbocker* XVIII. 216 The powder house, the pound, the poor-house and the county-house, are all objects of notice to the traveller. 1928 *Manch. Guardian Weekly* 7 Sept. 181/4 The spark that fired this powder-house was a letter protesting against the 'constant criticism' of the methods of Lancashire cricketers. 1789 *Rec. Early Hist. Boston* (1886) X. 183 Foster Thomas, powder-house-keeper. 1855 W. G. SIMMS *Forayers* iii. 39 Sinclair ..drew up an old powder-keg by a rope-hitch, which had been made about it. 1876 'MARK TWAIN' *Tom Sawyer* xxxii. 323 It was the treasure-box..along with an empty powder-keg, a couple of guns in leather cases,..a leather belt, and some other rubbish. 1893 W. K. POST *Harvard Stories* 6 One reason why they do it..is to make you flare up, you little powder keg. 1945 *Richmond* (Va.) *News-Leader* 17 Sept. 7/2 (*heading*) Argentina's militarist regime believed sitting on powderkeg. 1972 *Publishers' Weekly* 24 Jan. 21 His stories weeks before the revolt warned of the powderkeg inside the prison. 1975 *Publ. Amer. Dial. Soc.* 1973 LIX. 47 *Powder keg*..., a round, metal container for blasting powder, usually of twenty-five pounds capacity. 1890 KIPLING *Departmental Ditties* (ed. 4) 98 You shouldn't take a man from Canada And bid him smoke in powder-magazines. 1933 J. BUCHAN *Prince of Captivity* II. ii. 196 Birkpool is..becoming a powder magazine. 1979 G. LATTA tr. *Jacquemard-Sénécal's Eleventh Little Nigger* II. iv. 94 The reunion of the four of them..would constitute a sort of powder-magazine: one spark and the magazine blows up. 1980 ALEXANDER & ANAND *Queen Victoria's Maharajah* i. 11 A lucky hit on the powder magazine. 1949 *Electronic Engin.* XXI. 88/1 The production of intricate structures is due to the fact that moulding powders are now made by powder-metallurgical processes. 1975 BRAM & DOWNS *Manuf. Technol.* iii. 79 Cemented carbides are a typical powder metallurgical product. 1949 C. G. GOETZEL *Treat. Powder Metall.* I. p. vii, The final chapter of Part One covers briefly the many uses for metal powders that are somewhat beyond the sphere of interest of the powder metallurgist. 1954 H. UDIN et al. *Welding for Engineers* iii. 39 This process of spheroidization of pores or inclusions has been of great interest to powder metallurgists. 1933 *Engin. & Mining Jrnl.* CXXXIV. 373/1 What is frequently referred to as 'powder metallurgy' had its beginnings at the turn of the century when the metals tungsten and molybdenum first became commercial commodities. 1959 *Listener* 12 Mar. 453/1 Coolidge in America developed the process now known as powder metallurgy, by which a bar of compressed tungsten powder was sintered at a temperature below the melting point of the metal. 1970 *New Scientist* 12 Nov. 325/1 Powder metallurgy is commonly used to make precision components, and to fabricate exotic materials. 1939 L. DE LISSA *Life in Nursery School* ix. 158 Powder paints are cheap and suitable and can be obtained in good colours. 1955 E. BLISHEN *Roaring Boys* i. 33 How should I learn to distinguish between different types of brush, to mix powder paint? 1973 *Galt Toys Catal.* 49 *Powder paint set*..mix up a small quantity with water as required. 1934 *Physical Rev.* XLVI. 227 (*caption*) Powder patterns with H normal to the surface. 1951 L. F. BATES *Mod. Magn.* (ed. 3) xii. 457 The main method of studying domain structure is undoubtedly the Bitter figure or powder pattern method. 1965 CRAIK & TEBBLE *Ferromagnetism & Ferromagnetic Domains* 308 The first stage in the preparation of colloids or suspensions for the powder pattern or Bitter figure technique, is the production of magnetite (Fe_3O_4) in a very finely divided state. 1927 *Bull. U.S. Dept. Agric.* No. 1490, 7 Powder post is that class of defects in which the larvae of insects reduce the wood fibers of seasoned or partially seasoned wood to a powderlike condition. 1905 *Yearbk. U.S. Dept. Agric.* 1904 387 (*caption*) Work of powder post beetle.. in hickory poles. 1911 *Technical Ser. Bureau Entomol., U.S. Dept. Agric.* XX. III. 111 (*title*) A revision of the powder-post beetles of the family Lyctidæ of the United States and Europe. 1928 *Forestry* II. 42 The sapwood..has been reduced to a finely powdered, floury condition—the characteristic damage that gives the name of 'powder-post beetles' to the *Lyctus* species. 1963 *B.S.I. News* Mar. 6/1 The third British Standard will describe a test for determining the toxicity of wood preservatives to the powder post beetle, *Lyctus brunneus*. 1975 G. EVANS *Life of Beetles* iv. 94 Lyctidae, the powder post beetles (e.g. *Lyctus* spp.), have larvae which produce a very fine powdery dust. 1904 'O. HENRY' in *McClure's Mag.* Aug. 352/2 This stake comes in handier than a powder rag at a fat man's ball. 1906 —— *Four Million* (1916) 21 Delia finished her cry and attended to her cheeks with the powder rag. *a* 1911 D. G. PHILLIPS *Susan Lenox* (1917) II. ii. 33 Susan..safeguarded her nose against shine; she tucked the powder rag into the stocking. 1972 D. HASTON *In High Places* xi. 115, I could put this [failure] out of my mind swooping around the powder slopes.

c. With reference to the Debye-Scherrer method of X-ray crystallography (see *DEBYE), as *powder camera, diffraction, pattern, photograph, photography*.

1917 *Physical Rev.* X. 664 The powder photographs have an advantage..over ionization-chamber measurements, in that the intensities of reflection from different planes, as well as different orders, are directly comparable. 1924 R. W. G. WYCKOFF *Struct. Crystals* vi. 178 The spectrum lines which result from these reflections of monochromatic X-rays constitute a powder photograph. *Ibid.*, Such a powder pattern can be greatly simplified by filtering the X-rays to render them essentially monochromatic. *Ibid.*, The outstanding advantage of powder diffraction methods obviously lies in their ability to treat the many crystalline materials which do not grow large single crystals. *Ibid.* 185 A more extended description of these procedures is not justified because thus far they have found little application to powder photography. 1936 *Jrnl. R. Aeronaut. Soc.* XL. 411 The powder photograph is..a powerful means of recognising alloy phases. 1945 C. W. BUNN *Chem. Crystallogr.* v. 109 A powder camera consists essentially of an aperture system to define the X-ray beam, a holder for the specimen, and a framework for holding the photographic film. 1948 K. LONSDALE *Crystals & X-Rays* iii. 76 For powder photography monochromatic radiation is used..but the specimen is a mass of tiny crystals orientated in all directions. 1962 *Times* 4 Sept. 2/6 There are four X-ray generating sets and accessory equipment includes powder cameras. 1965 ADAMS & RAYNOR *Advanced Pract. Inorg. Chem.* xvii. 155 The intersection of a curved strip of film with these diffraction cones gives the familiar powder photograph, which consists of a series of curved lines. 1969 B. E. WARREN *X-Ray Diffraction* v. 67 Powder patterns are very often used for a precision measurement of the crystal axes.

powder, *v.*^{1} Add: **3. b.** (Further examples.) Phr. *to powder one's nose*, used also *euphem.* (with reference to a woman) for 'to go to the lavatory'.

1921 W. S. MAUGHAM *Circle* I. 28, I must powder my nose, Hughie. 1924 E. O'NEILL *Welded* II. 137 You'll want to go upstairs and powder your nose. 1927 S. ERTZ *Now East, now West* xvii. 261 She put no colour on her face,..which, if she powdered and didn't even redden her lips, always made people ask her if she were ill. 1930 A. BENNETT *Imperial Palace* lvi. 417 That's the bathroom and so on... You can hang your overcoat in there—and powder your nose. 1938 I. GOLDBERG *Wonder of Words* vi. 108 We are invited to wash our hands, or, if we wear dresses, to powder our noses. 1962 *Guardian* 5 Dec. 6/5 Useful information..about where to park..dine, stay overnight, and—for women—powder one's nose in comfort. 1969 R. T. WILCOX *Dict. Costume* (1970) 88/1 Venetian gentlemen also painted, powdered and patched. 1972 L. P. DAVIES *What did I do Tomorrow?* 72 I'll use your bathroom. To powder my nose, as nice girls say.

powder blue, *sb.* and *a.* Add: **1.** (Earlier example.)

1656 *Essex County, Mass. Probate Rec.* (1916) I. 233 Mace and Ribing, starch and poudarblu.

2. (Further examples.) Now also used of light blue shades.

1923 [see *CYCLAMEN c]. 1943 A. CHRISTIE *Moving Finger* xi. 130 A skin-tight powder-blue evening dress. 1967 *New Statesman* 28 July 110/3 Their powder-blue berets bob above helmets. 1970 *Cape Times* 28 Oct. 19/2 (Advt.), Rambler Rogue 1970, 12 800 miles, powder blue with black upholstery, executive's car. 1973 'S. HARVESTER' *Corner of Playground* III. i. 170 A waxed red hat with a powder-blue band. 1978 *Jrnl. R. Soc. Arts* CXXVI. 352/1 Just as there are many different shades of blue, for example powder blue,..wedgwood blue and pacific blue, so there are many different nuances of piano tone.

3. *attrib.* or as *adj.* Applied *spec.* to a type of Chinese porcelain of the Ch'ing period having a ground of powder blue. Also as **powdered blue.**

1900 F. LITCHFIELD *Pott. & Porc.* vii. 114 The date of what is called 'powdered blue' china is said to be the Kang-he period (1661–1722). 1906 S. W. BUSHELL *Chinese Art* II. viii. 36 The finely pounded pigment is blown upon the raw body to produce, when glazed, a 'powder blue', or *bleu fouetté* ground. 1911 [see *KANG-HAI, KANG-HE, KANG-HI]. 1960 R. G. HAGGAR *Conc. Encycl. Continental Pott. & Porc.* 352/2 A powdered-blue ground colour of great depth and richness was introduced at Meissen porcelain factory before 1725. 1965 [see *K'ANG-HSI]. 1971 L. A. BOGER *Dict. World Pott. & Porc.* 270/1 *Powder Blue Ware*, in Chinese ceramics; porcelain of the Ch'ing period, especially of the K'ang Hsi reign, which is characterized by a beautiful and rather even ground of bright soft blue with a slightly clouded appearance.

4. *Comb.*, as *powder-blue-grey* adj.

1952 S. SPENDER *Learning Laughter* 30 Fields and fields of powder-blue-grey cornflowers.

powder-box. Add: **c.** (Later examples.)

1895 *Army & Navy Co-op. Soc. Price List* 172/2 Powder Boxes—Puff, ea. 7/5, 9/4, [etc.]. 1925 *Heal & Son Catal.: Glass*, Powder Box. Height 6 in. Price 9/6. 1949 *Dædalian Q.* Fall 26 Powder box, powder box, powder your nose, How many petals are in a rose? 1979 G. LATTA tr. *Jacquemard-Sénécal's Eleventh Little Nigger* I. v. 46 He was still clutching his powder-box and puff in his clenched hands.

powdered, *ppl. a.* Add: **7.** Applied to foods that have been reduced to the form of a powder by dehydration.

1889 in A. Davis *Package & Print* (1967) Pl. 69 Pancake flour—self-rising mixture—..powdered milk without butter fat, corn sugar, [etc.]. 1909 *Chem. Abstr.* 933 Powdered milks are made to contain varying amounts of fat by the removal of more or less fat cream before evaporation. 1917 *Official Gaz.* (U.S. Patent Office) 2 Jan. 363/1 The Cabell Company, Baltimore, Md... Velvet *Particular description of goods.*—Powdered Egg. *Claims use* since Sept. 1, 1916. 1917 *Bakers Rev.* Nov. 13 (Advt.), *Velvet* Powdered Egg, made from selected fresh hen eggs, is entirely and instantly soluble. 1919 *N.Y. Times* 14 Dec. III. 5/1 Dry eggs from China..include the white of egg, the yolk of egg, and powdered whole egg. 1925 T. M. RECTOR *Sci. Preservation Food* xii. 172 Hermetic sealing is not used in the preservation of eggs and egg products though there is certainly a field for it in 'powdered' egg. 1935 *Economist* 3 Aug. 226/2 As for powdered and condensed milk,..these imported milk products are going to be made artificially scarce. 1943 *Industr. & Engin. Chem.* XXXV. 1204/2 The losses of vitamin D in powdered whole egg during storage were studied at two temperatures, 15° and 98·6°F. 1953 *Ann. Reg.* 1952 153 The provision of powdered milk every day for over a million children. 1972 G. DURRELL *Catch me a Colobus* v. 84 Complan, a sort of powdered milk which we found was very useful for rearing baby animals. 1980 *Sunday Times* (Colour Suppl.) 20 Jan. 70/1, I have orange juice..then..a cup of powdered coffee.

powderize (pau·dəraiz), *v.* [f. POWDER *sb.*^{1} + -IZE.] **a.** = POWDER *v.*^{1} 3 b. *rare.* **b.** = POWDER *v.*^{1} 6.

a 1800 S. PEGGE *Anecdotes Eng. Lang.* (1803) 259 Many words will admit *-ize* for the termination. A Hair-Dresser *powderizes*, while a Chemist..*pulverizes* [etc.]. 1903 *Sci. Amer. Suppl.* 18 Apr. 22818/1 Only one thing can be done to lighten the task, and that is to powderize the soap when the mixed materials are still warm. 1978 *Nature* 20 Apr. 715/2 The material was powderised and partially attacked by a mixture of hydrofluoric and hydrochloric acids.

powderless, *a.* Add: **b.** Not employing powder.

1953 *Photoengravers Bull.* XLIII. 171 The key to powderless etching in the Dow etching development lies in the chemical composition of the bath. 1959 *Times* 14 Jan. 12/4 Another precision etching system, the Lithotex powderless etcher, is being manufactured in Britain. 1967 V. STRAUSS *Printing Industry* v. 217/1 Though conventional etching methods are still used.., they are progressively replaced throughout the industry by powderless etching. 1978 *Penrose Ann.* LXXI. 162 Powderless etching is still rare [in India] and electronic engraving extremely so.

powder-monkey. Add: **a.** Now also a term for a member of a blasting crew; = POWDERMAN c. *U.S.*

1926 *Amer. Speech* II. 87/2 Dynamite is brought to the miner by the *powder monkey*. 1937 *Nat. Geogr. Mag.* Aug. 149/1 Hither and yon darted the 'powder monkeys', packing dynamite in the holes, shooting loose great masses of salt, which others broke up and loaded into trucks. 1939 W. FAULKNER *Wild Palms* 118 Suits from wop pick-and-shovel men and bohunk powder-monkeys and chink ore-trimmers. 1951 *Publ. Amer. Dial. Soc.* xv. 74 *Monkey* in various combinations has become a favorite term for handyman. There are *derrick monkeys, pump monkeys, boiler monkeys*, and *powder monkeys*. 1976 M. MACHLIN *Pipeline* xviii. 227 This time they wanted two powder monkeys, a couple of drillers..and four jobs for straight camp labor.

pow·der-room. [f. POWDER *sb.*^{1} + ROOM *sb.*^{1}] **a.** In Dict. s.v. POWDER *sb.*^{1} 5 b.

b. = *powder closet* s.v. *POWDER *sb.*^{1} 5 b.

1908 'F. DANBY' *Heart of Child* xv. 250 He liked to see.. his Staffordshire pottery en-niched in the quaint powder-room, opening out of the drawing-room. 1946 J. W. DAY *Harvest Adventure* xiv. 243 Look at the old drawings of Ockwells in Lysons' *Magna Britannica* [*sic*] and other works, and you will see that, a hundred years ago, the powder-room window was completely blocked up. 1966 M. M. PEGLER *Dict. Interior Design* (1967) 351 *Powder room*, originally a corner or small closet in the bedroom of an 18th-century house where one could go to have one's hair powdered.

c. A women's cloak-room or lavatory in a hotel or shop; *gen.* a lavatory. Also *attrib.*

1941 B. SCHULBERG *What makes Sammy Run?* xi. 272 She had just run into Laurette in the powder room. 1945 MENCKEN *Amer. Lang.* Suppl. I. 640 During the days of Prohibition some learned speak-easy proprietor in New York hit upon the happy device of calling his retiring-room for female boozers a *powder-room*. 1958 *Times Lit. Suppl.* 13 June 323/2 But there is too much

of the language of powder-room chatter ('When Terry asked me to marry him, as I thought I hoped he would..') and platitude for her tale to gain and hold sympathy. 1959 D. DU MAURIER *Breaking Point* 219 The call-box was just opposite the ladies' powder room. 1977 *Time* 5 Dec. 68/3 While the play is laced with affectionately bantering humor and a gamy ration of powder-room candor, the characters are stereotypical.

powder snow. [f. POWDER *sb.*^{1} + SNOW *sb.*^{1}] A newly fallen, light, dry snow. Also *attrib.*

1929 F. SMYTHE *Climbs & Ski Runs* xi. 169 The *staublawine* or powder-snow avalanche is of little danger to the ski-runner. 1946 P. BOTTOME *Lifeline* xxiii. 253 He himself moved over the light powder snow like a soundless ghost. 1955 E. HILLARY *High Adventure* 159 High on the mountain a cloud of powder snow was being blown off the ridge. 1972 D. HASTON *In High Places* xi. 118 It was obvious what was happening—a huge powder snow avalanche was passing. 1974 *Listener* 17 Jan. 76/1 Nice sunshine, powder-snow skiing.

powdery, *a.* Add: **4. powdery mildew,** a parasitic fungus belonging to the family Erysiphaceæ, or the disease it causes in plants, characterized by a white, floury covering of conidia on the parts attacked.

1889 *Jrnl. Mycol.* V. 214 (*heading*) Powdery mildew of the bean. 1913 G. & I. MASSEE *Mildews, Rusts & Smuts* 36 The entire group [*sc.* Erysiphaceæ] is often spoken of as powdery mildews, on account of the dense masses of conidia that are produced, and rest on the white patches of mycelium, giving them the appearance of having been sprinkled with flour. 1936 *Jrnl. Agric. Res.* LII. 645 (*title*) The diurnal cycle of the powdery mildew *Erysiphe polygoni*. 1978 P. P. PIRONE *Dis. & Pests Ornamental Plants* (ed. 5) i. 9/2 Powdery mildews are fungi that grow superficially on the leaves and stems of their hosts.

Powellism (pau·ĕliz'm). [f. the name of John Enoch *Powell* (born 1912) + -ISM.] The political and economic policies advocated by J. Enoch Powell; *spec.* one of restricting or terminating the immigration of coloured people into the United Kingdom.

1965 *Economist* 17 July 217/1 In the past few months a new word has found its way into British politics: Powellism. 1968 *Guardian* 9 Sept. 6/1 Mr Enoch Powell's alternative..is a drastic move towards laissez-faire economics and a stopping of immigration. Powellism is gaining ground within the Conservative ranks. 1972 M. WILLIAMS *Inside Number 10* x. 254 Although we realized only too well that to dissociate Labour from Powellism was not a winner in electoral terms, it was decided unhesitatingly that Harold should leave no doubt about where the Labour Party stood on matters of race. 1973 C. MULLARD *Black Brit.* II. vi. 66 It [*sc.* the Institute of Race Relations] has remained on safe, non-controversial ground, arousing as much black scorn as Powellism. 1977 M. WALKER *National Front* v. 127 The ghost of Powellism in the parliamentary party is still not laid... Powellism, Young's own ambition to lead the Club, and the bitter debates on the EEC in 1971 were.. enough to account for the crisis.

So **Pow·ellist** *a.* and *sb.*; **Pow·ellite** *a.* and *sb.*

1965 *Economist* 5 June 1129/2 The Conservatives should not turn Powellite in opposition to an incomes policy. 1966 [see *blue-water school* s.v. *BLUE *a.* 13]. 1968 *Guardian* 26 Apr. 22/2 (*heading*) MPs' temperatures soar as Powellite sparks fly. 1968 *Manch. Guardian Weekly* 5 Sept. 12 The Conservatives..are likely for a year or two to be more Powellist in economics. 1968 *Peace News* 29 Nov. 8 Working-class Powellists..are not so much thinking like colonialists, but rather as aborigines. 1977 M. WALKER *National Front* iv. 90 The NF was deeply concerned that it had missed the Powellite boat, and that it was not expanding as it should. *Ibid.* iv. 93 Stories in the press about a split in the movement between Powellites and the rest.

Powellize (pau·ĕlaiz), *v.* [f. the name of William *Powell*, of London, who invented the process + -IZE.] *trans.* To treat (timber) by boiling in a solution of sugar so as to preserve it and reduce shrinkage. So **Pow·ellized** *ppl. a.*; **Pow·ellizing** *vbl. sb.*

1903 *Sci. Amer. Suppl.* 5 Sept. 23139/1 The London city authorities..intend to repave the Strand thoroughfare with 15,000 Powellized blocks. 1913 *Chambers's Jrnl.* Aug. 621/1 Seeing that elm is plentiful and extremely low in price, Powellising should result in its more extended application. *Ibid.*, After being Powellised it becomes a very handsome and hard wood. 1929 *Encycl. Brit.* XXII. 222/1 Extensive tests carried out with Powellized sleepers on Indian railways give good results.

power, *sb.*^{1} Add: **4. f.** Used with preceding adj. or sb. to designate a movement to enhance the status of the group specified or the beliefs and activities of such a group. See also *black power, *flower power, etc.

1970 'J. MELVILLE' *New Kind of Killer* ii. 32 I'm working to establish Parent Power right now. 1972 *Pride of Lions* (Columbia Univ.) Apr. 2/2 What is important is that you come out, have gay pride and leave the dance with a sense of Gay power. 1972 *Guardian* 17 May 12/1 Pupil power flexes its muscles in London today. The organisers have called on all London secondary schoolchildren to join them in a one-day general strike. *Ibid.* 19 Sept. 12/3 Three different types of reformers: the pupil power movement: the egalitarians: and the orthodox educational reformers. 1974 *Howard Jrnl.* XIV. 38 The growth of 'pupil power' and the increase in truancy. 1974 N. BAGNALL (*title*) Parent power. 1974 *Times* 7 Dec. 5/4 (*caption*) Mr Narayan: his hope lies in village power. 1975 *Times* 30 Dec. 8/8 The old form a powerful group—'grey power' to adopt Professor Wilensky's phrase.

6. c. *the powers that be* (after Rom. xiii. 1): the authorities concerned; the elements exercising social or political control. Also in sing.

1526 [see Dict., sense 6 a]. 1793 W. B. STEVENS *Jrnl.* 24 Feb. (1965) 70 The Selfishness and Timidities essential to his nature..make him cling to the Powers that be. 1814 SCOTT *Waverley* I. xix. 281 The cautious Baillie justly observed, that..the tenantry and villagers might become riotous in expressing their joy, and give offence to the 'powers that be', a sort of persons for whom the Baillie always had unlimited respect. 1886 KIPLING *Departmental Ditties* (ed. 2) 5 Potiphar Gubbins, C. E., Is dear to the Powers that Be. 1909 *Westm. Gaz.* 11 Jan. 12/4 Perhaps next year the powers that be may take a little more trouble to discover the talent that lies outside London. 1924 LAWRENCE & SKINNER *Boy in Bush* 15 He had to hear the end of a story against the powers-that-be. 1930 *Times* 25 Mar. 23/7 One can only express the hope that the Power-that-be in Nanking will realize the desirability for proceeding slowly and gradually. 1956 A. WILSON *Anglo-Saxon Att.* II. ii. 331 Donald's audience was not so large as it had been for the first lectures, but even now there was a fair number..those who had hoped that they might trap Donald into some mistake and earn a reputation for standing no nonsense from the powers that be. 1976 *Equals* Oct./Nov. 2/3, I feel that the powers-that-be have expected too much to happen too quickly.

7. (Earlier and later examples.) *By the powers!* (earlier example).

1490 JOHANNES DE IRLANDIA *Meroure of Wyssdome* (1965) II. 49 He had overcummyn all their powaris of myrknes. 1803 G. COLMAN *John Bull* I. i. 12 By the powers she's well enough. 1974 B. & R. HILL *Spirit in Stone* iii. 39 When his training period was over, he secretly painted his face in the way directed by his 'power' (i.e. spirit helper) and set out.

10. a. (Later example.)

1905 J. M. SYNGE *Shadow of Glen* 25 I'm thinking it's a power of men you're after knowing.

b. Delete *vulgar* and add later examples.

1938 B. L. BURMAN *Blow for Landing* 161 There's a power of music in an anvil. 1947 *Daily Mail* 22 May 3/4 There's a power of difference between farming now and when I was a lad. 1958 *People* 4 May 16/1 (Advt.), Hungry children do themselves a power of good when they polish off the sandwiches. *a* 1974 R. CROSSMAN *Diaries* (1975) I. 400 Two days at Prescote have done me a power of good.

11. d. *Math.* A property of a set that is the same for any two sets whose elements can be placed in a one-to-one correspondence and that in the case of a finite set is equal to the number of elements it contains; = *POTENCY 5.

1903 B. RUSSELL *Princ. Math.* xliii. 364 Power is synonymous with cardinal number. *Ibid.*, To prove that there are powers higher than the continuum. 1953 A. A. FRAENKEL *Abstract Set Theory* i. 79 The cardinal of the continuum, often called the power of the continuum. *Ibid.* iii. 306 Cantor and many of his successors..call cardinals in general by the neutral name of 'powers' (Mächtigkeiten). 1961 L. F. BORON tr. *Kuratowski's Introd. Set Theory & Topology* v. 61 The set of all odd natural numbers has the same power as the set of all even natural numbers. *Ibid.*, The open interval $-\pi/2 < x < +\pi/2$ has the same power as the set of all real numbers. 1968 H. SHARP *Mod. Fund. Math.* vii. 240 The power of a finite set is simply the number of elements in the set.

13. d. Motive power or heat (as contrasted with light) obtained from an electricity supply. Usu. *attrib.*

1896 R. ROBB *Electric Wiring* v. 82 Wires that carry current for running motors, or for furnishing power in distinction from light, are commonly called 'power-wires'. 1904 W. R. BOWKER *Dynamo, Motor & Switchboard Circuits* v. 96 (*caption*) Low-tension system for power and lighting. 1931 *Electrician* 13 Nov. 654/1 Where it is desired to install power plugs in premises already lighted electrically a minimum of two 15A positions is demanded. 1941 E. WHITEHORNE *Elect. Wiring Specifications* vii. 98 (*caption*) Power wiring must serve specific loads and processes. 1958, 1975 [see *power point*, sense *18 below.]

14. (Further examples.) Also applied *concr.* to an engine that produces power.

1869 *Bradshaw's Railway Manual* XXI. 399 Indian Tramway..adapted according to local circumstances to cattle or locomotive power. 1953 BERREY & VAN DEN BARK *Amer. Thes. Slang* (1954) §82a/1 *Motor; engine*—1. chugger, coffee grinder, mill, percolator, power, stove. 1962 *Amer. Speech* XXXVII. 134 *Power*,..all the locomotives owned by a company. The expression is heard, 'The company has lots of power.' 1973 *Ibid.* 1969 XLIV. 245 A light engine crew moves *power* from one location to another. *Ibid.* 259 *Power*, 1: Number and type of locomotives on a train. 2: Locomotives available at a given time.

17. j. *more power to* (someone): good luck, may fortune favour (someone). See also *ELBOW *sb.* 4 g. Also *more power to one's arm*.

1842 S. LOVER *Handy Andy* ix. 89 'More power to you, Andy,' said the Squire. 1881 CARLYLE *Reminisc.* II. 321 More power to him! 1932 E. GLASGOW *Let.* 12 Jan. (1958) 112, I read and enjoyed and admired the articles by Allen Tate. They are fine and true. More power to him. 1948 'J. TEY' *Franchise Affair* xvi. 187 Hooray! More power to him! I begin to like the boy. 1973 P. MOYES *Curious Affair of Third Dog* viii. 107 'I'm trying to find Griselda, you see.' 'In that case, more power to your arm.'

k. *power behind the throne*: one who exercises power behind the scenes while appearing to have no authority to do so.

[1770 PITT *Speech* 2 Mar. in *London Museum* Apr. 249 A long train of such practices has at length unwillingly convinced me, that there is something within the court [in *Parl. Hist.* (1813) XVI. 843 something behind the throne] greater than the King himself.] 1866 MRS. LINCOLN in W. H. Herndon *Lincoln* (1889) III. 513, I told him [*sc.* Lincoln] once of the assertion I had heard coming from the friends of Seward, that the latter was the power behind the throne; that he could rule him. 1875 'MARK TWAIN' *Old Times Mississippi* in *Atlantic Monthly* June 728/1 A power behind the throne that was greater than the throne itself. It was the underwriters! 1905 H. A. VACHELL *Hill* ix. 198 It was his habit to consult his wife in emergencies. The chief cutter..said that Amelia was the power behind the throne. 1931 W. HOLTBY *Poor Caroline* vii. 277 I'd been..generally working in the *background*, but then I *liked* to be the power behind the throne. 1973 J. WAINWRIGHT *Devil you Don't* 8 She'd been blinded..at the possibility of controlling such a man. Of being the power behind such a throne. 1977 *Time* 8 Aug. 43/2 Anderson..will become the real power behind the throne.

18. a. *power absorption, company, group, -holder, hunger, -impulse, -instinct, logic(s, -loss, -lust, -mania, -maniac, -monger* (later examples), *motive, -producer, relation, -seeker, -soul, structure, struggle, turret, -urge, vacuum, -worship*. **b.** *power approach, cart, craft, ditching, drill, hoist, mower, saw, shovel, -vehicle* (examples), *wringer*. **c.** *power lever*. **d.** *power-carrying, -craving, -generating, -handling, -losing, -loving, -seeking* (examples), *-sharing* (all may be used as adjs. or sbs.); *power-greedy, -hungry, -lusting, -mad, -thirsty* adjs. **e.** *power-crazed, -driven* (later examples), *-obsessed* adjs.; *power-driving, -farming* sbs. **f. power amplifier,** an amplifier designed to deliver an output of appreciable power into a load; **power-assisted** *a.*, employing some inanimate source of power to assist manual operation; applied esp. to brakes and steering in cars where power from the engine is so used; so **power-assi·stance,** (the equipment for) the application of power to assist manual operation; **power bandwidth** *Electronics*, the range of frequencies over which a device can deliver a certain power or a signal with distortion less than a certain value; **power base,** a source of authority or support; **power block,** (*a*) a group of allied states, or a great power with its allies and dependencies; (*b*) *Naut.*, a power-driven pulley used to haul in a seine; **power board,** (*a*) a board or panel containing switches or meters for an electricity supply; (*b*) chiefly *N.Z.*, the controlling authority for the supply of electricity in an area; an electricity board; **power brake,** a power-assisted brake (in a motor vehicle); **power broker** orig. and chiefly *U.S.*, one who exerts influence or affects the distribution of political power by intrigue; hence **power-brokering, power-broking** *vbl. sbs.*; **power buzzer,** an electrical vibrator used in the war of 1914–18 to generate telegraphic earth currents; **power cable,** a cable transmitting electrical power; **power car,** a railway carriage incorporating an engine; **power centre,** a locus of political authority; a powerful person or institution; so **power-centred** *a.*, concerned with the study, acquisition, or exercise of political authority; **power cut,** a temporary withdrawal or failure of the electricity supply; also *fig.*; **power density** *Nuclear Physics*, the power produced per unit volume of a reactor core; **power dive,** a dive made by an aircraft with its engine(s) providing thrust; also *transf.* and *fig.*; so **power-dive** *v. intr.*; **power-egg,** on an airship (rarely, an aeroplane), an ovate housing for an engine; **power élite,** a social or political group that exercises power; **power factor** *Electr.*, (*a*) the ratio of the actual power delivery by an a.c. circuit (or a component in it) to the product of the r.m.s. values of current and voltage; (*b*) as a property of an insulating material: the power factor under specified conditions of a capacitor made with this material as dielectric; **power failure,** a failure of a power supply, esp. the electricity supply; **power frequency** *Electr.*, a frequency in the range used for alternating currents supplying power (commonly 50 or 60 Hz); **power game,** a contest for authority or influence, esp. in politics; **power law,** a relationship between two quantities such that the magnitude of one is pro-

portional to a fixed power of the magnitude of the other; **power level,** the amount of power being transmitted, produced, etc. (in some contexts measured relative to some reference level); **power line,** a conductor supplying electrical power, often spec. one supported by poles or pylons; **power-loader** *Mining*, a machine which loads coal on to a conveyor belt at the coal face; hence **power-loaded** *a.*; cf. *POWER LOADING *vbl. sb.*; **power-net,** a knitted stretch fabric used in women's underwear; **power oil,** oil brought up from a well and used on the spot as a source of power; **power-operated** *a.*, operated by power from an inanimate source; **power pack,** a unit for supplying power; *spec.* (*a*) one for converting an alternating current (from the mains) to a direct current at a different (usu. lower) voltage, and usu. comprising a transformer, rectifier, and capacitor; (*b*) (see quot. 1967); also *fig.*; **power package,** a self-contained source of power; **power pile** = *power reactor* below; **power play,** in sport, a concentration of players at a particular point; the style of play involving such concentrations; *spec.* in ice hockey, a group of players sent out against a depleted opposition; also *transf.*; **power point,** a point or socket (*POINT *sb.*^{1} A. 19 e) from which electrical machinery or heaters can be operated; also *fig.*; **power pole,** a pole used to support an overhead power line; **power reactor,** a nuclear reactor designed principally as a means of producing power; **power response** *Electronics*, the way the output power of a device depends on the signal frequency; *spec.* the power bandwidth; **power seat,** a power-assisted reclining seat; **power series** *Math.*, a series of the form $\ldots + a'_2x^{-2} + a'_1x^{-1} + a_0 + a_1x + a_2x^2 + \ldots$, where the *a*s are independent of *x*; also, a generalization of this for more than one variable; **power set** *Math.*, the set of all the subsets of a given set; **power spectrum,** the distribution of the energy of a wave-form among its different Fourier components; **power steering,** power-assisted steering (in a motor vehicle); **power stroke,** in a piston engine, a stroke during which the piston is moved by the expansion of the gases in the cylinder; **power-system,** a set of political beliefs or institutions founded or dependent upon coercion; **power take-off** (equipment for) the transmission of mechanical power from an engine, esp. that of a tractor or similar vehicle, to another piece of equipment; **power tool,** an electrically powered tool; **power-to-weight** (or **power–weight**) **ratio,** the ratio of the power an engine or motor can produce to its weight (or the weight of the vehicle, etc., containing it); **power train** *Mech.*, the mechanism that transmits the drive from the engine of a vehicle to its axle; also, this together with the engine and axle; **power transformer** *Electronics*, a transformer designed to accept a relatively large power, esp. one connected to a mains supply or power line to provide power at a lower voltage to a circuit or device; **power transistor** *Electronics*, a transistor designed to deliver a relatively high power; **power tube** *Electronics* = *power valve* below; **power unit,** a device supplying, or controlling the supply of, power; a power plant; **power valve** *Electronics*, a valve designed to deliver a relatively high power; **power wire,** a wire transmitting electrical power.

1901 *Jrnl. Inst. Electr. Engin.* XXX. 405 If the phase difference..had been but 3½° instead of 7°, the power absorption would have been 1 H.P. and not 2 H.P. 1920 *Ibid.* LVIII. 896/1 The grid of the power amplifier is given a negative potential. 1923 *Radio Times* 28 Sept. 36 (Advt.), Standard loud speaker..suitable for public purposes when used with a power amplifier. 1961 G. A. BRIGGS *A to Z in Audio* 16 The final stage is the power amplifier designed to feed the loudspeaker with a few watts of audio power. 1970 J. EARL *Tuners & Amplifiers* ii. 51 The majority of power amplifiers have the push-pull output transistors driven direct from a pair of driver transistors. 1938 *Jrnl. R. Aeronaut. Soc.* XLII. 416 We may conclude, therefore, that, for the average pilot, the power approach (or the undershoot technique) is a feasible method of approaching and landing. 1959 *Times* 1 Sept. 12/2 One of the technical improvements in cars that has taken place unobtrusively during the past four years has been the provision of power-assistance for the brakes. 1970 *Commercial Motor* 25 Sept. 65/1 The steering was light even though power-assistance is not fitted. 1928 *Punch* 21 Mar. (verso front cover) (Advt.), The power-assisted brakes give absolute control over the car under all conditions. 1950 *Gloss. Aeronaut. Terms* (*B.S.I.*) I. 38 *Power-assisted control*, a flying control in which the force needed to move the surface is provided partly by electrical or hydraulic means and partly by the pilot's physical effort. 1959 *Observer* 1 Mar. 21/5 The power-assisted steering is one of the best I have tried; it spins back swiftly after sharp corners..and does not transmit severe road shocks. 1975 C. NESBITT *Little Love & Good Company* xviii. 226 The size of the car I was learning in terrified me to begin with. It had power-assisted steering and I wasn't happy without gears. 1965 *Wireless World* Sept. 457/2 Power bandwidth is the curve of maximum output power (for the defined total distortion) versus frequency, plotted with logarithmic scales on both axes. 1977 *Gramophone* May 1773/2 The power bandwidth of the overall system (which is of course determined by that of the power amplifier) I found to be a little more restricted at its upper end than in some modern amplifiers. 1959 *Cambr. Rev.* 7 Feb. 311/2 This is clearly the altruistic formula for the egoistic power-base position just discussed. 1969 J. MANDER *Static Society* i. 70 Mexico.. became the prey of rival *caudillos*, each with his geographically inaccessible power-base. 1976 *Encounter* June 79/2 Franco passed on without ever heeding the advice of his more intelligent supporters, who urged him to prepare his successor a power-base at the centre of the political spectrum. 1958 *Times Lit. Suppl.* 21 Nov. 675/2 The frontispiece showing the geographical arrangement of world power blocks. 1960 M. SHARCOTT *Place of Many Winds* ii. 37 The power block has come into use so that the men no longer have to pull the net by hand. 1904 *Electr. Rev.* (N.Y.) XLV. 444 The power-board is a handsome marble panel equipped with Weston ammeter and voltmeter arranged for taking readings. [1918 *Statutes Dominion of N.Z.* 38 For every electric-power district there shall be an Electric-power Board.] 1938 R. FINLAYSON in D. M. Davin *N.Z. Short Stories* (1953) 242 The Power Board was brought to the pass at last of having to build a special concrete foundation for the poles. 1950 *N.Z. Jrnl. Agric.* Aug. 183/1 In New Zealand in the present [electricity] shortages there is an increasing tendency among power boards to adopt rationing. 1973 'D. HALLIDAY' *Dolly & Starry Bird* ii. 30 The power board is on the wall of the darkroom. 1977 *Daily Tel.* 26 May 19/1 (*heading*) Power board fined over fitter's death. 1896 G. RICHMOND tr. *Lieckfeld's Gas Engines* ii. 29 The simplest and oldest of the power brakes is that shown in fig. 3, 'The Prony Brake'. 1972 D. E. WESTLAKE *Bank Shot* ii. 16 Kelp..stomped on the brake. It was a power brake, and the car stopped on a dime. 1961 T. H. WHITE *Making of President 1960* ii. 52 None of the young men..could win the confidence of the aging backroom power brokers who wield such influence in the Democratic party. 1968 W. SAFIRE *New Lang. Politics* 349/1 During his mayoral campaign and often after he was elected, Republican John Lindsay spoke scornfully, and often despairingly, of the 'power brokers' (a phrase coined by Theodore White) who ran New York behind the scenes. 1972 *Village Voice* (N.Y.) 1 June 12/5 He's not going to get the nomination in Miami without coming to some kind of arrangement with the power brokers. 1977 *Listener* 21 Apr. 498/3 Chiang Ching..embarked on a series of intimate relationships with Yenan's power-brokers. 1975 *N.Y. Times* 10 Apr. 29/1 He..argued that conventions were 'undemocratic' because they were subject to power-brokering instead of the will of the people. 1977 *Guardian Weekly* 6 Nov. 9/2 The last opportunity for him [*sc.* Tito] to preside over a gathering of all the political clans and employ his power-broking to ensure an orderly succession. 1922 *Encycl. Brit.* XXXII. 489/1 The power buzzer..was a powerful vibrator worked by the current from a 10-volt accumulator, and connected to inconspicuous earths of insulated wire which could..be buried 6 ft. deep... Detachments of troops isolated by the enemy could send out code signals which could be picked up by listening sets..at ranges up to 3000 yd. 1923 R. STANLEY *Text-bk. Wireless Telegr.* (rev. ed.) II. xvii. 339 With a power buzzer of the 'Parleur' type on a 10-volt battery and a 100-yards earth base between the plates, in a fairly dry type of soil over chalk, good signals have been obtained over a range of 5 kilometres, the receiver being a 3-valve L.F. Amplifier connected to two earth plates 200 yards apart. 1905 HENRY & HORA *Mod. Electr.* xi. 219 Tough paper forms an excellent insulating material for electric light and power cables. 1959 E. H. CLEMENTS *High Tension* xi. 182 One of Douglas's tall poles bearing the power-cable across the open forest. 1974 E. AMBLER *Dr. Frigo* III. 241 The body was discovered..by workmen laying a power cable. 1977 *Mod. Railways* Dec. 492/1 Despite the apparently convincing case put forward for the positioning of the power car in the centre of the APT-P sets..it is reported that options are still open regarding the formation of any subsequent production trains. 1936 *Discovery* Nov. 356/1 The Union Pacific RR. has two twelve-car diesel-electric trains, of which the first two coaches are the power cars, each containing a 1,200 b.h.p. engine. 1946 *Nature* 5 Oct. 463/1 The 11,000 volt systems are in turn supplied by 33,000 volt or 66,000 volt systems, with proportionally higher power-carrying capacities. 1971 Power cart [see *green(s) fee* s.v. *GREEN *sb.* 17]. 1961 *Sunday Express* 19 Nov. 16 By one cunning dodge after another he has kept this one-time power centre of German militarism intact. 1962 M. McLUHAN *Gutenberg Galaxy* 12 The Renaissance prince tended to become an exclusive power centre surrounded by his individual subjects. 1977 F. YOUNG in J. Hick *Myth of God Incarnate* ii. 28 Cyril's attack on Nestorius is related to the political struggle between the ecclesiastical power-centres of Alexandria and Constantinople. 1960 *Encounter* Oct. 4/4 A party which deals with the real problems of real men and women will become immoral, power-centred. 1924 *Times Trade & Engin. Suppl.* 29 Nov. 239/1 With such an unusually large consumption of current,..the electric light, heat, and power companies can afford to sell power at a low cost. 1951 M. McLUHAN *Mech. Bride* (1967) 136/2 This is a good record for the thousands of people who work in power companies. 1961 BIGELOW & OTIS *Manchester, Vermont* xii. 113 Manchester..had two competing power companies until 1904. 1977 J. CLEARY *Vortex* i. 21 Get the power company out here... Shut off that power. 1936 *Discovery* Nov. 361/2 Part III is solely taken up by power craft. 1954 FISHER & LOCKLEY *Sea-Birds* v. 126 The oceanic sea-birds have solved these problems of mobility by becoming sailplanes as well as power-craft. 1924 P. RADIN tr. *Adler's Pract. & Theory Individual Psychol.* xix. 231 This, reduced to her line of power-craving, meant—'I must dominate everyone, draw everybody's attention to myself.' 1914 G. FRANKAU *Et Debellare Superbos* in *Poetical Wks.* (1923) I. 185 We grasped this sword for gain's sake, caste, nor king Power-crazed to his own people's ruining. 1952 M. ALLINGHAM *Tiger in Smoke* iv. 74 Teleprinters, radar..when we get a power cut the whole blessed police system is liable to go out of action. 1973 J. WAINWRIGHT *Touch of Malice* 97 It was a good clock—electric—and, apart from power cuts, it kept perfect time. 1953 W. E. UNBEHAUN *Hist. & Status Exper. Breeder Reactor* (U.S. Atomic Energy Comm. AECD-3712) 40 *Power density*, a measure of the power per unit of reactor core volume. 1955 *Proc. Internat. Conf. Peaceful Uses Atomic Energy* (United Nations) III. 238/2 The power density is highest in the seed with a value of better than 200 watts/cm^3. 1967 *Technology Week* 23 Jan. 28/1 The *Phoebus* 1 *B* reactor..uses a new hydrogen pump developed by North American's Rocketdyne Div. that can handle 150 lbs. of hydrogen per second for studies of operations in the higher power-density range. 1942 *Tee Emm* (Air Ministry) II. 95 The value of power ditching is so great that the pilot should always ditch before fuel is quite exhausted. 1930 R. DUNCAN *Stunt Flying* vi. 55 In a power dive, terrific drag is exerted on the main planes with a downward pressure on the tail-plane. 1937 *Times* 14 Dec. 17/2 During these evolutions they [*sc.* Japanese bombers] 'power-dived' to within a very short distance above the Panay's decks. 1941 STEINBECK & RICKETTS *Sea of Cortez* 163 The mosquitoes.. whooped and screamed and attacked, power-diving and wheeling up and diving again. 1954 *Ann. Reg.* 1953 373 Aeroplanes flew at increasing speeds, reaching the threshold of the speed of sound in level flight and much higher speeds in power dives from high altitudes. 1973 J. WAINWRIGHT *High-Class Kill* 24 Young Shaw had chucked himself over the guard-rail and power-dived into eternity. 1961 *Motor Cycle* 16 Mar. 334/1 Nowhere is a power drill more useful than in the garage. 1977 36 *Home Handyman Projects* (Austral. Home Jrnl.) 75/2 An economical way to buy a power drill and attachments is as a set by Black and Decker. 1898 A. G. ELLIOTT *H. de Graffigny's Gas & Petroleum Engines* iv. 73 This motor.. has become especially famous for its application to power-driven road vehicles. 1935 *Discovery* Nov. 326/2 A new power-driven spray painting outfit which can be carried by hand and can be run from an ordinary lighting socket has recently been produced. 1967 KARCH & BUBER *Offset Processes* i. 4 Early Power-Driven Presses. In 1814, R. Hoe and Company built the first steam-operated press to be used for printing. 1978 J. A. MICHENER *Chesapeake* 693 Four of their power-driven boats lay off the point of Tilghman Island. 1907 *Jrnl. R. Agric. Soc.* LXVIII. 130 Seed dresser, for hand or power driving. 1916 *Aeroplane* 1 Nov. 802/1 There is quite a long distance separating these gondolas—or 'power-eggs', as the Naval Air Service calls them—from the forward car. 1931 *Flight* 16 Jan. 49/1 The revolution counters and oil-thermometers for the outboard engines are mounted on their respective power-eggs, clearly visible for the pilots. 1961 F. K. MASON *Hawker Aircraft since 1920* 292 The two Griffon prototypes..were to be..replaced by Griffon 61 'power eggs' at a later date, so becoming the Tempest IV. 1953 C. W. MILLS *Veblen's Theory Leisure Class* p. xiv, The major change in national glamour, in which the debutante is replaced by the movie star, and the local society lady by the military and political and economic managers—the power elite. 1956 —— (*title*) The power elite. 1960 *News Chron.* 16 June 6/3 Behind the council lies the real power elite: The Federation of Economic Organisations. 1973 *Black Panther* 17 Mar. 7/2 The genocide, oppression, humiliation and repression this country's racist power elite has inflicted upon Native Americans, Black Americans also know. 1892 J. A. FLEMING in *Jrnl. Inst. Electr. Engin.* XXI. 606 The ninth column gives a number which it is convenient to call the power-factor of the transformer at no load—it is the ratio of the true to the apparent watts. If the currents and pressures were simple sine functions, then the power-factor in that case would be the cosine of the angle of lag of primary current behind the primary terminal potential difference. 1912 *Ibid.* XLIX. 323 The power factor and conductance of dielectrics under alternating electromotive force of low voltage. 1930 *Engineering* 24 Jan. 97/2 It [*sc.* a generator] is rated at 30,000 kv.-a., at 80 per cent. power factor, generating at 13,900 volts, three-phase and 60-cycles, and running at 200 r.p.m. 1950 *Ibid.* 28 July 80/2 Power-factor measurements up to 250 kV root-mean-square can be made with a Schering bridge. 1967 M. CHANDLER *Ceramics in Mod. World* iv. 133 If [*sc.* zircon porcelain] has not only a low power factor, but also good electrical properties in general. 1933 *Newnes Mod. Motor Repair* 684/2 When power failure at high revs. occurs, get someone to speed up the engine and listen at the tail pipe. 1961 *Providence* (Rhode Island) *Jrnl.* 29 Nov. 22 The impact with the utility pole caused a brief power failure in the immediate area of the accident. 1967 'T. WELLS' *Dead by Light of Moon* (1968) i. 8 Hey! The whole street is dark... Must be a power failure. 1974 M. BABSON *Stalking Lamb* xvii. 124 'Perhaps there is a power failure.' In the darkness, Sybilla sounded older. 1952 J. W. DAY *New Yeomen of Eng.* xvi. 185 Substituting cheap modern power-farming for expensive hand and animal labour. 1971 *Power Farming* Mar. 7/1 It is just 50 years since one of the most significant innovations in the world of power farming first saw the light of day. 1950 *Engineering* 28 July 80/2 Equipment which will provide impulse voltages up to 1,400 kV peak, power-frequency voltages up to 250 kV root-mean-square, and direct current voltages up to 200 kV. 1967 M. CHANDLER *Ceramics in Mod. World* iv. 130 Although excellent for insulation at power frequencies, porcelain is far from being the ideal insulating material where high frequencies..are involved. 1958 *Spectator* 1 Aug. 175/3 The landowners..the officials.. and the womanisers..are all playing an elaborate and stylised power game under the eyes of their neighbours. **1970** [see *Expo]. **1975** *Guardian* 20 Jan. 8/5 Britain's puny role in the East–West power game. **1956** H. Seligman in A. Pryce-Jones *New Outl. Mod. Knowl.* 159 The future will bring stream-lined power-generating units of the type described. **1945** W. S. Churchill *Victory* 26, I repulse those calumnies..that Britain..is a selfish, power-greedy..nation. **1941** Wyndham Lewis *Let.* 31 May (1963) 290, I mean that of the big power-groups..we are the inhabitants or controllers of the great ocean wastes. **1977** *Listener* 17 Feb. 223/2 Directors should be appointed for their competence, not because of the power group they represent. **1936** *Discovery* July 222/2 It is merely a question of the power-handling capacity of amplifier and loud-speaker. **1962** Simpson & Richards *Physical Princ. Junction Transistors* viii. 174 The mesa transistor should also be competitive as far as power-handling capability and ease of fabrication are concerned. **1972** 'G. Black' *Bitter Tea* (1973) ix. 139 A contractor's lorry..with a power hoist. **1927** A. Huxley *Proper Stud.* 29 Power-holders to whose material advantage it would have been to wield their power ruthlessly. **1971** T. W. Robinson *Cultural Revolution in China* i. 18 On the powerholders' side were those groups and individuals having a vested interest in maintaining the status quo. **1977** A. Giddens *Stud. in Social & Polit. Theory* x. 335 All power involves a certain 'mandate'..which gives power-holders certain rights and imposes on them certain obligations. **1977** *Dædalus* Fall 87 Schools produce the docile work habits the capitalists desire and serve to provide a facade of meritocracy for what is in fact a perpetual class of power holders. **1935** Huxley & Haddon *We Europeans* ii. 37 The power-hunger of potentates. **1946** 'G. Orwell' *James Burnham* 17 He seems to assume that power-hunger..is a natural instinct. **1946** —— *A. Koestler* in *Crit. Ess.* 134 Spartacus, however, is not represented as power-hungry, nor, on the other hand, as a visionary. **1960** F. Land *Lang. Math.* vi. 82 The power-hungry countries of the world are determined to remedy their situation. **1973** *Listener* 6 Sept. 307/3 There were ambitious, power-hungry men on both sides. **1936** Wirth & Shils tr. *Mannheim's Ideology & Utopia* iii. 124 Observing the mass-mind, especially its power-impulses and their functioning. **1938** B. Russell *Power* ii. 21 Their power-impulses..seemed to themselves indubitably righteous. **1944** 'G. Orwell' in *Horizon* X. 240 *No Orchids* is aimed at the power-instinct. **1919** *Phil. Mag.* XXXVIII. 637 We know of no theoretical reason for supposing that the power-law will give a better approximate representation than any other law, *e.g.* a sine-law. **1968** *Brit. Med. Bull.* XXIV. 257/1 Davis & Zerlin..have suggested that..the amplitude of the averaged vertex response varies with the loudness of the stimulus according to a power law with an exponent of 0·4. **1977** *Sci. Amer.* Jan. 90/2 The inverse-square power-law spectrum of crater sizes on the moon has the property that if there are enough primary craters between one kilometer and 10 kilometers in diameter to cover an entire planet, there are also enough craters between 100 and 1,000 kilometers across to cover it. **1934** A. L. Albert *Electr. Communication* iii. 50 If the amount of noise cannot be reduced, the only alternative is to raise the power level of the speech or music being transmitted. **1945** H. D. Smyth *Gen. Acct. Devel. Atomic Energy Mil. Purposes* viii. 85 The production goal..was set at a figure which meant that the pile should operate at a power level of 1000 kw. **1953** *Language* XXIX. 91 The human brain operates on a power-level of five watts. **1975** D. G. Fink *Electronics Engineers' Handbk.* i. 52 (*heading*) Power levels of speech. **1923** G. Collins *Valley of Eyes Unseen* xiv. 305 Luckily, I had just strength enough to reach up and touch the power-lever. **1951** M. McLuhan *Mech. Bride* (1967) 98/2 Her legs..are date-baited power levers for the management of the male audience. **1894** A. T. Snell *Electr. Motive Power* iii. 91 For very heavy power lines, in which copper cables of about 19/16 s.w.g. are used, the stalks are forged of cast steel, sometimes galvanised. **1956** *Nature* 17 Mar. 536/2 At this remote desert location, interference due to artificial signals from electromagnetic devices and power-lines was negligible. **1970** T. Hughes *Crow* 69 It was a naked power-line, 2000 volts. **1972** *Daily Tel.* 19 June 1/3 The aircraft narrowly missed plunging into a nearby reservoir and skimmed over power lines. **1943** *Trans. Inst. Mining Engineers* CII. 145 The point we are anxious about is whether there is mechanized machinery for taking big power-loaded outputs from such a dip as 1 in 6. **1963** *Economist* 12 Jan. 142/1 The industry raised its average percentage of power-loaded coal from 49 to 59 per cent between the two years. **1943** *Trans. Inst. Mining Engineers* CII. 40 The present development of the Meco-Moore power-loader differs from that of a few years ago. **1956** F. S. Atkinson in D. L. Linton *Sheffield* 269 Intensive efforts are being made to replace the hand-loading of coal on to conveyors by mechanical power-loaders. **1946** Koestler *Thieves in Night* 296 It startles me that its up-to-date, stream-lined power logics should be accompanied by all this maudlin opera stuff. **1959** *Times Lit. Suppl.* 9 Oct. 575/3 Over large tracts of their lives the winds of the Hobbesian power-logic blow unchecked. **1951** H. Arendt *Orig. Totalitarianism* i. 5 Persecution of powerless or power-losing groups may not be a very pleasant spectacle. **1922** Glazebrook *Dict. Appl. Physics* II. 109/1 If the condenser has internal power loss,..the power factor..will not be zero. **1960** *Farmer & Stockbreeder* 1 Mar. 129/2 It sharply cuts power-loss in transmission and hydraulics. **1935** *Mind* XLIV. 94 Plutarch and Lucian, experienced in the ways of power-loving Rome. **1923** D. H. Lawrence *Kangaroo* xvi. 344 The land..invites parasites now... What would happen if the power-lust came that way? **1959** S. Spender tr. *Schiller's Mary Stuart* iii. iv. 61 Your uncle, the power-lusting cardinal. **1962** *Times* 2 May 7/4 The power-mad politicians. **1976** B. Bova *Multiple Man* xviii. 201 He was insane, sir. Crazy. Power-mad. **1971** *New Scientist* 20 May 434/1 Psychology students studying power-mania. **1963** *Times* 26 Jan. 11/3 It would be hard to see why..even a power-maniac wanted to leave so charming an island. **1965** *Punch* 22 Sept. 443/2 Ropy little espionage agency, riddled with power-maniacs, sends courier to Europe to pick up important information. **1961** *Guardian* 16 Feb. 10/5 Intellectuals have been found flat on their faces for powermongers to walk over. **1977** *Rolling Stone* 19 May 41/2 Powermongers will undoubtedly find ways to obtain the requisite technology. **1955** P. Mullahy in H. S. Sullivan *Conceptions Mod. Psychiatry* 243 The energy of the infant, or rather its manifestations in the power motive, become quickly modified or transformed. **1940** *Sears, Roebuck Catal.* Spring/Summer 929 Power mower. **1957** D. Karp *Leave me Alone* xiv. 192 The young boys in the development cheerfully cut grass with power mowers but disdained raking. **1970** R. Lowell *Notebk.* 28 The lawns, the paths, the harbor—stitched with motors, Yawl-engine, outboard, power mower. **1978** J. A. Michener *Chesapeake* 836 In July he runs his power mower, pushing his lawn back year by year. **1950** *Vogue* (U.S.) 15 Nov. 150/2 (*caption*) Nylon power net pantie-girdle..for a smooth line under, say, tight velvet..pants. **1952** *Woman's Home Compan.* Nov. 3 (Advt.), Sheer powernet corselette with the smooth fit of a glove. **1963** *Times* 17 Apr. 16/7 The company will manufacture power-net, fish-net, marquisette curtaining and other fabrics in the Natal area. **1969** *Sears Catal.* Spring/Summer 2 This bra-slip... *The power-net frame*: cool and smooth..a stretchy blend of nylon and spandex. **1963** *Times* 20 Mar. 15/6 The power-obsessed journalist. **1957** Forbes & O'Beirne *Technical Devel. Royal Dutch/Shell* iii. 256 For correct operation it was essential that the 'power oil' should be free from impurities and it was therefore necessary to wash it thoroughly above ground. **1972** L. M. Harris *Introd. Deepwater Floating Drilling Operations* xi. 108 The direct system has individual power-oil lines from the drill vessel to the individual functions on the subsea stack. **1917** W. G. Raymond *Elem. Railroad Engin.* (ed. 3) x. 129 In the automatic block system the signal at the entrance of each block is power-operated, usually electric. **1940** *War Illustr.* 5 Jan. 554/3 The 'Wellingtons'..with five separate machine-gun positions, including two in power-operated turrets. **1962** *Lancet* 1 Dec. 1155/1 Conventional prosthetic limbs may in fact be incapable of substantial development, but power-operated ones seem certain to become much more practical and effective. **1937** *Jrnl. Inst. Electr. Engin.* LXXX. 194/2 This is derived from two grid-controlled rectifier valves operated from a 'power pack' of the usual construction. **1942** *Electronic Engin.* XV. 285/3 The experimental model described, together with a power pack to permit operation direct from the 50 cycle mains, can be assembled on a box chassis measuring only 6½in. × 6½in. **1958** *New Scientist* 1 May 20/2 Its thickness is controlled by plating time and current, indicated by a graduated ampere-hour meter on the power 'pack'. **1967** *Gloss. Mining Terms* (*B.S.I.*) viii. 21 *Power pack*, a motor-pump combination for producing power for hydraulic equipment. **1971** J. Z. Young *Introd. Study Man* iv. 69 The mitochondria carry the respiratory enzymes, and are hence called the power packs of the cell. **1972** F. Bradbury *Hydraulic Syst.* iii. 43 In a large plant..the control desk and the power pack may be separated by some distance or may be installed in separate machinery spaces. **1973** *Houston Chron.* 21 Oct. 3/1 (Advt.), Penske Road Race... It's big-time racing excitement in a box! With power pack, 2 cars and hand controls, 30-ft. of track. **1977** J. Hedgecoe *Photographer's Handbk.* 35 To produce more light a flash with a larger powerpack is needed. **1958** C. C. Adams *Space Flight* viii. 196 Auxiliaries. These include taxis and propulsion 'guns' for individual men in space suits, or reaction power packages attached like outboard motors to large objects. **1968** *Listener* 27 June 828/1 The power supply to this satellite was a 28 lb device called SNAP/9A. This power package contained as its energy source 17 kilocuries of Plutonium 238. **1945** H. D. Smyth *Gen. Acct. Devel. Atomic Energy Mil. Purposes* vii. 70 The whole of a power pile..has to be enclosed in very thick walls of concrete, steel, or other absorbing material. **1961** *Observer* 26 Feb. 18/1 This is now well known as power-play, a form developed by the All-Blacks and Springboks from the simple logic of the decisiveness of forward domination. **1961** J. S. Salak *Dict. Amer. Sports* 340 *Power play* (football), a play in which the offensive team concentrates blockers at one point. *Ibid.*, *Power-play* (ice hockey), launched when a team has an extra man or is trailing in the final minutes, all players rushing into the opponent's zone, and putting on the pressure. **1968** *Globe & Mail* (Toronto) 5 Feb. 17/6 The North Stars lead the NHL in power-play goals with 37. Balon scored on the power play. **1973** L. Mosley (*title*) Power-play: the tumultuous world of Middle East oil 1890–1973. **1976** *N.Y. Rev. Bks.* 15 Apr. 34/2 The Church was making a last desperate power play (for which it is now paying dear) to keep its immigrant children in line. **1976** *Washington Post* 19 Apr. D3/2 Chicago got the first goal of the game, by Cliff Koroll on a power play. **1978** *Dumfries Courier* 13 Oct. 4/2 At No. 3 Donald Bogie, with a variation of power play and finesse, proved too much for an experienced opponent. **1951** M. McLuhan *Mech. Bride* (1967) 98/2 To the mind of the modern girl, legs, like busts, are power points which she has been taught to tailor. **1953** K. Tennant *Joyful Condemned* xxxiii. 323 An iron was plugged in handily to a power point. **1958** C. Watson *Coffin scarcely Used* ix. 86 Purbright watched for the power points. There was one in each of the large bedrooms. **1975** G. Burdett *Do your Own Home Wiring* xiii. 124 An additional 13A power point can usually be supplied from a ring circuit in the form of a spur. **1978** *Listener* 2 Feb. 159/1 Surrealism..works that sprang into being simply by unplugging the power-points of the mind. **1959** M. Shadbolt *New Zealanders* 174 They found it [*sc.* a car], bashed and tangled, at the foot of a slanting power pole. **1973** *Black Panther* 20 Oct. 11/2 The United Boeing severed two power poles. **1906** *Chambers's Jrnl.* 27 Oct. 765/2 The internal-combustion engine is coming..rapidly into favour as a cheap power-producer for almost every kind of work. **1909** *Westm. Gaz.* 19 Oct. 4/1 The overhead valve system of this wonderful power-producer is not uncommon to the engine employed in this firm's racing cars. **1946** Power reactor [see *burn *v.*¹ 13 h]. **1962** [see *converter 3 f]. **1970** *IEEE Trans. Nuclear Sci.* XVII. i. 520 (*heading*) Silicon radiation detector monitors nitrogen 16 in a power reactor. **1958** *New Statesman* 9 Aug. 158/1 The meeting..symbolises the change in the power-relations of the Communist world. **1977** *Dædalus* Fall 66 In this approach it was stressed that the explanation of any institutional arrangement has to be attempted in terms of power relations and negotiations, power struggles and conflicts. **1962** R. F. Graf *Mod. Dict. Electronics* 234/2 *Power response*, the frequency-response capabilities of an amplifier running at or near its full rated power. **1963** *Wireless World* July 354/1 The power response at 10W output is −3dB at 15 c/s and 15 kc/s. **1970** J. Earl *Tuners & Amplifiers* iii. 68 The latest 'quality' amplifiers..boast a power response which is almost as good as the frequency response. **1975** *Gramophone* Nov. 953/3 Power response is taken to be the range of frequencies over which the amplifier can deliver at least half its rated power. **1960** *McGraw-Hill Encycl. Sci. & Technol.* XIV. 542/1 Bench or circular saws are the common woodworking type of power saw. **1969** in Halpert & Story *Christmas Mumming in Newfoundland* 107 A game of cards, the loan of a power-saw..can act as the favours that initially link two persons together. **1976** N. Thornburg *Cutter & Bone* iv. 108 The car, a late-model Buick Century, seemed to have every possible piece of optional equipment, including power seats. **1946** 'G. Orwell' *James Burnham* 4 The English Puritans, the Jacobins, the Bolsheviks, were in each case simply power-seekers. **1979** P. Alexander *Show me Hero* xxiii. 246 He was the perfect example of the power-seeker... He'd tread on anyone's face to get to the top. **1963** A. Heron *Towards Quaker View of Sex* i. 10 We see human energy that should be creative and loving deflected into activities that are coldly power-seeking. *Ibid.* iv. 41 Our tendencies toward aggression and power-seeking. **1977** P. Johnson *Enemies of Society* xii. 169 Marxism..has a methodology of power-seeking. **1893** A. R. Forsyth *Theory of Functions of Complex Variable* iii. 56 Any one of the continuations of a uniform function, represented by a power-series, can be derived from any other. **1938** F. E. Terman *Fund. Radio* v. 136 For electrode voltages such that the instantaneous plate current never became zero, the characteristics of a tube could be expressed in forms of the following power series. **1968** P. A. P. Moran *Introd. Probability Theory* iii. 135 $R(z)$ is a power series whose coefficients tend to zero. **1953** A. A. Fraenkel *Abstract Set Theory* ii. 96 The set of all subsets of S may..be called the power-set of S. **1963** Selby & Sweet *Sets, Relations, Functions* i. 18 Since A contained four elements, the power set 2^A was made up of 2^4 or 16 subsets, each subset an element of the power set. **1974** *Encycl. Brit. Macropædia* I. 521/1 The power set lattice.., defined by the inclusion relation on the power set $P(U)$ of all subsets of a set U, has important special properties not shared by lattices in general. **1972** *Times* 31 Oct. 15/3 It is safe to say that that kind of enforced power-sharing is practicable to the extent that the power to be exercised is circumscribed. **1973** *Hansard Commons* 20 Mar. 242/1 We have expressed positive views on many matters in the White Paper—for example,..on power sharing. **1978** P. Cosgrave *Margaret Thatcher* ii. 34 It was widely believed that if there had not been a general election in February 1974 his Northern Ireland 'power-sharing' executive would have worked. **1940** *Chambers's Techn. Dict.* 669/1 *Power shovel* (or *navvy*), an excavator consisting of a jib carrying a radial arm to the end of which a large bucket or scoop is attached. The bucket makes a radial cut. **1966** *McGraw-Hill Encycl. Sci. & Technol.* III. 416/2 Power shovels are made which can be rapidly converted into a dragline, crane, or backhoe by changing booms and modifying the rigging. **1971** P. O'Donnell *Impossible Virgin* iii. 57 Stacks of bricks, a concrete mixer, a power-shovel. **1922** D. H. Lawrence *Aaron's Rod* xxi. 311 Yield to the deep power-soul in the individual man, and obey implicitly. **1944** *Bell Syst. Techn. Jrnl.* XXIII. 282 The second part is devoted principally to the fundamental result that the power spectrum of a noise current is the Fourier transform of its correlation function. **1960** W. R. Bennett *Electr. Noise* x. 208 Many important phenomena are found to have a power spectrum even though the Fourier transform itself does not approach a limit. **1972** T. H. G. Megson *Aircraft Struct.* xii. 427 It is assumed that gust velocity is a random variable which may be regarded for analysis as consisting of a large number of sinusoidal components whose amplitudes vary with frequency. The power spectrum of such a function is then defined as the distribution of energy over the frequency range. **1932** *Automotive Industries* 10 Dec. 739/2 The greatest need for power steering exists undoubtedly in connection with heavy trucks and buses. **1976** M. Maguire *Scratchproof* xii. 182 She gunned the vehicle forward..as she swung the power-steering into a fierce lock. **1903** *Work* 9 May 218/2 Such engines have only one power stroke in every four. **1966** E. Rudinger *Consumer's Car Gloss.* 113 Every other stroke (the down-stroke) of the piston is a power stroke which drives the engine; whereas in a four-stroke engine, there is only one power stroke in every four. **1950** D. Riesman et al. *Lonely Crowd* xi. 255 Even those intellectuals..who feel themselves very much out of power and who are frightened of those who they think have the power, prefer to be scared by the power structures they conjure up. **1977** *Time* 4 July 6/1 Brezhnev already ranked No. 1 in the Kremlin power structure and was accorded the diplomatic status due a chief of state nearly everywhere he went. **1961** *Los Angeles Times* 4 Aug. III. 4 Bitter echoes of the 1960 power struggle..are still audible in party circles. **1969** *New Yorker* 20 Sept. 110/1 A serious power struggle could take place in the North. **1978** *Jrnl. R. Soc. Arts* CXXVI. 669/2 The detention of the four leaders had deprived one main group in the power struggle for the succession of its principal leaders. **1943** J. S. Huxley *Evolutionary Ethics* vii. 59 A naked power-system cannot tolerate tolerance or face even intellectual opposition. **1970** C. Furtado in I. L. Horowitz *Masses in Lat. Amer.* ii. 46 Allowing the landlord class to augment its share in aggregate income and to consolidate its position in the power system. **1929** *Sears, Roebuck Catal.* Spring/Summer 881 This power take-off and clutch pulley combined makes your Fordson tractor always ready for belt power. **1943** C. G. Barger *Automotive Mech.* I. iv. 123 Power take-offs are used on such vehicles as dump trucks, fire engines, wreckers, and..allow the accessories to be operated by the power of the engine. **1957** P. H. Wilkinson *Aircraft Engines of World* 20 (Advt.), The

Lycoming 825-h.p. T53, featuring front-or-rear power take-off, is the world's first turbine designed specifically for helicopters. **1958** *Times* 1 July (Agric. Suppl.) p. viii/3 The mechanism of the trailer is driven with the power-take-off from the tractor and unloads at the rate of three tons in seven minutes. **1972** *Proc. Inst. Mech. Engineers* CLXXXIV. iii. i. 284 Power take-off systems..supplying the power to operate various truck-mounted equipment on tipping gear, garbage packers,..and many more applications. *Ibid.*, Power take-offs (PTOs) are necessary on most of today's trucks. **1951** H. Arendt *Orig. Totalitarianism* v. 141 Hobbes..proceeded from this insight to a plan for a body politic best fitted for this power-thirsty animal. **1959** *Sears, Roebuck Catal.* Spring-Summer 1299 Make your Drill a new power tool every time you change accessories. **1961** *Times* 3 Oct. (Computer Suppl.) p. ii/5 Muscles are replaced by machines or power-tools. **1971** *New Scientist* 27 May 529/1 Other batteries still in the development stage..include the silver oxide/zinc cell and the titanium battery, both currently under evaluation..for application in TV, portable powertools and possibly electric vehicles. **1937** *Times* 13 Apr. (Brit. Motor Suppl.) p. vi/4 It is of 4½ litres capacity, and the power-to-weight ratio is given at 4·2lb. a brake horse-power. **1971** *Engineering* Apr. 72 (Advt.), Reyrolle Hydraulics axial piston pumps and motors. Fixed or variable displacement, with excellent power-to-weight ratio (up to 2·8 hp/lb). **1943** C. G. Barger *Automotive Mech.* I. iv. 107 (*heading*) Power train. **1946** W. H. Crouse *Automotive Mech.* i. 24 The power train consists of a series of gears and shafts, which mechanically connect the engine shaft with the car wheels. **1966** *Economist* 10 Sept. 1040/2 Chrysler has gained substantial sales in the United States in the past four years since it began offering a 5-year, 50,000-mile guarantee for the 'power train' (engine-transmission-rear-axle) of its American-built cars. **1976** *National Observer* (U.S.) 2 Oct. 8/2 Auto manufacturers are reluctant to offer special-engine versions of all their models and power-train combinations. **1929** K. Henney *Princ. Radio* xvi. 403 A.-c. voltages are likely to be picked up by the cores of audio transformers if they are near power transformers carrying a.-c. currents. **1975** D. G. Fink *Electronics Engineers' Handbk.* vii. 17 Electronic power transformers normally operate at a fixed frequency. The most popular frequencies are 50, 60, and 400 Hz. **1959** K. Henney *Radio Engin. Handbk.* (ed. 5) x. 34 The increased ruggedness of the supply due to the inherently stable physical structure of the power transistor. **1974** G. A. G. Bennett *Electricity & Mod. Physics* (ed. 2) xiv. 271/1 Power transistors are made by the same techniques as other transistors; but the collector, in which most of the heat is dissipated, is fused onto a thick metal mounting plate forming one face of the unit. **1924** Moyer & Wostrel *Pract. Radio* vii. 103 The volume of sound may be increased by using a power tube of, say, 5 watts of electric power, in the last stage of amplification. **1975** D. G. Fink *Electronics Engineers' Handbk.* vii. 21 Power tubes, in contrast to receiving-type tubes, handle relatively large amounts of power, and..a major emphasis is placed on efficiency. The traditional division between the two tube categories is at the 25-W plate dissipation rating level. **1942** F. H. Joseph *Lett. home from Brit. at War* 44 Machine guns, rear power turret having four, front power turret two. **1907** H. Allen *Gas & Oil Engines* xiv. 307 A 4-cylinder engine,..having cylinders 8in. diam. and 8in. stroke, capable of running up to 800 revs. per minute, gives a power unit, when using petrol, of 100 b.h.p. **1908** S. H. Moore *Mech. Engin. & Machine Shop Pract.* xix. 421 Synchronous motors are best adapted for power transmission plants and for large power units of high voltage. **1918** G. Sherwood *Farm Tractor Handbk.* vii. 110 The machine [*sc.* a Fordson tractor] is merely a power unit and transmission gear en bloc mounted on two pairs of wheels, together with the simplest of control and steering arrangements. **1963** *Amer. Speech* XXXVIII. 120 The auxiliary power unit supplying electrical power to the KC-97 when it is on the ground. **1967** C. J. Freezer *Model Railway Terminol.* 8/1 The power unit is a device which converts the high-voltage mains current into low-voltage currents, often with several outputs. **1922** D. H. Lawrence *Aaron's Rod* xxi. 311 But the deep power-urge is not conscious of its aims. **1947** Koestler in *Partisan Rev.* XIV. 345 The best way to prolong the power-vacuum to the west of the Russian bloc..is to proclaim an independent British policy. **1976** *Times* 10 Sept. 1/1 (*heading*) Mao succession struggle. Death leaves power vacuum in China. **1919** W. D. Owen *Guide Study Ionic Valve* x. 38 Power valves need to be very hard otherwise the plate voltage would cause a discharge across the space. **1944** *Wireless World* June 163/2 Although power valves are used, they are only lightly loaded and HT volts and current are quite low. **1909** *Chambers's Jrnl.* June 341/1 Thompson in Edinburgh introduced the first power-vehicle running on india-rubber tires. **1916** *Ibid.* Feb. 83/1 The power-vehicle is also invaluable for communication between commanders and their units. **1931** *Engineering* 9 Jan. 58/3 Volumetric efficiency is important..in that it affects the power-weight ratio of an engine. **1950** *Times Rev. Industry* Sept. 25/2 The Swiss Federal Railways has improved the power-weight ratio of single-phase main-line types [of locomotives]. **1902** H. A. Foster *Electr. Engineer's Pocket-Bk.* 766 Special precautions of this kind must be taken where sharp angles occur, or where any wires might possibly come in contact with electric light or power wires. **1911** W. Aitken *Man. Telephone* xx. 409 All power wires from the fuse board are now usually laid up in lead-covered cables. **1938** R. Finlayson *Brown Man's Burden* 41 The new power wires... Ten thousand volts, ehoa! **1941** 'G. Orwell' *England your England* in *Lion & Unicorn* i. 17 Power-worship..has never touched the common people. **1921** *Electrician* 11 Mar. 304/2 In the United States where an electric ironing machine..costs about the same as a washing machine, some women use their power wringer as a cold mangle. **1957** *Observer* 1 Dec. 10/5 There is no national test for power wringers so test the safety release yourself before buying.

g. Designating alcoholic liquids of a grade suitable for generating mechanical power.

1919 *Rep. Interdepartmental Comm. Alcohol for Power* 4 in *Parl. Papers* (Cmd. 218) X. 117 Some sections of the community believe that the words 'industrial alcohol' refer to an inferior spirit for drinking purposes. We recommend, therefore, that all alcohol for power or traction purposes should be described as 'power alcohol'... This description has already been adopted in Australia. *Ibid.* 7 All sales..of power alcohol should be made on the basis of a certified percentage by volume of absolute ethyl alcohol, with a minimum of 90 per cent. at a temperature of 62 deg. F. **1920** *Act* 10 & 11 *Geo. V* c. 18 § 11 In this section the expression 'power methylated spirits' means any methylated spirits (other than mineralised methylated spirits) which are intended to be used in generating mechanical power. **1922** G. W. Monier-Williams *Power Alcohol* vii. 198 The conversion of the alcohol into power methylated spirits may be carried out only by an authorised methylator. **1934** *Proc. World Petroleum Congr. 1933* II. 693/2 It is quite a recent innovation to market a 'power' kerosine. **1939** *Sun* (Baltimore) 17 Apr. 13/3 Senator Gurney..estimated that if farm surpluses were used to make 'power alcohol' to be mixed with gasoline, 840,000,000 bushels of grain would be diverted to that purpose annually. **1940** S. Miall *New Dict. Chem.* 330/2 Power methylated spirit is absolute ethyl alcohol, 92 volumes; benzol, 5 volumes; crude pyridine, 0·5 volume, and crude naphtha, 2·5 volumes. **1955** *Know Your Tractor* (Shell Guide) ii. 27 The cost to the farmer of gasoline and power kerosine is very variable throughout the world. **1957** *Encycl. Brit.* I. 543/1 The use of power methylated spirits practically ceased during World War II. **1961** L. M. Miall *New Dict. Chem.* (ed. 3) 446/2 *Power kerosine*, a volatile kerosine of high anti-knock value, essentially a blend of aromatic hydrocarbons, used in tractor engines and usually called tractor vaporizing oil (T.V.O.). **1975** E. M. Goodger *Hydrocarbon Fuels* vii. 134 'Power kerosine' or 'tractor vaporising-oil' denotes a kerosine blend prepared for spark-ignition engines used in agricultural tractors, which is taxed at a lower rate than that of gasoline.

h. Applied to a sportsman who applies great muscular power to his style of play; also of the style of play itself.

1958 *Observer* 15 June 24/1 The machine-like rhythm and efficiency of her power tennis. **1959** *Times* 29 May 4/7 A power player, he went for every shot. **1959** *Sunday Express* 12 July 12/1 His splendid piece of power-running. **1969** *New Yorker* 14 June 44/3 Graebner could now probably explode one. He has what is almost a setup on his power side. **1973** *Black Panther* 25 Aug. 13/1 Henry Aaron will..establish himself unquestionably as the greatest power hitter in baseball history.

power, *v.* Restrict †*Obs.* to sense in Dict. and add: **2.** *trans.* To supply with power, esp. for propulsion. Also *fig.*

1898 W. F. Durand *Resistance & Propulsion of Ships* v. 326 (*heading*) Powering ships. **1929** *Chicago Tribune* 31 Jan. 3/8 His plane is a Travelaire, powered with a whirlwind motor. **1954** *Essays in Crit.* IV. 313 Creative activity is often..powered by the drive to accomplish. **1959** *Times* 29 June 12/7 The big traction engines that had powered the carousels. **1962** *Times* 25 Apr. 16/6 The incident..could have powered strong conflict between faith and sex. **1973** *Sci. Amer.* Feb. 102/3 It is the gravitational energy from the falling material, rather than the rotation, that probably powers the X-ray sources. **1976** *Billings* (Montana) *Gaz.* 30 June 3-E/3 Larvell Blanks and George Hendrick each belted two-run, first-inning homers Tuesday to power the Cleveland Indians to a 4–1 victory over the Milwaukee Brewers. **1977** 'A. York' *Tallant for Trouble* iv. 53 The police launch..was powered by two big Perkins engines.

3. *intr.* **a.** To move or travel with great speed or force.

1972 J. Mosedale *Football* ix. 129 The key play sent Nagurski powering toward the line. **1973** 'D. Rutherford' *Kick Start* ii. 46 The big bike solved all traffic problems for me, whether I was powering to the head of a two mile traffic jam or pulling it on to its stand on a yellow line. **1974** *Oxford Times* 20 Sept. 19/3 Derek Clarke powered in from the right and unleashed a superb shot. **1977** *Navy News* Aug. 38/1 At Mount Wise the following day, Civil Service powered to 256–5 off their 55 overs. **1978** *Daily Tel.* 20 June 13/2 For Michael Marshall..diesel, electric and advanced passenger trains plainly have no appeal to compare with that of the majestic locomotives that powered along the track in the first half of the century.

b. To travel using an engine, esp. as an alternative or supplement to sail.

1975 *Daily Colonist* (Victoria, B.C.) 2 Apr. 21/2 We had to power most of the way, that's how little wind there was. **1976** 'F. Clifford' *Drummer in Dark* xv. 95 The Trident braked and powered round until it pointed down the long smeared runway. **1976** T. Heald *Let Sleeping Dogs Die* viii. 171 A seagoing cabin cruiser..was just beginning to power towards the narrow entrance to the cove.

So **pow·ering** *vbl. sb.*

1898 W. F. Durand *Resistance & Propulsion of Ships* v. 340 (*heading*) Powering by the law of comparison. **1899** *Engineering Mag.* Mar. 1011/1 It is in the powering of the two vessels that the great advance in marine engineering is most apparent. *Ibid.* 1011/2 The powering of the Oceanic is..about double that of the Great Eastern. **1976** R. Lewis *Witness my Death* i. 19 Coal and dust and slack for the powering of industrial furnaces.

pow·er-boat. Also **power boat, powerboat.** A motor-boat, esp. one with a powerful engine. Hence **pow·erboater,** one who travels by power-boat; **power boating,** travel by power-boat.

1908 R. W. Chambers *Firing Line* vii. 84 Every day..the swift power-boats sped northward to the Inlet. **1932** *World Today* LIX. 254 (*heading*) Power boating. *Ibid.* 258/2 Craft built by the British Power Boat Company have, to date, secured the championships of the world. **1953** *Richmond* (Va.) *News Leader* 2 Sept. 24/7 From the Elizabeth River at Norfolk to Gibson Island in Maryland there'll be..more races for sailors and power boaters alike. **1966** *Listener* 28 July 144/2 That horrible vehicle, the power boat. **1971** 'E. Fenwick' *Impeccable People* xxi. 115 The powerboat owners suddenly became invisible... The local powerboaters..remained the most unpopular minority conceivable. **1975** *Times* 17 June 14/6 Power-boat racing is the preserve of those who have a great deal of..money. **1977** *Modern Boating* (Austral.) Jan. 71/3 Warren has set out to produce a book which will give a potential power boat buyer a complete idea of what leisure powerboating is all about.

pow·er-drive, *sb.* [f. Power *sb.*¹ + Drive *sb.*] **1.** (Equipment for) the driving of machinery by mechanical or electrical power. Also *fig.*

1952 B. Ulanov *Hist. Jazz in Amer.* (1958) xvii. 203 Lionel Hampton..added..a power-drive on vibes and drums all his own. **1960** *Farmer & Stockbreeder* 15 Mar. (Suppl.) 13 The rugged, reliable AEI 'Stayrite' Single-phase power-drive has been specifically developed for arduous farming applications. **1971** P. J. McMahon *Aircraft Propulsion* iii. 78 The normal transmission efficiency of a power drive is expressed as the ratio of power output to power input.

2. The impulse to exercise power.

1954 R. F. C. Hull tr. *Jung's Pract. of Psychotherapy* in *Coll. Wks.* XVI. i. 19 The first corresponds to Freud's pleasure principle, the second to Adler's power-drive, the desire to be on top. **1964** Gould & Kolb *Dict. Social Sci.* 483/1 Some modern definitions stress the fact of the power drive of parties. **1969** J. Mander *Static Society* vi. 180 The power-drive implied in this process is an ugly thing. **1979** J. Sherwood *Hour of Hyenas* ii. 23 You are very striking-looking, but..my power drive is far stronger nowadays than my sex drive.

powered, *a.* Delete 'chiefly in parasynthetic combinations' and add: Utilizing mechanical power for propulsion. (Further examples.)

1935 C. G. Burge *Compl. Bk. Aviation* 377/2 The introduction of powered flight in 1903. **1960** *Which?* Mar. 48/1 There are two main categories of lawnmower—hand-operated and powered. **1971** *Engineering* Apr. 65 (Advt.), Straddle carriers, powered bogies,..level luffing cranes. **1974** 'D. Craig' *Whose Little Girl are You?* i. 5 A display of powered model aircraft.

Powerforming (pau·əɹfǭːɹmiŋ). Also with small initial. [f. Power *sb.*¹ + Re)forming *vbl. sb.*] The name of a process for reforming petroleum using a platinum catalyst. Hence **Pow·erformer,** an installation for this process.

1956 *Oil & Gas Jrnl.* 5 Mar. 62 A new catalytic-reforming process called Powerforming has been announced by Esso Research & Engineering Co. *Ibid.*, (*caption*) One of three Powerformers already operating is unit at left of picture at Carter's Billings refinery. **1958** W. L. Nelson *Petroleum Refinery Engin.* (ed. 4) xxi. 772 In catalytic reforming, only the platinum catalyst processes (Platforming,..Powerforming, Sovaforming, etc.) are sufficiently well defined to allow the prediction of yields. **1967** Bland & Davidson *Petroleum Processing Handbk.* iii. 32 The first commercial Powerforming installation was placed on-stream in July, 1955, at Baltimore, Md., by the Esso Standard Oil Co. **1969** *Daily Tel.* 30 Dec. 2/5 (*caption*) The fire..severely damaged one of the plant's two powerformers, used in the manufacture of top grade petrol.

powerful, *a.* Add: **A. 6.** (Earlier examples.) For *vulgar* read *colloq.*

1811 Byron *Let.* 25 June (1830) I. 249 For a long time I have been restricted to an entire vegetable diet,..so I expect a powerful stock of potatoes, greens, and biscuit. **1822** J. Woods *Two Years' Residence Eng. Prairie* 294, I also have got some beefs, and a powerful chance of corn.

7. *Comb.*, as *powerful-faced.*

1906 Rider Haggard *Benita* ix. 129 A clever powerful-faced man.

B. as *adv.* (Earlier and later examples.) For *vulgar* read *colloq.*

a **1822** in *Amer. Speech* (1956) XXXI. 270 *Powerful.* This word is much used by the middling and lower class of people in the interior of So: Carolina..: instead of saying a person is very strong, they would say, *he is powerful*; or *powerful strong.* **1832** W. Irving *Jrnl.* 10 Nov. (1919) III. 171 My gun is so powerful dirty. **1902** *Dialect Notes* II. 242 *Powerful*, adv. Very, as '*powerful much*'. **1927** *Amer. Speech* II. 362/1 Our lessons are all pow'eful hard to-day. **1942** *Morgantown* (W. Virginia) *Post* 26 Sept. 5 It does get powerful cold in those hills. **1977** F. Parrish *Fire in Barley* iii. 33 'Tes powerful hard t'temp me Mam's fancy. She d'play wi' her food, mos' times.

powerfulness. (Later examples.)

1961 *Chicago Rev.* Summer 94 Their art is an imitation of the inescapable powerfulness of this unknown and empty world. **1975** *Listener* 24 July 109/3 The powerfulness of Queenie Leavis's mind.

pow·erhouse. Also **power-house.** [f. Power *sb.*¹ + House *sb.*¹] **1.** A building in which power is produced on a large scale for driving machinery or for generating electricity for distribution.

Quot. 1881 refers to a building containing steam engines for pumping water.

1881 *Harper's Mag.* Mar. 597/1 He found himself in the end at that 'Power House' of which he had heard..for many a year. **1890, 1895** [in Dict. s.v. Power *sb.*¹ 18 c]. **1901** *Chambers's Jrnl.* Mar. 206/1 In the centre of the power-house there is a raised desk, upon which are a series of press-buttons or keys. **1922** Joyce *Ulysses* 238 The whirr of flapping leathern bands and hum of dynamos from the powerhouse urged Stephen to be on. **1922** D. H. Lawrence *England, my England* 175 Tall, ruined power-houses of tin-mines loomed in the darkness. **1964** Linsley & Franzini *Water Resources Engin.* xvi. 459 One of the world's largest underground plants is the Kemano powerhouse in Canada which has a generating capacity of 1,670,000 kw under a head of 2592 ft.

2. a. *fig.* A source of energy or inspiration; a strong person or animal; also a powerful group of people, e.g. a sports team. Also *attrib.*

1915 P. Geddes *Cities in Evolution* xiii. 312 Before long, then, the School of Civics..must become a familiar institution in every city, with its civic library in rapid growth and widening use, and all as a veritable power-house of civic thought and action. **1916** J. Buchan (*title*) The power-house. **1941** *Sat. Even. Post* 1 Feb. 54/3 A thresher is one of the few powerhouses in the shark family. **1941** *Time* (Air Exp. Ed.) 15 Sept. 25/3 In the final, Powerhouse Kovacs was too much for Riggs in the first set, 7–5. **1943** *Amer. Speech* XVIII. 105 A hard hitter is known as a *distance hitter, power house, power hitter, slugger*, or *heavy sticker*. **1943** *Sun* (Baltimore) 20 Sept. 14/5 In the first two Saturdays of the young football season,..the mid-West and Duke again have their power-house elevens. **1952** *N.Y. World-Telegram & Sun* 29 Sept. 16/3 Michigan State, Maryland and Georgia Tech were supposed powerhouses. **1961** *Ann. Reg. 1960* 423 There continued to be no national powerhouse for [drama]. **1970** *Nature* 6 June 982/1 Although almost unreadable, it is a powerhouse of condensed factual information. **1974** E. Tidyman *Dummy* iii. 33 Strong? Jesus, yes. Not big, but a powerhouse. **1979** *Harvard Mag.* May–June 41 In 1972 the M.I.T. team, which had become the real powerhouse, somehow convinced the student government to give them $2,000 to go to England to play the English champions.

b. A strong hand in a card-game.

1932 *Amer. Speech* VII. 335 *Power house*, a very strong card hand. **1953** G. S. Coffin *Acol & New Point Count* 34 The two clubs opening in Acol shows a rare powerhouse, the kind of hand you may expect to hold about once a month. **1958** *Listener* 11 Dec. 1012/2 A freak distribution resulted in South's power-house being useless against Four Spades doubled. **1967** P. Anderton *Play Bridge* iv. 28 An opening bid of 2 C. is reserved for the power-house and is forcing to game.

3. Applied *attrib.* or as quasi-*adj.* to loud or forceful popular music or performers of such music.

1942 Berrey & Van den Bark *Amer. Thes. Slang* §578/10 *Powerhouse coda*, a strong finale. **1946** R. Blesh *Shining Trumpets* xii. 282 The hot-riff band is called in musicians' parlance a *powerhouse* band. **1952** B. Ulanov *Hist. Jazz in Amer.* (1958) xiv. 163 His brass was beginning to sound like the powerhouse sections of the swing bands. **1955** P. Rossiter in A. J. McCarthy *Jazzbook 1955* 50 The first important revivalist group was formed..in the early 'forties..led by..Lu Watters... The approach was powerhouse with a two trumpet, trombone and clarinet front line. **1968** *Blues Unlimited* Sept. 26 The powerhouse singing of types like Blind Roosevelt Graves. **1975** *New Yorker* 19 May 8/2 (Advt.), Powerhouse drummer Elvin Jones finishes up with his quintet on Sunday. **1976** *Leicester Trader* 24 Nov. 4/7 Argent's powerhouse rhythm section..have joined with John Verity who saw out the last days with the old band.

power loa·ding, *vbl. sb.* [f. Power *sb.*¹ + Loading *vbl. sb.*] **1.** *Aeronaut.* The laden weight of an aeroplane divided by the total engine power.

1920 H. Woodhouse *Textbk. Appl. Aeronaut. Engin.* iii. 136 The lift loading of the machine per sq. ft. equals 8·3 lbs., and the power loading, 12·1 lbs. per h.p. **1927** *Observer* 12 June 17/3 The flights of the 'Horsley' have greater technical value than those wonderful American flights,..the 'Horsley' has an unprecedented high wing and power-loading. **1966** D. Stinton *Anal. Aeroplane* vii. 118 When power loading is used as a measure of merit for comparing different aircraft (W_0/P) lb/hp is used, where W_0 is the all-up-weight on take-off and P the sea level horsepower.

2. *Mining.* The loading of coal on to a conveyor belt at the coal face by means of a machine (cf. *power loader* s.v. *Power *sb.*¹ 18 f).

1943 *Trans. Inst. Mining Engin.* CII. 39 The films to be exhibited are shown by courtesy of the 'Power Loading' Committee, and illustrate the latest practice in power-loading in America. **1956** F. S. Atkinson in D. L. Linton *Sheffield* 269 With thinner seams power-loading presents a difficult problem and progress has been slower. **1962** *Economist* 20 Jan. 243/1 Powerloading equipment tends to upset the whole work cycle of mining.

power-loom. (Later and *fig.* examples.)

1843 *Ainsworth's Mag.* IV. 95 A stoker to the power-loom in which are weaving the future destinies of mankind. **1941** J. Masefield *In Mill* 6 More than a hundred power-looms were in full work. **1968** J. Arnold *Shell Bk. Country Crafts* xvi. 206 With the invention of the power-loom, in 1787, it became dominated by the machine.

pow·er plant. Also **power-plant, powerplant.** [f. Power *sb.*¹ + Plant *sb.*¹] (An) apparatus or an installation which provides power; those parts of a machine, vehicle, or aircraft which provide power; an engine; a power station. Also *transf.*

1890 *Jrnl. Franklin Inst.* CXXIX. 270 Power Plant... Street-car work imposes a very variable load on the engines and dynamos... We require stronger engines than are needed for constant loads. **1897** E. K. Scott *Electr. Power in Workshops* 65 Messrs. Willans and Robinson are also laying down an extensive power plant for their new works at Rugby. **1909** *Westm. Gaz.* 4 May 4/1 The rigid dirigible..cannot be made to lift the weight of the power-plant necessary to render it independent of all winds. **1933** *Newnes Compl. Wireless* 595 (*heading*) Radio power plant. Notes on the uses, installation and servicing of rotary transformers, rotary converters, motor generators, and hand-driven generators. **1935** R. H. Baker *Introd. Astron.* x. 209 Power derived from the waterfall, from the wind, and from fuel and food has its origin in the great power plant of the sun. **1942** W. Faulkner *Go down, Moses* 207 They..would be back once more in the little lost county seats as barbers and garage mechanics..and power-plant firemen. **1964** [see *mitochondrion]. **1967** *Jane's Surface Skimmer Systems 1967–68* 104/2 The powerplant is mounted at the aft end with an acoustic bulkhead and two crew rest rooms separating it from the main passenger cabin. **1973** *Nature* 16 Mar. 210/1 The remaining papers..covered gas turbine, steam and Stirling-cycle powerplants for vehicles.

pow·er-po:litics. [f. Power *sb.*¹ + Politic *sb.* 3, translating G. *machtpolitik* (f. *macht* might, power + *politik* politics).] Political action based on or backed by threats to use force. Hence **power-poli·tical** *a.*, pertaining to or characterized by power-politics; **power-politi·cian**, one who practises power-politics.

1937 G. M. Young *Daylight & Champaign* 135 We shall all have to..learn to talk of power-politics and art-form antecedents and the literary-critical approach, as if we had been cradled in Marburg and reared in Michigan. **1939** Wodehouse *Uncle Fred in Springtime* xiii. 179 The Duke's decision..to mobilize his nephew Ricky and plunge immediately into power politics was one which would have occasioned no surprise to anybody acquainted with the militant traditions of his proud family. **1940** A. Huxley *Let.* 9 Oct. (1969) 460, I am engaged at the moment on a strangely apposite study of Père Joseph, collaborator of Richelieu, the most astounding case of a power politician who was also a religious mystic. **1942** L. B. Namier *Conflicts* 21 For centuries Vienna and Paris had been the centres of European power-politics. **1942** Partridge *Usage & Abusage* (1947) 355/1 Hitler was the *protagonist* of Nazism, with its power-political ideology. **1959** *Oxf. Mag.* 4 June 448/1 The abolition of compulsory Latin was a power-political move intended to pacify the jealous monoliths of America and Russia. **1961** J. Wilson *Reason & Morals* ii. 125 If I do not use human beings as ends in themselves..it..results in that political tyranny which tends to corrupt the power-politician or the brain-washer. **1973** J. Burrows *Like an Evening Gone* ii. 31, I like that period... Besides—power politics of any age—it's a kind of adventure. **1977** *Church Times* 10 June 14/4 Even monasticism had been sucked into the power-political structure of the medieval Church.

Powhatan (pau-ˌătæn), *sb.* and *a.* [Native name.] **A.** *sb.* An Algonquian people of eastern Virginia; a member of this people; their language (now extinct). **B.** *adj.* Of, pertaining to, or characteristic of this people. Hence **Powhata·nic** *a.*

1608 J. Smith *True Relation of Occurrences in Virginia* sig. D, With a lowd oration he proclaimed me A werowanes of Powhaton, and that all his subjects should so esteeme vs, and no man account vs strangers nor Paspaheghans, but Powhatans. **1612** W. Strachey *Trav. Virginia* (1953) i. i. 35 There was ever Enmity..between the High-and Low Country, going by the names of Monacans, and Powhatans. **1785** T. Jefferson *Notes Virginia* xi. 167 We are told that the Powhatans, Mannahoacs, and Monacans spoke languages so radically different, that interpreters were necessary when they transacted business. *Ibid.* 168 The territories of the *Powhatan* confederacy, south of the Patowmac comprehended about 8000 square miles, 30 tribes, and 2400 warriors. **1860** H. R. Schoolcraft *Hist. & Stat. Information Indian Tribes* V. i. 35 Their language was so diverse from the Powhatanic dialects, which were of the Algonquin group, that not a word could be understood without interpreters. *Ibid.* 36 The older ones among them, preserve their language in a small degree, which are the last vestiges on earth, so far as we know, of the Powhatan language. *Ibid.* 37 All the sympathies of Virginians were with the Powhatanic tribes. **1907** *Amer. Anthropologist* IX. 135 When the English landed at Jamestown in 1607, the Powhatan confederacy was a thing of recent origin. **1912** T. Michelson in *28th Ann. Rep. U.S. Bureau Amer. Ethnol. 1906–7* (*map legend*) Languages the exact position of which is uncertain... Powhatan, Weapemeoc, Secotan, etc. **1915** J. Buchan *Salute to Adventurers* xxiv. 334 A voice..spoke the Powhatan language, which I knew. *Ibid.* xxvi. 355, I found some Indians..and told one who spoke Powhatan the issue of the fight. **1953** Wright & Freund *Strachey's Trav. Virginia* i. 35 The Powhatans were an Algonquian tribe and their name, which first came from the name of a town near the present Richmond, was later applied to the chief Wahunsonacock and his confederacy. **1974** *Encycl. Brit. Micropædia* VIII. 170/1 In the 1970s an estimated 3,000 Powhatan were reported, largely scattered along the Virginia coast. **1977** *Language* LIII. 259/2 Siebert places these data in the framework of Eastern Algonquian, as reconstructed by Bloomfield, and provides an English–Powhatan lexicon of 263 words.

Powindah (pau·ˌindă). Also **Povindah, Powandah,** ‖**pāvendeh.** [Pashto, f. Pers. *parvinda* merchandise.] A nomadic trading tribe of Afghanistan; a member of this tribe. Also *attrib.*

1851 H. B. Edwardes *Year on Punjab Frontier* I. ix. 454 The whole of the trade between India and Central Asia is carried on by periodical caravans, which cross and re-cross the soolimânee mountains every year. They are conducted by Afghan merchants, who are generally called Lohânees, but locally in the Dérajât Powinduhs, or Povindeuhs... Lohânee is not a name applicable to either the Kharotees or the Nâssurs, so I prefer calling them Powinduhs, a name which they all acknowledge. **1880** H. W. Bellew *Races of Afghanistan* xi. 104 During the cold weather, the Povinda is to be seen in most of the larger cities of India. *Ibid.* 105 These Povinda clans, though classed as subdivisions of the Ghilji people, differ from them in one or two important respects. **1885** E. Balfour *Cycl. India* (ed. 3) III. 275/2 Considering the wild and independent life the Povindahs lead, they are marvellously orderly and well-behaved when dispersed in British territory, travelling from one end of India to another. **1888** H. G. Raverty *Notes on Afghánistán* 489 The Náșirs or Náșirís, as they are also called, are about the most numerous of the Powandah tribes, and possess no land whatever of their own. **1895** [see *Mahsud]. **1920** *Blackw. Mag.* Oct. 445/1 Your car is halted at the boat bridge, to let the long Afghan powindah caravans pass. *Ibid.* 445/2 The powindahs are more often armed, each man's belt crammed with cartridges. **1934** Ahmad & Aziz *Afghanistan* iii. 17 The Gomal pass..is much used by Povindahs on their annual migration to their winter encampments on the Indus. **1953** J. Masters *Lotus & Wind* xii. 158 The man..was a Powindah horse-trader... When the Pushtu greetings were at last out of the way the Powindah said, 'Let us retire.' **1967** A. Swinson *N.-W. Frontier* v. 104 Now the Powindahs..are great clans of warrior merchants, Ghilzais and Kharotis as well as Powindahs proper. **1974** *Encycl. Brit. Macropædia* I. 169/1 A number of nomads (*pāvendehs*) cross the eastern frontier with their herds toward their summer pastures in Pakistan. **1978** 'M. M. Kaye' *Far Pavilions* xi. 177 They were only *Powindahs*—wandering, gipsy-like folk who live in tents and are always on the move.

powwow, pow-wow, pawaw, *sb.* Add: **3.** (Further examples.) *spec.*, a conference of senior officers. Also used occas. in general sense of 'bustle, activity'. For (Chiefly *U.S.*) read orig. *U.S.*

1893 'Mark Twain' in *Cosmopolitan* Sept. 629/1 Without the marring additions of human pow-wow and fuss and feathers and display. **1897** —— *Following Equator* xxxviii. 346 It all helps to keep up the liveliness and augment the general sense of swiftness and energy and confusion and pow-wow. **1925** Fraser & Gibbons *Soldier & Sailor Words* 229 *Pow-wow*, a senior officers' conference, as, for instance, of a General with the Commanding Officers of units after a field-day or an operation. **1926** J. C. Lincoln *Big Mogul* vi. 112 It seems there was a great pow wow, some wanted to wear one kind of thing and some another. **1930** G. MacMunn *Behind Scenes in Many Wars* xiii. 239, I visited him here several times, and attended his rather interminable pow-wows and conferences. **1936** F. Clune *Roaming round Darling* xxii. 221 O'Hea and the bandmaster..had a private pow-wow. **1944** *Sun* (Baltimore) 17 May 18/2 Governor O'Conor..is conceded to be in control of proceedings at the Democratic pow wow. **1954** *Manch. Guardian Weekly* 28 Jan. 3/2 The associated lobbies that oppose the [St. Lawrence] seaway, the railroads, coal-owners, and Eastern port authorities, went into a round of emergency pow-wows. **1962** *Press* (Vancouver) Nov. 9 He turned up in London at a periodical pow-wow initiated by one Cyrus Eaton. **1977** *Time* 30 May 33/2 What had brought the Mob chiefs together was a series of powwows with New York City Mafia bosses about the new Mob power structure.

powwowing, *vbl. sb.* Add: (Later examples.)

1905 J. C. Lincoln *Partners of Tide* i. 8, I cal'late there must have been some high old pow-wowin' in the old house, but..they fin'lly decided 'twas their duty to take the little feller to bring up. **1928** H. W. Shoemaker in *Publ. Pennsylvania Folk-Lore Soc.* I. iv. 15 He arrived at the Little Valley with the book and tried to raise dark Cathlin's spook by pow-wowing. **1938** A. Hark *Hex marks Spot* 52 Pow-wowing is a hidden but by no means secret art among the people of Pennsylvania. **1961** D. Waldo *Beat Drum Slowly* vii. 89 There's too much pow-wowing going on hereabouts.

Powysian (pōᵘˌi·siăn, pauˌi·siăn), *a.* and *sb.* Also † **Powisian.** [f. *Powys* (see below) + -ian.] **A.** *adj.* Of or pertaining to the historical principality of Powys in east-central Wales, or its inhabitants or dialect. **B.** *sb.* **a.** An inhabitant of Powys. **b.** The dialect of Powys.

Powys was established as the name of a Welsh county in 1974.

1868 *Coll. Hist. & Archæol. Montgomeryshire* I. 426 Powis, *Pow isa is the low country*, a name still given by the peasants of Montgomeryshire to the plain of Shropshire, and indeed derived from that, the only level portion of the ancient Powisian dominions. **1880** *Ibid.* XIII. 42 The men of Argoed..uphold the *brother's* right, and acknowledge not a sister's claim to inheritance... Hence we find no Boadiceas, no Cartismanduas, no queens, among the Powysians. **1897** P. Barbier *Age of Owain Gwynedd* (1908) i. 22 The death of Maredudd in 1132, and..the steadily growing power of Gwynedd..tended much to the diminution of Powysian influence. **1910** A. Jones *Hist. Gruffydd ap Cynan* 89 It was the activity of the Powysian princes, and especially of Maredudd,..that made Powys the object of attack. *Ibid.* 180 Gruffydd..threatened active hostility towards any Powysians who sought safety within his dominions. **1913** J. M. Jones *Welsh Gram.* 8 Powysian, the dialect of Powys, or North East and Mid Wales.

pox, *sb.* Add: **1. a.** (Later examples.) Now only *colloq.* and freq. used to designate any venereal disease.

1922 [see *lock *sb.*² 18]. **1930** *Amer. Speech* V. 392 *Pox*, any kind of venereal disease. **1930** Brophy & Partridge *Songs & Slang 1914–18* 88 But now she's standing in the gutter, selling matches penny-a-box: While he's riding in his carriage With an awful dose of — [awful dose of pox *in* **1965** —— *Long Trail* 70]. **1968** [see *dose *sb.* 2 d]. **1969** *Coast to Coast 1967–68* 191 'Yah! Luther stinking swine! He had the pox!' Simon..had yelled in return 'Old bugger Pope pokes the nuns!' **1976** J. O'Connor *Eleventh Commandment* viii. 101 Wally..strangled a prostitute for giving him a dose of the pox. **1977** *Sounds* 9 July 8/4 It's still potent diz bustin' rhino-stomp, a most touching lament to the pox, but it's not brash to an offensive level.

3. For † *Obs.* read *arch.* (chiefly *Lit.*) and add later examples.

1922 Joyce *Ulysses* 392 But they can go hang..for me with their bully beef, a pox on it. **1941** V. Woolf *Between Acts* 126 Hand me the mirror, girl. So. Now my wig... A pox on the girl—she's dreaming! **1963** *Sunday Times* 8 Sept. 29/3 Cool Shakespeare thrives in the sixth and phrases like 'Pox on't!' and 'Fie!' are in present usage. **1973** 'A. Hall' *Tango Briefing* xiv. 184 A pox on his grave repercussions, if he meant the whole thing'd blow up..why couldn't he bloody well say so.

4. *poxfiend*; **pox-doctor** *slang*, a doctor specializing in the treatment of venereal diseases; phr. *got up like a pox-doctor's clerk* (and varr.), dressed smartly but in bad taste, overdressed; **pox-fouled** *a.*, infected with syphilis.

1937 D. Jones *In Parenthesis* v. 118 You feel the pack of the Ox-blood Kid—it's as light as the Rig'mentals—there's a whole lot of them that work it: the Pox-Doctor's Clerk, for one, the chitties, and types of scullion bummers up. **1949** Partridge *Dict. Slang* (ed. 3) 1141/2 *Pox doctor's clerk, like*—or *got-up like*—*a*, in a very smart civilian suit: Naval: C. 20. **1965** E. Lambert *Long White Night* xv. 136 They was all dressed like they was at Buckingham Palace and Foran was done up like a pox doctor's clerk. **1965** P. Ferris *Doctors* iii. 57 They [*sc.* coloured doctors] can land a job in London as a pathologist or a pox-doctor, but that's about all. **1974** *Bulletin* (Sydney) 19 Jan. 13 Getting dressed up like Lord Muck and getting round kitted up like a flamin' pox doctor's clerk. **1922** Joyce *Ulysses* 419 And snares of the poxfiend. *a* **1915** —— *Giacomo Joyce* (1968) 9 The pox-fouled wenches and young wives that, gaily yielding to their ravishers, clip and clip again.

pox, *v.* Delete *Obs.* and add: (Later examples.) Also *transf.* and *fig.*

1784 Prince William *Let.* 23 July in P. Ziegler *King William IV* (1971) iii. 51 Oh, for..the pretty girls of Westminster..such as would not clap or pox me every time I fucked. **1802** G. Galloway in *Admirable Crichton* 70 Tho' we were pox'd wi' poverty and law. **1846** *Swell's Night Guide* 45 These kens are tenanted by a blackguard..school of pugging shakes, whose chief fame is in..poxing a swaddy. **1933** M. Lowry *Ultramarine* i. 51 That boy got all poxed up to the eyeballs, voyage before last... Yes, he was poxed all away to hell. **1935** in *Sc. Nat. Dict.* (1968) VII. 225/1 When I wuz away for my breakfast my mate oot o' pure duvilment poxt the stone on me. **1961** 'F. O'Brien' *Hard Life* xiv. 115 Conditions on the canal bank are nothing short of a scandal with men and women going about there poxed-up to the eyes. **1968** *Amer. Speech 1967* XLII. 295 *Pox*, a general colloquial term for 'ruining or damaging stone through faulty handling, milling, or carving'. **1968** *Sc. Nat. Dict.* VII. 225/1 *Pox*, *v.*,..to botch (a job), ruin (a piece of work). **1970** G. F. Newman *Sir, You Bastard* iii. 92 The car became expedient. He was poxed with running for trains, missing trains, and worse, catching trains crowded with sickly commuters. **1977** *Sat. Rev.* (U.S.) 23 July 15/2 Wilmington, Delaware, poxed at that time by 1,200 abandoned one- and two-story homes.

poxvirus (pǫ·ks,vəiᵊrə̆s). *Microbiol.* Also **pox virus.** [mod.L., f. Pox *sb.* + Virus.] Any of a group of large DNA viruses that cause smallpox and various other epidermal diseases in vertebrates.

1941 *Amer. Jrnl. Vet. Res.* II. 102/1 There are many different strains of pox virus, some of them less pathogenic than others. **1962** [see *arborvirus]. **1973** R. G. Krueger et al. *Introd. Microbiol.* xix. 522/1 Poxviruses propagate rapidly in a wide variety of mammalian and embryo tissues *in vitro* producing cytopathic effects in 24–48 hr.

poxy (pǫ·ksi) *a.* [f. Pox *sb.* + -y¹.] Infected with pox; spotty; *fig.*, trashy, worthless. Also as a general term of abuse.

1922 Joyce *Ulysses* 8 God knows what poxy bowsy left them off. **1950** M. Peake *Gormenghast* xxii. 149 Every poxy sunrise of the year, eh, that you burst out of the decent darkness in that plucked way? **1958** [see *moor *sb.*¹ 1 b]. **1959** I. & P. Opie *Lore & Lang. Schoolch.* ix. 171 Other unfortunates are 'Spotty Dicks'..their friends call them: Bumps,..Crater face,..Pimple bonce, Poxie, and Scabby guts. **1968** Bethell & Burg tr. *Solzhenitsyn's Cancer Ward* i. xix. 302 You know, that poxy-faced guy, the one with all the bandages. **1975** J. Pidgeon *Flame* i. 8 No more poxy weddings, no more bloody favours. **1976** J. O'Connor *Eleventh Commandment* ii. 31 The first tray..was full of poxy rings worth two or three quid. **1978** K. Bonfiglioli *All Tea in China* iii. xi. 153 My poxy friend Peter.

‖ **poya** (pōᵘ·yā). [Sinhalese *pōya*, f. Skr. *upavasatha* fast day.] In full, *poya day*: a day on which the moon enters one of its four phases, observed as a day of special religious observance by Buddhists in Sri Lanka.

1853 R. S. Hardy *Man. Budhism* i. 22 When the dark póya, or day of the new moon, has come, the sun moves in one day the distance of 100,000 yojanas from the moon. *Ibid.* ii. 51 Today so many men have observed the póya (or sacred day). *Ibid.* v. 116 It was the póya day, when the prince..went to the public alms-hall to distribute the royal bounty. **1889** M. Monier-Williams *Buddhism* xii. 316 These preachings are generally well-attended, especially on the four Poya days. **1910** tr. *Hackmann's Buddhism as Relig.* iii. ii. 116 Real preaching (not only reading of the sacred text) is done in some temples on the so-called *poya* days, the four quarters of the moon, when lay people assemble in the *vihāras* to present their gift. **1913** L. Woolf *Village in Jungle* ii. 40 If you had come last poya, Silindu, I could have given it. **1971** *Times Weekender* (Ceylon) 3 Oct. 9/2 He had obviously stoked up in anticipation of a dry Poya day ahead.

Poyning's Law. Delete 'See quot.' and add: Also (*correctly*) **Poynings' Law.** [f. the name of Sir Edward *Poynings* (1459–1521), Lord Deputy in Ireland, 1494–6.] The name given to a series of statutes, passed at Drogheda in 1494–5 and repealed in 1782, by which the Irish parliament was subordinated to the English Crown.

[**1612–13** in Coke *Fourth Part Institutes* (1644) lxxvi. 351 An Act made in the tenth year of H.7 called *Poynings Act*.] **1622** Bacon *Henry VII* 138 But Poynings (the better to make compensation of the Meagernesse of his Seruice in the Warres, by Acts of Peace) called a Parliament; where was made that memorable Act, which at this day is called Poynings Law, whereby all the Statutes of England were made to bee of force in Ireland. **1656** [in Dict.] **1797** *Encycl. Brit.* IX. 327/2 During his administration was enacted the law known by the name of *Poyning's Law*. **1827** H. Hallam *Const. Hist. Eng.* II. xviii. 719 This produced the famous statute of Drogheda in 1495, known by the name of Poyning's law. **1938** D. L. Keir *Const. Hist. Mod. Brit.* vii. 434 By Poyning's Law, in 1495, the Irish Parliament itself made applicable to Ireland all statutes lately made in England, and acknowledged the right of the King to be informed of the causes for its summons and to approve, in his Council, all bills to be introduced when it met. **1973** B. Bradshaw in B. Farrell *Irish Parliamentary Tradition* v. 69 Essentially Poynings' Law provided that a parliament could not be validly held in Ireland without the consent of the king..both to the convening of parliament and to the projected legislation.

Poynting (poi·ntiŋ). *Physics.* The name of John Henry *Poynting* (1852–1914), English physicist, used *attrib.* and in the possessive to designate concepts in electromagnetism, as **Poynting's theorem,** the theorem that the rate of flow of electromagnetic energy through a closed surface is equal to the integral over the surface of the Poynting vector; **Poynting('s) vector,** the vector product of the electric and magnetic field strengths at any point, which can often be interpreted as representing in magnitude and direction the rate of flow of electromagnetic energy.

1893 J. J. Thomson *Notes Recent Res. Electr. & Magn.* p. xiii, Poynting's theorem. **1913** L. Silberstein *Vectorial Mech.* i. 45 The flux of electromagnetic energy, per unit time and unit area, *i.e.* the so-called Poynting-vector. **1962** Corson & Lorrain *Introd. Electromagn. Fields* xi. 399 The Poynting vector for sunlight is approximately 1·4 kilowatts/meter² at the surface of the earth. **1971** D. A. Dunn *Models of Particles* iii. 74 The usual form of Poynting's vector, which appears in Poynting's theorem and expresses the amount and direction of power flow, is invariant to a Lorentz transformation. **1976** D. E. Soper *Classical Field Theory* x. 150 The energy current..and the momentum density..are given by the familiar Poynting's vector. **1977** *Daily Tel.* 14 Nov. 8 (Advt.), You'll find a complete range of ready-prepared programs for commonly-used calculations such as..field-strength and poynting vector due to electric dipole.

pozzolana, pozzuolana. Now also anglicized as **pozzolan (pozzulan, puzzolan).** Add to def.: **pozzolan cement,** a cement made from natural or artificial pozzolan, lime, and water.

1907 *Chem. Abstr.* I. 1078 (*heading*) Process of manufacturing limes, cements, and puzzolans. **1910** *Engineering News* LXIV. 597/2 The mixed Portland cement and puzzolan-cement is superior to Portland cement. **1918** R. Peele *Mining Engineers' Handbk.* xliii. 2217 Pozzulan cement proper is made by grinding hydrated lime and pozzulan, a volcanic material occurring near Naples. **1947** R. H. Bogue *Chem. of Portland Cement* xxix. 520 Only when the chemical action is completely understood will it be possible to design a 'pozzolan' of ideal composition for any particular purpose. **1951** *Econ. Geol.* XLVI. 311 Pozzolans are natural and artificial siliceous and aluminous substances which are not cementitious themselves but which react with lime in the presence of water at atmospheric temperatures to produce cementitious compounds. **1971** *Materials & Technol.* II. ii. 106 Pozzolan cement is made by grinding together a cement (usually Portland cement) and a powdered siliceous material.

pozzolanic *a.* (later examples); hence also **po:zzolani·city,** the property of combining with lime in the presence of water to form a cement.

1927 F. L. Brady *Introd. Building Sci.* xxi. 241 Mortars are used in England at the present day, which set by pozzolanic action. **1956** *Proc. Inst. Electr. Engin.* CIII. a. 225/1 Pulverized fuel fly ash from our power stations..has the pozzolanic property of setting with lime under water. **1960** *B.S.I. News* Jan. 15 'Pozzolanic portland cements' may contain up to twenty per cent. pozzolana but need not pass the test for pozzolanicity. **1972** *Daily Tel.* 29 Apr. 17 (Advt.), Manufacturers of pozzolanic cements and grouts.

pozzy¹: see *possie, pozzy.

pozzy² (pǫ·zi). *Army* and *Navy slang.* Also **possie, possy, pozi, pozzie.** [Origin unknown.] Jam, marmalade.

1916 *Daily Mail* 1 Nov. 4/4 'Pozzy' (jam). **1919** W. H. Downing *Digger Dial.* 59 *Pozi*, jam. 'Who puckeroed the pozi?' 'Who took the jam?' **1929** *Papers Mich. Acad. Sci., Arts & Lett.* X. 316/1 *Possie*, jam. This seems to be a variant of pozzi (or pozzy). **1962** Granville *Dict. Sailors' Slang* 91/1 *Pozzie* (*y*), jam or marmalade.

p-process: see *P III. 6.

practical, *a.* (*sb.*) Add: **A.** *adj.* **I. 1. b.** *spec.* of doors, windows, food, etc., forming parts of a theatrical or film set; operable; able to be used as in real life.

1933 P. Godfrey *Back-Stage* iv. 47 His [*sc.* the stage-carpenter's] doors and windows never open unless he has been told to make them 'practical'. **1960** O. Skilbeck *ABC of Film & TV* 98 *Practical.* Part, or fitting, of a Set which may have to operate exactly as though real; e.g. a door, tap, or light fitting. **1974** *Some Technical Terms & Slang* (Granada Television) s.v. *Practical*, Granada gave a 'practical' banquet in its play *The Dead*.

c. *Electr.* Applied to certain units (the ampere, volt, ohm, watt, coulomb, and farad) used for practical measurements, as contrasted with the absolute units of the C.G.S. system.

They are now part of the International System of Units.

1873 J. C. Maxwell *Treat. Electr. & Magn.* II. iv. x. 244 The practical unit of electromotive force is called the Volt. **1882** *Nature* 24 Aug. 391/2 Instead of expressing electrical quantities directly in absolute measure, the [International Electrical] Congress has embodied a consistent system, based on the Ohm, in which the units are of a value convenient for practical measurements. In this, which we must hereafter know as the 'practical system', as distinguished from the 'absolute system', the units are named after leading physicists, the Ohm, Ampère, Volt, Coulomb, and Farad. **1886** J. D. Everett *Units & Physical Constants* (ed. 2) xi. 151 The practical unit of capacity is the Farad. It is defined as 10^{-9} of the C.G.S. electro-magnetic unit of capacity. **1904** *Jrnl. Inst. Electr. Engin.* XXXIV. 172 He suggested that the prefix *ab* or *abs* should be used with the names of the practical units (Volt, Ampere, Ohm, etc.) to form names for the corresponding C.G.S. electro*magnetic* units. **1932** [see *international *a.* (*sb.*) A. 1 c]. **1963** Jerrard & McNeill *Dict. Sci. Units* 12 Units based on this definition..are known as electromagnetic units (e.m.u.). The quantities defined by these units are generally of an inconvenient size for practical work so units, known as practical units, are used. The latter may be obtained by multiplying the e.m. unit by a suitable conversion factor. *Ibid.* 13 The inconvenience of having three systems of electrical units, ab units, stat units and practical units has been overcome by the introduction of the metre, kilogramme, second, ampere units (M.K.S.). In this system, the practical units have the same value as the theoretical ones.

3. (Later examples.)

1970 P. Laurie *Scotland Yard* 292 *Practical*, the highest police compliment; particularly of a senior officer who can still distinguish between the formal processes of law and the realities of police work. **1972** *Police Rev.* 10 Nov. 1475/3 Dickens is on the side of the angels in applauding what we now loosely term 'good practical Police work'.

III. 6. *practical activity*, activity through which theory is realized and becomes actual; *practical attitude*, an attitude that is concerned with material facts and actual events; *practical criticism*, an analytical approach to literary criticism, advocated by I. A. Richards, which influenced and was further developed as *New Criticism; *practical nurse* chiefly *N. Amer.*, one who has completed a course of training in nursing practice but who is not a (state-)registered nurse; also *attrib.*; hence *practical nurse* v. trans., *practical nursing* vbl. sb.; *practical politics*, what actually takes place or is possible in political life, sometimes implying lack of moral principle (cf. *Realpolitik); hence *practical politician*.

1913 D. Ainslie tr. *B. Croce's Philos. of Practical* ii. i. 173 (*heading*) The practical activity in its dialectic. *Ibid.*, We shall no longer ask, therefore, whether the practical activity precede or follow knowledge. **1935** E. Burns *Handbk. Marxism* xiii. 213 There, where speculation ends, with real life, real positive science therefore begins, the representation of practical activity, of the practical process of the development of men. **1963** T. Bottomore tr. *Marx's Early Writings* 52 The criticism of the speculative philosophy of right does not remain within its own sphere, but leads on to *tasks* which can only be solved by *means of* *practical activity.* 1974 R. Stevens *James & Husserl* 177 The intentional continuity between projects of meaning and their fulfillment in practical activity. 1883 F. H. Bradley *Princ. Logic* 20 Not only are the genuine characteristics absent from a mere practical attitude, but we find present there a quality which is absent from real judgment. 1902 W. James *Var. Relig. Exper.* xi. 261 Our moral and practical attitude, at any given time, is always a resultant of two sets of forces within us. 1945 K. R. Popper *Open Society* II. xxiv. 213 It may be better to explain rationalism in terms of practical attitudes or behaviour. 1962 Macquarrie & Robinson tr. *Heidegger's Being & Time* I. vi. 238 So this phenomenon by no means expresses a priority of the 'practical' attitude over the theoretical. 1970 C. A. van Peursen in J. M. Edie et al. *Patterns of Life-World* 148 Theoretical truth is of importance, but in no case does it present the origin of truth as such. The practical attitude of man, 'shunning what is harmful and pursuing what is apt to promote well-being', is of more importance. 1929 I. A. Richards (*title*) Practical criticism: a study of literary judgment. 1958 *Oxf. Mag.* 13 Nov. 94/2 'Practical criticism' itself, as a term, is unsatisfactory... What on earth is the point of any sort of criticism if it isn't practical?.. In so far as it has a clear current sense, it means the analysis, apart from their literary context, of short passages of prose and verse. 1959 *Times Lit. Suppl.* 24 Apr. 241/3 It is not merely criticism; it is 'practical criticism' of a high order and of a kind which is too rarely found in France. 1972 *Ibid.* 3 Mar. 246/3 As for the assumptions involved in the identifying of me [*sc.* F. R. Leavis] with 'Practical Criticism', my first comment is that the formula isn't mine, and,..I have been known for my insistence, when having to use it, that 'Practical Criticism is criticism in practice'. 1977 *N.Y. Rev. Bks.* 15 Sept. 40/2 His practical criticism is not much concerned with the structure of an individual poem except as an embodiment of crisis. 1921 *Daily Colonist* (Victoria, B.C.) 1 Oct. 16/2 (Advt.), Situations Wanted—Female..experienced practical nurse, terms moderate. 1956 K. Hulme *Nun's Story* vii. 104 Our practical nurses ..can stand only a four-hour shift, but our sisters take unlimited duty. 1964 Mrs. L. B. Johnson *White House Diary* 24 Apr. (1970) 118 To my great delight we..saw a class in practical nursing, and I told them that the world was certainly waiting for their skills. So it is, as anybody who has tried to find a practical nurse for an elderly or ill member of the family can tell you. 1971 'A. Blaisdell' *Practice to Deceive* v. 68 Mrs. Carstairs would be back from practical-nursing her sister. 1979 W. Kienzle *Rosary Murders* 20 One of the orderlies and one of the practical nurses..had enjoyed a quick roll in bed. 1812 *Deb. Congress U.S.* (1853) 12th Congress 1 Sess., App. 2210 There were two circumstances, inherent in this system of coercing Great Britain by commercial restrictions, which ought to have made practical politicians very doubtful of its result. 1961 *Times* 10 Jan. 8 The impression remains that the Liberal leader is still the diplomatist, more at home in the chancery, or the corridors of the United Nations, not the father figure, so necessary in Canadian leadership, or the practical politician, able to talk about sewage problems. 1796 *Rep. Comm. House of Commons* (1803) XIV. 38 With a view to such a knowledge of practical Politics, as may be desired from the History of our experimental Legislation. 1826 Disraeli *Viv. Grey* I. II. xv. 217 'Hargrave', said his lordship, 'if you want any information upon points of *practical politics*'—that was his phrase..'there is *only one* man in the kingdom whom you should consult..and that's Stapylton Toad.' 1897 [see Practical *a.* 1 b]. 1919 F. Hamilton *Vanished Pomps of Yesterday* i. 30 As the inventor of 'Practical Politics' (*Real Politik*), Bismarck had a supreme contempt for fluent talkers and for words. 1939 I. Berlin *Karl Marx* iv. 74 It was his [*sc.* Marx's] first experience of practical politics: he conducted his paper with immense vigour and intolerance. 1961 N. P. Stallknecht in Stallknecht & Frenz *Compar. Lit.: Method & Perspective* vi. 121 We may..trace the notion of individual autonomy from its manifestation in religious practice and theological reflection through practical politics and political theory into literature and the arts.

B. *sb.* (Restrict 'in *pl.*' to senses in Dict.)

2. (Earlier example.)

a 1737 M. Green *Spleen* (1936) 15 That tribe, whose practicals decree Small-beer the deadliest heresy.

3. An examination, course, or lesson devoted to practice in a subject. Also *fig.*

1934 in Webster. 1955 *School Sci. Rev.* Nov. 38 For the practical paper, 93 per cent of candidates were successful —compared with 68 per cent who passed the theory, and of over 300 who passed the theory, only 7 failed the practical. 1961 *Times* 6 Nov. 14/2 Lieutenant Babington, straight out from England to take over a platoon..seems to have failed his practicals where discipline is concerned. 1966 *Rep. Comm. Inquiry Univ. Oxf.* II. 311 Nationally, the average for tutorials was much lower (1·7 hours against 6·3 hours at Oxford), while the national averages for lectures and practicals were about twice the Oxford figures. 1979 F. Olbrich *Sweet & Deadly* xi. 125 He would get through this damned exam if it was the last thing he did... There would still be the practicals, of course.

practicalism. Add: Also, = Pragmatism 4; also, in Communist usage, excessive attention paid to practical matters resulting in the disregard of theory.

1898 W. James *Philos. Concept. & Pract. Results* 5 The principle of practicalism.., as he [*sc.* C. S. Peirce] called it, when I first heard him enunciate it at Cambridge [Mass.] in the early '70s, is the clue..by following which ..we may keep our feet upon the proper trail. 1898 —— *Coll. Ess. & Rev.* (1920) 424 Now the principle of practicalism says that the very meaning of the conception of God lies in those differences which must be made in our experience if the conception be true. 1951 *Britannica Bk. of Year* 686/2 *Practicalism*, a Communist term for the fault of paying too much attention to practical problems of production, etc., and not enough to ideological propaganda. 1963 R. C. Tucker *Soviet Political Mind* ix. 181 Krushchev..has been criticized by certain elements within the Soviet Communist Party for being insufficiently so [*sc.* theory-oriented]. He has rejected their charge of 'practicalism'.

practicality. **1.** (Earlier and later examples.)

1828 Mill *Autobiogr.* (1924) 288 It is not for me to dispute the palm of practicality with these sage and cautious persons. 1961 E. Becker *Zen* iv. 107 The 'marvelous person' that is supposed to result from Zen exhibits more Chinese practicality than Indian speculation. 1972 *Physics Bull.* May 303/1 If the physicist's numeracy is combined with his practicality and technical competence one sees that he can present a formidable challenge in this market traditionally the preserve of the arts graduate.

practicant. Restrict †*Obs. rare* to first sense in Dict. and add later examples.

1952 *John o' London's* 2 May 434/1 In..nutrition, we have known him as a brake on the over-enthusiasm of the practicants in that novel subject. 1974 *Spectator* 9 Nov. 591/3 Practicants of Transcendental Meditation.

practice. Add: **1. c.** *Philos.* The active practical aspect as considered in contrast to or as the realization of the theoretical aspect.

c 1898 C. S. Peirce *Coll. Papers* (1934) V. 412 Science, when it comes to understand itself, regards facts as merely the vehicle of eternal truth, while for Practice they remain the obstacles which it has to turn, the enemy of which it is determined to get the better. 1907 F. C. S. Schiller *Studies in Humanism* iv. 130 It seems necessary, therefore, to conceive 'practice' more broadly as *the control of experience*... The aim of the doctrine of the 'subordination' of 'theory' to 'practice'..is merely *voluntarism*. 1937 C. Morris *Logical Positivism* v. 67 Science has integrated and utilized all of the dimensions of meaning, and may be said to walk on the three legs of theory, fact, and practice. 1969 D. Cairns tr. *Husserl's Formal & Transcendental Logic* 32 The distinction is after all a relative one; because even purely theoretical activity is indeed activity—that is to say, a practice (when the concept of practice is accorded its natural breadth).

d. A Marxist term for the social action which should result from and complement the theory of communism. Cf. *Praxis 1 c.

1899 *Social-Democrat* III. 358 (*title*) Social-democratic theory and practice. ? 1924 *Communist Internat.* I. 49 The practice of the class struggle is fertilised by theory, and in its turn becomes the fruitful soil for theoretical study. 1925 N. Bukharin *Lenin as Marxist* 17 If Leninism in practice is not the same as Marxism, then we get just that separation of theory from practice which is specially harmful for such an institution as the Institute of Red Professors. 1966 P. Heath tr. *Wetter's Soviet Ideology Today* I. vi. 139 It [*sc.* a theory] must issue, rather, in *correct* practice, i.e., in a practice that leads reality towards its *truth*. But the problem as to the criterion of truth is not thereby resolved. 1966 F. Schurmann *Ideology & Organization in Communist China* p. xlvi, By the time of Yenan, 'theory' had been canonized, and the Chinese Communists turned their attention to 'practice', namely organization and action. 1971 Z. A. Jordan *Karl Marx* I. vi. 67 Concepts such as those of progress, of the historically restricted scope of social laws, of ideology and social engineering (practice in the Marxian terminology).

12. (sense 2 b) *practice direction, master*; (sense 3) *practice-crew, -dress, -room* (examples); **practice bar** *Ballet* = *Bar *sb.*[1] 13 d; **practice-curve**, a curve or graph showing the relation of practice to progress; **practice pad**, a non-resonant pad, usu. circular and made of rubber or the like, on which to practise the art of drumming; **practice wicket**, (see quot. 1934).

1938 N. Streatfeild *Circus is Coming* viii. 132 Using the side of the steps as a practice bar and raising her right leg in an arabesque. 1946 —— *Party Frock* vi. 54 Sally was using the edge of the mantlepiece as a practice-bar. 1976 'M. Albrand' *Taste of Terror* xiv. 78 An empty space, a practice bar, one wall mirrored. 1887 *Century Mag.* XXXIV. 178/2 Freshmen formed a practice crew of their own. 1924 R. M. Ogden tr. *Koffka's Growth of Mind* v. 262 New configurations are also attributable to these lower centres; as is demonstrated by the fact that the practice curve improves by leaps which occur in learning new movements. 1956 B. R. Bugelski *Psychol. of Learning* xiv. 399 Learning curves are sometimes called 'practice' curves but about all that we can say for certain about such curves is that they will normally rise above a starting point if learning does take place. 1942 *Weekly Notes* 10 Jan. 19/1 (*heading*) Practice direction... The Judges of the Chancery Division have given the following direction: [etc.]. 1968 *Weekly Law Rep.* 24 May 815 (*heading*) Practice Direction (Divisional Court: Avoidance of Delay). 1977 C. Hampton *Criminal Procedure* (ed. 2) viii. 212 The Lord Chief Justice issued a practice direction in July 1967 explaining the procedure. 1934 A. P. Herbert *Holy Deadlock* 192 The young ladies of the chorus were in 'practice-dress', their plump legs naked from their 'trunks' to their ankles. 1937 M. Allingham *Dancers in Mourning* iii. 39 Her short white practice dress. 1979 K. O'Hara *Searchers of Dead* xi. 116 A scrawny girl in a practice dress..chasséed below him. 1885 Emden & Pearce-Edgcumbe *Compl. Coll. Practice Statutes, Orders & Rules* 1147 (*heading*) Office Rules settled by the Practice Masters, 1880, 1881, 1882. 1890 *Law Rep.: Queen's Bench Div.* XXV. 243 The practice masters..have at some period between 1880 and 1888 issued a direction that 'writs of summons before the Judicature Acts came into force may be renewed without an order'. 1937 *Encycl. Court Forms & Precedents* I. 13 There are eight King's Bench Masters... They have also control of the Central Office and one of them sits daily to exercise this control and to give directions with respect to questions of practice and procedure relating to the business of the Central Office...[*Note*] The Master discharging this duty (called 'the Practice Master') takes *ex parte* applications and also gives advice to solicitors on points of practice and procedure. 1966 *Masters' Practice Directions, Tables & Forms* (Supreme Court of Judicature) 1 These directions shall..supersede any Practice Masters' Rule or Direction dealing with the same subject. 1968 *New Yorker* 18 May 56, I started playing drums in junior high. I got a practice pad and sticks and a Paul Yoder method book. 1972 *Down Beat* 16 Mar. 19/3 First of all, get a drum set, not a practice pad. Then play records. 1921 W. de la Mare *Crossings* 17 No more scales in that musty-fusty old practice-room! 1922 [see *Bar *sb.*[1] 13 d]. 1963 *Times* 11 June 15/3, 14 sound-proof practice-rooms. 1977 *New Yorker* 19 Sept. 54/3 A friend of Robin's, Roman Markowicz, popped out of a practice room and stopped us. 1980 *Early Music* Jan. 43/1, I agreed to come to the choir practice-room. 1871 'Thomsonby' *Cricketers in Council* 23 Your first lessons may well be solely in hitting. Go to your practice wicket, and endeavour to hit hard..every ball that is bowled to you. 1934 W. J. Lewis *Lang. Cricket* 298 *Practice wicket*, a pitch with one wicket set up for the practice of batting or batting and bowling, usually in the nets.

practicism (præ·ktisiz'm). [f. Practic(e + -ism.] (See quot. 1957[1].)

1957 R. N. C. Hunt *Guide to Communist Jargon* xxxv. 121 Hence practicism is the tendency of party members to conduct their day-to-day work without regard to the theory which alone gives that work its justification and significance. 1957 *Economist* 26 Oct. 320/1 All the crimes in the jargon book of communist heresy—including such esoteric offences as 'practicism' (lack of clear revolutionary perspective).

pra·ctico-, combining form of Praxis or Practical *a.*: = practically.., practical and.., as in *practico-empirical, -social, -spiritual* adjs.; **practico-inert**, Sartre's term for his hypothesis that man's present freedom of action is limited as a result of the exercise of free choice, or praxis, by previous generations in their use of the material world.

1970 B. Brewster tr. *Althusser & Balibar's Reading Capital* (1975) II. iii. 83 All the visible phenomena and practico-empirical concepts produced by the economic world. 1966 A. Manser *Sartre* xiii. 207, I want to say.. that all men are slaves in so far as their experience of life takes place in the realm of the *practico-inert* and in the exact measure in which this realm is originally conditioned by scarcity. 1975 B. Cooper tr. *Aron's Hist. & Dialectic of Violence* ii. 41 Alienation presupposes the moment of original freedom and translucid *praxis*. Otherwise, it would but remain the experience of the practico-inert, the activity-passivity that we live out each day, and could not be recognized as the experience of bondage. 1975 F. Copleston *Hist. Philos.* IX. xvii. 379 Man thus falls under the domination of the 'practico-inert' which he himself has created. Man makes the machine; but the machine then reacts on man, reducing him to the level of the practico-inert, to what can be manipulated. 1976 A. S. Smith tr. *Sartre's Critique Dialectical Reason* 829 Practico-inert, matter in which past *praxis* is embodied. 1970 B. Brewster tr. *Althusser & Balibar's Reading Capital* III. 314 Ideology..is distinguished from a science not by its falsity, for it can be coherent and logical (for instance, theology), but by the fact that the practico-social predominates in it over the theoretical, over knowledge. *Ibid.* I. 54 In the 1857 *Introduction*, Marx writes.. 'practico-spiritual (*praktisch-geistig*) appropriation of this world'.

practicum (præ·ktikǝm). *N. Amer.* [a. late L. *practicum*, neut. of *practicus*, Gr. πρακτικός practical, concerned with action. (Cf. G. *praktikum* practical training.)] A practical exercise; a course of practical training.

1904 T. F. Hunt *Cereals in Amer.* p. vi, The ideal condition involves a study of the plant in the field. Unfortunately..no systematic course of instruction can be planned that will conform with the season of crop growth and meet the exigencies of the weather. Practicums should be supplied that will as far as possible remedy this defect. 1938 Kains & McQueston *Propagation of Plants* 458 To carry out these practicums it is desirable..to have available a plot of ground of not less than one-half acre. 1973 *Arithmetic Teacher* Apr. 244/1 (Advt.), The second week will be a practicum involving children. 1974 H. L. Foster *Ribbin'* vii. 301 Practicum experiences involve various aspects of the community life, tutoring students, working with teachers and students, and participating in school activities. 1978 *Daily Colonist* (Victoria, B.C.) 27 Aug. 17/5 His desire for a kindergarten practicum caused a few problems..but it was arranged through the Sooke school district.

practolol (præ·ktǝlǫl). *Pharm.* [Etym. unknown.] A white powder, $C_{14}H_{22}N_2O_3$, that is similar to propranolol in its effect on the heart but has less effect on respiratory functions.

1969 *Lancet* 14 June 1221/1 The British Pharmacopœia has issued the following supplementary list of approved names... Practolol, 4-(2-Hydroxy-3-isopropylaminopropoxy)-acetanilide. *Ibid.* 2 Aug. 227/1 Practolol..was given to forty-seven patients who had cardiac dysrhythmias after myocardial infarction. 1976 *Nature* 12 Aug. 595/2 Practolol (40 mg kg^{-1}) a β_1 receptor blocking agent that does not enter the brain in significant quantity did not inhibit hyperactivity. 1976 *Lancet* 6 Nov. 984/1 Practolol therapy may be associated with various adverse reactions. *Ibid.*, Practolol ('Eraldin') tablets were withdrawn from general use on Oct. 1, 1975.

prad. Add: (Further examples.)

The examples are all Australian.

1882 *Sydney Slang Dict.* 10/2 He blew on Sam who

frisked a lobb and the same day came it on Joe for fencing the prad got on the cross. 1916 C. J. Dennis *Moods of Ginger Mick* 27 'E sits there while I 'arness up me prad. 1930 *Bulletin* (Sydney) 17 Sept. 21/2 Our packhorses snorted suspiciously... The prads watched, all prick ears and snorts as the axe bit into the tree. 1933 *Ibid.* 27 Dec. 20 When it [*sc.* the rope] broke, the astonished prad plunged suddenly into the dam. Two doses of that remedy and he was a reformed animal. 1977 *Courier-Mail* (Brisbane) 31 Mar. 4/5 It would surely be more appropriate for the riding [for democracy] to be done on some business man rather than on a prad.

‖ **pradakshina** (prada·kʃinā). Also **pradakshna.** [Skr. *pradakṣiṇā*, f. *pra* in front + *dakṣiṇā* right.] In Hinduism and Buddhism, circumambulation of an object in a clockwise direction as a form of worship. Also *attrib.*

1810 E. Moor *Hindu Pantheon* 327 The respectful ceremony of *Pradakshna*, which consists in circumambulating several times the..object..to be reverenced, keeping, with closed palms, the right hand and the face towards it. 1883 M. Monier-Williams *Relig. Thought & Life in India* I. xii. 348 A pilgrim..sets out from the source of the Ganges,..and walks by the left bank of the river to its mouth..; then, turning round, he proceeds by the right side back to Gangotrī, whence he departed. This is called Pradakshiṇā. 1933 A. Stein *On Anc. Central-Asian Tracks* iv. 61 The ceremonial circumambulation or *pradakshina* prescribed by Indian custom. 1956 R. Pieris *Sinhalese Social Organisation* II. 84 He then.. reverentially performs *pradakshinā* three times to the diagram. 1976 *Jrnl. R. Soc. Arts* CXXIV. 677/2 The apsidal temple in question could have been a temple for the Nāga cult, with the four-sided Nāga image..standing in the apse, placed there to be worshipped in the *pradakṣiṇa* way.

præcocial, *a.* Add: Also **precocial.** Insert in etym. after *præcocēs*: (C. J. Sundevall 1836, in *Kungl. Svenska Vetenskapsakademien Handlingar 1835* 70). Add to def.: also extended to refer to the young of other animals which are independent soon after birth. (Later examples.)

1932 J. S. Huxley *Probl. Relative Growth* iii. 90 The same reasoning applies to Ocypoda, whose young are similarly precocial. 1937 Allee & Schmidt tr. *Hesse's Ecol. Animal Geogr.* xxiii. 484 The young [of water-birds] are praecocial and very soon learn to forage for themselves. 1949 A. Leopold *Sand County Almanac* I. 35 The hen plover is brooding the four large pointed eggs which will shortly hatch four precocial chicks. 1974 *Nature* 30 Aug. 732/2 The subject species, *Acomys cahirinus* (spiny mouse), is a murid rodent whose precocial infants possess functional motor and sensory capabilities within hours of birth. 1978 *Sci. Amer.* July 107/1 Reid has found that the contents of the kiwi egg are 61 percent yolk, a proportion half again as large as that found in the eggs of typical precocial birds.

prædella, erron. var. Predella.

1926 *Trans. Scottish Ecclesiol. Soc.* VIII. II. 71 The sanctuary is, perforce, very small and the Holy Table..is placed directly against the wall... The single step Prædella is semi-circular, and is enclosed by a beautiful tudor rail.

Praedesque (prī·de·sk), *a.* [f. the name of W. M. *Praed* + -esque.] In the manner or style of Winthrop Mackworth Praed (1802–39), poet, essayist, and writer of society verse. So **Prae·dian** *a.*; **Prae·dism**, the style of Praed's verse.

1865 *Dublin Univ. Mag.* II. 23 The best epigrams and Praedesque verses of the week. 1883 *Century Mag.* Feb. 595/1 Mr. Locker can write Praedesque poems. 1905 Mrs. H. Ward *Marriage of W. Ashe* I. ii. 29 Meanwhile the outer room gathered to hear the recitation of some *vers de société*, fondly believed by their author to be of a very pretty and Praedian make. 1927 *Observer* 15 May 6 What he was thinking of was polite badinage, Praedism, and Horatian levity.

prædormital (prīdǫ·ɹmităl), *a.* *rare*$^{-1}$. [? f. Præ- + stem of L. *dormītiō* sleep + -al.] = Hypnagogic *a.*

1947 V. Nabokov *Bend Sinister* 11 Suddenly, with the vividness of a prædormital image or of a bright-robed lady on stained glass, she drifted across his retina, in profile, carrying something.

præmunire, *sb.* **1.** (Later examples.)

1940 E. Pound *Cantos* lxx. 177 Treasons, felonies, new praemunires. 1961 E. F. Jacob *Fifteenth Cent.* vi. 253 In November, after consultation with the judges, writs under the statute of Praemunire were made out against Beaufort.

2. b. (Later hist. example.)

1902 J. Gairdner *Eng. Ch. 16th Cent.* viii. 141 Any subject henceforth bringing in bulls of excommunication was liable to a *præmunire*.

Prænestine (praine·stīn, prī-), *a.* and *sb.* [ad. L. *Prænestīnus* f. *Præneste* Palestrina: see -ine[1].] **A.** *adj.* Of or pertaining to the ancient city of Præneste or its inhabitants.

1880 tr. *Woltmann & Woermann's Hist. Painting* I. iv. 88 The engraved metal caskets of the kind commonly known as Prænestine cistæ, because they have been found for the most part at Prænestē, the modern Palestrina. 1885 *Encycl. Brit.* XIX. 654/2 Præneste was chiefly famed for its great temple of Fortune and for its oracle, in connexion with the temple, known as the 'Prænestine lots' (*sortes Prænestinæ*). 1937 *Oxf. Compan. Classical Lit.* 163/1 As regards Latin writing, in the inscription on the Praenestine *fibula*.., probably of the 6th c.,..the direction is from right to left. 1939 L. H. Gray *Foundations of Lang.* 332 The oldest record of Italic is a Praenestine fibula of the seventh century B.C., *Manios med fhefhaked Numasioi* 'Manius me fecit Numerio'. 1970 *Oxf. Classical Dict.* (ed. 2) 873/1 Praeneste has yielded the earliest specimen of Latin, whose peculiarities confirm Festus' statement..that Praenestine Latin was abnormal. 1976 *Archivum Linguisticum* VII. 60 Praenestine *fhfhaked* ..is an old reduplicated perfect remade into an aorist by the addition of *-t (> Early Latin -*d*).

B. *sb.* A native or inhabitant of Præneste.

1902 *Encycl. Brit.* XXXIII. 897/2 The Romans..were inclined to sneer at the pronunciation and idiom of the Prænestines. 1949 *Oxf. Classical Dict.* 726/1 Praenestines loyally resisted Pyrrhus..and Hannibal, and actually preferred their own status to that of Roman citizens.

Hence **Prænesti·nian**, the extinct Latin dialect spoken by the Prænestines.

1939 L. H. Gray *Foundations of Lang.* 333 To the [Latino-Faliscan] group also belonged the closely similar Hernician and Prænestinian. 1954 [see *Latino-Faliscan* s.v. *Latino-].

præses (prī·siz). [See Preses.] **a.** Var. Preses.

1763 Boswell *Jrnl.* 19 Jan. in *London Jrnl.* (1950) 155 It resembled a party's being worsted in the choice of praeses and clerk, at an election in a Scotch county. 1797, 1876 [in Dict. s.v. Preses]. 1898 P. S. Allen *Let.* 10 Oct. (1939) 16 When the Praeses introduced me, the Bishop said..'You mustn't leave me alone with a man, who makes such bold proposals, President.'

b. An academic moderator.

1841 *Rules Compilation of Catal.* in *Brit. Mus. Catal. Printed Bks.* I. p. v/1 The respondent or defender in a thesis to be considered its author, except when it unequivocally appears to be the work of the Praeses. 1853 C. C. Jewett *Smithsonian Rep. Construction of Catal. of Libr.* The Respondent or defender in a thesis, is to be considered its author, except when it unequivocally appears to be the work of the Praeses. 1931 G. S. Gordon *Let.* 19 Dec. (1943) 195 Is it possible that there is no such leisure for an academic official—whether Professor or Praeses—who has a conscience about the job he's paid for? 1967 *Anglo-Amer. Catal. Rules: Brit. Text* 27 Enter a dissertation written for defence in an academic disputation (according to the custom prevailing in European universities prior to the 19th century) under the praeses (the faculty moderator) unless the authorship can be well authenticated.

præsidial, var. Presidial *a.* and *sb.*

1918 C. G. Robertson *Bismarck* v. 285 Bismarck persuaded the King of Bavaria to write to the King of Prussia, inviting him..to take the Imperial Crown and exercise as Emperor his Praesidial rights in the Confederation.

Præsidium: see *Presidium.

pragmatic, *a.* and *sb.* Add: **A.** *adj.* **6.** (Further examples.)

1907 W. James *Pragmatism* iv. 136 The pragmatic value of the world's unity is that all these definite networks actually and practically exist. 1932 C. Morris *Six Theories of Mind* vi. 282 The pragmatic contribution to the theory of mind. 1948 *Mind* LVII. 358 These 'pragmatic paradoxes' as they have been called, are worth examination. 1964 A. W. Burks in Moore & Robin *Stud. in Philos. C. S. Peirce* 2nd Ser. viii. 143 Peirce's pragmatic principle of meaning. 1971 G. Petrović in R. Klibansky *Contemp. Philos.* IV. 393 To the uninformed the pragmatic theory of truth seems identical with that of Marx.

b. *spec.* Relating to the practical interpretation of political or social issues. Cf. *Pragmatism 4 b.

1961 *Mem. & Proc. Manch. Lit. & Philos. Soc.* CIII. 58 This was an explicit pragmatic democratic philosophy of an older generation. 1964 *Listener* 29 Oct. 654/2 Isn't there a danger that this kind of practical, pragmatic socialism, taking problems as they come, is going to rob you of a long-sighted view into the future. 1966 *Times* 11 Mar. 8/6 Mr. Wilson replied..that his 'policy was already very socialist and very pragmatic'. 1970 *Bull. Inst. for Study of U.S.S.R.* Aug. 17 The technocrat is more or less content with any ideology provided that it does not hamper pragmatic development. 1976 *Howard Jrnl.* XV. I. 3 Taking into account an admission of guilt or willingness to compensate for damage should only be done on the grounds that it is to the advantage of society to have the offender admit his guilt or pay for the damage. This is a very pragmatic attitude.

7. *Linguistics.* Of or pertaining to pragmatics. Cf. sense B. 4 below.

1935 B. Malinowski *Coral Gardens* II. iv. iv. 52 Since it is the function, the active and effective influence of a word within a given context which constitutes its meaning, let us examine such pragmatic utterances. 1953 C. E. Osgood *Method & Theory Exper. Psychol.* xvi. 699 The pragmatic dimension of semiotic, the relation between signs and their users or the effect of signs upon their users. 1957 C. Cherry *On Human Communication* vi. 226 Statistical communication theory abstracts from the semantic and pragmatic aspects of the set of signs used. 1964 E. A. Nida *Toward Sci. Transl.* iii. 36 There is a steady tendency for many terms to shift within the pragmatic area from an ethical response to an esthetic one. 1967 R. A. Waldron *Sense & Sense Devel.* iii. 49 The attitudes to language dealt with in the last chapter came into conflict with tradition in the stress they lay upon the pragmatic functions of language.

B. *sb.* **4.** *pl.* const. as *sing.* *Linguistics.* The study or analysis of linguistic signs as they relate to the human user and his behaviour (see quot. 1937). Also *attrib.*

1937 C. Morris *Logical Positivism* 4 Analysis reveals that linguistic signs sustain three types of relations (to other signs of the language, to objects that are signified, to persons by whom they are used and understood) which define three dimensions of meaning. These dimensions in turn are objects of investigation by syntactics, semantics, and pragmatics. 1952 *Mind* LXI. 205 The 'pragmatics' group would say that 'points of view' are the business of philosophers. 1954 *Mind* LXIII. 360 The step..from descriptive pragmatics to descriptive semantics. 1964 E. A. Nida *Toward Sci. Transl.* iii. 35 Pragmatics, in contrast to both semantics and syntactics, deals with the relation of symbols to behavior. 1969 I. I. Revzin in R. Klibansky *Contemp. Philos.* III. 332 In this domain there has been a general shift of interest from syntactics (the first set-theoretical and generative models) to semantics (and possibly pragmatics). 1971 *Language* XLVII. 522 The philosophical dichotomy is between 'semantics' and 'pragmatics', roughly corresponding to reference and inference respectively. 1975 *Ibid.* LI. 37 Partee..expresses reservations about the place of the referential/attributive distinction in natural language, and sees the possibility of assigning it to pragmatics. 1978 *Sci. Amer.* Nov. 82/2 The grammar of language includes rules of phonology, which describe how to put sounds together to form words; rules of syntax, which describe how to put words together to form sentences; rules of semantics, which describe how to interpret the meaning of words and sentences; and rules of pragmatics, which describe how to participate in a conversation, how to sequence sentences and how to anticipate the information needed by an interlocutor.

pragmatical, *a.* Add: **5.** (Later example.)

1938 C. Morris in *Internat. Encycl. Unified Sci.* I. I. 68 The pragmatical factor which complements and completes the formal and the empirical factors.

6. Of or pertaining to pragmatics. Cf. *Pragmatic *sb.* 4.

1939 *Mind* XLVIII. 480 It is to be noted that 'pragmatical' as it occurs throughout this paper designates the relations holding between signs and their users or interpreters, and is not to be confused with 'pragmatic' or 'pragmatist'. 1942 R. Carnap *Introd. Semantics* 10 Examples of pragmatical investigations are: a physiological analysis of the processes in the speaking organs..; a psychological analysis of the relations between speaking behavior and other behavior, [etc.]. 1946 C. Morris *Signs, Lang., & Behav.* 219 The present study has deliberately preferred to emphasize the unity of semiotic rather than break each problem into its pragmatical, semantical, and syntactical components. 1957 C. E. Osgood et al. in Saporta & Bastian *Psycholinguistics* (1961) 285/1 We may call the relation of signs to situations and behaviors..*pragmatical meaning*.

pragmatically, *adv.* Add to def.: Also, in a manner related to pragmatic philosophy, or to pragmatics.

1902 W. James *Var. Relig. Exper.* xviii. 448 Pragmatically, the most important attribute of God is his punitive justice. 1909 —— *Pluralistic Universe* viii. 321 Pragmatically interpreted, pluralism or the doctrine that it is many means only that the sundry parts of reality *may be externally related*. 1933 *Mind* XLII. 246 Prof. Schlick's 'Causality in Everyday Life and in Recent Science' sets itself to discover the meaning of causality pragmatically, from its use. 1948 *Mind* LVII. 358 The fault is not a fault of logic in the sense that the definition is formally self-contradictory. It is merely pragmatically self-refuting. 1964 E. A. Nida *Toward Sci. Transl.* iii. 46 Some Pentecostals respond 'pragmatically' to a passage of the Scriptures by engaging in shouting and dancing. 1969 C.-Y. Cheng *Peirce's & Lewis's Theories of Induction* xii. 129 Induction cannot be pragmatically significant in a *strong* sense.

pragmaticism. Add: **1.** (Further examples.)

1970 *Bull. Inst. for Study of U.S.S.R.* Aug. 17 Wiles also analyses the regimes in China and Cuba, where property relations are laid down by the Party: this, in his view, will inevitably lead to a kind of pragmaticism and finally to a bureaucracy bereft of ideological impetus. 1978 J. Updike *Coup* (1979) vi. 230 The development of a plausible pragmaticism.

2. Substitute for def.: The name given by C. S. Peirce to his pragmatic philosophy, esp. to the doctrine that concepts are to be understood in terms of their practical implications.

1905 C. S. Peirce in *Monist* XV. 166 So then, the writer, finding his bantling 'pragmatism' so promoted, feels that it is time to kiss his child good-by and relinquish it to its higher destiny; while to serve the precise purpose of expressing the original definition, he begs to announce the birth of the word 'pragmaticism', which is ugly enough to be safe from kidnappers. 1934 C. Hartshorne et al. *Coll. Papers C. S. Peirce* V. p. v, Pragmaticism (Peirce's term to indicate his divergencies from other pragmatists) was thus Peirce's way of insisting that abstractions must give an account of themselves. 1946 B. Russell in J. Feibleman *Introd. Peirce's Philos.* p. xv, Peirce's pragmatism (or pragmaticism, as he came to call it) is a very different doctrine from those of James and Schiller and Dewey. 1968 E. H. Madden in R. Klibansky *Contemp. Philos.* II. 33 Peirce's shift..to pragmaticism and the dispositional frequency theory.

pragmatism. Add: **4.** (Further examples.) Also, the philosophical method of inquiry of C. S. Peirce; = *Pragmaticism 2.

1902 C. S. Peirce in Baldwin *Dict. Philos. & Psychol.* II. 322/2 Synechism..is not opposed to pragmatism in the manner in which C. S. Peirce applied it, but includes that procedure as a step. 1928 T. S. Eliot *For Lancelot Andrewes* iv. 84 The great weakness of Pragmatism is that it ends by being of no *use* to anybody. 1946 B. Russell in J. Feibleman *Introd. Peirce's Philos.* p. xv, Pragmatism, for Peirce, was only a method; the truths which it sought to discover were absolute and eternal. 1967 *Encycl. Philos.* VI. 432/1 Pragmatism is a method of clarifying and determining the meaning of signs. 1971 G. Petrović in R. Klibansky *Contemp. Philos.* IV. 394 We have come a great distance..from the confusion of pragmatism with Marxism. 1974 K. R. Popper in P. A. Schilpp *Philos. K. Popper* I. 99 The tendency of English philosophers to flirt with nonrealistic epistemologies: phenomenalism, positivism,..sensationalism, pragmatism—these playthings of philosophers were in those days still more popular than realism.

b. *Politics.* Theory that advocates dealing with social and political problems primarily by practical methods adapted to the existing circumstances, rather than by methods which have been conformed to some ideology.

1951 A. B. Ulam *Philos. Found. Eng. Socialism* iii. 77 It is true that the Fabian movement and British socialism in general have been built upon foundations quite different from those of Marxism. We have here first of all a sturdy spirit of pragmatism. 1966 *New Left Rev.* Jan.–Feb. 25 Thompson at one point calls the Communist Party the 'alter ego' of the Labour and trade union Left. But he fails to see the implications of this; that the two are not so very different in nature, for they are united by a common pragmatism, which has led so often to a day-to-day accommodation. 1976 *Times* 2 Apr. 8/4 Your struggle with Adolf Hitler, when Britain cast overboard the philosophy of pragmatism, or utilitarianism—the philosophy of recognizing any group of gangsters, any puppets, as head of a country as long as they were in control of its territory. 1976 *Survey* Summer–Autumn 156 Now that Mao is dead there will no doubt be a sharp reaction towards 'pragmatism'.

pragmatist. Add: **2.** (Further examples.) Also, an adherent of historical pragmatism (cf. Pragmatism 3).

1892 W. Wallace tr. *Hegel's Logic* (ed. 2) 257 The motives which must be viewed by the pragmatist as really efficient. 1932 C. Morris *Six Theories of Mind* ii. 71 In this insistence that mind cannot be divorced from the world certain new realists and pragmatists are at one with the absolute idealists. 1937 —— *Logical Positivism* iv. 46 Two groups which at first sight might seem in opposition, namely: the pragmatists (or biological positivists), and the Wiener Kreis (the logical positivists).

b. (Later examples.)

1906 W. James in *Jrnl. Philos.* III. 337 (*heading*) Papini and the pragmatist movement in Italy. 1965 E. E. Evans-Pritchard *Theories Prim. Relig.* iii. 48 The emotionalist explanations of primitive religion which I have discussed have a strong pragmatist flavour.

pragmatistic, *a.* Add: (Earlier and later examples.) Hence **pragmati·stically** *adv.*

1906 W. James in *Jrnl. Philos.* III. 341 Subjective factors..are in *some* degree creative, then; and this carries with it..the admissibility of the entire Italian pragmatistic program. 1907 —— *Pragmatism* viii. 281 Concretely he means..just the pragmatistically unified and ameliorated world. *a* 1914 C. S. Peirce *Coll. Papers* (1958) VIII. II. ix. 247 But if this occasion did in actuality *not* arise, such habit of thought as the conditional proposition might produce would be a nullity pragmatistically and practically. 1961 *Proc. Aristotelian Soc.* LXI. 180 Basic concepts,..if chosen pragmatistically, would be like Quinian 'posits'. 1971 J. J. Shapiro tr. *Habermas's Toward Rational Society* v. 68 The successful transposition of technical and strategic recommendations into practice is, according to the pragmatistic model, increasingly dependent on mediation by the public as a political institution. 1974 R. A. Hocks (*title*) Henry James and pragmatistic thought.

pragmatize, *v.* Add: **2.** *intr.* To behave in accordance with, or give expression to, a doctrine of pragmatism.

1907 H. James *Let.* 17 Oct. in R. B. Perry *Tht. & Char. W. James* (1935) I. 428, I was lost in the wonder of the extent to which all my life I have (like M. Jourdain) unconsciously pragmatised. 1966 *New Statesman* 15 Apr. 537/1 The moment the election was won he [*sc.* Mr. Wilson] was 'pragmatising' around the place like a man possessed.

Hence **pragmatiza·tion**, the action or process of giving practical effect to theory.

1950 T. Wiesengrund-Adorno et al. *Authoritarian Personality* xvii. 726 It is precisely this pragmatization of politics which ultimately defines fascist philosophy.

‖ **Prägnanz** (pre·gnants). *Psychol.* [G., = conciseness, definiteness: orig. used in this sense by M. Wertheimer 1923, in *Psychol. Forschung* IV. 317.] The tendency, noticed in experiments with Gestalts, for configurations to be given their most concise and clearly definable interpretations.

1925 *Amer. Jrnl. Psychol.* XXXVI. 359 How the configuration conforms to laws of simplicity, pregnancy or precision (Prägnanz), symmetry and the like. 1935 K. Koffka *Princ. Gestalt Psychol.* iv. 110 The principle was introduced by Wertheimer, who called it the *Law of Prägnanz.* 1938 *Mind* XLVII. 91 In this connection the criticism of Gestalt, and particularly of the Law of Prägnanz, implicit in certain experiments carried out by Thouless is very relevant. 1963 J. Mann *Frontiers of Psychol.* iv. 131 The basic nature of human data-receiving and processing equipment, which organizes perceptual data without the conscious awareness of consent of the perceiver. An example of such a tendency is the law of *Prägnanz.* 1971 *Sci. Amer.* Dec. 70/2 It should be evident by now that some principle of *Prägnanz*, or minimum complexity, runs as a common thread through most of the cases.

Prague (prāg). [Name of capital of Czechoslovakia.] **1.** Used *attrib.* in *Prague School* and various associated Combs. to designate the linguistic theories, primarily in relation to phonology, developed by or associated with members of the Prague Linguistic Circle (*Cercle Linguistique de Prague*), especially during the 1920s and 1930s.

1935 [see *Morphonology]. 1936 *Eng. Stud.* XVIII. 159 Trnka is a member of the Cercle Linguistique de Prague, and his analysis may be described as an application to English of the principles of the so-called Prague phonological school. 1937 [see *Phonology]. 1939 L. H. Gray *Foundations of Lang.* 61 The phoneme..is..regarded as..a point in the psychological pattern ('phonology'; Edward Sapir and the Prague school). 1959 [see *Morpheme]. 1962 [see *Neutralization 3]. 1964 E. Bach *Introd. Transformational Gram.* vi. 135 The parallel between such incompletely specified segments and the archiphonemes of the Prague school is evident. *Ibid.* vii. 143 Such widely differing schools of linguists as the Prague circle, glossematicians, and American structuralists have all concurred in insisting that languages should be studied as structures. 1964 M. A. K. Halliday et al. *Linguistic Sci.* 148 Modern theories in phonology were first developed by the 'Prague circle' founded by the Russian linguist N. S. Trubetskoy, whose work has been followed up by Roman Jakobson and many others. 1968 J. Lyons *Introd. Theoret. Linguistics* 126 The Prague school phonologists would say..that it is not the phoneme /p/ which occurs in the word *spot*, but the archiphoneme /P/. 1976 *Archivum Linguisticum* VII. 128 The different surface realizations are due to a different functional sentence perspective. This concept has been central in the Prague School since its foundation.

2. **Prague ham**, a type of smoked ham (see quot. 1931).

1931 C. L. T. Beeching *Law's Grocer's Man.* (ed. 3) 219/1 *Prague* hams (Pragerschinken) are first salted in large vats and left in a mild brine for several months; they are smoked with beech wood and matured in cool cellars until marketed. 1959 W. Heptinstall *Hors d'Œuvre & Cold Table* II. 69 For a ham to be served hot, my preference goes to the Prague hams. They are..admirable for carving, also they have an exquisite flavour. 1961 *Harrods Food News* 4/2 Sliced Hams..Prague... lb. 12/-. 1976 T. Fitzgibbon *Food of Western World* 366/2 Prague ham..is served whole for a main course, either baked or boiled, and cold slices of it are often served as a first course.

So **Pra·guean, Pra·guian** *adjs.* and *sbs.*

1968 [see *Marked *ppl. a.* 1 c]. 1972 *Language* XLVIII. 385 To use Praguean terminology, there are interlingual archiphonemes which allow, within a broad category, wide language-specific (hence speaker-specific) phonological variation for each content unit. 1973 *Ibid.* XLIX. 193 As Prangueans would put it, both writing and speech are signaling systems of the first order; in other words, one cannot dismiss writing as being the signaling system of the second order, a signal of a signal. 1974 *Trans. Philol. Soc.* 1973 64 As a result of neutralization the mark is dropped; what is left is the complex of distinctive features characterizing both members of the opposition, the *archiphoneme* in Praguian terminology.

prahu. Freq. spelling of Proa.

1821 J. Leyden tr. *Malay Annals* 148 Tun Talani and the mantri Jana Petra returned to their prahus. 1850 [in Dict. s.v. Proa]. 1932 W. S. Maugham *Narrow Corner* xv. 114 The harbour was far from crowded: there were only two junks, three or four large prahus, a motor-boat and a derelict schooner. 1964 K. G. Tregonning *Hist. Mod. Malaya* 104 They each gave up all right to levy dues upon *prahus* and other local craft. 1968 *Punch* 4 Dec. 802/2 A fishing *prahu* picked me up. 1977 *Borneo Bull.* 7 May 4/1 The healthy state of the regatta was illustrated this year by the number of racing prahus taking part.

praia (prai,ă). Also **praya.** [Pg.] A beach, a sea-shore; a river-bank; a water-front.

a 1865 Smyth *Sailor's Word-bk.* (1867) 541 *Praia*.., the beach or strand on Portuguese coasts. 1890 *Engineer* 24 Jan. 65/2 A more practical scheme is the proposed building of the whole river front of the city,..and the construction of a broad praya suitable for wheeled conveyances. 1893 Kipling *Seven Seas* (1896) 13 Hail, Mother! Hold me fast; my Praya sleeps Under innumerable keels to-day. 1933 P. Fleming *Brazilian Adventure* I. xv. 127 We had meant to camp that night on the sandbank (from now on all sandbanks will be referred to as *praias*). *Ibid.* 129 We slept that night on the praia, under a million stars. 1947 J. Bertram *Shadow of War* I. 11 Flags..streamed from the..roof-tops of grey Victorian company offices along the *praya*. 1963 J. Fleming *Death of Sardine* iii. 41 One day, when Trigoso Praia..was 'discovered' the road might be an important promenade but now it was tarmac and potholes.

‖ **praire** (prę̄r). [Fr.] The European clam, *Venus verrucosa*, or the North American hard-shell clam, *Mercenaria* (formerly *Venus*) *mercenaria*, which has been introduced to parts of Europe.

1878 P. L. Simmonds *Commercial Products of Sea* I. xii. 147 The 'paires doubles' [*sic*] (*Venus verrucosa*) or clams of the Mediterranean..are never as delicate in flavour as when freshly caught.] 1929 A. E. Housman *Let.* 27 Feb. (1971) 278 In the interesting menu of a Paris restaurant..there are *praires*. These may be among the shell-fish which I have eaten at Marseilles, but I do not remember them. 1960 [see *Palourde]. 1971 M. McCarthy *Birds of America* 128 Sometimes I just have a dozen *praires* (which are cheaper than oysters)..for lunch. 1975 *Times* 31 May 7/3 A board..bearing praires, clovisses, Belan oysters, langoustines.

prairie. Add: Also 8, 9 **parara, pararie, praira,** 9 **praire, prairia. a.** (Further examples.) Also (*U.S. local*), a marsh, a swampy pond or lake.

1791 D. Bradley *Jrnl.* 19 Sept. (1935) 17 A prairia of two or three hundred acres where the grass or wild oats is 8 or 10 feet high and very thick. *Ibid.* 12 Oct. 22 Struck a large prairia in our course—found it impassable. 1794 W. Clark *Jrnl.* 1 Aug. in *Mississippi Valley Hist. Rev.* (1914) I. 421 An open..*Pararie*..handsomly interspersed with Small Copse of Trees. 1795 J. Smith in *Ohio Archaeol. & Hist. Q.* (1907) XVI. 380 We saw several pararas, as they are called. They are large tracts of fine, rich land, without trees and producing as fine grass as the best meadows. 1806 *New Eng. Republican* in *Massachusetts Spy* 16 July 1/5 A venerable Philosopher sitting in the middle of an immense Map, marked with vast praires, huge rivers, and mountains of salt. 1819 E. Dana *Geogr. Sk. Western Country* 37 The ore is dug from an open praira. *Ibid.* 108 There are two kinds of praira, the *river* and *upland*. 1834 D. Crockett *Narr. Life* xii. 85, I came to the edge of an open parara, and looking on before my dogs, I saw in and about the biggest bear that ever was seen in America. *c* 1834 H. Evans in *Chron. Oklahoma* (1925) III. 181 We could look and behold ..one continual large expanse of Pararie. 1916 *Dialect Notes* IV. 270 *Prairie, n.*, marsh. (Barataria Bay.) 1934 *Nat. Geogr. Mag.* LXV. 601 Shallow ponds, or 'prairies', with a tropical tangle of vegetation. 1942 M. K. Rawlings *Cross Creek* 51 We use the word 'prairie' in a special sense. We have no open plains, but around most of the larger lakes are wet flat areas thick with water grasses, and these we call our prairies. They are more nearly marshes, yet we save the word 'marsh' for the deep mucky edges of lake and river, dense with coontail and lily pads. 1951 *Collier's* 24 Nov. 16/3 The eastern half of the Okefenokee is open. There are 'prairies', or great fields of water from one to three feet deep, covered with white or yellow water lilies, purple bladderworts, [etc.]. 1958 S. A. Grau *Hard Blue Sky* iv. 234 'And they put the candles on squares of wood..and set 'em adrift.... And the candles they draw Anton up from the bottom.'.. 'Such a big prairie to find a man in.'

b. *prairie country* (examples), *farm, fire* (earlier and later examples; also *fig.* and *attrib.*), *flower* (examples), *hay, madness, town*; **prairie bottom**, a low-lying expanse of prairie land; **prairie-breaking**, the use of a prairie-breaker; also, an area of land ploughed or broken by this means; **prairie buffalo** = *plains buffalo* s.v. *Plain *sb.*[1] 10; **prairie-buster** = *prairie-breaker*; **prairie clover** (earlier and later examples); **prairie coal** *N. Amer.*, dried cattle or horse dung used as a fuel; = *buffalo-chips* s.v. Buffalo 5; **prairie cock** = Prairie-chicken (in Dict. and Suppl.) or *sage-grouse* s.v. Sage *sb.*[1] 5 c; **Prairie Cree** = *Plains Cree; **prairie crocus** *Canada*, a blue- or mauve-flowered anemone, *A. patens*, native to northern Europe but naturalized in parts of Canada; **prairie fox** (examples); **prairie hare**, either of two North American hares, the varying hare, *Lepus americanus*, or the jack-rabbit, *L. townsendii*; **prairie hawk** (earlier and later examples); **prairie marmot** (earlier and later examples); **prairie owl**, either of two North American owls, the burrowing owl, *Speotyto cunicularia*, or the short-eared owl, *Asio flammeus*; **prairie oyster**, (*a*) (earlier and later examples); (*b*) calves' testicles cooked and eaten as a delicacy; **prairie pea**, a milk vetch belonging to the genus *Astragalus*, esp. *A. crassicarpus*, or its fruit; **prairie pigeon** (later example); **prairie plough** (earlier examples); **prairie plover** (earlier and later examples); **prairie potato** = *prairie-turnip*; **prairie province** (also with capital initials) *Canad.*, (*a*) the province of Manitoba, *obs. exc. hist.*; (*b*) *pl.* the area consisting of the provinces of Manitoba, Saskatchewan and Alberta; **prairie rattler** (examples); **prairie rattlesnake** (earlier and later examples); **prairie rose** (earlier and later examples); **prairie smoke**, a North American name for *Anemone patens*, a small perennial herb with blue flowers which is widely naturalized in prairie regions; also called pasque flower or crocus; **prairie snake** (example); **prairie snipe** (examples); **prairie soil**, soil of the kind characteristic of the North American prairies; *spec.* in *Pedology*, a soil that is marked by a deep, dark-coloured surface horizon with a high organic content, is subject to moderate leaching, and occurs under long grass in subhumid temperate regions; **prairie squint**, a squint produced by exposure to the bright light of a prairie; also *fig.*; **Prairie State** (earlier and further examples); **prairie turnip** (later examples); **prairie wagon** (examples); **prairie warbler** (earlier and later examples); **prairie wolf** (earlier and later examples); **prairie wool**, in Canada, the natural, undisturbed plant cover of prairie land, predominantly composed of grasses.

1819 T. Say *Jrnl.* 24 Aug. in E. James *Acct. Expedition Rocky Mts.* (1823) I. vii. 131 Our party encamped.. in a..beautiful and level prairie bottom. 1834 A. Pike *Prose Sk. & Poems* 12 It is bordered by a strip of timber, ..and on the outside of this, a prairie bottom..of exceeding rich land. 1868 *Rep. Iowa Agric. Soc.* 1867 139 On strong prairie-bottom it [*sc.* the Rio Grande bearded wheat] is liable to get down. 1861 *Trans. Illinois Agric. Soc.* IV. 37 The plows were running..too deep for ordinary prairie breaking. 1879 *Scribner's Monthly* Nov. 132/2 It is only by resorting to figures that one can reach a comprehension of the aggregate extent of these long, narrow, black strips of 'prairie-breaking'. 1886 P. G. Ebbutt *Emigrant Life Kansas* 45 Will Hopkins..used to do a good deal of prairie-breaking, having a twenty-four inch plough and six yoke of oxen. 1806 Lewis & Clark *Orig. Jrnls. Lewis & Clark Expedition* (1905) V. 80 A species of Lizzard called by the French engages prairie buffalo are natives of these plains as well as those of the Missouri. I have called them the horned Lizzard. 1859 H. Y. Hind *North-West Territory* xii. 105/1 That there are two kinds of buffalo appears to be still a matter of doubt; they are stated to be the prairie buffalo and the buffalo of the woods. 1951 F. G. Roe *N. Amer. Buffalo* iii. 47 The Red River hunt had had to travel increasingly long distances westward to find the ordinary Prairie buffalo. 1952 J. W. Day *New Yeomen of Eng.* viii. 94 On one tract..which was recently broken up, a very good job was made by a three-furrow prairie-buster hauled by a tracklayer. 1961 *Guardian* 8 Mar. 5/3 Open land in parks should be under planning control as protection against what are termed 'prairie busters'. 1857 A. Gray *First Lessons in Botany* 95 *Petalostemon*, Prairie Clover... Chiefly perennial herbs,.. [with] small flowers. 1870 *Amer. Naturalist* IV. 581 The prairie clovers..are among the most interesting of the leguminose species. 1939 *Nat. Geogr. Mag.* Aug. 247/2 Prairie clovers may be white, pink, purple, or violet. 1968 Peterson & McKenny *Field Guide Wildflowers Northeastern & North-Central N. Amer.* 252 Prairie-clovers have pinnate leaves and dense, longish pink or white flower heads on wiry stems. 1939 C. L. Douglas *Cattle Kings of Texas* 324 He could not bring himself to relish food cooked with 'prairie coal'. 1948 *Southwest Rev.* Summer 238/1 When the permanent settlers and their families came, this 'prairie coal' became the standard fuel. 1972 J. Minifie *Homesteader* vii. 51 As he walked he collected horse-dung for his cook-fire. Weathered 'prairie coal' makes a quick, hot fire. 1805 Lewis & Clark *Orig. Jrnls. Lewis & Clark Expedition* (1905) III. xviii. 123 Send out Hunters to shute the Prairie Cock a large fowl which I have only Seen on this river. 1846 J. W. Webb *Altowan* I. ii. 31 The prairie cock (a large species of grouse, of a pepper-and-salt colour, and long, pointed tail)..rose at their feet. 1876 J. Burroughs *Winter Sunshine* v. 115 The prairie hens or prairie cocks set up that low musical cooing or crowing. 1900 H. Garland *Eagle's Heart* 107 A belated prairie cock began to boom. 1806 *Deb. Congress U.S.* (1852) 9th Congress 2 Sess., App. 1136 The quality of the land is supposed superior to that on Red river, until it ascends to the prairie country, where the lands on both rivers are probably similar. 1848 E. Bryant *California* iii. 34 Our march was..through an undulating prairie-country. 1853 *Trans. Lit. & Hist. Soc. Quebec* IV. 298 The prairie country of the Saskatchewan is roamed over by countless herds of buffalo, also by the reindeer and the beautiful antelope. 1907 W. O. Lillibridge *Where Trail Divides* 152 The darkness that precedes morning has the prairie country in its grip. 1922 *Beaver* Oct. 15/1 During that time there had been two half-breed rebellions in the prairie country. 1946 J. T. Adams *Album Amer. Hist.* III. 261 Hay Burning Stoves were useful in this prairie country. 1883 E. Petitot in *Proc. R. Geogr. Soc.* V. 649 They occupied the country between the Savanois Indians on the east and the Grandes-pagnes (also called Prairie-Crees), on the west. 1913 F. W. Hodge *Handbk. Indians of Canada* (1971) 621/1 (Index) Plain Crees = Paskwawininiwug... Prairie-Crees = Paskwawininiwug. 1922 A. J. A. Stringer *Prairie Child* 304 Prairie-crocuses [are] soft blue and lavender and sometimes mauve. 1951 *Chambers's Jrnl.* Aug. 505/1 The mauve prairie-crocus, really an anemone, which pushes out of grass knolls as soon as the snow melts, is the first source of pollen. 1969 N. W. Parsons *Upon Sagebrush Harp* iii. 16 The land was ..blue with furry pasqueflower, or prairie crocus. 1838 H. W. Ellsworth *Valley of Upper Wabash* v. 49 A late and lamented brother of the writer, who had just finished a prairie farm. 1884 'Mark Twain' *Huck. Finn* vii. 46 Hogs soon went wild in them bottoms after they had got away from the prairie-farms. 1886 P. G. Ebbutt *Emigrant Life Kansas* 198, I don't think Anderson had enough energy in him to start a prairie farm for himself. 1824 W. Owen in *Indiana Hist. Soc. Publ.* (1906) IV. I. 83 We then rode on to the prairie and rode twice through the prairie fire, which..moved very slowly. 1836 D. B. Edward *Hist. Texas* iv. 70 Why should there be any lack of timber, when by planting it.., and preserving it afterwards from the annual prairie fires..it would grow with such rapidity. 1852 A. Cary *Clovernook* 77 Stories of.. huge lights made by prairie fires. 1922 *Beaver* Oct. 17/2 One of the most..terrible sights of those early days was a stampede of hundreds of buffalo fleeing before a prairie fire. 1935 H. A. L. Fisher *Hist. Europe* I. xii. 144 The Moslem faith might have spread like a prairie fire through the Balkans. 1959 [see *Back-fire *sb.* 1]. 1963 A. Heron *Towards Quaker View of Sex* iii. 27 'The prairie fire' view of homosexual contact—that..if it were allowed legally and morally everyone would turn to it. 1836 J. Hall *Statistics of West* iv. 56 The prairie-flower displays its diversified hues. 1873 *Newton Kansan* 22 May 3/2 The wild prairie flowers..are beginning to look beautiful. 1894 *Harper's Mag.* Aug. 422/1 To be sure there were patches of orange prairie flowers all about. 1922 H. L. Wilson *Merton of Movies* 69 Ain't I the little prairie flower, growing wilder every hour? 1839 Marryat *Diary Amer.* 1st Ser. I. xvii. 206 [In statistical table of furs] Prairie fox..5,000. 1846 R. B. Sage *Scenes Rocky Mts.* xxviii. 241 For several nights I had a constant visitor in the shape of a prairie-fox,—a creature about twice the size of a large red squirrel. 1876 J. Burroughs *Winter Sunshine* iv. 108 The prairie fox, the cross fox, and the black or silver-gray fox, seem only varieties of the red fox. 1948 A. L. Rand *Mammals E. Rockies* 107 Kit Fox. *Vulpes velox* (Also called Prairie Fox). 1840 E. Emmons *Rep. Quadrupeds Mass.* 58 *Lepus Virginianus.* Harlan. Prairie Hare... This species is common throughout the New England States, and is known generally as the White Rabbit. 1866 W. R. King *Sportsman & Naturalist in Canada* 32 The Prairie Hare..is one of the largest hares of the continent, weighing from seven to eleven pounds, and is of a grey colour tinged with yellow. 1917 H. E. Anthony *Mammals Amer.* 280/1 Although called the Prairie Hare, this species is found also on mountain slopes. 1817 E. P. Fordham *Jrnl.* 17 Dec. in *Personal Narr. Trav.* (1906) viii. 143 Saw some prairie hawks, blue bodies, ash coloured belly and wings, tipped with black. 1898 H. S. Canfield *Maid of Frontier* 201 With a swoop like the swoop of the prairie hawk down swooping for the quail, the Paint Horse was away. 1907 W. O. Lillibridge *Where Trail Divides* 259 Swift as the swoop of a prairie hawk..the man's arms were about her. 1835 A. Brunson *Jrnl.* 28 Oct. in *Wisconsin State Hist. Soc. Coll.* (1900) XV. 283 Here I fed myself, but could get nothing but Prairie hay & pumpkins for my horse. 1845 *Cultivator* II. 93 Without any kind of..comfort, except what they may gather from a poor supply of prairie hay. 1867 *Harper's Mag.* July 138/2 A little stable, near which were great stacks of prairie hay. 1878 J. H. Beadle *Western Wilds* xxviii. 433 First rate prairie hay, on which stock will keep fat all winter. 1880 D. Currie *Lett. of Rusticus* 6/2 They [*sc.* horses] are being fed with prairie hay. 1949 *Daily Oklahoman* 13 Feb. D. 4/4 More than 2,500 tons of the prairie hay used in the recent haylift operations to save icebound livestock in the western states were supplied by hay growers around Vinita, Okla. 1912 J. Sandilands *Western Canad. Dict.* 35 *Prairie madness*, the melancholia which attacks the lonely homesteader. 1973 H. Robertson *Grass Roots* iii. 53 The loneliness and isolation which contributed to emotional breakdowns known in the West as 'prairie madness'. 1826 J. D. Godman *Amer. Nat. Hist.* II. 114 The Prairie Marmot... Commonly called Prairie-dog. 1888 *Ipswich* (Mass.) *Chron.* 15 Sept. 2/4 Usually a country that is inhabited by prairie dogs, or more properly by prairie marmots, has a dry, thin atmosphere. 1979 *Jrnl. R. Soc. Arts* CXXVII. 171/2 The prairie marmot at Whipsnade has been seen elsewhere in the downs. 1846 R. B. Sage *Scenes Rocky Mts.* xii. 110 The prairie-owl and rattlesnake maintain friendly relations with these inoffensive villagers [*sc.* prairie-dogs]. 1860 C. W. Wilson *Mapping Frontier* (1970) II. 108 Nothing to disturb me but the melancholy note of the prairie owl. 1907 W. O. Lillibridge *Where Trail Divides* 13 He would have watched the movement of a coyote or a prairie owl, for the simple reason that it was the only visible object endowed with life. 1917 T. G. Pearson *Birds Amer.* II. 101 Short-eared Owl... Other Names.—Marsh Owl; Swamp Owl; Prairie Owl. 1958 *Publ. Amer. Dial. Soc.* XXX. 9 Prairie Owl and prairie dog owl, equally frequent names for the burrowing owl,..are expressions used by 23% of the Nebraska informants. 1883 J. F. Keane *On Blue-Water* xii. 167 We all jumped up and agreed unanimously to propose the last toast at once in the shape of a prairie oyster—an egg broken into a cup without smashing the yolk, the toast poured in on the top of it, and the whole taken at a swallow. 1912 'Saki' *Chron. Clovis* 275 He hurriedly ordered another prairie oyster. 1920 [see *bromo-seltzer* s.v. *Bromo-]. 1939 C. Isherwood *Goodbye to Berlin* 51 Would you like a Prairie Oyster? 1941 *Amer. Speech* XVI. 181 English slang metaphor also has its place, as..*prairie oyster* for the testicles of a steer, a food morsel considered dainty. 1955 W. Foster-Harris *Look of Old West* viii. 234 Prairie or mountain oysters were an unmentionable part of a male animal. 1960 *Spectator* 25 Nov. 878 A Prairie Oyster, which is the raw yolk of an egg slipped whole into a glass containing a tablespoon of Worcester Sauce and a dash of sherry, with a flick of red pepper. 1979 A. Jute *Reverse Negative* 26 His eyes were bloodshot. His prairie oysters must have lost their potency. 1848 E. Bryant *California* ii. 28, I observed, also, a plant producing a fruit of the size of the walnut, called the prairie-pea. 1870 *Amer. Naturalist* III. 162 One of the earliest flowers [of the Kansas plains] is the Prairie-pea. 1943 B. A. De Voto *Year of Decision* 155 They..made spiced pickles of the 'prairie peas'. 1937 *Nat. Geogr. Mag.* Aug. 200/1 The Eskimo curlew, or 'dough bird' or 'prairie pigeon', as it was called by the gunners, apparently rivaled the passenger pigeon in numbers prior to 1885. 1831 W. Sewall *Diary* 30 Apr. (1930) 136 Sat off with the team, and a prairie plow which came on late last night with instructions, to commence breaking prairie. 1840 *Cultivator* VII. 33/1 It may be amusing to eastern readers, to hear a description of a 'prairie plow'. 1861 *Trans. Illinois Agric. Soc.* IV. 392 The sod should be broken with a prairie plow. 1851 W. Kelly *Excursion to California* I. v. 83 A stand of prairie plover most opportunely made their appearance as we pulled up. 1940 E. T. Seton *Trail of Artist-Naturalist* xxxii. 299 The white-tailed longspurs, the prairie plover, were all gone, wholly routed by the plough. 1828 J. C. Beltrami *Pilgrimage in Europe & Amer.* II. xvii. 321 Everything appeared to me delicious, even some roots which they call prairie-potatoes, and which I had before thought detestable. 1848 E. Bryant *California* iv. 54 A root or tuber, of an oval shape, about one and one-half-inch in length..is called the prairie potato. 1891 *Canadian Indian* Mar. 168 The prairie potato..yields when dry a light, starchy flour, and is often cut into thin slices and dried for winter. 1917 H. Kephart *Camping & Woodcraft* II. xxi. 379 Potato, Prairie. Prairie turnip.. Palatable in any form. 1876 J. C. Hamilton (*title*) The Prairie Province: sketches of travel from Lake Ontario to Lake Winnipeg. 1881 *Progress* (Rat Portage, Ontario) 12 Nov. 4/1 The editor of the Woodstock (Ont.) *Sentinel-Review*, proposes..to get up a huge excursion of marriageable girls in Ontario to proceed early next spring to the prairie province. 1908 M. A. Brown *My Lady of Snows* 221 The majority ruled, but the minority clamored from the prairie provinces. 1916 O. D. Skelton *Day of Sir W. Laurier* 97 The Winnipeg Board of Trade denounced the policy of 'crushing and trampling upon one hundred thousand struggling pioneers of the prairie province to secure a purely imaginary financial gain to one soulless corporation'. 1952 D. F. Putnam *Canad. Regions* 340/1 Manitoba, Saskatchewan and Alberta are known as the 'Prairie Provinces' because they include the Canadian section of the vast grass-covered interior plains of North America. 1959 *Manch. Guardian* 5 Aug. 4/4 Some of his [*sc.* Mr. John Diefenbaker's] friends in the prairie provinces think he has been clever to keep the Easterners out. 1965 *Globe & Mail* (Toronto) 5 Jan. B 5/1 All three Prairie provinces are particularly unhappy about the special incentives being offered to attract industrial enterprises. 1965 R. M. Hamilton *Canad. Quotations & Phrases* 130/2 The Prairie Province. From the title of the book which gave the phrase general circulation; after 1905 it was extended to include Saskatchewan and Alberta in 'The Prairie Provinces'. 1973 *Fisheries Fact Sheet* (Environment Canada Fisheries & Marine Service) No. 1. 4/3 The larger bodies of water in the Prairie Provinces. 1977 D. MacKenzie *Raven & Kamikaze* xi. 132 She'd come to England straight from a prairie-province university... Her father [was] a veterinarian in Saskatoon. 1878 J. H. Beadle *Western Wilds* 133 The only dangerous snakes are the little prairie rattlers, seldom over two feet long. 1948 *Chicago Tribune* 30 May 14 A prairie rattler coils to strike; its prey is a rabbit. 1977 *New Yorker* 6 June 47/2 One.. took up pentecostalism and died from shock while caressing a prairie rattler at a revival meeting back east in Tennessee. 1817 S. R. Brown *Western Gazetteer* 31 The only venomous serpents, are the common and prairie rattlesnake, and copper-heads. 1843 [see *Missisauga 2]. 1873 'Mark Twain' & Warner *Gilded Age* 125 Prairie-rattlesnakes..never strike above the knee. 1948 *Natural Hist.* Apr. 187/1 An extensive campaign was waged against the prairie rattlesnake. 1961 C. H. Pope *Giant Snakes* (1962) 152 In the northern United States the prairie rattlesnake may not give first birth until it is four or even five years old. 1822 J. Woods *Two Years' Residence Eng. Prairie* 303 The prairie-roses, balm.. and sassafras-wood..have all powerful scents. 1946 E. B. Thompson *Amer. Daughter* 36 We gazed in awe upon the prairie rose, a delicate pink flower growing close to the ground, whose thorny stem belied its tender beauty. 1963 *Canad. Geogr. Jrnl.* Aug. 54/2 Later the hardy prairie rose makes its appearance and fills the air with its sweet scent. 1893 *Jrnl. Amer. Folk-Lore* VI. 136 *Anemone patens*, var. *Nuttalliana*... gosling, prairie smoke, crocus. 1952 *Sun* (Baltimore) 26 Feb. 10/7 The Pasqueflower..is a bluish open bell shaped wild flower of the prairies... Patches of the flower at a distance give the impression of a bluish haze. This gives rise to its more familiar name 'prairie smoke'. 1958 *Weekend Mag.* (Montreal) 7 June 38/1 Earliest of spring flowers, Manitoba's crocus grows so profusely in places that it looks like a low-lying mist. Hence its nickname: the 'Prairie smoke'. 1845 J. C. Frémont *Rep. Exploring Expedition* 12 A large prairie snake..was occupied in eating the young birds. 1851 W. Kelly *Excursion to California* I. v. 80, I shot a brace of prairie snipe. 1917 T. G. Pearson *Birds Amer.* I. 247 Upland Plover. *Bartramia longicauda*... [Also called] Prairie Snipe. 1817 S. R. Brown *Western Gazetteer* 66 The common field near the town contains nearly 5000 acres, of excellent prairie soil. 1876 *Trans. Illinois Dept. Agric.* XIII. 288 The prairie soils are usually darker, more crude, coarser and wetter than the woodland. 1910 C. G. Hopkins *Soil Fertility* vi. 79 The undulating prairie soils vary from a gray silt loam on light clay in the older areas, to a dark brown silt loam, in the later formations, and the common flat prairie soils vary with age from drab silt loam to black clay loam. 1928 C. F. Marbut in *Proc. & Papers 1st Internat. Congr. Soil Sci.* IV. 21 The podsolic and lateritic soils of category VI have been subdivided into 8 sub-groups consisting of Tundra, Podsols, Brown Forest soils, Red soils, Yellow soils, Prairie soils (dark colored humid soils), Laterites and Ferruginous Laterites. 1974 E. A. Fitzpatrick *Introd. Soil Sci.* vii. 116 In the U.S.A. and elsewhere there are prairie soils or brunizems which are similar to chernozems but they have a middle horizon with a clay maximum and are slightly less fertile. 1946 Auden *Under Which Lyre* in *Harvard Alumni Bull.* 15 June 707/1 The sophomoric Who face the future's darkest hints With giggles or with prairie squints As stout as Cortez. 1963 R. D. Symons *Many Trails* ix. 92 He wears a grey felt hat, beneath which his tanned face is puckered in the 'prairie squint'. 1842 *People's Advocate* (Carrollton, Illinois) 6 Aug. 4/5 Federal Coon Whiggery extinct in the Prairie State! 1852 Mrs. Stowe *Uncle Tom's Cabin* II. xlv. 316 Farmers of rich and joyous Ohio, and ye of the wide prairie states. 1861 O. J. Victor *Hist. Southern Rebellion* I. 166 Illinois, the 'Prairie State', then proved that she was as rich in her patriotism as in her soil and exhaustless resources. 1868 *Harper's Mag.* June 123/2 When he pronounced 'good-by' to the Prairie State, at the State line, he said, 'Behind the cloud the sun is shining still.' 1949 J. Monaghan *This is Illinois* 138 The nation began to hum the wonders of the Prairie State. 1963 R. I. McDavid *Mencken's Amer. Lang.* 691 Illinois has had many nicknames..but *Prairie State* and *Sucker State* are the only ones surviving. 1970 *Daily Progress* (Charlottesville, Va.) 24 May 4/1 Illinois is the Prairie state. 1855 A. M. Murray *Let.* 5 Sept. (1856) II. 290 About forty miles from Chicago we passed the first prairie town of Joliet. 1867 *Atlantic Monthly* Mar. 326/1 Chicago, for fifteen years after it began its rapid increase, was perhaps of all prairie towns the most repulsive to every human sense. 1908 Kipling in *Collier's* 28 Mar. 11/1 'If you go as far as Winnipeg, you'll see the finest hotel in all the world.' 'Nonsense!' he said. 'You're pulling my leg! Winnipeg's a prairie-town.' 1977 H. Osborne *White Poppy* viii. 68 A prairie town in a cowboy film. 1857 J. Palliser *Jrnls.* (1863) 38 The root..receives the name of the Prairie Turnip by the half-breeds, who, with Indians, use it as food. 1941 D. McCowan *Naturalist in Canada* 246 The Crees and the Blackfeet were glad to make a meal from the edible root of the Prairie Turnip. 1956 D. Leechman *Native Tribes Canada* 110 The prairie Indians also ate service berries, wild cherries, red willow berries, prairie turnips, bitter root, and wild rose haws. 1856 Prairie wagon [see *AMBULANCE 3]. 1867 W. H. DIXON *New Amer.* I. iii. 37 We find that our big Concord coach has been exchanged for a light prairie waggon. 1948 *Chicago Daily News* 10 Apr. 6/2, I have an idea that too much of the squirrel rifle and prairie wagon tradition still runs in the bloodstream of most Americans. 1811 A. WILSON *Amer. Ornithol.* III. 87 [The] Prairie Warbler..I first discovered in that singular tract of country in Kentucky, commonly called the Barrens. 1917 T. G. PEARSON *Birds Amer.* III. 150/1 The Prairie Warbler is not very common on the prairies. 1960 R. T. PETERSON *Field Guide Birds of Texas* 217 Prairie Warbler... This warbler wags its tail. 1804 LEWIS & CLARK *Orig. Jrnls. Lewis & Clark Expedition* (1904) I. ii. 108 A Prarie Wolf come [*sic*] near the bank and Barked at us this evening. 1898 H. S. CANFIELD *Maid of Frontier* 39 The long howl of the prairie wolf rose on the air and hung tremulant. 1948 *Daily Ardmoreite* (Ardmore, Okla.) 18 Apr. 14/7 There are practically only two distinct kinds of wolves in America—the large gray timber wolf and the coyote or prairie wolf. 1963 R. D. SYMONS *Many Trails* 121 The grey jackal of the plains..we call in English the prairie wolf, but more often in corruption of its Aztec name, coyote. 1934 G. BETTANY *Valley of Lost Gold* 284 She loved..every blade of prairie wool. 1953 *Canad. Geogr. Jrnl.* June 245/1 The sheep crop the 'prairie wool'—that excellent hard forage composed of spear-grass, bunch-grass and buffalo-grass. 1970 [see *June grass]. 1973 R. D. SYMONS *Where Wagon Led* I. i. 13 The prairie grass was curled and dimpled—that's why they call it 'prairie wool'.

prairie-chicken. Substitute for def.: A North American grouse found in prairie regions and belonging to one of three species of the family Tetraonidæ, *Tympanuchus cupido, T. pallidicinctus*, or *Pediœcetes phasianellus*. Also *fig.* (Earlier and later examples.)

1840 *Picayune* (New Orleans) 13 Sept. 2/2 The travelling public will find..a fine table covered with white fish..and prairie chickens. 1949 *N. Dakota Hist.* Jan. 14 The 'coo' of the prairie chicken and the twittering of the meadowlark greet us. 1963 *Canadian Weekly* 30 Mar. 18/4 In 1945 the prairie chicken, or sharp-tailed grouse, by enactment was made an emblem of Saskatchewan. 1976 *National Observer* (U.S.) 3 July 10/6 He took Nicholson and his gang of cutthroats..and that cute little prairie chicken, Kathleen Lloyd, and made a damned good and entertaining movie. 1978 C. HARRISON *Field Guide Nests, Eggs & Nestlings N. Amer. Birds* 101 Greater Prairie Chicken... Nestling. Precocial and downy. Down pattern like that of the Lesser Prairie Chicken.

prairied, *a.* (in Dict. s.v. PRAIRIE.) (Later example.)

1930 H. N. SPALDING *From Youth to Age* 58 The happy cornlands of the prairied West.

prairie-dog. (Earlier and further examples.)

1774 J. R. PEYTON in J. L. Peyton *Adventures of my Grandfather* (1867) xii. 121 One of the singular and interesting sights on my route was the villages of the Prairie dogs. 1859 E. H. N. PATTERSON in L. Hafen *Overland Routes to Gold Fields* (1942) 110 Visited a prairie dog town this evening, which covers eighty acres. 1867 [see *LAY-OUT 2 b]. 1902 O. WISTER *Virginian* xvi. 176 There is a brown skunk down in Arkansaw. Kind of prairie-dog brown. 1914 B. M. BOWER *Flying U Ranch* 135 There ain't enough grass in our lower field to graze a prairie dog. 1932 S. ZUCKERMAN *Social Life Monkeys* ii. 23 Most writers describe the prairie dog of North America as an animal that lives in vast colonies. 1947 *Chicago Daily News* 20 Mar. 14/3 [They] make my book resemble a head of lettuce that has been gnawed by a pack of prairie dogs. 1961 *Maclean's Mag.* 29 July 23, I've seen me..lying at the edge of a field in Saskatchewan spying on the prairie dogs. 1976 *Billings* (Montana) *Gaz.* 1 July 2-A/4 Flath contends, however, that the endangered black-footed ferret lives within prairie dog towns.

prairie-hen. (Earlier and later examples.)

1804 LEWIS & CLARK *Orig. Jrnls. Lewis & Clark Expedition* (1904) I. iv. 181 Capt. Lewis..Saw great numbers of Prarie hens. 1909 G. PARKER *Northern Lights* 336 A prairie hen rustled by with a shrill cluck. 1933 *Sun* (Baltimore) 15 Apr. 4/2 Dr. Gross considered the advisability of mating the heath hen with a Wisconsin prairie hen.

prairie schooner. For *U.S.* read *N. Amer.* and add earlier and further examples. Also *Austral. colloq.* (see quot. 1911[1]).

1841 E. R. STEELE *Summer Journey in West* 134 So much is this appearance acknowledged by the country people that they call the stage coach, a prairie schooner. 1847 T. WEED *Let.* in T. W. Barnes *Mem. T. Weed* (1884) II. 149 We found the road..occupied with an almost unbroken line of wagons, drawn generally by two yokes of oxen, bringing wheat to the city. These teams are called 'prairie schooners'. 1867 [see *DOUBLE-DECKER b]. 1904 [see SCHOONER *sb.*[1] 2]. 1911 C. E. W. BEAN *'Dreadnought' of Darling* vii. 67 An old white-bearded patriarch of a fellow that had once appeared in one of the up-river towns with a 'prairie schooner'—one of those big white-hooded sort of ambulance waggons which the travelling hawkers drive from homestead to homestead over the plains in the West. 1911 *Daily Colonist* (Victoria, B.C.) 21 Apr. 4/2 Last summer one Sunday morning on the Cariboo road a prairie schooner stood by the roadside. 1949 *Amer. Speech* XXIV. 259 Hordes of treasure seekers from regions east of the mountains crossed the plains in covered wagons, the ships of the desert of the early Western emigrants. Such wagons in various regions were known as..*prairie schooners*. 1955 W. FOSTER-HARRIS *Look of Old West* vi. 159 The prairie schooners, developed from the Conestoga were smaller but still too heavy and clumsy for mountain work or badly broken country. 1957 L. EISELEY *Immense Journey* 19, I slid over shallows that had buried the broken axles of prairie schooners. 1961 [see *CONESTOGA 2]. 1977 *Time* 14 Feb. 54/1 There it will begin tests that will culminate in flights that could do for space colonization what the prairie schooner and the railroads did for the settling of America.

prairillon. For a 1860 *Scenes Rocky Mts.* 172 (Bartlett) read 1846 R. B. SAGE *Scenes Rocky Mts.* 172, and add earlier example.

1843 J. C. FRÉMONT *Rep. Exploration* 60 We were posted in a grove of beech,..with a narrow *prairillon* on the inner side.

praise, *sb.* Add: **1. c.** A laudatory utterance; *spec.* = *praise poem.

1861 tr. *Casalis's Basutos* II. xvii. 328 We often heard them recite, with very dramatic gestures, certain pieces... The natives called these recitations *praises*. 1901 G. M. THEAL *Rec. S.-E. Afr.* VII. 202 When the king goes out he is surrounded and encircled by these *marombes*, who recite these praises to him with loud cries, to the sound of small drums, iron and bells. 1929 *Bantu Stud.* (Johannesburg) July 201 We are concerned..with 'Izibongo' as the term denoting the 'Praises' of the Zulu Chiefs. 1937 G. P. LESTRADE in I. Schapera *Bantu-Speaking Tribes S. Afr.* xiii. 300 The tribal praise-poem reciter..makes a new praise from time to time. 1968 T. COPE *Izibongo: Zulu Praise-Poems* 51 The most primitive type of praise-poem is simply a collection of praises consisting for the most part of single lines or verses. 1970 R. FINNEGAN *Oral Lit. in Afr.* v. 111 The formalized praises which are directed publicly to kings, chiefs, and leaders, and which are composed and recited by members of a king's official entourage. 1979 G. FORTUNE in Hodza & Fortune *Shona Praise Poetry* 3 Fragments of the praises of individual kings of the Changamire dynasty have come down to us included in the clan praises of the Rozvi.

4. *praise-meeting* (examples), *-night*; **praise-house** *U.S.*, a small meeting-house for religious services; **praise-leader** *Sc.*, the leader of the singing in a church; **praise name,** in Africa, a name or title used in ceremonial contexts; a name applied to the subject of a praise poem; **praise poem,** a laudatory poem; *spec.* one of a genre belonging to the oral tradition of certain African peoples; so *praise poet, poetry*; **praise-reciter** = *praise poet* above; **praise song,** a laudatory song; *spec.* in Africa, = *praise poem* above; so *praise-singer, -singing*.

1862 H. WARE in E. W. Pearson *Lett. from Port Royal* (1906) 20, I went with him to the praise-house, where he has his school. 1867 *Nation* (N.Y.) 30 May 432/2 But the true 'shout' takes place on Sundays or on 'praise' nights..either in the praise-house or in some cabin. 1869 Praise-house [in Dict.]. 1920 C. JERDAN *Scottish Clerical Stories* xviii. 370 The minister..looked down over the side of the pulpit and said to the praise-leader, 'Is David ill?' 1862 J. M. MCKIM in *N. Amer. & U.S. Gaz.* 14 July 1/8 When dey come to de praise meeting dat night dey sing about it. 1862 H. WARE in E. W. Pearson *Lett. from Port Royal* (1906) 36 He had been up to the praise-meeting by Uncle Peter's invitation. 1863 H. G. SPAULDING in *Continental Monthly* Aug. 195/1 The present opportunities for religious worship which the freedmen enjoy consist of their 'praise meetings'—similar in most respects to our prayer meetings. 1904 D. KIDD *Essential Kafir* ii. 91 If the trouble does not vanish..the people..say to the spirits, 'When have we ceased to kill cattle for you, and when have we ever refused to praise you by your praise-names?' 1932 C. FULLER *Louis Trigardt's Trek* vii. 79 Molamoso ruled the country... This refers, however, to Legadimane, whose family or 'praise name' was Molamoso. 1935 *Critic* (Cape Town) Oct. 2 The Tswana-speaking clan called the *BaRaMoseki* has as its praise-name the name *Mokwena* (from *kwena*, 'crocodile', the 'totem' of the clan), and every member of that clan is addressed as *Mokwena* on suitable occasions. 1968 T. COPE *Izibongo: Zulu Praise Poems* I. ii. 26 A clan name is the personal name of its founder, and personal names are essentially praise-names. 1979 G. FORTUNE in Hodza & Fortune *Shona Praise Poetry* 71 The praise name is the most frequently used construction in praise poetry... Structurally the praise name is a single noun or a single complex nominal construction, one of whose constituents is a class affix. 1864 H. WARE in E. W. Pearson *Lett. from Port Royal* (1906) 253 It was not praise-night. 1935 *Critic* (Cape Town) Oct. 4 A praise-poem..consists of a number of stanzas, following each other in different order in different versions of the same poem. 1957 S. EINARSSON *Hist. Icelandic Lit.* 44 Most scholars assume that skaldic poetry originated at the courts of kings, the poems being praise poems to celebrate the deeds of these kings. 1965 I. SCHAPERA *Praise Poems of Tswana Chiefs* 6 It is still..common for someone to..recite praise-poems. 1977 *Amer. N. & Q.* XV. 148/2 The *Prothalamion* has not been sufficiently studied in the light of Horace's *Carmina*, several of which are praise-poems. 1979 G. FORTUNE in Hodza & Fortune *Shona Praise Poetry* p. x, The praise poems are written in the standard Shona orthography. 1935 *Critic* (Cape Town) Oct. 7 The old praise-poets had to compose in their heads, and had to remember as they went along. 1937 G. P. LESTRADE in I. Schapera *Bantu-Speaking Tribes S. Afr.* xiii. 296 Persons of but modest rank..compose their own praise-poems,..while those of higher status have theirs composed by..the praise-poets. 1965 I. SCHAPERA *Praise Poems of Tswana Chiefs* 5 There are..in every tribe some men who specialize in composing and reciting praises of chiefs... This they do not merely..to establish..a personal reputation as a *mmôki* (praise-poet, praise-reciter), but also in the hope of reward. 1970 R. FINNEGAN *Oral Lit. in Afr.* v. 111 The 'praise names'..often form the basis of formal praise poetry. 1971 *Listener* 2 Sept. 290/3 What is nowadays called 'bardic poetry' which is a genus of praise-poetry. 1977 *Westindian World* 3–9 June 13/4 It is sheer praise-poetry. 1979 G. FORTUNE in Hodza & Fortune *Shona Praise Poetry* 2 Praise poetry, especially of the more formal kind, is a mode of expression that is disappearing owing to urbanization and the replacement of traditional methods of education by schools. 1935 *Critic* (Cape Town) Oct. 7 A praise-reciter, whose business it is to know and remember praise-poems. 1965 Praise-reciter [see *praise poet* above]. 1954 M. F. SMITH *Baba of Karo* I. iii. 62 When you hear drumming, you hear the deep drum and you hear the praise-singers—you'll give them money! 1963 W. SOYINKA *Lion & Jewel* 61 And then I have to hire a praise-singer, And such a number of ceremonies Must firstly be performed. 1977 *Eastern Province Herald* (S. Afr.) 27 Apr., The installation of the new Chancellor..was a dignified affair but it is difficult to understand what relevance a Xhosa praise singer had to the function. 1957 *Africa* XXVII. 26 (*title*) The social functions and meaning of Hausa praise-singing. 1886 Praise song [in Dict.]. 1928 W. C. WILLOUGHBY *Soul of Bantu* iv. 368 Praise-songs, which make up in glory for all they lack in veracity, are chanted upon occasion by the men whom they extol. 1957 *Africa* XXVII. 29 The District Head..may request that the praise-songs of title-holders who are his particular friends..should also be sung. 1970 P. OLIVER *Savannah Syncopators* 65 Bussani tribesmen in Upper Volta singing praise songs for the chief of the village of Yarkatenga.

praise, *v.* Add: **3. c.** *absol.* To express approbation; to bestow praise.

c 1386 CHAUCER *Parson's Tale* (1877, Ellesmere MS.) 473 Certes, the commendacion of the peple is somtyme ful fals and ful brotel for to triste. this day they preyse tomorwe they blame. 1609 SHAKS. *Sonnet* cvi. 13 For we which now behold these present dayes, Haue eyes to wonder, but lack toungs to praise. 1879 *Fortn. Rev.* 1 Apr. 507 So Molière is read or witnessed; we laugh and we praise. 1896 *Forum* (N.Y.) Mar. 1 Whether we praise lavishly or venture to blame, two perils threaten us.

4. b. Catch-phrase *praise the Lord and pass the ammunition* (see quots.).

1942 F. LOESSER (*song-title*) Praise the Lord and pass the ammunition. 1942 *Life* 2 Nov. 43 On the cover and above are pictures of Captain William A. Maguire, the man who inspired the best of this war's songs, *Praise the Lord and Pass the Ammunition*... Legend and the song written by Frank Loesser have it that..up jumped the sky pilot, gave the boys a look And manned the gun himself as he laid aside the Book, shouting 'Praise the Lord and pass the ammunition!'. 1943 *Sun* (Baltimore) 17 Sept. 10/5 The navy..named a 35-year-old chaplain from nearby Haddonfield, N.J., as the man who first used the phrase 'Praise the Lord and pass the ammunition' during the attack on Pearl Harbor. The chaplain, Lieut. Com. Howell E. Forgy, was on his first visit home in three years. 1948 A. M. TAYLOR *Lang. World War II* (rev. ed.) 159 *Praise the Lord and Pass the Ammunition*: attributed to a minister at Pearl Harbor... Real author of the phrase seems to have been Naval Lieutenant Howell Forgy, Presbyterian chaplain.

praiseach (prǟʃa-χ). Also **praisseagh, prashack, prashagh, prashoge, prassia.** [Ir., f. L. *brassica* cabbage.] **a.** A porridge made from oatmeal, sometimes flavoured with vegetables. Also *fig.*, a mess, a collection of small pieces.

1698 J. DUNTON *Let.* in E. Maclysaght *Irish Life in 17th Cent.* (1969) 330 He chose rather to stay at home with Prashagh and Potatoes than hazard himself in France where he knew not that any such food grew. 1935 D. PIATT *Dialect in East & Mid-Leinster* 17/1 Praiseach... Secondary meaning: 'To make p. of a thing,' i.e., break in small pieces. 1969 C. CARFAX *Silence with Voices* viii. 51 Would I jam me wagon in the middle of a main road and wait to be made into *prashoge*?

b. The charlock, *Brassica arvensis*, or a related wild plant of the cabbage family.

1727 C. THRELKELD *Synopsis Stirpium Hibernicarum* s.v. Brassica, This is Praisseagh buigh in Irish, and grows plentifully in corn fields. 1859 *Ulster Jrnl. Archaeol.* VII. 278 In former times, when cabbages were not generally cultivated in Ireland, the wild kail (called in Irish *Praiseach*), was often made use of as a kitchen vegetable. 1880 T. MCGRATH *Pictures from Ireland* xi. 113 The growing oat crop struggles with the perennial thistle, dock, and prassia. 1904 N. COLGAN *Flora Co. Dublin* 22 *B[rassica] Sinapis*... Prashack. Yellow Weed. Charlock. 1943 D. A. WEBB *Irish Flora* 14 *B[rassica] arvensis*... Charlock, Praiseach... Tilled fields and waste places; common.

praiser. Add: **2. c.** *spec.* = *praise poet*. Cf. *MBONGO.

1904 D. KIDD *Essential Kafir* ii. 92 All chiefs keep a Court Praiser, whose business it is to go in front of the chief and sing his praises. 1937 G. P. LESTRADE in I. Schapera *Bantu-Speaking Tribes S. Afr.* xiii. 299 A praiser..may..alter the order..of stanzas. 1968 T. COPE *Izibongo: Zulu Praise Poems* I. ii. 26 When a man of distinction is rewarded for his services by the chief..he..establishes a great kraal and appoints a personal praiser, who will collect..and perfect his praises, so that they constitute what we call a 'praise-poem'.

Prakrit. Add: Also 8 **Pracort.** (Earlier and further examples.) Also *attrib.* or as *adj.*

1766 J. CLELAND *Way to Things by Words* 88 The Pracort is the vulgar language, so called in contradistinction to the Sanscort. 1880, etc. [see *MAHARASHTRI]. 1968 W. S. ALLEN *Vox Graeca* i. 14 In relatively ancient times this receives support from transcriptions into Prakrit (Middle Indian) on coins of the Greek kings of Bactria and India in the 1 and 2 c. B.C. 1971 [see *INDOLOGIST].

‖ **pralaya** (prala·yă). [Skr.] Dissolution, destruction of the world.

1922 JOYCE *Ulysses* 296 Questioned by his earthname as to his whereabouts in the heavenworld he stated that he was now on the path of pralaya or return but was still submitted to trial at the hands of certain bloodthirsty

entities on the lower astral levels. 1954 G. S. RAO *Indian Words in Eng.* 134/1 *Pralaya*, destruction, esp. the destruction of the whole world at the end of a Kalpa. 1970 V. MEHTA *Portrait of India* II. 62 We Hindus say that the universe has a *pralaya*, a death—a period of withdrawal for rest. We Madrasis have our *pralaya* all the time.

pralidoxime (prælidǫ·ksīm). *Pharm.* [f. *aldoxime* s.v. *ALDO- with arbitrary insertion of *p, r*, and *i* (from PYRIDINE).] (A salt of) the 2-hydroxyiminomethyl-1-methylpyridinium ion, $HO{\cdot}N{:}CH{\cdot}C_5H_4\overset{+}{N}{\cdot}CH_3$, which reactivates the enzyme cholinesterase and is used as an adjunct to atropine in the treatment of poisoning by certain cholinesterase inhibitors (as malathion and parathion).

1961 *Approved Names* (Brit. Pharmacopœia Comm.) 19 Pralidoxime Iodide, Picolinaldoxime methiodide, Protopam. 1965 *New Drugs* xlvi. 473/1 Pralidoxime chloride restores the depressed cholinesterase activity resulting from organophosphate poisoning. 1970 PASSMORE & ROBSON *Compan. Med. Stud.* II. xxxii. 2/2 Treatment of intoxication with carbamate or organophosphorus insecticides includes the use of atropine which antagonizes the muscarinic effects of acetylcholine and of pralidoxime, a cholinesterase reactivator. Pralidoxime is a quaternary ammonium compound (hydroxyiminomethyl-1-methylpyridinium) which is given slowly intravenously. 1974 M. C. GERALD *Pharmacol.* vii. 132 Normal cholinesterase activity is restored..when a cholinesterase reactivator, notably pralidoxime (2-PAM), is used as an antidote. 1978 *Daily Tel.* 23 Jan. 13/7 British soldiers and airmen are now equipped with pills which will enable them to survive three or four times the normal lethal dose of most nerve gases... The pills, produced from pralidoxime mesylate..are issued in batches for each 24 hours.

praline. Delete 'Chiefly *U.S.*' and add: The spelling *prawlin* is now *Obs.* (Earlier and further examples.)

1723 J. NOTT *Cook's & Confectioner's Dict.* sig. B4 Almonds Fry'd, or Prawlins. 1770 BORELLA *Court & Country Confectioner* 40 We beg leave to use the words *praline* [etc.]. 1906 *Mrs. Beeton's Bk. Househ. Managem.* 1079 (*heading*) Chocolate pralines. 1913 [see *NOYAU b]. 1951 *People* 3 June 2/1 (Advt.), Crunchy wafers sandwiched with chocolate praline. 1971 A. R. DANIEL *Bakers' Dict.* (ed. 2) 155/2 Praline sometimes consists of roasted blended whole almonds dipped in sugar boiled to the hard crack degree.

‖ **pralltriller** (pra·l,trilǝr). *Mus.* [G., f. *prallen* to bounce + *triller* trill.] (See quot. 1971.)

1841 J. BISHOP in *Hamilton's Dict.* (ed. 13) 114 *Pralltriller* (German), a transient shake. 1876 STAINER & BARRETT *Dict. Mus. Terms* 365/1 *Pralltriller*..a transient shake. 1928 *Daily Express* 23 Feb. 3 What is a pralltriller?.. A musical ornament, performed by trilling the ornamented note with the note above it. 1971 *Everyman's Dict. Mus.* (ed. 5) 529/1 *Pralltriller* (Ger.), the rapid repetition of a note, with a note a degree higher in between.

pram[1]. Add: **a.** Later example of spelling *prame*.

1834 G. CRABBE in *Poetical Wks.* I. i. 9 Vessels of all sorts, from the large heavy troll-boat to the yawl and prame.

c*. A small sailing-boat. *U.S.*

1937 *Sun* (Baltimore) 31 July 11/8 In the pram class, Bucky Wilson..scored a surprise victory. 1956 *Ibid.* 11 Oct. 21/4 Hard luck forced Mary Sullivan and Henry White out of the competition when a boom broke on one of their prams. 1966 *Amer. Speech* XLI. 237 The smallest [sailboats] are called *Prams*, and they measure up to about 10 feet long.

d. *praam bow* (example of spelling *pram bow*).

1902 *Rudder* Apr. 208 The fore overhang [of the Meteor] is neither the old clipper stem nor the new pram bow.

pram[2]. For *vulgar* or *colloq.* read *colloq.* and add further examples of sense 1.

1916 G. B. SHAW *Pygmalion* v. 173 When I was a poor man and had a solicitor once when they found a pram in the dust cart, he got me off. 1955 *Times* 4 June 7/4 There are women who would not exchange a familiar pram with a quirk in its steering for the best new one that money could buy. 1963 [see *pram-park* below]. 1970 [see *pram-pusher* below].

3. *attrib.* and *Comb.*, as (sense 1) *pramful, pram-handle, -load, race, rug*; **pram-park,** (*a*) (see quot. 1963); (*b*) a space, area, etc., where prams may be left; **pram-pusher,** one who pushes a pram; *spec.* a young mother; so **pram-pushing** *ppl. a.* and *vbl. sb.*

1957 M. FRAYN in *Granta* 9 Mar. 20/1 People said that an old woman had been arrested on the other side of the village, pushing a whole pramful of stolen goods along. 1977 F. BRANSTON *Up & Coming Man* xv. 168 A young mum and a pramful of kids. 1934 DYLAN THOMAS *Let.* 15 Apr. (1966) 102 Mothers are resting their bellies on pram-handles. 1972 *Where* Oct. 273/3 Staff took to the post office two pramloads of the report-and-appeal. 1973 *Times* 28 Feb. (Suppl. on Victoria Centre, Nottingham) p. iv/6 A flying squirrel pushes a pramload (which is a nest) of birds. 1963 *Times* 3 May 15/7 In the House of Commons on Monday Sir Robert Cary asked how one gets a pram on a bus. In New Zealand they are carried in special pram-parks on the front of the radiators, where they seem to be safe, but nothing can be left inside, as they are hung wheels foremost. 1965 R. RENDELL *To fear Painted Devil* xii. 136 We're going to have that extension done at last... A sun loggia... And a pram park! 1967 J. WILSON in L. Deighton *London Dossier* 35 Linguists wishing to meet *au pair* girls might do worse than to hang about the pram park inside Peter Jones department store. 1970 *Times* 23 Feb. 13/2 The scheme will include..seats and plants in the concourse, and a pram park. 1973 *Guardian* 3 Sept. 20 Pram parks should be provided inside shops. 1935 J. L. HODSON *Harvest in North* II. i. 39 Afe on yo' are nowt but skivvies and pram-pushers. 1963 *Guardian* 25 Jan. 8/7 The pram-pushers are always willing to discuss these, as a change from the inevitable baby-talk. 1970 A. PRICE *Labyrinth Makers* v. 64 Mothers bulldozing their way ahead with prams... Roskill adroitly slipped into the wake of one of the most aggressive pram-pushers. 1933 *Punch* 10 May 516/1 Possibly the pram-pushing girl's hat caught her eye. 1964 G. BUTLER *Coffin in Malta* vii. 198 Most husbands were competent nannies; he fully expected to do some pram-pushing himself. 1974 *Country Life* 7 Mar. 480/1 The pram-pushing Phil's anguished claim—'he's my son!' 1968 P. JENNINGS *Living Village* 123 In the scrapbook there is a very good colour photograph of a Boxing Day pram race. 1934 A. THIRKELL *Wild Strawberries* ix. 196 Ivy, run and get the pram rug and put it round her.

‖ **prana** (prā·nă). [Skr.] In Hindu religion, the 'breath of life'; hence in extended uses, a life-giving force or inspiration; the breath, breathing. Also **pranayama,** regulation of the breath; breathing-control.

1830 H. T. COLEBROOKE in *Trans. R. Asiatic Soc.* II. 11 The term *prâṇa*..properly and primarily signifies respiration, as well as certain other vital actions (inspiration, energy, expiration, digestion, or circulation of nourishment); and secondarily, the senses and organs. But, in the passages here referred to, it is employed for a different signification, intending the supreme Brahme. 1875 MONIER WILLIAMS *Indian Wisdom* ii. 40 Highest of all stands Prāṇa or Life. As the spokes of a wheel are attached to the nave, so are all things attached to Life. 1930 F. YEATS-BROWN *Bengal Lancer* v. 66 There was a *saddhu* at Puri who claimed to be able to resurrect sparrows..by breathing *prana* into them. 1938 S. BECKETT *Murphy* 196 He..trusted he would be granted Prana to finish a monograph. 1955 E. POUND *Section: Rock-Drill* (1957) xciv. 92 Above prana, the light, past light, the crystal. 1956 E. WOOD *Yoga Dict.* 25/1 The shorter unit of time often mentioned in Sanskrit philosophical literature, though not used in *prāṇāyāma* is called the second or moment (*kshaṇa*), and is often considered to be one quarter of the time taken up in shutting an eye. *Ibid.* 123/1 *Prana*, that Vital Air which is..concerned with the health and strength of the heart and its work in the body. 1959 [see *ORGONE]. 1960 J. HEWITT *Yoga* v. 70 Prana, to the Yogi, means much more than mere breath. Prana is actually the power behind and within breath. The power of the atom is Prana. Thought is Prana... It pervades the whole universe. 1960 KOESTLER *Lotus & Robot* I. iii. 117 The Yogi then demonstrated the extraordinary power of his chest muscles—the result of pranayama. 1970 *Man, Myth & Magic* v. 146/3 The idea of an astral body is very old. Ancient Indian writings describe the eight *siddhis* or supernormal powers which can be acquired through a type of yoga called *Pranayama*. 1971 'A. HALL' *Warsaw Document* xv. 185, I took a slow breath: the answer to panic is *prana*. 1979 W. H. CANAWAY *Solid Gold Buddha* xi. 76 He..did some *Pranayama*, and calmed himself, through the rhythmic breathing.

prance, *sb.* Add: **a.** (Further example.)

1898 F. P. DUNNE *Mr. Dooley in Peace & War* 184 He has th' gait proper f'r half-past six o'clock th' avenin' befure pay-day. But 'tis not th' prance iv an American citizen makin' a gloryous spectacle iv himsilf.

Hence **prancy** *a.*, resembling or suggestive of a prance.

1961 *New Statesman* 26 May 828/3 The 'Bohemian Jive', a prancy affair, is now an essential part of the repertoire, whether you wear points or sandals. 1963 *New Yorker* 22 June 4 The trumpeting band of Emil Coleman and the prancy one of Mark Monte.

Prandtl (præntʼl). [Name of Ludwig *Prandtl* (1875–1953), German physicist.] *Prandtl number*: a dimensionless parameter used in calculations of heat transfer between a moving fluid and a solid body, equal to $c_p\nu/k$, where c_p is the heat capacity of unit volume of the fluid, ν is its kinematic viscosity, and k its thermal conductivity.

1933 W. H. MCADAMS *Heat Transmission* iv. 96 Prandtl number = 1/Stanton group. 1954 R. STEPHENSON *Introd. Nucl. Engin.* vi. 245 The ratio of the quantity of heat transferred by convection to the quantity of heat transferred by conduction is given by the value of the Prandtl number for the fluid. 1958 [see *NUSSELT]. 1974 F. M. WHITE *Viscous Fluid Flow* ii. 84 Table 2-1 gives values of the Prandtl number for various fluids at 68°F. It shows that liquid metals have [a] very small Prandtl number, gases slightly less than unity, thin liquids somewhat higher than unity, and oils a very large value of *Pr*.

prang (præŋ), *sb.* *slang* (orig. *R.A.F.*). [etym. uncertain.] **1. a.** An accident in which an aircraft suffers damage; a crash-landing. **b.** A bombing-raid. Also *transf.* and *fig.*

1942 *Sun* (Baltimore) 7 Apr. 20/8 American flyers in the RAF Eagle Squadrons have introduced a new decoration. 'The Order of Prang'—but it never appears in official citations. 'Prang' is Eagle slang for crash. 1943 HUNT & PRINGLE *Service Slang* 53 'P/O Prune' is the title bestowed upon a pilot who has several 'prangs' on his record. 1945 PARTRIDGE *Dict. R.A.F. Slang* 45 *Prang*, a crash landing. (2) A bombing raid. 1946 G. GIBSON *Enemy Coast Ahead* 105, I like high-level attacks..or else it must be the very low-level prang. 1948 G. GREENE *Heart of Matter* III. i. 294 'There's no time like the present for a prang,' Bagster said, moving her firmly towards the bed. 1958 *Spectator* 16 May 614/1 The Prime Minister was questioned about the RAF's wizard prang on the Government's defence policy. 1979 N. SLATER *Falcon* ii. 36 Tell him about your wizard prangs in the war.

2. An accident or collision involving a road vehicle; a car-crash.

1959 *Sunday Times* 1 Nov. 23/2 The grisly enormities of American stock-car racing, with an hysterical ghoul of a commentator who revelled in every prang. 1971 A. DIMENT *Think Inc.* ii. 26 Might have had a bad prang before they re-sprayed her.

prang, *v.* *slang.* (orig. *R.A.F.*). [etym. uncertain.] **1.** *trans.* **a.** To crash or crash-land (an aircraft); to damage (part of an aircraft) during a crash-landing. Also const. *down*.

1941 *Tee Emm* (Air Ministry) July 6/1 Do they give a grateful sigh and shut up shop when the last serviceable aircraft has been pranged against a hangar because its pilot would land towards obstacles? 1942 *R.A.F. Jrnl.* 18 Apr. 1 Gremlins..run down the nose of the machine and tip you up and you prang a prop. 1942 *Tee Emm* (Air Ministry) II. 143 By now he didn't give a darn—He pranged her down beside the barn. 1944 'N. SHUTE' *Pastoral* v. 107 After so many operations it was an acute personal grief to him that he had pranged his Wimpey. 1977 *Belfast Tel.* 28 Feb. 9/1 (*caption*) The half of the propellor he is holding came off a Bristol fighter he 'pranged' in a schoolyard in 1925.

b. To bomb (a target) successfully from the air.

1942 [see *FINGER *sb.* 3 a]. 1943 B. J. HURREN *Eastern Med.* 27 One can picture the..rage of the German and Italian air commanders..each verging on apoplexy that their chosen pilots should not be able to 'prang' a ship which presented a clear, long, visible deck target area of some 600 feet by 90 feet wide. 1952 M. TRIPP *Faith is Windsock* v. 87 The Lancs broke off sharply at the last moment to prang Neuss. 1958 E. HYAMS *Taking it Easy* I. i. 16 The RAF said they didn't know how to, they just know about pranging the Luftwaffe and the railway yard at Ham.

c. To involve (a road vehicle or other object) in an accident; to crash or 'smash up'; to collide with.

1952 E. F. DAVIES *Illyrian Venture* iii. 50 'What height would you like to be dropped at?' 'Would 800 feet suit you?'.. 'I think I can manage that without pranging the mountain.' 1966 T. WISDOM *High-Performance Driving* ix. 97 The driver may well have left his 'flasher' on many corners ago and is happily oblivious of the fact until you move off on his signal and 'prang' him. 1971 *Daily Colonist* (Victoria, B.C.) 26 Feb. 2/1 Recently my rather ancient Chevvy II got pranged from behind—nothing serious, just a smashed tail light. 1973 A. MANN *Tiara* ix. 79 Most of them don't drive... If they prang a car, there's always plenty of witnesses to say it's the priest's fault. 1976 *Islander* (Victoria, B.C.) 22 Aug. 11/1 We had pranged a rock getting out of Oak Bay.

2. *intr.* To crash or crash-land an aircraft. Also *transf.*

1943 P. BRENNAN et al. *Spitfires over Malta* ii. 55 The upwind end of the landing-path was a maze of bomb-holes... I was too brassed off to worry whether I pranged or not. 1961 'J. ROSS' *Last August* iii. 31 A wasp was pranging against the window. 1968 *Daily Express* 26 Feb. 4/1, I knew we were going to prang, but all I wanted to do was to make sure that we weren't killed or seriously injured.

3. *trans.* In extended uses: to break, to smash; to hit, to strike heavily (*against*). Also *fig.*

1942 J. MOORE in *Observer* 4 Oct. 7/2 Now you talk..of pranging a date, meaning that you have left your popsy waiting outside the Unicorn while you continue to drink with the squadron in the Bull and Bush. 1943 HUNT & PRINGLE *Service Slang* 53 Jones pranged his arm at rugger to-day. 1946 *Slipstream* 38 Mind you don't prang yourselves against the table. 1948 PARTRIDGE *Dict. Forces' Slang* 147 He pranged the iron bedstead... He pranged his leg against the bedstead. 1977 F. PARRISH *Fire in Barley* x. 99 He was holding a pitchfork. 'I thought I'd prang a rabbit.'

Hence **pranged** *ppl. a.*; **pra·nging** *vbl. sb.*

1942 *Air News* Oct. 4/1 'Pranging', by the way, is a new R.A.F. expression which means smashing things up—including one's own aeroplane. 1946 BRICKHILL & NORTON *Escape to Danger* iii. 39 A couple of 109's hacked two Hurricane down near Montreuil on the 10th of June 1940, and Eric jumped from his pranged kite and ran for it. 1959 *Times Lit. Suppl.* 7 Aug. p. iii/3 Classic understating metaphors like 'having a party', 'falling in the drink', 'pranging', and so on, had their value in time of war. 1971 R. DENTRY *Encounter at Kharmel* vii. 117 Looking for the wreckage of a pranged aircraft.

prankster (præ·ŋkstǝɹ). orig. *U.S.* [f. PRANK *sb.*[2] + -STER.] One who plays pranks; a hoaxer, a practical joker.

1927 *Amer. Speech* II. 245/1, *-ster* also, for a time, gave signs of being moribund... It is, however, found in a number of new formations... *prankster*, [etc.]. 1940 O. NASH *Face is Familiar* (1942) 14 There is at least one thing I would less rather have in the neighbourhood than a gangster, And that one thing is a practical prankster. 1951 *Mind* LX. 468 Suppose a prankster laid out some railroad tracks which..diverged in such a manner as to make them look parallel. 1957 C. RICE *My Kingdom for Hearse* i. 6 Some practical prankster had evidently left a dead horse in his nose. 1969 *Daily Tel.* 14 Nov. 36/4 Police admitted that it was an almost impossible task..to

sift the bomb pranksters from the deadly serious ones. 1972 'M. YORKE' *Silent Witness* III. 51 The key..had been turned on her... 'Some prankster, I suppose... It was a very childish trick.' 1977 *Time* 24 Jan. 37/1 Ken Waller, a not-too-merry prankster who steals bits of his opponents' costumes in order to upset their concentration before they go onstage to face the judges.

p'raps (præps), *adv.* Also **praps, p'r'aps, p'rhaps.** Repr. colloq. pronuncs. of PERHAPS *adv.*

1835, 1837 [see PERHAPS *adv.* (*sb.*) 1]. 1898 G. B. SHAW *Candida* I. 93, I did think you a bit of a fool once; but I'm beginnin' to think that praps I was be'ind the times a bit. 1912 MASEFIELD *Widow in Bye St.* 24 We might go round one evening, p'raps. ?1912 R. FRY *Lett.* (1972) I. 358 Don't leave this letter about... P'raps you'll *want* to burn it. 1955 N. MARSH *Scales of Justice* iv. 75, I know what they'll say about me. Not you, p'r'aps, but the others. 1974 'P. B. YUILL' *Bornless Keeper* iv. 33 Praps we'd better both wait. 1976 D. CLARK *Dread & Water* vi. 126 P'raps you'll tell me what you're gunna do about it?

prasad, prasada (prasā·d, -a). *Hinduism.* [Skr. *prasāda* lit. clearness, kindness, grace, Hindi *prasād*.] **1.** A propitiatory offering of food made to a god; food which is offered to an idol and then shared among devotees.

1828 H. H. WILSON in *Asiatic Researches* XVI. 83 A *Chamár*, oh king, ministers to the *Sálagrám*, and poisons the town with his *Prasád*. *Ibid.* 96 At noon, he halted and bathed the god, and prepared his food, and presented it, and then took the *Prasád* and put it in a vessel, and fed upon what remained. 1855 —— *Gloss. Judicial & Revenue Terms* 424/2 All castes may partake of the *Prasád* of any image. 1875 MONIER WILLIAMS *Indian Wisdom* p. xxxvii, It is remarkable that the food offered to the gods, when appropriated and eaten by the priests, and the rice distributed by them to the people, are called *prasāda* (? = εὐχαριστία). 1913 J. N. FARQUHAR *Crown of Hinduism* ix. 381 In modern temples, the practice is to give every worshipper a portion of the food and of the water offered to the idol. The food is called *prasāda*, a grace-gift, and the water *tīrtha*, holy water. 1953 K. W. MORGAN *Relig. Hindus* vii. 296 The food is first offered to the Lord and what is eaten is His praśāda. 1965 'LAUCHMONEN' *Old Thom's Harvest* x. 132 East Indians..shared out prasad, mango..and rice. 1969 *Weekly Mail* (Madras) 26 July 7/5, 90 boys..became unconscious after taking 'prasad' at a religious function at Chakasigan village. 1979 D. QUINN *Fear of God* i. 52 The Indian said, 'Prasada, the remains are food offered to the Lord... The food is Krishna himself. You should eat it all.'

2. Divine grace or favour. Also *attrib.*

1895 E. W. HOPKINS *Relig. India* xv. 429 The *prasāda* doctrine (of special grace) belongs to a much earlier literature. 1921 R. E. HUME *Thirteen Princ. Upanishads* 59 As regards speculative knowledge of Ātman, its apprehension by means of human knowledge is opposed by the doctrine of prasāda, or 'Grace', in Kaṭha 2.20. 1964 R. ANTOINE et al. *Relig. Hinduism* xxiii. 247 This divine grace: *anugraha, prasāda, puṣṭi, kṛpā*, is *the* means which habilitates the bhakta to the obtainment and practice of the bhakti.

praseodymium (prēˈziǒdi·miǒm). *Chem.* [mod.L., f. G. *praseodym* (C. A. von Welsbach 1885, in *Monatshefte f. Chem.* VI. 490), f. Gr. πράσιος leek-green (f. πράσον leek) + G. *di*)*dym* DIDYMIUM: see -IUM. Named in allusion to the colour of its salts and its isolation, with neodymium, from the supposed element didymium.] A metallic element, similar to iron in appearance, which is a typical lanthanide and forms leek-green compounds in which it has a valency of three (rarely four), some of which are used to impart a yellow colour to glasses and ceramics. Atomic number 59; symbol Pr.

1885 [see *NEODYMIUM]. 1905 GOOCH & WALKER *Outl. Inorg. Chem.* xix. 493 For more than fifty years the elementary character of didymium was accepted, until Auer von Welsbach, by a most laborious process of fractional precipitation of the double nitrate of didymium and ammonium, succeeded in isolating two distinctly different double nitrates..from which were prepared two different series of salts, of different elements, which were now named praseodymium and neodymium. 1922 T. M. LOWRY *Inorg. Chem.* xxxiv. 672 Praseodymium also resembles cerium in forming a dioxide, PrO_2, when the nitrate is heated with potassium nitrate at 450°. 1950 *Thorpe's Dict. Appl. Chem.* (ed. 4) X. 183/2 All cerium-bearing minerals contain some praseodymium, e.g. cerite from Arendal (Switzerland) contains up to 8% of Pr_6O_{11}, monazite sand from Brazil, 5·5–6·2%. 1971 J. F. LIPTROT *Mod. Inorg. Chem.* xxvi. 438 Lanthanum, cerium, praseodymium, neodymium and gadolinium may be obtained by reduction of their trichlorides with calcium at about 1000 °C. 1974 *Encycl. Brit. Micropædia* VIII. 179/2 Praseodymium is about one-third as abundant as lead and about a thousand times more plentiful than gold in the igneous rocks of the Earth's crust... Natural praseodymium is all stable isotope praseodymium-141.

prashack, prashagh, etc., varr. *PRAISEACH.

prat, *sb.*[2] Delete *Rogues' Cant* and add: Also **pratt. 1. a.** Now usu. sing., the backside, rump. *slang* (orig. *Criminals'*).

1846 [see *NUT *sb.*[1] 7 a]. 1914 JACKSON & HELLYER *Vocab. Criminal Slang* 66 *Pratt*,..the human rear. 1952 R. STOUT *Prisoner's Base* i. 3, I have had to spend most of my time recently sitting on my prat. 1959 E. BORNEMAN *Tomorrow is Now* ix. 93 You gimme a pain in the royal pratt. 1972 D. DELMAN *Sudden Death* iii. 65 I'm a *shmo* about tennis, so if I fall on my prat a time or two you have to bear with me.

b. A hip-pocket. *U.S. Criminals' slang.*

1914 JACKSON & HELLYER *Vocab. Criminal Slang* 66 *Pratt*.., a hip pocket. 1927 [see *prat-digging* vbl. sb.] 1936 *Detective Fiction Weekly* 12 Sept. 93/1 In spite of the fact that a pocketbook may be removed most easily from a hip pocket known as a right or left 'pratt', the majority of men carry their money there.

2. A person of no account; a dolt, fool, 'jerk'. *slang.*

1968 M. BRAGG *Without City Wall* xii. 130 He had been looking for the exact word to describe David and now he found it: *prat*. 1973 J. WAINWRIGHT *Pride of Pigs* 32 Harris was a bit of a pompous prat. 1974 N. FREELING *Dressing of Diamond* 204 Want to get an eyeful, do you, dirty-minded prat that you are. 1980 J. WAINWRIGHT *Eye of Beholder* 18 The pompous prat. The I-know-people-in-high-places nut.

3. *attrib.* and *Comb.*, as *prat-faced* adj.; (in sense *1 b) **prat digger,** a pick-pocket; so **prat-digging** *vbl. sb.*; **prat frisk,** the theft of a wallet from a hip-pocket; **prat-kick,** a hip-pocket; **prat leather,** a wallet kept in the hip-pocket; **prat poke,** a wallet stolen from the hip-pocket.

1935 *Amer. Speech* X. 19/2 *Prat-digger*, a pickpocket, one who exploits the *prat kick*. 1955 *Publ. Amer. Dial. Soc.* xxiv. 69 Others specialize in hip pocket work and are called *prat diggers*. 1908 J. M. SULLIVAN *Criminal Slang* 19 *Pratt digging*, stealing from the hip pocket. 1916 G. A. ENGLAND *Pod, Bender & Co.* 291 It's a fact we've always been above such lays as pratt-digging. 1927 *Writer's Monthly* Nov. 390/1 The 'pratt'..is a trousers pocket. 'Pratt-digging' is stealing the 'pratt-leather' from the hip. 1976 U. HOLDEN *String Horses* i. 17 They liked to kiss each other lightly.., push each other with taunts. 'You prat-faced les. Get off.' 1924 Prat frisk [see *prat poke]. 1896 I. K. FRIEDMAN *Lucky Number* 154, 'I dipped it from yer prat-kick.'.. 'I means I took her from yer back pocket,' answered the rogue blandly. 1955 *Publ. Amer. Dial. Soc.* xxiv. 125 The hip pockets are *prat kicks*. 1908 J. M. SULLIVAN *Criminal Slang* 19 *Pratt leather*, a pocketbook in the hip pocket. 1927 Pratt leather [see *prat-digging vbl. sb.]. 1924 G. C. HENDERSON *Keys to Crookdom* 414 *Pratt poke*, purse kept in hip pocket. Pratt frisk—stealing such a purse, reefing a britch. 1955 *Publ. Amer. Dial. Soc.* xxiv. 115 When a wallet is taken from the hip pocket, it is known as a *prat poke*.

prat, *v.* Delete †*Obs.*, restrict *Sc.* to sense 1, and add: Also *Sc.* pret. **1. b.** To lark about; to trifle, romp. Freq. const. *with*.

1728 A. RAMSAY *Poems* (1953) II. 89 Some Beaus may snarl if we should prat. *a*1835 J. AFFLECK *Posthumous Poetical Wks.* (1836) 60, I never pretit onie where At midday, night or morn. 1851 A. MACLAGAN *Sk. from Nature* 153 As for his sons, their foes will find They're no to prat wi'! 1897 C.R. *Dunning Folk-Lore* 4 Thae brownies warna to prat wi'! They played gey pliskies whiles, an' did muckle mischeef.

2. To potter *about*; to fool around, to act in a silly or annoying manner. *slang.*

1961 PARTRIDGE *Dict. Slang* Suppl. 1231/2 *Prat about*, to potter, mess about. 1973 H. MILLER *Open City* xvii. 187 Sit down and stop pratting about.

3. a. *intr.* To simulate coyness. **b.** *trans.* To feign rejection of (someone). *U.S. Blacks.*

1970 C. MAJOR *Dict. Afro-Amer. Slang* 93 *Prat*, to play coy. 1972 'I. SLIM' in T. Kochman *Rappin' & Stylin' Out* 389 Pimping ain't no game of love, so prat 'em and keep your swipe outta 'em.

pratal, *a.* (Earlier and later examples.)

1847 H. C. WATSON *Cybele Britannica* I. 65 The proposed series of terms runs thus:—1. Pratal.—Plants of meadows, or rich and damp grass-land [etc.]. 1932 G. C. DRUCE *Comital Flora Brit. Isles* 69 Meadow Crane's-bill. Pratal. British. Meadows, grassy road-borders.

‖ **Prater**[2] (prā·tǝɹ). [Ger., ad. It. *prato* meadow.] The name of a large wooded park in Vienna.

1803 C. WILMOT *Let.* 2 Aug. in *Irish Peer* (1920) 207 The Prater..is esteem'd the most magnificent public walk and drive in Europe. 1819 M. WILMOT *More Lett.* (1935) 20 The turn out in the Prater of a Sunday Eve[g] is Magnificent. 1870 G. H. LEWES *Jrnl.* 10 Apr. in *Geo. Eliot Lett.* (1956) V. 89 Lytton then proposed that we should drive in the Prater. 1911 *Encycl. Brit.* XXVIII. 52/1 The Prater, a vast expanse (2000 acres) of wood and park on the east side of the city, between the Danube and the Danube Canal, is greatly frequented by all classes. 1938 W. J. TURNER *Mozart* xvi. 312 One fine autumn day when they were sitting in the beautiful Prater, Mozart spoke of his approaching death. 1945 C. ISHERWOOD *Prater Violet* 33 It is a warm spring evening in the Vienna Prater. 1974 A. GODDARD *Vienna Pursuit* II. 40, I..made a stately circuit on the Big Wheel in the Prater.

pratfall (præ·tfǫl), *sb.* Chiefly *N. Amer. slang.* Also **prat(t)-fall.** [f. PRAT *sb.*[2] + FALL *sb.*[1]] **a.** *Theatr.* A comedy fall; a fall on to the buttocks.

1939 N. COWARD *Play Parade* II. 108 Don't do a prat-fall in your first routine. 1941 L. ROSTEN *Hollywood* 316 The Hollywood writers—graduates of the westerns..masters of the chase, the 'pratt-fall'..—kept the movies moving. 1952 'E. BOX' *Death in Fifth Position* (1954) ii. 47 Some homicidal maniac..who enjoyed seeing ballerinas take fatal pratfalls. 1960 B. KEATON *Wonderful World of Slapstick* (1967) 96 Pop's pratfalls astonished Roscoe and everybody else. 1961 *Guardian* 27 Apr. 9/4 A more intelligent form of humour—away from the pratfall type of thing. 1977 *Time* 26 Dec. 49/1 Only Saturday-morning TV addicts could possibly endure the antics of *The World's Greatest Lover*, in which characters are forever shouting their lines, bulging their eyes and stumbling through pratfalls.

b. *transf.* and *fig.*

1953 R. BRADBURY *Fahrenheit 451* (1954) I. 56 Life becomes one big pratfall. 1956 D. KARP *All Honorable Men* 174 That gentleman is in for a rude surprise some morning soon. I understand he handles government contracts. Another pratfall soon. 1971 *Guardian* 25 Nov. 17/2 Performers who write their own material often take enormous pratfalls. 1977 *Rolling Stone* 7 Apr. 43/1 Why has an important investigation so quickly degenerated into a series of pratfalls?

Hence as *v. intr.*, to fall on to the buttocks. Hence **prat-fallen, -falling** *ppl. adjs.* Also *fig.*

1940 *Time* 29 Jan. 41/1 The sight of Sonja (for the fourth time in her professional career) pratt-fallen. *Ibid.* 30 Dec. 30/1 Opera at the Met has a way of prattfalling between two stools. 1942 *Ibid.* 3 Aug. 74/2 On the way to the plate he prat-falls on the carefully laid-out row of bats in front of the dugout. 1972 *Listener* 6 July 22/1 The eloquent gamey pratfalling scapegoat. 1973 D. LEES *Rape of Quiet Town* iv. 55 As the tension built up..it was a piece of prat-falling comedy that saved me.

‖ **pratiquant** (pratikaṅ), *a.* (*sb.*) [Fr.] Making a practice of religious duties or observances; practising. Also *absol.* as *sb.*

1902 G. ARTHUR *Let.* Mar. in *Some Lett. from Man of No Importance* (1928) 140 If the King is not religious in the 'pratiquant' sense of the word, he has a very strong sense of religion. 1956 S. BEDFORD *Legacy* III. iv. 154 Jules..says he couldn't marry someone who wasn't a Catholic... It isn't as though *he'd* ever shewed himself the least bit *pratiquant*. 1960 *Times* 27 Feb. 7/3 Fasting is an unhealthy habit; Communists tell the people of their Muslim lands. On the contrary, urge the new *pratiquants* in Turkey and elsewhere, it is astonishing how fitter you feel at the end of Ramadan. 1965 *New Statesman* 3 Dec. 890/2 After lapsing for 15 years, she had recently tried to become *pratiquant* again.

Pratt (præt). The name of Felix *Pratt* (1780–1859), Staffordshire pottery-manufacturer, used *attrib.* to designate a type of cream-coloured earthenware painted in high-temperature colours and widely produced in the late eighteenth and early nineteenth centuries. So *Pratt-type* adj., etc.

[1912 F. LITCHFIELD *Pott. & Porc.* (new ed.) vii. 195 The Pottery of Felix Pratt,..marked with the name *Pratt* and known as 'Pratt's ware', was in this district [*sc.* Fenton].] 1920 G. W. RHEAD *Earthenware Collector* xi. 197 One of the most successful pieces of Pratt ware is the mug of 'Midnight Conversation' in the Hanley Museum. *Ibid.*, A number of Pratt jugs have been known as Sunderland jugs. 1933 W. B. HONEY *Eng. Pott. & Porc.* vii. 100 The so-called 'Pratt ware' was made by other Staffordshire potters as well as by those of Sunderland and other places in the North. 1957 MANKOWITZ & HAGGAR *Conc. Encycl. Eng. Pott. & Porc.* 182/2 *'Pratt' type*, wares made at the end of the eighteenth and beginning of the nineteenth centuries, decorated with a distinctive palette of high temperature colours. 1963 *House & Garden* May 68/2 The pottery pieces..are mostly Pratt-ware made either just before or just after the beginning of the nineteenth century, and are part of a collection of Pratt. 1967 *Times* 14 Mar. 21/7 (Advt.), English pottery and porcelain..including..a Prattware grotto group. 1974 *Country Life* 21 Feb. (Suppl.) 56 Six Pratt Pot-lids and some Pratt Ware. *a*1977 *Harrison Mayer Ltd. Catal.* 99/1 Pratt Type and Peasant enamel Wares... 20 Slides.

praty, dial. corruption: see POTATO 2d δ in Dict. and Suppl.

‖ **pratyahara** (pratʸāhā·ra). *Yoga Philos.* [Skr.] (See quots.)

1882 E. B. COWELL tr. *Âchârya's Sarva-Darśana-Saṃgraha* xv. 267 Now in this way, having his mind purified by the 'forbearances'..the devotee is to attain 'self-mastery'..and 'restraint' (pratyáhára). 1899 MAX MÜLLER *Six Syst. Indian Philos.* vii. 458 We can hardly doubt that these postures and restraints of breathing..are helpful in producing complete abstraction (Pratyâhâra) of the senses from their objects, and a complete indifference of the Yogin towards pain and pleasure. 1942 D. D. RUNES *Dict. Philos.* 248/1 *Pratyāhāra*,..withdrawal of the senses from external objects, one of the psychophysical means for attaining the object of Yoga. 1957 *Encycl. Brit.* XII. 251/2 The next step of *pratyāhāra* or the withdrawal of the senses from their natural outward functioning answers to what modern psychology calls introversion. 1960 J. HEWITT *Yoga* viii. 114 Sense-Withdrawal, called by the Yogis Pratyahara. 1976 *Canberra Times* 23 Aug. 2/8 The ability to transcend thought is acquired by the application of concentrated mind to a single point, pratyahara.

prawn, *sb.* Add: **I. 1. c. prawn cocktail:** see *COCKTAIL 4; **prawn-fishing** = *PRAWNING *vbl. sb.* 2; so *prawn-fisherman*.

1921 *Chambers's Jrnl.* Sept. 590/1 Numerous are the adverse comments I've heard on the prawn fisherman and his ways. 1924 *Blackw. Mag.* Apr. 489/2 Neither the Lydons nor anybody else could make me enjoy prawn-fishing on that high walk at Galway. 1978 R. WADDINGTON *Catching Salmon* xi. 128 One cardinal principle..applies to prawn-fishing. If in a stocked pool no fish takes or follows your bait in the first few casts, it is..useless to continue.

II. *transf.* and *fig.* **2.** See sense b in Dict.

3. Applied to persons. **a.** Figuratively, or in a familiar manner. **b.** As a term of contempt: a fool, half-wit.

1845 DICKENS *Let.* 27 Jan. (1977) IV. 253 By the time he had finished this third dinner, his eyes protruded infinitely beyond the tip of that feature [*sc.* his nose]. You never saw such a human Prawn as he looked, in your life. 1895 W. P. RIDGE *Minor Dialogues* 207 Ah, I expect you're a saucy young prawn, Emma. 1937 PARTRIDGE *Dict. Slang* 657/1 *Prawn, silly*, a pejorative applied to persons. 1965 *Telegraph* (Brisbane) 5 July 8/4 Describing a fellow who was a bit eccentric, or one who was just plain nuts... Anyone would know what he [*sc.* an Australian] meant if he used the word..*prawn*.

c. In phr. *to come the raw prawn (over, with*, etc.): to attempt to deceive. *Austral. slang.*

1942 *Salt* 25 May 8 *Don't come the raw prawn*, don't try to put one over me. *c* 1948 S. L. ELLIOT in E. Hanger *Khaki, Bush & Bigotry* (1968) 36 The filthy rotten Crab, he'd better not come the raw prawn on us. 1951 CUSACK & JAMES *Come in Spinner* 306 Coupla bastards come the raw prawn over me on the last lap up from Melbourne and I done me last bob at Swy. 1959 E. LAMBERT *Glory thrown In* v. 41 Don't ever come the raw prawn with Doc, mate. He knows all the lurks. 1965 M. SHADBOLT *Among Cinders* xxi. 202 Don't you come the raw prawn with me. 1970 *Private Eye* 16 Jan. 16 Don't come the raw prawn with me. Ozzie Barry and me are just good friends.

prawning (prǭ·niŋ), *vbl. sb.* [f. PRAWN *v.* + -ING[1].] **1.** The action or process of fishing for prawn. Also *attrib.*

1886 [see PRAWN *v.* in Dict.]. 1973 [See *open go* s.v. *OPEN *a.* 22 c]. 1978 M. GILBERT *Empty House* xxiv. 219 Spades, buckets and prawning nets should be left in the porch. 1979 *Country Life* 24 May 1658/1 The Coquet..is a prawning and shrimping river.

2. Fishing for salmon using a prawn as bait.

1909 *Westm. Gaz.* 10 May 12/2 Prawning and spinning for salmon has begun on the Hampshire Avon. 1921 *Chambers's Jrnl.* Sept. 590/1 Prawning for salmon is looked down upon by many as being almost a form of poaching. 1931 E. TAVERNER et al. *Salmon Fishing* xxiii. 361 A rod, reel and line suitable for spinning will serve equally well for prawning.

praxeology (præksiǫ·lŏdʒi). Also **praxiology, praxo·logy.** [ad. F. *praxéologie* (L. Bourdeau 1882, in *Théorie des Sciences* II. VII. i. 463), or directly f. Gr. πρᾶξις action: see -OLOGY.] The study of such actions as are necessary in order to give practical effect to a theory or technique; the science of human conduct; the science of efficient action. So **praxeolo·gical, -iological** *a.*; **praxio·logist,** one who studies practical activity.

1904 W. R. B. GIBSON *Philos. Introd. Ethics* ix. 190, I say 'theory of experience' instead of theory of 'knowledge' or 'epistemology', in order to include the theory of action or 'praxology'. 1911 C. A. MERCIER *Conduct & its Disorders* p. viii, Apart from the general advantage..of having a systematic knowledge of conduct as a whole; there are certain special advantages to be derived from a study of Praxiology, if I may so term it. 1944 *Philos. & Phenomenol. Res.* IV. 527 The theoretical science of human action, praxeology, and especially its hitherto best developed part, economics or catallactics. *Ibid.* 533 Praxeology does not employ the term *rational*. It deals with purposive behavior, i.e., human action. *Ibid.* 537 The technological and the praxeological methods. 1945 Z. JORDAN *Devel. Math. Logic Poland* viii. 33/2 Kotarbinski's first published papers dealt with some problems of ethics and sociology, which were to supply the foundations of a general theory of action, called by him praxeology. 1961 T. KOTARBINSKI in *Methodos* XIII. 163 (*title*) The aspirations of praxiologists. 1962 E. NAGEL et al. *Logic, Methodol. & Philos. of Sci.* 211 (*title*) Praxiological sentences and how they are proved. 1965 D. WOJTASIEWICZ tr. *Kotarbinski's Praxiology* i. 1 Considerations included in the present work come within the scope of praxiology—the general theory of efficient action. *Ibid.*, The praxiologist concerns himself with finding the broadest possible generalizations of a technical nature. 1966 HOWARD & FOX tr. *Aron's Peace & War* p. xviii (*heading*) Praxeology, the antinomies of diplomatic-strategic conduct. *Ibid.* i. 4, I must first define international relations, then specify the characteristics of the four levels of conceptualization which we call *theory, sociology, history, praxiology*. 1973 *Times Lit. Suppl.* 6 July 787/3 The synthesis of these two modes of knowledge he labels 'praxeological'. 1973 B. B. WOLMAN *Dict. Behavioral Sci.* 286/1 *Praxiology*, psychology viewed as the study of actions, and overt behavior... Any normative science, such as, e.g. education, social philosophy, ethics, etc., that sets norms and goals for human actions.

-praxia (præ·ksiă), comb. form repr. Gr. πρᾶξις action in some mod.L. terms, as *APRAXIA, *ECHOPRAXIA, *PARAPRAXIA.

praxis. Add: **1. c.** A term used by A. von Cieszkowski in *Prolegomena zur Historiosophie* (Berlin, 1838), then adopted by Karl Marx *Zur Kritik der Hegelschen Rechtsphilosophie, Einleitung* in the *Deutsch-Französische Jahrbücher* (1844), to denote the willed action by which a theory or philosophy (esp. a Marxist one) becomes a social actuality. Also *attrib.* and *transf.*

This term, frequently translated as *practice, practical ability*, or *practical activity*, has been increasingly used since the 1960s, following the translation and availability of Marx's early writings.

1933 S. HOOK *Towards Understanding K. Marx* II. ix. 76 That is why Marx claimed that only in practice (*Praxis*) can problems be solved. 1936 —— *From Hegel to Marx* viii. 281 Practice (*Praxis*) was something much wider than *practicability*. It was selective behaviour... Marx's theory of the Praxis could explain what all other philosophers recognised but which they could not begin to account for, without writing fairy-tales, viz., how knowledge could give power. 1966 L. DUPRÉ *Found. Philos. Materialism* viii. 216 But for Marx, praxis is more than a principle of consciousness: it is a prereflective unity of nature and consciousness which can be explicated in thought but not initiated. 1969 D. MCLELLAN *Young Hegelians & Karl Marx* I. ii. 10 The main agent in this transformation was not to be thought, as in Hegel's philosophy, but will, which was the motive force for that synthesis of thought and action for which Cieszkowski coined the term, so influential later, of 'praxis'. 1971 R. J. BERNSTEIN *Praxis & Action* (1972) p. xi, Marx..went on to develop a thorough, systematic and comprehensive *theory* of *praxis*—a theory, which I shall argue, provides the key for understanding his basic outlook from his early speculations to his mature thought. 1974 *Times Lit. Suppl.* 31 May 582/5 'The embattled imagination' and 'maimed utopia', whose values are under threat in the praxis-obsessed intellectual climate of the Federal Republic. 1976 *Survey* Summer–Autumn 255 He ascribed to Marx, not a voluntarist doctrine as the negation of determinism, but a philosophy that conceived itself as historical praxis. 1978 *Daily Tel.* 23 Nov. 8/8 The new theology is seemingly based on the Marxian concept of praxis—the involvement of the oppressed in the historical processes of change.

d. Action that is entailed by theory or a function that results from a particular structure.

1953 E. L. ALLEN *Existentialism from Within* ii. 27 The Greeks did not speak of 'things' but of *pragmata*, implying that I have to do something (*praxis*) about them. 1962 MACQUARRIE & ROBINSON tr. *Heidegger's Being & Time* II. iv. 409 What is decisive in the 'emergence' of the theoretical attitude would then lie in the *disappearance* of *praxis*. *Ibid.*, And just as *praxis* has its own specific kind of sight ('theory'), theoretical research is not without a *praxis* of its own. 1966 B. HAIGH tr. *Luria's Human Brain & Psychol. Processes* i. 42 At first glance it may appear that lesions situated in very different parts of the brain may lead to a disturbance of praxis. 1968 J. M. HEATON *Eye* iii. 46 Thus, even at this early age, praxis has emerged, i.e. the activity of looking has become meaningful, an end in itself to the infant. 1970 E. PACI in J. M. Edie *Patterns of Life-World* 131 Since instruments are extensions either of the sensing body, or of the acting body, or of the body as organ of will and praxis, they represent a fusion of the self and nature in the body. 1972 PICCONE & HANSEN tr. *Paci's Function of Sci.* III. xix. 360 Science loses its function and society hides the meaning of praxis through technistic ideology.

e. (See quot.)

1950 [see *LEXIS 1].

pray, *v.* Add: **5. b.** *spec.* To make a formal petition *for* (something); to move a prayer (PRAYER[1] 5 in Dict. and Suppl.). Also *absol.*

1754 [in Dict.]. 1920 *Act* 10 & 11 *Geo. V* c. 67 §1 The Council shall..determine whether to issue the order as prayed for, or to issue the order with such modifications as may appear to be necessary. 1962 HANSON & WISEMAN *Parliament at Work* viii. 211 The need for such an Order arose from the attempts of a group of Conservative backbenchers to 'harry the life' out of the Labour Government of 1950–51..by 'praying' into the small hours of the morning.

praya, var. *PRAIA.

prayable, *a.* Restrict †*Obs.* to sense in Dict. and add: **b.** Of a prayer: that may be made.

1941 T. S. ELIOT *Dry Salvages* ii. 10 The hardly, barely prayable Prayer of the one Annunciation.

prayed (prēⁱd), *ppl. a.* [f. PRAY *v.* + -ED[1].] In *prayed-for* adj., that is sought in prayer.

1867 C. E. SMITH *Diary* 11 Mar. in C. E. S. Harris *From Deep of Sea* (1922) xvii. 224 Thank God for such a mercy! At last we are abreast of the longed-for, prayed-for Labrador. 1917 J. MASEFIELD *Lollingdon Downs* 75 In the lonely silence I may wait The prayed-for gleam—your hand upon the gate. 1952 DYLAN THOMAS *Let.* 21 July (1966) 376 I'll wait in all morning for your prayed-for call.

prayer[1]. Add: **1. d.** Slang phr. *not to have* (or *have got*) *a prayer*: to have no chance.

1941 B. SCHULBERG *What makes Sammy Run?* vi. 92 Get..back to New York. You won't have a prayer around here. 1957 R. A. HEINLEIN *Door into Summer* ii. 46 'I'm going to give you some advice.'.. 'Well?' 'Do nothing. You haven't got a prayer.' 1968 E. B. WHITE *Let.* 30 Dec. (1976) 574, I wish you luck. I don't think you have a prayer. 1973 A. ROSS *Dunfermline Affair* 113 He went for me... He was a big lad, and strong, but he didn't have a prayer. An amateur up against a professional almost never does. 1977 *Time* 10 Oct. 9/3 Mitterrand was prepared to sign anything back in 1972, when his party did not have a prayer of coming to power.

5. (Later examples of *spec.* sense.) See also quot. 1958.

1937 *Hansard Commons* 4 June 1307, I undertake, if the House will allow the remaining Regulations to be passed now, to amend No 95 immediately, and the notification of the Amendment will, of course, be subject to a Prayer, just as the Regulations themselves are. 1946 *May's Treat. Parliament* (ed. 14) xiv. 286 The last item of this group consists of motions for the disallowance of statutory orders or regulations... These motions are usually in the form of addresses to the Crown praying for the annulment of orders or regulations and are hence commonly called 'Prayers'. 1958 WILDING & LAUNDY *Encycl. Parliament* 431 *Prayer*, a motion to annul a Statutory Instrument... Such motions count as Exempted Business..and are taken at the end of the day's sitting. They must be moved during the forty days after the order is laid on the Table, at the expiration of which it automatically becomes law. 1968 *Observer* 21 Apr. 3/3 The British Medical Association..is arranging for a 'prayer' to be moved in Parliament. 1970 *Daily Tel.* 3 Nov. 2/6 Mr Enoch Powell, Conservative MP for Wolverhampton, South West, has a prayer down on the Commons Order Paper to annul the regulations. 1972 *Times* 23 Feb. 27/5 A..petitioner sought the direction of the court whether she might properly omit a prayer for costs from a petition which sought a decree of divorce. 1973 *N.Y. Law Jrnl.* 2 Aug. 5/3 Nowhere in this or any other document, has IBM denied the factual assertions, made by the United States, which are the basis for its prayer that IBM be held in contempt of court. 1975 J. P. MORGAN *House of Lords & Labour Govt.* ii. 63 Where affirmative resolution is required both Houses must give their approval before such Orders can be passed; where an Order becomes effective unless a Prayer for annulment is carried by either House (the negative resolution procedure), the Lords again enjoy the same rights as the Commons.

6. a. *prayer-attitude, -life, service.*

1953 R. KNOX *Off Record* xliv. 148 If one does a hop from Evangelicalism to the Church the difference is not so much one of doctrines as one of prayer-attitudes. *Ibid.* 150, I should find no difficulty in accepting the doctrine as doctrine, although it would make no addition to my own prayer-life. 1976 *Honolulu Star-Bull.* 21 Dec. F-2/3 Friends may call from 6 to 9 tonight at Dodo Mortuary, with prayer service scheduled for 7:30.

d. prayer bones *U.S.*, the knees; **prayer breakfast,** a breakfast during which prayers are offered; **prayer card,** a card used by a Member of Parliament for reserving a seat at prayers; **prayer chain,** a series of people each of whom receives a written prayer with an invitation to pass it or copies of it to others; **prayer circle,** a group of people who pray together; **prayer-cylinder** = PRAYER-WHEEL; **prayer day,** a day in Parliament on which prayers (see sense 5 above) are heard; **prayer-flag,** in Tibet, a flag on which prayers are inscribed; **prayer-gong,** a gong calling people to prayer; **prayer-niche,** in a mosque, a niche in the centre of a sanctuary wall indicating the direction of Mecca; **prayer-nut,** in a chaplet, a nut-shaped bead which opens to form a diptych with reliefs; **prayer plant,** a perennial herb, *Maranta leuconeura*, belonging to the family Marantaceæ, native to Brazil, bearing irregular, three-petalled, white flowers, and often cultivated as a house plant for the sake of its shiny, variegated leaves; **prayer ring** = *prayer circle; **prayer rug** (later examples); **prayer stool,** a stool for kneeling on while praying; **prayer ticket** = *prayer card; **prayer-value,** efficacy or worth for prayer; **prayer-wall,** a wall on which prayers are inscribed; = *MANI[2].

1926 *Amer. Speech* II. 362 *Prayer bones* (noun phrase), knees. 'Everyone get down on his prayer bones.' *a* 1944 J. CONROY in B. A. Botkin *Treas. Amer. Folklore* (1944) IV. 531 You've got to kneel down on your prayerbones... If you kneel down to save your poor old back, the little grains of sand eat into your prayerbones. 1970 C. MAJOR *Dict. Afro-Amer. Slang* 93 *Prayer bones*, the knees. 1966 *New Statesman* 4 Mar. 285/2 The Republican governor of Oregon, is in the vanguard of a movement that sponsors 'prayer breakfasts' for politicians all around the world. 1969 *Listener* 28 Aug. 271/1 (*caption*) Billy Graham speaks at a Honolulu prayer breakfast. 1959 P. G. RICHARDS *Hon. Members* iv. 75 No permanent reservation of seats is allowed... A Member who intends to be present at prayers at the start of a sitting can place a 'prayer card' on a bench; this card has to be obtained personally from an attendant at the House at any time after eight a.m. on the same day. 1975 *Daily Tel.* 16 Apr. 16 An interesting feature of the House of Commons before the Budget statement yesterday was the number of seats bearing Prayer Cards—reservations—on the Tory side. 1908 *Westm. Gaz.* 5 Oct. 4/1 Other ladies started prayer-chains to promote or defeat the different candidates' chances of victory. 1911 *Daily Colonist* (Victoria, B.C.) 11 Apr. 4/3 We have been requested to say something about the 'prayer chain' which is being worked again... We are told that in the time of Jesus it was said that whoever copied this prayer and sent it to nine persons would have great joy, and those who did not would have great sorrow. 1880 P. DEMING *Adirondack Stories* 25 As a preliminary to the sermon, a prayer-circle was formed. 1894 I. L. BISHOP *Among Tibetans* ii. 46 Prayer-cylinders which are turned by pulling ropes. 1897 *Geogr. Jrnl.* X. 35 A prayer-cylinder revolved by the wind. 1952 *Ann. Reg. 1951* 17 A motion..to cut off alcoholic refreshment after 10 p.m. on 'prayer days'. 1882 'SHWAY YOE' *Burman* I. xvii. 225 These prayer-flags..are made of paper, cut fancifully into figures of dragons, lizards, and the like, with embroidery-work round their edges. 1897 *Geogr. Jrnl.* X. 35 Groups of prayer-flags in memory of the dead are planted beside every village. 1936 [see *OBO]. 1952 [see *CHORTEN]. 1955 E. HILLARY *High Adventure* 62 We sat down wearily in the snow beside a clump of Tibetan prayer-flags. 1905 E. F. BENSON *Image in Sand* ix. 135, I adore theosophy, prayer-gongs, and letters from the ceiling. 1937 *Burlington Mag.* Oct. 193/2 The mihrab, or prayer-niche. 1971 *Country Life* 25 Feb. 426/1 The large construction..is an

Iranian prayer-niche in coloured tin enamel tiles (*faience*). 1937 *Burlington Mag.* Aug. 98/1 She holds a little silver chain, from which hangs..a 'prayer-nut' for a chaplet, in wrought silver. 1969 E. Wilkins *Rose-Garden Game* ii. 59 The..Chatsworth paternoster..has a terminal bead that is a little hinged box, which opens to show two miniature relief carvings... The prayer-nut is usually made of boxwood. 1953 J. Hersey *Garden in your Window* iv. 57 The Prayer Plant, while a bit rare, is simple to grow. 1956 Y. Field *House Plants* iv. 97 The small maranta is a very beautiful foliage plant. Since this plant closes its leaves at night or almost curls them together it is known as the Prayer Plant. 1977 Ward & Wellsted *Indoor Plants* 81/2 *Maranta leuconeura*... Prayer Plant, Rabbit's Tracks. This Brazilian plant has given rise to several spectacularly coloured foliage varieties. 1846 *Knickerbocker* XXVIII. 305 When a 'prayer ring' was to be formed, he announced it at the close of a sermon. 1904 Prayer-rug [see *Kulah[1]]. 1930 *Morning Post* 16 July 8/6 This fascinating old Koula Prayer Rug is believed to have been made for the Jewish Synagogue at Toledo. 1935 H. Edib *Clown & his Daughter* xlviii. 277 Pembeh touched him on the shoulder and pointed to a prayer-rug spread at the threshold of the room. 1962 C. W. Jacobsen *Oriental Rugs* 306 Tekke Prayer Rugs are available only from estates. 1979 *Guardian* 26 Oct. 15/2 The bearded Ayatollahs..[are] sweeping the most pressing problems under the prayer rug. 1908 *Daily Chron.* 6 Apr. 1/4 As they knelt upon the wooden prayer stool..they made no noise. 1924 J. E. Mills *From Back Benches* ii. 9 Lady Astor..staked out the second row corner seat below the gangway, and, attending regularly.., secured her ticket from the attendant which 'booked' the seat, providing she attended prayers. All went well until Mr. Joynson Hicks, returning..after nearly a year's absence, deposited his 'prayer ticket' in.. Lady Astor's seat. 1906 W. R. Inge *Truth & Falsehood in Relig.* iv. 102 It does not satisfy those who really believe in the supernatural occurrences, which it is proposed to maintain in consideration of their 'prayer-value'. 1953 R. Knox *Off Record* xliv. 149 It's no good contemplating becoming a Catholic unless you are prepared to accept doctrinal definitions which have, for you, no particular prayer-value. 1960 C. Winick *Dict. Anthropol.* 562/2 *Mani wall* or *prayer wall*, a low long wall of mud and stone, covered with flat rocks, on which Tibetan characters are carved. Devout Tibetans walk with the mani wall to their right to get benefit from it. Such walls are frequently more than a quarter of a mile long. 1974 *Listener* 17 Jan. 76/2 As you walk up the trail, prayer-walls bisect the paths. The act of walking past the wall is a prayer in itself.

prayer-bead. Add: **1.** Also *gen.*, one of a string of beads used in prayer.

1975 R. P. Jhabvala *Heat & Dust* 61 The white sadhu ..had all his possessions with him—a bundle, an umbrella, prayer-beads, and a begging bowl. 1976 'M. Delving' *China Expert* iv. 47 He wore a string of Tibetan prayer beads around his neck. 1979 E. Bercovici *Wolf Trap* 145 Cotton pointed to the beads. 'What are those?' he asked pleasantly. 'Prayer beads,' the man answered, holding the strand out, trying to smile. 'Arab beads.'

prayerfully, *adv.* (Further examples.)

1962 *Friend* 16 Mar. 326/2 The decision, like all moral decisions, can only be taken, prayerfully, on the merits of the case by the individuals concerned. 1971 'A. Garve' *Late Bill Smith* ii. 49 She sipped a dry martini.., raising her glass prayerfully to the success of the cruise. 1973 *Daily Tel.* 3 Mar. 3/1 Before he died of starvation, David wrote..a touching letter to his parents..talking hopefully and prayerfully about his own situation. 1977 J. F. Fixx *Compl. Bk. Running* i. 5, It is to be prayerfully hoped, but not reasonably expected, that some political leader will find the gumption to blurt out the melancholy truth.

prayer-meeting. (Earlier and later examples.)

1780 *Arminian Mag.* III. 155 Some of these coming over to the prayer meetings at Wednesbury..were utterly astonished... Presently a prayer meeting was set up at Darlaston. 1817 W. Sewall *Diary* 11 Jan. (1930) 7/1 Evening attended prayer meeting. 1928 J. Buchan *Runagates Club* xi. 298 It was the prayer-meeting, remember, which brought America into the War. 1954 D. S. Davis in *Ellery Queen's Mystery Mag.* June 27/1 Sue Thompson had..been..to Sunday prayer meeting. 1972 *News & Observer* (Raleigh, N. Carolina) 30 Dec. 4/3 Wednesday night prayer meeting..and the amen corner have gone with stewards who yelled out 'amen' during the sermon.

prayer-mill. (Earlier example.)

1832 J. Bell *Syst. Geogr.* V. 103 Prayer-mills, which are set in motion by wind or water.

pray-in: see *-in *suffix*[3].

praying, *vbl. sb.* Add: **b.** *praying ground*; **praying flag-staff,** a staff bearing a prayer flag (see *prayer[1] 6 d); **praying machine** (earlier and later examples); also *transf.*; **praying mat** (earlier example).

1877 T. W. R. Davids *Buddhism* 211 Everywhere in Tibet these praying flag-staffs meet the eye. 1935 Z. N. Hurston *Mules & Men* i. ii. 39 He went way down in de swamp behind a big plantation to de place they call de prayin' ground, and got down on his knees. 1967 W. Soyinka *Kongi's Harvest* 55 You must hurry or the confusion Will be worse than shoes before the Praying-ground at Greater Beiram. 1976 *Sunday Times* (Lagos) 26 Sept. 3/1 (*caption*) The scene at Obalende praying ground with the Lagos Chief..reading his address. 1826 S. Stallybrass *Jrnl.* 16 July in E. Stallybrass *Mem. Mrs. Stallybrass* (1836) iv. 203 An old man..was travelling sixty versts on foot, though not destitute of a horse, for the purpose of turning the *praying machine* for a week,.. in order to atone for past misconduct and drunkenness. 1879 *Good Words* 745/1 In the great temple there is a figure of Matreya, the coming Buddha... We..found ourselves before the top of the great praying machine, a revolving structure. 1972 C. Stephenson *Merrily on High* 185 Eastern [Orthodox] spirituality has tended to regard monks as primarily 'praying machines'. There has never been the tradition of 'scholar monks' which we have had in the west. 1869 'Mark Twain' *Innoc. Abr.* li. 543 To step rudely upon the sacred praying mats.

praying, *ppl. a.* Add: **a.** (Later example.)

1931 F. L. Allen *Only Yesterday* iv. 80 The 'praying Colonels' of Centre College.

b. praying band = *prayer circle* s.v. *prayer[1] 6 d; **praying mantis** = *praying-insect*.

1883 *Century Mag.* Sept. 788/2 The Woman's Christian Temperance Union is the lineal descendant of the Woman's Crusade of 1874, whose first 'praying band' was led..by Mrs. Thompson. 1900 C. W. Winchester *Victories of Wesley Castle* ii. 41 He had seen [him]..conducting a revival meeting with a praying-band, of which he was leader. 1937 *Sun* (Baltimore) (D ed.) 25 May 4/5 On the left sits a row of younger women—the 'praying band' or 'shout band'. 1895 Praying mantis [see Mantis]. 1899 [see *mule-killer* s.v. *mule[1] 5 c]. 1961 L. van der Post *Heart of Hunter* xii. 161 My old coloured nurse Klara, who had a Bushman mother..showed me my first praying mantis. 1973 M. R. Crowell *Greener Pastures* 40 They have established a cease fire, thanks to a praying mantis.

prazosin (prē̆·zosin). *Pharm.* [Arbitrary *pr-* (sometimes interpreted as representing *piperazine*), + Azo- + *-sin*.] A drug used (as the hydrochloride) in treating hypertension, being a vasodilator whose molecular structure incorporates quinazoline, piperazine, and furyl rings.

1970 *Jrnl. Clin. Pharmacol.* X. 417/1 These data suggest that prazosin hydrochloride is an efficacious agent for the therapy of ambulatory patients with arterial hypertension, particularly when a thiazide diuretic is given in concert. 1974 *Brit. Med. Jrnl.* 11 May 298/1 Prazosin is a new hypertensive drug thought to have a peripheral action involving direct relaxation of vascular smooth muscle and sympathetic blockade. 1978 A. S. Nies in Melmon & Morrelli *Clin. Pharmacol.* (ed. 2) vi. 200/1 Prazosin produces a number of side effects, the most prominent being weakness, dizziness, headache, palpitations and lack of energy... Postural hypotension occurring within two hours of the first few doses of prazosin has resulted in loss of consciousness in some patients.

pre-, *prefix.* Add: **A. I. 1.** With vbs., and ppl. adjs. and vbl. sbs. derived from them. *pre-address, -apprehend, -assemble, -audit, -book, -centrifuge, -clean, -coat, -compute, -cook, -decide, -dry, -film, -give* (*-given* ppl adj.; also *absol.* as sb.), *-grind, -incubate, -ionize, -know, -let, -lubricate, -machine, -own, -perceive, -plan, -polarize, -prepare, -pressurize, -publish, -qualify, -see, -separate, -soak, -think, -tune, -wash, -wear, -wrap, -write*; **prea·spirate** *Phonetics*, to aspirate (a sound) in advance of another sound; **pre-bai·ting,** the act or practice of accustoming vermin to harmless bait so that they will take poisoned bait more readily.

1912 W. Owen *Let.* Aug. (1967) 152, I didn't bring your pre-addressed envelope. 1964 J. Z. Young *Model of Brain* xii. 199 The particular classifying systems operating on any occasion are thus as it were pre-addressed. 1922 Joyce *Ulysses* 713 What..causes, before rising preapprehended,..did Bloom..recapitulate? 1934 *Jrnl. Eng. & Gmc. Philol.* XXXIII. 191 The preaspirated tenues in Scotland are due to the same Norse substratum. 1976 *Archivum Linguisticum* VII. 95 All geminates are held to have once been preaspirated. 1960 R. W. Marks *Dymaxion World of B. Fuller* 21/2 It is theoretically possible..to deliver a full-size, pre-assembled house by air. 1972 *Sci. Amer.* Oct. 118/2 The structure was prefabricated and preassembled in the carpentry shop complete with fans and electrical outlets. 1937 *Sun* (Baltimore) 2 Aug. 1/2 All pre-auditing shall be under the control of the executive branch of the Government. 1936 *Rep. Comm. Exper. Station* (Hawaiian Sugar Planters' Assoc.) 92 It would seem that a practice of prebaiting with unpoisoned cereal, followed by a heavy application of poisoned bait, should prove an effective means of control. 1944 J. S. Huxley *On Living in Revolution* x. 110 Careful study has now been made of the species [*sc.* the black rat], and this, with new methods of poisoning based on pre-baiting, is apparently providing the basis for effective control. 1973 *Times* 9 Mar. 14/1 In 1955 the anti-coagulant compounds, Warfarin and others, came on the market. No pre-baiting was needed with these. 1976 P. R. White *Planning for Public Transport* vii. 149 The amendment also stipulated that minibuses only would be permitted, not for hire or reward, and that passengers would be pre-booked. *Ibid.* viii. 183 The period stipulated for pre-booking by rail appears unrealistic. 1976 *Nature* 15 Jan. 114/2 The homogenate was squeezed through nylon cloth and precentrifuged at 500g for 10 min after adjusting the *p*H to 8.0. 1954 *Sun* (Baltimore) 13 Apr. B23/4 Instead of picking around through a pile at the vegetable counter,..she can buy precleaned fresh produce of a uniform quality. 1937 *Discovery* Sept. 283/2 The press is..precoated with fresh kieselguhr. *Ibid.*, This precoating is done from a small vat. 1973 *Metal Finishing Jrnl.* XIX. 353 About 500,000 tonnes per year of pre-coated steel sheet was used in the UK for buildings. 1948 *Amazing Sci. Fiction* Sept. 146/1 The luminous track on the radar screen had scarcely deviated from the pre-computed path. 1956 *Jrnl. Assoc. Computing Machinery* III. 284 The method used..is to precompute between pass 1 and pass 2 this adjustment based on the count of each duplicated region. 1959 *Proc. Eastern Joint Computer Conf.* 170/1 Prior to the start of a given shut-down, a pre-computed schedule most applicable to the current situation is abstracted from a library of typical schedules. 1946 *Fortune* Apr. 200/2 A high-priced Restaurant carrying a sideline of precooked quick-frozen meals on plastic plates. 1964 E. Bach *Introd. Transformational Gram.* v. 92 Except with carefully 'precooked' data, there will be many conflicting ways of drawing rules together. *a* 1974 R. Crossman *Diaries* (1975) I. 198 Very often the whole job is pre-cooked in the official committee to a point from which it is extremely difficult to reach any other conclusion than that already determined by the officials in advance. 1976 *Woman's Day* (N.Y.) Nov. 150/2 A step saver. No need to precook noodles. 1947 *Mind* LVI. 264 The meaning to be given to 'well-established' or to 'explanation', in history or the social sciences, cannot be pre-decided by a consideration of mathematics only. 1966 G. N. Leech *Eng. in Advertising* i. 4 It is patently false that he writes according to a predecided formula. 1961 *Dairy Industries* Sept. 652 Particles of the product to be dried fall in counter current to slowly rising pre-dried air flowing at a rate of 0·05 to 1 meter per second. 1962 J. T. Marsh *Self-Smoothing Fabrics* xi. 171 The majority of the finishing ranges for the crease-resisting process..increase production and reduce costs by some form of partial pre-drying. 1960 D. Wilson *Television Playwright* 16 Does insistence on cinematic grammar imply that all television drama should be pre-filmed? 1969 J. Elliot *Duel* i. ii. 38 It was to consist of three fifty-minute programmes, all prefilmed. 1943 M. Farber *Found. Phenomenology* xv. 506 The theory of pre-predicative experience, which 'pre-gives' the most primitive substrates in object-evidence, represents the first portion of the phenomenological theory of judgment. 1970 B. Brewster tr. *Althusser & Balibar's Reading Capital* (1975) iii. iv. 297 In Marx's theory..a synthetic concept of time can never be a pre-given, but only a *result*. 1974 *Sci. & Society* XXXVIII. 395 The specific structure of 'unevenness' of the 'ever-pregiven complex whole' which is its existence. 1976 *Brit. Jrnl. Sociol.* XXVII. 296 The Rankean identification of history with pre-given past events. 1973 R. Rendell *Some lie & Some Die* xiv. 121 He..smelt her grinding coffee beans—nothing pre-ground out of a packet for her. 1943 *Jrnl. Bacteriol.* XLVI. 383 One penicillin-containing set with and without glucose was allowed to preincubate uninoculated at 37°C., and a similar set at 2°C. 1977 *Lancet* 18 June 1310/1 The epithelial cells were not stained when tissue sections were preincubated in unlabelled α-B.T. before the standard reaction. 1940 *Jrnl. Appl. Physics* XI. 471/1 Then over the pre-ionized streamer channel the brilliant return stroke..follows at a speed of 10^{10} cm per second. 1979 *Nature* 7 June 477/3 For light ions, such long-distance propagation must be carefully arranged through a pre-ionised neutralising plasma channel, raising serious questions of possible propagation instabilities. 1867 J. S. Mill *Let.* 14 Feb. (1910) II. x. 76 Our freedom may be real though God preknows our actions. 1976 *Field* 30 Dec. 1292/2 (Advt.), Most of our beats are pre-let but we have one or two vacancies through late cancellation. 1961 *Motor Sport* Dec. 1003/1 In America Oldsmobile, Ford, Mercury, Lincoln, Plymouth, Dodge and Chrysler have adopted pre-lubricated chassis bearings. 1976 *Lebende Sprachen* XXI. 151/2 For further information about pre-lubricated bearings see Figure 1, Detail A. 1971 *Physics Bull.* July 406/3 Deep penetration welding using electron beams is becoming quite widely used for assembling pre-machined parts into complex assemblies as an economic alternative to forging, casting and mechanical fastening. 1977 *Offshore Engineer* Apr. 28/1 The system uses an internal clamp/welder to locate, clamp and make an inside root pass in about two minutes on a pre-machined pipe joint preparation. 1964 *Listener* 25 June 1030/3 Used cars are now referred to as pre-owned [in the U.S.A.]. 1970 M. Pei *Words in Sheep's Clothing* ii. 12 'Pre-owned' is described as the modern euphemism for 'second-hand'. 1977 *Caravan World* (Austral.) Jan. 3 (Advt.), 2 acres of pre-owned caravans and boats can be inspected. 1890 W. James *Princ. Psychol.* I. xi. 444 In short, *the only things which we commonly see are those which we preperceive*, and the only things which we preperceive are those which have been labelled for us, and the labels stamped into our mind. 1934 Webster, Preplan. 1948 *Times Rev. Industry* Aug. 18/3 Obviously continuous production must be pre-planned but thereafter is self-progressive. 1958 *Times* 11 Feb. 4/4 (Advt.), Preplanning the entire project ensures smooth continuity of operations and speedy completion. 1965 *Language* XLI. 92 Since this is not a preplanned book..there is a certain amount of repetition. 1976 *Daily Tel.* 13 Aug. 1/7 The positive anti-riot tactics, clearly pre-planned.., were introduced to counter any IRA organised trouble. 1977 J. M. Johnson in Douglas & Johnson *Existential Sociol.* viii. 231 None of the events described were anticipated or preplanned by the members in advance of their occurrence. 1949 *Jrnl. Acoustical Soc. Amer.* XXI. 199/1 The..temperature at which the output voltage falls to zero for a pre-polarized specimen. *Ibid.* 200/2 The voltage gradient necessary fully to pre-polarize varies with the time allotted to polarizing. 1957 E. G. Richardson *Technical Aspects Sound* II. ii. 70 The tube is metallized on the inner and outer surfaces, pre-polarized radially, and the alternating current applied in the same direction. 1968 *Guardian* 16 Feb. 3/2 Pre-prepared dishes, such as fish and chips. 1978 *Guardian Weekly* 29 Oct. 12/3 The President didn't deliver pre-prepared phrases, but stayed close to events. 1945 *Jrnl. Amer. Chem. Soc.* LXVII. 157/1 A current is passed through previously pressurized acid water, and bubbles form in prepressurized water when it is quickly frozen. 1971 *Arch. Biochem. & Biophysics* CXLII. 325/2 The enzyme was prepressurized for 10 min. 1973 W. H. Hallahan *Ross Forgery* vi. 120 'A Lodging for the Night' —first published in a collection..under the title *The New Arabian Nights* in 1882..was not known to have been pre-published separately. 1977 *Lancet* 25 June 1350/2 He asserted..that an article whose contents had already received detailed attention in the papers and on the air had..been prepublished and forfeited its claim to entry

as news in a medical journal. 1974 *Times* 27 Apr. 14/4 The drudgery of having to pre-qualify for [golf] tournaments. 1976 *Sunday Mail* (Glasgow) 28 Nov. 39/3 He finished in the top 25 last year, which ensured he doesn't have to prequalify this year for Turnberry in July. 1980 *Times* 29 Feb. 23 (Advt.), The Irrigation Department and the Electric Power Corporation invite qualified and experienced contractors..wishing to be pre-qualified as tenderers. 1931 Joyce *Let.* 22 Aug. (1966) III. 227 You advised me to proceed against Roth... I did though I presaw the result. 1967 E. Chambers *Photolitho-Offset* iii. 30 Pre-separated colour art work can be prepared using Bourges Colotone overlays. 1967 Karch & Buber *Offset Processes* iii. 64 Kits are useful in the preparation of pre-separated full-color process copy. 1919 *Science* 6 June 544/2 (*heading*) Presoaking as a means of preventing seed injury due to disinfectants and of increasing germicidal efficiency. *Ibid.* 545/1 The presoaked seeds are thoroughly wetted in the 1:80 solution..for ten minutes. 1974 *Indian Jrnl. Agric. Sci.* XLIII. 973/1 An experiment was laid out..in polythene bags with presoaked pumpkin seeds in different N solutions. 1960 *20th Cent.* Sept. 242 This kind of talk is not formalized or pre-thought. 1966 'A. Hall' *9th Directive* xii. 112 The Bureau..takes no action without the most serious pre-thinking. 1977 E. Leonard *Unknown Man No. 89* xvii. 155 Pre-think your options. 1964 J. Carnochan in D. Abercrombie et al. *Daniel Jones* 399 A series of pre-tuned reeds. 1974 Harvey & Bohlman *Stereo F.M. Radio Handbk.* iii. 41 Within this narrow band it is possible to pre-tune the r.f. stage to 98 MHz. 1977 Pre-wash [see *program, programme *sb.* 2 g (i)]. 1976 *New Musical Express* 12 Feb. 40/1 (Advt.), Genuine 'Levi & Levi' type jeans, preworn and shrunk, just need patches. 1934 Webster, Pre-wrap. 1959 *Times* 9 Mar. (Britain's Food Suppl.) p. vi/4 Pre-wrapped retail portions of natural cheese have been on the market for some years. 1963 *Economist* 29 June 1357/1 More and more cigars are..marketed pre-wrapped in large packs. 1951 Dylan Thomas *Let.* 12 Apr. (1966) 358, I would bring great packages of new poems to read, and much more pre-written prose to pad them in. 1969 *Computers & Humanities* IV. 106 This object code and a prewritten subroutine which searches the date item for the required elements becomes the Phase II control program.

2. With a sb. *pre-audit, -auditor, -censorship, -civilization, -coat, -evangelism, -excitation, -hearing, -incubation, -intimation, -knowledge* (later examples), *-negotiation, -oxygenation, -polarization, -pressurization, -publicity, -qualification, -rehearsal, -taste, -verbalization*; **pre-a·djunct** *Gram.*, an adjunct that precedes the word it modifies; also *attrib.*; **pre-aspira·tion** *Phonetics*, aspiration that precedes another sound; **preco·ntour** *Phonetics*, one or more unstressed syllables which precede the peak of a contour; **pre·pulse,** a preliminary pulse of electricity; **prerea·ction,** chemical reaction occurring before some other process; **pre·-rinse,** a preliminary rinse given to something before it is washed; **preseni·lity** *Med.*, premature senility; **pre·-soak,** (*a*) a soaking given prior to some subsequent process or treatment; (*b*) a liquid used for this; also *attrib.*; **pre·-wash,** a preliminary wash, used *spec.* as the name of a setting on an automatic washing-machine; also *attrib.*

1898 Pre-adjunct [see *head-word* s.v. *head *sb.* 66]. 1914 O. Jespersen *Mod. Eng. Gram.* II. xiv. 331 There are some adjectives that are hardly ever used predicatively, and on the other hand some that are hardly ever used as pre-adjuncts. *Ibid.* 333 The averseness to pre-adjunct employment..has been transferred to other words beginning with an *a-* of a different origin. 1957 *Publ. Amer. Dial. Soc.* xxviii. 123 *What, which, whose,* and *how* serve as pre-adjuncts: 1 What book?; 2 How far is it? 1945 S. Einarsson *Icelandic* i. 1 Aspiration, a breath (^{h}) following or preceding (preaspiration) the stops *p, t, k*, is indicated by an ^{h}. 1965 H. Wolter in *Proc. 5th Internat. Congr. Phonetic Sci. 1964* 595 The auditory impression of the pre-aspiration is rather like an [h], although in connection with [t] it may sound like an [f]. 1976 *Archivum Linguisticum* VII. 91 Preaspiration is not, according to Liberman, a voiceless vowel, although it does cause the end of a preceding vowel to be devoiced. 1938 *Sun* (Baltimore) 2 Apr. 6/2 It retains the principle of pre-audit but it makes the pre-auditor amenable to Presidential authority. 1962 *Guardian* 6 Nov. 9/4 Today's issue was subjected to pre-censorship by the State Prosecutor's office. *a* 1974 R. Crossman *Diaries* (1976) II. 663 The Lord Chancellor immediately replied that this would involve having pre-censorship all over again because if the licensees could be sued for libel then they would start controlling the plays. 1949 R. A. S. Macalister *Archæol. of Ireland* (ed. 2) p. x, It [*sc.* Ireland] has rendered to Anthropology the unique.. service of carrying a primitive European *Precivilization* down into late historic times. 1946 Precoat [see *filter aid* s.v. *filter *sb.* 5]. 1945 K. L. Pike *Intonation Amer. Eng.* iii. 29 Immediately preceding the stressed syllable of a primary contour there oftentimes will be one or more syllables which are pronounced in the same burst of speed with that primary contour but which themselves are unstressed. These syllables may be called precontours, and depend for their pronunciation upon the syllables which follow them. They may constitute grammatically independent words,..or they may be parts of a word. 1962 B. M. H. Strang *Mod. Eng. Struct.* 53 A contour may be preceded by one or more unstressed syllables forming a *pre-contour*; special meanings may be conveyed by varying the level of the pre-contour, but ordinarily it is spoken at pitch-level 3 unless the contour begins on level 3, which tends to lower the pre-contour. 1975 *Amer. Speech 1972* XLVII. 185 In many dialects, the medial consonant disappears in the precontour, as in *twenty-óne, United Státes*. 1968 F. A. Schaeffer *God who is There* v. ii. 143 Pre-evangelism must come before evangelism... The reason we have not been reaching many of these people is because we have not taken enough time with pre-evangelism. 1951 K. S. Lashley in Saporta & Bastian *Psycholinguistics* (1961) 186/2 Such contaminations might be ascribed to differences in the relative strength of associative bonds between the elements of the act, and thus not evidence for pre-excitation of the elements or for simultaneous pre-excitation. 1976 *Lancet* 30 Oct. 938/2 The electrocardiographic appearances in hypertrophic cardiomyopathy may erroneously suggest the presence of pre-excitation. 1934 Webster, Prehearing. 1968 *Listener* 5 Sept. 315/3 The programmes were recorded in July and at a pre-hearing the impression was of an extension of current permissiveness rather than musical revolution. 1977 *Gramophone* June 95/3 On this disc the Cathedral choir,.. have given an opulent preview (or pre-hearing) of some of the music for the great day. 1943 *Jrnl. Bacteriol.* XLVI. 383 In the presence of glucose, preincubation of the sterile solutions caused a clear-cut reduction in penicillin inhibition. 1977 *Nature* 6 Jan. 58/2 Preincubation with increasing numbers of suppressor cells yielded successively less active supernatants. 1923 J. W. Harvey tr. *Otto's Idea of Holy* xx. 171 This is nothing else than the pure *impulsion* to *redemption*, and the pre-intimation and anticipation of a boded 'good'. 1925 *Law Reports: Appeal Cases* 730 Expressing the matter in my own words, I would say that a threat is a pre-intimation of proposed action of some sort. 1962 *Science Survey* III. 290 For many of these fish, and certainly the majority of salmon, it is the first time the journey has been made and they can have no pre-knowledge of the route or the obstacles in front of them. 1979 G. Swarthout *Skeletons* 177 Guiding myself through the gloom as much by preknowledge as eyesight. 1960 *Guardian* 22 June 9/1 The pre-negotiations may last some days. 1967 *Economist* 25 Nov. 834/1 The idea..was to slide covertly into pre-negotiations with Britain by the device of asking the commission to talk with the British about devaluation. 1961 *Lancet* 19 Aug. 405/2 If a short-acting muscle relaxant is also used, conditions for intubation are obtained more quickly. Preoxygenation of the patient is a wise precaution. 1961 C. F. Gell in H. G. Armstrong *Aerospace Med.* x. 145/1 Preoxygenation, or the breathing of 100 per cent oxygen at sea level, is a procedure that was utilized prior to high level flights in airplanes without pressurized cockpits or in the absence of pressurized suits. 1965 *Gloss. Aeronaut. Terms* (*B.S.I.*) xvii. 3 *Preoxygenation*, a process of breathing pure oxygen before flight in order to give protection against decompression sickness by eliminating nitrogen from the body tissues and fluids. 1950 *Industr. & Engin. Chem.* Feb. 264/2 An alternating stress..is applied parallel to the direction of polarization, and conversely, expansion and contraction in the direction of an electric alternating current field superimposed parallel to the prepolarization. 1953 *Jrnl. Acoustical Soc. Amer.* XXV. 294/2 Although polycrystalline barium titanate is basically an electrostrictive material, by prepolarization it assumes properties which make it very similar to piezoelectric materials. 1971 *Arch. Biochem. & Biophysics* CXLII. 327/2 For most of the enzymic activities the restoration or stimulation of activity by pressure was closely related to the amount of activity remaining during the prepressurization. 1963 *Times* 19 Feb. 12/4 The sum includes pre-publicity, brochures, and the specialist 'after-care'. 1969 *New Scientist* 23 Oct. 172/1 An unusual amount of prepublicity hinting at a good show. 1979 *Times* 1 Dec. 12/5 No lavish pre-publicity on the free plug circuit. 1978 *Nature* 2 Feb. 474/1 A 25–50-ms conditioning prepulse was used to determine the relationship between inactivation and membrane potential before and after glyoxal treatment. 1969 *Jane's Freight Containers 1968–69* 302/3 Pre-qualification of tenderers on an international basis for the wharf construction has been processed in liaison with the World Bank. 1979 *Daily Tel.* 19 Oct. 24 (Advt.), Kingdom of Swaziland Ministry of Works, Power and Communications. International invitation for tender prequalification. 1975 *Nature* 2 Oct. 368/2 Prereaction did not enable him to go leaner than 3·4% methane. 1978 *Ibid.* 12 Jan. 165/2 Preincubation of the antiserum with purified β_1-μ reduced all peaks to background level, whereas pre-reaction with BSA had no effect. 1962 A. Nisbett *Technique Sound Studio* vi. 106 Pre-rehearsals and perhaps pre-recordings of the complicated parts will be necessary. 1972 *Listener* 6 July 3/1 A good deal of pre-rehearsal often takes place away from the station, but WGBH can only offer one hour of studio rehearsal time. 1950 J. G. Davis *Dict. Dairying* 71 A jetting pre-rinse which raises the temperature of the bottle at the same time. 1963 *Which?* Feb. 50/1 It was easy to give a cold pre-rinse and it improved the washing performance. 1970 *Ibid.* Oct. 293/1 For most, there was a pre-rinse and a choice of two washing programmes, depending on how dirty the dishes were. 1900 Dorland *Med. Dict.* 535/1 Presenility. 1902 *Brit. Med. Jrnl.* 16 Aug. 472/1 Another symptom of presenility is an early impairment of memory, especially of substantives. 1933 *Arch. Dermatol. & Syphilol.* XXVIII. 553 The cause of pseudoxanthoma elasticum is unknown... Jones, Alden and Bishop entertained the belief that since the disease is allied to the changes found in elastosis senilis, it is an evidence of presenility. 1972 *Albrecht v. Graefes Arch. f. klin. und exper. Ophthalm.* CLXXXIV. 314 Histologic changes in the corpus geniculatum laterale in older persons, presenility (Alzheimer's disease) and in senile dementia. 1919 *Science* 6 June 545/1 The disinfectant..must be applied at the end of the presoak period. 1920 *Jrnl. Agric. Res.* XIX. 371 A 6-hour presoak followed by a 6-hour treatment with formalin. 1969 *Chem. & Engin. News* 3 Feb. 16/1 Makers of enzyme-active detergents and presoaks compete strongly for a place in the..home laundry products market. 1976 *Chem. in Brit.* XII. 117/1 Proteases are incorporated into presoak and heavy duty washing powders and this is the major outlet for industrial enzymes. 1956 E. M. Forster *Marianne Thornton* 18 To look out through the high glass door, upon the magnificent Tulip Tree, became a ritual, and almost a pretaste of heaven. *a* 1974 R. Crossman *Diaries* (1975) I. 329 We were a typical Labour Party gathering, which gave us a pre-taste of the Blackpool hotels when Conference starts in a few days' time. 1959 J. C. Catford in Quirk & Smith *Teaching of English* vi. 188 Conscious preverbalisation in L_1, and translation into L_2, may be entirely suppressed, but errors due to interference from L_1 still keep breaking through. 1962 *Which?* Aug. 234/2 Up to 6 minutes pre-wash soak or 10 minutes washing. 1966 D. V. Davis *New Domestic Encycl.* iv. 131 If you have a fully-automatic washing machine, use the Pre-wash or Rinse programme. 1970 *Which?* May 143/2 The Bendix LTA had a pre-wash that could be included in the automatic cycle.

3. With an adj. *precombustible*.

1922 Joyce *Ulysses* 657 Fanned by a constant updraught of ventilation between the kitchen and the chimney-flue, ignition was communicated from the faggots of precombustible fuel to polyhedral masses of bituminous coal.

II. 4. a. In adverbial relation to an adj. **precommissu·ral,** anterior to a commissure of the brain; **preoptic** (examples).

1896 *Q. Jrnl. Microsc. Sci.* XXXIX. 184 The cerebrum of Platypus..conforms to the mammalian type, yet numerous features—such as..the 'precommissural area' —indicate its close Saceropsidean affinities. 1900 A. Hill tr. *Obersteiner's Anat. Central Nervous Organs* (ed. 2) 429 A small portion of the fornix which streams on to the septum pellucidum (præcommissural fibres of Huxley) ought to be reckoned as fibres of association. 1953 G. A. G. Mitchell *Anat. Autonomic Nervous System* viii. 99 A small fascicle of fornix fibres given off near the interventricular foramen passes downwards in front of the anterior commissure... In rabbits this precommissural fascicle ends in the medial and lateral septal nuclei and in the nucleus accumbens. 1971 J. Z. Young *Anat. Nervous System Octopus Vulgaris* iv. 69 There is a direct connection between the rather enigmatic precommissural region and the anterior pedal lobe. 1889 *Cent. Dict.* VI. 4694/3 *Preoptic*, anterior with respect to optic lobes; pregeminal: specifically noting the anterior pair of the optic lobes or corpora quadrigemina of the brain. 1907 J. B. Johnston *Nervous System Vertebr.* xviii. 297 Surrounding the preoptic recess is a layer of cells of very primitive character. 1970 *Jrnl. Anat.* CVII. 186 Evidence which indicates that the preoptic area is involved in the control of reproductive function.

B. I. 1. Also with (Eng. or other) sbs. directly forming sbs., as *pre-cancer, -climax, -delinquency, -menarche, -myelocyte*, etc. (below). Also *prehominid, *prepuberty, etc.

a. Formed on proper nouns (or their adjectives). *pre-Alfredian* (example), *-Arnoldian, -Augustinian, -Baconian* (examples), *-Caroline* [Charlemagne], *-Chaucerian* (example), *-Constantinian, -Copernican* (example), *-Darwinian* (examples), *-Galilean* (examples), *-Hitlerian, -Hitlerite, -Kantian, -Keynesian, -Linnæan, -Listerian, -Malthusian, -Mendelian, -Victorian* (examples); *pre-Celtic* (examples), *-European, -Fascist, -Greek* (example), *-Han, -Hispanic, -Nazi, -Saxon* (examples), *-Soviet, -Vedic*.

c. In pathological terms. *pre-cancerous* (earlier and later examples), *-epileptic, -malignant, -pathological, -symptomatic*.

d. Formed on other adjectives. *pre-capitalist, -capitalistic, -chemical, -cinematographic, -civilized, -coitional, -colonial, -conceptual, -conciliar, -copulative, -evolutionary, -experimental* (hence *-experimentally* adv.), *-feudal, -filmic, -industrial, -industrialist, -intellectual, -lexical, -literary, -matutinal, -modern, -orgasmic, -paroxysmal, -personal, -philosophical, -phonemic, -political* (examples), *-predicative, -prep, -preparatory, -primary, -rabbinical, -romanesque, -socialist, -telescopic, -theoretical, -tragic, -transformational*; freq. in *Gram.*, as *pre-accentual, -adjectival, -adverbial, -consonantal* (so *preconsonantally* adv.), *-infinitival, inflexional, -junctural, -pausal, suffixal, -syllabic*.

pre-a·dult, -adu·lt, prior to the attainment of adulthood; **pre-a·gonal, -ago·nic** *Med.*, occurring immediately before, or premonitory of, the death agony; **pre-agricul·tural** *Anthrop.*, that has not yet developed agriculture as a means of subsistence; **pre-analy·tic, -analy·tical,** preceding analysis; hence **pre-analy·tically** *adv.*; **pre-a·rticle** *sb. Gram.*, one of a set of words that can precede an article in a noun phrase; **pre-bacteriolo·gic, -lo·gical,** previous to the discovery of the relationship of bacteria to disease; **pre-ca·ncer** *sb. Med.*, a condition that implies an increased risk of the future development of cancer; cf. *pre-cancerous*, sense B. 1 c in Dict. and Suppl.; **prece·llular** *Biol.*, prior to the origin of cellular life; **pre-ci·vil,** prior to the development of social organization; **precli·max** *sb. Ecology*, the point in a plant succession at which development has ceased before the state of climax is reached; **preconvu·lsive** *Med.*, preceding a convulsion; **pre-cu·ltural,** pertaining to human existence prior to or independent of cultural development; **pre-deli·nquency** *sb.*, behaviour which is likely to result in (juvenile)

delinquency; hence **pre-deli·nquent** *a.* and *sb.*; **pre-emerge·nt** = *pre-emergence* (sense B. 2 below); **pre-erythrocy·tic** *Biol.*, occurring or existing in the period between the entry of a malaria parasite into the body as a sporozoite and the subsequent entry into red blood cells of schizonts descended from the sporozoites; **pre-expone·ntial** *Math.*, occurring as a non-exponential multiplier of an exponential quantity; **prega·mic** *Cytology*, prior to the formation of gametes; **prega·strular** *Biol.*, prior to gastrulation; **pregeolo·gical,** occurring in, or pertaining to, the period of the earth's history earlier than the time of formation of the oldest known rocks; hence **pregeolo·gically** *adv.*; **pregra·mmar** *sb. Linguistics* (see quots.); **programma·tical** *Linguistics*, applied to an assumed period in linguistic communication prior to the existence of grammatical structure; **pre-i·ctal** *Med.*, preceding a stroke or fit, esp. an epileptic fit; **pre-Inca·ic** = *pre-Incarial*; **pre-La·tin,** designating (any of) the Italic languages older than Latin; **prelo·gic** *sb.*, a mode of thought that does not yet conform to logical reasoning (cf. *prelogical *a.*); **pre-meio·tic** *Cytology*, occurring before meiosis; that has not yet undergone meiosis; **premena·rche** *sb. Med.*, the premenarchal period in a girl's life; **premena·rchal, -mena·rcheal, -mena·rchial,** of, pertaining to, or designating a girl in the few years before the onset of menstruation; **pre-mo·ral,** pertaining to a stage of development prior to the personal acceptance of moral responsibility; hence **pre-mora·lity** *sb.*; **premy·elocyte** *sb. Anat.* = *myeloblast; hence **premy:elocy·tic**; **preneopla·stic** *Med.*, before or prefatory to the development of a neoplasm; **pre·-noun** *sb. Gram.*, a word that generally precedes a noun and is in close syntactical relation to it; **pre-ovula·tory** *Med.*, before ovulation; **prepa·tent** *Med.*, applied to the period between parasitic infection of a host and the time when the parasite can be first detected; so **prepa·tency** *sb.*, the condition of an infected host during the prepatent period; **prepla·netary** *Astr.*, existing before the formation of planets; *spec.* constituting the material from which the planets were formed; **pre-pla·nting,** applied or performed before the crop is planted; also *absol.*; **preproinsulin** (prī̆·pro͵insiū̆lin) *sb. Biochem.*, a precursor of proinsulin; **prepu·beral** = *pre-pubertal*; so **prepu·berally** *adv.*; **pre-pubertal** (further examples); hence **prepu·bertally** *adv.*; **pre-refle·ctive, -refle·xive,** prior to reflection or reasoning thought; **pre-relativi·stic,** before the theory of relativity was published (in 1905 and 1915); **pre-reprodu·ctive,** prior to the time when an individual becomes capable of reproduction; **preschizophre·nic,** of or pertaining to the period prior to the onset of schizophrenia; as *sb.*, a person showing symptoms similar to those observed prior to schizophrenia; so **preschizophre·nia** *sb.*; **prese·nile** *Med.*, occurring in or characteristic of the period of life preceding old age, esp. the two or three decades immediately before; **preso·lar, preste·llar** *Astr.*, not (yet) having formed a star or stars; **pre-stru·cturalist,** prior to the development of structural linguistics; **pre-systema·tic,** prior to the development of a formal system; so **pre-systema·tically** *adv.*; **preve·rnal,** pertaining to a season before or very early in spring.

1965 W. S. Allen *Vox Latina* v. 85 Pre-accentual loss in *disciplina*. 1965 N. Chomsky *Aspects of Theory of Syntax* iv. 148 Verbs are strictly subcategorized into Intransitives, Transitives, pre-Adjectival, pre-Sentence. 1977 *Word 1972* XXVIII. 94 From an Old Irish system of separate comparative and superlative suffixes, the language will have changed to a system of separate comparative and superlative preadjectival particles. 1902 *Buck's Handbk. Med. Sci.* (rev. ed.) IV. 527/2 The condition.. becomes manifest during the growing or preadult period. 1974 *Sci. Amer.* Sept. 127/3 A husband, a wife and their preadult children. 1977 *Bull. Amer. Acad. Arts & Sci.* Oct. 25 Since the goals of adults tend to reflect their pre-adult formative experiences, the changing assessment of progress as it is traditionally understood may prove to be essentially an integrational phenomenon. 1976 *Archivum Linguisticum* VII. 32 Absolute-final position is the norm for final position, with pre-adverbial positions as optional variants in sentences where non-sentence adverbials occur. 1900 *Buck's Handbk. Med. Sci.* (rev. ed.) I. 563/1 Immediately preceding death there is an intense congestion of the viscera which frequently results in an outpour of serum. This condition, when involving the peritoneal cavity, is termed pre-agonal ascites. 1974 L. Watson *Romeo Error* iii. 66 Professor Negovskii of the Soviet Academy of Medical Sciences calls them shock, preagonal state, agony and chemical death. 1886 *Buck's Handbk. Med. Sci.* II. 49/1 A sudden high elevation of temperature..after a chill in a previously apyretic case, means a complication, and not a fatal issue, but a hyperpyrexia without chill, and with a profuse sweat, is pre-agonic. 1927 *Contemp. Rev.* July 85 The moon to pre-agricultural society was the real magic. 1947 *Sci. Amer.* Sept. 47/1 The preagricultural population..must have been vulnerable to changes in climate..and to the disappearance of species of prey. 1975 J. G. Evans *Environment Early Man Brit. Isles* iii. 55 The mixed deciduous woodlands of pre-agricultural Britain. 1962 H. R. Loyn *Anglo-Saxon Eng.* ii. 79 The pre-Alfredian period is rich in reference to Anglo-Saxon saints and scholars. 1929 C. I. Lewis *Mind & World-Order* ii. 54 The acceptance of such preanalytic data as an ultimate epistemological category would..put an end to all worthwhile investigation. 1939 *Mind* XLVIII. 89 He holds that philosophy must begin with the pre-analytic data of experience as these are expressed in common-sense judgments. 1927 *Jrnl. Philos.* XXIV. 9 The given is for science a continual challenge, always remaining 'pre-analytical', in the sense of never condemning as bootless the task of more searching analysis. 1965 A. C. Danto *Analytical Philos. Hist.* x. 207 From the Deduction Assumption, together with our pre-analytical notion of explanatory inadequacy, we may..elicit the remainder of Hempel's Analysis. 1934 Cohen & Nagel *Introd. Logic* xix. 385 Analysis..may reveal many complexities in the pre-analytically simple object or concept. 1951 *Mind* LX. 550 Something which we knew pre-analytically to be true. 1886 C. M. Yonge *Chantry House* I. iii. 22 In those pre-Arnoldian times no lofty code of honour was even ideal among school-boys. 1965 Pre-article [see *post-article* s.v. *post- B. 1 c]. 1971 Pre-article [see *non-lexical* s.v. *non- 3]. 1959 J. Blish *Clash of Cymbals* iii. 72 'Pre-Augustinean time' came to be something that a historian could know all about, but a physicist, by definition, nothing. 1964 P. F. Anson *Bishops at Large* i. 44 Restoring the pre-Augustinian tradition in Britain. 1865 Mill *Exam. Hamilton's Philos.* xxiv. 469 Generality of the pre-Baconian type. 1953 K. Britton *J. S. Mill* v. 152 The enquiry into microscopic conditions was distinguished in pre-Baconian logic as the enquiry into *material causes*. 1902 *Buck's Handbk. Med. Sci.* (rev. ed.) IV. 391/2 It..was long ago expressed by Alexander von Humboldt.., to be sure, in pre-bacteriologic language. 1892 *Pall Mall Gaz.* 6 Feb. 2/1 He was educated in the pre-bacteriological era, and had little sympathy with modern developments of medical science. 1965 S. Peller in Glass & Eversley *Population in Hist.* v. 94 In the pre-bacteriological era, the survival rates for the ruling families were far ahead of those of the general population. 1938 Dorland *Med. Dict.* (ed. 18) 1129/2 *Precancer*, an abnormal growth which is likely to develop into cancer. 1963 *New Statesman* 19 July 71 Pre-cancer means cells which, if left untreated, will eventually develop into invasive cancer. *Ibid.*, Microscopic studies of the cervical smears showed..five cases of pre-cancers. 1882 *Brit. Med. Jrnl.* 7 Jan. 5/1 (*heading*) The pre-cancerous stage of cancer, and the importance of early operations. 1975 *Sci. Amer.* Nov. 78/3 Screening programs have an additional shortcoming: for every cancer they detect they reveal perhaps 10 other abnormalities, many of which seem to be precancerous. 1949 I. Deutscher *Stalin* vi. 194 It was, however, a formidable task to adjust their pre-capitalist, often pre-feudal and even nomad ways of life to the Marxist, Communist policies of the central Government. 1952 V. A. Demant *Relig. & Decline of Capitalism* i. 28 Spontaneous social co-operation such as is universal in pre-capitalist societies. 1974 B. Pearce tr. *Amin's Accumulation on World Scale* I. ii. 141 Non-European precapitalist societies were not fundamentally different from those of Europe. 1979 *China Now* Jan.–Feb. 22/1 Abrupt transformation of the feudal and pre-capitalist organizational structures. 1940 Wirth & Shils tr. *Mannheim's Ideology & Utopia* iii. 108 Two types of irrationalism..on the one hand, precapitalistic, traditionalistic irrationalism. 1959 *Brno Studies in English* I. 73 It would be unjust..to explain this preference for pre-capitalistic epochs merely as a means of escape. 1897 Pre-Caroline [see *half-uncial* s.v. *half- II. n]. 1934 Priebsch & Collinson *German Lang.* ix. 356 The oldest German or rather Latin–German MS...is written in a pre-Caroline minuscule. 1948 D. Diringer *Alphabet* 545 The pre-Caroline book-hand in North Italy. 1946 *Nature* 21 Sept. 406/2 A new..theory: that they [*sc.* bacteriophages] were the direct descendants of precellular stages in the evolution of living forms. 1974 *Proc. Nat. Acad. Sci.* LXXI. 286/2 (*heading*) An hypothesis for the initiation of precellular evolution. 1934 S. Robertson *Devel. Mod. Eng.* (1936) ii. 18 The pre-Hellenic inhabitants of Greece..were not Indo-European, nor were the Etruscans in Italy, the pre-Celtic peoples who inhabited Britain, or the Basques in Spain. 1953 K. Jackson *Lang. & Hist. Early Brit.* 342 The name may very well be pre-Celtic. 1977 *Word 1972* XXVIII. 29 A pre-Celtic layer is now hardly identified with certainty through faint indices. 1909 R. Bridges *Let.* 29 Oct. (1940) 74, I have never studied the pre-Chaucerian verse. 1973 P. A. Colinvaux *Introd. Ecol.* xxix. 416 He was able to undo the mischief that contact poisons had done and leave the plantations in the security of the pre-chemical age. 1974 M. Taylor tr. *Metz's Film Lang.* iii. 32 The most obvious pictorial juxtaposition, the most properly literary effect of composition, were, to hear him, prophetically precinematographic. 1957 C. Vereker *Devel. Polit. Theory* iii. 98 The metaphorical language used by political thinkers who posit a pre-civil social condition is usually designated contractual. 1960 C. S. Lewis *Studies in Words* 62 That pre-civil condition was described as *nature* or 'the state of *nature*'. 1953 S. Spender *Creative Element* 30 It also includes the whole universe, the unexplored, or uncivilized, or pre-civilized areas of the map. 1956 R. Redfield *Peasant Society & Culture* iii. 77 Having developed out of the precivilized peoples of that very culture. 1916 F. E. Clements *Plant Succession* vi. 110 As a consequence [of reduced water-content], development would cease before reaching the climax proper, and the potential community ..may be called the preclimax. 1929 Weaver & Clements *Plant Ecol.* xviii. 424 The prairie itself is a preclimax to the deciduous forest with higher rainfall. 1960 N. Polunin *Introd. Plant. Geogr.* xi. 331 There are instances in which an apparent climax constitutes in reality a preclimax. 1959 K. E. L. Simmons in D. A. Bannerman *Birds Brit. Isles* VIII. 218 The ceremonies are not 'pre-coitional' in that they do not lead immediately to mating. 1961 *John o' London's* 18 May 543/2 The pre-colonial history of Africa. 1975 A. Drummond *Thames Jrnls. Vicesimus Lush* 18 No mention appears to have been made of potential goldfields in pre-colonial New Zealand. 1956 J. S. Bruner et al. *Study of Thinking* iii. 50 It is curiously difficult to recapture preconceptual innocence... It is as if the mastery of a conceptual distinction were able to mask the preconceptual memory of the things now distinguished. 1967 *Sunday Times* 22 Jan. 10 Everyone tends to exaggerate the mutual isolation of pre-conciliar days. 1976 *Times* 9 Aug. 11/1 The post-conciliar church polishes characteristically pre-conciliar weapons of censure, interdict, suspension and excommunication. 1978 *Times Lit. Suppl.* 25 Aug. 944/2 A spiritual tangle of casuistry and superstition entirely typical of pre-Conciliar Catholicism. 1951 Trager & Smith *Outl. Eng. Struct.* i. 28 Most Northern Middle Western speakers have simple vowel before final or pre-consonantal *r*. 1965 W. S. Allen *Vox Latina* i. 43 Before it [*sc.* h]..the articles take their preconsonantal rather than prevocalic form. 1976 *Archivum Linguisticum* VII. 163 He postulates the forms set down below, in which monophthongization would have been produced in pre-consonantal forms. 1953 K. Jackson *Lang. & Hist. Early Brit.* ii. 470 (*heading*) IE [*sc.* Indo-European] *gu*. This became *b* in CC. [*sc.* Common Celtic] initially except before *u*..; intervocally and preconsonantally, *g*. 1966 *Amer. Speech* XLI. 260 [R], as in SG [*sc.* Standard German], a lenis post-velar fricative, occurring preconsonantally and finally. 1977 D. Cupitt in J. Hick *Myth of God Incarnate* vii. 138 Pre-constantinian Christian art was scarce, unofficial, of very poor quality and often somewhat ambiguous. 1907 W. A. Turner *Epilepsy* ix. 195 He regarded the pre-convulsive fall in alkalinity as a 'biochemical aura'. 1972 M. Verzeano in Petsche & Brazier *Synchronization EEG Activities in Epilepsies* 178 When the preconvulsive state is reached (d), the passages of circulating activity through the network are very rhythmic, very rapid, and involve a large number of neurons discharging at a very high rate. 1964 C. S. Lewis *Discarded Image* iii. 22 Casual statements about pre- Copernican astronomy in modern scientists who are not historians are often unreliable. 1973 M. Amis *Rachel Papers* 24 During the long pre-copulative session I glanced downwards. 1927 B. Malinowski *Sex & Repression in Savage Society* iv. i. 179 In a pre-cultural condition there is no medium in which social institutions, morals, and religion could be moulded. 1963 Auden *Dyer's Hand* 87 Politics in every advanced society is..not concerned with human beings as persons and citizens but with human bodies, with the precultural, prepolitical human creature. 1876–7 W. James in R. B. Perry *Tht. & Char. W. James* (1935) I. xxviii. 478 Pre-Darwinians thought only of adaptation. They made organism plastic to its environment. 1880 *Atlantic Monthly* Oct. 444/1 Pre-Darwinian philosophers had also tried to establish the doctrine of descent with modification. 1899 A. H. Yapp *Cuckoo* iii. 189 In the case of the little birds and cuckoos we have had brought before us in the actions of the former what is decisively in the teeth of pre-Darwinian theory as of the Darwinian. 1921 G. B. Shaw *Back to Methuselah* p. viii, The pre-Darwinian age had come to be regarded as a Dark Age in which men still believed that the book of Genesis was a standard scientific treatise. 1932 Reckless & Smith *Juvenile Delinquency* vi. 168 (*heading*) Truancy as predelinquency. 1972 *N.Y. Law Jrnl.* 24 Oct. 4/8 California's Predelinquency Statute: A Case Study and Suggested Alternatives, by Robert L. Harris. 1951 Powers & Witmer *Experiment in Prevention of Delinquency* iii. 30 If the treatment group comprised only selected pre-delinquents..unhappy public relations might result. *Ibid.* v. 53 No boy who showed obvious pre-delinquent trends was excluded. 1977 M. Edelman *Polit. Lang.* iv. 69 Affluent adults may be 'predelinquent' or 'prepsychotic'; but it is not behavior that governs the connotations of these terms, but, rather, the statistical chances for a group and the belief that poor children are high risks. 1959 *Times* 21 Mar. 9/3 There is no small retail pack available yet of the pre-emergent weedkillers which are being tried extensively in commercial nurseries. 1928 L. J. J. Muskens *Epilepsy* vii. 257 The pre-epileptic headache..is little influenced by drugs. 1944 Huff & Coulston in *Jrnl. Infectious Dis.* LXXV. 237/2 For all the stages of the parasite [*sc.* *Plasmodium*] between sporozoite and erythrocytic trophozoite we shall use the term pre-erythrocytic stages. They are, of course, exoerythrocytic in the broadest meaning of the term. 1962 J. D. Smyth *Introd. Animal Parasitol.* vii. 87 (*caption*) Pre-erythrocytic schizonts of *Plasmodium cynomolgi* and *P. vivax* in parenchymal cells of rhesus monkey and man respectively; 7 days after mosquito inoculation. 1973 R. M. Pinder *Malaria* ii. 26 This process, which constitutes the pre-erythrocytic or primary tissue stage, takes from 8 to 12 days and is essential to allow the parasites to undergo the necessary metabolic adaptation for the change from life in a poikilothermic insect to that in a warm-blooded vertebrate. 1908 *Westm. Gaz.* 7 June 2/2 In India itself the idea and word 'Indian' hardly existed in pre-European times. 1971 *Black Scholar* June 35 We as black men are breaking loose..becoming once again the real black man in the full tradition of our pre-European forebears. 1890 W. James *Princ. Psychol.* II. xxviii. 647 By the pre-evolutionary naturalists, whose generation has hardly passed away, classifications were supposed to be ultimate insights into God's mind, filling us with adoration of his ways. 1923 *Proc. Soc. Exper. Biol. & Med.* XX. 371 (*heading*) The influence of nutrition during the pre-experimental period on the development of rickets in rats. 1971 *Jrnl. Gen. Psychol.* LXXXV. 157 During the pre-experimental period Ss were housed in group cages. 1970 *Ibid.* LXXXII. 47 It also suggests that pre-experimentally acquired tendencies..are important determinants of behavioral variability in experimental settings. 1966 D. G. Brandon *Mod. Techniques Metallogr.* iv. 182 Inclusion of the image term..leads to a correction factor,

a, in both the pre-exponential and exponential terms. 1970 *Nature* 12 Dec. 1086/1 The Arrhenius expression is a frequently erroneous semiempirical formula, with the temperature independence of the activation energy *E* and the pre-exponential factor *A* becoming increasingly questionable as the reaction temperature range broadens. 1938 *New Statesman* 19 Feb. 276/2 It is possible..that the present Italy willingly accepts that German domination in Central Europe which sons of the pre-Fascist Italy spilled their blood to prevent. 1934 Webster, Pre-feudal. 1949 I. Deutscher *Stalin* vi. 195 It was..a formidable task to adjust their..pre-feudal and even nomad ways of life to the Marxist, Communist policies of the central Government. 1962 H. R. Loyn *Anglo-Saxon Eng.* iii. 105 In one respect the techniques of warfare in pre-feudal England led to an inferior status on the part of the smith. 1974 M. Taylor tr. *Metz's Film Lang.* v. 114 The *iconology* (likewise prefilmic) that organizes the denotation of those same objects. 1880 W. James *Will to Believe* (1897) 254 The spencerian philosophy of 'Force', effacing all the previous distinctions between actual and potential energy, momentum, work, force, mass, etc., which physicists have with so much agony achieved, carried us back to a pre-galilean age. 1952 *Mind* LXI. 417 He thinks biology is in a 'pre-Galilean' stage. 1934 *Anatomical Rec.* LX (Suppl.) 92 The most distinctive spindles are the first pregamic figures. 1953 R. Wichterman *Biol. of Paramecium* ix. 273/1 The four micronuclear products of the first pregamic division enter quickly upon the second pregamic division without a resting stage between them.. to produce eight micronuclear products. 1894 *Biol. Lect.* (Marine Biol. Lab., Wood's Holl, Mass.) II. 2 This consideration led some morphologists to insist on the need of a more precise investigation of the præ-gastrular stages, and the desirability of taking as a starting-point not the two-layered gastrula but the undivided ovum. 1970 *Annales Embryol. & Morphogénèse* III. 133 (*heading*) Competence and induction in the pregastrular chick blastoderm. 1882 Pregeological [in Dict.]. 1899 *Geogr. Jrnl.* XIII. 233 A map of the world in early Cambrian times might show the influence of these pre-geological incidents. *Ibid.* 234 If the ocean basins were not formed pre-geologically, but have grown from the changes that have occurred during the long ages of geological time, then we must seek for a cause that has acted continuously, and is acting to-day. 1971 I. G. Gass et al. *Understanding Earth* ix. 137/2 The event which caused the loss of the primary atmosphere must have occurred fairly early in the history of the planet, presumably in pre-geological time (over 3600 million years ago). Otherwise we would expect to find some record of it in the rocks. 1949 *Archivum Linguisticum* I. 121 Is there anything against speaking of 'pre-grammar' as one speaks of prehistory alongside history? 1966 M. Pei *Gloss. Linguistic Terminol.* 216 *Pre-grammar*,..research into the state of the language in its prehistoric stages, and of the grammatical categories which cannot be accounted for by logical analysis. [1926 A. Sechehaye *Structure logique de la Phrase* i. 10 Il faut se souvenir de la différence..qui existe entre le langage prégrammatical et le langage organisé.] 1932 A. H. Gardiner *Theory of Speech & Lang.* ii. iv. 220 Ries takes up the strange position that they [*sc.* sentences without verbs] are 'pre-grammatical, or better still extra-grammatical phrases'. 1937 J. Orr tr. *Iordan's Introd. Romance Linguistics* iv. 331 Every form of grammar has an individual origin, and has its source in some pre-grammatical or extra-grammatical act which in process of time is transformed into grammar. 1940 A. H. Gardiner *Theory of Proper Names* 12 This term [*sc.* 'name'] belongs to the pre-grammatical stage of thought. 1960 E. H. Gombrich *Art & Illusion* iv. 134 It was the Greeks who taught us to ask '*How* does he stand?' or even 'Why does he stand like that?' Applied to a pre-Greek work of art, it may be senseless to ask this question. 1958 W. Willetts *Chinese Art* I. iv. 206 Criteria establishing the pre-Han date of objects on which they appear. 1972 *Trans. Oriental Ceramics Soc.* XXXVIII. 13 This long survival of the most rudimentary of techniques..contributed to the style of hard-fired pottery of the last pre-Han centuries. 1919 *Proc. Soc. Antiquaries Scotl.* LIII. 24 There is no doubt that they were used as such by the pre-hispanic tribes. 1931 *Times Lit. Suppl.* 9 Apr. 281/1 Pre-Hispanic Art in Argentina. 1963 *Times* 9 Feb. 9/7 One of the greatest obstacles to the dispersal of the thick mystery which hangs over Peru's prehispanic past is the fact that the later civilizations apparently had no form of writing. 1978 *Guardian Weekly* 26 Feb. 18/3 The pre-Hispanic Caribbean. 1959 *Encounter* July 68/1 With a broad pre-Hitlerian gesture Napoleon destroyed her book. 1934 C. Lambert *Music Ho!* iii. 224 The earnest and thoroughgoing sense of sin that gave such a peculiar flavour to pre-Hitlerite night life. 1958 *Electroencephalogr. & Clin. Neurophysiol.* X. 223/2 As a rule, toward the end of the attack, the record is characterized by the return of the pre-ictal spike complex. 1969 W. Pryse-Phillips *Epilepsy* v. 17 Grand mal fits..may be preceded by pre-ictal symptoms such as tiredness, irritability, etc. 1963 *Times* 9 Feb. 9/7 In the Machu Picchu area, at San Pedro de Cacha, an intact group of thousands of Incaic and pre-Incaic tombs, some dating from about 1200 B.C., was found only a few weeks ago by the Peruvian archaeologist, Dr. Manuel Chavez Ballón. 1934 A. Huxley *Beyond Mexique Bay* 251 Karl Marx went out imaginatively into the revolutionary future, Ruskin and William Morris into the pre-industrial past. 1956 R. Redfield *Peasant Society & Culture* 71 He saw chiefly what had come into those villages from preindustrial Europe. 1969 G. Greene *Trav. with my Aunt* i. xiv. 136 Horses moved slowly along, dragging harrows. We were back in the pre-industrial age. 1977 P. Johnson *Enemies of Society* vii. 93 If there had been a powerful ecological lobby in eighteenth-century England..the Western world would still be condemned to pre-industrial living standards. 1957 O. R. McGregor *Divorce in England* iii. 75 The peasant or domestic worker's family of pre-industrialist days..was largely a self-sustaining economic unit. 1977 *Language* LIII. 330, I will outline B's analysis of the variation in the form of the pre-infinitival complementizer in Guyanese decreolization. 1939 L. H. Gray *Foundations of Lang.* 202 In this absence of declension we may..see another survival of the pre-inflexional stage. 1976 *Archivum Linguisticum* VII. 163 In this paper, having accepted a pre-inflexional phase of IE[*sc.* Indo-European],..he tries to reduce the processes which give origin to the alternations I am studying. 1964 *Language* XL. 255 In speech development of the child, we can..establish a preintellectual stage. 1977 H. G. Burger in B. Bernardi *Concept & Dynamics of Culture* 433 Up to a certain point in time, then, a child exercises two separate mental stages: pre-intellectual and prelinguistic. 1968 *Language* XLIV. 84 The border between words constitutes a potential pause point. The features manifested at this prejunctural spot are correlated with the difference between rapid speech forms and deliberate speech forms. 1972 *Ibid.* XLVIII. 411 Prejunctural *-j* is subsequently vocalized along with the fronting of stressed *-ō* to *-æ-*, which is unrounded in West Saxon. 1874 W. Wallace tr. *Hegel's Logic* 53 It was..the main question of the pre-Kantian metaphysic. 1947 *Mind* LVI. 164 It is doubtful whether any reader hitherto sunk in dogmatic slumber pre- or post-Kantian would be awakened by these rambling and inconclusive pages. 1951 W. H. Walsh *Introd. Philos. Hist.* vii. 141 Hegel saw the way the abstract conception of reason favoured by Kant and (in general) the pre-Kantian rationalists had been countered by the many philosophies of feeling. 1979 *Studies in Eng. Lit.: Eng. Number* (Tokyo) 172 Dr. Beer's present work..is a valuable attempt to explore Coleridge's poetic intelligence with all its preoccupations coming from the poet's inborn interest in pre-Kantian Idealism and mysticism. 1959 *Ann. Reg. 1958* 88 The legislative assembly's rejection of measures to raise more local revenue illustrated some of the difficulties encountered in development—the pre-Keynesian thinking of many local leaders, [etc.]. 1960 *20th Cent.* Aug. 99 Old-fashioned, pre-Keynesian, *laissez-faire* liberalism. 1933 L. Bloomfield *Language* 373 Pre-Latin *[kolnis] 'hill' gives Latin *collis*. 1975 *Language* LI. 142 *S*-stem nouns like Pre-Latin **douk+es+i* are analogically re-analysed. 1957 C. La Drière in N. Frye *Sound & Poetry* ii. 103 The concord is of natural, or at least prelexical or paralexical, *suggestion* of the sound with its conventional reference. 1971 *Language* XLVII. 319 The Lexical Insertion Rule can insert a predicate into a prelexical terminal string only if the lexical marking of the predicate is not distinct from the feature specification of the prelexical terminal string. 1976 *Archivum Linguisticum* VII. 127 The lexical decomposition of McCawley and others yields several closely related items which differ only in one or more prelexical elements. 1903 *Pop. Sci. Monthly* Jan. 211 A correspondent..wrote to ask if the garden would accept as a gift the large and important collection of pre-Linnean books that it had been his pleasure to accumulate. 1936 *Discovery* Mar. 85/1 There is a rich collection of botanical incunabula, old herbals, and other pre-Linnaean items. 1962 H. R. Loyn *Anglo-Saxon Eng.* vi. 286 Botanists exercise themselves..to give post-Linnaean forms to strongly pre-Linnaean Anglo-Saxon generalized plant names. 1928 *Daily Express* 23 Nov. 10/2 The idea of shutting out extremism by barring its representatives at the ports seems to us as obsolete as pre-Listerian surgery. 1951 White & Hynes *Med. Bacteriol.* (ed. 5) v. 53 In pre-Listerian days the surgeon's scalpel frequently conveyed infection from one patient to another. 1931 G. Stern *Meaning & Change of Meaning* 12 Pre-literary developments are best left aside at first..and ..research should be restricted to periods represented by written texts. 1941 F. Klaeber *Beowulf* (ed. 3) p. lxvi, Of especial interest are the *gefrægn*-formulas, which unmistakably point to the 'preliterary' stage of poetry, when the poems lived on the lips of singers, and oral transmission was the only possible source of information. 1967 *N.Y. Rev. Bks.* 23 Feb. 33/2 The study of technology or town-planning is the same for preliterary or literary societies. 1937 R. H. Lowie *Hist. Ethnol. Theory* (1938) xii. 219 Into the notions about the dead this logical factor does not intrude, hence prelogic here runs riot untrammeled. 1957 H. J. Uldall in Hjelmslev & Uldall *Outl. Glossematics* i. 4 All languages..are based on this participative prelogic. 1897 *Lippincott's Med. Dict.* 824/2 Pre-malignant. 1961 *Lancet* 29 July 250/2 It is suggested that there is a premalignant defect in the genes which control the development of the reticulo-endothelial system, and that the abnormal stem cells derived from these genes undergo further changes and become malignant. 1974 R. M. Kirk et al. *Surgery* iv. 56 Hereditary polyposis..of the colon is pre-malignant. 1920 E. Pound in *Lett. J. Joyce* (1966) III. 9 His misspent premalthusian youth. 1965 P. Goubert in Glass & Eversley *Population in Hist.* xix. 467 Results obtained by the second method [in the study of demography] apply to all 'pre-Malthusian' times. 1963 V. Nabokov *Gift* ii. 119 The river runs into the murk of the prematutinal twilight that still hangs in the gorges. 1905 *Q. Jrnl. Microsc. Sci.* XLVIII. 490 It is evident..that we may group the cells that are produced in the life cycle of an animal or plant into three categories, viz. Premaiotic, Maiotic, and Post-Maiotic respectively. *Ibid.*, The synapsis represents that series of events which are concerned in causing the temporary union in pairs of pre-maiotic chromosomes, previously to their transverse separation and distribution, in their entirety, between two daughter nuclei. 1972 *Genetical Res.* XX. 201 The sensitive stage lay between the last premeiotic mitosis and the start of DNA synthesis. 1956 *Amer. Jrnl. Obstetr. & Gynecol.* LXXI. 1319 The premenarchal girl who produces estrogen but in insufficient quantities to cause bleeding. 1975 G. S. Richardson in J. J. Gold *Gynecologic Endocrinol.* (ed. 2) v. 56/1 The premenarchal ovary is a polycystic ovary with an unscarred 'porcelain' surface. 1956 *Obstetr. & Gynecol.* XIII. 724 (*caption*) Newborn. Childhood. Premenarche. 1958 E. & E. R. Novak *Gynecol. & Obstetr. Path.* (ed. 4) xxxv. 601 (*heading*) Pre-menarche. 1937 *Human Biol.* IX. 27 The positive slope of the regression of chest-width upon chronological age is greater in the case of premenarcheal girls than in the case of post-menarcheal girls. 1943 *Jrnl. Pediatrics* XXII. 529 An early menarche is associated with..greater than average height and greater than average weight during the premenarcheal years. 1942 Mazer & Israel *Diagn. & Treatm. Menstrual Disorders* ii. 35 A study of 175 post-menarchial and 175 premenarchial girls. 1968 M. R. Abell in J. J. Gold *Textbk. of Gynecologic Endocrinol.* ix. 193 (*heading*) Premenarchial endometrium. 1902 W. Bateson *Mendel's Princ. Heredity: A Defence* 7 The cases are all examples of discontinuous variation: that is to say, cases in which actual intermediates between the parent forms are not usually produced on crossing. [*Note*] This conception of discontinuity is of course pre-Mendelian. 1941 J. S. Huxley *Uniqueness of Man* iv. 121 The picture of the hereditary constitution of human groups which can now be drawn in the light of modern genetics is very different from any which could be framed in the pre-Mendelian era. 1977 Kruskal & Mosteller *Representative Sampling* (Univ. Chicago Dept. Statistics) 11 The notion is like the old pre-Mendelian genetic idea of the homunculus. 1966 F. Schurmann *Ideology & Organization in Communist China* 3 Some writers assert that the economy in premodern societies is a subsystem of a larger social system. 1970 I. L. Horowitz *Masses in Lat. Amer.* i. 7 This concept equates 'mass' to forces in the old society (old in the sense of pre-socialist, rather than pre-modern). 1977 *Sci. Amer.* Oct. 96/1 Premodern medicine blamed such failures on sepsis. 1898 L. F. Ward *Outl. Sociol.* v. 112 The evolving intellect throughout all this long pre-social and pre-moral period was exclusively devoted to the egoistic interests of individuals. 1963 L. Kohlberg in *Vita Humana* VI. 13 The six developmental types were grouped into three moral levels and labelled as follows: Level I. Pre-Moral Level. 1973 M. E. Wood *Children* 26 Children are first amoral..than enter a pre-moral stage, when social and authoritarian factors are the main restraints. 1943 *Mind* LII. 19 The pseudo-morality of sanctions lacks the inward reality of morality. It is in fact a sort of pre-morality. 1916 Jordan & Ferguson *Textbk. Histol.* viii. 221 Myeloblast (Premyelocyte; Hemoblast, Mesameboid Cell; Primitive Blood Cell; 'Lymphocyte').— This is the parent blood-cell of bone-marrow. 1931 M. G. Wohl *Bedside Interpretation of Lab. Findings* caption facing p. 88 Abnormal leucocytes:..premyelocyte with beginning neutrophilic granules. 1964 W. G. Smith *Allergy & Tissue Metabolism* iii. 44 It is accepted that eosinophils are differentiated from the premyelocytes of the bone marrow. 1963 *Jrnl. Clin. Path.* XVI. 319 Among the acute leukaemias of the granulocytic group, acute premyelocytic leukaemia is distinguished by the severity of its haemorrhages, the frequency of hypofibrinaemia, a rapidly fatal course, and an unusual cellular hyperplasia. 1938 *New Statesman* 21 May 890/1 A record whose pre-Nazi date can be guessed from the fact that Fritz Busch conducts the Berlin State Opera Orchestra. 1972 P. Black *Biggest Aspidistra* i. iv. 41 *Brigade Exchange*, a war play..created by the pre-Nazi German radio. 1941 *Cancer Res.* I. 45/1 The object of these investigations was the study of the influence of irritation on the first manifestation of neoplastic transformation, which, in the case of the skin, represents the conversion of the preneoplastic thickened epithelium into a papilloma. 1978 *Nature* 13 Apr. 635/2 Electron microscopy has shown that during the latent period only preneoplastic changes resembling those of Bowen's disease are present in rat epidermis. 1946 L. Bloomfield in C. F. Hockett *Leonard Bloomfield Anthol.* (1970) 460 Particles (*prenouns*) appear before nouns in less variety than before verbs [in Algonquian]. 1966 G. N. Leech *Eng. in Advertising* ii. 14 The pre-noun is broken down into four chief secondary classes, determiner.. numeral..adjective..and a certain range of nouns, including proper names and words denoting substances. 1968 R. Kyle *Love Lab.* xviii. 233 Have you ever known an orgasmic woman who wanted to go back to a pre-orgasmic condition? 1976 *Spare Rib* Nov. 16/1, I was going to start by telling you how I came to join the pre-orgasmic group, but I suppose a lot of things preceded that step. 1980 *Time* 28 Jan. 90/1 Psychologists persistently refer to unresponsive women as 'pre-orgasmic'. 1935 *Anat. Rec.* LXIV. Suppl. No. 1. 52 The beginning of the pre-ovulatory enlargement coincides with the beginning of oestrus. 1975 Franchimont & Burger *Human Growth Hormone* ii. iii. 175 During the first 10 days, FSH and LH fell and became stable at values similar to those encountered in the normally cycling female during the preovulatory period. 1899 Pre-paroxysmal [in Dict.]. 1907 A. W. Turner *Epilepsy* vi. 121 A form of pre-paroxysmal psychosis..is the feeling of good spirits and of exceptional well-being, which precedes the onset of an attack in some cases. 1977 *Lancet* 21 May 1095/1 Prepatency was documented in 35 patients who..started to excrete *Giardia*. The median prepatent period was 14 days. 1926 R. W. Hegner in *Q. Rev. Biol.* I. 399/1 The prepatent period extends from the time the infective parasites enter the body of the host until their offspring can be recovered by the usual laboratory methods. 1976 *Nature* 15 July 214/2 This highly virulent strain has a prepatent period of 8–14 d in mice. 1897 A. D. L. Napier *Menopause* iv. 78 For a variable period, usually about two years, the physiological economy is preparing for this pre-pathological change [*sc.* the climacteric]. 1977 M. Edelman *Polit. Lang.* v. 95 Ritualistic categorization further confuses feedback by defining a substantial proportion of the population as either pathological or prepathological. 1941 *Language* XVII. 225 A study of post-pausal and pre-pausal allophones reveals several recurrent differences between these and the corresponding allophones occurring elsewhere than at points of open juncture. 1973 A. H. Sommerstein *Sound Pattern Anc. Greek* v. 160 *Prepausal acute*: This phenomenon may well have nothing to do with the Acute-Grave rule; rather, it may be a feature of sentence intonation, viz. rise at end of phrase. 1948 J. L. Adams tr. *Tillich's Protestant Era* viii. 134 The dark ground of pre-personal being..is effective in every moment of our conscious existence. 1971 *Jrnl. Gen. Psychol.* LXXXIV. 257 The body, therefore, becomes man's expression in the world, the mirror of his being at the prepersonal and preobjective level. 1977 Fontana & van de Water in Douglas & Johnson *Existential Sociol.* iii. 126 For Merleau-Ponty, consciousness was basically the anonymous, prepersonal life of the body-subject. 1933 T. S. Eliot *Use of Poetry* 21 It is true also of the change from a pre-philosophical to a philosophical age. 1939 *Mind* XLVIII. 89 Mr. Loewenberg is a man with a mission: to rescue empiricism (*i.e.* pre-philosophical empiricism) from the empiricists. 1959 J. L. Austin *Sense & Sensibilia* (1962) vi. 55 Our ordinary, unamended, pre-philosophical manner of speaking. 1957 in *Amer. Speech 1972* (1975) XLVII. 222 To be safely pre-phonemic, let us begin with 1874, with Whitney's 'Elements of English Pronunciation', and observe what seemed important to him. 1960 H. M. Hoenigswald

Language Change viii. 73 The doctrine of gradual phonetic change may turn out to be a remnant from pre-phonemic days. **1968** R. A. Lyttleton *Mysteries Solar Syst.* ii. 54 Evidence seems to suggest that the meteorites do not represent original pre-planetary material but are a later product of the solar system. **1978** *Nature* 9 Feb. 504/1 The authors thus invoke another component—a pre-planetary disk which is accreting onto the central star. **1955** Pre-planting [see *post-emergence* s.v. *post- B. 1]. **1976** *Stillwater News* (Absarokee, Montana) 1 July 15/1 Whether pre-planting, post-plant or post-emergence herbicides are used depends upon type of crop, kinds of weeds and other conditions. **1963** Pre-political [see *pre-cultural* above]. **1977** *Jrnl. Politics* XXXIX. 7 Hobbes' resolutive-compositive method is an early and Rawls' original position a late example of the contractarian quest for political authority's pre-political foundations. **1943** Pre-predicative [see *pre-give*, sense *A. 1]. **1950** *Mind* LIX. 264 The first section of Husserl's book deals with what he calls 'pre-predicative' experience. **1971** *Jrnl. Gen. Psychol.* LXXXIV. 259 Merleau-Ponty..spoke more than previously of the interrelation of the predicative and prepredicative levels. **1975** Ld. Hailsham *Door wherein I Went* iv. 15 My formal education began at the age of five ..when I was sent to a pre-prep school in Rosary Gardens. **1960–1** *Where* Winter 16/1 *Pre-preparatory school*, an independent school for children under about 8. **1964** *Economist* 25 Apr. 356/1 Children do enjoy their pre-primary schooling. *Ibid.* 356/2 The pre-primary schools—voluntary to the age of five, then compulsory for one year. **1979** *Jrnl. R. Soc. Arts* CXXVII. 483/1 A task that continues uninterrupted from the pre-primary to the post-graduate class. **1975** *Nature* 11 Sept. 89/2 The existence of an analogous precursor to insulin, a preproinsulin, was strongly supported by the experiments of M. A. Permutt. **1979** *Ibid.* 21 June 675/1 The gene coding for preproinsulin II contained an additional intron of about 500 nucleotides between the region encoding amino acids 38 and 39 of the proinsulin II peptide chain. **1942** *Endocrinology* XXXI. 673 Prepuberal castration prevents the appearance of..physiological and behavioral reactions. **1949** *Radiology* LII. 112/1 The testes of normal chicks of this age are in the prepuberal state. **1977** *Yearbk. Obstetr. & Gynecol.* 329 Seventeen were prepuberal when leukemia was diagnosed. **1942** *Endocrinology* XXXI. 674 Male sexual behavior was shown by all 10 prepuberally castrated females. **1947** *Physiol. Rev.* XXVII. 275 Prepuberally castrated male chimpanzees show much more sexual activity than do similarly operated rodents. **1898** *Amer. Jrnl. Psychol.* IX. 257 The rate of growth.. decreases with fluctuations until about 10 years in girls and 12 years in boys, when the prepubertal acceleration sets in. **1932** S. Zuckerman *Social Life Monkeys* xvii. 267 Louttit holds that the nosing and circling activities of the prepubertal guinea-pig are part of its sexual responses. **1977** *Daily Colonist* (Victoria, B.C.) 19 Mar. 2/1 There may be some prepubertal girls using contraceptive pills 'to be sure'. **1978** *Bull. Amer. Acad. Arts & Sci.* Feb. 22 The eunuch or the prepubertal castrate does not develop the disease. **1937** *Nature* 3 Apr. 589/1 Prepubertally castrated male rats are not known to mate. **1942** *Psychosomatic Med.* IV. 190/1 (*heading*) Prepubertally castrated adults. **1977** *Sci. Amer.* Jan. 100/3 Samaritan religious tradition affords a kind of telescopic glimpse of the past: the ancient Judaism of pre-rabbinical times. **1978** S. H. Hodgson *Philos. of Reflection* I. i. ii. 107 The undistinguished unity of *primary*, pre-reflective, consciousness. **1932** *Mind* XLI. 116 We mean 'taking for granted', that state of pre-reflective unquestioning assurance. **1966** E. S. Casey tr. *Dufrenne's Notion of A Priori* 97 Pre-reflective thought experiences the *a priori* in the *a posteriori*. **1970** J. M. Edie *Patterns of Life-World* xviii. 339 Chomsky's.. own investigations lead us to what Merleau-Ponty termed the 'pre-reflexive' structures of experience. **1976** *Brit. Jrnl. Sociol.* XXVII. 360 Speaking can organize a separate mode of experience which is not simply a transmutation or decipherment of prepredicative thought. It has its own distinct phenomenological properties which exceed those of the prereflexive world. **1923** H. L. Brose tr. *Sommerfeld's Atomic Struct. & Spectral Lines* iv. 211 Coulomb's law is valid and likewise ordinary (pre-relativistic) mechanics. **1952** Koestler *Arrow in Blue* vi. 51 In pre-Relativistic days it was still just possible for the non-specialist to keep abreast of general developments in science. **1900** *Q. Jrnl. Microsc. Sci.* XLIV. 3 Reproductive period.—I have used this expression to denote the whole of that period in the life of a mammal, whether male or female, during which its generative organs are capable of the reproductive function; and in contrast to the Pre-reproductive and Post-reproductive periods which severally precede and follow it, during which the generative organs are either not fully developed or are degenerate. **1952** *New Biol.* XIII. 27 In ourselves the length of the prereproductive period of life has increased. **1977** J. L. Harper *Population Biol. of Plants* 231 The pre-reproductive period of trees is usually long. **1937** *N. & Q.* 6 Feb. 91/2 A pre-Romanesque church at Quintanilla de las Viñas. **1939** *Burlington Mag.* Mar. 110/1 Stone-sculptors of the pre-romanesque epoch lacked tradition and experience. **1907** H. M. Chadwick *Origin Eng. Nation* iv. 74 It is held that the remains..date from pre-Saxon times. **1962** H. R. Loyn *Anglo-Saxon Eng.* i. 7 Big rivers such as the Thames..preserve their pre-Saxon names. **1958** L. Bellak *Schizophrenia* i. 34 The pre-schizophrenic personality may be sociopathic, infantile,.. brilliantly highstrung. **1964** C. M. Thompson in M. R. Green *Interpersonal Psychoanal.* xxxiii. 315 So much for the outward picture of preschizophrenia. *Ibid.* 316 It is characteristic of the preschizophrenic that all true object relationships are impossible. **1965** A. F. Korner in B. I. Murstein *Handbk. Projective Techniques* ii. 29 Conversely, frank psychotics..often produce Rorschachs that reflect less of a schizophrenic process than do records of pre-schizophrenics. **1897** *Lippincott's Med. Dict.* 825/2 *Presenile*, occurring before old age: as, *presenile alopecia* or baldness. **1903** *Lancet* 22 Aug. 517/2 The patients in the severe cases are men as a rule in the pre-senile stage and they present well-marked cardio-vascular lesions. **1912** *Brit. Med. Jrnl.* 16 Nov. 1379/1 He has hitherto limited the term 'melancholia' to cases occurring at the climacteric and the pre-senile period of life. **1913** *Jrnl. Nervous & Mental Dis.* XL. 386 Further investigation of the types of mental makeup out of which an involutional depression may develop..may throw light on the peculiar combination of symptoms seen in the presenile psychosis. **1950** D. B. Kirby *Surg. of Cataract* x. 193/2 The presenile soft cataract up to the fourth decade may be handled by discission. **1976** F. Warner *Killing Time* i. vi. 17 Impaired mental and motor functions, presenile dementia, usually followed by death. **1962** E. Snow *Red China Today* (1963) lxxiii. 564 All three were born the 'year of the flood', 1949, and had no memory of a presocialist China. **1965** B. Pearce tr. *Preobrazhensky's New Econ.* 88 A country like the U.S.S.R...must pass through a period of primitive accumulation in which the sources provided by pre-socialist forms of economy are drawn upon very freely. **1973** *Sci. Amer.* Apr. 61/3 The molecules might be formed in the dense environs of a 'presolar nebula', that is, in the final phases of the collapse of a protostar into a self-luminous star. **1979** *Nature* 15 Feb. 556/1 According to the second model..the presolar grains condensed in a late supernova that triggered the collapse of the solar nebula. **1937** *Sci. & Society* I. 156 We have seen that pre-Soviet scholarship..had given a theoretical basis for the treatment of any language as potentially capable of developing any expressions required of it. **1949** M. Mead *Male & Female* xi. 230 In pre-Soviet Russia there seems to have been extraordinarily little valuation placed on women's child-bearing character. **1957** R. N. C. Hunt *Guide to Communist Jargon* xxiii. 81 Where pre-Soviet expansionist policy was concerned, the party line changed in the middle 'thirties. **1952** C. Payne-Gaposchkin *Stars in Making* (1953) ii. 43 Nebulae like the one in Orion must represent the primitive prestellar material. **1978** *Sci. Amer.* Apr. 112/1 The observational signpost for this prestellar stage is the emission by the cloud's many molecules of radio waves at wavelengths measured in millimeters. **1958** C. Rabin in *Aspects of Translation* 130 Pre-Structuralist works on grammar recognized this fact by separating the description of forms ('accidence') from the discussion of their meaning, which appeared as part of 'syntax'. **1961** *Brno Studies in English* III. 11 Cases.. were decidedly unknown to pre-structuralist study of language. **1933** L. Bloomfield *Language* 220 Some suffixes have *pre-suffixal stress*: the accent is on the syllable before the suffix. **1977** *Language* LIII. 10 If a German were to create a new word by adding the suffix *-ig* 'ish' to *Kind* 'child', the result would be [kindiç] with presuffixal *d* as in [kindər]. **1973** *Publ. Amer. Dial. Soc.* LX. 34 In such dialects, the /r/ is treated as if it were presyllabic, as an apical alveolar consonant. **1951** *Dorland's Med. Dict.* (ed. 22) 1212/1 *Presymptomatic*, existing before the appearance of symptoms. **1966** *New Society* 12 May 7/2 Donaldson's finding that one in six 'healthy' people has presymptomatic disease. **1978** *Nature* 9 Nov. 173/1 Bubbles moving in blood vessels have been successfully detected by the Doppler method in experiments confirming their significance and presymptomatic existence. **1951** N. Goodman *Struct. Appearance* I. i. 22, I use 'presystematically' for 'according to ordinary usage'. *Ibid.* 23 Care..must be exercised..when one word..has both a systematic and a presystematic use. For example, 'is a member of' is used in several different presystematic ways and also as the mere verbal reading of the systematic sign 'ε'. **1964** E. Bach *Introd. Transformational Gram.* v. 92 With..the free use of any presystematic knowledge of the language, we choose the analyses which maximize the independence of the classes set up. **1975** *Jrnl. Philos.* LXXII. 552 Presystematically, the physicalist ontological position is simply put: 'Everything is physical.' **1959** *Listener* 17 Sept. 429/1 The rings were unknown in pre-telescopic times. **1976** *Jrnl. R. Soc. Arts* CXXIV. 564/1 There were even observatories in pre-telescopic times. **1966** C. G. Hempel *Philos. Nat. Sci.* vi. 75 While the internal principles of a theory are couched in its characteristic *theoretical terms*.., the test implications must be formulated in terms..which are 'antecedently understood',..terms that have been introduced prior to the theory and can be used independently of it. Let us refer to them as *antecedently available* or *pretheoretical terms*. **1966** Y. Bar-Hillel in *Automatic Transl. of Lang.* (NATO Summer School, Venice, 1962) 8 So far, I have been using 'syntactic complexity' in its pretheoretical and unanalysed vague sense. **1968** A. J. Ayer *Origins of Pragmatism* 335 So far as anything can be, qualia are pre-theoretical. **1953** H. A. T. Reiche et al. tr. *Jaspers's Tragedy is not Enough* i. 31 Pre-tragic knowledge is rounded out, complete, and self-contained. **1957** N. Frye *Anat. Crit.* 210 The Greek *ananke* or *moira* is in its normal, or pre-tragic, form the internal balancing condition of life. **1960** H. Read *Forms of Things Unknown* xi. 177 From this point of view Stoicism is more complete; and above all that serene code of ethics achieved in ancient China—'the feeling of security without the shadow of tragedy, a natural and sublime humanity, a sense of being at home in this world, and a wealth of concrete insights', to quote Jaspers' own description of pre-tragic knowledge. **1965** N. Chomsky *Aspects of Theory of Syntax* 213 In the syntactic component of this (pretransformational) grammar, indices on category symbols were used to express agreement..but not subcategorization and selectional restrictions. **1973** *Word* 1970 XXVI. 396 Yet perhaps all we have to do is to rephrase Cohen's distinction between *langue bébé* and *langue adulte* by calling the former 'pre-transformational language'. Then, there would still remain the necessity of explaining how..a child suddenly employs transformations. **1908** *Westm. Gaz.* 8 July 12/2 Translated from Pre-Vedic Sanskrit. **1935** G. K. Zipf *Psycho-Biol. of Lang.* (1936) iv. 160 In pre-Vedic Sanskrit, the accent might be on the last syllable. **1905** F. E. Clements *Res. Methods Ecol.* 321 Prevernal, pertaining to early spring. **1908** *Science* 7 Feb. 207/1 Overtopped by the autumnal, the sublayers are successively those of the serotinal, estival, vernal and prevernal. **1926** [see *aspect *sb.* 14]. **1960** N. Polunin *Introd. Plant Geogr.* x. 285 In temperate forests..the seasonal aspects are apt to be important—in particular the prevernal (i.e. before spring) one of herbs which flower before the shading tree-leaves expand. **1933** Blunden *Charles Lamb* 193 It was Lamb's instinctive utterance of indignation against the spirit of the pre-Victorians, the tendency to make a boudoir or a Persian heaven. **1964** D. Owen *Eng. Philanthropy* (1965) 5 Victorians and pre-Victorians agitated..for more efficient employment of..charitable trusts. **1973** M. R. Booth *Eng. Plays of 19th Cent.* III. 25 The materialism of the age, its social ambition and self-seeking drive..are topics not really explored in pre-Victorian comedy. **1979** 'J. Gash' *Grail Tree* i. 17 Ceramics and pre-Victorian tapestries.

2. a. With sbs. or phrases forming *quasi*-adjs. or attrib. phrases. *pre-Broadway, -cession, -Christmas* (examples), *-Civil War, -coition, -college, -computer, -consonant, -crusade, -development, dinner, -employment, -enclosure, -examination, -game, -Inca, increase, -independence* (examples), *-invasion, -jazz, -launch, -legislation, -liberation, -life, -London, -lunch, -machine, -market, -marketing, -oidium, -ovulation, -pause, -phylloxera, -pottery, -qualificative, -radio, -recognition, -Reformation* (later example), *-relativity, -Renaissance* (examples), *-retirement, -revolution, -season, -seizure, -show, -sleep, -subject, -television, -theatre, -tour, -vowel, -work, -world.* **b.** With personal names. *pre-Hitler.* **c.** Adjs. of the type in 2 a, b above are sometimes used adverbially (cf. *pre *prep.*), as *pre-emergence* below, *pre-tax, *pre-war *advbs.*

pre-eme·rgence, occurring, performed, or applied before the emergence of seedlings from the soil; also *absol.* and as *adv.*; **pre-fla·me**, occurring in a gas flow before it reaches a flame.

The examples are arranged in alphabetical order and not, as in Dict., chronologically.

1977 *Times* 19 Nov. 13/3 The musical itself is definitely an oddity of the form, and comes here with less than triumphant pre-Broadway credits. **1920** *Chambers's Jrnl.* 13 Nov. 786/2 The natives obtained, individually or communally, land to which in the pre-cession days they could not have established a claim. **1925** T. Dreiser *Amer. Trag.* (1926) I. ii. xxvii. 338 Clyde..was invited by her to attend a pre-Christmas dance. **1976** *Morecambe Guardian* 7 Dec. 5/2 Getting party games organised is one of those pre-Christmas chores. **1961** *Georgia Rev.* Spring 10 In the pre-Civil War years, the South argued that the slave was not less humanely treated than the factory worker of the North. **1966** *Eng. Stud.* XLVII. 154 Professor Parry is at his best when he is dealing with New York, either the pre-Civil War capital or the tumultuous city of the 1920's. **1953** N. Tinbergen *Herring Gull's World* iv. 120 Head-tossing is the main part of the pre-coition behaviour. **1957** R. K. Merton *Student-Physician* 122 The greater intimacy between fathers and sons during the precollege years. **1960** *Farmer & Stockbreeder* 29 Mar. 39/1 (Advt.), Pre-College Student, not under 18,..to acquire sound agricultural background. **1976** *Sci. Amer.* Apr. 34/2 It was in this atmosphere that the National Science Foundation precollege curricula in biology and the social sciences became the focus of extended and bitter controversy. **1961** *Times* 3 Oct. (Computer Suppl.) p. viii/3 A most efficient system of manufacturing, restocking and transport had been devised in the pre-computer days. **1949** E. A. Nida *Morphol.* (ed. 2) ii. 20 The allomorphs are listed in a structurally corresponding fashion. First is given the preconsonant form and secondly the prevowel form. **1977** *Time* 7 Nov. 61/1 The message and the methods are modeled after those of Billy Graham, down to precrusade organization (by a staff of 27) and convert counseling. **1945** *Times* 29 June 5/6 Then there was at least a scheme in the pre-development stage to provide the V2 rocket with wings, which had great possibilities. **1942** C. Milburn *Diary* 25 Dec. (1979) 162 The sherry..was our pre-dinner appetiser. **1963** L. Deighton *Horse under Water* xiv. 60 We went back to H.K.'s for pre-dinner drinks. **1968** 'H. Pentecost' *Gilded Nightmare* (1969) i. iii. 46 The Trapeze Bar..is a predinner meeting place for the very rich. **1978** M. Gilbert *Empty House* x. 87 Roger ..was relaxing with his pre-dinner drink. **1940**, etc. Pre-emergence [see *post-emergence* s.v. *post- B. 1]. **1971** *Arable Farmer* Feb. 15/3 Tri-allate applied pre-emergence to wheat to control wild oat. **1977** J. L. Harper *Population Biol. of Plants* 112 In addition a 'safe site' is one from which specific hazards are absent—such as predators, competitors, toxic soil constituents and pre-emergence pathogens. **1942** *Amer. Rev. Tuberculosis* XLV. 643 The increasing adoption of preëmployment X-ray examination. **1949** H. C. Weston *Sight, Light & Efficiency* vii. 225 Whenever a pre-employment test..is applied the examinee should wear any glasses he is in the habit of wearing. **1971** *Flying* (N.Y.) Apr. 113/3 (Advt.), Airline employment test... Pre-employment tests. **1934** Webster, Pre-enclosure. **1945** H. J. Massingham in F. Thompson *Lark Rise to Candleford* p. ix, Intact from a pre-industrial and pre-Enclosure past. **1957** *Brit. Med. Jrnl.* 7 Sept. 551/2 Pre-examination strain can be defined as the condition wherein the nervous tension is of such a quality that it diminishes the efficiency of study and impairs the prospects of success. **1978** S. Allan *Inside Job* i. 17 She giggled nervously in the way Sheila remembered from pre-examination tension at school. **1924** *Colliery Guardian* CXXVII. 1443/2 Experiments show that contact with a heated surface may act in two ways: generally the ignition point is raised by the absorption of heat due to pre-flame combustion on a surface large enough to be only slightly affected itself. **1973** Boldt & Griffiths in J. P. Allinson *Criteria for Quality of Petroleum Products* v. 59 Amongst the main preflame products are the highly temperature sensitive peroxides and if these exceed a certain critical threshold concentration, the end gas will spontaneously ignite before the arrival of the flame front emanating from the sparking plug: this causes detonation or 'knocking'. **1951** *Time* 26 Feb. 78 C.C.N.Y., the heavy pre-game favorite each time, lost to Missouri (54–37), Arizona (41–38) and Boston College (63–59). **1976** *Billings* (Montana) *Gaz.* 17 June 1-H/5 Saturday's game at the Metra begins at 8:00 p.m., and pregame coverage goes on

the air at 7:30. **1977** J. F. Fixx *Compl. Bk. Running* xxiii. 258 If at some football training tables the pregame meal is still steak, it is only because common sense is too often no match for tradition. **1938** E. Waugh *Scoop* i. iv. 74 A volume of pre-Hitler German poetry. **1960** *News Chron.* 4 May 5/6 Old Berlin songs that recalled carefree pre-Hitler days. **1908** *Encycl. Relig. & Ethics* I. 469/2 (*heading*) The pre-Inca people. **1950** J. H. Steward *Handbk. S. Amer. Indians* VI. 533 In both *Inca* and pre-*Inca* Coastal sites there is found..a good deal of cotton in the seed. **1974** *Encycl. Brit. Macropædia* XIV. 133/1 The dryness of the central and southern coasts has preserved the remains of a long succession of pre-Inca peoples. **1976** *Evening Post* (Bristol) 23 Apr. 21/2 (Advt.), Mini 850, antique gold, at pre-increase price. **1977** *Horse & Hound* 14 Jan. 44/3 (Advt.), New Rice eventer at pre-increase price, £666 on the road. **1960** *Daily Tel.* 9 July 6 The example of the Congo strengthens the case.. for taking every possible precaution in our own territories to maintain law and order in the tense pre-independence days. **1977** *Time* 15 Aug. 15/1 In preindependence days, Makarios battled the British with the legendary Colonel George Grivas. **1944** *Hutchinson's Pict. Hist. War* 27 Oct. 1943–11 Apr. 1944. 413 (*caption*) A U.S. officer pointing out a target to general Eisenhower during pre-invasion manoeuvres by an American Armoured Unit in England. **1967** *Freedomways* VII. 111 He knows the bombing and shelling and mining we are doing are part of traditional pre-invasion strategy. **1926** Whiteman & McBride *Jazz* vii. 157 The foreign market for American music in pre-jazz times was poor. **1934** C. Lambert *Music Ho!* iii. 198 The melodic shape is clearly the most important factor in pre-jazz popular music. **1959** 'F. Newton' *Jazz Scene* iv. 68 Most of the material thus collected was 'pre-jazz'. **1963** *IEEE Trans. Product Engin. & Production* VII. iv. 39/1 The Surveyor spacecraft is subjected to several combinations of environment during prelaunch, boost, free flight, retro, landing and lunar operation. Vibration and shock are negligible during prelaunch operations because of careful handling. **1967** A. Battersby *Network Analysis* (ed. 2) 303 He then goes on to build up the pre-launch stock, advising M as soon as the required stock level has been reached. **1970** N. Armstrong et al. *First on Moon* ii. 45 The conversation..as always during a pre-launch breakfast, was studiedly casual. **1967** *Listener* 13 Apr. 481/1 What I should like to see is more consultation with Members of Parliament before legislation is prepared: what I call pre-legislation committees. *a* **1974** R. Crossman *Diaries* (1976) II. 528 A few weeks ago Roy Jenkins wrote me a long minute to say that he couldn't permit a pre-legislation committee on privacy. **1962** E. Snow *Red China Today* (1963) ii. 25 Near the Hsin Ch'iao I picked one [*sc.* a two-seater pedicab] up, pumped by a neatly dressed gray-haired gentleman who said he pulled a rickshaw at the old Peking Hotel in the 'pre-liberation' days. **1974** tr. *Wertheim's Evolution & Revolution* 287 In pre-liberation China religion in the Confucianist form was associated with the establishment. **1937** R. A. Wilson *Birth of Lang.* iv. 99 The formative energy which produced the tree must..have been latent in the pre-life matter. **1958** *Observer* 15 June 13/4 Scientists believe that pre-life processes may be occurring on the moon. **1962** F. I. Ordway et al. *Basic Astronautics* vi. 258 The search for evidence of extraterrestrial life, or at least prelife carbon compounds. **1967** J. B. Davis *Petroleum Microbiol.* ii. 19 There is no evidence of pre-life organic matter being incorporated in the sedimentary environment of geologic formations. **1959** P. Bull *I know Face* vi. 102 The Leeds incident occurred quite late in the pre-London tour. **1961** *Nottingham Even. Post* 28 July 12 The play is on its pre-London tour. **1962** G. Butler *Coffin in Oxford* iv. 64 They had..made a film..and had now moved.. to Oxford for the pre-London run. **1938** D. Kincaid *Brit. Social Life in India* xii. 276 The elderly gathered together in the clubs for pre-lunch drinks. **1955** A. Ross *Australia 55* 131 This was the first of four successive gloomy pre-lunch sessions for England. **1974** *Times* 4 Nov. 14/4 An appropriate pre-lunch appetizer. **1957** K. A. Wittfogel *Oriental Despotism* i. 13 Temperature and surface are the outstanding constant elements of the agricultural landscape. This was true for the pre-machine age; and it is still essentially true today. **1970** G. E. Evans *Where Beards wag All* xvi. 177 All, or nearly all, of the old terms connected with the pre-machine farming in the region are no longer used. **1963** *Wall St. Jrnl.* 9 Oct. 3/1 The council..proposed 'premarket' safety testing of cosmetics. **1977** *N.Y. Rev. Bks.* 14 Apr. 37/2 The modifications of law which constitute the subject of his book are elements of what Karl Polanya called the 'great transformation' from a pre-market to a market society. **1960** *Farmer & Stockbreeder* 5 Jan. 95/1 Results on farms throughout the country confirm the evidence of pre-marketing trials that Dictol will protect animals against husk. **1920** G. Saintsbury *Notes on Cellar-Bk.* i. 7 This was pre-oidium and pre-phylloxera wine. **1922** *Proc. Soc. Exper. Biol. & Med.* XIX. 380 Coincident with ovulation in the pigeon there occurs an increase of the blood sugar to 25 per cent. or more above the pre-ovulation value. **1975** *Ann. Human Biol.* II. 325 Variations in the pre-ovulation interval are also indicated by the timing of mid-cycle hormonal peaks. **1934** M. K. Pope *From Latin to Mod. French* ii. xvii. 222 The praepause form of the word, the one with sounded consonant, was retained very generally. **1953** *Language* XXIX. 419 There is agreement that there are pitch factors in at least two different kinds of pre-pause terminals ('terminal junctures'). **1920** Pre-phylloxera [see **pre-oidium*]. **1957** R. Campbell *Portugal* 53 We may never hope to taste again the crowning glories of the best pre-phylloxera vintages. **1972** *Country Life* 25 May 1309/3 Christie's will sell over 100 small lots of mid-19th century port and pre-phylloxera claret. **1949** W. F. Albright *Archaeol. Palestine* iii. 62 In the pre-pottery Neolithic Age man took an important forward step in the Near East. **1960** K. M. Kenyon *Archaeol. in Holy Land* ii. 45 It may be inferred with a high degree of probability that this Pre-Pottery Neolithic A settlement of Jericho was based on a successful system of agriculture. **1977** G. Clark *World Prehist.* (ed. 3) ii. 51 Phases II and III of the Mesolithic period in the Levant, commonly termed 'Pre-pottery Neolithic A and B' in the literature. **1924** H. E. Palmer *Gram. Spoken Eng.* 183 (*heading*) Adverbs in the pre-qualificative position... These immediately precede the qualificative... They also precede any other adverb they may modify. **1946** R. Blesh *Shining Trumpets* (1949) x. 220 The first hot records..sold by the millions and, in those preradio days, disseminated jazz more rapidly.. than a score of travelling bands. **1949** Bruner & Postman in Bruner & Krech *Perception & Personality* (1950) 26 Differential availability [of response systems]..leads to certain characteristic 'normalizing' prerecognition responses in our incongruity experiments. **1970** *Jrnl. Gen. Psychol.* LXXXIII. 24 As a corollary of lower recognition thresholds with increased information, we can expect fewer prerecognition responses. **1929** Pre-Reformation [see *pre-Renaissance* below]. **1920** A. S. Eddington *Space, Time & Gravitation* ix. 149 Action is one of the two terms in pre-relativity physics which survive unmodified in a description of the absolute world. **1946** *Mind* LV. 161 A pre-relativity physicist could use the figure..by interpreting AM and TM as curves of velocity. **1929** T. S. Eliot *Dante* i. 19 A directness of speech which Dante shares with other great poets of pre-Reformation and pre-Renaissance times. **1976** R. Pfeiffer *Hist. Classical Scholarship 1300–1850* i. 21 There seems to be a slight shifting of emphasis to the advantage of the classics, inconceivable in pre-Renaissance times. **1961** A. Heron *Solving New Probl.* 21 Does the evidence obtained support a rationale for adapting a pre-retirement planning and preparation programme to the needs of older employees of different occupational levels? **1965** J. Pollitt *Depression & its Treatment* vii. 91 Older patients have greater difficulty than those of pre-retirement age in re-adjusting their lives after illness. **1976** *Evening Post* (Nottingham) 16 Dec. 2/6 Support for a pre-retirement course run by Gedling Borough Council was so good that plans for a second session are already in the pipeline. *a* **1902** S. Butler *Way of All Flesh* (1903) xiv. 63 The pre-revolution French peasant. **1939** H. Nicolson *Diary* 3 Apr. (1966) 394 Apparently many of their [*sc.* the Polish army's] guns are pre-Revolution guns of the Russian Army. **1978** N. Marsh *Grave Mistake* iii. 91 A pre-revolution Russian stamp that was withdrawn on the day it was issued. **1961** *Dallas Morning News* 10 Oct. 2–2 It looks as if coach Hank Stram's men will meet the Bills just as they are developing into the kind of team they were expected to be in pre-season reckonings. **1970** N. Armstrong et al. *First on Moon* vii. 144 In sports, the Houston Oilers are showing plenty of enthusiasm in their early preseason workouts. **1975** *Cricketer* May 27/2 D. J. Insole will be giving all first-class umpires a pre-season briefing. **1979** *N.Y. Post* 10 Aug. 17 In one of our most bizarre pre-season presidential campaigns, an incumbent President is being dismissed by both the opposition and large sectors of his own party as a non-person. **1926** Rows & Bond *Epilepsy* iv. 87 In the pre-seizure period the disturbances of consciousness often commence with a slight difficulty in the power of attention and pass through the stages of dreamy states and fugues to complete unconsciousness. **1966** *Jrnl. Neurol., Neurosurg. & Psychiatry* XXIX. 253/2 The E.E.G. appeared normal on all the pre-seizure tracings. **1926** *Glasgow Herald* 19 Oct. 9 Before the contests, there are whispers of excellent pre-show performances of competing cows. **1960** *Farmer & Stockbreeder* 2 Feb. 84/1 Sons of several famous bulls were in competition in the showyard and their fortunes were the subject of considerable pre-show speculation. **1964** *Language* XL. 269 Nobody has made a thorough study of presleep soliloquies before. **1970** N. Armstrong et al. *First on Moon* xiii. 333 We're standing by for an exciting evening of TV and a presleep report. **1924** H. E. Palmer *Gram. Spoken Eng.* ii. 182 Adverbs in the pre-subject position. **1961** R. B. Long *Sentence & its Parts* xx. 471 No comma is used after pre-subject adjunct clauses functioning as clause markers in assertives. **1976** *Archivum Linguisticum* VII. 32 It looks like the same pre-subject position that we called I in the 'kernel' form of the sentence. **1965** *B.B.C. Handbk.* 28 Even though BBC radio's evening audience is much less than it was in pre-television days it is by no means inconsiderable. **1969** *Listener* 15 May 700/3 Sponsorship has entered into the scheme of things. So has advertising, which was never around in pre-television days. **1974** *Times* 17 Aug. 7/4 The older school of comedians, the pre-television comics. **1953** R. Fuller *Second Curtain* v. 74 The place was filling up: a few were eating pre-theatre meals. **1967** A. Bailey in L. Deighton *London Dossier* 49 This is really a lunch or pre-theatre restaurant, since it closes at 8.30 p.m. **1977** *Rolling Stone* 7 Apr. 30/2 Pre-tour jitters are an occupational hazard the McGarrigle sisters have avoided up to now. **1949** E. A. Nida *Morphol.* (ed. 2) ii. 16 Word-initial prevowel glottal stops. **1977** *Time* 26 Dec. 41/2 The message rings out, too, at the early morning pre-work prayer meetings held by businessmen. **1923** D. H. Lawrence *Birds, Beasts & Flowers* 99 Fishes, With.. their pre-world loneliness.

II. 3. In adjs. (also sometimes used as sbs.), chiefly *Anat.* and *Zool.* Also occas. with sbs. directly forming sbs., as *prealbumin. **pre-hy·oid**, in front of the hyoid bone; **premo·tor**, applied to the anterior part of the precentral area of the frontal lobe of the brain, which is concerned with the co-ordination of activities in the motor area immediately posterior to it; **pre-palatal** (further examples); *spec.* in *Phonetics*, of a consonant articulated with obstruction of the airstream immediately in front of the palate; **prepatellar** (earlier example); **prepatellar bursitis**, inflammation of the prepatellar bursa; = *housemaid's knee* s.v. Housemaid c in Dict. and Suppl.; **prespira·cular**, in front of a spiracle.

1949 I. F. & W. D. Henderson *Dict. Sci. Terms* (ed. 4) 351/1 *Prehyoid*, mandibulo-hyoid; *appl.* cleft between mandible and ventral parts of hyoid arch. **1974** D. & M. Webster *Compar. Vertebr. Morphol.* vii. 129 In living amphibians the hypobranchial muscles can be divided into a prehyoid and a posthyoid group. **1932** *Brain* LV. 534 In the baboon forced grasping appeared five to six days after removal of the motor and premotor areas. **1978** *Sci. Amer.* Oct. 52/2 (*caption*) The premotor area is involved in complex motor activity such as operating a typewriter. **1902**, etc. Pre-palatal [see *medio-palatal* adj. s.v. *medio- 2]. **1925** P. Radin tr. *Vendryès's Language* i. i. 23 We distinguish..the pre-palatals and the post-palatals. **1934** Priebsch & Collinson *German Lang.* ii. i. 88 The *s* was more prepalatal. **1958** J. Berry in J. A. Fishman *Readings Sociol. of Lang.* (1968) 741 Is it better..that all related languages of southern Ghana write the prepalatal affricate 'tʃ' uniformly so, or (under cultural pressure of the trade language), 'ch'? **1964** *Archivum Linguisticum* XVI. 22 The affricate *ch* [č] or the corresponding pre-palatal *x* [š]. **1973** *Amer. Speech* 1969 XLIV. 265 An *r* produced by passing the breath between the underside of the apex of the tongue and the postalveolar or prepalatal region. **1977** *Word* 1972 XXVIII. 248 Caballero..described the [ɾ] as a voiced prepalatal. **1882** C. B. Nancrede in J. Ashhurst *Internat. Encycl. Surg.* II. 717 (*heading*) Pre-patellar bursa. **1902** R. T. Frank tr. *Albert's Diagnosis Surg. Dis.* xxxiii. 370 Prepatellar bursitis requires but casual mention. A strictly circumscribed, elastic tense swelling directly in front of the patella is characteristic. **1927** W. C. Campbell *Orthopedics of Childhood* xii. 217 Pre-patellar bursitis is commonly due to excess kneeling. The symptoms are similar to those of bursitis elsewhere. **1964** *Australasian Post* 21 May 13, I rushed off in anguish and looked up prepatellar bursitis in a medical dictionary. It was.. housemaid's knee. **1902** *Nature* 16 Oct. 604/1 The last-mentioned [*sc.* the chorda tympani] is spoken of as..prespiracular in position. **1975** *Ibid.* 10 Apr. 483/3 Patterson confirms the homology between the spiracular groove of primitive actinopterygians..and the 'pre-spiracular groove' of rhipidistians.

pre (prī), *prep.* [A further development of *pre- B. 2 c; cf. *post *Lat. prep.* 6.] = Before *prep.* 8.

Usu. found in contexts where *before* would be equally appropriate and more agreeable.—Ed.

1973 G. Sims *Hunters Point* xiii. 119 'Have you tried phoning David's friends in Los Angeles?'.. 'They are all pre my era and I don't know their names.' **1975** H. Kissinger in *Dept. of State Bull.* 6 Oct. 532 Pre my being in office; those decisions were made in the previous Administration.

pre-accentual: see *pre- B. 1 d.

preach, *sb.*[2] Colloq. abbrev. of Preacher. *U.S.*

1968 D. Wilkerson *Hey, Preach—you're comin' Through!* 9 He grabbed my arm and blurted: 'Hey, Preach—you're comin' through!' **1969** C. F. Burke *God is Beautiful, Man* (1970) 96 Ananias..puttin' his hands on him like the preach down at the revival camp does.

preach, *v.* Add: **1. c.** Phr. *to preach to the converted*: to commend an opinion to those who already assent to it.

1867 Mill *Exam. Hamilton's Philos.* (ed. 3) xiv. 319 Dr. M'Cosh is preaching not only to a person already converted, but to an actual missionary of the same doctrine. **1916** G. Saintsbury *Peace of Augustans* iii. 144 One may be said to be preaching to the converted and kicking at open doors in praising..the four great novelists of the eighteenth century. **1971** *It* 2–16 June 14/4 The problem is as usual that one tends to be preaching to the converted —so the important thing is to make sure that people who don't know are informed.

preacher. Add: **4. preacher-man** *U.S. dial.*, = sense 1 a.

1899 in H. Wentworth *Amer. Dial. Dict.* (1944) 474/1. **1913** H. Kephart *Our Southern Highlanders* xiii. 286 Everywhere in the mountains we hear of biscuit-bread.. preacher-man, granny-woman. **1977** *Times* 23 May 5/1 A nice, homespun preacherman who spoke with a Southern drawl.

preachify, *v.* Add: Hence **prea·chifying** *ppl. a.*

1916 W. Owen *Let.* 13 July (1967) 399 His dogmatic, pig-headed, preachifying, self-sufficient manners and domineering tone. **1978** J. Anderson *Angel of Death* vii. 70 I'm not a great admirer of..paternalistic, preachifying Christianity.

preaching, *vbl. sb.* Add: **3.** *preaching-place* (later examples), *-stand, -stole*; **preaching-cross** (later examples).

1845 A. Wiley in *Indiana Mag. Hist.* (1927) XXIII. 37 Many new neighbors were taken in as preaching places. **1848** *Wesleyan Missionary Notices* VI. 164/1 In my last I expressed a desire..to open a preaching-place in a mountain district [of Jamaica]. **1856** Mrs. Stowe *Dred* I. xxiii. 314 The assembly poured in and arranged themselves before the preaching-stand. **1857** P. Cartwright *Autobiogr.* viii. 85 We took in a new preaching-place at a Mr. Moor's. **1953** M. Powys *Lace & Lace-Making* vi. 67 A small piece of lace like a straight collar about seven inches long and one and a half inches wide could be used as a 'protective' for a preaching stole. **1959** C. L. Wrenn *Word & Symbol* (1967) 23 The association of the saintly first preachers of Christianity in religious memory with the 'preaching crosses', which the missionary first set up at his oratory, is a well-known feature of the early Celtic Church. **1960** *Church & People* Nov.–Dec. 182 Africans flocking to our Mission churches and preaching places. **1970** M. Swanton *Dream of Rood* 13 It is clearly a preaching cross. Its message is evangelical, stating the role of Christ in the world of men. **1972** *Country Life* 17 Feb. 408/1 One such building at Sare dedicated to Saint Francis Xavier has a most unusual statue in coloured wood showing the saint in cassock, surplice and preaching stole.

preaching-house. (Earlier example.)

1747 J. Wesley *Jrnl.* 2 Nov. (1912) III. 321 Mr. J. Richards had just sent his brother word that he had hired a mob to pull down his preaching-house that night.

preachy, *a.* (Further examples.)

1955 A. Huxley *Let.* 29 May (1969) 748 If I seem to be smug and preachy, forgive me. **1966** *Word Study* Feb. 5/2 Getting so dogmatic, or preachy, or stuffy that our students rebel against us. **1978** *Time* 3 July 15/2 He is tiresomely preachy in his talks with non-Israeli leaders, repeating to the point of boredom his odd fact-and-fiction litany of Jewish biblical and legal rights, his self-justification for Irgun atrocities and his blend of self-righteous arrogance.

pre-Adamite, *sb.* and *a.* Add: **A.** *sb.* **4.** *N.Z.* An inhabitant of Canterbury Province before the settlement of 1850.

1930 L. G. D. Acland *Early Canterbury Runs* 1st Ser. i. 3 The old 'Pre-Adamites'..were those who had bought land from the New Zealand Company and settled here before the Canterbury settlers arrived. **1949** A. H. Reed *Story of Canterbury* iii. 55 To the Hays and Sinclairs and other 'pre-Adamites'—as those few who had arrived before the 'Pilgrims' came jocularly to be called—these ships represented shops, schools, churches, roads and other amenities of which they had for so long been deprived. **1977** *N.Z. Herald* 8 Jan. 1-6/6 Any reader whose antecedents were among these 'Pre-Adamites', as they are called, is invited to send their names..to [address given].

B. *adj.* **1.** (Further example.)

1916 *Nature* 25 May 259/2 For imitation, a pre-Adamite simian character, plays no small part in the ostensible development, mental, moral, and otherwise, of gregarious folk.

pre-adamitism (later example).

1880 A. Winchell *Preadamites* p. iii, The central idea of the work is human preadamitism.

preadapt (prī,ădæ·pt), *v.* [Pre- A. 1.] *trans.* To adapt beforehand; *spec.* in *Biol.*, to adapt (an organism) for life in conditions not yet available to it. Hence **preada·pted** *ppl. a.* [in *Biol.* tr. F. *préadapté* (L. Cuénot *La Genèse des Espèces Animales* (1911) iv. 291)].

1849 [in Dict. s.v. Pre- A. 1.] **1915** *Eugenics Rev.* VII. 56 By being warm-blooded, mammals and birds are enabled to maintain their normal activity throughout a wide range of temperature, and they may therefore be said to be preadapted to all temperatures within that range. **1947** *New Biol.* III. 90 For the most part cave animals are drawn from groups habitually living in damp, dim places, such as under stones or at the bottoms of streams; they are, so to speak, preadapted to life in caves before they enter them. **1952** *Ibid.* XIII. 25 An animal adapted to living in a small isolated volume would be preadapted to captivity. **1969** J. M. Weller *Course of Evol.* ix. 467 The evolutionary development..of these advanced and more complex feathers, whose original function was insulation, preadapted the very early birds for flight. **1970** T. H. Eaton *Evolution* viii. 121 In retrospect we can say that some thecodonts were 'preadapted' in certain ways for the life of birds. **1976** *Sci. Amer.* Aug. 38/2 Such a trend, may, however, have helped to preadapt the Egyptians to a ready acceptance of food production later.

Hence **pre,ada·ptive** *a.*, causing or characterized by preadaptation.

1915 *Eugenics Rev.* VII. 50 One can call indifferent or semi-useful, characters in a species which become evident adaptations on removal to a new habitat or on the acquirement of new habits, preadaptive or prophetic characters, or more briefly, preadaptations. **1944** G. G. Simpson *Tempo & Mode in Evol.* vi. 186 The direct development of adaptations in one environment may be preadaptive for another. **1969** J. M. Weller *Course of Evol.* i. 20 Potentially preadaptive mutations..are likely to accumulate within a population as recessives.

pre,adapta·tion. [Pre- A. 2.] Adaptation beforehand; *spec.* in *Biol.* [tr. F. *préadaptation* (L. Cuénot *La Genèse des Espèces Animales* (1911) iv. 306)], the possession or acquisition by an organism of heritable features which adapt it to an environment or mode of life which only later becomes available to it.

1886 [in Dict. s.v. Pre- A. 2]. **1915** *Eugenics Rev.* VII. 56 Versatility is an attempt at universal preadaptation, indeed at complete independence of particular circumstances. **1934** *Biol. Abstr.* VIII. 289/2 'Preadaptation': the occupation of empty regions is made by neighbouring spp. already prepared in the sense of having a necessary and sufficient adaptation. **1942** *Tee Emm* (Air Ministry) II. 144 Wear the special pre-adaptation goggles which will provide you with your hour's synthetic night before you tackle the real one. **1953** G. G. Simpson *Major Features Evol.* vi. 188 The term 'preadaptation' has been applied to a great variety of real or supposed evolutionary phenomena, from the appearance of a small mutation with selective value in the population in which it occurs to the sudden appearance of a form monstrous in its parental population but miraculously, one might almost say, adapted to some quite different way of life. **1978** *Sci. Amer.* Sept. 51/1 If a favored mutation does appear, it can be viewed as exhibiting a 'preadaptation' to that particular environment: it did not arise as an adaptive response but rather proved to be adaptive after it appeared.

Hence **pre,adapta·tional** *a.*, pertaining to or characterized by preadaptation.

1940 R. Goldschmidt *Material Basis Evol.* iii. 151 A return of the subspecies at one extreme end toward the starting point could only be accomplished by retracing the steps of preadaptational mutation to its original condition. **1944** G. G. Simpson *Tempo & Mode in Evol.* vi. 188 The field naturalist..is not likely to be satisfied in such cases with the preadaptational axiom that animals enter a new environment simply because they can.

pre-address, see *pre- A. 1. **pre-adjectival, -adjunct:** *pre- B. 1 d, A. 2.

preadmission. Add: **B.** *adj.* [Pre- B. 2.] Prior to admission.

1971 *Mod. Law Rev.* XXXIV. 642 The universities are to gain in both responsibility and autonomy..by assuming responsibility for pre-admission vocational training.

pre,adole·scent, *a.* and *sb.* [Pre- B. 1.] **A.** *adj.* **a.** Of a child: having nearly reached the beginning of adolescence. **b.** Of or pertaining to the two or three years before the beginning of adolescence.

1910 A. C. Perry *Probl. Elementary School* x. 201 It is probably true that the preadolescent girl can pursue her school work side by side with the boy without the slightest danger. **1925** *Arch. Neurol. & Psychiatry* XIV. 215 Since the patient is usually preadolescent, a separation of the cranial sutures..is likely to be present. **1935** E. Bowen *House in Paris* ii. i. 81 His pre-adolescent mind. **1949** M. Mead *Male & Female* xi. 232 The charming street-dance in which a little pre-adolescent girl dances to delight the men of the village. **1976** Deakin & Willis *Johnny go Home* vi. 85 Happier pre-adolescent family holidays.

B. *sb.* A preadolescent child.

1930 K. McHale *Pre-Adolescence* 3 Most people look upon the average pre-adolescent as one who has his second teeth..and who is not yet burdened with any difficult adjustments. **1951** *Child Devel.* XXII. 15 The opportunity to observe the play configurations of pre-adolescents was offered by the Guidance Study at the Institute of Child Welfare, University of California. **1960** *20th Cent.* Nov. 434 The idea of the unfallen, pre-adolescent has exercised an extraordinarily strong appeal. **1973** *Nature* 27 Apr. 582/3 A pabulum of romanticized science digested to gibberish for consumption by pre-adolescents.

Hence **pre,adole·scence**, the preadolescent period or stage of development.

1930 K. McHale *Pre-Adolescence* 3 The brain..reaches nearly maximum size during the years of pre-adolescence. **1949** E. B. Hurlock *Adolescent Devel.* i. 4 Because boys mature slightly later than girls, we may regard their preadolescence as extending from 11 to 12½ or 13, early adolescence from 13 to 17, and late adolescence from 18 to 21 years. **1972** *Sci. Amer.* July 76/2 She was able to follow the same child from birth to preadolescence.

pre-adult, -adverbial, -agonal, -agonic, -agricultural: see *pre- B. 1.

prealbumin (prī,æ·lbiŭmin, -ælbiū·min). *Biochem.* [Pre- B. 3 in Dict. and Suppl.; so called because it appears slightly in front of albumin during electrophoresis.] A plasma protein with an electrophoretic mobility slightly greater than that of albumin; *spec.* a tetramer in human blood which binds thyroxine and the retinol-binding protein.

1955 O. Smithies in *Biochem. Jrnl.* LXI. 634/2 The two components migrating more rapidly than the broad albumen zone are referred to as the pre-albumins, and, (, indicating the faster-moving component). **1959** *Nature* 3 Oct. 1067/2 (*heading*) Separation of prealbumins by starch gel electrophoresis. **1975** F. W. Putnam *Plasma Proteins* (ed. 2) I. ii. 72 Prealbumins have been described in other species such as the mouse; the molecular weight is only about 20,000 and the function is unknown. **1976** *Sci. Amer.* Sept. 58/2 When dietary protein intake is deficient, the two proteins that play a role in the transport of vitamin A (retinol-binding protein and prealbumin) are not made by the liver in adequate amounts.

pre-Alfredian: see *pre- B. 1 a.

preamble, *v.* Add: **II. 4.** (Later examples.) Also *transf.*

1951 W. Sansom *Face of Innocence* iv. 45 She might think this was a trick of Harry's to get her away with him, to preamble the marriage-bed. **1980** *Time* 28 Jan. 90/1 Nouns continue to be overrun by the jargonaut: the New York *Times* demands stronger sourcing, meetings are preambled, situations are impacted.

pre-amp (prī·æmp), abbrev. of *preamplifier.

1957 *Practical Wireless* XXXIII. 573/1 The pre-amp and output stages make use of highly efficient miniature valves. **1968** *Times* 29 Nov. (Sound of Leisure Suppl.) p. ii/2 The cost of stereo at its best is off-putting (a second matching speaker, stereo amplifier and pre-amp., stereo head, cartridge and stylus—your old turntable will still suffice). **1975** *Hi-Fi Answers* Feb. 81/1 (Advt.), Leak 12 watt mono valve amplifier, with Rogers pre-amp. £8. **1978** *Nature* 18 May 218/2 A preamp was mounted directly on the base of the detector (bias ~ 100V) and its output was fed to an amplifier.

pre,a·mplifier. [Pre- A. 2.] An amplifier designed to amplify a very weak signal (as from a microphone, pickup, or similar source) and deliver it to another amplifier for further amplification.

1935 *Television* VIII. 654/1 This pre-amplifier brought R4 signals up to consistent full loud-speaker strength. **1952** *Electronic Engin.* XXIV. 498/1 The low output signal obtained from a pick-up necessitates the use of a pre-amplifier. **1957** *New Yorker* 2 Nov. 95/2 Bogen RR550 DeLuxe FM-AM Receiver with built-in Preamplifier and Power Amplifier. **1965** *Wireless World* July 33 (Advt.), Transistor A.C. Microvoltmeter... Ideal as an oscilloscope preamplifier as it has a low noise level, low microphony, no hum and is independent of a mains supply. **1977** *Gramophone* June 117/1 Between the preamplifiers and the power amplifiers there are four Urei 527A ⅓-octave equalizers.

pre,a·mplify, *v.* [Pre- A. 1.] *trans.* To subject to a preliminary amplification; to amplify in a preamplifier. Hence **prea·mplified** *ppl. a.*

1959 *Geophysics* XXIV. 750 The reflected signal detected by the hydrophone is pre-amplified, passed through a variable passive filter, then amplified and printed on the recorder. **1978** *Nature* 13 July 135/1 After cell contact, lock-in amplifiers sample preamplified cell potentials..at selectable phase angles. *Ibid.* 14 Sept. 111/2 The signals were preamplified and detected by two automatic gain-controlled receivers.

Hence **pre,amplifica·tion**, the action or result of preamplifying.

1960 *Practical Wireless* XXXVI. 397/1 The amplifier has sufficient gain for many purposes with no pre-amplification. **1971** *Melody Maker* 13 Nov. 36/5 A new concept in sound engineering..is the Mark 3 Jam-Pak series, a unique idea in pre-amplification systems. **1977** *Rolling Stone* 13 Jan. 58/2 Most ribbon mikes produce lower output signals than most dynamic types and require more preamplification.

pre-anæsthe·tic, *a.* (*sb.*) *Med.* [Pre- B. 1.] **a.** Occurring before the introduction of anæsthetics into surgical practice.

1892 [in Dict. s.v. Pre- B. 1]. **1916** P. J. Flagg *Art of Anæsthesia* i. 1 The pre-anæsthetic period ends and the anæsthetic period begins with the discovery of ether in 1842 and its general introduction in 1846.

b. Used or carried out as a preliminary to the induction of anæsthesia. Also as *sb.*, a drug so used.

1930 [see *Nembutal]. **1934** *Current Res. Anesthesia & Analgesia* XIII. 169 (*heading*) Pernoston as a preanesthetic in surgery. **1957** S. M. Brooks *Basic Facts Pharmacol.* ii. 76 The most popular preanesthetic is a combination of morphine and atropine. **1974** A. Freeman in Lichtiger & Moya *Introd. Practice of Anesthesia* xxvii. 334 A barbiturate combined with a narcotic and a belladonna alkaloid produces suitable preanesthetic sedation in most children over 1 year of age.

Hence **pre-anæsthe·tically** *adv.*

1952 V. J. Collins *Princ. & Pract. Anesthesiol.* vii. 64 His work indicated one of the primary objects in the use of morphine preanesthetically namely, depression of basal metabolic rate and oxygen consumption.

pre-analytic(al, -ally: see *pre- B. 1.

pre-a·nimism. [Pre- B. 1 in Dict. and Suppl.] Primitive belief that certain powers exist in material objects. Cf. next.

1918 A. A. Brill tr. *Freud's Totem & Taboo* (1919) iii. 152 We have practically no further knowledge of pre-animism, as no race has yet been found without conceptions of spirits. **1937** *Nature* 27 Nov. 923/1 Dr. Robert Ranulph Marett..is best known as an anthropologist, the formulator of the theory of preanimism in the study of primitive religion. **1956** E. E. Evans-Pritchard *Nuer Relig.* xiii. 311 Many such origins have been propounded: magic, fetishism, manism, animism, pre-animism. **1963** S. Fuchs *Origin of Man & his Culture* xviii. 232 Andrew Lang maintained that pre-animism took two forms, that of magic and that of primitive monotheism.

pre-animi·stic, *a.* [Pre- B. 1.] Applied to a stage of religious culture, presumed to precede animism, at which the power or spirit attributed to a material object was believed to exist in the object.

1900 R. R. Marett in *Folk-Lore* XI. 162 (*title*) Pre-animistic religion. *Ibid.* 170, I propose that we examine a few typical cases of Powers, which, beneath the animistic colour that..has more or less completely overlaid them, show traces of having once of their own right possessed pre-animistic validity as objects and occasions of man's religious feeling. **1918** A. A. Brill tr. *Freud's Totem & Taboo* (1919) iii. 152 Our psychoanalytic view here coincides with a theory of R. R. Marett, according to which animism is preceded by a pre-animistic stage. **1949** Koestler *Insight & Outlook* xii. 175 Members of such very early, preanimistic types of society may be described as living in a state of original self-transcendence of consciousness. **1963** S. Fuchs *Origin of Man & his Culture* xviii. 232 It was soon necessary to assume a pre-animistic stage of religion.

pre-apprehend, -Arnoldian: see *pre- A. 1, B. 1 a.

pre-arrange, *v.* (Earlier example.)

1811 Jane Austen *Sense & Sens.* III. vii. 133 With.. the service pre-arranged in his mind, he offered himself as the messenger.

pre-article: see *pre- B. 1. **preaspirate:** *pre- A. 1. **preaspiration:** *pre- A. 2. **pre-assemble:** *pre- A. 1.

pre-asse·mbly, *sb.* and *a.* **A.** *sb.* [Pre- A. 2.] Preliminary assembly. Also *attrib.*

1958 *Engineering* 31 Jan. 140/2 From the fabrication shed the steel is moved over into the adjacent pre-assembly shed. *Ibid.* 140/3 The bulk of pre-assembly in the yard..is by welding.

B. *adj.* [Pre- B. 2.] Prior to assembly.

1977 *Design Engin.* July 37/3 This means that operations and parts are eliminated and that pre-assembly tolerances of components are relaxed.

preassure, *v.* (Earlier example.)

1697 J. Sergeant *Solid Philos.* 294 Being pre-assur'd the Thing has more Modes in it than we know.

pre-ato·mic, *a.* [Pre- B. 1.] Existing or occurring before the utilization of atomic energy or atomic weapons; characteristic of such a time.

1914 H. G. Wells *World set Free* iii. 141 Originally he had been something of a thinker upon international politics,..but the atomic bombs had taken him by surprise and he had still to recover completely from his pre-atomic opinions. **1945** R. A. Knox *God & Atom* ii. 28 The ladder that is meant to climb heaven from our front door-step climbs it, instead, from a period world which only history recaptures for us. It is definitely pre-Atomic. **1945** *N.Y. Times* 12 Aug. 8 E/4 To talk about limited, almost parochial pre-atomic subjects, there is the important war of 1914–18. **1956** A. Toynbee *Historian's Approach to Relig.* 215 It [*sc.* the world] has already experienced two devastating pre-atomic wars in one lifetime.

pre-audit: see *pre- A. 1, 2. **pre-auditor:** *pre- A. 2. **pre-Augustinian:** *pre- B. 1 a.

pre-auri·cular, *a. Anat.* [Pre- B. 3.] **1.** [ad. mod.L. (*sulcus*) *præ-auriculāris* (coined in Ger. by T. Zaaijer 1866, in *Natuurkundige Verhandelingen te Haarlem* XXIV. 28).] Designating a groove situated immediately anterior to the inferior margin of the auricular surface of the ilium, and better developed in the female than in the male.

[**1886** *Rep. Sci. Results Voy. H.M.S. Challenger* XVI. xlvii. 55 The most distinct examples of the sulcus præauricularis were found in the pelves of the Sandwich Island women.] **1897** *Lippincott's Med. Dict.* 822/2 Preauricular sulcus. **1911** *Anatomischer Anzeiger* XXXIX. 17 If now a dissection of the sacro-iliac joint is made in a woman, the praeauricular sulcus is seen to be entirely filled with ligamentous fibres. **1914** J. E. Frazer *Anat. Human Skeleton* v. 132 The pre-auricular groove..is as a rule only found on female bones. **1974** *Amer. Jrnl. Physical Anthropol.* XLI. 381/1 The pre-auricular groove of the ilium is usually cited as one of the morphological features to be noted when sexing human pelves.

2. Situated in front of the ear(s).

1901 *Amer. Anthropologist* III. 38 Preauricular. **1934** *Jrnl. Anat.* LXVIII. 533 Pre-auricular fistulae and pre-auricular appendages..lie along the line of the first pharyngeal depression. **1970** *Amer. Jrnl. Ophthalm.* LXXII. 798 (*caption*) Pre-auricular lymph node showing secondary malignant melanoma.

pre-bacteriologic(al), -baiting: see *pre- B. 1, A. 1.

prebiolo·gical, *a.* [Pre- B. 1.] Existing or occurring before the appearance of life; pertaining to the origin of life; = *prebiotic *a.*

1960 *Science* 22 July 200/3 Calvin's experiments have been criticized on the basis that the prebiological atmosphere contained only a small proportion of carbon dioxide. **1971** I. G. Gass et al. *Understanding Earth* ix. 123 Concentration of these compounds by geological processes established reservoirs of prebiological food. *Ibid.*, A prebiological environment. **1973** C. Sagan *Cosmic Connection* vii. 54 The search for prebiological organic chemicals on the Moon, on Mars, or on Jupiter is of great importance in understanding the steps leading to the origin of life.

So **prebio·logy**, the study of the origin of life, and of conditions before this; **prebio·logist**, a specialist in this.

1963 S. W. Fox in I. A. Breger *Org. Geochem.* ii. 40 The reaction mixture used represents..material which has significance in the context of prebiology, due particularly to the recent emphasis by Revelle..on carbon monoxide in the primitive atmosphere. **1971** I. G. Gass et al. *Understanding Earth* ix. 137/1 All the experimental evidence of prebiology points to the fact that the carbonaceous products were synthesized from gases. **1974** *Nature* 8 Mar. 180/1 Eight planetary astronomers..were assembled to meet ten practitioners interested in the origins of life (three exobiologists, two microbiologists, four chemical prebiologists and one palaeontologist).

prebio·tic, *a.* [Pre- B. 1.] = *prebiological *a.*

1958 *New Statesman* 23 Aug. 214/2 The dust is prebiotic, that is, before the formation of life, but with all the biochemical elements, from which life originated, waiting to be 'organised' into a chemistry of reproduction. **1962** F. I. Ordway et al. *Basic Astronautics* vi. 250 Lightning.. may have originally been responsible for the creation of the first organic molecules from prebiotic molecules. **1968** *New Scientist* 29 Aug. 437/1 Studies on prebiotic synthesis—the way in which the chemicals of life may have originated on the primitive earth—generally have an esoteric ring to them. **1973** B. J. Williams *Evolution & Human Origins* vii. 94/1 It was a closed system containing only water, the gases assumed to be present in the prebiotic atmosphere, and an energy source.

preboa·rd, *v.* [Pre- A. 1.] *trans.* To shape by heating on a form before dyeing rather than after.

1940 *Rayon Textile Monthly* Apr. 59/1 Nylon, the new synthetic yarn..has added a new operation to the manufacture of stockings... It has been found necessary to 'set' the shape of the stocking before it is dyed... At first, various methods of 'pre-boarding', or 'setting', the stockings were tried. **1950** B. E. Hartsuch *Introd. Textile Chem.* xi. 330 When full-fashioned nylon stockings are preboarded, they are mounted on metal forms of suitable shape, and these forms are heated externally with live steam for about 1 to 3 minutes. **1963** R. W. Moncrieff *Man-Made Fibres* (ed. 4) xliv. 660 The hosiery were seamed, pre-boarded 5 min. in 27 lb./in.² steam, scoured in 2 per cent sodium lauryl sulphate + 2 per cent caustic soda. They were dyed gold, finished with a standard resin emulsion hosiery finish, and postboarded dry at 200°C. **1970** *Ibid.* (ed. 5) xlvii. 809 The cycle of operations is normally (1) pre-board, (2) dye, (3) post-board, (4) dry.

Hence **preboa·rded** *ppl. a.*, **preboa·rding** *vbl. sb.*

1940 *Rayon Textile Monthly* Apr. 59/2 Various types of pre-boarding equipment. *Ibid.* 60/1 It is not necessary to lay the pre-boarded stocking out flat, as it still has to be dyed. **1953** K. H. Inderfurth *Nylon Technol.* viii. 208 Placing the gray stockings onto the preboarding form.. must be carefully and accurately performed in order to ensure satisfactory fit, wearing qualities, and sales appeal. **1964** E. R. Trotman *Dyeing & Chem. Technol. Textile Fibres* (ed. 3) xxiii. 512 In the case of pre-boarding, the most convenient time for the operation is before scouring.

pre-book: see *pre- A. 1.

Pre,bo·real, *a.* Also **pre-Boreal, preboreal.** [Pre- B. 1.] Applied to a European climatic period that followed the Arctic and preceded, or marked the transition to, the Boreal period, and was characterized by the spread of birch and pine forests. Also *absol.*

1924 *Jrnl. Linn. Soc.* (*Bot.*) XLVI. 497 (*caption*) Pollen-spectra... Holland (S.W. Sweden) (preboreal spectrum). **1934** *New Phytologist* XXXIII. 285 Erdtman suggests that it is partly on account of *P. montana* pollen that the pine maximum in many parts of the continent falls so much earlier than in England (i.e. in the Early Boreal or pre-Boreal). **1956** A. L. Armstrong in D. L. Linton *Sheffield* vi. 101 The known [Mesolithic] moorland sites are usually upon patches of old land surface.., or in the eroded banks of streams and cloughs, where their pre-Boreal age is testified by remains of the Pennine forest lying in the peat above them. **1976** *Sci. Amer.* Feb. 94/2 Sometime between 8300 and 8000 B.C. the cold Late Dryas period was superseded by a warmer period known as the Preboreal. Pine and birch forest reinvaded the European plain from the south, and the reindeer herd, following the retreating tundra, moved off to the north and northeast.

pre-Broadway: see *pre- B. 2 a.

Pre-Ca·mbrian, *a. Geol.* Also **pre-Cambrian, Precambrian.** [Pre- B. 1.] Of, pertaining to, or designating the time before the beginning of the Cambrian period and Palæozoic era, present-day rocks of which are marked by an almost complete absence of fossils. Also *ellipt.*, the Pre-Cambrian rocks or period.

1864 J. W. Salter in *Geol. Mag.* I. 290 The author suggests that the syenitic trap of St. David's is a part of the old pre-Cambrian land. **1875** [in Dict. s.v. Pre- B. 1]. **1910** Lake & Rastall *Text-bk. Geol.* xvii. 292 Everywhere in the British Isles there is a marked unconformity between the pre-Cambrian and the later deposits. **1915** C. Schuchert *Text-bk. Geol.* II. xxvii. 540 The result.. serves to make prominent the two most significant and distinctive features of the pre-Cambrian: (1) the wide-spread crustal revolutions,..and (2) the profound depth to which erosion has planed. **1921** A. W. Grabau *Textbk. Geol.* II. xxx. 198 (*heading*) Summary of the pre-Cambrian of the Canadian region. **1951** Auden *Nones* (1952) 71 That the red pre-Cambrian light Is gone like Imperial Rome Or myself at seventeen. **1959** *Times* 18 June (Queen in Canada Suppl.) p. xiii/2 The gnarled Precambrian surface of the Canadian Shield, the low hills of the Mackenzie valley are all foreign to the Yukon. **1969** E. W. Morse *Fur Trade Canoe Routes of Canada* v. 57 At the upper end of Lac des Alumettes the voyageurs came.. into close contact with the granite of the Precambrian Shield, rising straight from the water. **1971** *Nature* 25 June 498/1 The Pre-Cambrian covers 85% of the total length of geological time and is the longest geological division. **1977** *Jrnl. R. Soc. Arts* CXXV. 406/1 Uranium deposits..occur in well defined provinces mainly in Pre-cambrian terrain.

pre-cancer, -capitalist(ic): see *pre- B. 1.

precarial (prĭkēᵊ·riăl), *a. rare.* [f. Precary *sb.* + -al.] Of or pertaining to a precary.

1914 *Eng. Hist. Rev.* Jan. 137 It is also highly probable that precarial transactions were instrumental not only in the bringing together of ecclesiastical property, but also in utilizing it by means of dependent farms.

pre-Caroline: see *pre- B. 1 a.

precast (prī·kast), *a.* [Pre- A. 1.] **a.** Formed by casting before being placed in position; composed of units so made. **b.** Pertaining to or involving such a process.

1914 C. E. Fowler *Pract. Treat. Sub-Aqueous Foundations* xxvii. 553 The use of pre-cast piles for permanent construction at a reasonable cost is an ideal type. **1927** *Daily Express* 2 Mar. 3/6 The Concrete Products Association was formed yesterday to improve production..of 'pre-cast' units in building. **1932** Dowsett & Bartle *Pract. Formwork & Shuttering* ii. 28 A series of rolled steel filler joists arranged to carry a pre-cast floor. **1934** *Archit. Rev.* LXXV. 15/2 The floor is in pre-cast rose and green terrazzo tiles. **1960** *Times* 6 Oct. 7/4 It may be possible to make such roofs in plastic by precast methods. **1975** *Princ. Quality Concrete* (Portland Cement Assoc.) ii. 31 British engineers were experimenting with precast residential structures as early as 1905. Following World War II, precast construction became a permanent part of the European building trades.

precast (prīka·st), *v.* [Pre- A 1.] *trans.* To cast (an object, or concrete) before it is placed in position. Also *absol.*

Some passive uses are not distinguishable from a predicative use of *precast* adj.

[**1929** W. C. Huntington *Building Construction* iii. 71 Concrete piles may be precast or cast-in-place.] **1938** Wentworth-Shields & Gray *Reinforced Concrete Piling* ix. 81 The use of concrete piles which, instead of being pre-cast, are made by forming a hole in the ground and filling it with concrete..is steadily extending. **1950** *Archit. Rev.* CVII. 333/3 A large contract, where much of the work can be precast at the site. **1956** *Nature* 4 Feb. 200/1 The art of precasting concrete has given rise to an established industry. **1974** A. Hodgkinson *AJ Handbk. Building Struct.* v. 167/2 The more difficult the in situ formwork problem, the easier is the decision to precast. **1975** *Princ. Quality Concrete* (Portland Cement Assoc.) ii. 31 Since its discovery concrete has been used to precast specialty items.

So **preca·sting** *vbl. sb.*

1938 C. E. Reynolds *Concrete Construction* xi. 484 Where unskilled labour and only a minimum of supervision are available pre-casting may allow the manufacture of the members to be localised and carried out under more convenient conditions. **1974** A. Hodgkinson *AJ Handbk. Building Struct.* v. 167/2 A quite different reason for precasting in certain parts of the UK is the local attitude of carpenters and steel fixers which may cause the contractor to limit absolutely the number of employees on site.

precative, *a.* **a.** (Further examples.)

1845 [see *adhortative *a.*]. **1965** *Language* XLI. 12 *Jeṣam* and *yeṣam* represent full-grade precatives. *Ibid.* 520 A precative middle paradigm is given, but nothing is said about how middle precatives are formed.

precaution, *sb.* Add: **2.** *spec.* a precaution against conception in sexual intercourse; a contraceptive device. Usu. in *pl.*

1935 N. Mitchison *We have been Warned* iv. 419 What did he do to you? Was it—rape?.. Was he using any precautions? **1941** [see *it *pron.* 1 d]. **1968** B. Russell *Autobiogr.* II. ii. 97 From the first we used no precautions. **1969** G. Greene *Trav. with my Aunt* i. x. 98 If we didn't have a child together, it was purely owing to the fact that it was a late love. I took no precautions, none at all. **1975** T. Heald *Deadline* vii. 168 Neither had taken any precautions... Miss Morrison was pregnant.

precautiously, *adv.* Delete *rare* and add later examples.

1921 A. Dobson *Later Ess.* 163 A clever critic once observed of a popular novelist that few writers had better painted the inside of certain characters—adding *precautiously* 'so far as there is any inside'. **1922** Joyce *Ulysses* 534 Bloom. (Reflects precautiously.) **1964** M. Lane *Night at Sea* xii. 212 Ben put his injured hand precautiously behind him.

precedent, *sb.* Add: **5.** *precedent-setting* adj.

1967 *Economist* 14 Oct. 160/3 In Boston an outspoken lawyer, in a precedent-setting challenge to the constitutionality of the Massachusetts laws against marijuana, is asking whether all the fuss over pot is really worth it. **1977** *New Yorker* 19 Sept. 50/2 Some people feel that the conductor was challenging the soloist and that she more than met the challenge, with the result that the last movement is taken at a precedent-setting speed.

precellence. For † *Obs.* read *rare* and add later examples.

Quot. 1958 contains a rendering of Henri Estienne's title *La Précellence du langage françois* (1579).

1958 *Times* 6 May 12/4 When he delivers the Sir Basil Zaharoff lecture on May 20, Professor Alfred Ewert..will take as his subject 'The precellence of the French tongue'. **1978** R. Syme *Hist. in Ovid* vii. 132 There was another side to the oratorical precellence of Messalla Corvinus. Some found him prolix and lacking in bite.

precellular: see *pre- B. 1. **pre-censorship:** *pre- A. 2. **pre-centrifuge:** *pre- A. 1.

precept, *v.* Restrict † *Obs.* to senses in Dict. and add: **3.** *intr.* Of a local authority or similar body: to issue a precept; to make a demand *on* (a rating authority) for funds. Also *trans.*, to take by means of a precept. Cf. Precept *sb.* 4 d. So **prece·pting** *ppl. a.* and *vbl. sb.*

1911 *Encycl. Brit.* XXII. 915/2 To distinguish the rate the name of the precepting authority is frequently added or the purpose for which it is levied specified. **1961** *Times* 30 May (I.C.I. Suppl.) p. xiv/3 The county council precepts 72 per cent of the rates. **1962** L. Golding *Dict. Local Govt.* 310 Authorities which have no rating powers are known as precepting authorities as they issue precepts, i.e., demands on the rating authority or authorities, specifying the amount required in the £ of rateable value to meet their financial requirements. Examples of precepting authorities are county councils.., joint boards, parish councils, burial boards, river boards and port health authorities. **1974** *Daily Tel.* 26 Apr. 2/2 Where percentage rises are especially high one cause is said to be the practice of some old water boards of covering a large part of their expenditure by precepting on the general rates. **1979** *Kensington & Chelsea Newslet.* Oct. 1/3 These precepting authorities are all affected by substantial salary and wage increases themselves, and they therefore are bound to precept upon local authorities a very much greater amount next year than they have in the past.

preceptee (prīsepti·). *U.S.* [f. *PRECEPT(OR + -EE¹.] One who is being trained by a preceptor. Cf. *PRECEPTOR 3.

1974 *Med. Times* (N.Y.) Dec. 62/2 The benefits for the preceptee are substantial. **1975** *Jrnl. Med. Educ.* May 471/2 The conference is designed to accomplish the following objectives:..discuss monitoring and evaluation methodologies appropriate for either daily preceptor–preceptee interaction or faculty–preceptorship interaction.

preceptor. Add: **3.** *spec.* A physician or specialist who gives a medical student practical training. *U.S.*

1803 [in Dict., sense 1]. **1837** R. DUNGLISON *Med. Student* ii. 126 The question;—what subjects the office-student should peruse during his first year... Generally.. the preceptor gives himself but little trouble. **1864** S. CHEW *Lect. Med. Educ.* p. x, Is it necessary to pay attention to Medical Auscultation? My old preceptor considered it wholly useless. **1912** *Cycl. Amer. Med. Biogr.* II. 316/1 On the death of his preceptor, Dr. A. Torrence, [he] succeeded to his practice. **1925** A. FLEXNER *Med. Educ.* v. 107 A mere boy, fresh from school, he attended his preceptor in his office and on his visits. **1937** J. T. FLEXNER *Doctors on Horseback* i. 9 Morgan apprenticed himself to an experienced doctor; there was no other way of studying medicine... Preceptors were limited to repeating what they had learnt from their own preceptors. **1948** *Jrnl. Hist. Med.* Winter 96 He swept out the office, cleaned the instruments, kept the accounts... After three years of this he would, if he had his preceptor's recommendation, appear before three members of the Board of Censors of the County Medical Society. **1959** HAMMOND & KERN *Teaching Comprehensive Med. Care* vii. 82 Each General Medical Clinic student was assigned to a preceptorial group... Two staff physicians were assigned to each group as preceptors. **1976** *National Observer* (U.S.) 16 Oct. 10/3 A third-year Michigan State student who has served under two preceptors.

preceptorship. Add: **b.** The position of one who is being trained by a preceptor (cf. *PRECEPTOR 3). *U.S.*

1970 *Vital Speeches* 1 Aug. 634 In any new graduate education program we might be well advised to emphasize again a preceptorship method of training. **1972** *Science* 27 Oct. 380/2 D.O. (Osteopathy) students begin serious clinical exposure early under preceptorships with D.O.'s in family practice. **1974** *Med. Times* (N.Y.) Dec. 62/1 Students attended medical school and also went through a preceptorship with an experienced physician. **1976** *National Observer* (U.S.) 16 Oct. 10/3 In this preceptorship program, as it's called, medical students spend from 4 to 12 weeks working in a rural or community doctor's office.

precess, *v.* Restrict †*Obs.* to sense defined in Dict. and add: **2.** *intr.* To undergo precession.

1892 A. M. WORTHINGTON *Dynamics of Rotation* xiii. 135 The application of the couple is said to cause the spinning wheel to 'precess'. **1902** *Jrnl. Inst. Electr. Engin.* XXXII. 83 The pull of gravity on a spinning-top does not make it topple over, but makes it precess. **1942** SYNGE & GRIFFITH *Princ. Mech.* xiv. 418 A disk, 6 inches in diameter, is mounted on the end of a light rod 1 inch long and spins rapidly. It precesses once in 15 seconds. **1957** *Endeavour* Oct. 185/2 In each of these levels the nucleus precesses about the direction of H_0, but maintains its correct orientation in the field. **1971** I. G. GASS et al. *Understanding Earth* vi. 90/2 The axis of figure starts to precess about the axis of rotation. **1973** [see *PRECESSION 3c]. **1975** *Nature* 20 Feb. 590/3 When a single *He atom is placed in a magnetic field its nucleus.. precesses about the field direction in one of two permitted states, corresponding to two different energy levels. **1977** A. HALLAM *Planet Earth* 30/2 The satellites precess, or progressively change their orbiting path relative to the Earth's axis, due to these broad variations in the gravity field.

precession. Add: **3. a.** Now also used to denote the motion of the earth itself which manifests itself as the precession of the equinoxes. (Further examples.)

planetary precession, that part of the precession of the earth's axis caused by the gravitational attraction of the other planets.

1863 W. CHAUVENET *Man. Spherical & Pract. Astron.* I. xi. 604 The mutual attraction between the planets and the earth tends continually to draw the earth out of the plane in which it is revolving... The planetary precession is, then, the effect of a motion of the ecliptic upon the equator... The planetary precession does not affect the declination of stars, but changes their right ascensions, their longitudes, and their latitudes. **1913** S. E. SLOCUM *Theory & Pract. Mech.* vii. 430 (*heading*) Precession of the earth. *Ibid.*, An important case of regular precession is that furnished by the motion of the earth. **1926** H. N. RUSSELL et al. *Astron.* I. v. 141 The motion of the ecliptic pole produces..the planetary precession. **1939** SKILLING & RICHARDSON *Astron.* i. 16 Precession does not affect the position of the terrestrial poles upon the earth's surface. **1959** R. H. BAKER *Astron.* (ed. 7) ii. 59 The earth's precession is a slow conical movement of the earth's axis around a line joining the ecliptic poles, having a period of about 26,000 years. **1963** D. ALTER et al. *Pictorial Astron.* (ed. 2) xlvi. 211/2 This gradual north–south drift of the Southern Cross is a consequence of the precession of the earth, which produces a slow movement of the celestial pole among the stars on a circle with a radius of 23½°. **1971** BAKER & FREDRICK *Astron.* (ed. 9) ii. 49 It is the lunisolar precession that has been described... Planetary precession is the effect of other planets on the plane of the equator, so that its intersection with the ecliptic shifts slowly towards the east along the celestial equator. The result of the two precessions is the general precession.

b. The term is now applied to the motion of the body itself as well as to that of the axis of rotation. (Further examples.)

1907 FRANKLIN & MACNUTT *Elements Mech.* vii. 149 The torque required to produce precession of a spinning body depends upon the moment of inertia of the body and upon the angular acceleration which is involved in the continual change of direction of the axis of spin. **1942** *Tee Emm* (Air Ministry) June 56/1 He has a directional gyro—and should have some idea as to its rate of precession. **1958** *Engineering* 31 Jan. 132/3 If the weights are moved from one side of the point of balance to the other the direction of precession is reversed. **1962** F. I. ORDWAY et al. *Basic Astronautics* ix. 372 The antifriction motor..applies an additional torque in the direction of precession to compensate for friction in the bearings.

c. *spec.* The rotation of the spin axis of a nucleus, electron, etc., about the direction of a magnetic or an electric field.

1927 *Physical Rev.* XXIX. 395 Predicted and observed intensity relations for a number of band spectra are in agreement if we assume that σ is an electronic quantum number which is correlated with a precession about the internuclear axis. **1928** H. S. ALLEN *Quantum* xvi. 220 As the electron has a magnetic moment, its axis will experience a precession because of the couple due to its motion in the electric field. **1960** DICKE & WITTKE *Introd. Quantum Mech.* xii. 195 This torque produces a precession of the spin axis about the direction of the magnetic field; in other words, the particle acts like a gyroscope because of its spin angular momentum. **1965** *New Scientist* 1 July 36/3 This precession will alter the average area that the molecule presents to other molecules. **1973** O. HOWARTH *Theory of Spectroscopy* i. 14 The existence of precession explains why even a classical particle with a magnetic moment does not immediately align when put in a magnetic field. It precesses instead.

pre-cession: see *PRE- B. 2a. **pre-chemical:** *PRE- B. 1d.

‖**précieux** (presyö), *a.* [Fr.] = PRECIOUS *a.* 3. Also as *sb.* Cf. PRÉCIEUSE *sb.* (*a.*).

1891 M. S. VAN DE VELDE *French Fiction of To-day* I. iv. 109 A certain *précieux* hyper-refinement. **1939** *Burlington Mag.* Mar. p. xviii/1 The lives of other *précieux* in the stereotyped social and literary intercourse of the Salons. **1951** M. MCLUHAN *Mech. Bride* (1967) 63 Arno, Nash, and Thurber are brittle, wistful little *précieux* beside Capp. **1953** [see *BAROQUE *a.* (*sb.*)]. **1964** *Eng. Stud.* XLV. 111 As a *précieux* poet, the Duke [Orsino] is an accomplished master. **1969** *Listener* 8 May 637/1 There was point in A. C. Benson's defence in 1910 of 'The May Queen', that no *précieux* writer, with a care for his reputation, could have dared to write it... Certainly mid-19th-century literature was not *précieux*: it took risks.

precinct, *sb.* Add: **1. a.** (Further examples.)

1915 W. S. MAUGHAM *Of Human Bondage* xvi. 60 The precincts, with the exception of a house in which some of the masters lodged, were occupied by the cathedral clergy. **1956** *Newsweek* 9 Jan. 66/3 Just a few days before Christmas, nevertheless, the 230 tenants found eviction letters under their doors. In this way they learned that their proud precincts would be converted to house between 1,100 and 1,500 students by 1957. **1961** K. J. FRANKLIN *William Harvey* 59 He was offered an official residence in the precincts of Bart's.

b. (Further example.)

1921 L. STRACHEY *Queen Victoria* 415 For more than half a century no divorced lady had approached the precincts of the Court.

3. Also, a division of a city for the purpose of police control; *ellipt.*, = *precinct-house* (*b*), *-station* (see sense 5 below). *U.S.*

1864 [see *precinct station house*]. **1882** J. D. MCCABE *New York* xxiii. 374 The city is divided into thirty-five precincts, in each of which there is a station-house. **1894** P. L. FORD *Honorable Peter Stirling* 142, I had to go with them..to the precinct and speak to the superintendent. **1953** W. BURROUGHS *Junkie* (1972) ix. 90 They didn't find any junk on him so they took him to the third precinct to 'hold for investigation'. *Ibid.* x. 98 They drove back to the precinct and I was locked in. This time I was locked in a different cell. **1955** W. GADDIS *Recognitions* II. vi. 555 The case you reported to us as sadism and brutality reported by you to this precinct Tuesday December 20 at 10:17 A.M. resulted in false arrest. **1971** N. FREELING *Over High Side* III. 163 Watching..the cops from the ninety-ninth precinct, on the telly. **1974** *Amer. Speech* 1971 XLVI. 83 *Police station*, precinct.

4. A part of a town or community designated for a specific purpose; *spec.* one from which motor vehicles are excluded, esp. to allow pedestrians to shop in safety.

1942 H. A. TRIPP *Town Planning & Road Traffic* vii. 75 A great number of pockets will have been created, each of which will consist of a little local system of minor roads, devoted to industrial, business, shopping or residential purposes... Each pocket represents in its way a separate little community... The best term..seems to be 'precinct'. **1943** FORSHAW & ABERCROMBIE *County of London Plan* 51 Precincts are formed which can be maintained or re-planned as residential communities, business or industrial precincts. **1958** *Listener* 23 Oct. 643/1 The exclusion of wheeled traffic from the main shopping precinct. **1959** *Ibid.* 19 Mar. 509/2 The word 'precinct' implies an area free from through-traffic. *Ibid.* 22 Oct. 674/2 The Stevenage pedestrian precinct. **1961** L. MUMFORD *City in Hist.* ix. 276 In the original layout of the colleges in Oxford and Cambridge, medieval planning made its most original contributions to civic design: the superblock and the urban precinct divorced from the ancient network of alleys and streets.

5. *Comb.*, as (sense 3) *precinct caucus, level*; **precinct captain,** a leader of a political party in a precinct; **precinct court,** a court with jurisdiction over a precinct, subordinate to a county court; **precinct house,** (*a*) the headquarters of an election precinct; also *attrib.*; (*b*) a police station; **precinct sheet,** a register of eligible voters in a precinct; **precinct station** = *precinct house* (*b*) above; also **precinct station house; precinct worker,** one who promotes the interests of a political party in a precinct.

1954 B. & R. NORTH tr. *Duverger's Pol. Parties* I. i. 19 In the United States the caucuses formed at the county or city level co-ordinate the action of the *precinct-captains*. **1956** E. O'CONNOR *Last Hurrah* iii. 45 John, you'll see the precinct captains? **1977** *Time* 3 Jan. 38/2 Daley became a precinct captain at 21. **1976** *New Yorker* 24 May 118/2 In South Carolina precinct caucuses last night, the highest percentage of the votes—forty-seven per cent—was for 'Uncommitted'. **1704** in *N. Carolina Colonial Rec.* (1886) I. 605 Ordered that the Marshall bring forth the body of Tho: Evans to the next pr[e]cinct Court to answer the compl[aint]. **1943** L. E. PRICE in Boatright & Day *Backwoods to Border* 210 [Hooper] drove out to the precinct court in his rubber-tired carriage. **1863** *Rebellion Rec.* V. I. 77 The Mayor of Philadelphia..called upon all able-bodied men to assemble next morning at the precinct-houses of the election districts. **1899** T. HALL *Tales* 171 He did very well to copy off the entries in a precinct house register or to discover the important arrivals at the hotels. **1968** *Globe & Mail* (Toronto) 13 Jan. 25/5 Imagine committing a robbery half a block from a precinct house! **1972** *Village Voice* (N.Y.) 1 June 69/3 Up the street the team went, stapling on posters, casually strolling by the precinct house with the ladder. **1974** *Nation* (Barbados) 10 Mar. 3/5 Mr. Staines wrote his neighborhood precinct house asking that his thanks be passed on to the policeman. **1954** *Newsweek* 26 July 40/3 Latin America must be approached on the 'precinct level'. Labor leaders, teachers, local politicos must be convinced that inter-American cooperation will benefit the little groups. **1957** *Time* (Atlantic ed.) 20 May 13/2 Ike seems to find something distasteful in precinct-level party politics. **1974** *Union* (S. Carolina) *Daily Times* 23 Apr. 2/3 How do you catch the fraudulent? Would there ever be an updated precinct sheet to work from? **1936** J. STEINBECK *In Dubious Battle* ii. 12, I think I'll stop in at the precinct station. She might of got run over. **1975** *New Yorker* 16 June 114/2 The alleged beating two weeks earlier of a twenty-seven-year-old Chinese engineer inside the Fifth Precinct station. **1864** *N.Y. Herald* 4 Apr. 8/3 The body was removed to the Fourth precinct station house. **1952** *Time* 2 June 19/1 His deepest political instinct is party loyalty. From his start as a precinct worker and doorbell pusher in the wards of Cincinnati,.. he has been unmistakably Republican. **1976** *Washington Post* 19 Apr. B 2/3 More recently, Ray Krasnick, who headed the Tydings effort among precinct workers in Prince George's County, became Sarbanes' county co-ordinator.

precinctual (prīsi·ŋktiu,ăl), *a.* [f. PRECINCT *sb.* + *-ual*, perh. after *INSTINCTUAL *a.*] Of, pertaining to, or characteristic of a precinct.

1949 *Archit. Rev.* CVI. 144 The plan is clearly based on the precinctual theory, with interconnected squares, throughout which the pedestrian receives priority. **1960** *Guardian* 15 June 28/4 The road..'would be prejudicial to the precinctual character of the area'. **1965** *New Statesman* 30 July 146/1 There would be time to consider whether the Martin plan's obsession with 'enclosure' adds up to anything more than an unrevised residue of the *Architectural Review*'s 'Westminster Regained'—a precinctual proposal conceived at a time of positively prelapsarian innocence.

pre-cinematographic: see *PRE- B. 1d.

precious, *a.* (*sb.*, *adv.*) Add: **6. e.** *precious coral*, a coral belonging to the order Antipatharia, forming branching colonies resembling plants.

1906 S. J. HICKSON in Harmer & Shipley *Cambr. Nat. Hist.* I. xiii. 352 The 'precious coral' occurs in the Mediterranean. **1935** TWENHOFEL & SHROCK *Invertebr. Paleontol.* iv. 113 The compound corallum may be in the form of.. bushy growths as in the precious corals. **1979** *Sci. Amer.* Aug. 115/3 The detritus feeders include the true sponges, the antipatharians ('precious corals') and the gorgonians (sea fans).

precipit, *a.* For †*Obs. rare* read *rare* and add later example.

1922 JOYCE *Ulysses* 743 Not acting with precipit precipitancy.

precipitable, *a.* Add: **b.** *precipitable water* (Meteorol.): the quantity of water vapour in an atmospheric column of unit cross-section, expressed as the depth it would have if condensed to a liquid or as the mass.

1928 *Mem. R. Meteorol. Soc.* III. 2 If *m* is the mass of water vapour in grams over each square centimetre of the base of the stratosphere we have $mg = 33$, *i.e.* $m = \cdot034$

gram. Thus the water vapour in the stratospherei s equivalent to ·34 mm. of precipitable water. **1971** *Nature* 23 Apr. 503/1 Schorn *et al.* have observed average water contents in the northern hemisphere [of Mars] of up to 25 μm of precipitable water.

precipitate, *v.* Add: **III. 5.** Also, to cause (dust or other particulate matter in a gas) to be deposited on a surface.

1911 *Jrnl. Industr. & Engin. Chem.* Aug. 543/2 Cotton-covered wire when used as a discharge electrode..proved far more effective in precipitating the sulphuric acid mists. **1912** *Jrnl. Franklin Inst.* CLXXIV. 263 Plants.. built for precipitating the fumes from copper smelters and the dust from cement plants. **1938** *Trans. Inst. Chem. Engineers* XVI. 40/1 The gas is..passed through an electrofitter of the dry type where the greater part of the dust is precipitated. **1975** S. MASUDA in A. R. Blythe *Static Electrification 1975* iii. 154 Particles charged by collision with unipolar ions emitted from the discharge electrode are driven by the coulombic force on to the collecting electrode, where they are precipitated.

precipitated, *ppl. a.* Add: **2.** Also, deposited from a state of suspension in a gas.

1938 *Trans. Inst. Chem. Engineers* XVI. 38/1 The precipitated dust falls into hoppers below each section of the plant. **1971** M. ROBINSON in W. Strauss *Air Pollution Control* I. 267 The decline of [migration velocity] w_e at.. higher gas velocities is usually accounted for by reentrainment of precipitated dust.

precipitating, *ppl. a.* Add: **6.** Also, causing precipitation from a gas.

1912 *Jrnl. Franklin Inst.* CLXXIV. 262 The brush form of discharge from points is only a good precipitating agent when the current of gas is small.

precipitation. Add: **III. 5.** Also, the removal and deposition of particulate matter from suspension in a gas; the separation of crystals of a solute phase from a solid solution (see also *precipitation hardening* below).

1908 F. G. COTTRELL *U.S. Patent* 895,729 2/1 The gases or vapors containing the suspended particles enter the precipitation chamber A through pipe B. **1912, 1920** [see *ELECTROSTATIC *a.*]. **1926** *Trans. Amer. Soc. Steel Treating* X. 718 The idea that the hardness of an alloy may be increased by the precipitation of a soluble constituent from solid solution was first advanced by Merica and his associates in a hypothesis to account for the age-hardening of duralumin. **1938** *Trans. Inst. Chem. Engineers* XVI. 37/2 It is unlikely that any new cement works will be designed without provision for dust separation by electrical precipitation. **1958** A. D. MERRIMAN *Dict. Metallurgy* 3/1 *Ageing*, a precipitation process, often submicroscopic, which occurs when a supersaturated solid solution is allowed to rest at atmospheric temperature after quenching. **1967** A. H. COTTRELL *Introd. Metallurgy* xx. 372 Cu-Be alloys soften rapidly by discontinuous precipitation at temperatures above about 300°C, but this can be prevented by the addition of about 0·4 wt per cent cobalt. **1974** *Encycl. Brit. Macropædia* IV. 161/1 Electrostatic precipitation is a method for the precipitation of fogs..: a high voltage is applied across the gas phase to produce electrical charges on the particles. These charges cause the particles to be attracted to the oppositely charged walls of the separator.

8. Special Comb.: **precipitation hardening** *Metallurgy*, hardening of an alloy by heat treatment that causes the precipitation from solid solution of crystals of a solute phase; a strengthening process utilizing this phenomenon.

1926 R. S. ARCHER in *Trans. Amer. Soc. Steel Treating* X. 719 It is proposed in this paper..to develop the general theory of what may be called 'precipitation hardening'. **1931** *Jrnl. Iron & Steel Inst.* CXXIV. 671 The binary iron-boron alloys were incapable of hardening by quenching and the precipitation hardening was hardly noticeable. **1957** *Technology* Oct. 291/2 Parts made from it [*sc.* a new stainless steel] are subjected to 'precipitation hardening', a method of heat treatment designed to give increased strength by precipitating an inter-metallic compound between the metal particles. **1973** J. G. TWEEDDALE *Materials Technol.* I. vi. 169 Precipitation hardening..is a three-part treatment (1) a solution treatment at elevated temperature to dissolve the solute (2) a quenching operation to trap the solute..(3) a precipitation or ageing treatment to develop the required size of precipitate.

precipitator. Add: **2. b.** Also *spec.* an apparatus for removing particulate matter such as dust or smoke from a gas by passing it between electrodes so that the particles acquire an electric charge and are attracted to an oppositely charged surface.

1919 *Jrnl. Amer. Chem. Soc.* XLI. 587 (*heading*) An electrical precipitator for analyzing smokes. **1958** *Engineering* 28 Feb. 274/3 Dust is extracted from the 'used' air by electrostatic precipitators. **1971** *Time* 7 June 61/3 Equip the plants' stacks with electrostatic precipitators and wet scrubbers that would cut air pollution by 99%.

precipitin. Substitute for def.: An antibody that on reacting with its antigen produces a visible precipitate. (Later examples.) Also *fig.*

1912 [see *IMMUNOLOGIST]. **1931** *Syst. Bacteriol.* (Med. Res. Council) VI. xiii. 424 Kraus named the substance present in the bacterial filtrate precipitinogen while the term precipitin was used to denote the hypothetical substance or antibody formed or set free in the animal body in response to injections of an antigen or precipitinogen. **1966** *Lancet* 24 Dec. 1397/1 Some natives in the Territory of Papua and New Guinea who have chronic lung disease also have in their sera precipitins against a soluble substance in the roofs of their huts. **1971** 'D. HALLIDAY' *Dolly & Doctor Bird* xiii. 192, I don't think any of us felt anything: we carried our own precipitins, for the moment, against fear and danger.

b. *attrib.*, as *precipitin reaction, technique*; **precipitin test,** a means of establishing the identity of a substance by testing whether it reacts with a particular precipitin; hence *precipitin testing* vbl. sb.

1958 *Immunology* I. 87 Quantitative studies of the precipitin and agglutination reactions. **1962** M. RABAEY in A. Pirie *Lens Metabolism Rel. Cataract* 310 Generally, α-crystallin has been the first lens protein which has been detected by means of precipitin techniques. **1905** F. C. WOOD *Chem. & Microsc. Diagnosis* I. ix. 255 Before applying the precipitin test to a suspected stain, the presence of blood should, if possible, be determined either by the formation of Teichmann's crystals or by the spectroscope. **1952** M. E. FLOREY *Clin. Applic. Antibiotics* I. ii. 31 They..found precipitin tests inconclusive and considered in consequence that the urticaria was not the result of an antibody-antigen reaction. **1976** *Lancet* 6 Nov. 985/1 Antibodiesto thyroglobulin were detected by a precipitin test in an Ouchterlony plate. **1971** *E. Afr. Standard* (Nairobi) 13 Apr. 9/8 This will form the basis of establishing precipitin testing for East Africa.

precipitinogen (prisipiti·nŏdʒĕn). *Immunol.* Also † **preci·pitogen.** [f. prec. + *-OGEN.] A type of antigen which induces the production of a precipitin.

1904 G. H. F. NUTTALL *Blood Immunity & Blood Relationship* ii. 98 The immunifying substance, which they [*sc.* Obermayer and Pick] wrongly style 'precipitogen', as well as the precipitin they conclude are not albuminous. [*Note*] A misnomer..for the reason that it suggests a relation between precipitin and ('precipitogen') precipitable substances such as exists, for instance, between pepsin and pepsinogen, in other words that the 'precipitogen' is a forerunner of precipitin, which it is not. **1907** *Jrnl. Med. Res.* XVI. 173 Many or all of the other substances are assimilated more or less readily by the tissues, are neutralized in a relatively short space of time, and in some instances (precipitogens) definite reaction products to them may be formed. *Ibid.* XVII. 232 The complement, when incubated with antigen and low ($\frac{1}{10}$) dilution of immune serum, may..be absorbed through the union of precipitin and precipitinogen. **1931** [see prec.]. **1969** *Acta Path. & Microbiol. Scand.* LXXVII. 463 Some chemical properties of a group reactive precipitinogen from the *Fusobacterium* strain F1 have been investigated.

Hence **precipitinoge·nic** *a.*

1935 F. P. GAY *Agents of Dis. & Host Resistance* xxi. 418 Not all soluble proteins are precipitinogenic. **1970** B. G. F. WEITZ in H. W. Mulligan *Afr. Trypanosomiases* vi. 113 Cultivated trypanosomes are also agglutinogenic and precipitinogenic.

precipitron (prīsi·pitrǫn). [f. PRECIPI(TATOR + *-TRON.] A kind of electrostatic precipitator.

1941 *Iron & Steel Engineer* Sept. 78 (*heading*) Precipitrons for the steel industry. **1975** E. B. CEADEL in Barr & Line *Ess. Information & Libraries* 45 It should not take him much time to find out what..services consultants mean by 'electrostatic precipitrons'.

précis, *v. trans.* Add: (Earlier and later examples.) Hence **précised** (prē·sīd) *ppl. a.*

1856 LD. CANNING *Let.* 2 Apr. in E. Fitzmaurice *Life and Earl Granville* (1905) I. vii. 152 The lucid..way in which a heavy case is précis'd is admirable. **1916** *Daily Colonist* (Victoria, B.C.) 30 July 18/4 The main work of the Canadian War Records, that of compiling and precising the all-important diaries and appendices is, of course, in arrears. **1964** P. MACKESY *War for Amer.* ii. 60 The Ministerial viewpoint on the prospects offered by the advance from Canada is précised in CO 5/253, ff. 21–30. **1972** *Islander* (Victoria, B.C.) 9 Jan. 14/1 The new book Parsons on the Plains..is actually three books precis'd down and contained in one volume. **1977** C. HUSBAND in H. Giles *Lang., Ethnicity & Intergroup Relations* ix. 223 Headlines, as precised news values, represent an expression of 'constructed reality'.

precision. Add: **1. c.** *Statistics.* The reproducibility or reliability of a measurement or the like; used *spec.* to denote various measures or indices of this (see quots.). [The sense is due to W. Lexis, who used G. *präcision* (now *präzision*) (W. Lexis *Zur Theorie der Massenerscheinungen in der menschlichen Gesellschaft* (1877) ii. 25).]

1885 M. MERRIMAN *Textbk. Method of Least Squares* i. 1 The comparison of observations is necessary in order to determine the relative degrees of precision of different sets of measurements made under different circumstances. **1906** *Acta Univ. Lundensis* Ny Följd I. v. 7, *h* is called the measure of precision. **1911** G. U. YULE *Introd. Theory Statistics* xiii. 253 The reliability or precision of an observed proportion varies as the square root of the number of observations on which it is based. *Ibid.* xv. 304 The use of $\sqrt{2} \times \sigma$ (the 'modulus') as a measure of dispersion, of $1/\sqrt{2}\cdot\sigma$ as a measure of precision, and of $2\sigma^2$ as 'the fluctuation'. **1947** O. L. DAVIES *Statistical Methods in Res. & Production* 250 The normal distribution..is given by some writers the mathematical formulation: $he^{-h^2x^2}/\sqrt{\pi}$. The parameter *h* is then called the parameter of precision, but by comparison with the usual formula it is seen that this parameter is related to the standard deviation by the identity $h = 1/\sigma\sqrt{2}$. **1949** F. YATES *Sampling Methods for Censuses & Surveys* viii. 247 The relative precision of two different methods of sampling based on the same type of sampling unit may be defined as the reciprocal of the ratio of the sampling variances of the estimates given by the two methods when the same number of units are taken. **1957** KENDALL & BUCKLAND *Dict. Statistical Terms* 224 Precision is a quality associated with a class of measurements and refers to the way in which repeated observations conform to themselves; and in a somewhat narrower sense refers to the dispersion of the observations, or some measure of it, whether or not the mean value around which the dispersion is measured approximates to the 'true' value. In general the precision of an estimator varies with the square root of the number of observations upon which it is based. **1965** R. DEUTSCH *Estimation Theory* x. 154 Precision is a measure of how close the outcome of a measurement, or a sequence of observations, clusters about some estimated value of a specified parameter. **1965** D. V. LINDLEY *Introd. Probability & Statistics* II. v. 8 We shall call the inverse of the variance, the precision. The nomenclature is not standard but is useful. **1971** *Nature* 12 Feb. 484/1 An estimate of the precision (analytical reproducibility) of each K—Ar analysis is given as a ± value. **1974** *IEEE Trans. Instrumentation & Measurement* XXIII. 278/1 The desired precision of intercomparison was set at ±0·01μV (one part in 10⁸). **1974** *Nature* 8 Nov. 137/1 Radioactive isotopic dates invariably include their precision, that is, the repeatability, yet most earth scientists still take these figures as measures of accuracy.

d. In numerical work, the fineness of specification, as represented by the number of digits given and distinguished from *accuracy* (the nearness to the true value).

1948 *Math. Tables & Other Aids to Computation* III. 286 Numbers are stored to a precision of 35 binary digits. **1956** G. A. MONTGOMERIE *Digital Calculating Machines* vii. 12 Precision can be expressed in two ways: we may say that a number is correct to so many decimal places or to so many significant figures. **1962** *Gloss. Terms Automatic Data Processing* (*B.S.I.*) 14 A result may have more precision than it has accuracy, e.g. the true value of π to eight decimal figures is 3·1415927; the expression π = 3·1415249 is precise to eight figures but accurate only to about five. **1970** H. A. RODGERS *Dict. Data Processing Terms* 81/1 Strictly speaking there is a difference of precision between 1,000 and 1. × 10³; in the first case the low-order zeros are known to have that value while in the second case all that is known is the explicit digit and the multiplier. **1972** *Physics Bull.* Aug. 459/2 To precisions varying from 1 to 0·01 parts in a million, the value of $(2e/h)_J$ does not depend on whether the effects are observed in absorption or emission.

3. a. *attrib.* and *Comb.*, usu. implying an intended or actual precision of performance, execution, or construction.

1875 *Encycl. Brit.* III. 263/1 The theory of the common balance as we see it working in every grocer's shop, and.. of the modern precision balance. **1910** *Westm. Gaz.* 6 Jan. 4/2 Those wonderful American automatic precision tools that have played so conspicuous a part in almost every European factory. **1935** *Discovery* Jan. 9/1 Continuous knife-edges and continuous knife-edge bearings such as are now used in all precision balances. *Ibid.* Dec. 368/2 The optical outfits..consist of precision-made parts, the mere use of which teaches precision in working. **1937** *Ibid.* Apr. 112/1 These are real precision instruments,.. with lens of aperture F1·9 such as the old plate cameras never knew. **1939** *War Illustr.* 16 Dec. 440/3 The barrage ..gives London and other cities and vital points reasonable security from swooping raiders and precision-bombing. **1944** *Foundry* Feb. 116/1 Application of precision casting, utilizing either centrifugal or pressure methods, to the production of precision parts made from heat and corrosion resistant alloys is a recent development. **1950** *N.Y. Times Mag.* 27 Aug. 52/2 Strategic bombing as carried out by the American 8th and 15th Air Forces in Europe was 'precision bombing' directed, so far as operational accuracy permitted, against specific military targets. **1951** *People* 3 June 7/7 (Advt.), Just switch on that precision-built Arvin and discover for yourself how pleasant dry shaving can be. **1953** 'N. BLAKE' *Dreadful Hollow* 77 Yes, quite a precision-tool job. **1957** *Technology* Mar. 10/2 Other work includes..the repair of indicator gauges and other precision instruments. **1963** *New Yorker* 23 Nov. 23 (Advt.), What's low in upkeep, high in mileage,..precision-engineered with 42 hidden changes to date but looks the same every year? **1966** 'H. MACDIARMID' *Company I've Kept* viii. 187, I qualified as a precision fitter and obtained a job with a big general engineering firm. **1969** 'D. RUTHERFORD' *Gilt-Edged Cockpit* x. 172 We still think there's a market for individuality and a precision-built car. **1975** BRAM & DOWNS *Manuf. Technol.* i. 7 Gauge blocks are used as standards of measurement or reference in most precision-engineering works. **1976** J. VAN DE WETERING *Corpse on Dike* (1977) ii. 33 Mary kept her pistols in the drawer... 'Careful... They are precision instruments, both of them.'

b. Special Comb.: **precision approach radar,** a ground-based radar system used to follow accurately the approach of an aircraft and to enable landing to be supervised from the ground.

1950 *Electronics* Feb. 71/1 (*heading*) Airport surveillance and precision approach radar (GCA). **1956** *Electronic Engin.* XXVIII. 15/2 In practice, all airport radar requirements except precision approach radar..can be met. **1965** NAYLER & OWER *Aviation* xvii. 250/2 Ground Controlled Approach.., also known as Precision Approach Radar, gives the position of an approaching aircraft in elevation, azimuth, and range relative to the touch-down point on the runway.

precisionist, *sb.* and *a.* Add: **A.** *sb.* (Later examples.) Also *spec.* in *Art*: one of a group

of U.S. artists of the 1920s who employed a smooth, precise technique in their paintings.

1960 *Art in Amer.* III. 33/1 Its [*sc.* Cubism's] effects still pervade the most recent work of the Precisionists. **1974** *Encycl. Brit. Micropædia* VIII. 185/2 The Precisionists did not issue manifestos, and they were not a school or movement with a formal program. *Ibid.*, The Precisionists' style greatly influenced the American Magic Realists and the Pop artists. **1978** *Verbatim* Winter 6/2 And I am certain that that great precisionist intended a true rhyme.

B. *adj.* Employing or exhibiting precision as an artistic technique.

1960 *Art in Amer.* III. 32/1 The Precisionist painting process is one of continual distillation and editing. **1978** *Chicago* June 62/1 Precisionist paintings, ranging from realistic to abstract. **1979** *Jrnl. R. Soc. Arts* Nov. 747/2 Even American or German architecture looks a bit thrown together when compared with the precisionist craftsmanship of his high-gloss aluminium detailing.

Hence **preci·sionism,** the style or technique of the precisionist painters.

1960 *Art in Amer.* III. 47 When Jefferson consciously reacted against the light, impermanent and provincial qualities of American building..it was to another and more integral kind of 'precisionism' that he turned: to a precisionism of mass. **1978** S. F. YEH *Precisionist Painters* 10 The basis for Precisionism is to be found in Cubism.

pre-civil(ization, -ized: see *PRE- B. 1, A. 2. **pre-Civil War:** *PRE- B. 2a.

Pre-classic, *a. Archæol.* [PRE- B. 1.] Designating a period of Meso-American culture, about 1500 B.C. to A.D. 300.

1956 G. W. BRAINERD *Morley's Anc. Maya* (ed. 3) iii. 40 Maya history may be divided into three stages: (1) Pre-classic, extending from about 1500 B.C. to A.D. 317; (2) Classic, from A.D. 317 to 889; and (3) Postclassic, from 889 until 1697. **1967** L. DEUEL *Conquistadors without Swords* III. xv. 188 The emergence of an 'archaic age' in Mexican pre-history, now usually referred to as formative or pre-classic. **1970** BRAY & TRUMP *Dict. Archaeol.* 188/2 *Pre-classic* (*or Formative*) *period*, used in American archaeology for the period in which agriculture..formed the basis of settled village life. *Ibid.* 189/1 In the chronological sense..the Pre-classic period is usually taken to have ended c 300 AD. **1974** *Nature* 6 Dec. 472/1 The earliest occupation discovered belongs to the Real Xe phase of the Middle Preclassic Period... This cache was actually located beneath several preclassic floors.

pre-classical, *a.* Add: Also in extended uses.

1948 K. MALONE *Middle Ages* I. 28 The technic of adornment or elaboration was essentially the same in pre-classical and classical poetry. **1958** *Times* 15 Dec. 3/2 One of the first great pre-classical composers for whom general popularity seems surely destined is Monteverdi. **1976** *Observer* 16 May 1/2 He was the most brilliant and versatile of wind-players, starting as a classical bassoonist then branching out into the whole range of pre-classical wind instruments. **1978** *Early Music* Oct. 567/2 The cadenzas of the baroque and pre-classical eras were usually over a pedal point dominant rather than being the tonic six-four/dominant/tonic kind of the classical era.

preclean, -climax: see *PRE- A. 1, B. 1.

precli·nical, *a. Med.* [PRE- B. 1.] **1.** Of, pertaining to, or designating the first stage of a medical education, consisting chiefly of the necessary scientific studies without regular involvement with patients.

1930 A. FLEXNER *Universities* i. 14 Medicine stood almost still until the pre-clinical sciences were differentiated and set free—free to develop without regard to use and practice. **1948** F. ROBERTS *Med. Educ.* vii. 48 On entering the pre-clinical stage the student embarks on an intensive study of human anatomy and physiology. **1956** *Med. Press* 5 Sept. 225/1 In the more progressive medical schools it has been found of great benefit to give the pre-clinical student a series of specially selected clinical demonstrations. **1970** *Nature* 1 Aug. 431/2 In 1968–69 there were 6,017 pre-clinical and 7,024 clinical training places in medical schools in Britain. **1978** *Jrnl. R. Soc. Med.* LXXI. 373 The preclinical and paraclinical subjects that, it is widely accepted, comprise the early part of the European medical student curriculum.

2. Preceding the onset of recognizable symptoms that make a diagnosis possible.

1932 GAIGER & DAVIES *Vet. Path. & Bacteriol.* xvii. 277 The difficulty of diagnosing cases in the pre-clinical stage led to efforts to find a diagnostic agent for Johne's disease analogous to tuberculin for tuberculosis. **1943** *Brit. Jrnl. Tuberculosis* XXXVII. 98 How is the student of today going to prescribe tomorrow for the preclinical case diagnosed by mass-radiology? **1976** *Biol. Abstr.* LXI. 6389/1 (*heading*) Pulmonary cancer recognized in a preclinical phase of development.

3. Preceding clinical testing of a drug.

1962 *Folia Pharmacologica Japonica* 20 May 18* (*heading*) Consideration and some experiments of preclinical pharmacology of anti-cancer drugs. **1972** P. R. B. NOEL in Richards & Rondel *Adverse Drug Reactions* i. 3 The number of patients used..is not always very great and does not always exceed the number of animals used in the pre-clinical investigations.

precoat: see *PRE- A. 1, 2.

precocial, *a.* Now the usual spelling of PRÆCOCIAL *a.* (in Dict. and Suppl.).

precocious, *a.* Add: **3. b.** = PRÆCOCIAL *a.* (in Dict. and Suppl.).

1897 PARKER & HASWELL *Text-bk. Zool.* II. xiii. 382 The newly-hatched young may be..well covered with down and able to run or swim and to obtain their own food, in which case they are said to be precocious. **1970** R. A. & B. M. MAIER *Compar. Animal Behavior* ix. 193 Domestic chicks are precocious (well developed at hatching).

precog (prī·kǫg). Also **pre-cog.** [abbrev. of PRECOGNITION.] = PRECOGNITION 1. Also, one who predicts something; a person with precognition. Also *attrib.*, = PRECOGNITIVE *a.*

1966 *Listener* 19 May 727/2 It is generally recognized that 'pre-cog' dreams take place quite often; and with the knowledge which these dream researchers provide..it may be possible to obtain quite massive evidence of pre-cog dreaming. **1967** *Ibid.* 16 Mar. 359/3 Apart from the massive evidence of ESP in human life—what does the Professor say about 'pre-cog', I wonder—the abundance of ESP in nature must surely infuriate him. **1973** *Daily News* (N.Y.) 21 Aug. 53/1 Certain precogs prophesy the future with the buckshot approach, generalized predictions. **1977** *Sounds* 9 July 22/1 And as for the matter of whether the gent's armed with the sort of foresight Phillip K. Dick grants his 'precogs', you can just make up your own mind.

precognition. 1. (Later examples.)

1955 *Sci. Amer.* Oct. 116/3 The entire experimental series seemed to offer proof of some form of telepathy: 'pre-cognition' or 'post-cognition'. **1958, 1968** [see *EXTRA-SENSORY *a.*]. **1973** *Psychol. Abstr.* XLIX. 11/1 Telepathy and clairvoyance are seen as extensions of normal perceptual processes, precognition as the reverse of retrospective memory processes.

precognitive, *a.* (Later examples.)

1953 P. C. BERG *Dict. New Words* 127/2 *Precognitive telepathy*,..awareness by a percipient of images and ideas occurring at some future time in the mind of a subject or agent. **1974** *Sci. Amer.* June 118/2 One is more telepathic, more clairvoyant, more precognitive. **1975** *Physics Bull.* Mar. 125/2 The author has attempted to describe and briefly to discuss a number of cases claimed to show precognitive happenings.

precognize, *v.* Now usu. with pronunc. (prī·kǫgnəiz). Delete *rare* and add later examples.

1956 A. J. AYER *Probl. Knowl.* iv. 187 There is a tendency for them [*sc.* non-philosophers] to think that if future events were precognized, they would have to exist already... To precognize something is to know, not what *is* happening, but what *will* happen. *Ibid.*, Unless the event really were future there would be no question of one's *pre*-cognizing it. **1970** A. CAMERON et al. *Computers & Old Eng. Concordances* 29, I don't foresee the solution that you are precognizing for the simple reason that I don't think there is enough demand for any computer to be built.

preco·ital, *a.* [PRE- B. 1.] Occurring or performed as a preliminary to sexual intercourse.

1935 H. M. & A. STONE *Marriage Manual* viii. 265 A long period of precoital play and a considerable prolongation of the sexual act are unsuccessful in bringing about a culmination for the woman. **1953** A. C. KINSEY et al. *Sexual Behav. Human Female* ix. 361 The pre-coital techniques in marriage are..the same. **1963** A. HERON *Towards Quaker View of Sex* Inadequacies of the love-partner during pre-coital play. **1971** 'V. X. SCOTT' *Surrogate Wife* 98 They engaged in very little precoital play. **1973** M. AMIS *Rachel Papers* 23 In normal circumstances, with her embarrassment in any kind of pre-coital conversation,..the stiff-limbed movements: you were a plaything of her unease.

Hence **preco·itally** *adv.*

1971 *Nature* 16 Apr. 433/2 Apocrine glands contribute to total body odour, but a smegma pheromone would be exposed precoitally with exposure of the glans.

precoition(al): see *PRE- B. 2a, 1d. **pre-college:** *PRE- B. 2a. **precolonial:** *PRE- B. 1d. **precombustible:** *PRE- A. 3.

pre-combu·stion. [PRE- A. 2.] In certain diesel engines, commencement of combustion of the charge before it is drawn into the main cylinder, in an adjacent small chamber; usu. *attrib.*, denoting (an engine equipped with) a chamber for this purpose.

1923 L. H. MORRISON *Diesel Engines* xx. 472 One of the first of the precombustion engines developed in the United States was the Western... The precombustion chamber is located in the cylinder head almost in line with the cylinder bore. **1932** *Mod. Diesel* v. 43 On the Continent the line of development appears to have been directed towards securing a steady and progressive burning of the fuel by a system of pre-combustion, the charge being ignited in a partially separated chamber from which the more or less controlled expansion then passes to the working cylinder. **1972** S. H. HENSHALL *Medium & High Speed Diesel Engines for Marine Use* iv. 75 Pre-combustion chamber hot member engines are characterized by their smooth running, complete combustion with clean exhaust and a tendency to high fuel consumption, together with difficulties in starting in very cold conditions.

precommissural: see *PRE- A. 4a.

pre-compre·ss, *v. Building.* [PRE- A. 1.] *trans.* To compress prior to some other treatment. Hence **pre-compre·ssion,** *spec.* the compressive force exerted in prestressed concrete by the reinforcing rods.

1936 *Structural Engineer* XIV. 259/1 A sketch of the great 10,000 tons press constructed for the laboratory for Buildings and Public Works,..Paris... It was constructed in concrete with only a light reinforcement, but pre-compressed by the tension given to the hooped wire. **1940** *Ibid.* XVIII. 629 In a pre-stressed girder the fact that the part..has been given a pre-compression equal to or exceeding the said tension, renders it impossible for fissures to appear under normal loading or even higher loading. **1946** *Concrete & Constructional Engin.* XLI. 191 In a cracked section of a prestressed member, the pre-compression is either reduced or totally annulled. **1969** *Civil Engin.* (Easton, Pa.) June 45/1 In the arch section, over-compression of the upper flange was prevented by precompressing the arches: the span was made 1in. shorter at each end with hydraulic jacks. **1974** LIN & ZIA in B. Bresler *Reinforced Concrete Engin.* I. vi. 303 Prestressing imparts a precompression to a concrete member in its tension zone so as to increase its cracking resistance. **1975** KONG & EVANS *Reinforced & Prestressed Concrete* ix. 196 The precompression in the concrete tends to reduce the diagonal tension.

pre-compute: see *PRE- A. 1. **pre-computer:** *PRE- B. 2a.

preconceptional, *a.* Delete *rare*⁻¹ and add further examples.

1933 *Jrnl. Amer. Med. Assoc.* 25 Nov. 1703/1 (*heading*) Preconceptional and prenatal influences affecting the new-born. **1957** C. T. JAVERT *Spontaneous & Habitual Abortion* xvii. 377 The author prefers to begin prenatal care on a preconceptional basis and has found this approach to be particularly effective in the management of habitual abortion patients. **1975** *Amer. Jrnl. Obstetr. & Gynecol.* CXXIII. 717/2 Preconceptional irradiation.. may impair the female reproductive capacity.

preconceptual, -conciliar: see *PRE- B. 1d.

preconde·nse, *v.* [PRE- A. 1.] *trans.* To condense (a starting material for a polymer) so as to form a stable, low-molecular-weight intermediate which is convenient to handle and can be fully polymerized at a later stage in a process. Also **preco·ndensate** [after *distillate*, *filtrate*, etc.], an intermediate of this nature. So **preconde·nsed** *ppl. a.*

1950 R. W. MONCRIEFF *Artificial Fibres* xxvii. 261 The formaldehyde and urea are pre-condensed, being allowed to react at room temperature for about five hours until the viscosity is 6 centipoises at 20°C. **1953** *Jrnl. Soc. Dyers & Colourists* LXIX. 44/2 The loss will be further accentuated by differences in molecular size of the pre-condensate, particularly when highly precondensed resins are used, as in some proprietary products. **1962** J. T. MARSH *Self-Smoothing Fabrics* xi. 159 These pre-condensates are probably methanol ureas which may be formed by condensing urea and formaldehyde either at room temperatures or by refluxing for a short time. **1971** S. A. HEAP et al. in H. Mark et al. *Chem. Aftertreatm. of Textiles* vi. 272 There is..a growing tendency for finishers to purchase precondensed 'resin' in liquid form. *Ibid.* 279 The conventional process for the application of urea–formaldehyde precondensates consists of impregnating, drying, curing to bring about resinification, and washing.

precondition, *sb.* (Further examples.)

1912 A. H. POWLES tr. *Bernhardi's Germany & Next War* 4 The conscious increase of our armaments is not an inevitable evil, but the most necessary precondition of our national health. **1923** J. W. HARVEY tr. *Otto's Idea of Holy* xi. 86 It sheds a colour..upon the life and practice that are its precondition. **1948** *Sunday Pictorial* 18 July 7/6 He cannot accept the lifting of the blockade as a precondition. **1974** tr. *Wertheim's Evolution & Revolution* 211 Among the psychological preconditions for a revolution.. one should include a feeling among relatively large groups that avenues towards emancipation are consistently blocked.

precondi·tion, *v.* [PRE- A. 1.] *trans.* To bring into a desired state or condition beforehand. Hence **precondi·tioned** *ppl. a.*, **precondi·tioning** *ppl. a.* and *vbl. sb.*

1922 JOYCE *Ulysses* 494 Self which it itself was ineluctably preconditioned to become. **1961** L. P. V. JOHNSON *In Time of Thetans* vii. 54 Second stage. Bringing reverence and servility to preconditioned humanity. **1967** *Jrnl. Compar. & Physiol. Psychol.* LXIV. 360 Significantly greater CERs..were shown by the preconditioned groups than by control groups. *Ibid.* 360/1 Sensory preconditioning (SPC) involves the pairing of two neutral stimuli, S_1 and S_2. Following this procedure a response is conditioned to S_2. **1969** *Jane's Freight Containers 1968–69* 420/3 The fibre-board unit load device should be preconditioned for 48 hours under these conditions prior to testing. **1971** *Homes & Gardens* Aug. 19 Frying pans and baking tins should then be lightly greased with oil or cooking fat to pre-condition the non-stick coating. **1974** *Psychol. Abstr.* LI. 572/2 The strength of resultant saccharin aversions was inversely related to preconditioning saccharin familiarity.

pre-Conquest, *a.* (Further examples.)

1922 E. EKWALL *Place-Names Lancs.* 172 The example seems to indicate that Over Wyresdale was in pre-Conquest time common land to the townships round the lower Lune. **1927** J. J. HOGAN *Eng. Lang. in Ireland* 27 There is preliminary matter about pre-Conquest Ireland, then *Conquest* follows with some omissions. **1957** K. A. WITTFOGEL *Oriental Despotism* iii. 51 In pre-Conquest Mexico

the various forms of land and the obligations attached were carefully depicted in codices. **1960** S. CRUDEN *Scottish Abbeys* 54 To a small pre-Conquest church..there was added on the east a square choir with a rounded apse. **1975** 'S. MARLOWE' *Cawthorn Jrnls.* (1976) xi. 87 The inevitable merging of pre-Conquest and Spanish culture in Mexico.

preconscious, *a.* Add: (Earlier and later examples.) *spec.* in *Psychol.*, applied to memories and emotions existing at a deeper level than, or of a type different from, immediate memory or conscious thought, but which are accessible to and capable of being brought directly into consciousness; also *absol.* Cf. *FORE-CONSCIOUS *a.* (*sb.*) and quot. 1958.

1860 J. D. MORELL tr. *Fichte's Contrib. to Mental Philos.* iii. 43 It is not to be denied that all the apparently abnormal phenomena with which men are seized, in somnambulism, in vision,..and in ecstasy, spring out of the same spontaneous and preconscious region, from which all involuntary impulses and inspirations take their origin. **1867** H. MAUDSLEY *Physiol. & Path. of Mind* I. i. 15 The preconscious action of the mind, as certain metaphysical psychologists in Germany have called it. **1924** W. B. SELBIE *Psychol. Relig.* iv. 78 Consciousness, he [*sc.* Bergson] argues, only emerges when the individual becomes aware of his own mental states, and this allows for a preconscious or unconscious stage. **1925** J. RIVIERE tr. *Freud's Unconscious in Psycho-Anal.* in *Coll. Papers* IV. 25 We have now gained the conviction that there are some latent ideas which do not penetrate into consciousness... We may call the latent ideas of the first type preconscious. **1942, 1957** [see *FORECONSCIOUS *a.* (*sb.*)]. **1958** J. STRACHEY *Freud's Compl. Psychol. Wks.* XII. 262 In the 1925 English version, throughout the paper, 'foreconscious' was altered to 'preconscious', which has..become the regular translation of the German 'vorbewusst'. **1971** N. F. DIXON *Subliminal Perception* iv. 90 The problem is evidently one of confusing 'preconscious'—meaning antecedent physiological processes which do not have phenomenal representation—with the Freudian notion of 'a preconscious'. **1978** G. A. SHEEHAN *Running & Being* ix. 125 The preconscious..stores past preconceptions.

preco·nsciousness. [f. prec. + -NESS.] The state or condition of being preconscious; the preconscious part of the mind.

1930 W. EMPSON *Seven Types of Ambiguity* viii. 302 It is grasped in the pre-consciousness of the reader by a native effort of the mind. **1959** *Listener* 15 Oct. 624/1 This leads us to the idea of preconsciousness. That is the area of mental life in which everything we have experienced in the past, and can still remember, is stored. If consciousness is everything of which we are aware, then preconsciousness is everything which we can remember. **1970** *New Yorker* 26 Dec. 61 We learn of the 'hum' in the literal deeps of the poet's consciousness or pre-consciousness.

pre-consonant: see *PRE- B. 2a. **pre-consonantal(ly), -Constantinian:** *PRE- B. 1. **precontour:** *PRE- A. 2. **pre-convulsive:** *PRE- B. 1. **precook:** *PRE- A. 1.

precoo·l, *v.* [PRE- A. 1.] *trans.* To cool prior to use or some further treatment. So **precoo·led** *ppl. a.*, **precoo·ling** *vbl. sb.* Also **precoo·ler,** a device for precooling.

1904 *Physical Rev.* XIX. 330 From the compressor, the air passes successively through an aftercooler; a separator;..and finally through a precooler charged with broken ice or snow—reaching the liquefier at a temperature of about 2°. **1911** *Power* 14 Nov. 755/2 The Atchison, Topeka & Santa Fé Railway Company has erected at San Bernardino, Cal., a combined ice-manufacturing and precooling plant. *Ibid.* 759/1 For icing the cars, upon the completion of precooling, or for such cars as are not precooled, the ice is handled by the endless-chain conveyor. **1912** *Ice & Refrigeration* Jan. 14/1 Under some conditions the plant intended only to precool fruits before they are loaded is most advantageous while..under other conditions car precooling after loading is preferable. **1926** *Spectator* 18 Sept. 412/1 By having a hermetically sealed compartment with the commodity pre-cooled, a low temperature is maintained until the box is opened at destination. *Ibid.*, After the cleaning the pre-cooling takes place. **1936** *Discovery* Apr. 104/1 The compressed helium is first pre-cooled by liquid nitrogen. **1958** A. LAURIE et al. *Commercial Flower Forcing* (ed. 6) xii. 328 The cases containing the bulbs are placed in storage in early October at 32 to 34 °F, where they remain until shipping, which is usually in late November. Lily bulbs handled this way are referred to as precooled bulbs in contrast to bulbs sent direct to the florist immediately after digging and grading. **1958** *Times* 23 June 2/6 The more promising development is the use of precooling plants on the farms, the object being to cool the fruit quickly as soon as it is picked but before it is finally packed. **1970** *Times* 4 Sept. (Aviation Suppl.) p. ix/9 Imperial Metal Industries' Marston Excelsior subsidiary will display a range of heat exchangers, including..a precooler for the Hawker Siddeley Nimrod. **1973** *Times* 22 Sept. 13/1 We buy a quantity of double nosed Golden Harvest pre-cooled daffodils..which are delivered in the first week of October.

pre-copulative: see *PRE- B. 1d.

precordium² (prīkǭ·ɹdiŏm). *Anat.* Formerly also **præ-.** [Sing. of PRÆCORDIA.] = PRÆCORDIA.

1892 A. E. SANSOM *Diagn. Dis. Heart & Thoracic Aorta* xvi. 128 On placing the hand over the præcordium, the observer may be sensible of a peculiar vibration occurring over a certain area and at a certain period of the heart's action. **1900** DORLAND *Med. Dict.* 533/1 *Precordium*, same as *precordia*. **1934** *Arch. Internal Med.* LIV. 341 Pulsations of the heart were felt 3·5 cm. below the epigastric notch, and sometimes a systolic thrill was felt over the precordium. **1962** *Lancet* 13 Jan. 104/2 A pericardial friction-rub was audible over the precordium, but blood-pressure was 140/70 mm. Hg and there were no signs of cardiac failure. **1970** W. DRESSLER *Clin. Aids in Cardiac Diagn.* v. 66 Fig. 24 is from a 31 year old man who showed a heaving systolic impulse in the left half of the precordium between the 2nd intercostal space and the 5th rib. **1977** J. T. WILLERSON in Willerson & Sanders *Clin. Cardiol.* 101/2 Inspection and palpation of the precordium in the patient with cardiac disease are valuable and important parts of the physical examination.

pre-crusade: see *PRE- B. 2a. **pre-cultural:** *PRE- B. 1.

precu·re, *v.* [PRE- A. 1.] *intr.* Of a synthetic resin: to cure prematurely, making mechanical processing impossible. So **pre-cure** *sb.*, the premature curing of a synthetic resin; **pre-cu·ring** *vbl. sb.*

1935 C. ELLIS *Chem. Synthetic Resins* I. xxviii. 612 Avoidance of precuring during drying is essential if the molding properties of the resin are to remain unimpaired. **1936** H. W. ROWELL *Technol. Plastics* xxiii. 170 Another form of pre-cure is due to mixing a batch of quick curing powder or already cured powder with one having a normal rate of cure. **1943** *Brit. Plastics* XV. 235/2 Precuring.. shows up on the underside of the moulding as a slight chalkiness, in which the outlines of individual moulding powder particles can be seen. **1952** J. DELMONTE *Plastics Molding* viii. 208 If preheating is carried much beyond this temperature, it [*sc.* a thermosetting resin] will begin to precure excessively and will lose its flow qualities. **1962** J. T. MARSH *Self-Smoothing Fabrics* xi. 172 Where the material is batched or plaited in a warm condition immediately after the stenter, there are dangers of what has been termed 'pre-cure' which it is essential to avoid where mechanical processing takes place before the final condensation, and which it may possibly be wise to avoid generally.

precursor. Add: **3.** *Biochem.* and *Chem.* A compound which precedes another in a metabolic pathway or a chemical synthesis, esp. a naturally occurring one.

1889 C. A. MACMUNN *Outl. Clin. Chem. Urine* iii. 36 Although we know it [*sc.* urea] is formed from proteids, we cannot trace it back through its precursors—the intermediate products of metabolism. **1890** L. C. WOOLDRIDGE tr. *G. von Bunge's Text-bk. Physiol. & Path. Chem.* vi. 102 This compound is doubtless the precursor of haemoglobin, for there is no considerable quantity of any other compound of iron in the yolk. **1948** *Jrnl. Biol. Chem.* CLXXII. 651 (*heading*) Homoserine as a precursor of threonine and methionine in *Neurospora*. **1960** [see *CRYPTOXANTHIN]. **1971** *Nature* 30 July 304/1 An important feature of the process is the maintenance of the high orientation in the acrylic precursor throughout the carbonization processes. **1977** *Sci. Amer.* July 40/2 Ozone and its precursor, atomic oxygen, are destroyed by catalytic reactions that depend on H and OH.

precu·t, *v.* [PRE- A. 1.] *trans.* To cut prior to some other operation. Also *absol.* So **pre-cu·tting** *vbl. sb.*

1945 *Trans. Inst. Mining Engin.* CIV. 192 As with other types of loader which rely upon pre-cutting and blasting for the satisfactory preparation of the coal for loading, much difficulty was experienced. *Ibid.* 701 This roof can be a bad one if not properly controlled, and we felt we would not like to take the risk of pre-cutting and blowing coal, and thus removing all support. **1951** B. KELLY *Prefabrication of Houses* i. 4 There may be said to be various degrees of prefabrication, of which precutting might be one, the fabrication of panels another, [etc.]. **1960** J. SINCLAIR *Winning Coal* vii. 191 Two men pre-cut the face with an A.B. Fifteen Cutter. *Ibid.* 206 It may be preferable sometimes to pre-cut and load on the same shift with the separate cutter working about 40 yd ahead of the cutter-loader. **1961** M. BEADLE *These Ruins are Inhabited* (1963) xi. 160 The Co-ops are pre-cutting their meat and putting it in plastic film. **1964** A. NELSON *Dict. Mining* 343 *Pre-cutting*, a term used in machine mining where a coal-cutter makes a cut along the face in front of a cutter-loader.

Hence **precu·t** *ppl. a.*

1946 *Fortune* Apr. 244/2 Pease makes only wall, ceiling, and floor panels in the factory; roofs are built on the site from factory precut lumber by traditional methods. **1960** *Farmer & Stockbreeder* 29 Mar.-74/3 As the mower makes a new cut, the scatterer shakes up the pre-cut swath, thereby reducing the time necessary to make the hay. **1960** *Times* 29 Aug. 15/3 The Birmingham Co-operative Society has, for some time now, been selling pre-cut and prepacked meat. **1967** KARCH & BUBER *Offset Processes* iv. 112 (*caption*) Redi-Kut display letters are pre-cut and have a pressure-sensitive backing. **1974** *Encycl. Brit. Macropædia* XIV. 966/1 One of the earliest excursions into prefabrication was the adoption of precut framing lumber pieces for walls and partitions. **1976** *Sci. Amer.* Nov. 100/2 In the commonest form of stamping, a precut metal blank is formed in a mechanical press between a set of dies that have been carefully shaped to yield the desired part.

precut (prī·kɒt), *sb.* [f. the vb.] **a.** A cut made in something prior to some other operation on it. **b.** Something that has been precut.

1948 *Trans. Inst. Mining Engin.* CVII. 242 A long bottom jib was provided to give a pre-cut and so assist in the fracturing of the coal which is rather strong. **1960** J. SINCLAIR *Winning Coal* vii. 204 A middle pre-cut has been usefully employed where a floor tends to lift and break at the bottom of the coal. **1971** *Real Estate Rev.* Fall 49/1 They are to be distinguished from pre-cuts, which involve materials cut to proper size off-site and then conventionally constructed on site.

precy·stic, *a. Zool.* [PRE- B. 1.] Applied to protozoans that are preparing to encyst.

1926 C. M. WENYON *Protozool.* I. ii. 190 It may be that ..the precystic amœbæ..are produced by amœbæ living on the surface of the mucosa. **1938** *Archiv für Protistenkunde* XC. 405 The large majority of precystic *Colpoda* sink to the bottom to encyst. **1956** *New Biol.* XXI. 92 In preparation for this [encysting] the amoebae either digest or eject their inclusions and do not feed or grow, so that a generation of small amoebae with clear cytoplasm is produced—the precystic amoebae.

predacious, *a.* Add: **1.** Also, used of parasitic fungi which actually kill their hosts. Also *transf.* (Later examples.)

1908 C. ELIOT *Turkey in Europe* (ed. 2) iii. 73 The Turks never outgrew their ancestral character of predacious nomads. **1933** *Jrnl. Washington Acad. Sci.* XXIII. 140 Adhesion on hyphal tips..appears to be effective in the somewhat more feebly predacious activity of a fungus bearing solitary spores. **1936** *Mycologia* XXVIII. 307 (*heading*) A new predaceous fungus. **1946** *Ecology* XXVII. 257/1 The Chelonethida are of interest as predacious arachnids. **1964** R. M. & J. W. FOX *Introd. Compar. Entomol.* xiii. 372 Some predaceous species help to control other arthropods. **1971** R. C. W. BERKELEY in Hawker & Linton *Micro-Organisms* xii. 512 The activities of the predacious fungi have been much investigated in the hope that they could provide an effective means of controlling root parasitic eelworms.

predate, *v.*¹ **2.** (Later examples.)

1974 *Observer* (Colour Suppl.) 19 May 14/3 Houses that pre-date the introduction of noise controls and the airport itself. **1976** *Amer. N. & Q.* XIV. 98/1 Both poems predate the letters in the authorized edition of *The Letters of Junius* published in 1772. **1978** *N. & Q.* Dec. 532/2 Deeds of the Warren estates..provide instances in Lancashire which pre-date the above occurrence [of the name Diana] by at least three hundred years.

predate (prĭdē·t), *v.*² [Back-formation f. PREDATION.] **a.** *intr.* To seek prey. **b.** *trans.* Of a predator: to prey on, eat.

1974 *Trout & Salmon* Mar. 50/2 It is hoped that the stock of trout will predate sufficiently to minimise the problem [of coarse fish]. **1977** *Field* 13 Jan. 47/1 Man is a predator... To predate in person, instead of by proxy, is not unnatural. **1977** J. L. HARPER *Population Biol. of Plants* vi. 172 Wood pigeons (*Columba palumba*) cease to predate when the density of a food falls to a level at which the birds can no longer search quickly enough to pick up a sufficient quantity. **1977** *New Scientist* 27 Oct. 220/3 The eggs of many species of frogs are predated by many species of vertebrates and invertebrates.

predation. Restrict *Obs.* to sense in Dict. and add: **2.** *Zool.* The action of one animal preying upon another. Also *transf.* and *fig.*

1932 W. L. MCATEE in *Smithsonian Misc. Coll.* LXXV. No. 7. 144 Predation takes place much the same as if there were no such thing as protective adaptations. **1937** *Ann. Rep. Board of Regents Smithsonian Inst. 1936* 243 (*heading*) What is the meaning of predation? *Ibid.*, Predation has been shown..not to be, in a collective sense, an inexorable tax upon the luckless prey species. **1944** J. S. HUXLEY *On Living in Revolution* 61 The raids of the slave-making ants are not true war, but a curious combination of predation and parasitism. **1954** D. LACK *Nat. Regulation of Animal Numbers* xiv. 156 The small passerine birds cannot have been limited in numbers by predation. **1959** *Listener* 10 Dec. 1032/1 This predation of birds upon insects is of considerable practical importance. **1968** *Nature* 17 Aug. 694/1 Predation from vertebrates and the uncertainty associated with nests attached to palm leaves were certainly principal factors [in the death of *Scaphidura* chicks]. **1975** W. H. NESBITT in M. W. Fox *Wild Canids* xxvii. 395 They [*sc.* feral dogs] are a valuable part of the fauna by their sanitary predation activities. **1976** E. CURIO *Ethology of Predation* 1 Predation is an ecological factor of almost universal importance for the biologist. **1977** READER & CROZE *Pyramids of Life* i. 30/1 In its continuous 'predation' on plants, the elephant tears branches from trees, pulls great tufts of grass and roots from the earth. **1977** *Times* 11 Feb. 17/3 New entrants to the shipping business are said to have little chance of surviving outside the conference network because of..the threat of predation (the practice of selling below cost to destroy competition).

predatism (pre·dătiz'm). *Biol.* [f. PREDAT(ION + -ISM.] Predation; the mode of life of a predator.

1930 R. A. FISHER *Genetical Theory Nat. Selection* vii. 157 Though the principle of Bates is excluded when two species are actually equally acceptable or unacceptable, to demonstrate such equality with sufficient precision to exclude differential predatism, and to demonstrate it with respect to the effectual predatory population, would seem to require both natural knowledge and experimental refinement which we do not at present possess. **1946** C. T. BRUES *Insect Dietary* vi. 243 Predatism..is a commonplace mode of sustenance among animals, including man himself. **1964** *Biol. Abstr.* XLV. 5426/2 (*heading*) A new occurrence of predatism in polychaete worms.

predative (pre·dătiv), *a. rare.* [f. L. *prædāt-*, ppl. stem of *prædārī* to plunder + -IVE, after *native*, *passive*, etc.] = PREDATORY *a.*

c 1925 D. H. LAWRENCE *Virgin & Gipsy* (1930) iii. 50 She [*sc.* the gipsy-woman] was..just a bit wolfish... 'Good-morning, my ladies and gentlemen,' she said, eyeing the girls from her bold, predative eyes.

predator (pre·dătəɹ). *Zool.* [f. L. *prædātor* plunderer (see PREDATORY *a.*); cf. mod.L. *Predatores* (W. Swainson in Swainson & Shuckard *On Hist. & Nat. Arrangement of Insects* (1840) II. iii. 115).] An animal that preys upon another.

1922 W. M. WHEELER *Social Life Insects* ii. 46 Species that behave in this manner are not true parasites, but extremely economical predators, because they eventually kill their victims. 1931 W. C. ALLEE *Animal Aggregations* xiv. 246 The struggle for existence between predators and their prey. 1945 J. STEINBECK *Cannery Row* xvii. 69 The little octopi..prefer a bottom on which there are many caves and little crevices..where they may hide from predators. 1959 W. TRAVIS *Beyond Reefs* viii. 165 We all adopted the lazy movements and slow rhythms of these big predators [*sc.* sharks]. 1971 *Nature* 1 Oct. 345/1 Predators also show a tendency to continue to select a given type of prey, even though other types may be..more easily available.

2. *attrib.* **predator–prey** *adj. phr.*, concerning the ecological balance between a predator and its prey.

1946 *Q. Rev. Biol.* XXI. 235/2 In equations depicting predator–prey interactions in lower vertebrates, loss types may substitute naturally for each other. 1968 *Times* 2 Oct. 12/6 This clearly has considerable implications for understanding predator–prey relationships.

predatory, *a.* **4.** (Further examples.)

1925 *Jrnl. Mammalogy* VI. 29 The larger predatory mammals..require for proper sustenance animal food in large quantities. 1970 R. A. & B. M. MAIER *Compar. Animal Behavior* vii. 116 Defenses against predators are necessarily less than completely effective; otherwise, predatory animals could not survive.

predatoriness (examples).

1963 *Times* 7 Mar. 13/2 The techniques of power, of political manipulation, of the predatoriness of officialdom, become even more insidiously efficient. 1979 *Listener* 30 Aug. 284/2 Poverty..makes public predatoriness irresistibly attractive.

pre-daw·n, *sb.* and *a.* **A.** *sb.* [PRE- B. 1 in Dict. and Suppl.] The period before daybreak. Also *fig.* **B.** *adj.* [PRE- B. 2.] Occurring before dawn; of or pertaining to the period before daybreak.

1946 D. C. PEATTIE *Road of Naturalist* i. 13 So now I woke, in the pre-dawn of the desert. 1951 *Jrnl. Geophysical Res.* LVI. 325 Between the post-twilight and predawn, the intensity goes through a maximum which is partially localized to the north of our station. *Ibid.*, During the post-twilight period, an enhancement of intensity is observed in the western sky. In the predawn observations, no enhancement is noted in the eastern sky. 1963 J. LUSBY in B. James *Austral. Short Stories* 222 You felt the brittle pre-dawn tension of any war-time 'drome. 1965 G. MCINNES *Road to Gundagai* x. 178 The swirling pre-dawn mist lay wrapped about the feet of every illusion. 1973 G. HART *Right from Start* i. 30 On we went through October, 1970, the pre-dawn of the campaign. 1978 *Daily Tel.* 11 Nov. 11/1 A total eclipse of the sun at midday immersed the world..in semi-gloom, as in the pre-dawn.

predefine, *v.* Add: Hence **predefi·ned** *ppl. a.*

1929 R. BRIDGES *Testament of Beauty* IV. 124 But these philosophers..used the abstracted terms whereby they had pre-defined distinctions. 1976 *Brit. Jrnl. Sociol.* XXVII. 302 What is offered is not the theoretical construction of an object of study, but merely the systematization of predefined data. 1977 J. D. DOUGLAS in Douglas & Johnson *Existential Sociol.* i. 63 Even Goffman..has had little to say about the self except as a 'presenter' of predefined and learned social roles.

predelinquency, -delinquent: see *PRE- B. 1.

predeli·very, *a.* [PRE- B. 2.] Carried out in, or concerned with, the period preceding delivery of a baby.

1957 *Obstetr. & Gynecol.* IX. 633 (*heading*) Predelivery sedation with promazine. 1965 S. PELLER in Glass & Eversley *Population in Hist.* v. 91 Babies born out of wedlock had a perinatal mortality of 102, or 43 or 32 per 1,000.., depending on whether their mothers had been sheltered in the predelivery maternity division of the General Hospital for 0–7, 8–28, or 29–56 days.

predesignate, *v.* Add: Hence **prede·signated** *ppl. a.*

1961 J. B. WILSON *Reason & Morals* ii. 109 People do not consciously adopt a language-game for a deliberate and predesignated purpose. 1973 *Nation Rev.* (Melbourne) 31 Aug. 1449/3 Fortnightly..aficionados of the Australian pit fowl gather at the predesignated meeting place.

predestinating, *ppl. a.* (Later example.)

1956 R. MACAULAY *Towers of Trebizond* xxi. 242 The predestinating Calvinists in the Celtic mountains.

predestination. Add: Hence **predestina·tionism,** belief in predestination or the system of thought it entails.

1901 G. H. HOWISON *Let.* 21 July in R. B. Perry *Tht. & Char. W. James* (1935) II. 221 *Of course* I don't reconcile predestinationism and capricious free-will. 1937 *Mind* XLVI. 287 The unsatisfactory answer offered by Leibniz to readers frightened by *his* Predestinationism.

predestinator. **2.** Delete † *Obs.* and add later example.

1956 R. MACAULAY *Towers of Trebizond* xxi. 240 It is the Predestinators not the Pelagians who, as it says in the 9th Article, do vainly talk.

Hence **prede·stinatory** *a.* (*rare*) = PREDESTINATE *ppl. a.* 2.

1967 B. WRIGHT tr. *Queneau's Between Blue & Blue* xix. 203 If society gave me this predestinatory name, nature for her part provided me with peculiarly active grey matter.

predestine, *v.* Add: Hence **prede·stine** *ppl. a.* (*rare*).

1962 A. HUXLEY *Island* xiii. 204 These people are the propagandist's predestine victims.

predeterminative, *a.* Add: **2.** *Gram.* Having the quality of, or acting as, a predeterminer.

1961 R. B. LONG *Sentence & its Parts* ii. 42 In *only John* the adverb *only* is essentially predeterminative exactly as it is in *only the older children*.

predeterminer. Add: **2.** *Gram.* [f. PRE- B. 1 + *DETERMINER[1] 3.] One of a class of limiting expressions that precede the determiner. Also *attrib.*

1959 J. SLEDD *Short Introd. to Eng. Gram.* 116 The predeterminers, finally, are rather various but limited in number. 1961 *Amer. Speech* XXXVI. 159 It is customary to describe the English nominal as consisting of a sequence of constituents: predeterminers, determiners, adjectives, [etc.]. 1961 R. B. LONG *Sentence & its Parts* ii. 40 Predeterminer modifiers generally are adverbial in function, and mensurant, selectional, differential, conjunctive, or adjunct-like in force. 1964 [see *DETERMINER[1] 3]. 1965 O. THOMAS *Transformational Gram.* iv. 84 Like the other two classes of determiners, the predeterminers have a unique feature: they are invariably separated from the regular determiners by the word *of*. 1975 *Language* LI. 990 The articles are analysed with respect to..a class of modifiers sometimes known as predeterminers (*several, three*, etc.).

pre-development: see *PRE- B. 2a.

prediabe·tic, *a.* and *sb.* *Path.* [PRE- B. 1.] **A.** *adj.* Of, pertaining to, or designating a person in whom it appears that diabetes mellitus is likely to develop, but who does not exhibit its full symptoms. **B.** *sb.* A prediabetic person.

1921 *Jrnl. Amer. Med. Assoc.* 8 Jan. 79/2 A prediabetic stage in fat persons has been recognized. *Ibid.* 83/2 It is to the diabetic patient and his relatives that one can look most confidently for help in preventing diabetes... Such measures, however, are but general, and they will never suffice for the prediabetic. 1954 *Jrnl. Clin. Endocrinol.* XIV. 177 The high rates of fetal loss and the production of large babies are exemplified by the pregnancies of prediabetic women in the Cape Town region. 1962 H. ZAROWITZ in Ellenberg & Rifkin *Clin. Diabetes Mellitus* xi. 164 Because the hereditary potential for diabetes has been firmly established, this prediabetic state can occur at any time from birth to senescence. 1970 D. M. KIPNIS in Cerasi & Luft *Pathogenesis of Diabetes Mellitus* 46 Insulin sensitivity may be *increased* rather than decreased in prediabetics. 1976 *Billings* (Montana) *Gaz.* 27 June 1-B/5 Our baby will most likely be pre-diabetic (genetically disposed to diabetes, but not necessarily diabetic).

So **prediabe·tes,** the prediabetic state.

1937 *Endocrinol.* XXI. 195 'Prediabetes' in obesity is of interest because of its possible bearing on certain problems concerning the intermediate carbohydrate metabolism in obesity. 1964 *New Scientist* 27 Aug. 482/3 'Prediabetes', or the early stages, has been found in a surprisingly high percentage of the population.., and it is now known that definite changes can be detected before there is a noticeable rise in the blood sugar level. 1970 CAMERINI-DÁVALOS & COLE *Early Diabetes* IV. 256 Prediabetes is the condition of those persons who, because of a strong hereditary tendency, have a high probability for the later development of diabetes but in whom at the time of study, all tests of carbohydrate tolerance yield normal results.

predicability. (Later example.)

1965 F. SOMMERS in M. Black *Philos. in Amer.* 275, I shall use a reverse arrow to stand for the predicability relation between two terms.

predicant, *sb.* **1.** For 'Now *rare* or *Obs.*' read 'Now *Hist.*' and add later examples.

1816 T. J. HOWELL *Stranger in Shrewsbury* 130 The Dominicans, or Black Friars, were called in some places Jacobins, and in others Predicants. 1910 *Encycl. Relig. & Ethics* III. 176 The banishment of the pastors and the prohibition of public worship drove the people to private assemblies and the ministrations of lay preachers. Among the latter, who were known as 'predicants', François Vivens and Claude Brousson..were specially conspicuous. 1939 *Conc. Oxf. Dict. Eng. Lit.* 186/1 A monk of the order of the Predicants.

predicate, *sb.* Add: **1.** (Further examples.)

1903 B. RUSSELL *Princ. of Math.* I. iv. 45 We shall say that 'Socrates is human' is a proposition having only one term; of the remaining components of the proposition, one is the verb, the other is a predicate... Predicates.. are concepts, other than verbs, which occur in propositions having only one term or subject. 1962 A. MARTINET *Functional View of Lang.* ii. 44 *There was a riot, in the village, yesterday*... *There was* marks *the riot* as the predicate, i.e. as the element around which the others gravitate and in relation to which that function will be marked.

2. a. (Further examples.) Sometimes restricted to the main verb and its object or complement, to the exclusion of any adjunct. Also in *Logic* and *Math.*, freq. in wider use: an assertion or relation having one or more terms unspecified; a propositional function.

The generalization of *predicate* (G. *prädikat*) to include relations (many-place predicates) originated in Hilbert & Ackermann *Grundzüge der theoretischen Logik* (1928) 45: see quot. 1950.

1892 H. SWEET *New Eng. Gram.* I. 48 In language the logical connections between words extend over a wider area than the purely grammatical ones. Thus in such a sentence as *I came home yesterday morning*, the grammatical predicate to *I* is *came*, *home* and *yesterday* being grammatically connected with the predicate only, while *morning* is an adjunct to *yesterday* only. But in thought *yesterday* is as much part of the predicate as *came* itself, *came-home-yesterday-morning* being the logical predicate which, from a grammatical point of view may be regarded either as an extended predicate or a group-predicate. 1921 E. SAPIR *Language* ii. 37 The reduced sentence resolves itself into the subject of discourse—*the mayor*—and the predicate—*is going to deliver a speech*. It is customary to say that the true subject of such a sentence is *mayor*, the true predicate *is going* or even *is*. 1961 *Archivum Linguisticum* XIII. 81 The relative priority of such class concepts as noun and verb as against such as subject and predicate. 1968 J. LYONS *Introd. Theoret. Linguistics* viii. 334 *John killed Bill in Central Park on Sunday*. The subject is *John*; the predicate is *killed Bill*; and *in Central Park* and *on Sunday* are adjuncts.

1937 S. K. LANGER *Introd. Symbolic Logic* vii. 158 'Being white' has the properties of such a relation; any term, *a*, has it or does not have it, but since there is no second term we cannot say that *a* has this relation *to* any other. Such a relation of 'monadic' degree is called a predicate. *Ibid.* 159 The sole business of predicates in logic is to define classes. 1940 W. V. QUINE *Math. Logic* i. 27 'Is true' and 'is false'..are predicates by means of which we speak *about* statements. *Ibid.* 28 The verb 'implies'..is a binary predicate by means of which we talk *about* statements. 1943 *Trans. Amer. Math. Soc.* LIII. 42 Let us consider number-theoretic predicates, that is, propositional functions of natural numbers. 1950 tr. *Hilbert & Ackermann's Princ. Math. Logic* iii. 57 To the formula $x + y = z$ there corresponds a triadic predicate $S(x, y, z)$. The truth of $S(x, y, z)$ means that x, y, and z are connected by the relation $x + y = z$. [*Note*] Hitherto it has been customary in logic to call only functions with one argument place predicates, while functions with more than one place were called relations. Here we use the word 'predicate' in a quite general sense. 1965 HUGHES & LONDEY *Elem. Formal Logic* xxxix. 270 We shall..speak of the expressions, such as 'greater than' and 'between', which stand for two-place, three-place, etc., relations, as two-place, three-place, etc., predicates respectively. 1969 D. J. FOULIS *Fund. Concepts Math.* i. 14 Suppose that $P(x)$..becomes a proposition whenever x takes on any particular value in U. Then $P(x)$ is called a predicate or a propositional function, and the object variable x is called its argument. 1973 H. HERMES *Introd. Math. Logic* i. 40 In the statement *The crown jewels are kept in the Tower of London*, *The crown jewels* and *the Tower of London* can be understood as names for individuals and *are kept in* as a name for a predicate... *are kept in* is a name for a two-place predicate... *is tall* is a name for a one-place predicate.

3. *attrib.* and *Comb.*, as *predicate accusative, adjective, -centre* (*-centred* adj.), *clause, expression, marker, nominal, nominative, -part, -phrase, -position, -prefix, sentence, stress, -taking* adj., *term, variable, word*; **predicate calculus** [tr. G. *prädikatenkalkül* (Hilbert & Ackermann *Grundzüge der theoret. Logik* (1928) ii. 34)], any formal logic characterized by the use of existential quantifiers; cf. *propositional calculus* s.v. *PROPOSITIONAL *a.* b.

1887 W. W. GOODWIN *Greek Gram.* III. 194 The predicate nominative with the passive verbs of this class represents the predicate accusative of the active construction. *Ibid.* 196 The predicate adjective may be connected with its noun by the copula..or by a copulative verb. 1977 *Word* 1972 XXVIII. 79 In the following discussion I shall be concerned with predicate adjectives, except where otherwise noted. 1950 tr. *Hilbert & Ackermann's Princ. Math. Logic* p. ix, The terminology has been adapted to that of the *Grundlagen der Mathematik* by Hilbert and Bernays. For example, the term 'functional calculus' has been everywhere replaced by 'predicate calculus'. *Ibid.* iii. 67 We will now proceed, just as we did for the sentential calculus, to set up for the predicate calculus a system of axioms from which the remaining true sentences of the predicate calculus may be obtained by means of certain rules. 1955 A. N. PRIOR *Formal Logic* I. iv. 73 The calculus of predicational functions (often simply called the functional calculus, or the predicate calculus). 1966 *Mathematical Rev.* Jan. 7/1 (*heading*) Axiomatization of the infinite-valued predicate calculus. 1970 *Language* XLVI. 783 Whether grammatical or lexical-situational, all these relations are the linguistic counterpart of the predicate calculus. 1979 *Sci. Amer.* May 131/1 Could there exist an algorithm such that when it was given a statement written in precise mathematical language, it would report eventually whether the statement was true or false?.. For a powerful formalized language known as the predicate calculus it has been shown that no such algorithm exists. 1966 R. A. HALL *Pidgin & Creole Lang.* vi. 84 In the predicates of most pidgins and creoles, we find..virtually any type of free form or phrase, without any verb. Here are a few examples of nouns, pronouns, and adjectives as predicate-centers. *Ibid.* 85 Examples of other types of predicate-center include those containing adverbs or adverbial phrases. 1974 *Amer. Speech* 1970 XLV. 265 He differs from Becker in choosing a predicate-centered approach in which the verb is the centra lelement. 1966 *Eng. Stud.* XLVII. 257 Grammatically *that*-clauses.. may also function as predicate clauses. 1957 G. RYLE in M. Black *Importance of Lang.* (1962) 154 Predicate-expressions also denote what they are truly predicable of. 1966 R. A. HALL *Pidgin & Creole Lang.* vi. 83 The predicate in many pidgins and creoles..is often set apart from what goes before, by some special syntactic marker. South Seas Pidgin English..has a 'predicate-marker' /i-/, which is normally used when the subject..is not of the first or second person. 1965 N. CHOMSKY *Aspects of Theory of Syntax* iv. 181 'Bill is a lawyer.' The Predicate-Nominal of the latter is not singular, in the base structure. 1887 Predicate nominative [see **predicate accusative*]. 1957 D. L. BOLINGER in *Publ. Amer. Dial. Soc.* XXVIII. 150 This resolves the subject vs. predicate-nominative ambiguity in plain hwQs. 1924 O. JESPERSEN *Philos. Gram.* 145 We might also use the terms 'subject-part' and 'predicate-part' instead of 'primary' and 'adnex'. 1965 N. CHOMSKY *Aspects of Theory of Syntax* ii. 102 The Place and Time Adverbials that are associated with the full Predicate-Phrase. 1955 A. N. PRIOR *Formal Logic* II. iii. 160 The Schoolmen also very freely substitute singular terms for general ones, in the predicate- as well as the subject-position. 1966 R. A. HALL *Pidgin & Creole Lang.* vi. 83 Haitian and the other Central American French-based creoles have a series of predicate-prefixes, which indicate negation..and tense. 1964 *Language* XL. 46 The sentences to be discussed.. have *is* as their main verb. They will be referred to as predicate sentences. 1934 PRIEBSCH & COLLINSON *German Lang.* I. iii. 60 We might term such stresses predicate stresses, for they indicate what is the logical (if not the grammatical) predicate. 1974 *Canad. Jrnl. Linguistics* XIX. II. 153 Not only is there a problem for the analyst of knowing when to..assign a predicate-taking adjective to an *easy* or *eager* deep structure. 1901 A. SIDGWICK *Use of Words* 157 Predicate terms depend on artificial distinction. 1954 I. M. COPI *Symbolic Logic* iv. 67 We write the symbol for its predicate term to the left of the symbol for its subject term. 1937 A. SMEATON tr. *Carnap's Logical Syntax of Lang.* III. 84 In Language II, there are..not only numerical variables.., but also predicate-variables..and functor-variables. 1955 A. N. PRIOR *Formal Logic* II. iii. 158 The nearest thing to a general term-variable in the functional calculus is the predicate-variable (φ, etc.). 1932 A. H. GARDINER *Theory of Speech & Lang.* iv. 216 The subject-word places before the listener a thing to which he is to direct his attention, and the predicate-word tells him what he is to perceive or think about it.

predicate, *v.* Add: **3.** For *U.S.* read 'orig. *U.S.*' and add later *transf.* examples.

1968 *Globe & Mail* (Toronto) 5 Feb. 2/2 Mr. Diefenbaker said the federal Government had erred by predicating the conference on a bill of rights. 1973 *Times Lit. Suppl.* 20 July 836/1 A new conception of reality is demanded, predicated on dissatisfaction with formalist literature and rooted in the here and now. 1975 *High Times* Dec. 96/2 Some of the agents admitted 'they viewed routes of advancement within the DEA to be open to them predicated on the numbers of arrests they made and the amounts of narcotics they seized'. 1977 *Listener* 30 June 867/2 Crime predicated on sexual disorder I distrust.

predica·tional, *a.* [f. PREDICATION + -AL.] Of or pertaining to predication.

1894 J. VENN *Symbolic Logic* (ed. 2) ii. 59 It..concluded, in the predicational form,—using 'is' instead of 'is identical with'. 1921 W. E. JOHNSON *Logic* I. xiv. 237 Giving added significance to the predicational factor by bringing out the relation of an adjective to its determinable. 1922 *Ibid.* II. iii. 56 A function is called predicational when the component that determines its form is the characterising tie, which unites two variants related to one another as substantive to adjective. 1953 K. BRITTON *J. S. Mill* vi. 194, I say 'This snow is white'... The predicational form of sentence indicates this connexion [between whiteness and the other qualities of snow]: whereas 'There is snow and noise' does not—it is not asserted that the snow is noisy. 1955 [see *predicate calculus* s.v. *PREDICATE *sb.* 3]. 1961 *Brno Studies in English* III. 15 The unwarranted assumption that any word taken by itself must possess an independent predicational function. 1978 *Language* LIV. 90 Quantification is often predicational (i.e. with adjectives) in Japanese.

predicatival (predikātəi·văl), *a.* [f. PREDICATIV(E *a.* + -AL.] Of, pertaining to, or constituting a predicate. Also *ellipt.* as *sb.*

1891 H. A. STRONG et al. *Introd. Study Hist. Lang.* xvii. 290 A similar vacillation occurs in cases of the predicatival noun or predicatival attribute. 1937 A. H. GARDINER in *Mélanges de Linguistique et de Philologie offerts à J. van Ginneken* 310 Predicatival examples are not very frequent, e.g. *She is very Boston, Surely that knock* (i.e. at the front door) *is John*. 1958 A. A. HILL *Introd. Ling. Struct.* xvi. 274 A verb consisting of an *-ing* form..will be defined as a predicatival rather than as a predicator.

predicative, *a.* Add: **1. a.** (Further examples.) Also, of, pertaining to, or constituting a predicate. Also *ellipt.* as *sb.*

1914 O. JESPERSEN *Mod. Eng. Gram.* II. xiv. 330 As a rule words that can be used as adjuncts (pre-adjuncts) can also be used in the same form as predicatives. 1925 E. KRUISINGA *Handbk. Present-Day Eng.* (ed. 4) II. I. 235 Predicative Participles... The simple present participle is very frequently used with the copula *to be*, to form what is called the progressive. 1925 GRATTAN & GURREY *Our Living Lang.* xxi. 129 The following typical examples of Qualifiers and Predicatives. 1930 in J. T. Hatfield et al. *Curme Vol. Ling. Stud.* 46 A noun or pronoun in the subjective case may take a great variety of predicative cases. 1932 *Eng. Stud.* XIV. 129 By starting from the full meaning of the finite (or as it is now called: predicative) member of the group, the author compels us to look for a discussion of the progressive, perfect, etc., in the sections on *to be, to have*, etc. 1932 W. L. GRAFF *Lang.* ix. 328 If we emphasize the relationship of the referential parts to one another, it is noted that the Greenlandic sentence consists of a noun and its attributes, whereas the English one is formed by a subject and its predicate. Hence the further division into attributive or possessive languages and predicative ones. 1933 O. JESPERSEN *Syst. Gram.* 25 Some languages have a special case, or even two special cases in which predicatives are put: shall we say that 'a teacher' is in the 'predicative' case in 'he is a teacher' and in the 'illative' in 'he became a teacher'? 1942 R. W. ZANDVOORT in *Eng. Stud.* XXIV. 2 Only 10 of these [*sc.* forms of *to do*] are finite (predicative) forms..; the rest are non-finite (non-predicative). 1959 M. SCHLAUCH *Eng. Lang. in Mod. Times* viii. 230 He [*sc.* Deutschbein] contrasts especially the predicative sentence, which may be complicated but is apprehended as an organic whole..and the attributive sentence. 1966 *Eng. Stud.* XLVII. 50 A discussion of intransitive verbs combined with a predicative 'apposition' (e.g. *he died young*; *he died an admiral*).

b. In various special collocations, as *predicative clause, syntagm*, etc.; *predicative adjunct, appositive* = *object complement* s.v. *OBJECT *sb.* 7.

1963 Predicative adjunct [see *OBJECT *sb.* 7]. 1963 F. T. VISSER *Hist. Syntax Eng. Lang.* I. ii. 182 Syntactical units of the type..'he died *a martyr*' consist of a subject, a predicate in the form of an intransitive verb, and an adjunct (called 'predicative adjunct' in this study)... Jespersen uses the term 'quasi-predicate'; Curme the term 'predicative appositive' and F. T. Wood..the term 'pseudo-complement'. 1964 E. PALMER tr. *Martinet's Elem. Gen. Linguistics* iv. 116 The syntagm *il y avait*..is not autonomous but independent. We shall call it a *predicative syntagm*. *Ibid.* 119 The predicate comprises a predicative moneme, accompanied or not by modifiers. The predicative moneme is the element around which the sentence is organized, the other constituent elements marking their function by reference to it. 1965 *Language* XLI. 136 What Šaxmatov called 'predicative-attributive relations'. 1966 *Eng. Stud.* XLVII. 50 Intransitive verbs combined with a predicative 'apposition' (e.g. *he died young*..). I should prefer this term to Visser's 'predicative adjunct'. *Ibid.* 262 It is permissible to..speak of subject clause, predicate clause, predicative clause.

2. = PREDICATORY *a.* 1. *rare.*

1870 SWINBURNE *Let.* 19 Feb. (1959) II. 98, I trust you [*sc.* D. G. Rossetti] to 'cut close and deep'..if you find anything to pare away of the spouting or drawling, vociferous or predicative kind.

3. *Logic.* Of a function: of order only one greater than that of its argument of greatest order.

1906 B. RUSSELL in *Proc. London Math. Soc.* IV. 34 Norms..which do not define classes I propose to call *non-predicative*; those which do define classes I shall call *predicative*. 1910 WHITEHEAD & RUSSELL *Principia Math.* I. ii. 56 We will define a function of one variable as predicative when it is of the next order above that of its argument, *i.e.* of the lowest order compatible with its having that argument. 1936 *Mind* XLV. 498 The axiom of reducibility is adopted in *P*[*rincipia*] *M*[*athematica*]. This axiom moderates the second part of the theory by asserting that for every propositional function there is a formally equivalent one which is predicative, *i.e.*, has the lowest order compatible with its type. 1969 FEYS & FITCH *Dict. Symbols Math. Logic* v. 91 A property, of order only one greater than the order of what it applies to, is called by Russell a predicative property.

predicati·vity. [f. prec. + -ITY.] The fact or quality of predicating.

1963 W. V. QUINE *Set Theory* §36.265 This predicativity restriction obstructs Cantor's theorem. 1966 *Philos. Rev.* LXXV. 384 For each system in turn we examine the status of predicativity.

predicator. Restrict 'Now *rare*' to senses a and b and add further examples of sense c.

1958 [see *PREDICATIVAL *a.*]. 1961 [see *COMPLEMENT *sb.* 3 b]. 1966 *Amer. Speech* XLI. 204 All of the clauses here termed 'imperative' have as predicators verb forms which we can describe as present-tense subjunctives. 1966 G. N. LEECH *Eng. in Advertising* ii. 10 The elements of clause structure in English are:..P: Predicator (traditionally 'verb', but this term is needed for a class of word). 1969 *Eng. Stud.* L. 32 Clauses without a Predicator ..can assume the status of independent clauses.

predict, *v.* Add: **2. b.** *transf.* Of a theory, observation, etc.: to have as a deducible or inferable consequence; to imply.

1961 *Physical Rev.* CXXI. 1620 The theory predicts a linear dependence of M_{2p} on $[H_e/(T + \theta)]^2$, where θ is the experimentally determined Curie-Weiss constant. 1964 E. BACH *Introd. Transformational Gram.* viii. 186 General linguistic theory must provide a precise characterization of the way in which a theory can be said to 'predict' a given sentence. 1975 *Nature* 6 Feb. 442/1 Sensitivity to the taste of PTC predicts sensitivity to caffeine. 1976 *Sci. Amer.* July 39/3 The present isotopic ratios of neodymium therefore predict the total depletion in U-235. This calculation gives a result about 40 percent greater than the observed depletion. 1977 *Lancet* 24 Sept. 662/1 Running-water samples are perhaps closer to the water typically consumed in the home than are first-flush samples, and our results..indicate that they predict blood-lead more precisely.

4. To direct fire at with the aid of a predictor (sense *2).

1943 L. CHESHIRE *Bomber Pilot* iii. 57 They're predicting us now; looks like a barrage. 1952 M. TRIPP *Faith is Windsock* vi. 90 He saw a flak-burst below, then another, and another... 'Weave, Dig, the bastards are predicting us.'

predictability. (In Dict. s.v. PREDICTABLE *a.*) Add: (Further examples.) Also *attrib.*

1954 J. H. GREENBERG in H. Hoijer *Lang. in Culture* 4 Causality should not be confused with predictability. Perhaps only a predictability relation is discerned in some cases. 1955 *Bull. Atomic Sci.* June 227/3 The concept of predictability seems in some way related to the concept of pattern. 1972 *Archivum Linguisticum* III. 4 The imperfect subjunctive occurs consistently, with equal predictability, in the language of all characters, in all situations.

predictably (prĭdi·ktăbli), *adv.* [f. PREDICTABLE *a.* + -LY[2].] In a manner that can be or could have been predicted.

1914 J. H. SKRINE *Pastor Futurus* 88 The Pentecosts come back, as surely though not so predictably as the dawns. 1961 *Time* 13 Jan. 9/2 The British and French.. agreed to the setting up of a Communist state, North Viet Nam—which then, predictably, became a base for Communist operations. 1971 *Times* 25 Nov. (Canning Suppl.) p. ii/9 Metrication [of cans] is unlikely to be imposed, but economic pressures will probably hasten a voluntary change. Predictably, the International Organization for Standardization is busy on this and other metrication problems. 1975 *Times Lit. Suppl.* 24 Jan. 92/3 A full catalogue has been prepared by P. R. S. Moorey,..who has already made himself the undisputed master of this field... Predictably, the commentary..is authoritative and scholarly.

prediction, *sb.* Add: **3.** *attrib.* and *Comb.*, as *prediction paradox, study, table, value.*

1952 *Mind* LXI. 265, I hope Mr O'Connor will not mind my giving his paradox the new and somewhat more appropriate name of 'the prediction paradox'. 1950 S. A. STOUFFER *Measurement & Prediction* vi. 173 Each variable in a prediction study plays one of two possible roles. 1964 M. ARGYLE *Psychol. & Social Probl.* v. 70 Is it possible to work out a grand prediction table showing what treatment each individual should have? 1961 J. B. WILSON *Reason & Morals* iii. 144 Many scientists now use the prediction-value of scientific statements as virtually the only test of their truth.

predictionism (prĭdi·kʃəniz'm). [f. PREDICTION *sb.* + -ISM.] Belief in prediction or prophecy.

1919 P. H. OSMOND *Mystical Poets Eng. Church* vii. 215 He was a 'crank', dominated by extravagant notions—a victim of Predictionism and credulity. 1943 *Mind* LII. 200 The immediate issue is not Behaviourism but Predictionism, the doctrine which Professor C. I. Lewis so well sets out.

predictive, *a.* **a.** (Further examples.)

1908 *Westm. Gaz.* 9 May 4/4, I can see with prophetic eyes and hear with predictive ears a development of programme-music which may in the future militate somewhat against the dominant position of the opera. 1957 *Publ. Amer. Dial. Soc.* 1956 XXVI. 71 The results proved to be over 90% predictive. 1961 A. G. OETTINGER in *Proc. Symposia Appl. Math.* XII. 105 Predictive analysis yields a description of the syntactic structure of a sentence in terms consonant, although not identical, with old-fashioned parsing, immediate constituent theory.., or phrase-structure theory. 1964 E. A. NIDA *Toward Sci. Transl.* xii. 259 In contrast with this pass procedure is a 'predictive method'.., which more closely represents the mathematician's view of the language structure—one based on the expectations of what is to follow. 1966 I. RHODES in *Automatic Transl. of Lang.* (NATO Summer School, Venice, 1962) 206 Our method has become known as 'predictive analysis' and is based upon the universal habit on the part of the listener to *anticipate* the type of word which a speaker is about to utter. 1966 *Jrnl. Canad. Operational Res. Soc.* 117 *Predictive model*, a model used in a war game to predict the results of actions and interactions between opposing forces. 1972 *Jrnl. Social Psychol.* LXXXVIII. 145 Although the measures are required to compensate for educational disadvantage, they are also expected to have acceptable predictive validity in a system where all groups are competing for the limited further education places.

predictor. Add: **1. b.** *spec.* in *Statistics*, a variable whose value can be used in estimation; also *predictor variable*.

1950 S. A. STOUFFER *Measurement & Prediction* vi. 173 Each variable..can serve as a predictor. 1966 DRAPER & SMITH *Appl. Regression Analysis* iv. 104 There are many problems in which a knowledge of more than one independent (or 'predictor') variable is necessary in order to obtain better understanding and/or better prediction of a particular response. 1974 *Nature* 9 Aug. 466/1 Students' attitudes towards scientists were strongly related to only two of the predictor variables: deference and nurturance. 1975 *Sci. Amer.* May 97/1 Later species grew up in the shade of the pioneering species, and the numerical abundance of saplings in the understory proved to be a reasonable predictor of a species' success in reaching the canopy. 1977 *Canad. Jrnl. Linguistics* 1976 XXI. I. 21 Dashed lines are curves predicted on the basis of an additive model utilizing a multiple classification analysis.., which demonstrates the main effects of a given predictor (or independent variable).

2. *Mil.* An apparatus for automatically providing tracking information for an anti-aircraft gun from telescopic or radar observations.

1935 L. HART *When Britain goes to War* II. vi. 119 Greater progress has come through the invention of..the Vickers Predictor, whereby a combined calculation of the speed, course and height of the aeroplane is automatically made and electrically transmitted to the guns. 1936 *Sphere* 30 May 363 (*caption*) Operating the predictor, a delicate instrument for determining the range of enemy 'planes. 1941 *Ann. Reg.* 1940 69 The defenders adopted new tactics, gauging the path of the raiders by means of predictors instead of using searchlights. 1944 H. HAWTON *Night Bombing* 92 Cologne's defences were massive, but with the sky thick with aircraft the searchlights were bewildered and the predictors confused. 1962 S. PUGH *Fighting Vehicles & Weapons* II. 74 The most satisfactory solution so far is this British-designed combination of Swedish Bofors 40 mm. power-operated, automatic light anti-aircraft gun, L.70, with a radar/predictor fire control equipment known as Fire Control Equipment No. 27 'Yellow Fever'. 1974 *Encycl. Brit. Macropædia* IV. 1048/2 The original electronic analogue computers arose from the needs of anti-aircraft artillery 'predictors'.

predigested, *ppl. a.* (Later *lit.* and *fig.* examples.)

1902 *Brit. Med. Jrnl.* 17 May 1199/1 In the case of the premature infant that is unable to suck, it has been found to be of advantage to pass predigested food directly into the stomach by means of a tube. 1922 'K. MANSFIELD' *Let.* 17 July (1928) II. 229 What a relief it is to turn away from these little pre-digested books written by authors who have nothing to say! 1940 R. S. LAMBERT *Ariel & all his Quality* iv. 114 This paper [sc. *The Listener*] whose 'copy' comes to it predigested from other sources. 1975 *Country Life* 18 Dec. 1734/1 As early as 1789 Thomas Pitt, a Worcester chorister, published his pre-digested *Messiah* —ten anthems.

prediluvian, *a.* and *sb.* Restrict *rare* to the sb. and add further examples of the adj.

1928 V. G. CHILDE *Most Anc. East* i. 15 The Prediluvian kings' reigns are all incredibly long. 1931 C. WILLIAMS *Place of Lion* xvi. 277 Some incantation whereby the prediluvian magicians had controlled contentions among spirits. 1981 *Times Lit. Suppl.* 26 June 731/2 Their pre-diluvian fundamentalist faith.

pre-dinner: see *PRE- B. 2a.

prediscover, *v.* (Later example.)

1926 *Spectator* 3 July 18/2 The poet 'prediscovers' the Einstein theory.

pre-discovery. Add: **B.** *adj.* [PRE- B. 2.] Occurring or carried out before the discovery of something.

1946 *Nature* 9 Nov. 648/1 In the nomenclature of the time, these pre-discovery observations of Uranus are known as the 'ancient' observations. 1968 R. A. LYTTLETON *Mysteries Solar Syst.* vii. 234 LeVerrier, in his discussion, included the pre-discovery observations of Uranus going back to 1690.

predisposition. Add: **1.** Also, a tendency in a person to respond or react in a certain way. (Further examples.)

1936 *Discovery* Aug. 254/1 All these effects..can be shown to result from psychological inhibitions and predispositions. 1949 C. I. HOVLAND et al. *Exper. Mass Communication* vii. 192 A person soon 'forgets' the ideas he has learned which are not consonant with his predispositions. 1973 G. A. DAVIS *Psychol. of Problem Solving* ii. 18 Habit and conformity are implicit in such..personality concepts as rigidity,..predisposition,..fear of the unknown and, on occasion, pigheadedness. 1980 *Sci. Amer.* Apr. 112/1 It is generally accepted that most animal characteristics are the product of an interaction between inherited predispositions and the environment.

predissocia·tion. *Physics* and *Chem.* [PRE- A. 2.] The passage of a molecule between a quantized vibrational and rotational state (above its ground state) and a dissociated state of the same energy that is not quantized, the occurrence of which results in certain bands in the spectrum of the molecule being diffuse instead of having the normal rotational fine structure. Freq. *attrib.*

1924 HENRI & TEVES in *Nature* 20 Dec. 895/1 The molecule can be modified in its internal structure: the atoms are driven apart, the bonds are weakened, the molecule becomes more reactive, and the rotational movements are no longer quantified. This first modification is a preliminary preparation of the molecule for its total dissociation, and it is necessary to introduce a new term for this change. We propose to denote it by the term predissociation of the molecule. 1930 *Physical Rev.* XXXV. 1028 (*heading*) Predissociation of diatomic molecules from high rotational states. 1944 GLASSTONE *Theoret. Chem.* iv. 188 In some cases the predissociation spectrum is followed by a region of continuous absorption but, in other instances, bands with fine structure are found on both long and short wave sides of the predissociation bands. 1962 P. J. & B. DURRANT *Introd. Adv. Inorg. Chem.* vii. 222 Predissociation is commonly shown only in absorption, not in emission. 1966 BARNARD & MANSELL *Fund. Physical Chem.* ii. 96 A molecule undergoing pre-dissociation will dissociate within the time of one rotation and the rotational fine structure will tend to be lost in the gas phase spectrum. 1977 *Sci. Amer.* Feb. 95/1 Instead the molecule is disassembled at a lower energy through a phenomenon called predissociation.

Predmost (pre·dmǝ^u^st), *a.* Also **Předmost** (pə^ə^ʒe·dmost). [Anglicized form of the place-name *Předmost* near Brno (Brünn), in Moravia, Czechoslovakia.] Of, connected with, or relating to human or other remains, artefacts, etc., that evidence a Combe Capelle type of *Homo sapiens* of Upper Palæolithic culture, first excavated at Předmost between 1882 and 1894, and later at other sites, esp. in Central Europe and round the eastern Mediterranean; also *ellipt.* as *sb.* Hence **Predmo·stian.**

1916 H. F. OSBORN *Men Old Stone Age* iii. 257 Such very primitive forms as the Brünn or Předmost race of Upper Palæolithic times. *Ibid.* iv. 349 All these sculptures of the mammoth have in common the indication of a very small ear—similar to that in the Předmost model. 1921 M. C. BURKITT *Prehistory* x. 130 The male statuette of Brünn, if it be of the age of Předmost and not more ancient, appears to be a prolongation of the Aurignacian artistic technique. 1927 PEAKE & FLEURE *Hunters & Artists* v. 67 We may call it provisionally the Combe Capelle, or perhaps better the Predmost type by way of contrast with the Cro-Magnon type. 1931 *Times Lit. Suppl.* 23 Apr. 317/2 Sir Arthur Keith suggests that both the Cro-Magnons and the Predmostians are early Caucasians hailing ultimately from South-Western Asia. 1939 V. G. CHILDE *Dawn European Civilization* (ed. 3) i. 3 Mesolithic groups appear in general isolated and poorly equipped in contrast to Magdalenians or Předmostians. 1957 M. BULLOCK tr. *Boule & Vallois's Fossil Men* viii. 278 The long discussion of which these Predmost Men have been the object; their affinities are, above all, with the great Cro-Magnon race. 1960 W. HOWELLS *Mankind in Making* xiv. 210 There you have *Homo sapiens*. And there you have the men of the present, and the men of the Upper Paleolithic. Some of the latter, like the Předmost skulls, had rather strong brow ridges. 1977 BRACE & MONTAGU *Human Evol.* ix. 351 (*caption*) Předmost skull. Upper Paleolithic of Czechoslovakia.

prednisolone (predni·sŏlǝ^u^n). *Pharm.* [f. next with inserted *ol* (see -OL).] A synthetic steroid, $C_{21}H_{28}O_5$, which has similar properties and uses to prednisone, of which it is a reduced derivative.

1955 *Jrnl. Amer. Med. Assoc.* 21 May 166/1 Prednisolone (Meticortelone) and prednisone (Meticorten), two new synthetic steroids formerly known as metacortandralone and metacortandracin, have recently been recommended as effective in the treatment of rheumatoid arthritis. 1959 *Economist* 19 Dec. 1160/1 One of the five firms which produce the bulk of such drugs, charged chemists $17.90 for 100 tablets of prednisolone, under the brand name of Meticortelone. 1962 HARRIS & GRUBER in A. Pirie *Lens Metabolism Rel. Cataract* 379 Of the steroids, prednisolone, either as its phosphate or as the alcohol, caused the greatest reduction in cation recovery. 1966 *Lancet* 24 Dec. 1382/2 He was given the usual measures for heart-failure and antibiotics, oxygen, and prednisolone 10 mg. t.d.s. to 'buy time'. 1974 PASSMORE & ROBSON *Compan. Med. Stud.* III. xviii. 98/1 Until the prednisolone takes effect the patient's distress should be relieved by bronchodilator drugs.

prednisone (pre·dnisǝ^u^n). *Pharm.* [Prob. f. *pregnadiene* (f. *PREGNA(NE + *DIENE) + *-isone*, after *CORTISONE.] A colourless, crystalline, synthetic steroid, $C_{21}H_{26}O_5$, resembling cortisone but possessing greater glucocorticoid activity, which is used as an anti-inflammatory agent and to depress immune responses, esp. in the treatment of rheumatoid arthritis.

1955 *Dis. Chest* XXVII. 515 Metacortandracin (Metacortin) used in this study and furnished by Schering Corporation, Bloomfield, New Jersey has been re-named Prednisone. 1955 [see *PREDNISOLONE]. 1959 S. DUKE-ELDER *Parson's Dis. Eye* (ed. 13) xiv. 150 Prednisone and Prednisolone, are often preferable for systemic administration since they are less liable to excite the unfortunate side-effects associated with cortisone. 1965 *Spectator* 1 Jan. 12/3 Prednisone is especially useful for easing the pain of aged arthritics. 1974 R. M. KIRK et al. *Surgery* iv. 58 Generalised disease responds to chemotherapy, usually with a combination of vincristine, mustine, procarbazine and prednisone in pulsed doses.

predominately, *adv.* Delete 'Now *rare*' and add later examples.

1961 *Christian Sci. Monitor* 17 Oct. 4/7 WWRL's [*sc.* a radio station's] colorful mobile unit, cruising predominately Negro neighborhoods. 1965 C. WALSH in J. Gibb *Light on C. S. Lewis* 110 He quickly gained a wide audience...it was predominately high-brow and middle-brow. 1970 [see *NON-WORD]. 1973 *Yale Rev.* Spring 452 Other indications that the *Supplement* is predominately concerned with the modern..history of the English vocabulary. 1977 *Lancet* 6 Aug. 306/2 We were fascinated by the suggestion..that in San Francisco enteric diseases are predominately sexually transmitted.

pre-dry: see *PRE- A. 1.

pre-e·cho. [PRE- A. 2.] **1.** A faint copy of a louder sound occurring in a recording shortly before the original as a result of the accidental transfer of signals in a recording medium.

1935 *Gramophone* June 42/2 It appears to be not 'an echo' in the strict sense of the word, but a 'pre-echo', as in all the examples I list below it occurs *before the actual recording grooves* are reached by the needle. 1956 [see *POST-ECHO]. 1957 *N.Y. Times* 24 Feb. x. 15/1 Engineers say that a disk should not contain much more music than that;..the grooves will have to run too closely together with additional minutes;..there will be pre-echo, damage and results too ghastly to contemplate. 1962 *Times* 5 July 15/7 Prolonged storage [of tape] without rewinding.. can cause 'print-through' (detectable as pre-echo on some discs). 1967 A. L. LLOYD *Folk Song in England* i. 21 The phenomenon of pre-echo on magnetic tape. 1976 *Gramophone* Sept. 445/3 It certainly avoids pre-echoes in silent bars immediately followed by fortissimi.

2. A foreshadowing or anticipation.

1948 *Mind* LVII. 375 Professor Raphael..commends Price's refutation of the 'naturalistic fallacy' (a pre-echo of G. E. Moore's). 1961 *Times* 29 May 12/7 Is this a mere pre-echo of *My Fair Lady*? 1975 *Listener* 20 Nov. 674/1 The most fascinating political pre-echo since the boy Harold Wilson had his photo taken on the steps of Number Ten. 1977 *Gramophone* July 187/2 What an extraordinary pre-echo of Brahms this second piece becomes in this performance.

pre-ecla·mpsia. *Path.* [PRE- B. 1.] A condition of pregnancy characterized by high blood pressure and some other of the symptoms associated with eclampsia, and formerly thought to be associated with toxæmia.

1923 *Jrnl. Michigan State Med. Soc.* XXII. 144/1 (*heading*) Toxemias of pregnancy including pre-eclampsia, eclampsia and nephritis. 1929 H. J. STANDER *Toxemias of Pregnancy* 53 Pre-eclampsia is essentially eclampsia before the outbreak of convulsions and coma. 1955 H. M. CAREY in I. Donald *Pract. Obstetr. Probl.* ix. 130 Pre-eclampsia classically is defined as the appearance, after the 28th week of pregnancy, of œdema, hypertension and albuminuria. 1974 PASSMORE & ROBSON *Compan. Med. Stud.* III. II. xlii. 17/2 The term pre-eclampsia has also been used [for hypertension in pregnancy], on the basis that women who develop hypertension in late pregnancy may, unless treated, progress to the convulsive complication known as eclampsia. 1976 *Lancet* 18 Dec. 1341/1 Maternal pre-eclampsia, birth trauma, breech delivery, and disorders of hæmostasis have also been implicated in a disorder which predominantly affects immature male infants.

pre-ecla·mptic, *a.* and *sb.* *Path.* [PRE- B. 1.] **A.** *adj.* Characteristic of the state which precedes an eclamptic attack; of, exhibiting, or being pre-eclampsia. **B.** *sb.* A pre-eclamptic woman.

1899 J. C. EDGAR in C. Jewett *Pract. Obstetr.* xxiii. 517 Symptoms of eclampsia may be classified as those of the prodromal period, or pre-eclamptic state, and those of the attack. 1924 DE WESSELOW & WYATT *Mod. Views on Toxæmias of Pregnancy* viii. 90 It is practically certain that we are dealing with a pre-eclamptic. 1926 *Jrnl. Obstetr. & Gynæcol.* XXXIII. 21 The patient, 2-para, and seven months pregnant, was admitted for the pre-eclamptic signs of high blood-pressure, marked œdema and headache. 1960 LEVITT & ALTCHEK in Guttmacher & Rovinsky *Med., Surg., & Gynecol. Complications of Pregnancy* iv. 73/1 Clinically, the preeclamptic has more pronounced sodium and water retention than the normal gravida. 1962 *Lancet* 6 Jan. 7/1 Induction was performed in a patient with mild pre-eclamptic toxæmia. 1977 *Ibid.* 30 Apr. 923/1 Diabetic women in pregnancy often become pre-eclamptic.

pre-e·dit, *v.* [PRE- A. 1.] *trans.* To edit or sort as a preliminary to later editing; to prepare for computer processing by the addition or alteration of material. Hence **pre-e·dited** *ppl. a.*, **pre-e·diting** *vbl. sb.*

1934 WEBSTER, Pre-edit, *v.* 1938 *Amer. Speech* XIII. 36 In pre-editing, an editor may have to consult recent books in the subject-field concerning a particular word. 1958 *N.Y. Folklore Q.* Autumn 245 The valuable *Frank C. Brown Collection of North Carolina Folklore* would have been much more valuable had Brown spent more time in arranging and pre-editing while he was alive. 1960 [see *POST-EDIT *v.*]. 1968 *Amer. Documentation* Jan. 74/1 The documents are then returned to the DARE office for pre-editing and typing. 1970 A. CAMERON et al. *Computers & Old Eng. Concordances* 49 The not-so-friendly blue giant grudgingly yielded up a trial-run concordance to 1,100 pre-edited lines out of the approximately 5,500 lines of the complete gospel text. 1973 *Amer. Speech* 1969 XLIV. 192 The completed field records were..preedited and encoded. 1977 J. M. SMITH in P. G. J. van Sterkenburg et al. *Lexicologie* 242 There is a need for some 'intelligent' pre-editing of input... Such pre-editing is time-consuming.

Hence **pre-e·ditor,** one who carries out pre-editing.

1934 in WEBSTER. 1953 [see *POST-EDITOR]. 1958 *Aspects of Translation* 104 The pre-editor was to remove known ambiguities from the original text. 1960 E. DELAVENAY *Introd. Machine Transl.* 36 The role of the pre-editor would be to provide the machine with texts explicit from the graphio-semantic point of view.

pre-elect, *v.* **b.** Delete '*rare*' and add: Used *spec.* of the choice of heads of colleges and of certain classes of fellows in the universities of Oxford and Cambridge.

1977 *Daily Tel.* 28 Feb. 8/2 At Sidney Sussex, Cambridge, Miss A. P. Dowling, of Girton, has been pre-elected into a junior research fellowship. 1978 *Ibid.* 20 Jan. 14/4 At Christ's College, Prof. J. H. Plumb, Fellow of the College and Emeritus Professor of Modern English History, has been pre-elected to Mastership with effect from July 11, 1978.

pre-ele·ctric, *a.* [PRE- B. 1.] Occurring or pertaining to the time before the use of electricity, esp. in the making of gramophone records. Also *ellipt.*, a gramophone record not electrically recorded. Also **pre-ele·ctrical** *a.*

[1908 *Westm. Gaz.* 29 Feb. 12/2 It was in pre-electric-light days, and I couldn't find the matches.] 1934 C. LAMBERT *Music Ho!* IV. 257 Recording will have improved on the present methods as much as the present methods have improved on the old pre-electric horn recording. 1947 *Penguin Music Mag.* May 58 The connoisseur, who has his collection of rare pre-electrical recordings carefully card-indexed. 1960 *Guardian* 8 Mar. 7/1 One has to listen to old pre-electrics with a 'creative ear'. 1968 *Listener* 22 Feb. 250/1 It is scored for..stereophonic tape.., pre-electric gramophone, percussion, [etc.]. 1977 *Gramophone* Feb. 1321/1 The first five records concern themselves almost entirely with pre-1930 (mostly pre-electric) issues.

preem (prīm), *sb.*[2] and *v.*[2] U.S. slang abbrev. of PREMIÈRE *sb.* or *v.* in Dict. and Suppl.

1937 *Amer. Speech* XII. 317/2 *Preem*, first showing. 1937 [see *ORK]. 1942 BERREY & VAN DEN BARK *Amer. Thes. Slang* §590/30 *Preem*, to present a premiere performance. 1945 [see *HYPO *v.*]. 1945 [see *PREMIÈRE *v.*]. 1948 *Variety* 25 Aug. 1/2 The mother-daughter act..has been bought by ABC and set for an Oct. 4 preem. 1952 *N.Y. Daily News* 5 Aug. 23c/5 A new hour-long radio show..which preems via ABC [network] Sunday, Aug. 17. 1961 A. BERKMAN *Singers' Gloss. Show Business* 70 *Preem* (Var.), theatre premiere.

pre-emergence, -emergent: see *PRE- B. 2, 1.

preemie (prī·mi). *N. Amer. slang.* Also premie, premy. [f. *PREM(ATURE *sb.* + -Y[6], -IE.] A premature birth; a baby born prematurely. Also *attrib.*

1927 *Amer. Speech* II. 314/1 A baby delivered prematurely is a 'premy'. 1942 BERREY & VAN DEN BARK *Amer. Thes. Slang* § 534/6 *Premie*, a premature birth. 1949 *N.Y. Times* 25 Sept. I. 75/2 Saving 75 per cent of the 'preemies' born. *Ibid.*, The dread eye diseases said to afflict 'preemies' are unfounded. 1968 *Trans-Action* Oct. 7/2 The prematures were more likely to be below their proper grade in school. Among white children..19·4 percent of the preemies were..in special classes. 1975 *Time* (Canada ed.) 19 May 57/1 The preemie's sense of security is further heightened by the recorded sound of a pregnant mother's heartbeat piped into the artificial womb. 1976 *Word* 1971 XXVII. 61 The present-day premie nursery is the precursor of the prenatal assessment laboratory of tomorrow.

pre-e·mphasis. *Sound Recording* and *Broadcasting.* [PRE- A. 2.] A systematic distortion of a signal prior to transmission or recording, involving an increase in the relative strength of certain frequencies in anticipation of a corresponding decrease during reception or playback.

1940 *RCA Rev.* Jan. 359 The use of pre-emphasis circuit [*sic*] at the transmitter and a de-emphasis circuit at the receiver produces an overall gain in signal-noise ratio. 1942, etc. [see *DE-EMPHASIS]. 1959 K. HENNEY *Radio Engin. Handbk.* (ed. 5) xxi. 35 Below 100 cycles the characteristic of the [disk] recorder system is made constant-velocity by electric means. This tends to give pre-emphasis to the low frequencies... Above 500 cps a preemphasis above a constant velocity is given to the high frequencies, especially over the noise frequency range. The necessary characteristic for reproduction is the inverse of this curve. 1977 *Gramophone* Mar. 1476/3 The most important of these is the 'breathing' or 'pumping' effect as residual noise is heard rising and falling in level as the expander alters system gain in response to sudden changes in signal level... High frequency pre-emphasis, followed by mirror-image de-emphasis, reduces the effects of breathing.

So **pre-e·mphasize** *v. trans.*, to subject to pre-emphasis.

1968 COOK & LIFF *Frequency Modulation Receivers* xiv. 490 Both the left (*L*) and right (*R*) channels are fed to 75-μ sec high-pass filters, where they are preemphasized. 1974 M. MANDL *Mod. Television Syst.* ii. 29 The higher audio-frequency range at the transmitter is preemphasized. 1977 *Gramophone* Mar. 1476/1 A dodge of this kind is universally employed in VHF/FM broadcasting, where high frequencies are boosted (pre-emphasized) at the transmitter and attenuated (de-emphasized) at the receiver.

pre-employment: see *PRE- B. 2a.

pre-empt, *sb.* Restrict *Austral. colloq.* to sense in Dict. and add: **2.** *Bridge.* A pre-emptive bid.

1939 N. DE V. HART *Bridge Players' Bedside Bk.* iii. 34 Macleod's pre-empt showed an obvious fear of both major suits, from which he was trying to shut us out. 1959 *Listener* 8 Jan. 84/2 It [*sc.* the hand] could qualify for the bolder pre-empt of *Four Clubs*. 1962 *Times* 11 July 7/1 Few players would fancy a pre-empt with a two-suiter and two primary controls. 1972 R. MARKUS *Common-Sense Bridge* II. 65 If everybody knows the strength and weakness of your pre-empts they can easily take the right counteraction. 1977 *Homes & Gardens* Feb. 17 There are two types of hand where you should respond Three No-Trumps in reply to a pre-empt.

pre-empt, *v.* Add: **1. b.** (Earlier and later examples.)

1855 L. OLIPHANT *Minnesota & Far West* 162 Wal, I guess, if you can find a corner that's not pre-empted, you may spread your shavings there [for a bed]. 1913 J. LONDON *Valley of Moon* II Many [tables and benches].. were already pre-empted by family parties. 1944 AUDEN *Sea & Mirror* in *For Time Being* 15 Two wonders as one vow Pre-empting all.

2. *Bridge.* **a.** *intr.* To make a pre-emptive bid.

1914 M. C. WORK *Auction Devel.* 313 It is the exceptional case in which it is advisable to preëmpt with an original No Trump. 1920 —— *Auction Methods Up-to-Date* v. 65 His only chance is to preëmpt so strongly that his first bid will hold the declaration. 1947 S. HARRIS *Fund. Princ. Contract Bridge* I. i. 17 When North pre-empts but does not make a game bid, it is important for South to remember that he must not increase the contract unless he holds three quick tricks. 1964 *Official Encycl. Bridge* 435/1 The third player is best placed to pre-empt. 1972 R. MARKUS *Aces & Places* 35 South opened the bidding with 1 ♠, West doubled and North.. pre-empted to 4 ♠, which became the final contract.

b. *trans.* To thwart (a player) by making a pre-emptive bid.

1964 *Official Encycl. Bridge* 435/1 The third player.. knows that he cannot pre-empt his partner. 1972 R. MARKUS *Common-Sense Bridge* II. 65 Here is a hand..to show how easily you can be pre-empted into a ridiculous contract.

3. To set aside (one thing in favour of another); to preclude (something); to prevent (an occurrence); to forestall (someone).

1965 *Sun* 6 Dec. 7/7 In American TV you never, never say that a serial has been killed in favour of a new serial. It is always pre-empted. What they really mean is that it has been cancelled and a right established for the next one. 1968 *Listener* 5 Dec. 768/1, I think the Nazi regime by its own grotesque vileness pre-empted fictional effort. 1976 'A. HALL' *Kobra Manifesto* xvi. 217 He would kill me when the showdown came unless I could pre-empt him. 1977 *B.B.C. Radio 4 News 5 p.m.* 11 May (*recorded from oral evidence*) Federal rights pre-empt State rights. 1978 *Jrnl. R. Soc. Arts* CXXVI. 675/1 The targets serve to preempt such a situation arising.

Hence **pre-e·mpting** *vbl. sb.* and *ppl. a.*

1920 M. C. WORK *Auction Methods Up-to-Date* v. 61 With general strength, preëmpting is not necessary or advisable. *Ibid.* 63 A real preëmpting hand contains an unusual distribution of cards. 1965 H. KAHN *On Escalation* 287 It [*sc.* pre-emptive war] denotes an attack made because of a belief that the other side has determined to make an attack on the pre-empting party. 1967 *Listener* 2 Nov. 570/3 On the subject of 'pre-empting'—supplanting scheduled programmes in favour of special programmes of public interest—..he had some extremely interesting things to say.

pre-emption. Add: **1. b.** (Examples.) Also *concr.*, land so obtained or to be obtained.

1747 *First Rec. Baltimore Town* (1905) 21 Mr. Alexander Lawson applied also to enter his Preemption of making out Ground into the water. 1827 *United Empire Loyalist* (Toronto) 6 May 396/2 The first hundred purchasers of Town Lots, when they have erected a habitable house, will..be entitled to the pre-emption or privilege to purchase a Lot of Twenty-Five Acres..at..7s. 6d. per acre. *a* 1844 *Filson Club Hist. Q.* (1935) IX. 235 Each of these two men..had a pre-emption of 1400 acres. 1901 DUNCAN & SCOTT *Hist. Allen & Woodson Counties, Kansas* 582 Finding that the Indians would not settle on the Reserve, the Government, in 1860, had all of these lands offered for sale and opened to pre-emption. 1933 W. W. SPINKS *Tales Brit. Columbia Frontier* 110 Some of the land had already been pre-empted, and pre-emption amounted to an agreement by the government to sell the land to the pre-empter. 1968 R. H. PATTERSON *Finlay's River* 43, I see I have called it a homestead. Officially, in the books of the Land Registry, it is a pre-emption.

e. (Earlier examples.)

1780 in N. D. Mereness *Trav. Amer. Colonies* (1916) 643 Received a Letter and Preemption Warrant. 1784 J. FILSON *Discovery Kentucke* 37 The Settlement and pre-emption rights arise from occupation.

2. *Bridge.* The action of making a pre-emptive bid.

1961 *Times* 6 Dec. 8/3 A two-suiter is not built for pre-emption. 1962 *Times* 7 Mar. 3/6 A preemption in one of the minor suits is even less efficacious unless the hand has two tricks on the side. 1974 *Country Life* 26 Sept. 894/1 The hand is far too good for pre-emption. 1977 *Times* 17 June 12/1, I have been making notes of unsuccessful pre-emptions with their effect on subsequent bidding.

3. The action or an instance of setting aside or overriding something.

1978 *Nature* 20 Apr. 664/1 The issue of Federal pre-emption—Federal legislation that overrides state or local initiatives—lies at the heart of the current dispute. 1979 *Arizona Daily Star* 1 Apr. (Tucson T.V. Suppl.) 12/3 CBS hasn't treated this inspirational program very kindly. It's constantly being victimized by pre-emptions, time-slot changes and disappearances for up to three weeks at a stretch.

pre-emptioner (earlier examples).

1838 *Congress. Globe* 25th Congress 2 Sess. App. 142/3 Suppose a pre-emptioner was to go there and say, Mr. President, this house is too large for you; I..claim a pre-emption to part of this house. 1841 *Knickerbocker* XVII. 278 They amused themselves by calling the exclusives 'squatters', 'prëemptioners', etc. 1872 [see *HOMESTEADER].

pre-emptive, *a.* (*sb.*). Add: **2.** *Bridge.* Applied to a bid made with the expectation that it is high enough to prevent opponents from bidding normally and so obtaining adequate information.

1913 F. IRWIN *Auction High-Lights* 95 A preëmptive opening-bid in a major suit means that the bidder wants no information and wishes to play the hand at his own suit. 1916 'BASCULE' *Adv. Auction Bridge* I. 77 To what extent does it pay to make what are known as pre-emptive, or 'shut-out' bids? 1923 *Daily Mail* 5 May 8 The supporting bid,..the pre-emptive raise, and 'the switch' assume a new value. 1932 *Daily Tel.* 8 Oct. 15/5 In using the term 'pre-emptive' I am not in any way ascribing the meaning of 'shut-out' to that word. 1947 S. HARRIS *Fund. Princ. Contract Bridge* I. i. 17 The most valuable pre-emptive bid..is an opening bid of four of a major suit or five of a minor suit. 1952 I. MACLEOD *Bridge* v. 62 (*heading*) Pre-emptive responses. 1973 *Times* 20 Oct. 11/3 You will find plenty of opportunities for preemptive opening bids.

3. Designating an attack on an enemy who is thought to be about to make an attack himself (see also quots. 1966, 1971). Also *transf.*

1959 *Listener* 31 Dec. 1140/1 The American Strategic Air Command..might be prevented by a Russian pre-emptive strike from ever getting the Sword out of its scabbard. 1966 SCHWARZ & HADIK *Strategic Terminol.* 108 *Pre-emptive strike*, armed attack motivated by the conviction that an enemy attack is under way or is irreversibly imminent. Also called 'forestalling blow' or 'anticipatory attack', the pre-emptive strike differs from a so-called 'preventive' strike or war in that [etc.]. A strike or war..is preventive if the enemy still has the option of desisting from his planned aggression. 1970 *Times* 9 Oct. 15/2 December 7, 1941, when the Japanese stabbed America in the back at Pearl Harbour—or as we would say in these cooler, more euphemistic times, made their pre-emptive strike. 1971 E. LUTTWAK *Dict. Mod. War* 156/1 *Pre-emptive attack*, an attack launched in the belief that an enemy attack has already entered the executive phase, i.e. that the decision has already been made. Unless the attack actually reduces or eliminates the effect of the imminent attack, it cannot be called pre-emptive. 1976 LD. HOME *Way Wind Blows* xii. 167 There was no doubt at all about the most effective deterrent—it was the 'Polaris' submarine; which, because it was virtually undetectable, was a genuine second-strike weapon which robbed the pre-emptive attack of all its former attraction. 1978 *Times* 25 Jan. 17/2 It may well be that a guillotine is necessary... But if it is to be justified as a pre-emptive strike [etc.].

B. *sb.* (Later example.)

1930 L. G. D. ACLAND *Early Canterbury Runs* 1st Ser. ii. 26 In eighteen months nearly all the run except the pre-emptives had gone.

pre-e·mptively, *adv.* [f. PRE-EMPTIVE *a.* + -LY[2].] In a pre-emptive manner.

1917 E. BERGHOLT *Royal Auction Bridge* II. 148 By declaring 'pre-emptively', up to the full strength of his hand, Z. will no doubt be able to prevent B. from directing A. what to lead. 1952 I. MACLEOD *Bridge* vii. 84 A double is for penalties unless..(3) The double is of a suit not above the level of three (and not at the three level bid pre-emptively). 1959 *Encounter* Nov. 17/1 It is..easier to imagine the Soviet Union striking pre-emptively than to imagine the U.S...doing the same thing. 1968 G. JONES *Hist. Vikings* IV. iii. 392 After the death of Knut the ancient West Saxon dynasty pre-emptively reinherited England. 1975 *Sci. Amer.* Oct. 8/2 Our past practice of preemptively deploying the latest strategic weapons our technology affords us has neither forced nor persuaded the U.S.S.R. to stop deploying strategic weapons increasingly threatening to our security.

pre-emptor. Add: Also **pre-empter.** **1.** For *U.S.* read *N. Amer.* and add *Canad.* examples.

1860 *Brit. Colonist* (Victoria, B.C.) 12 Jan. 2/1 Pre-emptors run the risk of having to pay twice the amount required by the American government for wild land. 1933 [see *PRE-EMPTION 1 b]. 1962 G. NICHOLSON *Vancouver Island's West Coast* 265 A kindly Norwegian pre-empter.. assisted them in re-sawing and whittling the boards down to the proper dimensions by hand.

2. *Bridge.* One who makes pre-emptive bids.

1972 R. MARKUS *Common-Sense Bridge* III. 99 South.. overlooked the warning of the pre-emptor's bid and East's confident double.

preen, *v.*[2] Add: **2. b.** For (? *catachr.*) read *fig.* and add further examples.

1907 G. B. SHAW *John Bull's Other Island* p. liv, Not so pitiable as the virtuous indignation with which Judge Lynch, himself provable by his own judgment to be a prevaricator, hypocrite, tyrant and coward of the first water, preened himself at its expense. 1926 W. & E. MUIR tr. *Feuchtwanger's Jew Süss* I. 7 The Catholics were preening themselves on the probable extinction of the Protestant line in Swabia. 1943 A. CHRISTIE *Moving Finger* xi. 131 These schools..seem to take a delight in turning out girls who preen themselves on looking like nothing on earth. They call it being sweet and unsophisticated. 1948 O. WALKER *Kaffirs are Lively* xi. 164 South Africa..sometimes preens itself on its lack of lynch-law. 1972 'J. HERRIOT' *It shouldn't happen to Vet* i. 14 He had put one over on the young clever-pants vet and nobody could blame him for preening himself a little.

pre-enclosure: see *PRE- B. 2a.

pre-enginee·red, *ppl. a.* [PRE- A. 1.] Constructed from prefabricated units.

1958 *Times* 29 Mar. 5 (Advt.), APEE design to any specification pre-engineered buildings which can be erected speedily and economically by unskilled labour. 1974 *State* (Columbia, S. Carolina) 15 Feb. 9-B/7 (Advt.), Need men to erect pre-engineered buildings.

preen-gland (prī·n,glænd). *Zool.* [f. PREEN *v.*[2] + GLAND[2].] = *oil-gland* s.v. OIL *sb.*[1] 6e. Also called the uropygial gland.

1923 J. A. THOMSON *Biol. Birds* ii. 12 There is a striking paucity of skin-glands, for there is usually nothing but the preen-gland at the root of the tail. 1954 FISHER & LOCKLEY *Sea-Birds* viii. 190 This waxy oil..is similar in character to the oil from the preen-glands of birds. 1962 *Listener* 15 Nov. 807/2 It was noticed long ago that the bird rubs the beak on the preen gland which is situated on its back just in front of the tail. 1975 WALLACE & MAHAN *Introd. Ornith.* (ed. 3) iii. 84 The oil or preen gland (*uro-*

pygium) is a conical, bilobed structure, often with a tuft of tiny feathers that serve as a wick, located immediately in front of the tail.

pre-E·nglish, *a.* and *sb.* [PRE- B. 1 in Dict. and Suppl.] **A.** *adj.* **1. a.** Designating the period before settlement of English-speakers in the British Isles.

1922 E. EKWALL *Place-Names Lancs.* 26 We expect the name of such an important river (or at least its first el[ement]) to be of pre-English origin. 1922 F. KLAEBER *Beowulf* 190 The poet was interested in the old Anglian traditions—the only legends in *Beowulf* that are concerned with persons belonging to English (i.e., pre-English) stock. 1934 *Essays & Stud.* XIX. 157 *Ærgeweorc*..referring to constructions of the pre-English period. 1966 *Eng. Stud.* XLVII. 210 The oldest river-names are of pre-English origin.

b. Pertaining to a period before the adoption of a given word into English.

1960 C. S. LEWIS *Studies in Words* vi. 133 In modern English the two meanings are not at all related as parent and child. They can be explained only by the pre-English history of the word.

2. Prior to the emergence of the English language; *spec.* of or pertaining to the West Germanic or Anglo-Frisian dialect from which English developed.

1928 C. BERGENER *Contrib. Study Conversion of Adj. into Nouns in Eng.* 1 The conversion should have taken place in English, but for the sake of greater completeness also such cases have been included where the conversion was, or may have been, pre-English. 1933 *Mod. Lang. Notes* XLVIII. 383 For names not of English origin the authors..use..*pre-English*... The..term is a most unfortunate one, since it is ordinarily used in quite another sense, viz., to denote a word form in the hypothetical Germanic dialect out of which English developed. 1936 *Anglia* LX. 367 To the Langobardish *Laiamicho* answers a pre-English trisyllabic **Lāimikō* > **Laimikō*.

B. *sb.* The West Germanic or Anglo-Frisian dialect from which English developed. Also, English before written records.

1929 *Rev. Eng. Stud.* V. 179 A large and important group of writers and speakers..use *Anglo-Saxon* not in the sense 'Old English' but in the sense 'pre-English'. 1965 *Language* XLI. 34 The allophones of /g/..reveal..both [g] and [ʒ] in pre-English.

preening, *vbl. sb.* (In Dict. s.v. PREEN *v.*[2]) (Later examples.)

1890 E. COUES *Handbk. Field & Gen. Ornith.* II. iii. 129 Birds press out a drop of oil with the beak and dress the feathers with it, in the well-known operation called 'preening'. 1953 N. TINBERGEN *Herring Gull's World* iv. 41 Preening is a most vital occupation. 1975 J. A. G. BARNES *Titmice Brit. Isles* vii. 122 Preening must be considered almost as essential an activity as feeding.

preening (prī·niŋ), *ppl. a.* [f. PREEN *v.*[2] + -ING[2].] That preens (see PREEN *v.*[2] in Dict. and Suppl.); chiefly *fig.*, proud, self-confident.

1903 R. LANGBRIDGE *Flame & Flood* i. 2 The manner of Miss Lydia, as she nestled into repose upon the bench, was essentially that of the conquering fowl who, having winged her way out of the difficulties insurmountable until attained, looks back with preening self-congratulation on the terrors now safely left behind. 1959 *Times* 13 Jan. 10/6 The new, brightly preening *casa* built on the hillside by a wealthy Barcelona merchant. 1976 *Time* 27 Dec. 5/3 His preening charm and Irish good looks were also prominent in plays, films, television and supper clubs.

pre-e·ntry, *a.* [PRE- B. 2.] Prior to entry; *spec.* applied to a closed shop in which union membership is a prerequisite of appointment to a post. Cf. *POST-ENTRY *a.*

1941 [see *Air Training Corps* s.v. *AIR *sb.*[1] III. 3]. 1964 W. E. J. MCCARTHY *Closed Shop in Britain* ii. 52 On the most generous of estimates the number of workers affected by the pre-entry shop in all its forms is unlikely to be more than three-quarters of a million. This means that for every worker in a pre-entry shop there are four in post-entry shops. 1969 *Gloss. Aeronaut. & Astronaut. Terms* (*B.S.I.*) IV. 4 *Pre-entry streamtube*, the streamtube extending to the entry of a ducted body from infinity upstream. 1972 H. WILLIAMSON *Trade Unions* (ed. 2) ii. 22 In a pre-entry closed shop, membership of a union is essential before the man is appointed to a job. 1977 J. M. JOHNSON in Douglas & Johnson *Existential Sociol.* vii. 210 Because of my preentry preparations and personal contacts, I felt more or less at ease when I addressed a joint meeting of the social workers from the two CWS units to explain the purposes of the research. 1977 *Guardian Weekly* 25 Sept. 10/2 The pre-entry closed shop (that is the kind where you need a card to get a job).

pre-epileptic, -erythrocytic, -European: see *PRE- B. 1. **pre-evangelism:** *PRE- A. 2. **pre-evolutionary:** *PRE- B. 1d. **pre-examination:** *PRE- B. 2a. **pre-excitation:** *PRE- A. 2. **pre-experimental(ly, -exponential:** *PRE- B. 1.

pre-expo·sure. [PRE- A. 2.] A preliminary or premature exposure; *spec.* in *Photogr.*, one given uniformly to a sensitive film or plate in order to increase its sensitivity. So **pre-expo·se** *v. trans.*

1937 G. E. BROWN *Clerc's Photogr.* 582/1 (Index), Hypersensitizing, by pre-exposure. 1953 *Adv. Electronics* V. 76 In order to achieve high detectivity with photographic materials, it is necessary to pre-expose the negative uniformly in order to overcome the inertia of the material. *Ibid.* 77 Table VII shows..the density, *D*, of the film at optimum pre-exposure. 1967 E. CHAMBERS *Photolitho-Offset* v. 63 When double printing from positives, it is necessary to mask-out each positive in the areas where the other positive is required to print to prevent pre-exposure of the emulsion. 1972 P. PETZOLD *Effects & Exper. in Photogr.* (1973) 70/2 The film should be pre-exposed for only a few shots... Further frames could be pre-exposed at higher or lower levels. 1979 *Nature* 24 May 341/2 Pre-exposure of human red blood cells..to dilutions of *Bufo marinus* serum followed by washing and resuspension in fresh buffer, caused a dose-dependent inhibition of subsequent ³H-ouabain binding to these cells. 1979 *SLR Camera* June 73/2 The only trouble with pre-exposing the shadows in this way is that they tend to lose density in the print.

pref (pref). Abbrev. of *preference* (*share*) s.v. PREFERENCE 8.

1898 *Weekly Official Intelligence* 5 Mar. 145/2 Kinloch (Chas & Co.) Ord., 4/; ..Pref. 3/. 1927 *Financial Times* 10 May 1/3 Mexican Nat. 1st Pref. 1971 *Financial Mail* (Johannesburg) 26 Feb. 690/3 An alternative offer for MG's pref shares can be expected soon.

prefab (prī·fæb), *a.* and *sb. colloq.* Also **pre-fab.** [Abbrev. of *PREFABRICATED *ppl. a.*]

A. *adj.* Prefabricated. Also *transf.* and *fig.*

1937 *New Yorker* 27 Mar. 20 (*caption*) Darling, the Prefab Homes man was just here. 1944 *Archit. Rec.* Dec. 69/1 (*heading*) Expansible prefab house for postwar. 1958 *Times Lit. Suppl.* 21 Nov. 667/2 Wolsey's 'pre-fab.' chapel on the Field of the Cloth of Gold. 1962 J. PHILIPS *Dead Ending* (1963) II. iii. 77 That pre-fab bow tie. 1965 *New Scientist* 5 Aug. 326/2 Industrialized building is..a collection of developments. Some commentators make a division between 'light' and 'heavy', applying the former term to the progeny of the post-war 'prefab' systems, and the latter to such aids as pre-cast concrete sections of a ton or so each. 1966 T. PYNCHON *Crying of Lot 49* ii. 26 Barbed wire again gave way to the familiar parade of more beige, prefab, cinderblock office machine distributors, bottled gas works, [etc.]. 1973 C. BONINGTON *Next Horizon* xvii. 236 Neat rows of prefab huts, the homes of the Eskimoes. 1977 *Rolling Stone* 24 Mar. 58/2 The temptation to follow through with prefab notions of what that audience would like..was apparently too strong to resist.

B. *sb.* A prefabricated house or building; in Britain *spec.* a light, often single-storey house of the kind built in large numbers during and after the 1939–45 war when it was necessary to rehouse many people in a short time. Also *attrib.*

1942 *Time* 16 Mar. 77/3 This year 20% of all new houses may be prefabs. 1947 'N. SHUTE' *Chequer Board* ix. 250 Any young couple might live in a prefab when they start off first. 1949 G. COTTERELL *Randle in Springtime* 112 She continued with complaints about the people living in the new pre-fabs that had been put up where a V1 had demolished some houses in the road. 1958 *Spectator* 24 Jan. 109/2 The Crystal Palace, the first prefab in the world. 1958 U. BLOOM *Abiding City* xii. 200 England.. was rising from the ashes of the bombing, with the influx of pre-fabs springing up everywhere. 1959 *Times* 24 Sept. 15/2 Active youth clubs to keep the pre-fab element out of trouble. 1972 *Daily Tel.* 7 Dec. 16/2 The last 700 prefabs in London are to be demolished.

prefab (prī·fæb), *v. trans.* Colloq. abbrev. of *PREFABRICATE *v.* Hence **pre-fabbed** *ppl. a.*

1959 *Observer* 4 Oct. 21/6 Prefabbed to retail on both sides of the Atlantic, the Anglo-American telefilm serial is a celluloid bastard. 1959 *Encounter* Oct. 37/2 Pre-fabbed hand-crafts and papier-maché charm. 1973 'J. MARKS' *Mick Jagger* (1974) 67 You're still growing up in a blues environment whether you've prefabbed it or whether it's natural.

prefabricate (prīfæ·brikeɪt), *v.* [PRE- A. 1.] *trans.* To manufacture (sections of a building or similar structure) in a factory or yard prior to their assembly on a site, esp. when they are larger or more complex than those considered traditional; also with the building as obj. Also *absol.* and *fig.*

1932 W. H. HAM in *Architecture* Apr. 187/1 We can prefabricate 90 per cent of a house in the factory, assemble it, and make it a permanent, attractive, useful home. 1939 *Christian Sci. Monitor* 3 Mar. 4/1 Practically every steel bridge is prefabricated, or put together in the back yard of the bridge builders before the pieces are taken apart, labeled and shipped for erection on the site. *Ibid.* 4/2 Ironwork firm, by prefabricating, makes sure that parts will join. 1941 *Times* (Weekly ed.) 23 Apr. 2 Four new plants now being erected in Nebraska, Oklahoma and Texas..will assemble annually 3,600 heavy bombers from parts pre-fabricated in automobile factories. 1944 *Hansard Commons* 7 Mar. 1906 In the most recent class of frigates at least 80 per cent. of the structure has been prefabricated. 1947 *News Chron.* 8 Apr. 2/2 The political structure which is being pre-fabricated with some success cannot be placed in position until its economic foundations have been laid. 1960 E. DELAVENAY *Introd. Machine Transl.* vii. 108 How far will it be possible to 'pre-fabricate', so to speak, this vocabulary, when preparing a programme of automatic translation, by establishing in advance a mixed vocabulary peculiar to such a translation? 1964 *Times Rev. Industry* Feb. 3/1 Five Clyde shipbuilding firms are at present prefabricating houses. 1965 R. B. WHITE *Prefabrication* III. vi. 300 The overall tendency to prefabricate has continued to make headway, particularly for buildings which form part of a national programme of expansion or modernization. 1974 *Sci. Amer.* Feb. 94/3 The Crystal Palace was the first great iron-framed building... It was also..the first for which the structural units were prefabricated.

prefa·bricated, *ppl. a.* [PRE- A. 1.] **1. a.** Of a building or similar structure: constructed by assembling a relatively small number of components which have been made elsewhere. **b.** Of a component of such a structure: made in a factory or yard prior to use elsewhere in construction.

1933 *Archit. Rev.* LXXIV. 49/2 There are a number of houses, one among them being actually of the new 'pre-fabricated' type. 1935 *Economist* 23 Mar. 679/1 Even the 'pre-fabricated' house, which is assembled from sections, quickly constructed, completely fitted with air-conditioning and domestic equipment..is not regarded as cheap enough to initiate the revival which is so sorely needed. 1944 *Archit. Rev.* XCVI. 30 Mr. Churchill says that we need 500,000 prefabricated houses for temporary homes in the first two years of peace. 1945 in R. W. Zandvoort et al. *Wartime English* (1957) 190 Little more than 100 yards away lie several prefabricated U-boat sections, all in an advanced state towards completion. 1951 'J. WYNDHAM' *Day of Triffids* xvii. 298 The outbuildings are too small for our needs now—and I can't put up even prefabricated quarters singlehanded. 1959 *Listener* 17 Dec. 1072/1 Factories were built to manufacture prefabricated parts of buildings. 1963 N. MARSH *Dead Water* (1964) ii. 46 A large, prefabricated, multiple garage had been built. 1968 H. G. MILLER *Building Construction* iv. 34 These prefabricated sections, completely fitted with all doors and windows, are transported to the building site where the house can be quickly assembled on the prepared foundation. 1974 *Daily Tel.* 27 Nov. 9/1 So far about 10,000 tons of prefabricated sections of [the cruiser] Invincible have been assembled.

2. *transf.* and *fig.* Contrived, artificial.

1935 *Time* 7 Jan. 40/2 The youth is not having much success with his pre-fabricated recital. 1943 T. S. ELIOT *Reunion by Destruction* 12 The Church of South India is a pre-fabricated church. 1945 A. W. COYSH in *To start you Talking* ii. 30 The broadcast discussion is admittedly pre-fabricated. 1953 *Encounter* Oct. 14/1 This pre-fabricated public is made up not only of the more naïve party members..but also of fellow-travellers who read nothing but the pro-Communist press. 1963 [see *HAM *sb.*[1] B. 2].

prefabrica·tion. [PRE- A. 2.] The manufacture or use of prefabricated components.

1932 W. H. HAM in *Architecture* Apr. 195/1 Plaster.., down to ten years ago, prohibited any advanced thought along the line of prefabrication which would create economies worth while. 1946 *Sun* (Baltimore) 8 Feb. 4/2 Prefabrication seems to be the only solution to obtaining low-cost housing for veterans. 1952 J. B. SINGER *Plastics in Building* i. 25 Prefabrication, which at one time was only intended to bridge the gap in housing shortage, has been gradually extended to cover new fields. The use of factory-produced components is universally recognized as a means to rapid and economical building. 1960 I. CROSS *Backward Sex* 13 Albertville High School was a conglomeration of brick permanence and prefabrication. 1972 *Daily Tel.* (Colour Suppl.) 17 Nov. 79/2 For the upper ten storeys, standardisation and pre-fabrication were very extensively used.

prefabricator (prīfæ·brikeɪtəɹ). [f. *PREFABRICAT(E *v.* + -OR.] One who, or a business which, practises prefabrication.

1933 *Fortune* Apr. 54/3 Real-estate men offer house plus land for as little as $4,400. Against this new competition what have the prefabricators to show? 1940 *Reader's Digest* July 99 Gunnison Housing Corporation, largest of prefabricators, recently sold several factory-built houses in Springfield, Ill. 1949 *Archit. Rev.* CVI. 375/1 We find a degree and a habit of uniform standardization that no American prefabricator would even attempt to impose on his customers. 1965 R. B. WHITE *Prefabrication* I. i. 4 Foster Gunnison, pioneer prefabricator of New Albany, Indiana.

pre-fade (prī·feɪd), *a.* and *sb. Broadcasting.* [PRE- B. 2.] **A.** *adj.* Performed or occurring before programme material is faded up for transmission. Of apparatus: used for such monitoring. **B.** *sb.* Monitoring of programme material prior to fading it up for transmission; an instance of this; also, a technical facility for such monitoring.

1941 *B.B.C. Gloss. Broadcasting Terms* 24 *Pre-fade listening.* (1) Listening to a programme output before it is faded up for transmission. (2) Technical facilities provided for this purpose. 1949 F. FELTON *Radio-Play* ii. 19 There is also a 'pre-fade' apparatus by which the operator can listen to the record in advance. 1962 A. NISBETT *Technique Sound Studio* viii. 150 Listening to this disc on prefade it is brought into step with disc one. *Ibid.*, After a quick prefade check, the disc is once again faded up. *Ibid.* ix. 160 In the case of a typical prefade (that used for the BBC's Radio Newsreel) it is known that the duration from a certain easily recognizable point to the end of the record is exactly 1′ 17″. So the record is started exactly 1′ 17″ from the end of the programme, but not faded up. 1968 R. MILTON *Radio Programming* 313 While you are giving a talk from the studio, the technician may listen to a part of the recording he will use next, to be sure it is the right one. He will do this by using his pre-fade monitor. 1975 G. ALKIN *TV Sound Operations* 126 Gram desks are provided with a 'pre-fade' output so that the operator can listen for the cue without fading it up, so that the output is not heard by the audience. *Ibid.*, The method of cueing in discs is to play the recording on pre-fade until the cue is heard and then stop the turntable.

prefa·de, *v. Broadcasting.* [f. the sb.] *intr.* To employ pre-fade listening. Also *trans.*, to monitor (programme material) before fading it up for transmission. Hence **prefa·ded** *ppl. a.*, **prefa·ding** *vbl. sb.*

1962 A. NISBETT *Technique Sound Studio* viii. 149 For professional disc work..it is valuable to be able to 'prefade' (i.e. listen before fading up) while the studio loudspeaker is being used to monitor the rest of the programme. *Ibid.* ix. 160 For the close of many radio programmes prefaded music is used. *Ibid.* 161 Vocal music is not usually suitable for prefading. 1971 T. C. COLLOCOTT *Dict. Sci. & Technol.* 929/1 *Prefading*, listening to programme material and adjusting its level before it is faded up for transmission or recording.

pre-Fascist: see *PRE- B. 1 a.

prefectly (prī·fektli), *a. rare.* [f. PREFECT, PRAEFECT *sb.* + -LY[1].] Characteristic of or befitting a prefect.

1927 J. ELDER *Thomasina Toddy* xxii. 218 Anne recognised them with her most prefectly twitch of the lips.

prefectorial, *a.* Add: **2.** Of or pertaining to a prefecture (sense 3).

1942 E. PAUL *Narrow St.* iii. 24 Hours of squinting in the dingy misplaced prefectorial light. 1963 *Times* 18 Feb. 8/3 As incumbent of the Lyons prefecture, a massive grey building typical of the 'prefectorial baroque'..he represents the central power of Paris.

prefer, *v.* **7. a.** For † *above* read *above* and add later example.

1883 G. MOORE *Mod. Lover* II. vi. 105 There was one place he preferred above all others.

preferability. (Later examples.)

1962 L. J. COHEN *Diversity of Meaning* vi. 160 The preferability of one conceptual form to another. 1975 *New Yorker* 21 Apr. 96/2 The manual went on to stress the actuarial preferability of drivers leading a stable life.

preferable, *a.* Add: **3.** = PREFERENCE 8. *attrib.*

1913 *Act* 3 & 4 *Geo. V* c. 20 §97 (1) Such preferable securities as existed at the date of the sequestration, and are not null or reducible.

preferee. (Later example.)

1977 *Navy News* Sept. 4/3 Inevitably, there are more sea billets than preferees on the one hand while there is strong competition among the preferees for the comparatively small number of shore billets.

preference. Add: **1. b.** *spec.*, under the system of preferential voting (see *PREFERENTIAL *a.* c), the naming or numbering of candidates in the order desired by the voter; hence, the position in that order assigned to any candidate by the voter.

1908 *Westm. Gaz.* 20 Aug. 2/1 Some 272 of Haynes's supporters had not used their preference and so their votes were put aside as exhausted. 1955 E. LAKEMAN *How Democracies Vote* iv. 88 The Returning Officer either awards the appropriate number of points for each preference and adds them up, or, if each voter is obliged to number every candidate, adds up the preferences each candidate thus receives. 1965 *Austral. Encycl.* III. 367/1 In 1910 it [*sc.* Western Australia] made the marking of preferences necessary to a valid vote. 1975 *Irish Times* 10 May 1/5, I cannot dictate how my preferences should be distributed. In a democracy that is the right of the electorate.

7. b. (Earlier and later examples.)

1852 MRS. GASKELL *Cranford* (1853) vii. 133 We were six in number; four could play at Preference, and for the other two there was Cribbage. 1908 R. W. CHAMBERS *Firing Line* ii. 20 That kills our four at Bridge... We'll have to play Klondike and Preference now. 1977 V. S. PRITCHETT *Gentle Barbarian* v. 80 At Spasskoy he [*sc.* Turgenev]..played chess and draughts and games of Preference.

c. *Bridge.* A bid or pass by a responder indicating in which of two or more suits bid by his partner he wishes to play.

1919 R. F. FOSTER *Foster on Auction* I. 96 This bidding invariably shows a two-suiter... If he prefers the spades, he can bid two spades to indicate his preference. 1927 M. C. WORK *Contract Bridge* iii. 42 That bid would be a forced take-out and would not announce strength, merely a preference. 1958 *Listener* 4 Dec. 965/3 If..partner is two-suited it will be enough to give a spade preference when he is able to demand it.

8. preference bid = sense 7c above; **preference voting** = *preferential voting* s.v. *PREFERENTIAL *a.* c).

1927 M. C. WORK *Contract Bridge* iii. 148 *Preference bid*, a bid made to show preference for one suit over another, rather than strength (in case partner has bid a two-suiter). 1934 G. F. HERVEY *Contract Bridge Dict.* 117 Y's bid is a preference bid showing that he prefers the hand to be played in Spades. 1908 *Westm. Gaz.* 20 Aug. 2/1 The local Labour Party is inclined to boycott preference voting and advocate its members to plump.

preferential, *a.* Add: **c.** *preferential ballot, voting*, a form of voting found in various systems of proportional representation in which candidates are numbered in order of preference by the voter; the use of the alternative vote (see *ALTERNATIVE *a.* 6).

1870 *Putnam's Mag.* June 717/1 Mr. Hare's scheme is one which..may be called that of *preferential* voting. It ascertains the quota by dividing the whole number of voters by the whole number of representatives... This method, which we have called that of preferential voting, is also called by the Swiss reformers that of the electoral quotient. 1908 *Westm. Gaz.* 20 Aug. 2/1 The State of Western Australia..is now attempting..preferential voting in a simple form. 1911 *Ann. Amer. Acad. Pol. & Social Sci.* XXXVIII. 760 The preferential ballot for cities is a plan to restore majority elections and true representative government. 1926 [see *alternative vote* s.v. *ALTERNATIVE *a.* 6]. 1955 C. R. ADRIAN *Governing Urban Amer.* iii. 61 During their heyday—the first and second decades of the present century—reformers sponsored other organizational and procedural changes: preferential voting, such as the Bucklin and Ware systems, which did not catch on, [etc.]. 1976 J. ROGALY *Parliament for People* vi. 71 Two less satisfactory forms of preferential voting are the 'second ballot', used in France, and the 'Alternative vote'.

d. *Anthrop.* Esp. in phr. *preferential marriage, mating*: the preference within a tribe or kinship group for marriage to take place between persons standing in a particular relationship to each other, such as cross-cousins. Cf. *PRESCRIPTIVE *a.* 4 b.

1909 *Cent. Dict.* Suppl., Preferential marriage. 1920 R. H. LOWIE *Primitive Soc.* (1921) ii. 16 Among the Kariera of Western Australia the acquisition of a bride is complicated by certain rules of preferential mating. That is, a man is..practically obliged to mate with a particular type of cousin or some more remote relative. *Ibid.* 35 Cross-cousin marriage, levirate, and sororate are by no means the only terms of preferential mating. 1943 E. J. & J. D. KRIGE *Realm of Rain Queen* ix. 145 We shall turn first to the preferential marriages. A marriage is obligatory or approved, or discouraged..according as it strengthens..or conflicts with the..edifice erected by the [payment of] cattle. 1968 R. NEEDHAM tr. *Lévi-Strauss's Elem. Struct. Kinship* (1969) p. xxx, Societies which advocate marriage between certain types of kin adhere to the norm only in a small number of cases..hence the idea of calling such systems 'preferential', a name which.. expresses the reality. 1971 —— *Rethinking Kinship & Marriage* p. lxviii, One should not ask whether a tribe has a prescriptive as opposed to a preferential marriage system.

e. *transf.* = *PREFERRED *ppl. a.* 5.

1926 *Carnegie Scholarship Mem.* XV. 378 Since the smaller crystals grown in flat strips had no preferential orientation, none could be expected in the smaller round crystals. 1955 T. L. RICHARDS in H. S. Peiser et al. *X-Ray Diffraction by Polycrystalline Materials* xxi. 469 Preferential crystal growth in a definite crystal direction. 1968 R. RIEGER et al. *Gloss. Genetics & Cytogenetics* 161 *Selective* or *preferential fertilization*, fusion of germ cells of different genotypes from one or both sexes in combinations having nonrandom frequencies. 1977 *Lancet* 9 July 92/2 The preferential production of the IgE and IgG4 classes might..be due to structural peculiarities in the sensitising allergens.

preferentially, *adv.* Add: **2.** To a greater extent or degree.

1926 *Encycl. Brit.* II. 886/1 In the case of brass, steel and aluminium alloys, certain types of chemical reagents which act preferentially upon the material in the crystal boundaries contribute to the occurrence of such fractures. 1935 TIPSON & STILLER in Harrow & Sherwin *Textbk. Biochem.* ii. 61 The tendency is for the primary hydroxyl group to be preferentially esterified. 1957 G. E. HUTCHINSON *Treat. Limnol.* I. xvi. 851 Several crop plants assimilated ammonia preferentially above *p*H 7. 1971 *Physics Bull.* Mar. 141/3 In an equilibrium reaction these elements with greatest binding energy per nucleon are formed preferentially. 1977 *Sci. Amer.* Jan. 39/3 The black hole will therefore preferentially emit particles with charge of the same sign as itself and so will rapidly lose its charge.

preferred, *ppl. a.* Add: **3. b.** Applied to a set of numbers or values forming an approximate geometrical progression and used to determine the officially recommended values of a dimension or other characteristic with which standard components should be made, so as to cover most efficiently the range of possible requirements.

1922 *Mech. Engin.* XLIV. 791/1 A careful study of manufactured articles shows that even when sizes are determined by utility or use value, the choice of size is largely arbitrary. It is therefore obvious that if certain numerical values are universally accepted as preferred values, and if they are so spaced and of such extent as to fit in with all requirements met in deciding on sizes to be used, the arbitrary choices may be so made as to yield sizes expressible in terms of these preferred numbers. 1936 *Proc. IRE* XXIV. 159 Preferred numbers are certain numbers that have been selected to be used for standardization purposes in preference to any others. 1962 S. HANDEL *Dict. Electronics* 269 Manufacturers and users of electrical and electronic components such as fixed resistors and capacitors find that there are advantages in standardizing component values and adopt preferred values so that each value differs from the preceding one by a constant multiplier. 1963 JERRARD & MCNEILL *Dict. Sci. Units* 110 Preferred numbers are conventionally rounded off terms in a geometrical progression whereby a tenth multiple of the initial term is obtained after a predetermined number of terms, viz: $p, pq, pq^2, .. pq^{n-1}, pq^n$, where p is the initial number, n is the number of terms and q a factor such that $q^n = 10$.

5. *transf.* That is exhibited or adopted by a natural system, object, or substance more commonly than, or to the exclusion of, other apparently possible properties or modes of development; *preferred orientation*, an orientation which crystals in a material tend to adopt or in which they tend to form, usu. because of applied stress.

1929 *Chem. Abstr.* XXIII. 1051 The detn. of the presence and the quant. description of preferred orientations in cryst. masses. 1954 R. L. PARKER tr. *Niggli's Rocks & Min. Deposits* v. 153 If the variable attribute (grain diameter, for instance) can be characterized both by size and by frequency, a mean size must first be determined. To this end the central or preferred value (modal value) in the distribution must be sought..or the value of the arithmetic mean..calculated. 1956 *Jrnl. Iron & Steel Inst.* CLXXXIII. 99/2 The Zn coating on galvanized wire shows no preferred orientation. 1956 *Q. Jrnl. Geol. Soc.* CXII. 123 A specimen from Logie Head.. shows no apparent preferred orientation of quartzes and the measurements are evenly scattered over the fabric diagram. 1962 E. S. GOULD *Inorg. Reactions & Struct.* (rev. ed.) iii. 64 The configuration with bond directions towards the corners of a regular octahedron..may be shown to be preferred. 1968 M. S. LIVINGSTON *Particle Physics* x. 171 In such a plot, broad peaks are frequently observed centred about specific values of energy. Such a preferred value of energy indicates a state of the system of two particles for which decay into the final products is more probable than for lower or higher values. 1971 *Nature* 4 June 306/2 The wind has a strongly preferred direction up and down the length of the steep-sided Loch. 1974 D. M. ADAMS *Inorg. Solids* v. 97 In Chapter 4 we saw that the relative sizes of the ions in a crystal could be related to preferred coordination arrangements.

‖ **préfet** (prefɛ). [Fr.] The chief administrative officer of a department of France; = PREFECT, PRÆFECT *sb.* 1d.

1820 M. EDGEWORTH *Let.* 22 June (1979) 168 Mme Chéron whose son is Préfet I think of Toulouse or some Provincial town. 1861 TROLLOPE *Tales of All Countries* (ser. 1) 22 And the company are all talking to him as though he were the préfet. 1869 *Bradshaw's Railway Manual* XXI. 348 Directors: C. Bart, ex-Préfet, Paris. 1872 TROLLOPE *Golden Lion* xviii. 288 Colmar..has been accustomed to the presence of a préfet, and is no doubt important. 1908 O.E.D. s.v. *Prefecture* 3, A prefect or French *préfet*. 1942 'A. BRIDGE' *Frontier Passage* vi. 102 They went off together to the *Préfecture* at Bayonne. .. Crossman's prestige..made access to the *Préfet* easy. 1958 *Listener* 18 Sept. 431/1 While there might be scoundrels like Darnand, such could be balanced by the courage of *préfets* like Bousquet. 1974 S. COULTER *Château* I. viii. 49 'You are invited to the Préfet's... A reception.'.. She was impressed; the Préfet was the chief officer of the department.

pre-feudal: see *PRE- B. 1d. **pre-film(ic:** *PRE- A. 1, B. 1d.

pre--final, *sb.* and *a.* [PRE- B. 1.] **A.** *sb. Linguistics.* (See quot. 1933.) Cf. *POST-FINAL.

1933 L. BLOOMFIELD *Language* viii. 132 English final clusters consist of two, three, or four non-syllabics. One can describe the combinations most simply by saying that each cluster consists of a main final consonant, which may be preceded by a pre-final, which in turn may be preceded by a second prefinal; further, the main final may be followed by a post-final. 1965 *Amer. Speech* XL. 12 There is here no specification of finals with prefinals and post-finals.

B. *adj.* Preceding the final.

1957 *Publ. Amer. Dial. Soc.* XXVIII. 114 Each of the pre-final groups.

prefi·nished, *a.* [PRE- A. 1.] Of metal: coated or treated at the mill so as to make finishing by a subsequent manufacturer unnecessary.

1935 H. R. SIMONDS *Finishing Metal Products* v. 41 One manufacturer of tableware is..producing highly polished products from prefinished sheets. 1963 H. R. CLAUSER *Encycl. Engin. Materials* 550/2 Except for uniformity, a plain plated or painted surface looks the same whether it is made of prefinished metal or finished after fabrication. 1974 *Industr. Finishing* Oct. 10/1 Many types of coated or laminated strip product are in fact used in the construction of motor vehicles but in the present context a prefinished material is defined as one which only requires fabrication before it can be used in its final form.

So **prefi·nishing** *vbl. sb.*

1935 H. R. SIMONDS *Finishing Metal Products* v. 42 Lacquering and painting of sheets are an important part of the general prefinishing of raw materials. 1963 *Mech. World* CXLIII. 17/2 Rather than think in terms of functional or decorative coatings for pre-finishing,..it is probably more realistic to classify such pre-treatments as metallic or non-metallic.

prefi·re, *v.* [PRE- A. 1.] *trans.* To fire (pottery, clay, etc.) beforehand, *spec.* before glazing. Hence **prefi·red** *ppl. a.*, **prefi·ring** *vbl. sb.*

1944 E. ROSENTHAL *Porcelain & Other Ceramic Insulating Materials* I. viii. 203 Originally, pressed and other thin-walled articles were pre-fired at a temperature of between 800°–950°C. in order to give them the necessary mechanical strength so that they would not be deformed during the dipping process, but with the introduction of the aerograph for mass-production glazing pre-firing of the articles becomes superfluous. 1960 W. D. KINGERY *Introd. Ceramics* xiii. 428 The fine structure in the grog (pre-fired clay)..consists of fine mullite crystals in a siliceous matrix. 1961 M. FRANCIS tr. *Salmang's Ceramics* viii. 210 The grog should also be pre-fired at the temperature at which the brick will subsequently be used, so as to prevent the texture of the brick loosening owing to after-shrinkage of grog grains. 1965 G. J. WILLIAMS *Econ. Geol. N.Z.* xx. 366/1 Some use has been made of these materials for refractories at Kamo, where the shrinkage is controlled by prefiring part of the mix.

prefix, *sb.* Add: **2. b.** A word placed at the beginning of the registered name of a pedigree animal, esp. a dog, to indicate the establishment in which it was bred.

1893 [see *AFFIX *sb.* 4]. 1922 R. LEIGHTON *Compl. Bk. Dog* 367 A Prefix or Affix shall constitute part of a name. 1954 [see *AFFIX *sb.* 4]. 1961 C. H. D. TODD *Popular Whippet* 139, I remember some lovely dogs that sailed under the Poppy prefix. 1976 C. COOPER *Newfoundland* i. 31 New names were coming to join those of established breeders, among them Lt-Col. Reid-Kerr with his Gleborchd prefix.

prefix, *v.* **1.** Delete 'Now *rare*' and add later example.

1977 *Daily Tel.* 23 Feb. 32/3 Would-be exporters can 'pre-fix' the subsidy at the present level by arrangement with the European Commission in Brussels.

3. b. *Biol.* To fix with the first of two consecutively used fixatives.

1963 *Jrnl. Cell Biol.* XVII. 32 It was found that material prefixed in glutaraldehyde or acrolein..and refixed in osmium tetroxide contained numerous dense particles of diffuse contours..occupying the glycogen areas. 1971 *Nature* 29 Oct. 622/2 Even when the final suspension was prefixed by adding osmium tetroxide, damaged liposomes were not seen.

5. (Later example.)

1898 *Phil. Trans. R. Soc.* B. CXC. 85 The skin and musculature of the arm of Man are somewhat prefixed as compared with *Macacus*.

prefixal, *a.* Delete *rare*⁻¹ and add later examples. Hence **prefi·xally** *adv.*, in the manner of a prefix.

1922 S. GREW *Art of Player-Piano* 86 The shorter note may be affined prefixally to the note after it. 1962 D. C. SWANSON in Householder & Saporta *Probl. Lexicogr.* 74 'Prefixal' is self-explanatory; 'co-valent' is used here (instead of non-prefixal) to refer to all other types collectively. 1964 E. A. NIDA *Toward Sci. Transl.* vi. 135 Once the verb root has been given, the selection of possible suffixes is more strictly determined than in Navajo, in which the prefixal formations do not, to the same extent, help the reader to predict either the sequence of prefixes likely to occur or the root likely to follow. 1971 *Language* XLVII. 396 Reflexive prefixal *t-* metathesizes with an immediately following sibilant and assimilates to the latter with respect to voice and pharyngealization. 1975 *Ibid.* LI. 618 The prefixal vowel in *prenatal* (but not in *pregnant*) is lengthened.

prefixation. I. 1. Delete *rare*⁻¹ and add later examples.

1957 *Archivum Linguisticum* IX. 101 Hypercharacterization..of a person..is the common denominator of a wealth of processes:..the spreading prefixation of *i*(*l*) to the 3d sing. in modern substandard French. 1975 N. CHOMSKY *Logical Struct. Linguistic Theory* viii. 235 Any complete verb phrase can be turned into a noun phrase by prefixation of 'to'. 1978 *Amer. Speech* LIII. 64 This last observation prompts Samarin to suggest that glossolalia may provide evidence for a universal tendency to prefer suffixation to prefixation.

II. [f. PRE- + FIXATION.] **2.** *Anat.* [PRE- A. 4 b.] The state of a nerve of being prefixed (see *PREFIXED, PREFIXT *ppl. a.* 3).

1953 [see *POSTFIXATION *sb.* 1].

3. *Biol.* [PRE- A. 2.] The initial fixation of tissue that is subsequently to be fixed with a second fixative.

1963 *Jrnl. Cell Biol.* XVII. 32 The particles appearing when osmium tetroxide was used after glutaraldehyde or acrolein prefixation were similar to those appearing in rat liver after potassium permanganate..fixation. 1974 *Nature* 20 Dec. 722/1 The present study takes advantage of several circumstances: (i) cells could be frozen without prefixation or cryoprotectants.

prefixed, prefixt, *ppl. a.* Add: The form **prefixt** is *Obs.* **3.** *Anat.* Of a nerve: connected to the spinal cord relatively cranially. Cf. *POSTFIXED *ppl. a.* 2.

1892, etc. [see *POSTFIXED *ppl. a.* 2].

prefixial (prifi·ksiăl), *a.* [f. PREFIX *sb.*] = PREFIXAL *a.*

1975 *Verbatim* Dec. 13/2 'Bantu'..was coined in the 1850s by a philologist, Wilhelm Bleek, to characterize those peoples south of the Sahara speaking a group of related languages featuring prefixial concords, that is, a system in which the prefix is inflected to indicate grammatical case and number rather than, as in most western European languages, the suffix. 1976 *Archivum Linguisticum* VII. 136 Except for some prefixial combinations, the determinatum usually receives weak stress in English.

preflame: see *PRE- B. 2.

prefli·ght, *a.* [f. PRE- B. 2 + FLIGHT *sb.*[1]] **a.** Of or pertaining to the time before powered flight. *rare.* **b.** Of or pertaining to the preparations for a flight, or for flying in general.

1922 *Encycl. Brit.* XXX. 44/2 Almost all altimeters in use are based on the pre-flight aneroid in which the trade convention was to assume everywhere an atmospheric temperature of 10° C. 1942 *R.A.F. Jrnl.* 27 June 4 American boys..are to be given pre-flight aviation training. 1962 A. SHEPARD in *Into Orbit* 98 Then I went through my final pre-flight medical. 1968 J. SANGSTER *Touchfeather* xviii. 205, I could just see his head at the flight deck window as he made some of his pre-flight checks. 1970 N. ARMSTRONG et al. *First on Moon* ii. 35 She had left the Holiday Inn two hours earlier to drive to the preflight examination area.

prefli·ght, *v.* [f. prec.]. *trans.* To prepare (an aircraft) for a flight.

1971 *Flying* (N.Y.) Apr. 40/3 A moron could..preflight and satisfactorily steer the 172. 1975 L. D. KUSCHE *Bermuda Triangle Mystery Solved* 98 Each plane had been..preflighted and held a full load of fuel. 1975 *High Times* Dec. 70/2 So you load up and preflight the plane. 1976 B. LECOMBER *Dead Weight* vi. 75 Simon..did want a flying lesson... I told him to pre-flight the Cherokee.

prefo·cus, *a.* and *v.* [f. PRE- A. 1.] **A.** *adj.* Of a bulb: constructed so that the lamp is automatically focused upon fitting of the bulb; also applied to parts of such bulbs, esp. the cap, which make possible the necessary accurate positioning of the filament during manufacture.

1944 (*title*) Dimensions of prefocus lamp-caps and lampholders (British Standards Institution). 1950 YOUNG & GRIFFITHS *Automobile Electr. Equipment* (ed. 4) vi. 181 Fine adjustment for the filament position is.. essential in order to take care of slight manufacturing discrepancies in the filament distance from the bulb cap unless, of course, the bulb is of the pre-focus type now prevalent in British-made headlamps. 1967 L. HOLMES *Odhams New Motor Man.* x. 238/1 For Continental towing, British cars must have their headlamps converted to dip to the right instead of to the left, and for this purpose special pre-focus bulbs can be obtained. 1970 *A.A. Bk. of Car* 162/4 The cut-away in the flange of a pre-focus bulb ensures that it is correctly located and focused. 1972 *Gloss. Electrotechnical, Power Terms* (*B.S.I.*) IV. iii. 17 *Prefocus cap*, cap..which enables the luminous element to be brought into a specified position relative to the cap during the manufacture of the lamp.

B. *v. trans.* To make or adjust so that a lamp will be automatically focused when a bulb is fitted. So **prefo·cused** *ppl. a.*, **prefo·cusing** *vbl. sb.*

1951 *Philips Technical Rev.* XII. 309/1 It is of great importance that when a new bulb is fitted into a headlight its filament comes to lie exactly in the right position. This is ensured by a pre-focusing of the filament in the factory with the aid of a special adjusting device. *Ibid.* 312 (*caption*) On the base can be seen two of the three studs with which the lamp is focused: they ensure that the axis of the lamp coincides with that of the reflector, the filament being 'prefocused' with respect to the studs. 1954 A. W. JUDGE *Automobile Electr. Maintenance* (ed. 3) viii. 209 In the case of recent American-type headlamps, 'sealed-beam' units are employed. These consist of a lens, bulb, and reflector unit built into one water-and dust-proof pre-focused unit. 1970 *A.A. Bk. of Car* 162/4 *Quartz-halogen bulbs.* These are obtainable for fitting into pre-focused units in place of ordinary bulbs. 1977 *Nature* 6 Jan. 92/1 The built-in lamp turret keeps four pre-focussed lamps aligned, powered and ready to work.

preform, *v.* Add: *spec.* To form (plastic or other moulding material) into a shape, usu. one resembling a desired final shape, before some further processing. Also *absol.*

1936 H. W. ROWELL *Technol. Plastics* xx. 148 A 'tablet'..is made in a stock size of die and is not preformed to the approximate shape of the moulding. 1943 D. W. BROWN *Handbk. Engin. Plastics* i. 9 In order to reduce the size of the powder space required in moulds, the raw material is sometimes preformed into comparatively small pellets prior to being introduced into the mould. 1968 L. K. ARNOLD *Introd. Plastics* (1969) ii. 41 The powder may be preformed, that is, pressed into pellets or disks of convenient size to reduce bulk density and facilitate charging the mold. 1975 C. A. HARPER *Handbk. Plastics & Elastomers* xii. 36 There are..transfer-type presses that will automatically feed the powder, preform, preheat, transfer, and complete the molding cycle.

Hence **prefo·rming** *vbl. sb.*; also **prefo·rmer**, a press or similar device for preforming plastic.

1931 *Plastics & Molded Products* VII. 705 (Advt.), Trouble-free preforming and more economical production with Stokes single punch and rotary preform presses. 1952 J. DELMONTE *Plastics Molding* viii. 193 Standard single-punch preformers are illustrated in Figs. 2 and 3, and a rotary-type preforming machine is illustrated in Fig. 4. 1966 J. A. BRYDSON *Plastics Materials* x. 210 Preforming is carried out by compressing sieved powder that has been evenly loaded into a mould. 1968 *Encycl. Polymer Sci. & Technol.* IX. 26 Preformers are basically compacting presses; they may be mechanical, hydraulic, pneumatic, or rotary cam-type machines.

preform (prī·fǭɹm), *sb.* **1.** [f. the vb.] A moulded object which has to receive further processing to produce the final shape, which it usu. resembles.

1931 *Plastics & Molded Products* VII. 102 The test strip..is 7½ in. long by 1 in. wide and is molded from four preforms composed of a very soft grade of phenol-plastic compound. 1935 C. ELLIS *Chem. Synthetic Resins* II. lxviii. 1317 Tablets or preforms are made from the molding powder by applying high pressure..quickly and without heat. 1945 H. BARRON *Mod. Plastics* vii. 166 Preforms are employed where the use of fairly bulky moulding powder is a disadvantage. 1962 *Gloss. Terms Glass Industry* (*B.S.I.*) 33 *Preform*, a small glass product, normally employed in making glass-to-metal seals, formed by dry-pressing glass powder into shape in a die, the shape being fired to consolidate the glass particles into a non-porous component. 1968 *Encycl. Polymer Sci. & Technol.* IX. 26 It is easier for an operator to pick up a preform and place it in a mold cavity or transfer pot than to have to weigh a charge of granular material. 1970 *New Scientist* 12 Nov. 326/1 A precise quantity of [metal] powder is compacted and sintered so as to produce a preform which is approximately 85 per cent dense.

2. *Philol.* [PRE- A. 2.] A linguistic form reconstructed from later evidence.

1939 L. H. GRAY *Foundations of Lang.* i. 3 If..one has such a series as English (*he*) *bears*, Old Icelandic *berr*, Gothic *bairiþ*.., Old High German *birit*, Old Irish *berid*, Modern Irish *bheir*..., one may, by comparing and contrasting these forms in accordance with phonetic correspondences.., determine why they are here alike, and there unlike, and may perceive how they can all be derived from an hypothetical pre-form. 1972 *Language* XLVIII. 409 There are two possibilities for reconstructing the gen. pl. preforms of the two attested nouns.

preformationist. Add: (Later examples.) Also *attrib.* or as *adj.*

1936 *Mind* XLV. 221 Boodin begins by explaining why he prefers to class the cosmological theories of Plato and Aristotle, notwithstanding certain preformationist tendencies especially in the thought of Aristotle, under the heading of creation theories. 1960 *New Biol.* XXXI. 119 As development proceeds, suitable conditions start the synthesis of a succession of new substances. This, at least, is the contemporary opinion; the Preformationists in the eighteenth century thought otherwise, but their point of view is hard to square with the atomic hypothesis. 1973 *Jrnl. Genetic Psychol.* CXXII. 326 Preformationist prejudices were so powerful that early embryologists..'saw' nonexistent miniature adults when viewing embryos through their microscopes. 1975 *Nature* 5 June 449/1 These findings fit well the expectations of the Preformationists. 1978 *Ibid.* 9 Nov. 125/1 Until this metaphor became available, the only alternative to a crude preformationist kind of innatism was a mysterious force such as Driesch's 'entelechy', which would allow the developing organism to survive the vicissitudes of the environment and the embryologist's knife.

preformed, *ppl. a.* Add: *spec.* of plastic or other moulding material (cf. *PREFORM *v.*).

1918 H. ABRAHAM *Asphalts* xxvi. 453 (*heading*) Preformed joints and washers. 1935 C. ELLIS *Chem. Synthetic Resins* II. lxviii. 1319 The charge may be used in any of three forms, powder, preformed tablets or sheets. 1943 D. W. BROWN *Handbk. Engin. Plastics* i. 9 In some cases..the shape of the article renders it impossible to use preformed materials. 1971 E. W. DUCK *Plastics & Rubbers* iv. 60 A pre-formed sheet of the plastic is heated and then brought on to the surface of the mould and drawn tight..by the application of a vacuum. 1977 *Gramophone* June 122/1 The enclosure is very substantially made.., being braced..and filled with pre-formed foam fittings.

pre-Freu·dian, *a.* [PRE- B. 1.] Of or characterized by attitudes, etc., that were commonly accepted prior to Freud's pioneer work in psychoanalysis.

1937 *Harper's Mag.* Nov. 563/1 With the tools of semantic analysis, the authors laid in ruin the towering edifice of classical philosophy... Psychology (pre-Freudian) emerged in little better repair. 1938 J. M. KEYNES *Two Memoirs* (1949) 100 It was not only that intellectually we were pre-Freudian, but we had lost something which our predecessors had. 1959 P. TOWNEND *Died o' Wednesday* x. 169, I had..accomplished very little, merely my duty..a matter that caused our ancestors in..pre-Freudian days no perplexity. 1970 *Eng. Stud.* LI. 489 Ward seems quite untempted by the call to psychological speculation, and the book is positively pre-Freudian. 1976 *Jrnl. R. Soc. Arts* CXXIV. 625/1 The Angst of the pre-Freudian and Freudian era.

prefrontal, *a.* (*sb.*) Add: **A.** *adj.* **b.** (Earlier example.)

1888 *Phil. Trans. R. Soc.* B. CLXXIX. 3 (*heading*) Results of experiments upon the prefrontal region of the hemisphere.

c. *prefrontal leucotomy, lobotomy*, lobotomy of the prefrontal part of the brain.

1936, etc. [see *LOBOTOMY]. 1937, etc. [see *LEUCOTOMY].

prefulgence (prĭfʌ·ldʒĕns). [f. as PREFULGENT *a.*] = PREFULGENCY.

1892 G. GISSING *Born in Exile* II. IV. iii. 227 In his most presumptuous moments he had never claimed the sexual prefulgence which many a commonplace fellow so gloriously exhibits. 1916 F. SWINNERTON *Chaste Wife* xxiii. 317 Too stupid to understand anything but physical prefulgence or absolute social convention.

preg (preg), *a.* Colloq. abbrev. of PREGNANT *a.*[2] 1 a.

1955 W. GADDIS *Recognitions* I. v. 172 She's preg, baby. 1962 E. O'BRIEN *Lonely Girl* xvi. 188 Are you preg?.. 'Cos if you are, you won't be able to cycle. 1967 *London Mag.* Aug. 10 A bit of news which may just interest you. I am P-R-E-G and not by Roy. 1968 *New Society* 22 Aug. 266/1 'She's pregnant' is now used in many classes ('preg, preggy, preggers', whatever class they belonged to, are now not much used).

pre-game: see *PRE- B. 2 a. **pregamic:** *PRE- B. 1.

preganglio·nic, *a. Anat.* [PRE- B. 3.] Of a nerve of the autonomic nervous system: running from the central nervous system to a ganglion.

1895 *Jrnl. Physiol.* XVIII. 280 (*heading*) Note on regeneration of præ-ganglionic fibres of the sympathetic. 1897 *Ibid.* XXII. 223 The regeneration of post-ganglionic fibres presents a somewhat similar problem to that we have considered in relation to the pre-ganglionic fibres. 1903 *Brain* XXVI. 5 The nerve fibres which leave the central nervous system are all pre-ganglionic fibres; they end in connection with nerve cells in the ganglia, and these give off post-ganglionic nerve-fibres which end in the tissues. 1946 *Nature* 19 Oct. 556/1 Single shocks at low intensity delivered to the preganglionic nerve excite a single giant fibre therein. 1972 [see *INTERMEDIOLATERAL *a.*]. 1974 M. C. GERALD *Pharmacol.* v. 93 Conversely preganglionic sympathetic fibers are generally short, with long postganglionic fibers arising from the ganglion to the innervated tissues.

Hence **preganglio·nically** *adv.*

1937 *Jrnl. Physiol.* LXXXVIII. 6 Nicotine..shortens and diminishes the *P* waves set up both antidromically and preganglionically. 1967 *Jrnl. Pharmacol. & Exper. Therap.* CLV. 37/1 The difference was statistically significant whether stimulation was carried out preganglionically or postganglionically.

pregastrular, -geological(ly): see *PRE- B. 1.

pregermina·tion. [PRE- A. 2.] The treatment of seed to start the process of germination before planting. So **prege·rminated** *ppl. a.*, having been subjected to such treatment.

1942 H. I. BALDWIN *Forest Tree Seed* 230 Pregermination:—Germinative processes set in motion, but completion impossible because of external factors. Seed coat may or may not be broken. 1959 *Times* 7 Dec. (Agric. Suppl.) p. viii/3 Pregermination of conifer seed is practised before sowing. 1978 *Countryman* Winter 117 A new planting technique which is taking off in quite a big way is 'the fluid sowing of pre-germinated seeds'.

preggers (pre·gəɹz), *a. slang.* [f. PREG(NANT *a.*[2] + *-ers*, as in *bonkers, crackers*.] = PREGNANT *a.*[2] 1 a.

1942 M. DICKENS *One Pair of Feet* vii. 115 Let anyone mention in her hearing that they felt sick, and it would be all over the hospital that they were 'preggers'. 1960 F. RAPHAEL *Limits of Love* I. iii. 38 'I'm preggers,' Susan said. 1964 *Times* 4 Feb. 7/3, I would only offer my seat to a woman if she were carrying a baby, if she were preggers, or if she were obviously infirm. 1968 [see *PREG *a.*]. 1971 R. DENTRY *Encounter at Kharmel* vii. 115 'There was a strong suspicion that one of the women was preggers.' 'Eh?' 'Up the duff, sir.' 1980 C. FREMLIN *With no Crying* ix. 50 Preggers! Well, what do you know?

preggo (pre·go), *a. Austral. slang.* Also **prego.** [f. PREG(NANT *a.*[2] + *-O[2].] = PREGNANT *a.*[2] 1 a.

1951 CUSACK & JAMES *Come in Spinner* 226 Guinea's face lighted with unholy glee. 'A Parker prego? Did I hear right?' 1965 P. WHITE *Four Plays* 94 'Can't resist the bananas.' 'Yeah. They say you go for them like one thing when you're preggo.' 1971 *Guardian* 27 May 13/7 Uncommon in print, but familiar in speech, is preggo (pregnant).

preggy (pre·gi), *a. slang.* Also **preggie.** [f. PREG(NANT *a.*[2] + -Y[1].] = PREGNANT *a.*[2] 1 a. Also *transf.*

1938 N. MARSH *Death in White Tie* v. 63 There was your bag, simply preggy with banknotes. 1961 S. PRICE *Just for Record* vi. 57 If poor little preggy Emily Nugent had lived I might never be where I am today. 1968 [see *PREG *a.*]. 1970 K. GILES *Death in Church* iii. 69 She looks preggie, pretty, polite and a little sloshed. 1976 *Star* (Sheffield) 30 Nov. 2/7 Final fling for noisy Parkers shows Michael and preggie June back in England.

pre-give, -given: see *PRE- A. 1.

pre-glacial, *a.* Add: Hence **pregla·cially** *adv.*

1875 *Q. Jrnl. Geol. Soc.* XXXI. 61 The small thickness of preglacially weathered rock that the brief stay of the ice at these high elevations enabled it to remove. 1903 *Jrnl. Geol.* XI. 675 To whatever state of maturity the valley development had attained preglacially, it was directly modified by the advent of the glacial epoch. 1963 D. W. & E. E. HUMPHRIES tr. *Termier's Erosion & Sedimentation* 411 A smooth preglacially eroded surface.

preglo·ttalized, *ppl. a. Phonetics.* [PRE- A. 1.] Preceded by a glottalized sound. Hence **preglo:ttaliza·tion.**

1964 E. J. A. HENDERSON in D. Abercrombie et al. *Daniel Jones* 419 Vichintana Chantavibulya..discovered that her pronunciation of the palatal semi-vowel in initial position is preceded by a weak glottal plosive... There is a similar pre-glottalized articulation of the labio-velar semi-vowel. 1965 *Canad. Jrnl. Linguistics* XI. 1. 32 Phonological feats are possible on a paralinguistic level that seem impossible on a linguistic one,..as pre-glottalized stops and nasals. 1968 B. S. ANDRÉSEN *Pre-Glottalization in Eng. Stand. Pronunc.* 10, I shall regard the glottal stop and the following labial/alveolar/velar plosive as one phonological unit, and refer to this as a 'pre-glottalized' voiceless labial/alveolar/velar plosive. The use of these phones will be referred to as 'pre-glottalization'. 1969 *Eng. Stud.* L. 317 Preglottalization can be compared to other articulatory modifications of sounds, e.g. velarization of l, as in [ɫæl]. 1977 *Language* LIII. 317 Except in sentence-initial position, *b d* are either preglottalized or prenasalized, and the processes occur in complementary distribution—word-internally, pre-glottalization is the norm. 1977 *Trans. Philol. Soc.* 1975 222 For Oromo *d* represents either an alveolar implosive or a preglottalized voiced alveolar plosive.

pregnance. Delete †*Obs.* and add: **1.** (Later example.)

1959 I. JEFFERIES *Thirteen Days* xi. 171, I rode off early..into the pleasant pregnance of the day.

2. = *PRÄGNANZ. (But see quot. 1974.)

1948 *Brit. Jrnl. Psychol.* June 181 How little we still know of the details of this process of adaptation, of its conditions and limitations, of its relation to..'the law of pregnance'. 1969 G. N. SEAGRIM tr. *Piaget's Mechanisms of Perception* vi. 305 Pregnance..is only a coercive effect produced..by a form whose elements..succeed in compensating any deformations which are present. 1974 R. ARNHEIM *Art & Visual Perception* (rev. ed.) ii. 67 To compound the confusion, translators have rendered the German *Prägnanz* with the English *pregnance*, which means very nearly the opposite.

pregnancy[1]. Add: **1. a.** Also *transf.*, with reference to appearance: bigness, swollen shape.

1950 *Manch. Guardian Weekly* 4 May 3/4 Since Packard abolished its regal proboscis and succumbed to the epidemic pregnancy of current American models [of automobiles].

5. Special Comb.: **pregnancy test**, a test to establish whether a woman (or female animal) is pregnant; so *pregnancy testing*.

1929 *Jrnl. Amer. Med. Assoc.* 25 May 1746/1 Their 'pregnancy test'..is based on the injection of urine into immature white mice. 1962 L. DAVIDSON *Rose of Tibet* 313 Forcible mating begun... Army doctors followed..to make pregnancy tests. 1977 *Private Eye* 13 May 22/2 (Advt.), *Pregnancy Test Service.* Send small sample of urine & fee £3 for *reliable & strictly confidential* results by first class return post (plain sealed cover). 1977 *Times* 21 June 5/8 Scientists yesterday carried out pregnancy tests on a Colorado beetle found at the weekend on a rose bush in a garden at Peacehaven, Sussex. 1938 *Amer. Jrnl. Obstetr. & Gynecol.* XXXV. 362 A review of the most recent work in pregnancy testing is presented. 1971 *Guardian* 15 Apr. 22/4 An instant pregnancy-testing service is to be promoted in chemists' shops.

pregnane (pre·gnēin). *Chem.* [ad. G. *pregnan* (A. Butenandt 1930, in *Ber. d. Deut. Chem. Ges.* LXIII. 660), f. *pregnan-diol* *PREGNANEDIOL (cf. -ANE).] A synthetic, crystalline, saturated, tetracyclic hydrocarbon, $C_{21}H_{36}$, which has two stereoisomeric forms and from which a group of steroids, including progesterone and pregnanediol, is formally derived; *spec.* the 5β-isomer.

1932 *Ann. Rep. Progr. Chem.* XXVIII. 237 On oxidation there was formed the saturated diketone, $C_{21}H_{32}O_2$, pregnandione, and by Clemmensen reduction of the latter A. Butenandt, F. Hildebrandt, and H. Brücher have obtained the corresponding hydrocarbon, $C_{21}H_{36}$, pregnane. 1936 *Zeitschr. für Kristallogr.* XCIII. 478 Pregnane forms platy crystals elongated along the *b* axis and showing generally the *c* face only. 1959 I. L. FINAR *Org. Chem.* (ed. 2) II. xi. 456 The molecular rotation of any steroid is considered as the sum of the rotation of the fundamental structure (which is the parent hydrocarbon cholestane, androstane, or pregnane) and the rotations contributed by the functional groups. 1960 L. T. SAMUELS in D. M. Greenberg *Metabolic Pathways* I. xi. 432 The two 21-carbon compounds are 5α-pregnane (allopregnane)..and 5β-pregnane (pregnane). 1971 M. F. MALLETTE et al. *Introd. Biochem.* xx. 778 Steroids are named as derivatives of the polycyclic hydrocarbons gonane, estrane, androstane, and pregnane.

pregnanediol (pre:gnēin-, pre:gnăndai·ǫl). *Biochem.* and *Med.* Also **pregnandiol.** [ad. G. *pregnandiol* (A. Butenandt 1930, in *Ber. d. Deut. Chem. Ges.* LXIII. 660), f. L. *prægnan-s* pregnant + G. *di-* DI-[2] + *-ol* -OL; cf. *PREGNANE.] A crystalline steroid containing two hydroxyl groups, $C_{21}H_{36}O_2$, which is a product of the metabolism of progesterone and occurs in the urine during pregnancy.

1930 *Chem. Abstr.* XXIV. 2785 Pregnandiol is related to the sterols and gallic acid. 1934 *Jrnl. Biol. Chem.* CVII. 324 It also seems possible that our compounds may be related to pregnandiol. 1934 *Ann. Rep. Progr. Chem.* XXX. 216 Pregnanediol..is a physiologically inactive compound found in the urine during pregnancy. 1955 *Sci. Amer.* Jan. 55/3 In pregnancy urine there was found an inactive companion of estrone, called pregnanediol. Butenandt proved that pregnanediol was related to cholic acid, a component of bile, by breaking both down to a common product. 1961 *Lancet* 5 Aug. 277/2 The relationship between myometrial progesterone and pregnanediol in the urine is even less clearly defined. 1968 PASSMORE & ROBSON *Compan. Med. Stud.* I. xxxi. 13/1 Progesterone is reduced to pregnanediol, which is excreted as sulphate or glucuronide conjugates. 1976 *Path. Ann.* XI. 240 In 1946 Smith, Smith, and Hurwitz reported that stilbestrol administered to pregnant women increased urinary pregnanediol levels.

pregnenolone (pregnī·nǫlǝun). *Biochem.* and *Med.* [ad. G. *pregnenolon* (Butenandt & Westphal 1934, in *Ber. d. Deut. Chem. Ges.* LXVII. 2085), f. *pregn-an* *PREGNANE + *-en* -ENE + *-ol* -OL + *-on* -ONE.] A synthetic steroid, $C_{21}H_{32}O_2$, which is a reduced derivative of progesterone and was formerly used in the treatment of rheumatoid arthritis.

1936 *Chem. Abstr.* XXX. 3036 Similar mixts. of III and I or III and pregnenolone (IV) met with in the prepn. of the hormone from stigmasterol can be sepd. in the same manner. 1955 *Jrnl. Amer. Med. Assoc.* 21 May 166/1 All the patients had received, at one time or another, gold salts, pregnenolone, antibiotics, [etc.]..that produced, at the most, only palliative relief. 1964 E. J. W. BARRINGTON *Hormones & Evolution* ii. 54 In this way cholesterol is thought to be transformed into progesterone, through the intermediary pregnenolone, by partial degradation of its side chain. 1967 *Martindale's Extra Pharmacopoeia* (ed. 25) 1543/2 Pregnenolone and pregnenolone acetate have been used in the treatment of rheumatoid arthritis..but their value has not been substantiated.

prego, var. *PREGGO *a.*

pre-gra·duate, *a.* [f. PRE- B. 1, after POST-GRADUATE *a.* (*sb.*).] = UNDERGRADUATE *a.*

1937 *Discovery* Sept. p. lxxxi/1 (Advt.), The Pregraduate courses extend over three or four years and lead up to an Associateship of one of the colleges, and a B.Sc. Degree of the University of London. 1977 *Proc. R. Soc. Med.* LXX. 377/1 The knowledge and skills required to practise independently were expanding the pre-graduate curriculum to bursting point.

pre-grammar, -grammatical: see *PRE- B. 1. **pre-grind:** *PRE- A. 1. **pre-Han:** *PRE- B. 1 a.

preha·rvest, *a.* [PRE- B. 2.] Occurring before a crop is ready to be gathered.

1934 in WEBSTER. 1948 *New Biol.* V. 62 Some varieties of apple, Beauty of Bath for example, tend to fall before they are fully ripe or of good eating quality. This is known as preharvest fruit drop. 1951 [see *HORMONE 2]. 1971 T. T. KOZLOWSKI *Growth & Devel. Trees* II. viii. 359 In apple, three normal periods of fruit drop occur... These include 'early drop'..and the preharvest drop.

pre-hearing: see *PRE- A. 2.

prehea·t, *v.* [PRE- A. 1.] *trans.* To heat prior to use or to some other treatment.

1898 *Engineering Mag.* XVI. 245 The gas is usually not preheated, but the gas producers are set close to the furnace, so that the initial heat..is not lost. 1937 *Discovery* May 155/1 The tool having been evenly preheated is transferred to the high heat chamber. 1958 LAMBERMONT & PIRIE *Helicopters & Autogyros* 201 The ram-jets are run on propane, which is preheated before being ducted to them. 1958 *Times* 27 Oct. 11/5 Preheat the oven to 420 deg. F. (Regulo 7) and lay the hare directly on the grid. 1975 *Nature* 2 Oct. 368/2 The mixtures were preheated by an external source.

Hence **prehea·ted** *ppl. a.*, **prehea·ting** *vbl. sb.* and *ppl. a.*; also **prehea·ter**, a device for pre-heating.

1898 *Engineering Mag.* XVI. 245 This method of preheating may follow either the regenerative or the recuperative system. 1910 *Encycl. Brit.* I. 445/1 What is called a 'preheater' is used to warm up the compressed air before it enters in the motor cylinder. 1911 A. REYNOLDS tr. *Dichmann's Basic Open-Hearth Steel Process* vii. 59 The second way of producing a steam-air gas with high hydrogen content, consists in the employment of superheated steam, or preheated air supply. 1931 HOFFERT & CLAXTON *Motor Benzole* viii. 221 An efficient preheater should be capable of raising the maximum flow of oil from the temperature at which it leaves the heat-exchange system to a temperature of at least 135°C. 1952 FUCHS & BRADLEY *Welding Pract.* II. iii. 65 Preheating of the parts being welded is always a help towards crack prevention. 1960 *Farmer & Stockbreeder* 15 Mar. (Suppl.) 11/1 Put the mixture into a greased and floured 6in cake tin and bake in a pre-heated fairly hot oven..for 1 hour. 1967 KARCH & BUBER *Offset Processes* ii. 20 After one minute in an open preheater press, a reinforcement mat and backing sheet are added. Then the entire assembly is heated and later bonded. 1976 *Woman's Weekly* 6 Nov. 74/1 (Advt.), In this article she looks at pre-heating underblankets.

prehend, *v.* Restrict *Obs. rare* to sense in Dict. and add: **b.** *Philos.* To apprehend with or without conscious formulation of the perceived object; to interact in time and space with an object or event. Cf. *PREHENSION 3 b. Hence **prehe·nded, prehe·nding** *ppl. adjs.*

1925 A. N. WHITEHEAD *Sci. & Mod. World* (1926) vi. 153 Then the enduring pattern is a pattern of aspects within the complete pattern prehended into the unity of A. 1927 AUDEN & DAY-LEWIS in *Oxf. Poetry* p. vi, Emotion is no longer necessarily to be analysed by 'recollection in tranquillity': it is to be prehended emotionally and intellectually at once. 1929 A. N. WHITEHEAD *Process & Reality* 56 The essence of an actual entity consists solely in the fact that it is a prehending thing. 1933 —— *Adventures of Ideas* xiv. 268 There are the physical and the mental poles, and there are the objects prehended and the subjective forms of the prehensions. 1938 C. D. BROAD *Exam. McTaggart's Philos.* II. I. 4 When a person has repeatedly prehended certain particulars as having a certain characteristic *C* he may 'form an idea of' that characteristic. 1945 R. G. COLLINGWOOD *Idea of Nature* III. iii.

173 An iron filing prehends the magnetic field in which it lies, that is it converts that field into a mode of its own behaviour, responds to it. *Ibid.* 174 A plant prehends the sunlight. **1947** *Mind* LVI. 97 In certain circumstances, when a person 'sees' a physical object, he visually prehends that physical object... In other cases what he visually prehends is, not the physical object.., but a particular which stands in a certain relation to the visum. **1959** W. A. Christian *Interpretation of Whitehead's Metaphysics* i. 12 It has a subject (the prehending actual entity), an object or datum that is prehended, and a subjective form. **1971** J. B. Cobb in D. Brown et al. *Process Philos. & Christian Thought* xii. 220 The new occasion prehends all the entities in its past.

prehensible, *a.* (Later example in sense b of prec.)
1947 *Mind* LVI. 101 This line has parts of two different kinds, visually prehensible and not visually prehensible.

prehensile, *a.* Delete 'Chiefly *Zool.*' and add later *gen.* examples. Also *fig.*
1945 R. Hargreaves *Enemy at Gate* 23 Hungry, prehensile soldiers of fortune. **1959** *Listener* 15 Oct. 633/1 A prehensile readability. **1966** G. Greene in *New Statesman* 25 Feb. 254/2 Martha was the plump and prehensile wife of a German correspondent who was suspected of strong Nazi sympathies. She was said to look after men's needs with a simple and indiscriminate fervour. **1976** *Time* 27 Dec. 52/3 Retailers have been eying Kong's potential with prehensile enthusiasm.

prehension. Add: **3. b.** *Philos.* Apprehension of something perceived that may or may not involve cognition; the interaction that exists between a subject and an entity or event.
1925 A. N. Whitehead *Sci. & Mod. World* (1926) iv. 97 The word 'perceive' is, in our common usage, shot through and through with the notion of cognitive apprehension. So is the word 'apprehension', even with the adjective cognitive omitted. I will use the word 'prehension' for uncognitive apprehension: by this I mean apprehension which may or may not be cognitive. **1931** A. Wolf in W. Rose *Outl. Mod. Knowl.* xiii. 584 The 'interlockings' of actual occasions are called 'prehensions', and are conceived causally. Each actual occasion is generated from its prehensions of preceding occasions, and is prehended by succeeding occasions. **1938** C. D. Broad *Exam. McTaggart's Philos.* II. I. 4, I propose to substitute the artificial term *Prehension* for 'perception' when used in McTaggart's extended sense. I think that this word avoids the objections to 'perception' and to 'acquaintance', which I have pointed out. **1945** R. G. Collingwood *Idea of Nature* III. iii. 173 Everything enjoys what he calls 'prehensions', that is to say, somehow absorbs what is outside itself into its own being. **1959** W. A. Christian *Interpretation of Whitehead's Metaphysics* i. 12 A prehension is an operation in which an actual entity 'grasps' some other entity (actual or nonactual) and makes that entity an object of its experience. **1964** I. Leclerc in Reese & Freeman *Process & Divinity* 137 Form is the object of 'conceptual prehension', not of 'physical prehension'. **1971** V. Lowe in D. Brown et al. *Process Philos. & Christian Thought* i. 7 A prehension is not so much a relation as a relating, or transition, which carries the object into the makeup of the subject.

prehensive, *a.* Delete *rare* and add further examples. Also, pertaining to or involving prehension, esp. in sense 3b. Also *fig.*
1886 J. Sully *Teacher's Handbk. Psychol.* viii. 132 The discrimination and identification of the impression... This constitutes the first step in the process of perception. It may be marked off as the presentative or *prehensive* element. **1925** A. N. Whitehead *Sci. & Mod. World* (1926) iv. 90 Things are separated by space, and are separated by time: but they are also together in space, and together in time, even if they be not contemporaneous. I will call their characters the '*separative*' and the '*prehensive*' characters of space-time. *Ibid.* 98 For Berkeley's *mind*, I substitute a process of prehensive unification. **1932** D. Emmet *Whitehead's Philos. of Organism* iv. 87 Actual entities are..described as 'prehensive occasions', that is to say, events or concrete facts of becoming. **1941** P. Hughes in P. A. Schilpp *Philos. A. N. Whitehead* vi. 278 This activity of perceptual adaptation is a concrescence of prehensive processes, each of which has the quality of the act as a whole. **1966** *Punch* 13 Apr. 528/2, I shall propose to Longman to accept a work on the originality of Locke, Hobbes, and Hume, which will be as *pioneer* to a more prehensive work. **1974** *Nature* 6 Dec. 514/3 Limbless thalidomide children who carry out reaching and prehensive tasks with their mouths and teeth.
prehensiveness (further example).
1937 J. R. Firth *Tongues of Men* iii. 37 The very use of likeness and differences and the habitual comparison of ordered series of words assume the principle of 'inter-related prehensiveness' which may be called implication.

prehensorial (prīhĕnsōə·riăl), *a.* [f. Prehensori(um + -al.] = Prehensory *a.*
1903 *Proc. Zool. Soc.* I. 51 One cannot but wonder how the spider maintains a secure hold back downwards, especially when the powerful prehensorial legs of the first and second pairs are released.

pre-Hispanic: see *pre- B. 1a.

prehistorian. Delete *rare* and add further examples.
1936 *Discovery* Jan. 23/1 Its interest and importance are of equal..moment to prehistorian, archaeologist, and the student of religious beliefs. **1947** [see *distribution map* s.v. *distribution 9]. **1952** G. Sarton *Hist. Sci.* I. i. 6 Prehistorians have proved beyond doubt the existence of sophisticated cultures at very early times in many places. **1970** *Nature* 12 Dec. 1019/2 The very early carbon-14 dates..for the megalithic tombs of Brittany..have been regarded suspiciously by prehistorians. **1975** *Sci. Amer.* Feb. 41/3 A major French archaeological discovery that was declared fraudulent by many prehistorians in the 1920's has now regained credibility as a result of dating studies conducted at three independent laboratories. **1980** *Early Music* Jan. 85/1 Various groups of specialists: historians of theatre-décor, pre-historians of opera, analysts of court spectacle, experts on renaissance instruments.

prehistoric, *a.* Add: **a.** *prehistoric archæology*: the archæology of the prehistoric period.
1910 *Encycl. Brit.* I. 344/2 The more serious and cautious students of prehistoric archaeology. **1932** A. R. Radcliffe-Brown in *Rep. Brit. Assoc. Adv. Sci. 1931* (Centenary Meeting) 143 Another field that lies within the general field of Anthropology as now organised is that of Prehistoric Archæology. **1935** *Chambers's Encycl.* I. 387/1 Prehistoric archæology has no dates. **1948** A. L. Kroeber *Anthropol.* (rev. ed.) xix. 843 How about relations to history—with which prehistoric archaeology so obviously intergrades that no real line of demarcation can be drawn? **1974** *Encycl. Brit. Micropædia* II. 838/1 Childe was professor of prehistoric archaeology at the University of Edinburgh.
b. *transf.* and *fig.* (chiefly *joc.*).
1859 Geo. Eliot *Let.* 17 Aug. (1954) III. 133 Pug develops new charms... I think, in the pre-historic period of his existence, before he came to me, he had led a sort of Caspar Hauser life, shut up in a kennel in Bethnal Green. **1886** Kipling *Departmental Ditties* (1888) 15 Delilah Aberyswith was a lady..With..a little house in Simla in the Prehistoric Days. **1886** 'Mark Twain' *Speeches* (1910) 185, I can see that printing-office of prehistoric times yet, with its horse bills on the walls. **1924** J. Buchan *Three Hostages* vii. 105, I obediently sampled an old hock, an older port, and a most pre-historic brandy. **1968** M. Bragg *Without City Wall* x. 116 It's your success story which is jaded... It's been going on for centuries... Prehistoric! **1979** *Guardian* 12 June 23/2 Red and blue looked exactly the same to anyone watching on prehistoric black and white [television].
prehistorically *adv.* (further examples).
1974 *Verbatim* Dec. 1/1 Sulfur, iron,..and lead..were known to some peoples prehistorically. **1975** *Nature* 22 May 355/1 Wilkinson argues that 20th century studies on musk ox behaviour can be extrapolated back to predict the ways in which musk oxen could have been exploited prehistorically.

prehistory. Add: **b.** *transf.* Events or conditions leading up to a particular event, period, etc.
1931 N. Mitchison in *Time & Tide* 25 July 893/1 Psychologists have come nearer to discovering its causes [*sc.* the causes of unhappiness] than politicians have, but they are mostly bad historians, inventing—as Freud has done—their pre-history to suit their theories. **1958** *Times Lit. Suppl.* 26 Dec. 746/5 The pre-history of the Civil War will not be found in the new position of the mercantile and industrial classes. **1960** K. Amis *New Maps of Hell* (1961) i. 17 The prehistory of science fiction, up until 1914 or later, is admittedly as much British as American. **1974** *Encycl. Brit. Macropædia* X. 121/1 The latter third of the 19th century was a crucial point in the prehistory of jazz. **1977** *N.Y. Rev. Bks.* 27 Oct. 40/2 That is why the post-history of a work, the tradition it created, is as indispensable to the critic as its pre-history, its sources and the tradition it came from.

pre-Hitler: see *pre- B. 2b. **pre-Hitlerian, -Hitlerite:** *pre- B. 1a.

prehominid, *sb.* and *a.* [Pre- B. 1 in Dict. and Suppl.] **A.** *sb.* A creature belonging to an anthropoid genus that is considered to be an evolutionary ancestor of the hominids. **B.** *adj.* Of or pertaining to a creature of this kind.
1939 *Nature* 7 Jan. 18/1 The human affinities of his [*sc.* R. Broom's] recently discovered relics of new types of fossil prehominids. **1948** *Proc. Prehist. Soc.* XIV. 30 Several authorities refer to the Chopper-Chopping-tool Complex of the Far East as having been developed by the Prehominid stock of mankind—members of the *Pithecanthropus* group. **1959** B. Wall tr. *Teilhard de Chardin's Phenomenon of Man* III. ii. 194 To call *Pithecanthropus* and *Sinanthropus* pre-hominids might suggest that they were not yet quite man. *Ibid.* 195 Those creatures which (however pre-hominid in cranial structure) are already clearly situated *above* the point of origin..of our human race. **1968** A. S. Romer *Procession of Life* xviii. 290 Even if a pre-hominid wished to remain a tree-dweller, his living area would be much restricted unless he were able to venture out into the open. **1978** *Nature* 26 Oct. 744/1 Among living species, the pygmy chimpanzee..offers us the best prototype of the prehominid ancestor.

prehuman, *a.* Add: (Further examples.) Also as *sb.*
1883 A. Wilson in G. Allen et al. *Nature Studies* 105 That which the evolutionist and naturalist desire to know, is the nature of the forms which..must have connected the human root-stock with the pre-human root. **1932** S. Zuckerman *Social Life Monkeys* ii. 24 There is no clear reason why the social behaviour of the 'pre-humans' should be considered to have been like that of apes rather than like that of monkeys. **1958** A. R. Radcliffe-Brown *Method in Social Anthropol.* II. v. 181 It is reasonable to fix the real change from pre-human to human social life by reference to the beginnings of language. **1973** G. Olivier in M. H. Day *Human Evolution* 94 The possibility that man..may only have appeared very late, and that the other fossils, including *Pithecanthropus*, may be only pre-human hominids.

prehyoid: see *pre- B. 3. **pre-ictal:** *pre- B. 1.

pre-ignition. [Pre- A. 2.] Ignition of the fuel and air mixture in an internal-combustion engine before the passage of the spark.
1898 A. G. Elliott *Gas & Petroleum Engines* iv. 88 This third engine has a vaporizer which..is so constructed that even if it became red-hot there would be no risk of pre-ignition. **1909** *Westm. Gaz.* 1 Apr. 4/1 The compression has been increased to the highest point compatible with safety in regard to freedom from pre-ignition. **1951** 'S. Abbey' *Automobile Fault-Tracing* ii. 28 Pre-ignition and auto-ignition differ from detonation in that the incoming charge is ignited by an incandescent particle or surface before the sparking plug fires. **1970** *A.A. Bk. of Car* 50 Pre-ignition, like knocking, can cause extensive damage, as well as reduce engine power. **1977** I. M. Campbell *Energy & Atmosphere* v. 108 Autoignition (also referred to as preignition, knock, or pinking) arises in an internal combustion engine largely on account of the finite velocity of flame propagation from the spark zone.

preimage (prī,i·mĕdʒ). *Math.* [Pre- A. 2.] = *inverse image* s.v. *image *sb.* 7*.
1949 F. Blum tr. *B. L. Van der Waerden's Mod. Algebra* I. i. 3 The element $\phi(a)$ is called the image of a, while a is an inverse image (pre-image) of $\phi(a)$. **1958** [see *into *adj.*]. **1975** N. Chomsky *Logical Struct. Linguistic Theory* ix. 317 T carries the elements of its domain into strings of $\hat{P}$ which differ from one another and from their preimages in the sense of condition C6.

pre-implantation, *a.* *Biol.* [Pre- B. 2.] Occurring or existing between the fertilization of an ovum and its implantation in the wall of the uterus.
1945 W. J. Hamilton et al. *Human Embryol.* v. 49 (*heading*) Pre-implantation period. *Ibid.* 50 The pre-implantation stages of the human ovum must be essentially the same as those of the monkey. **1968** *Nature* 9 Nov. 596/1 Mosaicism, extending to most tissues of the body in mice, was achieved by fusing pre-implantation embryos together *in vitro.* **1972** R. L. Brinster in Balin & Glasser *Reproductive Biol.* xx. 751 In the human, where the gestation period is 270 days, the pre-implantation period is still only 8 days.

pre-impregnate, *v.* [Pre- A. 1.] *trans.* To impregnate (a material) with something prior to mechanical processing. So **pre-impregnated** *ppl. a.*, *spec.* (*a*) of paper insulation: impregnated with oil and resin before use in electric cables; of a cable: containing such insulation; (*b*) of reinforcing material for plastics: impregnated with synthetic resin before fabrication; **pre-impregnation.**
1933 *Jrnl. Inst. Electr. Engin.* LXXIII. 353/2 The design conditions are extremely simple as the dielectric depends solely on the use of the pre-impregnated paper, the selection of gas-space dimensions, and the pressure of the gas. **1937** *Ibid.* LXXXI. 634/1 With proper attention to the physical properties of the impregnating medium, this pre-impregnation ensures the absolute definition of the gas spaces up to the maximum working temperature of the cable. **1958** *Times* 1 Dec. 2/5 Aeroplastics Limited ..have commenced production of their new process of preimpregnating glass fabric. **1958** D. J. Duffin *Laminated Plastics* iv. 73 Either a dry or preimpregnated glass reinforcement is now laid in the mold. **1960** *Jrnl. Inst. Electr. Engin.* VI. 694/1 The pre-impregnated cable does not suffer from this kind of trouble because the paper is dried and impregnated in sheet form before application to the cable. **1965** *Mod. Plastics Encycl. 1966* 628/1 A pre-impregnation step is frequently used, in which the reinforcing web is passed through a resin bath before lay-up. **1970** *Materials & Technol.* III. xii. 871 Mats or fabrics are pre-impregnated with resin by passing through baths containing a solution of the resin. The solvents are removed by drying and the material is worked into mouldings. **1971** *Nature* 30 July 305/2 These preimpregnated layers are then moulded together in the normal way under heat and pressure, and cured.

pre-Inca: see *pre- B. 2a. **pre-Incaic:** *pre- B. 1. **pre-increase:** *pre- B. 2a. **pre-incubate:** *pre- A. 1. **pre-incubation:** *pre- A. 2. **pre-industrial** to **-intellectual:** *pre- B. 1. **pre-intimation:** *pre- A. 2. **pre-invasion:** *pre- B. 2a. **pre-ionize:** *pre- A. 1.

preiotation (prī,əiotēi·ʃən). [f. Pre- A. 2 + Iot(a (here standing for the palatal glide *y*) + -ation.] In the Slavonic languages, the development of a palatal glide before a vowel. So **preiotiza·tion**; **preio·tized** *a.*, preceded by a palatal glide.
1877 A. H. Keane tr. *Hovelacque's Sci. of Lang.* 281 The Lithuanian *este* becomes *jeste* in Church Slavonic; and this 'preiotation', as it is technically called, is a leading feature of all the Slavonic tongues. **1883** W. R. Morfill *Slavonic Lit.* i. 18 The difficulty of expressing the præiotised vowels is the same. **1887** —— in *Encycl. Brit.* XXII. 148/2 The addition of a *y* sound before vowels is one of the great characteristics of the Slavonic languages, called 'præiotization'. **1959** G. Nandriş *Old Church Slavonic Gram.* 6 The Glagolitic letters for preiotized ę and ǫ are ligatures.

pre-jazz: see *pre- B. 2a.

prejudice, *sb.* Add: **1. c.** *to terminate* (*dismiss,* etc.) *with extreme prejudice*: to kill, to assassinate. Hence *termination with extreme prejudice.* *U.S. slang.*
1972 B. F. Conners *Don't embarrass Bureau* (1973) ii. 99 'A few years ago when he wanted an agent..out of the organization he ended up dismissing him with extreme prejudice.' 'You mean he had him killed?'.. Ted nodded. **1974** W. Garner *Big enough Wreath* x. 123 'There is no question of anyone killing anybody.' 'There is. I'm asking it... Terminate with extreme prejudice?' **1974** F. Nolan *Oshawa Project* xvi. 105 Had he been taken out by his own people?.. He had seen some of those files with the brutal red block letters stamped diagonally across the page: *Terminate with extreme prejudice.* **1980** C. Pincher *Dirty Tricks* i. 10 A 'termination with extreme prejudice', as the CIA called its assassination projects in those days.

pre-junctural, -Kantian, -Keynesian: see *pre- B. 1. **preknow:** *pre- A. 1.

prelanguage, *sb.* and *a.* **A.** *sb.* (Stressed *pre·language.*) **1.** [Pre- B. 1 in Dict. and Suppl.] A form of communication preceding the emergence or acquisition of language.
1940 Bryant & Aiken *Psychol. of Eng.* iv. 33 What we may call 'pre-language' consisted of meaningful cries used exclusively in context to express emotions, messages, [etc.]. **1973** C. F. Hockett *Man's Place in Nature* xxv. 382 The lines of development we have described slowly gave rise to a vocal-auditory communicative system very different from a close call system: to an open system that we shall call *prelanguage.* *Ibid.*, Prelanguage was not language. **1978** *Verbatim* May 15/2 Language was supposed to begin with grammar, and prelanguage was in no way relevant to the development of speech.
2. [Pre- A. 2.] A hypothetical antecedent language.
1966 E. P. Hamp in Birnbaum & Puhvel *Anc. Indo-European Dial.* 107 This prelanguage would have arisen in Dacia.
B. *adj.* (Stressed *prela·nguage.*) [Pre- B. 2.] Prior to the emergence of language.
1964 *Language* XL. 240 Various prelanguage stages of development.

prelapsarian, *a.* (Further examples.)
Quot. 1934 is taken by 'H. MacDiarmid' from an unknown source.
1934 'H. MacDiarmid' *Stony Limits* 65 'The buoyant Prelapsarian naturalness of a country girl.' **1949** Auden *Nones* (1952) 36 Her pallid affected heroes Began their hectic quest for the prelapsarian man. **1972** *Time* 22 May 32/2 Nature unspoiled, inhabited by prelapsarian man. **1977** *Daily Tel.* 24 Mar. 15/5 Glenda Jackson's evocation of poet Stevie Smith suggests a woman of prelapsarian innocence.

prelatial, *a.* Add: Also, that is a prelate.
1886 F. G. Lee *King Edward VI* iii. 142 Both as regards what the prelatial preacher said, and what he did not say, it appeared..unsatisfactory and..inadequate.

pre-Latin: see *pre- B. 1. **pre-launch:** *pre- B. 2a.

pre-law (prīlǭ·), *a.* (*sb.*) *U.S.* [Pre- B. 2.] Of or pertaining to subjects studied in preparation for a course in law. Also *ellipt.* as *sb.*
1961 in Webster. **1971** *Mod. Law Rev.* XXXIV. 650 Their North American counterparts who have undertaken significant pre-law work in another discipline. **1976** *National Observer* (U.S.) 21 Feb. 16/1 (Advt.), Undergraduate prelaw program, premed, and other preprofessional offerings. **1976** *Billings* (Montana) *Gaz.* 17 June 3-B/1 (Advt.), [She]..plans to attend the University of Montana in Missoula to study pre-law. **1978** *Detroit Free Press* 5 Mar. 9/1 Career training offered at Madonna for deaf students includes..pre-law and education.

pre-legislation: see *pre- B. 2a. **pre-let:** *pre- A. 1. **prelexical:** *pre- B. 1d. **pre-liberation, -life:** *pre- B. 2a.

prelim (prīli·m, prī·lim), colloq. abbrev. of Preliminary *sb.* and *a.* **a.** (See note *s.v.* Preliminary *sb.* b in Dict.) A preliminary practice, examination, contest, inquiry, or report. Also *transf.*, a student in a preliminary class. Also *attrib.*
1891 C. Dawson *Let.* 19 Feb. in R. S. Churchill *Winston S. Churchill* (1967) I. Compan. I. v. 228, I was so glad to hear you had successfully passed the 'Army Prelim', allow me to congratulate you. **1901** *Daily News* 1 Apr. 5/6 We arrived at Putney, just in time to see Oxford come out for their 'prelim'. **1902** *Daily Chron.* 19 Dec. 5/2 The English public school boy goes north for months of special tutoring for his 'prelim.'; thereat, probably, to fail in his English paper. **1904** *Daily News* 28 Dec. 6 While yet in the preliminary class,..she..said, 'I want a canvas six feet long.' 'What does she want with a six-foot canvas?.. She's only a "prelim"!' **1923** L. J. Vance *Baroque* xxvii. 173 A fight that'd make the Dempsey-Carpenteeyay bout look like a cooked prelim. **1928** *Collier's* 18 Aug. 25/2 You're nothin' but a has-been, staggerin' around like some prelim boy. **1958** *Times* 17 Oct. 3/5 The general college rule which stipulates that undergraduates who twice fail 'prelims' should be sent down or rusticated. **1965** D. Francis *Odds Against* vi. 86 I'll put one of the boys on to it and let you have a prelim. Is it urgent? **1974** N. Freeling *Dressing of Diamond* 41 I've been waiting to hear from you to set a prelim afoot. **1977** P. Cosgrave *Cheyney's Law* ii. 17 Tommy went to Cambridge... His prelims were only fair.
b. Usu. in *pl.* The preliminary matter of a book. Cf. *preliminary *sb.* c.
1927 *Observer* 18 Dec. 4 He tells..about signatures, prelims, end-papers, uncut and unopened pages, issues and imperfections. **1932** *Times Lit. Suppl.* 7 Jan. 13/2 When, with the 'prelims' of 'The Painted Veil', he [*sc.* W. S. Maugham] comes up against a more important bibliographical problem, his description is confused. **1957** *Ibid.* 1 Nov. 664/3 Besides a few corrections in the text there are some notes in pencil among the prelims..and in ink on the back endpaper. **1960** [see *blad *sb.*[2] 2c]. **1976** *Indexer* Oct. 93/2 The index..should be provided by the publisher much as he provides prelims and jacket copy.

prelimen (prīlai·mən). [f. L. *præ* before + *līmen* threshold.] A preliminary step.
1898 C. S. Sherrington in *Phil. Trans. R. Soc.* B. CXC. 50 The requisite prelimen to the original aim of the inquiry [having been] carried through, the examination of certain spinal reflexes has been proceeded to.

preliminary, *sb.* Add: **c.** Usu. in *pl.* The preliminary matter of a book. Cf. *prelim b.
1888 C. T. Jacobi *Printers' Vocab.* 103 *Preliminary*, any matter coming before the main text of a work—title, preface, contents, etc. **1903** A. E. Housman *Let.* 12 Feb. (1971) 64, I don't quite know the meaning of 'the preliminary', but I enclose the dedication which is to follow the title page. **1977** *N. & Q. for Somerset & Dorset* Sept. 301 The list of abbreviations on p. 70 would have been more usefully placed in the preliminaries.

prelingual, *a.* Add: **a.** (Further examples.)
1924 R. M. Ogden tr. *Koffka's Growth of Mind* v. 322 The behaviour of the child during his pre-lingual period. **1976** *Word* 1971 XXVII. 132 What constitutes semantic salience for the pre-lingual child?
b. [Pre- B. 3.] Located in front of the tongue.
1953 *Archivum Linguisticum* V. 69 According to their articulating points Marr's consonants fall into labial, 'prelingual' (dentialveolar and palatal), and 'postlingual' (velar and ultravelar).

prelinguistic, *a.* and *sb.* [Pre- B. 1.] **A.** *adj.* = Prelingual *a.* a.
1900 H. Sweet *Hist. Lang.* iv. 39 Even in the prelinguistic stage in which gesture predominated, there must have been some principles of order. **1901** [see *inner *a.* 2b]. **1919** M. K. Bradby *Psycho-Anal.* x. 133 A dream does..sometimes belong to a pre-linguistic stage of mental experience. **1941** *Mind* L. 421 Mathematics will vanish with the rest of our intellectual heritage if we revert to our pre-linguistic apehood. **1951** Trager & Smith *Outl. Eng. Struct.* 13 Prelinguistic data are now available. **1963** Ervin & Miller in J. A. Fishman *Readings Sociol. of Lang.* (1968) 70 The prelinguistic sounds of deaf and hearing children are indistinguishable in the first three months. **1972** *Language* XLVIII. 487 The identification of a prelinguistic stage has for decades rested on the intuition that everything which precedes the first utterance identifiable as a word is non-linguistic. **1979** N. Lash *Theol. on Dover Beach* ii. 29, I am not suggesting that there is..any such thing as pre-linguistic or non-linguistic human experience.
B. *sb. pl.* [-ic 2.] The study of biological and physiological aspects of speech.
1949 [see *microlinguistics *sb. pl.*]. **1953** *Internat. Jrnl. Amer. Linguistics Memoir* VIII. 28 'Prelinguistics' studies the physical and biological aspects of speech. **1959** M. Joos in J. A. Fishman *Readings Sociol. of Lang.* (1968) 186 This line, along which he [*sc.* the linguistic analyst] can shift items into the 'past' or the 'future' as he moves from pre-linguistics towards metalinguistics. **1964** Crystal & Quirk *Syst. Prosodic & Paralinguistic Features in Eng.* ii. 27 Speech results from activities which create a background of voice set ('the idiosyncratic, including the specific physiology of the speakers, and the total physical setting'). This is in the area of 'prelinguistics'.

pre-Linnean, -Listerian, -literary: see *pre- B. 1.

preliterate, *a.* [Pre- B. 1.] Applied to social groups or cultures which have not acquired a form of writing. Also as *sb.*, a person belonging to such a group or culture. Cf. *non-literate *a.*
1925 E. Faris in *Amer. Jrnl. Sociol.* XXX. 710 For some time the writer has been using..the term 'preliterate' to designate the peoples of the *sociétiés inférieures*, as Lévy-Bruhl calls them. *Ibid.* 712 Pre-literate man is... one in whose culture there is no written literature. *Ibid.*, Pre-literates do not have cities. **1933** A. R. Radcliffe-Brown in *Encycl. Social Sci.* IX. 202/1 The confusion which has resulted in the attempt to apply to preliterate societies the modern distinction between criminal law and civil law can be avoided. **1957** *Antiquity* XXXI. 211 Rubbish deposited in superimposed layers by preliterate men. **1962** E. R. Service *Primitive Social Organization* i. 8 Where an Arunta-like way of life is not yet significantly altered by modern influences it is a culture that is primitive, ancient, and preliterate. **1966** E. G. Stanley *Continuations & Beginnings* 127 To understand the use of tags and set phrases, whole half-lines of verse used repeatedly, it is useful to know about some kinds of preliterate composition. **1967** *N.Y. Rev. Bks.* 23 Feb. 33/3 He writes about the last 8000 years of preliterate Europe, which means that he excludes Greeks and Romans. **1970** [see *non-literate *a.*].
Hence **preli·teracy,** the quality or state of being preliterate.
1957 G. Clark *Archaeol. & Society* (ed. 3) i. 22 Clearly.. some difference of opinion is likely to exist as to precisely at what stage preliteracy gives way to literacy. **1967** *N.Y. Rev. Bks.* 23 Feb. 33/1, I wish I knew who invented the word 'preliteracy' to indicate the illiteracy of certain extinct or living cultures... Preliteracy points to literacy as the next step in human evolution... In the latter part of the nineteenth century illiteracy was still illiteracy—not preliteracy.

pre-load, *sb.* [Pre- A. 2.] A load applied beforehand; *spec.* (*a*) one in a bearing or machine part (see *preload *v.*); (*b*) the tension in heart muscle at the end of diastole.
1941 *Motor Commerce* July 25/3 Adjust the pinion assembly to the correct bearing pre-load of 12 to 17 in.-lbs., and lock the adjusting nuts in position by a lock-washer. **1954** *Sun* (Baltimore) (B ed.) 3 Nov. 36/4 This earth pile acts as a preload or surcharge on the subsoil... The preload or surcharge is to get a settlement of the earth before the towers are constructed. **1962** *Amer. Jrnl. Physiol.* CCII. 936/1 Theoretically, with no load on the muscle, the velocity of shortening is maximal... Experimentally, the closest one can approach such a condition with heart muscle is to use the smallest preload from which a contraction occurs that produces analysable data. **1970** *Circulation Res.* XXVI. 114/1 The muscle is prestretched to a certain initial length using a 'preload', P, and temporarily fixed at that length. **1971** J. J. Gregory in Ayres & Gregory *Cardiol.* xii. 225 Just as ventricular function may be evaluated by change in afterload, interventions that alter preload may also be utilized to test myocardial function. **1971** *Power Farming* Mar. 71/1 Dynamic loading in a tightened bolt may vary from no stress at all to that exceeding the bolt's preload. **1972** H. E. Ellinger *Automechanics* xxiii. 400 Pinion preload is sufficient to eliminate any end play in the pinion shaft and still low enough to prevent bearing damage. **1976** *Circulation* LIII. 298/2 In the intact heart it is impossible to completely separate preload from afterload.

preload, *v.* [Pre- A. 1.] *trans.* To load beforehand; *spec.* in *Engin.*, to design or make (a bearing or machine part) in such a way that there are internal loads independent of any working load (e.g. to reduce noise in operation).
1945 R. K. Allan *Rolling Bearings* x. 253 Another method..is to preload the bearings by axial adjustment. **1950** O. J. Horger in M. Hetényi *Handbk. Exper. Stress Analysis* xi. 556 Seeger investigated the fatigue resistance of hollow shafts which were preloaded axially so as to produce compressive prestress below yield-strength value of the steel. **1952** C. H. Dix *Seismic Profiling for Oil* iii. 44 In some cases it is desirable to drill several shot holes, preload each one of them, and not plan to reshoot any of them. **1969** W. Wrigley et al. *Gyroscopic Theory* xiv. 324 When two ball bearings are preloaded against each other, the result is a much stiffer assembly. **1969** *Jane's Freight Containers 1968–69* 439 (*caption*) They are pre-loaded and sealed by shippers.
So **preloa·ded** *ppl. a.*, **preloa·ding** *vbl. sb.*
1936 *Brit. Pat.* 471,989, The object of our invention is to provide improved brake-operating mechanism which permits a pre-loaded spring to be used in conjunction with compensation between the brakes on the front and rear wheels of a vehicle for normal braking. **1941** *Automobile Engineer* XXXI. 180/3 Owing to the substantial design of the housing it has not been found necessary to employ pre-loading of the bearings. **1945** R. K. Allan *Rolling Bearings* x. 235 The primary object of preloading is to eliminate..radial and/or axial movements in a bearing when it is functioning in a machine. **1962** *John o' London's* 22 Feb. 182/2 The dice are, as it were, already pre-loaded. **1969** W. Wrigley et al. *Gyroscopic Theory* xiv. 324 Under the preloaded condition, the pair of bearings have axial and radial stiffnesses vs. load which are very nearly constant. **1971** B. Scharf *Engin. & its Lang.* xii. 137 Preloading is frequently used in high precision work and in order to ensure noiseless running. **1976** Attewell & Farmer *Princ. Engin. Geol.* xi. 826 Any pre-loading or the application of consolidation pressure must..start with the condition that the initial stresses applied to the soil are less than the existing shear strength of the soil.

pre-logic: see *pre- B. 1.

prelogical, *a.* [Pre- B. 1.] Preceding or prior to logic or logical reasoning; chiefly in *Anthrop.*, applied to the thinking of persons or cultural groups which is based on myth, magic, etc. Also *transf.* and *absol.*
1893 C. S. Peirce *Coll. Papers* (1933) IV. I. iv. 62 The whole of the theory of numbers belongs to logic; or rather, it would do so, were it not, as pure mathematics, *prelogical*, that is, even more abstract than logic. **1923** L. A. Clare tr. *Lévy-Bruhl's Primitive Mentality* ii. 91 To prelogical mentality, cause and effect present themselves in two forms, not essentially different from one another. **1926** —— tr. *Lévy-Bruhl's How Natives Think* I. ii. 78 By designating it [*sc.* the mentality of primitives] 'prelogical' I merely wish to state that it does not bind itself down, as our thought does, to avoiding contradiction. **1933** H. Read *Art Now* I. 34 Art conceived as a stage in the ideal history of mankind, as a pre-logical mode of expression, as something necessary and inevitable and organic. **1935** *Mind* XLIV. 544 Insistence on the alogical, or prelogical, character of the aesthetic consciousness. **1943** *Amer. Speech* XVIII. 220 The first stage of human development ..is that of the savage, prelogical mentality, with a one-valued semantics (or system of evaluations), in which.. 'everything is everything else' by 'mystic participation'. **1959** *Spectator* 11 Sept. 339/3 If we seek the pre-logical and oppose the march of intellect, we are the enemies of science..and the worshippers of myth. **1967** C. L. Markmann tr. *Fanon's Black Skin, White Masks* (1968) vi. 159 The prelogical thought of the phobic has decided that such is the case. **1977** G. W. Hewes in D. M. Rumbaugh *Lang. Learning by Chimpanzee* i. 28 Early language was also characteristically 'prelogical'.

pre-London: see *pre- B. 2a. **pre-lubricate:** *pre- A. 1.

prelude, *sb.* Add: **2.** (Later examples of sense 'an introduction, preface (to a literary work)'.)
1889 *Mod. Lang. Notes* IV. 350 Grein's sixteenth Canto of the 'Christ'.., is a transitional passage... The whole passage forms a kind of interlude, while it is also a prelude to Part III. **1892** J. Earle *Deeds of Beowulf* p. xxiv, It is not easy to account for this Prelude, which really throws no light on the poem, nor in any way helps the narrative.

prelude, *v.* Add: **1. b.** (Further example.)
1915 J. Buchan *Nelson's Hist. War* II. ix. 34 Von Kluck preluded it [*sc.* an enveloping movement] by a heavy bombardment of Binche and Bray.
3. a. (Later example.)
a **1945** E. R. Eddison *Mezentian Gate* (1958) xxxix. 214 The musicians tuned their instruments, preluded and, when the murmur of talk was stilled.., struck up a cavatina.
Hence **prelu·dingly** *adv.*, in a prelusive manner. *rare.*
1932 J. Joyce in *New Statesman* 27 Feb. 261/1 Preludingly he conspews a portugaese into the gutter, recitativing.

Preludin (prīliū·din). *Pharm.* Also **preludin.** A proprietary name for phenmetrazine hydrochloride.
1954 *Trade Marks Jrnl.* 17 Mar. 270/1 *Preludin*... All goods included in class 1 [*i.e.* chemical products used in industry, science, etc.]. Albert Boehringer, [*et al.*].., trading as C. H. Boehringer Sohn, Ingelheim am Rhein.., Germany; manufacturers. **1955** *Official Gaz.* (U.S. Patent Office) 13 Sept. TM59/2 C. H. Boehringer Sohn, Ingelheim am Rhein, Germany..*Preludin.* For anti-depressant and anti-obesity drug. Use since April 1, 1954. **1955** *Chem. Abstr.* XLIX. 8566 (*heading*) Identification of Preludin (2-phenyl-3-methylmorpholine) and Ritalin..before and after passage through the body. **1957** *Brit. Med. Jrnl.* 5 Jan. 30/2 'Preludin', a new drug for controlling the appetite, is, I believe, gaining popularity among those who.. try to slim. It was discovered in 1953 and first marketed in Germany in 1954. **1959** [see *phenmetrazine]. **1960** *Spectator* 22 July 120/3 A few weeks ago we criticised the irrationality of the campaign then being waged by some newspapers against the pep-pill Preludin. **1970** Passmore & Robson *Compan. Med. Stud.* II. v. 39/1 Appetite suppressant drugs include phenmetrazine (preludin), chlorphentermine and diethylpropion (tenuate). Their value in this respect is limited and emotional dependence has been reported. **1974** M. C. Gerald *Pharmacol.* xv. 286 Phenmetrazine (Preludin) swept the market in Sweden after its introduction..in 1955. This drug..continues to be the abused drug of choice in Sweden. **1976** D. Hare *Teeth 'n' Smiles* I. 20 *Anson*: Is that heroin? *Peyote*: Preludin. *Anson*: Ah. Preludin is a fuck-pump... It enlarges your sexual capacity.

preludize, *v.* Add: (Earlier and later examples.)
1842 J. R. Planché *White Cat* I. 9 The leader preludizes on the violin. **1978** *Gramophone* Feb. 1389/3 When at a recital Hans von Bülow had to follow a singer of whom he thought little he would sometimes sit at the piano and preludize for a moment or two on the passage from Beethoven's Ninth Symphony.

prelumirhodopsin. *Biochem.* [Pre- A. 2.] An isomer of rhodopsin, stable only at very low temperatures, which is formed by the action of light on rhodopsin and changes spontaneously to lumirhodopsin.
1963 Yoshizawa & Wald in *Nature* 30 Mar. 1280/1 The intermediate was shown to be stable to about −140°, above which it forms lumirhodopsin. We shall call it prelumirhodopsin. **1968** A. White et al. *Princ. Biochem.* (ed. 4) xl. 904 After pre-lumirhodopsin is formed by the isomerization of one double bond, all subsequent steps proceed spontaneously. **1976** *Nature* 5 Feb. 424/2 The results of a picosecond study indicated that the formation of prelumirhodopsin when rhodopsin is irradiated probably involves a restricted change in the geometry of retinal rather than a complete isomerisation. *Ibid.* 22 Apr. 726/2 In the photolysis of the visual pigment rhodopsin the intermediate first observed is bathorhodopsin (formerly called prelumirhodopsin).

pre-lunch: see *pre- B. 2a.

pre-luncheon, *sb.* and *a.* **A.** *sb.* [Pre- A. 2.] A light mid-day meal preceding luncheon.
1873 C. M. Yonge *Pillars of House* II. xvi. 110 A preluncheon or nooning of cake and wine within an hour of the meal of the day.
B. *adj.* [Pre- B. 2.] Held or occurring before luncheon.
1975 *N.Y. Times* 29 Aug. 29/1 The four actors..did a jungle scream in unison for the press at a preluncheon cocktail party.

prem (prem), *sb.* and *a.* [Abbrev. of *premature *a.* or *sb.*] **A.** *sb.* A premature baby. **B.** *adj.* Premature; of or pertaining to premature babies.
1953 Baker *Australia Speaks* vi. 142 *Prem*, a hospital term for a premature baby. **1960** *News Chron.* 18 Mar. 8/6 Children's specialists are improving the outlook for 'prems'—frail, under-developed babies, some born too soon. **1961** *News of World* 23 Apr. 13 My last was three months prem. **1962** G. Butler *Coffin in Oxford* ix. 126 Father Mahoney was standing..at the end of the corridor by the nursery and 'prems' block. **1963** L. Diack *Labrador Nurse* xxix. 142 I've got a lovely prem. Four pounds. **1972** R. Lewis *Fool for Client* ii. 44 My daughter..was took bad in the night and they think the baby is going to be a prem. **1976** 'D. Halliday' *Dolly & Nanny Bird* iii. 41 The last time I heard her whisper like that, the incubator lights had cut out in a prem ward.

pre-machine: see *pre- A. 1, B. 2a.

pre-make-ready. *Printing.* [Pre- A. 2.] (See quots.)
1948 R. R. Karch *Graphic Arts Procedures* x. 256 Premakeready includes those operations done in advance of placing the form on the printing press. **1964** *Gloss. Letterpress Rotary Printing Terms (B.S.I.)* 16 *Pre-make ready*, the operations relating to the obtaining of a good printing result which take place before the printing plates or formes go to the press. **1967** V. Strauss *Printing Industry* vii. 423/2 Pre-makeready is..that part of quality control in letterpress which deals with the printing-image carrier and has the purpose of reducing to a minimum the time spent for makeready of type forms on the press.

premalignant, -Malthusian: see *pre- B. 1.

pre-man, *sb.* and *a.* **A.** *sb.* [Pre- B. 1 in Dict. and Suppl.] (Stressed *pre-·man.*) A hominid or man-like creature that lived before the appearance of man, *Homo erectus* and *H. sapiens.* **B.** *adj.* [Pre- B. 2.] (Stressed *pre-ma·n.*) Occurring or belonging to the time before the appearance of man.
1921 H. G. Wells *Short Stories* (1927) 687 Men, our ancestors, had their first glimpse of the pre-men of the wilderness. **1947** *Sci. Illustr.* Sept. 60/1 Desert so devoid of life that it takes the visitor back to pre-man time on earth. **1953** J. S. Huxley *Evolution in Action* vi. 136 We can distinguish..three stages in the physical evolution of man. First, the deployment of the pre-men. **1971** J. Z. Young *Introd. Study Man* xxxiii. 457 Up to the stage that may be called pre-man (*Australopithecus*) the changes were gradual. **1977** *Time* 7 Nov. 48/1 Leakey has found more and better pre-man and early man fossils than any other anthropologist.

Premarin (prĭmĕə·rin). *Pharm.* Also **premarin.** [f. Pre(gnant *a.*[2] + Mar(e[1] + -in[1].] A proprietary name for a mixture of œstrogenic compounds obtained from the urine of pregnant mares and used in the treatment of various conditions, esp. those caused by or involving œstrogen deficiency.
1942 *Official Gaz.* (U.S. Patent Office) 21 July 498/1 Ayerst, McKenna & Harrison (United States) Limited, Rouses Point, N.Y...*Premarin.* For pharmaceutical preparations for the treatment of ovarian deficiencies. Claims use since Feb. 24, 1942. **1944** *Jrnl. Amer. Med. Assoc.* 19 Aug. 1098/2 Premarin.—An amorphous preparation containing the naturally occurring, water soluble, conjugated forms of the mixed estrogens obtained from the urine of pregnant mares. *Ibid.*, Premarin Tablets. **1956** *Trade Marks Jrnl.* 2 May 312/1 *Premarin*... Estrogenic preparations. Ayerst, McKenna & Harrison, Limited.., St. Laurent, Quebec, Canada; manufacturers and merchants. **1961** *Lancet* 2 Sept. 504/2 In these patients continued administration of premarin resulted in a suggestive, though inconclusive, improvement in mortality as compared with a comparable control group. **1977** *Ibid.* 19 Nov. 1063/1 A two to three fold increase in the incidence of ovarian cancer was recorded in women treated for menopausal symptoms with 'Premarin' (conjugated equine œstrogens) usually combined with stilbœstrol.

premarital, *a.* [Pre- B. 1.] Occurring before marriage.
1886 [in Dict. s.v. Pre- B. 1]. **1915** T. F. A. Smith *Soul of Germany* v. 97 During his pre-marital years he may form many such irregular acquaintanceships. **1937** A. Huxley *Let.* 17 Dec. (1969) 430 As to pre-marital continence, there is a great deal of evidence that this is important if there is to be higher education. **1951** M. McLuhan *Mech. Bride* (1967) 32/1 The tendency of the modern housewife, after a premarital spell in the business world, to embrace marriage and children but not housework. **1963** A. Heron *Towards Quaker View of Sex* i. 6 An increase in transient pre-marital sexual intimacies. **1976** *Drum* (E. Afr. ed.) Nov. 26/1, I think your girl is one of those sensible ones who do not like to indulge in pre-marital sex. **1980** 'R. Deacon' *Spy!* iii. 70 It was essential to cover up this pre-marital affair.
Hence **prema·ritally** *adv.*
1973 S. Fisher *Understanding Female Orgasm* ii. 34 Religiosity plays a role in how sexually active women are premaritally. **1975** R. H. Rimmer *Premar Experiments* (1976) i. 108 They aren't fully aware of the impact that freer and open sexuality, premaritally and postmaritally, will have on human goals and values.

pre-market(ing): see *pre- B. 2a.

prematuration (premătiūrēi·ʃən, prīmætiurēi·ʃən), *sb.* [f. Prematur(e *a.* (*adv.*) + -ation.]
1. The fact of making or becoming mature unnaturally early.
1909 *Westm. Gaz.* 3 Feb. 2/1 The systems followed in the schools of the leading civilised races of the world make for prematuration.
2. [after F. *prématuration* premature birth.] = Prematureness. *rare*[−1].
1977 A. Sheridan tr. *Lacan's Écrits* iv. 137 This is a point that I think I have myself helped to elucidate by conceiving the dynamics of the so-called *mirror stage* as a consequence of a prematuration at birth.

prematuration (prīmætiurēi·ʃən), *a.* [f. Pre- B. 2 + Maturation.] Occurring before maturation.
1919 T. H. Morgan *Physical Basis Heredity* xii. 142 Most of the eggs pass through this early prematuration stage in the larvæ and some of them may reach the maturation stage in the pupa. **1974** *Austral. Jrnl. Agric. Res.* XXV. 883 Only in one cultivar..could a difference in rate of development between samples be attributed with any confidence to pre-maturation cold acquisition.

premature, *a.* (*adv.*). Add: Also pronounced (premătiūə·ɹ) in predicative use. **2. b.** *Obstetrics.* Born or occurring before full term (but usu. after the stage when the fœtus normally becomes viable).
1754 W. Smellie *Coll. Cases & Observations in Midwifery* II. xiii. 213 (*heading*) On the situation of the child during pregnancy, the signs of conception and premature labour. **1775** A. Hamilton *Elem. Pract. Midwifery* 122 When a woman miscarries in early Gestation, *this* they consider as an Abortion; but, if in the later Months, *that* they term a Premature Birth. **1800** *Med. Facts & Observations* VIII. 190 She has since borne six children by premature labour. **1840** [see *induction 9]. **1878** *Obstetr. Jrnl.* VI. 163 (*heading*) Case illustrating the viability of extremely small premature children. **1923** J. H. Hess *Premature & Congenitally Diseased Infants* iii. 40 Heat regulation is one of the least developed functions of the premature infants. **1924** C. Mackenzie *Heavenly Ladder* xviii. 244 The shock brought on a premature travail, and she was delivered of a boy in the Vicarage. **1969** D. Baird *Combined Textbk. Obstetr. & Gynæcol.* (ed. 8) xxxiii. 544 By international agreement a 'premature' infant has been defined as one weighing 2,500 g. (5½ lb.) or less at birth. **1973** *Sci. Amer.* May 27/2 Most of the mothers in the experimental group had had a premature baby (gestation period less than 38 weeks).
c. *premature ejaculation* (see quot. 1974); = *ejaculatio præcox.
1910 A. Abrams *Diagnostic Therapeutics* iii. 230 Occasionally onanism is followed by various grades of impotency (usually psychic) and premature ejaculation. **1925, 1928** [see *ejaculatio præcox]. **1942** T. P. Wolfe tr. *Reich's Function of Orgasm* v. 138 Hysterical men suffer either from erective impotence or premature ejaculation. **1968** T. Wiseman *Quick & Dead* 140, I with my quick grin and premature ejaculations. **1974** Passmore & Robson *Compan. Med. Stud.* III. xxxv. 34/1 Partial impotence is common and may take the form of a failure to ejaculate or ejaculation before entry into the vagina or before orgasm is reached, i.e. premature ejaculation.
B. *sb.* *Obstetrics.* A child born before full term.
1900 in Dorland *Med. Dict.* **1923** J. H. Hess *Premature & Congenitally Diseased Infants* xiv. 313 In the premature especially the skin is delicate, lacking the horny layer. **1960** A. K. Geddes *Premature Babies* iii. 18 An irregular respiratory rhythm is normal for prematures.

premature (pre·mătiūəɹ), *v.* *Mil.* [f. Premature *a.*] *intr.* Of a shell or other projectile: to explode prematurely. Of a gun: to fire a shell that explodes prematurely. Hence **pre·maturing** *ppl. a.*
1916 'Boyd Cable' *Doing their Bit* v. 83 A shrapnel prematuring at the muzzle, and the bullets that should have gone lifting high and clear inside the case smashing, perhaps, into the open rear of a gun-emplacement or a battery a few hundred yards in front of the prematuring gun. **1918** G. Frankau *Judgement of Valhalla* 49 Behind, a cratered slope, with batteries Crashing and flashing, violet in the dusk, And prematuring every now and then.

prematurely, *adv.* Add: **b.** *spec.* in *Obstetrics.*
1812 *Medico-Chirug. Trans.* III. 137, I have however now before me, a list of *seventy-eight* labours occurring prematurely, either from the spontaneous action of the womb, or from accidental violence. **1902** *Brit. Med. Jrnl.* 17 May 1197/2 The dry mouth and the weak digestion, and the frequency of gastro-intestinal disorders in the prematurely born are matters of every-day observation. **1943** *Lancet* 9 Dec. 320/1 An infant, prematurely born, is, although in a normal stage of development, inadequately prepared to contend against the operation of external agents.

prematurity. Add: **3.** (Later examples.)
1927 *Times Lit. Suppl.* 10 Feb. 90/3 Our advice is to save this book for a dismally wet afternoon: tea will arrive with a startling prematurity. **1961** E. A. Powdrill *Vocab. Land Planning* iii. 45 The term 'prematurity' is used by planners, very often as a sort of delaying action in grounds of refusal. *Ibid.* 46 Thus, the term 'prematurity', used in the sense of preventing something from happening in advance of the proper time, also implies that at some future date the proposal could be approved.
4. *Obstetrics.* The birth of a baby before full term.
1875 *Trans. Edin. Obstetr. Soc.* III. 260, I have seen nothing to warrant me viewing prematurity—that is, at and after the seventh month—as necessarily convertible with debility. **1902** *Brit. Med. Jrnl.* 17 May 1196/2 In this contribution the type of prematurity which is considered is that of the infant expelled from the uterus at the seventh month of intrauterine life. **1937** A. Tow *Dis. of Newborn* iii. 63 More careful antepartum care has definitely lowered the incidence of prematurity. **1971** *Sci. Amer.* Oct. 118/2 Prematurity from spontaneous abortion, affecting approximately one pregnancy in 10, is the main source of mortality.

pre-matutinal: see *pre- B. 1d.

pre-med (prī·me·d), *a.* (*sb.*[1]) Chiefly *U.S.* Colloq. abbrev. of *premedical *a.* (*sb.*).
1962 E. Snow *Red China Today* (1963) xxxv. 263, I finished my pre-med work in three years and won a scholarship to the American University in Beirut. **1971** 'S. Ransome' *Trap 6* (1972) iii. 27 He's about to begin his premed courses and he's not sure whether he'll make it all the way through to his M.D. **1972** W. P. McGivern *Caprifoil* (1973) xiii. 212 London told us you did a year of pre-med at Birmingham University. **1977** *Time* 28 Mar. 70/3 After high school Lily entered Detroit's Wayne State University as a premed student because 'I wanted to be a doctor'. **1978** *Detroit Free Press* 5 Mar. 9/1 Career training offered at Madonna for deaf students includes art, journalism, public relations, pre-med and pre-dentistry [etc.].

pre-med (prī·me·d, prī·med), *sb.*[2] Colloq. abbrev. of *premedication. Also *attrib.*
1964 D. Francis *Nerve* vi. 64 A brisk nurse told us he was going to the operating theatre within minutes and not to disturb the patient, as he had been given his pre-med. **1974** C. Fremlin *By Horror Haunted* 110 The shaving, the marking-up, the pre-med injections. **1977** *Observer* 4 Sept. 22/1 You'll get a pre-med, a jab to make you drowsy, at about ten.

premedical, *a.* (*sb.*) Chiefly *U.S.* [Pre- B. 1.] Of, pertaining to, or designating subjects studied in preparation for a medical course. Also *ellipt.*, a premedical course of studies. Cf. *pre-med *a.* (*sb.*[1]).
1904 *Bot. Gaz.* XXXVII. 225 This general text-book of botany is written for premedical and pharmaceutical students in particular. **1928** *Brit. Med. Jrnl.* 1 Sept. 363/2 The elementary sciences, the pre-medical subjects of chemistry, physics, biology, and botany, are in the curriculum to familiarize the student with the structure and behaviour of the materials with which, and upon which, he will have to work. **1940** A. Huxley *Let.* 7 July (1969) 455 Matthew has got through his second year of pre-medical quite well. **1961** *Lancet* 26 Aug. 484/1 Major Titov's first question, on returning from Space recently, concerned his wife's premedical examination results. **1976** *New Yorker* 16 Feb. 41/3 They reserved their serious efforts for the medievalists, true scientists, linguists, prelaw and pre-medical students, other scholars.

premedicant. [Pre- A. 2.] A drug given as premedication. Also *attrib.*
1960 *Proc. R. Soc. Med.* LIII. 673/1 The sister in charge administered the premedicant combinations in rotation, leaving the anaesthetist and the assessor in ignorance of the drugs given to particular patients. **1964** *Brit. Jrnl. Anaesthesia* XXXVI. 703/1 When evaluating the effects of various premedicants, the authors were impressed with the very high incidence of pre-operative vomiting and nausea which occurred when pethidine 100 mg was given alone. **1971** Pryor & Macalister *Gen. Anaesthetic & Sedation Techniques for Dentistry* ix. 60 The description of premedicant drugs earlier in this chapter should enable the anaesthetist to select appropriate combinations for any particular circumstance. **1977** *Lancet* 10 Dec. 1220/2 Anæsthetic premedicants include the minor tranquillisers (e.g., diazepam), major tranquillisers (e.g., droperidol), barbiturates, and opiates, but all possess undesirable side-effects.

premedicate, *v.* [Pre- A. 1.] *trans.* To give preparatory medication to, now esp. before anæsthesia. Hence **premedicating** *vbl. sb.*
1846 [in Dict. s.v. Pre- A. 1]. **1940** MacIntosh & Pratt *Essent. Gen. Anæsthesia* x. 91 The dose of premedicating drug a patient will require can often be roughly gauged by his resistance to alcohol. **1972** *Nature* 15 Dec. 411/1 Before the injection..patients were premedicated with 100 mg of pethidine.
Hence **premedica·tion,** medication given prior to or in preparation for the main treatment; *spec.* a pre-anæsthetic. Cf. *pre-med *sb.*[2]
1926 *Surg., Gynecol. & Obstetr.* XLIII. 103/2 All patients received as a premedication half a gram of veronal and 2 centigrams of morphine. **1932** *Brit. Jrnl. Anaesthesia* IX. 41 For the purpose of this discussion, by *premedication* is understood a new conception of pre-anaesthetic medication, whereby the patient is rendered unconscious in his bed before the administration of the anaesthetic. **1965** J. Pollitt *Depression & its Treatment* iv. 57 If an emergency operation must be performed.., the combination of chlorpromazine..and a barbiturate..is effective as premedication.

premeditate, *v.* **1.** (Later examples.)
1929 S. Leslie *Anglo-Catholic* xvi. 231 Your Aquin often premeditated modern theories, but he is generally truest..when his followers or commentators try their hardest to explain him away. **1965** K. Sisam *Struct. Beowulf* 3 *Beowulf*, with more lapses and more use of devices that help an improviser, has many of the marks of premeditated art.

pre-meiotic to **-Mendelian:** see *pre- B. 1.

premenopausal, *a.* *Med.* [Pre- B. 1.] Of or pertaining to the years preceding the menopause.
1939 *Jrnl. Clin. Investigation* XVIII. 177/2 The pituitaries from premenopausal individuals were low in gonadotropic potency. **1944** *Jrnl. Clin. Endocrinol.* IV. 577/1 While anovulatory cycles may occur at any age, it seems certain that they are far more common in the premenopausal years than they are in younger women. **1956** C. F. Fluhmann *Managem. Menstrual Disorders* xxv. 322 A series of 173 hospital patients over forty years of age with various types of abnormal uterine bleeding are illustrative of the 'premenopausal' period. **1975** *Lancet* 5 July 7/2, 6 women were premenopausal, with ages ranging from thirty-two to forty-four years.
So **preme·nopause,** the stage of a woman's life immediately preceding the menopause.
1941 Mazer & Israel *Diagn. & Treatm. Menstrual Disorders* xxi. 308 A number of clinicians reported favorably on the use of testosterone propionate in cases of dysfunctional uterine bleeding, especially that of the premenopause. **1957** *Amer. Jrnl. Obstetr. & Gynecol.* LXXIII. 985 Since the average age of menopause in the American white woman is 46, it appears that age 40 would be the beginning of the premenopause. **1968** R. W. Kistner in Astwood & Cassidy *Clin. Endocrinol.* II. vi. ix. 697 Effective treatment of the premenopause should.. produce regular, but not excessive, uterine bleeding.

preme·nstrual, *a.* [Pre- B. 1.] Occurring before menstruation; *premenstrual tension*, tension felt prior to menstruation. Also *transf.*
1885 [see *postmenstrual *a.*]. **1928** *Jrnl. Amer. Med. Assoc.* 14 Jan. 109/1 With Premenstrual Tension.—This was shown by a considerable group of women and manifesting [*sic*] itself by extreme nervousness, symptoms of autonomic imbalance, irritability, psychic changes and a feeling of tremendous tension. **1943** *Amer. Jrnl. Dis. Children* LXV. 302 The premenstrual state persisted, and fifteen months later the child menstruated. **1954** G. I. M. Swyer *Reproduction & Sex* iv. 43 The 'premenstrual tension' of which some women complain. **1970** R. Lowell *Notebk.* 72 Revolution, Drugging her terrible premenstrual cramps, Marches..to storm the city. **1974** J. Cooper *Women & Super Women* 11 Other occupations are.. smashing crockery from pre-menstrual tension. **1978** F. Weldon *Praxis* xxii. 203 You're hysterical. I expect you're pre-menstrual.
Hence **preme·nstrually** *adv.*
1931 *Arch. Neurol. & Psychiatry* XXVI. 1054 The blood of this patient showed twice the amount of female sex hormone that is normally found premenstrually. **1973** J. Zubin *Contemp. Sexual Behavior* viii. 160 Mean levels of hostility are highest premenstrually.

preme·nstruum. *Med.* [f. Pre- B. 1 in Dict. and Suppl. + Menstruum.] The stage of the menstrual cycle which precedes menstruation.
1910 *Trans. N.Y. Obstetr. Soc. 1909–11* 229 In the third stage called by them [*sc.* Hitchmann & Adler] the premenstruum, the mucous membrane which is now thick and velvety, can be divided into a superficial, compact, and deeper, spongy layer. **1938** *Jrnl. Amer. Med. Assoc* 21 May 1722/1 During the premenstruum, the concentration of estrogen in the blood rises and affects the sympathetic nervous system. **1969** *Sunday Times* 14 Sept. 54/3 Crimes of violence by women, most often involving their own families, are more often committed during the premenstruum. **1977** *Lancet* 24/31 Dec. 1330/2 It is common knowledge that exacerbations of acne and skin allergies occur during the premenstruum.

preme·tallize, *v.* [Pre- A. 1.] *trans.* To convert (a dye) before use into a metal chelate form by treatment with a metal salt, usu. in order to improve fastness properties. Hence **preme·tallized** *ppl. a.*
1948 Kirk & Othmer *Encycl. Chem. Technol.* II. 252 The formation of the metallic complex can be accomplished as an after-treatment, or the dye can be premetallized and applied to the fiber in the form of its soluble alkali salt. **1949** *Jrnl. Soc. Dyers & Colourists* LXV. 490/2 Ultralan Orange RS—This is a premetallised dye giving bright oranges when applied from strongly acid dye-baths, under which conditions the most level dyeings are obtained. **1963** A. J. Hall *Textile Sci.* ii. 82 Nylon can also be dyed with wool and cotton dyes (notably the acid wool dyes, pre-metallised dyes and direct cotton dyes). **1963** *Times* 31 May 19/6 The use of premetallised dyes for wool has also increased..despite the poor trading conditions in the textile printing industry.

premiation (prīmiēi·ʃən). *rare.* [f. as Premiate *v.*] Reward; the act of rewarding, a prize-giving.
a **1490** Johannes de Irlandia *Meroure of Wyssdome* (1965) II. 53 And þocht euirilk man and ressonable creatur incontinent eftir þar deid and partyn furth of the wauirld have certane knaulage of þar dampnacioun or premiacioun [etc.]. **1930** J. Ritchie in *Scots College, Rome* iii. 93 We witnessed two great functions. The first was a premiation at the Gregorian University when..we saw John Joseph Dyer..marching up for his gold medals.

premie, var. *preemie.

premier, *a.* and *sb.* Add: **B.** *sb.* **b.** (Further Canad. examples.)
This sense is now *obs.* in *Austral.* and *Canad.* usage: cf. sense B. c below and note *s.v.* Prime Minister 3 b in Dict.
1883 *Brandon* (Manitoba) *Daily Mail* 29 Jan. 2/1 It says that several of those roughly classed as Ministerialists will in all probability vote 'no confidence' in the present Premier. **1916** A. Bridle *Sons of Canada* 14 It is of prime importance to remember how..so impersonal a figure ever came to be Premier of Canada.
c. *Austral.* and *Canad.* The chief minister of a State or Province.
1853 *Hamilton* (Ontario) *Gaz.* 3 Oct. 2/6 In the prosecution of this singularly dignified scheme—we shall say nothing of its abstract honesty—the Premier scruples not to employ the influence which his position invests him with. **1902** *Parl. Debates Austral.* 1901–2 XI. 14528/2 Is it the case, as stated by the Premier of Queensland, that he (the Premier) has made repeated applications for a detailed statement of the receipts and expenditure of the departments transferred to the Commonwealth, and that such statement has not yet been supplied to him. **1917** N. McNeil in J. O. Miller *New Era in Canada* 197 Why did Honoré Mercier, as Premier of Quebec, place a reference to the Pope in the preamble of his Jesuits Estates Bill? **1929** M. de la Roche *Whiteoaks* xi. 151 Look at the situation in the Province of Quebec! There the women have no vote. 'We are Latins!' their Premier exclaims. **1930** W. K. Hancock *Australia* x. 209 In 1916 a Labour Premier of New South Wales..handed his resignation, not to the official head of the State, but to caucus. **1969** T. Jenkins *We came to Australia* I. ii. 30 Australia has a Prime Minister in the capital, Canberra, but..each of its six States has its own 'local' Prime Minister, known as a Premier. **1972** *Ann. Reg. 1971* 79 On 21 October the Progressive Conservative Party in Ontario under a new Premier, Mr William Davis, retained power with an increased majority winning 78 seats.
d. *U.S.* The Secretary of State. ? *Obs.*
1855 *N.Y. Herald* 22 Nov. 4/4 The casting vote between the Premier and the Kitchen is subject to the caprices and vacillations of the President, whose official position makes him supreme over both the action of the premier and the counsels of the Kitchen. **1878** *Harper's Mag.* Mar. 490/2 The diplomatic anteroom, where foreign dignitaries await audience with the Premier, is handsome in its appointments. **1886** E. Alton *Among Law-Makers* vii. 68 The Secretary of State..is sometimes (though not accurately) referred to as 'The Premier'. **1905** *Washington Post* 21 Mar. 4 Elihu Root..is ideally equipped for the duties of the Department of State, but it is considered unlikely that he could be induced to return to the Cabinet, even as premier. **1925** W. H. Smith *Hist. Cabinet U.S.A.* 28 He [*sc.* the Secretary of State] is frequently spoken of as the 'premier' of the cabinet, but there is no such title or designation known to our laws.
e. The Prime Minister of a country other than Great Britain or one of its colonies or a nation belonging to the British Commonwealth. Also used as a title prefixed to the surname of a premier.
1936 [see *Meiji]. **1942** W. S. Churchill *End of Beginning* (1943) 14 We sent Premier Stalin—for that I gather is how he wishes to be addressed..—exactly what he asked for. **1961** *N.Y. Times* 21 May IV. 1 Premier Khrushchev has made propaganda capital out of that fact. **1976** *Daily Tel.* 20 July 4/1 This is assumed to refer to some sort of demonstration similar to April's Peking riot by supporters of Teng and the late Premier Chou En-lai.

‖**premier cru** (prəmye krü). Also **premier crû.** Pl. **premier(s) crus.** [Fr., lit. 'first growth'.] A wine of the best quality. Also *transf.*, *fig.*, and *attrib.* Cf. *cru, *growth[1] 1d.
1868 E. L. Beckwith *Pract. Notes Wine* x. 47 The old, well-known *premiers crûs*, or first growths, retain their ancient and honoured places at the head of French wines. **1875** H. Vizetelly *Wines of World* i. 13 Branne-Mouton, next-door neighbour to Château Lafite, and noted for its nutty aroma,..is deserving..of being ranked among the *premiers crûs*. **1928** P. M. Shand *Bk. French Wines* ii. 58 Château Haut-Brion ranks..as the peer of the three great *Premiers Crus* of the Médoc. **1951** [see *cru]. **1965** P. O'Donnell *Modesty Blaise* i. 16 He has a wonderfully varied list of girl-friends. From *premier cru* to honest *vin du pays*. **1970** *Guardian* 21 May 13/1 The American demand begins at the top with the five Premiers Crus, and moves steadily down the 1855 classification of the Médoc wines. **1976** *Time* 20 Dec. 27 (Advt.), The best cognacs come only from the Grande and the Petite Champagne districts, the 'premiers crus' of the Cognac region. **1978** L. Pryor *Viper* (1979) ii. 31 The Long Beach race..the United States' only true road race..lacks the *premier cru* quality of Monaco.

‖**premier danseur** (prəmye dañsör). Pl. **premiers danseurs.** [Fr., lit. 'first dancer'.] A leading male dancer in a ballet company. Cf. *première danseuse.
1828 [see *danseur]. **1860** Thackeray *Roundabout Papers* iii, in *Cornh. Mag.* May 634 Sir Alcide Flicflac (premier danseur of H.M. Theatre)! **1930** C. W. Beaumont *Hist. Ballet in Russia* VII. 49 Dutac was honoured with the position of *premier danseur* during the reign of Alexander I. **1930** —— tr. *Noverre's Lett. on Dancing & Ballets* 21 If he [*sc.* the *maître de ballet*] concentrate his attention on the *premières danseuses* and *premiers danseurs*, the action becomes tedious. **1938** A. L. Haskell *Ballet* ii. 20 The field-marshal de Bassompière, [was] a *premier danseur* between campaigns. **1969** *Times* 11 Nov. 9/3 The vicissitudes of a *premier danseur* only half the size of his ballerina partner are worthy of consideration. **1973** R. Hayes *Hungarian Game* xxxv. 211, I watched a traffic cop directing sluggish cars like a *premier danseur* in a cattle pen. **1978** *Chicago* June 26/3 Jacques d'Amboise, premier danseur of the New York City Ballet, returns to choreograph a special number for the evenings.

première, *sb.* Add: Now also with pronunc. (pre·miēəɹ, *U.S.* prĭmi·əɹ). Also **premier, premeer.** (Earlier and later examples.) Also *transf.*
1889 'F. Leslie' *Let.* 9 Feb. in W. T. Vincent *Recoll. Fred Leslie* (1894) II. xxii. 81 It upset all of us and made us more nervous than a *première*. **1890** G. B. Shaw *Let.* 28 Feb. (1965) I. 244 This does not..include the expenses of the *première* at Amsterdam. **1915** *Sat. Even. Post* 9 Oct. 62/2 She always accompanies me to our premières. **1930** E. Mannin *Confessions & Impressions* II. iv. 137 He complimented me on my literary *première* and told me to keep

on writing. 1937 *New Republic* 19 May 48/1 Miss Gaynor [arrives] at another preemeer in smart mourning. 1941 *Commonweal* 10 Jan. 294/1 The movie première—pronounced pre-meer, with heavy emphasis on the second syllable—is a national phenomenon. 1957 *Times* 9 Sept. 11/4 Each season, when Balmain finally settles down to the production of a new collection of some 150 models, Ginette Spanier is sent away for three weeks' holiday, returning only a few days before the all-important *première*. 1968 S. Challis *Death on Quiet Beach* viii. 115 Fane was due to attend a late premier of his current movie. 1978 J. Anderson *Angel of Death* xiii. 144 'I'll be the only actress in the world who could do it justice.'.. 'I'll come to the premiere.'

première (pre·mieᵊɹ, prĭmiᵊ·ɹ), *v.* Also **premier.** [f. prec.] *trans.* To present or perform (a play, film, programme, or the like) for the first time; to reveal (a new product). Also *absol.* and *intr.* for *pass.* Hence **premie·red, premiering** *ppl. adjs.*

1940 *Winchell Coll.* (Topeka, Kansas) *Jrnl.* 11 Dec. 4/6 There's irony in the request of Grinnell college's alumni that Frank Capra should premiere 'Meet John Doe' there. 1941 W. C. Handy *Father of Blues* v. 70 With..Gordon Collins, Lew Hall, all-time end men, premiering on the flanks, you'd feel a strange enchantment creeping over you. 1943 *Newsweek* 13 Sept. 101 Keepsakes, a new Sunday program..premièred on Blue Sep. 5, 8–8:30. 1945 G. Antheil *Bad Boy of Music* ii. 16 This symphony, my 'First Symphony', was later to be premiered in Berlin. 1945 Mencken *Amer. Lang.* Suppl. I. 387 A few of its [sc. *Variety's*] characteristic inventions will suffice:..*to premier* (often shortened to *to preem*). 1952 'E. Box' *Death in Fifth Position* (1954) i. 2 My company is going to première an important new ballet tonight. *Ibid.* vii. 181 By the time *Eclipse* was to be premièred, Ella had infuriated Miles..by threatening to leave the company. 1955 L. Feather *Encycl. Jazz* 141/1 He [sc. W. G. Fuller]..was co-composer and arr. of..*The Swedish Suite* premiered at Carnegie Hall '48. 1967 *N.Y. Herald Tribune Internat.* 11–12 Feb. 5/4 André Kostelanetz, who commissioned the work and premiered it 25 years ago, asked Mr. Lindsay to do the reading and will conduct the performance. 1973 *Times* 11 Apr. 12/7 In Frankfurt the Theater am Turm, which has premiered most of Peter Handke's plays, is run on similarly cooperative lines. 1975 *Publishers Weekly* 1 Dec. 62/3 He managed to keep the title in the public eye until the film premiered in December 1939. 1976 *National Observer* (U.S.) 10 Apr. 20/2 The premiere of a bizarre Scots drama, Menzies Mc Killop's *Future Pit*, now joined in repertory by a premiering trio of one-acters: Frank B. Ford's *Waterman*, Gladden Schrock's *Glutt*, and Michael Casales' *Cold. Ibid.* 16 Oct. 10/3 The ABC Evening News..premiered last week..and was notable in at least two respects. 1977 *Custom Car* Nov. 13/4 Saab premiered their long-rumoured Turbo 99.

‖**première danseuse** (prəmyĕr daṅsöz). Also *ellipt.* **première.** Pl. **premières danseuses.** [Fr., fem. of *PREMIER DANSEUR.] A leading female dancer in a ballet company; a ballerina.

1828 [see *DANSEUSE]. 1846 R. Ford *Gatherings from Spain* xxiii. 327 Egyptians, whose women are the premières danseuses on these occasions, in which [gipsy] men never take a part. 1867 *Galaxy* Aug. 441 The dancer who has passed the chrysalis ballet-girl stage, and is now a full-fledged, butterfly *première*. 1887 J. Payn *Holiday Tasks* 13 But here his eye wanders..from the photo of the *première danseuse* at the Frivolity Music Hall. 1890 G. B. Shaw *London Music 1888–89* (1937) 314 Many a *première danseuse* holds her position in spite of a neck and wrists which are, dancingly considered, dead as doornails. 1911 [see *BALLERINA]. 1930 [see *PREMIER DANSEUR]. 1942 L. Kirstein *Bk. of Dance* xiv. 319 The great..artist Vaganova,..frequently shines as its *première* [in the *Lac des Cygnes*]. 1974 *Sat. Rev. World* (U.S.) 19 Oct. 40/2 The *première danseuse étoile* of the Opéra in Paris. 1978 Ld. Drogheda *Double Harness* xx. 261 In 1919, she [sc. Ninette de Valois] was engaged as *première danseuse* for the season of international opera at Covent Garden.

premiership. Add: **1. b.** *U.S.* The office of Secretary of State. Cf. *PREMIER *sb.* d. ? *Obs.*

1928 H. Minor *Story Democr. Party* 69 Madison had cabinet troubles, too. Monroe accepted the premiership in March 1811.

pre-mi·lking. [Pre- A. 1.] The removal of milk from a cow's udder before the birth of her calf.

1953 K. Russell *Princ. Dairy Farming* xiii. 157 Premilking can be done either by hand or machine. 1960 *Farmer & Stockbreeder* 5 Jan. 101/3 Too many concentrates before calving can result in pre-milking being necessary. 1970 W. H. Parker *Health & Dis. in Farm Animals* xiii. 167 Premilking..is a severe handicap to the calf.

premisal. Delete ? *Obs.* and add later example.

1912 *Catholic Encycl.* XIV. 75/1 Ethics may not be divided from psychology and theodicy, any more than from deductive logic. With the proper premisals then from the one and the other here assumed, we say that the Creator could not have given man a fixed nature, as He has, without willing man to work out the purpose for which that nature is framed.

premium. For 'Pl. -iums, formerly -ia' read 'Pl. -iums, -ia' and add: **3. a.** (Further examples.) *spec.*, a sum paid in addition to the rent on a leased property.

1859 Geo. Eliot *Let.* 19 Feb. (1954) III. 14 There was a house after my own heart at Mortlake..but it turned out to have a premium affixed to the lease, which made it too expensive. 1924 A. Christie *Poirot Investigates* iii. 71 'We've got a flat—at last!.. It's dirt cheap. Eighty pounds a year!'.. 'Big premium, I suppose?' 1966 *New Statesman* 21 Jan. 71/2 If railwaymen work genuinely longer or more difficult hours, and get overtime or shift premia in compensation, this is fair enough. 1966 *Economist* 29 Jan. 386/1 The case for higher night premia would be 'examined' in a later report, but he most definitely did not recommend them now. 1970 M. Greener *Penguin Dict. Commerce* 263 Very often when property is leased, the lessee, in addition to paying a rent for an agreed period, pays a lump sum. This is known as a premium, or sometimes as 'key money', and was once intended to avoid taxation and disguise the true rent. 1974 M. B. Brown *Econ. of Imperialism* viii. 177 Some foreign issues [of stocks] were certainly made more attractive because of the premiums at which they were issued.

b. *Comm.* (See quot. 1928.)

1928 *Funk's Stand. Dict.* II. 1956/3 *Premium*,..any object offered free to those who purchase goods to a certain value, as a set of books given free as an inducement to subscribe to a magazine. 1930 Lucas & Benson *Psychol. for Advertisers* xii. 204 $1,502,000,000 is spent annually on advertising. This is divided as follows: Newspapers..$690,000,000..Premiums, programs and directories..25,000,000. 1954 R. J. Schwartz *Dict. Business & Industry* 392/1 *Premium*, something given free or at a nominal price to induce an actual sale or to promote interest in a product. 1963 *Sunday Times* 17 Nov. 11/1 A rapidly-growing little specialist industry is growing round the 'take-a-plastic-daffodil-madam' school of retailing... A premium, in their jargon, can be anything given away or sold cheap to persuade people..to buy, stock, sample or re-order a product. 1974 *Encycl. Brit. Micropædia* VIII. 191/1 Until the 1900s the most popular premiums were pictures and trade cards.., which were collected and exchanged by enthusiastic consumers whose collections became quite valuable.

c. *Finance.* The excess of the forward price of a currency or a commodity over the spot price.

1933 B. Ellinger *This Money Business* x. 101 In normal times the difference between 'spot'—i.e. the rate for immediate delivery—and 'forward' rates depends on the rates of interest in the respective countries, but in abnormal times merchants may find a growing premium or discount on the forward rate over the spot rate. 1957 [see *FORWARD *a.* 4]. 1971 R. F. Pither *Man. Foreign Exchange* (ed. 7) x. 138 Forward rates of exchange are quoted as a 'margin' or 'difference' against the 'spot' rate of the currency concerned, or as a 'premium' or 'discount' on the 'spot' rate, or they may be quoted 'outright'. 1978 R. G. F. Coninx *Foreign Exchange Today* viii. 111 Forward margins are referred to as *premiums* or *discounts*. *Ibid.* 113 With indirect quotations, premiums indicate that the home currency enjoys higher interest rates than the quoted currency.

6. a. (Further examples.)

1906 Galsworthy *Man of Property* xxiv. 295 When Mrs. MacAnder dined at Timothy's, the conversation.. took that wider, man-of-the-world tone current among Forsytes at large, and this, no doubt, was what put her at a premium there. 1932 *Time* 28 Mar. 30/2 The news put Philharmonic subscriptions back at a premium last week. 1974 *Times* 14 Mar. 11/2 Sadly, space is at a premium in most department stores.

b. *fig. to put* (or *place*) *a premium on* (something) and varr., to put a high value on something esp. as an inducement or incentive.

1907 G. B. Shaw *John Bull's Other Island* p. xvi, In short, our circumstances place a premium on political ability whilst the circumstances of England discount it; and the quality of the supply naturally follows the demand. 1911 —— *Getting Married* 142 Our democratic and matrimonial institutions..put a premium on want of self-respect in certain very important matters. 1933 J. W. N. Sullivan *Limitations of Sci.* iv. 132 The struggle for existence takes the place of the human breeder. Nature sets a premium upon certain varieties as compared with others. 1939 A. Huxley *After Many a Summer* i. xi. 147 He's been greedy and domineering, among other reasons, because the present system puts a premium on those qualities. 1959 [see *PEARL HARBOUR].

7. a. (sense *3 b) *premium promotion, selling*; **premium apprentice,** an apprentice who has paid a premium for instruction in his intended trade.

1927 F. H. Shaw *Knocking Around* vi. 54 My greatest efforts of all should be expended in an endeavour to ameliorate the lot of that hard-lying ocean Ishmael, the premium apprentice. 1979 *Jrnl. R. Soc. Arts* Dec. 36/2 When I left school,..I put in a happy period as a premium apprentice at the Sentinel Waggon Works at Shrewsbury. 1962 S. Strand *Marketing Dict.* 562 *Premium promotion*, the use of premiums (inexpensive gifts) in the promotion of the sale of products or services. 1974 *Encycl. Brit. Micropædia* VIII. 191/2 *Premium promotion*, an advertisement, often part of the product package, that induces prospective purchasers to buy the product by offering a free gift or a reduced price. 1966 *Lebende Sprachen* XI. 109/1 *Premium selling*, offering an item with the purchase of another product, either free or for a nominal additional payment, as an inducement to buy the product.

b. Passing into *adj.* Of a commodity, etc., esp. petrol: superior in quality and therefore commanding a higher price; of a price: such as befits an article of superior quality; higher than usual. orig. *U.S.*

1928 *National Petroleum News* 24 Oct. 115 (Advt.), This is our anti-knock gasoline, a premium motor fuel. 1931 *Economist* 5 Sept. 422/2 The profit to the garage on the sale of petrol..is now 2d a gallon on national 'commercial' grades and 2½d on national 'premium' grades. 1945 H. S. Bell *Amer. Petroleum Refining* (ed. 3) xviii. 278 The refiner cannot approach the desired knock rating of 80 for premium motor fuels..by simple skimming and thermal cracking except by a material reduction of the end point of his product. 1961 I. L. Horowitz *Philos., Sci. & Sociol. of Knowledge* v. 54 A world which pays a premium price for technological manipulation. 1965 *New Statesman* 23 Apr. 634/1 There were the garages selling the well-known, branded petrols, each in three main grades—Super, Premium and Regular. 1970 *Daily Tel.* 30 Jan. 19/1 All supersonic travellers would fly 'premium class' at a slightly lower rate than that paid at present by first-class passengers, but with the same comfort. 1977 *Listener* 1 Dec. 708 Qube (sc. U.S. cable television) has ten 'premium' channels where you pay per programme. 1979 *Guardian* 22 June 9/8 Trout will for some time still be a premium fish, selling at about £1 each.

premium bond. Also **Premium Bond.** [f. Premium + Bond *sb.*[1]] A debenture earning no interest but eligible for lotteries; *spec.* (in full *Premium Savings Bond*) since 1956, a British government bond not bearing interest but with the periodic chance of a cash prize. Also *attrib.* (See also *ERNIE.)

[1882 R. Bithell *Counting-House Dict.* 237 A number of Lottery Loans of the worst class have been started in some of the German States, and also in Austria... It would be impossible to get subscriptions to them to any great extent in this country if called by their proper name. The name of Premium-Loans..has therefore been substituted.., and the money that has been extracted from the pockets of unfortunate dupes by these means is enormous. 1908 *Westm. Gaz.* 29 Aug. 2/2 Two of the largest of these lotteries, the Panama and Congo premium-bearing loans, are two of the most scandalous pieces of finance which Europe has ever witnessed. Here is an exciting chance of winning a fortune by gambling; let us get the money somehow to buy half a dozen of the bonds, and work no more!] 1908 *Economist* 12 Sept. 477/2 The practical man in the street who knows anything about premium bonds is quite aware that they are in their nature and intention lotteries. 1918 *Ibid.* 19 Jan. 79/2 The report of the Select Committee on Premium Bonds.. concludes with the following paragraphs:—'..We do not, therefore, advise that an issue of Premium Bonds be made at the present time.' 1931 *Star* 8 May 6/3 Every trick—from premium bonds to guessing the number of beans in a bottle—seems to have been tried. 1940 Graves & Hodge *Long Week-End* v. 77 He [sc. Horatio Bottomley] was then launching new prize schemes—the Premium Bond Scheme of 1918, for example, to which his readers subscribed £90,000. Out of this he had agreed to pay £10,000 in prizes. 1956 H. Macmillan in *Times* 18 Apr. 5/2 Finally, I have something completely new for the saver in Great Britain—a premium bond. 1957, etc. [see *ERNIE]. 1957 *Observer* 25 Aug. 9/3 New National Savings reported last week totalled £26,689,000 (including £1,200,000 Premium Bonds and £2,566,000 accrued interest). 1958 *Times Lit. Suppl.* 15 Aug. p. x/3 He also prefers pools to premium bond gambling—in which a bloke can't choose his own combination of numbers, so how does one know that it's on the level? 1962 H. O. Beecheno *Introd. Business Stud.* xiv. 140 The Premium Savings Bond scheme has taken advantage of our national love of a gamble, holding out the possibility of a reward for the few much higher than other savings methods would give. 1974 *Guardian* 27 Mar. 1/1 Another £500,000 a month added to Premium Bond prizes.

premiumed, *a.* (in Dict. s.v. Premium). Add: Also, that has paid a premium.

1927 *Daily Express* 5 July 5/5 The trade may also be entered as a premiumed apprentice or as a beginner at a nominal wage.

premi·x, *v.* [Pre- A. 1.] *trans.* To mix beforehand.

1934 in Webster. 1966 *Gloss. Terms Internal Plastering* (B.S.I.) 17 A plaster in which a lightweight aggregate has been pre-mixed dry with a gypsum plaster to give a low density. 1966 *McGraw-Hill Encycl. Sci. & Technol.* V. 292/1 Gaseous fuels can be premixed with air or oxygen, in which case the mixture can be fed to a flame holder and burned in a very efficient manner. 1972 [see *PREMIX *sb.*]. 1976 *Nature* 20 May 259/2 Rattlesnake venoms are neutralised also when premixed *in vitro* with either rattlesnake plasma or commercial antivenin.

Hence **premi·xed** *ppl. a.*, **premi·xing** *vbl. sb.*

1941 *Engineers' Digest* II. 417 (*heading*) Premixed combustion of gaseous fuel for steel finishing operations. *Ibid.*, Premixing provides accuracy of air to fuel-gas proportioning over the widest conceivable operating range. 1945 H. Barron *Mod. Plastics* vii. 163 (*caption*) Rotary premixing machines in which powdered phenolic resins (Novolak) is mixed with hexamethylenetetramine, wood flour, and colouring material. 1959 *Economist* 31 Jan. 431/2 A 'flexible' type of road carriageway base incorporating premixed water-bound macadam. 1960 *Farmer & Stockbreeder* 8 Mar. 75/1 The idea being that two different fertilizers can be applied at the same time without pre-mixing. 1963 A. M. Neville *Properties of Concrete* iv. 205 If instead of being batched and mixed on the site concrete is delivered ready for placing from a central plant it is referred to as ready-mixed or pre-mixed concrete. 1978 *Sci. Amer.* Apr. 155 In a premixed flame the gases are mixed prior to the burning and the rate of combustion depends on the flow rate.

premix (prī·miks), *sb.* and *a.* [Pre- A. 2.] **A.** *sb.* A mixture prepared beforehand; *spec.* (*a*) *Agric.*, a powder or granular preparation into which a drug or the like has been incorporated and which is mixed with animal feed to introduce the drug, etc., into it in suitably low concentrations; (*b*) synthetic resin to

which various substances have been added to make it suitable for moulding; (*c*) (see quot. 1976).

1957 *Times* 2 Dec. (Agric. Suppl.) p. vi/4 A source of greater fear to those familiar with oestrogen effects is the inhalation of dust from a concentrated premix, which would present a definite hazard. 1960 O. Skilbeck *ABC of Film & TV* 98 *Premix*, when dubbing is likely to prove especially difficult, or when insufficient heads are available for the number of tracks involved, some tracks may be combined at a first premix stage and added to the remainder later. 1963 H. R. Clauser *Encycl. Engin. Materials* 519/1 The unsaturated resin is first mixed with fillers, fibers, and catalyst to provide a nontacky compound... The premix is molded at pressures of 150 to 500 psi and at temperatures ranging from 250–310 F. 1963 *Poultry Sci.* XLII. 1264/2 Premixes containing menadione sodium bisulfite complex (16 gm./lb.) were made employing soybean meal or corn as the carrier. 1967 Simonds & Church *Encycl. Basic Materials for Plastics* 58/2 Premix is a damp, sometimes sticky variation of molding compound differing technically from so-called 'molding compound' only in the ratio of monomer which is used. 1971 *Farmer & Stockbreeder* 23 Feb. 25/1 Emtryl is available in two forms, as a soluble powder and as a premix. 1972 Quick & LaBau *Handbk. Film Production* xviii. 201 In the event of an extremely complex mix, pre-mixes may be desirable. This is where a few of the tracks are mixed to a desired level and then that master track mixed with the remaining channels of information. The director makes decisions concerning whether or not to pre-mix his film. 1976 B. Armstrong *Gloss. TV Terms* 71 *Premix*, a preliminary dub of certain sound tracks, usually music and effects, before the final mix.

B. *adj.* Premixed.

1963 H. R. Clauser *Encycl. Engin. Materials* 537/2 Premix molding materials are physical mixtures of a reactive thermosetting resin.., chopped fibrous reinforcement..and powdered fillers (usually carbonates or clays). 1968 P. I. Smith *Plastics as Metal Replacements* i. 45 The manufacturer makes available to the moulder either conventional pre-mix compounds or pre-impregnated chopped strand glass mat all ready for moulding. 1975 *Petroleum Rev.* XXIX. 96/1 A semi-closed circuit deep sea diving breathing set, using air or a pre-mix gas. 1976 P. Hill *Hunters* xii. 176 A huge pre-mix concrete lorry was disgorging its load. 1977 *Evening News* 11 June 11/6 Producer and artist then live with these 'pre-mix' tapes for about two months.

premi·xture. [Pre- A. 2.] A mixture prepared beforehand.

1934 in Webster. 1972 *Physics Bull.* Jan. 20/2 Two separate arrays of injector tubes are used to introduce premixtures of fluorine with helium and nitric oxide with carbon dioxide at the upstream end.

pre-modern: see *PRE- B. 1 d.

premo·dify, *v.* *Linguistics.* [Pre- A. 1.] *trans.* To modify (a word or phrase) by an immediately preceding word or phrase. So **premo·difying** *ppl. a.*; **premodifica·tion, premo·difier.**

1962 R. Quirk *Use of English* x. 164 The premodification of nouns by nouns was a common feature of English before Germans studied science or America was discovered. 1966 G. N. Leech *English in Advertising* xiv. 127 In advertising language, the interesting part of the noun group is the pre-modifying part... Noun groups with lengthy pre-modifications are italicised. *Ibid.* 128 Premodifiers which can have the designative, or categorising function are nouns, adjectives and compounds. 1972 *Language* XLVIII. 456 The relatively empty *do* [in the ungrammatical sentence *He had well done it*] does not permit a premodifier *well*, but with a richer verb premodification is normal, e.g. *He has well revealed the causes.* 1973 G. W. Turner *Stylistics* iii. 81 Words preceding the head word in a group are conveniently called 'modifiers' (sometimes 'premodifiers'). 1976 *Amer. Speech* 1974 XLIX. 82 It [sc. *much*] collocates with *like* in an affirmative sentence if it is premodified, hence *I like him very much*.

premonitory. B. (Earlier example.)

1834 *Knickerbocker* IV. 307 The premonitories seize me before I have time to run to the doctors for relief.

premonstrate, *v.* For † *Obs.* read *rare* and add later example.

1857 A. Mathews *Tea-Table Talk* I. 251 Marks, natural or acquired, premonstrate a talent for locomotion.

premonstration. For † *Obs.* read *rare* and add later example.

1920 E. H. Begbie *Mirrors of Downing St.* i. 9 His intuitions are amazing. He astonished great soldiers in the war by his premonstrations.

pre-moral(ity): see *PRE- B. 1.

premo·rbid, *a.* *Med.* [Pre- B. 1.] Preceding the occurrence of symptoms or disease.

1939 *Amer. Jrnl. Psychiatry* XCV. 1041 A number of factors were included: age, sex, physical build, premorbid personality..and permeability quotients. 1953 *Jrnl. Nerv. & Mental Dis.* CXVII. 516 The scale was used to evaluate each patient in..the following three areas: (a) the premorbid history; (b) possible precipitating factors; (c) signs of the disorder. 1969 *Jrnl. Amer. Med. Assoc.* 18 Aug. 1085/2 Marihuana may have a psychotogenic effect even in an individual with a healthy premorbid personality.

pre-mortem, *a.* Add: Also as *sb.* in *fig.* use.

1971 *Listener* 7 Jan. 18/1 'The death of the symphony orchestra' was discussed on Radio 4, in a kind of pre-mortem. 1972 F. Warner *Lying Figures* iii. 33 What of love? *Guppy*. A post-prandial pre-mortem.

premotor: see *PRE- B. 3.

premou·lt, *a.* and *sb.* *Zool.* **A.** *adj.* [Pre- B. 2.] Existing or occurring just before a change of plumage in birds or the shedding and replacement of the integument of insects, crustaceans, or reptiles. **B.** *sb.* [Pre- B. 1 in Dict. and Suppl.] A premoult stage or period.

1957 R. A. H. Coombes in D. A. Bannerman *Birds Brit. Isles* VI. 312 (*heading*) The pre-moult migration of the sheld-duck. 1964 *Oceanogr. & Marine Biol.* II. 303 At moult [of the crab, *Carcinus maenas*], uptake of water, averaging 66·3% of the premoult weight, takes place. 1967 P. A. Meglitsch *Invertebr. Zool.* xvi. 681/1 The physiology of most of the body parts [of arthropods] is affected by premolt. 1973 *Nature* 9 Mar. 133/2 An insect does not enter premoult if its thoracic glands have been removed.

premultiplica·tion. *Math.* [Pre- A. 2.] Multiplication by a prefactor.

1862 *Phil. Trans. R. Soc.* CLI. 316 Let that matrix be reduced by premultiplication with a unit-matrix. 1972 *Computer Jrnl.* XV. 250/2 Pre- and post-multiplication is preserved.

premultiply, *v.* (Examples.)

1862 [see *post-multiply* vb. s.v. *POST- A. 1 a]. 1972 Roberts & Shipman *Two-Point Boundary Value Probl.* viii. 223 If each side of (8.8.12) is premultiplied by the square partitioned matrix of order $n(m+1)$,..the following partitioned matrix is obtained. 1978 [see *post-multiply* vb. s.v. *POST- A. 1 a].

premunition. Restrict 'Now *rare*' to senses in Dict. and add: **3.** *Med.* [ad. F. *prémunition* (E. Sergent et al. 1924, in *Bull. de la Soc. de Path. exotique* XVII. 38).] (The production of) a resistance to disease due to the presence of the causative agent in the host in a harmless or tolerated state.

[1924 *Tropical Dis. Bull.* XXI. 492 For absolute immunity the distinguishing term proposed is 'immunity'.. and for relative immunity, 'premunition' brought about by a process of 'premunition', with corresponding verb. (Unfortunately English words corresponding with 'premunition' and 'premunir' do not at present exist.)] 1934 T. W. M. Cameron *Internal Parasites of Domestic Animals* iv. 218 In..premunition, removal of the latent infection may permit of re-infection. 1951 G. Lapage *Parasitic Animals* vii. 205 Some experts believe that the greater resistance of the negro race to human malarial parasites..is really premunition. 1971 P. C. C. Garnham *Progress in Parasitol.* vi. 101 Host and parasite eventually settle down together in the state of premunition. 1975 *Tropical Animal Health & Production* VII. 125 (*heading*) The premunition of adult cattle against Babesiosis.

premunity (primiū·niti). *Med.* [f. Premun(ition + -ity, after *immunity*.] (See quot. 1938.) So **premu·ne** *a.*, exhibiting premunity.

1938 G. O. Davies *Vet. Path. & Bacteriol.* (ed. 2) vii. 145 Premunity denotes a state of resistance or tolerance to infection which only lasts so long as the infecting organism is present in the tissues of the host. 1948 U. F. Richardson *Vet. Protozool.* i. 7 The more uniform 'premunity' of animals to protozoa in an infected area is mainly due to the more uniform exposure to infection and reinfection. *Ibid.* iv. 86 Infection derived from a premune animal is less severe than that from an active case.

premunization (prī:miunəizēi·ʃən). *Med.* [f. Premun(ition + -ization, after *immunization*.] The action or result of premunizing; premunition (sense *3).

1941 J. T. Cuthbertson *Immunity against Animal Parasites* iv. 49 Resistance to reinfection with the malarias of monkeys, birds, and dogs have all been shown to depend in part on premunization. 1975 *Tropical Animal Health & Production* VII. 126 Premunisation was effected by the injection of infected blood and control of the subsequent infection with drugs.

premunize (prī·miunəiz), *v.* *Med.* [f. Premun(ition + -ize, after *immunize*, as anglicization of F. *prémunir* (E. Sergent et al. 1924, in *Bull. de la Soc. de Path. exotique* XVII. 37).] *trans.* To introduce pathogens into (a host) so as to produce premunition. Hence **pre·munized, pre·munizing** *ppl. adjs.*

1925 E. Sergent et al. in *Trans. R. Soc. Tropical Med.* XVIII. 384 We ask our British colleagues to..consider whether it is possible to give currency to the verb 'to premunize' and to Anglicize the word 'premunition'. 1934 T. W. M. Cameron *Internal Parasites of Domestic Animals* ii. 45 Immune animals..are premunized and may act as carriers. 1938 *Proc. R. Soc. Med.* XXXI. 1301 When a premunized host is superinfected..little apparent effect is produced. 1963 E. Sergent in P. C. C. Garnham et al. *Immunity to Protozoa* iii. 44 A typical example of a premunizing vaccine is the antitubercular vaccine, BCG. 1975 *Tropical Animal Health & Production* VII. 125 Twenty-five cattle..were premunised with virulent *Babesia bigemina*. *Ibid.*, The results of haematological and serological immune responses of premunised cattle.. are reported.

premy, var. *PREEMIE.

premyelocyte, -cytic: see *PRE- B. 1.

prena·sal, *a.* and *sb.* **A.** *adj.* **1.** *Anat.* and *Zool.* [f. Pre- B. 3 + Nasal *a.*] (In Dict. s.v. Pre- B. 3.)

2. *Linguistics.* [f. Pre- B. 2 + Nasal *sb.*] Occurring before a nasal consonant.

1973 J. M. Anderson *Struct. Aspects Lang. Change* 137 Modifications in French, revolving around nasalization of prenasal vowels.

B. *sb.* *Linguistics.* [f. Pre- A. 2 + Nasal *sb.*] A prenasalized consonant.

1948 R. A. D. Forrest *Chinese Lang.* v. 93 The Heh-Miao..have rid their language of all compound consonants (except the prenasals). *Ibid.* 94 The irregular representation of the initials with prefixed homorganic nasals ('prenasals') is puzzling.

Hence **prenasa·lity,** the quality or state of being prenasalized.

1976 *Language* LII. 332 Another class of consonants involving nasality..argues against any solution to the problem of prenasality in which a single feature has the entire segment as its domain.

prena·salize, *v.* *Linguistics.* [Pre- A. 1.] *trans.* To pronounce (a consonant) with initial nasalization. Chiefly as pa. pple. or ppl. adj. Hence **prenasaliza·tion.**

1956 Jakobson & Halle *Fund. of Lang.* iv. 43 Such relatively rare phonemes as the discontinuous nasals (the so-called prenasalized stops). 1961 Webster, Prenasalization. 1973 J. M. Anderson *Struct. Aspects Lang. Change* 119 Some consonants..may become aspirated, pre-aspirated, glottalized, prenasalized, [etc.]. 1976 *Language* LII. 331 Another class of segment is much more common..: this is the type generally described as 'prenasalized stops'. 1977 *Ibid.* LIII. 317 Prenasalization of these consonants, while common, is limited in the following ways: except in sentence-initial position, *b d* are either preglottalized or prenasalized, and the processes occur in complementary distribution.

prenatal, *a.* Add: (Further examples.) Also = *ANTENATAL *a.* 2.

1909 Chesterton *Orthodoxy* iv. 94 This proves that even nursery tales only echo an almost pre-natal leap of interest and amazement. 1909 *Westm. Gaz.* 19 Apr. 2/3 In some forgotten strange pre-natal world, Where rose-crowned summer smiled on placid seas. 1938 *New Statesman* 19 Feb. 298/2 Pre-natal clinics are increasing. 1960 C. MacInnes *Mr. Love & Justice* 207 Step up to the pre-natal clinic, darling. See what they have to say.

prenatally, *adv.* (further examples).

1953 R. Lehmann *Echoing Grove* ii. 73 Already midsummer dawn fainted pre-natally in the high uncurtained windows. 1965 *Science* 15 Jan. 306/3 This period of neural differentiation occurs prenatally in guinea pigs.. and during the 1st week of postnatal life in rats. 1976 *Lancet* 18 Dec. 1352/2 Of interest was the occurrence of measles prenatally in the mother of 1 patient..during the third trimester of gestation.

pre-Nazi: see *PRE- B. 1 a. **prenegotiation:** *PRE- A. 2. **preneoplastic:** *PRE- B. 1.

preneu·ral, *a.* and *sb.* *Zool.* [Pre- B. 3.] **A.** *adj.* In chelonians, applied to a skeletal element that lies between the nuchal bone and the neural bones. **B.** *sb.* A bone in this position.

1904 *Amer. Jrnl. Sci.* CLXVIII. 274 There is a preneural bone, whose anterior border has occupied a notch in the hinder border of the nuchal. 1957 *Bull. Mus. Compar. Zool. Harvard* CXV. 171 The very similar term 'pre-neural' has long been in use for an element that is immediately posterior to the nuchal. 1969 R. Zangerl in C. Gans *Biol. Reptilia* I. vi. 333 Most of the Mesozoic genera retain some amphichelydian characters,..but.. the occurrence of a preneural is erratic. *Ibid.* 334 Preneural elements occur occasionally.

prenex (prī·neks), *a.* *Logic.* [ad. late L. *prae-nex(us* tied or bound up in front: see Pre- A. 1 and Nexus.] Of or relating to a quantifier placed initially in a formula whose scope affects the whole formula; *spec.* in phr. *prenex normal form* (see quot. 1944).

1944 A. Church in *Ann. Math. Stud.* xiii. 60 Thus we have that a w.f.f. is in prenex normal form if and only if all its quantifiers are initially placed, no two quantifiers are upon the same variable, and every variable occurring in a quantifier occurs at least once within the scope of that quantifier. *Ibid.* 61 Use of the prenex normal form was introduced by C. S. Peirce, although in a different terminology and notation. 1950 W. V. Quine *Methods of Logic* (1952) iv. 226 Let us speak of a quantifier as prenex in a sentence when..it is initial..and its scope reaches to the end of the sentence. *Ibid.* 243 The prenex universal quantifiers..are dropped for ease in reading. 1951 *Jrnl. Symbolic Logic* XVI. 32 For any formula *F* of our *m*-valued formalization there is a formula *G* in prenex normal form which is weakly equivalent to *F*. 1965 Hughes & Londey *Elem. Formal Logic* xli. 297 A wff in which all the quantifiers occur at the beginning, all are affirmative, and in which their scope extends to the end of the whole wff, is said to be in Prenex Normal Form. 1974 Boolos & Jeffrey *Computability & Logic* ix. 112 Thus..F_1 is in prenex form but F_1 is not; and since F_2 is a prenex formula logically equivalent to F_1, F_2 is a prenex form of F_1.

prenominal, *a.* Add: **b.** Preceding a substantive. Also as *sb.*

1961 *Amer. Speech* XXXVI. 163 Roughly speaking, the prenominal adjectivals will be only single, simplex (descriptive) adjectives. 1964 *Language* XL. 45 As a prenominal the genitive has two unique characteristics. 1965 *Ibid.* XLI. 283 Prenominal and postnominal modifiers. 1978 *Ibid.* LIV. 26 A prenominal numeral like *cinq* 'five' may be pronounced with a final consonant in all positions.

pre-noun: see *PRE- B. 1.

prenova (prīnōᵘ·vă), *a.* and *sb.* *Astr.* **A.** *adj.* [Pre- B. 2.] Preceding development of a star into a nova. **B.** *sb.* [Pre- B. 1 in Dict. and Suppl.] A star prior to its becoming a nova.

1939 D. B. McLaughlin in *Pop. Astron.* XLVII. 418 The pre-nova stage. During this portion of its life, the star is either constant or irregularly variable through a small range. 1943 *Publ. Observatory Univ. Michigan* VIII. 188 In the pre-nova state the star is very hot, probably 50,000° to 60,000° K. 1956 Z. Kopal in A. Beer *Vistas in Astron.* II. 1499 (*heading*) Internal constitution of the pre-novae. 1957 C. Payne-Gaposchkin *Galactic Novae* xi. 313 The pre-nova is regarded as a hydrogen-poor subdwarf with contraction the main source of energy. 1976 *Nature* 22 Jan. 172/3 First indications that the nova was unusual came when a search for the prenova on the Palomar Sky Survey plates showed no star at the nova's position.

prentice, *sb.* Add: **4.** *prentice-boy* (later examples), *-player*; *prentice-hand, -work*: later examples.

1839 C. J. Lever *Confessions H. Lorrequer* vi. 47 A red-hot orangeman,..vice-chairman of the ''Prentice Boys'. 1852 Dickens *Bleak Ho.* (1853) i. 1 Fog cruelly pinching the toes and fingers of his shivering little 'prentice boy on deck. 1898 Prentice hand [see *ELEVEN *a.* 2 c]. 1907 'Mark Twain' *Christian Sci.* ii. iii. 127 They seem to me to prove the presence of the 'prentice hand. 1980 *Times* 12 Jan. 10/6 Breaded scallops and fragile wun-tun came from an expert, not a prentice, hand. 1963 *Times* 7 May 8/7 How fine it would be to hear Sir John Gielgud on a stage where Edmund Kean was a prentice player. 1881 'Mark Twain' *Prince & Pauper* xxii. 269 His frantic and lubberly 'prentice-work found but a poor market for itself. 1975 *Times Lit. Suppl.* 29 Aug. 963/1 It is unfair to print such academic prentice-work cheek by jowl with the work of experienced professionals.

pre-nu·clear, *a.* [Pre- B. 1.] **1.** *Phonetics.* That occurs before a nucleus.

1952 W. Jassem *Intonation of Conversational Eng.* vi. 61 The distinctive feature of the prenuclear tunes is..the relation of the final tone of the tune to the initial tone of the following tune. 1961 [see *INTERNUCLEAR *a.* 3]. 1966 J. E. Buse in C. E. Bazell *In Memory of J. R. Firth* 54 Of the four pre-nuclear classes, prepositions, determinatives and number particles are mutually exclusive with..the tense particles. 1976 *Archivum Linguisticum* VII. 38 One feature both subclasses share is their insistence that any non-nuclear occurrence be in prenuclear position rather than in tail (postnuclear) position.

2. Preceding the development of nuclear weapons.

1960 *Guardian* 14 Sept. 10/5 In the pre-nuclear age power was physical violence. 1965 *Economist* 11 Dec. 1181/1 Mr Kosygin's attitude is frightening because it is so stunningly out of date. It is pre-nuclear, pre-coexistence. 1968 *Punch* 10 Apr. 543/3 A strange, pre-nuclear atmosphere prevailed.

preobrazhenskite (prī,obrāʒe·nskəit). *Min.* [ad. Russ. *preobrazhenskit* (Ya. Ya. Yarzhensky 1956, in *Doklady Akad. Nauk SSSR* CXI. 1087), f. the name of P. I. *Preobrazhensky* (1874–1944), investigator of Russian salt deposits: see -ite[1].] A hydrated magnesium borate found as nodules in salt deposits in Kazakhstan.

1957 *Amer. Mineralogist* XLII. 704 *Preobrazhenskite*... It occurs in colorless, lemon-yellow, and dark gray nodules in fine-grained halite-polyhalite rock. 1972 *Soviet Physics: Doklady* XVI. 519/1 The equation of preobrazhenskite so obtained corresponded to the chemical composition $3MgO \cdot 5{\cdot}5B_2O_3 \cdot 4{\cdot}5H_2O$, but not $3MgO.5B_2O_3.4{\cdot}5H_2O$ as originally proposed. *Ibid.*, The most important crystal-chemical complex in the structure of preobrazhenskite comprises a set of infinite boron-oxygen chains perpendicular to the b axis with a new type of radical; the repeated link in each chain comprises four triple boron-oxygen rings..and two additional $BO_2(OH)_2$ tetrahedra.

preoccupied, *ppl. a.* **b.** (Examples.)

1842 *Rep. Brit. Assoc. Adv. Sci.* 113 The genus of birds, *Plectorhynchus*, being preoccupied in Ichthyology, is changed to *Plectorhamphus*. 1898 [see *BENTONITE]. 1913 [see *BALUCHITHERIUM]. 1967 J. R. & P. H. Napier *Handbk. Living Primates* iii. 377 The subgeneric name *Lyonogale* is substituted for the preoccupied *Tana* Lyon.

pre-Œdipal (prī,ī·dipăl), *a.* *Psychoanalysis.* Also **preœdipal.** [f. Pre- B. 1 + *ŒDIPAL *a.*] That is prior to the onset of the Œdipal phase of development. Also **pre-Œ·dipus** *attrib.*

1932 *Internat. Jrnl. Psycho-Anal.* XIII. 282 Our insight into this early, pre-Oedipal, phase in the little girl's development comes to us as a surprise. 1958 *Ibid.* XXXIX. 516 (*title*) The preoedipal attachment to the mother. 1961 J. Strachey et al. tr. *Freud's Compl. Psychol. Wks.* XXI. 230 The phase of exclusive attachment to the mother, which may be called the pre-Oedipus phase, possesses a far greater importance in women than it can have in men. 1974 G. & R. Blanck *Ego Psychol.* ii. 36 The beginning introjections, incorporations, and identifications are variously described in the literature as primitive, archaic forms of superego, or as..preoedipal super-ego.

pre-oïdium: see *PRE- B. 2 a.

pre-op (prī,ǫ·p), colloq. abbrev. of *PRE-OPERATIVE *a.* Also as *sb.*, ellipt. for *preoperative preparation* or the like.

1934 S. Kingsley in *Famous Plays of 1934* 159, I was kind of worried about that preop insulin. 1956 K. Hulme *Nun's Story* x. 162 All pre-op medication given. 1972 M. Crichton *Terminal Man* i. v. 44 Pre-op patients..often didn't want to see people. 1976 'R. Gordon' *Doctor on Job* iv. 34 The razor's a bit ropey. It's the one they use for the ward preops. 1977 *Lancet* 5 Feb. 301/2 His details are entered in the book for the agreed date, with comments such as 'very fat' or 'needs pre-op physio'.

pre-operation, *sb.* and *a.* Add: **A.** *sb.* **2.** A preoperational activity. **B.** *adj.* [Pre- B. 2.] Prior to a surgical operation.

1971 *Nature* 13 Aug. 456/1 For some time, it has been believed that young children are unable to form transitive inferences about quantity until they pass the stage of logical preoperations at about 7 yr old. 1976 J. Snow *Cricket Rebel* 107, I relaxed as the pre-operation drugs took effect and I moved into another world. 1978 *Detroit Free Press* 16 Apr. D4/2 This way, except for a pre-operation visit, 'I just see them unconscious.'

preopera·tional, *a.* *Psychol.* [Pre- B. 1.] That precedes operational thought, usu. typified by the mental processes of children aged between 2 and 7. (Cf. *OPERATION 4 b.)

1953 Mays & Whitehead tr. *Piaget's Logic & Psychol.* ii. 12 Starting from the postulate that all logical problems arise in the first place from manipulations of objects, we can now say that this period is pre-operational. 1960 J. S. Bruner *Process of Educ.* iii. 34 In this so-called preoperational stage, the principal symbolic achievement is that the child learns how to represent the external world through symbols established by simple generalization; things are represented as equivalent in terms of sharing some common property. 1964 Lunzer & Papert tr. *Inhelder & Piaget's Early Growth of Logic in Child* 291 Whether or not the co-ordination is complete, operations and pre-operational co-ordinations enter into the most diverse kinds of behaviour. 1975 M. D. Smith *Educ. Psychol.* ii. 40 Pre-operational two to seven uses egocentric speech.

preoperative (pri,ǫ·pĕrătiv), *a.* *Med.* [Pre- B. 1.] Given or occurring before a surgical operation.

1904 [in Dict. s.v. Pre- B. 1]. 1954 Martin & Hynes *Clin. Endocrinol.* (ed. 2) iii. 69 The use of pre-operative iodine or anti-thyroid drugs..adds further difficulty in the histological diagnosis of thyroidectomy specimens. 1957 *New Biol.* XXIV. 54 Without preoperative treatment with heparin there would be every likelihood of the blood clotting within the operated blood vessel. 1977 *Lancet* 25 June 1352/1 In preoperative preparation of the bowel most surgeons used neomycin for 1–3 days, or phthalylsulphathiazole for 3–10 days.

Hence **preo·peratively** *adv.*, before an, or the, operation.

1931 [see *POSTOPERATIVELY *adv.*]. 1957 *Ann. R. Coll. Surgeons* XXI. 368 The visual defect pre-operatively affected both eyes nearly equally. 1976 *Lancet* 27 Nov. 1205/2 All the patients feel well, and there have been no occlusive vascular episodes postoperatively. However, in the only patient (no. 2) whose electrocardiogram was abnormal preoperatively, no change has occurred.

pre-ordination. (Later example.)

1948 *Mind* LVII. 180 The well-worn antagonisms of the immanent and transcendent, of finite sinfulness and divine perfection and preordination, which centuries of theological brooding have failed to dissipate.

pre-orgasmic, -ovulatory: see *PRE- B. 1. **pre-ovulation:** *PRE- B. 2 a. **pre-own:** *PRE- A. 1. **pre-oxygenation:** *PRE- A. 2.

prep, *sb.* and *a.* For '*School and College slang*' read '*slang* (orig. *School* and *College*)' and add: **1. a.** (Further examples.) Also, lessons and exercises to be done by a pupil after school hours, either at school or as homework (see *HOMEWORK 2). Also *attrib.*

1900 *Dialect Notes* II. 51 (College Words and Phrases) *Prep*... Preparation. 1911 Beerbohm *Zuleika D.* xxii. 313 With his elbows on the kitchen table..sat Clarence, intent on belated 'prep'. 1939 R. C. Woodthorpe *Rope for Convict* v. 51 I've just remembered I haven't done the prep. he sent me. 1961 E. S. Turner *Phoney War* vii. 28 At prep time, they were not allowed to use ink, for fear of damaging the art treasures on the walls. 1972 *Where* Sept. 237/3 The standard half-hour homework, or 'prep' as it is called in some schools, is purely notional. 1976 *Daily Times* (Lagos) 4 Sept. 19/2 (Advt.), Boarding with daily coaching, strict attention, and Prep Supervision for: Forms One to Five male and female in Lagos. 1977 J. I. M. Stewart *Madonna of Astrolabe* i. 8 Two of her boys.. must be got home in time to be calmed down and persuaded to do their prep. 1979 *Homes & Gardens* June 97/2 There is a general shuffling, an air of impatience, boredom even, of the sort you find at prep. time in public schools.

b. Short for Preparation in various other senses. Also *attrib.*

1925 D. H. Lawrence *Let.* ? 17 Dec. (1962) II. 870 Tell Achsah, lest she make any preps for me. 1934 *Amer. Speech* IX. 237/2 The curtailed word or back-shortening *prep*... In this sentence, *The team had an intensive prep yesterday afternoon*, it has the same connotation as *drill* or *practice*. 1976 M. Millar *Ask for me Tomorrow* (1977) xvi. 127 A little too perfectly groomed, as if he'd just been given the full treatment in..a mortician's prep room. 1976 *Amer. Speech* 1973 XLVIII. 204 He is given a *prep* 'surgical preparation'. 1977 *Hot Car* Oct. 69/1 (*caption*) Persevere with this as it is an important prep stage for the paint.

c. *Horse-racing.* A race that is a preparation for a more important event. *U.S.*

1944 *Sun* (Baltimore) 4 Mar. 9/4 A better-than-fair horse, which he..guided to second place..in the $7,500-McClennan, the widener prep won by Sun Again. 1975 *New Yorker* 15 Sept. 110/2 It isn't often that a hundred-thousand-dollar race is a prep for a two-hundred-and-fifty-thousand-dollar one. 1977 *Ibid.* 3 Oct. 112/1 Quiet Little Table..won the prep, running head-and-head down the stretch with Wise Philip and Jatski.

2. Restrict '*U.S.*' to sense b. **a.** (Earlier and later examples.) See also *PREPARATORY *a.* 2 a.

1895 J. L. Williams *Princeton Stories* 128 After awhile he found himself walking with the freshman way out toward the Prep. school. 1903 *Chicago Record-Herald* 7 June iii. 1/1 A crowd of nearly 4,000 university, 'prep' school and grammar school rooters cheered from the bleachers. 1930 [see *FACT *sb.* 4 f]. 1934 C. Lambert *Music Ho!* v. 298 They are..as childish as the hidden rivers and prep school puns that adorn Joyce's *Anna Livia Plurabelle*. 1943 *Scrutiny* XI. 287 His [sc. Wodehouse's] humour is a cross of Prep-school and *Punch*. 1959 T. S. Eliot *Elder Statesman* ii. 63 You were expelled from your prep school for stealing. 1971 *New Yorker* 15 May 51/2 Dulwich College—what the English call a public school and we call a prep school. 1974 *Listener* 14 Mar. 339/3 The prep-school language and the note of petulance nicely convey the immaturity..of this band of heroes. 1976 *National Observer* (U.S.) 17 Apr. 10/4 It was not the Vietnam War that pinched the adolescent rush to prep schools.

b. (Examples.)

1899 A. H. Quinn *Pennsylvania Stories* 117 He was going to tell all those people, from the Governor down to the prep in the gallery, who came from his own old school, just what the College had done for *him*. 1948 *Chicago Daily News* 6 Dec. 23/4 (*caption*), 2 preps die in Oregon bush crash. 1978 *Maledicta* 1977 I. 223 A stout prep who wore saucer-shaped glasses.

c. Short for *preparatory school*. Also *attrib.*

1895 J. L. Williams *Princeton Stories* 244 Charlie Symington was a well-built prep. boy who had been known to strike out three men with the bases full. 1924 H. de Sélincourt *Cricket Match* v. 158 To know whose call it is..was driven into me at the prep. 1927 W. E. Collinson *Contemp. Eng.* 21 My attendance at Dulwich College Preparatory School (the Prep.) coincided with the South African war. 1934 *Amer. Speech* IX. 237/2 In the following sentence the term refers to high schools, *Nebraska Preps Have Major Encounters this Week*. 1969 *Eugene* (Oregon) *Register-Guard* 3 Dec. 2D/1 The Eagles thought, to win the state B prep football title, they'd have to throw more than they did. 1976 *Honolulu Star-Bull.* 21 Dec. H-2/1 Going back to last Saturday's game, it was great to see just how far Russ Francis has come since his prep days at Kailua High.

prep (prep), *v.*[1] *U.S. slang.* [f. prec., sense 2.] *intr.* To attend a preparatory school, be a preparatory school student.

1915 *Dialect Notes* IV. 236 *Prep*, preparatory: used as a verb 'to attend a preparatory school'. 1920 F. Scott Fitzgerald *This Side of Paradise* ii. 43 Where'd you prep? 1936 L. C. Douglas *White Banners* xiv. 305 Thomas and this Colonel Livingstone had prepped together at this academy. 1967 *Boston Sunday Herald Mag.* 16 Apr. 8/2 What school do you go to?.. Where did *you* prep? 1977 *New Yorker* 23 May 91/1 A native of Peoria, Illinois, who prepped at Lawrenceville, Davis graduated from Princeton with highest honors in history.

prep (prep), *v.*[2] *slang* (orig. *U.S.*). [f. Prep *sb.* 1, or shortening of Prepare *v.*] **a.** *trans.* To prepare (someone or something); to train (an animal) for racing; to prime (a witness); *spec.*, in hospital terminology, to prepare (a patient) for an operation. Also *absol.*

1927 *Amer. Speech* II. 313/1 Ask whether the 'tenthirty appendectomy has been prepped yet?' For some reason a patient's abdomen is not shaved, it is 'prepped', that is, prepared for the surgeon. 1936 *Esquire* Sept. 160/3 *Little Lord Fauntleroy* and *Dodsworth* [sc. two films] have been 'prepping prod.' (preparing production) for some time. 1937 'J. Bell' *Murder in Hospital* vii. 133 Macdonald started to prep him. *Ibid.*, She gave him his atropine for the operation when she had finished prep'ing the leg. 1943 *Sun* (Baltimore) 12 Oct. 16/5 (*heading*) Attention [sc. a horse] being prepped for New Orleans 'Cap. 1961 'K. Norway' *Waterfront Hosp.* i. 19, I told Nurse David, 'Five minutes—we'll have to prep on the table.' 1965 *Eng. Stud.* XLVI. 461 *Prepster* 'one who is being prepped; trainee'. 1967 *Boston Sunday Herald* 14 May ii. 3/1 Anyone planning to enter greyhound racing should know it costs close to $600 to prep each dog for the races. 1968 J. D. MacDonald *Pale Grey for Guilt* (1969) xii. 145 Somebody prepped her pretty good, Sheriff. I might even have thought she saw somebody she sincerely mistook for me. 1969 I. Kemp *Brit. G.I. in Vietnam* v. 110 The gun-ships must have done a thorough job of 'prepping' the L.Z. because, apart from sporadic sniper fire,.. we met no opposition there. 1972 M. Goldberg *Karamanov Equations* xxiii. 226 Have the nurse prep him... Neck, chest, and groin. 1975 *Globe & Mail* (Toronto) 18 Sept. 5/1 Looking as Tory as the advance men who precede the Premier to prep the crowd, he strolled through the market.

b. *intr.* for *refl.* To prepare oneself (for an event); to practise, to train (esp. in sport). *U.S.*

1934 *Amer. Speech* IX. 237/2 Verbal use of the word appears in *Beavers Arrive in Omaha to Prep for Husker Fray*. 1937 *Sun* (Baltimore) 22 Apr. 17/1 (*heading*) Track preps for Ky. Derby. 1941 *Ibid.* 3 July 16/1 The latter.. was prepping for the New Castle Handicap to be run on Saturday. 1949 *N.Y. Times* 4 Sept. v. 2/6 A pitcher, who had prepped earnestly for many years in the minors..was cut from the roster. 1972 *Newsweek* 10 Jan. 24/3 Mrs. Nixon has been prepping for the trip for weeks. 1977 *Time* 28 Nov. 66/1 Akers had prepped as a Royal assistant before moving into the head coaching job at Wyoming in 1975.

pre-pa·ck, *v.* [Pre- A. 1.] *trans.* To pack or wrap (an article, usu. of food) on the site of production or before retail. Also *fig.* So **pre-packed** *ppl. a.*, **pre-pa·cking** *vbl. sb.* Also **pre-pa·cker.**

1928 *Daily Express* 23 Mar. 3/1 The public..would abandon bread altogether in favour of pre-packed foods, all of them comparatively expensive. 1931 J. W. Wingate *Man. Retail Terms* xv. 344 *Prepacking*, merchandise packed by the store in advance of sale. 1952 *Times* 6 Aug. 2/2 Describing the method of pre-packing butter in most machines, the report states that..the rate of delivery is between 60 and 80 packets a minute. 1957 *Times* 2 July (Agric. Suppl.) p. v/4 These requirements..have done much to encourage prepacking on the farm in units ready for retail sale without further wrapping or weighing. 1962 H. O. Beecheno *Introd. Business Stud.* ii. 14 Now we have the ability to pre-pack, preserve and store the vast majority of the goods available on the market. 1974 'E. Anthony' *Malaspiga Exit* i. 14 The average pre-packed American beauty. 1976 C. Bermant *Coming Home* ii. iii. 149 The reasoning seemed to be pre-packed and clogged with clichés and slogans. 1976 *Milton Keynes Express* 23 July 23/3 (Advt.), Wholesale fruit and vegetable merchants and pre-packers of quality produce. 1977 *Oxf. Diocesan Mag.* July 14/2 We have exciting new concepts of mission—no longer seeing the Church as going out with a prepacked Gospel to sell.

pre·-pack, *sb.* [Pre- A. 2.] A container or wrapper in which an article (usu. of food) is enclosed on the site of production or before retail.

1957 *Daily Mail* 26 Sept. 8/5 Business is growing so fast that the sale of pre-packs is expected to increase by 70,000,000 a year. 1973 *Times* 1 Feb. 4/2 Prepacks containing 'A' grade eggs may be decorated with a red band. 1976 *Oxf. Consumer* Mar. 8/2 Prepacks of biscuits and shortbread must be marked with their weights if they weigh more than 50g.

pre-pa·ckaged, *a.* [Pre- A. 1.] Packaged on the site of production or before retail. Also *fig.* So **pre-pa·ckaging** *vbl. sb.* Also **pre-pa·ckage** = *PRE-PACK *sb.*

1944 R. E. Lee *Television* 179 Local stations will transmit pre-packaged variety shows. 1945 *Business Week* 18 Aug. 91 An ill-timed effort to market prepackaged frozen meat. 1947 *Printers' Ink* 3 Jan. 70 (*heading*) Produce marketers survey pre-packaging to stem competition from frozen foods. 1957 *Times* 2 July (Agric. Suppl.) p. vii/2 The growing popularity of pre-packaged foods, of self-service shopping, of refrigerators. 1960 *News Chron.* 22 Sept. 13/1 The pre-package industry was born. 1963 *Supermarket & Self-Service* (Johannesburg) Nov./Dec. 9/1 Place a mass display of pre-packaged cheese on a dump table in the middle of a side aisle. 1965 *Wireless World* July 345/2 (*heading*) Pre-packaged semiconductors for the retail market. 1966 *Rep. Comm. Inquiry Univ. Oxf.* II. 470 To establish a system of required lectures would.. place a disastrous emphasis on prepackaged instruction. 1976 *Times* 1 May (Food Suppl.) p. iii/6 Prepackaging.. has..been used for cuts of fresh meat.

preparation, *sb.* Add: **1. c.** (Further examples.)

1914 'I. Hay' *Lighter Side School Life* iv. 114 A prefect.. awarded both signallers fifty lines for creating a disturbance in Preparation. 1971 *Black Scholar* Jan. 64/1 (Advt.), Teaching load each semester is 7 to 8 hours with two preparations.

6. (Further examples.)

1960 *Harper's Bazaar* Oct. 117/2 The..unfussy packaging of men's preparations. 1964 *New Statesman* 6 Mar. 354/1 In this country a doctor can prescribe from a list of about 5,000 'preparations', mostly mixtures of drugs. 1972 *Guardian* 7 Nov. 11/3 The Vichy preparations include complete ranges of cleansers, toners, moisturisers, night creams.

10. (sense 1 c) *preparation book*; (sense 8) *preparation sermon.*

1896 E. Turner *Little Larrikin* viii. 76 The diminution of the pile of preparation books. 1843 *Knickerbocker* XXI. 261 On the very day of the preparation sermon at Tinnecum, a number of young persons were assembled.

preparationist (prepărēi·ʃənist). *temporary.* [f. Preparation *sb.* + -ist.] One who favours naval and military preparedness.

1915 A. L. Lowell in *World's Work* (N.Y.) XXX. 719/1 The preparationists..fix their attention primarily on the means of securing the safety of our own land from injury by war. *Ibid.*, To the preparationists..the suggestion of a league to enforce peace ought to appeal as a means of doing on an international scale the thing they are seeking to do for the United States.

preparator. Delete *rare* and add further examples.

1884 *Science* 11 Apr. 443 While, however, the use of the photograph for outlines diminishes the labor of the artist about one-half, it increases that of the preparator. 1931 A. A. Morris *Digging in Yucatan* xvii. 267 A phenomenally skillful Japanese artist preparator who was then working for the American Museum of Natural History. 1937 *Nature* 2 Jan. 16/1 The [preservation] process was repeated once, and the bones were then ready for the preparators, who mended cracks and other deficiencies. 1938 *Times* 18 Jan. 15/5 In addition the Museum preparators have made use of skeletons already in the collection.

preparatory, *a.* **2. a.** (Earlier and further examples.) Also *U.S.*, applied to a (usu. private) school in which pupils are prepared for college entrance.

1822 M. Edgeworth *Let.* 23 Jan. (1971) 328, I have asked all your questions my dear mother about the preparatory school for Pakenham... Mr. Malthus and Dr. Batten declared that they should prefer having a boy sent to them from the Charter-House to having him from any lesser preparatory school. 1848 *Indiana Gen. Assembly Doc.* (1849) II. 279 Connected with the Institution is a flourishing Grammar School, which serves the double purpose of a Normal School and a Preparatory Department. 1851 C. Cist *Sk. Cincinnati in 1851* iii. 69 The Classes in the course of study in the Preparatory Department, are divided among the Adjunct Professors of Mathematics and Languages and the Professor of Modern Languages. 1879 *Scribner's Monthly* Dec. 207 The Johns Hopkins is seeking..to penetrate downward into the preparatory schools. 1903 *World's Work* (N.Y.) Sept. 3884/1 The preparatory school..take[s] boys from twelve to fourteen years of age to fit them in from three to six years for entrance to our best colleges. 1924 *Granta* 25 Apr. 361/2 At the age of eight, he arrived at his Preparatory School, 'The Wick', in Sussex. 1949 *Manch. Guardian Weekly* 20 Jan. 5/2 The son of a bishop, he went through a fashionable preparatory (that is, public) school. 1954 A. S. C. Ross in *Neuphilol. Mitt.* LV. 26 School-boys at their preparatory school..should be addressed as *Master*. 1963 [see *junior college* s.v. *JUNIOR *a.* (*sb.*) 5]. 1969 *Listener* 9 Jan. 41/3 Such a young man..can often command a tiny handout to tide him over by entering the profession of preparatory schoolmaster, purely as a temporary measure. 1972 *Lebende Sprachen* XVII. 35/2 US preparatory school—BE private secondary school. 1976 *Southern Even. Echo* (Southampton) (Advt. Suppl.) 6 Nov. 7/6 Resident matron required for January for Boys' Preparatory School.

prepare, *v.* Add: **1. e.** (Further examples.) Also, in extended sense, to be willing or determined *to do* something. *Be prepared*, the motto of the Scout and Guide organizations.

1902 G. B. Shaw *Mrs Warren's Profession* Pref. p. ix, Nor am I prepared to accept the verdict of the medical gentlemen who would compulsorily sanitate and register Mrs Warren. 1908 R. S. S. Baden-Powell *Scouting for Boys* i. 20 The badge..of the first class scout consists of a brass arrow head with the motto on it '*be prepared*'. *Ibid.* 48 The scouts' motto is founded on my initials, it is: *be prepared*. *Ibid.* 49 A scout..must Be Prepared at any time to save life. 1939 G. B. Shaw *Geneva* i. 13, I came here to place it before a body of persons of European distinction. I am not prepared to discuss it with an irresponsible young woman. 1948 E. Gowers *Plain Words* vi. 40 The recipient of a letter may feel better..if he is told that the Minister 'is not prepared to approve' than he would have done if the letter had said 'the Minister does not approve'. *Ibid.* 41 'The Board have examined your application and they are prepared to allocate 60 coupons for this production. I am accordingly to enclose this number of coupons.'.. *Prepared to allocate* should be *have allocated*. Since the coupons are enclosed, the preparatory stage is clearly over. 1961 *New Eng. Bible Acts* x. 47 Is anyone prepared to withhold the water for baptism from these persons? 1963 A. Christie *Clocks* xxvii. 229 What I say to you is: 'Be prepared.' And I don't mean it in the Boy Scout sense. 1972 J. Pover *Chinese Agenda* (1973) xiv. 190 It's better to be prepared..that's what they taught us in the Boy Scouts. 1976 *Daily Tel.* 20 July 2/1 The Government was not prepared to fight for realistic exclusive zones for British fishermen.

prepared, *ppl. a.* Add: **b.** spec. *prepared core* Archæol. (see Core *sb.*[1] 5), *food*, *piano* (see quot. 1960).

1918 A. Huxley *Let.* 12 Aug. (1969) 160 Thirty-two quintals of sugar and prepared foods. 1946 *Mod. Music* Summer 205 *Four Sonatas* for prepared piano by John Cage were also heard. 1952 *Musical Q.* XXXVIII. 124 Cage's pieces for what he calls the 'prepared piano' offer an array of tightly organized little sounds of many colors. 1959 J. D. Clark *Prehist. S. Afr.* vi. 142 The characteristic technique of the Middle Stone Age times is the prepared core and specially prepared flake with thin section and faceted butt. *Ibid.* 157 The same basic prepared-core technique. 1960 *20th Cent.* Apr. 348 Cage, an American, is the originator of the 'prepared piano',..in which the pitch and timbre of certain notes are altered by attaching..metal, rubber, wood, etc., to the strings at various distances from the point at which the hammer strikes. 1964 H. Hodges *Artifacts* vii. 102 The core may be very carefully flaked in preparation so that ultimately one final blow will detach a flake tool of the required shape, and these prepared cores are generally the result of hammer, punch or pressure flaking. 1972 *Listener* 7 Sept. 292/3 Further rapid growth [is] expected in the prepared-foods trade. 1977 *Belfast Tel.* 24 Jan. 9/3 The same professionalism marked John Cage's sonata for prepared piano. For this another small grand piano had been fitted out in advance with an assortment of bolts, nuts and rubber wedges according to a carefully specified plan.

preparing, *ppl. a.* (Later example.)

1864 'E. Wetherell' *Old Helmet* I. i. 21 The other figures, the dark walls and ivy, the servants and the preparing collation, were only a rich mosaic of background for those two.

pre-paroxysmal: see *PRE- B. 1 d.

pre-partum (prī,pā·ɹtəm), *a.* [L. phr., 'before birth', used *attrib.*] = *ANTE-PARTUM *a.*

1858 R. Barnes *Physiol. & Treatm. Placenta Prævia* ii. 71, I believe these considerations present a rational explanation of a multitude of cases of præ-partum hæmorrhage, [etc.]. 1950 *Amer. Jrnl. Obstetr. & Gynecol.* LIX. 1116 In cases of prepartum bleeding, cesarean section to facilitate abdominal exploration was practically always necessary. 1978 J. Updike *Coup* (1979) v. 186 'My goodness, you're touchy these days.' 'Pre-partum blues,' he suggested.

prepatency, -patent, -pathological, -pausal: see *PRE- B. 1. **pre-pause:** *PRE- B. 2 a.

prepay, *v.* Add: (Further examples.)

1899 A. E. W. Mason *Miranda of Balcony* xv. 216 He wires me..and prepays the reply. 1973 *Philadelphia Inquirer* (Today Suppl.) 7 Oct. 42/2 He'll be in to prepay.

prepaid *ppl. a.* (further examples); **prepayment** (further *attrib.* examples).

1926 C. Connolly *Let.* Sept. in *Romantic Friendship* (1975) 172, I..sent a prepaid wire. 1977 *Modern Railways* Dec. 465/2 Urgent introduction of a wide-ranging system of pre-paid bus tickets. 1906 Prepayment [see *coin-box*]. 1970 *Which?* Mar. 90/1 Fire meters had defective prepayment mechanisms. 1977 *Wandsworth Borough News* 16 Sept. 14/3 Cash from pre-payment meters, believed to be £107, was stolen from the home.

prepense, *a.* Add: **a.** (Further example of *malice prepense*.)

1923 D. H. Lawrence *Birds, Beasts & Flowers* 128 Full of malice prepense, and overweening.

c. (Further example.)

1919 W. B. Yeats *If I were Four & Twenty* in *Irish Statesman* 23 Aug. 212/1 For he [sc. Claudel] is prepense, deliberate.

preperceive: see *PRE- A. 1. **prepersonal** to **-phonemic:** *PRE- B. 1. **pre-phylloxera:** *PRE- B. 2 a. **preplan:** *PRE- A. 1. **pre-planetary, -planting:** *PRE- B. 1. **prepolarization:** *PRE- A. 2. **prepolarize:** *PRE- A. 1.

prepo·lymer. [Pre- B. 1 in Dict. and Suppl.] An intermediate in a process of polymerization which is convenient for manipulation and can be fully polymerized at a later stage in the process.

1956 *Mod. Plastics* July 111/3 If this thickening process is interrupted at the appropriate point, the prepolymer or sirup has about the same consistency as molasses (in June, not January!). 1962 *New Scientist* 25 Oct. 205/1 The state of polymerization of special synthetic fluids such as aviation turbine oils or polypropylene glycol (the latter a prepolymer used in the manufacture of polyurethane foams). 1971 *Penrose Ann.* LXIV. 23 The second [solventless ink] system uses liquid photo-sensitive pre-polymers which are polymerized and solidified immediately by exposure to high doses of ultra-violet light. 1978 *Sci. Amer.* June 107/1 The third method of consolidating stone is to treat it with organic monomers and prepolymers.

So **prepo:lymeriza·tion; prepo·lymerize** *v. trans.*, to convert (a monomer) into a prepolymer.

1949 B. L. Davies *Technol. Plastics* ix. 151 If the prepolymerization of the resin is carried too far, the flow in the mould is restricted so that only a limited range of mouldings can be produced. 1956 *Mod. Plastics* July 111/3 (*heading*) Prepolymerization. *Ibid.* 114/1 Thermosetting resins cannot be prepolymerized to any useful degree in this manner, since they set up to a firm gel at a relatively low degree of cure. 1967 Margerison & East *Introd. Polymer Chem.* iv. 200 If all traces of inhibitors.. are not removed from the monomer, irreproducible rates of polymerization are often found; for this reason it is the practice to 'pre-polymerize' 1 or 2% of the monomer before use.

preponderously, *adv. rare.* [f. Preponderous *a.* + -ly[2].] To a preponderous degree; excessively.

1921 *Public Opinion* 5 Aug. 133/1 Is it a city or merely a village preponderously overgrown?

prepose, *v.* Restrict † *Obs.* to senses 1 and 3 and add: **2. a.** (Later examples.)

1946 O. Jespersen *Mod. Eng. Gram.* V. xv. 220 *Well to do* = 'well off, living in easy circumstances' is often preposed, generally written with hyphens: *a well-to-do farmer*. 1951 W. K. Matthews *Languages U.S.S.R.* iv. 53 Syntactically Altaic follows the rule of subordinating, in this case preposing, secondary to principal categories. 1971 *Language* XLVII. 276 The former [example] would result if the NSR preceded the postcyclic transformation which preposes *away*. 1975 *Ibid.* LI. 386 Truncated passives can be generated transformationally..by a rule which obligatorily preposes the underlying object..into underlying subject position. 1978 *Studies in Eng. Lit.: Eng. Number* (Tokyo) 106 It is obvious that there are similarities between the rule which preposes *higher into the heavens* in (46a) and the one which preposes, for example, *up into the clouds* in (42b).

preposed *ppl. a.*: delete † and add later examples. Also **prepo·sing** *vbl. sb.*
1888 *Trans. & Proc. Mod. Lang. Assoc. Amer.* 1887 III. 39 It is a characteristic of Anglo-Saxon poetry ..to introduce an idea with a pronoun... This preposed pronoun is noticed by all writers upon A.-S. style as frequently standing at the head of the sentence. 1928 O. Jespersen *Internat. Lang.* ii. 153 Word-order with preposed subject (as in English) 'Are you ill?') cannot well be used in an I.A.L. 1970 B. M. H. Strang *Hist. English* v. 290 A strongly falling pattern overall, which has been less general in later English because of the increase in preposed weak particles. 1975 *Language* LI. 83 Passivization...may involve not one but two transformational operations—subject postposing and object preposing. 1976 J. S. Gruber *Lexical Struct. Syntax & Semantics* i. ii. 70 The preposing..serves to add emphasis to the phrase, changing the meaning slightly. 1978 *Language* LIV. 774 Bernard Mohan...presents some formulas and the results of a 'grammaticality' test..for various sentences with adverb-preposing or topicalization below and above a set of predicates. *Ibid.* 283 These failures are also significant facts about the speech of Achilles:..(d) Syntax: sentence length; clause length; preposed relative clauses.

preposition. 3. Delete *rare* and add further examples.
Now usu. hyphened in this sense.
1930 T. Sasaki *On Lang. R. Bridges' Poetry* 26 In French, where postposition, and then pre-position, of the adj. attrib. is the general rule. 1946 O. Jespersen *Mod. Eng. Gram.* V. xxi. 392 Historically, however, *for* was a subordinating conjunction, as shown (1) by the possibility of pre-position. 1961 *Moderna Språk* LV. 243 Another theory advanced by Jespersen is that some words cannot be freely used in pre-position because their signification demands a complement: thus *ashamed* (of something, to do something). 1963 F. T. Visser *Hist. Syntax Eng. Lang.* I. i. 19 The extreme scarcity, however, of examples, in Old English with pre-position of the clause..renders the correctness of this interpretation doubtful.

pre-position (prīpǝzi·ʃǝn), *v.* [Pre- A. 1.] *trans.* To position (esp. military equipment) in advance. So **pre-posi·tioning** *vbl. sb.*
1962 *Daily Tel.* 7 May 12/6 The area was selected as one where climatic conditions would provide the maximum test for seaborne military 'pre-positioning'... Stockpiling of military equipment, or as military experts call it, pre-positioning, has been accelerated in Europe. 1974 C. Johnston *Brink of Jordan* i. iv. 23 This gave the unsettling impression that they [sc. Saudi and Syrian forces] were simply pre-positioned there so as to create the best results for their countries in an eventual carve-up of Jordanian territory. 1979 *Sci. Amer.* Jan. 39/3 Indeed, our own proposals improve the quick-reaction defense of Europe by prepositioning more armor in Europe and rebasing most of the powerful combat air strength now held on the carriers to land bases in support of NATO.

prepositional, *a.* Add to def.: Formed with a preposition; serving as, or having the function of, a preposition. (Further examples.)
1940 C. C. Fries *Amer. Eng. Gram.* 130 The prepositional infinitive was made up of the preposition or function word *to* and the dative case of a verbal noun. 1961 R. B. Long *Sentence & its Parts* iii. 77 The prepositional adverbs *but* and *like* sometimes enter into subordinate clauses (then best regarded as interrogative in type) instead of preceding them and functioning as prepositions with declarative-clause objects. *Ibid.* vii. 185 Prepositional units used as subjects are assigned singular force. *Over the fence is out. To admit the truth would be to endanger the whole enterprise.* 1963 F. T. Visser *Hist. Syntax Eng. Lang.* I. iv. 390 This object is traditionally called the 'prepositional object',..it would be more correct to call it a 'direct object dependent on a prepositional group-verb'... Other terms for 'prepositional group-verbs' are 'phrasal verbs', 'group verbs', 'compound verbs'. 1965 N. Chomsky *Aspects of Theory of Syntax* ii. 101 It is well known that in Verb-Prepositional-Phrase constructions one can distinguish various degrees of 'cohesion' between the Verb and the accompanying Prepositional-Phrase. 1965 *Language* XLI. 156 Prepositional object: *John is looking at her* (actually this is not what is usually called a prepositional object, but the object of a prepositional verb-unit *look at*). 1966 G. N. Leech *Eng. in Advertising* ii. 12 Prepositions are a class of words which occur in initial position in a type of adverbial group, the 'prepositional group'.

preposi·tionless, *a.* [f. Preposition + -less.] Lacking or without a preposition.
1956 A. H. Smith *Eng. Place-Name Elements* i. 6 The normal syntax demanded a prepositionless nominative. 1963 F. T. Visser *Hist. Syntax Eng. Lang.* I. iv. 328 To the general replacement of the prepositionless complements by prepositional complements there are a few exceptions. 1965 *Language* XLI. 133 The subchapter on direct (i.e. prepositionless) verbal complements contains 'The accusative of price, measure and quantity'. 1966 *Eng. Stud.* XLVII. 54 The traditional and vague term *direct object*..which corresponds to an OE complement in the accusative form and, later on, to a postverbal prepositionless stem-form.

prepotency. Add: **1. b.** *Psychol.* The quality inherent in a particular stimulus or response that makes it prepotent (cf. next).
1928 *Psychol. Rev.* XXXV. 420 What response the animal will make at a given moment depends upon the prepotency of the stimuli. 1953 B. F. Skinner *Sci. & Human Behav.* xiv. 220 When two responses are strong at the same time, only one can be emitted. The appearance of one response is called 'prepotency'.

prepotent, *a.* Add: **1. c.** *Psychol.* Applied to the effective stimulus and its response when stimuli with different, conflicting, responses occur together.
1906 C. S. Sherrington *Integrative Action Nervous Syst.* vi. 228 It is those stimuli which..are most fitted to excite pain which, as a general rule, excite in the 'spinal' animal..the prepotent reflexes. 1928 *Psychol. Rev.* XXXV. 420 The animal behaves as it does because a certain prepotent stimulus in the environment has forced it that way. 1948 W. McDougall *Introd. Social Psychol.* (ed. 29) 439 The 'prepotent reflexes' of sex, fear, and rage. 1953 B. F. Skinner *Sci. & Human Behav.* xiv. 220 The prepotent response does not, merely by virtue of its having been emitted, alter the strength of the dispossessed response. 1960 Hinsie & Campbell *Psychiatric Dict.* (ed. 3) 572/2 In general, nociceptive reflexes, such as the flexion reflex, are prepotent to all other types of reflex competing for the final common pathway.

pre-pottery: see *pre- B. 2a.

prepper (pre·pǝɹ). *School and College slang.* [f. Prep *sb.* and *a.* 2 + *-er¹.] **a.** *U.S.* A preparatory sports team or a member of such a team. **b.** A preparatory school (Preparatory *a.* 2 a).
1945 *Richmond (Virginia) Times-Dispatch* 10 Oct. 18/4 (*heading*) Thomas [Jefferson High School] heads off heavy slate for preppers. 1956 'M. Innes' *Appleby plays Chicken* i. 1. 16 My public school was..nothing to my prepper. 1962 'R. Gordon' *Doctor in Swim* xi. 73 'Actually, I'm a stinks beak in a prepper,' he confessed. 1974 *Anderson (S. Carolina) Independent* 19 Apr. 3a/3 Audie Mathews, 6-4, of Chicago Heights, Ill., one of the nation's most coveted preppers, is reported to be considering North Carolina State, Illinois, Oregon, Purdue and UCLA. 1974 *Times* 29 Oct. 16/4 Cheam is dominantly the oldest prepper in the business, having started in the Surrey suburb in 1645.

preppy (pre·pi), *a. U.S. School and College slang.* Also **preppie.** [f. Prep *sb.* and *a.* + -y¹.] Of, pertaining to, or characteristic of a pupil at a preparatory school (see *preparatory *a.* 2a); immature; (see also quot. 1980). Also as *sb.*, a pupil at a preparatory school. Hence **pre·ppiness.**
1900 *Dialect Notes* II. 51 *Preppy*, silly, immature. 1970 *New Yorker* 16 Nov. 52/3 His first year as a prep had left Junius feeling like a pound of plaster of Paris. 1971 M. McCarthy *Birds of Amer.* 10 When he finally did ask,..it was in a casual preppy voice. 1975 B. Meggs *Matter of Paradise* i. ii. 30 All you preppies had those funny names. *Ibid.* vi. vi. 96 Bubbling along now in his keen preppie way. 1977 *New Yorker* 11 July 80/2 They are washable, stylish, practical translations of the all-American look—of preppiness and of L. L. Bean. 1978 *N.Y. Times* 6 May 12/4 The pair of loafers my sons refuse to own because they're too preppy. 1980 W. Safire in *N.Y. Times Mag.* 30 Mar. 9 The word that sums up the rage of the fashion world is 'preppie'... Suddenly, sweater counts, the buttons are down, the sweaters and skirts are back.

pre-precipita·tion, *a. Metallurgy.* [Pre- B. 2.] Applied to phenomena occurring at or immediately before the onset of precipitation from solid solution in alloys, esp. the separation of submicroscopic particles from which crystals later develop.
1939 M. Cohen in *Metals Technol.* Oct. 14 (*heading*) Pre-precipitation behavior of the alloy. *Ibid.* 15 The resistance and dilation curves indicate knot formation even at 200°, although there is no pre-precipitation hardening at that temperature. 1949 J. E. Garside *Process & Physical Metall.* xiii. 312 The process of age-hardening consists of two stages—(i) Pre-precipitation stage resulting in the formation of nuclei; (ii) Growth of the nuclei, first into particles of ultramicroscopic size, and then of microscopic size. 1959 B. Chalmers *Physical Metall.* viii. 391 The desirable results of precipitation are almost always obtained when a very large number of precipitate particles or pre-precipitation zones are formed. 1965 Hunsicker & Stumpf in C. S. Smith *Sorby Centennial Symposium Hist. Metall.* xvii. 287 In the face of mounting evidence the conviction became widespread that some sort of 'pre-precipitation' structure was developed at low temperatures and possibly at higher temperatures prior to or during and contributory to actual precipitation.

pre-predicative: see *pre- B. 1.

pre-preference, *a.* (Earlier example.)
1869 *Bradshaw's Railway Manual* XXI. 128 New capital is created as a 4½ per cent. pre-preference or debenture stock, the first preference being made second.

pre-preg, *sb.* (*a.*) [f. *pre-(im)preg(nated) *ppl. a.*] A fibrous material (e.g. glass or carbon fibres) that is pre-impregnated with synthetic resin for use in the manufacture of reinforced plastics. Freq. *attrib.* or as *adj.*
1954 R. H. Sonneborn *Fiberglas Reinforced Plastics* iii. 63 The pre-preg materials are fabrics or mats that are pre-loaded, using resin mixtures that are essentially the same as are used in standard molding operations. 1958 D. J. Duffin *Laminated Plastics* iv. 72 Although more expensive than standard materials, prepregs are widely used in the aircraft industry and elsewhere because of the considerable savings in time—and therefore labor—costs. 1965 *Mod. Plastics Encycl.* 1966 167/3 Pre-preg laminates are made by pre-impregnating glass cloth with a resin solution. 1969 *Sci. Jrnl.* Feb. 43/3 Prepreg is made by dipping numerous tows or groups of fibres in a dilute solution of resin in acetone and then laying them down side by side..on a firm flat surface. 1972 *Physics Bull.* Nov. 664/1 Fibre-polymer systems for the fabrication of composite parts are used either as prepregs, in which fibre and polymer are precombined and are ready for the fabrication operation, or as wet systems in which the dry fibre and the polymer are brought together during the moulding process.
Hence **pre-preg** *v. trans.*, to pre-impregnate with synthetic resin; **pre-pregged** *ppl. a.*; **pre-pregging** *vbl. sb.*
1964 Oleesky & Mohr *Handbk. Reinforced Plastics* ii. iii. 77/2 Epoxy resins may be readily adapted to prepregging using glass fabrics or roving. *Ibid.* ix. i. 545/1 The general specifications governing roving for filament winding also apply to roving to be prepregged. 1970 *Encycl. Polymer Sci. & Technol.* XII. 26 Sheet molding compounds is the term used by the Society of the Plastics Industry to identify these new reinforced, basically prepregged sheet molding products.

pre-preparatory: see *pre- B. 1d. **pre-prepare:** *pre- A. 1. **pre-pressurization:** *pre- A. 2. **pre-pressurize:** *pre- A. 1. **pre-primary:** *pre- B. 1d.

prepri·mate. [Pre- B. 1 in Dict. and Suppl.] An evolutionary ancestor of the primates, or an animal showing characteristics that are more highly developed in the primates.
1931 E. A. Hooton *Up from Ape* ii. 67 Our arboreal pre-primate ancestors must have been very closely similar to the existing tree shrews of the Order Insectivora. *Ibid.* 69 When the insectivorous pre-primate became a hand-feeder and a manipulator of objects, it opened up for its descendants a new route of evolutionary progress. 1971 G. H. Bourne *Ape People* xii. 322 The shape of the jaw seemed to be more like that of a modern preprimate known as the tarsier.

preprint, *sb.* Add: (Further examples.) Also *attrib.*
1939 E. C. Bingham *Some Defs. Rheology* 1 This paper is issued in preprint form primarily to stimulate discussion. 1951 J. A. Wheeler in *Niels Bohr* 174 A paper... seen in preprint form through the kindness of the authors. 1961 D. J. Boorstin *Image* iv. 141 They [sc. scientists] now use the device of the 'preprint'. This is a version of an article made available before its 'publication'. 1970 *Nature* 26 Dec. 1356/1 We can show how the preprint system was set up by giving references to a large number of preprints.

prepri·nt, *v.* [Pre- A. 1.] *trans.* To print in advance, *spec.* to print and issue (part of a work) before publication of the whole. Hence **pre-pri·nted** *ppl. a.*
1898 E. D. P. Evans (*title*) Meaning of minister in Anglo-Saxon. (Preprinted from the Philological Society's Transactions, 1925–26, part II.) 1958 *Practical Wireless* XXXIV. 72/2 It is provided with interchangeable paper on plastic shields which can be supplied either pre-printed or blank. 1964 *Economist* 11 July 165/1 The answer has been preprinted colour gravure. 1965 L. W. Beck *Stud. Philos. Kant* vii. 108 This paper..is preprinted here by kind permission. 1970 A. Davidson *Returns of Love* ii. 21 You pick up the little card which pops out of the slot, knowing that it's been pre-printed. 1977 It May 6/1 You can purchase books of 25 pre-printed slips at two pence per book.

pre-process, *v.* [f. Pre- A. 1 + Process *v.*²] *trans.* To subject to a preliminary processing. Hence **prepro·cessing** *vbl. sb.* Also **prepro·cessor,** a machine for preprocessing.
1964 *Ann. N.Y. Acad. Sci.* CXV. 568 The use of some analog equipment, such as filters, to preprocess data. 1965 *Proc. Internat. Fed. Information Processing Congr.* II. 329/1 (*heading*) ISODATA—a..self-organizing computer program for the design of pattern recognition preprocessing. 1966 [see *post-processing* s.v. *post- A. 2 a]. 1967 E. H. Lennon in Cox & Grose *Organiz. Bibliogr. Rec. by Computer* iv. 83 In order to reduce errors..the system provides for a preprocessing and storage of data from a data description..in secondary storage. 1967 *Computer Jrnl.* IX. 360/1 The preprocessor was..designed to translate these two dialects into a form acceptable to the Atlas compiler. 1970 A. Cameron et al. *Computers & Old Eng. Concordances* 68 Some preprocessing of the concordances is a desirable prerequisite to editing. 1976 *Nature* 29 Jan. 294/2 The three detected voltages were then digitised, preprocessed and recorded on magnetic tape for later computer processing.

prepro·cessed, *ppl. a.* [Pre- A. 2.] Of food: processed before being offered for sale; needing little preparation.
1961 M. Beadle *These Ruins are Inhabited* (1963) viii. 107 Englishwomen..can't afford the extra cost of preprocessed food or commercial laundry charges. 1962 F. I. Ordway et al. *Basic Astronautics* xiii. 501 Fresh meat and vegetables under conventional refrigeration obviously cannot be considered. However, some preprocessed, small bulk, frozen foods are a possibility.

pre-production, *sb.* and *a.* **A.** *sb.* **1.** [Pre- A. 2.] (A) preliminary or trial production.
1938 *New Statesman* 20 Aug. 282/1, I have seen pre-productions, for the Festival, of this week's naturalistic plays. 1947 Crowther & Whiddington *Science at War* i. 49 The General Electric Company worked out a method of pre-production, by which small quantities of new valves could be produced by formerly unskilled women workers, while the problems of mass-production were being worked out.
2. [Pre- B. 1 in Dict. and Suppl.] Work prior to production; preparation for production.
1976 *Times* 20 Dec. 63/1 We're here in New York doing preproduction. 1979 D. Lowden *Boudapesta 3* xviii. 100 In films, 80% of scripts written never reached the stage of preproduction.
B. *adj.* [Pre- B. 2.] Prior or preliminary to (a) production.
1946 *Nature* 21 Dec. 897/2 An extremely active development department using a large number of pre-production machines. 1959 *Times* 19 Feb. 3/4 Testing of valves up to pre-production stage. 1959 P. Bull *I Know Face* iv. 66 Other pre-production costs included advance payment to producer. 1967 *Jane's Surface Skimmers Systems* 1967–68 133/1 The design team..is based in a factory at Zaporozhe,..where all prototypes and pre-production engines..are developed and built. 1970 M. Torme *Other Side of Rainbow* (1971) iii. 34 Nearly every television variety show operates in a three-phase pattern. First, there is a preproduction period in which songs are chosen, scripts are written, [etc.]. 1973 *Times* 14 Nov. 2/5 Noise measurements on the 20 preproduction aircraft had been disappointing. 1977 *Engin. Materials & Design* Aug. 17/1 The results obtained were comparable with those expected from a pre-production run of conventionally designed radiators.

pre-profe·ssional, *a.* and *sb.* [Pre- B. 1 in Dict. and Suppl.] **A.** *adj.* Prior or preliminary to professional training.
1948 *Mind* LVII. 387 University instruction in psychology should serve..as a preparation for other fields (e.g. law, medicine, engineering), in the form of pre-professional courses. 1975 *N.Y. Times* 21 Dec. IV. 12/4 New professional (a currently popular term in education) programs, such as Mills's prelaw 'Administration and the Legal Process'.
B. *sb.* One who is training for a profession.
1970 *Jrnl. Gen. Psychol.* LXXXIII. 140 When I speak to audiences, I find that the myth is still believed firmly, even by preprofessionals. 1977 *Early Music* Apr. 263/1 The conference was designed to draw together professionals and college-age pre-professionals.

prepro·gram, *v.* Also **preprogramme.** [Pre- A. 1.] *trans.* To program (a computer or calculator) beforehand. Also *transf.* and *fig.* So **prepro·grammed** *ppl. a.*, **prepro·gramming** *vbl. sb.* (*U.S.* also -gramed, -graming).
1964 *Economist* 16 May 746/2 Ships operated entirely by pre-programmed computer are practicable. 1965 H. Kahn *On Escalation* viii. 166 Deterioration in international relations will provide a pre-programmed crash defense program. 1970 *Computers & Humanities* IV. 355 Produce a number of specific programs, which use the experience gained to develop appropriate software for combined interactive and preprogrammed modes of operation. 1971 J. Z. Young *Introd. Study Man* p. viii, His brain may indeed be pre-programmed to operate in this way. 1973 P. Evans *Bodyguard Man* xvii. 110 It was as though his thought-processes had been pre-programmed, as though this was a situation that he had foreseen. 1974 Hawkey & Bingham *Wild Card* xv. 129 Because of the preprogramming..all I have to do is transmit the starting instructions. 1977 J. D. Douglas in Douglas & Johnson *Existential Sociol.* i. 60 Even the simplest of human activities, such as walking down the street, cannot be preprogrammed without danger of catastrophe. 1977 *Sci. Amer.* Apr. 94/1 (Advt.), A series of preprogrammed hand-held calculators that virtually revolutionized numerical data processing.
Hence as *sb.*, an already existing program.
1971 J. Z. Young *Introd. Study Man* p. vii, The whole structure of our language and thought is limited by a preprogramme in the organization of the brain.

preproinsulin: see *pre- B. 1.

prepster (pre·pstǝɹ). *U.S.* [f. Prep *sb.* and *a.* + -ster.] (See quot. 1965.)
1965 *Amer. Speech* XL. 164 A *Prepster* not only denotes a preparatory student in collegiate slang but also a trainee. 1974 *Sumter (S. Carolina) Daily Item* 22 Apr. 14 His credentials—23 points per game and a fantastic field goal average of 60 per cent, plus nine rebounds per game and seven assists per outing as a senior—as a prepster.

prepsycho·tic, *a.* (*sb.*) *Psychol.* [Pre- B. 1.] Of or relating to symptoms, or to the period of time, prior to the onset of a psychosis. Also as *sb.* Hence **prepsycho·tically** *adv.*
1927 Henderson & Gillespie *Text-bk. Psychiatry* ix. 218 Are there certain pre-psychotic traits which indicate a severe grade of disturbance in the event of a psychosis developing? 1931 *Internat. Jrnl. Psycho-Anal.* XII. 298 (*title*) Ego defence and the mechanism of oral ejection in schizophrenia: the psycho-analysis of a pre-psychotic infant. 1935 *Brit. Jrnl. Med. Psychol.* XV. 140 By 'prepsychotic' I mean someone who is showing ominous signs of an impending psychosis but who is not yet frankly psychotic. 1941 *Jrnl. Amer. Med. Assoc.* 16 Aug. 522/1 A severe hypotension, which existed prepsychotically as well as during the course of the psychosis, has been improved. 1959 H. Nielsen *Fifth Caller* iii. 48 There was a tragic case—a man, prepsychotic, who needed expert attention. 1977 *Time* 4 Apr. 41/2 Tom Verlaine..delivers raw, jabbing vocals in a declamatory, prepsychotic style similar to Patti Smith's. 1977 *Lancet* 27 Aug. 449/1 There are no doubt prepsychotics who abuse alcohol.

prepuberal(ly, -pubertally: see *pre- B. 1.

prepu·berty. [Pre- B. 1 in Dict. and Suppl.] The period of life preceding puberty, esp. the two or three years immediately before. Also *attrib.*
1922 R. T. Frank *Gynecol. & Obstetr. Path.* iv. 74 Prepuberty.—Practically it begins with the time at which budding into womanhood is first noticed and ends with the onset of the first menstruation. *Ibid.* 75 In some instances..in which the menstruation is delayed or remains in abeyance, although ovulation occurs, the prepuberty and the puberty stage cannot be differentiated clinically. 1933 *Amer. Jrnl. Dis. Children* XLIII. 329 We firmly believe that there is a prepuberty rise in basal metabolism. 1941 *Jrnl. Pediatrics* XIX. 291 Adolescence lasts from the time of puberty until about 21 years. Today we are not discussing this whole subject, only certain phases of it and perhaps particularly that part of adolescence which we might call prepuberty and puberty. 1949 M. Mead *Male & Female* xiii. 280 For the old institutionalized hostilities of..girls' teen and boyish pranks, the new pre-puberty dating pattern is being substituted. 1976 *Times* 1 Sept. (Fashion Suppl.) p. vii/1 All three categories of clothing—male, female and pre-puberty.

prepube·scent, *a.* [Pre- B. 1.] = *pre-pubertal* adj. s.v. Pre- B. 1 in Dict. and Suppl.
1904 G. S. Hall *Adolescence* I. p. x, Rousseau would leave prepubescent years to nature..and allow the fundamental traits of savagery their fling till twelve. 1932 R. F. Fortune *Sorcerers of Dobu* ix. 276 The pre-pubescent girls of his own age have lain with boys older than he. 1965 F. Sargeson *Mem. Peon* vi. 174 A pre-pubescent male was..demanding..'that you leave my mother alone'. 1978 J. Hyams *Pool* v. 61 This prepubescent midget has got the hots for me.
Hence **prepube·scence** = *prepuberty.
1916 *Arch. Internal Med.* XVII. 887 We may consider boys in the period of prepubescence as individuals of adult form but of small size, growing rapidly, and as yet scarcely influenced by the internal secretions of the sex glands. 1950 *Psychiatry Q.* XXIV. 495 It would appear that any study of attitudes towards death during prepubescence or adult life, would throw further light on the psychodynamics of human behavior. 1977 *N.Y. Rev. Bks.* 29 Sept. 13/1 During Baum's prepubescence the Civil War took place.

prepubic, *a.* (Further examples.)
1918 *Bull. Amer. Mus. Nat. Hist.* XXXVIII. 331 In the ornithischian dinosaurs the expanded prepubic process appears to have served as a base for a forward extension of the pubic-ischio-femoralis. 1934 *Anatomischer Anzeiger* LXXVIII. 44 The prepubic skeletal element [of *Ascaphus truei*] is cartilaginous. 1954 D. & M. Webster *Compar. Vertebr. Morphol.* v. 102 In prototherian and metatherian (but not eutherian) mammals there is a fourth skeletal element on each side [of the pelvic girdle], called the marsupial or prepubic bone.

prepubis. Add: **1.** (Later example.)
1956 A. S. Romer *Osteol. Reptiles* vii. 308 In this case [sc. the pterosaur skeleton] the majority opinion is that the true pubis is included in the plate and that this anterior element is a neomorph, a prepubis.
2. = *epipubis.
1931 G. K. Noble *Biol. Amphibia* x. 240 A prepubis may have been a primitive character of modern Amphibia. 1969 *Nature* 14 June 1091/1 Here I report the presence of the so-called marsupial bone (prepubis or epipubis) in the skeleton of *K[ryptobaatar] dashzevegi*. This bone has never before been found in fossil Multituberculata material. 1975 *Nature* 26 June 698/1 Marsupial bones (the epipubes, prepubes) occur in Marsupialia, Monotremata and have also been found in Cretaceous Multituberculata.

prepublica·tion, *a.* and *sb.* [Pre- B. 2.] **A.** *adj.* Produced, issued, or occurring in advance of publication.
1932 F. Scott Fitzgerald *Let.* 6 Feb. (1964) 332 A pre-publication review which would..generate information destined..to hurt the sale of my book. 1936 *Amer. Speech* XI. 372/1 The prepublication announcement and the blurb on the jacket of his recent book. 1962 F. Bowers *Bibliogr. & Textual Crit.* iii. iv. 78 It constituted a pre-publication state decided on before public sale could have offered any opportunity for objections. 1977 *Time* 13 June 62/1 Seldom has an anthology of critical essays aroused so much prepublication anxiety as Diana Trilling's *We Must March My Darlings*.
B. *sb.* [Pre- A. 2.] The action or fact of publishing beforehand; publication in advance.
1958 *Nature* 30 Apr. 547/1 In the opinion of the editor... an article about the paper that had appeared in *Medical World News* contained so much of its substance..as to constitute prepublication and disqualify its claims for space in a journal that aims to publish original material. 1975 P. Harcourt *Fair Exchange* ii. iii. 105 This book... where did you get it?..you realize it's pre-publication?

prepublicity, -publish: see *pre- A. 2, 1. **prepulse:** *pre- A. 2.

prepu·nched, *ppl. a.* [Pre- A. 1.] Of a card or the like: having holes already punched in it. Of information: already stored or represented as a pattern of holes. So **prepu·nch** *v. trans.*
1953 *Proc. IRE* XLI. 1274/1 While at his desk the programmer may then assemble into a single deck of cards some prepunched library programs together with cards especially punched for the problem to be solved. 1957 *Practical Wireless* XXXIII. 705/1 Pre-punched holder plates take such focal components as valves, etc. 1964 T. W. McRae *Impact of Computers on Accounting* ii. 47 We are likely to see a great deal more of inter-business documentation in the form of machine-sensible input, such as sending a purchase order as a pre-punched card. 1965 *New England Jrnl. Med.* CCLXXII. 1221/2 The clerk..obtained a deck of data-processing cards that had been prepunched with the patient's file number.

prepupa (prīpiū·pă). *Ent.* [Pre- B. 1 in Dict. and Suppl.] An insect late in its larval development, during a relatively quiescent phase in which preparations for the transformation into a pupa take place; also, in certain beetles, a distinct instar preceding the pupa stage.
1925 A. D. Imms *Gen. Textbk. Entomol.* 186 The prepupa represents a greatly abbreviated instar. 1955 P. A. Buxton *Nat. Hist. Tsetse Flies* xi. 380 Generally speaking one measures the duration of life of the puparium in an inclusive way, from the time when the larva becomes immobile, through the life of prepupa and pupa, to the emergence of the adult. 1959 E. F. Linssen *Beetles Brit. Isles* II. 85 After this third moult..it [sc. a mould beetle] turns into what is called a pre-pupa. 1975 *Nature* 17 Apr. 592/2 In such specimens..there is no visible response to background, whether the caterpillar or prepupae are exposed to long or short hours of daylight.

prepupal (prīpiū·păl), *a. Ent.* [Pre- B. 1.] Immediately preceding a larval insect's change into a pupa.
1906 J. B. Smith *Explanation Terms Entomol.* 108 Prepupal: that stage in the larva just preceding change to the pupa. 1925 A. D. Imms *Gen. Textbk. Entomol.* 185 A brief period of quiescence follows which represents the prepupal instar. 1933 *Jrnl. Ecol.* XXI. 131 The excellent photographs..illustrate the adult sawflies..and the larval, prepupal and pupal stages. 1971 Borror & DeLong *Introd. Study Insects* (ed. 3) iv. 69/1 The other changes begin in the prepupal stage of the last larval instar.

preputial, *a.* (Later examples.)
1971 [see *fixative *sb.* 3]. 1976 *Lancet* 20 Nov. 1107/2 Culture of swabs from the preputial sac..and from the lesions of two patients, suggested that the source of infection in boys is the prepuce or urethra rather than the bowel as in girls.

prequalification, -ative: see *pre- A. 2, B. 2a. **prequalify:** *pre- A. 1.

prequel (prī·kwĕl). [f. Pre- A. 2 + Se)quel *sb.*] A book, film, etc., the events portrayed in which or the concerns of which precede those of an existing completed work.
1973 *Britannica Bk. of Year* 1973 732/3 *Prequel*, a literary work whose narrative sequentially precedes that of an earlier work. 1977 *National Observer* (U.S.) 1 Jan. 1/4 Cammer..his plan to write a book, *Freedom from Compulsion*... He calls it a 'prequel' to his earlier book, *Up from Depression*. ''Prequel' is a word I coined', he explains. 'It's a sequel except it's on a subject that comes before.' 1977 *Globe & Mail* (Toronto) 17 Sept. 37/5 The Silmarillion, for which Tolkien coined the term Prequel, describes not only the creation of Middle Earth, but of the universe. 1979 *Films & Filming* Mar. 11 In this 'prequel' Tom Berenger stars as Butch Cassidy and William Katt as Sundance.

prerabbinical, -radio: see *pre- B. 1d, 2a.

pre-Raph. (prīræ·f), *a.* (*sb.*) *Colloq.* abbrev. of Pre-Raphaelite *sb.* and *a.* (senses A. 1 and B. 1). Hence **pre-Ra·phly** *adv.*
1874 L. Troubridge *Life amongst Troubridges* (1966) 80 A new rage..painting the panels of the shutters of our bedrooms,..you can't think how pre-Raph. they look. Pale blue ground..and in each panel drops a flower, of course very pre-Raphly done. 1944 J. Lees-Milne *Prophesying Peace* (1977) 54 [illegible] William [Morris] planted these? 1960 *Listener* 27 Aug. 312/1 The charms of Laura Troubridge written when she was 'a madly Pre-Raph' young Victorian. 1975 *Times* 25 Sept. 13/5 William Gaunt's brilliant little 1940s book at that time contained all that most of us knew..about the Pre-Raphs.

pre-Raphaelism. (Earlier example.)
1852 Geo. Eliot *Let.* 24–25 July (1954) II. 48 The British Quarterly has one subject of which I am jealous—'Pre-Raphaelism in Painting and Literature'.

pre:-Raphaeli·tically, *adv.* [f. Pre-Raphaelitic *a.* + -al + -ly².] In a manner suggestive of the pre-Raphaelites.
1893 G. B. Shaw *Let.* 1 Mar. (1965) I. 490 You are glad to..come back PreRaphaelitically to Giotto again. 1927 *Glasgow Herald* 1 July 4/6 The drabness of Arnold Bennett's pre-Raphaelitically accurate Five Towns.

pre-Raphaelitish, *a.* (Earlier example.)
1854 A. Thackeray *Jrnl.* Feb. in H. Ritchie *Lett. A. T. Ritchie* (1924) v. 61 Mr. Millais was there, a tall good-looking Pre-Raphaeliteish young man.

pre-Raphaelitism. (Further examples.)
1974 J. Christian *Pre-Raphaelites in Oxford* 5 In the history of Pre-Raphaelitism nowhere played a more important part than Oxford. 1978 *Bodl. Libr. Rec.* X. 52 In view of Oxford's strong connections with Pre-Raphaelitism, it is doubly gratifying to be able to record that the Bodleian Library has proved a comparatively rich quarry.

pre-ra·tional, *a.* [Pre- B. 1.] Intuitive, instinctive, based on mental processes more primitive than reason.
1903 C. A. Strong *Why Mind has Body* xi. 274 Not reasoning, but some deep pre-rational instinct..is the basis of our belief in other minds. 1919 M. K. Bradby *Psycho-Anal.* iii. 39 In our intuitive pre-rational unconsciously based convictions we shall rediscover the beliefs of primitive man. 1921 E. Sapir *Language* ii. 39 We have been assuming that the material of language reflects merely the world of concepts and, on what I have ventured to call the 'pre-rational' plane, of images, which are the raw material of concepts. 1936 Wirth & Shils tr. *Mannheim's Ideology & Utopia* iii. 108 A mode of thought is thus created which conceives of history as the reign of pre- and super-rational forces. 1941 *Mind* L. 378 It is, however, pertinent to one central topic, morality as pre-rational, rational, and post-rational. 1948 P. Tillich *Protestant Era* p. xxiv, The trend of the younger generation in Europe toward the vital and prerational side of the individual and social life. 1957 J. S. Huxley *Relig. without Revelation* (rev. ed.) iii. 52 This pre-rational phase of individual mental life.

prereaction: see *pre- A. 2.

pre-rea·der. [Pre- B. 1 in Dict. and Suppl.] **a.** A book designed for students who cannot yet read. **b.** A person who cannot yet read. Also **pre-rea·ding** *ppl. a.* and *vbl. sb.*
1965 *Language* XLI. 548 There is no reason to believe that prereading training on Latin phonology would be of any benefit. 1966 J. Derrick *Teaching Eng. to Immigrants* ii. 104 A set of pictures is published for use at this pre-reading stage. *Ibid.* v. 182 There is preparatory work even to this—work which may be called 'pre-reading' and which is aimed at familiarizing the pupils with the notion of reading and with some of the musculo-sensory activities that are involved in the whole process. *Ibid.* 183 Many elementary courses do in fact include a pre-reader or picture book which can be used with the class if pupils' copies are available. 1972 *Sci. Amer.* Dec. 114/3 Prereaders will like the action just fine; small sophisticates in the early grades will find the challenge interesting. 1976 *Word* 1971 XXVII. 501 This spontaneously developed ability of the prereader has been noted only sporadically.

prerecognition: see *pre- B. 2a.

pre-reco·rd, *v.* [Pre- A. 1.] *trans.* To record for subsequent use, esp. in film-making and broadcasting. So **pre-reco·rded** *ppl. a.* (*prerecorded tape*, magnetic tape on which sound has been recorded prior to its sale); **prereco·rding** *vbl. sb.*
1937 M. Steiner in N. Naumburg *We make Movies* xiv. 233 Pre-recording means pre-scoring, pre-playing with an orchestra, piano, or whatever is required of the song or dance number to be used in the picture. 1941 *B.B.C. Gloss. Broadcasting Terms* 24 *Pre-Record* (*v. trans.*), to record a programme on a closed circuit for subsequent reproduction. 1954 *Newsweek* 11 Oct. 55/2 In TV, as in the movies, it is not unusual to pre-record musical numbers, but this is generally done a few days before the performance. 1958 *Sunday Times* 3 Aug. 3/6 Now that the station has bought a new record-player and gets all its music in prerecorded tapes, the personal touch is all but gone. 1962 *Times* 5 July 15/4 Pre-recorded tapes have been issued in some quantity in Great Britain. 1965 *Listener* 30 Dec. 1087/3, I can only say that pre-recording again ruined what might have been a delightful three-quarters of an hour. 1972 *Daily Tel.* 6 Jan. 1/2 Mr Wilson, Leader of the Opposition, visited a television studio yesterday morning to pre-record his contribution to the programme. 1978 *N.Y. Times* 30 Mar. B22/6 (Advt.), The synchronized cassette player lets you make perfect prerecorded presentations. 1978 *Lancashire Life* Sept. 131/1 Few people feel that the pre-recorded cassette is of comparable quality to the long playing record.

preredu·ction. *Genetics.* [ad. G. *praereduction* (Korschelt & Heider *Lehrb. der Vergleichenden Entwicklungsgeschichte der Wirbellosen Thiere* (Allgemeiner Theil) (1903) II. vi. 586): see Pre- A. 2 and Reduction.] **a.** Reduction of chromosome number at the first of the two meiotic cell divisions, rather than at the second. Opp. *postreduction a.
1905 *Proc. Acad. Nat. Sci. Philadelphia* LVII. 186 He has to assume a complex axial metamorphosis, which is wholly unnecessary on the basis of a prereduction. 1915, 1921 [see *postreduction a].
b. Separation of homologous chromatids or genes at the first of the two meiotic cell divisions, rather than at the second. Opp. *postreduction b.
1934, etc. [see *postreduction b].
Hence **preredu·ctional** *a.*, involving or pertaining to prereduction; **preredu·ctionally** *adv.*
1905 *Proc. Acad. Nat. Sci. Philadelphia* LVII. 187 He describes a prereductional division of the bivalent chromatin nucleolus. *Ibid.* 195 Whenever the heterochromosomes occur in pairs in the spermatogonia.. their univalent components become separated in the first maturation mitosis, *i.e.*, divide prereductionally. 1950 [see *postreduction b]. 1950 [see *postreductionally *adv.*].

pre-reflective, -reflexive: see *pre- B. 1.

pre-re·gistered, *ppl. a.* [Pre- A. 1.] Registered in advance, *spec.* in *Printing*, brought into alignment or coincidence beforehand (cf. Register *v.* 4 b).
1967 Karch & Buber *Offset Processes* vi. 219 For multicolor work, cross-hair register marks sight the original material in precision-notched and pre-registered frames. 1973 *LSA Bull.* Mar. 24 The 13 countries with the largest numbers of pre-registered members (all with 20 or more)... The U.S.A...had 189 pre-registered participants.

preregistra·tion, *a.* and *sb.* **A.** *adj.* [Pre- B. 2.] Of or pertaining to the period of a doctor's training between qualification and registration (cf. *houseman 6).
1922 *Brit. Med. Jrnl.* 2 Sept. 423/2 After January 1st, 1923, prospective medical students will also be required to pass a pre-registration examination in chemistry and physics. 1962 'D. Margerson' *Med. as Career* x. 67 House officer posts, including the two obligatory pre-registration appointments, will usually take up eighteen months to two years. 1964 *Lancet* 5 Sept. 519/2 If..the two-year clinical course were followed by graduation and two years' preregistration house-appointments, hospitals in the 'network' would get preregistration doctors a year earlier than they do now. 1977 *Daily Tel.* 22 July 6 (Advt.), You'll get £2,636 to continue your normal medical studies. And £4,429 during your pre-registration period.
B. *sb.* [Pre- A. 2.] Registration in advance, *spec.* in *Printing*, the action of bringing into register in advance.
1967 E. Chambers *Photolitho-Offset* vi. 68 Loading and pre-registration are carried out on a separate appliance, thereby enabling subjects to be prepared in advance. 1973 *LSA Bull.* Mar. 24 Pre-registrations for the Congress.

pre-rehearsal: see *pre- A. 2. **pre-relativity:** *pre- B. 2a. **pre-relativistic:** *pre- B. 1.

pre-relea·se, *a.* (*sb.*) [Pre- B. 2.] **1. a.** Designating a period before the date fixed for release of a film.
1927 *Glasgow Herald* 15 Nov. 9/7 An amendment..providing that pre-release cinema shows should take place in provincial centres as well as in London was agreed to without a division. 1928 *Daily Express* 9 July 9 There is a pre-release presentation of Dolores del Rio in 'The Gateway of the Moon', at the Kensington Kinema. 1973 *Times* 11 May 2/6 The film..had only been booked for five prerelease test runs at cinemas in Leeds. 1976 L. St. Clair *Fortune in Death* vii. 68 He's made..a Near East thriller. They wanted some pre-release publicity.
b. Of or pertaining to the period before release of a suspect or prisoner.
1958 F. Norman *Bang to Rights* iii. 106 You have to atend [*sic*] what they call a prerelease course. 1959 *New Statesman* 7 Mar. 335/2, I am very glad that Critic called attention to the scheme for pre-release leave to selected prisoners. 1961 *Lancet* 22 July 204/1 Such measures as the open prison and the pre-release hostel were proving invaluable. 1971 R. Cross *Punishment, Prison & Public* ii. 68 Attached to some prisons are pre-release hostels from which the inmates go out to daily regular work. 1976 *Newmarket Jrnl.* 16 Dec., The first mention of any possible charge relating to drink-driving came after he had given a blood sample and a pre-release breath test.
2. *ellipt.* as *sb.* A film or record given restricted availability before being generally released.
1929 *Sunday Dispatch* 13 Jan. 16/3 We, in London, have been privileged to view many pre-releases. 1978 *Sunday Times* 29 Jan. 43/3 'Top Ranking' started life in Britain as a status single, available only as an imported 'pre-release'.

pre-relea·se, *v.* [Pre- A. 1.] *trans.* To release beforehand.
1968 *Punch* 24 Apr. 589/1 There's a lot to be said for pre-releasing the decimal coinage at the lower end of the scale. *a* 1974 R. Crossman *Diaries* (1975) I. 147 We had pre-released the news to the *Daily Express*, the *Guardian* and the *Evening Standard*. 1976 *Church Times* 2 Apr. 10/1 He persistently declines to extend to the Press that assistance (such as circulating in advance scripts of major speeches, or sticking to the text of speeches thus pre-released) which so greatly facilitates newspaper production.

pre-reproductive: see *pre- B. 1.

prerequire, *v.* (Later example.)
1975 *Christian* II. 228 There are two levels of psychotherapy, the second prerequiring the first.

pre-retirement, -revolution: see *pre- B. 2a.

pre-revolu·tionary, *a.* and *sb.* [Pre- B. 1 in Dict. and Suppl.] **A.** *adj.* **1.** Existing before a (particular) revolution.
1861 [in Dict. s.v. Pre- B. 1]. 1867 H. W. Beecher in *N.Y. Ledger* 7 Sept. 2/3 Planted in 1646, it was more than a hundred years old when the pre-revolutionary excitements were taking place in Boston. 1874 T. B. Aldrich *Prudence Palfrey* x. 166 Since the hanging of a witch or two in pre-revolutionary days, the office of sheriff there had been virtually a sinecure. 1936 *Burlington Mag.* Mar. 130/2 Baroque paintings of the pre-Revolutionary decades in France. 1961 *Times* 29 Dec. 11/1 Pre-revolutionary St. Petersburg. 1976 *New Yorker* 15 Nov. 213/1 Some extraordinarily rich evocations of pre-revolutionary village life in China.
2. Of a society or its condition: verging on social or political revolution.
1964 R. D. Hopper in I. L. Horowitz *New Sociol.* xix. 313 In pre-revolutionary societies, there is formed a group that is marginal to the structure of political power and social prestige. 1972 'H. Buckmaster' *Walking Trip* 85 It is a prerevolutionary situation I'm told by a white politician in Salisbury, and he thinks the next step is violence. 1974 *Listener* 24 Jan. 108/3 An artist in a pre-revolutionary situation concentrates upon actually producing art. 1975 A. Beevor *Violent Brink* vii. 165 England, the despair of Marx, seemed at last about to move into a pre-revolutionary situation.
B. *sb.* One who prepares the way for a revolution.
1937 *Times Lit. Suppl.* 1 May 323/2 Mr. Stephen Spender's 'The Destructive Element', with its presentation of Henry James as a master pre-revolutionary.

pre-rinse: see *pre- A. 2. **pre-romanesque:** *pre- B. 1d.

pre-Romantic, *a.* (*sb.*) *Mus.* and *Lit.* Also **pre-romantic.** [Pre- B. 1 in Dict. and Suppl.] Pertaining to or characteristic of the period before the Romantic Movement. As *sb.*, a composer or writer of that period.
1934 C. Lambert *Music Ho!* i. 57 Purcell, the most picturesque of the pre-Romantic composers. 1938 C. Connolly in *New Statesman* 6 Aug. 223/1 The English pre-romantics..balanced their love of childhood by their hope of heaven. 1947 A. Einstein *Mus. Romantic Era* xi. 134 One can find examples of it, particularly in the opera, even in pre-Romantic opera. 1959 *Brno Studies in English* I. 104 Bulwer attempted to follow in the steps of the great representatives of the English pre-romantic and romantic period, especially of William Godwin and Lord Byron. 1962 *Times* 16 Feb. 15/3 Mozart is observed..as the clever pre-romantic tune-spinner. 1963 N. Frye *Romanticism Reconsidered* 9 It is obvious that in pre-Romantic poetry there is a strong affinity with the attitude that we have called sense... But the pre-Romantic structure of imagery belonged to a nature which was the work of God. 1978 D. Grylls *Guardians & Angels* iv. 112 In her treatment of parent-child relations Jane Austen is pre-Romantic. 1980 *Church Times* 25 July 6/4 The enormous revolution in literary taste which began in the 'twenties..demoted Spenser, the tribe of Ben Jonson, and the eighteenth-century pre-romantics.

pres: see *prez¹˒ ².

‖ **pré salé** (pre sale). Also **pré-salé.** [Fr.] A salt meadow, *spec.* one on which sheep are reared; freq. used *attrib.* or *ellipt.* to designate the flesh of sheep reared on a salt meadow.
1839 F. A. Kemble *Let.* 21 July in *Rec. Later Life* (1882) I. 255 That peculiar close short turf which creates South Down and Pré Salé mutton. 1903 J. M. Falkner *Nebuly Coat* i. 3 The low-lying meadows..where as good-tasting mutton is bred as on any *pré-salé* on the other side of the Channel. 1930 A. Bennett *Imperial Palace* xxiii. 145 It was a man's menu... Turtle soup. Sole *Palace*. Pré-salé with two vegetables. 1935 M. Morphy *Recipes of All Nations* 54 It is an error to imagine that French mutton and lamb are inferior to English meat. Their *pré-salé* is equal to any. 1966 P. V. Price *France: Food & Wine Guide* 48 The most vaunted type of lamb comes from flocks that feed on pastures by the sea, which gives the meat a special delicacy; this type is known as *pré salé* (salty meadow) and has an *Appellation Contrôlée*.

presanctify, *v.* Add to def.: Said also in some Anglican churches on Good Friday. (Earlier and later examples.)
1758 S. Redford *Important Inquiry* (ed. 2) App. 397 They offer up and shew the people the Sacrament reserved on those two solemn days, which they call the *imperfect* Mass, or the Mass of the *presanctified*. *a* 1773 A. Butler *Feasts Catholic Church* (1774) vi. iv. 355 This is called the 'Mass of the Pre-sanctified Mysteries, *Missa præsanctificatorum*'. 1839 *Penny Cycl.* XIV. 57/2 In the Greek or Constantinopolitan church three Liturgies are in use, those of Basil, Chrysostom, and the Liturgy of the Præ-sanctified. 1909 *Daily Chron.* 10 Apr. 5/4 At Westminster Cathedral yesterday morning the Mass of the Presanctified was solemnised in the presence of Archbishop Bourne. 1957 *Oxf. Dict. Chr. Ch.* 1101/1 In the Middle Ages all present at the Mass of the Presanctified received Holy Communion. 1965 C. E. Pocknee *Parson's Handbk.* (ed. 13) 156 In some churches the Mass of the Pre-sanctified has been reintroduced with a general Communion from the reserved sacrament. 1978 C. Jones et al. *Study of Liturgy* 409 Communion from the reserved sacrament, the so-called Mass of the Pre-Sanctified.

presbycousis (prezbikū·sis). *Med.* Also presby(a)cusis, -ac(o)usia (-ăkū·ziă), -kousis. [mod.L., f. Gr. πρέσβυς an old man (cf. Presbyopia) + ἄκουσις hearing (ἀκούειν to hear).] Loss of acuteness of hearing due to age.
1890 Billings *Med. Dict.* II. 386/1 Presbykousis. 1892 F. P. Foster *Med. Dict.* IV. 2640/2 Presbycusis. 1896 *Syd. Soc. Lex.*, Presbycousis. 1900 *Lancet* 18 Aug. 538/2 To this last belonged the progressive deafness which came on between the ages of 40 and 50 years, and which had been called presbyacousia. 1911 Stedman *Med. Dict.* 703/1 Presbyacusia. 1958 *Times Rev. Industry* Aug. 13/1 Gradual blunting of auditory acuity with the increasing age of the worker..is known as 'presbycusis'. 1970 *Brit. Med. Jrnl.* 30 May 524/2 It is possible that the high level of everyday noise in industrial countries plays some part in the development of presbyacusis. 1976 *Listener* 2 Sept. 279/1 Presbycousis has not yet set in. Most people over 30, said Jeremy Bugler,..lose some of their hearing.

presbyope. Delete *rare*⁻⁰ and add examples.
1880 L. Owen tr. *Giraud-Teulon's Elem. Treat. Function of Vision* ii. ii. 30 The presbyope presents himself generally under the following aspect: he has always enjoyed excellent distant vision... But he is drawing near to forty-five or fifty and begins to experience a certain difficulty in reading small type. 1900 C. H. May *Man. Dis. Eye* xxiv. 332 The presbyope is compelled to hold reading, writing, sewing, and other forms of near work farther away than the usual distance, making such efforts uncomfortable. 1937 *Jrnl. Optical Soc. Amer.* XXVII. 332/1 Presbyopes should invariably wear their reading correction (reading spectacles). 1974 *Nature* 25 Oct. 729/2 It is known that convergence can bring about accommodation; and it seems likely that in a presbyope, devoid of accommodative ability, convergence would still be accompanied by a central command for accommodation.

presbyteress. Restrict † *Obs.* to sense 1 in Dict. and add: **2.** (Later example.)
1901 J. Wordsworth *Ministry of Grace* v. 271 In these [*sc.* the 'Didascalia' and the 'Constitutions'] and similar books the elders Widows are sometimes mentioned under the title of πρεσβύτιδες, a name for which we have no nearer equivalent than the somewhat ambiguous and inexact 'Presbyteresses'... The Virgins, Widows and Presbyteresses have the first place among the women in church.

presca·ler. *Electronics.* [Pre- A. 2.] A scaling circuit employed to scale down the input to a counting or other scaling circuit so that it can deal with high counting rates.
1954 E. H. W. Banner *Electronic Measuring Instruments* xi. 277 When very short resolving times are essential, it is necessary to use a hard-valve pre-scaler in front of a scaler using these new tubes. 1967 *Electronics* 6 Mar. 152 (*caption*) The counter is used in pairs as a prescaler. 1972 *Physics Bull.* Feb. 114/1 A new high performance, low priced scaler timer for use in schools and technical colleges... It features a fast (1 μs) prescaler of integrated circuit construction with a bold two digit read out of counts. 1976 *CB Mag.* June 72/2 Because the VHF prescaler is built-in, the decimal point is placed properly on the display. 1977 *Design Engin.* July 81/3 A prescaler extends this range to 800 MHz.

preschizophrenia, -ic: see *pre- B. 1.

preschool, *a.* and *sb.* **A.** *adj.* [Pre- B. 2.] (Stress variable.) Of or pertaining to the time before a child is old enough for school.
1924 *Jrnl. Amer. Med. Assoc.* 5 Jan. 1/1 We weigh our preschool children nude on their birthdays. 1934 [see *attentive *a.* 1 b]. 1946 *Nature* 23 Nov. 737/1 They [*sc.* defects] develop during the pre-school age, and accurate knowledge concerning them is lacking. 1958 *Word* XIV. 170 In this experiment, preschool and first grade children ..were presented with a number of nonsense words. 1960 *Guardian* 29 Apr. 6/4 Dr Tizard suggested the opening of pre-school centres for mentally handicapped children. 1979 *Dædalus* Spring 76 Participation rates for mothers with children of preschool age increased from 18 to 27 percent between 1966 and 1976.
B. *sb.* [Pre- B. 1 in Dict. and Suppl.] (Stressed *pre-school.*) A kindergarten or nursery school for children of preschool age.
1934 in Webster. 1937 S. V. Benét *Thirteen o' Clock* (1938) 275 Nope, that doesn't work any more, what with pre-schools, automats and movies. 1958 *Word* XIV. 159 The oldest children at the Preschool were five years old. 1966 Bereiter & Engelmann *Teaching Disadvantaged Children in the Preschool* i. 3 This is a 'successful' preschool for disadvantaged children. 1977 *Caravan World* (Austral.) Jan. 67/3 Sue drops their young son at pre-school every morning.
Hence **pre-schoo·ler,** a child who is too young to attend school; a child who attends preschool; **pre-schoo·ling** *vbl. sb.*, the education of a preschool child.
1954 [see *-schooler]. 1958 *Word* XIV. 159 First graders did significantly better than preschoolers. 1965 *Economist* 4 Sept. 883/1 Even as pre-schoolers, these children do not lack experiences. 1971 *Daily Colonist* (Victoria, B.C.) 28 Aug. 18/1 Sesame Street, for pre-schoolers, to date has gotten few brickbats. 1972 *Guardian* 11 Aug. 12/1 Preschooling is the outstanding economical and effective device in the general approach to raising educational standards in EPAs. 1974 *Ibid.* 19 Mar. 20/2 The schools would be there to offer gipsy children preschooling opportunities. 1979 *Guardian* 5 June 11/3 Pre-schoolers should know their full name by 2½.

presciencelessness (pre·siĕnslĕsnĕs). *rare*⁻¹. [f. Prescience + -less + -ness.] The state or condition of lacking prescience.
1928 Hardy *Winter Words* 5 Led by sheer senselessness And presciencelessness Into unreason.

pre-score, *v.* [Pre- A. 1.] *trans.* To score or inscribe beforehand.
1977 *Engin. Materials & Design* Aug. 29/3 A diamond tool to scribe the fired material, or a system that prescores the ceramic alumina in its green state.

pre-sco·ring, *vbl. sb. Cinemat.* [Pre- A. 1.] The recording of a sound track in advance of the shooting of the film it is to accompany.
1937 *Jrnl. Soc. Motion Picture Engineers* XXIX. 356 We do not record songs or orchestras on the set... We record them in advance, usually before the picture goes into production. This we call pre-scoring. 1937 [see *pre-record *v.*]. 1948 E. Lindgren *Art of Film* i. ii. 31 The flexibility of sound-recording methods is well illustrated by the device known as pre-scoring. 1949 B. Woodhouse *From Script to Screen* vii. 111 The principle known as pre-scoring is used extensively.

prescree·n, *v.* [Pre- A. 1.] *trans.* To screen (in any sense) beforehand. Hence **prescree·ned** *ppl. a.*, **prescree·ning** *vbl. sb.*
1967 Karch & Buber *Offset Processes* v. 167 Although no claim is made that the use of this screen will improve the quality of the picture, this new technique will reduce the cost of prescreening the photographs and then stripping the half-tone negative into the flat. 1967 E. Chambers *Photolitho-Offset* xi. 174 With a prescreened film the contact screen is absent, and..the sensitivity of the film itself is varied in the form of a dot pattern. 1977 *Time* 20 June 48/2 The microwave detector could at the very least be used for pre-screening women—especially those under 35 who are ordinarily not encouraged to have mammograms unless they have a family history of breast cancer or symptoms of the disease. 1977 *Jrnl. R. Soc. Arts* CXXV. 241/1 Careful prescreening for human health hazards will have taken place. 1978 *New York* 3 Apr. 84/1 College Background Singles Only—Newsletter of pre-screened, higher quality singles events plus our own tennis and theatre parties.

prescribable (prĭskrai·băb'l), *a.* [f. Prescribe *v.* + -able.] That can or may be prescribed; capable of being prescribed.
1967 *Economist* 25 Feb. 708/1 The Central Health Services Council's committee on the classification of proprietary preparations has laid down for doctors' guidance its views on when a food is a food and not a drug and therefore not prescribable. 1977 *Lancet* 24/31 Dec. 1360/1 Only a fraction of these are prescribable.

prescribe, *v.* Add: **2. c.** Delete † *Obs.* and add later examples.
1961 *Parthenon* (Marshall Univ., W. Va.) 10 Nov. 3/3 The '*Third Unabridged*' does not, of course, pretend to prescribe. It seeks, rather, to describe. 1978 *Amer. Speech* LIII. 70 Conceived as a modern dictionary that describes but does not prescribe, *Webster's Third New International Dictionary of the English Language* brought forth a deluge of adverse criticism and scathing reviews.
4. Delete † *Obs.* and add later example.
1919 K. Routledge *Mystery of Easter Island* viii. 116 As both the lifeboat and the cutter were carried in the waist of the ship when we were at sea, the space available for 'constitutionals' was prescribed.
prescribing *vbl. sb.* (earlier example).
1542 N. Udall *Erasmus's Apophthegmes* sig. N5ᵛ, Signifying, not to bee any prescribyng to the Romaines, how ferre thei ought to extend their empier.

prescription. Add: **I. 1.** (Earlier and later examples.)
1542 N. Udall *Erasmus's Apophthegmes* sig. f4ᵛ, The moste parte of people is barred from offendyng, onely by prestripcions [*sic*] of lawes, but a philosopher accumpteth and vseth reason in stede of lawes. 1960 J. O. Urmson *Conc. Encycl. Western Philos.* 143/1 Moral judgements, on this view [*sc.* the prescriptivist's], share with imperatives the characteristic that to utter one is to commit oneself, directly or indirectly, to some sort of precept or prescription about actual or conceivable decisions or choices. 1963 *English Jrnl.* May 337/2 Note that this statement [from Sir James Murray's preface to the Dict.] contains not one word about fixing the language, about proscription or prescription of any kind. 1968 J. Lyons *Introd. Theoret. Linguistics* i. 43 It should be stressed that in distinguishing between description and prescription, the linguist is not saying that there is no place for prescriptive studies of language.
III. 6. *prescription charge, pad.*
1928 E. O'Neill *Strange Interlude* ii. 61 He..goes to the table and taking a prescription pad from his pocket, hastily scratches on it. 1961 *Daily Herald* 9 Feb. 9 Of the doubled prescription charge his argument was: 'It is ludicrous exaggeration to say that by and large a 2/os. charge is any more of a burden than a 1/os. charge was in 1949.' 1965 *Ann. Reg.* 1964 47 Medical prescription charges would be abolished (at a cost of £25 million a year) and pensions would be increased. *a* 1974 R. Crossman *Diaries* (1975) I. 35 Kenneth Robinson gave the case for abolishing prescription charges. 1975 M. Simpson *Chrome Connection* iii. 61 The indentations on the prescription pad bore witness to his complicity.

prescriptionist. (In Dict. s.v. Prescription.) Add: **b.** (Further example.)
1906 *Dialect Notes* III. 151 (*Arkansas*) Mr. H. B. Mayes has accepted a position as *prescriptionist* for James S. Robinson, one of the most prominent druggists of Memphis.
c. = *prescriptivist *sb.*
1954 *Mind* LXIII. 258 It would be correct to call him an ethical 'prescriptionist'. 1964 C. Barber *Ling. Change Present-Day Eng.* i. 8 In fact they become moralists or prescriptionists, intent on telling us how we ought to talk.
Hence **prescri·ptionism** = *prescriptivism 2.
1962 *Amer. Speech* XXXVII. 215 Long's expressed dissatisfaction with school grammars..as well as his impatience with common proscriptions like *everybody they* indicate his rejection of traditional prescriptionism.

prescriptive, *a.* Add: **1. a.** Now also *spec.* in *Linguistics.* (Opp. *descriptive *a.* 3 b: cf. *normative grammar* s.v. *normative *a.* b.)
1933 O. Jespersen *Essent. Eng. Gram.* i. 19 Of greater value, however, than this *prescriptive* grammar is a purely *descriptive* grammar. 1963 *English Jrnl.* May 338/1 An accurate description of the language as it is actually used, kept simple by the relative absence of variants.. will in itself serve prescriptive purposes. 1968 J. Lyons *Introd. Theoret. Linguistics* i. 43 Linguistics..is *descriptive*, not *prescriptive* (or normative). 1977 *Time* 4 Apr. 5/3 The point to be made to Ms. Spaak, the Académie Française and all other prescriptive-normative institutions that would like to see language spoken in a certain way: c'est impossible.
b. *Philos.* Having or implying an imperative force. Also *absol.*
1946 *Jrnl. Philos.* XLIII. 35 The issue is whether a definition shall be taken as prescriptive in empirical enquiry or used as a convenient tool constantly responsible to facts. A nominal definition is by definition prescriptive. 1951 [see *descriptive *a.* 3 a]. 1952 R. M. Hare *Lang. Morals* i. 2 If moral language belongs to the genus 'prescriptive language', we shall most easily understand its nature if we compare and contrast first of all prescriptive language with other sorts of language. 1961 I. L. Horowitz *Philos., Sci. & Sociol. of Knowl.* vii. 88 Whatever the ratio of descriptive and prescriptive elements in an ideology, it is clearly a different qualitative entity than either religion or science. 1963 R. M. Hare *Freedom & Reason* v. 72 If moral judgements were *singular* prescriptives.., there would be less difficulty. 1967 *Encycl. Philos.* II. 314/2 All the views of definition that have been proposed can be subsumed under three general types of positions... These three general positions will be called 'essentialist', 'prescriptive', and 'linguistic' types. 1976 T. D. Perry *Moral Reasoning & Truth* 176 Every moral statement is prescriptive in the sense that it entails a certain imperative.
4. (Earlier example.)
1765 Johnson *Preface* in *Plays of Shakespeare* I. p. vii, The Poet, of whose works I have undertaken the revision, may now begin to assume the dignity of an ancient, and claim the privilege of established fame and prescriptive veneration.
b. *Anthrop.* Applied to marriage traditionally considered obligatory between persons in certain categories of relationship to each other within a tribe or kinship group. (Cf. *preferential *a.* d.)
1958 *Amer. Anthropologist* LX. 75 Prescriptive marriage rules entail enduring affinal ties between groups. 1961 E. Leach *Rethinking Anthropol.* iii. 54 Needham.. claims to have demonstrated that a rule of prescriptive patrilineal cross-cousin marriage is an impossibility. 1968 R. Needham tr. *Lévi-Strauss's Elem. Struct. Kinship* (1969) p. xxxi, Exceptional cases apart, they do what they say they must, hence the reason for calling their marriage system 'prescriptive'. 1971 F. Korn in R. Needham *Rethinking Kinship & Marriage* 113 The terms can be consistently arranged in an asymmetric prescriptive terminology.

prescriptivism (prĭskri·ptiviz'm). [f. Prescriptive *a.* + -ism.] **1.** *Linguistics.* The practice or advocacy of prescriptive grammar; the belief that the grammar of a language should lay down rules to which usage must conform.
1954 *College English* XV. 395/1 Professor Bloomfield comes to the conclusion that what is taught in an English class must be some form of..prescriptivism, checked by the limits of fact as established by linguistics. *Ibid.* 395/2 Bloomfield defends prescriptivism first because it has social utility. That is, the public judges..our students by the language they use. 1957 *Eng. Lang. Teaching* XII. i. 10 It is not for their prescriptivism as such that the older teaching grammars stand condemned. 1964 *Word* XX. 289 The charge of prescriptivism is also made against Chomsky. 1971 *Archivum Linguisticum* II. 54 We are probably all aware of the operation of even weaker collocational constraints as we search for the 'right' choice among, say, *achieve, accomplish, effect*,..etc. to associate with *plan* or *project* or *proposal*.., and a certain inescapable 'prescriptivism' informing language choices is perhaps worthy of note in passing. 1976 *Amer. Speech* 1973 XLVIII. 264 Prescriptivism is wrong, the reader is told again.
2. *Philos.* The theory that (moral) judgements have prescriptive force akin to that of imperatives; freq. contrasted with *descriptivism 1.
1963 R. M. Hare *Freedom & Reason* ii. 16 Let me refer to the type of doctrine..as 'universal prescriptivism'. 1963 I. L. Horowitz *Power, Politics & People* 15 His [*sc.* Mills's] role in sociology as a contributor to its debates on descriptivism and prescriptivism. 1967 *Encycl. Philos.* II. 317/2 There are two main varieties of prescriptivism. The nominalist variety explains definitions as semantic rules for assigning names to objects, while the formalist variety regards definitions as syntactic rules for abbreviating strings of symbols. 1973 *Heythrop Jrnl.* XIV. 136 (*title*) Prescriptivism in theory and practice. 1976 T. D. Perry *Moral Reasoning & Truth* i. 33 Moore's famous doctrine of the 'naturalistic fallacy' which has been accepted in principle by three of the four major tendencies in analytical ethics: intuitionism, emotivism, prescriptivism.

prescriptivist (prĭskri·ptivist), *a.* and *sb.* [f. as prec. + -ist.] **A.** *sb.* An adherent or advocate of prescriptivism. **B.** *adj.* Of, pertaining to, or characteristic of prescriptivism.
1952 T. Pyles *Words & Ways Amer. Eng.* xi. 272 But he is likely not to see any reason why absolute uniformity, the desideratum of the prescriptivist, should be any particular concern of the student of language even if it were possible of attainment. 1959 *Aristotelian Soc. Suppl. Vol.* XXXIII. 167 It seems to me that only prescriptivist prejudice can deny that we have here a morality. 1960 J. O. Urmson *Conc. Encycl. Western Philos.* 143 To call a thing good is thereby to offer guidance about choices; and the same might be said of the other moral terms. Descriptivists, however, refuse to admit that this feature is part of the *meaning* of moral terms. Their principal

opponents, who may be called 'prescriptivists', hold that it *is* part of the meaning. **1964** E. Bach *Introd. Transformational Gram.* v. 90 But the decision to edit..has nothing in common with the prescriptivist's zeal. **1967** *Encycl. Philos.* II. 317/2 The prescriptivist assimilates definitions to imperative sentences rather than to declarative sentences. **1973** *Heythrop Jrnl.* XIV. 139 His normative views are no more abitrary or relativist than those of any utilitarian, despite his non-naturalist and prescriptivist theory of meta-ethics. **1976** T. Eagleton *Crit. & Ideology* v. 174 It is this purely prescriptivist morality.. which finds a later echo in the moral ideology of Kant. **1977** *Publ. Amer. Dial. Soc. 1974* LXI/LXII. 8 The English teacher who..is suddenly bereft of her prescriptivist techniques and her substitution drills.

prescriptivity (prīːskripti·viti). [f. Prescriptive *a.* + -ity.] = Prescriptiveness.
1963 R. M. Hare *Freedom & Reason* i. 6 The prescriptivity of moral judgements explains both why there should be thought to be a problem about moral freedom, and how to approach its solution. **1966** *Amer. Philos. Q.* III. 305/2 We questioned the 'prescriptivity' of the deduced conclusion of normative syllogisms. **1976** T. D. Perry *Moral Reasoning & Truth* 9 When we point out to our interlocutor what other moral judgements his present judgement commits him to in view of its universalizability and 'prescriptivity' we shall often be able to force him to withdraw it.

prescriptorial (prīːskriptōə·riăl), *a. rare.* [f. Pre- B. 1 + Scriptorial *a.*] Existing before the use of writing.
1897 J. W. Powell in *16th Ann. Rep. U.S. Bureau Amer. Ethnol. 1894–95* p. xcvi, The names are associative or symbolic in the vague fashion characteristic of prescriptorial ideation.

Presdwood (pre·stwud). Also **presdwood.** [Alteration of *pressed wood.*] (See quot. 1940.)
A proprietary name in the U.S.
1940 *Chambers's Techn. Dict.* 671/1 *Presdwood*.., tradename for a strong building-board having water-resisting properties. **1946** *Amer. Jrnl. Roentgenol.* LV. 198/2 The procedure was repeated using tempered presdwood phantoms. **1949** *Official Gaz.* (U.S. Patent Office) 5 Apr. 36/1 Masonite Corporation, Chicago... *Presdwood*... For fiberboard, insulating board, composite board, [etc.]... Claims use since Oct. 6, 1926. **1951** R. Mayer *Artist's Handbk.* v. 193 In the majority of instances Presdwood is superior to wooden panels.

pre-season, -seizure: see *pre- B. 2a. **pre-see:** *pre- A. 1.

presele·ct, *v.* [Pre- A. 1.] *trans.* To select in advance.
1864 [in Dict. s.v. Pre- A. 1.] **1910** *Proc. Amer. Inst. Electr. Engin.* XXIX. 189 The secondary switches are inserted between the line-switches and the first selectors in such a way that the primary line switches pre-select idle secondaries and the secondaries pre-select idle first selectors. **1941** J. S. Huxley *Uniqueness of Man* ii. 56 Immigrants were pre-selected for..the qualities making up the pioneer spirit. **1961** *Which? Reports on Cars* 7 The cars tested..are in no way pre-selected by the dealer or especially inspected by the manufacturer. **1974** *Country Life* 28 Nov. 1673/1 Four pushbuttons..can be set to pre-select any given four FM stations.
Hence **presele·cted, -sele·cting** *ppl. adjs.*
1910 *Proc. Amer. Inst. Electr. Engin.* XXIX. 184 A line switch always uses a pre-selected idle trunk instead of making a selection after a subscriber starts to call as the Strowger selector switches do. **1924** H. H. Harrison *Introd. Strowger Syst. Autom. Telephony* i. 16 The combination of such pre-selecting switches with the trunking methods..made the large capacity exchange system a commercial possibility. **1933** Auden in *Rev. Eng. Stud.* (1978) Aug. 305 The pre-selected gear adjusted. **1962** J. Riordan *Stochastic Service Syst.* vi. 126 A. B. Clarke.. proposed..that the process be observed, in the stationary condition, for a preselected busy time τ. **1974** D. Kyle *Raft of Swords* iii. 19 A short-range missile..capable of being fired at pre-selected targets. **1977** *Offshore Engineer* June 60/3 A computer program..carried out a complete fatigue analysis of pre-selected joints of the tower.

presele·ction, *sb.* and *a.* **A.** *sb.* [Pre- A. 2.] Selection in advance; *spec.* the operation or use of a preselector.
1924 H. H. Harrison *Introd. Strowger Syst. Autom. Telephony* 145/2 (Index), Pre-selection. **1930** W. K. Hancock *Australia* x. 207 Members of branches join with the unionists who live in the same area to choose by a pre-selection ballot the local party candidate, and to elect delegates to attend the State conference of the party. **1941** J. S. Huxley *Uniqueness of Man* ii. 55 Pre-selection was at work on the pioneers. The human cargo of the *Mayflower* was certainly not a random sample of the English population. **1948** R. V. Pound *Microwave Mixers* i. 28 Since most of the circuits..were designed for use in pulse-radar systems, preselection is achieved by means of the resonant TR cavity of the duplexer that precedes the converter. **1950** J. Atkinson *Herbert & Procter's Telephony* (new ed.) II. xviii. 572/1 The trunking between the subscribers' lines and the 1st group selectors makes use of two stages of preselection by means of 10-outlet unidirectional mechanisms. **1957** *Railway Mag.* Nov. 758/2 Route-setting is used, with pre-selection facility.., the controls remaining stored until conditions allow of their becoming effective. **1962** *Lancet* 6 Jan. 23/2 All cases with chorioretinitis or cerebral calcification, were excluded. There was no other preselection. **1976** [see *preselector a]. **1979** *Daily Tel.* 24 Sept. 4/8 Mr Bob Hawke, president of the Australian Council of Trade Unions and former President of the Australian Labour party, announced yesterday that he would be a contender for pre-selection for the safe Labour seat of Wills at the next Federal election.
B. *adj.* [Pre- B. 2.] Occurring before selection.
1977 *Daily Tel.* 7 Nov. 2 The Service's Ground Branch is most seriously affected, with one in three group captains nominated for command making it known at preselection stage that they are not interested in taking over their own stations.

presele·ctive, *a.* [Pre- A. 3.] That preselects or permits preselection.
1925 *Jrnl. Inst. Electr. Engin.* LXIII. 660/2 If there are switches with a large number of outlets the problem does not arise; neither would it arise if one could use preselecting outgoing secondary switches, i.e. switches which themselves found the line before it was wanted... Such a circuit will no doubt arrive and, when it does, outgoing switches of a pre-selective type will for 10-point switches ..completely sweep the board. **1930** *Engineering* 17 Oct. 498/3 The pre-selective device consists of an arrangement whereby the gear-control lever can be set for any gear, but the selected gear will not actually be engaged until a pedal is depressed. **1941** J. S. Huxley *Uniqueness of Man* ii. 54 Pre-selective influences are those which attract certain types into an environment and discourage others. **1955** *Times* 16 Aug. 2/6 The Conquest Century has the characteristic Daimler transmission, which comprises a fluid flywheel and a preselective epi-cyclic gearbox. **1971** B. Scharf *Engin. & its Lang.* xvi. 234 More sophisticated overhead chain conveyors are provided with a mechanism by means of which any one of a number of discharge points can be preselected at any loading station. The material will then be automatically discharged or moved on to a side line... These conveyors are termed preselective overhead chain conveyors.

presele·ctor. [Pre- A. 2.] **a.** *Teleph.* A switch which when a subscriber lifts his receiver automatically connects the calling line to an idle trunk by a hunting action, independently of impulses produced by dialling; formerly also = *line finder* s.v. *line *sb.*[2] 32.
1912 J. Poole *Pract. Telephone Handbk.* (ed. 5) xxxii. 535 The line-switch used by Siemens is a specially neat arrangement... It is called a 'pre-selector' by Messrs Siemens, and each switch is complete in itself. *Ibid.* 536 Secondary line-switches or pre-selectors are used in both systems to facilitate and economise the connections. **1921** W. Aitken *Autom. Telephone Syst.* I. 3 A preselector is a switch that automatically selects an idle line of a group when the receiver is lifted. **1924** [see *hunt *v.* 8*]. **1950** J. Atkinson *Herbert & Procter's Telephony* (new ed.) II. xviii. 572/2 The 1st preselectors are arranged in groups, so that each group carries..an equal volume of traffic. The preselectors of one group are trunked via 2nd preselectors to a maximum of 100 1st selectors. **1976** T. H. Flowers *Introd. Exchange Syst.* iv. 89 The choice between preselection and line-finding is mostly a question of economics. The first needs one exchange switch, or exchange line, and the second one switch per cord circuit. The quantity of switches needed as line finders is thus much less than the quantity as pre-selectors, but whereas line finders must be full-sized exchange switches to achieve satisfactory traffic loading, pre-selector switches may have as few as ten contacts in the banks.
b. *Telecommunication.* A tuned circuit preceding the first mixer in a superheterodyne receiver; an analogous filter in a microwave receiver.
1930 *Electronics* Sept. 279/1 (*heading*) An improved preselector circuit for radio receivers. *Ibid.* 308/1 The essential features of this preselector are shown in Fig. 6. **1951** A. Sheingold *Fund. Radio Communications* xv. 307 The preselector, when present, helps to maintain a favorable signal-to-noise ratio and minimizes interference effects. **1971** M. G. Scroggie *Found. Wireless & Electronics* xxii. 396 Because the i.f. amplifier is relied upon for most of the selectivity, the preselector tuning circuits do not have to be very sharp, so slight errors in gauging are not serious. **1975** D. G. Fink *Electronics Engineers' Handbk.* xxv. 72 Narrow-band filters in the receive path, often called preselectors, are built using mechanically tuned cavity resonators or electrically tuned YIG resonators. Preselectors can provide up to 80 dB suppression of signals from other radar transmitters in the same rf band but at a different operating frequency.
c. A gearbox that enables a driver to select the next gear at any time before the change is actually made (by means of a separate pedal). Usu. *attrib.*
1930 *Engineering* 17 Oct. 498/2 There is one cam for each gear, all mounted on a common shaft coupled to the pre-selector lever. **1935** *Economist* 7 Dec. 1144/1 It is a natural step from the power unit to the transmission, where most important developments have centred round such features as the fluid flywheel and the pre-selector gear. **1969** *Driving* (Ministry of Transport) xvi. 198 'Pre-selector' transmissions, mostly found on buses and coaches, have a lever by which the driver can select gears in advance, ready for later changes... No gear change takes place until a gear-change pedals is pressed and released. **1979** J. Leasor *Love & Land Beyond* i. 13 An electrical gear change which could be used as an ordinary box or as a preselector.

pre-se·ll, *v. Comm.* [Pre- A. 1.] *trans.* To promote (a product) before it is available to the consumer; to persuade (the consumer) in advance to buy a product. Also *transf.* So **pre-se·lling** *vbl. sb.* and *ppl. a.*, **pre-so·ld** *ppl. a.*
1950 in Webster Add. **1958** *Washington Post* 22 Sept. A2/1 Campaign organizers and the American Heritage Foundation public services advertising campaign has already done much to 'pre-sell' the public. **1959** *Times* 7 Apr. 14/4 It is the turn of the television programme to provide pre-sold material for cinema films. **1959** P. Wood in S. Spender tr. *Schiller's Mary Stuart* 8 English audiences are far less indulgent to her than foreign ones, who are pre-sold on the pathos of her situation. **1960** *Times* 26 Oct. 4/2 Plays expanded from television originals are also, in a sense, pre-sold. **1961** *Economist* 11 Mar. 984/1 Others believed..that the 'pre-selling' of the major products by advertising direct to the consumer would have a much more potent effect when the barrier of the counter had been removed and she had nothing to do but pick them up. **1962** E. Godfrey *Retail Selling & Organization* iv. 33 Pre-sold goods. In some cases the preliminary stages of the sale will have been completed before the customer comes into the department, through advertising. **1967** *Guardian* 21 July 3/2 We deliberately avoided preselling the film to America. **1973** *Publishers Weekly* 30 Apr. 50/3 The recent January issue..was so relatively 'with it' that the entire issue was presold. **1977** *Daily Tel.* 2 Dec. 19/1 The interviews..have been pre-sold to the United States and other foreign countries.

presence. Add: **1. d.** The quality in reproduced sound that gives a listener the impression that the recorded activity is occurring in his presence (see also quot. 1950).
1950 *Audio Engin.* Sept. 33 In motion picture work presence refers to the lack of localization of the reproduced sound, so that the eye is beguiled into believing that the sound issues from the location the eye follows... A second use of the term *presence* indicates the degree of intimacy achieved... A third type of presence is detail presence, in which an auditor is able to pick out an individual instrument or soloist, and more or less easily follow its melodic line throughout the changing mass of sound. **1952** H. F. Olson *Musical Engin.* vii. 262 The reverberation-frequency characteristic has a marked effect upon presence. Excessive reverberation in the low-frequency range reduces presence. A uniform directional pattern in the directivity characteristic of a loudspeaker enhances the presence. **1957** *IRE Trans. Audio* V. 106/2 If the need for great 'presence' calls for a very close microphone position, the reproduction may cause a solo instrument to sound much too large, and this can be corrected by attenuating the difference channel relative to the sum. **1958** *Proc. Inst. Electr. Engin.* CV. B. 609/1 The second observation concerns the critical nature of the frequency band in the region 2–4 kc/s... Deficiency in this band gives a distant impression; slight excess gives a forward quality, sometimes referred to as 'presence'. **1974** Harvey & Bohlman *Stereo F.M. Radio Handbk.* v. 127 (*caption*) Curves showing prominence given to mid-range and bass frequencies by the presence control. **1976** G. Alkin in J. Borwick *Sound Recording Pract.* xxiv. 364 Some types of Lavalier microphone have a non-linear frequency response which peaks in the 'presence' region (between about 4 and 6 kHz) to restore clarity of diction.
e. *Politics.* The maintenance by a nation of political interests and influence in another country or region; *spec.* the maintenance of personnel, esp. armed forces, on the soil of an allied or friendly state; *concr.*, armed forces stationed in this way. Also *transf.*, denoting the representation of a nation's interests at an event.
Cf. Fr. *présence*, in same sense as in quot. 1955.
[**1955** *Times* 4 Aug. 5/3 Times had changed, he said, and there was no longer any need for outmoded oriflamme to guarantee the *présence française*, or rather the *permanence française*, which could only exist 'if we respond to the wishes of the peoples oversea'.] **1958** *Spectator* 7 Feb. 176/2 The 'presence of France' must be maintained. **1961** *Listener* 21 Dec. 1058/1 As Britain and France step back on to the side-lines [in Africa], the United States steps forward to join them there. This new presence..was not at first easy for Britain to accept. **1963** *Ann. Reg. 1962* 319 An effective United Nations 'presence' in South West Africa. **1966** *Punch* 22 June 898/1 How small can a 'presence' be, of the sort we are going to maintain East of Suez?.. The Americans have a presence of 380,000 men in Vietnam alone, and regard that as barely enough. **1972** *Times* 18 Mar. 12/4 Setbacks in the Arab world that followed his liquidation of the guerrilla presence in Jordan. **1975** *Listener* 25 Sept. 390/1 They were known as the Trucial States. When the British presence was withdrawn in 1971, they became a federation called the United Arab Emirates. **1977** *Time* 10 Oct. 11/3 Working out a formula that would allow some Palestinian presence at Geneva was the focus all week long of intense bargaining.
5. a. Also *transf.*
1959 *Sunday Times* 18 Jan. 16/8 For a painter to have a presence is already an achievement. By 'presence' I mean the variously-definable something that bids a visitor pause and is one of the signs of greatness. **1977** 'E. Anthony' *Silver Falcon* vii. 135 The chestnut..had that indefinable quality known in the horse world as presence.

presenile, -senility: see *pre- B. 1, A. 2.

presenium (prisī·niŏm). *Med.* [f. Pre- B. 1 in Dict. and Suppl. + L. *senium* feebleness of age.] The period of life preceding old age.
1926 *Lancet* 16 Oct. 820/2 Presenile Mental Disorders. In this article the term presenile is applied to mental disorders arising in the period of life beginning in the late 'forties and extending to the early 'sixties. This period includes the climacterium in women, and certain common mental disorders met with in both sexes during the years which precede the actual period of old age or senility. The more frequent clinical types of mental disorders that are encountered during the presenium are as follows. **1976** Smythies & Corbett *Psychiatry* vii. 132 Dementia occurring in the presenium demands thorough and complete investigation.

presensitiza·tion, *a.* and *sb.* **A.** *adj.* [Pre- B. 2.] Existing or occurring before sensitization.
1964 W. G. Smith *Allergy & Tissue Metabolism* ii. 28 These responses rapidly returned to pre-sensitisation levels.
B. *sb.* [Pre- A. 2.] Sensitization beforehand.
1977 *Lancet* 27 Aug. 419/1 Data from patients whose serum had been tested for the presence of lymphocytotoxic antibodies before transplantation against a panel of at least 40 random lymphocyte donors were used in analyses of the effect of humoral presensitisation.
So **prese·nsitize** *v. trans.*, to sensitize beforehand; **prese·nsitized** *ppl. a.*
1963 *Lancet* 5 Jan. 45/1 In 2-month-old neonatally thymectomised C3H mice, an established Ak skin graft broke down within 12 days after the injection of lymphoid cells from C3H donors presensitised against Ak. **1967** Karch & Buber *Offset Processes* vi. 230 Presensitized plates are surface coated by the manufacturer and ready for exposure when removed from the original package. **1977** *Lancet* 27 Aug. 419/1 Recipients whose serum reacted against more than 5% of the panel members were regarded as being presensitised.

present, *a.* (*adv.*) Add: **I. 1.** *present company excepted* (and varr.), phr. used to indicate that a generalization does not apply to the hearers of it; (*to be*) *among those present*: to be present (at a function, etc.); to be in the vicinity; (orig. used in reports of social gatherings, etc.; hence in jocular use).
1793 J. O'Keeffe *London Hermit* i. ii. 25 Sir, you should always except the present company. **1832** *Reg. Deb. Congress U.S.* 14 June 3530 Mr. C[layton] observed that the gentleman ought to remember that the present company is always excepted. **1846** Dickens *Dombey* (1848) iii. 20 There's a Tartar within a hundred miles of where we're now in conversation, I can tell you, Mrs. Richards, present company always excepted too. **1913** F. L. Barclay *Broken Halo* vii. 92 'Present company excepted' is always understood, without being expressed, when sweeping generalities are being made. **1925** Wodehouse *Carry on, Jeeves!* iv. 84, I hopped out of bed pretty early next morning, so as to be among those present when the old boy should arrive. **1947** —— *Full Moon* vi. 111 There had unquestionably been mosquitoes among those present. **1975** G. Moffat *Miss Pink* iii. 54 Women never strike out for themselves... Present company excepted, of course.

present, *sb.*[2] Add: **2. e.** *a present from* (*Brighton* etc.): an inscription on a piece of souvenir pottery etc., bearing the name of the town in which it is sold; hence, a piece of pottery etc. so inscribed, a souvenir.
1852 Dickens *Bleak Ho.* (1853) iv. 28 We found a mug, with 'A Present from Tunbridge Wells' on it. **1890** Kipling *Courting of Dinah Shadd* 125 She gave me a drink out of a china mug wi' gold letters—'A Present from Leeds'. **1921** W. de la Mare *Mem. Midget* viii. 49 A gay little bumper of milk gilded with the enwreathed letters, 'A Present from Dover'. **1962** N. Mitford *Water Beetle* 113 The china cabinet will contain Rose Pompadour Sèvres cheek by jowl with A Present from Bexhill. **1964** F. Sinclair *Three Slips to Noose* vii. 61 A small square room furnished with..china shepherds and presents from Clacton. **1974** J. Stubbs *Painted Face* i. 32 A small ashtray..inscribed *A Present from Brighton.*

present, *v.* Add: **I. 1. c.** Also *fig.*
1923 *Adelphi* Aug. 236 Osbert is a born impresario... Osbert 'presents' the [Sitwell] family, and does it with originality.
2. a. (Further example: cf. sense *9 b.)
1880 [see *presbyope].
4. c. Of a radio or television producer or broadcaster: to bring (a broadcast item) before the listening or watching audience, to introduce or announce (a programme). Of a performer: to perform (a song etc.).
1933 *Radio Times* 14 Apr. 120/1 Tonight Mr. James Agate 'presents' perhaps the greatest..star of this glittering series. **1961** *N.Y. Times* 30 Jan. 19 Ellie Mao, soprano, and Frederick Fuller, baritone, presented a program of folksongs entitled 'East meets West' in Carnegie Recital Hall last night. **1972** *Listener* 10 Aug. 187/2 The need felt by..the Presentation Department at Television Centre..to do something more than merely 'present' the evening's programmes. 'Presenting' means seeing that the programmes get on the air tidily and on time, smoothing over or explaining disaster, making announcements and trailing coming events. **1976** *Laurel* (Montana) *Outlook* 9 June 6/3 A solo, presented by Marilyn Parker, was accompanied by Mrs. Markegard.
5. (Later examples.) Also *absol.*
1976 *Daily Tel.* 20 July 3/2 How can anyone, any lawyer, present any case that is acceptable in common sense. **1976** *Dallas Morning News* 22 Sept. 10B/5 Learn to read, learn to listen, learn to think, learn to write, learn to present. *Ibid.*, Businesses have botched sales efforts mainly because their people..could not present their information in clear and 'selling' English.
9. b. Also more widely in *Med.*: of a condition: to be manifest, to occur. Of a patient: to present himself or appear for an initial medical examination. Cf. sense 2b in Dict., and quot. 1880 in sense *2a.
1925 *Boston Med. & Surg. Jrnl.* 23 July 179/1 A rather marked purplish hemorrhagic area presented about the wound. **1960** *Lancet* 16 Jan. 138/2 A patient presenting with an exacerbation of bronchitis was initially assigned by the doctor to one of three categories. **1972** *Nature* 8 Sept. 102/2 These complications may present as hypersensitivity reactions.., but most often they take the form of gastric erosions. **1976** *Lancet* 20 Nov. 1107/2, 73 boys who presented to their general practitioners..with symptoms of urinary-tract infection..were referred to a three-year prospective study. **1977** *Proc. R. Soc. Med.* LXX. 262/1 It is not unusual for patients to present in an eye department with symptoms associated with either poor accommodation or poor convergence.

presentable, *a.* Add: **4.** (Earlier and later examples.) Also *fig.*
1800 M. Edgeworth *Parent's Assistant* (ed. 3) VI. 147 Do send my *ooman* to me to make me *presentable.* **1801** —— *Belinda* I. iv. 132 Excuse me for showing you the simple truth; well dressed falsehood is a personage much more *presentable.* **1907** J. M. Synge *Lett. to Molly* (1971) 172, I will show it to you tomorrow if it is presentable enough. **1925** C. Connolly *Let.* 23 Apr. in *Romantic Friendship* (1975) 72 He [*sc.* Maurice Bowra] is extremely presentable in any society. **1965** F. Raphael *Darling* xix. 87 It was enough to be socially presentable. **1974** *Times* 19 Oct. 6/7 Bowra remarked that he had had his hair cut—'makes one more presentable'. The word 'presentable'.. was a very important epithet in the Bowra system of social terminology... Those who had 'unpresentable' pinned on them were remorselessly barred. **1979** A. Morice *Murder in Outline* ii. 12 A group of more presentable-looking seniors granted the privilege of handing round the apéritifs.

presentation. Add: **III. 5. a.** (Later examples.) Also, a display or show (e.g. of slides) used esp. in advertising.
1972 G. Bromley *In Absence of Body* i. 13 'We've got the OOO-Frooty presentation tomorrow.'.. The presentation was to show the client proposals for a new [advertising] campaign. **1976** *National Observer* (U.S.) 12 June 3/4 Picnic lunches, public speeches, and presentations about Proposition 15 were to be the order of the day. **1976** J. H. Spencer *Surgenor Campaign* i. 16 Cusack taking him through a slide presentation on their international capability: twenty-seven offices in fourteen countries.
c. In Broadcasting, the action or an instance of presenting a programme; also *ellipt.* for *presentation department* (see sense 10 below).
1941 *B.B.C. Gloss. Broadcasting Terms* 26 *Programme presentation.* (1) Action of presenting a sequence of programmes by means of a framework of microphone announcements... (2) Framework of microphone announcements in a sequence of programmes, its purpose being to supply continuity, to link programmes together, and to attract listeners. **1963** [see *Newspeak]. **1968** *Listener* 22 Aug. 252/1 In bad periods of radio..language is usually what they [*sc.* programmes] are about, or, to call it by its new, pompous name, Presentation. **1968** *Radio Times* 28 Nov. 23/3 Television presentation by Nick Hunter. **1974** *Some Technical Terms & Slang* (Granada Television), *Presentation*, the department within Granada responsible for shape and coordination of the daily programme schedule. **1978** *Listener* 7 Dec. 762/4 The business of neat, informative presentation.
IV. 10. (in sense 4) *presentation binding, bowl, box, clock* (example), *copy* (earlier and later examples), *cup, drawing, pack, plate, silver, watch*; (in sense 1 b) *presentation dress*; (in sense 5c) *presentation assistant, studio, suite*; **presentation department** (see quot. 1978).
1941 *B.B.C. Gloss. Broadcasting Terms* 24 *Presentation Assistant* (*abbrev.* P.A.), Broadcasting official immediately responsible for the smooth running of a sequence of programmes, and hence for co-ordinating the activities of programme producers, announcers, and engineers directed to that end. **1939–40** *Army & Navy Stores Catal.* 841 Books in presentation bindings. **1952** J. Carter *ABC for Bk. Collectors* 139 *Presentation binding*, used variously for *gift binding* or *author's binding.* **1907** *Yesterday's Shopping* (1969) p. li/2 Presentation bowls. **1973** L. Cooper *Tea on Sunday* ii. 30 Silver, some of it..presentation cups and bowls, shining behind shining glass doors. **1908** *Sears, Roebuck Catal.* 333/2 Alaska metal tableware set in a fancy presentation box. **1976** *Sunday Mail* (Glasgow) 28 Nov. 31/4 The two bigger boxes have ribbons and bows on them. We know there is a demand for the presentation box. That's why they are dearer. **1935** D. L. Sayers *Gaudy Night* i. 17 A Presentation Clock was to be unveiled. **1803** Scott *Lett.* (1932) I. 182 Be so good as to disperse the following presentation copies with 'From the Editor', on each. **1938** [see *cancellandum]. **1978** A. Waugh *Best Wine Last* xxiv. 304 There were a great many presentation copies, signed by brother and sister writers. **1973** Presentation cup [see **presentation bowl*]. **1978** *A–Z of BBC* (ed. 2) 163/1 Presentation Department is editorially responsible for supervising the transmission operation; for promoting programmes on the screen; for network identification..; for programme announcements and public service information,..; and for running the Television Duty Office. **1975** *Country Life* 20 Feb. 428 Among the drawings there are those, aptly christened by Johannes Wilde 'presentation drawings'. **1896** *Girl's Own Paper* 12 Dec. 161/1, I was borne off to the Court Dressmaker to choose the..presentation dresses. **1938** N. Marsh *Death in White Tie* viii. 83 He looked at the two photographs... One was of the Lady Mildred Potter in the presentation dress of her girlhood. **1976** *Shooting Times & Country Mag.* 9–15 Dec. 11/1 (Advt.), Supplied in presentation pack with supply of BB shot. **1867** C. L. Eastlake in *Queen* 15 June 470/1 If the pieces of 'presentation plate'..were only entrusted to art-workmen of sound education, we might hope for something better than the everlasting palm trees, camels and equestrian groups. **1967** N. Freeling *Strike Out* 40 Here on shelves was presentation silver..for Rob was the best bicycle champion Holland had had. **1960** *B.B.C. Handbk.* 40 It (*sc.* the Television Centre] will provide the service with..seven major production and two presentation studios. **1974** *B.B.C. Handbk. 1975* 264/2 The Television Centre houses separate presentation suites incorporating network control rooms, and studios for announcements and weather forecasts. **1931** M. Allingham *Police at Funeral* iv. 51 Presentation watch... The company gave him this watch.

presentational, *a.* Add: Also, of or pertaining to presentation in other senses. Also *absol.* (Further examples.)
1928 *Daily Tel.* 19 July 18/3 The intelligent theatres of New York..show an admirable sympathy both for good European drama and new forms of presentational art. **1929** A. N. Whitehead *Process & Reality* II. iv. 170 Perception which merely, by means of a sensum, rescues from vagueness a contemporary spatial region, in respect to its spatial shape and its spatial perspective from the percipient, will be called 'perception in the mode of presentational immediacy'. **1942** S. K. Langer *Philos. in New Key* iv. 97 The meanings given through language are successively understood, and gathered into a whole by the process called discourse; the meanings of all other symbolic elements that compose a larger, articulate symbol are understood only through the meaning of the whole... This kind of semantic may be called 'presentational symbolism', to characterize its essential distinction from discursive symbolism, or 'language' proper. **1958** C. H. Whitman *Homer & Heroic Tradition* vi. 107 It is by means of the image and the poetic symbol..that language is made presentational. *Ibid.* 127 If metaphor flags and facts prevail, the presentational is lost, and the poem sinks to prose. **1959** W. A. Christian *Interpr. Whitehead's Metaphysics* iii. 54 For occasions of a relatively high grade of complexity, a contemporary region may be given in presentational immediacy. That is, the region may be perceived as the locus of sensa, and as the subject to mathematical relations. **1962** *Listener* 3 May 765/1 A personal style, an effective or original rhetoric, a brilliant presentational technique are cheerfully accepted in lieu of any real human adequacy. **1975** *Nature* 23 Oct. 723/1 An author with the technical knowledge and presentational skill of John Maddox needs no introduction. **1980** *Times* 24 May 14/3 It all was done for 'presentational' reasons..not merely to make the Americans feel good, but..to show solidarity.

presenta·tionally, *adv.* [f. prec. + -ly[2].] In a presentational manner; by means of or as regards presentation.
1934 *Mind* XLIII. 390 We have a number of unconscious sensations as elements in wholes in which they do not figure presentationally, but only as ingredients. **1954** *Mind* LXIII. 196 How does this internal occupation of the visual field itself come to occupy that field presentationally? **1958** C. H. Whitman *Homer & Heroic Tradition* vi. 105 Yet when we speak of the 'direct appeal' of an image or symbol, we mean that it appeals as an image of painting or sculpture appeals to the mind, that is, presentationally. **1979** *Guardian* 30 June 8/4 Presentationally speaking, as they say in the public relations business.

presentatively, *adv.* (Later example.)
1878 S. H. Hodgson *Philos. of Reflection* I. 172, I represent to myself what *I imagine* you to be feeling presentatively.

pre:sent-day·, *a.* [Present *a.* (*adv.*) 10.] Current, contemporary; now in existence or in use; prevalent; living at the present time.
1887 [in Dict. s.v. Present *a.* (*adv.*) 10]. **1902** in C. W. Cunnington *Eng. Women's Clothing* (1952) ii. 47 Present-day fashions require for the ideal figure an upright poise of the shoulders. **1925** I. A. Richards *Princ. Lit. Crit.* 222 That Dante is neglected is due only indirectly to his present-day obscurity. **1926** D. L. Sayers *Clouds of Witness* vi. 142 A present-day girl, who rushes about bareheaded in all weathers. **1930** *Times Educ. Suppl.* 31 May 245/3 Much of present-day British India never was under Mogul rule. **1934** *Amer. Speech* IX. 83/1 An invitation from Edward C. Ehrensperger to address the Present-Day English Section of the Modern Language Association of America. **1934** *Discovery* Nov. 309/1 There is a reversal back to negative phototropism, that is, they [*sc.* the termites] revert to the normal conditions of their present-day life. **1946** 'S. Russell' *To Bed with Grand Music* v. 77 A pigskin bag at present-day prices. **1959** *Universities & Left Rev.* Spring 54/1 The passivity of the present-day working-class reader. **1967** E. Short *Embroidery & Fabric Collage* iii. 63 In present-day furnishing also there is a feeling for function as well as decoration. **1974** P. Erdman *Silver Bears* iii. 35 My family has an obligation not only to present-day Iran, but also to the Persia of the past. **1977** G. W. H. Lampe *God as Spirit* ii. 51 Saul's sudden possession by the Spirit when he met a group of ecstatics coming down from a high place, preceded, like some of their present-day counterparts, by a musical group.
Hence **present-day·ness,** actuality, contemporaneity.
1963 *Times* 9 Jan. 11/3, 1960 doesn't view the eighteenth century as 1920 did—fashion and social economic state of society affect our attitude to the past, and it's the designer's job to bring out this present-dayness.

prese·ntedness. *Philos.* [f. Presented *ppl. a.* + -ness.] (See quot. 1951.)
1925 C. D. Broad *Mind & its Place* v. 255 The premise ..might be derived from an uncritical jump from 'presentedness' to 'presentness'. **1933** *Mind* XLII. 307 It is part of the notion of a specious present that a certain characteristic, which we will call 'presentedness', has a maximum value at a certain point in it, and tails off to nothing in two opposite directions within it, *viz.* towards the point where the perceived past merges into the remembered past, and towards the point where the anticipated future merges into the perceived future. **1951** *Mind* LX. 162 'Presentedness' seems to be Broad's name for the characteristic Hume called 'force and vivacity'. **1954** *Mind* LXIII. 33 Broad does not explain what he means by 'presentedness' beyond saying 'This is meant to denote a psychological characteristic.'

presentee[2] (prezĕnti·). *joc.* [f. Present *a.* (*adv.*) in imitation of Absentee.] One who is present. Hence **presentee·ism.**
1892 'Mark Twain' *Amer. Claimant* xxi. 211 There was an absentee who ought to be a presentee—a word which she meant to look out in the dictionary. **1931** H. Withers *Everybody's Business* ix. 161 Certainly he is an absentee.. —if he adopted the habit of dropping in at the works and making well-meant suggestions.., is it likely that his presenteeism would be helpful? **1943** *Nat. Liquor Rev.* July 4/2 The Kaiser Company's public relation officials discovered that the term 'absenteeism' irked the people who read it... The Kaiser Company..changed its policy and praised those who were on the job by using the term 'presenteeism'.

pre-se·ntence, *sb.* and *a.* **A.** *sb. Linguistics.* [Pre- B. 1 in Dict. and Suppl.] A construct that precedes or underlies the formation of a sentence.
1940 Bryant & Aiken *Psychol. of Eng.* iv. 33 These primitive 'pre-sentences' came to be broken up. **1965** *Language* XLI. 459 Let us call these sequences of morphemes presentences when they label nonsurface P-markers.
B. *adj.* [Pre- B. 2.] **a.** That occurs before a judicial sentence.
1957 *Encycl. Brit.* VI. 719/2 Under the federal rules of criminal procedure and the law of a few states, a pre-sentence investigation by the probation service and a report to the trial judge must be made. **1974** *Guidelines to Volunteer Services* (N.Y. State Dept. Correctional Services) 36 *Pre-sentence report*, a background investigation conducted by a probation department following an individual's conviction of a crime. **1979** *Arizona Daily Star* 1 Apr. A8/2 As Raymond said in a pre-sentence interview: 'My closest friends are my mother and my brothers. They're the only people I can trust.'
b. *Linguistics.* That occurs before a spoken or written sentence.
1965 N. Chomsky *Aspects of Theory of Syntax* ii. 102 Sentence Adverbials which form a 'pre-Sentence' unit in the underlying structure. *Ibid.* iv. 148 Verbs are strictly subcategorized into Intransitives, Transitives, pre-Adjectival, pre-Sentence, etc.

presenter. Add: **7.** One who presents or introduces a programme on radio or television.
1967 *Listener* 24 Aug. 249/2 A few words spoken into a camera by a presenter can smooth..an awkward script. **1974** *Radio Times* 16 Feb. 17/3 You and Yours. Presenter Lyn MacDonald. **1976** *Evening Times* (Glasgow) 1 Dec. 6/1 It's the fact that the Nationwide presenter made a quick dash by air from London to Abbotsinch and then on to Paisley.

presentiality. a. (Later examples.)
1911 tr. *Aquinas's Summa Theol.* I. xiv. 205 His glance is carried from eternity over all things, as they are in their presentiality. **1969** G. Leff *Hist. & Social Theory* i. 18 Its presentiality takes precedence over all its other attributes.

presentiate, *v.* Add: (Later examples.) Hence **prese·ntiated** *ppl. a.*; **presentia·tion,** the act of rendering present.
1974 *Southern Calif. Law Rev.* XLVII. 800 If we say that the squirrel is futurizing the present, *i.e.*, preparing for the future, or presentiating the future, *i.e.*, bringing the future into the present, we are expressing an inaccurate anthropomorphism. *Ibid.* 802 The entire credit structure, including the monetary system itself, is founded on presentiation. Virtually no aspect of life in a modern society is left untouched by presentiation related to exchange. *Ibid.* 805 Because trouble is expected in a relation, efforts may be made in advance to deal with it transactionally, i.e., to eliminate it before it occurs through resolving the conflicts in advance, thereby turning what would have been relational into what is simply an allocated (presentiated) cost. **1976** *Jrnl. Econ. Issues* X. 15 Contracts 'presentiate' the future, or bring the future into the present where it is arranged. *Ibid.* 47 The transaction would be fully presentiated.

presentic (prīze·ntik), *a. Gram.* [f. Present *a.* (*adv.*) + -ic.] Pertaining to or characteristic of the present tense.
1964 H. S. Sørensen in *Eng. Stud.* XLV. (Suppl.) 81 As in the case of the incomplete perfect, grammarians have overlooked the fact that what is 'presentic' about the perfect is the point of reference, and nothing else. **1965** W. Winter *Evidence for Laryngeals* 210 In B nonpresentic forms with an /a/ suffix were found in one and the same paradigm with presentic forms with an /n/ infix and no /a/ suffix.

presenting, *ppl. a.* (Further examples in *Med.*: cf. *present *v.* 9 b.)
1911 R. C. Cabot *Differential Diagnosis* 17 A 'presenting symptom', comparable to the 'presenting part' in obstetrics, may turn out to be of minor importance when we have studied the whole case. **1948** Martin & Hynes *Clin. Endocrinol.* iv. 96 Increasing weight may be the presenting symptom. **1960** R. D. Laing *Divided Self* vi. 111 All that could not find direct expression and open acknowledgement in her was condensed in her presenting symptom. **1973** *Times* 17 Oct. 14/4 The temptation to shoplift may be a presenting complaint in depressive illness or melancholia.

presentist, *sb.* (*a.*) Delete *rare* and add further examples. Also, one who has a bias towards the present or is influenced by present-day attitudes. Hence **pre·sentism.**
1927 [see *passéisme]. **1956** *N.Y. Times Bk. Rev.* 8 Jan. 22/3, I think Mr. Nevins' review underscores the danger of 'presentism'; I suggest historians would strengthen their position by applying the chief test of their profession—perspective and caution in contemporary analyses. **1975** *Nature* 24 Apr. 729/3 Such history as is dealt with reads soundly, but it is often drawn from secondary sources and professional historians of science would judge it presentist and Whiggish. **1976** T. Stoianovich *French Hist. Method* i. 36 Even the attempt to understand the past in its own terms is 'presentist' to the extent that it is founded on what contemporary science and bias lead us to believe to have been its own terms. **1977** *Times Lit. Suppl.* 27 May 655/1 The author wants to explain how the world got to be the way it is at the start of the last quarter of the twentieth century. 'Presentism' accordingly governs the way he distributes attention.

presently, *adv.* Add: **2. a.** Revived in U.S. and to some extent in Great Britain.
1939 *Topeka* (Kansas) *State Jrnl.* 20 Feb. 12/1 Sunner is presently minister of interior and one of the outstanding leaders of the Falangists. **1943** *Time* 20 Sept. 25 They said Mussolini assured them he would return to power and re-establish the Fascist regime, comparing himself presently with Napoleon—the parallel being Napoleon's exile on Elba. **1945** *Richmond* (Virginia) *Times-Dispatch* 21 June 1/3 The one class of cadets presently at the academy. **1949** *Sun* (Baltimore) 9 Apr. 6/1 The members of the presently major coalition can hardly refuse to meet with the Mayor. **1957** G. Marx *Let.* 12 Apr. (1967) 213, I am presently building a house and doing my own show, but sometime within the next two months I'll make it. **1958** *Economist* 9 Aug. 433/1 It is entirely possible that Mr Macmillan..may now be getting greater commendation from the commentators of his generation than he will eventually get from historians; certainly the praise presently being heaped upon him seems to be..a consequence of the recent recovery in the Conservatives' fortunes. **1968** *Globe & Mail* (Toronto) 17 Feb. 52 (Advt.), We want a go-getter who is well established and presently calls on machinery, tool, and equipment supply firms. **1968** B. Foster *Changing Eng. Lang.* v. 215 This meaning of 'at present'..is one which once again has been reintroduced from across the Atlantic where it had also lingered on, with the result that it is now in good use in England. 'Warm air is presently moving north-east' reported a B.B.C. weather bulletin (20 May 1963). **1969** *Daily Tel.* (Colour Suppl.) 24 Jan. 8/4 Ivan Cooper, a Protestant and former Unionist,..and presently chairman of the Derry Labour Party, was elected chairman. **1971** *Nature* 2 July 23/1 The Caribbean area is a subplate presently attached to the South American plate. **1978** *N.Y. Times* 30 Mar. D14/5 (Advt.), GTE Sylvania is presently engaged in the research, design, development and production of high energy Lithium Battery power sources for use in highly specialized applications. **1978** *Dumfries Courier* 13 Oct. 2/5 Mr. William O'Brien, solicitor, Dumfries, for the accused, said Mr. Savage was presently unemployed, his last employment being a year ago.

pre-separate: see *pre- A. 1.

‖ **presepio** (prese·pio). [It., f. L. *præsæpe* enclosure, stall: cf. Præsepe.] A crib; a model of the manger in which Christ was laid.
1759 M. W. Montagu *Let.* 19 July (1967) III. 220 The devout people who spend 20 years in making a magnificent presepia [*sic*] at Naples. **1958** *Listener* 25 Dec. 1067/1 Every Italian family has its own *presepio*, as the crib is called. **1969** E. H. Pinto *Treen* 173/2 Presepio or Nativity figure tableaux have been and still are used in most parts of Christendom.

preservability. (In Dict. s.v. Preservable *a.*) Delete *rare* and add later examples.
1959 *Brno Studies in English* I. 11 This feature [*sc.* the documentary, preservable character of written utterances], which one may perhaps term 'preservability', has been appreciated by men since time immemorial. **1972** *Science* 22 Sept. 1067/1 Some biologic groups show fossil diversities closer to their actual diversities than do other groups because of inherent differences in preservability.

preservation. Add: **1. b. preservation order,** a legal obligation on an owner to preserve a building of historic interest or value.
1947 *Act* 10 & 11 *Geo. VI* c. 51 § 29 If it appears to a local planning authority that it is expedient to make provision for the preservation of any building of special architectural or historic interest in their area, they may for that purpose make an order (in this Act referred to as a 'building preservation order') restricting the demolition, alteration or extension of the building. **1953** *Act* 1 & 2 *Eliz. II* c. 49 § 11 The Minister may..make an order (in this Act referred to as a 'preservation order') placing under the more lasting protection of the Minister a monument with respect to which an interim preservation notice is in force. **1968** *Guardian* 13 Aug. 5/8 Residents of..Bridgnorth have won a battle to preserve the Crown Hotel..from being knocked down... A preservation order on the building has been confirmed by the Minister of Housing and Local Government. **1971** 'J. Ashford' *Bent Copper* iv. 27 Parkham Green village was justly famous for its ancient architectural beauty... The whole village was covered by a special preservation order. **1978** N. J. Crisp *London Deal* iv. 83 Those houses are the subject of preservation orders now.

preservationist (prezəɹvē·ʃənist). [f. Preservation + -ist.] An advocate of preservation, *esp.* one who advocates the preservation of historic buildings or antiquities. Also *attrib.* or as *adj.* Hence **preserva·tionism,** the practice or advocacy of preservation.
1927 *Blackw. Mag.* Sept. 314/1 The excuses made for her [*sc.* the peregrine falcon] by modern 'Preservationists' are altogether paltry. **1937** *Archit. Rev.* LXXXII. 50/1 We would not..do more than note in passing Raphael's plea for the preservation of antiquities and his appointment as controller of monuments by Pope Leo X. Distance has dimmed too much for us the true nature of Raphael and his kind as constructive artists, as opposed to the 'preservationists' of today, for us to risk such an appointment. **1957** *Observer* 10 Nov. 3/4 Town planning councils, preservationist maniacs and 'good taste' committees all came in for a drubbing. **1959** *Archit. Rev.* CXXVI. 205/2 Old houses in New Zealand have not yet acquired a period value. Preservationism is completely absent; too much so, one is inclined to say. **1960** *Guardian* 25 Feb. 1/2 A preservationist undergraduate..threatened to send 'Save Cowley Vicarage' telegrams. **1961** *Architect & Building News* 21 June 815/1, I hope the preservationists will not let up for a minute in their struggle to protect our dwindling countryside. **1973** *Times* 16 Oct. 2/5 In 1964 the London County Council wanted to demolish it, but a public inquiry ended in a victory for preservationists led by Sir John Betjeman. *a* **1974** R. Crossman *Diaries* (1975) I. 623 She regarded it as pure sentimentalism and called it 'preservationism', a word of abuse. **1978** *Courier-Jrnl.* (Louisville, Kentucky) 16 Apr. D-4/1 Announcement of the discovery was delayed until recently ..because the mining concern..feared there would be a preservationist outcry.

preserve, *sb.* Add: **1. a.** (Later example.)
1839 J. D. Hooker in L. Huxley *Life J. D. Hooker* (1918) I. 43 That Capt. Ross did *not* intend to treat me thus..I am sure, from his asking me to tell the quantity of preserves for animals required.
5. *attrib.* and *Comb.*, as (sense 2) *preserve-can, -dish, -jar, -pot.*
1882 W. D. Hay *Brighter Britain!* I. v. 138 There were empty preserve-cans, gallipots, and oyster-shells! **1856** M. J. Holmes '*Lena Rivers* 108 The big preserve dish got broken. **1867** G. W. Harris *Sut Lovingood* 92 Preserve jars, vinegar jugs, seed bags, yarb bunches..all mix'd. **1885** E. P. Roe *Nature's Serial Story* xliii. 307 Racoons.. will uncover preserve-jars.., and with the certainty of a toper uncork a bottle and get drunk on its contents. **1969** R. & D. De Sola *Dict. Cooking* 183/2 *Preserve jar*, jar or pot for holding homemade preserves. **1854** Thoreau *Walden* 235 He goes to the mill-pond, she to her preserve-pot.

preserver. Add: **2. c.** = Life-preserver 2.
1912 *Chambers's Jrnl.* Aug. 636/1 In the panic which is certain to ensue after a wreck even the handling of this preserver would be awkward in the narrow passages and gangways.

pre-se·rvice, *a.* [Pre- B. 2.] Of or relating to a period before a person or thing is ready for service (*spec.* national service) or use.
1928 [see *in-service* s.v. *in *prep.* 17*]. **1944** W. Temple *Church looks Forward* xix. 135 During the war, with the Pre-Service Units of various kinds, still more has to be done. **1948** *News Chron.* 24 Aug. 2/3 Britain's self-supporting Boy Scouts still find it easier to make recruits than do the Treasury-backed pre-Service cadet organisations. **1963** F. F. Laidler *Gloss. Home Econ. Educ.* 56 *Preservice training*, training given to a person before he/she enters an occupation. **1967** M. Chandler *Ceramics in Mod. World* v. 150 Short-term pre-service protection can be provided by coating such bricks with tar. **1974** H. L. Foster *Ribbin'* iv. 164 A few years ago I participated as a guest lecturer in a preservice education program for teacher aides in a large city.

preserving, *vbl. sb.* Add: **b.** (Further examples.) **preserving sugar,** a coarse kind of sugar used in the preserving of fruits.
1861 Mrs. Beeton *Bk. Househ. Managem.* xxx. 758 In all the operations for preserve-making, when the preserving-pan is used, it should not be placed on the fire, but on a trivet. **1909** M. Little *Cookery Up-to-Date* xi. 235 Weigh the apricots and to every pound of fruit add three-quarters of a pound of preserving sugar. **1916** *Daily Colonist* (Victoria, B.C.) 13 July 7/1 (Advt.), Preserving Kettles. Best grey enamel, 12 quarts, strong bail handle. **1921** *Ibid.* 30 Oct. 3/1 (Advt.), Preserving Quinces, 2 lbs. for 25¢. **1926** *Ibid.* 13 July 7/5 (Advt.), Preserving Crocks. For butter or eggs; four-gallon size, with cover. **1948** *Good Housek. Cookery Bk.* 667 If you intend to make jams, jellies and marmalades regularly, you would be well advised to invest in a good strong preserving pan. **1949** *Nat. Geogr. Mag.* XLVI. 193/2 [Watermelon] will cross with the so-called preserving melon, or citron, which is simply a hard, white-fleshed watermelon, good only for preserving. **1970** *Canad. Antiques Collector* July–Aug. 23/2 One of the first commercial potteries..to make.. preserving jars. **1972** K. Stewart *Times Cookery Bk.* xx. 265 Strawberry Jam... 4 lb strawberries 3½ lb granulated or preserving sugar. **1977** *Western Morning News* 30 Aug. 2/3 Attractive Sale of Furniture and Effects, including:..preserving pan.

pre-se·t, *a.* [Pre- A. 1.] Decided or determined in advance; (of apparatus, etc.) set in advance of its operation.
1934 *Practical Wireless* 10 Mar. 1127/2 (*heading*) Bank of pre-set condensers. **1946** *Jrnl. Inst. Electr. Engin.* XCIII. II. 426/2 Pressure switches..which close circuit on the pressure falling to a pre-set point, may be installed. **1954** *N.Y. Times Mag.* 29 Aug. 49/1 The pre-set missile is not guided; it is fired, like a shell, at a pre-determined trajectory. **1961** G. Millerson *Technique Television Production* iii. 23 (*caption*) Height is pre-set and not readily adjusted. **1966** D. G. Brandon *Mod. Techniques Metallogr.* 102 The specimen stage..will remain in a pre-set position sufficiently long to permit any area to be photographed with the full resolution of the instrument. **1973** *Country Life* 29 Nov. 1796 (Advt.), This ultra-modern..oven..switches off at a pre-set time. **1978** R. V. Jones *Most Secret War* xlv. 447 Graphite rudders that were placed in the main jet so as to deflect the stream of incandescent gases and thus turn the rocket on to a pre-set trajectory both in bearing and in elevation.

pre-se·t, *v.* [Pre- A. 1.] *trans.* To set or fix (apparatus) in advance of its operation; to settle or decide beforehand.
1946 *Sun* (Baltimore) 18 July 13/2 A pressure gauge.. can be preset to open at an altitude where there is enough oxygen for a man to breathe. **1958** *Engineering* 11 Apr. 468/3 Temperature controllers which can be remotely pre-set from the control room. **1960** [see *bias *a.*, *sb.*, and *adv.* B. 7]. **1962** D. Slayton in *Into Orbit* 68 You could preset the trainer for any kind of mission you wanted to fly. **1962** *Which?* May 148/1 The grinder was pre-set to give a certain particle size. **1977** *Gramophone* Feb. 1343/2 Under a sliding lid..are four similar controls for pre-setting your choice of 'instant' programmes.

prese·xual, *a.* [Pre- B. 1.] Preceding or not yet influenced by sexual activity or sexual awareness; pre-pubertal; also *Gram.*, not (yet) differentiated by natural gender.
1919 M. K. Bradby *Psycho-Anal.* iv. xii. 164 Because our own sexuality is associated with sense of guilt, to us innocence implies a pre-sexual state of mind. **1925** D. H. Lawrence *St. Mawr* 204 They [*sc.* pine-trees] hedged one in with the power..of the pre-sexual primeval world. **1927** B. Malinowski *Sex & Repression in Savage Society* I. ix. 77 The development of pre-sexual life at this stage also differs in Europe and Melanesia. **1949** *Archivum Linguisticum* I. 168 A syntactic peculiarity of old Bulgarian declension..discloses the existence of a presexual division of nouns into animate and inanimate. **1961** R. F. C. Hull tr. *Jung's Coll. Wks.* IV. 118 The necessity for this becomes really urgent when we ask ourselves whether the intense joys and sorrows of a child in the first years of his life, that is, *at the presexual stage*, are conditioned solely by his sexual libido. **1971** G. H. Bourne *Ape People* xi. 254 Presexual play in humans often involves the nibbling or biting of the earlobes.

preshow: see *pre- B. 2a.

preshri·nk, *v.* [Pre- A. 1.] *trans.* To shrink (fabric) prior to cutting or (a garment) prior to sale, so as to prevent shrinkage following washing or cleaning.
1936 G. G. Denny *Fabrics* (ed. 4) I. 104 A process for completely pre-shrinking cotton and linen fabrics. **1963** *Home Dressmaking* (B.B.C.) i. 4 Pre-shrink wool, cotton, silk fabrics, linings, interfacings, tapings, etc., before cutting. **1975** J. Labarthe *Elements of Textiles* vii. 301 There is no economy to any customer to buy cotton garments..unless these have been preshrunk by a dependable process. **1978** *Detroit Free Press* 5 Mar. D 9/1 Preshrink fabric and fringe, or it may shrink and pucker from steam.
Hence **preshru·nk** *ppl. a.*; also **preshri·nkage,** the process of preshrinking.
1942 G. G. Denny *Fabrics* (ed. 5) v. 191 Use of terms 'Full shrunk', 'Preshrunk', 'Shrunk'..prohibited if there is residual shrinkage left in the goods. **1951** *Good Housek. Home Encycl.* 251/1 All cloth..should be shrunk before cutting, unless it is guaranteed pre-shrunk. **1960** *Guardian* 9 May 4/6 Pre-shrinkage is one of the things the Irish are rather good about. *a* **1963** L. MacNeice *Astrol.* (1964) vii. 232 Astrology has been used to sell anything from pre-shrunk shirts to alcoholic drinks. **1975** J. Labarthe *Elements of Textiles* vii. 301 An..advantage given by a satisfactory preshrinkage process is that the strength of the fabric has been increased. **1978** *Detroit Free Press* 16 Apr. A14/4 (Advt.), Water-repellent nylon taffeta with pre-shrunk cotton flannel lining.

preside, *v.* **3.** (Later examples.)
1967 *Decision & Decision-Makers in Mod. State* (Unesco) 82 The council is presided by the President or the Vice-President. **1974** *Amer. Speech 1971* XLVI. 113 The meeting was presided..by Dean Pinero.

president, *sb.* Add: **2. f.** The priest or minister who presides at the Eucharist; the celebrant.
[**1867** M. Dods et al. tr. *Writings of Justin Martyr & Athenagoras* 63 There is brought to the president of the brethren bread and a cup of wine mixed with water. (*Note*) This expression may quite legitimately be translated 'to that one of the brethren who was presiding'.] **1945** G. Dix *Shape of Liturgy* v. 111 Justin says: '... Then the bread is 'offered' to the president and a cup of water mingled with wine.' **1971** *Order for Holy Communion* (Alternative Services Series 3) 30 The Breaking of the Bread. The president breaks the consecrated bread, saying [etc.]. *Ibid.* 31 The president and the other communicants receive the holy communion. At the administration the ministers say to each communicant, [etc.]. **1973** in *Mod. Eucharistic Agreement* 63 The eucharistic gathering and its president live their dependence on the one Lord and great High Priest. **1977** *Oxf. Diocesan Mag.* Aug. 17/2 The building now consecrated, the Eucharist began, with the Bishop of Oxford as president, and the Bishop of Reading, the Archdeacon of Berkshire, the Vicar and the Curate..as concelebrants.
g. At some sporting events, a referee, judge, or official in charge.
1961 F. C. Avis *Sportsman's Gloss.* 285/2 *President*, the senior judge in a group, as required at international show jumping competitions. **1971** L. Koppett *N.Y. Times Guide Spectator Sports* xiv. 210 The official in charge of [fencing] competition is called the 'president'. **1975** *New Society* 10 July 81/2 The 'president' (ie, ref) of a fencing match. **1976** *Sunday Tel.* 13 Mar. 36/6 Too few countries trouble to train presidents—officials who take charge of bouts.

presidential, *a.* Add: Hence **preside·ntialism,** the system or practice of presidential government; **preside·ntialist,** a supporter or advocate of such government.
1964 J. E. S. Hayward in *Parliamentary Affairs* XVIII. 35 *L'Express* drew the conclusion that the Opposition must accept Presidentialism and find a candidate for the next election. **1965** *Economist* 28 Aug. 787/3 Professor Burns himself is a convinced Presidentialist (and Democrat), and offers some concluding recommendations for reducing the four-party competition to two. **1973** W. G. Andrews in *Political Stud.* XXI. 311 The French constitutional structure has undergone radical change from parliamentarism towards presidentialism since 1958. **1974** *Times* 6 Nov. 14/6 If one begins to indulge in presidentialism after the South American pattern, then we shall have changed republics. **1975** *Government & Opposition* X. 28 The bipolarizing pressures inherent in a system of presidentialism based on election by universal suffrage have led to changes.

presiding, *ppl. a.* Add: **b.** *presiding judge* (U.S.); **presiding elder,** an elder who has charge of a district in the U.S. Methodist Church; **presiding officer,** an official in charge of a polling-station at an election.
1831 J. M. Peck *Guide for Emigrants* 258 There are three [Methodist] districts, over each of which is a presiding Elder. **1844** I. D. Rupp *He Pasa Ekklesia* 447 A presiding elder, though no higher as to order than an elder, has charge of several circuits and stations, called collectively a district. **1904** G. H. Lorimer *Old Gorgon Graham* ix. 186 The Doc...knew more Scripture when he was sixteen than the presiding elder. **1961** W. E. B. DuBois *Worlds of Color* ix. 137 He emphasized to the Presiding Elder the plan of giving up the old church and moving across the river. **1802** *Deb. Congress U.S.* 19 Jan. (1851) 117 The constant change of presiding judges..hung up the business. **1874** 'H. Churton' *Toinette* xxiii. 245 Geoffrey's counsel called the attention of His Honor, the Presiding Judge of the Court, to the fact. **1745** *Life & Adventures B.-M. Carew* vii. 86 By this Means no presiding Officer has it in his Power to make one more than two, which sometimes happens in the Elections amongst other Communities. **1872** *Act* 35 & 36 *Vict.* c. 33 sched. I. § 21 The returning officer shall appoint a presiding officer to preside at each station. **1978** D. Devine *Sunk without Trace* xviii. 170 Voting papers, to be valid, must have the official stamp embossed on them by the Presiding Officer at the time of issue to the individual voter.

Presidium (prĭsi·diŏm, -z-). Also **Præsidium.** [Russ. *prezidium*, ad. L. *præsidium*, garrison, f. *præsidēre* (see Preside *v.*).] The presiding body or standing committee in a Communist organization, esp. in the Supreme Soviet. Also *attrib.*
1924 *Observer* 23 Mar. 13/5 In a second decree the Presidium of the Union C.E.C. decided to replace the sentence of ten years strict isolation passed on the Catholic Archbishop Ciepliak by the All-Russian C.E.C. by expulsion from the territories of the Union of Socialist and Soviet Republics. **1927** *Glasgow Herald* 10 Oct. 11 Mr Arthur Horner (South Wales), a member of the National Executive of the Miners' Federation, presided, and was supported by a presidium of 11. **1930** *Economist* 1 Nov. (Russian Suppl.) 1/2 The Central Executive Committee meets ordinarily about three times a year and, when it is not sitting, is represented by a small elected committee or Presidium, and by the Council of People's Commissaries, which is the executive organ. **1931** G. D. H. Cole in W. Rose *Outl. Mod. Knowl.* 727 The Congress [of Soviets]..is ..an occasional gathering of delegates, represented between sessions by a Præsidium. **1955** *Times* 9 May 10/3 The presidium of the Supreme Soviet yesterday annulled the Anglo-Soviet and Franco-Soviet treaties of alliance. *Ibid.* 14 July 6/6 The Praesidium, known in Stalin's day as the Politburo, is the highest policy-making body in the Soviet Union. **1958** *Spectator* 20 June 791/3 Khrushchev's last execution of a Præsidium member, in 1956, was kept quiet for a month or so. **1960** *Evening Bull.* (Philadelphia) 14 Dec. 13/6 A Russian official who made the mistake of referring to a member of the Presidium (the 14-member body that rules the Soviet Union) as a fool. **1968** *Listener* 1 Aug. 133/3 Dubcek started at the top with his bloodless revolution. First, he infiltrated the Party Praesidium, effectively the Party's ruling body, and excluded Novotny supporters in separate reshuffles. **1974** L. Deighton *Spy Story* xiii. 129 Madame Furtseva, the first woman to reach the Presidium of the Central Committee.

presie, var. *prezzie.

pre-sleep: see *pre- B. 2 a. **pre-soak:** *pre- A. 1 and 2. **pre-socialist:** *pre- B. 1 d.

pre-Socra·tic, *a.* and *sb. Philos.* Also **Presocratic, presocratic.** [Pre- B. 1.] **A.** *adj.* Of or relating to the period before Socrates (chiefly the sixth and early fifth centuries B.C.) when, in Greece, systematic enquiry into things and their causes began.
1871 [in Dict. s.v. Pre- B. 1]. **1892** J. Burnet *Early Greek Philos.* 2 The common practice of treating this younger contemporary of Sokrates [*sc.* Demokritos] along with the 'pre-Socratic philosophers' has obscured the true course of historical development. **1913** P. V. Cohn *Nietzsche's Compl. Wks.* (Index) XVIII. 117 The real philosophers of Greece pre-Socratic. **1957** Kirk & Raven *Presocratic Philosophers* p. vii, We have limited our scope to the chief Presocratic 'physicists' and their forerunners, whose main preoccupation was with the nature (physis) and coherence of things as a whole. **1964** C. S. Lewis *Discarded Image* iii. 37 The pre-Socratic philosophers of Greece invented Nature. **1974** *Nature* 8 Nov. 130/2 In the treatment of Greek science emphasis is laid upon the importance of the presocratic belief that causal relationships existed between natural phenomena. **1977** I. Murdoch *Fire & Sun* 33 Sexual love (Aphrodite) as cosmic power had already appeared in Presocratic thought in the doctrines of Empedocles.
B. *sb.* Any of the Greek philosophers of the sixth and fifth centuries B.C. who preceded Socrates (d. 399 B.C.).
1945 B. Russell *Hist. Western Philos.* (1946) I. xiii. 126 'The Good' dominated his [*sc.* Plato's] thought more than that of the pre-Socratics. **1957** Kirk & Raven *Presocratic Philosophers* 1 The Neoplatonist Simplicius,..who lived a whole millennium after the Presocratics, made long and evidently accurate quotations, in particular from Parmenides, Empedocles, [etc.]. **1972** E. Hussey *Presocratics* i. 1 What gives the group of Presocratics such unity as it possesses is..that all these men were involved in the movement of thought which led to the separation of science and philosophy from one another and from other ways of thinking. **1977** I. Murdoch *Fire & Sun* 51 Nor is he at all like the cosmic 'gods' of the Presocratics.

presolar: see *pre- B. 1. **pre-sold:** *pre-sell *v.* **pre-Soviet:** *pre- B. 1 a. **pre-spiracular:** *pre- B. 3.

press, *sb.*[1] Add: **I. 5*.** *Psychol.* Something in the environment to which (a need in) the organism reacts (see quot. 1938).
1938 H. A. Murray *Explorations in Personality* ii. 40 A tendency or 'potency' in the environment may be called a *press*... For example, a press may be nourishing, or coercing, or injuring, or chilling,..or amusing or belittling to the organism. *Ibid.* 42 The endurance of a certain kind of press in conjunction with a certain kind of need defines the duration of a single episode. **1953** *Jrnl. Abnormal & Social Psychol.* XLVIII. 532/2 So we know *two* things about his narrators: their ambition and their most recent press. That press, as our hypothesis predicts, they projected directly into their..Tests. **1969** J. W. Getzels in Lindzey & Aronson *Handbk. Social Psychol.* (ed. 2) V. xlii. 501 There was no evidence that student press influenced the level of aspiration, at least so far as Merit students are concerned. **1973** *Jrnl. Genetic Psychol.* CXXIII. 87 Four slides were used to test for the presence of hostile press.
II. 6. b. In Gymnastics, a raising of the body by continuous muscular effort.
1901 *Health & Strength* Apr. 36/2 (*heading*) One arm body press... Lie flat on the ground..and with hand beneath centre of chest press the body up to arm's length. **1956** Kunzle & Thomas *Freestanding* i. 22 The presses to handstand are one of the best forms of strength training because at the same time the gymnast learns how to fight for and maintain a hand balance when the arms feel extremely tired.
c. *Weight-lifting.* A raising of a weight from the floor to shoulder-height followed by its gradual extension above the head.
1908 *Health & Strength Ann.* 93 Continental lifts differ considerably from those in practice in this country... The Continental 'Press' is a cross between the above [*sc.* the 'Push'] and the English 'Press'... The Continental 'Press' can only be distinguished from our 'Arm Press' by a slight side wriggle. **1914** *Ibid.* 83 Thomas Inch lifted 304½ lbs. (bent press) at Scarborough in December. **1925** F. G. L. Fairlie *Official Rep. VIIIth Olympiad, 1924* 255 Middle-weights... Two hands, Military Press: Galimberti (Italy), 214½ lb. **1928** *Health & Strength Ann.* 77 Lifters are urged to maintain themselves in a state of readiness on the three Olympic lifts, viz: 'Two Hands Clean and Military Press with Barbell', 'Two Hands Snatch', and the 'Two Hands Clean and Jerk with Barbell'. **1935** *Encycl. Sports* 704/2 There are swings, presses, snatches, jerks, all made with one hand, as well as two-hand and shoulder lifts. **1975** *Oxf. Compan. Sports & Games* 1099/1 At the 1924 Olympic Games the lifts were one hand snatch, opposite one hand jerk, two hands clean and press, two hands snatch, and two hands clean and jerk.
d. The action of pressing clothes.
1932 D. C. Minter *Mod. Needlecraft* 145/2 Muslin and lawn dresses usually require a final all-over press. **1957** J. Osborne *Look Back in Anger* I. 16 I'll give them a press while I've got the iron on. **1962** M. Duffy *That's how it Was* iii. 33 The girls would..run up something new..to wear the same evening with a quick press before they went out. **1975** Byfield & Tedeschi *Solemn High Murder* i. 6 'These things could do with a press if that's possible.' The smell of tropical mildew clung to the rumpled winter-weight clericals he handed the man.
e. In Basketball, any of various forms of close marking by the defending team. Also *transf.*
1961 J. S. Salak *Dict. Amer. Sports* 341 *Press* (basketball), a maneuver designed to hamper the offensive team's ability to move the ball toward their basket. There are many types of 'presses'. **1971** L. Koppett *N.Y. Times Guide Spectator Sports* iii. 86 The press itself creates openings for the offense. **1976** *Honolulu Star-Bull.* 21 Dec. H-1/5 A full-court press enabled Kalani to wipe out a 13-point third quarter lead. **1978** W. Safire *Political Dict.* 248 'Full-court press' became White House lingo in the late sixties... In politics, the term has come to mean a strenuous effort to get legislation passed probably because of its resemblance to 'all-out pressure'. In basketball, however, the phrase is used only to describe a defense.

PRESS

13. e. Also *to go to press* (also *fig.*), *to read for press.*

1715 T. Hearne *Let.* 2 Feb. (MS.), I find Mr. Urry's Chaucer advertised as being to go to yᵉ Press in a little time. **1810** *Irish Mag.* III. 279/2, I shall, therefore..go immediately to press, be squeezed into the genteelest form I can. **1846** G. Dodd *Brit. Manuf.* 6th Ser. 57 To read for press—that is, to search for the minutest errors. **1929** Yeats *Let.* 13 Sept. (1954) 768, I will work at it here and there... I should go to press with it next spring. **1933** [see *hope *sb.*^{1} 4 a]. **1951** [see *bed *sb.* 6 c]. **1961** *Financial Times* 11 July 6 At the time of going to press..it is not possible to determine any very definite trend of trading at the present time.

h. *a good press*: see *good *a.* 13. Hence *to have* (*receive*, etc.) *a good* (or *bad, mixed*, etc.) *press*: to be favourably (or unfavourably, divergently, etc.) commented on or criticized in current newspapers, journals, etc. Also *transf.*, to receive (favourable, etc.) publicity, to be (favourably, etc.) appraised in conversation or in literature.

1908 [see *good *a.* 13]. **1913** R. Fry *Let.* Oct. (1972) II. 373 Has it [*sc.* an exhibition] been a success, and has there been any decent Press on it? **1915** [see *good *a.* 13]. **1920** *Sat. Rev.* 10 July 26 Mr Austen Chamberlain has a very bad press. **1928** [see *good *a.* 13]. **1932** *Statesman* (Calcutta) 2 Aug., It was the clearest case, for years, of how county cricket should not be conducted. Allom had a lively Press last Wednesday! **1934** H. G. Wells *Exper. Autobiogr.* II. vii. 501, I wish I could hear at times of people still reading these three stories: they got, I think, a rather dull press. **1958** *Listener* 13 Nov. 769/1 Cromwell had rather a mixed press for his great day. **1961** P. Kemp *Alms for Oblivion* 1 In Britain General Franco had not enjoyed a good Press. **1967** *Observer* 26 Nov. 8/3 The Phoenicians had a largely hostile press from the Bible and from their rivals the Greeks and Romans. **1976** *Women's Report* Sept./Oct. 4/1 Chiswick Women's Aid has had a good press recently because the DHSS has withdrawn some of its grant money. **1977** *Sunday Times* 30 Jan. 38/1 Rape is enjoying a very educative Press from TV dramatists at the moment.

i. Usu. with *the*: used collectively for journalists, esp. reporters; also, of an individual reporter.

1926 in S. Bent *Ballyhoo* (1927) ii. 55 At least a half dozen times since the wedding the unfortunate composer has been badgered by the press until some such statement as 'we are very happy' has been wrung from him. **1949** 'J. Tey' *Brat Farrar* xii. 102 'He says he's a reporter,' Lana said... 'Oh, *no*!' Bee said. 'Not the Press. Not already.' **1951** M. Dickens *My Turn to make Tea* iv. 45 'Here's the Press, Waldo,' his wife told him, 'come to put Marjorie in the *Post*.' *Ibid.* vii. 122 Sister..said that if I was The Press, Matron had deputed her to show me round. **1956** C. Mackenzie *Thin Ice* x. 129 The dinners of the East Indiamen were held once a quarter without excessive formality and, what was more important for the speaker, without the Press. **1973** A. S. Neill *Neill! Neill! Orange Peel!* II. 235 The Salvation Army damsel..came to a young man sitting alone. 'Are you saved?' 'Press,' he said. 'Oh, I beg your pardon,' and she moved hastily away. **1974** P. N. Walker *Major Incident* viii. 95 As the police were desperately trying to clear the streets, the first of the press were trying to drive in. **1978** M. Butterworth *X marks Spot* II. i. 73 Arrange for the exhumation forthwith. Seal off Highgate Cemetery... No Press. No television.

IV. 14. (Further *attrib.* examples.)

1952 J. Gloag *Short Dict. Furnit.* 374 *Press cupboard*, a large cupboard with a superstructure consisting of a shelf with smaller cupboards behind it..introduced during the second half of the 16th century. **1959** L. A. Boger *Compl. Guide Furnit. Styles* xxii. 384 The name *press cupboard* was given in America to a form of cup-board resembling the English hall and parlor cupboard. **1970** *Canad. Antiques Collector* Jan. 29/1 A further kind of cupboard..was called a press, or press-cupboard, and was about the same general size and shape as a modern wardrobe. **1975** *Oxf. Compan. Decorative Arts* 651 *Press cupboard*, a large cupboard, sometimes confused with a court cupboard, which came into use in the latter half of the 16th c. and remained in fashion until the 18th c. It had the upper part recessed with contained cupboards and a shelf running in front of them.

V. 15. a. (*a*) *press-shop* (later examples), *-table*; (*b*) *press advertising, boss, camera, campaign, censor, censorship* (further examples), *club, freedom, interview, pass, photo, photograph, photographer, photography, ticket.*

1961 *Travel Topics* June 41/1 When one first thinks of press advertising, it conjures up the thought of taking space in the national dailies or Sunday papers. **1932** E. Pound *Let.* 18 Feb. (1971) 239 There is no reason why young England shd. pardon the ineffable polluters and saboteurs. What they have done to stifle literature in Eng., tho not so important as the press-bosses' stifling of economic discussion, is all of piece [*sic*]. **1948** A. L. M. Sowerby *Dict. Photogr.* (ed. 17) 89 The typical Press camera consists of a frame containing the shutter, fitted at the back for plates in dark slides and with the lens carried on a flat panel supported at the four corners by struts and connected with the camera body by bellows. **1964** M. McLuhan *Understanding Media* xx. 200 The press camera contributed to radical changes in the game of football. **1974** *Encycl. Brit. Macropædia* XIV. 330/2 Press cameras are loaded with sheet film..for fast, hand-held shooting; they are traditionally of folding-bellows design with a lens standard on an extendable baseboard. **1903** 'Vigilans sed Æquus' *German Ambitions* vi. 86 The German press campaign against our army in South Africa. **1951** M. McLuhan *Mech. Bride* (1967) 40/1 The working woman was put into adolescent short skirts and told in big press campaigns that the age-old tyranny of man was at an end. **1900** W. S. Churchill *Let.* 1 May in R. S. Churchill *Winston S. Churchill* (1967) I. Compan. II. 1174 Wolverton is here, one of the press censors. **1940** L. Durrell *Spirit of Place* (1969) 65 George Seferiades..chief foreign press censor, who is a remarkable poet and person. **1939** 'G. Orwell' in *New English Weekly* 12 Jan. 203/2 The radio, press-censorship, standardised education and the secret police have altered everything. **1978** 'A. York' *Tallant for Disaster* vi. 93 Even the British have press censorship... What about all those D-notices and things? **1896** *Peterson Mag.* Mar. 311/1 The Pittsburgh Women's Press Club made a wise choice in selecting for a secretary Miss Marie de Sayles Coyle. **1967** L. T. Braun *Cat who ate Danish Modern* ii. 20 Why don't we meet for drinks at the Press Club? **1974** *Times* 18 Nov. 15/1 Advertisers threaten press freedom if they try to use their advertising power as a form of censorship. **1923** *Radio Times* 23 Sept. 18/3 Mr. J. W. Reith, the General Manager of the B.B.C...has managed to avoid..the usual press interviews. **1976** L. Henderson *Major Enquiry* viii. 47 The report of Shenton's press interview was given great prominence by the *Evening News*. **1914** *Automobile Topics* 6 June 303/1 Primary cause for protest was the method adopted by the Speedway management of distributing press passes. **1977** H. Innes *Big Footprints* II. i. 103 They weren't interested in my press pass or the fact that I was an American TV man. **1964** M. McLuhan *Understanding Media* xx. 200 A press photo of battered players in a 1905 game. *Ibid.*, The press photo coverage of the lives of the rich. **1980** R. McCrum *In Secret State* xviii. 168 The dashing whizz-kid of the press photos. **1944** M. Laski *Love on Supertax* ii. 31 Suppose you wanted a really flattering press photograph and I knew someone who'd fake it up. **1974** 'J. Le Carré' *Tinker, Tailor* xxiii. 196, I had with me the American press photographs of the arrest. **1922** M. Arlen *Piracy* 7 Those young women of patrician and careless intelligence, whom it is the pet mistake of bishops, diarists, press-photographers, and Americans, to take as representing the 'state' of modern society. **1974** 'M. Innes' *Appleby's Other Story* v. 44 One has to think of the reporters and press-photographers. **1922** L. Warren *Journalism* xxi. 230 In a book such as this it is quite out of the question to go into details concerning press photography. **1980** *Times* 3 Mar. 14/6 *Life*..was press photography for the press photographer at its most splendid. **1958** *Engineering* 11 Apr. 461/1 The current expansion programme, which includes the opening of a new press shop later this year and a new assembly building early in 1959. **1959** *Motor Manual* (ed. 36) i. 8 In the latest press shops, all the presses engaged in the production of one component are arranged in a long line, and are linked by roller conveyors. **1971** *Engineering* Apr. 20/2 Mounting of the equipment on the movable press-table is also easy. **1851** J. Chapman *Diary* 10 July in G. S. Haight *Geo. Eliot & J. Chapman* (1940) 191 Spencer gave me a ticket for the Opera..and might have had an excellent place but for the vexing regulation that 'press tickets' must be exchanged which destroyed my chance of admittance. **1976** 'D. Fletcher' *Don't whistle 'Macbeth'* 17 Some idiot in the box office had allocated press tickets for the first matinée instead of the first night.

b. (*b*) Operated by pressing, as *press-cock, switch*; (see also sense 16d below); also *press-button *sb.* and *a.*, *press-fastener, etc.

1932 *Jrnl. R. Aeronaut. Soc.* XXXVI. 854 The ideal starter..was a self-contained unit in which only one simple operation, such as pressing a press-cock, was required. **1892** E. J. Houston *Dict. Electr. Words* (ed. 2) 424/2 *Pressel*, a press switch or push connected to the end of a flexible pendant conductor.

16. a. press-mould (see quot. 1974); so **press-mould** *v.*, **press-moulding** *vbl. sb.* **press-moulded** *ppl. a.*

1971 *Country Life* 27 May 1303/1 The [Staffordshire slipware] dish is press-moulded and is signed 'I.S.'. **1974** Savage & Newman *Illustr. Dict. Ceramics* 233 *Press-mould*, an absorbent mould made of lightly fired clay or plaster of Paris, and into which clay is pressed by hand to make such objects as small ornaments for relief or sprigged decoration. *a* **1977** *Harrison Mayer Ltd. Catal.* 95/2 A range of simple Press Moulds in 5 basic shapes. **1969** Speck & Sutherland *Eng. Antiques* 190/2 Press-moulded glass. **1958** H. Wakefield in Edwards & Ramsey *Connoisseur Period Guides: Early Victorian Period* 100/2 It was the period in which the process of press-moulding was first developed for the production of dishes and other open shapes.

b. press attaché, a diplomat responsible for the dealings of an embassy with the press; **press baron**, a powerful newspaper owner, a newspaper magnate, esp. one who is a member of the peerage (see *baron 2 b); **press boat**, a boat reserved for the use of reporters at a boat race or similar event; **press book**, (*a*) a volume of press cuttings; (*b*) a book printed at a private press, a type of fine book (see *fine *a.* 12 d); **press-box** (earlier and later examples); **press card**, a document that authorizes a reporter to practise journalism, or one that gains him admission; **press clipping** orig. *U.S.* = *press cutting*; also *attrib.*; hence *press-clipper*; **press conference**, a meeting at which journalists and other representatives of the news media are given an opportunity to put questions to a politician, writer, etc.; also (*rare*) (with hyphens) as *v. trans.*; **press corps**, a group of reporters (usu. in a specified place); **press-corrected** *a.*, designating a text of which the proof sheets have been corrected before publication; **press correction**, (*a*) the act or process of correcting errors in a text during preparation for publication; (*b*) an error marked for correction; **press-corrector**, a proof-reader; **Press Council**, a body established in the U.K. in 1953 to raise and maintain professional standards among journalists; **press coverage**, the reporting (of an event) by the press; **press-cutter** = *press cutting agency*; **press cutting** (earlier and later examples); *press-cutting agency* (further examples), *album, book, bureau, people*; **press day**, (*a*) a day on which journalists are invited to an exhibition, a performance, etc.; (*b*) the day on which a journal goes to press; **press digest**, a digest or summary of press reports; **press kit**, a dossier prepared for journalists; **press notice**, a review in a newspaper or other periodical of a book, play, or the like; **press number**, a number at the foot of the page of an early printed book showing on which press or by which printer the page was printed (see quot. 1961); **press office**, an office within an organization or government department responsible for dealings with the press; **press-proof** (examples); **press release**, an official statement offered to newspapers for publication; **press-revise** (examples); **press run**, a spell of allowing a printing-press to run; the amount of printed material produced as a result; **press secretary**, a secretary who deals with publicity and public relations; **press show**, a performance given for the press, esp. a film shown to journalists before general release; also *attrib.*; so **press-show** *v. trans.*; **press stand**, a section of the tiered seats for spectators at racing or field events reserved for reporters; also *attrib.*; **press table**, a table reserved for journalists esp. in a court of law; **press time**, the time at which a newspaper goes to press; **press view**, a viewing of an exhibition by journalists before it is open to the general public.

1938 A. Barmine *Mem. Soviet Diplomat* i. 16 When Krestinsky was at the Berlin Embassy, Stern had served for many years as his Press Attaché. **1980** 'R. Deacon' *Spy!* iii. 86 She had made a favourable impression with the press attaché. **1958** *Spectator* 20 June 794/3 The history of the rise in the peerage of the press barons..is one of the shoddiest episodes in the whole story of the press. **1975** *Times* 3 July 14/3 (*caption*) Press barons together; Lord Thomson shares a smile with..Lord Beaverbrook. **1870** D. J. Kirwan *Palace & Hovel* xxiv. 363 By the side of the Press boat, the Umpire's boat..was anchored, many of the passengers wearing the rival colors. **1901** R. H. Davis in *Scribner's Mag.* Aug. 131/1 The press-boats buried their bows in the waters of the Florida Straits and raced for the cable-station at Port Antonio. **1897** A. Beardsley *Let.* 6 Jan. (1971) 240, I quite forgot to return you the cuttings for your press book. I enclose them now. **1930** *Publishers' Weekly* 19 Apr. 2116/2 The past five years has seen keen collecting interest in Press books both early and modern. **1976** *Times Lit. Suppl.* 5 Mar. 271/3 There is also a large output of less sumptuous..books..produced by a host of part-time private presses, small publishers who commission fine books, and trade printers who..take time off to print a worthwhile book... It is to cover these books that the term 'press books' has been coined. **1889** *Sporting Life* (Philadelphia) 10 July 5/5 The upper stand ..will contain the seats for ladies and their escorts and the private boxes, not forgetting the press box. **1976** Dexter & Makins *Testkill* 61 Festing followed me to the Press box and sat..in silence until the end of the game. **1934** *N.Y. Times* 20 Feb. 18/3 The number of press cards has been cut by 55 per cent. **1951** 'A. Garve' *Murder in Moscow* iii. 41, I went on to see the head of the Soviet Press Department and collect my press card. **1976** *Times* 27 Feb. 15/2 The use of fake press cards by soldiers in Ulster puts genuine journalists in danger. **1903** *Everybody's Mag.* July 127/1 The press-clippers caught every reprint. **1903** *Christendom* Apr. p. ii (Advt.), United States Press Clipping Bureau. **1904** G. B. Shaw *Let.* 6 Apr. (1972) II. 416 PPS I subscribe to an American press clipping agency. **1942** D. Powell *Time to be Born* (1943) i. 20 Julian fussed with some press clippings. **1975** *Language for Life* (Dept. Educ. & Sci.) xv. 232 The same is no less necessary for English, the 'materials' of which are duplicated sheets, press-clippings, files, photographs, and so on. [**1923** A. Cecil in *Cambr. Hist. Brit. Foreign Policy* III. viii. 628 [During the 1914–18 war] Lord Robert Cecil used to hold a kind of weekly reception for American journalists, when they were at liberty to question him on Foreign Affairs.] **1937** *Time* 1 Mar. 9/3 One afternoon Mrs. Roosevelt stole into the President's regular semi-weekly press conference to say good-by to her husband. **1953** *Manch. Guardian Weekly* 2 Apr. 7/4 Another general was soon to press-conference himself into the Presidency. **1958** *New Statesman* 15 Mar. 332/3 This programme..takes one of two forms: either it is a press-conference in which an eminent person is questioned by journalists in several countries, or it is a straight discussion between those taking part. **1976** *Eastern Even. News* (Norwich) 9 Dec. 1/5 'I don't believe anyone in this industry wants a dispute,' Sir Derek said at a Press conference during a visit to Bedlay Colliery, Lanarkshire. **1940** G. Seldes *Witch Hunt* i. 6 He came to Trier and used the American press corps. **1974** *Sunday Times* 21 July 1/3 A 200-strong international Press corps confined to the hotel by the island's [*sc.* Cyprus's] 24-hour curfew. **1964** F. Bowers *Bibliogr. & Textual Crit.* v. ii. 139 Editors should choose the First Folio press-corrected reading..instead of the quarto and the uncorrected Folio reading. *Ibid.* I. iii. 19 A brief look at some problems of press-correction will illustrate with suitably neutral examples. *Ibid.*, Press-correctors do not deliberately introduce typographical errors in the copy. **1947** *Minutes of Evidence R. Comm. on Press* 12 Nov. 23/2 in *Parl. Papers 1947–8* (Cmd. 7330) XIV. 533 The proposal is that there should be a Press Council, something..approximating to the General Medical Council.., and that there should be punishments and rewards instituted in order to raise and preserve the standards of professional behaviour within the newspaper profession. **1953** *Times* 5 Nov. 4/2 The new Press Council had proclaimed deep concern at the unwholesome exploitation of sex by certain newspapers and periodicals. **1977** *Evening Post* (Nottingham) 27 Jan. 6/1 If a newspaper were genuinely hostile to the Labour Party and decided, as a result, that in future no reference would be made to it or its troubles and triumphs, there would be an excellent case for reporting the newspaper to the Press Council for failing to do its duty. **1957** J. Mitford *Poison Penmanship* (1979) 34 These examples represent only a very tiny sampling of press coverage of this part of the case. **1961** C. Willock *Death in Covert* iii. 71 All goes down to advertising. Whynne says we'll get it back twice over in press coverage. **1976** *Times* 27 Feb. 15/1 Documents from army sources critical of press coverage in Northern Ireland. **1901** G. Gissing *Let.* 30 Nov. in *G. Gissing & H. G. Wells* (1961) 200, I have never dared to subscribe to the press-cutters, for I remember..the day when a press notice meant a sneer which disturbed my work. **1898** G. B. Shaw *Let.* 24 Mar. (1972) II. 22 A sheaf of pamphlets & press cuttings. *a* **1916** 'Saki' *Infernal Parliament* in *Square Egg* (1924) 148 Pasting notices of modern British plays into a huge press-cutting book. **1922** A. E. Housman *Let.* 26 Oct. (1971) 206 The press-cutting agency sends me..more notices than I want to see. **1929** T. S. Eliot *Dante* iii. 63 He, Dante Alighieri, was an important person who kept press-cutting bureaux busy. **1936** 'G. Orwell' *Let.* 26 Aug. in *Coll. Ess.* (1968) I. 228, I don't know what sort of reviews it got in France—I only saw about two..the press-cutting people didn't get them. **1941** V. Nabokov *Real Life S. Knight* xi. 102 A press-cutting agency began to pepper him with samples of praise. **1942** 'M. Innes' *Daffodil Affair* i. 37 He has consulted his colleagues; assistants have been turning over press cuttings. **1967** 'E. Peters' *Black is Colour* iii. 53 Things like the press-cutting book and the photographs get into arrears very easily. **1967** J. B. Priestley *It's an Old Country* vii. 84 Magazines and paperbacks, jigsaw puzzles, photograph and press-cutting albums. **1923** A. Huxley *Antic Hay* vii. 103 It was Press Day. The critics had begun to arrive. **1956** J. Symons *Paper Chase* xiii. 99 'Press day. Very busy.' He waved the galleys. **1972** C. Fremlin *Appointment with Yesterday* xiv. 113 The Editor ringing up, more and more irate, as press day drew near. **1958** *New Statesman* 20 Sept. 368/3 The press-digest which the President and Mr Dulles receive from the US embassy in London. **1977** G. Markstein *Chance Awakening* xxv. 76 The press digest was lying on his desk. **1968** *Globe & Mail* (Toronto) 17 Feb. 1/8 The ad hoc committee of five had already quietly rented space in a downtown Ottawa office building and prepared a slick press kit. **1977** *New Yorker* 3 Oct. 36/2 Our advance word on this event [*sc.* the publication of a new encyclopaedia] came to us in the form of a fat press kit, stuffed with fact sheets and kind words about the work. **1888** 'Mark Twain' *Let.* 1 Oct. in C. Clemens *Mark Twain* (1932) iii. 49, I thank you ever so much for not forgetting to remember to send me the press notice. **1977** J. Aiken *Last Movement* i. 37 'What about your opening?'.. 'Big success. I'll show you our press notices.' **1895** *Funk's Stand. Dict.*, Press-number. **1949** *Harvard Library Bull.* III. II. 198 (*title*) Press numbers as a bibliographical tool. **1961** T. Landau *Encycl. Librarianship* (ed. 2) 283/2 *Press number*, small figures which in books printed between 1680 and *c.* 1823 often appear at the foot of a page, sometimes twice in a gathering. The figures indicate on which press in the printer's workshop the sheet was printed or perhaps the identity of the worker. **1937** L. Hellman *Diary* 17 Oct. in *Unfinished Woman* (1969) viii. 87, I have been to the Press Office [in Valencia]..and paid a visit to Rubio, the Press Chief. *a* **1974** R. Crossman *Diaries* (1976) II. 269, I must send it straightaway across to the Press Office in Transport House. **1841** W. Savage *Dict. Art of Printing* 597 *Press proof*, a good impression of a sheet of a work, or of a job, to read it carefully by, and to mark the errors, previous to its being put to press. **1972** P. Gaskell *New Introd. Bibliogr.* 115 The third and final stage of proof correction was the press proof, when a forme or sheet was read for residual blemishes..just before the actual printing run was about to begin. **1958** M. H. Saringulian *Eng.-Russ. Dict. Libr. & Bibliogr. Terms* 148/1 Press release. **1964** W. Markfield *To Early Grave* (1965) ii. 29 He sent out press releases, and the *Brooklyn Eagle* ran a small story. *a* **1974** R. Crossman *Diaries* (1975) I. 67, I therefore gave instructions that for one month all the press releases and all the actual letters to authorities written in my name on planning permissions and compulsory purchase orders should be sent to me. **1976** *Oxf. Diocesan Mag.* July 14/2 There must be the news angle to the press release which, of course, should be factual and not based on rumours or hearsay. **1888** C. T. Jacobi *Printers' Vocab.* 103 *Press revise*, the final proof for press or machine. **1960** G. A. Glaister *Gloss. Bk.* 324/1 *Press revise*, an extra proof from the corrected type when ready for machining. **1958** *New Statesman* 15 Mar. 328/2 Since there is no 'preventive censorship', a paper which incurs the wrath of the government risks losing its entire press-run, which is simply impounded and placed in the Reuilly Barracks. **1976** M. Ierley *Year that tried Men's Souls* III. 198 (*caption*) At the left is the page as it appeared when Publisher Benjamin Towne began his press run. **1959** J. Ludwig in *Tamarack Rev.* Summer 20 Eisenhower with that puzzled look which meant if his press secretary didn't say something fast he was a goner. **1967** H. P. Levy *Press Council* p. xiii, Sir Richard Colville, the Press Secretary to Her Majesty the Queen, kindly read the chapter on the Royal Family in typescript. **1976** P. Alexander *Death of Thin-skinned Animal* xv. 150 He..announced himself as the London correspondent of *Paris Match* and said he'd like to speak to Colonel Njala's press secretary. **1958** *Vogue* July 44 American horror films..are never press-shown and are a disappointment to connoisseurs. **1961** *John o' London's*

15 June 671/1 A hard-boiled press-show audience. **1962** *Ibid.* 2 Aug. 115/1 On my way to the press-show of *The Lion*. **1963** *Movie* Jan. 20/3, I don't think there are any plans for press-showing it. **1972** *Times* 3 June 7/3 In Rome..I started going to press shows. **1914** *Automobile Topics* 6 June 303/1 Incidentally each applicant was put through the third degree in order to establish his complete identity and right to the press stand privileges. **1915** G. Patten *Courtney of Center Garden* 53 Passing the press stand, Whip caught Chatterton's eye again. **1937** E. Rickman *On & off Racecourse* vi. 137 He would usually watch the racing from the press-stand. **1922** Joyce *Ulysses* 454 From the presstable, coughs and calls. **1974** F. Nolan *Oshawa Project* i. 1 By the time the speeches started, the general was drunk... Every correspondent at the press table..could see the signs. **1927** S. Bent *Ballyhoo* ix. 240 It may be timely..but the reasons for printing it are that there is a glut of space to be filled in advance of news press-time, and that it must be filled with bait which will give the paper 'attention value'. **1978** *Rugby World* Apr. 19/1 At press-time, Royal High were level on 14 points with Glasgow Academicals and Madras, each club having two games to play, but with only one of the sides to go up alongside Leith. **1890** G. B. Shaw *London Music 1888–89* (1937) 284 My ticket for the Press view at the Old Masters on Friday! *Ibid.* 368, I have been at the Royal Academy all day, 'Press-viewing' it. **1929** R. Fry *Let.* 27 Dec. (1972) II. 646, I may be able to wangle you one [ticket] for the Press view on Monday.

c. (sense *6d) **press cloth**, a piece of cloth placed between the fabric and iron while pressing; **press line**, a crease made by a pressing iron; **press mark**, a mark left on fabric by the impress of an iron; hence **press-mark** *v. trans.*; **press-pad**, a soft pad used in pressing clothes.

1918 M. J. Rhoe *Dress you Wear* xi. 127 Nearly all pressing is done over a damp press cloth. **1933** A. M. Miall *Home Dressmaking* vii. 51 You should have a second wet press-cloth ready, and change to it as soon as the first dries, to avoid scorching. **1964** *McCall's Sewing* viii. 118/1 Always use a press cloth to prevent shine when necessary to press on the right side of the garment. **1979** *Tucson* (Arizona) *Citizen* 20 Sept. 2B/3 A final pressing (with press cloth) from the right side will give your coat (and its hem) a brand-new look. **1947** C. Talbot *Compl. Bk. Sewing* xxxi. 208/1 Remove the sharp line by moving the seam back and pressing the sleeve under the seam, removing the press lines from the sleeve. **1948** H. Hall *Home Dress-Making Simplified* vii. 64 It is important to press-mark all the side seams of the waist, the shirt, and the sleeves, as these seams will be the fitting lines of the dress. **1948** E. L. Towers *Standard Processes in Dressmaking* xvi. 116 If press marks appear on the right side of the garment, hold the fabric in the steam of a kettle to remove them. **1974** J. Robinson *Penguin Bk. Sewing* I. ii. 44 If during making a few press marks..do show then remove these by steaming. **1924** W. D. F. Vincent *Cutters' Pract. Guide Overcoats* 73/1 A good plan when damping fronts, lapels and collar is to damp through a double piece of cloth from the back, the silk being face down on the soft cloth press-pad.

d. From the vb. stem (cf. sense 15 b in Dict. and Suppl.): **press fit** *Engin.*, an interference fit between two parts in which one is forced under pressure into a slightly smaller hole in the other; cf. *shrink fit* s.v. *shrink *v.* 17; hence **press-fitted** *a.*; **press-key**, a control or switch similar to a piano key, operated by pressing the end with the finger.

1888 *Lockwood's Dict. Mech. Engin.* 265 *Press fit*, a fitting of contiguous parts slightly tighter than a sliding fit.., to allow of the sliding parts being pressed together with a hydraulic press. **1902** *Internat. Libr. Technol.* III. § 22. 33 In a press fit, the internal piece..must be enough larger than the hole to insure the development of enough friction between the two pieces to hold it there securely when pressed home. **1971** B. Scharf *Engin. & its Language* xi. 111 Considerable effort is required to assemble the parts: this is reflected in the use of terms such as force, drive or press fit. **1970** K. Ball *Fiat 600, 600D Autobook* vi. 53/2 The side bevels embody the axle shaft slip joint cavities, the free bevels being mounted on a shaft press-fitted into the differential casing. **1976** *Gramophone* Dec. 1092/1 Tape transport is controlled by an array of press-keys all fitted with a non-slip tread to prevent finger slip.

press, *v.*^{1} Add: **I. 1. a.** *to press the button*: see *button *sb.* 4b and cf. *press-button *sb.* and *a.*

c. *to press the flesh*: to greet by physical contact; *spec.* to shake hands. *U.S. slang.*

1926 Maines & Grant *Wise-Crack Dict.* 8/1 *Press the flesh*, shake hands. **1933** A. E. W. Mason *Sapphire* ii. 16 'Press the flesh,' said I, extending my hand. **1975** W. Safire *Before the Fall* vi. v. 436 The Soviet leader [*sc.* Brezhnev] surprised Kissinger..with his American political habit of 'pressing the flesh'—punching an arm, squeezing, back-patting. **1977** *National Observer* (U.S.) 22 Jan. 14/3 After the assassination of John Kennedy, some said no future President would be able to 'press the flesh'. But both Lyndon Johnson and Gerald Ford felt that personal appearances were integral to campaigning. **1977** *Time* 7 Nov. 31/2 Aides had to coax him into playing fewer tennis matches with celebrities..and spending more time pressing the flesh.

d. Also in Gymnastics, with various prepositions.

1956 Kunzle & Thomas *Freestanding* i. 25 From prone support jump up to a knee and elbow balance... From there learn to press up to handstand and then lower again. *Ibid.* 26 Use the ankles to bounce the body into the air again, pressing through with the toes to get the maximum impulse. *Ibid.* ii. 32 Straighten out with the knees, press off on to one leg and lower the trunk sideways. **1964** G. C. Kunzle *Parallel Bars* iii. 83 Do not neglect specific strength training, such as..pressing to handstand against the wall bars.

2. c. *intr.* In Golf and Tennis. (See quots. 1975, 1977.)

1910 *Encycl. Brit.* XII. 223/2 *Press*, to strive to hit harder than you can hit with accuracy. **1922** Wodehouse *Clicking of Cuthbert* vi. 132 Keep the head still.. don't press. **1975** *Oxf. Compan. Sports & Games* 423/1 To 'press' is to try to hit the ball too hard, usually with a resultant mis-hit. **1977** *Tennis World* Sept. 17/2 'Pressing' is trying too hard: a player is said to be pressing if his shots are over-eager or impatient.

4. a. *spec.* To smooth or flatten (fabric or clothes) with an iron or clothes press. Also with *out.*

1901 A. H. Rice *Mrs. Wiggs of Cabbage Patch* vi. 92 She pressed out Asia's best dress. **1908** M. H. Morgan *How to dress Doll* viii. 67 Sew the tucks firmly, then press them open. *a* **1911** D. G. Phillips *Susan Lenox* (1917) I. iii. 37 I'm going to wear my white dress with embroidery, and it's got to be pressed. **1928** A. Christie *Mystery of Blue Train* xxiv. 195 He found the imperturbable George pressing trousers. **1949** D. Smith *I capture Castle* xiv. 257 Your frock's quite a bit creased, miss... I could press it, if you like. **1957** C. MacInnes *City of Spades* I. ix. 68 At one time I pressed suits by day and worked in the Post Office by night. **1976** C. Dexter *Last seen Wearing* xvi. 123 The little woman at home cooking a meal for you and probably pressing your pants or something.

b. To dry and flatten (leaves, flowers, etc.) in order to preserve them.

1785 T. Martyn tr. *Rousseau's Lett. Elements Bot.* viii. 82 Your pile of plants and papers thus arranged, must be put into the press, without which your plants will not be flat and even; some are for pressing them more, others less. **1840** C. Fox *Jrnl.* 22 Mar. in *Memories Old Friends* (1882) vi. 75 Clara has been collecting flowers, and they have been together pressing many of them. **1930** R. Macaulay *Staying with Relations* ii. 20 'You see, I press... Do you enjoy pressing, Catherine?' 'Flowers, she means,' Benet explained. 'Isie likes to keep her verbs intransitive.' **1974** W. C. Cartner *Fun with Botany* 23 Plant specimens can be pressed and dried for further study... Lay out the fresh specimens between sheets of newspaper, and press the sandwich between two boards.

c. To make (a gramophone record), to record (a song, etc.). *colloq.*

1918 [see *matrix 4 c]. **1929** Wilson & Webb *Mod. Gramophones* xi. 253 The stampers which press records have to be kept at a certain temperature in order that the record material will flow properly. **1954** W. W. Johnson *Gramophone Bk.* 55 By 1929 one record manufacturer alone was pressing records at the rate of a million a week. **1968** P. Oliver *Screening Blues* 5 In the ensuing months more stores carried Race records, specially pressed for the Negro market. **1976** *Sunday Times* 21 Mar. 58/3 Island is coy about how many albums it is pressing.

II. 8. d. (Later example.)

1968 *Globe & Mail* (Toronto) 3 Feb. 10/1 About 2,000 enemy troops had pressed an attack there since Tuesday against a U.S.-advised Vietnamese garrison of about the same size.

III. 15. a. (Further examples.) Also with *on.* Freq. *fig.* and in colloq. phr. *to press on regardless*, to persevere despite the dangers or difficulties (see *regardless *a.* 1 c). Also as *adj. phr.*

a **1599**, **1870** [in Dict.]. **1916** Joyce *Portrait of Artist* (1969) v. 243 And why were you shocked, Cranley pressed on in the same tone, if you feel sure that our religion is false and that Jesus was not the son of God? **1921** G. B. Shaw *Back to Methuselah* v. 266. After passing a million goals they press on to the goal of redemption from the flesh. **1930** *Flight* 23 Oct. 1177/2 Lord Thomson and his gallant crew would still have said press on, instead of crying halt, in airship development. **1948** Partridge *Dict. Forces' Slang* 147 *Press on, regardless*—or merely *press on*, to act keenly, to be efficiently busy. Hence, *press-on type*, an almost too keen person—applied mostly to 'operational types'. They press on, regardless of fog, flak, fighter opposition. **1950** G. Hackforth-Jones *Worst Enemy* iii. 212 Action was needed to stem this tide of defeatism. Head down was the way to progress through the blizzard. 'When in doubt, press on.' A good motto that. **1952** M. Tripp *Faith is Windsock* xiv. 209 The Vicar was laudatory: 'A magnificent press-on effort, old chap.' **1958** *Times* 18 Dec. 11/4 A few colourful wartime metaphors survive... A third and uncouth example, to press on regardless, stands for a dashing and stoical, if disillusioned, perseverance which continues to find a place in life today just as it did in the early days of the war. **1959** *Listener* 5 Mar. 428/2 While the scientists press on regardless, the humanists go on worrying. **1960** *Times Lit. Suppl.* 18 Mar. 182/1 What vitality the man must have had! And it is this vitality which Mr. Coulter's press-on-regardless manner succeeds very well in conveying. **1961** J. Dawson *Ha-Ha* i. 7 The other students..used to wave as they passed and cry: 'How goes it?' or 'Press on regardless.' **1968** *Listener* 15 Aug. 203/2 That kind of Irishman—admirable rather than safe: the kind I'd heard junior RAF men in the war refer to as 'a press-on type'. **1977** *Drive* May–June 54/2 Covering 40 miles for every gallon of 4-star fuel (even press-on drivers could manage at least 35 mpg).

17. (Later example.)

1811 J. Love *Let.* 29 Oct. (1840) 349 To press after attaining and communicating to others more of the beginnings and pledges of that glorious life which now we view at a distance.

press, *v.*^{2} Add: **2. d.** Also in phr. *to press into service.*

1926 *Discovery* June 191/2 Bait, such as a meal-worm, may be pressed into service [by the bird-photographer] to entice a bird on to some particular twig. **1935** *Yachting* Dec. 82/3 *Press into service*, a reminiscence of the press-gangs which caused the War of 1812 by stopping American merchantmen on the high seas and 'pressing' members of their crews into service in the British navy. **1961** *New Eng. Bible Mark* xv. 21 Simon, from Cyrene,..was passing by..and they pressed him into service to carry his cross. **1978** K. J. Dover *Greek Homosexuality* ii. 97 They masturbate constantly..if no living being with a suitable orifice is available, but prefer horses, mules, or deer..; even the neck of a jar may be pressed into service.

press-. The stem of Press *v.*^{1}, used in combination with advbs. to form adjs. designating things that can be pressed *down, in, on*, etc. (See also *press *sb.*^{1} 15 b (*b*), 16 d.)

1903 *Work* XXV. 218/2 A treacle tin, washed out and dried, with a burner soldered in the press-in lid, will serve quite well if the experiments are conducted outside the house. **1936** A. Ransome *Pigeon Post* xviii. 189 It was an ordinary tin of paint with a wire handle..and a press-in lid. **1962** L. S. Sasieni *Princ. & Pract. Optical Dispensing* viii. 203 The third type [*sc.* of bridge lining] (press-on or snap-on) is shaped roughly in the form of a half tube which presses on to, and snaps over, the metal bridge. **1963** *Rep. Comm. Inquiry Decimal Currency* viii. 68 in *Parl. Papers 1962–3* (Cmnd. 2145) XI. 195 The two main groups of cash registers are the 'press-in' key type and the 'press-down' key type. **1975** B. Wood *Killing Gift* (1976) II. i. 48 A vacuum jar with a press-on lid.

press agency. [f. Press *sb.*^{1} + Agency.] = *news agency* (*b*) s.v. *news *sb.* (*pl.*) 6 c.

1897 H. Maxwell *Sixty Years a Queen* VIII. 190 The British Government has no official or semi-official organ in the press. Official pronouncements are communicated.. to press agencies, and through them find their way into journals of all shades of politics. **1973** D. May *Laughter in Djakarta* ii. 35 He ought to make a further check on the news by going to the Indonesian press agency.

press agent. (In Dict. s.v. Press *sb.*^{1} 16 b.) Add: (Examples.) Also, more widely, one employed by any person or organization to handle publicity.

1883 *Railway Age* 25 Jan. 46/3 On general principles.. we desire to observe that the associate press agent, or some one who makes Wichita conspicuous in the dispatches is an ass. **1902** Wh. H. Chantrey *Theatre Accounts* ii. 28 Salaries... Press Agent and Bill Inspector 5-10-0. **1917** Wodehouse *Uneasy Money* x. 114 Roscoe Sherriff, her press agent. **1949** *Chicago Tribune* 9 Dec. 18/3 This was the first time that a press agent had hit on a truthful first page story in a month of Sundays. **1964** M. McLuhan *Understanding Media* xxi. 213 Today's press agent regards the newspaper as a ventriloquist does his dummy. **1977** *Time* 10 Oct. 61/1 A former pressagent, Condon, 62, boasts average book sales of 1.3 million.

Hence **pre·ss-agent** *v. trans.*, to advertise in the manner of or by means of press agents; **pre·ss-agented** *ppl. a.*; **press-agenting** *vbl. sb.*; **pre·ss-agentry**, the employment or activities of press agents.

1909 Wodehouse *Swoop* II. ii. 68 Come now, your Grand Grace, is it a deal? Four hundred and fifty chinking o'Goblins a week for one hall a night, and press-agented at eight hundred and seventy-five. **1913** *Writer's Mag.* Nov. 172/1 There is no 'side line' open to the young writer better than press agentry. **1920** W. T. Tilden *Art of Lawn Tennis* 3, I shall be accused of 'press-agenting' my own book by this statement. **1926** *Daily Express* 6 Aug. 3/5 Even the Hohenzollerns know something about Press agentry. **1930** P. W. Slosson *Great Crusade & After* x. 271 The same press-agenting which helped make the reputation of a grand-opera star..was also at the service of a pugilist. **1933** *Nation* (N.Y.) 11 Jan. 43 He press-agented most of the striking new theories, from those of the Lombrosian criminology..on down to the neo-Nietzchean doctrines of Elie Faure. **1939** *Sun* (Baltimore) 22 Nov. 13/1 Mr. Frank Murphy, the present highly press-agented Attorney General. **1947** M. Berger in R. de Toledano *Frontiers of Jazz* 100 Bunk indulged in some personal press-agenting. **1948** *Archit. Rev.* CIV. 89 The same press-agentry that ballooned the popularity of other stars in the field of jazz. **1959** *Time* (Atlantic ed.) 24 Aug. 45 A longstanding and well pressagented public 'feud'. **1973** E. Bullins *Theme is Blackness* 9 Whether they will regard Black dramatic criticism seriously and not degenerate into professional press agentry. **1977** *Irish Press* 29 Sept. 10/2 Meanwhile back in Ireland we are faced with the usual problem of press agentry.

pre·ss-board. [f. Press *v.*^{1} or *sb.*^{1} + Board *sb.*] **1.** An ironing-board; *spec.* (see quot. 1939).

1849 G. G. Foster *New York in Slices* i. 14 The press-board has been placed across the back corner of the shop. **1896** J. C. Harris *Sister Jane* i. 17 I've got this press-board on my lap, or I'd fetch it myself. **1924** W. D. F. Vincent et al. *Cutters' Pract. Guide Body Coats* 33/1 The seam should be placed straight on the press-board in front of you. **1939** M. B. Picken *Lang. Fashion* 116/2 *Press-board*, padded board, a small ironing board, used for pressing fabrics when sewing.

2. *Electr. Engin.* (Written **pressboard.**) A material consisting of compressed laminations of paper, used as a separator or insulator in electrical equipment; a piece of this.

1910 H. M. Hobart *Dict. Electr. Engin.* II. 415/1 *Pressboard*, sometimes termed *pressed board*, a fibrous material closely resembling press-spahn. **1926** A. P. M. Fleming in J. A. Fleming *Electr. Educator* II. 1354/2 The manufacture of pressboards..is the same as for paper making so far as the production of pulp. **1952** J. P. Casey *Pulp & Paper* II. xvi. 957 Pressboards made..in thicknesses ranging from 0·005 to 0·125 in. are used as a spacing and insulating medium. **1973** R. W. Sillars *Electr. Insulating Materials* vii. 123 Pressboard is prepared from cotton rag fibres or from pulp processed like other electrical papers, but instead of drying out as a single layer, a number of wet layers are placed together, pressed in a hydraulic press and dried by heat.

pre·ss-button, *sb.* and *a.* [f. Press *v.*^{1} or *sb.*^{1} + Button *sb.*] **A.** *sb.* **a.** = *push-button *sb.*

1892 [see *pressel]. **1977** *Gramophone* Nov. 959/3 To the right of the main tuning knob..are two press-buttons.

b. A fastener similar to a press-stud.

1907 *Yesterday's Shopping* (1969) 404/1 Pocket for powder, lined white silk, with puff, and press button fastening. **1917** M. A. Souder *Notion Department* xv. 122 There are two types of snap fasteners: those built upon the principle of the ball and pocket reinforced with a wire spring, properly designated as snap fastener, and those of a flatter and structurally weaker design of a ball and socket without this wire spring, called press buttons. **1933** *Archit. Rev.* LXXIV. 30/1 (*caption*) The upholstery is easily removable on the motor car press-button principle.

B. *adj.* **a.** = *push-button *a.* a.

1958 *Oxford Mail* 23 Aug. 3/6 Very neat press-button catches are fitted on all doors. **1965** *Wireless World* July 34/1 (Advt.), Press-button operation.

b. = *push-button *a.* b.

1948 *Daily Tel.* 23 Apr. 5/2 Lord Montgomery said although we heard much talk of 'press-button' warfare, scientists had not so far produced any new weapon that could justify the discarding of the present-day technique of land warfare. **1958** *Listener* 12 June 990/3 A press-button world. **1965** M. McIntyre *Place of Quiet Waters* i. 5 All this high-powered, press-button living is wrong. **1971** *Daily Tel.* 30 Jan. 3/4 It was joked about as obsolete and useless in the age of press-button warfare.

pressed, *ppl. a.*^{1} Add: **1.** (Further examples.)

1850 E. Dobson *Rudimentary Treat. Manuf. Bricks & Tiles* I. iii. 83 Pressed bricks..are prepared by putting the raw bricks one at a time, when nearly dry, into a metal mould, in which they are forcibly compressed by the action of a powerful lever which forces up the piston forming the bottom of the mould. This gives a very beautiful face to the brick. **1869** *Our Young Folks* V. 86 We are making pressed glass nowadays that is almost as clear and beautiful as blown. **1895** *Army & Navy Co-op Soc. Price List* 111/2 Pressed veal & ham (Blanchflower's). **1912** J. Armstrong *Motor* 266, I expect to see the clutch and flywheel as pressed parts in the near future, while pressed-steel pistons and connecting-rods are most likely to become common. **1926** F. Hurst *Appassionata* I. 10 Two empty pressed-glass perfume-bottles that had stood equidistant on that dressing-table ever since you could remember. **1935** H. C. Bryson *Gramophone Record* ix. 212 As soon as the press is opened, the steam commences to circulate in the dies, so that the operator has to remove the pressed record speedily or it will adhere to the rapidly-warming die. **1955** *Sci. Amer.* Jan. 68/1 One result of these studies has been the discovery of two new kinds of magnetic materials, the ferrites and the pressed-powder magnets. **1955** *Railway Mag.* June 388/2 The floor..is built up of 1½ in. thick boards..bolted to dove-tailed galvanised steel sheeting carried on pressed-steel floor members. **1968** *Radio Times* 28 Nov. 20/1 The week's 'Newly Pressed' pop records. **1976** *Country Life* 1 Apr. 814/1 Apart from the addition of a 'bib' spoiler, the Mexico has the normal pressed-steel front of the rest of the range.

2. *U.S. slang.* Well dressed.

1970 H. E. Roberts *Third Ear* 11/1 *Pressed*, to be very well dressed. **1972** T. Kochman *Rappin' & Stylin' Out* 165 Being well dressed is..expressed kinetically ('pressed'), and..the term refers to a favored norm.

pressel (pres'l). [f. Press *v.*^{1}]. A press-button switch; orig., one attached to a flexible pendant conductor. Also *pressel-switch.*

1892 T. O'C. Sloane *Stand. Electr. Dict.* 434 *Pressel*, a press-button often contained in a pear-shaped handle, arranged for attachment to the end of a flexible conductor, so as to hang thereby. **1911** W. P. Maycock *Electr. Wiring* (ed. 4) ii. 159 Instead of a cord pull-switch, one might connect an ordinary or two-plate ceiling-rose to the switch leads..and hang therefrom a 'pressel' or suspension or pendant switch..which is a convenient pattern for operating with one hand. **1916** G. Frankau *Guns* 21 And he hears, as he plays with the pressel-switch, the strapped receiver click on his ear that listens, listens. **1971** B. W. Aldiss *Soldier Erect* 79, I handed the microphone to Gor-Blimey. After some frigging about with the pressel-switch until he got things right, he spoke to Blue Spot. **1973** B. Callison *Web of Salvage* ii. 25 He released the pressel switch and the static came back.

presse-pâte (pre·s,pat). [Fr.] The section of a paper-making machine in which superfluous water is extracted from the pulp before it is formed into sheets or rolls. Also *attrib.*

1888 Cross & Bevan *Text-bk. Paper-Making* vi. 96 The presse-pâte system, originally adopted for the treatment of straw, has of late years been extensively applied to esparto. The presse-pâte consists of the wet end of a paper machine, and is furnished with sand-tables and strainers. **1937** E. J. Labarre *Dict. Paper* 194/2 Presse-pâte..is a machine practically identical with the wet end of the paper-machine... It serves to extract 'loose' water from the (wood) pulp, which is allowed to accumulate on a press roll or round an iron rod, until a sheet or roll of wet board of sufficient thickness is obtained. **1963** R. R. A. Higham *Handbk. Papermaking* ii. 62 The pulp may either be concentrated on deckers, or presse-pâtes, or left in slush form.

presser. Add: **1. a.** Also, one who presses wool into bales.

1911 W. H. Koebel *In Maoriland Bush* viii. 122 The 'presser' climbs inside the high, square, wooden structure that rises in the centre of the floor, in readiness to receive the fleeces. **1955** G. Bowen *Wool Away!* vii. 95 Pressing wool is a simple straightforward job, but good pressers work without waste movement and without getting in each other's way. **1965** J. S. Gunn *Terminol. Shearing Industry* II. 9 *Presser*, a skilled man who presses the wool into bales so that they are not 'light on' (short in weight).

c. Also in *Glass-making.*

1962 *Gloss. Terms Glass Industry* (*B.S.I.*) 45 *Presser*, a worker who shapes glass by pressing in a mould by hand or by machine.

3. a. (Further example.)

1873 *Young Englishwoman* Mar. 150/2, I get the stitching as close as the width of space between the needle hole and the edge of presser.

5. presser-bar (*b*) (examples); **presser-eye** *Spinning*, an aperture or eye through which cotton yarn passes before being wound on the spindle; **presser-foot** (examples); also *attrib.*

1908 *Sears, Roebuck Catal.* 41/2 The presser bar is round and fitted with a presser bar adjuster by which the pressure on the goods is regulated. **1974** J. Robinson *Penguin Bk. Sewing* ii. 36/1 Presser Bar..Stitch Length Regulator. **1892** J. Nasmith *Students' Cotton Spinning* ix. 340 In short, the traveller performs the same function as the flyer eye in the throstle or the presser eye in the roving frame. **1895** *Montgomery Ward Catal.* 262/1 Parts for Old Style Low Arm Singer..Presser Foot. **1908** *Sears, Roebuck Catal.* 41/1 The presser foot has a very large under surface, which extends on both sides of the needle and holds any weight goods firmly in place over the feed. The forward part of the presser foot nearest the operator is curved upward so that foot will not catch in seams of fleecy materials. **1932** Presser-foot [see *hemming *vbl. sb.*^{1} b]. **1961** *Which?* Nov. 277 (*caption*) Presser-foot screw..presser-foot lever. **1964** A. Butler *Teaching Children Embroidery* 29 Stitches with the presser foot, on ..a [sewing] machine with a zigzag attachment.

press-fastener. [f. Press *v.*^{1} + Fastener.] = *press-stud.

1926–7 *Army & Navy Stores Catal.* 664/3 Press fasteners ..in black or white, 1 dozen on card. **1956** *Good Housek. Home Encycl.* (ed. 4) 185/2 One side must be left open.., press fasteners or hooks and eyes being used to close it. **1960** *Mrs. Beeton's Cookery & Househ. Managem.* 145 Hooks and Eyes, Press Fasteners. **1976** J. Wainwright *Who goes Next?* 95 Racing gloves..fastened at the back of the wrist, with a good press-fastener.

press-gang, *sb.* Add: **1.** Also *transf.*

1771 C. Burney *Present State of Mus. France & Italy* 119 These boys are a kind of *press-gang*, who seize all other boys they can find in their way to the church, in order to be catechised. **1969** I. & P. Opie *Children's Games* i. 18 In some places the press-gang think they will be successful if they demand, 'Join the ring or tell us your sweetheart's name.'

2. *joc.* A group of journalists; the press (Press *sb.*^{1} 13g).

1840 *Spirit of Times* 11 Jan. 535/1 In compliment to the 'Press gang' Messr. Prentice and Weissinger..were invited to occupy seats over the Judge's stand. **1859** L. Wilmer *Our Press Gang* xxvi. 353 Our newspapers, in general are the organs of the mob, and..the Press Gang *itself* is..a mob of the worst kind. **1859** G. H. Lewes *Let.* 20 Apr. in *Geo. Eliot Lett.* (1954) III. 54 Nor have I any relations with the pressgang here..the Edinburgh papers might advantageously be employed in this matter. **1869** 'Mark Twain' *Lett. to Publishers* (1967) 27, I will attend to the Buffalo books for the press..did I tell you that I took dinner with the whole press gang yesterday? **1889** *Pall Mall Gaz.* 21 Oct. 4/3 Ask what you like, my good sir; don't you know I am one of the press gang myself? **1941** F. L. Mott *Amer. Journalism* xxxv. 603 Many stories are told of the conviviality of the Chicago 'press gang' of the nineties, and from the legends which grew up..sprang the concept of the romantic sot of the newspaper office. **1975** H. Waugh *Bride for Hampton House* (1976) i. 1 She..waved at the press gang when she took the big elevator to the newsroom floor.

pressie, var. *prezzie.

pressing, *vbl. sb.*^{1} Add: **1.** (Further examples.)

a **1911** D. G. Phillips *Susan Lenox* in *Hearst's Mag.* (1915) June 538/1 Susan finished her pressing and started to dress. **1960** *Vogue Pattern Bk.* Autumn 61 Careful pressing, at every stage in making clothes, helps to give your work that smooth professional finish. **1969** T. C. Thorstensen *Pract. Leather Technol.* xii. 192 This pressing of the oil removes some of the high melting point components and gives the oil a lower cold test. **1976** *Southern Even. Echo* (Southampton) 15 Nov. 15/5 The match was one of two halves, with Basingstoke doing all the pressing in the first.

2. b. An article formed or shaped in a press; *spec.* a gramophone record made by stamping a blank with a matrix.

1922 *Metal Industry* XX. 273/1 Many parts are now being manufactured as pressings which a few years ago would have been thought impossible. **1927** *Gramophone* Sept. 139/1 The new white label pressings arrived just in time for me to take them to Paris. **1952** Godfrey & Amos *Sound Recording & Reproduction* v. 143 A vinyl pressing is superior to one made in shellac in that it is far less fragile. **1959** *Motor Manual* (ed. 36) 7 A pressing is produced by squeezing the sheet of steel between a die, securely anchored in the base of the press, and the punch which is forced into it at a pressure of 750 tons or more. **1959** *Times Rev. Industry* Nov. 46/2 Quantities of pressings and castings are required by makers of household equipment. **1962** *Times* 5 July 15/6 A master tape of a recent recording of music by Prokofiev and an ordinary commercial pressing of a disc made from it were started simultaneously. **1973** D. Westheimer *Going Public* v. 77 Lee..turned on the stereo. It was an LP pressing of some old John Kirby RPM singles.

3. *pressing cloth, machine* (earlier and later examples), *pad, plant, rag, room*; **pressing board** (*b*), an ironing board.

1969 E. H. Pinto *Treen* 151/1 (*heading*) Ironing and Pressing Boards. **1917** E. R. Hambridge *Simple Dressmaking* iii. 65/2 Wring out the pressing cloth *very* tightly, and lay it on the fabric. **1974** Lippman & Erskine *Dressmaking made Simple* iv. 59 A pressing cloth can be obtained pre-treated. **1825** *Austin Papers* (1924) II. 1028 We intend to send a gin and probably a pressing machine. **1940** *Chambers's Techn. Dict.* 671/2 *Pressing machine*, a machine in which the whole forming operation is carried out by pressing the plastic glass by a plunger forced into a die or mould. The machine may be operated by hand or it may be fully automatic. **1947** C. Talbot *Compl. Bk. Sewing* xxxi. 208/1 The rounded shaping of armhole and shoulder..must be protected in pressing. A tailor's cushion, a pressing pad, or a sleeve board will help. **1974** Lippman & Erskine *Dressmaking made Simple* vi. 87 Press..shaped parts over the tailor's pressing cushion/pad. **1958** *Manch. Guardian* 21 Jan. 6/6 Oriole Records Ltd... has its own pressing plant and..presses all Mr. Lonsdale's records for him. **1934** A. L. Hird *Princ. & Pract. Needlework & Dressmaking* iv. 41 *Do not apply water directly* to any materials except wool and wool mixtures, but always by way of pressing rag or as steam. **1922** O. Mitchell *Talking Machines* vi. 70 When the discs pass into the pressing room the steel backing is laid upon a heated table with the mould upwards. **1937** F. Stark *Baghdad Sketches* 36 He was a tailor in his spare time, and he used the roof as a pressing room.

pressive, *a.* Add: **4. b.** *Psychol.* That pertains or relates to environmental press (see *press *sb.*^{1} 5*).

1938 H. A. Murray *Explorations in Personality* ii. 96 In emotional action it is the sudden, close, pressive situation that seems to 'do the work' by releasing energy in the motor centres of the interbrain.

press lord. Also with capital initials. [f. Press *sb.*^{1} + Lord *sb.*] A powerful newspaper owner, a newspaper magnate, esp. one who is a member of the peerage. Cf. *press baron* s.v. *press *sb.*^{1} 16b.

1930 *Economist* 30 Aug. 396/2 At the Labour Party Conference..resolutions will be discussed..putting forward claims, rather reminiscent of those of the Press Lords, that certain Labour interests should dictate how the Cabinet should be composed. **1932** *Ann. Reg. 1931* 23 The 'press lords', Lord Rothermere and Lord Beaverbrook, who had never abated their hostility to Mr. Baldwin, had once more opened a campaign against him, and in London at any rate they were not without a considerable following. **1947** *Sat. Rev. Lit.* (U.S.) 22 Feb. 15/2 Roy Howard still dresses and acts among the press lords like a police reporter. **1955** T. H. Pear *Eng. Social Differences* 22 Some Press-lords are wounded..or amused at the suggestion that increased circulation is one of their first aims. **1965** Auden *About House* (1966) 15 Only a press lord Could have built San Simeon. **1977** *Time* 17 Jan. 5/2 But the press lord from Down Under is no stranger to *Time* readers—or *Time* staffers.

pressman^2. Restrict † *Obs.* to sense 2 in Dict. and add **1.** Also *fig. rare.*

1978 *Church Times* 17 Feb. 11/4 In religious education there has been a constant battle waged to rise to the challenge of the 1944 Act and ensure that the subject is taught by qualified specialists rather than by willing amateurs or, worse, by reluctant press-men.

pressmanship. Add: **1.** Also, skill as a pressman.

1923 H. A. Maddox *Printing* ix. 106 Most printing is in the indifferent class and much good typography is marred by careless pressmanship. *Ibid.* x. 125 Printing from three-colour half-tone blocks demands a high standard of pressmanship.

press-mark. Add: (Further examples.) Now chiefly with reference to manuscripts and early books in old libraries. Also *attrib.*

1941 N. R. Ker *Medieval Libraries* p. xviii, Pressmarks are useful indications of provenance just in so far as they are distinctive. **1952** J. Carter *ABC for Bk.-Collectors* 141 Seymour de Ricci, in his *English Collectors of Books and Manuscripts*, shows how much can be learned from the study of press-marks by anyone concerned with the provenance of books. **1963** *Times Lit. Suppl.* 1 Mar. 160/3 A stock of the press-mark tickets used by Louis-Henri. **1971** *Eng. Stud.* LII. 351 The press-mark of the Leiden manuscript is not *Voss 106*, but *Voss. 106*, in which *Voss.* is an abbreviation of *Vossianus.*

Hence **press-mark** *v. trans.* and *intr.* (see quots. 1889 and 1895).

1889 *Cent. Dict.*, *Press-mark*.., *v.t.* and *i.* To place a press-mark on; also, to use press-marks. **1895** *Funk's Stand. Dict.*, *Press-mark*,..*vt.* & *vi.* To mark (a book) with characters showing the proper place in a bookcase. **1915** *Trans. Bibliogr. Soc.* XIV. 5 The Society's library..has been rearranged and re-pressmarked.

press officer. Also with capital initials. [f. Press *sb.*^{1} + Officer *sb.*] An official appointed by an individual or institution to handle publicity and public relations.

1919 J. Buchan *Mr. Standfast* xx. 366, I was about to make a rush for..one of the Press officers, who would.. be in the way of knowing things. **1941** *Whitaker's Almanack* 321/2 Colonial Office... *Press Officer*, A. Ridgway.. £800. **1949** H. Nicolson *Diary* 14 Jan. (1968) 163 One man..had asked Philip Jordan, as Attlee's Press Officer, to describe what the P.M. had had for breakfast that morning. **1959–60** A. Butler in *Parliamentary Affairs* XIII. 57 Journalists under great pressure make use of guidance from Press officers in Government departments when handling reports and Bills. **1961** *Times* 25 July 2/2 (Advt.), 'English Electric' require a Press Officer..to..be responsible for all press and public relations functions. **1969** A. G. Thomas in L. Durrell *Spirit of Place* 57 In 1939 Durrell..moved to Athens working first for the Embassy as an unestablished press officer. **1972** 'H. Carmichael' *Naked to Grave* v. 71 You would get all the information from the press officer which was currently available.

pressoreceptor (pre·sorise·ptəɹ). *Physiol.* Also with hyphen. [f. *presso-*, taken as comb. form of *pressure* + Receptor.] A proprioceptor which responds to changes in blood pressure.

1941 P. Bard *Macleod's Physiol. in Mod. Med.* xl. 576 (*caption*) Effects of electrical stimulation of nerves from pressoreceptors and chemoreceptors in the dog. **1943** *Physiol. Rev.* XXIII. 244 In man the diastolic pressure has been shown to rise during standing. This is a sign of increasing vasomotor tone. It is compensatory in nature until it encroaches too far upon pulse pressure. The reflexes originating in the vascular presso-receptors are responsible for this control. **1975** *Investigative Urology* XII. 465 We studied the influence of pressoreceptor stimulation on micturition reflex and urethral pressure profile.

presspahn (pre·s,ʃpān, pre·spān). *Electr. Engin.* Also **press-spahn.** [a. G. *preßspahn* (now *-span*) pressboard, orig. pieces of card for pressing clothes f. *preß-* (in comb.) pressed, pressing + *span* shred, splinter.] Pressboard (sense *2).

1904 *Electr. Engineer* 14 Sept. 412/2 Creasing 'fibre, presspahn, etc.,' destroys the glazed surface..and is.. likely to reduce its insulating value. **1913** Barr & Archibald *Design Alternating Current Machinery* ii. 39 Press-spahn, in thin sheets, is largely used for the insulation of low-voltage windings. **1938** W. T. MacCall *Electr. Engin.* I. iii. 38 Many special forms of paper and cardboard, such as presspahn,..are used for moderate pressures.

press roll^1. *Papermaking.* [f. Press *sb.*^{1} + Roll *sb.*^{1}] (See quot. 1940.)

1881 J. Dunbar *Pract. Papermaker* (ed. 2) 47 The author has tried a contrivance which effectually prevents the paper breaking at the press rolls. **1937** E. J. Labarre *Dict. Paper* 195/1 *Press rolls* are pairs of heavy rolls, termed 'first', 'second', and 'third' press rolls according to their order in the paper-machine, serving to press out the water from the web of paper. **1940** *Chambers's Techn. Dict.* 671/1 *Press rolls*.., heavy cylinders of the paper-making machine which press out moisture from the wet web. Before the last pressing, the web is reversed in order to remove felt-marks.

press roll^2. *Jazz.* [f. Press *sb.*^{1} + Roll *sb.*^{2}] A drum-roll (see Roll *sb.*^{2} 2) in which the sticks are pressed against the drum-head.

1934 [see *drag *sb.* 7 i]. **1939** W. Hobson *Amer. Jazz Music* iii. 53 A 'press roll', one of the many rhythmic patterns which have been used by jazz drummers for years. **1956** S. Longstreet *Real Jazz* 148 A press roll is played on snare drums. **1966** *New Yorker* 11 June 153/1 Barbarin, in addition to a marvellous press roll, in which his sticks come at one another low from opposite edges of the snare in fat blurs, like gulls after the same clam, uses a high hat [etc.]. **1977** *Ibid.* 9 May 51/1 He principally used..powerful, accented press rolls on his snare drum.

press-room^1. Add: **2.** (Later examples.)

1878 *Rep. Vermont Board Agric.* V. 79 At the end from the road were the press and wash rooms. **1966** P. V. Price *France: Food & Wine Guide* 164 Visitors see the pressroom..and the huge stone container for 13,000 litres of wine.

3. A room reserved for the use of reporters.

1902 *Evening Star* (Washington, D.C.) 16 Dec. III. 10/3 The press room at the District building is quite a center of interest. **1941** F. L. Mott *Amer. Journalism* xxxv. 607 President Theodore Roosevelt..provided a press room and telephones for the reporters. **1943** L. C. Wilson in F. L. Mott *Journalism in Wartime* 99 Merriman Smith, on the White House,..fills a notebook..and then he literally runs to the United Press telephone booth in the Press Room. **1952** F. L. Mott *News in Amer.* xiii. 137 The doors were opened, and there was a grand rush for the telephones in the press-room and for taxis and cars. **1973** *Guardian* 19 Apr. 1/6 The President's sudden appearance last night in the White House press rooms. **1976** *National Observer* (U.S.) 31 Jan. 1/1 The press room..is a bit boggling. It is really a complex of rooms, encompassing a Western Union setup, a Sports Comm., Inc., service.., a bank of dictation phones, a wall of releases and handouts, and eight long rows of tables..stacked with 125 typewriters. **1977** *Cleethorpes News* 6 May 7/2 She spent most of her time in the Press room before leaving through a small but enthusiastic crowd.

press-spahn, var. *presspahn.

press-stud. [f. Press *v.*^{1} + Stud *sb.*^{1}] A fastener made of metal, plastic, etc., used for joining two parts of a garment etc. together and consisting of two components, one with a short shank which is pressed into a corresponding hollow in the other. Also *attrib.*

1917 *Harrods Gen. Catal.* 1425/2 Press Studs, Black and White..2/9 per gross. **1928** *Daily Express* 19 Mar. 5/5 The chalk will leave an impression at exactly the correct place for the other half of the press-stud to be sewn. **1955** *Times* 29 June 12/4 A beautifully tucked skirt, complete even with a pink and white striped nylon petticoat kept in place by press studs attached to the dress. **1966** *Price List* (Olney Amsden & Sons Ltd.) 31 Press stud tape.. 22/6 doz. yards. **1974** *Drive* Autumn 32/2 Floor coverings used to be fixed by press studs. **1977** *Offshore Engineer* Apr. 74/2 The boot's normal strap and buckle fastening is backed up with a press-stud arrangement which allows the wearer to release it in two seconds.

pre·ss-up. [f. Press *v.*^{1} + Up *adv.*^{1}] An exercise in which the body is raised from a prone position by straightening the arms while keeping the hands and feet on the ground and the legs straight; (see also quot. 1961).

1947 J. Bertram *Shadow of War* 208 Press-ups are a fairly strenuous exercise at the best of times. **1955** M. E. B. Banks *Commando Climber* iii. 33 We went through the usual climbing exercises such as press-ups with the finger tips and not the palms of the hands. **1956** Kunzle & Thomas *Freestanding* i. 22 Start with the feet on the third or fourth wall bar and do press-ups as before. **1961** J. S. Salak *Dict. Amer. Sports* 341 *Press Up* (mountain climbing), an upward movement on rock completed by pressing down on the palms of the hands on large flat holds and ledges. **1967** *New Scientist* 28 Dec. 766/1 When we say that an athlete is 'fit', we generally mean that he can perform some arbitrary feat, like 50 press-ups, without getting unduly puffed. **1978** *Rugby World* Apr. 33/3 It's hard on the club coach who has to motivate players with whom he probably played and has to try to get one of his best friends to do another half-a-dozen sprints or extra press-ups.

pressure, *sb.* Add: **II. 6. b.** (Further examples.) Also stress, strain; in *Finance*, forces (*on* a currency) tending towards a change in its value.

1961 *N.Y. Times* 5 July 31 Throughout 'The Making of a President' Mr. White shows wonderfully well how the pressures pile up on candidates. **1964** *Ann. Reg. 1963* 203 Pressure on the peso..became so strong in May that the authorities could no longer resist it. **1976** *Times* 30 Mar. 4/4 Not that they do not want freedom; but it brings pressures and choices with which they find it hard to cope. **1976** *Howard Jrnl.* XV. I. 13 There may also be cases in which a period of detention is necessary as a respite from problems or pressures that will otherwise entrap the offender in greater trouble. **1977** A. Ecclestone *Staircase for Silence* v. 95 As a sensitive man he registered the pressures which were..shaping men to make choices which would carry them to such lengths.

c. (Earlier and later examples.)

1812 *Q. Rev.* May 159 At the end of the same session, the third bill, from the pressure of business, was given up without having come to a final hearing. **1911** D. H. Lawrence *White Peacock* III. v. 442 In spite of his pressure of business he had become a County Councillor. **1926** S. Jameson *Three Kingdoms* x. 292 She worked on with an aching heart on the evenings when pressure of business kept her in the office until it was too late even to see Sandy before he was in bed and asleep. **1938** R. C. Hutchinson *Testament* I. vi. 60 They were supposed to undergo examination every week, but that, from the pressure on the doctors' time, was often omitted. **1960** E. Stopp tr. *St. François de Sales's Sel. Lett.* 238, I can see that it is hopeless to wait for a better opportunity, since continual pressure of affairs seems to be my fate.

7. a. (Further examples.)

1949 *Sun* (Baltimore) 7 Feb. 8/2 There is no doubting the fortitude the Norwegians show thus far against Russian pressure. **1964** Gould & Kolb *Dict. Social Sci.* 530/2 [S. E. Finer] reserves the term pressure for those activities ..which amount to the 'application or threatened application of a sanction should a demand be refused'. **1966** *Listener* 19 May 713/1 The effects of the sort of social pressures described by Professor Sprott depend largely on the age of the person, the length of time they last, and their intensity. **1976** *Times* 30 Mar. 4/2 The Conservative Party should resist well intentioned pressures to spell out in detail what it would do when it won a general election.

b. *to bring pressure* (*to bear*: to exert influence to a specific end; *to bring* (or *put*) *pressure on* (someone): to urge or press (someone) strongly in order to persuade.

1864 W. Hardman *Let.* 21 Apr. in S. M. Ellis *Lett. & Mem. Sir W. Hardman* (1925) 172 Some pressure had evidently been brought to bear. **1897** Put pressure on [in Dict.]. **1908** A. F. Bentley *Process of Govt.* x. 208 We frequently talk of 'bringing pressure to bear' upon someone, and we can use the word here with but slight extension beyond this common meaning. Pressure, as we shall use it, is always a group phenomenon. **1912** T. Dreiser *Financier* xlv. 489 He thought once of going to Mrs. Cowperwood and having her bring pressure to bear on her husband. **1934** *Amer. Speech* IX. 11/2 *To put on the pressure* is to run the cards of a long suit in an effort to force the discard of cards which might otherwise win tricks. **1937** *Sun* (Baltimore) 22 July 1/5 Republic Steel Corporation officials and leaders of a local back-to-work movement and of a law and order league brought sustained 'pressure' on city officials of an Ohio town, seeking the use of force in reopening strike-bound mills. **1960** L. P. Hartley *Facial Justice* iii. 25 These dissidents brought pressure to bear on their Governments to leave the upper air alone. **1961** *Times-Picayune* (New Orleans) 1 Jan. II. 3 This might be done to arouse those who have been squeezed out by the trims to exert pressure on the Legislature, so it would be more receptive to a tax proposal later in the year.

IV. 9. **a.** *pressure drop, gradient.* **b.** *pressure transducer.* **d.** *pressure sensation.* **f.** *pressure-reducing, -retaining* adjs. **g.** other Combs.: *pressure-sensitive* adj.

1949 O. G. Sutton *Sci. of Flight* i. 14 The solution of practical problems, such as the determination of the pressure-drop in pipes. **1964** J. C. Catford in D. Abercrombie et al. *Daniel Jones* 31 For normal voice the liminal pressure-drop across the glottis is of the order of 3 cm of water. **1918** *Meteorol. Gloss.* (Met. Office), *Gradient wind*, the flow of air which is necessary to balance the pressure-gradient. **1968** R. A. Lyttleton *Mysteries Solar Syst.* i. 27 For a heated gas-cloud rotating round the sun, there can be a pressure-gradient perpendicularly away from the general equatorial plane of the distribution. **1934** Webster, Pressure-reducing, adj. **1950** *Sci. News Let.* XV. 79 The liquids are displaced from the tanks by an inert gas. Nitrogen, stored under a high pressure and fed to the tanks through pressure-reducing valves, is commonly used for this purpose. **1971** B. Scharf *Engin. & its Lang.* xii. 177 Pressure control valves may be pressure-reducing valves which maintain reduced pressure on the downstream side (i.e. after the valve), pressure-retaining valves which maintain the pressure on the upstream side, and indirect pressure control valves to maintain the pressure at a point other than in the line in which the valve is located. **1895** *Amer. Jrnl. Psychol.* VII. 81 *Druckempfindung*, pressure sensation. **1901** [see *contact *sb.* 1 d]. **1932** *Mind* XLI. 363 Whenever I touch anything I have pressure-sensations with a characteristic local sign. **1937** *Jrnl. Exper. Psychol.* XX. 458 There were relatively more pressure sensitive spots on the dorsal side of the arm than on the ventral side. **1970** *New Yorker* 3 Oct. 108/3 Repair the damage with pressure-sensitive tape. **1949** *Jrnl. Sci. Instruments* XXVI. 327/2 (*caption*) Final form of electronic pressure transducer. **1956** *Nature* 25 Feb. 380/1 The time it takes for the wave to cover a known distance was measured by means of two barium titanate crystal pressure-transducers. **1963** H. K. P. Neubert *Instrument Transducers* iv. 348 Through the year the main application of piezoelectric pressure transducers has been the 'engine indicator' for use with internal-combustion engines, which employs quartz disks or piles of disks.

10. pressure arch *Mining*, a distribution of pressure over an excavation resembling that in a structural arch, caused by increased pressure on the side walls of the excavation, which act as abutments supporting the strata forming the roof; **pressure-boiler**, a boiler designed to withstand great pressure, for heating liquids above the normal boiling point; **pressure breathing** (see quot. 1965); **pressure broadening** *Physics*, pressure-dependent broadening of spectral lines caused by collisions of emitting molecules with their neighbours in a fluid: so **pressure-broadened** *a.*; **pressure cabin**, in an aircraft, an airtight cabin in which the air is maintained at a pressure safe and comfortable for the occupants; **pressure cable** *Electr. Engin.*, a paper-insulated cable that contains gas or oil under pressure within the outer sheath or pipe, in order to counteract the tendency of the oil to move away from the conductors in operating conditions and enable higher voltages to be used; **pressure chamber**, a chamber designed to hold material under pressure, or in which pressure can be applied; **pressure (die-)casting**, die-casting in which the metal is forced into the mould under pressure; a casting so made; so **pressure-cast** *a.*; **pressure drag** *Aeronaut.*, the drag on a moving body which results from the aerodynamic pressure distribution over its surface; form drag; **pressure flaking** *Archæol.*, the flaking of flint tools by applying pressure with a hard point; hence **pressure-flaked** *a.*, shaped in this way; **pressure-flaker**, a pointed bone tool used for pressure-flaking; **pressure flask**, a flask designed to withstand pressure greater than that of the atmosphere; **pressure hold** *Mountaineering*, a hold maintained by the exertion of sideways or downward pressure; **pressure hull**, the hull (or part of the hull) of a submarine which is designed to withstand the pressure of the sea when the vessel is submerged; **pressure jump** *Meteorol.*, a mobile zone of atmospheric disturbance, characterized by a steep pressure gradient and usu. marking a discontinuity in the height of an inversion layer; so *pressure-jump line*; **pressure lamp**, a portable oil or paraffin lamp in which the fuel is forced up into the mantle or burner by the air pressure in an enclosed reservoir, which is increased by pumping with a built-in plunger; **pressure line** = *pressure ridge*; **pressure microphone**, a microphone which responds to the instantaneous pressure of sound waves; **pressure mine**, a mine designed to be activated by the temporary reduction in hydrostatic pressure caused by a passing ship; **pressure pack, package**, a dispenser containing a substance, freq. an aerosol, under pressure; so **pressure-packaged** *ppl. a.*; **pressure-packaging** *vbl. sb.*; **pressure pad**, a pad designed to transmit or absorb pressure; **pressure pattern**, a pattern of prevailing atmospheric pressures; usu. *attrib.*, as *pressure-pattern flying*, denoting the use of air routes which enable aircraft to take advantage of the air currents associated with such patterns to economize on fuel or time; **pressure ridge** (further examples); **pressure saucepan** = *pressure cooker a; **pressure sore** *Med.*, a sore produced by continued pressure on a part of the body; **pressure stove**, a portable stove supplied with oil or paraffin under pressure; **pressure suit**, a garment that can be made airtight and inflated to protect the wearer against low ambient pressure (as in high-altitude flight); **pressure tank**, a tank in which a fluid, esp. fuel, is held under pressure; **pressure tendency** *Meteorol.* = **barometric tendency*; **pressure-tight** *a.*, (of a joint, container, or the like), tightly enough constructed to prevent the passage of a fluid under pressure; hence **pressure-tightness**; **pressure vessel**, a vessel designed to contain material at high pressures; esp. in a nuclear reactor, a vessel containing the reactor core immersed in the pressurized coolant; **pressure wave**, a wave consisting of a sudden change in pressure propagated through a medium; **pressure welding**, welding in which pressure is applied to the parts to be joined; welding brought about by pressure. Also *pressure cooker, *pressure(-)feed *v.* and *sb.*, etc.

1950 Ferrari & Wardell in E. Mason *Pract. Coal Mining* I. ix. 145/2 Props and bars, chocks and cutter nogs, all of which can be withdrawn and reset, are used for carrying the weight of stone inside the pressure arch. **1958** A. Nelson *Methods of Working* iv. 32 A true pressure arch can only exist underground where two side abutments exist, each being strong enough to support its share of the load on the arch and also to provide the lateral thrust necessary for its stability. **1973** L. J. Thomas *Introd. Mining* viii. 340 The recommendation of the committee was that stalls should be limited to three-quarters of the width of the pressure arch for stability, and that pillar width should be equal to stall width. **1891** S. P. Sadtler *Hand-bk. Industr. Org. Chem.* v. 179 Three grammes of substance are placed in a small beaker (preferably of metal), which is placed as one of several in a Soxhlet pressure-boiler, or the test is carried out in the Lintner pressure-flask,—and heated to the temperature of boiling water. **1952** A. Huxley *Let.* 20 May (1969) 644, I have had no return of my iritis and, thanks to the newly invented pressure-breathing treatment,..have practically eliminated the slight chronic bronchitis. **1965** *Gloss. Aeronaut. Terms (B.S.I.)* xvii. 3 *Pressure breathing*, a technique in which oxygen is supplied to the lungs at a pressure higher than the ambient barometric pressure. **1936** *Rev. Mod. Physics* VIII. 48/2 These pressure-broadened lines show the expected larger shifts and half-widths. **1970** A. F. Harvey *Coherent Light* xxv. 1095 For most atmospheric phenomena the pressure-broadened line is appropriate. **1932** *Physical Rev.* XXXIX. 860 The pressure shift of spectral lines, unexplained by the usual theories of pressure broadening. **1967** W. R. Hindmarsh *Atomic Spectra* viii. 86 The three main causes of line-broadening are: the natural width of atomic energy levels..; Doppler width..; and collision, or pressure broadening. **1935** *Jrnl. R. Aeronaut. Soc.* XXXIX. 1045 Pressure cabins and/or free oxygen in the cabins are both being experimented with today. **1948** 'N. Shute' *No Highway* iv. 109 They..went into the rear fuselage, behind the pressure cabin. **1965** J. D. Storer *Behind Scenes in Aircraft Factory* iii. 33 A door in a normal pressure cabin would have to withstand an outward load of some seven tons. **1931** M. Hochstadter et al. in *Jrnl. R. Soc. Arts* LXXX. 95 The utilisation of an impregnated paper cable..could be greatly increased if it were possible to get rid of the heterogeneity in the dielectric, or to render it harmless so far as the time-voltage curve and stability are concerned... It is possible to do this (a) By the use of a very thin impregnating oil and the provision of channels along the cable... (b) By radial compression of the cable in such a way that the radial 'breathing' is reversible at all temperatures, such vacuous spaces as tend to form being closed by the compression or the pressure in them raised to such an extent that no ionisation takes place. The latter alternative leads to the 'Pressure Cable'. **1966** *IEEE Trans. Power Apparatus & Syst.* LXXXV. 375 (*heading*) A few aspects of the general problem concerning tightness of connections formed by pressure cables. **1973** J. G. Tweeddale *Materials Technol.* II. ii. 41 A pressure-cast material is likely to be..more uniformly consistent in structure than other cast materials. **1979** Mills & Mansfield *Genuine Article* iv. 71 'Beirut Sovereigns.' Untold numbers of these superb pressure cast forgeries now adulterate the market. **1922** *Proc. Inst. Mech. Engin.* I. 27 Pressure castings in iron though not yet out of the experimental stage, would..tend to eliminate holes due to occluded air or shrinkage. **1933** *Iron Age* 30 Nov. 18/1 Brass pressure castings are subject to porosity in the same manner that zinc and aluminum die castings are subject to porosity. **1973** J. G. Tweeddale *Materials Technol.* II. ii. 42 (*caption*) Systems for pressure casting. (a) Gravity pressure. (b) Gas pressure. (c) Mechanical pressure. (d) Centrifugal pressure. (e) Another system for using centrifugal pressure. **1915** *Jrnl. Chem. Soc.* CVIII. ii. 820 A press is described..which by means of a lower cylinder with narrow holes bored vertically in it forces a molten solid out of the pressure chamber and so causes a sudden drop in pressure. **1934** *Jrnl. Cellular & Compar. Physiol.* V. 335 A pressure chamber was constructed which permits of viscosity measurements by the 'centrifuge method' at high hydrostatic pressures. **1966** *Lancet* 24 Dec. 1406/1 A small brass pressure chamber was constructed of about 20 ml. capacity. **1919** *Bull. Amer. Inst. Mining Engin.* Feb. 240 On fracture, the pressure die casting will be found to consist of a dense closely grained outer stratum and a porous inner stratum. **1933** *Machinery* XXXIX. 781/1 Pressure die-castings do not have a homogeneous structure, but, upon fracture, exhibit a dense fine-grained exterior and a coarser grained interior. **1973** J. G. Tweeddale *Materials Technol.* II. ii. 8 We have..injection moulding technologists concerned only with pressure-die-casting of thermo-plastic polymers. **1950** Kuethe & Schetzer *Found. Aerodynamics* xii. 212 Two types of drag, form or pressure drag and skin friction, are evident in the flow of a viscous incompressible fluid past a body. **1959** F. D. Adams *Aeronaut. Dict.* 79/1 Inasmuch as pressure drag is a function of form, form drag and pressure drag are sometimes considered synonymous. **1961** H. H. Koelle *Handbk. Astronaut. Engin.* v. 23 In ideal, inviscid, and incompressible two-dimensional flow there is no pressure drag. The sum of the pressure-force components in the free-stream direction is zero. **1934** *Geogr. Jrnl.* LXXXIII. 302 A knife of neolithic age, pressure-flaked from tabular chert. **1959** J. D. Clark *Prehist. S. Afr.* vi. 161 Small pressure-flaked points shaped like equilateral triangles. **1954** S. Piggott *Neolithic Cultures Brit. Isles* ii. 43 The same floor produced three antler tines, and Floor 58..another, associated with a heap of minute flakes, indicating their use as pressure-flakers. **1927** Peake & Fleure *Hunters & Artists* 49 The new technique..includes a high finish by the process of pressure-flaking, that is to say the removal of small thin flakes by pressing near the edge with a bone tool rather than by striking with another stone. **1949** K. P. Oakley *Man the Tool-Maker* v. 29 Some of the tribes..dress spear-heads by pressure-flaking. **1959** J. D. Clark *Prehist. S. Afr.* vi. 157 The finish is much finer and pressure flaking is frequently used. **1891** Pressure flask [see *pressure boiler* above]. **1967** *Oceanogr. & Marine Biol.* V. 188 Pressure flask for studying plants under a few atm. **1941** T. A. H. Peacocke *Mountaineering* iv. 46 Sideways pressure-holds, with the palms of the hands pointing down, should be used as much as possible. **1955** J. E. B. Wright *Technique of Mountaineering* Pl. 10 (*caption*) Last man on Kern Knotts Crack using plimsolls for pressure holds for feet. **1975** W. Unsworth *Encycl. Mountaineering* 120/1 Pressure holds are holds where there is no grip as such and one relies on the friction of the rock. **1923** *Man. Seamanship* (Admiralty) II. 171 Situated on the pressure hull..is what is known as the 'diver's connection'. **1966** *McGraw-Hill Encycl. Sci. & Technol.* XIII. 211/2 The pressure hull, comprising all of the inner and part of the outer hull, is the strong hull that resists external sea pressure. **1974** L. Deighton *Spy Story* xx. 210 The crash came like a sledgehammer pounded against the hollow steel of the pressure hull... Obviously some dire damage had been done to the submarine. **1950** M. Tepper in *Jrnl. Meteorol.* VII. 21 It is proposed that a squall line might be considered as a disturbance generated by accelerations along the cold front and which travels along the warm sector inversion as a gravitational wave. It is recommended that any series of meteorological events similar to this mechanism be called a Pressure Jump Line. *Ibid.* 23/1 The leading edge of this pressure gradient, which shall be referred to as the pressure jump, is clearly defined on the maps, seems to undulate rather violently, and moves in a non-uniform manner. **1955** *Sci. News Let.* 12 Mar. 170/2 This squall line is also known as a pressure jump line, since a sudden rise in barometric pressure always accompanies it. **1963** E. R. Reiter *Jet-Stream Meteorol.* iv. 271 We may..point out some..research work..which makes a pressure jump that travels as a gravity wave along the inversion between moist and dry air responsible for the formation of squall lines and tornadoes. **1967** *Oceanogr. & Marine Biol.* V. 42 Sudden rises in sea level..have been experienced occasionally at coastal locations with the passage over the sea of a moving squall line or pressure jump. **1939–40** *Army & Navy Stores Catal.* 279 'Tilley' lamps... A 'pressure' lamp. **1958** L. Durrell *Balthazar* vii. 154 A whole encampment..had sprung up in the darkness, fitfully lit by oil and paraffin stoves, by pressure lamps and braziers. **1974** J. Wainwright *Hard Hit* 150 The white-hot mantle of the pressure-lamp. **1909** *Daily Chron.* 3 Sept. 1/2 Much of our hard work was lost in circuitous twists around troublesome pressure lines and high, irregular fields of very old ice. **1934** Olson & Massa *Appl. Acoustics* v. 93 If the response corresponds to the variations in pressure of the medium, it is termed a pressure microphone. **1966** *McGraw-Hill Encycl. Sci. & Technol.* VIII. 360/1 Pressure microphones are inherently nondirectional (omnidirectional), because pressure is a scalar and not a vector quantity. **1949** J. S. Cowie *Mines, Minelayers & Minelaying* viii. 162 The Germans, meanwhile, had played their last card, the 'Oyster' or pressure mine. **1957** *Encycl. Brit.* XV. 535/1 Since pressure mines are fired by the reduced water pressure produced by a ship passing over them, the best method of sweeping them is to simulate the passage of a ship by towing a large, expendable target over them. **1969** *New Scientist* 28 Aug. 421/2 In fairways that are not more than 100 or 200 feet deep, the pressure mines are a special hazard. **1958** Herzka & Pickthall *Pressurized Packaging (Aerosols)* iv. 78 The first pressure packs marketed in Great Britain were packed in aluminium dispensers which monopolized the British market until the advent of the all-tinplate dispenser in 1955. **1959** *News Chron.* 30 June 6/5 The pressure-pack has no screw-cap and the dispensing valve automatically seals itself after use. **1966** Harris & Platt in A. Herzka *Internat. Encycl. Pressurized Packaging* vi. 81 In the case of the pressure pack..the basic source of energy is provided by the propellant, which may be either compressed or liquefied gas. **1958** *Food Technol.* XII. 331/1 Results of one or both of the above mentioned methods will indicate whether the

existing product is applicable to the pressure package. **1957** *Mod. Packaging* Dec. 156/2 Many foods that are pressure packaged cannot be subjected to heat without quality loss and thus must be refrigerated. **1959** *News Chron.* 30 June 6/3 A familiar product..is the pressure-packaged insecticide. **1957** *Mod. Packaging* Dec. 156/2 Most foods that have application to pressure packaging will require some sort of preservation treatment. **1912** *Machinery* 31 Oct. 148/2 The shell stripper..also acts as the spring pressure pad during the drawing operation. **1941** C. W. Hinman *Pressworking Metals* vi. 75 The power of compression is adjusted by nuts beneath a pressure pad at the lower end of the casing. **1969** *Times* 7 Mar. 15/1 A group of research workers in California claims to have discovered a way of presenting visual information to blind people through an array of pressure pads which vibrate against the skin of their backs. **1979** B. Freemantle *Charlie Muffin's Uncle Sam* xvi. 146 Battery-operated bells..rang if..anyone..stepped on one of the pressure pads..around the display cases. **1946** *Sci. News Let.* 2 Nov. 278/2 A new technique, 'pressure pattern flying', is now available to air pilots on the Atlantic route from Europe... This new technique consists in determining the shortest flight-time path to the destination by a series of late accurate reports..which locates pressure areas and enables a pilot to take advantage of the airflow circulating around them. **1954** *N.Y. Times* 6 June II. 31/2 During the past eight years Trans World Airlines pilots flying over the Atlantic..have mastered the techniques of getting maximum range from their planes by flying 'pressure patterns'. **1962** G. D. P. Worthington *Flight Planning* iv. 53 The study of pressure pattern flying and the increasing knowledge being obtained of jet streams has made it apparent that the shortest route is not always the quickest or most economical. **1913** I. Cowie *Company of Adventurers* xv. 264, I..was aroused every now and again by the cracking, rumbling and thunderous resounding of the ice as the cold took a firmer grip on it and upheaved it into pressure ridges. **1951** *Beaver* June 13 The entire ice-cover is criss-crossed by a network of pressure ridges. **1975** E. Iglauer *Denison's Ice Road* ix. 230 Denison pointed to a pressure ridge, a wide band of broken ice, several feet high. **1951** *Good Housek. Home Encycl.* 229/1 For cooking a single dish for four or five persons, or a complete meal for one or two, a pressure saucepan will serve. **1889** *Buck's Handbk. Med. Sci.* VIII. 748/3 (Index), Pressure-sores. **1905** R. Howard *Surg. Nursing* ii. 23 In applying back splints to the leg and foot..the heel itself only rests lightly on the splint, otherwise a pressure-sore may occur. **1977** *Lancet* 10 Sept. 548/1 Chairfast patients consistently had a higher pressure-sore frequency than bedfast patients of a similar degree of helplessness. **1914** *Handbk. Amat. Camping Club* 51 The increasing popularity of the paraffin pressure stove, the best-known form of which is perhaps the 'Primus', is an indication that this form of kitchen range probably best fits the camper's bill. **1956** C. Evans *On Climbing* viii. 129 For high altitudes, and extreme cold, it is possible to have a pressure-stove made with an extra large cup to hold more priming fuel. **1966** B. Kimenye *Kalasanda Revisited* 31 The self-pitying thought, 'I might as well be dead', kept recurring in his mind as he pumped his pressure stove to boil a kettle of tea. **1936** *Flight* 1 Oct. 340/2 To enable the pilot to stand the extremely low pressure encountered at about 50,000ft..a special 'pressure-suit' has been produced. **1949** *Startling Stories* Sept. 125/2 The multiple layers of my pressure suit had made movement very difficult. **1962** J. Glenn et al. *Into Orbit* 244 G-suits are not to be confused with pressure suits (or, now, spacesuits) which the Astronaut wears during space flight to maintain atmospheric pressure at high altitudes. **1977** P. Way *Super-Celeste* 58 The sudden expansion of his pressure suit turned his body into a heavy, rigid block... He was catapulted..into..the sky. **1862** *Electrician* 10 Jan. 115/2 (*heading*) Mr. Reid's pressure tank, used in testing cables during manufacture. **1917** 'Contact' *Airman's Outings* viii. 225 A small gravity tank for his machine, to be used when the pressure tank is ventilated by a bullet. **1929** F. P. Gibbons *Red Napoleon* xii. 403 'Pressure tank hit,' Binney shouted. 'I'm dumping her.' **1962** *Sci. Survey* III. 85 Modern machines [*sc.* electrostatic generators] of this type, enclosed in steel pressure-tanks, have produced about 10 MeV. **1939** R. C. Sutcliffe *Meteorol. for Aviators* xvii. 218 The closest attention should always be given to the pressure tendencies. **1946** W. L. Donn *Meteorol.* vii. 118 The pressure tendency is the net change in pressure for the preceding 3 hours. **1970** F. W. Cole *Introd. Meteorol.* xiv. 323 While pressure, as such, is not a useful weather parameter, the change in pressure and the pressure tendency are both helpful in developing a forecast. **1946** *Nature* 21 Dec. 897/2 Pressurization of cabins for high-altitude flying now appears to be essential... This creates a fresh outlook on the body structure, which now has to be a pressure-tight shell. **1963** R. Hammond *Automatic Welding* ii. 85 Eight spot welds fix the spider to the rim, eliminating rivet holes and ensuring a pressure-tight joint. **1951** H. H. Doehler *Die Casting* xi. 458 (*heading*) Inspecting for pressure tightness. **1970** tr. *Zoebl's Fund. Hydraulic Circuitry* viii. 156 This type of plant is used for checking the pressure tightness of welded pipes. **1915** Haven & Swett (*title*) The design of steam boilers and pressure vessels. **1960** *Practical Wireless* XXXVI. 298/1 X-ray photographs of the welds in the pressure vessel of a nuclear power station. **1962** *Newnes Conc. Encycl. Nucl. Energy* 707/2 For safety reasons the reactor is located inside a steel pressure vessel about 70 ft in diameter and 70 ft high. **1975** Bram & Downs *Manuf. Technol.* ii. 60 If the product is a pressure vessel, very high-integrity welds are essential. **1949** O. G. Sutton *Sci. of Flight* vi. 140 (*heading*) Pressure waves caused by moving bodies. **1956** A. H. Compton *Atomic Quest* iii. 212 The pressure waves in a metal strained far beyond the elastic limit. **1962** A. Nisbett *Technique Sound Studio* xi. 191 The water-tank artificial reverberation machine.., in which the sound is modulated on to a 80-kc/s carrier and fed via a piezo-electric element to produce pressure waves in the water-tank. **1975** T. Allbeury *Special Collection* iv. 18 He could feel the vibration in his ears, the pressure waves of the guns and bombs. **1926** Stoughton & Butts *Engin. Metall.* vii. 137 (*heading*) Electric heating for pressure welding. **1954** H. Udin et al. *Welding for Engineers* iii. 34 Oxyacetylene pressure welding is accomplished by butting together under pressure the two pieces of metal to be joined and heating the junction by oxyacetylene torches. **1967** A. H. Cottrell *Introd. Metall.* xxii. 435 In extrusion, the great pressure developed between the metal and the die can lead to sticking, due to pressure welding.

pre·ssure, *v.* orig. *U.S.* [f. the sb.] **1. a.** *trans.* To exert pressure on. Chiefly *fig.*, to urge or impel (someone *to do* something or *into* a situation or course of action); to drive or force (someone *out of* something). Also *absol.* (const. *for*), to exert pressure, to press. Hence **pre·ssured** *ppl. a.*, of work, affairs, etc.: urgent, pressing; of people: under pressure.

1939 R. Chandler in *Dime Detective Mag.* Jan. 103/2 I'm not trying to pressure you. **1944** *Sun* (Baltimore) 6 Oct. 7/3 You can't pressure the War Labor Board into action through strikes. **1951** L. Z. Hobson *Celebrity* (1953) x. 140 It's too bad Gregory Johns is so set against public appearances, but even the studio isn't trying to pressure him. **1957** J. F. Horner *Summary of Scientology* i. 10 Finally he was pressured into writing a popular treatise on Scientology. **1960** *Daily Mail* 12 May 10/6 Baldies can be 'pressured' into paying more than £200 for a course of treatment. **1961** S. Raven *Eng. Gentleman* iii. iv. 170 Was Rufus trying to pressure him? asked Henry. Certainly not. Just trying to make him see things in a sensible way. **1963** *Economist* 12 Jan. 111/2 Preachers.. 'were pressured out of the pulpits they held'. **1968** P. Oliver *Screening Blues* vi. 248 The imputation that the singers were pressured by the record companies which is frequently made, though probably having some measure of truth, is probably much overstated. **1971** H. Cheetham *Portrait of Oxford* xiii. 202 The trouble about an Oxford education..is that no-one..pressures you into working. **1971** C. Bonington *Annapurna South Face* iii. 32, I..was near to exhaustion from weeks of pressured work and worry. **1973** J. Goodfield *Courier to Peking* i. iii. 45 You personally have never pressured for unlimited resources. **1976** *Times* 8 July 2/5 He said the personalities of Joseph Markham and later Clive Mildoon began as fantasies, providing relief for the pressured and overburdened public figure of Mr John Stonehouse.

b. *trans.* To gain by bringing pressure to bear. *U.S.*

1944 *Sun* (Baltimore) 7 May 8/3 He intervened himself and pressured a better settlement for the unions. **1952** *Ibid.* 22 Mar. 6/4 Other strong unions will now immediately pressure comparable or greater gains for their own people.

2. *trans.* = *pressurize *v.* 1.

1961 in Webster. **1979** *Daily Tel.* 8 June 2/1 The engine on the right would have continued to pressure the No. 3 [hydraulic] system under normal circumstances.

pressure cooker. [f. Pressure *sb.* + Cooker.] **a.** An airtight vessel in which food can be cooked in steam under pressure, so that a higher temperature is reached and the food is cooked more quickly.

1915 *Jrnl. Home Econ.* VII. 375 Why should the modern household more than the modern factory reject a tool of value? This question might well be asked concerning pressure cookers. **1919** *Delineator* Nov. 53/1 Of all modern household saving devices on the market to-day.. there is no one article which does more toward lessening household burdens..than the pressure cooker. **1937** H. W. Tilman *Ascent of Nanda Devi* x. 111 Weight was saved by the abandonment of two pressure cookers..a cooker that cooks by steam under pressure and in which, I think, even a pair of boots would be made edible. **1950** F. Swinnerton *Flower for Catherine* 57 Probably some tough old sheep, with lambs of her own, that I shall have to tenderize in the pressure cooker. **1951** *Good Housek. Home Encycl.* 229/1 Some pressure cookers have lids which seal internally, fitting under a rim. **1969** *Islander* (Victoria, B.C.) 5 Oct. 5/2, I had brought along my pressure cooker for speed, and soon I had this filled with the tasty ingredients for a stew.

b. *fig.* (freq. *attrib.*).

1954 Koestler *Invis. Writing* iv. 54, I had acquired it [*sc.* Russian]..by the same pressure-cooker method by which I had learnt modern Hebrew. **1958** *Spectator* 13 June 759/1 Strict curfews and a huge concentration of troops restored order, but the valve of the pressure-cooker was seen to be under enormous strain. **1968** Mrs. L. B. Johnson *White House Diary* 18 Dec. (1970) 758 Every day of the last four years it seems to me she [*sc.* Mrs. Hubert Humphrey] has grown..and especially in the pressure cooker of a campaign. **1974** *Spartanburg* (S. Carolina) *Herald* 18 Apr. c4/1 With the season now almost four months old and the pressure cooker of the Masters just behind them, most of the game's top guns are taking a break. **1976** L. Sanders *Hamlet Warning* (1977) iv. 36 Santo Domingo was a pressure cooker, ready to explode. **1976** *National Observer* (U.S.) 25 Dec. 6/5 An auction's pressure cooker atmosphere is no place for split second decisions involving tens of thousands of dollars.

Hence (*lit.* and *fig.*) **pressure-cook** *v. trans.* (also *absol.*), **pressure-cooked** *ppl. a.*, **pressure-cooking** *vbl. sb.*; also **pressure cookery**.

1940 *Sears, Roebuck Catal.* Spring/Summer 612 Your meals will taste better, too, because pressure cooking retains all the delicate flavors. **1950** *Mrs. Beeton's Bk. Househ. Managem.* 1155 (*heading*) Hay-box and pressure cookery. **1951** *Good Housek. Home Encycl.* 616/2 Pressure-cook for 25–30 minutes. *Ibid.* 618/2 Pressure cookery is the only safe method for the home sterilising of bottled and canned vegetables. **1958** *Listener* 2 Jan. 13/2 The Russians..avoid early specialisation or pressure-cooking in education. **1958** *Woman* 27 Sept. 4/3, I pressure-cooked it until the meat left the bone. **1960** *Farmer & Stockbreeder* 9 Feb. (Suppl.) 2/1 Nut and raisin pressure-cooked loaf cake is covered with a double thickness of greased paper. **1968** *Time* 11 Oct. 28 We concocted Chicago from one Bat for peace, Numerous Democratic toads, And a pressure-cooked American flag.

pressure-feed, *v.* and *sb.* [See Feed *v.* 7, 8 c and Feed *sb.* 5.] **A.** *v. trans.* To supply (material, esp. fuel) by means of applied pressure. **B.** *sb.* Also **pressure feed**. A supply system in which the flow of material is maintained by applied pressure; the supplying of material in this way. So **pre·ssure-fed** *ppl. a.*, supplied with material in this way; utilizing a pressure feed.

1904 *Autocar* 6 Aug. 164/1 (*heading*) Carburetter. Pressure feed from the exhaust. *Ibid.*, The lubricator..is also pressure fed. **1906** W. W. Beaumont *Motor Vehicles* II. viii. 158 Pressure-fed lubrication is adopted for the engine and for the petrol feed. *Ibid.* xii. 206 Pressure feed is used, obtained by a shunt from the exhaust pipe to a distributing box. **1909** D. Leechman *Carburetters* ix. 66 Petrol is maintained at a certain level..by means of a float and needle valve when the petrol is pressure fed. **1914, 1925** [see *gravity(-)feed *sb.*]. **1936** *Discovery* Sept. 299/2 The rocket motor of today with its..complicated system of pressure feeds, safety valves and dual liquid-fuel storage tanks, little resembles the old-style cardboard tube and gunpowder. **1961** H. H. Koelle *Handbk. Astronaut. Engin.* xx. 16 Pump-fed systems are lighter than pressure-fed systems for all but the very shortest-duration requirements.

pressure group. [f. Pressure *sb.* + Group *sb.*] A group or association of people representing some special interest, who bring concerted pressure to bear on public policy. Also *attrib.* Hence **pressure groupism**, activity characteristic of a pressure group.

1928 P. H. Odegard *Pressure Politics* vii. 202 The character of pressure groups as of individuals can frequently be understood from the manner in which they spend their money. **1934** W. Lippmann *Method of Freedom* iii. 97 We come then to the conclusion that it is not the pressure groups as such which make it impossible for the state to act in the general interest..but pressure groups attached to and reinforced by political machines. **1936** A. Huxley *Eyeless in Gaza* xxii. 316 One joined the Party, one distributed literature, one financed pressure-groups. **1937** B. Zeller *Pressure Politics in N.Y.* viii. 229 The outstanding feature of modern pressure group technique is the widespread use of propaganda channels. *Ibid.* ix. 263 This frequent and flagrant abuse of pressure group politics. **1941** W. Temple *Citizen & Churchman* iii. 56 One American professor opened his reply with the words: 'It is obvious that the Church is a pressure-group.' **1953** Wodehouse *Performing Flea* 154 You can no longer put a negro on the stage unless you make him very dignified. Owing to the activities of the Negro pressure group, comic negro characters are absolutely taboo. **1954** *Encounter* Mar. 58/2 The illusion..that US opposition to the admission of Communist China to the UN is..an ephemeral, emotional attitude dictated by a nebulous pressure-group called 'The China Lobby'. **1959** *Oxf. Univ. Gaz.* 16 Mar. 795/1 It is an interesting aspect of the workings of pressure-group politics to see that in Australia and New Zealand it is the interests of the manufacturers which prevail, with both political parties, against the general good of those countries, which urgently require greater production and export of farm products. **1960** *Encounter* Jan. 39/2 Those literal-minded pressure-groups which haunt all public organs of opinion. **1966** *New Statesman* 13 May 683/1 Comment on this affair included, not surprisingly, accusations of pressure groupism and intolerance. **1971** P. Gresswell *Environment* 18 In a literate society, such pressure groups [*sc.* amenity societies] are a necessary adjunct to parliamentary democracy. **1978** *Dædalus* Spring 75 Almost equally annoying to scientists is the change in political attitudes; members of Congress now tend to look on them as just another selfish pressure group, and not as the wizards of perpetual progress.

pressure head. [f. Pressure *sb.* + Head *sb.*] **1.** The pressure exerted by a fluid, expressed as the height of a column of fluid which would produce that pressure by virtue of its weight (cf. Head *sb.* 17 a and b).

1907 Woodward & Preston tr. *Sorel's Carbureting & Combustion in Alcohol Engines* viii. 160 Let..*H* = pressure head on the liquid. **1974** J. A. Fox *Introd. Engin. Fluid Mech.* viii. 283 A velocity change of 1 ft/s can generate a pressure head of approximately 125 ft if it occurs rapidly enough.

2. *Aeronaut.* A pitot-static tube.

1930 *Aircraft Engin.* II. 95/2 The standard pressure head..consists of pressure and static tubes mounted parallel and near to each other. **1964** E. H. J. Pallett *Aircraft Instrument Man.* ii. 11 Pressure heads now in general use are..electrically heated.

pressure jet. [f. Pressure *sb.* + Jet *sb.*[2]] **a.** Used *attrib.* with reference to a type of oil burner in which the fuel is burned at a fine nozzle through which it is passed under pressure.

1911 W. H. Booth *Liquid Fuel* xiii. 212 The pressure-jet system will recover from 70 per cent. to 75 per cent. of the theoretical calorific value of the oil fuel used in actual practice. **1920** E. C. Bowden-Smith *Oil Firing* iv. 90 In the pressure-jet burner great care must be taken to maintain a high temperature, or the burner may fail entirely. **1968** J. Slome *Domestic Oil-Fired Central Heating* iii. 61 Pressure jet burners can make quite a considerable amount of noise.

b. A small jet engine mounted at the tip of a helicopter rotor blade and supplied with compressed air through a duct in the blade. Freq. *attrib.*

1950 *Jrnl. Helicopter Assoc. Gt. Brit.* III. 153 The pressure jet helicopter is..suitable for longer ranges. **1958** Lambermont & Pirie *Helicopters & Autogyros* 201 The XV-1 has a reciprocating engine..which acts by feeding air through tubes to small pressure jets at the tips of the three-bladed rotor during vertical flight.

pressure plate. [f. Pressure *sb.* + Plate *sb.*] **a.** A plate for detecting or receiving pressure, *spec.* in an anemometer. **b.** A plate for applying pressure, *spec.* in a clutch.

1845 *Rep. Brit. Assoc. Adv. Sci. 1844* 24 A contrivance ..is affixed to the pressure plate, by means of which the fluid is deposited at a variable rate, but always depending on the force on the pressure plate at the time. **1856** *Ibid.* *1855* 127 The force of the wind is ascertained by means of a circular plate having an area of four square feet, which is kept by the vane at right angles to the current of the wind... To this pressure-plate is attached a wire which communicates with a recording pencil. **1892** *Q. Jrnl. R. Meteorol. Soc.* XVIII. 174 It was not possible to obtain a perfectly distinct line from the pen of the Pressure Plate during a gale unless the paper moved at least 2 inches in each minute. **1921** J. V. Woodworth *Amer. Tool Making* xxvii. 445 The pressure on the foot treadle, which causes the pressure plate to clamp the can and lid against the chuck, also throws in the friction clutch which starts the work. **1958** *Newnes Compl. Amat. Photogr.* iv. 71 In a 35 mm. camera, the pressure plate should be examined carefully. **1963** D. H. McIntosh *Meteorol. Gloss.* (ed. 4) 16 In the pressure-plate anemometer, the deflexion of a flat plate placed in the wind is measured; its use is confined mainly to atmospheric turbulence measurement. **1976** *Horse & Hound* 3 Dec. 66 (Advt.), Superb rebuilt Land-Rovers fitted with..new clutch and pressure plate.

pressure point. [f. Pressure *sb.* + Point *sb.*[1]] **1. a.** A point where pressure is supposed to stimulate or inhibit convulsions.

1876 [in Dict. s.v. Pressure *sb.* 10]. **1885** J. Ross *Handbk. Dis. Nervous Syst.* 167 Pressure points are frequently observed in spasmodic affections. Pressure upon certain points puts a stop at times to the convulsion when present, and consequently these points may be called pressure-arresting points. In other cases the convulsions are brought on by pressure on particular points, and these may..be called pressure-exciting points. **1896** J. M. Da Costa *Med. Diag.* (ed. 8) ii. 233 There are 'pressure-points' which when acted on will cause the convulsive movements to be arrested. **1910** J. L. Salinger tr. T. Ziehen in A. Church *Dis. Nervous Syst.* 1082 The relation of the pressure-points is also noteworthy. Pressure upon these occasionally increases the severity of the attack, or may produce a new attack [of hysteria].

b. One of numerous small areas on the skin that are specially sensitive to pressure; also, a point at which pain is felt on pressure.

1882 *Amer. Jrnl. Med. Sci.* LXXXIV. 589 Dr. Meyer discovered a painful pressure-point at the upper part of the brachial plexus. **1891** W. Stirling tr. *Landois's Text-bk. Human Physiol.* (ed. 4) II. xiv. 1018 The 'pressure-points'..lie much closer together, and are more numerous than the temperature-points. **1906** C. P. Flint et al. tr. *Sahli's Diagnostic Methods* 758 The pressure sense is not scattered diffusely in the skin, but..depends upon localized organs. The projections of the latter upon the surface of the skin are called 'pressure points'. **1940** *Jrnl. Exper. Psychol.* XXVI. 516 This parallelism of the two kinds of sensation [*sc.* pressure and vibration] could not be proved for the pressure points. **1958** R. Wartenberg *Neuritis, Sensory Neuritis, Neuralgia* xli. 406 In cases of neuralgia of the last intercostal nerve, typical pressure points, especially on the back, could always be found.

c. A point where an artery can easily be pressed against a bone to inhibit bleeding.

1909 R. Howard in *Sci. & Art of Nursing* III. xxii. 5 (*heading*) The main arteries and their pressure points. **1933** Bailey & Love *Surg. for Nurses* xxix. 308 Pressure points..are situations in which large arteries are adjacent to bones, and so are easily compressed against the rigid underlying part. **1954** Diehl & Laton *Health & Safety for You* iv. 46/1 (*caption*) Find the pressure point between the wound and the heart; press against the bone. **1973** *Guardian* 11 Apr. 11/4 He..sliced the top off his thumb. But he knew about pressure points.

d. A pressure sore, or a point where one is apt to develop owing to the pressure on it.

1929 E. L. Eliason et al. *Surg. Nursing* xvi. 389 After the immediate effects of the operation have disappeared, a vigilant watch must be kept for the development of pressure points. **1941** K. D. Keele *Mod. Home Nursing* iii. 43 Pressure points of the body have to be learnt. They are those parts which, being exposed to the brunt of bearing the weight of the body, get most wear. **1964** M. C. T. Morrison *Basic Princ. Accident Surg.* xvi. 89 The patient should be nursed on pillows or foam rubber pads to distribute the pressure evenly over the whole of his back or side rather than on his 'pressure points'. **1969** Brain & Walton *Dis. Nervous Syst.* (ed. 7) xiv. 630 The dressing should be well covered with adhesive plaster attached to skin some distance from the pressure points and changed every day.

2. *fig.* A person or thing that can be used by someone as a means of exerting pressure on another.

1975 T. Allbeury *Palomino Blonde* xv. 91 The girl.. is being used as a pressure point on him to give the details of his discovery to the Soviets. **1977** R. Ludlum *Chancellor Manuscript* xv. 167 Hoover..will soon control the pressure points of the country. He'll be running it. **1978** *Internat. Relations Dict.* (U.S. Dept. State Library) 25/1 An international political strategy relating two or more issues in negotiations, and then using them as tradeoffs or pressure points, much as in a 'carrots and stick' technique.

pressure-test, *sb.* and *v. trans.* [f. Pressure *sb.* + Test *sb.*[1]] **A.** *sb.* A test for pressure of any kind, or of ability to withstand or sustain pressure. **B.** *v. trans.* To subject to a test of this nature.

[**1882** R. Sennett *Marine Steam Engine* xxvii. 570 The water-pressure test should be double the working steam pressure, provided that during the examination no indication of weakness is observed.] **1888** W. C. Unwin *Testing of Materials of Construction* xv. 479 It may be useful to examine if the rate of hardening in pressure tests can be expressed as simply as that in tension tests. **1897** [in Dict. s.v. Pressure *sb.* 9 b]. **1941** Wyndham Lewis *Let.* 17 Oct. (1963) 300 No. 1 doctor said that if the [ocular] pressure-test gave a negative result, then another cause must be looked for. **1957** *Sun* (Baltimore) 19 June 40/3 They were pressure-testing it with oxygen, after welding it, when the tank exploded. **1977** *Offshore Engineer* June 50/3 We pressure-tested flotation chambers to check water-tightness.

pressure treatment. [f. Pressure *sb.* + Treatment.] **a.** *Timber.* Impregnation of timber with a preservative fluid, such as creosote, under applied pressure.

1914 Moon & Brown *Elem. Forestry* xii. 235 The Rueping, Card, Lowry and other more or less important processes are in common use but they are all variations of the same pressure treatment. **1942** H. D. Tiemann *Wood Technol.* xvii. 258 Wood preservation against decay consists in impregnating wood, either by soakage or by pressure treatment, with antiseptic liquids. **1950** A. J. Panshin et al. *Forest Products* iv. 65 Practically all southern pine poles are given a full-length pressure treatment with creosote oil before being placed in service.

b. *Biol.* Subjection (of cells, organisms, etc.) to increased pressure.

1940 *Biol. Bull.* LXXVIII. 106 It is clear that the time relationships between nuclear and cytoplasmic division may be somewhat disturbed by pressure treatment. **1956** *Internat. Rev. Cytol.* V. 215 The cell is rendered incapable of performing the work of cleavage by any combination of temperature and pressure treatments that jointly weakens the gel structure to approximately the same degree. **1970** S. B. & A. M. Zimmerman in A. M. Zimmerman *High Pressure Effects on Cellular Processes* viii. 183 Numerous vacuoles appear in the cell cytoplasm following decompression and may be representative of some damage caused by the pressure treatment.

Hence (as a back-formation) **pre·ssure-treat** *v. trans.* So **pre·ssure-treated** *ppl. a.*; **pre·ssure-treating** *vbl. sb.*

1936 *Jrnl. Cellular & Compar. Physiol.* VIII. 159 The Amoebae are placed in a centrifuge-pressure chamber which is so designed that the control and pressure-treated specimens are simultaneously centrifuged for a suitable period. **1938** Hunt & Garratt *Wood Preservation* vi. 221 Untreated wood, exposed when pile heads are cut off to grade, may be pressure treated by a method developed by E. F. Hartman. **1950** A. J. Panshin et al. *Forest Products* iv. 66 One of the empty-cell processes is usually employed, unless customer specification for total retention of creosote is such that a full-cell pressure-treating process is required. **1960** *Farmer & Stockbreeder* 2 Feb. (Suppl.) 9/1 Concrete stanchions and pressure-treated timber trusses and purlins. **1963** H. R. Clauser *Encycl. Engin. Materials* 738/1 Pressure-treating operations must be conducted in plants with considerable equipment. *Ibid.*, Some of the American species are easy to pressure-treat, others are difficult to penetrate with any chemical. **1970** S. B. & A. M. Zimmerman in A. M. Zimmerman *High Pressure Effects on Cellular Processes* viii. 187 The pressure-treated cells revealed a loss of Golgi complex and pinocytotic channels.

pressure tube. [f. Pressure *sb.* + Tube *sb.*] **1.** A tube open at one or more points to a surrounding fluid whose velocity or pressure it is used to measure. Usu. *attrib.* in *pressure tube anemometer*.

1894 *Q. Jrnl. Meteorol. Soc.* XX. 186 Lately he had spent a good many hours by the side of Mr. Dines's pressure tube anemometer, watching the action of the pen in squalls of wind. **1920** G. Taylor *Austral. Meteorol.* viii. 77 The pressure and suction tubes..act together to move the float. **1970** R. W. Longley *Elements Meteorol.* vii. 140 The record..came from a Dines pressure-tube anemometer which is able to measure rapid fluctuations in the wind.

2. A tube in which pressurized coolant or moderator is passed through the core in certain types of nuclear reactor.

1961 J. K. Pickard et al. *Power Reactor Technol.* iv. 214 A graphite reflector can be used..which compensates for the loss of neutrons in the pressure tubes. **1968** Moore & Holmes in *Steam Generating & Other Heavy Water Reactors* (Brit. Nuclear Energy Soc.) 3/2 One of the important features of a pressure tube reactor is that not only can different materials be used for the moderator and coolant, but the operating conditions of these media can be selected quite independently to suit their functions.

pressurization (preʃəraizēi·ʃən). [f. as next + -ation.] The action or result of pressurizing (*lit.* or *fig.*).

1937 *Jrnl. Aeronaut. Sci.* IV. 99/1 The problems of cabin pressurization will increase rapidly as wall pressure differentials of over 1300 lbs. per sq. ft. will maintain. **1946** *Nature* 21 Dec. 897/2 Pressurization of cabins for high-altitude flying now appears to be essential with the adoption of the gas turbine. **1958** 'P. Bryant' *Two Hours to Doom* 86 Have the pressurization..set for full. **1963** *Daily Tel.* 26 Apr. 14 A luncheon appointment between a publicity expert and an M.P. to discuss a particular interest appears suspect to Mr. Edelman as 'pressurization'. **1969** *Ibid.* 21 Nov. 2/3 A decompression explosion can also be ruled out as the plane was too low to need pressurisation. **1975** F. R. Palmer in W. F. Bolton *Eng. Lang.* ii. 44 The stop (or plosive) consonants..involve the pressurization of air pushed up from the lungs into the vocal tract. **1979** *Daily Tel.* 2 July 22/2 Cruel and critical but sympathetic pressurisation each day from the organising consultants.

pressurize (pre·ʃəraiz), *v.* [f. Pressure *sb.* + -ize.] **1.** *trans.* To produce or maintain pressure artificially in (a container, closed space, etc., esp. an aircraft); to apply pressure to.

1944 *Aeronautics* Sept. 56/2 The fuselage will be pressurized so that at all altitudes cabin conditions will be equivalent to a height of 8,000 ft. **1958** *Times* 25 Jan. 9/2 This machine consists of a small electric compressor pump, which pressurizes the container to a pressure of 80 lb. to the square inch. **1958** *Times* 14 Aug. 9/7 This means she [*sc.* an undersea cargo ship] would have an empty space which must either be pressurized to balance the pressure of the sea, or made strong enough to withstand that pressure. **1970** *Nature* 18 Apr. 249/2 Before re-entry, the camera and payload section were sealed and pressurized to two atmospheres. **1972** *Daily Tel.* 11 Dec. 15/6 We returned from our first drive, pressurised the lunar module cabin and took our helmets off. **1975** *Sci. Amer.* July 52/3 The surface of the journal and the surface of the bearing are separated by a film of lubricant when the journal is turning rapidly enough to pressurize the wedge with lubricant.

2. To subject to moral or mental pressure or suasion; to urge or constrain.

1956 *Essays in Crit.* VI. 238 The best poems..have all these qualities together with a strict sense of form that pressurizes the colloquial idiom. **1964** *Listener* 26 Nov. 859/2 This is the move which concedes the initiative, since White is now able to gain space and pressurize the black squares on the king's side. **1970** N. Bawden *Birds on Trees* ii. 32 Charlie said we didn't want to pressurize him. **1973** 'M. Innes' *Appleby's Answer* xiii. 116 Perhaps Bulkington has developed a quiet pressurising line on his young charges. **1978** *Guardian Weekly* 30 Apr. 15/3 U.S. officials are pressurising the Saudis to increase production capacity.

Hence **pre·ssurizing** *vbl. sb.*

1946 *Sun* (Baltimore) 16 May 6/3 Pressurizing enables passengers to enjoy the comfortable flying conditions of 8,000 feet while at a ceiling of 25,000. **1976** J. M. Kelly *Stud. Civil Judicature Roman Republ.* v. 130 The pressurizing of the defendant into surrendering the plaintiff's property.

pressurized (pre·ʃəraizd), *ppl. a.* [f. as prec. + -ed[1].] **1.** Containing, or made to contain, fluid under pressure. **a.** Of an aircraft cabin, spacesuit, etc.: designed to maintain an interior air pressure close to normal atmospheric pressure in a low-pressure environment.

1938 *Time* 23 May 33 Without pressurized cabins, planes now fly as high as 14,000 feet. **1945** *Times* 2 Oct. 2/4 It has a pressurized cabin which, up to a height of 20,000 ft., maintains a pressure inside the passengers' cabin equivalent to that at 8,000 ft. **1949** *Archit. Rev.* CV. 237/2 A very highly streamlined low-wing, four-motor monoplane, with tricycle undercarriage and pressurized cabin. **1951** A. C. Clarke *Across Sea of Stars* (1959) 4 We could live comfortable for a month in our pressurised tractors. **1958** *Times* 28 Aug. 9/4 'Colonists' [on the moon] ..unable to quit their pressurized suits for a moment. **1962** F. I. Ordway et al. *Basic Astronautics* xiii. 516 (*caption*) First pressurized suit..was developed for Wiley Post and worn by him in 1934. **1975** D. Lodge *Changing Places* i. 3 They were protected from the thin, cold air by the pressurized cabins of two Boeing 707s.

b. Of an aerosol container or spray.

1955 *Industr. & Engin. Chem.* June 1198/1 Aerosol products are pressurized, self-spraying products that at the press of a valve button deliver an active ingredient in a fine spray (insecticides and room deodorants), a heavier spray (paints and enamels), a foam (shave creams), and newest among the applications, a dry powder. **1958** *Times* 24 Nov. 11/5 Helena Rubinstein has a new pressurized scent spray. **1961** *Lancet* 2 Sept. 506/1 It was decided to recommend that the surgical staff should use.. a pressurised powder spray of polybactrin. **1976** *Which?* Feb. 37/1 Technically, an aerosol is simply a fine spray. Most people use the word to describe the pressurised can that produces a spray.

2. Of a fluid: increased in pressure; *pressurized-water reactor*, a nuclear reactor in which the coolant is water at high pressure.

1953 *Chem. & Engin. News* 3 Aug. 3187/3 The AEC is continuing research and development work on the pressurized water reactor. **1957** E. Hyams *Into Dream* iii. vi. 236 As if the pressurized air contained some poison of the mind. **1958** *Times Rev. Industry* July 32/2 Pressurized fluid is fed to the power valve which..actuates the brake. **1960** *Economist* 22 Oct. 392/2 One important aspect of *Dreadnought* is that it gives British technicians a first experience of a pressurised water reactor. **1966** *McGraw-Hill Encycl. Sci. & Technol.* XI. 363/1 The sodium-cooled reactor originally installed on the submarine USS *Seawolf* has been replaced by a pressurized-water reactor. **1968** M. Woodhouse *Rock Baby* xviii. 182 Find out if there's still pressurized gas within the casing. **1977** *Sci. Amer.* June 46/1 The lower piston rests against the dividing plate, supported by a cushion of pressurized nitrogen gas.

3. *fig.* Of a person or situation: subject to pressure; under moral, mental, or social pressure or constraint.

1959 *Times* 30 Nov. 4/1 This age of pressurized competition. **1965** *Punch* 22 Sept. 425/2 This going out of London has made the job very pressurised. Weekends are hectic.

press-work. Add: **3. b.** The pressing or drawing of metal into a shaped hollow die; a piece of metal shaped by such means.

1896 O. Smith *Press-Working of Metals* i. 14 In press-work the metal is sometimes heated as in forging, but in the great majority of cases it is handled cold. **1903** *Engineering* 16 Jan. p. v (Index), Press work for sheet metal. **1904** *Ibid.* 22 Jan. 132/3 We illustrate below a very remarkable specimen of 'press' work... The barrel shown ..has been drawn out of a steel-plate 1/16th of an inch in thickness. **1941** C. W. Hinman *Pressworking of Metals* i. 1 Today, it is a wonder how the past generation ever produced any satisfactory presswork. **1963** Bird & Hutton-Stott *Veteran Motor Car* 82 This was a most remarkable piece of presswork and comprised a complete chassis..of deep-section side girders, upswept over the rear axle.

Hence as *v. trans.*, to shape (metal) in this way; usu. as *pres. pple.*; so **pre·ss-working** *vbl. sb.*

1896 O. Smith (*title*) Press-working of metals. *Ibid.* v. 120 It is taken for granted that the materials employed for press-working..must be to some considerable degree in a malleable or ductile physical condition. **1941** C. W. Hinman *Pressworking of Metals* vi. 62 (*heading*) Press-working nonmetallic materials. **1949** *Tool Engineers Handbk.* (Amer. Soc. Tool Engineers) lxxiii. 1058 Heated dies are used in the pressworking of brass and magnesium. **1958** Eary & Reed (*title*) Techniques of pressworking sheet metal.

prestation. Add: **a.** (Further example.)

1973 *Proc. Gen. Board of Faculties Oxf. Univ.* CXXXIII. 568 The directive also lays down that in the case of provision ('prestation') of services [etc.]... This expression refers to a short visit to another country in order to provide services on a temporary or transient basis.

c. *Anthrop.* A gift, payment, or service that forms part of some traditional function in a society, given or due either to specific persons or to the group.

1889 W. R. Smith *Lect. Relig. of Semites* xi. 403 The very idea of an execution implies a public function, and not a private prestation. *Ibid.* 413 Even in the theology of the Rabbins penitence atones only for light offences, all grave offences demanding also a material prestation. **1935** B. Malinowski *Coral Gardens* I. vi. 204 Since the English language has a really unaccountable and intolerable gap, I am deliberately introducing here the word 'prestation' in the French sense, that is, of legally defined services to be tendered by one individual or group to another. **1951** *Jrnl. R. Anthrop. Inst.* LXXXI. 35/2 In Kachin type systems it is an exchange of women for gifts (prestations). *Ibid.* 51/1 The 'prestations'..may not only take on a variety of forms, they may have several quite different structural functions. **1954** I. Cunnison in *Mauss's Gift* p. xi, There is no convenient English word to translate the French *prestation* so this word itself is used to mean any thing or series of things given freely or obligatorily as a gift or in exchange; and includes services, entertainments, etc., as well as material things. **1957** M. Fortes in R. Firth *Man & Culture* 178 Exogamy is evidently enforced without exception, as we should expect with a jural obligation that..is validated by prestations on both sides. **1967** F. Barth in R. Firth *Themes in Econ. Anthropol.* 152 The rights of the cultivator as *user* as distinct from *owner* are expressed in the symbolic prestation of one pot of beer to the title holder after each harvest. **1968** R. Needham et al. tr. *Lévi-Strauss's Elem. Struct. Kinship* (1969) vi. 77 The *ufuapie* exchange prestations which are economic, legal, matrimonial, [etc.]. **1973** *Sci. Amer.* July 74/1 In general anthropologists have argued that the goods are a 'prestation', which has been defined as the act of paying in money or service what is due by law or custom.

Prestel (preste·l). The proprietary name of a computerized visual information system operated by British Telecommunications, by which data selected from one or more data bases may be made to appear on a television screen by dialling an appropriate telephone number.

1978 *Times* 28 July 17/7 A Post Office brain-child, originally called Viewdata and now known as Prestel.., is now operating in a test version in preparation for its full public launch. **1978** *Trade Marks Jrnl.* 6 Dec. 2689/1 *Prestel*... Electrical, electronic and electro-mechanical apparatus, instruments and installations; monitoring, control and data storage apparatus and instruments... The Post Office. **1979** *Observer* 11 Feb. 42 (Advt.), Even at this early stage, there are thousands of pages of information available to Prestel subscribers. It's a sign of the way television is moving from being a simple means of entertainment to a much more complex domestic information medium.

prestellar: see *pre- B. 1.

prestige. Add: Also with pronunc. (prestī·dʒ). The pronunc. (pre·stidʒ) is no longer heard.

3. *attrib.* or quasi-*adj.* (not clearly distinguishable from some of the examples listed in sense 4 below). Cf. *prestigeful *a.*, *prestigious *a.* 2.

1934 R. Benedict *Patterns of Culture* (1935) iv. 85 The Dionysian bent in the North American vision quest..did not usually have to make compromise with prestige groups and their privileges. **1937** *Time* 16 Aug. 34/2 The cinema has a special category for what it calls 'prestige pictures'. **1944** W. S. Maugham *Razor's Edge* vii. 325 Though she didn't much care for them [*sc.* some modern paintings] she thought quite rightly that they would be a prestige item in their future home. **1949** L. P. Hartley *Boat* xi. 156 If only they could all put off their company manners and change into their old clothes! But no; this was a prestige occasion. **1953** *Time* 23 Mar. 104/2 A 'prestige production', in broadcasting circles, is a show that abounds in a specific type of intelligence. **1957** *Times Lit. Suppl.* 8 Nov. 674/1 Serious books, normally prestige ware, had overnight changed into consumer goods, so that both conscience and bank-balance slept tight. **1958** M. Argyle *Relig. Behaviour* viii. 85 Prestige suggestion, in which people change their opinion after being told that a prestige person holds a different one. **1961** D. Jenkins *Equality & Excellence* viii. 175 Too many expensively educated young women aspire to careers as secretaries or receptionists in 'prestige offices'. **1962** *Rep. Comm. Broadcasting 1960* 92 in *Parl. Papers 1961–2* (Cmnd. 1753) IX. 259 The occasional, highly-advertised prestige programme put on for the occasion of a Christian festival. **1967** *Word Study* Mar. 3/2, I do find it difficult not to blame Miss Prouty at least a little bit..for rejecting usages so widely current in the prestige dialect that they are sanctioned even by her own textbook. **1968** *Globe & Mail* (Toronto) 13 Jan. 37/2 (Advt.), Accommodation available in Toronto's finest small nursing home, central prestige location, single or double occupancy, with or without private bath. **1969** *Nature* 29 Nov. 840/1 The Soviet Union still..seems to treat all scientific and technological progress (from sputniks to fish-spotting and from television sets to trans-continental pipelines) as primarily 'prestige' achievements. **1971** E. Jones in J. Spencer *Eng. Lang. W. Afr.* 84 Yams are a kind of prestige crop and item of food in rural West Africa. **1974** *Times* 20 Sept. 1/4 Aston Martin, one of Britain's prestige car companies. **1977** *Irish Times* 8 June 13/6 (Advt.), Superb town residence in prestige location.

4. *Comb.*, as *prestige-object, -principle, -product, -structure, -value, -word*; *prestige-hunting, -ranking, -rating* vbl. sbs.; *prestige-bearing, -building, -conferring, -conscious, -marking* adjs.; **prestige advertising**, advertising with the principal aim of furthering the prestige of the advertiser (rather than increasing sales, etc.).

1958 P. Shore in N. Mackenzie *Conviction* 39 I.C.I. is not alone in this kind of prestige advertising. **1959** *Manch. Guardian* 2 July 6/5, I doubt whether prestige advertising is important in recruiting university graduates. **1972** *Lebende Sprachen* XVII. 46/2 Prestige advertising. **1949** R. K. Merton *Social Theory* ii. v. 201 Pickpockets who.. delight in mastering the prestige-bearing feat of 'beating a left breech'. **1964** R. A. Hall *Introd. Linguistics* 21 Prestige-bearing persons. **1965** *Economist* 13 Nov. 723/1 A very different and very prestige-building new activity. **1961** D. Jenkins *Equality & Excellence* viii. 151 Those prestige-conferring occupations which used to be reserved for those 'of good family'. **1971** *Guardian* 25 Sept. 8/1 Prestige-conscious companies like IBM and Alcan. **1930** M. Mead *Growing up in New Guinea* iii. 29 But this is neither child labour nor idle prestige hunting on the part of the parents. **1957** M. Joos *Readings in Linguistics* 376/2 The dialects and idiolects of higher prestige were more advanced in this direction [of phonetic drift], and their speakers carried the drift farther along so as to maintain the prestige-marking difference against their pursuers. **1955** D. Chapman *Home & Social Status* iii. 42 The piano, which was formerly the principal prestige-object. **1939** *Brit. Jrnl. Psychol.* Jan. 220 Concrete manifestations of the 'prestige-principle' at work. **1958** *Observer* 25 May 16/2 To-day 'culture' is being marketed as a prestige-product. **1955** T. H. Pear *Eng. Social Differences* i. 29 How far does their [*sc.* adolescents'] prestige-ranking of occupations resemble that made by adults? **1957** Young & Willmott in 'C. H. Rolph' *Human Sum* vii. 140 An earlier national study of the prestige-ranking given to occupations by people who were predominantly non-manual workers. **1960** *New Left Rev.* Sept.–Oct. 3/1 The changing patterns of prestige-ranking. **1954** J. A. C. Brown *Social Psychol. of Industry* v. 140 The worker may be..upset when he is moved to another job at the same pay, but with a lower prestige-rating. **1949** R. K. Merton *Patterns of Influence* iv, in Lazarsfeld & Stanton *Communications Res.* ii. 198 He begins his climb in the prestige-structure at a relatively high level. **1929** L. D. White (*title*) The prestige value of public employment in Chicago. **1942** *Mind* LI. 170 Mathematics has indeed, a tremendous prestige value. **1958** *Listener* 21 Aug. 283/2 The two- or three-garage house, even the monster car itself, looked like losing its prestige value. **1967** E. Short *Embroidery & Fabric Collage* iii. 68 A good patchwork quilt has a prestige value in keeping with the labour that goes into the making of it. **1972** J. L. Dillard *Black English* vi. 233 The prestige value of more expensive toys, bicycles, and athletic equipment. **1964** C. Barber *Ling. Change Present-Day Eng.* ii. 25 There are many men of the professional classes who, far from practising the sounds of R[eceived] P[ronunciation] and the prestige-words of R[eceived] S[tandard], are deliberately refusing to do so. **1964** *Eng. Stud.* XLV. (Suppl.) 22 But very often there is a marked difference in tone between the foreign and the native terms, the former being felt as prestige-words, the latter as the plain terms.

prestigeful (prestī·ʒfŭl), *a.* [f. Prestige + -ful.] = *prestigious *a.* 2.

1956 C. W. Mills *Power Elite* iii. 53 There is..an appreciation of the new for its own sake: that which is new is prestigeful. **1959** *Encounter* Aug. 71/1 The more or less prestigeful 'pure fields'. **1961** S. R. Herman in J. A. Fishman *Readings Sociol. of Lang.* (1968) 507 In the new environment he [*sc.* an immigrant] is often without the prestigeful status he enjoyed in his country of origin and he is very much in need of recognition. **1967** M. Argyle *Psychol. Interpersonal Behaviour* v. 93 The experimenter should be prestigeful, an attractive person of the opposite sex, or at any rate compatible. **1971** *Times Lit. Suppl.* 22 Oct. 1310/4 Bateson had his contributions refused publication by the prestigeful periodical *Nature*. **1974** R. A. Hall *External Hist. Romance Lang.* 22 Many scholars.. are inclined to follow the folk-lore of our Western European culture in ascribing the status of 'language' only to those types of speech which manifest the prestigeful features just mentioned.

prestigey (prestī·ʒi, prestī·dʒi), *a. colloq.* [f. Prestige + -y[1].] = *prestigious *a.* 2.

1963 *Spectator* 27 Sept. 385 Desires for prestigey bigness. **1968** J. Bingham *I Love, I Kill* xi. 133 What you want is the serious actor bit. Something more prestigey. **1968** M. Richler *Cocksure* vi. 36 We're no more than a bauble..a prestige-y trinket.

prestiginous (presti·dʒĭnəs), *a. rare*[−1]. [f. Prestige + *-in-* + -ous, irreg. after *multitudinous*, etc.] = *prestigious *a.* 2.

1896 G. B. Shaw *Let.* 16 Mar. (1965) I. 614 A commercial and prestiginous success for Janet.

prestigious, *a.* Restrict 'Now *rare*' to sense in Dict. and add: Also with pronunc. (prestī·dʒəs). **1.** (Further examples.)

1957 *Eng. Lang. Teaching* XII. i. 5 Ogden, whose prestigious virtuosity in paraphrase had enabled him to work Basic English out. **1974** *Times Lit. Suppl.* 11 Jan. 32/3 For the period of nearly five years during which he remained as Prime Minister after the war he was..engaged in promoting policies which were actively disliked, or accepted reluctantly, by a majority of his supporters. This was the essential nature of the prestigious balancing act which he was constantly obliged to perform.

2. Having, showing, or conferring prestige (sense 2).

In this sense many prefer to use *prestigeful *a.* or some other adjective.

1913 Conrad *Chance* i. iii. 76 'You have had all these immense sums.. *What* have I had out of them?' It was perfectly true. He had had nothing out of them—nothing of the prestigious or the desirable things of the earth. **1958** *Economist* 25 Oct. (Suppl.) 19/1 But then came a form of competition that the American automobile industry had never envisaged—a competition from other industries for the consumer's dollar spent on prestigious purchases. **1960** *Time & Tide* 8 Oct. 1179/1 The commercial [television] companies agreed—..to give ITN enough cash for its extremely prestigious and worthy coverage of the United Nations. **1963** *Listener* 18 Apr. 656/2 Once established in these prestigious places men leave only if they have to. **1967** G. Steiner *Lang. & Silence* 72 Recent French linguistic philosophy also assigns a special function and prestigious authority to silence. **1969** *Daily Tel.* (Colour Suppl.) 5 Sept. 32/1 Those hotels such as every prestigious capital needs. **1970** B. M. H. Strang *Hist. English* 75 Of course, before 1770, not everyone was confined to the English of his town or village unless he hiked or hacked to another; many were exposed to the highly prestigious and influential written form. **1973** *Oxf. Univ. Gaz.* CIII. Suppl. 5. 33 The small but prestigious collection of German drawings in the Department. **1974** *Times* 27 Apr. 15/5 A career in pure science is still more socially prestigious, in Britain, than one in engineering or in applied science. **1975** *Physics Bull.* May 219/3 Halley was already quite distinguished, established as the Savilian Professor of Geometry at Oxford (a prestigious position).

prestigiously *adv.*, **prestigiousness**: further examples, corresponding to *prestigious *a.* 2.

1962 *Listener* 27 Dec. 1098/1 Art has become a commodity, albeit a highly prestigious one. But it is its very 'prestigiousness' that has brought upon its none too sturdy back the hordes of P.R.O.s and promoters. **1968** J. M. Ziman *Public Knowl.* vi. 118 He uses the standard technical words and phrases of the subject, not..to associate himself prestigiously with his would-be colleagues.

presto, *a.*[1], *adv.*[1], *sb.*[1] Add: **A.** *adj.* Also *transf.*

1952 A. Christie *They do it with Mirrors* i. 9 Everyone's life has a tempo. Ruth's was *presto* whereas Miss Marple's was..*adagio*. **1976** C. Bermant *Coming Home* ii. vii. 215, I was an andante being in a *presto* setting.

pre-stress, *sb.* and *a.* **A.** *sb.* [Pre- A. 2.] Tension applied to an object during manufacture or prior to some other treatment, usu. in order to counteract applied compressive loads (as in prestressed concrete).

1934 *Engineering News-Record* 13 Sept. 345/1 A prestress of 8,392 lb. per sq. in. **1940** *Structural Engineer* XVIII. 642/1 (*heading*) Diminution of the preliminary tensile pre-stress in steel by shrinkage and creeping. **1956** *Archit. Rev.* CXIX. 146/3 Strips of shuttering supported on props are necessary under the transverse diaphragms and these also support the precast units before the prestress is applied. **1967** *New Scientist* 10 Aug. 295/1 The higher the prestress applied, the higher the fatigue strength of the section. **1977** *Design Engin.* July 64/1 Forces depend on prestress, i.e. initial deflection of spring, as well as direction of motion.

B. *adj.* [Pre- B. 2.] Occurring before a stressed syllable.

1973 *Word 1970* XXVI. 98 In the traditional analysis long vowels occur only under stress... Therefore, no prestress vowel may be long. **1975** *Amer. Speech 1972* XLVII. 171 The sets of intersyllabic consonants and consonant clusters differ remarkably in different slots; for instance prestress /bh bt nk/ (as in *abhór, obtáin, enquíre*).

prestress (prīstre·s), *v.* [Pre- A. 1.] *trans.* To apply stress to (an object or material) prior to some other treatment; to introduce stress into (an object) during manufacture, so as to enable it more successfully to withstand applied loads; *spec.* with reference to reinforced concrete (cf. *prestressed *ppl. a.*).

1934 *Engineering News-Record* 13 Sept. 345/1 The idea of destroying the bond between steel rods inserted in concrete, and prestressing the rods in tension and the concrete in compression, is not new. 1936 *Structural Engineer* XIV. 252/1 The concreting operation is carried out in the usual manner, the only difference being that the longitudinal rods are pre-stressed. 1940 *Concrete & Constructional Engin.* XXXV. 330/1 Thin piano wires, of a strength of 350,000 lb. to 450,000 lb. per square inch, are prestressed to a stress equivalent to half that of yield point. 1967 *New Scientist* 10 Aug. 295/1 When the concrete is prestressed the steel is dynamically opposed to the applied load. *Ibid.*, Any series of units can be cast separately and then prestressed together to convert them into a monolithic whole. 1971 *Materials & Technol.* II. iv. 113 (*caption*) The vertical outer wall..was prestressed with Freyssinet cables.

So **prestre·ssing** *vbl. sb.*

1934 *Engineering News-Record* 13 Sept. 345/1 One advantage of the prestressing is to postpone the formation of cracks. 1940 *Structural Engineer* XVIII. 629 Notwithstanding the cost of the pre-stressing operations, this great saving in materials renders also pre-stressed designs very economical. 1953 *Sci. News Let.* 24 Jan. 63/2 In one phase of the study, it was found that prestressing doubled the ability of one aluminum alloy, used in the aircraft industry, to carry an external load. 1964 C. W. Glover *Structural Precast Concrete* xix. 328 The basic idea of pre-stressing is to induce in the unloaded members stresses that are contrary to the stress normally produced by loading. 1974 *Encycl. Brit. Macropædia* IV. 1078/1 The calculation of the initial tensile force required in the prestressing tendons to produce compressive stresses that will counteract the tensile stresses in the concrete.

prestre·ssed, *ppl. a.* [f. prec. + -ED[1].] Previously subjected to stressing; into which stress has been deliberately introduced during manufacture; *spec.* of concrete: reinforced by steel rods or wires which have been tensioned while the concrete is setting, so that after setting they tend to compress the concrete and thereby strengthen it.

1936 *Structural Engineer* XIV. 251/1 Two telegraph posts 40 ft. long,..one..in pre-stressed concrete, and the other..in ordinary reinforced concrete, were subjected to alternate stressing. 1948 *Concrete & Constructional Engin.* XLIII. 260 The resiliency and freedom from cracks of prestressed concrete make the material very suitable for railway sleepers and runways. 1955 *Times* 19 July 4/7 Huge pipes in cast iron, or spun iron, or prestressed concrete. 1963 Simonds & Church *Conc. Guide Plastics* (ed. 2) vii. 172 These prestressed laminated structures show strengths three to four times those of the unstressed fabric laminates. 1969 *Jane's Freight Containers 1968–69* 68/1 Crane—18-inch pre-cast, pre-stressed concrete piles; pre-stressed concrete deck. 1973 *Times* 14 Mar. 4/7 As Lancashire County Council's chief assistant in charge of bridges he designed about fifty, including..its first prestressed bridge.

pre-stre·tch, *v. Building.* [Pre- A. 1.] = *pre-tension *v.* Hence **pre-stre·tched** *ppl. a.*; **pre-stre·tching** *vbl. sb.*

1936 *Structural Engineer* XIV. 251/1 By inducing tension in the reinforcement one secures..a decrease of the tension produced in the concrete by the shearing process, or even its total suppression if the reinforcement is pre-stretched in two directions. 1941 *Concrete & Constructional Engin.* XXXVI. 93/1 M. Freyssinet's device with pre-stretched wires needs no anchorage for the manufacture of a long row of articles after one stretching operation. 1946 *Ibid.* XLI. 147 With pre-stretching the member has to remain in the mould until the stretching force, produced by tensioning of the reinforcement.., can safely be transmitted to the concrete. 1949 [see *post-stretching *vbl. sb.*]. 1965 E. C. Hiscock *Cruising under Sail* (ed. 2) v. 79 Terylene..is..more suitable for halyards and headsail sheets; but for those purposes Marlow Ropes Ltd. make a pre-stretched three-strand rope.

pre-structuralist, -suffixal: see *pre- B. 1.
pre-subject: *pre- B. 2a.

presumptive, *a.* Add: **3. b.** *Embryol.* Applied to undifferentiated tissue that becomes a specified part in the normal course of development.

1935 *Jrnl. Morphol.* LVIII. 432 Each presumptive mid-gut epithelial nucleus appropriates a portion of the vitellophage cytoplasm and thus are formed the definitive mid-gut epithelial cells. 1950 B. M. Patten *Early Embryol. of Chick* iv. 66 The sharpness of the boundaries between different presumptive areas..is an entirely artificial device. 1977 *Sci. Amer.* July 76/3 In the 1910's and 1920's Harrison and his colleagues demonstrated that in amphibian embryos the presumptive limb region (the region that will later produce a leg) behaves in many ways like a typical epimorphic field.

presupposition. Add: **3.** Comb., as *presupposition-free* adj.; **presuppositionless** *a.* (further examples); hence **presupposi·tionlessness.**

1966 *Jrnl. Philos.* LXIII. 699 (*heading*) Completeness theorems for some presupposition-free logics. 1972 *Jrnl. Symbolic Logic* XXXVII. 424 'Presupposition-free' here refers to the absence of presuppositions that there are individuals in the domain over which individual variables range. 1906 *Mind* XV. 281 There is no absolutely presuppositionless psychology. 1940 *Philos. Rev.* XLIX. 285 The idea of a presuppositionless philosophy. 1974 *Jrnl. Ecumenical Stud.* XI. 140 Presuppositionless appreciation of..convictions. 1976 *Word 1971* XXVII. 191 A presuppositionless analysis of a child's corpus will not have any theoretical import. 1940 M. Farber *Philos. Ess. in Memory E. Husserl* 44 The claim of presuppositionlessness has been made at various times.

presuppositional (prīsʊpǝzi·ʃǝnǎl), *a.* [f. Presupposition + -al.] Of or pertaining to presuppositions.

1909 W. M. Urban *Valuation* i. 14 The method of psychological worth analysis we may..characterise as the Presuppositional Method. It begins with analysis of presuppositions. 1954 *Mind* LXIII. 154 Banishing all this presuppositional meaning. 1975 R. M. Kempson *Presupposition* iv. 74 Anomalies..arise for a presuppositional account with each logical connective. 1978 *Language* LIV. 494 McClaran's paper, 'Presuppositional aspects of Yucatec sentences', investigates the semantic and syntactic properties associated with Yucatec verbs bearing the suffixes *ik*, *Ah*, *il*, and *Al*.

pre-syllabic, -symptomatic: see *pre- B. 1.

presyna·ptic, *a.* **1.** *Cytology.* [Pre- B. 1.] Prior to meiotic synapsis.

1909 *Ann. Bot.* XXIII. 21 In common with Grégoire ('07), we may adopt, provisionally at least, the following scheme of phases for convenience of clearness in description. The prophases of division naturally fall into two periods, the pre-synaptic and the post-synaptic phases. 1912 [see *leptotene]. 1921 *Ann. Bot.* XXXV. 367 Fig. 5 represents a presynaptic pollen mother-cell.

2. *Physiol.* [Pre- B. 3.] Of, pertaining to, or designating a neurone that transmits a nerve impulse across a synapse. Opp. *postsynaptic *a.* 2.

1937 *Proc. R. Soc.* B. CXXII. 113 The response is erratic in that by no means every pre-synaptic stimulus yields a post-synaptic response. 1950 [see *hyperpolarize *v.*]. 1965 G. H. Bell et al. *Textbk. Physiol. & Biochem.* (ed. 6) xxxix. 796 In general the presynaptic fibre divides up into numerous fine branches which then end in greatly expanded terminals, presynaptic knobs, which make intimate contact with part of the membrane of the cell body or dendrites of the postsynaptic cell. 1979 *Internat. Rehabilit. Med.* I. 45/1 New evidence of the blocking of pain specific receptors..by morphine-like substances produced by presynaptic dendrites.

Hence **presyna·ptically** *adv.*

1971 *Nature* 12 Nov. 102/1 In the central nervous system, amphetamine releases presynaptically bound NE [*sc.* norepinephrine] or DA [*sc.* dopamine] and blocks their re-uptake. 1976 *Ibid.* 3 June 418/1 The fact that chlorpromazine also blocks α-adrenoceptors, possibly presynaptically located, may also contribute to the enhanced presence of catecholamines at the synaptic cleft.

pre-systematic(ally): see *pre- B. 1.

preta·pe, *v.* [Pre- A. 1.] *trans.* To prerecord using magnetic tape. So **preta·ped** *ppl. a.*, **preta·ping** *vbl. sb.*

1968 J. Philips *Hot Summer Killing* (1969) iii. ii. 151 The networks will be giving up commercial time. Pretaped shows are already prepared. 1972 *Listener* 6 July 3/1 Becton insists on live transmission... 'You are not really trusting people with the airtime if you insist on pre-taping.' 1972 W. P. McGivern *Caprifoil* (1973) viii. 144 He pre-taped a series of television speeches to camouflage his absence. 1973 *Sociometry* XXXVI. 103 Subjects listened to pre-taped instructions. 1976 *National Observer* (U.S.) 16 Oct. 6/2 They were a poor substitute for journalism, pretaped and all but tensionless.

‖ **prêt-à-porter** (prɛtapɔrte). [Fr., 'ready to wear'.] Phr. used *attrib.* and *absol.* to denote clothes that are sold in standard sizes ready for wear.

1957 *Punch* 16 Jan. 136/3 Gloves, scarves, jewellery, and *prêt-à-porter* clothes...all the fleeting frivolities..of the passing mode. 1958 M. Stewart *Nine Coaches Waiting* vi. 71 The young and lovely buy dresses *prêtes à porter*... Off the peg. 1959 *Guardian* 4 Dec. 6/3 The prêt-à-porter spring and summer shows are in full swing in Paris. 1967 *Times* 21 Feb. 9/2 As at the Paris prêt-à-porter fair, the Mary Quant stand was jampacked for the parades. 1973 *Sat. Rev. Arts* (U.S.) Jan. 84/3 The last two pieces contain an onslaught of information about the vigorous young designers and the boom of *prêt à porter*. There is worry that the heyday of French couture is over. 1977 *New Yorker* 11 July 79/1 The feverish search on Seventh Avenue for novelty and for sure profits has come to rival the showings of the Paris prêt-à-porter.

pre-taste: see *pre- A. 2.

pre-·tax, *a.* and *adv.* **A.** *adj.* [Pre- B. 2.] Designating gross assets, earnings, funds, or profits considered before the deduction of tax.

1963 *Times* 7 June 17/2 Group pre-tax profit is £124,000, £14,000 more than forecast, and after tax of £64,000 there is available to the holding company nearly £60,000. 1968 *N.Y. Times* 12 Jan. 38 They forecast a sales gain of 8 to 10 per cent but see an almost dramatic improvement in margins as pretax earnings rise 15 to 20 per cent above those of 1967. 1969 *Times* 2 May 28/5 Pre-tax profits are up from £810,000 to £930,000. 1977 *New Yorker* 29 Aug. 47/1 They'd give me five per cent of the pretax profit.

B. *adv.* [*pre- B. 2c.] Before the deduction of tax.

1976 *Daily Tel.* 17 Feb. 19 (*heading*) Lonrho advances 35pc pre-tax. *Ibid.* 16 July 17/1 It is encouraging to see a £4·57 million turnround to interim profits of £3·61 million pre-tax.

prete·ctal, *a. Anat.* [ad. mod.L. *prætectālis*, f. Pre- B. 3 + *tect(um + -al.] Lying in front of the tectum; of or pertaining to the pretectum.

1925 *Jrnl. Compar. Neurol.* XXXIX. 195 (*caption*) Horizontal section through the posterior commissure, showing the pretectal nucleus in relation to the other parts of the thalamus and midbrain. 1959 *Folia Psychiatrica et Neurologica Japonica* XIII. 268 The pretectal region is located between the posterior part of the thalamus and the tectum, equipped with admixture of cells of both thalamic and mesencephalic origin. 1973 *Brain Res.* LXIII. 360 The fact that two or three optic tract volleys were needed to evoke a response in the short ciliary nerve..indicates a low excitability of the pretectal neurones concerned.

Hence (as a back-formation) **prete·ctum**, the pretectal region of the brain.

1961 *Lancet* 9 Sept. 568/2 The disturbances in ocular motility and the radiological findings, led to the diagnosis of a small lesion in the pretectum. 1973 *Nature* 1 June 295/1 The control of pupil size..is known to be mediated by a reflex pathway through the pretectum.

pre-tee·n, *a.* orig. *U.S.* [f. Pre- B. 2 + Teen *sb.*[2]] Prior to one's teens; denoting the years of a child's life (usu. immediately) before the age of thirteen. Also *absol.* as *sb.* Hence **pre-tee·nager, pretee·ner**, a child (just) before the age of thirteen.

1960 V. Packard *Waste Makers* vii. 76 Even pre-teen boys' shoes were slated for obsoleting. They were being designed away from their 'sexless' look to a real 'nervous' look of flashy casualness. 1966 *N.Y. Times* 6 Jan. 33 Darlene [is] mother of two pre-teen children. 1966 *Economist* 10 Dec. 1144/2 One feature of nightlife on the Strip which the casual visitor is most likely to notice are the regular contingents of pre-teenagers, especially young girls of twelve and even less. 1967 *Atlantic Monthly* Jan. 77 The texts of many popular songs are so obviously coital that one wonders how they get on the radio and are sold openly to pre-teens. 1967 *Punch* 15 Mar. 377/2 Close behind the teen-age revolution, the emancipation of the pre-teens is gathering momentum... In North America the female pre-teen is already 'a knacky, switched-on dolly'. 1969 *Punch* 12 Mar. 384/3 By the time that the American child has reached the age of three, he is known as a sub pre-teener. 1970 *Daily Tel.* 8 May 17 The tendency is for children to experiment with cigarettes from as early as eight years old, and pre-teenagers are often regular smokers. 1972 J. L. Dillard *Black English* vi. 260 Twelve- and thirteen-year-old boys revealed the same type of half-funny, half-pathetic misinformation about sex which practically all preteens seem to have. 1972 *N.Y. Times* 3 Nov. 3/5 (Advt.), Both for preteen sizes 6 to 14. 1977 *Maclean's Mag.* 2 May 23/3 A mini crime wave involving gangs of teen-agers and pre-teens. 1978 *Church Times* 25 Aug. 6/2 An imaginative collection of prayers and poems for pre-teenagers.

pre-telescopic, -television: see *pre- B. 1d, 2a.

pretence, pretense, *sb.* Add: **7.** *attrib.* (in sense 5), passing into *adj.*, denoting something that is imitative or 'phoney'.

1941 *Punch* 17 Sept. 256/3 That lorry buzzing along High Street has got some pretence bombs and it's going to strew them about and we've got to pretend they have been dropped by the Blen. 1953 *Mind* LXII. 209 If I dream of a snake my dream must contain, if not a snake then an illusory or pretence snake.

pretend, *sb.* Restrict †*Obs. rare* to sense in Dict. and add: **2.** In (imitation of) children's use: the act of pretending in imagination or play (cf. Pretend *v.* 15b). Also *attrib.* passing into *adj.*, denoting a thing or action that is imitative or imaginary.

1888 F. H. Burnett *Sara Crewe* 1. 28 One of her 'pretends' was that Emily was a kind of good witch, and could protect her. Poor little Sara! 1911 G. Stratton-Porter *Harvester* iii. 48 Not so indifferent after all... That was all 'pretend!' But she waited just a trifle too long. 1928 Barrie *Peter Pan* ii. 70 in *Plays*, Now that they know it is pretend they acclaim her greedily. *Ibid.* iv. 97 It is a pretend meal this evening, with nothing whatever on the table. *a* 1936 Kipling *Something of Myself* (1937) i. 10, I have learned since from children who play much alone that this rule of 'beginning again in a pretend game' is not uncommon. 1955 J. Masters *Coromandel!* 31 It's all pretend, Jason, isn't it? 1959 J. L. Austin *Sense & Sensibilia* (1962) vii. 72 The water in toy beer-bottles is not toy beer, but *pretend* beer. 1960 *Guardian* 3 May 2/1 All 'pretend' space outfits can be dangerous and should be banned. 1962 *Listener* 4 Jan. 20/2 A diminutive, waif-like figure, dressed in rags, with his pretend sword and his pretend gun. 1965 G. McInnes *Road to Gundagai* iii. 54 'It's only pretend,' she kept on saying. 'You mustn't be afraid of pretend.' 1974 W. Rees-Mogg *Reigning Error* 109 Gold is real money and paper is pretend money.

pretending, *ppl. a.* Add: Also (in senses 3d, 15b of the vb.), of a thing or action: imitative, imaginary; of a game, etc.: that involves pretence or imitation. Cf. *pretend *sb.* 2 above.

a 1901 C. M. Yonge *Autobiogr.* in C. Coleridge *C. M. Yonge* (1903) iii. 95 They were not perfect playmates, for they called all 'pretending games' falsehood. 1960 *Times* 27 Apr. 1/3 Only a proper castle, not an 18th/19th-century Gothic pretending one. 1965 *Vogue* Aug. 64 Pretendin' racoon, pretty as a picture.

‖ **prétendu** (pretandü). ? *Obs.* Also fem. **prétendue.** [Fr.] An intended husband or wife; a fiancé(e).

1847 Thackeray *Van. Fair* (1848) xxxiii. 295 In reply to the exhortation of her daughter's *prétendu*, Mr. Pitt

Crawley. 1850 —— *Pendennis* II. i. 9 Lady Ann Milton, Mr. Foker's cousin and *prétendue*. *Ibid.* xx. 201 She has her mamma on one side, her *prétendu* on the other.

pre-tension (prī·te·nʃǝn), *sb.*[2] [f. Pre- A. 2 + Tension *sb.*] Tension in an object applied previously or at an early stage of a process, e.g. that applied to the reinforcing steel in the manufacture of prestressed concrete.

1936 *Structural Engineer* XIV. 251/2 It is necessary to produce the pretension of the steel and to manufacture concrete of high resistance at a cost sufficiently low to allow one to preserve the greater part of the savings effected on the materials. 1941 *Concrete & Constructional Engin.* XXXVI. 74/1 Shrinkage, elastic deformation, and creep under stress reduce the pre-tension by 10 tons to 20 tons per square inch. 1976 G. S. Ramaswamy *Mod. Prestressed Concrete Design* i. 5 Inexpensive end anchorages are produced at the two ends of the bar to retain the pretension.

pre-tension (prī·te·nʃǝn), *v.* [f. Pre- A. 1 + Tension *v.*] *trans.* To apply tension to (an object) prior to some other treatment, esp. incorporation in a structure.

1937 *Rep. Building Res. Board 1936* (Dept. Sci. & Industr. Res.) 99 In connection with some tests on a particular lightweight aggregate, four beams have been tested to determine the effect of pre-tensioning the tension reinforcement. 1949 P. W. Abeles *Princ. & Pract. Prestressed Concrete* ix. 63/2 A special process has been developed for the manufacture of hollow slabs in approximately 100 yd. long production lines, in which the wires are pre-tensioned in order to avoid any sag. 1973 *Sci. Amer.* Apr. 114/1 The output of work per cycle can be increased somewhat by pretensioning the fiber before it is immersed in the brine.

Hence **pre-te·nsioned** *ppl. a.*, **pre-te·nsioning** *vbl. sb.* (freq. *attrib.*).

1936 *Structural Engineer* XIV. 261/2 Thus, with two different steels, an effect was produced similar to that produced by M. Freyssinet's method of pre-tensioning. 1937 *Rep. Building Res. Board 1936* (Dept. Sci. & Industr. Res.) 98 The pre-tensioning apparatus is left in position until the concrete has hardened sufficiently to take the stresses induced in it. 1949 P. W. Abeles *Princ. & Pract. Prestressed Concrete* ix. 58/2 Precast articles having pre-tensioned steel are mainly applicable to slabs, sleepers, beams, [etc.]..where mass production is possible. 1964 C. W. Dunham *Adv. Reinforced Concrete* viii. 400 Pretensioned members can be used for other parts of a structure besides the floors and roof. 1965 *Economist* 5 June 1176/1 The first multi-storey prestressed concrete building in the world combining pretensioned, prestressed roof beams manufactured in the factory with post-tensioning on site using the Magnel system. 1971 J. R. Libby *Mod. Prestressed Concrete* xiv. 451 Pre-tensioning benches that can be moved from job site to job site have been used to a limited degree. 1975 Kong & Evans *Reinforced & Prestressed Concrete* ix. 196 In pre-tensioning, the tendons pass through the mould, or moulds for a number of similar members arranged end to end, and are tensioned between external end anchorages, by which the tension is maintained while the concrete is placed.

preter-, præter-, *prefix.* Add: **2. preterse·nsuous** *a.* = *pretersensual* adj.

1963 V. Nabokov *Gift* iii. 172 If..he had had to answer before some pretersensuous court..he would scarcely have decided to say that he loved her.

preterist. 1. (Later example.)

1962 V. Nabokov *Pale Fire* 35 A preterist: one who collects cold nests.

prete·rm, *a.* and *adv. Obstetrics.* **A.** *adj.* [Pre- B. 2.] Born or occurring after a pregnancy that lasted significantly less than the normal time; *spec.* (see quot. 1977[2]).

1928 A. Gesell *Infancy & Human Growth* xv. 300 The pre-term child is viable even though he may have completed but three-quarters of his allotted uterine life-period. 1933, 1971 [see *post-term *a.*]. 1977 *Lancet* 11 June 1255/1 Cigarette smoking during pregnancy is associated with an increased proportion of pre-term deliveries. *Ibid.* 30 July 246/1 The terms 'prematurity' and 'immaturity', with their vague and multiple meanings, have been replaced by the precise terms 'low birthweight' (under 2500 g) and 'preterm' (less than 37 completed weeks).

B. *adv.* [*pre- B. 2c.] Before the end of the normal period of pregnancy.

1977 *Lancet* 9 July 87/1 We gave pregnant Sprague-Dawley rats 4 mg/kg indomethacin..and killed the fetuses at cæsarean section shortly pre-term.

prete·rminal, *a.* [Pre- B. 1.] Preceding that which is terminal.

1947 *Radiology* XLIX. 311/2 A similar preterminal course..is found with such toxic agents as the nitrogen mustards. 1965 N. Chomsky *Aspects of Theory of Syntax* ii. 84 A terminal string is formed from a preterminal string by insertion of a lexical formative. 1976 *Lancet* 4 Dec. 1253/2 Children with severe ketoacidosis in association with viral infections have symptoms and signs in the preterminal stage very similar to those of Reye's syndrome. 1977 *Navy News* June 39 (Advt.), Why not spend your pre-terminal leave with us and be introduced to the company.

prete·st, *v.* [Pre- A. 1.] *trans.* To test beforehand; *spec.* in *Psychol.*, to test in advance (the efficacy of questions or the methods of administration for use in a projected test). Hence **prete·sted** *ppl. a.*, **prete·sting** *vbl. sb.*

1949 C. I. Hovland et al. *Exper. Mass Communication* ii. 26 Qualitative pretesting consisted of face-to-face interviewing... After the first few interviews..the items were revised and pretested again. 1949 R. K. Merton *Social Theory* vi. 163 'Pre-testing' in social affairs is only a rough approximation. 1951 in M. McLuhan *Mech. Bride* (1967) 41/2 Columbia Broadcasting System uses Dr. Flesch's findings to pre-test radio scripts. 1969 N. A. Rosen *Leadership Change & Work-Group Dynamics* (1970) i. 20 Pre-testing, even on a small number of workers, would have led to communication among them. 1970 I. L. Horowitz *Masses in Lat. Amer.* i. 7 It further assumes that the masses have an historical mission and a pre-tested political direction. 1970 D. Goldrich et al. in *Ibid.* v. 176 The interview schedule had been pretested in Santiago. 1977 *Sci. Amer.* Apr. 25/1 The requirement does not apply to spherical tanks because they are built under conditions where the welds can be pretested for integrity.

pre-test, *sb.* and *a.* **A.** *sb. Psychol.* [Pre- A. 2.] An experimental test designed to assess the efficacy of questions or methods of administration intended for use in a projected test. Also occas., a preliminary or qualifying test. Also *attrib.*

1949 C. I. Hovland et al. *Exper. Mass Communication* ii. 26 The purpose of the quantitative pretest was the advance determination of the approximate distribution of answers to each question and the relationship between questions. 1966 J. S. Bruner *Beyond Information Given* (1974) xviii. 321 The experiment was carried out with six- and seven-year-olds and began with a pretest. 1970 *Jrnl. Gen. Psychol.* LXXXIII. 240 After the pretest, all rats were given Richter-type tests for 30 consecutive days. 1971 *Ibid.* LXXXIV. 99 Gross differences in motor performance..were reduced by training all Ss to asymptote during a pretest session. 1972 *Jrnl. Social Psychol.* LXXXVI. 14 The first, a pretest phase, was used to select 'high conformers'. 1973 *Jrnl. Genetic Psychol.* CXXII. 101 Their..pretest scores dropped significantly. 1976 *Columbus* (Montana) *News* 10 June 1/3 You have to have passed your 15th birthday and a pretest.

B. *adj.* (With hyphen.) [Pre- B. 2.] Existing before a test.

1960 *Farmer & Stockbreeder* 2 Feb. 76/1 The Board explains that daily gain on test has been introduced as a measurement of growth which is not complicated either by pre-test environment or the weekly slaughtering routine.

pretheatre, -theoretical: see *pre- B. 2a, 1d. **prethink:** *pre- A. 1.

pretonic, *a.* Add: Also *absol.* as *sb.*, = Pretone.

1953 K. Jackson *Lang. & Hist. Early Brit.* ii. 634 There is also..an *h*- prefixed to vowels; but in Brittonic there was not the same extension of this to pretonics not originally ending in -*s* in British that there was in Pr.I. 1973 *Archivum Linguisticum* IV. 24 This kind of system may be dealt with in the same way as the different pretonics of the individual tones.

Hence **preto·nically** *adv.*, as regards a pretone.

1953 K. Jackson *Lang. & Hist. Early Brit.* ii. 322 It [*sc.* a Latin pronunciation (au)] is not reduced pretonically to *ō*; e.g. *awdur*, *cawlai*.

pre-tour, -tragic: see *pre- B. 2a, 1d.

pretrai·ning, *vbl. sb. Psychol.* [Pre- A. 1.] Training which takes place in advance of an experiment or test; also *attrib.* Hence (as a back-formation) **pretrai·n** *v. trans.*

1955 *Jrnl. Exper. Psychol.* L. 180 This pretraining has generally consisted of verbal paired-associates learning in which stimuli are the same as, or substitutes for, those of the motor task. 1957 J. S. Bruner *Beyond Information Given* (1974) i. 35 A subject is first given some pretraining, in one of four pretraining groups. 1959 *Psychol. Abstr.* XXXIII. 764/1 The stability of generalized expectancies (GEs) developed under 2 pretraining conditions and with differing frequencies of past reinforcement. 1971 *Nature* 9 July 124/2 Cats were pre-trained, using classical conditioning. 1973 *Jrnl. Genetic Psychol.* CXXII. 17 Children aged 2-½ to 5-½ were pretrained on two three-choice simple discrimination problems. 1974 *Psychol. Abstr.* LII. 1244/1 In Exp II..the type of response in pretraining..was varied.

pre-transformational: see *pre- B. 1d.

pretrea·t, *v.* [Pre- A. 1.] *trans.* To treat beforehand. Hence **pretrea·ted** *ppl. a.*

1934 *Jrnl. Physical Chem.* XXXVIII. 795 Comparing the form of the characteristic curve of the normally developed strip S with those of the strips that were pretreated with the iron citrate solutions..the latter are steeper. 1950 *Nucleonics* Mar. 48/1 It may be necessary to pretreat and purify the water before using it in the pile. 1956 *Nature* 21 Jan. 136/1 A variation of the relative concentrations..may account for the variation..of the shape of light curves observed in differently pretreated algae. 1963 *Mechanical World* CXLIII. 17/1 Of recent years the production of pre-treated metal surfaces has increased enormously. 1975 *Jrnl. Immunol. Methods* VIII. 383 Peritoneal macrophages..cultivated for 48 hr on glass pretreated with poly-L-lysine.

pretrea·tment, *sb.* and *a.* **A.** *sb.* [Pre- A. 2.] Treatment given beforehand. Also *attrib.*

1925 *Jrnl. Forestry* XXIII. 921 Within each pretreatment the seed experienced variations of that treatment. 1946 *Nature* 23 Nov. 748/2 Effective control was measured six weeks after treatment, all plots in this case having received pre-treatment with nitro-chalk seven days in advance of the weed-killer application. 1955 *New Biol.* XVIII. 88 Freshly excised grafts, after pre-treatment with 15 per cent glycerol in Ringer, were sealed off in glass tubes and frozen. 1961 *Times* 12 Apr. 17/6 The coil of steel is passed through chemical pre-treatment baths. 1973 *Times* 29 Oct. 20/7 Liquid wastes from all trade and industrial sources, broadly speaking, drain without pretreatment in to the River Tees or into Tees Bay. 1978 *Jrnl. R. Soc. Arts* CXXVI. 686/1 The aim of pre-treatments is to increase the action of the adhesive bonding forces by eliminating greases, dirt and the oxides which exist on the surface.

B. *adj.* [Pre- B. 2.] Existing before treatment.

1961 *Lancet* 2 Sept. 499/2 This..gave an opportunity for estimation of the pretreatment level of serum-cholesterol. 1962 *Ibid.* 12 May 989/2 In 1 other patient infected with a proteus strain, organisms isolated during and after treatment were more resistant than the pretreatment cultures. 1972 *Jrnl. Social Psychol.* LXXXVI. 84 The..pretreatment mean attitude score for the nine treatments is approximately equal.

pretrial, *sb.* and *a.* **A.** *sb.* (Stressed *pre·trial.*) [Pre- A. 2.] A preliminary hearing before a trial. Also *attrib. U.S.*

1938 E. J. Ellison in *Christian Science Monitor* 15 June (Weekly Mag.) 3/1 Why not have a special judge to clear out the legal underbrush, and call it a 'pre-trial'. *Ibid.* 3/2 Some two weeks before a case is scheduled for trial, the opposing parties appear before a pre-trial judge. 1938 *Daily Progress* (Charlottesville, Va.) 18 Aug. 4/1 The 'pre-trial' system was introduced in Detroit six years ago. 1970 *Daily Colonist* (Victoria, B.C.) 23 Apr. 10/1 Black Panther chairman Bobby Seale was accused at his pretrial hearing of ordering the death of a Panther suspected of turning informer. 1971 *N.Y. Law Jrnl.* 23 Nov. 17/5 If case cannot be settled the Part I judge will assign the case to a pre-trial examiner, who will, with the aid of the attorneys, prepare the pre-trial order. 1976 *National Observer* (U.S.) 17 Jan. 2/2 The judge had ordered the papers not to print stories about a pretrial hearing for a murder defendant because the judge did not want prospective jurors to be influenced by arguments made at the hearing. 1978 *Chicago* June 116/3 Tom Sullivan sat at the defense table in the Hanrahan case, and he handled pretrial matters for Kemer's co-defendant.

B. *adj.* (Stressed *pretri·al.*) [Pre- B. 2.] Of or pertaining to the period before a trial or trials.

1948 B. Vesey-Fitzgerald *Bk. Dog* 749 Whereas in pre-trial days the range [of sheepdog] was severely restricted, nowadays, as a result of trials, the scope for selection is nation-wide. 1971 *Times* 20 Mar. 11 Most have already spent more than a year in jail awaiting trial, and pretrial detention will be deducted from their sentences. 1978 *Times* 3 Nov. 17 (*heading*) Barristers' immunity from claims in negligence in pre-trial work narrowed. *Ibid.* 13 Nov. 7/8 Mr. Nazaryan, who has been in pre-trial detention for almost a year, is charged with anti-Soviet agitation and propaganda.

prettied: see *pretty *v.*

prettification (pri:tɪfikē̆i·ʃǝn). [f. Pretty *a.* + -fication.] The fact or process of making pretty; prettifying.

1909 in Webster. 1920 *Times Lit. Suppl.* 23 Sept. 617/1 Such work is..the counterfeit of romance. It gives us, not a celebration of life, but a prettification of it. 1930 A. I. Nazaroff *Tolstoy* vi. 97 He is described very realistically, without the slightest trace of prettification or sugar-coating. 1966 *New Statesman* 23 Dec. 935/2 The sanctification of emotional impotence. The prettification of stultified tragedy. 1969 *Daily Tel.* 10 July 21/3 This [manner], together with some prettification of the action, deprived the play of the savagery that is surely there. 1978 *Listener* 20 July 76/3 The writer, Georgina Masson.. is up in arms over the prettification of the cemetery.

prettify, *v.* Add: Also **prettyfy.** Delete *colloq.* and add further examples. Also *fig.* Hence **pre·ttified** *ppl. a.*; **prettifying** *vbl. sb.* (further examples). Also **pre·ttifier**, one who prettifies.

1867 G. du Maurier *Let.* in D. du Maurier *Young George du Maurier* (1951) 273 Then D and I walked through the lovely Bois de Boulogne to the Mare d'Auteuil which has been brutally modernised and prettyfied. 1889 *Cent. Dict.*, Prettified. 1919 B. Tarkington *Let.* 14 June in *On Plays* (1959) 13 You know..why all the magazines *haf* to have the prettified girl cover. 1934 *Sun* (Baltimore) 2 Apr. 8/1 (*heading*) Prettifying war. *Ibid.*, These endeavors [to outlaw certain forms of warfare] are based on the assumption that we are making progress if somehow we can manage to prettify war. 1936 L. C. Douglas *White Banners* iii. 63 A man doesn't try to prettify himself very much, or make himself over to look different. He wants to be important for owning something rather than being something. 1955 *Times* 24 Aug. 7/4 To anyone who once heard a chanty at sea, such prettified verses, however musical, will always seem a travesty of their originals. 1960 W. Miller *Canticle for Leibowitz* xiv. 151 A place of majesty that overawed the would-be prettifiers. 1970 *Daily Tel.* 21 Feb. 8/5 The 19th-century weakness for prettifying the lives of great men. 1971 P. Gresswell *Environment* 89 Too much prettyfying is damaging enough but the bleak 'serviceable' attitude to details causes the more widespread damage... For this reason, many housing estates..are depressing. 1973 *Times* 30 Oct. 14/8 The Cubist works are prettified exercises in taste. 1976 *Early Music* Oct. 402/2 The *Dido* gathering should be suave and elegant but not prettified or frivolous.

pretty, *a.* (*sb.*) Add: **A. II. 3. a.** (Further examples.) Also *U.S.*

1878 J. H. Beadle *Western Wilds* xxiv. 387 A half-breed squaw, about as 'pretty' as a wild-cat struck with a club. 1891 'Mark Twain' tr. *Hoffmann's Slovenly Peter* (1935, Ltd. Ed.) 25 'Try how pretty you can be Till I come again,' said she. 'Docile be, and good and mild.' 1938 *Amer. Speech* XIII. 6/2 *Pretty*,..good; fine; excellent. 'He was a real pretty ball player.'

b. (Further examples.) Freq. in negative contexts. Also in phr. *to say pretty things*, to speak consolingly or in a condescending manner.

1811 A. Constable *Let.* 28 Apr. in J. Constable *Corr.* (1962) I. 63 Uncle D.P.W. here for a pretty week. 1811 Jane Austen *Sense & Sens.* II. v. 80 It was not very pretty of him, not to give you the meeting. 1898 G. B. Shaw *Philanderer* iv. 140 *Paramore.* I can only admire you, and feel how pleasant it is to have you here. *Julia*... And pet me, and say pretty things to me! I wonder you dont offer me a saucer of milk at once! 1931 E. O'Neill *The Haunted* iv, in *Mourning becomes Electra* 246 Peter is coming, and I want everything to be pretty and cheerful. 1937 M. Allingham *Dancers in Mourning* iii. 43 Go out and say pretty things... We'll all back you up. 1957 P. Kemp *Mine were of Trouble* ii. 28, I have learnt something of that frantic advance on Toledo and the final battle. It is not a pretty story. 1973 *Black Panther* 8 Sept. 17/1, I slipped back..and observed some of these same officers... Their tactics weren't very pretty.

c. (*to come to*) *a pretty pass*: see *pass *sb.*[2] 7a.

4. e. In phrases: (*as*) *pretty as paint*, *as a picture* (cf. *picture *sb.* 2h), *as a speckled pup*, etc.

Most of the phrases are more or less restricted to the U.S.

1906 *Dialect Notes* III. 151 *Pretty as a speckled pup*,.. exceedingly pretty. 1909 *Ibid.* 359 *Pretty as a picture*,.. very pretty: often used of a fine specimen of fruit. 1922 E. V. Lucas *Genevra's Money* xvi. 112 Now, there's that girl—she's as pretty as paint. If I were the kind of feller that does these things I could make a fool of myself over her. 1926 M. J. Atkinson in J. F. Dobie *Rainbow in Morning* (1965) 88 As pretty as a speckled pup under a new-painted buggy; as pretty as a speckled hen; as pretty as a picture. 1927 *Amer. Speech* Dec. 169 To him 'pretty as a heart flush' was the supremely beautiful. 1927 E. O'Neill *Marco Millions* iii. i. 141 Here! Let me get a good look at you! Why, you're still as pretty as a picture and you don't look a day older! 1936 N. Streatfeild *Ballet Shoes* vi. 77 Cook said it was as pretty as a picture, and Clara that it put her in mind of something off a Christmas card. 1976 *Time* 27 Sept. 39/2 Girls are variously 'ugly as homemade soap' or 'pretty as a speckled pup'.

5. b. (Later examples.)

1930 E. B. White *Let.* 4 July (1976) 93 The Pierce, after some brilliant road work, burned out a generator—which will cost me a pretty penny. 1978 L. Block *Burglar in Closet* i. 4 The attaché case..was a slim model in cocoa Ultrasuede that had cost someone a pretty penny.

B. *sb.* **a.** (Further examples.) Also as a form of address, with ellipsis of *my*.

1934 *Amer. Speech* IX. 288/2 *A pretty*, any good-looking girl. 1952 M. Allingham *Tiger in Smoke* xiv. 203 He's all right, pretty. He's all right now. 1972 [see LA, L.A. s.v. *L 7].

b. (Earlier and later examples.)

1736 *Boston News-Let.* 15 Apr. 2/2 (Advt.), Just arrived, and to be sold cheap, a choice variety of Haberdashery,.. Dutch Prettys, Silk Cane and Watch Strings, [etc.]. 1895 *Dialect Notes* I. 392 *Pretty*, a picture or similar article; a toy. 1927 W. E. Collinson *Contemp. Eng.* 54, I well remember the disgust we children felt at a lady (an English-woman) who..called a fancy cake a pretty! 1952 M. Allingham *Tiger in Smoke* viii. 130, I ought to keep that [miniature]. It must go down among the other pretties in the show table in the drawing-room. 1957 H. Croome *Forgotten Place* xx. 229 Scarves, handkerchiefs, nylons, pretties, were pushed aside, or sent flying. 1977 *Daily Mirror* 10 May 12/5 Perhaps the brisk sales of pretties in the shops now shows which way the wind is beginning to blow. 1977 J. Wambaugh *Black Marble* (1978) x. 225 Probably buying her pretties with what little money the kennel took in.

d. *ellipt.*, a pretty good sum (of money), a *pretty penny* (see sense A. 5b in Dict.). *U.S.*

1851 G. Thompson *Diary* 28 Jan. in N. E. Eliason *Tarheel Talk* (1956) 138, I would not send her an ugly [valentine] for a pirty. 1909 G. Stratton-Porter *Girl of Limberlost* xxi. 393 I'd give a pretty to know that secret thing you say you don't. 1927 *Amer. Speech* II. 277/1 *I'll bet you a pretty*, I'll bet you a good deal. 1935 H. Davis *Honey in Horn* v. 46 I'll bet you a pretty he ain't got any [money]. 1941 W. C. Handy *Father of Blues* v. 69 I'd give a pretty to the ear that could forget them.

e. *Golf.* The fairway.

1907 *Westm. Gaz.* 13 Sept. 3/1 Often..he will get just as far as if he had been lying on the 'pretty'. 1909 *Ibid.* 11 Sept. 7/2, I happened upon Daniel Lambert..wielding a heavy mashie among the thistles that flourish along the pretty to the tenth. 1927 *Daily Tel.* 12 Feb. 10/5 When the ball went sailing down the pretty, straight and true, what a satisfaction it was to both of them.

D. b. pretty-by-night *U.S.* = *marvel of Peru* (Marvel *sb.* 6); **pretty-face (kangaroo, wallaby)**, the whip-tailed wallaby, *Macropus parryi*, which is found in southern Queensland and northern New South Wales and has white markings on its head; **pretty please**, an emphatic or affected colloq. form of request.

1872 E. Eggleston *End of World* xxv. 169 She planted some pretty-by-nights in an old tea-pot. 1890 *Harper's Mag.* Jan. 282/1 Hollyhocks and larkspur and pretty-by-nights blossomed in the door-yard. 1911 C. Harris *Eve's Second Husband* 275 The 'pretty-by-nights' under the window..refused to consider the tragedy of Adam's unfaithfulness. 1931 W. N. Clute *Common Names of Plants* 135 The four o'clock..bears the name of pretty-by-night and lives up to it. 1947 M. Henry *Misty of Chincoteague* ix. 92 You kin cut a few of them purty-by-nights and some bouncin' Bess fer a centerpiece. 1887 W. S. S. Tyrwhitt *New Chum in Queensland Bush* viii. 145 The smaller kind [of kangaroo], known as pretty faces or whip tails,..are rather smaller and of a grey colour, with black and white on the face. 1911 W. H. D. Le Souëf *Wild Life Austral.* vii. 215 The most graceful of the Kangaroos..are locally called Pretty-face or Whip-tailed Kangaroos. 1943 C. Barrett *Austral. Animal Bk.* xi. 93 Its [*sc.* the whip-tail wallaby's] southern ally, often called 'pretty-face'..is among the most beautiful of all marsupials, with its slender, graceful body, its very long and slender tail, and white-and-grey face markings. 1970 W. D. L. Ride *Guide Native Mammals Austral.* v. 47 Whiptail, Pretty-face Wallaby... tail very long and slender, very marked white face-stripe. [1891 R. T. Cooke *Huckleberries* 169 Say 'please' now—real pretty.] 1959 A. Sinclair *Breaking of Bumbo* v. 74 She was saying, *Please. Pretty please.* 1964 *Time* 28 Feb. 28/3 Can I, pretty please? 1966 'T. Wells' *Matter of Love or Death* xii. 122 'I really can't.' I squeezed her hand. 'Not even if I say pretty please?' 1968 J. D. MacDonald *Pale Grey for Guilt* (1969) xv. 180, I guess you're not going to give it back just because I say pretty please with sugar. 1970 W. Smith *Gold Mine* xxxix. 105 Please, please, pretty please times three. 1973 C. Mason *Hostage* vii. 106 Say...'pretty please, with sugar on it.'

pretty, *adv.* Add: **1. b.** *pretty much*, almost, very nearly; approximately.

1806 D. Roe *Diary* 27 May (1904) 30 They got sum horsfish & that was pretty much all. 1861 [in Dict.]. 1873 [see *account *sb.* 9c]. 1937 E. C. Vivian *Tramp's Evidence* vii. 90 Crandon goes to bed with the dicky-birds, pretty much. 1961 'S. Gillespie' *Neighbour* vi. 93 Her flat was pretty much what he had expected. 1976 *Southern Even. Echo* 13 Nov. 12/7 The defendant.. 'pretty much on impulse' took the television.

2. a. For now *rare* and *illiterate* read *colloq.* and add further examples. *Spec.* in phr. *to sit* (or *be sitting*) *pretty*, to be comfortably placed or well situated; to be in a fortunate or advantageous position.

1864 Trollope *Can you forgive Her?* I. xxii. 173, I must go down. The Duchess of St. Bungay is here, and Mr. Palliser will be angry if I don't do pretty to her. 1891 J. Newman *Scamping Tricks* i. 2 We can talk pretty to each other. *Ibid.* vi. 46, I saw they were started on the road of mutual admiration, and travelling pretty, and that he meant calling again. 1902 *Free Lance* 5 Apr. 8/2 They must be spoken 'pretty' to, caressed, humoured, coaxed. 1921 M. Moore (*title of musical comedy*) Sittin' Pretty. 1924 Bolton & Wodehouse (*title of musical comedy*) Sitting Pretty. 1925 Wodehouse *Sam the Sudden* xv. 106 If you're American, we're sitting pretty, because it's only us Americans that's got real sentiment in them. 1932 —— *Hot Water* i. 32 We're sitting pretty. The thing's in the bag. 1932 S. Gibbons *Cold Comfort Farm* xvi. 223 It was nearly half past two, and everybody seemed sitting pretty for the sunrise. 1937 *Times Lit. Suppl.* 25 Dec. 970/2 This submerged Dickens, who would not 'play pretty' to any orthodoxy old or new, comes nearer to raising his head in the Christmas Books than in his longer works. 1939 'N. Blake' *Smiler with Knife* xviii. 256 I'm sitting pretty for the moment, she thought; but [they]..will go over this district with a fine-tooth comb. 1947 *Times* 18 Nov. 2/3 Did he think the country was 'sitting pretty'? 1957 L. P. Hartley *Hireling* xi. 88, I shouldn't be sitting where I am, sitting pretty, to coin a phrase, if it wasn't for you. 1959 *Listener* 13 Aug. 239/1 At the moment the motor industry is 'sitting pretty'. 1967 O. Wynd *Walk Softly* xi. 182 Toba was still sitting pretty, at the most pausing for reassessment. 1972 *Driving* (Dept. of Environment) (ed. 2) 119 Always try to 'park pretty'; that is, squared up in the middle of the marked space. 1976 *Washington Post* 23 May G 2 (*heading*) In some cases, they're sitting pretty.

pretty (pri·ti), *v.* [f. the adj.] Freq. const. *up*. **a.** *refl.* To make (oneself) pretty; to make or dress (oneself) *up* to look attractive. **b.** *trans.* To make (something or someone) pretty or attractive; also used ironically, to spoil or injure. Also *absol.* Hence **prettied** (up) *ppl. a.*; **prettying** (up) *vbl. sb.*

1916 H. L. Wilson *Somewhere in Red Gap* ii. 70 All I think is that he's trying to pretty himself up for Nettie. 1932 *Sun* (Baltimore) 23 Aug. 4/4 The women [pilots] were sent up a new supply of cold cream today, which enabled them to 'pretty up' for their landing. 1935 M. M. Atwater *Murder in Midsummer* viii. 72 The nurse..thought her patient should have waked up to tears and moans, and here she was fussing about the set of the lavender knitted thing about her thin shoulders. Prettying up for company! 1939 R. Chandler *Big Sleep* xxii. 182 A low-voiced prettied-up rhumba. 1943 —— *Lady in Lake* (1944) viii. 50 She was gone a week and came back all prettied up. 1950 D. D. Paige in *Lett. E. Pound* p. xxv, The general aim has been to present a volume that can be read consecutively with as little eye fatigue as possible. The editor alone is responsible for these prettyings up. 1953 K. Tennant *Joyful Condemned* xxi. 203 They wanted ..plenty of time to pretty their hair. 1953 *Here & Now* (N.Z.) Oct. 5/2 Again, isn't it rather a sham to seal the roads along which she will travel and pretty up the buildings lining them. 1959 D. Niland *Big Smoke* ii. vii. 160 He took the bottle from the bar and..smashed its end into a jagged, terrible weapon... He said, 'I'll pretty up your face, boy.' 1960 *Farmer & Stockbreeder* 19 Jan. (Suppl.) 1/3 I had made up my mind to marry Sue. The next night was our last night ashore and I had spent an hour and a half readying myself and prettying up. 1961 *John o' London's* 6 July 57/4 Even Jaques Becker.. can't resist both tarting and prettying up the Modigliani legend. 1969 *Sears Catal.* Spring/Summer 30 Smocking pretties the yoke and sleeves of this broadcloth dress with its ruffled stand-up collar. 1972 *Times* 29 Aug. 10/2 People pay a great deal of attention..to prettying their houses. 1974 'M. Yorke' *Mortal Remains* v. iii. 154 Elsie's still prettying herself... She's been in the beauty shop all afternoon. 1979 J. Scott *Clutch of Vipers* vi. 91 The Chief Constable came in; looking..bull-shouldered in these prettied surroundings.

pretty-boy (pri·tiboi). Also without hyphen. [f. Pretty *a.* 4 + Boy *sb.*[1]] A foppish or effeminate man; a male homosexual. Also used ironically, a 'tough', a thug. Also *attrib.*

1885 *Daily News* 26 Jan. 3/7 The style termed by irreverent mashers the pretty-boy clip, the style sometimes called the upward drag, and the whim which ranges from a delicate fringe to furze-bush proportions, at first amazed and amused the neat Japanese damsels. 1898 R. Hughes *Lakerim Athletic Club* 241 Sawed-Off had sniffed scornfully that lawn-tennis was a game fit for nobody but girls and pretty boys. 1931 *Amer. Mercury* Nov. 353/2 *Pretty boys*,..the circus bouncers; strong-arm men. 1941 Baker *Dict. Austral. Slang* 57 *Pretty-boy*, an effeminate young man. 1946 G. Millar *Horned Pigeon* ix. 117 A pretty boy with wavy, brown hair flowing glossily over his round head... 'Who is it?' 'The *colonnello*'s bum-boy. A shit.' 1955 M. Allingham *Beckoning Lady* v. 71 The middle-aged pretty-boy face, complete with protuberant blue eyes and corrugated dark brown hair. 1956 'E. McBain' *Cop Hater* (1958) xix. 157 Scar tissue hooded his eyes. He owned cauliflower ears and hardly any teeth. His name, of course, was 'Pretty-Boy Krajak'. 1968 M. Woodhouse *Rock Baby* xi. 111, I walked past pretty-boy... Close up, pretty-boy smelled worse than ever. 1970 E. R. Johnson *God Keepers* (1971) iv. 46 A man named Al Brunning, pretty-boy greaseball. 1973 M. Amis *Rachel Papers* 98 She was referring to the Beatles record (late-middle period—between pretty-boy rock and bleared occult) which had just come to an end. 1974 P. De Vries *Glory of Hummingbird* (1975) xi. 153 You're not cross-eyed..and your ears are pasted on straight. Not any pretty-boy, but probably photogenic.

prettyism. (Earlier example.)

1776 T. Anburey *Let.* 20 Nov. in *Trav. Interior Parts Amer.* (1789) I. 109 These *Enfant du Diable* [*sc.* skunks] differ from your *Enfant du Diable*, the London beaux, who have all their prettyisms perhaps, but are externally exhaling their pestiferous odours.

pretty-pre·ttiness. [f. Pretty-pretty *a.* + -ness.] The state or quality of being pretty-pretty; excess of prettiness.

1901 M. Beerbohm in *Sat. Rev.* 20 Apr. 500/2 For prettiness—even pretty-prettiness—in the right place I have as great a taste as anyone else. 1924 'L. Malet' *Dogs of Want* i. 26 The coquettish little Cities of the Plain..and their cheap pretty-prettiness of countless hotels. 1931 *Observer* 6 Sept. 6/4 The revulsions into Sunday School pretty-prettiness are equally surprising. 1945 E. Bowen *Ivy Gripped Steps* in *Demon Lover* 130 His elder brother's jibes at his pretty-prettiness. 1960 Koestler *Lotus & Robot* ii. vi. 165 A culture with a surface polish of utterly refined pretty-prettiness. 1965 *Observer* 17 Jan. 28/1 A yearning for the pretty-prettiness of the days when women were women.

pretty-pretty, *a.* and *sb.* Add: **A.** *adj.* (Earlier and later examples.) Also as *adv.*

1877 *Punch* 3 Feb. 47/2 To paint pretty-pretty, to compose namby-pamby, and perpetuate the modish and the monstrous. 1928 Galsworthy *Swan Song* i. xi. 80 Nothing pretty-pretty about that memorial—no angels' wings there! 1937 [see *arty-and-crafty *a.*]. 1952 L. T Stanley *Woman Golfer* 61 A 'pretty-pretty' swing may look nice, but it doesn't get you far. 1961 [see *bitchy *a.* 1]. 1962 'K. Orvis' *Damned & Destroyed* xxi. 157 He just might sing pretty-pretty and tell you exactly what's cooking in Moss's fat head. 1973 *Times* 5 Oct. 13/1, I love the baby dresses and suits which are pretty without being pretty-pretty. 1980 A. Alpers *Life K. Mansfield* vi. 114 Some flabby fiction, and some pretty-pretty verse.

B. *sb.* (Further examples.) Also *sing.*, in absol. use.

1899 [see *chance *v.* 4c]. 1929 W. Deeping *Roper's Row* xxxv. 401 But that was a monstrous argument to use, mush, the pretty-pretty, a kitten-faced sentimentality. 1934 C. Lambert *Music Ho!* v. 327 Prokofieff's third piano concerto..is curiously lacking in any sense of direction, oscillating disturbingly between the pretty-pretty and the ugly-ugly.

pre-tune: see *pre- A. 1.

pretzel. Add: **1.** (Earlier and later examples.) Also *fig.* and *attrib.*

1856 [see *blutwurst]. 1857 C. Kingsley *Two Yrs. Ago* III. ix. 271 After him came..like in *Struwelpeter*, Caspar, bretzel in hand. 1858 *Harper's Mag.* Aug. 327/1 Eating *pretzels* and drinking what is here called *bière de Bavière*. 1915 *Lit. Digest* 12 June 1443/1 Prunella's painting pretzels in Przemysl. 1932 E. Wilson *Devil take Hindmost* vii. 45 The pretzel man with his basket and the roast-chestnut man have come out again. 1933 R. L. Sutton *Arctic Safari* 24 The booby prize was a string of pretzels at least four feet long. 1933 *Sun* (Baltimore) 15 Aug. 8/7 In the old days your newest reporter would have been sufficiently cultured to know that one doesn't twist pretzels; one bends them. Hence the artisan or craftsman who performs the work is a pretzel bender. 1945 *Finito! Po Valley Campaign* (15th Army Group) 5 The air forces twisted the enemy's rail lines into pretzels of steel. 1961 Wodehouse *Ice in Bedroom* xvi. 128 You know as well as I do that Chimp Twist is as crooked as a pretzel. 1968 Mrs. L. B. Johnson *White House Diary* 7 Feb. (1970) 627

Little round tables with cokes and pizza, peanuts and pretzels. 1973 *Time* 25 June 11/3 Ervin Committee Member Lowell Weicker, dropped in for beer and pretzels. 1975 *New Yorker* 31 Mar. 29/2 She dumps an armful of immense pretzel stick cuttings into the pail. 1976 R. Cowper *Paradise Beach* in *Custodians* 71 He snatched her hands to his lips and set about them as if they were a pair of pretzels.

2. *Mus. slang.* (See quots.) Also *pretzel bender.*

1936 *Metronome* Feb. 61/2 *Pretzel*—French horn. 1936 *Amer. Mercury* May p. x/2 *Pretzel bender*—one who favors the French horn. 1945 L. Shelly *Jive Talk Dict.* 16 *Pretzel*, French horn. *Ibid.* 31 *Pretzel bender*, French horn player.

prevalence. 3. (Further examples.)

1839 *Ann. Rep. Registrar-Gen. England* 87 The prevalence of a disease..is expressed by the deaths in a given time out of a given number of living. 1857 T. W. Grimshaw et al. *Man. Public Health Ireland* xxvii. 298 From statistics [of small-pox]..it appears that its greatest prevalence is observed in May, the cases in that month being 13·7 per cent. of the total cases occurring in the year. 1961 M. Schorer *Sinclair Lewis* iv. viii. 471 He talked..about the prevalence of American slang in British speech. 1975 *Nature* 20 Mar. 168/3 Any successful preventative measure against leprosy will be shown by a fall in the number of new cases or in the incidence rate: the total number of cases (or 'prevalence' rate) will change much more slowly, because of the inclusion of patients who are already crippled.

pre-Vedic: see *pre- B. 1a.

preveniently, *adv.* (Further examples.)

1974 *Times Lit. Suppl.* 29 Nov. 1343/1 Are they those misconceptions of the nature and role of poetry which this most preveniently resourceful essay in metaphysics would correct? 1977 G. W. H. Lampe *God as Spirit* iii. 74 God's Spirit works preveniently to bring a person to conversion. 1977 *Theology* LXXX. 192 Whenever and wherever there is response to the divine love, *there* that love is preveniently at work.

preventative, *a.* and *sb.* Add: **A.** *adj.* (Later examples.)

1936 W. H. S. Smith *Let.* 18 July in *Young Man's Country* (1977) ii. 16 My principal occupation was trying 'Bad Livelihood' cases. These are preventative cases for securing the good behaviour of criminals by taking a bond. 1939 *Ann. Reg. 1938* 93 Two new types of prison sentences were proposed. One was called 'corrective training'... The other was called 'preventative detention', and was to be for not less than two and not more than four years for persons over 30, but up to ten years on certain types of offenders with long criminal records. 1968 *Globe & Mail* (Toronto) 5 Feb. 10/8 A Democratic administration had gotten the United States deeply involved in Vietnam..by failing to apply 'preventative diplomacy'. 1973 *Black Panther* 21 July 8/3 The People's Free Health Clinic.. provides medical treatment and preventative care on a clinical level. 1976 *Oxf. Mission Q. Paper* July/Sept. 15 Many of the people are illiterate and it takes a lot of patient talking and convincing before they can see the benefits of preventative medicine.

B. *sb.* **c.** A contraceptive; = *preventive *sb.* c.

1901 J. A. Godfrey *Science of Sex* ii. vi. 257 The checks employed by women are more diverse in mechanical detail... Opinions differ greatly as to both the reliability and the physiological harmlessness of these forms of preventative. *Ibid.* 256 So long as the sheath remains whole, it is an absolute preventative. 1918 R. B. Armitage *Private Sex Advice to Women* x. 130 The use of 'contraceptives' or 'preventatives' is considered justified in certain cases. 1934 Dylan Thomas *Let.* Oct. (1966) 143 Do you believe in preventatives?

preventer. Add: **2.** (Further example.)

1920 *Chambers's Jrnl.* Mar. 208/1 A single set of hydrofoils under the bow, known as a preventer, helps to lift the boat when getting up speed, while checking any tendency to nose-dive.

3. b. (*a*) *preventer backstay* (later examples), *brace* (later example), *gasket* (later example), *guy* (later example), *sheet*.

1912 W. I. Downie *Reminisc. Blackwall Midshipman* ii. 22, I expect preventer backstays were practically a permanent part of her equipment during the trip. 1939 H. Hughes *Through Mighty Seas* x. 264 For the first time I saw preventer back-stays being rigged from the main top-gallant mast. 1867 G. E. Clark *Seven Years of Sailor's Life* xx. 203 We had the barque under a close-reefed main-topsail, and with preventer braces on the yard, flew on the waves. 1926 T. M. Hemy *Deep Sea Days* ii. 64 It was 'all hands wear ship'; then up aloft and change the preventer braces from one side to another. 1907 M. Roberts *Flying Cloud* xxxii. 304 Budd went aloft on the cro' jack-yard and passed a couple of preventer gaskets. 1923 *Man. Seamanship* (Admiralty) II. xi. 188 In the recent America Cup Races, both craft tried taking a preventer guy out to the weather crosstrees if the wind was light enough. 1867 G. E. Clark *Seven Years of Sailor's Life* xix. 191 The mainsail was furled, and preventer sheets put on the fore boom.

4. In full *blow-out preventer*. A heavy valve or assembly ('stack') of valves usu. fitted at the top of an oil well during drilling and closed in the event of a blow-out.

1916 A. B. Thompson *Oil-Field Devel. & Petroleum Mining* vii. 367 An apparatus which is largely employed with rotaries is what is called a 'Blow-out Preventer'. 1934 *Proc. World Petroleum Congr. 1933* I. 370/2 The main feature is a heavy rubber sleeve packer held in a container which is free to revolve in ball bearings inside the body of the preventer. 1962 *Economist* 15 Sept. 1046/2 The blow-out preventers and the drill bit itself can be lowered. 1972 L. M. Harris *Introd. Deepwater Floating Drilling Operations* x. 98 All preventers should have a pressure rating in excess of the maximum that could be expected on the wellhead. 1976 *Offshore Engineer* July 6/4 Shell has recovered, from 130m of water, the blow-out preventer (bop) stack which fell to the seabed while being lowered into position from the rig *Chris Chenery*. 1977 *Daily Tel.* 29 Apr. 2/1 They went to the gang boss and it was decided to rectify it when another driller noticed a small stream of mud running from the preventer's outlet.

preventionism (prīve·nʃǝniz'm). [f. Prevention + -ism.] A policy of prevention. So **preve·ntionist**, one who favours such a policy.

1918 A. Gray tr. *Grelling's Crime* II. ii. 109 All these questions..must simultaneously be answered in the affirmative, if the preventionists wish to justify their point of view. *Ibid.* 118 When preventionism suits their purpose, they speak of the right and the duty of the anticipated defence against future attack. 1978 D. Gryll's *Guardians & Angels* ii. 66 The child-cruelty furore.. promoted sympathy for children's sufferings, and preventionists tended to deprecate not only cruelty but strictness.

preventive, *a.* and *sb.* Add: **A.** *adj.* **2.** spec. *preventive arrest, detention, maintenance, war.*

1639 Preventive war [in Dict.]. 1908 *Act 8 Edw. VII* c. 59 § 10 (1) Where a person is convicted on indictment of a crime,..and subsequently the offender admits that he is or is found by the jury to be a habitual criminal, and the court passes a sentence of penal servitude,..the court ..may pass a further sentence ordering that on the determination of the sentence of penal servitude he be detained for such period not exceeding ten nor less than five years,..and such detention is herein-after referred to as preventive detention. 1918 E. Pound *Let.* 15 Nov. (1971) 140 The wholesale preventive arrests surely prevented another rising... Similar preventive arrests would have prevented the Easter rising. 1932 *Rep. Dept. Comm. on Persistent Offenders* iv. 14 in *Parl. Papers 1931–2* (Cmd. 4090) XII. 553 Preventive Detention is a sentence intended for 'professional' criminals or criminals who definitely give themselves up to a career of serious crime. 1945 *Facts on File* 28 Feb. 70/2 Paris reports that the French police have released from 'preventive detention' P. G. Wodehouse, British novelist. 1948 H. Nicolson *Diary* 29 Nov. (1968) 155 A preventive war is always evil. 1953 B. V. Bowden *Faster than Thought* iv. 86 The technique of preventive maintenance..has considerably reduced the number of valve failures which occur while the machines are running. 1959 *Listener* 16 Apr. 674/1 Southern Rhodesian Government publishes a new preventive detention bill. 1963 *New Statesman* 22 Feb. 260/1 Preventive detention, a relatively humane experiment begun in 1908 and modified 40 years later, has been finally adjudged a failure and abolished. 1966 Schwarz & Hadik *Strategic Terminol.* 130 *Preventive war*, war initiated to prevent the enemy from making gains he might be expected to make if he were allowed to initiate the war, or that he might make without resorting to war if not forcibly opposed in good time. 1967 E. Duckworth in Wills & Yearsley *Handbk. Managem. Technol.* vi. 119 *Preventive maintenance*, a scheme for regular overhaul of equipment to prevent unexpected breakdown.

b. spec. *preventive medicine.*

[1769 W. Buchan *Domestic Med.* ii. 531 Dr Mead recommends a preventive medicine.] 1870 *Food Jrnl.* I. i. 22 Preventive medicine received a great impulse by the labours of Coleman in relation to ventilation. 1881 [in Dict.]. 1926 *Eng. Rev.* Sept. 315 What preventive medicine can do is illustrated by the clearing up of the infective diseases of Panama. 1957 A. Huxley *Let.* 22 Feb. (1969) 818 If we could combine Krishnamurti with old Dr Vittoz's brand of psychotherapy..and a sensible diet, we would have solved the problem of preventive medicine. 1963 *Lancet* 19 Jan. 145/2 (*heading*) Statistical methods in clinical and preventive medicine.

B. *sb.* **c.** A contraceptive; = *preventative *sb.* c.

1822 F. Place *Illustr. & Proofs Princ. Population* v. 150 The proposals of Mr. Malthus, to persuade the poor that they have no right to eat.., as well as Mr. Godwin's infanticide, are..proposals to commence at the wrong end. The remedy alone can be found in preventives. 1901 E. B. Foote *Home Cycl.* iv. vii. 1137 To lessen the 'vicious employment of preventives' outside of marriage. 1911 R. Brooke *Let.* Mar. (1968) 292 The Church'll declare that God gave every man the Right-not-to-use-Preventives. 1943 R. Malkin *Marriage, Morals & War* 159 In 1940, when Lieutenant Colonel Gardner of the United States Army Medical Corps gave happy birth to the idea of a pamphlet explaining venereal diseases, which..would be sold together with a set of preventives and prophylactics. 1961 A. H. Nethercot *First Five Lives A. Besant* vii. 117 Before marriage, pure will power and moral determination were to be the only preventives. Physical preventives were..against the will of God.

preventorium (prīventōǝ·riǝm). orig. and chiefly *U.S.* [f. Prevent *v.*, after Sanatorium.] An institution where preventive care is given to people at risk from tuberculosis or other diseases.

1907 W. Ewart in *Jrnl. Balneol. & Climatol.* XI. 155 The place for the 'Prevention-Sanatorium', or 'Preventorium', is the sea-coast. 1909 *Boston Even. Transcript* 10 Nov. (*heading*) To fight tuberculosis New York will have $700,000 preventorium. 1929 W. B. Tomson *Prevention of Tuberculosis* vii. 51 The preventorium stands out as the predominating instrument for prophylaxis and early treatment in childhood in America. 1930 *Aberdeen Press & Jrnl.* 21 Oct. 6/1 There are two preventoria in Aberdeen. 1936 *Dict. Amer. Biogr.* XVIII. 130/1 In 1909 he [*sc.* Nathan Straus] established in his cottage in Lakewood, N.J., the pioneer tuberculosis preventorium for children. 1953 H. R. Leavell in Leavell & Clark *Textbk. Preventive Med.* xvi. 497 'Preventoriums', where children from families with tuberculosis were sent and cared for..did not prove useful enough to justify their cost, and they were not taken over by government. 1968 *Awake!* 8 May 12/2 When I was eighteen, I had to spend some months in a preventorium.

preverb, *sb.* and *a. Gram.* **A.** *sb.* With pronunc. (prī·vǝ̄ɹb). **1.** [Pre- B. 1 in Dict. and Suppl.] A particle or prefix preceding the stem of the verb.

1930 [see *Alfredian *a.*]. 1939 L. H. Gray *Foundations of Lang.* iii. 62 In English, many compound verbs borrowed from French consisting of a preverb (commonly, but erroneously, called a preposition..) and a base-word, and serving either as a noun..or as a verb, are distinguished in their use..by a difference of stress-accent. 1946 L. Bloomfield in C. F. Hockett *Leonard Bloomfield Anthol.* (1970) 460 Certain particles, *preverbs*, freely precede verb stems. 1951 *Archivum Linguisticum* III. 28 The early loss in NGmc of unstressed prefixes (*i.e.* mainly preverbs). 1967 C. J. Fillmore in *Glossa* I. 91 Many of the syntactic properties of the positive and negative adverbial elements called 'preverbs' have already been discussed.

2. [Pre- A. 2.] A verb which precedes another verb; an auxiliary verb.

1965 F. Behre in *Eng. Stud.* XLVI. 91 It is the content of the pre-verb that is qualified by *if* and not the pre-verb (*would*, *should*, *could*, *might*, etc.).

B. *adj.* With pronunc. (prīvǝ̄·ɹb). [Pre- B. 2.] Occurring before a verb. Also as *adv.*

1976 *Archivum Linguisticum* VII. 33 Essentially, then, medial position means non-initial, pre-verb position. 1976 *Amer. Speech 1974* XLIX. 82 The collocation of *much* with *prefer* applies only when *much* is preverb, as in *I much prefer a dry wine.*

preverbal (prīvǝ̄·ɹbǎl), *a.* [Pre- B. 1.] **1.** Preceding the formulation of an utterance; prior to or present before the development of speech.

1931 G. Stern *Meaning & Change of Meaning* v. 115 A pre-verbal phase of 'gedankliche Gliederung' follows on the preliminary adjustment. 1938 I. Goldberg *Wonder of Words* ii. 28 Obviously there was what may be called a preverbal intelligence. 1957 C. E. Osgood et al. in Saporta & Bastian *Psycholinguistics* (1961) 287/1 The auditory effects of hearing 'hammer' do not produce behavior in any way relevant to *hammer* object in the pre-verbal child. 1978 Russell & Dewey in P. Moore *Man, Woman, & Priesthood* vii. 93 Mother-figures reign in the pre-verbal layers of personality laid down in infancy. 1979 *Nature* 13 Dec. 724/1 The stimulus comparison task..appears to have been carried out at a post-perceptual, pre-verbal level.

2. *Gram.* Preceding the verb. Also *absol.* as *sb.*

1948 [see *complementation]. 1958 *Archivum Linguisticum* X. 32 Or a post-verbal, counter to appearances, may vie with a preverbal. 1959 M. A. K. Halliday *Lang. of Chinese 'Secret Hist. Mongols'* vi. 88 The conditional is marked lexico-grammatically, by a system of adverbs; these, designated by position as 'preverbal' and 'final', occur either alone or in combination one with the other. 1972 J. L. Dillard *Black English* iv. 152 Abram.. is represented as using the pre-verbal durative particle *a* (*Your a gwine*). *Ibid.* v. 220 A preverbal *done* can be found as far back as Dunbar. 1977 *Word 1972* XXVIII. 147 The use of a preverbal particle in sentence initial position has recently become more common in the literary language.

preverbalization, -vernal: see *pre- A. 2, B. 1.

previable (prīvai·ǎb'l), *a.* [Pre- B. 1.] Before the stage when a fœtus has developed sufficiently to survive outside the womb.

1910 F. J. Taussig *Prevention & Treatm. Abortion* 2 Abortion is the pre-viable expulsion of the human ovum. 1936 —— *Abortion* iv. 81 The previable stage of pregnancy. 1945 *Jrnl. Obstetr. & Gynæcol.* LII. 35/1 The term 'pre-viable' has been proposed for the separate category of liveborn premature infants with a birth-weight of less than 2½ pounds. 1972 *Daily Tel.* 24 May 17/4 Research on pre-viable foetuses (those which may show some signs of life, but which have no hope of an independent existence) should be safeguarded further by limiting such work only to foetuses weighing less than 300 grammes (about two-thirds of a pound). 1978 *Church Times* 7 July 12/5 Difficult questions also arise over baptism, when a fully formed but pre-viable foetus is delivered by a legally procured abortion, still living but inevitably destined to die.

preview, *sb.* Restrict *rare* to sense 1 in Dict. and add: **2. a.** (Later examples.) Also, a foretaste, a glimpse.

1935 *Sun* (Baltimore) 4 Apr. 1/5 Voting,..held this week in Michigan, Illinois and Maryland, has left a somewhat mottled political picture as a 'preview' of the important Presidential contest which will take place next year. 1938 *Ibid.* 24 Jan. 6/3 His preview of the budget probabilities for the fiscal year now current. 1946 *War Report* (B.B.C.) 263 They were taking with them into captivity a preview of the wreckage of Hitler's Deutschland. 1956 W. H. Whyte *Organization Man* i. 10 The best place to get a preview of the direction the Social Ethic is likely to take in the future. 1959 *New Statesman* 31 Jan. 136/3 Mr Dene..was to be given a pre-view of army life on a special advance visit to Winchester Barracks. 1978 LaRosa & Tanenbaum *Random Factor* xii. 179 The morning of November 18 brought a preview of the winter to come.

b. (Occas. **prevue.**) *spec.* A showing or presentation of films, books, exhibitions, etc., before they are available to the public. Also *attrib.*

1922 *Opportunities in Motion Picture Industry* (Photoplay Research Soc.) 76 Where the studio employs a number of directors usually all of them sit in on the 'previews' that are given a film before it is actually ready for the final release. 1928 L. North *Parasites* 84 He attended a preview of a picture made by a small independent group of players. 1931 *Amer. Speech* VII. 74 Why should the word *trailer* be used to apply to a prevue of a motion picture? 1936 W. de la Mare in *J. Freeman's Lett.* p. xvii, That dubious puff, the pre-view, was not as yet in fashion. 1940 *Times* (Weekly ed.) 7 Aug. 17/3 A pre-view and mannequin parade of women's sportswear, coats, and costumes was held in London on Thursday. 1955 *Radio Times* 22 Apr. 42/2 (*heading*) Sport. Today's results and weekend pre-view. 1958 *Photoplay* Oct. 15 The first studio preview of *Stage Struck.* 1961 G. Millerson *Technique Television Production* i. 15 Picture monitors. These preview screens give a continuous view of what the three or more studio cameras and other video sources are seeing. 1977 *Rolling Stone* 13 Jan. 22/5 They finally came to a preview theater to see the 'Stairway to Heaven' segment.

preview, *v.* Restrict *rare* to sense in Dict. and add: **2. a.** *trans.* To show or present (a film, etc.) before its public presentation; to give a preview or foretaste of (something).

1928 L. North *Parasites* v. 66 We pre-view a picture every week. 1939 *Sun* (Baltimore) 26 Sept. 10/3 In some respects it promises to preview the World Series. 1950 Blesh & Janis *They all played Ragtime* i. 16 Such was the Negro's position in our society that it was inevitable that this rich new vein of music should be previewed for white America in whorehouses. 1951 *Newsweek* 27 Sept. 74/3 Euclid previewed a new line of vehicles which it is counting on to send its sales volume soaring. 1965 *Observer* 31 Jan. 23/5 The BBC did preview 'Culloden', but only on the same day that it was shown. 1966 *Listener* 15 Sept. 397/1 The first edition,..which previewed the Commonwealth Conference, merits only subdued congratulation. 1968 *Radio Times* 28 Nov. 33/1 William Douglas Home introduces his first television play—and previews his own edgy performance. 1977 *New Yorker* 6 June 108/1 His full-length opera 'The Voyage of Edgar Allan Poe', was 'previewed' in Minneapolis.

b. *intr.* Of a production, performance, etc.: to be previewed.

1978 *Tucson Mag.* Dec. 99/1 On the same afternoon, the TMA League's annual 'Christmas Fair' previews to members only. 1980 *Times* 11 June 9 Yet another massive stage project, now previewing at the Aldwych, where it officially opens on June 19. 1981 *Times* 26 Aug. 9/3 *Two Gentlemen of Verona* and *Titus Andronicus,* which start previewing tonight.

Hence **pre·viewer**; **pre·viewing** *ppl. a.*

1970 *Guardian* 6 Aug. 8/2 The previewing critics. 1970 *Globe & Mail* (Toronto) 25 Sept. 12/5 Some of the previewers yesterday were the photographer's grandmother, Mrs. Arthur S. King, Mr. and Mrs. J. A. Blackey, [etc.]. 1976 *National Observer* (U.S.) 23 Oct. 22/1 The previewers all have a common goal: to latch on to the money and fame that can be made on the billion-dollar-a-year lecture circuit.

previous, *a.* Add: **2. d.** (Further examples, including examples as *sb.*)

1885 *Ordinances Univ. Cambr.* 6 By Grace, 4 Apr. 1878, selected Candidates for the Civil Service of India who are Candidates for Honours are excused the Previous Examination. 1905 *Abol. Compulsory Greek* (Cambridge Univ.) 6 As regards the smaller and the local schools it may be pointed out that these are practically unaffected by the requirements of the Previous... In such schools the inclusion or exclusion of Greek in the regular curriculum does not depend on the Previous Examination. 1950 M. Marples *University Slang* 81 Soon after this date [*sc.* 1863] *Little-go* died out at Oxford, leaving the field to *Smalls,* and retired to Cambridge, where its official title, corresponding to *Responsions,* is *Previous.* 1979 *Jrnl. R. Soc. Arts* Oct. 706/2 As late as 1861, the subjects were little more than a repetition of the 'Previous' on a slightly larger scale.

c. *ellipt.* as *sb.* Previous convictions. *slang.*

1935 G. Ingram *Cockney Cavalcade* x. 168 He ain't got no 'previous', so you ought-a get bound over, didn't yer, Jack? 1970 G. F. Newman *Sir, You Bastard* i. 34 'Neither has any previous, Terry,' Burgess said. 'I thought perhaps the fella might have had a little bit,' he shrugged. 1974 E. Jones *Barlow comes to Judgement* 14 Sitting on the benches..were ten men... Nine of them had previous. The tenth had a clean record. 1977 'M. Underwood' *Murder with Malice* v. 56 Anthony Rivings..five convictions for dishonesty..three other Rivings..all with previous.

prevoc·alic, *a.* Also (occas.) **prae-vocalic.** [Pre- B. 1.] Before a vowel; of or pertaining to the position before a vowel. Hence **prevoca·lically** *adv.*

1909 in Webster. 1934 M. K. Pope *From Latin to Mod. French* II. xvii. 223 In educated speech throughout the sixteenth century final consonants, whether single or supported, were maintained when in prae-vocalic position in the phrase. 1943 A. L. Kroeber in *Univ. Calif. Publ. Linguistics* I. 30 The stops p, t, k are universal and stable. In the Arizona dialects they are less aspirated, prevocalically, than in other languages and have often been written b, d, g. 1949 *Language* XXV. 400 It is..improbable that *l* [in Umbrian] was treated differently before *j* and prevocalic *l.* 1957 N. Frye *Sound & Poetry* 114 The two methods are even more sharply contrasted in the line endings of the two poets, where the frequencies of Spenser's prevocalic and Milton's postvocalic clusters are more than doubled. 1964 R. H. Robins *Gen. Linguistics* 101 In Scots English, /r/ occurs both prevocalically and postvocalically. 1966 [see *chest register* s.v. *chest 10 b]. 1970 B. M. H. Strang *Hist. English* I. ii. 51 Dark *l* post-vocalically, and clear *l* pre- or inter-vocalically. 1976 *Archivum Linguisticum* VII. 164 Both to postulate an analogical extension of the pre-vocalic forms, since in the majority of the cases they do not even exist, and to assume lost vowels, is to beg the question.

pre-vowel: see *Pre- B. 2a.

prevue, var. *Preview *sb.* 2b.

pre-wa·r, *a.* and *adv.* **A.** *adj.* [Pre- B. 2.] Pertaining to or characteristic of the period before a war, esp. the wars of 1914–18 and of 1939–45.

1908 [O.E.D. s.v. Pre- B. 2]. 1908 *Daily Chron.* 24 Apr. 7/3 The Transvaal Government..are thoroughly honest—a great difference from the pre-war days. 1915 *Political Q.* May 90 The relations of the engineers (employers and workmen) were governed by two separate pre-war agreements. The earlier..made in 1911,..the second ..dated April 1914. 1917 [see *foggy *a.* 5 b]. 1926 Galsworthy *Silver Spoon* III. vii. 277 How stupid and pre-war! Why couldn't he, like her, be free, be supple, take life as it came? 1928 *Publishers' Weekly* 30 June 2617 The government of France has succeeded in stabilizing the franc at 25·52 to the dollar, approximately one-fifth of its pre-war ratio. 1938 *Encycl. Brit. Bk. of Year* 20/1 Tricycle undercarriages familiar in pre-war days, have been re-introduced in modernized form for greater safety in high-speed landings. 1942 E. Paul *Narrow St.* xxix. 260 What dealt our pre-war world its mortal blow was the supine cowardice and hypocrisy of so-called democrats. 1946 *R.A.F. Jrnl.* May 172 All the women were free..to resume their pre-war occupation... Already they have resumed their pre-war studies. 1958 *Listener* 6 Nov. 719/1 He [*sc.* Lord Montgomery] has severe and deserved strictures on the pre-war governments for their failure to bring the army up to date. 1975 *New Yorker* 28 Apr. 112/2 The Thai Nguyen steelworks now produces only pig iron, but planners hope that in two or three years it will regain its prewar capacity of a hundred and seventy thousand metric tons of steel annually. 1977 *Times* 22 Apr. 2/8 The pre-war, Spanish-style house has.. indoor swimming pool.

B. *adv.* [*Pre- B. 2 c.] Before a war, esp. the wars of 1914–18 and of 1939–45.

1920 *Econ. Conditions Central Europe* I. 12 Four million tons of coal were imported annually pre-war, mainly to Petrograd and Baltic ports. 1923 *Westm. Gaz.* 25 Aug. 4/5 The new tourist hails from districts and from classes which, pre-war, never dreamed of leaving England. 1928 *Daily Tel.* 4 Sept. 9/6 Some time pre-war there was a large contract out for tender from a foreign Government for water tanks. 1959 *Times Lit. Suppl.* 24 Apr. 288/4 The turnover of land is low, indicating a considerable decrease in distress sales as compared with pre-war. 1974 *Daily Tel.* 17 July 18 Pre-war, the 1933 Wimbledon final ..bears a similarity to this last championship.

pre-wash: see *Pre- A. 1, 2. **pre-wear:** *Pre- A. 1. **pre-work, -world:** *Pre- B. 2a. **pre-wrap, -write:** *Pre- A. 1.

Prex. Add: (Earlier and later examples.) Also *transf.*

1828 *Yankee* (Portland, Maine) 16 July 232/1 Our Frex says this: You surely miss [etc.]. 1906 *N.Y. Even. Post* 11 June 6 If the various unpopular 'Prexes' would study the grounds of their unpopularity. 1942 Berrey & Van den Bark *Amer. Thes. Slang* § 854/10 Government officials and employees... Prex, Prexy, the President. 1967 H. Kemelman *Nine Mile Walk* 150, I was still officially a member of the faculty [of history] and as such was invited to the President's annual Christmas reception for the faculty. I accepted..because of past favours from Prex.

Prexy (pre·ksi). *U.S. slang.* Also **Prexie, p-.** [f. prec. + -y[6].] = Prex in Dict. and Suppl.

1871 L. H. Bagg *4 Years at Yale* 655 The title 'Prex'.. is oftener used alone to designate him [*sc.* the President] among the Seniors, the modified form of 'Prexy' is somewhat in vogue, in familiar talk. 1905 *N.Y. Even. Post* 1 Sept. 7 Scores of entering classes are lined up in chapel to listen to good advice from the dean or 'Prexie'. 1909 O. D. von Engeln *At Cornell* 58 The avenue is still, for Prexy is delivering his annual address in the Armory. 1929 *Publishers' Weekly* 22 June 2859/1 Professor Charles E. Merriam..has hobnobbed with politicians as well as prexies. 1948 *Variety* 25 Aug. 1/4 Madison Sq. Garden 'prexy' Gen. John Reed Kilpatrick. 1973 *Center City Office Weekly* (Philadelphia) 9 Oct. 3 Dr. Richard N. Harner, prexy of the Epilepsy Foundation, will appear in a taped interview with Frank Ford. 1974 *Cleveland* (Ohio) *Plain Dealer* 13 Oct. c. 13/1 While the NHL is controlled basically by the board of governors..the silver-haired prexy still wields a powerful stick when it comes to meting out fines and suspensions. 1979 *Honolulu Advertiser* 8 Jan. A. 3/1 Brokers at Stapleton Assoc. think they had more than their share of the breaks in '78. Prexy John Stapleton started things off by busting his leg on a ski trip.

prey, *sb.* Add: **4. b.** *bird of prey*: esp. a bird belonging to the order Falconiformes or Strigiformes. Also *attrib.* and *fig.* (Earlier and further examples.)

a 1398 Trevisa tr. *Bartholomæus Anglicus' De Proprietatibus Rerum* (1975) II. 1288 Most hote briddes of complexioun and colerik, as briddes of pray, haueþ þe vtter partyes ȝelowe. 1603 Shakes. *Meas. for M.* II. i. 1 We must not make a scar-crow of the Law, Setting it up to feare the Birds of prey. 1899 W. E. H. Lecky *Democracy & Liberty* (ed. 2) I. p. xxii, He [*sc.* W. E. Gladstone] had a wonderful eye—a bird of prey eye—fierce, luminous and restless. 1920 H. E. Howard *Territory in Bird Life* vii. 269 A bird of prey would have more difficulty in approaching a flock unawares than it would have in approaching a single individual. 1956 D. A. Bannerman *Birds Brit. Isles* V. p. v, It is a sad fact that several of our most noble birds of prey can no longer be studied in what were once their native haunts. 1974 M. Bijleveld *Birds of Prey in Europe* i. 1 During the last two hundred years, the European continent has seen a period of intense persecution of the diurnal birds of prey.

prez[1] (prez). Also **pres** and with capital initial. Colloq. abbrev. of President *sb.*

1892 [see J.C.R. (*J III)]. 1936 *Esquire* Sept. 64/2 Mr. Roosevelt may be Mr. President to statesmen but he's the Prez to *Variety.* So is Harry Cohn, prez of Columbia Picts. 1942 Berrey & Van den Bark *Amer. Thes. Slang* § 183/4 *Pres,..prez,* president. 1956 *Washington Star* 7 Nov. A. 47 We should give the President our full support... Let's give three rousing cheers for the Pres! 1969 C. Burke *God is Beautiful, Man* (1970) 88 So this derty rat fink he says to the prez of the gang, Caiaphas, 'What's in it for me?' 1973 *Philadelphia Inquirer* (Today Suppl.) 14 Oct. 29/2 Reuben Malonado, 'prez' of the Royal Javelins, picked up an easy $150 a week. 1975 *N.Y. Times* 27 Feb. 20/5 'Look, there's the Prez,' one shouted when Mr. Ford came into view.

prez[2], **pres,** colloq. abbrevs. Present *sb.*[2]

1922 Joyce *Ulysses* 270 Accept my little prez. 1967 *She* Dec. 9/2 Perfect tree prez. for the husband of any wife who's fed up with having her best kitchen knives whisked away to the toolshed.

prezzie (pre·zi). *colloq.* Also **pres(s)ie,** (*rare*) **presee.** [f. prec. + -ie.] = prec.

1937 E. D. Metcalfe *Let.* 27 Jan. in F. Donaldson *Edward VIII* (1974) xxv. 312 The rest of the time has been spent shopping (buying presees for Wallis). 1961 J. Rose *At Cross* 141 Bella said 'I brought you quite a lot prezzies.' 1967 A. Diment *Dolly Dolly Spy* xiv. 184 We'll have the pressies first. 1975 *Australian* 24 Apr. 13 From endeavours yesterday to discover what presents the Whitlams were taking overseas with them, we can infer the rest of the following piece of government policy: From this day forth no public announcements will be made about the nature of prime ministerial pressies. 1977 *H. &* ... *Queen* Dec. 164/2 A presie from Mummy. 1980 *Times* 12 Mar. 21/3 Beswick is chuffed by the reception from the shop floor. He is getting little prezzies from them.

‖**priamel** (pri,ā·məl). Also **Priamel.** Pl. **priameln.** [G., ad. L. *praeambulum*: see Preamble *sb.*] A kind of epigrammatic verse cultivated in Germany in the fifteenth and sixteenth centuries; also applied to a similar literary form in ancient Greek poetry.

1950 *Chambers's Encycl.* XI. 838/2 *Rosenplüt,* ..recited epigrammatic poems (in a form known as the Priamel) at public ceremonies in honour of towns, princes or noblemen. [1953 T. G. Rosenmeyer tr. *Snell's Discovery of Mind* iii. 48 Sappho makes use of the 'preamble', a species of folk poetry emphasizing one thing above the rest.] 1962 R. W. B. Burton *Pindar's Pythian Odes* viii. 106 The last sentence of the epode..is in the form of a Priamel or *praeambulum,* a series of parallel statements leading by stages to a climax, an extended form of paratactic simile of a type frequent in archaic poetry. 1976 *Oxf. Compan. German Lit.* 683/2 *Priamel,* a minor poetic form, cultivated in the 15th and 16th c., in which, after a preparatory cumulative build-up, a comic or witty *pointe* forms the final line... The chief exponent of Priameln is Hans Rosenplüt. 1976 *Classical Q.* XXVI. 194 The first line is a Priamel, the three terms of which are all applicable to Peleus, who is to be the subject of the myth that follows.

priapulid (prəi,æ·piulid), *sb.* and *a.* [f. mod.L. name of class or phylum *Priapulida,* f. generic name *Priapulus* (J. B. P. A. de M. de Lamarck *Hist. Nat. Animaux sans Vertèbres* (1816) III. 76), diminutive form of Priapus: see -id[3].] **A.** *sb.* A marine unsegmented worm belonging to the class Priapulida, found in mud at the bottom of cold seas. **B.** *adj.* Of, pertaining to, or designating an animal of this kind.

1906 H. Theel in *Kungl. Svenska Vetenskapsakad. Handlingar* XL. iv. 5, I am now going to give a short review of those Priapulids..which pass their lives in northern and arctic seas. 1916 A. Huxley *Burning Wheel* 36 Your heaven's so, With a path leading up to it past a row Of votary Priapulids. 1951 L. H. Hyman *Invertebrates* III. xiii. 184 The priapulids are animals of modest size, up to 8 cm., and drab coloration. *Ibid.* 194 The priapulid larvae live in the bottom muck along with the adults. 1967 *New Scientist* 14 Sept. 547/2 In the case of the priapulids..a plausible explanation for their pentamery seems hard to furnish. 1979 *Sci. Amer.* July 114/1 The mud supported an active group of burrowing invertebrates, with priapulid worms predominant.

pricasse, var. *Prikaz.

price, *sb.* Add: **I. 1. f.** *colloq.* A high price.

1920 'K. Mansfield' *Bank Holiday* in *Athenæum* 6 Aug. 166/1 He likes to watch..her puzzled eyes lifted to his: 'Aren't they a *price*!'

5. (Further examples.) Freq. in phr. *at a price.*

c 1647 Clarendon *Hist. Rebellion* (1703) II. vii. 189 So much enamoured on Peace, that he would have been glad, the King should have bought it at any price. 1849 Thackeray *Pendennis* I. xiii. 118 He's too young to marry ..and..poor as Job. Can't have him at no price, I say, Mr. Bo? 1859 B. Jerrold *Wit & Opinions D. Jerrold* 155 We love peace, as we abhor pusillanimity; but not peace at any price. 1873 C. M. Yonge *Pillars of House* I. xi. 230 Mr. Froggatt says he would not go at any price. 1923 R. Fry *Let.* 29 Apr. (1972) II. 533 The British public won't have me at any price. 1928 A. Christie *Mystery of Blue Train* xxi. 172 I'm going to leave you... I can't stand my father-in-law at any price. 1934 G. B. Shaw *On Rocks* Pref. 177, I am not offering you the truth at a price for my own profit. 1961 L. Mumford *City in Hist.* xvii. 544 The machines..that would lend themselves to decentralization in a life-centered order, here become either a means to increase congestion or afford some slight temporary palliation—at a price. 1974 J. Pope-Hennessy *R. L. Stevenson* viii. 151 Louis Stevenson had stipulated..that he would not at any price stay in a hotel..but wished to live in a house. 1978 'W. Haggard' *Poison People* iv. 144 It's..illegal to hold it [*sc.* gold] in quantity. I don't say bullion can't be found at a price.

5*. *what price —?*: what is the value or use of —?, what is the likehood of —? Freq. merely an expression of contempt: 'so much for —'.

1893 P. H. Emerson *Signor Lippo* xiv. 52 What price you, when you fell off the scaffold? 1895 H. W. Nevinson *Neighbours of Ours* iii. 73 What price the little backstairs Dook? 1899 R. Whiteing *No. 5 John St.* ix. 94 What price grammar? It don't seem to teach people to keep a civil tongue in their 'ead. 1905 E. Nesbit *Oswald Bastable* 93 Oswald now thought that politeness was satisfied..so he said: 'What price treasures?' 1907 G. B. Shaw *Major Barbara* II. 245 *Bill* (cynically..) Wot prawce Selvytion nah? 1914 C. Mackenzie *Sinister St.* II. iv. ix. 1114 It's all very nice for you to be so calm. But what price its being my watch that's lost, not yours, old sport? 1920 D. H. Lawrence *Women in Love* i. 10 'What price the stockings?' said a voice at the back of Gudrun. 1930 R. Lehmann *Note in Music* vii. 301 But what price jaunts on Sundays—eh? 1959 M. Gilbert *Blood & Judgement* xvi. 164 Quick work... What price the law's delays. 1973 'B. Graeme' *Two & Two make Five* iv. 31 What price himself to replace Perkins, he asked himself with cynical amusement. 1977 *New Scientist* 12 May 336 (*heading*) What price Australian uranium?

V. 14. *price-boom, -boost, control* (so *price-controlled* adj.), *freeze, hike, -level, -maintenance* (examples) (so *price-maintain* vb. trans., *-maintained* ppl. adj.) *raiser, range, -regulation, review, rise, -wave*; *price-conscious, -sensitive* adjs.; **price-cutting** (further examples); hence (as a back-formation) *price-cut* vb. intr. and trans.; also *price cut* sb.; *price-cutter* (later examples); **price discrimination,** the action of charging different prices to different customers for the same goods or services; **price-earnings ratio** (see quot. 1965); **price elasticity** (see quot. 1971); hence *price-elastic* adj.; **price-fixing,** the action of introducing a fixed or standard price for something esp. by agreement between manufacturers; also *attrib.*; hence (as a back-formation) *price-fix* vb. trans., *price-fixed* ppl. adj.; **price-gouging,** the action of increasing prices by large amounts at once; **price index,** an index showing the variation in the prices of a set of goods, etc., since a chosen base period; also (with hyphen) *attrib.*; **price leader** orig. *U.S.,* a dominant firm that determines the prices within an industry; hence *price leadership*; also *price-leading* ppl. adj.; **price-list** (examples); **price movement,** a fluctuation in price; **price ring,** an association of traders formed to control certain prices; **price-slashing,** price-cutting by large amounts; so *price-slasher*; **price stop,** a ban on price increases; **price support,** assistance in maintaining price levels regardless of supply or demand; **price system** (see quot. 1968); **price tab** *U.S.,* a bill; **price-tag** (earlier and later examples); also *fig.*; hence as *v. trans.* and *price-tagging* vbl. sb.; **price ticket** = *price-tag*; **price war,** intense competition among traders by price-cutting.

1928 *Britain's Industr. Future* (Liberal Industr. Inquiry) IV. xx. 268 The rapid industrial slump which followed the price-boom of 1919–20. 1961 *Wall St. Jrnl.* 23 Jan. 2/2 A price-boost might well be delayed until midsummer. 1974 *News & Press* (Darlington, S. Carolina) 25 Apr. 8/6 Both business men and consumers are fearful that the lifting of economic controls will set off new waves of price and wage boosts. 1961 *Wall St. Jrnl.* 4 Oct. 1 Farmers aren't as price conscious as last year, so we can get more money on a sale. 1963 *Economist* 20 July 281/1 The price-conscious professional classes. 1974 *Country Life* 28 Nov. 1662/1 From the price-conscious north I have news of good stocking fillers. 1914 *Automobile Topics* 12 Dec. 321 (*caption*) Ford loses price control suit. 1936 *Discovery* Apr. 128/1 He waxes..mildly indignant over price-control of new metals by monopolies. 1944 *Sun* (Baltimore) 6 Oct. 13/5 Price control clinics, manned by officers and enlisted men, to hear reports from GIs on instances of overcharging, were ordered established. 1955 T. H. Pear *Eng. Social Differences* 184 The freeing of both tea and coffee from price-control. 1974 *Listener* 3 Oct. 422/3 The attempted price controls..have been far too severe. 1948 *Hansard Commons* 5 Mar. (Written Answers) 105 Mr. J. Morrison asked..What are the wholesalers' and retailers' margin of profit allowed on ..such household goods as are price controlled. 1976 *Sci. Amer.* Nov. 138/2 Of its two million citizens more than 12,000 were dead of the flu and its concomitant pneumonia by the middle of November, against a macabre background of military embalmers and a price-controlled quick-coffin industry. 1925 *Wireless Dealer* I. ii. 259/1 The retailer who is given a big discount must not price-cut to the public. 1928 *Publishers' Weekly* 30 June 2596 If turnover is secured by price cuts which decrease the normal profit [etc.]. 1957 *Chem. & Engin. News* 1 Apr. 28/2 In the chemical industry it is impossible to tell in advance whether a price cut may at some time in the future 'tend substantially to reduce competition'. 1964 *Financial Times* 31 Jan. 1/2 The..Adsega supermarket chain..has been price-cutting cigarettes. 1965 *Mod. Law Rev.* XXVIII. v. 554 The assumption of the publishers was that 'best sellers' would be price-cut. 1967 *Economist* 4 Mar. 845/1 The steady erosion of prices that followed the arrival of the price-cutters. 1969 D. C. Hague *Managerial Econ.* iv. 89 Even if several firms do follow the price cutter..price cutting may still be attractive. 1929 *Times* 2 Nov. 7/5 That could only be done with the abolition of the suicidal policy of price-cutting and competition. 1962 E. Godfrey *Retail Selling & Organization* i. 6 Price-cutting and the widespread introduction of supermarkets have made competition very difficult to meet. 1974 'G. Black' *Golden Cockatrice* ii. 32 I'll fight a price-cutting war by matched price-cutting. 1957 Clark & Gottfried *University Dict. Business & Finance* (1967) 276 When the purpose or result of such *price discrimination* is to reduce competition or to injure competitors, either of the seller or the buyer, it is illegal under the anti-trust laws. 1969 D. C. Hague *Managerial Econ.* iv. 83 We were considering a special case of price discrimination... We supposed that the producer was in the most favourable of all situations and could charge a different price to each individual consumer. 1974 *News & Courier* (Charleston, S. Carolina) 25 Apr. 17-C/2 White failed to show International was guilty of breach of contract and price discrimination. 1961 *Dallas Morning News* 9 Apr. iv. 1 Foods, which long had been considered 'recession resistant' but hardly dynamic stocks, have been acting like growth stocks, going to higher price-earnings ratios. 1965 *McGraw-Hill Dict. Mod. Econ.* 390 *Price-earnings ratio,* the current market price of a company's stock expressed as a multiple of the company's per-share earnings. It is computed by dividing the annual per-share earnings of a company into the market-value of its stock. For example, if company A's stock is selling at $100 per share and the company earned $5 per share, the price-earnings ratio is 20. 1968 *Newsweek* 25 Nov. 91/2 Other stocks may continue to show solid earnings growth but then they become overexploited. Investors simply bid too high for them. This shows up in the price-earnings ratio. 1972 *Observer* 22 Oct. 15/3 A couple of dark clouds... One is the sky-high price-earnings ratio of your stock. 1964 *Economist* 15 Feb. 620/2 The argument that books are price-elastic. 1967 *Times Rev. Industry* Mar. 16/1 This company clearly believes that shoe demand is price elastic. 1976 P. R. White *Planning for Public Transport* vi. 127 Weekend and day return fares..relate to trips such as shopping and visiting friends, demand for which is more price-elastic. 1952 T. W. Hutchinson tr. *Schneider's Pricing & Equilibrium* i. 23 We can measure the reaction of the quantity demanded to changes in price, when all other prices and income remain constant, by the price elasticity. 1971 J. A. Perrow *Econ.* i. 17 Price elasticity may be defined as the responsiveness of demand for a good to a small change in its price. 1976 P. R. White *Planning for Public Transport* viii. 157 The non-business market. Here, the time and price elasticities are almost the reverse. 1949 *Time* 25 July 24/2 In the past, prices had been held down by a combination of price fixing and subsidies. Bread was price fixed, so were cooking oils [etc.]. 1933 K. T. Langguth *Financial Dict.: Eng.–German* 186 Price-fixed. 1949 *Consumer Reports* Aug. 344/1 Places to buy price-fixed merchandise at less than the established price. 1971 'E. McBain' *Hail, Hail, Gang's all Here* ii. 183 That apartment's price-fixed... If he gets out, they can put a new tenant in and legally raise the rent. 1920 *Argus* (Melbourne) 4 June 6 Competition will reduce prices in time, but price-fixing..will only arrest the tendency to cheapness. 1930 *Economist* 15 Feb. 352/2 Rationalisation must also be distinguished from price-fixing associations or cartels. 1965 *Spectator* 26 Feb. 251/1 The price of whisky and gin was slashed following the end of price-fixing. 1973 *Country Life* 29 Nov. 1773/3 This 20-year-old price-fixing procedure [for Champagne] comes to an end in 1975. 1958 *Times Rev. Industry* Feb. 106/2 Those whose task it was to determine relaxations of the price-freeze. 1978 Ld. Hailsham *Dilemma of Democracy* xix. 122 Inflation has led to a demand either for a price freeze, or a wage freeze or both. 1967 *Guardian* 5 Aug. 7 Negro housewives..are the victims of price-gouging in the neighbourhood shops..owned by whites. 1974 *Aiken* (S. Carolina) *Standard* 22 Apr. 1-B/2 Each wage demand that is not balanced against productivity and each incident of price gouging motivated by greed help perpetuate the inflation cycle. 1977 *Times* 8 Aug. 42/3 This week the agency will open an investigation of alleged price gouging on fuel oil used in home heating; one consumer group is claiming the FEA permitted oil companies to overcharge by $2 billion last year. 1948, 1968 Price hike [see *hike *sb.* 2]. 1977 *Rolling Stone* 24 Mar. 16/2 Industry spokespersons tend to cite increased costs on every level when explaining the dollar price-hike. 1886 Price index [see *index *sb.* 9 e]. 1930 *Economist* 5 Apr. 763/1 The Economist price index has fallen during the past two years much more heavily than price indices in certain other countries on the gold standard. 1930 W. K. Hancock *Australia* ix. 184 The Statistician's price-index numbers. 1954 M. Beresford *Lost Villages* vi. 183 It cannot be said that our price-indexes are yet near perfection. 1954 E. H. Carr *Interregnum* 77 A price-index issued by the labour section of the Moscow Soviet for the calculation of wages in Moscow. 1973 Price index [see *index *sb.* 9 e]. 1936 A. R. Burns *Decline of Competition* iii. 77 The United States Steel Corporation is more frequently classified as a price leader than any other American corporation. 1962 *Economist* 13 Jan. 151/2 The International Nickel Company of Canada..is the acknowledged price leader. 1936 A. R. Burns *Decline of Competition* iii. 76 Price leadership exists when the price at which most of the units in an industry offer to sell is determined by adopting the price announced by one of their number. 1979 *Internat. Jrnl. Sociol. of Law* May 133 For price leadership to work, the price leader must have close to the largest share of the market. 1961 *New Left Rev.* July–Aug. 5/2 In Italy, each of the industrial sectors is dominated by a single, price-leading firm. 1927 Bowley & Stamp *Nat. Income 1924* 58 On account of the change in price-level, we should substitute a comparative level of £9,500, [etc.]. 1940 *Economist* 13 Jan. 53/1 Voluntary negotiation..can only result in a welter of independent wage decisions in different industries, each bearing a different relation to the general price level. 1972 *Accountant* 19 Oct. 485/2 Differences between financial statements prepared along the alternative bases of current-value and price-level accounting. 1872 *Young Englishwoman* Dec. 662/3 Will you be so kind as to send a price-list of the combs and hair-pins. 1915 W. Owen *Let.* 8 Jan. (1967) 313 What my friend advised me to do is to get price-lists and samples from England immediately. 1973 *Sat. Rev. Society* (U.S.) May 68/1 A list of catalogs or 'price lists', of items stocked by the Government Printing Office. 1960 *Guardian* 10 Dec. 9/6 The intending signatories agreed..that they would price maintain their vehicles. 1964 *Financial Times* 31 Jan. 1/2 More price-maintained lines would be added to their lists of reductions. 1968 *Times* 29 Nov. p. iv/4 Since they are still price-maintained it is not possible for retailers to cut prices, though in recent years budget labels have emerged. 1930 *Economist* 13 Sept. 483/1 The price-maintenance scheme is ultimately financed by the Reich. 1965 *Mod. Law Rev.* XXVIII. v. 552 Fifty years have now passed since the head of a well-known department store, in opposing enactment of a general price-maintenance law, told a Congressional Committee [etc.]. 1969 D. C. Hague *Managerial Econ.* xiv. 297 Is there a well-established (and legal) tradition of price maintenance? 1934 *Discovery* Sept. 245/2 They [*sc.* the farmers] have the advantage of being able to hear things that should be known at once to them: such as price movements, weather reports, harvest conditions and prospects. 1948 G. Crowther *Outl. Money* (rev. ed.) iii. 95 This is one way in which price movements have a direct causal effect on the level of production and employment. 1965 J. Meuvret in Glass & Eversley *Population in Hist.* xxi. 517 Here again, however, price-movements can afford some illumination. 1906 'Mark Twain' *Autobiogr.* (1924) II. 24 That congregation's real estate stands at a low figure. What they are anxious to have now..is a price-raiser. 1965 *Punch* 7 July 2/1 George Brown's challenge to price-raisers to justify themselves. 1925 *Ladies' Home Jrnl.* May 146/2 Price ranges from 25c to 45c. 1937 M. Hillis *Orchids on your Budget* (1938) iii. 46 Another good rule is not to attempt to have everything come within the same price range. 1973 D. Westheimer *Going Public* i. 15 The fact that so much of your business is in the lowest price-range has its positive side... You'll find your bread-and-butter business is in the lower price-range. 1919 J. M. Keynes *Econ. Consequences Peace* vi. 225 The effect on foreign trade of price-regulation and profiteer-hunting as cures for inflation is even worse. 1935 *Economist* 12 Oct. 704/1 This shortage..is largely a consequence of planning and price-regulation. 1959 *Chambers's Encycl.* XI. 195/2 Price-regulation was in existence in Babylonia as early as the middle of the third millennium B.C. 1960 *Farmer & Stockbreeder* 22 Mar. 78/3 On Wednesday, representatives of the Branch met seven Conservative M.P.s at the House of Commons..when there was a long and useful discussion about many aspects of the White Paper and the Price Review. 1969 *Times* 6 Jan. 7/7 In spite of occasional controversy, especially before annual price reviews, the overall impression one gets of the past 10 years is one of fair stability. 1928 *Britain's Industr. Future* (Liberal Industr. Inquiry) II. viii. 97 The majority of cartels and price rings fall under the category of Trade Associations.. and not under that of Public Companies or Corporations. 1957 *Observer* 1 Dec. 1/4 This is no moment..for a price ring designed to keep prices up, or restrictive principles to prevent them going down. 1965 M. Hilton tr. J. Meuvret in Glass & Eversley *Pop. in Hist.* xxi. 511 The price-rise..can be explained..by bad harvests. 1977 *Times* 4 Oct. 15/2 Price rises are still at an unacceptable level. 1966 *Economist* 3 Dec. 1046/1 In the price-sensitive group of semi-manufactures, the bigger impact was on imports of textiles, paper and, above all, iron and steel. 1976 *Scotsman* 24 Dec. 3/7 Dunford and Prudential were justified in passing on information of a 'price-sensitive' character to their institutional shareholders, as potential underwriters. 1964 *Punch* 11 Mar. 377/3 John Bloom.. is a notorious price-slasher. 1930 *Publisher's Circular* 14 June 793/3 The economic and cultural consequences of reckless price-slashing. 1940 *Economist* 28 Dec. 799/1 The 'price-stop order', designed to prevent war-time increases in prices, has been reinforced several times by stricter penalties; but it has not been possible to prevent some rise in prices. 1950 *Ann. Reg. 1949* 244 A price stop was placed on certain essential commodities. 1949 *Sun* (Baltimore) 10 Sept. 11/4 Corn from this year's crop is expected to move into Government hands under price-support programs to join more than 400,000,000 bushels remaining there from the 1948 crop. 1957 M. Swan *Brit. Guiana* 95 Price support has come in the form of the Commonwealth Sugar Agreement. 1965 J. L. Hanson *Dict. Econ.* 326/1 *Price Support.* The U.S. Government's method of giving assistance to farmers. Prices are fixed well above the equilibrium level and so output cannot be completely disposed of on the market, the U.S. Government agreeing to purchase at the fixed prices any surpluses resulting from this policy. 1962 M. McLuhan *Gutenberg Galaxy* 118 Complex markets, price-systems, and commercial empires. 1968 P. A. S. Taylor *Dict. Econ. Terms* (ed. 4) 85 *Price system.* This is an economic system in which prices are determined by the forces of the market. 1974 *Encycl. Brit. Macropædia* XIV. 1004/2 A price system weighs the desires of consumers in terms of the prices they are willing to pay for various quantities of each commodity or service. 1949 *Sun* (Baltimore) 12 Sept. 1/8 A big victory for labor in the board's belief that companies should pick up the price tab on pensions. 1881 *Harper's Mag.* Sept. 587/1 Untying a little green price tag from the handle of the umbrella. 1942 D. Powell *Time to be Born* (1943) iv. 97 Vicky was uncomfortably aware of Miss Finkelstein's eagle eye putting price tags on her suit, her hair, her shoes. 1951

Sport 30 Mar. 7/2 When I remember what a record transfer-fee price-tag did to Bryn Jones.., I can only sympathise with you. 1961 L. van der Post *Heart of Hunter* II. viii. 123 Though no price-tag could be put on them [*sc.* protected animals], we knew our lives would be immeasurably poorer without them. 1971 C. Fick *Danziger Transcript* (1973) 20 Your uniform smells as though the price tags are still on it. 1972 *Countryman* Winter 61 These faceless experts make an attempt to price-tag the social benefit of forestry. 1974 W. Foley *Child in Forest* II. 159 She price-tagged them by instinct. 1977 *Offshore Engineer* July 14/2 The NEB says that 'the project is likely to incur a 20–30% cost overrun' on the $8,000 million price tag it currently sees as realistic. 1972 *Straits Times* (Malaysian ed.) 23 Nov. 6/5 Encik Khir chaired the meeting which was held to resolve the problems facing shopkeepers over price-tagging. 1977 *Daily Times* (Lagos) 11 Jan. 17/2 Given the low level of enlightenment in the country, the intransigence of Nigerians to any government directive, price tagging was born with a lot of problems which have so far retarded its success. 1934 *Archit. Rev.* LXXVI. 27/2 A well-lettered price-ticket, decorative value apart, is more desirable than a label covered with hieroglyphics. 1957 P. Worsley *Trumpet shall Sound* viii. 159 Natives..tore the price-tickets off the goods. 1930 *Economist* 22 Mar. 652/2 Experience shows that this group invariably emerges from a price-war with a stronger hold on the oil markets than before. 1969 D. C. Hague *Managerial Econ.* iv. 74 A market which is free from the dangers of occasional, or continual, price cutting or even major price wars. 1977 *Times* 6 Aug. 3/1 A tea price war began yesterday as packers ordered cuts after auction prices fell. 1891 G. Clare *Money-Market Primer* 89 At all times some semblance of agreement is traceable between the respective price-waves.

price, *v.* Add: **1. d.** *to price out of the market*: to eliminate (oneself or another) from commercial competition through prohibitive prices; to charge a prohibitive price for (goods or services) or to (the customer). Also simply *to price out*: to charge a prohibitive price to.

1938 *Sun* (Baltimore) 3 Jan. 8/3 Building material dealers and manufacturers, and to a less extent building labor, not only priced themselves out of the market but also priced the country out of an anticipated increase of $2,000,000,000 of national income. 1946 *Ibid.* 10 Aug. 4/1 Our price policies in the past..have had a tendency to price our export commodities out of the world market. 1946 *Your Investments* Sept. 9 Many consumers were being priced right out of the market..by the accelerated rise in living costs. 1947 *Daily Progress* (Charlottesville, Va.) 4 June 1/3 Government support prices for peanuts are so high that 'it forces peanut butter up so far as we are being priced out of the market'. 1949 *Sun* (Baltimore) 28 Jan. 10/1 Earlier support plans have simply priced cotton out of the world market. 1955 *Times* 15 June 3/1 The country should realize that we could be easily priced out of international markets. 1958 *Spectator* 14 Feb. 201/1 As for the story that we should have been 'priced out of our export markets', time has shown that this does not happen so easily as pessimists predict. 1971 *Guardian* 6 Sept. 9/8 Swiss exports may be pricing themselves out of world markets. 1975 *Times* 4 Sept. 2/1 In an effort to price out [football] hooligans..most Saturday concessionary fares are being ended. 1977 *Guernsey Weekly Press* 21 July 1/6 His members were very concerned about the risk of being priced out of the market.

e. *to price up*: to increase the price of.

1943 *Our Towns* (Women's Group on Public Welfare) ii. 58 The shop then prices up the goods in order to cover.. the commission. 1976 N. Roberts *Face of France* xxv. 227 The [champagne] trade started pricing up its wares to restrain demand.

price-current. Add: Also *pl.* **prices-current** (occas. used for the sing.).

1815 *Niles' Reg.* IX. 3/2 This account of the *selling prices* of the several stocks mentioned, is taken from the *public prices current* of the two places. 1839 *Southern Lit. Messenger* V. 38/2 There are no daily papers..no prices current—no reports from the stock market. 1856 *Trans. Mich. Agric. Soc.* VII. 533 A glance at our 'prices current' might suffice to satisfy the most incredulous. 1965 J. L. Hanson *Dict. Econ.* 326/1 *Prices Current,* a price list showing the prices ruling at a certain date.

priceless, *a.* Add: **3.** *colloq.* Amusing, absurd, ludicrous; delightful.

1907 *Punch* 23 Jan. 59 Lady Bountiful: Oh, dear Miss Smith, *do* send me some of your priceless little sketches for my rummage sale on the 26th. 1914 D. O. Barnett *Let.* 19 Nov. (1915) 11 There was a priceless 'drunk' here the other day when I was on guard... He made the most magnificent remarks en route and so did the chaps who were carrying him. 1921 G. B. Shaw *Back to Methuselah* II. 87 What a priceless humbug old Lubin is! 1924 D. Moore *Fen's First Term* xii. 127 She had been a 'priceless idiot'. 1925 'R. Crompton' *Still—William* xi. 201 'I do hope I remember all this when I wake up,' said the Toreador, 'it's too priceless.' 1978 S. Naipaul *North of South* I. i. 29 The European..burst out laughing... 'Can you imagine how they must have..rolled their eyes? Absolutely priceless.'

pricelessness (earlier example); **pri·celessly** *adv.*

1879 Trollope *Eye for Eye* II. i. 13 There came a day in which the pricelessness of the girl he loved sank to nothing. 1910 G. Murray tr. *Euripides' Iphigenia in Tauris* 62 Brother, and home, and sister pricelessly Beloved. 1934 G. B. Shaw *On Rocks* I. 222 You see, what makes your diagnosis so pricelessly funny to me is that as a matter of fact my life has been a completely intellectual life, and my training the finest intellectual training in the world. 1977 J. B. Hilton *Dead-Nettle* vi. 61 Frank, you are pricelessly sweet.

pricey (prəi·si), *a. colloq.* Also **pricy.** [f. Price *sb.* + -y[1].] Expensive, high-priced. Also *comb.*

1932 'C. L. Anthony' *Service* III. ii. 101 I've got the day off to-day—been up to a sale to see about a show-case. But I couldn't touch it. It was a very pricey article. 1944 *World's Press News* 31 Aug. p. iii/1 (*heading*) 'Pricy' at second hand... The advertiser offered 4s. 6d. for each issue of *Vogue* ten days after publication. The price of *Vogue* is 3s. 1953 D. Whipple *Someone at Distance* xxvii. 243 'Pricey, I know,' continued Mr. Pye. 'But worth it, Madam.' 1957 *Economist* 19 Oct. (Suppl.) 10/1 The 'pricier' models like the Ford Zodiac or the Vauxhall Cresta. 1962 M. Procter *Devil in Moonlight* xi. 114 It's pricey... It might cost you a lot of money. 1971 *Daily Tel.* 19 Aug. 2/5 Meat has become a very pricey business for most households. 1976 W. H. Canaway *Willow-Pattern War* xvii. 174 A pricey-looking transistor radio. 1978 *SLR Camera* Aug. 88/1 It can..be fitted with a motor drive unit, but not with the wide variety of viewing heads and viewing screens available for the more pricy sisters in the catalogue.

prick, *sb.* Add: **14.** Also, an early kind of knitting-needle. Cf. *knitting-prick* (Knitting *vbl. sb.* 3).

[1707 *Rec. Convention R. Burghs Scotl.* (1880) I. 431 For the better improvement of stocking manufactures it is thought fitt that for hereafter all prick stockings may be made of three plyed wosten and of due proportione.] 1838 W. Howitt *Rural Life Eng.* I. III. iii. 309 They knit with crooked pins called pricks. 1969 E. H. Pinto *Treen* 304 Bow curved needles, made from wire sharpened both ends..and known as pricks, were commonly used for 'bump' or coarse knitting.

17. a. (Further examples.) For (Now *low.*) read *coarse slang.*

1680 Rochester *Poems* 14 But though St. James has the honor on't, 'Tis consecrate to Prick and Cunt. 1744, 1763 [see *bollock 1]. 1896 A. Beardsley *Let.* c3 Dec. (1970) 223 Yes *everything* is phallic shaped except Symons's prick. 1922 Joyce *Ulysses* 424 Trinity medicals... All prick and no pence. 1965 W. Young *Eros Denied* xiv. 132 You know, the young men's pricks seem to be getting bigger and bigger. It must be the Welfare State. 1971 'A. Burgess' *MF* iii. 39 His nakedness and limp prick..were now properties of the changing room. 1976 'E. McBain' *Guns* (1977) ii. 38 Jocko had..a very small pecker... Blood on the bulging pectorals, tiny contradictory prick.

c. As a vulgar term of abuse for a man.

1929 *Amer. Speech* IV. 343 *Prick,* one in authority who is abusive or unjust. 1934 H. Miller *Tropic of Cancer* 110 Jesus, what I'd like is to find some rich cunt—like that cute little prick, Carl. 1935 J. T. Farrell *Guillotine Party* 193 That's what I think of you, Merton..you're a p...k! 1937 Partridge *Dict. Slang* 659/1 *Prick*... An offensive or contemptuous term (applied to men only). 1961 P. Kemp *Alms for Oblivion* ii. 40 Winn drafted a bitter reply, concluding with the *cri-de-coeur*: 'Uncomplaining gravest difficulties here but how long oh how long must we continue to kick against the pricks in your office.' 1967 'E. Trevor' *Freebooters* xi. 124 We don't like bein' pushed around by an incompetent prick of a commanding officer. 1971 B. W. Aldiss *Soldier Erect* 52 Don't you call *me* a cunt, you Midland prick, you, or I'll sort you out! 1973 J. Wainwright *Devil you Don't* 25 John Smith said: 'Some men have big pricks.' 'Some men *are* big pricks.' 1978 M. Puzo *Fools Die* xi. 115 They have good jobs, big futures. And the pricks won't even do their service.

18. Delete ?*Obs.* and add later examples.

1975 B. Meyrick *Behind Light* iv. 57 Normally Pa had thirty or so good leaves left to make rolled 'pricks' as a change from pressing into wads. *Ibid.,* Soon our back pantry was full of thick hanging 'pricks' of twisted and rolled tobacco. 1977 *Navy News* Feb. 6/6 The hair on the nape of the neck was bound in yarns..and called a perique. In my days we rolled leaf tobacco in a similar way and called the result a 'prick', just modern spelling of an old word.

21. prick-farrier *Services' slang,* a medical officer; **prick punch** (later example); **prick-shooting,** shooting at the 'prick' or target; **prick-stitch** (see quots.); so **prick-stitch** *v. trans.* and *intr.,* **prick-stitching** *vbl. sb.*; **prick-sucker** *coarse slang,* a fellator or fellatrix; **prick-teaser** *coarse slang* = *cock-teaser* (*cock *sb.*[1] 23); also **prick-tease** *sb.*; hence (as back-formations) **prick-teased, prick-teasing** *ppl. adjs.*

1961 Partridge *Dict. Slang* Suppl. 1232/1 *Prick(-)farrier,* a medical officer: R.A.F. regulars': since ca. 1928. 1971 S. Kerry *Doctor's Cabin* iv. 48 'Meet Doc Kerry, our prick farrier.' They both laughed. 'No offence meant,' said Johnny. 'It's just a vulgar Naval term for a surgeon.' 1921 *Daily Colonist* (Victoria, B.C.) 2 Oct. 26/5 A chalk mark is good, but a prick punch makes a mark that will not rub out. 1801 T. Roberts *Eng. Bowman* 241 Of Prick-shooting..the marks used in this kind of shooting have..consisted either of a small circular piece of white paper, fixed to a post..; or of a target. 1887 W. Butt *Ford's Archery* (rev. ed.) 138 This prick-shooting next became known as the paper-game. 1924 W. D. F. Vincent et al. *Cutters' Pract. Guide Body Coats* 62/1 Hunting Coats are generally finished with plain seams, and have the front edges seamed and prick-stitched. 1928 A. S. Bridgland *Mod. Tailor* II. xviii. 242 *Prick-stitch.*—This stitch is employed to give either strength or appearance, and consists in alternately passing the needle straight up and down through the material, the stitch itself being either a back or a side-stitch. 1933 J. E. Liberty *Pract. Tailoring* iii. 18 *Prick-stitch.* This is exactly similar to side-stitch but is made by two actions, one upward, the other downward, the stitch actually being pricked alternately through the material which would be too thick for side-stitching. *Ibid.* v. 54 Prick back over the felling and along the seam for not quite ¼ in., then prickstitch parallel to the felling up to the top of the welt and to the same width. *Ibid.* vii. 100 To obtain the desired result, it will need to be prick stitched. 1955 —— *Ibid.* (ed. 2) v. 56 The..bar tack, with a little addition of a short row of prickstitching midway between the flap seam and the top edge of the jetting. 1964 *McCall's Sewing* xiii. 239/2 *Prick stitch.* Take short half-back-stitches in which only two or three threads of the fabric are picked up. Pull each stitch..taut. 1968 J. Ironside *Fashion Alphabet* 83 *Glove-stitching,* known as 'prick' stitch. Stitch used for hand glove-making and sometimes for sewing very heavy materials and leather. The thread is taken through one layer of material and then the other. 1868 *Index Expurgatorius of Martial* 21 Cotilus, the prick-sucker,..is shown to be the effeminate of men. 1974 *New Direction* IV. iv. 5/4 From then onward she became an ardent prick-sucker. 1977 E. J. Trimmer et al. *Visual Dict. Sex* (1978) ii. 31 A girl who works her way through several partners without..ever having intercourse will become known as a '*prick-tease*'. 1975 D. Durrant *With my Little Eye* viii. 'Prick-teased boys had up for rape. 1961 Partridge *Dict. Slang* Suppl. 1232/1 *Prick-teaser,* a late C. 19–20 variant of *cock-teaser.* 1971 R. Busby *Deadlock* i. 8 He laughed and pulled her roughly across the seat. 'A prick teaser, are we?' 1978 F. Norman *Dead Butler Caper* v. 32 That Gloria's a right prick teaser. She'll con 'im something rotten. 1967 'P. Loraine' *W.I.L. One to Curtis* i. 16, I supplied..an empty house..for whatever prick-teasing kind of a party they wanted to throw. 1972 J. Mann *Mrs. Knox's Profession* iv. 24 He shouted after her, 'Prick-teasing bitch.'

prick, *v.* Add: **1. a.** Also, to wound or disable (a game bird) by shooting. Cf. *pricked *ppl. a.* 1 c.

1900 'Blagdon' *Shooting* 89 Too often, when cover is deficient and birds are consequently difficult to approach, there is a tendency to take long shots at birds which are really beyond sporting range, with the result that a large number of birds are 'pricked', or slightly wounded, without being brought to bag. 1916 *Shooting Don'ts* 13 Don't 'brown' into a covey. To be continually killing more than one bird at a shot will make you suspected. It results in waste of birds, on account of the number that get 'pricked', and die.

22. (Further examples.)

1789 *Ann. Agric.* XI. 51 My first parcel [of seeds] was pricked in upon a small garden bed. 1913 J. Weathers *Twentieth-Cent. Gardening* vii. 67 Annuals sown under glass are first of all 'pricked out' into other pots or boxes when large enough to handle. 1935 A. G. L. Hellyer *Pract. Gardening* v. 46 Seeds..should not be very close together unless it is certain that time will be available to 'prick' them out. 1952 C. E. L. Phillips *Small Garden* vi. 55 When the youngsters have developed their first pair of true leaves, prick them off into other boxes or pots. 1978 'E. Peters' *Morbid Taste for Bones* i. 8 He was only to help Brother Cadfael prick out early lettuces.

28. b. *prick out,* to come into view as specks or points.

1930 R. Macaulay *Staying with Relations* xx. 305 By two o'clock a few stars had pricked out, tiny candles shaking between the drifting gloom of clouds.

prick-eared, *a.* Add: **1.** (Later examples.) Also, of corn or wheat.

1922 Blunden *Shepherd* 81 From the young corn the prick-eared leverets stare At strangers come to their land. 1940 C. Day Lewis tr. *Virgil's Georgics* i. 25 The dangers of showery spring, When the prick-eared crop already bristles along the plains. 1946 L. B. Lyon *Rough Walk Home* 11 Prick-eared, he lurks To leeward, patiently bold.

pricked, *ppl. a.* Add: **I. 1. c.** Of a game bird or part of a bird: wounded or disabled by shooting. Also *transf.*

1937 *Discovery* Dec. 385/1 Wounded or pricked birds left about the place are numerous. 1940 N. M. Sedgwick *Young Shot* v. 45 The guns should carefully mark where game falls,.. where..a bird with a pricked wing comes down. 1952 *Chambers's Jrnl.* Apr. 212/1 Thereafter, for days, a salmon was seen constantly to move in the same place, and gradually I became certain that it was one and the same fish that had twice been lost. Never did any of us visit the Island but there it was, rolling every few minutes as so often a pricked fish will. 1958 R. Waddington *Grouse* ix. 105 Very high grouse mean more pricked birds and the moor in general is the sufferer.

3. (Later examples.)

1820 M. Edgeworth *Let.* 20 Dec. (1971) 226, I think the pricked map upon the whole better out and I have sent the proof sheets and left it out. 1880 in L. Higgin *Handbk. Embroidery* 107 Designs..on pricked paper. 1900 F. Jackson *Hist. Hand-Made Lace* 216 *Pricked,* the term used in pillow lace-making to denote the special pricking out of the pattern upon parchment. 1927 'R. Crompton' *William—in Trouble* v. 119 Half a dozen Italian stamps.. turned out..to be 'pricked' and useless for collections. 1933 *Burlington Mag.* Jan. 43/1 There are pricked designs for figures which can be identified on vestments still existing. 1967 *Daily Tel.* 1 Feb. 13/2 Making lace borders, using a pricked parchment pattern, and placing pins in the holes, which control the completed stitches.

II. 8. *pricked-out,* of seedlings, planted out in a bed after being moved from the trays or boxes in which the seeds were germinated.

1938 G. Greene *Brighton Rock* iv. 154 The small pricked-out plants irritated him like ignorance. 1975 *Country Life* 13 Feb. 388/3 Space the pricked-out seedlings at seven by four.

pricker. Add: **4. a.** Phr. *to get* (or *have*) *the pricker*: to become (or be) angry. *Austral.* and *N.Z. slang.*

1945 Baker *Austral. Lang.* vi. 121 A man in a temper is said..*to have the dingbats, the pricker or the stirks.* 1955 D. Niland *Shiralee* 102 You've got the pricker properly, eh? You'll knock him into next week, will ya? 1959 G. Slatter *Gun in my Hand* viii. 91 You'll come a gutzer son I ses... Got the pricker with me.

pricking, *vbl. sb.* Add: **1. b.** Phr. *pricking of* (or *in*) *one's thumbs,* used in various constructions with allusion to quot. 1605 in Dict.: an intuitive feeling or hunch; a premonition, a foreboding.

1935 'G. Orwell' *Clergyman's Daughter* i. 43 'I had a feeling I was going to meet you to-day.' 'By the pricking of your thumbs, I presume?' 1946 D. C. Peattie *Road of Naturalist* v. 52, I stood then on the back platform of the flying *Overland* with the knowledge that I had got into new terrain, not easily to be mastered, pricking in my thumbs. 1966 E. Palmer *Plains of Camdeboo* vi. 92 There should have been a pricking in our thumbs the morning we went to see the fossils, but we had no warning at all.

6. b. In *Palæography,* the piercing of a series of holes on a leaf to assist with the ruling of lines; a set of such holes.

1908 E. Johnston *Writing & Illuminating* (ed. 2) vi. 110 The writing line dots are pricked through all the sheets by means of a fine awl or needle... See also *methods of ruling without pricking.* 1912 E. M. Thompson *Introd. Gk. & Lat. Palaeogr.* 55 In earlier MSS. these prickings are often found near the middle of the leaf. 1971 T. A. M. Bishop *English Caroline Minuscule* p. xii, In English MSS. written after c. 900 the prickings for horizontal ruling are found only in outer lateral margins. 1973 *Bodl. Libr. Rec.* IX. 12 Prickings with an awl had been made along outer bounding-lines of four bifolia simultaneously with hair-sides up. 1976 *Codicologica* I. 78 Other aspects of the medieval book: the nature of parchment, ink, pricking, and ruling.

8. (Later examples.)

1935 A. G. L. Hellyer *Pract. Gardening* v. 47 Pricking out should be done as soon as it becomes necessary. 1952 C. E. L. Phillips *Small Garden* vi. 53 Pricking out simply means lifting the seedlings from this nursery bed..and replanting them more widely somewhere else. 1976 *Abingdon Herald* 9 Dec. 5/2 A heating cable will enable you to carry on plants after pricking out.

10. pricking-iron *Saddlery* (see quot. 1960); **pricking-up** *Basketry* (see quot. 1912).

1904 P. N. Hasluck *Saddlery & Harness-Making* (1962) II. i. 18 Fig. 32 shows a tool used in stamping the lines preparatory to stitching. These tools vary in width from three teeth, which are used only for round points and scalloped work, to twenty-four teeth for straight lines... (*caption*) Fig. 32.—Pricking-iron. 1946 N. Wymer *Eng. Country Crafts* v. 47 From the start he was taught to keep his needle-holes as equidistant as possible, the length he must make each stitch being marked off for him on his leather by means of a pricking-iron. 1960 G. E. Evans *Horse in Furrow* xvi. 206 On the cart-trace *back* decorative sewing..was the rule. A *pricking iron*—a chisel-shaped implement with points or teeth at regular intervals on the blade—was first used to mark out the pattern and to ensure that the stitches were uniformly placed. 1975 J. H. L. Shields *To handmake a Saddle* x. 42 The end to be stitched is then pricked with a No. 4 pricking iron. 1912 T. Okey *Introd. Art of Basket-Making* 153 *Pricking up,* turning up the stakes after their insertion in the bottom with the point of the shop-knife to form the framework of the sides of a basket. 1959 D. Wright *Baskets & Basketry* vi. 136 *Pricking-up*: turning up of willow stakes over the point of the knife after they have been inserted into the base. 1960 E. Legg *Country Baskets* 93 The turning up of the stakes is done rather differently than with cane, the *pricking up* method being adopted.

prickle, *sb.*[1] Add: **6. b.** (Earlier example.)

1875 *Encycl. Brit.* I. 897/1 The cells..next in order are polygonal, and not unfrequently possess pointed processes or prickles projecting from them, hence the name, *prickle cells,* employed by Schultze.

9. prickle-cell (earlier and later examples).

1875 Prickle-cell [see *6 b]. 1962 Blake & Trott *Periodontology* ii. 17 This epithelium consists of a few layers of prickle cells. 1974 R. M. Kirk et al. *Surgery* v. 72 Histologically the tumour is of the prickle-cell layer, invading the deeper tissues, and later spreading to the regional lymph glands.

prickly, *a.* Add: **1. b.** (Earlier and later examples.) Also, of persons: quick to react angrily, touchy.

c 1862 E. Dickinson *Poems* (1955) II. 490 His pretty estimates Of Prickly Things. 1935 N. Mitchison *We have been Warned* iv. 340 Would you mind particularly if the C.P. were involved..? They're a nasty, prickly lot. 1943 A. Ransome *Picts & Martyrs* i. 11 It's Mother she's getting at, not us... She's prickly with disapproval. 1950 *Listener* 9 Nov. 482/1 Hence the prickly suspicions of the new China's relations with the Western Powers. 1957 R. Watson-Watt *Three Steps to Victory* cxxiv. 453, I was probably being needlessly prickly. 1975 *N.Y. Times* 2 Apr. 37/2 They were self-conscious gentry, prickly of their privileges and independence. 1980 T. Morgan *Somerset Maugham* iii. 168 Janet Vale of the *Morning Telegraph* found him prickly.

3. prickly ash, substitute for def.: one of several North American shrubs or trees whose aromatic bark is used medicinally, including those belonging to the genus *Zanthoxylum,* esp. *Z. americanum,* of the family Rutaceæ, and the angelica tree, *Aralia spinosa,* of the family Araliaceæ; (earlier and later examples); **prickly Moses** *Austral.,* one of several species of *Acacia* bearing prickles, esp. *A. verticillata, A. juniperina,* or *A. pulchella*; **prickly poppy,** substitute for def. s.v. Poppy *sb.* 3: an annual or perennial herb belonging to the genus *Argemone* of the family Papaveraceæ, native to North or Central America, esp. *Argemone mexicana,* a widespread weed of tropical and subtropical regions; **prickly rhubarb** = *Gunnera.

1709 J. Lawson *New Voy. Carolina* 101 Prickly-Ash grows up like a Pole. 1743 J. F. Gronovius *Flora Virginica* II. 150 *Aralia*... Gambriar and Prickly-ash. 1778 J. Carver *Trav. N.-Amer.* 393 The chief...prepared for him a decoction of the bark of the roots of the prickly Ash. 1860 M. A. Curtis *Bot.* 91 Prickly ash. (*Aralia spinosa*)... The berries..are thought by some to be also a valuable remedy for the bite of a rattlesnake. 1899 M. Going *Field, Forest, & Wayside Flowers* 74 Prickly-ash, and hackberry..are thus unsystematic in their mode of conducting their affairs. 1931 M. Grieve *Mod. Herbal* I. 70/1 The Prickly Ash..is a small North American tree. *Ibid.* 71/2 The name Prickly Ash has also been given to *Aralia spinosa*.., the Prickly Elder, or Angelica Tree. 1938 M. K. Rawlings *Yearling* xxix. 379 She..made him a tonic of prickly ash and poke-root and potassium. 1975 M. C. Davis *Near Woods* ii. 27 A clumped prickly ash had found a home among sandstone boulders. 1887 *Australian* Apr. 9/3 An expedition was now made into the scrub for fishing rods... I cannot recommend 'snap-scrub' for a rod, nor that awful thing which our philosopher called 'prickly moses'. 1965 *Austral. Encycl.* VII. 276/1 Prickly Moses, a corruption of 'prickly mimosa' applied to several species of wattle. 1724 P. Miller *Gardeners & Florists Dict.* I. s.v. *Argemone,* Argemone is a sort of Poppy, and some call it the prickly Poppy. 1760 [see Poppy *sb.* 3]. 1869 *Amer. Naturalist* III. 163 The Prickly Poppy (*Argemone*) looks now like a common thistle. 1898 A. M. Davidson *Calif. Plants* 112 The prickly poppy will send out great white flowers with crumpled petals and a great many yellow stamens. 1977 Lewis & Elvin-Lewis *Med. Bot.* ii. 31/2 Prickly poppies are widely distributed in weedy habitats in temperate and tropic regions. 1895 W. Robinson *Eng. Flower Garden* (ed. 4) 501/2 Gunnera (Prickly Rhubarb).—South American plants remarkable for large and handsome foliage, somewhat resembling that of gigantic Rhubarb. 1900 *Century Bk. Gardening* 98/2 Gunneras are called 'Prickly Rhubarbs', and the big leaves are not unlike those of a large Rhubarb. 1952 A. G. L. Hellyer *Sanders' Encycl. Gardening* (ed. 22) 217 Gunnera (Prickly Rhubarb)... Hardy herbaceous perennials. First introduced mid-nineteenth century.

prickly pear. (Earlier and further examples.)

1612 W. Strachey *Trav. Virginia* (1849) I. x. 119 Here is a cherry-redd fruict both within and without..which wee call the prickle peare;..they beare a broad, thick, spungeous leafe, full of kernells. 1672 W. Hughes *Amer. Physitian* 38 Most call it the Prickle-Pear Bush, and the fruit the Prickle-Pear. 1725 H. Sloane *Voy. Jamaica* II. p. vi, Tab. VIII Shews..the sort of Prickly Pear, thought in Jamaica to be that particular kind of Opuntia, whereon feeds the small Worm or Beatle, from whence comes in Cochineel. 1739 P. Delegal in *Georgia Hist. Soc. Coll.* (1840) I. 188 The islands in Georgia are full of the prickly pear shrubs which feed flies. 1836 J. Hildreth *Dragoon Campaigns Rocky Mts.* xvi. 141 It was covered with the prickle-pear. 1917 *Nature* 20 Sept. 57/2 The prickly pear cactus (*Opuntia tuna*) has become extensively naturalised [in Hawaii]. 1925 T. S. Eliot *Poems 1909–1925* 98 Here we go round the prickly pear Prickly pear prickly pear... At five o'clock in the morning. 1956 C. Mackenzie *Thin Ice* iii. 36 We left Tangier about an hour before sunrise, riding through plantations of prickly pear. 1974 V. Nabokov *Look at Harlequins* (1975) I. iii. 12 We walked round the house, skirting prickly-pear shrubs. 1978 G. D. Rowley in V. H. Heywood *Flowering Plants of World* 65/1 Opuntias (prickly pears) are grown commercially in parts of Mexico and California for their large juicy fruits.

prick-madam. (Later examples.)

1955 G. Grigson *Englishman's Flora* 182 Prick-madam was the name used [for the yellow stonecrop] in the sixteenth and seventeenth centuries. 1978 *Verbatim* May 2/2, I have never grown *stonecrop*; now that I know it as prickmadam I am tempted to try.

pricy, var. *Pricey *a.*

pride *sb.*[1] Add. **I. 3. b.** Phr. *pride and prejudice*; occas. *prejudice and pride.* Cf. Prejudice *sb.* 3.

1610 J. Hall *Sixt Decade of Epistles* v. 42 Lay downe first, all pride and prejudice, and I cannot fear you. 1647 J. Taylor *Liberty of Prophesying* xii. 185 Epiphanius makes pride to be the onely cause of heresies..Pride and Prejudice cause them all, the one criminally, the other innocently. 1650 —— *Holy Living* iv. 323 There is in it [*sc.* anger] envy and sorrow, fear and scorn, pride and prejudice, [etc.]. 1758 *Idler* 13 May 41/2 The prejudices and pride of man. 1758 C. Lennox *Henrietta* II. 48 The triumph of virtue over pride and prejudice. 1769 H. Brooke *Fool of Quality* IV. 292 Reason, and the workings of nature had begun to get the better of pride, and prejudice, in the peer. 1782 F. Burney *Cecilia* V. x. 379 The whole of this unfortunate business..has been the result of Pride and Prejudice. 1782 Cowper *Hope* in *Poems* I. 170 Now truth perform thine office, waft aside The curtain drawn by prejudice and pride. 1796 R. Bage *Hermsprong* I. xxxi. 204 But the tender interest they had in each other was torn asunder by pride and prejudice; and this pride and this prejudice, she feared, had been infused into the tender mind of Miss Campinet. 1813 Jane Austen (*title*) Pride and prejudice.

5. b. pride-of-California, a perennial wild pea with pink or violet flowers, *Lathyrus splendens,* native to California; **pride of China, pride of India,** add: = Margosa, Neem (in Dict. and Suppl.); (earlier and later examples).

1785 G. Washington *Diary* 13 June (1925) II. 383 Next 3 rows of the Seed of the Pride of China. 1803 J. Davis *Trav. U.S.A.* 79 The mocking-bird..was warbling, close to my window, from a tree called by some the Pride of India. 1834 J. J. Audubon *Ornith. Biogr.* II. 191 They ..feed voraciously on..the berries of the pride of India. 1835 J. H. Ingraham *South-West* II. 101 The 'pride of China',—the universal shade-tree in the south-west. 1893 *Harper's Mag.* Apr. 756/2 This causeway broadened into a sandy street under huge pride-of-India trees, whose branches met overhead. 1895 'F. Franceschi' *Santa Barbara Exotic Flora* 64 *Lathyrus Splendens,* appropriately called 'the pride of California',..has made its appearance in our gardens quite lately. 1949 *Bull. Hist. & Philos. Soc. Ohio* VII. 71 A tall conical envelope of straw..protected the Pride of China, a tree brought from New Orleans. 1970 W. Smith *Gold Mine* xxvii. 63 The moonlight came in through the window, playing shadow pictures through the branches of the Pride of India tree onto the wall. 1976 *Hortus Third* (L. H. Bailey Hortorium) 638/2 *Lathyrus...splendens* Kellogg. Pride-of-California...somewhat shrubby.

II. 12. Delete † and substitute for def.: A group of lions forming a social unit. (Later examples.)

1929 *Times* 30 Sept. 12/6 Owing to the dry weather a pride of 16 lions, including females and cubs, concentrated on the Kajiado road..less than 20 miles from Nairobi. 1940 V. Pohl *Bushveld Adventures* x. 218 Presently we distinguished outlines of several other forms beyond the one we now knew to be the leader of the pride. 1964 C. Willock *Enormous Zoo* v. 75 We found the pug-marks of a pride of lion. 1975 *Sci. Amer.* May 54/2 The social unit of the lion—the pride—is a long-lasting entity. *Ibid.* 55 (*caption*) A typical pride usually includes two or three adult males, from five to ten adult females and a number of cubs.

15. *pride-swollen* adj. (earlier and later examples).

1598 Marston *Scourge of Villanie* x. sig. H3, These pride-swolne dayes. *a* 1846 B. R. Haydon *Autobiogr.* (1927) i. 10 His large, red, pride-swollen, big-featured face.

pride, *v.* **4.** (Later example with *for.*)

1850 D. M. Craik *Olive* I. v. 71 How Elspie then prided herself for the continual tutoring which had made the image..an image of love.

prideful, *a.* Add: Also *N. Amer.* **a.** (Later examples.)

1900 *Century Mag.* Dec. 293 The doctor's stately and prideful wife. 1945 R. Hargreaves *Enemy at Gate* 64 A prideful, unbending spirit was in no mood to bow a compulsory knee without a fight for it. 1956 B. Chute *Greenwillow* v. 64 He's prideful, and that's a sin, but he's been good to me. 1974 G. M. Fraser *McAuslan in Rough* 159 When that veteran has not only learned his political science at Govan Cross but is also a member of an independent and prideful race. 1974 R. Helms *Tolkien's World* iv. 73 The true heroism in this situation was..the endurance of his men, forced by his prideful act to exhibit their loyalty to the death. 1977 *Time* 28 Feb. 23/1 He also was continuing to have his problems with prickly, prideful Senator Robert Byrd.

b. (Later examples.) Also, meriting a feeling of pride.

1939 *Sun* (Baltimore) 19 Oct. 11/3 She was very prideful of this, and when she finished Bill..went over and congratulated her. *a* 1967 A. Ransome *Autobiogr.* (1976) i. 19, I was practising day in day out the simpler conjuring tricks that were to lead me to the prideful moments of a professional magician who, before vast audiences, should produce rabbits out of a hat. 1968 *Globe & Mail* (Toronto) 17 Feb. 28 They find they can now choose from a prideful list for entertaining..or just revel in that greatest joy of all..a leisurely candlelit dinner with fine wines and matching service. 1978 J. Carroll *Mortal Friends* I. v. 56 Collins was aware of the pleasure, the prideful pleasure, Brady was taking in his words.

pridefully, *adv.* (Later examples.)

1947 S. J. Perelman *Westward Ha!* (1949) x. 123 His new ball-point fountain pen, which he had been exhibiting pridefully all morning. 1977 *Time* 14 Feb. 56/3 A play that will most appeal to people of the sort he has so wickedly satirized—the pridefully literate.

priderite (prəi·dərəit). *Min.* [f. the name of R. T. *Prider,* 20th-c. Australian geologist, who made a study of the suite of rocks from which the first identified sample was taken: see -ite[1].] A lustrous black titanate of potassium and barium that is optically similar to rutile and occurs as rectangular prisms and plates.

1951 K. Norrish in *Mineral. Mag.* XXIX. 496 (*heading*) Priderite, a new mineral from the leucite-lamproites of the West Kimberley area, Western Australia. *Ibid.* 500 The formula of priderite is approximately $(K, Ba)_{1.33}(Ti,Fe)_8O_{16}$. 1960 *Jrnl. Geol. Soc. Austral.* VI. 72 Priderite, a potassium titanate.., is the mineral which was earlier referred to as rutile. 1968 *Mineral. Mag.* XXXVI. 869 The priderite bearing rocks are alumina-deficient and alkali-rich but sodium-poor.

priest, *sb.* Add: **I. 2. b.** *priest-in-charge* (see quot. 1977).

1941 A. Thirkell *Northbridge Rectory* iii. 59 At St. Sycorax, where he was priest-in-charge, a title which gave him deep pleasure, he indulged in a perfect orgy of incense and vestments. 1963 *Times* 4 Feb. 12/4 In the summer..he will become priest-in-charge of Titsey..in the Southwark diocese. 1976 *Oxf. Diocesan Mag.* July 18/2

PRIESTED

Stanford Dingley is not typical of rural parishes, in that, though very small, it has had its own priest, though not actually resident, for the last 20 years—a rector, a curate, and now a priest-in-charge. **1977** Macmorran & Elphinstone *Handbk. Churchwardens* (new ed.) vi. 61 The unbeneficed clergy..fall into two classes: first, ministers in charge of benefices which for the time being lack the services of any incumbent (generally called 'curates-in-charge' or 'priests-in-charge'): and secondly, assistant curates, viz. clergymen appointed to assist incumbents within their parishes. **1979** *Guardian* 31 Oct. 10/7 Actually he isn't even a vicar. He's just a priest in charge who's been there about eighteen months.

III. 10. a. *priest-king* (later examples), *-ruler*.

1898 R. Brown in R. M. Dorson *Peasant Customs* (1968) I. 168 The majestic figure of the Priest-king of Uru-salim. **1920** H. G. Wells *Outl. Hist.* III. xix. 124/2 The beginnings of organized war, first as a bickering between villages, and then as a more disciplined struggle between the priest-king and god of one city and those of another [in Mesopotamia]. **1928** A. Evans *Palace of Minos* II. II. 774 The remains of the remarkable painted relief of the personage wearing a plumed lily crown and collar, in whom we may with good reason recognize one of the actual Priest-Kings of Knossos. **1939** J. D. S. Pendlebury *Archaeol. Crete* iv. 249 One [chariot] drawn by winged griffins contains two female figures, one of whom.. has a headdress resembling in some ways that of the Priest King. **1958** *Times Lit. Suppl.* 24 Jan. 42/2 But he [sc. Charlemagne] was not a priest-king. **1978** *Listener* 28 Sept. 402/4 Were not priest-kings adored just because they were victims, in the sense that they were sacrifices for the people? **1920** H. G. Wells *Outl. Hist.* III. xvi. 94/2 There [in the Euphrates-Tigris valley] flourished the first temples and the first priest-rulers that we know of among mankind.

priested, *ppl. a.* (In Dict. s.v. Priest *v.*) **a.** (Later example.)

1916 Joyce *Portrait of Artist* (1969) v. 221 A priested peasant, with a brother a policeman in Dublin and a brother a potboy in Moycullen.

priesteen (prīstī·n). [f. Priest *sb.*; see *-een[2].] Anglo-Irish diminutive of Priest *sb.*

1907 J. M. Synge *Playboy of Western World* I. 23 'It isn't fitting,' says the priesteen, 'to have his likeness lodging with an orphaned girl.' **1912** Joyce *Let.* 19 Aug. (1966) II. 304 Can your friend in the sodawater factory or the priesteen write my verses? **1922** —— *Ulysses* 212 The quaker's pate godlily with a priesteen in booktalk.

priestess. Add: **3.** *Comb.*, as *priestess-queen* (after *priest-king*).

1920 H. G. Wells *Outl. Hist.* III. xix. 114/1 The Sumerians allowed much more freedom and authority to women than the Semites. They had priestess-queens, and one of their great divinities was a goddess, Ishtar.

‖ **prikaz** (prika·z). Also **pricasse, prikas.** Pl. **prikazy.** [Russ.] In Russia: an office or a department, esp. in the central administration (now only *hist.*); an order or a command.

a **1725** C. Whitworth *Acct. Russia* (1758) 61 The Court..was very numerous and magnificent, being filled on solemn occasions by the *Bojars*, or privy Counsellors, with all the officers of each Pricasse. **1854** R. G. Latham *Native Races Russ. Empire* xii. 165 The *Prikas* is a kind of *Divan* or *Council*, consisting of two Kirghiz and two Russian assessors. **1886** *Encycl. Brit.* XXI. 105/1 He [sc. Kotoshikhin] served in the ambassador's office (*posolski prikaz*), and when called upon to give information against his colleagues fled to Poland about 1664. **1905** *Contemp. Rev.* Feb. 155 No law, ukaz, prikaz, can be decided upon except on the report of a Minister. **1917** [see *collegium]. **1952** S. Harcave *Russia* iv. 58 Administrative offices (*prikazy*) were set up as new problems arose. **1963** N. V. Riasanovsky *Hist. Russia* xviii. 212 The authority of a prikaz extended over a certain type of affairs, such as foreign policy in the case of the ambassadorial prikaz. **1971** J. S. Reshetar *Soviet Polity* vi. 217 Collegium decisions can be carried out only by an order (*prikaz*) issued by the minister because the collegium as such has no authority.

prim, *a.* Add: **c.** *prim-lipped* adj.

1926 J. Masefield *Odtaa* xix. 318 A prim-lipped man, with the look of a 'spoiled priest',..seemed to be in charge of the guard. **1953** E. S. Grenfell in C. K. Stead *N.Z. Short Stories* (1966) 74 The old man turned a..prim-lipped face to the parson.

prima[2]. Add: **b.** Short for *prima ballerina, Prima donna.

1930 T. Karsavina *Theatre St.* xiv. 183 From a group in the first wing, a sanctum reserved only for primas, an infuriated figure rushed up to me. **1951** Green & Laurie *Show Biz* 571/1 *Prima*, prima donna. **1968** J. M. White *Nightclimber* xv. 97 Her voice lacked the sheer power needed to make a real impression in the bigger opera-houses. In central and eastern Europe, nevertheless, she always enjoyed the reputation of an undisputed prima.

‖ **prima ballerina** (prī·mă balərī·nă, bælər-). Pl. **prime ballerine, prima ballerinas.** [It., lit. = first female dancer.] **1. a.** A ballerina of the highest rank; the leading ballerina of a ballet company. Cf. *ballerina, *première danseuse.

1900 J. T. Grein in *Sunday Special* 23 Sept. 2/2 Curiously enough for many years the London public has failed to do justice to the prima ballerina. **1912** J. E. C. Flitch *Mod. Dancing & Dancers* iv. 62 At the Alhambra ..the dancing of the *prime ballerine*, almost all of whom were foreign, left little to be desired. **1918** D. H. Lawrence *Let.* 12 Jan. (1962) I. 536, I dreamed you were a sort of *prima ballerina*—which is the translation of a cinema star, I suppose. **1921** *Dancing Times* June 709 Lydia Lopokova has made a welcome re-appearance..as prima ballerina of the Diaghileff Company. **1955** *Times* 19 May 3/7 The transformation of Mephistopheles into the *prima ballerina* Mephistophela is thoroughly Heinesque. **1958** *Times* 13 Sept. 10/5 Her Giselle was a fragile, vulnerable creature, excessively shy, and bearing no mark of the *prima ballerina*. **1976** P. Harcourt *Dance for Diplomats* iii. 30 He paid homage to his *prima ballerina* and the *corps de ballet*.

b. prima ballerina assoluta [It., lit. = absolute], a prima ballerina of outstanding excellence.

1870 T. A. Brown *Hist. Amer. Stage* 33/2 She made her *debut* as prima ballerina assoluta in the Teatro Regio in Ancona. **1904** A. Bennett *Great Man* xvii. 181 The glimpse which Henry had of the *prima ballerina assoluta* in her final pose..caused him to turn..to Geraldine to see whether she was not shocked. **1915** M. E. Perugini *Art of Ballet* xxxii. 272 With Signorina Maria Bordin..as *prima ballerina assoluta*..the production achieved instant success. **1928** *New Statesman* 31 Mar. 793/1 Tilly Losch (*prima ballerina assoluta* of the Vienna Opera-house). **1957** G. B. L. Wilson *Dict. Ballet* 37 Only twice in the history of the Imperial Th. was the additional title of 'prima ballerina assoluta' bestowed—on Pierina Legnani and on Mathilde Kschessinska. **1975** *New Yorker* 26 May 31/1 She..said that the Bolshoi's former prima ballerina assoluta, Galina Ulanova, was now one of its most valued teachers. **1978** Ld. Drogheda *Double Harness* xxi. 300 She would inherit the position of *prima ballerina assoluta*, with her name standing alone at the top of the list of dancers.

2. *transf.* and *fig.* *spec.* an important or self-important person; something which has leading status in its particular field.

1954 *Economist* 11 Sept. 1/1 A prolonged attack of stage fright has repeatedly postponed the entrance of the new fighters..into squadron service with the RAF and the new *prima ballerina*, the supersonic P.1 fighter, has not turned up this week at rehearsals. **1964** M. Clive *Day of Reckoning* viii. 71 Any old lady..could turn..her daily round into a sort of sacred ballet, herself the prima ballerina. **1975** N. Luard *Robespierre Serial* xvi. 146 I'm right with you however our prima ballerina responds. But what the hell are you going to tell him?

Primacord (prəi·măkǫɹd). *Mil.* Also **primacord.** [f. Primer *sb.*[2] + Cord *sb.*[1]] A proprietary name in the U.S. for a type of detonating fuse consisting of a core of high explosive in a textile and plastic sheath.

1937 *Official Gaz.* (U.S. Patent Office) 23 Feb. 699/2 The Ensign-Bickford Co., Simsbury, Conn... *Primacord*. For detonating fuse. Claims use since Dec. 11, 1936. **1950** [see *cordtex]. **1959** *N.Y. Times Mag.* 31 May 38/2 One crew has its primacord all set to blow a big obstacle. **1972** *Daily Colonist* (Victoria, B.C.) 27 Feb. 30/1 Police.. warned parents..that 1,000 feet of primacord—an explosive detonating cord used to trigger dynamite—had gone missing.

primacy. Add: **1. b.** *Psychol.* The predominance of certain impressions, esp. first impressions, over subsequent or derived ones, in the mind or memory; also *attrib.*, as *primacy effect, principle*.

1896 M. W. Calkins in *Psychol. Rev. Monogr. Suppl.* I. II. 35 Ordinary self-observation has..enumerated frequency, recency, vividness.., and primacy (the earliest position in a definite series of events) as the factors of interest. **1913** C. E. Seashore *Psychol. in Daily Life* v. 151 Familiar illustrations of the secondary or quantitative laws are (1) the law of primacy: other things being equal, the first impression will be the most effective. **1926** R. M. Ogden *Psychol. & Educ.* xii. 199 Primacy is popularly expressed by the statement that 'first impressions are lasting'. **1931** *Psychol. Rev.* XXXVIII. 217 The law of primacy has also been suggested on the basis of experimentation, and this has been contrasted with recency. **1953** C. I. Hovland et al. *Communication & Persuasion* iv. 117 Experimental psychology for a long time postulated a Law of Primacy and a Law of Recency. **1959** Lambert & Fillenbaum in Saporta & Bastian *Psycholinguistics* (1961) 457/1 The European cases in most instances fail to support either a primacy or a habit strength principle. **1971** *Sci. Amer.* Aug. 85/1 The increased probability of recall for the first few words in the list is called the primacy effect.

prima donna. Add: **1.** (Further examples.) Also **prima donna assoluta** [It., lit. = absolute], a *prima donna* of outstanding excellence.

1782 W. Beckford *Let.* 5 Apr. in J. W. Oliver *Life William Beckford* (1932) v. 110 Our Prima Donna, Miss Fawkener..has real talent. **1855** Geo. Eliot in *Fraser's Mag.* LII. 50/2 He will..interpolate no *cantata* to show off the powers of a *prima donna assoluta*. **1938** *Oxf. Compan. Mus.* 749/2 The term *Prima Donna Assoluta* ('absolute first lady') is sometimes used to make perfectly clear the position of the *very* most important woman member of an opera company. **1958** *Listener* 14 Aug. 250/2 A singer who is hailed as a *prima donna assoluta*. **1976** S. Galatopoulos (*title*) Callas: prima donna assoluta.

2. *transf.* and *fig.* A person of the highest standing in a particular field or activity; one who behaves in a self-important or temperamental manner. Also as *v. intr.*

1834 [see *Angrian *a.*]. **1846** *Swell's Night Guide* 36 Here also hang out some of the prima donnas of the flags and curbs, some of the small fry of 80, Quadrant. **1861** B. Hemying in H. Mayhew *London Labour* (1862) Extra vol. 215/1 Two classes of prostitutes come under this denomination—first, kept mistresses, and secondly, prima donnas or those who live in a superior style. **1877** A. Macmillan *Let.* in C. Morgan *House of Macmillan* (1943) vii. 117 It is clear that our *Prima Donna* must be paid on a different scale from the others. **1936** *Amer. Mercury* May p. x/2 *Prima donna*, the first-class gripe artist; a temperamental [jazz] musician. **1938** *Times Lit. Suppl.* 639/3 We see her [sc. Madame de Staël]..as the 'prima donna', exacting, torrential and exasperating. **1940** E. Hemingway *For whom Bell Tolls* xiv. 181 Stop prima-donnaing and accept the fact. **1943** *Sun* (Baltimore) 24 Sept. 14/2 A willingness to merge his identity with that of the journal of which he was a part. He was no prima donna. **1948** D. Cecil *Two Quiet Lives* II. 146 The most trivial points..were enough to produce a violent explosion of prima donna temperament. **1970** S. Ellin *Bind* iv. 22 You've been putting on a prima donna act for the last hour. What's it all about? **1970** 'B. Mather' *Break in Line* ix. 117, I felt no resentment... It was going to be hairy enough without any prima-donnaing on my part. **1973** C. Bonington *Next Horizon* ix. 140 He had invited Royal Robbins..to be chief [climbing] instructor, and the two men, both prima donnas in their own right, could not have offered a greater contrast. **1976** Botham & Donnelly *Valentino* iv. 34 Di Valentina was rapidly becoming the prima donna of the Manhattan cabaret set. **1978** *Jrnl. R. Soc. Arts* CXXVI. 537/2 The industrial designer tends often to adopt the rôle of catalyst rather than that of a prima-donna in his relationship with his colleagues in the development team.

Hence **prima do·nna-ish** *a.*, **prima do·nna-ism, prima-do·nnaship.**

1889 *Scottish Art Rev.* II. 114 Miss Macintyre..is still too young and amateurish to make it possible to predict whether she will be..spoiled by her early prima-donnaship. **1961** A. Wilson *Old Men at Zoo* i. 25 It..served to increase my dislike for their unusual touchy, prima donna-ish relationship. **1969** P. Dickinson *Pride of Heroes* i. 16 Pibble had taken a prima-donna-ish dislike to the stationmaster. **1970** C. F. Hockett *Leonard Bloomfield Anthol.* p. xiv, We can still know..his reaction to the pettishness, the prima-donnaism, the neglect of already accumulated experience, and the antiscientific bias that have all too often characterized our discussions. **1973** 'B. Mather' *Snowline* vi. 73 He is apt to get prima donna-ish when he is out of temper. **1980** *Daily Tel.* 14 Jan. 12 We hope that he will go on being equally modest and lazy, cocking a snook at the prima-donna-ish antics of some chess masters while continuing from strength to strength.

prima facie. Add: **A.** *adv.* (Further examples.)

a **1676, 1797** [see *doli capax]. **1900** [see *en bloc]. **1955** *Times* 5 May 4/1 If the right given to the tenant was an asset, it was something which *prima facie* would on his death pass to his personal representative. **1971** *Mod. Law Rev.* XXXIV. 693 The recommendation was prima facie *unlawful*.

B. *adj.* (Further examples.) Also *ellipt.* as *sb.*

1916 G. B. Shaw *Androcles & Lion* p. lx, An objection from an average stockbroker constitutes in itself a *prima facie* case for any social reform. **1955** *Times* 1 July 6/5 The Magistrate..said that on the girl's evidence there was plainly a *prima facie* case. **1978** P. Lovesey *Waxwork* 153 If anyone can supply the *prima facie* evidence that her husband was involved, it's her. **1980** N. Freeling *Castang's City* xxv. 169 You've nothing for a *prima facie*... Any lawyer could knock it down.

primal, *a.* Add: **1. b.** *Psychol.* Relating or pertaining to such needs, fears, behaviour, etc., as form the origins of emotional life, esp. as in Freud's theory that, in the hypothesized murder of the dominant father who possesses the females in a primal horde, lies the unconscious origin of the Oedipus complex and the beginning of conscious emotions.

1918 A. A. Brill tr. *Freud's Totem & Taboo* iv. 218 If the totem animal is the father, then the two main commandments of totemism..agree in content with the two crimes of Oedipus..and also with the child's two primal wishes whose insufficient repression..forms the nucleus of perhaps all neuroses. **1934** R. Money-Kyrle tr. *Róheim's Riddle of Sphinx* i. 81 All this forms a 'religion' in which the infantile primal fantasies recur in a projected form. **1950** J. Strachey tr. *Freud's Coll. Papers* V. 229 Parricide, according to a well-known view, is the principal and primal crime of humanity. **1961** R. Fliess *Ego & Body Ego* III. v. 301 (*heading*) Primal hate against the eldest brother. **1968** M. Harris *Rise of Anthropol. Theory* xvi. 425 In this fashion, the primal patricide, helped along by hereditary memory traces in the 'racial unconscious', gave rise to the Oedipus complex, nuclear family, incest taboo, [etc.]. **1969** P. A. Robinson *Freudian Left* 109 Freud had in fact asserted that the primal-crime hypothesis was not to be understood as a simple statement of fact. **1970** A. Janov *Primal Scream* v. 55 It is when we force the neurotic patient to feel, rather than act out his primal fears that we can help him understand the feelings that are terrorizing him. **1973** D. Nicholson-Smith tr. *Laplanche & Pontalis's Lang. of Psycho-Anal.* 332 If we consider the themes which can be recognised in primal phantasies..the striking thing is that they all have one trait in common: they are all related to origins. *Ibid.* 334 This anticathexis..is unlikely to derive from the super-ego, whose formation is subsequent to primal repression. **1976** N. Thornburg *Cutter & Bone* i. 7 He was.. indistinguishable from the evangelists and fire-worshipers, the pornographers and primal screamers. **1977** *Undercurrents* June–July 16/1 Another group is that at Atlantis near Burtonport, Co. Donegal, who practice self-sufficiency and primal therapy. *Ibid.* 26/1 Tim Elloart discusses the development of the treatment, its basic methods, and the attitude of primal therapists to other forms of psychiatric treatment. **1978** *Listener* 19 Oct.

499/1 In America, one of the best-known of the therapies which claim to help you relive those early traumas is primal therapy.

c. Special collocations (sense *1 b): *primal father*, the dominant male, possessing all the females, assumed to have existed by some theories of the origins of social life; *primal horde*, a conjectured original form of human group; *primal law*, the conjectured law of nature whereby human beings originally lived, esp. under the dominance of the male; *primal scene*, a Freudian term for the first time that a child is emotionally aware of his parents copulating.

1918 A. A. Brill tr. *Freud's Totem & Taboo* iv. 245 An ideal could arise having as a content the fullness of power and the freedom from restriction of the conquered primal father, as well as the willingness to subject themselves to him. **1934** R. Money-Kyrle tr. *Róheim's Riddle of Sphinx* iv. 179 All human institutions are regarded as foundations of the primal father. **1918** A. A. Brill tr. *Freud's Totem & Taboo* iv. 246 The family was a reconstruction of the former primal horde and also restored a great part of their former rights to the fathers. **1934** R. Money-Kyrle tr. *Róheim's Riddle of Sphinx* iii. 171 What is the relation between the ontogenic conception of culture and the primal horde theory. **1969** P. A. Robinson *Freudian Left* 98 Verbal communication is the only vehicle of traditional continuity we know of; but, *ex hypothesi*, the primal-horde epoch must have been over before the development of speech. **1903** J. J. Atkinson (*title*) Primal law. *Ibid.* i. 210 The following thesis, however, on the genesis of primal law in human marriage, treats of a *conjectural* series of events in the ascent of man. **1925** J. Strachey tr. *Freud's Infantile Neurosis* in *Coll. Papers* III. v. 510 We will proceed with the study of the relations between this 'primal scene' and the patient's dream. **1955** M. Klein et al. *New Directions in Psychoanal.* xiii. 327 When, in this emotional state, he covers his eyes with his hand he is, I think, reviving the young infant's wish never to have seen and taken in the primal scene. **1957** M. McCarthy *Memories Catholic Girlhood* viii. 204, I conceived an aversion to apricots..from having watched her with them, just as though I had witnessed what Freud calls the primal scene. **1973** D. Nicholson-Smith tr. *Laplanche & Pontalis's Lang. of Psycho-Anal.* 335 Should we look upon the primal scene as the memory of an actually experienced event or as a pure phantasy? **1977** A. Sheridan tr. *Lacan's Écrits* iii. 96 The Wolf Man never managed..to integrate his recollection of his primal scene into his history.

primalism (prəi·măliz'm). *rare.* = Primality.

1904 Chesterton *G. F. Watts* 145 This indescribable primalism, which we have noted as coming out in the designs, in the titles and in Watts' very oil-colours.

primality. Restrict *rare* to sense in Dict. and add: **2.** *Math.* [f. Prime *a.* + -al + -ity.] The property of being a prime number.

1919 L. E. Dickson *Hist. Theory Numbers* I. xvii. 397 A test for the primality of $2^n \pm 1$. **1958** *Computer Jrnl.* I. 101/1 It proved possible to test for primality a series of numbers of the form $2^K - 1$. **1975** *Nature* 16 Oct. 544/1 They added that A. Ferrier, using a desk machine, had just demonstrated the primality of another large number.

‖ **prima materia** (prəi·mă matiᵊ·riă). [L., = first matter. Cf. Gr. ἡ πρώτη ὕλη.] = *materia prima.

1906 W. B. Yeats *Poems, 1899–1905* p. xii, I know I have been busy with the Great Work, no lesser thing than that, although it may be the Athanor has burned too fiercely, or too faintly and fitfully, or that the *prima materia* has been ill-chosen. **1919** D. H. Lawrence *Phoenix II* (1968) 232 From the conjunction of fire and water within the living plasm arose the first matter, the Prima Materia of a living body, which, in its dead state, is the alchemistic Earth. **1954** R. F. C. Hull tr. *Jung's Coll. Wks.* XVI. II. iii. 189 The refining of the *prima materia*, the unconscious content, demands endless patience, perseverance..and ability on the part of the doctor. **1969** K. Minogue in Ionescu & Gellner *Populism* 209 Nationalism has supplied them with a vocabulary suitable to their self-assertion expressing their claim to be unique and valuable, rather than simply indistinguishable *prima materia* fit for the inevitable process of industrialization.

primaquine (prəi·mă-, prī·măkwīn). *Pharm.* [f. Prima[2] + Quin(olin)e.] A synthetic quinoline derivative which is used (in the form of an orange-red crystalline phosphate) in the treatment of malaria; 8-(4-amino-1-methylbutylamino)-6-methoxyquinoline, $C_{15}H_{21}N_3O$.

1949 *Jrnl. Amer. Med. Assoc.* 3 Sept. 26/2 The Malaria Study Section, Laboratory of Tropical Disease, National Institutes of Health ..suggested the term 'primaquine', to distinguish the compound from 'pentaquine' and 'isopentaquine'. **1960** *Times* 11 Nov. 17/1 Primaquine, one of the newer and most efficient drugs for the treatment of malaria. **1962** *Lancet* 1 Dec. 1133/1 An erythrocyte defect similar to that found in primaquine sensitivity. **1965** *New Scientist* 19 Aug. 440/1 Medical officers at the Walter Reed Army Research Institute in Washington..report that the anti-malarial drugs, chloroquine and primaquine..are proving less and less effective. **1973** B. J. Williams *Evolution & Human Origins* iv. 65/2 Such hemolytic reactions also occur in persons with G6PD deficiency ion response to certain drugs, including some antimalarials (receiving the terms 'primaquine sensitivity', etc.). **1974** M. C. Gerald *Pharmacol.* iv. 77 Of the 15,000 compounds ..screened as substitute antimalarials, only two, chloroquine and primaquine, were found to be superior to quinine. **1978** *Nature* 22 June 627/1 The antimalarial drugs selected for the WHO list are chloroquine, primaquine, pyrimethamine and quinine.

primarily, *adv.* Add: Also increasingly, following Amer. usage, with pronunc. (prəimeᵊ·rĭli).

primary, *a.* and *sb.* Add: **A.** *adj.* **I. 3. a.** (Further examples.)

1899 Middleton & Chadwick *Treat. Surveying* I. v. 170 The methods..are not so complicated, or so minutely accurate, as those employed in dividing up a 'grand' or primary triangulation. **1920** W. N. Thomas *Surveying* xiii. 382 On the Ordnance Survey the first framework of triangles set out over the country constituted the 'Principal' or 'Primary' triangulation. **1923** Glazebrook *Dict. Appl. Physics* III. 571/2 For precision of definition, it is essential that there shall be one, and only one, material standard to represent each of the fundamental units. This is called the primary standard, and is preserved under the strictest conditions of custody, used only at very rare intervals, and then solely for purposes of comparison with the corresponding secondary standards. **1945** R. A. Knox *God & Atom* iii. 41 St. Thomas..distinguished God as the Primary Cause from those secondary causes to which we attribute this or that effect in our daily experience, and taught that the influence of the Primary Cause was present everywhere, conspiring with the secondary cause to produce the effect. **1966** *McGraw-Hill Encycl. Sci. & Technol.* XIV. 441/1 In 1907, the International Union for Cooperation in Solar Research adopted the value 6438·4696 Å as the primary standard for the wavelength of red radiation from cadmium measured relative to the meter. **1973** *Nature* 12 Jan. 146/3 The time interval between publication in a primary journal and the appearance of the corresponding abstract. **1973** *Sci. Amer.* Dec. 65/2 The tangle that springs up where the forest has been felled is the first stage in the growth of a 'substitute forest'. In more formal terms it is an early stage in the development of a 'secondary forest', which will replace the cleared 'primary forest'. **1975** J. B. Harley *O.S. Maps* ii. 18 A succession of local central meridians were as a result brought into use before the associated tertiary triangulations could be adjusted to the primary triangulation of Great Britain.

II. 4. b. (Later examples.)

1914 M. Drummond tr. *Haberlandt's Physiol. Plant Anat.* i. 61 More often..it [sc. the middle lamella] also comprises the primary thickening layers. **1943** *Bot. Rev.* IX. 125 Current concepts of the origin of primary vascular tissues..are much confused. *Ibid.* 129 'Procambium' has come to mean specifically the vascular meristem from which primary xylem and phloem are derived. **1953** K. Esau *Plant Anat.* iii. 39 The primary walls have primary pit fields. *Ibid.* iv. 77 If these cells [that give origin to the meristem] are the direct descendants of the embryonic cells..the meristems are called primary. **1971** F. C. Ford-Robertson *Terminol. Forest Sci.* 41/1 The primary (cell) wall..is the wall of the meristematic cell modified during differentiation.

d. *primary age*, used *attrib.* of children receiving or ready to receive primary education; also *ellipt.* as *primary*; *primary education*, add to def.: now formally applied in Great Britain to the education of children between the ages of five and eleven years; *primary school* (later examples).

1908 A. Ruhl *Other Americans* x. 173 In the gymnasium four little primary girls were imitating..the gestures of the elocution teacher. **1944** *Act 7 & 8 Geo. VI* c. 31 §8 It shall be the duty of every local education authority to secure that there shall be available for their area sufficient schools—(*a*) for providing primary education, that is to say, full-time education suitable to the requirements of junior pupils. *Ibid.* §7 The statutory system of public education shall be organised in three progressive stages to be known as primary education, secondary education, and further education. *Ibid.* §9 For the purpose of fulfilling their duties under this Act, a local education authority shall have power to establish primary and secondary schools. **1956** H. M. Pollard *Pioneers Popular Educ. 1760–1850* xxi. 265 Kay-Shuttleworth recommended that particular attention be paid to the large primary schools which had grown up..in The Hague [etc.]. **1958** K. Lovell *Educ. Psychol. & Children* xvi. 198 The primary school child like the pre-school child has his fears and anxieties. **1963** Barnard & Lauwerys *Handbk. Brit. Educ. Terms* 151 *Primary education*,..comprises full-time education suitable to the requirements of junior pupils (*i.e.* pupils under twelve). **1964** Curtis & Boultwood *Introd. Hist. Eng. Educ.* (ed. 3) viii. 184 All normal children..may be transferred at the age of 'eleven plus' from the primary or preparatory school to one type or another of secondary school. **1964** D. Holbrook *Eng. for Rejected* 54 How inefficient of the primary school to suppose that Joan has 'no imagination'. **1972** *Jrnl. Social Psychol.* LXXXVI. 167 The first study cited presents highly similar results for black and white primary-age subjects. **1976** *Times* 27 Apr. 14/7 It is impossible to generalize about English primary schools: some are progressive, informal, unauthoritarian, while others are strictly traditional. **1978** *Nagel's Encycl.-Guide: China* 316 *Primary Schools* take children from 7 to 13. For these six years, they learn little more than Chinese and arithmetic.

e. *primary assembly, meeting* (earlier examples); *primary election* (examples); also *primary caucus*.

1792 *Ann. Reg.* 1789 [214/2] The primary elections had for some days been carried on in the different districts of Paris. *Ibid.* [215/1] The inhabitants of every district in France, preparatory to the election of delegates, hold what is called a primary assembly, where they choose a prescribed number of electors who are to act for the whole in the choice of a representative. **1801** *Spirit of Farmers' Museum* 61 The Editor of the Gazette of the United States..notices the 'Primary Assemblies' of our towns. **1821** *Massachusetts Spy* 11 Apr. 3/3 This was all the *hocus-pocus* of a primary caucus. **1829** *Niles' Reg.* XXXVI. 363/2 The battle is in reality fought in the primary meetings, and not on the day appointed by law for the election. **1835** C. P. Bradley *Biogr. Isaac Hill* 54 The freemen of the State were called upon to give at their primary elections, an expression of their opinion. *a* **1850** T. Ford *Hist. Illinois* (1854) 88 Personal politics..were carried from the primary elections into the legislature. **1885** *Century Mag.* Apr. 825 Nine out of ten of our wealthy and educated men..are really ignorant of the nature of a caucus, or a primary meeting, and never attend either. **1961** *Atlanta Constitution* 4 Nov. 1 An investigation of Atlanta's recent primary election produced 'no evidence' that any irregularities took place. **1974** *Hartsville* (S. Carolina) *Messenger* 22 Apr. 6-A/1 This is the first time in the state's history that the Republican Party decided to nominate its candidates in a primary election, rather than by the convention method.

f. *primary spermatocyte* (Zool.), a spermatocyte which will undergo meiosis to yield further spermatocytes.

1896 E. B. Wilson *Cell* iii. 122 The primary spermatocyte first divides to form two daughter-cells known as spermatocytes of the second order or sperm-mother-cells. Each of these divides again..to form two spermatids or sperm-cells. **1927** *Jrnl. Exper. Zool.* XLIX. 463 The darkly colored pycnotic primary spermatocytes..are the most conspicuous cells in the germinal epithelium. **1960** W. B. Crow *Synopsis of Biol.* viii. 40 In the grasshopper each primordial germ cell divides eight times producing $2^8 = 256$ cells... The cells finally formed by such division are called primary spermatocytes; they undergo the reduction division.

g. *primary endosperm nucleus* (Bot.), the (usu. diploid) nucleus formed in an ovule by fusion of the two polar nuclei; also, the (usu. triploid) nucleus formed by fusion of a sperm nucleus with these nuclei.

1899 *Bot. Gaz.* XXVII. 58 The polar nuclei may fuse to form the primary endosperm nucleus. **1950** Robbins & Weier *Bot.* ix. 205/2 The nucleus resulting from this triple fusion is called the primary endosperm nucleus. **1960** W. B. Crow *Synopsis of Biol.* viii. 42 The immotile male nuclei..are carried to the embryo-sac wherein one fertilizes the ovum... The other nucleus usually in angiosperms fuses with the primary endosperm nucleus, itself the product of fusion of two polar nuclei, so that triple fusion occurs, the triploid nucleus provided dividing up to form the nuclei of the endosperm. **1974** G. W. Burns *Plant Kingdom* xx. 487/1 One sperm passes to the egg, uniting with it to form the zygote, and the other combines with the two polars, producing the primary endosperm nucleus.

h. Designating an earthquake P wave (see *P III. 3).

[**1912** *Nature* 5 Sept. 4/2 The usual seismographic record shows three chief groups of disturbances, due respectively to the longitudinal and transverse waves through the core..and to the superficial waves round the crust. These ..are complicated and supplemented by reflections of the deep waves at the surface, and sometimes by twin earthquakes caused by the primary.] **1919** *Proc. R. Soc. Edin.* XXXIX. 161 Tables familiar to all seismologists, in which times of transit of the primary and secondary waves are expressed in terms of the arcual distances of the stations of observation from the..epicentre. **1955** *Sci. Amer.* Sept. 56/3 In 1897 R. D. Oldham of England identified on seismograms three main types of seismic waves: (1) primary (P) waves, which are compression-and-expansion waves like those of sound; (2) secondary (S) waves, which vibrate at right angles to the direction of travel, as light waves do; (3) surface waves, which appear in the upper 20 miles or so near the earth's surface. **1968** R. A. Lyttleton *Mysteries Solar Syst.* ii. 56 The velocity of the primary waves.. is always essentially faster than that of the secondary waves.

i. *primary road* = *main road* s.v. Main *a.* 8 b.

1956 R. Braddon *Nancy Wake* xviii. 215 They had frequently to cross primary roads. **1974** *State* (Columbia, S. Carolina) 27 Feb. 18-A/1 There are roads (even primary roads) which look impressive on a map but which fade away into mystery on the ground.

6. f. Add defs.: (i) Orig. applied to compounds regarded as being derived from any of four molecules (water, ammonia, hydrogen chloride, and hydrogen, by replacement of one hydrogen atom by an organic radical. This sense survives in mod. use with respect to ammonia derivatives, *spec.* amides (but see quots. 1965), amines, and ammonium salts, and is extended to analogous derivatives of other elements, esp. phosphorus. [The sense is due to Gerhardt & Chiozza, who used F. *primaire* (*Compt. Rend.* (1853) XXXVII. 88).]

1854 *Q. Jrnl. Chem. Soc.* VI. 195 The amides thus produced, which we shall call primary amides, represent a molecule of ammonia in which 1 atom of hydrogen is replaced by the negative radicals. **1888** Bloxam *Chem.* (ed. 6) 586 The amides, like the amines..may be primary, secondary, or tertiary accordingly as one, two, or three atoms of H in the NH_2 group has been replaced. **1889** G. M'Gowan tr. *Bernthsen's Text-bk. Org. Chem.* 119 Just as amines are derived from ammonia, so from phosphuretted hydrogen, PH_3, are derived primary, secondary, and tertiary phosphines by the exchange of hydrogen for alcoholic radicals. **1938** G. H. Richter *Textbk. Org. Chem.* xiv. 230 The reduction of compounds that contain a $>C=N-$ linkage also produces primary amines. **1962** Cotton & Wilkinson *Adv. Inorg. Chem.* xx. 392 The phosphines are less basic than amines of the same type, but for phosphines the order is tertiary > secondary >

primary, whereas for amines it is commonly irregular but usually with primary > tertiary. **1965** *Nomencl. Org. Chem.* (I.U.P.A.C.) C. 176 The generic name 'amine' is applied to compounds NH_2R, NHR^1R^2, and $NR^1R^2R^3$, which are called primary, secondary, and tertiary amines, respectively. *Ibid.* 188 Compounds containing one, two, or three acyl groups attached to nitrogen bear the generic name 'amide'. When only one acyl group is attached to a nitrogen atom, the generic name 'primary amide' may be used; when two acyl groups are so attached, the generic name 'secondary amide' may be used; and when three acyl groups are so attached, the generic name 'tertiary amide' may be used. *Ibid.* 190 *N*-substituted primary amides $R^1{-}CO{-}NHR^2$ and $R^1{-}CO{-}NR^2R^3$. [*Note*] These compounds have been called, respectively, secondary and tertiary amides, but this usage is not recommended.

(ii) Subsequently applied to organic compounds (except amines, etc.: see (i) above) in which the characteristic functional group is located on a saturated carbon atom which is itself bonded to not more than one other carbon atom. [Applied orig. to alcohols by H. Kolbe, who used G. *primär* (*Ann. d. Chem. u. Pharm.* (1864) CXXXII. 102).]

1864 *Chem. News* 26 Nov. 260/1 Primary alcohol. **1888** Morley & Muir *Watts's Dict. Chem.* (rev. ed.) I. 100/1 In the primary alcohols the carbon-atom joined to the hydroxyl is connected immediately with only one other carbon atom. **1929** L. A. Coles *Introd. Mod. Org. Chem.* xii. 140 Methyl and ethyl alcohol are the two most important members of the primary alcohol series. **1968** J. March *Adv. Org. Chem.* ix. 866 Primary alcohols or aldehydes can be converted directly to nitriles by air oxidation in the presence of ammonia, a strong base..and a copper complex.

(iii) Applied to a saturated carbon atom which is bonded to only one other carbon atom; also, bonded to or involving a primary carbon atom. Of an ion or a free radical: having (respectively) the electric charge or the unpaired electron located on a primary carbon atom.

1903 Walker & Mott tr. *Holleman's Text-bk. Org. Chem.* I. 46 A carbon atom which is only linked to one other carbon atom is called primary. **1926** H. G. Rule tr. *J. Schmidt's Text-bk. Org. Chem.* 70 When a carbon atom is combined in such a manner that only one of its four valencies is satisfied by carbon, it is termed a primary carbon atom. **1951** I. L. Finar *Org. Chem.* iii. 31 A primary carbon atom is one that is joined to one other carbon atom. **1968** J. March *Adv. Org. Chem.* ix. 866 Primary amines at a primary carbon can be dehydrogenated to nitriles. **1972** DePuy & Chapman *Molec. Reactions & Photochem.* iv. 46, 2,2-Dimethylcyclohexanone..cleaves to give the tertiary alkyl radical rather than the primary alkyl radical. **1972** Norman & Waddington *Mod. Org. Chem.* ix. 116 A tertiary carbonium ion is a relatively more stable species than a primary carbonium ion..and is formed much faster. **1972** S. J. Weininger *Contemp. Org. Chem.* v. 106 The secondary ($-CH_2-$) carbon-hydrogen bonds..of propane are more easily broken than the primary ($-CH_3$) carbon-hydrogen bonds of ethane.

g. *Electr.* Substitute for entry: (i) Orig. of an electric current: supplied directly by a cell or battery, as opposed to an induced current. Now, with reference to any device utilizing electromagnetic induction, esp. a transformer: of, pertaining to, or carrying the input electrical power.

1837 M. Faraday in *Ann. Electr., Magn., & Chem.* I. 176 The conducting power of the connecting system A B D was sufficient to carry all the primary current. *Ibid.* 177 These experiments establishing..a distinction between the primary or generating current and the extra current, led me to conclude that the latter was identical with the induced current described..in the first series of these researches. **1862** *Electrician* 21 Feb. 183/2 In the primary wire of the induction coil, the 'return' or 'extra' current is an effect which is equally objectionable. *c* **1865** [in Dict.]. **1893** G. Kapp *Dynamos* xvii. 435 Such an apparatus is known as a transformer, the coil through which we send the alternating current being called the primary or driving coil. **1896** F. Bedell *Princ. Transformer* i. 3 The primary electromotive force..is equal to the product of the number of primary turns, and the rate at which the magnetic flux in the magnetic circuit is changing. **1929** A. T. Dover *Electr. Traction* (ed. 2) x. 281 The primary winding..has tappings to give 1000, 800, and 220 volts. **1938** Kerchner & Corcoran *Alternating-Current Circuits* vii. 193 The primary current could thus be made to lead or lag the primary voltage by adjusting the degree of coupling between the two transformer windings. **1963** Williams & Prigmore *Electr. Engin.* x. 333 In the open-circuit test ..the rated voltage..is applied to the primary terminals of the transformer. **1971** H. A. Romanowitz *Introd. Electr. Circuits* xxii. 491 When a transformer is used to deliver energy at a higher voltage than that at which it is received, the primary winding is the one with the small number of turns and the larger wire in its coils.

(ii) Of a cell or battery: in which the chemical reaction that generates the current is irreversible, and which therefore cannot store electrical energy applied to it.

1882 *Engineer* 19 May 365/2 The distinction between a primary and a secondary battery is in no sense an important one when we are considering either as the producer of a current. **1886** R. Wormell *Electr. in Service of Man* 427 The primary battery gives out a current of electricity, and the secondary acquires a condition which gives it also in turn the power of producing an electric current. **1922** Glazebrook *Dict. Appl. Physics* II. 59/2 Polarisation is one of the difficulties encountered in all primary cells. **1971** H. A. Romanowitz *Introd. Electr. Circuits* ii. 36 The wet cells and dry cells just described, which are not rechargeable, are classified as primary cells.

h. Applied to the testes and ovaries as sexual characters essential to reproduction and determined directly by the genetic sex, without hormonal intervention; sometimes the sexual ducts and organs are included, as essential to reproduction though developing as a result of hormonal influence (these are otherwise classed as either accessory or secondary characters). Cf. *secondary *a.*

1871 Darwin *Descent of Man* I. viii. 253 With animals which have their sexes separated, the males necessarily differ from the females in their organs of reproduction; and these are the primary sexual characters. *Ibid.* 254 Unless..we confine the term 'primary' to the reproductive glands, it is scarcely possible to decide which ought to be called primary and which secondary. **1894** H. Ellis *Man & Woman* ii. 18 When we are dealing with Man it is perhaps most convenient to set aside as primary the sexual glands..and the organs for emission and reception in immediate connection with these glands. **1926** J. R. Baker *Sex in Man & Animals* ii. 26 The primary sexual characters are..the testes and ovaries. The accessory sexual characters are the obviously useful sex characters other than the testes and ovaries, such as the vas deferens ..and the vagina... The secondary sexual characters are those which seem not to be directly concerned in reproduction, such as beards, antlers, and crests. **1948** C. D. Turner *Gen. Endocrinol.* viii. 263 The primary sex characters are the gonads... Ducts and glands involved in the transmission of gametes or developing zygotes are known as the sex accessories. **1960** B. I. Balinsky *Human Embryol.* xvi. 432 The primary sex characters distinguishing a male from a female animal are the sex glands—the testis and the ovary, respectively. By secondary sex characters are meant the distinctions between the sexes other than the presence of sex glands. These..fall into two groups: the organs which are essentially necessary for reproduction,..and organs or characters..such as the spurs and comb in the cock, the beard in man, [etc.]. **1977** E. J. Trimmer et al. *Visual Dict. Sex* (1978) xxii. 262/1 A boy castrated before puberty will fail to develop the primary and secondary sexual characteristics of a normal male that would otherwise be stimulated by hormones produced by the testes.

i. *Geol.* Of a mineral: that is an original constituent of the rock. Of a rock: whose constituents have undergone no alteration since formation. Usu. applied *spec.* to minerals and rocks that have crystallized from magma.

1886 J. Geikie *Outl. Geol.* xiii. 151 It is the primary or original constituents of a rock which ought to determine its species, but these are often replaced by secondary minerals, and thus it is not in all cases possible to say what were the primary minerals. **1905** —— *Struct. & Field Geol.* iii. 37 Those rock-constituents which crystallised out from the magma are termed primary or original, to distinguish them from another group of minerals which are of later origin than the rocks in which they occur. **1914** J. Park *Text-bk. Geol.* xii. 191 A primary mineral or rock constituent is one that is developed during the cooling of the molten magma, or, in the case of sedimentary rock, that appeared among the original constituents. **1921** *Trans. Geol. Soc. S. Afr.* XXIV. 116 In some varieties..a titaniferous lime-iron garnet comes in, together with primary calcite. **1931** A. Johannsen *Descr. Petrogr. Igneous Rocks* I. ii. 28 The primary minerals of the igneous rocks are comparatively few in number... The feldspars, the pyriboles, the micas, quartz, olivine, and the feldspathoids are practically all. **1966** Read & Watson *Beginning Geol.* xii. 147/1 In this environment, the primary minerals..are no longer stable, and they give place to new minerals and fabrics more in harmony with the new conditions. **1971** B. W. Sparks *Rocks & Relief* iv. 132 A relatively small class of primary calcareous rocks, known as carbonatites, are [*sic*] associated in some areas with alkaline igneous rocks.

j. *Sociol.* Esp. as *primary group*: a term for the sort of direct and informal relationships that an individual forms by reason of family or environmental associations which are considered basic to social life and culture (see esp. quots. 1933 and 1971).

1894 Small & Vincent *Introd. Study of Society* III. ii. 183 (*heading*) The primary social group: the family. **1909** C. H. Cooley *Social Organization* iii. 23 By primary groups I mean those characterized by intimate face-to-face association and coöperation... They are primary..chiefly in that they are fundamental in forming the social nature and ideals of the individual. **1933** —— et al. *Introd. Sociol.* iv. 55 If human nature belongs, then, to men in association what kind or degree of association is required to develop it?.. Are there simple forms of association..? It appears that there are, and we shall call them *primary groups*. *Ibid.*, A primary group may be defined as a group of from two to possibly fifty or sixty people..who are in relatively face-to-face association for no single purpose, but merely as persons rather than as specialized functionaries. **1950** E. A. Shils in Merton & Lazarsfeld *Continuities in Social Res.* 25 The primary group, they say, 'served two principal functions in combat motivation: it set and emphasized group standards of behavior and it supported and sustained the individual in stresses he would not otherwise have been able to withstand.' **1971** Z. Barbu *Society, Culture & Personality* ii. 34 E. Farris rejects the criterion of face-to-face association and specifies that intimate relationships, together with group consciousness, *esprit de corps*, and the feeling of 'we' constitute the main characteristics of a primary group. On the other hand E. A. Shils considers primary groups as identical with what other writers call informal groups, i.e. more or less spontaneous gatherings based on some kind of mental affinity.

k. *Physics* and *Astr.* Of, pertaining to, or designating radiation that is not produced by other radiation but may itself produce other (secondary) radiation; of cosmic rays: originating outside the earth's atmosphere.

1900 J. S. Townsend in *Proc. Cambr. Philos. Soc.* X. 218 The apparatus shown..was used to determine the relative intensities of the secondary radiations given out by different bodies, and the intensity of the secondary radiation compared with that of the primary radiation which excites it. **1921** J. Scott-Taggart *Thermionic Tubes* i. 23 When the primary or original electron attains a velocity sufficiently high to break off electrons from the gas molecule, it will leave the latter positively charged. **1938** R. W. Lawson tr. *Hevesy & Paneth's Man. Radioactivity* (ed. 2) v. 61 The number of secondary electrons corresponding to each incident primary electron depends on the velocity of the primary electron. **1944** *Ann. Reg. 1943* 361 Results which they regarded as..confirming that the main part of the primary cosmic ray radiation does not consist of electrons. **1946** *Electronic Engin.* XVIII. 75/1 An electrode will emit primary electrons when its temperature is raised sufficiently to overcome the work function of the material of which it is made. **1959** K. Henney *Radio Engin. Handbk.* (ed. 5) vii. 15 Secondary emission may be obtained by electron bombardment of pure metals... In this case a part of the energy of the bombarding or primary electron is transferred to one or several conduction electrons of the solid. **1973** Smith & Jacobs *Introd. Astron. & Astrophysics* xviii. 459 The fact that positrons..are only one-tenth as numerous indicates that these electrons are primary particles, and not secondaries like the light nuclei. This conclusion is based on the fact that more positrons than electrons are produced when primary cosmic rays collide with interstellar atoms.

l. *primary poverty*, lack of means to buy the basic necessities of life.

1901 B. S. Rowntree *Poverty* p. viii, Families whose total earnings are insufficient to obtain the minimum necessaries... Poverty falling under this head I have described as 'primary' poverty. **1909** M. F. Davies *Life in Eng. Village* II. xii. 140 An estimate has been made of the minimum cost at which food, fuel, dress, household sundries, and house-room..can be obtained..and it has then been seen how many families were below this standard, or in primary poverty. **1936** R. C. K. Ensor *England 1870–1914* xiv. 515 The number of people found by Rowntree in 'primary' poverty in 1901 was 15·46 per cent. of the wage-earning class in York. **1960** *Guardian* 25 Feb. 3/6 Primary poverty was uncommon..although 130 fathers were out of work. **1964** M. Laski in S. Nowell-Smith *Edwardian England* iv. 173 Primary poverty, that is to say,..conditions where the family income..is insufficient to maintain health and working efficiency.

m. *primary succession* (Ecol.) = *prisere.

1905 F. E. Clements *Res. Methods Ecol.* iv. 241 Primary successions..arise on newly formed soils, or upon surfaces exposed for the first time, which have in consequence never borne vegetation before. **1932** Fuller & Conard tr. *Braun-Blanquet's Plant Sociol.* xiii. 235 The Anglo-American school distinguishes between primary successions or sequences of communities which originate independently of men and secondary successions. **1952** P. W. Richards *Trop. Rain Forest* xii. 269 The successions or seres leading to the establishment of stable climax Rain forest are classified..into primary successions or priseres starting on soil not previously occupied by plants ..and secondary successions or subseres. **1961** Hanson & Churchill *Plant Community* v. 151 (*caption*) Primary succession on sand and gravel. **1973** P. A. Colinvaux *Introd. Ecol.* vi. 77 The successions so far discussed are all primary successions, that is to say they are supposed to proceed by pioneering new sites.

n. *primary industry, production*, the husbandry or use of raw materials, as in agriculture, forestry, fishing, mining, etc.; *primary produce, products*, the fruits of these activities; *primary producer*, one engaged in such industries; *primary-producing* adj., that produces or is the source of raw materials.

1930 *Economist* 8 Feb. 290/1 Our exporting manufacturers will be faced with lessened purchasing power in the hands of primary producers overseas. **1935** *Ibid.* 12 Jan. 57/2 Our own index of the dollar prices of primary products shows a rise of fully 20 per cent. **1941** Baker *N.Z. Slang* v. 38 We were gaining [1880–1900] a footing in world markets for our primary produce. *Ibid.* vii. 60 [The] social and economic existence [of Australia and New Zealand] is largely dependent upon primary production—upon the soil. **1950** *N.Z. Jrnl. Agric.* Jan. 3/1 The basis of New Zealand's economic standards is the country's primary industries. There has been an inclination to overlook the fact that in the economic structure of New Zealand agriculture must always be the cornerstone. *Ibid.* Aug. 99/1 It has been a particularly interesting experience to attend in my capacity as Minister of Agriculture..and to see in action the machinery of the several bodies that administer the interests of different sections of primary producers. **1956** T. Balogh in A. Pryce-Jones *New Outl. Mod. Knowl.* iv. 506 Political influence in primary-producing countries was secured by concluding bulk-buying agreements. **1959** A. H. McLintock *Descr. Atlas N.Z.* 44 In common with other exporters of primary products, New Zealand representatives have, at international conferences, drawn attention to the serious effects produced by protective barriers established at unrealistically high levels. **1965** S. T. Ollivier *Petticoat Farm* xi. 160 The primary produce from New Zealand went to feed other countries. **1966** G. W. Turner *Eng. Lang. Austral. & N.Z.* iv. 85 In both Australia and New Zealand, a land naturally infertile by English standards has been made to yield products which have made the term 'primary industry' synonymous with farming, by the application of scientific agriculture. **1974** M. B. Brown *Econ. of Imperialism* iv. 94 The extraction by the United States and by other developed countries of minerals and primary products from the whole world would not be regarded by classical economists as any sort of plunder. **1974** *Globe & Mail* (Toronto) 12 Oct. 8/3, 37 per cent of Quebec's labor force is employed in primary industry (mining, forestry, etc.) and manufacturing. **1975** *Listener* 11 Sept. 322/3 Indexing of the export prices of primary-producing countries against the rise in prices of the goods *they* need to import. **1976** *Oxford Times* 12 Mar. 11/1 The role of a University Department of Agricultural Science is..to provide graduates who can..become leaders in the field of primary production. **1977** *Herald* (Melbourne) 18 Jan. 4/1 On the talks so far, the Primary Industry Minister, Mr. Sinclair, probably sums up best by saying that he is 'relatively happy'.

o. *primary air*, air admitted to the fuel in a furnace or burner at or before the earliest stage in its burning.

1931 *Engineering* 9 Jan. 39/3 The primary air and coal enter at the side of the burner through a long, narrow port. **1932** *Discovery* Aug. 248/1 The 'primary air' or 'bottom air' entering from beneath the fuel bed..meets the incandescent coke, and immediately the oxygen disappears and is converted into carbon dioxide. **1971** E. R. Norster *Combustion & Heat Transfer in Gas Turbine Syst.* 91 The earlier Python engine had a 'vaporizing' tubular-type chamber, the fuel being pre-mixed with some primary air before being fed tangentially into a swirl chamber.

p. *primary constriction* (Cytology), a chromosomal constriction associated with the centromere.

1932 C. D. Darlington *Rec. Adv. Cytol.* 495 *Primary* or *attachment constriction*, that always associated with the spindle attachment. **1937** *Ibid.* (ed. 2) 575 *Primary* or *centric constriction*, that always associated with the centromere. **1957** C. P. Swanson *Cytol. & Cytogenetics* v. 112 The localized centromere produces the primary constriction. **1969** Brown & Bertke *Textbk. Cytol.* xviii. 344/1 This 'constriction' associated with the centromere is often loosely called the centromere, or kinetochore, but more exactly it is the primary constriction.

q. *Ecol.* Forming part of the lowest trophic level in a community, either as a producer or as a consumer that feeds on a producer; of or pertaining to a producer.

[**1893** W. K. Brooks in *Mem. Biol. Lab. Johns Hopkins Univ.* II. 148 A few forms are so predominant that..we may regard the great primary food-supply as made up of two simple protozoa, Globigerina and the Radiolarians, and some five or six unicellular plants.] **1934** *Q. Rev. Biol.* IX. 163/1 In connection with..his phrase ('primary food supply') which I have undertaken to use, Brooks gives..a meaning somewhat more comprehensive than that which I prefer... It suits my purpose better to confine the usage to the microscopic plants equipped with chlorophyll and capable of manufacturing carbohydrates from raw materials. **1940** *Ibid.* XV. 48/2 The various primary animals (phytophages) of a formation are often independent of each other. **1953** E. P. Odum *Fund. Ecol.* iv. 82 In speaking of productivity, it is important to distinguish between the basic or primary productivity..and consumer or secondary productivity. **1956** *Limnology & Oceanogr.* I. 116/1 The diagram shows how some streams may be fertile in having high total respiratory metabolism and yet possess little primary productivity. **1969** B. K. Sladen in Sladen & Bang *Biol. of Populations* vii. 92 The knapweed plants were the producers, and nine animal species were primary consumers eating the knapweed plant. *Ibid.* 93 The productivity of the producers, the photosynthetic plants, must be greater than that of the primary consumers. **1971** M. Alexander *Microbial Ecol.* xvi. 410 Aquatic algae are typically primary organisms in that their energy is obtained from sunlight. **1976** S. B. Chapman in *Methods in Plant Ecol.* iv. 161 Primary production is the production of organic matter by photosynthesis. *Ibid.*, Gross primary production, the total amount of organic matter produced (including that lost in respiration) over a given period of time. *Ibid.*, Net primary production, the amount of organic matter incorporated by a plant or an area of vegetation (gross primary production minus the loss due to respiration) over a given period of time.

r. *Psychol.* Relating to abilities or traits which, through factor analysis, appear basic to other aspects of intelligence or personality.

1938 L. L. Thurstone *Primary Mental Abilities* p. vi, These subsequent studies have had the advantage of some orientation about the first seven primary factors as landmarks... It is probably better to find the principal landmarks in the cognitive and conative primary traits by means of group procedures. *Ibid.* vi. 92 The tests will be improved by making them relatively pure measures of the primary abilities. **1958** K. Lovell *Educ. Psychol. & Children* iii. 58 Thurstone has termed these primary mental abilities and claims that they are found both in very young children and in adults. **1965** R. B. Cattell *Sci. Analysis of Personality* iii. 60 Several 'primary abilities' were also involved, such as numerical, spatial, verbal, and logical abilities. **1969** H. J. & S. B. Eysenck *Personality Struct. & Measurement* 328 We have seen that while primary factors emerge in considerable profusion, these seldom if ever agree precisely with those postulated by Cattell and Guilford. **1970** L. J. Bischof *Interpreting Personality Theories* (ed. 2) vi. xii. 464 In his current work, Cattell calls the source traits Primary Personality Factors. With few exceptions, the Primary Personality Factors are bipolar or dichotomized. **1972** *Jrnl. Social Psychol.* LXXXVI. 187 Disputes about the number and nature of primary personality factors in questionnaire data are still rife.

s. *primary structure* (Aeronaut.), those parts of an aircraft whose failure would seriously endanger safety.

1939 *Aircraft Engin.* Feb. 66/1 The primary structure of the main plane is erected in a wall jig. **1959** F. D. Adams *Aeronaut. Dict.* 132/1 Elements not a part of the primary structure include cowlings, fairings, windshields, etc. **1964** J. E. D. Williams *Operation of Airliners* ix.

133 To ensure that throughout the operational life of the aircraft the possibility of disastrous fatigue failure is remote, the primary structure must either have an extremely long probable life or be designed to fail safely.

t. Of radar: transmitting radiation for targets to reflect, not requiring any generation of signals by targets.

1945 R. Watson-Watt in *Nature* 15 Sept. 323/2 Primary radar is that form of radar which 'does not require the co-operation of the object to be located'. **1960** T. J. Morgan *Radar* xii. 137 The measurement of the wind direction and velocity was carried out by means of direction finding stations, then later by primary radar. **1963** R. S. H. Boulding *Princ. & Pract. Radar* (ed. 7) xxii. 470 In secondary radar as distinct from normal (or primary) radar, the target plays an active part in the operation of ranging and position finding.

u. *primary stress*, the principal stress in a word (see Stress *sb.* 8). Also *attrib.* So *primary-stressed* adj., (of a word or syllable) carrying a primary stress.

1951 Trager & Smith *Outl. Eng. Struct.* i. 36 There must be a stress phoneme whose characteristic is maximum normal loudness, which we may call *primary stress*. **1964** G. L. Trager in D. Abercrombie et al. *Daniel Jones* 267 When a clause begins with the primary-stressed syllabic, there are only two pitches. **1968** Chomsky & Halle *Sound Pattern Eng.* 34 We note that primary stress falls on the prefix if the stem is monosyllabic. *Ibid.* 114 The other..vowels..have never received primary stress at any stage of their derivation. **1971** *Eng. Stud.* LII. 349 From recorded readings we know that the unnatural stress-patterns which result can be emended by a prolongation of the first of two primary-stressed syllables. **1972** *Language* XLVIII. 328 In simple declarative sentences ending in a predicate, primary stress is often most naturally placed on the subject. *Ibid.* 331 Surface structure alone is insufficient to determine primary-stress placement.

v. *primary structure* (Biochem.), the sequence of amino-acids forming the chain structure of a protein or polypeptide, as opposed to the three-dimensional configuration and arrangement of the chains.

1952 K. U. Linderstrøm-Lang *Proteins & Enzymes* 58 The presence of intrahelix as well as interhelix bonds may justify a classification into secondary (intrahelix) and tertiary (interhelix) structures, as distinct from the primary structure of the simple β-chain. **1964** G. H. Haggis et al. *Introd. Molecular Biol.* iii. 48 The peptide chains coil up, and fold back on themselves, to form a complex three-dimensional molecular structure, and..the determination of sequence, or primary structure, is only the beginning of a description of the protein molecule. **1970** R. W. McGilvery *Biochem.* viii. 150 Any disruption of protein structure that does not involve the primary structure ought to be self-healing once the cause of the disruption is removed. **1974** *Nature* 29 Nov. 351/2 Let us now discuss the effect of natural selection on secondary or tertiary structure, as natural selection acts through these higher order structures and not on primary structure.

B. *sb.* **4.** (Further examples.)

1967 E. Short *Embroidery & Fabric Collage* i. 14 If the three primaries are placed on a white background the reverse will happen. **1972** *House & Garden* Feb. 70/2 The most successful rooms are those using vivid primaries. **1979** *Guardian* 13 June 12/5 Cellular cotton shirts come in all pastels and primaries.

5. (Earlier and later examples.) Also, a primary circuit, current, etc.

1837 M. Faraday in *Ann. Electr., Magn., & Chem.* I. 200 The renitency encountered in the conductors will necessarily exercise a due influence in lessening the force of secondary currents, but cannot be made available as a cause of the comparative atony which these currents, by the initial impulses of the primary, invariably display. **1896** F. Bedell *Princ. Transformer* i. 2 The alternating current transformer..consists simply of two independent circuits, a primary and a secondary, wound independently upon a common core of laminated soft iron. **1931** B. Brown *Talking Pictures* iii. 36 Modulation of the voltage is accomplished through a transformer, the primary of which is supplied with the speech recording current coming from the microphone, via the amplifiers. **1938** Kerchner & Corcoran *Alternating-Current Circuits* vii. 182 Circuit 1, energized by means of an alternating potential difference, is called the primary. **1967** M. F. Buchan *Electr. Supply* vi. 156 Consider now what happens if the secondaries are connected in delta, whilst the primaries are star-connected to a 3-wire supply.

6. (Earlier and later examples.) Add to def.: now usu. = *primary election*, one at which candidates for political office in the U.S.A. are chosen.

a **1861** T. Winthrop *Life in Open Air* (1863) 147 'Boys,' said he,..'when I accepted the office of Orator of the Day at our primary and promised to bring forward our resolutions in honor of Mr. Wade.' **1868** *All Year Round* 19 Sept. 351/2 He is 'powerful' in 'primaries', where he votes early and often for his favourite candidates. **1900** B. C. Clark in *Mod. Eloquence* X. p. xvii, Those of you who remember as I do the times that tried men's souls will not, I hope, forget their humble servant when the primaries shall be held. **1908** *Contemp. Rev.* Apr. 404 Other Western States have passed similar laws for direct primaries. **1930** *Economist* 7 June 1267/1 They have nominated Mr. Franklin Fort, a 'dry' Congressman, as his opponent in the Republican 'primaries'. **1966** *Listener* 25 Aug. 289/3 He could be freed from this bondage either by a system of primaries in which the field of choice were extended to others beyond the narrow range of committed party workers or through the introduction of the alternative vote. **1967** *Boston Herald* 1 Apr. 1/1 George C. Wallace, Jr., said Friday that he was thinking of entering the New Hampshire presidential preferential primary next March 12. **1976** *Times* 26 Feb. 15/2 Sitting Presidents need to do far more than escape disaster in the primaries, which are popularity contests among their own party's voters.

8. *Physics* and *Astr.* A primary ray or particle, esp. a primary cosmic ray.

1923 *Physical Rev.* XXII. 243 The emission is comprised in part of electrons whose speeds are not appreciably less than that of the primaries. **1932** *Ibid.* XLI. 545 The average number of secondaries per primary is about 100 in iron and 230 in lead. **1942** J. D. Stranathan *'Particles' of Mod. Physics* xii. 488 Many of the primaries and many of the large number of secondaries formed high in the atmosphere are unable to penetrate the entire atmosphere. **1956** *Spaceflight* Oct. 27/1 The charged cosmic ray primaries..consist of approximately 80 per cent. protons, 18 or 19 per cent. alpha particles, and the remainder heavier nuclei. **1959** K. Henney *Radio Engin. Handbk.* (ed. 5) vii. 15 If all the energy brought into the body by the bombarding primary could be transformed into the energy required for the emission of secondary electrons, the ratio δ would be very high. **1974** *Encycl. Brit. Macropædia* V. 203/1 The discovery in meteoritic crystals of particle tracks produced by primaries heavier than iron.

9. *Gram.* In Jespersen's terminology, a word or group of words (normally a noun or a noun-phrase) of primary importance in a phrase or sentence. Cf. *adjunct *sb.* 5 b, *subjunct.

1924 O. Jespersen *Philos. Gram.* vii. 97 We may, of course, have two or more coordinate adjuncts to the same primary: thus, in *a nice young lady*, the words *a*, *nice*, and *young* equally define *lady*. **1928** —— *Internat. Lang.* II. 97 When adjectives are made into primaries, we have the endings already considered. **1928** *Mod. Lang. Rev.* XXIII. 143 After a chapter on clauses as 'primaries', about 150 pages are devoted to a thorough survey of relative clauses. **1935** [see *noun-equivalent]. **1940** *Eng. Stud.* XXII. 88 *Primary, secondary* and *tertiary*, intended to connote.. *headwords, attributes* and *adjuncts*, the terms representing their relative importance or ranks within the sentence, headwords coming first. **1959** M. Schlauch *Eng. Lang. in Mod. Times* viii. 221 In this system [of Otto Jespersen's] a leading term, for instance a noun subject, is a primary.

primate, *sb.*[1] Add: **4.** (Usually with pronunciation prəi·mēⁱt.) Substitute for def.: A mammal belonging to the order Primates, which includes man, apes, monkeys, and several groups of prosimians. Also *attrib.* (Examples.)

1898, 1899 [see Primates *sb. pl.*]. **1906** E. Ingersoll *Life of Animals* 7 The higher the Primate in the scale of organization the more perfectly are its fore limbs and hands adapted to seizing and handling objects. **1929** R. M. & A. W. Yerkes *Great Apes* i. 2/1 No infrahuman primate, least of all the great ape, has been thoroughly domesticated. **1967** J. R. & P. H. Napier *Handbk. Living Primates* p. v, Animal behaviour, ecology and genetic biology..today dominate basic research trends in primate biology. **1977** Rainier III & Bourne *Primate Conservation* p. xviii, All authors of this book agreed that the royalties earned should be used to further primate conservation.

primatology (prəimătǫ·lŏdʒi). [f. Primate *sb.*[1] 4 + -ology.] The study of primates.

1941 J. F. Fulton in T. C. Ruch *Bibliographia Primatologica* I. p. xi, Dr. Ruch has followed Linnaeus in adopting the term 'primate', and in the title of his bibliography he has introduced a new and useful derivative 'Primatology'. **1942** Ruch & Fulton in *Science* 9 Jan. 47/1 It is only within the past decade or two that the study of the primates as a distinct zoological group has come into sufficient stature to warrant separate designation—'primate biology' or if you will, 'primatology'. **1956** *Nature* 17 Mar. 505/2 Duckworth had very wide interests in anatomy, teratology, primatology, anthropology, archaeology and general natural history. **1967** *Guardian* 20 Oct. 6/3 With the growth of primatology we are increasingly aware how unusual are man's problems in dealing with his fellows. **1973** *Sci. Amer.* Jan. 11/1 Geza Teleki ('The Omnivorous Chimpanzee') is completing work for his Ph.D. in primatology at the University of Georgia. **1976** *Primate Eye* VII. 1 In this issue of Primate Eye, we are including a number of reviews of books in the field of primatology.

Hence **primatolo·gical** *a.*, of or pertaining to primatology; **primato·logist**, one who studies primates.

1945 M. F. A. Montagu *Introd. Physical Anthropol.* ii. 11 Dead primates enable the primatologist to make detailed studies of their anatomy. **1949** *Antiquity* XXIII. 126 One of the few technical treatises on man's embryology that is anthropological and primatological in its orientation. **1957** L. Eiseley *Immense Journey* 105 Primatologists may therefore be forgiven their fumblings over great gaps of millions of years from which we do not possess a single complete monkey skeleton. **1973** *Nature* 18 May 175/2 A general section on the radiation of the placental mammals, placing primate evolution in its proper perspective—a perspective which is far too readily forgotten in many modern primatological studies. **1977** *Ibid.* 20 Oct. 654/3 He launched an attack on primatologists..for the way they have generally discussed the adaptive significance of social behaviour.

primavera. (Later example.)

1953 R. Chandler *Long Good-Bye* xv. 95 The table of polished primavera.

prima vista. 2. (Examples.)

1845 [see Transpose *v.* 7]. **1974** *Listener* 14 Feb. 218/2 The London Symphony Orchestra..is still full of excellent players who can read almost everything *prima vista*.

prime, *sb.*[2] Add: **I. 1. b.** *Linguistics.* A simple, indivisible linguistic unit.

1959 F. W. Householder in *Word* XV. 231 (*title*) On linguistic primes. **1961** —— in Saporta & Bastian *Psycholinguistics* 19/1 We must recognize at least two kinds of linguistic units: (1) *ultimate units*, or *primes*, out of which other more complex units may be constructed. **1963** J. Lyons *Structural Semantics* ii. 11 The lexeme is a formal unit of grammatical analysis, established distributionally. It may or may not be a 'prime'..though it frequently is for Greek. **1964** E. Bach *Introd. Transformational Gram.* v. 58 Each *level* of a linguistic theory comprises..a set of *primes* (i.e., atoms or indivisible elements). **1965** N. Chomsky *Aspects of Theory of Syntax* 222 Each level *L* is a system based on a set of primes (minimal elements—i.e. an alphabet). **1975** —— *Logical Struct. Linguistic Theory* iii. 105 If *a* and *b* are (not necessarily distinct) primes of L, we can form a⌢b and b⌢a as new elements of L.

2. d. (Further examples.)

1917 D. W. Payne *Founder's Man.* p. xi, The prime mark ' above a number means minutes or linear feet. **1964** *Amer. Jrnl. Physics* XXXII. 264/2 The prime (′) here indicates ordinary differentiation of a function of a single variable. **1973** A. H. Sommerstein *Sound Pattern Anc. Greek* iii. 91 A Roman numeral followed by a prime indicates that the convention in question is intended to *replace* the Chomsky-Halle convention of the same number. **1976** *Physics Bull.* May 191/2 We would like to draw the attention of readers to the equation $(B_\theta)_i = B_\theta - D'M$ (our prime) appearing in the last column.

5*. Colloq. abbrev. of Prime Minister (sense 3). Cf. Prime *sb.*[1] 10.

1916 A. Huxley *Let.* May (1969) 99 The Prime received suddenly one morning a letter..'Dear Mr. A[squith]'. **1924** Galsworthy *White Monkey* i. ii. 14 Didn't he think that the cubic called 'Still Life'—of the Government, too frightfully funny—especially the 'old bean' representing the Prime?

5.** *Cycling.* An especially difficult stage in a long-distance cycle race.

1959 *Observer* 31 May 32/4 Weatherlaw was the first 'prime'—a specially marked stretch of hilly road which gives the first three men to the summit a bonus of money and time. **1961** *Times* 7 June 5/6 Pewter tankards being offered as prizes at the town primes along the route. **1975** *Oxf. Compan. Sports & Games* 235/1 On mountainous stretches certain summits are designated as *primes*.

5*.** Short for **prime rate*.

1973 *Business Week* 10 Feb. 19 The all-out struggle over the prime. **1977** *Offshore Engineer* July 19/3 Citibank's current prime is 6¾%. **1978** *Daily Tel.* 25 Nov. 19/2 At 11·5 p.c., the prime now stands at its highest level since October 1974.

prime, *a.* Add: **3. b.** *Broadcasting.* Pertaining to or associated with the largest audience of the day. See also *prime-time 3.

1959 *Times Lit. Suppl.* 6 Nov. p. xxxi/5 John Fischer, the editor of *Harpers*,..asks for an autonomous authority empowered to produce programmes of exceptional merit, financed by a levy on the income of the broadcasters, who will also be under an obligation to transmit these programmes in the cherished prime-viewing hours. **1961** Webster s.v., Prime television time. **1976** *Broadcast* Dec. 15/3 We have to go through the routine again slap in the middle of prime listening time on a Saturday morning. **1977** *Times* 7 Dec. 19/1 It is seldom that anyone gets handed such a quantity of prime television time to do what he likes with.

4. a. *prime rib*, best rib of beef, i.e. one of the first two ribs in the forequarter. Also *attrib.*

1960 E. David *French Provincial Cooking* 335 To satisfy customers, butchers bone, trim and tie up secondary cuts of meat..and sell them at a small amount less than, say, sirloin or prime ribs. **1973** D. Barnes *See the Woman* (1974) ii. 195 I'll buy us a prime-rib dinner. **1973** *Listener* 19 Apr. 501/1 A landscape of luscious rib roasts, lamb chops, shell steaks, T-bone steaks, sirloin steaks, fillet mignon, prime ribs, veal piccata and so on. **1978** *Chicago Tribune Mag.* 2 Oct. 8/2 The 25-cent hot beef special is now a $2.75 prime rib sandwich, served with salad and potato.

b. *spec.* used of land, the position of real estate, etc.

1634 [in Dict.]. **1850** *Househ. Words* 3 Aug. 433/1 Sir Roger Rockville..was the last of a very long line... His first known ancestor came over with William, and must have been a man of some mark,..for he obtained what the Americans would call a prime location. **1961** J. D. Adams in *Webster* s.v., Prime farming land. **1976** *Western Mail* (Cardiff) 27 Nov. 17/2 (Advt.), A charming detached Freehold four-bedroom residence built in 1935,..in a prime residential area. **1977** *Grimsby Even. Tel.* 24 May 4/9 (Advt.), A thriving newsagents, tobacconist, sweets and general business situated in a prime position in a growing village close to Grimsby. **1978** *Church Times* 1 Sept. 4/1 It may still occupy a prime site on a busy thoroughfare.

8*. *Astronautics.* Originally designated to take part or be used in a space mission, esp. in *prime crew*, the original person or persons selected to man a spacecraft.

1965 *Life* 3 Dec. 48 Pete Conrad, who spent a week in space, is the prime crew. **1970** N. Armstrong et al. *First on Moon* iii. 63 At 4:30 A.M. the transfer vans arrived—one prime, one backup. **1970** R. Turnill *Language of Space* 13 Every manned space flight, American or Russian, has a back-up crew, to replace the prime crew in the event of illness or death.

9. a. prime cost, the direct cost of something, not including discounts, expenses involved, etc.; **prime (interest, lending) rate**, the lowest rate of interest offered on bank loans at a given time and place.

1718 C. Hitchin *Receivers & Thief-Takers* 11 For instance, suppose you steal Goods to the value of twenty Pounds prime Cost. **1732** R. Johnson *Let.* 6 Oct. in *Cal. State Papers, Amer. & W. Indies* (1939) 231, I have examined what dutys are payd in this Province on English European ships or goods, and there is only 2½ p.c. our currency upon the prime cost of goods from Europe in general. **1775** in *15th Rep. R. Comm. Hist. Manuscripts* App. vi. 297 in *Parl. Papers 1897* (C. 8551) LI. 1 Vessels have come from Hispaniola, and sold gunpowder to the Provincials at prime cost. **1890** A. Marshall *Princ. Econ.* I. vi. vi. 519 This is the Prime cost which a manufacturer has commonly in view when, trade being slack, he is calculating the lowest price at which it would be worth his while to accept an order. **1925** S. E. Thomas *Elem. Econ.* xiii. 168 The excess is 'dumped' on foreign markets at a price just sufficient to cover prime cost, or even below prime cost. **1938** Bowers & Rowntree *Econ. for Engineers* (ed. 2) xiii. 258 Prime costs, according to accounting procedure, are direct labor and materials expenses alone... Many economists, however, define prime costs as all *avoidable* costs, so that a plant shutdown would eliminate all prime costs. **1944** A. Cairncross *Introd. Econ.* xvi. 199 Normally, some surplus over prime or variable costs will be earned. **1953** Stonier & Hague *Textbk. Econ. Theory* v. 110 'Variable', 'prime' and 'direct' costs represent all those costs which can be altered in the short run as output alters. **1962** S. Strand *Marketing Dict.* 573 *Prime cost*, the cost of a product involving labor and all the parts that go into making it. **1978** J. Kellock *Elements of Accounting* x. 175 The value of rejected materials or by-products will be credited to the cost of raw materials purchased or deleted from the prime cost. **1958** *Wall St. Jrnl.* 29 Dec. 8/3 The 'prime' rate was thus brought back to within a half percentage point of the 4½% rate which was in effect from August, 1957... The 'prime' rate is the interest banks charge their biggest borrowers with the best credit status. **1970** *Daily Tel.* 4 Mar. 19 Several reductions in 'prime lending rate' by small American banks, are not likely to lead to cuts by major American banks for at least a fortnight. **1972** *Bankers Mag.* (Boston, Mass.) Winter 45/1 Many bankers have turned critical on the concept of a prime rate. **1973** *N.Y. Law Jrnl.* 26 July 3/3 Better-quality construction loans with a takeout were being made at the beginning of the year at a spread of about three points over the prime rate. **1978** S. Brill *Teamsters* vi. 252 Glick's Las Vegas loans were given at a time when he was far from being a candidate for a prime rate.

prime, *v.*[1] **4.** Delete *local* and add later examples. Also *fig.*

1930 *Engineering* 11 Apr. 473/1 Special arrangements for priming the pumps are not required, as the latter themselves exhaust all the air automatically during the first few revolutions, when the engine is being started on compressed air. **1973** L. Russell *Everyday Life Colonial Canada* v. 64 The..shaft had a piston with a leather diaphragm, which had to be wetted ('primed') by pouring a little water into the pump. **1977** T. Sharpe *Gt. Pursuit* xiii. 124 Significance is all... Prime the pump with meaningful hogwash.

b. *Aeronaut.* To inject fuel into (the cylinders of an aircraft engine) to facilitate starting. Also *intr.*, or with the engine as *obj.*

1915 G. A. Burls *Aero Engines* i. 20 L is a cock, or tap, communicating with the 'combustion chamber'..and may be used to 'prime' the chamber with a few drops of petrol. **1927** V. W. Pagé *Mod. Aircraft* (1928) xiv. 571 Prime engine by injecting a small quantity..of gasoline through each priming cock. **1931** M. M. Farleigh *Princ. & Probl. Aircraft Engines* x. 166 The cylinders should then be primed with fresh gasoline in the case of extreme sub-zero climatic conditions. **1939** *Aero Engines* II. 256 Do not prime excessively,..one stroke of the priming pump usually being found sufficient for a hot engine. **1941** A. W. Judge *Aircraft Engines* II. xi. 395 If the engine had been properly primed with mixture it was possible to start it by means of the starter magneto alone. **1977** D. Beaty *Excellency* vi. 80 He primed the engines, pressed the starter button, heard the propeller creak round.

c. *fig.* Cf. *pump priming *vbl. sb.*

1959 *Conferences & Exhibitions* Mar. 23/2 The aim is to introduce a trade to a fair in which it has not exhibited before by 'priming the pump'. **1963** *Times* 26 Jan. 12/1 He was understandably slow to prime the pump when a quick success in the Brussels negotiations might have given enough extra, uncontrollable impetus to start an inflation.

7. *Biol.* and *Med.* To treat (an animal or tissue) so as to induce a desired susceptibility or proclivity.

1943 *Jrnl. Endocrinol.* III. 273 Excellent results were obtained with rabbits..primed by five daily injections of 2 mg. of AP61B or AP118B. *Ibid.*, Of eleven rabbits primed with five daily doses of 1 mg. only four accepted the buck. **1963** *Rec. Progress Hormone Res.* XIX. 674 In the latter effect, the body is in a true sense 'primed' for new biological activity. **1967** *Science* 17 Nov. 939/3 The 30 mice anesthetized with ether were exposed for 30 seconds to atmosphere containing 5 ml of ether per 1 liter of air; half of these were acoustically primed 10 seconds later. **1971** *Nature* 24 Dec. 456/1 In guinea-pigs primed with DNP-OA, injection of allogeneic lymphoid cells stimulates synthesis of antibodies to both hapten and carrier. **1975** *Jrnl. Compar. & Physiol. Psychol.* LXXXIX. 214/2 At 16 days postnatally, 18..mice were acoustically primed by a 30-sec exposure to the sound produced by an electric bell.

primed *ppl. a.*, (*b*) *Biol.* and *Med.*, rendered susceptible; prepared; (cf. sense 7 above).

1943 *Jrnl. Endocrinol.* III. 273 In considering the question of inducing superovulation from the primed ovary, the first question arising is whether the ovulation-producing act of mating is sufficient to cause ovulation in a greater than normal number of follicles. **1960** *Proc. Soc. Exper. Biol. & Med.* CIV. 589/2 The primed mechanism responds to the second stimulus of the protein, but the resulting antibody is apparently specific for certain particular loci on the protein surface. **1967** *Science* 17 Nov. 940/1 Biochemical examinations of primed and nonprimed mice may reveal whether differences in oxidative phosphorylation..are associated with changes in audiogenic seizure susceptibility. **1975** *Behavior Genetics* V. 328 In an attempt to determine the increase in susceptibility due exclusively to acoustic priming, the seizure severity scores from the nonprimed..mice of a concurrently run study.. were subtracted from the severity scores of the primed, non-cross-fostered mice of the present study.

prime, *v.*[2] Delete 'Now only *dial.*' and add: **2.** *U.S.* To pull off the lower leaves of tobacco plants.

1792 J. Pope *Tour S. & W. Terr. U.S.* 63 [The Creeks] scarcely ever weed, hill, prime, top or succour their Tobacco. **1963** H. Garner *Best Stories* 167 Taking suckers first make [*sic*] it better to prime after. *Ibid.* 168, I thought of the rows upon rows still to be primed of sand leaves, the lowest leaves on the plant.

primed (praimd), *a.* [f. Prime *sb.*[2] + -ed[2].] Having the symbol ′ as a superscript.

1927 *Proc. R. Soc.* A. CXIII. 630 We shall make it a rule always to use..primed or multiply primed letters such as ξ′ and α″ to denote parameters, representing matrix rows and columns. **1944** C. Palache et al. *Dana's Syst. Min.* (ed. 7) I. 13 Projection values are primed, and polar values unprimed. **1968** *Amer. Jrnl. Physics* XXXVI. 1105/1 He obtained $F' = l^2(1, \gamma, \gamma)F$ and $a' = l(\gamma^2, \gamma^3, \gamma^3)a$, for the force and acceleration transformed to the instantaneous (primed) rest frame of the electron.

prime mo·ver. [f. Prime *a.* (*adv.*) + Mover[1].] **1.** (See Mover[1] 3. Further examples.)

1972 *Science* 26 May 892/3 Jamie L. Whitten,..a prime mover in the passage 18 years ago of Public Law 566. **1973** *Amer. Speech 1969* XLIV. 293 Concerning the origin of the jargon, Adams..relies on the theories of his informants... Their views focus on the probable ages of the originators: teenagers or adults or an intermediate group of 'prime movers'. **1973** *Nature* 17 Aug. 467/1 Both of them have been prime movers, in the period of forty years covered, in securing the establishment of ethology as a separate, respectable and inevitable branch of animal biology. **1977** *Jrnl. R. Soc. Arts* CXXVI. 40/2 Maria Grey, 'prime mover' of this enterprise, would have been the first to wish to have associated with her the men and women who helped in the work.

2. (See Mover[1] 2b. Further examples.)

1967 R. Whitehead in Wills & Yearsley *Handbk. Managem. Technol.* iv. 55 Up to almost the present time engineering has been concerned with the study of prime movers—heat converted into usable energy. **1971** *Sci. Amer.* Sept. 37/3 By the 16th century the waterwheel was by far the most important prime mover. *Ibid.* 152/3 The piston engines in the nation's more than 100 million motor vehicles have a rated capacity in excess of 17 billion horsepower, or more than 95 percent of the capacity of all prime movers (defined as engines for converting fuel to mechanical energy). **1974** *Petroleum Rev.* XXVIII. 783/2 Primary production equipment, comprising gas/oil separators and forwarding pumps with gas turbine prime movers.

3. A towing vehicle; *spec.* (see quot. 1963).

1938 T. J. Hayes *Elem. Ordnance* xix. 677 Artillery prime movers are used to tow artillery. **1945** *Finito! Po Valley Campaign* (15th Army Group) 51 A German convoy of two 170 mm cannon pulled by prime movers. **1962** *Exhib. Brit. Military Vehicles* 128 This vehicle is a development of the tractor 30 ton 6×4 G.S...for semi-trailer F.V. 12002 which it will replace in the service as prime mover for the 50 and 60-ton semi-trailer tank transporters. **1963** *Dict. U.S. Mil. Terms* (U.S. Dept. Defense) 171 *Prime mover*, a vehicle, including heavy construction equipment, possessing military characteristics, designed primarily for towing heavy, wheeled weapons and frequently providing facilities for the transportation of the crew of, and ammunition for, the weapon. **1969** *Age* (Melbourne) 24 May 61/9 (Advt.), Commer semi trailer outfit complete, 61 mod., petrol motor, prime mover and 34 ft. semi trailer... $2000. **1976** *Daily Times* (Lagos) 22 Sept. 15/1 Those above eight tonnes but not articulated prime movers i.e. with trailers now cost ₦35. **1977** 'D. Rutherford' *Return Load* ii. 28 He had..invested a legacy..in a Leyland prime mover... Like most owner-drivers he made a practice of hiring the semi-trailers which, when hitched to the tractor, made up the complete articulated vehicle.

primer, *sb.*[1] Add: **2. d.** *N.Z.* (With pronunc. pri·maɪ.) One of the primer classes, covering the first years of instruction in a primary school; also, a child in a primer class.

1928 *Syllabus of Instruction for Primary Schools* (N.Z. Dept. Educ.) 55 In all schools teachers of Primer classes will use 'Physical Exercises and Games for Infants'. *Ibid.*, Where there are three teachers: Primers do infant work; Squad I, Tables 1–36; [etc.]. **1947** 'A. P. Gaskell' *Big Game* 92 There was Micky, her [*sc.* the teacher's] smallest primer, a little wizened creature with sad eyes. **1957** J. Frame *Owls do Cry* xxviii. 125 He is in primer three at school. **1963** N. Hilliard *Piece of Land* 191 It seemed no time since he'd been in the primers. **1963** B. Pearson *Coal Flat* i. 8 She had taught in the primers of his school when he was in Standard six.

primer, *sb.*[2] Add: **2. b.** *Biochem.* A molecule that serves as a starting material for a polymerization (see quot. 1976). Freq. *attrib.*

1954 Cantarow & Schepartz *Biochem.* xvii. 391 A 'primer' of branched polysaccharide, the main linkages of which are α-1,4, is essential for the action of animal phosphorylase. **1963** *Proc. Nat. Acad. Sci.* XLIX. 533 (*heading*) Formation of DNA-RNA hybrids with single-stranded DNA as primer. **1965** M. W. Neil *Vertebr. Biochem.* (ed. 2) xii. 182 Polysaccharide synthesis involving the addition of uridine diphosphate-bound units to a primer chain is widespread in nature, and is the mechanism whereby such macromolecules as cellulose, chitin, starch and the mucopolysaccharides, in addition to glycogen, are elaborated. **1976** Conn & Stumpf *Outl. Biochem.* (ed. 4) xviii. 507 Primer, in biochemistry, refers to the initial terminus of a molecule onto which additional units are added to produce the final product. **1977** D. E. Metzler *Biochem.* xv. 903/2 The enzyme displays many of the properties expected of a DNA-synthesizing enzyme. It requires a template strand of DNA as well as a shorter primer strand.

c. *Physiol.* (See quots. 1963, 1975.) Freq. *attrib.*

1963 Wilson & Bossert in *Recent Progress Hormone Res.* XIX. 674 We propose to distinguish the releaser effect, involving the classical stimulus-response mediated wholly by the central nervous system, from the primer effect, in which the endocrine and reproductive..systems are altered physiologically. *Ibid.* (*caption*) The pheromone may be the primary stimulus causing a quick behavioral response (releaser effect), or it may act more slowly and indirectly by altering the physiology and 'priming' the animal for a different behavioral repertory (primer effect). **1971** *Nature* 16 Apr. 432/2 Pheromonal primer effects are near-universal in social mammals, including primates. **1975** *Ibid.* 20 Nov. 194/2 The action of pheromones is commonly divided into two classes..: chemical 'releasers' of specific acts of behaviour, and 'primers' which seem to act initially on the endocrine system.

3. Delete † *Obs.* and add later examples.

1937 *Times* 13 Apr. (Brit. Motor Suppl.) p. xiii/1 Before colour can be applied the body undergoes a number of preparatory stages, being thoroughly washed down with an acid cleaner and afterwards with hot and cold water, and dried off in preparation for the first coat of primer. **1958** *Listener* 14 Aug. 251/2 If the patches are touched in with primer and undercoat you will not run into any trouble. **1969** W. R. R. Park *Plastics Film Technol.* vi. 156 An effective primer works better with a thick coating. **1976** *Southern Even. Echo* (Southampton) 11 Nov. 8/3 These were then given a 'primer' coat of lime plaster—almost like white-wash—and the geometrical designs painted on.

5. *Aeronaut.* = *priming pump* s.v. *priming *vbl. sb.*[1] 8.

1923 *Gloss. Aeronaut. Terms* (Brit. Engin. Standards Assoc.) 48 *Engine primer*, a device for supplying fuel to the induction pipe or combustion chambers to facilitate starting. **1932** Chatfield & Taylor *Airplane & Engine* (ed. 2) x. 225 In automobiles this temporary excess of fuel for starting is supplied by means of the choke but for airplane engines a primer is usually used. **1939** *Aero Engines* II. 256 Always turn off the primer after use.

prime-time. Restrict † *Obs.* to senses in Dict. and add: **3.** *Broadcasting.* (Except in *attrib.* use usu. as **prime time.**) The time of day when an audience is expected to be at its largest; a peak listening- or viewing-period. Also *attrib.* and *absol.*, prime-time television. Also *transf.*

1964 *Variety* 2 Dec. 31/3 For the first time in years, WNBC-TV has copped the number one rating position in prime time, in the highly competitive N.Y. market. **1966** [see E.S.T. s.v. *E. III]. **1971** *Daily Tel.* 13 Feb. 15/7 A 2p coin will buy three minutes time for local calls in prime time and six minutes at night and weekends. *Ibid.* 17 Apr. 19/3 The average [commercial local radio] station should aim to sell some 17,500 minutes of prime time in an average year at an average rate of £10 per minute. **1973** R. Stout *Please pass Guilt* (1974) xiv. 143 That ad would have made a wonderful five-minute spot... She would have been glad to pay for prime time—say ten o'clock. **1976** *Billings* (Montana) *Gaz.* 18 June 12-D Jaclyn Smith is one of the gals who's huckstered in TV commercials a committee studied along with prime-time programs to determine the image given women on the small screen. **1977** *New Yorker* 10 Oct. 124/2 The Grand Central Racquet Club..charges the highest fee I know of for renting either of its two courts—forty-five dollars an hour in prime time. **1978** G. Vidal *Kalki* vii. 179 Wasn't Kalki blown to bits before our very eyes on prime-time?

primeur. Add: **a.** (Later examples.) **b.** New wine.

1913 E. Wharton *Custom of Country* ii. xii. 172 A bill burdened by Undine's reckless choice of *primeurs*. **1924** R. Fry *Let.* 2 July (1972) II. 555 They raise three crops a year of *primeurs*. **1937** W. Fortescue *Sunset House* vi. 118 She prides herself upon her *primeurs*, being a scientific gardener. **1950** *Vogue* Aug. 100/4 Intellectuals..spend a lot on:..Exotic food (but not *primeurs*). **1968** A. & G. Sainsbury *France & her People* i. 19 In Brittany are places..with a mild climate which has made them famous for the production of *primeurs*, the early fruit and vegetables. **1973** *Times* 15 Dec. 11/3 A wine can be called 'Primeur' if it is offered for drinking before the date when the wines made in the normal way and bearing a vintage date are put on the market. **1975** *Harpers & Queen* May 34/2 Beaujolais is the success of the century,..even the new *primeur*..now brought over to be tasted at two months old. **1978** *Chicago* June 206/2 The Wassermans discuss the continuing tendency to produce Rhône reds—wines to be aged for as long as 30 years—in the *primeur* fashion.

primevalness. (Example.)

1971 D. Crystal *Linguistics* ii. 49 What evidence there was about language-history..militated against acceptance of even the most basic assumptions used in the arguments about primevalness.

primidone (pri·midōun). *Pharm.* [f. *p(y)rimid(ine + *-di)one.] A white, crystalline pyrimidine derivative, $C_{12}H_{14}N_2O_2$, which is an anticonvulsant used esp. to treat *grand mal* and psychomotor epilepsy. Cf. *Mysoline.

1953 *Brit. Med. Jrnl.* 5 Sept. 540/1 The introduction of primidone ('mysoline'; 5-phenyl-5-ethylhexahydropyrimidine-4:6-dione) as an anticonvulsant drug some two years ago. **1958** [see *Mysoline]. **1961** *Lancet* 9 Sept. 569/1 He was treated with primidone..and benzhexol hydrochloride..with striking reduction in the frequency of the seizures. *Ibid.* 569/2 He was discharged on primidone therapy. **1974** M. C. Gerald *Pharmacol.* xi. 214 Among the safest and most effective drugs are..primidone (Mysoline) and diphenylhydantoin for psychomotor seizures.

priming, *vbl. sb.*[1] Add: **7*.** In the sense of Prime *v.*[1] 4 (in Dict. and Suppl.). Cf. *pump priming *vbl. sb.*

1888 *Lockwood's Dict. Mech. Engin.* 266 *Priming*,.. (2) the priming of a force pump is the expulsion of the air from the water space, in order that the water shall enter into the partial vacuum thus produced... (3) The fetching of a lift pump by pouring liquid into the bucket in order to produce sufficient vacuum to enable it to draw. **1928** A. L. Dyke *Aircraft Engine Instructor* 217 The idling system also contains an air bleed which serves the.. purpose of..contributing to the operation of the priming device. **1931** M. M. Farleigh *Princ. & Probl. Aircraft Engines* x. 166 When..continued priming of the cylinders fails to bring about any combustion, the ignition should be checked carefully both for quality of spark and the time of its occurrence. **1969** W. T. Ingram et al. *Gloss. Water & Wastewater Control Engin.* 247 *Priming*,..the action of starting the flow in a pump or siphon.

7.** *Biol.* and *Med.* (See *prime *v.*[1] 7.)

1943 *Jrnl. Endocrinol.* III. 270 Pituitary extracts were administered by a series of subcutaneous injections for the purpose of stimulating the follicles ('priming'). **1963** *Recent Progress Hormone Res.* XIX. 673 New external stimuli following the priming are required to release the altered behavior patterns. **1967** *Science* 17 Nov. 939/2 Acoustic priming appears to be ineffective before the age of 14 days, corresponding to the normal onset of hearing in mice. **1975** *Behavior Genetics* V. 324 This failure of the 17-day-old albino mice to exhibit as great a change in seizure severity as a result of acoustic priming might have been due to their innately elevated auditory thresholds. **1978** *Nature* 5 Jan. 10/1 Production of interferon can also be modulated in other ways; pretreatment of cells with small amounts of homologous interferon before addition of an interferon inducer often increases the yield, a phenomenon termed 'priming'.

8. priming pump *Aeronaut.*, a small pump in an aircraft for priming its engine.

1932 R. Mahachek *Airplane Pilot's Man.* vi. 49 On large engines the choke is replaced by a priming pump which injects fuel directly into the intake system. **1942** D. M. Crook *Spitfire Pilot* 79, I..gave the priming pump a couple of strokes, and pressed the starter button.

priming (prəi·miŋ), *vbl. sb.*[3] *U.S.* [f. Prime *v.*[2] + -ing[1].] The action of removing the lowest leaves, or other layers of leaves, from a tobacco plant; also, the leaves removed.

1899 M. L. Floyd *Cultivation of Cigar-Leaf Tobacco* 14 The first priming, which means the first four leaves taken from the stalk, also the last priming, which means the last four or six leaves taken from the top of the stalk, are kept separate. **1904** E. Glasgow *Deliverance* 166 The very primings ought to be as good as some top leaves. **1938** *Daily Progress* (Charlottesville, Va.) 21 Oct. 5/1 Following the change from 'stalk cutting' to 'priming' (cutting of separate leaves for curing in bundles), less heat was required.

priming (prəi·miŋ), *ppl. a. Biol.* and *Med.* [f. Prime *v.*[1] + -ing[2].] That primes (see *prime *v.*[1] 7).

1930 *Amer. Jrnl. Physiol.* XCII. 129 The first test consisted of 'priming' injections of two rat units of purified extract into all animals. **1940** *Anat. Rec.* LXXVII. 1 Four to 6 month old rabbit does of medium-sized strains were injected subcutaneously..with a priming dose of the gonadotropic material. **1975** *Behavior Genetics* V. 324, 24 hr after the priming exposure the pigmented mice show an identically large increase in audiogenic seizures, whereas the albino mice have a lesser increase.

primitive, *a.* and *sb.* Add. **A.** *adj.* **I. 1. b.** Applied to behaviour or mental processes that apparently originate in unconscious needs or desires and have not been affected by objective logical reasoning.

1910 *Amer. Jrnl. Psychol.* XXI. 115 The following investigation of children's spontaneous constructions and primitive activities is made in the hope..that a clearer, saner insight into the child's nature and needs may follow. **1919** M. K. Bradby *Psycho-Anal.* iii. 28 The mind is unevenly developed, and what is relatively primitive coexists with what is advanced without completely harmonising with it. **1923** L. A. Clare tr. *Lévy-Bruhl's Primitive Mentality* 32 If then, primitive mentality avoids and ignores logical thought, if it refrains from reasoning and reflecting, it is not from incapacity to surmount what is evident to sense. **1924** *Brit. Jrnl. Med. Psychol.* IV. 32 Synthetic or intuitional conceptions of the unconscious, based on analogies with primitive notions and behaviour. **1962** M. Gabain tr. *Piaget's Moral Judgment of Child* ii. 189 It is not nearly so natural as one would think for primitive thought to take intentions into account.

2. b. *Anthrop.* That relates to a group, or to persons comprising such groups, whose culture, through isolation, has remained at a simple level of social and economic organization.

[**1781** Gibbon *Decl. & F.* III. xxxviii. 638 From this abject condition, perhaps the primitive and universal state of man.] **1903** C. S. Myers in *Rep. Cambr. Anthropol. Exped. Torres Straits* II. ii. 143 Stories which travellers relate about the remarkable capacity possessed by primitive peoples for distinguishing faint sounds amid familiar surroundings, cannot be accepted as evidence of an unusually acute hearing. **1920** R. H. Lowie *Primitive Society* (1921) i. 12 The knowledge of primitive society has an educational value that should recommend its study. **1938** R. Bunzel in F. Boas *Gen. Anthropol.* 333 There are ..certain primitive societies where the accumulation of wealth is considered undesirable. **1954** R. Firth in *Inst. Primitive Society* ii. 15 As I (and I think most of my colleagues) use it, 'primitive' is little more than a technological index—a shorthand term for a type of economic life in which the tool system and level of material achievement is fairly simple: little use of metals; no complex mechanical apparatus; no indigenous system of writing. **1963** *Brit. Jrnl. Sociol.* XIV. 21 Many books by social anthropologists have titles which include the word *primitive*. When we use this word..we refer to a low level of technology which limits social relationships to a narrow range. **1976** J. Friedl *Cultural Anthropol.* viii. 316 The primitive economy is one that is controlled exclusively by the local community.

II. 4. b. Applied to a parent language at an early, unrecorded, or reconstructed stage of its development into a group of dialects or languages.

1878 T. L. K. Oliphant *Old & Middle Eng.* i. 13 The Primitive Aryan *katvar* changes to the Gothic *fidwor* (our *four*). **1895** Kellner & Bradley *Morris's Hist. Outl. Eng. Accidence* (rev. ed.) iii. 30 The Teutonic languages differ much more from Primitive Aryan in the consonants than in the vowels. **1898** Primitive Germanic [used s.v. Germanic *a.* 2]. **1914** H. C. Wyld *Short Hist. Eng.* ii. 32 Parent, or Primitive Germanic, was divided into three great branches. **1920** *Trans. Philol. Soc. 1916–20* 129 (*heading*) Primitive Slavonic. **1933** L. Bloomfield *Language* i. 13 If a language is spoken over a large area,.. the result will be a set of related languages... We infer that..the Germanic (or the Slavic or the Celtic)..have arisen in the same way; it is only an accident of history that for these groups we have no written records of the language, as it was spoken before the differentiation set in. To these unrecorded parent languages we give names like *Primitive Germanic* (*Primitive Slavic*, *Primitive Celtic*, and so on). [*Note*] The word *primitive* is here poorly chosen, since it is intended to mean only that we happen to have no written records of the language. German scholars have a better device in their prefix *ur-* 'primeval'. **1972** M. L. Samuels *Linguistic Evol.* 2 The alternation corresponding to *stand–stood* was regular in the Indo-European system, and so with that corresponding to *seek–sought* in Primitive Germanic.

5. b. (Further examples.) *primitive cell*, the smallest unit cell of any particular lattice, having lattice points at each of its eight corners only; *primitive lattice*, a lattice generated by the repeated translation of a primitive cell.

1931 *Zeitschr. f. Kristallogr.* LXXIX. 501 The cell chosen is..not necessarily the primitive, i.e. smallest cell, as such a cell would often demand a description in oblique and inconvenient axes. But it is always either the primitive cell or a one- or three-face-centred or a body-centred cell. **1932** *Ann. Rep. Progr. Chem.* XXVIII. 263 *P* stands for primitive lattice. *Ibid.*, The rhombohedral lattice is designated by *R*, and the hexagonal by *C* or *H* according as the crystallographic axes coincide with or are perpendicular to the primitive translations of the lattice. **1945** C. W. Bunn *Chem. Crystallogr.* vii. 223 In a set of symbols characterizing a space-group, the first is always a capital letter which indicates whether the lattice is simple (*P* for primitive), body-centred (*I* for inner), side-centred (*A*, *B*, or *C*), or centred on all faces (*F*). **1966** *McGraw-Hill Encycl. Sci. & Technol.* III. 595/1 The three primitive cells of the cubic lattices are, respectively, a cube, a rhombohedron with a plane angle of 109° 28′, and a rhombohedron with an angle of 60°. The two rhombohedra are extremely inconvenient to handle; consequently, the body-centered and face-centered cubes are adopted in their stead. **1974** D. M. Adams *Inorg. Solids* ii. 12 In general it is convenient to work with the cell of highest symmetry and this is not necessarily primitive.

c. Applied to any root of an integer *n* such that the least power to which the root can be raised to yield unity modulo *n* is the totient of *n*.

1837 J. Hymers *Treat. Theory Algebraical Equations* x. 193 If *r* be one of the roots and *a* be a primitive root of the prime number *n*..it is proved..that all the roots of this equation may be represented by *r*, *ra*, [etc.]. **1916** G. A. Miller et al. *Theory & Applic. Finite Groups* xv. 308 For any prime *p*, it is shown in the theory of numbers that there exists a primitive root *g* of *p* such that $1, g, g^2, \ldots, g^{p-2}$, when divided by *p*, give in some order the remainders $1, 2, 3, \ldots, p-1$. **1972** J. E. & M. W. Maxfield *Discovering Number Theory* viii. 65 A primitive root (mod *m*) exists for $m = 2, 4, p^a$, and $2p^a$, where *p* is an odd prime and *a* is a positive integer. There is no primitive root for other values of *m*.

d. *Group Theory.* [tr. G. *primitiv* (S. Lie *Theorie der Transformationsgruppen* (1888) I. xiii. 221).] Applied to a substitution group whose letters cannot be partitioned into disjoint proper subsets in a way that is preserved by every element of the group.

1888 *Amer. Jrnl. Math.* X. 300 A group in the plane is primitive when with each ordinary point which we hold, no invariant direction is connected. **1897** W. Burnside *Theory of Groups of Finite Order* ix. 177 A simple group can always be represented in primitive form. **1933** L. P. Eisenhart *Continuous Groups of Transformations* ii. 80 The group of motions in the euclidean plane is primitive. **1968** D. Passman *Permutation Groups* i. 14 Let G be a transitive permutation group of prime degree. Then G is primitive.

e. *Logic* and *Math.* [tr. It. *primitive* (G. Peano 1897, in *Atti della R. Accad. delle Sci. di Torino* XXXII. 568).] Applied to concepts and propositions that serve as the basis of a deductive system and are not further defined or demonstrated; *primitive recursive* (see *recursive *a.* 2a).

1903 B. Russell *Princ. Math.* p. xi (*heading*) Two indefinables and ten primitive propositions in this calculus. **1910** Whitehead & Russell *Principia Math.* I. i. 95 Following Peano, we shall call the undefined ideas and the undemonstrated propositions primitive ideas and primitive propositions respectively. **1922** tr. *Wittgenstein's Tractatus* 121 The possibility of crosswise definition of the logical 'primitive signs' of Frege and Russell shows by itself that these are not primitive signs and that they signify no relations. **1932** Lewis & Langford *Symbolic Logic* i. 23 Thus it is proved that these primitive ideas and postulates for logic are the only assumptions required for the whole of mathematics. **1952** P. Geach tr. *Frege's Philos. Writings* 161 The same happens for the formula *a*=*b*. In some cases its meaning can be assumed as a primitive idea, in others it is defined. **1959** M. Bunge *Causality* ix. 233 Neither Aristotle nor his followers seem to have been aware of the *logical* necessity of admitting.. a set of unexplained or primitive concepts and ideas in order to avoid reasoning in a circle. **1970** E. Duckworth tr. *Piaget's Genetic Epistemol.* 7 Simultaneity, then, is not a primitive intuition; it is an intellectual construction.

f. Applied to those *n*th roots of unity of which the *n*th power, but no lower power, is unity.

1916 G. A. Miller et al. *Theory & Applic. Finite Groups* xvii. 325 For $p^2 = 9$, the six primitive ninth roots of unity are $\rho, \rho^2, \rho^4, \rho^5, \rho^7, \rho^8$ and are the roots of $x^6 + x^3 + 1 = 0$. **1971** E. C. Dade in Powell & Higman *Finite Simple Groups* viii. 274 We conclude that *F* contains a primitive *e*th root of unity.

10. *Primitive Baptist*: in the U.S.A., a member of a loosely organized secession of conservative character from the Baptist Church; also *attrib.*

The *Primitive Methodist Connexion* (after 1902 known as *Primitive Methodist Church*) united in 1932 with the United Methodists and Wesleyan Methodists to form the Methodist Church. The *Primitive Methodist Church, U.S.A.*, remains however a separate denomination.

1851 T. A. Burke *Polly Peablossom's Wedding* 143 Brethren Crump and Noel were both members of the Primitive Baptist Church. **1856** in N. E. Eliason *Tarheel Talk* (1956) 288 Was recived by examinytion on the primitive baptis faith. **1872** Z. N. Morrell *Flowers & Fruits* vi. 72 There was also an organization calling themselves 'Primitive Baptists', on the Colorado River. **1933** *Sun* (Baltimore) 24 Aug. 6/4 Elder A. J. Harrison..was elected head of the Ketockin Association, Old School, Primitive Baptists. **1948** *Daily Ardmoreite* (Ardmore, Okla.) 15 July 14/1 The Washita Valley Primitive Baptist association will meet at the Primitive Baptist church here, July 22. **1972** J. S. Hall *Sayings from Old Smoky* 144 Dave Reagan said of the Primitive Baptists, 'They are just like yellow jackets. They'd 'cruit [recruit] up in summer, and in winter they'd all die out'.

11. *Art.* **a.** Applied to the art and artists of pre-Renaissance western Europe.

[**1843** A. de Montor (*title*) Peintres primitifs.] **1847** Ld. Lindsay *Sk. Hist. Christian Art* II. ii. 93, I strongly suspect an ancestral relation between them [*sc.* the frescoes of the Baptistery at Parma] and the primitive and interesting school of Bologna. **1857** G. Scharf *Handbk. Paintings by Anc. Masters* (Art Treasures Exhib., Manchester) 5 Ottley,..an earnest student of the earlier periods of Italian art, had formed a small, but very authentic, collection of primitive works. **1923** J. Gordon *Mod. French Painters* ix. 94 In the early Italian primitive painters, and, indeed, in primitives of every order, we find beneath the artists' learning the foundations laid upon what may be called folk painting. **1927** R. Fry *Flemish Art* i. 24 This realization of space implies a sense of colour as a plastic function which is also almost entirely absent in primitive Flemish art. **1932** Konody & Lathom *Introd. French Painters* i. 3 What is known as primitive French painting is a hybrid art, composed of Italian, French, Spanish and German elements in varying proportions. **1970** *Oxf. Compan. Art* 925/1 Within the European context art historians and connoisseurs have used the term 'primitive' for early phases within the historical development of painting or sculpture in the various European countries.

b. Executed by one who has not been trained in a formal manner. Also, imitative of an early style suggesting lack of formal training. Of an artist: without formal training. Cf. *naïf *a.* 1b, *naïve *a.* 1c.

1942 J. Lipman *Amer. Primitive Painting* 5 The critic.. has come..to evaluate primitive art positively rather than negatively. *Ibid.* 7 The primitive artist typically allowed himself free rein in depicting pose, gesture..and background. **1952** M. McCarthy *Groves of Academe* (1953) viii. 148 On the walls were dark paintings of the first presidents, clergymen and theologians, a primitive engraving showing William Penn and the Indians. **1957** Primitive painting [see *naïve *a.* 1c]. **1962** W. Gaunt *Everyman's Dict. Pictorial Art* I. 12 A native development [in U.S.A.] of great interest in the eighteenth and nineteenth centuries was that of a 'primitive' or folk art, practised by sign painters and other craftsmen and amateurs. **1964** J. Summerson *Classical Lang. Archit.* v. 39 Laugier's primitivism..certainly appealed to him [*sc.* Sir John Soane] but he was prepared to go much further than Laugier in..inventing a 'primitive' order of his own. **1967** Primitive portrait [see *mourning-piece* s.v. *mourning *vbl. sb.*[1] 5]. **1976** *Sunday Times* (Colour Suppl.) 8 Feb. 7/3 Beryl Cook, seaside landlady and primitive painter, talks to Allen Saddler. **1978** I. Murdoch *Sea* 126 Hartley and Fitch were sitting stiff and upright, like a married pair rendered by a primitive painter.

12. *primitive accumulation* (Econ.): in Marxist theory, the original accumulation of capital, supposedly derived from the expropriation of small producers or smallholders, from which capitalist production was able to start; hence *primitive socialist accumulation*: the accumulation of capital which would be needed to start socialist production, also to be derived from the expropriation of small producers, smallholders, or peasants.

1887 Moore & Aveling tr. *Marx's Capital* II. viii. xxvi. 736 The whole movement..seems to turn in a vicious circle, out of which we can only get by supposing a primitive accumulation..preceding capitalistic accumulation. *Ibid.*, This primitive accumulation plays in Political Economy about the same part as original sin in theology. *Ibid.* 738 The so-called primitive accumulation ..is nothing else than the historical process of divorcing the producer from the means of production. **1935** E. Burns *Handbk. Marxism* xvi. 258 The so-called primitive accumulation of capital consisted in this case in the expropriation of these immediate producers. **1950** A. Erlich in *Q. Jrnl. Econ.* LXIV. 69 This formative period of modern capitalism..had now to find its counterpart in 'primitive socialist accumulation' which was assumed to serve as midwife in the same way for the socialist society of the future. **1959** *Listener* 29 Oct. 726/1 Trotsky proposed to carry through this Draconian programme, of what he called 'primitive socialist accumulation' without Stalin's terrible methods. **1965** B. Pearce tr. *Preobrazhensky's New Econ.* 67 If we partly exclude the operation of the law of value..we must accordingly replace its regulatory action by another law, inherent in planned economy at its present stage of development—the law of primitive socialist accumulation. **1967** I. Deutscher *Marxism in our Time* (1972) 242 It was out of the question that a country like this should be able to achieve socialism in such circumstances. It had to devote all its energies to 'primitive accumulation', that is, to the creation under state ownership of the most essential economic preliminaries to any genuine building of socialism.

B. *sb.* **I. 1. b.** (Later examples.) Also *transf.*, someone uncivilized, uncultured.

1924 *Brit. Jrnl. Med. Psychol.* IV. 35 The primitive has in many ways a contact with his environment of a refinement and subtlety that is more than a match for civilized brains. **1926** L. A. Clare tr. *Lévy-Bruhl's How Natives Think* 13 Primitives... By this term, an incorrect one, yet rendered almost indispensable through common usage, we simply mean members of the most elementary social aggregates with which we are acquainted. *a* **1936** Kipling *Something of Myself* (1937) vii. 184 Out of the woods..came two dark and mysterious Primitives. **1967** [see *Charley, Charlie 8]. **1972** *Buenos Aires Herald* 2 Feb. 7/1 The primitives fight for their territories and economic planners insist that the vast region must be opened. **1977** M. Cohen *Sensible Words* iii. 122 The newly emphasized methods of linguistic analysis include studying the language of children and 'primitives'.

4. a. (Further examples.) Also, a naïve painter; also *transf.* of artists working in another medium. **b.** (Further examples.) Also *attrib.* Also *transf.* of other art forms.

1907 R. Fry *Let.* 5 Mar. (1972) I. 282 A great Ferrarese altarpiece... The effect will be fine in our Primitive room. **1910** E. Singleton *Art of Belgian Galleries* i. 17 The Last Supper is one of the most profound and best-painted works of the Fifteenth Century; and if one were to make a list of five or six supreme masterpieces of the Flemish Primitives, this would have to be included. **1922** C. Bell *Since Cézanne* 51 One definitely artistic gift..many children do possess..is a sense of the decorative possibilities of their medium. This gift they have in common with the Primitives; and this the *douanier* possessed in an extraordinary degree. **1923** [see *primitive *a.* 11a]. **1932** F. F. Sherman *Early Amer. Painting* p. xv, Numerous dealers in antiques..offer them for sale..as 'primitives'. Primitives they certainly are not... They are worthless as works of art or of antiquity. **1934** *Musical Q.* Apr. 214 The Primitives stem from Moussorgsky, through Debussy and the *Sacre*. **1947** G. Greene *19 Stories* 155 The first season of 'primitives' [*sc.* films] was announced (a highbrow phrase). **1951** R. Firth *Elem. Social Organis.* v. 163 When we talk..of the Italian primitives..we are referring ..to art that is distinguished primarily by being earlier in time, though it..also bears the character of lack of sophistication. **1952** O. Kallir in A. M. Moses *Grandma Moses* p. xv, Grandma Moses is called a 'primitive'. Each of her pictures shows plainly that its author has had no art training. **1958** *Listener* 21 Aug. 269/2 The school of the 'primitives', represented by John Osborne, Sheelagh Delaney, and..Bernard Kops. **1959** E. Pound *Thrones* ci. 78 *Hs'uan Tsung*, 1389 natus, painted kittens, and Joey said, 'are they for real' before primitives in the Mellon Gallery. **1964** Mrs. L. B. Johnson *White House Diary* 20 Jan. (1970) 56 There was also a little American Primitive —just made you merry to look at it. **1974** P. De Vries *Glory of Hummingbird* (1975) iii. 39 We respected the artist's [*sc.* a writer's] reluctance to show portions of work not in sufficiently polished form because we felt..that here was a true primitive. **1976** *Sunday Times* (Colour Suppl.) 8 Feb. 22/1 Most of all, her paintings are funny. You can't say that about many primitives. **1977** *Jrnl. R. Soc. Arts* CXXVI. 35/2 The Flemish Primitives..would superimpose the dark colours and leave the pale colours transparent.

II. 6. (Later examples.) Also, = *phonetic *sb.*

1814 J. Marshman *Elem. Chinese Gram.* 36 If we then add the 214 elements to the 1689 primitives, we shall have one thousand nine hundred and three characters producing nearly the whole language. **1874**, **1907** [see *phonetic *sb.*]. **1909–10** L. Bloomfield in C. F. Hockett *Leonard Bloomfield Anthol.* (1970) 1 Derivative nouns and verbs also stand..in a definite ablaut relation to their primitives. **1975** *Language* LI. 969 It. *bozz-ello*..is an authentic derivative from *bozza*; while *bosel*, *bossel*, *bozel* in Renaissance French is a cluster of completely isolated forms lacking a primitive—a situation which reflects on the grammatical status of *-el*.

8. *spec.* a complete primitive (see Dict., this sense). (Examples.)

1885 A. R. Forsyth *Treat. Differential Equations* i. 8 The relation, which exists between the variables themselves without their differential coefficients and which is the most general one possible, is called sometimes the general solution, and sometimes the primitive, of the differential equation. **1929** T. C. Fry *Elem. Differential Equations* ii. 27 This relation includes every possible solution of the differential equation. It is called the general solution or primitive. **1969** B. Spain *Ordinary Differential Equations* i. 9 Obtain the differential equations corresponding to the primitives..$y = c \log x$..[etc.].

9. *Logic* and *Math.* A primitive concept or proposition (see *A. 5e). Also in extended use.

1950 *Jrnl. Symbolic Logic* XV. 130 Hence ϕ and μ as defined above will suffice as the sole primitives for the arithmetic of positive integers. **1960** G. Bergman *Meaning & Existence* ii. 44 It is not required that an improved language be interpreted by interpreting separately all, or even any, of its primitives. **1964** M. Black *Compan. Wittgenstein's Tractatus* 25 We find Wittgenstein..constantly returning to the theme of the 'logical indefinables' or the 'logical primitives'. **1964** R. H. Robins *Gen. Linguistics* iv. 133 Many linguists are prepared to accept these terms [*sc. contrast* and *distinctive*] as primitives, i.e. as requiring no further definition within linguistics. **1975** *Language* LI. 621 We are not yet in a position to characterize seriously the semantic representation of roots. My guess is that we are not yet aware of the majority of semantic primitives. **1975** M. A. Slote *Metaphysics & Essence* iii. 41 This notion of (an) experience, like the other notions we have been using as primitives, is not just an arbitrary primitive with which to attempt the definition of the concepts we wish to define. **1976** J. S. Gruber *Lexical Struct. Syntax & Semantics* ii. i. 260 Interpretive semantics is valuable only for those functions which a logical calculus entails, and for this it must operate on trees of semantic primitives.

primitivism. Add: **1.** (Further examples.) Also, a belief in the desirability of a 'return to nature'; an exaltation of simplicity or of irrationalism; the practice of primitive art.

1934 A. Huxley *Beyond Mexique Bay* 257 To introduce a salutary element of primitivism into our civilized and industrialized way of life. **1938** *Burlington Mag.* June 302/2 Far more primitivism has been attributed to him [*sc.* Rousseau] than is actually justified. **1939** J. Charlot *Art from Mayans to Disney* xi. 86 Political cartoons reminiscent of Monnier and Grandville, which are not flavoured at all by the 'primitivism' of his [*sc.* Posada's] later work. **1947** A. Einstein *Mus. Romantic Era* ix. 95 What formal primitivism, after Beethoven! **1950** E. H. Gombrich *Story of Art* xxvii. 440 This Primitivism advocated by Gauguin became perhaps an even more lasting influence on modern art than either Van Gogh's Expressionism or Cézanne's way to Cubism. **1951** *Essays in Crit.* I. 97 The term 'Back to Nature' covers the many and varied forms of primitivism. **1952** J. Summerson *Sir John Soane* 17 Soane's 'primitivism'..is formalized at Bentley Priory (1798). **1958** H. R. Hitchcock *Archit. in 19th & 20th Cent.* 450/2 'Primitivism' in painting and sculpture has been of recurrent importance since the days of the Fauves and the Expressionists; a comparable primitivism in architecture has been much rarer, except for Gaudí. **1969** *Daily Tel.* 10 Feb. 10/4 He [*sc.* Thomas Mann] puts the German character on the operating table..: the loneliness, the smug provincialism, the Wagnerian primitivism, the eroticism. **1976** *Survey* Summer–Autumn 107 Khrushchev was only comprehensible if one began by accepting his Marxist primitivism.

primitivist (pri·mitivist), *sb.* and *a.* [f. Primitive *a.* + -ist.] **A.** *sb.* A believer in primitivism (sense 1); an advocate of the superiority of primitive customs or of primitive art; a person who uses obsolete methods or techniques. **B.** *adj.* Of or pertaining to primitivism or to the primitive, esp. in art; irrational, opposing scientific development.

1926 W. R. Inge *Lay Thoughts* 204 So the Utopians are usually primitivists. They glorify the noble savage, who runs wild in woods. **1934** *Musical Q.* Apr. 213 Three currents are left in the wake of the Modern Movement—Primitivist, Classicist, Popularist. **1949** B. Willey in *Ideas & Beliefs of Victorians* (B.B.C.) 43 Perhaps as Rousseau and other primitivists had urged, civilisation was a monstrous aberration, and men were happier and better when fresh from the hands of God or Nature in some primeval Eden. **1952** J. Summerson *Sir John Soane* 33, I mentioned earlier the 'primitivist' element which is so important a factor in the Soane style. **1961** *Times* 7 June 17/3 All the rather flashy vitality and 'primitivist' imagery of his old manner have been discarded. **1975** *Nature* 20 Mar. 219/1 Attacked by the new school of linguistics and cognitive epistemology as an ignorant primitivist, Skinner not only maintains his position but makes it more dogmatic. **1977** *Times Lit. Suppl.* 15 July 874/3 Nothing but primitive commonplaces: without laws, private property or rulers, the Indians 'live according to nature'. It is a primitivists' fantasy world characterised by the observance of natural moral practices. **1977** D. Watkin *Morality & Archit.* i. ii. 25 He [*sc.* Viollet-le-Duc] has..a related 'primitivist' notion that Roman and Renaissance architecture lost contact with the pure fount of Greek truth, and is thus morally and stylistically in questionable taste.

Hence **primitivi·stic** *a.*

1943 [see *isolationistic *a.*]. **1948** L. Spitzer *Linguistics & Lit. Hist.* 210 Claudel can sing..not only 'Georgica', as did Vergil in a primitivistic mood. **1958** H. R. Hitchcock *Archit. 19th & 20th Cent.* iv. 60 Soane's Dulwich Gallery of 1811–14, outside London, is likewise built of common brick and has similarly primitivistic detailing. **1959** *Encounter* Nov. 76/1 A kind of atavism, an inability to think..in any but primitivistic terms. **1972** M. Bradbury in Cox & Dyson *20th-Cent. Mind* III. xii. 343 Golding's universe is normally a-social or perhaps presocial, primitivistic at its core and yet also conscious that it is only through *knowing* our primitivism that we will find our innocence.

primitivize (pri·mitivaiz), *v.* [f. Primitive *a.* + -ize.] *trans.* and *intr.* To render primitive; to impute primitiveness to; to simplify; to return to an earlier stage. So **primitiviza·tion**, **pri·mitivizing** *vbl. sb.* and *ppl. a.*

1942 Primitivisation [see *overinclusion]. **1955** *Times* 18 May 8/6 It does not bring about a primitivizing or animalizing of the human, but rather it celebrates man at his human best. **1959** *Encounter* May 50/2 Mr. Logue is a primitivizing poet. **1968** D. Lawton *Social Class, Lang. & Educ.* iii. 23 'Cultural deprivation' or absence of external stimulation resulted in a 'primitivization' of an individual's behaviour. **1969** D. Daube *Roman Law* iii. 168 A common failing of modern research into ancient law is the inclination to primitivize the sources, to press the naive side of any statement or custom and overlook the element of sophistication which is often quite strong. **1971** *Jrnl. Gen. Psychol.* LXXXIV. 208 In psychology the term 'regression' refers to a primitivization of behavior, a 'going back' to a less mature way of behaving which the individual has 'outgrown'. **1976** T. Stoianovich *French Hist. Method* 146 Since the impairment of an existing superstructure provokes economic primitivization, staunch support develops in favor of a viable new superstructure.

primmy (pri·mi), *a. rare.* [f. Prim *a.* + -y[1].] Tending to primness.

1879 [see *governessy *a.*].

‖**primo** (prī·mo), *a.* and *sb.* [It., = first: cf. Prima[2].] **A.** *adj.* Used in some phrases, chiefly musical, as **primo basso**, chief bass singer (example *fig.*); **primo buffo**, chief male comic singer or actor; **primo tenore (assoluto)**, chief tenor singer (of outstanding ability); **primo uomo**, singer of the chief male part; (see also quots.).

1740 J. Grassineau *Mus. Dict.* 183 *Primo*, the first; this word is often abridg'd, P°, I°..and added to other words, as *Primo canto*,—*the first treble.* **1801** Busby *Dict. Mus.*, *Primo* (Ital.). First: as *Primo Violono* [sic], first violin; *Primo Flauto*, first flute. **1826** M. Kelly *Reminisc.* I. 48 The celebrated Genaro Luzzio was the primo buffo, and the principal female, La Coltellini, was delightful, both as a comic actress and singer. **1848** Geo. Eliot *Let.* 8 June (1954) I. 266 Dear Quartett of Friends—I may still say so, though I fear your primo basso has departed. **1855** —— in *Fraser's Mag.* July 50/2 He will write no *part* to suit a *primo tenore.* **1876** Stainer & Barrett *Dict. Mus. Terms* 366/1 *Primo*, (*It.*) First (masc.), as *tempo primo*, at the original pace or time; *violino primo*, first fiddle; *primo buffo*, chief comic actor or singer; *primo musico* and *primo uomo*, principal male singer in the opera. **1880** Grove *Dict. Mus.* II. 509/1 It was *de rigueur* that the First Man (*Primo uomo*) should be an artificial Soprano. *Ibid.* 514/1 The chief, or *Buffo* group, consisted of two Female Performers..and three Men, distinguished as the *Primo Buffo*, [etc.]. **1889** G. B. Shaw *London Music 1888–89* (1937) 166 His dignity as *primo tenore assoluto.* **1938** *Times Lit. Suppl.* 28 May 370/3 What the *primo uomo* was to the eighteenth-century opera..the man with the stick is to the symphony concerts. **1955** E. Dent in H. Van Thal *Fanfare for E. Newman* 86 The first half of the nineteenth century..was the triumph of the *prima donna* over the *primo uomo*, the *castrato* hero of the eighteenth century, though she soon had to face her rival in the *primo tenore.*

B. as *sb.* **1.** *Mus.* In a pianoforte duet, the upper part; the pianist who plays this part.

1792 J. A. K. Colizzi *Three Duets for Two Performers on the Harpsichord or Piano Forté* 3 Primo. **1883** Grove *Dict. Mus.* III. 30/2 In pianoforte duets, *Primo* or 1mo is generally put over the right-hand page, and then means the part taken by the 'treble' player. **1954** K. Dale *19th-Cent. Piano Mus.* xii. 284 Swirling cadenzas of scales in demisemiquavers for performance by *primo.* *Ibid.* 286 The playing of six quavers in a bar by *primo* against four by *secondo* throughout the whole of the trio section cannot have been altogether easy for the young performer. **1965** *Listener* 1 July 33/3 Britten's rhythmic control as *secondo* balancing and counteracting Richter's natural genius as *primo.* **1964** G. Read *Mus. Notation* xviii. 298 For four-hand music on one instrument, the two parts are placed on facing pages..the first player (*primo*; top part) reads from the right-hand page.

2. (With capital initial.) A title given to an official of the Royal Antediluvian Order of Buffaloes.

1879 *Buffalo* 16 Jan. 3/3 Yours fraternally, Primo James Dewsbury. *Ibid.* 5/1 Primos who..have been elevated to the position of Knights of the Order of Merit. **1928** *Daily Express* 2 Aug. 9/5 Mr. J. C. E. Cartwright.., Grand Primo of England, inaugurated the Croydon and District Provincial Grand Lodge... The Grand Primo of the new lodge is Mr. L. R. N. Percy. **1966** *R.A.O.B. Centenary, South Yorkshire, 1866–1966* (Royal Antediluvian Order of Buffaloes), Of the 17 Provincial Grand Primos..only two have been taken from us by the call of the reaper.

primoge·niturist. *rare.* [f. Primogeniture + -ist.] One who believes that the right of succession or inheritance belongs to the first-born.

1976 I. Murdoch *Henry & Cato* i. 7 His father, a rigid primogeniturist, had left everything to..the elder son.

primordial, *a.* Add: **1. b.** *primordial soup*: see *soup *sb.*

4. a. *primordial meristem* = *promeristem* (Pro-[2] 1 in Dict. and Suppl.).

1925 Eames & MacDaniels *Introd. Plant Anat.* iii. 41 The youngest cells in a region of growing plant body in which the formation of new organs or parts of organs is taking place constitute a promeristem, or primordial meristem. **1943** *Bot. Rev.* IX. 142 The limits between primordial meristem cells at the apex and procambium cells below are vague.

primordiality. (Earlier and later examples.)

1874 W. Wallace tr. *Hegel's Logic* 297 The cause therefore appears as passing into its correlative, and to be losing its primordiality in the latter. **1977** J. A. Fishman in H. Giles *Lang., Ethnicity & Intergroup Relations* i. 17 Primordiality denotes both primacy, in the sense of a presumably original essence, as well as primitivism or irreducibility.

primordium. Add: **b.** (Examples.) In *Embryol.* = *anlage.

1875 Bennett & Dyer tr. *Sachs's Text-bk. Bot.* ii. 531 In Primulaceæ..five protuberances (primordia) appear on the receptacle above the calyx, each of which grows up into a stamen. **1898** A. Willey in *Nature* 25 Aug. 390/1 The word that commends itself to me [for the German 'Anlage']..is primordium. **1908** F. R. Lillie *Devel. of Chick* 8 The ovum is the primordium of the individual, the ectoderm the primordium of all ectodermal structures,.. the first thickening of the ectoderm over the optic cup the primordium of the lens, etc. **1935** *Jrnl. Morphol.* LVIII. 425 The primitive mesenteron..consists of a single layer of squamous epithelium dorsal to the attachments of the cardiac primordia, and two layers ventral to them. **1965** Bell & Coombe tr. *Strasburger's Textbk. Bot.* 115 The primordia of the leaves arise at the apex from the outer layers of cells. **1978** M. J. T. FitzGerald *Human Embryol.* ix. 75 At the end of the fifth week the primordia of the hands and feet are already apparent.

primp (primp), *a.* [f. the vb.] Smart, neat, prim.

1835 *Fraser's Mag.* July 17 Your primp wizand faces. **1903** *N.Y. Times* 26 Sept. 4 (Advt.), All-weather coats they are—just as primp, good-fitting and handsome as a man could wish to wear. **1931** *Aberdeen Press & Jrnl.* 19 Feb., Scotia's leed has mony a kin', Tae fit baith primp an' pliskie. **1966** J. S. Cox *Illustr. Dict. Hairdressing* 122/1 *Primp*, smart, neat.

primp, *v.* Delete *dial.* and add: **1.** (Further examples.) Also *refl.*

1914 R. Frost *North of Boston* 103 Lord, if I were to dream of everyone Whose shoes I primped to dance in! **1945** J. Steinbeck *Cannery Row* viii. 47 A Lee cousin primped up slightly wilted heads of lettuce the way a girl primps a loose finger wave. **1959** *Numbers* Feb. 30 Primping yourself up like a damned quean. **1965** F. Knebel *Night of Camp David* xiv. 232 She came willingly enough, after primping her hair and smoothing her charcoal linen dress. **1974** 'A. Haig' *Peruvian Printout* 45 When Heinrich..came back..Shirley even forgot to primp herself.

b. (Earlier and later examples.) Also const. *up.*

1887 *Harper's Mag.* Mar. 544/1 When you was primping so, I thought all the time it was for Mrs. Rainwater. **1901** W. N. Harben *Westerfelt* iv. 49 Ef you want to primp up a little an' bresh that hoss-hair off'n yore pants, go in yore room. **1937** *Daily Tel.* 31 Aug. 12/4 It [*sc.* the women's dressing-room of an American flying-boat] is described as containing 'mirrors and leather-covered stools for primping'—which I take to mean such running repairs as passengers find necessary. **1939** N. Coward *Words & Music* in *Play Parade* II. 120 In tropical heat Nobody who's sweet, survives We powder and primp And try to be sympathetic. **1977** *Daily News* (Perth, Austral.) 19 Jan. 6/4 (*caption*) Dorothy Hamill, 1976 Olympic figure skating champion, primps before making her New York debut with the Ice Capades at Madison Square Garden.

c. *intr.* To make tidying or smoothing movements. *rare.*

1881 I. M. Rittenhouse *Maud* (1939) 1 Eva..pulled down her basque, 'primped' at her hair,..and looked expectantly towards the door.

3. *trans.* and *intr.* To move (oneself or another person) fussily or mincingly.

1951 W. Sansom *Face of Innocence* xiii. 184 She primped us over to Roddy with all the posturing, like a dove stamping out its love-dance, of one person meeting another. **1953** J. Masters *Lotus & Wind* vii. 89 She opened the door..and primped along the passage. The skirt clung so tightly around her thighs that she had to hobble. **1977** *N.Z. Listener* 15 Jan. 46/4 The comedians pranced and primped with Ronnie Corbett.

primped *ppl. a.*: also **primped-up**; **primping** *vbl. sb.* (later examples).

1935 Z. N. Hurston *Mules & Men* (1970) i. vi. 126 Tain't no use in you gittin' yo' mouf all primped up for no hoein' and rakin' out of me, Bertha. Call yo' grandson and let him do it. **1959** *News Chron.* 11 Aug. 6/4 Adolescence..is early enough to begin real primping. **1963** *Listener* 28 Mar. 570/1 One had the primped-up stage, gorgeous to the eye. **1977** *New Yorker* 16 May 108/2 When the course is closed to play and the green-keeping staff is giving it a final primping for the opening round.

primrose, *sb.* (*a.*) Add: **6. b.** A commercial soap of a yellowish colour. In full, *primrose soap*.

1796 M. Edgeworth *Parent's Assistant* (ed. 2) II. 127 A fresh assortment of..Primrose Soap. **1907** *Yesterday's Shopping* (1969) 39/2 *Soaps* (*Plain*). Pale Primrose (Army & Navy) bar about 3 lb. 0/9. The Royal Primrose (J. Knight's) bar about 3 lb. 0/9½. **1909** H. G. Wells *Tono-Bungay* iii. i. 265 We had added to the original Moggs' Primrose several varieties of scented and super-fatted. *c* **1938** *Fortnum & Mason Catal.* 56/2 *Soaps, Household*.. Primrose Royal..per bar 1/3.

7. *primrose yellow* (further examples); *primrose-coloured* (earlier example), *-vested* adjs.; **primrose soap**: see sense 6b above.

1788 *Gazetteer* 12 May 2/3 The train was a primrose coloured goffree'd crape spotted with blue crape in relief. **1922** Joyce *Ulysses* 195 Primrosevested he greeted gaily with his doffed Panama as with a bauble. **1907** *Yesterday's Shopping* (1969) 478/2 Oil Colours..Primrose yellow..Raw sienna. **1954** T. S. Eliot *Confid. Clerk* i. 32, I thought a primrose yellow would be cheerful. **1978** *Vogue* 1 Mar. 128 Shirtdress..primrose yellow with Peter Pan collar.

8. Primrose Day (earlier example); **Primrose League** (further examples); so **Primrose Leaguer.**

1884 E. W. Hamilton *Diary* 21 Mar. (1972) II. 581 Mrs. G...carried a splendid bouquet of primroses,..to show that the 'Primrose Leaguers' have no title to appropriate the flower to themselves. *Ibid.* 19 Apr. 597 'Primrose Day'. Were I an admirer of Lord Beaconsfield, I should be furious that his memory should be so ridiculed. **1912** Chesterton *Manalive* ii. ii. 240, I have faced many a political crisis in the old Primrose League days at Herne Bay. **1923** J. M. Murry *Pencillings* 146 Disraeli..was a far more remarkable man than the most enthusiastic Primrose Leaguer has ever imagined. **1959** B. & R. North tr. *Duverger's Pol. Parties* (ed. 2) i. i. 66 The Primrose League, an organization distinct from the party proper, aimed at social mixing. **1975** R. Taylor *Lord Salisbury* viii. 134 He told the Primrose League in a memorable speech: 'You may roughly divide the nations of the world as the living and the dying.'

B. as *adj.* (Earlier and later examples.)

1788 *Gazetteer* 12 May 2/3 An immense panache of white, blue and primrose feathers. **1815** in R. W. Chapman *Jane Austen's Pride and Prejudice* (1923) 398 Primrose sandals, and white kid gloves. **1931** [see *jumper *sb.*[2] 3c]. **1976** *S. Wales Echo* 25 Nov. 27/4 (Advt.), Bathroom/w.c., half-tiled in Primrose, matching Primrose suite.

primrose, *v.* **a.** (Further examples.)

1928 *Daily Express* 10 Mar. 5/3 There are few of the many who enjoy the country who will be able to resist primrosing. **1941** E. Bowen *Look at Roses* 122 This afternoon..we'll go primrosing. **1967** 'L. Bruce' *Death of Commuter* viii. 82 'I'm going to take her primrosing tomorrow,' he told Carolus. 'In Langley Wood'. **1973** J. Thomson *Death Cap* vi. 88 To go bird's-nesting, or blackberrying or primrosing.

primrosy, *a.* Add: **c.** *fig.* (Cf. *primrose path* s.v. Primrose *sb.* (*a.*) 7.)

1908 E. V. Lucas *Over Bemerton's* xx. 202 His duty always lies along the primrosiest path.

primuline (pri·miulīn). *Chem.* Also **Primuline.** [f. Primul(a + -ine[5].] A synthetic yellow dyestuff which is the sodium salt of the sulphonic acid derivative of primuline base (see below) and is used in the dyeing of cotton.

The dye may also contain other analogous salts.

1887 *Dyer & Calico Printer* VII. 101/1 A new series of colours..are just now being brought out by Brooke, Simpson, & Spiller (Limited). These colours, the discovery of Mr. A. G. Green, promise to compete successfully with the direct cotton colours at present in the market. The basis of the series is a compound, to which the name of primuline has been given. **1919** E. de B. Barnett *Coal Tar Dyes* 108 Primuline dyes cotton in bright yellow shades, but these are too fugitive to be of any value. **1950** *Thorpe's Dict. Appl. Chem.* (ed. 4) X. 215/1 The sodium salts of the sulphonic acids of the higher thionated *p*-toluidines (primuline bases) were first manufactured..in 1887 under the name of 'Primuline' and employed for the production of so-called 'ingrain colours' by ..diazotising and developing upon the fibre. **1968** E. N. Abrahart *Dyes* v. 135 Among the direct dyes C.I. Direct Red 70..employs Primuline as diazo component and Schäffer's acid as coupling component.

2. Special Combs.: **primuline base**, a yellow thiazole derivative, $C_{21}H_{15}N_3S_2$, which is obtained when *p*-toluidine is heated with sulphur and is an intermediate in dye manufacture; *loosely*, any of the related compounds also formed by this process; **primuline red**, a red dyestuff obtained from primuline base by diazotization followed by coupling with β-naphthol.

1889 *Jrnl. Chem. Soc.* LV. 228 The product is not homogeneous, but consists of about 50 per cent. of a base, $C_{14}H_{12}N_2S$,..40 per cent. of primuline-base, and 10 per cent. of unaltered paratoluidine. **1913** *Thorpe's Dict. Appl. Chem.* (ed. 2) IV. 385/1 These more condensed bases ('primuline bases') were obtained in larger amount by increasing the proportion of sulphur to 4½–5 atoms to 1 mol. amine. **1961** Cockett & Hilton *Dyeing of Cellulosic Fibres* v. 156 Primuline Base may be sulphonated to give the monosulphonic acid, the sodium salt of which is the yellow dye Primuline, a direct dye for cellulosic fibres. **1968** E. N. Abrahart *Dyes* v. 135 The sulphurization products of *p*-toluidine, primuline base and dehydrothio-*p*-toluidine are used..in the manufacture of some valuable cotton dyes. **1900** *Jrnl. Soc. Chem. Industry* 30 Apr. 345/1 Primuline Red is obtained by dyeing with Primuline, diazotising, and developing with β-naphthol. **1917** *Jrnl. Soc. Dyers & Colourists* XXXIII. 140/1, I replied by pointing out the greater fastness of Primuline Red compared to Benzopurpurine.

primus, *a.* and *sb.* Add: **A.** *adj.* **1.** *primus inter pares* (further examples); also fem. *prima inter pares*.
1909 Webster, Prima inter pares. **1919** M. Beer *Hist. Brit. Socialism* I. ii. v. 162 He could be their self-sacrificing father and teacher, their authoritative adviser and leader, but never the *primus inter pares*. **1961** *Times* 12 Oct. 18/1 But this is a ballet in which Dame Margot is but *prima inter pares*. **1973** *Times* 20 Oct. 13/5 Herr Schneiderhan did not attempt to stand out as a virtuoso but attacked his solos as *primus inter pares*, playing with and to his accompanying strings. **1979** *Guardian* 5 May 21/3 Mrs Thatcher..becomes..*primus inter pares* in that quaint and English system known as Cabinet Government.

B. *sb.* **2.** Also **Primus**. The proprietary name of a make of pressure stove or lamp, usu. burning paraffin; *loosely*, any pressure stove. Freq. *attrib.*
1904 *Outing* Mar. 698/1 At last we found and packed with rucksacks, small kerosene cans, Primus stove, etc. **1904** *Railway Mag.* XIV. 45/1 A ½ in. scale locomotive is more expensive to construct, and needs a 'Primus' burner. **1907** *Athenæum* 12 Oct. 436/2 Robinson with great efforts made the 'Primus' work, and then burnt the stew with it. **1910** *Trade Marks Jrnl.* 22 June 989 Primus... Stoves. Aktiebolaget B.A. Hjorth & Co..., Stockholm, Sweden; manufacturers. **1910** *Official Gaz.* (U.S. Patent Office) 5 July 246/2 Aktiebolaget B.A. Hjorth & Co., Stockholm... *Primus*. **1933** E. A. Robertson *Ordinary Families* iii. 53 It was asking too much of anyone's stomach to expect primus cookery. **1944** M. Laski *Love on Supertax* i. 11 Have you ever tried..to light a primus stove?.. The methylated spirits flare up... You must frantically pump paraffin through to the burner. **1951** G. Millar *White Boat from England* ii. 16 We had several gadgets for the primus, including a pyramidal toaster. **1973** J. Stranger *Walk Lonely Road* xiv. 108 Coffee, thick and strong and sweet, brewed over a Primus. **1974** O. Manning *Rain Forest* iii. i. 254 Simon folded back the flaps of a large, square tent... A primus lamp hung from the roof.

prince, *sb.* Add: **I. 3. b.** A person with power or influence; a magnate. *U.S.*
1841 J. S. Buckingham *America* III. 427 Capitalists and merchants [of Boston]..are here called 'princes'. **1884** *Century Mag.* Sept. 796 At a shady end of the veranda, are seen the railroad king,..the bonanza mine owner, the Texas rancher, and the Pennsylvania iron prince. **1904** [see *baron 2b]. **1976** T. Gifford *Cavanaugh Quest* (1977) viii. 137 He was a perfect reflection of the typical Minneapolis power broker, though somewhat better dressed than the grain barons and the department store princes and computer tycoons.

c. An admirable or generous person. *colloq.* (chiefly *U.S.*).
1911 H. B. Wright *Winning of Barbara Worth* xvi. 252 Yes sir, gents, I'm here to tell you that that there man, Jefferson Worth, is a prince—a prince. Let me tell you what he done for me. **1939** I. Baird *Waste Heritage* v. 69 Hep ain't like other guys, he's a prince. **1951** J. D. Salinger *Catcher in Rye* iii. 31 He's crazy about *you*. He told me he thinks you're a goddam prince. **1966** J. Cleary *High Commissioner* viii. 164 'You have a lot of time for him, haven't you?' 'They don't come any better. He's a prince, you know?'

IV. 11. prince's pine, (*b*) (earlier and later examples).
1807 F. Pursh *Jrnl. Bot. Excursion* (1869) 15 Pyrola umbellata calld here Princess [*sic*] pine. **1818** A. Eaton *Man. Bot.* (ed. 2) 203 *Chimaphila*..*umbellata*, (prince's pine, bitter wintergreen). **1884** [see Pipsissewa]. **1954** C. J. Hylander *Macmillan Wild Flower Bk.* 280 Pipsissewa... Also known as Prince's Pine, this is a trailing and somewhat woody perennial.

12. *Prince of Wales check*, a large check pattern; *Prince of Wales('s) feathers*: (*a*) see Feather *sb.* 8; also *ellipt.* and *fig.*; (*b*) = *crape-fern* (*crape *sb.* 3 b); *Prince of Wales knot* (see also quot. 1978); *Prince Rupert's drop* (further example).
1958 P. Mortimer *Daddy's gone a-Hunting* vii. 35 A tall, thin man in Prince of Wales check. **1959** *Sunday Express* 21 June 14/3 Prince of Wales check trousers. **1960** *News Chron.* 11 July 6/5 The Prince of Wales check Sudan cotton in grey and black. **1972** *Vogue* Feb. 73 Prince of Wales check wool tent coat. **1882** T. H. Potts *Out in Open* 108 T[odea] *superba*, 'the glory of the west'. How great the impression made by its marvellous beauty, may be assumed from the number of familiar names.. bestowed upon it, as the Royal fern, the King's fern, Prince of Wales' feather. **1919** T. Wright *Romance of Lace Pillow* ix. 82 Other patterns were the *Prince of Wales's Feathers*. **1933** *Flight* 29 June 626/2 And a final break-up in a 'Prince of Wales Feathers', were other manoeuvres which held the spectators literally spellbound. **1944, 1951** [see *crape-fern]. **1958** C. Freeman *Pillow Lace in E. Midlands* 46 The names given to the various patterns often refer to some element of the design. .. Other favourites were..Prince of Wales's Feathers [etc.]. **1966** H. Sheppard *Dict. Railway Slang* (ed. 2) 9 *Prince of Wales*, blowing off steam by engine. **1971** D. J. Smith *Discovering Railwayana* x. 58 *Prince of Wales*, short for Prince of Wales' feathers, a plume-like emission of steam. **1977** Binney & Burman *Change & Decay* 143/2 (*caption*) Lea, Wiltshire; St. Giles. Bell of 1622, with Prince of Wales feathers. **1977** *R.A.F. News* 22 June–5 July 1/1 The manœuvre takes the nine Gnats up into a Prince of Wales feathers, with eight aircraft trailing white smoke and the leader trailing red. **1897** *Sears, Roebuck Catal.* 222/2 Illustration No. 2205 shows the De Joinnile [necktie] as worn with an ordinary finger ring. No. 2206 shows it tied in a Prince of Wales knot. **1971** *Guardian* 3 Aug. 9/3 Broad kipper ties..tied in loose Prince of Wales knots, were featured by many Paris houses. **1978** 'K. Blake' *Professionals 1: Where Jungle Ends* iii. 39 His Prince of Wales knotted tie in rich brown. **1862** Ruskin *Unto this Last* iv. 145 Nay, boiled bulbs they might have been—glass bulbs—Prince Rupert's drops, consummated in powder..for any end or meaning.

Prince Albert. **1.** The name of Prince Albert Edward, afterwards King Edward VII, used *attrib.* and *absol.* to designate a kind of frock coat or suit made fashionable by him. orig. *U.S.*
1884 I. M. Rittenhouse *Maud* (1939) 270, I ran out and ushered Mr Lyons in, gotten up to kill in his Prince Albert coat. **1890** [see *jimswinger]. **1895** *Montgomery Ward Catal.* 273/1 Prince Albert Suits. **1897** Kipling *Day's Work* (1898) 221 A man in a black Prince Albert, without a collar, came up. **1903** S. Clapin *New Dict. Amer.* p. viii, If a Londoner is fortunate enough to cross the Atlantic,.. but is unfortunate enough to have to buy a frock-coat..he must call it a 'Prince Albert'. **1919** *Ladies' Home Jrnl.* July 19/2 Instinctively I looked about him for revolvers. There were none, not even the slightest bulge at the hips of the Prince Albert coat he wore. **1927** *Scribner's Mag.* Feb. 164/1 Antone was dressed for a call, having donned a shiny Prince Albert coat over his collarless shirt. **1941** W. C. Handy *Father of Blues* vii. 91 Glittering young devils in silk toppers and Prince Alberts. **1967** C. O. Skinner *Madame Sarah* viii. 147 Amid the..city coats, Prince Alberts and pin-striped trousers, Sarah looked vainly for a single woman. **1972** H. Kemelman *Monday the Rabbi took Off* xv. 103 Others [*sc.* Chassidim]..favored a Prince Albert, which because it was warm, they kept open.

2. The name of Prince Albert, the Prince Consort, used in *pl.* to designate foot- or toe-wraps worn by tramps, sailors, etc., inside boots; the boots themselves. *Austral.*
1893 K. Mackay *Out Back* (ed. 2) ii. v. 191 With bent shoulders..they 'mouched' along,..showing glimpses of brown, unwashed skin above the frayed edges of their 'Prince Alberts', the toes of their bluchers gaping wide. **1903** 'T. Collins' *Such is Life* (1937) i. 52 Unlapping from his feet the inexpensive substitute for socks known as 'prince-alberts'. **1924** *Truth* (Sydney) 27 Apr. 6 *Prince Alberts*, rags or bandages used by a swagman or sundowner around his feet in place of socks. **1945** Baker *Austral. Lang.* 105 *Prince Alfreds* or *Prince Alberts* as synonyms for *toe-rags*. These terms developed from the malign suggestion that the Prince Consort was so poor when he came to England to marry Queen Victoria that he wore toe-rags instead of socks. *Ibid.*, Rough lace-up boots were also known as *Prince Alberts* in Queensland in the closing years of last century. **1958** J. Bisset *Sail Ho!* v. 48 These foot-wraps were known in British ships as 'Prince Alberts'.

Prince Charming. [Partial tr. F. *Roi Charmant*, the name of the hero of the Comtesse d'Aulnoy's *L'Oiseau Bleu* (1697). In English the name first appears as that of the hero of Planché's *King Charming* or *Prince Charming*, and was later adopted for the hero of various fairy-tale pantomimes, *esp.* the *Sleeping Beauty* and *Cinderella*.] A fairy-tale hero; the lover every girl dreams of; a perfect young man.
[**1850** *Times* 27 Dec. 5/2 Lyceum... Then followed the principal attraction of the evening, in the shape of a new and original fairy extravaganza, in two acts, entitled *King Charming; or, the Blue Bird of Paradise*. *Ibid.*, We would particularly notice..the final scene of the restoration of King Charming.] **1855** *N.Y. Daily Times* 25 Dec. 4/3 The burlesque of 'King Charming'..The legend is taken from the Countess D'Aulnoy's fairy tale of 'L'Oiseau Bleu', and relates to the adventures of the wonderful *Prince Charming*, an immortal monarch, gifted with perpetual youth. *Ibid.* 4/4 Mrs. H. C. Watson as *Prince Charming* scarcely did herself justice. **1862** *Welcome Guest* 4 Jan. 152/3 They tell me I am good looking, but I don't believe it; although the young ladies of my acquaintance, who are too bold for me, call me 'Prince Charming'. **1913** W. J. Locke *Stella Maris* xv. 205 Love she had heard of, the love of Prince Charming for Princess Rose. **1920** [see *backfisch]. **1929** E. Wilson *I thought of Daisy* v. 286 When Prince Charming comes along, he's always just a great, big, strong, clean-limbed American. **1931** G. C. D. Odell *Ann. N.Y. Stage* VI. 427 A holiday frippery, Planché's King Charming, or, the Blue Bird of Paradise, ushered in, on Christmas Eve [1855], a success that lasted, without interruption, until January 12th. *Ibid.* 450 [Laura Keene] hired the unlucky Metropolitan theatre..and advertised the opening for December 24th, with Two Can Play at that Game and Prince Charming... The Broadway Theatre on the same evening produced King Charming, as it called Planché's extravaganza... Prince Charming was not played at Miss Keene's..some miscreant had so slashed the chief scene of Prince Charming that the play could not be given. **1936** R. C. K. Ensor *England 1870–1914* vii. 215 He [*sc.* Rosebery] had come to the front as the Prince Charming of politics—young, handsome, rich, eloquent, candid, and popular. **1939** J. Curtis *What Immortal Hand* iii. 36 She had not the slightest wish to marry anybody from Lowdham Street... No, she wanted some kind of Prince Charming like they had in the pictures. **1939** Wodehouse *Uncle Fred in Springtime* xv. 225 There was a look in her eyes that made me think right away that she was feeling he was her Prince Charming. **1945** G. Endore *Methinks the Lady* (1947) ii. 27 And your Prince Charming? You met him at Mona's? **1960** E. Eliot *They all married Well* xiii. 201 Virginia Bonynge was too pretty..to be forever scorned by *all* the Prince Charmings who abounded in London. **1961** *Guardian* 17 May 6/1 Like most other English schoolgirls, I spent many an hour dreaming... Prince Charmings used to come and carry me off on white chargers. **1975** *Radio Times* 9 Jan. 43/2 They are destined to find their Prince Charmings at a commercial fair in Rochefort. **1976** *Daily Mail* 25 Mar. 22/1 We all know.. the story of Cinderella... Our hero is a progressive Prince Charming who doesn't mind the privilege and protocol provided he can marry the girl of his choice. *Ibid.*, Way back in the days of Dr Kildare, Richard Chamberlain must have been every girl's idea of Prince Charming. **1977** *Christian* IV. 124 The beast has by the kiss of love become the Prince Charming who, far from having to be shut away, is now the very life and soul of what I am.

prince consort: see Consort *sb.*[1] 3, Prince *sb.* 6.

princely, *a.* Add: **1. b.** *princely states*, the states of India that were ruled by native princes before the Indian Independence Act of 1947. Also (*rare*) in *sing.*
1952 *Columbia Lippincott Gazetteer of World* 835/2 By Jan. 26, 1950, when India became a sovereign republic, all acceding princely states had been brought within the constitutional framework. **1959** *Listener* 19 Mar. 497/2 The Princely States..have disappeared. **1967** Singha & Massey *Indian Dances* iv. 59 Both the South Indian Princely States and the Madras High Court recognized their rights and status. **1975** *Times* 29 Aug. 12/1 The map of India..was redrawn to incorporate the former princely states. **1980** H. R. F. Keating *Murder of Maharajah* xviii. 225 Here is a simple District Superintendent of Police..invited to a princely state because a murder has been committed.

prince's feather. **b.** (Earlier and later examples.)
1712 J. Mortimer *Whole Art Husb.* II. 166 [see quot. 1721 in Dict.]. **1883** W. Robinson *Eng. Flower Garden* 14/1 Prince's Feather, Love-lies-bleeding.. Among annuals none are more in want of judicious use and appreciation than these. **1925** E. Glasgow *Barren Ground* i. xi. 134 A narrow path led between rows of log cabins, each with its patchwork square of garden, and its clump of gaudy prince's feather or coxcomb by the doorstep. **1974** M. Allan *Plants that changed our Gardens* i. 41 Prince's Feather..grows up to 5 ft tall and has flower plumes of deep crimson.

princess. Add: **5. b.** Used as a form of address to a woman or girl. *colloq.*
1924 J. Buchan *Three Hostages* xviii. 254, I have waked you from sleep, my princess. Therefore so far it is good. **1968** J. Symons *Man whose Dreams came True* i. v. 42 'This is celebration night, Princess.' He had called her Princess the whole evening. **1969** A. Laski *Dominant Fifth* iii. 120 Ah come on, princess, you're being morbid. **1970** J. Aiken *Embroidered Sunset* i. 7 It was one of Uncle Wilbie's pleasantries to address Lucy as Princess. **1972** R. Ludlum *Osterman Weekend* i. 12 Hey, Princess—get your brothers out and help your mother with the smaller bags.

7. *princess ring, princess* (*tele*)*phone*.
1962 M. & G. Gordon *Journey with Stranger* (1963) iii. 28 If you want a good bargain in princess rings, go down New Road in Bangkok. **1973** —— *Informant* iii. 13 She was..constantly twisting a princess ring from Thailand. **1966** B. Glemser *Dear Hungarian Friend* viii. 139 She.. picked up the azure blue Princess telephone, and said, 'Hello?' **1969** D. E. Westlake *Up your Banners* (1970) xxxi. 210 We had to go next door and use Mrs. Lupowitz' Princess phone, a skittering pink beast that traveled all over its waxed table..whenever anybody tried to dial it. **1972** G. Baxt *Burning Sappho* ii. 42 He'd been watching Pat circling her gold princess phone for ten minutes. **1973** *Sat. Rev. Society* (U.S.) Mar. 70/1 Colored phones and such models as the lighter and smaller 'Princess' and the 'Trimline' (with the dial mechanism in the handset). **1976** M. Machlin *Pipeline* ii. 29 Brandon opened the chest and removed a brown princess telephone from it.

8. princess dress (earlier and later examples); also *princess cut, line, tunic*.
1867 in A. Adburgham *Shops & Shopping* (1964) xi. 119 Princess breakfast dress. **1872** *Young Englishwoman* Oct. 541/1 A long princess-tunic. *Ibid.* 546/1 A lovely Princess dress, with cape. **1877** in A. Adburgham *Shops & Shopping* (1964) xvi. 178 The Princess cut. **1960** C. W. Cunnington et al. *Dict. Eng. Costume* 172/1 *Princess dress*,..a style popularly associated with the Princess of Wales when *c.* 1878–80 it was very fashionable. **1964** *McCall's Sewing* i. 5/1 A narrow centre panel, often seen in a princess-line dress, will add height. **1968** J. Ironside *Fashion Alphabet* 25 *Princess*, a line which follows the curves of the body..darted for shape but with no seam at the waistline. **1973** *Times* 15 Nov. 1/3 The wedding dress was in pure white silk in the traditional *Princess* line.

‖ **princesse lointaine** (prænses lwantɛ̃n). [Fr., lit. 'distant princess', title of a play by E. Rostand (1868–1918), based on a theme of the poetry of the 12th-cent. troubadour Jaufré Rudel.] An ideal but unattainable woman. Also as *attrib. phr.*, aloof, unapproachable.
1921 W. J. Locke *Mountebank* viii. 96 The woman who could satisfy all his romantic imaginings was the Princesse Lointaine. **1934** A. Thirkell *Wild Strawberries* xi. 238 Pierre had sat with Agnes, feeling like Geoffroy Rudel with the Princesse Lointaine. **1940** A. Christie *Sad Cypress* i. i. 17 That little air of yours—aloof—untouchable—*la Princesse Lointaine*. **1949** R. Harvey *Curtain Time* 127 Papa was the romantic and the mystic, following the gleam of the fiercer bass, the more golden walleyed pike, and sailing in search of a muskelunge, as if for some *princesse lointaine*. **1957** L. Durrell *Bitter Lemons* 97, I had seen her..driving about the hills with the same *princesse lointaine* expression. **1962** I. Murdoch *Unofficial Rose* xxxiii. 314 He had loved Lindsay as the enticing but untouchable *princesse lointaine*. **1969** M. Drabble *Waterfall* 42 She lay there, the unachievable *princesse lointaine*. **1976** P. Quennell *Marble Foot* iv. 162 Constant's *princesse lointaine* was the Chinese film-actress Anna May Wong.

Princeton-First-Year. Applied to a form of male homosexual activity in which partners achieve orgasm by intercrural friction.
1969, 1971 [see *plain sewing* s.v. *plain *a.*[1] and *adv.* C.c]. **1980** *Times Lit. Suppl.* 21 Mar. 324/5 'Princeton-First-Year' is a more condescending version of the term 'Princeton Rub'; that is, *coitus contra ventrem*.

Princetonian (pri:ns,təʊ·niăn), *sb.* and *a.* [f. *Princeton* (see below) + -ian.] **A.** *sb.* A student or graduate of Princeton University, New Jersey, U.S.A. **B.** *adj.* Of or pertaining to Princeton University.
1876 (*title of newspaper*) The Princetonian. **1896** J. Barnes (*title*) A Princetonian. A Story of undergraduate life at the College of New Jersey. **1898** J. W. Alexander *Princeton* 5 These two renowned and useful organizations ..[are] exclusively Princetonian. *Ibid.* 26 Some cynical Princetonian has said that nothing less than two armies and a revolution could drive a son of Old Nassau to a New England college. **1928** [see *old-line* s.v. *old D. 3]. **1949** *Cavalier Daily* (Univ. of Virginia) 22 Oct. 4/2 Staffers of the Daily Pennsylvanian visited the Princetonian offices following the Penn-Princeton football game. **1971** M. McCarthy *Birds of Amer.* 192 The languid voice of a Princetonian major in government studies. *Ibid.* 193 The ultra-wasp Princetonian, who bore the curious name of Silvanus Platt. **1977** *New Yorker* 23 May 88/1 Thus ringingly begins an 'Alumni Primer' that is distributed to each Princetonian during the summer following his.. graduation from that hallowed institution.

principal, *a.* and *sb.* (*adv.*) Add: **A.** *adj.* **1. a.** *principal boy*, the leading male role in a pantomime, usu. played by a woman; *principal girl*, the leading female role in a pantomime.
1893 H. E. McLelland *Jack & Beanstalk* 11 She's Jill, our 'principal girl', the gallery's joy. *Ibid.*, So it's only natural, as I'm the principal boy! **1897** G. B. Shaw *Our Theatres in Nineties* (1932) III. 24 Why..is the 'principal boy' expected to be more vulgar than the principal girl? **1900** [in Dict.]. **1901** R. J. Broadbent *Hist. Pantomime* xxi. 224 Towards the close of the 'fifties ..the character of Harlequin began to be played by women, the origin of what is now known as the 'principal boy'. **1910** Baroness Orczy *Lady Molly* ix. 236 The little actress looked ready to cry... 'I am principal boy at the Grand,' she explained. **1925** M. W. Disher *Clowns & Pantomimes* xvii. 317 This obsession with barbaric splendour was shown in Harris's choice of 'principal boys'. **1932** D. L. Sayers *Have his Carcase* xv. 201 Airy-fairy-Lilian they used to call me when I was principal boy in old Rosenbaum's shows. **1962** *Oxford Mail* 24 Dec. 6/6 In 1947 he met his wife in pantomime at Lincoln. 'She was the principal girl, Maid Marion. I was just one of the bad robbers.' **1969** *Listener* 2 Jan. 18/1 Gone are the good days when principal boys were girls. **1971** *Petticoat* 17 July 28/1 He..was wearing a pink voile shirt with principal-boy sleeves. **1975** *Times* 2 May 11/3 The greatest of pantomime principal boys, Dorothy Ward.

11. *principal component*, one of the components of a set of statistical data (regarded as points in a multi-dimensional space) which contribute most strongly to its variance; freq. *attrib.* (in *sing.* or *pl.*), designating a method of analysis which involves finding the principal component and removing the variance due to it, and repeating this successively; *principal ideal*, an ideal whose elements can be generated by multiplying some particular member of the ring by each member (including itself) in turn; *principal quantum number* (Physics), the quantum number symbolized by *n* (see *N I. 4 b); *principal stress*, each of the three purely tensile or compressive stresses acting in mutually perpendicular directions into which any combination of stresses acting at a point can be resolved.
The examples are arranged in alphabetical order (in Dict. they are arranged chronologically).
1933 H. Hotelling in *Jrnl. Educ. Psychol.* XXIV. 421 We..determine the components, not exceeding *n* in number, and perhaps neglecting those whose contributions to the total variance are small. This we shall call the method of principal components. **1963** Sokal & Sneath *Princ. Numerical Taxon.* vii. 195 Two different methods of factor analysis are customarily practiced: the principal components method is largely employed by British factor analysts. **1968** *Brit. Med. Bull.* XXIV. 236/2 Examples of the use of the older established techniques of principal components analysis..and factor analysis..rather than taxonomic analysis, predominate in the literature. **1969** A. P. Dempster *Elem. Contin. Multivariate Anal.* vii. 136 The principal component analysis of a given sample relative to a given reference inner product over variable-space consists of finding the eigenvalues and eigenvectors of the sample covariance inner product relative to the reference inner product. The eigenvalues found in this way will be called sample principal components of total variance relative to the chosen reference inner product or, more briefly, principal components. **1937** A. A. Albert *Mod. Higher Algebra* (1938) xi. 255 Every two quantities of a principal ideal ring have a greatest common divisor. **1965** J. J. Rotman *Theory of Groups* iv. 66 A principal ideal domain is a domain in which every ideal is a principal ideal. **1970** D. M. Burton *First Course in Rings & Ideals* ii. 19 An ideal (*a*) generated by just one ring element is termed a principal ideal. **1922**, etc. Principal quantum number [see *N I. 4 b]. **1973** J. G. Tweeddale *Materials Technol.* I. ii. 23 A specific maximum possible number of orientations of orbital pattern, each pattern being completely occupied if two electrons are in it, is possible for each mode, the number being determined by the relevant Principal Quantum Number. **1858** W. J. M. Rankine *Man. Appl. Mech.* i. v. 94 The three conjugate normal stresses are called principal stresses, and their directions, principal axes of stress. **1922** Glazebrook *Dict. Appl. Physics* I. 803/2 The intensity of greatest shearing stress at any point is equal..to one-half the algebraic difference of the greatest and least principal stresses. **1944** A. Holmes *Princ. Physical Geol.* vi. 78 The various types of faults depend on the relationships between the three principal stresses.., assuming of course, that..the stress difference is sufficient to bring about fracture and movement. **1971** I. G. Gass et al. *Understanding Earth* xix. 272/1 Making certain other reasonable assumptions about the shear fracture character of the fault, it is possible to derive..the direction of the maximum principal stress.

12. *Surveying.* = Primary *a.* 3a in Dict. and Suppl.
1790 *Phil. Trans. R. Soc.* LXXX. 248 The first set [of Secondary triangles] consists of thirty-five, whereby the relative distances of so many points have been determined from certain stations of the principal series. **1795** *Ibid.* LXXXV. 490 (*heading*) Of the selection of the angles constituting the principal triangles, and the manner of reducing them for computation. **1847** W. Yolland *Acct. Measurement of Lough Foyle Base* vii. 113 From the commencement of the Irish Survey, in 1825, secondary objects required for breaking up the principal triangulation into a smaller network..were regularly observed at the same time as observations were made of the principal stations. **1920** [see *primary *a.* 3 a].

13. *Seismology.* Applied to the most intense shock or earthquake occurring in a sequence.
1899 C. Davison *Hereford Earthquake Dec. 17, 1896* iv. 199 The principal shock was also registered by magnetographs at Kew. *Ibid.* 200 The principal earthquake was felt by several persons in the Isle of Wight. **1902** *Q. Jrnl. Geol. Soc.* LVIII. 374 The focus of this shock was evidently close to the northern end of the focus of the first and principal shock. **1938** L. D. Leet *Pract. Seismol.* ix. 295 Davison has cited several well-studied cases..where the range of audibility of after-shocks increased progressively and systematically, giving evidence of decreasing depth of foci following the principal shock. **1965** A. Holmes *Princ. Physical Geol.* (ed. 2) xxv. 893 The principal shock, which generally lasts only a few seconds, or at most, and rarely, a few minutes, may be preceded by fore-shocks and is invariably followed by a series of after-shocks.

B. *sb.* **1. a.** (Further examples.)
1827 S. Rodman in B. Swan *New Bedford in 1827* (1935) 3/2, I visited the High School, Wm. Johnson, principal. **1833** *Century Mag.* XXX. 780/1, I am, sir.. permitted to be the Principal of the Canterbury, (Conn.) Female Boarding School. **1949** *Lubbock* (Texas) *Morning Avalanche* 23 Feb. i. 10/6 Price was named to the position of principal of the new school. **1973** *N.Y. Law Jrnl.* 30 July 13/7 Petitioner, a probationary assistant principal in James Wilson Young High School, resigned in writing on Feb. 13, 1973 (effective June 29, 1973) after being pressed to do so by both the District Principal (Mr. Covell) and the Assistant District Principal (Mr. Pecorale) of the School District. **1974** *Hartsville* (S. Carolina) *Messenger* 22 Apr. 1-A/4 Dr. Black pointed out that students who presented certificates to their principals in September, 1973 will not need another one.

e. A rank in the Civil Service.
1890 A. E. Housman *Let.* 9 Oct. (1971) 27 The Administrative Principal, Mr Webb, has to-day taken up the comparison of Trade Mark applications. *Ibid.* 28 The position as Principal held by Mr Webb would give him a status not justified..by length of service. **1915** F. G. Heath *Brit. Civil Service* xxii. 230 The salaries are as under: Twelve Analysts (Second Class), £160 by £15 annually to £350;..one Deputy Principal, £700 by £25 to £800; and one Principal Chemist, £1200, and after five years.., £1500 per annum. **1951** T. A. Critchley *Civil Service Today* ii. 38 Under these high officials..are the assistant secretaries, in charge of divisions, and principals, in charge of sections. **1951** *Posts in Civil Service for University Graduates* 9 The Principal's day to day work is not confined to his desk. **1967** *Times Rev. Industry* Feb. 100/2 The full time staff is small... Of these five are 'principals' —as the Civil Service calls executives.

f. A fully-qualified practitioner or partner in a professional business.
1968 *Economist* 13 Apr. 41/1 A young man can..become a principal in general practice..after four to five years' training..plus a year's pre-registration service—in hospital. **1972** *Accountant* 26 Oct. 502 (Advt.), Principals with the responsibility for training newly articled clerks should ensure that the best course of action for their newly articled clerks is to enrol with the Metropolitan College. **1975** *Law* (Employment Service Agency: Careers and Occupational Information Centre) 8 The principal with whom he has served his articles or apprenticeship may offer him an opening.

2. a. (Later examples.)
1962 H. O. Beecheno *Introd. Business Stud.* xiii. 117 Whereas an agent is not normally allowed to relend his principal's money at interest..a bank is allowed to do this. **1976** *Times* 22 Apr. (Baltic Exchange Suppl.) p. i/9 The Baltic is unusual in being open both to middle men and principals.

f. A leading performer in a drama or entertainment.
1936 N. Streatfield *Ballet Shoes* xiv. 212 The production was on a very large scale... The principals became unduly important. **1961** *Times* 20 June 16 The three principals are admirable: as Danila, Mr. Yuri Soloviev gives a tremendous performance; he has a prodigious technique in leaps and turns. **1971** *Morning Star* 28 June 4 Steve Hodson and Gillian Blake, the two principals in Yorkshire TV's new children's series. **1976** *National Observer* (U.S.) 8 May 20/3 To add to my discomfort, I found the voices of the principals, Michael Cristofer and Tyne Daly, distinctly unpleasant.

‖ **principessa** (printʃipe·sä). [It., f. med.L. *principissa* (see Princess *sb.*)] An Italian princess. Also *attrib.*
1823 Lady Blessington *Jrnl.* 27 July in E. Clay *Lady Blessington at Naples* (1979) 38 The Principessa Partanno is..no longer in her *première jeunesse*. **1861** C. M. Yonge *Young Step-Mother* xxiii. 333 He should go and marry the Principessa Bianca, a foreigner and Papist. **1945** E. Waugh *Brideshead Revisited* i. ii. 50 The Principessa Fogliere gave a ball and Lord Malton was not asked. **1963** *Times Lit. Suppl.* 11 Jan. 21/3 The increasingly angular principessa who adores him. **1967** P. E. H. Durston *Mortissimo* (1968) iii. 27 The most elegant little *principessa* in the Villa Borghese. **1973** 'D. Jordan' *Nile Green* xxxiii. 159 The princess..smiled, a weary principessa smile. **1979** N. Slater *Falcon* i. 22 The pretty principessas would be storming the Embassy gates.

principium. Add: **1. c.** *principium individuationis*: the principle through which an entity is differentiated from matter, or being from non-being. (Cf. Individuation 1 *note*.)
1694 Locke *Essay Hum. Und.* (ed. 2) ii. xxvii. 179 'Tis easie to discover, what is so much enquired after, the *principium Individuationis*, and that 'tis plain is Existence it self, which determines any sort of Being to a particular time and place incommunicable to two Beings of the same kind. **1739** Hume *Treat. Hum. Nature* I. iv. 349 In order to justify this system, there are four things requisite. *First*, To explain the *principium individuationis*. **1883** F. H. Bradley *Princ. Logic* 265 It was shown above.. that space and time-relations are no *principium individuationis*; for they fall within the *what*, and do not make the *this*. **1947** *Horizon* Feb. 150 But above this layer of all-embracing 'identification' Sorge reappears as *principium individuationis*, isolating a person and stimulating his intellect into the frightful awareness of his nakedness and his fate. **1965** *Listener* 2 Sept. 344/1 Something.. must divide us from other people, there must be some kind of *principium individuationis*.

principle, *sb.* Add: **5. c.** *first principle*: a primary proposition, considered self-evident, upon which further reasoning or belief is based; freq. in *pl.*
In some quots. influenced by sense 3 in Dict.
1638 [see Principle *sb.* 10 a]. **1690** Locke *Essay Hum. Und.* i. ii. 10 Those (as they are called) first Principles. **1701** J. Norris *Ideal World* I. ii. 75 As much above the Possibility as the Necessity of Demonstration, in one Word, a very first Principle. **1785** [see Principle *sb.* 7 a]. **1817** Coleridge *Biog. Lit.* I. xii. 253 Philosophy in its first principles must have a practical or moral as well as a theoretical or speculative side. *Ibid.* 260 Those original and innate prejudices..which to all but the philosopher are the first principles of knowledge. **1871** B. Jowett tr. *Plato's Dialogues* I. 711 And this is the reason why every man should expend his chief thought and attention on the consideration of his first principles: —are they or are they not rightly laid down? **1934** *Times Lit. Suppl.* 19 July 497/2 It is to Coleridge's search for first principles in literature that appeal is made. **1961** *Cambr. Daily News* 10 Feb. 7 Once you have absorbed the first principles in art.

7. c. Also in phr. *on general principles*, freq. in weakened sense: in general, for no specified reason, from a settled motive.
1894 Somerville & 'Ross' *Real Charlotte* II. xxi. 90 She had no particular dislike for Francie..but on general principles she was pleased that discomfiture had come to Miss Fitzpatrick. **1898** R. Hughes *Lakerim Athletic Club* 246 Pretty wanted to punch his head on general principles, but decided it would be better to beat him at tennis. **1914** *New Republic* 26 Dec. 15/1 In one Spanish village he was locked up on general principles, because the King happened to be passing through town that day. **1930** J. C. Ransom *God without Thunder* ii. viii. 173 It is like the flattery of the man who flatters us on general principles. **1938** 'G. Graham' *Swiss Sonata* 87 Vicky will be held responsible for it just on general principles, and that will be that.

d. Phr. *in principle*: theoretically; in general but not necessarily in individual cases.
1820 G. Canning in C'tess of Airlie *Lady Palmerston* (1922) I. vi. 102 So objectionable does it appear to them in principle as well as in practice. **1859** Palmerston in P. Guedalla *Palmerston Papers* (1928) 117 The First Method would evidently be the best in Principle. **1874** Geo. Eliot *Let.* 15 July (1956) VI. 67, I am thoroughly opposed in principle (quite apart from any personal reference to myself) to the system of *contemporary* biography. **1932** *Ann. Reg. 1931* 295 These [proposals] were rejected by Washington, but the latter made a counter-proposal: that France accept the scheme 'in principle', and leave to a conference of technical experts those practical modifications which she desired. **1951** J. Cornish *Provincials* 102 Still, we were loggers; we had won in principle. **1963**

Richardson & Toynbee *Thanatos* 87 When we say that we know a thing it involves us in saying that it is, at least in principle, verifiable by the senses.

principled, *ppl. a.* Add: **4.** Based on or guided by (technical) principles or rules; not arbitrary or *ad hoc*.
1968 P. M. Postal *Aspects Phonol. Theory* iii. 47 The one principled way to make such a choice independently of the grammar..is to pick one of the several representations which eliminate *all* phonetically predictable features. **1970** *Canad. Jrnl. Linguistics* XVI. i. 3 Both factions agree on what constitutes the natural domain of linguistics:..the principled explanation of sound–meaning relations in languages. **1972** *Language* XLVIII. 301 It is interesting that there is a principled basis for choosing, in certain types of circumstances, a rule with a global environment over a corresponding non-global rule plus an ordering statement. **1978** *Ibid.* LIV. 410 As there is no principled way to prevent base rules from generating intransitive prepositions in any PP position, 5c could be base-generated, given E's base rules.

pring (priŋ), *sb.* [Echoic.] The sound made by a bell.
c **1921** D. H. Lawrence *Phoenix II* (1968) 183 She heard a loud prrring-prrring of a bicycle-bell.

pring (priŋ), *v.* [Echoic.] *intr.* To make a sound like a bell.
1927 *Scots Observer* 30 Apr. 10/4 The bell of no. 13 was pringing lustily.

pringling (pri·ŋgliŋ), *ppl. a.* [f. Pringle *v.* + -ing[2].] That pringles, or causes a prickly sensation.
1896 A. Conan Doyle *Uncle Bernac* i. 7, I..pressed my lips upon the wet and pringling gravel. **1923** C. Morley *Where Blue Begins* i. 5 In the golden light and pringling air he felt excitable and high-strung.

pri·ngling, *vbl. sb.* [f. the vb.] A prickly and tingling sensation.
1956 P. Wentworth *Silent Pool* xli. 214 Young Watson felt a pringling at the back of his neck.

prink, *v.*[3] (Later examples.) Also, to walk daintily or with precise movements.
1962 M. Baldwin *Death on Live Wire & On stepping from Sixth Storey Window* 11 Uncle Cyclops Had one eye To bulge at ankles prinking past. **1962** J. Onslow *Bowler-Hatted Cowboy* xiii. 124 In the morning a doe with her twin fawns had passed us, stopping to nibble at the willow bushes as she prinked down the hill.

prinkle (pri·ŋk'l), *sb. Sc.* [Origin obscure.] A young coal-fish, *Pollachius virens*.
1832 P. Buchan *Secret Songs of Silence* (MS.) 177 The laddie and the lassie, Gaed out to gather prinkle, O. **1903** G. Sim *Vertebrate Fauna of 'Dee'* 238 In the young stages, the names of 'Prinkle', 'Gerrick', 'Poodlie', are given [to the coal fish]. **1943** W. S. Forsyth *Guff o' Waur* 54 Wupp it weel wi' curly 'oo', The prinkles to confoun'. **1972** *Which?* May 135/1 Saithe..may be called coal fish, coley, and a whole host of local names from cooth to prinkle.

Prins (prins). *Chem.* The name of H. J. *Prins* (1889–1958), Dutch chemist, used *attrib.* to designate the condensation of an olefin with formaldehyde or other aliphatic aldehydes in the presence of a dilute mineral acid.
1944 *Jrnl. Chem. Soc.* 296 Under the conditions of the Prins reaction, a normal, acid-catalysed addition to the olefinic linking occurs. **1970** *Tetrahedron Lett.* i. 37 (*heading*) The isolation and synthesis of a novel tetracyclic ether from East Indian sandalwood oil. A facile intramolecular Prins reaction.

print, *sb.* Add: **I. 3.** *esp.* = *finger-print* s.v. *finger *sb.* 15.
1924 P. MacDonald *Rasp* vi. 88 But how to explain the finger-prints? And Deacon did not know of those prints. **1929** 'G. Daviot' *Man in Queue* ii. 18 Attached..was a report from the finger-print department. There was no trace of these prints in their records. **1936** A. Christie *Cards on Table* xxii. 215 Handled it with gloves,..and.. the last prints would be those of Mrs. Benson herself. **1938** N. Marsh *Artists in Crime* v. 66 If you come across any keys, try them for prints. **1952** E. Grierson *Reputation for Song* xxi. 170 They *will* take his prints. **1957** F. & R. Lockridge *Tangled Cord* (1959) xiii. 170 Harry here gets to thinking maybe he touched something and left prints. **1975** J. McClure *Snake* viii. 113 The drinking vessels had been cleaned of any prints, and the wash basin..given a thorough rub-over. **1980** P. G. Winslow *Counsellor Heart* ii. 32 While the print man was working Capricorn had a few words with..the Divisional Surgeon.

5. (Later examples.) Also *attrib.*
1909 [see *creamery 1 b]. **1955** J. G. Davis *Dict. Dairying* (ed. 2) 153 The cream..goes through either a box moulder for packing into 56 lb. boxes or a pat moulder for wrapping 1 lb. and ½ lb. prints. **1963** M. McCarthy *Group* ix. 194 Libby was scandalized by the amount of fresh print butter Polly mixed in..*plus* brandy and sherry. **1972** E. Wigginton *Foxfire Bk.* 188 When it [*sc.* a mould] is filled, push down on the handle of the mold, which acts like a piston, thus releasing the 'print' of butter.

6. (Earlier and later examples.) Also, a pattern printed on fabric.
1756 E. Holyoke in G. F. Dow *Holyoke Diaries* (1911) 16 Put Prints out to whiten. **1825** E. Weeton *Let.* 22 Apr. in *Jrnl. of Governess* (1969) II. 352 When you open the parcel..you will find the print which I have procured for you: there are two patterns, a yard each. **1917** *Harrods Gen. Catal.* 1409/2 Best English Prints, for Servants' Dresses. **1957** M. B. Picken *Fashion Dict.* 264/1 *Print*, fabric stamped with design by means of paste dyes used on engraved rollers, wood blocks or screens. **1964** *McCall's Sewing* i. 7/1 Like stripes, prints also can be used to create desirable effects. *Ibid.* vii. 108/1 First check to see if the print has a regular directional pattern. **1972** *Vogue* June 94 Mandarin coat and slit dress of matching print. **1976** *Times* 25 Mar. 11/3 Made in pure silk chiffon, it was chosen from a range of evening dresses in many prints and colours. **1976** *Guardian* 2 Apr. 4/1 Detail from Nympheus: a hand blocked print on linen copied from an 11th century Chinese painted silk, now in the British Museum.

II. 7. Used without *in* or *into*.
1932 E. V. Lucas *Reading, Writing & Remembering* iii. 69, I have no recollection that the article ever reached print. *Ibid.* vi. 121, I publish it here for the first time—it has waited only forty-five years for print. **1934** H. G. Wells *Exper. Autobiogr.* I. vi. 356 This success whetted my appetite for print and I sent Harris a further article.. which he packed off to the printers at once. **1950** *Science News* XV. 7 There is one fundamental reason why free-lance articles so rarely see print. **1970** G. F. Newman *Sir, You Bastard* viii. 204 He knew that the fray with James would make some print. **1977** *Time* 14 Mar. 31/1 Nor is it likely that a British version of the Pentagon papers or the Watergate scandals would ever have seen the light of print.

9. Restrict †*Obs.* to sense in Dict. and add: **b.** *in the print*, in the printing trade.
1973 L. Heren *Growing up Poor in London* ii. 39 For our mother, only a minimum of education was required to ensure a good safe job in the print. *Ibid.* viii. 193 For her [*sc.* his mother] a good job in the print and a house in Bromley or Beckenham meant security, respectability and keeping yourself to yourself.

11. a. (Later examples.)
1942 D. Powell *Time to be Born* (1943) iv. 80 A few names, if sufficiently in the public prints, naturally did stick. **1961** R. M. Williams in D. N. Barrett *Values in America* iii. 74 The criticisms levelled against higher education in the public prints. **1973** *Daily Tel.* (Colour Suppl.) 30 Nov. 7/2 The popular prints are interested in Parliament only when something dramatic blows up, or when an MP, possibly overworked, makes an ass of himself.

b. Also in general use.
1928 *Daily Mail* 25 July 18/5 Prints of the Memorandum and Articles of Association can be inspected at any time.

13. In mod. use applied to (*a*) a (usu. positive) photographic picture produced on an opaque medium for direct viewing (as opposed to a transparency); (*b*) a positive copy of a motion picture (on a transparent medium).
(*a*) **1915** [see *colour transparency* s.v. *colour *sb.*[1] 17 b]. **1939** Mack & Martin *Photogr. Process* ix. 304 It is the final objective of the photographic process to produce a positive image (either a transparency or a paper print). **1958** C. L. Thomson *Colour Films* 39 Transparencies as a basis for prints have the advantage that the printer knows in advance that the colour photograph is a successful one, and will therefore produce an acceptable print. **1970** C. B. Neblette *Fund. Photogr.* xxi. 299 Ektachrome paper is designed for making color prints from transparencies on Ektachrome and Kodachrome film. **1978** *Sci. Amer.* Apr. 110 (*caption*) In this negative print made with the 48-inch Schmidt telescope on Palomar Mountain the nebulas appear black.
(*b*) **1912** F. A. Talbot *Moving Pictures* viii. 87 The majority of cinematograph manufacturing establishments undertake to develop negatives, and to supply positive prints ready for projection. **1914** J. B. Rathbun *Motion Picture Making* ii. 29 The light of the projector passes through the transparent positive print and traces the image on the screen. **1942** *Sun* (Baltimore) 23 Feb. 10/7 Lou Fonseca, director of the American League's motion picture, 'The Ninth Inning', said today that a number of prints of the film are now available for clubs, schools and other organizations. **1973** H. Gruppe *Truxton Cipher* (1974) xiv. 141 The movie was a..bad Western... The print was on its sixth tour through the Atlantic Fleet, and ..the worse for wear. **1978** R. Hill *Pinch of Snuff* xii. 125 They made a film I'm interested in... I'd like to find out how many prints there were.

13*. A signal on magnetic tape produced by print-through.
1950–1 *B.B.C. Quarterly* V. 250/2 On playback..the comparatively high-level prints resulting from storage are replayed at their initial level. **1958** H. G. M. Spratt *Magn. Tape Recording* iii. 110 The strength of the print rises rapidly immediately the reel has been wound up and then, with the rate of increase falling, tends towards an ultimate limiting value. **1962** A. Nisbett *Technique Sound Studio* iv. 84 The erasure is more marked on the small printed signal than on the main body of the recording (the print being reduced by perhaps 16 dB, as against 3 dB off the main signal).

15. a. print chain, an endless chain of printing types in some printers; **print-maker, -making** (as sense 12); **print order**, an order for a certain number (of an issue of a book, paper, etc.) to be printed; **print room** (examples); **print run** = *run *sb.*[1] 20 d; **print-script**, a style of handwriting that imitates typography; **print train** = *print chain* above; **print wheel**, a disc having printing types round its rim that can be brought into position by rotation of the disc.
1967 R. Bregzis in Cox & Grose *Organiz. Bibliogr. Rec. by Computer* V. 119 The cards are printed out by an IBM 1401 computer using a print chain with an expanded set of 101 characters. **1972** *Computers & Humanities* VII. 97 High quality printout, suitable for publication purposes and from an extended print chain with a large character set, is highly desirable. **1928** M. Dobson *Block-Cutting & Print-Making by Hand* ii. 4 There are..many terms peculiar to the craft of print-making. *Ibid.* xx. 176 He who would be his own print-maker will really find no great obstacle in the way. **1961** *Times* 18 May 17/5 The work of 27 Soviet print-makers..and..almost our first view of current artistic activity in Russia. **1965** Zigrosser & Gaehde *Guide to Collecting Orig. Prints* i. 4 Print-making is a democratic form of art, for it enables not one but many persons to own and enjoy the same work of art... Every medium has both utilitarian and aesthetic functions. This is especially true of printmaking. **1977** J. Trevalyan in S. Turner *Handbk. Printmaking Supplies* 8 Today printmaking has grown into a vast and profitable business... When several printmakers meet together their talk is almost always where can this or that be bought. **1953** Pohl & Kornbluth *Space Merchants* (1955) ii. 13 The first issue comes out in the fall, with a print order of twenty million. **1971** M. Russell *Deadline* vii. 75 There's a 15,000 extra print order. Some newsagents sold out by ten yesterday. **1979** J. Sherwood *Hour of Hyenas* iii. 32 The only subject discussed was the money value of various contracts and the size of the resulting print order. **1849** *Index to Add. MSS. Brit. Mus. 1783–1835* p. iv, A certain number also of manuscripts..have long since been transferred to the Print Room. **1862** Geo. Eliot *Let.* 17 May (1956) IV. 34 We..went to the Printroom of the British Museum to see Italian portraits of 15th cent. **1901** A. Whitman *Print-Collector's Handbk.* x. 132 Six officials have presided over the destinies of the Print Room. **1921** E. J. Sullivan *Art of Illustration* xxxvii. 252 Push past the unpretentious and silent swing door to the Print Room. **1956** Hayden & Bunt *Old Prints* p. viii, He will pass from our modest book to the masterpieces in the Print Rooms of the British and the Victoria and Albert Museums. **1970** B. Allen *Print Collecting* vi. 97 The little blue card..entitles him to be a regular visitor to the Print Room. **1975** *Language for Life* (Dept. Educ. & Sci.) xxi. 311 These books..cannot always command the large print runs of text books. **1979** *Bookseller* 23 June 2836/1 This compilation..sold out of its first print-run rapidly. **1979** *Times* 19 Dec. 12/4 Increasingly..books with a strong American end are being entirely printed in the United States, with a proportion of the print run bought for the United Kingdom. **1922** *Print Script* (Board of Educ.) 5 During the last few years the movement in favour of 'print-script' has spread so widely in the schools of this country that a wish has been expressed that the experience of the Board's Inspectors should be made available to the general public. **1932** A. J. Fairbank *Handwriting Man.* Introd., Of the problems arising in the schools today..., one..is how to adapt the handwriting which has been named 'print-script'..so that there may remain in the revised model nothing to hinder the tempo at which an adult writes. **1955** P. Rudland *From Scribble to Script* 6 A comparatively recent innovation in the teaching of handwriting is what is known as *print-script*... Print-script originated from an address on penmanship given by Edward Johnston at the Annual Conference of Teachers in 1913. **1959** J. C. Gagg *Beginning Three R's* xii. 82 It would seem that print-script is.. the most useful first style for slower children. **1975** *Language for Life* (Dept. Educ. & Sci.) xi. 185 Some teachers believe that a print-script should be used. **1970** *Computers & Humanities* IV. 247 Lest one despair of limitation in the number of possible escape codes, it should be pointed out that the standard IBM print train for the 360 line of computers permits 240 characters and to date (August 1969) only 174 are used on the extended set now planned by the Library of Congress for catalog card printing in all Roman alphabets. **1941** T. J. Rhodes *Industr. Instruments for Measurement & Control* iv. 144 The multiple-point recording potentiometer is achieved by substituting a print wheel for the recording pen and by equipping the potentiometer with a motor-driven selector switch synchronized with the print wheel. **1961** L. W. Hein *Introd. Electronic Data Processing* xiii. 255 Print wheel 1 will be used to print the first letter of the name, and wheel 120 will be used for the cents position of the net pay figures on the stub. **1970** A. Cameron et al. *Computers & Old Eng. Concordances* 39 We had print wheels made; we never even got to fitting them on the machine.

b. print journalism, writing, reporting or writing for newspapers (as opp. television); so *print journalist*; **print medium** (usu. pl., media), newspapers (as opp. broadcasting); **print union**, a trade union for printers (also *printing union* s.v. *printing *vbl. sb.* d).
1975 *Listener* 1 May 578/3 Michael Barratt can be taken to task for unwittingly imposing the techniques of popular print journalism on television reportage. **1977** *Listener* 1 Dec. 708/1 Eric Sevareid..has done for television what Walter Lippman did..for print journalism. **1971** *Ibid.* 25 Nov. 711/3 Some other print journalists—the editor of the *Evening Standard* is one—have chosen to break the solidarity of press and television to suggest that television's standards have been lower. **1973** *Time* 25 June 15/3 It will be the print journalists' and historians' task to review and criticize. **1975** *Listener* 10 Apr. 463/3 The camera simply recorded..the cycle of a diplomatic issue in a way no print journalist could ever hope to describe. **1968** *Globe & Mail* (Toronto) 13 Feb. b3/1 The message..is accentuated in print media advertising. **1972** *Guardian* 29 Jan. 11/5 The print media..claim..that they are being discriminated against. **1978** *Verbatim* Feb. 1/1 Part of the responsibility for our bent language rests with the print media. **1959** *Daily Tel.* 22 July 1/5 Print unions reject hours and pay offer. **1975** *Times* 7 Aug. 1/4 (*heading*) Print unions' ultimatum to 'Observer'. **1980** *Times* 13 Aug. 1/7 The unexpectedly strong reaction among other print unions brings a new dimension into the dispute over machine managers' pay. **1976** *New Yorker* 13 Sept. 103/3 (Advt.), Print writing is tougher than television writing.

print, *v.* Add: **II. 6. b.** *to print out*: to produce in or as a print-out.
See also 14 c below.
a **1884** Knight *Dict. Mech.* Suppl. 722/1 The sending operator prints out his message in plain letters at the dis-

tant end of the line, whether the receiving operator is at the instrument or not. **1953** *IRE Trans. Instrumentation* June 68 This Binary Outscriber indicated that it would adequately perform its required task: that of rapidly printing-out the memory contents. **1955** *IRE Trans. Electronic Computers* IV. 2/2 The typewriter can operate directly upon information received from the accumulator and alphanumeric translator and print out a completely general format. **1957** *IRE Trans. Instrumentation* Sept. 194/1 A Flexowriter prints out the character that was stored in the interim flip-flop storage registers. **1969** *Listener* 10 July 44/3 In the case of the two computers, of course, the actual difference could be shown by printing out their programmes. **1977** *Sci. Amer.* Sept. 23/1 (Advt.), Results are displayed to hospital personnel and printed out as reports.

8. b. (Later examples.)
1912 *Englishwoman* July 73 The line block will print well on paper on which the delicate shades of the half-tone would be lost. **1971** D. Potter *Brit. Eliz. Stamps* ix. 95 On rare occasions the embossed stamp appears albino, when the colour has failed to print through lack of ink. **1979** *SLR Camera* June 41/1 For the other cases which will not print on these medium grade papers you'll have to use harder or softer grades to produce a result that even approaches the satisfactory.

d. With pass. force: to appear in print; to be printed.
See also quot. 1775 in sense 6 in Dict.
1930 *Sat. Rev. Lit.* (U.S.) 2 Aug. 21/2 The *Return* began printing in *Collier's*. **1953** *Northampton Dioc. Mag.* Autumn 11 The spoken word rarely prints satisfactorily. **1973** M. Russell *Double Hit* xviii. 132 The newsvendors' stands stood untended: the first editions were still printing.

11*. Of magnetic tape or a recorded signal: to give rise to print-through. Also *trans.*, to transfer (a signal) as a result of print-through.
1950–1 *B.B.C. Quarterly* V. 246/2 After a certain time, ..measure the level of the signal printed on to the erased slip. **1952** *Appl. Electronics Ann. 1951* 43/2 Trouble.. occurs sometimes with the programme 'printing' from one layer to another of the reel. **1958** J. Tall *Techniques Magn. Recording* iii. 33 Homogeneous tape was..favored until a few years ago in some parts of Europe and is still used occasionally there. It is subject to one major fault..: it 'prints' excessively.

III. 14. a. Also, to produce (a print) of a motion picture or from a transparency.
1915 J. B. Rathbun *Motion Picture Making* ii. 36 In printing the positive film from the negative, the teeth of the sprockets in the printing machine pass through both films, holding them in perfect register. **1931** B. Brown *Talking Pictures* x. 229 The usual method in printing is to mask the sound track space on the unexposed film and then print off the picture. *Ibid.* 243 While printing cinema films is similar in principle to ordinary photographic work, it is carried out on entirely different lines, due to the enormous length of the negative. **1974** *Encycl. Brit. Macropædia* XII. 549/2 From this optical negative the sound track can be printed photographically on the exhibition release prints. Sometimes the release prints are printed as positives directly from the final magnetic track, rather than from a negative made from this track.

b. (Further example.)
1929 R. H. Goodsall *Beginner's Guide Photogr.* vii. 37 A piece of printing paper is placed in contact with the negative and the light allowed to pass through the latter... Where there is little or no deposit on the negative it prints dark.

c. *to print out* (intr.): to produce an image (or, of an image, to appear) without chemical development. Also *trans.*, to print without development.
See also 6 b above.
1882 *Rep. Brit. Assoc. Adv. Sci. 1881* 595 The author has 'printed out' the spectrum on chloride of silver. **1902** *Encycl. Brit.* XXXI. 703/2 Considerable use is also made of ready-prepared platinotype paper, sensitized with salts of platinum and iron, which can be printed out entirely or only partly printed and developed with potassic oxalate. **1906** R. C. Bayley *Compl. Photographer* xiv. 175 In order that P.O.P. may be sufficiently sensitive to be usable at all, and give a rich image by printing out, it is not sufficient that it should contain silver chloride only. **1913** Hind & Randles *Handbk. Photomicrogr.* xiv. 230 The paper is printed-out in daylight until all detail is visible, then developed in potassium oxalate solution. **1948** *Rep. Progress Physics* XI. 255 When exposures are increased by a factor of 10^7 to 10^8 the photographic material 'prints out'; it darkens visibly due to a process of reduction of the emulsion grains. **1953** *Phil. Mag.* XLIV. 223 This [exposure to radiation] caused silver to print-out internally both in strained crystals and in crystals which had been annealed. **1963** John & Field *Textbk. Photogr. Chem.* ii. 26 Emulsions of this type..were coated on paper, to give 'printing-out papers', so called because the image printed out directly. **1965** *Photogr. Jrnl.* CV. 285/2 A positive image is 'printed out', that is, it becomes visible without chemical development.

d. *trans.* To produce a positive print from (a negative or transparency).
1913 F. A. Talbot *Pract. Cinematogr.* vii. 93 An enterprising amateur who had an excellent negative handed it over to a topical-film firm to print and circulate. **1929** R. H. Goodsall *Beginner's Guide Photogr.* vii. 37 When the negative has been developed, washed and dried..it is ready to be printed. **1940** G. G. Quarles *Elem. Photogr.* x. 133 When a number of negatives are to be printed at one time, it is well to sort them into piles according to contrast. **1974** C. Swedlund *Photogr.* ix. 221/2 Contact printing was the original method of printing negatives, and it remains the process most typically used for negatives of such size..that their images do not require enlargement.

e. *to print down* (trans.): to transfer a photographic image from (a negative) to a printing plate.
1923 F. T. Cockett *Photo-Litho. & Offset Printing* 30 The seccotined worsted will adhere to the base glass and to the cut edges of the negatives so that the whole series can be printed-down in one operation. **1944** J. C. Tarr *Printing To-day* ix. 106 The negatives are printed down on to the metal vacuum frames. **1967** E. Chambers *Photolitho-Offset* iv. 48 Further treatment would result in a grey dot formation which would prove unsatisfactory for printing-down to metal.

f. *to print in* (trans.): to transfer (an image on a negative) to another negative that has already been exposed once; to produce an additional image on (an exposed negative); also *absol.*
1929 R. H. Goodsall *Beginner's Guide Photogr.* viii. 51 This is one method by which clouds may be printed-in. The foreground negative is exposed first, partially developed, and then returned to the easel and the sky portion printed in from another negative while the foreground is screened by a card. **1956** *Focal Encycl. Photogr.* 910/2 To print in large areas, such as a sky, a plain card is used to shade the remainder of the image. **1958** *Newnes Compl. Amat. Photogr.* 1 The man who spends joyful evenings printing in clouds. **1976** M. J. Rosen *Introd. Photogr.* v. 119/2 For printing in small areas of the print, the typical tool is an opaque mask of light cardboard or plastic with a small hole in it. To print in, the print first is given its normal exposure.

15*. [f. *print *sb.* 3.] **a.** *trans.* To test (an object) for finger-prints.
1938 N. Marsh *Death in White Tie* xv. 163 We'd better print the brandy-glass. **1951** A. Hocking *Death disturbs Mr. Jefferson* ii. 24 Austen said to the policeman: 'Print all the rest of the stuff, will you?' **1967** 'D. Shannon' *Rain with Violence* (1969) iii. 39 The lab men had printed the patent leather tote bag. **1971** 'L. Egan' *Malicious Mischief* (1972) i. 9 Dick Hunter, who had just been made Detective again..was printing the kitchen door.

b. To record the finger-prints of (a person).
1952 J. Steinbeck *East of Eden* I. 484 Ever been mugged or printed? **1955** D. W. Maurer in *Publ. Amer. Dial. Soc.* xxiv. 147 He is *printed* (his fingerprints taken) and *mugged* (photographed). **1957** C. MacInnes *City of Spades* ii. xiii. 192 The screws can print you in the nick at Brixton. **1970** G. F. Newman *Sir, You Bastard* viii. 232 He had been charged, printed, and provisionally questioned.

15.** To make (a printed circuit or component).
1946 *Business Week* 23 Feb. 19/2 (*caption*) Developed for proximity fuses, radio circuits 'printed' on ceramic plates are space savers adaptable to miniature pocket receivers. **1946** *Wireless World* Oct. 349/1 The connections are printed on the panels, the 'ink' being a solution of silver, which is dried, baked on and finally varnished over. *Ibid.*, Even the resistors are printed by the use of appropriate solutions. **1956** *Appl. Electronics Ann.* 1955–56 46/1 It is now possible to print a complete piece of electronic equipment except for the larger components. **1958** *Daily Mail* 8 Sept. 8/2 Using modern techniques of etching and engraving, a wiring circuit is actually 'printed' on to a flat base. **1966** *McGraw-Hill Encycl. Sci. & Technol.* X. 596/1 Resistive inks are composed of various forms of carbon with a resin binder..and a solvent vehicle. This mixture is applied through a stencil to form a rectangular pattern..and then baked... Resistive elements printed in this manner have wide tolerance limits. **1973** Dokter & Steinhauer *Digital Electronics* vi. 227 What is printed is not generally the whole circuit but merely the wiring connections.

printability (pri:ntăbi·liti). [f. Printable *a.* + -ity.] **a.** Of type, a block, a plate, etc.: capacity to produce print. **b.** Of paper: capacity to take print. **c.** Of language, statements, etc.: suitability or fitness to be printed. Also *attrib.*
1967 E. Chambers *Photolitho-Offset* iv. 43 The positives are printed-down by the deep-etch method of plate-making which creates a plate with firm hand dot fringes of good printability, sharpness and long press-life. **1969** *R. & E. Coordinator* (Res. & Engin. Council Graphic Arts Industry) Apr. 8/1 A meter and a modified IGT Printability Tester are used to measure the drying time..of quick drying inks into paper. **1971** *Scholarly Publishing* II. 361 The first experimental paper, both sized and unsized, substantially exceeded the minimum specifications ..and was found to have good printability as well. **1979** M. Russell *Touchdown* i. 18 His only difficulty would be to stay within the limits of printability.

‖ **printanier** (prǣntanye), *a.* (*sb.*) Also **printanière**. [F. *printanier*, lit. 'of springtime' (*printemps*), f. L. *primus* first + *tempus* time.] Made from or garnished with spring vegetables. Also as *sb.*
1861 Mrs. Beeton *Bk. Househ. Managem.* 78 (*heading*) Potage printanier, or spring soup. **1867** Trollope *Claverings* I. xix. 237 There's just a little soup, printanière; yes, they can make soup here. **1897** *Sears, Roebuck Catal.* 15/1 Soups..French Bouillon, Julienne, Printanier, Vegetable [etc.]. **1907** *Yesterday's Shopping* (1969) 19/1 *Soups*..Printanière—½ lb. tin 0/5¼. *c* **1938** *Fortnum & Mason Price List* 57/1 *Soups*..Printanier—per tin 1/–. **1965** *House & Garden* Dec. 84/2 *Printanier*, neatly.. diced very young vegetables, sometimes cooked in with the dish, but mainly used for garnishing.

printed, *ppl. a.* Add: **2. b.** Used of a writer.
1893 W. G. Collingwood *Life & Work J. Ruskin* I. v. 56 He was quite an artist; and a printed poet!

c. printed matter, paper, leaflets, papers, cards, circulars, etc., that are printed, not written. So **printed-matter mail**, a cheaper rate of postage than that for ordinary mail.
1876 in *Jrnl. R. Soc. Arts* (1976) May 343/1, I am directed by the Council of this Society respectfully to draw your Lordship's attention to the anomaly which exists with reference to the conveyance of printed matter by the post, owing to the distinction which the Post-office makes between the book post and the newspaper post. **1897** *Post Office Guide* July 375 The Articles which are entitled to be sent at the rate applicable to *Printed Papers* are mostly impressions or copies obtained upon paper, parchment, or cardboard, by means of printing, lithography, engraving, photography, or any other mechanical process easy to recognize... Besides these articles, there are some others which are admitted, though not really printed matter, as, for instance, manuscripts intended for the press (when sent with the proofs of the same), papers impressed for the use of the blind, and cardboard drawing models stamped in relief. **1918** *Ibid.* July 9 The expression 'Printed Paper' means a packet not exceeding 2 oz. in weight which consists of or contains one or more of the following articles or documents, that is to say:—..Books..Sketches..Maps.. [etc.]. **1929** H. Crane *Let.* 6 Sept. (1965) 345 My old landlady..had thought they were 'printed matter' and had failed to forward them. **1929** D. H. Lawrence *Let.* 9 July (1962) II. 1163, I will address them..by registered printed-matter mail. **1957** C. Brooke-Rose *Languages of Love* 32 Two newspapers and three letters for her lay on the hall table. One was only printed matter, and one an unsealed invitation. **1959** *Spectator* 14 Aug. 181/1 Printed matter is not much read by the non-Communists or the uncommitted. **1967** *Economist* 15 July 238/3 With the inevitable rise in the price of newspapers..and the probable ending of resale price maintenance on confectionery and even tobacco, newsagents will be keener than before to sell printed matter.

4. Also *printed-out* (see *print *v.* 14 c).
1934 *Jrnl. Optical Soc. Amer.* XXIV. 316/1 Weigert found that if shorter exposures than were required to produce a printed-out image were given..the image rendered visible by suitable development was dichroic. **1976** K. I. & R. E. Jacobson *Imaging Syst.* ii. 55 The colour of the printed out silver is usually reddish purple.

5. *printed circuit*, an electric circuit in the form of a flat sheet of insulating material bearing thin conducting strips and components, usu. mass-produced by a method that involves printing the circuit design on the sheet using a stencil or photograph of it; so *printed circuitry, wiring*; also applied to individual components made by such processes.
1946 *Business Week* 23 Feb. 19/2 On the back of the plate are solder spots that connect through holes in the plate with the 'printed' circuit. **1946** *Sci. News Let.* 2 Mar. 133 'Printed wire', the new development that reduces wiring radio circuits to a two-dimensional lithographic process. **1946** *Wireless World* Oct. 349/1 (*heading*) Printed 'wiring'. **1952** *Electronic Engin.* XXIV. 129/1 A major disadvantage of printed circuits has been the difficulty of incorporating satisfactory resistors. **1956** *Mod. Plastics* XXXIII. 223/2 Use of printed circuitry in 1955..increased in electronic computers, industrial control units, servo-mechanisms, and similar equipment. **1958** *Daily Mail* 8 Sept. 8/2 Research men working with printed circuits have evolved a way of reducing the thickness of the copper foil which carries the current from five-thousandths of an inch..to half a thousandth of an inch. **1966** *McGraw-Hill Encycl. Sci. & Technol.* X. 596/2 Printed capacitors are fabricated as part of the conductor circuit pattern when a high dielectric constant ceramic..is used as the circuit base material. Conductive patterns are screened on opposite sides of the circuit base to form the capacitor. **1970** J. Earl *Tuners & Amplifiers* vi. 141 The vast majority of tuner-amplifiers are now transistored, the designs being based on printed circuit boards..and sometimes integrated circuits. **1971** *Engineering* Apr. 30/1 The introduction of solid-state electronics and printed circuitry to vending machinery has minimized the number of separate electrical relays required. **1977** *Gramophone* Nov. 960/3 Normally these would flex the board during transit, with the possibility of cracking the printed wiring.

6. Produced by print-through (sense *1).
1950–1 *B.B.C. Quarterly* V. 245/1 In an investigation of accidental printing, one of the main difficulties is in measuring printed signals which are more than 50 db lower in level than the recorded signal, but are closely spaced about it on the tape. **1962** A. Nisbett *Technique Sound Studio* iv. 84 (*caption*) Wavelength of original and printed signal.

printer. Add: **1. a.** (Further example.)
1966 J. & P. Dixon *Photography* i. 18 A good printer is of tremendous value to a studio. Without seeing the original subject he can take the negative and produce prints of superb quality.

2. a. (Earlier and later examples.) Now usu. short for *teleprinter.
1859 T. P. Shaffner *Telegr. Manual* xviii. 273 The apparatus comprises two essential mechanisms, the 'Transmitter' or 'Compositor', and the 'Receiver' or 'Printer'. **1928** A. Williams *Telegraphy & Telephony* ii. 36 (*caption*) A specimen of the printing done with the Creed printer. **1960** I. Fleming *For your Eyes Only* 26 Anything you have to say I'll put straight on the printer to London. **1972** R. Busby *Reasonable Man* xviii. 161 Williams flicked the telex message... 'When this came up on the printer it jogged my memory.'

d. *Photogr.* and *Cinemat.* An apparatus for producing positive prints from negatives.
1912 F. A. Talbot *Moving Pictures* viii. 82 This is the Williamson printer. **1940** G. G. Quarles *Elem. Photogr.* x. 135 The actual darkroom technique of making the exposure will depend upon the type of printer available.

The simplest..is the simple hinged-back printing frame. **1951** R. Spottiswoode *Film & its Techniques* viii. 196 The step printer (whether of the contact or optical type) is much like a camera or projector. The printing and printed films are pulled down one frame at a time..and held in front of a frame-sized aperture while exposure takes place. **1974** L. Lipton *Independent Filmmaking* ix. 371 The highest quality masters are made with optical and contact step printers, although continuous contact machines are also used.

e. *Computers.* A device which produces a printed record of the input or output of a computer of which it is part or to which it is connected.

1946 *Math. Tables & Other Aids to Computation* II. 103 The static outputs of a total of 80 decade counters and 16 PM counters are connected to the printer. **1949** E. C. Berkeley *Giant Brains* viii. 137 The recorder consists of a printer, a reperforator, and a tape transmitter... The printer is a regular teletypewriter connected to the machine. **1962** *Communications Assoc. Computing Machinery* V. 477/2 Routines intended for line-at-a-time printers do not yield optimal output for typewriters. **1974** 'A. Haig' *Peruvian Printout* 37 The high-speed printer in the New York computer room was still for a moment... Howard and Sam walked round the backs of the printers lifting the boxes of printout on to a rubber-wheeled trolley. **1975** T. Allbeury *Special Collection* xix. 135 Their addresses were typed up on the fast printer in 3·4 seconds.

4. a. printer-slotter, a machine used for printing on cardboard or other packaging materials.

1954 *TAPPI* Feb. 144A/1 Another..development which appears very near is a printer-slotter arranged to handle aniline inks. **1957** *Ibid.* Nov. 190A/2 A completely new design Langston printer-slotter is currently available in three sizes... This machine is of a split or opening construction so arranged that a two-color press can be easily converted..into a press of three, four, or even more stations. **1968** *Globe & Mail* (Toronto) 13 Jan. 48/2 (Advt.), Corrugated box printer-slotter operator. Call Mr. Cherry.

printergram (pri·ntəɪgræm). [f. *TELE)PRINTER + TELE)GRAM.] A telegram transmitted by telex.

1932 *Telegraph & Telephone Jrnl.* Oct. 2/2 Printergrams are charged for at the same rate as phonograms. **1942** Crooks & Dawson *Dict. Typewriting* (ed. 4) 268 Telex subscribers..have the advantage of being able to send and receive inland or overseas telegrams directly to and from the Post Office and/or the cable companies who are also on Telex. Telegrams handled in this manner are known as 'Printergrams'. **1976** [see *PHONOGRAM 3].

printery. Add: Also *Austral.* and *Afr.* **1.** (Later examples.)

1921 Mencken *Amer. Lang.* (rev. ed.) 187 Printery..appeared very early, and..has been reinforced by many analogues, e.g., beanery, bootery [etc.]. **1943** K. Tennant *Ride on Stranger* x. 102 The Order owned..a share in a printery. **1969** *Sydney Morning Herald* 24 May 1/7 Mr Newton..watched the search for several minutes before announcing: 'I'm bored with this. I'm off to the factory (a printery he operates in Canberra).' **1973** *New Journalist* (Austral.) July–Aug. 6/2 The licence of 3KZ is owned by Industrial Printing and Publicity Co. Ltd. This began as a Labor printery. **1975** B. Garfield *Hopscotch* xxii. 226 An *Evening Standard* van..returning to the printery from its last delivery. **1979** V. S. Naipaul *Bend in River* xiv. 229 We didn't have many printeries in the town.

printing, *vbl. sb.* Add: **a.** (Later examples.)

1966 *Listener* 1 Sept. 301/2 In the United States they have far greater fingerprinting output than we have.. though they have not quite universal printing. **1970** O. Dopping *Computers & Data Processing* xv. 234 In programming editing operations for printing, it is essential to consider how the form will look to the user.

b. Also, the total number of copies (of a book) printed at one time; an impression.

1928 *Publishers' Weekly* 26 May 2117 A best selling novel... Four large printings were necessary before publication. **1933** *Morning Post* 7 July 14/7 (Advt.), 7 printings in 5 weeks. *Peter Abelard.*

c. *printing out*: the production of an image without chemical development (cf. *PRINT *v.* 14c); *printing-out paper* (abbrev. *P.O.P.), a printing paper capable of being used for this.

1891 W. E. Woodbury *Gelatino-Chloride of Silver Printing-Out Process* ii. 6 One of the principal advantages of the chloride emulsion paper is its ability to give good prints from weak negatives—superior, in fact, to any that could be obtained by other printing-out processes. **1902** *Encycl. Brit.* XXXI. 703/2 The most notable change in recent years is the supersession of albuminized papers by papers coated by machinery with emulsions of silver haloids in gelatine, the chloride being used for most of the printing-out papers,..while the bromide forms the basis of most of the developable papers. **1904** [in Dict., sense d]. **1918** J. R. Roebuck *Sci. & Pract. of Photogr.* vi. 127 Photographers generally still rank prints on printing out paper ahead of prints on developing paper, but the former requires more time and labor, so that they have gone largely out of general use. **1939** [see *printing paper* (sense d below)]. **1968** H. Asher *Photogr. Princ. & Pract.* (1970) viii. 230 Printing-out papers are not normally stocked by dealers but they are still obtainable. **1976** K. I. & R. E. Jacobson *Imaging Syst.* ii. 54 One of the oldest systems.. uses printing out papers which produce visible images of good photographic quality directly on exposure to a strong light source..without the need for development or any other amplification process. *Ibid.* 55 Exposure to light of lower intensity for longer periods..causes these internal latent images to act as centres for the printing out of silver.

c*. = *PRINT-THROUGH 1.

1949 S. J. Begun *Magn. Recording* v. 98 Impregnated mediums exhibit 'printing' to an objectionable extent. **1950–1** *B.B.C. Quarterly* V. 248 (*caption*) Records of printing at various frequencies. **1962** A. Nisbett *Technique Sound Studio* iv. 84 (*caption*) Printing depends on the thickness of the tape base layer, as well as temperature and physical shock.

d. *printing industry*; **printing-frame** (further example); **printing paper** (*a*) (earlier example); (*b*), delete '(also *printing-out paper*, abbrev. *P.O.P.*)' and see sense c above; (further examples); **printing union** = *print union* s.v. *PRINT *sb.* 15b.

1858 Printing-frame [see *ACCELERATOR b]. **1976** *Times* 23 Mar. 1/5 The union said in a letter to the printing industries committee that the question had not arisen because of the dispute at Barnsley. **1806** R. Sutcliff *Trav. N. Amer.* (1811) xiv. 258 The mill..is..employed in making writing and printing paper. **1939** Mack & Martin *Photogr. Process* ix. 313 Silver halide printing papers may be divided roughly into two classes, viz.: printing-out papers ('P.O.P.') in which the reduction of the halide to metallic silver is completed by the action of the light, no development being required; and developing-out papers ('D.O.P.') in which the latent image formed by exposure is subsequently developed. **1968** G. L. Wakefield *Introd. Photogr.* viii. 146 A printing paper is coated with a sensitive emulsion similar to that on a film but much slower and generally sensitive only to blue light. **1976** M. J. Rosen *Introd. Photogr.* v. 117/1 Photographic printing papers are manufactured in a variety of contrast grades. **1976** *Times* 23 Mar. 1/5 The union asked other printing unions to follow its example and refuse to take part in joint meetings or federated chapels (office branches) where the institute was represented.

printing, *ppl. a.* (Earlier and later examples.)

1841 Wright & Bain *Brit. Pat.* 9204 In Sheet 3 we exhibit a side view,..of an electro-magnetic printing telegraph. **1849** *Rep. Brit. Assoc. Adv. Sci.* Notices & Abstr. 133 A colloquial and also a printing telegraph are used. **1929** *Bell Syst. Techn. Jrnl.* VIII. 267 Commercial telegraph operation..is carried on almost exclusively by two well known methods, manual morse and printing telegraph. **1940** *Chambers's Techn. Dict.* 674/2 *Printing telegraph*, a telegraph system in which the received signals are translated and operate a printing machine, giving a readable message. **1968** *Gloss. Terms Offset Lithogr. Printing* (B.S.I.) 28 *Printing pressure*, the pressure applied at the point of contact between two printing surfaces to transfer ink from one surface to the other. **1971** *Gloss. Electrotechnical, Power Terms* (B.S.I.) III. iii. 7 *Printing telegraphy*, any method of telegraph operation in which the received signals are automatically recorded as printed characters.

printing-house. Delete 'Now only *Hist.*' and add later examples. *Printing House Square* (examples in allusive use).

1861 B. Moran *Jrnl.* 3 Dec. (1949) II. 917 *The Times*, is filled with such slatternly abuse of us and ours, that it is fair to conclude that all the Fishwives of Billingsgate have been transferred to Printing House Square. **a 1910** 'Mark Twain' *Autobiogr.* (1924) II. 285 Orion severed his connection with the printing-house in St. Louis. **1938** H. Nicolson *Diary* 9 Sept. (1966) 358 Colin Coote.. says that *The Times* leader urging the Czechs to surrender their fringes was written by Leo Kennedy and merely glanced at by Geoffrey Dawson. He is appalled by the lack of responsible guidance in Printing House Square. **1951** S. Jennett *Making of Bks.* iii. 48 In some printing houses one of the boxes in the case is set aside for the reception of defective or battered types. **1956** C. Cockburn *In Time of Trouble* vi. 88 Is, indeed, anyone, anywhere, truly worthy of *The Times*? This was the awfully solemn thought which..sometimes oppressed Printing House Square. **1964** F. Bowers *Bibliogr. & Textual Crit.* III. i. 64 McKerrow..once remarked that some contradictory pieces of evidence could be reconciled most easily by the hypothesis that the entire printing-house had adjourned to the nearest pub and got drunk. **1978** *Times* 18 Apr. 17/7 Surely you cannot be so short-sighted in Printing House Square as to..believe..as you grandly state in a recent leading article. **1979** *N.Y. Rev. Bks.* 25 Oct. 44/4 The great printing houses became nodal points in a semi-secret network of cultural communications.

printing-office. (Earlier examples.)

1733 B. Franklin *Poor Richard* (title-page), Printed and sold by B. Franklin, at the New Printing-Office. **1802** *Monthly Mag.* XIV. 347/2 This portrait is done by the letter-engraver who executed in wood for the printing-office of Fust.

pri·nt-out. [f. vbl. phr. *to print out* (*PRINT *v.* 14c, 6b).] **1.** *Photogr.* Used *attrib.*, = *printing out* (*PRINTING *vbl. sb.* c).

1899 P. N. Hasluck *Bk. Photogr.* 184/1 Something may be done in the development of print-out papers. **1929** *Proc. 7th Internat. Congr. Photogr.* 1928 23 (*heading*) Parallelism between photo-electric conductivity effects and direct print-out effects. **1930** O. Wheeler *Photogr. Printing Processes* iii. 28 While daylight is still commonly employed for exposure in print-out processes, artificial light..is often used commercially. **1939** Mack & Martin *Photogr. Process* v. 175 If an emulsion receives a very great exposure, the halide is reduced directly by the light without the aid of a developer. This print-out effect was the only means of producing a photograph in the very earliest days of photography. **1965** *Photogr. Jrnl.* CV. 285/2 A special kind of print-out paper has been in use for data recording for some years. **1973** W. Thomas *SPSE Handbk. Photogr. Sci. & Engin.* vi. 407 Sufficient exposure of an emulsion to light causes visible darkening without development (a print-out image).

2. (A sheet or strip of) printed matter produced by a computer or other automatic apparatus; the production of such matter. Also *fig.*

1953 *IRE Trans. Instrumentation* June 68 An hour's operation includes some 40 print-outs of the complete memory. **1957** *Ibid.* Sept. 193/2 The reading system consists of a film reader, control and decoding section, and print-out equipment. *Ibid.* 194/1 A stepping switch is caused to step to the next position at the end of each character print-out. **1961** *Aeroplane* CI. 573/2 In addition to the automatic print-out unit for the recording of positional information, an optional addition to the data presentation unit is a computer to convert the slant range and elevation co-ordinates into ground range and height data. **1966** 'C. E. Maine' *B.E.A.S.T.* vi. 79 Synøve and..Wetherby..were standing by one of the print-out machines, reading a long sheet of paper as it emerged. **1969** *New Scientist* 1 May 238/2 Everyone should be entitled to a print-out of the information in the data bank in regard to him. **1971** K. Gottschalk in B. de Ferranti *Living with Computer* iv. 31 The drafting and print-out of leases, wills and forms used in lawyers' offices to eliminate repetitive work. **1974** *Ellery Queen's Mystery Mag.* Nov. 148/1 We could use computers at this end too, to run through possible letter combinations and produce the necessary printouts almost at once. **1979** M. Babson *Twelve Deaths of Christmas* xvii. 87 Sod your computers, *I'm* getting the printout from the marrow of my bones.

print-shop. Add: **1.** (Further examples.)

1778 *English Mag.* Feb. 59/1 Notwithstanding the many satirical exhibitions at the print-shops, of grown gentlemen learning to dance..there are many arguments that may be used in defence of this genteel exercise. **1897** A. Beardsley *Let.* 13 Apr. (1970) 302 The book and print shops [in Paris] are an evergreen joy to me.

2. *U.S.* A printing-office or printery.

1921 *Amer. Printer* 5 Nov. (*heading*) Visit to an old Oxford printshop. **1961** R. L. Duncan *Voice of Strangers* IV. iii. 246 He went into the print shop, where Fletcher had just finished cleaning the press. **1970** *Eng. Stud.* LI. 164 His quarrel with Sir John Cheke.., the exegetic controversy with Edward Lee, the question of Henry VIII's divorce were all reflected in the books pouring out of Basle's printshops. **1977** C. McCarry *Secret Lovers* xiii. 170 Each afternoon he sent Joëlle to collect the typescript from the printer; she took it back to the print shop every morning.

print-through (pri·ntþrū). Also **print through.** [f. vbl. phr. *to print through*.] **1.** The accidental transfer of recorded signals to adjacent layers in a reel of magnetic tape. Freq. *attrib.*

1956 R. E. B. Hickman *Magn. Recording Handbk.* ii. 24 The print-through is..dependent upon the output level of the original signal. **1958** H. G. M. Spratt *Magn. Tape Recording* iii. 110 The ratio of the level of the original signal to that of the last echo before and the first echo after that signal is termed the print-through or transfer ratio. **1962** *Times* 5 July 15/7 Dangers to tape..arise from prolonged storage without rewinding, which can cause 'print-through' (detectable as pre-echo on some discs). **1975** *Hi-Fi Answers* Feb. 76/2 To be sure no print through is occurring it would be a good idea if you were to rewind each tape in your collection at least once in six months to redistribute the tape in the reel and minimise any print through effect.

2. *Printing.* (See quot. 1961.)

1961 R. F. Bowles *Printing Ink Manual* xi. 349 Another factor which has a marked effect on the quality of the print is the degree of 'print through' which is visible. 'Print through' is the degree to which the print is visible on the reverse side of the sheet, and is the combination of 'show-through' and 'strike-through'. **1973** L. C. Young *Materials in Printing Processes* ix. 121 When the printing on one side of a paper can be seen through the paper, this print-through may either be due to show-through or strike-through... Normally the vehicle only penetrates a short distance into the sheet, but even this will have the effect of reducing the printing opacity, so making print-through more likely.

prion. Insert in etym. after 'mod.L.': (Comte B. G. E. de la V. Lacépède *Tableaux Méthodiques des Mammifères et des Oiseaux* (1799) 14). Substitute for def.: A small saw-billed petrel belonging to the genus once so called, now included in the genus *Pachyptila* of the family Procellariidæ and found in southern seas. (Examples.)

1848 J. Gould *Birds Austral.* VII. 54 (*heading*) Dove-like Prion. **1862** *Proc. Zool. Soc.* 125 In form and colouring it is precisely similar to the other Priones. **1901** A. J. Campbell *Nests & Eggs Austral. Birds* II. 917 Mr Travers frequently found these Prions caught in the branches of scrubby trees. **1937** *Discovery* May 141/2 The prion is a ghost. A fluttering thing of pale grey-blue and white. **1959** *New Biol.* XXIX. 112 Some small bird species such as the prions owe their survival to the fact that they are only active at night, thereby avoiding the skuas' attack. **1972** K. Simpson *Birds in Bass Strait* 71/2 The Dove or Antarctic Prion is perhaps one of the most numerous of all southern seabirds.

prior, *a.* Add: **a.** *prior charge*, in *Finance*: see quots. 1968, 1974. Also (with hyphen) as *attrib. phr.*

1877 *Encycl. Brit.* VII. 15/1 The Companies Clauses Act, 1863, part iii., which makes debenture stock a prior charge on the undertaking, and gives the interest thereon priority of payment over all dividends or interest on any shares or stock of the company. **1930** *Economist* 22 Mar. 653/1 Foreign bonds, industrial prior-charge stocks and

even industrial preference shares shared in the general tendency, though to a less conspicuous extent. *Ibid.* 29 Mar. 695/2 Gilt-edged stocks and well-secured industrial debentures and prior charges. **1968** Johannsen & Robertson *Managem. Gloss.* 105 *Prior charges*, all types of debentures, preference shares and other stocks ranking for payment of interest or dividend in precedence to the ordinary shares. **1974** *Terminol. Managem. & Financial Accountancy* (Inst. Cost & Managem. Accountants) 62 *Prior charge capital*, those classes of share and loan capital, the holders of which have a claim on the profits and assets of a business before the ordinary shareholders.

c. *Statistics.* Applied to the result of a calculation made in ignorance of, or previously to, some observation(s); *prior probability*, the probability that a hypothesis is true calculated without reference to certain relevant observations. Opp. *POSTERIOR *a.* 1b.

1921, etc. [see *POSTERIOR *a.* 1b]. **1977** *Sci. Amer.* May 126/3 With this valuable extra information, which statisticians call a 'prior distribution,' it is possible to construct a superior estimate of each player's true batting ability.

priorate. Add: **1. a.** Also, the (term of) office of a prioress.

1925 C. S. Durrant *Link betw. Flemish Mystics & Eng. Martyrs* I. x. 150 The Priorate of Mother Salome has ever been looked back to as a time when [etc.].

priorite (prəi·ŏrəit). *Min.* [ad. G. *priorit* (W. C. Brögger 1906, in *Skr. udgivne af Vidensk.-Selsk. i Christiania (Mat.-Nat. Kl.)* I. vi. 111), f. the name of G. T. Prior (1862–1936), British mineralogist: see -ITE[1].] A mixed oxide, chiefly of niobium, titanium, and yttrium, with traces of several other elements, which occurs as black or dark brown orthorhombic crystals.

1907 *Jrnl. Chem. Soc.* XCII. II. 886 Another mineral of this series is one from Swaziland, South Africa, analysed by G. T. Prior..; for this, the name priorite is proposed. Blomstrandine and priorite are isomorphous and are respectively dimorphous with polycrase and euxenite. **1944** C. Palache et al. *Dana's Syst. Min.* (ed. 7) I. 796 Priorite was found originally at Urstad on the island of Hitterö in southwest Norway. **1966** *Amer. Mineralogist* LI. 156 Only two well-established Ce–Y rare-earth mineral series are known for which names have been assigned to the end members: 3. aeschynite–priorite series [;] 4. britholite–abukumalite series. **1968** I. Kostov *Mineral.* 254 Euxenite group... The chief members of the group are euxenite.., priorite ($(Y, Th)(Nb, Ti)_2O_6$).

prioritize (prəiˌǫ·ritəiz), *v.* orig. *U.S.* [f. PRIORITY + -IZE.] **a.** *trans.* To designate as worthy of prior attention, to give priority to (in the sense of *PRIORITY 2). **b.** *trans.* To determine the order in which (items) are to be dealt with, to establish priorities for (a set of items). Also *absol.*

A word that at present sits uneasily in the language.—Ed.

1973 T. H. White *Making of President 1972* xii. 325 The storefront operators in the counties that Malek had 'prioritized' had identified independents, wavering Democrats and 'don't knows'. **1975** R. Burns *Alvarez Jrnl.* 47 But in the meantime I've got to prioritize the operations, and the priority standard is the probability of conviction. **1977** *Time* 14 Mar. 28/2 From then on toward midnight, he tries, in his own words, 'to prioritize'. **1977** *Daily Colonist* (Victoria, B.C.) 15 May 33/5 A special committee had been struck..to prioritize their recommendations and to report. **1981** *Times* 3 Feb. 13/6 In the Nato headquarters..we are well used to prioritizing our targets.

Hence **prio:ritiza·tion**; **prio·ritized**, **prio·ritizing** *ppl. adjs.*

1977 *Financial Times* 24 Dec. 3/6 It has two meanings, depending on whether one is doing the prioritization, or having it thrust upon one. **1977** *Time* 14 Mar. 28/2 Prioritizing takes him into the Oval Office to talk each day with the President and to drop in on..Vice President Mondale. **1978** *Verbatim* Feb. 1/2 A teacher in Mill Valley has drawn up a 'prioritized list of all components of the school program'.

priority. Add: **1. b.** *Taxonomy.* The claim of the first validly published Latin name to be taken as the correct one for any given organism.

1842 *Rep. Brit. Assoc. Adv. Sci.* 109 We have no hesitation in adopting as our fundamental maxim, the 'law of priority', viz... The name originally given by the founder of a group or the describer of a species should be permanently retained. **1928** D. B. Swingle *Textbk. Systematic Bot.* vii. 68 By agreement botanists do not go back of Linnaeus' 'Species Plantarum' (1753) to establish priority in the publication of names. **1953** E. Mayr et al. *Methods & Princ. Systematic Zool.* xi. 213 It would be unfair..to blame all name changes on the law of priority. **1963** Davis & Heywood *Princ. Angiosperm Taxon.* viii. 291 Enough information should be given to indicate why a name has not been adopted if it appears to have priority over the accepted name.

2. Also, the right to precede others or to receive attention, supplies, etc., before others. Hence *transf.*, an interest having a prior claim to consideration; often in *pl.* or preceded by a qualifying word, as *first, high, top priority.*

1917 *Times* 10 Mar. 6/4 The Minister of Munitions..has issued an important Order under the Defence of the Realm Act as to priority of war work... During the last 12 months the Ministry of Munitions has been administering a scheme which ensured for war work and for work of national importance priority over all other work in regard to labour and materials. **1922** *Encycl. Brit.* XXXII. 147/1 The Priorities Committee undertook whenever necessary to administer priorities in the production of all raw materials and finished products. *Ibid.* 835/1 The labour needs of employers in war industries were graded as entitled to 'Super-Priority', 'First' or 'Second Class Priority', or as not deserving special treatment. **1940** *Economist* 18 May 893/1 How far can we tap reserves of skill for war work by the full mobilisation of this class of man power and its allocation, according to an infrangible schedule of priorities, exclusively to war and export manufacture? *Ibid.* 24 Aug. 236/2 There was no priority at all until June of this year, and since then there has only been a general Priority of Production Direction which went no further than to notify to industry two short lists of very broad categories of munitions. **1941** *New Statesman* 26 Apr. 429/2 First priority is being given to dairy cattle. **1944** *Daily Tel.* 23 Sept. 2/2 The obvious remedy for that would be to make civil aviation priority No. 1 at the Ministry. **1948** 'N. Shute' *No Highway* ii. 49, I think this trip to Canada is top priority of anything that's going on at Farnborough today. **1949** G. Cotterell *Randle in Springtime* iv. iii. 213 I'm going into furnishing, see. Anything you've got to have dockets and priorities for. Lino, stair carpets. **1958** *Listener* 11 Sept. 368/1 The Minister had to explain that their area did not have early priority. **1960** M. Spark *Ballad of Peckham Rye* iii. 42 She came up with an estimate and said 'priority'... I said, 'Excuse me, Miss Coverdale, but I've got two priorities already.' **1968** *Highway Code* 37 (*caption*) Give priority to vehicles from opposite direction. **1969** *Morning Star* 29 Jan. 1/1 Improvements..are much less than could be achieved if the Government got its priorities right. **1970** G. F. Newman *Sir, You Bastard* iv. 122 There was a priority on at the Yard, all detectives were being called back. **1972** A. Ulam *Fall of Amer. Univ.* v. 211 He would couple this frankness with a plea to the young not to give up, to work within the system, for with reordered priorities this country might still be saved and might even be worth saving. **1977** *Listener* 26 May 682/2 There will be questions of social priorities involved.

5. In sense 2, passing into *adj.* (Further examples.)

1917 *Times* 10 Mar. 6/4 (*heading*) National work. Important priority scheme. **1922** *Encycl. Brit.* XXX. 818/2 Trades specified in a priority list drawn up with reference to the relative urgency of the industrial requirements of the country. **1934** T. E. Lawrence *Let.* 8 Apr. (1938) 795 She..has no one aboard now to get her priority treatment. **1942** *Times* (Weekly ed.) 2 Dec. 15 Various priority and freight rationing schemes are in operation. **1946** K. Tennant *Lost Haven* (1947) xvi. 248 Young Len's working for the mill, ain't he? And that's a priority job. **1960** O. Manning *Great Fortune* xiii. 153 This young man might have been granted a priority flight over Europe. **1967** *Guardian* 10 Jan. 4/4 Measures should be taken to increase the ratio of teachers to children in educational priority areas. **1976** *Broadcast* Dec. 1/3 The achievement of a shorter working week for weekly paid staff is a priority objective.

priorly, *adv.* Delete *rare* and add earlier and later examples.

1779 R. Baker *Remarks Eng. Lang.* (ed. 2) 94 It seems a wonder that we have no such word as *priorly*. It would be naturally formed from *prior*, and would be very useful. **1965** *Amer. Psychologist* XX. 1007/2 After a certain point in human evolution, the only means whereby man could fill his evolutionary niche was through the cultural transmission of the skills necessary for the use of priorly invented techniques, implements, and devices. **1970** *Jrnl. Gen. Psychol.* LXXXII. 207 It states that organisms respond to discrepancies between contemporaneous stimuli and some internalized average of other inputs, experienced priorly or as context.

Priscol (pri·skǫl). *Pharm.* Also **priscol.** A proprietary name for *TOLAZOLINE.

The trade-mark registration in the U.S. is no longer active (see next, quot. 1949).

1938 *Trade Marks Jrnl.* 18 May 604/2 Priscol... All goods included in class 3 [i.e. chemical substances prepared for use in medicine and pharmacy]. Society of Chemical Industry in Basle.., Basle, Switzerland; manufacturers and merchants. **1938** *Official Gaz.* (U.S. Patent Office) 14 June 247/2 Society of Chemical Industry in Basle... Priscol for preparation having an action on the circulation of the blood. Claims use since May 6, 1936. **1945** *Federation Proc.* IV. 114/1 Although priscol, 15 mgm., was adrenolytic in relation to blood pressure, it was not adrenolytic or sympatholytic in respect to the cervical sympathetic functions studied. **1956** Ld. Amulree in A. Pryce-Jones *New Outl. Mod. Knowl.* 222 On the theory that the confusion [in senile dementia] is increased by a shortage of oxygen within the brain, priscol and other vaso-dilators have been used without any definite improvement in the patient's condition. **1968** W. C. Bowman et al. *Textbk. Pharmacol.* xxix. 763 Tolazoline (Priscol) exhibits a wide range of pharmacological activity in addition to its antiadrenaline action.

Priscoline (pri·skǫlīn). *Pharm.* Also **priscoline.** [f. *PRISCOL + -INE[5].] A proprietary name in the U.S. for *TOLAZOLINE.

1949 *Jrnl. Amer. Med. Assoc.* 21 May 272/2 Since this paper was written, the name priscol has been changed to priscoline. **1950** *Official Gaz.* (U.S. Patent Office) 22 Aug. 996/2 Ciba Pharmaceutical Products, Inc., Summit, N.J... *Priscoline*... For preparations having an action on the circulation of the blood. Claims use since Mar. 15, 1949. **1966** W. F. Barker *Peripheral Arterial Dis.* vi. 85 Priscoline is perhaps one of the most widely used drugs in peripheral vascular occlusions.

prisere (prəi·sīəɪ). *Ecol.* [f. PRI(MARY *a.* + *SERE *sb.*[2]] A sere that began on an area not previously occupied.

1916 F. E. Clements *Plant Succession* ix. 182 Within the same climax, seres are classified as primary and secondary, i.e., as priseres and subseres. **1926** Tansley & Chipp *Study of Vegetation* ii. 19 We distinguish a sere beginning on bare soil..as a primary sere, or prisere. **1939** A. G. Tansley *Brit. Islands & their Vegetation* x. 218 Any concrete example of succession, i.e. any definite series of communities leading to a climax, is called a sere, and the complete succession from bare habitat to climatic climax is a primary sere or prisere. **1960** N. Polunin *Introd. Plant Geogr.* xi. 324 The complete sere just indicated is a primary sere (prisere), beginning on a bare substratum without organic material. **1977** A. S. Collinson *Introd. World Vegetation* vi. 103 Where plants begin their succession on virgin territory..the process is termed a primary succession or prisere.

‖ **prisiadka** (prisya·tkă). Also **prisjádka, prisyadka.** [Russ.] A dance-step in which the male dancer squats on his heels and kicks out each leg alternately to the front. Also used for the dance itself. Cf. *KAZACHOK.

1938 B. Schönberg tr. *Sachs' World Hist. of Dance* i. 28 Wilder still are the Bavarian *Schuhplattler* and the Ukrainian *prisjádka* with their heel stamping. *Ibid.* 30 The squat-fling dance is still the possession of many European peoples... In the Ukraine it is known by the term *prisjádka*. **1972** *Daily Tel.* 14 Aug. 6/9 They..burst out into wild Ukrainian dancing with every possible variation of the squatting step 'prisyadka'. **1977** J. Wambaugh *Black Marble* (1978) xii. 292 'I don't care if I'm six feet tall,' Valnikov said, squatting on his haunches, trying some prisiadka kicks that put him temporarily on his ass.

prism. Add: **7. prism-binocular(s)**, binoculars containing two pairs of triangular prisms, introduced so as to shorten the apparatus and improve the stereoscopic effect.

1901 *Brit. Optical Jrnl.* Sept. p. iv (Advt.), Busch's Prism Binoculars... The lightest and most portable Prism Binocular on the market. **1919** *Jane's Fighting Ships* 59 adv., Bausch & Lomb Optical Co... Field Glasses (Stereo Prism Binocular). **1957** *Encycl. Brit.* III. 583/2 Ernst Abbe took the matter up *de novo* in 1893 when he designed prism binoculars and telescopes. His constructions were the forerunners of the modern prism binocular.

prison, *sb.* Add: **1. d.** In *Roulette* and related board-games: a position on the board where bets are held in abeyance until the next round of play; *spec.* in phr. *to put* (a stake) *in prison.*

1867 *Bohn's Hand-Bk. Games* 346 The punters may..have their stake moved into the middle semicircles of the colour they then choose, called 'la première prison', the first prison, to be determined by the next event, whether they lose all or are set at liberty. **1940** Wodehouse *Eggs, Beans & Crumpets* 32 When Zero turns up..stakes on the even chances aren't scooped up—they are what is called put in prison. **1977** P. Arnold *Encycl. Gambling* 247/1 *Prison*, a convention whereby a stake on the even-money chances at roulette is left on the table, or 'put in prison' when zero appears, to be either retained by the bettor or lost according to the next spin.

c. *prison-without-bars* (colloq.): an open prison (*OPEN *a.* 2c).

1948 *Manch. Even. News* 10 Nov., The former governor of Britain's 'prison-without-bars' at Loudham Grange. **1952** 'J. Henry' *Who lie in Gaol* v. 69, I heard a great deal of the many advantages I would enjoy at the prison-without-bars at York; in fact it was looked upon as a form of heaven by most of the prisoners [at Holloway]. **1959** 'H. Carmichael' *Stranglehold* vi. 58 A solicitor who was doing time at the prison-without-bars.

3. a. (*a*) *prison chaplain, -clock, -discipline* (earlier example), *guard, library, officer, pallor, reform, -wall* (later examples; also *fig.*), *warder, yard.* (*c*) *prison camp* (examples), *farm, hospital.* **b.** *prison-keeper* (earlier example), *visitor*; *prison-visiting* sb. **c.** *prison-grey, -like* (later examples) adjs. **d. prison haircut** = *prison-crop.*

1925 *Scribner's Mag.* Oct. 386/1 The scene is a Turkish prison-camp during the recent war. **1978** *Lancashire Life* Nov. 150/1 (Advt.), Mr. P——, a Pole who arrived in England in 1947 after..escaping from a German prison camp. **a 1902** S. Butler *Way of All Flesh* (1903) lxv. 293 He might experimentalise advantageously upon the viler soul of the prison chaplain. **1910** *Encycl. Brit.* V. 851/2 Prison chaplains are appointed by the home secretary. **1972** N. Marsh *Tied up in Tinsel* iii. 78 The prison chaplain gave a short, civilized sermon. **1898** O. Wilde *Ballad of Reading Gaol* 18 The prison-clock Smote on the shivering air. **1818** T. F. Buxton *Inquiry Prison Discipline* 137 Having..described two..opposite modes of prison discipline, I would suggest.., that a comparison of these is the most certain criterion of their respective merits. **1834** J. S. Mill in *Monthly Repos.* VIII. 590 Has not a notion grown up within a few years, (we believe a very false one), that the increased mildness of prison-discipline has made our gaols..places where the prisoner is actually too comfortable, and too well off? **1961** *Atlanta Constitution* 4 Nov. 1 The jury praised the administration and operation of the Atlanta Police Department, the Fulton Tax Commissioner's Office, the Bellwood and Alpharetta prison farms, [etc.]. **1968** *Listener* 15 Feb. 210/1 As remarkable..is the improvement he has brought about in his year in charge of the smaller prison farm, Tucker. **1975** C. Weston *Susannah Screaming* (1976) iii. 39 Delgado made a break from the

prison farm where he had been sent after a period of good behavior in a federal cellblock. **1956** 'H. MacDiarmid' *Stony Limits & Scots Unbound* 90 A flash of sun in a country all prison-grey. **1961** W. T. Ballard *Night Riders* i. 15 Two wore the uniform of prison guards, three the striped suits of convicts. **1970** G. Jackson *Let.* 10 June in *Soledad Brother* (1971) 40, I am being tried in court right now..for the alleged slaying of a prison guard. **1977** *Time* 12 Dec. 47/3 With only good time remaining as a route to early release, the potential for abuse by prison guards would be heightened as well. **1974** *Times* 17 Aug. 7/1 A snotty little nervous kid with a prison haircut. **1933** J. Buchan *Prince of Captivity* II. i. 178 You would spend some weeks in a prison hospital till they patched you up. **1943** F. Thompson *Candleford Green* ix. 142 Such a journey..and a prison hospital..at the end of it. **1978** P. G. Winslow *Coppergold* 48 He fell a victim to influenza..was taken to the prison hospital. **1855** Dickens *Dorrit* (1857) I. i. 4 The prison-keeper appeared carrying..a basket. **1967** H. Pinter *Night School* in *Tea Party & Other Plays* 101, I was running the prison library. **1979** K. Bonfiglioli *After You with Pistol* vi. 31 He gets a nice job in the prison library but *horrid* things happen to him in the showers. **1839** E. A. Poe in *Burton's Gentleman's Mag.* Oct. 206 This prison-like rampart formed the limits of our domain. **1916** D. H. Lawrence *Amores* 77 The town Glimmers with subtle ghosts Going up and down In a common, prison-like dress. **1944** A. L. Rowse *Eng. Spirit* xxxv. 244 That sepulchral, prison-like building. **1970** P. Dickinson *Seals* ii. 53 Many criminals..are really only happy..when..their day is shaped by a prison-like discipline. **1907** B. Thomson *Story of Dartmoor Prison* xxi. 260 The better class of men came to realize that prison officers were their friends rather than their enemies. **1961** *Observer* 9 Apr. 22/8 He refers to prison officers as prison warders, a title abandoned something like thirty years ago. **1978** P. Lovesey *Waxwork* 79 It is quite impossible to conduct a conversation through an iron grille with two prison officers at my client's shoulder. **1935** A. J. Cronin *Stars look Down* II. xx. 446 He sat there with his prison pallor upon him. **1977** *New Yorker* 24 Oct. 141/1 He squints into the unaccustomed sunlight..and..suffers from a case of prison pallor. **1890** W. Booth *In Darkest Eng.* I. ix. 74 Once the work of Prison Reform is taken in hand by men..who are in full sympathy with the class for whose benefit they labour. **1972** A. Roudybush *Sybaritic Death* (1974) vii. 67 His original project had been to devote his activities to the cause of prison reform. **1838** H. Martineau *Retrospect of Western Travel* I. 224, I trust that the practice of prison-visiting will gain ground. **1973** L. Cooper *Tea on Sunday* i. 21 Barry Slater, the unfortunate legacy of Alberta's spell of prison visiting. **1837** H. Martineau *Society in Amer.* II. III. iv. 285 Every prison visitor has been conscious, on first conversing privately with a criminal, of a feeling of surprise at finding him so human. **1975** N. Freeling *What are Bugles blowing For?* xv. 88 Vera made a good prison visitor. **1855** Trollope *Warden* xvi. 248 No convict, slipping down from a prison wall, ever feared to see the gaoler more entirely than Mr. Harding did to see his son-in-law. **1898** O. Wilde *Ballad of Reading Gaol* 16 The weeping prison-wall. **1951** M. Kennedy *Lucy Carmichael* I. vii. 62 Rickie peeped for a moment over the prison walls of his own depression. **a 1902** S. Butler *Way of All Flesh* (1903) lxiv. 286 The prison warder..sent for the doctor. **1914** *Prison Officers Mag.* Nov. 450/2 For the past four years the majority of the Irish Prison Warders have favoured us with their confidence and support. **1928** [see *WARDERING *vbl. sb.*]. **1961** [see *prison officer* above]. **1978** M. Butterworth *X marks Spot* 179 With two escorting prison warders as witnesses. **1642** in *Rec. Early Hist. Boston* (1877) II. 70 The Constables are appointed..to take care for the building a salt peter howse in the prison yarde. **1776** *Jrnls. Continental Congress U.S.* (1906) IV. 121 Resolved, That the said J. Connolly be allowed..to walk in the prison yard or hall. **1851** J. J. Lancaster in *Rep. Sel. Comm. Passengers' Act* 142 in *Parl. Papers* XIX. 1 Those in Millbank [*sc.* a London military hospital] are drawn up in the prison-yard or wards. **1856** Dickens *Dorrit* (1857) II. vi. 383 They prowled about..in the old, dreary, prison-yard manner. **1963** N. Marsh *Dead Water* (1964) v. 126 She ..walked aimlessly..as if the garden were a prison yard.

prisoner[1]. Add: **1. a.** *prisoner of conscience*, one who is detained or imprisoned because of his or her political or religious beliefs.

1961 *Amnesty* 11 July 2/1 What are the facts we need? There are thousands of them and each one is a human being, a prisoner of conscience behind bars because of his political views or religious beliefs. **1962** *Time to keep Silence* (Amnesty International), Prisoners of Conscience are of two distinct sorts—those prepared to suffer persecution because they have the courage of their convictions, and those who are prisoners of their own conscience because they lack the courage of their convictions. When the latter become the former, all of us will be free. **1970** *Times* 20 Apr. 6/5 A great many prisoners of conscience, it claims, are sent with or without trial 'to the so-called special psychiatric hospitals'. **1977** *Guernsey Weekly Press* 21 July 6/6 A prisoner of conscience is a person who is detained because of his political, racial or religious beliefs.

b. prisoner's dilemma (see quot. 1957); **prisoner's friend** *Armed services*, an officer who represents a defendant at a court martial.

1957 Luce & Raiffa *Games & Decisions* v. 95 We turn now to a different example of a non-zero-sum game. This one is attributed to A. W. Tucker... The following interpretation, known as the prisoner's dilemma is popular: Two suspects are taken into custody and separated. The district attorney is certain that they are guilty of a specific crime, but he does not have adequate evidence to convict them at a trial. He points out to each prisoner that each has two alternatives: to confess to the crime the police are sure they have done, or not to confess. If they both do not confess..he will book them on some very minor trumped-up charge..and they will both receive minor punishment; if they both confess they will be prosecuted, but he will recommend less than the most severe sentence; but if one confesses and the other does not, then the confessor will receive lenient treatment for turning state's evidence whereas the latter will get 'the book' slapped at him. **1963** *Jrnl. Abnormal Psychol.* LXVI. 308/2 Trust of the other person plays a critical role in determining choices made in Prisoner's Dilemma games. **1977** A. W. Tucker *MS. letter* (*copy in O.E.D. files*), The Prisoner's Dilemma is my brain child. I concocted it at Stanford in early 1950 as a catchy example to enliven a semi-popular talk on Game Theory... My example became known by the 'grapevine', but I did not publish it. **1914** 'Bartimeus' *Naval Occasions* xxi. 198 The Prisoner's Friend then gave evidence. **1972** J. Potter *Going West* 191 He was prisoner's friend to you in that spot of unpleasantness during the war.

2. a. *prisoner of war* (further examples). Freq. abbrev. *P.O.W., POW* (see P II). Also *attrib.*, esp. in *prisoner(s) of war camp.* Hence *prisoner-of-wardom* nonce-word.

1922 C. E. Montague *Disenchantment* x. 146 To 'take it out of' German prisoners of war. **1922** *Encycl. Brit.* XXXII. 163/1 The inspection of prisoners-of-war camps by the accredited representatives of the protecting State. **1944** [see *KRIEGIE]. **1946** *Encycl. Brit. Bk. of Yr.* 605/1 (*caption*) Japanese inmates of a prisoner of war camp on Guam. **1961** *Times* 7 June 17/1 A Union prisoner-of-war camp. **1974** *Times* 4 Mar. 9/4 (*heading*) Prisoner-of-wardom. **1974** [see *OFLAG].

pri·sonful. [-FUL.] As much or as many as a prison will hold.

1911 G. B. Shaw *Getting Married* 173 If a prisonful of thieves were asked what induced them to take to thieving, [etc.]. **1922** A. Bennett *Lilian* II. ii. 66 Only the malice of a prisonful of women could have seriously asserted her to be older than Felix.

prisonize (pri·z'nəiz), *v.* [f. PRISON *sb.* + -IZE.] *trans.* To cause (a person) to adapt himself to prison life. Chiefly in *pass.*: to adapt to the attitudes and social behaviour of prison life, esp. at the expense of one's 'normal' personality. Hence **pri:soniza·tion**, the fact or process of becoming adapted to prison life and unfitted for the outside world.

1940 D. Clemmer *Prison Community* xii. 299 As we use the term Americanization to describe a greater or less degree of the immigrant's integration into the American scheme of life, we may use the term *prisonization* to indicate the taking on in greater or less degree of the folkways, mores, customs, and general culture of the penitentiary. *Ibid.* 300 First offenders..'wise up', as the inmates say, or in other words, by association they become prisonized. **1963** T. & P. Morris *Pentonville* vii. 169 Prisonization may be defined as the continuous and systematic destruction of the psyche in consequence of the experience of imprisonment, and the adoption of new attitudes and ways of behaving..which may frequently make it impossible for the individual to act successfully in any normal social role. *Ibid.* 170 While the majority of inmates become 'prisonized' in some aspects, few are wholly prisonized. *Ibid.*, A certain type of prisonized man whose behaviour forms the hard core of inmate subculture. **1972** E. Heffernan *Making it in Prison* i. 7 The 'outlaw' has been found to be the most highly 'prisonized' in terms of loyalty to the inmate code. *Ibid.* ii. 32 If any category may be considered 'new' to prisonization, it is the professional, with shorter sentence length..and fewer total years in an institutional atmosphere. **1973** Cull & Hardy *Fund. Criminal Behavior & Correctional Syst.* viii. 147 He found that those inmates who were farthest from release..were the most prisonized. **1975** D. Duffee *Correctional Policy & Prison Organization* iii. 35 The counter-suggestion was made that prisonization is just an institutional form of deprivation felt by the lower class, uneducated, and black offenders most of their lives.

priss (pris). *U.S. colloq.* [Back-formation from next.] One who is prissy; a prim girl; an effeminate man, a 'pansy'.

1923 G. McKnight *Eng. Words* iv. 61 Youthful impatience with anything or anybody that interferes with a good time is expressed by such names as *kill-joy, frost, wet-blanket*... To these may be added, from the language of girls,..*poor potato,..cuckoo, old priss, old Jane. Ibid.* 62 If she [*sc.* a girl] is unpopular, she is..a *priss*, a *tomato*, a *chunk of lead*, a *drag*. **1942** Berrey & Van den Bark *Amer. Thes. Slang* §825/32 *Priss*, a girl who objects to 'necking'. **1975** A. Bergman *Hollywood & LeVine* (1976) ii. 23, I..was led to my room by an elderly priss named Roy. He told me I looked the rugged type. **1976** *National Observer* (U.S.) 16 Oct. 10/3 Randall..is television's consummate comedy priss, his overelocution and self-righteousness maddeningly funny.

prissy (pri·si), *a.* (and *sb.*) *colloq.* (orig. *U.S.*). [Perh. blend of PRIM *a.* and *SISSY.] Precise and over-particular; prim, priggish, or prudish, esp. in a supposedly effeminate way. Also *Comb.* and as *sb.*

1895 J. C. Harris *Mr. Rabbit at Home* iv. 40 Once, when I was courting, I spoke of a sitting hen, but the young lady said I was too prissy for anything. **1905** *Dialect Notes* III. 91 *Prissy, adj.* Precise, nice, over-particular. 'She's awful *prissy*.' Rare. **1925** A. Woollcott in 'L. Carroll' *Alice's Adventures in Wonderland* p. viii, The extraordinary contrast between the cautious, prissy pace of the man and the mad, gay gait of the tale he told. **1927** *Amer. Speech* II. 362/1 *Prissey*.., a boy who acts like a girl. 'Don't be such a prissey, Jim.' **1927** D. Marquis *Archy & Mehitabel* xxiv. 107 Some strait laced prune faced bunch of prissy mouthed sisters of uncharity. **1929** W. Faulkner *Sound & Fury* 49 He don't like that prissy dress. **1932** E. Hemingway *Death in Afternoon* xvii. 205 He should redeem..the prissy exhibitionistic, aunt-like, withered old maid moral arrogance of a Gide. **1948** M. Gilbert *They never looked Inside* xii. 177 She reminded him instantly of one of Walt Disney's prissy little rabbits. **1952** A. Wilson *Hemlock & After* I. ii. 36 He was disgusted at the precise, prissy tones in which he heard himself saying, 'It's lovely to be in the country.' **1957** *Listener* 2 May 722/1 His work is haunted by a disagreeable, sickly, prissy *art nouveau* rhythm. **1963** *Guardian* 28 Mar. 1/3 The rather prissy manner in which Dr Beeching has chosen to announce his determination to make us turn into a new economic man. **1968** *Times Lit. Suppl.* 28 Nov. 1327/1 The prissy, petit-bourgeois level of Robespierre. **1973** B. Broadfoot *Ten Lost Years* xxi. 244 We weren't quite as prissy as this New Generation thinks we were. **1975** L. Gillen *Return to Deepwater* viii. 140 Good grief, you little prissy, you've been kissed before, certainly!

Hence **pri·ssified** *a.*, **pri·ssily** *adv.*; **pri·ssiness.**

1934 Webster, Prissily,.. *adv.*—prissiness, *n.* **1957** *Observer* 15 Sept. 13/2 When it aspires to epigram, the dialogue falls into a quaint, soggy prissiness. **1958** *Spectator* 22 Aug. 246/2 Little girls prissily painted a still-life of oranges. **1963** P. M. Hubbard in *Mag. of Fantasy & Sci. Fiction* Jan. 8/1 He had a slightly prissified voice. **1976** *Listener* 15 July 58/1 The blackcurrant eyes, the prissily pursed mouth. **1976** 'J. Ross' *I know what it's like to Die* xxvi. 161 An accountant's clerk of terrifying prissiness.

pristane (pri·stēin). *Chem.* [ad. G. *pristan* (Y. Toyama 1923, in *Chem. Umschau auf d. Gebiete d. Fette, Oele, Wachse u. Harze* XXX. 186/1), f. L. *prist-is*, Gr. πρίστ-ις saw-fish, (*loosely*) shark; see -ANE.] A saturated hydrocarbon, now known to be 2,6,10,14-tetramethyl pentadecane, $C_{19}H_{40}$, which occurs in the liver oils of certain sharks and related species and is a colourless oil solidifying below about 30° C.

1923 *Jrnl. Chem. Soc.* CXXIV. I. 890 The pure hydrocarbon, which is named pristane, has the following characters:..b.p. 158°/10 mm.,..296°/760 mm., without decomposition. **1963** *Nature* 20 July 284/1 It is estimated that pristane constitutes about 1·1 per cent of the total paraffin fraction of crude wool wax. **1965, 1971** [see *PHYTANE]. **1975** *Sci. Amer.* June 94/3 Pristane is also found in some marine organisms, whereas phytane almost never appears except in organisms that have been contaminated by petroleum.

pristine, *a.* Add: Now usu. with pronunc. (pri·stīn).

2. In various *transf.* and extended senses: having its original condition; unmarred, unspoilt. Of a natural object, physical feature, or the like: unspoilt by human interference, untouched; virginal, pure. Of a manufactured object: spotless, pure in colour; fresh, good as new. Hence, in weakened sense: brand-new, newly-made. orig. *U.S.*

These transferred uses, though now increasingly common, are regarded with disfavour by many educated speakers.

1923 W. Stevens *Harmonium* 59 The responsive man, Planting his pristine cores in Florida, Should prick thereof. **1940** W. Faulkner *Hamlet* III. i. 180 The furious cold rain..galloped on in tearful and golden laughter across the glittering and pristine earth. *Ibid.* IV. i. 298 This time the Justice raised one hand, in its enormous pristine cuff, toward her. **1942** —— *Go down, Moses* 166 He felt the old lift of the heart, as pristine as ever, as on the first day. **1951** *Everywoman* May 68/2 (Advt.), Has a slim waist, full skirt and a pristine white detachable collar. **1955** *Bull. Atomic Sci.* Apr. 119/2 The endeavor to retain the pristine secrecy of atomic energy information. **1967** S. Attanasio tr. *Hohendorf's Life & Times of Goethe* 44 Christiane was a short, attractive brunette, with a pristine mouth and round cheeks. **1974** *BP Shield Internat.* Oct. 28/4 The value of Lake Sibaya as a pristine system cannot be overemphasised. **1975** *Times* 11 Aug. 4/5 Gone the cluttered spike, the chatter of teleprinters; his habitat now is the pristine, air-conditioned new building on the Barbican promenade of Lazard Brothers. **1976** *Milton Keynes Express* 25 June 44/1 (Advt.), The quality of the coachwork can only be described as in pristine condition. **1977** *Gramophone* May 1725/2 (Advt.), All records and tapes obtained through DLR, including budget labels, are factory fresh, unconditionally guaranteed, pristine products.

Hence **pri·stinely** *adv.*

1899 *Westm. Gaz.* 16 May 2/3 This indignant Tory thinks that what would be pristinely beautiful as Dollis Hill would be newly ugly as Gladstone Park. **1972** *Bookseller* 3 June 2456/2 Roughly half the books mentioned are not published within that week. It is only the populars who insist on having something pristinely fresh, published that day. **1975** F. King *Needle* ix. 48 A Bentley like that one, though it would be a shame to damage and stain something so pristinely beautiful.

‖ **Prisunic** (prizūnik). [Fr., lit. 'sole price'.] One of a chain of multiple stores (in France) in which a cheap class of goods is sold (orig. all at the same price).

1965 R. Postgate *Plain Man's Guide to Wine* (ed. 2) ii. 39 The '*Prisunic*' and other French shops which sell every-day wines. **1967** J. Porter *Chinks in Curtain* ii. 23 A few messy looking shops. A Prisunic. Oh dear! I plodded on. **1970** *Guardian* 19 Dec. 3/5 The old couple set up house in the doorway of the prisunic down the road. **1971** *Ibid.* 30 Oct. 3/7 The..grocer says he is still feeling the draught from the *prisunic* and..may decide to go.

prisyadka, var. *PRISIADKA.

privacy. Add: Also with pronunc. (pri·vǎsi). **1. b.** The state or condition of being alone, undisturbed, or free from public attention, as a matter of choice or right; freedom from interference or intrusion. Also *attrib.*, designating that which affords a privacy of this kind.

1814 J. Campbell *Rep. Cases King's Bench* III. 81 Though the defendant might not object to a small window looking into his yard, a larger one might be very inconvenient to him, by disturbing his privacy, and enabling people to come through to trespass upon his property. **1890** Warren & Brandeis in *Harvard Law Rev.* IV. 193 (*title*) The right to privacy. *Ibid.* 196 The question whether our law will recognize and protect the right to privacy..must soon come before our courts for consideration. **1901** G. B. Shaw *Capt. Brassbound's Conversion* II. 252 Well, I am afraid I want a little privacy, and, if you will allow me to say so, a little civility. **1933** *Post Office Electr. Engineers' Jrnl.* XXVI. 224/1 Overseas radio telephone services operated by the Post Office are provided with privacy equipment on all channels where the necessary deciphering equipment is provided at the distant end. **1940** *Chambers's Techn. Dict.* 752/2 *Secrecy* (or *privacy*) *system*, modification of speech-frequencies within the speech-band, so that during transmission from a radio transmitter to a receiver the signal is unintelligible and cannot be tapped. **a 1953** E. O'Neill *More Stately Mansions* (1964) II. iii. 130 Can I never have a moment's privacy in my own home? **1965** [see *PRIVATE *a.* (*sb.*) 2 f (*a*)]. **1970** R. K. Kent *Lang. Journalism* 106 *Privacy, right of*, the right of a citizen not to have details of his life explored in the press... The right of privacy also prevents the use of a person's name or picture in an advertisement without his permission. **1975** R. H. Rimmer *Premar Experiments* III. 233 In the meantime, you can live in one of Premar's privacy rooms. **1976** *Billings* (Montana) *Gaz.* 27 June 5-D/5 (Advt.), There's also a large patio with privacy fence and a double attached garage, all on a nicely landscaped half acre. **1977** *Chicago Tribune* 2 Oct. XII. 21/1 (Advt.), Huge patio deck with privacy fence and decorator touches. **1978** I. Murdoch *Sea* 375 When Titus appeared I decided to go outside to avoid interruption and ensure privacy.

private, *a.* (*sb.*) Add: **2. a.** *private individual, person* (further examples).

1673 J. Ray *Observations Journey Low-Countries* 305 When the Gallies are at home those [slaves] that belong to private persons are permitted to lodge in their Masters houses. **1885** *List of Subscribers, Classified* (United Telephone Co.) (ed. 6) 233 (Advt.), The Birkbeck Bank opens Drawing Accounts with trading firms and private individuals. **1930** G. B. Shaw *Apple Cart* p. xviii, Socialists have said to me that they were converted by seeing that the nation had to choose, not between governmental control of industry and control by separate private individuals [etc.]. *Ibid.* p. xix, We cannot do this as private persons. It must be done by the Government or not at all. **1931** M. Allingham *Look to Lady* xiv. 150 When publicity is fatal..then the private individual has to get busy on his own account. **1960** N. Mitford *Don't tell Alfred* iii. 41 If my husband were a private person, none of this would matter. **1975** N. Freeling *What are Bugles blowing For?* xi. 67, I find it miserable. Everyone so callous... But I'm just a private individual. **1978** *Verbatim* Sept. 1/1 In certain circles, it has become popular to the point of irritation to characterize every eventual, dyspeptic, close-mouthed selfish crank as a *private person*.

c. *private member* (earlier and later examples). Hence *private member's bill*, a bill introduced in Parliament by a private member.

In quot. 1606 *private member* means 'not a member of the Privy Council'.

1606 *House of Commons Jrnl.* 13 May I. 308/2 Petitions heretofore delivered by the Privy Council:... 28 *et* 43 *Eliz.*... Last Session, by Mr. Hare, a private Member. **1835** *Mirror of Parliament* 25 Feb. 69/2 In the last Session, Wednesday was usually devoted to Bills brought in by private Members of Parliament... The Chancellor of the Exchequer.—Nobody is more sensible than I am of the extent to which the country is often indebted to private Members, who undertake the management of public Bills. **1852** Disraeli *Ld. George Bentinck* xxvii. 580 Instead of experiencing the usual and almost inevitable doom of private members of parliament and having his statements shattered by official information, Lord George Bentinck on the contrary was the assailant and the successful assailant of an administration on these very heads. **1908** J. Redlich *Procedure House of Commons* I. II. iii. 173 This, it was complained, was done systematically, and the result had been to destroy all chance of private members' bills being carried. **1930** *Daily Express* 6 Nov. 19/2 It was a private members' day, on which neither Government policy nor the fate of the Government came up for discussion. **1939** W. I. Jennings *Parliament* vii. 180 Many private members' Bills are 'inspired' by interests outside. **1950** *Erskine May's Law of Parl.* (ed. 15) xiv. 287 Private Members' bills have precedence on Friday up to the Friday before Good Friday. **1964** Abraham & Hawtrey *Parl. Dict.* (ed. 2) 31 *Private member's bill*, a public bill introduced by a private member... It must be carefully distinguished from a private bill. **1969** *Listener* 10 July 37/2 Seldom is a government prepared to offend many of its supporters in Parliament and the country by making itself responsible for contentious reforming measures. Private members' legislation is therefore a necessity. **1976** H. Wilson *Governance of Britain* x. 171 Any Private Member's Bill, such as those which under Standing Orders can be introduced for a limited period of the year under a procedure of balloting for priority, must if a penny of public expenditure is involved carry with it the cachet of 'Queen's recommendation signified'.

d. (Later example.) Also *private trade, trading.*

1671 in *Publ. Hudson's Bay Rec. Soc.* (1942) V. 5 That Capta. Guillam & all others..bee examined what private trade hath bin by them. **1821** G. Simpson *Jrnl.* 8 Jan. in *Ibid.* (1938) I. 212 Chastellan & Lamallice..are renewing their old practice of carrying on Private Trade with the Indians. **1929** *Times* 26 Feb. 17/5 He courageously scrapped his own Bolshevist economic theories in 1921 and reinaugurated private trading. **1965** B. Pearce tr. *Preobrazhensky's New Econ.* 97 At the present moment the State Bank hardly grants any credits to private trade and industry. **1979** *Guardian* 12 Nov. 7/1 Most trading skills, shops and trucks remain in the hands of private traders. Private trading is not illegal.

f. *private detective, investigator*, a detective who is engaged privately and is not a member of an official police force. Also (orig. *U.S.*) in *colloq.* and *slang* collocations, as *private dick* (*DICK *sb.*[6]); *private eye* (*EYE *sb.*[1] 3d); also (with hyphen) *attrib.*; hence as vb. intr. and *private-eyeing* vbl. sb.

(*a*) **1868** Trollope *He knew he was Right* (1869) I. xix. 150 'The man was a policeman once.' 'What we call a private detective.' **1873** G. Lening *Dark Side N.Y. Life* 59 A jealous wife engages a private detective to watch her husband. **1898** F. Remington *Crooked Trails* i. 19 He rode a Spanish pony..and arrested Polk, his guide, and two private detectives, whom Polk had bribed to set him over the Rio Grande. **1905** Chesterton *Club of Queer Trades* i. 40 Though only a private detective myself, I will take the responsibility. **1936** A. Christie *ABC Murders* v. 38 'Then you're not—anything to do with the police, sir?' 'I am a private detective.' **1940** R. Chandler *Farewell, my Lovely* iii. 21 Philip Marlowe, Private Investigator. One of those guys, huh? **1965** D. Francis *Odds Against* v. 71 The one thing people want when they employ private investigators is privacy. **1974** V. Gielgud *In Such a Night* ii. 17 Giacomo told me of your reputation as a private investigator. **1975** J. Wainwright *Square Dance* 173 You'd be surprised how professional..some of the better provincial private detective agencies are.

(*b*) **1912** A. H. Lewis *Apaches N.Y.* vi. 128 But w'at wit' th' stores full of private dicks a booster can't do much. **1938** R. Chandler in *Dime Detective* June 23/1 We don't use any private eyes in here. So sorry. **1939** —— *Big Sleep* xviii. 127 Ohls pulled a chair up and sat down and said: 'Evening, Cronjager. Meet Phil Marlowe, a private eye who's in a jam.' **1946** E. O'Neill *Iceman Cometh* I. 14 Yuh remember dey used to send down a private dick to give him the rush to a cure, but de lawyer tells Harry nix, de old lady's off of Willie for keeps dis time and he can go to hell. **1952** Wodehouse *Pigs have Wings* i. 20 'You mean she's a sleuth?..' 'Substantially that, miss. I gather that she leaves the rougher work to her subordinates.' 'Still she's a genuine private eye.' **1962** [see *gum-shoe* s.v. *GUM *sb.*[2] 9]. **1964** Wodehouse *Frozen Assets* vi. 119, I imagine private-eye-ing is one of those things where you've either got the knack or you haven't. **1971** B. Malamud *Tenants* 25 He felt in the house,..a presence other than himself. Nothing new but who now? Private Eye snooping for one cause or another? **1974** E. Ambler *Dr. Frigo* II. 98 Isn't all research private-eye work, Doctor? **1975** J. Hone *Sixth Directorate* IV. v. 176 That's not what I'm here for—to carry on your private eyeing for you. **1979** G. Swarthout *Skeletons* 231 She had offered to marry me again..if I would private-eye for her.

g. *private army*, an army not recruited by the State; a mercenary force. Also *transf.* and *fig.*

1941 W. Temple *Citizen & Churchman* ii. 25 Anything like a 'private army' is a contradiction of the civilized state. **1950** V. Peniakoff *Private Army* Part IV (*title*) Popski's private army. **1959** M. Gilbert *Blood & Judgement* ix. 95 The police were a private army. **1964** Gould & Kolb *Dict. Social Sci.* 482/2 *Factions, cliques..private armies, lobbies, pressure groups*..are terms which, like *party*, denote voluntary associations to influence government. **1968** *N.Y. Times* 23 July 41 (*heading*) Norman Mailer enlists his private army to act in film. **1968** 'J. Welcome' *Hell is where you find It* i. 15 The security and secret services, so called, had recently..been shaken up and amalgamated... We were an off-shoot, a semi-amateur show, a private army. **1969** D. Bagley *Spoilers* v. 60 'Some of my patients had been cutting up ructions at the Howard Club. Johnny didn't like it.' 'And you had to take your own private army to back you up?' **1979** J. Rathbone *Euro-Killers* ii. 27 Their riot sticks.. infuriated him—he hated private armies.

h. *private developer* (see *DEVELOPER f).

1961 *Kentish Times* 28 July 10 There is little land left in the urban district, with its Green Belt setting, for either Council or private developer. **1965** *New Society* 26 Aug. 6/1 A modern civic centre to be built by a private developer under council guidance. **1970** *Guardian* 17 Aug. 5/2 The private developer is in a better position to judge the demands for development than the planner. **1972** *Country Life* 25 May 1330/1 Berkshire has given planning permission for some 18,000 houses, of which private developers build less than 3,000 new houses a year. **1975** *Times* 30 Aug. 13/5 How many private developers are interested in urban renewal?

3. b. *private parts* (earlier and later examples). Also *transf.* and *fig.*

1785 Grose *Dict. Vulgar T.*, *Commodity*, a woman's commodity; the private parts of a modest woman, and the public parts of a prostitute. **1853** *Law Jrnl. Rep.* XXXI. III. 123/1 What do you mean in law by exposing his *person*? The indictment should have been for exposing his *private parts*. **a 1930** D. H. Lawrence *Last Poems* (1932) 157 The reddened limbs..and the half-hidden private parts. **1959** I. & P. Opie *Lore & Lang. Schoolch.* vi. 96 It may be a verse in which the private parts are mentioned, as in the baby-washing songs of Tiny Tim. **1969** *Listener* 27 Feb. 282/1 He has an objection to 'showing his private parts' to the reader; and to this modesty, as much as to his dreadful loquacity, must, alas, be ascribed his failure to produce the great autobiography that one might have expected from him. **1971** *Farmer & Stockbreeder* 23 Feb. 30/1 Major Ogilvie recalls some mothers feeling embarrassed at having to see the 'private parts' of an animal's body—like teats and udders—being handled by a man.

4. a. Also, as a sign or notice indicating that a room or the like is private. (Further examples.)

a 1911 D. G. Phillips *Susan Lenox* (1917) II. xi. 285 The frosted glass door marked 'Private'. **1973** 'D. Craig' *Bolthole* i. 15 He saw a wide staircase with a tasselled rope across it and a Private sign. **1973** G. Mitchell *Murder of Busy Lizzie* xv. 177 My sitting-room is the one marked PRIVATE.

c. *private play, theatre* (earlier examples), *theatrical* (earlier and later examples); also *attrib.* or as *adj.*; **private bar** = *lounge bar* s.v. *LOUNGE *sb.* 4; also *ellipt.*; **private bath(room)**, a bath(room) set aside for private use, usu. one attached to a room in a hotel or guest house; **private beach**, a beach that is privately owned, esp. by a hotel for the use of guests; **private box** (see Box *sb.*[2] 8); **private business** *Eton College slang*, extra tuition; **private collection**, a collection (of paintings or the like) in private possession; **private development**, development (sense *3 d) undertaken by a private individual or company; **private hotel**, a residential hotel or boarding house which receives guests only by private arrangement; **private inquiry**, work undertaken by a private detective (see sense 2 f above); hence **private inquiry agency, agent**; **private joke**, a joke understood only by oneself or a privileged few; **private motoring**, motoring in a privately owned vehicle; so **private motorist**; **private view** (earlier examples).

1909 G. B. Shaw in *Nation* 28 Aug. 787/2 Mr. Chesterton..sees in every public-house a temple... He enters ostentatiously, throws down all the shields and partitions that make the private bar furtive. **1910** H. G. Wells *Hist. Mr. Polly* viii. 259 The policeman...put his head inside the Private Bar, to the horror of every one there. **1953** K. Tennant *Joyful Condemned* xxxiv. 340 The gossip of the Private Bar. **1963** N. Marsh *Dead Water* (1964) i. 15 There was only one other woman in the private beside Jenny. *Ibid.* ii. 55 'I want another drink. Anyone join me?'.. He made towards the old private bar. **1972** M. Gilbert *Body of Girl* xii. 107 She was in here.. just after we opened. She came into the private bar. **1975** A. Hunter *Gently with Love* xxxiii. 132 Come into the private—I would not have you leave without a crack. **1825** E. Weeton *Jrnl.* 14 June (1969) II. 384, I like to bathe alone, and a private bath is just to my taste. **1906** 'O. Henry' *Four Million* 47 The double front room with private bath. **1910** *Bradshaw's Railway Guide* Apr. 1148 Bedrooms with private bath and telephone. **1974** *Country Life* 21 Mar. 692/3 (Advt.), My bedroom, with its own tv and private bath. **1910** *Bradshaw's Railway Guide* Apr. 1148 Suites of rooms with private Bathrooms. **1961** *Sphere* 6 May 212 A new 1st-class hotel, the Hibiscus, with private beach, opens this summer. **1975** S. Brett *Cast* iii. 24 Marius's got a villa down the South... It's a lovely place. Private beach. **1632**, etc. Private box [see Box *sb.*[2] 8]. **1829** [see *FAMILY *sb.* 9c]. **1897** Kipling *Let.* 1 June in C. E. Carrington *Rudyard Kipling* (1955) x. 254 We went to the Lyceum... Irving put a private box at our disposal. **1979** *Country Life* 9 Aug. 393/2 The top floor will contain 30 private boxes. **1900** J. S. Farmer *Public School Word-Bk.* 158 *Private-business*,.. extra work with the tutor. **1979** D. Newsome *On Edge of Paradise* ii. 87 Half-an-hour's preparation for his Private Business lecture on Napoleon. **1899** R. Fry *Let.* Oct. (1972) I. 174 He took me to an amazing private collection, full of marvellous drawings and sculptures all looking far better for being in a private place. **1979** R. Cox *Auction* i. 24 There were several Memlings in Austrian private collections. Stefan Zweig owned one. **1961** *Recreation* Dec. 531/1 Areas should..have room around the edges to protect the values of the area from encroachment by private developments. **1971** 'D. Halliday' *Dolly & Doctor Bird* ii. 19 Coral Harbour is a private development in one of the moneyed quarters of New Providence Island. **1975** *Times* 30 Aug. 13/5 Impressions of the results of the last boom period for private development are still..fresh. **1857** G. H. Lewes *Jrnl.* 11–18 May in *Geo. Eliot Lett.* (1954) II. 326 We went to Dingley's Private Hotel—very comfortable. **1910** *Bradshaw's Railway Guide* Apr. 1012/2 Cullen's private hotel and family boarding house. **1936** [see *CATCH *v.* 40]. **1960** L. Davidson *Night of Wenceslas* i. 8, I had really borrowed the money from old Imre, who lived in the same private hotel. **1962** Bull & Richardson *Hotel & Catering Law* (rev. ed.) iii. 37 This chapter is concerned with premises which are not conducted as 'inns', that is with private premises as distinct from public premises. It is convenient to refer to such premises as 'private hotels', this term also including guest houses, boarding-houses, apartment houses, and similar places. *Ibid.*, The private hotel proprietor reserves to himself the right to pick and choose his guests, and does not hold himself out as willing to receive anyone who calls. He makes a separate contract, either written or verbal, with his guests. **1970** C. Whitman *Death out of Focus* x. 153 It was a typical private hotel bedroom. **1972** *Times* 1 Apr. 14/1 The very words 'boarding house' are out. Now it is 'guest house' at the very least, and possibly even 'private hotel'. **1874** M. Clarke *His Natural Life* III. xxii. 331, I dabbled a little in the Private Inquiry line of business. **1892** Private inquiry agency [see *INQUIRY, ENQUIRY 4]. 1897 A. Morrison Dorrington Deed-Box iii. 98 Your respectable talents will be devoted to the service of Dorrington & Hicks, private inquiry agents. 1923 Private inquiry agent [see *INQUIRY, ENQUIRY 4]. 1948 'J. Tey' Franchise Affair x. 106 We cannot expect you to turn yourself into a private inquiry agent on our behalf. 1973 R. Lewis Of Singular Purpose v. 113 All solicitors use private enquiry agents... They are often ex-policemen. 1974 'M. Innes' Appleby's Other Story xiii. 181 Miss Kentwell works for a private enquiry agency..of the highest repute. 1940 E. Coxhead Wind in West i. 19 She answered it [sc. an advertisement]..for a sort of private joke. 1978 J. McNeil Consultant v. 68 Susan had learned..not to ask for explanations of Webb's private jokes. 1974 Country Life 2 May 1050/1 The restrictions on private motoring caused by the higher cost of petrol. 1976 Times 3 Aug. 2/1 Between 1964 and 1974 passenger traffic increased by 40 per cent, private motoring growing by nearly 65 per cent, rail staying level and bus traffic declining. 1949 Daily Chron. 13 May 4 (caption) Private motorists made themselves popular by giving lifts to people who would otherwise have had to walk long distances. 1979 Times 18 June 4/5 The Egon Ronay Guide to Transport Cafés..should also be of help to the economy-minded private motorist. 1790 F. Reynolds Dramatist i. 12 Where arises the pleasure at an Opera, a private Play, or a Speech in Parliament? 1794 C. Mathews Let. 3 Aug. in A. Mathews Mem. Charles Mathews (1838) I. 106, I left England without calling on Wayte, to whom I am indebted for a few articles; among which are the dresses for the private play. 1868 P. Fitzgerald Life David Garrick I. vi. 158 It was once determined to get up a private play..and the parts were cast in a moment. 1784 W. Hayley (title) Plays of three acts written for a private theatre. 1807 E. Weeton Let. 18 Nov. (1969) 50 She..was never outdone in elegance of movement at a Ball, out-performed at a private Theatre. 1787 J. Powell (title) The romantic & private theatricals. 1818 Keats Let. 23 Jan. (1931) I. 96, I began an account of a private theatrical—Well it was of the lowest order, all greasy and oily. 1914 G. B. Shaw Fanny's First Play 153 The end of a saloon in an old-fashioned country house..has been curtained off to form a stage for a private theatrical performance. 1837 Dickens Pickw. xxviii. 289 A select two or three,..were being honoured with a private view of the bride and bridesmaids, up stairs. 1840 —— Old C. Shop xxix. 254 Miss Monflathers,..at the head of the head Boarding and Day Establishment..condescended to take a Private View with eight chosen young ladies. 1847 E. Gray Let. 18 Apr. in M. Lutyens Ruskins & Grays (1972) v. 33 John [Ruskin] is going to a private view of the Royal Academy.

d. Used with reference to medical treatment and facilities for which fees are charged to the patient instead of being provided by the state or a public body; from 1946 in the United Kingdom *spec.* of treatment and facilities outside the *National Health Service* (see *NATIONAL *a.* 5), as *private bed, nursing, patient*, etc. Also *ellipt.*

1754 W. Smellie Midwifery II. xxvi. 437, I attended a private patient. 1801, 1843 [see sense 1 b in Dict.] 1860 F. Nightingale Notes on Nursing vi. 38, I have often seen the private nurse go on dusting..while the patient is eating... The above remarks apply much more to private nursing than to hospitals. Ibid. 39 Generally, the only rule of the private patient's diet is what the nurse has to give. 1924 A. Bennett Price of Love xii. 256 In those days of State health insurance all doctors were too [illegible] patients. 1934 P. Bottome Private Worlds xii. 113 They stood in a small private room off the ward, and looked down at the moaning woman on the bed. 1936 D. L. Sayers Gaudy Night ix. 191 He's in a private ward, so you can get in any time. 1942 M. Dickens One Pair of Feet vii. 110 Sister Adams..told me that I was to..go on day duty on the Private wards. Ibid. 115 I went on cutting bread savagely and the Private Nurse stirred milk with pursed lips. 1944 [see *DOC'S BODY 2]. 1943 G. Greene Ministry of Fear [illegible] home and [illegible]. 1946 Act 9 & 10 Geo. VI c. 81 §5 The Minister may allow any medical practitioner..on the staff of a hospital providing hospital and specialist services to make arrangements for the treatment of his private patients either at the hospital or at any other such hospital. 1946 P. Bottome Lifeline ii. 33 Ours is not a state-run affair..but a private Hospital. 1956 P. Scott Male Child i. 26, I spent most of April in a private nursing home. 1960 C. Watson Bump in the Night i. 15 He lay in a small private ward of Chalmsbury General Hospital. 1961 Ann. Reg. 1960 9 There had been a demand that 'private patients' who relieved the finances of the health service by paying their own doctor should get free access to N.H.S. drugs. 1967 M. Sharman Face of Danger i. 7 They..walked out of the private wing of the hospital. 1967 P. Willmott Consumer's Guide Brit. Social Services vi. 158 Private beds amount to little over one per cent of the total number of beds in use. Ibid., Financial help towards the cost of private treatment is provided by several provident associations. 1969 B. Turner Circle of Squares iv. 27 Poor Flisch had a private room. 1977 Guardian 1 July 6/5 Hospital laboratory technicians..will refuse to carry out tests on private patients. 1977 P. Johnson Offshore Islanders vi. 400 Labour ministers lacked the will to impose a salaried service on the medical profession... Private practice, private beds in hospital, private health insurance were permitted. 1976 W. J. Burley Wycliffe & Schoolgirls iii. 67 She was a staff nurse, and chucked up her job to go into private nursing. 1976 N. Leigh-Taylor Doctors & Law iv. 35 The Government has announced that it intends..to abolish private treatment in N.H.S. hospitals.

e. *Teleph.* and *Telegr.* (i) Applied to (*a*) a line that is permanently for the exclusive use of the subscriber or is not connected to the public network; (*b*) a number that is ex-directory; (*c*) a number at a private address rather than business premises.

1878 Telegr. Jrnl. VI. 51/1 The regulations concerning the despatch and receipt of telegrams, the tariffs for the same, and for the renting of private wires. 1885 List of Subscribers (United Telephone Co.) p. vii, The Charge for Private Lines is at a fixed annual rental, payable in advance, varying with the situation and the distance apart of the points connected. 1911 W. Aitken Man. Telephone xxiii. 476 Private Lines..are lines not having exchange service. 1924 J. Buchan Three Hostages xvi. 235 This must be a private telephone..of which only his special friends knew the number. 1933 D. L. Sayers Murder must Advertise viii. 129 He was not in the telephone-book, but his private number would doubtless be on the telephone-clerk's desk. 1940 War Illustr. 16 Feb. p. ii/1 Taking the final proof of his commentary on the foreign news of the day to the 'private wire' room, to be telegraphed or telephoned to Manchester. 1942 A. Christie Body in Library vi. 59, I had a private line put in connecting my bedroom with my office. 1969 N. Freeling Tsing-Boum xiii. 95 Good morning. Police Judiciaire!.. I'm at a private number in Marseilles; will you..clear me a direct line. 1972 L. Moir tr. Simenon's Maigret & Flea ii. 34 You'll know where to find me. My private number's in the book. 1974 Encycl. Brit. Macropædia XVIII. 95/1 Private-line systems for data communications have come into widespread use in the past decade. 1974 D. Gray Dead Give Away vii. 72 Cyril decided on a nap in his study, where he had a private line, on which he could ring Nina. 1976 H. MacInnes Agent in Place xxiii. 241 Tony..dialled Bill's number—not his private line, just the ordinary one. 1976 T. H. Flowers Introd. Exchange Syst. i 11 Picture telegraphy..is possible over the telephone service lines but difficulties discourage small users and encourage large users of such services to rent private circuits not subject to switching.

(ii) *spec.* of an exchange: serving private lines; *private branch exchange*, an exchange on private premises by which private lines may be connected to the public network.

1891 J. Poole Pract. Telephone Handbk. vii. 124 Fig. 102 represents a type of switch-board which was designed by the writer in 1881 for the use of private telephone exchanges. 1905 Ann. Rep. Amer. Telephone & Telegr. Co. 1904 6 There is an enormous increase in the number of private branch exchanges in hotels. 1911 W. Aitken Man. Telephone xxi. 416 No hotel or warehouse of any standing is now considered complete without a private branch exchange connected to the 'Central' by a number of circuits. 1943 A. L. Albert Fund. Telephony viii. 170 A private branch exchange or *PBX* in a large store, hotel, manufacturing plant, or to serve a college campus. 1974 Encycl. Brit. Macropædia XVIII. 94/1 Typical automatic switching systems in operation include the step-by-step system, used..for local exchanges and private branch exchanges.

(iii) Describing components in a telephone exchange which belong to a circuit whose potential indicates the condition of a particular subscriber's line and enables its condition to be ascertained without interfering with calls in progress; esp. in *private wire* (see quot. 1969).

1906 J. Poole Pract. Telephone Handbk. (ed. 3) xxx. 486 When a current is started and stopped through the 'private' magnet, the end of the side-switch arm slips under the outer tooth. 1919 R. Mordin Strowger Automatic Telephone Exchange i. 23 The whole arrangement of fixed contacts is called the connector bank; the upper half the private bank, and the lower the line bank. The moveable contacts are termed respectively the private wiper and the line wiper. 1927 C. W. Wilman Man. Automatic Telephony vi. 55 This wire is comparable with the test wire in a manual system inasmuch as it indicates whether a particular line is free or busy... It is..known as the private line (because it prevents intrusion on a busy trunk). 1969 S. F. Smith Telephony & Telegr. A vi. 153 A third wire is therefore provided on all connexions through the exchange, the potential of which indicates the condition of the circuit. This avoids intrusion on calls in progress and is called the private wire, usually abbreviated to '*P-wire*'.

5. b. *private bank* (see Bank *sb.*[3] 7 a); hence *private banker, banking*; *private car*, (*a*) *U.S.*, a privately owned and used railway carriage; (*b*) a motor car owned and used privately, as distinct from a commercial vehicle; *private family* (earlier example); *private income*, an income derived from private sources, as investments, property, etc; an unearned income; *private means*, income derived from private sources (cf. *private income* above); *private press*, a printing and publishing house of limited resources and output, often operated for the owner's personal satisfaction rather than profit, and usually issuing small editions of books designed to meet higher standards of production than those of commercial publishers; *private residence* = *private house*; *private room*, a room in a club, hotel, etc., that may be hired for private use (see also sense 5 d below); *private school* (further examples); also, a school that is independent of a State system of education (see *INDEPENDENT *a.* 5 d); *private service*, domestic service in a private house.

1714 in A. McF. Davis Tracts Currency Massachusetts Bay (1902) 115 Which does most of all import them, the Publick or the Private Bank? 1802 M. Edgeworth Let. 1 Dec. (1979) 43 Private banks never issue any notes. 1978 M. Birmingham Sleep in Ditch 120 My mother wanted me to be a banker..in one of the small, distinguished private banks. 1837 in W. L. Mackenzie Life & Times M. Van Buren (1846) 178 The Bills of the banks of this State only shall be circulated as Money by private bankers. 1978 P. Noyes Who is Simon Warwick? viii. 104 A house which only a private banker could possibly have described as a cottage. 1836 in W. L. Mackenzie Life & Times M. Van Buren (1846) 176 If the fetters are knocked off by the repeal of the Restraining Law, private banking associations may be formed. 1954 Econ. Hist. Rev. VII. 167 A tentative sketch of some developments in London private banking is offered. 1897 Kipling Capt. Cour. ix. 186 Send 'Constance', private car, here, and arrange for special [train]. 1926 Brit. Gaz. 12 May 1/3 There were few private cars on the roads and nearly every vehicle was labelled 'Food only'. 1938 E. Ambler Cause for Alarm xiv. 232 We had to wait for a private car and a van to pass. 1979 B. Peterson Peripheral Spy ii. 34 It really must be important when Col. Petrovich waited for you with his private car. 1873 C. M. Yonge Pillars of House IV. xlix. 385 Between his private income and the endowment he would be able to keep up..a staff of Curates. 1910 A. Bennett Clayhanger iv. ii. 470 It's a good thing she has a private income of her own. 1923 J. M. Murry Pencillings 86 No one really pays much attention now to the subtle problems which tormented Henry James, simply because no one would earn any gratitude by solving them. Even the attempt to solve them calls for a private income. 1941 'G. Orwell' in Partisan Rev. VIII. 496 Nearly all [Home Guard] commands are held by retired colonels, people with 'private' incomes or, at best, wealthy business men. 1952 M. Laski Village iii. 65 Because she's got a private income no one ever expected her to go out and take a job. 1971 G. Household Doom's Caravan ii. 53 He was a bachelor with a private income. 1862 Mrs. J. B. Speid Our Last Years in India vi. 149 In the case of married military men, under the grade of field officer, I cannot see, unless they have private means..how they can escape involvement. 1976 C. Bermant Coming Home ii. ii. 131, I was without connections and without private means. 1834 J. Martin Bibliogr. Catal. Bks. Privately Printed p. v, The second portion of the work, consisting of an account of the publications from literary clubs, and private presses. 1900 Library I. 407 Since the days when Horace Walpole started as a master-printer at Strawberry Hill quite a number of book-lovers have amused themselves with the management, and occasionally with the actual working, of a private press. 1922 D. B. Updike Printing Types II. xxii. 215 The types of the Kelmscott, Doves, and other English private presses were from his [sc. E. P. Prince's] hand. 1934 H. Waddell Let. in M. Blackett Mark of Maker (1973) xii. 112 The man who is secretary of the Pilgrim Trust (Tom Jones)..runs a very luxurious private press, for which he wants me to do a translation. 1955 S. H. Steinberg Five Hundred Years of Printing iii. 217 It was pleasure in fine printing, or at least in printing according to personal taste, rather than commercial success that made kings and nobles set up private presses in the seventeenth and eighteenth centuries. 1968 Times 30 Jan. 13/3 (Advt.), First editions and private press books, including the Shakespeare Head Press Homer, 1930–31. 1978 Times Lit. Suppl. 15 Sept. 1024/1 Some private presses eventually extended their operations so much that their work is more properly described as commercial. 1885 List of Subscribers, Classified (United Telephone Co.) (ed. 6) 8 Any Subscriber who pays £20 a year for an Exchange connection can..have his Private Residence joined up with the system. 1974 P. Lovesey Invitation to Dynamite Party iii. 34 'There was a second explosion..at Sir Watkin Wynn's residence.' 'A private residence? What have they got against Sir Watkin Wynn?' 1824 Scott Redgauntlet III. vii. 197 Walking into the inn, [he] demanded from the landlord breakfast and a private room. 1847 C. Brontë Jane Eyre I. xi. 172 When I asked a waiter if any one had been to inquire after a Miss Eyre, I was answered in the negative; so I had no recourse but to request to be shown into a private room. 1879 Trollope John Caldigate II. xviii. 251 'I suppose I can have a private room here, at noon tomorrow?' asked Caldigate, turning to the woman at the bar. 1920 'Sapper' Bull-Dog Drummond 7 Have we ever had staying in the hotel a man called le Comte de Guy?.. Has he ever fed here, or taken a private room? 1974 J. Gardner Return of Moriarty 87 A private room had been booked for Moriarty and his guest,..at the Café Royal. 1875 Trollope Prime Minister (1876) I. i. 6 He had been at a good English private school. 1914 C. Mackenzie Sinister St. II. iii. iii. 547, I don't think it is snobbishness... It's a throw back to primitive life in a private school. 1944 Private school [see *INDEPENDENT a. 5 d]. 1945 Guide to Educ. Syst. Eng. & Wales (Min. of Educ.: Pamphlet No. 2) 59 Private school, independent school owned by a private individual or group of individuals. 1969 T. Jenkins We came to Australia i. iii. 41 There are both State and private schools. 1976 C. Bermant Coming Home ii. iii. 154, I was a teacher in one of those private schools, which was basically a very expensive..crammer for the sons of oil sheiks. 1934 D. L. Sayers Nine Tailors 139 Deacon was a waiter in some club... He wanted to try private service. 1953 A. Christie Pocket Full of Rye v. 36 She'd been in private service first and after that in various cafés. 1978 M. & N. Ward Home in Twenties & Thirties 38/1 There was..an inexorable reduction in the number of people engaged in private service.

6. b. *private account*, a bank account relating to one's personal (as opposed to business) assets; *private call*, a personal telephone call to or from one's place of work; *private life*, a person's domestic or personal (way of) life, as distinct from that relating to employment, official position, etc. (freq. with a notion of sense 3).

1924 'Sapper' Third Round i. 34 [The cheque] is drawn on my private account. 1973 A. Behrend Samarai Affair viii. 84 He compiled a list of every individual connected with the case... Thus:..Bank Manager, Hartley's private account. 1942 E. Waugh Put out More Flags i. 52 There's a ridiculous woman up the line saying is this a private call? 1974 'J. Le Carré' Tinker, Tailor xi. 84 The misuse of unlisted Circus telephones for private calls. 1526 R. Whytford Martiloge f. cxxiiv, He resigned his crowne, & lyued a holy priuate lyfe. 1843 Dickens Mart. Chuzz. (1844) xvi. 193 A full account of the Ball, with the Server's own particulars of the private lives of all the ladies that was there! [illegible] Some interesting details of the general's private life. 1943 J. B. Priestley Daylight on Saturday xxii. 209 Her private life, now in ruins, insisted upon claiming her attention, and she could not pretend to herself that it was less important than the private lives of all the other women in the factory. 1973 A. Behrend Samarai Affair viii. 81, I was speaking of Mr Gosling's private life... I know next to nothing of Mr Gosling's private life.

7. b. *private notice question* (see quot. 1964).

1871 Hansard Commons 27 Feb. 941, I wish to ask some questions of the Prime Minister, of which circumstances prevented me from giving any other than a Private Notice to him. 1913 Ibid. 21 Jan. 225 Private notice [illegible] May I ask the Chancellor of the Exchequer a question of which I have given him private notice. 1939 G. F. M. Campion Introd. Procedure of House of Commons iv. 126 Private Notice Questions are of two kinds: (a) those of an urgent character, and (b) non-urgent. 1952 Daily Express 13 Oct. 12/2 A private notice question can cause more flutter in the Civil Service than any other of the few instruments of torture left in the hands of back-bench members. 1964 Abraham & Hawtrey Parl. Dict. (ed. 2) 168 On specially urgent matters, 'private notice questions' may be asked after the end of the time allotted by the standing orders to questions for oral answer. A member who wishes to avail himself of this privilege must give notice of the terms of his question to the minister and to the Speaker not later than twelve o'clock on the day on which he seeks to ask it. 1976 S. Lloyd Mr Speaker, Sir iii. 88 The Speaker also has power under the Standing Orders to allow what are called Private Notice Questions (I will refer to them from now on as P.N.Q.s), ones which in his opinion are of an urgent character and relate either to matters of public importance or to the arrangement of business.

c. *private secretary*: see Secretary *sb.*[1] 2; *private secretaryship*, the office or post of private secretary.

1773 R. Jephson Let. 2 Mar. in D. Garrick Private Corr. (1831) I. 530 Our friend Tighe is much engaged in his office of Private Secretary to the Lord Lieutenant, but is getting better health and more strength every day. 1814 Jane Austen Mansf. Park II. xiii. 255, I would rather have been private secretary to the first Lord than any thing else. 1869 Trollope Phineas Finn II. lxv. 232 The Duke of St. Bungay was at work as a Private Secretary when he was three-and-twenty. 1880 E. W. Hamilton Diary 25 Apr. (1972) I. 3 Horace Seymour and Henry Primrose are the two [illegible]. 1891 W. Fraser Disraeli & his Day (ed. 2) 43 Mr Algernon Greville became, some years afterwards, Private Secretary to the Duke. 1930 J. B. Priestley Angel Pavement [illegible] the private secretary jobs. Yours is one of them, isn't it? 1954 K. Amis Lucky Jim iv. 48 Our influential friend will shortly be declaring his private secretaryship vacant. 1974 R. Ingham Yorke i. 5 You'll be my new private secretary, so you'll get to know everything.

d. *private law* (see quot. 1923).

1773 E. Erskine Inst. of Law of Scotl. i. 9 Public law is that which hath more immediately in view the public weal... Private law is that which is chiefly intended for ascertaining the civil rights of individuals. The former part of law of Scotland is to be the proper subject of this treatise. 1923 W. J. Byrne Dict. Eng. Law 104/2 Private or civil law deals with those relations between individuals with which the State is not directly concerned; as in the relations between husband and wife, parent and child,..contracts, torts, trusts, legacies. 1932 J. F. Jolowicz Hist. Introd. Roman Law i. 5 The change from republic to empire did not make any immediate difference to private law. 1969 D. Daube Roman Law iii. 131 The basic structure..is still largely dominated by the criminal trial; but the cases discussed have shifted to lesser crimes and even near to private law. 1973 I. M. Sinclair Vienna Convention on Law of Treaties vi. 86 The potentially misleading nature of private law analogies.

e. *private international law* (see quots.).

1834 J. Story Commentaries Conflict of Laws i. 9 The jurisprudence, then, arising from the conflict of the laws of different nations, in their actual application to modern commerce and intercourse, is a most interesting and important branch of public law... This branch of public law may be fitly denominated private international law, since it is chiefly seen and felt in its application to the common business of private persons. 1861 R. Phillimore Commentaries Internat. Law IV. p. iii, This volume is devoted to the consideration of Jus Gentium—Private International Law, or Comity: that is, strictly speaking, the law which ought to govern the legal relations of individuals not being the subject of the State which administers the law. 1938 G. C. Cheshire Private Internat. Law (ed. 2) i. 21 The expression 'Private International Law', coined by Story in 1834,..and used on the Continent by Foelix in 1843, has been adopted by Westlake and Foote and most English authors. The chief criticism directed against its use is its implication that the subject forms a branch of International Law. There is, of course, no affinity between Private and Public International Law. The latter comprises those universally accepted customs which are recognized by States in their public relations with each other; the former consists of rules which the Courts of each territorial jurisdiction follow when a dispute containing some foreign element arises between private persons. 1962 J. F. McMahon in Brit. Year Bk. Internat. Law 1961 326 The European Economic Community Treaty..devotes two of its articles to what it calls 'approximation of laws'... This gradual approximation of law, however, will have to take place both in the field of legislation and in the jurisprudence of the courts. The Court of the European Communities is not itself concerned with private international law and that question has only been invoked on one occasion before the Court.

f. *private treaty* (see quot. 1973).

1858 Estates Gaz. 16 Aug. 16/1 (Advt.), To be sold, by private treaty, a substantial and well-built house. 1922 V. Sackville-West Heir i. 19 Are we to try for auction or private treaty? Personally I think the house at any rate will go by private treaty. 1937 D. M. D. Alexander [illegible] Auctioneers & Estate Agents (Chartered Auctioneers' & Estate Agents' Inst.) 19 Almost all general urban practices depend..upon the commissions earned on the sale of houses and properties whether such sales are by private treaty or auction. 1973 Westland & Rodway Place of your Own i. 11/2 In Scotland..houses are more often sold 'by private treaty'. This way, the owner places a reserve, or 'upset' price on the property and invites those interested to make offers, in writing, by a specified date. On that date, the offers are examined, and the property will usually go to the highest bidder. An offer made this way is binding by law, unless you withdraw it before it is formally accepted... Some properties in England and Wales are offered for sale on these terms. 1979 Irish Times 18 Sept. 13/1 One of the very few [houses] that would justify its private treaty price tag of over three-quarters of a million pounds.

g. *private war*: a war fought by a restricted number of participants from personal or private motives. Also *transf.*

1866 C. M. Yonge Dove in Eagle's Nest I. p. vi, An offended nobleman, having sent a Fehdebrief to his adversary, was thenceforward at liberty to revenge himself by a private war. 1894 Kipling Jungle Bk. 85 A wolf who obeyed the orders of this boy who had private wars with man-eating tigers was not a common animal. 1948 G. V. Galwey Lift & Drop vii. 198 Operating a war of his own against the gang and the Law. 1973 J. R. L. Anderson Death on Rocks x. 175 [illegible] You and Simon may feel that you have a private war against Potterton, but this is more than a private war. 1974 'G. Black' Golden Cockatrice xi. 194 A killing that was one incident in the continuing private war the Russians and the Chinese have been waging against each other.

h. *private company*: a company whose membership and transfer of shares are limited by law.

1908 Act 8 Edw. VII c. 69 §121 For the purposes of this Act the expression 'private company' means a company which by its articles—(a) Restricts the right to transfer its shares; and (b) Limits the number of its members..to fifty; and (c) Prohibits any invitation to the public to subscribe for any shares or debentures of the company. 1928 Act 18 & 19 Geo. V c. 45 §55 If any company, being a private company, alters its articles in such manner that they no longer include the provisions which..are required to be included in the articles of a company in order to constitute it a private company, the company shall, as on the date of the alteration, cease to be a private company. 1928 Britain's Industr. Future (Liberal Industr. Inquiry) vi. 84 The most important existing legal distinction is between Public Companies..and Private Companies, limited to not more than 50 shareholders. 1948 Act 11 & 12 Geo. VI c. 38 §31 If at any time the number of members of a company is reduced, in the case of a private company, below two,..and it carries on business for more than six months while the number is so reduced, every person who is a member of the company during the time that it so carries on business..shall be severally liable for the payment of the whole debts of the company contracted during that time. 1961 T. E. Utley Occasion for Ombudsman ii. 18 The recent case of the Esso Petroleum Company, which, being a private company, [illegible].

i. *private world*: a private 'realm' within which one moves or lives; cf. World *sb.* 10.

1922 A. Huxley Crome Yellow xiii. 128 He determined to retire absolutely from it [sc. the great world] and to create..at Crome a private world of his own. 1968 Listener 19 June 1024/1 Never has the private world of the thwarted male been so shamelessly exposed to view. 1976 S. Hynes Auden Generation ix. 296 The private world of love is threatened by public violence.

j. *private sector*: that part of an economy, industry, etc., which is free from direct state control. Usu. with *the*.

1952 E. Sukkett-Unger Compar. Econ. Syst. iii. 59 Coordination of freedom and planning obviously influences the formation of private and public sectors within the whole economic structure... The composite [illegible] countries still reveal larger private than public sectors. Their private sectors are relatively much larger than those of the countries under Eastern planning. 1965 J. L. Hanson Dict. Econ. 327/2 Private sector. The part of the economy that is not under State control. [illegible] 1980 Illustr. London News Mar. 19/1 The extension of the steel strike into the private sector.

k. *private language*: a language which can be understood by the speaker only, esp. in *Logic* involving the query whether such a concept can have meaning. Also *loosely*, a language shared by a privileged few.

1953 G. E. M. Anscombe tr. Wittgenstein's Philos. Investigations i. 94 Sounds which no one else understands but which I 'appear to understand' might be called a 'private language'. 1955 L. P. Hartley Perfect Woman xiii. 121 Why should you understand my private language? 1964 Amer. Philos. Q. I. 20/1 A private language is one of which it is not merely the case that it is not understood by anyone other than the speaker, but more that it is logically impossible that it should be understood by anyone other than the speaker. 1979 D. Francis Whip Hand xiv. 173 The reins felt alive, carrying messages... A private language, shared, understood.

l. *private-label*: used *attrib.* to denote a product manufactured by a particular company for sale through its own retail markets; cf. *own-label* s.v. *OWN *a.* 4 a.

1961 Economist 11 Mar. 964/1 There are the usual 'private-label' teas, flour, butter, and dried cereals, fruit and pulses; besides these, private label jams and biscuits are quite common and several companies market their own canned peas, soups, canned fruit and canned vegetables; there is even a private-label pine essence. 1971 Guardian 9 June 13/2 Supermarkets' private label brands were selling at around [illegible],..while the major manufacturers were sometimes cutting their prices by as much as 15.

10. b. (Further examples.)

1859 Geo. Eliot Let. 10 Apr. (1954) III. 43 The letter is marked 'private'. 1971 A. Price Alamut Ambush x. 125 The letter?.. He said it'd be a bit much to open it because it was marked 'private'.

16. Delete †*Obs.* and add later examples.

1876 'Mark Twain' Tom Sawyer xxv. 272 I'll smoke private and cuss private. 1883 —— Life on Mississippi iii. 54 They all drank more than usual—not together, but each man sidled off and took it private, by himself. 1905 [see *CASE *sb.*[1] 6 c].

B. *sb.* **7.** (Examples.)

1940 C. McCullers Heart is Lonely Hunter II. iv. 155 He's so fat he hasn't seen his privates for twenty years. 1955 P. Bowles tr. Beckett's Molloy 77 She..thrust her stick between my legs and began to titillate my privates. 1979 'E. McBain' Calypso v. 49 The dancer..wiped the black man's glances over what the Vice Squad would have called her 'privates'.

9. Short for *private school*.

1932 C. Connolly Let. 6 Apr. in Romantic Friendship (1975) 64, I met quite a nice small boy who is at my private. 1959 M. Mitford Christmas Pudding v. 81 At my private..we had a most handy little cemetery for the fathers, just behind the cricket pav. 1940 —— Pigeon Pie iv. 80 It is exactly like one's private. 1966 Listener 21 July 128/1 What private were you at?

private enterprise. [Private *a.* (*sb.*) 4 c, 5.] A business or other commercial activity that is privately owned and free of direct state control; such concerns collectively. Also *attrib.*

1844 [see enterprise *sb.* 1 b]. 1859 Mill On Liberty v. 191 When private enterprise, in a shape fitted for undertaking great works of industry, does not exist in the country. 1888 E. Bellamy Looking Backward xxii. 331 Credit..was the only means you had for concentrating and directing it [sc. capital] upon industrial enterprises. It was in this way a most potent means for exaggerating the chief peril of the private enterprise system of industry by enabling particular industries to absorb disproportionate amounts of..capital. [illegible]

So **private enterpriser**, an advocate of or participant in private enterprise.

1896 G. B. Shaw in Labour Leader 19 Dec. 443/2 [illegible]

privateer, *sb.* Add: **2.** (Later *fig.* example.)

1836 Dickens Let. 1 Nov. (1965) I. 185, I perceive that 'Bells Life'—The Carlton Chronicle, and some other Weekly papers, are in the habit of re-publishing my sketches from the Chronicle verbatim... Some remonstrance in the paper might have the effect of inducing the Privateers at all events to acknowledge the source from which they derive the Articles.

3*. An advocate or exponent of private enterprise.

1940 Amer. Guardian 5 Apr. 4/3 It is the general policy of the privateers never to reduce their rate unless forced to do so by public competition in the shape of municipally or federally owned [electric power] plants. 1965 Spectator 19 Feb. 223/1 These two engaging privateers

..are concerned not so much with steel as with the general pattern of British politics. 1979 Arizona Daily Star 22 July c 10/1 As a privateer, Serrano drove himself to the races and for the last half of the series even served as his own mechanic.

privateness. Delete 'Now *rare*' and add later examples.

1922 E. R. Eddison Worm Ouroboros xv. 212 I'll walk apart, madam,..if thou wouldst have privateness to deliver thy mind. 1939 'G. Orwell' Coming up for Air iv. vi. 276 The privateness of all those lives! 1941 —— Lion & Unicorn 15 What it does link up with..is the addiction to hobbies and spare-time occupations, the *privateness* of English life. 1941 E. R. Eddison Fish Dinner vi. 92 He let go her hands and stood..as if withdrawn for the moment into some inside privateness of deliberation.

privatism (prəi·vătiz'm). [f. Private *a.* + -ism.] An inclination or tendency to be private (in various senses); the use or advocacy of personal or private ideas, institutions, etc. Hence **privati·stic** *a.*, **privati·stically** *adv.*

1948 C. S. Lewis in Williams & Lewis Arthurian Torso ii. vi. 188 'Privatism.' This occurs when the poet writes what the reader, however sensitive and generally cultivated he may be, could not possibly understand unless the poet chose to tell him something more than he has told. 1970 Time 28 Dec. 6 Few observers of the U.S. scene foresaw that political passions on the campuses would become muted in a new emphasis on 'privatism'. 1970 C. A. Reich Greening of Amer. x. 284 'Human nature' was not necessarily always privatistic, grasping, competitive, materialistic. 1971 J. J. Shapiro tr. Habermas's Toward Rational Society vi. 121 Student activists are less privatistically oriented to professional careers and future families than other students. 1971 Atlantic Monthly Nov. 119, I cannot say in blanket fashion whether this Mao myth is 'good or bad'... The 'privatistic' alternative, anyway, in a country with *per capita* income perhaps one twentieth of America's, is not a glittering one. 1977 Times Lit. Suppl. 2 Sept. 1054/2 Self-interest—what Dr Kammen euphemistically calls 'privatism'—and secularism were strong. 1978 T. Honoré Tribonian 37 They were trained largely in private law..and imperial constitutions, let alone criminal and public law, had little part to play in legal education. The privatistic outlook of these men is likely to be reflected in the output of constitutions.

privatization (prəi:vătəizēⁱ·ʃən). [f. Private *a.* + -ization.] **1.** The policy or process of making private as opposed to public, *spec.* the advocacy or exploitation by the State of *PRIVATE ENTERPRISE; = *DENATIONALIZATION 2.

1959 News Chron. 28 July 2/6 Erhard selected the rich Preussag mining concern for his first experiment in privatisation. 1960 Ibid. 22 Apr. 11/5 Complete privatisation was opposed by the Socialists..because they feared..the little man selling out his shares to the big capitalists. 1970 Observer 25 Jan. 1/6 He foresaw 'privatisation' of many sectors of industry now in public ownership. 1970 J. Cotler in I. L. Horowitz Masses in Lat. Amer. xii. 440 If rural marginality allows for the..privatization of State power, the political sphere demands..a new line of social integration. 1976 National Observer (U.S.) 1 May B6/3 The contrast between then and now measures the tendency toward privatization and withdrawal of our commitments from the open, public arena that has occurred during the course of the Twentieth Century. 1976 Globe & Mail (Toronto) 12 Dec. 5/7 Privatization in the handing over of elements of the public service to the private sector is threatening the livelihoods of thousands of public servants. 1977 Ibid. 20 Jan. 6/1 The Government published a working paper..which set out some possibilities..including this: 'The possibility of the private sector providing goods or services that are now provided through government enterprise and programs.' The government, it seemed was toying with the idea of 'privatization'. 1979 New Statesman 6 July 14/3 This political formula of controlled privatisation depends on not too many people finding the stringent limits on expression spiritually intolerable.

2. The act or process of regarding as personal or separate, *spec.* the concept of an institution, activity, discipline, etc., seen in terms of its relation to the individual rather than to society generally or to a part of society.

1968 Listener 6 June 720/1 On these two points, the privatisation of death, and the loss of any sense of an appropriate length of life, I have contrasted contemporary English society with alien cultures—Celtic or African. 1969 J. H. Goldthorpe et al. Affluent Worker in Class Struct. iv. 96 Our findings would indicate as the most probable concomitant of these workers' orientation to work and of their present type of employment what we have earlier referred to as *privatisation*—a process, that is, manifested in a pattern of social life which is centred on, and indeed largely restricted to, the home and the conjugal family. 1972 Clergy Rev. Mar. 209 Despite our privatization of God, we are all sharers in spiritual matters. 1973 Times Lit. Suppl. 26 Oct. 1307/2 To buy things like that secondhand car which initiates the simultaneous privatization and social extension of their life-style. 1974 J. I. M. Stewart Gaudy iii. 40 He had carried with him all the privatization of experience that characterizes bourgeois life. 1975 Amer. Anthropologist LXXVII. 260 While the stress on language and on intersubjectivity is fine, the privatization of anthropology, its ultimate inability to issue in a set of explanatory propositions, is disastrous. 1976 Nichols & Armstrong Workers Divided 20 We also seek to indicate..what is sometimes called 'privatization'..: a nuclear family-based separation from community. 1976 Jrnl. Church & State XVIII. 209 (title) Does Church–State separation necessarily mean the privatization of religion? 1978 Times Lit. Suppl. 17 Feb. 194/4 The modern 'privatization' of religion owes more to constitutional developments than to Christianity having become 'controversial'. 1979 E. Norman Christianity & World Order vi. 80 To regard Christianity as being..concerned primarily with the relationship of the soul to eternity, is these days denounced within Christian opinion as a 'privatization' of religion.

privatize (prəi·vătəiz), *v.* [f. Private *a.* + -ize.] **1.** *trans.* To regard as personal or separate, *spec.* in the sense of *PRIVATIZATION 2.

1969 W. Glen-Doepel tr. Metz's Theol. of World v. 109 The societal dimension of the Christian message was not given its proper importance but, implicitly or explicitly, treated as a secondary matter. In short, the message was 'privatized' and the practice of faith reduced to the timeless decision of the person. Ibid. 114 It is impossible to privatize the eschatological promises of biblical tradition: liberty, peace, justice, reconciliation. Again and again they force us to assume our responsibilities towards society. 1972 Biblical Theol. Bull. Feb. 47 Because of their transcendental, existentialistic, personalistic drift these theologies have privatized and presentialized the Christian message. 1977 Times Lit. Suppl. 8 Apr. 441/4 Few of his dockers were 'privatized' in the sense in which that term was used by Goldthorpe and his colleagues in their Luton study. 1979 E. Norman Christianity & World Order vi. 80 The modern politicized Christians also 'privatize' religion.

2. To make private as opposed to public, *spec.* of the State, to assign (services, industries, etc.) to *PRIVATE ENTERPRISE; = *DENATIONALIZE *v.* 2 b.

1970 New Society 5 Feb. 222/3 Is the Office of Health Economics trying to hint that the best place to start totally privatising the National Health Service is at eye level? 1972 Daily Tel. 10 Feb. 16 Some local government services (water, refuse collection, fire-fighting, sea-side amenities, art galleries, museums) could be 'privatised', as well as national health services and State schools. 1976 Globe & Mail (Toronto) 12 Dec. 7/2 The House might want to address itself to the question whether we want the Crown corporations to continue to be involved in business or whether they should all be privatized. 1979 Ibid. 24 May 7/5 Mr. Clark intends to proceed with his promise to 'privatize' Petro Canada.

Hence **pri·vatized** *ppl. a.*

1973 Goldthorpe & Lockwood in Sociol. Rev. II. 154 At this point we may return to our earlier distinction between the 'privatised' and the 'socially aspiring' worker. 1968 Brit. Universities Ann. 95 Until recently, the values of the entertainment industry, and the demand for purely 'privatised' standards of morality gave the young little to act collectively about. 1969 R. Blackburn in Cockburn & Blackburn Student Power 195 Condemned to a trapped existence in anonymous private or public bureaucracies, 'industrial' man is promised the domestic joys of a 'privatized' existence. 1971 J. J. Shapiro tr. Habermas's Toward Rational Society ii. 13 The adult role anticipated at the university..is therefore unsuited for supporting a privatized orientation bound to career and advancement. 1976 Spare Rib Dec. 4/3 As long as most people live in nuclear family-type set-ups..babysitting will remain a 'privatised', individual act. 1978 A. Brittan (title) The privatised world. Ibid. v. 121 The 'privatised self' is explicable in terms of developments in the class structures of western societies.

privilege, *sb.* Add: **2.** (Later N. Amer. examples.)

1968 Globe & Mail (Toronto) 13 Feb. 32/7 (Advt.), Kennedy Rd.—Steeles, home privileges, parking, 1–2 girls. Ibid., $20. wk. Gentlemen 18–25. Good meals, lunches, privileges. Nova Scotian family. 1972 New Yorker 22 July 48/1 Hillside homesites..with ocean beach privileges. 1976 J. Lee Ninth Man 57 Thirty dollars a week... But you get icebox privileges. All my tenants get icebox privileges.

8. *privilege leave* (earlier and later examples.)

1883 Kipling Let. 14 Aug. in C. E. Carrington Rudyard Kipling (1955) iv. 53 Privilege leave..gives you the pleasant duty of enjoying yourself in a cool climate for thirty days and being paid £20 for that duty. 1980 J. Ditton Copley's Hunch ii. ii. 138 Anybody who escapes from enemy hands is entitled to leave—over and above the ordinary ration of privilege leave.

privilege, *v.* Add: **1. c.** *trans.* *R. C. Ch.* To make (an altar) privileged.

1844 Orthodox Jrnl. 6 Jan. 3/2 The high altar was privileged by Gregory XIII.

d. In *pa. pple.* Entitled *to* (a special right).

1856 Mrs. B. G. Ferris Mormons at Home xii. 199 A few who call themselves physicians..are privileged to a seat in this important assemblage.

privileged, *ppl. a.* Add: **d.** *Eccl.* Applied to days in the Church's calendar which are placed in the highest category of importance, or one of the higher categories (e.g. as regards the precedence they take when two feasts or observances coincide).

1877 J. D. Chambers Divine Worship in Eng. v. 85 Sundays..are distinguished..into Principal Privileged, Greater Privileged, Minor Privileged, Inferior Semi-privileged or ordinary Sundays. Ibid. 87 The Privileged Sundays, according to the present Anglican Rite, appear, beside the Principal Double Festivals and their Octaves, to be the First Sunday in Advent, Passion Sunday and Palm Sunday, and Sunday within the Octave of the Ascension. The Privileged Ferials: Ash Wednesday, the Four Days before Easter, the Vigils, Fasts and days of Abstinence above enumerated. 1953 Anglican Services v. 56 Ordinary (or lesser) Sundays..give way..to feasts or the privileged Octave days of feasts of Our Lord. Ibid. 57 Ferias are of three classes, Privileged, greater, and ordinary. Ibid. 61 Octaves are of three kinds, privileged, common, and simple.

Privy-Councilship. [-ship.] = Privy-Counsellorship.

1910 Blackw. Mag. Sept. 422/1 Even Privy Councilship does not turn nonsense into sound argument.

‖**Prix de Rome.** (prī də rɔm). [Fr., = prize of Rome]. In full *Grand Prix de Rome*. One of a group of prizes awarded annually by the French Government, established by Louis XIV in 1666 for competition by young painters and sculptors, extended in 1720 to include architects, and in 1803 to include musicians and engravers. The winner of the first prize in each category is entitled to a period of study in Rome; also, the winner of a *prix de Rome*.

1879 Grove Dict. Mus. I. 233/2 In 1828 he [sc. Berlioz] took the second, and at last, in 1830,..the first prize—the 'Prix de Rome'. 1884 R. & E. Holmes tr. Berlioz' Autobiogr. I. xxii. 113 The intention of the Government, in establishing the Prix de Rome, was, first, to bring forward year by year the most promising among the young French composers; secondly, to enable them, by means of a pension, to devote themselves entirely for five years to the study of music. 1889 F. F. Buffen Musical Celebrities 53 At the age of nineteen Gounod succeeded in gaining the second 'Prix de Rome' for his cantata, 'Marie Stuart et Rizzio', and in 1839 took the 'Grand Prix' with his composition of 'Fernand'. 1905 J. Webster Wheat Princess i. 10 Allow me to present Monsieur Benoit, the last *Prix de Rome*—he is the man to paint your ghost. 1906 W. J. Locke Beloved Vagabond (1907) xvi. 204 'You a *Prix de Rome*, Master?' 'Yes, my son, in Architecture.' He was clothed in a new and sudden radiance. To a Paris art student a *Prix de Rome* is what a Field Marshal is to a private soldier. 1957 Observer 29 Dec. 11/5 The Master had trained as an architect, won the Prix de Rome. 1968 'S. Jay' Sleepers can Kill xv. 153 Oh, well, there goes the Prix de Rome. He put the charcoal down. 1972 Guardian 22 July 9/6 John Skeaping..was living in Rome, having won the Prix de Rome for sculpture.

‖**prix fixe.** (prī fīks). [Fr., lit. = fixed price.] A meal served in a hotel or restaurant at a fixed price, a table d'hôte meal (cf. *À LA CARTE); the menu offered at such a meal. Also *attrib.*

1883 R. L. Stevenson in Magazine of Art VI. 274/2 You taste the food of all nations in the various restaurants; passing from a French *prix-fixe*, where every one is French, to a roaring German ordinary where every one is German. 1930 A. Bennett Imperial Palace xiv. 82 Prevent customers who prefer the *prix-fixe* from choosing more expensive things than the price will stand. 1933 'G. Orwell' Down & Out vii. 59 A *prix fixe* restaurant where we went for dinner. 1966 P. V. Price France: Food & Wine Guide 33 In the majority of restaurants there will be at least one *prix fixe*, a fixed price menu, offering several courses. 1973 Guardian 28 May 3/4 In Chantilly..there is *crème chantilly* with everything. But ..you wouldn't believe how the *prix fixe* can expand. 1975 Times 20 Dec. 10/7 The *prix-fixe* menus..averaged now 25 francs.

prize, *sb.*[1] Add: **3. c.** *glittering prizes.*

1875 F. Arnold Our Bishops & Deans I. v. 286 There are certain glittering prizes which are the great attractions to these. 1923 Ld. Birkenhead in Times 8 Nov. 7/4 The world continues to offer glittering prizes to those who have stout hearts and sharp swords. 1976 F. Raphael (title) Glittering prizes. 1977 A. Clarke Let. from Dead ix. 104 Just keep your trap shut..and remember the glittering prizes. 1978 Broadcast 3 Apr. 9/3 Party political broadcasts are not the glittering prizes that once they seemed to be... The public are bored by them.

4. a. (*a*) *prize essay, poem* (later examples); now also *fig.* (as adj.) describing undesirable qualities: outstanding, unrivalled, complete, utter. (*b*) *prize pig* (later example).

1831 Edin. Rev. LIII. 556 The world..is pretty well agreed in thinking that the shorter a prize-poem is, the better. 1856 C. M. Yonge Daisy Chain ii. xviii. 548 He..had written the best prize poem ever heard at Oxford. 1857 Geo. Eliot Let. 22 May (1954) II. 329 Meditations about a new book..when the Prize Essay has reached a second edition. 1933 Blunden Charles Lamb 21 George Richards, whose Oxford prize-poem delighted Byron. 1952 E. O'Neill Moon for Misbegotten i. 63 Hogan. All prize pigs, too! I was offered two hundred dollars apiece for them. 1956 K. Tillotson Matthew Arnold & Carlyle 139 Arnold opens non-committally, using a technique of evasion common in prize-poems. 1976 Southern Even. Echo (Southampton) 13 Nov. (Suppl.) 5/3 The final episode finds Katy..accused of writing to a young man regarded as a prize flirt. 1978 M. Tripp Wife-Smuggler v. 58 I've been made a fool, a prize bloody fool. 1980 E. G. Wilson John Clarkson iv. 46 A Cambridge prize essay was bound to have a good circulation.

b. *prize-giving* (examples); **prize-book** (earlier and later examples); **prize-roll**, a roll or list of prize-winners.

1839 C. Sinclair Holiday House xii. 274 Being the best scholar there [sc. at school], he might..receive a whole library of prize-books. *c* 1909 D. H. Lawrence Collier's Friday Night (1934) i. 4 Then on the next shelf prize-books in calf and gold. 1905 E. M. Forster Where Angels fear to Tread v. 124 Fortunately the school prize-giving was at hand. 1955 E. Blishen Roaring Boys ii. 100 Prize-giving..didn't flow naturally out of what had gone before, as it does in a grammar school. 1973 R. Parkes Guardians vii. 124 There they all were, droning away..as though at some Kafkaesque prize-giving. 1912 Chambers's Jrnl. May 329/1 A medal can be verified occasionally if the prize-roll or some other collateral document is extant.

prize, *sb.*[3] Add: **3.** *prize ship* (earlier example); **prize-master** (earlier and later examples).

1760 in Essex Inst. Hist. Coll. (1911) XLVII. 125 He put a Prize Master (as he called him) and three more of his Hands on board the Sloop. 1916 in Outlook (N.Y.) 9 Aug. 823/2 Prizes cannot be brought into the waters of the United States for the purpose of laying up by a prize master. 1931 Times Lit. Suppl. 16 July 555/3 For his conduct as a prizemaster in the captured Généreux..he was advanced to commander. 1937 C. S. Forester Happy Return i. vii. 80 Gerard, whom he had left on board as prizemaster, had served in a Liverpool slaver. 1710 Boston News-Let. 26 June 2/2 On the said day arrived Her Majesties Ship the Feversham..with Col. Hunter, our Governour, and with him a Prize ship of 300 Tons.

prize, *sb.*[4] Add: **3. prize-beam,** a beam used in packing tobacco.

1800 W. Tatham Hist. & Pract. Ess. Tobacco 52 As all tobacco must be in due case when it is put into the hogshead, so must the prize-beam retain its depressed position until two distinct ends are attained, to wit, that of giving a compact consistency to the cake [etc.].

prize, *v.*[3] Add: **2.** (Earlier and later examples.)

1724 H. Jones Present State Virginia 40 [They] by Degrees *prize* or press it with proper Engines into great Hogsheads. 1902 U.S. Dept. Agric. Farmers' Bull. No. 60. 17 The leaves..are tied into hands and bulked down for a short time, after which they are 'prized' into hogsheads.

b. To pack (persons) into a narrow space.

1799 W. Beckford Let. 16 Aug. in J. W. Oliver Life W. Beckford (1932) x. 269 Assure Lady Heard that she shall not be worn to death with seeing Sights,..nor prysed into rumbling Carriages.

prizeman. Add: Hence **pri·zewoman.**

1940 Horizon Jan. 61 Miss Pitter, a Hawthornden prizewoman.

prize-money. Add: **a.** (Earlier example.)

1749 New Hampsh. Probate Rec. (1916) III. 733, I give to Doctor Robert Ratsey all my Waidges, Prize money, [etc.].

b. [f. Prize *sb.*[1]] Money awarded as a prize or as prizes.

1934 in Webster. 1961 N.Y. Times 24 Jan. 23 He won the Masters, the United States Open and a record $80,738 in prize money. 1973 M. Amis Rachel Papers 194, I was entering a national under-21 short-story competition, sponsored by one of the colour magazines. With the prize-money we might just have a few days in Paris ourselves.

prizer[2]. Add: **b.** A prize-winner.

1846 E. Copleston Let. 9 Dec. in W. J. Copleston Mem. E. Copleston (1851) 188 My delight was not a little heightened, by seeing my horned countrymen of North Devon among the 'prizers'.

prize-ring. (Earlier example.)

1822 Sunday Times 20 Oct. 4/2 Bill Cropley, one of the heroes of the prize-ring, but now a hard-working coal-heaver.

prizewoman: see *PRIZEMAN.

pro, *Latin prep.* Add: **A.** *prep.* **2. pro bono publico.** (Later examples.) Now freq. used as a signature to an open letter (as to a newspaper).

1914 'I. Hay' Lighter Side School Life vii. 194 Fiery old gentlemen write..to say that in their young days boys were boys and not molly-coddles. Old friends like *Materfamilias*, *Pro Bono Publico*,..rush into the fray... There is quite a riot of pseudonyms. 1922 Joyce Ulysses 306 Someone..ought to write a letter *pro bono publico* to the papers about the muzzling order for a dog the like of that. 1973 G. Beare Snake on Grave viii. 40 He would..write a letter to *The Times* which he would sign 'Pro Bono Publico'. 1977 New Yorker 12 Sept. 133/1 A politician who speaks for an important industry is considered very much pro bono publico.

4. pro forma. Also with hyphen and as one word. (Further simple and *attrib.* examples.) *pro-forma invoice*, an invoice sent to a purchaser in advance of the ordered goods, so that formalities may be completed (see also quot. 1965); also *absol.* as *sb.*, an official form for completion; a pro-forma invoice.

1827 W. Bolling in Virginia Mag. Hist. & Biogr. (1935) XLIII. 240 Then called proforma at Mr. Robertson's to see my sister. 1858 P. L. Simmonds Dict. Trade Products 303/2 *Pro-forma-account*, a model or sketch account; a pattern bill of particulars. 1895 Funk's Stand. Dict., *Pro forma*,.. as a matter of form; as, a *pro forma* invoice. 1928 Blunden Undertones of War ii. 19 He rejoiced in inventing new Army Forms, which he called 'pro forma's'... Some of them were such that one's best information could not find a heading in them. 1930 M. Clark Home Trade 100 An order may be received from an unknown person or firm... In such cases a pro forma invoice may be dispatched. 1945 Ann. Trop. Med. & Parasitol. XXXIX. 226 A senior member of the nursing staff..checked that the patient took the tablet and recorded each dose given and taken on a *pro-forma*. 1959 Punch 27 May 705/1, I do not know in precisely what form the Department of Meteorology of the Imperial College of Science is asking for information about hail, but at the very least there should be a 'proforma' with columns headed. 1965 J. L. Hanson Dict. Econ. 329/2 *Pro-forma invoice*, a commercial document with three main uses: (i) A polite request for payment when a supplier is unwilling to allow his customer credit; (ii) With goods sent on approval, becoming an ordinary invoice if the goods are retained; (iii) When goods are sent to an agent to be sold; (iv) In foreign trade when goods are exported on consignment, informing the importer of the expected prices of the goods. 1977 Wandsworth Borough News 16 Sept. 2/2 A pro-forma for all organisations..should contain a sentence encouraging applicants to think of any aspect of their work which might not be covered by the form. 1978 Jrnl. R. Soc. Med. LXXI. 413 Details of the illness were recorded on a proforma.

10. pro tempore. (Further examples of the abbreviated form *pro tem.*)

1835 Dickens Let. ?30 Oct. (1965) I. 85 Through the stupidity of Frisby who was in attendance pro: tem: Frank Ross 'dropped in' to my writing room. 1886 [see *GADGET]. 1913 Sat. Even. Post 4 Oct. 47 It was proper that Sergeant Bagby, in his capacity as host *pro tem.* should do the..explaining. 1955 Times 12 May 11/6 One feels that this is only a capital *pro tem.*, making do until the kingdom reaches some new turning-point in its fortunes. 1974 'E. Lathen' Sweet & Low xv. 146 Would you be willing to take charge of our cocoa trading—on a pro-tem basis?

B. *sb.* **1. a.** (Later examples.)

1835 E. FitzGerald Let. July in FitzGerald to his Friends (1979) 18 But then I get a settled home, a good companion, and the other usual pro's that desperate people talk of. 1969 V. E. Frankl in Koestler & Smythies Beyond Reductionism 419 All the protesters are actually anti-testers, they have no 'pro', no positive alternative to offer, but they are fighting against, rather than struggling for something.

C. *prep.* For, in favour of. **D.** *adj.* or quasi-*adj.* **a.** Favourable, positive, supportive; favourably disposed.

1837 H. Martineau Diary in Autobiogr. (1877) II. iv. 109 In the morning I am *pro*, and at night..*con* the scheme. 1961 Dallas Morning News 17 Feb. i. 5 We're getting more 'pro' letters than 'con' on horse race betting. 1966 'W. Cooper' Mem. New Man i. iv. 49 'In touch' was a phrase everybody used... They used it in a pro sense; being in touch was most desirable. 1974 R. Harris Double Snare v. 32 It's nice of you to be so pro the idea—*I* don't feel pro or against.

b. *pro-attitude* (Philos.), an attitude such as approval, pleasure, satisfaction, etc., which is the normal reaction to all things considered ethically good.

1935 C. A. Campbell in Mind XLIV. 298 All usages of the term 'good' signify at least this common feature in that to which goodness is attributed, viz., that it is the object of what may perhaps least misleadingly be called a *pro*-attitude. 1939 W. D. Ross Foundations of Ethics xi. 284 This attractive character, or..the fact that we have a pro-attitude towards them, seems to be all that is common to these three kinds of thing that are habitually called good. 1947 A. C. Ewing Def. of Good ii. 68 A slightly different view would be that to call anything 'good' is to say that it is the object of some 'pro attitude' on the part of most people... I use 'pro attitude' to cover desiring, liking, seeking, choosing, approving, admiring, etc. 1949 Mind LVIII. 90 The *ground* of a pro-attitude lies..in the concrete factual characteristics of what we pronounce good. 1964 A. Edel in I. L. Horowitz New Sociol. 224 We may omit here the internal operation of general values, in the sense of obligations and pro-attitudes not peculiar to social science—regard for truth, objectivity and impartiality, [etc.]. 1967 G. R. Grice Grounds of Moral Judgement i. 9 Wanting, for Nowell-Smith is one of many pro-attitudes, and..it is fair to say that his use of the term 'pro-attitude' conceals distinctions of importance.

pro, pro., *abbrev.* Add: **2. a.** In the sense of Professional *sb.* 2.

1866 Sporting Life 17 Oct. 4/4 County matches..are also the true source of our supply of professionals of ability, for you rarely hear of a good 'pro' until he has played for his county. 1885, etc. [in Dict.]. 1902 C. J. C. Hyne Mr. Horrocks, Purser 124 'I tell you the man's not a theatrical.'.. 'Never knew any pro. yet bring either honour or profit to any boat,' said the Purser. 1932 John o' London's 25 June 426/1, I spent all my holidays practising in tournaments and having coaching from a pro. 1951 'J. Tey' Daughter of Time xvii. 209 One wouldn't expect an amateur to walk into the Yard and solve a case that had defeated the pro's. 1960 [see *GOV]. 1965 New Statesman 7 May 712/1 Randall is essentially a pro in the tradition of technician editors, able, almost apolitical. 1975 J. Symons Three Pipe Problem xvi. 158 They're not pros, how long do you think they'll stand up under questioning?

b. *attrib.* or as *adj.* in the sense of Professional *a.* 4, esp. in sporting uses.

1932 A. J. Morrison New Way to Better Golf ii. 25 The pupil..did not recognize me as the handy man of the pro shop and a former caddy. 1949 Times Digest (Richmond, Va.) 26 Nov. 10/4 Riggs was enthusiastic about the crowds which his pro tennis stars have been drawing. 1961 Boxing News 20 Oct. 10 Next live pro item will be the Maurice Cullen–Guy Gracia bout from Newcastle on November 13. 1970 Washington Post 30 Sept. D1/1 Baseball needs such finishes in order to provide some counter-interest to college and pro football, already moving into high gear. 1975 C. James Fate of Felicity Fark ii. 20 These *Krauts* are all pro athletes in disguise. 1978 S. Brill Teamsters vi. 236 Three former pro football players were partners. 1978 Rugby World Apr. 51/3 New Zealander Ken Bousfield, a Sydney player, rejected a league offer of over £17,000 for three years at the end of last season. But this year, he has turned pro with Penrith.

c. *pro('s) shop*, a (work)shop run by the resident professional at a golf club.

[1905 H. Vardon Compl. Golfer iv. 39 The proper place for him [sc. the beginner] to go is the professional's shop which is attached to the club of which he has become a member. Nearly all clubs have their own professionals, who are makers and sellers of clubs.] 1932 [see sense 2 b above]. 1937 H. Longhurst Golf 3 An extremely high standard of business morality obtains among professional golfers. A novice may enter a pro's shop [to buy clubs], a chicken ready for the plucking, and yet come out with all his feathers on. 1953 J. Turnesa Low Score Golf ii. 13 Let's leave the pro shop and go over to the lesson tee. 1964 D. Langdon How to play Golf & stay Happy iii. 27 Palmer..rushed back to the pro's shop after a disastrous round of 76, slammed his driver into the vice and filed away the club face. 1976 T. Gifford Cavanaugh Quest ii. 31 The members' golf committee had allowed Billy to live in the room over the pro shop.

3. Abbrev. of (*professional*) *prostitute*. Cf. *PROFESSIONAL *sb.* 2 b.

1937 in Partridge Dict. Slang. 1941 B. Schulberg What makes Sammy Run? ix. 247 He treats all women like pros. 1950 [see *BAG *sb.* 16*]. 1968 H. C. Rae Few Small Bones ii. i. 79 She's a semi-pro actually... She works in a garage..during the day, but at night she ..entertains. 1976 'E. McBain' Guns iv. 95 Benny already had himself two girls..experienced pros who were bringing in enough cash every week to keep him living pretty good.

pro-, *prefix*[1]. Add: **II. 4. b.** *pro-element, -infinitive, -name, -sentence, -syllable, -verb* (further examples), *-vicariate, -word*; **pro-form** *Linguistics*, a pronoun or other lexical unit substituted for a longer expression. **c.** *pro-infinitival, -syllabic* adjs.

1975 N. Chomsky Logical Struct. Linguistic Theory x. 560 'So' is introduced as a pro-element standing for the verb phrase in such sentences as 'John saw him and so did I'. 1964 Katz & Postal Integrated Theory Ling. Descr. iv. 83 We stipulate in the general theory of linguistic descriptions that the dictionary entry of every pro-form (i.e., every form dominated by the constituent Pro) must contain the semantic marker (Selector). 1969 Canad. Jrnl. Linguistics XIV. 49 In this structure the head noun is a pro-form which is of the *du* class. 1976 Analysis XXXVI. 80 Proforms and modified proforms, upon given occasions of their use, get their semantic content from their antecedents. 1905 O. Jespersen Growth & Struct. Eng. Lang. viii. 208 Another recent innovation is the use of *to* as what might be called a pro-infinitive instead of the clumsy *to do so*: 'Will you play?' 'Yes, I intend to.' 'I am going to.' 1940 —— S.P.E. Tract liv. 153 This leads to the possibility of using an isolated *to* as a 'pro-infinitive': Will he sing? Yes, he wants to. 1934 J. J. Hogan Outl. Eng. Philol. iii. xiv. 136 The modern 'pro-infinitival' *to*, as in *I want to*. 1964 Eng. Stud. XLV. 88 *To* here stands..for *to tell me*, or *to do so*, for which reason Jespersen calls it a 'pro-infinitive'. 1765 J. Elphinston Princ. Eng. Lang. Digested II. viii. 184 Instead of a name or noun repeated, a pro-name or pronoun. 1972 Language XLVIII. 461 To answer the question *Did he do it?*, one requires an element that will affirm or deny. It may appear alone, and in that case will carry the sentence intonation, accent and all. Such a single answer-word might be *Unquestionably*. It may also appear at the end of a pro-sentence, e.g. *He did it unquestionably*. 1976 Analysis XXXVI. 83 The antecedent of the modified prosentence 'it is false', is the quoted sentence together with the quotes. 1948 J. R. Firth in E. P. Hamp et al. Readings in Linguistics II (1966) 185 In certain of its prosodic functions the neutral vowel might be described temporarily as a pro-syllable. However obscure or neutral or unstressed, it is essential in *a bitter for me* to distinguish it from *a bit for me*. In contemporary Southern English many 'sounds' may be pro-syllabic. 1956 Archivum Linguisticum VIII. 123 Structurally the difference [between ij and ɐr] is that of a vocalic phoneme followed by a junctional prosody and a consonantal phoneme preceded by a prosyllabic prosody. 1924 O. Jespersen Philos. Gram. vi. 83 We should get a class of substitute words which might be divided into pro-nouns, pro-adjectives, pro-adverbs, pro-infinitives, pro-verbs..but it could hardly be called a real grammatical class. 1976 J. S. Gruber Lexical Struct. Syntax & Semantics ii. i. 267 For the other occurrences of pro-verbs, the nature of the semantic category common to both the pro-verbs and the verbs embedded in them is more obscure. 1881 Dublin Rev. July 173 The districts of Lake Tanganyika, and the Victoria Nyanza have already been created Pro-Vicariates Apostolic. 1965 Language XLI. 393 We can define a set of proword substitutions which are similar to various types of zeroing.

5. a. *pro-abortion, -abortionist, -Allied, -Ally, -American* (further example), *-Arab, -Axis, -Boche, -Boer* (earlier example), *-British* (examples), *-business, -Communist, -Fascist, -German, -Israeli, -Nazi, -Soviet, -West, -Western, -Zionist*; **pro-kno·ck** *a.* and *sb.*, (a substance) tending to cause knocking when present in the fuel burnt in an internal-combustion engine; **pro-li·fe** *a.*, in favour of the maintenance of life; *spec.* against inducing abortion; hence **pro-li·fer**, someone with these views. **b.** Also in comb. with a sb. (or verb-stem) + -EER; *pro-Britisher*, *-marketeer* [*MARKETEER 3]. **c.** *pro-Arabism*, *-Germanism*, *-Sovietism*.

1976 Pro-abortionist [see *pro-lifer* below]. **1976** *National Observer* (U.S.) 31 Jan. 5/3 If Carter can appeal successfully to both proabortion and right-to-life Democrats, he might as well hand him the nomination now. **1977** *Lancet* 2 July 48/2 Pro-abortionists are left, therefore, with two familiar Parliamentary tactics of filibustering to prevent progress. **1919** W. S. CHURCHILL in M. Gilbert *Winston S. Churchill* (1977) IV. Compan. i. 536 None of the pro-Allied Russian Governments would meet them. **1915** H. MUENSTERBERG in *Fatherland* 22 Dec. 341/1 The psychological equation of his personality makes him a pro-German in all that is best in him, and only his temper and his perpetual desire to be with the masses made him a pro-Ally. **1916** *Lit. Digest* (N.Y.) 1 Jan. 3/2 He has been trapt into the nets of those who wove the pro-Ally newspaper opinion in this country. **1916** MRS. BELLOC LOWNDES *Let.* 29 May (1971) 72 In America it has been rejected by one set of publishers as pro-German—by another as pro-Ally. **1971** D. E. WESTLAKE *I gave at the Office* (1972) 159 They were revolutionaries who were pro-American, which is very rare in the world today. **1920** G. BELL *Let.* 12 Sept. (1927) II. xviii. 499 It is only quite recently that I have realized how prominent a place I have occupied in the public mind here as the pro-Arab member of the administration. **1973** 'D. RUTHERFORD' *Kick Start* ix. 186 The Russians..are openly pro-Arab. **1959** *Daily Tel.* 2 Nov. 10/2 Traditional pro-Arabism influenced us. **1938** *New Statesman* 25 June 1054/1 After he [*sc.* M. Imrédy] had made a speech to this effect, M. de Kánya hastened to deliver his pro-Axis speech of June 1st. **1942** *Times Rev. Year 1941* 3 Jan. p. i/2 One result of their [*sc.* the Germans'] intrigues was a *coup d'état* in Iraq, where on April 3 Raschid Ali and a group of military malcontents expelled the Regent and set up a pro-Axis government. **1915** *National Rev.* Apr. 169 A pro-Boche Government would have been bundled out 'neck and crop' last August. **1923** KIPLING *Irish Guards in Gt. War.* II. 160 Some pro-Boche agent in the far-off lands where it was purchased. **1896** *Daily News* 22 Apr. 5/1 If it were indeed a necessity of the situation to be pro-Boer or pro-British..then as Britons we should be for the British, we admit. **1927** H. DOBBS in *Lett. Gertrude Bell* II. 543 The so-called pro-British sections of the populations. **1980** P. VAN GREENAWAY *Dissident* vii. 148 A man, if he is anti-Soviet, must therefore be pro-British..and *vice versa*. **1927** *Leader* 31 Dec. 517/1 That was unexpected talk to what in Ireland is called a Chamber of Commerce—Chambers whose members are mostly pro-Britishers and Shoneens. **1975** HUNT & SHERMAN *Econ.* (ed. 2) xii. 155 A large amount of probusiness propaganda. **1978** *New York* 3 Apr. 34/3 It might also provide Kennedy with a pro-business image for a presidential campaign. **1950** *New Yorker* 6 May 96/2 The pro-Communist *Lettres Françaises* is the best literary weekly in Paris. **1976** 'M. BARAK' *Secret List H. Roehm* iii. 35 The establishment of a pro-communist spy ring. **1937** DUCHESS OF ATHOLL in Koestler *Spanish Testament* 6 A so-called 'Radical Government', reinforced later by members of Señor Robles' pro-Fascist party. **1940** 'G. ORWELL' in *World Rev.* (1950) June 28 The government..are subjectively pro-Fascist. **1914** R. BROOKE *Let.* 3 Sept. (1968) 613 The intellectuals..are mostly pacifists and pro-Germans. **1915** KIPLING *Let.* 12 Aug. in C. E. Carrington *Rudyard Kipling* (1955) xvii. 433 Munthe..tells me of all his grief..to find that Sweden is so—not pro-German but afraid of Russia. **1938** E. PHIPPS *Let.* 9 Jan. in M. Gilbert *Winston S. Churchill* (1976) V. xliv. 894 Van's displacement.. would be represented as a victory for the pro-Germans in England. **1964** *Times Lit. Suppl.* 12 Nov. 1018 He [*sc.* Frank Harris] spent the 1914–18 War in the United States, where he wrote pro-German propaganda. **1914** W. B. YEATS *Tribute to Thomas Davis* (1947) 12, I am not more vehemently opposed to the Unionism of Professor Mahaffy than I am to the pro-Germanism of Mr. Pearse. **1940** *Tablet* 4 May 421/1 The pro-Germanism of expedience which was once practised by certain politicians. **1975** J. CROSBY *Affair of Strangers* iv. 33 French policy.. is far more pro-Arab than the French people are. The French people..would be pro-Israeli. **1927** *Jrnl. Inst. Petroleum Technologists* XIII. 301 Amyl nitrate and nitrite..according to Midgley are pro-knock. **1928** *Ibid.* XIV. 188 They might have some indications..as to how the pro-knock worked as against the anti-knock in that particular type of flame propagation. **1953** E. M. GOODYER *Petroleum & Performance in Internal Combustion Engines* viii. 189 Ignition accelerating materials are those which act as pro-knocks in spark-ignition engine fuels, and include organic peroxides, nitrates, nitrites, and various sulphur compounds. **1973** K. OWEN in Hobson & Pohl *Mod. Petroleum Technol.* (ed. 4) xv. 596 Monomethylaniline..acts as an anti-knock agent in its undecomposed form but..if it is decomposed too early in the combustion cycle it can even have a pro-knock effect. **1961** G. SMITH *Business of Loving* xvii. 284 Benny divides people into.. the life-enhancers and the life-diminishers; he is a believer in the first, a leader of the pro-life party. **1978** *Dædalus* Spring 155 Its eleven members include only five scientists, one of whom was known for his prolife position. **1979** *Time* 30 July 6 As the oldest of eleven children (all married), I'd like to point out our combined family numbers more than 100 who vote only for pro-life candidates. **1976** *National Observer* (U.S.) 31 Jan. 5/2 Carter.. had misled proabortionists and prolifers. **1976** *Observer* 24 Oct. 9/1 Anti-abortion forces have been organising to overturn the decision [of the U.S. Supreme Court]... The pro-lifers, as they prefer to call themselves, can no longer be written off as a lunatic fringe. **1979** *Time* 30 July 6 Pro-lifers have children, pro-choicers do not. **1961** *Economist* 2 Dec. 877/1 Mr Heath's speech may legitimately be criticised by pro-marketeers. **1976** H. WILSON *Governance of Britain* 10, I had to be aware of.. the balance between committed pro-marketeers and committed anti-marketeers. **1936** *New Yorker* 29 Feb. 24/3 Ernst Röhm, who in 1928 had written a pro-Nazi autobiography. **1974** G. JENKINS *Bridge of Magpies* vii. 113 Her husband was the boss of the pro-Nazi underground movement. **1950** L. FISCHER in Koestler et al. *God that Failed* 215 This I did not understand in the years when I was pro-Soviet. **1977** *Listener* 21 Apr. 499/3 Mao's pro-Soviet opponents. **1950** L. FISCHER in Koestler et al. *God that Failed* 224 My years of pro-Sovietism have taught me that no one who loves people and peace should favour a dictatorship. **1952** *Sun* (Baltimore) 6 Feb. 1/5 The Soviet claims..treated such 'neutralism' on the part of United Nations members as pro-Sovietism. **1968** 'A. BRIDGE' *Portuguese Escape* iii. 44 One of the real stars is pro-West, and arranged with our man in Hungary to bring him away. **1976** *Billings* (Montana) *Gaz.* 17 June 2-E/1 Kenya, the only pro-West nation on the East African coast. **1934** WEBSTER, Pro-Western. **1969** H. KAHN *On Escalation* i. 24 If [*sc.* the United States] could have invaded Iraq in 1958 to restore a pro-Western government. **1960** P. VAN GREENAWAY *Dissident* v. 123 Is it possible for a man to be pro-Western *without* being anti-Soviet? **1949** KOESTLER *Promise & Fulfilment* i. ii. 13 'Somehow we are like the Arabs'..confessed a sincere and pro-Zionist Englishman. **1971** D. MEIRING *Wall of Glass* viii. 65 The idea that Jew could kill Jew in Palestine for political reasons was the more intolerable the more pro-Zionist you were.

pro-, *prefix*². Add: **I. proaccelerin** (-ǣkse·lĕrin) *Biochem.*, a relatively labile procoagulant present in the blood; **pro-activator** *Biochem.*, a precursor of the activator of a compound; **probasi·dium** *Bot.* [ad. F. *probaside* (P. Van Tieghem 1893, in *Jrnl. de Bot.* VII. 80)], in some fungi, a part of a basidium, or an early stage in its development, in which nuclear fusion takes place; **probio·tic** *a.* = *PRE-BIOLOGICAL, *PREBIOTIC *adjs.*; **proca·rcinogen**, a substance that is not directly carcinogenic itself but is converted in the body into one that is; so **procar:cinoge·nic** *a.*; **procoa·gulant** *sb.* and *a. Biochem.*, (of or pertaining to) any substance that promotes the conversion of the inactive prothrombin to the clotting enzyme thrombin; **pro-conve·rtin** *Biochem.* [CONVERT *v.*], a relatively stable procoagulant present in the blood; **proery·throblast** *Med.* [ad. It. *proeritroblasti* (A. Ferrata *Morfologia del Sangue* (1912) v. 232)], the earliest recognizable precursor of the red-cell series, characterized by a large nucleus with nucleoli and by basophilic cytoplasm; **proestrus**, var. *pro-œstrum* below in Dict. and Suppl.; **pro-ethnic** *a.*: see as *main entry; **profibrinoly·sin** *Biochem.* = *PLASMINOGEN; †**proga·mete** *Biol.*, a structure able to give rise to one or more gametes; **prohe·terocyst** *Biol.*, an incipient heterocyst; **proho·rmone** *Physiol.*, a natural precursor of a hormone; **proi·nsulin** (pro,i·n-) *Biochem.*, the natural precursor of insulin; **promeristem** (later examples); **pro-mitocho·ndrion** *Cytology*, an inactive form of mitochondrion; **promy·elocyte** *Med.*, a cell intermediate in development between a myeloblast and a mature myelocyte; so **promy·elocytic** *a.*; **pro-œstrus**, also **proestrus**, **procestrus**; (further examples); **propla·stid** *Cytology*, a small unspecialized plastid, able to differentiate into a plastid of any type characteristic of the species; **prosecre·tin** *Physiol.*, a supposed precursor of secretin; **prothe·ca** [THECA], in Foraminifera, the primary wall; **protri·chocyst** *Zool.* [ad. G. *protrichocyste* (B. M. Klein 1928, in *Arch. f. Protistenkunde* LXII. 210)], an undeveloped trichocyst.

1951 P. A. OWREN in *Proc. 3rd Internat. Congr. Internat. Soc. Hematol.* 379, I wish to propose the terms proaccelerin and accelerin instead of Factor V and Factor VI, because ..these factors constitute the system which is responsible for the acceleration of prothrombin conversion. **1966** *McGraw-Hill Encycl. Sci. & Technol.* II. 266/1 The interactions of tissue thromboplastin, calcium ions, and several proteins of plasma, including proaccelerin..and proconvertin, result in the conversion of prothrombin into a proteolytic enzyme, thrombin. **1956** T. ASTRUP in *Blood* XI. 783 In blood, human milk, tears, and in other body fluids enzymatically acting activators of plasminogen are also found, or can be produced. The production of activating agents in these cases is caused by the transformation of a precursor (a proactivator). **1973** *Jrnl. Clin. Invest.* LII. 2591/2 Conversion of highly purified plasminogen proactivator to plasminogen activator was shown to result in the generation of chemotactic activity. **1928** C. W. DODGE tr. *Gäumann's Compar. Morphol. Fungi* xxv. 415 This enlarged hyphal cell which..forms the first stage of the basidium..is called [the] probasidium. **1979** I. K. ROSS *Biol. Fungi* vi. 156 In the spring, each cell of the teliospore [in *Puccinia graminis*] functions as a probasidium and produces a thin-walled metabasidium. **1954** *New Biol.* XVI. 44 We have as yet no basis for confidence about the probiotic state. **1971** J. Z. YOUNG *Introd. Study Man* xxvi. 372 A probiotic soup of amino-acids, ribose, four purine and pyrimidine bases, and a source of high-energy phosphate. **1963** *Clin. Pharmacol. & Therapeutics* IV. 111/1 A compound requiring metabolic activation is one which when administered to animals is very likely not carcinogenic by itself ('procarcinogen') but requires transformation in the host to become a 'proximate' carcinogen—a sort of lethal synthesis. **1975** *Pharmacol. Basis of Cancer Chemotherapy* 129 (*heading*) Procarcinogens and their bioactivation. **1944** *Jrnl. Exper. Med.* LXXX. 121 The papers dealing with the 'procarcinogens' show clearly that the substances thus designated do not cause neoplastic changes but act either by enabling the real carcinogens to reach susceptible cells or by promoting the formation of growths. They are in other words procarcinogenic. **1976** *New Scientist* 9 Dec. 586/2 Cigarette smoke..contains procarcinogenic polynuclear aromatic hydrocarbons which are broken down by enzymes in the lungs. **1958** LANDABURU & SEEGERS in *Amer. Jrnl. Physiol.* CXCIII. 178/1 Other factors support the production and early conversion of thrombin, and these we call procoagulants. **1960** *Nature* 16 Mar. 930/2 The control of prothrombin activation is by way of anticoagulants and procoagulants functioning in dynamic equilibrium. **1964** W. H. SEEGERS *Prothrombin* ii. 200 There is a procoagulant effect noticeable in whole blood or plasma following the alimentary intake of certain kinds of fats. **1971** R. S. SHERIDAN *Human Physiol.* xiv. 323/2 (*caption*) Intermediates of prothrombin activation may result in the formation of a number of other procoagulants as well as anticoagulants. **1976** *Nature* 12 Apr. 711/1 It has been shown that human fibroblasts contain a potent procoagulant activity called 'tissue factor' (TF). **1951** P. A. OWREN in *Proc. 3rd Internat. Congr. Internat. Soc. Hematol.* 383 This substance acts as the limiting factor in prothrombin conversion and I have thus chosen to give it the name proconvertin. **1975** *Nature* 17 June 621/2 The coagulation of blood is envisaged as a complex but ordered succession of processes, and at least four of the many factors (prothrombin, proconvertin, Christmas factor and Stuart factor) are known to be dependent on vitamin K. **1927** A. PINEY *Rec. Adv. Hæmatol.* ii. 19 It is obvious that the adherents of the monophyletic school will be of opinion that the red corpuscle is derived from the primitive stem cell (hæmocytoblast). They conclude that all sorts of transitions can be found between the large non-hæmoglobiniferous cells (pro-erythroblasts) and the mature, fully hæmoglobiniferous corpuscle. **1962** *Lancet* 27 Jan. 208/1 A continuous morphological spectrum of cells was evident, indicating many transitional forms between proerythroblasts and primitive blast cells. **1969** RAYNOR *Fairbairn's Atlas Hæmatol. Cytol.* (1970) 7 The proerythroblast is not itself the functional stem cell serving as a self-maintaining progenitor of the normoblast series. **1967** E. C. LOOMIS et al. in *Arch. Biochem.* XII. 1 We suggest the following names for the components: (1) Fibrinolysin ...; 2). Profibrinolysin—the inactive form or precursor of fibrinolysin. This compound is the proenzyme form from serum or plasma activated by streptokinase, organic solvents and other enzyme activators. **1958** *Observer* 14 Dec. 4/3 A precursor, profibrinolysin, is present in the blood and is changed to fibrinolysin by natural agents released when needed. **1968** A. WHITE et al. *Princ. Biochem.* (ed. 4) xxxi. 733 The proteolytic enzyme, plasmin (fibrinolysin), ordinarily exists in plasma as the inactive precursor..plasminogen (profibrinolysin). **1892** *Q. Jrnl. Microsc. Sci.* XXXIII. 6 In my terminology I have used the words 'gametogonium and progamete to express, from slightly different points of view, a cell which divides to form gametes, or (rarely) passes into the state of a gamete. *Ibid.* 34 In this sense we may call the 'parthenogenesis' of Metazoa only one polar body is formed, and the ovum, rather a progamete than an oosphere, segments and develops directly. **1904** *Proc. Amer. Acad. Arts & Sci.* XL. 231 The zygospores are formed between the gills of the host, and the progametes arise at times from branches of the same hypha. **1970** *Nature* 14 Nov. 686/1 A close pattern of heterocysts and presumptive heterocysts ('proheterocysts') is apparent. **1973** *Jrnl. Cell Sci.* XII. 641 In the presence of ammonia, heterocyst development is affected, so that a pattern consisting largely of proheterocysts, rather than mature heterocysts, is formed. **1935** *Amer. Jrnl. Physiol.* CXII. 311 Many of the published opinions concerning prohormones have been more or less in the nature of speculations, rather than from directed experiments planned to give information concerning its existence or properties. **1970** *Proc. Nat. Acad. Sci.* LXVII. 1637 Unlike the islet insulin, the parathyroid may store its hormone as the prohormone, with conversion taking place when the gland is stimulated. **1977** *Lancet* 25 June 1341/2 Vitamin D is a pro-hormone which only becomes active on transformation to its 25-hydroxy derivative, a process that is subject to pronounced but poorly understood constraints. **1916** A. E. SCHÄFER *Endocrine Organs* xvii. 128 Provisionally, it will be convenient to refer to this hypothetical autacoid as 'proinsulin'. It must, however, be stated that it has yet to be determined whether the active substance is present as such in the pancreas or whether it exists there as pro-insulin, which becomes elsewhere converted into the active autacoid. **1967** D. F. STEINER et al. in *Science* 26 Apr. 700/2 The labeling data reported here support our earlier interpretation that component *b* is a precursor in the biosynthesis of insulin. It might be less cumbersome, therefore, to designate this material 'proinsulin'. **1969** *Nature* 15 Nov. 696/1 Proinsulin has little or no biological activity, but is present in the circulation and produces insulin-like effects when injected into normal animals. **1970** *Jrnl. Clin. Investigation* XLIX. 506/2 At present data concerning the biological activity of human proinsulin are not available. **1925** EAMES & MACDANIELS *Introd. Plant Anat.* iii. 41 Promeristems gradually become differentiated. **1953** K. ESAU *Plant Anat.* iv. 78 The initiating cells and their most recent derivatives are often distinguished, under the name of promeristem. **1976** BELL & COOMBE tr. *Strasburger's Textbk. Bot.* (rev. ed.) 89 Primary embryonic tissues are those which are derived ontogenetically directly from the tissue of the embryo, and they are referred to as primordia or promeristems. **1969** CRIDDLE & SCHATZ in *Biochem.* VIII. 323/2 Since the term 'proplastid' is well established.., the mitochondria-like particles from anaerobic yeast cells were correspondingly termed 'promitochondria'. **1974** *Nature* 15 Mar. 258/2 Such mitochondria as yeast promitochondria do not contain all the carriers of the respiratory

chain and possess an enhanced resistance to anaerobiosis. **1925** STRONG & ELWYN *Bailey's Textbk. Histol.* (ed. 7) vi. 142 The myelocytes are the most abundant developmental forms of marrow... The most immature are known as promyelocytes, the fully matured as metamyelocytes. **1977** *Lancet* 15 Oct. 806/2 Cytoplasmic vacuolation, similar to that in erythroblasts, occurs in promyelocytes in the bone-marrow of alcoholics. **1957** L. K. HILLESTAD in *Acta Medica Scand.* CLIX. 189 This paper deals with three cases of a special type of acute myelogenous leukemia... The white blood cell picture in the peripheral blood resembles that of the more chronic forms of leukemia, as it is dominated by promyelocytes and myelocytes with very few myeloblasts. A logical name for this type of leukemia is *acute promyelocytic leukemia*. **1973** *Brit. Jrnl. Hæmatol.* XXV. 255 Acute promyelocytic leukemia..is now recognized as a distinct clinical and pathological entity, classically characterised by..replacement of bone marrow by abnormal promyeloblasts and promyelocytes. **1923** *Amer. Jrnl. Anat.* XXXII. 306 Through its action on procestrous and ovulation the corpus luteum indirectly inhibits those growth processes which are initiated by the maturing follicles. **1923** Proestrum [see *METŒSTRUS]. **1927** *Nature* 4 Dec. 790/1 It can no longer be affirmed that the procestrus of the lower mammal corresponds simply to the menstrual flow of the human female. **1966** **1973** Proestrus, procestrus [see *METŒSTRUS]. **1976** *Sci. Amer.* July 32/2 In the normal estrous cycle of the rat the pituitary secretes large amounts of luteinizing hormone...in the afternoon of proestrus, approximately 30 hours after the initial increase in estradiol secretion by the ovaries. **1934** I. F. RANDOLPH in *Bot. Gaz.* LXXIII. 345 Since these bodies have been found to occur as plastid primordia in the cytoplasm of meristematic cells in maize, and since they are here found to be definitely concerned with the formation of chloroplasts, the term 'proplastid' will be used for such bodies. **1934** L. W. SHARP *Introd. Cytol.* (ed. 3) vi. 89 The differentiated plastids seen in mature tissues may be traced back to plastid primordia, or proplastids in the young cells of the meristem or embryo. **1967** KIRK & TILNEY-BASSETT *Plastids* xiv. 497 It may be generally true that the proplastid on dividing, the chloroplast's precursors plant cell has to start dividing, the chloroplasts revert to proplastids to facilitate the plastid division that leads to the plastid types in the cell are to be maintained. **1902** BAYLISS & STARLING in *Jrnl. Physiol.* XXVIII. 331 The distribution of 'prosecretin', as we have proposed to call the mother-substance, corresponds ..precisely with the region from which acid is introduced into the lumen excites secretion from the pancreas. **1935** *Amer. Jrnl. Physiol.* CXII. 511 In this study we have attempted to obtain concrete evidence concerning the existence of prosecretin. **1962** R. A. GREGORY *Secretory Mech. Gastro-Intestinal Tract* xii. 157 Bayliss & Starling originally supposed that it [*sc.* secretin] might exist in the form of an active precursor 'prosecretin' from which secretin was liberated by acid hydrolysis. This view was later abandoned. **1964** M. F. GLAESSNER *Princ. Micropalæont.* vi. 108 The protheca or primary wall consists of a layer of clear transparent calcite (diaphanotheca), and a thin dark outer rind-like film (tectum). **1963** K. A. ALLEN tr. *Pokorny's Princ. Zool. Micropalæont.* I. vi. 236 In some of these forms [of Foraminifera] there is only a single undifferentiated layer, the protheca. **1933** G. N. CALKINS *Biol. Protozoa* (ed. 2) iv. 135 The Hoxocysts at rest..are capsules filled with a densely staining..substance. They appear to be connected with the silver line system and..are here represented by granules when the trichocysts are undeveloped. In such granular form they are sometimes called 'protrichocysts'. **1969** *Jrnl. Cell Sci.* XXVIII. 67 Overwintering the trichocysts consisting of amorphous material are variously referred to as protrichocysts, mucoid trichocysts, mucigenic bodies, or secretory ampules. **1972** M. S. GARDINER *Biol. Invertebrates* xix. 850/2 Electron micrographs reveal that the stripes contain refringent granules, considered protrichocysts, which are..blue and in *Stentor cœruleus*, giving this species its characteristic colour.

pro-abortion(ist): see *PRO-¹ 5 a. **pro-accelerin, -activator**: see *PRO-² I.

proa·ctive, *a. Psychol.* [f. PRO-¹ I + ACTIVE *a.* 2] Of a mental effect from a previous situation which is active in a subsequent activity, esp. in learning theory, as *proactive inhibition*, *interference*, the inhibition of or interference with learning caused by effects that remain active from conditions preceding that learning.

1933 WHITELY & BLANKENOP in *Jrnl. Exper. Psychol.* XVI. 852 Objective results..indicate..a somewhat inhibitive influence of the various sets upon learning, and similar conditions of prior and later learning, they suggest what might be called a *proactive inhibition*. However, perhaps the term *proactive inhibition* might better be reserved to designate the detrimental influence of conditions introduced prior to learning upon a subsequent recall, thus differentiating it from..retroactive inhibition. **1940** *Amer. Jrnl. Psychol.* LIII. 174 By proactive inhibition is meant rather [*sc.* the retardation of the learning of an activity when some other activity has occurred as a prior condition (sometimes called negative transfer)], or (b) the fact of poorer retention of an activity when some other activity has occurred as a prior condition to the original learning of that activity, than when a period of comparative rest preceded the original learning. **1963** *Mind* LII. 360 Then we find that remembering is functionally dependent, on the one hand, on learning that is also its proactive and on that which occurs afterwards (retroactive inhibition). **1951** *Jrnl. Psychol.* Mar. May 93 It seems unlikely that proactive inhibition would have any permanent effect on the learning of the interpolated series.. retroactive and pro-active processes are sometimes selectively dissociated. **1967** HILGARD & ATKINSON *Introd. Psychol.* (ed. 4) xii. 324/2 Proactive inhibition plays a very significant role when 'experienced' subjects are used in a formal experiment. **1971** E. SALTZ *Cognitive Bases of Human Learning* vi. 204 Most studies of interference are some variation of one of the following three experimental paradigms: (a) retroactive interference, (b) proactive inhibition, or (c) interference between.

pro-Allied, -Ally: see *PRO-¹ 5 a.

pro-am, *a.* (*sb.*) [f. PRO *abbrev.* + AM(ATEUR 2.] Of a sport or other activity: practised by or open to both professionals and amateurs (see also quot. 1951). Also *ellipt.* as *sb.*, one who takes part in such activities; a pro-am event.

[**1931** *N.Y. Times* 26 May 34/2 (*heading*) Kinder-Revenor..lead the Pro-Amateurs with 71.] **1949** *Sun* (Baltimore) 12 Mar. 22/1 In the pro-am division, Cross went with Al Jamison, professional from Quantico, Va., for one of the 63's. **1951** GREEN & LAURIE *Show Biz* 571/1 Pro ams, professional amateurs, those pseudotypes who constantly appeared on so-called amateur radio and vaudeville programs. **1968** *N.Y. Times* 11 July 48 They send helicopters to take the pro's and pro-am to another. **1970** *Times* 13 Feb. 13/3 Since turning professional in 1968 Oosterhuis has won only..one pro-am event. **1976** *Billings* (Montana) *Gaz.* 14 June 1-0/7 Pate..practiced Tuesday in light rain and played in a proam Wednesday.

2. prode·lta *a.* and *sb.*, *Geol.*, (the part of a delta) lying underneath and beyond the sloping front of a delta; so **prodeltaic** *a.*; **profi·lmic** *a. Semiotics* [ad. F. *profilmique*: cf. E. Souriau in *Revue Internationale de Filmologie* (1951) II. VII–VIII], happening or situated in front of a camera; **proneu·ral** *a.*, of the first bone in a turtle's carapace, situated in front of the neural bones; also *absol.*

1940 E. S. HILLS *Outl. Structural Geol.* i. 4 The bottom-sets or prodelta clays represent the finest material carried over the floor of the sea or lake in which the delta was built out. **1959** *Amer. Assoc. Petroleum Geologists' Bull.* XLIII. 2495 This subgroup may be..the prodelta and frontal zones. **1961** W. & E. E. HUMPHRIES tr. *Termier's Erosion & Sedimentation* xi. 227 This bed is, perhaps, comparable to that formed on a prodelta. **1969** BENNISON & WRIGHT *Geol. Hist. Brit. Isles* xiv. 319 The high percentage of silt in the clays has led to a comparison with some modern pro-delta sediments. **1975** HOBSON & TIRATSOO *Introd. Petroleum Geol.* ii. 32 The sediments of the delta front, pro-delta and continental shelf are organically fairly rich. **1968** MURCHISON & WESTOLL *Coal* v. 89 The seaward advance of delta-fronts and prodeltaic muds, silts and sands. **1974** *Nature* 8 Feb. 344/2 Interbedded sheets and lenses of moderately well sorted prodeltaic and littoral sands. **1973** P. WILLEMEN in *Screen* Spring/Summer 13 Profilmic events should be divided into signifying reality and into non-signifying reality (eg on one level, a city is a signifying reality, a mountain range is not). *Ibid.*, In the cinema one 'sections' the profilmic reality. **1974** M. TAYLOR tr. *Metz's Film Lang.* iii. 33 That great artist..manages to have beauty, which has been pitilessly rejected from every 'profilmic' occasion. **1952** A. CARR *Handbk. Turtles* i. 36 Along the mid-line twelve of the bones of the carapace are arranged in a row. In front is the proneural bone (usually known as the nuchal). **1967** P. C. H. PRITCHARD *Living Turtles of World* 10 The foremost bone in the turtle shell..is large; it is called the proneural or nuchal bone. Behind the proneural comes a midline row of eleven or fewer bones, called neurals.

pros, prahu. Add: (Further examples.)

The length of these boats is now often in excess of thirty feet [see quot. 1977].

1923 [see *BALLAHOU, BALLAHOO]. **1957** P. WORSLEY *Trumpet shall Sound* vii. 133 The officer..stayed put, only to see the prophet himself arrive in a bedraggled prau. **1971** *Walkabout* (Austral.) Nov. 55/1 On Groote Island.. there are many cave paintings. Some depict Indonesian praus from Macassar. **1973** *Daily Tel.* 2 Oct. 19/6 Crato II, a 31ft proa with a 15ft outrigger and carrying 250 sq. ft of sail was also being repaired. **1977** *Austral. Sailing* Jan. 26 For your reporter and for the new Crossbow II, described as a twin masted proa 73 ft overall, this was the highpoint of the fifth annual week of sailing speed trials.

B. *sb. phr.* (Further example of *sing.*)

1923 GALSWORTHY *Captures* 59 The house rocked with pro and con.

D. Used as *prep. phr.*

1895 W. STEVENS *Let.* 4 Aug. (1967) 6 Nor have I any suggestions pro and con anything in particular.

pro-Arab: see *PRO-¹ 5 a. **pro-Arabism**: *PRO-¹ 5 c. **pro-attitude**: *PRO *Latin prep.* D. 1 b.

proavis (pro,ē̆ⁱ·vis). Pl. **-aves** (-ē̆ⁱ·vīz). [f. PRO-² + L. *avis* bird.] A hypothetical animal forming an evolutionary link between fossil reptiles and fossil birds. So **proa·vian** *a.*, of or pertaining to an animal of this kind; also as *sb.*

1907 F. NOPCSA in *Proc. Zool. Soc.* I. 235 An effort to condense these hypothetical changes into a drawing is given in text-fig. 82, which might..be called a 'Pro-Avis'. **1910** W. P. PYCRAFT *Hist. Birds* ii. 37 It is to the earlier Jurassic formations..that we must look for traces of the pro-avian types. *Ibid.* 38 What these 'pro-aves' were like we can only dimly surmise. **1926** G. HEILMANN *Orig. Birds* iv. 191 The term Proavian or Proavis covers a form intermediate between reptile and bird. *Ibid.* 192 Before starting to reconstruct the proavian skeleton, we must try to gain a clear understanding of its particular peculiarities..that would lead on to the bird. **1962** J. C. WELTY *Life of Birds* xxii. 481/2 Heilmann attempted the reconstruction of a hypothetical, missing-link 'proavian'. **1974** I. C. J. GALBRAITH tr. *Dorst's Life of Birds* I. xvi. 531 An intermediate hypothetical being [between reptiles and birds] has had to be erected and named Proavis. *Ibid.* 532 It is really more appropriate to imagine a series of proaves.

pro-Axis: see *PRO-¹ 5 a.

probabilify (prǫbăbi·lifai), *v.* [f. L. *probābil-is* PROBABLE *a.* + -IFY.] *trans.* To give probability to; to give (some proposition) reasonable grounds for being true. So **probabilifica·tion**, the action or process of rendering probable; hence **probabilifi·able**, **probabilifica·tory** *adjs.*

1936 H. H. PRICE *Truth & Corrigibility* 9 It is probabilification which will chiefly concern us. *Ibid.* 16 Some parts of the evidence will greatly increase the probability of other parts. *Ibid.* 18 A system where the mutual support..is only of the probabilificatory kind. **1949** W. KNEALE *Probability & Induction* i. 11 Some conjunction containing A would not probabilify B to the same degree or even at all. *Ibid.*, Admittedly the notion of probabilification requires further elucidation. **1953** *Mind* LXII. 455 Probabilifiable by sense-perception. **1966** A. FLEW *God & Philos.* vii. 148 A traveller's tale not probabilified by any promising theory. **1973** J. L. MACKIE *Truth, Probability & Paradox* v. 213 Equivalent contraposition does not hold even for probabilification between propositions.

probabilism. Add: **3.** The name given to theories in various fields, freq. contrasted with deterministic or possibilistic theories, which claim that the governing laws are not invariant, but state only probabilities or tendencies.

1952 O. H. K. SPATE in *Geogr. Jrnl.* CXVIII. 419 It does not seem certain that 'probabilism' is our ultimate..is the automatic alternative to a vigorous environmentalism... There may be a middle third, one which might call 'probabilism'. **1955** *Psychol. Rev.* LXII. 209/2 As Mises has pointed out in dismissing probabilism in physics, macrolaws have their origin in different microsituations. **1956** E. BRUNSWIK in K. R. HAMMOND *Psychol. E. Brunswik* (1966) xvii. 509 The statistical mechanics and quantum theory, being of a microscopic character, has little to do with the probabilism of functional psychology. **1969** H. & M. SPROUT *Ecol. Perspective* vi. 107 A familiar version of 'behavioral model, derived largely, one suspects, from classical economics, might be called 'common-sense probabilism'. **1970** L. J. COHEN *Implications of Induction* i. 17 Some of Popper's arguments against Carnap are examples of anti-probabilism at this level. **1970** *Jrnl. Gen. Psychol.* LXXXIII. 108 Heider..focused on this region and his assumptions are similar to Brunswik's with regard to probabilism, multiple mediation, and objective reference.

probabilist. Add: **2.** More generally, one who holds any theory of probabilism (cf. *PROBABILISM 3). (Example.)

1965 H. & M. SPROUT *Ecol. Perspective* vi. 107 The common-sense probabilist assumes that the individual applies his environmental knowledge rationally to the choice of ends achievable.

3. An expert or specialist in probability theory.

1973 *Nature* 1 June p. i/1 (Advt.), Forty-six leading probabilists are represented in..

B. *adj.* = *PROBABILISTIC *a.* 2.

1960 L. DELAVENAY *Introd. Machine Transl.* vi. 93 Linguists who become automatic translation programmers will have to be trained on probabilist methods. **1970** L. J. COHEN *Implications of Induction* i. 7 The attack mounted here against probabilist theories of inductive syntax.

probabilistic, *a.* Add: **2.** Pertaining to or expressing probability; subject to or involving chance variations or uncertainties.

1951 *Philos. of Sci.* XVIII. 216 The recognition of the probabilistic character of environmental cue-object and means–end relationships through replacement of the traditional absolute right–wrong alternatives..by cues or means of lower statistical validity. **1957** N. CHOMSKY *Syntactic Struct.* ii. 17 The development of probabilistic models for the use of language (as distinct from the syntactic structure of language) can be quite rewarding. **1965** C. H. SPRINGER et al. *Adv. Methods & Models* i. 11 Models which are based on the mathematics of statistics and probability, into which we introduce the uncertainties which usually accompany our observations of real events, are called probabilistic models. **1966** C. G. HEMPEL *Philos. Nat. Sci.* v. 65 Many important laws and theoretical principles in the natural sciences are of probabilistic character. **1972** *Computers & Humanities* VII. 17 His lengthier treatment of two stochastic models notes that probabilistic stylistics is somewhat more advanced than deterministic approaches. **1978** *Sci. Amer.* Feb. 131/3 The strict determinism of classical mechanics is abandoned in the quantum theory and is replaced by a probabilistic interpretation of measurements at the microscopic level.

Hence **probabili·stically** *adv.*, in a probabilistic manner; in terms of probabilities.

1955 *Science* 11 Nov. 910/1 Not only perception but also thinking and valuing are fruitfully conceived as only in some degree probabilistically valid achievements. **1965**

Language XLI. 201 Between the total workings of such a determinate system..and the sound a speaker produces there is a layer of indeterminacy that can only be handled probabilistically. **1975** *Nature* 17 July 166/2 Equations such as (2) and (3) can provide very simple examples of fully deterministic systems whose dynamics are best described probabilistically. **1978** *Sci. Amer.* June 99/2 The processes that govern the placing of telephone calls are so complicated that it is more fruitful to view them probabilistically than to do so deterministically.

probability. Add: **1. b.** (Further example.)

1835 DICKENS *Let.* 14 Dec. (1965) I. 105, I shall not in all human probability be home before Wednesday Week.

3. Also, used of quantities which are derived logically by inferential or inductive reasoning, when mathematical concepts may be inapplicable or insufficient. (Further examples.)

1764 T. BAYES in *Phil. Trans. R. Soc.* LIII. 376 The probability of any event is the ratio between the value at which an expectation depending on the happening of the event ought to be computed, and the value of the thing expected upon it's [*sic*] happening. **1939** *Internat. Encycl. Unified Sci.* I. vi. 48 (*heading*) Probability as a unique logical relation. **1949** HUTTEN & REICHENBACH tr. *H. Reichenbach's Theory of Probability* p. v, Philosophical analysis was the starting point for a new mathematical construction of the calculus of probability. **1951** S. S. STEVENS *Handbk. Exper. Psychol.* 44/2 Russell once told his lecture audience that 'probability is the most important concept in modern science, especially as nobody has the slightest notion what it means'. **1975** A. R. WHITE *Modal Thinking* iv. 68 The common philosophical confusion of probability with estimates of probability.

4. *attrib.* and *Comb.*, as *probability amplitude*, *calculus*, *field*, *function*, *generating function*, *judgement*, *measure*, *proposition*, *relation*(*ship*), *statement*, *value*, *wave*; **probability curve**, a graph of a probability distribution; **probability density**, a probability distribution that is a continuous function; **probability distribution**, a function whose integral over any interval is the probability that the variate specified by it will lie within that interval; **probability paper** (see quot. 1933); **probability sample**, a sample whose members are chosen randomly; **probability space**, a space each point of which is an outcome and has a probability associated with it; **probability theory**, a branch of mathematics that deals with quantities having random distributions.

1936 *Physical Rev.* XLIX. 520/2 Let a_s denote the probability amplitudes of states in which the neutron is free and in a state *s*. **1944** H. REICHENBACH *Philos. Found. Quantum Mech.* xvii. 84 Since the probabilities are derived always as the squares of complex functions, these latter functions are sometimes called probability amplitudes. **1970** I. E. MCCARTHY *Nuclear Reactions* I. iii. 74 The probability amplitude description of a quantum process is a very simple one. **1940** *Mind* XLIX. 265 While scientists apply the probability-calculus with great success to, *e.g.*, microphysics, biology and vital statistics, philosophers are still not unanimous as to the right interpretation and 'meaning' of probability. **1971** *Times Lit. Suppl.* 1 Oct. 1180/4 The mathematics of Xenakis's highly sophisticated techniques derived from the probability calculus are no more discernible by the human ear than are the mathematics of the rudimentary chance operations favoured by John Cage. **1893** Probability curve [see *NORMAL *a.* 2 e]. **1914** [see *probability paper* below]. **1964** M. MCLUHAN *Understanding Media* v. 52 Some computer that translates our least gesture into a new probability curve. **1939** H. JEFFREYS *Theory of Probability* i. 24 We shall usually write this briefly $P(dx|p) = f'(x)dx$, *dx* on the left meaning the proposition that *x* lies in a *particular* range *dx*. $f'(x)$ is called the probability density. **1961** POWELL & CRASEMANN *Quantum Mech.* ii. 59 The fundamental postulate..states that the quantity $\psi^*\psi = |\psi|^2$ is to be interpreted as a probability density for a particle in the state ψ. More precisely,..in a measurement of the position of the particle, the probability $P(\mathbf{r})d\mathbf{r}$ of finding it in a volume element $d\mathbf{r} = dxdydz$ at a point **r** is proportional to $|\psi(\mathbf{r})|^2 d\mathbf{r}$. **1968** P. A. P. MORAN *Introd. Probability Theory* ii. 68 When $f(t)$ is continuous, $f(t)dt$ can be regarded as the probability that *X* lies in the range $(x, x + dt)$ when *dt* becomes small. $f(t)$ is then known as the probability density of the distribution. **1937** H. CRAMÉR *Random Variables & Probability Distributions* ii. 11 The use here made of the terms *probability function* and *distribution function* corresponds to the terminology of Kolmogoroff. **1944** H. REICHENBACH *Philos. Found. Quantum Mech.* xxiv. 111 The specification of these values is therefore replaced by the statement of their probability distributions. **1970** G. A. & A. G. THEODORSON *Mod. Dict. Sociol.* 314 A probability distribution gives the probable frequency of occurrence of each category or class interval..of a given variable. **1940** *Mind* XLIX. 272 The definition of randomness as 'Nachwirkungsfreiheit' is wholly inadequate for the purpose of characterising a probability-field. **1965** R. C. JEFFREY *Logic of Decision* vii. 103 A probability field is a collection of propositions which contains the denial of any proposition that it contains, and which contains the conjunction and the disjunction of any pair of propositions that it contains. **1906** *Acta Univ. Lundensis* I. v. 8 The values of the probability function $\phi(x)$ are most conveniently tabulated by Sheppard. **1974** H. FRANK *Introd. Probability & Statistics* ii. 40 A probability function is defined as a function $Y = f(X)$ such that *X* is a random variable and *Y* is the set of probabilities associated with *X*. **1949** *Jrnl. R. Statistical Soc.* B. XI. 217 The recurrence relation for the probability-generating function $\Pi_{r+1}(s)$ for the entire population in the $r + 1$th generation is obtained by substituting $G(s)$ for *s* in $\Pi_r(s)$. **1968** P. A. P. MORAN *Introd. Probability Theory* ii. 79 We consider the probability generating function obtained by multiplying (2.62) by s^k and summing over the possible values. **1914** C. D. BROAD *Perception* ii. 150 The correct probability is always that relative to the knowledge of the person who makes the probability-judgment. **1934** *Philos. Rev.* XLIII. 133 The given experience of the moment of knowing is the basis of a probability-judgment concerning the experience. **1954** L. J. SAVAGE *Found. Statistics* iii. 33 Let me say precisely what is meant..by a probability measure. **1971** *Sci. Amer.* Aug. 95/1 Mathematical probability is based on a special function that assigns to each subset *A* of a given set Ω a positive real number that represents the probability that a point selected 'at random' from the set Ω will actually be in *A*. This function is called a 'probability measure' on the set Ω, and we shall denote it by *m*. **1914** A. HAZEN in *Trans. Amer. Soc. Civil Engineers* LXXVII. 1549 Probability Paper.— The practical difficulty with the plotting on Fig. 1 is the great curvature of the lines showing the required storage. This difficulty is so great as to make the method unsatisfactory in most cases; but it has been removed by using paper ruled with lines spaced in accordance with a probability curve, or, as it is otherwise called, the normal law of error. **1933** *Med. Res. Council Special Rep. Ser.* No. 183. 9 A special form of graph paper known as 'probability paper' has been prepared, on which the ordinates represent a scale of percentages so spaced that the actual distances on the paper are proportional to the corresponding values of the normal equivalent deviation. On 'logarithmic probability paper' the scale of ordinates is identical with that on ordinary probability paper, but the scale of abscissae is logarithmic. **1958** E. J. GUMBEL *Statistics of Extremes* i. 29 If the theory holds, the observations plotted on probability paper ought to be scattered closely about the straight line. **1914** C. D. BROAD *Perception* ii. 150 This is not as much objectivity as is wanted for probability-propositions. **1922** tr. *Wittgenstein's Tractatus* 111 There is no special object peculiar to probability propositions. **1921** J. M. KEYNES *Treat. Probability* i. 4 If a knowledge of *h* justifies a rational belief in *a* of degree *α*, we say that there is a probability-relation of degree *α* between *a* and *h*. **1965** P. CAWS *Philos. of Sci.* xxxiv. 261 If premise and conclusion are both known, some probability relation may be established between them. **1955** O. KLEIN in W. Pauli *Niels Bohr* 102 The basic ideas of quantum theory, where the causal relationship between events is replaced by a probability relationship. **1955** F. C. MILLS *Statistical Methods* (ed. 3) xix. 659 A probability sample is one for which the inclusion or exclusion of any individual element of the population depends on the application of probability methods, not on personal judgment, and which is so designed and drawn that the probability of inclusion of any individual element is known. **1972** *Jrnl. Social Psychol.* LXXXVIII. 208 Field interviewers.. administered the Rokeach Value Survey to a national probability sample of 1489 American adults. **1968** J. B. JOHNSTON et al. *Sets, Functions, & Probability* v. 163 We shall now describe how a probability space can be associated with certain chance phenomena in the real world. **1975** I. STEWART *Concepts Mod. Math.* xvii. 247 Axiomatic probability theory works entirely in terms of probability spaces. **1930** J. LAIRD *Knowl., Belief & Opinion* xvii. 376 According to this view, probability-statements are statements of proportions in very large series. **1939** E. NAGEL *Princ. Theory of Probability* iv. 23 Probability statements are on a par with statements which specify the density of a substance; they are not formulations of the degree of our ignorance or uncertainty. **1941** *Mind* L. 48, I shall here refer to the contrast between two very well-known types of probability-theory. **1962** *Listener* 15 Nov. 793/1 In psychological research, too, probability theory plays an essential part because many of the variables can be measured only approximately. **1974** P. ERDMAN *Silver Bears* ii. 16 At M.I.T... he had become fascinated with the probability theories of..John von Neumann and Oskar Morganstern [*sic*]. **1922** W. E. JOHNSON *Logic* II. xi. 251 Mill's position is paradoxical, since he apparently attributes a higher probability-value to a law, merely on the ground of its width. **1940** *Mind* XLIX. 267 From a purely mathematical point of view there would be no obstacle to the choice of series of the above kind as fields of measurement for probability-values. **1942** *Electronic Engin.* XV. 149/1 The term wave is commonly used in a very wide sense, to cover almost everything from a heat wave to the probability waves of modern physics. **1956** E. H. HUTTEN *Lang. Mod. Physics* v. 186 Physicists have sometimes spoken of the 'probability wave'; but this phrase must not be taken literally.

probable, *a.* (*sb.*) Add: **A.** *adj.* **3. c.** Now chiefly U.S. Law. *probable cause*, reasonable cause or grounds (for making a search or preferring a charge).

a **1676** M. HALE *Historia Placitorum Coronæ* (1736) II. xviii. 150 They are not to be granted without oath made before the justice of a felony committed, and that the party complaining hath probable cause to suspect they are in such a house or place, and do shew his reasons of such suspicion. And therefore I do take it, that a general warrant to search in all suspected places is not good, but only to search in such particular places, where the party assigns before the justice his suspicion and the probable cause thereof, for these warrants are judicial acts, and must be granted upon examination of the fact. **1789** J. MADISON in *Congress. Reg.* I. 428 The rights of the people to be secured in their persons, their houses, their papers, and their other property from all unreasonable searches and seizures, shall not be violated by warrants issued without probable cause, supported by oath or affirmation, or not particularly describing the places to be searched, or the persons or things to be seized. **1811** *U.S. Circuit Court Rep.* (1827) 37 What, then, is the meaning of the term 'probable cause'? We answer, a reasonable ground of suspicion, supported by circumstances sufficiently strong in themselves to warrant a cautious man in the belief, that the person accused is guilty of the offence with which he is charged. **1850** *Calif. Sup. Court* I. 11 The offence which.. there is probable cause to suppose he has committed. **1878** *U.S. Rep.* XCVII. 646 If there was probable cause of seizure, there was a reasonable cause. If there was a reasonable cause of seizure, there was a probable cause. **1927** *Ibid.* CCLXXV. 106 The liquor was obtained by a search and seizure instituted without warrant or probable cause. **1937** N. B. LASSON in *Johns Hopkins Stud. Hist. & Pol. Sci.* LV. II. iv. 125 In the searching of automobiles.. no warrant but only the existence of probable cause is necessary to constitute the search a reasonable one. **1954** *Fed. Reporter* (1955) CCXXII. 556/1 An 'anonymous tip' to officers..was not 'probable cause' for arrest without warrant. **1976** *Washington Post* 14 Dec. A18/2 Why should the United States, without even a whisper of probable cause, be bugging a dependent ward that had been formally delivered into its care by the United Nations?

f. *probable error*: the difference between the mean of a distribution and the first or third quartiles, i.e. an error of such a magnitude that larger and smaller errors are equally likely.

It has now been largely superseded as a measure of accuracy or consistency by the standard error.

1812 *Phil. Mag.* XXXIX. 241 All that can be gained is, that the errors are as trifling as possible—that they are equally distributed—and that none of them exceed the probable errors of the observation. **1854** *Amer. Jrnl. Sci. & Arts* XVII. 396, I have calculated..the amount of probable error in the determinations of many of the atomic weights. **1872** THOMSON & TAIT *Elem. Nat. Philos.* I. iii. 113 The probable error of the sum or difference of two quantities, affected by independent errors, is the square root of the sum of the squares of their separate probable errors. **1886** *Proc. R. Soc.* XL. 66 Throughout this discussion the technical term 'probable error' has been used; it may in every instance be replaced by Mr. Galton's very apt name 'quartile'. **1889** F. GALTON *Natural Inheritance* v. 58 The term Probable Error is absurd when applied to the subjects now in hand, such as Stature, Eye-colour, Artistic Faculty, or Disease. I shall therefore usually speak of Prob. Deviation. **1903** *Biometrika* II. 273 Unfortunately custom has not taken this standard deviation as the measure of the goodness of the sample, but the whole theory having developed from the normal curve, the probable error instead of the standard deviation has been chosen, i.e. ·67449 × standard deviation. **1938** D. C. BARTON in A. E. Dunstan et al. *Sci. of Petroleum* I. VIII. 369/2 Determinations of relative gravity with a probable error of ±10 or even ±5 tenth milligal are of value in geophysical prospecting. **1968** R. A. LYTTLETON *Mysteries Solar Syst.* iii. 104 The value of approximately $0{\cdot}08 \times 10^{12}$ dyn cm^{-2}..suggests a radius as large as 2770 km, which exceeds the quoted observed value by a little more than twice the probable error.

B. *sb.* Restrict † *Obs.* to sense in Dict. and add: **b.** One who will probably, though not certainly, be successful; a likely candidate, competitor, etc.; *spec.* a member of the supposedly stronger team in a trial match (opp. to *POSSIBLE *sb.* 1 d).

1906 *Pall Mall Gaz.* 23 Jan. 2 The last two 'probables' are untried men as far as Parliament is concerned. **1909** *Westm. Gaz.* 28 May 12/3 All probables ran. **1976** *West Lancs. Evening Gaz.* 8 Dec. 18/1 Fylde lock Bill Beaumont is one of four Lancashire players included in the Probables team for the first England trial at Twickenham on December 18. **1977** R. LUDLUM *Chancellor Manuscript* xxxii. 349 'What have you found out about our four candidates?'..'One..I'd say a probable.'

c. *Mil.* An aircraft recorded as probably shot down. Also, a submarine probably destroyed.

1940 in *Winged Words* (1941) 27 They were the new Heinkel 113s... We got..three or four of what we call 'probables'. **1940** W. S. CHURCHILL in D. McLachlan *Room 39* (1968) vi. 131 At least 5 probables [*sc.* submarines] have occurred since the beginning of new year. **1944** *Sat. Even. Post* 22 July 73/3, I chalked him up with only a probable, because I did not see it crash. **1955** C. S. FORESTER *Good Shepherd* II. 278 Out of his escort force he had lost a destroyer... But he had sunk two probables and a possible. **1977** L. DEIGHTON *Fighter* v. 278 The RAF announced that..584 aircraft 'were probably destroyed'... One assumes that half the probables were downed.

probableness. Delete † *Obs.* and add later example.

1951 J. S. BRUNER in Blake & Ramsey *Perception* v. 128 The higher the 'probableness' or likeness to English of our nonsense words, the less the amount of stimulus information (in terms of length of exposure) necessary for recognizing them correctly.

proband (prǭᵘ·bǣnd). [ad. L. *proband-us*, gerundive of *probāre* to test, examine.] An individual chosen as a propositus because of the presence of some trait whose inheritance is to be studied. Also *attrib.*, as *proband method*, *test*.

1929 *Resumptio Genetica* IV. 296/1 (Index), Proband. **1931** E. & C. PAUL tr. *Baur's Human Heredity* xi. 501 A graphic representation of the parents, grandparents, great-grandparents, etc., of the individual from whom an investigation starts (such a person is called a proband) is known as that person's ancestral table. *Ibid.* 508 The proband method is wrongly supposed to be difficult to apply. **1940** HINSIE & SHATZKY *Psychiatric Dict.* 431/2 The practicable statistical method of probands in the study of selective population groups..is called the proband method. **1962** *Lancet* 29 Dec. 1341/2 Seven probands had a first-degree relative who had been treated with an antidepressant drug while under hospital care for a depressive illness. **1967** *Economist* 9 Dec. 1065/4 It is said that the Proband test effectively screens disease

mals. **1977** *Jrnl. Med. Genetics* XIV. 125/2 The same balanced translocation was found in the proband's sister, in 5 of 6 sibs of the mother and in 2 children of one of them.

Pro-Banthine (prǭᵘbæ·nþīn). *Pharm.* Also **Probanthine**, **probanthine**. [f. PRO(PYL + *Banthine*, a proprietary name for methantheline bromide, a related compound (see quot. 1954).] A proprietary name for *PROPANTHELINE.

1953 *Official Gaz.* (U.S. Patent Office) 17 Mar. 571/1 G. S. Searle & Co., Skokie, Ill... *Pro-Banthine*... For medicinal agent for the treatment of abnormal conditions of the gastro-intestinal system in tablet and ampoule form. Claims use since July 31, 1952. **1953** *Trade Marks Jrnl.* 29 Apr. 358/2 Pro-Banthine... Pharmaceutical preparations in the form of tablets or ampoules for the treatment of gastro-intestinal disorders. G. D. Searle & Co..., Village of Skokie, State of Illinois, United States of America; manufacturers. **1954** [see *PROPANTHELINE]. **1955** *Radiology* LXIV. 331/1 Probanthine (15 mg.) should be given intramuscularly one hour before the examination to reduce pancreatic secretion to a minimum and to lower the concentration of enzymes in the juice. **1958** J. H. BURN *Lect. Notes Pharmacol.* (ed. 5) 28 Methantheline (Banthine) and propantheline (Probanthine) are used to relieve the pain of gastric and duodenal ulcers. **1967** H. BECKMAN *Dilemmas in Drug Therapy* 300 In Pro-Banthine you have pure anticholinergic action and in Donnatal there is this action plus the sedative effect of the contained phenobarbital. **1969** I. KEMP *Brit. G.I. in Vietnam* vii. 154, I was told I had a 'pyloric spasm'—in the upper intestine—and given tablets of probanthine, which at last eased the pain.

probasidium: see *PRO-² I.

probate, *v.* Add: **3.** *U.S.* To place (a convicted person) on probation. Hence **pro·bated** *ppl. a.*, of or pertaining to a sentence of probation.

1961 in WEBSTER. **1972** *N.Y. Times* 3 Nov. 42/1 [He] was given a 10-year probated sentence. **1977** *Time* 19 Sept. 38/2 That was when he announced that 'whether women like it or not, they are sex objects' as he set free on a probated sentence a 15-year-old youth who had raped a 16-year-old coed in a high school stairwell.

probation. Add: **I. 3.** Used in the U.K. by Criminal Courts for certain adult offenders. (Further examples.)

1921 E. RUGGLES-BRISE *Eng. Prison Syst.* ix. 108 The principle of conditional conviction is common to most penal codes... It may take the form..of judicial reprimand, or of being bound over to be of good behaviour, or of probation. **1969** F. FINLAY *Boy in Blue Jeans* xix. 218 I'm amazed to hear you got probation. *Ibid.* 219 Within weeks of being put on probation Christopher left the hostel and he was sacked from his job. **1971** R. CROSS *Punishment, Prison & Public* i. 20 Probation is essentially the suspension of punishment conditional on there being no further offence for a period during which the offender is placed under personal supervision. *Ibid.* iii. 110 It leads one to wonder whether magistrates might not make more use of probation and less of imprisonment. **1973** *Howard Jrnl.* XIII. 346 He applies crisis theory to probation, and rightly suggests that there is scope for short-term probation.

III. 5. *probation officer* (examples); **probation order**, a court order committing an offender to a period of probation; **probation report**, a probation officer's report on an accused person submitted to a court before sentence is passed, a social inquiry report; **probation service**, probation and after-care service, a function which carries responsibility for the oversight of probationers and the care of accused persons and discharged prisoners.

1906 *Rep. N.Y. Probation Comm.* i. 8 The duties of the probation officer were to inquire into the previous history of any defendant when so directed by the court. **1909** *Westm. Gaz.* 1 Apr. 2/1 Section 107 of the Act contains a most useful conspectus of the substitutes for imprisonment which are at the disposal of the Court. They are.. discharging the offender and placing him under the supervision of a probation officer. **1922** H. H. GODDARD *Juvenile Delinquency* 78 Finally she was brought to the Bureau [in Ohio] September 1918, by the probation officer to see if we could give any advice. **1930** *Morning Post* 8 Aug. 10 The Home Secretary has decided to introduce an experimental scheme for training full-time probation officers. **1975** *Howard Jrnl.* XIV. II. 28 Such work should also be familiar to the reporting probation officer if he is to be of real value to the sentencing court. **1921** E. RUGGLES-BRISE *Eng. Prison Syst.* ix. 109 The extent to which Probation Orders are applied varies to a great extent in different parts of the country. **1948** *Act* 11 & 12 *Geo. VI* c. 58 §3 Where a court by or before which a person is convicted of an offence..is of opinion that having regard to the circumstances, including the nature of the offence and the character of the offender, it is expedient to do so, the court may, instead of sentencing him, make a probation order, that is to say, an order requiring him to be under the supervision of a probation officer for a period to be specified in the order of not less than one year nor more than three years. **1969** *Listener* 26 June 908/3 Each probation officer has around 60 cases at a time (most in Wales and fewest in London; women probation officers, dealing with female offenders, have a smaller case-load). Out of these, he can expect about six or seven to be so successful that the probation order can be cut short. **1978** J. B. HILTON *Some run Crooked* ix. 91 The Bench..made a probation order: and Harbutt listened like a model penitent. **1973** J. PATRICK *Glasgow Gang Observed* iv. 39 The probation report offered this summary of Dick at the age of fourteen. **1977** *Wandsworth Borough News* 16 Sept. 17/2 [He] was remanded in custody for probation reports. **1958** *New Statesman* 11 Oct. 479/3 It is fair to say..that even if the probation service can ever carry this added burden, it is likely to be found..that the rehabilitative part of the work calls for welfare workers with a special kind of training not hitherto given to probation officers in this country. **1972** *Times* 19 Dec. 3/6 For the first time since the probation service was set up in 1907, officers in inner London yesterday went on strike.

probationary, *a.* Add: **1.** (Further examples.)

1922 *Act* 10 & 11 *Geo. V* c. 30 §40 Such of those persons as..are serving a probationary period preliminary to establishment. **1963** T. & P. MORRIS *Pentonville* iv. 78 Once he has successfully completed his one-year period of probationary service, the prison officer is in effect a prison officer for life. **1964** A. SWINSON *Six Minutes to Sunset* iv. 54 He passed the Indian Civil Service examinations from Wren's and like most of his class spent two probationary years at Balliol and a third to get his degree. **1971** *Guardian* 8 Feb. 7/7 Introduction of a probationary licence for a period of one year after passing the test.

2. (Further examples.)

1926 *N.Z. Educ. Gaz.* 1 Dec. 202/2 Teachers selected for service as probationary assistants must understand they are still under training. **1963** B. PEARSON *Coal Flat* iii. 47 There was her routine of school, and her relations..with Miss James the probationary assistant. **1964** M. BANTON *Policeman in Community* ii. 17 On the Friday he has instructed a number of new recruits (probationary constables) to parade at the station at 6 a.m.

probationer. Add: **b.** (*e*) *Lord Probationer*, a newly appointed Scottish judge before he undergoes his trial and takes the oath. *Obs.* (*f*) *N.Z.* A teacher during his first year in a school after training at a teachers' training college.

e. 1799 *Edin. Weekly Jrnl.* 22 May, William Macleod Bannatyne, Esq. having gone through his trials as Lord Probationer, took the oaths and his seat on the Bench by the title of *Lord Bannatyne*. **1838** W. BELL *Dict. Law Scotl.* 176 The form of trial [for new judges]..consists in the presentee, or Lord Probationer as he is called, hearing and reporting, and delivering an opinion on certain of the causes depending in court. **1910** *Pall Mall Gaz.* 26 Apr. 3/5 He appears again in the First Division, and the junior judge reports to the judges of that court the judgments the Lord Probationer has pronounced.

f. 1921 *N.Z. Educ. Gaz.* 1 Dec. 21/1 Central classes for the instruction of pupil-teachers, probationers, and uncertificated teachers in science and in drawing and handwork may..be established by an Education Board. **1922** *Ibid.* 1 June 62/2 Pupil-teachers and probationers may not attend any classes in hygiene established for uncertificated teachers. **1963** B. PEARSON *Coal Flat* i. 8 You were here [*sc.* at this school] before, as probationer, weren't you?

probative, *a.* Add: **2.** (Further examples.)

1971 *Jrnl. Gen. Psychol.* LXXXIV. 222 A theorem deducible only from the conjunction of axioms as their only implicate is, therefore, most probative of the theory that contains them. **1972** *Mod. Law Rev.* XXXV. I. 73 The views of the majority have been treated as the most probative, if not the dispositive, factor. **1975** *Sci. Amer.* May 118/1 A new argument for credibility seemed to have emerged when one of the troubled men underwent a polygraph test, which at best is hardly fully probative.

probatively *adv.* (further example); also **pro·bativeness**, the quality of affording proof.

1971 *Jrnl. Gen. Psychol.* LXXXIV. 222 Theorems differ in their degree of probativeness of a theory. **1973** *N.Y. Law Jrnl.* 20 July 2/2 Movant has failed to probatively demonstrate that respondent has willfully failed and refused to comply with the judgment of the court.

probatory, *a.* Add: **1.** (Further example.)

1970 *Internat. Jrnl. Cancer* V. 311/1 Samples of tumours or normal tissue were obtained by probatory excision.

3. (See quot.)

1924 P. S. ALLEN in *Library* Mar. 255 The manuscripts are identified in the catalogue by the first words of the second leaf, the 'probatory words'.

probatum. 2. probatum est (further examples).

1831 M. EDGEWORTH *Let.* 20 Jan. (1971) 474 Mother in law gave Fanny..the best receipt for a poultice that ever was (probatum est). **1884** G. MEREDITH *Let.* 17 Sept. (1970) II. 745 All material conquest follows self-conquest... *Probatum est.*

probe, *sb.* Add: **2. c.** Any small device, esp. an electrode, which can penetrate or be placed in or on something for the purpose of obtaining and relaying information or measurements about it, or of exciting radiation in it.

1924 *Physical Rev.* XXIV. 597 Potential distribution and ion concentration were investigated by Langmuir's modified probe method. **1938** *Proc. IRE* XXVI. 1534 The electric field intensity was measured by a small probe with a crystal detector, followed by an audio-frequency amplifier and a copper-oxide meter. **1943** F. E. TERMAN *Radio Engineer's Handbk.* iii. 260 Just as waves can be set up in space by straight wires and loops, so can the wave-guide modes be excited by electric probes and loops. **1965** *Wireless World* July 31 (Advt.), A completely new transducer..utilizes the variation in capacitance between its probe and the object under investigation to provide an electrical signal. **1971** *Sci. Amer.* Dec. 76/1 W. L. Bretz in our laboratory designed and built a small probe that could record the direction of airflow at strategic points in the respiratory system of ducks. **1972** *Physics Bull.* Jan. 23/3 In its basic form the pulse echo apparatus..comprises a heavily damped piezoelectric transducer source, often called a probe, which is placed on the surface of the sample under test. **1977** *Sci. Amer.* Aug. 63/1 [On the ocean floor] temperature gradients are determined by plunging a long cylindrical probe several meters into the soft sediment and measuring the temperature at one-metre intervals with fixed thermistors.

d. *Aeronaut.* and *Astronautics.* (i) A tube fitted to the nose or wing of an aircraft in order to fit into a drogue towed by another and convey fuel from it in aerial refuelling.

1949 *Flight* 11 Aug. 178/2 Either the tanker or the aircraft to be refuelled..could be fitted with the 'probe'. **1950** C. H. LATIMER-NEEDHAM *Refuelling in Flight* 186 On the nose of the fighter aircraft, a horizontal tubular member, or probe, approximately 4 ft. in length, is fitted, and this is aimed by the pilot at the trailing drogue so that as the fighter closes with the tanker the probe enters the drogue and thus makes contact. **1966** [see *DROGUE 3]. **1978** *Aeroplane Monthly* Jan. 35/1 This sub-variant incorporated the necessary 'plumbing' to permit in-flight refuelling, achieved with the aid of a probe of gigantic proportions.

(ii) A projecting device on a spacecraft designed to engage with the drogue of another craft during docking.

1969 *Times* 23 May 1/3 Ground control told the astronauts that it suspected that the ring, which serves as a mount for the docking probe, had slipped by about three degrees. **1970** N. ARMSTRONG et al. *First on Moon* iv. 80 The command module had at its top a 'probe', a triangularly shaped assembly with a pencil-like point. **1970** R. TURNILL *Lang. of Space* 34 Three tiny capture latches on the nose of the probe provide the first steadying link, and then the command module crewman fires a gas bottle which thrusts the two together so that 12 docking latches snap shut to complete the process.

e. A small, usu. unmanned, exploratory spacecraft (other than an earth satellite) for transmitting information about its environment; also, a rocket or an instrument capsule for obtaining measurements in the upper atmosphere.

1953 *Jrnl. Brit. Interplanetary Soc.* XII. 73 The probe will arrive at Mars nine months after opposition. **1958** *Observer* 17 Aug. 1/6 From then on the probe will be on its own for about 59 hours, coasting through space, gradually slowing down under the pull of the earth's gravity. **1959** F. D. ADAMS *Aeronaut. Dict.* 132/1 *Probe*,..3. An instrumented research rocket, or its payload, for penetrating the upper atmosphere or beyond. **1967** *Technol. Week* 23 Jan. 2/2 (Advt.), Motorola command receivers are ready to prove themselves again and again..in high altitude probes ..and in a multitude of new tactical applications. **1968** *Times* 15 Nov. 8/5 The Russian probe was not able to measure the lower 25 kilometres of the Venusian atmosphere. **1970** R. TURNILL *Lang. of Space* 122 A second spacecraft..is intended to go into orbit around Saturn, and drop off at least one probe. **1977** *Nature* 8 Sept. 98/3 The 11 instruments on board include television cameras, infrared and ultraviolet spectrometers, charged particle detectors, magnetometer and plasma wave detectors and, for the first time on an interplanetary probe, radio wave detectors.

4. b. *fig.* A penetrating investigation. Also in other transferred senses of the vb.

1903 *Christendom* 9 May 151/1 Few words are commoner in newspaper headlines than 'probe', which is newspaper English for an investigation of alleged abuses. **1930** *Amer. Speech* VI. 119 Probe started in junk yard blaze. **1945** *Ann. Reg. 1944* 307 With an obbligato of court injunctions, Congressional 'probes', Gallup polls,..the case dragged on. **1948** I. BROWN *No Idle Words* 99, I have just seen an inquiry into a fatal explosion in a factory described as a Blast Probe. **1959** *Listener* 31 Dec. 1140/2 Such conventional forces should be capable of performing all the functions for which the troops of Nato are at present organized: to deal with frontier incidents, to distinguish between local probes and deliberate sustained attack. **1962** A. HUXLEY *Island* xi. 177 Slanting down through chinks in the green vaulting overhead, the long probes of sunlight picked out here a row of black and yellow water jars, there a silver bracelet. **1962** *Listener* 15 Mar. 477/3 The results of the so-called Berlin probe—the recent meetings between Mr Gromyko and the American Ambassador in Moscow—have been 'no wickets and no runs'. **1971** J. B. CARROLL et al. *Word Frequency Bk.* p. vii, The AHI Corpus is..a highly informative probe of what might be called the American school lexicon. **1980** R. MCCRUM *In Secret State* iv. 26 Would Hayter start an internal probe into the background to the Lister business?

4*. *fig. Spec.* in *Nuclear Physics*, applied to a particle which can be used to penetrate nuclei, atoms, etc., and reveal their internal structure.

1955 C. G. DARWIN in W. Pauli *Niels Bohr* i. 5 The α-particle was always Rutherford's favourite. He could see that its great mass and its great energy made it the most effective of all probes to show what was in the atom. **1971** S. KAUFMAN in L. C. L. Yuan *Elem. Particles* iv. 160 Extremely high-energy..projectiles are required to produce the mass equivalent of these strange particles and to provide probes of short enough wavelength to 'see' any internal structure. **1972** DEPUY & CHAPMAN *Molec. Reactions & Photochem.* v. 78 Quenching is another useful probe for determination of mechanism. **1974** I. E. MCCARTHY *Nuclear Reactions* I. i. 6 The invention of accelerating machines promised new probes, for example protons, deuterons, and even heavier ions. **1975** SPIRO & LOEHR in Clark & Hester *Adv. Infrared & Raman Spectroscopy* I. iii. 135 To manipulate the composition of the sample, Oseroff and Callender employed a 'pump' laser beam, to establish the photo-stationary state, coaxial with a 'probe' beam, which produced the Raman spectrum. **1975** *Nature* 5 June 459/1 This effect in α-phenylethylamine was first noticed by Hug *et al.*, who realised that it originated in the two degenerate asymmetric deformations of the methyl group and could function as a new probe of chirality.

5. probe-and-drogue, used *attrib.* with reference to (i) a method of aerial refuelling (see 2d (i) above), or (ii) a method of docking spacecraft (see 2d (ii) above); **probe microphone** (see quot. 1955); also (*colloq.*) **probe mike.**

1951 *Engineering* 27 Apr. 491/1 In the probe-and-drogue system, the tanker trails a hose..to the end of which is attached a conical metal drogue..with the open end facing rearward. **1959** *Times* 8 Sept. 4/2 The range of the Vulcan V bomber will be increased significantly by the use of the probe and drogue aerial refuelling system. **1970** R. Turnell *Lang. of Space* 34 *Docking tunnel*... So called because it contains the interlocking probe and drogue system for linking up the two craft in space. **1955** *Gloss. Acoustical Terms* (*B.S.I.*) 24 *Probe microphone*, a microphone or device incorporating a microphone for measuring sound pressure at a point in a sound field without significantly altering by its presence the sound field in the neighbourhood of the point. **1976** K. Benton *Single Monstrous Act* iii. 37 'What is it?.. A probe mike?' 'That's it... It's shaped like a spike.' **1979** 'J. Le Carré' *Smiley's People* (1980) xxi. 257 They'd like to run a couple of probe mikes into the ground floor.

probe, *v.* Add: **2.** (Further examples.) Cf. *probe *sb.* 4b.

1884 *N.Y. Weekly Tribune* 12 Mar. 1/2 The Senate Committee did not probe the Public Works Department in vain. **1915** C. Mackenzie *Guy & Pauline* 228 If he could only probe by some remark a generous impulse. **1953** *Manch. Guardian Weekly* 5 Mar. 3/1 The press exhaustively probed the unpublished agenda and was then kept..firmly out of earshot. **1977** F. Branston *Up & Coming Man* xiv. 150 Headlines were mostly variations of 'CID probe M-Way Rolls death mystery'.

4. (Further examples.)

1887 M. Corelli *Thelma* II. iv. 66 Lady Winsleigh.. had..the cleverness to probe into Thelma's nature and find out how translucently clear and pure it was. **1906** G. Meredith *Let.* 5 Apr. in *Amer. N. & Q.* (1973) XI. 69/2 'Beauchamp's Career' does not probe so deeply, but is better work on the surface. **1923** *Times Lit. Suppl.* 4 Jan. 9/3 The only instrument by which his *fin-de-siècle* soul.. could probe to something solid to live by. **1959** *Listener* 14 May 827/1 If an aggressor were to try a probing action it's just as likely that he would probe on the sea, or even under the sea, as on land or in the air. **1962** *Ibid.* 5 July 3 (*heading*) Anthony Crosland and Donald MacRae probe into the state of the nation.

probing *vbl. sb.*, **prober** (further examples).

1948 I. Brown *No Idle Words* 99 They [*sc.* sub-editors] are probers to a man. **1954** L. MacNeice *Autumn Sequel* 75 The probing mind begins to fail the prober. **1958** *Listener* 20 Nov. 822/2 If the probing [of the moon] is carried out recklessly..then the extra-terrestrial bodies will be contaminated. **1970** *Times* 26 Feb. 4/6 Rescue workers today delved with probing rods into a mass of snow. **1974** *State* (Columbia, S. Carolina) 15 Feb. 1-A/4 (*heading*) White House refuses material for probers.

probenecid (prōᵘbe·nĭsid). *Pharm.* [f. Pro(pyl + Ben(zoic *a.* + -*e*- + A)cid *a.* and *sb.*] A white, crystalline, bitter-tasting powder which is a uricosuric agent used esp. to treat gout; *p*-(di-*n*-propylsulphamoyl)benzoic acid, $(C_3H_7)_2NSO_2 \cdot C_6H_4COOH$.

1950 *Ann. Internal Med.* XXXIII. 18 Benemid. [*Note*] Sharp and Dohme's trademark for p-(di-n-propylsulfamyl)-benzoic acid. This drug has been tentatively given the generic designation 'probenecid'. **1953** *Proc. Soc. Exper. Biol. & Med.* LXXXII. 604/1 Probenecid has been shown to inhibit reversibly the renal tubular secretion of a number of organic acids, such as penicillin.., phenolsulfonphthalein.., and para-aminohippuric acid. **1963** *Lancet* 5 Jan. 54/2 People with these vague aches associated with uric-acid blood-level elevation have responded well to colchicine intravenously as initial treatment and probenecid daily. **1974** A. Henry in R. M. Kirk et al. *Surgery* xv. 305 Acute attacks [of gout] are controlled by colchicine or phenylbutazone, while uricosuric agents such as probenecid keep the plasma uric level down and must be continued throughout life.

probertite (prōᵘ·bəɹtait). *Min.* [f. the name of Frank H. *Probert* (1876–1940), U.S. mining engineer + -ite¹.] A hydrated borate of sodium and calcium, $NaCaB_5O_9.5H_2O$, which is found as colourless, monoclinic crystals at a number of locations in California.

1929 A. S. Eakle in *Amer. Mineralogist* XIV. 427 The new borate..occurs as one of the minerals of the kernite deposit in the Kramer District, Kern County, California, and the name 'probertite' is proposed for the mineral, in honor of Frank H. Probert, Dean of the Mining College, University of California, to whom the writer is indebted for specimens, photos and notes of its occurrence. **1949** *Amer. Mineralogist* XXXIV. 19 Probertite..is monoclinic, has a radiating prismatic habit, perfect (110) cleavage and specific gravity of 2·141. **1964** Nies & Campbell in R. M. Adams *Boron, Metallo-boron Compounds & Boranes* iii. 93 Hot concentrated borax liquors can be prepared in which the calcium content frequently is manyfold its equilibrium value, and when it finally deposits from these hot solutions it is usually in the form of probertite, $NaCaB_5O_9 \cdot 5H_2O$.

probing, *ppl. a.* (Further examples.)

1909 *Daily Chron.* 10 Aug. 7/2 He answered probing, keenly-put questions with dogged determination not to betray himself. **1962** F. I. Ordway et al. *Basic Astronautics* v. 187 Contact devices..either require direct contact with the surface in making the measurements or are located in a probing craft on, or in the atmosphere of, the world under examination. **1962** *Daily Tel.* 11 June 12/5 The probing talks between Russia and the United States on Berlin. **1972** *Jrnl. Social Psychol.* LXXXVI. 158 His reasoning is assessed by a series of predetermined probing questions that are administered with each situation.

probiotic: see *pro-² 1.

probit (prōᵘ·bit). *Statistics.* [f. Prob(ability + Un)it.] The unit which forms the scale into which percentages may be transformed so that data evenly distributed between 0 and 100 per cent become normally distributed with a standard deviation of one probit.

1934 C. I. Bliss in *Science* 12 Jan. 38/1 These arbitrary probability units have been called 'probits'. **1947** D. J. Finney *Probit Analysis* iii. 20 The probit of the proportion *P* is defined as the abscissa which corresponds to a probability *P* in a normal distribution with mean 5 and variance 1. **1967** J. M. Rendel *Canalisation & Gene Control* ii. 27 This distance from the mean measured in standard deviations is called a 'probit'.

b. *attrib.*, as *probit line, unit*; **probit analysis**, the technique of using probits in statistical analysis.

1947 D. J. Finney *Probit Analysis* i. 6 The statistical treatment of quantal assay data has been much aided by the development of probit analysis. **1956** *Nature* 25 Feb. 356/2 In general, high resistance is associated with flat probit/regression lines. The probit lines for resistant strains are nearly always flatter than for normal strains. **1958** *Immunology* I. 225 On transforming the per cent haemolysis into probit units and plotting these results against the reciprocal of the corresponding serum dilution, a linear relationship was obtained. **1968** *Brit. Med. Bull.* XXIV. 248/1 An analysis of variance is given, from which it can be seen, for instance, whether a fit by a set of parallel probit lines is justified. **1975** *Jrnl. R. Statistical Soc.* C. XXIV. 259 Probit analysis of dose-response curves and surfaces.

problem. Add: **3. c.** As the second element in various *Combs.* and collocations describing (*a*) a supposedly insoluble quandary affecting a specified group of people or a nation; (*b*) a real or imagined chronic personal difficulty, as *credibility, drink, health, weight problem.*

1950 M. Hay *Foot of Pride* vi. 161 The ship struck a mine..and all on board, save one, were drowned. A senior official of the British Immigration Office..impulsively expressed his relief that this particular Jewish problem had been solved. **1957** [see *Jewish *a.* 1]. **1965** L. Hughes in *Negro Digest* Sept. 57/1, I know I am The Negro Problem. **1969** 'J. Morris' *Fever Grass* iv. 44 She had the body of a ballet dancer with a weight problem. **1970** D. Bagley *Running Blind* iv. 83 He had a drinking problem at one time and decided to cut it out. **1971** 'A. Garve' *Late Bill Smith* i. 15 'Sugar?' 'No, thanks. I've a waistline problem.' **1974** E. Ambler *Dr. Frigo* i. 41 If Villegas had a health problem which could be helped by a change of climate [etc.]. **1977** *Grimsby Even. Tel.* 24 May 7/1 [He] told the court he had a drink problem and asked to be given a chance. **1978** S. Brill *Teamsters* ii. 48 As a convicted bank robber, the inmate has a credibility problem.

d. In various colloq. phrases, as *no problem*, simple, easy, 'the question does not arise'; *that's your* (*his*, etc.) *problem*, used to disclaim responsibility or connection.

1963 *Amer. Speech* XXXVIII. 271 *No sweat* means 'no problem'. **1967** M. Kenyon *Whole Hog* xxii. 217 'Don't you think he just might bring out the acid and the humane killer again? For me?' 'That's your problem.' **1973** M. Amis *Rachel Papers* 117 Finally, every time I emptied my glass, he took it, put more whisky in it, and gave it back to me, saying 'No problem' again through his nose. **1976** L. Sanders *Hamlet Warning* (1977) xxiii. 207 'Shouldn't we tell the hotel people what to do with the debris?'.. 'That's their problem.' **1977** C. Forbes *Avalanche Express* xi. 116 'If I catch you fooling around I'll break your arm.' 'No problem,' John assured him easily.

7. (*a*) *problem analysis, -game, paper, -programmer, -situation, skin, -solution, -tackler*; *problem-free, -ridden* adjs.; (*b*) *problem book, column, letter, page, picture, play* (further examples); (*d*) 'in which problems of a personal or social character are manifested', as *problem case* (Case *sb.*¹ 8), *child, family, parent*; **problem-oriented** *a. Computers*, (of a computer language) devised in the light of the requirements of a certain class of problem; **problem-solver**, one who finds solutions to difficult or perplexing questions or situations; hence **problem-solving** *sb.*, the action of finding solutions to such problems; also as *adj.*, applied to behaviour, mental processes, equipment, etc., involved in or related to this activity; **problem tape** *Computers*, a magnetic tape containing the numerical information for a problem.

1969 J. Argenti *Managem. Techniques* 200 (*heading*) Problem analysis. **1931** F. M. Ford *Let.* 14 Mar. (1965) 200 It might have an enormous sale as a problem book. **1937** 'L. Q. Ross' *Educ. Hyman Kaplan* 2 Here was a student who might, unchecked, develop into a 'problem case'. **1949** Koestler *Promise & Fulfilment* iii. iv. 328 A large number of the immigrants of recent years are psychological problem cases. **1920** J. Taft in *Proc. Nat. Conf. Social Work* 63 The placing and replacing of a problematic child..is also costly... The problem child is such a costly, nagging, persistent proposition that..we are forced to bring intelligence to bear upon his case. **1944** H. G. Wells *'42 to '44* 83 What can one forecast from America, the great problem-child of humanity? **1964** M. Argyle *Psychol. & Social Probl.* ix. 123 Another group, which includes problem children and psychopathic delinquents, have had discipline that was too strict and harsh. **1977** D. Beaty *Excellency* iv. 50 We're the bankrupt problem child of the E.E.C. **1974** M. Cecil *Heroines in Love* ix. 213 The problem columns of all the new magazines. **1937** W. de B. Hubert in C. P. Blacker *Social Problem Group?* vi. 122 It is not at present known with certainty what proportion of families showing *both* mental defect and social problems..contribute to the total number of problem families. **1958** *Sunday Times* 26 Jan. 18/6 The number of hospital admissions is six times greater in the case of children from problem families. **1977** P. Johnson *Enemies of Society* xiv. 191 It means more 'problem families' and so more crime. **1964** P. Worsley in I. L. Horowitz *New Sociol.* 385 There is no 'problem-free' solution. **1921** H. E. Palmer *Princ. Language-Study* xiii. 145 Many types of puzzles and problem-games are practically identical with mathematical problems. **1970** O. Norton *Dead on Prediction* i. 7 Whenever a man picks up a woman's magazine he *always* turns to the problem letters. **1961** *Computer Jrnl.* IV. 217/1 In most if not all of the current computer program languages—called 'problem-oriented languages'—the programmer must be concerned in some degree with how his program will be handled either on a specific computer or on a class of computers with specific characteristics. **1967, 1970** [see *machine-oriented* adj. s.v. *machine *sb.* 10]. **1973** C. W. Gear *Introd. Computer Sci.* viii. 319 These statements are part of what we call a Problem Oriented Language because they provide a language in which the problem can be described, but in which the method of solution is not described. **1974** M. Cecil *Heroines in Love* viii. 192 The problem page ('Why when I have this beautiful home..do I feel the need for something more?'). **1919** *Granta* 1 May 4/2 The solutions of the problem paper in the Mathematical Tripos. **1961** *Economist* 2 Dec. 909/3 Short 'problem papers' on points of specific difficulty may be either inspired by the staff or requested by a government office or an outside group. **1932** A. S. Neill (*title*) The problem parent. *Ibid.* i. 9 There is never a problem child; there is only a problem parent. **1956** A. G. McRae *Hill called Grazing* x. 104 Problem children, if you like, though I prefer to think of them as the offspring of Problem Parents. **1962** *Listener* 7 June 976/2 The issue of the problem-parent and the uncomprehending home. **1910** *Punch* 30 Mar. 219 (*caption*) The problem picture. **1979** G. MacDonald *Camera* xiii. 180/2 As the [nineteenth] century advanced social realism in Britain did not mature... Instead middle-class dilemmas were explored in 'problem pictures' with titles like 'The Confession',.. 'The Prodigal Daughter', etc. **1941** G. Heyer *Envious Casca* iv. 53 A problem-play, is it? **1957** V. Brittain *Testament of Experience* (1979) iii. 96 My chief passion was for work..and my fourth for intellectual drama and 'problem' plays. **1970** R. E. C. Houghton *Shakespeare's Measure for Measure* 5 The very term 'problem plays' is loose and ambiguous—were they problems to their author, or are they only problems to the modern critic?.. The most intelligible use of the term would be for a play primarily concerned to present a moral problem. **1970** O. Dopping *Computers & Data Processing* xix. 305 Those who design programming systems..are sometimes called system programmers. In contrast, the user's normal programmers are sometimes called problem programmers or application programmers. **1924** R. Graves *Mock Beggar Hall* 59 A disturbing problem-ridden affair demanding the comments of a moralist. **1950** *Mind* LIX. 385 The problem-situation involving rational argument and discussion. **1978** J. Dunn in Hookway & Pettit *Action & Interpretation* 169 There is no doubt much redundancy and not a little error of one kind and another in agents' characterisations of their problem situations. **1970** *Cape Times* 28 Oct. 3/2 (Advt.), A daily cleanser for problem skins. **1956** J. Klein *Study of Groups* 192 (*heading*) Cooperative versus solitary problem-solution. **1899** J. Milne *Romance of Pro-Consul* xvi. 173 He [*sc.* Sir George Grey] was the problem-solver called in late. **1929** R. Frost *Let.* 6 Jan. (1964) 194, I don't believe in myself as a problem-solver. **1974** *Times* 31 Aug. 4/6 Practical problem-solvers can contribute much to education. **1979** *Dædalus* Summer 148 Pragmatism..is a fairy tale of energies magically released..into what Dewey called the 'situation', jointly apprehended by the problem-solvers involved. **1931** *Psychol. Rev.* XXXVIII. 337 Problem-solving by insight is regarded..as qualitatively different from problem-solving on the basis of trial and error. **1964** *Language* XL. 237 Most problem-solving situations involve concept evocation rather than the formation of new concepts. **1966** A. Battersby *Math. in Managem.* i. 15 A computer is said to have applied a problem-solving programme to the proposition that the base angles of an isosceles triangle are equal. **1978** *Amer. Poetry Rev.* July/Aug. 38/4 Every once in a while the problem-solving yields to an almost basking, sunny Calvinism. **1979** *Yale* Apr. 4/2 Man is a problem-solving animal. **1963** *Times* 29 Jan. 9/4 Contributing to a Forum, colliding politely at a Meeting Point, joining in the domestic Parliament of a Woman's Hour, onward the problem-tacklers go. **1948** *Math. Tables & Other Aids to Computation* III. 9 It was intended that the routine tapes should contain all the orders, the table tapes should contain numerical information of a general nature, comparable to function tables used in manual computing, and the problem tape should contain numerical information specific to the problem being solved. **1956** G. A. Montgomerie *Digital Calculating Machines* x. 213 The input tapes are of three kinds. First the problem tape containing the numerical information for a set of data and also, usually, some instructions.

problematic (prǫblĕmæ·tik), *sb. Sociol.* [f. Problematic *a.* 1.] Something that constitutes a problem, or an area of difficulty in a particular field of study.

1957 R. K. Merton *Social Theory* (rev. ed.) ii. 127 Working out its problematics, i.e., the principal problems (conceptual, substantive and procedural). **1969** R. Blackburn in Cockburn & Blackburn *Student Power* 194 The dialectical approach to the same problematic adopted by such writers as Isaac Deutscher and Herbert Marcuse enabled them to obtain a more lasting insight into the dynamic of Soviet society. **1971** *Catholic Q.* XXXIII. 439 Essays..sharing the common problematic of attempting to situate the language of faith within language as a whole along the lines indicated by modern linguistic analysis. **1977** R. H. Brown in Douglas & Johnson *Existential Sociol.* ii. 77 A humanistic sociology investigates the problematics of feeling and meaning.

pro·blemless, *a.* [f. Problem + -less.] Devoid of or unaffected by problems; presenting no problems.

1924 *Public Opinion* 3 Oct. 320/1 Jowett went serenely on his way—apparently problemless. **1967** *Lingua* XVII. 55 As far as word classes are concerned, problemless or 'regular' hypotheses about little-known languages are always very suspicious! *a***1970** E. Starkie in J. Richardson *Enid Starkie* (1973) iv. x. 80 The advantages... The problemless life. It was sacrificing nothing.

pro-Boche: see *pro-¹ 5a.

proboscidal, *a.* Add: Also *fig.*

1922 Joyce *Ulysses* 456 He assumes the avine head, foxy moustache and proboscidal eloquence of Seymour Bushe.

pro-Britisher: see *pro-¹ 5b. **pro-business:** *pro-¹ 5a.

proby, probie (prōᵘ·bi). *colloq.* [f. Probationer + -y⁶, -ie.] A probationer; *spec.* U.S., a fireman undergoing probation.

1899 L. Becke *Old Convict Days* ii. 42 For a proby (probationer) you're a plucky one. I won't report you. **1946** *Richmond* (Va.) *News Leader* 25 Jan. 11/4 There are times when the 'probies' (students at the probationary firemen's school) must think [etc.]. **1969** *Publ. Amer. Dial. Soc.* lii. 30 *Proby*, a man on probation because he is new on the force [*sc.* the Denver Fire Department]... 'John is still a proby.'

procainamide (prokē̆ⁱ·nămaid). *Pharm.* Also procaineamide. [f. **procaine amide*.] An amide, $NH_2C_6H_4 \cdot CONH \cdot CH_2CH_2N(C_2H_5)_2$, which is formally derived from procaine (an ester), and is used in cardiac therapy (esp. to control arrhythmia) in the form of a hydrochloride, a white hygroscopic solid.

1954 *Lancet* 8 May 957/1 It is now established that procainamide is superior to quinidine in the treatment of ventricular tachycardia. **1954** *Brit. Pharmaceut. Codex* 610 Procainamide Hydrochloride is *p*-amino-N-(2-diethylaminoethyl)-benzamide hydrochloride and may be prepared by treating the NN-diethylethylenediamine with *p*-nitrobenzoyl chloride, and reducing the nitro-compound obtained. **1971** L. Schamroth *Disorders Cardiac Rhythm* lvii. 326/2 Follow-up therapy may be carried out with oral procaineamide: 250 to 500 mgm 4 to 6-hourly. **1979** *Sci. Amer.* Dec. 52/1 Procainamide, which is administered to counteract irregular rhythms of the heart, must be given to most patients every three hours in order to provide blood levels near the therapeutic range.

procaine (prōᵘ·kē̆ⁱn). *Pharm.* Also procain. [f. Pro-¹ + Co)caine.] The synthetic compound 2-diethylaminoethyl *p*-aminobenzoate, $NH_2C_6H_4 \cdot COO \cdot CH_2CH_2N(C_2H_5)_2$, which is used as a local anæsthetic, usu. in the form of its hydrochloride, a white, crystalline solid. Cf. *novocain.

1918 *Jrnl. Amer. Med. Assoc.* 23 Feb. 537/2 Procaine is the official name of the product introduced as novocaine. **1919** *Ibid.* 6 Sept. 757/2 Procain is employed largely in infiltration anesthesia. **1940** H. A. McGuigan *Appl. Pharmacol.* 536 The activity of procaine is increased by the addition of 0·25 per cent sodium bicarbonate or 0·50 per cent potassium sulfate to the solution injected. **1951** A. Grollman *Pharmacol. & Therapeutics* xvii. 324 Procaine is relatively non-toxic, being destroyed rapidly by the liver. **1958** *Daily Mail* 24 July 5/7 He thought it contained procaine to help her fibrositis. **1962** H. Heath in A. Pirie *Lens Metabolism Rel. Cataract* 364 The actions of some drugs, e.g. zoxazolamine, pentobarbital and procaine, are potentiated in the scorbutic state. **1976** H. Ferguson *Confessions of Long Distance Acid Head* 7 Apart from cannabis, I have used barbiturates,..morphine, cocaine, procaine, ritalin, even apomorphine once.

b. Special Combs.: **procaine amide** = *procainamide; **procaine penicillin**, an insoluble salt of procaine and benzylpenicillin, which is an antibiotic used in the form of a suspension in oil and which releases penicillin slowly after intramuscular injection.

1950 *Jrnl. Pharmacol. & Exper. Therapeutics* XCVIII. 21/2 (*heading*) The action of procaine amide..on ventricular arrhythmias. **1963** V. Schrire *Clinical Cardiol.* xix. 348 Procaine amide should be used in preference to intravenous quinidine. **1968** I. L. Rubin et al. *Treatm. Heart Dis. in Adult* x. 240 Procaine amide is administered intravenously as the drug of choice in the treatment of ventricular tachycardia. **1947** *Proc. Staff Meetings Mayo Clinic* XXII. 567 The preparation is a procaine salt of penicillin G (duracillin) which was prepared in the research laboratories of Eli Lilly and Company. Procaine penicillin G is a crystalline, nonpyrogenic substance which is prepared by combining one molecule of procaine base..with a molecule of penicillin. **1949** Florey & Jennings in H. W. Florey et al. *Antibiotics* II. xxxvii. 1220 The first accounts of trials with procaine penicillin in man all mentioned the freedom of patients from local or general toxic effects. **1961** *Times* 27 Mar. 5/4 The fungicidal activity of the antibiotic procain penicillin. **1967** *Martindale's Extra Pharmacopœia* (ed. 25) 1007/2 Procaine penicillin is administered by intramuscular injection to create a depot from which penicillin is slowly liberated.

procaineamide: see *procainamide.

procarbazine (prokā·ɹbăzīn). *Pharm.* [f. Pro(pyl + Carb(amic *a.* + Hydr)azine, formative elements of the systematic name.] A hydrazine derivative whose hydrochloride is used in the treatment of some neoplastic diseases, esp. Hodgkin's disease; *N*-*p*-isopropylcarbamoylbenzyl-*N'*-methylhydrazine, $C_{12}H_{19}N_3O$.

1965 *Brit. Med. Jrnl.* 18 Dec. 1473/1 (*heading*) Natulan (procarbazine) combined with radiotherapy in management of inoperable malignant melanoma. **1974** R. M. Kirk et al. *Surgery* iv. 58 Hodgkin's disease (Lymphadenoma)... Generalized disease responds to chemotherapy, usually with a combination of vincristine, mustine, procarbazine and prednisone in pulsed doses. **1977** *Lancet* 14 May 1041/2 Of all cytotoxic drugs the teratogenicity of the alkylating agents is particularly well-documented; these, and procarbazine too, are best avoided at all stages of pregnancy if possible.

procarcinogen(ic): see *pro-² 1.

procaryon, etc., varr. prokaryon, etc.

‖ **procédé** (prosede), *sb.* [Fr.] Manner of proceeding; method, procedure, process.

1872 Geo. Eliot *Let.* 14 Mar. (1956) V. 256, I altogether abominate that *procédé* of M. Forgues and others who undertake to trim and abridge. **1935** *Scrutiny* IV. iii. 285 The characteristic *procédé* may be seen in this extract from the second section of his [*sc.* Santayana's] essay on Russell. **1962** *Listener* 15 Nov. 832/3 All the familiar *procédés* of opera have vanished, whether set numbers or Wagnerian leading-motives.

procedural (prosī·diūrăl), *a.* [Procedure + -al.] Of or pertaining to procedure. Hence **proce·durally** *adv.*

1889 F. W. Maitland in *Pol. Sci. Q.* IV. 506 Our collections include a few documents which bear no legislative authority, namely..a few procedual [*sic*] formulas. **1908** R. Pound in *Illinois Law Rev.* Nov. 232 The necessity of patient cutting away by the courts of an *abatis* of procedural obstacles in order to attack the substantive points before them. **1919** H. A. L. Fisher *Stud. Hist. & Politics* 52 It includes a reform of the civil, penal, and procedural codes. **1938** *Mind* XLVII. 529 Prof. Schilpp is capable of lapsing into revolting phrases like 'procedurally formal'; but this gives no proper indication of his style, which is on the whole vigorous and beautifully clear. **1947** *Daily Tel.* 13 Dec. 1/1 Mr. Marshall, although sure that M. Molotov's words were 'not seriously designed for consumption here, but for another audience', said that such procedural methods 'do not inspire respect for the Soviet Government'. **1963** J. Prescott *Case for Hearing* vi. 89 We'll make the girl a ward of court... Any quick and procedurally easy application to the magistrates' court for permission to marry would be out. **1968** *Brit. Med. Bull.* XXIV. 252/2 A computer can be used to take procedural decisions during the test itself. **1978** *Nature* 13 July 104/3 Now at least it seems as if the commission is leaning over backwards to be procedurally fair.

procedure. Add: **1. e.** *Computers.* A set of instructions for performing a specific task.

1946 *Ann. Computation Lab. Harvard Univ.* I. iv. 98 There are many coding routines..which occur so frequently as to make standard coding procedures of real value. This chapter includes..certain of the longer procedures. **1954** *IRE Trans. Electronic Computers* III. 15/1 Specialized procedures, formed from combinations of basic procedures,..are needed to achieve higher speeds of operation for special purposes. **1965** *Data Processor* Oct. 22/3 A procedure is a block of instructions designed to perform a specific function such as the calculation of overtime pay in a payroll application... Procedures share common elements with different programs. Seldom-used procedures can be held in auxiliary storage and called into the main storage only when required. **1970** O. Dopping *Computers & Data Processing* xix. 315 Subroutines are called procedures in Algol. A number of standard functions are obligatory among these procedures, ..but in addition to these the programmer may write his own procedures.

process, *sb.* Add: **1. c.** Used in *Philos.*, esp. in and with reference to the work of A. N. Whitehead (1861–1947), to designate the course of becoming rather than being.

1926 A. N. Whitehead *Relig. in Making* iii. 114 In this fusion of ground with consequent, the creative process brings together something which is actual and something which, at its entry into that process, is not actual. The process is the achievement of actuality by the ideal consequent, in virtue of its union with the actual ground. In the phrase of Aristotle, the process is the fusion of being with not-being. **1949** O. Lee *Existence & Inquiry* 11 Because the world seen as process was very different from what it had been before, a new theory of inquiry was needed to deal with it. Dialectic was the answer first proposed—a logic of process. **1964** E. E. Harris *Found. of Metaphysics in Sci.* xxii. 451 Samuel Alexander, J. C. Smuts, Lloyd Morgan and Henri Bergson..expounded theories of process and evolutionary pluralism. Whitehead..like Hegel, attempted to reconcile pluralism with monism and process with holism. **1977** *Theology* LXXX. 187 The world is a dynamic totality of events..hence it is a process, *from* the given past *through* the present.. and *towards* a future.

5. b. *Social Sciences.* The continuing interaction of human groups and institutions, esp. as observed and studied through its effects in social, political, cultural, etc., life, with the aim of finding underlying patterns of behaviour in the data available, freq. contrasted with the study of such aspects of society through its structures. Also *attrib.*

1887 Moore & Aveling tr. *Marx's Capital* I. i. i. 12 The different proportions in which different sorts of labour are reduced to unskilled labour as their standard, are established by a social process that goes on behind the backs of the producers, and, consequently, appear to be fixed by custom. **1898** E. A. Ross in *Amer. Jrnl. Sociol.* III. 860 Everything that is being done to bring to light the processes of socialization and control contradicts the easy-going theory that actual society is a spontaneous product due to the social instincts of man. **1902** L. F. Ward in *Ibid.* VII. 761 Ratzenhofer shows the precise *modus operandi* of the whole process of social assimilation. **1928** [see *educational *a.* 2]. **1939** *Jrnl. Psychol.* VIII. 389 The question may be raised whether any light upon this situation can be obtained by examining the *process* of personality development for leads to..more satisfactory methods and procedures. **1951** R. F. Bales *Interaction Process Anal.* p. iii, An attempt to formulate some of the basic structural characteristics and dynamic processes one would expect to find in small groups. **1954** *Amer. Anthropologist* LVI. 398 In *synchronic* studies of national character, we are discussing not the origins of the culture or the society, but the process of learning of identifiable human beings living within that society at a given period. **1958** *Pol. Stud.* VI. 243 The term 'process' seems to enter social and political discourse today in two different ways: it can be used widely or it can be used more specifically. *Ibid.* 248 Here I think we probably find a plausible separate use for this process notion—to refer to an isolable complex of interactions between procedural rules..and the internal and external relations of various kinds of social groups. **1960** *Amer. Anthropologist* LXII. 18 The recent advances in the structural-functional approach are impressive..compared to the lag and the disagreements over how to conceptualize cultural processes. **1969** K. Cauthen *Science, Secularization & God* v. 165, I have already declared myself in favor of a metaphysical philosophy based on the process model argued so persuasively by Whitehead and Hartshorne. **1971** R. F. Murphy *Dialectics of Social Life* i. 31 The study of 'process' and 'dynamics' is thus not as processual and dynamic as we would like to believe, for it commonly approximates a seriation of structures through time.

6. c. (*d*) A method of straightening and styling the hair by chemical means; *transf.* hair thus treated; the chemicals which effect this. *U.S. Blacks.*

1964 L. Hairston in J. H. Clarke *Harlem* 288 Sonny rubbed the process in so thick with his rubber gloves, it started stingin' a little t'rough the heavy layer of grease he packed in my scalp. *Ibid.* 293 By Friday my process 'd need retouchin'. **1967** *Trans-Action* Apr. 8/1 Time may pick up when a familiar car cruises by and a few dudes drive down to Johnny's for a 'process' (hair straightening and styling). **1970** E. Ofari in *Black Scholar* Oct. 49/2 Draper apparently has never heard of cadillacs, processes, chitterlings, the blues. **1972** B. G. Cooke in T. Kochman *Rappin' & Stylin' Out* 64 The 'process' of hair straightening is now considered demeaning; most black brothers have abandoned it.

d. A linguistic operation or change.

1954, etc. [see *item *sb.* 2d]. **1964** R. H. Robins *Gen. Linguistics* v. 212 Different forms of the paradigms are then described as the result of processes, vowel change.. etc. applied to the root form. (*Process* in this use is a descriptive term; it has nothing to do with historical processes in time or with changes in the forms of the language through the years.) **1974** R. Quirk *Linguist & Eng. Lang.* v. 92 Educated opinion here is well-informed about the 'existing processes' of English.

12. a. (Examples referring to individual cells.)

1893, etc. [see *dendron]. **1977** *Sci. Amer.* Aug. 108/2 In the nervous system a network of nerve cells with elongated processes communicate with one another by secreting neurotransmitters, which traverse the tiny gap between two nerve cells.

13. a. (sense *1c) *process motif, theism, -thinker, thought, view*; (sense 6c) *process block* (further examples), *department, embossing, -engraver* (examples), *engraving, lens, -maker, photography, plate* (examples), *print, -printer, reproduction*; also with reference to a kind of colour printing in which a continuous and wide range of colours is produced by superimposing half-tones in each of three or four different colours; as *process colour, ink, printing*; with reference to industrial processes, esp. continuous ones, as *process cost, engineer, engineering, industry, operation, plant, sheet, work* (further example); *process-type* attrib.; (sense *6d) *process approach, model, morpho-*

phonemic; **b. process annealing** *Metallurgy*, heat treatment applied to an alloy after cold working to prepare it for further cold working; **process black**, a black ink suitable for use in process work; **process butter** (see quots. 1902, 1906); **process camera**, a camera specially designed for taking photographs for use in process work; **process chart**, a diagram showing the sequence and sometimes the time and place of the different stages in an industrial or commercial process, or the different activities performed by an employee; **process cheese**, cheese made by melting and blending (and often emulsifying) other cheeses; also *fig.*; cf. **processed cheese*; **process control**, the regulation and control of the physical aspects of an industrial process, esp. automatically by instruments; freq. *attrib.*; hence **process controller**; **process heat**, heat supplied or required for an industrial process; so **process heating** *vbl. sb.*; **process philosophy**, philosophy based on the theory of process (sense *1c); **process projection** *Cinemat.*, projection on to the back of a translucent screen, the other side of which is used as a background for ordinary filming; **process schizophrenia**, endogenous schizophrenia that does not seem connected with environmental causes; hence **process schizophrenic**; **process shot** *Cinemat.* (see quot. 1973); **process steam**, steam supplied or required for an industrial process other than power generation; **process theology**, theological theory based on the concept of process (sense *1c); hence **process theologian**; **process water**, water used in an industrial process; **process worker**, one who works in process printing or in an industrial process.

1936 *Metals Handbk.* (Amer. Soc. Metals) 211 *Process annealing*, heating iron base alloys to a temperature below or close to the lower limit of the critical temperature range followed by cooling as desired. **1977** R. B. Ross *Handbk. Metal Treatm.* 322 The purpose of Process annealing is to remove work hardening prior to further cold work. **1972** W. P. Lehmann in *Language* XLVIII. 266 Recent grammatical study has led to a preference for a process approach in linguistic analysis. **1907** *Yesterday's Shopping* (1969) 471/1 *Process black*..for use in drawings intended for process reproduction. **1964** E. Chambers *Camera & Process Work* xv. 208 The ink manufacturers make inks suitable for proofing, these are usually sold under such names as half-tone black, process black or press black. **1888** C. T. Jacobi *Printers' Vocab.* 104 *Process blocks*, illustrations in relief produced by any mechanical process. **1925** Process block [see *letter-press 1]. **1899** *Jrnl. Franklin Inst.* CXLVII. 94 (*heading*) Renovated or process butter. **1902** Leffmann & Beam *Select Methods of Food Analysis* 370 So-called 'process' or 'renovated' butter, made by rendering old or inferior samples, purifying the fat, coloring, salting, and molding it, is now a familiar commercial article. **1906** L. L. Van Slyke *Mod. Methods of testing Milk & Milk Products* i. 18 Renovated or process butter is the product made by melting butter and reworking, without the addition or use of chemicals or any substances except milk, cream or salt. **1911** Simmons & Mitchell *Edible Fats & Oils* iv. 44 The preparation of 'process' butter from stale or unsaleable genuine butter. **1895** *Photogr. Jrnl.* XIX. 313 In the construction of a process camera rigidity and parallelism and ability to stand wear and tear have to be carefully studied. **1967** Karch & Buber *Offset Processes* v. 141 The lithographic plate is usually produced from a negative which is made with the process camera. **1974** J. Craig *Production for Graphic Designer* 72 The first step in making a printing plate is to photograph the copy..using a special camera, called a process camera. **1941** Colvin & Stanley *Running Machine Shop* vi. 256 Process charts can also be made to follow the operator instead of the part or product. **1968** B. Yuill *Supervision Princ. & Techniques* xxi. 240 The layout must be carefully planned by using such techniques as materials and man process charts, which show the proposed courses of materials and the movement of manpower through the plant. **1977** P. E. Hicks *Introd. Industr. Engin. & Managem. Sci.* iv. 63 Whereas analysis was restricted only to operations and inspections in using the operation process chart, the flow process chart includes additional consideration of moves, delays, and storages. **1926** T. R. Pirtle *Hist. Dairy Industry* i. 110 The development of the process or packaging cheese business is one of the outstanding accomplishments of the cheese industry in recent years. **1951** M. McLuhan *Mech. Bride* (1967) 24/1 These wondrous totalitarian techniques for making the public into process cheese. **1972** *Federal Register* XXXVII. 11722/3 The amendments..will have the effect of providing for optional use of buttermilk in pasteurized process cheese food. **1926** F. B. Wiborg *Printing Ink* xx. 241 *Process color inks.* These inks are made exclusively for the purpose of printing pictorial subjects... Special process inks are made for this class of printing. **1951** R. G. Radford *Letterpress Machine Work* II. xii. 134 The majority of experienced process colour printers prefer the Miehle two-revolution machine. **1968** *Heidelberg News* Sept. 4/3 Only use process colour when it is justified. Four colour work is seen at its best, has most impact, alongside black and white. **1931** *Electronics* Oct. 144 (*heading*) Electronic oscillators for industrial process control. **1945** D. P. Eckman *Princ. Industr. Process Control* x. 202 Many simple controllers serve in industrial process control as safety devices to protect process equipment from overloads of temperature or pressure. **1967** *Economist* 8 Apr. 162/2 It would make a great difference to the market for, and the cost of, industrial process control if engineers could make headway in designing standardised sections of process control instruments, etc., that could be assembled, building-block fashion, at the client's plant. **1977** *Sci. Amer.* Sept. 122/3 A small process-control computer monitors the temperature, directs the insertion and withdrawal of the wafers and controls the internal environment of the furnace. **1951** *Proc. Inst. Electr. Engineers* XCVIII. ii. 609/1 The output signal of the process controller is transmitted to the regulating unit, which adjusts the physical quantity upon which the controlled quantity depends in order to restore it to the desired value. **1955** *Automatic Control* (1957) ii. 26 Process controllers supervise the manufacture of plastics, synthetic fibers, drugs, the whole range of products of the chemical industry. **1926** S. I. Levy *Introd. Industrial Chem.* ii. 47 The process cost sheets reveal clearly the great importance of chemical efficiency. **1974** *Terminol. Managem. & Financial Accountancy* (Inst. Cost and Managem. Accountants) 21 *Process cost centre*, a cost centre in which a specific process or a continuous sequence of operations is carried out. **1967** *Times* 18 Jan. 16/4 There are restrictive practices in the industry, particularly in the machine room and process department. *Ibid.* 16/7 *Process department*, prepares photographs and line drawings as plates. **1968** *Guardian* 29 Feb. 11/2 Joe Balfour's work wasn't good enough for process department, so he was transferred to assembly. **1931** R. R. Karch *Printing & Allied Trades* xvii. 166 Thermography is known as 'raised letter' printing, and 'process embossing'. **1948** —— *Graphic Arts Procedure* i. 7 Imitation engraving is known by several names, among which are *raised-letter printing, process embossing.* **1935** *Proc. Inst. Production Engineers* XIV. 158 Process and rate department (under chief process engineer). **1948** W. H. Schutt *Process Engin.* i. 1 The process engineer must visualize exactly how the article should be made and what equipment, tools, and floor space are required. **1960** *McGraw-Hill Encycl. Sci. & Technol.* X. 642/2 Process specifications are set by process engineers (as distinguished from product engineers) and cover just how processes are to be controlled. **1948** W. H. Schutt (*title*) Process engineering. **1980** *Jrnl. R. Soc. Arts* May 326/2 'Design for Production' is a discipline which involves both product and process engineering. **1923** H. A. Maddox *Printing* x. 126 Process engravers usually adapt their filters to certain ink standards. **1951** R. G. Radford *Letterpress Machine Work* II. xii. 127 Discoveries..have made it possible for the process engraver to make a set of three- or four-colour half-tone plates by photo-mechanical means. **1894** *Amer. Dict. Printing & Bookmaking* 464/1 *Process printing or engraving*, a method by which engravings are made by the aid of photography. **1965** *Listener* 23 Sept. 462/3 The year was 1872, when..*Punch's* tentative introduction of process engraving first heralded the disappearance of the laborious procedure of reproducing line drawings by wood engraving. **1933** E. Molloy *Newnes Engin. Pract.* III. 794/2 The faster the machine..ran, the greater the amount of steam that would be available for process heat. **1947** O. Lyle *Efficient Use of Steam* xx. 597 The two principal uses of process heat are for the heating of water or watery solutions and for the evaporation of water. **1971** *Materials & Technol.* II. xii. 751 Steam generators which raise steam for process heat operate from 15 to about 150 p.s.i. **1926** S. I. Levy *Introd. Industr. Chem.* iii. 65 (*heading*) Process heating by steam. **1971** *Materials & Technol.* II. xii. 751 Steam has found considerable employment as a heat carrier for process heating. It can easily be ducted from a boiler to the vessel or column wherein the processes take place. **1951** *Industr. & Engin. Chem.* Dec. 2695 (*heading*) Quality control in the process industries. **1926** Process ink [see *process colour* above]. **1974** J. Craig *Production for Graphic Designer* 109 Because process inks are transparent, it is the light reflected from the paper's surface that supplies the light to the ink. **1902** *Encycl. Brit.* XXXI. 696/2 In the 'Process' lens, Series V. *f*/8, the combination is adjusted to secure identical size and sharpness of each colour-image in three-colour process work. **1961** Process lens [see *colour correction* s.v. *colour *sb.*¹ 18]. **1900** *Fortn. Rev.* Jan. 65 Engraving..as a profession, and as a means of obtaining fame, has entirely died out; the engraver nowadays is a process-maker. **1972** W. P. Lehmann in *Language* XLVIII. 267, I should like to propose that support for a process model of language has been provided by recent typological studies. *Ibid.* 269 In a process-model grammar of language, nominal modifiers are introduced by embedding. **1977** *Trans. Philol. Soc. 1975* 23 The earliest generative ('process morphophonemic') solution I know of to this problem is that of Bloomfield (1933, §13.9). **1967** C. Michalson *Worldly Theol.* i. 19 Daniel Day Williams..was the first theologian to bring the process motifs into combination with other theological traditions. **1958** *IRE Trans. Industr. Electronics* VII. 23/1 Process operations are characterized by the continuous and cyclic handling of large liquid, gas, and bulk flow streams. **1941** W. M. Urban in P. A. Schilpp *Philos. Whitehead* 319 The general group of modern philosophies which are called process philosophies, philosophies, which, in Bergson's terms, find more of reality in becoming than in that which becomes. **1949** B. M. Loomer in D. Brown et al. *Process Philos. & Christian Thought* (1971) 76 The second criticism..runs to the effect that process philosophy, being a kind of naturalism and consequently predisposed in favour of continuity of explanation, neglects the discontinuous qualities of existence. **1960** *Times Lit. Suppl.* 15 Apr. p. xv/2 Dr. Pittenger's Christology seems to be the product of a combination of panentheism, process philosophy, and Christian existentialism. **1971** D. Brown et al. *Process Philos. & Christian Thought* p. v, In recent years, however, process philosophy has come to mean especially, though not exclusively, the philosophy of Alfred North Whitehead and his intellectual descendants, most notably Charles Hartshorne. **1940** *Jrnl. Soc. Motion Picture Engineers* XXXIV. 252 The origination of a combination of projectors superimposing identical prints of the same background on the screen simultaneously compounded the light delivery of a single machine and therefore greatly expanded the scope of background process photography. **1970** C. C. Ammonds *Printing: Basic Sci.* x. 158 Correction for two colours in the achromatic lens is adequate for the simpler forms of color photography, but for process photography..much greater correction is necessary. **1928** C. S. Darling *Exhaust Steam Engin.* ix. 188 It is..possible in a process plant to obtain useful heat from the condenser of a turbine. **1894** *Amer. Dict. Printing & Bookmaking* 460/2 A number of photographers and printers.. hope that they can print direct from process plates. **1931** R. R. Karch *Printing & Allied Trades* xvi. 162 Four-color process plates are used in printing most of the magazine advertisements in color. **1965** Zigrosser & Gaehde *Guide to Collecting Orig. Prints* iv. 71 All prints made by photomechanical methods are called process prints. **1901** *Edin. Rev.* Apr. 551 A few..were found ready to submit their work to the uncertainties and vagaries of the process-printer. **1931** R. R. Karch *Printing & Allied Trades* xvi. 161 *Process printing.* By the use of three transparent colors, red, yellow, and blue, illustrations may be printed that contain all the colours of the rainbow. **1962** L. M. Larsen *Industr. Printing Inks* ii. 33 With the coming of process printing..it has been necessary to use high color strength inks. **1974** J. Craig *Production for Graphic Designer* 105 Four-color process printing is the method used to reproduce full-color continuous-tone copy. **1939** *Jrnl. Soc. Motion Picture Engineers* XXXII. 589 Developments in process projection equipment and technology. **1951** A. Cornwell-Clyne *Colour Cinematogr.* (ed. 3) vii. 583 Process projection, called in Britain 'background projection', a somewhat more precise description, assumed great importance economically in ratio to the continuous rise in the cost of film production. *Ibid.*, Process projection, as a technique, was evolved for the purpose of dispensing with the necessity of location photography. **1977** *Times Lit. Suppl.* 29 Apr. 525/3 There are eight steel engravings after Turner, reproduced absolutely facsimile—line for line. This is close to the edge of possibility in process reproduction. **1962** *Psychol. Bull.* LIX. 329/1 Process schizophrenia involves a long-term progressive deterioration..with little chance of recovery. **1967** Hilgard & Atkinson *Introd. Psychol.* (ed. 4) xxi. 536/2 They hypothesized that process schizophrenia..might be caused by some sort of brain damage..; consequently process schizophrenics might respond..in a manner similar to patients with diagnosed brain damage. **1935** *Proc. Inst. Production Engineers* XIV. 165 Process sheets are drawn up showing the operations to be performed on each component in their correct sequence. **1953** J. J. Rose *Amer. Cinematographer Hand-bk. & Ref. Guide* (ed. 8) 150 Process shots have been the means of saving studios considerable production time and expense in filming scenes for pictures having a foreign locale. **1960** K. Amis *New Maps of Hell* ii. 61 Slow-motion process shots of newts. **1973** D. A. Spencer *Focal Dict. Photographic Technol.* 492 *Process shot*, studio film shot in which a still or moving background is rear projected on to a translucent screen in front of which the action takes place—also called transparency process. **1924** *Power Engineer* XIX. 454/2 In view of the scattered nature of the works, it is not advantageous to attempt to collect condensed process steam. **1954** E. Molloy *Power & Process-Steam Plant* i. 4 Breweries require large amounts of process steam. **1963** A. Jaganmohan tr. *Shlyakhin's Steam Turbines* xi. 160/1 Back-pressure turbines are used..where both electrical energy and process steam are required at the same time. **1972** D. A. Pailin in Cox & Dyson *20th-Cent. Mind* iii. iv. 130 Process theism cannot deal adequately with the nature of God's actuality. **1974** M. Wiles *Remaking of Christian Doctrine* vi. 110 For the process theologian there is no essential problem about the transcendent God's activity in the world. **1977** *Theology* LXXX. 187 The North American process theologians would say..the world is a dynamic totality of events..and not of things. **1971** D. Brown et al. *Process Philos. & Christian Thought* p. vi, When process theology is talked about in American (and to some extent British) theological schools today, Bergson, Berdyaev and Teilhard may be in the background, but the work of Whitehead, Hartshorne, Ogden and Cobb is primarily in mind. **1975** *Times Lit. Suppl.* 15 Aug. 926/2 Process theology, in which God is in a state of *becoming*, has been shown to be anything but Christian. **1977** F. Young in J. Hick *Myth of God Incarnate* ii. 42 Evolutionary theology and process theology are not foreign to the Christian tradition. **1977** *Theology* LXXX. 189 There is, in Whitehead, Hartshorne, and other process thinkers, a full recognition of the reality of natural evil as well as of moral evil and man's sinfulness. **1972** D. A. Pailin in Cox & Dyson *20th-Cent. Mind* iii. iv. 123 Charles Hartshorne..has been the leading exponent of the 'process' view of God... Whitehead is the father of process thought. **1966** *McGraw-Hill Encycl. Sci. & Technol.* X. 642/1 Processes have..been classified into continuous or process-type operations, as in an oil refinery, and intermittent (or repetitive) or manufacturing-type operations. **1972** Process view [see *process thought* above]. **1928** *Rep. Water Pollution Res. Board 1927–8* 9 In some factories.. alternative methods are employed by which the production of process water is avoided. **1949** G. E. H. Lewis *Factory Steam Plant* iii. 46 Where process water heating is not feasible the air preheater merits consideration. **1978** *Environmental Conservation: Chemicals* (Shell Internat. Petroleum Co.) 2 This has been achieved mainly by segregating contaminated process water from the usually much larger volume of uncontaminated storm and cooling water, and..by re-using and recycling process water. **1924** C. A. Suckan *Supervision & Maintenance of Steam-Raising Plant* v. 77 Where steam is used for process work such as boiling. **1908** C. T. Jacobi *Printing* (ed. 4) xxiv. 251 The process worker will have arranged his screens at.. angular distances for the respective colours. **1974** *Nature* 15 Feb. 421/1 In England the death of a 71 year-old former process worker at ICI is being investigated in order to establish whether or not it was caused by long exposure to the fumes of the monomer.

process, *v.*¹ Add: **3. a.** (Further examples.) Also, to operate on (data) (cf. *data processing* s.v. *datum 3). Also *fig.*

1948 A. Toynbee *Civilization on Trial* 84 The form in which this culture has been 'processed' for export. **1957** *B.B.C. Handbk.* 47 The News Bureau..selects and processes news and other items of urgent information for transmission by teleprinter to the news departments. **1958** *Newnes Compl. Amat. Photogr.* 283 Pakolor film can be processed by the user. **1959** *Times Lit. Suppl.* 27 Feb. 109/3 Mr. Morgan presents a deliberately narrowed vision of life, where every detail is 'processed' to fit. **1960** E. Delavenay *Introd. Machine Translation* 122 Language data are indeed processed not only with translation in mind but with the aim of obtaining the widest and deepest penetration of such facts as the relationships between words. **1968** *Brit. Med. Bull.* XXIV. 189/2 Only data which can be explicitly formulated..can be processed by a computer. **1968** *Listener* 4 July 17/3 Rock music is the most efficient medium of creative expression. A song can be composed, processed and broadcast round the world in a week. **1970** *Daily Tel.* 20 May 2/4 The heart of the system is a computer which processes radio signals and continually plots the airliner's position on a moving chart. **1971** *Nature* 11 June 344/1 It may take two months for this volume of vaccine to be processed. **1972** *Language* XLVIII. 271 The right hemisphere of the human brain can also process oral symbols for concrete nouns. But only the hemisphere with a specialized speech center can process verbs. **1976** P. Hill *Hunters* v. 43 We're processing the statements that have already been taken.

b. To subject (a person) to a process, as of registration, examination, or analysis. orig. *U.S.*

1935 *Sun* (Baltimore) 16 Apr. 4/1, 900 applicants were put through medical examinations and transported to army camps to be 'processed'. **1945** H. L. Mencken *Amer. Lang.* Suppl. I. 417 *To process*, now threatens to take its place in the language alongside *to contact*... The New Dealers gave it a much wider range..widening it to include human beings among its objects. It has since been adopted..both in its older sense of doing something to inanimate materials and in its new sense of mauling and manipulating God's creatures. **1948** D. Soibelman *Therapeutic & Industr. Uses of Music* vi. 132 One physician has reported that, since installing music in his waiting room, he has found the average time taken to process a patient reduced by..one-half. **1954** *Manch. Guardian Weekly* 23 Dec. 15/3 All Chinese students..including those whose cases are still being processed, are completely free to travel anywhere in the United States. **1959** *Times Lit. Suppl.* 27 Mar. 173/2 Maupassant, Lautrec, Gauguin—one by one the wild boys are being expertly processed, attractively jacketed, to emerge as items suitable for ticking off on library lists. **1977** *Detroit Free Press* 11 Dec. 21-A/2 Officials at the center said 12 victims were processed there.

processed *ppl. a* (further examples); *processed cheese* = *process cheese* s.v. *process *sb.* 13b; **processing** *vbl. sb.* (further examples); also *attrib.*

1912 *U.S. Dept. Agric. Yearbk. 1911* 387 Processing consists in heating the cans to a sufficiently high temperature to insure the preservation of their contents. **1918** Thom & Fisk *Bk. Cheese* vi. 84 *Processed cheeses.* Cheese of any group may be run through mixing and molding machines and re-packaged in very different form from that characteristic of the variety... The possible variations are numerous. **1933** *Sun* (Baltimore) 15 July 1/6 He accordingly proclaimed August 1 as..the date upon which the processing tax should become operative. **1936** *Discovery* May 157/1 A well-prepared wood or esparto paper can be more permanent than a carelessly processed rag paper. **1958** *Newnes Compl. Amat. Photogr.* 280 Use a film for which processing kits are available. **1958** *Times Lit. Suppl.* 29 Aug. 478/4 It is instructive, if unedifying, to follow the tergiversations and admire the polemical acrobatics of various practitioners of the art of literary processing, notably of that arch-processor V. Yermilov. **1960** Processed pea [see *garden-pea* s.v. *garden *sb.* 6]. **1964** L. Deighton *Funeral in Berlin* xxxviii. 231 His processed cheese sandwiches. **1966** A. Young in *Spero* I. ii. 19 He sat behind me in Homeroom, sportshirt, creased pants, shiny black pointy-toed stetsons, jacket, processed hair. **1972** *Daily Tel.* 29 Apr. 12/1 The fully-automated, processed tourist rarely visits Genoa. **1977** B. Pym *Quartet in Autumn* vii. 63 Fresh vegetables..would be better than processed peas. **1977** J. Hedgecoe *Photographer's Handbk.* 70 (*caption*) Various types of processing drum are made for color prints. **1979** *SLR Camera* Jan. 43/1 The catalogue almost swells at the seams with such goodies as processing drums, colour analysers, printing filters, [etc.].

process, *v.*² Add: **1.** (Further examples.)

1902 *To-Day* 20 Aug. 113/1 Neither Barnum nor the new Lord Mayor will be able to process this year. **1912** A. Huxley *Let.* 23 June (1969) 46 On the Bismarck Tower a bonfire was lighted and 1000 odd students processed from the tower to the University. **1953** H. Nicolson *Diary* 4 July (1968) 242 We process in robes to the City Hall where there are many graduands. **1962** G. Moore *Am I too Loud?* xxiii. 254 The vision of our young and beautiful Queen processing slowly up the aisle in her gorgeous robes is never to be forgotten. **1971** K. Thomas *Relig. & Decline of Magic* iii. 63 They also involved processing across the field with cross, banners and bells to drive away evil spirits and bless the crops.

2. *trans.* To lead or carry (a person, etc.) in procession; to traverse (an area) in procession.

1959 *Times* 10 Dec. 14/7 The Lord of Miracles is solemnly processed all round the city. **1968** D. M. Smith *Mod. Sicily* iii. 484 The flagellants then processed the streets as they had done in 1647 and 1773. **1974** D. Avery *Not on Queen Victoria's Birthday* vii. 117 Most of the families left their feasting to attend the sermon in the church under the impression that the saint was to be processed afterwards.

Hence **proce·ssing** *vbl. sb.* and *ppl. a.*

1920 *Blackw. Mag.* Dec. 712/2 The bowings and curtseyings and processings and workings of the Puddispor congregation could have told anybody *that*! **1959** *Antiquity* XXXIII. 19 A single row of processing animals. **1977** *Gramophone* May 1724/2 So we must be prepared for ..the shuffling of processing feet, the coughing and the tramping of the congregation.

processable (prōᵘ·sesăb'l), *a.* [f. Process *v.*¹ + -able.] That can be processed. Hence **pro:cessabi·lity**, the capacity to be processed.

1956 *Industr. & Engin. Chem.* May 930/1 Processing studies of Adiprene B urethane rubber have resulted in an understanding of a number of the factors required for processability. *Ibid.* 932/1 They become progressively tougher and more viscous while remaining 'processable' even in an advanced state of scorch. **1965** J. R. Scott *Physical Testing of Rubbers* ii. 45 Examples of this empirical approach to the problems of measuring processability are the procedures developed in the U.S.A. for the wartime synthetic rubber program. **1967** Cox & Grose *Organiz. Bibliogr. Rec. by Computer* vii. 183 This may help us in problems of transmission of machine processable cataloguing data. **1971** P. M. Hubbard *High Tide* ii. 24, I was a sort of raw material, which had not proved as processable as they had hoped.

processer, var. *processor.

procession, *sb.* Add: **1. c.** (See also quot. 1937.)

1937 Partridge *Dict. Slang* 661/1 *Procession*, as applied to a race, esp. a boat-race (above all, one in which there are only two crews), implies 'an ignominious defeat'. **1958** *Times* 22 Sept. 14/2 Although she [*sc.* the British yacht] made up half a minute..it was obvious that..it could not now be anything more than a procession.

d. *Cricket.* A rapid succession of batsmen; a batting collapse.

1891 W. G. Grace *Cricket* iii. 76 West Gloucestershire [scored] 6 only. Only nine overs were bowled, and it was a most inglorious procession. **1927** M. A. Noble *Those 'Ashes'* 210 The Civil Service first innings was almost a procession. They were able to make only 59. **1977** *Times* 17 Jan. 7/1 The Australian procession started when Turner was caught by Majid off Sarfraz for 11.

processional, *a.* Add: **b.** (Later example.) Also, traversed by a procession or processions.

1906 H. Begbie *Priest* viii. 124 The sound of a heavy step approaching from the processional aisle on the south caught her ear. **1942** *Country Life* 9 Oct. 695/1 (*caption*) Processional way to Buckingham Palace from Victoria Station. **1973** *Times* 15 Oct. 4 Thieves broke into St Albans Abbey, Hertfordshire, on Saturday night and stole..the silver gilt top of a processional cross.

processionally, *adv.* (Further examples.)

1851 G. B. Pagani *Life of Rev. A. Gentili* iii. 190 Father Gentili commenced the custom of going processionally on Sundays and singing the Litany of the Holy Name of Jesus, from the chapel at Gracedieu all the way to Osgothorpe. **1895** 'Mark Twain' in *N. Amer. Rev.* July 10 There were now three bullets in that one hole—three bullets imbedded processionally. **1936** G. B. Shaw *Simpleton* ii. 59 (*stage directions*) Kanchin and Janga enter processionally, reading newspapers. **1960** R. W. Marks *Dymaxion World of B. Fuller* 8/1 In Fuller's special argot, however, 'regenerative' means 'multi-orbital, cyclic, processionally concentric'—a definition which itself requires definition.

processioner. 4. (Earlier examples.)

1731 *Bristol Parish* (Va.) *Vestry Bk.* (1898) 59 Order'd that George Tucker be Prosessioner in the Stead of Robert Tucker junr who is Lame and cannott Officiate as prosessioner. **1795** in L. P. Summers *Ann. Southwest Va.* (1929) 463 The said Processioners to examine their business the first day of February next. **1828** W. Bolling in *Va. Mag. Hist. & Biogr.* (1938) XLVI. 321 Attended the processioners around my lines..to the upper line between Dr. Watkins and myself.

processor (prōᵘ·sesəɹ, *U.S.* prǫ·s-). Also processer. [f. Process *sb.* + -or.] **a.** A person who performs a process. **b.** A machine or system which performs a process. Cf. *microprocessor* s.v. *micro- 1.

1909 M. B. Saunders *Litany Lane* ii. 10 Her tragic face..was already being 'blocked' for the night's press in many a rushing 'processor's' den. **1934** *Planning* I. xx. 5 Various industries handling agricultural products between the first processor and the consumer have recently been turned over to the N R A. **1948** *Times* 2 Mar. 2/3 [The] film processor..was sentenced to 21 months' imprisonment. **1959** *Times* 9 Mar. (Britain's Food Suppl.) p. xviii/1 Processors of Vegetable Oils for the Biscuit and Margarine Industries. **1960** *Farmer & Stockbreeder* 5 Jan. 53/2 It would provide the foundation on which producers and processers could develop quality and reduce costs. **1962** *Times* 9 Oct. (Uganda Suppl.) p. v/4 African processers have invested considerable sums in permanent salting vats. **1977** *New Yorker* 29 Aug. 62/2 Any program that runs on the system can access any information physically accessible to the (central) processor. **1978** *Homes & Gardens* Oct. 140/1 Food processors, or kitchen machines..are food choppers and slicers with some abilities at mixing.

processual, *a.* Restrict *Roman Law* to sense in Dict. and add: **b.** Pertaining to a social or linguistic process.

1957 R. K. Merton *Social Theory* (rev. ed.) ix. 316 An instructive processual analysis of the formation of subgroups. **1958** Willey & Phillips *Method & Theory in Amer. Anthropol.* 5 In the context of archaeology, processual interpretation is the study of the nature of what is vaguely referred to as the culture-historical process... It implies an attempt to discover regularities in the relationships given by the methods of culture-historical integration. **1960** *Amer. Anthropologist* LXII. 19 How processual analysis might be more systematically related to structural analysis. **1970** *Antiquity* XLIV. 28/1 In both these works..historical activities tend to be viewed as being essentially descriptive, while the ultimate aims of archaeology are characterized as being processual, that is to say, concerned with the formulation of general rules of cultural behaviour. **1971** [see *process *sb.* 5b]. **1977** *Word 1972* XXVIII. 295 Since understanding sentences, and generating others, in the language described implies constant switches from process to system and from system to process..he will have to find the most reasonable compromise between the requirements of systematic and processual simplicity.

prochlorperazine (proklǫɹpe·răzīn). *Pharm.* [f. Pro(pyl- + Chlor- + *pi)perazine.] A pale yellow viscous liquid which is used, usu. in the form of one of its salts, as a tranquillizer; 2-chloro-10-[3-(4-methylpiperazin-1-yl)-propyl]phenothiazine, $C_{20}H_{24}ClN_3S$.

1958 *Psychiatric Res. Rep.* IX. 23 Proc[h]lorperazine treatment of psychotic patients was begun at Longview Hospital in November, 1955. **1959** *Jrnl. Amer. Med. Assoc.* 16 May 361/1 Caffeine and sodium benzoate seems to be an effective and rapidly acting antagonist to the toxic manifestations of prochlorperazine. **1977** *Lancet* 9 July 94/2 Dizziness and vomiting can be avoided by intravenous prochlorperazine 12·5 mg given 5 min before the mexiletine injection.

prochromosome (prōᵘkrōᵘ·mǒsōᵘm). *Cytology.* Also with hyphen. [ad. G. *prochromosom* (J. B. Overton 1905, in *Jahrb. f. wissensch. Bot.* XLII. 126): see Pro-² and *chromosome.] One of the densely staining heterochromatic masses seen in certain interphase nuclei, frequently associated with centromeres; = *chromocentre.

1906 *Proc. R. Soc.* B. LXXVII. 557 These bodies correspond in fact exactly to what in 1904 we have already described as the Anlagen of the premeiotic chromosomes in the corresponding cells in the testes of Periplaneta, and there can be no doubt that they represent also the structures subsequently alluded to as prochromosomes by Overton, Miyake, and Strasburger in the same stage in certain mono- and dicotyledonous plants. **1907** *Ann. Bot.* XXI. 335 In certain plants there is a tendency of the chromatin to form lumps or masses in the resting condition, in which there is often a general uniformity in size, and when the number of such lumps approaches the number of somatic chromosomes, each mass has been looked upon by some as representing a prochromosome, or the centre of organization of a chromosome. **1934** [see *euchromocentre]. **1965** A. K. & A. Sharma *Chromosome Techniques* viii. 205 The number of prochromosomes may be equal to, or if fused, less than, the number of chromosomes in the complement. **1969** Brown & Bertke *Textbk. Cytol.* i. 8/2 Flemming (1882) illustrated stained nodes in the nuclear network... Overton named them *prochromosomes* in 1905 and 3 years later Baccarini called them *chromocenters*; both names are still in use.

procidence. Add: Now usu. as mod.L. procidentia. Freq. distinguished from *prolapse* or restricted to the more severe kinds (see quots.).

1607 Topsell *Four-f. Beasts* 388 Another disease called *Procidentia ani*, that is to say, the falling out of the fundament. **1829** S. Cooper *Good's Study of Med.* (ed. 3) V. 146 If the descent [of the uterus] be only to the middle of the vagina, it is called *relaxatio uteri*; if to the labiæ, *procidentia*; if lower than the labiæ, *prolapsus*. The distinction is of trifling importance. **1888** A. H. N. Lewers *Pract. Text-bk. Dis. Women* viii. 113 When the uterus has partly, or wholly passed the orifice of the vulva, the case is called one of 'procidentia'. *Ibid.* 114 In extreme cases of procidentia the whole uterus lies outside. **1903** J. P. Tuttle *Treat. Dis. Anus* xvii. 667 Procidentia..is practically always applied to those cases in which all the coats of the bowel descend. **1956** H. E. Bacon et al. *Proctology* xviii. 215 The difference between prolapse and procidentia lies in the coats of the bowel involved. *Ibid.* 218 Protruding through the anal orifice, procidentia is diagnosed from a series of circular folds irregularly placed. **1974** Passmore & Robson *Compan. Med. Stud.* III. xxviii. 62/2 In third degree prolapse (procidentia) the whole uterus lies outside the introitus.

procidentia: see prec.

Procion (prōᵘ·siǫn). A proprietary name for any of a large class of reactive dyestuffs based on 1,3,5-triazine and covering a wide range of colours. Usu. *attrib.*

1956 *Trade Marks Jrnl.* 23 May 424/1 *Procion*... Dyes, dyestuffs and colouring matters, none being for laundry or toilet purposes. Imperial Chemical Industries Limited, ..London, S.W.1; Manufacturers and Merchants—10th February 1956. **1957** *Listener* 18 July 73 (Advt.), Polythene, for instance,..bright and fast new dyes like the 'Procion' range... They're all I.C.I. discoveries, you know. **1957** *Official Gaz.* (U.S. Patent Office) 23 July TM 129/1 *Procion*... For dyes, dyestuffs, and colouring matters. **1971** R. L. M. Allen *Colour Chem.* xiii. 205 Procion dyes provide a complete range of shades in both the M and H series. **1974** *Sci. Amer.* Jan. 41/1 The branching patterns of a cell can be traced by injecting the fluorescent dye Procion yellow or by stimulating and then recording from its axons in roots and connectives. **1976** *Nature* 25 Mar. 338/2 Procion dye was injected intracellularly by iontophoresis to determine axonal morphology of one or two cells in a particular brain.

Proclian (prǫ·kliăn), *a.* Also **Pro·cline** (-lain). Of or relating to Proclus (A.D. ?410–85), a neo-Platonist philosopher and head of the Athenian school after Plutarch and Syrianus, his views, or works.

1912 F. von Hügel *Eternal Life* vii. 118 We find in Eckhart a..scientific, still predominantly Proclian, thirst for intellectual utter simplicity and clearness. **1951** *Mind* LX. 417 The Proclian *Liber de causis*. **1967** I. P. Sheldon-Williams in A. H. Armstrong *Camb. Hist. Later Greek & Early Med. Philos.* 477 The Procline Neoplatonism had worked out those implications. *Ibid.*, The enemies were..the Procline theology based on polytheism, and the Procline theurgy deriving from the belief in a supernatural power inherent in the phenomenal world.

procli·max. *Ecol.* [f. Pro-² 1 + *climax *sb.* 4 b.] (See quot. 1938.) Also *attrib.*

1934 F. E. Clements in *Jrnl. Ecol.* XXII. 45 For those cases in which the community is modified and held for a more or less indefinite period in some other condition, the term 'proclimax' is suggested. **1938** Weaver & Clements *Plant Ecol.* (ed. 2) xviii. 480 Proclimax is a general term which includes all the communities that simulate the climax to some extent..but lack the proper sanction of the existing climate. It thus includes subclimax, preclimax, and postclimax, as well as disclimax. **1951** *Jrnl. Ecol.* XXXIX. 81 The pro-climax communities of Mediterranean cliffs..develop directly from the crustose-lichen stage. **1975** R. H. Whittaker *Communities & Ecosystems* (ed. 2) iv. 182 These may be termed 'subclimaxes', or 'proclimaxes' of various sorts, in Clements' treatment.

Procline: see *Proclian *a.*

proclisis (prō·klisis). *Gram.* [mod.L., f. Gr. πρό forward + κλίσις, f. κλίνειν to lean.] Pronunciation as a proclitic; the transference of stress to a following word.

1893 J. Clark *Man. Linguistics* vi. 158 The existence of proclisis, which naturally is lifted in emphatic positions, or freeing itself at the end of a sentence. **1955** *Archivum Linguisticum* VII. ii. 135 The facts of enclisis and proclisis in Polish are generally on the same lines as in Czech. **1964** *Language* XL. 276 Since he objects to the application of the idea of proclisis as well as enclisis, he keeps to the terms 'anteposition' and 'postposition' in referring to the placing of the pronoun with respect to the verb.

proclitic, *a.* and *sb.* Add: **A.** *adj.* (Further example.)

1973 A. H. Sommerstein *Sound Pattern Anc. Greek* ii. 11 Prepositions, being proclitic, are separated from the following word by a single # boundary.

Procne: see *Progne.

procoagulant: see *pro-² 1. **pro-Communist:** *pro-¹ 5a.

proconsul. Add: **3.** (See quot.)

1939 *Sunday Times* 8 Jan. 13/1 A pro-Consul..is a resident member of a British trading community abroad, generally a shipping agent or merchant of repute.

4. Usu. with initial capital. [A. T. Hopwood 1933, in *Ann. Mag. Nat. Hist.* 10th Ser. XI. 98.] An extinct ape belonging to the genus so called, known from Miocene fossil remains discovered in East Africa by Louis Leakey (1903–72) in 1932.

1933 *Jrnl. Linn. Soc.: Zool.* XXXVIII. 457 It would seem that the dentition of *Proconsul* is more primitive than that of the chimpanzee. **1954** *New Biol.* XVII. 12 The remains of fossil apes are especially abundant [on Rusinga Island in Lake Victoria]..of which the most famous has been *Proconsul*. **1962** *Listener* 5 Apr. 589/1 This animal, which has been named Proconsul, existed approximately 25,000,000 years ago. **1973** J. Bronowski *Ascent of Man* i. 38 A classical find made by Louis Leakey goes by the dignified name of *Proconsul*... (The name *Proconsul*..was coined to suggest he was an ancestor of a famous chimpanzee at the London Zoo in 1931 whose nickname was Consul.) **1977** A. Hallam *Planet Earth* 284 Proconsul, an animal about the size of a small baboon, is known from a good skull, jaws and some limb bones.

proconsulate. b. (Later example.)

1933 G. Arthur *Septuagenarian's Scrap Bk.* 243 The soldier whose proconsulate was extended for over two years and who was then wistfully asked if he would not come back again.

proconsulship. (Later example.)

1976 *Church Times* 25 June 11/4 The only absolutely firm date in the whole of the New Testament is the proconsulship of Gallio, shown by an inscription to have begun in the early summer of 51.

proconvertin: see *pro-² 1.

procreator. (Later example.)

1975 *Way* Suppl. xxv. 12 In the sexual union, man and woman under God become procreators.

procrypsis (prokri·psis). *Zool.* [f. Pro- (cf. Procryptic *a.*) + Gr. κρύψις concealment.] Protective colouring in animals.

1920 G. D. H. Carpenter *Naturalist on L. Victoria* 196 Procryptic colouring conceals its wearer from danger, causing it to resemble either the general surroundings or some particular part thereof (Special Procrypsis). **1933** *Discovery* Nov. 357/2 The resemblance of creatures to their environment..technically termed Procrypsis. **1946** *Nature* 24 Aug. 278/1 Is resemblance to part of a dead ant to be sharply separated from resemblance to a whole, living ant? To a Darwinian, one is an example of special procrypsis, while the other is pseudoaposematic. **1977** M. Tweedie *Insect Life* ii. 74 We see among them [*sc.* insects] the first stages of the type of adaptation commonly known as camouflage, but among biologists as cryptic adaptation or procrypsis.

procto-. Add: **proctoscope, -scopic** *a.* (earlier examples); **procto·scopy**, use of, or examination with, a proctoscope.

1896 *Mathews' Med. Q.* III. 203 A little practice in Kelly's method teaches the operator how to carry the end of the proctoscope away from the prostate. *Ibid.* 208 The proctoscopic mirror faces the operator. *Ibid.* 332 A central aperture, which..escaped my own digital perception..and was not discovered until subsequently revealed by proctoscopy. **1977** *Lancet* 21 May 1085/1 The diagnosis of inflammatory bowel disease was..confirmed at proctoscopy.

proctology (prǫktǫ·lŏdʒi). [f. Procto- + -logy.] The branch of medicine concerned with the anus and rectum or (with some writers) with the anus and the whole colon.

1899 *Trans. Ohio State Med. Soc.* 257 Thos. Chas. Martin M.D. Cleveland. Teacher of Proctology in the Cleveland College of Physicians and Surgeons; Proctologist to the Cleveland General Hospital. **1929** W. E. Minor *Clin. Proctology* iv. 37 Some years ago I adopted a terminology which recognizes only one type of hemorrhoid. I find this solution very acceptable and helpful to the student of proctology. **1956** H. E. Bacon et al. *Proctology* p. ix, The late Joseph M. Mathews defined proctology as the science that treats of surgical diseases of colon, rectum and anal canal. **1959** K. Zimmerman in R. Turell *Dis. Colon & Anorectum* II. lviii. 1195 The practice of 'office proctology', or 'ambulant proctology', was for many years in the hands of charlatans. **1976** *Times* 29 Apr. 18/4 The section of proctology of the Royal Society of Medicine held their annual dinner at 1 Wimpole Street, yesterday evening.

Hence **proctolo·gic, proctolo·gical** *adjs.*; **procto·logist**, an expert or specialist in proctology.

1899 *Trans. Amer. Proctologic Soc.* I. p. v, Article I. The name of this Association shall be the *American Proctologic Society.* Article II. Its object shall be the cultivation and promotion of knowledge in whatever relates to Disease of the Rectum and colon. **1899** *Ohio State Jrnl.* 8 June 3/1 A new national medical association to be known as the American Proctological society was organized yesterday. **1899** Proctologist [see above]. **1907** *Proctologist* I. 1 No medical work..is complete, today, that does not deal fully and explicitly with proctological diseases. **1926** L. J. Hirschman *Handbk. Dis. Rectum* (ed. 4) 7 As efficient diagnostic and therapeutic service should be rendered to the proctologic patient as to any other. **1950** J. P. Nesselrod *Proctology* ii. 54 Proctologists are not yet in full accord..with regard to the extent of their field. **1959** K. Zimmerman in R. Turell *Dis. Colon & Anorectum* II. lviii. 1195 Many..proctologic conditions may be treated in the office. **1964** *Punch* 26 Aug. 307/2 Proctologists, who are admired as working in a gold mine. **1971** *Dis. Colon & Rectum* XIV. 8/2 Barium-enema examination and proctologic evaluation were requested. **1979** *Guardian* 9 Jan. 8/5 Mr Carter's own proctological history. **1980** S. Stein *Resort* i. 14 Politicians are assholes attended by proctologists.

procto·rially, *adv.* [f. Proctorial *a.* + -ly².] In a proctorial capacity; in the manner of a proctor.

1883 H. S. Holland in S. Paget *H. S. Holland* (1921) III. iv. 290 You speak of a deeper sense of the power of evil—I have felt it proctorially. **1971** F. R. Leavis in *Human World* Aug. 5 As for myself and Cambridge, I haven't to complain that I suffered..proctorially enforced oppression.

Proctor (prǫ·ktəɹ). The name of Robert George Collier *Proctor* (1863–1903), English bibliographer, used *attrib.* or in the possessive in **Proctor method** = *Proctor('s) order*; **Proctor number**, the number assigned to an early printed book in Proctor's *Index*; **Proctor('s) order**, a system of classifying early printed books geographically and chronologically, first used in the *Index to the Early Printed Books in the British Museum* (1898–1938) begun by Proctor.

[**1903** *Library* IV. 195 Proctor's 'Index'..is arranged under towns and countries in chronological order.] **1904** A. W. Pollard in *Library* V. 22, I think it was Baer of Frankfurt who first gave the author the pleasure of seeing a 'Proctor number' quoted side by side with that of Hain. **1931** M. R. Stillwell *Incunabula & Americana* 22 In the majority of instances, it will be found that Proctor's order prevails in whatever place-entry bibliography one may wish to consult. **1934** A. Esdaile *National Libraries of World* i. 18 Incunabula..gathered by Robert Proctor into one room..and arranged by order of countries, towns, presses, and date, an arrangement now often called 'Proctor order'. **1952** J. Carter *ABC for Bk.-Collectors* 143 *Proctor's order*, the classification of early printed books, on scientific typographical principles, by country, town and printer. **1955** *N. & Q.* May 229/1 Henry Bradshaw..arranged the books in what he called his natural history method in bibliography; it meant the arrangement and classification of them as natural objects are classified. This arrangement is now popularly known as the Proctor method. **1961** T. Landau *Encycl. Librarianship* (ed. 2) 291/1 *Proctor order*, system of classification of incunabula named after the arrangement in R. G. C. Proctor's Index of early printed books in the British Museum. **1967** Cox & Grose *Organiz. & Handling Bibliogr. Rec. by Computer* 137 Since the end of the last century it has been customary to catalogue incunables in 'Proctor Order'.

proctotrupid (prǫktotrū·pid), *a.* and *sb. Ent.* Also **Procto-, -trypid.** [f. mod. L. family name *Proctotrupidæ*, f. generic name *Proctotrupes* (P. A. Latreille *Précis des Caractères génériques des Insectes* (1796) 108), f. Procto- + Gr. τρυπ-ᾶν to bore: see -id³.] **A.** *adj.* Of, pertaining to, or designating a proctotrupid. **B.** *sb.* A small wasp belonging to the family Proctotrupidæ or the superfamily Proctotrupoidea, which include parasitoids of insects and spiders.

1869 [in Dict. s.v. Procto-]. **1891** [see *chalcid *a.* and *sb.*]. **1932** Riley & Johannsen *Med. Entomol.* xxiv. 422 In California a little proctotrupid..invades houses in numbers in the fall. **1932** E. Step *Bees, Wasps, Ants Brit. Isles* 184 The wings are much simpler..in the Chalcids and Proctotrypids. **1972** L. E. Chadwick tr. *Linsenmaier's Insects of World* 300/2 Equally small and also at home in the water are a few proctotrupids.

proctotrupoid (prǫktotrū·poid). *Ent.* [f. mod.L. name of superfamily *Proctotrupoidea*: see prec. and -oid.] = *proctotrupid *sb.*

1954 Borror & DeLong *Introd. Study of Insects* xxix. 716 The proctotrupoids are not as common as the chalcids or ichneumons. **1971** R. R. Askew *Parasitic Insects* viii. 155 The great majority of proctotrupoids are endoparasites.

procumbent, *a.* Add: **3.** *Zool.* Of a tooth: lying along the jaw.

1874 T. C. Jerdon *Mammals of India* 62 Upper middle incisors distant; lower ones procumbent. **1902** *Encycl. Brit.* XXX. 506/1 In the lower jaw there is a single pair of procumbent incisors, followed by several small teeth representing the canine and early premolars. **1977** Roonwal & Mohnot *Primates S. Asia* 41 It holds small fruit in both hands while chewing, and larger food, such as an unpeeled banana, is chipped with its procumbent lower incisors. **1978** *Nature* 17 Aug. 663/1 Diplodocids had elongated tapering snouts with delicate, procumbent teeth for selecting smaller plant parts.

procuracy. Restrict †*Obs.* to other senses in Dict. and add: **1. a.** (Later example.)

1978 *Jrnl. R. Soc. Arts* CXXVI. 672/2 The rôle, function and powers of public security organs, the procuracy and the courts are more closely defined.

procurator. 2. a. For †*Obs.* read '*Obs.* except *Hist.*' and add later examples.

1897 Addis & Arnold *Cath. Dict.* (ed. 5) 761/2 The procurators or official agents of monasteries of nuns should not hold office more than three years. **1909** B. Ward *Dawn Cath. Revival* I. 55 The procurator [at Douay] was Rev. Gregory Stapleton, who had held that office since 1773. **1931** J. Clayton *St. Hugh of Lincoln* v. 35 The procurator was guest master... Other visitors to the Grande Chartreuse claimed the procurator's time... They loved him, Hugh the procurator, for the gracious speech and courteous treatment.

procurement. Add: **3.** *Mil.* The action or process of procuring equipment and supplies. Freq. *attrib.*

1957 [see *logistical *a.* 4]. **1958** *Times Rev. Industry* Mar. 9/2 Strategic materials on the active stockpiling list had reached their procurement priority levels by the middle of last year. **1966** *Amer. Speech* XLI. 300 It receives the plans and decisions of the Secretary of Defense and Secretary of the Air Force, and its own Chief of Staff, and translates them into training, logistic, and procurement programs. **1966** *Electronics* 17 Oct. 103 It is this growth in avionics complexity that is making military procurement officers insist on built-in test capability. **1977** *R.A.F. News* 30 Mar.–12 Apr. 10/4 It is then the task of the operational requirements staff to define the parameters in a detailed operational requirement which..is passed to the Procurement Executive who put it to industry to see how it can best be met.

procureur. Add: **c.** = Procurer 4.

1910 *Times* 29 Apr. 14/1 The procureurs (the cant name is 'ponce') at work in this country are mostly foreigners. **1979** W. J. Fishman *Streets East London* 52/2 Lodging houses infested by thieves, *procureurs* and prostitutes.

‖ **procureuse** (prokürö̂z). [Fr.] = Procuress 2.

1930 E. Waugh *Vile Bodies* vi. 105 What a coarse face..she looks like a *procureuse*. **1968** C. Cooper *Thunder & Lightning Man* iv. 52 Does she condone it?.. Is she just an old *procureuse*? **1977** 'R. Player' *Month of Mangled Models* vii. 123 She's only a common procureuse dressed like a duchess.

procurrent (prokʊ·rĕnt), *a. Ichthyol.* [ad. L. *procurrent-*, *procurrens*, pres. pple. of *procurrere* to run forward.] Of a fish's fin: having rays that are almost parallel.

1902 Jordan & Evermann *Amer. Food & Game Fishes* 538 Procurrent (fin). With the lower rays inserted progressively further forward. **1931** J. R. Norman *Hist. Fishes* iii. 72 True spines are never developed in this [caudal] fin, but rudimentary or procurrent rays resembling spines may be found at the base of the lobes.

procurvature. [f. Pro-² + Curvature.] = Procurvation.

1903 R. I. Pocock in *Ann. & Mag. Nat. Hist.* XI. 411 The species described by Keyserling as *Trechona pantherina* appears to me to be the female of *auromitens*, in spite of a less procurvature of the anterior line of eyes.

procuticle (prokiū·tik'l). [f. Pro-² + Cuticle.] The inner, thicker layer of the cuticle of an arthropod, situated below the epicuticle and comprising the exocuticle (if present) and the endocuticle.

1951 A. G. Richards *Integument of Arthropods* xvi. 148 The term procuticle is proposed for the embryologically original (parent) chitin-protein fraction. *Ibid.* 149 The procuticle may remain seemingly unchanged in soft transparent cuticle and soft areas.., in which case the fully formed cuticle is said to consist of epicuticle and endocuticle... Or the outer portion of the procuticle may become hardened and darkened by sclerotization, giving an outer dark exocuticle and an inner transparent endocuticle. **1959** W. Andrew *Textbk. Compar. Histol.* iii. 91 It [*sc.* the cuticle of arthropods] presents a great diversity of structure but in general may be divided into an outer part without chitin, the 'epicuticle', and an inner part with chitin, the 'procuticle' or 'endocuticle'. **1962** Gordon & Lavoipierre *Entomol.* ix. 53 The procuticle confers on the integument amongst other properties those of hardness and strength and is the real skeletal support of the body of the insect. **1967** [see *exocuticle a]. **1976** C. P. Friedlander *Biol. Insects* i. 16 When first secreted the entire procuticle is in the endocuticle condition; subsequently a large proportion of it is hardened.. to form the exocuticle.

procyonid (prō̆u·si,ǫnid), *sb.* and *a.* [f. mod.L. family name *Procyonidæ*, f. generic name *Procyon* (G.C.C. Storr *Prodomus Methodi Mammalium* (1780) 35): see Procyon.] A mammal belonging to the family Procyonidæ, which includes racoons and pandas. Also as *adj.*, of, pertaining to, or resembling an animal of this kind.

1909 *O.E.D.*, Procyonid. **1910** H. F. Osborn *Age of Mammals* iv. 288 This [*sc.* the Lower Miocene] is the first geological appearance of the characteristically American family of raccoons, or procyonids. **1921** *Proc. Zool. Soc.* 419 The genus [*Ailuropoda*] is neither Ursid nor Procyonid. **1941** *Geol. Ser. Field Mus. Nat. Hist.* VIII. 33 (*title*) A new procyonid from the Miocene of Nebraska. **1964** E. P. Walker *Mammals of World* II. 1179/2 Most procyonids travel in pairs or family groups. **1973** *Nature* 28 Sept. 218/2 Its [*sc.* the giant panda's] closest affinities are with the ursids (bears) or procyonids (raccoons). **1978** T. A. Vaughan *Mammalogy* (ed. 2) xii. 217/1 Procyonids are of modest size, weighing from less than a kilogram to about 20 kg.

procyonine (prō̆u·si,onīn), *a.* [f. mod.L. subfamily name *Procyoninæ*: see prec.] (Earlier and later examples.)

In Dict. s.v. Procyon.

1869 [see *Arctoid *a.* (*sb.*)]. **1921** *Proc. Zool. Soc.* 418 If the tooth in *Ailuropoda* is not Ailurine or Procyonine, it is certainly not Ursine.

prod, *sb.*¹ Add: **1. b.** *N. Amer. colloq.* (Further examples.)

1910 B. Edwards in H. A. Dempsey *Best of Bob Edwards* (1975) v. 96 The old man was on the prod. **1947** B. A. De Voto *Across Wide Missouri* 26 Not only the Arikaras but the Blackfeet were on the prod. **1962** [see *onery, onnery, o'n'ry].

2. b. *Founding.* Any of a number of pointed projections, intended to hold the loam, on the flat metal base used for preparing a loam mould.

1888 *Lockwood's Dict. Mech. Engin.* 267 The pyramidal or conical points cast on loam and core plates for the retention of the loam are termed prods. **1889** J. G. Horner *Pract. Iron Founding* viii. 103 A plate..is cast, studded over with 'prods' to hold the loam which is swept over its face. **1923** — *Mod. Ironfoundry* vii. 65 Prods are cast on many loam mould plates. Generally, they occur on one side only, and the pattern prods are mounted in a strip of wood, provided with a handle.

Prod (prǫd), *sb.*³ and *a.* Also **prod.** An Anglo-Irish colloq. abbrev. of Protestant *sb.* 2 a and *a.* 1. Cf. *Prot *sb.* and *a.*

1942 E. Bowen *Seven Winters* 51 She spoke of 'Prods' (or, extreme, unctuous Protestants) with a flighty detachment that might have offended many. **1961** *Spectator* 28 Apr. 599 He was a 'Mick', I was a 'Prod' but we found no difficulty in being friends although we differed in faith. **1970** M. Kenyon *100,000 Welcomes* ii. 14 A long-hair student, or a Prod, or similar riff-raff. **1974** *Irish Democrat* Dec. 7/2 This is..about O'Brienism, which is based on the fearful symmetry that taig is taig and prod is prod and never the twain shall meet. **1977** P. Carter *Under Goliath* iii. 15 Most of the kids were in tough Prod gangs, like the Tartans... They always seemed to..tell if you were as hard-line Prod as they were.

Proddy (prǫ·di), *a. colloq.* (chiefly *Anglo-Irish*). Also **proddy.** [f. *Prod *sb.*³ and *a.* + -y⁶.] Protestant. Also *Comb.*, as (children's slang) *Proddy-hopper, -woddy*; **Proddy Dog** (opp. *Cat*: Catholic).

1954 W. K. Hancock *Country & Calling* i. 50 And they would sing: *Proddy Dog, Proddy Dog, Sitting on a well, Up comes the Devil And pulls him down to hell.* Then we and the Catholic boys would pelt each other with cow-dung. **1958** I. Cross *God Boy* 165 Proddy-hopper, proddy-hopper, go to hell. **1959** I. & P. Opie *Lore & Lang. Schoolch.* xvi. 344 In Ireland, both north and south, Catholics are 'Cathies' and Protestants 'Proddy-woddys'. *Ibid.*, In Staines..R.C. children call the Protestants 'Old Proddy Dogs'... They still call them 'Proddy Dogs' at Ilford. **1961** *Spectator* 28 Apr. 603 In other streets Papist and Proddy schoolboys could pass in peace. **1968** T. Parker *People of Sheets* 60 We always divided up into the same two sides, The Cats and The Proddy Dogs. **1975** G. Seymour *Harry's Game* v. 76 Nice safe little billet.. in a nice Proddy area... I'm not going..to sit on my arse in Proddyland.

prodelision. (Further examples.)

1933 *Trans. Philol. Soc. 1931–32* 32 They would point to the patent phenomena of adaptation or assimilation of the two sounds to one another..and also to the facts of elision, prodelision, and crasis (in the case of vowels), that is, of coalescence. **1968** W. S. Allen *Vox Græca* iv. 96 Much rarer than elision is the process of 'prodelision' in which it is the short initial vowel of the second word that is lost after a final long vowel or diphthong.

prodelta(ic): see *pro-² 2.

prodigiosin (prodidʒiō̆u·sin). *Biochem.* [a. G. *prodigiosin* (E. Kraft *Beiträge zur Biol. des B. Prodigiosus* (Dissertation, Würzburg, 1902) 37, following suggestion by K. B. Lehmann), f. mod.L. *prōdigiōs-us* marvellous, Prodigious (former specific epithet of the bacterium now called *Serratia marcescens*): see -in¹.] A dark red crystalline pigment with antibiotic properties which is produced by certain bacteria of the genus *serratia* and has a molecule ($C_{20}H_{25}N_3O$) consisting essentially of three pyrrole rings linked to a central carbon atom.

1914 *Chem. Abstr.* VIII. 2894 Prodigiosin, the pigment of *B[acillus] prodigiosus*, is more sol. in alc. than in water. **1950** A. H. Corwin in R. C. Elderfield *Heterocyclic Compounds* I. vi. 320 Tripyrrylmethanes are also of some interest because of the fact that they are leuco bases of tripyrrylmethane dyes, one of which, prodigiosin, has been isolated from a natural source. **1968** A. Albert *Heterocyclic Chem.* (ed. 2) v. 234 Pyrroles are well represented in Nature. Apart from the many porphins and related tetrapyrroles.., there is the bacterial pigment prodigiosin, and the antibiotic netropsin. **1971** *Jrnl. Antibiotics* XXIV. 636 Prodigiosin..is the bright red pigment of *Serratia marcescens* and was probably responsible for many medieval 'miracles' involving the appearance of blood stains on the Holy Host.

prodigiosity (prodidʒiǫ·sĭti). [f. L. *prōdigiōs-us* Prodigious + -ity.] **1.** A person or thing of enormous size; a monster.

1895 G. Meredith *Amazing Marriage* II. xxxvi. 407 We're none of us 'fifty feet high, with phosphorous heads', as your friend..says of the prodigiosities.

2. A marvellous quality or performance.

1910 W. J. Locke *Simon* vi. 77 He had fallen in love with her when she had first taken Marseilles captive with the prodigiosities of her horse Sultan.

prodnose (prǫ·dnō̆uz), *sb.* [f. Prod *v.* + Nose *sb.*] An inquisitive person, a nosey-parker; *spec.* a detective.

1934 Dylan Thomas *Let.* 11 May (1966) 126 Singing as loudly as Beachcomber in a world rid of Prodnose. **1965** *Spectator* 12 Feb. 213/2, I shall be greatly disappointed if some prodnose does not get a PhD thesis out of these pages in 2065. **1968** V. C. Clinton-Baddeley *My Foe Outstretch'd* vi. 103 He was sensitive about his reputation as an amateur prodnose. **1973** D. Robinson *Rotten with Honour* 97 I'll tell you why, you squalid prodnose. **1976** *Listener* 5 Aug. 135/2 Were the other lonely prodnoses with clipboards, operating in the dark, copying me? Or was I, daunting thought, the only pollster operating in the entire nation?

prodnose (prǫ·dnō̆uz), *v.* [f. prec.] *intr.* To pry; to be inquisitive. So **pro·dnosing** *vbl. sb.*

1958 *Spectator* 3 Oct. 430/1 At this time [*sc.* the 1940s] the social virtue of prodnosing..was still at a fairly harmless stage of development. **1969** *Daily Tel.* (Colour Suppl.) 31 Oct. 20/1 It is perhaps high time that the industrial psychologists who are encouraged to prodnose into most things got to work on the Press.

prodromal, *a.* Add: **2.** Now usu. with pronunc. (prodrō̆u·mal).

produce, *sb.* Add: **6.** *produce broker* (earlier example), *trade* (earlier example).

1851 C. Cist *Sk. Cincinnati in 1851* 143 Forwarding and Commission merchants and Produce brokers. **1872** *Rep. Vermont Board Agric.* I. 161 B. F. Rugg, who was then engaged in the produce trade,..undertook to carry out a plan for controlling the Boston butter market.

produce, *v.* Add: **1. b.** (Earlier and later examples in the sense 'to bring (a performer or performance) before the public'.) Now also *spec.* to administer and supervise the production of (a film or broadcast programme). Also *absol.*

1836 Dickens *Let.* 25 Aug. (1965) I. 171 A farce in two acts..to be produced at the Saint James's Theatre on the first of October. **1897** G. B. Shaw in *Sat. Rev.* 13 Feb. 170/1 Like all plays under Mr. Barrett's management, 'The Daughter of Babylon' is excellently produced. **1912** F. A. Talbot *Moving Pictures* 329 The Hepworth Manufacturing Company..recently has produced several powerful and excellent film-plays. **1923** *Radio Times* 28 Sept. 23/3 The whole production produced and directed by Mr. R. E. Jeffrey, who has adapted this well-known play for wireless transmission. **1935** E. F. Dyer *Producing School Plays* ii. 24 If he [*sc.* the producer] finds no inner meaning there is either something wrong with him or with the play which he has chosen to present; either he is not a suitable person to produce the play, or the play is unworthy of production. **1937** 'M. Innes' *Hamlet, Revenge!* I. i. 17 I'm producing. And I've built a sort of Elizabethan stage. **1940** C. P. Purdom *Producing Plays* i. 2 Any one who knew anything of the theatre could recognize a play produced by Mr. Granville-Barker. **1966** *Listener* 6 Oct. 515/2, I think it was over-ambitious of Mr Wheeler to produce and write the script, yet one cannot belittle his success in presenting very clearly the broad scope of his subject. **1971** N. K. Parrott in J. R. Brown *Drama & Theatre* iv. 87 *Othello* got produced, mainly because somebody wanted to do it and convinced enough other people to join him in presenting it.

3. b. Also *absol.*

1976 'A. Garve' *Home to Roost* ii. 26 She had naturally expected to start a family... There was no apparent physical reason why we shouldn't produce.

c. *absol.* To produce the goods, money, results. *slang.*

1970 G. F. Newman *Sir, You Bastard* viii. 226 Ring me. And you'd better produce. **1977** *New Yorker* 24 Oct. 64/3 One queen's 'husband' asked her to 'produce' for four of his friends and stabbed her when she declined.

producer. Add: **1. b.** The person who produces a dramatic performance, film, or broadcast programme.

1891 Scott & Howard *Life E. L. Blanchard* I. 213 Though he was a clever actor, he rose to greater fame as what we should now call a stage-manager or producer of plays. **1896** G. B. Shaw in *Labour Leader* 19 Dec. 443/4 Our extraordinary clever producer of Ibsen would get enormously rich. **1909** *Westm. Gaz.* 30 Apr. 5/2 Mr. Louis Calvert's appointment as stage manager, or, according to the more modern term, 'producer' of the so-called Millionaires' Theatre in New York, may be reckoned a high compliment. **1911** D. S. Hulfish *Cycl. Motion-Picture Work* II. 95 The producer is in charge of the studio. **1912** F. A. Talbot *Moving Pictures* 329 There is every indication that the British producers are making up headway. **1915** *Times* 26 Nov. 11/4 The English film-producers of all branches are rapidly proving that..the hustle of the American 'producer', and the mobile features..of the foreign actor are not essential. **1925** *Scribner's Mag.* Sept. 283/2 The great Delando, most resourceful of the Broadway producers, put down the last act of 'The Republic'. **1933** *Radio Times* 14 Apr. 72/2 It will be interesting to see how the producers handle a 'spectacular' show of this kind. **1938**, **1944** [see *director 1 g]. **1949** A. Huxley *Ape & Essence* 24 Titles, credits and finally..the name of the *producer*. **1961** G. Millerson *Technique Television Production* 190 The producer may be the business-head of the programme, responsible for organization, finance, policy, etc., while his director is concerned with interpretation, staging and directing its production. **1973** *Radio Times* 26 July 55/1 The producer [in TV] has overall charge of the production.. and it is he who marshals the resources of the BBC to make sure the production is appropriate to the play and to the series of which the play is part. **1976** M. Maguire *Scratchproof* iv. 51 Sam Goldwyn used to say that a producer shouldn't get ulcers, he should give them.

c. *producer-in-chief.*

1939 M. Spring Rice *Working-Class Wives* i. 13 Men ..are the recognised producers-in-chief. **1976** *S. Wales Echo* 23 Nov. 6/6 He is also producer-in-chief of a series of plays for Granada which will be shown in Britain.

2. (Earlier example.)

1784 Adam Smith *Wealth of Nations* (ed. 3) II. iv. viii. 515 The interest of the consumer is almost constantly sacrificed to that of the producer.

3. *gas producer*, substitute for def.: a furnace for producing fuel gas by passing a current of air and usually steam through hot solid fuel so that incomplete combustion occurs; *producer gas*, gas so produced, used as a low-grade but inexpensive fuel and consisting chiefly of nitrogen and carbon monoxide with smaller amounts of hydrogen and carbon dioxide. (Further examples.)

1902 *Encycl. Brit.* XXVIII. 595/1 In all the attempts to make water gas up to that date the incandescence of the fuel had been obtained by 'blowing' so deep a bed of fuel that carbon monoxide and the residual nitrogen of the air formed the chief products, this mixture being known as 'producer' gas. **1939** *Times* 28 Mar. 11/2 Fuel costs favoured the producer gas engine compared with the oil or petrol engine. **1941** *Thorpe's Dict. Appl. Chem.* (ed. 4) V. 368/1 In the Thwaite cupola producer.., where a dry-air blast..is employed, it is usual to add a certain proportion of limestone to the fuel charge in order to form a liquid slag with the ashes. **1967** M. Chandler *Ceramics in Mod. World* ii. 81 If producer gas or heavy oils are used..it is necessary to have a muffle throughout the length of a kiln.

4. *Ecol.* Any organism or part of an organism that produces the organic compounds it needs from simple substances such as water, carbon dioxide, or nitrogen. Freq. *attrib.*

1941 *Q. Rev. Biol.* XVI. 395/1 In the marine plankton the plant type of life is sometimes called 'producer' and the animal type 'consumer' plankton. *Ibid.* 397/1 Certain important groups of producer plankton..were absent from the neighbourhood of the poles. *Ibid.*, The hotter regions..favor the more catabolic types of metabolism among the producers by a general increase of vital velocities. **1942** *Ecology* XXIII. 400/2 In the language of

community economics introduced by Thienemann ('26), autotrophic plants are producer organisms, employing the energy obtained by photosynthesis to synthesize complex organic substances from simple inorganic substances. **1953** E. P. Odum *Fund. Ecol.* ix. 223 Considering the fresh-water environment as a whole, the algae are the most important producers. **1976** T. C. Emmel *Population Biol.* i. 21 The first trophic level in ecosystems is represented by green plants and comprises the producers (or autotrophs). **1978** *Sci. Amer.* Mar. 102/2 These rodents derive energy directly from the primary producers (leaves, shoots, seeds and buds), from other consumers (invertebrates in the forest litter and occasionally birds' eggs and young) and from decomposers (fungi).

5. *attrib.* and *Comb.*, as *producer-exhibitor, -novelist, -retailer*; **producer goods** (see quot. 1956); **producer-oriented** *a.*, interested in or favouring the producer of goods rather than the consumer.

1920 *Stage Year Bk.* 52 It became increasingly certain, however, that the main body of exhibitors was opposed to the producer-exhibitor. **1951** *Manch. Guardian* 14 May 4/2 It is, of course, extremely difficult to give a coherent picture when you are showing objects..that include producer goods, consumer goods, and transport. **1956** J. C. Swayne *Conc. Gloss Geogr. Terms* 114 *Producer goods*, goods used to make other goods. **1969** *Listener* 13 Feb. 219/2 The producer-novelist David Thomson gave us a good example in *From Oblivion to Obscurity* (Third Programme), by another novelist, with previous radio successes, F. C. Ball. **1962** *Times* 18 Oct. (Walter Thompson Suppl.) p. ii/3 Monopolists and critics of advertising..are fundamentally producer-oriented. **1964** M. McLuhan *Understanding Media* (1967) II. xxxi. 333 Nearly all of our technologies since Gutenberg have been ..not producer-oriented, but consumer-oriented. **1938** *Daily Tel.* 16 Feb. 14/6 Finally, it is said that compulsory pasteurisation would threaten the economic existence of the small producer-retailer. **1960** *Farmer & Stockbreeder* 5 Jan. 87/2 Producer-retailers with over 50 head of poultry will again be required to contribute ½d per dozen.

produ·cership. [f. Producer + -ship.] The position or function of a producer.

?**1924** G. B. Shaw *To a Young Actress* (1960) 65 You are not within five years hard work of being good enough for Comisarjevsky, whose producership I have made a condition of my consent to the Parisian Pygmalion. **1926** *Spectator* 1 May 796/1 By a judicious system of African producership and land ownership peace has prevailed. **1933** V. A. Demant *God & Man & Society* iii. 70 The only unassailable standard of producership is the fulfilment of the real demands of the whole body of consumers. **1960** *Times* 16 Mar. 10/4 By 1944 [Dr. Rennert] had risen to the chief producership of the Berlin Städtische Oper.

producing, *ppl. a.* (Further examples.)

1884 *U.S. Tenth Census* X. 13 'Shale oil' was found at a depth of 751 feet, and in November, 1871, producing sand was struck at 1,110 feet. **1920** *Stage Year Bk.* 51 One or two of the bigger and better producing firms had begun to look higher. **1927** *Petroleum Devel. & Technol. in 1926* (A.I.M.E. Petroleum Div.) 202 Where it occurs, naturally no complete travel from the injection well to the producing wells is had. **1932** *New Yorker* 23 July 2/2 (Advt.), Rockland Producing Company 'Death Takes a Holiday': Fri., July 22. **1976** *Daily Tel.* 3 Dec. 5/5 So far..most of Mexico's producing wells have been on land or in the relatively calm and shallow waters of the Gulf of Mexico.

product, *sb.*¹ Add: **1. a.** Also, more widely, applied to other mathematical entities (as events, matrices, permutations, sets, tensors, vectors, etc.) obtained by certain defined processes of combination of two or more entities, the processes not necessarily being commutative and the entities combined not necessarily being of the same kind (cf. *sum *sb.*¹). (Further examples.) Cf. *inner product* s.v. *inner *a.* (*sb.*²) 1 k, *outer product* s.v. *outer *a.* (*sb.*¹) 3.

1892 F. N. Cole tr. *Netto's Theory of Substitutions* ii. 23 The substitution which results from the successive application of two or more substitutions we call their product. **1913** C. E. Cullis *Matrices & Determinoids* vi. 158 If *A* and *B* are any two matrices, the product AB is defined below to be a certain third matrix which is completely known when *A* and *B* are known. **1941** Courant & Robbins *What is Math.?* ii. 110 By the 'intersection' or 'logical product' of *A* and *B* we mean the set consisting only of those elements which are in *both A* and *B*. **1962** B. D. Seckler tr. *Gnedenko's Theory of Probability* i. 22 The event consisting in the simultaneous occurrence of *A* and *B* will be called the product, or intersection, of the events *A* and *B* and will be denoted by *AB*. **1965** Birkhoff & Mac Lane *Survey Mod. Algebra* (ed. 3) vii. 188 Show that tensor products are distributive on direct sums. **1965** Patterson & Rutherford *Elem. Abstract Algebra* ii. 35 Defining the product of two permutations in this way, we obtain a binary operation in the set of all permutations... It is not commutative. **1972** A. G. Howson *Handbk. Terms Algebra & Anal.* ii. 11 The Cartesian product, $A \times B$, of two sets *A* and *B* is defined to be the set of all ordered pairs (*a*, *b*) of elements from *A* and *B* respectively. *Ibid.* xxxiv. 172 The vector product is *not* associative.

3. a. Now freq. that which is produced commercially for sale. Also *collect.*, merchandise, esp. gramophone records.

1928 S. R. Hall *Mail-Order & Direct Mail Selling* xx. 373 To build up a successful mail-order business, one must first have or create the article or products that have mail-order possibilities. **1950** A. Tack *Sell your Way to Success* viii. 113 Work out the sales points around your product. **1976** *Street Life* 7–20 Feb. 22/3 Some records can be sold like soapflakes, but the majority are not purely 'product'. **1977** *Time* 12 Dec. 68/2 More product, to borrow the record-company jargon, from the pianist who burst out of Russia two years ago and has been a one-man industry ever since.

b. The value of goods produced, esp. *gross national* (or *total annual*) *product* (see *gross *a.* 6c).

1888 E. Bellamy *Looking Backward* xxii. 314 The total annual product of the nation..would not have come to more than three or four hundred dollars per head. **1962** *Listener* 29 Mar. 548/1 There is a fixed proportion of the gross national product which can be spent on defence. **1966** *Ibid.* 17 Mar. 374/1 [We could] keep our defence expenditure between five and six per cent of gross national product.

6. *attrib.* and *Comb.*: (in sense 1) *product event, integral, measure, space*; (senses 2–4) *product design, division, group, launch, line, mix, morpheme, nucleus.*

1959 *Listener* 21 May 885/1 A blackboard in the product design section. **1980** *Jrnl. R. Soc. Arts* May 335/2 There is no doubt that the four companies..have put a lot of effort into product design. **1970** *Financial Times* 13 Apr. 14/3 The new structure of the British Steel Corporation consisting of product divisions is generally considered to be one of two logical management systems for a modern business of international proportions. **1968** P. A. P. Moran *Introd. Probability Theory* i. 2 Given any two events, A_1 and A_2, we define a 'product' event denoted by A_1A_2 which occurs if any events E_i occur which belong to both A_1 and A_2. **1957** *Which?* Autumn 2 In the United States and the Scandinavian countries, there are independent organisations that issue reports, regularly or occasionally, each dealing with the brands widely available in one product-group. **1962** E. Godfrey *Retail Selling & Organization* i. 7 The Nielsen Survey..showed that in three product groups—coffees, soft drinks and toilet soaps—50 per cent of all grocery shops offered one or more items below list price. **1968** P. A. P. Moran *Introd. Probability Theory* vi. 290 Expressions such as (6.40) are known as 'product integrals'. This is a mathematical concept which is related to a product of a number of factors in the same way that an integral is related to a sum. **1976** *Times* 22 Apr. 11/4 They were for use in explaining to salesmen, dealers and executives a product launch, a new marketing philosophy or some new twist in internal communications. **1969** *Time* 17 Jan. 52 The company has greatly broadened its product line, introducing seven new models in the past two years. **1980** *Jrnl. R. Soc. Arts* May 325/2 It has over 400 product lines and thousands of variants. **1950** W. Feller *Introd. Probability Theory* I. v. 91 Independence of trials means product measure. **1968** P. A. P. Moran *Introd. Probability Theory* iv. 194 ϕ is the 'product measure' generated by the separate measures on the n one-dimensional spaces. **1953** F. G. Moore *Manuf. Managem.* v. 85 Valid comparisons of over-all figures sometimes become almost meaningless because variations in the 'product mix'—the quantities of different items produced—occur continually. **1965** H. I. Ansoff *Corporate Strategy* (1968) i. 18 Strategic decisions are..concerned.. specifically with selection of the product-mix which the firm will produce. **1970** O. Dopping *Computers & Data Processing* xxii. 365 Product mix is an economically important problem which can in many cases be handled by computer. **1965** *Language* XLI. 365 The sentences would have no string structure at all if transformations combined morphemes from separate elementary sentences into novel product-morphemes (portmanteau blends). **1931** G. Gamow *Constitution of Atomic Nuclei* iii. 63 The particle (α or β) ejected from the nucleus No. 1 may leave the product-nucleus No. 2 in an excited state. **1963** Product nucleus [see *delayed neutron]. **1968** E. T. Copson *Metric Spaces* viii. 134 The product space $X \times T$ is an $(m+n)$ dimensional Euclidean space. **1968** P. A. P. Moran *Introd. Probability Theory* i. 10 The space of all such events is known as the product space of the two spaces.

7. Special Combs.: **product champion**, a person entrusted with the promotion of a product or idea; **products liability**, a manufacturer's legal responsibility to the consumer for his product or products; **product moment**, of a set of pairs of statistical data, the sum of the products of the elements of each pair; freq. *attrib.* (with hyphen), designating a correlation coefficient (symbol *r*) calculated from this, equal to the covariance divided by the geometric mean of the variance; cf. *Pearson b, *correlation 1c.

1969 *Observer* 2 Nov. 12/7 'Product champions' are appointed to help push new ideas through the natural opposition within a big company. **1976** *Jrnl. R. Soc. Arts* CXXIV. 725/1 We welcome ideas from serious-minded enthusiasts or 'product champions'. **1972** *Guardian* 12 Aug. 10/5 American judges and legislators have created..a completely new set of rights known as product liability law. The dramatic effect of this is to make the manufacturer..strictly liable to the ultimate consumer. **1976** *National Observer* (U.S.) 23 Oct. 11/2 Kerry Choi, head of product-liability research for the Association of Trial Lawyers of America..says most of the cases he handles for member attorneys involve inadequate warnings. **1978** *Rep. R. Comm. Civil Liability & Compensation for Personal Injury* I. v. xxii. 255 Products liability must be considered in the context of public concern to protect the interests of the consumer. **1904** *Drapers' Company Res. Mem.* (Biometric Ser.) I. 32 We shall obtain by the method of mean square contingency satisfactory results, *i.e.*, values close to the coefficient of correlation as found by product moment or four-fold division methods. **1904** *Amer. Jrnl. Psychol.* XV. 78 The method of 'product moments' is valid, whether or not the distribution follow the normal law of frequency, so long as the 'regression' is linear. **1918** *Biometrika* XII. 87 We are now in a position to set down the algebraical values of the product-moment coefficients. **1925** R. A. Fisher *Statistical Methods* vi. 146 Such an estimate is called the correlation coefficient, or the product moment correlation, the latter term referring to the summation of the product terms, *xy*, in the last equation. **1930** M. Ezekiel *Methods Correlation Anal.* viii. 127 The value $\Sigma(xy)$ is sometimes called the product moment. **1951** Paden & Lindquist *Statistics for Econ. & Business* xiv. 231 The use of the mean x/σ_X, y/σ_Y product for this purpose was first proposed by the English statistician Karl Pearson and is therefore called the Pearson product-moment coefficient of correlation. **1972** *Jrnl. Social Psychol.* LXXXVII. 33 Pearson product-moment correlations were calculated between fear ratings and intensity of escape attempts. **1972** E. Lukacs *Probability & Math. Statistics* iv. 90 In the bivariate case one also has a mixed moment (product-moment) of second order.

production. Add: **I. 1. b.** (Earlier and later examples.)

1784 Adam Smith *Wealth of Nations* (ed. 3) II. iv. viii. 515 Consumption is the sole end and purpose of production... The mercantile system..seems to consider production and not consumption, as the ultimate end and object of all industry and commerce. **1817** D. Ricardo *Princ. Pol. Econ.* xx. 444 The cost of production, and therefore the prices of various manufactured commodities, are raised to the consumer by one error in legislation. **1887** Moore & Aveling tr. *Marx's Capital* I. ii. vi. 147 In order that a man may be able to sell commodities other than labour-power, he must of course have the means of production, as raw material, implements, &c. **1933** S. Hook *Towards Understanding K. Marx* xi. 120 For Marx it is the relations of production, not the forces of production and not the conditions of production, which are the basis of the cultural superstructure. **1964** S. M. Miller in I. L. Horowitz *New Sociol.* 300 We need sustained economic growth, high production, and high employment in order to solve many of the problems of the unemployed and the poor today in America. **1977** *Undercurrents* June–July 23/1 First, to infiltrate the key institutions, including 'the means of production' (The People's Warehouse).

II. 3. c. The action or process of producing a stage play, film, or other performance. Also, the performance itself.

1894 [in Dict., sense 3 a]. **1925** *Scribner's Mag.* July 7/1 Jesse Lynch Williams has been..preparing a play for production in New York in the fall. **1928** Barrie *Peter Pan* p. viii, I remember writing the story of *Peter and Wendy* many years after the production of the play. **1932** *New Yorker* 11 June 52/2 Wherever she appears.., as a telephone operator in some big, showy production—a breath of humanity sweeps over the screen. **1937** *Printers' Ink Monthly* May 40/2 *Production*, the building, organizing and presenting of a radio program. **1942** *N. & Q.* 12 Sept. 161/1 'Business of the Stage' denotes the movements, groupings, vocal inflections, etc., of the players, which are settled at rehearsal. The modern term is 'Production'. **1949** *Radio Times* 15 July 15/2 '*The Dilettanti*' by Thomas Love Peacock. Adapted for broadcasting... Production by Noel Iliff. **1952** Granville *Dict. Theatr. Terms* 101 *In production* (of a play), in rehearsal and general preparation for *production*. A *dark* theatre sometimes has a notice stating that 'this theatre is closed; a new play is *in production*'. **1962** A. Nisbett *Technique Sound Studio* 267 *Production*, compilation, and studio direction, etc., of a programme. **1976** M. Maguire *Scratchproof* ii. 30 Loose talk..can ruin a production before it even gets off the ground.

d. *fig.* An unnecessarily elaborate performance; a fuss, commotion, drama. Freq. in phr. *to make a production* (*of, out of*, etc., something).

1941 B. Schulberg *What makes Sammy Run?* vii. 128 Something tells me that when our blast comes, it will really be a production. **1959** R. Condon *Manchurian Candidate* (1960) ix. 131 You make a production out of it like I was involved somehow in your life. **1962** M. & G. Gordon *Journey with Stranger* (1963) xi. 74 The simplest tasks at home became productions when travelling. **1967** S. Woods *Case is Altered* ix. 107 You've made rather a production of this, Inspector. **1974** R. Butler *Buffalo Hook* ii. 15 Why should there be this big production over a cargo..that's covered by insurance anyway.

V. 7. a. (Further examples.)

1929 T. H. Burnham *Engin. Econ.* xv. 192 Production control necessitates a system of records and charts which indicate at a glance whether the planned production is being adhered to, or if departure therefrom is occurring, at what stage the divergence is arising. **1938** E. Ambler *Cause for Alarm* i. 17 He's afraid of the production figures falling off. **1941** B. Schulberg *What makes Sammy Run?* xi. 283 Production costs have been too high. **1943** J. B. Priestley *Daylight on Saturday* viii. 52 If our lads was fightin' like 'ell..yer'd see them production figures take a high jump. **1951** R. Firth *Elem. Social Organization* iv. 136 The organization of production tends to be based not merely on a system of cash rewards... A production relationship is often only one facet of a social relationship. **1957** *Technology* Mar. 8/1 Then he goes into the production shops, where he gains experience of the many aspects of aircraft construction. **1958** J. F. Magee *Production Planning & Inventory Control* i. 1 A manager necessarily thinks of problems in production planning in terms of people and their responsibilities. **1962** A. Battersby *Guide to Stock Control* x. 89 In striving for the shop-floor efficiency associated with long manufacturing runs, the Production Manager will always be tending to drive stock levels upwards. **1966** *New Statesman* 20 May 753/1 (Advt.), Book Publishers invite applications..for the post of production controller to supervise the production of a section of

their Home Education list, from manuscript to bound copy. **1970** O. Dopping *Computers & Data Processing* xxii. 364 The most common problem in production planning in a workshop is to determine the sequence in which different operations, pertaining to different orders, should be placed in the different machine groups. **1975** *North Sea Background Notes* (Brit. Petroleum Co.) 7 Production licences..give the licensee exclusive rights over a specific area to explore for and produce hydrocarbons. **1975** *Petroleum Economist* Aug. 309/1 The Hamilton Brothers facility involved developing a production riser as a technical innovation. **1976** *Scotsman* 24 Dec. 13/7 (Advt.), Experience of production control within a high volume fabrication and pressing shop would be a distinct advantage. **1977** *Observer* 24 Apr. 1/6 The rig crew was about to install a safety valve on the top of a production pipe. **1978** P. Sutcliffe *Oxf. Univ. Press* iv. x. 166 Frowde..acquired great expertise and used it to good effect in producing his own books. He was his own production manager, firm and clear in his instructions to Hart.

b. In *Broadcasting* and *Cinemat.* (sense 3 c), *production assistant, clerk, director, editor, manager, staff, team*; also *production control, control room.*

1960 D. Davis *Gram. Television Production* 77 *Production Assistant* (*P.A.*), the director's personal assistant on a programme. **1969** W. Rutherford *Gallows Set* i. 18 In one group were the members of the film crew. In the other..were..the most senior director..[and] Anne, his production assistant. **1972** *Listener* 21 Dec. 852/1 Sequence of calls before a shot. Production Assistant: 'Quiet. Going for a take. Standing by'. Director: 'Right'. **1963** *Movie* Apr. 11/1 Production clerk at R.K.O., then second assistant director. **1961** G. Millerson *Technique Television Production* i. 15 Some networks prefer to have separate rooms for production control. *Ibid.*, Through the window of the production control room we can see the studio below. **1937** *Printers' Ink Monthly* May 40/2 *Production director*, individual in charge of the radio studio programs. **1961** G. Millerson *Technique Television Production* i. 15 Facing these monitors sit the production director and his assistant. **1972** Production editor [see *make-up editor* s.v. *make-up 6]. **1938** *Times* 7 Jan. 13/6 Beside him [*sc.* the producer] sits the production manager, whose functions are similar to those of a stage manager in a theatre. **1959** W. S. Sharps *Dict. Cinematogr.* 121/1 *Production manager*, in filming this post is often held by the Assistant Director and he is responsible for ensuring that everything and every person concerned in production is available at the right place at the right time. **1973** *Listener* 22 Nov. 727/1 In..the film-within-the-film, the wife of the production manager..sits sourly knitting on set. **1962** A. Nisbett *Technique Sound Studio* 246 *Control cubicle* (*BBC*), the soundproof room equipped with control desk..which is occupied by production and operational staff. *Ibid.* 10 Although each member of the production team is concentrating on his own job, this is geared to that of the team as a whole. **1974** *Listener* 27 June 820/2 My wanting to make this series of documentaries..aided by my production team's talents as film makers.

c. Special Combs.: **production brigade**, in communist countries, a unit within a commune required to meet specified agricultural production figures; **production engineering**, the planning and control of the manufacturing processes, plant, and equipment involved in the production of any manufactured product; so **production engineer**; **production line**, a line (*line *sb.*² 19c) along which things undergo successive stages of production; **production number**, a spectacular song or dance in a musical show or revue; **production platform**, a platform (sense *8d) used in the production of oil or gas from the sea bed; **production reactor**, a nuclear reactor designed to produce fissile material; **production relations**, in Marxist theory the social relations arising from and essential to the process of production, as those between the controllers of the means of production (raw materials, land, machinery, etc.) and the labour force; **production run**, a run (Run *sb.*¹ 19c) for the purpose of the routine production of a product; **production-sharing** (see quot. 1963); **production testing**, testing under the conditions that would prevail during production; hence **production-test** *v. trans.*; **production well**, a well from which oil or gas is actively being produced.

1962 E. Snow *Red China Today* (1963) lix. 453 About management? It is in the hands of the production brigade. **1978** HuaKuo-Feng in *Peking Rev.* 10 Mar. 22/1 In some cases the cadres arbitrarily demand grain and money from the commune, production brigade, production team or commune members or even requisition labour power. **1920** *Engin. Production* Mar. 128/3 As a production engineer it has frequently been necessary for me to select machines suitable for the intensive manufacture of various components. **1921–2** (*title*) Proceedings of the Institution of Production Engineers. **1940** E. J. H. Jones *Production Engin.* i. 3 From the schedules thus compiled, it is possible for the production engineer to ascertain whether sufficient plant will be available for the work in hand. **1946** G. Galle in *Philips Resistance Welding Handbk.* ii. 41 Production engineers should be careful to watch that the seam welder is not overloaded. **1966** S. Beer *Decision & Control* i. 7 Tests have been run in the works to see whether one kind of machine tool is better than another. The purchasing department says that *A* is better than *B* because it is cheaper to buy. The production engineers say that *B* is better than *A* because it produces a better job. **1920** *Engin. Production* Nov. 467/2, I look forward to the time when the curriculum of our universities shall include lectures on production engineering. **1921–2** *Proc. Inst. Production Engineers* I. 39 The name calls up visions of conveyors, elevators, and all such gear, which is by no means the only part of production engineering. **1956** *Nature* 4 Feb. 200/1 The art of precasting concrete has given rise to an established industry producing staple things like blocks, tiles, floor beams and parts for small structures, where problems are in the nature of factory production-engineering. **1966** *McGraw-Hill Encycl. Sci. & Technol.* X. 640/1 Production engineering as a planning activity takes place between product design and the planning of the over-all manufacturing process. **1978** R. V. Jones *Most Secret War* xl. 377 The superiority of American production engineering was often a powerful—even vital—aid. **1935** T. H. Burnham *Engin. Econ.* (ed. 3) II. viii. 245 Many nice problems arise as to when it is economic to break up a group system and lay down a special production line. **1943** J. S. Huxley *TVA* xii. 112 One, two, and three bedroom houses were built on outdoor production lines and distributed to various communities. **1958** *Listener* 27 Nov. 903/1, I suggest one of those 'production-line' chickens, which is big enough for four. **1964** M. Argyle *Psychol. & Social Probl.* viii. 107 A production line must go at the speed of the slowest man. **1971** *Daily Tel.* 14 Aug. 7/4 The nauseating 'production line' feeling experienced in so many crematoria. **1975** *N.Y. Times* 8 Oct. 3/1 The new telescope is to be a production-line 25-meter model..and has been chosen to keep costs to a minimum. **1936** *Metronome* Feb. 61/2 *Production number*, show tune. **1959** *Listener* 12 Nov. 845/3 One production number, danced in silhouette against a changing background, marked an exciting advance in presentation. **1967** Wodehouse *Company for Henry* v. 98 One of those big production numbers so popular in revue, where the whole strength of the company let themselves go in uninhibited dance **1964** *Oil & Gas Jrnl.* 12 Oct. 104/1 Offshore Louisiana suffered its worst battering in history from Hurricane Hilda... The industry must decide what to do about multiwell production platforms that have been sheared off and sunk. **1976** *Scotsman* 27 Dec. 6/1 The order from Brazil for a production platform from the McDermott yard at Ardersier. **1956** Production reactor [see *converter 3 f]. **1966** *McGraw-Hill Encycl. Sci. & Technol.* XI. 359/1 Water is used as a coolant in the United States production reactors, whereas in the United Kingdom gas cooling has been the basis for most designs. **1950** T. H. Marshall *Citizenship & Social Class* ii. 108 The essential factor in this theory is the conception of these production-relations as forces determining the life-situation of individuals. **1973** C. D. Kernig *Marxism, Communism & Western Society* VII. 36/1 In the historico-materialist view of history..it is assumed that revolutions arise out of a state of conflict between production and production relations. **1979** G. A. Cohen *Karl Marx's Theory of Hist.* ii. 31 A production relation binds at least one person(s)-term and at most one productive force(s)-term, and no other type of term. **1967** D. Goch in Wills & Yearsley *Handbk. Managem. Technol.* 146 Compared with the standard usage of 3 lb of moulding powder per unit, the production run required an additional 30 lb. **1973** J. Leasor *Host of Extras* i. 22 Many people think that the Ford Model T had the longest production run of any car—eighteen years. Rolls beat them on this..by being in production with the Silver Ghost for nineteen consecutive years. **1963** *Economist* 8 June 1046/1 The concept that the Indonesians prefer is one of 'production-sharing', under which the contractor takes over existing or potential development, brings in his capital equipment to carry it out, and has as his reward a share of what is produced, processed and sold. **1975** *Petroleum Economist* Sept. 348/1 The Lebanese government has invited bids from interested oil companies for offshore acreage on a production-sharing basis. **1960** *Farmer & Stockbreeder* 19 Jan. Suppl. 41/1 It should provide a means of production-testing individual hens. **1975** *Petroleum Economist* Aug. 286/1 An exploratory well was production tested in June last year at rates of 4 200 b/d of condensate and 23 million cubic feet per day of gas. **1975** *Offshore* Sept. 75/2 While the well was not production tested, Shell indicates it could be a major discovery and Amoco calls it 'potentially significant'. **1934** *Proc. World Petroleum Congr. 1933* I. 359/1 (*heading*) The tubing of production wells under pressure. **1976** *Offshore Engineer* Mar. 6/4 At present only one production well is operating; depending on final assessment of reserves, from four to six wells will be used.

d. Designating a vehicle or appliance made in the ordinary course of production, as opposed to one made for testing or other special purposes.

1961 *Motor Sport* Dec. 1002 This talented designer has shown quite outstanding genius in placing another production B.M.C. engine across the front of his Mini. **1971** 'D. Rutherford' *Clear the Fast Lane* 39 The usual changes needed before a production car is ready for rallying. **1972** *Lebende Sprachen* XVII. 136/2 Six Jet-stream prototypes are to be built before production aircraft begin to emerge within the next few months. **1974** *Encycl. Brit. Macropædia* XII. 567/2 Stock-car racing, limited to American production-model passenger cars with suitable modifications, began at Langhorne, Pennsylvania, on July 4, 1939. **1974** *Guardian* 26 Mar. 32/4 Disabled drivers' tricycles should be replaced with modified production cars. **1978** *Gramophone* Mar. 1642/2 A raffle for the first ever production model Quad Electrostatic loudspeaker.

productional (prodʊ·kʃənăl), *a.* [f. Production + -al.] Of, pertaining to, or resulting from production.

1931 *Economist* 7 Feb. 285/2 In comparison with its 'productional' programme the Government's 'distributional' programme has been very modest. **1961** G. Millerson *Technique Television Production* 11 You will find here..several controversial hypotheses, for which the author must be held responsible (dynamic composition; aural composition; productional rhetoric). **1961** E. Wilson *Shaw on Shakespeare* 255 Shaw denounced script cuts and productional schemes which distorted the plays and destroyed their integrity.

productionism (prodʊ·kʃəniz'm). [f. Production + -ism.] A doctrine based upon the importance of production.

1930 *Times Lit. Suppl.* 23 Oct. 858/1 They [*sc.* some plays] may be unreadable..because the author was obsessed by the theory of 'productionism' and..contributed nothing of substance or of design but only a few wisps of straw. **1963** J. S. Huxley in *New Scientist* 27 June 713/2 In the USSR we have an analogous situation that may be called 'productionism' largely to keep up with and beat the USA in industrial efficiency.

productionize (prodʊ·kʃənaiz), *v.* [f. Production + -ize.] *trans.* To produce for general use; to put into production. So **produ·ctionized** *ppl. a.*; **produ·ctionizing** *vbl. sb.*

1937 *R. Air Force Q.* VIII. 102 Generally, an experimental military aircraft..attains speeds on tests which are not equalled by the subsequent '*productionized*' craft in the service. **1939** *Neuphilol. Mitt.* 137 Productionize, vb. **1957** *Times Survey Brit. Aviation* Sept. 15/7 There may well be less chance of reducing costs by 'productionizing'. **1961** *Flight* LXXX. 431/2 Ferranti are 'productionizing' the MoA moving-map display. **1970** *Sci. Jrnl.* May 37/2 A parallel programme inside the factory was devoted to productionizing the equipment.

productive, *a.* (*sb.*) Add: **A.** *adj.* **1. c.** *Med.* Of a cough: that raises mucus or sputum.

1923 *Radiology* I. 168/2 At the time of examination ..she had a persistent and slightly productive cough. **1965** *Brit. Jrnl. Industr. Med.* XXII. 194/1 Those who smoked, and more particularly those who had a productive cough, had lower ventilatory capacities and lower forced expiratory ratios than the remainder.

3. Also, esp. in Marxist theory, as *productive forces.*

1907 L. Boudin *Theoretical System K. Marx* ii. 26 The basis of the structure [of society] is a given state of the development of the productive forces of society. **1909** E. Untermann tr. *Marx's Capital* III. xv. 293 The means, this unconditional development of the productive forces of society, comes continually into conflict with the limited end, the self-expansion of the existing capital. **1927** E. & C. Paul tr. Lenin in D. Ryazanoff *Karl Marx* 123 At a certain stage of development, the material productive forces of society come into conflict with the ..property relationships within which they have hitherto moved. **1973** C. D. Kernig *Marxism, Communism & Western Society* VII. 36/1 The concept 'productive forces' is very important for the Marxist interpretation of revolution... Designated as 'productive forces' are working people, means of production..developed and employed by people and, in addition, the means of labour (raw materials, natural resources) that are consumed.

productivity. Add: **1. b.** *Ecol.* The rate of production of biomass.

1908 J. Johnstone *Conditions of Life in Sea* ix. 179 It is much more difficult to attempt..estimations of the productivity of a sea area, than merely to attempt to ascertain the mass of life at one particular time. **1934** *Q. Rev. Biol.* IX. 175/2 Along the Southern California coast the greatest productivity is..within fifty miles of shore. **1953** E. P. Odum *Fund. Ecol.* iv. 82 It is important to distinguish between the basic or primary productivity on the one hand and consumer or secondary productivity on the other. **1960** N. Polunin *Introd. Plant Geogr.* xv. 478 There is insufficient data as yet to compare the vegetational productivity of different climatic zones. Thus although some tropical inland waters may be more prolific as producers of plant or animal life than some extra-tropical ones, others are practically barren. *Ibid.* xvi. 520 The average productivity of similar land and ocean *areas* appear to be roughly comparable over the year, being said to be of the order of three tons of dry material per acre. **1970** W. D. Russell-Hunter *Aquatic Productivity* xii. 226 This primary productivity of the oceans..amounts to considerably more than half of the primary productivity of the entire world.

c. *Econ.* The rate of output per unit of input, used esp. in measuring capital growth, and in assessing the effective use of labour, materials, and equipment. Also *transf.*

1899 J. B. Clark *Distrib. Wealth* iv. 49 We have said that the *specific* productivity of labor fixes wages... In like manner, the *specific* productivity of capital fixes interest. **1930** *Economist* 18 Jan. 107/2 Still, if productivity has risen high, profits have not followed suit. **1936** J. M. Keynes *Gen. Theory Employment* xi. 138 Many discussions of this subject seem to be mainly concerned with the physical productivity of capital in some sense. **1947** *Amer. Econ. Rev.* May 402 We must try to distinguish as sharply as possible between increase and decrease of productivity caused by larger or smaller output volume—volume productivity increase—and the increase and decrease of productivity caused by real improvement of production or organization—real productivity increase. **1955** *Sci. Amer.* July 35/1 Machinery prices being similar for the managements of all the plants, productivity was a direct function of the average hourly earnings of production workers. **1957** *Introd. Work Study* (Internat. Labour Office) i. 5 *Productivity*..is..the arithmetical ratio between the amount produced and the amount of the resources used in the course of production. **1962** *Daily Tel.* 15 Nov. 1/4 The pace at which our productivity is increasing is too slow, and we hope this National Productivity Year will ginger things up a bit. **1965** H. Wilson in *Oxf. Times* 3 Dec. 16/2, I think I may claim the responsibility for introducing the word 'productivity' to Whitehall, and I'm not proud of it. **1970** *Physics Bull.* July 291/1 The 'productivity' of the Council has also edged up, with the subject boards considering more courses at fewer meetings. **1976** A. W. A. Peterson in T. S. Barker *Econ. Struct. & Policy* v. 107 The model of production..assumes that industrial investment brings about higher labour productivity as a result of the introduction of more modern equipment.

2. *attrib.* and *Comb.*, as (sense *1c) *productivity agreement, deal, measure, team.*

1978 *Cornish Guardian* 27 Apr. 12/7 An active participant in the introduction of the Company Productivity Agreements. **1970** G. Greer *Female Eunuch* 243 By playing upon..discontent with wage freezes and productivity deals, an adroit Tory can convert the working class to the most arrant conservatism. **1977** *Navy News* Dec. 18/1 The opinion that in industry some 'spurious productivity deals' are now being drawn up has been voiced in a letter received from a PO. **1959** J. W. Kendrick *Wages, Prices, Profits & Productivity* (Amer. Assembly, Columbia Univ.) ii. 39 The most commonly used productivity measure is 'output per man-hour'. **1953** *Britannica Bk. of Year* 638/2 A team of experts appointed to study methods of increasing production assumed the title of Productivity Team.

‖ **produit net** (prodwi net). [Fr., lit. 'net product'.] In the politico-economical doctrine of the Physiocrats, the amount of the excess of the value of agricultural products over the cost of their production. (Cf. Physiocrat.)

1792 A. Young *Trav. France* I. xxii. 559 By pursuing the jargon of the *produit net*, and making it variable, instead of fixed, every species of inconvenience and uncertainty has arisen. **1885** *Encycl. Brit.* XIX. 360/1 The real annual addition to the wealth of the community consists of the excess of the mass of agricultural products (including..metals) over their cost of production. On the amount of this 'produit net' depends the wellbeing of the community. **1931** M. Dobb in W. Rose *Outl. Mod. Knowl.* xvi. 598 The essential definition of 'productive' as creative of surplus or *produit net*. **1965** Seldon & Pennance *Everyman's Dict. Econ.* 53 Cantillon's contention that only agricultural enterprises yield a surplus over the costs of production gave rise to the Physiocrats' concept of 'produit net'.

pro-element: see *pro-¹ 4b.

proembryo. Insert at beginning of etym.: [a. G. (M. J. Schleiden *Grundzüge der Wissenschaftlichen Botanik* (1843) II. iii. 52).] Delete 'the *suspensor* of Phanerogams' and 'Now little used'. Add: In seed plants, the group of cells formed by the early divisions of the zygote after fertilization; also, a young embryo. Also *attrib.* (Further examples.)

1862 F. Currey tr. *Hofmeister's On Germination of Higher Cryptogamia & Fructification of Coniferæ* xvi. 441 The breaking up of the pro-embryo of the Coniferæ into a number of independent suspensors is a phenomenon of the most peculiar kind. **1919** F. O. Bower *Bot. Living Plant* xvii. 274 The very first division of the zygote stamps the polarity of the pro-embryo. **1950** D. A. Johansen *Plant Embryol.* xii. 76 The young proembryo completely fills the archegonium in *Torreya taxifolia.* **1964** H. J. Dittmer *Phylogeny & Form in Plant Kingdom* xxiv. 587 Embryonic development in this plant [*sc.* tansy mustard] begins with a series of transverse divisions of the zygote forming a chain of proembryo cells. **1978** *Nature* 2 Feb. 441/1 Frequently only two seedlings germinate from a single seed and this suggests that only a small proportion of proembryos reach maturity.

proembryonic *a.* (later example).

1957 H. C. Bold *Morphol. Plants* xxvii. 522 All of the proembryonic cells descended from the zygote may begin to develop into embryos.

proenzyme (pro,e·nzaim). *Biochem.* [f. Pro-² 1 + *enzyme.] The inactive precursor of an enzyme; = Zymogen.

1900 in Dorland *Med. Dict.* **1902** C. A. Herter *Lect. Chem. Path.* viii. 254 The glandular cells from which they [*sc.* enzymes] come apparently do not hold them in any considerable quantity, but contain substances called proënzymes, from which they are produced. *Ibid.*, These proënzymes, pepsinagen and rennet zymogen. **1976** *Nature* 22 Jan. 235/2 Plasminogen is the plasma proenzyme which, on conversion to its active form, plasmin, is considered responsible for lysis of fibrin deposits resulting from physiological or pathological activation of the coagulation cascade.

proerythroblast: see *pro-² 1. **pro-estrus,** var. *pro-œstrum* s.v. Pro-² 1 in Dict. and Suppl.

pro-e·thnic, *a.* [Pro-¹ and ².] **1.** (See Pro-² 1.) (Further examples.)

1906 J. H. Moulton *Gram. N.T. Gk.* I. ix. 221 The Greek participle..represents the proëthnic participle. **1935** G. K. Zipf *Psycho-Biol. of Lang.* iv. 162 The shift in accent which took place in almost every Indo-European dialect after the pro-ethnic parent language broke up into its dialects.

2. Favouring the Gentiles, as opp. to the Jews. *rare.*

1920 R. Harris *Testimonies* II. ii. 13 Propagating by testimonies a Gospel which is at once pro-ethnic and anti-Judaic.

Hence **pro-e·thnically** *adv.*

1920 R. Harris *Testimonies* II. ii. 16 As it is written: Father of many nations (ἐθνῶν) have I set thee... The extract from Genesis turns on the use of the word ἐθνῶν, and the words are used pro-ethnically.

proette (prōᵘe·t). [f. Pro *abbrev.* + *-ette.] A female professional golfer.

1968 *Maclean's Mag.* Sept. 39 For obvious reasons, the LPGA objects to its members being called 'pro' golfers, and is trying to popularize the description 'proettes'. **1969** *Sunday Times* 5 Oct. 20 Even a lady 'proette' has lost the US Women's Open through signing for a correct total but a 5 and a 4 when she meant a 4 and a 5. **1971** *Time* 28 June 41 And while no proette has ever topped $50,000 for a season, Jack Nicklaus for one has picked up that much in a single tournament. **1975** *Auckland* (N.Z.) *Star* 18 Jan. 35 Two Australian proettes started the second round of the Benson & Hedges $7000 women's golf classic..today in fine style... New Zealand proette Marilyn Smith..turned in 37.

pro:-Europe·an, *sb.* and *a.* [pro-¹ 5a.] **A.** *sb.* One who favours or supports Europe or other European countries; *spec.*, a supporter of (British membership of) the European Economic Community. **B.** *adj.* Favouring or supporting Europe or other European countries; *spec.*, supporting (British membership of) the European Economic Community.

1944 *Sun* (Baltimore) 8 July 6/4 An even more rhetorical appeal was issued on the same day by the *PPF*. ('pro-European'). **1962** *Guardian* 10 July 18/7 The Liberals alone have a consistently pro-European tradition. **1963** *Ann. Reg. 1962* 35 The pro-Europeans in the country were a majority of the under-45s, of the over-£25 a week men.., and of those educated past the age of 16. **1969** *Time* 4 July 23 Maurice Schumann, 58, Minister of Foreign Affairs, combines impeccable Gaullist credentials with a pro-European outlook. **1970** *Manch. Guardian Weekly* 4 Apr. 8 A strongly pro-European speech from Mr Roy Hattersley, the Minister of Defence. **1971** *New Yorker* 3 July 64 The highly uncertain future if Britain does not go into Europe has not yet been adequately explained or understood, according to some anxious pro-Europeans. **1979** *Guardian* 28 Apr. 32/2 A tougher line on Europe may upset the more pro-European wing of her party.

prof (prǫf). Also † *U.S.* **proff.** (Colloq.) abbrev. of Professor 4. Also *attrib.*

1838 *Yale Lit. Mag.* Feb. 144 For Proffs and Tutors too, Who steer our big canoe, Prepare their lays. **1859** G. H. Lewes *Jrnl.* 14 July in *Geo. Eliot Lett.* (1954) III. 116 At Berne I called on Prof. Schiff. **1888** [see Professor 4 b]. **1916** H. L. Wilson *Somewhere in Red Gap* ii. 74, I bet Wilbur thinks the prof is awful old-fashioned, playing with his fingers that way. **1933** Auden *Dance of Death* 8 With profs. from Germany. **1949** H. Wadman *Life Sentence* 16 Prof, how are you? You don't look a day older. **1967** O. Wynd *Walk Softly* iv. 42 It's certainly not the local practice at prof level. Heads of departments come near to being living gods. **1975** C. Fremlin *Long Shadow* x. 79 You were at the reception desk, phoning up to the Prof.'s room. **1979** L. Meynell *Hooky & Villainous Chauffeur* xi. 150 Don't call me *Prof*; it's an abbreviation I find particularly distasteful.

pro-Fascist: see *pro-¹ 5a.

‖ **proferens** (profe·renz). *Law.* Pl. **proferentes** (profere·ntiz). [L., pres. pple. of *proferre* to offer, adduce.] The party which proposes or adduces a contract or a condition in a contract. Also in phr. *contra proferentem*, or *-es*, used of legal decisions made 'against the proposer(s)', with reference to the maxim *verba cartarum fortius accipiuntur contra proferentem,* the words of contracts should be interpreted most forcibly against him who adduces them'.

[*a* **1626** Bacon *Elem. Common Lawes Eng.* (1630) i. 11 *Verba fortius accipiuntur contra proferentem.* This rule that a mans deedes and his words shall be taken strongliest against himselfe,..is..a rule drawn out of the depth of reason. **1766** Blackstone *Comm.* II. xxiii. 380 That the deed be taken most strongly against him that is the agent or contractor, and in favour of the other party. '*Verba fortius accipiuntur contra proferentem.*'] **1927** *Times Law Reports* 3 June 528/2 At the least the expression was ambiguous and must be construed *contra proferentem.* **1935** *Lloyd's List Law Rep.* 9 May 306/1 In this case the underwriters were not the *proferentes.* **1947** J. Charlesworth *Law of Negligence* (ed. 2) xxix. 617 The Court of Appeal held that the word..was ambiguous and must be construed *contra proferentem.* **1971** R. A. Percy *Charlesworth on Negligence* (ed. 5) xvi. 670 On the first point, Mackinnon J. and the Court of Appeal held that the word 'indemnify' was inapt to exclude a claim by the plaintiffs, and at least was ambiguous and must be construed *contra proferentes.* **1974** E. R. H. Ivamy *Marine Insurance* (ed. 2) xxiii. 349 He saw no reason for coming to the conclusion that the insurers were the *proferentes.*

profesh (profe·ʃ). *slang.* Abbrev. of Profession 6; applied *spec.* to the theatrical profession. Also *U.S.*, the body of professional tramps.

1901 J. London *Let.* 6 Dec. (1966) 126 Wyckoff is not a tramp authority... Wyckoff only knows the working-man... The profesh are unknown to him. **1907** —— *Road* 236 The profesh are the aristocracy of The Road. **1914** E. Pugh *Cockney at Home* 192 'Mr. Alexander,..being a hartist in his profesh, which there's only one thing as keeps him off the London stage at this present moment, and that is—' 'Eggs!' **1936** Wodehouse *Laughing Gas* xviii. 192 We're most of us in the profesh downstairs.

profess, *v.* **1. 1. a.** (Later examples.)

1939 A. Clarke *Sister Eucharia* i. 8 The day she was professed a year Ago. **1975** *Anglo-Saxon Eng.* IV. 140 If the manuscript was a gift to William of St Calais around 1083 when the first monks were professed at Durham, it is written in a script..that would have been familiar to the new bishop.

profession. Add: **III. 6. e.** Applied allusively and *euphem.* to Prostitution 1.

1888 Kipling *In Black & White* 78 Lalun is a member of the most ancient profession in the world. **1914** C. Mackenzie *Sinister St.* iv. ii. 862 There's only Miss Carlyle who's in the profession and comes in sometimes a little late. **1922** A. Woollcott *Shouts & Murmurs* ii. 57 The Actor and the Streetwalker... The two oldest professions in the world—ruined by amateurs. **1936** *Times Lit. Suppl.* 18 Apr. 338/4 Blackham has attempted a comprehensive survey of the activities of womankind from 'the oldest profession' to the magistracy.

professional, *a.* (*sb.*) Add: **A.** *adj.* **II. 3.** *professional* (*middle*) *class*, members of the learned and skilled professions regarded collectively. Freq. (with hyphen) *attrib.*

1888 [in Dict.]. **1919** G. B. Shaw *Heartbreak House* p. viii, Just as Ibsen's intensely Norwegian plays exactly fitted every middle and professional class suburb in Europe. **1960** C. Day Lewis *Buried Day* 131 The professional-class families. **1965** M. Morse *Unattached* ii. 74 Social class (professional-middle to lower working class). **1979** G. St. Aubyn *Edward VII* i. 29 Gibbs had been brought up as a member of the professional Middle Class.

4. a. (Earlier and further examples.)

1798 in *Deb. Congress U.S.* (1852) 10th Congress 1 Sess., App. 2741 The solemn air and dictatorial manner of a professional schoolmaster. **1836** *New Sporting Mag.* July 198 On this point I heard a remark from one of the professional [cricket] players. **1883** W. James *Let.* 13 Jan. in R. B. Perry *Tht. & Char. W. James* (1935) I. 611 [S. H.] Hodgson is..a *gentleman* to his finger tips and a professional philosopher as well. **1946** *Mind* LV. 149 But is this work to bear fruit only in the narrow and specialised fields that professional philosophers inhabit? **1968** P. McKellar *Experience & Behav.* xi. 277 Such a person might..be a professional philosopher who gets on well with other people in parties and other social situations.

b. (Earlier example.)

1851 J. Pycroft *Cricket Field* iv. 56 The chief patronage..was..in London. There the play was nearly all professional: even the gentlemen made a profession of it.

c. Disparagingly applied to one who pursues relentlessly an activity or belief that is regarded with disfavour; inveterate, habitual, ruthless.

1879 *Cornh. Mag.* Oct. 414 It is one of the misfortunes of the professional Don Juan that his honour forbids him to refuse battle. **1937** *Time* 18 Jan. 75/2 Chekhov was a strong supporter of Zola and the Dreyfusards, Suvorin was a professional anti-Semite. **1978** J. Krantz *Scruples* vi. 167 The 'extra man' invited to sit next to her at dinner was either a homosexual or a professional leech who dined out every night by mere virtue of being unmarried and mildly presentable.

d. Reaching a standard or having the quality expected of a professional person or his work; competent in the manner of a professional.

1926 C. Connolly *Let.* 8 May in *Romantic Friendship* (1975) 124, I think one must be pretty professional to succeed [as a writer]. **1945** 'A. Gilbert' *Black Stage* iv. 56 'This chap's got his head screwed on all right,' exclaimed Goodier. 'Looks like the professional touch to me.' **1969** M. Butterworth *Vanishing Act* xi. 124 The old and tried method of bulk-carrying—by crew members of ships and aircraft who do it on a regular, professional basis, and know all the angles—is safest and best. **1973** D. Francis *Slay-Ride* ix. 100 The cutting edges had been sharpened like razors and the point would have been good as a needle. A professional job: no amateur could have produced that result with a few passes over a carborundum. **1979** *Daily Mail* 31 Jan. 9/6 The average [career] adviser in schools was strictly an amateur... But those responsible for guiding university students into the right jobs were highly professional.

e. Of technical equipment: of a type or standard used by professionals.

1955 *Brit. Communications & Electronics* II. 48/2 Professional recording equipment ranges from the fixed studio equipment, through mobile apparatus, down to portable machines. **1965** *Wireless World* Sept. 460/1 An important thing to check, especially in professional audio amplifiers, is the stability margin. **1971** *Hi-Fi Sound* Feb. 67/2 A professional pickup—a studio or lab component—may well be robust and dependable, but not a candidate for top hi-fi systems, in which tracking weights are at their lowest. **1975** G. J. King *Audio Handbk.* x. 222 Professional machines operating at 38 or 76 cm/s may adopt essentially constant-current recording over the primary bandwidth.

B. *sb.* **2. b.** *spec.* A prostitute. Cf. *profession 6 e.

1861 [see *amateur 2]. **1973** 'D. Jordan' *Nile Green* xxvii. 121 'I'm not a professional,' she said, too coolly, 'and he's not precisely my type.' **1977** *Listener* 17 Feb. 215/1 The girl he offers a client in order to clinch a deal ..is already a professional.

C. *Comb.*, as *professional-looking* adj.

1936 A. Huxley *Eyeless in Gaza* vi. 64 She [*sc.* a model ship] was so professional-looking. **1980** J. Cartwright *Horse of Darius* v. 67 Dieter had..prepared a very professional-looking salad.

professionality. (Earlier example.)

1861 *Economist* 27 Apr. 456/1 The pungency is given.. by that additional flavour of professionality.

professionalize, *v.* Add: **1.** (Further examples.)

1947 *Mind* LVI. 393 In the third period..philosophy has been professionalized. **1954** [see *lubritorium]. **1974** *Nature* 11 Jan. 122/3 Steeds 'professionalises' the subject so that a research student with the aid of this book should be able to make intelligent use of anisotropic elasticity.

Hence **profe·ssionalized** *ppl. a.*; **professionalizing** *vbl. sb.* (further examples); also as *ppl. a.*; **professionalization** (further examples).

1923 G. B. Shaw *Perfect Wagnerite* (ed. 4) 152 Wagner was not only the highly professionalized royal conductor of Dresden..: he was also the author of the saying that music is kept alive..on the cottage piano of the amateur. **1958** *Oxford Mag.* 20 Feb. 290/1 The increased professionalisation of sport. **1959** B. Wootton *Social Sci. & Social Path.* ix. 287 The history of this rapid growth of professionalization, and of the splintering of generalized welfare work into numerous highly specialized professions, is an interesting story. **1969** H. Perkin *Key Profession* i. 20 The professionalization of university teaching..turned even more on the reform of Oxford and Cambridge. **1972** *Science* 12 May 645/1 The professionalizing of forestry created a community of interest between private and public policy makers. **1973** L. Holcombe *Victorian Ladies at Work* i. 19 There was a raising of the status of the workers in teaching and nursing, this 'professionalization' being..distinctive..of the period. **1975** *Language for Life* (Dept. Educ. & Sci.) xxiii. 341 Language should become a well-established option in Dip. H.E. courses and..institutions selecting for a professionalising year should look upon it as an important qualification for acceptance. **1977** *N.Y. Rev. Bks.* 14 Apr. 38/3 The bar became increasingly professionalized—which meant that lawyers and judges were drawn from higher social classes. **1977** R. Holland *Self & Social Context* ix. 265 The relation between sociology and psychology in the United States and Britain, both as disciplines and as *professionalising* disciplines. **1979** *Dædalus* Spring 15 They also believe in the 'professionalization' of sociology.

professionize, *v.* For *rare*⁻¹ in Dict. read *rare* and add: **2.** *trans.* To turn (an activity) into a profession.

1920 *Christian World* 23 Sept. 14/1 They professionized the study and diverted literature from its true and best purpose.

professor. Add: **II. 4. d.** A schoolmaster, a personal tutor; *spec.* a secondary school headmaster. Chiefly *U.S.*

1903 *Dialect Notes* II. v. 326 *Professor*, a male teacher. This abuse of the word 'professor' seems to have grown up in the country districts recently. It is now applied indiscriminately to any schoolmaster. **1940** W. Faulkner *Hamlet* i. iii. 65 He's going to be the new school professor next year... Or so they claim. **1972** *Buenos Aires Herald* 4 Feb. 13/5 (Advt.), Spanish. Perfect accent, very clear pronunciation with experienced professor.

5. a. (Further examples.)

1816 *Sussex Weekly Advertiser* 22 July, Mr. Lambert, professor of Cricket, has published the whole art of playing. **1954** F. C. Avis *Boxing Ref. Dict.* 87 *Professor*, a familiar name for a boxing coach.

b. (Earlier and later examples.)

1774 in C. S. R. Hildeburn *Century of Printing* (1886) II. 182 Catalogue of New and Old Books, to be sold by Auction, by Robert Bell, Bookseller and Professor of Book-Auctioneering, on Monday, the Seventh of February, 1774. **1848** W. C. Macready *Diary* 9 Dec. (1912) II. 415 At James's Hotel, where I dined, the landlord introduced me to *Professor* (!) *Risley*—the balancer and posture-master; *of course* I shook hands with him, etc.! **1896** C. H. Shinn *Story of Mine* 56 They were never out of sight of pilgrims—Irishmen with wheelbarrows,..'professors' with divining rods and electric 'silver detectors'. **1927** *Amer. Speech* III. 27 Most of those who insist on being given the title 'professor' are quacks or fakers of some kind... The title 'Professor' is now applied more often jocularly than seriously. **1972** *Times* 30 May 2/8 Their [*sc.* performing fleas] trainer, 'Professor' Len Tomlin, was 'too upset' last night to speak about the tragedy which had struck his troupe of 15 performers.

professor (profe·səɪ), *v.* Also with initial capital. [f. the sb.] *trans.* To address (a person) as 'professor'.

1893 W. James *Let.* 8 July (1920) I. 345 Both you and Angell, being now colleagues and not students, had better stop Mistering or Professoring me. **1901** —— *Let.* 16 June (1920) II. 148, I professor-ed you because I had read your name printed with that title in a newspaper letter. **1908** —— *Let.* 28 July (1920) II. 308 Dear Bergson,—(can't we cease 'Professor'-ing each other?)

professoriat (prǫfesōᵊ·riăt). [f. L. *professōrius* belonging to a public teacher (see Professor) + *-at.] **a.** = Professoriate 1. **b.** = Professoriate 2. Also *fig.*

1860 [see Professoriate 2]. **1933** *Sun* (Baltimore) 19 Dec. 1/2 Privately, they express complete distaste for the program, contempt for the 'professoriat'. **1978** *Encounter* Feb. 62/1 In this essay Davie's learning, verbal intelligence, scrupulousness, and his sense of how a poet actually works, combine to do credit to a text and, incidentally, to the Professoriat.

profibrinolysin: see *pro-² 1.

proficiency. Add: **3.** *attrib.* and *Comb.*, as *proficiency certificate, level, test*; **proficiency badge** *Scouting and Guiding*, a badge worn to mark achievement in a given test of skill or endurance; **proficiency pay** *Mil.*, increased pay given in respect of proficiency.

1921 *Daily Colonist* (Victoria, B.C.) 10 Apr. 9/4 It is proposed to further consider the proficiency badges which a Two Star Wolf Cub can obtain. **1970** *Policy, Organisation & Rules of Scout Assoc.* ii. 25 The Cub Scout proficiency badges are designed to give advanced training in the twelve areas referred to. **1970** G. F. Newman *Sir, You Bastard* 259 Sneed still held the proficiency certificate for using people. **1977** P. Strevens *New Orientations Teaching of English* viii. 91 *Proficiency level*, particularly the distinction between *beginners* and *non-beginners.* **1909** *Westm. Gaz.* 24 Apr. 2/3 £450,000 is the charge for the service or proficiency pay of British soldiers. **1918** E. S. Farrow *Dict. Mil. Terms* 473 *Proficiency pay*, in the British service, extra pay, varying from 3d. to 6d. daily, issuable to soldiers of cavalry, artillery, infantry and school of musketry, according to conditions laid down in the Royal Warrant. **1960** D. D. Eisenhower in *Public Papers of Presidents of U.S. 1960–1961* 49 Additional longevity pay of career personnel,..an increased number of men drawing proficiency pay. **1918** E. S. Farrow *Dict. Mil. Terms* 473 *Proficiency test*, in target practice, the annual test conducted to determine the proficiency of organizations in collective marksmanship.

profile, *sb.* Add: The pronunc. (prōᵘ·fail) is now usual. **1. b.** A biographical sketch or character study (common in journalistic use since *c* 1920); a summary description or report.

a **1734** R. North in J. L. Clifford *Biogr. as an Art* (1962) 31 As for the many sketches or profiles of great men's lives, pretended to be synoptical or multum in parvo, we are sure there is nothing we look for in them. **1840** Dickens *Let.* 26 Nov. (1969) II. 158, I have gone through your two profiles, and marked them in pencil here and there. **1925** *New Yorker* 21 Feb. 9 (*heading*) Profiles. **1927** *Observer* 23 Oct. 6/2 No man can better give a thumbnail sketch of..a personality intimately known. In this volume we have glimpses of a few political personages... But novel 'profiles' of writers whom he has known are not to be found. **1930** H. Crane *Let.* 21 Nov. (1965) 357 One assignment is a 'profile' of Walter Teagle, president of Standard Oil (N.J.). **1942** *Observer* 29 Nov. 1/4 M. Maisky, the Russian Ambassador, is the subject of the 'Profile' on page 7. **1952** M. Steen *Phoenix Rising* v. 112 He's the big guy..who does the profiles in the Saturday edition. **1959** *Economist* 20 June 1096/2 He is the author of 'Profiles in Courage', a prize-winning and best-selling work of popular history. **1962** *Listener* 7 June 1004/1 A film profile of Julian Bream. **1975** *Language for Life* (Dept. Educ. & Sci.) xiv. 217 We have in mind a profile which would include diagnostic information and examples of written work. *Ibid.* xvii. 249 More reliable and productive would be a detailed profile of every child's strengths and weaknesses. **1976** *Liverpool Echo* 6 Dec. 1/5 United States President-elect Jimmy Carter was today examining profiles of 70 possible candidates for Cabinet and other senior posts.

2. Also, *in lost profile* (see *profil perdu).

1967 W. Ames *Prince Albert & Victorian Taste* xi. 139 The Duke..is seen in lost profile, with just enough of his nose and chin showing to be unmistakable.

3. a. (Further examples.) In *Physical Geogr.*, the outline of part of the earth's surface as seen in a vertical section along a straight line or a line following the course of a valley or river; *profile of equilibrium*, the profile of a graded river or stream; a profile such that the velocity is just sufficient to transport all the load supplied to it from above; also, an analogous profile of a beach, such that the amount of sediment deposited is balanced by the amount removed.

[**1841** A. Surell *Étude sur les Torrents des Hautes-Alpes* i. 2 Quand on relève le profil en long du thalweg, on obtient..une courbe sensiblement continue, dont la pente s'élève, ou, si l'on aime mieux, dont la tangente s'approche de la verticale, à mesure qu'on approche du col. *Ibid.* iv. 18 Le profil longitudinal forme une courbe continue, convexe vers le centre de la terre.] **1868** *Min. Proc. Inst. Civil Engineers* XXVII. 549 The longitudinal profile of the irrigation canal, leading out of such a river should be..a regular inclined plane. *Ibid.*, The profile course of a river ..described a curve, concave to the horizon throughout, but more inclined near its source than elsewhere. **1894** *Jrnl. Geol.* II. 77 The profile of a consequent stream may for a time possess unequal slopes at its subsequent falls, but it soon attains a tolerably systematic curve of descent. *Ibid.*, Following certain French writers, the profile of the stream when this balanced condition has been reached has been called the profile of equilibrium. **1902** *Jrnl. Geol.* X. 1 (*heading*) Development of the profile of equilibrium of the subaqueous shore terrace. **1924** *Q. Jrnl. Geol. Soc.* LXXX. 581 There is little indication of the U-shaped transverse profile which is so characteristic of the Towy valley near Nant Stalwyn. **1944** A. Holmes *Princ. Physical Geol.* xiv. 292 Along a shore of submergence the slope of the initial surface may be either steeper or gentler than that of the ideal profile of equilibrium. **1950** *Geol. Mag.* LXXXVII. 430 Fig. 1 represents the profile (perpendicular to the fold-axis) of part of the quarry face. **1952** M. L. Begeman *Manuf. Process* (ed. 3) xix. 488 (*caption*) Milling profile of a locomotive side rod. **1954** W. D. Thornbury *Princ. Geomorphol.* xviii. 476 (*caption*) Comparison of topographic profiles across North America and the South Atlantic basin. **1961** L. E. Doyle *Manuf. Processes & Materials* xxxiv. 755 An optical comparator.. offers one way of checking the profiles and positions of gear teeth. **1964** V. J. Chapman *Coastal Veg.* viii. 194 The beach profile is often such that it allows the waves to break close in-shore at high tide. **1968** R. W. Fairbridge *Encycl. Geomorphol.* 871/2 During the last glacial period of the Pleistocene, mean sea level was lowered by about 100 meters or more. River valleys cut down towards a new profile of equilibrium. **1969** D. J. Easterbrook *Princ. Geomorphol.* vi. 131 Below its junction with the Platte River, the profile of the Missouri River steepens because of the entry of gravel from the Platte. **1972** M. G. Cross *Oceanogr.* i. 14 A profile of the ocean bottom under the ship's track is drawn by a recorder. **1976** S. Judson et al. *Physical Geol.* xiii. 311/2 The long profile of a stream from its headwaters to its mouth is..steepest in its upper reaches.

c. The shape of a wave.

1902 *Encycl. Brit.* XXXII. 579/2 Mr Froude made the assumption that the profile of the wave was a curve of sines. **1952** R. W. Ditchburn *Light* iv. 81 A wave of irregular profile may always be regarded as the sum of a series of simple harmonic waves. **1959** E. Pulgram *Introd. Spectrogr. of Speech* ii. 33 The profiles of waves A and C are so different from one another that one cannot help wondering whether, apart from pitch, they really represent the same tone. **1975** E. Hecht *Schaum's Outl. Theory & Probl. Optics* i. 1, $g(x+vt)$ is a solution of the wave equation corresponding to an arbitrary profile $g(x)$ propagating in the negative x-direction. *Ibid.* 2 Show that $f(x-vt)$ is a progressive wave moving in the positive x-direction with an unchanging profile.

e. *transf.* A characteristic personal manner; an attitude, a policy (of a country, government, etc.). *low profile*: see *low *a.* 23.

1961 *Musical Amer.* May 14/1 In all of Prokofieff's music..we find his profile—his 'signature'—his craftsman's attitude. **1962** *Listener* 11 Jan. 105/2 Marschner's application of a powerful declamatory style lends Heiling an extraordinary dramatic profile. **1970** *Guardian* 16 Dec. 10/2 The United States..has repeatedly committed itself to keeping its profile low. **1972** *Times* 30 Mar. 2/1 The most complicated question is the profile the Army should adopt at the start of Ulster's 'marching season'. **1972** *Guardian* 12 Apr. 14/1 The British profile during the present crisis in Vietnam has been as low as could be conceived. *Ibid.* 24 May 13/6 There is only one realistic way to deal any sort of worthwhile blow at the IRA... This is to 'raise the profile' very briefly in one..area. **1978** S. Brill *Teamsters* ix. 323 Jackie expanded his base and his profile by joining civic groups.

4. d. *Soil Sci.* The set of horizons of which a soil is composed, as displayed in a vertical cross-section down to the parent material.

1906 E. W. Hilgard *Soils* x. 165 (*caption*) Soil profiles illustrating differences in soils of humid and arid regions. **1923** *Soil Sci.* XVI. 95 In this scheme of soil classification, the soil profile includes the whole thickness, upon which the soil-forming processes have operated, from the surface down to the parent rock or geologic substratum. The importance of a separation of a profile into its natural divisions is emphasized. **1927** N. C. Comber *Introd. Sci. Study of Soil* xiii. 144 Areas in which the profiles are essentially alike are grouped together and the characteristic profile is given a definite name. **1946** L. D. Stamp *Britain's Struct. & Scenery* xi. 94 Over a large part of Highland Britain..there has been insufficient time for the weathering of rocks and the formation of a complete profile. **1954** W. D. Thornbury *Princ. Geomorphol.* iv. 80 Some soils have profiles that could not have developed under a single set of soil-developing controls, but consist of a younger profile developed under existing topographic and climatic conditions superposed upon an older profile formed under different conditions. **1976** M. D. Gidigasu *Laterite Soil Engin.* xx. 512 In well-developed laterite profiles, the laterite horizon ranges in texture from gravelly soils to laterite rock.

e. *Geol.* A representation of the form of the interface between strata obtained from measurements made at points lying on a straight line; also, the line itself; *to shoot a profile*, to make such measurements.

1929 *Colorado School of Mines Q.* Mar. 108 For a number of potential profiles covering the ground, lines of equal resistivity may be drawn in plan view, instead of equipotential lines. **1929** *Trans. Amer. Inst. Mining & Metalling Engineers* LXXXI. 597 To delimit a newly discovered dome and determine the depth to the top of the cap..profiles are shot by the refraction method. **1931** [see *profiling *vbl. sb.* 3]. **1940** C. A. Heiland *Geophysical Explor.* ix. 499 In a new area a profile is first shot to determine the normal sequence of beds. *Ibid.* x. 736 The survey..is of interest because of the excellent correlation possible between resistivity profiles. *Ibid.* 739 Fig. 10-71 shows resistivity-depth curves for a dipping vein taken along three profiles, laid out 15° off strike, at increasing distances from the outcrop. **1949** *Geophysics* XIV. 57 (*heading*) Airborne magnetic profile above 40th parallel, eastern Colorado to western Indiana. **1950** *Bull. Amer. Assoc. Petroleum Geologists* XXXIV. 1384 The slope of a time-distance curve plotted from a profile across the dome usually showed salt velocity. **1960** M. B. Dobrin *Introd. Geophysical Prospecting* (ed. 2) v. 86 Figure 5-13 shows two sample refraction records shot from opposite directions, each made with a spread of 22 detectors spaced 300 ft apart along a profile in line with the shots. **1977** *Nature* 3 Nov. 23/2 (*caption*) Seismic and magnetic profiles from locations indicated in Fig. 1. Vertical scale for the seismic profiles is in seconds of two-way travel time.

f. The outline formed on a graph or chart by joining the scores that a person has obtained in tests for various personality traits, esp. in order to provide a quantified result easily comparable with the results of others or of the norm; a similar type of diagrammatic repre-

sentation of measured individual attributes for purposes of comparison. Also *transf.*

1932 Darrow & Heath in K. S. Lashley *Stud. Dynamics of Behavior* 68 By repetition and comparison of tests on the same person, we find a tendency for the shape of the profile to be characteristic of a given individual. **1940** T. L. Kelley *Talents & Tasks* 28 The full line is the profile of an individual, and the dotted line that of the average participant in the type of job being considered. **1946** *Jrnl. Clin. Psychol.* II. 23/1 Several recent articles have discussed the usefulness of the Minnesota Multiphasic Personality Inventory in various clinical situations ..and one has dealt with the problem of test profiles. **1948** *Eng. Stud.* XXIX. 109 Two variables are considered of prime importance in identifying specific patterns of tone in speech: the shape of the curve of speech melody, and the position of the principal stress on that curve. A given combination of shape and stress-position will be referred to as a profile. **1957** R. B. Cattell *Personality* ix. 366 The definition of a profile as a set of ordered measurements (corresponding to the mathematical definition of a vector quantity) applying to a single case. **1960** J. B. Carroll in Saporta & Bastian *Psycholinguistics* (1961) 342/1 Métraux's..'profiles' of speech development include descriptions of the tendency of the child to verbalize with regard to his own and others' behavior. **1973** *N.Y. Law Jrnl.* 24 July 4/4 To be sure, Ruiz-Estrella did fit the hijacking profile, but no one contends that this statistical survey..can come close to supplying traditional probable cause for a search. **1973** *Times* 26 July 8/2 Mr. Ehrlichman said that the need for a psychiatric profile of Dr. Ellsberg had prompted the decision to break into the office. **1974** *Physics Bull.* Nov. 505/1 Abstracts would be sent selectively to subscribers..according to their interest 'profiles'. **1977** *Language* LIII. 186 Given that we have participants of particular socio-economic profiles—but not taking their individuality into account—and given the specific situation, these are the choices which are most expected.

g. *Astr.* (A diagram of) the way the intensity of radiation varies with wavelength from one side of a line in a stellar spectrum to the other.

1933 *Proc. Nat. Acad. Sci.* XIX. 642 (*caption*) Schematic profiles of lines in the spectrum of a nova expanding with constant high velocity. *Ibid.*, The predicted profile of a star with an effectively transparent radiation shell, ejecting matter symmetrically with respect to two hemispheres. **1953** L. H. Aller *Astrophysics* p. v, The abundance of calcium may be determined from the profile of the '*K*' line in the solar spectrum. **1957** *Encycl. Brit.* XXI. 195l/1 The profiles of faint lines are strongly affected by the resolving power of the spectrograph. **1971** *Nature* 15 Jan. 214/1 He was one of the first to measure intensity profiles of the Fraunhofer lines in the solar spectrum and to attempt to explain these by theoretical models.

h. (A diagram of) the way a quantity varies along a line, esp. a vertical line through the earth or atmosphere; more widely, any graph in the form of a line.

1953 *Jrnl. Geophysical Res.* LVIII. 519 Five temperature profiles are obtained which represent stratospheric conditions over New Mexico during October 1952. **1955** *Sci. Amer.* Sept. 168/2 If we took gravity readings all over the earth and corrected them to sea level, we would have a gravity profile of the geoid. This profile undulates. **1963** G. L. Pickard *Descriptive Physical Oceanogr.* vi. 97 Temperature/depth or salinity/depth profiles..are usually drawn as the first stage in examining oceanographic data. **1970** *Sci. Jrnl.* Apr. 50/3 The measurement of temperature profiles within the atmosphere from satellite heights. **1971** I. G. Gass et al. *Understanding Earth* v. 85/2 (*caption*) Crustal temperature–depth profiles, as a function of heat flow and surface radioactivity. **1972** *Nature* 25 Feb. 417/1 Oxygen isotope profiles through the entire depth of the ice sheets of Greenland and Antarctica..provide an excellent record of climatic changes. **1977** *Lancet* 29 Oct. 932/1 The area irradiated by the transducer was calculated from beam profiles obtained by scanning a piezo-electric probe hydrophone through the centre of the beam, parallel to the transducer face. **1978** *Nature* 20 Apr. 725/1 PAS staining of the modified bovine receptors revealed a radically different profile and showed the presence of at least four major peaks with molecular weights in the region of $3{\cdot}9 \times 10^4$, 5×10^4, $1{\cdot}25 \times 10^5$ and $1{\cdot}8 \times 10^5$.

i. *Astronautics.* A particular sequence of accelerations undergone by a space rocket in flight; the plan of a space flight as regards the nature and duration of successive trajectories.

1962 K. A. Ehricke *Princ. Guided Missile Design* I. vii. 774 (*caption*) Constant thrust acceleration profile with and without intermediate coast period. **1962** R. C. Duncan *Dynamics Atmospheric Entry* i. 16 The functional phases of the direct-entry profile are: 1. Orbital phase... 2. Departure phase... 3. Free-fall phase... 4. Approach phase... 5. Landing phase. **1966** *Electronics* 3 Oct. 131 In the Mercury and Gemini programs, and in hypothetical mission profiles, the crew's time usually limits the number of adjustments that can be made in an experiment. **1972** *Jrnl. Spacecraft & Rockets* IX. 259 (*heading*) Thrust profile shaping for spin-stabilized vehicles.

7. (Earlier example.)

1824 J. Decastro *Mem.* 43 The master carpenter had forgot to saw off one of the unpainted pieces of profile belonging to a wing.

8. *profile picture, study, writer*; **profile chart** *Ecol.* = *profile diagram*; **profile cut**, a method of cutting a diamond in which it is sliced into thin plates that are polished on one side, finely grooved on the other, and bevelled on the edge; also *attrib.* as *adj.*; **profile diagram** *Ecol.*, a representation of a vertical section through a forest, showing the outlines of the individual components of the vegetation; **profile drag** *Aeronaut.*, that part of the drag on an aerofoil or aircraft which arises directly from its profile and from skin friction, i.e. those parts of the drag which are not attributable to lift; **profile grinding** *Engin.*, grinding in which the wheel extends the whole width of the work and is given a profile which when viewed at right angles to the axis of rotation is the negative of the one it is desired to produce on the work; so **profile-grind** *v. trans.*, **-ground** *ppl. a.*; also **profile grinder**, a machine for this; **profile paper** (earlier example); **profile shot**, a photograph or view of the human face in profile; **profile stage property** = sense 7.

1926 Tansley & Chipp *Aims & Methods in Study of Vegetation* iv. 65 Profile charts record diagrammatically the vertical relations of the vegetation..as seen in profile or 'elevation'. **1965** P. J. Fisher *Jewels* vi. 83 Normally, for the conventional brilliant cut, an octahedral diamond crystal is sawn into two halves... For the new profile cut the same crystal is sawn into four plates. **1970** R. Webster *Gems* (ed. 2) xx. 378 The Profile cut allows considerably greater area of visible diamond than a brilliant cut of similar size. Viewed from above a Profile cut diamond resembles a row of baguettes joined by a common table facet. **1976** 'D. Craig' *Faith, Hope & Death* xxii. 156 Good stuff, like profile-cut diamonds. [**1933** Davis & Richards in *Jrnl. Ecol.* XXI. 369 The stratification is very irregular and ill-marked, as can also be seen from the diagrammatic profile in Fig. 6.] **1952** P. W. Richards *Trop. Rain Forest* ii. 24 Because the direct observation of the stratification of the Rain forest usually offers insuperable difficulties, Davis & Richards..adopted the device of constructing profile diagrams to scale from accurate measurements of the position height and width and depth of crown of all the trees on narrow sample strips of forest. **1974** Mueller-Dombois & Ellenberg *Aims & Methods Vegetation Ecol.* viii. 148 Profile diagrams can be used to illustrate details in vertical spacing of species. **1922** *Flight* XIV. 692/2 Prandtl calls this increment of the drag at given lift by the trailing vortex system the 'induced drag', and the drag of the wing of infinite aspect ratio and of the same section he calls the 'profile drag'. **1936** *Jrnl. Aeronaut. Sci.* IV. 13/2 The covering of cellulose acetate may be highly polished to lower the profile drag. **1979** *Nature* 20–27 Dec. 778/3 These calculations take into account the effects of lift (or, in aeronautical jargon, 'induced drag') and of the power needed to overcome direct air resistance ('profile drag'). **1941** *Automobile Engineer* XXXI. 169/3 (*caption*) Profile grinding a helical gear by the Maag gear grinding process. **1956** *Ibid.* XLVI. 348/2 The floor-to-floor time was 12 minutes, or 6 minutes per gear. To profile grind at such rates a reasonable standard of gear preparation is necessary. **1950** C. R. Hines *Machine Tools for Engineers* xi. 234 Profile or contour grinders. These grinders are similar to pantograph milling machines. **1968** S. Tolansky *Strategic Diamond* viii. 67 (*caption*) Shaped profile grinder roller for making ceramic spark plug. **1975** Bram & Downs *Manuf. Technol.* i. 31 Microscopes are incorporated in suitable machine tools such as profile grinders. **1917** T. R. Shaw *Precision Grinding Machines* x. 155 (*heading*) Profile and form grinding. **1956** *Automobile Engineer* XLVI. 347 (*heading*) Faster profile grinding of spur gears. **1968** S. Tolansky *Strategic Diamond* viii. 66 Profile grinding is used extensively and ubiquitously for both large and small components which require to be of exact size or have complex shape. *Ibid.* 67 In spite of the hardness of the carbide profile grinding wheels and rollers they soon wear and lose both accurate outline and..accuracy in dimensions. **1941** *Automobile Engineer* XXXI. 168/3 In some cases the gears are hobbed and in others pinion type profile ground cutters are employed. **1874** 'Mark Twain' *Gilded Age* xvii. 160 He plotted the line on the profile paper. **1793** J. Woodforde *Diary* 25 Nov. (1929) IV. 80 They were so kind as to bring us a profile Picture of our late worthy friend Mr. DuQuesne. **1967** P. A. Whitney *Silverhill* iii. 48 He could not photograph me properly. Someone less skilled had taken over, but profile shots were not the same. **1968** L. Deighton *Only when I Larf* ii. 24 He swung round to give me a profile shot. **1972** I. Hamilton *Thrill Machine* viii. 34 She always managed to turn slightly this way or that to give Joe a profile shot. **1854** A. C. Mowatt *Autobiogr.* xvii. 308, I suppose you will send some profile stage properties to my room. **1959** *N. & Q.* Feb. 84/1 An initial profile-study gives some biographical facts. **1978** *Rugby World* Apr. 36/2 Being wholly inarticulate on any aspect of Rugby, I decided it might be wiser to confine my questioning to Gareth's second enthusiasm which, according to a generation of profile-writers, is fishing.

profile, *v.* Add: **1. b.** To compose or present a profile (sense *1 b) of (a person). Also *transf.* orig. *U.S.*

1948 *Word Study* Apr. 3/1 A student publication at Wayne wrote: 'Pan Profiles Russell Beggs' *Panorama*.' **1959** J. Thurber *Years with Ross* v. 85 Ross..took..the flagpole sitter..and profiled him. **1967** *Times Rev. Industry* Oct. 12/2 In February when the *Review* profiled Mr. Len Neal..he spoke enthusiastically of his department. **1970** T. Lupton *Managem. & Social Sci.* (ed. 2) iv. 98 Ways of measuring and 'profiling' the many structural characteristics of organizations. **1974** *Observer* 24 Mar. 29 (*caption*) Anthony Sampson profiles Mrs Katherine Graham, whose newspaper exposed the Watergate scandal. **1979** *Tucson* (Arizona) *Citizen* 20 Sept. 7B/3 Hugh Downs hosts a magazine format show. Tonight, disco star Donna Summer is profiled.

c. To summarize or register (information).

1971 *Nature* 19 Mar. 153/2 The user constructs a list of words and phrases (search terms) that summarize (profile) his information requirements. **1975** *New Yorker* 12 May 93/1 Electrical wires from the model power plants..ran over to a control room, where electronic equipment could absorb the findings of a hundred and twenty instruments that profiled, among other things, hull pressures, mooring forces, and six degrees of freedom of motion.

2. (Examples in *Engin.*) Cf. *profiling *vbl. sb.* 2.

1905 J. Horner *Engineers' Turning* xv. 294 Fig. 368 is a tool in its holder used for profiling ball handles, as used on lathes, and other machine tools. **1953** G. S. Schaller *Engin. Manufacturing Methods* xix. 340 (*caption*) Vertical rotary-head milling machine profiling a vertical surface. **1973** J. G. Tweeddale *Materials Technol.* II. vi. 146 Side-cutters are often profiled axially to cut a specific shape.

3. *intr.* To present one's profile to view; *spec.* in *Bullfighting*, to stand in profile in preparation for a charge.

1932 E. Hemingway *Death in Afternoon* 347 Profiling with more style, his kills would gain greatly in emotion. **1957** A. MacNab *Bulls of Iberia* xv. 205 To get it to charge he has to profile on the contrary horn, making it feel sure it will catch him each time. **1973** *Black World* Sept. 84 Ever get tired of people posturing, Posing and profiling? **1974** F. Nolan *Oshawa Project* ii. 14 He'll be over here.. profiling for the newsreels.

4. *trans.* To measure or investigate the profile (sub-senses of *4) of. Cf. *profiling *vbl. sb.* 3.

1932 *Physics* II. 174 One of the earliest applications of the seismograph was its use in profiling salt domes. **1960** *Econ. Geol.* LV. 204 Sometimes the geologist wants to profile a particular stratigraphic horizon..instead of the land surface. **1972** *Physics Bull.* Feb. 85/1 The Clarendon Laboratory at Oxford..has improved still further its original method of profiling the atmosphere from a satellite. **1978** *Nature* 5 Jan. 49/2 Side-scan sonar has been used..to profile icebergs, by lowering a sonar transducer vertically..from the side of a boat.

profiler (prōᵘ·failəɪ). [f. Profile *v.* + -er¹.]

1. A profile machine.

1904 *Electr. World & Engin.* 19 Mar. 581/2 (*heading*) Motor driven profiler. **1927** *Daily Tel.* 11 May 18/3 (Advt.), Gear cutters, Profilers, Radials, &c. **1957** W. H. Armstrong *Machine Tools for Metal Cutting* vii. 156 The distinguishing feature of a profiler is a profiling unit that is mounted on a spindlehead which may be fed transversely. **1963** Jones & Schubert *Engin. Encycl.* (ed. 3) 1007 Most of the profilers used at the present time are hand-operated, so far as the feeding movements are concerned.

2. An instrument for measuring profiles, esp. of strata of rock or the sea bed.

1959 *World Oil* Apr. 107/2 A versatile new marine exploration device..has been applied recently to problems in connection with petroleum exploration... Known as the..continuous seismic profiler, the technique essentially is a continuous sound reflection device. **1969** J. W. Mavor *Voyage to Atlantis* v. 104 We carried five basic instrument systems. First, a sonar system..to measure depth. Three instrument systems, the seismic profiler, the magnetometer and the gravimeter. Finally, [etc.]. **1972** J. G. Dennis *Structural Geol.* xvi. 368 Most present-day profilers work with sound sources such as high voltage sparks or air guns. **1973** *Nature* 22 June 455/2 Seismic reflexion profiler data reveal a distinct basement ridge. **1975** *McGraw-Hill Yearbk. Sci. & Technol.* 291/2 Velocity shears top to bottom are given by a free-falling electromagnetic profiler... The instrument measures minute voltages induced by the flow of sea water in the Earth's weak magnetic field. **1978** *Nature* 7 Dec. 601/2 Temperature profiles were taken from the RV Oceanographer with the microstructure profiler (MSP), a winged instrument which falls freely through the water measuring temperature, pressure, and conductivity.

profiling (prōᵘ·failiŋ), *vbl. sb.* [f. Profile *v.* + -ing¹.] **1.** The drawing of profiles.

1888 [in Dict. s.v. Profile *v.*].

2. *Engin.* The shaping of a part, orig. by means of a tool guided by a template or pattern. Freq. *attrib.*, esp. in *profiling machine* (= *profile machine* s.v. Profile *sb.* 8).

1892 [in Dict. s.v. Profile *v.*]. **1950** C. R. Hine *Machine Tools for Engineers* ix. 155 (*caption*) Four-spindle vertical Hydrotel milling machine with automatic 360-deg profiling attachment. **1957** [see *profiler 1]. **1967** A. Battersby *Network Analysis* (ed. 2) 377 This measure is likely to utilize the drilling machine inefficiently, because of the odd bits of time spent in waiting for successive small batches from the profiling machine.

3. *Geol.* and *Physical Geogr.* [f. Profile *sb.*] The measurement or investigation of profiles, esp. of strata; *spec.* by means of measurements made at points lying on a straight line.

1929 *Trans. Amer. Inst. Mining & Metall. Engineers* LXXXI. 598 Mapping structures..by means of refraction profiling. **1931** F. H. Lahee *Field Geol.* (ed. 3) xxiii. 680 Field work may be conducted according to one of three main plans: (1) fan-shooting or fanning by the refraction method; (2) profile shooting, or profiling by the refraction method; and (3) profiling by the reflection method. **1938** B. McCollum in A. E. Dunstan et al. *Sci. of Petroleum* I. viii. 396/2 In most cases..the profiling of this very shallow boundary cannot be successfully carried out at the present time by reflection. **1963** J. B. Hersey in M. N. Hill *Sea* III. iv. 65 Continuous refraction profiling should prove especially valuable in the study of unconsolidated sediments in deep water. **1968** R. W. Fairbridge *Encycl. Geomorphol.* 1227/2 Sub-bottom acoustic profiling..has demonstrated many wave-cut terraces partly hidden beneath a thin veneer of late Holocene sediments. **1977** R. J. Rice *Fund. Geol.* viii. 141 Profiling. A third approach to the analysis of hillslope forms is by measurement of representative profiles.

profilmic: see *pro-² 2.

profilograph. Add to def.: = *profilometer 2. (Examples.)

1941 *Jrnl. Amer. Ceramic Soc.* XXIV. 229 The closest approach to this was a profilograph which had been constructed by Abbott at the University of Michigan and a somewhat similar device which had been built in Germany by Schmaltz. These instruments used a diamond stylus mechanically mounted on a mirror in such a way that movements of the diamond point in a direction normal to the surface of the material could be magnified by an optical lever system and the results recorded on a photographic film. **1963** *Engineering* 20 Sept. 360/1 The Bump-cutter employs 110 12 in diameter saw blades impregnated with natural diamonds. The design principle is based on a profilograph with the cutting head mounted on a 16 ft long frame supported by drive wheels at the rear and steering wheels at the front. **1976** Attewell & Farmer *Princ. Engin. Geol.* x. 751 Unless joint surfaces are sufficiently exposed..to lend themselves to profilograph examination, it is difficult to see how a reduction factor can be applied.

profilometer. 1. Substitute for def.: Any instrument or device for measuring the profile of the face. (Example.)

1939 *Jrnl. Amer. Med. Assoc.* 18 Nov. 1903/2 A description of Dr. Joseph Safian's profilometer is found in his book 'Rhinoplastic Surgery'.

2. Any instrument for measuring or recording the roughness of a surface; *spec.* (*a*) one in which a fine stylus is drawn over a metal surface; (*b*) one consisting of a wheeled frame for travelling along a road.

1937 *Metal Cleaning & Finishing* May 426 During the past two years, a new instrument called the 'Profilometer' has been developed. This instrument supplements the earlier work in this field, and enables roughness measurements to be made rapidly with a portable instrument. **1938** *Times* 26 Nov. 9/4 The profilometer—designed and constructed at the Road Research Laboratory—has been used to measure the riding qualities of a number of experimental sections. This 16-wheeled machine compares irregularities in the surface by integrating the vertical rise and fall in profile of the road above a given datum, the results being recorded as 'inches per mile'. **1949** G. Schlesinger *Factory* xi. 219/1 Generally the Profilometer provides only 'average' readings..determining the roughness of the surface without producing conclusions as regards the waviness. **1958** H. M. Sherrard *Austral. Road Pract.* xvi. 302 The profilometer consists of a wheel resting freely on the road surface whose vertical movement is traced on a ribbon of paper, thus giving a profile of the road. **1966** R. Ashworth *Highway Engin.* xii. 254 The profilometer consists of a 16-wheeled articulated carriage arranged so as to support the recording gear at a constant height above the continuously arranged level of the road surface at these 16 wheel points. **1976** [see profilometry below].

Hence **profilo·metry**, the use of such an instrument; **profilome·tric** *a.*

1971 T. F. J. Quinn *Applic. Mod. Physical Techniques Tribol.* i. 32 This 'microscopic' approach must involve the use of as many methods of surface examination as possible. The more conventional methods of optical microscopy..and surface profilometry..are often too coarse for this approach. **1975** D. F. Moore *Princ. & Applic. Tribol.* ii. 15 The profilometric, cartographic, and photogrammetric measurement techniques..deal with a complete representation of surface roughness. **1976** J. Halling *Introd. Tribol.* ii. 21 The most usual method for the study of surface geometry is profilometry. In the profilometer a very fine diamond stylus..is drawn over the surface irregularities... The vertical movement of the stylus..is measured and amplified, usually electronically, so that the recorded output provides a picture of the actual surface.

‖ **profil perdu** (profil pɛrdü). Also **profile perdu.** [Fr.] (See quot. 1959.) Also *attrib.*

1959 P. & L. Murray *Dict. Art* 257 *Profil perdu*, (Fr. lost profile) is that view of a head in which the profile is lost because the whole head is turned so far away that only the outline of the cheek is visible. By extension, the *profil perdu* of any object is what is seen of it when it is more than half turned away from the spectator. **1961** *Times* 14 Feb. 6/6 Rubens and Velasquez repeat the idea with back views, profile or *profile-perdu* heads. **1967** E. Wymark *As Good as Gold* xv. 214 She..stood gazing down the street, her figure a *profil perdu* against a grey sky.

Profintern (prǫ·fintəɪn). [Russ. *Profintérn*, f. *Krásnyĭ Internatsionál Profsoyúzov* Red International of Trade Unions, after *Komintérn* *Comintern.] An international organization of left-wing Trade Unions, founded in 1921 and dissolved in 1937.

1928 R. W. Dunn *Soviet Trade Unions* vi. 65 Through its delegates in the Red Trade Union International (Profintern) to participate in the international labor movement. **1938** *Encycl. Brit. Bk. of Yr.* 335/1 The Red International of Labour Unions ('Profintern') was founded at Moscow in 1921 under the auspices of the Third International to work for the reorganization of the Trade Union movement throughout the world on militant and revolutionary lines. **1949** I. Deutscher *Stalin* 401 Parallel with the Comintern, the *Profintern* (the International of the Red trade unions) had opposed itself to the so-called Amsterdam International. **1958** *Economist* 1 Nov. 424/2 The fact that A. Lozovsky, who had vanished previously, is also reported as dead in 1952 suggests that the former head of the International of Red Trade Unions (Profintern) may also have been a victim of the anti-Jewish drive. **1977** *N.Y. Rev. Bks.* 26 May 26/4 Nikolsky was a representative of the Profintern, the Trade Unions International.

profit, *sb.* Add: **7. a.** *profit economy, -monger* (further example), *-mongering* (example), *plan, -planning*; *profit-bearing, -conscious* (hence *-consciously* adv.), *-hungry, -linked* adjs.; (obj. genitive) *profit-maximizer*; *profit-cashing, -generating, -making* (further examples), *-maximizing, -seeking,* sbs. and adjs.; (instrumental, etc.) *profit-motivated, -oriented* adjs.

1918 W. S. Churchill *Let.* 10 Sept. in M. Gilbert *W. S. Churchill* (1975) IV. vii. 145 The lives they have saved and the prisoners they have taken have made these 18,000 men the most profit-bearing we have in the army. **1945** *Richmond* (Va.) *Times-Dispatch* 5 May 12/3 Profit-cashing by those who desired a clean slate over the week-end stalled numerous leaders. **1960** *Farmer & Stockbreeder* 1 Mar. 149/2 (Advt.), Every profit-conscious egg producer must have Evans Maxilay and Topscore strains. **1976** *Western Mail* (Cardiff) 22 Nov. 8/4 A number of their most profit conscious private sector industries. **1972** *Physics Bull.* June 366/3 Profit-consciously,..the company will normally supply the crawler only as part of its contract inspection service. **1943** *Sun* (Baltimore) 8 Feb. 3/2 The 'profit economy' has not always been equal to the demands of war. **1970** R. Stavenhagen in I. L. Horowitz *Masses in Lat. Amer.* vii. 242 Sol Tax describes the Panajackel Indians' economy as being a 'penny capitalism'.. because they are oriented towards a profit economy. **1976** *National Observer* (U.S.) 14 Aug. 4/1 Profit-generating factors seem to have existed on Sundays in the Big Apple. **1939** *Sun* (Baltimore) 18 Aug. 6/1 The profit-hungry Celanese Corporation of America..is creating a public resentment. **1972** *Accountant* 12 Oct. 447/3 (*heading*) Profit-linked share incentive schemes. **1973** *Times* 31 Oct. 1/1 (*heading*) Government puts limit on profit-linked rises in Phase Three changes. **1953** E. Smith *Guide Eng. Traditions & Public Life* 76 'Public Schools'..are not conducted for the purpose of profit-making. If there should be any excess of income over expenditure it is used for improvements in the school. **1974** *Times* 15 Oct. 5/2 A straightforward profit-making job with very few public benefits. **1968** *Listener* 28 Mar. 403/1 A pure profit-maximiser would already be attacking these hindrances to profit if he could. **1961** *Southern Econ. Jrnl.* Oct. 163/1 Where the industry's product price has been kept below the 'profit-maximizing' and 'entry-limiting' prices due to fears of public reaction. **1968** *Listener* 4 Apr. 437/2 So much for constraints of profit-maximising. **1977** *Dædalus* Fall 92 Larger and larger fractions of the GNP are being produced in sectors..that are clearly not competitive profit-maximizing sectors. **1961** *Spectator* 2 June 808 The profitmongers have an uncanny nose for threats. **1884** W. Morris in *Justice* 17 May 2/2 Ugliness is but a part of the bestial waste of the whole system of profit-mongering, which refuses cultivation and refinement to the workers. **1973** *Listener* 14 June 805/1 The public's distrust of..endorsements, knowing them to be profit-motivated and not spontaneous. **1978** W. Garner *Möbius Trip* ii. 57 The tatty commercialism of profit-motivated research. **1976** *Nigerian Chron.* 18 Aug. 7/4 Government has recently decided to grant full autonomy in personnel matters to all profit-oriented parastatals. **1979** *Arizona Daily Star* 5 Aug. E3/4 'I think we need a president who is more profit-oriented,' he said. **1967** D. Goch in Wills & Yearsley *Handbk. Managem. Technol.* 158 *Profit plan*, profit target based on a predetermined rate of return required from the invested funds represented by the fixed assets and working capital employed in carrying on a business activity. **1964** E. C. D. Evans (*title*) Profit planning and the measurement of return on capital employed. **1927** M. Sadleir *Trollope: a Comm.* 148 The profit-seeking [of the Great Exhibition] ..lay behind the pious ejaculations of an inspired Press. **1949** I. Deutscher *Stalin* ii. 27 The evils of modern profit-seeking industrialism. **1965** H. I. Ansoff *Corporate Strategy* (1968) iii. 37 It would seem..that profit-seeking, or maximization of profit, would be the natural single business objective.

b. **profit foul** *U.S. Basketball*, an intentional or 'professional' foul committed to prevent one's opponents from scoring (? *Obs.*); **profit margin**, the margin that remains in a business operation when the costs involved are deducted from profits, usu. considered as a percentage of the capital employed (cf. Margin *sb.* 2b, quot. 1866); **profit(s) motive** (usu. with *the*), the incentive that the possibility of making profits gives to individual or free enterprise; **profit-sharing** (further examples); also as adj.; **profit squeeze**, the diminishment in profit margins due to costs rising relatively faster than selling prices with insufficient compensation from increased sales; **profit-taking** (further examples); so *profit-taker*.

1952 *Sun* (Baltimore) 15 Jan. 17/3 Veteran Coach Murray Greason..criticized today what he termed widespread use of the 'profit foul' in basketball. **1926** *Encycl. Brit.* III. 225/2 Various measures were taken during the War to restrict profiteering, especially in belligerent countries. These included the fixing of maximum prices, and in some cases of profit margins at each stage of production and distribution. **1974** *Guardian* 25 Mar. 22/6 The discount stores operate on narrow profit margins. **1976** B. Williams *Making of Manchester Jewry* iii. 67 The general move was in the direction of mass sales at a low profit margin. **1931** Patterson & Scholz *Econ. Probl. Mod. Life* (ed. 2) ii. 37 The profits motive represents the modern crystallization of the economic force of self-interest. Hence, production is guided by market value and turns toward the production of luxuries for which there is an effective demand. **1936** J. M. Keynes *Gen. Theory Employment* xxiii. 335 In conditions in which the quantity of aggregate investment is determined by the profit motive alone, the opportunities for home investment will be governed..by the domestic rate of interest. **1947** A. E. Waugh *Princ. Econ.* xxxvi. 863 A society that used coercion instead of the profit motive would not need to establish such institutions as those of free enterprise and free contract. **1975** *Verbatim* May 3/2 Never forget that the profit motive can sometimes be used to a customer's advantage. **1900** *Econ. Rev.* X. 239 (*heading*) Two profit-sharing concerns. **1920** M. Beer *Hist. Brit. Socialism* II. iv. xiv. 292 Profit-sharing and Industrial Co-partnership schemes have been re-examined. **1949** *Here & Now* (N.Z.) Oct. 11/2 Far from being the predatory capitalist, he offers himself as the profit-sharing employer. **1975** *Times* 14 Jan. 14/1 Only 22 per cent of the sample belonged to profit sharing or bonus schemes related to the profits the company makes. **1979** *West Lancs. Even. Gaz.* 28 May 11 (Advt.), The company offers an attractive starting salary, operates a profit-sharing scheme and other benefits. **1958** *Wall St. Jrnl.* 8 Dec. 1/6 The outstanding feature of 1959 may be a further profit squeeze. **1969** J. Argenti *Managem. Techniques* iii. 11 If margins fall, profits fall, unless the company can somehow increase sales volume—and that is not easy. This phenomenon is known as 'the Profit Squeeze' and most companies today are feeling its effects. **1552** Profit taker [see sense 7a in Dict.]. **1968** *Economist* 5 Oct. 78/1 What does seem to have arrived is the day of the profit-taker. **1980** *Times* 5 Jan. 19 The profit takers pulled out of gold. **1928** *Daily Mail* 25 July 18/3 In the Electrical group Bournemouth and Pooles eased on profit-taking. **1976** *Birmingham Post* 16 Dec. 9/10 Stocks settled for a small gain after two rally attempts were stalled by profit-taking.

c. *pl.*, as **profits tax**, a tax on business profits, *spec.* that levied on company profits, as *excess profits tax* (*excess *sb.* 6b).

1920 [see *corporation tax* s.v. *corporation 7]. **1938** *Ann. Reg. 1937* 38 The Chancellor..was urged from many quarters to withdraw the profits tax and meet his requirements from a further increase in the income tax. **1947** *Western Daily Express* 14 June 1 The increased profits tax was criticized at the annual meeting of the Association of British Chambers of Commerce. **1958** *Times* 7 Jan. 15/1 The effect on capital allowances for profits tax purposes of new plants coming into operation. **1974** *Times* 17 Apr. (Ontario, Manitoba & Saskatchewan Suppl.) p. iv, Both provinces have imposed a special 'windfall profits' tax to prevent developing companies from profiting unduly from share increases in the world price of their products.

profitability. Delete *rare* and add: (Further examples.) Also, the state of being profitable; the capacity to make a profit. Also *attrib.*

1924 J. Stamp *Current Probl. Finance & Govt.* 11, I may venture to say that we have almost reached a limit of profitability along the old lines of deductive reasoning. **1931** *Economist* 26 Sept. 548/2 In so far as an industry's profitability is improved the weight of its fixed interest charges is lessened. **1961** *Listener* 2 Nov. 691/2, I am not frightened by profits; no one should ever be frightened of profits. Profitability is the only measure of success. **1964** *Daily Tel.* 15 Feb. 9/2 (*heading*) Guthrie starts BOAC drive for profitability. **1969** *Times* 6 Jan. 7/8 But for foot-and-mouth disease, production would certainly have risen considerably and the price position would already have caused alarm among producers near the profitability margin. **1979** *Jrnl. R. Soc. Arts* CXXVII. 617/1 There is a direct connection between profitability and survival.

profiteer (prǫfitiᵊ·ɪ), *sb.* [f. Profit *sb.* + -eer; cf. Fr. *profiteur* (*de guerre*).] One who profits; *spec.* one who seeks to make excessive gain, as by the extortionate sale of necessary goods. Also as second element in *war profiteer*.

1912 *Athenæum* 21 Dec. 756/3 The fundamental unfairness of the relations between the wage-earner and the 'profiteer'. **1914** *Englishwoman* Nov. 94 The tricks of the armament profiteers are fresh in the public mind. **1918** W. Owen *Let.* 10 Aug. (1967) 568 All the stinking Leeds & Bradford War-profiteers. **1942** W. S. Churchill *End of Beginning* (1943) 130 If there are any would-be profiteers of disaster who feel able to paint the picture in darker colours, they are certainly at liberty to do so. **1952** A. Powell *Buyer's Market* ii. 137 In the twilight world of undergraduate conversation..a kind of stage 'profiteer' or 'tycoon': a man of Big Business and professionally strong will. **1975** *New Yorker* 21 Apr. 134/3 There have already been reports of some killings in several cities, where government police, tax collectors, war profiteers,..have been among the targets previously announced. **1976** *Economist* 16 Oct. 15/2 The radicals also had..strong support among ex-Red Guards and other profiteers of the 1960s cultural revolution. **1980** *Times* 9 May 15/4 Dante ..puts profiteers next door to sodomites in the Seventh Circle of Hell.

profiteer (prǫfitiᵊ·ɪ), *v.* [f. prec.] **a.** *intr.* To practise profiteering; to be a profiteer.

1916 *New Age* 17 Feb. 361/1 The companies are..not only removed from the common temptation to profiteer, but are guaranteed a practically fixed income. **1920** R. Macaulay *Potterism* ii. iii. 94 She had merely profiteered out of it all, and had a good time. **1928** R. Campbell *Wayzgoose* i. 20 Journalists are..profiteering on the brains of sheep.

b. *trans.* To obtain (money) by profiteering; to exploit (a person) financially. *rare.*

1923 S. Kaye-Smith *End of House of Alard* i. 12 He wouldn't lend us any of the money he profiteered out of those collapsible huts. **1928** *Sunday Express* 3 June 13/1 Aren't we being profiteered here?

profiteering (prǫfitiᵊ·riŋ), *vbl. sb.* [f. as prec. + -ing¹ (see note).] The action or fact of seeking to make an excessive profit, as by providing necessities at extortionate prices. Also *attrib.*

Quot. 1814 is apparently an independent and isolated formation. The word was revived in the early twentieth century by A. R. Orage and others.

1814 *Guernsey Star & Gaz.* in *New Age* (1919) 21 Aug. 278/2 The extortionate profiteering that is being practised by the tradesmen in the public market. 1914 *New Age* 27 Aug. 391/2 England is at war upon profiteering. *Ibid.* 15 Oct. 561/2 The profiteering braggadocio..of 'City Man' and his confederates. 1919 *Act* 9 & 10 *Geo. V* c. 66 (*title*) An act to check profiteering. *Ibid.* §8 This Act may be cited as the Profiteering Act, 1919. 1922 W. J. Locke *Tale of Triona* vi. 56 'A dog and a rose and a glass of wine,' said she, 'are a woman's due for amusing a man. But a motor-car is profiteering.' 1939 A. Thirkell *Before Lunch* xii. 307 He said he'd take a hundred more for it than he gave. No, no, I said... No profiteering. I'll give what you gave. 1976 F. Zweig *New Acquisitive Society* ii. v. 112 Profiteering could also cover excessive or illegitimate rents. 1978 P. Boardman *Worlds of Patrick Geddes* ix. 307 The mainsprings of the Financial Age were ..the perfection of profiteering-techniques.

profiterole. Delete † *Obs.* and add: Also **profiterolle.** Now *spec.* a small hollow case of choux pastry usu. filled with cream and served with chocolate sauce.

1884 F. J. Delize *Franco-Amer. Cookery Bk.* 131 Range the profiteroles in pyramid form in the centre. 1889 A. B. Marshall *Cookery Bk.* xiv. 315 *Chocolate profiteroles*... Make a choux pastry..and force it out from the bag on to a dry baking tin in shapes about the size of a small button mushroom. 1906 *Mrs. Beeton's Bk. Househ. Managem.* lxii. 1667 *Profiteroles* (Fr.), a kind of light cake, baked in hot ashes, and filled with cream or custard. 1949 N. Mitford *Love in Cold Climate* ii. vi. 264 Chocolate profiterolles with real cream. 1960 F. Raphael *Limits of Love* i. x. 129 Between dances, Andrew and Julia ate.. chocolate profiteroles and hot sausages. 1972 *Daily Tel.* (Colour Suppl.) 25 Aug. 42/3, I had three puddings (40p plus each); first profiteroles which were a credit to the pâtissier, Patrice.

profiting, *ppl. a.* (s.v. Profit *v.*) (Later example.)

1908 *Daily Chron.* 3 Oct. 5/4 So many profiting interests are concerned that there can be little doubt as to the ultimate formation of a syndicate.

proflavine (proflēⁱ·vin). *Pharm.* Also -in (-in). [f. Pro-2 + Flavine in Dict. and Suppl.] A yellowish-brown crystalline solid, 3,6-diaminoacridine, $C_{13}H_{11}N_3$, which is used, in the form of an orange-red hydrated sulphate, as an antiseptic.

1917 C. H. Browning in *Brit. Med. Jrnl.* 16 June 825/1 The Medical Research Committee's Department of Biochemistry and Pharmacology has..continued to give us valuable aid, especially by providing an experimental supply of 'proflavine'. *Ibid.* 21 July 71/1 The name 'proflavine' has been suggested to us by the Medical Research Committee. 1917 *Lancet* 3 Nov. 676/1 Proflavine..is..a preliminary product in the manufacture of acriflavine or flavine. 1945 *Times* 7 Aug. 10/2 Other interesting items included 24,000,000 hypodermic tablets, 40,000,000 acriflavine and proflavine tablets. 1958 *Nature* 11 Oct. 983/1 Proflavin, although not interfering with synthesis of deoxyribonucleic acid in phage-infected cells, inhibits the maturation of phage progeny particles. 1970 Passmore & Robson *Compan. Med. Stud.* II. xviii. 41/1 Acridine dyes, e.g. proflavine, acriflavine, are more active than aniline dyes against Gram-positive bacteria, and are also active against Gram-negative bacilli. 1971 D. J. Cove *Genetics* x. 137 The starting point of these studies was a 'phage strain which carried an *rII* mutation induced by the acridine proflavin.

profluence. 2. Restrict † *Obs.* to sense a in Dict. and add: **b.** (Later example.)

1950 M. Peake *Gormenghast* lxxi. 392 The windows.. appeared to be sprinkled over the green façades..with an indiscriminate and wayward profluence that gave no clue as to how the inner structures held together.

pro-form: see *pro-1 4b.

profundal (profʊ·ndăl), *a.* and *sb. Ecology.* [ad. G. *profund* Profound: see -al.] **A.** *adj.* Applied to the region of the bed of a lake lying below the thermocline. **B.** *sb.* The profundal region of a lake bed.

[1928 K. E. Carpenter *Life in Inland Waters* viii. 180 Modern workers usually prefer to recognise a 'sub-littoral zone'.., ending at about 50 metres in lakes of the plain type.., and below this a 'profound' or 'deep-water' region.] 1931 *Ecol. Monogr.* I. 233 In fresh-water lakes few animals have proved themselves capable of living in that unusual habitat, the anaerobic profundal zone of the lake bottom. *Ibid.* 245 In Third Sister Lake the three major benthic zones have the following approximate extent: littoral, 0–3 m.; sublittoral, 3–10 m.; profundal, 10–18 m. 1957 G. E. Hutchinson *Treat. Limnol.* I. xv. 818 Enabling the animals of the profundal benthos to live not merely under conditions of low oxygen and high CO_2 tension. 1961 *Ekol. Polska* A. IX. 352 A total of 2300 samples was taken from the profundal of Lake Tajty. 1965 B. E. Freeman tr. *Vandel's Biospeleol.* i. 9 The profundal regions of lakes..constitute an environment similar to the subterranean media. 1972 *Oikos* Suppl. No. 14. 5 In the profundal of a eutrophic lake the environment is relatively homogeneous and the species diversity low compared with the littoral.

prog, *sb.*3 Add: (Earlier and later examples.) **prog** *v.*3 (earlier and later examples).

1890 Barrère & Leland *Dict. Slang* II. 152/2 *Proggins* (university), proctor. 1892 *Granta* 13 Feb. 196/1 [Proctor] What do you mean by this, Sir? You have been following me about for the last ten minutes... [Freshman] Oh—er—I only wanted to see you proggins some one! 1935 D. Sayers *Gaudy Night* xii. 255 The Proggins was just coming..round the corner of Broad Street. 1935 N. Mitchison *We have been Warned* iv. 428 'I warned him he might get progged.'.. 'We might make the progs feel a bit awkward!' 1945 G. B. Grundy *Fifty-Five Years at Oxford* iv. 55 He did not care a — for all the — proggins in the kingdom. 1965 *Guardian* 6 May 5/3 This evening may be the last..on which undergraduates can be progged. *Ibid.* 5/5 The progs have chased and chased us Up and down the town.

prog (prɒg), *a.* and *sb.*4 *slang.* Also **Prog.** Abbrev. of Progressive *a.* (*sb.*) in Dict. and Suppl.

1958 'N. Blake' *Penknife in my Heart* vii. 91 The Lanes, his hosts, were a prog. couple. 1965 *N.Y. Times* 1 June 33 The 'progs' or progressives believe Tewkesbury lives too much in the past. 1968 *Listener* 29 Aug. 280/2 Chaps like us..who don't believe in change, do far more for the Church than a thousand bloody progs like Pope John. 1971 *Progress* (Cape Town) May 1/1 (*heading*) Prog. expansion programme. *Ibid.* 1/2 (*heading*) Swing to Progs in North Rand. 1977 *Guardian Weekly* 11 Dec. 7/1 Liberal-minded South Africans cheered their favoured Progressive Federal Party... Much applause for the gains of the 'progs', as they are locally termed.

prog (prɒg), *sb.*5 *slang.* Abbrev. of *programme *sb.* 2 e.

1975 *Listener* 11 Dec. 790/1 Nice to have you with us on the prog, we say, don't we, fans?

prog, *v.*1 2. (Further U.S. dial. examples.)

1935 Z. N. Hurston *Mules & Men* i. vi. 128 We progged thru the woods that was full of magnolia, pine..and many kinds of trees whose name I do not know. 1949 'J. Nelson' *Backwoods Teacher* vi. 63 He took a stick and progued around in the hole. 1949 *Sun* (Baltimore) 9 June 10/6 A progger..is a fellow that goes progging for frogs.

progamete: see *pro-2 1.

progeny. Add: **6.** *attrib.* and *Comb.*, as **progeny test,** an assessment of the genetic value of an individual made by examining its progeny; so **progeny-test** *v. trans.*, to assess in this way; **progeny-tested** *ppl. a.*, **progeny-testing** *vbl. sb.*

1918 Babcock & Clausen *Genetics in Relation to Agric.* xv. 293 We find that the progeny test of individual plants was first used by Le Couteur and Shirreff. But it was Louis de Vilmorin who first gave special attention to the value of the progeny test (1856). 1932 [see **performance test*]. 1953 Srb & Owen *Gen. Genetics* xxiii. 501 The wisdom of using sires proved good by progeny tests and by careful observations on relatives has become evident to almost all breeders. 1960 *Farmer & Stockbreeder* 8 Mar. 101/2 A.I. organizations are in a favoured position, because they deal with large numbers, and can progeny-test many bulls. 1971 *Ibid.* 23 Feb. 13/3 The Milk Marketing Board's Warren Farm progeny tests..could hardly be called a tremendous success. 1972 *Country Life* 6 Jan. 53/1 Extensive progeny testing of A1 bulls is carried out... By the use of egg transfer it should be possible to produce large numbers of calves from potential bull-breeding females and to progeny test them. 1944 *Jrnl. Agric. Res.* LXIX. 471 Use of progeny-tested dairy sires would be a little more likely to increase the rate of improvement. 1974 *Country Life* 12 Dec. 1853/1 This not only offers the advantage of the progeny-tested sire as compared with the crossing bull, which is what the farmer has to use at present, but a shorter and more concentrated calving season for the stockman. 1933 *Amer. Naturalist* LXVII. 502 Progeny testing in poultry breeding can be used in evaluating the breeding potentiality of either sire or dam. 1970 Progeny testing [see **performance testing*]. 1977 *Jrnl. Agric. Sci.* LXXXVIII. 129/1 We can calculate the expected response to selection for various types of progeny... We may also be able to choose between the use of different types of family or between, say, family selection and a progeny-testing programme.

progeria (prodʒɪə·riă). *Path.* [mod.L., f. Gr. προγήρ-ως prematurely old + -ia1.] A fatal disease of children characterized by symptoms usually associated with senility.

1904 H. Gilford in *Practitioner* Aug. 210 The name progeria, for which I am indebted to Mr. James Rhoades and Professor Arthur Sidgwick, is not only a far better word [than micromegaly], but is a true description of the distinguishing features of the two cases. *Ibid.* 217 The name progeria..has been given in recognition of the senile characters which form such a conspicuous feature of the disease from the beginning. 1927 *Times* (Weekly ed.) 28 Apr. 475/2 Cases of premature senility in children (goblins) described as progeria, the persistence in an adult (a-teleioses) of child characters (elves). 1957 L. Eiseley *Immense Journey* 108 The cause of this curious disease, known as progeria, or premature aging, is totally unknown. 1969 *Guardian* 13 Jan. 2/3 A post-mortem examination will be carried out on..a 9-year-old girl who died of a disease that gave her the physical characteristics of a 90-year-old woman. Norma..was the second member of her family to be suffering from progeria.

Hence † **proge·rian** *sb.*, a person with progeria; also *attrib.*; **proge·ric** *a.*, of or being a person with progeria.

1913 *Lancet* 1 Feb. 305/1 Progerians pass from a delayed childhood into a premature old age. *Ibid.* 306/1 The total length of the progerian face from nasion to chin is only 84 mm. 1914 *Boston Med. & Surg. Jrnl.* 16 July 110/2 Progerians are usually dwarfs. 1933 R. W. B. Ellis tr. *Apert's Infantilism* vii. 73 (*caption*) Mould of the upper and lower jaws in the same progeric patient as in figs. 11 and 12. 1945 *Amer. Jrnl. Dis. Children* LXIX. 276/2 The conditions..are postulated for the progeric patient..during the period from 2 to 6 years of age, when most of his subcutaneous fat vanished. 1976 *Nature* 22 Apr. 713/1 With factor VII-deficient plasma, both normal and progeric cells showed a markedly prolonged clotting time.

pro-German, -ism: see *pro-1 5a, c.

progestational (prodʒestēⁱ·ʃənal), *a.* [f. Pro-2 1 + Gestation + -al.] Relating to, promoting, or being part of the physiological preparations for pregnancy; applied *esp.* to substances whose physiological effects resemble those of progesterone; *progestational proliferation*, proliferation of the endometrium in preparation for pregnancy.

1923 G. W. Corner in *Physiol. Rev.* III. 467 It seems preferable to avoid the suggestion of falsity or imitation inherent in the prefix *pseudo* [in *pseudopregnancy*], by using the terms progravid, progestational. 1928 *Amer. Jrnl. Physiol.* LXXXVI. 78 Sections taken through the middle of the uterine horn showed that no progestational proliferation had taken place in these animals. 1929 *Ibid.* LXXXVIII. 326 It is possible that there is an antagonistic relation between oestrin and the progestational substance. 1932 *Ibid.* C. 111 The production of a progestational endometrium in the uterus of a castrate rabbit. 1944 [see *progestin]. 1948, 1949 [see *gestagen]. 1954 [see *hydroxyprogesterone]. 1960 *Times* 19 Jan. 3/5 (Advt.), Applicants should have had considerable experience in steroid chemistry, and be well acquainted with recent work on gonadotropic, oestrogenic and progestational hormones. 1970 *Sci. Jrnl.* June 26/2 The ovaries produce ova and the oestrogenic and progestational hormones, mainly oestradiol and progesterone, which are concerned with the development and maintenance of the attributes of femaleness, including sexual receptivity and the inception of pregnancy. 1974 *Fertility & Sterility* XXV. 575/1 A special uterine reaction to the luteal hormone (progestational proliferation) was identified in the adult rabbit.

Hence **progesta·tionally** *adv.*, as regards progestational activity or state.

1948 W. H. Pearlman in Pincus & Thimann *Hormones* I. 441 Corticosterone..(progestationally inactive in the Clauberg test). 1958 *Jrnl. Clin. Endocrinol. & Metabolism* XVIII. 350 The highest point of formation of progestationally effective substances in the corpus luteum appears to be between the seventh and eighth day following ovulation. 1968 R. W. Kistner in Astwood & Cassidy *Clin. Endocrinol.* II. vi. ix. 680 If an endometrial biopsy..reveals 'progestationally immature endometrium',..one tablet daily..can be prescribed.

progesterone (prodʒe·stĕrōᵘn). *Physiol.* [ad. G. *progesteron* (W. M. Allen et al. 1935, in *Klin. Wochenschr.* 17 Aug. 1182/1), blend of *progestin and its G. synonym *luteosteron* (K. H. Slotta et al. 1934, in *Ber. d. Deut. Chem. Ges.* LXVII. 1271), f. Luteo-, repr. *corpus luteum*: see *-sterone.] A female steroid sex hormone, $C_{21}H_{30}O_2$, which is secreted by the *corpus luteum* and also made synthetically, and is responsible for the cyclical changes in the uterus in the latter part of the menstrual cycle and also necessary for the maintenance of pregnancy. Also *attrib.* and *Comb.*

1935, etc. [see *progestin]. 1949 [see *gestagen]. 1957 *Times* 2 Dec. p. vi/2 When P.M.S. is used in conjunction with a steroid hormone progesterone it will bring maiden sheep into season to produce an extra crop of lambs in their second summer (six months earlier than normal). 1958 [see *gestagen]. 1961 L. Martin *Clin. Endocrinol.* (ed. 3) ix. 229 Owing to its feeble progesterone-like action,..large doses must be given. 1965 Lee & Knowles *Animal Hormones* ii. 20 As the corpus luteum matures, increasing amounts of progesterone are produced which depress the output of LH and LTH. 1966 [see *diosgenin]. 1968 *Times* 11 Nov. 10/8 Udry and Morris suggest that of the two ovarian hormones, oestrogen tends to increase the likelihood of human sexual activity, and progesterone to decrease it. 1974 *Daily Colonist* (Victoria, B.C.) 25 Aug. 26/2 By 1954, Chang and Pincus had found two progesterone hormones that worked. The birth-control pill was born. 1974 *Fertility & Sterility* XXV. 575/2 The following trivial names are used in this paper:.. progesterone (pregn-4-ene-3,20-dione). 1979 *Jrnl. R. Soc. Arts* CXXVII. 417/2 It was possible to inhibit ovulation in experimental animals with progesterone.

progestin (prodʒe·stin). *Physiol.* [f. Pro-1 5 + Gest(ation + -in1.] **a.** Progesterone, esp. an unpurified preparation of it. **b.** = *progestogen, *gestagen.

1930 W. M. Allen in *Amer. Jrnl. Physiol.* XCII. 174 We have as yet proposed no name for this hormone of the corpus luteum... In so far as we are acquainted with its physiological behavior, its chief action lies in its ability, by alteration of the endometrium, to aid gestation in the castrated rabbit; and for this reason we wish to propose for it the name *progestin*, i.e., a substance which favors gestation. 1935 [see *androsterone]. 1935 W. M. Allen et al. in *Nature* 24 Aug. 303/2 Heretofore, two different names have been used in the literature for this

hormone (progestin, luteosterone). For the sake of international uniformity, we agree to use..the name *progesterone* for the pure hormone. 1936 *Jrnl. Amer. Med. Assoc.* 23 May 1809/1 The Council adopted the following terms: (1) *progesterone* to indicate the chemically pure substance..(2) *progestin* as a general term to indicate the substance (and other chemically allied substances..in case any such compounds are subsequently discovered) without reference to the state of chemical purity. 1944 J. Hoffman *Female Endocrinol.* xli. 711 The term *progestin* is applied to crude extracts of corpus luteum tissue which possess progestational activity. The name progesterone describes the crystalline form of the luteal hormone. 1945 H. Burrows *Biol. Actions of Sex Hormones* xxi. 389 More than one substance having the same kind of action as progesterone is known, and a generic name is therefore convenient. The author has used the term 'progestin' to denote chemical compounds having a biological action comparable with that of progesterone. [*Note*] 'Progestogen' has been suggested as an alternative generic name. *Ibid.* 390 The corpora lutea of the ovary are the chief natural source of progestin... The placenta is another source. 1961 J. Rock in C. A. Villee *Control of Ovulation* xii. 228 The two artificial steroids are called 'progestins' because of the fact that their action resembles that of progesterone. 1966 *New Scientist* 15 Dec. 620/1 The synthetic progestins which are contained in all oral contraceptive tablets were in fact introduced for the wrong reason. 1976 *Sci. Amer.* Feb. 32/2 The steroid hormones..include the male sex hormones (collectively called the androgens), the female sex hormones (the estrogens and the progestins) and the hormones secreted by the cortex.

progestogen (prodʒe·stŏdʒen). *Physiol.* Also **progestagen.** [f. as prec. + -o + -gen; the variant spelling may reflect the influence of *gestagen.] = *gestagen.

1941 Dorland & Miller *Med. Dict.* (ed. 19) 1171/2 *Progestogen*, a general term for any substance possessing progestational activity. 1945 [see *progestin]. 1962 *Lancet* 12 May 1012/1 In those premature labours where relaxation of the cervix precludes uterine contractions, such measures as cervical suture and the administration of progestogens can prevent disaster. 1968 Passmore & Robson *Compan. Med. Stud.* I. xxxvii. 11/1 The principal ovarian hormones..fall into three broad functional categories, oestrogens, progestagens and androgens. *Ibid.* 14/2 A more useful pharmacological definition of a progestagen is a substance which induces secretory changes in an oestrogen-primed endometrium. 1968 *Times* 28 Nov. 14/4 Most oral contraceptive pills contain both a progestogen and an oestrogen. 1977 *Lancet* 21 May 1101/2, 7 women.. used low-dose progestagen pills.

Hence **progestoge·nic** *a.* = *progestational *a.*

1949 H. E. Nieburgs *Hormones in Clin. Pract.* viii. 176 The progestogenic preparations are mainly employed for the prevention and treatment of habitual and threatened abortion. 1969 *Sunday Times* 14 Sept. 54/3 The remedy may be a change to a more 'progestogenic brand' [of oral contraceptive]. 1973 *Nature* 9 Mar. 88/1 Administration of sex hormones or their analogues, whether androgenic, oestrogenic or progestogenic, can produce temporary remission in about one-quarter to one-third of patients with advanced breast cancer. 1977 *Lancet* 19 Nov. 1085/1 Evidence is accumulating to indicate that it is principally the œstrogenic and not the progestagenic component of combined oral-contraceptive preparations which causes the acceleration of blood-clotting.

proglacial (proglēⁱ·sĭal, -ʃăl), *a. Geomorphol.* [f. Pro-2 2 + Glacial *a.*] Situated or occurring just beyond the edge of an ice-sheet or glacier.

1937 Wooldridge & Morgan *Physical Basis Geogr.* xxii. 393 The term 'extra-morainic' has been proposed to cover all such lakes, but they are evidently not literally extra-morainic in every case, and the term 'pro-glacial' is preferable. 1957 G. E. Hutchinson *Treat. Limnol.* I. i. 89 During the early stages of deglaciation, proglacial lakes collected between the Cary Moraine and the receding ice. 1970 R. J. Small *Study of Landforms* xi. 395 His reconstruction of pro-glacial drainage conditions. *Ibid.*, The waters of the lakes escaped, either into adjacent lakes or away from the pro-glacial zone altogether. 1972 R. J. Price *Glacial & Fluvioglacial Landforms* (1973) vii. 185 As the glacier front moves down-valley the former proglacial fluvioglacial deposits are overridden by the ice.

prognathously (prɒgnæ·pəsli). *adv.* [f. Prognathous *a.* + -ly2.] In a prognathous manner; with the jaw prominent or protruding.

1974 N. Gordimer *Conservationist* 226 A jaw of fine teeth... Set rather prognathously, in the forward-jutting rounded arc that, in life, would make a wide white-toothed smile. 1976 T. Sharpe *Wilt* xiv. 146 He smiled prognathously.

Progne. Add: Also in Gr. form **Procne.** 1. (Later examples.)

1956 E. Pound tr. *Sophocles' Women of Trachis* 41 As Progne shrill upon the weeping air, 'Tis no great sound. 1980 'A. T. Ellis' *Birds of Air* 125 'You seem very cheerful,' she snapped... 'Oh, I am,' said Mary... Procne to mute Philomela grieving for her Itys, she thought.

prognostical, *a.* (*sb.*) Restrict † *Obs.* to B and add: **A.** *adj.* (Later examples.) *rare.*

1894 H. Latham *Service of Angels* 40 Is it prognostical? Is it proleptic? 1924 Keynes in *Econ. Jrnl.* XXXIV. 316 We come to something more prognostical of Alfred in a little device of William Marshall's latter days.

prognostically, *adv.* (Later examples.)

1911 *Amer. Jrnl. Med. Sci.* CXLI. 644 Studying one's cases prognostically. 1977 *Lancet* 4 June 1199/2 H.D.L. is much more important prognostically.

progradation (progrădēⁱ·ʃən). *Physical Geogr.* [f. next + -ation.] The seaward advance of a beach or coastline as a result of the accumulation of river-borne sediment or beach material.

1909 W. M. Davis in *Geogr. Jrnl.* XXXIV. 303 There is good reason for regarding the action of the streams in terracing or degrading the former valley floors as the cause of the progradation of the strand-plain. 1937 Wooldridge & Morgan *Physical Basis Geogr.* xxi. 332 Progradation may result from the extensive deposition of river alluvium, as in deltas. 1967 *Oceanogr. & Marine Biol.* V. 130 Brothers (1954) concluded that dune formation during subsequent periods of shoreline progradation added great quantities of sand to the foreland around Auckland. 1971 *Nature* 10 Sept. 91/2 The virtual elimination of shelf seas, during a prolonged phase of tectonic stability and peneplanation following rapid build-up of evaporites and progradation of coastal plain sediments.

prograde (prōᵘ·grēⁱd), *a.* [f. Pro-1 1 + Retro)grade *a.* and *sb.*] **1.** *Petrol.* Of a metamorphic change: resulting from an increase in temperature or pressure. Opp. *retrograde *a.* 3e.

1967 K. G. Cox et al. *Introd. Pract. Study Crystals* x. 200 The volatile constituents, H_2O and CO_2, are expelled from the rocks with rising temperature and cease to be available for the formation of low-temperature minerals (usually rich in H_2O and CO_2) when the temperature declines at the end of prograde metamorphism. 1977 A. Hallam *Planet Earth* 174 The metamorphism of sedimentary rocks..involves the production of water vapor, carbon dioxide and other gaseous substances... This type of metamorphism is called prograde metamorphism, and takes place principally in response to increasing temperature.

2. *Astr.* From west to east; anticlockwise as seen from north of the ecliptic; = Direct *a.* 3. Opp. Retrograde *a.* 1a.

1969 *Nature* 19 July 243/2 Once in a prograde orbit, the tides on the Earth would begin to push the Moon outwards to its present position. 1977 *Ibid.* 3 Mar. 15/3 The overhead motion of the Sun relative to an observer fixed on the surface [of Venus] is prograde with a speed of about 4 m s−1.

prograde (progrēⁱ·d), *v. Physical Geogr.* [f. Pro-1 1 + Retro)grade *v.*] **a.** *intr.* Of a shore or shoreline: to undergo progradation.

1909 W. M. Davis in *Geogr. Jrnl.* XXXIV. 303 After having maturely retrograded the cliffs, the waves have prograded the strand-plain. The strand-plain broadens a little opposite each valley, for now that the shore-line is prograding, the rivers have opportunity of building their deposits forward. 1929 H. Meredith *East Anglia* ii. 63 Its beach progrades in inverse proportion to the retrograding of the Ness itself. 1954 W. D. Thornbury *Princ. Geomorphol.* xvii. 443 Beach features are particularly ephemeral forms along a retrograding shore line, but along a shore line that is advancing seaward or is prograding they may be semipermanent. 1967 *Oceanogr. & Marine Biol.* V. 123 Thus valley train material extending seaward off Alaska has prograded almost to the headlands. 1978 *Nature* 17 Aug. 655/1 During the Campanian, fluvial, deltaic and coastal plain systems prograded eastwards across marine strata such that by the end of the Campanian (∼69 Myr bp) the area of marine deposition was greatly restricted.

b. *trans.* To cause to prograde.

1909 [see sense a]. 1919 D. W. Johnson *Shore Processes* v. 223 Just so long as the current aggrades (builds up) the seabottom offshore, the waves will prograde (build forward) the shore. 1939 Revelle & Shepard in P. D. Trask *Rec. Marine Sediments* iv. 279 The coarse material carried to the sea by rivers may temporarily prograde the shore, forming deltas. 1968 R. W. Fairbridge *Encycl. Geomorphol.* 133/1 Marine Deposition Coasts. These are another type of Secondary coasts that have been prograded by waves and currents.

Hence **progra·ded** *ppl. a.*, **progra·ding** *vbl. sb.* and *ppl. a.*

1910 *Jrnl. Geol.* XVIII. 166 The prograding of the shore beyond the headland. 1918 *Trans. & Proc. N.Z. Inst.* L. 215 Some of the material is thrown up on the beach, so that the shore-line advances seawards, leaving a prograded strip of new land. 1919 D. W. Johnson *Shore Processes* v. 223 Following Davis we may call any shore which is experiencing such a long-continued advance into the sea, a prograding shore, and distinguish it from the more usual retreating or retrograding shore. The prograding of a shoreline..may continue for a few years, a few centuries, or many thousands of years. 1940 *Geogr. Jrnl.* XCVI. 261 Occasional deep lows, locally called pits and containing water, are scattered about amongst the shingle. They were usually found during rapid prograding to the lee of bends. 1968 R. W. Fairbridge *Encycl. Geomorphol.* 895/2 (*caption*) Prograded spit, Whananaki Inlet, Northland, New Zealand. 1978 *Nature* 12 July 131/1 The environment was interpreted as a prograding tidal estuary where shelf deposits pass upwards into fields of migrating dunes and megaripples.

program, programme, *sb.* Add: These two forms have become established as the standard N. Amer. and British spellings respectively, with the exception that *program* is usual everywhere in connection with *Computers.* This latter distinction is followed in this article and throughout the Suppl. in editorial matter. **2. a.** (Further examples.)

1936 G. Greene *Gun for Sale* iii. 107 They sat two programmes round at the cinema. 1975 *Cricketer* May 41/3 (Advt.), Immediate cash paid for all programmes up to 1960. £5 minimum pre-war cup finals. 1976 *Globe & Mail* (Toronto) 16 Feb. 17/1 This was the season's final program in the Toronto Symphony's Young People's Concerts series.

b. Also *transf.*, a planned series of activities or events.

1941 *Bull. Amer. Assoc. Petroleum Geologists* XXV. 1256 Past successes enable us now to look ahead to a difficult but orderly exploration program, rather than a frenzied, inefficient scramble for immediately needed oil. 1949 Shurr & Yocom *Mod. Dance* 3 The fallacy of this statement became increasingly apparent as personal contact with, and participation in, teacher education programs in the colleges and universities increased. 1955 *Science* 25 Nov. 1003/2 The satellite program..is already underway. 1961 Berkner & Odishaw *Sci. in Space* i. 15 An orderly scientific program [of space research] cannot be conducted if the unreliability associated with new vehicles is always present. 1976 *Liverpool Echo* 7 Dec. 17/9 On the Wirral the West Cheshire programme was reduced to three games. 1977 *Nature* 11 Aug. 487/2 A joint programme to provide a set of soil structure standards..has been under way for five years at the Universities of Warsaw and Moscow.

c. *Mus.* A sequence of objects, scenes, or events intended to be suggested by a musical composition or used to determine its structure.

1854 H. F. Chorley *Mod. German Mus.* II. 306 There is no parroting such a programme..to an opera as the overture to 'Leonora'. 1883 Grove *Dict. Mus.* III. 34/2 There is a growing tendency amongst critics and educated musicians to invent imaginary 'programmes' where composers have mentioned none. 1944 W. Apel *Harvard Dict. Mus.* 605/1 In the final analysis, there are two types of program music: that which is good music regardless of the program; and that which is poor music even with a 'good' program. 1962 *Listener* 15 Mar. 489/3 While the majority of his piano pieces bear explanatory titles, in his symphonies he was far more reticent as to the underlying programme. 1974 *Encycl. Brit. Micropædia* VIII. 231/3 Only in the so-called Romantic era, from Beethoven to Richard Strauss, is the program an essential concept.

d. = *dance programme* s.v. *dance *sb.* 7.

1899 A. E. W. Mason *Miranda of Balcony* iv. 40 He compared programmes with Miranda... Four dances must intervene before he could claim her. 1913 Mrs G. de H. Vaizey *College Girl* xxvii. 369 The three programmes were filled to the last extra. 1949 N. Marsh *Swing, Brother, Swing* iii. 42 Her coming-out ball had been here... She felt the cord of her programme grow glossy under the nervous pressure of her gloved fingers. 1976 *Times* 11 June 14/6 Guests carry around little programmes to remind them who they are dancing with.

e. *Broadcasting.* (i) A broadcast presentation treated as a single item for scheduling purposes, being broadcast between stated times and without interruption except perhaps for news bulletins or advertisements.

1923 *Radio Times* 28 Sept. 1 From November 14th last year..we have..transmitted roughly 1,700 distinct evening programmes. 1933 *Ibid.* 14 Apr. 72/2 In the programmes for Friday next, you will find particulars of *Looking In*, first television review. 1946 *B.B.C. Year-bk.* 62 About 120 new programmes..are put on the air every week. *Ibid.*, Such famous programmes as 'Itma', 'Music Hall', and 'Workers' Playtime'. 1962 A. Nisbett *Technique Sound Studio* i. 17 Certain topical and miscellany programmes..go on the air in the form of a mixture of live and recorded segments. 1976 *Times* 21 May 2/8 Since the programme I have had about a dozen other nasty telephone calls.

(ii) A radio service providing a regular succession of programmes on a particular frequency.

1939 R. Macaulay *Lett. to Sister* (1964) 96 I've just read the debate on B.B.C. in Hansard... The fact is we *can't* get on without 2 programmes. 1939 [see *home *a.* 2e]. 1945, etc. [see *light *a.*1 19b]. 1946 *B.B.C. Year-bk.* 51 The Director-General promised that within ninety days of the end of hostilities in the West, the BBC would provide its listeners in the United Kingdom with two full-scale alternative programmes. *Ibid.*, It is the aim of the Home Service to provide the home programme of the people of the United Kingdom. 1968 *B.B.C. Handbk.* 48 The popular music programme on 247 metres became Radio 1. The Light Programme, on 1500 metres and VHF, became Radio 2. The Third Network, which embraces the Music Programme, the Third Programme, Study Session and the Sports Service, became Radio 3. 1980 *Times* 31 July 15/3 Radio 3 used to be the most civilized and broad-ranging programme in the world.

f. *Electronics.* Also *programme signal.* A signal corresponding to music, speech, or other activity.

1935 Nilson & Hornung *Pract. Radio Communication* viii. 356 The program fed into the mixer does not always come directly from a microphone. 1948 A. L. Albert *Radio Fund.* xiv. 569 The frequency deviations of a program signal in frequency-modulation can be made, and is made, quite large... If the program has a wide frequency deviation, but noise does not, the signal-to-noise ratio of the output of a discriminator will be high. 1954 Molloy & Pannett *Radio & Television Engineers' Ref. Bk.* iv. 12 It is necessary at times to compress the programme from a range of 50 to 22 db. 1959 K. Henney *Radio Engin. Handbk.* (ed. 5) xxi. 7 Program signals have very complex wave shapes. 1970 J. Earl *Tuners & Amplifiers* iv. 77 The amplifier..must process the programme signal (e.g. the signal from the pickup, radio tuner, tape recorder or whatever). 1977 *Gramophone* Nov. 937/1 The 2760 copier unit completes the system..and has automatic end-of-tape sensing and erasure of programme.

g. (i) A sequence of operations that a machine can be set to perform automatically.

1945 J. P. Eckert et al. *Description of ENIAC* (PB 86242) (Moore School of Electr. Engin., Univ. of Pennsylvania) 1 The intended use of the ENIAC is to compute large families of solutions all based on the same program of operations. 1954 *Amer. Machinist* 25 Oct. 136/1 The operator..sets a combination of switches calling for table movements equivalent to blueprint dimensions, or a 'program', then presses a starting button. 1962 E. Bruton *Automation* vi. 74 An automatic washing machine may be designed to wash for four minutes, empty, and spin-dry for ten. This is its programme. It can be 'programmed' in other ways. 1970 *Which?* Oct. 293/1 For most, there was a pre-rinse and a choice of two washing programmes, depending on how dirty the dishes were. 1972 *Daily Tel.* 11 Jan. 11/2 There's a Westinghouse electric clothes-dryer..which takes 12 lb of clothes and has five drying programmes: auto-dry, wash 'n 'wear, time dry, air fluff and low heat. 1977 *Times* 9 July 21/4 The ability of modern machines to offer merely rinse and dry programmes for clothes that have been prewashed by hand.

(ii) *Computers.* A series of coded instructions which when fed into a computer will automatically direct its operation in carrying out a specific task. Also *transf.*

1946 *Nature* 20 Apr. 527/2 Control of the programme of the operation of the machine [sc. ENIAC] is also through electrical circuits. 1947 *Math. Tables & Other Aids to Computation* II. 358 An important limitation upon programming is that the machine must adhere to a prescribed linear course of operation. It cannot at any point choose between two subsequent programs on the basis of results already obtained. 1950 *Phil. Mag.* XLI. 256 The problem of constructing a computing routine or 'program' for a modern general purpose computer which will enable it to play chess. 1953 *Proc. IRE* XLI. 1245/1 A large family of high-speed, large-scale, stored-program, digital computers have been built. *Ibid.* 1247/1 This conditional instruction makes it possible for the programmer to write programs which take different courses of action depending upon the results of previous computation. 1960 *Times* 4 Oct. (Computer Suppl.) p. v/3 To prepare this sequence of instructions, or program (a spelling now adopted in computer terminology), the programmer will have broken down an operation into its simplest elements. 1971 *Times Lit. Suppl.* 4 June 635/2 Were accurate estimation of the merits of such positions possible, the next world chess champion could quite conceivably be a computer programme. 1972 *Sci. Amer.* Mar. 42/3 Computer instructions are so complicated that programmers are often baffled when they look at programs they have written but have not seen for several months, and a third party usually finds them inscrutable. 1972 R. M. Lee *Short Course Basic Fortran IV Programming* i. 8 Programs are written in one of the many user-oriented languages, such as FORTRAN, and then translated into machine language... The translation is done by the computer. 1974 *Sci. Amer.* Oct. 105/1, I have described the timing and the characteristics of the coordination of the eye-head movements that are elicited by the appearance of a visual target, and have presented our evidence for the conclusion that the programs for eye-head coordination are not present in the central nervous system in their entirety. 1977 W. S. Davis *Operating Syst.* v. 58 Before any program can be run, it must first be set up (cards loaded in the card reader, the printer loaded with..paper.., tapes and disk packs mounted, and so on).

h. *Psychol.* and *Educ.* In human and animal learning, a series of step-by-step questions or tests (freq. designed to be used in a teaching machine operated by the learner) aimed at the establishment of learning patterns through the stimulus of rewarding correct responses or behaviour at each step.

1950 B. F. Skinner in *Psychol. Rev.* LVII. 207/2 Such a set was randomized in a program of reinforcement repeated every hour. In changing to this program from the arithmetic series..the pigeons were soon able to sustain a constant rate of responding under it. 1958 —— in *Science* 24 Oct. 971/2 The machines themselves cannot be adequately described without giving a few examples of programs. 1961, etc. [see *linear *a.* 3c]. 1962 *Listener* 17 May 855/2 The drawback to a multiple choice programme..is that plausible wrong answers must be presented to the student, and he may remember these instead of the correct ones. 1967 Coulthard & Smith in Wills & Yearsley *Handbk. Managem. Technol.* xi. 204 Two types of programme are currently in use: 1. The linear programme—which repeats a statement just made, omitting a key word or words, and requires the trainee to remedy the omission... 2. The intrinsic programme—which provides an explanation of a key point in the subject and asks the trainee to select the correct answer to a question from several alternative answers. 1976 W. B. Kolesnik *Learning* x. 226 The program can be used in grades one through twelve in the area of language, arts, mathematics, [etc.].

4. *programme-book, -card* (further examples), *-making* (examples; also in senses other than 2b), *note, -seller* (examples), *vendor*; (sense *2e) *programme content, director, editor, engineer, item, -maker* (see also sense in Dict.), *planner, planning, staff, time*; (sense *2g) *program step, tape, testing*; **programme boy, girl,** a boy or girl employed to sell programmes at a place of entertainment; **programme-building,** the selection of items for a concert or for a period of broadcasting; so **programme-builder; programme chairman** *U.S.*, one who arranges the programme of events or the agenda for a particular event for a society, etc.; **programme company,** a company authorized to make programmes and advertisements for broadcasting on British commercial television; **programme contractor** = **programme company*, so **programme-contracting** *a.*; **program(me) control,** (*a*) = *programmer 3; (*b*) control of or by a program(me); **program(me) controller** = *programmer 1c, 3; **program counter** *Computers*, a register in the control unit of a computer which contains the address of the next instruction to be executed, this number being increased by one each time unless an instruction to do otherwise occurs; **programme girl**: see **programme boy*; **programme junction** *Broadcasting* (see quot. 1941); **program library** *Computers* = *library1 2*; **programme line** *Telecommunications* (see quot. 1940); **programme movie** = **programme picture*; **programme music** (earlier and later examples); **programme pencil,** a small pencil for filling in a programme-card at a dance, etc.; **programme picture,** a cinema film made relatively cheaply and intended to be shown as part of a programme that includes another film as the main feature; **program register** *Computers* = *control register* s.v. *control *sb.* 5; **programme service** *Broadcasting*, a service consisting in the regular broadcasting of radio or television programmes for reception by the public; **programme symphony,** a symphony with a programme (sense *2c).

1954 *Grove's Dict. Mus.* (ed. 5) VI. 943/2 Philip Hale's long series of notes for the Boston Symphony Orchestra made the programme-books of that orchestra valuable historic documents. 1976 *New Yorker* 9 Feb. 102/3 The wretched program book gave no texts—only excerpts from the Woolf cycle, and translations of the rest. 1921 *Dict. Occup. Terms* (1927) § 889 Programme boy, girl, seller. 1928 *B.B.C. Handbk.* 73/2 The programme builders believe that..the 60,000 hours of programmes will receive the liveliest and most general approval..by the application of a common-sense policy. 1947 *Penguin Music Mag.* Dec. 27 Every programme-builder should know the symphonic repertoire from A to Z. 1961 *Listener* 21 Dec. 1088/3, I don't know what was in the mind of the programme-builder of the concert given by the B.B.C. Symphony Orchestra. 1928 *B.B.C. Handbk.* 74/1 The best method of explaining the details of programme building is to follow a week's programmes from their first beginnings to the day on which they are broadcast. 1935 *Discovery* Sept. 277/1 It may be left safely to the B.B.C., whose experience and standards of programme building..may be relied upon to result in presentations in line with public approbation. 1948 *Penguin Music Mag.* Feb. 93 Mr. Barbirolli's extraordinary skill in programme-building. 1957 *Encycl. Brit.* IV. 208/2 It is principally under the headings (5) and (6).. that radio has created expression forms peculiar to itself, and in most other respects program-building is creative only in the sense that the program-builders can build combinations of suitable music and speech around one or another central idea. 1948 *Penguin Music Mag.* June 135 The Orchestra's first programme-card was several degrees more adventurous than any before it. 1961 *U.S. Nat. Bureau Standards Rep. Fiscal Year 1961* 57 The Bureau contributed to the planning and success of the Symposium through the efforts of..Dr. C. M. Herzfeld, Program Chairman. 1976 *Billings* (Montana) *Gaz.* 27 June 5-E/3 She is a past president, program chairman and secretary of the chapter. 1958 *New Statesman* 22 Mar. 375/3 As each programme-company in turn began broadcasting in the regions, the tale was always the same: of viewers with a choice, most chose the ITV channel most of the time. 1962 *Rep. Comm. Broadcasting 1960* 1 in *Parl. Papers 1961–2* (Cmnd. 1753) IX. 259 Independent television comprises not only the ITA but also the programme companies, which have at various dates since 1954 been appointed by the ITA. 1968 *Listener* 11 July 84/3 What are called the programme companies, in ITV, are set up in the first instance as contractors on the basis of the advertising franchise in their area. 1958 *New Statesman* 20 Dec. 880/1, I am not sure that I agree that the effect of the competition on the programme-content is as superficial as they seem then to have found. 1964 M. McLuhan *Understanding Media* (1967) xxviii. 293 Like the radio that it still provides with program content, the phonograph is a hot medium. 1968 *Listener* 29 Aug. 285/3 There is now hardly a significant publication, from the weekly reviews to the mass-selling dailies, which does not have equity in one or other of the programme-contracting companies. 1954 *Act* 2 & 3 *Eliz. II* c. 55. § 2 The programmes broadcast by the Authority shall..be provided not by the Authority but by persons (hereafter in this Act referred to as 'programme contractors') who, under contracts with the Authority, have, in consideration of payments to the Authority.., the right and the duty to provide programmes or parts of programmes to be broadcast by the Authority. 1958 *New Statesman* 5 July 1/2 With commercial success and popular support behind them the existing programme contractors, and those who would like to become programme contractors, are in a much more powerful position than they were when ITV was an untried gamble. 1962 *Rep. Comm. Broadcasting 1960* 166 in *Parl. Papers 1961–2* (Cmnd. 1753) IX. 259 The [Independent Television] Authority's power to control the companies, once they are appointed programme contractors, is illusory and negligible. 1945 J. P. Eckert et al. *Description of ENIAC* (PB 86242) (Moore School of Electr. Engin., Univ. of Pennsylvania) B-5 The simplest procedure for handling the problem is to devote one multiplier program control to each of the n multiplications. 1951 M. McLuhan *Mech. Bride* (1967) 21/1 The president of the National Broadcasting Corporation ridiculed the proposal to separate business control from program control. 1953 *Proc. IRE* XLI. 1271/2 (*heading*) Program control of external units. 1957 D. M. Considine *Process Instruments & Controls Handbk.* ix. 78 The operation of a tire vulcanizer, on a completely automatic timed basis, is an example of program control. 1977 *Design Engin.* July 15/3 Eight parallel latched outputs are available as binary, BCD, or as a 7-segment-plus-decimal-point output under program control. 1957 D. M. Considine *Process Instruments & Controls Handbk.* ix. 78 By controlling sequence, intervals, and rates of change, a program controller may encompass all the operations in a complete industrial process. 1961 *Times* 4 Aug. 2/3 Independent Television in Wales..now invite application for the following key posts:.. Programme Controller. In this position, experience of Television production, a knowledge of Welsh, and broad interests over the whole field of entertainment are essential. The Controller in Cardiff will plan and budget all programmes, and must know how and where to produce or acquire material. 1967 F. W. Clarke *Installing Small Pipe Central Heating* vi. 37 Where a boiler supplies hot water and serves the heating, the various jobs required of it can be simply co-ordinated by a programme controller. This turns the heating or the hot water on and off at selected times, as set on the clock. 1976 C. Bermant *Coming Home* ii. iv. 161, I returned to Granada and sent a memo to the programme controller. 1946 *Math. Tables & Other Aids to Computation* II. 102 Counters are used not only for arithmetic purposes, but also as a part of the programming circuits which determine when and how a given unit shall perform. Each unit whose operations consume more than one addition time has such a program counter. 1962 Huskey & Korn *Computer Handbk.* xvi. 29 In the case of a jump instruction, the address for the next instruction to be fetched comes from the address part of the present instruction... The state of the program counter must be changed to agree with this address so that it will count on from the new starting point. 1977 *Design Engin.* July 15/3 There are four parallel inputs, a testable sense input, three bi-directional control flags for use as inputs or outputs, a program counter, a two-word stack for nested subroutine calls, and an instruction-decode programmable logic array. 1961 G. Millerson *Television Production* i. 15 He may operate the buttons and faders for video switching himself, but most networks consider the programme director too preoccupied with the many other aspects of production, and delegate this job to another person. 1972 *Listener* 6 July 2/1 Michael Rice, the Programme Director, asked one of his people..to lead the production team. 1929 *Radio Times* 8 Nov. 387/3 Broadcast reading has only been tried half-heartedly. The programme editors still suffer from ..fear of not pleasing everybody all day long. 1949 *Ibid.* 15 July 41/3 He returned to the BBC as a programme engineer. 1962 Programme engineer [see *balance *sb.* 14 b]. 1905 *Daily Chron.* 13 Feb. 9/3 An interesting story of a medical student's love for a programme girl..was told. 1918 [see *cloak-room b]. 1921 [see **programme boy*]. 1979 G. Latta tr. *Jacquemard-Sénécal's Eleventh Little Nigger* i. v. 47 The programme girls..persuaded her to swallow a considerable amount of whisky. 1962 Programme item [see **programme planning*]. 1941 *B.B.C. Gloss. Broadcasting Terms* 25 *Programme junction*, brief interval between the end of one programme and the beginning of the next, used for switching operations whereby transmitters are linked to, or detached from, the network concerned. 1975 *Listener* 23 Oct. 532/2 There was internal machinery to see that there were common programme junctions. 1960 *Ann. Rev. Automatic Programming* I. 93 (*heading*) MERCURY autocode: principles of the program library. 1977 R. E. Harrington *Quintain* xii. 139 Sanderson gave me the constants and I just ran the program. I didn't even write it. He got it from the program library. 1940 *Chambers's Techn. Dict.* 676/2 *Programme* (or *program*) *line*, a transmission line, of superior propagation characteristics, for relaying broadcasting programmes. 1944 *Proc. IRE* XXXII. 601/1 A key located to the left of the VU meter should be used to connect this meter to the outgoing program line. 1929 *Radio Times* 8 Nov. 393/1 There are people who..abuse the programme-makers! 1977 *Broadcast* 13 June 5/2 Programme makers could..put their ideas to the empirical test by means of a pilot programme. 1904 W. James *Let.* 1 Jan. in R. B. Perry *Tht. & Char. W. James* (1935) II. 201 Münsterberg has the most extraordinary power of schematization and program-making. 1949 *Penguin Music Mag.* Feb. 19 That almost perfect example of programme-making, *Music in Miniature.* 1980 *Listener* 3 Jan. 21/1 Industrial disputes [in broadcasting] took up more time than programme-making. 1935 *Movie Mirror* Dec. 106/1 Dropping into the theater, prepared for a regular program movie, my interest was caught after the first few feet and worked up to a fever pitch at the final reel. 1879 Grove *Dict. Mus.* I. 232/2 Berlioz was one of the most uncompromising champions of what, for want of a better name, has been dubbed 'programme music'. 1954 C. S. Lewis *Eng. Lit. in Sixteenth Cent.* i. ii. 139 Disorder in life rendered by disorder in art. This is in poetry what 'programme music' is in music. 1923 M. R. Werner *Barnum* 319 And then in Barnum's program notes each year appeared this notice. 1942 E. Blom *Music in England* ix. 149 Ella also wrote his own programme notes. 1958 'E. Dundy' *Dud Avocado* i. vii. 136 I'd never seen a ballet whose story I was able to follow even when the programme-notes were in English. 1965 *Listener* 25 Nov. 874/1 The relevance of the play to the 'thirties in Britain and also to Radio in Europe Week was so admirably condensed in a programme note in *Radio Times* by the producer, Douglas Cleverdon, that it does not seem worth labouring. 1978 *Daily Tel.* 20 Jan. 13/1 To 25 years' experience before the camera can be added (I learn from the programme note) some recent experience as a director in the theatre. 1895 *Montgomery Ward & Co. Catal.* Spring & Summer 115/2 'Programme' Pencils, round, enameled in colors with gilt tip and ring. Suitable for use in lady's memorandum book. 1921 E. N. Hull *Sheik* i. 9 She hesitated, tapping her programme-pencil against her teenth. 1928 *Sunday Dispatch* 19 Aug. 14/2 A 'programme' picture is a film which costs from £6,000 to £8,000 or thereabouts, and cannot be called a 'super'. 1935 *Movie Mirror* Dec. 38/3 Your Reviewer Says: An average program picture, but Velez fans will want to see it for sure. 1956 Programme planner [see *b.b.c.]. 1961 A. Wilson *Old Men at Zoo* ii. 92 The television engineers and programme planners with whom the office now seemed filled. 1974 *Guardian* 23 Jan. 1/5 If the [TV] close-down had been at 10.30 there would have been more room for manoeuvre by the programme planners. 1940 R. S. Lambert *Ariel & all his Quality* iii. 77 Charles Siepmann..was promoted to..Director of Programme Planning. 1962 *Rep. Comm. Broadcasting 1960* 159 in *Parl. Papers 1961–2* (Cmnd. 1753) IX. 259 What particular 'time slots' each [TV company] is to occupy, and with what programme items... That is to say, the overall programme planning. 1948 *Gloss. Computer Terms* (Mass. Inst. Technol. Servomechanisms Lab. Rep. R–138) 8 *Program register*, the part of the computer used for holding orders after they are extracted from storage but before they are carried out. 1956 Program register [see **control register*]. 1962 R. V. Oakford *Introd. Electronic Data Processing Equipment* iii. 37 Information (normally instructions) can be transferred to the program register in the control unit from general memory or from the arithmetic (process) unit. 1921 Programme seller [see **programme boy*]. 1977 M. Babson *Lord Mayor of Death* v. 45 Here comes the programme seller. 1940 L. R. Lohr *Television Broadcasting* ii. 23 Television transmitters have been in operation from time to time in Philadelphia, Schenectady,.. and Bridgeport, but none of these had established a program service for the public at the time of writing. 1962 A. Nisbett *Technique Sound Studio* i. 17 The next link is a continuity suite where the entire programme service is assembled. 1940 R. S. Lambert *Ariel & all his Quality* ii. 43 Programme and administrative staff had not been divided into watertight..categories. 1977 *Listener* 28 Apr. 540/2 There has been a planned increase of programme staff, facilities and output. 1950 *High-Speed Computing Devices* ix. 157 The Type 604 can perform 60 program steps, or operations, per card; a program step includes any one of the four arithmetic operations, or a number transfer. 1956 [see *command *sb.* 1 d]. 1978 *Sci. Amer.* Feb. 29/2 (Advt.), Because of this dual capability, it can..identify which program step a system was executing at the time of malfunction. 1934 C. Lambert *Music Ho!* iii. 162 Nationalism..destroys both the aristocratic quality of the eighteenth-century abstract symphony and the individualist quality of the nineteenth-century programme symphony. 1962 *Listener* 29 Nov. 941/2 The third symphony..was suggested by visits to Mycenae and Venice. Programme symphonies are even more out of fashion than the normal type, but I can only say that I tried to express the emotions aroused in me by the places rather than to paint pictures of them. 1948 *Math. Tables & Other Aids to Computation* III. 126 We can have on the first section of the program tape.. the program for arranging the data in order by age. 1964 C. Dent *Quantity Surveying by Computer* iii. 26 In other cases the data tape is read in under control of the instructions (stored in the memory by the program tape), the data being worked on as it is read in. 1959 J. Jeenel *Programming for Digital Computers* viii. 393 In program testing one usually employs certain techniques especially developed for this purpose. 1964 F. L. Westwater *Electronic Computers* ix. 140 The coding is tested on the computer... (This is called 'program testing'.) 1957 *Practical Wireless* XXXIII. 529/2 Fewer programmes.. would enable the BBC to..reject the dross which is still allowed programme time. 1977 M. Babson *Lord Mayor of Death* xii. 82 Programme vendors..were..shaking their heads regretfully at would-be customers.

program, programme, *v.* (The note s.v. *program, programme *sb.* applies equally to the vb.) Add: **1.** (Further examples.)

1912 A. Bennett *Jrnl.* 16 Feb. (1932) II. 44 On Wednesday morning at 7 a.m. as 'programmed' a week ago, I began 'The Regent'. 1949 *Archit. Rev.* CVI. 375/1 Let us.. consider a country like Japan where, after wholesale destruction, four million minimum dwellings are now being programmed. 1956 *Sun* (Baltimore) (B ed.) 24 Sept. 10/2 Senator Scott found..that 75 per cent of the soil bank outlay of $261,000,000 programmed for 1956 was to be spent in twelve mid-Western farm states. 1970 *Daily Tel.* 30 Jan. 2/4 He tried to programme her day into housework and study, but with four children and a pile of nappies it did not work. 1977 *Gramophone* Feb. 1308/1 The items are programmed in a quite interesting way, the fireworks of Liszt's *Hungarian Rhapsody* No. 2 being followed by the cool renunciation of Satie's *Gymnopédie* No. 1. 1979 *Church Times* 25 May 14/5 When the ceremonies and speeches were over, the General..was programmed to leave the gathering and walk along the red carpet to his car.

† **2.** *intr.* To write programme notes. *Obs. rare.*

1889 G. B. Shaw *London Music in 1888–89* (1937) 243 He programmed in a pat-the-young-man-on-the-back style.

3. *trans.* and *intr.* To broadcast. *U.S.*

1937 *Amer. Speech* XII. 101 *To program* means..to broadcast. 1967 *Boston Sunday Herald* 26 Mar. ii. 8/2 (Advt.), Personalities are an important ingredient in today's radio, and WCOP provides warm, personable, well established people—they program 24 hours a day with your listening pleasures in mind. 1969 *N.Y. Rev. Books* 2 Jan. 17/3 We can program twenty more hours of TV in South Africa next week to cool down the tribal temperature raised by radio last week. 1978 *Chicago* June 22/1 CSO does not program enough contemporary music.

4. a. To express (a task or operation) in terms appropriate to its performance by a computer or other automatic device; to cause (an activity or property) to be automatically regulated in a prescribed way.

1945 J. P. Eckert et al. *Description of ENIAC* (PB 86242) (Moore School of Electr. Engin., Univ. of Pennsylvania) B-4 In this fashion, problems involving numbers of multiplications far in excess of 24 can be programmed. 1949 *Nature* 22 Oct. 684/2 The problem must be programmed, that is, it must be split up into a series of simple operations which the machine can perform. 1952 *Phil. Mag.* XLIII. 1245 When a mathematician assembles the set of orders required to work out the solution of a problem he is said to be programming this problem for the machine. 1955 *IRE Trans. Industr. Electronics* II. 3/1 Industry needs more flexible methods of programming machine cycles to achieve automatic operation of machine tools in limited-quantity production. 1958 *Times Rev. Industry* Feb. 52/2 The Burroughs Typing Sensimatic has unmatched flexibility. The control unit permits each job to be individually programmed. 1962 G. A. T. Burdett *Automatic Control Handbk.* xix. 2 The engineer must programme the operations which the machine is to carry out. Preferably this programming, i.e. planning in advance and in sequence all the steps of the required operations, should involve the minimum of human effort and..a computer may be used. 1971 *Sci. Amer.* Apr. 71 Evidently an annual cycle of feeding and fasting is also programmed in the animal. 1973 A. Parrish *Mech. Engineer's Ref. Bk.* xix. 15 If feeds and speeds are programmed such that the spindle motor is producing its maximum horse power, any hard spots in the work piece can result in stall. 1977 *Sci. Amer.* Sept. 187 (*caption*) Typical task for a traveling-wire EDM is cutting gear teeth... When a slow and complex series of cuts is programmed, it can run unattended for 60 hours.

b. To incorporate (a property) *into* a computer or other device by programming.

1972 Carr & Mize *MOS/LSI Design & Application* viii. 233 The uniqueness desired within the master PLA chip is often programmed into the master chip by changing only the gate mask. 1977 D. Bagley *Enemy* xv. 121 He's installed a scad of microprocessors in that control board... He could program his timetables into them. 1977 *Nature* 11 Aug. 571/3 This book..deals with the problem of programming 'common sense' into a computer.

5. a. To cause (a computer or other device) automatically to do a prescribed task or perform in a prescribed way; to supply with a program. Also *absol.*

1945 J. P. Eckert et al. *Description of ENIAC* (PB 86242) (Moore School of Electr. Engin., Univ. of Pennsylvania) B-1 The problem of programming the ENIAC. *Ibid.*, We then wish to program the first accumulator to transmit its contents twice into the second one. 1947 *Proc. IRE* XXXV. 761/1 When an accumulator is programmed to transmit subtractively, it will transmit, not the number it holds, but the complement of the number it holds. 1950 *Phil. Mag.* XLI. 256 (*heading*) Programming a computer for playing chess. 1961 K. Amis *New Maps of Hell* i. 33 Here an airborne device, programmed to detect and forestall aggressive intentions, ends by prohibiting most kinds of human action. 1962 *Lancet* 8 Dec. 1215/2 The Pegasus computer can be programmed to punch out the desired results on standard teleprinter tape. 1973 *Daily Tel.* 8 Jan. 13 (Advt.), What you get for £505. 1. Full central heating and domestic hot water... 5. Time switch to programme the boiler. 1973 *Sci. Amer.* Sept. 87/1 Cardiographic instruments can be programmed to sound an alarm if an alarming event occurs. 1976 M. M. Mano *Computer System Archit.* ii. 74 For small quantities it is more convenient to use a programmable ROM, referred to as PROM... Each cell in a PROM incorporates a link that can be fused by application of a high current pulse. A broken link in a cell defines one binary state and an unbroken link represents the other state... This allows the user to program the unit in his own laboratory..to achieve the desired relationship between input address and output data. 1976 *Physics Bull.* Dec. 535/3 The operation is computer controlled so that the mirrors can be programmed to follow a particular source round the sky.

absol. 1954 *Amer. Machinist* 25 Oct. 134/2 Tool Engineers..will have to learn a new approach to tooling. Instead of designing massive fixtures or intricate mechanical controls, they will 'program'. 1958 *Oxf. Mag.* 29 May 470/1 It is not difficult to learn to program..backed by regular university lectures in numerical analysis and computing. 1966 *Sci. Amer.* Sept. 72 The ability to write a computer program will become as widespread as the ability to drive a car. Not knowing how to program will be like living in a house full of servants and not speaking their language. 1977 *Daily Tel.* 14 Nov. 8 (Advt.), To program, just read down the column, making the appropriate keyboard entries as you go!

b. *fig.* To train to behave in a predetermined way.

1963 *Language* XXXIX. 455 He succeeded in programming the live bees that crowded around the imitation insect to head in a prescribed direction to seek and find nectar. 1966 L. Jones in A. Chapman *New Black Voices* (1972) 459 We have always been separate, except in our tranced desire to be the thing that oppressed us, after some generations of having been 'programmed'.. into believing that our greatest destiny was to become white people! 1967 *Freedomways* VII. 131 The black student is being educated in this country as if he were being programmed in white supremacy and self-hatred. 1968 *New Scientist* 19 Dec. 653/1 To what extent can astronauts, environmentally be-suited, rigidly programmed, and electrically guided to their destination, be said to resemble the courageous explorers of the past? 1975 A. Price *Our Man in Camelot* iv. 71 Your cover is perfect... You were trained and programmed for just such an operation as this. 1976 J. Ross *I know what it's like to Die* ii. 12 Violent death programmed him to action in a predetermined routine; his reflexes conditioned by his training.

6. *Psychol.* and *Educ.* To form into a teaching programme (*program, programme *sb.* 2 h).

1958 B. F. Skinner in *Science* 24 Oct. 976/2 When material is adequately programmed, adjacent steps are often so similar that one frame reveals the response to another. 1971 Pittenger & Gooding *Learning Theories in Educ. Practice* iii. 91 Programming complex behavior requires careful planning and sequencing of material.

7. *intr. Astronaut.* Of a spacecraft: To perform a scheduled and automatically controlled manœuvre.

1958 *Daily Progress* (Charlottesville, Va.) 11 Oct. 1/5 He said the first stage appeared to have 'programmed'—started curving on its trajectory to the northeast—higher than it should. 1962 M. Caidin *Man-in-Space Dict.* 156/2 *Programming*, movement of a booster vehicle through assigned trajectory maneuvers in flight, as when a booster launches from a vertical position, then programs over toward horizontal flight. 1962 J. Glenn in *Into Orbit* 189 'We're programming in roll OK,' I said.

programmable (prōᵘ·græmăb'l), *a.* and *sb.* Also (*rare*) **programable.** [f. Program, programme *v.* + -able.] **A.** *adj.* Of an apparatus or an operation: capable of being programmed. Also *fig.*

1959 *Times Rev. Industry* May 36/2 The investigating team..designed a 'push-button office'... Such an arrangement would..be a data processing system..not based on a ..programmable computer. 1965 M. Frayn *Tin Men* ix. 51 Filling up a football coupon is another job which a computer could easily be programmed to do... We have a range of variables which can be identified in advance and manipulated according to predetermined rules. It's programmable. 1967 Cox & Grose *Organization & Handling Bibliogr. Rec. by Computer* ii. 46 With a system such as I have described all the elaborate mechanical and studio devices used to compose pages and flats are now replaced by programmable operations generated by algorithm from simple input instructions. 1971 *Sci. Amer.* Aug. 100/1 (Advt.), Others know us as a computer company: more than 10,000 own our programmable calculators and computers. 1972 Carr & Mize *MOS/LSI Design & Application* viii. 232 (*caption*) Programable logic array (PLA) contains multiple ROMs and flip-flop feedback elements for sequential logic elements. 1979 *Nature* 13 Sept. 131/2 The laser frequency..and data recording were controlled by a programable calculator. 1979 B. Peterson *Peripheral Spy* vi. 157 Predictable, hell! I'm programmable, that's what. Just like a godamned computer.

B. *sb.* A programmable calculator.

1975 *New Scientist* 27 Feb. 506 New handheld programmables will appear in 1975–76 not unlike the HP-65 at as little as £85. 1977 *Sci. Amer.* Apr. 94/2 (Advt.), All our hand-held and portable programmables incorporate HP's special RPN logic system.

Hence **pro:grammabi·lity,** the property of being programmable.

1966 *Jrnl. Assoc. Computing Machinery* XIII. 369 (*heading*) Use of multiwrite for general programmability of search memories. 1975 *Daily Tel.* 17 July 7 (Advt.), Programmability overcomes both limitations—and makes a calculator vastly more powerful. 1977 *Sci. Amer.* June 79/1 Programmability transformed the slick slide-rule calculator into an advanced scientific machine.

programmatic, *a.* Delete *rare* and add further examples (corresponding to various senses of 'programme').

1904 G. S. Gordon *Let.* 25 Oct. (1943) 4 Pardon this very egotistical and programmatic letter dear Molly. 1935 *B.B.C. Ann.* 81/2 Announcements relating to B.B.C. policy in every respect—programmatic, engineering, or other—are issued. 1937 *Proc. Prehistoric Soc.* III. 265 There is no shortage of programmatic declarations, of resumés of problems and aims. 1941 *Mind* L. 395, I may fitly use these programmatic statements as my text on which to hang the main reflections. 1947 *Proc. IRE* XXXV. 757/1 Both digital and programmatic information must be stored: the machine must be able to remember both the numbers that are operated on and the instructions for performing the operations. 1958 *Listener* 27 Nov. 861/1 No new Pope ever makes what might be called a programmatic speech giving the policies he intends to follow. 1971 S. Herrick *Astrodynamics* I. vii. 176 Programmatic or program-assisted singularities are destructive bits of sub-programming.. of which two are noteworthy. 1974 *Listener* 24 Jan. 121/3 A symphonic poem along Straussian programmatic lines. 1977 M. Goulder in J. Hick *Myth of God Incarnate* iv. 65 In the programmatic opening paragraph of Acts, Luke designates the advance of the church as being in four stages. 1979 *Dædalus* Summer 105 Even the most apparently blithe comedies are far more programmatic than a haphazard jumble of anecdotal detail.

Hence **programma·tically** *adv.*, in the manner of a programme or programme music; in accordance with a programme; with regard to a programme.

1947 A. Einstein *Music in Romantic Era* xi. 126 Spohr wrote a 'Characteristic Tone-Painting in the Form of a Symphony'..which makes use of a poem as the starting-point for revivifying programmatically the traditional form of the symphony. 1952 *Word* VIII. 97 Still other American linguists are programmatically rejecting the strict asseverance of meaning from sound and the exclusion of meaning from linguistic science. 1971 S. Herrick *Astrodynamics* I. vii. 179 Programmatically or computationally, especially with automatic, automaton machinery, there is a great difference. 1974 *Times Lit. Suppl.* 29 Mar. 344/5 Beardsley's admirable drawing and programmatically tortured lettering. 1978 *Nature* 23 Feb. 785/3 The book..offers to approach flow phenomena programmatically in the framework of a general dynamic theory.

programmed (prōᵘ·græmd), *ppl. a.* [f. PROGRAM, PROGRAMME *v.* + -ED¹.] **1.** Predetermined or controlled by a program (see *PROGRAM, PROGRAMME *v.* 4a, 5a, *sb.* 2g). Also *transf.*

1947 *New Republic* 23 June 13/3 The machine, having been properly briefed.., will perform its programmed task. 1953 *Proc. IRE* XLI. 1235/1 A significant new concept in non-numerical computation is the idea of a general-purpose programmed computer—a device capable of carrying out a long sequence of elementary orders analogous to those of a numerical computer. 1958 *Engineering* 28 Feb. 265/2 After taking off vertically..the vehicle veered to the south-east along its programmed trajectory. 1959 E. M. Grabbe et al. *Handbk. Automation, Computation & Control* II. iv. 13 Programmed checks..consist of the verification of the correctness of the system operation by means of special procedures introduced by the system user. 1964 M. McLuhan *Understanding Media* (1967) iii. 46 In the new electric Age of Information and programmed production. 1966 P. O'Donnell *Sabre-Tooth* xviii. 241 Willie nodded, taking the cue and moving into his role like a programmed robot. 1967 D. Wilson in Wills & Yearsley *Handbk. Managem. Technol.* 45 In addition to the validation of input data, internal programmed controls are usually included during computer processing. 1973 M. Crichton *Terminal Man* 106 The programmed machine could exceed the capabilities of the programmer. 1974 E. Ambler *Dr. Frigo* iii. 155, I am not referring to the immediate tactical success, but to..the programmed success of the future. 1978 G. A. Sheehan *Running & Being* ii. 27 It stripped off those layers of programmed activity and thinking.

2. *Psychol.* and *Educ.* Presented in the form of a teaching programme (*PROGRAM, PROGRAMME *sb.* 2h), or employing such a programme.

1958 B. F. Skinner in *Science* 24 Oct. 975/3 Immediate feedback encourages a more careful reading of programmed material. 1962 *Listener* 23 Aug. 273/2 Hundreds of thousands of Americans..are now being taught by machine, and by what are termed programmed texts. 1963 *Guardian* 15 Jan. 6/1 'Earth in Orbit'..is a geography book with a difference. It is the first programmed textbook of English origin. 1967 *Punch* 22 Mar. 407/2 The shades of Mr. Chips, let alone Dr. Arnold, have long ago fled before an influx of black-masked technicians panning stealthy TV cameras over programmed learning machines. 1967 Coulthard & Smith in Wills & Yearsley *Handbk. Managem. Technol.* 204 Programmed learning is already extensively used in this country in schools (teaching mathematics and the sciences), in the armed services, by the airlines, and in companies with extensive training programmes for salesmen, technicians, etc. 1968 *Brit. Universities Ann.* 91 Experiments in programmed teaching have been going on in several institutions notably at Bradford. 1970 W. C. Sahakian *Psychol. of Learning* x. 187 Programmed instruction is designed so that maximal reinforcement ensues by successfully controlling environment. 1974 *Nature* 22 Mar. 300/1 (Advt.), The reader will find that in his understanding of the concepts of the subject and his confidence in using them are both developed rapidly as he works through the programmed problems.

programmer (prōᵘ·græməɹ). Also (*rare*) programer. [f. PROGRAM, PROGRAMME *sb.* or *v.* + -ER¹.] **1.** One who programmes, in various senses, as: **a.** (In Dict. s.v. PROGRAMME *sb.*) **b.** One who devises a course of programmed instruction. **c.** One who plans or chooses programmes for broadcasting. **d.** One who arranges something according to a programme.

1890 [in Dict. s.v. PROGRAM, PROGRAMME *sb.*]. 1958 *Science* 24 Oct. 971/1 The machine itself..is a labor-saving device because it can bring one programmer into contact with an indefinite number of students. 1966 *Listener* 27 Jan. 147/1 It is one of the unwritten laws of programming that even if the same-only-different is to be served up it must be loudly proclaimed to be different. 1966 *Punch* 27 July 146/1 It was Baden Powell who revolutionized the process of growing up... Even his most grudging admirers must admit that he was an improvement on witch doctors and tribal elders as a programmer of boys' spare time. 1975 J. De Bres tr. *Mandel's Late Capitalism* vii. 236 The only means at the disposal of late capitalist economic programmers for the correction of actual development when they deviate from predictions, is State intervention in the economy. 1977 *Daily Tel.* 2 Dec. 16/6 Using a computer device..they can let the programmers know, while the show is in progress, that they are unhappy with what they are viewing as television entertainment. If they are in the majority, it will be taken off the air and replaced by something else. 1978 *Detroit Free Press* 2 Apr. 1C/1 One programer at ABC believes that the show will help ABC's image among those who for religious or other reasons, hold ABC and TV in general responsible for moral decay.

e. *spec.* One who writes computer programs.

1948 *Math. Tables & Other Aids to Computation* III. 45 Magnitudes of numbers at each step have to be studied by the programmer..to avoid exceeding capacity in other operations. 1951 *Electronic Engin.* XXIII. 140/1 Most machines print the results of computations at the will of the programmer by means of some electromechanical device. 1958 *Oxf. Mag.* 29 May 469/1 Numerical analysts and programmers are in great demand in the laboratories of government and industry. 1968 N. Chapin 360 *Programing in Assembly Lang.* vi. 139 A compiler program causes the computer to accept as input the source program the programer has written... The computer translates the source program..and produces the object program. 1974 *Maclean's Mag.* (Toronto) Jan. 44/3 As any programmer will tell you, a computer is merely as good as the data fed into it. 1980 R. McCrum *In Secret State* iii. 12 Quitman was not a trained programmer, but he had been given the standard course in basic computer access methods.

2. = *programme picture* s.v. *PROGRAM, PROGRAMME *sb.* 4.

1936 *Movie Mirror* Feb. 118/2 *Frisco Waterfront*.. A better-than-average programmer that brings Rod La Rocque back to American films. 1939 *Motion Picture Herald* 11 Nov. 56/1 Very ordinary programmer. It is so simply told as to fail to arouse much for or against criticism. 1974 *Radio Times* 14 Feb. 9/3 When..Montgomery directed this Guadalcanal actioner, everybody hoped it would certainly be a handsome cut above its other gun-toting rival *Guadalcanal Diary* made in 1943 without the benefit of real locations or scriptwriters. Instead the result was a routine action programmer with Cagney in rather restrained mood.

3. A device that automatically controls the operation of something in accordance with a prescribed programme; in quot. 1945 a part of an early electronic computer analogous to the control unit of later ones.

1945 J. P. Eckert et al. *Description of ENIAC* (PB 86242) (Moore School of Electr. Engin., Univ. of Pennsylvania) B-2 The use of the master programmer is being stressed since it is the mechanism in the ENIAC which enables one to link the simple sequences of instructions given the other units of the computer into a complete whole. 1962 J. Glenn in *Into Orbit* 143 During the powered phase of flight..an electronic programmer inside the Atlas would guide it along the prescribed path. 1968 *McGraw-Hill Yearbk. Sci. & Technol.* 59 Until the mid-1950s this automation was in the form of special machines with mechanical programmers (cams, levers, and stops, operating in conjunction with powered lead screws) or tracer mechanisms, which cause the cutting tool to follow a path described by simultaneously moving a 'feeler' over a model of the part to be made. 1968 *Which? Guide to Central Heating* 73/1 A programmer is basically a more versatile time switch, which gives you independent control over your room heating and water heating. 1973 *Daily Tel.* 4 Dec. 11/4 If you go out for a few hours turn down the room thermostat or adjust your boiler programmer.

programming (prōᵘ·græmiŋ), *vbl. sb.* Also (*rare*) programing. [f. PROGRAM, PROGRAMME *sb.* or *v.* + -ING¹.] †**1.** The writing of programme notes. *Obs. rare.*

1889 G. B. Shaw *London Music in 1888–89* (1937) 243 Sir George patronized everyone in his programming.

2. *Broadcasting.* The choice, arrangement, or broadcasting of radio or television programmes.

1940 R. Lohr *Television Broadcasting* p. ix, The first (part) deals with television in relation to the public, that is, the programming. 1955 *Broadcasting* 15 Oct. 84/2 ZIV is delighted that NBC has taken this step..because better programming is good for the entire industry. 1958 B. Ulanov *Hist. Jazz in Amer.* xviii. 226 WNEW, the New York radio station that has proved most adventurous in its programming of popular music. 1960 *News Chron.* 21 Sept. 6/5 Peak-time programming. 1964 M. McLuhan *Understanding Media* (1967) v. 65 TV caused drastic changes in radio programming. 1973 *Guardian* 16 Mar. 10/1 'Complementary programming', the principle on which BBC-2 now gives you a serious programme while BBC-1 has a light one. 1976 *Broadcast* 29 Mar. 16/2 The five station types are defined by size and programming capability.

3. Planning carried out for purposes of control, management, or administration, esp. in economics.

1943 *Sun* (Baltimore) 1 July 14/2 The President transferred from Mr. Jones' RFC to Mr. Wallace's BEW full control over the programming of imported strategic materials. 1959 *Listener* 21 May 884/2 The design of controls, the programming of production methods, and so forth. 1967 E. Duckworth in Wills & Yearsley *Handbk. Managem. Technol.* 117 Operational research in general has to do with the programming function in industry and all the methods described will be found to apply to the programming problems. 1975 J. De Bres tr. *Mandel's Late Capitalism* vii. 237 There is undoubtedly a certain reciprocal effect, of a both technical and economic character between planning of production and accumulation within individual companies and programming of the economy as a whole.

4. The operation of programming a computer; the writing or preparation of programs; *programming language* = *LANGUAGE *sb.* 1d.

1945 J. P. Eckert et al. *Description of ENIAC* (PB 86242) (Moore School of Electr. Engin., Univ. of Pennsylvania) B-1 An elementary programming procedure. 1947 *Electronic Engin.* XIX. 107/1 It remains to show how the automatic sequencing of operations—or programming—is achieved. 1949 *Math. Tables & Other Aids to Computation* III. 376 The most time-consuming factor occurring in the use of the ENIAC is the 'programming' (i.e. setting up the machine for a specific problem). 1954 *Sci. News* XXXIV. 60 The 'programming' of calculations to be done by computors is lengthy and tedious. 1959 *Computer Bull.* II. 81/1 This committee held three meetings starting on 24 January 1958 and discussed many technical details of programming language. 1962 *Lancet* 8 Dec. 1215/1 The final stage of programming must therefore consist in the translation of the flow diagram into actual coded orders which the machine can understand. 1967 *Economist* 8 Apr. 163/1 Programming schools are springing up everywhere. 1968 N. Chapin (*title*) 360 programming in assembly language. 1977 *Sci. Amer.* Sept. 236/3 Most of the programming languages in service today were developed as symbolic ways to deal with the hardware-level concepts of the 1950's.

5. *Psychol.* and *Educ.* The preparation and organization of the material necessary to a course of programmed instruction.

1954 B. F. Skinner in *Harvard Educ. Rev.* XXIV. 96 In addition to the advantages which can be gained from precise reinforcement and careful programming, the device will teach reading at the same time. 1969 [see *LINEAR *a.* 3c]. 1976 W. B. Kolesnik *Learning* vi. 124 The method of instruction most closely identified with the concept of learning-as-conditioning is commonly referred to as programming. *Ibid.* 125 While programming does sometimes involve the use of mechanical or electronic equipment, it need not do so.

programmist (prōᵘ·græmist). Also programist. [f. PROGRAM, PROGRAMME *sb.* + -IST.] One who writes programme notes.

Found only in the writings of G. B. Shaw.

1888 G. B. Shaw in *Star* 16 Nov. 2/3 The usual Dvorakian dressing of Bohemian rhythms and intervals which gives the musical programmist an opportunity of writing about 'national traits'. 1889 — *How to become Mus. Critic* (1960) 144 Mr Joseph Bennett, the programist, says of the symphony—'It has been described by the composer as a "little" work.' 1896 — *Our Theatres in Nineties* (1932) II. 94 The average programmist would unblushingly write, 'Here the composer..has abruptly introduced the dominant seventh of the key of C major into the key of A flat.'

progress, *sb.* Add: **7.** *progress clerk, committee, department, man, manager, payment*; **progress chaser**, an employee responsible for ensuring that work is done efficiently and to schedule (cf. *CHASER¹ 7); hence **progress-chasing**; **progress report**, an interim report on progress made to date.

1930 *Daily Tel.* 18 Dec. 12/6 (Advt.), Progress chasers wanted for aircraft work. 1943 J. B. Priestley *Daylight on Saturday* vii. 64 Even my friend Moss takes to it (sc. factory work) better than I do. You know she's a progress chaser now, don't you? 1977 *Lancashire Life* Mar. 46/1 The imminent threat of war found Arthur Lowe, progress chaser with Fairey Aviation, joining the Duke of Lancaster's Own Yeomanry in Manchester. 1943 J. B. Priestley *Daylight on Saturday* viii. 52 Doesn't matter what we do here—progress chasing and all the rest of it. 1971 R. Lewis *Fenokee Project* i. 10 Pete dealt with the progress chasing on outstanding contracts. 1921 *Dict. Occup. Terms* (1927) § 939 *Progress clerk*..traces and pushes forward work in its various stages from operation to operation until it is ready for delivery... Keeps a record of output, etc. in works. 1942 R. P. Jeffcott *Girls growing Up* iv. 77 A progress clerk..is conscious of the superiority of her life over that of a girl who does a petty routine job all day. 1974 *Evening News* (Edinburgh) 8 Oct. 11/4 (Advt.), Progress clerk, £1600 plus, have you previous experience in electronics? 1914 E. T. Elbourne *Factory Admin. & Accounts* iii. 39 (*heading*) Progress Committee—under Works Manager. 1925 C. L. Bolling *Commerc. Managem.* x. 148 The managerial staff or 'progress department' should prepare diagram of the plans of the advisory staff. 1932 S. E. Thomas *Commerce* xvi. 248 It is the duty of the Progress Department to follow up the various processes, to prepare job cards, material charts, etc. 1922 Progress man [see *CHASER¹ 7]. 1957 C. Smith *Case of Torches* ii. 26 The small offices..where the engineers and draughtsmen and progress men spent their time. 1925 C. L. Bolling *Commerc. Managem.* x. 150 Each job undertaken is notified to the progress manager. 1953 S. E. Thomas *Commerce* xxvii. 388 The executives under the works manager may include..a Progress Manager, who sees that the operations..take place in the order required. 1959 *Wall St. Jrnl.* 12 June 5/1 The Army..doesn't require 'progress payments'—funds advanced by the Government, usually monthly, to help offset a company's costs as it carries out major contracts. 1977 *Herald* (Melbourne) 17 Jan. 2/6 The company had obtained nearly $120,000 in progress payments but had completed only one contract in that period it was alleged. 1929 Saunders & Anderson *Business Reports* iv. 43 Progress report is another report which may be limited to time, place, and handling of data similar to the informational report. 1943 J. B. Priestley *Daylight on Saturday* xxv. 189 Cheviot was now frowning hard at the progress reports on his desk. 1972 'E. Ferrars' *Breath of Suspicion* iii. 38 So, if I see her again, you want a sort of progress report.

progress, *v.* Add: **6.** (Further examples.) Also, *spec.* to cause (work, etc.) to make regular progress towards completion.

1954 'N. Shute' *Slide Rule* 184, I was chiefly occupied..in progressing the design and construction of the factory at Portsmouth. 1965 E. Gowers *Fowler's Mod. Eng. Usage* (ed. 2) 479/2 *Progress*... *Prŏ'grĕss* is usual for the transitive verb, now much used in the manufacturing and building industries in the sense of pushing a job forward by regular stages. 1976 *Southern Even. Echo* (Southampton) 16 Nov. 13/2 (Advt.), Purchasing Administrator to work in busy spares section. Placing and progressing orders. 1978 *Observer* 12 Feb. 12/3 Welders to be trained to make more tack items to allow them to progress their own work to completion.

progression. Add: **10.** *Spectroscopy.* A series of regularly spaced lines or bands in a spectrum which arise from transitions to or from a series of energy levels having consecutive quantum numbers.

1926 *Physical Rev.* XXVIII. 638 In other words we suppose that the absorption bands whose stimulation is associated with these series all belong to a single n″ progression. 1940 P. Pringsheim *Fluorescence & Phosphorescence* ii. 156 If excited molecules of a vapor are in one definite vibrational level v′ of an electronic state T′, they can return from there to all existing levels v″ of the ground state and thus produce an emission spectrum in which the lines corresponding to v″ = 0, 1, 2.. form a regular 'progression'. 1965 R. N. Dixon *Spectroscopy & Struct.* vi. 129 When an electronic spectrum is studied in absorption at moderate temperatures most of the molecules in the lower state will be in the vibrational level with v″ = 0, and the vibrational structure will consist of one series of bands, corresponding to consecutive values of v′. Such a series is called a progression. The bands are labelled v′–v″. Thus the absorption series 0–0, 1–0, 2–0, 3–0.. is a progression in the upper state vibration frequency. 1976 *Chem. Physics Lett.* XLI. 189/2 The Raman spectrum is completely dominated by an intense band at 316 cm⁻¹ and its associated overtone progression.

progressism (progre·siz'm). [f. PROGRESS *sb.* + -ISM.] = PROGRESSIONISM.

1892 B. Bosanquet *Meeting of Extremes in Contemp. Phil.* 206 Men do not, under the influence of such progressism, abstain from one or more climaxes of the finite may have been attained in the past. 1923 W. R. Inge *Outspoken Ess.* 2nd Ser. 26 The 'Progressism' of much modern thought is a poor substitute for this belief in the substantial reality of the eternal values.

progressive, *a.* (*sb.*) Add: **A.** *adj.* **2. b.** (Earlier and later examples.) Also of games other than cards.

1870 W. B. Dick *Amer. Hoyle* (ed. 10) 56 There is another variety to be met with occasionally, which may be styled 'Progressive Jack-Pots'. 1885 C. M. Seaver (*title*) Standard guide to progressive euchre. 1886 I. M. Rittenhouse *Maud* (1939) 366 So we had a jolly little time, playing progressive euchre, and indulging in some music. 1888 A. Randall-Diehl *Two Thousand Words* 169 *Progressive euchre*, a game of cards in which a player starts at one table, among the several scattered about the room. If he wins at the first table, he passes on to the next, and is credited with one game. If he wins again, he moves forward one table. 1903 R. Brooke *Let.* 27 Dec. (1968) 5, I am going to have quite a gay time this week—progressive whist, fancy dress-balls etc. 1904 *Bridge & Progressive Bridge* 19 Progressive Bridge.. The Rules of Bridge apply except that no account is taken of games or rubbers. 1906 *Daily Colonist* (Victoria, B.C.) 1 Jan. 15/1 Mrs. Piggott entertained a number of her lady friends..progressive five hundred being the amusement. 1907 *Yesterday's Shopping* (1969) 361/1 At Home. Progressive Hearts—o'clock. R.S.V.P. Progressive Bridge..Progressive Whist. 1926 R. Macaulay *Crewe Train* II. ii. 118 Do you enjoy whist?.. We have very nice progressive drives on Wednesday nights. 1963 M. Kendon *Ladies' College Goudhurst* 20 In 1905 the girls were allowed 'a fun evening of Progressive games.

c. *progressive assimilation*, in Philol. the process whereby a sound is modified by or harmonized with one closely preceding it.

1915 G. Noël-Armfield *Gen. Phonetics* ix. 91 If the first sound carries its influence forward the assimilation is said to be progressive. 1934 M. K. Pope *From Latin to Mod. French* ii. i. 64 Assimilations and dissimilations may be either regressive or progressive, i.e. a sound may be modified in anticipation of a sound following, or the articulatory position of one sound may modify the pronunciation of a later one. 1939 L. H. Gray *Foundations of Lang.* iii. 66 Assimilation may..be either progressive, when the first phoneme modifies the second,..or regressive, when the second modifies the first. 1964 C. Barber *Ling. Change Present-Day Eng.* iii. 63 A historical example of progressive assimilation is seen in words like *watch*..where the rounded vowel is seen to result from the influence of the preceding w. 1977 *Canad. Jrnl. Linguistics* XXII. 114 Chapter 5 shows..the action of progressive and regressive assimilation.

d. *progressive proofs* (see quot. 1960).

1931 Plack & Clunes *Letterpress Printing* xiv. 247 *Block-Maker's Progressive Proofs.* The colour sheets of 'progressives' provided by the process block-maker must be followed with absolute exactitude. 1948 *Graphic Arts Procedures* vii. 221 The two preceding pages show 'progressive proofs' and the final result of four-color process printing. 1960 G. A. Glaister *Gloss. of Bk.* 331/1 *Progressive proofs*, the proofs made in colour-printing as a guide to shade and registration. Each colour is shown separately and imposed on the preceding ones.

e. Psychol. *Progressive Matrices* (see *RAVEN).

1939 *Brit. Jrnl. Med. Psychol.* XVIII. 16 Progressive Matrices (Sets A, B, C, D and E 1938) obtainable from Messrs H. K. Lewis & Co. 1948 *Psychometrika* XIII. 36 In Sets E and D of the Progressive Matrices, the speed with which certain details are perceived seems to be important for the solution. *Ibid.* 41 The Progressive Matrices are loaded in several factors, plus a factor common to all of them. 1954 A. Anastasi *Psychol. Testing* x. 261 The Progressive Matrices Test, developed in England by Raven..should also be included... This test, designed as a measure of Spearman's *g* factor, requires primarily the eduction of relationships within abstract material.

3. e. *Educ.* Of teaching methods, types of schools, etc.: aiming to develop individual capability and character in children rather than to achieve standardized results.

1839 tr. A. Necker de Saussure (*title*) Progressive education. 1910 Chesterton *G. B. Shaw* 185 Shaw has always made this one immense mistake (arising out of that bad progressive education of his), the mistake of treating convention as a dead thing. 1924 *Progressive Education* Apr. 3 The Progressive Schools are increasing rapidly. 1943 [see *ESSENTIALIST *adj.*]. 1946 E. Hodgins *Mr. Blandings builds his Dream House* (1947) 22 The creative, anarchistic, and sexual freedoms of a progressive school. 1959 *Listener* 5 Feb. 244/1 A man who tries to exercise authority in the manner of a sergeant-major will get short shrift in a progressive school. 1967 *Guardian* 14 Oct. 8/4 The Progressive Education Association founded in 1919, had done its work so well that it was dissolved in 1955. 1976 *Listener* 29 Apr. 536/1 Anxious children did particularly poorly in a progressive classroom.

f. Of taxation: (see quot. 1902).

1889 G. B. Shaw *Fabian Ess. Socialism* 193 The Radical progressive income taxes singing together, and the ratepaying tenants shouting for joy. 1902 *Encycl. Brit.* XXXIII. 127/2 The question whether the burden of taxation should not be *progressive*—the proportion of the sum taken by the State from the tax-payers increasing with the wealth of the individual. *Ibid.* 199/1 A general system of progressive taxation. 1976 *National Observer* (U.S.) 21 Aug. 3/2 By using the progressive income-tax structure, state and local governments are placing the burden back on those hardest hit by raising regressive local taxes. 1978 *Times* 18 July 15/5 Progressive taxation is imposed in this country in order to moderate the unequal distribution of income.

g. *progressive kiln*, a long kiln through which timber to be dried is slowly passed on trucks.

1920 A. L. Howard *Manual of Timbers of World* 397 In all progressive kilns the timber is piled on trucks, and moved at regular intervals through zones of varying temperature and humidity. 1971 *Timber Trades Jrnl.* 14 Aug. 26/1 Small mills naturally use compartment kilns for export drying, but a progressive kiln will now pay off at 60m³ per day and offers lower consumption of heating fuel and electricity.

h. *Gram.* = *EXPANDED *ppl. a.* 2b.

1924 H. E. Palmer *Gram. Spoken Eng.* ii. 149 Tenses composed of the verb (be) and the -ing-form are called *Progressive Tenses*. 1932 *Jrnl. Eng. & Gmc. Philol.* XXXI. 253 The creation of the progressive form resulted from the desire to express the idea of progressive action, action going on: 'He is writing a letter to his mother.' 1945 *Trans. Philol. Soc.* 130 The distinction between the 'terminate' and the 'progressive' aspects of the verb, e.g. *I go*/*am going*. 1957 R. W. Zandvoort *Handbk. Eng. Gram.* i. ii. 36 What was said..of the imperfective or durative aspect of the present participle also applies to its predicative use, as in *The ships were sailing out of the harbour*. This construction is known as the *progressive*.

i. Of music: modern, experimental, innovatory, avant-garde; used with reference to several distinct musical developments, as *progressive jazz, pop, rock*, etc.

1947 *Down Beat* 23 Aug. 1/4 Stan Kenton next month returns to the band business... First recordings will be for an album to be titled Concert in Progressive Jazz. *Ibid.* 19 Nov. 1/3 Stan..abhorred references to his music as 'jazz', himself using the descriptive phrase 'progressive jazz'. 1959 F. Newton *Jazz Scene* vi. 117 Leading 'progressive' players like the pianist Lennie Tristano. 1963 R. I. McDavid *Mencken's Amer. Lang.* 744 It [sc. funk] also designates progressive bop containing a strong blues element which marks its Negro origin. 1970 E. Lee *Music of People* vii. 147 A new style has arisen, usually called 'progressive pop', which is of such musicality that it has been heralded by some critics as a new art form of music. 1975 *New Yorker* 21 Apr. 71/1 (Advt.), Headhunters, a progressive disco-jazz quintet created in Herbie Hancock's image. 1977 *It* May 26/3 Patti wrote this heavy condemnation of 'progressive' rock radio as we hear it now.

4. a. (Further examples.)

More recently in South Africa, designating several political parties committed to a policy of multi-racialism; also, freq. a name or term adopted by radical, left-wing, or communist parties.

1930 W. K. Hancock *Australia* viii. 163 The area to be developed was also within constituencies held by the Country Progressive party, on whose support the Victorian Government was dependent. 1954 *Manch. Guardian Weekly* 16 Sept. 3/3 The ludicrous and largely Communist-dominated 'Progressive Party' campaign of 1946. 1955 *Treatm. Brit. P.O.W.'s in Korea* (H.M.S.O.) 4 The 'progressive' view—the Communist view—was the only one allowed. 1969 A. G. Frank *Latin Amer.* xxii. 269 Concerned and progressive people everywhere scrutinize these..laws, and often criticize them. 1971 *Progress* (Cape Town) May 5/4 There had been talk of a split for a long time; the Press had even coined the term the 'Progressive Group' of the United Party. 1976 R. Williams *Keywords* 207 Nearly all political tendencies now wish to be described as progressive, but..it is more frequently now a persuasive than a descriptive term.

b. Characterized by (the desire to promote) change, innovation, or experiment; avant-garde, advanced, 'liberal'.

1908 H. G. Wells *War in Air* ii. 35 It was always a very rhetorical and often trying affair, but in those progressive times you have to make a noise to get a living. 1949 'J. Tey' *Brat Farrar* i. 11 The great house in the park was a boarding-school for the unmanageable children of parents with progressive ideas and large bank accounts. 1953 M. McCarthy *Groves of Academe* i. 4 In a progressive community where the casserole and the cocktail and the disposable diaper reigned. 1956 [see *LABOUR *sb.* 6c]. 1974 *Howard Jrnl.* XIV. 99 The Rev. W. D. Morrison, whose outspoken views in the 1890s led to the most progressive document on prison reform since the writing of John Howard himself. 1976 *Church Times* 27 Aug. 6/5 There was something either in those particular temperaments or in the 'progressive' ethos, that militated against contentment.

5. *Comb.*, as *progressive-minded* adj.

1955 Koestler *Trail of Dinosaur* 206 The worshippers of tyranny and terror usurp the rightful place of a truly progressive party and cunningly direct the energies of the progressive-minded into 'anti-Fascist' crusades. 1975 A. Bergman *Hollywood & Le Vine* v. 60 A bad time for progressive-minded people.

B. *sb.* **1.** (Further examples.)

1921 J. R. Hornady *Bk. of Birmingham* ii. 27 In the North we divide; there you will find Republicans, Democrats, Progressives, Independents, and so forth. 1954 *Britannica Bk. of Year* 658/1 Awareness of Communist thought produced the word..Progressive, a Communist sympathizer. 1955 *Treatm. Brit. P.O.W.'s in Korea* (H.M.S.O.) 35 These 'Progressives' were given a major role to play as mouthpieces of Communist propaganda. 1958 *Listener* 4 Dec. 941/1 The true Progressive was essentially urban and middle class. 1970 *Cape Times* 28 Oct. 1/1 The United Party has been a bit bitter about the Progressives, believing that they should not exist. 1971 *Rand Daily Mail* 4 Dec. 1/1 Mr. Bill Carr's decision to seek election to the Johannesburg City Council as a Progressive is more than just a coup for that particular party. 1976 *Times* 7 Aug. 13/1 The progressives seem to behave like theological scrap merchants, swopping bits and pieces from the ecumenical marketplace according to their fancy, clucking noises of disapproval are heard from the Vatican.

2. *pl.* Shortened from *progressive proofs* (sense *A 2d).

1923 H. A. Maddox *Printing* x. 125 A final set of colour proofs (progressives) is prepared for the guidance of the printer. 1932 [see sense *A 2d].

3. One who favours, advocates, or practises progressive (sense *A. 3e) educational methods.

1908 H. G. Wells *Anal. Frustration* vii. 73 The 'natural virtue' schools of such educational 'progressives' as Neill and his associates. 1941 H. Croome *You've gone Astray* xvi. 168 You may know the true progressive..by the fact that he calls children not boys and girls, but 'kids'. 1961 Curtis & Boultwood *Short Hist. Educ. Ideas* (ed. 3) xx. 578 Pleas by the Progressives for activity methods, informal learning, and the encouragement of self-discipline and initiative. 1969 M. Ash *Who are the Progressives Now?* 2, 23 Of still greater gall to the old-style progressive at the Dartington Colloquy..was the pervasive belief of some others present, that by a well-meant concession to the independent progressive schools a niche could be found for them in the State system as recipients of..difficult children. 1976 *Times* 10 Apr. 13/3 In the past the debate between traditionalists and progressives in education has often taken place on the abstract and abstract plane of educational philosophy.

4. *Gram.* Shortened from *progressive form, tense*, etc. (sense *A. 3h).

1961 R. B. Long *Sentence & its Parts* v. 127 The point of view is internal, as with all progressives. 1965 N. Chomsky *Aspects of Theory of Syntax* 216 Such Verbs as *own*..occur freely with or without Progressive. 1978 *Language* LIV. 418 Half of its discussion of imperfectivity (30–40) is devoted largely to an examination of the English progressive in which it is clear that he is arguing from form to meaning, rather than the reverse.

progressivism. (Further examples.)

1961 Curtis & Boultwood *Short Hist. Educ. Ideas* (ed. 3) xx. 581, 1958..marked the final withdrawal of the pragmatist group from identification with progressivism. 1978 *Church Times* 17 Mar. 9/1 One might cite William Golding as representative of those intellectuals who were brought up to Wellsian rationalism and progressivism..and then rejected it. 1979 *Daily Tel.* 12 Oct. 5/1 Mr Amin [President and Prime Minister of Afghanistan] did not refer to Communism or Socialism in his speech but stressed that the constitution would be based on 'progressivism', the word commonly used by Communists in parts of Asia where Communism is still frowned upon.

progressivist. Add: **a.** (Further examples.)

1945 K. R. Popper *Open Society* I. iv. 39 Had he [sc. Plato] been a progressivist, he might have hit at the idea of a classless, equalitarian society. 1959 *Times* 23 Sept. 13/6 The younger composers..are indistinguishable in outlook from the vanguard of our English progressivists.

b. *attrib.* or as *adj.*

1969 *Daily Tel.* 13 Nov. 14/3 A favourite progressivist argument..is that people who are most vehemently against something are often the very same people who are unconsciously most inclined to it themselves. 1977 D. Watkin *Morality & Archit.* III. ii. 94 According to the progressivist view, anything which reminds one 'of the past' is regarded as a vice.

proguanil (progwā·nil). *Pharm.* [f. PRO(PYL + *BI)GUAN(IDE + -*il*.] A bitter-tasting synthetic compound, 1-*p*-chlorophenyl-5-isopropylbiguanide, $C_{11}H_{16}ClN_5$, which is used, in the form of its white crystalline hydrochloride, in the prevention and treatment of malaria. Cf. *PALUDRINE.

1949 *Brit. Med. Jrnl.* 15 Jan. 88/2 (*heading*) 'Paludrine' (proguanil) in prophylaxis and treatment of malarial infections caused by a West African strain of *P. falciparum*. 1956 *Nature* 25 Feb. 368/1 Further mosquito infectivity experiments have confirmed that, in patients carrying gametocytes of a proguanil-resistant strain of *P. falciparum*, anopheline mosquitoes can be infected from a person taking 0·4 gm. proguanil daily. 1961 *New Scientist* 20 Apr. 119/1 It was found that if a strain of *P. gallinaceum* was transmitted serially from chick to chick, and the birds were treated with gradually increasing doses of proguanil, the organisms became highly resistant to its action. 1970 Passmore & Robson *Compan. Med. Stud.* II. xx. 20/1 Proguanil (paludrine) acts on the malaria parasite both in the blood and in the liver, probably by interfering with the reduction of folic acid. 1977 *Times* 5 July 7/6 Protection is easy: one of the antimalarial drugs such as proguanil should be taken.

proheterocyst: see *PRO-² I.

prohibited, *ppl. a.* Add: *prohibited area*, a region which only authorized persons may enter.

1940 *Hansard Commons* 5 Mar. 195 *Mr. Woodburn* asked the Secretary of State for War whether, in his designation of a part of Scotland as a prohibited area, he has considered the repercussions of this decision on the Scottish tourist industry. 1948 E. Pound *Pisan Cantos* (1949) lxxxiii. 126 Bein' aliens in prohibited area. 1961 F. H. Burgess *Dict. Sailing* 164 *Prohibited area*, an area marked on a chart, where anchoring, trawling, or fishing etc., may be forbidden by authority. 1964 G. Lyall *Most Dangerous Game* vi. 48 The convictions for flying in prohibited areas.

prohibition. Add: **4.** Now usu. with reference to the restrictions on the manufacture and sale of intoxicating drinks in the United States (1920–33) under the Volstead Act.

1925 W. J. Bryan *Mem.* 186 His views on the initiative and referendum and prohibition had not altered. 1927 *New Republic* 21 Sept. 109/1 The Republicans have been laboring night and day to keep prohibition from becoming an issue in the campaign. 1931 M. F. Furness *Mem. Sixty Years* xix. 246 We of course heard much talk of Prohibition, and also saw a good deal of its absence. 1966 C. M. Bowra *Mem. 1898–1939* viii. 212 Maurice had been in the United States during the fantastic boom of the twenties. This was also the period of prohibition, which made it a matter of prestige and honour to drink too much, especially hard liquor. 1970 *Nature* 19 Sept. 1186/2 For that matter the laws [in the U.S.A.] against marihuana have become so widely abused as to appear to many as a new Prohibition.

6. *prohibition amendment, candidate, era, law, state, system*; **prohibition party** (examples).

1883 *Harper's Mag.* Dec. 162/1 The prohibition amendment was defeated [in Ohio]. 1877 *Ibid.* Dec. 146/1 R. Pitman..Prohibition candidate for Governor. 1909 G. F. Parker *Recoll. Grover Cleveland* (1911) 71 In addition, something over 300,000 votes had been cast for the Prohibition and the Greenback candidates. 1922 S. Lewis *Babbitt* xiii. 174 He was blind and deaf from prohibition-era alcohol. 1968 *N.Y. City* (Michelin Tire Corp.) 32 The area later became famous as the 'speakeasy belt' during the Prohibition era. 1884 *N.Y. Weekly Tribune* 20 Aug. 7/1 The platform..means specifically that the prohibition law shall be enforced. 1949 *Daily Oklahoman* 13 Feb. D.2/2 A petition will be circulated calling for repeal of the state's liquor prohibition law. 1869 in D. L. Colvin *Prohibition in U.S.* (1926) iv. 73 We adopt the name of the National Prohibition Party, as expressive of our primary object. 1948 *Time* 12 Jan. 13/1 The Prohibition Party has nominated a candidate for President ever since its formation in 1869. 1892 *Outing* Dec. 209/1 They have no beer here: North and South Dakota are prohibition states. 1949 *Time* 10 Oct. 27/3 The state's church-going United Drys..were fiercely proud of living in a prohibition state. 1877 *Harper's Mag.* Dec. 146/2 He argues..the prohibition system a success.

prohibitionism. (Further example.)

1915 *N. Amer. Rev.* Dec. 948 All the speakers agreed that the bad saloon did more harm to the liquor trade than prohibitionism.

prohibitiveness. (Earlier example.)

1891 T. Hardy *Let.* 15 July (1978) I. 240 The late Lord Lytton..abolished prohibitiveness as between himself & the mass of the thinking public..by consenting to paper covers at 25/-.

prohibitory, *a.* Add: **3.** *Gram.* = PROHIBITIVE *a.* 3.

1925 G. R. Driver in A. S. Peake *People & Bk.* 97 Since *lā* and *l̄ā* implied a precative and a prohibitory sense respectively, there was no risk of confusion between these forms.

prohormone: see *PRO-² I. **pro-infinitive, -infinitival:** *PRO-¹ 4b, c. **proinsulin:** *PRO-² I. **pro-Israeli:** *PRO-¹ 5a.

project, *sb.* Add: **5. b.** *Educ.* An exercise in which pupils are set to study a topic, either independently or in co-operation, from observation and experiment as well as from books, over a period of time.

1916 D. Snedden in *School & Society* 16 Sept. 420/2 Some of us began using the word 'project' to describe a unit of educative work in which the most prominent feature was some form of positive and concrete achievement. 1919 J. A. Stevenson in *School Sci. & Math.* Jan. 57 A project is a problematic act carried to completion in its natural setting. 1924 *Progressive Education* I. 72 A distinguishing earmark of a project, then, is the whole child responding to a situation; it is child activity. 1938 *New Statesman* 8 Jan. 46/2 *New Schools for Old* shows us the changes now being introduced into American public school methods of education. Children are encouraged to cope with the practical problems of life, and emphasis laid upon the 'project' or collective enterprise. 1942 B. Clements et al. *Projects for Junior School: Teachers' Bk.* i. 5 When working out a project the teacher gives help only when and where necessary, since the basic principle of modern teaching is child activity and teacher guidance. 1959 *Housewife* June 16 Cristy, who in one crowded summer, enjoys a library reading project, a visit to a Kansas farm and a course in baby care. 1961 Curtis & Boultwood *Short Hist. Educ. Ideas* (ed. 3) xx. 580 Not only were large-scale projects on such topics as 'Conservation' and 'Pan-Americanism' undertaken by many schools—often all the schools of an area—as part of the curriculum, but, in addition, community service by school children became common. 1965 *Nursing Times* 5 Feb. 191/1 By etymological definition, a project is a plan, scheme or design. Educationally, it refers to any exercise in which students gather their own information on a subject, arrange it and present it in an interesting form. 1976 P. Dickinson *King & Joker* i. 19 Can I do next term's project on it [sc. a toad]?

c. *N. Amer.* A government-subsidized block of houses or apartments available at low rents. *housing project*: see *HOUSING *sb.*¹ 7.

1932 *Amer. City* Aug. 82/1 (*heading*) Federal Aid Now Offered for Low-Cost Housing and Slum-Clearance Projects. *Ibid.* 82/2 All housing projects should be large-scale developments. 1958 *Hearings Housing Act 1958* (U.S. Congress, Comm. Banking & Currency) 7 July 78 There is new thinking now to break up the projects and even though they are slum sites or urban redevelopment sites, to build scattered on the site. 1966 *Listener* 29 Sept. 454/1 Jim lives in one of a group of fifteen-storey-high buildings that make up a project, a city-owned housing estate. A family can rent an apartment cheaply in a project if it has a low income. Officially intended to replace the slum neighbourhoods of ten years ago, the projects are stark, anonymous, all-brick slums now. 1968 *Globe & Mail* (Toronto) 3 Feb. 3/4 A 2,000-unit high-rise and low-rise project. 1975 *New Yorker* 29 Sept. 43/3, I lived in a project then. The floors were so new they didn't have to be covered with anything.

d. A co-operative enterprise, often with a social or scientific purpose, but also in industry, etc.

1952 D. Riesman in *Antioch Rev.* Dec. 426 Professor John R. Seeley is now engaged in directing a large research project. 1952 Auden *Nones* 61 Thou shalt not worship projects nor Shalt thou or thine bow down before Administration. 1965 H. I. Ansoff *Corporate Strategy* ii. 17 CIT uses long-term profitability over the lifetime of the project as the yardstick for evaluation. 1966 T. Pynchon *Crying of Lot 49* iv. 88 When they grew up they..got stuck on some 'project' or 'task force' or 'team' and started being ground into anonymity. 1969 J. Argenti *Managem. Techniques* v. 26 Major projects call for the co-ordination of large numbers of machines, sub-contractors, finance, local authorities, design staff and so on. 1974 *Howard Jrnl.* XIV. 37 One aspect of a larger collective project being undertaken by members of the Centre for Contemporary Cultural Studies on 'mugging'.

8. *attrib.* and *Comb.* (sense 5), as *project approach, area, book, engineer, house, housing, manager, method, officer, work.*

1973 *N.Y. Law Jrnl.* 4 Sept. 5/6 We are endeavoring to try appropriation cases on a project approach. 1961 *Economist* 2 Dec. 935/1 It means accepting—as in India—the idea of a 'project area' in which all the means of modernisation will be made to converge. 1976 *Billings* (Montana) *Gaz.* 20 June 3-B/1 We joined Montana Federation of Women's Clubs because it offers prizes in all sorts of project areas. 1947 A. Einstein *Mus. Romantic Era* xiii. 181 The scheme for the composition of such a work is found in Schumann's 'project-book'. 1976 *Columbus* (Montana) *News* (Fair Bk. Suppl.) 10 June 18/1 Those exhibiting must have the project books up to date and be enrolled in the project and unit in which they are exhibiting. 1931 F. L. Eidmann *Econ. Control Engin. & Manuf.* iv. 45 In plants where the engineering projects are large,..it has been found that a very good way of handling the work is to assign one engineer to the task of 'living' with the job from start to finish... This engineer is known as a 'project engineer' in some organizations. 1973 *Times* 12 Nov. 28/8 Let us see the professionals all take a greater share of responsibility if a project is to be a success instead of leaving it to the 'jack of all trades'—the project engineer. 1967 G. Jackson *Let.* Nov. in *Soledad Brother* (1971) 139, I thought most blacks..understood..that these places were built with us in mind, just as were the project houses, unemployment offices, and bible schools. 1970 D. Goldrich et al. in I. L. Horowitz *Masses in Lat. Amer.* v. 182 Those invaders who could qualify by 'normal' criteria for project housing would receive it. 1973 *Black Panther* 14 Apr. 7/3 Her application for project housing was refused by the Chattanooga Housing Authority. 1965 *Guardian* 31 Mar. 11/3 The truth lies..in the person of what is usually known as the project manager. 1976 *Southern Even. Echo* (Southampton) 10 Nov. 19/6 Conder, based at Winchester, are also acting as the project managers. 1916 J. C. Moore in *School Sci. & Math.* Nov. 688 The project method in science is nothing new, though the name often calls forth an attack... The story of every great invention is the story of a project. 1925 W. H. Kilpatrick *Foundations of Method* xxi. 346 'You defend then the term 'project method'?'.. 'If it be thought of as a purposeful way of treating children in order to stir the best in them and then to trust them to themselves as much as possible, yes, I approve it.' 1943 H. Read *Educ. through Art* vii. 233, I am aware that serious criticisms have been made of the project method of teaching, but they seem to be based on a formless type of project. 1953 Curtis & Boultwood *Short Hist. Educ. Ideas* xvii. 476 After the Great War of 1914–18 Dewey's problem method was reinterpreted by W. H. Kilpatrick as the project method. 1968 H. I. Ansoff *Corporate Strategy* 1 Dr Ansoff worked for the RAND Corporation as a project officer. 1973 R. Hayes *Hungarian Game* xxxi. 183 Generally speaking, sir, agents administer *things*, while Project Officers administer *people*. 1941 *Manch. Guardian Weekly* 14 Mar. 214/4, I..discover 'underlying object of fostering project work on citizenship'. 1958 *Sunday Times* 15 June 4/8, I want more, not less, practical mathematics in junior schools..and suitably graded mathematical project work in secondary schools.

project, *v.* Add: Also *South. U.S. dial.* projeck, projick. **II. 6. d.** *South. U.S. dial.* To wander, saunter, stroll (*around*); to trifle, mess, play *with*.

1828 J. Hall *Lett. from West* 290 A man who goes into the woods..has a..great deal of *projecking* to do, as well as hard work. 1845 W. T. Thompson *Chron. Pineville* 107 You see what comes of your projectin' about town, when you ought to be gwine home. 1848 — *Major Jones's Sk. Trav.* 62, I didn't know whether he was projeckin' with me or not. 1891 'O. Thanet' *Otto the Knight* 66 Quality liked projeckin' roun' de kitchin. 1893 H. A. Shands *Some Peculiarities of Speech in Mississippi* 51 *Projicking*, a word used by negroes and illiterate whites to mean *fooling, trifling*; as, 'If you don't stop your projickin' with me, I'll lick you.' 1906 F. Lynde *Quickening* 135 Don't you know you oughtn't to go projecting around in the woods all alone? 1929 W. Faulkner *Sound & Fury* 10 Don't you start no projecking with Queenie. *Ibid.* 67 Is you been projecking with his graveyard? 1957 *Daily Progress* (Charlottesville, Va.) 5 Feb. 8/1 It beats all get-out how some people are always tinkering and 'projecking' on how to do things different from the way most people do them.

9. c. To cause (an image) to be visible on a screen situated at a distance. Also *absol.*

1865 *Rep. Brit. Assoc. Adv. Sci. 1864* II. 98 The impressive character of the image projected [by a magic lantern], being often stereoscopic in aspect. 1902 *Encycl. Brit.* XXVII. 95/2 In the magic lantern an electric lamp or limelight..projects, through an objective lens, the successive images of the film upon a distant screen. 1935 *Television Today* II. 599/2 (*heading*) Projecting the cathode-ray oscillograph picture. 1946 R. Lehmann *Gipsy's Baby* 57 Leisure employs me..as a kind of screen upon which are projected the images of persons. 1964 *Photogr. Jrnl.* CIV. 152/1 Microfilm images..can be projected directly on to printing plates. 1969 *Focal Encycl. Film & Television Techniques* 378/1 The projection cathode ray tube..produces a very bright picture which can be projected with a suitable optical system. 1979 D. Meiring *Foreign Body* ii. 30 Now it's film time... We'll project on the wall, just to the left of the bar.

d. To cause the image(s) on (a photograph, film, or slide) to be visible on a screen.

1896 R. W. Paul *Brit. Pat. 4686*, My invention relates to an improved apparatus for producing representations of moving scenes, figures or objects by projecting onto a screen..by means of..suitable projecting apparatus, a series of photographic pictures of such scenes... In order to give a definite position to the picture on the film which is to be projected I prefer to employ the following mechanism. 1912 F. A. Talbot *Moving Pictures* ix. 91 The film to be projected is carried upon a spool mounted on an arm or bracket above the mechanism. 1949 Kidd & Long *Filmstrip & Slide Projection* 8 Often miniature slides can be projected in the standard projector (some of which project both miniature slides and filmstrips). 1962 L. Deighton *Ipcress File* xv. 87 If they had to have a major here to project the film it might just be worth watching. 1971 L. B. Happé *Basic Motion Picture Technol.* i. 37 Corresponding prints could..be projected by similar anamorphic lenses to show a picture filling a very wide screen. *Ibid.* 39 The proportions of the new Cinemascope format when projected were 8 units wide by 3 high.

10. b. To attribute (an emotion, state of mind, etc.) to an external object or person, esp. unconsciously. Also *absol.*

1923 J. S. Huxley *Ess. Biologist* iv. 167 Certain neurotic types project their depression so as to colour everything that comes into their cognizance a gloomy black. 1925 A. & J. Strachey tr. *Freud's Coll. Papers* III. 458 It was incorrect of us to say that the perception which was suppressed internally was projected outwards; the truth is rather..that what was abolished internally returns from without. 1939 *Jrnl. Psychol.* VIII. 409 It appears that the subject projects similar patterns or configurations upon widely different materials. 1960 E. E. Cummings *Let.* 30 Jan. (1969) 266 Indeed this correspondent can't help suspecting yourself of what the psychoanalysts call 'projecting'. 1966 I. G. Sarason *Personality* xii. 181 A defense mechanism through which an individual unconsciously projects his own undesirable characteristics to others than himself. 1975 K. R. Scherer et al. *Human Aggression & Conflict* iv. 116 The subjects were projecting their needs into their imagery.

c. To convey to others, esp. by one's manner and actions (a positive image of one's personality or attributes). Usu. *absol.*

1955 *Psychiatry* XVIII. 217/2 The self-evaluation projected by his clothes and manner. 1957 *Economist* 12 Oct. 130/1 This matter of 'projection' is taken very seriously. 'He simply doesn't project' can be as final a dismissal of political aspiration as the fact that a man is known to have beaten a whole series of wives. 1959 *Encounter* Sept. 50/2 Competing with the roar of the machines..the actors struggle to project. 1960 *News Chron.* 28 July 4/5 Unable to 'project' publicly, in private he deploys considerable private charm. 1967 M. Argyle *Psychol. Interpersonal Behaviour* iii. 55 If a person behaves unpleasantly, or in some other way fails to live up to the image he has projected, equilibrium is disturbed.

III. 12. a. *pass.* = next sense.

1902 J. M. Baldwin *Dict. Philos. & Psychol.* II. 414/2 The radiations taken together are called the 'projection system', the lower centres being projected upon the cortex. 1925 *Jrnl. Neurol. & Psychopath.* VI. 5 We draw the conclusion that the upper part of the retina in apes is projected on the medial side of the corpus geniculatum externum. 1926 *Brain* XLIX. 2 The nasal part of the retina is always projected laterally to the temporal half [of the external geniculate body].

b. *intr.* *Physiol.* Of an area or organ of the body, or its nerves: to have or be nerve fibres extending *to* an area. Also const. *upon*.

1936 *Jrnl. Compar. Neurol.* LXIV. 37 The anterior thalamic nuclei project to a small part of the orbital surface of the frontal lobe. *Ibid.*, The nucleus ventralis posterior, projects entirely upon the cortex of the central sulcus and the postcentral convolution. 1951 T. C. Ruch in S. S. Stevens *Handbk. Exper. Psychol.* iv. 125/2 The ablation of a cortical area truncates the axons running to it, and the locus of the resulting retrograde degeneration establishes which thalamic nucleus projects to the particular area ablated. 1975 D. H. Ford *Anat. Central Nerv. Syst.* vii. 83 The white matter of the [spinal] cord..consists of association fibers..which connect adjacent levels of the cord, longer association bundles which project up or down the various funiculi to interconnect further segments of the cord, or very long projection systems which project to supracord levels or which enter the cord from higher levels. 1978 *Sci. Amer.* July 38/1 These fibers usually branch repeatedly and may project to distant parts of the nervous system or leave the nervous system to innervate effector tissues.

projectile, *a.* and *sb.* Add: **A.** *adj.* **5.** *Literary Criticism.* (See quots.)

1929 I. A. Richards *Pract. Criticism* 357 Aesthetic or 'projectile' adjectives..raise several extraordinarily interesting questions... In so far as they register the projection of a feeling into an object they carry a double function. 1949 Brooks & Warren *Fund. Good Writing* x. 351 What I. A. Richards calls 'projectile' adjectives: that is, adjectives which function, not so much to give an objective description, as to express the writer's or speaker's feelings... The '*miserable* wretch' may actually be smiling happily. The woman who has just been called 'a *great little* wife' may be large or small.

projecting, *vbl. sb.* (Later examples.)

1959 *Times Lit. Suppl.* 16 Jan. 39/1 *Introduction to Cine*..starts off the complete beginner and takes him as far as editing and projecting. 1960 E. H. Gombrich *Art & Illusion* xi. 385 We prefer suggestion to representation, we have adjusted our expectations to enjoy the very act of guessing, of projecting.

projection, *sb.* Add: **IV. 6. b.** *Math.* Any homomorphism from a vector space or the like into a part of itself such that each element of the part is mapped on to itself; also, a homomorphism from a group into a quotient group.

1942 *Amer. Jrnl. Math.* LXIV. 115 The study of groups which have projections on abelian groups. 1950 *Bull. Amer. Math. Soc.* LVI. 488 The systematic use of these injection and projection homomorphisms is at the heart of our formulation of the duality phenomena. 1976 D. E. Christie *Basic Topology* vii. 191 An indispensable tool for products is the projection, a function from the product to one of the factors.

7. b.

I. *Orthomorphic*: also called *CONFORMAL; IV. *Zenithal*: also called *azimuthal*; VII. Another conventional projection is the *two-point equidistant*, showing accurately the distances from every point to each of two chosen points. Other projections named after their inventors are *Albers'*, a conical equal-area projection with two standard parallels; and *Clarke's*, projected from a centre outside the globe onto a diametrically opposite plane. Projections may be *interrupted*, so that the representation is not convex but lobate or partly dissected; and *transverse*, representing a globe rotated through a right or other angle from its conventional orientation.

1910 J. I. Craig *Theory Map-Projections* v. 45 (*heading*) Transverse conical projection... This projection is of no practical importance. *Ibid.* 53 (*heading*) Zenithal or azimuthal projections. *Ibid.* 55 Lambert's equivalent azimuthal projection. 1910 *Encycl. Brit.* XVII. 656/1 (*caption*) Clarke's perspective projection for a spherical radius of 108°. 1912 A. R. Hinks *Map Projections* i. 6 There is a class of projections sometimes named azimuthal, from the fact that the azimuths, or true bearings, from the centre of the map, of all points, are shown correctly. *Ibid.* 7 The objection to the term *azimuthal* is that it is hard to pronounce, and several writers have followed German in calling always this class of projection *zenithal*. *Ibid.* iii. 26 Albers' conical equal area projection. 1922 C. Close in *Ordnance Survey Prof. Papers* No. 5. 6 (*heading*) Two-point equidistant projection. 1927 J. A. Steers *Introd. Study Map Projections* vi. 152 (*heading*) 'Interrupted' projections. 1969 G. C. Dickinson *Maps & Air Photographs* i. 9 Distances..can be shown correctly from one, or two, but no more, chosen points on certain special projections. [*Note*] The zenithal equidistant and the two-point equidistant projections respectively. *Ibid.* 17 If a map can be broken in some areas that do not matter—the oceans if map is needed mainly for continental areas, or vice versa—and the meridians gathered together at several 'central' meridians the good qualities of the 'central' areas are more widely spread. Fig. 4F shows such an interrupted sinusoidal projection. *Ibid.* 20 Let us try projecting it [sc. the globe] onto a flat sheet of paper touching the globe, at the north pole to begin with, and furthermore see what happens when we move the source of light... The results from a group of projections known collectively as zenithal or azimuthal projections. *Ibid.* 24 If in relation to the parallels, the words 'their true distance apart' in the preceding specification is [*sic*] replaced by 'spaced so that the area between them is the same as that on the globe' we get Alber's projection. *Ibid.* 25 Suppose we want to make a transverse Mercator's projection based on (say) 90°W. and not the equator.

d. *Econ.* A forecast based on present trends.

1952 *Economist* 30 Aug. 526/1 The FBI's figure..amounts almost exactly in total to a direct projection of the sharp upward trend in consumption during 1950 and 1951. 1962 *Listener* 16 Aug. 235/1 When this work has reached the stage of placing the various national projections alongside one another, the foundations of planning will have been laid. 1969 *Times* 4 Sept. 7 The eminent thinker acknowledges that economic performance is not conclusive but insists that it furnishes the basic structure and framework of power. Here is his G.N.P. projection for 1980. 1976 *Time* 27 Dec. 48/2 Among those who doubt the Carter projections are the members of Time's Board of Economists.

8. a. *spec.* The process of projecting (an image on) a film or transparency on to a screen for viewing.

1896 R. W. Paul *Brit. Pat. 4686*, I prefer to employ the following mechanism,..causing the film to be propelled instantaneously a small amount, after which it remains still for projection of the picture. 1912 F. A. Talbot *Moving Pictures* ix. 99 This second lens is used for the projection of lantern slides. 1953 L. J. Wheeler *Princ. Cinematogr.* vi. 193 Both lanterns must be accurately trained on the screen to give the appearance of continuous projection. 1976 *Times* 22 Apr. 11/3 Amplified sound, music, lighting, slide projection, videotape, all the tools of the professional chatterbox will be deployed to tempt conference organizers.

b. *Mus.* The projective quality of sound; acoustic penetration. Also *transf.* of an instrument.

1977 *Gramophone* Dec. 1045/1 So fine was the earlier recording that the later one..is not necessarily an improvement, even if the sound has slightly more clarity and projection. 1977 *Oxf. Times* 16 Dec. 16 The Allegri Quartet..tested the viola and cello in exchanged positions. The increased projection of the viola was remarkable.

9. b. *Psychoanal.* The unconscious process or fact of projecting one's fears, feelings, desires, or fantasies on to other persons, things, or situations, in order to avoid recognizing them as one's own and so as to justify one's behaviour. Also in more general use.

1909 Peterson & Brill tr. *Jung's Psychol. of Dementia Praecox* iv. 87 By the method of outward projection they frequently place the responsibility on some foreign agency. 1923 J. S. Huxley *Ess. Biologist* iv. 167 This projection, or interpretation of external reality in terms of one's self, is a curious and almost universal attribute of the human mind. 1924 J. Riviere tr. *Freud's Defence of Neuro-Psychoses* in *Coll. Papers* I. ix. 180 In paranoia the reproach is repressed in a manner which may be described as *projection*; by the defence-symptoms of distrust directed against others being erected. 1938 G. W. Allport *Personality* vi. 173 There is likewise a complementary form of projection whereby a person does not attribute his own frame of mind to others but rather one that justifies and explains his own frame of mind to himself. 1944 *Horizon* IX. 169 His [sc. Lenin's] fanatical hatred of the Bourgeoisie of which, in analytical terms, the Russian revolution was merely a 'projection'. 1950 T. S. Eliot *Cocktail Party* I. ii. 59 The man I saw before, he was only a projection—I see that now—of something that I wanted. 1966 R. D. Laing et al. *Interpersonal Perception* ii. 16 Projection refers to a mode of experiencing the other in which one experiences one's outer world in terms of one's inner world. 1966 *Listener* 5 May 653/2 Rogozhin..who tries to win Nastasya with money, and ends by murdering her, can be seen to be a projection of Myshkin's urge to power and destruction. 1975 K. R. Scherer et al. *Human Aggression & Conflict* iv. 117 Perhaps through the mechanism of complementary projection, they perceived the students to be particularly hostile, dangerous, and intent on overpowering the soldiers. 1976 Smythies & Corbett *Psychiatry* xv. 271 Projection means the attribution to external agencies of one's own psychological conflicts.

c. The conveying of a positive image of one's personality to others by one's manner and actions.

1955 *Times* 10 May 3/7 Attack, boldness, and what actors call 'projection' of the artist's personality, are undeniably all there. 1957 [see *PROJECT *v.* 10 c].

9*. *Physiol.* and *Psychol.* The process whereby a stimulus is perceived as being located at a point other than where the sensation or perception occurs (see quots.).

1887 G. T. Ladd *Elem. Physiol. Psychol.* II. vi. 387 The law of eccentric projection is generally stated thus: Objects are perceived in space as situated in a right line off the ends of the nerve-fibres which they irritate. 1890 W. James *Princ. Psychol.* II. xvii. 41 The other cases of translocation of our sensations are equally easily interpreted without supposing any 'projection' from a centre at which they are originally perceived. *Ibid.* 42, I conclude..that there is no truth in the 'eccentric projection' theory. 1892 Van Liew & Beyer tr. *Ziehen's Introd. Physiol. Psychol.* iv. 77 By 'eccentric projection' we understand the fact that a sensation produced by the stimulation of the nerve-*trunk* instead of the nerve-*ends* is regularly attributed to irritation of the peripheral ramifications of the nerve. 1902 J. M. Baldwin *Dict. Philos. & Psychol.* II. 358/2 *Projection*, the spatial objectivation of objects in sense perception... This usage is vague and descriptive, varying from the mere recognition of a spatial datum to the hypothesis of the spatial projection of states at first purely 'inner' and unspatial. It is also complicated with the hypothesis..that nervous projection.., to the periphery, sometimes extends out in lines at right angles to the sensitive surface. *Ibid.*, *Projection* (*nervous*, or '*eccentric*'), the property of the nervous system whereby stimulations are referred to the periphery of the body or to the end-organs. 1972 *Encycl. Psychol.* III. 47/1 *Projection, eccentric*, the introspective observation that sensory experiences are usually localised outside the body at the same position as the stimulus object... Thus the blue is seen as on the sky rather than in the retina.

9.** *Physiol.* The spatial distribution, in the brain, or other parts of the central nervous system, of the points to which nerves or nerve impulses go from any given area or organ; const. *on, upon, to* the receiving part; also *concr.*, a tract of projection fibres.

1924 *Scand. Sci. Rev.* X. 18 [This case verified my supposition..that every limited lesion of the calcarine cortex causes a corresponding limited blind spot in the visual field, or, that there exists a mathematical projection of the peripheral retina in the calcarine cortex.] *Ibid.* 37 The projection of the retina on the calcarine cortex. 1925 *Jrnl. Neurol. & Psychopath.* VI. 3 It is very probable that the projection of the retina on the primary centres in the ape is similar to that in man. 1934 *Proc. R. Soc.* B. CXV. 504 Although the existence of the cortico-pontine fibres has long been recognized.., there has as yet been no solution of the problem of their distribution in the pons and projection on the cerebellum. 1936 *Jrnl. Compar. Neurol.* LXIV. 7 The thalamic projection to the frontal cortex has occasioned much discussion. 1938 J. F. Fulton *Physiol. Nerv. System* xv. 335 In addition to the corticospinal projections, the cerebral cortex in the higher forms gives rise to a vast extrapyramidal projection passing to many subcortical levels. 1951 T. C. Ruch in S. S. Stevens *Handbk. Exper. Psychol.* iv. 136/1 The projection of the body surface upon the posteroventral nucleus of the thalamus was worked out in greater detail. 1973 W. J. S. Krieg *Synoptic Functional Neuroanat.* 4/2 In the pons..the cortical projections are broken into bundles, and many fibers form connections to the cerebellum here.

V. 10. *projection dynamics, phenomenon, screen, surface*; **projection booth, box** = *projection room* below; **projection-fibre**, substitute for def.: a nerve fibre connecting one part of the central nervous system with another, esp. the cortex with the brain stem or spinal cord; (further examples); **projection lens**, the objective lens in a film or slide projector, which projects an enlarged image into space; **projection printing** *Photogr.*, printing in which an optical system is placed between the negative and the printing paper, so that enlargement or reduction of image size is possible; hence **projection printer**, an apparatus for this; **projection room**, a room in a cinema or film studio designed to contain the projector and its operators, through windows in the wall of which the film is projected; **projection rule**, in *Transformational Grammar*, a rule, based on underlying phrase markers, for combining lexical senses of words so as to predict their semantic role in a given sentence; **projection system**, add: in mod. use, a system of projection fibres; (earlier and later examples); **projection test** = *projective test* (see *PROJECTIVE *a.* 5b); **projection welding**, resistance welding in which welding is effected at one or more projecting points of contact previously formed in the components by pressing; so **projection weld** *sb.* and (with hyphen) *v. trans.*, **-welded** *ppl. a.*; also **projection welder**, an apparatus for projection welding.

1929 F. Green *Film finds its Tongue* xviii. 249 Out in the theatre, sitting in the audience, is an Observer. He has a telephone that leads to the projection booth. 1968 *Globe & Mail* (Toronto) 17 Feb. 45 (Advt.), Recreation area consisting of large family room with..projection booth and screen for home movie entertainment. 1934 S. Chesmore *Behind Cinema Screen* ix. 85 In modern theatres the projection box is roomy..and well lit. 1966 P. O'Donnell *Sabre-Tooth* iv. 67 The projection box was equipped with a kershaw filmstrip and slide projector. 1953 C. E. Osgood *Method & Theory in Exper. Psychol.* vi. 229 (*heading*) Projection dynamics in perception. 1920 S. W. Ranson *Anat. Nerv. System* xviii. 297 Many of the fibers of the medullary white center connect the cerebral cortex with the thalamus and lower lying portions of the nervous system. These are known as projection fibers, and may be divided into two groups according as they convey impulses to or from the cerebral cortex. 1970 L. J. A. DiDio *Synopsis of Anat.* xix. 437/2 White matter of the cerebral hemispheres... It is composed of myelinated nerve fibers that may be divided into three groups: projection fibers, commissural fibers, and association fibers. *Ibid.*, Projection fibers are those that establish either ascending or descending connections between the cerebral cortex and structures outside the telencephalon. 1917 C. N. Bennett *Guide to Kinematogr.* ix. 131 A secondary effect is often produced through the additional length of focus of the projection lens. 1962 *Which?* Mar. 68/1 The slide is put into a slide carrier in the projector and slid in front of a lamp and behind a projection lens.

1962 Henderson & Gillespie *Text-bk. Psychiatry* xii. 294 When a failure of repression occurs the paranoid symptoms develop as projection phenomena. **1940** Lucas & Dudley *Making your Photographs Effective* xi. 168 The apparatus required for projection printing consists of the projection printer, or enlarger,..and the easel. **1965** M. J. Langford *Basic Photogr.* xviii. 324 The term 'enlarger' although common usage, is deceptive. 'Projection printer' is the more accurate description of an optical device to give prints both larger and *smaller* than the original negative. **1923** *Brit. Jrnl. Photogr.* LXX. 350 The remaining factor in contact printing is the distance between the light and the negative... The question of printing distance operates equally in projection printing. **1974** A. Feininger *Darkroom Techniques* II. 51 Unlike contact printing, ..projection printing allows a photographer a considerable amount of control as far as the final appearance of the print is concerned. **1914** R. Grau *Theatre of Science* iii. 48 The fixture and office furniture are of massive mahogany and plate glass and the projection room is the last word in luxurious splendor. **1930** *Aberdeen Press & Jrnl.* 29 Mar. 7/4 A fire occurred in the projection room of the Swan Cinema. **1975** *Language for Life* (Dept. Educ. & Sci.) xxv. 425 Almost a quarter of the schools had a projection room. **1962** Katz & Postal *Integrated Theory Ling. Descr.* iii. 64 The set of projection rules of a semantic component is..an unordered set. Each rule applies when the conditions of its application are met, and no two rules apply in the same case because no two rules have the same conditions of application. **1964** Fodor & Katz *Struct. Lang.* xix. 493 A semantic theory must contain two components: a dictionary of the lexical items of the language and a system of rules (which we shall call *projection rules*). **1965** N. Chomsky *Aspects of Theory of Syntax* iv. 154 The projection rules must now be adapted to detect and interpret conflicts in feature composition. **1966** J. J. Katz *Philos. Lang.* iv. 153 A system of *projection rules* that provide the combinatorial machinery for projecting the semantic representation for all supraword constituents in a sentence. **1977** *Language* LIII. 93, I assume..that semantic representations are complex objects, related to different aspects of syntactic structure by means of 'projection rules', or 'interpretive rules', of different types. **1946** Koestler *Thieves in Night* 170 You are fond of people..as projection-screens for your own feelings. **1954** —— *Invisible Writing* xxvi. 276 My emotions were self-centred, and those who inspired them served merely as projection-screens. **1890** W. James *Princ. Psychol.* I. ii. 59 The entire cortex being, according to him [*sc.* Munk], nothing but a projection-surface for sensations, with no exclusively or essentially motor part. **1876** *Quain's Elem. Anat.* (ed. 8) II. 565 First projection system..between the convolutions above and the cerebral ganglia..corresponds for the most part to the corona radiata. **1890** A. Hill tr. *Obersteiner's Anat. Central Nerv. Organs* 168 Through the fibres of this system sense-pictures are projected on the perceptive cortex, and..the cortex..reflects outwards again the states of stimulation, information with regard to which is transferred to it by means of sensory nerves. The whole of these conducting paths Meynert, therefore, terms a 'projection system'. **1958** M. Argyle *Relig. Behaviour* ix. 104 The orthodox [church members] scored higher on ego-defensiveness and dependency measured by various projection tests. **1962** *Listener* 11 Jan. 62/2 Achievement motivation as measured by the projection test in which children were asked to write stories about pictures. **1967** M. Argyle *Psychol. Interpersonal Behaviour* i. 18 In 'projection tests' subjects are asked to tell a story about people shown in rather vaguely-drawn pictures... There is considerable doubt over the validity of such projection tests, and they cannot be said to provide very good predictions. **1950** Hipperson & Watson *Resistance Welding* iii. 86 Projection welds may be made with a great variety of projection shapes and sizes. *Ibid.* 88 Unequal thicknesses of sheet may be projection welded. **1961** J. A. Oates *Welding Engineer's Handbk.* xxiii. 249 In cases where the projection welds have to be made on a narrow flange it is an advantage to use an elongated projection. **1980** L. M. Gourd *Princ. Welding Technol.* xi. 167 Reinforcing rings are frequently projection-welded around holes in sheet-metal tanks. **1950** Hipperson & Watson *Resistance Welding* i. 25 A few typical projection welded applications..are shown. **1980** L. M. Gourd *Princ. Welding Technol.* xi. 166 (*caption*) Examples of projection-welded details. **1946** *Philips Resistance Welding Handbk.* i. 15 Owing to the number of spots, projection welders are of a higher kVA. than normal spot welders. **1968** Romans & Simons *Welding Processes & Technol.* v. 39 The majority of projection welders are operated by compressed air. **1918** Hamilton & Oberg *Electric Welding* iii. 119 The welding of sheet metal is not restricted to one spot at a time, as any reasonable number of welds can be made at one operation by the method known as 'point-' or 'projection-welding'. **1975** Bram & Downs *Manuf. Technol.* ii. 63 In projection welding the component is shaped to provide localised current flow, concentrating the welding heat at the areas of projection.

proje·ctional, *a.* [f. Projection *sb.* + -al.] Of, pertaining to, or connected with projection (in various senses).
1899 *Phil. Trans.* B. CXCI. 298 The large system of fibres just described above is probably both an associational and projectional system. **1949** *Mind* LVIII. 76 If the term 'God' is really non-significant, Findlay's earlier description of the religious attitude should be translated into emotive or projectional terms.

proje·ctionist. [f. as prec. + -ist.] One who operates a film projector.
1922 A. C. Lescarboura *Cinema Handbk.* vii. 285 One reel must serve for a large number of projectionists. **1938** *Times* 23 May 11/1 A suggestion that the cinema projectionists' dispute in London and the home counties might be extended. **1958** X. Fielding *Corsair Country* viii. 167 The projectionist is making the final adjustments to his Heath Robinson apparatus. **1969** B. Patten *Notes to Hurrying Man* 14 This is the projectionist's nightmare: A bird finds its way into the cinema..smashes into a screen. **1973** J. Wainwright *Pride of Pigs* 158 They usually employed a lad—the 'second projectionist', so-called—to carry the reels to and from the projection room. **1977** *Western Morning News* 30 Aug. 11 (Advt.), Classic Entertainment Centre. We have an immediate vacancy for a Senior projectionist. Experience on 'Kalee 21' essential.

projective, *a.* Add: **2. d.** *projective plane*, that two-dimensional manifold which may be regarded as a spherical shell with all pairs of antipodal points identified; it is an example of a *projective space*, a space which may be regarded as obtained by taking a vector space of the next higher dimension, identifying all vectors which are multiples of one another, and omitting the origin.
1900 Projective plane [in Dict., sense 2 a]. **1910** Veblen & Young *Projective Geom.* I. iv. 97 Any such space we call a properly projective space. **1942** *Amer. Jrnl. Math.* LXIV. 137 A satisfactory analytic theory may be developed for every projective plane in which Desargues' Theorem is valid. **1960** Hilton & Wylie *Homology Theory* iii. 133 The real projective space P^n may be defined as the image of the *n*-sphere S^n under identification of all pairs of antipodal points. **1962** B. H. Arnold *Intuitive Concepts Elem. Topology* iii. 71 A projective plane can be considered as a disk and a Möbius strip whose edges are joined. **1964** C. E. Springer *Geom. & Anal. Projective Spaces* vi. 150 A projective space is orientable if the dimensionality of the space is odd and nonorientable if it is even. **1975** I. Stewart *Concepts Mod. Math.* xiii. 199 This is exactly what is happening in the projective plane: going round once things get twisted; going round twice brings them back to normal.

5. b. *Psychol.* Of or pertaining to the projection of unconscious feelings, fears, fantasies or desires; esp. of tests designed to reveal unconscious elements of personality by responses to words, pictures, etc. Also *ellipt.* as *sb.*
1895 J. M. Baldwin *Mental Devel. in Child* vi. 120 All of them [*sc.* stages of attitude] belong in the 'projective' stage of the child's sense of self, *i.e.*, they all go to furnish data which he afterwards appropriates to himself as 'subject'. **1939** *Jrnl. Psychol.* VIII. 404 No attempt has been made to provide a complete review of all the projective methods now being used. **1954** L. Bellak *TAT & CAT in Clin. Use* p. x, The T.A.T. in common with all the other projective tests, is still far from being a properly established instrument. **1956** A. I. Hallowell in B. Klopfer et al. *Devel. Rorschach Technique* II. xiv. 476 Rorschach theory, as well as that underlying other projective tests, has been based on the general assumption..that 'every subject's responses..are *determined* by psychological attributes of that subject'. **1966** I. G. Sarason *Personality* xii. 180 The ideas behind projective techniques have been largely psychoanalytic. **1971** *Jrnl. Gen. Psychol.* LXXXIV. 321 The clinician using the auditory method is now able to consider stimulus properties when evaluating projective material. **1976** L. R. Aiken *Psychol. Testing & Assessment* (rev. ed.) viii. 223 Questionnaires and projectives are useful, but the most popular psychometric device for determining attitudes is an attitude scale.

7. *projective verse*, a term invented by C. Olson (1910–70), American poet and poetical theorist, to describe a brand of verse propelled by its inherent energy and composed according to a system of poetic values in which structure, lay-out, and breathing have an importance not accorded them in traditional forms. Hence *projective poet*, etc. Also *ellipt.*
1950 C. Olson in *Poetry New York* III. 15 *Projective verse* teaches, is, this lesson, that that verse will only do in which a poet manages to register both the acquisitions of his ear *and* the pressures of his breath. *Ibid.* 20 Which gets us to..the degree to which the projective involves a stance toward reality outside a poem [etc.] *Ibid.* 22 Eliot ..has only gone from his fine ear outward rather than, as I say a projective poet will, down through the workings of his own throat to that place where breath comes from, where breath has its beginnings, where drama has to come from, where, the coincidence is, all art springs. **1962** E. Mottram in *London Mag.* Dec. 71/1 Projective or open verse is a 'stance towards reality' as it brings the verse into being—'some simplicities that a man learns if he works in *open*, or what has been called *composition by field*, as opposed to inherited line, stanza, over-all form, what is the "old" base of the non-projective'. **1962** *Listener* 27 Dec. 1102/1 Of the 'projective verse' school, Ginsberg and..Edward Dorn seem to me remarkable talents. **1963** *Ibid.* 7 Mar. 435/3 A poet I liked very much is Robert Bly. In versification he is not 'projective', but in tone and attitude he is. **1967** *Book Week* (Washington Post) 19 Mar. 6/1 Here what he [*sc.* Olson] calls the Projective Open or Field verse (as opposed to the systematic Closed Forms of the past) is put to work, using *line, syllable, breath*, as principles he has preached. His one theme is energy—how a man's energy is expended in history and in space.

projector. 4. b. Substitute for def.: An apparatus containing a source of light and a system of lenses for projecting on to a screen an enlargement of an image or a slide, film, or opaque surface. (Examples.)
1912 F. A. Talbot *Moving Pictures* 135 In colour work the projector requires a special type of shutter. **1915** W. H. Chantrey *Theatre Accounts* (ed. 2) 78 Cinematograph projectors should be fitted with two metal film-boxes of substantial construction. **1926** *Encycl. Brit.* II. 961/2 In June 1895 Thomas Armat of Washington..arrived at the principle of the modern projector, a device in which the film, moving intermittently, has periods of rest and illumination in large excess of the period of movement. **1962** *Movie* Oct. 12/3 The flickering blue light of the projector in the viewing theatre sequence. **1964** M. McLuhan *Understanding Media* (1967) xxix. 311 The present dissociation of projector and screen is a vestige of our older mechanical world of..separation of functions. **1977** J. Hedgecoe *Photographer's Handbk.* 312 Projectors for 35 mm transparencies and smaller are now usually magazine loaded.

5. *Comb.*, as *projector lamp, -man.*
1962 *Which?* Mar. 74/1 Projector lamps are expensive. **1972** *Gloss. Electrotechnical, Power Terms (B.S.I.)* IV. iii. 15 *Projector lamp*, lamp in which the luminous element is so mounted that the lamp may be used with an optical system projecting the light in chosen directions. **1927** *Observer* 17 Apr. 3 The picture is..'ridden in'—that is, the orchestra work up to an appropriate climax, and at a given bar the projector-man 'makes his throw'.

projicient. Restrict †*Obs.* to sense in Dict. and add: (prodʒi·ʃi̇ĕnt). **B.** *adj.* Concerned with an individual's perception of his surroundings.
1904 *Nature* 8 Sept. 465/1 [Reporting C. S. Sherrington.] In presence of the arcs of the great projicient receptors and the brain there can be few receptive points in the body the activities of which are totally indifferent to one another. **1927** J. H. Parsons *Introd. Theory Perception* vii. 143 The projicient senses—vision.., hearing.., and smell—in the head segments provide those sensations which occupy the focus of the perceptual pattern, the field of attention. **1954** *A.M.A. Arch. Neurol. & Psychiatry* LXXII. 472 On the other hand, the admonitions from the group representative are apprehended largely by visual and auditory means; the authority figure is very much more clearly defined by 'projicient' modalities, and, we guess, is always clearer to the individual.
So **projicience** (prodʒi·ʃi̇ĕns), projicient activity or ability.
1906 C. S. Sherrington *Integrative Action Nervous Syst.* ix. 324 It is in the leading segments that we find the 'distance-receptors'. For so may be called the receptors which, acting as sense-organs, initiate sensations having the psychical quality termed projicience. **1927** J. M. Parsons *Introd. Theory Perception* ii. 7 At a somewhat higher level there is evidence of response to radiation of shorter wave-length—light, and perhaps ultra-violet radiation. As soon as this occurs the germ of projicience is found. **1931** *Brit. Jrnl. Psychol.* XXII. 143 Many, if not most, of the stimuli, to which man and the higher mammals respond through projicience, cannot be adequately described without using qualitative terms. **1949** A. Gesell et al. *Vision* xii. 196 This ability is a topographic discrimination, an elementary form of projicience.

prokaryon (prŏᵘkæ·ri,ǫn). *Biol.* Pl. **prokarya.** [f. Gr. προ- Pro-² + κάρυον nut, kernel; cf. *prokaryote.] The structure in a prokaryote which contains the genetic material; the prokaryotic 'nucleus'.
1957 E. C. Dougherty in *Jrnl. Protozool.* IV. Suppl. 14/1 For the moneran nucleus I propose *prokaryon*.., and for that of higher organisms, *eukaryon*... From these derive..the nouns *prokaryosis* and *eukaryosis* and their corresponding adjectives, *prokaryotic* and *eukaryotic*.., denoting, respectively, 'the condition of possessing prokarya or eukarya'. **1969** Brown & Bertke *Textbk. Cytol.* vi. 96/1 The more or less central region of indeterminate shape and no bounding membrane is sometimes called the nucleus but more properly the prokaryon.

prokaryote (prŏᵘkæ·riǫt). *Biol.* Also **-caryote.** [a. F. *procaryote* (É. Chatton 1925, in *Ann. des Sci. Nat.: Zool.* VIII. 76), f. as prec. + Gr. -ώτ-ης.] A prokaryotic organism. Opp. *eukaryote*.
1963 *Cold Spring Harbor Symp. Quantitative Biol.* XXVIII. 1/1 The distinction of *eukaryotes* which possess a characteristic chromosome nucleus and *prokaryotes* where the nuclear equivalent does not show any chromosome-like structures, is not bounded by a nuclear envelope, and does not divide by mitosis. **1967** Kirk & Tilney-Bassett *Plastids* xi. 364 This theory..makes it unnecessary to explain the separate evolution of photosynthetic ability in the prokaryotes (organisms such as bacteria and blue-green algae, with no separate membrane-bounded nucleus, or other organelles) and the eukaryotes (higher organisms, including algae and higher plants, with chromosomes inside a membrane-bounded nucleus, and other membrane-bounded organelles). **1969** Loewy & Siekevitz *Cell Struct. & Function* (ed. 2) i. 4 The genetic information in procaryotes, at least in the organisms studied so far, is located on a single chromosome that consists of a circular double strand of DNA and that lacks the basic proteins called histones. **1976** W. C. Schefler *Biol.* xiii. 242/1 Procaryotes therefore have a very simple structure, which leads biologists to believe that a division of living organisms into these two basic cellular types occurred long before the evolutionary development of plants and animals as distinct groups. **1976** *Sci. Amer.* Sept. 167/2 No plant or animal is able to fix nitrogen, only prokaryotes: organisms, including bacteria and blue-green algae, that have no cell nuclei.

prokaryotic (prŏᵘkæri,ǫ·tik), *a.* *Biol.* Also **-caryotic.** [f. as prec. + -ic.] Having no nuclear membrane in its cell; belonging to the group of organisms so characterized, which comprises bacteria and blue-green algæ. Opp. *eukaryotic*.
1957 [see *prokaryon *a.*]. **1969** [see *conidium]. **1974** *Taxon* XXIII. 246 Cells with a general organization between entire prokaryotic and eukaryotic cells have not been found. **1976** *Sci. Amer.* Feb. 35/2 Bacteria are prokaryotic cells, that is, their genetic material is distributed throughout the cytoplasm. **1976** W. C. Schefler *Biol.* xiii. 242/1 Procaryotic cells..lack mitochondria, chloroplasts, an endoplasmic reticulum, Golgi bodies, and lysosomes.

proke, *v.*¹ **2.** (Later examples.)
1843 N. Macleod *Crack aboot Kirk* (ed. 2) I. 2 If a man has a sair leg or a sick body ye needna keep prokin' at him and roarin' in his lug a' day that he's no weel. **1914** P. MacGill *Children of Dead End* xxi. 149 I'm sick of prokin' in the gutters here. **1941** E. R. Eddison *Fish Dinner* xiv. 261 The plague that sat dozing in her mouth's corner proked at him swiftly.

prokinesis (prŏᵘkəini̇·sis, -kɪn-). *Zool.* [ad. G. *prokinetik*, f. Gr. πρό Pro-² + κινητικός moving: see Kinetic *a.* (*sb.*) and *kinesis.] A process, found in some birds and lizards, by which the upper bill or jaw may be raised relative to the cranium by rotation about a hinge anterior to the eyes.
1962 T. H. Frazzetta in *Jrnl. Morphol.* CXI. 287/2 It is here advocated that..Hofer's 'prokinesis' be modified to designate any kinetic joint anterior to the eyes. **1964** *Ibid.* CXIV. 3/1 The basic avian condition is prokinesis from which rhynchokinesis evolved. **1964**, etc. [see *rhynchokinesis]. **1973** *Nature* 11 May 73/1 Prokinesis and rhynchokinesis may have evolved independently in neognathous and palaeognathous birds, or rhynchokinesis may have evolved from some form of prokinesis.
So **prokine·tic** *a.*
1960 *Q. Rev. Biol.* XXXV. 219/2 In view of the fact that there are so many differences in the actual operation of kinesis in the various orders [of birds], I doubt if the terms pro- and rhyncho-kinetic actually are morphologically meaningful. **1964** *Jrnl. Morphol.* CXIV. 4/2 The prokinetic condition is the most widespread and apparently the primitive condition in the class Aves. **1974** P. J. K. Burton *Feeding & Feeding Apparatus in Waders* II. 35 An important feature of rhynchokinesis is that the pivots about which the two jaws move are relatively much further apart than in prokinetic species.

pro-knock: see *pro-¹ 5a.

prolactin (prolæ·ktin). *Physiol.* [f. Pro-¹ + Lact(ation + -in¹.] A gonadotrophic polypeptide hormone which promotes lactation.
1932 O. Riddle et al. in *Proc. Soc. Exper. Biol. & Med.* XXIX. 1211 We..have identified this same hormone, which we shall here call 'Prolactin' as the hitherto undefined pituitary principle which is essential for lactation in mammals. **1941** *Nature* 11 Jan. 44/2 The activity was at first ascribed to a single hormone of the pituitary which was called prolactin. Very recent work has gone to show that this activity is, in fact, due to the co-operation of two separate hormones of the anterior pituitary—prolactin and glycotropin. **1952** [see *lactogen]. **1974** *Sci. Amer.* Sept. 53/3 There is probably a role for a third gonadotropic hormone of the pituitary, prolactin, in maintaining the steroid-producing function of the corpus luteum for its usual 14-day life span, but there is some doubt whether this is true in humans. **1979** *Jrnl. R. Soc. Arts* CXXVII. 416/2 We know that the blood prolactin level is directly related to suckling frequency.

prolactinoma (prolæktinŏᵘ·mă). *Med.* [f. prec. + *-oma.] A tumour that produces excessive quantities of prolactin.
1975 S. Franks et al. in *Hormone Res.* VI. 273 Pituitary tumours causing disorders of reproduction are almost without exception 'prolactinomas'. **1977** *Lancet* 9 Apr. 779/2 The incidence of prolactinoma in patients with pituitary tumours is likely to vary in different endocrine clinics.

prolamine (prŏᵘ·lămīn, -in). *Biochem.* Formerly also **prolamin.** [f. *proline with inserted *-am* (f. Amide).] Any of a class of proteins which occur in the seeds of cereals and are characterized by solubility in a 70–90 per cent solution of alcohol and insolubility in water.
1908 T. B. Osborne in *Science* 2 Oct. 422/1 Prolamins..form a unique and sharply differentiated group of proteins which occur in quantity in the seeds of cereals... I propose this name for the group which heretofore has been simply called alcohol-soluble proteins. The name refers to the relatively large proportion of proline and amide nitrogen which they yield on hydrolysis. **1921** *Monthly Bull. Agric. Intelligence & Plant Dis.* XII. 400 The only well-defined proteins found in oats are a venaline,..and a prolamine soluble in alcohol, and belonging to the same group as the gliadin of wheat. **1931** E. C. Miller *Plant Physiol.* ix. 518 The prolamins..are known to occur only in the seeds of cereals. *Ibid.*, The glutelins and prolamins are collectively termed 'glutens'. **1938** *Thorpe's Dict. Appl. Chem.* (ed. 4) II. 493/1 Prolamines (gliadins) are present only in small quantities in the inner endosperm of rice. **1971** *Sci. Amer.* Aug. 36/3 The nutritional proteins in corn kernels are classified in four categories according to solubility: (1) albumins, soluble in water; (2) globulins, soluble in saline solutions; (3) prolamines, soluble in moderately strong alcohol.

prolan (prŏᵘ·læn). *Physiol.* [a. G. *prolan* (B. Zondek 1929, in *Zeitschr. f. Geburtshülfe u. Gynäkologie* XCV. 363), f. L. *prōl-ēs* progeny: see -an.] The name given to what was formerly thought to be one female sex hormone but is now known to comprise both follicle-stimulating hormone (*prolan A*) and luteinizing hormone (*prolan B*).
1931 *Biol. Abstr.* V. 1357/2 The hormone of the anterior lobe of the pituitary body, called prolan, is an activator of the ♀ sex gland. **1936** *Brit. Med. Jrnl.* 28 Mar. 628/2 Prolan preparations are also found to be effective here [*sc.* in habitual miscarriage from no apparent cause]. **1966** Rowlands & Parkes in A. S. Parkes *Marshall's Physiol. of Reproduction* (ed. 3) III. xxv. 51 This gonadotrophic material, called Prolan by Aschheim and Zondek, was at first thought to be of hypophysal origin. **1971** *Path. Biol.* XIX. 1119 Estimation of..prolan B at concentrations lower than the threshold of immunological activity during various pathological conditions is rendered possible.

prolapsed, *ppl. a.* Add: Also *fig.*
1926 S. Baldwin *On England* 111 We see the sentences of the ancients clean run like athletes and fit for their work as compared with the prolapsed and slovenly figures of so much of our own diction.

prolately, *adv.* (Earlier example.)
1866 B. H. Kennedy *Public School Latin Primer* 110 The Infinitive stands... 4. Prolately, after Prolative Verbs and Adjectives.

prolating (prŏᵘlēⁱ·tiŋ), *vbl. sb.* [f. Prolate *v.* + -ing¹.] Increase or extension.
1919 *Empire Rev.* 256 The loss of wealth, high taxation, the dislocation of trade and industry with their attendant evils, labour unrest and the prolating of unemployment.

prola·tively, *adv.* [f. Prolative *a.* + -ly².] As a prolative infinitive.
1888 B. H. Kennedy *Rev. Latin Primer* 163 The Infinitive of a Copulative Verb used Prolatively is followed by a Complement in the Nominative.

prole (prŏᵘl), *sb.* and *a.* Freq. *derogatory.*
A. *sb.* Abbrev. of Proletarian *sb.*
1887 G. B. Shaw *Let.* 21 Oct. (1965) I. 176 We call the working men proles because that is exactly what they are. **1939** 'G. Orwell' *Coming up for Air* I. ii. 18 There's a lot of rot talked about the sufferings of the working class. I'm not so sorry for the proles myself. **1939** Joyce *Finnegans Wake* (1964) 39 The doubles of Perkin and Paullock, peer and prole. **1949** 'G. Orwell' *Nineteen Eighty-Four* II. 209 The dumb masses whom we habitually refer to as 'the proles', numbering perhaps 85 per cent of the population. **1954** Koestler *Invisible Writing* III. xxi. 238, I know that these little Proles can't help being what they are—dumb, callous, primitive. **1958** *Observer* 11 May 15/2 Even the boatswain, a 'good prole' who doesn't believe in unions or democracy, is afraid to take the helm. **1967** J. Potter *Foul Play* i. 11 Make way there, you proles. **1975** I. Murdoch *Word Child* 10 'I'm fed up with hearing the proles binding about the price of meat,' said Freddie. **1977** J. I. M. Stewart *Madonna of Astrolabe* x. 138 In the Blunderville heyday the Mumfords were the next things to proles.
B. *adj.* Abbrev. of Proletarian *a.* c.
1938 'G. Orwell' *Let.* 20 Apr. in *Coll. Ess.* (1968) I. 314 As to the great proletarian novel, I really don't see how it's to come into existence. The stuff in *Seven Shifts* is written from a prole point of view, but of course as literature it's bourgeois literature. **1949** —— *Nineteen Eighty-Four* I. 12 A woman down in the prole part of the house suddenly started kicking up a fuss. **1959** *News Chron.* 21 Oct. 3/1 She came from prole roots in St. Pancras. **1963** *Listener* 7 Feb. 263/1 Excellent radio actresses like Gladys Young, for whom our prole dramatists write no parts nowadays. **1969** 'H. Calvin' *Chosen Instrument* x. 132 Superior journalists..would think it Victorian and prole to know shorthand. **1977** *Times* 20 Apr. 12/2 What has happened in England is that..the proles, or to be exact the prole élite, have seized the power.

prolet- (prŏᵘle·t). Abbrev. [after Russ. *prolet-* in *proletkul't* for *proletárskaya kul'túra* proletarian culture] of Proletarian *a.* and *sb.*, as in *prolet-art, -cult,* (*-kult*), *-cultist, -cultural* adj., used to designate cultural activities (esp. such as were started in Russia after 1917) which supposedly reflect or encourage a purely proletarian ethos.
1921 E. & C. Paul *Proletcult* ii. 19 Proletarians who are alive to their class interest..will insist upon Independent Working-Class Education, upon proletarian culture, upon Proletcult. **1921** *Glasgow Herald* 28 Dec. 11 There is little proof that there has been general misrepresentation, and it is general misrepresentation that the prolet-cultist wants. **1922** *Ibid.* 4 Jan. 5 One cannot imagine any greater stimulus to the 'Prolet-Cultural' movement. **1931** H. G. Wells *Work, Wealth & Happiness of Mankind* (1932) xv. 732 From the Proletarian springs 'Prolet-art', for example, among the first fruits of the new spirit. It is art without individuality. **1943** E. M. Almedingen *Frossia* II. 84 Frossia told her about *Coppelia* at the neighbouring Proletcult. **1961** *New Left Rev.* May–June 27/2 The follies of *proletcult*, the stridency and crude class reductionism which passed for Marxist criticism. **1963** A. Hartley *State of England* ii. 26 The *Proletkult* of a philosopher such as Sartre is purely intellectual. **1967** *Oxf. Compan. Theatre* (ed. 3) 952/2 *Trades Unions Theatre*, Moscow..was created in 1932 from the Proletcult Theatre, started after the October Revolution by Sergei Eisenstein. **1974** Moore & Parry *Twentieth-Cent. Russ. Lit.* (1976) ii. 24 The magazine *Proletarskaya kul'tura* incorporated the name of the movement, which itself became, in shorthand, *Proletkul't*. **1976** T. Eagleton *Crit. & Ideology* v. 165 Such purely gestural, shamefaced materialism will provoke..the reaction of those who press their questioning of the intrinsic élitism of literature and its aesthetics to neo-*proletkult* limits.

proletaire. b. (Earlier example.)
1833 J. S. Mill *Let.* 2 Feb. in *Wks.* (1963) XII. 140 Those of the St. Simonians who retain their connection with the Pere Suprême and with each other, have made themselves *prolètaires*.

proletarian, *a.* and *sb.* Add: **A.** *adj.* **c.** (Further examples.) Also as *proletarian revolution*: the stage of political development predicted by Marx when the proletarians would overthrow capitalism.
1903 G. B. Shaw *Man & Superman* v. 196 We have been driven to Proletarian Democracy by the failure of all the alternative systems. **1925** tr. *Lenin's Proletarian Revolution* i. 11 Almost a third of this pamphlet..is devoted by this windbag to a twaddle which must be very agreeable to the bourgeoisie, as it..obscures the question of the proletarian revolution. **1934** C. Lambert *Music Ho!* IV. 248 The sleeves themselves are rolled up in the most approved proletarian fashion. **1935** W. Empson *Some Versions of Pastoral* 3 It is hard for an Englishman to talk definitely about proletarian art. **1966** *Guardian* 13 Dec. 8/2 [China] The example of the great proletarian cultural revolution. **1973** C. D. Kernig *Marxism, Communism & Western Society* VII. 238/1 The proletarian revolution becomes an act of human emancipation or the self-realization of man after his self-alienation. **1975** *Chinese Econ. Stud.* VIII. IV. 3 Countless proletarian heroes have suddenly emerged. **1976** *Times* 28 Sept. (China Suppl.) 10/3 Under the banner of 'proletarian internationalism' ideological motives seem to be distinguishable features of Chinese aid. **1977** *China Now* June 4/1 Tachai is an expression of the proletarian revolution China is in the midst of today (in contrast to the democratic revolution between 1949 and 1966). **1978** *Times Lit. Suppl.* 1 Dec. 1402/5 Sociologists and realistic novelists—including proletarian novelists—find it difficult if not impossible to describe the texture of this world.
B. *sb.* (Earlier and later examples of mod. sense.)
1870 *On the War* (Internat. Working-Men's Assoc.) 4 Mindful of the watchword of the International Working-men's Association: *Proletarians of all countries unite*, we shall never forget that the workmen of *all* countries are our *friends* and the despots of *all* countries our *enemies*. **1966** D. Wilson *Quarter of Mankind* i. 7 The peasant, not the urban proletarian, is the central character in China's drama. **1969** A. G. Frank *Latin Amer.* (1970) xxiii. 360 The rural and slum proletarians..tend to be quite near-sighted and to see only the land and the jobs which they want but don't have.

proletarianism. (Later examples.)
1918 *Nation* (N.Y.) 7 Feb. 131/2 Fidelity to the cause of international proletarianism..must stay the hands of the Bolshevik peace negotiators. **1975** D. Francis *High Stakes* v. 75 If politicians..searched diligently amongst their antecedents for proletarianism and denied aristocratic contacts..who was I to spoil the fun?

proleta:rianiza·tion. [f. Proletarianize *v.* + -ation.] The fact or process of rendering or becoming proletarian (sense c).
1920 *19th Cent.* Sept. 445 If State agriculture in Russia comes to be on a larger scale, will there not be a sort of proletarianisation of the peasants? **1936** 'H. MacDiarmid' *Lucky Poet* (1943) iii. 145 The line we advocate ..will..greatly speed up the proletarianization of Scottish Arts and Letters. **1948** C. S. Fox tr. *Röpke's Civitas Humana* III. vi. 140 Proletarianisation means..that human beings have got into a highly dangerous sociological and anthropological state which is characterised by lack of property, lack of reserves of every kind.., by economic servitude, uprooting, massed living quarters, militarisation of work..; in short, by a general devitalisation and loss of personality. **1961** L. P. Hartley *Two for River* 45 The appalling vulgarity of that town! Nowhere has the proletarianization of the English race gone so fast, or so far. **1966** F. Schurmann *Ideology & Organization in Communist China* i. 40 Given the unilinear development of history, the process of world-wide proletarianization is inevitable. **1974** *Daily Tel.* 24 May 3/2 The initiative for getting rid of the oak tables came from one or two students keen to promote the 'proletarianisation' of the college. **1977** P. Johnson *Enemies of Society* xiv. 190 The proletarianization of the British middle class, that leading creator and custodian of western civilization, is one of the most significant social changes of our times.

proletarianize, *v.* Add: (Further examples.) Hence **proleta·rianized** *ppl. a.*; **proleta·rianizing** *vbl. sb.*
1921 tr. *Rathenau's New Society* 60 To many it is not agreeable to picture to themselves the aspect of a thoroughly proletarianized country. **1937** 'C. Caudwell' *Illusion & Reality* vi. 115 Proletarianise the craftsman to the level of a labourer or machine-minder. **1949** 'M. Innes' *Journeying Boy* i. 6 The lower stratum of the intellectual class was being proletarianised. **1957** M. McCarthy *Mem. Catholic Girlhood* i. 40 The peculiarly fatigued, dusty, proletarianized character of American municipal entertainment. **1962** *Guardian* 9 Mar. 7/2 Many of the most important American cities..have become increasingly proletarianised in the worst sense. **1965** D. E. C. Eversley in Glass & Eversley *Population in Hist.* ii. 66 A fall in the death rate..accelerated the proletarianizing process. **1966** D. Wilson *Quarter of Mankind* iii. 37 The middle class has become a captive collaborator with the Communist Party, having to work for its own downfall in the interests of a reconstructed, proletarianized nation. **1974** *Daily Tel.* 15 Jan. 14/3 Now officially referred to..as 'Mr Tony Benn'. So the..process by which the man who was once called Viscount Stansgate has tried so determinedly to proletarianise himself is almost complete

1974 B. Pearce tr. *Amin's Accumulation on World Scale* I. 27 Analogous to this process is the mobilizing of the internal colonial reserves, as with the proletarianizing of blacks in the United States. **1979** *Jrnl. R. Soc. Arts* CXXVII. 354/2 The proletarianizing of the north-east coast.

proleta·rianly, *adv.* [f. Proletarian *a.* and *sb.* + -ly².] According to proletarian views.
1931 *Time & Tide* 3 Oct. 1130 This *rentier* has been unfeeling enough to practice what is capitalistically called virtue, but is proletarianly known as vice; *he has saved money!*

proletariate, -at. Add: The usual spelling is now *proletariat*. **2. a.** *fig.* (Further example.)
1881 *Nature* 24 Feb. 387/1 First..was the Sparrow, the most impudent proletariat—I had almost said Social democrat, because the whole world today has that bad word in the mouth.
b. (Earlier and later examples.) *dictatorship of the proletariat*: the Communist ideal of proletarian supremacy following the overthrow of capitalism and preceding the classless state.
1856 Geo. Eliot in *Westm. Rev.* X. 75 The Proletariat, or those who are dependent on daily wages. **1886** tr. *Marx's Manifesto of Communists* 8 The small middle class, the artisans, merchants, mechanics, shopkeepers, and farmers, are all doomed to fall into the ranks of the Proletariat, because their small capital cannot compete with that of the millionaire, and..their skill is depreciated by new modes of production. Thus the Proletariat recruits from all classes of population. **1920** M. Beer *Hist. Brit. Socialism* II. III. vi. 120 O'Brien's was the policy of a relentless class war by a proletariate that was absolutely unable to convert its votes into political power. **1927** H. Laski *Let.* 20 Nov. in *Holmes–Laski Lett.* (1953) II. 998 Did I tell you that I have traced the origins of the famous 'Dictatorship of the Proletariat' to Babeuf? **1937** E. St. V. Millay *Conversation at Midnight* ii. 69 The dictatorship of the proletariat, though not yet present and in this room, is a fact! **1941** *New Statesman* 19 Apr. 407/1 Unluckily the proletariat are even more conservative in their food than the bourgeoisie. **1964** P. G. Casanova in I. L. Horowitz *New Sociol.* 71 Mills..could not believe that Marx's view that the proletariat was the force of history could be applied to the United States of 1960. **1972** W. Leonhard in C. D. Kernig *Marxism, Communism & Western Society* II. 429/2 The term 'dictatorship of the proletariat' was probably coined in 1837 by Auguste Blanqui. *Ibid.* 434/1 After Stalin's death..a change again took place in the Soviet presentation of the dictatorship of the proletariat. **1976** T. Eagleton *Crit. & Ideology* ii. 58 In nineteenth-century England..there were sound political reasons why the proletariat should be excluded from literacy, but sound religious reasons why it should not be.

proletariza·tion. [f. F. *prolétaire* Proletaire + -ize + -ation.] = *proletarianization. Also **pro·letarize** *v. trans.*; **pro·letarized** *ppl. a.*
1918 *Times* 19 Aug. 5/6 We are drifting towards the complete proletarization of the official classes. **1920** E. Antonelli *Bolshevist Russia* I. iii. 80 The theory of 'proletarizing' the peasants was never abandoned. **1952** Gerth & Martindale tr. *Weber's Anc. Judaism* I. ii. 56 Indebted or landless, hence, proletarized Israelites. **1968** P. B. Austin *On being Swedish* xiii. 86 The sinister cleavage between highbrow and lowbrow, the proletarization of the soul which lies at the root of so many of the western world's spiritual and even material ills, is here virtually unknown. **1975** *Daily Tel.* 15 Aug. 12 The real aim is the proletarisation of society.

prolidase (prŏᵘ·lidēⁱz, -s). *Biochem.* [f. *prol(ine + Im)id(o- + *-ase.] A proteolytic enzyme which hydrolyses peptide bonds formed with the nitrogen atom of proline or 4-hydroxyproline.
1937 Bergmann & Fruton in *Jrnl. Biol. Chem.* CXVII. 191 We should like to call this enzyme *prolidase* in order to indicate that in the substrates of this enzyme the proline nitrogen is present as imido nitrogen. **1951** *Adv. Enzymol.* XII. 220 Prolidase is specifically activated by Mn⁺⁺. **1976** *Metabolism* XXV. 504 Prolidase is an important enzyme in completely degrading collagen to free amino acids. Prolidase is present in human intestinal mucosa.

pro-life, -lifer: see *pro-¹ 5a.

proliferate, *v.* Add: **1. b.** Also more widely: to give rise to an increasing number of offspring, to reproduce prolifically.
1926 *Socialist Rev.* Apr. 33 Those who..have neither the time to care for their children's moral well-being, nor the space which is necessary to ensure their physical welfare, proliferate, unchecked. **1955** *Sci. Amer.* May 32/3 Normal man carries throughout life a host of microbes which now and then start proliferating and cause disease.
d. *gen.* Of things, happenings, etc.: to increase greatly in numbers; to be(come) rife.
1961 *Daily Tel.* 19 Jan. 12/3 There was more than a suspicion that the system which had gradually evolved over the years was not able to produce the highest quality in British football—a quality which becomes more and more indispensable as international matches proliferate. **1972** *Sci. Amer.* Feb. 13/3 Indeed, microwave devices may proliferate as much as television sets have proliferated. **1976** *National Observer* (U.S.) 21 Aug. 6/6 In recent years, cut-rate sporting goods have proliferated to appeal to the less fussy adventurer.
2. b. To produce (esp. nuclear weapons) in large quantities.
1971 *Nature* 24 Dec. 493/2 Both superpowers continue to proliferate nuclear weapons. **1974** *Daily Tel.* 24 June 1/2 The Government strongly supports international agreement not to 'proliferate' nuclear weapons.

proliferating, *ppl. a.* (Further examples.)
1964 *Time* 4 Dec. 21/1 In Europe, U.S. diplomats are still trying to promote a multilateral nuclear fleet (MLF) as an alternative to proliferating national forces. **1967** *Economist* 25 Feb. 698/2 Indeed, the case for Britain (and France) going non-nuclear would become more persuasive after the signing of the treaty than it would in a treatyless, proliferating world. **1968** *Ibid.* 13 Apr. 41/1 Those who struggle on in the hospital service, taking the necessary diplomas of the proliferating specialist colleges.

proliferation. Add: **1. c.** *transf.* Enlargement or extension; an increase in number (*of*); now esp. of nuclear weapons.
1920 H. G. Wells *Outl. Hist.* II. 641/2 The British and French were at first the leading peoples in this great proliferation of knowledge. **1955** *Sci. Amer.* May 88/2 With the proliferation of cults came a tendency to worship privately without the intercession of priests. **1966** *Listener* 4 Aug. 177/3 One of the most noticeable results of setting up a Royal Commission has been the proliferation of proposals for legal solutions to industrial relations problems. **1966** Schwarz & Hadix *Strategic Terminol.* 89 *Proliferation*, in the context of nuclear strategy the increase of the number of states possessing independent national control over nuclear weapons, by the acquisition and development of the necessary technological capacity and industrial capabilities. **1967** *Observer* 5 Mar. 13/5 The issue is whether proliferation of bombs can be prevented without fettering peaceful nuclear developments. **1972** *Sci. Amer.* Feb. 15/3 Most of the problems are related to the fact that a large proliferation of microwave devices would make heavy demands on part of the electromagnetic spectrum. **1976** *National Observer* (U.S.) 21 Aug. 5/3 Is the United States' role..so dominant in the nuclear field that we could stop proliferation if our own policies were strict and consistent?

proliferative, *a.* (Examples of general use.)
1952 W. D. Jacobs *William Barnes* ii. 16 Words of the Anglo-Saxon vocabulary and their proliferative quality. **1969** E. Bishop *Compl. Poems* 83 Our problems Becoming helplessly proliferative.

proliferator (proli·fĕrēⁱtəɹ). [f. Proliferate *v.* + -or.] One that proliferates; *esp.* one that advocates or engages in the production of nuclear weapons.
1974 *Nature* 31 May 397/2 If the present nuclear countries had busied themselves more over nuclear arms control in the past 5 years, they would have a better moral platform from which they could harangue proliferators.

proliferent (proli·fĕrĕnt), *a.* *nonce-wd.* [f. as Proliferous *a.*: see -ent.] Prolific.
1922 Joyce *Ulysses* 377 Solicitude for that proliferent continuance.

prolificacy. (Later examples.)
1926 V. A. Rice *Breeding & Improvement of Farm Animals* v. 61 'Prolificacy' implies especially frequent or numerous production. **1962** G. MacEwan *Blazing Old Cattle Trail* i. 3 A strong point about the longhorns was their prolificacy. **1971** *Farmer & Stockbreeder* 23 Feb. 24/2 Mr. Doane's recipe for increasing prolificacy of a flock included selection of rams with sex drive and fertility.

prolifically, *adv.* (Further examples.)
1915 C. S. Jones *Hohenzollerns* 167 He had for many years sought to win the favour of the great Frederick by writing prolifically on agriculture. **1947** O. Barfield in *Ess. presented to Charles Williams* 113 We owe them all to tarning, a process which we find prolifically at work wherever there is poetry.

proline (prŏᵘ·līn). *Biochem.* Formerly also **prolin.** [ad. G. *prolin* (Fischer & Suzuki 1904, in *Ber. d. Deut. Chem. Ges.* XXXVII. 2843), contraction of *pyrrolidin* *pyrrolidine.] A colourless, crystalline amino-acid, pyrrolidine-2-carboxylic acid, $C_5H_9NO_2$, the lævorotatory form of which is a constituent of most proteins.
1904 *Jrnl. Chem. Soc.* LXXXVI. I. 917 (*heading*) Derivatives of prolin. *Ibid.*, The proline used was derived from gelatin. **1911** *Jrnl. Biol. Chem.* IX. 205 Prolin is ordinarily determined in the ester hydrolysis by alcoholic extraction of the amino-acids whose esters boil below 90° at less than 1 mm. pressure. **1913** *Ibid.* XIII. 513 Proline, when added to blood used for perfusing a surviving dog's liver, leads to no increase in the normal formation of acetoacetic acid, nor is the acetoacetic acid excretion of glycosuric animals markedly increased by the administration of proline. **1946** *Nature* 5 Oct. 474/2 All the proline residues appear to be situated at points roughly one third and two thirds along the length of the molecule. **1954** A. White et al. *Princ. Biochem.* vii. 109 Proline is present in all proteins which have been studied. **1973** Biggs & Woodson *Clin. Biochem.* iii. 38 All amino acids—with the exception of proline, which has an imino group—have an amino group on the α carbon.

prolix, *a.* **1. a.** Delete †*Obs.* and add later example.
1973 M. Amis *Rachel Papers* 8 Mother's was a prolix and generally rather inelegant parturition.

Prolixin (proli·ksin). *Pharm.* A proprietary name in the U.S. for fluphenazine hydrochloride, $C_{22}H_{26}F_3N_3OS.HCl$, a phenothiazine derivative used as a tranquillizer.
1959 *Dis. Nervous System* XX. 170/2 The fluphenazine used in this study was supplied..under the trade name Prolixine [*sic*]. **1960** *Official Gaz.* (U.S. Patent Office) 1 Mar. TM 8/2 Olin Mathieson Chemical Corporation, New York... *Prolixin.* For central nervous system depressant preparations. First use Sept. 18, 1959. **1972** A. Goth *Med. Pharmacol.* (ed. 6) xix. 217 The phenothiazine antiemetics include chlorpromazine (Thorazine), fluphenazine dihydrochloride (Prolixin; Permitil), perphenazine (Trilafon), [etc.]. **1973** *Black Panther* 23 June 13/2 Do these drugs include Prolixin, a powerful mood stabilizer which includes side effects in from 50 to 100% of the cases that include pseudo-Parkinson's Disease—painful tremors and rigidity, restlessness ('the prolixin shuffle'), muscle contractions, spasms of the back, neck and throat, [etc.]?

prolly (prǫ·li), *adv.* Representation of a colloq. pronunciation of 'probably'.
1962 J. D. MacDonald *Key to Suite* vii. 112 The girls have dresses on today. I guess you prolly noticed. **1969** K. Giles *Death cracks Bottle* iv. 37, I don't know wot 'appen to it. The mice prolly. **1976** N. Thornburg *Cutter & Bone* i. 29 'Some big old buck nigger prolly do it,' Ronnie said.

prolly, var. *proly *sb.* and *a.*

proloculus (prolǫ·kiŭləs). *Zool.* Also *erron.* **proloculum.** [f. Pro-² 1 + Loculus.] In Foraminifera, the first chamber formed by the zygote.
1928 J. A. Cushman *Foraminifera* ii. 12 The initial chamber in the foraminifera is known as the proloculum. **1935** Twenhofel & Shrock *Invertebr. Paleontol.* ii. 36 The zygote secretes a small shell around itself, the proloculum of the microspheric shell. **1952** R. C. Moore et al. *Invertebr. Fossils* ii. 40/2 The size of the proloculus of megalospheric forms varies considerably in certain species. **1975** *Nature* 18 Sept. 208/2 In random thin sections these seem to arise from a cluster of small chambers or from a large initial chamber which may be the proloculus.

prolong, *sb.* Restrict †*Obs.* to sense in Dict. and add: **2.** A prolongation.
1905 *Electrochem. & Metall. Industry* III. 9/1 This product..is a by-product with the European smelters, who use sheet-iron 'prolongs' on the condensers to collect it.

prolongability (prolǫŋăbi·lĭti). [f. Prolongable *a.* + -ity.] Capacity to be prolonged or lengthened.
1950 D. Jones *Phoneme* 183 'Prolongability'..is an infallible criterion for determining which type of length a vowel or diphthong has.

prolongedly (prolǫ·ŋėdli), *adv.* [f. Prolonged *ppl. a.* + -ly².] At length, extensively, over a long period.
1934 F. Rolfe *Desire & Pursuit of Whole* xxiv. 268 Crabbe plastered a double-handful of the glutinous rice, firmly and prolongedly, over the sot's face. **1972** G. Jones *Kings, Beasts, & Heroes* p. xvii, The *Beowulf* poet's business is with a non-divine hero of wondertale descent whom he associates closely, constantly, and prolongedly with the antecedents of northern tribal history.

prolongingly (prolǫ·ŋiŋli), *adv.* *rare*⁻¹. [f. Prolonging *vbl. sb.* + -ly².] At length, prolongedly.
1851 H. Melville *Moby Dick* III. xlii. 243 A voice that prolongingly moulded every word.

proluse (prolⁱū·z), *v.* *nonce-wd.* [Back-formation from Prolusion.] *intr.* To give an introductory discourse; to prolusionize.
1917 Kipling *Diversity of Creatures* 330 This they permitting, he, emboldened thus, Prolused of humankind promiscuous.

proly (prŏᵘ·li), *sb.* and *a.* Usu. *derogatory.* Also **prolly.** [f. Proletarian *a.* and *sb.*] **A.** *sb.* = *prole *sb.*
1959 J. Cary *Captive & Free* 84 Hooper, son of a lorry driver, who worked his way through the grammar school to a varsity, had, like many of his type, a great contempt for the class from which he had sprung. He called it the prolies. He thought of a proly as a born slave and parasite. **1969** E. McGirr *Entry of Death* iii. 64 The politicians say that the prollies prefer to sit back and wait..rather than take direct haction. **1970** K. Giles *Murder Pluperfect* viii. 166 Our Lady might have been..loving the prolies.
B. *adj.* Of working-class origin; = *prole *a.* Also in phr. *prolier-than-thou*, after *holier-than-thou* (see *holy *a.* 5c).
1971 *Listener* 2 Sept. 311/3 This 'prolier than thou' account. **1972** C. Drummond *Death at Bar* iii. 69 It was one thing..for people of decent family and quite another for prolly sergeants. **1977** *Time Out* 28 Jan.–3 Feb. 3/3 The prolier-than-thou sectarianism of the new convert to Marxism is all too evident in her letter.

prom (prǫm). *colloq.* [Abbrev. of Promenade *sb.*] **1.** *U.S.* = *promenade *sb.* 2c.
1894 D. Morrow in H. Nicolson *Dwight Morrow* (1935) 41, I have..been invited over to the Smith Junior Prom. **1899** A. H. Quinn *Pennsylvania Stories* 170 All you children can get tickets of me for the Senior Prom right now. **1914** G. Atherton *Perch of Devil* i. 74 The Prom is anything but an exclusive affair. **1924** [see *drag *v.* 1 e]. **1930** P. W. Slosson *Great Crusade* (1931) xii. 342 Dancing was a universal convention, and the formal balls, 'proms' and 'hops' became extremely expensive affairs. **1956** [see *jump *sb.*¹ 1 d]. **1972** T. P. McMahon *Issue of Bishop's Blood* (1973) vii. 88 A girl..who went into a decline because she hadn't been invited to the Junior Prom at Santa Clara University. **1977** I. Shaw *Beggarman, Thief* I. vi. 76 Girls who..went..to the proms at which he played the trumpet in the band.
2. = *promenade concert* (s.v. Promenade *sb.* 4b); *the Proms*, the Henry Wood Promenade Concerts, now given annually at the Royal Albert Hall, London (also in *sing.*).
1902 *Free Lance* 4 Jan. 358/1 There is never one of the programmes at the Proms..unworthy of the..most cultured music lover. **1913** H. Walpole *Fortitude* II. i. 180 A walk or two and going into the gallery at Covent Garden once or twice and the Proms sometimes. **1927** *Morning Post* 16 Aug. 6/7 (*heading*) Mozart's music at the 'Proms'. **1930** *Daily Express* 8 Sept. 6/2 Faces are certainly funny things—as you remarked when we went to the 'Prom.' at the Queen's Hall the other night. **1944** *Times* 10 June 5/4 While the L.S.O...has continued in the path it designed for itself..the Proms have continuously evolved. **1945** *Ann. Reg. 1944* 369 A concert in celebration of the fiftieth anniversary of the first 'Prom'. **1955** *Times* 19 Aug. 2/5 During his lifetime Henry Wood used the Proms as a stamping ground for new music. **1958** *Times* 20 June 3/6 This year is Puccini's centenary and, for the first time in a Prom, a whole act from one of his operas is included. **1973** *Listener* 14 June 810/1 Henry Wood decreed that the nine symphonies of Beethoven should be the backbone of the Proms. **1976** *Ibid.* 8 Apr. 440/1 The word 'Prom' no longer has anything to do with promenading: it means something like the Albert Hall Proms—young people of the student class enjoying high culture in a spirited, informal manner.
3. = Promenade *sb.* 2a in Dict. and Suppl.
1909 J. A. Glover-Kind *I do like to be beside Seaside* (1970) (song) 4, I do like to stroll upon the Prom, Prom, Prom. **1910** *Bradshaw's Railway Guide* Apr. 1142/2 Charmingly situated on the West Prom., directly facing the Sea. **1925** *Glasgow Herald* 18 July 8/7 A scrap of conversation overheard on the 'prom' of a well-known resort. **1953** *Manch. Guardian* 21 Dec. 7/8 (Advt.), Southport.—Clifton Hotel, Prom.: A.A., R.A.C.: 86 bedrooms, lift, garage. **1973** P. Lovesey *Mad Hatter's Holiday* i. 7 Moscrop joined the general movement in the direction of the West Pier..one of a long parade of freshly-arrived visitors taking that first bracing turn along the prom.
4. *attrib.* and *Comb.*, as (sense 1) *prom dress, girl, night*; (sense 2) *prom concert, -goer.*
1973 'S. Harvester' *Corner of Playground* II. iii. 93 She had queued to see ballet at Covent Garden, gone to Prom concerts, visited art galleries. **1976** *National Observer* (U.S.) 26 June 17/4 Prom dresses, cheers, put-downs, yearbook inscriptions...we remember them all. **1978** *Detroit Free Press* 5 Mar. (Spring fashion Suppl.) 19 (Advt.), The prom dress is back... In a big, beautiful way. **1894** *Outing* XXIV. 68/2 For two days..in January the room is crowded with 'Prom' girls and their escorts. **1947** *Penguin Mus. Mag.* II. 37 There is one body of Prom-goers who..are seen and heard too much,..a source of continual irritation to the serious Promenader. **1971** *Guardian* 26 Aug. 10/1 Tonight promises to be an adventurous evening for Prom-goers. **1974** *State* (Columbia, S. Carolina) 3 Mar. 6-F (Advt.), Sugarplum styles—beribboned with satin or velvet, strewn with flowers..to make your prom night unforgettable.

promarketeer: see *pro-¹ 5b.

promastigote (promæ·stigŏᵘt), *a.* and *sb.* *Zool.* [erron. f. Gr. πρό (? taken as opp. Gr. α-, in *amastigote*) + μαστιγ-, μάστιξ whip (used to render Flagellum) + -ωτης (see -ot² and -ote).] (Applied to) the flagellated form assumed by parasitic protozoans of the genus *Leishmania* when carried by arthropods.
1971 *Jrnl. Parasitol.* LVII. 626/2 Promastigote (leptomonad) flagellates..were from 2 strains [of *Leishmania hertigi*]... Body width frequently tapering rather abruptly.., similar to some promastigotes of *L. adleri*. **1974** *Nature* 7 June 588/1 Examination of the promastigotes of *L. hertigi* immediately revealed cytoplasmic inclusions of a type which had not been observed in any of our previous studies. **1976** *Ibid.* 19 Aug. 689/1 We present evidence that transformation from the mammalian (amastigote) stage of *L. donovani* to the insect and/or culture (promastigote) stage is inhibited by host hamster spleen homogenates.

promazine (prŏᵘ·măzīn). *Pharm.* [f. Pro(pyl + M(ethyl + *-azine* (f. Az(o- + -ine⁵).] The compound 10-(3-dimethylaminopropyl)phenothiazine, $C_{17}H_{20}N_2S$, which is used (in the form of its white crystalline hydrochloride) as a tranquillizer.
1956 *Jrnl. Amer. Med. Assoc.* 5 May 45/1 Treatment with promazine satisfactorily relieved the withdrawal symptoms of acute alcoholic intoxication in all patients who remained in the hospital for the complete course of therapy. **1958** *Martindale's Extra Pharmacopœia* (ed. 24) 398 Promazine has actions similar to those of chlorpromazine. It has been used for its tranquillising effect in psychotic conditions, especially those associated with hyperactivity, in acute alcoholism and delirium tremens, and in the withdrawal symptoms of drug addiction. **1965** J. Pollitt *Depression & its Treatment* iv. 46 Admission to hospital for continuous narcosis may be necessary. This may be induced by four-hourly doses of barbiturates, but these should give way as soon as possible to phenothiazine derivatives such as chlorpromazine or promazine. **1971** [see *promethazine].

promenade, *sb.* Add: **2. a.** Now most freq. a paved walk raised alongside the beach at a seaside resort.
1892 B. Potter *Jrnl.* Mar. (1966) 234, I was disappointed with the Hoe. It is exactly like the grounds of the Naval Exhibition, broad asphalt promenades, cigar kiosks, and even the lighthouse all complete. **1899** P. E. Amy *Beautiful Jersey* 69/2 *Promenades.* The chief of these is the Esplanade, a marine promenade, practically a continuation of Albert Quay. **1938** G. Greene *Brighton Rock* I. i. 5 The ghost train diving between the grinning skeleton [*sic*] under the Aquarium promenade, the sticks of Brighton rock, the paper sailors' caps. **1939** *Blackpool Official Guide*, Blackpool's promenades cover the whole seaboard of seven miles in length. *Ibid.*, There is no other seafront which combines such a variety of scenery as Blackpool's famous promenade. **1958** J. Betjeman *Coll. Poems* 117 Prepare for an evening of dancing and cards And forget the sea-breeze on the dry promenades. **1977** *Lancashire Life* Dec. 57/1 Along the promenade at that point there were cannons at intervals.
b. *spec.* A gallery at a music-hall, frequented by demi-mondaines and their followers.
The two most notorious promenades were at the Empire and the Alhambra music-halls in Leicester Square.
1863 *Observer* 18 Jan. 6/2 The Alhambra, of all the music halls, is the one least entitled to use the name... The balcony is converted into a promenade for loose women and the simpletons who run after them. **1899** Beerbohm *Around Theatres* (1953) 32 We bought two seats... We passed, on our way to them, into the far-famed Promenade. **1906** A. Bennett *Whom God hath Joined* i. 20, I saw the great Charlie the other night..in the promenade at the Empire. **1914** C. Mackenzie *Sinister St.* II. iv. ii. 870 Michael reached the Orient Palace of Varieties, and..joined the throng of the Promenade. *Ibid.* 872 On the Promenade where it was quite certain that every woman had a history to account for her presence there, how utterly living had quenched life. **1915** Kipling in *Nash's Mag.* Oct. 133/1 He was as communicative as—as a lady in the Promenade. **1918** A. Bennett *Pretty Lady* i. 2 Behind the audience came the restless Promenade, where was the reality which the stage reflected. **1964** C. Mackenzie *My Life & Times* III. 236 A flash tart who frequented the promenades of the Empire or the Alhambra. **1979** R. Blythe *View in Winter* iv. 182 The wonderful Empire in Leicester Square..the promenade there..was frequented by the first-class ladies of the town.
c. A ball or dance at a school or college. *U.S.*
1887 *Lippincott's Mag.* Aug. 298 The most important society event of the year is the Junior Promenade. **1905** *N.Y. Herald* 22 Jan. 10 The fair guests invited to the Junior Promenade, the great event of the college year. **1933** *Fortune* Aug. 90/3 True jazz..is even losing its great popularity at college promenades. **1972** *Lebende Sprachen* XVII. 35/2 US promenade—BE school dance, college dance.
d. *Dancing.* (See quots.) Also *promenade position*.
1953 K. Ambrose *Beginners, Please!* vi. 43 When a pivot is made on one leg (as in Figs. 5–6–7..) the movement is termed a *promenade*. **1956** J. C. Milligan *101 Scottish Country Dances* 28 *Promenade for three couples* is a formation, not a method of progression. **1957** G. B. L. Wilson *Penguin Dict. Ballet* 222 *Promenade*, (1) a slow turn on one foot while the body is held in a set position, such as an arabesque; (2) a slow turn in a pas de deux when the danseuse, on point, is turned round by her partner. **1967** Chujoy & Manchester *Dance Encycl.* 747/1 *Promenade position*, in ballroom dance, a couple in closed position moving sideways to the left, with the side of the foot leading. **1968** J. C. Milligan *Introd. Scottish Country Dancing* 51 Promenade can be done by two, three or four couples. It is done in both reel and strathspey time and takes eight travelling steps.
3. Short for *promenade deck*: see sense 4a.
1845 *Knickerbocker* XXV. 61 On the upper deck the engineers and sailors, ladies, emigrants and gentlemen, sat side by side upon the single seat which ran all round the promenade. **1873** W. D. Howells *Chance Acquaintance* i. 1 On the forward promenade of the Saguenay boat ..Miss Kitty Ellison sat. **1974** 'G. Black' *Golden Cockatrice* iv. 74, I..went up one deck..going out on to the open promenade which was empty.
4. a. *promenade deck* (earlier and later examples); **b.** **promenade band** (further example); **promenade concert**, add to def.: or at which a proportion of the audience stands; (earlier and later examples).
1893 *World* 11 Oct. 23/2 Long before the run of a successful Savoy opera is over Sir Arthur's melodies are dinned into our ears by every promenade band and street piano. **1839** Dickens *Let.* 9 Feb. (1965) I. 640 Kate at the Promenade Concert. **1921** *Daily Colonist* (Victoria, B.C.) 24 Mar. 4/1 It should combine the features of a convention hall with the facilities for holding promenade concerts at which at least 3,000 persons can be accommodated. **1954** *Grove's Dict. Mus.* (ed. 5) IX. 357/1 On 30 June 1944 the Promenade Concerts were closed down..on account of flying bombs, but they were replaced on 1 July by broadcasts from Bedford. It was there that Wood conducted his last performance on 28 July. **1962** *Listener* 2 Aug. 189/3 The programme engineer at the Promenade Concerts noticeably reduced the volume of cheers at the conclusion of Mahler's Third Symphony. **1829** *Amer. Traveller* (Boston) 14 Apr. 2 The engraving above, exhibits what may emphatically be termed a *Land Barge*.. with a cabin, berths, &c. below; a promenade deck, awning, seats, &c. above. **1931** M. de la Roche *Finch's Fortune* vi. 114 He strode out with the best of them on the promenade deck. **1973** 'D. Mariner' *Beaufort Dossier* x. 177 The promenade deck was really a broad passage... Its roof was the..boat-deck overhead.

promenade, *v.* Add: **2.** (Earlier and later examples.)
1818 T. Brown *Brighton* II. i. 22 Their more fortunate comrades *promenade* Hyde Park, or the Mall. **1977** *Lancashire Life* Aug. 33/2 Promenade Blackburn's bustling modern centre today, and you might find anxiety over a new town rival puzzling.

promenader. Add: **b.** *spec.* One who attends a promenade concert; *esp.* one who stands.
1889 G. B. Shaw in *Star* 19 Aug. 2/4 With the floor.. hidden by a crowd of promenaders too closely packed to promenade. **1918** A. Bennett *Roll-Call* I. ii. 36 Promenaders promenaded in and out of the corridor. **1959** *Listener* 29 Oct. 716/1 Even the Promenaders..do not respond joyfully to a reduction in the large share of familiar classics. **1966** K. S. Sorabji in 'H. MacDiarmid' *Company I've Kept* ii. 65 That archetype of the 'democratic' audiences, the Promenaders. **1977** *Times* 23 July 14/6 There was general enthusiasm among the young promenaders for the English flavour of this year's jubilee season.

prometaphase (prŏᵘme·tăfēⁱz). *Cytology.* [f. Pro-² + *metaphase* s.v. *meta- 4.] The stage in mitotic or meiotic nuclear division, following prophase and preceding metaphase, during which the spindle is formed and the chromosomes become oriented towards it.
1931 W. J. C. Lawrence in *Cytologia* II. 361 The degree of repulsion gradually diminishes and diakinesis is abruptly terminated by the sudden converging of the chromosomes on the centre of the nucleus. This stage (pro-metaphase) is very brief and is characterised by the secondary association of a number of the bivalents. **1969** Brown & Bertke *Textbk. Cytol.* xix. 404/1 Evidence at prometaphase and anaphase seems to indicate that the centromere is 'pulled' toward the pole or poles, the arms of the chromosome apparently passively following along.

promethazine (prome·þăzin). *Pharm.* [f. Pro(pyl + Meth(yl + *-azine* (f. Az(o- + -ine⁵).] A bitter-tasting antihistamine compound, 10-(2-dimethylaminopropyl)phenothiazine, $C_{17}H_{20}N_2S$, which is used chiefly as an anti-emetic and a sedative, usu. in the form of its hydrochloride, a white powder.
1951 *Lancet* 26 May 1143/2 'Avomine' (promethazine 8-chlorotheophyllinate) is a safe and effective remedy for seasickness. **1959** *Which?* July 69/2 All that is certain is that hyoscine, meclozine, cyclizine and promethazine do act to prevent travel sickness, and all have some side-effects **1961** *Lancet* 23 Sept. 688/1 To control shivering a mixture of promethazine hydrochloride 50 mg., pethidine 50 mg., and chlorpromazine 50 mg. was made up to 20 ml. 2–4 ml. of this mixture was administered intravenously as required. **1971** B. R. Jones *Pharmacol.* iii. 15 Post-operative vomiting is reduced by including in the premedication an intramuscular injection of either an antihistamine such as Promethazine ('Phenergan') or a tranquillizer of the phenothiazine type such as Promazine.

promethea (promī·þiă). [L. *Promēthea*, fem. of adj. *Promētheus*: see Prometheus.] In full, *promethea moth*. The North American silk moth, *Callosamia promethea*, of the family Saturniidæ.
1889 *Cent. Dict.* s.v. Prometheus. **1901** M. C. Dickerson *Moths & Butterflies* 119 Late in fall and winter we may find brown cocoons containing the chrysalides of the Promethea moth swinging from the branches of the wild cherry. **1905** V. L. Kellogg *Amer. Insects* xiv. 422 The promethea-moth.., light reddish brown in female, and blackish and clay colour in male.., is perhaps the most abundant of all these giant moths. **1909** G. Stratton-Porter *Girl of Limberlost* xv. 300 He found a splendid Promethea on a lilac in a corner. **1949** *Nat. Hist.* Feb. 80/2 The Cecropia and Promethea have their wings wrapped around their bodies. **1954** Borror & DeLong *Introd. Study of Insects* xxvi. 507 The promethea moth, *Callosamia promethea* (Drury), is sometimes called the spice-bush silk moth. **1972** Swan & Papp *Common Insects N. Amer.* 268 Promethea moth... Canada to Florida, west to the Great Plains.

Prometheanism (promī·þiăniz'm). [f. Promethean *a.* + -ism.] Conduct or policy resembling that traditionally ascribed to Prometheus.
1957 N. Frye *Anat. Crit.* 157 Patterns of the 'Romantic agony', chiefly sadism, Prometheanism. **1976** M. J. Lasky *Utopia & Revolution* (1977) xi. 412 He [*sc.* Locke] disdained the incendiaries of the age with their fiery images, and deplored the whole tradition of Prometheanism.

prometheus. 2. a. Substitute for def.: = *promethea. (Examples.)
1889 in *Cent. Dict.* **1972** Swan & Papp *Common Insects N. Amer.* 265 Efforts to reel or card the fibers of the Prometheus and Polyphemus cocoons have not been successful.

promethium (promī·þiŏm). *Chem.* orig. **-eum.** [mod.L.: see Prometheus and -ium.] An artificially produced metallic element (traces of which have subsequently been found in nature) which is a lanthanide whose longest-lived isotope has a half-life of about 18 years. Atomic number 61; symbol Pm. Cf. *illinium.
1948 Marinsky & Glendenin in *Chem. & Engin. News* 9 Aug. 2348/2 We propose, therefore, the name 'prometheum' (symbol Pm) for element 61 after Prometheus, ..who stole fire from heaven for the use of mankind... This name..symbolizes the dramatic way in which the element may be produced in quantity as a result of man's harnessing of the energy of nuclear fission **1949** *Sun*

(Baltimore) 22 Sept. 21/4 An element 500 times 'hotter' than radium is being studied... It is promethium, one of the rare earth metals. **1955** T. D. O'Brien et al. in Sneed & Brasted *Comprehensive Inorg. Chem.* IV. vi. 184 Promethium comes between neodymium and samarium in the lanthanide series. It has an oxidation number of +3 and possibly +4. **1959** Seaborg & Valens *Elements of Universe* 126 Promethium was discovered as a result of its artificial production in the atomic reactor. *Ibid.*, Promethium nitrate is a rather undistinguished-looking powder. **1976** Cotton & Wilkinson *Basic Inorg. Chem.* xxvi. 450 Promethium occurs only in traces in U ores as a spontaneous fission fragment of ^{238}U. Milligram quantities of pink $^{147}Pm^{3+}$ salt can be isolated..from fission products in spent fuel of nuclear reactors.

prometryne (prō^u·mĕtrəin). *Agric.* Also **-tryn** (-trin). [f. Pro(pyl + Me(thyl + *-tryne* (f. Tri(azi)ne).] A herbicide, 2,4-bis (isopropylamino)-6-methylthio-1,3,5-triazine, $C_{10}H_{19}N_5S$, which is usu. employed in the form of a wettable powder against annual grasses and broad-leaved weeds.

1961 *B.S.I. News* July 28 Common names for pesticides... Prometryne. 4,6-bisisopropylamino-2-methylthio-1,3,5-triazine. **1968** *Weed Control* (Nat. Acad. Sci., Washington) x. 181 Prometryne..has..a melting point of 118°C, and a solubility of 48 ppm in water. It is highly soluble in many organic solvents. **1971** *Arable Farmer* Feb. 15/3 Resistance of..peas to prometryne..occurs as a result of such metabolic conversions within the plant. **1976** J. R. Plimmer in Kearney & Kaufman *Herbicides* (ed. 2) II. xix. 925 Granular formations of prometryn did not cause vapor damage to cotton plants exposed to their vapor.

Promin (prō^u·min). *Pharm.* Also **promin.** A proprietary name in the U.S. for a glucoside derivative of 4,4′-diaminodiphenyl sulphone, $C_{12}H_{12}N_2O_2S$, which has a bacteriostatic action and has been used esp. to treat leprosy.

1937 *Official Gaz.* (U.S. Patent Office) 29 June 1008/2 Parke, Davis & Company, Detroit... *Promin.* For p-amino-benzene-sulfonamide used in treatment of streptococcic infections. Claims use since April 16, 1937. **1939** *Canad. Med. Assoc. Jrnl.* XL. 319/1 The following compounds have been investigated: p-aminobenzene sulphonamide (sulphanilamide),..a glucoside derivative of 4:4′ diamino diphenyl sulphone (promin). **1948** *Call-Bulletin* (San Francisco) 12 Apr. 3/6 Diasone and two companion drugs, promin and promizole, have shown promise against both tuberculosis and leprosy but they have some disadvantages. **1958** *New Scientist* 11 Sept. 798/2 Eventually it was found by Dr. John Lowe in 1948 that the simpler parent compound—diamino diphenyl sulphone..(now called dapsone) was more active than promin. **1971** W. H. Jopling *Handbk. Leprosy* vi. 52 Dapsone..had the advantage over Promin of being much cheaper and of being suitable for oral administration (Promin has to be given intravenously).

prominence, *sb.* Add: **3. b.** *Phonetics.* The degree to which a sound or syllable stands out from its phonetic environment.

1929 I. C. Ward *Phonetics of Eng.* xiv. 135 The effect of prominence is produced by the very intimate combination of length, stress, pitch, and inherent sonority of sounds. **1949** C. E. Bazell in *Travaux du Cercle Linguistique de Copenhague* V. 83 When two features..are simultaneous in overt order, no distinctions of prominence (e.g. syllabic/asyllabic) can be made. **1950** D. Jones *Phoneme* 137 Prominence is an effect perceived objectively by the hearer. It is thus quite a different thing from stress which is a subjective activity on the part of the speaker. *Ibid.* 144 It is a natural tendency of English people and speakers of other stress languages to attribute all prominences to stress alone. **1962** A. C. Gimson *Introd. Pronunc. Eng.* 219 Among the vowels prominence increases as the vowel becomes more open. *Ibid.* 221 Sound qualities also contribute to an impression of prominence but mainly through the characteristic relationship of certain qualities with unaccented syllables and others with accented syllables. **1973** *Word 1970* XXVI. 62 Not every syllable given accentual prominence in the sentence is a rhythmic accent with temporal prominence.

prominent, *a.* (*sb.*) **B.** *sb.* **1. b.** Delete †*Obs. rare* and add later N. Amer. examples.

1948 *Sun* (Baltimore) 19 June 3/2 Hillcrest Country Club, a favorite among movie prominents, was blown up today by an accumulation of gas. **1975** *Maclean's Mag.* (Toronto) 15 Dec. 62/2 Several prominents such as John Diefenbaker and Joey Smallwood pushed for a reprieve.

promiscuous. Add: **2. b.** Now esp.: indiscriminate in sexual relations (cf. Promiscuity 2).

1900 [in Dict.]. **1924** C. Connolly *Let.* Dec. in *Romantic Friendship* (1975) 32, I am not promiscuous but I can't be loyal to an icicle. **1937** A. Huxley *Let.* 17 Dec. (1969) 430 Where it is 'done' to attach a great deal of importance to the achievement of promiscuous satisfaction..there..will the amount of attention given to other matters decline. **1949** *Times Lit. Suppl.* 21 Oct. 680/2 She was promiscuous in her favours, and did not at once become his alone. **1955** G. Greene *Quiet American* II. ii. 130 One starts promiscuous and ends like one's grandfather, faithful to one woman. **1978** S. Herzel in P. Moore *Man, Woman, & Priesthood* viii. 119 It is precisely because men *can* compartmentalize that they are more easily promiscuous than women.

6. b. (Earlier examples.)

1826 Disraeli *Vivian Grey* II. xii. 170, I do wish you'd come in some day quite promiscuous. **1827** *Ibid.* III. viii. 115 It's remarkable wrong to tax 'em all promiscuous. **1875** Swinburne *Let.* 5 Nov. (1960) III. 82 It turned up promiscuous when at last wanted after ten years.

promise, *sb.* Add: **4. a.** (Further examples.)

1886 E. G. White *Hist. Sk. Foreign Missions Seventh-Day Adventists* 281/2 They should be ready to counsel and instruct those who have newly come to the faith, and who give promise of possessing ability to work for the Master. **1919** T. S. Eliot *Poems*, Grishkin is nice.. her friendly bust Gives promise of pneumatic bliss. **1971** *Daily Tel.* 17 June 3/3 To police he showed 'promise' of becoming a sophisticated criminal.

5. *promise-breaking* (examples), *-keeping* (later example).

1842 Dickens *Let.* 22 Mar. (1974) III. 146 You.. have set me down, I know, as a neglectful, erratic, promise-breaking and most unworthy person. **1940** *Mind* XLIX. 231 Utilitarians have tended rather to overestimate the disastrous consequences of promise-breaking. *a* **1974** R. Crossman *Diaries* (1975) I. 249 From the point of view of the electorate this technical promise-keeping is quite unimportant.

promise, *v.* Add: **3. a.** (Later examples.)

1904 L. T. Meade *Love Triumphant* II. ix. 192 If anything could induce me to promise myself to a man..it would be to Cedric Vershoyle. **1967** C. Potok *Chosen* xii. 200 My father promised my sister to the son of one of his followers when she was two years old.

5. b. Delete 'Now *arch.* or *dial.*' and add later examples.

1963 J. Kennaway *Bells of Shoreditch* I. i. 12 He said, 'I promise you, you're looking very well.' **1969** [see *ghastly *a.* 1 b].

promised, *ppl. a.* Add: *promised land* (later examples).

1920 Wodehouse *Jill the Reckless* (1922) x. 142 The rank and file of the profession were greeted, like Moses on Pisgah, with a fleeting glimpse of the promised land, consisting of a large desk [etc.]. **1965** *Amer. N. & Q.* Apr. 128/2 Nearly 100 songs tell about travel around the Horn and across the Plains to the promised land, life in California..and the great joy of going home. **1972** D. Dakin *Unification of Greece* 268 Just when the promised land seemed to be within their grasp, the Greeks... suffered ignominious defeat. **1975** J. Cleary *Safe House* vi. 250 The farmers..had headed west to the Promised Land of California.

promissory, *a.* **1. b.** (Further examples.)

1882 [see Agent *sb.* 4]. **1911** *Oklahoma Session Laws* (3rd Legislature) 216 If any such promissory note or assessment is not paid when due, action may be brought thereon. **1960** G. Durrell *Zoo in my Luggage* iii. 79, I paid him the two shillings, and then wrote out a promissory note for the other five shillings. **1972** *Times* 18 Feb. 20/5 In 1963 UDT..accepted a number of promissory notes.

promitochondrion: see *pro-^{2} 1.

promizole (prō^u·mizō^{u}l). *Pharm.* Also **Promizole.** [f. *Promi(n + *thia)zole.] A bacteriostatic agent, 2-amino-5-*p*-aminobenzenesulphonylthiazole, $C_9H_9N_3O_2S_2$, formerly used to treat leprosy and tuberculosis. Formerly a proprietary name in the U.S.

1944 *Proc. Staff Meetings Mayo Clinic* XIX. 26 Promizole is the trade name of 4,2′-diaminophenyl-5′-thiazole-sulfone [*sic*]. **1944** *Official Gaz.* (U.S. Patent Office) 11 Apr. 191/1 Parke, Davis & Company, Detroit... *Promizole.* For chemotherapeutic agent for the treatment of bacterial infections. Claims use since Dec. 23, 1943. **1948** [see *Promin]. **1964** P. Feeny *Fight against Leprosy* xvi. 147 From every continent there was coming a steady stream of literature about the effects of promin, diasone, sulphetrone and one more weapon that had been added to the arsenal, promizole.

Prommer (prǫ·məɹ), *colloq.* Also **prommer.** [f. *prom 2 + -er^1.] One who attends a promenade concert (*esp.* at the Royal Albert Hall); a promenader.

1947 *Penguin Music Mag.* May 36 Even in 1932 a Prommer could follow a score without digging his elbow into someone every time he turned a page. **1960** *Times* 30 July 9/1 (*heading*) Prommers Stay for Gerhard. **1969** *Listener* 12 June 836/3 Traditionally, of course, your average Prommer is a catholic music-lover..who jolly well doesn't care who sees him enjoying himself on Saturdays. **1972** *Guardian* 14 Aug. 8/1 A regular Prommer who couldn't make it this particular night and is..listening at home. **1975** *Times* 19 Sept. 9/8 The finale was a young man's reading, and seemingly much to the liking of the even younger prommers.

promo (prō^u·mo), *a.* and *sb. colloq.* [Abbrev. of *promotional *a.*, *promotion 2 a.] **A.** *adj.* = *promotional *a.* **B.** *sb.* Publicity, advertising; *spec.* a promotional trailer for a television programme. *Comb.*, as **promo man,** a promotion man, publicity organizer.

1963 *Amer. Speech* XXXVIII. 156 Other stump words or clipped forms such as *info*..and *promo.* **1966** *Sat. Rev.* (U.S.) 12 Feb. 8/3 'Will Robin escape?... Will Batman arrive in time to save him? The worst is yet to come!' And sure enough, minutes later on came the second of this trilogy—*The Blue Light.* What a promo! **1971** D. E. Westlake *I gave at the Office* (1972) 11, I did promo voice-overs for the new TV shows. **1972** *Village Voice* (N.Y.) 1 June 46/1 A slightly agitated promo-man-about-town stopped me one day recently as I was crossing Sheridan Square. **1973** *Publishers Weekly* 10 Sept. 54 (Advt.), With big national TV promo by the author.. Robert Rosefsky is really doing a terrific job of promoting his book on TV. **1974** *Some Technical Terms & Slang* (Granada Television), *Promo*, a short promotional trailer for a programme. **1976** *New Musical Express* 17 Apr. 22/4 Despite promo pics that make them look like 12-year-olds, their 'charisma' is more David Cassidy style. **1978** *Advertising Age* (Chicago) 16 Oct. 19 (*heading*) Oxford promo is publicity happening.

promote, *v.* Add: **I. 1. c.** *Sport* (chiefly *Assoc. Football*). To transfer (a team) to a higher division of a league (see *promotion 1 d).

1924 *Times* 5 May 6/6 Bristol City,..promoted a year ago, return to a lower division. **1949** *Times* 25 Apr. 6/2 (*heading*) Swansea Town promoted.

d. *Curling.* To move (another stone) forward by striking.

1937 T. Henderson *Lockerbie* 58 He left the stone alone..deeming it safer play to promote the Minister's stone.

e. *Bridge.* To establish (a relatively low card) as a winner; to secure (a trick) by this action.

1959 *Listener* 31 Dec. 1178/3 A further spade lead will promote the nine of diamonds. *Ibid.*, The fifth heart will promote one of North's trumps. **1962** *Ibid.* 12 Apr. 662/2 The defence would take two rounds of clubs and play a third club, promoting a trick for West's nine of hearts.

2. a. *spec.* To further the sale of (an article) by advertising or other modes of publicity; to publicize (a venture, person, etc.). Also *absol.*

1930 *Publishers' Weekly* 31 May 2732/2 The books all to be individualized in appearance and fully promoted. **1965** *Melody Maker* 3 Apr. 7/3 With the group over here to promote their latest recording,..they could well make the chart. **1971** D. Potter *Brit. Eliz. Stamps* x. 117 These packs are heavily promoted, with full-page colour advertisements in the national press. **1976** *National Observer* (U.S.) 30 Oct. 9/3, I love chocolate-chip cookies, and I love to promote.

c. *Chem.* To increase the activity or effectiveness of (a catalyst) by addition of another substance; to act as a promoter of (a catalyst) or in (a catalytic reaction). *Loosely* (passing into 2 a), to initiate, catalyse.

[**1920** *Jrnl. Physical Chem.* XXIV. 243 When more than one of the components are themselves catalysts a difficulty presents itself in choosing between 'promoter' and 'promoted'.] **1930** N. K. Adam *Physics & Chem. of Surfaces* viii. 280 Many reactions go on at the surface of charcoal. It is a good catalyst for promoting halogenations. **1936** R. H. Griffith *Mechanism of Contact Catalysis* iii. 82 The fact that a substance may act as a poison to a catalyst, and yet itself be promoted by that catalyst, is obviously quite possible. **1940** Glasstone *Textbk. Physical Chem.* xiii. 1128 On an ordinary iron catalyst one atom only in 2,000 appears to be able to catalyze the reaction between nitrogen and hydrogen, but when suitably promoted the proportion of active points is increased ten-fold. **1946** *Chem. Abstr.* XL. 4876 The catalytic action is promoted by a smaller quantity of BF_3. **1947** *Jrnl. Polymer Sci.* II. 41 The presence of small amounts of relatively high molecular weight mercaptans greatly promotes the copolymerization reaction. **1967** R. W. Lenz *Org. Chem. Synthetic High Polymers* x. 270 *N,N*-Dimethylaniline promotes the spontaneous decomposition of benzoyl peroxide, and this combination can be used to initiate polymerization reactions at low temperatures. **1975** P. H. Emmett in Drauglis & Jaffee *Physical Basis for Heterogeneous Catalysis* 21 Why then is a K_2O-Al_2O_3 promoter better than Al_2O_3 alone in promoting an iron synthetic ammonia catalyst?

IV. 8. *slang* (orig. *U.S.*). To borrow or obtain (usu. illicitly). Also to exploit (someone) *for* material advantage.

1930 *Amer. Mercury* Dec. 457/1 *Promote*, to steal. 'We got to promote a boat to run the stuff in.' **1934** J. M. Cain *Postman always rings Twice* 97 If I hadn't been there, and begun promoting him for something to drink that afternoon, maybe he'd be here now. **1941** *Argus* (Melbourne) *Week-End Mag.* 15 Nov. 1/4 In Army parlance to arrange something is always to 'tee up'; just as to borrow something is to 'promote' it. **1942** Z. N. Hurston in A. Dundes *Mother Wit* (1973) 226/1 You skillets is trying to promote a meal on me.

promoted (promō^u·tėd), *ppl. a.* [f. Promote *v.* + -ed^1.] That has been promoted; furthered, advanced. **1.** *Chem.* Of a catalyst: containing or influenced by a promoter (sense *1 d).

[**1920**: see *promote *v.* 2 c.] **1927** *Jrnl. Amer. Chem. Soc.* XLVIII. 2821 In the case of pure iron and promoted iron catalysts for ammonia synthesis, oxygen is the type of poison which seemed best suited for such a study. **1968** E. K. Rideal *Concepts in Catalysis* vi. 120 Promoted iron catalysts used in the ammonia synthesis are generally prepared by reduction in hydrogen at 500°C of a mixture 95% Fe_3O_4 plus 4–5% Al_2O_3 plus 0–1% K_2O. **1975** P. H. Emmett in Drauglis & Jaffee *Physical Basis for Heterogeneous Catalysis* 18 The carbon monoxide chemisorption indicates that about 50 percent of the singly promoted (Al_2O_3 the only promoter) catalyst is usually covered with promoter.

2. In other senses of the vb.

1962 *Listener* 6 Dec. 975/3 A splendid, and splendidly promoted, mixture of Baedeker, gazetteer, and travellers' tales. **1977** *Cleethorpes News* 6 May 31/1 Two fine individual performances destroyed promoted Cromwell-Drewery on their Premier debut.

promotee (promō^{u}tī·). [f. Promote *v.* + -ee^1.] One who is or has been promoted.

1958 E. H. Carr *Socialism in One Country* I. I. iii. 110 The railways, whose staff adopted..an unwelcoming.. attitude to promotees. **1966** *Punch* 2 Feb. 174/3 Crimes of local importance have to be handled at higher and higher levels to provide something for the promotees from *Z Cars* to do. **1977** *Hongkong Standard* 12 Apr. 8/1 This promotion causes the promotee to lose money on the deal.

promoter. Add: **I. 1. d.** *Chem.* A less active additive which increases the activity of a catalyst; more generally, a substance which improves a catalyst in some way. Also, a substance used as an initiator in a catalytic polymerization reaction.

1911 J. Y. Johnson *Brit. Pat.* 19,249 6 Very much better yields can be obtained in the synthetical production of ammonia from its elements if there be employed, as the catalytic agent, iron in admixture with certain bodies as hereinafter explained... These bodies my foreign correspondents [*sc.* Badische Anilin & Soda Fabrik, Germany], for the sake of brevity, term 'promoters'. **1927** *Jrnl. Amer. Chem. Soc.* XLVIII. 2826 The results show that the predominant effect of promoters of the type alumina and potassium aluminate is to increase the number of catalytically active atoms relative to the total number of metal atoms present. **1930** N. K. Adam *Physics & Chem. of Surfaces* viii. 241 Many promoters are simply refractory supports for a metallic catalyst. **1961** J. N. Anderson *Appl. Dental Materials* (ed. 2) xxiv. 251 When these promoters meet the benzoyl peroxide in the polymer, they start a chain of events similar to that which occurs when heat is applied. **1963** P. H. Plesch *Chem. Cationic Polymerization* iv. 149 The patent literature contains a great many combinations of a metal halide and a co-catalyst (often called promoter, especially in American patents), most of which are substances which can combine with the metal halide to form a protonic acid. **1970** G. Odian *Princ. Polymerization* vii. 464 The reactive cyclic ether used as a component of the catalyst system is referred to as a promoter (or a cocatalyst). **1971** *Sci. Amer.* Dec. 52/3 It has been found that iron plus a few percent of the oxides of potassium and aluminum, which are known as promoters, give a longer-lived catalyst and one with greater resistance to impurities in the feed stream.

e. One who organizes or actively supports a sporting event, entertainment, etc., esp. for profit.

1936 [see *football pool* s.v. *football 4]. **1950** *Sport* 7–11 Apr. 22/3 This would involve the full co-operation of sports promoters and the B.B.C. **1951** *Sunday Pictorial* 21 Jan. 16/3 He's a promoter's dream and has sold £450 worth of tickets for this outing. **1956** B. Holiday *Lady sings Blues* (1973) xxii. 180 If it had been left to the management and promoters, I could have shot myself long ago. **1964** *Melody Maker* 28 Nov. 3 Deejays and promoters must stop being idiots. **1971** *Daily Tel.* 27 May 2/6 The Isle of Wight County Council last night rejected all three farm sites proposed by Richard Roscoe, a promoter, for staging a pop festival.

f. *Med.* An agent that causes tumour promotion (*promotion 2 d) (see quot. 1978).

1947 *Brit. Jrnl. Cancer* I. 390 Dibenzanthracene is undoubtedly a potent Initiator, but a weak Promotor; benzpyrene is moderately potent both as Initiator and Promotor; croton oil, on the other hand, is exceptionally potent as a Promotor, but quite useless as an Initiator. **1969** *Progress Exper. Tumor Res.* XI. 50 When the initiating agent is given..to mice and this is followed by repeated skin application of the promoter after a lapse of as much as 380 days, tumors arise within 32–42 days from the beginning of promoting treatment. **1970** [see *promoting *ppl. a.* 2 b]. **1976** *Maclean's Mag.* (Toronto) 27 Dec. 22/3 These carcinogens don't usually cause cancer unless they mix with other agents called promoters. **1978** *Nature* 20 July 271/1 Tumour promoters are compounds which are not carcinogens but which can induce tumours in mice treated with a subcarcinogenic dose of a chemical carcinogen.

g. *Genetics.* An essential part of an operon, situated between the operator and the structural gene(s), at which transcription starts. [The sense is due to F. Jacob et al., who used F. *promoteur* (*Compt. Rend.* (1964) CCLVIII. 3128).]

1967 C. R. Woese *Genetic Code* iv. 93 The operator locus comprises both the modulation sequence and the punctuation for tape-reader attachment. Jacob and his associates..have more recently adduced evidence for the existence of the latter, which they call *promoter*, by a study of deletions. **1971** D. J. Cove *Genetics* xi. 163 Mutations in the promoter region lead to a reduction in the rate of messenger synthesis from the whole operon. The promoter is thought to be the region for the attachment of the DNA-dependent RNA polymerase, the enzyme responsible for transcription. **1975** *Nature* 13 Mar. 118/1 The promoter has been loosely defined as the site on the DNA where the RNA polymerase recognises some signal which allows it to bind tightly and initiate transcription.

II. 3. a. For †*Obs.* read '*Obs.* exc. *Hist.*' and add: *β.* (Later example.)

1955 W. W. Greg *Shakespeare First Folio* iv. 150 The Act may well have been a dead letter except for action by professional promooters.

III. 4. (Earlier and later examples.)

1699 *Edin. Gaz.* 26–29 June, Munday last being the Day appointed for the publick Graduation..Mr. William Scot..Promoter for this year declam'd an Elegant Harangue. **1858** *Min. Univ. St. Andrews* (MS.), XVII. 415 The Senatus appoint the ex-Rector to act in the meantime as pro-Rector and Promotor. **1962** *Aberdeen Univ. Rev.* Autumn 313 The other graduands who have obtained First Class Honours are called up, one by one, by the Promotor.

promoting, *vbl. sb.* Add: **2. b.** *Chem.* The action of *promote *v.* 2 c.

1936 R. H. Griffith *Mechanism of Contact Catalysis* iii. 82 The most useful oxides, to give the promoting effect with ion-oxide catalysts, were of the unreducible type, and would give spinels of the same cubic symmetry as Fe_3O_4. **1947** *Jrnl. Polymer Sci.* II. 42 Aliphatic mercaptans in the buradiene-styrene recipe with persulfate as catalyst exert a strong promoting effect upon the copolymerization.

promoting, *ppl. a.* Add: **2. b.** *Med.* That causes tumour promotion (*promotion 2 d).

1944 *Jrnl. Exper. Med.* LXXX. 101 (*heading*) The initiating and promoting elements in tumor production. **1966** *Lancet* 31 Dec. 1457/2 Nevertheless, should such initiated skin be subsequently treated with a promoting agent (not itself carcinogenic), the train of events leading to cancer is again accelerated so that a visible tumour appears at the initiated site. **1970** *Cancer Res.* XXX. 312/1 It is generally agreed that promoting agents cause epidermal hyperplasia.., but not all hyperplastic agents have been shown to be promoters. **1976** [see *promotion 2 d].

promotion. Add: **1. c.** *Curling.* The positional advancement of a stone by striking. Cf. *promote *v.* 1 d.

1897 *Encycl. Sport* I. 261/1 A canny, or quiet, draw is expected from the lead, and it is better that his stone, when spent, should be short of the tee, where it is in the way of promotion. **1937** T. Henderson *Lockerbie* ix. 57 Ye're a graun' curler. That yin's in the wey o' promotion.

d. *Sport* (chiefly *Assoc. Football*). Reallocation of a team to a higher division of a league on the basis of final position in a league table after a season's play.

1907 *Times* 15 Apr. 11/3 Notts Forest and Chelsea have now made themselves certain of promotion. **1949** *Times* 25 Apr. 6/2 Swansea Town assured themselves of promotion to the Second Division. **1965** [see *inject *v.* 2]. **1967** *Listener* 17 Aug. 223/3 'Millwall must be certainties for promotion': thus Bill Nicholson, one of the two or three most analytic club managers in England.

e. *Boxing.* The organization or staging of a contest; a contest staged.

1951 *Sport* 7–13 Jan. 14/4 This show is but the first of the season at Streatham but will be the forerunner of many more value-for-money promotions at the South London hall. **1958** F. C. Avis *Boxing Ref. Dict.* 87 *Promotion*, the organization of a boxing contest. **1962** *Listener* 27 Sept. 466/2 The odd-looking circular building would be just the place for boxing promotions.

f. *Phonetics.* The intensification of normal stress levels in verse; an instance of this.

1956 S. Chatman in *Kenyon Rev.* XVIII. 424 What I have called 'metrical tension' can conveniently be described as 'promotions' or 'suppressions' of the stress levels of normal non-verse speech under the pressures of the abstract metrical pattern. **1973** *Word 1970* XXVI. 55 These examples, from Prator, illustrate suppression and promotion respectively.

g. *Bridge.* The action of promoting a card or trick (see *promote *v.* 1 e).

1959 *Listener* 31 Dec. 1178/2 The problem is to prevent North..from winning a second trump trick by promotion. **1964** *Official Encycl. Bridge* 633/2 *Trump promotion*, the creation of trump tricks through forcing the premature use of the trump cards of the opposition.

h. *Transformational Gram.* The translation of material from an embedded to a main sentence.

1968 R. W. Langacker *Lang. & Struct.* II. v. 128 A permutation rule that we will call Not Promotion is an interesting example that accounts for a subtle type of ambiguity in negative sentences. **1973** P. Schachter in *Language* XLIX. 19 The derivation of both constructions involves the *promotion* of material from an embedded into a matrix sentence.

2. a. *spec.* The furtherance of the sale of something by advertisement or other modes of publicity. Also *transf.*

1925 C. Morley *Thunder on Left* ix. 112 He spoke of the Elevated Railroad's limited appropriation for promotion, of the peculiar problems of transportation publicity. **1928** *Publishers' Weekly* 26 May 2169 Promotion cannot be done without waste... But the idea back of the new mergers is the idea of outlets, of promotion, of selling more goods. **1932** *Ibid.* 7 May 1945/2 A free gift offer in a full page Book-of-the-Month Club promotion. **1958** *Listener* 23 Jan. 148/1 Sir Miles [Thomas]..began his business career as an adviser to Mr. W. R. Morris..on 'sales promotion'. **1962** *Advertisers' Rev.* 26 Oct. 17/2 Price-reductions were easily preferred to all other types of promotion... Other types of promotions considered effective were sample offers, banded offers and free gifts. **1969** *Morning Star* 20 Nov. 5 Heather..had to show a good knowledge of BEA as well as poise and charm. She will represent the airline in promotions and conferences. **1976** *National Observer* (U.S.) 13 Nov. 9/4 Some industry observers say one of those promotions could be trading stamps, although they cost four times as much as a game promotion and require a long-term commitment from a store. **1980** *Bookseller* 12 Jan. 194/1 (Advt.), Evans Brothers Limited require a Publicity Controller to deal with the promotion of their Adult and Children's Trade Books. Applicants should..be capable of dealing with all aspects of book promotion and advertising. *Ibid.* 194/3 The successful candidate..will have experience in organising direct mail promotions, preparing catalogues, leaflets and adverts, [etc.]

c. *Chem.* The action of *promote *v.* 2 c.

[**1920** *Jrnl. Physical Chem.* XXIV. 263 Such an example of auto-catalysis..is to be distinguished from simple auto-catalysis, in which a reactant or product has a direct catalytic effect on the reaction. Here the specific effect of the product is on the catalyst... From the point of view of promoter action such a term as 'auto-activation' or 'auto-promotion' would be more appropriate.] **1926** *Jrnl. Chem. Soc.* 1817 (*heading*) Promotion with iron. **1940** Glasstone *Text-bk. Physical Chem.* xiii. 1128, 1 per cent. of ceria gives the optimum promotion for the catalytic action of thoria on the combination of oxygen and hydrogen. **1968** E. K. Rideal *Concepts in Catalysis* v. 97 The V–O stretching frequency ν = 1025 cm^{-1} is shifted to longer wavelengths by promoters such as alkaline sulphates, where the order of activity in promotion is $Cs > Rb > K > Na$.

d. *Med.* The furtherance of neoplastic growth following its initiation by a carcinogen; the conversion of latent tumour cells into active, malignant ones; (see quot. 1964^1).

1944 *Jrnl. Exper. Med.* LXXX. 121 Non-Specific Promotion.—It seems certain that many agents and influences which have no actual carcinogenicity will be found to stimulate the multiplication of latent neoplastic cells. **1964** *Progress Exper. Tumor Res.* IV. 209 Cocarcinogenesis is not synonymous with tumor promotion, since promotion inherently denotes a specific step in the sequence of events leading to skin tumor formation. Conversely, cocarcinogenesis describes a situation in which response to a carcinogen is increased by a second factor introduced concurrently with the carcinogen, with no implication of impact at a specific step in tumor formation. *Ibid.* 226 Apparently, there is a threshold for the amount of croton oil that must be used in each application in order to achieve promotion. **1976** *Brit. Jrnl. Cancer* XXXIV. 660/1 The events related to tumor promotion are understood better since the isolation and chemical characterization of active promoting agents by Hecker (1968) and Van Duuren (1969).

6. (sense *1 d) *promotion contender, rival*; *promotion-challenging, -chasing, -seeking* sbs. and adjs.; (sense *2 a) *promotion campaign, kit, man, scheme, stunt*; **promotion bar,** a notional barrier imposed to restrict a person's promotion above a certain level unless certain specified requirements are met; **promotion sheet,** a record of an employee's service showing his claim to promotion or increased pay.

1973 C. Mullard *Black Britain* II. iv. 43 Besides overt racism employers operated a promotion bar. **1962** *Listener* 19 July 116/1 The distributors were rash in putting it [*sc.* a film] on without a promotion campaign. **1972** *Oxf. Mail* 15 Feb. 12/6 Carlisle are in their customary promotion-challenging position. **1977** *Sunday Express* 30 Jan. 31/8 Big Brian Joicey grabbed two goals to earn Barnsley a well-deserved victory over promotion-chasing rivals Bournemouth. **1978** *Rugby World* Apr. 19/1 At the time of writing they still had three matches left to play, including games against fellow promotion contenders Jedforest and Selkirk. **1964** *Bookseller* 2 May 1783/1 The Society of Young Publishers reports that the half-way mark has been reached in their appeal for £1,000 towards the production of 'promotion kits' for mailing to 2,000 booksellers and 1,000 libraries throughout the country. **1958** *Spectator* 25 July 138/1 Given a really good labour-saving idea, the British are not so reluctant to try it as many high-powered promotion men would have us believe. **1977** *Rolling Stone* 19 May 14/4 Bennett formerly worked as a Beatles promotion man. **1977** *News of World* 17 Apr. 24/7 It was just as well that most of Brighton's promotion rivals faltered yesterday. **1925** C. Morley *Thunder on Left* xx. 250 The children had found some deceptive promotion scheme advertised in a cheap magazine. **1958** *New Statesman* 26 Apr. 548/1 Some people seem to think that there is something bogus about these promotion schemes and that their cost would be better applied in making a permanent price reduction. **1976** *Norwich Mercury* 10 Dec., The mark of a promotion-seeking team is to win when playing badly. **1909** *Daily Chron.* 23 Mar. 1/4 The manipulation of their promotion sheets, with the object of postponing the payment of justly earned increased salaries, was one of the grievances of the strikers. **1955** W. Gaddis *Recognitions* II. ii. 372 It was all fixed up about this guy who jumped out a window... It's something for a TV promotion stunt.

promotional (promō^u·ʃənăl), *a.* [f. Promotion + -al.] Promotive; of or pertaining to promotion (usu. sense 2 in Dict. and Suppl.) or promoters; relating to advertising.

1922 *Universalist Leader* 13 May 4 Experience has disclosed that emotional contributions are promotional of the very poverty we had felt moved to alleviate. **1926** *Publishers' Weekly* 22 May 1675/1 As a further promotional step the..Company is displaying a letter commenting on its service. **1935** *Sun* (Baltimore) 10 Jan. 11/5 Consolidated Gas picked up more than a point on hopes of an amicable adjustment of its rate controversies with the regulatory authority through adoption of a 'promotional' schedule of charges. **1950** *Sport* 22–28 Sept. 14/1 A promotional 'war' seems certain over the privilege of staging a fight with Lee Savold. **1959** *Sociometry* Sept. 273 Some effects of promotional frustration on employees' understanding of, and attitudes toward, management. **1960** W. Taplin *Advertising* i. 12 Some of the ideas involved in advertising..the complex of promotional activities of which they form part. **1963** *Guardian* 2 Feb. 12/7 What about that for promotional literature? **1978** *Incorporated Linguist* Summer 60/3 Nearly twenty years ago, a totally unknown mineral water gained national fame in France through a relatively inexpensive and brilliant promotional campaign relying solely on the slogan '*Bébé aime Charrier*' originally spelt 'B.B.' thus capitalizing instantaneously on the tempestuous Bardot-Charrier romance. **1980** *Bookseller* 12 Jan. 165/2 His publishers' promotional machinery was faultlessly efficient.

promotor. Restrict †*Obs.* to sense in Dict. and add: **2.** *Chem.* = *promoter 1 d.

1931 *Chem. Abstr.* XXV. 428 The influence of various other promotors and protectors in hydrogenation is studied. **1976** J. Petró in Szabó & Kalló *Contact Catalysis* II. v. 60 Additives which have a favourable effect on the operation of catalysts are termed promotors (activators).

3. Var. *promoter 1 f. *rare.*

promovable (prō^{u}mū·văb'l), *a.* [f. Promove *v.* + -able.] = Promotable *a.*; esp. in phr. *removable and promovable* (cf. Promotable *a.*, quot. 1887), of Sc. and Ir. magistrates.

19th- and 20th-century examples of Promove *v.*, etc., may be found in *S.N.D.*

1920 W. O'Brien *Evening Memories* xiii. 216 Two paid magistrates removable and promovable at the caprice of Dublin Castle.

prompt, *sb.* Add: **II. 2. b.** **prompt corner,** the prompter's corner off-stage; **prompt entrance** (see quot. 1952); **prompt script** = Prompt-book; **prompt-side,** substitute for def.: the side of the stage where the prompter takes up his position, usu. on the actor's left (*U.S.*, on the right); (earlier and later examples); **prompt table,** the table on which the prompter rests his book; **prompt-word,** a word spoken by a prompter (in quot. *transf.*).

1933 P. Godfrey *Back-Stage* i. 17 A good stage-manager is never far from the prompt corner. **1967** *Listener* 21 Sept. 370/3 When the moment came, I was standing in the prompt corner with..the stage-director. **1886** M. Mackintosh *Stage Reminisc.* ii. 25 It would give him a better chance with the audience to sing it from the 'prompt' entrance. **1890** B. Hall *Turnover Club* xx. 187 One evening he was standing in the 'prompt entrance' with a prominent actress who was starring at the house then. **1952** Granville *Dict. Theatr. Terms* 143 *Prompt entrance*, that way on to the stage from the prompt corner which is for the use of the stage management. **1920** Wodehouse *Jill the Reckless* (1922) xiv. 210 The assistant stage-director bent sedulously over the footlights.. shading his eyes with the prompt script. **1824** R. Humphreys *Mem. J. Decastro* 30 A thunder-drum, which now stands on the prompt side of the theatre. **1963** C. Mackenzie *My Life & Times* I. 90 The next thing I remember is sitting on my nurse's knee in the front of the dress circle of the Cork Opera House on the right-hand side, or as an actor would say, on the prompt side. **1844** J. Cowell *Thirty Years passed among Players* II. ii. 59/2 Every book or manuscript they have an opportunity to place upon a prompt-table. **1967** *Listener* 21 Sept. 369/2 He caught sight of Trebel..lurking by the prompt table. **1918** A. Quiller-Couch *Foe-Farrell* 176, I knew..that I must break his fate to him. I even gave him the prompt-word. 'Homelike', I suggested.

prompt, *a.* (*adv.*) Add: **A.** *adj.* **4.** *Nucl. Physics.* Of a neutron or gamma ray: emitted within a small fraction of a second as the direct result of a fission, as distinct from radiation due to the decay of fission products.

1947 M. Deutsch in C. Goodman *Sci. & Engin. of Nucl. Power* I. ii. 84 The average total kinetic energy of the two fission fragments from U^{235} is about 160 Mev. In addition, the prompt neutrons have kinetic energies totaling about 5 Mev. If the energy of excitation of the fission fragment immediately after its formation is insufficient to cause neutron emission, the nucleus may lose energy by the emission of prompt gamma rays. **1950** [see *delayed neutron*]. **1962** H. D. Bush *Atomic & Nuclear Physics* vii. 140 'Prompt' γ-rays are also observed due to the excitation of the primary fission products. **1973** *Physical Rev.* VII. C. 1180/1 In Table 1 the energy emitted per fission event in the form of prompt-γ radiation and the number of γ rays per fission are given for a number of energy intervals above 0·14 MeV. **1974** S. E. Hunt *Fission, Fusion & Energy Crisis* vi. 53 Reactors are operated so that if they relied on the neutrons emitted instantaneously on fission, on so-called 'prompt' neutrons, alone, the fission chain reaction would be subcritical, i.e. the reaction would die away.

B. *adv.* **b.** Promptly; soon.

1910 W. M. Raine *Bucky O'Connor* ii. 23 The reverend gentleman..had this diverting experience so prompt after he was wishing for it.

C. **prompt-critical** *a. Nucl. Physics*, critical even when the effect of delayed neutrons is neglected and prompt neutrons alone are considered; hence **prompt-criticality.**

1954 R. Stephenson *Introd. Nucl. Engin.* ii. 64 When the *k* for a U^{235} reactor is exactly equal to 1·0073, there is sufficient reactivity to maintain the chain reaction by means of the prompt neutrons alone, and the reactor is said to be prompt critical. **1956** S. Glasstone *Princ. Nucl. Reactor Engin.* iv. 242 The rapid increase in the flux when the reactivity exceeds the prompt-critical value would make the reactor difficult to control, and special precautions must be taken that this condition does not arise in reactor operation. **1973** D. R. Inglis *Nucl. Energy* IV. 120 Plutonium has only about one-third as many delayed neutrons as does uranium (about 0·23 percent of neutrons are delayed), and this provides only a narrow margin for adjustments below a prompt critical condition. **1954** R. Stephenson *Introd. Nucl. Engin.* vii. 269 This is the condition for prompt criticality. **1974** S. E. Hunt *Fission, Fusion & Energy Crisis* vi. 53 Prompt criticality would be even more serious in a fast reactor of the Dounreay type than in the more usual thermal neutron reactors, since the neutrons do not need to be slowed down between one generation of fissions and the next.

prompter. Add: **2. b.** Also *Comb.* in possessive, as **prompter's bell, box, copy, table** = *prompt-bell, -box, -copy, *table* s.v. Prompt *sb.* 2 b.

1812 C. Mathews *Let.* 20 Jan. in A. Mathews *Mem. Charles Mathews* (1838) II. viii. 187, I..never did believe (nor will I, till I hear the prompter's bell) that Drury would be played in next winter. **1895** W. Archer *Theatr. World 1894* 369 Where, at the stroke of the prompter's bell, a new world is revealed to the delighted sense. **1775** F. Abington *Let.* in *Private Corr. D. Garrick* (1835) II. 32 Begging leave to sit in the prompter's box. **1870** E. L. Blanchard *Diary* 23 Mar. in Scott & Howard *Life E. L. Blanchard* (1891) II. 381 Adelphi; Byron's new four-act drama of *The Prompter's Box: A Story of the Footlights and the Fireside.* **1770** A. Murphy *Let.* 20 Sept. in *Private Corr. D. Garrick* (1831) I. 399 Having the prompter's copy in my drawer for above two years past. **1834** W. C. Macready *Diary* 10 June (1912) I. 152, I came up and was first at rehearsal; from the prompter's table I wrote a hasty note to R. Price. **1889** J. L. Toole *Reminisc.* I. i. 29, I was sitting at the prompter's table, when I heard a voice at my elbow.

promptive, *a.* (Further example.)

1955 W. Faulkner *Fable* 384 The third man..said,.. just diffident and promptive, as you address someone.. who may have temporarily forgotten your need or forgotten you: 'Paris'.

promptuary, *sb.* (*a.*) **A.** *sb.* **1.** Delete ?*Obs.* and add later example.

1977 'E. Crispin' *Glimpses of Moon* xi. 222 He looked.. like a gorilla at large in an unguarded banana promptuary.

promyelocyte, -cytic: see *pro-^{2} 1. **proname:** *pro-^{1} 4 b.

pronase (prō^u·nē^{i}z). *Biochem.* [f. *pro(tei)nase.] A purified preparation of proteinase from cultures of the bacterium *Streptomyces griseus.* (A proprietary name in the U.S.)

1960 *Jrnl. Biochem.* (Japan) XLVIII. 600 The above mentioned, excellent proteolytic activity of *Streptomyces griseus* protease is expected to develop many new industrial and scientific applications, in cooperation with the success in commercial production of this enzyme. [*Note*] Partially purified *Streptomyces griseus* protease is put on the market from the Kaken Chemical Co., Ltd., in the trademark of '*Pronase*'. **1962** *Science* 2 Nov. 594/1 The zona pellucida may be removed from all stages of the mouse egg by digestion with pronase. **1963** *Official Gaz.* (U.S. Patent Office) 22 Jan. TM132/2 California Corporation for Biomedical Research, Los Angeles, Calif... *Pronase* for highly purified proteinase prepared from *Streptomyces griseus* culture broth. First use May 9, 1961. **1976** *Ann. Rev. Microbiol.* XXX. 163 Ribonuclease digestion of a phenolic extract containing the antiviral factor resulted in complete loss of activity, whereas treatments with trypsin or pronase did not lessen antiviral activity.

pronatalist (pronē^i·tălist), *a.* and *sb.* Also **pro-natalist.** [f. Pro-^{1} 5 a + Natal *a.*^{1} and *sb.* + -ist.] **A.** *adj.* Of or pertaining to the encouragement of large families, esp. by the state; in favour of or advocating large families. **B.** *sb.* A pronatalist person.

1938 *Mod. Law Rev.* II. 102 The suppression of abortion was part of the policy of the *Alliance Nationale pour l'Accroissement de la Population Française*, founded by J. Bertillon in 1896. (Three months after this had been founded, Paul Robin set up his counter organisation, '*La Ligue de la Régénération Humaine*', for advocating birth-control and attacking the pro-natalists. *Ibid.* 105 In most countries and at most periods there have been pro-natalist policies directed at encouraging marriage and the raising of large families.) **1953** *Population Stud.* VII. 39 Even the most ambitious of French pro-natalists admit to-day that policy must be directed to encouraging the birth of a third and possibly a fourth child, rather than concentrating on the large family. **1974** *Encycl. Brit. Macropædia* XIV. 818/2 Taken as a whole, Europeans in the Middle Ages were pronatalists. **1974** *Sci. Amer.* Sept. 125/3 Between the 1930's and the 1950's there was..an increase in pronatalist intervention by the [U.S.] Government, particularly through the tax structure. **1976** *Nature* 4 Nov. 7/1 More and more governments are reversing pro-natalist policies and instituting family planning programmes. **1979** *Jrnl. R. Soc. Arts* CXXVII. 415/1 Some countries are still avowedly pronatalist.

So **prona·talism,** advocacy or encouragement, esp. by the state, of large families.

1938 *Mod. Law Rev.* II. 105 The semi-official (e.g. municipal) encouragement of birth-control in the first half of the nineteenth century was severely shaken by the Franco-Prussian war, which was followed by an intensive pro-natalist campaign. Similarly, a further impetus to pro-natalism was given during the last war, and the policy is now an official one. **1974** D. V. Glass in H. B. Parry *Population* I. v. 73 Japan, too, in spite of its high Gross Reproduction Rate, included pronatalism in its imperialist policy.

pronate, *ppl. a.* (Further *literary* example.)

1938 S. Beckett *Murphy* ii. 24 He raised his left hand..and seated it pronate on the crown of his skull.

pro-Nazi: see *pro-^{1} 5 a.

prone, *sb.* Delete 'Now rare' and add: Also ‖**prône. 2.** (Further examples.) Also prayers, exhortations, etc., attached to the sermon.

[**1763** C. Cordell *Divine Office for Use of Laity* I. p. vi, The prayers, publications, and familiar instructions used at the Parish-Mass, on Sundays..either immediately after the Gospel, or before the Lavabo are called in France the Prône, from a Greek word signifying the Nave of the Church.] *a* **1773** A. Butler *Moveable Feasts Catholic Church* (1774) I. v. 78 A Person who besides Morning and Evening Prayers has devoutly attended High-mass with a Prone or Sermon. **1860** F. C. Husenbeth *Life Monsignor Weedall* iii. 50 These sermons were called by the French name of *Prones*... In these *Prones*, however, no one surpassed Mr. Weedall. **1912** A. Fortescue *Mass* vii. 295 The Prayers of the Faithful..became the *prône*, commands to pray for all classes of people, living and dead, which are still given out before the sermon. **1915** F. E. Brightman *Eng. Rite* II. 1037 The Bidding of the Bedes has not stood alone, but has formed part of a group of vernacular devotions, instructions, and notifications, attached to the Sermon, and known as the 'Prone'. **1937** W. Douglas *Church Mus. in Hist. & Pract.* v. 120 The most powerful [urge] was that toward the use of the vernacular. Both in England and in Northern Europe..this tendency had brought about a series of vernacular public devotions called the Prone, in connection with the Sermon at High Mass. It contained a bidding prayer for intercessions, a confession and absolution, the Creed, the Lord's Prayer, and the Ten Commandments with explanation, and the Church notices. **1968** F. E. Vokes in *Studia Evangelica* V. 146 The so-called 'Prone', the vernacular instruction in the Mass. **1972** J. G. Davies *Dict. Liturgy & Worship* 76/2 The association between the sermon and the intercessions which is found in some of the Reformed churches is probably derived from their association in the prone. **1978** D. H. Tripp in C. Jones et al. *Study of Liturgy* II. III. xi. 255 The normal Sunday worship was to be a preaching-service, based on the medieval prone.

prone, *a.* Add: **6. c.** Const. *absol.* with preceding sb. (usu. with hyphen).

1926, etc. [see *accident *sb.* 10 b]. **1973** J. M. White *Garden Game* 47 They were fundamentally good boys..but they were also violence-prone. **1974** *Times* 18 Oct. 16/4 The traditionally drought-prone areas. **1975** *Publishers Weekly* 25 Aug. 287/2 This tale of a wayward bus line founded by the author's failure-prone father.

proneural: see *pro-^{2} 2.

prong, *sb.*^{2} Add: **2. c.** (Earlier examples of U.S. sense.)

1725 in *Amer. Speech* (1940) XV. 300 To a Gum on the sound side of the north prong of the Spring Swamp. **1784** G. Washington *Diary* 25 Sept. (1925) II. 311 Carpenters Creek, a branch of Jackson's..which is the principal prong of James River. **1834** J. M. Peck *Gazetteer Illinois* III. 217 It [*sc.* Crawford's Creek] enters the south prong of Bear creek.

pronghorn, *sb.* Add: **2.** [f. Prong *sb.*^{2} + Horn *sb.*] A projecting stabbing implement (in context a gas lamp). *nonce-wd.*

1922 Joyce *Ulysses* 439 The navvy..gores him with his flaming pronghorn.

pronk (prǫŋk), *v. S. Afr.* [Afrikaans, to show off, strut, prance, ad. Du. *pronken* to strut.] *intr.* Of springbok: to leap in the air, to buck, esp. as an alarm-signal. Hence **pro·nking** *vbl. sb.*

1896 F. V. Kirby *In Haunts of Wild Game* ii. 49 He quickly settles down into a long 'rocking-horse' canter, or else goes 'pronking' away, as the Boers style it. **1915** *Chambers's Jrnl.* Nov. 703/1 When a whole troop of these antelopes are thus leaping..'pronking', or 'pranking', as the Boers call it. **1957** R. Campbell *Portugal* v. 90 Both the mules..began to rise off the ground as if the earth were red-hot, like springboks 'pronking', with all four feet at once, arching their backs. **1966** E. Palmer *Plains of Camdeboo* ix. 154 Every hunter in the past had his theory as to why springbuck pronk, just as every Karoo farmer has today. **1971** *Sunday Mail Family Section* (Brisbane) 10 June 6 The beautiful springbok.. gives a spectacular alarm signal... It springs into the air, back arched, displaying a crest of pure white hairs. This is called 'pronking'.

pronk (prǫŋk), *sb. slang.* [Origin uncertain: cf. Du. *pronker* fop.] A weak or effeminate person, a softie; a crank, fool, mug.

1959 C. MacInnes *Absolute Beginners* 25 Here the pronk half rose in his ballet tights and saluted. *Ibid.* 33 No one is going to..try to blackmail me with that crazy old mixture of threats and congratulations that a pronk like you falls for. **1972** L. Henderson *Cage until Tame* iv. 28 Whoever this pronk Durant was he had a lot to learn. **1976** —— *Major Enquiry* xv. 102 This pronk reckons he can..point out the right one [*sc.* car] with a hazel twig.

pronograde (prō^u·nogrē^{i}d), *a.* [f. L. *prōn-us* Prone *a.* + *-gradus* going, walking: see Grade *sb.*] Moving on all fours.

1902 [see *orthograde *a.*]. **1918** F. Wood-Jones *Probl. Man's Ancestry* 22 The likeness [to man] still further diminished in the lemurs, and in the general run of pronograde quadrupedal mammals it reached a minimum. **1971** F. S. Hulse *Human Species* (ed. 2) vii. 168 For moving or standing in a pronograde position, that is, on all fours, a somewhat tube-shaped pelvis is quite adequate. **1978** *Nature* 14 Dec. 706/1 A narrow trochlea with pronounced anteromedial and posterolateral borders such as is characteristic of cercopithecoid monkeys (and many other pronograde quadrupedal mammals).

pronominal, *a.* (*sb.*) Add: **A.** *adj.* **2.** (Further examples.) Also, characterized by the presence of a pronoun.

1924 O. Jespersen *Philos. Gram.* 303 There are two kinds of questions; 'Did he say that?' is an example of one kind, and 'What did he say?' and 'Who said that?' are examples of the other. Many names have been proposed for these two kinds: yes-or-no question or categorical question *v.* pronominal question [etc.]. **1928** H. Poutsma *Gram. Late Mod. Eng.* (ed. 2) I. vii. 381 Sweet..distinguishes them as general and special questions; Kruisinga..as disjunctive and pronominal questions. **1931** G. Stern *Meaning & Change of Meaning* xi. 332 German *etcetera*..may mean 'podex, crepitus ventris, devil, cacare', etc. This might be called a pronominal use. **1960** E. Delavenay *Introd. Machine Transl.* 120 Prepositions require further programming..as do pronominal verbs (s'améliorer). **1965** B. Collinder in Bessinger & Creed *Medieval & Linguistic Stud.* 28 Adolf Noreen distinguished between *expressive* and *pronominal* semantemes. **1978** *Language* LIV. 369, I also argue that the pro-predicates have an especially close resemblance to 'pronominal determiners' like the italicized portions of the following.

B. *sb.* (Further example.)

1971 *Language* XLVII. 169 The prefixes of the inflectional category include aspects such as *si-*..and *yi-*.., modals such as *ni-*..and *di-*.., and pronominals such as *yi-*..and *i-*.

pronominalize *v.* (further examples); hence **prono·minalized, prono·minalizing** *ppl. adjs.*

1961 *Amer. Speech* XXXVI. 163 These are reducible identically to the pronominalized *each one, this one*..or finally all the way to *it.* **1971** *Language* XLVII. 527 The indefinite article, when truly indefinite (when pronominalized by *one*), introduces a noun in a qualitative rather than an identifying sense. **1975** *Archivum Linguisticum* VI. 78 The subject of *ofereode* is pretty clearly the (unhappy) result of the situation outlined in the stanza; the genitive, which has been pronominalized, referring to the situation itself. **1977** *Language* LIII. 97 The pronominalizing languages present in surface more of the logical structure of restrictive relative clauses than do the non-pronominalizing languages. **1978** *Amer. Speech* LIII. 31 In colloquial English, *they* can be said to pronominalize a [+ human, + III] antecedent of undetermined sex, regardless of number.

pronominalization (pronǫ:minăləizē^i·ʃən). [f. Pronominalize *v.* + -ation.] The process or fact of replacing (a noun or noun phrase) by a pronoun. Also *attrib.*

1961 *Amer. Speech* XXXVI. 163 The optional deletion of the modifier..occurring in the grammar after the pronominalization. **1965** N. Chomsky *Aspects of Theory of Syntax* iii. 145 Sameness of reference requires reflexivization of the second Noun Phrase (this is also true of pronominalization). **1970** *Language* XLVI. 172 The essence of the pronominalization rule adopted..is that any noun phrase..may be expressed as a pronoun. **1971** J. P. Thorne in A. J. Aitken et al. *Edin. Stud. Eng. & Scots* 59 The rules governing pronominalisation require either that the pronoun occurs in the deep structure of the sentence or that the deep structure contains two identical noun phrases, in which case a transformational rule re-writes one of them as a pronoun in the surface structure. **1979** *Trans. Philol. Soc.* 58 Hankamer (1971)..has argued that pronominalization is a special instance of deletion.

pronosophical (prō^{u}nosǫ·fikăl), *a. nonce-wd.* [f. Gr. προνοέω to foresee + *-sophical* as in *philosophical.*] Having the wisdom of foresight; previsionary.

1922 Joyce *Ulysses* 425 Stephen: Gesture..would be a universal language, the gift of tongues rendering visible.. the first entelechy, the structural rhythm. Lynch: Pronosophical theology. Metaphysics in Mecklenburg street!

pronotum. Add: (Later examples.) **pronotal** *a.* (examples).

1904 *Proc. U.S. Nat. Mus.* XXVII. 389 The pronotum [of *Ceratophyllus multispinosus*] has two rows of few weak bristles. *Ibid.* 395 *Ceratophyllus alaskensis*... Male: Head flattened on top as usual. Pronotal ctenidium of 22 spines. **1925** A. D. Imms *Gen. Textbk. Entomol.* II. 665 In many species [of flea] the pronotum carries a row of stout spines forming the pronotal comb. **1947** C. A. Hubbard *Fleas Western N. Amer.* iv. 38 Taxonomically the only portion which has any great significance is the pronotum and the absence or presence on it of the pronotal comb or ctenidium. **1959** Southwood & Leston *Land & Water Bugs Brit. Isles* iii. 17 Head and pronotum [of the juniper shieldbug] with concolorous puncturation except sometimes near the pronotal angles. **1972** Swan & Papp *Common Insects N. Amer.* xii. 123 The head [of lace bugs] is often covered with a hood, the pronotum extended on the sides like another pair of wings. *Ibid.* xxiii. 655 Genal and pronotal spines absent [in the human flea].

pronoun. Add: **b.** *Comb.*, as *pronoun-form, -object.*

1933 L. Bloomfield *Language* xvi. 269 Among the substantives are some pronoun-forms which, by over-differentiation, do not serve as actors: *me, us, him, her, them, whom.* **1973** *Word 1970* XXVI. 81 In the first-person plural, the subject pronoun-form functions as an indirect object: *They give me the money.* **1957** R. W. Zandvoort *Handbk. Eng. Gram.* I. v. 78 *To do* is used as a notional verb, chiefly in its non-finite forms and with a neuter pronoun-object (*something, nothing, this, that, it, what?* etc.). **1963** F. T. Visser *Hist. Syntax* I. iv. 425 The period of transition from Old to Middle English is characterised by a considerable number of remarkable changes in the form of most of the reflexive pronoun-objects.

pronounal, *a.* (Further example.)

1961 R. B. Long *Sentence & its Parts* ii. 39 Inflection for the possessive is fundamentally a nounal or pronounal variety of inflection.

pronto (prǫ·nto), *a. Mus.* [a. It. *pronto* ready, prompt, f. L. *promptus* quick: cf. next.] (See quots.)

Quot. 1740 represents quasi-*adv.* use.

1740 J. Grassineau *Mus. Dict.* 184 *Pronto*, readily, quick, nimbly, without loss of time. **1908** L. J. De Bekker *Stokes' Encycl. Mus. & Musicians* 517/2 *Pronto, It.* Quick, ready. **1938** *Oxf. Compan. Mus.* 759/1 *Pronto* (It.) 'Ready', 'prompt'. So *Prontamente*, 'promptly'.

pronto (prǫ·nto), *adv. colloq.* (orig. *U.S.*). [a. Sp. *pronto*, f. as prec.] Quickly; promptly, straight away.

1850 L. H. Garrard *Wah-to-Yah* ix. 134 Bent told him to *vamos, prento!* (go quick). *Ibid.* xi. 162 Me be off prento. *Ibid.* xvii. 225 Esta—'here'—said she... Vamos—prento, por el rancheros—'go quick for the rancheros.' **1911** H. B. Wright *Winning of Barbara Worth* v. 96 All we have to do with it is to push for Rubio City pronto. **1926** J. Black *You can't Win* vi. 66 If we was in the city I'd take fifty cents of it purty pronto. **1934** R. Macaulay *Going Abroad* ii. 26, I bring zem wizout fail; pronto, right away. **1938** E. Bowen *Death of Heart* III. v. 405 I'm going to take you right back—now, pronto, at once. **1948** 'N. Shute' *No Highway* v. 128 Will you.. get Honey back here, pronto? **1952** M. Allingham *Tiger in Smoke* x. 165 When the war ended you were slung out pronto. **1966** *Listener* 22 Dec. 927/3 He finished up by saying that there would be a punitive expedition pretty pronto if the stuff was not returned. **1976** P. Cave *High Flying Birds* iii. 33 You tell that bastard to come and see me... Pronto.

Prontosil (prǫ·ntosil). *Pharm.* Also **prontosil.** [G., proprietary name.]. **1.** A proprietary name for a reddish-brown crystalline bacteriostatic dye, 2′,4′-diaminoazobenzene-4-sulphonamide, $C_{12}H_{13}N_5O_2S$, which was the first of the sulphonamide drugs to be known and was given (usu. orally) in the treatment of a wide range of bacterial infections; formerly also called *Red Prontosil.*

1936 *Proc. R. Soc. Med.* XXIX. 313 Advice will be given of a new class of substances, which have an actual chemotherapeutic action in streptococcal infections (prontosil, prontosil S). **1936** *Trade Marks Jrnl.* 15 July 874/2 Prontosil... A medicated dye preparation for human use in the prevention and treatment of Streptococcus infections. Bayer Products, Limited. **1937** *Lancet* 13 Mar. 626/1 Trefouël and others..have shown that the soluble red azo-dye Prontosil Soluble (for injection) and the almost insoluble orange azo-dye Red Prontosil (oral) are bactericidal only after reduction to sulphanilamide. **1938** F. B. Young *Dr. Bradley Remembers* viii. 437 Some German, it seemed—it was always the Germans in those days—had compounded a new synthetic drug..marketed under the name of Prontosil. **1938** *Times* 8 Dec. 18/5 The new drug, prontosil, the value of which in treating human puerperal fever was clearly established at the Queen Charlotte's Hospital isolation blocks and research laboratories. **1942** *Endeavour* July 122/2 The origin of these and other developments of bacterial chemotherapy was the synthesis by Mietzsch and Klarer of the drug known as prontosil, and the disclosure of its therapeutic properties by Domagk. **1965** *Listener* 10 June 847/1, I can remember helping to make some quinine ampoules for the treatment of a young wife who was dying in her first pregnancy. They were useless. A few years later a proprietary drug called 'Prontosil' came to us, and by giving tablets of that by mouth we began to cure puerperal fever. **1972** 'J. Herriot' *It shouldn't happen to Vet* (1973) xxi. 143 Just a suggestion, James..I honestly think that the situation calls for a little Prontosil.

2. Special Combs. (none now current): **Prontosil Album** [L. *album*, neut. of *albus* white], a name for sulphanilamide; **Prontosil Red** or **Rubrum** [L. *rubrum*, neut. of *ruber, rubrus* red] = sense 1; **Prontosil soluble** (see quot. 1958).

1937 *Lancet* 27 Feb. 510/1 From the 12th to the 29th the patient also received one tablet of Prontosil Album by mouth every six hours. **1958** *Martindale's Extra Pharmacopœia* (ed. 24) 1250 Proprietary preparations of sulphanilamide were formerly marketed in Great Britain under the names Prontosil Album (Bayer Prod.) and Streptocide (Evans Medical Supplies). **1938** *Brit. Med. Jrnl.* 8 Jan. 104/2, I have obtained successful results in the treatment of erysipelas and acute dermatitis by using as a local application prontosil red, 7½ grains being dissolved in 1 oz. of distilled water. **1938** *Lancet* 10 Sept. 647/1 The new rule..will apply not only to sulphanilamide but to related substances including products sold under the following names:..Prontosil rubrum.., Prontosil Soluble, [etc.]. **1959** Evers & Caldwell *Chem. of Drugs* (ed. 3) ix. 114 Prontosil rubrum is broken down in the body to 4-aminobenzenesulphonamide. **1964** Ariëns & Simonis in E. J. Ariëns *Molecular Pharmacol.* I. 77 One of the bioactivations first recognized as such was that of Prontosil Rubrum.., the compound found by Domagk..to be effective in the protection of mice against bacterial infections. **1936** *Lancet* 11 July 107/2 There seems to be some risk of toxic reactions associated with the administration of Prontosil (soluble) intravenously. *Ibid.* 5 Dec. 1323/2 When prontosil soluble was substituted for the sulphonamide, there was no bactericidal effect. **1958** *Chambers's Techn. Dict.* 1017/2 Prontosil-soluble, which has greater antibacterial effect [than Prontosil] and is more soluble, is the disodium salt of 4′-sulphonamido phenylazo-7-acetylamino-1-hydroxy naphthalene-3·6-disulphonic acid.

pronuba. Add: **2.** [C. V. Riley 1872, in *Nature* 26 Sept. 444/1.] In full, *pronuba moth.* A small white North American moth of the genus formerly so called, now usually included in the genus *Tegeticula*, esp. the yucca moth, *T. yuccasella*, which pollinates yucca plants.

1872 *Nature* 26 Sept. 444/1 The larva of *Pronuba* eats through the Yucca capsule in which it fed, enters the ground and hibernates there. **1905** V. L. Kellogg *Amer. Insects* xvi. 577 (*caption*) Pronuba-moth depositing eggs in ovary of Yucca. **1929** Robbins & Rickett *Botany* xxvi. 437 Without the *Pronuba* moth the *Yucca* would not produce seeds. **1946** D. C. Peattie *Road of Naturalist* ii. 24 The smoky little Pronuba moths..come to the blossoms, mate there, hover and hide and perform small extraordinary rites without which no Joshua tree would stand and brandish arms at heaven.

pronunciamento. Add: (Earlier and later examples.) Also *attrib.*

1835 *Morning Courier & N.Y. Enquirer* 23 Nov. 2/2 It is not..a question of a *pronounciamiento* in favor of federalism or centralism, or who shall govern. **1906** R. Fry *Let.* 20 Nov. (1972) I. 273 The Trustees..will pass some sort of resolution accepting my *pronunciamento.* **1929** *Times* 1 Nov. 17/4 The worst result of the Spanish doctor's pronunciamento is likely to be a regular cult of conscientious and laboured irregularity. **1939** H. G. Wells *Holy Terror* IV. iii. 439 The pronunciamento bosses of those little old republics there used to be in South America. **1951** E. Paul *Springtime in Paris* xi. 202 Then came Zhdanov's pronunciamento in Russia condemning jazz as degenerate and unfit for proletarian amusement. **1968** R. Hargreaves *Bloodybacks* x. 265 Although there was anything but unanimity of sentiment and opinion among the inhabitants of the Thirteen Colonies, the Congressional *pronunciamento* had committed them to continued resistance. **1972** *Daily Tel.* (Colour Suppl.) 10 Nov. 18/1 The political revolution consisted largely of intimidation, arbitrary rule..and windy *pronunciamientos.* **1975** *New Yorker* 20 Jan. 61/3, I wish he had prevailed on the author to drop a few samples of the writer-hero's more high-flown, phony-eloquent pronunciamentos.

pronunciation. Add: **6.** *attrib.* and *Comb.*, as **pronunciation key,** a list of symbols providing a guide to pronunciation; **pronunciation-spelling,** the spelling of words in accordance with their usual pronunciation; an instance of this.

1962 C. L. Barnhart in Householder & Saporta *Probl. in Lexicogr.* 174 Two great scholars..said that the pronunciation key was of no importance whatsoever; they felt that any key that used symbols consistently was adequate. **1966** *Random House Dict. Eng. Lang.*, (*heading*) Pronunciation Key. **1944** H. J. Uldall in E. P. Hamp et al. *Readings in Linguistics II* (1966) 149 If there have been cases of spelling-pronunciation, there have also been cases of pronunciation-spelling. **1953** K. Jackson *Lang. & Hist. in Early Brit.* I. ii. 70 If the spelling..depended on tradition alone..there would be far more cases of pronunciation-spellings betraying lenition than the few which do exist. **1979** *Amer. Speech* LIV. 33 There are no spelling pronunciations in the corpus, although there are pronunciation spellings like the British *cuppa* and *pinta*, as in *cup of tea* and *pint of milk.*

procemium. (Later examples of var. form.)

1826 *Hansard Commons* 25 Apr. 590 The whole of the authorities..formed a very dry proemium to the proposal with which it [*sc.* the bill] terminated. **1933** R. Tuve *Seasons & Months* i. 11 Both analogues and influences may be found for the 'alma Venus' of Lucretius' proemium.

pro-œstrus, var. *pro-œstrum* s.v. Pro-^{2} 1 in Dict. and Suppl.

proof, *sb.* Add: **I. 1. a.** (Further examples.)

1927 A. H. McNeile *Introd. to N.T.* 13 He therefore makes no use of proof-texts, and no suggestions that Christianity is the real and 'fulfilled' Judaism. His sole 'proofs' are the actual words and deeds of the Master and the effects which they produced. **1945** R. Knox *God & Atom* iii. 42 In the long run, you felt, the first three Proofs stood or fell by the value of the causality argument.

II. 10. b. For †*Obs.* read *Hist.* and add later example.

1956 W. S. Churchill *Hist. Eng.-Speaking Peoples* I. II. i. 126 They..were clad in proof, but..they cast aside their ring-mail.

11. c. The aeration of dough by leaven before baking. Cf. *prove *v.* 1 g.

1903 Kipling *Five Nations* 23 There is no proof in the bread we eat or rest in the toil we ply. **1961** *Sunday Times* 5 Feb. 30/4 You knead it [*sc.* the dough] again, but this time for a few minutes only—just to 'knock out the proof', as the bakers say.

III. 13. b. Delete † *Obs.* and add further examples.

1948 C. Abel *Business of Photogr.* xxxix. 357 Ownership of the proofs submitted by a portrait photographer to a customer has long been a matter of argument in the profession. **1960** MORTENSEN & DUNHAM *How to pose Model* (ed. 3) xii. 152 As soon as you have picked out the proofs that you want to make into prints, destroy *all* the others.

14. (Earlier and later examples of current sense.)

1889 J. ATKINS *Coins & Tokens of Possessions & Colonies of Brit. Empire* 5 Proofs exist of both these pieces. **1920** *Brit. Numismatics Jrnl. 1918* IV. 131 The same artist issued proofs either for a sixpence or a farthing. **1969** *Times* 21 July p. v/4 Proofs (special coins struck from highly polished dies mainly for collectors) have been issued for this denomination. **1977** *Times* 5 May 20/2 Jubilee crown proofs are in silver.

IV. 18. a. (sense 4) *proof-test* vb.; (senses 12–14) *proof coin, copy, -correcting* (earlier example), *correction* (further examples), *-corrector, impression* (later example), *stage, state* (example).

1949 W. H. SHELDON *Early Amer. Cents 1793–1814* i. 39 *Proof* coins were never struck for circulation... Proofs were first used as presentation pieces. **1969** *Coin Investor* 18 Jan. 3/1 The difference between a proof coin and a brilliant uncirculated one is that a proof coin has been specially struck on a highly polished die. **1976** *National Observer* (U.S.) 17 Jan. 17/4 (Advt.), Highly sought after by collectors, Proof coins are the ultimate expression of the minting art. Dies and solid gold blanks are polished by hand to a mirror finish. **1806** SCOTT *Let.* Oct. (1932) I. 327, I would like to see them in the proof copy in case any minute alterations may yet occur to me. **1975** A. POWELL *Hearing Secret Harmonies* ii. 66 He held under his arm what looked like the proof copy of a book. **1850** W. M. THACKERAY *Let.* June in H. Ritchie *Lett. A. T. Ritchie* (1924) iii. 29 Then comes printing and proof correcting & so forth. **1940** *Chambers's Techn. Dict.* 677/2 *Proof corrections*, additions or emendations to a proof. They should be made in ink, and clearly indicated in the margin. **1978** *Hart's Rules for Compositors & Readers* (ed. 38) 34 (*heading*) Proof-correction marks. **1928** L. P. SMITH *Words & Idioms* 162 We are becoming more and more the slaves of schoolmasters and proof-correctors. **1932** A. E. HOUSMAN *Let.* 19 June (1971) 322 On p. 32 the proof-corrector has again given directions for making a change which ought not to be made. **1910** *Connoisseur* Oct. 93/1 How to distinguish Proof Impressions. **1895** F. M. FORD *Let.* (1965) 8 You may tell Longmans that W.M.R. had offered to revise my chapters as they go to Press wh. is better than in the proof stage I shd. think. **1966** N. NICOLSON in H. Nicolson *Diaries & Lett., 1930–1939* 118 The name had to be altered again..at proof-stage. **1910** *Connoisseur* Oct. 93/2 To such a man even the names of the various proof states must be a source of bewilderment. **1951** *Sun* (Baltimore) 30 Jan. 5/3 Their major purpose..is to proof-test some slide-rule work being done by nuclear physicists and weapons experts.

b. **proof-glass** (examples); **proof load** *Mech.*, a load which a structure must be able to bear without exceeding specified limits of deformation; *loosely*, proof stress; **proof-read** *v. trans.*, to read (printer's proofs) and mark errors for correction; hence **proof-read** *ppl. a.*; **proof-reader** (earlier example); **proof-reading** *vbl. sb.* and *ppl. a.* (examples); **proof-slip** *Typog.* = PROOF-SHEET; **proof strain** *Mech.*, the strain produced by the proof stress; *loosely*, proof stress; **proof strength** (earlier example); **proof stress** *Mech.*, the stress required to produce a specified permanent deformation of a material or structure; **proof theory** (see quot. 1942); hence **proof-theoretic** *a.*, of or pertaining to proof theory; **proof-theoretically** *adv.*, in a proof-theoretic manner.

1765 H. JACKSON *Ess. Brit. Isinglass* 73 We likewise advise them to a serious Perusal of our new-invented *Proof-glasses.* **1848** *Knickerbocker* XVIII. 380 With what profound deliberation he drew his proof-glass from the bung-hole of a brandy-pipe. **1858** W. J. M. RANKINE *Man. Appl. Mech.* II. iii. 287 The toughness of the bar, or the extension corresponding to the proof load. **1930** *Engineering* 21 Feb. 241/2 This working load is equivalent to one-half the proof load, which again is almost one-half the ultimate strength of the chain. **1973** A. PARRISH *Mech. Engineer's Ref. Bk.* xii. 5 A 'proof load' is a specified load which a lifting appliance shall withstand without showing permanent set exceeding a specified amount or showing other defect. **1933** G. STEIN *Autobiogr. Alice B. Toklas* v. 139 A good many years later Jane Heap said that she had never appreciated the quality of Gertrude Stein's work until she proof-read it. **1937** W. FOLLETT in *Atlantic Monthly* Jan. 49/1 *The New York Times*..is..the most nearly proofread of the larger metropolitan dailies. **1951** L. HELLMAN *Autumn Garden* I. 14 His publishers..want the manuscript... I'll have to proofread it with him tonight. **1964** F. BOWERS *Bibliogr. & Textual Crit.* IV. iv. 126 The automatic assumption is surely wrong that every forme of cheap commercial printing was necessarily proof-read. **1966** 'H. B. TAYLOR' *Triumvirate* xx. 115 The galleys of proofread type. **1976** *Times Lit. Suppl.* 2 July 813/4 It is smoothly written, although no better illustrated than Professor Donohue's survey..and far less well proofread. **1855** I. C. PRAY *Mem. J. G. Bennet* 41 From this post he was transferred to that of a proof-reader in the printing-house of Wells & Lilly. **1852** GEO. ELIOT *Let.* 15 June (1954) II. 36 Between theatre-going and proof-reading, my spiritual eyes are burning as dim and bleared as gas-lights. **1937** W. FOLLETT in *Atlantic Monthly* Jan. 49/2 Such a thing it is to be equipped, or cursed, with the proofreading eye. **1977** *Early Music* Oct. 470/2 Manuscripts..*A* and *B* show more evidence of proofreading. **1883** 'MARK TWAIN' *Lett. to Publishers* (1967) 162 You must glance through all the proof-slips..for..I have added footnotes and other stuff which you have not seen. **1967** COX & GROSE *Organiz. & Handling Bibliogr. Rec. by Computer* v. 118 The material along with the worksheet and the LC proof slip or the Title II card..is then routed to the Catalogue Department... All books for which there are proof slips or Title II depository cards are processed. **1858** W. J. M. RANKINE *Man. Appl. Mech.* II. iii. 273 Resilience or Spring is the quantity of mechanical work required to produce the proof strain. **1862** [see *proof stress* below]. **1888** J. G. HORNER *Lockwood's Dict. Mech. Engin. Terms* 268 A proof strain would in all cases be short of that which would have a crippling effect. **1811** *Niles' Reg.* I. 311/1 The same process repeated until the ley has acquired proof strength. **1862** W. J. M. RANKINE *Man. Civil Engin.* II. i. 226 Resilience, or Spring..is the quantity of mechanical work required to produce the proof-stress on a given piece of material, and is equal to the product of the proof strain, or alteration of figure, into the mean load which acts during the production of that strain; that is to say, in general, very nearly one-half of the proof load. **1935** *Discovery* Apr. 112/2 By the use of aluminium alloy with a ·1 per cent. proof stress of 17 tons per square inch instead of the 15 tons per square inch alloy now used, the hull weight of such a [flying] boat could be kept as low as 12½ per cent. of the total weight. **1962** *BSI News* June 12/1 Qualities of steels determined by the ratio of the 0·2 per cent proof tensile stress at elevated temperatures to the minimum tensile strength at room temperature. **1952** S. C. KLEENE *Introd. Metamath.* xiv. 213 The proof-theoretic equivalents provability and irrefutability refer only to the enumerable infinity of formal proofs... The set-theoretic notions are actually equivalent to the proof-theoretic ones. **1967** —— *Math. Logic* 118 The proof-theoretic approach to the predicate calculus. **1952** —— *Introd. Metamath.* xiv. 425 The axioms are 'consistent' proof-theoretically. **1967** —— *Math. Logic* 118 We now develop some of the further results proof-theoretically. **1942** D. D. RUNES *Dict. Philos.* 255/2 *Proof theory.* The formalization of mathematical proof by means of a logistic system..makes possible an objective theory of proofs and provability, in which proofs are treated as concrete manipulations of formulas (and no use is made of meanings of formulas). **1969** *Listener* 10 July 44/2 The analogy he was using was one with proof theory in logic, where from initial axioms and rules of inference you can produce true theorems—rather like doing geometry exercises in school. **1979** *Sci. Amer.* Oct. 138/3 For this purpose Hilbert introduced a new theory called proof theory, or metamathematics, in which meaningful statements about the meaningless signs and configurations of the axiomatic system could be formulated.

proof, *v.* Add: **1. c.** To aerate (dough) by the action of yeast before baking. Cf. *PROVE *v.* 1 g.

1875 *Encycl. Brit.* III. 253/2 After this laborious process the finished dough is covered over for some time.. during which fermentation again begins, and the mass is 'proofed'. **1972** *Countryman* Autumn 45 There were three ovens, one above the other, in our bakehouse. The bottom one was small and used only for proofing buns and certain cakes.

d. = *proof-read* vb. s.v. *PROOF *sb.* 18 b.

1960 *Times* 16 Feb. 6/1 Bristol set its own papers and printed and proofed them in the city. **1974** R. C. DENNIS *Conversations with Corpse* viii. 70 Arenas was reading a pink report, making occasional corrections of grammar or punctuation... He handed me the pages he had already proofed. **1979** A. EASSON *Elizabeth Gaskell* i. 44 William also proofed much of her work.

Hence **proofed** *ppl. a.*; **proo·fing** *vbl. sb.*

1902 Proofed, proofing [see PROOF *v.* 2]. **1909** *Westm. Gaz.* 3 Sept. 5/2 Dr. Bartsch, of the Royal Proofing Office at Great Lichterfelde, communicates the result of experiments he has made in disinfecting large quantities of books with hot air. **1940** *Chambers's Techn. Dict.* 678/1 *Proofed tape*, cotton cloth coated with a rubber compound, wrapped round rubber-insulated cables. **1953** *News Chron.* 2 June 2/2 Squatting..in the light-weight (six pound) proofed cotton nylon tents. **1957** *Times* 20 Dec. 19/2 The moth-proofing of wool, the rot-proofing of jute, the mould-proofing of paint. **1958** *Times Lit. Suppl.* 18 Apr. 214/5 We are told nothing about the special proofings of the *Second World War* volumes. **1976** H. WILSON *Governance of Britain* viii. 157 The Conservative leader is fairly well proofed against trouble with the [party] machinery. **1977** *Broadcast* 13 June 7/2 There were..only ten days between the signature of the Report and its proofing and publication.

Prooshian (prū·ʃǎn), *a.* and *sb.* Also Prooshan, Prooshun, Proosian. Joc. var. PRUSSIAN *a.* and *sb.*

1837 [see PRUSSIAN *a.* 2 b]. **1843** DICKENS *Mart. Chuz.* (1844) xix. 239 Some people..may be Rooshans, and some may be Prooshans; they are born so, and will please themselves. **1871** F. C. BURNAND *More Happy Thoughts* (ed. 2) xxix. 214 The Gay Prooshians have no end of ships. **1878** W. S. GILBERT *H.M.S. Pinafore* II. 28 For he might have been a Roosian, A French, or Turk, or Proosian, Or perhaps Itali-an! **1899** KIPLING *Stalky & Co.* 128 'My word!' said M'Turk,.. 'The Prooshian Bates has an infernal straight eye.' **1914** R. BROOKE *Let.* 24 Aug. (1968) 611 To Hell with the Prooshians. **1922** JOYCE *Ulysses* 324 The Prooshians and the Hanoverians.

proot (prūt), *int.* [Etym. obscure.] A command to a donkey to move faster. Hence **proot** *v. intr.* To cry proot. Cf. PROO *int.*

1879 R. L. STEVENSON *Trav. with Donkey* 18 [He] taught me the true cry or masonic word of donkey-drivers, 'Proot!' *Ibid.* 20 'Proot!' seemed to have lost its virtue. I prooted like a lion, I prooted mellifluously like a sucking-dove; but *Modestine* would be neither softened nor intimidated. **1950** L. G. GREEN *In Land of Afternoon* ix. 132 One expert driver..encourages his mules by yelling the principal parts of Greek verbs... The word of command '*Proot!*', which appears to have come from France with the Huguenots, is more commonly heard.

prop, *sb.*[1] Add: **1. g.** *Rugby Football.* One of two outside front-row forwards who support the hooker in a scrummage.

1950 B. H. TRAVERS *Let's talk Rugger* iii. 50 In a 3-4-1 scrum the wing forwards have to push the front-row props towards the hooker all the time. **1959** *Sunday Times* 15 Mar. 40/7 Later in the half, Wood the Irish prop, was hauled back for a five-yard scrum when many people thought he had forced his way over for a try. **1960** E. S. & W. J. HIGHAM *High Speed Rugby* 154 In order to achieve a well-balanced and fairly comfortable scrum, it is desirable to pair off the two props and the two locks so that they are, as nearly as possible, of the same length of body and the same length of leg. **1971** [see *LOCK *sb.*[2] 11 c]. **1977** *Western Morning News* 1 Sept. 10/7 Perhaps the most significant move, however, is the inclusion of Nigel Redgrave, another to rejoin Albion..at loose head prop.

4. **prop forward** *Rugby Football*, = sense 1 g above; **prop-free front** *Coal-mining* (see quot. 1967); **prop-root** [tr. G. *stützwurzel* (K. Goebel *Organographie der Pflanzen* (1901) II. 479)], a root springing from the base of a plant above ground level, providing extra support; **prop-word** (see quot. 1892).

1951 *Sport* 30 Mar.–5 Apr. 6/3 The greatest surprise of the whole 26 is the omission of Bill Hopper, the young Leeds prop forward. **1978** *Rugby World* Apr. 33/2 Mayer belonged to a by-gone age in that, despite his size, he always was scrupulously fair on the field, and the revolution in prop-forward scrummage techniques over the past few years, to some extent, passed him by. **1956** F. S. ATKINSON in D. L. Linton *Sheffield* xiv. 270 Much pioneer work has been done in the coalfield to the north-east of Sheffield and in Nottinghamshire to develop a new method of mining at the coalface, known as the 'prop-free front' system. With this method a strong, flexible scraper-chain conveyor, called an armoured conveyor, is installed in the space between the nearest roof supports and the face of coal being worked... To maintain intact the roof between the vertical supports and the wall of coal, cantilever bars are used which are supported by the props behind the conveyor. **1967** *Gloss. Mining Terms (B.S.I.)* XI. 11 *Prop-free front*, a system of supports in a longwall face in which props are not normally set between the conveyor and the coal. The roof above and in advance of the conveyor is supported by cantilever bars set on props on the goaf side of the conveyor. **1905** I. B. BALFOUR tr. *Goebel's Organogr. Plants* II. 277 A series of transitions..leads us from the soil-roots to those which spring from the base of the stem of many Monocotyledones, and which soon entering the soil serve as prop-roots. **1938** FRITSCH & SALISBURY *Plant Form & Function* x. 109 These prop-roots serve the purpose of augmenting the somewhat feeble primary root-system. **1953** K. ESAU *Plant Anat.* xvii. 474 Others [*sc.* roots] serve mainly as supporting organs, such as the prop roots in the mangrove plants and, on a smaller scale, in the grasses and sedges. **1976** NORSTOG & LONG *Plant Biol.* vi. 165 A conspicuous feature of the majority of the trees [of the tropical rain forest]..is their smooth, thin, lichen-covered bark and their flaring buttresses or, in the case of some smaller trees, stilt-like prop roots. **1892** H. SWEET *New Eng. Gram.* I. 66 Another way of using the adjective without its noun in English is to substitute the unmeaning noun-pronoun *one* for the noun, the inflection of the noun being transferred to the prop-word, as we may call it. **1914** O. JESPERSEN *Mod. Eng. Gram.* II. 248 The reason why the word *one* has been chosen to fulfil the role of a prop-word is chiefly to be sought in the frequent and quite natural use of *one* (by itself) to take the place of a substantive just mentioned. **1934** *Language* X. 370 We call *drinking* in one employment a participle, in another a gerund, *one* now a numeral, now a prop-word, now a pronoun. **1965** *Eng. Stud.* XLVI. 59 A careful survey of the propword 'one' question.

prop, *sb.*[2] Add: **2. b.** A diamond; a valuable piece of jewellery. *Criminals' slang.*

1914 JACKSON & HELLYER *Vocab. Criminal Slang* 66 *Prop*... General circulation amongst pickpockets and looters. A diamond stud originally, now comprehending diamonds in any sense... Example: 'Any heel gun can get a breech poke, but it takes an A1 claw to grab a prop.' **1925** *Flynn's Mag.* 7 Mar. 191 *Prop*, a large diamond. **1931** [see *GROIN *sb.*[1] 3*]. **1971** S. HOUGHTON *Current Prison Slang* (MS.) 17 *Prop*, nice piece of jewellery.

3. *Comb.* (all ? *Obs.*), as **prop-getter, man, -nailer** *Criminals' slang*, one who steals props (sense 2); a pickpocket.

1901 'J. FLYNT' *World of Graft* 220/2 *Prop-getters*, thieves who make a specialty of 'lifting' scarf-pins. **1931** *Police Jrnl.* Oct. 505 *A prop getter*, a thief who steals scarf-pins. **1935** *Amer. Speech* X. 19/2 *Prop man*, a pickpocket or *snatcher* who lifts stickpins containing valuable stones. **1862** H. MAYHEW *London Labour* Extra vol. 25 'Prop-nailers', those who steal pins and brooches. **1886** H. BAUMANN *Londinismen* 146/1 Prop-nailer.

prop, *sb.*[5] Abbrev. of PROPRIETOR.

? **1880** in W. Whitman *Daybks. & Notebks.* (1978) I. 157 Herman Beckurts—prop: Denver *Tribune*. **1913** W. T. ROGERS *Dict. Abbrev.* 157/1 *Prop*..., proprietor. **1956** H. GOLD *Man who was not with It* III. xxviii. 264 The prop. on this busy corner was with it and for it. **1974** A. ROSS *Bradford Business* 15 A painted board..said *Redlands Hotel—props. K. & G. Lyall.* **1980** *Guardian* 25 July 12/8 At Porters..(the Viscount Newport prop.—and head waiter on a busy Thursday evening).

prop, *sb.*[6] *colloq.* [f. PROP(ELLER.] **1.** A propeller, esp. on an aircraft.

1914 *Flight* 10 Jan. 43/2 He made a fine glide from 650ft., making a perfect landing with the 'prop' stationary. **1918** E. M. ROBERTS *Flying Fighter* 239, I crashed into a hedge, smashed my prop to bits, and then the machine landed on its nose in the next field. **1931** *Daily Express* 13 Oct. 15/3 A smiling young man jumped back from the roaring prop. **1935** C. S. FORESTER *Afr. Queen* viii. 139 Must 'a' just 'it a rock with the tip of the prop. **1969** G. MACBETH *War Quartet* 33 We ran, Clumsy in fleece and leather, to the field, Hearing the props whirl. **1974** L. DEIGHTON *Spy Story* xix. 202 The propellers came to a standstill... For a moment the sub became unstable... Then the props picked up speed.

2. *attrib.* and *Comb.* **a.** Simple *attrib.*, as *prop swinger*; in the sense 'propeller-driven', as *prop bomber, plane, trainer.*

1975 *New Yorker* 8 Sept. 25/2 Man's most sophisticated machines of war were sent to hover..over the towns and villages of the Plain of Jars: light spotter planes at 2,000 feet; prop bombers, gunships, and flareships at 5,000 feet; [etc.]. **1965** J. V. PACILIO *Discovering Aerospace* 50 The aircraft engineer would say that a prop plane 'loses its efficiency' at this speed. **1973** *Black Panther* 13 Oct. 14/2 All prop planes were..supposed to land on Runway 31-Right. **1936** F. CLUNE *Roaming round Darling* xiii. 117 The quartet of 'prop swingers' and 'contracters' departed hopefully for the aerodrome. **1974** *Daily Tel.* (Colour Suppl.) 19 Apr. 15 They make jet trainers and prop trainers.

b. Special Combs.: **prop-fan,** (an aircraft engine incorporating) an airscrew having broad blades swept back from a direction perpendicular to the rotation axis; **prop-jet** = *TURBOPROP; **prop-shaft,** a propeller shaft, esp. of a motor vehicle; **prop wash,** a surge or wash of air created by the action of a propeller; also *transf.*

1970 METZGER & GANGER *Results of Initial Prop-Fan Model Acoustic Testing I* (NASA N 71-25785) 1 In order to explore the low noise potential of a Prop-Fan as an aircraft propulsion system, a noise survey was conducted. *Ibid.* 3 The test item consisted of a 21-inch diameter, 12-bladed, manually adjustable pitch, shrouded Prop-Fan model with 22 fixed pitch recovering vanes. **1977** *Jrnl. R. Soc. Arts* CXXV. 352/1 Propeller developments in the guise of so-called prop fans are being studied in the USA. Uninstalled performances as good as those for low pressure ratio fans are estimated at high subsonic speeds. **1977** *New Scientist* 1 Dec. 567/1 The propfan..has eight thin, swept-back blades which allow it to turn at high speeds without encountering the compressibility problems associated with conventional propellers. *Ibid.* 567/2 On a flight from New York to Miami the propfan airliner would take several minutes less than its turbofan counterpart, because of reduced climb and descent times. **1946** P. H. WILKINSON *Aircraft Engines of World* II. 264 Most of the atmospheric jet engines now in production or under development for use in piloted aircraft are turbojets or propjets. **1963** *Engineering* 1 Nov. 560/3 The Dart was the first propjet engine to go into commercial service. **1971** *Flying* Apr. 2/1 (Advt.), The MU-2F and MU-2G outperform competitive prop-jets simply because they are the only propjets in their category designed to utilize a low-drag high-speed wing. **1972** *Daily Colonist* (Victoria, B.C.) 5 Mar. 16/1 The pilot of a Mohawk Airlines propjet which crashed..apparently knew he would not make it to the airport runway. **1964** C. BARBER *Ling. Change Present-Day Eng.* iv. 87 *Prop-shaft*..'propellor shaft'. **1965** *Listener* 17 June 914/3 Telemetering devices which can measure the vibration in a wheel or a prop-shaft. **1976** *Drive* Jan.–Feb. 78/1 Propshaft vibration is an occasional nuisance cured only by diligent garage detective work. **1941** *Amer. Speech* XVI. 168/1 *Prop wash*, an expression of disbelief. (Air Corps.) **1944** *U.S. Air Services* May 16/1 All we could feel was the breathing of tightly packed men..and the animal shudder of the glider as it swung into the prop wash and swung out again. **1958** L. WOLFF *Low Level Mission* 19 The prop-wash blasts of the four engines would interfere with other planes to the rear. **1977** *Observer* 28 Aug. 21/4 The pilot climbed aboard... The diminutive figure.. waved bravely in the prop-wash.

prop, *sb.*[7] *Criminals' slang.* Abbrev. of PROPERTY *sb.* used in *Comb.*, as **prop game,** the practice of defrauding householders into paying exorbitant prices for unnecessary house repairs; **prop man,** one who perpetrates such a fraud. (See also PROPS *sb. pl.* in Dict. and Suppl.)

1966 *Evening Echo* (Bournemouth) 20 Apr. 15/1 The 'prop game' appeared to be..unique to Leeds, something that has cropped up in the last two and a half years. **1966** *Guardian* 13 Dec. 5/4 Gangs operating from Leeds are known as 'the prop men' because the racket began in Leeds when so-called property repairers made exorbitant charges after the gales of February, 1961. **1967** N. LUCAS *CID* viii. 97 The 'prop game'..was a method by which men obtained money from old people by posing as officials.

prop, *v.*[1] Add: **1. b.** In various humorous and ironic phrases. Usu. with *up.*

1908 WODEHOUSE & WESTBROOK *Globe By The Way Bk.* 34 As regards Boarding-The-'Bus and Propping-The-Public-House-Wall, the issue is perhaps more open. But here again I look to see the representatives of the old country well to the fore. **1938** G. HEYER *Blunt Instrument* x. 196 There's a couple propping the wall up at the end of the street. You know the style: kissing and canoodling for the past hour. **1950** 'J. TEY' *To love & be Wise* xi. 133 You'll find him propping up the counter of the post-office. **1965** *Listener* 7 Oct. 539/1 He was to be seen almost every night propping up the left end of the bar in the Wheatsheaf. **1973** J. PATTINSON *Search Warrant* ii. 37 A solitary man was propping up the bar. **1978** B. PRIESTLEY *Island Emperor* iii. 25, I lay in the sun and..I propped up Niki's bar.

4. For *Australian* read 'orig. *Austral.*' and add further examples. Also *transf.*

1928 'BRENT OF BIN BIN' *Up Country* 171 How they raced and propped and wheeled on desperate courses bristling with pitfalls. **1946** *Sun* (Baltimore) 4 Oct. 16/4 Tacato Briar was unprepared for break and propped coming out of gate. **1954** *Ibid.* 10 July 9/3 Sans Egal went to the front at once and opened a lead of some six lengths along the backstretch. However, when he entered the final straightaway, he attempted to 'prop' and lost much of his lead. **1969** T. KENEALLY *Survivor* 70 Seconds later a university sedan, driven by George the university guard, wheeled fast in through the gate and propped at the front of the house. **1970** P. WHITE *Vivisector* 602 The present mob might have trampled Rhoda underfoot if it hadn't suddenly realized she was something beyond its experience, so it propped, and divided.

propædeutic, *a.* and *sb.* **B.** *sb.* **1.** (Earlier example.)

1798 A. F. M. WILLICH *Elem. Crit. Philos.* 19 In the mean time Kant's system, or rather his elementary *Propedeutic* for a system, acquired still greater reputation.

propædia (prōᵘpī·diă). [ad. Gr. προπαιδεία: see PROPAIDEIA.] An introductory volume of the 15th edition of the *Encyclopædia Britannica* (published in 1974) in which information is presented in the form of short outlines. (Cf. *MACROPÆDIA, *MICROPÆDIA.)

1974 [see *MICROPÆDIA]. **1974** *Times* 12 Jan. 12/1 The first volume, propaedia, will be introductory, setting forth a classification of all knowledge into 10 parts, each with a long essay.

propagand, *sb.* (Earlier and further examples.)

1795 W. COBBETT *Bone to gnaw for Democrats* 13 Citizen David, painter to the Propagande, has represented Liberty under the form of a Dragon. **1806** 'C. CAUSTIC' *Democracy Unveiled* (ed. 3) I. 75 Vile propagands in every city Make smooth the path of French banditti.

propagand (prǫpăgæ·nd), *v.* [Back-formation from PROPAGANDA.] *trans.* and *intr.* = PROPAGANDIZE *v.* in Dict. and Suppl.

1901 *Westm. Gaz.* 11 Jan. 2/2 Being free to 'propagand' he has not hesitated to do so. **1923** *Ibid.* 16 May 8/1 Russia was spending large sums out of her Secret Service in order to propagand in the East against British interests. **1935** N. MITCHISON *We have been Warned* II. 208, I expect he'll propagand me a lot. **1938** E. HEMINGWAY *Fifth Column* (1939) 171 That typical French *ivresse* that you were propaganded to believe did not exist. **1948** W. FORTESCUE *Beauty for Ashes* xii. 77 Would I consent to be a voluntary speaker and 'propagand' on platforms?

Hence **propaga·nded** *ppl. a.*; **propaga·nding** *vbl. sb.* and *ppl. a.*

1920 R. FROST *Let.* 19 Sept. (1972) 94 Good luck with the propaganding. **1937** F. P. CROZIER *Men I Killed* xii. 268 They discuss it in awed whispers, well away from the propaganding microphones of the B.B.C. **1958** *Times Lit. Suppl.* 24 Oct. 604/2 Faulty reasoning..and unfelicities like 'propaganding' are a few of the obstacles it [*sc.* a book] presents. **1968** *Economist* 31 Aug. 45/1 The fear-bound and self-propaganded Kremlin leaders. **1971** F. R. LEAVIS in *Human World* Aug. 9 To see it [*sc.* the present government] replaced by one that has an alternative party-backing, representing a proclaimed and propaganded different policy and programme.

propaganda. Add: **2. a.** (Earlier examples.)

1790 J. MACPHERSON *Let.* 27 Sept. in A. Aspinall *Corresp. George, Prince of Wales* (1964) II. 98 All Kings have..a new race of Pretenders to contend with, the disciples of the propaganda at Paris or, as they call themselves, Les Ambassadeurs de genre humain. **1797** *Gentl. Mag.* Aug. 687 The Propaganda, a society whose members are bound, by solemn engagements, to stir up subjects against their lawful rulers.

3. The systematic propagation of information or ideas by an interested party, esp. in a tendentious way in order to encourage or instil a particular attitude or response. Also, the ideas, doctrines, etc., disseminated thus; the vehicle of such propagation.

1908 LILLEY & TYRRELL tr. *Programme of Modernism* 102 The Church..soon felt a need of new methods of propaganda and government. **1911** G. B. SHAW *Blanco Posnet* 324 Though we tolerate..the propaganda of Anarchism as a political theory..we clearly cannot.. tolerate assassination of rulers on the ground that it is 'propaganda by deed' or sociological experiment. **1929** G. SELDES *You can't print That!* 427 The term propaganda has not the sinister meaning in Europe which it has acquired in America... In European business offices the word means advertising or boosting generally. **1938** R. G. COLLINGWOOD *Princ. Art* ii. 32 Where a certain practical activity is stimulated as expedient, that which stimulates it is advertisement or (in the current modern sense, not the old sense) propaganda. **1957** R. N. C. HUNT *Guide to Communist Jargon* 132 The Soviet Government not only has an elaborate machinery for conducting such propaganda abroad..but also does the same at home through the press, radio, films etc. **1974** *Anderson* (S. Carolina) *Independent* 23 Apr. 4A/6 CIA went on employing propaganda fronts long after anybody except professionals on both sides was paying any attention to the propaganda. **1976** A. J. RUSSELL *Pour Hemlock* xiv. 166 White propaganda, the truth; gray, a composition of half-truths and distortions; or black, a pack of lies.

4. *attrib.* and *Comb.*, as (sense *3) *propaganda campaign, chief, film, fund, leaflet, poster, raid, technique, war, warfare.* (See also sense 2 b in Dict.) **propaganda machine,** an organization responsible for the dissemination of propaganda.

1937 KOESTLER *Spanish Testament* vi. 133 One of the most effective propaganda campaigns launched by the rebels was that relating to the alleged shooting of hostages by the Madrid Government. **1974** *Encycl. Brit. Macropædia* XV. 39/2 Today several hundred more or less scholarly books and thousands of articles shed substantial light on the psychology, techniques, and effects of propaganda campaigns, major and minor. **1942** *Short Guide Gr. Brit.* (U.S. War Dept.) 1 The first and major duty Hitler has given his propaganda chiefs is to separate Britain and America. **1950** KOESTLER in *God that Failed* 27 The absurdity of a propaganda-chief who only reads his own paper. **1973** D. MAY *Laughter in Djakarta* x. 161 It's a propaganda film..anti-neo-colonialism. **1978** CADOGAN & CRAIG *Women & Children First* x. 224 According to widely shown propaganda films, the most adept German spy was bound to give himself away eventually through mispronunciation. **1842** *Communist Chronicle & Communitarian Apostle* I. v. 77 The propaganda fund shall be devoted to the propagation of the doctrines of communism. **1947** F. FRENAYE tr. *C. Levi's Christ stopped at Eboli* (1948) xvii. 162 When their ship came back to Trieste from Odessa, Communist propaganda leaflets were found on board. **1978** A. WAUGH *Best Wine Last* xv. 178 The propaganda leaflets that our aeroplanes scattered behind the German lines. **1948** Propaganda machine [see *MACHINE *sb.* 8]. **1972** H. MACINNES *Message from Malaga* ii. 35 A propaganda machine is only as effective as people are stupid. **1978** F. MACLEAN *Take Nine Spies* iv. 153 The Russians only responded with counter-blasts from their own propaganda machine. **1945** *New Yorker* 31 Mar. 52/1 A propaganda poster pasted on the remaining walls. **1979** *Listener* 1 Nov. 604/3 English [war] propaganda posters are bland alongside those of America, Holland or Sweden. **1934** *Ann. Reg. 1933* I. 181 On July 1 occurred the first of a series of propaganda raids by German aeroplanes over the Austrian frontier, when leaflets abusing the Dollfuss Government were dropped. **1927** H. D. LASSWELL (*title*) Propaganda technique in the World War. **1975** *New Yorker* 21 Apr. 133/1 The Communists are taking full advantage of their highly developed propaganda techniques. **1838** tr. *Recoll. Caulincourt, Duke of Vicenza* I. iv. 74 The English Cabinet was well aware that a propaganda war was impossible as long as Russia should continue allied to France. **1974** D. SEAMAN *Bomb that could Lip-Read* vi. 49 The I.R.A...are already winning the propaganda war, the one that finally matters. **1979** *Guardian* 22 Feb. 6/1 The [Russian] propaganda war against China continues to intensify. **1942** *R.A.F. Jrnl.* 18 Apr. 31 Propaganda warfare in the field was used.

propaganda (prǫpăgæ·ndă), *v.* [f. the sb.] *trans.* = PROPAGANDIZE *v.*

1921 J. F. PORTER *Sir Edward Elgar* 10 Elgar..never attempted to propaganda his work. **1949** H. L. MENCKEN in *Philologica: Malone Anniversary Stud.* 317 There is..a desire to get rid of circumlocution and the waste of words, as in..to propaganda, to steam-roller, to belly-ache.

propagandic. For *rare*$^{-0}$ in Dict. read *rare* and add examples.

1939 W. FORTESCUE *There's Rosemary* xxxix. 242 Next morning I studied my newspaper to see if my propagandic statistics had been faithfully recorded and was aghast to see, in 'leaded caps', a paragraph headed *The Baby fizzled.* **1946** L. KREY in W. S. Knickerbocker *20th Cent. English* 409 Just such grossly indecent or propagandic publications as often yield fabulous returns.

propagandism. Add: (Earlier examples.)

1800 *Aurora* (Philadelphia) 8 May 3 A war undertaken by the coalesced power against the system of *Propagandism* by which they have been menaced. **1807** *Weekly Inspector* (N.Y.) 28 Mar. 75/1 We have ever been disposed to attribute the wonderful success of the French, since their revolutionary era, to *Propagandism* or, in other words, to the poison of their principles, circulated by their emissaries; and corrupting the *mind* of the nations they proposed to attack.

propagandist, *sb.* (*a.*) Add: **A.** *sb.* **1.** (Earlier and later examples.)

1797 BURKE *Two Lett. on Conduct of our Domestick Parties* 109 How can I help it if this Royal propagandist will preach the doctrine of the rights of men? **1929** J. FINEBERG tr. *Lenin's What is to be Done?* in *Coll. Wks.* IV. II. 147 A propagandist..must present 'many ideas', so many indeed that they will be understood as a whole only by a (comparatively) few persons. **1942** *Sun* (Baltimore) 12 Jan. 8/1 The rumors have been fostered by Hitler's propagandists. **1976** *Daily Tel.* 20 July 3/1 Dr Goebbels, the Nazi propagandist, could not have invented a more vilifying tag than that of the Black Panther.

B. *adj.* (Earlier examples.)

1824 D. WEBSTER *Speech on Greek Revolution* 5 It may be easy to call this resolution *Quixotic*, the emanation of a crusading or propagandist spirit. **1833** *Blackw. Mag.* June 933 Portugal..has been abandoned..to the revolutionary spoliation and propagandist arts of France.

propagandistic *a.* (further examples); **propagandistically** *adv.* (further example).

1941 G. G. SCHOLEM *Major Trends in Jewish Mysticism* viii. 306 Men of tireless literary and propagandistic activity. **1957** I. ASIMOV *Earth is Room Enough* (1960) 19 Such propagandistic lies were not uncommon. **1976** *New Yorker* 26 Jan. 66/3 Kim Il Sung figures he has made enough headway, even propagandistically, at the United Nations for the moment. **1977** *N.Y. Rev. Bks.* 23 June 4/2 There was no such propagandistic cause as anti-communism to impel those peach-cheeked youngsters to wage a war against an enemy caught up in the thrall of a fanatical, even suicidal nationalism.

propagandize, *v.* Add: **a.** (Further example.) Also, to subject (a person) to propaganda; to encourage to a belief thus.

1928 *Observer* 11 Mar. 13/4 A crowd of the peasants.. tries to 'propagandize' an American soldier. **1933** *Sun* (Baltimore) 3 Oct. 12/1 Those who have only money enough..should not be propagandized into spending beyond their means. **1938** *Daily Tel.* 7 Jan. 12/4 They had too much common-sense to be propagandized. **1969** *Wall St. Jrnl.* 14 Feb. 1/6 South Korea..propagandizes its citizens heavily. **1974** tr. *Sniečkus's Soviet Lithuania* 31 The Party..propagandised Marxist-Leninist theory and intensified the ideological training of its members.

b. (Further examples.) Also, to disseminate propaganda.

1967 *Cold Spring Harbor Symp. Quant. Biol.* XXXII. 7/2 Being honest, I always have to mention when I am propagandizing against cigarettes—that 80% of heavy cigarette smokers do *not* die of lung cancer. **1974** K. MILLETT *Flying* III. x. 335, I propagandize and make coffee. **1977** *New Yorker* 6 June 137/1 He propagandized for wilderness preservation as well as urban amenity.

Hence **propaga·ndizing** *ppl. a.* and *vbl. sb.*

1855 J. D. HOWARD in *N. Amer. Rev.* LXXX. 5 The early conquests of the Saracens, then, are to be ascribed ..to the propagandizing spirit of their new faith. **1860** *Even. Jrnl. Tract* No. 13. 2 What class of men north and south did Mr. Fillmore represent? The old Whig or Clay party south who had no sympathies with the slave propagandising element of the Democratic party. **1927** S. BENT *Ballyhoo* iii. 87 The propagandizing of screened officials. **1978** *N.Y. Times* 30 Mar. B2/1 Here is a perfectly legal industry..based on the commercial exploitation and propagandizing of something that is illegal. **1979** *Country Life* 8 Nov. 1687/4 'Folk song' is a silly, sentimental, misleading and propagandising concept.

propagate, *v.* Add: **4.** (Examples of the active voice.)

1973 *Sci. Amer.* Feb. 73/2 Internal waves propagate energy through the body of a stratified fluid in much the same way that energy is propagated by waves at the surface. **1975** *Nature* 8 May 157/1 The flagellum of this organism propagates waves both distally and proximally in common with other trypanosomes.

b. *refl.* for *passive.*

1880 [see *cathode ray* s.v. *CATHODE c]. **1908** tr. *Suess' Face of Earth* III. IV. i. 4 Some kind of wave propagating itself freely through the crust of the earth.

c. *intr.* for *passive.* To be propagated, to travel.

1943 F. E. TERMAN *Radio Engineers' Handbk.* III. 255 Waves of a variety of types may propagate down a wave guide. **1957** J. J. STOKER *Water Waves* x. 374 In practically all of this book we assume that the medium in which waves propagate is water. **1966** C. R. TOTTLE *Sci. Engin. Materials* vii. 168 Griffith in 1920..proposed that glass possessed many fine cracks in the surface, which could propagate through the material and cause failure. **1967** *Oceanogr. & Marine Biol.* V. 31 The elevation ζ may be specified as a function of position and time, representing an external surge propagating into the sea area. **1969** *Sci. Jrnl.* Dec. 44/3 In a waveguide the microwave energy propagates down the inside of a hollow conductor. **1971** *Nature* 3 Dec. 292/2 When a laser beam propagates through a mixture of gases. **1974** *Sci. Amer.* Jan. 38/3 In the nerves of both higher and lower animals it is the cell membrane that..enables the nerve impulse to be set up and to propagate. **1977** *Nature* 21 July 203/2 The proton is knocked out of the nucleus and could be observed propagating freely after the collision.

propagating *ppl. a.* (examples of the sense 'travelling').

1971 *Sci. Amer.* June 22/1 The energy stored in the inverted population is then available to amplify a propagating light wave at a particular frequency. **1972** *Ibid.* Jan. 18/3 The muffling of seismic signals might be achieved by conducting the test in an underground material where a comparatively small fraction of the energy of the explosion would appear in a propagating seismic wave. **1973** *Physics Bull.* Nov. 657/3 It may be helpful to consider first the ways in which propagating gravitational waves are like electromagnetic waves.

propagation. Add: **7.** *Chem.* In a chain reaction, the step or series of steps in which product molecules are formed or polymer chains lengthened, but which is self-perpetuating by virtue of the regeneration or relocation of reactive centres; e.g. in polymerization, reaction of a radical with a molecule of monomer to form a longer radical. Freq. *attrib.*

1928 *Proc. R. Soc.* A. CXXII. 621 The conditions which govern the propagation of the stable reaction chains. **1940** *Ann. Rep. Progr. Chem.* XXXVI. 74 When an active molecule possesses such a long lifetime, the rate of formation and destruction of these centres plays no part in the rate of polymerisation, which is then solely determined by the velocity of the propagation reaction. **1950** *Thorpe's Dict. Appl. Chem.* (ed. 4) X. 88/1 This is followed by the propagation stage, in which successive additions of monomer molecules occur, maintaining an odd electron, characteristic of a free radical, at the end of the growing chain. **1964** J. M. CRABTREE et al. tr. *V. N. Kondrat'ev's Chem. Kinetics of Gas Reactions* ix. 613 The chain propagation processes involving HO_2 radicals are important in the so-called slow oxidation of hydrogen which occurs.. at relatively high pressures. **1973** K. J. SAUNDERS *Org. Polymer Chem.* i. 10 This new radical then adds further monomer molecules in rapid succession to form a polymer chain. In this propagation the active centre remains, being continuously relocated at the end of the chain.

8. *attrib.* and *Comb.*, as *propagation tray*; **propagation coefficient, constant,** or **factor** *Physics*, the coefficient of the distance in an equation representing the propagation of a wave (quot. 1943 represents an equivalent def.); **propagation function** *Physics*, = *PROPAGATOR 3.

1943 *Gloss. Terms Telecommunications (B.S.I.)* 7 *Propagation coefficient, propagation constant*, the natural logarithm of the vector ratio of the steady-state amplitudes of a wave at a specified frequency, at points in the direction of propagation separated by unit length. **1911** J. A. FLEMING *Propagation Electr. Currents* ii. 68, *P* is a complex quantity and therefore may be written in the form $\alpha + j\beta$. It is called the Propagation constant of the line. **1963** R. W. DITCHBURN *Light* (ed. 2) ii. 28 The wavelength..is denoted by λ. An associated constant $\kappa = 2\pi/\lambda$ is called the wavelength constant (or propagation constant). **1976** G. R. OLHOEFT in R. G. J. Strens *Physics & Chem. of Minerals & Rocks* 262 All of the material properties (other than geometric boundary conditions) which are important to the description of the propagation of an electromagnetic wave (or lack of propagation) are described in the propagation constant by the quantities μ, ϵ, and σ. **1958** CONDON & ODISHAW *Handbk. Physics* IV. vii. 108/2 The solution is a plane wave..varying periodically in time with the frequency $\nu = \omega/2\pi$ and advancing in the $+x$ direction through space with a complex propagation factor $\gamma = j\omega\sqrt{\epsilon^*\mu^*} = \alpha + j\beta$. **1949** *Physical Rev.* LXXVI. 770/1 Many of the properties of the integrals are analyzed using formal properties of invariant propagation functions. **1970** J. SCHWINGER *Particles, Sources, & Fields* iii. 145 Considering spin 0 particles and their real scalar sources,..we examine the effect of adding an additional weak source $\delta K(x)$. It is given by $\delta W(k) = \int (dx)\delta K(x)\phi(x)$, where $\phi(x) = \int (dx')\times\Delta_+(x-x')K(x')$. This combination of source and propagation function, measuring the effect of pre-existing sources on a weak test source, is the *field* of the sources. **1977** *Grimsby Even. Tel.* 31 May 9/5 A hundred small cannabis plants in propagation trays, plus smoking pipes and an LSD tablet were found.

propagator. Add: **1. c.** (Later examples.) Also, a small box with a transparent lid and a base that can be heated, for germinating seeds or raising seedlings.

1914 W. F. ROWLES *Garden under Glass* i. 16 The propagator itself may consist of a box..covered by loose sheets of glass. **1950** E. J. KING *Propagation of Plants* iii. 29 A propagator is a small frame built inside a heated greenhouse in such a way as to give even higher temperatures than those in the main part of the house. **1971** *Daily Tel.* 13 Feb. 7/3 A week or two lost in raising seeds in a heated greenhouse, or electric propagator in the home, is not a matter of extreme urgency. **1976** *Abingdon Herald* 9 Dec. 5/2 A small propagator..enables one to raise seeds at any time. **1979** *Garden* CIV (Advt.), Humex Mono-top Propagator. Pick of the bunch—the electrically heated propatray base surmounted by an.. enclosure with sliding doors for easy access.

d. The male copulative organ; the penis. Now *arch.*

1670 J. OGILBY *Africa* 451 Lastly, they have little Bellies, broad Feet, long Toes, and furnish'd, as most of the Blacks upon the Guinee Coast, with large Propagators. **1971** *Black Scholar* June 12/1 Setting foot on the shores of West Africa in 1550, the Englishman was struck by the African's religion..and his 'Propagator', which he perceived monstrous.

3. *Physics.* An algebraic function that is taken as representing the propagation of a particle on the sub-atomic scale, esp. between its space–time points of creation and annihilation.

1951 *Physical Rev.* LXXXIV. 1233/1 G' (1,2), the 'quantum propagator', is a function of $(x_{\mu 1}x_{\mu 1})$ not containing any Dirac operators. **1958** *Ibid.* CXII. 1417/1 A coincidence arrangement appears capable of testing the electron propagator at distances approaching the nucleon Compton wavelength, which is comparable with our present limit on the photon propagator. **1973** J. A. REISSLAND *Physics of Photons* vi. 186 If the effect of a perturbation $H'(t)$ is to create a particle at position $\mathbf{r}'$ at time t', the function (6.19) traces the propagation of that particle until it is removed at $\mathbf{r}$ at a time t. Hence the term 'propagator' for the one-particle Green's function $G(\mathbf{r},t;\ \mathbf{r}',t') \equiv i\eta(t-t')\langle a(\mathbf{r},t)a^+(\mathbf{r}',t')\rangle$ where $a^+(\mathbf{r}',t')$ creates a particle at $\mathbf{r}',t'$ and $a(\mathbf{r},t)$ destroys it at $(\mathbf{r},t)$. **1974** *Physics Bull.* May 191/2 Propagators (or Green's functions) have been much used by physicists.

propagule. Restrict *rare* to sense in Dict. and add: **2.** A seed, spore, or other product of a plant which is disseminated to form a new individual; also, occasionally used as a name for the products of asexual reproduction in certain lower animals. Hence **propa·gular** *a.*, of or pertaining to a propagule.

1905 F. E. CLEMENTS *Res. Methods Ecol.* iv. 241 Dormant seeds and propagules are abundant. **1959** *New Biol.* XXVIII. 80 Extrinsic barriers between breeding populations..effectively prevent or check gene flow because they are not crossed by adults, or by seeds, pollen, or other 'propagules'. **1965** BELL & COOMBE tr. *Strasburger's Textbk. Bot.* 204 Propagules may consist of single cells (spores), groups of cells (gemmae), or entire organs or complexes of organs (bulbils). **1967** *Oceanogr. & Marine Biol.* V. 494 Several hydroids are particularly adapted.. to ensure asexual reproduction by means of propagular stolons whose brittle and curved extremities can catch a neighbouring leaf. **1969** *New Scientist* 20 Nov. 402/1 Any fungal propagules landing on its surface will be killed. **1974** *New Phytologist* LXXIII. 981 Wallace pointed out that the small propagules of Compositae..allowed that family to gain an early footing on islands. **1977** *Nature* 3 Nov. 48/2, I report here that the marine red alga *Centroceras* produces missile-shaped propagules which are carried away by ocean currents.

propamidine (propæ·midīn). *Pharm.* [f. PROP(ANE + AMID(E + -INE[5].] A diamidine with bactericidal and fungicidal properties which is used, usu. in the form of its isethionate (a white, hygroscopic powder), in dressing minor wounds or burns; 1,3-di(4-amidinephenoxy)propane, $CH_2[CH_2{\cdot}O{\cdot}C_6H_4.C(NH){\cdot}(NH_2)]_2$.

1941, **1951** [see *PENTAMIDINE]. **1964** W. G. SMITH *Allergy & Tissue Metabolism* iv. 56 Mast cell degranulation will account for the histamine liberating activity of compound 48/80. This is also true of the histamine releasing properties of..propamidine. **1976** 'J. HERRIOT' *Vets might Fly* (1977) ii. 20 Lord Hulton was a devotee of May and Baker's Propamidine Cream and used it for all minor cuts and grazes in his cattle.

propane. Add: **2.** *attrib.*, as *propane gas.*

1974 *Janet Frazer Catal.* Spring/Summer 458/4 Portable refrigerator... Suitable for use with Butane or Propane gas. **1979** 'J. ROSS' *Rattling of Old Bones* iii. 25 The propane-gas lamps..were flooding the cupboard with hissing white light.

propanediol (prōᵘ:pē¹ndəi·ǫl). *Chem.* [f. PROPANE + DI-[2] + -OL.] = *propylene glycol* s.v. *PROPYLENE 2; also, a derivative of this.

1894 G. M'GOWAN tr. *Bernthsen's Text-bk. Org. Chem.* (ed. 2) 206 Propylene glycol is known in two isomeric forms, viz.: (*a*) Tri-methylene glycol† or β-Propylene glycol,..(*b*) α-Propylene glycol‡. [*Note*] † 1-3-Propanediol. ‡ 1-2-Propane-diol. **1952** *Jrnl. Pharmacol. & Exper. Therapeutics* CV. 452 The 1,3-propanediol ethers have greater protective action in electroshock seizures than the corresponding position isomers of the 1,2-propanediol series. **1966** *McGraw-Hill Encycl. Sci. & Technol.* XIV. 16/2 According to basic chemical structure, four categories of tranquilizers are delineated: phenothiazines..; rauwolfia derivatives..; propanediols, for example, meprobamate or Miltown; diphenylurethanes. **1976** *Ann. Rev. Microbiol.* XXX. 421 A clear example of evolution to reverse function in the laboratory is the evolution of propanediol catabolism through recruitment of enzymes normally involved in L-fucose and L-lactate metabolism.

propanidid (propæ·nidid). *Pharm.* [Arbitrarily f. PROPAN(E.] A colourless or yellowish oily liquid which is given intravenously in solution as a short-acting anæsthetic; propyl-4-diethyl-carbamoylmethoxy-3-methoxy-phenylacetate, $C_{18}H_{27}NO_5$.

1964 *Brit. Jrnl. Anæsthesia* XXXVI. 655/1 The new intravenous narcotic propanidid..displays a biphasic action on ventilation consisting of stimulation followed by depression. **1965** *Med. Jrnl. Austral.* 21 Aug. 329 Propanidid is a unique, short-acting intravenous anæsthetic agent with many interesting features—in particular, its mode of inactivation. **1971** PRYOR & MACALISTER *Gen. Anæsthetic & Sedation Techniques Dentistry* x. 63 Propanidid has made a tremendous contribution towards making the induction of anaesthesia pleasanter for the patient and working conditions easier for the dental surgeon.

propanol (prōᵘ·pănǫl). *Chem.* [f. PROPAN(E + -OL.] Either of the isomers of propyl alcohol; sometimes *spec.* normal propyl alcohol, 1-propanol (cf. *isopropanol* s.v. *ISO- b).

[**1892** *Proc. Chem. Soc.* 16 June 128 Citric acid, on this system, would be named propanoltrioic acid or simply propanoltri-acid.] **1894** G. M'GOWAN tr. *Bernthsen's Text-bk. Org. Chem.* (ed. 2) 92 Normal propyl alcohol. [*Note*] 1-Propanol. *Ibid.* 93 Secondary propyl alcohol. [*Note*] 2-Propanol. **1926** H. G. RULE tr. *Schmidt's Text-bk. Org. Chem.* I. 111 Propyl alcohol, propanol, $C_3H_7.OH$. Both of the theoretically possible structural isomerides of this formula are known. **1951** I. L. FINAR *Org. Chem.* I. iv. 62 Propylene, propene, C_3H_6, may be prepared by heating propanol or *iso*-propanol with sulphuric acid. **1964** N. G. CLARK *Mod. Org. Chem.* viii. 140 Two isomeric propanols exist, of which the primary alcohol is less important. **1964** J. C. MATTHEWS *Mod. Chem. Course* xxv. 302 Add 20–25 drops of ethanol or propanol slowly with shaking.

propantheline (propæ·nþēlīn). *Pharm.* [f. PROP(YL + *X)ANTH(ENE + *-el* (f. E(THY)L) + -INE[5].] A white, bitter-tasting powder, $C_{23}H_{30}NO_3Br$, which is a quaternary ammonium bromide derived from xanthene, and is a parasympatholytic agent used esp. for treating the symptoms of peptic ulceration. Also *propantheline bromide.* Cf. *PRO-BANTHINE.

1954 *Brit. Jrnl. Pharmacol.* IX. 218 (*heading*) A comparison of the peripheral parasympatholytic and autonomic ganglion blocking activities of methantheline ('Banthine') and propantheline ('Pro-Banthine') with atropine and hexamethonium. *Ibid.* Propantheline (2'-di*iso*propylaminoethyl xanthene-9-carboxylate methobromide) is the isopropyl analogue of methantheline and differs from it only in degree of activity.., its actions being qualitatively similar. **1955** *Lancet* 10 Sept. 526/2 Propantheline bromide can be used to relieve colic, whether due to spasm of the sphincter or associated with biliary reflux and pancreatitis. **1958** [see *PRO-BANTHINE]. **1970** O. L. WADE *Adverse Reaction to Drugs* ii. 13 Atropine or propantheline in the doses required to reduce gastric motility and gastric secretion in the treatment of peptic ulcer may cause unpleasant dryness of the mouth or blurring of vision. **1977** *Lancet* 26 Nov. 1134/1 The effects of propantheline bromide on the eyes should be considered when the drug is used in the treatment of hyperhidrosis.

propassion. (Later examples.)

1875 H. E. MANNING *Internal Mission of Holy Ghost* xiv. 392 In our Divine Lord there were no passions... We say indeed, that there were in Him pro-passions. **1876** —— *Glories of Sacred Heart* ix. 266 All those pro-passions, as they are called—because the Church never speaks of passions when it speaks of the Sacred Heart.

propellent, *a.* and *sb.* Add: Now usu. spelt **-ant.** **A.** *adj.* (Further examples.)

1919 R. H. GODDARD in *Smithsonian Misc. Coll.* LXXI. II. 6 This enables high chamber pressures to be employed..and also permits most of the mass of the rocket to consist of propellant material. **1945** *Soap & Sanitary Chemicals* Apr. 125/3 Methyl chloride is an excellent propellant gas for aerosols to be used against insects out of doors.

B. *sb.* **a.** (Further examples.)

1917 W. S. CHURCHILL in M. Gilbert *Winston S. Churchill* (1977) IV. Compan. I. 125 What proportion of our total explosive or propellant output could be based upon 10 million gallons of whiskey? **1931** L. H. MORRISON *Amer. Diesel Engines* v. 129 The explosion of these lighter parts of the fuel provides the propellant whereby the remainder of the fuel is injected into the engine cylinder in a finely atomized condition. **1936** W. S. CHURCHILL in *Second World War* (1948) I. 538 By ammunition is meant projectiles (both bombs and shells) and cartridge-cases containing propellent. **1939** W. BEVERIDGE *Blockade & Civilian Population* 8 Fats, broadly speaking, are all directly convertible into munitions, because they can be used, and are very largely used, in making propellants. **1966** *McGraw-Hill Encycl. Sci. & Technol.* V. 153/1 Besides black powder, which is mainly used in sporting rifles, the common gun propellants are either nitrocellulose or a mixture of nitrocellulose and nitroglycerine.

b. A substance that is used (alone, or reacting with another) as a source of the hot gases that provide the thrust in a rocket engine. Cf. *FUEL *sb.* 2*c, *OXIDIZER 1, *MONOPROPELLANT *sb.* and *a.*

1919 R. H. GODDARD in *Smithsonian Misc. Coll.* LXXI. II. 67 Let us assume, for case (a) (many small secondary rockets), as well as for case (b) (large secondary rockets), that the ratio of mass of metal to mass of propellant is the minimum reasonable amount that can be expected. **1948** M. J. ZUCROW *Princ. Jet Propulsion & Gas Turbines* xii. 467 The propellants employed in a rocket motor may be a solid, two liquids (fuel plus oxidizer), or materials containing an adequate supply of available oxygen in their chemical composition (mono-propellants). **1955** *Times* 29 June 16/3 Some five years ago we faced a difficult decision as to the most suitable propellents for the type of rockets which we were developing. **1957** *Spaceflight* I. 51/1 A propellant can be either fuel or oxidizer. **1967** *New Scientist* 21 Sept. 594/2 Project officials ordered a series of critical tests to determine just how much propellant would be available for use after the main retrorocket had been fired. **1974** *Sci. Amer.* Aug. 7/1 (Advt.), Launch weight of the spacecraft was only 1,108 pounds, including 66 pounds of propellant and 122 pounds of science instruments.

c. The compressed fluid in an aerosol container or the like that causes its contents to be ejected.

1945 *Soap & Sanitary Chemicals* Apr. 127/1 Nitrous oxide also has a pressure too high for it to be practical as a propellant. **1957** H. R. SHEPHERD in E. Sagarin *Cosmetics Sci. & Technol.* xxxvi. 804 While some aerosol cosmetics are made with a single propellent, it is customary to use a mixture or a solution of propellents. **1966** *McGraw-Hill Encycl. Sci. & Technol.* V. 279/1 In the small first-aid water fire extinguisher, a propellant must be provided. Usually this is carbon dioxide. **1973** *Daily Tel.* 21 Aug. 15/7 The propellent in the anti-perspirant caused freezing in the lung tissues resulting in loss of oxygen and death. *Ibid.* 15/8 A variety of compressed or liquified bases are used in aerosols and technically known as propellents. Most are halogenated hydrocarbons, which are also refrigerants, and sniffing of such propellents is a recognised form of drug abuse in Britain and the United States. **1978** *N.Y. Times* 30 Mar. 62/3 The nitrous-oxide cartridges are made for use as the propellant in restaurant-sized equipment for whipping cream and non-dairy toppings.

propeller. Add: Also propellor. **3. a.** (Further example.)

1980 *Daily Tel.* 22 Aug. 2/3 SMM Engineering..has won an £800,000 order..for a propellor and a bow steering unit for an 80,000-ton tanker.

c. A device on an aeroplane analogous to a ship's propeller in form and function; now always, an airscrew (formerly, a pusher as opposed to a tractor).

1842 W. S. HENSON *Brit. Patent* 9478 2 In the place of the movement or power for onward progress being obtained by movement of the extended surface or plane, as is the case with the wings of birds, I apply suitable paddle wheels or other proper mechanical propellers. **1843** G. CAYLEY in *Mechanics' Mag.* 8 Apr. 278/1 The broad horizontal rudder, or tail, H, capable of being turned on its hinge to any angle,..gives the power of ascent and descent when the propellers are used. **1853** —— *Let.* 22 June in J. L. Pritchard *Sir George Cayley* (1961) 263 When perfected this toy [*sc.* Chinese flying top] furnishes a very beautiful specimen of the action of the screw propeller in air. **1871** *English Mechanic* 27 Jan. 448/2 Through the instrumentality of the hollow bladed screw-propeller.. shall we..be enabled to sustain and propel the mechanism which we shall construct, in and through the atmosphere. **1888** H. MIDDLETON *Brit. Patent* 9725 5 Propellors and supporters for flying machines consisting of curved flexible wings with ribs. **1894**, etc. [see *airscrew* s.v. *AIR *sb.*[1] B. III. 6]. **1908** H. G. WELLS *War in Air* viii. 271 She..swept down to the water..and came..rolling and..writhing.., halting and then coming on again, with her torn and bent propeller still beating the air. **1917** R. B. MATTHEWS *Aviation Pocket-bk. 1917* iv. 93 An airscrew is described as a tractor when placed in front of the main planes, and as a propeller when fitted behind. **1928** *Trans. Inst. Mining Engineers* LXXVI. 110 A considerable amount of research-work has been carried out on screw propellors in wind-tunnels. **1937** J. H. YOUNGER et al. *Airplane Maintenance* xv. 298 A propeller develops thrust by its reaction on a mass of air, which it pushes backward. **1968** MILLER & SAWYERS *Technical Devel. Mod. Aviation* vi. 205 The advantage of the turboprop stems essentially from the greater efficiency of the propeller than the jet as a means of propulsion at low speeds and for take-off. **1977** D. ANTHONY *Stud Game* vi. 38 A lounge decorated with..an ancient propeller Lindbergh might have used.

5. *propeller efficiency*; *propeller-driven* adj.; **propeller fan,** a fan that produces a flow of air parallel to the axis of rotation of its impeller; *esp.* one with the impeller unenclosed or in a very short casing that does not restrict the air flow; **propeller shaft,** a shaft transmitting power from an engine to a propeller or to the driven wheels of a motor vehicle; **propeller turbine** = *TURBOJET; usu. *attrib.*

1953 D. O. DOMMASCH *Elem. Propeller & Helicopter Aerodynamics* iii. 69 The development of the turbo-prop engine has made flight at high subsonic Mach numbers possible for propeller-driven aircraft. **1973** *Black Panther* 13 Oct. 14/2 They had steered 553 in behind a small propeller driven plane. **1909** *Westm. Gaz.* 4 Mar. 4/2 Propeller efficiency also constitutes one of the most serious drawbacks to the experimenter. **1897** *Proc. Inst. Mech. Engineers* 469 The propeller fan was for volume of air without compression, the centrifugal for volume with compression. **1937** C. KELLER et al. *Theory & Performance Axial-Flow Fans* i. 1 In axial-flow or propeller fans the air or gas flows essentially in an axial direction through the fan. **1950** T. H. F. HOLMAN et al. *Textbk. Heating & Ventilating* vi. 77 Fans are of two main types, the propellor fan..and the centrifugal fan. **1966** W. C. OSBORNE *Fans* ii. 47 Larger propeller fans of up to 24 ft (7·5 m) or more in diameter are often really axial flow fans in very short casings, since such a design is mechanically more satisfactory. *Ibid.*, Large 'propeller' fans are commonly used on cooling towers. **1970** M. G. LUFF *Air Conditioning* viii. 261 In terms of ventilation and air conditioning there are three main types of fan, the propeller fan, the axial flow fan and the centrifugal fan. **1839**, etc. Propeller shaft [in Dict.]. **1913** A. E. BERRIMAN *Aviation* iii. 24 In monoplanes the propeller-shaft is ordinarily on a level with the middle of the body. **1945** P. H. WILKINSON *Aircraft Engines of World* 345 The drive shaft..drives the two concentric propeller shafts for the two contra-rotating propellers through planetary spur reduction gears. **1962** *Which? Car Suppl.* Oct. 143/1 New propellor shaft coupling. **1973** *Daily Tel.* 13 July 13 (Advt.), There's no ugly propeller shaft tunnel in the floor of the car to spoil the stretch of deep-pile carpets. **1945** *Jrnl. R. Aeronaut. Soc.* XLIX. 196/2 Perhaps the large blower of the jet or propeller turbine unit may encourage the study of boundary layer control possibilities of higher maximum lift coefficients. **1951** *Engineering* 6 July 8/1 The propeller-turbine engine could be expected to improve appreciably in power output and specific weight as combustion temperatures were increased from current values. **1968** MILLER & SAWYERS *Technical Devel. Mod. Aviation* vi. 183 (*heading*) Propeller turbines succeed with the Viscount. *Ibid.*, The propeller-turbine engine was of more immediate interest to the airlines than the jet, and by 1950 the idea of fitting it to existing airplanes like the DC-6 and Constellation was becoming widely accepted in the United States.

propelling, *vbl. sb.* and *ppl. a.* Add: **b.** *ppl. a.* Also in *Comb.*, as *propelling pencil*, a mechanical pencil containing a screw by which the lead may be projected and retracted.

1895 *Montgomery Ward Catal.* 185/1 Men's Propelling and Repelling Pencil..takes full size and length of lead. **1933** D. L. SAYERS *Murder must Advertise* x. 169 If Bredon really wanted a propelling pencil he ought to get an Eversharp. **1955** S. BECKETT *Molloy* 136 Between my fountain-pen and my propelling-pencil. **1962** L. DEIGHTON *Ipcress File* xi. 71 Murray used a propelling pencil with changeable coloured leads. **1976** M. BUTTERWORTH *Remains to be Seen* ii. 15 Silver propelling-pencil poised over the page.

propensely, *adv.* Add: **3.** Favourably, readily. *nonce-use.*

1922 JOYCE *Ulysses* 409 If I had poor luck with Bass's mare, perhaps this draught of his may serve me more propensely.

proper, *a.* (*adv.*, *sb.*) Add: **I.** *adj.* **2. c.** *Physics.* (See quot. 1924.) [tr. G. *eigen*(*zeit*) proper (time) (H. Minkowski 1908, in *Nachr. von der k. Ges. der Wissensch. zu Gottingen* (Math.-phys. Klasse) 103).]

1916 *Monthly Notices R. Astron. Soc.* LXXVI. 704 The element *ds* integrated along the geodetic line gives $s = \int ds$. This is what Minkowski calls the proper-time of the material particle. **1923** *Proc. R. Soc.* A. CII. 530 Where *m* and ($-e$) are the 'proper mass' and charge of the electron respectively. **1924** A. S. EDDINGTON *Math. Theory Relativity* i. 34 Quantities referred to the space-time system of an observer moving with the body considered are often distinguished by the prefix proper- (German, *Eigen-*), e.g. proper-length, proper-volume, proper-density, proper-mass = invariant mass. **1942** P. G. BERMAN *Introd. Theory Relativity* iv. 41 Whenever the two events can be just connected by a light ray which leaves the site of one event at the time it occurs and arrives at the site of the other event as it takes place, the proper time interval τ_{12} between them vanishes. **1952** C. MØLLER *Theory of Relativity* iv. 137 While q^0/c^2 expresses the source density of proper mass, we see that the source density for relativistic mass is $((\mathbf{f}.\mathbf{u}) + q)/c^2$. **1964** W. G. V. ROSSER *Introd. Theory Relativity* iii. 102, l_0..is the length of the fish measured in the coordinate system in which it is at rest; l_0 is called the proper length of the fish, whilst l..is the length of the fish measured in an inertial reference frame relative to which the fish is moving with uniform velocity. **1970** *Nature* 17 Oct. 272/1 Proper mass (equivalently, rest-mass, proper energy, or rest-energy) is the most important Lorentz-invariant scalar associated with any system.

3*. *Math.* and *Physics.* Used in collocations as a translation of G. *eigen* own, proper, characteristic.

a. Applied to a vibration or oscillation: = *NORMAL *a.* 2 c.

1873 *Proc. London Math. Soc.* IV. 258 The problem of determining the proper tones of any spherical cavity bounded by rigid walls. **1909** *Westm. Gaz.* 4 Sept. 10/1 All elastic bodies, including metals, when made fast at one end, vibrate when subjected to a shock from outside. The vibrations so caused are what are known as proper vibrations. **1962** S.-I. TOMONAGA *Quantum Mech.* I. i. 16 It was possible to arrange the proper oscillations in order by giving each of them an integral number *s* which has the physical meaning that $(s-1)$ is the number of nodes of the oscillation.

b. = *EIGEN-, as *proper function* = *EIGENFUNCTION; *proper value* = *EIGENVALUE.

1930 RUARK & UREY *Atoms, Molecules & Quanta* xv. 526 Such an aggregate of E values is often referred to as a 'spectrum of characteristic values', or 'proper values'. **1935** PAULING & WILSON *Introd. Quantum Mech.* iii. 58 The functions $\psi_n(x)$ which satisfy Equation 9-8 and also certain auxiliary conditions..are variously called wave functions or eigenfunctions (Eigenfunktionen), or sometimes amplitude functions, characteristic functions, or proper functions. **1938** [see *EIGENVALUE]. **1958** [see *LATENT *a.* i]. **1975** GRAY & ISAACS *New Dict. Physics* 574/1, ψ must always be finite... The integral of $|\psi|^2$ over all space must be equal to 1... Wave functions obtained when these conditions are applied are called proper wave functions and form a set of characteristic functions of the Schrödinger wave equation. These are often called eigenfunctions and correspond to a set of fixed energy values in which the system may exist, called eigenvalues (proper values).

II. 5. c. *Math.* (i) Applied to any subset (subgroup, etc.) that does not constitute the entire set.

1906 W. H. & G. C. YOUNG *Theory of Sets of Points* iii. 16 A set which is contained entirely in another set is called a component of the latter set, and, if there are points of the latter set not belonging to the former set, it is said to be a proper component of the other. **1937** R. D. CARMICHAEL *Introd. Theory of Groups of Finite Order* i. 28 A subgroup of *G* which is not identical with *G* is called a proper subgroup of *G*. **1953** A. A. FRAENKEL *Abstract Set Theory* i. 21 A subset of *S* which is different from *S*, is called a proper subset. **1965** B. MITCHELL *Theory of Categories* i. 6 If the monomorphism $\alpha: A' \to A$ is not an isomorphism, we shall call A' a proper subobject of *A*.

(ii) Applied to a subgroup (subring, etc.) that does not constitute the entire group, and has more than one element.

1953 W. LEDERMANN *Introd. Theory of Finite Groups* ii. 31 Every group *G* has two trivial or improper subgroups, namely, *G* itself and the group which consists of the unit element by itself ($I^2 = I$); all other subgroups are called proper subgroups. **1965** PATTERSON & RUTHERFORD *Elem. Abstract Algebra* ii. 39 A subgroup of *G* other than *G* or *E* is called a proper subgroup. *Ibid.* iii. 100 Every non-zero ring has two ideals, namely the ring itself and the subset consisting of 0 alone. Any ideal which is not one of these two is called a proper ideal.

III. 10. c. *proper Bostonian*: = *BRAHMIN b. Also *attrib.* or as *adj.*

1947 C. AMORY (*title*) The Proper Bostonians. *Ibid.* i. 12 Outside observers have claimed to be able to tell the Proper Bostonian male by waistcoat, and the Proper Bostonian female by hat. **1956** C. W. MILLS *Power Elite* iii. 58 Proper Bostonians and proper San Franciscans.. would be genuinely embarrassed..[by] cheap publicity. **1969** A. LASKI *Dominant Fifth* v. 180 Daughter of a not particularly wealthy and certainly not Proper Bostonian American. **1973** R. L. SIMON *Big Fix* i. 10 Her proper Bostonian background, the old shipping family back on Lewisburg Square. **1977** J. CLEARY *High Road to China* vii. 231 She was only a mild rebel: there was still too much of the Proper Bostonian in her.

B. *adv.* **1.** Also, correctly, in a genteel manner (of speech).

1915 *Dialect Notes* IV. 188 Talk proper before your teacher. **1952** M. ALLINGHAM *Tiger in Smoke* ii. 41 Perhaps she'll 'ave another go at teachin' me to speak proper, pore soul. **1966** F. Shaw et al. (*title*) Lern yerself Scouse. How to talk proper in Liverpool. **1980** *Listener* 22 May 665/1 He has not learnt how to talk proper.

properdin (prǫpǝ·ɹdin). *Phys.* [f. L. *prō-* Pro-¹ + *perd-ere* to destroy (see Perdition) + -in¹.] A protein found in the blood and concerned with the body's response to infection.

1954 *Science* 20 Aug. 279/1 This protein, tentatively named *properdin*.., acts only in conjunction with complement and Mg⁺⁺ and participates in such diverse activities as the destruction of bacteria, the neutralization of viruses, and the lysis of certain red cells. Properdin is a normal serum constituent and differs from antibody in many respects, particularly in its lack of specificity and in its exact requirements for its interactions. [*Note*] Hans Hirschmann..suggested this name. **1963** C. J. C. Britton *Whitby & Britton's Disorders of Blood* (ed. 9) xiv. 356 Properdin levels are low in acute leukæmia. **1975** Spragg & Gigli in Mathieu & Kahan *Immunol. Aspects Anesthetic & Surg. Pract.* vi. 116 Properdin has been highly purified as a distinct protein with a molecular weight of 223,000 daltons.

properly, *adv.* Add: **6.** *Math.* So as to form a proper subset or a proper subgroup (see *proper *a.* 5c).

1965 E. Schenkman *Group Theory* iv. 125 If Π is a set of primes, a Sylow Π-subgroup of a group G is a Π-subgroup of G not properly contained in any Π-subgroup of G. **1968** E. T. Copson *Metric Spaces* i. 6 The set A is..properly contained in B if every member of A belongs to B and there is at least one member of B which does not belong to A. **1971** E. C. Dade in Powell & Higman *Finite Simple Groups* viii. 326 The group H..is properly contained in G.

Propertian (prǫpǝ·ɹʃ¹ăn), *a.* [f. L. *Propertius* (see below) + -an.] Belonging to or characteristic of Sextus Aurelius Propertius, Latin elegiac poet of the first century B.C., or his poetry.

1871 J. R. Lowell *My Study Windows* 217 Goethe, who was classic..in his 'Hermann and Dorothea', and at least Propertian in his 'Roman Idyls', wasted his time..on the mechanical mock-antique of an unreadable 'Achilleis'. **1918** E. Pound *Let.* 22 Nov. in *Lett. J. Joyce* (1966) II. 424, I hope my Propertian ravings will amuse you. **1930** W. S. Maugham *Cakes & Ale* xiv. 161 Love lyrics and elegies in the Propertian manner. **1974** *Classical Q.* XXIV. 96 The verb of the first clause is rendered in English by the pluperfect. The Propertian passage displays a structural similarity.

propertied, *a.* **3.** Delete *nonce-use* and add later example.

1909 M. E. Albright *Shakesperian Stage* 147 The Elizabethan stage..was little more than a union of the old *sedes* and *plateæ* of the moralities, or the propertied and unpropertied stages of the interludes.

property, *sb.* Add: **2. d.** *ellipt.*, shares or investments in property.

1964 *Financial Times* 3 Mar. 19/5 There was a little more interest in Properties, sentiment being helped by last Friday's Gallup poll. **1977** *Evening Post* (Nottingham) 24 Jan. 16/9 Properties ran into profit-taking with Haslemere 176p, MEPC 62p, Land Securities 161p, and Stock Conversion 161p on offer.

e. A person (esp. one engaged in show business) regarded as a commercial asset, esp. in phr. *hot property* (cf. *hot *a.* 6a, 9b), a success or sensation, a 'hit'. *colloq.*

1958 J. Blish *Case of Conscience* xiv. 153 Signor Egtverchi is now a hot property... Suddenly..he is worth a lot of money. **1969** *Rolling Stone* 28 June 22 The Hagers, potentially hot property, now have Record One. **1980** M. Gilbert *Death of Favourite Girl* xii. 114 Katie was a big property by then and..naturally I was ready to talk about her.

5. e. *Linguistics.* An intrinsic aspect or function.

1953 C. E. Bazell *Linguistic Form* iii. 38 The acoustic and articulatory property-complexes are 'genuine' aspects of the phonemes. **1962** E. F. Haden et al. *Resonance-Theory for Linguistics* iv. 49 Every entity in language has a Property and a Form... The Property of each entity is internal to it, corresponding to its function in the complex of which it is a part. **1965** N. Chomsky *Aspects of Theory of Syntax* iv. 160 In any given linguistic system lexical entries enter into intrinsic semantic relations of a much more systematic sort than is suggested by what has been said so far. We might use the term 'field properties' to refer to these..aspects.

8. a. *property account, developer, -holder* (earlier example), *-owner* (earlier and later examples), *right, speculator, value*; *property-based, -owning* adjs.; **property bond,** a share or bond in property; **property mark,** a mark indicating ownership; **property qualification** (earlier example); **property tax** (earlier and later examples).

1869 *Bradshaw's Railway Man.* XXI. 417 Expended... Property accounts—materials..[$]139,463. **1974** *Terminol. Managem. & Financial Accountancy* (Inst. Cost & Managem. Accountants) 58 *Real or property account*, the record of an asset, (e.g. buildings, plant and equipment, cash, etc.). **1957** K. A. Wittfogel *Oriental Despotism* 2 The modern property-based system of industry. **1974** tr. *Wertheim's Evolution & Revolution* 27 Only one of the filiation lines leads to social progress, namely the one passing through the 'property-based', 'multicentered', or ..'open' society. **1970** *Daily Tel.* 17 Jan. 22/3 The considerable expansion of property values since the war..is the great selling point for property bonds, compared with other investment plans. **1972** *Accountant* 12 Oct., Property bonds..reflect the value of the property owned by the property fund without being subject to the vagaries of the stock exchange. **1970** *Harper's Bazaar* Oct. 76/1 Property developers..wreaked vandalism upon the cities and countryside of England. **1977** M. Walker *National Front* v. 125 A property developer called Roy Bramwell. **1824** *Deb. Congress U.S.* (1856) 18th Congress 1 Sess., App. II. 3129 The memorial of the..property-holders of the city of Baltimore. **1899** *Amer. Anthropologist* Oct. 601 Property marks are used very frequently by the Eskimo tribes of Alaska. They occur almost exclusively on weapons used in hunting, which, after being dispatched, remain in the bodies of large game. **1865** *Harper's Mag.* July 154/2 It is the nightmare of property-owners. **1941** W. Temple *Citizen & Churchman* v. 75 What are the rights of property-owners in respect of the property which they own? **1979** C. E. Schorske *Fin-de-Siècle Vienna* ii. 46 The property owners of the inner city..feared the competition of vast new housing construction. **1923** *Spectator* 19 May 837/2 It remains to state as clearly as may be what means lie ready to develop a property-owning democracy. **1978** *Countryman* LXXXIII. 37 (*title of poem*) Towards a property-owning democracy. **1807** *Deb. Congress U.S.* 16 Nov. (1852) 916 The Constitution of the United States requires no property qualification in the elected. **1942** W. Temple *Christianity & Social Order* ii. 27 Men are sinful, so property-rights are needed, not so much for the satisfaction of the rich as for the protection of the poor. **1968** *Listener* 27 June 847/1 A friend likened it [*sc.* a vacuum cleaner] to a property speculator's cocktail cabinet. *a* **1974** R. Crossman *Diaries* (1975) I. 46 Nobody in a Labour Cabinet is going to object to an action which is extremely popular outside London and which will only ruin property speculators. **1808** in *57th Rep. R. Comm. Hist. Manuscripts* 139 in *Parl. Papers 1902* (Cd. 931) LIII. 1 How do the farmers with you talk of the *Property Tax*? **1978** *N.Y. Times* 30 Mar. B6/3 The legislative leaders and Governor Carey agreed today to..offer low-income taxpayers, particularly the elderly, a property-tax protection program. **1914** *Proc. 6th Nat. Conf. City Planning* (U.S.) 102 Suddenly he finds his property values injured..because someone has chosen to construct a small retail store. **1979** V. S. Naipaul *Bend in River* vi. 109 The big recent rise in property values in the town.

b. Now also *Cinemat.* (*b*) *property manager, truck, wagon, woman, workshop*; **property-plot** (example); **property-room** (earlier examples). See also *Property box 2.

1959 W. S. Sharps *Dict. Cinematogr.* 121/2 *Property manager*, the person responsible for obtaining, storing and supplying all the inanimate items required for a set. **1933** P. Godfrey *Back-Stage* iv. 44 The property-plot is a detailed list of every article of furniture and every other stage accessory used throughout the play. **1784** J. Byng *Torrington Diaries* (1934) I. 176 In the property room is a profusion of wigs, and truncheons. **1829** H. Foote *Compan. Theatres* 38 In the line with the flies, over the auditory, are carpenters' shops, property-rooms, store-rooms, &c. **1961** Bowman & Ball *Theatre Lang.* 280 *Property truck*, a wagon offstage on which properties can be placed until needed. **1963** *Movie* May 19/3 He sat in the back of the property truck writing the ending. **1895** *McClure's Mag.* V. 55/1 The baggage-wagons and the property-wagons have stopped near the dressing-rooms. **1808** *Monthly Pantheon* I. 692/2 His wife was (in the technical language of the theatre) a dresser and property-woman. **1829** H. Foote *Compan. Theatres* 38 Beneath it is the *printing-office*; and over it are property workshops.

pro·perty box. *Theatre.* [f. Property *sb.* + Box *sb.*²] †**1.** A seated compartment (Box *sb.*² 8) which may be privately rented. *Obs.*

1812 *Dramatic Censor 1811* 53 And Ladies Carlisle and Jersey rent Property boxes, which reduces the number of Ladies seceders to fourteen. **1828** J. Ebers *Seven Yrs. King's Theatre* ix. 243 In this year (1825) and the preceding one, the terms existing in the property-boxes expired.

2. A box (Box *sb.*² 1) in which stage properties are stored.

1864 P. Paterson *Glimpses Real Life* 3 Richard's truncheon we knew was in the property-box. **1890** B. Hall *Turnover Club* ii. 28, I met on the rocks, with a 'property-box', A gloomy theatrical man.

propertyless, *a.* Add: (Earlier and later examples.) Also *ellipt.* as *sb.*

1822 W. Cobbett *Cottage Econ.* 107 They were formerly the sons and daughters of small farmers; they are now the progeny of miserable property-less labourers. **1886** [in Dict.]. **1912** Belloc *Servile State* i. 16 Two classes of free citizens, the one capitalist or owning, the other propertyless or proletarian. **1941** R. Humphreys *Latin Amer.* 20 For the Indian and the propertyless, independence meant not new freedom but new masters. **1977** *Dædalus* Summer 72 They predicted, in Hegelian fashion, an abolition of private property and class ownership of the means of production by a regenerated communism among the propertyless masses.

Hence **pro·pertylessness,** the state of being propertyless.

1964 M. McLuhan *Understanding Media* xxxiii. 357 Hence the specter of joblessness and propertylessness in the electric age.

prophage (prōᵘ·fēⁱdʒ). *Biol.* [Contraction of F. *probactériophage* (Lwoff & Gutmann 1950, in *Ann. de l'Inst. Pasteur* LXXVIII. 734): see Pro-² and *phage.] The form which a temperate phage has in a lysogenic bacterium: it is incorporated into and replicates with the bacterial genome, and is only potentially lytic.

1951 [see *induce *v.* 4 e]. **1955** *Sci. Amer.* Apr. 93/1 Some prophages control the production of substances by their hosts (e.g., diphtheria toxin) or have other important effects on them. **1969** A. M. Campbell *Episomes* vi. 81 Prophage insertion is a special kind of recombination process. **1973** R. G. Krueger et al. *Introd. Microbiol.* xviii. 507/1 The evidence suggested that the virus resided in a novel state as a prophage and that the presence of the prophage or phage genome in the cell rendered the cell immune to vegetative replication of other similar temperate phage particles by the same mechanism that limited its own vegetative replication.

propham (prōᵘ·fæm). [f. Pro(pyl + Ph(enyl + Carb)am(ate.] Isopropyl *N*-phenylcarbamate, $C_6H_5·NH·CO·O·CH(CH_3)_2$, a white crystalline substance used as a selective herbicide to control weeds among crops, esp. during germination.

1955 *Proc. Brit. Weed Control Conf. 1954* I. 183 The almost complete failure of the more insoluble propham, CIPC and CMU to give control of wild oats may also be connected with the inability of the material to reach the wild oats at the critical time at Site I, where TCA was successful. **1958** *Weed Control Handbk.* (Brit. Weed Control Council) i. 19 For the control of wild oats in sugar beet and peas, propham has given results comparable with those obtained with trichloroacetic acid when applied in the same way—before the final seed-bed cultivations. **1973** Ashton & Crafts *Mode of Action of Herbicides* xiii. 202 Propham is rapidly broken down by micro-organisms in the soil; such degradation is promoted by warmth and moisture. For this reason propham has proved most useful in cool season crops.

prophase. Substitute for entry in Dict.: **prophase** (prōᵘ·fēⁱz). *Cytology.* [ad. G. *prophase* (E. Strasburger 1884, in *Arch. f. mikrosk. Anat.* XXIII. 250): see Pro-² and Phase.] The first stage in a mitotic or meiotic nuclear division, preceding prometaphase, during which the chromosomes become visible and shorten and the nuclear envelope disappears. Also *attrib.* or as *adj.*

Prophase of the first meiotic division is divided into *leptotene, *zygotene, *pachytene, *diplotene, and *diakinesis, in that order.

1884 *Jrnl. R. Microsc. Soc.* IV. II. 714 In both [plants and animals] we find that in the 'prophases' of nuclear division cytoplasm is collected at the future poles of the cell-nucleus. **1887, 1898** [in Dict.]. **1903** *Bot. Gaz.* XXXV. 250 The first division in Salamandra is characterized by a long period of growth of the cell and nucleus during the prophase..; the first longitudinal fission taking place during the prophase and the second during the metaphase or anaphase. **1911** *Q. Jrnl. Microsc. Sci.* LVII. 27 The most important difference between the pre-meiotic and meiotic prophases is the entire absence in the former of any appearance of fusion of chromatin threads, such as takes place in zygonema. **1931** E. B. Ford *Mendelism & Evolution* I. i. 18 Blocks of genes cross over together, as would be expected if, in fact, this phenomenon is due to the twisting of the chromosome threads round each other during the prophase of the reduction division. **1948** R. A. R. Gresson *Essent. Gen. Cytol.* v. 33 (*caption*) Prophase chromosome to show arrangement of chromosomes. **1957** C. P. Swanson *Cytol. & Cytogenetics* iii. 49 Prophase is said to be initiated at the moment when the chromosomes become visibly distinct. **1970** Ambrose & Easty *Cell Biol.* ix. 296 In late prophase..the spindle develops and the asters move apart.

Hence **propha·sic** *a.*

1913 *La Cellule* XXIX. 311 The first indication that the prophasic changes have begun is seen in the breaking down of the anastomoses along the lines between the heavier portions of the reticulum. **1929** *Bot. Gaz.* LXXXVIII. 376 In the latter case the visible division of the chromonema would always be prophasic. **1951** H. C. Bold in G. M. Smith *Man. Phycol.* xi. 225 The Florideae furnish a conspicuous exception in that the spermatium nucleus is usually described as in a prophasic condition at the moment of union.

prophet, *sb.* Add: **II. 5. c.** (Earlier examples.)

1843 *Ainsworth's Mag.* III. 220 What's to win the Derby?.. What say the prophets? **1862** *Times* 31 Dec., Prophets, tipsters and welshers—the parasites of the ring.

III. 6. c. *prophet-like* (example).

1857 G. H. Lewes *Biogr. Hist. Philos.* (ed. 2) II. 319 Now grave, prophetlike, and impassioned.

propho (prǫ·fo). *slang* (orig. *U.S.*). [f. Proph(ylaxis + *-o².] Prophylaxis of venereal disease. Also *attrib.*

1919 L. L. Lincoln *Company C, 11th Engineers* 37 In his efforts to get away with it he paid many visits to 'Doc. Propho'. **1921** J. Dos Passos *Three Soldiers* IV. 202 That's one thing you guys are lucky in, don't have to worry about propho. **1925** —— *Manhattan Transfer* III. i. 281 Just my propho kit. **1959** J. Braine *Vodi* vii. 115 Jack..had pointed out that there were no propho stations in Civvy Street.

prophylactic, *a.* and *sb.* Add: **B.** *sb.* **b.** A condom.

Condoms were formerly used more for their prophylactic properties against venereal disease than for their contraceptive properties.

1943 [see *preventive *sb.* c]. **1950** 'D. Divine' *King of Fassarai* xviii. 143 'What were you doing before this?' 'Handin' out prophylactics in Baltimore.' **1964** G. McDonald *Running Scared* v. 73 Prophylactics made them [*sc.* women] more willing... People could have intercourse..without restraint. **1972** C. Potok *My Name is Asher Lev* iii. x. 259 Along the..beach lay..beer cans, bits of paper, a prophylactic. **1975** *Listener* 27 Nov. 728/3 The GI wore his packet of prophylactics in his cap and propositioned every woman in sight.

prophyll (prōᵘ·fil). *Bot.* [f. Pro-² + Gr. φύλλ-ον leaf.] **1.** Now the usual form of *prophyllon* (Pro-² 1). (Examples.)

1905 I. B. Balfour tr. *Goebel's Organogr. Plants* II. 382 Prophylls are characterized first of all by their position. We find them..usually in pairs at the base of the lateral shoots. **1921** J. Small *Textbk. Bot.* vii. 75 The first few leaves on a lateral branch are described as prophylls. There are commonly two, and they may be spiny..or just small and simpler than the other leaves... The latter prophylls are usually called bracteoles. **1953** K. Esau *Plant Anat.* xvi. 413 The first cataphylls on a lateral shoot are called prophylls. **1976** Bell & Coombe tr. *Strasburger's Textbk. Bot.* (rev. ed.) 131 These first leaves of the side shoot, which are frequently of simple form.., are termed prophylls.

2. = *protophyll.

1971 D. W. Bierhorst *Morphol. Vascular Plants* ii. 24/2 This [*sc.* the protocorm] is a parenchymatous moss bearing..a number of avascular, leaflike structures ('prophylls').

propinquitous, *a.* Add: For *nonce-wd.* read *rare* and add further examples.

1941 *Cold Spring Harbor Symp. Quant. Biol.* IX. 155/1 The tendency to propinquitous union. **1974** R. Adams *Shardik* xxviii. 234 Let us just step outside for a stroll in some nice, lovely place with no propinquitous walls or bushes.

propio-, propion-. Add: **propiola·ctone,** a pungent, colourless, liquid β-lactone, $CH_2·CO·O·CH_2$, which is used as a disinfectant; also **β-propiolactone**; **propionyl** (earlier example).

1917 *Chem. Abstr.* XI. 2577 Propiolactone..was prepd. by treating an aq. soln. of $CH_2ICH_2CO_2Na$ with $AgNO_3$ and stirring the resulting ppt. for 15–25 min. **1952** *Chambers's Jrnl.* Apr. 255/2 Treating natural wool with a chemical called propiolactone increases its diameter and density without appreciable effect upon its surface properties. **1966** *McGraw-Hill Encycl. Sci. & Technol.* I. 463/2 β-Propiolactone works best at high humidities and is active even below 10° C. **1850** *Phil. Trans. R. Soc.* CXL. 129 The existence of a series of bases of the formula $C_nH_{n+1}N$, *i.e.* bases containing formyl, acetyl, propionyl (metacetyl), butyryl, &c., appears to be still doubtful.

propionic, *a.* **1.** (Earlier example.)

1850 [see *propyl].

propitiative (propi·ʃi‚ătiv), *a.* *rare.* [f. Propitiate *v.* + -ive.] Tending to propitiate; propitiatory, conciliatory.

1928 *Observer* 19 Feb. 17/2 Where the majority of passengers have to travel in a brutalising congestion, the sight of half-empty 'firsts' next door is not propitiative.

proplastid: see *pro-² 1.

propodus (prǫ·podǝ̆s). *Zool.* [f. Pro-² + Gr. πούς, ποδ- foot.] = Propodite.

1945 T. H. Savory *Spiders Brit. Isles* (ed. 2) 196 The propodus ends in a large claw and often a pair of smaller accessory claws. **1976** *Nature* 11 Mar. 136/1 Among the factors affecting the strength of a crab's master claw is the cross-sectional area of muscle in the manus (that part of the propodus behind the apposing fingers).

proportional, *a.* and *sb.* Add: **A.** *adj.* **1.** *proportional counter*, an ionization chamber in which the voltage between the electrodes is great enough to produce gas amplification but not so great that the output pulse ceases to be proportional to the initial ionization; so *proportional counting*; *proportional limit* (Mech.), the maximum stress to which a body or material can be subjected without a departure from the proportionality of stress and strain.

1906 *Proc. R. Soc.* XII. 427 Bars tested in torsion to some stress below the elastic limit (proportional limit of bar) and the direction of torsion reversed immediately. **1937** *Rev. Sci. Instruments* VIII. 254 (*heading*) Properties of the proportional (Geiger-Klemperer) counter. *Ibid.* 255/1 The proportional or Geiger-Klemperer counter which is operated in a range of potential below discharge but above the region of ionization by collision has been used in several series of experiments, notably where the detection of protons in the presence of gamma-radiation is required. **1939** *Rev. Mod. Physics* XI. 213/2 With the gases mentioned, stable proportional counting and freedom from extremely critical voltage control can only be achieved with the aid of high gain amplifiers. **1950** D. H. Wilkinson *Ionization Chambers* vi. 157 With simple precautions as to the allowed region of initial ionization the proportional counter becomes a precision instrument, and may be used as an accurate measure of particle energy. **1950** E. E. Wahlstrom *Introd. Theoret. Igneous Petrol.* v. 115 Both the elastic limit and the proportional limit vary with the time through which the stress is applied. **1958** Faires & Parks *Radioisotope Lab. Techniques* xi. 111 In order to obtain the high electric fields required to produce gas amplification, the proportional counter usually takes the form of a metal cylinder, having a fine wire, insulated at its ends, stretched along its length. **1961** G. R. Choppin *Exper. Nuclear Chem.* xi. 186 This fact and the sensitivity to impurities make Geiger-Müller counting less attractive than proportional counting. **1971** *Nature* 16 Apr. 448/1 Several investigators have measured the diffuse cosmic X-ray intensity in the 1/4 keV band by means of rocket-borne gas proportional counters. **1972** Mallows & Pickering *Stress Anal. Probl. S.I. Units* ii. 37 It is common..to use the proportional limit, elastic limit, and yield stress as alternatives for one another in design calculations: since they are usually close to one another, this is quite acceptable.

2. b. *proportional representation*, a system of parliamentary representation based on numerical (rather than regional) divisions of the electorate, *spec.* one by which each party is represented in proportion to the numerical strength of the vote it receives, usually by means of a method of transferable vote (see Transferable *a.* in Dict. and Suppl.).

1870 *Putnam's Mag.* June 720/1 When once the theory of proportional representation is reduced to practice..it will assert its superiority. **1873** Mill *Autobiogr.* vii. 302 The two greatest improvements which remain to be made in Representative Government... One of them was Personal, or, as it is called with equal propriety, Proportional Representation. **1884** [see sense 2 in Dict.]. **1909** [see Transferable *a.*]. **1917** H. G. Wells in *Times* 30 Mar. 7/5 The essential point to grasp is that Proportional Representation is not a novel scheme, but a carefully worked-out remedy for universally recognized ills. **1940** F. A. Hermens *Democracy & Proportional Representation* 1 There are few devices of democratic government on which opinions differ so sharply as on proportional representation. **1952** [see *Gaullist *a.* and *sb.*]. **1963** J. Grimond *Liberal Challenge* xi. 314, I believe ..that the best solution might be to keep the present system for the Commons but introduce proportional representation for a second chamber. **1976** *Scotsman* 24 Dec. 4/3 A series of amendments to provide for the Scottish and Welsh assemblies to be elected by proportional representation were tabled yesterday.

proportionalism. **2.** (Further example.)

1954 B. & R. North tr. *Duverger's Pol. Parties* ii. iii. 377 In all three P.R. replaced a modified majority system (second ballot in Switzerland and Norway, some features of proportionalism in Denmark).

proportionality. Add: **1. b.** *limit of proportionality* (Mech.) = *proportional limit.

1888 W. C. Unwin *Testing of Materials of Construction* iii. 65 Between A and B a sensible, but slight, curvature appears in the diagram, and a sensible, though small, deviation from proportionality begins to appear in the stresses and strains. Bauschinger calls the point A the limit of proportionality. **1930** *Engineering* 3 Jan. 31/1 The figures for limit of proportionality..were obtained from stress-strain diagrams. **1971** B. Scharf *Engin. & its Lang.* iv. 22 The following values may be obtained by means of a tensile test: limit of proportionality..; yield stress..; proof stress [etc.].

c. *constant* or *factor of proportionality* (also *proportionality constant, factor*): the constant ratio of one variable to a second to which it is proportional.

1919 *Smithsonian Misc. Coll.* LXXI. ii. 9 The cross section, S, should obviously be as small as possible; and this condition will be satisfied at all times, provided it is the following function of the mass of the rocket (M − m), $S = A(M - m)^{2/3}$, where A is a constant of proportionality. **1937** *Rev. Sci. Instruments* VIII. 255/1 These are all connected with..proportionality between resultant electrical impulse and initial ionization per particle. The proportionality factor should not depend much upon the kind of particle entering the counter. **1955** J. Lindhard in W. Pauli *Niels Bohr* 188 Accordingly, ω and the average excitation potential become proportional to Z, i.e. $\hbar\omega = I = Z.I_0$. However, it was found difficult to calculate the proportionality constant I_0. **1962** F. I. Ordway et al. *Basic Astronautics* x. 396 The proportionality factor C_F, called thrust coefficient, is a result of the expansion of the combustion gases through the nozzle.

2. A formula affirming the proportionality of two or more quantities.

1954 A. Dresden tr. *van der Waerden's Science Awakening* vi. 176 The proportionality a:b::c:d means that na > mb implies nc > md, and na = mb implies nc = md, and na < mb implies nc < md, no matter how the integers m and n are chosen. **1976** *Nature* 27 May 301/2 The luminosity of a star of solar mass satisfies the approximate proportionality $L \propto G^7M^5 \propto t^2$.

proposal. **1.** Restrict †*Obs.* to sense in Dict. and add: **b.** *Philos.* (See quots.) *rare.*

1932 W. E. Johnson in *Mind* XLI. 1 (*title*) Probability: the relations of proposal to supposal. *Ibid.* 5 Any statement of implication, whether formal or material, may be translated into the language of probability by taking the implicans as supposal and the implicate as proposal. **1944** *Mind* LIII. 98 Following Johnson I shall call the proposition on the left of the solidus the 'proposal' and that on the right of it the 'supposal'.

3. c. An offer or tender. Now *U.S.*

1748 [in Dict., sense 3 a]. **1914** *Chicago Tribune* 8 May 14 Sealed proposals plainly marked on the outside 'Proposals for coal'..will be received at the Indian Office. **1935** H. W. Horwill *Dict. Mod. Amer. Usage* 244/1 *Proposal* is sometimes used in Am. in the sense of *tender*.

propose, *v.* Add: **3. e.** Also with *for* and (more usually) *to*. (Further examples.)

1855 Thackeray *Newcomes* II. xiii. 127 Perhaps neither of them will propose for her. **1872** A. C. Steele *Broken Toys* II. xxv. 167, I am going to Vere Court tomorrow to propose for Nella Vere. **1911** G. B. Shaw *Getting Married* 227 Sykes. When her blood boils about it ..she doesnt care what she says. *Reginald.* Well: you knew that when you proposed to her. **1928** E. O'Neill *Strange Interlude* ix. 341 Ned's just proposed to me. I refused him, Charlie. **1931** H. Walpole *Judith Paris* ii. 311 She had been amazed when the handsome young Pomfret Herries had proposed for her in marriage. **1978** T. Sharpe *Throwback* iii. 19 Mr Flawse..took a swig of brandy to steady his nerves. The bloody woman was proposing to him.

proposita (prōᵘpǫ·zitā). [L., fem. of *prōpositus*: see *propositus.] A female propositus.

1970 *Jrnl. Med. Genetics* VII. 180/2 The proposita.. presented as a profoundly retarded, 21-year-old girl with poor posture. **1976** *Nature* 15 Jan. 139/1 A study of 12 different blood group systems in the proposita and her parents failed to exclude paternity.

proposition, *sb.* Add **6. a.** *spec.* in *U.S.*, a constitutional proposal.

1921 *Congress. Rec.* 21 Feb. 3537/1 If this particular proposition were a law, and the Federal Trade Commission were given the authority that is herein provided, they would protect the country and the people in it by proper license. **1979** *Tucson Mag.* Jan. 25/1 Today's mad, mad world of Proposition 13 and other horrors. **1979** *Time* 13 Aug. 24/1 County officials hope the mistake will work in their favor, on the somewhat shaky grounds that the absurdity of Proposition .004 will defeat it.

7. a. An enterprise or project submitted for consideration or action; a matter, problem, or undertaking which requires attention; also with respect to ease or difficulty of performance, etc., as an *easy, serious, tough proposition* and with regard to likelihood of (commercial) success, as *business, mining proposition*. orig. *U.S.*

1877 R. J. Burdette *Rise & Fall of Mustache* 258 For a long time the good lady held out stoutly against the chicken proposition. **1893** *Scribner's Mag.* June 756/1 'Arn't you ashamed to tell me this?' 'Of course I am, but that isn't the proposition just now.' **1896** Ade *Artie* xviii. 168 I'm goin' against a tough proposition. **1902** O. Wister *Virginian* ii. 19 The biggest tobacco proposition for five cents got out yet. *Ibid.* xviii. 214, I saw over in a fenced meadow..what he was pleased to call 'the proposition'. Proposition in the West does, in fact, mean whatever you at the moment please. **1906** *Daily News* 24 Jan. 12/3 Later on it was found that the Main Reef series of gold-bearing rock dipped into this property, which became an attractive mining 'proposition', as the slang phrase goes. **1909** S. E. White *Rules of Game* iii. ix. 181 We're the only two business propositions in this country. **1929** *Daily Express* 7 Nov. 2/5 Every industry I want to nationalize must be a business proposition. **1932** E. Wilson *Devil take Hindmost* viii. 70 Care of the worker is a money-saving proposition. **1941** *Strand Mag.* June 140/2 All I know is that I've got to find her. The question is, how? And it won't surprise you to hear that it appears to me to be a pretty stiff proposition. **1958** *Economist* 26 July 271 Once the basic capital expenditure has been made on a machine and microphone, this is definitely an economic proposition.

b. *transf.* of persons.

1901 *Tit-Bits* 27 July 416/2 He was a pretty smooth proposition himself. **1908** C. E. Mulford *Orphan* xiv. 178 I'd rather have him with me in a mix-up than against me. He's the coolest proposition loose in this part of the country at any game. **1915** T. Burke *Nights in Town* 19 He is educated..to regard himself as, in the Broadway phrase, a serious proposition. **1925** C. E. Mulford *Cottonwood Gulch* vii. 92 Knife fighters are bad propositions. **1979** 'H. Howard' *Sealed Envelope* v. 74 Soon as I discovered she was an easy proposition I dropped out. I don't go for a twist who sleeps in anybody's bed.

8. *Comb.*, as (sense 4) *proposition-forming* adj.

1955 A. N. Prior *Formal Logic* iii. iii. 261 We may define in terms of 'ε' another proposition-forming operator. **1968** Hughes & Cresswell *Introd. Modal Logic* ii. 23 'It is necessary that' is thus a (monadic) proposition-forming operator on propositions.

proposition (prǫpozi·ʃǝn), *v.* orig. *U.S.* [f. the sb.] **1.** *trans.* To make or present a proposition to (a person). (Sometimes with unfavourable connotations.)

1924 H. C. Witwer *Love & Learn* x. 275, I finally got Ike sold on the idea that my plan was a good thought and he departed to proposition Hershel. **1927** *Collier's* 24 Dec. 36/4 He propositioned her to use his lodge at Big Bear [Lake] for her party. **1935** A. J. Pollock *Underworld Speaks* 91/2 *Propositioned*, asked to join in an unlawful undertaking. **1938** *Amer. Speech* XIII. 194 One need be very little of a purist to recoil from the current expression 'I *propositioned* him', meaning, of course, 'I made a proposition to him'. **1947** E. Andrews *Hist. Scientific Eng.* xiv. 252 It makes us more tolerant today when we hear a businessman speak of 'contacting' a customer or, worse yet, of 'propositioning' him. **1949** *Chicago Tribune* 10 Dec. 12/5 Count that day gained in which A sofomore sonny Won't proposition pop For movie money! **1949** 'J. Tey' *Brat Farrar* iii. 20 I'm propositioning you... What is wrong with the proposition? **1967** *Punch* 28 June 949/2 While being propositioned by Lord Beaverbrook about becoming the film critic of the *Evening Standard*, I nervously filled in a yawning silence by telling this anecdote. **1978** J. A. Michener *Chesapeake* 660 [They] sailed up

river to the landing of a farm owned by an old man.., and there they propositioned him: 'You ain't gonna have much more use for your long gun... We aim to buy it.' **1979** *Dædalus* Summer 46 The deputy..proceeds to proposition her with a debased contract.

2. To request sexual favours from (a person); to solicit.

1936 J. G. Cozzens *Men & Brethren* i. 139 There's no real reason to be embarrassed because your clerical collar keeps you from feeling free to proposition every woman you meet. **1946** 'P. Quentin' *Puzzle for Fiends* iii. 23 You ..had him proposition you in a canoe? **1949** K. Malone in Kirby & Woolf *Philologica* 317 *To proposition*, at least in the field of amour, is not only distinct from *to propose*, but in a sense antithetical to *propose*. **1953** W. Burroughs *Junkie* vi. 60, I remember once he told me how he'd been propositioned by a queer who offered him twenty dollars. **1963** T. Pynchon *V.* xi. 328 The girls were professional and tried for a while to proposition Fausto and Dnubietna. **1974** M. Gilbert *Flash Point* xii. 105 This girl stopped me, by asking me for a cigarette... There was no question of me propositioning her or annoying her. **1975** *New Yorker* 4 Aug. 25/1 In Hyde Park, that black whore had propositioned him as he walked from work toward the Tube.

propositional, *a.* Add: **a.** (Further examples.) *spec.* applied to speech and language in which statements and assertions occur.

1874 J. H. Jackson in *Med. Press & Circ.* 14 Jan. 21/1 But if *I* have to say, 'Gold is yellow', I have to revive the words, and I have to put them in propositional order. **1879** —— in *Brain* I. 312 A speechless patient may retain the word 'no', and yet have only the interjectional or emotional, not the propositional, use of it; he utters it in various tones as signs of feeling only. **1892** *Mind* I. 10 There are *five* independent laws, which are necessary and sufficient for propositional synthesis. **1922** tr. *Wittgenstein's Tractatus* 45 The sign through which we express the thought I call the propositional sign. *Ibid.* 51 An expression is thus presented by a variable whose values are the propositions which contain the expression... I call such a variable a 'propositional variable'. **1926** H. Head *Aphasia* i. iii. 39 Occasionally he can not only use 'yes' and 'no' correctly, but can even repeat them. Here, then, we have propositional speech and voluntary utterance. **1932** Lewis & Langford *Symbolic Logic* ix. 267 In elementary functions only propositional variables occur. **1935** Weisenburg & McBride *Aphasia* x. 277 When propositional speech is negligible, spoken expression may nevertheless be frequent. **1943** W. G. Hardy in *Cornell Univ. Abstr. of Theses* (1944) 57 Richards..now concerns himself with the educative values of propositional analysis. **1947** H. Reichenbach *Elem. Symbolic Logic* 179 The recursive definition of the term 'propositional expression'. **1955** A. N. Prior *Formal Logic* i. iii. 50 We introduce the symbol 'o' as a propositional *constant*, to stand..for some arbitrarily chosen proposition. **1957** J. Eisenson in L. E. Travis *Handbk. Speech Path.* (1959) xii. 438 When linguistic symbols are used to communicate a specific idea or to elicit a specific response, we are dealing with propositional speech. **1968** N. Rescher *Topics in Philos. Logic* v. 51 The most satisfactory course is to base the logic of belief (and of assertion) upon a propositional analysis, rather than one articulated in terms of sentences, inscriptions, utterances, or the like. **1973** J. J. Zeman *Modal Logic* p. v, The book..contains a detailed development of non-modal propositional logic. **1976** *Listener* 9 Dec. 743/1 The capacity for propositional speech—the synthesis of words into statements that, by their form, give extra meaning to those words. **1977** *Word 1972* XXVIII. 223 In propositional speech (both in forming constructions and in understanding them), the patient had difficulty sorting out relationships between more than two critical items.

b. In special collocations, as *propositional attitude*, an attitude which can vary expressed towards a proposition that does not vary; *propositional calculus*, a calculus which formalizes the basic truth-functional operations possible in logical propositions and gives notation, indicating conjunction, disjunction, negation, etc., to their sentential connectives; *propositional connective*, a connective which is used as a logical operator in propositions; *propositional function* (see quot. 1910 and cf. *predicate *sb.* 2a).

[**1904** B. Russell in *Mind* XIII. 509 Belief is a certain attitude towards propositions, which is called knowledge when they are true, error when they are false.] **1939** *Mind* XLVIII. 479 When we disbelieve, then doubt, and finally believe a proposition, it must be the *same* proposition toward which we have these different attitudes; that is, propositions must be invariant under change of propositional attitude. **1940** B. Russell *Inquiry into Meaning & Truth* xi. 204 A negative basic proposition thus requires a propositional attitude, in which the proposition concerned is the one which, on the basis of perception, is denied. **1966** W. V. Quine *Ways of Paradox* xv. 189 Striving and wishing, like believing, are propositional attitudes and referentially opaque. **1969** J. Hintikka *Models for Modalities* iii. 87 (*heading*) Semantics for propositional attitudes. **1903** B. Russell *Princ. Math.* ii. 12 It is not with such entities that we are concerned in the propositional calculus, but with genuine propositions. **1938** *Jrnl. Symbolic Logic* III. 83 Chapter 2 supplements the propositional calculus, in effect, with the Boolean algebra of one-place predicates. **1959** *Ibid.* XXIV. 97 Gödel proves the non-existence of a finite matrix characteristic for the intuitionist propositional calculus IC. **1966** *Mathematical Rev.* XXXI. 4/1 Since the main results are existential, a really conclusive solution requires a narrow meaning of 'propositional calculus'. **1973** J. J. Zeman *Modal Logic* ii. 7 Our first task will be..to set down some propositional calculi. Many systems discussed later will be presented as extensions of the propositional calculus. **1938** *Jrnl. Symbolic Logic* III. 84 A deductive system is presented which involves..the propositional connectives, and prediction and quantification with respect to individuals. **1952** S. C. Kleene *Introd. Metamath.* iv. 73 In particular, ⊃, &, ∨, ¬ are propositional connectives, and operators of the forms ∀x and ∃x are quantifiers..these six are logical operators. **1974** *Jrnl. Philos. Logic* III. 202 The propositional connectives concern statements. **1903** B. Russell *Princ. Math.* ii. 13 Where there are one or more real variables, and for all values of the variables the expression involved is a proposition, I shall call the expression a *propositional function*. **1910** Whitehead & Russell *Principia Math.* I. i. 15 Let φx be a statement containing a variable x such that it becomes a proposition when x is given any fixed determined meaning. Then φx is called a 'propositional function'; it is not a proposition, since owing to the ambiguity of x it really makes no assertion at all. **1943, 1969** [see *predicate *sb.* 2 a]. **1974** *Jrnl. Philos. Logic* III. 196 A predicate, or propositional function, is a statement-valued function, i.e. a function that maps objects of some kind into statements (propositions).

Hence **proposi·tionalist,** someone who is concerned with the logic of propositions; **propositionally** *adv.* (earlier example).

1879 *Brain* II. 215 Then propositionally 'yes' and 'no' give assent or dissent to anything whatever. **1952** *Mind* LXI. 61 The spurious character of our propositionalist's argument becomes evident. **1970** W. V. Quine *Philos. of Logic* i. 3 The propositionalist by-passes differences between languages.

propositionize, *v.* Delete *rare* and add further examples.

1878 *Brain* I. 312 Loss of speech is therefore the loss of power to propositionise. It is not only loss of power to propositionise aloud (to talk), but to propositionise either internally or externally. **1920** *Ibid.* XLIII. 119, I believe that under the uncouth word 'propositionizing' is included what I understand by 'symbolic thinking'. **1921** *Brit. Jrnl. Psychol.* XI. 185 There are some aspects of the loss of function in aphasia which are not strictly comprised under the heading of 'propositionising'. **1955** R. Jakobson in H. Werner *On Expressive Lang.* 77 The patient fails to operate with contiguity, while operations based on similarity remain intact. Thus he [*sc.* an aphasic] loses the ability to propositionize. **1956** —— in Jakobson & Halle *Fund. Lang.* ii. iv. 71 The impairment of the ability to propositionize, or generally speaking, to combine simpler linguistic entities into more complex units, is actually confined to one type of aphasia.

‖**propositum** (prōᵘpǫ·zitǝ̆m). *Philos.* [L., neut. of *prōpositus*: see next.] The first premise of a syllogism; an argument, principal theme or subject propounded.

1858 A. De Morgan *On Syllogism* (1966) 83, I see great difference in the *propositum* between 'This house was built by Jack' and 'This is the [or even *a*] house that Jack built'. **1913** [see *judicatum]. **1920** S. Alexander *Space, Time, & Deity* II. 249 'Proposition' contains a reference to language, and 'propositum' would be a better, though a pedantic name.

propositus (prōᵘpǫ·zitǝs). Pl. **propositi.** [a. L. *prōpositus*, pa. pple. of *prōpōnere* (see Propone *v.*).] An individual who was the first member of a family to come to the notice of a researcher, and through whom investigation of a pedigree began. Cf. *proband, *proposita.

1926 *Eugenics Rev.* XVIII. 248 ☛ Points to the Propositus or central figure in the pedigree. **1939** *Brit. Jrnl. Psychol.* XXX. 9 When propositi are separated into groups of comparable mental grade, defect is seen to be more common among relatives of simpletons than among relatives of idiots. **1956** *Nature* 7 Jan. 40/1 The factor was transmitted to them by the paternal grandmother (generation II) of the propositus. **1961** *Lancet* 19 Aug. 437/2 We have collected the details of 107 sibships; the propositi attended our clinics. **1977** *Ibid.* 3 Sept. 504/1 The propositus (family C) presented with pituitary insufficiency.

propoxur (propǫ·ksūᵊɹ). [f. Prop(yl + Ox(y + Ur(ethane.] An insecticide having a long-lasting ability to produce rapid incapacitation of affected insects; *o*-isopropoxyphenyl-*N*-methylcarbamate, $CH_3·NH·CO·O·C_6H_4·O·CH(CH_3)_2$.

1964 *Zeitschr. f. Angewandte Zool.* LI. 332 Deposits on filter paper of..carbamates (carbaryl and propoxur), prepared up to 105 days ago, were repeatedly applied to imagines of five different strains of house-flies (Musca domestica). **1977** *Time* 12 Sept. 56/3 Health authorities are now using more of other insecticides, such as Malathion and propoxur to kill DDT-resistant mosquitoes.

propoxyphene (propǫ·ksifīn). *Pharm.* [f. Prop(io- + Oxy- + *-phene* (f. Phen-, pheno-).] A mild narcotic analgesic, chemically related to methadone, which is given orally (usu. as the hydrochloride, a whitish powder) esp. in cases of chronic or recurrent pain, its effects being similar to but weaker than those of codeine; (+)-α-4-dimethylamino-3-methyl-1,2-diphenylbut-2-yl propionate, $(CH_3)_2N·CH_2·CH(CH_3)·C(C_6H_5)(CH_2·C_6H_5)·O·CO·CH_2CH_3$.

1955 *Arch. Internat. de Pharmacodyn.* CIV. 165 The analgesic activity of Propoxyphene has been demonstrated in patients with chronic pain. **1957** *Jrnl. Amer. Med. Assoc.* 29 June 966/2 Propoxyphene is a new synthetic analgesic with the effectiveness of codeine but having less undesirable gastrointestinal side-effects. *Ibid.* 969/2 Since the preparation of this article, the Council on Drugs of the American Medical Association has changed the generic name of propoxyphene hydrochloride to dextro propoxyphene hydrochloride. **1974** M. C. Gerald *Pharmacol.* xiv. 272 In recent years propoxyphene has been among the most frequently prescribed drugs. **1976** *National Observer* (U.S.) 28 Aug. 3/1 The drug agency has suggested that the Justice Department's Drug Enforcement Administration (DEA) rule that propoxyphene-containing products be added to the list of the Controlled Substances Act's 'Schedule IV' drugs. This would mean that physicians must renew prescriptions for Darvon and similar drugs every six months.

proppy (prǫ·pi), *a.*² *Austral. colloq.* [f. Prop *v.*¹ + -y¹.] Of a horse: tending to prop (Prop *v.*¹ 4) or stop suddenly in mid-stride, faltering; also of other animals. Hence **pro·ppily** *adv.*

1945 Baker *Austral. Lang.* 70 Another extension is the adjectival use of *proppy* for a horse that jibs and plays up when ridden or driven. **1951** H. G. Lamond in Murdoch & Drake-Brockman *Austral. Short Stories* 213 Both [dogs] walked proppily on tiptoes. **1969** *Australian* 24 May 35/5 King's Delight had a bruised sole on the near fore, and Clare said the horse was proppy in his action.

propraetorship. (Later example.)

1824 J. H. Newman in *Encycl. Metrop.* (1845) X. 280/1 From the period of his Consulate to his Proprætorship in Cilicia.

propranolol (propræ·nǝlǫl). *Pharm.* [f. Pro(pyl + *pr(op)anol with reduplication of final -ol.] The compound 1-isopropylamino-3-(1-naphthyloxy)-2-propanol, $C_{16}H_{21}NO_2$, which is a β-adrenergic blocking agent used mainly (in the form of a colourless crystalline hydrochloride) in the treatment of cardiac arrhythmia.

1964 *Brit. Med. Jrnl.* 19 Sept. 720 (*heading*) Effect of propranolol (Inderal) in angina pectoris: preliminary report. **1965** J. H. Burn *Lect. Notes Pharmacol.* (ed. 8) 10 When propranolol is given, a patient performs a given amount of work at a lower heart rate and he can perform a greater amount of work without feeling pain. **1972** *Lancet* 16 Sept. 565/2 In our clinical experience, propranolol is effective in the treatment of heroin dependence. **1974** *Sci. Amer.* June 62/3 This complex series of events can be turned off by propranolol, a drug that prevents the noradrenaline from combining with the beta adrenergic receptor. **1980** *Brit. Med. Jrnl.* 29 Mar. 885/1 We have..investigated the effect of giving higher doses of a non-selective beta-blocker (propranolol)..to patients with suspected myocardial infarction.

proprietariat (proprǝiĕtēᵊ·riăt). *nonce-wd.* [f. Proprietary *a.*, after *proletariat*.] The propertied class.

1896 G. B. Shaw *Rep. Fabian Policy* 4 It [*sc.* the Fabian Society]..does not believe that the moment will ever come when the whole of Socialism will be staked on the issue of a single General Election or a single Bill in the House of Commons, as between the proletariat on one side and the proprietariat on the other. **1928** —— *Intelligent Woman's Guide Socialism* 223 The Proletariat and the Proprietariat face each other. **1950** —— *Farfetched Fables* 86 The feudal proprietariat is all for well policed private property.

proprietary, *sb.* and *a.* Add: **B.** *adj.* **1. a.** (Further examples.) *proprietary name* or *term*, a word or phrase over which a person or company has some legal rights, esp. in connection with trade (as a trade mark).

1911 *T. Eaton & Co. Catal.* Spring & Summer 175 (*heading*) Proprietary Medicines. **1921** W. A. Craigie *Let.* 18 Feb. (Oxf. Dict. files), We have expressly recognized that it [*sc.* *Vaseline*] is a proprietary term. **1924** *Pocket Oxf. Dict.* 932/1 *Vaseline*... Proprietary term introduced in 1872 by R. A. Chesebrough. **1930** *Engineering* 7 Mar. 304/1 These stock, or proprietary, engines are made by..specialists. **1930** *Economist* 22 Nov. 957/1 The Economic Council was unable to agree as to whether the undertaking by retailers selling proprietary articles to charge the price fixed by the manufacturers..should be prohibited. **1933** *O.E.D.* (new impr.) s.v. *Vaseline*... A proprietary term, introduced by R. A. Chesebrough in 1872. *Ibid.* Suppl. s.v. *Ferozone*... Proprietary name. **1958** *New Statesman* 28 June 822/1 Many [doctors]..tend to prescribe a well-advertised proprietary brand because they have no time to consult their list for a cheaper standard preparation. **1972** *Physics Bull.* Aug. 489/2 'Freon' is the proprietary name for Du Pont's brand of the fluorinated derivatives of hydrocarbons used as refrigerants and aerosol propellants. **1974** *Islander* (Victoria, B.C.) 4 Aug. 10/1 By the mid-18th century, more than 200 so-called 'proprietary medicines' were being sold in Britain and in the American colonies.

b. *proprietary company* = *private company* s.v. *private *a.* 7 h. *Austral.*

1896 *Companies Act* (Victoria, Austral.) § 2 'Proprietary Company' means a company..which..(a) has not more than twenty-five members or shareholders; (b) does not receive deposits, except from its members or shareholders..; (c) does not use its title without the addition thereto immediately before the word 'limited' of the word 'proprietary'. **1973** R. N. Purvis *Purvis on Proprietary Companies* i. 8 A public company must have at least three directors, whereas a proprietary company need have only two.

proprietous (pro‚prǝi·ĕtǝs), *a.* [f. Propriety + -ous.] Characterized by (extreme) propriety or punctilious behaviour. Also *Comb.* Hence **propri·etously** *adv.*

1844 *Ainsworth's Mag.* VI. 228 An elderly person, whose proprietous grey silk dress..conveyed the idea of a distant relative. **1845** *Ibid.* VII. 498 The rows of canvass-covered proprietous-looking [bathing] machines. **1882** 'L. Malet' *Mrs. Lorimer* I. v. 120 My dear, I feel a little stifled when I think of you guarded by these proprietous and unimaginative dragons. *Ibid.* vi. 132 Mrs. Mainwaring had lost for a moment that proprietous self-command and calm dignity of demeanour, which..were certainly liable to keep most people at a very respectful distance from her. **1913** D. H. Lawrence *Sons & Lovers* viii. 181 He must see a girl home from the skating rink—quite proprietously—and so can't get home. **1974** R. B. Parker *Godwulf MS* i. 9 The elevator that took me to the fourth floor was covered with obscene graffiti that some proprietous soul had tried to doctor into acceptability. **1979** *Verbatim* Autumn 1/1 In Marvin Pope's exhaustive scholarly study of *The Song of Songs*, the Bride's vivid description of her passion for the Bridegroom as 'my bowels were moved for him'—striking in its proprietous King James context (5:4)—is rendered 'my inwards seethed for him.'

propriety, *sb.* Add: **7.** Also, *to play propriety*, to ensure correct moral behaviour, act as a chaperone.

1836 [in Dict.]. **1877** V. Lush *Jrnl.* 13 Feb. (1975) 187, I invited a few of the Choir here after practice, with Mrs Scott and Mrs Lloyd to play propriety and we had a very jolly evening. **1925** I. Smith *Marriage in Ceylon* 137 Angela..had felt it would not be 'the thing' to accept hospitality of the man who in a few days' time would be her husband without someone to 'play propriety'.

proprioceptor (prōᵘ·priosepto̧ɹ). *Physiol.* [f. L. *propri-us* own, Proper + -o + Re)ceptor.] Any sensory structure which receives stimuli arising within the tissues (other, usually, than the viscera); *esp.* one concerned with the sense of position and movement of a part of the body. Cf. *exteroceptor, *interoceptor.

1906 [see *exteroceptor]. **1927** Haldane & Huxley *Animal Biol.* i. 24 The receptor organs are those parts of the living organism which are specially sensitive to the changes going on around them. Some of them are affected by the changes going on inside the body in muscles and joints and in the organ of balance (proprioceptors), others by the changes taking place in the world outside (exteroceptors). **1938** *Jrnl. Exper. Biol.* XV. 112 The campaniform sensilla will act as proprioceptors for the palps. **1940** Fraenkel & Gunn *Orientation of Animals* iv. 37 In the knee-jerk the stimulus is translated by the proprioceptors in the leg, goes to the central nervous system and is there, so to say, reflected back to the leg, where it is translated into effect or action. **1964** J. Z. Young *Model of Brain* xv. 242 Proprioceptors are probably present [in octopus arms] but it is not clear how they could appropriately signal the position of the arms. **1974** D. & M. Webster *Compar. Vertebr. Morphol.* x. 200 The most studied proprioceptor is the muscle spindle.

So **proprioce·ption,** the reception of information by proprioceptors and its interpretation; **proprioce·ptive** *a.*; **proprioce·ptively** *adv.* (*rare*).

1906 C. S. Sherrington *Integrative Action of Nervous Syst.* iv. 130 Since in this field the stimuli to the receptors are given by the organism itself, their field may be called the proprio-ceptive field. *Ibid.* ix. 349 The supposition that the organ [*sc.* the cerebellum] is the chief co-ordinative centre or rather group of centres of the reflex system of proprio-ception. **1927** Haldane & Huxley *Animal Biol.* v. 123 Proprioceptive organs may affect the consciousness. Thus we can tell how much our knee is bent even with our eyes shut, owing to the joint-organs, or how great a weight we are holding, owing to the muscle-organs. **1958** S. H. Bartley *Princ. Perception* xiv. 315 Kinesthesis and the vestibular sense have been called proprioception, since they have something to do with end results originated by the activity of the body itself. **1961** L. F. Brosnahan *Sounds of Language* i. 15 Articulations which are similar..must require greater precision of execution.. in order to keep the resulting speech sounds distinct, both proprioceptively and auditorily. **1968** *New Scientist* 7 Nov. 315/1 Proprioceptive information..describes such things as the tension in the muscles and the location and movement of the parts of the limb. **1977** *Lancet* 5 Nov. 977/1 He had..absent vibration sense, and very reduced proprioception in the lower limbs.

proprio-spinal (prōᵘ-prio‚spǝinăl), *a.* *Anat.* Also as one word. [f. as prec. + Spinal *a.*] Situated wholly within the spinal cord.

1904 C. S. Sherrington in *Nature* 8 Sept. 461 (*caption*) Proprio-spinal neurones. *Ibid.* 463/1 We..arrive at the following reflex chain for the scratch reflex: (i.) The receptive neurone... (ii.) The long descending proprio-spinal neurone.., from the shoulder segment to the grey matter of leg segments. **1920** H. Head *Studies in Neurol.* II. iii. 504 If these proprio-spinal paths have been destroyed, or their functions abolished, the mass-reflex may appear in its complete form even though some sensation and voluntary movement are still present below the level of the lesion. **1959** *Brain* LXXXII. 614 We have retained the name 'fasciculi proprii' for the fibres under review, as there is no suitable English equivalent. Sherrington's name 'proprio-spinal fibres'..might be suggested. But it seems to us that a name which includes the term 'spinal' is undesirable. **1974** *Exper. Brain Res.* XXI. 188 Our results show that long propriospinal fibres stimulated a few segments caudal to the cervical enlargement.

props, *sb. pl.* Add: **1. a.** (Earlier examples.)

1841 *Spirit of Times* 16 Oct. 396/2 There we subsisted by *spouting*, not Shakespeare, but our dresses and props. **1854** E. L. Blanchard *Diary* 22 Nov. in Scott & Howard *Life E. L. Blanchard* (1891) I. 125 Go to Drury; see props.

b. A familiar name for a property-man or the props department.

Quot. 1831 may be an example of *prop *sb.*⁸

1831 P. Egan *Show Folks* 23 'Good Houses' now to make him right, The *Treasury* to swell: The Actors [*sic*] need—the *Props* delight—And '*All's well, that ends well!*' **1889** [in Dict.]. **1902** Patterson & Bateman *By Stage Door* 192 While he was 'Props' he was discharged..for not yelling 'fire!' at the right time. **1921** Galsworthy *Six Short Plays* 128, I want 'Props'. *Ibid.* (*stage-direction*) 'Props' goes out through the French windows. **1933** P. Godfrey *Back-Stage* iv. 48 No self-respecting 'Props' will spend a penny on new materials if odd scraps will serve. **1976** M. Maguire *Scratchproof* ii. 23 What do you think of our tack-room interior? Have props done a good job?

c. as *sing.* a stage property.

1911 C. Pollock *Footlights* 257 By-play with small articles, rehearsed twenty times, is blundered over when the player finds the 'prop' actually in his hands. **1961** M. Catto *Mister Moses* iii. 80 The stage had been set—it awaited the last theatrical prop. **1976** *Early Music* Oct. 394/1 Each tableau, step, gesture, prop and lighting-cue must arise from necessity and have its effect at once. **1978** *Listener* 23 Mar. 366/3 Ronnie Barker's face..is,.. as with all true comedians, his best prop.

2. *transf.*

1898 A. M. Binstead *Pink 'Un & Pelican* vi. 146 And when at last the day came for him to go, he 'collected his props', as he called getting his belongings together, most reluctantly. **1926** *Publishers' Weekly* 10 July 120/1 Woodard-Clarke's [window-display]..took a middle course between the painted background route and the 'props' of nature. The clear blue sky was conveyed by blue cloth of chiffon-like texture. **1948** M. Gilbert *They never looked Inside* iii. 35 Have a cigarette? They are part of the office props. **1976** S. Barstow *Right True End* iii. xiv. 224 A shot of me standing on a stone pier beside a whelk-stall, holding a huge crab by one claw. It must have been dead, lent to me as a prop by some kindly stall-holder.

3. *attrib.* and *Comb.* **a.** *props department, girl, man*; **props room** = *property room* s.v. Property *sb.* 8b. **b.** Formed with the *sing.*, as *prop boy, girl, man*; **prop basket** = *property box 2; **prop table** (see quot. 1964); **prop wagon,** a property wagon (see quot.).

1952 Granville *Dict. Theatr. Terms* 144 It is the traditional boast of an old actor that he was born in a *prop basket* in the prompt corner. **1935** *Motion Picture* Nov. 38 Colman, who was off in a corner talking with a prop boy, heard her. **1960** K. A. Ommanney *Stage & School* v. xv. 413 All movable articles are stored in the props department. **1970** R. Leach *Theatre for Youth* i. v. 43 Masks properly belong to the props department. **1964** E. Crampton *Handbk. Theatre* 260 These tables are the direct responsibility of the prop-girl. **1977** S. Brett *Star Trap* iii. 31 There's always the stage staff... Nothing like a warm little props girl to comfort a chap. **1942** Berrey & Van den Bark *Amer. Thes. Slang* § 605/12 *Property man*,..prop man. **1951** in H. Downs *Theatre & Stage* II. 818/2, I once knew a 'props' man who was..a marvel. **1971** *Esquire* July 88/3 We were watching the prop men lug the elephant tubs and the chimp- and lion-act furniture. **1978** M. Puzo *Fools Die* xxviii. 323 Script-girls, secretaries, studio accountants, cameramen, propmen, the technical crews, the actors and actresses, the directors and even the producers. **1957** Props room [see *balloon *sb.*¹ 6c]. **1977** C. Wood *James Bond* xii. 103 This place..was like the props room of a folded theatrical company. **1939** Burris-Meyer & Cole *Scenery for Theatre* xiii. 390 Prop table, left. **1964** E. Crampton *Handbk. Theatre* 260 *Prop tables*, tables set in the wings ..for properties to be taken on and brought off-stage during a performance. **1927** *Hollis St. Theatre Progr.* (Boston) 19 Sept. Gloss., *Prop wagon*, wagon for transporting paraphernalia and equipment used on carnival.

propugnaculum. (Earlier example.)

1773 Boswell *Jrnl. Tour Hebr.* 10 Sept. (1785) 198 Before it the ocean roars, being dashed against monstrous broken rocks; grand and awful *propugnacula*.

propulsion. Add: **3.** *attrib.*, as *propulsion-jet, -system*; **propulsion gun,** a hand-held device that an astronaut can cause to eject a jet of compressed gas so as to propel him in space.

1958 C. C. Adams *Space Flight* viii. 196 Auxiliaries. These include taxis and propulsion 'guns' for individual men in space suits, or reaction power packages attached like outboard motors to large objects. **1965** *Life* 18 June 26/2 White himself used a camera attached to his propulsion gun, and McDivitt operated another at his window inside. **1935** Balmer & Wylie *After Worlds Collide* i. 14 The earth around the huge metal cylinder had been melted by the blasts of its atomic propulsion-jets. **1961** *Amer. Speech* XXXVI. 170 In the years after 1919, he [*sc.* Robert Esnault-Pelterie] had taken the initiative in stimulating studies on propulsion systems for interplanetary travel. **1966** *Electronics* 14 Nov. 16/3 In addition, four panel discussions will bring together the nation's top men in the fields of space policy and launch and propulsion systems.

propyl. Now also with pronunc. (prōᵘ·pǝil). Delete first sentence of def. and substitute: Either of two isomeric radicals, $CH_3·CH_2·CH_2—$ (1-*propyl*, *primary* or *normal propyl*) and $(CH_3)_2CH—$ (2-*propyl*, *secondary propyl* or *isopropyl*); sometimes *spec.* the normal form. (Earlier and later examples.)

1850 *Phil. Trans. R. Soc.* CXL. 127 Substances of the formula $C_{12}H_{11}N$ will also be obtained..by fixing upon aniline the radical (propyl) belonging to the missing alcohol of propionic acid* (metacetic acid). [*Note*] *A more appropriate name for metacetic acid, proposed by Dumas, Malaguti and Leblanc.., as it is the *first* acid of the series $C_nH_nO_4$ that exhibits the character of a *fatty* acid. **1887** Moore & Aveling tr. *Marx's Capital* I. i. 18 Butyric acid is a different substance from propyl formate. Yet both are made up of the same chemical substances, carbon (C), hydrogen (H), and oxygen (O), and that, too, in like proportions—namely, $C_4H_8O_2$. **1950** *Thorpe's Dict. Appl. Chem.* (ed. 4) X. 223/1 *n*-Propyl radical is formed by exposing the vapour of di-*n*-propyl ketone to ultra-violet light at 2 mm. pressure. **1951** I. L. Finar *Org. Chem.* I. viii. 122 All aldehydes can be made to undergo the Cannizaro reaction by treatment with aluminium ethoxide. Under these conditions the acid and alcohol are combined as the ester,..*e.g.*..propionaldehyde gives propyl propionate.

pro:pylthiou·racil, any propyl derivative of a thiouracil; *spec.* 6-*n*-propyl-2-thiouracil, $C_7H_{10}N_2OS$, an antithyroid substance used to combat thyrotoxicosis.

1945 *Jrnl. Clin. Endocrinol.* V. 424/1 The compounds 6-ethylthiouracil and 6-*n*-propylthiouracil are among a group which are approximately ten times as active as thiouracil when tested in rats. *Ibid.* 424/2 It seemed fitting to test the effectiveness of ethylthiouracil and propylthiouracil in human beings suffering from hyperthyroidism. **1961** *Lancet* 23 Sept. 688/2 On 2 of these patients there was subsequent difficulty in management with a lack of the usual responsiveness to propylthiouracil. **1975** B. Catz *Thyroid Case Stud.* 13 The dose of propylthiouracil taken daily was 600 mg.

propylene. Add: **2.** Used *attrib.* or in *Comb.* in the names of derivatives: **propylene glycol,** either of two isomeric liquids, $CH_2OH·CHOH·CH_3$ and $CH_2OH·CH_2·CH_2OH$; *spec.* the 1,2-glycol (the former), which has a wide variety of uses, chiefly as a solvent or carrier, as a constituent of antifreeze, and in the food and perfume industries; **propylene imine** (also **pro:pylen(e)i·mine**) [ad. G. *propylenimin* (Gabriel & Ohle 1917, in *Ber. d. Deut. Chem. Ges.* L. 815)], a synthetic, colourless, highly inflammable liquid, $CH_3·CH·CH_2·NH$, which is widely used industrially, freq. in polymerized form and esp. as a binding agent with dyes and adhesives, and in the manufacture of plastics, paper, etc.

1885 Morley & Green in *Jrnl. Chem. Soc.* XLVII. 132 From the aqueous portion of the distillate 140 grams of propylene glycol (boiling at 185–195°) can be obtained by fractional distillation. **1926** *Jrnl. Biol. Chem.* LXVIII. 416 It was..desired to convert propylene glycol into β-hydroxybutyric acid by a set of reactions which would not involve carbon atom (2). **1951** I. L. Finar *Org. Chem.* I. xvii. 329 Lactic acid..may be prepared: (i) By oxidising propylene glycol with dilute nitric acid. **1966** *Kirk-Othmer Encycl. Chem. Technol.* (ed. 2) X. 649, 1,2-Propylene glycol, $CH_3CHOHCH_2OH$, is a colorless and odorless liquid with a slightly sweet taste. **1917** *Jrnl. Chem. Soc.* CXII. i. 564 β-Bromopropylamine and β-bromo*iso*propylamine.. both yield β-bromo*iso*propylamine, this result being explicable by the intermediate compound being propylenimine. **1944** *Jrnl. Org. Chem.* IX. 133 Propylenimine was prepared from 2-amino-1-propanol. **1966** *Kirk-Othmer Encycl. Chem. Technol.* (ed. 2) XI. 527 Ethylenimine and propylenimine are colorless mobile liquids with a strong ammoniacal odor. **1971** *Nature* 16 Apr. 460/2 Propylene imine is an important chemical intermediate with a variety of applications in the production of polymers, coatings, adhesives, textiles and paper finishes.

propylidene (propi·l-, propǝi·lidīn). *Chem.* [ad. F. *propylidène* (E. Reboul 1876, in *Compt. Rend.* LXXXII. 30): see Propyl and *-idene.] The bivalent radical $CH_3CH_2CH=$; usu. *attrib.*, esp. in names of derivatives of this.

1876 *Jrnl. Chem. Soc.* XXIX. i. 894 Propionic aldehyde is treated with phosphoric chloride, and forms a propylidene dichloride homolog[o]us with ethylidene dichloride; its formula is $CH_3—CH_2—CHCl_2$, and it boils at 85°–87°. **1903** A. J. Walker tr. *Holleman's Text-Bk. Org. Chem.* i. 146 Propylene chloride, $C_3H_6Cl_2$,..is not identical with the reaction-product obtained by treating acetone with phosphorus pentachloride..nor with that from propionaldehyde, $CH_3·CH_2·CHCl_2$ (propylidene chloride). **1951** I. L. Finar *Org. Chem.* I. iv. 49 General methods of preparation of the olefins... By the action of zinc dust on methanolic solutions of *gem*-dihalogen derivatives of the paraffins.., *e.g.*, propylene from propylidene bromide. **1976** *Chem. Physics Lett.* XXXVII. 220/2 The barrier to conversion of propylidene to propylene is greater than in C_2H_4.

propylitization (prǫ:pilitǝizēⁱ·ʃǝn). *Petrol.* [f. Propylit(e + -ization.] The hydrothermal alteration of an igneous rock to propylite.

1903 A. Geikie *Text-bk. Geol.* (ed. 4) II. 812 The solutions..in their progress..induce chemical and mineralogical changes in the surrounding rocks which thus undergo various transformations, being sometimes weakened by the removal of certain constituents, as in propylitisation and kaolinisation. **1965** G. V. Williams *Econ. Geol. N.Z.*

viii. 87/1 The mineralization is a typical Tertiary andesitic propylitization of which there is evidence from Te Puke..in the south to Great Barrier Island in the north. *Ibid.*, They found..that zones of intense propylitization (containing known gold-bearing veins) are detectable electrically.

propyne (prōᵘ·pain). *Chem.* [f. PROP(ANE + *-YNE.] A gaseous, unsaturated hydrocarbon, CH_3—C⋮CH, which resembles acetylene, from which it is formally derived by replacing a hydrogen atom by methyl; allylene.

1931 *Jrnl. Chem. Soc.* 1610 The names of hydrocarbons containing the triple linking will end in *yne, diyne*, etc... Examples: Propyne, heptyne. **1935** *Jrnl. Amer. Chem. Soc.* LVII. 1089/2 The gaseous product of the reaction, propyne, was not determined quantitatively, but was identified through the fact that it formed a silver salt with ammoniacal silver nitrate solution. **1969** M. JULIA in H. G. Viehe *Chem. of Acetylenes* v. 341 Acetylene itself with photolyzed diazomethane..gave propyne and allene. **1975** GUTSCHE & PASTO *Fund. Org. Chem.* viii. 208 Treatment of propyne with water (in the presence of mercuric ion..) yields 2-propenol, which immediately rearranges to..acetone.

pro-rate, *v.* Add: Also **prorate**. **a.** (Earlier and later examples.)

1860 *Congress. Globe* 21 Dec. 180/1 The amendment.. requires this company to pro-rate passenger fare with all railroad companies or lines which terminate either at Alexandria, Washington or Baltimore. **1921** *Oil & Gas Jrnl.* 1 July 3/1 *(heading)* Are runs to be prorated? **1926** J. ISE *U.S. Oil Policy* xii. 116 California oil was cutting deeper into the eastern markets all the time, and in July [1923] some of the pipe-line companies in the eastern fields started to prorate runs. **1963** J. MITFORD *Amer. Way of Death* v. 66 In all likelihood, idle time of employees is figured in and prorated as part of the 'man-hours'. **1978** *Time* 3 July 37/3 San Jose businessman Larry Whitaker ..said he would pro-rate his own $18,921 property tax cut among his 150 California employees.

Hence also **pro-ra·ted** *ppl. a.*, **pro-ra·ting** *vbl. sb.*

1911 F. E. WEBNER *Factory Costs* 212 On the other hand, there is no possible way of entirely avoiding a prorating or averaging of expense. **1921** *Oil & Gas Jrnl.* 1 July 3/1 In order to conserve space in the containers there may be instituted in the near future a prorating program. **1967** *Boston Sunday Globe* 23 Apr. 24 (Advt.), A pro-rated portion of the purchase price.

proration (prorēⁱ·ʃən). [f. PRO-RATE *v.* + -ION.] The action or an instance of prorating; *spec.* allocation of the permitted production of oil or gas between competing operators, fields, etc.

1923 *Oil Weekly* 22 Sept. 12/1 The Eastern fields and those of the Middle West are without proration problems. **1931** *Economist* 20 June 1324/2 Oil proration, copper curtailment agreements, and railroad mergers indicate the swing of the pendulum. **1954** R. CASSADY *Price Making in Petroleum Industry* vii. 114 Under a proration scheme.. a sort of floor is placed under prices by limiting the amount of oil to be withdrawn on a basis of the needs of the market. **1957** *Times* 11 Dec. 16/5 The Pembina Oilfield is subject to strict proration; but all the wells have produced the maximum allowable since they were completed. **1960** *Economist* 15 Oct. 264/1 The ideas of a petroleum exporters' cartel, miscalled 'international proration'..are not inherently unfeasible. **1971** *Nature* 28 May p. viii/1 (Advt.), Only a Wang 100 can do side calculations, analyses, prorations and serve as an adding machine.

Hence **prora·tioning** *vbl. sb.* in the same sense.

1948 E. V. ROSTOW *National Policy for Oil Industry* iii. 21 In 1932, the Supreme Court declared prorationing legal. **1959** DE CHAZEAU & KAHN *Integration & Competition in Petroleum Industry* vii. 163 Prorationing to market demand has clearly brought about orderly marketing of crude oil with much fewer and less extreme price changes. **1960** *Guardian* 15 Oct. 7/5 The Organisation of Petroleum Exporting Countries..would..try to keep prices stable by regulating production (the system known to the industry as 'pro-rationing'). **1962** W. A. LEEMAN *Price of Middle East Oil* ix. 230 In effect the Venezuelans proposed a system of world-wide 'prorationing', similar to the arrangements already established in Texas, Louisiana, Oklahoma, and a number of other states.

proroguing, *vbl. sb.* (Later example.)

1937 G. FRANKAU *More of Us* vi. 69 And, as he donned those shoes Shoemaker Lobb webs From toe to heel with best bespoken broguing, This house of lords seemed ripe for his proroguing.

pros (prɒs), var. *PROSS.

1905 *Sessions Paper* 8 Feb. 556 She is only a *pros.*; you know her. **1972** J. MILLS *Report to Commissioner* 104 I'm a pros, man, I shoot up in my thighs.

‖**prosa** (prōᵘ·ză). *Eccl.* Pl. **prosae**. [L.] = PROSE *sb.* 2. Cf. *PROSULA.

1801 T. BUSBY *Dict. Mus.*, *Prosæ*, certain hymns used in the Romish church consisting of rhyme without measure. **1907** [see *PROSULA]. **1929** *Exultet Roll* (Brit. Mus.) 5 In the South of Italy, at least from the early part of the tenth century till the thirteenth, a custom existed of writing out this prosa..on a separate roll distinct from the other services of the day. **1957** N. FRYE *Anat. Crit.* 275 The emergence of the 'prosa' out of the sequence in medieval music. **1970** P. EVANS *Early Trope Repertory of St. Martial de Limoges* i. 9 Both the prosa and the prosula are basically literary in their conception. The prosa is created by adding a text to the pre-existent melismatic sequentia which follows the Alleluia.

prosauropod (prosǭ·ropod), *sb.* and *a.* *Palæont.* [ad. mod.L. name of infraorder *Prosauropoda* (F. von Huene 1920, in *Zeitschr. für Induktive Abstammungs- und Vererbungslehre* XXII. 211), f. PRO-² 1 + mod.L. *Sauropoda*: see SAUROPOD *a.* and *sb.*] **A.** *sb.* A dinosaur belonging to the infraorder Prosauropoda, which includes herbivorous, usually bipedal, saurischians. **B.** *adj.* Of or pertaining to an animal of this kind.

1951 C. C. YOUNG in *Palaeontologia Sinica* CXXXIV. 88 For the sake of convenience we may discuss the prosauropods together. **1962** E. H. COLBERT *Dinosaurs* iv. 80 The prosauropods quickly developed to become the dinosaur giants of their day. **1965** *Proc. Linn. Soc.* CLXXVI. 211 Each of the two prosauropod families transferred from the Carnosauria..bears..a striking resemblance to one of two existing prosauropod families. **1971** E. C. OLSON *Vertebr. Paleozool.* II. viii. 354 Some genera..among animals often classed as prosauropods were intermediate in structures related to gait, being only partially quadrupedal. **1978** *Nature* 17 Aug. 662/1 Prosauropods probably could feed tripodally—supporting their weight on the hindlimbs and stout tail.

proscenium. Add: **3.** *proscenium box* (earlier examples), *curtain, drop, opening*.

1828 J. R. PLANCHÉ *Paris & London* (1830) I. v. 24 *(stage direction)* A Diagonal View of the Stage of the Odeon is seen through the wings—the proscenium boxes, L. **1849** THACKERAY *Pendennis* I. xiv. 124 One of the illustrious patrons of the Museum Theatre, and occupant of the great proscenium-box, was..the Marquis of Steyne. **1829** *Harlequin* 20 June 46 The only drop below the proscenium curtain was the very fine pierced forest limbs, which every frequenter of Drury Lane Theatre must recollect. **1975** C. HOGGET *Stage Crafts* i. 14 Pelmet for proscenium curtains should overhang reveals to allow curtains to open fully. **1827** T. DIBDIN *Reminisc.* II. 115 One artist offered to paint me a proscenium drop (as we call the painted cloth which falls between the acts). **1889** *Theatre* XIII. 292 The proscenium opening is formed by groups of columns on either side of the first proscenium box. **1974** *Encycl. Brit. Micropædia* VIII. 244/1 The proscenium opening was of particular importance to the realistic playwrights of the 19th century.

prosciutto (proʃū·to). Also (erron.) **prosciuto**. [It., 'ham'.] Italian spiced ham. Also (pleonastically) *prosciutto ham*.

c **1938** *Fortnum & Mason Price List* 39/2 Proscuitto [*sic*] (Hors d'Œuvre Sliced Ham)..3/9. **1945** E. WAUGH *Brideshead Revisited* I. iv. 90 Melon and prosciuto on the balcony. **1952** V. CANNING *House of Seven Flies* iii. 53 Charlie had ventured too far in search of black market *vino, prosciutto* and anything else he could lay his hands on. **1960** I. FLEMING *For Your Eyes Only* 153, I shall have melon with prosciutto ham. **1964** MRS. L. B. JOHNSON *White House Diary* 6 May (1970) 131 We had a gourmet lunch, beginning with prosciutto and melon. **1965** *Guardian* 4 Aug. 6/4 *Prosciutto* ham was among the delicacies on the buffet table. **1967** 'J. CROSS' *To Hell for Half-a-Crown* xv. 190 She had fixed melon with *prosciutto*. **1977** C. McFADDEN *Serial* xx. 46/2 Kate..sauntered up to a vantage point in front of the prosciutto.

prose, *sb.* Add: **4. a.** Also, a dull, prosy person. *colloq.*

1844 DICKENS *Mart. Chuz.* xxxvii. 439, I verily believe you have said that fifty thousand times, in my hearing. What a Prose you are!

6. *prose book, work*; **prose fiction**, the genre of fictional narratives written in prose; **prose-poem** (earlier examples); **prose sense**, the meaning of a poem as it can be paraphrased in prose; **prose style**, characteristic manner of writing in prose.

1940 DYLAN THOMAS *Let.* 13 May (1966) 248, I do not want to write another straight prosebook yet. **1841** Prose-fiction [see PERFECT *a.* 4 a]. **1848** MILL *Pol. Econ.* I. II. xiv. 467 The most successful writer of prose fiction (Scott). **1919** V. WOOLF in *Times Lit. Suppl.* 10 Apr. 189/2 It is for..[the historian of literature] to ascertain whether we are now at the beginning, or middle, or end, of a great period of prose fiction. **1957** *Encycl. Brit.* XVI. 573/2 Dickens, perhaps the most remarkable genius in the history of English prose fiction. **1842** POE in *Graham's Mag.* Jan. 69/1 Criticism is *not*..an essay, nor a sermon, nor an oration,..nor a prose-poem. **1850** C. KINGSLEY *Alton Locke* I. ix. 139 That great prose poem, the single epic of modern days, Thomas Carlyle's 'French Revolution'. **1947** C. BROOKS *Well Wrought Urn* xi. 182 The 'prose-sense' of the poem is not a rack on which the stuff of the poem is hung. **1852** THACKERAY *Esmond* III. iii. 88 His [*sc.* Addison's] prose style I think is altogether inimitable. **1906** R. BROOKE *Let.* 10 May (1968) 51 This effort has..worked..havoc in my carefully elaborated prose-style. **1959** G. D. PAINTER *Proust* I. viii. 115 His [*sc.* Proust's] prose style was..faded and artificial. **1976** N. FREELING *Lake Isle* xiii. 120 Someone, presumably..has been phoning somebody. A prefect to judge from the prose style. *c* **1827** MILL *Speech* in *Adelphi* (1924) I. 692 The very small number of good prose works which have been published for many years past, except indeed novels. **1934** J. JOYCE *Let.* 1 June (1966) III. 306, I work every day alone at my big long wide high deep dense prosework. **1978** W. WHITE *Whitman's Daybks. & Notebks.* I. p. xxiii, Every name of a person, place, book, poem, prose work, or a 'situation' in Whitman's life and times that seemed to me to call for annotation, I have annotated.

prosecretin: see *PRO-² 1.

prosecute, *v.* **6. d.** (Further examples.)

1865 *Chambers's Encycl.* VII. 799/1 If a person is murdered, some one of the relatives naturally prosecutes. **1901** G. B. SHAW *Capt. Brassbound's Conversion* III. 286 The counsel for the prosecution can proceed to prosecute. The floor is yours, Lady Waynflete. **1966** *Listener* 9 June 828/1 Even when the police prosecute, committal for trial cannot be left entirely to their discretion. **1971** *Reader's Digest Family Guide to Law* 743/1 A private individual has the right in most cases to follow the same procedure, even if the police have decided not to prosecute.

prosecuting *ppl. a.* (earlier and later examples).

1832 *Jrnl. Indiana Ho. Representatives* 6 Dec. 33 Duly elected Prosecuting Attorney of the 2d Judicial Circuit. **1912** M. NICHOLSON *Hoosier Chron.* 180 The Republican prosecuting attorney of Ranger County joined with the local bank in certifying Miles's probity. **1959** *Granta* 6 June 33/2 'Where were you?' shouted the bluff prosecuting counsel. **1976** *Daily Mirror* 16 July 9/2 Prosecuting authorities are now more sensitive to the need to investigate suspicions of corruption.

prosecutorial (prɒsikiŭtǭᵊ·riăl), *a.* [f. PROSECUTOR + -IAL.] Of or pertaining to a prosecuting official or his duties. Also *transf.*

1973 *N.Y. Law Jrnl.* 19 July 1/8 To our minds the participants' attempt to set up a federal crime for which these defendants stand convicted went beyond any proper prosecutorial role. **1975** *Columbia Law Rev.* LXXV. 130 Prosecutorial discretion is the power held by an agency or official charged with enforcement of the law to exercise selectivity in the choice of occasions for the law's enforcement. **1978** *N.Y. Times* 30 Mar. B3/5 The Midtown Enforcement Project has received a Federal grant..to bring innovative enforcement and prosecutorial methods to the area from 30th Street to 60th Street from river to river. **1978** *Listener* 29 June 848/1 His closely-argued, mercilessly prosecutorial book.

proselyting, *vbl. sb.* (Later examples.)

1931 H. F. PRINGLE *Theodore Roosevelt* II. xiii. 456 Mrs. Storer went on with her proselyting. **1948** *Richmond* (Va.) *Times-Dispatch* 13 Feb. 30/1 San Francisco..was guilty of one of the worst cases of out-and-out..proselyting and subsidizing yet seen.

pro-sentence: see *PRO-¹ 4b.

proseology (prozǫ·lŏdʒi). *colloq. rare.* [f. PROSE *sb.* + -OLOGY.] Prolix, turgid, or confusing prose.

1909 *Daily Chron.* 19 July 3/4 To plough through all the extracts from journals, letters, &c., which form the book needs not a little patience and sticking power. Doubtless, those who can get beneath all this proseology will find much to excite them. **1968** *Economist* 29 June p. xi/2 The usual, easy proseology reviling the accursed folk or crying the beloved country.

pro shop: see PRO* *abbrev.* 2c.

prosimian, *a.* and *sb.* (Examples.) Substitute for defs.: **A.** *adj.* Of, belonging to, or designating prosimians. **B.** *sb.* A mammal belonging to the suborder Prosimii, a group of primitive, mostly arboreal primates, which includes lemurs, lorises, galagos, and tarsiers.

1925 *Bull. Geol. Soc. China* IV. 142 In some area intermediate between the Oriental and Ethiopian regions the centre of prosimian dispersal was located. **1945** *Bull. Amer. Mus. Nat. Hist.* LXXXV. 181/2 It seems quite possible that New World and Old World monkeys arose independently from Eocene prosimians... Their prosimian ancestors, if distinct, must have been closely allied. **1959** *New Scientist* 10 Dec. 1175/1 Professor Schultz considers that the very important exploratory function of the hands of simians superseded the tactile sense of the prosimians (such as the lemurs). **1966** [see *HIGH *a.* 6d]. **1972** T. A. VAUGHAN *Mammalogy* vii. 117 *(caption)* Hands and feet of some prosimian primates. **1974** *Sci. Amer.* Apr. 127/1 The prosimians and the tenrecs, unlike any other mammals, respond sharply to external temperature. **1977** P. NAPIER *Lemurs, Lorises & Bushbabies* i. 9 The lemurs, lorises and bushbabies are called prosimians.

prosiopesis (prɒsiopī·sis). *Gram.* [f. Gr. προ- before + σιώπησις taciturnity, f. σιωπᾶν to be silent.] Jespersen's term for ellipsis of the beginning of a grammatical structure in speech.

1924 O. JESPERSEN *Philos. Gram.* x. 142 The subject must generally be expressed, and those few cases in which it is omitted, may be explained through prosiopesis, which sometimes becomes habitual in certain stock exclamations like *Thank you*. **1927** —— *Mod. Eng. Gram.* III. 226 *It* is occasionally left out through prosiopesis. **1935** H. STRAUMANN *Newspaper Headlines* 41 Whether such sentences should be considered as incomplete or elliptic or as aposiopesis and prosiopesis is a question of terminology. **1966** G. N. LEECH *Eng. in Advertising* viii. 78 One marginally casual feature of advertising language is the occurrence of what..has been termed prosiopesis.

prosiphonate, *a.* (Examples.)

1935 TWENHOFEL & SHROCK *Invertebr. Paleont.* ix. 378 Prosiphonate ones [*sc.* septa] are found in Mesozoic forms [of cephalopod]. **1952** R. C. MOORE et al. *Invertebr. Fossils* ix. 371/1 In late Paleozoic and Mesozoic ammonoids, this structure [*sc.* the siphuncle] is mostly confined to immature stages, and forward-pointing septal necks appear in submature whorls. This condition is described as prosiphonate. **1970** R. M. BLACK *Elements Palaeont.* viii. 86 Short septal necks encircle the siphuncle where it passes

a variety of cognitive and intellectual *prosthetic* devices to the brain—a kind of eyeglasses for the mind.

3. *Biochem.* Applied to a non-protein group forming part of or combined with a protein, e.g. in an enzyme. [This sense is due to A. Kossel (1892), who used G. *prosthetisch* (*Arch. f. Anat. u. Physiol.* (*Physiol. Abt.*) (1893) 157).]

1898 J. A. MANDEL tr. *Hammarsten's Text-bk. Physiol. Chem.* (ed. 2) ii. 48 The nucleoproteids..may be considered as combinations of a proteid nucleus with a side chain, which Kossel calls the prostetic [*changed to* prosthetic *in ed.* 4 (1904)] group. **1932** *Science* 10 June 615/2 It is of fundamental importance to decide whether the activity of insulin is a function of the whole molecule or of a prosthetic group, non-proteid in composition, which is attached to the protein complex. **1939** *Nature* 25 Nov. 886/2 It is now known also that vitamin B_1, including its pyrophosphate derivative, is identical with the prosthetic group of the enzyme carboxylase. **1954** K. J. LAIDLER *Introd. Chem. Enzymes* v. 94 The cytochromes are therefore complete enzymes in themselves, their prosthetic groups being successively reduced and oxidized during the course of a biological oxidation. **1964** G. H. HAGGIS et al. *Introd. Molecular Biol.* iii. 42 The total structure of a protein molecule often includes, in addition to its peptide chains a component of different chemical constitution, known as a prosthetic group. The haem component of haemoglobin is an example of a group of this kind. **1973** R. G. KRUEGER et al. *Introd. Microbiol.* vii. 238/1 Prosthetic groups are usually permanently attached to the enzyme, having been built into the enzyme when it was synthesized.

B. *sb. pl.* The branch of surgery concerned with the replacement of defective or absent parts of the body by artificial substitutes.

1894 in G. M. GOULD *Illustr. Dict. Med. & Biol.* **1911** G. H. WILSON *Man. Dental Prosthetics* Pref., This book has been written in response to the oft-repeated request by teachers and members of the dental profession for a concise modern text-book on Dental Prosthetics. **1963** *New Scientist* 28 Nov. 525 Here is only one more instance of a remarkable flowering of techniques..ranging from prosthetics to chromosomal manipulations, which have tremendous potential for good, but also allow a mockery of nature.

prosthetist (prǫ·sþītist). *Surg.* [f. PROSTHET(IC *a.* + -IST.] One who designs and fits prostheses.

1902 *Buck's Ref. Handbk. Med. Sci.* (ed. 2) V. 513/2 Napoleon..certainly made many cripples and should be hailed as the patron saint of prosthetists. **1924** D. D. CAMPBELL *Full Denture Prosthesis* x. 235 It scarcely need be observed that no prosthetist, sure of his method of securing central occlusion, will avail himself of such a mechanical aid. **1953** H. R. B. FENN et al. *Clin. Dental Prosthetics* xix. 453 *(caption)* If this type of denture is to be successful it requires absolute accuracy on the part of both the prosthetist and the technician. **1970** M. VITALI in G. Murdoch *Prosthetic & Orthotic Pract.* xiii. 531 The prosthetist..could not say when the pylon is going to be ready for delivery but it has helped a lot to understand some of the problems of the amputee.

prosthion (prǫ·sþiǫn). *Anat.* [Neut. of Gr. πρόσθιος foremost (f. πρόσθεν before, in front): cf. *-ION².] The lowest or the most forward point of the maxilla between the two central incisors.

1925 *Biometrika* XVII. 55 If there be two points used, the prosthion and the alveolar point..then the computation of the angles and sides of the fundamental triangle of the skull becomes impossible. **1933** *Jrnl. R. Anthrop. Inst.* LXIII. 42 Some modern craniometricians speak of the prosthion and the alveolar point as if they were coincident. **1937** *Amer. Jrnl. Physical Anthrop.* XXII. 486 Prosthion has..been generally used to designate both the most forward and the lowest points, in accordance with the measurement being taken. The distinction is made, however, by Buxton and Morant, who call the most forward point prosthion and the lowest the alveolar point. **1937** [see *NASION]. **1974** MOORE & LAVELLE *Growth of Facial Skeleton in Hominoidea* iv. 196 Up to the stage of eruption of the first permanent molar, the face of the orang utan is growing strongly downwards and forwards. Subsequently, the face begins to tilt so that the prosthion, instead of moving downwards and forwards, moves forwards and somewhat upwards.

prosthodontia (prǫsþodǫ·ntiă). *Dentistry.* [f. as next + -IA¹.] Prosthodontics.

1917 F. A. PEESO *Crown & Bridge-Work* viii. 151 *(heading)* Relations of prosthodontia and orthodontia. **1934** F. W. FRAHM *Princ. & Technics Full Denture Construction*, p. xiv, The term 'prosthetic dentistry', or 'prosthodontia', may..be defined as an account of methods for the replacement of any lost or missing parts of the dental organism by artificial substitutes. **1947** H. H. HORNER *Dental Educ. Today* ix. 266 The School of Dentistry..offers graduate courses in orthodontia, prosthodontia, and oral surgery.

prosthodontics (prǫsþodǫ·ntiks), *sb. pl.* *Dentistry.* [f. as next + -IC; cf. *ORTHODONTICS *sb. pl.*] The branch of dentistry concerned with the design, manufacture, and fitting of artificial replacements for teeth and other parts of the mouth; prosthetic dentistry.

1947 H. H. HORNER *Dental Educ. Today* ix. 266 Postgraduate and refresher courses are offered in orthodontics.., prosthodontics.., [etc.]. **1977** M. M. HUDIS *Dental Lab. Prosthodontics* p. xv, The basic concepts will be developed so that they may be integrated into the total concept of prosthodontics. **1978** *Who's Who* 1495/2 Formerly Professor and Chairman of the Prosthodontics Department, University of Southern California.

prosthodontist (prǫsþodǫ·ntist). *Dentistry.* [f. PROSTH(ESIS + Gr. ὀδοντ-, ὀδούς tooth + -IST.] One who practises prosthodontics.

1917 F. A. PEESO *Crown & Bridge-Work* viii. 151 Quite frequently when the services of an orthodontist has been employed the coöperation of the prosthodontist is necessary to complete and render his work permanent. **1934** F. W. FRAHM *Princ. & Technics of Full Denture Construction* vi. 74 The prosthodontist should avail himself of the use of the very best appliances and technics that will help him to develop and adapt the very finest that is possible in the construction of dentures. **1977** *Daily Colonist* (Victoria, B.C.) 21 May 34/3 'I don't like to go to the dentist myself,' says Dr..Tregaskis..dental professor and prosthodontist.

prostie, var. *PROSTY.

Prostigmin (prosti·gmin). *Pharm.* Also -ine (-in), and with small initial. A proprietary name for neostigmine (s.v. *NEO- 1 b).

1931 *Trade Marks Jrnl.* 11 Nov. 1500/2 Prostigmin... Chemical substances prepared for use in medicine and pharmacy. The Hoffmann–La Roche Chemical Works Limited... London. **1935** *Times* 9 Mar. 9/4 Later, using a drug known as prostigmin, she obtained much better results in the treatment of a severe case of myasthenia. **1946** *Official Gaz.* (U.S. Patent Office) 19 Mar. 327/1 *Prostigmin.* For medicinal preparation... Claims use since Apr. 29, 1931, in the form 'Prostigmine'; and since Nov. 8, 1931, on the mark as shown. **1956** LD. AMULREE in A. Pryce-Jones *New Outl. Mod. Knowl.* II. 217 The use of prostigmine in the treatment of myasthenia gravis and of the extract of thyroid and of its synthetic substitute thyroxin in the treatment of diseases of the thyroid, are further examples of the way in which modern therapy can overcome hitherto irremediable diseases. **1961** *Lancet* 2 Sept. 512/1 Various unsuccessful regimens were tried, including a diabetic diet, acetylcholine derivatives, and prostigmine. **1967** H. BECKMAN *Dilemmas in Drug Therapy* 47/1 In sinus and nodal tachycardia, neostigmine (Prostigmin) may occasionally effect a conversion to normal rhythm. **1974** M. C. GERALD *Pharmacol.* vii. 132 Such reversible inhibitors,..neostigmine (Prostigmine), for example, act for several hours after a single dose.

prostisciutto (prǫ:stiʃū·to). *nonce-wd.* [Blend of PROSTITUTE *sb.* and *PROSCIUTTO.] A female prostitute regarded metaphorically as an item on a menu.

1930 S. BECKETT *Whoroscope* 1 What's that? A little green fry or a mushrooming one? Two lashed ovaries with prostisciutto?

prostitute, *ppl. a.* and *sb.* Add: **B.** *sb.* **1. c.** A man who undertakes male homosexual acts for payment; usu. *male prostitute*.

1948 [see *male prostitute* s.v. *MALE *a.* 1 e]. **1958** L. DURRELL *Balthazar* vii. 157 A magnificent-looking male prostitute whose oiled curls hung down his back and whose eyes and lips were heavily painted. **1967** *Listener* 1 June 718/2 Few of them ever told me what they were in for, though in the case of Ralph, a male prostitute known to the wing as Suzanne, it was only too obvious. **1975** *Daily Tel.* 24 July 3/6 Many of the boys became male prostitutes.

2. b. Delete 'Now *rare*' and add further examples.

1889 G. B. SHAW *Let.* 31 Aug. (1965) 223 The radical who writes conservative articles is considered a prostitute. **1980** C. FITZGIBBON *Rat Report* vi. 122 You damned us.. for turning scientists into military prostitutes.

prostitution. Add: **1. d.** Of men: the undertaking of homosexual acts for payment.

1886 R. F. BURTON *Terminal Ess.* in *Arabian Nights' Entertainments* X. 242 According to Gomara there were at Tamalipas houses of male prostitution. **1975** P. McCUTCHAN *Very Big Bang* xiii. 122 Porn, poncing, male prostitution—you name it, the court'll send you down for it. **1975** *Times* 21 June 2/3 Runaway boys..were procured for a male prostitution ring by offers of food and shelter.

prosty (prǫ·sti). *U.S. slang.* Also **prostie**. Abbrev. of PROSTITUTE *sb.* 1 a. Cf. *PROSSIE.

1930 *Bookman* Dec. 397/2 These song writers exploit motherhood into a mania, and go Rotary about any old half-baked, klux-ridden Dixie backwash. They make love in kindergarten terms, turn Sunday-school superintendent about prosties whom they term 'faded roses' or 'butterflies', and otherwise trade on primitive mass ideas. **1941** [see *HOPPY *sb.*²]. **1951** GREEN & LAURIE *Show Biz* 571/1 *Prostie*, prostitute. Variety's sensitized way of describing female characters comparable to those in early Mae West plays. **1972** G. BAXT *Burning Sappho* vii. 128 The prostie in the orange wig. **1976** J. HAYES *Missing* i. 29 If she was a prostie, he couldn't afford her fee.

‖**prosula** (prǫ·sulă). *Eccl.* Pl. **-æ**. [mod.L., dim. of L. *prosa* *PROSA; see -ULE.] (See quots.)

1907 ORME & WYATT tr. *Wagner's Introd. Gregorian Melodies* xiv. 245 A second kind of trope, which also goes back to Tutilo, resembles the Sequences, and is often like them called *Prosa*, or, when of lesser extent, *Prosula*. **1958** W. APEL *Gregorian Chant* iii. 433 Twenty-five of the Offertoires..are further amplified by the addition of a *prosula*..that is, a new text appended to the end of a verse, usually the last. *Ibid.*, The music for the *prosula* or, as we would say, for the trope, is identical with the closing passage of the verse. **1970** [see *PROSA]. **1975** *Anglo-Saxon England* IV. 134 A Kyriale, beginning imperfectly with part of the prosula to the *Kyrie eleison* entitled *Clemens rector*. *Ibid.* 135 A number of Kyries do not have prosulae.

prosy, *a.* **1.** (Earlier examples.)

1814 JANE AUSTEN *Let.* 9 Sept. (1952) 402 The scene with Mrs. Mellish, I should condemn; it is prosy & nothing to the purpose. **1823** SCOTT in *Ballantyne's Novelist's Library* V. p. lxxxvi, Perhaps, to be circumstantial and abundant in minute detail, and in one word, though an unauthorized one, to be somewhat *prosy*, is one mode of securing a certain necessary degree of credulity in hearing a ghost-story.

pro-syllable, -syllabic: see *PRO-¹ 4 b, c.

Prot (prǫt), *sb.* and *a.* Also **prot.** A colloq. abbrev. of PROTESTANT *sb.* 2 a and *a.* 1; *spec.* opp. to *Catholic* (freq. in derog. or contemptuous use). Cf. *PROD *sb.*³ and *a.*

1725 J. THORNTON *Let.* in *Dublin Rev.* (1916) Apr. 318 Sir George Brown, I hear, is got to Gant, there be-moaning his folly in having tied himself up to an old Prot, who cunningly settled all she had out of his reach. **1737** R. CHALLONER *Let.* 15 Sept. in *Recusant Hist.* (1970) X. 351 Our Prelate here, has..absolutely refused to consent to the Parties being married first by a Priest and then by a Prot. minister. **1843** M. EDGEWORTH *Let.* 3 Dec. (1971) 599 The average salary of the Irish priest is £290 per Annum and average of Prot—£120 or £130. **1900** C. M. YONGE *Mod. Broods* v. 50 Oh, she is a regular old Prot.. almost a Dissenter. **1900** M. CREIGHTON *Let.* 16 Nov. (1904) II. 454 The position was 'I would meet you if I could: but I am not going to be bullied by a handful of Prots.' **1937** AUDEN & MACNEICE *Lett. from Iceland* xiii. 204, I know a Prot Will never really kneel, but only squat. **1955** E. POUND *Section: Rock-Drill* (1957) lxxxvi. 24 Yes yr/ Holiness, they are all of them prots. **1971** B. SLEIGH *Smell of Privet* x. 87 'You must never say cup, always chalice. Don't be such a Prot, my dear!' I had no idea what a 'Prot' was, but vowed to myself I would never be one again. **1977** P. WAY *Super-Celeste* 100 Back in Belfast..there were Prot bombs and Catholic bombs and SAS bombs.

protactinium (prōᵘtækti·niŏm). *Chem.* Also **protoactinium** (prōᵘ:to,ækti·niŏm). [mod.L., coined (as *protactinium*) in Ger. (Hahn & Meitner 1918, in *Physik. Zeitschr.* XIX. 211/1) see PROTO- and *ACTINIUM 2. So named because the principal isotope produces actinium by radioactive decay.] A radioactive metallic element of the actinide series, which occurs in small quantities as a decay product in uranium ores, and whose longest-lived isotope has a half-life of about 33,000 years. Atomic number 91; symbol Pa.

The spelling *protactinium* is that adopted by the International Union of Pure and Applied Chemistry.

1918 *Jrnl. Chem. Soc.* CXIV. II. 346 Assuming that 8% of the uranium atoms disintegrating produce 'protoactinium', the quantity in the 73 mg. is that in equilibrium with 86 grams of uranium. **1919** *Chem. Abstr.* XIII. 1181 Protactinium is one of the 5 new radioactive elements occupying a place in the periodic table hitherto vacant. **1934** *Nature* 8 Sept. 386/1 The protoactinium was precipitated with zirconium as phosphate in the earlier stages, partly freed from zirconium by fractional crystallization and precipitated together with tantalum. **1934** *Times* 13 Sept. 9/4 The successful isolation of protactinium, the parent element in the actinium series of radioactive changes, was announced yesterday. **1959** *New Scientist* 17 Dec. 1264/3 From 60 tons of waste material from the production of uranium from its ores, chemists at Windscale have with some difficulty extracted 100 grammes of protactinium. **1962** COTTON & WILKINSON *Adv. Inorg. Chem.* xxxii. 906 Protactinium as ²³¹Pa.. occurs in pitchblende, but even the richest ores contain only about 1 part Pa in 10⁷. The isolation of protactinium..is difficult, as indeed is the study of protactinium chemistry generally, owing to the extreme tendency of the compounds to hydrolyze. **1970** J. W. GARDNER *Atoms Today & Tomorrow* vi. 99 The decay chain proceeds from thorium (atomic number 90) via protactinium (atomic number 91) to uranium. **1973** PHILLIPS & MILNER in J. C. Bailar et al. *Comprehensive Inorg. Chem.* V. 79 Protactinium is the third member of the actinide series, but it possesses many of the properties of the Group V elements in its chemical reactions... The stable valency state in solution is Pa(V), and the existence of Pa(IV) is also possible.

protagonist. Add: **1, 2.** Also *pl.*, the leading characters in a play, story, contest, etc.; the most prominent or most important individuals in a situation or course of events.

Fowler's classification of the plural as an absurd use (*Dict. Modern English Usage* p. 471; maintained in Sir E. Gowers' second edition, p. 489) may be challenged on the grounds that derivation from Greek πρῶτος *first*, does not preclude a plural form, and limitation to the singular is strictly relevant only in the context of ancient Greek drama.

1671 [in Dict.]. **1930** D. L. SAYERS *Strong Poison* viii. 109 I'm getting a certain amount of light on the central figures in the problem—what journalists like to call the protagonists. **1950** G. B. SHAW *Shakes versus Shav* 135 Living actors have to learn that they too must be invisible while the protagonists are conversing, and therefore must not move a muscle nor change their expression. **1952** *Sunday Times* 6 July 4/8 The soliloquy is a special problem on the screen but in 'Julius Caesar' there are

PROTAGOREAN

only two of them, both short. Mr. Mankiewicz has told me that he hopes to use an intimate technique, giving importance above all to the characters of the protagonists. **1962** L. Auchincloss in I. Howe *Edith Wharton* 33 The change..comes with the infiltration of the other protagonists of the drama, the Spraggs, the Wellington Boys, the Gormers, etc.]. **1976** *Times* 29 Oct. 1/1 Strong opposition to more cuts in public expenditure were voiced at a meeting of the Cabinet on Tuesday. The protagonists were Mr Crosland..Mr Shore..and Mr Benn.

¶ **3.** [Through confusion of sense 2 with Pro-² 5a.] A proponent, advocate, or supporter (of a cause, idea, etc.).

In this use the notion of 'a leading personage' is not implied. In some contexts there is ambiguity between this sense and sense 2.

1935 *Hansard Commons* 4 June 1718/2 My right hon. Friend the Member for Epping and others on the right have come out in this House as the protagonists of self-determination. **1938** A. P. Herbert *What a Word!* 15 99, I heard with horror..that the word 'protagonist' is being used as if it were pro-tagonist—one who is for something, and opposed to ant-agonist, one who is against it. **1952** *Times* 2 Apr. 5/4 As a protagonist, in and out of Parliament and especially in the county of Durham, of the fullest use of land for food production, I feel impelled to reply to my friend Miss Nancy Price's letter..and to defend the town council of Worthing. **1961** E. K. Klein et al. *Economics: Model of U.K.* iv. 121 Professor Robbins, a firm protagonist of the importance of the influence of demand over the period. **1972** *Observer* 15 Oct. 39/6 In practical terms, I wish I thought that, in 1975, we shall read..alongside *protagonist*, not just 'also: proponent' but 'also, *improperly*: proponent'. **1974** *BSI News* Sept. 12/3 Over the last few years the relative merits of the pascal and the bar have been discussed interminably.. Protagonists of the pascal do not think its magnitude is of any relevance to its choice. **1975** *Times Lit. Suppl.* 22 Aug. 939/2 Köhan was..an unabashed protagonist of the technical superiority of Western civilization. **1979** *Jrnl. R. Soc. Arts* CXXVII. 334/1 A protagonist of and expert on the Added Value concept.

Hence ¶ **protagonism** *rare*, the defence or advocacy of a cause, idea, etc.

1909 *N.Y. Even. Post* 27 Nov. 6 The principal character..is gradually drawn into a protagonism of common sense, candour, and progress. **1937** *Mind* XLVI. 311 This method, the validity of which depends upon the progressive series of experiments developed in the *Phenomenology*, is dramatic; it requires the experimenter to be protagonist in his factitious protagonism.

Protagorean (protægǭri·ən), *a.* and *sb.* [f. *Protagoras* (Gr. Πρωταγόρας) the name of a Greek philosopher of the 5th century B.C. + -an.] **A.** *adj.* Of or pertaining to Protagoras or his philosophy. **B.** *sb.* An adherent or admirer of the philosophy of Protagoras. Hence **Protagore·anism**, the Protagorean philosophy.

1678 R. Cudworth *True Intell. Syst.* i. 10 The Protagorean Philosophy made all things to consist of a Commixture of Parts (or Atoms) and Local Motion. **1845** *Encycl. Metrop.* II. 614/1 The need of which a reason, he asserts, in opposition to the Protagorean notion of man being the measure of all things, which he treats as a silly truism. **1887** *Encycl. Brit.* XXII. 236/1 Socrates rested his scepticism upon the Protagorean doctrine that man is the measure of his own sensations and feelings. *Ibid.*, In the review of theories of knowledge which has come down to us in Plato's *Theætetus* mention is made..of certain 'incomplete Protagoreans'. **1907** *Hibbert Jrnl.* Jan. 439 A Protagorean treatise of the fifth century B.C. **1921** T. R. Glover *Pilgrim* 176 The idea of Christian charity has been perverted,..to mean a Protagorean acceptance of the equal value of all opinions. **1932** *Times Lit. Suppl.* 21 Apr. 282/2 It supplies the key to the interpretation of the refined Protagoreanism which the author avows. **1954** *Essays in Crit.* IV. 55 There Arnold figures as a Protagorean sceptic. **1958** *Times Lit. Suppl.* 17 Jan. 26/2 For Professor Guthrie this is a genuine outline of Protagorean thought.

protalus (protē̆ⁱ·lɒ̆s). *Physical Geogr.* Also **pro-talus.** [f. Pro-² + Talus¹.] A rocky ridge or lobe on the lower edge of an existing or former snow-bank, composed of frost-fractured boulders and other debris that have slid or rolled over the snow from a talus or scree higher up the slope, or been transported by solifluction from the talus. Usu. *attrib.*, as *protalus lobe, rampart.*

1934 K. Bryan in *Geogr. Rev.* XXIV. 656 The word 'nivation' is the name of a process of excavation around snowbanks described by Matthes. The use of the same word for these ramparts of blocks is likely to prove misleading, and the reviewer suggests that 'protalus rampart' would be appropriate for the features. **1962** *Prof. Papers U.S. Geol. Survey* No. 324. 61/2 Many deposits thin downslope to a featheredge, but in the mountains some terminate in a lobate mass bearing one or more arcuate ridges in which the fragments tend to be oriented imbricate to the slope... Such protalus lobes appear to have flowed slowly forward as a unit, presumably by solifluction. **1970** R. J. Small *Study of Landforms* xi. 378 The disintegrated material may slide across the bank of firn below, and accumulate to give moraine-like piles of debris ('pro-talus') running approximately parallel to the headwall. **1976** *Scottish Geogr. Mag.* XCII. 182 An excellent example of a protalus rampart occurs in Wester Ross 10 km south-east of Gairloch. *Ibid.* 184 The protalus complex lies at the foot of a scree slope that rises through a vertical height interval of 100–200 m.

protamine. Substitute for entry in Dict.: **protamine** (prō·tămīn). *Biochem.* and *Med.* Also † -in (-in). [ad. G. *protamin* (F. Miescher 1874, in *Verhandl. d. Naturforsch. Ges. in Basel* VI. 153): see Proto- and Amine.] **1.** Any of a class of basic proteins of relatively low molecular weight which occur combined with nucleic acids in the sperm of many species of fish, and which have the property of countering the anti-coagulant action of heparin; orig. *spec.* that obtained from the salmon.

1874 *Jrnl. Chem. Soc.* XXVII. 794 Nuclein occurs in the salmon roe in the form of an insoluble compound with the new base protamine, the latter constituting no less than one-fourth of the weight of the roe. **1896** *Ibid.* LXX. i. 582 The names *salmine* and *sturine* are suggested by [sic] the two protamines. **1902** *Encycl. Brit.* XXXI. 724/1 These Protamines..take up water and yield the bases above referred to. **1928** W. V. Thorpe tr. *Kossel's Protamines & Histones* ii. 1. 4 Increasing knowledge of the hydrolysis products of the protamines confirmed Kossel's view that the protamines belonged to the protein group and were the most elementary type of this class of compounds. **1954** A. White et al. *Princ. Biochem.* ix. 195 The protamines are relatively small and simple proteins, deficient in many amino-acids and extremely rich in arginine. **1970** R. W. McGilvery *Biochem.* xxvi. 653 The effects of heparin can be reversed by giving protamine, the highly basic small protein associated with nucleic acids in fish sperm. **1971** D. R. Williams *Metals of Life* iii. 27 The threshold between polypeptides and proteins is arbitrarily set at MW 5000 since the smallest physiologically active polymers, the protamines (found in spermatozoa), have molecular weights commencing at this value. **1973** B. A. Brown *Hematol.* iv. 132/1 At the completion of the surgery, protamine is administered in order to neutralize the effects of the heparin.

2. *attrib.* and *Comb.* **a.** In various combs., as *protamine-insulin, protamine-zinc(-insulin)*, denoting suspensions of insulin with a protamine, and usu. also zinc chloride, which have greater stability and a more prolonged hypoglycæmic action than insulin alone.

1935 C. L. Hill tr. N. B. Krarup (*title*) Clinical investigations into the action of protamine insulinate. **1936** *Canad. Public Health Jrnl.* XXVII. 137/2 In the preparation of protamine insulin for injection, a suitable quantity of protamine, buffered with sodium phosphate, is added to regular insulin. **1936** *Canad. Med. Assoc. Jrnl.* XXXV. 240/1 The other solution contained a buffer phosphate, 1 c.c. of which, when added to the protamine-zinc-insulin complex, adjusted the reaction so that the hydrogen-ion concentration..was identical with that of blood. **1956** *Nature* 1 Feb. 223/2 Robinson and Fehr were able to estimate the percentage of insulin in protamine insulin by using the upper phase from a butanol/acetic/water mixture as a chromatographic solvent. **1960** M. Spark *Bachelors* vii. 100 Protamine zinc for more prolonged coverage throughout the evening and the following night. She needs to suits. **1962** H. Burn *Drugs, Med. & Man* xiv. 148 While the maximum time of action of insulin alone was six hours, that of protamine-insulin was 12–18 hours, and finally that of zinc-protamine-insulin was 24–30 hours.

b. protamine sulphate, a salt of a protamine and sulphuric acid, given as an aqueous solution to neutralize the anti-coagulant effect of heparin; **protamine titration**, a test of the clotting ability of blood in which blood is first made incoagulable with heparin and then titrated against protamine sulphate until clotting occurs; a value so obtained.

1936 *Jrnl. Pharmacol. & Exper. Therapeutics* LVIII. 80 After standing 4 or 5 days the precipitated protamine sulfate has settled and is removed by centrifugation. **1962** *Listener* 3 May 769/2 Certain mixtures of large molecules—for example, mixtures of starch, gelatin, and protamine sulphate—will form drops which have great stability. **1971** S. I. Rapaport *Introd. Hematol.* xxiv. 311 One-tenth volume of 1 per cent protamine sulfate is added to plasma, and the plasma is examined for a precipitate after 15 minutes at 37°C. **1949** J. G. Allen et al. *Jrnl. Lab. & Clin. Med.* XXXIV. 473 (*heading*) A protamine titration as an indication of clotting defect in certain hemorrhagic states. **1949** *Jrnl. Amer. Med. Assoc.* 30 Apr. 1251/1 The blood..showed an increased protamine titration when the prothrombin level was normal or near normal, when fibrinogen levels were not abnormal and when fibrinolysin was not grossly disturbed. **1973** B. A. Brown *Hematol.* iv. 132/1 The protamine titration..is used to estimate the minimum required dose of protamine.

protan (prō̆ᵘ·tæn), *sb.* (*a*). *Ophthalm.* [f. *protan-* in *protanomaly, *protanopia, etc.] A protanomalous or protanopic person. Also as *adj.*

1944 D. Farnsworth in *Inter-Society Color Council News Let.* lvi. 8 There are 3 or more types of Anomaly (chromatic imbalance). 1. n. Protan; adj. Protanous: reduction of red-bluegreen discrimination relative to interesting axis. **1962** *Lancet* 15 Dec. 1269/1 The former includes protan and deutan defects with 3 or more alleles at each locus, and the latter includes true hæmophilia and Christmas disease. **1968** [see *Ishihara]. **1974** *Nature* 23 Aug. 653/1 It is highly probable that there are at least two gene loci involved in each of four X-linked diseases: colour blindness (protan and deutan), haemophilia (A and B), muscular dystrophy (Duchenne and Becker), and retinal degeneration (retinoschisis and Norrie's disease).

protandrous, *a.* Add: [ad. G. *protandrisch* (F. H. G. Hildebrand *Die Geschlechter-Vertheilung bei den Pflanzen* (1867) 17).] **1.** (Earlier and later examples.)

1870 A. W. Bennett in *Jrnl. Bot.* VIII. 317 (*heading*) Protandrous. **1965** Bell & Coombe tr. *Strasburger's Textbk. Bot.* 602 Many apparently protandrous flowers..were borne in the axils of the dead leaves.

2. *Zool.* = Proterandrous *a.* 2.

1897 Parker & Haswell *Textbk. Zool.* I. vi. 285 A few forms [of Nematodes] are hermaphrodite, but, instead of having a double set of reproductive organs, as in Platodes, organs of the ordinary female nematode type are present, and the gonads produce first sperms and afterwards ova. Such animals are said to be protandrous (male products ripe first). **1929** *Amer. Naturalist* LXIII. 577 The lengths of males [of *Pandalus danae* Stimpson] indicated that they were in two year-groups... On the other hand the females seemed to belong to a third year-group.. This appeared to suggest that the form under consideration was protandrous. **1973** *Nature* 5 Oct. 265/1 Both protandrous and protogynous sequential hermaphroditism have been reported in fishes, and there is some evidence that smaller and younger specimens of [*Limea*] species are predominantly males, while larger and older individuals are mostly females.

protandric *a.* (examples in the sense of *protandrous *a.* 2); **protandry** (examples in *Zool.*).

Delete †1882 *Nature* (Annandale).

1870 [see *protogyny]. **1887** *Bergens Museums Aarsberetning* VII. 29 It may not be amiss to draw a comparison between the protandric hermaphroditism of Myxine and the hermaphroditism of the few other hermaphroditic vertebrates known. **1892** J. A. Thomson *Outl. Zool.* 632 (Index), Protandry of Myxine. **1932** *Proc. 6th Internat. Congr. Genetics* II. 26 Several of the ovoparous species show a large percentage of intersexuality when young, with a strong tendency toward protandry. *Ibid.* 27 This species is regularly protandric, each young animal producing many thousands of sperm balls. **1951** [see *gonochoric adj.* s.v. *gono-]. **1970** *Nature* 11 July 184/2 Gross examination of gonads in larger males and smaller females showed no evidence of protandry. **1975** *Ibid.* 15 May 221/2 He considered it likely that *A. equina* is a protandric hermaphrodite which mainly self-fertilizes and retains larvae within the parent.

protanomal (prō̆ᵘ·tănǫ-măl). *Ophthalm.* [ad. G. *protanomal* (W. A. Nagel 1907, in *Zeitschr. f. Psychol. und Physiol. d. Sinnesorgane: Abt. II* XLII. 67), f. Gr. πρωτ- Prot- + G. *anomal* anomalous.] A person with protanomaly.

1925 J. H. Parsons *Introd. Study Colour Vision* ii. iii. 183 Of the partial protanopes (protanomal, Nagel) Donders and König do not record any case, v. Kries one only, whereas Nagel, Guttman, and Abney and Watson record a considerable number. **1973** *Jrnl. Optical Soc. Amer.* LXIII. 236/1 In Fig. 2, the red primary is assigned a value of 637 nm for normals and deuteranomals, and 625 nm for protanomals.

Hence **protano·malous** *a.*, having protanomaly; also *absol.*

1951 *Amer. Jrnl. Psychol.* XXII. 370 It is now customary..to distinguish two groups of anomalous trichromates, upon the analogy of the two groups of dichromates,—the red-anomalous or protanomalous trichromates, whose sensitiveness to red is below normal, and the green-anomalous or deuteranomalous trichromates, whose sensitiveness to green is below normal. **1958** [see *protanomaly]. **1965** *Science* 9 July 186/1 The protanomalous subject can match all colors of the spectrum with mixtures of three hues but requires more red in each mixture than the normal subject. **1975** *Sci. Amer.* Mar. 172/3 Some have considered that the protanomalous need more red in their mixture because, although they have normal cone pigments, the red signals are too weak.

protanomaly (prō̆ᵘtănǫ·măli). *Ophthalm.* [f. prec. + -y³.] A form of anomalous trichromatism marked by a reduced sensitivity to red and subnormal discrimination between red, yellow, and green hues.

1938 *Proc. Physical Soc.* L. 674 As the results for only one protanomalous observer are given, it is impossible to talk of any stages of protanomaly except by comparison with the results for the deuteranomalous observers. **1946** W. D. Wright *Res. Normal & Defective Colour Vision* xxiv. 297 Anomalous trichromatism is subdivided into three corresponding groups—protanomaly, deuteranomaly and tritanomaly—from the fact that each group has characteristics intermediate between those of the trichromat and the corresponding dichromat. **1973** *Vision Res.* XIII. 2033 The condition of anomalous trichromacy thus includes the categories of simple and extreme protanomaly and simple and extreme deuteranomaly.

protanope (prō̆ᵘ·tănō̆ᵘp). *Ophthalm.* [ad. G. *protanop* (J. von Kries 1897, in *Zeitschr. f. Psychol. und Physiol. d. Sinnesorgane* XIII. 248), f. Prot- + Gr. priv. ἀν- + ὤψ, ὠπ- eye, face.] A person with protanopia.

1908 *Psychol. Bull.* V. 298 When red and blue are mixed to match blue-green the protanope requires a relative excess of red in his purple mixture; and in similar determinations with red and yellow the protanope may demand five times as much red as the deuteranope. **1924** tr. J. von Kries in *Helmholtz's Treat. Physiol. Optics* II. 402 In order to have brief descriptive terms for the relation that has been found to exist here, without expressing any theoretical bias, the writer suggests the names *protanopes* and *deuteranopes* to describe the two kinds of dichromats, that is, persons who lack the first com-

ponent or the second component, respectively, of the normal visual organ. **1953** *Jrnl. Physiol.* CXXI. 565 The protanopes confuse red with green and are characterized by a very low sensitivity to light of long wavelengths. **1959** [see *dichromat, dichromate *sb.*²]. **1971** *Vision Res.* XI. 1034 Protanopes are deficient in some red-sensitive cone pigment that is present in normal eyes.

protanopia (prō̆ᵘtănō̆ᵘ·piă). *Ophthalm.* [mod. L., f. prec. + -ia¹.] A form of dichromatic colour-blindness marked by an insensitivity to red and an inability to distinguish between red, yellow, and green hues.

1903 J. M. Baldwin *Dict. Philos. & Psychol.* II. 371/1 In..the so-called red-blindness, the red end of the spectrum is shortened, and the maximum brightness is further towards the green [protanopia]. **1923** L. C. Martin *Colour* x. 146 In complete protanopia or deuteranopia we have the spectrum consisting [broadly speaking] of two parts of differing hue separated by a grey or white. **1950** *Jrnl. Optical Soc. Amer.* XL. 46/2 If the scores at either + 5 or − 5 are incorrect the subject has complete red-green color deficiency, i.e. either protanopia or deuteranopia. **1974** *Ophthalmic Res.* VI. 381 Colour discrimination was tested with the Ishihara plates and Nagel anomaloscope..was typical for protanopia.

Hence **protano·pic** *a.*

1908 *Psychol. Bull.* V. 298 Certain individuals..make their mixtures much redder, and other individuals much greener than the average. Here again we must distinguish between a protanopic and a deuteranopic subtype. Investigations since Rayleigh's pioneer publication have shown that the former sub-type is much more numerous than the latter. **1973** *Vision Res.* XIII. 1767 The reduction in gradient of the deuteranopic isochromatic line through the blue-green and purple region suggests an explanation of why some deuteranopes cannot distinguish either the protanopic or the deuteranopic diagnostic figures in the Ishihara test.

protargol (prō̆ᵘ·tă-jǫl). *Med.* [a. G. *protargol* (A. Neisser 1897, in *Dermatologisches Centralblatt* I. 5), f. *prot-ein* Protein + L. *arg-entum* silver; cf. -ol.] A substance made from protein and various compounds of silver, used as a mild antiseptic and a stain.

1898 *Boston Med. & Surg. Jrnl.* 25 Aug. 194/2 In the few cases of purulent ophthalmia that have been treated with protargol the length of the course of the disease has apparently only been slightly shortened, but the severity of the attack has been decidedly lessened. **1907** J. H. Parsons *Dis. Eye* xxix. 580 In more severe cases..protargol, 15 to 20 per cent., should be rubbed into the margins of the lids with a stumpy camel's hair brush. **1938** *Stain Technol.* XIII. 154 Protargol exhibits an isoelectric point roughly about pH 4, at which acidity it is precipitated, hence it might be expected to behave somewhat like a dye by having its selectivity altered by variation of pH of the staining mixture. **1976** *Acta Zool.* LVII. 117 The figures demonstrate the close connection between the surface cilia as shown by SEM [*sc.* scanning electron microscopy] and the subsurface kinetosomes revealed by the protargol impregnations.

protarsal (protā·rsăl), *a.* *Ent.* [f. Protars(us + -al.] Of or pertaining to the protarsus.

1902 R. I. Pocock in *Proc. Zool. Soc.* II. 391, 2nd leg with superior basal and anterior apical femoral spine, three inferior apical protarsal spines..and one inferior median tarsal spine.

protasis. Add: **2.** (Earlier and later examples.)

a **1568** Ascham *Scholemaster* (1570) 57 He began the Protasis with *Trochaijs Octonarijs.* **1961** *Listener* 5 Oct. 527/2 For a good deal of his new novel one might as well be reading the protasis of a fair-to-middling detective story.

3. (Earlier and later examples.) Also *fig.*

1588 W. Kempe *Educ. Children* sig. G.4ᵛ, Only the protasis or first part of our similitude is attributed but to Cato, for want of a like similitude garnished with like authoritie. **1904** [see *if-clause* s.v. *if *conj.* (*sb.*) 10]. **1922** Joyce *Ulysses* 704 Positing what protasis would the contraction for such several schemes become a natural and necessary apodosis? **1971** *Language* XLVII. 81, I use the term 'conditional sentence' to cover the entire complex sentence consisting of a protasis and an apodosis.

protea. In etym., for 'Linnæus 1737' substitute 'Linnæus *Hortus Cliffortianus* (1737) 29'. Substitute for def.: An evergreen shrub or small tree of the genus so called, belonging to the family Proteaceæ, usually native to southern Africa or Australia, and bearing cone-like heads of small flowers with prominent bracts. Also *attrib.* (Further examples.)

1770 R. Weston *Universal Botanist* I. 221 Cape Protea or Silver-tree. **1804** H. Andrews *Botanists Repository* V. tab. cccxlix, From the great number of the divided leaved Proteas, we are led to conjecture, that they are as numerous as those with entire leaves. **1901** L. H. Bailey *Cycl. Amer. Hort.* III. 1438/2 Proteas are tender shrubs which are among the most attractive and characteristic plants of the Cape of Good Hope. **1951** [see *disa]. **1972** J. Burmeister *Running Scared* 13 Their daughter had been chosen to present a protea, South Africa's unlovely national flower, to the President. **1972** Palmer & Pitman *Trees S. Afr.* I. xxiv. 503 The close-set protea leaves may yield as important a source of food to sugarbirds and sunbirds as do the flowers.

Protean, *a.* (*sb.*) Add: Now freq. with pronunc. (prō̆ᵘtī·ăn). Also **protean. A.** *adj.* **1. c.** Of a theatrical performer: characterized by the ability to take several parts in the same piece; quick-change; also *transf.* of such a performance. orig. *U.S.* Cf. sense B 1b in Dict.

1897 *Daily Tel.* 10 Mar. 4/5 Few will deny that Leopoldi Fregoli..is..alert, versatile, neat in his business, quick as lightning in his changes, and..the best 'protean' entertainer that the oldest playgoer has ever seen. **1909** Webster, *Prot(e)an*.. 3. *Theat.* Noting an actor who plays different parts in a play; hence, noting a performance of this kind. *Slang.* **1952** Granville *Dict. Theatr. Terms* 145 *Protean act*, an act performed by a: *Protean entertainer*, a lightning-change artist. An impersonator. **1961** Bowman & Ball *Theatre Lang.* 282 *Protean actor, protean artist.* Hence *protean act, protean drama*, etc.

d. Of animal behaviour: unpredictable, following no obvious pattern.

1959 Chance & Russell in *Proc. Zool. Soc.* CXXXII. 67 We therefore propose the term 'Protean Displays'. The mythical Proteus frustrated would-be captors by constantly changing his shape, so that they had nothing systematic to which to react. *Ibid.* 68 Protean displays involve rapid, sudden transitions, as one obvious component of their confusing effect. **1967** *New Scientist* 13 July 96/2 The term protean behaviour was coined to cover behaviour that is sufficiently unsystematic to prevent the predator from predicting in detail the positions, actions or both, of the prey. **1979** *Nature* 6 June 368/1 We have surveyed the occurrence of such protean behaviours and defined them as behaviours which are sufficiently unsystematic in appearance to prevent a reactor predicting in detail the position and/or actions of the actor. *Ibid.*, The erratic nature of a protean display defeats anticipation by the predator.

protease. Add to def.: a proteinase or peptidase. (Earlier and later examples.) [First formed as F. *protéase* (G. Malfitano 1900, in *Ann. de l'Inst. Pasteur* XIV. 420).]

1903 *Ann. Bot.* XVII. 237 There is at present evidence that enzymes which digest proteids (proteases) occur..in certain lowly Algæ, in some fungi, and in various Phanerogams. **1923** [see *peptidase]. **1931** [see *proteinase]. **1949** Abraham & Heatley in H. W. Florey et al. *Antibiotics* I. ii. 85 This can happen..by the secretion of destructive exocellular enzymes such as penicillinase, protease, or peptidase. **1966** A. Suggs in J. Pryse Lem *Metabolism Rel. Cataract* 330 An endopeptidase called β-protease..will attack lens protein in acid pH and a second enzyme, the α-protease,..attacks the breakdown products produced by the endopeptidase. **1973** Zeffren & Hall *Study of Enzyme Mechanisms* ix. 188 Chymotrypsin is one of several proteolytic enzymes or proteases which function collectively in the mammalian intestine. **1976** *Nature* 12 Feb. 580 This may be related in part to the presence of protease inhibitors in connective tissues.

protect, *v.* Add: **1. a.** Also *absol.*

1894 E. Fawcett *New Nero* ii. 26 Music..was always an expression of..that soulless and mysterious will-to-live, which for ever creates, protects, and perpetuates. **1914** W. B. Yeats tr. *Sophocles' Oedipus at Colonus* in *Coll. Plays* 343 *Theseus.* If God sent you hither, you need no protection of mine, but God or no God my mere name will protect.

c. To attempt to preserve (a threatened plant or animal species) by preventing collecting, hunting, etc.; to restrict access to (land valued for its wild life or its undisturbed state).

1893 *Zoologist* XVII. 390 If a particular species were declining, and were known to frequent a particular place, the County Council should..be called upon to protect that restricted area. **1935** *Discovery* Oct. 304/1 To protect a bird proved..to be noxious simply brings bird-protection into contempt. **1969** F. N. Hepper in J. Fisher et al. *Red Bk.* 360/1 The need to protect plants for their own sake is becoming increasingly accepted by those in authority.

2. (Earlier examples.)

1789 *Deb. Congress U.S.* 9 Apr. (1834) 106 [Measures] calculated to encourage the productions of our country, and protect our infant manufactures. **1825** J. S. Mill in *Westm. Rev.* III. 415 The various classes of manufactures are protected from foreign competition.

4. c. To provide (an electrical device or machine) with safeguards against too high a current or voltage.

1875 *Telegr. Jrnl.* III. 60/2 Lightning protectors invented to protect telegraph lines. **1888** D. Salomons *Managem. of Accumulators & Private Electric Light Installations* (ed. 3) II. ii. 97 Put a safety fuse in every switch and wall plug... Every lamp is protected in this way..against accidental short-circuits. **1935** Monseth & Robinson *Relay Syst.* xiii. 458 The first zone..is protected by an instantaneous balanced-beam impedance element. **1975** D. G. Fink *Electronics Engineers' Handbk.* VII. 27 Two types of fast-blow fuses are used to protect power-tubes.

5. *Chem.* **a.** To prevent the alteration or removal of (a particular group or part of a molecule) in a reaction, by first causing it to form an unreactive derivative from which the original structure can later be regenerated.

1889 G. M'Gowan tr. *Bernthsen's Text-bk. Org. Chem.* 352 When it is wished to prepare the mono-nitro-compounds, the aniline must again be 'protected', either by using its acetyl compound or by nitrating in presence of excess of concentrated sulphuric acid. **1929** Mitchell & Hamilton *Biochem. Amino Acids* i. 90 In other words, the chloracetyl group, introduced to protect the amino group of the amino acid is, after it has performed its protective function, itself transformed into an amino acid group. **1951** I. L. Finar *Org. Chem.* I. xi. 203 If the synthesis requires reaction with one halogen atom only, the most satisfactory procedure is to 'protect' the other halogen atom by ether formation and subsequently decompose the ether with concentrated hydrobromic acid. **1964** N. G. Clark *Mod. Org. Chem.* vi. 90 By forming the dibromide..the double bond is 'protected' from the ensuing reaction, and may be restored later to the compound by zinc treatment. **1971** D. R. Williams *Metals of Life* ix. 134 Freeman et al. have found that the side-chains in simple peptides which are usually said to be 'protected' give amino acids again are protected by transition metal ions, the best protectors being copper (II) and nickel (II).

b. To render (a hydrophobic sol) inert to the flocculating action of small concentrations of an electrolyte.

1903 *Jrnl. Chem. Soc.* LXXXIV. ii. 135 The capacity of colloidal solutions to protect a colloidal solution of gold against the precipitating action of an estimated quantity of sodium chloride is expressed as the gold number. **1909** J. Alexander tr. *Zsigmondy's Colloids & Ultramicroscope* iii. 77 Another colloid which protects the nascent colloidal gold was discovered by Faraday, and called by him 'jelly'. **1939** *Thorpe's Dict. Appl. Chem.* (ed. 4) III. 287/2 A quartz suspension protected by gelatine will possess the cataphoretic velocity and isoelectric point of gelatine. **1966** Guckel & Seifert *Physical Chem.* xxii. 685 A lyophobic sol is often stabilized by addition of a lyophilic sol, which is then termed a protective colloid. An example is gelatin, which protects the silver bromide sol used in photographic emulsions.

protectant, *a.* Delete †*Obs. rare*⁻¹ and add examples of the sense: protecting (esp. plants) against disease.

1943 *Phytopathology* XXXIII. 627 (*heading*) The slide-germination method of evaluating protectant fungicides. **1954** *Jrnl. Econ. Entomol.* XLVII. 462 (*heading*) Protection of stored shelled corn with a protectant dust in Indiana. **1977** *Protecting World's Crops* (Shell Internat. Petroleum Co.) 2 Protectant fungicides, if applied before the disease occurs, kill or inhibit the development of fungal spores or mycelia (strands of fungus) before they can penetrate the host.

B. *sb.* An agent that protects, esp. a plant against disease.

1935 *Jrnl. Pomol. & Hort. Sci.* XIII. 262 The effects of the wetting agents..upon the biological activities of the protectant have to be determined. **1940** *Phytopathology* XXX. 2 Synthetic organic chemicals developed by the Crop Protection Institute were efficient, non-injurious seed protectants for combating damping-off in Lima beans. **1960** *New Scientist* 4 Aug. 344/1 Removal of the dark green colouring matter present in crude preparations of pyrethrum markedly increased its stability in sunlight, and further improvement was achieved by the addition of..a number of antioxidants... The data suggest that these compounds act primarily as protectants for pyrethrin II and cinerin II, whereas the absence of 'chlorophyll' pigments enhances the stability of pyrethrin I and cinerin I. **1975** *Nature* 22 May 329/1 It is clear that as a protectant, sclareol is highly specific for rust fungi.

protected, *ppl. a.* Add: **a.** (Further examples.) Now also, receiving legal immunity or exemption.

1926 *Brit. Gaz.* 11 May 3/6 The London Central Meat Market, at Smithfield, is now a protected area, and barriers have been drawn across all the approaches. **1942** *Ann. Reg. 1941* 27 Now they [*sc.* men above the reservation age] were to be reserved only if engaged on what was called 'protected work', i.e. work which the Government recognised as of national importance. **1944** *Manpower* (Ministry of Information) 112 Firms were divided into 'protected' and 'unprotected' establishments, according to the urgency of the work they were doing. **1956** J. C. Swayne *Conc. Gloss. Geogr. Terms* 114 *Protected state*, a territory, e.g. Brunei, Kuwait, under a ruler who receives the protection of another state. The protecting state controls foreign affairs, but has no jurisdiction over internal matters. **1968** 'C. Aird' *Henrietta Who?* vii. 64 You're a protected tenant... No one can make you leave. **1975** 'J. Bell' *Victim* i. 12, I took in my second lot of tenants. Protected tenure. **1978** *Country Life* 20 July 148/3 Jersey has no statutory list of protected buildings as has existed in Britain for the last three decades.

b. Of a species whose survival is threatened: affected by laws preventing collecting, hunting, etc.

1930 J. S. Huxley *Bird-Watching & Bird Behaviour* vi. 115 The National Trust and the Royal Society for the Protection of Birds..are..paying watchers to see that protected birds are not shot or robbed of their eggs. **1936** *Discovery* Feb. 34/2 Any local authority may now, on application to the Ministry [of Agriculture] take the bird [*sc.* the little owl] off the protected list. **1959** E. F. Linssen *Beetles Brit. Isles* I. 156 In an effort not to bring about its [*sc.* the Great Silver Beetle's] total destruction, naturalists refrain from publishing records regarding its distribution... It should be treated as a 'protected' species and left alone. **1976** *Hortus Third* (L. H. Bailey Hortorium) 798/1 In most cases it is better to leave these plants to live where they are native and protected than to move them to gardens. **1978** *Vole* Dec. 18/1 It seemed like every damn animal on this planet should be a protected species.

protecting, *ppl. a.* Add: **a.** (Earlier examples of *protecting duty*.)

1785 *Daily Universal Reg.* 1 Jan. 2/3 The Protecting Duties, so generally called for in Ireland. **1820** *Ann. Reg.* 73/2 The American timber being of an inferior quality to that from the Baltic, required a protecting duty.

b. *spec.* in *Chem.*, applied to a group introduced into a molecule in order to protect a feature of that molecule in a reaction. (Cf. *protect *v.* 5a.)

1947 *Nature* 12 Apr. 500/1 The use of another protecting unit easily removed by hydrolysis in combination with the carbobenzoxy method would considerably extend the use of the latter. **1952** L. J. Desha *Org. Chem.* (ed. 2) xiii. 246 Finally, the protecting acetyl group is removed by hydrolysis. **1968** I. L. Finar *Org. Chem.* (ed. 4) II. xiii. 584 This enamine can react with another amino-acid, and the protecting group is removed by mild bromination.

protection. Add: **1. c.** Freedom from molestation obtained by paying money to a person who threatens violence or retribution if payment is not made; hence protection money itself. Also in other extended uses.

1860 [implied at *protection-rent* below]. **1903** *Independent* 15 Jan. 148/2 I'm sure that no one man knows all ends of this business of 'protection'. **1930** *Economist* 25 Oct. 754/1 A gangster would take it upon himself, say, to organise the selling of fish in one district in Chicago... The fishmonger who did not care for protection would find his shop bombed. **1938** G. Greene *Brighton Rock* II. i. 78 I've got protection. You be careful. **1938** F. D. Sharpe *Sharpe of Flying Squad* xix. 202 A man offered to sell 'protection' to the bookmakers—at a price. **1962** D. Francis *Dead Cert* ix. 106, I..asked the owners straight out if they were paying protection. **1977** 'W. Haggard' *Poison People* III. iv. 123 A man whose prosperous business had kicked at increasing demands for protection and had therefore finally gone to the wall.

d. An attempt to preserve certain animals, plants, or undisturbed areas of land by enforcing rules governing access, collecting, hunting, etc.

1880 *Act* 43 & 44 *Vict.* c. 35 It is expedient to provide for the protection of wild birds of the United Kingdom during the breeding season. **1894** W. H. Hudson *Lost Brit. Birds* 1 It was thought best to leave out any species represented by at least three or four pairs that have some measure of protection afforded to them when breeding. **1895** G. S. Anderson in Roosevelt & Grinsell *Hunting in Many Lands* 377 (*heading*) Protection of the Yellowstone National Park. **1930** J. S. Huxley *Bird-Watching & Bird Behaviour* vi. 115 Protection has brought the bittern back to breed and boom in Norfolk. **1936** *Discovery* Sept. 293/1 From time to time the moose is over-hunted in some districts, but after a few years' protection they come back again. **1952** H. L. Edlin *Changing Wild Life of Brit.* v. 71 The Harriers, typical hawks of the marshes, became very rare, but under protection a few continue to nest. **1969** F. N. Hepper in J. Fisher et al. *Red. Bk.* 360/2 The I.U.C.N. itself has taken the principal lead in this field by initiating a scheme for the protection of plant species.

e. *Electr. Engin.* The action or result of *protect *v.* 4c.

1890 Slingo & Brooker *Electr. Engin.* xvii. 725 The way in which it [*sc.* the cut-out] affords this protection is by automatically disconnecting the circuit when the current..exceeds a certain predetermined limit. **1920** *Whittaker's Electr. Engineer's Pocket-Bk* (ed. 4) 428 Merz-Price protection may operate by a balance of voltages or a balance of currents. The former is used for the protection of cables, and the latter for the protection of transformers and alternators. **1962** *Newnes Conc. Encycl. Electr. Engin.* 612/1 The fuse forms the basis of most small, simple distribution-system protection, combining overcurrent protection and fault isolation. **1977** R. W. Smeaton *Switchgear & Control Handbk.* xxvi. 27 The proper choice of protection is based on equipment size, application reliability, shutdowns, probability of faults, and economics.

f. *Chem.* The action of *protect *v.* 5a and b.

1909 J. Alexander tr. *Zsigmondy's Colloids & Ultramicroscope* xviii. 185 The origin of the protection of the gold can be most simply explained by the assumption that specific attractive forces bring about a union of the ultramicrons of metal and protective colloid. **1939** *Thorpe's Dict. Appl. Chem.* (ed. 4) III. 287/1 The protection of sols is of great importance and has been practised empirically since ancient times. **1947** *Nature* 12 Apr. 500/1 The use of the carbobenzoxy reagent for protection of amino-groups in the course of peptide synthesis has..some limitations. **1951** I. L. Finar *Org. Chem.* I. xii. 215 'Protection' of the double bond is conveniently carried out by the addition of bromine which is subsequently removed by zinc dust in methanolic solution. **1958** J. W. Mullin in Cremer & Davies *Chem. Engin. Pract.* VI. xi. 459 Protection is effected by a number of lyophilic molecules which envelop a lyophobic particle and cover it with a monomolecular layer.

g. *Bridge.* (See quot. 1967.)

1952 I. Macleod *Bridge* vii. 88 Naturally, if there is an element of protection about your bid,..partner will realize that you may be quite a bit weaker. **1958** *Listener* 30 Oct. 709/3 This is a situation in which I cannot look for protection. **1967** *Bridge Players' Encycl.* 393/2 *Protection*, reopening with a bid or double when the opposing bidding has stopped at a low level.

h. *Mountaineering.* (See quot. 1971.)

1966 C. Bonnington *I chose to Climb* iii. 46 There was no protection and it was now necessary to pivot round on one's toes to grasp the smooth, square-cut edge of the bulge. **1971** —— *Annapurna South Face* Gloss. 323 *Protection*, quantity and quality of running belays used to make a pitch safe to lead.

4. (Earlier examples.)

1789 *Deb. Congress U.S.* 15 Apr. (1834) 150 He conceived it the duty of the committee to pay as much respect to the encouragement and protection of husbandry ..as they did to manufactures. **1820** *Hansard Lords* 26 May 579 Let your lordships consider..what would be the effect..if the existing system of protection were abolished, and a fixed duty..were substituted.

5. protection forest, a forest whose purpose is to provide a dense cover of vegetation which helps to inhibit erosion and conserve water; **protection money**, money paid to secure protection (sense *1c); **protection racket**, an illegal scheme for the levying of protection money; † **protection rent** = *protection money.*

1889 W. Schlich *Man. Forestry* I. i. 47 Already in the middle ages so-called 'Protection Forests' existed. **1928** R. S. Troup *Silvicultural Syst.* ix. 116 To afford protection against erosion, landslips, and avalanches in mountainous regions, to conserve the water-supply in catchment areas and to prevent floods; forests maintained for such purposes are termed 'protection forests'. **1974** Longman & Janík *Trop. Forest* vi. 123 Forest reserves are also particularly appropriate in steep terrain to prevent erosion and rapid run-off of water in catchment areas, and these have been recognised for many years as 'protection forests'. **1923** *Nation* (N.Y.) 24 Oct. 449 The men that help unload get $1 a case, and the revenue officers $2 protection money. **1934** R. Graves *I, Claudius* xx. 289 Shopkeepers in the town and farmers in the country had to pay secret 'protection-money' to the local captains; if they refused to pay there would be a raid at night by masked men, their house would be burned down and their families murdered. **1972** T. Lilley *'K' Section* vi. 29 The opium dens and brothels closed. The coffee-shops that now refused to pay protection money. **1937** E. Ambler *Uncommon Danger* x. 149 His business then was intimidating shopkeepers—the 'protection racket', as it is called now in America. **1954** T. S. Eliot *Confidential Clerk* II. 62 Colby doesn't need your protection racket So far as I'm concerned. **1976** D. Daiches in D. Villiers *Next Year in Jerusalem* 275 The characteristics of a Chicago gangster tale: a leader organizing a protection racket, violent measures taken against those who refuse protection money. **1860** *Leisure Hour* 19 July 460/2 In return for black-mail or protection-rent, they shared the property of those who paid it, and engaged to defend it from aggressions.

protectionism. Add: (Earlier and later examples.)

1852 *Punch* 31 July 53/1 If a steam-boat does accidentally 'put in' with a few voyagers, it is met, in the first place, by a spirit of Protectionism and high prices in the shape of pier dues. **1945** K. R. Popper *Open Society* I. vi. 97 What I demand from the state is protection..for my own freedom and for other people's... [This] view..may be called 'protectionism'. **1955** *Times* 4 June 5/1 The Canadian Government's decision to amend the Customs tariff—the changes are due to come into effect today—has a suggestion of protectionism. **1969** *Listener* 14 Aug. 201/1 This protectionism..was originally intended to stave off the intrusions of the American cinema. **1977** *Time* 30 May 51/1 In times of recession, nations inevitably turn toward protectionism as a means of shielding jobs from the threat of foreign goods.

protective, *a.* (*sb.*) Add: **A.** *adj.* **1. a.** (Later examples.)

1938 *Encycl. Brit. Bk. of Year* 467/1 The term 'protective foods' was originally applied to milk and green leaf vegetables, because they made good the deficiencies commonly found in human diets... It has come to include.. foodstuffs..which protect the body against disturbance in structure or function of its organs and parts..which protect, in short, against 'disease'. **1940** *Topeka* (Kansas) *State Jrnl.* 19 Apr. 1/8 De Geer in his broadcast declared The Netherlands would resist with arms any attempt by a foreign power to extend protective help to her. **1944** J. S. Huxley *On living in Revol.* xiii. 139 Agriculture..can be devoted mainly to providing protective foodstuffs. **1955** *Gloss. Terms Radiol.* (*B.S.I.*) 67 *Protective material*, material which is used to provide protection against ionizing radiation. **1974** *Times* 2 Feb. 18/7 The Inland Revenue accepts that in certain trades the employee has to supply his own tools, protective clothing, etc. *a* **1977** *Harrison Mayer Ltd. Catal.* 4/1 Always wear suitable protective clothing.

b. *protective coloration, colouring*, an animal's colouring that blends with its habitat, enabling it to conceal itself. Also *fig.*

1892 F. E. Beddard *Animal Coloration* iii. 86 A South American bittern..affords an excellent instance of the advantages which result from a protective coloration. **1918** G. H. Thayer *Concealing-Coloration in Animal Kingdom* ii. 25 We have..Obliterative Coloration, and Mimicry, as the two main principles of Protective Coloration. **1934** *Proc. Nat. Acad. Sci. Washington* XX. 559 (*heading*) Does protective coloration protect? **1937** Koestler *Spanish Testament* II. 372, I was able to observe..what direct biological forces this process of protective coloration exerts. Guilty or innocent, the prisoner changes form and colour, and assumes the mould that most easily enables him to secure a maximum of those minimal advantages possible within the framework of the prison system. **1941** A. Christie *Evil under Sun* x. 183 Protective colouring is your line. Remain rigidly non-active and fade into the background! **1949** N. Mitford *Love in Cold Climate* iii. 30 As far as my fellow guests were concerned, I was clearly endowed with protective colouring;..I might just as well not have been there at all. **1957** T. W. Kirkpatrick *Insect Life in Tropics* viii. 214 This instance of protective coloration is unlike most others. **1977** B. Colloms *Victorian Country Parsons* xi. 211 To avoid trouble he became over-anxious to please and grew adept at assuming protective colouration [*sic*]. **1979** *Books & Bookmen* Jan. 16/1 One strand of English (and American) poetry, the strand that reflects nineteenth-century bourgeois values at their most unequivocal, can only survive by adopting the protective colouring of a game.

c. *Electr. Engin.* Providing protection against too high a current or voltage.

1896 R. Robb *Electric Wiring* v. 171 All conductors.. must be provided near the point of entrance to the building with some protective device which will operate to shunt the instruments in case of a dangerous rise in potential, and will open the circuit and arrest an abnormal current flow. **1922** [see *bias *v.* 5]. **1962** *Newnes Conc. Encycl. Electr. Engin.* 612/1 A great variety of protective equipment is marketed for distribution systems because of its influence on capital expenditure on such items as switchgear. **1976** *Billings* (Montana) *Gaz.* 6 July 3-D/1 A protective relay adjacent to a Kemmerer, Wyo., power plant was blamed Monday for the Fourth of July electrical black-out that darkened most of Utah. **1978** *Gramophone* Apr. 1796/3 There is also a mains voltage adjuster and twin 2.5A protective mains fuses.

2. a. (Earlier example.)

1820 *Ann. Reg.* 771/1 The protective or restrictive system.

c. *protective arrest, custody*, the detention (of a person) either allegedly or truly for his own protection.

[**1933** R. Bernays *Special Correspondent* xliii. 222 Jews, Socialists, pacifists, Liberals—anyone who has engaged in political agitation or is believed to be hostile to the New Germany—are incarcerated without trial and for an indefinite period. The German name is *Schutzhaft* (protective detention), the idea being that they are asylums..for men who otherwise might suffer grievous bodily harm for their political opinions at the hands of their infuriated countrymen.] **1935** S. Lewis *It can't happen Here* xv. 154 It was blandly explained..that they were merely being safeguarded. Sarason did not use the phrase 'protective arrest'. **1936** *Sun* (Baltimore) 17 Feb. 8/3 He is declared to have been placed under protective custody. Now this phrase is a deliberate steal from the vocabulary of Nazi Germany, its purpose being to cast a pall of dignity around the proceeding when Brown Shirts take an opponent into a Brown House for the purpose of beating him with a rubber hose. *Ibid.*, 'Protective arrest' sounds better than 'ganging up' and gives the impression that the State in its beneficent wisdom is protecting somebody when the only need for protection is protection against the protectors. **1940** *Topeka* (Kansas) *State Jrnl.* 8 May 3/1 When I was asked to take some letters..I agreed readily thinking they might be an open sesame for sleeping quarters. They were—under British 'protective arrest' in Spillum. **1940** C. V. Wedgwood *William the Silent* vii. 192 He was even forced to take monks and priests into protective custody—and it really was protective custody, though the catholics represented it as plain imprisonment. **1947** P. Woodruff *Wild Sweet Witch* 7 The Deputy Commissioner of the day did take a man into protective custody to prove he was not the panther. **1964** N. Marsh *Dead Water* iii. 79, I wish I could put you under protective custody. **1973** 'D. Shannon' *No Holiday for Crime* viii. 128 Pat's reformed pusher ready to tell all sitting safe in protective custody.

3. *Chem.* **a.** Having the quality or property of protecting a sol (cf. *protect *v.* 5b); *spec.* in *protective colloid*, a lyophilic colloid whose presence in small quantities protects a lyophobic sol.

1906 *Jrnl. Soc. Chem. Industry* 31 May 484/1 To insure the satisfactory production of bright deposits it is in all cases essential to employ clear, well filtered solutions. The authors explain the observed phenomena by supposing that the bright deposit is formed by causing the metal to retain its amorphous condition and preventing it from becoming crystalline. The mutual protective effect of colloids upon one another..is probably the chief factor. **1909** J. Alexander tr. *Zsigmondy's Colloids & Ultramicroscope* iii. 77 Lobry de Bruyn (1898), characterized gelatin jelly as a protective colloid (Schutzkolloid). **1939** *Thorpe's Dict. Appl. Chem.* (ed. 4) III. 287/1 Zsigmondy showed that the sharp colour change from red to blue displayed by gold sols under the influence of electrolytes could be used as a means of obtaining a quantitative comparison of the protective action of different colloids. **1950** E. K. Fischer *Colloidal Dispersions* vi. 247 In the manufacture of colored pigments, protective colloids aid in keeping the particle size small. **1960** [see *peptization]. **1967** G. P. A. Turner *Introd. Paint Chem.* xi. 160 If the protective colloids are not truly compatible with the film-former, gloss will be reduced and the film weakened.

b. = *protecting *ppl. a.* b.

1932 *Chem. Abstr.* XXVI. 5072 The method of synthesizing peptides which consists in stabilizing the amino group of 1 acid with a protective group R, then so altering the CO_2H group as to enable it to couple with a 2nd amino acid and removing the group R after the coupling has been effected. **1968** I. L. Finar *Org. Chem.* (ed. 4) II. xiii. 582 Bergmann (1932) introduced carbobenzoxy chloride as an amino protective group, and this appears to be the most widely used method of protection.

B. *sb.* Also, a contraceptive sheath.

1971 *It* 2–16 June 23/3 (Advt.), Protectives by post:.. Durex Gossamer Doz. 50p. **1977** *Lancet* 15 Oct. 811/1 Although the condom, or male protective, is marketed primarily as a method of contraception, some stress is laid both by the manufacturers and by the medical profession on its value as a prophylactic against sexually transmitted diseases.

protector, *sb.* **1. a.** Add to small-type note: Also, *Protector of the Poor*, a term of respect formerly used in British India, as by Indian servants to their masters.

1890 Kipling in *Macmillan's Mag.* June 149/2 The news does not come from my mouth, Protector of the Poor. **1894** —— *Jungle Bk.* 165 'Was it to help thee steal green corn?..' 'Not green corn, Protector of the Poor—melons', said Little Toomai. **1910** *Blackw. Mag.* Feb. 167 They sent him to Lord Caerlaverock, for the ex-viceroy loved to be treated as a kind of consul-general for India. But this Protector of the Poor proved a broken reed. **1911** F. H. Burnett *Secret Garden* iv. 25 The native servants.. in India..called them [*sc.* their masters] 'protector of the poor' and names of that sort. **1952** J. Masters *Deceivers* ix. 99 Protector of the Poor, at first cockcrow the villain called for a lotah for purposes of nature.

b. (Further examples.)

1867 G. H. Selkirk *Guide to Cricket Ground* ii. 33 *Pads*, leg guards and protectors for the abdomen. **1898** *Sci. Abstr.* I. 240 (*heading*) Telephone line protectors. **1922** *Lillywhites' Sports Requisites* 7 Wicket-keeping Sundries. Palmer's Abdominal Protectors. **1934** *Jrnl. Inst. Electr. Engineers* LXXIV. 236/2 Protectors are connected to every open-wire Post Office line.

d. (*a*) A man who keeps a mistress; (*b*) a man who looks after a prostitute in return for her earnings, a pimp. Cf. Protection 1b.

1905 [see *man's woman* s.v. *man *sb.* 21]. **1938** F. D. Sharpe *Sharpe of Flying Squad* x. 116 Prostitutes and their protectors were roped into the stations by the dozen. **1954** *Britannica Bk. of Year* 637/2 A group of criminals making a living from organized prostitution was referred to as a Vice-Ring, the leader of such a group being a Vice-Chief..or—with reference to the prostitutes controlled by him—a Protector.

e. One by whom protection from harassment is assured or who collects protection money. Cf. *protection 1c.

1933 H. G. Wells *Shape of Things to Come* II. 153 The man who wanted to be left alone in peace..was pressed to pay his tribute to the gang. Or he would not be left in peace. And even if his particular 'protectors' left him in peace, there might still be other gangs about for whom they disavowed responsibility and with whom he had to make a separate deal. **1977** J. Wainwright *Nest of Rats* iii. 15, I was wise enough to choose my own 'protector'.

protectorate, *sb.* Add: **2. c.** (Earlier example.)

1844 *Times* 30 July 4/5 Queen Pomane [of Tahiti] had been forced to accept the 'Protectorate' of the French flag.

3. (Earlier and later examples.)

1860 E. B. Andrews in G. E. Metcalfe *Gt. Brit. & Ghana* (1964) 285/1 The Protectorate on this side of the Volta. **1871** *Act* 34 *Vict.* c. 8 Whereas the inhabitants of certain territories in Africa adjoining Her Majesty's settlements of Sierra Leone, Gambia, Gold Coast, and Lagos, and the adjacent protectorates. **1921** *Brit. Year Bk. Internat. Law 1921–2* 114 Virtually colonies; constitutionally foreign soil—that is the definition of 'protectorates': juridical monsters. **1923** *Publ. Permanent Court Internat. Justice* B. iv. 27 The extent of the powers of a protecting State in the territory of a protected State depends, first, upon the Treaties between the protecting State and the protected State establishing the Protectorate, and, secondly, upon the conditions under which the Protectorate has been recognised by third Powers as against whom there is an intention to rely on the provisions of these Treaties. **1955** *Sci. Amer.* Mar. 60/2 The protectorate borders on the vast jungle belt of central Africa in which yellow fever is endemic. **1961** L. van der Post *Heart of Hunter* I. vii. 110 Although Lobatsi was in a British Protectorate the railway itself belonged to Southern Rhodesia.

4. *protectorate land, law.*

1936 *Discovery* June 189/1 Nigeria is in the peculiar state of having both mandated and protectorate lands within its boundaries. **1961** L. van der Post *Heart of Hunter* I. vii. 112 The provincial commissioner..would have up the station-master and draft instructions to ensure that the station was run in the spirit as well as the letter of Protectorate Law.

protectorist (prote·ktŏrist). *Hist.* [f. Protector *sb.* + -ist.] = Protectorian *sb.*

1913 J. Willcock *Life Sir H. Vane* xvi. 275 About half the members of the Commons were Protectorists or supporters of the constitution prescribed in *The Petition and Advice.*

protectrice. For †*Obs.* in Dict. read 'Now *rare*' and add later example.

1974 J. Flint *Cecil Rhodes* v. 105 Lobengula...had imagined that Queen Victoria was now his ally and protectrice.

protegulum (prote·giulɒ̆m). *Zool.* [mod.L., f. Pro-² 1 + L. *tegulum* covering.] In brachiopods, the embryonic form of the shell.

1891 C. E. Beecher in *Amer. Jrnl. Sci.* XLI. 344 All brachiopods, so far as studied by the writer, have a common form of embryonic shell, which may be termed the protegulum. **1904** *Amer. Jrnl. Sci.* CLXVII. 283 The protegulum of this species [*sc. Stropheodonta perplana*] is nearly circular. **1935** Twenhofel & Shrock *Invertebr. Paleont.* viii. 269 Growth proceeds around the protegulum, mainly along the anterior and lateral margins of the two valves. **1959** L. H. Hyman *Invertebrates* V. xxi. 531 Each valve begins as a minute plate, the protegulum, presumably composed of periostracum. **1973** P. Tasch *Paleobiol. Invertebrates* vii. 263/2 The outer surface of the mantle flaps (lobes) became white and smooth, indicating initial shell formation (protegulum).

proteiform, *a.* (Earlier and later examples.)

1793 *Critical Rev.* IX. 183 Pathologists have, within these last years, differed greatly respecting the cause of this Proteiform disease [*sc.* gout]. **1944** S. Putnam tr. *E. da Cunha's Rebellion in Backlands* ii. 84 The proteiform mestizo of the seaboard.

protein. Add: Now pronounced (prō̆ᵘ·tīn).

It seems likely that in proposing the word *protéine* Mulder was adopting a suggestion made to him by Berzelius: see *Nature* (1951) 11 Aug. 244.

c. (Further examples.)

1917 *Nature* 15 Feb. 471/2 Just as the protein supply in meat may be compensated for by the greater utilisation of the protein-rich pulses. **1925** C. H. Browning *Bacteriol.* iii. 46 Instead of adding peptone a useful procedure is to digest the minced meat at the commencement by pancreatic extracts containing the protein-splitting ferment trypsin, which produces peptone bodies in the mixture and yields a very suitable basis for nutritive medium. **1928** *Physiol. Rev.* VIII. 418 If, as is stated by Mathews.., Fischer was induced to turn his attention to protein chemistry by Kossel, the debt science owes this great biochemist is beyond estimation. **1946** *Nature* 19 Oct. 556/2 Estimations are made on protein-free filtrates, prepared by adding to 0·2 c.c. serum, 11 c.c. water and 0·5 c.c. of each of the Folin-Wu reagents. **1953** *Amer. Naturalist* LXXXVII. 255 The heterogeneity which is the dismay of the protein chemist attempting to solve purification problems may be the very basis for his existence as a human being. **1956** *Nature* 28 Jan. 190/1 Although reticulocytes have practically their full complement of hæmoglobin, evidence from amino-acid incorporation studies suggests that these cells..still have protein-synthesizing capacity. **1960** *Farmer & Stockbreeder* 9 Feb. 110/2 (Advt.), Lobolettes are protein-packed, rich in energy and fully fortified with vitamins, minerals and trace elements. **1961** *Lancet* 29 July 258/1 These figures may be explained by the presence of one or more insulin antagonists..or by protein-binding of the insulin in the blood. *Ibid.* 5 Aug. 284/1 Rona and Takahaski first demonstrated..that.. the plasma-calcium consisted of a diffusible fraction and a non-diffusible protein-bound fraction. **1972** J. Maddox *Doomsday Syndrome* iii. 75 Protein deficiency is still a serious cause of stunted development..among poor people even in advanced societies. **1979** *Arizona Daily Star* 5 Aug. (Parade Suppl.) 14/3 Protein-rich foods such as meat, dairy products and nuts should be eaten.

d. Special Combs.: **protein plastic,** a plastic in which protein is the chief component; *esp.* a casein plastic; **protein shock** *Med.*, a disturbed state produced by the parenteral introduction into the body of a foreign protein; also, protein therapy; **protein therapy** *Med.*, the production of protein shock for therapeutic purposes.

1936 Sturken & Woodruff *U.S. Patent 2,040,033* 1/1 Our invention relates to the production of protein plastics which may be cured in the mold without the necessity of a prolonged cure in formaldehyde solution or formaldehyde vapor. In the past, protein plastics such as casein have found many uses in the light plastics field. **1943** H. R. Fleck *Plastics* iv. 82 The two most widely known and industrially important protein plastics are those formed from casein..and those formed from the proteins present in soya beans. **1969** *Encycl. Polymer Sci. & Technol.* XI. 696 The term 'protein plastic' is specifically interpreted commercially to mean casein plastic. **1917** *Jrnl. Exper. Med.* XXVI. 699 The mechanism of recovery following the so called 'protein shock therapy'. *Ibid.* 705 By means of the protein shock, antibody-rich fluids (serum) are forced into the lymph channels. **1935** F. P. Gay *Agents of Dis. & Host Resistance* lxiii. 1506 Symptomatic disturbances during protein shock are an increased pulse rate, sweating, decreased blood pressure, increased peristalsis, increased lymph flow and lymph volume, and a mobilization of the serum enzymes. **1964** S. Duke-Elder *Parsons' Dis. of Eye* (ed. 14) xiv. 149 In certain aspects the response to cortisone resembles that to fever therapy by 'protein shock'. **1917** *Jrnl. Amer. Med. Assoc.* 8 Sept. 766/2 The application of foreign protein therapy to the acute, subacute and chronic arthritides. **1940** B. I. Comroe *Arthritis* xv. 195 Substances..now employed for non-specific protein therapy include bacterial vaccines. **1967** *Biol. Abstr.* XLVIII. 7122/1 (*heading*) The use of protein therapy for gastric and duodenal ulcers.

proteinase (prō̆ᵘ·tīnēⁱz, -s). *Biochem.* [a. G. *proteinase* (Grassmann & Dyckerhoff 1928, in *Zeitschr. f. physiol. Chem.* CLXXIX. 41): see Protein and *-ase.] Any enzyme that hydrolyses proteins to smaller polypeptides.

1929 *Chem. Abstr.* XXIII. 615 Plant proteases... The proteinase and the polypeptidase of yeast. **1931** *Biochem. Jrnl.* XXV. 256 In their main conclusions these authors agreed that green malt contains (1) a protease or proteinase (to adopt the nomenclature of Grassmann) which appears to attack..crystalline egg-albumin..and (2) at least one peptidase which attacks the dipeptide leucylglycine. **1941** *Adv. Enzymol.* I. 76 Pepsin, trypsin, and chymotrypsin represent the three best recognized proteinases. **1963** *Biochem. Jrnl.* LXXXVI. 100/1 The purified proteinase probably liberated peptides from a boiled sample of α_2-crystallin, and it seems probable that the lens proteinase is an endopeptidase. **1970** Passmore & Robson *Compan. Med. Stud.* II. xviii. 33/1 Potent bacterial proteinases and collagenases decompose muscle tissue and collagen.

proteinoid (prō̆ᵘ·tīnoid), *sb.* (*a.*) *Biochem.* [f. Protein + -oid.] A protein-like polypeptide or mixture of polypeptides obtained by heating a mixture of amino-acids. Also *attrib.* or as *adj.*

1956 S. W. Fox in *Amer. Scientist* XLIV. 353 Evidence that proteinoids can be formed by heating one or two amino acids has now accumulated. *Ibid.* 354 The appearance of aspartic acid after hydrolysis is part of the evidence for a proteinoid product. **1968** *New Scientist* 4 Apr. 41/1 When Sidney Fox first discovered this form of amino acid condensation ten years ago, he was immediately attracted by the idea that these quasi-proteins—or proteinoids,..might represent the first evolutionary step ..that led to the true proteins. *Ibid.* 41/2 Perhaps the most striking similarity that proteinoids bear to true proteins..is that..they act like enzymes. **1971** *Nature* 7 May 42/1 In contrast to the plausible explanations for proteinoid formation, there seems to be no satisfactory concept for the genesis of polynucleotide templates in the presumed conditions of the primitive Earth. **1977** A. Hallam *Planet Earth* 236 (*caption*) Cell-like structures called 'proteinoid microspheres' have been produced by evaporating organic chemicals on hot lava beds.

proteinosis (prō̆ᵘtīnō̆ᵘ·sis). *Path.* [mod.L., ad. G. *lipoid*)*proteinose* (E. Uhrbach in J. Jadassohn *Handb. der Haut- und Geschlechtskrankheiten* (1932) XII. 336): see Protein and -osis.] The abnormal accumulation or deposition of protein in tissue.

1937 *Arch. Dermatol. & Syphilol.* XXXV. 357 A subsequent microscopic examination revealed the tinctorial and cellular features of lipoid proteinosis. **1954** *Ann. Internal Med.* XLI. 163 Lipoid proteinosis is a peculiar abnormality of fat deposition characterized by the appearance of white or yellow plaques and nodules in the skin and mucous membranes, producing hoarseness due to vocal chord involvement. **1958** *New England Jrnl. Med.* 5 June 1123 (*heading*) Pulmonary alveolar proteinosis. **1961** *Lancet* 30 Sept. 733/2 A man aged 36 gave a 2-year history of increasing breathlessness with diffuse shadowing on the chest radiograph. 15 months after the onset of symptoms a lung biopsy..showed that he had pulmonary alveolar proteinosis. **1974** *Arch. Dermatol.* CX. 594/2 This finding suggests that lipid accumulation in lipoid proteinosis lesions is not due to a primary defect in lipid metabolism, but could be due to secondary adherence to glycoproteins.

proteinuria (prō̆ᵘtīnⁱū̆ə·riă). *Med.* [mod.L., ad. F. *protéinurie* (L. Hugounenq 1901, in *Lyon Médical* CXVI. 87): see Protein and -uria.] The presence of abnormal quantities of protein in the urine.

1911 *Jrnl. Physiol.* XLII. 238 It seemed important to determine how far an individual suffering from so grave a disturbance of protein metabolism as this condition of proteinuria betokens, could be in nitrogenous equilibrium. *Ibid.* 241 Of some significance is the fact that the excretion of creatin appears to be characteristic of Bence-Jones proteinuria. **1976** *Acta Med. Biol.* XXIV. 9 These patients with symptomless persistent proteinuria appear to constitute a distinct clinical group which is characterized by normal renal function..and a more favorable prognosis.

Hence **proteinu·ric** *a.*, of, pertaining to, or suffering from proteinuria.

1932 *Dorland's Med. Dict.* (ed. 16) 1042/2 Proteinuric. **1969** *Metabolism* XVIII. 556/1 A polypeptide exhibiting diabetogenic and anti-insulin properties has been isolated from the urine of 33 of the 35 proteinuric diabetic patients. **1977** *Lancet* 21 May 1108/2 Groups divided up according to whether they remained normotensive or developed mild pre-eclampsia or proteinuric pre-eclampsia as defined by Nelson.

protension. Delete *rare* and add: **2.** (Later example.)

1890 W. James *Princ. Psychol.* II. xx. 222 In the case of protension or mere farness it [*sc.* the neural process] is more complicated.

3. (Further example.)

1935 *Philos. of Sci.* II. 236 In a similar reaction against Wundtian atomism Külpe also added duration (protension) to the list.

b. (With spelling **protention,** after G. *protention* (E. Husserl 1922 in *Jahrb. f. Philos. u. phänomenol. Forsch.* I. III. ii. § 77.145), perh. infl. by *retention 2c.) In Phenomenology, extension of the consciousness of some present act or event into the future. Hence **prote·ntional** *a.* Cf. *retention 2c.

1931 W. B. Gibson tr. *Husserl's Ideas* III. ii. 216 The same holds good, according to the naïvely natural view, in respect of *anticipation*.., or previsional expectation. At first there comes in the immediate '*protention*' (as we might put it). *Ibid.* 237 Continuous changes in an opposite direction: 'after' corresponding to 'before', a protentional continuum corresponding to the retentional. **1941** *Philos. & Phenomenol. Res.* II. 341 These expectations—Husserl calls them..'protentions'—belong, of course, to our present acting. **1966** A. Gurwitsch *Phenomenol. & Psychol.* ix. 149 The notions of protention and particularly of retention are at the center of the Husserlian theory concerning the experience of time. **1974** D. Carr *Phenomenol. & Probl. of Hist.* iv. 103 It is the act conceived as the living present, with its horizons of retention and protention, which is 'responsible' for the original or primary givenness of anything.

protensity. Delete *rare*⁻⁰ and add examples.

1915 G. F. Stout *Man. Psychol.* (ed. 3) II. i. 212 In all sense presentations we can discern Quality, Intensity, and Protensity or Duration. **1920** S. Alexander *Space, Time, & Deity* II. 130 The 'protensity' of sensation is nothing but its continuance, that is, again, a continuous repetition of the sensation in time. **1935** *Philos. of Sci.* II. 236 The four classical attributes of sensation are quality, intensity, extensity and protensity. **1964** Jakobson & Halle in D. Abercrombie et al. *Daniel Jones* 99 The tense/lax opposition should..be..viewed as a separate, 'protensity' feature which..corresponds to the quantity features in the prosodic field. **1972** *Language* XLVIII. 31 This difference between laxness diphthongization and all other types is a simple consequence of the fact that laxness is a protensity feature.

protention. See *protension 3b.

proteo- (prō̆ᵘtiˌo-), comb. form of Protein; **proteocla·stic** *a.* [see Clastic *a.*] = Proteolytic *a.*; **proteoglycan** (-glai·kæn) (see quot. 1969); **proteoli·pid, -ide**, a complex that contains protein and lipid moieties and is insoluble in aqueous media but soluble in organic solvents; cf. *lipoprotein* s.v. *lipo-. Also Proteolysis, etc.

1929 I. F. & W. D. Henderson *Dict. Sci. Terms.* (ed. 2) 260/2 Proteoclastic. 1959 W. Andrew *Textbk. Compar. Histol.* v. 195 The granular amoebocytes or granulocytes of the oyster carry out intracellular digestion by means of sucroclastic, lipoclastic, and proteoclastic enzymes. 1969 Hascall & Sajdera in *Jrnl. Biol. Chem.* CXLIV. 2384/2 We refer to this fraction, which is a basic building block for the cartilage matrix, as proteoglycan subunit. Because of the selectivity of the techniques used in its purification we presume that it contains only covalently bound protein. The second component is a glycoprotein fraction.. [*Note*] The term 'proteoglycan' is used to describe macromolecules which consist primarily of polysaccharide which is presumed to be bound covalently to the small amount of protein present. 'Glycoprotein' is used to indicate a macromolecule which is primarily protein, but which contains covalently bound saccharide. 1974 *Sci. Amer.* May 67/1 It does appear, however, that the wet surfactants on the ocean, which are known to be glycoproteins and proteoglycans, are reasonably good carriers of phosphate, of various organic molecules, of the scarcer ions of seawater and of heavy metals. 1977 *Lancet* 4 June 1190/2 The major proteins of cartilage are proteoglycan..and type-II collagen. 1950 Folch & Lees in *Federation Proc.* IX. 171/2 (*heading*) Brain proteolipides, a new group of protein-lipide substances soluble in organic solvents and insoluble in water. 1971 [see *lipoprotein* s.v. *lipo-]. 1976 *Nature* 25 Mar. 348/1 The L-glutamate binding proteins of insect and crustacean muscle are hydrophobic proteins (that is, proteolipids) extractable with organic solvents.

proteolytic, *a.* Add: Hence **proteoly·tically** *adv.*, as regards or by means of proteolysis or proteolytic enzymes.

1903 *Ann. Bot.* XVII. 613, I have further succeeded in preparing a proteolytically active glycerin-extract from the roots. 1970 *Nature* 12 Dec. 1097/1 Plasmin..proteolytically degrades fibrin and fibrinogen preferentially to other substrates. 1978 *Ibid.* 21 Sept. 182/2 One current speculation is that the two forms of fibronectin may be the same gene product, but the surface form is proteolytically processed to the slightly smaller plasma form.

proter (prō^u·tǝɹ). *Biol.* [a. F. *proter* (Chatton & Lwoff 1936, in *Arch. Zool. expér. et gen.* LXXVIII. 85), f. Gr. πρότερος in front.] In ciliate protozoa, the anterior of the two organisms formed by transverse fission. Cf. *opisthe.

1950 A. Lwoff *Probl. Morphogenesis in Ciliates* xi. 73 In ciliates of the *Leucophrys* type, the proter and the opisthe are modelled in the anterior and posterior parts of the parent. 1961 Mackinnon & Hawes *Protozool.* iv. 222 At binary fission..the oral structures then go to the anterior daughter, or proter.

protero-. Add: **Proterozoic** *a.* (examples of *absol.* use).

1905 Chamberlin & Salisbury *Geol.* I. i. 17 In these four great series of sedimentary rocks there are, here and there, intrusions of igneous rocks, and in some places the sedimentary beds have been metamorphosed into crystalline rocks by heat and pressure. This is particularly true in the lowest of these series, the Proterozoic, where a large part of the sediment is metamorphosed, and where there is much igneous rock. 1971 *Nature* 25 June 498/1 The establishment of global subdivisions for the Upper (Late) Pre-Cambrian, or the Proterozoic, is particularly important. 1977 A. Hallam *Planet Earth* 189 The Proterozoic was, compared with the Archaean, dominantly a period of crustal reworking.

proterogenesis (prǫtĕrodʒe·nėsis). *Biol.* [mod.L. (coined in Ger. by O. H. Schindewolf 1925, in *Neues Jahrb. f. Min., Geol. und Paläont.* LII. B. 337): see Protero- and -genesis.] The anticipation of future evolutionary development in the early stages of an organism's life. Hence **pro:terogene·tic** *a.*, **-gene·tically** *adv.*

1938 *Rep. Brit. Assoc. Adv. Sci.* 79 Among the examples quoted by Schindewolf in support of the principle of proterogenesis is one drawn from the cephalopod family the *Clymendiæ* which lived during the Devonian period. It consists of a number of genera and species in which, at one end of the series, the shell has the normal type of spiral coil..with an almost circular outline throughout development. In the next member of the series the innermost portion of the spiral has a triangular outline. In other members of the series the latter form of outline finds every degree of expression up to one in which it prevails at all stages of growth, including the adult. The series as it stands may be quoted in support of either the proterogenetic or the tachygenetic view, according to which end of the series is taken as the starting point. Schindewolf adopts the former. *Ibid.* 82 Trend characters, on the other hand, arise either cœnogenetically or deuterogenetically and proceed proterogenetically or tachygenetically towards later or earlier stages in life-history respectively in successive generations. 1947 A. M. Davies *Introd. Palæont.* (ed. 2) iv. 139 In some cases this change is contrary to the rule of palingenesis, since the youthful whorl-shape foreshadows the adult whorl-shape of forms that come later in time... This reverse sequence to that of palingenesis is termed proterogenesis or cænogenesis, and has been observed in other animal phyla, particularly in the Graptolites. 1947 H. H. Swinnerton *Outl. Palæont.* (ed. 3) x. 195 The significance of the inner capricorn whorls in alimorphs may accordingly become anticipatory or recapitulatory, proterogenetic or palingenetic according to which view is adopted. 1966 Davis & Langerl tr. *Hennig's Phylogenetic Systematics* iii. 228 The theory of the 'early ontogenetic origin of types' assumes a special place in discussions of the origin of higher categories or new types. Equivalent or nearly equivalent are the concepts of 'proterogenesis' (Schindewolf), paedomorphism (Garstang), and—as Wettstein 1942 emphasizes—the designations diametagenesis.., fetalization.., and neomorphosis.

proteroglyph (prǫ·tĕroglif). *Zool.* [ad. F. *protéroglyphe*, mod.L. *Proteroglypha* (A. H. A. Duméril 1853, in *Mem. Acad. Sci.* XXIII. 415), f. Protero- + Gr. γλυφή carving.] A venomous snake belonging to a group characterized by grooved fangs in the front of the mouth. So **proterogly·phous** *a.*

1895 Proteroglyphous [in Dict. s.v. Protero-]. 1896 *Proc. Zool. Soc.* 616 In the Proteroglyphs adapted to life in the sea, a similar series of modifications takes place. 1956 L. M. Klauber *Rattlesnakes* II. xi. 715 The front-fanged snakes whose fangs are permanently erect are referred to as proteroglyphs. 1965 R. & D. Morris *Men & Snakes* viii. 177 With the sea-snakes and the cobras, we come to a condition known as Proteroglyphous. Here there has been a reduction in the structure of the upper jaw, the front region having disappeared, bringing the poison fangs to the fore... These are the so-called 'fixed-front-fang' snakes. 1969 A. Bellairs *Life of Reptiles* I. v. 193 According to this hypothesis the viperids would have had a separate ancestry from the proteroglyphs (elapids and sea snakes).

protest, *sb.* Add: **4. c.** In Adlerian psychology, a personal, perhaps unconscious, dissent or attempted dissociation from one's self or circumstances; esp. *masculine protest* (see quots. 1917 and 1972).

1917 Glueck & Lind tr. *Adler's Neurotic Constitution* (1921) iii. 49 The dynamics of the neurosis can therefore be regarded (and is often so understood by the neurotic because of its irradiation upon his psyche) as if the patient wished to change from a woman to a man. This effect yields in its most highly colored form the picture of that which I have called the 'masculine protest'. 1939 H. Orgler *A. Adler* v. 128 The second little theft was carried out as a protest against his being released on parole. 1972 H. Papanek in Freedman & Kaplan *Interpreting Personality* iii. 127 The term 'masculine protest' refers to the attitude of a boy or girl who is raised in a patriarchal culture, in which the real man is respected and admired and the feminine role connotes submissiveness and immaturity.

d. The expressing of dissent from, or rejection of, the prevailing social, political, or cultural mores.

1953 S. A. Brown in A. Dundes *Mother Wit* (1973) 40/2 We go then to what is called the New Negro Movement, then to..Social Protest. 1967 *Listener* 8 June 752/3 Mr Woodcock..traces the development of protest from the first tramps of *Down and Out in Paris and London* to the final achievement of Big Brother. 1968 *Ibid.* 4 July 22/3 Unlike many American authors of his generation, he has [not]..the rather breathy enthusiasm of those who have jumped alongside the youthful millions on the band-wagon..of Protest. 1975 A. Powell *Hearing Secret Harmonies* ii. 72, I was watching a programme..dealing with protest, counterculture, alternative societies.

5. *attrib.* and *Comb.* **a.** Demonstrating or representing a protest against a specific action or proceeding, as *protest banner, button, camp, group, meeting* (further examples), *movement, rally, strike*; designating a literary or artistic medium which seeks to register or portray dissatisfaction with a given event, style, etc., as *protest art, literature, music, poetry, song*; also *protest-singer, -singing*; (sense *4 c) *protest mechanism*; also *protest-oriented* adj. **b.** Special combs.: **protest march** = *march *sb.*[4] 1a; hence as *v. intr*; also **protest marcher**; **protest vote,** a vote placed with a minor faction and considered to represent a protest against the policies of a greater; so **protest voting.**

1973 S. Henderson *Understanding New Black Poetry* 16 Not 'protest' art but essentially an art of liberating vision. 1976 *Milton Keynes Express* 18 June 3/3 Their threat to swamp the area with protest banners had been lifted at the last minute. 1972 *Sat. Rev.* (U.S.) 27 May 6/2 A large protest button, reading: 'Memorial Day, 1969, 35,000 GI's dead in Vain. No More.' 1968 'O. Mills' *Sundry Fell Designs* i. 9 This..must be her eleventh protest camp, not counting non-overnight demonstrations in Trafalgar Square but counting the Aldermaston marches. 1961 B. R. Wilson *Sects & Society* 1 The sect, as a protest group, has always developed its own distinctive ethic. 1973 *Freedom* 1 Sept. 4/1 The various 'protest' groups had lost interest in The Bomb. 1960 *Times Lit. Suppl.* 5 Feb. 77/1 Mr. Klaus Roehler's stories..invite automatic comparison with other outcrops of post-war protest literature. 1975 *Listener* 16 Jan. 69/1 A..flood of protest literature which circulated through the underground channels of *samizdat*, or clandestine publishing. 1959 'M. Derby' *Tigress* iv. 151 What was he..doing in Ceylon? Leading a hydrogen bomb protest march? 1963 *Economist* 9 Nov. 550/2 The people were protest-marching. 1966 C. Achebe *Man of People* i. 4 Protest marches and demonstrations were staged up and down the land. 1967 *Punch* 8 Nov. 699/1, I see an army with banners, protest-marching up and down Charing Cross Road. 1976 *Eastern Even. News* (Norwich) 9 Dec. 3/1 The meeting held at the Hippodrome theatre after a protest march with banners through the town centre. 1960 *Guardian* 12 Oct. 8/5 Sir Edgar Whitehead..failed to..speak to the patient thousands of protest marchers. 1976 J. Wainwright *Bastard* v. 74 The leather-stampers [*sc.* policemen] who stroll alongside the protest marchers. 1920 *Challenge* 21 May 45/1 Adler..has shown how this protest mechanism is responsible for neurotic manifestations of another kind. 1939 L. MacNeice *Autumn Jrnl.* vii. 30 In the sodden park on Sunday protest Meetings assemble. 1965 S. T. Ollivier *Petticoat Farm* x. 137 The Richards brothers.. called a protest meeting of all suppliers. 1909 *Westm. Gaz.* 5 Nov. 5/2 A protest movement is being organised in Belgium against the interference of England in the internal policy of Belgium, especially in regard to the Congo question. 1974 tr. *Wertheim's Evolution & Revolution* iv. 114 In such a case, it should not be called a counterpoint any more, but a social protest movement. 1969 *Listener* 5 June 806/1 Can he [*sc.* Bob Dylan] have forgotten entirely the horrors that gave such a fine edge to his protest music? *Ibid.* 6 Feb. 163/3 Mr Desmond Bird spoke of our 'protest-oriented' society. 1973 S. Henderson *Understanding New Black Poetry* 25 There has been, despite denials, some protest poetry in the sixties. 1960 *Guardian* 11 July 5/3 A protest rally was held in Trafalgar Square. 1977 W. H. Manville *Good-bye* ii. 16 A lot of show-biz people were going to sing and tap-dance at a last-ditch protest rally. 1968 *Guardian* 19 Sept. 9/4 Brave new causes for brave new protest singers. 1969 *Listener* 5 June 805/3 Bob Dylan's *Nashville Skyline* LP completes his recent renunciation of the rebellious, CND-orientated protest-singer image in favour of that of a fun-loving country boy. 1966 *Punch* 19 Jan. 70/2 Anyone tired of protest singing must have been cheered to learn that a group in California ..is rapidly climbing the charts with seventeenth-century songs. 1953 J. Greenway *Amer. Folksong of Protest* 3 Protest songs are unpleasant and disturbing. 1966 *Punch* 9 Feb. 208/2 The rise of the protest songs seems to be doing something for the audibility of lyrics. 1979 *Oxford Times* 21 Dec. 15 What happened to protest songs? Well, here's one, asking: 'Do you find it attractive to be radio-active?' 1974 T. Allbeury *Snowball* xxii. 138 A million workers were due to vote on protest strikes. 1973 *Irish Times* 2 Mar. 9/3 In the event, West Mayo threatened a protest vote. 1976 *Times* 3 Feb. 7/5 A substantial part of the [French] communist vote is a protest one. 1948 in M. McLuhan *Mech. Bride* (1967) 6/6 A blank ballot is the only means of protest voting.

protest, *v.* Add: **1. d.** *trans.* With direct speech as obj.

1903 E. Childers *Riddle of Sands* v. 48 'I'm not boring you, am I?' he said suddenly. 'I should think not,' I protested. 1919 V. Woolf *Night & Day* xii. 154 'But I do read De Quincey,' Ralph protested 'more than Belloc and Chesterton.' 1952 M. Laski *Village* xvi. 218 'But it's quite a good idea,' protested Martha. 1976 B. Freemantle *November Man* iii. 36 'And why the hell not?' he protested.

7. a. Also const. *at.*

1947 Partridge *Usage & Abusage* 46/1 The following from a newspaper placard, *The Daily Worker*, Feb. 6, 1938,—40,000 protest at food prices. 1969 *Daily Tel.* 22 Apr. 29 Conservatives protested angrily..at the Government's failure to announce new contribution rates.

b. *trans.* To protest against (an action or event); to make the subject of a protest. Chiefly *U.S.* Cf. 2c in Dict.

1904 *Brooklyn Eagle* 5 June **5/6 Many of the students are much incensed at the judges and will probably protest the decision. 1927 E. G. Mears *Resident Orientals on Pacific Coast* i. 6 The Peking Foreign Office has regularly protested acts of injustice and violence. 1930 C. Johnson *Negro in Amer. Civilisation* xx. 297 They are protesting the disposition of public school officials to ignore vocational training for Negro youth. 1944 *Sun* (Baltimore) 22 July 2/1 For Hitler it was sufficient that this former chief of staff resigned in 1938 to protest Hitler's march into Austria. 1951 *Newsweek* 27 Sept. 74/3 The residents of Follanshee..have protested the sale, claiming it would throw 2,441 persons out of work. 1956 [see *heel *sb.*[1] 11]. 1966 H. Kemelman *Saturday Rabbi went Hungry* ii. 15 For one thing, I protest their having been singled out. They were pushed and one of them fell. 1977 H. Fast *Immigrants* ii. 82 Dan protested naming the child after Jean's mother. 1978 *Dædalus* Summer 188 They protest the brutal simplicities, the unilinearity and determinism of the great cruel myths of modernization.

Protestant, *sb.* and *a.* Add: **A.** *sb.* **I. 2. a.** (Further examples.)

1864 J. H. Newman *Apologia pro Vita Sua* vii. 425 If Protestants wish to know what our real teaching is,..let them look, not at our books of casuistry, but at our catechisms. 1900 C. M. Yonge *Mod. Broods* v. 50 You seem to me like the Roman Catholic child, who said there were five sacraments, there ought to be seven, but the Protestants had got two of them. 1938 O. C. Quick *Doctrines of Creed* II. xiii. 134 Neither Thomistic orthodoxy nor the modernism of the Liberal Protestants can take such an interpretation seriously. 1955 R. Macaulay *Let.* 5 Feb. in *Last Lett. to Friend* (1962) 190 It is this tendency to rule out Protestants (including Anglicans) from the Church of Christ that is so tiresome and silly. 1962 C. Quin tr. *E. Amand de Mendieta's Rome & Canterbury* viii. 183 After the Reformation movement had led to a vigorous reaction among Protestants against excessive devotion to the Virgin Mary, Roman Catholic theologians have always regarded the defence of the legitimacy of such devotion as one of their chief tasks. 1966 D. E. Jenkins *Guide to Debate about God* iii. 65 Protestants have tended not to be very concerned about the collapse of reason in relation to awareness of God. 1973 *Ann. Reg.* 1972 376 In Northern Ireland the most sinister development was sectarian assassination; 81 Catholics and 40 Protestants.

b. (Further examples.) Also, a Low Church member of the Church of England.

The name continues to be disfavoured by many Anglicans (see also sense c below).

1834 J. H. Newman *Let.* 30 July (1891) II. 59 The word Protestant does not, as far as I know, occur in our formularies. It is an uncomfortable, perplexing word, intended to connect us..with the Protestants abroad. We are a 'Reformed' Church, not a 'Protestant'. 1874 J. H. Blunt *Dict. Sects* 447/2 High Churchmen of modern times.. have..objected to the designation of Protestant as being (1) one of too negative a character to express at all justly

the principle of Catholic resistance to the uncatholic pretensions and practices of Rome: and (2) as being a name which is used by so many sects as to be inclusive even of heresy. 1913 C. Mackenzie *Sinister St.* I. ii. vi. 239 'Finding out for yourself,' echoed Chator with a look of alarm. 'I say, you're an absolute Protestant.' 'Oh, no I'm not,' contradicted Michael. 'I'm a Catholic.' 1933 G. Faber *Oxf. Apostles* iii. 73 They [*sc.* the Tractarians] were hostile to Roman pretensions..but they claimed the same title of Catholic..and they loathed the title of Protestant only less than that of Dissenter. 1960 *Daily Tel.* 15 Nov. 12/8 Surely Canon Lionel Lydekker is mistaken when he writes that the original meaning of Protestant was 'protesting against any tampering with the Holy Catholic Faith'. *Ibid.*, Canon Lydekker's definition of 'Protestant' is in line with 18th-century usage when, in this country, it meant Anglican, as distinct from Nonconformist, just as on the Continent it still means Lutheran as distinct from Calvinist (or 'reformed').

c. A member of a nonconformist or non-episcopal Church.

1958 M. Argyle *Relig. Behaviour* xii. 157 By Protestants we mean to refer to the main 'nonconformist' denominations such as Methodists, Presbyterians and Baptists. 1963 Auden *Dyer's Hand* 350 In New England Protestants of Anglo-Scotch stock consider themselves a cut above Roman Catholics and those of a Latin race. 1977 R. L. Wolff *Gains & Losses* 8 Dissenters or Nonconformists, Protestants of many varying sects, who dissent from the Church of England.

B. *adj.* **1. a.** (Further examples.) *spec.* designating (*a*) Christian belief and principles outside the Communion of Rome (now the most common sense); (*b*) a nonconformist or non-episcopal Church (cf. sense A. 2c above).

(*a*) 1861 C. M. Yonge *Young Stepmother* xxix. 448, I wonder if the omnibus is too protestant to leave a parcel at the convent. 1930 T. Parsons tr. *Weber's Protestant Ethic* i. 35 Business leaders and owners of capital, as well as the higher grades of skilled labour, and even more the higher technically and commercially trained personnel of modern enterprises, are overwhelmingly Protestant. *Ibid.* iii. 80 The conception of the calling thus brings out that central dogma of all Protestant denominations which the Catholic division of ethical precepts into *præceptia* and *consilia* discards. 1935 E. Gill *Lett.* (1947) 342 Prudishness is more typically the vice of the protestant puritan. 1940 R. Niebuhr *Christianity & Power Politics* i. 18 This is the issue upon which the Protestant Reformation separated itself from classical Catholicism. 1954 B. Griffiths *Golden String* vii. 114 When his mother heard that it would be necessary for him to attend a Protestant service, she had replied that she would rather go to the workhouse with her eleven children than submit to that. 1958 B. Pym *Glass of Blessings* xiv. 164 It was absurd to have this suspicious Protestant attitude towards convents. 1960 *Daily Tel.* 10 Nov. 14/7 The Rev. John Castle asks: 'Where in her formularies is the Church of England described as "Protestant"?' The answer is (albeit singly and solely) in the Coronation Service. Whether or no the contention that the oath then taken by the Sovereign regarding the 'Protestant Religion' is merely one imposed by the State without the fiat of the Church, is beside the point. 1965 C. E. Pocknee *Parson's Handbk.* (ed. 13) i. 7 These quotations from well-known Roman Catholic writers are sufficient to disprove the idea that there is something 'protestant' or anti-Roman in the custom of bowing as the normal act of reverence. 1976 *Listener* 29 Apr. 526/2 'The Good Lord' may have Protestant connotations, which would be inappropriate in translating ['le Bon Dieu' in] a Catholic poem. 1976 *Times* 28 Sept. 2/1 Mr. Mason, Secretary of State for Northern Ireland, yesterday made a determined attempt to win the hearts and minds of Ulster's disgruntled population, Roman Catholic and Protestant, by pledging himself to tackle their serious economic ills.

(*b*) 1864 J. H. Newman *Apologia pro Vita Sua* v. 248 They [*sc.* Anglican Bishops] were..fraternizing..with Protestant bodies, and allowing them to put themselves under an Anglican Bishop, without any renunciation of their errors. 1942 J. Baillie *Invitation to Pilgrimage* x. 71 Some of the lesser Protestant sects, Quaker, Methodist, and others,..tended to be 'perfectionist'—sometimes even to a greater degree than Mediaeval Catholicism has ever been. 1954 R. Macaulay *Let.* 19 Aug. in *Last Lett. to Friend* (1962) 166 Conversions.. from the unlovely Protestant churches. 1963 Auden *Dyer's Hand* 350 In New England..the most respectable Protestant denominations are the Congregationalists and the Unitarians. 1973 *Times* 16 Apr. 1/7 He will meet Roman Catholic, Anglican and Protestant church leaders, and Mr Whitelaw, Secretary of State for Northern Ireland. 1977 R. L. Wolff *Gains & Losses* 13 Those who were more Protestant than the legally established Church of England had their own troubled history.

b. Protestant ascendancy, the Anglo-Irish ruling class in Ireland, which has been Protestant since the Reformation, as opp. the native Irish who remained Roman Catholic; also *transf.*; **Protestant Episcopal** (later examples); hence **Protestant Episcopalianism, Episcopalism; Protestant ethic,** the ethical outlook towards business enterprise which, according to Max Weber's analysis, first evolved in protestant Europe through the teachings of Calvin that to be successful through hard work was a person's duty and responsibility; also **Protestant work ethic.**

1827 Protestant ascendancy [see sense 1 a in Dict.]. 1875 F. Arnold *Our Bishops & Deans* I. iii. 148 What idea of Protestant truth was conveyed to the Roman Catholics by the favourite phrase 'Protestant ascendancy'? 1922 R. Dunlop *Ireland from Earliest Times* v. 129 The Treaty of Limerick marks the beginning of a new period known as that of the Protestant Ascendancy. 1936 R. M. Douglas *Irish Bk.* 138 He [*sc.* the Catholic Irishman] was 'loyal', and..became an informer, a policeman, or a soldier serving the Protestant ascendancy. 1966 C. M. Bowra *Memories 1898–1939* xii. 272 His mother, an O'Reilly, was a true daughter of the Protestant Ascendancy. 1977 *Herald* (Melbourne) 17 Jan. 7/3 The neglect of the Castle Hill uprising reflects the prejudices of the Protestant ascendancy in Australia. 1857 *Church Rev.* Jan. 562 The Protestant Episcopal is representative—republican; but not democratic. 1920 *Catholic World* Sept. 777 The best they can offer now as a way out of the hated 'Protestant Episcopal' is 'Protestant Catholic' church! 1961 *N. & Q.* June 236/2 His thesis is that the name 'Protestant Episcopal', though historically justifiable, is a liability to the missionary efforts of American Anglicans and should be replaced by 'The American Episcopal Church'. 1977 *Time* 10 Oct. 37/1 Social relations executive of the Protestant Episcopal Diocese of Massachusetts. 1956 R. Macaulay *Towers of Trebizond* i. 14 My aunt had inherited..strong prejudices against..all American religious bodies except Protestant Episcopalianism. 1836 *Southern Lit. Messenger* II. 282 In regard to Protestant Episcopalism in America, it may be safely said that, prior to this publication of Dr. Hawks, there were no written memorials extant. [1904 M. Weber in *Archiv. f. Sozialwissenschaft und Sozialpolitik* XX. 1 (*title*) Die protestantische Ethik und der 'Geist' des Kapitalismus.] 1926 R. H. Tawney *Relig. & Rise of Capitalism* 320 Both 'the capitalist spirit' and 'Protestant ethics'..were a good deal more complex than Weber seems to imply. 1930 T. Parsons tr. *Weber's Protestant Ethic* iii. 89 We thus take as our starting-point in the investigation of the relationship between the old Protestant ethic and the spirit of capitalism the works of Calvin, of Calvinism, and the other Puritan sects. 1944 *Social Res.* Feb. 61 Weber's original intention in *The Protestant Ethic* must be seen against the background of his time. 1956 W. H. Whyte *Organization Man* i. i. 6 The organization man.. needs..something that will do for him what the Protestant ethic did once. 1968 C. Armstrong *Balloon Man* xii. 146 She was tough. In her own way, Protestant ethic or whatever, she was. 1977 *Listener* 7 Apr. 455/1 Infant prodigies..are the ultimate denial of the Protestant Ethic —hard work can produce lesser rewards than sheer talent. 1980 *Country Life* 24 Apr. 1283/3 Mrs Smith had the Protestant work-ethic in the very marrow of her bones.

Protestantism. Add: **1.** (Further examples.)

1862 J. H. Newman *Let.* 28 June (1970) XX. 216, I do hereby profess *ex animo*..that Protestantism is the dreariest of possible religions; that the thought of the Anglican service makes me shiver. 1864 —— *Apologia pro Vita Sua* v. 181, I held a large bold system of religion, very unlike the Protestantism of the day, but it was the concentration and adjustment of the statements of great Anglican authorities. 1930 T. Parsons tr. *Weber's Protestant Ethic* 27 We are dealing with the connection of the spirit of modern economic life with the rational ethics of ascetic Protestantism. 1938 O. C. Quick *Doctrines of Creed* ii. 12 While the error of scholasticism was to tie down Christianity to a particular philosophy, the error of much modern Protestantism has been to disparage philosophy altogether. 1952 P. Tillich *Courage to Be* v. 114 Protestantism..was established as a strictly authoritarian and conformist system, similar to that of its adversary, the Roman Church of the Counter-reformation. 1964 A. Watts *Beyond Theol.* vi. 155, I am not speaking of modern 'liberal Protestantism', but of that 'old-time Bible religion'. 1974 *Encycl. Brit. Macropædia* XV. 136/2 Providence..may also be experienced as personal guidance. This latter phenomenon is common..in some forms of Protestantism in which generally each person is expected to have a private experience of divine guidance. 1976 B. Griffiths *Return to Centre* xv. 106 Protestantism opened the Bible to the private interpretation of every man. 1976 *Times* 13 Aug. 13/6 What gives them their weight and validity is their promulgation by the *magisterium* and surely it is this authority that the traditionalists are ready to defy. Is not this nearer to the heart of Protestantism? 1977 B. Colloms *Victorian Country Parsons* xii. 228 William Kingsley was a broadminded parson whose sermons and services were suitable for most brands of Protestantism.

2. (Later example.)

1948 F. W. Dillistone in M. Warren *Triumph of God* iii. 80 The energy of young Protestantism was bound to be poured out in the effort to consolidate its own position within the hostile environment by which it was surrounded.

Protestantization (prǫ:tėstǎntaizēɪ·ʃǝn). [f. Protestantize *v.* + -ation.] The action or fact of rendering Protestant; conversion to Protestantism.

1880 W. James in *Atlantic Monthly* Oct. 447/2 After Charles IX. and Louis XIV., no general protestantization of France. 1977 S. Schoenbaum *Shakespeare* v. 54 Further forays into Protestantization followed.

protester. Add: **2. c.** An opponent of the established order, esp. one who actively remonstrates over an issue of public importance; a demonstrator.

1960 *Times Lit. Suppl.* 1 Jan. 3/2 Those cold-water apartments where hipsters and protesters..try to go their own way. 1968 *Listener* 4 July 22/3 Absolutely and actively in sympathy with the protesters over most issues, he can ..trace out the roots of their discontents. 1971 'E. Lathen' *Longer the Thread* vii. 65 Thatcher ..would not believe..that homicide investigations came to a halt whenever youthful protesters threatened to act up. 1976 'R. B. Dominic' *Murder out of Commission* v. 41 Protesters picked up some pretty strange habits during the Vietnam War.

proteus. Add: **3. b.** [Adopted as the name of a genus by J. N. Laurenti in his *Synopsis Reptilium* (1768) 35.] = *olm. (Earlier and later examples.)

a 1829 H. Davy *Consolations in Travel* (1830) iv. 190 The same infinite power..has given the Proteus to the deep and dark subterraneous lakes of Illyria,—an animal to whom the presence of light is not essential. 1902 Baskett & Ditmars *Story of Amphibians & Reptiles* vii. 48 *Proteus*..lives in a cave in Austria. 1965 B. E. Freeman tr. *Vandel's Biospeleol.* iii. 23 *Proteus* represents a veritable giant amongst the European cave fauna, because it reaches a length of 30 cm.

prothallial, *a.* Add: (Later examples.) *spec.* applied to a small cell formed at an early stage in the development of the male gametophyte of certain gymnosperms, or a similar structure in certain pteridophytes.

1892 *Ann. Bot.* VI. 214 Judging from the obvious continuity of the dividing wall with the lateral walls of the prothallial chambers..there can hardly exist a doubt that the two cavities result from a primary transverse division of the cell. 1910 Coulter & Chamberlain *Morphol. Gymnosperms* vi. 277 The appearance of two evanescent vegetative (prothallial) cells is a feature only of the Abietineae. 1929 Robbins & Rickett *Botany* xxii. 326 Two are very small cells (prothallial cells) which die and disintegrate almost as soon as formed. 1965 Bell & Coombe tr. *Strasburger's Textbk. Bot.* 593 In the originally unicellular microspore (pollen grain) [of *Pinus*] one or two (or sometimes more) small cells are formed at a particular spot just inside the wall; these prothallial cells often degenerate after a time. 1969 F. E. Round *Introd. Lower Plants* xi. 132 The microspore divides within the spore wall to form a single 'prothallial' cell and an antheridial initial cell. 1973 A. Cronquist *Basic Bot.* xvi. 262 The prothallial cell [of *Selaginella*] is considered to be an evolutionary vestige of the body of the gametophyte.

prothallium. Add: (Later examples.) Also, a homologous structure in the development of certain gymnosperms.

1892 J. B. Farmer in *Ann. Bot.* VI. 213 (*title*) On the occurrence of two prothallia in the embryo sac of *Pinus*. 1935 C. J. Chamberlain *Gymnosperms* xiii. 309 In some heterosporous pteridophyte ancestor [of the conifers], the prothallium (gametophyte) became included within the spore..and finally disappeared entirely.

prothallus. Add examples referring to gymnosperms.

1940 *Chambers's Techn. Dict.* 680/1 The term prothallus is extended to cover homologous stages in the life-cycle of Gymnosperms. 1965 K. R. Sporne *Morphol. Gymnosperms* i. 19 Cell-walls are being laid down between the nuclei of the female prothallus, a process which, in most gymnosperms, continues until the whole of the prothallus becomes circular.

protheca: see *pro-[2] 1.

prothesis. 3. (Later examples.)

1968 P. M. Postal *Aspects Phonol. Theory* vii. 144 The Prothesis rule is that which inserts an [i] in the front of verbs containing less than one vowel. 1976 *Language* LII. 307 Evidence for prothesis in Spain dates from the Vulgar Latin of the 2nd century.

prothetely (proþe·tǝli). *Ent.* [ad. G. *prothetelie* (H. J. Kolbe 1903), in *Allgemeine Zeitschr. f. Entom.* VIII. 1), f. Gr. προθεῖν to run before + τέλος end: see -y[3].] In certain insect larvæ, the development of one part of the body, esp. the wings, at a faster rate than that of the rest. Hence **prothete·lic** *a.*, of or pertaining to this type of development.

1934 Folsom & Wardle *Entomol.* (ed. 4) iii. 169 As a rare abnormality a holometabolous larva may possess two pairs of true external wing-pads... The phenomenon is termed prothetely. 1940 R. Goldschmidt *Material Basis of Evolution* iv. 283 The actual working of this timing mechanism can be inferred from cases of so-called prothetely, where a single larval organ, e.g. wings or antennae, metamorphoses alone. *Ibid.* 284 (*caption*) Head of prothetelic caterpillar of *Lymantria dispar* with pupal antennae. 1960 H. Oldroyd tr. *Jeannel's Introd. Entomol.* iv. 83 (*caption*) Prothetelic larva of *Tenebrio molitor*.. showing rudiments of wings. *Ibid.*, Prothetely. Quite often when bred artificially, but more rarely in nature, insect larvae may have rudiments of wings, which have matured at a greater rate than the rest of the body. 1978 *Nature* 23 Mar. 350/2 They..obtained morphogenetic effects (precocious-prothetelic adults after administration to young instars) only in Heteroptera.

prothrombin (proþrǫ·mbin). *Phys.* [a. G. *prothrombin* (A. Schmidt *Zur Blutlehre* (1892) xii. 202), f. Gr. πρό Pro-[2] + θρόμβ-ος clot, Thrombus: see -in[1].] A protein formed in the liver and normally present in blood, whose conversion into thrombin is an essential part of the clotting process.

1898 E. A. Schäfer *Text-bk. Physiol.* I. 160 A fibrin ferment (thrombin), or its precursor (prothrombin), producing the formation of fibrin from fibrinogen. *Ibid.*, The fibrin ferment is sometimes spoken of as 'thrombin', and the nucleo-proteid material in the plasma from which it is produced is then termed 'prothrombin'. 1912 J. G. McKendrick *Princ. Physiol.* viii. 117 The theory at present in vogue is that when blood is shed there is at once the death of many colourless cells. These contain a protein called pro-thrombin, which, in turn, produces an enzyme known as thrombin. 1936 [see *hypoprothrombinæmia* s.v. *hypo- II]. 1946 [see *oxalated* ppl. a. s.v. *oxalate *v.*]. 1957 *Sci. News* XLIV. 31 In the presence of ionized calcium, of a further globulin component of the plasma, and probably of other factors, thromboplastin converts

prothrombin, by what must be a relatively minor change in its molecule, to thrombin. This last substance now reacts directly with fibrinogen, changing it to fibrin. 1974 R. M. Kirk et al. *Surgery* ii. 10 Vitamin K analogue, 10 mg intramuscularly each day, replaces the deficiency but if there is concomitant liver damage, prothrombin is not synthesised. 1976 [see *proconvertin* s.v. *pro-[1] 1].

b. *attrib.*, as **prothrombin time,** the time taken for blood or plasma to clot when an excess of a calcium salt and possibly other natural components of the clotting mechanism (besides prothrombin) are added.

1935 *Amer. Jrnl. Med. Sci.* CXC. 505 By omitting the addition of excess thromboplastin, but otherwise following the directions of the method outlined for the determination of prothrombin, one determines the clotting time of recalcified plasma, which for normal plasma is from 90 to 130 seconds. This method is still commonly called Howell's Prothrombin Time. 1957 A. J. Quick *Hemorrhagic Dis.* ii. 44 The prothrombin time of normal human plasma is consistently 12 seconds, whereas when measured by the two-stage test, it varies from 244 to 452 units. 1972 *Daily Colonist* (Victoria, B.C.) 1 Feb. 2/1, I am on anticoagulants and have a prothrombin time test taken weekly to determine dosage.

protic (prō^u·tik), *a.* *Chem.* [f. Prot(on in Dict. and Suppl. + -ic.] Of a liquid, esp. a solvent: possessing protons whose binding is sufficiently loose for them to participate in protonation; hydrogen-bonded.

1944 *Jrnl. Physical Chem.* XLVIII. 53 In all protic solvents, protons enter strongly into any consideration of acid-base properties. 1965 Phillips & Williams *Inorg. Chem.* I. xv. 556 Similarly, in other 'protic' solvents a special role is played by substances producing H^+ ions, or the characteristic anion of the solvent (e.g. NH_2^- in NH_3, F^- in HF, and HSO_4^- in H_2SO_4). 1973 E. J. King in Covington & Dickinson *Phys. Chem. Org. Solvent Syst.* iii. 333 We first distinguish two broad classes [of solvent] based on dielectric constant. In solvents of high dielectric constant, often referred to loosely as polar solvents, ion-pairing is minimal, even negligible in dilute solutions... By contrast, in solvents of low dielectric constant, loosely called non-polar solvents, ion-pairing is important and acid strength depends on the choice of standard base... Each broad class in turn is sub-divided into hydrogen-bonded and non-hydrogen-bonded solvents. The term *protic* is frequently used for the first sub-division, *aprotic* for the second. 1975 *Nature* 3 Jan. 40/1 These reactions show that molecular nitrogen can be reduced at a single metal site in a protic medium with negligible discharge of dihydrogen or displacement of dinitrogen by hydride ligands.

protide (prō^u·taid). *Biochem.* Also **-id** (-id). [a. F. *protide* (G. Bertrand 1923, in *Bull. de la Soc. de Chim. biol.* V. 102), f. *protéine* Protein: see -ide.] A generic term for a protein, peptide, or amino-acid. Hence **proti·dic** *a.*

The scheme of nomenclature proposed by Bertrand met with little favour among English-speaking scientists (see *Chem. & Engin. News* (1952) 5 May 1910). *Protide* is now to be found almost exclusively in translations or abstracts from French.

1936 A. P. Mathews *Princ. Biochem.* xxii. 215 The International Union of Pure and Applied Chemistry suggested that the proteins be called 'protides' in consonance with the 'glucides' and 'lipides'. 1958 (*title*) Protides of the biological fluids. *Ibid.* p. v, Some biochemical constituents of protidic nature or origin in the living animal are considered in a further nine original papers. 1962 *Biol. Abstr.* XXXIX. 948/1 (*heading*) Study of the protidic fraction of Digitaria exilis. 1973 *Compar. Biochem. & Physiol.* B. XLV. 225 (*heading*) Protides of the Mustelidae: comparative study of plasma lactate dehydrogenases. 1975 *Biol. Abstr.* LX. 4700/2 Some data..suggest that the initiation of growth is strictly dependent in [*sic*] nutritional factors (mainly protids). 1977 *Lancet* 10 Dec. 1242/2 The 26th colloquium on protides of the biological fluids will be held in Bruges on May 1–5.

protist. Add etym.: [f. mod.L. *Protista*, f. G. *protisten* (E. Haeckel *Generelle Morphologie der Organismen* (1866) I. 203): see Protista *sb. pl.*] (Later examples.)

1905 J. McCabe tr. *Haeckel's Evol. Man* I. vi. 98 In the case of the protists, the entire organism usually consists of a single autonomous cell throughout life. 1926 C. M. Wenyon *Protozool.* I. 4 A typical Protist consists of a small portion of cytoplasm and a nucleus. 1965 B. E. Freeman tr. *Vandel's Biospeleol.* vi. 62 Not a single free-living aquatic protist can be considered as a true cavernicole. 1975 *Nature* 7 Aug. 467/2 These [phytoflagellates] were the ancestors of all plants, and..of non-photosynthetic protists and animals.

protistan, *sb.* (Example.)

1940 [see *metachronal *a.*].

protistology (prōtistǫ·lŏdʒi). [f. Protist, Protista *sb. pl.* + -ology.] The study of organisms included in the Protista. Hence **protisto·logist,** one who studies protists.

1911 J. A. Thomson *Introd. Sci.* iv. 110 It might also be convenient to have a special science of Protistology for the minute and simple organisms which seem to hesitate between plant and animal life. 1911 *Q. Jrnl. Microsc. Sci.* LVI. 396 Cytologists and protistologists alike have been content..with assuming that the Bacteria are a group of simple organisms. 1926 C. M. Wenyon *Protozool.* I. i. 4 It is safer to regard them all as one large group, the Protista, the study of which is known as Protistology. 1951 *John o' London's Weekly* 17 Aug. 501/2 A scholar..may spend his days at the microscope and read only treatises on protistology. 1965 B. E. Freeman tr *Vandel's Biospeleol.* vi. 61 The disappearance of pigments in *Euglena* placed in darkness is a common occurrence, well known to protistologists. 1973 *Microscopy* XXXII. 325 He [*sc.* Georges Deflandre]..was soon deep in studies of protistology (the biology of unicellular organisms).

protium (prō^u·tiǝm). *Chem.* [mod.L., f. Gr. πρῶτ-ος first + -ium.] The 'normal', most abundant isotope of hydrogen, having only a proton in the nucleus and forming at least 99·98 per cent (by volume) of naturally occurring hydrogen; symbol ^1H (also H^1). Cf. *deuterium, *tritium.

1933 [see *deuterium]. 1936 *Nature* 12 Dec. 1021/1 Several attempts have been made to determine the ratio protium–deuterium (^1H:^2H) in ordinary water, and the results mostly fall into two groups, either near 5500 or near 9000. 1957 G. E. Hutchinson *Treat. Limnol.* I. iii. 211 Hydrogen has two stable isotopes, H^1, sometimes called protium, and H^2 or D, usually called deuterium. 1972 *Nature* 31 Mar. 202/1 Of late, both Cornforth and Arigoni have developed techniques for the solution of this problem based in the chemical synthesis of CH_3 – groups containing one atom each of the three hydrogen isotopes, protium, deuterium and tritium. 1975 *Physics Bull.* May 211/1 In the case of the neutron diffraction experiments D_2O was used instead of H_2O as deuterium is a better coherent scatterer than protium.

proto-. Add: **1.** (*a*) *proto-chemistry* (later example), *-culture, -history, -music, -novelist, -phoneme, -poet, -scientist*; **protocu·ltural** *a.*, belonging to such origins as can be surmised of human cultural development; † **pro·togram** *Obs.*, an acronym; **protograph** (further example); † **protogra·phic** *a. Obs.*, acronymic; **protohisto·rian,** (*a*) (see sense 1 (*a*) in Dict.); (*b*) one who studies proto-history; **proto-histo·rical** *a.* = *proto-historic* adj.; **pro·to-literate** *a.*, characterized by the most primitive kind of writing; **proto-scienti·fic** *a.*, belonging or relating to primitive science, or to an early stage in scientific development; **pro·tosyntax** (see quot. 1940); hence **protosynta·ctical** *a.*; **protosynta·ctically** *adv.*; **pro·totheme** (see quot. 1897); **proto-typographer** (further examples).

1976 *Times Lit. Suppl.* 12 Nov. 1418/2 The development of alchemy and proto-chemistry [in China]. 1961 A. I. Hallowell in S. L. Washburn *Social Life Early Man* 237, I suggested that the level of development represented by cultural adaptation can be focused more sharply in evolutionary perspective if we hypothecate a protocultural phase in hominid evolution. 1976 *Sci. Amer.* Oct. 104/2 The difference is not necessarily related to the confinement of our troop but may simply reflect protocultural differences. 1971 R. M. & F. M. Keesing *New Perspectives Cultural Anthropol.* 48 There must have been 'protomen' with 'protoculture'. 1924 *Glasgow Herald* 27 Sept. 4 'Anzac' is one of the first protograms to which the war gave birth. It is used..to describe anything pertaining to the 'Australian and New Zealand Army Corps'. 1933 H. Wentworth *Blend-Words in Eng.* 3 Words formed from the initials of other words—called *letter words* ..and *protograms* (F. H. Vizetelly)—are fewer. 1974 *Bible Translator* July 317 According to Russian biblical scholarship these basic sections are..(1) the protographs of the Septuagint and the New Testament [etc.]. 1924 *Glasgow Herald* 27 Sept. 4 The great majority of words of the protographic type have been coined within the last decade. 1949 *Proc. Prehist. Soc.* XV. 196 That difficult problem so often shirked by prehistorian and protohistorian—the mechanics of cultural diffusion. 1928 V. G. Childe *Most Anc. East* viii. 176 The implements of the protohistorical period were almost entirely of metal. 1950 A. Huxley *Themes & Variations* 54 That Golden Age of Peace, which not long since was regarded as a mere myth, but is now revealed by the light of archaeology as a proto- and pre-historical reality. 1920 R. R. Marett *Psychol. & Folk-lore* xi. 249 The value of proto-history, as it is sometimes termed. 1947 H. C. E. Zacharias (*title*) Protohistory. An explicative account of the development of human thought from Palaeolithic times to the Persian monarchy. 1980 *Encounter* May 66/1 We—the workers in British protohistory during the last 30 years—have suffered corporately from inadequate preliminary education, leading to the mental counterparts of asthma, myopia and strabismus. 1942 Delougaz & Lloyd *Pre-Sargonid Temples* i. 123 The architectural history of the Sin Temple bears out..the subdivision of the Early Dynastic period into three and the Proto-literate into at least two distinct phases. 1971 Proto-literate [see *pictographic* adj. s.v. *pictograph]. 1963 Auden *Dyer's Hand* 474 A music which sounds remarkably like primitive proto-music. 1977 *Rolling Stone* 21 Apr. 41/3 Ultimately, *14 Canons* is a unique type of protomusic—a series of potentially extendible alchemical exercises. 1976 *Times Lit. Suppl.* 19 Nov. 1459/4 Bunyan's humanity and his raciness and his humour and everything that makes him a proto-novelist. 1960 H. M. Hoenigswald *Lang. Change & Ling. Reconstruction* xii. 132 If a split affects the same proto-phoneme in each daughter language, the partial likeness between the sets of correspondences is impaired. 1974 R. W. Wescott in *Language Origins* 116 Only eight proto-phonemes (which are more nearly equivalent to contemporary morphophonemes than to contemporary Phonemes) appear in all five of their formulations. The eight are p, t, k; m, n; y, w; e. 1963 Auden *Dyer's Hand* 34 Whatsoever Adam called every living creature, that was..its Proper Name. Here Adam plays the role of the Proto-poet. 1934 Webster, *Protoscientific*, adj. 1968 M. Bunge in R. Klibansky *Contemp. Philos.* II. 4 In the underdeveloped (protoscientific) disciplines, fact-collecting passes for the sole respectable occupation. 1978 *Sci. Amer.* Jan. 69/1 Overshadowed by scholasticism, the work of the protoscientists was ignored or treated as heresy, and its proponents endured ridicule and some persecution. 1940 W. V. Quine *Math. Logic* vii. 292 The part of syntax which omits membership will be called protosyntax... Protosyntactical definability is intended not as an approximation to constructivity, but as something more inclusive. The notion of a non-theorem, e.g., is protosyntactically definable, yet presumably not constructive. 1943 *Mind* LII. 272 A restricted portion of the syntax (that which omits membership) is distinguished by the label 'protosyntax'. 1964 *Amer. Philos. Q.* I. 265/1 The entire construction is done..within Quine's protosyntax. 1897 Prototheme [see *deuterotheme* s.v. *deutero-]. 1905 *N. & Q.* III. 176/1 These protothemes in familiar intercourse, or even on more serious occasions, often received the termination *-a, Seax*, for instance, becoming *Seaxa*. 1931 *Library* XII. 109 This volume is printed with the type of Johannes de Salsburga and Paulus de Constantia, the prototypographers of Barcelona. 1976 *Times Lit. Suppl.* 22 Oct. 1328/2 Thanks to Caxton, England had a native prototypographer who worked with patriotic gusto in the national language.

2. a. (Further examples.) Also in *Philol.*, forming sbs. and adjs. designating the earliest attested or hypothetically-reconstructed form of a language or family of languages (cf. *primitive *a.* 4b). (The list that follows represents only a selection of possible items.) *proto-Algonquian, -Aryan, -Athapaskan, -Australian, -Australoid, -Austronesian, -Corinthian* (further examples), *-Gallo-Romance* (also *-Romanic*), *-Germanic, -Greek* (further examples), *-Hattic, -Indo-European, -Italic, -Malay, -Medic, -Polynesian, -Romance, -Semitic* (examples), *-Slavonic.* Also with geographical names and sbs., as *proto-Atlantic, -Nile, -ocean, -Thames*; and with astronomical names, as *proto-earth, -Jupiter, -sun* (hence *protosolar* adj.). Also *protocloud, -cluster* in *2b, *proto-continent, *protogalaxy, *protoplanet, *protostar. **proto-Hi·ttite,** the language of the Hattian people, philologically unrelated to Hittite.

1939 L. Bloomfield in C. Hockett *Bloomfield Anthol.* (1970) 352 Our basic forms are not ancient forms, say of the Proto-Algonquian parent language. 1974 *Canad. Jrnl. Linguistics* XIX. 145 As was mentioned above, Proto-Algonquian palatalization came down into Fox pretty much unscathed. 1904 G. S. Hall *Adolescence* II. xviii. 657 The Todas of India, whom some call proto-Aryans. 1938 Partridge *World of Words* iv. 126 The Latin may be traced to an Aryan original; but the proto-Aryan form.. was caused by some accidental circumstance. 1964 M. E. Krauss in *Internat. Jrnl. Amer. Linguistics* XXX. 118 (*title*) Proto-Athapaskan-Eyak and the problem of Na-Dene: the phonology. 1966 J. T. Wilson in *Nature* 13 Aug. 676/1 It is proposed that, in Lower Palaeozoic time, a proto-Atlantic Ocean existed so as to form the boundary between the two realms, and that during Middle and Upper Palaeozoic time the ocean closed by stages. 1972 *Sci. Amer.* Nov. 62/3 In Devonian times an order of jawless freshwater fishes, cousins to the orders that once flourished on opposite sides of the proto-Atlantic, inhabited the streams of the region that is now the European and Asiatic flanks of the Urals. 1918 *Phil. Trans. R. Soc.* B. CVIII. 382 This fossil human skull of a not yet adult Proto-Australian presents..the general picture of a cranium similar in all respects to the cranium of the Australian of to-day. *Ibid.*, The Proto-Australian is, in some very important features, to be sharply differentiated from Neanderthal man. This is nowhere more clearly seen than in the palate and teeth. 1923 R. B. Dixon *Racial Hist. Man* iv. ii. 374 The Australian population thus appears to be made up almost entirely of two types, the Proto-Negroid and Proto-Australoid, of which the former is concentrated in the north and northwest, the latter in the south and southeast. 1959 Proto-Australoid [see *gerontomorphic *a.*]. 1963 G. B. Milner in C. Mohrmann et al. *Trends in Mod. Linguistics* 68 Dempwolff had found a sufficient body of evidence to justify his setting up a Proto-Melanesian language..as he had reconstructed a Proto-Polynesian language, both of which he regarded as ultimately descended from Proto-Austronesian. 1976 *Language* LII. 221 The systematic reconstruction of Proto-Austronesian..phonology and lexicon was first attempted by Otto Dempwolff. 1932 *Times Lit. Suppl.* 8 Sept. 622/4 Corinth, where the Protocorinthian style forms a natural transition between the Geometric and the Orientalizing. 1973 *Univ. Oxf. Ann. Rep.* 1970–71 8 Publications: 'A Protocorinthian Dinos and Stand'. 1969 *Times* 18 July 6/4 The proto-earth may have swept up from the dust cloud much more silicate material than it now possesses. 1977 A. Hallam *Planet Earth* 18/2 Most probably, the Moon formed from a dense atmosphere, generated by the high temperatures of solid-particle accretion at the surface of the proto-Earth. 1950 *Language* XXVI. 9 A concrete example of how this type of intermediate reconstruction can be done and what it gives us can be seen in the phonological system of Proto-Gallo-Romance. 1964 *Ibid.* XL. 32 If Provençal should turn out to belong to it, 'Proto-Gallo-Romance' is the obvious choice. 1946 *Stud. in Philol.* XLIII. 463 Then with a similarly acquired statement of Proto-Provençal, we can formulate Proto-Gallo-Romanic. 1934 Priebsch & Collinson *German Lang.* iv. 236 The Proto-Italic and Proto-Germanic peoples. 1960 *Amer. Speech* XXXV. 227 From pre-Scandinavian or Proto-Germanic to Old and Modern Icelandic. 1964 *Language* XL. 294 Next he reviews the history of the problem of the Proto-Germanic long stops. 1972 *Ibid.* XLVIII. 407 Proto-Germanic, which should be based on the internal reconstructions of the individual dialects. 1959 T. Burton-Brown *Early Medit. Migrations* iii. 66 There were established, from at least as early as the end of the Third Millennium, some

kind of 'proto-Greek' people. 1964 E. Palmer tr. *Martinet's Elem. Gen. Linguistics* v. 149 Tsakonian is a proto-Greek dialect. 1968 W. S. Allen *Vox Graeca* i. 30 It may be mentioned that in Proto-Greek, and still preserved in Mycenaean, there was a series of 'labio-velars'. 1933 E. H. Sturtevant *Compar. Gram. Hittite Lang.* i. 29 There seems to be no need for the cumbrous terms 'Proto-Hattic' or 'Proto-Hittite'. 1948 D. Diringer *Alphabet* v. 89 Some scholars call them 'Proto-Hattic' or 'Proto-Hittite'. 1924 A. H. Sayce in *Jrnl. R. Asiatic Soc.* 245 Proto-Hittite is the name given by Dr. Forrer to the prefixal language, examples of which are found in the cuneiform texts of Boghaz Keui. 1952 O. R. Gurney *Hittites* vi. 122 The name Proto-Hittite has been widely adopted in order to avoid confusion with the official Hittite, but is somewhat misleading, since it suggests an earlier stage of Hittite, whereas it is a language totally unrelated to the latter. The name Hattian is preferable. 1947 R. S. Wells in *Word* III. 15 Linguists have reconstructed large parts of the vocabulary of Proto-Indo-European. 1955 W. P. Lehmann in *Language* XXXI. 355 (*title*) Proto-Indo-European Resonants in Germanic. 1960 *Amer. Speech* XXXV. 227 Specific laryngeal problems in Proto-Indo-European phonology. 1979 *Amer. Speech* 1978 LIII. 266 We have virtually no evidence for the earlier history of Proto-Indo-European forms. 1968 *Language* XLIV. 269 On the evidence of Latin, Oscan, and Umbrian, Proto-Italic still had the phrasally prior final consonants that have disappeared in Proto-Romance. 1976 *Archivum Linguisticum* VII. 62 Oscan *-tt-* represents a proto-Italic cluster *-ky-. 1976 *Sci. Amer.* May 113/1 James B. Pollack and his co-workers..suggest that exactly the same process would have taken place within the miniature solar system of the Jovian satellites, with the proto-Jupiter the source of the heat. 1909 A. C. Haddon *Races of Man* 18 *Indo-Chinese, Parcæans* or *Southern Mongols*:..Those members who spread into the East Indian Archipelago are often called Oceanic Mongols, but a better term is Proto-Malays; and it is from these the true Malay is derived. *Ibid.* 14 The broadening of the head is probably due to an early mixture with a Proto-Malay stock. 1947, 1958 Proto-Malay [see *Jakun]. 1964 W. A. Hamid in W. Gungwu *Malaysia* iii. xii. 179 The Proto-Malays are the tribes to be found in the interior forests among the foothills of the Malay archipelago. 1877 A. H. Sayce in *Trans. Philol. Soc.* 1875–6 136 In Protomedic and Susianian..the initial is similarly always dropped in the plural of the verb. 1880 —— *Introd. Sci. Lang.* ii. x. 321 The Protomedic group of languages to which Accadian belongs, in the Ural-Altaic family. 1894 Protomedic [see *Medic *sb.*[3]]. 1972 *Sci. Amer.* Apr. 116/1 Primate forms found in fossil forest beds deposited 35 million years ago beside the proto-Nile. 1975 *Nature* 29 May 376/1 Young rift oceans (proto-oceans) are commonly the site of large scale evaporite deposition. 1930 R. Paget *Human Speech* vii. 145 Several other gesture-words from Proto-Polynesian. 1973 *Amer. Speech* 1970 XLV. 118 The reconstruction of some proto-Polynesian forms. 1949 *Archivum Linguisticum* I. 151 The Proto-Romance consonant clusters... We use this term, instead of the vague..'Vulgar Latin'. 1978 *Language* LIV. 182 There is no discussion of proto-morphophonemics, which might conceivably have raised the issue of umlaut in Proto-Romance. 1948 D. Diringer *Alphabet* 214 The proto-Semitic alphabet. 1969 *Word* XXV. 115 The Proto-Semitic consonant system is generally assumed to have had a voiced velar stop phoneme */g/ as established by a set of correspondences throughout the Semitic family. 1920 *Trans. Philol. Soc.* 1916–20 128 In Proto-Slavonic all final consonants fell out. *Ibid.* 130 Beside the palatalization there is another sweeping tendency in Proto-Slavonic phonology. 1951 *Archivum Linguisticum* III. 205 The work is a succinct presentation of Protoslavonic morphology. 1975 *Nature* 11 Sept. 91/1 S. Ramadurai.. argued that carbonaceous chondrites contain interstellar graphite grains from the protosolar nebula. 1978 *Ibid.* 16 Mar. 239/2 Further conditions which must be satisfied are ..penetration of this element into the protosolar cloud. 1974 *Sci. Amer.* Mar. 51/3 At a distance of perhaps 20 million miles from the protosun, a fifth of the way to the present orbit of the earth, a very few nonvolatile materials could have condensed into solid particles. 1969 Bennison & Wright *Geol. Hist. Brit. Isles* xv. 346 The consequent drainage pattern developed on the eastwards-tilted Mesozoic rocks included the proto-Thames.

b. proto-bi·face *Archæol.*, an early form of biface; **pro·tocell,** a body postulated as ancestral to the cell; **protocerebrum,** (*b*) the anterior segment of the brain of an arthropod; **pro·tocloud** *Astr.*, a protogalactic cloud; **pro·tocluster** *Astr.* = prec.; **protocneme** (prō^u·toknīm) *Zool.* [Gr. κνήμα tibia; cf. Cnemial *a.*], one of six pairs of primary mesenteries which are found in corals of the order Scleractinia; **protodo·lomite** *Min.*, a mineral with a composition near that of dolomite, $CaMg(CO_3)_2$, but an imperfect crystal structure; **protody·nastic** *a.* (further examples); **protoe·nstatite** *Min.*, an artificial, high-temperature form of the magnesium silicate $MgSiO_3$; **protofi·bril** *Biol.*, a filament of protein that is a component structural element of a fibril or *spec.* of a microfibril; **protofi·lament** *Biol.*, a filament of protein, about 5 nanometres in diameter, a group of which constitute a microtubule; **protohu·man** *sb.* and *a.* *Anthrop.*, (pertaining to or being) one of the man-like prehistoric creatures from which man is held to have evolved; **pro·tolife,** inanimate existence representing a late stage in the evolution of life; **pro·tolith** *Petrol.* [Gr. λίθ-ος stone] (see quot. 1972); **protoli·thionite** *Min.* [ad. G. *protolithionit* (F. Sandberger *Untersuchungen über Erzgänge* (1885) ii. 169): cf. Lithionite], a variety of zinnwaldite containing a higher proportion of lithium and a lower proportion of iron; **protomerite** (earlier and later examples); **proto-Neoli·thic** *a. Archæol.*, belonging to or characteristic of the earliest Neolithic period; also *absol.* as *sb.*; **protonephri·dium** *Zool.* [mod.L., coined in Ger. by B. Hatschek *Lehrb. der Zool.* (1889) II. 160: see Nephridium], in certain invertebrates, esp. flatworms, an excretory system made up of solenocytes opening into ducts leading to pores in the exterior surface; also, a larval nephridium of this type; so **protonephri·dial** *a.*; **pro:toperithe·cium,** in some fungi, an ascogonium from which have grown out one or more trichogynes, some of which develop into part of a perithecium if spermatization occurs; hence **pro:toperithe·cial** *a.*; **protopetro·leum** [mod.L., coined in Ger. by C. Engler 1897, in *Ber. d. Deut. Chem. Ges.* XXX. 2360], an intermediate product in the formation of petroleum from organic debris; **pro·tophenomenon** *Philos.*, a primary phenomenon; **protophloem,** add: [a. G. *protophloëm* (E. Russow 1873, in *Mém. Acad. Impér. Sci. St.-Pétersbourg* 7 sér. XIX. 4]; (further examples); **pro·topod** *a. Ent.* [ad. It. *protopodo* (A. Berlese 1913, in *Redia* IX. 127)], of an insect larva, lacking abdominal segmentation and limbs; **protopro·teose** *Biochem.*, any of a class of proteoses that are soluble in water and dilute salt solutions and are formed during gastric secretion; **protosco·lex** *Zool.*, a vesicle formed from the germinal layer of a hydatid cyst and capable of development into a scolex or a secondary cyst; **pro·tostele** *Bot.* [Stele 2], a simple type of stele in which a central core of xylem is surrounded by a cylinder of phloem: hence **protostelic** (-stī·lik) *a.*; **pro·totheca** *Zool.* [Theca], a cup-shaped basal plate which is formed at the start of the development of a colony of stony corals; **pro·totroch** *Zool.* [Gr. τροχός wheel], a pre-oral ciliated ridge encircling the body of the trochosphere larva of certain invertebrates, including polychæte worms and some annelids and molluscs; **protovermi·culite** *Min.*, a mineral similar in composition to vermiculite but containing more water, and found as large yellow or brown scales; **protoxylem,** add: [a. G. *protoxylem* (E. Russow 1873, in *Mém. Acad. Impér. Sci. St.-Pétersbourg* 7th Ser. XIX. 3)]; (further examples).

1975 *Nature* 7 Aug. 470/2 Some protohandaxes may have one face made entirely of cortex, so not all of these artefacts can be called protobifaces and the term protohandaxes is preferable. 1976 *Ibid.* 8 July 104/2 Other tool forms, such as..protobifaces..occur with less frequency. 1965 S. W. Fox *Orig. Prebiol. Syst.* 372 The explanation has been extended to permit us to visualize a spontaneous synthesis of protein-like material sufficiently similar to yield a protocell which could spontaneously include ATP-splitting ability. 1974 Ponnamperuma & Gabel in Carlile & Skehel *Evolution in Microbial World* 407 Oparin does not in any way imply that the coacervates he and his associates have studied were the actual precursors of the protocell. 1977 A. Hallam *Planet Earth* 236 Carbonaceous meteorites also contain organic spheres, and mineral grains coated with organic sheaths, that have been likened to 'protocells'. 1897 *Q. Jrnl. Microsc. Sci.* XL. 261 Viallanes has shown, by his very careful researches on the structure of the adult brain.., that it consists in insects of three segments... The first or protocerebrum, including the optic centres, corresponds to the first segment in Peripatus. 1969 *New Scientist* 10 July 56/2 It is generally known that the regulating clock mechanism of insects lies in the protocerebrum. 1970 *Nature* 31 Oct. 412/1 A protocloud formed at that time would initially expand with the Universe, but at a reduced rate. 1971 *Proc. Internat. School of Physics 'Enrico Fermi'* XLVII. 336 The density fluctuations associated with protoclusters—and *a fortiori* protogalaxies—would be too small to be detected. 1976 *Nature* 11 Nov. 114/2 Four stages might usefully be distinguished: (1) the creation of a massive protocluster cloud; (2) the separation of individual protostars from such a protocluster cloud; [etc.]. 1900 Protoeneme [see *metacneme* s.v. *meta- 4]. 1916 H. S. Pratt *Man. Common Invertebr. Animals* 138 The gullet is joined with the body wall by all of the protocnemes. 1940 L. H. Hyman *Invertebrates* I. vii. 589 These original twelve septa, which arise as couples, are called protocnemes. 1956 J. W. Wells in R. C. Moore *Treat. Invertebr. Paleont.* F 333/2 When the first 6 mesenteric pairs (comprising 12 protocnemes) have developed the embryonic period [of scleractinians] is terminated. 1955 Graf & Goldsmith in *Bull. Geol. Soc. Amer.* LXVI. 1566 These poorly ordered near dolomites, or protodolomites, also have been observed to form during the rapid cooling of dry periclase-calcite assemblages through the dolomite stability field. *Ibid.* 1567 In view of their relatively narrow compositional range, it appears probable that protodolomites have a relatively high degree of short-range Ca–Mg order, rather than being merely metastable, disordered, high-magnesium calcites. 1967 *Oceanogr. & Marine Biol.* V. 151 The deficiency of magnesium in the Red Sea brine might be caused by dolomitization of the carbonate rocks (some evidence of which is found in the presence of crystals of protodolomite in the core from the Discovery Deep). 1902 *Nature* 6 Nov. 14/2 [Professor E. Smith] intends to give a full account of the structure of the brain in the predynastic and protodynastic Egyptians. 1962 S. E. Finer *Man on Horseback* vii. 89 These are the traditional monarchies where the ideals of nationality, liberty, equality and popular sovereignty have not yet penetrated. Another and better description is perhaps the proto-dynastic societies, societies where allegiance is owed to the dynasty. 1977 G. Clark *World Prehist.* (ed. 3) v. 236 There can be no doubt of the existence precisely at the period of transition from the Predynastic to the Protodynastic or Archaic period of Egyptian history of innovations that stemmed from Mesopotamian sources. 1939 *Jrnl. Amer. Ceramic Soc.* XVIII. 110/1 Constitution of steatite... On heating to 800°, talc lost its H_2O and was transformed into protoenstatite. 1962 *Ibid.* XLV. 156/2 The rate of the metastable inversion of protoenstatite to clinoenstatite during cooling is very sensitive to particle size. 1965 L. Bragg et al. *Crystal Struct. Minerals* xii. 236 The detailed shape and relative positions of the silicate chains depend on the relative positions of the Mg atoms and their surrounding, octahedrally coordinated, oxygens. In protoenstatite the chains are fully extended, whereas in enstatite and clinoenstatite they are slightly different in shape and not fully extended. 1961 Filshie & Rogers in *Jrnl. Molecular Biol.* III. 785 It can be observed..that a high concentration of lead has entered each microfibril and become bound to preferred sites, revealing a composite structure consisting of filamentous subunits relatively unstained by lead (henceforth to be referred to as protofibrils) each of the order of 20 Å in diameter. 1966 *New Scientist* 24 Feb. 480/2 Protofibrils, some 20 angstroms wide, may be observed to occur in a regular array, and it is widely accepted that they aggregate around an annulus, with perhaps nine outer protofibrils and two further protofibrils inside. 1971 *Nature* 22 Jan. 253/1 Wood fibres are hollow tubes composed of layers of cellulosic protofibrils embedded in a matrix of hemicellulose and lignin. 1971 *Proc. Nat. Acad. Sci.* LXVIII. 1766 If we assume a protofilament arrangement of monomeric subunits..in a microtubule, it becomes apparent that a homofilament microtubule can be constructed only if the number of protofilaments is even (i.e. 12 or 14 in the usual model) while a heterofilament microtubule always results if an odd number (11 or 13) of protofilaments are assembled. 1977 *Jrnl. Protozool.* XXIV. 4/1 Microtubules..can be thought of as protofilaments that are end-to-end polymers of dimers which are then bound together to form a tube with an open lumen. 1910 *Daily Chron.* 9 Apr. 6/2 The Oceanic negro is far removed from primitive man, but.. he inherits, as we all do, but happily in a lesser degree, the savage instincts of the proto-human. 1954 L. C. Eiseley in W. L. Thomas *Current Anthropol.* 69/1 We have.. stumbled into the world of essentially cultureless or almost cultureless proto-human types which are diverse in form because they represent evolution still at work upon the parts of the body. 1954 W. La Barre *Human Animal* iv. 83 The linearity of man, his relative hairlessness, his clothing, and his culture-based carnivorousness suggest that the proto-humans, like the anthropoids, were warm-climate-adapted animals. 1971 R. M. & F. M. Keesing *New Perspectives in Cultural Anthropol.* 45 Sharing must be viewed as a crucial protohuman innovation. 1978 *Sci. Amer.* Apr. 94/1 Excavation of these protohuman sites has revealed evidence suggesting that two million years ago some elements that now distinguish man from apes were already party of a novel adaptative strategy. 1966 *Palaeogeogr., Palaeoclimatol., Palaeoecol.* II. 54 The formation of cobionts and protolife through inorganic photosynthesis stopped at the beginning of this period of transition. 1977 A. Hallam *Planet Earth* 236/1 The stromatolites are universally regarded as the remains of true life: the earlier microscopic fossils may well also represent the remains of blue-green algae, but it is perfectly probable that they represent some form of primitive protolife. 1972 *Gloss. Geol.* (Amer. Geol. Inst.) 571/2 *Protolith*, the unmetamorphosed rock from which a given metamorphic rock was formed by metamorphism. Syn: *parent rock*. 1974 *Nature* 15 Mar. 199/1 This investigation attempts to decipher the premetamorphic age of the protolith of a recrystallised breccia from Apollo 16. 1892 *Dana's Syst. Min.* (ed. 6) vi. 627 *Protolithionite*, a lithium-iron mica from the granite of the Erzgebirge, Fichtelgebirge, etc. Color dark. Optically nearly uniaxial... Sandberger regards it as the source of the zinnwaldite, hence the name. 1959 *Amer. Mineralogist* XLIV. 1297 It is a lithium-iron mica, closely related to zinnwaldite and containing a large amount of the protolithionite component of the lepidolite series. 1885, 1921 Protomerite [see *epimerite]. 1962 J. D. Smyth *Introd. Animal Parasitol.* vi. 73 In some forms [of gregarines], the protomerite is drawn out into a specialised region for attachment. 1921 R. A. S. Macalister *Text-bk. European Archaeol.* I. x. 549 A culture independent of any of those which we have now considered, namely the 'Protoneolithic' Campignian. 1924 [see *Asturian *a.* and *sb.*]. 1931 *Antiquity* V. 520 Menghin distinguishes a Protoneolithic, and a Mixoneolithic, in the latter of which the Neolithic arts found their full expression. 1960 C. Winick *Dict. Anthropol.* 440/2 *Protoneolithic*, in some classifications, the lower, or early, Neolithic era, consisting of the Campignian and Ertebole cultures. 1895 Protonephridial [see *protonephridium* below]. 1963 R. P. Dales *Annelids* v.98 The metanephridial funnels or protonephridial solenocytes lie in the coelomic fluid. 1895 E. S. Goodrich in *Q. Jrnl. Microsc. Sci.* XXXVII. 479 The nephridia of the Planarians..are formed of a main duct, which branches out into fine tubules ending blindly internally in flame-cells; they do not develop beyond this 'protonephridial' condition—protonephridium of Hatschek. 1900 *Ibid.* XLIII. 742 For its [*sc.* the nephridium's] closed representation..and for closed 'head-kidneys', the term Protonephridium might, perhaps, be used with advantage. It is the name proposed by Hatschek for the closed nephridia of the Platyhelminths. 1930, 1967 [see *metanephridium* s.v. *meta- 4]. 1949 A. S. Romer *Vertebrate Body* ii. 19 The excretory organs [of Amphioxus]..are tiny tubes (protonephridia) of a type found in certain invertebrates. **1978** L. C. Oglesby in P. J. Mill *Physiol. Annelids* xiv. 619 In only one group, the Rotifera, is there direct evidence that the protonephridia serve an osmoregulatory role. **1955** G. M. Smith *Cryptogamic Bot.* (ed. 2) I. xii. 450 If appropriate spermatidia or conidia are not available for the trichogynes, there is no further development beyond the protoperithecial stage. **1976** *Ann. Rev. Microbiol.* XXX. 98 Nutritional control is important for the initiation of protoperithecial development and conidiogenesis. **1941** *Bot. Rev.* VII. 396 A haploid mycelium or a multicellular trichogyne of *Neurospora sitophila*.., through which nuclei of opposite sex are passing *en route* to the ascogonium in a proto-perithecium, where they are destined to take part in the formation of the first pair or pairs of conjugate nuclei. **1974** *Nature* 24 May 383/1 In *N[eurospora] crassa*, protoperithecia have been induced to develop into fruiting bodies, albeit sterile. **1909** *Econ. Geol.* IV. 625 Engler thus enumerates the various stages which in his opinion occur in the formation of petroleum from organic matter:..4. Formation of liquid hydrocarbons and violent reaction with 'cracking' into light or gaseous products = formation of protopetroleum. **1938** B. T. Brooks in A. E. Dunstan et al. *Sci. of Petroleum* I. 52/1 Accordingly it might be expected that protopetroleums in transition stages will be found in geologically recent strata in the form of solid or semi-solid material. **1973** R. E. Chapman *Petroleum Geol.* ii. 32 There is general agreement that the main source of petroleum is the organic matter buried with a fine-grained sediment (usually a clay); and that diagenesis of this organic matter leads to a 'protopetroleum' which, before or during migration, becomes modified by the physical and chemical environment—particularly by increasing temperature during burial—until it eventually becomes petroleum. **1953** G. E. M. Anscombe tr. *Wittgenstein's Philos. Investigations* I. 167 Our mistake is to look for an explanation where we ought to look at what happens as a 'proto-phenomenon'. That is, where we ought to have said: *this language-game is played*. **1966** *Amer. Philos. Q.* III. 7/1 We should look simply at what is said as a proto-phenomenon. **1902** Protophloem [see *metaphloem* s.v. *meta- 4]. **1953** K. Esau *Plant Anat.* xii. 286 The primary phloem may be divided into protophloem and metaphloem. **1965** Protophloem [see *metaphloem* s.v. *meta- 4]. **1925** A. D. Imms *Gen. Textbk. Entomol.* 179 In the protopod phase metamerism is incomplete, the abdomen being imperfectly differentiated. **1934** Folsom & Wardle *Entomol.* (ed. 4) iii. 172 The protopod larva is characterized by a lack of differentiation of the internal and external organs. **1969** R. F. Chapman *Insects* xx. 400 Among the parasitic Hymenoptera the first instar larva hatches as a type known as a protopod larva. **1891, 1916** Protoproteose [see *heteroproteose* s.v. *hetero-]. **1936** A. P. Mathews *Princ. Biochem.* xxii. 221 'Proto-proteoses', precipitated by half saturation of their solutions by ammonium sulphate. **1971** *Exper. Parasitol.* XXX. 233/1 Protein synthesis in larval *Echinococcus granulosus* protoscolices occurs by the pathway involving amino acyl-adenylates and amino acyl-tRNA as intermediates. **1976** *Lancet* 9 Oct. 811/2 Since the Lebanese also eat raw liver a hydatid of the tonsil might arise from implantation of a protoscolex in a tonsillar crypt. **1901** L. A. Boodle in *Ann. Bot.* XV. 705 A centrally placed solid stele (protostele), consisting of a central mass of xylem..surrounded by a continuous ring of phloem. **1919** F. O. Bower *Bot. Living Plant* xxi. 330 Generally in young sporelings there is a simple stele of a type called a 'protostele', having a solid xylem-core, and phloem surrounding it. **1957** H. C. Bold *Morphol. Plants* xxiii. 446 The most primitive genera and the juvenile stages of most others have stems that contain protosteles. **1975** J. D. Haynes *Botany* xxii. 331 The stele of most members of this group [*sc.* lycopods] is a protostele. **1902** *Encycl. Brit.* XXV. 413/2 There is good reason to suppose that the protostelic condition is primitive in evolution. **1957** H. C. Bold *Morphol. Plants* xxiii. 447 Internally the roots are exarch and protostelic. **1904** H. M. Bernard in *Ann. Mag. Nat. Hist.* XIII. 4 The parent colony of a calicle rises out of a basal cup—the Prototheca... The term 'prototheca' was suggested to me in conversation by my friend Prof. Jeffrey Bell. **1906** S. J. Hickson in Harmer & Shipley *Cambr. Nat. Hist.* I. xiv. 386 The calicoblasts form..a skeletal plate at the aboral end of the coral embryo, which becomes turned up at the edges to form a shallow saucer or cup. This cup is called the 'prototheca'. **1935** Twenhofel & Shrock *Invertebr. Paleont.* iv. 78 The embryonic skeleton of a typical coelenterate has the shape and appearance of a little, hollow, conical cup and is known as the prototheca. **1897** A. T. Masterman in *Q. Jrnl. Microsc. Sci.* XL. 291 There are three prominent ciliated bands, the preoral (or prototroch), the collar-band..and the trunk band. **1904** *Amer. Naturalist* XXXVIII. 500 Cells arising from the first quartette..make up a cell row which very probably forms at least a part of the second ciliated band on the head of the adult, in a position corresponding with that of the prototroch of the annelid larva. **1932** Borradaile & Potts *Invertebrata* vii. 207 A band of cilia round the base [of the Pilidium larva] constitutes the prototroch. **1959** *Q. Jrnl. Microsc. Sci.* C. 89 The ectoderm of the mouth region [of the pre-adult *Scoloplos armiger*]..includes transitorily the prototroch cells. **1978** K. S. Richards in P. J. Mill *Physiol. Annelids* ii. 48 In the prototroch the compounding of cilia helps to eliminate such lateral stresses. **1877** *Proc. Acad. Nat. Sci. Philadelphia* 269 Professor Geo. A. König described a micaceous mineral from Magnet Cove, Ark., to which he gave the name Protovermiculite... The mineral occurs in large foliated plates, loose in the soil. **1948** *Amer. Mineralogist* XXXIII. 656 Protovermiculite from Magnet Cove, Arkansas. Large golden yellow scales. **1902** *Phil. Trans. R. Soc.* B. CXCV. 135 The protoxylem is separated from it [*sc.* the pith] by a large mass of primary metaxylem. **1974** *New Phytologist* LXXIII. 979 The helically thickened protoxylem..is stretched during the elongation of the axis.

c. More widely, prefixed to adjs. and sbs. designating an original, early, or undeveloped form of an artistic or political movement, as *proto-Baroque, -Cubist* adjs., *-Fascism* (hence *-Fascist* sb. and adj.), *-Marxian* adj., *-Renaissance* (also *attrib.*), *-romantic* adj.

1935 *Burlington Mag.* Apr. 159/1 Why are the artists working about the year 1800 gothic-manneristic, classicistic, proto-Baroque, high Baroque? **1977** *Dædalus* Summer 2 Titian's removal of the Virgin..to the right side of the worshipers in his Pesaro Madonna—once considered a protobaroque stylistic invention. **1979** *Jrnl. R. Soc. Arts* Nov. 767/2 If I can use the historical analogy again, with sixteenth-century Italy, the proto-Baroque or Mannerist period, which seems to have had the same kind of doubts and plu[r]alism characteristic of our age. **1959** H. Read *Conc. Hist. Mod. Painting* v. 156 The metamorphic *Three Dancers* is in fact a turning point in Picasso's art almost as radical as was the proto-cubist *Demoiselles d'Avignon*. **1938** *New Statesman* 19 Feb. 302/1 It [*sc.* racism] is proto-Fascism, based on mysticism on the one hand, and pseudo-biology on the other. **1945** H. Read *Coat of Many Colours* i. 3 Lucian, one of those romantic exiles who brought some light and liberty into a proto-fascist world. **1959** —— *Conc. Hist. Mod. Painting* iv. 119 Anarchists.., proto-fascists in some cases, the Dadaists, adopted Bakunin's slogan: destruction is also creation! **1973** *Black Panther* 28 Apr. 8/3 The danger in a 'professional' national police force is the same as that of a volunteer army. In both we find an elitist, racist, proto-fascist orientation and esprit. **1977** M. Walker *National Front* i. 15 Proto-fascist, crypto-fascist,..quasi-fascist; the sub-groups..multiply and do little to impose meaning on the confusion. **1969** P. A. Robinson *Freudian Left* 168 Portrait of Hegel as loyal son of the Enlightenment and proto-Marxian critic of the European social order. **1909** *Cent. Dict. Suppl.*, *Proto-Renaissance*.., a revival movement in art and literature preceding the Renaissance proper, especially that which began in the reign of the Emperor Frederick II. (1194–1250). **1911** *Encycl. Brit.* XX. 468/1 A 'Proto-Renaissance', the characteristic of which was a fresh interest in surviving remains of classical antiquity. **1942** N. Pevsner *Outl. Europ. Archit.* iv. 61 The Tuscan Proto-Renaissance of S. Miniato..i.e., the architecture of Florence in the 11th century, and nothing else. **1945** *Burlington Mag.* Jan. 23/2 Such, however, was the popular 'proto-Renaissance' in our country. **1948** N. Pevsner *Outl. Europ. Archit.* (rev. ed.) v. 82 Of Romanesque or Proto-Renaissance connections there are here none left. **1963** *Times Lit. Suppl.* 26 Apr. 312/4 'Proto-Renaissance' Romanesque 'antique' models. **1947** A. Einstein *Mus. Romantic Era* viii. 81 Many of the traits in Mozart's works can be considered 'Romantic' or proto-romantic. **1971** *Country Life* 12 Aug. 392/1 The drawings are large in scale..and far removed from the experimental, proto-romantic work that we associate with Brown.

protoactinium, var. *Protactinium.

proto-Algonquian: see *proto- 2 a.

protoanemonin (prǭᵘto,āne·mǒnin). *Chem.* [Coined in Japanese as *purotoanemonin* (Asahina & Fujita 1920, in *Jrnl. Pharmaceut. Soc. Japan* XL. 3): see Proto- and Anemonin (cf. quot. 1920).] A poisonous, vesicant, pale yellow oil, which is isolated from many plants of the family Ranunculaceæ, and is an unsaturated lactone, $CH_2=C\cdot CH=CH\cdot CO\cdot O$, having bacteriostatic and fungistatic properties.

1920 *Chem. Abstr.* XIV. 1384 The sharp, oily substance volatile with steam obtained from *Ranunculus scleratus* L., consists in the main of the mother substance of anemonin, and has been given the name *protoanemonin*, which under spontaneous union of 2 mols. passes into anemonin. **1949** H. W. Florey et al. *Antibiotics* I. xiv. 605 Protoanemonin ($C_5H_4O_2$) was obtained as a pale yellow irritating oil. It solidified to give anemonin within a few hours at room temperature but was stable at 5°C. in 1 per cent. aqueous solution. **1958** *New Biol.* XXVI. 44 All three species [of buttercup] contain a glycoside 'ranunculin' which is readily broken down in damaged tissues to give glucose and an unsaturated lactone—protoanemonin. This lactone is poisonous and blistering, and makes *R. bulbosus* and *R. acris* very unpalatable to stock. **1972** *Science* 5 May 512/2 Lesser in toxicity are the vitamin antagonists such as protoanemonin..and amino acid antagonists such as mimosine.

proto-Aryan to **-Austronesian:** see *proto- 2a. **proto-Baroque, -biface:** *proto- 2c, b.

protobiont (-bəi·ǫnt). *Biol.* [f. Gr. πρωτο- Proto- + βιοντ-, pres. pple. stem of βιοῦν to live, f. βίος life.] A small drop of fluid surrounded by a membrane, hypothesized as ancestral to living cells.

1964 A. Synge tr. *Oparin's Chem. Origin of Life* iii. 61 The droplets would come to contain a constantly increasing concentration of the corresponding catalysts when the mass of the droplet increased as it grew by polymerising the monomers of the surrounding medium. Such coacervate droplets with an improved organisation are still only hypothetical but, for convenience of discussion, we shall refer to them in what follows by the provisional name of 'protobionts'. **1970** A. L. Lehninger *Biochem.* xxxiv. 782 Oparin..has suggested that the first cells, which he called protobionts, arose when a boundary or membrane formed around one or more macromolecules possessing catalytic activity, presumably proteins. **1971** *Sci. Amer.* May 30/1 Judging from the various forms of life we know today, the first protobionts were probably microscopic in size and single-celled in structure. **1978** *Ibid.* Sept. 65/1 To Oparin the reproductive machinery and DNA are only the ultimate biochemical subtleties that turned metabolically competing protobionts into living cells.

protocell, -cerebrum: see *proto- 2b. **protochemistry:** Proto- 1 in Dict. and Suppl.

protochlo·rophyll. *Biochem.* [a. G. *protochlorophyll* (N. A. Monteverde 1893, in *Acta Horti Petropolitana* XIII. 210): see Proto- and Chlorophyll.] A naturally occurring photoactive precursor of chlorophyll.

1894 *Jrnl. R. Microsc. Soc.* 702 The same author [*sc.* Monteverde] finds, in etiolated leaves, besides xanthophyll and carotin, a pigment to which he gives the name *protochlorophyll*. It displays a distinct red fluorescence. **1928** *Science* 7 Dec. 570/2 Protochlorophyll is not a decomposition product of some other organic substance, as leucophyll, but is a pigment which develops without the influence of light and changes photochemically into chlorophyll upon exposure to light. **1951** *Ann. Rev. Plant Physiol.* II. 131 Protochlorophyll is a pale greenish pigment containing two hydrogen atoms less than chlorophyll. **1956** E. I. Rabinowitch *Photosynthesis* II. xxxvii. 1759 Up to 90% of the protochlorophyll, accumulated in the dark, are quantitatively converted, within a minute or less of moderately strong illumination, into chlorophyll *a*. **1976** *Photochem. & Photobiol.* XXIV. 555 We now extract cells with acetone... Protochlorophyll is obtained by extracting this solution with petroleum ether..and extracting this petroleum ether fraction with 80% acetone to remove substances which interfere with subsequent chromatography.

protocho·rdate, *sb.* and *a.* *Zool.* [ad. mod.L. *Protochordata* (F. M. Balfour *Treat. Compar. Embryol.* (1881) II. xii. 27), f. Proto- + *chordate *a.* and *sb.*] **A.** *sb.* A small marine animal belonging to one of the subphyla Hemichordata or Cephalochordata, which form a group considered to be related to ancestors of the vertebrates, and are characterized by a dorsal nerve cord, a notochord, and gill slits. **B.** *adj.* Of, pertaining to, or resembling an animal of this kind.

1894 A. Willey *Amphioxus & Ancestry of Vertebrates* v. 242 Of the free-living protochordates, the lowest type of organisation is undoubtedly presented by the *Enteropneusta* (Hemichorda). *Ibid.* 282 There is no *a priori* reason for doubting that the Vertebrate mouth is completely homologous with the Protochordate mouth. **1918** H. F. Osborn *Orig. & Evol. Life* viii. 246 The principal.. secretory glands..doubtless had their beginnings among the ancestors (protochordates) of the vertebrated animals. **1933** L. A. Adams *Introd. Vertebrates* i. 17 All these modern protochordates are small and live in the sea. **1951** C. K. Weichert *Anat. Chordates* ii. 12 The animals included in this category [*sc.* Acrania] are believed to show similarities to the ancestors of the chordates and for this reason are sometimes called the protochordates. **1978** *Nature* 5 Jan. 61/2 Attention has focused on the protochordate endostyle since the demonstration that certain of its cells could bind iodine.

protocloud to **-cneme:** see *proto- 2b.

protococcoid, *a.* Add: (Example.) Also **protococcoi·dal** *a.*

1965 F. E. Round *Biol. Algae* i. 3 The unicells may be sub-divided into non-motile (Protococcoidal), 'amoeba'-like (Rhizopodial) and motile cells (Flagellate). **1967** M. E. Hale *Biol. Lichens* iii. 45 The common protococcoid algae are soon killed off by the fungus.

protocol, *sb.* Add to etym.:

The word has now entered the general vocabulary of English in senses *5 b, c.

2. a. Also *fig.*

1923 A. Huxley *Antic Hay* xii. 184 The parting kiss.. was already in the protocol, as signed and sealed before her departure by giggling Molly.

c. [In Gr. sense.] The first sheet of a roll of papyrus, bearing the manufacturer's official mark; this mark itself.

1885 *Encycl. Brit.* XVIII. 233/1 The first sheet of a roll was named πρωτόκολλον... On the Arab conquest of Egypt in the 7th century, the manufacture was continued, with the substitution of Arabic in marking the protocol. **1905** W. E. Crum *Catal. Coptic MSS. Brit. Mus.* 181 Upon the 1st *selis*, above the Coptic text, is part of the protocol in large Kufic characters. **1912** E. M. Thompson *Introd. Gk. & Lat. Palaeogr.* 25 After their conquest of Egypt in the seventh century, the Arabs continued the manufacture of papyrus and also affixed protocols to their rolls.

d. *Protocols of the (Learned) Elders of Zion*: a spurious publication of Russian origin purporting to describe Jewish plans for the domination of the world. Also, *Protocols of Zion*.

1920 (*title*) The Jewish peril: protocols of the Learned Elders of Zion. **1921** *Times* 16 Aug. 9/6 The so-called 'Protocols of the Elders of Sion' were published in London last year under the title of 'The Jewish Peril'. **1937** H. G. Wells *Star-Begotten* vii. 128 These Reds—Moscow—Bernard Shaw—New Dealers—Atheists—Protocols of Zion, all of that—mere agents. **1941** —— *You can't be too Careful* v. i. 238 You can study how the new pogromism was revived in that curious and impudent forgery, *The Protocols of the Elders of Zion*. **1974** *Jewish Chron.* 20 Dec. 15/4 The antisemitic Protocols of the Elders of Zion have gone through more editions in Arabic than any other language. **1979** O. Sela *Petrograd Consignment* 142 Wasn't he desperate to read the Protocols of the Elders of Zion again; wasn't another pogrom all he lived for.

e. *Philos.* A statement which forms an essential part of a person's description of

something experienced or perceived; a basic statement that can be verified or assessed.

1936 *Mind* XLV. 275 N[eurath] expresses Protocol-propositions in the form 'Charles's protocol (there is a table in the room perceived by Charles)'. **1956** J. O. Urmson *Philos. Analysis* viii. 121 Protocols are direct reports of the given and are justified with reference to the given... I choose between the possible protocols 'This is red' and 'This is green' by seeing which correctly reports experience. **1965** P. Caws *Philos. Sci.* xi. 74 A set of protocol sentences..constitutes a *protocol*.

3. (Further examples.) *spec.* the detailed record of the procedure and results in a scientific experiment; hence, experimental procedure.

1887 *Amer. Jrnl. Psychol.* I. 136 The protocol here is admirable, taken on the spot by Mr. Birchall and printed in full, and Mr. Guthrie is very positive in stating that there were a large number of 'complete successes'. **1910** *Amer. Jrnl. Physiol.* XXVII. 36 The protocols of the experiments above discussed follow. **1923** [see *intratracheally* adv. s.v. *intra-]. **1929** I. A. Richards *Pract. Crit.* 4, I lectured the following week partly upon the poems, but rather more upon the comments, or protocols, as I call them... I asked each writer to record on his protocol the number of 'readings' made of each poem. **1947** *Ann. Rev. Microbiol.* I. 357 The data are not given, but from the protocols it appears probable that the hemagglutinin persists while toxicity and infectiousness are lost. **1961** *Lancet* 22 July 213/2 Not only the urinary but also the fæcal porphyrin excretion must be considered. Protocols of published cases [of porphyria] frequently lack this vital information. **1973** *Jrnl. Genetic Psychol.* CXXII. 192 Each S's protocol was scored in terms of the number of responses to critical items which indicated a preference for balance. **1976** *Amer. Speech 1974* XLIX. 12 The phonetic transcriptions and marginal notes made from that record and entered in the workbooks are a protocol, and they are regarded as working notations. **1977** *Lancet* 9 Apr. 805/1 The most common treatment protocol appears to have been a series of 6 treatments (range 1–30) at a dose of 250 (range 100–400) roentgens per treatment.

5. a. Also used of analogous departments in other countries.

1975 M. Sinclair *Long Time Sleeping* xii. 150 'I wonder when was the last time we refused to accept an Ambassador?' 'I'll ask Protocol in the morning.' **1980** J. Hone *Flowers of Forest* I. 13 His job..had been in Protocol... It was his function to control liaison..between our own and other allied intelligence services.

b. An official form of procedure and etiquette in affairs of state and diplomatic relations; the observance of this.

1945 in Webster Add. **1949** *Washington Post* 22 Mar. 1/3 He [*sc.* President Truman] felt that it would not be good protocol for him to be away hobnobbing with an Englishman out of office at the very time that the man now in charge of Britain's foreign affairs is in Washington. **1953** *Economist* 12 Dec. 799/1 Alleged breaches of diplomatic protocol at the Bermuda conference. **1957** *Listener* 5 Sept. 337/2 Dr Adenauer..chatted gaily and frankly... Protocol and affairs of state were set aside. **1980** J. Cartwright *Horse of Darius* iii. 46 Our sovereign is coming here... The protocol is very important. Who comes to visit. How we keep in touch with Tehran.

c. In extended and general uses, any code of conventional or proper conduct; formally correct behaviour.

1952 *Daily Express* 27 Mar. 4/3 He punctuates his work with a gay laugh..and a first-name informality with colleagues... Not that protocol is in danger. For behind his exuberance is a tough grasp of ceremonial and what is due. **1954** W. Faulkner *Fable* 18 The old generalissimo turned, his two confreres..flanking him in rigid protocol. **1954** *Times* 27 Nov. 6 The ceremony was taking its course in accordance with academic protocol. **1959** *Woman's Own* 20 June 37/2 Prince Edward and Princess Alexandra had a childhood as free from protocol as their mother's had been. **1971** B. W. Aldiss *Soldier Erect* 20 That sort of American approach was even harder to master than the Ancient British protocol but, once mastered, it gave positive results.

7. a. (sense *2c) *protocol sheet*; (sense *5a) *protocol department, section*.

1912 E. M. Thompson *Introd. Gk. & Lat. Palaeogr.* 24 Among the Romans the protocol-sheet was inscribed with the name of the Comes largitionum..and with the date and the name of the place where it was made. **1958** L. Durrell *Mountolive* vi. 133 Then, turning, he completed his devoirs to the Protocol section. **1977** 'S. Leys' *Chinese Shadows* (1978) viii. 182 Masses can always be arranged by appointment: one should apply at the Protocol Department in the Ministry of Foreign Affairs.

b. *spec.* in *Philos.* (sense *2e), as *protocol language, proposition, sentence, speech, statement*.

1933 *Philosophy* VIII. 98 Carnap then considers the 'protocol language'. Scientific evidence is derived from protocol propositions that describe our perceptions, feelings, thoughts, etc. *Ibid.*, Finally, Carnap develops the thesis that protocol speech is part of physical speech. **1935** *Analysis* II. 59 The form of protocol statements cannot be found, but must be fixed by a convention. **1937** A. Smeaton tr. *Carnap's Logical Syntax of Lang.* v. 317 The statement of the protocol-sentences is the affair of the physicist who is observing and making protocols. **1965** P. Caws *Philos. of Sci.* xi. 73 A protocol sentence must be such that a decision as to its empirical truth or falsity can be reached after a finite number of observations.

‖**protocolaire** (prǫtokǫlę̄r), *a.* [Fr.] Characterized by a strict regard for protocol; formal, ceremonial.

1934 'A. Bridge' *Ginger Griffin* xviii. 232 But of course James wouldn't—he's too *protocolaire*. **1958** *Spectator* 24 Jan. 101/1 Less *protocolaire* than Trooping the Colour. **1962** T. Zinkin *Caste Today* 29 Occasionally there are festivals or private ceremonies... On those occasions, as in the most protocolaire of royal banquets, the place and part to be taken by each are known to all. **1975** N. Freeling *What are Bugles blowing For?* xxiii. 137 A list..of engagements..ranging from the unavoidable to the purely *protocolaire*. **1979** H. Wilson *Final Term* viii. 161 Duncan Sandys..said that even in pre-Amin days he had always insisted on the title 'Britain' despite protocolaire objections.

protocolar (prǭᵘ·tokǫlǎɹ), *a.* *rare.* [f. Protocol *sb.* + -ar[1]: cf. prec.] Of, pertaining to, or characterized by (a) protocol; formal, *protocolaire*.

1905 *Truth* 22 June 1589/1 To some extent it [sc. *The Mikado*] is protocolar, but one must not for that think it an empty form. **1960** *News Chron.* 12 Apr. 6/7 The Canadian Prime Minister, found himself in the peculiar protocolar position of sending a message of sympathy..at the same time as he was sending a vigorous protest.

protocolic, *a.* For *nonce-wd.* read *rare* and add later example.

1969 *Punch* 8 Jan. 60/3 It..cuts out all your time-consuming Ambassadorial summonses, stern notes..and general argy-bargy in protocolic triplicate.

pro·tocone. *Zool.* [f. Proto- + Cone *sb.*[1]] An inner cusp on the front corner of a mammalian upper molar tooth.

1888 H. F. Osborn in *Amer. Naturalist* XXII. 1072 (*table*) Proposed terms... Protocone... Protoconule... Protoconid. **1896** *Proc. Zool. Soc.* 573 In the deciduous 4th premolar [of *Centetes*] likewise the protocone develops first. **1922** [see *deuterocone* s.v. *deutero-]. **1933** A. S. Romer *Vertebr. Paleont.* xii. 248 A single cusp is found at the inner apex; this was originally believed to represent the original reptilian cone and hence is called the protocone. **1968, 1971** [see *metaconule]. **1976** *Nature* 5 Aug. 464/2 The upper first deciduous molar displays spatial dominance of the protocone.

protoconid (-kǭᵘ·nid). *Zool.* [f. prec. + *-id[5].] A cusp on a mammalian lower molar tooth corresponding to the protocone on an upper molar.

1888 [see prec.]. **1896** *Proc. Zool. Soc.* 568 In the lower molar [of *Gymnura*] the protoconid evidently develops fast. **1907** [see *deuteroconid* s.v. *deutero-]. **1933** [see *metaconid]. **1971** W. D. Turnbull in A. A. Dahlberg *Dental Morphol. & Evolution* ix. 168 (*caption*) The eocristid and its associated primary cusp, the protoconid or eoconid..are shown. **1976** *Nature* 5 Aug. 464/2 There is a spatially dominant protoconid with a large flat buccal face, a lingually facing anterior fovea, and an inferiorly projecting mesiobuccal enamel line.

pro·tocontinent. *Geol.* [f. Proto- + Continent *sb.*] = *supercontinent. Hence **pro:tocontine·ntal** *a.*

1958 L. King in *Continental Drift* (Univ. of Tasmania) 13 Recent studies..suggest once more the validity of continental drift, with two proto-continents Laurasia and Gondwana. **1968** *Sat. Rev.* (U.S.) 2 Mar. 48 (*caption*) The hegira of the Indian subcontinent about 60,000,000 years ago is suggested on the above map of earth's proto-continent. **1977** *Sci. Amer.* Mar. 104/2 As long as the fairly coherent supercontinent could move in relation to its adjacent ocean floor, the ocean floor could have been subducted, partially melted and chemically differentiated to manufacture typical continental crust at the leading edge of the protocontinent. **1978** *Nature* 10 Aug. 547/1 The protocontinental tectosphere produced in island-arc environments or along the active margins is probably thin, chemically heterogeneous and poorly consolidated.

protoconule (-kǭᵘ·niul). *Zool.* [f. *protocon(e + -ule.] An intermediate cusp between the protocone and the paracone of a mammalian upper molar tooth.

1888 [see *protocone]. **1905** [see *metaconule]. **1933** A. S. Romer *Vertebr. Paleont.* xii. 248 Between protocone and paracone there is often a smaller cusp, the protoconule. **1968** R. Zangerl tr. *Peyer's Compar. Odontogr.* 187 Intermediate cusps occurred along the trigon, a protoconule between protocone and paracone and a metaconule between protocone and metacone.

pro·tocorm. *Bot.* [ad. F. *protocorme* (M. Treub 1890, in *Ann. Jard. Bot. Buitenzorg* VIII. 30), f. Proto- + Corm[2].] A tuber-like body produced in the seedling stage of certain pteridophytes and orchids which grow in association with mycorrhiza. Also *attrib.*

1891 F. O. Bower in *Proc. R. Soc.* L. 267 The sporophyte [of *Phylloglossum*] consists of two parts:—(i) the protocorm, with its protophylls and roots, and (ii) the strobilus. **1905** I. B. Balfour tr. *Goebel's Organogr. Plants* II. 232 The chief mass of the seedling [of orchids] is formed of the 'protocorm'. **1938** G. M. Smith *Cryptogamic Bot.* II. vii. 180 A massive globose structure (the protocorm)..grows through the gametophyte. **1959** S. Shushan in C. L. Withner *Orchids* iii. 53 Continued enlargement of the embryo, which can henceforth be called the protocorm, results in either a smooth or irregular globular mass of cells. **1962** K. R. Sporne *Morphol. Pteridophytes* iv. 64 The protocorm might well be regarded as a derivative and retrograde development. **1967** *New Scientist* 14 Sept. 551/1 By subculturing every few weeks, protocorm formation may be kept going indefinitely. **1977** J. Arditti *Orchid Biol.* 211 Within 4–6 months, a mass of protocorms is formed.

proto-Cubist: see *proto- 2c. **proto-cultural, -culture:** *proto- 1.

pro:to-di·asystem. *Linguistics.* [f. Proto- + *diasystem* (1954), f. Gr. διά through + System.] A hypothetical reconstruction of the system of linguistic relationships in a proto-language (see also quot. 1969).

1964 *Word* XX. 376 If I use the term proto-language for a reconstruction and proto-dialect for the real speech, my implication is that the reconstruction is in fact a proto-diasystem. **1968** *Language* XLIV. 485 Proto-languages are to be understood as proto-dia-systems, with over-all sets of correspondences between related linguistic structures. **1969** S. P. Durham *Computer in Reconstruction of Proto-Diasystem: Franco-Provençal* I. i. 14 Since a diasystem is a combination into one single descriptive statement of the features of several dialects viewed horizontally, a proto-diasystem can be considered a combination into one single statement of the features of several dialects viewed vertically, that is, historically. **1972** *Computers & Humanities* VII. 6 A corpus of about 1000 words, chosen from eleven villages, provided a concordance to suggest a reconstruction of the Franco-Provençal proto-diasystem and a tentative phonology.

protodolomite to **-filament:** see *proto- 2a, b, c.

pro·to-form, *sb.* (and *a.*) *Linguistics.* [f. Proto- + Form *sb.* 5c.] A hypothetical form of a word or part of a word from which actual words have been derived. Also *attrib.* or as *adj.*

1964 *Language* XL. 145 We must use different symbols to represent the proto-forms. **1965** *Canad. Jrnl. Linguistics* X. 94 Reconstruction of proto-form phonological shapes. **1965** *Language* XLI. 305 There is no reason to assume stress on a first syllable of the proto-form. **1976** *Archivum Linguisticum* VII. 63 Both of these assumed proto-forms might appear surprising in view of the dearth of direct evidence in Latin texts and inscriptions.

pro·togalaxy. *Astr.* Also with hyphen. [f. Proto- + Galaxy *sb.*] A vast mass of gas, not yet formed into stars, postulated as a preliminary stage in the evolution of a galaxy.

1959 *Astrophysical Jrnl.* CXXX. 43 These nuclei had already originated before the protogalaxy condensed into stars. **1969** *Monthly Notices R. Astron. Soc.* CXLV. 417 The central part of the cloud begins to collapse, and the collapsing region gradually grows until it eventually includes almost the entire mass of the proto-galaxy. **1977** *Sci. Amer.* Oct. 43/3 Shapley's work laid the foundation for the present view that the globular clusters were formed 10 to 13 billion years ago during the gravitational collapse of the protogalaxy, a vast cloud of gas consisting of hydrogen and helium.

Hence **protogala·ctic** *a.*, of, pertaining to, or being a protogalaxy.

1969 *Monthly Notices R. Astron. Soc.* CXLV. 407 We assume for simplicity that the proto-galactic material consists of pure hydrogen. **1977** *Sci. Amer.* Oct. 44/2 As the protogalactic cloud contracted local regions of higher density became self-gravitating and condensed within a relatively short time into globular star clusters.

proto-Gallo-Romance, -Romanic: see *proto- 2 a.

protogenic (-dʒe·nik), *a.*[2] *Chem.* [f. Proto(n in Dict. and Suppl. + *-genic.] Of a solvent (or solute): having a tendency to protonate most solutes (or solvents). Opp. *protophilic *a.*

1931 N. F. Hall in *Chem. Rev.* VIII. 194 The protogenic and protophilic character of very weak acids and bases is largely masked by the overwhelming prominence of the similar properties of water. **1940** Glasstone *Textbk. Physical Chem.* xii. 958 For convenience solvents are divided roughly into three categories according as the molecules are (*a*) proton acceptors, i.e., basic, or protophilic, (*b*) proton donors, i.e., acidic, or protogenic, or (*c*) neither donors nor acceptors, i.e., aprotic. **1969** R. G. Bates in Coetzee & Ritchie *Solute-Solvent Interactions* ii. 51 The four classes of solvents, 'amphiprotic', 'protogenic', 'protophilic', and 'aprotic', are not clearly restrictive and the proper classification of many solvents is in doubt. **1973** E. J. King in Covington & Dickinson *Physical Chem. Org. Solvent Syst.* iii. 333 Formic acid is an obvious example of an acidic or protogenic solvent, but it has weak basic properties too.

pro:to-geome·tric, *a.* Also **proto-Geometric** and as one word. [f. Proto- + *geometric *a.* d.] Designating the period preceding the Geometric Age in Greece, or the pottery attributed to this period, contemporaneous with the collapse of Mycenæan civilization on the mainland and the period of cultural decline that followed it (*c* 1100–*c* 900 B.C.).

1926 *Cambr. Anc. Hist.* IV. xvi. 580 Between the flourishing of the Creto-Mycenaean civilization, and the geometric period proper, there lies a long period which has been named, not very happily, the proto-geometric: a period of cultural decay. **1933** *Jrnl. Hellenic Stud.* LIII. 162 Cremation first appears in connexion with the proto-Geometric style in pottery. **1939** J. D. S. Pendlebury *Archaeol. Crete* iv. 243 This type of tomb, square with a circular vault above, continues in Lasithi and the

neighbourhood into Proto-geometric times. **1950** H. L. Lorimer *Homer & Monuments* i. 42 Simple as the proto-Geometric culture is, it marks a period, not of decline, but of renascence. **1953** *Antiquity* XXVII. 75 The Protogeometric iron sword of about 30 inches long, when new, was quite as effective for cutting as for thrusting. **1961** *Oxf. Mag.* 16 Feb. 232/1 The earliest Proto-geometric wares found at Smyrna can be dated about 1000 B.C. **1973** J. Boardman *Greeks Overseas* (ed. 2) i. 3 The finds in Athens cemeteries show that after a very short while, probably by about 1050 B.C., the new 'Protogeometric' style of vase-painting had been evolved from the debased Mycenæan forms.

proto-Germanic to **-Greek:** see *proto- 1, 2a.

protogynous, *a.* Add: [ad. G. *protogynisch* (F. H. G. Hildebrand *Die Geschlechter-Vertheilung bei den Pflanzen* (1867) 17).] (Earlier and later examples.)

1870 A. W. Bennett in *Jrnl. Bot.* VIII. 320 *Chlora perfoliata*..is protogynous. **1909** J. R. A. Davis tr. *Knuth's Handbk. Flower Pollination* III. 221 The flowers [of *Bartsia alpina*] in Greenland were found by Warming to be feebly protogynous, the anthers dehiscing soon after the maturation of the stigma. **1973** *Nature* 23 Mar. 275/2 Some cross-pollination may occur because the flowers are protogynous and have some insect visitors.

2. *Zool.* = Proterogynous *a.* 2.

1931 F. A. E. Crew in W. Rose *Outl. Mod. Knowl.* 285 In quite a number of forms protandrous..or protogynous..hermaphroditism is the rule, it being usual for an individual first to function as a female or male, and then, later, as a male or female. **1973** *Nature* 12 Oct. 333/1 This system is an illustration of the phenomenon of social determination of phenotype, which includes social control of sex reversal in protogynous fishes.

protogyny. (Earlier and later examples.)

1870 A. W. Bennett in *Jrnl. Bot.* VIII. 315 The terms Protandry and Protogyny used by Hildebrand to express, in the one case the development of the stamens before the pistils, in the other case the development of the pistil before the stamens, are so convenient and expressive that I have adopted them in this paper. **1940** *Nature* 30 Mar. 485/2 Protandry, protogyny, special floral arrangements and other devices could also be listed. **1972** *Science* 12 May 601/2 In these plants, dioecism, protandry, protogyny..are common.

protohæm, -heme (prǭᵘ·tohīm). *Biochem.* [a. G. *protohäm* (H. Fischer et al. 1931, in *Zeitschr. f. physiol. Chem.* CXCV. 21): see Proto- and *hæm, heme.] A ferrous chelate derivative of a protoporphyrin; *spec.* = *hæm, heme a.

1931 *Chem. Abstr.* XXV. 2157 By using a technic which excludes the possibility of oxidation, etioheme, protoheme, mesoheme and meso-ester heme were obtained cryst. from the corresponding porphyrins. **1963** R. P. Dales *Annelids* iv. 88 In haemoglobin the haem is the red protohaem, in chlorocruorin it is the green chlorocruorohaem. **1963** C. H. Doering tr. *Karlson's Introd. Mod. Biochem.* ix. 183 Peroxidases oxidize substrates by employing H_2O_2 as an oxidizing agent. Depending on their origin, they may contain either a red heme (protoheme or closely related ones) or a green heme. **1970** R. W. McGilvery *Biochem.* xxi. 499 The normal product of porphyrin synthesis is protoheme IX, which is quickly bound by globin peptides and protected from oxidation.

proto-Hattic to **-Jupiter:** see *proto- 1, 2a, b.

pro·to-language. *Linguistics.* [f. Proto- + Language *sb.*] A hypothetical parent language from which actual languages or dialects have been derived.

1948 W. F. Twaddell in *Language* XXIV. 139 Our reconstruction of a proto-language is theoretical and partial. **1950** H. M. Hoenigswald in *Ibid.* XXVI. 357 We may refer to Meillet's rule that in reconstructing the vocabulary of a proto-language we need the testimony of three, rather than two, independent witnesses. **1955** *Orbis* IV. 428 Many distinguished linguists have expressly declared that the assumption of a uniform proto-language conflicts with all we know about languages actually observed. **1964** *Word* XX. 376 The reconstructed proto-language is a formula, a statement on relationships, albeit in a diachronic rather than synchronic direction, and does not therefore represent, at least not necessarily, a genuine speech form that ever existed. **1977** R. Williams *Marxism & Lit.* I. ii. 25 In one area this movement was 'evolutionary' in a particular sense: in its postulate of a proto-language (proto-Indo-European) from which the major 'family' had developed.

protolife: see *proto- 2b.

protolingui·stic, *a.* [f. Proto- + Linguistic *a.*] Descriptive of communication or signs which are understood without the use of verbal language (see quot. 1964); of communication, etc., from which language is presumed to have developed. Also, relating to the study of proto-language. Hence **protolingui·stics** *sb. pl.*

1964 *Discovery* Oct. 32/3 Those [*sc.* symbols]..which are commonly translated into or from ordinary language; and those which are not so translated—like clothes, architectural design and furniture... We are concerned here with symbols of this latter kind and with behaviour to which a meaning is attached when it occurs in a specific setting. In both these kinds of strictly non-verbal communication we are dealing with messages which are not para-linguistic, but proto-linguistic. **1975** M. Critchley in E. H. & E. Lenneberg *Foundations Lang. Devel.* I. i. 6 One point must..be kept well to the fore when ruminating upon this topic of protolinguistics. From an anatomic, physiological angle, speech is a parasitic function. **1976** R. W. Wescott in *Ann. N.Y. Acad. Sci.* CCLXXX. 104 As used here, the term 'protolinguistics' will mean the analytic and comparative study of protolanguages. *Ibid.* 115 Some protolinguistic problems, finally, are due almost exclusively to lack of information. **1978** *Sci. Amer.* Apr. 104/3 If, as I suppose, the hominids under observation communicated only as chimpanzees do or perhaps by means of very rudimentary protolinguistic signals, then the observer might feel he was witnessing the activities of some kind of fascinating bi-pedal ape.

proto-literate, -lith: see *proto- 1, 2b.

protoli·thic, *a.* *Archæol.* [f. Proto- + Gr. λίθος stone, after *neolithic*, etc.] **1.** A term introduced by the American ethnologist W. J. McGee to designate a type of primitive stone implements formerly in use amongst the Seri Indians of eastern Mexico (see quots.).

1897 *Amer. Anthropologist* X. 326 In this stage of development (called *protolithic* after McGee) stone implements come into more or less extended use in connection with implements of shell, tooth, etc. **1898** W. J. McGee in *17th Ann. Rep. Bur. Amer. Ethnol. 1895–6* I. 295 None other so well represents protolithic culture.

2. Belonging to the earliest stone age: eolithic.

1931 *Antiquity* V. 518 In the new terminology three major divisions are recognised, the old Lower and Middle Palaeolithic being grouped together as 'Protolithic', the Upper Palaeolithic and Mesolithic as 'Miolithic', the 'Neolithic' continuing as usual though now including the old Aeneolithic. **1962** A. D. Krieger in Jennings & Norbeck *Prehist. Man in New World* 29 Menghin..employs six culture stages applied to both Old and New World, namely, (1) Protolithic (2) Epiprotolithic, (3) Miolithic, (4) Epimiolithic, (5) Neolithic, and (6) Chalcolithic. The first two are more or less equivalent to Lower Paleolithic as used by Old World archeologists... Menghin is careful to avoid any suggestion that such stages in the New World are of an age equal to those of the Old World; even his Protolithic in America may be no more than twenty thousand years old.

protolithionite: see *proto- 2b.

protologue (prǭᵘ·tolǫg). *Taxonomy.* Also **protolog.** [f. Proto- + -logue.] The description and other details accompanying the first publication of the taxonomic name of a plant or animal.

1905 Schuchert & Buckman in *Science* 9 June 900/1 For the sake of accuracy we suggest that the original description by words (type-description) be called the protolog. **1939** *Jrnl. Bot.* LXXVII. 206 A 'protologue' is the printed matter (description etc.) accompanying the original publication of a name or epithet. **1957** W. T. Stearn *Linnaeus's Species Plantarum* xiii. 126 For each Linnaean species all the constituent elements in Linnaeus's protologue must be taken into consideration. **1970** *Watsonia* VIII. 43 The entire protologue reads as follows.

protology. Restrict †*Obs. rare*⁻⁰ to sense in Dict. and add: **2.** (protǫ·lǒdʒi). The study of or enquiry into origins. Hence **protolo·gical** *a.*, that pertains to what is original or primitive.

1903 *21st Ann. Rep. Bur. Amer. Ethnol. 1899–1900* 138 In the quaint protology, or science of first things, of the Iroquois things are derived from things through transformation and evolution. **1937** *Proc. Prehist. Soc.* III. 188 When man began to control his environment to a greater extent, 'things' became for him imbued with 'force', and in this way arose the magical and protological conception of the world. **1974** *Archiv für Rechts- und Sozialphilosophie* LX. 385 We have found that the learning of basic logical principles and methods via protological calculus has also a didactic advantage. **1977** M. Goulder in J. Hick *Myth of God Incarnate* iv. 75 In place of the primitive eschatology, the stress would now fall on protology.

protolytic (-li·tik), *a.* *Chem.* [f. Proto(n in Dict. and Suppl. + *-lytic.] Applied to a reaction or process in solution which consists in the transfer of a proton from one molecule to another, one of the molecules usu. being of the solvent; also applied to the solvent itself. Also **proto·lysis**, a protolytic reaction; proton transfer in solution.

1931 N. F. Hall in *Chem. Rev.* VIII. 191 Acid-base or protolytic reaction may be defined by the general equation..A⇌B + ⊕ (acid⇌base + proton). **1934** *Chem. Abstr.* XXVIII. 5740 Acids and bases are designated [by J. N. Brønsted] 'protolyte' and their reaction 'protolysis'. **1959** I. M. Kolthoff in Kolthoff & Elving *Treat. Analytical Chem.* I. I. xi. 411 It is clear that a base can be of any charge type, its charge being one less positive than that of its conjugate acid, for example, $Al(OH)_2^+$, NH_3, HCO_3^-, CO_3^{-2}. Its reaction with an acid solvent is called a protolysis reaction and no distinction is made between a dissociation (NH_3 in water) and hydrolysis (cyanide in water). **1968** V. Gutmann *Coordination Chem. in Non-Aqueous Solutions* i. 6 Acids and bases are defined as proton donors and proton acceptors respectively and acidbase reactions are regarded as being due to proton transfer reactions (protolysis). **1969** T. R. Blackburn *Equilibrium* ii. 66 If a solvent is both appreciably acid and appreciably basic (as water is), it will have a measurable autoprotolysis constant. Such solvents are called protolytic.

‖**protoma** (prǫ·tǒma). Pl. **-æ, -as.** [mod.L., ad. Gr. προτομή Protome.] = Protome in Dict. and Suppl.

1931 *Antiquity* V. 330 Fragments of horns and ears of the stone protomas of the bulls forming the capitals. **1953** *Proc. Prehist. Soc.* XIX. 45 From Ghar Dalam there are two protomae of animals, in the same ware, and having a similar decoration to the pottery described above.

proto-Malay, -Marxian: see *proto- 2a, c.

protome. Delete *rare* and add: Now *spec.* the forepart of an animal represented decoratively, as in (ancient) sculpture.

1886 A. J. Evans in *Jrnl. Hellenic Stud.* VII. 14 The horse's head, or protomé, as is well known, is introduced generally in a sunken square. **1933** C. Seltman *Greek Coins* v. 60 The device of the foreparts of two lions facing one another. The son..replaced one of the half-lions by the forepart of a bull, and at first the two *protomes* were back to back. **1942** *Antiquity* XVI. 356 But there is one motive, the bull's protome, that does seem really distinctive of the Halaf style. **1972** *Scripta Hierosolymitana* XXIV. 41 A Greek artist conceived the strange idea of.. combining two traditional elements..by uniting a lion *protome* with a swinging handle. **1978** W. A. P. Childs *City-Reliefs of Lycia* 13 The lintel was decorated with four winged bull-protomes.

proto-Medic: see *proto- 2a.

protomer (prǭᵘ·tǒməɹ). [ult. f. Gr. πρῶτος first + *-mer.] **1.** *Chem.* [f. Proto(n or *proto(tropy.] Any prototropic tautomer.

1923 [see *prototropy]. **1968** *Jrnl. Amer. Chem. Soc.* XC. 1575/2 Analysis of the protomer stabilities in terms of relative chemical binding energies is risky. **1979** C. Roussel et al. in J. V. Metzger *Thiazole & its Derivatives* II. vii. 379 Protic solvents stabilize more the form giving the stronger hydrogen bond: so the mercapto protomer.. should become apparent in the ultraviolet spectra recorded in ethanol.

2. *Biochem.* [f. Prot(ein.] Any of the protein subunits of which an oligomeric protein is built up.

1965 J. Monod et al. in *Jrnl. Molecular Biol.* XII. 89 The identical subunits associated within an oligomeric protein are designated as protomers. *Ibid.* 106 Within oligomeric proteins the protomers are in general linked by a multiplicity of non-covalent bonds, conferring both specificity and stability on the association. **1970** *Nature* 28 Nov. 828/1 By definition a protomer constitutes one primitive cell. **1980** *Biophysical Chem.* XI. 49/1 The treatment has been extended to include aggregated structures containing from two to five protomers, thus encompassing almost the whole spectrum of oligomeric proteins likely to be encountered.

Hence **protome·ric** *a.*

1923 *Jrnl. Chem. Soc.* CXXIII. 828 The two forms of the ion are tautomeric..; but they yield protomeric hydrides, and isomeric derivatives with radicles such as methyl and ethyl. **1968** *Jrnl. Amer. Chem. Soc.* XC. 1575/1 In the cases of the heteroaromatic isomers, the protomeric equilibrium has been..found to favor the amide in solution. **1974** *Nature* 24 May 316/1 The accumulation of data on the protomeric units of several synthetases. **1979** R. Barone et al. in J. V. Metzger *Thiazole & its Derivatives* II. vi. 20 When the electronegativity of the substituent borne by the amino group increases, the protomeric equilibrium is expected to be shifted towards the imino structure.

protomerite: see *proto- 2b. **protomusic:** *proto- 1.

proton. Add: **2.** *Physics.* A stable sub-atomic particle which has a positive charge numerically equal to that of the electron, forms a part (or in the commonest isotope of hydrogen the whole) of all atomic nuclei, and is a baryon with a mass of 938·3 MeV (1836 times that of the electron), spin of ½, and isospin of ½; it is now usu. regarded as a particular state of a nucleon. [Perh. suggested by, if not derived from, the name of William *Prout* (1785–1850), English chemist and physician, who suggested that hydrogen was a constituent of all the elements.]

1920 *Engineering* 17 Sept. 382/3 Sir Ernest [Rutherford], replying, said that..a clear nomenclature was certainly wanted; the term 'prouton' for [*read* or] 'proton' might be suitable for the H nucleus. **1922** J. Mills *Within Atom* ii. 13 The hydrogen atom is composed of only one proton and one electron. **1926** *Nature* 9 Oct. 526/2 Recourse was had to passing fairly large amounts of hydrogen—up to one litre—through heated palladium, in the hope that at the moment of exit a fraction of the protons and electrons would combine to form the helium nucleus. **1942** J. D. Stranathan *'Particles' Mod. Physics* xi. 416 By firing alpha particles through a cloud chamber filled with nitrogen, it has been possible to photograph the track of the proton leaving the nitrogen nucleus. **1955** *Sci. News Let.* 10 Sept. 170/1 The number of protons in the nucleus determines the kind of matter the atom forms, whether hydrogen, uranium, iron, carbon or any other. **1962** H. D. Bush *Atomic & Nucl. Physics* iii. 62 Isotopes will occur

where atomic nuclei have the same number of protons but differing numbers of neutrons. **1968, 1972** [see *neutron]. **1979** *Nature* 7 June 483/2 A novel prediction of many such grand unified models is that baryon number is not absolutely conserved; so the proton may actually decay!

3. Special Comb. (all for sense *2): **proton accelerator**, a particle accelerator designed to accelerate protons; **proton acceptor, donor**, a substance or species which is able to accept protons from, or give them up to, a substrate; also *proton donator*; so *proton-accepting, -donating* ppl. adjs.; **proton gradiometer**, a rod with a proton magnetometer at each end, which may be stood on the ground in a vertical position to measure local variations in field strength at ground level due to features just below the surface; **proton magnetometer**, a magnetometer in which the magnetic field strength is determined from the frequency of the voltage induced by the precession of protons in hydrogen atoms (e.g. in water) following the removal of a stronger magnetic field applied to orient the protons; **proton-precession magnetometer** = prec.; **proton synchrotron**, a synchrotron designed to accelerate protons.

1947 Proton accelerator [see *bevatron]. **1961** Livingston & Blewett *Particle Accelerators* xiii. 438 The synchrocyclotron, which has been so successful as a proton accelerator in the 100- to 700-Mev range, requires a solid-core magnet. **1969** *Times* 5 Feb. 13/6 Four new particles of matter have been discovered in two independent experiments using the proton accelerators at Brookhaven..and at the European Nuclear Research Centre. **1925** *Jrnl. Chem. Soc.* CXXVII. I. 1383 All prototropic changes appear to involve..(i) the removal of a proton from one part of a molecule to some outside basic or proton-accepting component of the system, and (ii) the addition to another part of the molecule of a proton..from some acidic or proton-donating component of the system. **1966** Gucker & Seifert *Physical Chem.* xix. 537 Many a neutral molecule is amphiprotic because it has an ionizable proton and a proton-accepting oxygen or nitrogen atom. **1925** *Jrnl. Chem. Soc.* CXXVII. I. 1378 We..regard the acid as a proton donator and the base as a proton acceptor. **1940** Glasstone *Text-bk. Physical Chem.* xii. 958 It is unlikely that free protons exist to any extent in solution, and so the acidic or basic functions of any species cannot become manifest unless the solvent molecules are themselves able to act as proton acceptors or donors, respectively: that is to say, the medium itself must have basic or acidic properties. **1953** L. C. Jackson tr. *Ketelaar's Chem. Constitution* 85 A base, according to Brönsted, is a proton-acceptor just as an acid is a proton-donor. **1973** A. W. Adamson *Textbk. Physical Chem.* xii. 544 NH_3 is a much better proton acceptor than is H_2O. **1925** Proton-donating, donator [see *proton-accepting, acceptor* above]. **1940, 1953** Proton donor [see *proton acceptor* above]. **1977** Helmprecht & Friedman *Basic Chem.* vii. 149, H_2SO_4 is a more effective proton donor and a stronger acid than HSO_4^-. **1960** M. J. Aitken in *Archaeometry* III. 38 Previous articles..have described the use of the proton *magnetometer* for archaeological prospecting. The proton *gradiometer* is a development which not only has advantages in operation, but is also less complex and therefore cheaper to construct. *Ibid.* 39 With the two-bottle system of the proton gradiometer all these large-scale effects are avoided for they affect both bottles equally. **1970** *Oxf. Univ. Gaz.* C. Suppl. vi. 25 Comparative surveys using proton gradiometer, pulsed magnetic induction, and soil conductivity meter. **1959** Proton magnetometer [see *eye *sb.*[1] 3 g]. **1961** *Antiquaries' Jrnl.* XLI. 44 In an attempt to locate these burials..a survey with proton magnetometer followed by selective excavation was organized. **1975** *New Yorker* 12 May 58/3 With a proton magnetometer, the geologists began aeromagnetic surveys to develop a general view of the region's geological structure. **1976** W. M. Telford et al. *Appl. Geophysics* iii. 136 The important advantages of the proton magnetometer are that it measures absolute field strength and that its sensitivity (~1γ) is higher than any of the instruments considered so far. The fact that it requires no orientation or levelling makes it very attractive for marine and even more for airborne operations. **1958** *Jrnl. Geophysical Res.* LX. 880 The proton-precession magnetometer has been successfully adapted..for measuring the horizontal and vertical components of the earth's magnetic field. **1971** I. G. Gass et al. *Understanding Earth* xvi. 236/1 Fluxgate and proton-precession magnetometers were developed to measure magnetic fields to one part in 10⁵. **1947** *Proc. Physical Soc.* LIX. 677 (*heading*) Theory of the proton synchrotron. **1966** *Daily Tel.* 25 Nov. 26/3 Nimrod, a proton-synchrotron, is essentially a circular racetrack, round which bursts of sub-atomic particles are repeatedly accelerated until they approach the speed of light. **1975** *McGraw-Hill Yearbk. Sci. & Technol.* 208/2 The collision of protons of energy 28 GeV from the CERN proton synchrotron with a metal target yielded..charged pions and kaons.

protonate (prǭᵘ·tǫnēit), *v.* *Chem.* [f. Proton in Dict. and Suppl. + -ate[3].] **a.** *trans.* To transfer a proton to (a molecule, group, atom, etc.), a co-ordinate bond being formed to the proton. **b.** *intr.* To receive a proton in this way. So **pro·tonated** *ppl. a.*, having received a proton; bonded to an additional proton; **pro·tonating** *vbl. sb.*

As used by Pitzer (quot. 1945), *protonated* had a slightly different, more restricted sense, being applied to the two bonds (each with a bridging hydrogen atom) which link the boron atoms in the electron-deficient compound diborane, B_2H_6.

1945 K. S. Pitzer in *Jrnl. Amer. Chem. Soc.* LXVII. 1127/1 Let us call it a proton containing double bond or, shorter, a protonated double bond. **1951** *Jrnl. Amer. Chem. Soc.* LXXIII. 3647/2 With the carbinols..the fact that one *p*-amino grouping was never protonated is consistent with the view that only one of the rings is involved in resonance interactions. **1958** *Ibid.* LXXX. 3715/1 Compounds such as tri-*p*-aminophenylmethanol..when dissolved in sulfuric acid form stable carbonium ions in spite of the strong electron-withdrawing groups present in the form of protonated amino substituents. **1966** Smith & Cristol *Org. Chem.* xiii. 795 Pyrrole..does protonate in the presence of very strong acids with the destruction of the aromatic sextet. **1969** T. R. Blackburn *Equilibrium* ii. 66 Any acid stronger than the hydronium ion will simply protonate water molecules to produce an equal quantity of H_3O^+. **1972** Cotton & Wilkinson *Adv. Inorg. Chem.* (ed. 3) v. 176 Even protonated carbonic acid, or more properly, the trihydroxycarbonium ion, $C(OH)_3^+$..has been observed in solutions of carbonates or bicarbonates in FSO_3H–SbF_5–SO_2 solutions at −78°. *Ibid.* 182 Fluorosulfuric is one of the strongest of pure liquid acids. It is commonly used in presence of SbF_5 as a protonating system. **1975** *Nature* 30 Oct. 823/2 Calculations..show that this spectral shift can be associated with the formation of the protonated $C=N^+$ bond.

protonation (prǭᵘtǫnē̆i·ʃən). *Chem.* [f. prec. + -ion.] The action or result of protonating.

1948 *Jrnl. Biol. Chem.* CLXXV. 249 The effectiveness of the positively charged ammonium group in preventing appreciable protonation of the carboxyl group of the L-leucine cation in 100 per cent sulfuric acid can be appreciated when it is remembered that both acetic acid and monochloracetic acid are completely ionized in this solvent. **1958** *Jrnl. Amer. Chem. Soc.* LXXX. 3715/1 If one of the benzene rings in diphenyl- or triphenylmethanol is replaced by a pyridine ring, protonation of the nitrogen atom would be expected to decrease the stability of the carbonium ion produced in sulfuric acid. **1970** *Nature* 25 July 370/1 The double helix denatures abruptly at both low and high *p*H because of protonation of amine groups near *p*H 3 and ionization of hydroxyl groups near *p*H 11·5. **1976** *Sci. Amer.* June 43/3 The release and uptake of protons by intact cells was clearly different from the change in protonation observed during the photoreaction of isolated purple membrane.

†**protone** (prǭᵘ·tǭᵘn). *Biochem. Obs.* [ad. G. *proton* (A. Kossel 1898, in *Zeitschr. f. physiol. Chem.* XXV. 174), f. *pro(tamin* Protamine + *pep)ton* Peptone.] Any of various peptone-like substances produced as the primary products of hydrolysis of protamines.

1898 [see *hexone 1]. **1916** A. P. Mathews *Physiol. Chem.* iv. 136 It has been suggested by Taylor that the protamine, salmin, may be made up of these tri-peptides, or protones, united as follows. **1928** A. T. Cameron *Textbk. Biochem.* viii. 115 According to Kossel protamines are hydrolysed first into protones which are compounds containing two radicals of arginine..united with one of alanine, or serine, or proline, or valine.

protonemal, *a.* (examples), **protonematal** *a.* (further examples).

1875 Bennett & Dyer tr. *Sachs's Text-bk. Bot.* II. 318 These latter [leaves] then put forth protonemal filaments, which produce first of all a flat pro-embryo; and upon this finally new leaf-buds arise. **1938** G. M. Smith *Cryptogamic Bot.* I. ii. 135 The protonematal initial develops into a green filament (the primary protonema), also differentiated into nodes and internodes. **1958** *Nature* 19 Apr. 1139/2 In the course of studies on protonematal regeneration and growth in the moss *Splachnum ampullaceum* (L.) Hedw., the effect of gibberellic acid on the growth of the protonemata was also tested. **1969** F. E. Round *Introd. Lower Plants* viii. 101 The protonemal stage is..generally ignored in the description of liverworts.

proto-Neolithic to **-nephridium:** see *proto- 2b.

protonic (protǫ·nik), *a.*[2] [f. Proton in Dict. and Suppl. + -ic.] **a.** Of, pertaining to, or characteristic of, a proton or protons.

1929 A. N. Whitehead *Process & Reality* 127 Each proton is a society of protonic occasions. **1932** E. Mollov *Compl. Wireless* 113/1 If..an atom happens to lose one or more of its planetary electrons..some of the protons become unbalanced, and there is a surplus of protonic force. **1953** R. B. Braithwaite *Scientific Explanation* 93 Hydrogen atoms behave..as if they were solar systems each with an electronic planet revolving round a protonic sun. **1961** G. R. Choppin *Exper. Nuclear Chem.* i. 1 The existence of nuclei means, therefore, that there must also be another force present which is strong enough to counterbalance the protonic repulsion and hold the nucleons together. **1976** *Nature* 23 Sept. 298/2 It is important at each step to vary the *p*H of the aqueous buffered component to keep the protonic activity of the solvent constant.

b. *Chem.* Of an acid, solvent, etc.: possessing a proton which can be used in protonation. Of a hydrogen atom in a molecule: available for use in protonation; possessing some positive charge.

1951 *Jrnl. Polymer Sci.* VI. 518 If we use this value for both the methyl and butyl salts in the protonic solvents, we obtain the *B* values given in..Table VIII. **1953** Audrieth & Kleinberg *Non-Aqueous Solvents* ii. 28 Why should not any protonic solvent, capable of undergoing limited self-ionization into hydrogen ion and some base-analog ion, serve as a parent substance of a system of acids, bases, and salts? **1966** Phillips & Williams *Inorg. Chem.* II. xxxiii. 563 Organometallic compounds undergo typical reactions with compounds which contain protonic hydrogen, i.e. which act as acids towards them... Of these reactions, hydrolysis is one of the most important. **1968** [see *hydridic *a.*]. **1969** T. C. Waddington *Non-Aqueous Solvents* i. 8 In terms of Brønsted-Lowry or protonic acids, the strongest acid in a solvent is the protonated form of that solvent and the strongest base the deprotonated form.

Hence **proto·nically** *adv.*

1979 *Science* 7 Dec. 1157/2 The semifluid bimolecular lipid membrane and the plug-through complexes form a condensed, continuous nonaqueous (protonically insulating) sheet that acts as the osmotic barrier and separates the aqueous proton conductors on either side.

proto-Nile: see *proto- 2a.

protonmotive (prǭᵘ:tǫnmǭᵘ·tiv), *a.* *Physics* and *Biochem.* Also **proton motive.** [f. Proton in Dict. and Suppl. + Motive *a.*] Of, pertaining to, or characterized by the movement of protons in response to an electric potential gradient; *protonmotive force*: a force analogous to the electromotive force, which acts on the proton gradient across cell membranes and comprises the sum of the electric potential difference and the pH gradient across the membrane.

1966 P. Mitchell in *Biol. Rev.* XLI. 494 The operation of the proton-translocating ATPase and o/r chain systems in an ion-tight membrane would create both a pH differential and a membrane potential, conveniently described together as a protonmotive force..by analogy with electromotive force. **1977** Hall & Baker *Cell Membranes & Ion Transport* iv. 79 The energy derived from the proton gradient (the proton motive force) may be used to drive oxidative phosphorylation. **1978** *Nature* 7 Sept. 14/3 Bonner suggested that a modified protonmotive ubiquinone cycle satisfies the kinetics of cytochrome interactions. **1979** *Science* 7 Dec. 1153/3 The first protonmotive device conceived by man was the electromotive hydrogen-burning fuel cell, invented by William Grove in 1839. **1980** *Federation Proc.* XXXIX. 1706/1 According to Mitchell's chemiosmotic hypothesis, a protonmotive force..energizes the synthesis of ATP by a proton-translocating ATPase.

protonosphere (protǫ·nǒsfīəɹ). [f. as prec. + -o + Sphere *sb.*] (See quot. 1960.)

1960 F. S. Johnson in *Jrnl. Geophysical Res.* LXV. 578/1 The name protonosphere is adopted here to describe the ionized medium above about 1800 km where protons are the principal ionized constituent, the name ionosphere being reserved for the lower region consisting of heavier atmospheric ions, such as atomic oxygen. **1973** *Q. Jrnl. R. Astr. Soc.* XIV. 197 Estimates of the flux of ionization into and out of the protonosphere. **1975** *Physics Bull.* July 338/3 (Advt.), There are vacancies for three Research Students, to commence in October 1975 and work for the degree of PhD... 2. Ionosphere and protonosphere by satellite radio transmissions.

Hence **pro:tonosphe·ric** *a.*

1971 *Radio Science* VI. 849/1 The observed fluxes must be lower limits to the total protonospheric fluxes, since H^+ ions may also be differing from the protonosphere and contributing to the *F*-region via charge exchange below 800 km. **1978** *Nature* 2 Feb. 428/1 Temporal changes of protonospheric content are indicative of the filling and draining of this region.

protonotary, prothonotary. Add: **5.** *Comb.* **prothonotary warbler**, an American warbler, *Protonotarius citrea*, of the family Parulidæ, distinguished by a deep yellow head and breast, green back, and blue-grey wings.

1783 J. Latham *Gen. Synopsis Birds* II. II. 494 Prothonotary W[arbler]... This inhabits Louisiana, where it has obtained the name of *Protonotaire*. **1811** A. Wilson *Amer. Ornithol.* III. 72 Prothonotary Warbler... They are abundant in the Mississippi and New Orleans territories, near the river. **1874** E. Coues *Birds of Northwest* 47 Prothonotary Warbler... This species..only reaches the lowermost Missouri. **1977** *Daily Tel.* 24 Jan. 10/4 Bird-watchers..already know what a prothonotary warbler is.

protonovelist, -ocean: see *proto- 1, 2a.

protopathic, *a.* Add: **2.** *Neurology.* In the theory that there are two (or three) sets of nerves and sensory receptors supplying the skin, the epithet of the coarser and more primitive sensibility (involving pain and temperature) and of the parts of the nervous system on which it is based.

1905 H. Head et al. in *Brain* XXVIII. 106 The position of the point stimulated cannot be recognised and each stimulus causes a widespread radiating sensation... To this form of sensibility we propose to give the name 'protopathic'. **1912** J. G. McKendrick *Princ. Physiol.* xiii. 224 If a sensory nerve to an area of skin is divided, sensibility may return if the ends unite. The sensations that return first have been termed protopathic, and depend on heat, cold, and pain spots. **1920** [see *epicritic *a.*]. **1942** *Brain* LXV. 110 Head's protopathic and epicritic fibres.. come to grief when they make contact with the hard facts of anatomy. **1951** W. L. Jenkins in S. S. Stevens *Handbk. Exper. Psychol.* xxx. 1172/1 Also abandoned for the lack of experimental confirmation are Head's proposal of a dual system of four epicritic and four protopathic skin senses [etc.]. **1958** *Ann. N.Y. Acad. Sci.* LXXIV. 30 Return of protopathic sensibility commences about six weeks following section of the sensory nerves and is completed within six months. **1974** M. & D. Web-

STER *Compar. Vertebr. Morphol.* xii. 278 The often unmyelinated protopathic fibres..arise from small cell bodies in the dorsal root ganglia. **1977** *Lancet* 11 June 1271/2 Head, in his model of the protopathic and epicritic nervous system,..introduced the notion of processing of sensory input at the level of entry into the central nervous system.

protope·ctin. *Biochem.* [ad. G. *protopektin* (A. Tschirch-Bern 1907, in *Ber. d. Deut. Pharm. Ges.* XVII. 242): see PROTO- and PECTIN.] = PECTOSE.
1908 *Chem. Abstr.* II. 431 (*heading*) On pectin and protopectin. **1922** *Biochemical Jrnl.* XVI. 704 The soluble pectin probably develops from an insoluble pectic substance contained in the cell wall.. This insoluble pectin corresponds to the protopectin of Fellenberg, and to the pectose of earlier investigators. **1951** Z. I. Kertesz *Pectic Substances* iii. 55 Protopectin is now believed by some to represent very large and therefore water-insoluble) pectinic acid molecules. **1962** S. M. Siegel *Plant Cell Wall* i. 16 The constitutional differences between protopectin and the other pectic substances remain unclear. Empirically, the protopectins are distinguished by their insolubility, and, in general, by a higher molecular weight. **1973** J. J. Doesburg in L. P. Miller *Phytochemistry* I. x. 272 In most plant tissues, the pectic substances are present in the form of water-insoluble protopectin. Ripe fruits are the main exceptions having a part present in soluble form, which is formed from protopectin during the ripening process. **1976** Bell & Coombe tr. *Strasburger's Textbk. Bot.* (ed. 30) 64 Protopectin can..be regarded as a cementing substance holding the cells of a tissue together.

Hence **protope·ctinase** [*-ASE] (see quots.).
1937 *Nat. Gaz.* LXXXIII. 331 Protopectinase,—The term applied to the enzyme which hydrolyses or dissolves protopectin, with the resultant separation of the plant cells from each other, usually spoken of as maceration. *Ibid.* 338 Various strains of potatoes, of a wide range of mealiness, were used as test tissue for protopectinase activity, with the idea that the most mealy might have middle lamellae most easily hydrolyzed by the enzyme. **1951** Z. I. Kertesz *Pectic Substances* xiv. 335 Until a few years ago it was believed that protopectinase is distinct from the enzyme which hydrolyzes the 1,4 glycosidic linkages in pectinic acids... However, there is now a growing tendency toward the view that the two enzymes (protopectinase and pectin-polygalacturonase) are identical. **1973** J. J. Doesburg in L. P. Miller *Phytochemistry* I. x. 281 In the past a specific enzyme, protopectinase, was thought to produce soluble pectic material from protopectin... Since there is no adequate proof of the existence of such a separate enzyme..it is believed that the action of 'macerating' enzymes is exerted by one or more of the pectic enzymes mentioned hereafter.

protoperithecium to **-phenomenon**: see *PROTO- 2b.

protophi·lic, *a. Chem.* [f. PROTO(N in Dict. and Suppl. + *-PHILIC.] Of a solvent (or solute): having a tendency to remove a proton from most solutes (or solvents). Opp. *PROTOGENIC *a.*[2] Also **pro·tophile** (*rare*), such a substance.
1930 N. F. Hall in *Jrnl. Chem. Educ.* VII. 787 The terms *protophilic* and *hydrophilic* have been proposed to describe the tendency of a molecule to unite with protons, and it would seem that some such word as *protophile*, forbidding as it is, would arouse less prejudice than the term *base* used in such a broad and subversive manner. *Ibid.* 792 Next there is the basic strength, or protophilic tendency of the solvent. **1931**, **1940** [see *PROTOGENIC *a.*[2]]. **1953** Audrieth & Kleinberg *Non-Aqueous Solvents* ii. 33 Amphiprotic solvents occupy a position intermediate between those of marked protophilic (*sic*) character, such as ammonia and the amines, and those of distinct protophobic character, such as acetic and hydrogen fluoride. **1969** [see *PROTOGENIC *a.*[2]]. **1973** E. J. King in Covington & Dickinson *Phys. Chem. Org. Solvent Syst.* iii. 334 Ethanolamine and dimethyl sulphoxide are basic or protophilic solvents.

protophloem, -phoneme: see *PROTO- 2b, 1 *a.*

pro·tophyll. *Bot.* [f. PROTO- + Gr. φύλλ-ον leaf.] In club-mosses, a structure resembling a leaf produced on the upper surface of the protocorm or tuber.
1891 F. O. Bower in *Proc. R. Soc.* L. 267 The sporophyte [of *Phylloglossum*] consists of two parts:—(i) the protocorm, with its protophylls and roots, and (ii) the strobilus. **1902** *Encycl. Brit.* XXXII. 74/1 The plant [sc. *Phylloglossum*] is produced by tubers, which resemble the protocorm in bearing first a number of protophylls. **1938** G. M. Smith *Cryptogamic Bot.* II. vii. 180 From the upper surface of the protocorm arise a few to many erect, conical outgrowths (protophylls) which are leaf-like in function. **1962** K. R. Sporne *Morphol. Pteridophytes* iv. 64 Further protophylls appear in an irregular manner.

pro-toplanet. *Astr.* [f. PROTO- + PLANET *sb.*[1]] A large diffuse body of matter in a solar or stellar orbit, postulated as a preliminary stage in the evolution of a planet.
1949 *Astrophysical Jrnl.* CIX. 309 A simple model is therefore considered first, consisting of two spherical masses ('protoplanets') in near contact, located inside the gaseous disk surrounding the sun. **1952** H. C. Urey *Planets* i. 13 First, a spherical or irregular cloud must rapidly collapse to a flat disk... Second, the disk of gas would break up into a Kolmogoroff spectrum of turbulent eddies... Finally a system of protoplanets, one for each of the planets, would be left at the appropriate distance from the sun. **1971** I. G. Gass et al. *Understanding Earth* ix. 135/2 This conclusion suggests..that a considerable degree of fractionation had already taken place in the protoplanet before it condensed into a solid body. **1974** *Sci. Amer.* Mar. 27/3 The second compositional class would have consisted of protoplanets formed just after the metallic iron-nickel alloy condensed out of the solar nebula.

Hence **protopla·netary** *a.*, of, pertaining to, or being a protoplanet.
1962 *Nature* 13 Jan. 89/1 Meteorites are generally assumed to have originated by the disruption of protoplanetary bodies in the region now occupied by the asteroids. **1977** *Nature* 13 Oct. 584/1 When the cloud collapses and a new hot star is created in its centre, the flattened protoplanetary disk formed from the remnants of the cloud continues to be cold.

protoplasm. Add: Before von Mohl coined the word in this sense it had been used (also in Ger.) with a slightly different meaning by J. E. Purkinje (*Uebersicht der Arbeiten und Veränderungen der schlesischen Ges. für vaterländische Kultur 1839* 82).

protoplasmal, *a.* (Example.)
1885 W. S. Gilbert *Mikado* i. 7, I can trace my ancestry back to a protoplasmal primordial atomic globule.

protoplast[1]. Add: **2. b.** The living contents of a cell; *esp.* in recent usage, a living cell whose cell wall has been removed or destroyed.
1884 *Rep. Brit. Assoc. 1883* 536 When the protoplast is in its normal position lining the cell wall, this core of protoplasm filling the pore would offer great resistance to a bodily passage of the protoplast. **1930** *Jrnl. Microsc. Soc.* 365 For the protoplast of the Cyanophyceae and Schizomycetes the author proposes the term *archiplast*. **1941** E. B. Wilson *Cell* (ed. 3) ii. 23 Cytosome and nucleus taken together form a living unit or protoplasmic system that is often spoken of as the protoplast (Hanstein) or 'cell-unit'... **1961** W. Weibull in *Jrnl. Bacteriol.* LXVI. 696 [Bacterial] Protoplasts obtained by lysis in sucrose will be designated as 'protoplasts'. **1970** Ambrose & Easty *Cell Biol.* viii. 271 The outer wall and capsule in many bacteria can be digested away by enzyme treatment leaving the protoplast, which is surrounded by a membrane still retaining the main permeability characteristics of the original bacterium.

protopod to **-Polynesian:** see *PROTO- 2b, 1 *a*, 2b.

protopo·rphyrin. *Chem.* [a. G. *protoporphyrin* (Fischer & Lindner 1925, in *Zeitschr. f. physiol. Chem.* CXLII. 147): see PROTO- and *PORPHYRIN.] Any of a group of fifteen isomeric porphyrins, $C_{34}H_{34}N_4O_4$, in which the porphin nucleus has four methyl, two vinyl, and two propionic acid substituents; one isomer (*protoporphyrin IX*) occurs widely in living organisms, notably as hæm (its ferrous chelate derivative).
1925 *Chem. Abstr.* XIX. 2714 The esters of ooporphyrin, Kämmerer's porphyrin, Papendieck's porphyrin and CO₂- and HCl-porphyrins are identical... The name 'protoporphyrin' is proposed for all of these. **1937** *Jrnl. Biol. Chem.* CXVIII. 531 Probably the most important porphyrin in nature is protoporphyrin IX... There are fifteen possible isomers of this compound, but this one..is the only one so far demonstrated in nature. **1961** *Lancet* 26 Aug. 450/2 This man..produced no abnormal quantity of uroporphyrin but excessive amounts of protoporphyrin which were localised mainly within his circulating red cells. **1963** A. White et al. *Princ. Biochem.* (ed. 3) xli. 792 A common pathway exists for the synthesis of heme and of chlorophyll leading to formation of protoporphyrin IX... Insertion of iron into the latter results in heme formation. In plants, in addition to synthesis of heme, magnesium is inserted into protoporphyrin IX to form magnesium protoporphyrin IX, which is converted in plastids to chlorophyll. **1975** *Nature* 26 June 706/1 The prosthetic group of haemoglobin has a protoporphyrin structure (ferrohaem) in which the iron atom is ionically bound.

Hence **pro:toporphyri·a** *Med.*, the presence of protoporphyrin in the red blood cells.
1956 in *New Gould Med. Dict.* (ed. 2) 975/1. **1961** I. A. Magnus et al. in *Lancet* 26 Aug. 451/2 The absence of uroporphyrin clearly distinguishes this syndrome from congenital porphyria. It seems to be a hitherto undescribed erythropoietic condition for which we suggest the name 'erythropoietic protoporphyria'. **1975** *Sci. Amer.* July 73/1 Investigators can easily induce these typical symptoms without serious consequences in patients suffering from mild forms of erythropoietic protoporphyria, so that the disease is one of the few of its kind where the action spectrum for a direct effect of light has been studied in detail. **1977** *Proc. R. Soc. Med.* LXX. 572/2 Erythropoietic protoporphyria (EPP) is a disorder, usually of autosomal dominant inheritance, in which large amounts of protoporphyrin are found in the plasma.

protoproteose to **-solar:** see *PROTO- 1, 2a, b, c.

pro-tosome. *Genetics.* [f. PROTO- + *-SOME[4].] The larger of two particles which together were postulated to constitute a gene; cf. *EPISOME. (No longer current.)
1931, 1966 [see *EPISOME a].

pro-tostar. *Astr.* [f. PROTO- + STAR *sb.*[1]] A contracting mass of gas in which nucleosynthesis has not yet begun, representing an early stage in the formation of a star.
1954 H. Alfvén *Orig. Solar Syst.* xii. 188 This condensation may have taken place from a 'protostar' of the type considered by Spitzer and others. **1970** *Nature* 15 Aug. 49/3 It has been suggested that protostars are formed when some of the gas and dust associated with the spiral arms of the galaxy piles up into clouds. **1976** [see *OBSERVATIONALLY *adv.*]. **1977** J. Narlikar *Struct. Universe* ii. 27 Such a cloud contracts as a whole, but subsequently breaks up into smaller subunits or 'protostars' when instability develops in the system.

Also **protoste·llar** *a.*, of, pertaining to, or being a protostar or protostars.
1973 *Nature* 17 Aug. 425/1 Conditions are therefore appropriate for the formation of solid planets if the nature of the protostellar body is such that dispersal of the particulate matter does not take place soon after its formation. **1975** *Ibid.* 6 Feb. 393/2 The initially low metal abundance could possibly arise from the properties of the gas required for protostellar collapse and the mechanisms which limit the greatest mass which can condense into a star. **1976** *Astron. Jrnl.* LXXXI. 1092/2 There is..a cluster of protostellar objects..southwest of M17 at the apparent edge of fragment B.

protostele to **-syntax:** see *PROTO- 1, 2a, b.

protota·xic, *a. Psychol.* [f. PROTO- + TAX(IS + -IC.] Applied to a hypothetical first or basic stage of experiencing or receiving impressions; also, related to a primal type of experience. See also *PARATAXIC, *SYNTAXIC *adjs.*
1945 P. Mullahy in *Psychiatry* VIII. 183/2 Prototaxic symbolization seems without reference to an ego, to 'I' or 'me' because the infant has no, or only a rudimentary, sense of self. **1953** H. S. Sullivan *Interpersonal Theory of Psychiatry* (1955) ii. 28 These modes are: the prototaxic, the parataxic, and the syntaxic. I shall offer the thesis that these modes are primarily matters of 'inner' elaboration of events. *Ibid.* 29 The prototaxic mode, which seems to be the rough basis of memory, is the crudest, the earliest, and possibly the most abundant mode of experience. **1969** A. Nesi *Theories of Psychol.* xx. 247 The first stage he [sc. Sullivan] called the prototaxic. **1972** L. Saltzman in Freedman & Kaplan *Interpreting Personality* vi. 176 (*heading*) Prototaxic mode of experience. **1975** J. C. Gowan *Trance, Art & Creativity* ii. 24 The prototaxic mode is notable for the scary, hair-raising aspect of the numinous.

proto-Thames to **-theme:** see *PROTO- 1, 2a, b.

protothe·tic (-þe·tik). *Logic.* [ad. G. *protothetik* (S. Leśniewski 1929, in *Fundamenta Math.* XIV. 4), f. Gr. πρωτο- PROTO- + θετικός fit for placing, positive, f. θετός, ppl. adj. of τιθέναι to set, place.] A type of propositional calculus on the basis of which Leśniewski developed his system of logic (see quots. 1945, 1955). Also **protothe·tics** *sb. pl.* in the same sense.
1940 *Jrnl. Symbolic Logic* V. 83 Protothetic involves not only propositional variables..but also truth function variables. **1945** Z. Jordan in *Polish Sci. & Learning* vi. 24/2 Leśniewski's system consists of three parts. The first of them, called Protothetic, corresponds to what is known as the 'calculus of equivalent statements'..or the 'theory of deduction', together with that of the apparent variables. It makes use of one axiom and of one logical constant (implicatively only). **1946** [see *MEREOLOGY]. **1955** A. N. Prior *Formal Logic* III. iii. 293 The basis of Leśniewski's logic is the 'protothetic', i.e. propositional calculus enriched with functional variables and quantifiers,—and on this he builds two further disciplines called 'ontology' and 'mereology'. **1963** O. Wojtasiewicz tr. *Łukasiewicz's Elem. Math. Logic* iv. 97 The sentential calculus can be extended by the introduction of variable functors and what are called *quantifiers*. One such system, containing the sentential calculus, is S. Leśniewski's *protothetics*. **1974** *Jrnl. Philos. Logic* III. 231 The extended propositional calculus which serves as a basis for this theory is not full protothetic.

prototroch: see *PROTO- 2b.

prototroph (prō·totrǭf, -trǫf). *Genetics.* [f. as next.] A strain (usu. of bacteria or fungi) which can grow on the simplest medium necessary for the growth of its species, without supplementary nutrients.
1946 Ryan & Lederberg in *Proc. Nat. Acad. Sci.* XXXII. 172 We propose to designate as a prototroph any strain which has the nutritional requirements of the 'wild type' from which it was derived irrespective of how it became prototrophic. **1952** *Genetics* XXXVII. 720 The occurrence of prototrophs, thus selected, from platings of thoroughly investigated auxotroph parents has been taken as *prima facie* evidence of crossing. **1958** *Heredity* XII. 169 Bursts of white, prototrophic cells frequently arise from colonies of certain strains of red, adenine-requiring yeasts. The occurrence of prototrophs appears to result from back-mutation at the locus for adenine

requirement. **1975** J. B. Jenkins *Genetics* viii. 311 Prototrophs are cells that can grow on minimal medium.

Hence **pro·totrophy**, the state of being a prototroph.
1952 *Genetics* XXXVII. 720 The methionineless stock is, fortunately, so stable that back mutations to prototrophy are undetectable under the conditions of crossing experiments. **1974** *Nature* 8 Feb. 387/1 *E. coli* strain H/r 30 R requiring arginine was used for measurement of mutation to prototrophy. **1976** *Ibid.* 24 June 736/2 It was impossible to grow the population for long periods without reversion to prototrophy.

prototrophic (-trǭ·fik, -trǫ·fik), *a.* [f. Gr. πρωτο- (see PROTO-) + -τροφ-ή nourishment + -IC.] **1.** *Bot.* [ad. G. *prototroph* (A. Fischer *Vorlesungen über Bakterien* (1897) v. 47).] = *autotrophic* adj. s.v. *AUTO-[1].
1900 A. C. Jones tr. *Fischer's Struct. Bacteria* v. 48 A better classification would be to divide the bacteria, according to their mode of life, into three biological groups, prototrophic, metatrophic, and paratrophic. *Ibid.* Prototrophic species are those which either require no organic compounds at all for their nutrition (nitrifying bacterium), or which, given the smallest quantity of organic carbon, can derive all their nitrogen from the atmosphere (bacteria of root-nodules). With them may be classed..sulphur and iron bacteria. **1935** F. O. Bower *Bot. Living Plant* (ed. 2) xxviii. 430 On the basis of nutrition Bacteria have been classified into three groups: (i) Prototrophic, those which require no organic compounds at all for their nutrition. These are represented by the nitrifying Bacteria, which live in open nature, in the soil, and are never parasitic. (ii) Metatrophic... (iii) Paratrophic. **1940** *Nature* 26 Oct. 541/1 If virus was the first form of life, it ought to be possible to find prototrophic viruses which feed on inorganic materials.

2. *Genetics.* Being a prototroph.
1946 [see *PROTOTROPH]. **1952** *Genetics* XXXVII. 721 Prototrophic stereomycin-sensitive (S^s) stocks were crossed to S^r (streptomycin-resistant) auxotrophs by plating the parents on minimal-streptomycin agar. **1958** [see *PROTOTROPH]. **1978** *Nature* 29 June 753/2 When complementary auxotrophs of opposite mating type are conjugated and plated on minimal medium (MM), several types of prototrophic colonies grow up.

Hence **prototro·phically** *adv.*
1960 *Nature* 30 Apr. 731/1 These cells..now grow prototrophically at a reduced rate identical to that of the spermidine-deficient cells.

prototropy (protǫ·trŏpi, prǭ·totrǫpi). *Chem.* [f. PROTON in Dict. and Suppl. + Gr. τροπή turn, turning.] Tautomerism in which the forms differ only in the position of a proton; migration of a proton from one part of a molecule to another.
1923 T. M. Lowry in *Jrnl. Chem. Soc.* CXXIII. 828 Prototropy, or the reversible change of protomers, which forms the subject of the present communication, is..a change which involves the migration of a proton or hydrogen nucleus. **1953** C. K. Ingold *Structure & Mechanism in Org. Chem.* x. 529 When migration is considered always to depend on proton-transfer processes..and it is such outstanding importance, that it is usually designated by the special name prototropy. **1964** N. G. Clark *Mod. Org. Chem.* xv. 296 The behaviour of ethyl acetoacetate is another example of the phenomenon of tautomerism, in particular of prototropy.

Hence **prototro·pic** *a.*, of, pertaining to, or exhibiting prototropy.
1925 *Jrnl. Chem. Soc.* CXXVII. 1382 Prototropic changes, which involve only the migration of a proton, are catalysed both by acids and by alkalis, like the hydrolysis of an ester. **1947** *Nature* 11 Jan. 86/1 Dislocation of a-hydrogen in glutamic acid can be interpreted as due to reversible condensation of the amino-group with the carbonyl in the prosthetic component of aminopherase to form a prototropic system. **1953** C. K. Ingold *Structure & Mechanism in Org. Chem.* x. 579 When a hydroxypyridine, or a-pyridone, as it is usually named after its prototropic tautomer, is alkylated, the formed quaternary ammonium ion passes into its anhydro-base, the N-alkyl-a-pyridone. **1976** *Nature* 6 May 15/1 Canellin showed that dynamic structure-activity relationships are capable of analysing quite complex situations, such as relative ionic populations in a prototropic equilibrium mixture.

prototype. Add: **b.** *spec.* That of which a model is a copy on a reduced scale.
1920 *Flight* 8 Jan. 57/2 Anyone can make a model resembling a full-size prototype that won't fly; this is not a scientific model. **1924** H. Greenly *Model Railways* i. 6 *Tonnage coefficient* is the scale equivalent of the weight of a train or loco. in tons actual of the prototype. **1942** *Model Railway News* Jan. 24/2 Were the Americans to enter the British market (i.e. the market based on British prototypes) the story would..be different. **1955** E. A. Steel *Model Mech. Engin.* i. 1 The working model must be an engineering job. A model locomotive is built to the scale of 1¾ in. to the foot or one-eighth the size of a prototype of similar design. **1967** G. C. Frazer *Model Railway Terminol.* No. 18. 2/2 On the prototype the gauge is always expressed in feet and inches, or metric measurements. In the model it can be similarly expressed, but..it is customary to use numbers or letters to describe the gauges. **1975** *Railway Modeller* Jan. 18/1 An hour or two beside a main line can provide one with quite a collection of short trains suitable for modelling. A study of prototype railway magazines is another prolific source.

2. *Electr.* A basic filter network (usu. having series and shunt reactances in inverse proportion) with specified cut-off frequencies, from which other networks may be derived to obtain sharper cut-offs, constancy of characteristic impedance with frequency, etc. Freq. *attrib.*
1923 *Bell Syst. Technical Jrnl.* II. 28 Mid-series and mid-shunt sections derived from prototypes other than the 'constant-*k*' wave-filter..are other possible units. **1932** W. L. Everitt *Communication Engin.* vii. 170 The fundamental data required for a filter are the pass band and the impedance into which it is to work. From these data, the prototype is designed. **1962** *Newnes Conc. Encycl. Electr. Engin.* 298/1, *m*-Derived Sections. These are 'derived' from the prototype sections above... The result is to introduce resonance into the shunt arm and/or antiresonance into the series arm, so that the attenuation rises much more steeply. **1977** M. H. Klayton *Fund. Electr. Technol.* xvii. 378 This band-pass filter is another form of the constant-*k* prototype circuit. The same elements can be rearranged to form a constant-*k* prototype band-suppression filter, sometimes called a wave trap.

3. The first full-size working version of a new vehicle, machine, etc., or a preliminary one made in small numbers so that its performance and methods of mass-production can be evaluated. Freq. *attrib.*
1932 *Flight* 26 Feb. 170/1 The A.B. 20 was actually begun as a three-engined machine, like its prototype the D.B. 70. **1935** C. G. Burge *Compl. Bk. Aviation* 261/1 The first experimental machine of a new type is usually made by a special department in the factory... This first or 'prototype' machine is, to a large extent, hand-made. **1949** *Flight* 28 Dec. 590/1 Of these one or two, such as the Ha 138 flying boat, may have been built as prototypes, but in the main the machines are 'prototypes only'. **1948** J. S. Murphy *Production Engin.* ii. 20 The previous model forms an ideal basis for the prototype, and experimental and development work can be carried out under practical conditions. **1948** N. Shute *No Highway* i. 16 That was the prototype Reindeer, the one we had seen for the trials. **1955** *Times* 10 Aug. 8/4 The American figures assumed the large-scale development in 15 to 20 years' time of types of reactor now at the prototype stage. **1964** *Jrnl. Geophysical Res.* LXIX. 2991/1 The development of a prototype lunar transponder..demonstrates the feasibility of designing future transponders for hard landings on the moon. **1970** P. H. Hill *Sci. Engin. Design* iii. 47 When entering the experimental stage of the design process..one should first deal with the mock-up, then the model, and finally the prototype after the mock-up and model have proven the real worth of the design. **1978** *Daily Tel.* 7 Apr. 2/5 Three camels from Longleat Wildlife Park..were driven to miles to Reading, where the prototype of the camel-milking machine..had been set up.

prototypic, *a.* (Further examples.)
1926 *Physiol. Rev.* VI. 322 The author..advanced the view that the blood plasma of Vertebrates and Invertebrates with a closed circulatory system is..but a reproduction of the sea water of the remote geological period in their prototypic representatives of such animal forms first made their appearance. **1963** A. Farrer in Mascall & Box *Blessed Virgin Mary* 28 A change which is prototypic of our own adoption, and the breaking of our bondage. **1965** L. Nortbourne tr. *Schuon's Light on Anc. Worlds* iv. 76 Semi-divine beings, the prototypic and normative personages whom every man has to imitate in all things. **1974** H. Ashley *Engin. Anal. of Flight Vehicles* i. 7 During the first decade of powered flight, 1903–1913, configuration was going through rapid, somewhat haphazard evolution. There were therefore many deviations from the prototypic arrangement outlined in Section 1.2, e.g., the Wright Flyers..employed antisymmetric warping of the wing structure for roll control.

prototypical, *a.* (Later examples.)
1964 Gould & Kolb *Dict. Social Sci.* 257/1 Fads, unlike merely 'prototypical' class systems. **1967** D. Cooper *Psychiatry & Anti-Psychiatry* i. 19 In the popular mind the schizophrenic is the prototypical madman. **1978** J. Sacks in P. Moore *Man, Woman, & Priesthood* iii. 30 The claim of Korah was the prototypical denunciation of the priesthood in the name of equality. **1979** *Amer. Speech* 1978 LIII. 281 The prototypical image of interruption..is perhaps those in which the interrupting material is simultaneous or overlapping; that is, the interrupting material is uttered..concurrently to what is interrupted.

prototypically, *adv.* (Further example.)
1957 F. Read *Tenth Muse* xxx. 281 An attempt to find architecture of a new universal art..represented prototypically by Greek architecture and later by Byzantine architecture.

prototyping (prǭ·totaipiŋ), *vbl. sb.* [f. PROTOTYPE + -ING[1].] The design, construction, or use of a prototype. Freq. *attrib.*
1951 *Chem. Age* LXV. 833/1 Prototyping of machines, instruments, and of chemicals is an essential step in this development. *Ibid.* 833/2 Prototyping may also prove of value in the estimation of costs. **1963** *Economist* 13 Nov. 726/1 Means for using the accumulated technical knowledge developed through such techniques as experimentation, prototyping, intervention and micromodelling. **1976** Hilburn & Julich *Microcomputers/Microprocessors* i. 4 Most microprocessor manufacturers have microcomputer systems, sometimes called prototyping systems, available to the designer. **1977** *Sci. Amer.* Sept. 90/2 (Advt.), There's room for eight plug-in boards..using the prototyping board for experimenting with interfaces to other equipment.

prototypographer to **-xylem:** see *PROTO- 1, 2b.

protozoology. Add: (Further example.) Hence **pro:tozoolo·gical** *a.*, of or pertaining to protozoology; **protozoo·logist**, an expert or specialist in protozoology.
1906 *Nature* 29 Nov. 117/2 When the protozoologist has worked out his life-histories and obtained his results, then the medical man steps in and carries off the honey to the medical hive. **1922** *Daily Mail* 17 Nov. (Advt.), Botanical, zoological and protozoological work. **1944** L. E. H. Whitby *Med. Bacteriol.* (ed. 4) vii. 90 The most commonly used preparations in protozoological work are Leishman's stain and Giemsa's stain. **1966** *Year Bk. Med.* XXI. 88 Probably best known of *E. histolytica* than of any other amoeba, and what is revealed is not only of practical importance to medical students but of only greatest interest to protozoologists. **1971** *Daily Tel.* 18 June 25 (Advt.), A Protozoologist..is required to join a team responsible for the maintenance and supply of free living Protozoa and to assist in the research programme of the Centre. **1972** *Nature* 3 Mar. 44/1 Most of his 500 publications deal with protozoological problems. **1974** *Ibid.* 3 May 89/3 In former times protozoology was considered to be a suitable subject by which students could be introduced to biology.

protracted, *ppl. a.* **1. a.** (Earlier and later examples of *protracted meeting*.)
1832 *Patriot & Farmer's Monitor* (Kingston, Ontario) 10 Apr. 2/6 It is now required of the Episcopal Methodists preachers, to make the public acquainted with their motives for establishing Protracted Meetings. **1935** W. J. Bryan *Mem.* 17 A protracted meeting held in a Christian Church. **1948** *Chicago Tribune* 21 Nov. vii. 17 Loafers' Glory..acquired that name years ago when a protracted meeting was held in the little log schoolhouse.

protrichocyst: see *PRO-[1] 1.

protriptyline (protri·ptilīn). *Pharm.* Also (*erron.*) **protryptiline, protryptyline.** [f. PRO-(PYL + TRI- + HE)*ptyl* (s.v. HEPTANE) + -INE[5].] A tricyclic antidepressant, 5-(3-methylaminopropyl)dibenzo[*a*, *e*] cycloheptatriene, $C_{19}H_{21}N$, given as the hydrochloride, a white bitter-tasting powder.
1963 *Amer. Jrnl. Psychiatry* CXX. 594/1 In an attempt to evaluate a derivative of amitriptyline, MK-240 (protriptyline), where the 'sleepiness' was removed by a change in the chemical formula, the drug was studied in 61 patients with various types of depression. **1966** *Psychosomatics* VI. 346/1 Protriptyline is at least five times as active as amitriptyline against depressive reactions. **1966** *Canad. Med. Assoc. Jrnl.* XCIV. 1220/1 Protriptyline gave chemical results which compared favourably with those of antidepressants of proven value. **1978** *Nature* 23 Mar. 320/1 The tricyclic antidepressant drugs which are tertiary amines..were more potent inhibitors than those which are secondary amines (desmethylimipramine, nortriptyline, protriptyline).

protrusively, *adv.* (Later example.)
1975 H. Thomson *Occlusion* ix. 156 The whole procedure is repeated for the outgoing movements beginning at intercuspal position and moving laterally on each side and then protrusively.

protuberance. 2. b. (Earlier example.)
1869 *Jrnl. Franklin Inst.* LVII. 317 The first protuberance which I observed, is represented by Fig. 1. Above an intensely luminous peak-shaped mass, rising from the border of the sun, is spread a cloud-like formation.

protu·beranceless, *a.* [f. PROTUBERANCE + -LESS.] Without a protuberance; flat, regular.
1954 W. Faulkner *Fable* 153 The entire earth one unbroken machined de-mountained dis-rivered expanse of concrete paving protuberanceless by tree or bush or house.

proud, *a.* (*sb.*, *adv.*) Add: **I.** *adj.* **2.** (Further U.S. examples.)
1938 M. K. Rawlings *Yearling* viii. 69 Be proud things come so bountiful. *Ibid.* xx. 195 I'd be proud to eat breakfast before I go. **1949** H. Hornsby *Lonesome Valley* xxiii. 302 I'm just as proud to see you, as if you was one of my own young 'uns! **1951** H. Giles *Harbin's Ridge* x. 99, I was sure proud Granny was there that day.

II. 7. d. (Later example.)
1970 *Country Life* 1 Oct. 856/1 Your case is the same as that of the farmer who sows his winter wheat too early; by the time the cold weather arrives the crop is 'proud'—too lush and forward.

9. (Further examples.)
1960 R. C. Bell *Board & Table Games* vii. 172 The inlay pieces..were fitted into them [sc. recesses], leaving an excess standing proud. **1961** P. Audemars *Stolen like Magic Away* vi. 92 The exposed rails of the track..stand proud and quite high above the flints. **1974** *Good Motoring* July/Aug. 18/2 The horn push, sited right across the central spoke of the steering wheel, is well proud of the spoke and this gives rise to occasional, accidental blasts.

E. *Comb.* **a.** *proud-arsed*, *-crested* (later examples), *-lidded*, *-necked*, *-plumed*, *-visaged*. **b.** *proud-arching*, *-fed*, *-pied* (later allusive example).
1919 A. Huxley in *Coterie* Sept. 61 The swan's proud-arching opulent loveliness. **1925** — *Those Barren Leaves* ii. 101 The proud-arsed broad-shouldered break and run. **1944** Blunden *Shells by Stream* 57 And look, those birds with perfect ease, Proud-crested. **1939** — *Note & Far* 37 A proud-fed but a puny hill. **1946** *Day Lewis Poems* 1943–47 88 Falsified palm, visible reticence, Proud-lidded mask. **1934** T. S. Eliot *Rock* ii. 75 Yet they walk in the

street proudnecked, like thoroughbreds ready for races. **1912** C. Mackenzie *Carnival* xlii. 373 When April pauses to survey her handiwork, assuming in the contemplation of the proud-pied earth the warmth and maturity of midsummer. **1949** Blunden *After Bombing* 9 With a sounding proud-plumed company. By a glittering sea. **1844** J. R. Lowell in *Graham's Mag.* July 15 Though these proud-visaged hopes, once turned to fly, Hurl backward many a deadly Parthian dart.

Proustian (prū·stiăn), *sb.* and *a.* [f. the name of *Marcel Proust*, French writer (1871–1922). + -IAN.] **A.** *sb.* An admirer or imitator of Proust.
1919 R. Fry *Let.* 28 Oct. (1972) II. 464 I've..got Proust's second volume... I forget whether you are a Proustian or not. **1928** *Sunday Express* 26 Feb. 3/1 If you desire to be a Jurgenist you must join the Proustians or..that other great clique of..Joyceans. **1936** A. Huxley *Eyeless in Gaza* xiv. 183 You're a Proustian, I take it? **1958** *Times* 6 Mar. 13/5 Proustians will be glad to have these 'moves on the chess-board of Time' in a volume that..stimulates endless new ideas. **1973** *Listener* 2 Aug. 155/3 An ardent Proustian who was trying to persuade me to..plunge into the full 12-volume *Remembrance of Things Past*.

B. *adj.* Of, pertaining to, or characteristic of Proust, his writings, or his style.
1926 A. Huxley *Jesting Pilate* i. 139 The decaying relics of feudalism..form the stormy background to the Proustian comedy. **1929** [see *IMITATION *sb.*]. **1932** *Times Lit. Suppl.* 2 Apr. 274/2 The Proustian distinction between 'involuntary memory' and 'voluntary memory'. **1936** L. P. Smith *Reperusals & Re-Collections* ii. 13 An immense leisurely, true novel, written with a Tolstoyan or Proustian amplitude, which allows space for an immense copiousness of detail and for infinite digressions. **1943** J. Lees-Milne *Ancestral Voices* (1975) 186 Lady Crewe believes no relationship, no emotion, no motive to be straightforward, and suspects everything and everyone. This is truly Proustian. **1958** *Spectator* 10 Jan. 51/1 Too often the adjective 'Proustian' evokes a kind of decadent Baroque. **1976** A. Powell *Infants of Spring* viii. 123 A lack of interest for individuals in what might be called the Proustian sense was perhaps characteristic, too, of the whole of the Arts Society.

Hence **Prou·stery** *nonce-wd.*, a Proustian manner; **Proustia·na**, memorabilia of Proust.
1928 Proustery [see PROUSTIAN *sb.* above]. **1965** *Guardian* 10 Dec. 6/5 She invited me to her home to see her Proustiana. **1973** *Times* 4 Oct. 17/2 Considering the volume of Proustiana..it requires an artist's skill to pick out the white stones of this life and work.

prove, *v.* Add: **A. 2.** *pa. pple.* (Further examples of the form *proven*.) Now common in the U.S.
1916 *Nat. Real Estate Jrnl.* (U.S.) 15 Mar. 110/2 Time has proven the jurisdiction of our creation for we have advanced with a knowledge and better understanding of ourselves and our profession. **1927** E. O'Neill *Marco Millions* ii. i. 106, I believe that what can be proven cannot be true. **1932** C. Kelly *U.S. Postal Policy* iv. 81 How little he understood the will of the American people is proven by their unyielding demand for this service. **1934** *Dial. Amer. Biogr.* XIV. 531/1 He had already proven himself a bold operator in various speculative fields. **1943** W. Rougheap *Art of Murder* 131 Our national nostrum, 'Not Proven'..a verdict which has been construed by the profane to mean 'Not Guilty, but don't do it again'. **1944** *Scots Mag.* Aug. 387 That his volume had a very large circulation for some generations is proven from a statement quoted by Stevenson. **1957** B. & C. Evans *Dict. Contemp. Amer. Usage* 399/1 The participle *proven* is respectable literary English. In the United States it is used more often than the form *proved*. In Great Britain *proved* is used more often and *proven* sounds affected to many people. **1959** I. Epstein *Judaism* xvii. 302 This doctrine of resurrection..cannot be proven philosophically. **1963** L. Hughes in *Liberator* July 4/1 Savages have proven a point. **1971** *Sci. Amer.* July 5 (Advt.), The MGB's 1798 c.c. twin-carb engine is proven time and again in competition. **1976** *Brit. Jrnl. Social.* XXVII. 313 Many of its specific predictions have been proven wrong by the course of events over the last century. **1979** *Economist* 11 Aug. 74/1 If it means a form (or rather, forms) of involvement that seem profitable to employees and employers alike, both can be proven right. **1980** *Daily Tel.* 18 Mar. 15/3 As yet..chloracne is the only disorder proven to result from human contact with dioxin.

B. I. 1. a. (Further example.)
1881 'Mark Twain' *Prince & Pauper* xxv. 198 He seized Miles by the arm, dragged him to the window, and began to devour him from head to foot with his eyes..stepping briskly around him and about him to prove him from all points of view.

4. For *Coal-mining* read *Mining* and add further examples.
1869 *Trans. N.Z. Inst.* II. 368 At Coromandel..the [gold-bearing] lodes have been 'proved' to a depth of over 300 feet from the surface. **1883** J. Bradshaw *N.Z. as it is* xiii. 178 In the Coromandel and Thames gold-fields, reefs have been 'proved' to a depth of over 600 feet below the sea level. **1978** *Nature* 6 Apr. p. xix/1 The Steeple Aston Borehole (1970–1972) proved Jurassic, Upper Coal Measures, Upper Devonian and igneous intrusive rocks. **1979** *Jrnl. R. Soc. Arts* Jan. 84/2 Figure 1 shows the extent of the coalfield under the land and also shows that about 10 kilometres of coal measures under the sea.

f. *Homœopathy.* [tr. G. *prüfen* (Hahnemann).] To give (a drug) to healthy persons to ascertain the symptoms it produces.
[**1833** C. H. Devrient tr. *Hahnemann's Organon of Healing* 26 115 Employ these medicines whose pure effects have been proved upon a healthy person in the manner that is useful to the cure of diseases homœopathically.] **1843** *Brit. Jrnl. Homœopathy* I. 160 It is essential that a preparation precisely similar to that proved should be always employed. **1920** *Encycl. Brit.* XIII. 645/2 To ascertain the curative virtues of any drug it must be 'proved' upon healthy persons—that is, taken by healthy individuals of both sexes in a state of health in gradually increasing doses. **1974** *Homoeopathy* June/July 93, I got these proved during the 1938/1939 period..and I wanted Nelsons to potentise (the drug)..but no one would cooperate.

g. *intr.* Of dough: to become aerated by the fermentation of yeast prior to baking; to rise. Also of yeast: to cause such fermentation.
1854 C. Tomlinson *Cycl. Useful Arts* I. 181/1 The whole of the dough..is left about an hour..to prove. **1854** A. Baker *Gloss. Northamptonshire Words & Phrases* II. 139 In making a cake, if it rises well, it is said to prove well. Yeast will often say 'It is good yeast, it proves so well'. **1909** Mrs. Beeton *Cookery Bk.* (new ed.) 1305 Knead well, and leave dry, cover over with a clean cloth, and let it prove for (1 hour). **1912** Jack & Strauss *Woman's Bk.* 214/2 After shaping the loaves, place them on a baking-sheet..and set them to prove for fifteen or twenty minutes. **1923** W. G. R. Francillon *Good Cookery* (ed. 2) xvi. 286 *Currant loaf*...Set to prove. Then bake in a quick oven for about fifteen minutes. **1959** *Woman* 14 Mar. 21/1 Place the dough in a tin and prove in a warm but not too hot place. **1972** *Guardian* 30 June 11/3 Form the dough into a ball... Cover with a clean cloth and stand in a warm place to prove.

II. 7. c. *refl.* To evince proof of one's abilities or prowess.
1961 *Spec. Ed.* 38 Mar. 38 But no one could help him, in his agonising struggle to save them—and prove himself. **1964** *Harper's Mag.* Dec. 55/1 A third such student, Caesar, came from a Latin-American country where a man is usually felt to prove himself by having many affairs, and thus to have a mistress after marriage. **1973** *Times* 1 Dec. 13/6 You have to prove yourself to your father, don't you? We all need to prove ourselves to our fathers.

8. d. *to prove too much*, to pursue an argument too far; to establish by argument a proposition so inclusive as to yield unhelpful results.
1791 J. Mackintosh *Vindiciae Gallicae* iv. 215 To this I answer, *first*, that such reasoning will prove too much, as it tends, in its proper extent, it impeaches the great system of morals, of which political principles form only a part. **1801** T. Jefferson *Let.* 23 Sept. in Koch & Peden *Life T. Jefferson* (1944) 565 It may be objected that this proves too much, as it proves you cannot enter the ship of a friend to search for contraband of war. **1870** J. H. Newman *Gram. of Assent* vi. 133 The theory to which I have referred cannot be carried out in practice. It may be rightly said to prove too much. **1903** 'L. Carroll' *Sylvie & Bruno Concluded* iii. 41 But surely that involves the logical fallacy of *proving too much*?

12. a. For *U.S.* read *N. Amer.* and add earlier and later examples.
1867 A. D. Richardson *Beyond Mississippi* xi. 138 He does not see the land again until ready to 'prove up' which he may do after thirty days. Then he revisits his claim. **1878** J. H. Beadle *Western Wilds* ii. 43 My wife 'proved up' on her Cherokee blood. **1912** *Daily Colonist* (Victoria, B.C.) 12 Mar. 12/5 Proved up, with hard work, and with special license to shoot from lands. **1938** F. G. Macgregor *North-West of 16* v. 67 They also had to bring fifteen acres under cultivation and to erect some sort of abode. Carrying out these obligations and getting title to the land was termed 'proving up'. **1963** G. H. Thomson *Crocus Country* xix. 124 The homesteaders, however, were obliged to fence so many acres before they could 'prove up' or be given the patent to them. **1967** *Islander* (Victoria, B.C.) 16 Nov. 12/2 Mr. James Foulds, however, proved up on his homestead and left in 1911. **1977** *New Yorker* 11 July 37/2 He tried to educate people to the responsibilities of living on public lands—to prove up, so that they could stay within the rules, legalize their occupancy, prove up.

b. = sense 1 d. Also *absol.* orig. *N. Amer.*
1921 *Daily Colonist* (Victoria, B.C.) 13 Oct. 2/4 Several claims staked up and down the channel from the discovery have proved up in a similar manner. **1926** *Ibid.* 17 Jan. 35/5 The richest mineralization that has been proved up in the whole district. **1975** *Offshore Aug.* 68/1 The company is aiming to prove up about 20 trillion cu feet of the fuel to justify an initial transmission line to northern markets. **1977** *Bulletin* (Sydney) 22 Jan. 42/1 It frequently takes as much as 10 to 12 years from an initial discovery to prove up a viable mine, arrange finance and forward sales-contracts.

13. prove out, to establish (something) as correct or workable; to test (a system or process) exhaustively. Also *intr.* (*or refl.*, orig. *U.S.*)
1959 *Wall St. Jrnl.* 13 May 14/2 Fishermen are pondering the commercial possibilities of pollock... If pollock proves out, however, it has a great future. **1964** D. F. Galouye *Counterfeit World* xii. 98 Of all the empirically-conceived..systems, this one had to go on to final verification. It could be proved out conclusively. **1967** *Electronics* 6 Mar. 28 (Advt.), We'll run a sufficient number of samples to prove out the process. **1969** *Daily Mail* 13 Jan. 5/7 The Moon is a convenient body on which to prove out some of our systems and programmes. **1972** *Daily Tel.* (Colour Suppl.) 10 Nov. 10/1 A duration of 50 to 60 days must be proved out. **1976** *Fortune* 1 Mar. 2/4 By making a few copies of Buddhist Scriptures... **1976** N. Thornburg *Cutter & Bone* iv. 102 If your hundred-to-one chance proves out, and Wolfe actually is the one.

proved, *ppl. a.* Add: **2.** (Further examples.)
1941 A. Tarski *Introd. Logic* i. vi. 118 Statements established in this manner are called *proved statements* or *theorems*. **1965** B. Mates *Elem. Logic* vi. 97 We may wish to use previously proved theorems.

proven, *ppl. a.* Add: **1.** (Later examples.) In some uses passing into sense 2 ('tested, approved, shown to be successful').
1963 *Automobile Engineer* LIII. 231/1 One..well known manufacturer is experimenting with an automatic stepped ratio transmission based upon proven principles. **1966** *Archivum Linguisticum* XVI. 125 A few relatively recent historians of the language..continue to repeat seemingly contradictory, debatable and even erroneous statements on this subject as if they were proven facts. **1968** *Globe & Mail* (Toronto) 17 Feb. 36 (Advt.), We have an opportunity for a mature, ambitious man..with proven sales ability. **1975** *Daily Tel.* 27 Nov. 29/1 (Advt.), Applicants should possess..a proven success record within a national consumer goods environment. **1976** *Sunday Times* 8 Feb. 55/3 (Advt.), The successful candidate will have a proven profitable track record as a manager. **1977** T. M. Bernstein *Dos, Don'ts & Maybes of Eng. Usage* 181 The form *proven* is used as an attributive adjective..and particularly in certain technical locutions, such as 'a *proven* oil field'.

2. Also (of a person, etc.) in wider use.
1942 W. Faulkner *Go down, Moses* 144 Boss and the negroes..the coons and cats, because the proven hunters..scorned such other than shooting..to test their little dogs. **1950** P. Stern (*title*) The proven marksmanship. **1961** *Sunday Express* 8 Oct. 30/6 Little elegant is a proven mudlark. **1968** *Globe & Mail* (Toronto) 17 Feb. 26 (Advt.), An outstanding opportunity exists for a proven manager to join a new management team.

provenance. Add: **a.** (Earlier example.)
1785 E. Sheridan *Jrnl.* (1960) 61 Miss Anstruther as I before mention'd Elegant, fashionable in her appearance, but nothing of her provenance in her manner that caught me much in Miss Brook.

b. The history or pedigree of a work of art, manuscript, rare book, etc.; *concr.*, a record of the ultimate derivation and passage of an item through its various owners.
1883 J. Dickson *Dancing Floor* i. vi. 112 If I knew the provenance of the manuscript, I might be able to understand it better. **1946** 'M. Innes' *From London Far* ii. x. 146 That aspect of the history of a valuable picture they call provenance? Who owned the picture last..and who owned it before that. **1948** A. Christie *Labours of Hercules* xi. 220 Sir Reuben would have purchased a Renaissance goblet, provenance unspecified. **1960** E. A. Lowe *Eng. Uncial* 21 A Canterbury origin is probable, Canterbury provenance is certain. **1966** *Listener* 10 Feb. (2017) 182 Both pictures..have the same provenance..I found a number of three on Madison Avenue I was introduced to a man who asked if I had a provenance of the picture. **1970** J. N. Barron *Law, Painting* 136 *Provenance*, a history or pedigree of a painting; the establishment of the identity of successive owners since its execution. Also included would be all published documents, catalogues, and journals that contain references to the painting, along with reproductions, exhibitions, and sales records, as well as correspondence, especially of the artist, in which mention of it may be made. **1974** A. Price *Other Paths to Glory* ii. ix. 203 He was only interested in where it came from, eh? *Provenance*.

c. *Forestry.* The location in which tree seed is collected. Also, seed from a specific location.
1933 *Empire Forestry Jrnl.* XII. 198 The problem of races in trees which..is about the geographical location and where the seed collected..and the genetic character of the mother trees. The author prefers the term 'provenance'..has come to regard the real key to the whole of the problem. **1942** H. J. Baldwin *Forest Tree Seed* iii. 29 Provenance is of tremendous significance to the success of plantations. **1956** M. L. Anderson tr. *Köstler's Silviculture* 86 It is the seed of the most different provenances, thirty different provenances of pine, had been carried out as early as 1821. **1970** H. L. Edlin *Collins Guide to Tree Planting & Cultivation* vi. 100 The true origin of each stock is called its provenance. *Ibid.* 101 A second generation of the same stock would still have that provenance.

Hence **pro·venanced** *a.*, provided with a record of provenance; established as to origin.
1939 *Nature* 20 May 848/2 Only adequately provenanced objects should be collected; only so can we hope to map, ultimately, the cultural regions of pre-industrial Britain. **1971** J. Mann *Charitable End* iii. 27 Most of it's [sc. an art collection's] not provenanced, you know. **1975** *Numismatic Chron.* 198 (*heading*) Some provenanced finds of Crusader bezants.

Provençal, A. *adj.* **1.** (Further examples.)
1650 [see *LANGUEDOCIAN *a.* and *sb.*]. **1936** A. W. Clapham *Romanesque Archit.* iv. 78 The Provençal school..is not the most closely akin to the global external form. **1971** *Guardian* 11 Feb. 9/3 A quilted skirt in fine wool Provençal print.

2. Designating a style of cookery characteristic of Provence, typically containing rich savoury ingredients.
1841 Thackeray in *Fraser's Mag.* June 720/1 You know what a *Provençale sauce* is?..A rich, savoury mixture, of garlic and oil. **1966** *Harrod's Food News* Sept. 2/2 Individual Scampi Provençal—per carton 9/6. **1974** *Times* 23 May 7/6 Try my favourite Provençal mixture with tomatoes, herbs and anchovies.

B. *sb.* **2.** (Earlier and further examples.)
1841 Thackeray in *Fraser's Mag.* xix. 124 The French have three dialects, the Wallon..the Provençal..and the Norman. **1916** *Chambers's Jrnl.* Jan. 35 The Provençal chant. **1961** *Archivum Linguisticum* XIII. 33 An older Provençal term. **1964** *Archivum Linguisticum* XVI. 5 Occitan—or Provençal, as it is more usually known—was one of the major European literary languages... Provençal is analogous, referring as it does both to the language as a whole and to the one particular dialect of Provence. **1977** *Listener* 28 July 118/1 The verbal texts..are sung in the original Provençal.

3. *Cookery.* (See quots.)
1974 *Times* 5 Sept. 13/5 *Provençale*, a dish which contains tomatoes, garlic and peppers. **1977** *Time* 28 Nov. 31/3 Particularly recommended are..*provençales* with garlic, tomatoes and herbs, served somewhat like a pizza.

Hence **Provença·lism**, an idiom or mode of expression typical of the Provençal language; **Provençalist**, a student of Provençal language and literature.
1934 *Times Lit. Suppl.* 30 Aug. 587/1 Mr. W. P. Shepard, himself no mean Provençalist. **1958** *Provençalism* [see *CATALANIST]. **1965** *Times Lit. Suppl.* 16 Sept. 1023/4 A brilliant essay..that Provençalists ignore at their peril. **1978** *Language* LIV. 473 The Berlin school was spearheaded by K. A. F. Mahn, who, as a Provençalist, became something of a rival of Raynouard and Diez.

Provence. Add: *Provence rose*, the cabbage rose, *Rosa centifolia*, or a variety of it, esp. one bearing fragrant red flowers, or a hybrid produced by crossing *R. centifolia* and *R. gallica*; also, a flower of one of these plants. (Further examples.)
Substitute for note: There is a long-established confusion between Provence and Provins roses, reflected in quots. 1597 and 1903 (in Dict.). Quot. 1578 in Dict. probably refers to the Provins rose: cf. *provence rose* s.v. PROVINCE *sb.* 10.
1597 Gerard *Herball* ii. i. 801 The Damaske Rose is called..of some *Rosa provincialis*, or Rose of Provence. **1707** J. Mortimer *Whole Art Husbandry* I. xviii. 477 The Red Provence Rose, whose Branches and Leaves are bigger and greater than those of the common Red Rose. **1817** *Repository of Arts* (Ackermann) Jan. 53/1 A..bonnet..ornamented with Provence roses and fancy flowers. **a 1821** Keats *Cap & Bells* in R. Monckton Milnes *Life Keats* (1848) II. 219 She..wetted three or four White Provence rose-leaves with her fairy tears. **1837** T. Rivers *Rose Amateur's Guide* 2 It is therefore very probable that it [sc. *Rosa centifolia*] was called the Provence rose from growing more abundantly in that province. *Ibid.*, Hybrid roses, between this [sc. *R. centifolia*] and *Rosa gallica*, are called Provence Roses by the French amateurs of the present day. **1847** C. M. Yonge *Scenes & Characters* xxv. 301 Those are some of the Provence roses for Miss Weston. **1848**, **1860** [see *CALLICO]. **1909** R. G. Kingsley *Roses* iii. 39 The Moss Rose,..originally a sport from the common Provence or Cabbage Rose, was also introduced into England from Holland in 1596. **1955** G. S. Thomas *Old Shrub Roses* xii. 136 *Rosa gallica*..(the 'French Rose' or 'Rose of Provins'). The name 'Provins' has been used in the old names, *Rosa provincialis*, but great confusion arose with the Provence Rose, *Rosa centifolia*. **1978** J. Harkness *Roses* xiii. 180 These old roses [sc. centifolias] have mostly been known by three different names from their country of origin: Provence rose, from their common names: Cabbage roses, Provence roses and Cent-feuilles.

provender, *v.* Add: **3.** *intr.* To partake of provender; to feed *on*. *rare*.
1819 Keats *Let.* 12 June (1931) II. 376 Infidel Rooks do not provender with Elisha's Ravens. **1891** C. Graves *Field of Tares* iv. 45 Last Leaving the iron horse provendering on coal and water..we follow the footsteps of the man with the black valise.

provenience. Add: Now chiefly *U.S.* (and to some extent *Canad.*). Elsewhere *provenance* is the more usual form. (Further examples.)
1955 J. R. Hulbert *Dict.*, *Brit. & Amer.* 54 It would be redundant and space consuming to give two pronunciations and spellings of *provenience* every time a word containing an *r* in this position turned up. **1968** P. M. Postal *Aspects Phonol. Theory* iv. 71 Thus in English the grossest divisions seem to correspond to the Germanic, Romance, and Greek provenience of forms. **1978** *Maledicta* 1977 Winter 133, I am concerned here particularly with lexical reflections of the diddle of foreigners, as they show linguistic provenience in English, although they are certainly not all of English provenience. **1978** *New York* 3 Apr. 81/2 Why should the hero's hypertrophic sense of nobility be of Dutch provenience? Why should someone who can ferret out olfactorily the point where, blindfolded, he is..writing thickness of a ballpoint pen—be the means for his blowing up the world?

pro·venly, *adv.* [f. PROVEN *ppl. a.* + -LY[2].] In a proven manner.
1887 O. B. Shaw *Let.* 27 May (1965) 170 The provenly heroic Annie Besant. **1967** *Times Rev. Industry* July 33/2 The uncommitted clearing banks are staying firmly on the sidelines until the industry has provenly made its way. **1972** *Daily Tel.* 30 Dec. 18 The otter is provenly beneficial to fishing. **1975** *Ibid.* 30 Aug. 11/2 The doctor concerned was fully trained and provenly competent.

prover. Add: **I. 1. d.** *Homœopathy.* A healthy person on whom the effect of a drug is tested. Cf. *PROVE *v.* 1 f.
1843 *Brit. Jrnl. Homœopathy* I. 162 The prover should choose a period when he is in the best of health. **1848** C. J. Hempel tr. C. Hering in *Jahr's New Manual* I. 29 The prover should ..commence at the first with..the most unbounded praise and admiration on the Austrian provers. **1929** *Encycl. Brit.* XII. 645/2 The manifestations of a drug can thus be obtained..by a number of provers... This..reaction..can of course be only produced by persons on 'provers', constitutes the distinguishing feature of the homœopathic materia medica. **1931** J. E. Barker *Miracles of Healing* vii. 101 It may seem easy to match the symptoms of a patient with a drug producing the same symptoms in healthy provers who have experimented with it. In reality this is exceedingly difficult.

1974 *Homœopathy* June/July 89 After taking the thirty powders the provers have a rest for a month, and then have a further thirty powders.

proverb, *sb.* Add: **1. c.** (Earlier example.)
1743 in A. D. Candler et al. *Georgia Rec.* (1913–16) XXIII. 513 He had then recourse to his Usual Salve, (well known, to all persons at Savannah with whom he converses, even to a proverb) That He was Seventy Years of Age, His Memory decayed, etc.

5. (Examples.)
1859 *Fireside Games for People* 104 Proverbs. One of the party is sent out of the room: the rest busying themselves with thinking of a proverb..to be discovered by him on his return. **1867** 'Aunt Carrie' *Popular Pastimes for Field & Fireside* 168 Proverbs. The interpreter selects some one to leave the room; those remaining agree upon a proverb (etc.). **1870** 'L. Hoffmann' *Drawing-Room Amusements & Evening Party Entertainments* ii. 90 Proverbs. This is another 'guessing' game. **1895** *Montgomery Ward & Co. Catal.* Spring & Summer 236/3 Proverbs, The old standard game revised, consisting of 100 cards comprising 250 proverbs. **1954** C. M. Yonge *Womankind*..(1907) 66 We have been playing games (e.g. Proverbs, Memory Test, etc.) this evening. **1975** *Word in Play* 257/2 *Proverbs*... [It is sometimes called hidden proverbs, the proverbs are hidden.]

6. *proverb card* (later examples).
1966 S. Mann *Collecting Playing Cards* vii. 141 *Proverb Cards*..a complete pack of proverbs, each illustrated..with English Proverbs. Made c. 1700... The value is indicated by a single suit-mark at top right. **1973** *Jrnl. Playing-Card Soc.* Nov. 44 There seem to be two packs of Proverb cards that could conceivably be the one referred to by Lenthall.

proverbial, *a.* (*sb.*) Add: **A.** *adj.* **1. a.** Also *absol.*
1858 *N. & Q.* Feb. 76/2 The description given certainly enlightens the reader as to a number of characteristics of the proverb, but equally certainly leaves him doubtful about where the proverbial leaves off.

2. (Later examples.) Also used with allusive force to introduce a word or expression that is familiar as (part of) a proverb or catch-phrase.
1924 *Argosy* 27 Sept. 463/1 In the more than proverbial four-leaf clover. **1938** S. Wolff *Trial of Mary Dugan* xxiv. 282 Like the proverbial man from Missouri, I have to be shown. **1951** H. Ashford *Murder of Marcus Keiser* xv. 225 The whole thing is rather ideal—the proverbial house party, a murder, everyone under suspicion. **1937** M. Allingham *Dancers in Mourning* viii. 108 White Walls were normally contained an excitable household... This morning..the proverbial donkey-wrench had landed squarely in the middle of the brittle machinery. **1976** J. Snow *Cricket Rebel* vi.) Having bowled a short ball at a batsman during one match he sarcastically inquired of Spectators whether that was the proverbial red rag to a bull. This is the proverbial red rag to a fast bowler. **1976** 'D. Fletcher' *Don't whistle 'Macbeth'* 211 The proverbial penny had dropped... It certainly made me very uncomfortable.

proverbialism (earlier and later examples); also **PROVERBIALITY** *n.*
1832 W. Motherwell in A. Henderson *Scottish Proverbs* p. xiv, Zachary Boyd, Rector of Glasgow University, has left..quite a cycle of common proverbialisms. **1935** *Times Lit. Suppl.* 14 Dec. 854/4 We pick them [sc. proverbs] up..and recognize them as part..the great gun men's epigrammatic sayings and above all from books... This aspect of modern proverbialism should not be ignored. **1976** *Hiroshima Stud. Eng. Lang. and Lit.* XXI. 11 The effort to such proverbialisms indicates his dependence on the world of experience.

pro-vicariate: see *PRO-[1] 4b.

provident, *a.* Add: **1.** *Provident Society* (examples). Also *Provident Club*.
1858 M. Tuckett *Diary* 12 Nov. (c 1975) 18 We betook ourselves to the Polytechnic where a stall awaited us, in the sale for the Provident Society. **1869** *Bradshaw's Railway Manual* XXI. App. 98 The United Kingdom Railway Officers' and Servants' Association, and Railway Provident Society. **1968** A. Bryant *Hist. Brit. United Provident Assoc.* p. 3 During the 'twenties and 'thirties many Provident Clubs became linked with particular hospitals. **1939** F. Gomersall *Self-Help* vii. 49 During the first half of May, 1836, petitions were received by the Government from a number of societies in South Lancashire... These included..the Provident Society of Salford. **1978** P. Sutcliffe *Oxf. Univ. Press* ii. xii. 63 He started a provident club for medical aid and a clothing club.

provide·ntialism. [f. PROVIDENTIAL *a.* + -ISM.] The belief that events are predestined, whether by God or by fate.
1927 J. S. Huxley *Relig. without Revelation* 18 The release of God from the anthropomorphic disguise of personality..also provides release from that vice which may be termed Providentialism. **1934** H. G. Wells *Exper. Autobiog.* I. v. 264 The ultimate adoption of the Five Year Plan and its successor has been the completest change over from the providentialism of Marx to the one-hated and despised method of the Utopists. **1954** C. S. Lewis *Eng. Lit. in 16th Cent.* i. ii. 148 His [sc. Fabyan's] philosophy of history is a simple Providentialism which leaves him completely agnostic about second causes.

providentialness. (Later example.)
1903 H. Whiting *Sanctuary* i. 10 The sense of general providentialness on which Mrs. Peyton reposed.

province. Add: **I. 2. a.** *spec.* in recent use, Northern Ireland.
1972 *Ann. Regist. 1971* 26 A horrifying escalation of violence in the Province. **1977** [see sense 10 below].

5. (Earlier example.)
1789 *Ann. Agric.* XI. 293 All the animation, vigour, life, and energy of luxury, consumption, and industry, which flow with a full tide through this kingdom, wherever there is a free communication between the capital and the provinces.

6. a. Also, an area containing a distinctive group of animal or plant communities. (Earlier and later examples.)
1847 H. C. Watson *Cybele Britannica* I. 14 Eighteen 'Provinces', or groups of counties, have been marked out on the map; each of these provinces..in natural districts..were grouped into provinces. **1856** *Q. Jrnl. Geol. Soc.* XVI. p. xxxv, Thus natural provinces are constituted, each including a group of characteristic forms peculiar to itself. **1932** Fuller & Conard tr. *Braun-Blanquet's Plant Sociol.* xiv. 365 The floristic province is characterized by at least one climax community. **1947** R. Good *Geogr. Flowering Plants* iii. 38 This classification divides the floras and floristic units of the world first into kingdoms, then into regions..and finally into provinces. **1957** P. Dansereau *Biogeography* i. 14 Each area (the provinces here) holds a more or less homogeneous portion of its total history. **1973** J. W. Valentine *Evolutionary Paleoecol. Marine Biosphere* ii. 74 The regions that constitute distributional units of organisms are called biogeographical regions or provinces (biotic provinces, faunal provinces, floral provinces, and so on). **1974** *Sci. Amer.* Apr. 83/3 The marine faunas today are partitioned into more than 30 provinces, each of which has a substantial roster of endemic species that inhabit only that province and a only low percentage of common species.

b. In full *petrographic* or *petrographical province*. An area of igneous rocks that appear to have been formed during the same period of igneous activity, presumably from the same magma. [cf. G. *geognostisch bezirk* (H. Vogelsang 1872, in *Zeitschr. d. deutsch. geol. Ges.* XXIV. 525).]
1886 J. W. Judd in *Q. Jrnl. Geol. Soc.* XLII. 54 There are distinct petrographical provinces, within which the rocks erupted during any particular geological period present certain well-marked peculiarities in mineralogical composition and microscopical structure, serving at once to distinguish them from the rocks belonging to the same general group, which were simultaneously erupted in other petrographical provinces. **1886** F. Rutley in *Ibid.* XLIII. 96 Lavas of totally distinct characters have been poured out..from the same vent, so that the use of the term 'petrographic province' seemed to be of rather doubtful propriety. **1910** Lake & Rastall *Text-bk Geol.* xiii. 230 The occurrence of chemical peculiarities running through the rocks of a series which were erupted at the same time..and not only that, but that the rocks..so long as they were not all brought together by chance, but that there must be some real relationship between the different types. **1941** *Amer. Jrnl. Sci.* CCXXXIX. 542 (*heading*) Compositions (less ionized) of the Chemical-Compositions of residual magmas in New Zealand petrographical provinces. **1965** H. Williams et al. *Petrography* i. 10 The markedly potassic, leucitic lavas of the region around Rome and Naples..form a petrographic province. **1962** P. T. Bronner tr. *Belousov's Basic Probl. in Geotectonics* xxxii. 657 At first it was thought..that different geographical areas are characterized by different 'petrographic provinces'... This was the origin of the study of petrographic provinces.

c. In full now *physiographic province*. An extensive region all parts of which have a broadly similar geology and topography and which differs significantly from adjacent regions.
1893 J. W. Powell in *14th Ann. Rep. U.S. Geol. Survey, 1892–3* I. 71 One of the results of this interpretation is the recognition of geologic provinces... The geologic province is the unit of past geography: throughout each the successive deposits represent a definite chronologic sequence, and throughout each there was generally a history of continuous, consistent, and mutually corroborative series of geologic changes. **1896** *Ibid.* 75 Willis *Northern Appalachians* (Nat. Geogr. Monogr. I. No. 6) 197 The plains were the homes of the most populous Indian tribes... The ranges of the mountain were a barrier to intercourse long after the several topographic provinces had come under one national government. **1914** *Ann. Assoc. Amer. Geographers* IV. 83 The confusion will be worse when the plotting of census and other statistics by physiographic provinces has become common. **1936** *Bull. Amer. Assoc. Petroleum Geologists* XX. 1378 (*caption*) Outline map of Mexico showing principal physiographic and structural provinces. **1940** [illegible] **1974** *Physics Bull.* Oct. 430/1 Another striking feature of the Mercurian surface is the asymmetry in distribution of the major physiographic provinces (also a characteristic of the moon and Mars).

† d. *Soil Science.* In full *soil province*. (See quots.) *Obs.*
1909 *Bull. Bureau of Soils* (U.S. Dept. Agric.) No. 55. 26 The complete scheme of classification, so far as perfected by the Bureau of Soils, also provides for the grouping of these series..into thirteen great soil provinces, as shown in the map. **1913** *Ibid.* No. 96. 7 A soil province is an area having the same general physiographic expression, in which the soils have been produced by the same forces or groups of forces and throughout which each rock or soil material yields to equal forces. **1924** F. E. Bear *Soil Managem.* ix. 76 Province refers to a large land area in which either the mode or the source of origin, or both, of the surface soil are quite similar throughout. Thus the Glacial and Loessial Province of the Bureau of Soils includes the en-

tire land area in the United States over which the glacial processes were most important in the formation of the original soil.

e. = *oil province* s.v. *OIL *sb.*¹ 6e.

1926 [see *oil province* s.v. *OIL *sb.*¹ 6e]. **1933** *Bull. Amer. Assoc. Petroleum Geologists* XVII. 1107 The earliest..trap to form in many American petroleum provinces was a reservoir rock which was wedged out and overlapped by an impervious cap rock. **1966** *McGraw-Hill Encycl. Sci. & Technol.* X. 61/2 Underground occurrences of petroleum may be classified as pools, fields, and provinces. **1971** *Daily Tel.* 29 Dec. 5/3 This huge oil yield from the northern 'province' of the North Sea will have important consequences for this country.

III. 8. (Earlier example.)

1690 LOCKE *Essay Hum. Und.* IV. xx. 362 They seemed to me to be the three great Provinces of the intellectual World, wholly separate and distinct one from another.

IV. 10. province rose = *Provins rose* s.v. *PROVINS or *Provence rose* s.v. PROVENCE in Dict. and Suppl.; also *absol.*; **province-wide** *a.*, extending throughout or pertaining to a whole province.

1597 GERARD *Herball* III. i. 1802 The greate Rose..is generally called the greate Province Rose. **1629** J. PARKINSON *Paradisi in Sole* cix. 413 Some Gentlewomen have caused all their damaske stockes to be grafted with province Roses, hoping to have as good water, and more store of them. *Ibid.*, The flowers are..of a sent not so sweete as the damaske Province. **1731** P. MILLER *Gard. Dict.* s.v. Rosa, The Damask, Province, and Frankfort Roses grow to the Height of seven or eight Feet. **1964** P. WORSLEY in I. L. Horowitz *New Sociol.* 380 Government intervention in province-wide infrastructural fields, such as air-ways, bus-lines, insurance etc. **1977** *Belfast Tel.* 22 Feb. 8/8 The old Loyalist merry-go-round of.. province-wide protests and rallies for the converted are discarded.

provincial, *a.* and *sb.* Add: **A.** *adj.* **2. a.** Specifically of Canada.

1795 *Quebec Gaz.* 8 Jan. 3/1 Clerk of the Provincial Court for the District of Three Rivers. **1849** J. E. ALEXANDER *L'Acadie* I. 35 It was found necessary to intermingle the newly arrived regulars with the Glengarry light infantry, a provincial corps. **1878** *Herald* (Ottawa) 24 Jan. 1/4 Two whiskey informers..were under the protection of the Provincial Police. **1965** *Globe & Mail* (Toronto) 10 Mar. 1/6 Provincial police said the single-engined plane..struck the lines with its undercarriage. **1968** *Ibid.* 3 Feb. 11/1 He said his department is seeking to have provincial services extended to Indians. **1976** *Telegraph-Jrnl.* (St. John, New Brunswick) 12 Aug. 1/1 He will recommend a *provincial* tax hike.

4. a. Also *provincial theatre*.

1867 *Harper's Mag.* Dec. 96/1 The provincial theatres compare favorably with those of London. **1952** GRANVILLE *Dict. Theatr. Terms* 145 *Provincial theatre*, the stage outside London.

c. *spec.* Of a university other than the older universities of Oxford and Cambridge (or other than that of Oxford only).

1914 C. MACKENZIE *Sinister St.* II. III. ix. 688 It was still natural to regard Cambridge as a provincial university, and to take pleasure in shocking the earnest young Cambridge man with the metropolitan humours and airy self-assurance of Oxford. **1955** *Ann. Reg. 1954* 351 *Lucky Jim*..was an example of the work of the new 'provincial school' about which there was much talk in the year. **1958** *Times Lit. Suppl.* 17 Jan. 30/4 Talk of.. 'the red-brick intellectuals', though no Movement founder-member had done more than *teach* at one of the provincial universities. **1966** C. M. BOWRA *Memories 1898–1939* xiii. 320 In the United States the academic profession had ties all over the country and was not divided as in England into Oxford and Cambridge on the one side and 'provincial' universities on the other. **1978** *Encounter* July 8/1, I studied at an English provincial university.

5. (Further examples.)

1813 M. EDGEWORTH *Let.* 6 Apr. (1971) 10 He.. speaks excellent language but with a strong provincial accent which at once destroys all idea of elegance. **1863** TROLLOPE *Rachel Ray* I. vi. 118 Mrs. Rowan perceived at once that Mrs. Tappitt was provincial,..but she was a good motherly woman. **1899** J. MCCARTHY *Reminisc.* II. xxxv. 252 Rather tall, very angular, surprisingly awkward..with a rough provincial accent and an uncouth way of speaking. **1909** A. W. EVANS tr. *A. France's Penguin Island* VII. ix. 312 Provincial women, since they wear low heels, are not very attractive, and preserve their virtue with ease. **1954** C. S. LEWIS *Eng. Lit. in 16th Cent.* I. i. 68 Scotch poetry had already a considerable achievement behind it and was by no means a local or provincial department of English poetry.

7. b. Delete *erron.* Substitute for def.: Epithet of the Provins rose (see *PROVINS).

B. *sb.* **2. a.** (Further examples.)

1916 JOYCE *Portrait of Artist* (1969) 48 If the minister did it he would go to the rector: and the rector to the provincial: and the provincial to the general of the jesuits. **1960** [see *DEFINITORY *sb.*]. **1973** *Franciscan* XV. 168 The Community Retreat conducted by Brother Luke, the American Provincial.

4. c. In Canada: a member of a provincial police force.

1936 W. B. MOWERY *Paradise Trail* 4 On his flight across the provinces he had..slipped out of several tight squeezes with the Provincials. **1952** H. GARNER *Yellow Sweater* 143 One of the Provincials took me upstairs. **1963** J. N. HARRIS *Weird World Wes Beattie* xi. 137 The provincials were extremely dubious about trying to find a weapon in the depths of Lake Muskoka in March.

5. (Later examples.)

1913 C. MACKENZIE *Sinister St.* I. II. v. 210 She used to laugh and tell him he was a regular old 'provincial'. **1954** C. S. LEWIS *Eng. Lit. in 16th Cent.* I. i. 83 Until we have trained ourselves to feel that 'gudeman' is no more rustic or homely than 'husband' we are no judges of Douglas as a translator of Virgil. If we fail in the training, then it is we and not the poet who are provincials.

9. (Further examples.)

1961 *Listener* 31 Aug. 325/1 His thoughts about Beckford and Beckett, Jouhandeau and Camus, the *anti-roman* and the English provincials. **1973** *Times* 2 July 15/6 The London papers stood out for a long time after the provincials had joined with him.

provincialate. (Further examples.)

1911 A. BRENNAN *Life St. Lawrence of Brindisi* xviii. 179 During his Provincialate the Friars of Piedmont.. renewed their petitions. **1930** T. S. WESTBROOK *Glimpses Catholic Eng.* 70 During the Provincialates of Agnellus.. and of his successor Albert of Pisa..the brethren at Oxford lived in the strictest poverty. **1960** J. B. DOCKERY *Christopher Davenport* viii. 121 During the Provincialate of Sancta Clara the community were well clothed. **1976** *Oxford Times* 3 Dec. 13/3 For the Jesuit student, as later for all his religious brethren during his Provincialate, he came as a breath of fresh air.

provincialism. Add: **3.** (Earlier example.)

1770 *Monthly Rev.* XLII. 180 His language..is, moreover, frequently debased with certain provincialisms.

4. *Ecol.* The development of biogeographical provinces. Cf. *PROVINCE 6a.

1969 *Spec. Papers Geol. Soc. Amer.* No. 119. 1 Provincialism increased by the addition of the Malvinokaffric Province. **1975** *Nature* 22 May 353/2 Why should the early Devonian faunas exhibit more provincialism, for instance, than those of the late Silurian.

provinciality. Add: **2.** *Ecol.* The restriction of the distribution of a plant or animal community to a particular province or group of provinces. Cf. *PROVINCE 6a.

1969 *Spec. Papers Geol. Soc. Amer.* No. 119. 3 The waxing and waning of provinciality displayed by Lower Devonian invertebrate faunas can be viewed in another way. **1971** J. G. JOHNSON in *Amer. Jrnl. Sci.* CCLXX. 257 Degrees of faunal resemblance..can be measured by a Provinciality Index (PI) consisting of a weighted ratio of common and endemic genera. **1976** *Nature* 24 June 695/1 It is noteworthy that an abrupt increase in phyletic rate commonly coincides initially with the brief upsurge in cladogenetic rate accompanying major provinciality increase.

provincializa·tion. [f. PROVINCIALIZE *v.* + -ATION.] A making or becoming provincial; conversion into a province.

1924 *Glasgow Herald* 16 Sept. 7 In a vigorous speech [he] emphasised the provincialisation and Indianisation aspect of the Report. **1967** A. N. SHERWIN-WHITE *Racial Prejudice in Imperial Rome* I. 24 Critognatus contrasts Roman imperialism, as revealed by the permanent exploitation and provincialization of southern Gaul (Gallia Narbonensis) unfavourably with the former temporary devastation of the country by the Cimbri. **1976** *Northern Miner* (Toronto) 19 Aug. 6/1 The government of Saskatchewan took the first step in the provincialization of the potash industry.

provi·ncialized, *ppl. a.* [f. PROVINCIALIZE *v.* + -ED¹.] Made, or having become, provincial; having become a province.

1955 *Times* 23 June 11/6 A country which had felt itself neglected and provincialized under the Danish and, especially, the Swedish connexion. **1974** *Sci. Amer.* Apr. 85/3 A widespread, soft-bodied fauna of low diversity gave way to a slightly provincialized, skeletonized fauna of somewhat higher diversity.

proving, *vbl. sb.* Add: **1. b.** *Homœopathy.* The testing of a drug (see *PROVE *v.* 1 f).

1843 *Brit. Jrnl. Homœopathy* I. 291 In the provings of the insoluble substances, such as calcarea, silica, &c., the symptoms produced by the first doses are rarely experienced..in the subsequent ones. **1881** *Encycl. Brit.* XII. 126/2 The record of such provings constitutes a large part of the literature of homœopathy. **1905** J. H. CLARKE *Homœopathy Explained* xvi. 122 There is always this check in homœopathy—the provings can be tested in practice. **1936** H. A. ROBERTS *Princ. & Art of Cure by Homœopathy* i. 16 The results of such investigation would enrich the homœopathic materia medica by completing provings of some of the older remedies, and by bringing out provings of new remedies. **1975** C. H. SHARMA *Man, Homoeopathy & Natural Med.* i. 16 Before a homoeopathic remedy can be used by a physician, it has to go through a series of 'provings'.

4. b. *N. Amer.* The action of establishing a claim. Also with *up*. Cf. PROVE *v.* 12.

1958 J. G. MACGREGOR *Northwest of 16* v. 67 They also had to bring fifteen acres under cultivation and to erect some sort of abode. (Carrying out these obligations and getting title to the land was termed 'proving up'.)

6. *attrib.* (Further examples.)

1944 *Air News Yearbk.* II. 188 Poland and Norway represented the 'proving ground'. **1948** 'N. SHUTE' *No Highway* i. 28 They did a whole lot of proving flights over the route before they put it into regular operation. **1959** *Listener* 15 Jan. 147/2 And, finally, proving time: once you have reconstituted the yeast and made a dough, carry on with your normal timing. **1971** M. LEE *Dying for Fun* ix. 55 All over his desk were scattered invitations —art galleries, press conferences..air trips and proving flights. **1975** *Harpers & Queen* May 27/1 The other day, I and my dough came to be separated at a crucial point in the 'proving' process. **1979** *Nature* 8 Feb. 430/1 Scientists have used astronomy as a proving ground for theories of gravity ever since Newton explained the sizes and shapes of the planetary orbits.

Provins (prǫvæ̃ns). The name of a French town, thirty miles east of Melun, used attrib. to designate a variety of *Rosa gallica*, formerly known as *Rosa provincialis*, the apothecary's red rose, which has long been cultivated there. Cf. *province rose* s.v. *PROVINCE 10.

1837 T. RIVERS *Rose Amateur's Guide* 11 In France, this [sc. *Rosa gallica*] is called the 'Provins Rose'. **1902** JEKYLL & MAWLEY *Roses for Eng. Gardens* ii. 13 These two names, Provence and Provins, for two classes of garden rose..are so much alike... Provence is the Cabbage Rose (*R. centifolia*): Provins is *Rosa gallica*, the garden kinds being mostly striped. **1955** C. C. HURST in G. S. Thomas *Old Shrub Roses* ix. 61 The Provins Roses were also much appreciated in India and in England. **1978** J. HARKNESS *Roses* xiii. 173 It [sc. *Rosa gallica officinalis*] is also known as the Apothecary's Rose, a reference to its uses in medicine, and as the Provins rose, because that French town specialized in making conserves and medicine from it.

provirus (provəi³·rʊs). *Biol.* [f. PRO-² + VIRUS, after *PROPHAGE.] The form which a DNA or RNA virus has when incorporated into, and able to replicate with, the DNA of a host cell.

1952 *Physiol. Rev.* XXXII. 419 Most of the cells perpetuate the potentiality of producing virus, although the virus itself is rarely detectable in them. For this reason, such cells are considered as infected with a provirus, a perpetuating, but immature and nonlytic agent. **1953** S. E. LURIA *Gen. Virol.* xiv. 277 We may suppose that in the recovered plant the virus is mainly in a condition (provirus) similar to the prophage. **1964** *Proc. Nat. Acad. Sci.* LII. 323 It has been suggested that the provirus of Rous sarcoma virus-infected cells is composed of DNA. **1970** *Nature* 5 Sept. 1023/1 It is widely believed that cells transformed with Rous sarcoma virus (RSV) contain a DNA transcript of the viral RNA, the so-called 'provirus'.

Hence **provi·ral** *a.*

1969 C. D. DARLINGTON in C. W. M. Whitty et al. *Virus Dis. & Nervous Syst.* 137 Diseases such as Kuru and Scrapie having combined genetic, cytoplasmic and pro-viral components. **1976** *Nature* 15 July 190 (*heading*) Proviral sequences of baboon endogenous type C RNA virus in DNA of human leukaemic tissues.

provision, *sb.* Add: **7.** (Earlier and later examples of West Indian sense.)

1808 J. STEWART *Acct. Jamaica* 100 Ground provisions (as they are called), or roots... These roots, or ground provisions, are so productive (particularly the yam), [etc.]. **1827** [see *ground-provisions* s.v. GROUND *sb.* 18 a]. **1955** *Caribbean Q.* IV. 1. 51 A large number of the contractors used these payments to acquire small plots of land in which they planted cocoa, provisions, and later, nutmeg trees. **1965** 'LAUCHMONEN' *Old Thom's Harvest* i. 11 Bet we can grow some whopping good crop of provision on that piece of land, Pa.

10. *provision basket* (example), *book*, *dealer* (further example), *farm*, *farmer*, *house*, *importer*, *man*, *pit*, *shop*, *store*, *trade* (earlier example), *train*, *wagon*; **provision pay** (earlier example).

1876 'MARK TWAIN' *Tom Sawyer* xxviii. 268 The gay throng filed up the main street laden with provision baskets. **1922** *Beaver* Apr. 9/2 A record of the provisions stocked, with their weight or quantities, was entered as they were received in the 'Provision Book', in which was also entered the allowances as they were given out. **1877** *Harper's Mag.* Jan. 284/2 They sold some grapes and apples and pears to the provision dealer in exchange for beef and chicken. **1958** J. CAREW *Black Midas* i. 9 At the back of the village were rice-fields, small provision farms..and wild-cane reeds. **1953** E. MITTELHOLZER in *Caribbean Anthol. Short Stories* 41 Hoolcharran had begun as a provision farmer, and lived in a mudhouse. **1798** W. TOMISON *Jrnl.* 2 Feb. in A. M. Johnson *Saskatchewan Jrnls. & Corr.* (1967) 108 The rest employed bringing ice for the provision house. **1804** J. ORDWAY in *Jrnls. Lewis & Ordway* (1916) vi. 166 We continued building, raised a provision & Smoak house 24 feet by 14 f. **1903** *N.Y. Times* 15 Oct. 1 Deacon Cotten..was dickering with representatives of meat and provision houses for supplies. **1885** *List of Subscribers, Classified* (United Telephone Co.) (ed. 6) 174 Provision importers. **1872** *Boston* (Mass.) *Ordin.* (1873) 193 The vehicles of market or provision men. **1683** *Rec. East Hampton, N.Y.* (1887) II. 131 For his Wages hee is to be payd the some of thirty five pound in probision pay. **1887** *Courier-Jrnl.* (Louisville, Kentucky) 3 Feb. 7/4 Within a very few minutes after the opening the crowd in the provision pit increased. **1854** M. S. CUMMINS *Lamplighter* xv. 115 Willie accompanied them as far as the provision-shop. **1796** *Boston* (Mass.) *Directory* s.v. *Fletcher*, Provision store. **1830** *Reg. Deb. Congress U.S.* 11 May 429/2 The provision trade of the West. **1896** *Harper's Mag.* Apr. 764/1 Blücher..found that he had captured..all the enemy's hospital outfit, his field-smithies, and his provision-train. **1765** R. ROGERS *Jrnls.* p. viii, I tarried till August 26th, and was then ordered with 100 men to escort the provision-waggons. **1925** G. STUART *40 Yrs. on Frontier* I. 97 Three days were consumed in getting together the equipment of men and horses with provision wagons and everything necessary.

provision, *v.* Add: **a.** (Further examples.) Also *refl.* **b.** (Further examples.) Also with *up*.

1903 R. BEDFORD *True Eyes* viii. 48 Why didn't you provision from home? **1928** *Daily Express* 11 Aug. 4/6 The main thing to remember in going to the islands is to provision-up for your stay well ahead. **1941** *Pitman's Business Educ.* Oct. 152 Without access to overseas supplies of oil, Germany has attempted to provision herself by the seizure of Rumania and by the invasion of Russia. **1973** *Animal Behaviour* XXI. 306/2 We suspect that the females were provisioning separate cells.

provisional, *a.* (*sb.*) Add: **A.** *adj.* **1. a.** (Further examples.) *Provisional Government*: now *spec.* a government set up to rule until constitutional self-government can be established; *Provisional I.R.A.*: the unofficial wing of the Irish Republican Army instituted in 1970; *provisional (driving-)licence*: a licence issued to a learner-driver; *provisional order*: (see quot. 1963).

1848 *Act* 11 & 12 *Vict.* c. 63 s. x They shall make a Provisional Order under their Hands and Seal of Office. **1870** *Act* 33 & 34 *Vict.* c. 1 §2 Any Select Committee of the House of Commons to which any Bill for confirming Provisional Orders has been referred in relation to any Provisional Order therein contained may examine witnesses upon oath. **1916** WELLS & MARLOWE *Hist. Irish Rebellion of 1916* ix. 47 At the Post Office was established the Headquarters of the 'Provisional Government of the Irish Republic'. **1931** *R.A.C. Guide 1931–32* 34 To enable an applicant suffering from a disability to learn to drive a motor vehicle of any special construction ..the Licensing Authority may..grant him a provisional licence for a period of three months. **1963** J. F. GARNER *Administrative Law* iii. 42 Provisional orders are made by a Minister of the Crown under the authority of a statute, and they are therefore sometimes described as a form of delegated or subordinate legislation, but they have no legal force until they have been included (usually by way of reference in a schedule) in a Provisional Orders Confirmation Act. **1965** J. CH'ÊN *Mao & Chinese Revolution* (1967) I. viii. 172 Under the Constitution, the Provisional Soviet Government was elected with Mao as its chairman. **1970** TIERNEY & MACCURTAIN *Birth Mod. Ireland* 131 Pearse then stepped out on to the portico and read the Proclamation of the Provisional Government of the Irish Republic. **1970** *Times* 9 Apr. 12/2 The recent formation of a 'provisional' I.R.A. Council. **1971** S. A. DE SMITH *Constitutional & Administrative Law* xv. 342 Provisional orders, which do not have legal effect till confirmed by Act of Parliament and are therefore not a form of delegated legislation at all. **1973** *Times* 11 Oct. 2/5 Mr McMorrow had been active in the Provisional IRA in Londonderry. **1974** *Guardian* 22 Mar. 8/7 The Environment Department has turned down a plea for stricter eyesight tests for people applying for their first provisional driving licence. **1976** *Burnham-on-Sea Gaz.* 20 Apr. 24/4 Mrs —— told the court that she only held a provisional licence and this had now expired. **1978** *Times* 6 Mar. 2/6 Under the Provisional IRA's new structure, each active service unit is largely self-contained, and in contact only with the central command.

B. *sb.* **3. a.** One whose tenure of office is of a temporary nature; a provisional governor.

1848 A. H. CLOUGH *Let.* 26 Feb. in J. Bertram *N.Z. Lett. of T. Arnold* (1966) 78 Will the army and Nationals rally around this government, or allow the people to set up their Provisionals. Inasmuch as the Provisionals are all in the Ministry, I suppose they may please themselves.

b. A member of the Provisional I.R.A.

1971 *Guardian* 11 Aug. 1/5 Some senior members of the IRA Provisionals, known to have been in Belfast recently, have..arrived. **1974** *Listener* 14 Mar. 323/1 The Provisionals' traditional method of discipline: putting a gun barrel behind a man's knee and blowing off his knee cap.

‖ **provisorium** (provizō³·riʊm). [Ger.] A provisional or interim measure or condition.

1957 *Listener* 28 Nov. 867/1 Since it has not been possible to reach such understandings subsequently.. the provisorium flowing from these circumstances has endured. **1963** *Economist* 3 Aug. 428/1 Bonn was not a 'provisorium' but a 'transitorium'.

provitamin (prōᵘ-vitămin). *Biol.* Also **pro-vitamin.** [a. G. *provitamin* (Windhaus & Hess 1926, in *Nachr. von der K. Ges. d. Wissensch. zu Göttingen* (1927) 175): see PRO-² and *VITAMIN.] A substance which is converted into a vitamin within an organism. (Freq. with following capital letter indicating relationship to a specific vitamin.)

1927 ROSENHEIM & WEBSTER in *Lancet* 5 Feb. 306/2 These observations suggest that the provitamin (we propose to use this convenient term, suggested by Prof. Windhaus, for the parent substance of vitamin D) is destroyed by bromine. **1943** *Endeavour* Apr. 73/2 It became evident that, though the diets of the tropical natives were often deficient in the calcifying vitamin,.. they really had ample supplies because of the action of sunlight on the provitamin. **1952** *New Biol.* XIII. 40 Doubling the number of chromosomes in pure yellow corn caused a 40% increase in the carotenoid pigment content, including the active provitamin A fraction of the carotenoids. **1971** *Nature* 22 Jan. 255/2 Vitamin D..is produced in the skin when ultraviolet radiation is absorbed by the pro-vitamin 7-dehydrocholesterol. **1976** H. CAMPION et al. in B. E. C. Nordin *Calcium, Phosphate & Mineral Metabolism* xii. 445 The two principal provitamins D, ergosterol..and..7-dehydrocholesterol..are formed in vivo by two very similar routes.

provo¹, provoe. Add: Also with capital initial. **1.** (Later example.) Also *transf.*, a provost-cell.

1779 *New-Jersey Jrnl.* (Chatham, N.J.) 13 Apr. 3/1 The other two are safely lodged in the provo of the continental troops. **1832** W. DUNLAP *Hist. Amer. Theatre* iv. 43 The Jail, then called the provo, where American prisoners suffered for asserting the rights of their country, scowled on the east. **1865** W. REID in *Cincinnati Daily Gaz.* 13 Dec. 1/3 He was boasting of his success with the 'cussed free niggers'. We've got a Provo' in our town that settles their hash mighty quick. He's a downright high-toned man, that Provo', if he is a Yankee.

2. *Comb.*, as **provo-marshal**: cf. PROVOST-MARSHAL.

1919 G. B. SHAW *Peace Conference Hints* vii. 102 The estimate of military crime which any statistician can give..without consulting a provo-marshal. **1934** —— *Too True to be Good* II. 50 Offences which cannot be stated on a charge sheet and dealt with by the provo-marshal.

provo, Provo² (prōᵘ·vo). [a. Du. *provo*, abbrev. of F. *provocateur*.] A member of a group of young Dutch agitators of anarchist persuasion, whose policy was to provoke the authorities; the Dutch anarchist group or movement. Also *attrib.*

1966 *Times* 15 June 1/5 For several weeks there has been unrest in Amsterdam. Young men and women calling themselves 'provos', from the French *provocateur*, who reject any authority or discipline, have gathered in certain parts of the city to provoke police intervention. **1967** *Listener* 19 Jan. 83/2 A somewhat riotous group of youngsters, who called themselves Provos, organized themselves and started to prove the validity of their organization's name by provoking the authorities. **1967** J. EASTWOOD *Little Dragon from Peking* x. 97 Hitch-hikers, *autostops*, *Blousons noirs*, provos from Amsterdam. **1968** *Listener* 22 Feb. 233/1 Police action against a Provo demonstrator when Princess Beatrice of the Netherlands was married in 1966. **1970** *New Yorker* 8 Aug. 50/3 One of the most interesting aspects of Provo, the Dutch movement that was among the first and brightest of the radical movements of the last decade, was that it blossomed forth with a number of responsible civic ideas. **1976** J. VAN DE WETERING *Corpse on Dike* v. 58 You look funny..but you don't look like a hippie or a provo or a bird-of-protest.

provo, Provo³ (prōᵘ·vo, prǫ·vo). *colloq.* [abbrev. of PROVISIONAL *a.* (*sb.*).] A member of the Provisional I.R.A. Also *attrib.* or as *adj.*

1971 *Guardian* 14 Aug. 9/7 In their bombing campaign the Provos seem to have hit on a policy..described as being the best way to bring down Stormont. **1972** *New Yorker* 19 Feb. 52/2 There are still no more than a few thousand I.R.A. men, Provo or Official, in the Six Counties. The Officials have less than half as many members as the Provos. **1973** *Daily Tel.* 27 Jan. 1/2 IRA men who recognise courts are automatically disowned by the Provos. **1976** *Church Times* 26 Nov. 5/2 The march squelched on to a new rallying point as a mob of Provo IRA thugs had barred the way into Falls Park. **1977** *Cork Examiner* 8 June 16/2 The Provos also claim that two soldiers were killed in a bomb explosion in West Belfast.

provocable, *a.* (Later example.)

1850 A. H. CLOUGH *Let.* 3 Jan. in J. Bertram *N.Z. Lett. of T. Arnold* (1966) 188 There is a great blessing..in being set down among uncongenial people—for me at least who am over provocable.

‖ **provocateur** (provǫkatōr). [Fr., = 'instigator, provoker'.] One who provokes a disturbance; an agitator; an *agent provocateur*. Also *attrib.*

1922 U. SINCLAIR *They call me Carpenter* xxvii. 94 The poor devils who went on strike were locked out of the factories..and their policies bedevilled by provocateurs. **1925** L. TROTSKY *Whither England?* v. 99 It must also thoroughly understand that the strike will fail to be immediately defeated only if it is able to offer the necessary resistance to the strike-breakers, provocateurs, Fascisti, etc. **1934** C. STEAD *Seven Poor Men of Sydney* iv. 112 What y' raisin' 'ell for; where y' come from? You're a *provocateur*. **1940** 'G. ORWELL' *Inside Whale* 142 To say 'I accept' in an age like our own is to say that you accept..submarines, spies, provocateurs, press censorship, [etc.]. **1956** A. L. GOODHARD in A. Pryce-Jones *New Outl. Mod. Knowl.* 581 The most important task..is the final extirpation..of all the remnants of these *provocateur* fabrications. **1961** C. COCKBURN *View from West* vi. 67 It looked much as though there might have been some *provocateurs* at work. **1974** T. P. WHITNEY tr. *Solzhenitsyn's Gulag Archipelago* I. I. viii. 319 The trial of the provocateur R. Malinovsky. **1976** 'J. DAVEY' *Treasury Alarm* i. 13 So you want me to..tell you if he's a genuine bloated capitalist or some sort of provocateur.

provocation. Add: **III. b.** *attrib.* **provocation test** *Med.*, a test to ascertain whether or not a person is alive.

1966 *Lancet* 31 Dec. 1466/2 On Oct. 12, 1965, patient was anæsthetized with halothane for a few minutes as a provocation test. **1971** *Essentials from Rep. Organtranspl.* (Netherlands Red Cross) 12 Provocation-tests and the best possible recording techniques should be used.

provocative, *a.* and *sb.* Add: **A.** *adj.* **2.** (Later examples.) Now limited to sexual contexts.

1933 [see *EXOTIC A. 2 b]. **1960** [see *BEEHIVE 1 c]. **1980** I. ST. JAMES *Money Stones* I. vii. 24 Her provocative teasing looks.

provocator (prǫ·vǒkeⁱtəɹ). [ad. Fr. *provocateur*.] A provoker or challenger; = *PROVOCATEUR.

1896 W. LE QUEUX *Secret Service* iv. 79 From Paris 'flying brigades' of spies and provocators are sent out. **1913** *Amer. Yearbk. 1912* 392/2 This..has caused a reawakening of the revolutionary movement..the old Terrorist wing having practically disappeared..on account of the exposure of the activity of Eugene Azeff and his staff of police spies and provocators within its ranks. **1918** A. GRAY tr. *Grelling's Crime* II. ii. 132 If even the creator of the defensive Entente of 1904 was regarded as a dangerous provocator, [etc.].

‖ **provodnik** (prəvpdnʸi·k). [Russ.] In the U.S.S.R.: **a.** A guide. **b.** An attendant or guard on a train.

1888 J. C. MURRAY tr. *S. Maimon's Autobiogr.* xviii. 148, I was once seized as a prowodnik myself. **1927** *Contemp. Rev.* June 729 Two *provodniks*, or train attendants, looked after our coach. **1936** P. FLEMING *News from Tartary* ii. 23, I went back to my compartment and found the provodnik. **1976** *National Observer* (U.S.) 21 Feb. 7/2 A provodnik is shaving in one of the two lavatories at the height of the morning rush.

provolone (prǫvolōᵘ·ni). [It., f. *provola* cheese made from buffalo's milk.] An Italian smoked cheese, often made in a variety of shapes, as spherical, pear-shaped, etc. Also *attrib.*

1946 A. SIMON *Conc. Encycl. Gastron.* IX. 22/1 *Provolone*, an all-the-year-round Italian cheese. **1952** S. KAUFFMANN *Philanderer* (1953) xii. 198 Madeline had gone shopping..to get him the anise and java ring and provolone that he loved. **1967** *Boston Sunday Globe* 23 Apr. (Advt. Section) 7/2 For color and flavor contrast add some sliced Swiss cheese, Italian Provolone. This last is light in color, sharp, tangy, cuts without crumbling, and has an agreeable flavor. **1968** V. & M. PETTITT *Len Deighton's Continental Dossier* 25 *The Basilicata*—a rather remote wild area where Romans caught bears for the Colosseum... Specialities: Provolone cheese, Aglianico di Vulture—a full red wine. **1975** *New Yorker* 4 Aug. 20/3 Authentic Philadelphia hoagies, which are sort of submarines made of Genoa salami, cooked salami, provolone, capicola, lettuce, tomato, olive oil, and assorted spices. **1978** *Detroit Free Press* 16 Apr. (Detroit Suppl.) 7/1 The house specialty ($1.25–$6.50) is made from eight kinds of Italian luncheon meat, salamis, provolone cheese, Italian bread and sweet peppers.

provost guard. *U.S.* A body of soldiers acting as military police under a provost-marshal; also, the quarters used by these.

1778 *Jrnls. U.S. Continental Congress* (1908) X. 74 About thirty [officers] who have been confined in the provost guard and in the most loathsome gaols. **1864** O. W. NORTON *Army Lett.* (1903) 212 Company K is provost guard and river patrol. **1883** SWEET & KNOX *On Mexican Mustang through Texas* xlii. 595 We may be caught by the provost-guard, and put in the bull-pen. **1887** G. B. MCCLELLAN *McClellan's Own Story* iv. 69 These..I at once brought to the city and employed as a provost-guard.

pro-West, -Western: see *PRO-¹ 5a.

prowl, *v.* Add: **2. d.** *Criminals' slang* (in U.S.). To examine, search, or inspect (a place or person), esp. before committing a robbery; to 'case'; to rob.

1914 JACKSON & HELLYER *Vocab. Criminal Slang* 67 *Prowl*, noun... An expeditionary investigation; a survey in transit; a search of the person or of a place in the sense of '*frisk*'; a burglary; a sneak; a saunter. Also used as a verb in the same senses. **1926** J. BLACK *You can't Win* xi. 136 I'd rather 'prowl' one of them than any business man. *Ibid.* xx. 318 He magnanimously suggested that I 'prowl the joint' he lived in. **1938** in *Amer. Speech* XIII. 158/1 Store is prowled. **1943** R. CHANDLER *Lady in Lake* xii. 71, I went back to the kitchen and prowled the open shelves above and behind the sink. **1977** 'M. INNES' *Honeybath's Haven* xv. 137 Some sort of sneak-thief had conceivably been prowling the dead man's property.

prowl, *sb.* Add: **a.** *on the prowl* (further examples); now freq. in search of an amorous partner.

1922 JOYCE *Ulysses* 600 A figure of middle height on the prowl, evidently, under the arches saluted again, calling: *Night!* **1946** *Sun* (Baltimore) 3 July 4/5 That big cat..is reputedly on the prowl again. **1959** W. BROWN *Cry Kill* iii. 31 Not a beauty like Lola Stuart, but good enough to catch the eye of any guy on the prowl. **1966** *N.Y. Times Bk. Rev.* 27 Mar. 35/1 Including his memorable encounters with an emancipated American college girl on the prowl. **1972** F. WARNER *Lying Figures* II. 9 Out on the prowl tonight, lover-boy? **1973** 'E. PETERS' *City of Gold & Shadows* iii. 45 A normal minor wolf on the prowl, with..an eye cocked for congenial company.

b. *Comb.*, as **prowl car** orig. *U.S.*, a police patrol car having a radio link with headquarters; **prowl dog** = *guard dog* s.v. GUARD *sb.* 18 in Dict. and Suppl.

1937 *Sun* (Baltimore) 6 Sept. 2/7 The man..climbed into the rear seat of our prowl car. **1953** H. CLEVELY *Public Enemy* xxix. 229 There's a prowl car outside... You were followed here. **1963** J. JOESTEN *They call it Intelligence* IV. xix. 188 A prowl car, manned by Western police, providentially arrived on the scene. **1967** N. LUCAS *C.I.D.* xi. 169 The presence of one of the Austin vans in the area had not passed unnoticed by the alert crew of a Berkshire County Police wireless prowl car.

1971 *Islander* (Victoria, B.C.) 16 May 11/1 Meantime another prowl car pulled into the yards. **1971** *Southerly* XXXI. 71 A prowl car told us to switch our parkers on. **1974** W. GARNER *Big enough Wreath* xii. 163 We got patrols. We got prowl dogs.

prowler. Add: (Further examples.) Also, a burglar, a sneak thief.

1912 D. LOWRIE *My Life in Prison* i. 5 Inadvertently we had left the back door open one night and a nocturnal prowler had taken advantage of it. **1926** J. BLACK *You can't Win* xix. 295 What a fox he is, to roll his money up in the curtain... What chance would a prowler have of finding his money? **1955** H. KURNITZ *Invasion of Privacy* (1956) xv. 99 It was Zorn's first ride in a police car. The radio chattered endlessly..of prowlers, burglars, rapists. **1976** *Flintshire Leader* 10 Dec. 1/7 Many of them are elderly or handicapped and live in fear of prowlers, car accidents and falls on uneven pathways.

proword: see *PRO-¹ 4b.

prox., abbrev. of PROXIMO.

1881 G. B. SHAW *Let.* 14 July (1965) 39 After the 1st prox. my address will be 37 Fitzroy Street W. **1935** A. P. HERBERT *What a Word!* iii. 64 There must be millions of our citizens who have not the least notion what is meant by your *inst.*, *prox.*, and *ult.* **1962** *Daily Tel.* 10 Dec. 12/7 'Inst', 'prox' and 'ult', which even today are scattered broadcast.

proxemics (prǫksī·miks). *Sociol.* [f. PROX(IMITY + *-emics*; cf. *EMIC *a.*] The study of the spaces that people feel it necessary to set between themselves and others as they vary in different social settings, or between different social groups or cultures; also the study of the feeling for space between people as it is manifested in aspects of culture such as the planning of houses or towns, in language, etc.

1963 E. T. HALL in *Amer. Anthropologist* LXV. 1003 Proxemics, the study of how man unconsciously structures microspace—the distance between men in the conduct of daily transactions, the organization of space in his houses and buildings, and ultimately the layout of his towns. **1969** *Guardian* 29 Sept. 7/2 Though territoriality and its effects has been studied for many years now in connection with animal life, Dr Hall is..the first person to link the concept direct with human beings and..has coined a purely human word for it: proxemics. **1974** *Language Sciences* Aug. 32/3 Ever since Edward Hall.. made public the results of his research on man's use of space, the interrelated observations and theories of which he calls proxemics, [etc.]. **1976** U. ECO *Theory of Semiotics* 10 *Kinesics and proxemics*: the idea that gesturing depends on cultural codes is now an acquired notion of cultural anthropology.

Hence **proxe·mic** *a.*, of, relating or pertaining to proxemics.

1963 E. T. HALL in *Amer. Anthropologist* LXV. 1003 (*title*) A system for the notation of proxemic behavior. **1965** [see *KINESIC *a.*]. **1971** *Times Lit. Suppl.* 4 June 653/4 In man 'proxemic' behaviour ranges from the distance two people maintain while engaged in conversation or the way a group of people arrange themselves, to architecture and city planning. **1976** J. F. KESS *Psycholinguistics* vi. 145 A detailed notational system for proxemic behaviors along a number of dimensions.

proximal, *a.* (*sb.*) Add: **A.** *adj.* **2. c.** *Dentistry.* Of, pertaining to, or being opposing surfaces of adjacent teeth in the same arch.

1908 G. V. BLACK *Work on Operative Dentistry* II. 3 Cavities occurring in the proximal surfaces of the teeth are called proximal cavities. **1944** S. HEMLEY *Fund. Occlusion* vi. 150 The teeth in the same arch in the adult dentition are normally in proximal contact with each other on both the mesial and the distal surfaces. **1963** C. R. COWELL et al. *Inlays, Crowns, & Bridges* iv. 35 The restoration covers the incisal edge of the tooth as well as the affected proximal surface. **1975** G. T. CHARBENEAU *Princ. & Pract. Operative Dentistry* xi. 262/1 The annoyance of food impaction between such teeth with an open proximal contact will be the initial concern of the patient.

3. *Psychol.* Applied to the stimuli immediately responsible for a perception or sensation.

1935 K. KOFFKA *Princ. Gestalt Psychol.* iii. 80 The table..can be called a stimulus for our perception of a table;..the excitations to which the light rays coming from the table give rise are called the stimuli for our perception. Let us call the first the distant stimulus, the second the proximal stimuli. **1955** F. H. ALLPORT *Theories of Perception* v. 147 The gestaltists point out the necessary differences between the proximal stimulus-pattern and the percept,..and attribute the 'way things look' to the organizing forces of the brain-field. **1971** *Jrnl. Gen. Psychol.* LXXXV. 3 A relation between the various aspects of perception and the proximal and distal stimuli is clearly revealed in almost all experiments.

proximate, *a.* **2.** (Examples of *proximate analysis*.)

1831 [see ANALYSIS 3]. **1857** W. A. MILLER *Elem. Chem.* III. i. 6 The separation of wheat flour into starch, sugar, gluten, ligneous fibre, and oily matter, affords an instance of proximate analysis. **1951** CAMPBELL & GIBB *Methods of Analysis of Fuels & Oils* i. 1 The proximate analysis of coal, which is carried out on coal ground to pass a 72 B.S. sieve and air-dried, involves the direct determination of (*a*) moisture, (*b*) volatile matter, and (*c*) ash, the remainder, the so-called 'fixed carbon', being obtained by difference. **1971** M. F. MALLETTE et al. *Introd. Biochem.* ix. 314 The crude lipid of the proximate analysis found on..certain food labels refers to the nonvolatile material derived by weighing the residue after evaporation of the extraction solvent.

proximity. Add: **2.** *attrib.* **proximity fuse,** a detonator in a missile that employs radar to operate it automatically when within a predetermined distance of a target; so **proximity-fused** *a.*

1945 *Sci. News Let.* 6 Oct. 214/1 The proximity fuze, a tiny radio set device in the nose of the projectile, is rated as the U.S.A. No. 2 secret weapon. **1956** A. H. COMPTON *Atomic Quest* 53 Applications of radar in the form of the 'proximity fuse' were critically important in bringing victory [in World War II]. **1972** *Guardian* 11 Jan. 11/2 In the Chinese campaign in Hongkong..the guerrillas eventually produced bombs equipped with photo cells, magnetic and proximity fuses, and vibration detectors that would set off the charge if you so much as looked at it. **1978** R. V. JONES *Most Secret War* xliv. 427 The new proximity-fused shells.., although originally a British invention, had been developed and engineered in America.

proxy, *sb.* Add: **III. 6. a. proxy form,** a form on which a proxy vote is registered; **proxy sitting** *Spiritualism*, a sitting arranged with a medium and attended by one person at the request of another, usu. unknown, person who hopes for news of someone recently dead; **proxy war** *U.S.*, a war limited in scale or area, instigated by a major power which does not itself become involved.

1930 *Economist* 6 Sept. 453/2 Accordingly they may, and should, use the company's money for the printing and postage of proxy forms. **1927** N. WALKER *Bridge* III. iv. 139, I had known about practically nothing that was mentioned in the two previous proxy Leonard sittings. **1933** *Proc. Soc. Psychical Res.* XLI. cxxx. 139 (*title*) A consideration of a series of proxy sittings. **1948** *Mind* LVII. 393 Telepathy is again invoked as the source of supernormal material: a well-worn hypothesis, which fails to cover the data obtained by proxy-sittings, cross-correspondences and so forth. **1962** C. D. BROAD *Lect. Psychical Res.* xv. 352 The essentials of a proxy-sitting are the following. The experimenter..receives in writing from some person, often a complete stranger to him, a few distinctive facts about a certain recently deceased individual... The specified facts are such as would suffice to enable the experimenter to recognize with some probability that the medium was referring to the individual in question. **1955** *N.Y. Times* 9 Jan. 8E/5 A threat that the United States would instantly retaliate with atomic weapons against the heart of the Communist world if the Commies started another proxy or brush-fire war. **1978** *Amer. Polit. Sci. Rev.* Sept. 971/2 Proxy wars, as the Athenians discovered in trying to rule their empire indirectly, are extremely costly.

b. *Petrol.* and *Min.* Applied to a mineral that proxies another.

1931 [see *PROXY *v.* 2]. **1949** F. H. HATCH et al. *Petrol. Igneous Rocks* (ed. 10) I. ii. 69 These Al''' atoms which function as silicons are 'proxy Al's', and each unit contains two of these. **1965** A. W. G. WHITTLE in G. J. Williams *Econ. Geol. N.Z.* x. 150/2 It is probable that an appreciable amount of 'proxy-nickel' was leached during the hydrothermal alteration of the peridotites to serpentinites.

proxy, *v.* Restrict *rare* to sense in Dict. (s.v. PROXY *sb.*) and add: **2.** *Petrol.* and *Min. trans.* To occur in place of, esp. in a crystal lattice. Also *intr.*, const. *for*. Hence **pro·xying** *vbl. sb.*

1922 A. JOHANNSEN in *Jrnl. Geol.* XXX. 640 The [German] author says a 'gabbro tendency' is shown, and that diallage is proxied by an amphibole of similar chemical composition. **1925** *Amer. Jrnl. Sci.* CCIX. 313 It is not possible for Br atoms to proxy for Na atoms in halite; they can occupy only places of Cl atoms. **1931** A. JOHANNSEN *Descr. Petrogr. Igneous Rocks* I. 189 'Proxy-minerals' is a translation of the German words *stellvertretende Gemengteile*, used by von Leonhard [in 1823] for minerals which take the place of other minerals in a rock; i.e., proxy them but do not replace them in the sense of molecular replacement... Thus if a certain type rock contains biotite, and another is like it in every way except that the dark mineral is hornblende, then in the latter the hornblende proxies biotite, and hornblende is the proxy-mineral. **1946** *Amer. Mineralogist* XXXI. 423 In tetrahedral positions Al proxies part of the Si. *Ibid.* 424 Other variables in the chemical composition [of the montmorillonite group of minerals] are..the proxying of OH by F as in the micas. **1963** W. A. DEER et al. *Rock-Forming Minerals* II. 353 The richterite..has an unusually high content of titanium some of which may proxy for silicon in tetrahedral positions.

pro-Zionist: see *PRO-¹ 5a.

prozone (prōᵘ·zōᵘn). *Immunol.* [Contraction of *pro-agglutinoid zone*, f. PRO-¹ + *AGGLUTINOID + ZONE *sb.*] The range of relative quantities of precipitin (or agglutinin) and antigen within which the expected precipitation (or agglutination) fails to occur when they are mixed; the mixture so produced, usu. containing antibody in excess. Freq. *attrib.*

[**1914** H. ZINSSER *Infection & Resistance* vi. 162 In the study of agglutinin and precipitin reactions, phenomena exactly analogous to the Neisser-Wechsberg effect have been noticed, in the case of the agglutinins, the so-called 'pro-agglutinoid' zone being a case in point.] **1916** *Jrnl. Immunol.* I. 6 The fourth line in this table represents the so-called prozone in which excess of precipitinogen inhibits precipitation. **1934** ZINSSER & BAYNE-JONES *Textbk. Bacteriol.* (ed. 7) xvi. 222 The specificity of the prozones is demonstrable in two ways. In the first place, bacteria that have been subjected to the action of serum showing such a prozone, without being agglutinated, will no longer agglutinate when subsequently emulsified in a potent agglutinating serum. Again, absorption of a prozone serum with the homologous bacteria will remove the prozone. **1964** D. F. GRAY *Immunol.* xi. 114 The 'constant antibody' precipitin reaction illustrates a phenomenon of diagnostic importance that may occur in agglutination tests, viz. the prozone, in which antigen: antibody aggregation is interfered with in the presence of an excess of antigen or of antibody. **1970** PASSMORE & ROBSON *Compan. Med. Stud.* II. xxii. 15/2 The prozone phenomenon..is probably due to the stabilizing effects of high protein concentration on the particles; the protein coating increases the net charge of the particles and brings about increased electrostatic repulsion between individual particles, thus opposing the efforts of the antibody molecules to link them together.

prozzy, var. *PROSSIE.

Pru (prū). Colloq. abbrev. of *Prudential Assurance Company*; *esp.* in phr. (*the*) *man from the Pru*, a representative of the Company who calls regularly at private houses to collect life insurance premiums (see quot. 1963).

1927 W. E. COLLINSON *Contemp. Eng.* 111 The best known English company, the Prudential, is often called the Pru. **1961** C. COCKBURN *View from West* ii. 19 To what extent..is our entire view of life..determined by what is told us by the Men from the Pru? **1963** *Times* 24 Apr. 12/3 The Prudential Assurance Company has discontinued the issue of life insurance policies paid by weekly premiums. This does not mean that 'the man from the Pru' will no longer be calling from door to door: all new 'industrial' life policy premiums will ordinarily be collected every four weeks instead of weekly. **1973** A. BEHREND *Samarai Affair* xiii. 141 She..said with no more emotion than if addressing a man from the Pru, 'My husband's told me what you've come for.' **1978** *Guardian Weekly* 15 Jan. 21/3 Next year I hope the Cottesloe [Theatre] will offer us, like the man from the Pru, a definite policy.

prude (prūd), *v.* [f. the sb.] *intr.* To conduct oneself in the manner of a prude; to act prudishly. Hence **pru·ding** *vbl. sb.*

1737 H. CAREY *Musical Century* 13 Crowds of coxcombs thus deluding, Cringing, chatt'ring, oggling, flatt'ring By coquetting and by pruding, All are victims to my art. **1850** C. M. YONGE *Henrietta's Wish* x. 151 'Pruding,' said Beatrice, 'showing openly that you like it to be observed how prudent and proper you are.' **1923** V. L. SILBERRAD *Lett. Jean Armiter* ix. 194 Girls aren't brought up in cotton wool nowadays as you were. We do as we jolly well like! It's no good preaching and pruding.

prun- (prūn). *Chem.* [f. PRUN(US.] A formative element used in the names of several substances which occur in trees of the genus *Prunus*, as **pru·nase** [*-ASE], an enzyme which hydrolyses β-glucosides (notably prunasin), liberating glucose, and occurs chiefly in bitter almonds and yeast, as well as in the fruit of several *Prunus* species; **pru·nasin** [f. prec. + -IN¹], a crystalline substance found in a number of trees, notably the bird cherry, *P. padus*; the racemic form of a glucoside of the nitrile of mandelic acid, $C_6H_5{\cdot}CH(CN){\cdot}OC_6H_{11}O_5$; **pru·netin** [*-ETIN], a colourless crystalline isoflavone derivative, $C_{16}H_{12}O_5$, which is the monomethyl ether of prunetol and occurs combined as glycosides in the wood and bark of several trees; **pru·netol** [f. prec. + -OL] = *GENISTEIN; **pru·nitrin** [-*trin* prob. after DEXTRIN], a colourless crystalline glucoside of prunetin.

The quotations follow in chronological order.

1910 H. FINNEMORE in *Pharmaceutical Jrnl.* XXXI. 604/1 This aqueous solution was shaken with ether, which at once precipitated a nearly colourless, semi-crystalline product, consisting mainly of a new dihydric phenol, $C_{16}H_{12}O_5$, to which the name prunetin is assigned. *Ibid.*, Prunetin contains a single methoxy group, and when demethylated by boiling with hydriodic acid yields the corresponding trihydric phenol, prunetol, $C_{15}H_{10}O_5$. *Ibid.*, Further treatment of the aqueous solution yielded a colourless glucoside which when hydrolysed gave prunetin, and was obviously the mother substance of that phenol, and to which the name prunitrin is accordingly given. **1912** H. E. ARMSTRONG et al. in *Proc. R. Soc.* B. LXXXV. 360 It appears to be desirable to assign a distinct name in future to the enzyme in 'emulsin' by which the resolution of the simple cyanophoric glucoside is effected; as it occurs very generally in the various species of Prunus, we propose to term it Prunase; also it will be convenient to use the name Prunasin in speaking of the glucoside (d-mandelonitrile glucoside) which, hitherto, we have termed Fischer's glucoside. **1918** PERKIN & EVEREST *Natural Org. Colouring Matters* vii. 205 Prunetin,..colourless needles, melting-point 242°, dissolves in alkalis with a slight yellow colour. **1936** W. STILES *Introd. Princ. Plant Physiol.* v. 106 The prunasin is now hydrolysed by means of prunase into glucose and mandelonitrile. **1940** *Thorpe's Dict. Appl. Chem.* (ed. 4) IV. 283/1 This β-glucosidase is also called prunase and latterly, β-phenylglucosidase. **1956** I. L. FINAR *Org. Chem.* II. vii. 244 The enzyme zymase hydrolyses amygdalin into one molecule of glucose and a glucoside of (+)-mandelonitrile (this compound is

identical with prunasin, a naturally occurring glucoside). **1959** N. CAMPBELL in E. H. Rodd *Chem. Carbon Compounds* IVB. viii. 925 Genistein (prunetol), 5:7:4'-trihydroxyisoflavone, $C_{15}H_{10}O_5$, colourless needles,.. occurs along with luteolin in dyers broom. *Ibid.*, Prunetin, ..genistein 7-methyl ether,..is isolated from the bark of a wild cherry related to *Prunus emarginata*..and the commercial timber muninga (*Pterocarpus angolensis*). *Ibid.* 926 Prunitrin, prunetin 4'-glucoside..occurs in *Prunus Serotina* L. and is synthesised by methylating sophoricoside. **1976** *Nature* 15 Apr. 604/1 The ecological success of bracken is partly..because of its ability to synthesise various secondary compounds which deter predators and phytopathogens. These compounds include the cyanogenic glycoside, prunasin, toxic because on enzymatic hydrolysis HCN is released.

prune, *sb.* Add: **1. d.** *slang* (orig. *U.S.*). A disagreeable or disliked person; a simpleton; spec., *Royal Air Force*, the personification of stupidity and incompetence (also, as a fictitious title, *P.O. Prune*). Hence **pru·nery**; **pru·nish** *a.*

1895 W. C. GORE in *Inlander* Dec. 112 *Prune*, one who is disagreeable, and irritable. **1941** BAKER *Dict. Austral. Slang* 57 *Prune*, a simpleton, fool. **1942** *Tee Emm* (Air Ministry) II. 67 All because the Prunes of the Air Force *will* ignore the existence of A.A. Danger Areas. **1942** *Observer* 4 Oct. 7/2 The Royal Air Force has adopted him [*sc.* P.O. Prune] now, and an official magazine is devoted to the purpose of trying to cure him of his prunery! *Ibid.*, One day, I think, in some solemn dictionary Prune will become immortal. **1942** T. RATTIGAN *Flare Path* I. 30 They call me P.O. Prune—he's a character in The Training Manual—sort of crazy, good-tempered, half-witted sort of bloke..and I—well, I kind of act P.O. Prune for them. *Ibid.* II. i. 53 He's not quite so prunish as he lets on. **1943** C. H. WARD-JACKSON *Piece of Cake* 49 *Prune, Pilot Officer*, a fictitious character who behaves as every officer should not, created by Squadron Leader Anthony Armstrong and the artist RAFF (L.A.C.W. Hooper)... Prune was created to teach pupils and other flying personnel how things should not be done. **1944** 'N. SHUTE' *Pastoral* ii. 35 He wished..that he knew what it was that worried her, whether it was some prune that she had left at her last station. **1961** G. SMITH *Business of Loving* viii. 201 Snap out of it, you moonstruck old prune. **1963** *Listener* 28 Feb. 392/3 No horse-play, no gremlins: in Mr Barr's script the Prunes have all been turned into serious-minded Prisms. **1970** *Women Speaking* Apr. 5/1 If a man doesn't like a girl's looks or personality, she's a.. prune, lemon. **1978** J. KRANTZ *Scruples* xiii. 368, I think she's a bit of a prune.

3. (Further examples.)

1922 JOYCE *Ulysses* 690 A sofa upholstered in a prune plush. **1976** *Vogue* Jan. 74/1 Prune silk crepe de chine with tiny white print.

4. (Further examples.) Also adj. phrs. *prunes and prismy*, *pruny and prismy*.

c **1909** D. H. LAWRENCE *Collier's Friday Night* (1934) i. 8 She says this in a very quaint 'prunes-and-prisms' manner, with her chin in the air and her hand extended. **1922** JOYCE *Ulysses* 365 Say prunes and prisms forty times every morning, cure for fat lips. **1931** *Time & Tide* 4 July 802/2 A tougher-minded generation than ours may find it altogether too prunes-and-prismy. **1940** G. D. H. & M. COLE *Counterpoint Murder* v. 50 She's forty if she's a day, and all pruny and prismy. **1979** *Daily Tel.* 22 Nov. 14/7 She regales us with an amusing chronicle of Lady Lytton's attempts to find a congenial companion among the straightlaced Indian Civil Service wives, whose 'prunes and prisms' expressions she found most off-putting.

5. *prune-orchard*, *-rancher*, *-whip*; *prune-coloured*, *-dark* adjs.; **prune-brandy,** an alcoholic beverage prepared from prunes; **prune-juice,** also, nonsense *U.S. slang*; **prune picker** *U.S. colloq.*, a Californian.

1895 M. PEMBERTON *Impregnable City* vi. 41 Drink that, and when you've drained the bumper, we'll have some prune brandy. **1872** *Young Englishwoman* Nov. 595/1 A hat of duck-green turquoise is trimmed with prune-coloured velvet. **1923** *Blackw. Mag.* Oct. 499/2 The foothills..were covered with a shadow over which prune-coloured clouds hung. **1941** L. MACNEICE *Plant & Phantom* 64 With prune-dark eyes, thick lips, jostling each other. **1957** J. KEROUAC *On Road* (1958) III. iv. 199 Here we were dealing with the pit and prunejuice of poor beat life. **1965** WODEHOUSE *Galahad at Blandings* x. 170 We decided that a big Society wedding was a lot of prune juice and we wanted no piece of it. **1921** *Chambers's Jrnl.* Mar. 173/1 Prune-orchards do not need irrigating. **1918** L. E. RUGGLES *Navy Explained* 112 *Prune picker*, a native of California. So called because of the abundant prune crops. **1929** *Papers Mich. Acad. Sci., Arts & Lett.* X. 316 *Prune picker*, a Californian. **1921** *Chambers's Jrnl.* Mar. 174/2 With proper pruning and cultivation the prune-rancher has an assured..living. **1942** 'R. WEST' *Black Lamb & Grey Falcon* II. 26 Their coffee-brown beauty which fastidious nostrils, secretive lips and eyes like prune-whip made refined and romantic.

prune, *v.*² Add: **3.** (Further examples.)

1836 J. H. NEWMAN in *Brit. Mag.* X. 137 Prune thou thy words. **1925** B. BEETHAM in E. F. Norton *Fight for Everest, 1924* 368 Bow to the inevitable, countenance what you deem to be reasonable, but prune early with a firm pencil anything excessive. **1970** *Railway Mag.* Oct. 546/1 The locomotive-hauled stock of British Railways has been drastically pruned in recent years.

b. Also with *out*.

1955 *Bull. Atomic Sci.* Mar. 94/2 Undoubtedly, these will be pruned out in the next edition. **1973** *Sci. Amer.* June 93/3 Shannon therefore proposed that the computer should not consider all possible moves from each position but should prune out the most obvious of bad moves.

pruner. Add: **2.** A tool used for pruning trees or shrubs.

1895 *Montgomery Ward Catal.* Spring & Summer 391/3 This pruner, being made with the shear cut, will work with double the ease of any other pruner. *Ibid.*, Waters' Improved Tree Pruner. **1916** L. H. BAILEY *Pruning Man.* vi. 192 (*caption*) Double-lever and single-lever pole pruner. **1949** E. HYAMS *Not in Our Stars* xiii. 160 The long-arm pruner had tired his arms and shoulders. **1971** *Country Life* 14 Oct. 955/3 My arm aches from using the sickle, and the heavy pruner. **1975** E. WIGGINTON *Foxfire 3* 26 Got to where I couldn't press the pruners enough to cut a big limb.

pruning, *vbl. sb.*² Add: **1.** (Later examples.)

1941 P. P. PIRONE *Maintenance of Shade & Ornamental Trees* v. 58 Proper and systematic pruning helps trees better to withstand adverse environmental conditions. **1972** G. E. BROWN (*title*) The pruning of trees, shrubs and conifers.

2. (Later examples.)

a **1930** D. H. LAWRENCE *Last Poems* (1932) 289 Humanity needs pruning. **1969** *Listener* 6 Mar. 322/1 It is a surprise to find the *Lyric Symphony*..a highly charged, expansive outpouring in seven longish movements of which the first three could themselves do with some pruning. **1970** T. LUPTON *Managem. & Social Sci.* (ed. 2) ii. 47 A drastic alteration of working practices and some pruning of manpower. **1971** *Nature* 16 July 206/2 If a second edition is prepared, the editors would do well to perform some judicious pruning.

prunus. Add to etym.: adopted by Linnæus (*Hortus Cliffortianus* (1737) 186) as the name of a genus. **1.** Also, a tree or shrub belonging to this genus, esp. one of many varieties of cherry cultivated for the sake of their ornamental, pink or white flowers. (Later examples.)

1901 L. H. BAILEY *Cycl. Amer. Hort.* III. 1445/2 It is an important point in the growing of these grafted Prunuses to remove all sprouts from the stock as soon as they appear. **1945** J. BETJEMAN *New Bats in Old Belfries* 22 Pinkly bursts the spray Of prunus and forsythia across the public way. **1966** *New Statesman* 7 Jan. 25/2 The evergreen prunus we call cherry-laurel. **1972** *Countryman* Summer 48 The old prunus on the lawn..immediately caught my eye.

prurience. 3. (Later examples.)

1926 H. W. FOWLER *Mod. Eng. Usage* 473/1 *Prurience, -cy.* There is no differentiation; *-ence* is recommended. **1974** C. RICKS *Keats & Embarrassment* i. 15 Prurience, pruriency, and prurient came fully into their modern meaning (from 'itching') in the eighteenth century.

prurient, *a.* Add: **3.** (Further example.) Also *absol.* or as *sb.*

1911 G. B. SHAW *Blanco Posnet* 334 The farcical comedy which has scandalized the critics in London..is played to the respectable dress circle of Northampton with these same jests slurred over so as to be imperceptible by even the most prurient spectator. **1969** *Punch* 29 Jan. 159/1 We've had the prudes and the prurients, the 'Love-Outs' and the love-ins, sex without marriage and marriage without sex. **1974** C. RICKS *Keats & Embarrassment* i. 15 The prurient is characterized by a particular attitude..of cherishing, fondling or slyly watching.

prurit. For †*Obs. rare*⁻¹ read *rare* and add later example.

1953 S. BECKETT *Watt* 182 A diffuse ano-scrotal prurit.

prurition (prūri·ʃən). *rare*⁻¹. [f. PRURIT(US + -ION.] = PRURITUS.

1922 JOYCE *Ulysses* 695 He scratched imprecisely with his left hand, though insensible of prurition, various points and surfaces of his partly exposed..skin.

prushun (prʊ·ʃən). *U.S. slang.* [Origin obscure.] A boy who travels with a tramp and begs for him.

1893 *Century Mag.* Nov. 106/1, I once knew a kid, or prushun, who averaged in Denver nearly three dollars a day. **1899** 'J. FLYNT' *Tramping with Tramps* 396 *Prushun*, a tramp boy. An 'ex-prushun' is one who has served his apprenticeship as a 'kid' and is 'looking for revenge', *i.e.*, for a lad that he can 'snare' and 'jocker', as he himself was 'snared' and 'jockered'. **1907** J. LONDON *Road* (1914) 235 If he travels with a 'profesh', he [*sc.* 'a road-kid'] is known possessively as a 'prushun'. **1927** *Dialect Notes* V. 459 The tramp lives in idleness while the boy goes about begging food for both. Many continue as *prushuns* until middle life, and when their master dies are left helpless.

prusik (prʊ·sik), *a.* *Mountaineering.* Also (*erron.*) **prussik**, and with initial capital. [f. the name of Karl *Prusik*, Austrian mountaineer.] Used to designate a method invented by Dr. Prusik of ascending a climbing rope by means of two separate continuous loops, each attached to it by a special knot (*Prusik knot*) which tightens when weight is applied and slackens when it is removed, thus enabling the loop to be moved up the rope. Also *ellipt.* as *sb.* Hence as *vb.*, to climb with the aid of prusik loops or similar devices; **pru·siking** *vbl. sb.*

1937 E. A. M. WEDDERBURN *Alpine Climbing* vi. 101 By employing the Prusik method he [*sc.* a man who has fallen] may be able to get himself out of a crevasse unaided... He first attaches the middle-sized loop of cord to the climbing rope as high as he can reach with a Prusik knot.., passes it through his waist loop, [etc.]. **1946** J. E. Q. BARFORD *Climbing in Brit.* v. 68 The Prusik Knot or Friction Hitch. This is a new and very useful hitch which is used for attaching a subsidiary rope or sling to the main rope. **1955** M. BANKS *Commando Climber* v. 83 We.. moved on carefully, realising that, despite Prussik slings, rescue operations are extremely difficult. **1956** C. EVANS *On Climbing* vii. 105 Each person in the party should have two Prusik loops (nine-foot lengths of Italian hemp cord spliced to make a rope ring). **1959** H. MERRICK tr. *Harrer's White Spider* 199 It is..obvious that Longhi was unfamiliar with the 'Prusik-knot' technique. *Ibid.* 200 It would not have taken Longhi half an hour to 'Prusik' himself by his own efforts up to the overhang. **1968** [see *JUMAR]. **1972** D. HASTON *In High Places* i. 8 He can use a special wrap-around knot to attach himself to the rope. This is called a prussik knot, after its German inventor. *Ibid.* ix. 103 Mike and I prussiked up in two hours to the high point. *Ibid.* xi. 120 A carry from Camp IV, up the overhanging prussiks, along the horizontal horrors of the ice ridge and up the never-ending icefields above. **1977** *Guardian Weekly* 20 Mar. 19/2 Two of the instructors.. help to provide a wide variety of experience in the arts of prusiking, abseiling, and all the necessary techniques, to both the advanced specialist and to schoolchildren.

Prussian, *a.* and *sb.* Add: **A.** *adj.* **1.** (Earlier example.)

Prussian binding (see quot. 1882), *collar* (see quot. 1955).

1565 R. SHACKLOCK tr. *Hosius's Hatchet of Heresies* (*title-page verso*), Fixing his eye on Prussian grounde, He sawe holy Hosius makyng this boke. **1880** *Queen* 19 June (Advt.), Sun umbrella cover..bound Self, Scarlet, Green, or Blue Prussian Binding. **1882** CAULFIELD & SAWARD *Dict. Needlework* 412/2 *Prussian bindings.* These are designed for the binding of mantles, dressing-gowns, and waterproofs... They consist of a silk face and cotton back, having a diagonal twill. **1932** D. C. MINTER *Mod. Needlecraft* 113/1 Neaten the bottom of a fitting sleeve with lute ribbon or Prussian binding. **1955** J. E. LIBERTY *Pract. Tailoring* (ed. 2) xi. 211 The Prussian or Double collar, is made to button up to the neck and has a small stand. There are variations of this type, from the style very like the double linen collar worn with a tie, to that where the fall lies flatter on the fronts and does not fit so closely to the stand, 2 to 3½ in. fall, as on uniforms. **1968** J. IRONSIDE *Fashion Alphabet* 52 *Prussian collar*, a fairly high-standing turned-down collar, as on military great-coats.

2. a. Prussian blue. (Further examples.) Also *ellipt.*

1835 [see *ANTWERP]. **1911** O. ONIONS *Widdershins* iv. 154 The daylight had gone, but I knew that 'Prussian' would be about the colour for the eyes. **1940** [see *ANTWERP].

d. A variety of pea with large, bluish seeds. Also *ellipt.*

1804 J. GARDINER *Amer. Gardener* 43 Spanish morottos, rouncivals, prussians, green and white, marrowfats, and other large late peas are the kinds to sow this month. **1824** J. C. LOUDON *Encycl. Gardening* (ed. 2) III. viii. 618 The egg, the Moratto, the Prussian blue, and the Rouncivals, ..are all very fine eating peas. **1832** J. TOD *Annals Rajast'han* II. 765, I never saw finer crops of Prussian-blues,..cauliflowers, celery, and all that belongs to the kitchen-garden. **1915** *N. & Q.* 6 Nov. 370/1 Prussian blues are a particular description of peas. Those and Marrowfats were in my early days considered the best varieties of that vegetable.

B. *sb.* **1.** (Earlier examples.)

1554 W. PRAT tr. *Discr. Aphrique* sig. B6v, The Germaynes, Italyons, Spanyardes, Frenchemen, Scottes, Iryshmen, the Danes, Liuones Prussiens. **1565** R. SHACKLOCK tr. *Hosius's Hatchet of Heresies* f.2ᵛ This agrement of fayth..floryshed among..the Germanes.. Catholyke Russians, Prussians, or Masouians.

2. = *OLD PRUSSIAN *sb.* b.

1888 J. WRIGHT tr. *Brugmann's Elem. Compar. Gram. Indo-Gmc. Lang.* I. 11 The Baltic division consists of Prussian, Lithuanian, and Lettic. **1972** W. B. LOCKWOOD *Panorama Indo-European Lang.* 139 A catechism, in Prussian and German, was twice printed in 1545.

Prussianism. Add: (Further examples.) Also, the militaristic concepts and disciplinary methods regarded as typical of the Prussian system. Hence **Pru·ssianist** *a.*

1915 *Chambers's Jrnl.* Oct. 664/2 Then we are getting.. 'Prussianism', 'Prussianisation',.., with all their meanings. **1916** A. HUXLEY *Let.* 31 Mar. (1969) 95 One discusses too the collapse of English civilisation, whose rapid decay under the sinister influence of Prussianism is everywhere apparent. **1922** Prussianist [see *GOOSE-STEP]. **1942** *R.A.F. Jrnl.* 2 May 6 Yet there is discipline... It is all done without Prussianism. **1944** F. A. HAYEK *Road to Serfdom* 7 It was the prevalence of socialist views and not Prussianism that Germany had in common with Italy and Russia. **1945** K. R. POPPER *Open Society* II. xii. 27 Hegel..became the first official philosopher of Prussianism. **1978** *Christian* V. 18 Pope Leo XIII decided..to come to terms with both French Republicanism and Prussianism.

Prussianize, *v.* Add: (Later example.) Also *intr.*, to act in a manner regarded as typical of Prussians. **Prussianized** *ppl. a.*, **Prussianizing** *vbl. sb.*, **Prussianization**: later examples.

1909 *Q. Rev.* CCX. 664 A similar policy pursued in a neighbouring territory—the attempted prussianisation of the inhabitants of Prussian Poland. *Ibid.*, Since his [*sc.* Bismarck's] time there has been a steady attempt to prussianise them. *Ibid.*, So far, the prussianising policy had wholly failed. **1915** [see *PRUSSIANISM]. **1927** 'IXION' *Further Motor Cycle Reminisc.* 82 The victim [of the practical joke] occupied a minor official position, by dint of which he Prussianized rather too freely. **1963** *Times Lit. Suppl.* 18 Jan 45/2 The Poles in Prussia's eastern provinces were not only devout Catholics but also Polish patriots who..offered stubborn resistance to Bismarck's Prussianization policy. **1976** *Listener* 19 Feb. 202/1 After 60 years, some Hanoverians, especially the military, had become Prussianised.

Pru·ssianly, *adv.* [f. PRUSSIAN *a.* + -LY².] In a manner regarded as typical of Prussians.
1917 *Daily Chron.* 26 July 2/5 People don't shove quite so selfishly, don't scowl at each other so Prussianly. **1932** BLUNDEN *Face of England* 72 He stumped, yet more Prussianly, into the pavilion. **1979** F. MORTON *Nervous Splendour* i. 7 Helmet Prussianly spiked.

prussic, *a.* (Examples of *prussic acid* used *attrib.* and *fig.*)
1860 MRS. GASKELL *Let.* ?28 June (1966) 912 The thunder yesterday gave me a dreadful headache so I had to..have my prussic acid medicine made up. **1927** D. H. LAWRENCE *Mornings in Mexico* 12 Such a suave, prussic-acid sarcasm.

Prussification (prɒsifikēⁱ·ʃən). [f. PRUSSI(AN *a.* + -FICATION.] = PRUSSIANIZATION. So **Pru·ssify** *v. trans.* = PRUSSIANIZE *v.*
1898 *Daily News* 21 Jan. 4/5 The Bill to increase the fund for the Prussification of the Polish provinces of the kingdom. **1904** *Daily Chron.* 23 Nov. 4/6 His proposal was successfully opposed by the other kingdoms as tending to the Prussification of the new empire. **1924** *Contemp. Rev.* Mar. 301 The Russians were trying their hardest to russify, and the Prussians were trying their hardest to prussify their Polish provinces.

pry, *v.*² Add: (Further examples.) Also *fig.*
1896 C. M. SHELDON *His Brother's Keeper* iii. 66 We managed to pry out of him that he had seen you and Eric go down the ladders. **1921** E. O'NEILL *Diff'rent* II. 245 It was always like pryin' open a safe for me to separate him from a cent. **1926** *Harper's Mag.* Feb. 363/1, I stood rooted to the spot and you could not have pried me away. **1927** S. ERTZ *Now East, Now West* ii. 21 He walked about the decks..hand in hand with Cleve, whenever that friendly child could be pried loose from some new and fascinating acquaintance. **1933** D. GARNETT *Pocahontas* v. 46 Holding out his clenched fist for her to pry his fingers open. **1947** S. BELLOW *Victim* (1948) i. 14 Philip pried off the caps on the handle of a metal cabinet in the kitchen. **1954** T. S. ELIOT *Confidential Clerk* I. 18 She's come to pry some cash from the money-box. **1968** J. AIKEN *Whispering Mountain* iii. 64 Owen's teeth were pried open and the neck of the bottle forced between them. **1976** *Time* 20 Dec. 1/2 When Watergate raised questions about the integrity of the Executive Branch, Congress appointed an independent prosecutor to pry out all the facts. **1978** C. TOMLINSON *Shaft* 14 As if this place could be pried out of now.

Przewalski (prεyəvæ·lski). Also **Prejevalsky, Przevalsky.** The name of N. M. *Przewalski* (1839–1888), Russian explorer, used *attrib.*, *ellipt.*, and in the possessive to designate a wild horse, *Equus przevalskii*, collected by him in 1876 in Central Asia, its native land, and named after him by I. S. Poliakov in 1881 (*Izvestiya Imper. Russ. Geogr. Obshchestva* XVII. 1).
1881 *Ann. Mag. Nat. Hist.* VIII. 16 (*heading*) Prejevalsky's Horse. **1884** *Nature* 21 Aug. 391/2 Przevalsky's wild horse has warts on its hind-legs as well as on its fore-legs. *Ibid.*, Nor has Przevalsky's horse any dorsal stripe. **1928** *Daily Tel.* 3 Jan. 5/1 Of the numerous arrivals at the zoo during the past year, the following are amongst the rarest or most valuable:..a Prejevalski's wild horse,..and two Komodo monitor lizards. **1951** G. G. SIMPSON *Horses* iii. 17 It is usually said that the only living true wild horse..is the central Asiatic race or subspecies called Przewalski's horse. **1969** J. FISHER et al. *Red Bk.* 101/1 Przewalski's horse..is the sole surviving species of wild horse. *Ibid.*, 101/2 The Przewalski horse can be mistaken for the kulan, or Mongolian wild ass. **1972** *Guardian* 22 May 7/3 (*caption*) A young Przewalski filly. Only about 30 of these Mongolian wild horses are..left in the mountains of China. **1973** *Daily Tel.* 23 Aug. 14 The Przewalski, whose primitive characteristics include a hog mane and mealy muzzle, is the only true horse now living in a wild state in the Gobi Desert, where it is almost extinct.

ps-. In words beginning thus the only pronunc. current is that with initial (s); the indication of an alternative (ps) in the following main entries would be misleading and is accordingly not shown.

psalm, *sb.* Add: **3.** *psalm-singing* adj. (earlier and later examples); **psalm-singer** (later examples). Both words frequently have somewhat derogatory connotations.
1818 'A. BURTON' *Adventures J. Newcome* 254 *Psalm-singer*, an epithet of the greatest possible contempt. **1908** J. M. SULLIVAN *Criminal Slang* 17 *Psalm singer*, a prison trusty; an informer. **1909** *Daily Chron.* 15 Dec. 5/5 Cromwell, the greatest ruler England ever had, was, with his glorious Ironsides, a Psalm-singer. **1818** 'A. BURTON' *Adventures J. Newcome* 59 Ye skulking, d—d psalmsinging crew! **1909** *Daily Chron.* 15 Dec. 5/5 In our war with the Boers we found Psalm-singing Dutchmen more than a match for our troops. **1964** Psalm-singing [see *Bible-banging* ppl. adj. s.v. *BIBLE III].

psaltery, *sb.* Add: **1.** Also, a modern imitation of this.
1901 W. B. YEATS *Let.* 20 July (1954) III. 354 Dolmetsch has interested himself in the chanting..and has made a psaltery for Miss Farr. It has 12 strings. **1975** *Gramophone* Oct. 709/1 The psaltery heard here is a bigger instrument than those we see in old paintings... It has seventy-three strings..and is played with both hands, the instrument lying flat on a table.

psaltery, *v.* For † *Obs. rare*⁻¹ read *rare* and add later example.
1903 G. B. SHAW *Let.* 11 June in *Florence Farr, Shaw, Yeats* (1946) 17 Are you too busy psalterying to copyright my new play at the Bayswater theatre?

psammic (sæ·mik), *a.* *Ecol.* [f. Gr. ψάμμ-ος sand + -IC 1.] Inhabiting areas of sand or gravel.
1938 J. R. CARPENTER *Ecol. Gloss.* 222 *Psammic*, concerning communities on sand or gravel. **1965** B. E. FREEMAN tr. *Vandel's Biospeleol.* vii. 69 The banks of rivers and lakes contain numerous psammic Rotifera.

psammite. Delete *rare* and add: In later use, a sediment or sedimentary rock composed of medium-sized particles (now commonly defined to be between 1/16 mm and either 2 or 4 mm diameter); also, a metamorphic derivative of a sandstone. (Further examples.)
1882 A. GEIKIE *Text-bk. Geol.* 154 (*heading*) Gravel and sand rocks (psammites). **1933** *Geogr. Jrnl.* LXXXI. 158 If the water table approaches the surface evaporation takes place, causing a concentration of the salts carried down. These salt-sand crusts are psammites. **1962** READ & WATSON *Introd. Geol.* I. v. 260 (*heading*) The psammites: sand, sandstone, greywacke, arkose. **1977** A. HALLAM *Planet Earth* 174 Meta-sandstones are sometimes called psammites.

psammitic *a.* (further examples); also, derived by metamorphism from a sandstone.
1882 A. GEIKIE *Text-bk. Geol.* 87 *Psammitic*, or sandstone-like, composed of rounded grains, as in ordinary sandstone: when the grains are larger (often sharp and somewhat angular) the rock is *gritty*, or a grit. **1910** *Q. Jrnl. Geol. Soc.* LXVI. 377 They may be conveniently divided into (1) a Psammitic Group, consisting chiefly of flaggy quartz-felspar granulites..; and (2) a Pelitic Group, consisting chiefly of garnetiferous mica-schists (or gneisses) which are often coarse and quartzose. **1921** G. W. TYRRELL in *Geol. Mag.* LVIII. 501 It is suggested that the terms *psammitic* and *pelitic* might usefully be restricted to the hard metamorphic rocks which have been changed beyond the limits implied by the corresponding Latin terms [sc. *arenaceous* and *argillaceous*]. **1956** E. W. HEINRICH *Microsc. Petrogr.* iv. 99 Psammitic and pelitic rocks may be further subdivided on the basis of mineralogical composition. **1959** *Trans. R. Soc. Edin.* LXIII. 554 In a series of psammitic and slightly calcareous pelitic schists, such as those of Morar, there is great difficulty in discovering the metamorphic grade in relation to that of other areas. **1962** READ & WATSON *Introd. Geol.* I. ix. 539 Psammitic gneisses commonly form massive pinkish rocks composed principally of quartz and feldspar.

psammo-. Add: **psa·mmophile** *sb.* and *a.* *Bot.* [a. F. *psammophile* (J. Thurmann *Essai de Phytostatique* (1849) I. xiii. 268): see *-PHIL, -PHILE], (a plant) thriving best in sandy soil; **psammophi·lic** *a.* = *psammophilous* adj.; **psammophilous** *a.* (further examples); **psa·mmophyte,** a plant characteristic of sandy habitats; **psa·mmosere** *Ecol.* [*SERE *sb.*²], a plant succession having its origin on sand.
1888 F. A. LEES *Flora W. Yorkshire* 69 Others [*sc.* plants]..are quite as much psammophiles, if their stations be any guide. **1901** C. MOHR *Plant Life Alabama* 131 The slender, wiry culms of this grass..render the species one of the most striking types of psammophile plants. **1973** M. A. SLEIGH *Biol. Protozoa* xi. 264 These psammophilic forms [of Protozoa] have been studied by J. Dragesco. **1888** F. A. LEES *Flora W. Yorkshire* 78 (*heading*) Arenaceous soils and psammophilous species. **1909** E. WARMING *Oecol. Plants* lxviii. 263 The fourth formation, composed of large dune-grasses, is much more psammophilous than halophilous. **1961** R. D. MANWELL *Introd. Protozool.* xv. 305 Some species [of Sarcodina] appear to be restricted to beach sands... Such species are said to be 'psammophilous'. **1903** W. R. FISHER tr. *Schimper's Plant-Geogr.* II. i. 80 The formations of the sandy sea-shore and of dunes serve as excellent examples of the vegetation of psammophytes. **1909** E. WARMING *Oecol. Plants* lxix. 264 *Triticum junceum*..is one of the halophilous psammophytes that begin the production of dune. **1973** F. DI CASTRI in di Castri & Mooney *Mediterranean Type Ecosystems* ii. 22 The psammophyte *Ambrosia chamissonis* was remarkably similar in the coastal fringe of Chile and western North America. **1916** Psammosere [see *lithosere* s.v. *LITHO-]. **1929** WEAVER & CLEMENTS *Plant Ecol.* iv. 74 The differences of hardness and stability result in very dissimilar seres. These are distinguished as (rock) lithoseres and (sand) psammoseres, respectively. **1964** V. J. CHAPMAN *Coastal Vegetation* vi. 150 The dune succession..is quite clearly a prisere, and since it develops on sand it is often known as a Psammosere.

psammoma. Add: Hence **psammo·matous** *a.*, of or pertaining to a psammoma.
1919 J. EWING *Neoplastic Dis.* xx. 309 In other tumors of nerve-trunks the structure is that of typical perivascular endothelioma as occurring in the dura mater, and psammomatous changes may appear in them. **1974** *Radiology* CXIII. 34/2 The second characteristic of psammomatous calcification is its diffuse extent throughout the abdomen due to serosal and omental implants.

psepho-, comb. form of Gr. ψῆφος pebble, used esp. in terms relating to voting (cf. PSEPHISM).

psephocracy (sīfǫ·krăsi). *rare.* [f. prec. + -CRACY.] The form of government which results from election by ballot; representative government. So **pse·phocrat,** an elected ruler, or an adherent or supporter of government by election.
1966 *New Statesman* 15 Apr. 531/1 How then did Britain..become a democracy?.. It never did... What we do have is representative government, or the rule of the ballot-box, or (in one word) psephocracy. *Ibid.*, 531/2 Psephocracy on the British model has been extended, thanks to the advice of British psephocrats, to a couple of dozen nations. **1970** *Sci. Jrnl.* Feb. 27/1 The present system [of government] is more of a leadership than a referred system—the so called 'psephocracy'.

† psephograph (sī·fograf). *Obs.* [f. *PSEPHO- + -GRAPH.] A machine for the automatic recording of votes.
1906 *Westm. Gaz.* 18 Dec. 7/2 The machine, of which a young Italian, Signor Boggiano, is the inventor, and which is known as the 'Psephograph', or vote-recorder, has the appearance of an upright box. **1907** *Daily Chron.* 11 Mar. 6/2 The plebiscite taken by means of the vote recording psephograph..has resulted as follows:—For the Tunnel, 3,212; against, 812.

psephology (sīfǫ·lŏdʒi, sef-). [f. *PSEPHO- + -OLOGY.] The study of public elections, and statistical analysis of trends in voting; *loosely*, the prediction of electoral results.
1952 D. E. BUTLER *Brit. Gen. Election of 1951* 1 It..seems appropriate to preface this book with a discussion of why elections merit study and an examination of how much has been..learnt from psephology... I am indebted to Mr. R. B. McCallum for the invention of this word to describe the field of research in which he is so eminent a pioneer. It is derived from ψῆφος—the pebble which the Athenians dropped into an urn to vote. **1952** *Economist* 4 Oct. 18 (*heading*) British psephology. **1957** *Ibid.* 21 Sept. 917/2 Even with the present degree of Liberal revival in the country, prudent (as distinct from roseate) psephology suggests that the only new seats the Liberals might pick up [etc.]. **1958** *Times* 6 Nov. 6/5 (*heading*) Material for psephology—Chichester figures anxiously awaited. **1973** *Guardian* 25 May 12/2 Even with the aid of psephology, it remains difficult to detect what precisely turns votes.

So **psepholo·gical** *a.*, **psepholo·gically** *adv.*; **psepho·logist,** a political scientist who specializes in the study of elections; an electoral analyst or commentator.
1952 *N.Y. Herald Tribune* 8 Aug. (Late City Ed.) 12/3 He [*sc.* R. B. McCallum] suggested I [*sc.* D. E. Butler] call myself a psephologist. **1952** *Daily Express* 30 Sept. 4/3 (*heading*) Psephologically speaking: you may vote for a good-looking party. **1955** D. E. BUTLER *Brit. Gen. Election of 1955* p. v, Acknowledgements..Mr. R. B. McCallum and Mr. H. G. Nicholas, my psephological mentors, gave me much valuable help. **1958** *New Scientist* 27 Mar. 32/1 The next General Election should provide the first opportunity modern Britain will have seen of testing a little-known and much neglected psephological law. **1958** *Times* 6 Nov. 6/5 The psephologists at Central Office and Transport House will be..ready to read into the results of Chichester, and all the other by-elections pending, sinister or cheering evidence. **1962** *Times* 16 Mar. 13/2 Psephologically, and even more psychologically, it [*sc.* the Orpington by-election] is in a class by itself. **1964** *Time* 4 Nov. 3 By 9 p.m., when Vermont had plopped into the Democratic column and television's psephologists flatly declared Johnson the winner, the answer became obvious: very, very, big. **1969** *Daily Tel.* 9 Sept. 17/6 If psephological calculations mean anything, enough votes will be mustered by Spain to secure an adverse vote against Britain. **1977** *Oxford Times* 29 Apr. 10/2 Whatever happens next Thursday will be psephologically interesting. **1977** *Times* 25 July 8/4 The repercussion of this estrangement (the psephological importance of which has yet to be fathomed) was felt in the economy. **1977** *Time* 21 Nov. 28/2 Psephologists will be sorting out the particulars for months to come, but one trend was clear in last week's off-year election returns: a solid vote for sanity.

pseud (siūd), *a.* (*sb.*) *colloq.* [f. the Gr. stem ψευδ- false, or as a shortening of PSEUDO quasi-*adj.* or PSEUDO-.] = *PSEUDO quasi-*adj.* B. Also *absol.* as *sb.* *Pseuds'* or *Pseud's Corner*, used of the pretentious or insincere generally (with allusion to the use in quot. 1968²).
1962 *Spectator* 26 Oct. 656 Present-day trend-setters, pseud as they come. **1964** *Ibid.* 20 Mar. 379/1 The pseuds and intellectual craze-mongers seem to have dropped *cinéma-vérité* almost as quickly as they took it up. **1968** *Jazz Monthly* Apr. 28/2 As well as being the creator of an avant-garde film on human buttocks, Miss Ono has a long list of other achievements which must put her in the running for the title of Pseud of the Century. **1968** *Private Eye* 22 Nov. 3/3 (*heading*) Pseuds' Corner. **1971** *Guardian* 21 Oct. 14/2 Woodstock, the drug scene, the race war, Vietnam... The genre is familiar, and the path through it can verge dangerously close to Pseud's Corner. **1973** P. DICKINSON *Green Gene* I. i. 20 A few big firms..which..don't mind spending a bit on a pseud paper which makes people..think it's all not so bad as all that... It's a real pseud. **1976** *Listener* 25 Mar. 362/2 His ability to ad lib flowery prose with panoramic views overlooking Pseud's

Corner. **1977** *Ibid.* 7 Apr. 447/1 A dreamy piano solo, recalling both Beiderbecke's 'In a Mist' and (I know this sounds pseud) early Schoenberg. **1980** *Oxford Times* 11 Jan. 13/1 Both strike me as the slightly pseud face of rock 'n' roll.

pseudandry: see *PSEUDO- 2a. **pseudentity:** see *pseudo-entity* s.v. *PSEUDO- 2a.

pseudepigram (siūde·pigræm). *nonce-word.* [f. *pseud-* (see PSEUDO-) + EPIGRAM, punning on *pseudepigraph*.] A pretended epigram.
c **1905** F. ROLFE *Nicholas Crabbe* (1958) viii. 60 The frolicsome Thorah screamed pseudepigrams everywhere.

pseudergate: see *PSEUDO- 2a.

pseudery (siū·dəri). *colloq.* [f. *PSEUD *a.* (*sb.*) + -ERY.] An affected or pompous manner of expression, usu. with intellectual pretensions; an example of this.
1972 *Guardian* 24 Feb. 10/3 There's nothing like an overt piece of pseudery to make one feel all lilywhite. **1975** *Daily Tel.* 30 Aug. 6/2 In another paper, a psychiatrist solemnly reported his finding that 'in general fat people do not go to university'... These farragoes of improbable pseudery..tend to be prefaced by an enormously portentous address. **1976** *Broadcast* 12 Jan. 9/3 Capital's Sunday pseudery gets the boot. **1978** *Daily Tel.* 19 Jan. 12/6 Best of all is Mr Ward's pithy dismissal of the sort of supernatural pseudery that has enthralled the credulous since the beginning of time.

pseud-idea: see *pseudo-idea* s.v. *PSEUDO-2a.

pseudish (siū·diʃ), *a.* *colloq.* [f. *pseud-* (see PSEUDO-) or *PSEUD *a.* (*sb.*) + -ISH¹.] Of architecture: imitative and exaggerated. Of other arts: affected, spurious. Also *ellipt.* as *sb.* Hence **pseu·dishness** = *PSEUDERY.
1938 O. LANCASTER *Pillar to Post* 66 Pseudish. This style which attained great popularity both in this country and in America (where it was generally known as Spanish-colonial), is actually our old friend Pont Street Dutch with a few Stockholm trimmings and a more daring use of colour. **1945** *Archit. Rev.* XCVII. 165/1 The Georgian Movement slid into Pseudish, but the ideal—of chaste simplicity—remained. **1972** *Jazz & Blues* Nov. 30/3 Other contributions are getting dangerously pseudish. **1975** *Times Lit. Suppl.* 28 Nov. 1429/2 This style, which surely earns Betjeman's label 'pseudish'. **1976** *Listener* 23 Dec. 814/1 Better, perhaps, than the pseudish silences that have been creeping over telly art in the past year. **1978** *Punch* 6 Sept. 374/1 We're accustomed to pseudishness in Arts Council catalogues.

pseudo, quasi-*adj.* (*sb.*, *adv.*) Restrict † *Obs. rare* to senses in Dict. and add: Now usu. with pronunc. (siū·do). **B.** *adj.* Pretentious, insincere, sham, affected, meaningless; having aspirations beyond true worth. Also *absol.* as *sb.*, a pretentious or insincere person. Pl. **pseudoes, pseudos.**
Now used independently of its relation to the combining form PSEUDO- which characterizes the senses in Dict.
1945 *Archit. Rev.* XCVII. 110/1 The flamboyant, the vulgar, the monumental and the 'pseudo' have had their day. **1958** *Times* 11 Dec. 6/2 The whole conception was 'pseudo'. **1959** K. TYNAN in J. Feiffer *Sick Sick Sick* (ed. 2) Introd., The real Bohemians of Greenwich Village as well as the pseudoes. **1959** K. R. POPPER *Logic Sci. Discovery* ii. 51 Nothing is easier than to unmask a problem as 'meaningless' or 'pseudo'. **1962** *John o' London's* 19 Apr. 371/1 So all the pseudos flocked in. **1964** J. SYMONS *End of Solomon Grundy* III. ii. 150 That's the trouble with the country to-day, too many pseudo people in it... This garage trouble, now, it comes from the pseudos like that man Grundy. **1967** *Observer* (Colour Suppl.) 28 May 14/2 The undiscriminating, arty chat of a campus pseudo. **1977** *Time* 18 Apr. 1/1 Intellectuals all over the world, real or pseudo, proudly proclaim that 'democracy' won.

pseudo-. Add: Now usu. with pronunc. (siū·do). **1. a.** *pseudo-antithesis, -argument, -art, -artist, -book, -communism, -communist, -conversation, -criticism, -definition, -democracy, -difficulty, -emotion, -fact, -folk* (in examples *attrib.*), *-Freud, -grammar, -historicity, -history, -intellectual, -knowledge, -language, -life, -linguistics, -literature, -logic, -moralist, -morality, -mystic, -mysticism, -need, -objectivity, -passive, -passivization, -philosopher, -philosophy* (earlier and later examples), *-principle, -procedure, -proverb, -question, -religion, -simplicity, -theology, -thesis, -word.*
The examples are given here in an alphabetical sequence for convenience of location.
1949 KOESTLER *Insight & Outlook* xi. 169 The biological approach..makes these appear..as a typical pseudoantithesis. **1943** *Mind* LII. 139 The methodological unification here attempted..helps to eliminate pseudo-arguments. **1977** *Theology* May 173 That 'many people would prefer' to do something is an alarmingly Benthamite pseudo-argument. **1960** *Encounter* Mar. 83/1 The dull air of *acedia* that hangs over the mass pseudo-arts. **1934** DYLAN THOMAS *Let.* Dec. (1966) 147 This is the quarter of the pseudo-artists. **1928** D. H. LAWRENCE *Let.* 1 Apr. (1932) 718 That was very nice of you, to send me that little pseudo-book full of red gold. **1945** KOESTLER *Yogi & Commissar* III. iii. 225 The spreading of Russian pseudo-communism over Europe can be stopped only by a true socialist movement. **1948** *Civil & Mil. Gaz.* (Lahore) 11 Apr. 1/1 Nineteen workers of the Lahore Mint, suspected to be Communists or pseudo-Communists, were arrested. **1926** D. H. LAWRENCE *Glad Ghosts* 30 The pseudo-conversation was interrupted. **1951** N. FRYE in D. Lodge *20th Cent. Lit. Crit.* (1972) 423 The literary chit-chat which makes the reputations of poets boom and crash in an imaginary stock exchange is pseudo-criticism. **1956** J. H. WOODGER tr. *Tarski's Logic, Semantics, Metamath.* 285 The sentences under (2), clearly related to the axiom of reducibility of *Principia Mathematica*, can be called pseudodefinitions in accordance with the proposal of S. Leśniewski. **1965** *Language* XLI. 37 Looked on as a theory, traditional grammar consists of primitive notions only. Its definitions are pseudodefinitions, mere embellishments. **1960** KOESTLER *Lotus & Robot* I. v. 161 The result is a pseudo-democracy in a political vacuum. **1977** M. EDELMAN *Polit. Lang.* vii. 130 One of the few psychiatrists to examine such meetings as political phenomena concludes that the self-government is in fact 'pseudodemocracy'. **1905** W. JAMES *Ess. Radical Empiricism* (1912) xi. 254 Closely connected with this pseudo-difficulty is another one of wider scope and greater complication. **1963** J. LYONS *Structural Semantics* ii. 18 A pseudo-difficulty created by posing a pseudo-question. **1949** KOESTLER *Insight & Outlook* xv. 206 The scientist..dismissed them with a shrug as pseudoemotions and purely conventional attitudes. **1938** R. G. COLLINGWOOD *Princ. Art* iv. 61 He would not have based his theory on a pseudo-fact. **1972** S. FISHER *Female Orgasm* (1973) xv. 390 A tremendous volume of pseudofact is being transmitted to people of all age levels about the nature of sexual response. **1962** *Times* 25 July 13/1 The reach-me-down pseudo-folk poetry. **1976** A. MURRAY *Stomping Blues* xi. 212 No less pretentious..are those pseudo-folk blues musicians. **1951** M. LOWRY *Let.* 25 Aug. (1967) 252 You might call it pseudo-Freud and the philosophy of 'nothing but'. **1963** *Times Lit. Suppl.* 31 May 391/2 The mystique is Victorian home-life made..intellectually respectable by pseudo-Freud. **1927** L. BLOOMFIELD in C. F. Hockett *Leonard Bloomfield Anthol.* (1970) 190 Prescientific notions about language, with the dismal study of pseudo-grammar, still prevail in our schools. **1935** *Mind* XLIV. 407 For the general public of Goethe's day (including Goethe himself and other imaginative writers) the concept 'Hellenic' was as little historical as is that of 'Aryan' for the modern Nazis;..like it, it entailed a terrific parade of pseudo-historicity. **1958** T. F. T. PLUCKNETT *Early Eng. Legal Lit.* i. 11 Then there is the pseudo-historicity of our law. **1880** A. H. SAYCE *Introd. Sci. of Lang.* II. ix. 239 An attempt was made to extract a pseudo-history from the Greek myths. **1946** R. G. COLLINGWOOD *Idea of Hist.* 180 Meyer's great merit lies in his effective criticism of the openly positivistic sociological pseudo-history fashionable in his time. **1973** B. J. WILLIAMS *Evolution & Human Origins* iv. 48/2 These taxonomies were used, in turn, to construct pseudo-histories. **1938** *Sun* (Baltimore) 16 Apr. 8 Attacks on the profit motive by many of the pseudo-intellectuals who have supported and colored so much of the Administration's policy in the past. **1977** P. JOHNSON *Enemies of Society* xvi. 218 The fatuous Mary Wimbush, the pseudo-intellectual. **1842** W. NEWNHAM *Reciprocal Influence of Body & Mind* ii. 24 That pseudo-knowledge..would leave its possessor without a single ray of duty. **1957** C. DAY LEWIS *Poet's Way of Knowledge* 16 If you like to think of science as 'knowledge', and poetry as at best some kind of 'pseudo-knowledge', no one can stop you, but you will be thinking in terms unacceptable to many scientists today. **1979** *Dædalus* Spring 15 The critics accuse the positivists of surreptitiously..transforming a pseudo-knowledge into a power which..can only be exerted 'in the interests of the dominant class'. **1960** Pseudolanguage [see *INTERLANGUAGE *sb.*]. **1978** *Amer. Speech* LIII. 61 Samarin suggests that glossolalia is on the same continuum as actual language and is closely related to other kinds of 'pseudolanguage' which can be produced by any sufficiently uninhibited person who is merely playing with language-like sounds. **1942** F. BROWN in *Unknown Worlds* Mar. 6/2 A formula for giving pseudolife to inanimate objects. **1978** G. A. SHEEHAN *Running & Being* ii. 32 His solution..is not to impersonate the achiever... This would be a pseudolife. **1962** H. A. GLEASON in Householder & Saporta *Probl. in Lexicogr.* 86 The reaction to popular pseudo-linguistics. **1964** M. A. K. HALLIDAY et al. *Linguistic Sci.* i. 6 Workers in other fields have gone on working with their own do-it-yourself pseudo-linguistics, being content..with inexact observations and *ad hoc* categories. **1972** *Language* XLVIII. 438 It is indicative of Arens' conservatism and his appreciation of neo-Humboldtian pseudo-linguistics in Germany that he should expand this last chapter. **1944** *Mind* LIII. 185 He therefore can only distinguish 'true literature' (the expression and communication of an experience) from 'pseudo-literature' (*e.g.* advertisement, propaganda, pot-boiling, Collingwood's 'magic' and 'entertainment'). **1964** *Listener* 9 Jan. 61/1 How can John Raymond in *The Sunday Times* deal with seven volumes of verse in just over 1,000 words?.. He can afford to generalize and chat, and in this way he provides a kind of pseudo-literature. **1960** K. AMIS *New Maps of Hell* i. 21 Time travel..is inconceivable, but..an apparatus of pseudo-logic..is set up to support it. **1964** A. WYKES *Gambling* ii. 50 The Victorian pseudo-moralists who screamed..of the dangers of drink and gambling were for the most part unthinking pleasure-stiflers. **1943** *Mind* LII. 19 This is the morality of obedience at its best and surest. Doubtless it is easily confused with the pseudo-morality of sanctions. **1961** *Encounter* Feb. 78 Hugh Kingsmill described Lawrence as 'a pseudo-mystic'. **1964** P. F. ANSON *Bishops at Large* ix. 344 The pseudo-mysticism propagated by Mrs Besant and leading members of the Theosophical Society. **1960** *Commentary* June 472/2 The continual creation of pseudo-needs as a basis for production. **1979** *Time* 8 Jan. 72/3 Lasch detects narcissism nearly everywhere, in the buzz words of the 'human potential' movements, in the 'pseudo needs' created by advertisers for restless consumers. **1946** KOESTLER in *New Writing & Daylight* VII. 82 Novels date more than drama and poetry. The reason for this is the novel's pseudo-objectivity. **1973** MATIAS & WILLEMEN tr. M. Cegarra in *Screen* Spring/Summer 185 The pseudo-objectivity of the film. **1964** *Language* XL. 77 Smx, marks indefinite voice, or pseudopassive. **1965** N. CHOMSKY *Aspects of Theory of Syntax* ii. 104 It is now possible to account for 'pseudo-passives', such as 'the proposal was vehemently argued against',..by a slight generalization of the ordinary passive transformation. *Ibid.* 106 Where 'on the boat' is a V[erb]-Complement in 'John decided on the boat' (meaning 'John chose the boat'), it is subject to pseudopassivization by the passive transformation. **1828** DISRAELI *Voy. Capt. Popanilla* iv. 35 A state of existence which has puzzled many pseudo philosophers. **1842** W. NEWNHAM *Reciprocal Influence of Body & Mind* 159 It is also employed by many pseudo-philosophers as a convenient term. **1966** *Eng. Stud.* XLVII. 154 In the mid-twentieth century the typical Bohemian has become the beatnik poet or pseudo-philosopher. **1817** JANE AUSTEN *Sanditon* (1925) vii. 92 It were Hyper-criticism, it were Pseudo-philosophy to expect from the soul of high toned Genius, the grovellings of a common mind. **1897** Pseudo-philosophy [see *IRRATIONALIST]. **1879** W. JAMES in *Mind* IV. 337 Illusory simplification..is made by invoking some sham term, some pseudo-principle, and conglomerating it and the data into one. **1964** M. A. K. HALLIDAY et al. *Linguistic Sci.* vii. 218 Where he has learned..such a sheer quantity of linguistic material..testing all of it becomes a 'pseudo-procedure'; it just cannot be done. **1965** *Language* XLI. 206 Scholastic pseudo-procedures of discovery. **1949** KOESTLER *Insight & Outlook* vii. 101 A similar pseudo-proverb is, 'He never works between meals'. **1934** R. CARNAP *Unity of Sci.* 40 The danger may arise of being diverted by the material mode of speech into considering pseudo-questions. **1947** D. RYNIN *Johnson's Treat Lang.* 353 On this assumption, 'Is everything known to God?' is for us not a genuine question, but a pseudo-question. **1963** Pseudo-question [see *pseudo-difficulty* above]. **1927** A. HUXLEY *Proper Stud.* 220 There is a powerful religion, or rather pseudo-religion, of sexual purity. **1956** R. C. ZAEHNER in A. Pryce-Jones *New Outl. Mod. Knowl.* 66 The modern pseudo-religions the most obvious of which was Hitlerism. **1969** *Daily Tel.* 24 Apr. 20/3 The pseudo-religion of Social Justice. **1931** E. SAPIR in *Amer. Mercury* XXII. 205/2 The simplicity of English in its formal aspect is..really a pseudo-simplicity or a masked complexity. **1951** M. MCLUHAN *Mech. Bride* (1967) 141/2 It is the weak and confused who worship the pseudosimplicities of brutal directness. **1940** C. S. LEWIS *Let.* 17 Jan. (1966) 176 You will presently see both a Leftist and a Rightist pseudo-theology developing. **1961** 'F. O'BRIEN' *Hard Life* x. 75 His talk is always full of 'ifs' and 'buts', rawmaish and pseudo-theology. **1977** *Rolling Stone* 30 June 62/2 There has also been the startling growth of psychological technology: all the encounter groups and how-to manuals, all the new therapies and analyses, all the pseudotheologies and gurus and disciplines. **1935** R. CARNAP *Philos. & Log. Syntax* 21 All these philosophical theses are deprived of empirical content, of theoretical sense; they are pseudo-theses. **1963** —— in P. A. Schilpp *Philos. of R. Carnap* 51, I argued in detail that the thesis of materialism was just as much a pseudo-thesis as that of idealism. **1951** S. ULLMANN *Princ. Semantics* II. 59 In view of the hybrid nature of particles, it might be convenient to label them *pseudo-words*. **1954** *Archivum Linguisticum* VI. 18 Ullmann..speaks of 'pseudo-words' which would have no full semantic status.

b. *pseudo-American, -antique, -divine, -Elizabethan, -existing, -Georgian, -historic(al), -infantile, -intellectual, -Marxist, -medical, -medieval, -mystical, -philosophic(al), -psychological, -religious* (later examples), *-revolutionary, -romantic* (later examples), *-sophisticated, -Spanish, -technical* adjs.
The examples are also given here in an alphabetical sequence for convenience of location.
1938 Pseudo-American [see *KIDDO]. **1964** C. BARBER *Present-Day Eng.* ii. 20 English pop-singers have developed a special pseudo-American accent of their own. **1936** *Burlington Mag.* May 219/2 The Byzantine pseudo-antique character. **1959** Pseudo-antique [see *KATHAREVOUSA]. **1950** D. GASCOYNE *Vagrant* 33 To be with God, and not pseudo-divine Scorn-inspired self-deceivers. **1946** BLUNDEN *Shelley* 205 One of the endless pseudo-Elizabethan or at least post-Elizabethan compositions. **1956** Pseudo-Elizabethan [see *MIMSEY *a.*]. **1904** B. RUSSELL in *Mind* XIII. 353 False propositions, according to Meinong, are the non-subsisting, merely pseudo-existing objectives of erroneous judgments. **1905** E. WHARTON *House of Mirth* I. i. 7 Its marble porch and pseudo-Georgian façade. **1936** J. BUCHAN *Island of Sheep* vii. 125 The house was..a pseudo-Georgian edifice of red brick with stone facings. **1919** M. BEER *Hist. Brit. Socialism* I. I. v. 51 The soul of moral philosophy was *ius naturale*, which is..pure ethics in a pseudo-historic guise. **1905** O. JESPERSEN *Growth & Struct. Eng. Lang.* x. 246 That pseudo-historical and anti-educational abomination, the English spelling. **1927** W. E. COLLINSON *Contemp. Eng.* 7 Pseudo-infantile forms like pinny (pinafore). **1977** D. MORRIS *Manwatching* 185 The pseudo-infantile woman displays pouted lips, wide-open eyes, and child-like body postures. **1944** KOESTLER in *Horizon* Mar. 173 The pseudo-intellectual hangers-on whose primary motive is..neurosis pure and simple. **1956** A. S. C. ROSS in M. Black *Importance of Lang.* (1962) 99 To say *Miss Austen* instead of *Jane Austen* is either precious or pseudo-intellectual. **1938** *Ann. Reg. 1937* 197 M. N. Pokrovsky... His 'school' is now persecuted as holding pseudo-Marxist, anti-Leninist and therefore unscientific conceptions. **1945** H. READ *Coat of Many Colours* xliv. 219 If Balzac had followed the advice of our pseudo-Marxist critics, he would have made his works subservient to his political theories. **1908** *Jrnl. Amer. Med. Assoc.* 28 Nov. 1860/2 Among the pseudo-medical institutions that have been investigated and closed through fraud orders by the Post-office Department was a Cincinnati concern known as the Epileptic Institute. **1977** *Gay News* 24 Mar. 18/3 Old pseudo-medical myths die hard. **1978** J. UPDIKE *Coup* (1979) iii. 108 Some..calibrated diet whose pseudo-medical niceties were catered to even in the depths of our famine. **1883** Pseudo-medieval [see *HIGH *a.* 6a]. **1967** E. SHORT *Embroidery & Fabric Collage* iv. 117 Too often a designer who happily experiments with plant forms and

animals will, when confronted with the human figure, resort to the hackneyed, pseudo-medieval figure in the nebulous draped garment which is so often seen in church work. **1933** *Mind* XLII. 184 It is indeed to the Greeks, or at any rate to Plato, that this argument, not inconsistently with its quasi-Kantian, Christian or Hebraic (and some will say, pseudo-mystical) flavour, harks back. **1960** KOESTLER *Lotus & Robot* II. xi. 245 The rest is pseudo-mystical verbiage. **1961** *Encounter* Feb. 78 This credo is expressed in pseudo-mystical terms. **1914** J. LONDON *Let.* 26 Mar (1966) 418, I have played with philosophy, expositing the power of mind over matter... While this is..pseudo-philosophic, nevertheless it will make it most palatable to ..the folk who will read it. **1922** C. BELL *Since Cézanne* 82 We shall then be armed..against the portentous 'Ist', whose parthenogenetic masterpiece we are not in a state to relish till we have sucked down the pseudo-philosophic bolus that embodies his eponymous 'Ism'. **1933** M. OAKESHOTT *Experience* IV. 243 A psychology which is not scientific certainly exists, and certainly would not be superseded by a science of psychology... Nevertheless, such a psychology..would be a pseudo-philosophical form of experience. **1940** *Mind* XLIX. 99 The encumbrance of these largely parasitic philosophical and pseudo-philosophical ideas. **1946** *Mind* LV. 360 When we ask what in fact constitutes the order or form of the vegetative realm or any below the highest one, the answer is given in terms derived from human experience and purpose. Or else it is a worse answer in pseudo-mechanical, pseudo-psychological terms, like 'vital force'. **1964** *Eng. Stud.* XLV. 419 Professor Bodelsen's work..eschews..the flights of pseudo-psychological fancy of the gaudier school of criticism. **1938** *Burlington Mag.* Jan. 44/1 Even in old Siena..there were, not only pagan painters, but also pseudo-religious artists ..who substituted prettiness for piety. **1943** K. MANNHEIM *Diagnosis of our Time* vii. 102 It is not a matter of chance that both Communism and Fascism try to..superimpose a pseudo-religious integration. **1957** J. S. HUXLEY *Relig. without Revelation* iii. 59 Only in high civilisations does art become emancipated from religious or pseudo-religious domination. **1978** *China Reconstructs* Nov. 5/1 Their pseudo-revolutionary line and its counter-revolutionary aims are being thoroughly criticized. **1979** *Dædalus* Winter 132 This leads them sometimes..to become unexpected allies..of pseudorevolutionary, violent minorities. **1927** R. H. WILENSKI *Mod. Movement in Art* 29 The degenerate romantic and pseudo-romantic art of the nineteenth century. **1961** D. G. JAMES *Matthew Arnold* i. 26 We may call it 'Romantic' if we like; it is better to call it pseudo-romantic or even sentimental. **1977** *Time* 19 Sept. 11/1 Against such alienation, the pseudo-romantic exploits of Andreas Baader and Ulrike Meinhof hardly needed ideological underpinnings to strike a responsive chord. **1935** W. G. HARDY *Father Abraham* 134 Abraham saw a quiet, controlled, secretive man with that indefinable air of assurance which travel imparts—considerably different, indeed, from the pseudo-sophisticated Lugal of argumentative days but yet recognizable. **1960** A. L. HENCH in *Amer. Speech* XXXV. 73 One of my brighter students recently asked me..why I had used the word *sophisticated* to praise a literary critic... The student said he understood the word meant 'artificial', 'adulterated', 'tarnished', even 'slightly corrupt'... I told him that of course the word sometimes was used to mean 'artificial' or 'false' but that I thought persons who used it this way actually meant *pseudosophisticated*. **1928** H. CRANE *Let.* 31 Jan. (1965) 315 The perfect labyrinth of 'villas'—some pseudo-Spanish, some a la Maya. **1964** H. KÖKERITZ in D. Abercrombie et al. *Daniel Jones* 143 In America, *Don Quixote* and *Don Juan* have assumed a pseudo-Spanish pronunciation. **1945** *Mind* LIV. 185 The philosopher..must be particularly conscious of the pitfalls of language, and especially of all pseudo-technical and emotive language. **1964** P. STREVENS in D. Abercrombie et al. *Daniel Jones* 120 They all commonly bear the same pseudo-technical label.

2. a. pseuda·ndry [Gr. ἀνήρ, ἀνδρ- man], the use by a woman of a male pseudonym; **pseude·rgate** *Zool.* [a. F. *pseudergate* (Grassé & Noirot 1947, in *Compt. Rend.* CCXXIV. 219): see *ERGATE], in certain genera of termites, a blind, wingless member of the colony, carrying out some of the functions of the workers; **pseud-idea:** see *pseudo-idea* below; **pseu·do-a·cid** *Chem.* [tr. G. *pseudosäure* (A. Hantzsch 1899, in *Ber. d. Deut. Chem. Ges.* XXXII. 577)], a compound which is not itself an acid but which exists in equilibrium with, or is easily converted into, an acidic form and thus undergoes some typical reactions of acids; hence **pseudo-aci·dic** *a.*; **pseudo-aci·dity**; **pseu:do-asy·mmetry** *Chem.*, the property of an atom of being bonded to two enantiomorphic groups; hence **pseu:do-asymme·tric** *a.*; **pseu·do-base** *Chem.* [a. G. *pseudobase* (A. Hantzsch 1899, in *Ber. d. Deut. Chem. Ges.* XXXII. 595)], a compound which is not itself a base but which exists in equilibrium with, or is easily converted into, a basic form and thus undergoes some typical reactions of bases; hence **pseudo-ba·sic** *a.*; **pseu:do-basi·city**; **pseudobe·dding** *Geol.*, a structure in which an appearance of stratification has been produced by a cause other than deposition in the apparent planes of stratification; **pseudo-boléite, -boleite** (-bōᵘ·lǝit) *Min.* [a. F. *pseudoboléite* (A. Lacroix 1895, in *Bull. du Muséum d'Hist. naturelle* I. 40), f. *boléite*, similar mineral named after *Boleo*, near Santa Rosalia, Lower California, Mexico, where both were first found], a hydroxide-chloride of lead and copper occurring as translucent blue crystals; **pseudo-bulb** (later examples); **pseudo-bulbous** *a.* (later examples); **pseu:docholine·sterase** *Biochem.*, an enzyme present in the blood and in the liver, brain, and certain other organs which acts on the same esters as cholinesterase and some others besides; **† pseudocirrhosis** *Path.* [after G. *pseudolebercirrhose* (F. Pick 1896, in *Zeitschr. f. klin. Med.* XXIX. 395)] = *PICK'S DISEASE I; **pseu·docode** *Computers*, a programming language that is not a machine language and has to be translated by a computer before it can be executed; **pseu:do-compatibi·lity** *Bot.*, the fertilization of flowers by pollen which would normally be incompatible (*INCOMPATIBLE *a.* 6b); **pseudoconglo·merate** *Geol.* (see quot. 1972); **pseudocotu·nnite** *Min.* [ad. G. *pseudocotunnit* (G. vom Rath 1877), alteration of It. *pseudocotunnia* (A. Scacchi 1873, in *Atti della R. Accad. delle Sci., Fis. e Matem. di Napoli* VI. IX. 38), f. *cottunia*, similar mineral named after D. *Cotungo* (1736–1822), It. anatomist], a potassium lead chloride, K_2PbCl_4, found as dull yellowish or whitish crystals on Vesuvius; **pseudocyesis** (examples); cf. *PSEUDOPREGNANCY I; **pseudocyphe·lla** *Bot.* [CYPHELLA], a small pore in the lower surface of certain lichens; **pseudodo·minance** *Genetics*, the expression of a recessive allele as a result of the deletion of the part of the chromosome bearing the corresponding dominant allele; so **pseudodo·minant** *a.*; **pseu·do-entity, pseude·ntity** *Philos.*, something falsely called or regarded as an entity; **pseu·do-existence** *Philos.* (see quots.); **pseudofæ·ces** *Zool.*, a mixture of mucus and particulate matter from the water that collects in the mantle cavity of a mollusc and is expelled without passing through the digestive system; hence **pseudofæ·cal** *a.*; **pseudofo·vea** *Ophthalm.*, a point of maximum sensitivity on the retina other than the fovea, such as may develop in a squinting eye; hence **pseudofo·veal** *a.*; **pseu·dofracture** *Med.*, a defect in bone that appears on a radiograph as one of a series of narrow, well-defined lines of translucence; **pseu·dogene** *Genetics*, a section of a chromosome that is an imperfect copy of a functional gene; **pseu·dogley** *Soil Science* [a. G. *pseudogley* (W. L. Kubiëna *Bestimmungsbuch und Systematik der Böden Europas* (1953) 295)], a gley resulting from temporary or seasonal waterlogging due to poor drainage of surface water, rather than from the permanent existence of a high groundwater table; **pseudoglio·ma** *Ophthalm.*, any condition that gives rise to signs similar to those of retinoblastoma; **pseudoglo·bulin** *Biochem.* [a. G. *pseudoglobulin* (F. Hofmeister 1899: see *Zeitschr. f. physiol. Chem.* (1900) XXXI. 140, *Beiträge z. chem. Physiol.* (1901) 361)], the fraction of serum globulin that is soluble in pure water and saline solutions but precipitated by half-saturation with ammonium sulphate; **pseu·dogout** *Path.*, a joint disorder resembling gout but produced by deposits of crystals of calcium pyrophosphate rather than sodium urate and occurring most often in the knee; **pseudo-gyny** *Ent.*, pseudogynous condition; **pseu:do-hallucina·tion,** an isolated, brief, and vivid sensory experience, commonly auditory or visual, occurring in clear consciousness and in the absence of any external stimulus; so **pseu:do-hallucina·tory** *a.*; **pseu:dohomose·xual** *a.*, pertaining to or designating homosexual behaviour which acts as an outlet for the expression of fear, aggression, dependency, dominance, etc., rather than being genuinely sexual; hence **pseu:do-homosexua·lity**; **pseudo-hypertrophy** (earlier example); **pseu:dohy:poparathy·roidism** *Med.*, a familial disorder in which the features of hypoparathydroidism are accompanied by skeletal and developmental abnormalities and which is caused by a failure of tissues to respond to parathyroid hormone; so **pseu:do-hy:poparathy·roid** *a.*; **pseu·do-idea, pseu·d-idea,** a meaningless or false idea; **pseu·do-instruction** (also *pseudo instruction*) *Computers*, an instruction similar to a computer instruction in form that is not executed as an instruction by hardware but used to control a compiler or assembler; **pseu:do-isochroma·tic** *a.* *Ophthalm.*, composed of different colours that appear the same to a colour-blind person; **pseudoi·xiolite** *Min.*, an oxide of iron, manganese, tantalum, and niobium that has a highly disordered orthorhombic structure and is similar to ixiolite, differing in changing to tantalite or columbite on heating and in lacking tin; **pseu·dokarst** *Geomorphol.* [cf. It. *pseudocarsico* adj., *-carsismo* sb. (G. B. Floridia 1941, in *Boll. della Soc. di Sci. nat. ed econ. di Palermo* XXIII. 12), and see quot. 1960¹], karst-like topography in ground other than limestone produced by subterranean erosion rather than solution; hence **pseudoka·rstic** *a.*; **pseudo-lympho·ma** *Path.*, any of various conditions involving enlargement of lymph nodes which bear some resemblance to lymphoma but are not malignant; hence **pseudolympho·matous** *a.*; **pseudomembranous** *a.*, add *Path.* and substitute for def.: applied to conditions in which mucous membranes are covered with a sheet formed of exudate; (further examples); **pseudomu·cin** *Med.* [a. G. *pseudomucin* (O. Hammarsten 1882, in *Zeitschr. f. physiol. Chem.* VI. 209)], the thick, tenacious, semi-opaque liquid present in pseudomucinous cysts; hence **pseudomu·cinous** *a.*, epithet of the commonest kind of ovarian cyst, containing mucin; **pseu:domycorrhi·za** *Biol.* [a. Sw. *pseudomykorrhiza* (E. Melin *Studier över de Norrländska Myrmarkernas Vegetation* (1917) II. v. 358)], an association of tree roots and fungi, often mildly pathogenic ones, in the absence of true mycorrhiza; so **pseu:domycorrhi·zal** *a.*; **pseudoneuro·tic** *a.* *Psychol.*, of or pertaining to types of mental illness in which superficial symptoms of neurosis are found in conjunction with underlying symptoms of psychosis, esp. of schizophrenia; **pseu·do-object** *Gram.*, a noun or pronoun that appears to be, but actually is not, an object; **pseu·do-operation** (also *pseudo operation*) *Computers* = *pseudo-instruction* above; **pseu·do-order** *Computers* = *pseudo-instruction* above (see also quot. 1955); **pseudo-pela·de** *Path.* [a. F. *pseudo-pelade* (L. Brocq. et al. 1905, in *Annales Dermatol. et Syph.* VI. 1): cf. PELADA], the appearance of multiple bald patches on the scalp; **pseudoperidium** (later examples); **pseu:doperithe·cium** *Biol.*, in fungi belonging to the order Laboulbeniales, a structure resembling a perithecium in which the vestigial walls have degenerated; **pseudo-pla·nkton** *Biol.*, organisms attached to drifting debris or vegetation; so **pseudoplankto·nic** *a.*; **pseudo-plasmo·dium** *Biol.*, a structureless aggregate of distinct unicellular organisms; **pseudopla·stic** *a.* and *sb.*, (a liquid) that is non-Newtonian, esp. in having a viscosity that decreases with increasing shearing stress; hence **pseu:doplasti·city**; **pseu:dopseu:dohypoparathy·roidism** *Med.*, a familial disorder in which the skeletal and developmental abnormalities of pseudohypoparathyroidism are present without the biochemical abnormalities common to hypoparathyroidism and pseudohypoparathyroidism; **pseu·dopupil** *Ent.* [ad. G. *pseudopupille* (S. Exner *Physiol. d. Facettirten Augen v. Krebsen u. Insecten* (1891) ii. 18)] (see quot. 1977); **pseudo-quadraphony** (-ǫ·fǒni), sound reproduction in which signals from two sources are fed to four speakers in such a way as to give a partial effect of quadraphony; **pseudora·bies,** (*a*) (see quots. 1897, 1912) (?*obs.*); (*b*) *Vet. Sci.*, an infectious viral disease of the central nervous system that causes intense pruritus and usu. death in cattle and affects other domestic animals in varying ways; **pseudorace·mic** *a.* *Chem.*, applied to a racemic substance consisting of mixed or intergrown crystals of the optically active isomers; so **pseudora·cemate**; **pseu·doreaction** *Physiol.*, a spurious positive response to a test; **†pseudoredu·ction** *Cytology*, the apparent halving of the number of chromosomes through synapsis; **pseudorota·tion** *Chem.*, a change in molecular configuration involving concerted displacements of the

constituent atoms, which is equivalent to a rotation of the molecule coupled with a permutation of some of the constituents, but does not involve any actual rotation of the molecule; hence **pseudorota·tional** *a.*; also **pseudorota·te** *v. intr.*, to undergo such a change; **pseudorota·ted** *ppl. a.*; **pseudoru·tile** *Min.*, an oxide of iron and titanium, $Fe_2Ti_3O_9$, that is formed as an intermediate stage in the change of ilmenite to rutile; **pseu·do-salt** *Chem.*, a compound which is normally covalent but which under certain conditions exists as or in equilibrium with an ionized, salt-like form; **pseu·dosocial** *a.*, exhibiting or designating seemingly social behaviour that arises from individual reactions to a need or external stimulus rather than from genuinely social reasons; **pseu·dospecies,** a term used for the different national or racial groups to denote the illusory nature of the belief that they have evolved genetically into different and separate species (inferior to one's own); so **pseu:dospecia·tion,** a false division into species following these lines; **pseudosphere, pseudospherical,** substitute: **pseudosphe·rical** *a.* *Geom.*, being or pertaining to a surface or space whose curvature is everywhere equal and negative; (further examples); hence **pseu·dosphere,** the pseudospherical surface that is generated by rotating a tractrix about its asymptote; (examples); **pseu·do-squeeze** *Bridge*, play whereby an opponent is, or may be, misled into discarding or unguarding a potentially winning card, although he has alternative discards; cf. *SQUEEZE *sb.*; **pseu·do-statement,** an expression that formally resembles a statement but does not refer or correspond to an objective fact, being rather used for its subjective effect on the hearer or reader; **pseu·dostem,** the apparent trunk of a banana plant or a closely related species, which is made up of closely packed leaf sheaths enclosing a stem; **pseu:dostratifica·tion** *Geol.* = *pseudobedding* above; **pseudostra·tified** *a.* (*a*) *Geol.*, exhibiting or of pseudostratification; (*b*) *Histology* (see main entry below); **pseudota·chylyte, -ite** *Petrogr.*, a dark glassy rock resembling tachylyte that results from vitrification by frictional heat generated during dynamic metamorphism; **pseu·dotill** *Geol.*, a deposit similar to a till but non-glacial in origin; **pseudoti·llite** *Geol.* [ad. G. *pseudotillit* (M. Schwarzbach *Das Klima der Vorzeit* (ed. 2, 1961) v. 34], a deposit similar to a tillite but non-glacial in origin; **pseudo-tra·chea** *Ent.*, a fine food-channel in the mouthparts of many flies; also, an organ found in certain woodlice which resembles an insect trachea; **pseu:dotuberculo·sis** *Vet. Sci.* [mod.L., coined in Ger. as the specific epithet of the *Pasteurella* species causing the disease (A. Pfeiffer *Ueber die bacilläre Pseudotuberculose bei Nagethieren* (1889) 5)], any of several diseases clinically and anatomically similar to tuberculosis that occur chiefly in rodents, birds, and warm-blooded animals, esp. sheep, and are caused by species of *Pasteurella* or *Corynebacterium*; **pseudotube·rculous** *a.*, resembling (that of) tuberculosis; **pseu·dotumour** *Path.*, a swelling or other condition that gives rise to the clinical signs of a neoplasm but is not neoplastic, i.e. is not characterized by the persistent proliferation of cells having no physiological function; **pseudou·racil** *Biochem.*, the uracil residue in pseudouridine; **pseudou·ridine** *Biochem.*, a nucleoside, 5-ribosyluracil, found in transfer RNA and differing from uridine in that the sugar residue is attached to the base at a carbon rather than a nitrogen atom; **pseu·dovitamin** *Biochem.*, a compound that is not a vitamin but closely resembles some particular vitamin in molecular structure; **pseudovite·llus** *Ent.* = *MYCETOME; **pseudo-wa·vellite** *Min.* [ad. G. *pseudo-wavellit* (F. Henrich, after Laubmann, 1922, in *Ber. d. Deut. Chem. Ges.* LVB. 3016)], a hydrated basic phosphate of calcium and aluminium, $CaAl_3(PO_4)_2(OH)_5.H_2O$, found as white, grey, or yellow crystals (see quot. 1951); **pseudo-wo·llastonite** *Min.* [a. F. *pseudowollastonite* (A. Lacroix *Minéral. de la France* (1893–5) I. 624)], the high-temperature form of the calcium silicate $CaSiO_3$, which normally changes to wollastonite at ordinary temperatures; **pseudoxantho·ma (ela·sticum)** *Path.* [mod.L., coined in Ger. (J. F. Darier 1896, in *Monatschr. für prakt. Dermatol.* XXIII. 616): cf. XANTHOMA], a congenital disease in which there is widespread disturbance of connective tissue formation, leading to soft, yellowish papules and plaques in the skin, cardiovascular disorder, and ultimately death.
1928 H. M. PAULL *Literary Ethics* xvii. 189 All sorts of devices are adopted by the author, who does not wish his name to be known. He may..disguise it in various ways, using initials, asterisks, a reversed name, or one of the opposite sex,..respectively initialism, asterism, boustrophedon and pseudandry. **1961** T. LANDAU *Encycl. Librarianship* (ed. 2) 294/1 *Pseudandry*, a woman author writing under a masculine pseudonym. **1957** RICHARDS & DAVIES *Imms's Gen. Textbk. Entomol.* (ed. 9) III. 380 Pseudergates occur in *Zootermopsis* and..the so-called workers of *Mastotermes* are probably also of this form. **1969** R. F. CHAPMAN *Insects* xxxiv. 706 The pseudergate is a central form from which various others can be derived. **1979** R. M. ALEXANDER *Invertebrates* xix. 436 Pseudergates also remove the eggs as the queen lays them. **1899** *Jrnl. Chem. Soc.* LXXVI. I. 399 The author [*sc.* Hantzsch] describes as pseudo-acids those substances which do not contain a hydrogen atom directly displaceable by metals, but which are capable of changing into a salt-forming isomeride. **1910** [see *pseudo-acidity* below]. **1929** P. WALDEN *Salts, Acids, & Bases* v. 127 In the homogeneous condition pseudo-acids are associated, usually in the dimolecular form. The carboxylic acids are typical pseudo-acids. **1952** TURNER & HARRIS *Org. Chem.* ix. 135 Primary and secondary nitroparaffins are pseudo-acids, dissolving in aqueous alkali..to give the ions of the *aci*-nitroparaffin or true (nitronic) acid. **1953** C. K. INGOLD *Struct. & Mechanism in Org. Chem.* x. 576 The pseudo-basic carbinol corresponds to the pseudo-acidic phenylnitromethane. **1910** N. V. SIDGWICK *Org. Chem. Nitrogen* vii. 144 The nitroparaffins are typical instances of pseudo-acids. In fact it was in this connexion that Hantzsch developed the theory of pseudo-acidity. **1927** Pseudo-acidity [see *pseudo-basicity* below]. **1907** J. B. COHEN *Org. Chem.* ii. 93 Pseudo-asymmetry of the character of the trihydroxyglutaric acids..is afforded by cyclopropane derivatives. *Ibid.* 94 The cyclic carbon atoms 2 and 4 are asymmetrical *per se*; 3 is symmetrical though structurally identical with 2 and 4. It follows..that 1 and 3 are pseudo-asymmetric. **1962** E. L. ELIEL *Stereochem. Carbon Compounds* iii. 28 Such a pseudoasymmetric atom does not give rise to dissymmetry in the molecule as a whole. **1975** *Nature* 13 Nov. 96/3 Some of these ideas have now appeared in a very general paper..which was followed by two detailed papers on compounds showing axes and planes of pseudo-asymmetry. **1899** *Jrnl. Chem. Soc.* LXXVI. I. 400 Pseudo-bases are substances which, by isomeric change, are capable of giving a true base of the ammonium hydroxide type from which the salts are derived. **1951** C. R. NOLLER *Chem. Org. Compounds* xxx. 591 The *N*-alkyl-α-pyridones are obtained by the oxidation of the quaternary hydroxides. These strong bases appear to be in equilibrium with the α-hydroxydihydropyridines, which are called pseudo bases. **1960** K. W. BENTLEY *Natural Pigments* i. 14 The α- and γ-pyranols are, in fact, pseudo bases, forming a salt and water on treatment with acid. **1921** *Jrnl. Chem. Soc.* CXIX. 1470 One of the more characteristic reactions of pseudo-basic carbinols is the formation of ethers by simple treatment with alcohols. **1927** *Ann. Rep. Progr. Chem.* XXIV. 115 In the same way that in mobile cation tautomerism the hydrogen ion forms a more or less stable covalent link with (negative) carbon, whereas the sodium ion tends to remain in the electrovalent state (thus giving rise to the phenomena of pseudo-acidity), so..the hydroxide ion tends to co-ordinate with positive carbon, whereas very stable anions like chloride..tend to retain their ionic condition..: thus arises the phenomenon of pseudo-basicity. **1850** D. T. ANSTED *Elem. Course Geol., Mineral., & Physical Geogr.* 579/1 (Index), Pseudo-bedding. **1893** *Q. Jrnl. Geol. Soc.* XLIX. 395 Cracks, small at first, have little by little grown into deep joints, and so the pseudo-bedding has been gradually produced. **1939** *Jrnl. Geol.* XLVII. 72 The result of this nonuniformity in distribution of sand is that a type of pseudobedding is developed by concentration or combining of laminae representing the approach slopes of the ripple deposits. **1970** R. J. SMALL *Study of Landforms* xi. 365 Many of the polished surfaces observable in glaciated valleys represent pseudo-bedding planes which have been revealed by removal of overlying sheets of rock by a quite different mechanism. **1897** *Mineral. Mag.* XI. 333 Pseudoboleite... Between boleite and cumengeite. Boleo, Lower California. **1951** C. PALACHE et al. *Dana's Syst. Mineral.* (ed. 7) II. x. 81 Pseudoboleite and boleite have been considered to be identical but the evidence favors the individual character and tetragonal symmetry of the former species. **1964** *Mineral. Abstr.* XVI. 457/2 The mineralogy of the metalliferous areas of Iran is systematically described. Rarer species include..pseudoboléite. **1890** W. WATSON *Orchids* ii. 18 Usually only one pseudo-bulb is developed at the apex or growing point of each rhizome yearly. **1934** R. STOUT *Fer-de-Lance* xvii. 292 Check..the shipment of pseudo-bulbs. **1959** T. B. MORRIS *Death among Orchids* x. 77 Fat pseudo-bulbs and leaves had been torn and bruised by the fall. **1979** B. & W. RITTERSHAUSEN *Orchids in Colour* 14 The plants [*sc.* cymbidiums] produce a number of well rounded pseudo bulbs. **1901** L. H. BAILEY *Cycl. Amer. Hort.* III. 1166/2 The pseudo-bulbous species..should be hosed over thoroughly. **1976** *Hortus Third* (L. H. Bailey Hortorium) 798/2 The rhizomes of such pseudobulbous genera, when well grown, regularly develop one or more new pseudobulbs each season. **1943** MENDEL & RUDNEY in *Biochem. Jrnl.* XXXVII. 59/1 It is the purpose of this paper to show that there exist in the animal body two esterases capable of hydrolysing acetylcholine: a true cholinesterase acting exclusively on choline esters, and a non-specific enzyme, which hydrolyses not only esters of choline but a variety of non-choline esters as well. Moreover, experiments with both enzymes at high and low concentrations of acetylcholine, have revealed a decisive difference between the two esterases, calling for a sharp distinction of the true cholinesterase from the non-specific enzyme, for which we venture to suggest the name pseudo-cholinesterase. **1974** M. C. GERALD *Pharmacol.* iii. 52 Alcohol..and the skeletal muscle relaxant succinylcholine are broken down by the enzymes alcohol dehydrogenase and pseudo-cholinesterase, respectively. **1900, 1940** Pseudocirrhosis [see *PICK'S DISEASE I]. **1953** *Proc. IRE* XLI. 1252/1 Problems are submitted to the computer expressed in pseudo code. An 'interpretive' routine then decodes the input information and calls the subroutines into play as required. **1954, 1958** [see *INTERPRETER 5 b]. **1959** M. H. WRUBEL *Primer of Programming for Digital Computers* ii. 24 Pseudo-codes are often easier to learn than the machine language; only a single pseudo-code instruction is needed to generate frequently used functions such as square root..and log *x*. **1969** J. E. SAMMET *Programming Languages* iv. 129 The early compiling work done in the United States by Dr. Grace Hopper initially involved very artificial pseudo-codes rather than mathematical notation. **1979** *Personal Computer World* Nov. 61/2 Some [programmers] use pseudocode, a written problem definition language that looks like PL/1 or Pascal. **1943** *Nature* 16 Jan. 70/1 *N*[*icotiana*] *Forgetiana* and *N. alama* show pseudo-compatibility only in exceptional circumstances, that is, the actions of the various allelomorphs of the switch gene are always distinctive... Pseudo-compatibility marks the breakdown of the distinction between the types produced by the *S* [*sc.* sterility] allelomorphs. **1977** *Jrnl. Hort. Sci.* LII. 475 Pseudocompatibility was maximized by pollinating old flowers with large quantities of pollen. **1896** C. R. VAN HISE in *16th Ann. Rep. U.S. Geol. Survey* I. 679 The autoclastic rocks which readily show their origin may be called dynamic breccias, and those which resemble ordinary conglomerates may be called pseudo-conglomerates. **1957** F. J. PETTIJOHN *Sedimentary Rocks* (ed. 2) viii. 367 (*caption*) Brecciated siltstone... Brecciation was contemporaneous with sedimentation; a pseudoconglomerate. **1972** *Gloss. Geol.* (Amer. Geol. Inst.) 574/1 *Pseudoconglomerate*, a rock that resembles, or may easily be mistaken for, a true or normal (sedimentary) conglomerate; e.g...a sandstone packed with many rounded concretionary bodies, or an aggregate of rounded boulders produced in place by spheroidal weathering and surrounded by clayey material. **1889** J. L. LOBLEY *Mount Vesuvius* x. 313 Pseudocotunnite... Chloride of Lead with Chloride of Potassium... This mineral, which has the composition of a combination of Cotunnite and Sylvine, was obtained from the sublimations of the crater of Vesuvius following the great eruption of 1872. **1933** *Mineral. Abstr.* V. 269 New records of minerals from Vesuvius are thenardite, pseudocotunnite,..and hieratite. **1817** J. M. GOOD *Physiol. System Nosology* v. iii. 415 Pseudocyesis. Symptoms of pregnancy without impregnation... Spurious Pregnancy. **1859** *Med. Times & Gaz.* 3 Sept. 225/1 There are two varieties of pseudo-cyesis or spurious pregnancy, a *local* and a *constitutional*. **1950** H. B. FRIEDGOOD in R. H. Williams *Textbk. Endocrinol.* x. 657 The sensation of fetal movements is reported usually during the fourth or fifth months of pseudocyesis and seems to be caused by intermittent contractions of the abdominal musculature. **1975** S. L. ROMNEY et al. *Gynecol. & Obstetr.* x. 152/2 Special forms of psychogenic amenorrhea include pseudocyesis or false pregnancy..and anorexia nervosa. **1980** *Daily Tel.* 15 Mar. 3/3 Pseudocyesis, or spurious or phantom pregnancy as it is variously known is a psychological disorder in which a woman has the false but fixed belief that she is pregnant. It is a not uncommon condition, often around the menopause. **1882** *Encycl. Brit.* XIV. 554/1 They [*sc.* cyphellæ] are generally naked, but are often also pulverulent or sorediiferous, in which latter case they are called pseudo-cyphellæ. **1964** *Oxf. Bk. Flowerless Plants* 64/1 The inner layers of the plant [*sc.* *Pseudocyphellaria crocata*]..show as conspicuous spots through small holes (pseudocyphellae) scattered over the lower surface. **1976** M. E. HALE in D. H. Brown et al. *Lichenology* i. 6 They [*sc.* foliose and fruticose lichens] have pores, pseudocyphellae or cyphellae, for gas exchange. **1930** *Amer. Naturalist* LXIV. 561 An actual loss or inactivation of chromatin material, leading to exaggeration and pseudo-dominance. **1938** *Yearbk. Carnegie Inst.* XXXVII. 306 The 'Minute' effect is lethal when homozygous..and gives pseudo-dominance to straw in hybrids. **1969** W. D. STANSFIELD *Schaum's Outl. Theory & Probl. Genetics* viii. 157 When an organism heterozygous for a pair of alleles, *A* and *a*, loses a small portion of the chromosome bearing the dominant allele, the recessive allele on the other chromosome will become expressed phenotypically. This is called pseudodominance, but it is a misnomer because the condition is hemizygous rather than dizygous at the locus. **1965** *Science* 26 Nov. 1123/3 This pseudodominant expression of *fa* can be understood by considering *N* as a deficiency for salivary band 3C7, wherein lies the wild-type allele of *fa*; the facet phenotype is expressed because the wild-type allele is missing. **1975** J. B. JENKINS *Genetics* iv. 165 The white-eye mutant allele was pseudodominant in deletions numbered 258-11..and 264-31. **1896** W. CALDWELL *Schopenhauer's System* iii. 149 A pseudo-entity ..like 'mere matter' or a mere Epicurean god in the interstellar spaces. **1912** *Mind* XXI. 214 'Matter' is..a pseudentity. **1937** B. RUSSELL *Princ. Math.* (ed. 2) p. xi, I do not mean that statements apparently about points or instants or numbers, or any of the other entities which Occam's razor abolishes, are false, but only that they need interpretation which shows that their linguistic form is misleading, and that, when they are rightly analysed, the pseudo-entities in question are found to be not mentioned in them. **1944** M. BLACK in P. A. Schilpp *Philos. B. Russell* 231 Vocabulary, by promoting the hypostatization of pseudo-entities, encourages false beliefs concerning the *contents* of the world. **1956** J. HOLLOWAY in A. Pryce-Jones *New Outl. Mod. Knowl.* 31 Metaphysical jargon could be abandoned once for all, and so could all pretended references to pseudo-entities which could never conceivably be observed. **1904** B. RUSSELL in *Mind* XIII. 207 What is called the existence of an object in presentation is not really existence at all: it may be called pseudo-

existence. 1934 *Mind* XLIII. 375 Confronted with Meinong's obscure and tentative utterances about immanence and pseudo-existence, Mr. Russell reasonably protested. 1953 *Phil. Trans. R. Soc.* B. CCXXXVII. 360 The pseudofaecal strings are conveyed to the end of the waste canals and so into the angle between the inner lobe of the mantle edge and the ctenidial membrane.., above which they are caught in the exhalant current and carried away. 1967 J. H. Day in G. H. Lauff *Estuaries* 401/2 It feeds at low tide levels, spooning up the surface silt with its chelae, sucking out the detritus, and discarding the silt as pseudofecal pellets. 1975 *Jrnl. Marine Biol. & Ecol.* XVII. 4 Since filtration rate and pseudofaecal production determine the amounts of material ingested they are important in studying the efficiency with which bivalves control their rates of ingestion. 1936 *Q. Jrnl. Microsc. Sci.* LXXIX. 207 It was mostly through this aperture ventral to the siphons that collections from the mantle (pseudofaeces) were expelled on sudden closure of the valves. 1976 B. L. Bayne et al. in *Marine Mussels* v. 145 All material to be removed is transferred to the rejection tracts of the mantle edge. These tracts convey the material to the posterior margin of the inhalant siphon, adjacent to the exhalant opening, where it is deposited as pseudofaeces and is carried away by the exhalant current. 1937 *Mind* XLVI. 252 A patient suffering from hemianopia developed a 'pseudo-fovea' in the sound halves of his eyes. 1967 Lyle & Wybar *Lyle & Jackson's Pract. Orthoptics in Treatm. of Squint* (ed. 5) ix. 208 The definition of the image (visual acuity) at the 'pseudo-fovea' will be less clear than at the 'suppressed' fovea. 1946 N. A. Stutterheim *Squint & Convergence* xii. 37 The technique is..to have the non-squinting eye occluded in order to encourage the true fovea and to discourage pseudo-foveal inclinations. 1930 *Amer. Jrnl. Roentgenol.* XXIV. 31/1 Fromme considers the point of involvement of those spontaneous pseudofractures to be about 1 inch below the epiphyseal line. 1950 Shank & Kerley *Textbk. X-Ray Diagnosis* (ed. 2) xi. 136 Pseudo-fractures or *umbauzonen* are always associated with systemic disease or malacic processes, such as Paget's diseases, adolescent rickets.., fibrous dysplasia and osteomalacia. 1976 Gordan & Vaughan *Clin. Managem. Osteoporoses* vii. 80 The pathognomonic x-ray finding of osteomalacia is the presence of bilateral symmetrical pseudofractures, described by Looser in 1920 and therefore often called Looser zones. 1977 C. Jacq et al. in *Cell* XII. 109/1 The 5S DNA of Xenopus laevis, coding for oocyte-type 5S RNA, consists of many copies of a tandemly repeated unit of about 700 base pairs. Each unit contains a 'pseudogene' in addition to the gene. The pseudogene has been partly sequenced and appears to be an almost perfect repeat of 101 residues of the gene. *Ibid.* 109/2 This homologous structure was nearly as long as, and almost an exact repeat of, the gene itself; hence the name—pseudogene. 1978 *Nature* 19 Jan. 205/2 The GC-rich region of 5S DNA also contains a pseudogene which is identical to the first 101 base pairs of the gene at all but 10 sites. 1953 W. L. Kubiëna *Soils of Europe* 242 The modification of 'gley-like soil' to *pseudogley* has been made here to make it conform to the rules of nomenclature, whereby the type designation should be expressed by a noun. *Ibid.* 244 In typical pseudogley the concretions are extremely numerous and are readily visible to the naked eye. 1965 B. T. Bunting *Geogr. of Soil* xiv. 163 Soils on elevated sites on clayey parent materials develop the mottling typical of gleization at some depth beneath a dark brown, granular or blocky structured surface horizon. Such soils have been termed pseudogleys. 1973 J. Mulqueen in Schlichting & Schwertmann *Pseudogley & Gley* 713 The pseudogley soils at Ballinamore are stratified into essentially two layers. 1884 H. R. Swanzy *Handbk. Dis. Eye* xvii. 307 Purulent inflammation of the vitreous humour (to which unfortunately the name pseudo-glioma is sometimes applied). 1946 Berens & Zuckerman *Diagnostic Exam. Eye* ix. 258 In children retinoblastoma (glioma) should be differentiated from an abscess of the vitreous (pseudoglioma). 1962 *New Scientist* 5 Apr. 801/1 Ophthalmologists had been used to examining babies with an abnormal mass of organized tissue behind the lens of the eye, and the non-committal and portmanteau term of 'pseudoglioma' had served as a diagnosis in many of their cases. 1905 *Jrnl. Physiol.* XXXII. 329 A portion of the globulin solution was dialysed... The euglobulin which fell out was removed... The fluid remaining after dialysis contained the pseudo-globulin. 1964 W. G. Smith *Allergy & Tissue Metabolism* vi. 69 Bradykinin is present in normal blood as an inactive precursor, bradykininogen, which is a component of the pseudoglobulin fraction of plasma. 1962 D. J. McCarty et al. in *Ann. Internal Med.* LVI. 711 (*heading*) The significance of calcium phosphate crystals in the synovial fluid of arthritic patients: the 'pseudogout syndrome'. *Ibid.* 712/1 It is suggested that these patients represent a discrete type of arthritis, labeled 'pseudogout' because in some respects it resembles classical gouty arthritis. 1972 *Daily Colonist* (Victoria, B.C.) 2 Mar. 2/2 Pseudo-gout is an attack that resembles an attack of gout, in that it strikes at a joint (usually the knee) with dramatic suddenness and with just as severe pain. 1975 *Amer. Jrnl. Roentgenol.* CXXIII. 532/1 Acute attacks of pseudogout have occurred following surgery, trauma, and injection of mercurial diuretics. 1903 *Jrnl. R. Microsc. Soc.* 172 E. Wasmann returns with fresh light to a discussion of 'pseudogyny' in *Formica sanguinea*. 1890 W. James *Princ. Psychol.* II. xix. 116 From ordinary images of memory and fancy, pseudo-hallucinations differ in being much more vivid, minute, detailed..abrupt and spontaneous, in the sense that all feeling of our own activity in producing them is lacking. 1902 A. R. Defendorf *Kraepelin's Clin. Psychiatry* 7 This group of hallucinations, which has been variously designated as psychic hallucinations.., pseudohallucinations (Hagen), and apprehension hallucinations,..involves several or all of the sensory fields, and..stands in close relation to the other contents of consciousness. 1968 P. McKellar *Experience & Behav.* iv. 120 What is sometimes called 'pseudo-hallucination'..involves a projected perceptual-like image, but one in which the person concerned recognizes the subjective nature of the occurrence. 1902 W. James *Var. Relig. Exper.* x. 251, I refer to hallucinatory or pseudo-hallucinatory luminous phenomena, *photisms*, to use the term of the psychologists. 1955 L. Ovesey in *Psychiatry* XVIII. 17 The dependency and power components..seek completely different, non-sexual goals, but make use of the genital organs to achieve them... For this reason, I have designated these two components as pseudohomosexual. *Ibid.*, This paper consists of a case study that provides clinical documentation for the concept of pseudohomosexuality. 1962 I. Bieber et al. *Homosexuality* i. 10 A neurosis divisible into true and pseudohomosexual types. 1966 I. B. Weiner *Psychodiagnosis in Schizophrenia* xiv. 309 Pseudohomosexual concerns often emerge in the course of these reflections because the thought, 'I am a failure', readily translates into ideas of being something less than an adequate man and hence a homosexual. 1878 D. F. Lincoln tr. Eulenburg in *Ziemssen's Cycl. Pract. Med.* XIV. 155 More and more stress is laid on the connection between pseudo-hypertrophy and progressive muscular atrophy. 1954 *Arch. Dis. in Childhood* XXIX. 404/1 Any effect of chronic hypocalcaemia on physique is likely to be more evident if the disease is present early, and the preponderance of dwarfism among pseudo-hypoparathyroid cases may be in part a reflection of the fact that in them the disorder is probably present from birth. 1942 F. Albright et al. in *Endocrinology* XXX. 922 (*heading*) Pseudo-hypoparathyroidism—an example of 'Seabright-Bartan Syndrome'. 1966 Wright & Symmers *Systemic Path.* II. xxxii. 1126/2 In pseudohypoparathyroidism, because hypocalcaemia is present from birth, the epiphyses and sutures close early, with the result that the patient is short and thickset in stature and has a characteristic round face. 1976 *Lancet* 20 Nov. 1106/2 Among suggested causes [of neonatal hypocalcæmia] are..decreased responsiveness of end-organs to parathyroid hormone (a form of pseudohypoparathyroidism), and vitamin-D deficiency. 1863 H. Spencer *First Princ.* ii. 36 We can entertain them [*sc.* hypotheses] only as we entertain such pseud-ideas as a square fluid and a moral substance. 1879 W. James *Coll. Ess. & Rev.* (1920) 130 Professor Bain would no doubt say that nonentity was a pseud-idea not derived from experience and therefore meaningless. 1911 —— *Some Probl. of Philos.* xii. 197 The pseudo-idea of a connection which we have, Hume then goes on to show, is nothing but the misinterpretation of a mental custom. 1957 D. D. McCracken *Digital Computer Programming* xv. 182 The very first order of business on jumping into the interpretive routine is to increase index 1 by 1 so that it contains the location of the first pseudo instruction. 1967 P. A. Stark *Digital Computer Programming* xii. 198 In addition to all the arithmetic and input-output instructions..the symbolic language has a number of pseudo-instructions to the symbolic assembler. 1975 R. M. Graham *Princ. Systems Programming* 388 Pseudo instructions are used to define symbols, define constants, reserve space in the object segment, and provide the assembler with other information. 1879 T. J. Dills tr. J. Stilling in *Arch. Ophthalm.* VIII. 182 If we intermix the different shades of both interchangeable colors in smaller or larger squares, in such manner that the squares of the one color form letters and figures, and those of the other the groundwork, so that the different intensities alternate in ground and letter..the question as to judgment of colors is rendered unnecessary, the inquiry being merely about letters, numbers, figures. This is the principle of the pseudo-isochromatic plates. 1949 H. C. Weston *Sight, Light & Efficiency* vii. 241 For distinguishing between the varieties of colour-sense deficiency,..what are called pseudo-isochromatic plates are available. 1970 R. A. Moses *Adler's Physiol. of Eye* (ed. 5) xxi. 638/1 Pseudoisochromatic color plates are patterns of colored and gray dots that reveal one pattern to the normal, another to the color deficient. 1963 E. H. Nickel et al. in *Amer. Mineralogist* XLVIII. 976 The ixiolite-like minerals that convert to columbite-tantalite on heating cannot be considered as true ixiolites. For want of a better name, it is suggested that they be referred to as disordered columbite-tantalite, or as 'pseudo-ixiolite'. 1971 *Canad. Mineralogist* X. 758 In all other pegmatites, pseudo-ixiolite forms either tabular grains in a medium-grained albite + quartz + muscovite + garnet assemblage in the internal parts of the pegmatite bodies. 1954 W. R. Halliday *Pseudokarst* (Technical Note No. 25, Nat. Speleological Soc., U.S.) 1 Because the non-solugenic processes are analogous rather than homologous, it seems preferable that they should be considered as independent phenomena, i.e. pseudokarst. *Ibid.*, Pseudokarst may be recognized in three major non-calcareous realms: basalt flows, glaciers, and certain soils. 1960 *Bull. Nat. Speleol. Soc.* XXII. ii. 109/1 About 25 years ago, European geologists began to discuss features of non-solutional origin which are analogous to those of areas of karstic geomorphology. These they termed pseudokarst... Hans Peter Kosack, noted German geomorphologist, believes that the term *pseudokarst* was first employed in print in an Italian publication in 1941... However, as Dr. Kosack has pointed out (pers. comm.),..H. Cramer employed the term in an unpublished study of the karst of the British Isles which was prepared in 1936, and believes that the term was in use in Europe as early as 1930. *Ibid.*, Areas showing pseudokarstic features are distributed quite widely throughout the Western United States. 1968 R. W. Fairbridge *Encycl. Geomorphol.* 849/2 Pseudokarsts produced by piping display disappearing streams, sinkholes,..residual hills and caves. 1971 J. N. Jennings *Karst* i. 5 More obviously pseudokarstic are larger caves..and collapse depressions. 1963 *Cancer* XVI. 928/2 A primary lymphocytic tumour of the lung will be defined as either a malignant lymphoma of the lymphocytic type or an inflammatory pseudolymphoma that originally involves only the lung, or the lung and its regional lymph nodes, and in which there is no evidence of dissemination of the tumor for at least 3 months after the diagnosis is established. 1976 *National Observer* (U.S.) 25 Sept. 14/3 'What did you find?' I asked. 'A lesion, very rare,' he exulted. 'It is called pseudolymphoma.'.. 'Is it cancer?' 'No.' 1972 *Clin. & Exper. Immunol.* X. 202 The diagnosis was localized pseudolymphomatous changes in the parotid glands and regional lymph nodes. 1976 *Chest* LXX. 358/1 Apparently these 'pseudolymphomatous' lesions have the potential to regress with appropriate therapy..or may progress to frank neoplasia. 1924 *Arch. Pediatrics* XLI. 565 (*heading*) Report of a case of pseudomembranous ileocolitis. 1952 *Gastroenterol.* XXI. 212 Acute pseudomembranous inflammation involving portions of the intestinal tract after abdominal operations has been a subject of major importance within the past decade. 1977 *Lancet* 16 Apr. 839/1 Pseudomembranous enterocolitis may arise ..as a complication of colonic obstruction. 1883 *Jrnl. Chem. Soc.* XLIV. 875 Paralbumin is only a mixture of a mucoid substance, pseudomucin, with varying proportions of albumin. 1901 J. L. Rothrock in C. A. L. Reed *Text-bk. Gynecol.* xl. 603 Pseudomucinous (Proliferating) Cysts.—To this group belong the greater proportion of ovarian cysts. 1968 J. W. Huffman *Gynecol. Childhood & Adolescence* xiv. 279/2 Pseudomucinous Cystadenomas. These tumors owe their name to their contents, a gel-like fluid, pseudomucin, secreted by the cells of the epithelium lining the locules within the cysts. 1927 M. C. Rayner *Mycorrhiza* viii. 142 These pseudomycorrhizas are usually simple and unbranched. 1934 *Forestry* VIII. 102 These pseudomycorrhizas show aberrant structure in many respects. 1952 S. A. Waksman *Soil Microbiol.* iii. 88 Pseudomycorrhiza..are endotrophic in nature. 1959 J. L. Harley *Biol. Mycorrhiza* iv. 69 The name pseudomycorrhiza must be used with some care... Those who held a pseudoreligious belief that mycorrhizas 'benefited' their hosts..would refute, by the suggestion that the state being observed was pseudomycorrhizal rather than mycorrhizal, any observation which looked contrary to their own belief. 1949 *Psychiatric Q.* XXIII. 249 In establishing the diagnosis of the pseudoneurotic form of schizophrenia, it will be necessary to demonstrate the presence of the basic mechanisms of schizophrenia. 1966 I. B. Weiner *Psychodiagnosis in Schizophrenia* xvi. 398 Borderline and pseudoneurotic are two of many nosological terms that have been proposed to identify fairly stable personality states in which schizophrenic features are implied but not overtly manifest. 1971 *Brit. Med. Bull.* XXVII. 77/1 There are schools which diagnose simple or pseudo-neurotic forms of schizophrenia. 1965 *Language* XLI. 399 Both reject the same pseudo-objects: From.. *The candidate spoke two hours*:.. (*They reported*) *the speaking of two hours by the candidate*... *Two hours were spoken by the candidate*. 1966 *Eng. Stud.* XLVII. 54 *It* (as a kind of pseudo-object) appears with transitive and intransitive verbs, and finally with original nouns and adjectives (to lord it, to queen it, to rough it) indicating the verbal function of these nominal parts of speech. 1956 *Jrnl. Assoc. Computing Machinery* III. 299 One of the principal objectives of the PACT I compiler has been to eliminate as much as was immediately feasible of the bookkeeping and rudimentary thinking which is involved in the preparation of a computational problem for a large-scale, high-speed digital computer. To this end, a set of PACT pseudo-operations was developed. These operations are more closely related to the computational problem than the machine operands. 1976 W. G. Rudd *Assembly Lang. Programming* v. 62 The START pseudo-operation orders the assembler to begin the assembly process. 1951 M. V. Wilkes et al. *Preparation of Programs for Electronic Digital Computer* 17 A converse of the fact that orders are represented in the machine by numbers is that numbers may be represented outside the machine by 'pseudo-orders', that is, tape entries which are punched in the same form as orders but which are merely intended to be used as constants and are never to be obeyed as orders. 1955 *Jrnl. Assoc. Computing Machinery* II. 1 It is well known that programming may be simplified by the use of pseudoorders. We define these as additional orders (like square root..) which the machine cannot perform. [*Note*] In Wilkes, Wheeler and Gill [1951: see prec. quot.]..what we call pseudoorder is denoted there by 'order', whilst the word pseudoorder is used in a different context. 1964 F. L. Westwater *Electronic Computers* ix. 143 Each subroutine was allocated a code number, and 'pseudo-orders' were written using these numbers as functions. 1909 *Brit. Jrnl. Dermatol.* XXI. 27 Dr. J. M. H. Macleod showed a case of pseudo-pelade (Brocq) or cicatricial alopecia in a man, aged 34 years, affecting chiefly the vertex of the scalp. 1975 S. L. Moschella et al. *Dermatol.* II. xxv. 1206 The etiology of pseudopelade is a mystery. 1887 H. E. F. Garnsey tr. *A. de Bary's Compar. Morphol. & Biol. Fungi* v. 275 The hymenium [of Uredineae] and the rows of spores which proceed from it are enclosed in a membranous envelope composed of a single layer of cells (the peridium, pseudo-peridium or paraphyses-envelope). 1965 Bell & Coombe tr. *Strasburger's Textbk. Bot.* 512 In some genera..all the spores of the peripheral chains and the terminal spores of the other chains lose their spore-like character before breaking through the epidermis and cohere as a firm investment (pseudoperidium). 1832 J. Lindley *Introd. Bot.* i. iii. 207 Pseudoperithecium; Pseudohymenium; Pseudoperidium; terms used by Fries to express such coverings of Sporidia as resemble in figure the parts named perithecium, hymenium, and peridium in other plants. 1895 [see *imperfect *a.* 8 b]. 1903 *Bot. Gaz.* XXXV. 154 The head, which strongly suggests the pseudoperithecium, if it may so be termed, of the more highly differentiated species of Gymnoascus, is thus a remarkable combination of two elements of independent origin. 1928 C. W. Dodge tr. *Gäumann's Compar. Morphol. Fungi* xxiv. 365 At maturity, the vestigial walls of the perithecium [of *Coreomyces*] degenerate, leaving the developing ascogonium surrounded only by the walls of the original cells of the distal region, a pseudoperithecium. 1916 B. D. Jackson *Gloss. Bot. Terms* (ed. 3) 312/1 Pseudoplankton..organisms accidentally found floating. 1935 P. S. Welch *Limnology* ix. 208 Pseudoplankton—debris mingled in plankton. 1947 R. Ruedemann *Graptolites N. Amer.* i. 19 The majority of the typical graptolites lived as pseudoplankton. 1898 *Amer. Naturalist* XXXII. 14 It is highly probable that many graptolites were indeed pseudo-planktonic. 1969 Bennison & Wright *Geol. Hist. Brit. Isles* v. 105 The graptolites were planktonic or pseudoplanktonic. 1892 R. Thaxter in *Bot. Gaz.* XVII. 392 The essential characters of a pseudo-plasmodium are common to both groups. 1966 *McGraw-Hill Encycl. Sci. & Technol.* XI. 44/1 Labyrinthulidae..are uninucleate marine organisms... An aggregate of many individuals may form a motile pseudoplasmodium. 1929 R. V. Williamson in *Industr. & Engin. Chem.* Nov. 1108/1 Certain types of dispersions do not flow in accordance with the laws of either ideal fluids or ideal plastics. The flowing properties of such dispersions are similar in many respects to the flowing properties of ideal plastics; therefore we

shall refer to them as pseudoplastics. The primary difference between pseudoplastic flow and ideal plastic flow is the absence of a real yield value in pseudoplastic flow. 1958 *Ibid.* Jan. 10/2 Aqueous solutions of the poly(ethylene oxide) with intrinsic viscosity of 9 exhibit a mucous-like stringiness, and when observed in a variable shear-rate viscometer can be classed rheologically as pseudoplastic. 1962 *Lancet* 15 Dec. 1263/1 They suggest that the plastic or pseudoplastic flow of abnormally viscous bile caused by metallic ions is converted into a newtonian flow by the action of chelating agents. 1979 A. L. Lydersen *Fluid Flow & Heat Transfer* i. 2 (*caption*) Velocity gradient du/dy as a function of the shear stress $\tau = F/A$. a, dilatant fluid; b, Newtonian fluid; c, pseudo-plastic fluid; d, Bingham plastic fluid. 1938 Weiss & Louis in A. E. Dunstan et al. *Sci. of Petroleum* II. xx. 1128/2 The phenomenon of pseudo-plasticity is of importance in the physico-chemical examination of the large molecules present in petroleum. 1958 *Industr. & Engin. Chem.* Jan. 11/2 The initially high viscosity and the low shear rate pseudoplasticity are reduced. 1967 *New Scientist* 9 Feb. 334/3 Pseudoplasticity describes the fall of viscosity with increasing shear rate. 1952 F. Albright et al. in *Trans. Assoc. Amer. Physicians* LXV. 339 We wish to present today a case with all of the characteristics of pseudohypoparathyroidism except that she has no manifestations suggesting hypoparathyroidism—no hyperphosphatemia, no hypocalcemia. Thus she might be said to have..a 'pseudo-pseudo-hypoparathyroidism'. 1962 *Lancet* 19 May 1075/1 (*heading*) Chromosomal analysis in gonadal dysgenesis with pseudopseudohypoparathyroidism. 1975 *Arch. Dermatol.* CXI. 90/1 A 31-year-old woman with the characteristic features of pseudopseudohypoparathyroidism, such as shortened metacarpals and metatarsals, round facies, and normal serum calcium values, was studied. 1971 *Kybernetik* IX. 159/1 The *deep pseudopupil* of Dipterans is not to be confused with the *corneal pseudopupil*..and especially not with the *reduced corneal pseudopupil*.., in spite of the remarkable similarity of these phenomena. 1977 *Sci. Amer.* July 108/2 Looking at the eye of an insect, we frequently see a black spot in the center of the eye. As the insect rotates its head the black spot always points in the direction of the observer. The spot is known as the pseudopupil: the facets in it look black because they reflect less light in the direction of the observer than the facets in the rest of the eye. 1975 G. J. King *Audio Handbk.* 2 Pseudo-quadraphony..is designated 2–2–4, which implies that the four loudspeakers obtain their signals from the two-channel source. 1976 *Which?* May 99/3 Apart from these, there is a sort of 'fake' quad called, variously, Hafler (after its inventor), ambiophony, or pseudoquadrophony. With this system, you can derive some quadrophonic effect from ordinary stereo recordings and broadcasts. 1897 *Lippincott's Med. Dict.* 840/2 *Pseudorabies*, hysteria resembling rabies, or a condition in animals resembling rabies. 1906 *Jrnl. Nerv. & Mental Dis.* XXXIII. 741 Pseudo-Rabies.—Five cases are cited of pathological alcoholic intoxication in which the patients were wild, making murderous attacks, attempting to bite persons or even trees, bed clothes, etc. 1912 J. J. Walsh *Psychotherapy* xx. ii. 753 There seems no doubt that..pseudo-rabies occurs; that is, persons are bitten by a dog, become seriously disturbed over the possibility of rabies developing, and..there is either a neurosis simulating many symptoms of true rabies, or..even death may take place. 1931 *Jrnl. Exper. Med.* LIV. 246 Among the laboratory animals, rabbits are stated to be more susceptible to pseudorabies than guinea pigs. 1957 Smith & Jones *Veterinary Path.* ix. 258 Pseudorabies may be suspected in disease outbreaks in which animals die shortly after showing very severe pruritics limited to a segment of the skin. 1926 *Jrnl. Chem. Soc.* ii. 2779 Whether he is dealing with a racemate or a pseudo-racemate. 1973 S. F. Mason in Ciardelli & Salvadori *Fund. Aspects & Rec. Devel. Optical Rotatory Dispersion & Circular Dichroism* iii. 212 A pseudoracemate of *lel* and *ob* diastereoisomers is not optically inactive. 1897 Kipping & Pope in *Jrnl. Chem. Soc.* LXXI. 991 Whilst retaining the name racemic compound for a substance belonging to class (*a*), we propose..to call those belonging to class (*b*) pseudoracemic, in order to distinguish them from mere mixtures of the two antipodes on the one hand, and from racemic compounds on the other. 1951 S. Coffey tr. *Wibaut's Org. Chem.* viii. 218 Cases are also known in which mixed crystals of the optical antipodes separate (pseudo-racemic mixed crystals). 1972 R. A. Jackson *Mechanism* v. 84 If the two compounds have opposite configurations no such solid solution will in general be possible, and either a simple eutectic mixture or a 'pseudo-racemic' compound will be formed. 1900 Dorland *Med. Dict.* 544/2 *Pseudoreaction*, a clumping or other bacterial reaction not due to the presence of the typhoid bacillus. 1928 L. E. H. Whitby *Med. Bacteriol.* xxiii. 238 A reaction occurs in both arms, that on the control being a pseudo-reaction whereas that on the test arm is a combination of a pseudo and a positive reaction. 1977 *Compar. Biochem. & Physiol.* B. LVI. 272/2 The nature and origin of the pseudo-reaction of the marine molluscan tissues remains obscure. 1899 *Jrnl. Morphol.* XV. Suppl. 71 It may be stated..in regard to the number of chromosomes, that it is plainly greater than in the first spermatocyte division, which is known to be post-synaptic, *i.e.*, after the pseudo-reduction. 1931 W. Shumway *Textbk. Gen. Biol.* vi. 149 The split..develops so that when the metaphase occurs only half the diploid number of chromosomes may be counted *but* each of these has four parts. There has been no reduction in the sense that any chromosomes have been lost; they have merely united in pairs so that the change in the number of visible chromosomes is called pseudo-reduction. 1972 Cotton & Wilkinson *Adv. Inorg. Chem.* (ed. 3) xiii. 400 A cyclic 5-coordinate intermediate is formed which then pseudorotates. 1960 *Rev. Mod. Physics* XXXII. 451/1 (*caption*) The pseudorotated figure (b) corresponds to sending (a) through a clockwise rotation of $\frac{2}{5}\pi$ and permuting nuclei $1 \rightarrow 3$, $3 \rightarrow 2$, and $2 \rightarrow 1$. 1947 *Jrnl. Amer. Chem. Soc.* LXIX. 2484/2 (*heading*) Pseudo-rotation of ring puckering. 1960 *Jrnl. Chem. Physics* XXXII. 937/1 One..attractive mechanism for the exchange process in PF_5 and PCl_5 is a purely internal pseudorotation. 1974 *Nature* 31 May 474/2 By dynamic reversal of the H_2O binding step before pseudorotation, the two oxygens in the pair of oxoniums could equilibrate with the H_2O oxygens of the medium. 1976 Emsley & Hall *Chem. of Phosphorus* ii. 60 A shorthand notation is required for the *tbp* structures and their pseudorotational transformations. 1966 Teufer & Temple in *Nature* 9 July 180/1 As a result of an investigation of several altered ilmenite concentrates by X-ray techniques hitherto not applied to this problem, a new crystalline phase has been identified as a major constituent of altered ilmenite. The new phase crystallizes in a disordered structure of hexagonal symmetry and has the theoretical composition $Fe_2O_3 \cdot 3TiO_2$. We propose the name 'pseudorutile' for this new mineral. 1975 *Amer. Mineralogist* LX. 905/2 The electrochemical corrosion model is consistent with the pseudorutile composition being a stable alteration product of ilmenite in which all the iron is in the ferric state. 1910 N. V. Sidgwick *Org. Chem. Nitrogen* vii. 152 The mercury must migrate from one position to the other, according to the solvent, just as the hydrogen atom does with a pseudo-acid, and hence mercuric nitroform should be called a pseudo-salt. 1930 *Chem. Abstr.* XXIV. 4021 In alc. soln. there can exist an equil. between the pseudo-salts and the true salts. 1953 C. K. Ingold *Struct. & Mechanism in Org. Chem.* x. 577 Pseudo-salt formation can be seen in the reactions of methylquinolinium salts with various sources of carbanions, for instance Grignard reagents, or pseudo-acidic carbonyl or nitro-compounds. 1964 M. Argyle *Psychol. & Social Probl.* v. 60 One very common type of juvenile delinquent is the 'pseudo-social' delinquent, so called because he is perfectly well behaved towards other members of his gang, but not to people outside of it. 1968 Hebb & Thompson in Lindzey & Aronson *Handbk. Social Psychol.* II. 734 It is important also to exclude pseudosocial behavior, in which grouping occurs only because of some stimulus external to the group: examples are animals running from a forest fire. 1974 *Black Panther* 16 Mar. 11/3 Without new insights, we must really fear new outbreaks of that reactionary pseudospeciation which found (we hope) its climax in Hitler. 1975 *N.Y. Times* 26 Jan. x. 1/1 Erikson calls this 'pseudospeciation'..meaning that man has falsely created divisions where there are none. 1965 E. Erikson in *Amer. Jrnl. Psychiatry* CXXII. 246/1 Sociogenetic evolution has split mankind into pseudo-species, into tribes, nations and religions..which bind their members into a pattern of individual and collective identity, but alas, reinforce that pattern by a mortal fear of and a murderous hatred for other pseudo-species. 1968 —— *Identity* i. 41 Man as a species has survived by being divided into what I have called *pseudospecies*. 1974 *Black Panther* 23 Feb. 10/2 That means we have a common faith ..that each pseudospecies and each empire in some dialectical way added new elements to a more universal sense of humanity. 1889 *Cent. Dict.*, Pseudosphere. 1909 L. P. Eisenhart *Treat. Differential Geom. Curves & Surfaces* viii. 274 The surface of revolution of a tractrix about its asymptote is called the pseudosphere, or the pseudospherical surface of the parabolic type. 1926 J. E. Campbell *Differential Geom.* ii. 28 The formulae of spherical trigonometry or of pseudospherical trigonometry will apply to any surfaces which have the same ground form as the sphere or the pseudosphere. 1947 L. P. Eisenhart *Introd. Differential Geom.* iv. 284 The length of the segment of a tangent to a meridian from the point of contact to the axis of rotation is *a*, and consequently the meridian curve is a tractrix... These pseudospherical surfaces are said to be of the parabolic type. They are called pseudospheres. 1965 J. D. North *Measure of Universe* iv. 60 In 1868 E. Beltrami..showed that Lobachevsky's plane geometry holds in Euclidean space on certain surfaces of constant negative curvature (the pseudospheres) and that these could be conformally represented on a plane. 1909, 1926 Pseudospherical [see *pseudosphere* above]. 1956 J. R. Newman *World of Math.* I. iv. 645 All we know about space, he [*sc.* von Helmholtz] said, is what we have learned from experience. If we lived in a spherical or pseudo-spherical space our sensible impressions of the world would dictate the adoption of the non-Euclidian geometries of Riemann or Lobachevsky; nothing in our intuition would require us to adopt a 'flat-space' Euclidean system. 1939 N. de V. Hart *Bridge Players' Bedside Bk.* xi. 54 The term pseudo-squeeze should be reserved for those occasions when a player by ruse or subterfuge deliberately creates in the mind of an opponent the illusion that it is safe to discard from a certain suit and fatal to discard from another, when in fact the reverse is the case. 1975 *Times* 22 July 5/1 Careful defence was needed by the British pair..to avoid being caught in a pseudo-squeeze. 1926 I. A. Richards *Sci. & Poetry* vi. 56 We must confine ourselves to the other function of words, or rather..to one form of that function, let me call it pseudo-statement. *Ibid.* 59 A pseudo-statement is a form of words which is justified entirely by its effect in releasing or organizing our impulses and attitudes..; a statement, on the other hand, is justified by..its correspondence..with the fact to which it points. 1933–5 Wittgenstein *Blue & Brown Bks.* (1958) 71 One of the reasons why we are tempted to make our pseudo-statement is its similarity with the statement 'I only see this.' 1940 *Kenyon Rev.* 271 Poetry consists essentially of pseudo-statements. 1947 D. Rynin *Johnson's Treat. Lang.* 333 Only if the expression has statement meaning shall we consider it a genuine statement; otherwise we shall call it a pseudo-statement, provided it satisfies the purely grammatical requirements. 1894 *Kew Bull.* 231 The stem (pseudo-stem) in Musas usually arises from a perennial rootstock. 1927 *Bot. Gaz.* LXXXIV. 337 If..the trunk is cut across..it is found to be a pseudostem composed of the overlapping close-fitting leaf sheaths alone. 1957 *New Biol.* XI. 71 The sappy, leafy banana trunks..are really pseudostems consisting of the overlapping leaf-bases. 1972 J. W. Purseglove *Trop. Crops: Monocotyledons* II. 357 The last leaves are produced at nodes on the flowering stem in the centre of the pseudostem. 1874 *Q. Jrnl. Geol. Soc.* XXX. 253 The great masses of gabbro in Rum often exhibit that pseudo-stratification so often observed in igneous rocks. 1941 *Amer. Jrnl. Sci.* CCXXXIX. 1 (*heading*) The development of pseudo-stratification by metamorphic differentiation in the schists of Otago, New Zealand. 1959 *Econ. Geol.* LIV. 1161 Even, regular layering is the commonest type of pseudostratification in the section. 1874 *Q. Jrnl. Geol. Soc.* XXX. 245 When the northern face of this mountain is viewed from a little distance, the whole mass presents the appearance of being made up of a number of concentrically curved beds. This pseudo-stratified appearance is.. very frequently presented by masses of undoubted igneous origin. 1940 *Trans. Geol. Soc. S. Afr.* XLII. 59 The rocks in this area are very clearly pseudostratified and form hilly country. 1917 S. J. Shand in *Q. Jrnl. Geol. Soc.* LXXII. 199 The name pseudotachylyte has been adopted in recognition of the fact that these rocks have a great similarity to tachylyte. 1954 H. Williams et al. *Petrography* xi. 202 X-ray investigation and measurement of the refractive index have shown some pseudotachylytes to be cryptocrystalline products of extreme crushing of rocks such as granite, without actual melting. 1971 *Nature* 17 Sept. 189/1 More than half of the rocks consist of a very irregular pattern of small black veins embracing gneiss fragments—'pseudotachylytes'. 1977 A. Hallam *Planet Earth* 177 Very rarely temperature may rise enough for melting, and such frictionally-produced melts are called pseudotachylites. 1957 J. K. Charlesworth *Quaternary Era* I. xxvii. 569 Solifluxion..produces stony clays or pseudo-tills. 1966 *Earth-Sci. Rev.* II. 249 None of these thickness criteria, alone, can distinguish tills..from pseudo-tills. 1963 R. O. Muir tr. *Schwarbach's Climates of Past* v. 39, I would recommend that the term tillite be applied not only to undoubted moraines but to all moraine-like sediments of probable or possible glacial or glacio-marine origin. Those later shown to be..of non-glacial origin, may then, more properly, be called pseudo-tillites. 1963 D. W. & E. E. Humphries tr. *Termier's Erosion & Sedimentation* x. 205 These graywackes, which must be deposited in deep water.., consist of grains of angular sand set in an argillaceous groundmass, sometimes with boulders or fragments (pseudo-tillites) not unlike the material of glacial moraines. 1968 R. W. Fairbridge *Encycl. Geomorphol.* 473/2 Any sort of 'accidental' mixture such as is caused by a gravitational flow..can easily be taken for a glacial till, i.e., it is a pseudotillite. 1890 B. T. Lowne *Anat., Physiol., Morphol., & Devel. of Blow-Fly* I. iv. 146 The Pseudotracheæ are cylindrical channels on the oral surface of the disc. 1925 A. D. Imms *Gen. Textbk. Entomol.* iii. 595 Fine trachea-like food channels or pseudotracheæ become evident [in Diptera]. 1954 *New Biol.* XVII. 44 Those species [of woodlice] which can withstand drier conditions are also those which possess these 'pseudotracheae'. 1975 *Nature* 27 Mar. 325/1 On each half labellum of *Drosophila* there are..some 25 taste pegs between the pseudotracheae. 1896 G. M. Sternberg *Text-bk. Bacteriol.* ii. xvi. 608 Preisz (1894) has compared the bacillus of pseudo-tuberculosis described by Nocard with that of Pfeiffer, of Parietti, and of Zagari, and finds them identical. 1899 E. O. Jordan tr. *Hueppe's Princ. Bacteriol.* iv. 201 If tubercles occur in which, instead of the tubercle bacillus, other bacteria are found the affection is called pseudo-tuberculosis. 1959 R. Lovell in Stableforth & Galloway *Infectious Dis. Animals* I. vi. 250 Caseous lymphadenitis of sheep. This is a chronic disease frequently referred to as pseudo-tuberculosis and widely distributed in South America, Australia and New Zealand. 1977 Andrewes & Walton *Viral & Bacterial Zoonoses* xxviii. 145 The numerous cases of pseudotuberculosis in zoos and research establishments are most probably due to the contamination of feeding-stuffs and water by wild birds and rodents. 1907 *Jrnl. Compar. Path. & Therapeutics* XX. 53 Mainly on account of the acid-fast character of the bacilli, the disease has been referred to as a pseudo-tuberculosis, and Bang has suggested that it should be called 'chronic bovine pseudo-tuberculous enteritis'. 1957 S. L. Robbins *Textbk. Path.* xxix. 1101/1 Pseudotuberculous (giant cell) thyroiditis. This form of inflammation of unknown etiology is so named because of its histologic appearance. 1962 *Lancet* 19 May 1042/1 We describe here a further case of pseudotuberculous mesenteric adenitis,..together with evidence that 3 other children in the same family and a pet dog had also been infected with *Past. pseudotuberculosis*. 1901 A. P. Ohlmacher in Hektoen & Riesman *Text-bk. Path.* I. 288 Of the various pathogenic blastomycetic species obtained from morbid processes in man or under saprophytic conditions, all failed to produce anything else than 'pseudotumors'. 1938 Smith & Gault *Essent. of Path.* xxiii. 232 This resemblance [to true tumors] ceases on microscopic study when the histologic evidence of neoplastic change of the cells is found to be lacking. The cells in such pseudotumors will show evidence of inflammatory hyperplasia. 1944 J. F. Brailsford *Radiol. Bones & Joints* (ed. 3) xvii. 241 A pseudo-tumour of the spinal cord was recorded by W. E. Dandy—this was found to be a shell of bone surrounding the posterior half of the spinal cord. 1974 Passmore & Robson *Compan. Med. Stud.* III. xxxiv. 12/1 A pseudotumour is one of the names given to a syndrome of raised intracranial pressure unassociated with a space-occupying lesion. 1964 G. H. Haggis et al. *Introd. Molecular Biol.* ix. 218 In 5-ribosyl-uracil ('pseudouracil') the uracil ring is attached to the backbone ribose through the carbon at position 5 in the uracil ring, rather than through the nitrogen at position 3. 1970 Ambrose & Easty *Cell Biol.* iv. 133 It contained, like most tRNA molecules, a number of unusual bases, such as inosine... and pseudouracil.., and certain methylated forms of normal bases. 1978 *Sci. Amer.* Jan. 62/3 In some systems the control function is associated with a particular modified nucleotide in the tRNA molecule, for example a uracil that has been converted into a pseudouracil. 1959 *Biochimica et Biophysica Acta* XXXII. 571 It is proposed (by Dr. A. Michelson) that this substance be called pseudouridine, with the symbol Ψ for the prefix 'pseudo' in abbreviations. 1964 A. White et al. *Princ. Biochem.* (ed. 3) xxx. 607 Formation of pseudouridine may also occur at the polynucleotide level, but the mode of synthesis is unknown. 1978 *Sci. Amer.* Jan. 55/1 (*caption*) Other structural modifications also occur. For example, the nucleoside pseudouridine (ψ) has its base attached to the ribose through a carbon atom instead of a nitrogen atom. 1951 *Abstr. of Papers 120th Meeting Amer. Chem. Soc.* 22C (*heading*) Crystalline pseudovitamin B_{12}. 1956 *Nature* 28 Jan. 188/1 Several substances, such as factor *A* and pseudo-vitamin B_{12}, are closely related to cobalamin. 1967 *Oceanogr. & Marine Biol.* V. 383 As in many other crustaceans, vitamin B_{12} is found in *Nephrops*... It occurs as the analogs Factor B and pseudovitamin B_{12}. 1858 T. H. Huxley in

Trans. Linn. Soc. XXII. 208 The central mass..completely simulates the vitellus of an impregnated ovum; and I will therefore term it a 'pseudovitellus'. 1899 D. Sharp in Harmer & Shipley *Cambr. Nat. Hist.* VI. viii. 588 There exists [in aphids]..a peculiar structure, the pseudovitellus, a sort of cellular, double string. 1924 *Philippine Jrnl. Sci.* XXIV. 150 Henneguy..also described the origin of the 'pseudovitellus' from the follicular epithelium. 1946 Pseudovitellus [see *mycetome]. 1925 *Mineral. Mag.* XX. 463 Pseudowavellite... Hydrated phosphate of aluminium with lime, ferric iron, and rare-earths; occurring as white encrustations (trigonal needles) on limonite and wavellite at Amberg, Bavaria. So named because of its resemblance to wavellite, of which it is perhaps an alteration product. 1942 [see *lewistonite]. 1951 C. Palache et al. *Dana's Syst. Min.* (ed. 7) II. 837 Available evidence indicates that crandallite and pseudowavellite are best considered as a single species with some variation of composition, the name crandallite having priority. 1906 *Amer. Jrnl. Sci.* CLXXI. 105 Pseudo-wollastonite appears either in the form of small irregular grains often tabular in shape or in short prisms or fibers arranged in parallel or divergent groups. 1942 *Ibid.* CCXL. 729 It is a most remarkable fact that, if powdered glass of the composition $CaSiO_3$ is crystallized at any temperature, the product always consists almost exclusively of pseudo-wollastonite, but lumps of the same glass will crystallize readily to wollastonite with only a trace of pseudo-wollastonite present at temperatures between 800° and 1100°. 1970 R. W. Andrews *Wollastonite* 2 There are two polymorphs of calcium monosilicate: wollastonite the low temperature form, and pseudowollastonite..the high temperature form. *Ibid.* 5 Natural pseudowollastonite has been reported from only one locality, in Iran, near the head of the Persian Gulf. 1900 Dorland *Med. Dict.* 545/1 *Pseudoxanthoma*, a disease resembling xanthoma. 1901 *Brit. Jrnl. Dermatol.* XIII. 232 The author ranges himself with Darier, and considers the condition to be due to a degeneration of elastic tissue... The qualification 'pseudo'-xanthoma should be insisted upon. 1933 *Arch. Dermatol. & Syphilol.* XXVIII. 553 The histologic evidences of pseudoxanthoma elasticum are fragmentation and degeneration of the elastic tissue. 1961 *Lancet* 12 Aug. 356/2 Although the lesions attributable to pseudoxanthoma elasticum may be widely distributed through the tissues and organs of the body, it is more commonly recognised by dermatologists and ophthalmologists than by general physicians. 1977 *Proc. R. Soc. Med.* LXX. 569/1 Instead of the well ordered wavy collagen bundles of normal skin, the middle and lower dermis in pseudoxanthoma elasticum shows an irregular network of tangled, curled and branching fibres.

b. *Cytology.* **pseudodi·ploid** (**-te·traploid,** etc.) *adjs.*, having a chromosome complement which differs from the normad diploid (tetraploid, etc.) complement in constitution but not in number. So **pseudodi·ploidy,** etc.

1923 *Bot. Gaz.* LXXVI. 330 Of two plants from our cultures, each of which had a total of 48 chromosomes in their somatic cells.., one appears to have been a chromosomal mutant of the type $(4n+1-1)$ and the other a mutant of the type $(4n+1+1-1-1)$. Such forms obviously cannot properly be called $4n$ or tetraploid. They ..may be classified as modified tetraploids, or at most as 'pseudotetraploids'. 1977 *Lancet* 30 Apr. 961/1 Pseudodiploidy, 46,XX,D,+,17q−, was observed in 14 marrow and in 10 blood-cells by the Giemsa staining method. 1978 *Nature* 16 Mar. 262/1 Stable diploid, or occasionally pseudodiploid transformed cell lines have been obtained after transformation by various tumour viruses.

pseudoallele (siūdo,æ·līl). *Genetics.* [f. Pseudo- + *allele, or a back-formation from *pseudoallelism.] Each of two or more mutations that resemble alleles of a single gene functionally, in affecting the same process or property, but differ structurally, in that crossing-over is possible between them.

1948 *Genetics* XXXIII. 113 The bithorax mutants, *bx* and *bx*[3] (locus, 3–58·8), are pseudo-alleles of bithoraxoid-dominant, *bxd*^D. 1956 *Nature* 17 Mar. 504/2 Phenomena such as position effect and pseudoalleles make the classical corpuscular picture of the gene obsolete in modern genetics. 1962 W. R. Singleton *Elem. Genetics* xiv. 235 In any organism susceptible of precise analysis, pseudo-alleles have turned out to be the rule rather than the exception. 1975 V. Grant *Genetics of Flowering Plants* iv. 75 The pseudoalleles are so close on the genetic map that crossovers between them occur only very rarely.

So **pseudoalle·lic** *a.*, behaving as or consisting of pseudoalleles; **pseudoalle·lism,** pseudoallelic state or property.

1938 *Yearbk. Carnegie Inst.* XXXVII. 305 Any dominant mutant which is lethal when homozygous and which shows pseudo-allelism to a dissimilar, non-allelomorphic but neighboring mutant is probably a deficiency. 1948 *Genetics* XXXIII. 113 Pseudo-allelism is characterized by the presence of closely linked genes, which seem to act developmentally like one. 1953 *Adv. Genetics* V. 208 The most completely investigated pseudo-allelic series is the case of 'lozenge'. 1975 J. B. Jenkins *Genetics* ix. 388 Primarily as a result of the analysis of pseudoallelic series, the gene emerged conceptually as a unit of function, or cistron. 1975 V. Grant *Genetics in Flowering Plants* iv. 74 The phenomenon of crossing-over within the limits of what had been considered a gene became known as pseudoallelism.

pseudo-American to **-book:** see *pseudo- 1 a, b, 2 a.

pseudobre·ccia. *Geol.* Also with hyphen. [f. Pseudo- + Breccia.] A limestone in which partial and irregular dolomitization has produced a texture similar to that of a breccia. So **pseudobre·cciated** *a.*; **pseu:dobreccia·tion,** the structure or state of a pseudobreccia.

1907 A. Strahan et al. *Geol. S. Wales Coal-Field* VIII. (Mem. Geol. Survey No. 247) 10 Mr. [R. H.] Tiddeman notes that some of the limestones of Mumbles Head..and Oxwich exhibit pitted bedding-surfaces and a pseudo-brecciated structure. *Ibid.*, The structure referred to as pseudo-brecciation gives the rock the appearance of having been crushed into small angular fragments and re-cemented. *Ibid.* 12 In Callencroft Quarry a pseudo-breccia..is exposed, but dolomitisation appears to have taken place along a fault also. 1913 *Jrnl. Geol.* XXI. 407 That this is actually a case of pseudobrecciation, and not a brecciated structure due to the cementation of a dolomite breccia in a calcareous matrix, is evident from a microscopical examination. 1963 D. W. & E. E. Humphries tr. *Termier's Erosion & Sedimentation* XIV. 301 This bed is a pseudo-brecciated layer of Early Malm (Late Jurassic) age. *Ibid.*, The compaction of the sediment was due to migration of 'imbibed' water... This compaction is probably associated with the pseudo-brecciation. 1969 Bennison & Wright *Geol. Hist. Brit. Isles* ix. 189 The limestones of the D Zone are oolitic or rubbly with pseudobreccias and are highly fossiliferous.

pseudo-bulb(ous, -cholinesterase: see *pseudo- 2 a.

pseudocide (siū·dosaid). Also with hyphen. [f. Pseudo- + Sui)cide *sb.*[1] and *sb.*[2]] **a.** A pretended attempt at suicide, undertaken with the intention of failure. **b.** One who makes such an attempt. Hence **pseu·docidal** *a.*

1959 Lennard-Jones & Asher in *Lancet* 30 May 1138 (*heading*) Why do they do it? A study of pseudocide. *Ibid.* 1140/1 People are described who deliberately harm themselves or take an overdose of tablets without wishing to die. We have termed these actions 'pseudocide'. 1970 *Hospital Tribune* 23 Mar. 22/1 Another characteristic of the pseudocidal persons was found to be 'their understandable propensity to leave suicide notes'. *Ibid.* 22/2 The typical pseudocide is remarkably different from the usual picture associated with actual suicidal persons. 1976 *New Society* 22 Jan. 147/1 The reason this attempt is called 'pseudo-cide' is that it is invariably a gesture and is not meant to succeed—witness the fall in the number of attempts during the junior doctors' strike, when relief was not necessarily at hand.

pseudocirrhosis to **-compatibility:** see *pseudo- 1 a, 2 a.

pseu·do-concept. *Philos.* Also without hyphen. [f. Pseudo + Concept *sb.*] A notion which is sometimes treated as a concept though it cannot be properly conceptualized or grasped by the mind.

1866 H. L. Mansel *Philos. of Conditioned* 93 It is not to be wondered at..that our positive conception of God as a Person cannot be included under this pseudo-concept of the Infinite. 1901 A. E. Taylor *Probl. Conduct* viii. 439 The religious experience..may for all we know prove to be itself a mere illusion, and the Absolute a mere pseudo-concept. 1917 D. Ainslie tr. *Croce's Logic* ii. 37 'Conceptual fictions' is a manner of speech... For brevity's sake we shall call them pseudoconcepts. 1937 *Mind* XLVI. 228 The consequence is rigidly deduced that the 'transcendent' One cannot even without contradiction be said to be *One*; thus it is apparently a 'pseudo-concept'. 1956 J. O. Urmson *Philos. Analysis* 91 There are two objects with a certain relation between them, in spite of the fact that the concept 'object', thus used, is a metaphysical pseudo-concept. 1967 *Encycl. Philos.* II. 266/2 The work of economists, like that of all other scientists, belongs to the category of utility itself, not to that of truth. 'Economic man' is a paradigm case of a pseudo concept.

pseudoconglomerate to **-entity:** see *pseudo- 1 a, b, 2 a, b.

pseu·do-event. orig. *U.S.* [Pseudo- 1.] An event arranged or brought about merely for the sake of the publicity which it generates. Hence **pseudo-eve·ntful** *a.*

1962 D. J. Boorstin *Image* i. 9 A pseudo-event..is not spontaneous, but comes about because someone has planned, planted, or incited it. Typically, it is not a train wreck or an earthquake, but an interview. *Ibid.* iv. 161 A pseudo-eventful by-product of the [celebrity] star system is what *Time* magazine has accurately described as 'non-books.' These are printed matter between covers, usually put together by someone other than the ostensible autobiographer. 1962 *Spectator* 22 June 823/2 Another pseudo-event three thousand miles away. On March 17, I had stood on Fifth Avenue, barred from crossing Manhattan by a quarter of a million people..parading..in ..rosettes and badges saying 'Kiss Me, I'm Irish.' 1963 *Guardian* 21 Jan. 8/2 By enthroning the pseudo-event at the heart of an £800 million industry [*sc.* football pools], the promotors have made clear..that nothing..is what it seems. 1966 D. Jenkins *Educated Society* i. 36 The centre of the public stage will be occupied by performers, whether politicians or 'celebrities', who live out a life of highly publicised pseudo-events for the delectation of multitudes. 1969 *Daily Tel.* (Colour Suppl.) 1 Aug. 10/4 Beauty Festivals are pseudo-events of almost clinical purity. 1976 *Listener* 28 Oct. 524/3 The [US] election campaign became a perfect pseudo-event..a game played on..trains, in..planes, and in shopping precincts.

pseudo-existence to **-Freud:** see *pseudo- 1 a, b, 2 a.

pseudogamy (siūdǫ·gămi). *Biol.* [ad. G. *pseudogamie* (W. O. Focke *Pflanzen-Mischlinge* (1881) vii. 510): see Pseudo- and *-gamy.] **a.** In an apomictic plant, development of an embryo following pollination without fertilization.

1900 B. D. Jackson *Gloss. Bot. Terms* 213/2 *Pseudogamy*, parthenogenetic fruiting, as pollination without impregnation of ovules. 1908 *Ann. Bot.* XXII. 42 *Humaria rutilans* is..an example of the so-called apogamous development of the ascocarp, or pseudogamy. 1956 *Nature* 21 Jan. 141/2 A third group of populations [of *Culex pipiens*] is interfertile with the others in one direction only. Parthenogenesis, with pseudogamy, predetermination and multiple compatibility genes can be excluded as causative mechanisms. 1974 *New Phytol.* LXXIII. 1243 Pseudogamy is a type of agamospermic reproduction in which seed development is stimulated by pollination although the male nucleus does not fuse with the egg. 1976 Bell & Coombe tr. *Strasburger's Textbk. Bot.* (rev. ed.) 403 In many angiosperms embryos develop from diploid unfertilized egg-cells (parthenogenesis and angiospermy..), but sometimes this development requires the stimulus of pollination (pseudogamy).

b. The fusion of two vegetative nuclei.

1907 *Annales Mycologica* V. 422 Fusion of two vegetative nuclei: pseudogamy. 1928 C. W. Dodge tr. *Gäumann's Compar. Morphol. of Fungi* iv. 13 The sexual processes occurring outside in the thallus between two sexually differentiated vegetative cells..called pseudomixis (pseudogamy of Hartmann). *Ibid.*, Since the copulating cells are not morphologically distinguished from other vegetative cells and since only the release of specific developmental stimuli..marks this anastomosis of two vegetative cells as a sexual process, pseudogamy is often distinguished with difficulty from the usual pseudo-sexual anastomoses which are brought about by food relations.

Hence **pseudo·gamous** *a.*

1932 C. D. Darlington *Rec. Adv. Cytol.* xv. 434 Moderate or even high pollen fertility is sometimes found in the pseudogamous *Potentilla* species. 1964 W. Williams *Genetic Princ. & Plant Breeding* viii. 290 Hybrid endosperms in pseudogamous apomicts derived from crosses between unrelated parents promote stronger growth in apomictic seedlings. 1974 *New Phytol.* LXXIII. 1246 The three pseudogamous species..would set seed when pollinated by each other or intraspecifically, but not when pollinated by any of quite a large selection of different *Potentilla* species.

pseudogene to **-hallucinatory:** see *pseudo- 1 a, b, 2 a.

pseudoha·logen. *Chem.* Also with hyphen. [a. G. *pseudohalogen* (Birckenbach & Kellermann 1925, in *Ber. d. Deut. Chem. Ges.* LVIII. 786): see Pseudo- and Halogen.] Any of a class of compounds (in some cases hypothetical) which have small molecules built up from atoms of electronegative elements and which closely resemble the halogens in many respects.

1925 *Chem. Abstr.* XIX. 1996 (*heading*) Pseudo halogens. 1954 R. C. Brasted in M. C. Sneed et al. *Comprehensive Inorg. Chem.* III. ix. 223 The similarities between the pseudohalogens and the halogens are as follows: 1. They are in general fairly volatile. 2. They show affinity for metals, with which they combine directly to form salts. 3. The silver, lead, and mercury (I) salts are insoluble in water. [Etc.] 1962 Cotton & Wilkinson *Adv. Inorg. Chem.* xxii. 465 The most important pseudohalogens are $(CN)_2$, cyanogen; $(OCN)_2$, oxycyanogen (existence in free state uncertain); $(SCN)_2$, thiocyanogen; $Se(CN)_2$, selenocyanogen; and $(SCSN_3)_2$, azido-carbon disulfide. 1965 Phillips & Williams *Inorg. Chem.* I. xii. 467 All the pseudo-halogen anions enter into simple acid-base reactions with the proton in aqueous solution.

Hence **pseudoha·lide** [*halide], a compound, ion, or radical formed by a pseudohalogen.

1925 *Chem. Abstr.* XIX. 1996 An attempt was made to prep. $(SeCN)_2$, $(CNO)_2$ and $(TeCN)_2$ by electrolysis of K pseudo halides in alc. solns. 1965 Phillips & Williams *Inorg. Chem.* I. xii. 467 The difference between the two series of anions, halides and pseudo-halides, lies in the unsaturation of a pseudo-halide which is most marked in CN^-. The unsaturation places the pseudo-halides high relative to the halides in the spectrochemical series. 1967 *New Scientist* 28 Dec. 766/1 A novel method of making carbonyl pseudohalides by simple substitution. In this way they could obtain from potassium cyanide the pseudohalide carbonyl cyanide.

pseu:doherma·phroditism. *Med.* Also with hyphen. [f. Pseudo- + Hermaphroditism.] The condition of having the gonads and chromosomes of one sex and some anatomical and secondary characteristics of the other sex.

Contrasted with true hermaphroditism, in which there are gonads or gonadal tissue of both sexes.

1881 *Amer. Jrnl. Obstetr.* XIV. 105 The case is..an interesting and rare one, whether it be a case of double congenital ovarian hernia..or whether it proves to be that equally rare malformation, masculine pseudo-hermaphroditism with feminine external genitals. 1924 R. Muir *Text-bk. Path.* xix. 705 General hypoplasia of the uterus is observed in ovarian defect and may be attended by other abnormalities, e.g. pseudo-hermaphroditism. 1950 H. B. Friedgood in R. H. Williams *Textbk. Endo-*

crinol. x. 633 Another example of the importance of considering functional as well as structural changes in making a diagnosis is seen in the differentiation of intersexuality of genetic origin from female pseudohermaphroditism due to congenital adrenal hyperplasia. 1974 Passmore & Robson *Compan. Med. Stud.* III. xxiii. 70/2 In males the resulting lack of testicular androgen during intrauterine development leads to..pseudohermaphroditism, the baby being born with apparently female genitalia despite a normal male chromosome complement.

So **pseu:doherma·phrodite,** a person with this condition.

1895 *Amer. Jrnl. Obstetr.* XXXII. 528 The sexual instincts of this anomalous male being were directed toward men, not women—a fact which leads the author to remark that hermaphrodites and pseudo-hermaphrodites are *degenerates*. 1928 L. Loeb in E. V. Cowdry *Special Cytol.* II. xxxiv. 1184 In male pseudo-hermaphrodites..a development of the mammary gland may occur in certain cases. 1975 Fraser & Nora *Genetics of Man* iv. 53/1 A 46,XY male with a chromatin-negative buccal smear who is female in external appearance is a male pseudohermaphrodite.

pseudo-historic to **-literature:** see *pseudo- 1 a, b, 2 a.

‖**pseudologia fantastica** (siūdolōu·dʒia fæ̆ntæ·stikă). *Psychol.* Also **pseudologia phantastica.** [mod.L., ad. Gr. ψευδολογία falsehood + Gr. fem. φανταστική or med.Lat. fem. *phantastica* imaginary.] A condition, often associated with other abnormal traits, in which a person fabricates stories about himself in order to inflate his importance but readily changes or abandons them when challenged. Also *ellipt.* as *pseudologia*.

1909 *Westm. Hospital Rep.* XVI. 68 There are some cases of moral imbecility in which this [lying] is the chief..feature, yet so pronounced that it has been looked upon by some as a distinct disease, and has been given a name all to itself: 'Pseudologia Fantastica'. 1917 C. E. Long tr. *Jung's Coll. Papers Analytical Psychol.* i. 71 Our case has another analogy with *pseudologia phantastica*: The development of the phantasies during the attacks. 1934 Owen & Zilboorg tr. *Fenichel's Outl. Clin. Psychoanal.* xi. 443 The most remarkable feature in pseudologia is that the patient really is speaking the truth..and that unwittingly the phantastic lies are distorted expressions of his repressed infantile sexual history. 1949 *Horizon* Mar. 216 A hysteric, suffering from *pseudologia phantastica*. 1960 I. Bennett *Delinquent & Neurotic Children* iii. 87 Imaginative or pathological lying (pseudologia fantastica).

pseudologic: see *pseudo- 1 a.

pseudologue (siū·dolǫg). [f. as Pseudologer: see -logue.] A compulsive liar; someone suffering from *pseudologia fantastica*; a pseudologer.

1949 *Psychiatric Q.* XXIII. 19 Pseudologues have little or no insight into real behavior or into the harm they may cause others.

pseudolymphoma to **-membranous:** see *pseudo- 1 b, 2 a.

pseudomonas (siūdomōu·năs). *Biol.* [mod.L. (W. Migula 1897, in *Arbeiten aus dem Bakteriol. Inst. der Technischen Hochschule zu Karlsruhe* I. 237), f. Pseudo- + Gr. μονάς unit.] A bacterium of the genus *Pseudomonas*, which comprises aerobic Gram-negative species that occur chiefly in soil and water, are generally rod-shaped, frequently produce soluble pigments, and include many plant pathogens but few animal pathogens. Freq. *attrib.*

1917 *Jrnl. Bacteriol.* II. 174 The vibrios are..closely akin to the fluorescent (or pseudomonas) bacteria. 1950 C. J. Witton *Microbiol.* xxvi. 366 Pseudomonas intestinal infections of infants and pseudomonas bronchopneumonias have also been reported. 1961 *Lancet* 22 July 179 Recurrence of Pseudomonas infection while on I.M. polymixin after sterile culture for 16 days. 1966 *New Scientist* 21 July 150/1 Currently, research is centred on fish bacteria which cause disease, such as vibrio and pseudomonas. 1977 *Time* 24 Jan. 55/3 These have sharply reduced infection from pseudomonas bacteria, which once killed nearly a third of all burn victims.

So **pseudomo·nad,** a bacterium of the genus *Pseudomonas*, or one of the family Pseudomonadaceæ that includes it, or one of the order Pseudomonadales that includes the family.

1921 R. E. Buchanan *Agric. & Industr. Bacteriol.* xxxiv. 407 The organism is a typical yellow pseudomonad in morphology, culture and physiology. 1958 Pelczar & Reid *Microbiol.* xii. 125/1 Many *Pseudomonas* species, or 'pseudomonads', produce water-soluble pigments. 1966 *McGraw-Hill Encycl. Sci. & Technol.* XI. 63/1 Pseudomonads, as members of the genus are familiarly known. *Ibid.* 64/1 *Aeromonas* is composed of pseudomonads which physiologically resemble bacteria of the genus *Aerobacter*. 1973 J. Levy et al. *Introd. Microbiol.* xiii. 334 Order Pseudomonadales... The pseudomonads are ubiquitous in nature, are largely free-living.., and have achieved economic importance in that they are the most common cause of food spoilage. 1975 *Nature* 24 Apr. 671/1 The most striking metabolic capability of pseudomonad organisms is their versatile utilisation of organic compounds as sole sources of carbon, nitrogen and energy.

pseudomoralist, -morality: see *pseudo- 1 a.

pseudomorph. Add: Hence **pseudomo·rphically, pseudomo·rphously** *advs.* Also **pseu·domorph** *v. trans.* = Pseudomorphose *v.*; **pseu·domorphed** *ppl. a.*, **pseu·domorphing, pseudomo·rphosing** *vbl. sbs.*

1876 F. Zirkel *Microsc. Petrogr.* vi. 114 The alteration did not happen in a proper pseudomorphosing manner, since the contours of the hornblende are..no longer recognizable. 1923 *Trans. R. Soc. Edin.* LIII. 371 Olivine, pseudomorphed by brownish-green serpentine, is quite common. 1931 *Econ. Geol.* XXVI. 595 Features ..preserved pseudomorphically. 1943 *Amer. Jrnl. Sci.* CCXLIII A. 539 The early soft mineral, franckeite, was found strongly replaced, pseudomorphously, by minerals of higher-temperature significance, such as pyrrhotite. 1959 *Trans. R. Soc. Edin.* LXIII. 554 The schists exhibit distinct retrogressive characteristics, notably in the common pseudomorphing of once-idiomorphic garnet by chlorite-free assemblages. 1960 *Jrnl. Petrology* I. 211 In the aureole of the Glen Doll diorite the rocks of high oxidation ratio..have been reduced, hematite being pseudomorphed by magnetite. 1962 W. A. Deer et al. *Rock-forming Minerals* V. 50 A mineral occurring pseudomorphously after loparite..has been named metaloparite. 1971 *Canad. Jrnl. Earth Sci.* VIII. 634/1 Some of the large embayed chromites are pseudomorphically replaced and rimmed by a..garnetiferous aggregate. 1975 *Nature* 17 Jan. 183/2 The large amount of pseudomorphed olivine..suggests that the older dykes were very basic olivine basalts.

pseudomucin to **-neurotic:** see *pseudo- 1 a, b, 2 a.

pseudonym. a. (Earlier example.)

1833 J. S. Mill in *Tait's Edin. Mag.* III. 347 *Junius Redivivus* is the (somewhat inappropriate) pseudonyme of a writer who is one and indivisible.

pseudo-object to **-passivization:** see *pseudo- 1 a, 2 a.

pseu·dopatient. [f. Pseudo- + Patient *a.* and *sb.*] Someone who pretends to have the signs, symptoms, and history of a medical case in order to gain admission to a hospital as a patient. Cf. *Munchausen b.

1973 *Science* 19 Jan. 251/1 If the sanity of..pseudopatients were always detected, there would be prima facie evidence that a sane individual can be distinguished from the insane context in which he is found. 1974 *Med. Jrnl. Austral.* II. 385/2 The fact that studies using pseudopatients as a means of evaluating health services have until the present been poorly conceived should not blind us to the possible values of the technique. 1977 *Times Lit. Suppl.* 4 Feb. 125/1 Research workers and journalists are increasingly lying their way into hospitals with the intention of writing up their experiences. It is high time that the World Health Organisation added the diagnostic category of 'pseudo-patient' to its International Classification of Diseases.

pseudo-pelade to **-plasticity:** see *pseudo- 1 a, b, 2 a.

pseudopod. Add: **1.** Also *fig.*

1951 V. Nabokov *Speak, Memory* xi. 162 The undulating plump shadows of older foliage on the water..were rhythmically palpitating, extending and drawing in dark pseudopods. 1975 *N.Y. Times Bk. Rev.* 8 June 21/2 Gifts of memory and mimicry, and pseudopods of learning, extend what is only an essay into a thick 'Anatomy' of cosmic..speculation.

4. *Spiritualism.* (See quot. 1920.)

1920 E. E. Fournier d'Albe tr. *Schrenk Notzing's Phenom. Materialisation* 25 The recent investigations by W. J. Crawford have shown that white light acts destructively on the pseudopods or psychic projections from the medium's body. 1945 N. Collins *London belongs to Me* i. viii. 88 The medium was invited to materialise the ectoplasmic hand inside the wax. Then when the séance was over a cast of the pseudopod could be cast in plaster.

pseudopodium. Add: **1.** Also *fig.*

a 1902 S. Butler *Note-bks.* (1912) xii. 196 My reviewers felt no sense of need to understand me... When the time comes that they want to do so they will throw out a little mental pseudopodium without much difficulty.

pseu·dopotential. *Physics.* [f. Pseudo- + Potential *a.* and *sb.*] A potential distribution assumed for the purposes of calculation as an approximation to the actual potential.

1956 C. Herring in R. Breckenridge et al. *Proc. Conf. Photoconductivity, Atlantic City, 1954* 86 The idea of using such a repulsive pseudopotential is due to Hellman..and has been applied with some success to molecules..and metals. 1965 *Rev. Mod. Physics* XXXVII. 388/2 Since we are making the Born approximation, the best choice for $v(r)$ is a pseudopotential that fits free (p, p) scattering. 1968 *Times* 2 Dec. 17/1 The theory, known as the 'pseudopotential method', does not allow the properties of the metal to be calculated without guessing either the shape of the wave or the force exerted by the atoms on the electrons. 1976 Coles & Caplin *Electronic Struct. Solids* iv. 93 Pseudopotential methods are now sufficiently finely developed that not only can accurate band structures be calculated for metals (and alloys) but also meaningful estimates can be made of the difference in energy of different crystallographic structures for an element.

pseudopre·gnancy. [f. Pseudo- + Pregnancy[1].] **1.** *Med.* An abnormal condition in which many of the signs and symptoms of pregnancy are present in a woman who is not pregnant; = *pseudocyesis* s.v. Pseudo- 2 a in Dict. and Suppl.

1860 T. H. Tanner *Signs & Dis. Pregnancy* i. 7 The term *pseudo-pregnancy* is also sometimes applied to diseases which simulate pregnancy. 1893 T. M. Madden *Clin. Gynæcol.* xl. 463 Symptoms of Pseudocyesis.—In cases of pseudopregnancy we frequently find all the general symptoms of pregnancy counterfeited with an exactitude that might well seem marvellous if we did not take into consideration the circumstances. 1972 M. M. Garrey et al. *Gynæcol. Illustr.* v. 92 Amenorrhœa... Causes. 1. Pregnancy and missed abortion. 2. Psychological. (a) Stress and emotional disturbance. (b) Pseudopregnancy. (c) Anorexia nervosa.

2. *Zool.* A state marked by changes in the reproductive organs and mammary glands similar to those of early pregnancy, occurring naturally in many female mammals after ovulation when fertilization has not occurred and also capable of being induced experimentally.

1913 Hill & O'Donoghue in *Q. Jrnl. Microsc. Sci.* LIX. 135 As the œstral cycle in Dasyurus differs considerably from that of the Eutherian mammal, it has been found necessary to introduce two new terms, viz. Post-œstrus, to designate the period which intervenes between œstrus and ovulation; and Pseudopregnancy, to designate the period which, in the non-pregnant animal, follows ovulation, and in which the changes in the ovary, mammary glands and uteri are essentially similar to those in the pregnant female. 1940 *Endocrinology* XXVII. 125 We produced pseudopregnancy [in mice] by three methods. (*a*) Females obtained from a low fertility group were allowed to copulate frequently but did not become pregnant. (*b*) Females were mated with vasectomised males. (*c*) Females were stimulated daily on the cervix uteri with a glass rod. 1970 W. B. Yapp *Introd. Animal Physiol.* (ed. 3) viii. 293 In those animals, such as mice, in which there is scarcely any pseudopregnancy, the corpora lutea develop very little and soon degenerate unless copulation occurs.

So **pseudopre·gnant** *a.*, of, in, or characteristic of the state of pseudopregnancy.

1913 *Q. Jrnl. Microsc. Sci.* LIX. 159 Comparison of our preparations of pseudo-pregnant uteri with those of normal post-partum uteri demonstrates..that the regressive changes in the glands are identical in the two. 1932 S. Zuckerman *Social Life Monkeys & Apes* v. 73 The pseudo-pregnant state is generally shorter than the pregnant, and varies in degree from animal to animal. 1972 *Theol. Stud.* XXXIII. 432 The resulting embryos could be transplanted to a dozen or a hundred pseudo-pregnant healthy but genetically nondescript surrogate mother cows.

pseudo-principle: see *pseudo- 1 a.

pseu·do-problem. [f. Pseudo- + Problem.] A problem which is unreal either because it has no possible solution or because there exists a confusion in the elements of which it is composed.

1911 W. James *Some Probl. Philos.* x. 156 There is a pseudo-problem, 'How can the finite know the infinite?' which has troubled some English heads. 1923 Ogden & Richards *Meaning of Meaning* vii. 268 When the pseudo-problems due to cross vocabularies are removed. 1933 *Mind* XLII. 339 The difficulty which, I think, led Russell to his theory of descriptions had its source in a mistake which logicians are prone to make, and which has given rise to many pseudo-problems. 1938 C. W. Morris in *Internat. Encycl. Unified Sci.* I. ii. 57 Current scientific formulations embody many pseudo problems which arise from the confusion of statements in the language of semiotic and the thing-language. 1956 F. Copleston *Contemp. Philos.* xii. 209 Some would wish to define a pseudo-problem as a question which we are unable to answer, not simply because we here and now lack the means of answering it,.. but because no way of answering it is conceivable. 1979 *Trans. Philol. Soc.* 209 The pseudo-problems introduced by prematurely formed classificatory schemes or prejudiced reactions to them.

pseudo-procedure: see *pseudo- 1 a.

pseu·do-proposition. *Philos.* [f. Pseudo- + Proposition.] An apparent proposition which is unreal because it does not have intelligible meaning.

1883 *Mind* VIII. 24 It is unnecessary here to occupy space with examples of these three familar kinds of pseudo-proposition. 1934 *Mind* XLIII. 335 A pseudo-proposition being a series of words that may seem to have the structure of a sentence but is in fact meaningless. 1966 *Philos. Rev.* LXXV. 315 In the pseudo-propositions of the *Tractatus* we see how things really are.

pseudo-proverb to **-racemic:** see *pseudo- 1 a, b, 2 a.

pseudora·ndom, *a. Math.* Also with hyphen. [f. Pseudo- + Random *a.*] Satisfying one or more statistical tests for randomness but produced by a definite mathematical procedure.

1949 *Seminar on Sci. Computation, Nov.* (Internat. Business Machines) 104/2 A random number c lying between 0 and 1 is selected from a store, or a pseudo-random number c lying between 0 and 1 is computed arithmetically. 1954 *Jrnl. Assoc. Computing Machinery* I. 88 Modern-day usage of high-speed electronic digital computing machines frequently involves the consumption of a very large quantity of random numbers. It is desirable that the machine..be able to manufacture its own random or pseudo-random numbers. 1966 A. Battersby *Math. in Managem.* vii. 170 For convenience, genuinely random numbers are replaced by 'pseudo-random' numbers; these are generated by fairly simple calculations which are known to produce strings of numbers indistinguishable in the short run from truly random ones. 1973 *Sci. Amer.* May 19/1 The recipient of a coded message can then be provided with a generator that operates exactly like the one used to add pseudorandom digits to the original message.

Hence **pseudora·ndomly** *adv.*

1963 *IBM Jrnl. Res. & Devel.* VII. 233/2 A periodic pseudorandomly generated 256-bit pattern. 1976 *New Scientist* 3 June 529 Each alarm is interrogated pseudo-randomly every 2-3 seconds and a small PDP8 computer identifies the change of state on a visual display unit.

pseudo-ra·tional, *a.* [f. Pseudo- + Rational *a.*] Assumed to be, or treated as, rational although beyond experience or proof. So **pseudo-ra·tionalism,** a theory or system based on pseudo-rational arguments or assumptions; **pseudo-ra·tionalist,** an adherent or advocate of such a theory; **pseu:do-rationa·lity,** pseudo-rational quality or nature; **pseu:do-rationaliza·tion,** unjustified or spurious rationalization.

1909 W. James *Pluralistic Universe* v. 211 Hegel was the first non-mystical writer to face the dilemma squarely and throw away the ordinary logic, saving a pseudo-rationality for the universe by inventing the higher logic of the 'dialectic process'. 1927 A. Huxley *Proper Stud.* 207 And though the earliest philosophies and religions may seem intellectually very remote from ourselves, we feel, none the less, that the emotions and intuitions to which they give rational, or pseudo-rational, expression are recognizably akin to our own. 1929 N. K. Smith tr. *Kant's Critique Pure Reason* 394 If in employing the principles of understanding we do not merely apply our reason to objects of experience, but venture to extend these principles beyond the limits of experience, there arise pseudo-rational doctrines which can neither hope for confirmation in experience nor fear refutation by it. 1936 *Mind* XLV. 268 It is the mark of a pseudo-rationalist to refer to *the* real world, or to speak of the *certainty* of any propositions, or to assume that sense-data support only one system. 1952 K. R. Popper *Open Society* (ed. 2) xxiv. 227 What I shall call 'pseudo-rationalism' is the intellectual intuitionism of Plato. 1960 Koestler *Lotus & Robot* I. i. 51 This combination of mystic assertion and pseudo-rational proof is as old as the world. *Ibid.* ii. 100 There is a tendency in the human mind never to leave a symbol alone, an itch to debase it by pseudo-rationalizations. 1976 *Contemp. Psychoanal.* XII. 93 Any dialogue of ideological infrastructures through overt discussions.. should..reflect the irretrievable ambivalence of ideas.. which..are pseudo-rational in that they unify human experience according to semiotic rules but conflict with strictly scientific statements, because they cannot be experimentally demonstrated.

pseudoreaction to **-salt:** see *pseudo- 1 a, b, 2 a.

pseu·doscalar, *sb.* and *a.* *Math.* and *Physics.* [f. Pseudo- + Scalar *a.* and *sb.*] **A.** *sb.* **a.** A quantity that transforms as a scalar under rotation but changes sign under reflection. **b.** A sub-atomic particle whose wave function is such a quantity, the particle having zero spin and odd parity.

1938 N. Kemmer in *Proc. R. Soc.* CLXVI. 137 The quantities occurring in our possible fundamental equations are one scalar (ϕ), two 4-vectors (χ_α and ϕ_α), two anti-symmetrical tensors of the second order ($\chi_{\alpha\beta}$ and $\phi_{\alpha\beta}$), two 'pseudovectors', that is, totally antisymmetrical tensors of the third order ($\chi_{\alpha\beta\gamma}$ and $\phi_{\alpha\beta\gamma}$), and finally one 'pseudoscalar' (χ_{0123}). 1950 *Physical Rev.* LXXVIII. 805/2 It then follows..that the π-meson is a pseudoscalar. 1966 *McGraw-Hill Encycl. Sci. & Technol.* IX. 566/2 Thus **S·p** is a pseudoscalar, and so such a term cannot occur in the angular distribution of a parity conserving process.

B. *adj.* Involving or being a pseudoscalar.

1941 *Physical Rev.* LX. 151/2 Pseudo-scalar theories can give a scattering small enough to agree with that observed. 1947 *Rev. Mod. Physics* XIX. 5/1 The initial vector meson is decomposed into a pseudoscalar meson and a photon. 1949 [see *coupling constant* s.v. *coupling *vbl. sb.* 8]. 1974 *Physics Bull.* Dec. 579/2 Pions are pseudoscalar objects and cannot condense out in states of zero momentum (in infinite nuclear matter) because they do not have the quantum numbers of the vacuum.

pseu·do-science. Also **pseudoscience.** [f. Pseudo- + Science.] A pretended or spurious science; a collection of related beliefs about the world mistakenly regarded as being based on scientific method or as having the status that scientific truths now have.

1844 *Northern Jrnl. Med.* I. 387 That opposite kind of innovation which pronounces what had been before recognised as a branch of science, to have been a pseudo-science, composed merely of so-called facts, connected together by misapprehensions under the disguise of principles. 1859 [in Dict. s.v. Pseudo- 1 a]. 1911 G. B. Shaw *Doctor's Dilemma* p. xcii, The pseudo science of the commercial general practitioner, who foolishly clamors for the prosecution..of the Christian Scientists when their patients die. 1911 J. G. Frazer *Golden Bough: Magic Art* (ed. 3) I. iii. 113 Magic as a pseudo-science. 1912 J. J. Walsh *Psychotherapy* I. v. 38 Astrology, is the typical example of pseudo-science in medicine. 1928 C. Dawson *Age of Gods* vi. 134 These pseudo-sciences were held in higher honour by the Babylonians themselves than the more utilitarian branches of knowledge. 1937 *Brit. Jrnl. Psychol.* XXVII. 246 We may..consider..psychoanalytical theory as illustrative of the manner in which various influences combine to produce what we may call pseudo-science. 1957 J. S. Huxley *Relig. without Revelation* iii. 47 Theology has been, as my grandfather T. H. Huxley said, only a pseudo-science. 1960 *Guardian* 9 Dec. 5/3 The pseudo-science of the academic pollster. 1977 A. Giddens *Stud. in Social & Polit. Theory* i. 58 His [sc. K. Popper's] endeavour to establish clear criteria of demarcation between science and pseudo-science shares much of the same impetus as the concern of the logical positivists to free science from mystifying, empty word-play.

Hence **pseudo-scienti·fic** *a.*, **pseu·do-scientist.**

1873 *Q. Rev.* CXXXV. 192 The pseudo-scientific teachers of what has..been termed..the Agnostic Philosophy. 1898 [see *collective *a.* (*sb.*) A. 2 d]. 1902 A. Machen *Hieroglyphics* v. 126 The only people who have always a plain answer to a plain question are the pseudo-scientists..who think that one can solve the enigma of the universe with a box of chemicals. 1902 *Encycl. Brit.* XXV. 472/1 This was the pseudo-scientific note of the new anti-Semitism, the theory which differentiated it from the old religious Jew-hatred. 1914 R. A. S. Macalister *Philistines* ii. 44 The pseudo-scientific hypothesis that Samson (like Achilles, Heracles, Max Müller, Gladstone, and other demonstrated characters of mythology) was a solar myth. 1926 Fowler *Mod. Eng. Usage* 348/1 *Mentality*... Some like it because it is longer than *mind*;..and some because it has a pseudo-scientific sound about it that may impress the reader. 1928 R. Macaulay *Keeping up Appearances* xix. 213 The pseudo-scientists..like Freud, poor old man, who's hypnotised himself with observing diseased eroto-maniacs and thinking them normal. 1960 Hanrahan & Bushnell *Space Biol.* vii. 109 Not only fiction-writers but also some pseudo-scientists have dabbled in anti-gravity. 1964 M. A. K. Halliday et al. in J. A. Fishman *Readings Sociol. of Lang.* (1968) 164 The spurious rigour of some pseudo-scientific 'measurements' of the 'efficiency' of language. 1973 C. Sagan *Cosmic Connection* (1975) viii. 59 An enormous interest is apparent in a range of pseudo-scientific or borderline-scientific topics—astrology, scientology, the study of unidentified flying objects, [etc.]. 1975 *Times* 26 Sept. 19 (*heading*) Lord Zuckerman deplores pseudo-scientists.

pseudoscopy. (Examples.)

1951 L. P. Dudley *Stereoptics* i. 21 If the two components of a stereogram be transposed so that the 'left-eye' view is seen by the right eye, and the 'right-eye' view by the left eye, the result is no longer a true stereoscopic effect, but is what is termed pseudoscopy. 1960 *New Scientist* 14 July 142/3 In 1953, when interest in 3-D had a temporary revivial, I took up experiments..on a new line. These were based on the observation that 'pseudoscopy'—that is to say, the effects of the left eye seeing what the right eye ought to have seen, and *vice versa*—were not at all as bad as one would expect.

pseu·dosex. [f. Pseudo- + Sex *sb.*] Pseudo-sexual activity; also, perverted sexual activity.

1951 E. Bergler *Neurotic Counterfeit-Sex* iii. 104 In analyzing this type of pseudo-sex, one is surprised at the seeming contradiction in the mechanical performance. 1964 *Listener* 12 Nov. 754/2 Pseudosex..takes up much of the zoo monkey's time; like violence itself, it is a symptom and a reliable index of stress in society. 1972 *Ibid.* 9 Mar. 320/2 He related the present craze for pseudo-sex to an equally current repressed hostility in society—pointing out that the kinky aim is to put us off the pleasures of real sex. 1978 M. Puzo *Fools Die* xii. 129 This was a group of publications that drowned the American public with information, pseudoinformation, sex and pseudosex: culture and hard-hat philosophy.

pseudose·xual, *a.* Also **pseudo-sexual.** [f. Pseudo- + Sexual *a.*] **1.** *Zool.* In certain crustaceans (see quots.).

1925 A. M. Banta in *Zeitschr. für Indukt. Abstammungs- und Vererbungslehre* XL. 28 (*heading*) A thelytokous race of Cladocera in which pseudo-sexual reproduction occurs. *Ibid.* 37 Since these eggs are apparently in every way like sexual eggs in other Cladocera, except that they develop without fertilization, they will be referred to as pseudo-sexual eggs and this type of reproduction will be referred to as pseudo-sexual reproduction as contrasted with the usual type of parthenogenesis and the normal sexual reproduction in Cladocera. 1957 *New Biol.* XXIII. 56 The story of reproduction in *Daphnia* is made more complex by certain races, particularly in arctic regions, which have a process of pseudo-sexual reproduction. 1967 G. E. Hutchinson *Treat. Limnol.* II. xxiv. 510 Ruttner-Kolisko (1946)..found that in *Keratella hiemalis* pseudo-sexual resting eggs are produced parthenogenetically at the height of the development of the population..in July.

2. Applied to sexual behaviour motivated by the fear of aggression, desire for dominance, etc., that frequently results from overcrowded social conditions, rather than by genuinely sexual aims.

1951 E. Bergler *Neurotic Counterfeit-Sex* iii. 141 All their sexual (one should say, rather, *pseudo-sexual*) activity is but a blind for totally unrelated baby-conflicts. 1960 *Jrnl. Nerv. & Mental Dis.* CXXXI 203/1 The difference in status shows itself very soon by a pseudo-sexual act. 1961 C. & W. Russell *Human Behav.* iii. 175 A real sexual relationship between two people lies in the fact that each is proportionately freed from pseudosexual involvement. 1969 *Sci. Jrnl.* Apr. 86/2 To the extent that such behaviour looks sexual, but in fact relates almost entirely to agonism, it can be described as pseudosexual.

Hence **pseudose·xually** *adv.*

1964 *Listener* 12 Nov. 755/1 Neglectful mothering.. produces juvenile delinquent monkeys, precociously aggressive and pseudosexually active.

pseudo-simplicity, -social: see *pseudo- 1 a, 2 a.

pseu:do-solariza·tion. *Photogr.* [f. Pseudo- + Solarization.] = *Sabatier effect* s.v. *Sabatier, *solarization 1 b.

1889 R. Meldola *Chem. of Photogr.* vi. 219 The observation of Sabatier, that a collodion wet plate becomes reversed if, towards the end of development, daylight is suddenly admitted to the room, relates to cases of what may be called pseudo-solarisation. 1969 M. J. Langford *Adv. Photogr.* xi. 228 The effect discussed here, although described by photographers as solarisation, is the Sabattier effect or 'pseudo-solarisation'. The effect is easily distinguished—the reversal of weakest densities, and the formation of a thin contour line around strong tone boundaries. 1972 *Exper. Mechanics* XII. 423 A method is described for obtaining well-defined fringes..in photoelastic stress analysis. It is based on an edge effect which occurs during the pseudo-solarization of films during development.

pseudo-sophisticated to **-stratification:** see *pseudo- 1 b, 2 a.

pseudostra·tified, *a.* [Pseudo- 2.] **1.** (See s.v. *pseudo- 2 a.)

2. *Histology.* Applied to epithelium composed of a single layer of columnar cells each in contact with the basal lamina but whose nuclei are at varying distances from it, so that they give the appearance of several layers of cells.

1905 J. S. Ferguson *Normal Histol. & Microsc. Anat.* ii. 30 Diagram showing the manner in which all the epithelial cells of pseudo-stratified ciliated epithelium reach the basement membrane. 1974 J. G. A. Rhodin *Histology* iii. 72/1 Non-ciliated pseudo-stratified columnar epithelium occurs in parts of the male urethra, excretory ducts of many glands, the ductus epididymis.., and the ductus deferens.

pseudosuchian (siudosū·kiăn), *sb.* and *a.* *Palæont.* [f. mod.L. *Pseudosuchia* (K. A. von Zittel *Handbuch der Palæont.* (1890) Abth. I. III. 637), f. Pseudo- + Gr. σοῦχ-ος crocodile: see -ian.] **A.** *sb.* A small fossil reptile belonging to the suborder Pseudosuchia of the order Thecodontia, which includes primitive, often bipedal, carnivores, whose remains have been found in Triassic formations. **B.** *adj.* Of or pertaining to an animal of this kind.

1913 R. Broom in *Proc. Zool. Soc.* II. 624 The type of this Pseudosuchian is a fairly complete skeleton from Elgin. *Ibid.* 631 The Pterodactyl and Pseudosuchian skulls are almost exactly similar in essentials. 1926 *Glasgow Herald* 14 Aug. 4/2 The Pseudosuchians have long since passed, but they fulfilled their promise and gave us birds. 1971 *Nature* 12 Nov. 75/2 The pseudosuchians and the ornithischians would seem, in fact, to be quite irreconcilable in terms of pelvis and ankle structure. *Ibid.* 76/1 It may be deduced that the ornithischians arose from pseudosuchian reptiles which are at present unknown.

pseudotachylite to **-uridine:** see *pseudo- 1 a, b, 2 a, b.

pseudovary. (Earlier examples of **pseudo-varium, pseudovum.**)

1858 T. H. Huxley in *Trans. Linn. Soc.* XXII. 208 The number of chambers..is necessarily regulated by the.. rate at which new pseudova are detached from the pseudo-varium. *Ibid.*, The germ increases in size, and gradually becomes separated from the terminal chamber by the successive development..of new pseudova.

pseu·dovector, *sb.* and *a.* *Math.* and *Physics.* [f. Pseudo- + Vector.] **A.** *sb.* A vector whose sign is unchanged when the signs of all its components are changed.

1923 A. S. Eddington *Math. Theory of Relativity* vi. 179 R^μ is the pseudo-vector representing the displacement from the charge (ξ, η, ζ, τ) to the point (x, y, z, t) where x^μ is reckoned... We call it a pseudo-vector because it behaves as a vector for Galilean coordinates and Lorentz transformations. 1938 [see *pseudoscalar *sb.*] 1966 *McGraw-Hill Encycl. Sci. & Technol.* II. 413/1 In general, vectors associated with rotations belong to the category of pseudovectors. In particular, the angular velocity vector associated with the motion of a rigid body is a pseudovector. 1972 *Jrnl. Physics* B. V. 992 It is convenient to express in the standard way the (real) antisymmetric tensor $\mathbf{A}^\nu(t)$ in terms of the pseudovector $V(t)$ which it uniquely defines.

B. *adj.* Involving or being a pseudovector.

1947 *Rev. Mod. Physics* XIX. 2/2 We are left with mesons of spin 0..or possibly spin 1 (vector or pseudo-vector fields). *Ibid.* 3/1 Little attention has been shown to the pseudovector theory. 1964 *Cambr. Rev.* 24 Oct. 51/1 The assumed conservation of the pseudovector current.

pseudovitamin to **-xanthoma:** see *pseudo- 1 a, 2 a.

pshent (pʃent). = Pschent, p-skhent.

1922 Joyce *Ulysses* 500 On his head is perched an Egyptian pshent.

psi[1] (psai, sai). [Gr. ψεῖ.] **1. a.** The name of Ψ, ψ, the 23rd letter of the Greek alphabet.

c 1400 Mandeville *Trav.* (1725) iii. 25 What Lettres thei ben,..with the Names..α Alpha, β Betha,..χ Chi, ψ Psi, [etc.] 1848 [see *chi]. 1955 W. Pauli *Niels Bohr* 30 A general agreement was reached following the substitution of abstract mathematical symbols, as for instance psi, for concrete pictures.

b. *Nuclear Physics.* A neutral, strongly interacting particle that is distinguished by an exceptionally long lifetime in relation to its mass of 3·1 MeV, has a spin of +1, zero hypercharge, zero isospin, and negative parity, and is produced by the collision either of protons or of electrons and positrons at high energies; freq. written ψ. Also (*psi prime* or ψ′), a similar particle of mass 3·7 MeV that decays into a psi and two pions.

Also designated J.

1974 *Times* 18 Nov. 1/8 American physicists yesterday announced the discovery of a new kind of elementary particle which has been given the name Psi. 1974 J.-E. Augustin et al. in *Physical Rev. Lett.* 2 Dec. 1406/2 We have observed a very sharp peak in the cross section for $e^+e^- \to$ hadrons, e^+e^-, and possibly $\mu^+\mu^-$... The resonance has the parameters $E = 3{\cdot}105 \pm 0{\cdot}003$ GeV, $\Gamma \leq 1{\cdot}3$ MeV... We suggest naming this structure ψ(3105). 1974 G. S. Abrams et al. in *Ibid.* 9 Dec. 1453/2 The recent discovery of a very narrow resonant state coupled to leptons and hadrons has raised the obvious question of the existence of other narrow resonances also coupled to leptons and hadrons. We therefore began a systematic search..and quickly found a second narrow resonance decaying to hadrons. The parameters of this new state (which we suggest calling ψ(3695)) are $M = 3{\cdot}695 \pm 0{\cdot}004$ GeV, $\Gamma < 2{\cdot}7$ MeV. 1975 *Physics Bull.* Jan. 12/3 Discussion between the two groups soon established the authenticity of the 'J' particle, as Brookhaven called it (Stanford suggests calling it ψ). 1975 *Physical Rev. Lett.* 7 July 1/1 Although the newly discovered narrow boson resonances at $M = 3{\cdot}095$ GeV (ψ or J) and at $M' = 3{\cdot}684$ GeV (ψ′) have yet to be fully explored, the data so far indicate that these particles are of a hadronic nature with quantum numbers $J^P = 1^-$ and $IG^C = 0^{--}$. 1978 *Sci. Amer.* Mar. 50/2 The main significance of the discovery of the psi particle was that it provided compelling evidence for the existence of a fourth kind of quark, which had earlier been named the 'charmed' quark.

2. Paranormal phenomena or faculties collectively; the psychic force supposed to be manifested by these. Freq. *attrib.* and *Comb.*

1942 R. H. Thouless in *Proc. Soc. Psychical Res.* XLVII. 5, I suggest that we should use a term proposed by Dr Wiesner, and call this group of effects [sc. extra-sensory perception] the 'psi phenomena', a term which has the important negative merit that it implies no theory as to their nature. *Ibid.* 8 Accepting the reality of *psi*, we may seek to fit it into the existing framework of scientific explanation. 1946 *Jrnl. Parapsychol.* X. 146 *Psi*, a general term to identify personal factors or processes in nature which transcend accepted laws. It approximates the popular use of the word 'psychic' and the technical one, 'parapsychical'. 1948 J. B. Rhine *Reach of Mind* ix. 112 How general psi capacities are we do not yet know. 1950 *Times Lit. Suppl.* 9 June 361/2 The whole field of 'the Unseen', survival after death, ghosts, mediums, precognition, gypsies, astrology, spiritual healing, in short, what it is now fashionable to call 'psi'. 1955 A. Huxley *Let.* 27 Aug. (1969) 761 The effects [of amanita]..are quite alarmingly powerful, and it will obviously take a lot of very cautious experimentation to determine the right psi-enhancing dose of the mushroom. 1968 *New Scientist* 10 Oct. 77/1 He was even starting to produce scores that were significantly *below* chance (a well-known effect which parapsychologists call 'psi-missing'). 1969 J. I. M. Stewart *Cucumber Sandwiches* 31 'Are individuals any longer described as psychic, Arthur?'.. 'The jargon changes. The 'psi-factor' is all the go now.' 1976 *Jrnl. Soc. Psychical Res.* XLVIII. 267 Empirical evidence exists to the contrary; i.e. that psi can act independently of the brain without following the limiting characteristics of space, time and physical causality. 1977 *N.Y. Rev. Bks.* 13 Oct. 45/3 The very essence of sound experimental design in parapsychology is to close all cheating loopholes. Until they are closed, no experiment indicating sensational psi powers is worth publishing.

psi[2], var. *p.s.i.* s.v. *P II.

psilocin (sai·losin). *Chem.* [a. G. *psilocin* (A. Hofmann et al. 1958, in *Experientia* XIV. 108): f. as next.] The alkaloid 3-(2-dimethylaminoethyl)-4-hydroxyindole, $C_{12}H_{16}N_2O$, which is the active hallucinogenic metabolite of psilocybin and is found in traces in psilocybin-containing mushrooms.

1958 A. Hofmann et al. in *Experientia* XIV. 109/1 A second substance, closely related to *Psilocybin* but found only in traces, has been called *Psilocin*. 1963 *Listener* 7 Feb. 238/2 Psilocin, which occurs as a compound in a hallucination-causing Mexican mushroom, has also been recently synthesized by military scientists. 1974 M. C. Gerald *Pharmacol.* xvii. 324 Cross-tolerance has been demonstrated between LSD and mescaline, psilocybin, and psilocin. 1975 Brimblecombe & Pinder *Hallucinogenic Agents* iv. 108 The active principles of teonanacatl, the sacred mushroom of Central America, are the alkaloids psilocin..and psilocybin... Most species of *Psilocybe* and a number of *Stropharia* and *Conocybe* species contain these compounds.

psilocybin (sailosai·bin). *Chem.* [a. G. *psilocybin* (A. Hofmann et al. 1958, in *Experientia* XIV. 108), f. mod. L. *psilocybe* (see below), f. Gr. ψιλός bare, smooth + κύβη head: see -in[1].] An alkaloid, $C_{12}H_{15}N_2O \cdot H_2PO_3$, which is the phosphate ester of psilocin and is the hallucinogen present in several Central American species of mushroom (notably *Psilocybe mexicana*), producing effects similar to those of LSD but less strongly and for a shorter time.

1958 A. Hofmann et al. in *Experientia* XIV. 109/1 The compound has been given the name *Psilocybin*; it possesses indole characteristics and contains phosphorus. 1962 A. Huxley *Let.* 18 Sept. (1969) 939 Mescalin, LSD and psilocybin all produce a state of affairs in which verbalizing and conceptualizing are in some sort by-passed. One can talk about the experience—but always with the knowledge that 'the rest is silence'. 1966 T. Pynchon *Crying of Lot 49* i. 17 Effects of LSD-25, mescaline, psilocybin, and related drugs on a large sample of suburban housewives. 1970 K. Platt *Push-button Butterfly* (1971) iv. 43 He would be selling grass, meth, acid..psilocybin-coated grass, peyote buttons. 1975 [see *psilocin].

psionic (sai,ǫ·nik), *a.* [f. *psi + *-onic*, as in *electronic*.] Pertaining to or involving 'psi'. So **psio·nics** *sb. pl.* [-ic 2], (the study of) the paranormal; **psio·nically** *adv.*

1952 *Astounding Sci. Fiction* XLIX. 119/2 The psionic translator in his belt would have brought him the sense of every syllable, and enabled even these psionic illiterates to understand him. 1953 T. Sturgeon *More than Human* iii. 207 A gravity generator, to increase and decrease..weight... Gravitics is the key to everything. It would lead to the addition of one more item to the Unified Field—what we now call psychic energy, or 'psionics'. 1960 P. Anderson in 'E. Crispin' *Best SF Five* (1963) 228 Research has taught us just enough about psionics to show we can't imagine its potentialities. 1966 *Analog Science Fact/Fiction* Nov. 29/1 I'm going to have to do the real end of the work—the psionics end. 1975 *Homes & Gardens* Nov. 63/3 In the years since Lakhowski first talked of fundamental, or psionic, energy in cells, radiesthesia has not come very far. 1976 *Psionic Med.* xi. 6 Dr. Wright wondered whether the case histories of patients treated psionically would throw any light on these questions. 1978 C. Humphreys *Both Sides Circle* xix. 202 Dr Lawrence, at the age of ninety, founded The Psionic Medical Society.... This drew together homoeopathy and radiaesthesia.

psittacine, *a.* and *sb.* Now usually pronounced (si·tăkəin). Substitute for etym.: [a. mod.L. subfamily name *Psittacinæ*, f. generic name *Psittacus* (Linnæus *Systema Naturæ* (1735)), a. L. *psittacus* parrot: see -ine[1].] (Earlier and later examples.)

1874 *Proc. Zool. Soc.* 592 Many deductions can be made as to the mutual relations of the several genera of the Psittacine suborder. 1901 *Trans. Linn. Soc. Zool.* VIII. 257 It is possible to derive the Psittacine type as a very apocentric modification of this metacentre. 1973 *Observer* 4 Nov. 5/3 Medical evidence and a growing suspicion..are adding urgency to demands..that import restrictions should be placed on the 'psittacines'. 1976 *Amer. Speech* 1973 XLVIII. 265 The psittacine linguistic and cultural relativity with which the argument begins is a very dead duck at the argument's conclusion.

psittacism (si·tăsiz'm). [ad. F. *psittacisme* (Leibnitz *Nouveaux Essais sur l'entendement humain* (1765) ii. 145) or G. *psittazismus*, f. Gr. ψιττακός parrot: see -ism.] The mechanical repetition of previously received ideas or images that reflects neither true reasoning nor feeling; repetition of words or phrases parrot-fashion, without reflection, automatically. Hence **psi·ttacist; psittaci·stically** *adv.*

1896 A. G. Langley tr. *Leibnitz' New Ess. conc. Hum. Und.* ii. xxi. 196 All that they do think about it [sc. the future life] is but a *psittacism*, or gross and vain images after the Mahometan fashion, in which they themselves see little likelihood. 1901 J. Wilson in *N. & Q.* 9th Ser. VIII. 183/1 The words *in nomine Domini nostri J.C.* being, I fear, too often repeated psittacistically. 1902 *Amer. Jrnl. Sociol.* VII. 751 Leibnitz characterized human progress as a psittacism. 1904 *Amer. Jrnl. Religious Psychol. & Educ.* May 107 Then followed monographs on psittacism and symbolic thought, heredity, and laughter. 1923 Ogden & Richards *Meaning of Meaning* vi. 230 There will be some to whom a word is merely a stimulus to the utterance of other words without the occurrence of any reference—the psittacists, that is to say, who respond to words much as they might respond to the first notes of a tune which they proceed almost automatically to complete. *Ibid.* x. 349 Psittacism is the use of words without reference. 1936 *Amer. Speech* XI. 173/2 This procedure would reduce a science to the level of a fastidious psittacism—a blind alley leading to the dead end of a circumscribed habit. 1938 S. Chase *Tyranny of Words* iv. 35 Speaking without knowing is called 'psittacism', but is a practice not confined to parrots. 1975 *Jrnl. Roman Stud.* LXV. 187 Military matters are also the subjects of contributions from two younger British scholars, which are remarkable not least for being unencumbered by the needlessly repetitive anthologies of modern literature, a healthy immunity from that pernicious psittacism too widespread in this volume.

psittacosis. Add: Usu. with pronunc. (s-). (Further examples.) Cf. *ornithosis.

1930 *Aberdeen Press & Jrnl.* 10 Jan. 5 Recently three persons were reported to have died from 'parrot plague' at Berlin and two at Prague. The disease is rare and is known to medical science as psittacosis. 1955 *Times* 16 Aug. 4/2 Psittacosis, or 'parrot disease', may attack or be carried by birds other than parrots. It is not yet known how the zoo budgerigars caught the disease. 1966 *Listener* 3 Nov. 652/3, I remember as a child that my father bought a pair of brightly coloured and rather handsome parakeets, almost at the same time as the great psittacosis scare started, so we got rid of our nice new parakeets. 1970 I. Murdoch *Fairly Honourable Defeat* i. v. 52 It was called psittacosis because people thought you could only get it from parrots, but in fact you can get it from any bird. Pigeons are notorious carriers of psittacosis. 1973 *Times* 29 Oct. 15/8, I have at the moment under my care five patients with psittacosis all infected by a recently imported parrot which was ill when purchased and died a week later.

Hence **psittaco·tic** *a.*

1947 W. P. Blount *Dis. Poultry* xliii. 389 Mice inoculated with psittacotic material often die about the seventh day.

psocid (sōᵘ·kid, -sid). *Ent.* [f. mod.L. family name *Psocidæ*, f. generic name *Psocus* (J. C. Fabricius *Supplementum Entomologiæ Systematicæ* (1798) 198), f. Gr. ψώχ-ειν to grind: see -id[3].] A small winged or wingless insect with long, segmented antennæ, belonging to the family Psocidæ or the order Psocoptera, which includes book-lice and other pests feeding on fungi, algæ, cereal products, or decaying vegetable or animal matter.

1891 *Insect Life* IV. 188 The correspondence is less striking than that of a Mallophagan with a Psocid. 1922 *Entomol. Monthly Mag.* LVIII. 104 The occurrence of various species of Psocids..inside houses, has been frequently observed. 1959 [see *bark-louse*]. 1967 M. E. Hale *Biol. Lichens* xi. 159 Psocids, lice-like insects that infest larch trees, also eat lichens. 1975 R. D. Hughes *Living Insects* ix. 226 Psocids are omnivorous, feeding on fragments of plant or animal materials.

psophometer (sofǫ·mitəɪ). *Electr.* [f. Gr. ψόφος noise + -meter.] An instrument for giving a reading approximately proportional to the subjective aural effect of the noise in a communication circuit.

1938 *Jrnl. Inst. Electr. Engineers* LXXXIII. 261/1 The circuit noise meter (or psophometer) is an instrument which has been designed for measuring the disturbing effect of power induction on telephone communication. 1953 *P.O. Electr. Engineers' Jrnl.* XLVI. 112/1 The use of the psophometer has extended in recent years from the function of power circuit noise interference measurement to the measurement of overall noise on [telephone] transmission lines and equipment. 1975 J. E. Flood *Telecommunication Networks* ii. 48 Noise is usually measured by a meter having a frequency-weighting network which has been standardised by the CCITT. Such instruments are called psophometers.

Hence **psophome·tric** *a.* (see quot. 1943); **psophome·trically** *adv.*

1943 *Gloss. Terms Telecommunications (B.S.I.)* 15 *Psophometric voltage*, the voltage at 800 c/s between two points in a telephone system which, if it replaced the disturbing voltage, would produce the same degree of interference with a telephone conversation as the disturbing voltage. 1951 *Electronic Engin.* XXIII. 35 Typical figures permitted are 2 mV of weighted (psophometric) ripple. 1975 J. E. Flood *Telecommunication Networks* ii. 48 For white noise in the band 0–4 kHz, the effect of psophometric weighting is to reduce the noise level by 3·6 dB. *Ibid.*, Noise power is..measured in units of pWp (picowatts psophometrically weighted) or dBmp (decibels relative to 1 mW psophometrically weighted).

psoralen (sōᵊ·rălen). *Chem.* and *Pharm.* Also **-ene** (-īn). [f. mod.L. *Psoral-ea* (f. Gr. ψωραλέος itchy, mangy), generic name of an Indian leguminous herb, *P. corylifolia*, from seeds of which it was first isolated + *-en* (cf. -ene).] A crystalline tricyclic lactone, $C_{11}H_6O_3$, containing fused coumarin and furan ring systems, which occurs in certain plants and is taken orally or applied in ointments to treat certain skin disorders; any derivative of this compound.

1933 H. S. Jois et al. in *Jrnl. Indian Chem. Soc.* X. 46 A petroleum ether extract of the seeds of *Psoralea corylifolia* gave a dark reddish-brown oil and a crystalline solid $C_{11}H_6O_3$ now named Psoralen, melting at 162°. 1959 N. Campbell in E. H. Rodd *Chem. Carbon Compounds* IVb. viii. 883 Examples of the furocoumarins are found in psoralene and angelicin, which differ from each other in the points of attachment of the furan ring to the aromatic nucleus of coumarin. 1969 *Observer* 26 Jan. 5/5 Patients who are prescribed psoralens—drugs sensitive to ultra-violet light and taken to bring a healthy glow back to unnaturally white patches of skin—should resist the temptation to use them as 'sun-tan pills'. 1978 *Lancet* 11 Mar. 538/1 Psoralens form photoadducts and interstrand cross-links with D.N.A. in the presence of U.V.-A. and mounting evidence indicates that these events are mutagenic.

psoriasiform (sorəi,æ·sifǭɪm, sorəi·ăsifǭɪm). *Med.* [f. Psoriasi(s + -form.] = Psoriatiform *a.*

This is the more usual adj. of the two.

1897 *Brit. Jrnl. Dermatol.* IX. 477 Dr. Fox brought the case again to demonstrate a curious psoriasiform eruption disseminated about the wrists. 1977 *Lancet* 3 Sept. 475/2 Exfoliative dermatitis and psoriasiform, eczematous, lichenoid, erythematous, and mixed rashes were observed.

psst, *int.* An onomatopoeic sound expressing a hiss, often to attract attention.

1922 Joyce *Ulysses* 285 What is she? Hope she. Psst! 1938 M. K. Rawlings *Yearling* xvi. 188 Buck hissed at him. 'Psst. You got him. Leave him lay.' 1963 V. Nabokov *Gift* v. 302 One could already hear the energetic '*psst, psst*' of Shahmatov, who had been served the wrong order. 1972 W. M. Estes *Streetful of People* i. 63 '*Psst!* You're going the wrong way,' hissed Calvin Turnbough. 1975 *New Yorker* 13 Jan. 28/3 So now everybody is responding. Bresson is like that. *Psst, psst, psst*—the steam gathers, then the lid blows off. 1976 *Times* 21 May 16/4 Psst! Have you heard the latest?

psy- (sai). orig. *U.S.* Abbrev. of Psychological *a.*, used esp. in *psy-war*: see *psychological warfare* s.v. *psychological *a.* 3.

1954 *Britannica Bk. of Year* 638/1 Mental tensions were reflected in *psy-war* (psychological warfare). 1965 *Wall St. Jrnl.* 27 July 1/1 In the local bureaucratic jargon, this can come under the heading of 'civic action' or 'rural reconstruction' or 'psy war'. 1966 *New Statesman* 14 Oct. 549/1 As psy-op cadres take the field And start to reconsolidate. 1971 *Times Lit. Suppl.* 7 May 519/1 In one sense, 'psywar' is not the same as propaganda—is its opposite even: while propaganda aims to influence belief, psywar influences action, and above all persuades those it is directed at into inaction. 1974 *Jrnl. R. United Services Inst.* June 38/1 Good Psy-Ops must be addressed to the need level in force and this need level is determined by multiple factors, not simply by the efficiency of an army's logistical services or the level of combat losses. 1974 *Black Panther* 16 Mar. 13/3 Such tactics amount to nothing less than psychological warfare, or as they say in Vietnam—'psy-war'.

psych (saik), *sb.* *colloq.* Also **psyche.** [f. Psych(ology, Psych(iatry, etc.] **1.** Psychology or psychiatry. Freq. *attrib.*

1895 W. C. Gore in *Inlander* Nov. 64 *Psych.* n., psychology. 1910 in *Dialect Notes* (1914) IV. 129 He was feeling sadly as he thought of Psych and Chem. 1946 P. Carter in Aldiss & Harrison *Decade 1940s* (1975) 111 Shut up or I'll have the psych corpsman go over you. 1951 *Galaxy Sci. Fiction* May 138/1 He had put the entire student body through interrogation and a psych check. 1953 'T. Sturgeon' *More than Human* iii. 204 He went through medical school too, and psych. 1960 *Analog Science Fact/Fiction* Nov. 12/1, I checked with one of our own psych men... Lefferts has definite paranoid tendencies, he says. 1975 R. Rimmer *Premar Experiments* (1976) i. 78 Since I didn't feel like watching TV in the sitting room, I concentrated on my psych book. 1976 *Amer. Speech* 1973 XLVIII. 297 In most large, metropolitan hospitals, there are customarily two or more units devoted to psych.

2. a. *pl.* Psychical research.

1927 *Observer* 2 Oct. 7 The story of his magic and his mysticism is good, but it is not half such satisfying spookery as is going out..from the offices of the S.P.R. Beginners in 'Psychs' may get a thrill or two from the book.

b. A psychic person.

1975 *Publishers Weekly* 7 Apr. 81/3 He has great ESP powers, so he volunteers to help his friend Ahmed of the Rescue Squad trace psyches in distress.

3. A psychologist, psychiatrist, or psychoanalyst.

1946 P. Carter in Aldiss & Harrison *Decade 1940s* (1975) 113 The psychs probably have a spy or two planted in this room. 1947 L. MacNeice *Dark Tower* 181 You don't mean a psycho-analyst?.. We do not believe in the psych. 1962 P. Mortimer *Pumpkin Eater* xvi. 146 It's only that the doctor, that psyche, did say that I shouldn't have another child. I'm in the middle of treatment, Jake says, for depression. 1964 R. Petrie *Murder by Precedent* x. 156 He'd better see a psych—no, his family doctor. 1968 *Listener* 19 Dec. 810/1 'That would be very foolish, but also of some inconvenience to me,' the psych said.

4. *Bridge.* (Usu. as **psyche.**) A psychic bid.

1965 [see *psych *v.* 2]. 1969 A. Truscott *Gt. Bridge Scandal* 272 His rather puerile psyche clearly indicates that he knew nothing about the hearts opposite. 1973 L. Meynell *Fatal Flaw* iv. 33 Vyvyan already knew Nancy's play..very dependable; no 'psyches' or fancy tricks. 1980 *Oxford Times* (City ed.) 25 Jan. 11/6 Barry Rigal reports on the two methods of dealing with psyches—the successful and the unsuccessful—against Surrey.

psych (saik), *v.* *colloq.* Also **psyche.** [f. *psych(oanalyse *v.*, etc.: cf. prec.] **I. 1.** *trans.* To subject to psychoanalysis.

1917 *Metropolitan Mag.* Jan. 20/1 Well, she went to this psychoanalyzer; she was 'psyched'. 1928 *Daily Express* 31 Dec. 2/5 While for some patients being 'psyched' may be a step towards being cured, to others it may amount to being infected. 1943 F. Brown *Angels & Spaceships* (1955) 178 It isn't *fair* to psych a guy when he doesn't know what he's talking about. 1965 P. Wylie *They both were Naked* i. i. 20 I've been psyched so much I can hardly daydream as I used to. 1973 K. Giles *File on Death* i. 10 He's been psyched to the best of our skill.

II. 2. *intr.* In *Bridge*, to make a bid that misrepresents one's hand in order to deceive one's opponents.

1952 [implied in *psyching *vbl. sb.*]. 1965 *Sunday Times* (Colour Suppl.) 31 Jan. 38/2 'Obviously Jones has psyched.. Don't tell me you wouldn't try five-spades.'.. You catch them out in a full-blooded psyche. 1969 A. Truscott *Gt. Bridge Scandal* 271 He had no technical reason to suppose that his partner had psyched. 1977 *Detroit Free Press* 11 Dec. 22-c/1 The psychic bid was invented by Dorothy Rice Sims in the early 1930s. It soon became the vogue to 'psyche', usually with hit and miss results.

III. 3. *trans.* To influence (someone) psychologically; to excite, stimulate; (usu. with *up*) to prepare (oneself or another) mentally for a special effort or the like; (usu. with *out*) to gain a psychological advantage over, to intimidate, to demoralize. Freq. as pa. pple. or ppl. adj. orig. and chiefly *U.S.*

1957 *Venture Sci. Fiction* Jan. 18/2 A growing moodiness had driven her..to get out alone... She couldn't understand the pull she felt... Ever since those moments in the Monster's cage... *Damn* the Monster! Had the thing psyched her? 1961 *Milwaukee Jrnl.* 8 Nov. II. 16/3 'We didn't think we could beat Maine with an orthodox offense,' Hatch said. 'We hoped this way to provoke some defensive miscues and also to "psych" our own kids into believing they had something extra going for them.' 1963 *Amer. Speech* XXXVIII. 205 *Get psyched out*, v. phr., slang term applied to losing one's nerve while skiing downhill. 1966 *Time* 29 Apr. 35 Having discovered psychology, the cops induce 'truth' by psyching the subject. 1967 J. Severson *Great Surfing* Gloss., *Psyched out*, mentally incapacitated; generally referring to a surfer's reaction to the big surf. To become frightened, shook up. 1968–70 *Current Slang* (Univ. S. Dakota) III–IV. 96 *Psyched up*, adj., excited. 1968 *New Yorker* 10 Aug. 78 He's never tried to psych us, or insult us with a pep talk. 1969 *Ibid.* 14 June 72/3 It's not that I'm psyched out by him, but I'm playing great and he hits three all-time winners. 1970 N. Armstrong et al. *First on Moon* xiii. 318 He always likes to get you psyched up for tragedy. 1971 E. Bullins in W. King *Black Short Story Anthol.* (1972) 63 Dandy thought that the way she managed things and worked the love and affection from people was like a pimp who psychs out his whores. 1972 *N.Y. Times* 4 June 4/8 At tiring moments she tried to psyche herself up by muttering, 'McGovern, McGovern, I've got to win for McGovern.' 1973 P. A. Whitney *Snowfire* vii. 131 He was absolutely without fear. Nothing ever psyched him out before a race. 1973 *Massachusetts Daily Collegian* 26 Apr. 8/1 The states get psyched... New England..has begun to get just a little excited about the upcoming bicentennial of the American Revolution. 1974 H. L. Foster *Ribbin'* vi. 252 The teacher psychs himself—that is, puts himself in a certain frame of mind so that he can deal with the realities of his teaching assignment. 1974 *Canad. Mag.* 21 Sept. 27/1 It's harder for me to get myself up for practices but I still get myself up for a game. Only you have to psych yourself harder as time goes by. *a* 1976 J. Quarry in *6,000 Words* 165 Pressure doesn't psych me. 1977 I. Shaw *Beggarman, Thief* iii. viii. 313, I had no business being at the net. I was trying to psyche you into missing the shot. *Ibid.* ix. 318, I could see something was psyching him out and it worried me. 1978 *Telegraph* (Brisbane) 28 Sept. 23/4 For months we had been psyching ourselves up for this very rare entertainment delight. 1979 *Chatelaine* (Canada) Jan. 22/3 Psych yourself into ignoring your pet's emotional pleas for more food. 1979 *Tucson* (Arizona) *Citizen* 20 Sept. 6D/1 It's hard for our kids to get psyched up for a dual meet, especially this early in the season.

4. *trans.* With *out.* To analyse in psychological terms; to work out.

a 1961 D. Hulburd in *Webster* s.v., I psyched it all out by myself. 1973 *Daily Tel.* (Colour Suppl.) 30 Nov. 38/1, I would have come away from his [sc. Geller's] various feats as from any others I could not psych out—certain there was a simple, logical, rather ordinary explanation that escaped me. 1974 K. Millett *Flying* (1975) iv. 435 Mother's X-ray eyes met Celia once, had it all psyched out in three minutes. 1978 S. Brill *Teamsters* iii. 88 Most others could never approach Hoffa's ability..to psyche out the opposition's thinking so consistently.

5. *intr.* With *out.* To break down mentally; to become confused or deranged.

1970 *Atlantic Monthly* Feb. 84, I psyched out. I'll be damned if I know how. 1971 J. Mandelkau *Buttons* xiv. 155, I psyched out! 1972 R. Bloch *Night-World* (1974) xiv. 90 It had been a real rip-off, and at first Tony had psyched out on the whole scene. 1973 *To our Returned Prisoners of War* (U.S. Secretary of Defense, Public Affairs) 8 *Psych out*,..to become confused or disturbed.

Hence **psy·ching** *vbl. sb.*

1952 I. Macleod *Bridge* vii. 95 The other main rule for intelligent psyching is to read your partner's bid as phoney before the opponents find out, and to thicken the smoke screen. 1974 *Times Lit. Suppl.* 22 Feb. 182/5 The reading and psyching of an opponent is quite as fascinating as chess. 1974 H. L. Foster *Ribbin'* vi. 250 The psyching function helps them overcome their middle class, nonphysical, open personality. 1975 *Time Out* 16 May 13/2 None of the heavy 'psyching' or banal superstitions that I'd expected. 1977 *Washington Post* 7 Sept. E9 In time the players realized that 'psyching' was not a dividend payer.

psychagogue. Add: Also **psychogogue. 2.** (Further example.)

1928 Auden *Poems* 20 Eyes Look in the glass, confess The tightening of the mouth; Know the receding face A blemished psychogogue; But symmetry will please.

Psyche. Add: **1. c.** *Psyche-knot* (earlier and later examples).

1888 A. R. Diehl *Two Thousand Words* 170 *Psyche knot*, the style of wearing the hair in a projecting coil in the middle of the back of the head. 1968 J. Updike *Couples* v. 404 Her hair was pinned up in a psyche knot.

d. *Psychol.* The conscious and unconscious mind and emotions, esp. as influencing and affecting the whole person. Also *Comb.*

1910 C. G. Jung in *Amer. Jrnl. Psychol.* XXI. 226 Disease is an imperfect adaptation; hence in this case we are dealing with something morbid in the psyche. *Ibid.* 254 This explains a part of the conflict in the child's psyche. 1940 H. G. Baynes *Mythol. of Soul* v. 154 Split off from the psychic hierarchy as an infantile *idée fixe*, it resisted the decisive transition from the infantile to the cultural psyche. 1949 J. Strachey tr. *Freud's Outl. Psycho-Anal.* i. 1 We know two things concerning what we call our psyche or mental life: firstly, its bodily organ and scene of action, the brain (or nervous system), and secondly, our acts of consciousness. 1958 *Times Lit. Suppl.* 23 May p. xii/2 The transformation and re-birth of the psyche in the individual's development towards maturity and integration. 1959 *Ibid.* 23 Jan. 44/5 If the stability of the self is threatened by too much division the psyche asserts itself by projecting an image of wholeness upon the sphere of consciousness. 1961 *Times* 4 Sept. 5/7 This democratic and psyche-conscious age. 1976 *Jrnl. Analytical Psychol.* XXI. 193 A heart ailment..need not arise from the heart only; it can also arise from the psyche of the sufferer.

2. a. (Earlier example.)

1820 M. Edgeworth *Let.* 19 Aug. (1979) 224 You know the prints of the Berne Costume. Pray look at the *butterfly wing* caps—Brobdignag butterflies... This picturesque Psyche costume.

psyche, var. *psych.

psychedelia (saikĭdī·liă). [Back-formation from next: see -ia[2].] Psychedelic articles or phenomena collectively; the subculture associated with psychedelic drugs.

1967 *Melody Maker* 27 May 9 Apparently today's hippie must be expanded and experienced in the whys and wherefores of psychedelia but it cannot be said that the products of this society are all 'junkie'. 1967 *Listener* 10 Aug. 169/2, I am unfriendly to the whole idea of psychedelia and the very notion of a chemical paradise with all those tangerine trees and marmalade skies seems absurd and slightly tatty. 1968 *Rat* 13–16 May 17/2 Buttons, posters, trip glasses, zodiac pendants, psychedelia. 1972 D. Sale *Love Bite* xiv. 172 The crazy jumble of sights and sounds and smells... Here was psychedelia that could never be achieved artificially in a discothèque; a relaxed and happy sense of awareness without the use of pills or pot. 1976 *Maclean's Mag.* 15 Nov. 4/1 In the Sixties his [sc. Dr. Timothy Leary's] name was synonymous with psychedelia.

psychedelic (saikĭde·lik, -dĭ·lik), *a.* and *sb.* Also occas. **psychodelic.** [Irreg. f. Gr. ψυχή (see Psyche) + δηλ-οῦν to make manifest, reveal (f. δῆλος manifest, visible) + -ic.

Proposed by H. Osmond in a letter to Aldous Huxley early in 1956: see G. Smith *Lett. of Aldous Huxley* (1969) 795.]

A. *adj.* **1. a.** Of a drug: producing an expansion of consciousness through greater awareness of the senses and emotional feelings and the revealing of unconscious motivations (freq. symbolically); usu., = *psychotomimetic *a.*

1957 H. Osmond in *Ann. N.Y. Acad. Sci.* LXVI. 429, I have tried to find an appropriate name for the agents under discussion: a name that will include the concepts of enriching the mind and enlarging the vision... My choice, because it is clear, euphonious, and uncontaminated by other associations, is psychedelic, mind-manifesting. 1959 *Times Lit. Suppl.* 13 Nov. 665/3 He is so far from condemning the use of psychedelic drugs as to believe that, if wisely directed, they may help to open closed minds to dimensions of experience which would otherwise remain closed to them. 1965 *Jrnl. Amer. Med. Assoc.* 11 Jan. 104/1 The use of hallucinogenic (psychotomimetic, dysleptic, psychedelic) substances to produce altered states of consciousness is not new. 1965 *Brit. Jrnl. Philos. Sci.* XVI. 150 The popular issue of consciousness-expanding, or psychedelic drugs. 1967 *New Statesman* 3 Feb. 154/3 Mr. Andrews stands for many poets..who are trying to reach beyond ordinary experience, in his case through the 'mental voyages' of psychedelic drugs. 'This LSD is pure hero food.' 1970 R. C. Zaehner *Concordant Discord* iii. 42 If psychedelic drugs attest the existence of a timeless heaven, they none the less point to the existence of a timeless hell. 1970 H. Perry *Human Be-In* 111 The core society's preoccupation with psychedelic drugs has been a way of avoiding their real hang-up: the attack of the young on middle-class values, particularly status and property. 1974 *Howard Jrnl.* XIV. 99 He compares religious experience to the chemical experience induced by psychedelic drugs. 1975 Brimblecombe & Pinder *Hallucinogenic Agents* i. 4 The psychedelic drug is said to enrich the mind, to enlarge the vision, and to create a mystic insight, but the term has achieved its maximum use and notoriety in the lay rather than the scientific literature.

b. Of, pertaining to, or produced by such a drug.

1963 (*title of periodical*) The psychedelic review. 1965 G. Cummins *Swan on Black Sea* 116 He said that yours was possibly a psychedelic condition. 1966 *New Statesman* 16 Sept. 387/1 LSD-takers, or acidheads, look upon Aldous Huxley as a sort of John the Baptist... Huxley baptised with mescalin, but now there is this larger psychedelic vision. 1967 *Punch* 22 Feb. 280/1 How.. would one set about 'integrating LSD into the fabric of American society', as one sober supporter of the psychedelic experience suggested? 1970 G. Greer *Female Eunuch* 172 The state induced by the kiss is actually self-induced, of course, for few lips are so gifted with electric and psychedelic possibilities.

c. Concerned with or characterized by the use of such drugs.

1966 *New Statesman* 4 Mar. 305/2 The work done by the 'psychedelic' specialists Timothy Leary, R. Alpert and R. Metzner in the United States has explored the possibilities of expanding awareness by the use of hallucinogenic drugs, in particular LSD-25. 1967 *Times* 3 July 7/4 Since the drug aspect of the psychedelic cult attracted a great deal of unfavourable publicity and a number of unsavoury hangers-on, Dr. Timothy Leary.. was asked to resign his appointment as Professor of Clinical Psychology at Harvard Medical School. 1967 *Los Angeles Free Press* 10 Nov. 8/3 If you want a big picture of Brando on his bike, why travel to a psychedelic shop when you can get it at the drug store. 1967 *Amer. Jrnl. Psychiatry* CXXIII. 1202/1 The rationale of psychedelic therapy with alcoholic patients is focused on the alienation-breaking potential of 'peak' or psychedelic experiences induced with the aid of LSD. 1972 M. D. de Rios *Visionary Vine* ii. 26 While some societies such as those in Peru have thousands of years of psychedelic tradition behind them, advanced industrial societies often find themselves in deep trouble as segments of their society are suddenly discovering the use of powerful mind-altering substances. 1976 *New Musical Express* 31 July 8/2 This obsession with the bizarrity came out of the psychedelic '60s.

2. Producing an effect or sensation held to resemble that produced by a psychedelic drug; *spec.* having vivid colours, often in bold abstract designs or in motion.

1965 *Los Angeles Free Press* 5 Nov. 4/1 (Advt.), The record stores won't sell and the radio banned..The Psychedelic Sound of 'The Trip' (original version). 1966 *Life* Sept. 61/1 The world of art is 'turning on'. It is getting hooked on psychedelic art. 1966 *Melody Maker* 12 Nov. 9/1 Bobby Darin..has come back after years of absence..right in the middle of psychedelic pop, freak-outs and happenings. 1967 *Wall St. Jrnl.* 9 Feb 1/4 Psychedelic fabrics are becoming the rage. 1967 *Daily Tel.* 24 Oct. 19 Dupont showed its spring news in the new Royal Lancaster, where a batch of whirling lights and psychedelic patterns enlivened the theatre. 1968 *Globe & Mail* (Toronto) 3 Feb. 23/4 'Topless' dancers gyrating in the glow of psychedelic slides and lights. 1968 *Southerly* XXVIII. 279 He is wearing his psychedelic union jack trousers and tee-shirt. 1969 *Observer* 12 Jan. 8/3 The very latest psychedelic colours, electric purples and greens. 1971 *Hi-Fi Sound* Feb. 71/1 A discotheque with a psychedelic lighting display making you virtually blind. 1973 C. & R. Milner *Black Players* v. 139 He drives a secondhand Volkswagen van, which had been gaily painted in psychedelic designs by its previous owners. 1977 B. Pym *Quartet in Autumn* ii. 22 He..proceeded to check the items in his shopping-bag—a 'psychedelic' plastic carrier, patterned in vivid colours, hinting at some unexpected aspect of his character.

B. *sb.* **1.** A psychedelic drug.

1956 H. Osmond *Let.* in G. Smith *Lett. Aldous Huxley* (1969) 795 To fathom Hell or soar angelic, Just take a pinch of psychedelic (Delos to manifest). 1957 —— in *Ann. N.Y. Acad. Sci.* LXVI. 429 The psychedelics help us to explore and fathom our own nature. 1959 R. C. Johnson *Watcher on Hills* x. 162 Experience under the psychedelics may have an important contribution to make to Art. 1965 A. Huxley *Let.* 22 July (1969) 803 As you say in your letter, we still know very little about the psychodelics. 1974 M. C. Gerald *Pharmacol.* xvii. 318 Drugs in this same category are also termed hallucinogens, psychedelics, psychotogens, psychodysleptics, and so forth. 1977 *Martindale's Extra Pharmacopoeia* (ed. 27) 880/2 The group of drugs termed variously psychodysleptics, psychotomimetics, in some cases psychedelics, or, more usually but often inappropriately, hallucinogens are substances which as their principal action provoke abnormal mental changes, particularly in cognitive and perceptual spheres. 1977 *Rolling Stone* 16 June 24/3 Ergot also contains a powerful psychedelic: lysergic acid amide, a close relative of LSD, with about ten percent of that drug's mind-altering potency.

2. A person who takes a psychedelic drug or has a psychedelic life-style.

1966 *Time* 11 Mar. 43 Such dangers do not deter the acid heads or 'psychedelics'—even though some users are willing to admit that they found no great 'show', or had a 'freak trip' (a bad one) or 'tripped out' (the worst kind). 1967 *Economist* 17 June 1240/1 The East Village has supplanted Greenwich Village as a new meeting ground for poets, beats, psychedelics and plain old-fashioned bohemians.

Hence **psychede·lically** *adv.*, in psychedelic colours; also various *nonce-words*, as **psy·chedel, psychedeliac** (-dĭ·liæk) = *psychedelic *sb.* 2; **psy:chedelicate·ssen,** a shop selling psychedelic articles.

1966 *Life* 9 Sept. 68/3 True 'acid rock' goes deeper psychedelically than just lyrics. 1967 *Time* 24 Feb. 55 In Los Angeles, the leading psychedelicatessen is the Headquarters. 1967 *Listener* 24 Aug. 252/1 Those bells which announce the approach of psychedels like medieval lepers. 1970 V. Canning *Great Affair* xvi. 300 Troops.. wearing combat helmets—some psychedelically painted. 1975 *Time Out* 9 May 55 Zig Zag badges are optional but retired psychedeliacs requesting 'Andmoreagain' are asking for disappointment and maybe even a bunch of fives. 1976 *Homes & Gardens* July 39/1 The front door of their old rectory is painted psychedelically and Shirley opens it wearing a comfortable kaftan.

psychedelicize (saikĭde·lisaiz), *v.* *colloq.* [f. prec. + -ize.] *trans.* To make psychedelic; to render more colourful and lively.

1966 *Life* 9 Sept. 68/4 Buttons with the slogan 'Psychedelicize Suburbia'. 1975 *Time Out* 15 Aug. 12/4 Reviewing style became a bizarre psychedelicised variant of the traditional 'Good beat. Chart chance' mode.

psychiatric, *a.* Add: (Further examples.) Also, connected with or affected by mental illness that can be treated medically; *psychiatric social work*, social work designed to support and supplement psychiatric treatment; so *psychiatric social worker*.

1919 M. C. Jarrett in *Mental Hygiene* Apr. 215 There is a misconception..that the psychiatric social worker has a different function from other social workers. *Ibid.* 219 The future social worker..will have included in her professional education some knowledge of all the different branches of social work—psychiatric social work, medical social work, family rehabilitation, child welfare, community service. 1940 Hinsie & Shatzky *Psychiatric Dict.* 13/1 Affectation... It is perhaps more commonly observed among those who are not, strictly speaking, psychiatric, though it may..appear prominently in hysteria. 1957 *Times* 15 Oct. 14/5 Britain's first psychiatric prison, which is to be built at Grendon Underwood in Buckinghamshire, is expected to be ready for occupation in 1962. 1962 N. E. Whitten in A. Dundes *Mother Wit* (1973) 408/1 Cases such as the following have been given..by a white psychiatric social worker. 1965 J. Pollitt *Depression & its Treatment* i. 6 It is rarely necessary for the patient to realise that evidence for a psychiatric reaction is being sought. 1971 E. D. Smigel *Handbk. Study of Social Probl.* 12 Beginning in the 1920s psychiatric sociology has been a developing subdiscipline. 1976 S. B. Guze (*title*) Criminality and psychiatric disorders. 1977 E. Ambler *Send no more Roses* xi. 253 The book..may..be of some sociological interest to specialists, particularly in the field of psychiatric social work.

psychiatrically *adv.* (further examples).

1921 'M. B., Oxon' *Cosmic Anat.* ix. 122 It is quite possible that *psychiatrically* Jung's deduction was right. 1965 J. Pollitt *Depression & its Treatment* vi. 85 Outpatient treatment and short-stay admission is now widely practised for all psychiatrically ill patients. 1975 *Times* 5 Aug. 2/8 The drugs would be administered only to people who had recovered from a period of depression and were psychiatrically well.

psychiatrist. Add: (Further examples.) Now usu. a practitioner of psychiatry.

1922 R. S. Woodworth *Psychology* i. 16 According to the psychiatrists, mental disturbance is primarily an affair of emotion and desire rather than of intellect. 1931 F. L. Allen *Only Yesterday* viii. 198 Psychiatrists were installed in business houses to hire and fire employees. 1959 *Daily Tel.* 9 Apr. 1/3 There would be psychiatrists, social psychologists, and penologists on the staff so that they could approach the problem of crime from all points of view. 1971 *Lancet* 29 May 1124/1 A radiologist to the N.H.S. is medical and a radiographer non-medical, and similarly with..psychiatrist and psychologist, ophthalmologist and optician.

psychiatrize (saikai·ătraiz), *v.* [f. Psychiatr(y, -(ist + -ize.] *trans.* To treat psychiatrically; *psychiatrize away* (nonce-use), to do away with by means of psychiatry or its concepts. Hence **psychi·atrized** *ppl.a.*; **psychi:atriza·tion.**

1929 *Sunday Dispatch* 6 Jan. 3/5 Parents may also be psychiatrized to study their traits and home-life. 1954 E. Jenkins *Tortoise & Hare* vi. 59 He couldn't be psychiatrized against his will. 1964 P. Meadows in I. L. Horowitz *New Sociol.* 451 Psychiatrized conformity masked as true individuality. 1977 *Times Lit. Suppl.* 6 May 565/3 The psychiatrization of 'perverse' pleasures. 1978 *Church Times* 10 Feb. 11/1 We have psychiatrised away the Seven Deadly Sins. Pride, anger, avarice, envy, sloth, gluttony and lust have been made respectable. They have become self-fulfilment, stress, incentive, insecurity, inertia, defective metabolism and emotional tension.

psychic, *a.* (*sb.*) Add: **1. a.** (Further examples.) Also, having a psychical rather than a physical or physiological origin (cf. Psychical *a.* 1 a).

psychic blindness = *psychical blindness* s.v. Psychical *a.* 1 a; *psychic determinism* = *psychical determinism* s.v. *psychical *a.* 1 a; *psychic energizer*, an antidepressant drug, esp. one effective against psychotic states; *psychic unity*, a supposed similarity of the mental make-up of all mankind.

1890 W. James *Princ. Psychol.* I. ii. 41 Munk..was the first to distinguish in these vivisections between sensorial and *psychic* blindness... Psychic blindness is inability to recognize the *meaning* of the optical impressions, as when we see a page of Chinese print but it suggests nothing to us. 1896 *Alienist & Neurologist* XVII. 520 Hysteria is a constitutional psycho-neuropathy with morbid impulsions, caprices, delusions, hallucinations, and illusions, psychic and sensory. We see it displayed..in men with psychical impotency and in women after the menopause. 1910 *Jrnl. Abnormal Psychol.* V. 68, I have successfully treated by Freud's psychoanalytic method cases of homosexuality, psychic impotence..and many other so-called perversions. 1924 E. & C. Paul tr. *Wittels's S. Freud* 267 By psychoanalysts the term 'sexual' is used with wide connotations, so that 'libido' becomes almost synonymous with 'psychic energy', with conation and also with what Bergson terms the 'vital impetus'. 1925 J. Laird *Our Minds & their Bodies* ii. 32 'Psychic' tumours or false pregnancies have deceived skilled observers. 1931 W. B. Gibson tr. *Husserl's Ideas* i. ii. 89 Blindness to ideas is a kind of psychic blindness. 1943 *Jrnl. Nerv. & Mental Dis.* XCVIII. 184 Freud's theory of psychic determinism does not ignore human values. 1953 G. Devereux *Psychoanal. & Occult* ii. 38 Due partly to the psychic unity of mankind..and partly to the limitation imposed upon the general direction of thought processes by the ethos of the culture area [etc.]. 1955 *Internat. Jrnl. Psycho-Anal.* XXXVI. 355 Although the concept of psychic determinism is generally accepted without qualification as an aspect of scientific causality, it sometimes appears difficult to reconcile it with the feeling of free will. 1957 N. S. Kline in *Congr. Rep. 2nd World Congr. Psychiatry* i. 212 Psychic Energisers. We have found that iproniazid (Marsilid) may represent a new principle of drug action since it is capable of increasing psychic energy. 1961 Psychic energizer [see *psychostimulant* sb. and adj. s.v. *psycho-]. 1963 *Listener* 7 Feb. 238/2 Such 'psychic poisons' as lysergic and diethylamide, LSD-25, which produces extreme mental confusion. 1967 J. J. Honigman *Personality in Culture* iv. 97/2 The assumption of psychic unity permits an investigator to apply the same principles of psychology to many people and to use his experience in his own culture (or with other cultures he has studied) as controls. 1968 *New Scientist* 2 May 226/1 The so-called 'psychic poisons', capable of inducing temporary or even permanent insanity. 1970 Aguilera & Messick *Crisis Intervention* i. 2 Psychic determinism is the theoretical foundation of psychotherapy and psychoanalysis. 1974 S. Arieti *Amer. Handbk. Psychiatry* I. iii. 67/1 Regardless of national boundaries, great rapidity has characterized the use of these new drugs, be these 'tranquilizers'..or 'psychic energizers'. 1974 M. Mendelson *Psychoanalytic Concepts of Depression* (ed. 2) vii. 254 Unlike the energy of science..psychic energy is directional.

d. *psychic income* (Econ.): the non-monetary or non-material satisfactions that accompany an occupation or economic activity.

1904 F. A. Feller *Princ. Econ.* xlii. 402 It is well to recall also the distinction between wealth income, money income, and psychic income... The money expression of psychic income can be only approximately attained. 1937 Gemmill & Blodgett *Economics* I. iii. 56 Psychic income means the actual enjoyment or gratification which comes to a person through the consumption of commodities and services. 1948 M. H. Umbreit et al. *Fund. Econ.* xxii. 346 The fact that individuals desire to work in..pleasant occupations has led to the concept of psychic income. 1975 *New Society* 3 July 3/2 Views of metal rooftops have been replaced by grass and flowers or attractive paving, giving office workers an inflation-proof bonus in what economists call psychic income.

3. a. (Earlier and later examples.)

psychic research = *psychical research* s.v. Psychical *a.* 3; so *psychic researcher*; *psychic surgery*, surgery that is ostensibly performed by psychic or paranormal means; so *psychic surgeon*.

1880 *Spiritualist* XVI. 18/2 (*heading*) Psychic action from a distance. 1836 *Discovery* June 185/1 A curious outbreak of what the African calls *kupagawa na pepo*, i.e., to be 'ridden by demons' has occurred recently in Mombasa..and other East African towns, almost in the form of a psychic epidemic. 1939 'N. Blake' *Smiler with Knife* iii. 54 He'll write up the Yarnold Cross ghost, and that'll bring a horde of sightseers and psychic researchers up to the farm. 1968 S. Hynes *Edwardian Turn of Mind* v. 145 His spectrum of interests—biology, psychic research, and socialism—make him an Edwardian radical in spirit. *Ibid.*, The psychic researchers' arguments against scientific scepticism. 1975 Milner & Smart *Loom of Creation* iv. 250 The greatest problem I have experienced in describing and substantiating psychic surgery is that previously most authors have reported something akin to conventional surgery in hospitals. 1975 W. & M. Uphoff *New Psychic Frontiers* iii. 165 (*caption*) [He] witnessed eleven and filmed ten 'psychic surgeons' in the Philippines. 1977 *Time* 12 Dec. 46/2 One part reported on 'psychic surgery', in which Filipino healers supposedly diagnose tumors and other problems, and then use psychic forces—not scalpels—to make incisions and treat them.

4. *Bridge.* Of a bid, bidder, or bidding: deliberately misrepresenting the player's hand so as to deceive the opponents.

1932 D. R. Sims *Psychic Bidding* ii. 18, I shall attempt to outline a few types of psychic bids. *Ibid.* 22 A clever psychic bidder will now employ the barricade bid of two or even three No Trumps. 1936 E. Culbertson *Contract Bridge Complete* i. iii. 55 How do the experts distinguish a bona fide bid from a psychic one? 1952 I. Macleod *Bridge* i. 13 In the years 1932–5 two notable teams were pre-eminent in duplicate Bridge in this country. One captained by Harry Ingram..devastated their opponents with their psychic bidding. 1975 *Times* 20 Dec. 10/8 A player has made a psychic opening bid and does not hold a possible trick. 1977 *Washington Post* 7 Sept. E 9 In the early 1930s, just after the birth of psychic bidding, many players indulged in this mania of fabricating bids.

B. *sb.* **1. b.** The realm of perceptual, mental, or physical phenomena that seem to transcend known physical laws (see Psychical *a.* 3).

1909 [see *oil *v.* 1 b]. 1960 R. F. C. Hull tr. *Jung's Nature of Psyche* in *Coll. Wks.* (1969) VIII. 181 It appears the psychic is an emancipation of function from its instinctual form and so from the compulsiveness which ..causes it to harden into a mechanism.

3. *Bridge.* A psychic bid (see sense A. 4 above); *controlled psychic*: (see quots. 1959, 1962).

1932 D. R. Sims *Psychic Bidding* i. 15 The strategical bids which, under the name of 'psychics', are being extensively misused. 1936 *Punch* 2 Dec. 639/3 Unless

PSYCHICAL

North's last bid was a pure psychic, he should certainly hold the King of Spades himself. **1959** T. Reese *Bridge Player's Dict.* 40 A controlled psychic, as opposed to an ordinary psychic, is one made in accordance with a prearranged system. **1962** *Listener* 8 Nov. 786/1 The British pair in the open room were playing what are known as 'controlled psychics'. That is to say, a player would sometimes open the bidding on very slight values without taking a great risk, for there would be a built-in mechanism to prevent the partnership going too high.

psychical, *a.* Add: **1. a.** (Further examples.) *psychical determinism*, the theory that an individual's mental responses and actions are determined by his previous mental actions or his unconscious mind; *psychical distance*, the mental distance from subjective emotions or involvement supposed necessary for the appreciation of the aesthetic qualities inherent in some kinds of experience (see quot. 1976); *psychical unity* = *psychic unity* s.v. *psychic *a.* 1 a.

1863 J. F. Collingwood tr. *Waitz's Introd. Anthropol.* I. ii. i. 273 If theology feared that an original difference of language..would involve the original unity of the human species..the science of language restores to theology the psychical unity of mankind. **1876** W. James *Coll. Ess. & Rev.* (1920) 29 We have no space to discuss the sources of the English prejudice in favor of psychical determinism. **1897** C. H. Judd tr. *Wundt's Outl. Psychol.* v. 323 The ability to produce purely qualitative effects..which we designate as psychical energy. **1902** D. G. Brinton *Basis of Social Relations* i. 19 When we have such evidence as this for the psychical unity of the human species [etc.]. **1912** E. Bullough in *Brit. Jrnl. Psychol.* V. 87 (*heading*) 'Psychical distance' as a factor in art and aesthetic principle. **1913** E. Jones *Papers on Psycho-Anal.* ii. 21 When the sublimation process is not sufficiently potent to provide an outlet for the accompanying psychical energy, other paths of discharge have to be forged. **1938** R. M. Ogden *Psychol. of Art* vii. 142 The illusion of 'psychical distance' is destroyed when the actors appear in the audience. **1960** J. Strachey *Freud's Compl. Wks.* VI. p. xiv, This is the truth which he [*sc.* Freud] insists upon..: it should be possible in theory to discover the psychical determinants of every smallest detail of the processes of the mind. **1976** Rader & Jessup *Art & Human Values* iii. 54 The phrase 'psychical distance' was employed..to denote 'the marvelling unconcern of the mere spectator' in the moment of aesthetic contemplation. Because the word 'psychical' has a misleading connotation..some recent aestheticians have preferred the term 'aesthetic distance'.

3. *psychical researcher*, one who studies or investigates psychical phenomena.

1892 W. James *Coll. Ess. & Rev.* (1920) 320 The 'psychical researchers', though kept at present somewhat out in the cold, will inevitably conquer the recognition which their labors also deserve. **1912** J. Buchan *Moon Endureth* iv. 135 The Presences might be..spirits..behaving as psychical researchers think they do. **1931** A. Huxley *Music at Night* 102 There is the scientific Psychical Researcher, whose views on the future life..seem to be almost indistinguishable from those held by Homer and the author of Ecclesiastes. **1965** *Jrnl. Soc. Psychical Res.* XLIII. 32 Mr Priestley..has..a number of interesting and original suggestions to offer that are well worth the serious consideration of psychical researchers.

psychics. **2.** (Examples.)

1942 'M. Innes' *Daffodil Affair* iii. i. 80 She represents a rare but fairly well-understood morbid condition—that of one individual split up into several personalities... I should have imagined it to be pretty well off the slate of serious psychical inquiry. Lucy is psychopathology, not psychics. **1977** *Gramophone* June 31/2 In a quite separate field, that of psychics, Percy Wilson's passing will have saddened many friends and associates.

psychism. **3.** (Further examples.)

1872 Geo. Eliot *Let.* 4 Mar. (1956) V. 253 Ideas of spirit-intercourse, 'psychism' and so on, have come before me in the painful form of the lowest charlatanerie. **1895** *Folk-Lore* VI. 79 Analysed under the dry light of anthropology, its psychism [*sc.* that of psychical research] is seen to be only the 'other self' of barbaric spiritual philosophy 'writ large'. **1974** *Christian* I. 322 They want to know more about apparently unusual goings-on, and this interest is not far removed from the curiosity about psychism and the occult.

4. (Examples.)

1962 *Times* 4 May 9/6 Psychism is, of course, a function of the brain and..when psychism is disturbed the brain function is affected. **1970** G. Ordish tr. *Chauvin's World of Ants* vii. 171 *Myrmica*..is an ant relatively low on the ladder of ant psychism.

psychist. Add: **c.** A bridge-player who practises psychic bidding.

1952 I. Macleod *Bridge* vii. 95 The stupid psychist.. would perhaps bid Three No Trumps.

psycho (saɪ·ko), *sb.* and *a. colloq.* [Abbrev. of various words beginning with this element.]

A. *sb.* **1.** Psychoanalysis or psychology. Also *attrib.* or as *adj.*

1921 R. Macaulay *Dangerous Ages* v. 102 'Psychoanalysis, I mean.' 'Oh, psycho... Not that insomnia is always a case for psycho, you know.' **1938** N. Marsh *Artists in Crime* x. 148 The psycho people say one shouldn't repress things. **1939** 'J. Bell' *Death at Half-Term* vii. 133 Now don't you let loose any of your high-faluting psycho-stuff on me. **1946** J. Cary *Moonlight* xix. 138 She gave you a guilty complex... My psycho man says that's the worst kind, it's so unreasonable. **1960** *Times* 25 June 12/3 The mind was so important in this work [*sc.* radionics] that it excluded all physical considerations. They were coming down solidly on the psycho side, but not the psychic side.

2. A psychologist.

1925 A. Huxley *Let.* 5 Mar. (1969) 243 The psychos imagine that they have shed some light on art by affirming that the origin of art is an infantile coprophily.

3. A psychopath.

1942 [see *mental *sb.*]. **1947** *Sat. Rev. Lit.* (U.S.) 18 Jan. 19/3 A large percentage of 'psychos' were exposed to unwholesome mother-influence. **1959** C. MacInnes *Absolute Beginners* 11 Wiz has for all oldies..the same kind of hatred psychos have for Jews or foreigners or coloureds. **1973** R. C. Dennis *Sweat of Fear* vii. 45 He's some kind of psycho. He gets freak vibes—you know, like pictures in the head. **1980** *Daily Tel.* 7 Nov. 15/4 He finally runs down the psycho in a morgue, of all appropriate places, where he is pursuing a girl called Amy.

B. *adj.* **1.** Psychological.

1927 *Variety* 1 June 314 Psycho drama flops... The Compagnie des Jonchets, a private club, was over its head with the psychological drama 'Le Souffle sur la Flamme'. **1976** *Denbighshire Free Press* 8 Dec. 6/2 The programme is completed by the psycho thriller 'Night Caller', AA certificate film.

2. Psychopathic.

1936 R. Chandler *Man who liked Dogs* in *Black Mask* Mar. 19/1 Since when can a cop sign as complaining witness on a psycho case? **1957** J. D. MacDonald *Executioners* ii. 22 Maybe I didn't act worried enough... I think he's psycho. **1958** A. Wilson *Middle Age of Mrs. Eliot* ii. 149 Honestly I think she's a bit psycho at times. **1976** R. Barnard *Little Local Murder* ix. 109 That sort of bloke ought to be locked away. They're psycho, that's what they are.

psycho (saɪ·ko), *v. colloq.* [f. *psychoanalyse *v.* or prec. sb.] *trans.* = *psych *v.* 1.

1925 [see *intense *a.* 4 b]. **1925** *Christian World* 4 June 7/2 How many of us spend twenty minutes a day in consciously 'psychoing' ourselves..? **1928** C. Mackenzie *Extraordinary Women* xviii. 343 A friend of mine took me to be psychoed last spring... It's the latest thing since the war. **1937** A. Thirkell *Before Lunch* xi. 293 He has had every inhibition psychoed and is perfectly free. **1946** J. Cary *Moonlight* xix. 138 'It's a complex,' said Kathy. 'We ought to get her psychoed.' **1960** N. Marsh *False Scent* ii. 66 You'd better get yourself psychoed, my poor Charles.

psycho-. Add: **psychæsthetic**, var. *psycho-æsthetic* below; **psychasthenia** (earlier and later examples); [ad. F. *psychasthénie* (P. M. F. Janet 1893, in *Rev. gén. des Sci. pures et appliquées* IV. 176)]; **psychasthenic** *a.* (earlier and later examples); also as *sb.*, a person with psychasthenia; **psychoa·ctive** *a.* = *psychotropic *a.*; hence **psychoacti·vity**; **psycho-æsthe·tics**, the study of the psychological aspects of æsthetic perception; hence **psycho-æsthetic** (also **psychæsthetic**) *a.*; **psy·chobabble** *colloq.* (orig. *U.S.*), jargon that is much influenced by the concepts and terminology of psychology and is used esp. by laymen in referring to their own personality or relationships; hence **psy·chobabbler**, one who uses such jargon; **psychoce·ntric** *a. Psychol.*, treating the psyche or mind, rather than the body, as the important factor in human behaviour; **psychoche·mical** *a.*, pertaining to the relationship between chemicals and the mind, esp. the way the former can be used to modify the latter; also (of a chemical), psychotropic; also as *sb.*, a psychotropic chemical; **psychoche·mistry**, the chemistry of the mind; **psychocu·ltural** *a.*, relating to the interaction of the culture in which individuals live and their psychological characteristics; **psychocu·rative** *a.*, of or pertaining to the healing of mental or psychological disorders; **psy:chodiagno·sis, -diagno·stics** *Psychol.* [after Ger. (H. Rorschach *Psychodiagnostik* (1921))], the investigation of a subject's personality, esp. by means of Rorschach and other projective tests; hence **psy:chodiagno·stic** *a.*; **psychodysle·ptic** [Gr. δύσληπτος hard to take hold of] = *psychotomimetic *sb.*; **psychoe·ndocrine** *a.*, relating to or involving both the endocrine glands and mood and behaviour; **psy:choendocrino·logy**, the branch of science concerned with the relationship between the secretions of the endocrine glands and a person's mood and behaviour; hence **psy:choendocrinolo·gic** *a.*, **-endocrinolo·gically** *adv.*, **-endocrino·logist**; **psychogeo·graphy**, that branch of psychological speculation or investigation which is concerned with the effects on the psyche of the geographical environment; so **psychogeogra·phic, -ical** *adjs.*; **psycholeptic** *a.*, (*b*) characterized by a sudden fall in psychic tension; (*c*) (of a drug) sedative; **psycholytic** (-li·tik) *a.* [*-lytic], applied to a drug such as LSD which can disturb or disrupt certain emotional reactions that have become fixed in the unconscious or can block normal channels of response; chiefly in *psycholytic therapy*, therapy that combines controlled use of low dosages of such drugs with psychotherapeutic instruction for the patient and subsequent discussion; **psychomime·tic** *a.* and *sb.* = *psychotomimetic *a.* and *sb.*; **psychomoti·lity** *Psychol.*, physical movement which reflects or is evidence of mental activity; **psychoneu·ral**, of or pertaining to the relationship or interaction between the mind and the nervous system; **psy:choneuroe·ndocrine, -neu:roe:ndocrinolo·gic, -lo·gical** *adjs.*, of or pertaining to the joint or mutual action of the nervous system, the endocrine system, and behaviour; so **psy:choneu:roendocrino·logy**, the branch of science concerned with this; **psychoneurology** (further example; cf. *neuropsychology); hence **psy:choneurolo·gical** *a.*; **psychoneurosis** (delete and see as main entry below); **psycho-optic** *a.* (examples); so **psycho-o·ptical** *a.*; **psy:chopharmaceu·tical** *a.* and *sb.*, (a drug) that is psychotropic; **psy·cho-philosophy**, philosophical reasoning based on subjective criteria, or on subjective psychic criteria; hence **psy·cho-philosopher**; **psy:chophy:sicotherapeu·tics** *nonce-wd.*, remedial treatment of mind and body; **psycho-poli·tical** *a.*, characterized by the interaction of politics or political events and behaviour; so **psycho-po·litics**; **psycho-pri·smatism** [cf. Prismatic *a.* 2] (see quot.); **psycho-se·nsory** *a.*, pertaining to the conscious perception of sensory impulses; **psycho-sociological** *a.* (further examples); so **psycho-socio·logist, -socio·logy**; **psychosomatic** (delete and see as main entry below); **psy·chosphere**, the sphere or realm of consciousness; cf. *noosphere; **psychosti·mulant** *sb.* and *a.*, (a drug that is) antidepressant; **psy·chosyndrome**, a syndrome in which the symptoms are psychological; **psycho-sy·nthesis**, the integration of disjoint elements of the psyche or personality by means of psychoanalysis; hence (*nonce-wds.*) **psycho-sy·nthesist**, one who practises or advocates this; **psycho-synthe·tic** *a.*; **psycho-vi·sual** *a.*, pertaining to psychological factors associated with vision, such as the emotive connotations of particular colours, and to the centre in the brain associated with such processes; see also *visuo-psychic* s.v. Visuo- in Dict. and Suppl.

1900 S. B. Collins tr. *M. de Fleury's Medicine & Mind* v. 206 Psychasthenia..seemed to be modified..parallel with the oscillations of the blood pressure. **1926** [see *extrovert *sb.* (and *a.*)]. **1968** *New Scientist* 5 Sept. 500/1 Rupp suffered from a psychasthenia which led him to ascribe fictional properties to positrons. **1901** C. R. Corson tr. *Janet's Mental State of Hystericals* vi. 520 It is very rare to meet a psychasthenic patient who is, if we may so speak, a pure type of this affection. *Ibid.* 521 Abulia is a common characteristic with hystericals and psychasthenics. **1906** W. James *Let.* 6 May (1920) II. 254 Pierre Janet discussed lately some cases of pathological impulsion or obsession in what he has called the 'psychasthenic' type of individual. **1977** A. Sheridan tr. *J. Lacan's Écrits* ii. 16 States as diverse as phantasmatic fear, anger, active sorrow, or psychasthenic fatigue. **1961** *Perspectives in Biol. & Med.* IV. 428 Asynchrony [prevails] after application of analeptic and psychoactive drugs. **1967** *New Scientist* 19 Jan. 128/1 Glossy magazines and sombre journals of opinion alike have discovered an intense interest in psychoactive drugs, drugs which affect the way people behave and feel. **1974** M. C. Gerald *Pharmacol.* xviii. 350 These medicines are most often tranquilizers, sedatives, and other psychoactive agents. **1975** *Daily Colonist* (Victoria, B.C.) 19 Oct. 17/2 Advisers..avoided taking any position on the two most commonly used 'psychoactive or mood-altering' drugs—alcohol and tobacco. **1977** *Rolling Stone* 30 June 123/2 (Advt.), Psychoactive mushrooms... Chart, illustrations—tests for chemicals—105 alkaloid mushrooms, 42 psilocybin. **1971** *McGraw-Hill Yearbk. Sci. & Technol.* 357/2 These tribesmen, having discovered that the narcotic constituent of the mushroom is excreted with almost undiminished psychoactivity, incorporated a ritual urine-drinking ceremony. **1973** *Nature* 6 Apr. 367/3 These two amphetamine derivatives, which show profound psychoactivity in man. **1925** W. H. J. Sprott tr. *Kretschmer's Physique & Character* ii. xiv. 258 An indefinite number of individual temperamental shades emerge from the psychæsthetic and diathetic proportions. **1943** [see *psychomotility* below]. **1951** *Jrnl. Aesthetics* X. 2 Our discussion will have to go in two directions: (1) What are the specific attributes of the art of the blind? (2) What psycho-aesthetic implications result from it for the world of the normal-sighted? **1973** *Screen* Spring/Summer 65 While there is perhaps no 'eternal and immutable essence' of the cinema as opposed to the theatre.., there is at least a psycho-aesthetic conditioning of each art by the technical constraints which define and constitute them. **1909** *Encycl. Relig. & Ethics* II. 448/2 *Psycho-æsthetics*,..the application of psycho-physiology to the study of æsthetic states... Helmholtz in Germany, and Grant Allen in England, tried to determine the physiological concomitants of certain phenomena of the Beautiful. **1939** *Time & Tide* 8 Apr. 454/1 Your temperament told from your taste in Old Masters. A stimulating essay in psycho-æsthetics. **1976** *National Observer* (U.S.) 8 May 15/1 For the consumer who doesn't understand psycho-babble, trying to sort out the various specialties can be downright mind-boggling: Gestalt, TA (Transactional Analysis), bioenergetics, sex therapy, behavior modification, [etc.]. **1977** R. D. Rosen (*title*) Psychobabble. **1977** *Proc. R. Soc. Med.* LXX. 806/1 This was yet another American death book, full of psychobabble and journalistic cuttings from every other American death book. **1980** *Times Lit. Suppl.* 16 May 544/3 The book is written in colloquial American spliced with psychobabble, a language in which the highest commendation is to say of someone 'She was a person.' **1977** *N.Y. Times Mag.* 20 Nov. 124/4 The psychobabblers not only outnumber the rest of us, but..they have The Force on their side. **1978** *Guardian Weekly* 22 Jan. 19/1 She mocked the manners and morals and especially the 'mindless prattle' of the psychobabblers among whom she lives [in California]. **1936** J. O. Wisdom in *Proc. Aristotelian Soc.* XXXVI. 62, I shall try to establish my psychocentric analysis of *right*. **1949** *Mind* LVIII. 390 There is the traditional 'psychocentric' conception..: the dualistic conception, which regards the human being as a compound of two distinct but interacting entities, mind and body. **1956** J. B. Rhine in A. Pryce-Jones *New Outl. Mod. Knowl.* 205 There have been psychocentric schools of psychology ..but none of these psychocentric views has ever prevailed widely in academic psychology. **1958** Psychochemical [see *psychotomimetic *a.*]. **1959** *New Scientist* 20 Aug. 222 The Committee appears to have been particularly impressed by what the US Army's chemists told it about the so-called psychochemical weapons. *Ibid.*, Whatever the intrinsic power of 'psychochemicals' may prove to be, the picture of a bloodless war painted by the Congressional committee is hard to believe. **1965** B. Inglis *Drugs, Doctors & Disease* iii. 110 Nowhere has the evidence of the power of placebo effect been more striking than in the new market for psycho-chemicals: pep pills and tranquillisers. **1972** G. Watson (*title*) Nutrition and your mind: the psychochemical response. **1973** 'A. Hall' *Tango Briefing* xiv. 176 Obviously psychochemicals but not related to mescaline or lysergic acid. **1977** *Rolling Stone* 21 Apr. 46/4 The Soviet Union was hard at work in psychochemical research. **1900** *Amer. Jrnl. Psychol.* XI. 600 The writer takes up..passive and then active sadness, morbid joy, their original mechanism, their psychophysiology, psycho-chemistry, [etc.]. **1931** *Chem. News* 23 Jan. 51/1 Colloidal and physiological chemistry have advanced to the extent that we should now be able to envisage a Psychochemistry, or Chemical Psychology. **1951** M. A. Straus in *Amer. Sociol. Rev.* June 374 One is led to what Frank has called a 'psychocultural' rather than a purely psychological explanation of the phenomena of bilingual inferiority. **1977** *Canad. Jrnl. Linguistics* 1976 XXI. 226 The wider implications of language as a psychocultural, evolutionary phenomenon. **1901** A. C. Halphide *Psychic & Psychism* i. 21 There are many schools of Psycho-curative systems, all of which might be classified under the title Mental Medicine. **1953** *Cape Times* 14 Feb. 5/2 The doctors believe that the installation of a pigeon loft at the hospital may have a psycho-curative effect. **1940** *Proc. R. Soc. Med.* XXXIII. 173 (*heading*) Myokinetic psychodiagnosis: a new technique of exploring the conative trends of personality. **1969** J. E. Exner *Rorschach Systems* i. 5 Rorschach might well be appalled were he to perceive how the technique is utilized in contemporary psychodiagnosis. [**1930** *Amer. Jrnl. Psychiatry* X. 50 The Rorschach 'Psychodiagnostik' test, consisting of ten symmetrical ink-blots.] **1937** *Amer. Jrnl. Orthopsychiatry* VII. 320 With this paper we want to introduce a new concept in the theory and a new tool in the practice of Rorschach's psychodiagnostic ink blot test. **1949** S. Rosenzweig *Psychodiagnosis* i. 1 As a psychodiagnostic art clinical psychology derives historically from two chief sources—the psychometric and the psychodynamic. **1932** *Character & Personality* I. 2 The proper aim of a quarterly for psychodiagnostics and allied studies seems to us to be to establish an organic connection among the numerous specialized branches of psychology. **1960** H. J. Eysenck *Exper. Personality* I. p. ix, Experiments in psychogenetics, psychopharmacology, psychodiagnostics [etc.], ..all..form part of the programme of research. **1970** *Jrnl. Aesthetics* XXIX. 105 (*heading*) The ink blot test, 'psychodiagnostics' and Hermann Rorschach's aesthetic views. **1961** Kalinowsky & Hoch *Somatic Treatments in Psychiatry* ii. 8 Psychodysleptics or psychotomimetics. This refers to a group of drugs which can produce the so-called 'model psychoses' and which have characteristically hallucinogenic and mildly stimulant properties. **1967** *WHO Chron.* XXI. 465/2 Three classes of psychotropic drugs are particularly dependence-producing: the anxiolytic sedatives, the psychodysleptics (hallucinogens), and the psychostimulants. **1974** *Nature* 27 Sept. 314/1 Some psychodysleptics (mescaline sulphate and LSD), when injected during the photosensitive larval period, suppress diapause induction as if the larvae were subjected to a long 16-h photophase. **1946** *Psychosomatic Med.* VIII. 176 (*heading*) Psychoendocrine relationships in pseudocyesis. **1958** M. Reiss *Psychoendocrinol.* i. 13 The psychoendocrine concept is based on the discovery that the activity of the pituitary is related to the function of the hypothalamus. **1977** *Proc. R. Soc. Med.* LXX. 513/2 Psychoendocrine relationships in affective disorders. **1953** M. Reiss in *Internat. Rec. Med.* CLXVI. 196 Psychoendocrinology will become in psychiatric research a much more recognized branch than it is at present. **1958** —— *Psychoendocrinol.* i. 20 The responsibility of the psychoendocrinologist in such a case has become very grave indeed. *Ibid.* 27 No doubt people like Tamerlane..who had undescended testicles..have seemed brilliant just because they were psychoendocrinologically not completely mature, and therefore more accessible to new impressions and situations. **1961** *Psychosomatic Med.* XXIII. 449/1 (*heading*) Psycho-endocrinologic studies in a male with cyclic changes in sexuality. **1975** S. Arieti *Amer. Handbk. Psychiatry* IV. 554/2 Psychoendocrinologists have been mainly preoccupied with the basic psychophysiological exploration of the significance of hormonal responses as reflections of intrapsychic processes. *Ibid.* 555/1 By the late 1950s, then, it was generally recognized not only that psychoendocrinology rested on a solid experimental foundation, but that psychological stimuli were, in fact, among the most potent of all natural stimuli to the pituitary-adrenal cortical system. **1953** J. L. Moreno *Who shall Survive?* iii. 436 There is also in psychogeography in respect to a certain criterion either a yes or a no, whatever the motivation of this yes or no may be. *Ibid.* 440 The psychogeographic mapping of the community shows..the relationship of local geography to psychological processes. **1958** *Archit. Rev.* CXXIV. 1/1 This microclimatology of the psyche is something to which every town-dweller can testify, and in a city like Paris..it is a more than personal affair—that document of psychogeography, André Breton's *Nuit du Tournesol*, which ought on the face of it to be an entirely private exercise in erotic topography, can be read with understanding, even by those who have never visited Paris. *Ibid.*, It shows 'quartiers d'états d'âme' and 'gradients of psychogeographical drift'—factors not generally taken into account by the average planning authority. **1963** *Listener* 14 Feb. 299/1 The kind of psychogeographic studies made by the Situationists are all very well in communicating a feeling about man/environment relationships. **1974** *Times Lit. Suppl.* 14 June 630/2 The book promises to become a midwife's guide to the birth of a new discipline, which one expects will be inelegantly dubbed 'psychogeography', rather than 'geopsychology'. **1925** E. & C. Paul tr. *Janet's Psychol. Healing* I. x. 558 Individuals in whom psychological tension is unstable, suffer from sudden relaxations of this tension, succumb to psycholeptic crises. **1940** H. G. Baynes *Mythol. of Soul* xi. 882 We could then regard the whole drama as a psycholeptic crisis with its characteristic feeling-symptom of the end of the world. **1961** Kalinowsky & Hoch *Somatic Treatments in Psychiatry* ii. 8 Delay has proposed the following classification of the new drugs based upon their predominant action. (A) Psycholeptics or Sedatives... (C) Psychodysleptics or psychotomimetics. **1971** Zirkle & Kaiser in A. Burger *Med. Chem.* (ed. 3) II. lv. 1412/1 The antipsychotic agents were originally given such names as tranquilizers.., ataraxics.., psycholeptics, and psychosedatives. **1962** D. D. Jackson in *Jrnl. Nerv. & Mental Dis.* CXXXV. 436/1 More accurately, perhaps, we should speak of psycholytic drugs given by psychosogenic therapists. **1963** R. A. Sandison in R. Crocket et al. *Hallucinogenic Drugs* 34 This total experience of the unconscious, brought about by the power of LSD to loosen the psyche, has led to a feeling that the hallucinogenic drugs should be renamed the psycholytic drugs. This name, which is free from the many objections attached to the word 'hallucinogenic' was first suggested and adopted at Göttingen last year. **1964** D. F. Downing in M. Gordon *Psychopharm. Agents* I. xiii. 606 In this context they [*sc.* psychotomimetic agents] are frequently known as psycholytic drugs because of their power to loosen the psyche. **1974** Arieti & Brody *Amer. Handbk. Psychiatry* (ed. 2) III. 425/1 *Psycholytic therapy*, this technique consists of a series of drug sessions in which small doses of LSD..are given to a number of patients in an outpatient setting. These sessions are associated with individual or group therapy. **1964** M. McLuhan *Understanding Media* xxxi. 308 Our children are striving to carry over to the printed page the all-involving sensory mandate of the TV image. With perfect psycho-mimetic skill, they carry out the commands of the TV image. **1967** *WHO Chron.* XXI. 464/2 Psychodysleptics, also called 'hallucinogens', 'psychomimetics', or 'psychodelics', are compounds that produce abnormal mental phenomena, particularly in the cognitive and perceptual spheres. **1969** *Listener* 28 Aug. 295/2 *Grand Hotel, The Age of Innocence, Dr. Finlay's Casebook*..all inextricably jumbled together into a deliriously psychomimetic paradise. **1974** S. Arieti *Amer. Handbk. Psychiatry* (ed. 2) I. 67/2 Exaggerated expectations about the uncritical use of 'psychomimetics' (mainly LSD 25) in the treatment of mental disorders and especially about the power of some drugs to enlarge the field of consciousness and provide new philosophical and religious insights are unrealistic. **1925** W. J. H. Sprott tr. *Kretschmer's Physique & Character* ix. 134 The psychomotility of the cycloid is even and adequate to the stimulus, and motor expressions and movements are well rounded, fluid, and natural. **1934** E. B. Strauss tr. *Kretschmer's Text-bk. Med. Psychol.* iv. 43 In a mild degree traces of these Parkinsonian features often typify the psychomotility of advanced old age. **1943** H. Read *Educ. through Art* iv. 79 Within the main cycloid and schizoid groups, there are a considerable number of psychaesthetic variants and a considerable degree of psychomotility. **1969** H. E. King in Zubin & Shagass *Neurobiol. Aspects of Psychopathol.* vi. 99 (*heading*) Psychomotility: a dimension of behavior disorder. **1890** W. James *Princ. Psychol.* II. xx. 164 Thus we should escape the responsibility of explaining, by falling back on the everlasting inscrutability of the psycho-neural nexus. **1923** J. S. Huxley *Ess. Biologist* iv. 134 The mind, or shall we say the psycho-neural organization. **1949** [see *hypnoanalysis]. **1969** *Word* 1967 XXIII. 469 Nor can we yet identify all of the psychoneural factors which enter into the final stage of speech perception, 'understanding'. **1954** M. Reiss in *Jrnl. Mental Sci.* C. 687 Such efforts are unavoidable if progress is to be made in psycho-neuro-endocrinologic problems. *Ibid.* 701 The influence of the various treatments on a psycho-neuro-endocrine cycle. **1971** —— in D. H. Ford *Influence of Hormones on Nervous System* p. xix, Our association was named Psychoneuroendocrinological because it emphasizes not only the beginning but also the end of the most important pathophysiological vicious circle in the body. **1972** *Science* 9 June 1115/2 A group of scientists..have organized an International Society of Psychoneuroendocrinology. *Ibid.* 1115/3 The relationships of such psychoneuroendocrine studies to the clinical observations on man are dealt with in papers by Abrams. **1978** *Nature* 14 Dec. p. xii. (Advt.) Psychoneuroendocrinology is an attempt to provide the essential interdisciplinary approach to research in human reproduction. **1921** *Edin. Rev.* Jan. 61 In London the Psycho-Neurological Society has been formed..for the study and discussion of problems in psychotherapy. **1928** H. P. Weld *Psychol. as Science* viii. 156 (*heading*) Psychoneurological theories. **1943** *Amer. Jrnl. Psychiatry* C. 181/2 Apart from the therapeutic implications of the transsection of the anterior thalamic radiations, the method has important implications for experimentation in clinical neurology and in psychoneurology. **1885** Landois & Stirling *Text-bk. Human Physiol.* II. xiii. 921 The psycho-optic centre,..according to Munk, embraces the outer convex part of the occipital lobe of the dog's brain. **1937** *Arch. Neurol. & Psychiatry* XXXVII. 1173 Both kinds of movement belong to the so-called psycho-optic reflexes because, being produced by visual stimuli they are performed more or less instinctively. **1954** S. Duke-Elder *Parsons' Dis. Eye* (ed. 12) xxvii. 462 The involuntary reflexes which depend on vision (fixation, fusional movements, convergence, etc.)—the psycho-optical reflexes—are centred in the visual cortex of the occipital lobe. **1964** *Dis. Nerv. System* XXV. 233/2 The effects of discontinuing psychopharmaceuticals in a large group of long-term schizophrenic patients. **1965** *New Scientist* 18 Mar. 719/1 So far medical researchers have largely had to rely on a patient's behaviour pattern to assess the effects of the so-called psychopharmaceutical drugs—those which can be used nowadays with considerable effect against certain types of nervous disorder. **1969** *Sci. News* 20 Dec. 581 One advantage of doxepin is its apparently low toxicity compared to other psychopharmaceuticals. **1966** *New Statesman* 18 Feb. 243/2 (Advt.), ESP Psycho-philosopher, having evolved new theory concerning influence on environment at a distance and through thought, seeks volunteers to co-operate in test. **1960** *IRE Trans. Electronic Computers* IX. 524/1 The pragmatic philosophy of C. S. Peirce helped save much of philosophy from the sterilizing effect of psychophilosophy. **1922** Joyce *Ulysses* 659 Heliotherapy, psychophysicotherapeutics, osteopathic surgery. **1921** *Q. Rev.* Oct. 397 The exaggeration..would..have made its dogmatic definition sooner or later inevitable, but Manning's championship of it assisted its appearance at the psycho-political moment. **1934** H. G. Wells *Exper. Autobiogr.* II. ix. 798 This psycho-political autobiography. **1948** J. Towster *Polit. Power in U.S.S.R.* iv. 53 While the unification of nations is the goal *ne plus ultra* there are enormous psycho-political obstacles in the way. **1971** K. Millett *Sexual Politics* II. iii. 73 The psycho-political tactic here is a pretence that the indolence and luxury of the upper-class woman's role..was the happy lot of all women. **1961** *Guardian* 2 Nov. 8/2 Robert Jungk..wanted to do his thesis on what he called 'psychopolitics', the interaction between mass psychology and mass psychology movements and politics. **1980** *Boston Globe* 3 Feb. B1 Kantor claims that people's current patterns of interaction, or 'psychopolitics', are based on 'critical identity images'. **1934** H. Hiler *Notes Technique Painting* 332 *Psycho-prismatism*, the affective psychology of colour. The study of the reactions of human beings or animals to the various colours. **1910** W. A. Turner *Three Lect. Epilepsy* 25 He has described the psycho-motor, psycho-sensory, psycho-visual, and psycho-auditory centre in close relation to the motor, sensory, visual, and auditory centres. **1947** H. C. Elliott *Textbk. Nervous Syst.* xix. 238/1 Caudal to its upper part [*sc.* that of the sensory cortex] lies..the psychosensory region. **1959** S. Duke-Elder *Parsons' Dis. Eye* (ed. 13) iv. 38 The pupils participate in several reflexes, three of which are of clinical importance:.. 3. The psychosensory reflex, whereby a dilatation occurs on psychic and sensory stimuli. **1928** *Amer. Jrnl. Sociol.* Nov. 447 Knowing these [laws of suggestion], we could follow their lead into greater knowledge of phenomena of a psychosociological nature. **1970** Touraine & Pécaut in I. L. Horowitz *Masses in Lat. Amer.* iii. 67 The resulting normative and psychosociological changes [in the social system] are analyzed..as a function of the change in values judged necessary to attain the industrialized state. **1966** *Punch* 9 Mar. 332/3 Those big black career advertisements and their rich esoterica about openings for crystallographers, systems analysts and industrial psycho-sociologists. **1908** *Science* 10 July 54 Psycho-sociology. **1957** R. K. Merton *Student-Physician* 53 A middle ground which has been described as social psychology (or, by some, psycho-sociology). **1973** *Screen* Spring/Summer 151 Various revelatory and therapeutic methods belonging to modern psycho-sociology. **1913** J. Murray *Ocean* x. 228 We may say that within the biosphere a sphere of reason and intelligence has been evolved in man, who attempts to interpret and explain the cosmos; this may be called the psycho-sphere. **1957** P. B. Sears *Ecology of Man* 10 To these might be added *Mind*—the Psychosphere, studied by psychologists, anthropologists and other social scientists. **1975** O. L. Reiser *Cosmic Humanism & World Unity* iii. 97 The Psychosphere may be regarded as a psychic-magnetic environment, an 'auric field', beyond the Van Allen Radiation Belt. **1961** Musser & O'Neill *Mod. Pharmacol. & Therapeutics* (ed. 2) xix. 361 Prior to the development of these newer drugs, called psycho-stimulants or psychic energizers, apathetic and depressed patients were treated with caffeine and the amphetamines. **1963** *Wall St. Jrnl.* 21 Jan. 12/4 The three areas in which the largest number of new agents are being investigated were psycho-stimulants, broad spectrum antibiotics and cholesterol-reducing agents. **1966** J. D. P. Graham *Pharmacol.* iv. 26/2 This drug [*sc.* dexamphetamine sulphate] is given to mentally depressed patients. It may therefore be termed a psychostimulant drug. **1967** [see *psychodysleptic* adj. above]. **1971** T. A. Ban in O. Vinar et al. *Advances in Neuropsychopharmacol.* 212 Conflict tolerance in humans may also increase under the influence of psychostimulants in general. **1973** *Proc. R. Soc. Med.* LXVI. 359/2 Would-be psychiatrists are taught to describe, define and treat disembodied psychosyndromes instead of learning to apply modern investigative science to finding causes. **1976** Smythies & Corbett *Psychiatry* vii. 113 Non-specific endocrine psychosyndromes occur with apathy, depression and lability of mood. **1919** C. Long in M. K. Bradby *Psycho-Anal.* p. vi, We aim at a reconstruction of life which can only be conceived as a psycho-synthesis. **1924** J. Riviere tr. *Freud's Coll. Papers* II. xxxiv. 395 The neurotic human being brings us his mind racked and rent by resistances; whilst we are working at analysis of it and removing the resistances

this mind of his begins to grow together; that great unity which we call his ego fuses into one all the instinctual trends which before had been split off and barred away from it. The psycho-synthesis is thus achieved during analytic treatment without our intervention. **1940** H. G. Wells *Babes in Darkling Wood* 9 The mental break-down of Gemini..bring [*sic*] the methods of a leading psycho-analyst and modern psychosynthesis into the story. **1975** M. & N. Samuels *Seeing with Mind's Eye* iii. 37 Currently, visualization is being used in a number of different psychotherapeutic techniques—including..directed daydreams, Psychosynthesis, and behaviorist desensitization. **1944** H. G. Wells *'42 to '44* 172 What a psycho-analyst calls the Unconscious, but which, according to the psycho-synthesist, is merely a multitude of reaction systems out of contact with the main directive system. **1940** —— *All Aboard for Ararat* ii. 80 The core of the new world must be (listen to these words!) Atheist, Creative, Psycho-synthetic. **1910** Psycho-visual [see *psycho-sensory* adj. above]. **1969** G. C. Dickinson *Maps & Air Photographs* iv. 63 The conventional colour sequence, which follows spectrum order from violet through shades of blue, green, yellow and orange to red (or more commonly brown), accords well with the psycho-visual properties of colours—blues for submarine areas are 'recessive', reds for hills 'stand out'—but there can be unfortunate suggestive overtones. **1971** *Nature* 19 Mar. 180/1 It is hoped that a laboratory equipped for psychovisual studies will..report on the degree to which descriptions of 'artificial' ball lightning resemble those of the natural phenomenon that are recorded in the scientific literature.

psycho-acou·stic, *a.* [f. Psycho- + Acoustic *a.*] Pertaining to the perception of sound and the production of speech or to the study of these. Also **psycho-acou·stical** *a.*, **-acou·stically** *adv.*

1885 W. Stirling tr. *Landois's Text-bk. Human Physiol.* II. xiii. 922 The psycho-acoustic centre..lies, in the dog,..in the region of the second primary convolution. **1946** *Jrnl. Aeronaut. Sci.* XIII. 255 The Psycho-Acoustic Laboratory, Harvard University. **1953** *Electronic Engin.* XXV. 24/2 Acoustic and psycho-acoustic distortion. These forms of distortion occur through imperfections in the acoustics of studios and listening rooms, the use of mon-aural (single channel) reproducing chains, [etc.]. **1953** *Language* XXIX. 85 Measurement of fidelity has to be made psycho-acoustically in terms of the whole speech signal, not just in terms of its linguistic content. **1958** *New Scientist* 30 Oct. 1165/1 A great deal of psycho-acoustical work has concerned simple sounds, like pure tones, or sharp clicks. **1971** *Computers & Humanities* V. 312 Use of the computer in investigating psycho-acoustical phenomena. **1975** G. J. King *Audio Handbk.* vii. 163 Several psycho-acoustical principles govern the subjective assimilation of quadraphony. **1976** *Gramophone* Jan. 1269/2 Now the physical and psycho-acoustic conditions for correct perception of direction of sound can be expressed by equations. **1977** *New Scientist* 1 Dec. 574/2 Ways and means of psychoacoustically handling the retrieved signals so that they produce a realistic surround of sound.

So **psycho-acou·stics**, the science of the perception of sound and the production of speech; **psy:cho-acousti·cian**, an expert or specialist in psycho-acoustics.

1948 *Mind* LVII. 388 Harvard divided psychology into two Departments—the Department of Social Relations.. and the Department of General Psychology, with three.. laboratories of General Psychology, Physiological Psychology and Psycho-Acoustics. **1958** *New Scientist* 30 Oct. 1165/3 The science of acoustics..has branched out into two fields which are essentially modern in idea and purpose... The first field is ultrasonics... The second field is psycho-acoustics, which is, essentially, the study of hearing, but more especially the whole speech-hearing process. **1963** Ervin & Miller in *Child Psychol.* (*62nd Yearbk. Nat. Soc. Study of Educ.* i) iii. 111 Information on the actual phonetic cues used by the child could be obtained by using artificially constructed vocalic stimuli. Such studies have been conducted by psychoacousticians on adults, but not on children. **1976** *Canad. Jrnl. Linguistics* XXI. i. 2 It would be pointless, e.g., to try to incorporate all of the practically infinite variations of the acoustic wave form into a descriptive system for psycho-acoustics when what is required is an adequate description of those specific properties of the sound pattern which significantly affect the perception of speech. **1977** *Rolling Stone* 30 June 41/2 Next I met David Wessel, a psychoacoustician (one who studies human perception of sound) from Michigan State University.

psychoactive, -activity, -æsthetic(s: see *psycho-.

psychoa·nalyse, *v.* Also with hyphen. [Back-formation from next, after *analysis, analyse*.] *trans.* To subject to or treat by psychoanalysis; = *analyse *v.* 3 b. Also *absol.*

1911 *Amer. Jrnl. Psychol.* July 423 It is..hoped that Freud will..psychoanalyze Goethe. **1922** J. Macy in D. H. Lawrence *Sons & Lovers* p. ix, Let whoever cares to try analyze or psychoanalyze. **1924** C. Mackenzie *Heavenly Ladder* xxiii. 288, I could psycho-analyse all Bloomsbury now. They all suffer from an inferiority complex. **1969** *Listener* 14 Aug. 219/2 Professor Baker scrupulously does not diagram, hypothesise or psycho-analyse. **1973** *Amer. N. & Q.* XI. 78/1 To psychoanalyze a dramatic hero is to treat him as a human being.

Hence **psycho-a·nalysed** *ppl. a.*

1928 'R. West' *Strange Necessity* 240 A psycho-analysed person who has made the realization that all persons he dreams of are disguised versions of himself.

psychoana·lysis. Also with hyphen and (*rare*) as **psychanalysis.** [ad. F. *psycho-analyse* (S. Freud 1896, in *Rev. Neurologique* IV. 166): see Psycho- and Analysis.

Freud earlier used *psychische analyse* and *klinisch-psychologische analyse* (*Neurol. Centralbl.* (1894) XIII. 364).]

a. A therapeutic method originated by Freud for treating disorders of the personality or behaviour by bringing into a patient's consciousness his unconscious conflicts and fantasies (which are attributed chiefly to the development of the sexual instinct) through the free association of ideas, analysis and interpretation of dreams and parapraxes, etc., and allowing him to relive them by transference. **b.** A theory of personality and psychical life derived from this, based on concepts of the ego, id, and super-ego, the conscious, pre-conscious, and unconscious levels of the mind, and the repression of the sexual instinct; more widely, a branch of psychology dealing with the unconscious.

[**1898** H. Ellis in *Alienist & Neurologist* XIX. 610 The influence of fear is not denied by Breuer and Freud, but they have found that careful psychic analysis frequently reveals that the shock of a commonplace 'fear' is really rooted in a lesion of the sexual emotions.] **1906** *Jrnl. Abnormal Psychol.* I. 28 Their importance with relation to treatment (by the method of 'psycho-analysis') is made clear. **1913** *Q. Rev.* Jan. 143 'Psycho-analysis' has been strongly advocated by some men in the medical profession... It consists in carefully and systematically resuscitating the patient's past memories, thus making him aware of his buried and unconscious mental processes, when those are brought before his present consciousness. **1924** W. B. Selbie *Psychol. of Relig.* 286 Psycho-analysis is the name given to the process by which the hidden depths of the individual consciousness can be revealed. **1932** *Sun* (Baltimore) 12 Sept. 6/3 Psychanalysis for the majority of informed Marylanders is not..a new kind of gypsy dreambook. **1938** *Internat. Jrnl. Psycho-Anal.* XIX. 1 Everything of importance that we know concerning the play of instinctual forces and the course they follow in the homosexual we have derived from psycho-analysis. **1958** Thorpe & Schmuller *Personality* ix. 215 According to psychoanalysis, all of us are characterized by primitive impulses, but our parents and the culture have required us to develop standards of behavior which lead to the repression of these impulses, forcing them into the unconscious. **1964** A. Anastasi *Fields of Applied Psychol.* xiv. 383 It is apparent that active interpretation by the therapist is a major feature of psychoanalysis. **1964** M. Argyle *Psychol. & Social Probl.* ii. 26 Neither of the two major theoretical approaches in psychology—psychoanalysis or learning theory—can explain the phenomena of childhood socialization satisfactorily. **1966** *Times* 13 Jan. 11/4 Freudian and Jungian psychanalysis.. has had its day. **1974** J. Mitchell *Psychoanal. & Feminism* II. ii. 153 His theories of the nature of sexuality and the evidence of its repression..were developed in a double intellectual context, that of Marxisim and of psychoanalysis. **1975** A. Ryle *Frames & Cages* iii. 19 In psychoanalysis, the therapist will..embark on a relationship with the patient, the understanding and evolution of which is the central process of therapy... It is through this process that unconscious primitive fantasies are replaced by freer and less idiosyncratic apprehension of the self and the world.

psychoa·nalyst. Also with hyphen. [f. prec., after *analysis, analyst*.] One who practises or has training in psychoanalysis.

1911 *Amer. Jrnl. Psychol.* July 434 The business of the psychoanalyst is to provide a means by which the emotion attached to a repressed complex may find expression, by being transformed. **1918** *Jrnl. Educ.* Mar. 153/1 Dr. Pfister devotes a couple of pages to an exposition of the need for the psychoanalyst to be himself 'free from complexes'. **1921** R. Macaulay *Dangerous Ages* v. 88 The psycho-analyst doctor would really want to hear details. **1947** A. Huxley *Let.* 9 Mar. (1969) 567 Marlow is one of those classical cases, so dear to psychoanalysts, with a fixation on his mother. **1977** A. Sheridan tr. *J. Lacan's Écrits* iii. 105 Of all the undertakings that have been proposed in this century, that of the psychoanalyst is perhaps the loftiest.

psy:choanaly·tic, *a.* Also with hyphen and (*rare*) as **psychanalytic.** [f. as prec., after *analysis, analytic*, or ad. G. *psychoanalytisch*.] Of, pertaining to, or employing psychoanalysis.

1906 *Jrnl. Abnormal Psychol.* I. 31 The strict 'cathartic', psycho-analytic method advocated by Freud. **1932** *Sun* (Baltimore) 12 Sept. 6/3 In the twenties the works of Freud, Jung, Jones and others of the psychanalytic schools sold like novels. **1979** *N.Y. Rev. Bks.* 25 Oct. 23/2 *Patricide*..draws on some of the best recent psychoanalytic literature.

So **psy:choanaly·tical** *a.*, in the same sense; **psy:choanaly·tically** *adv.*, by means of, in respect of, or towards psychoanalysis.

[**1857** *Russell's Mag.* Nov. 163/2 [Poe] chose..the psycho-analytical. His heroes are monstrous reflections of his own heart in its despair, not in its peace.] **1908** *Jrnl. Abnormal Psychol.* III. 209 It would have added greatly to the interest of this question if a psycho-analytical investigation had been resorted to. **1919** *N.Y. Tribune* 22 Dec. 10/6 'Psychoanalytically speaking,' he writes, 'I think you were wrong about H. third and the carving knife.' **1927** *Observer* 12 June 12/4 The interesting question of whether medical men may treat a patient psycho-analytically was dealt with in Court the other day. **1931** R. Campbell *Georgiad* i. 13 Nor would he dogmatise his pet perversions With psycho-analytical assertions. **1959** B. Wootton *Social Sci. & Social Path.* iv. 145 Even the psychoanalytically inclined are beginning to allow a rather less rigid, and therefore less dismal attitude. **1962** *Lancet* 22 Dec. 1311/1 He begins his book with an account of Freud's philosophical theories as a background to the later development of psychoanalytical concepts by the neo-Freudians. **1971** *Times* 1 Feb. 20/4 (Advt.), Psycho-analytically Orientated Group counters inhibition, anxiety, depression. **1977** C. Storr *Tales from Psychiatrist's Couch* x. 104 The strict, psycho-analytical view that it is utterly improper for the analyst to have anything to do with the husbands, wives, parents, children.. of his patients.

psychobabble, -er: see *psycho-.

psychobio·graphy. [f. Psycho- + Biography.] **a.** A biography dealing esp. with the psychology of the subject.

1931 *Brit. Jrnl. Psychol.* XXII. 96 The majority of 'psychobiographies' and pseudo-scientific works on the psychology of character. **1969** *Daily Tel.* 20 Mar. 22/2 Frank E. Manuel, the American historian,..in this 'psycho-biography' attempts to analyse aspects of Newton's conduct. **1974** *Publishers Weekly* 13 May 56/2 To understand Mishima's death, Scott-Stokes thinks, one must understand his life. With this useful psychobiography he would seem to have made his case. **1977** *N.Y. Rev. Bks.* 23 June 23 (Advt.), An iconoclastic psychobiography of Freud's disciple.

b. The art of writing psychobiographies; the interpretation of life histories in psychological terms, or the psychological analysis of a historical person.

1965 *Hist. & Theory* IV. 357 This is still psychobiography, but the psychopathological model has given way to a model of man in history. **1975** *Times Lit. Suppl.* 24 Jan. 90/2 The problem..is how to read the personality into the action, and it is the claim of psychobiography that this can not only be done but that it can be done systematically. The classic exercise in psychobiography is Erik Erikson's *Young Man Luther*.

Hence **psychobio·grapher**, a writer of psychobiography; **psy:chobiogra·phic, -ical** *adjs.*; **psy:chobiogra·phically** *adv.*

1972 *N.Y. Times Bk. Rev.* 10 Dec. 36 It [*sc.* a book] makes no large historiographical, mythological or psychobiographical assumptions. **1975** *Times Lit. Suppl.* 24 Jan. 90/2 A psychobiographer would naturally reply that this is to take too literal and simplistic a view of the information that is contained in the sources. **1977** *Oxf. Lit. Rev.* II. iii. 8/1 This touches on processes which are not consonant either with the communal norms of cultural expectations or with the detective work of psychobiographers. **1977** *N.Y. Rev. Bks.* 15 Sept. 40/4 He is a psychobiographical critic.., interested in the poem not as artifact but as evidence. **1978** *Dædalus* Spring 229 It is ..contrary to the psychobiographic drive of most of the younger scientists. **1979** *N.Y. Times* 15 July II. 19 Leopold cuts a rather poor figure, psychobiographically, emerging as the standard narcissistic parent who sees his child only as an extension of his own ego.

psychobio·logy. Also with hyphen. [f. Psycho- + Biology.] The study of the biological basis of behaviour or mental phenomena; the interaction of mental and biological factors in an organism.

1902 *Encycl. Brit.* XXXII. 65/2 This connection of vegetal and animal functions remains one of the obscurest in all psycho-biology. **1923** A. Meyer in E. Winters *Coll. Papers* (1952) IV. 244 In connection with psychiatric work, but from an angle quite different from Freudism, there had developed during the last twenty-five years a less spectacular objective psychobiology. **1946** [see *sociobiology]. **1966** *New Scientist* 24 Feb. 464/3 Dr Allan Jacobson..reported that, by taking extracts from the brains of rats that had been taught to perform certain tasks, and injecting them into untrained rats, the latter seemed to acquire a degree of memory for these tasks... New experiments, however, undertaken by five workers from the departments of psychobiology, and of molecular and cell biology, University of California, Irvine, appear to cast considerable doubts on the validity of Dr Jacobson's conclusions. **1975** *Sci. Amer.* Dec. 7/2 (Advt.), *A Primer of Psychobiology* is a brief, informative introduction to what is known about the structure and function of the nervous system and how these relate to behavior.

So **psy:chobiolo·gic, -lo·gical** *adjs.*, of or pertaining to psychobiology; both psychological and biological; **psy:chobiolo·gically** *adv.*, in a psychobiological manner; in relation to psychobiology; **psychobio·logist**, an expert or specialist in psychobiology.

1901 *Amer. Jrnl. Psychol.* XII. 206 The experiments must conform to the psycho-biological character of an animal if sane results are to be obtained. **1934**, **1935** [see *field *sb.* 17 d]. **1935** Adams & Zener tr. *Lewin's Dynamic Theory of Personality* iii. 79 The environment is..to be defined not physically but *psychobiologically*, that is, according to its quasi-physical, quasi-social, and quasi-mental structure. **1941** *Amer. Speech* XVI. 216 All human behavior is a social psychobiological continuum in which there is no real dichotomy. **1946** *Nature* 24 Aug. 252/2 This is unfortunate at a time when a lot of young men are coming out of the medical services with the idea that there is something in psychosomatic medicine... It is high time we began to try to find out what this is instead of mouthing big phrases such as 'psychobiologic unit'. **1961** Webster, Psychobiologist. **1971** *Amer. Jrnl. Psychiatry* CXXVIII. 706/1 The psychobiologic changes occurring during the [menstrual] cycle do not appear to produce a specific behavioral effect since the form and severity of the symptoms differ among the various studies. **1973** *Sci. Amer.* Sept. 25/1 We strike a balance..in graduate medical education between the preparation of technologically based specialists and psychobiologically trained physicians. **1977** *Proc. R. Soc. Med.* LXX. 690/1 The underlying pathology of anorexia nervosa may comprise one extreme of the psychobiological nutritional disturbances possibly common to many cases of 'periodic oedema'. **1977** D. M. Rumbaugh [illegible] in *Language Learning* [illegible] as a very insightful, [illegible] psychobiologist. **1979** *Time* 2 Apr. 47/2 [illegible] psychobiologist, [illegible] psychopharmacologist are..replacing [illegible] chairmen of hospital [illegible]

psychocentric, -chemical, -chemistry, -cultural, -curative: see *psycho-.

psychodelic, var. *psychedelic *a.* and *sb.*

psychodiagnosis, -diagnostic(s: see *psycho-.

psychodrama. Also with hyphen. [f. Psycho- + Drama.] **1.** A form of psychotherapy in which a patient acts or performs extempore with or in front of fellow patients and therapists in a way that dramatizes the patient's problems or difficulties; an extempore psychotherapeutic play of this kind. Also *fig.*

1937 J. L. Moreno in *Sociometry* I. 9 The psychodrama is human society in miniature, the simplest possible setup for a methodical study of its psychological structure. **1952** W. J. H. Sprott *Social Psychol.* xi. 241 The meetings of [illegible] in effect..psycho- [illegible] currents came to the [illegible] **1965** *New Statesman* 9 [illegible] last February at the [illegible] which BG's [*sc.* Ben [illegible] years of resentment [illegible] the old leader. **1968** [illegible] Its methods centre [illegible] (Stanislavsky for [illegible] Oriental philosophy. [illegible] (Advt.), Community. [illegible] encounter, gestalt, [illegible] -enactment psycho- [illegible]. **1978** *Listener* 19 [illegible] believes the answer [illegible] pains, repressed [illegible] treatment known as 'psychodrama'.

2. A play or film in which psychological elements are the main interest.

[**1927**: see *psycho *a.* 1.] **1963** *Movie* Apr. 22/2 A kind of psychodrama about racial relationships. **1975** C. [illegible] *Flick's Wedding*..was [illegible] **1977** *Time* 5 Sept. 26/2 [illegible] film of the long-running Broadway psychodrama.

psychodrama·tic, *a.* [f. prec., after *drama, dramatic*.] **1.** Of or by means of therapeutic psychodrama.

1937 J. L. Moreno in *Sociometry* I. 25 When we apply psychodramatic principles to art, especially in the theatre, one notes that the presentation of the role is often interrupted by foreign elements, foreign to the role, betraying the private personality of the actor. **1977** A. Sheridan tr. *J. Lacan's Écrits* ii. 9 Psychodramatic treatment seeks its efficacy in the abreaction that it tries to exhaust on the level of play.

2. Pertaining to or of the nature of a psychodrama (sense 2).

1943 *Sewanee Rev.* LI. 309 It could just as easily be accidental in the psychodramatic reverie of an insane and uneducated son grieving at the loss of his mother. **1946** *N.Y. Times* 15 Oct. 29/1 'Frontiers of the Mind', a collection of six dramatized case histories, will be presented..under the auspices of the [illegible] Psychodramatic Theatre.

So **psychodrama·tics**, (*a*) the use of psychodrama as therapy; (*b*) psychological dramatics; **psychodrama·tist**, (*a*) one who directs or takes part in therapeutic psychodrama; (*b*) one who writes psychodramas (sense *2).

1937 J. L. Moreno in *Sociometry* I. 10 But in interpersonal therapy, especially in one of its forms which can be called psychodramatics, the situation is enormously more complex. **1953** J. L. Moreno *Who shall Survive?* p. xvi, [illegible] psychodramatist and roleplayer. **1957** V. Nabokov *Pnin* ii. 45 He went on with his Slavic [illegible] **1973** *Nation Rev.* [illegible] uncertainties..are the [illegible] series produced by [illegible] psychodramatists. **1979** C. E. Schorske *Fin-de-Siècle Vienna* vii. 345 The harsh [illegible] *Murderer, Hope of Women* [illegible] [*sc.* Kokoschka] to liberate himself.

psychodynamics. Add: (The study of) the activity of and interrelation between the various parts of an individual's personality or psyche. (Further examples.)

[illegible] Prior to the emergence of psychodynamics, interest was focused in the [illegible] factors in their relationship to organic disease. **1957** 'T. Sturgeon' [illegible] Psychodynamics has come a long way, but it hasn't begun to alter the fact that human beings are the most feral, vicious..and self-destructive creatures God ever made. **1960** H. J. Eysenck *Exper. in Personality* II. 105 (*heading*) Experiments in psychodynamics. The excitation-inhibition balance in neurotics and in normals. **1963** J. A. Johnson *Group Therapy* v. 164 The therapist will have formed an opinion of the psychodynamics of each member from previous meetings and should now use this information as a helpful guide in anticipating as well as evaluating behavior associated with separation. **1975** Rosenbaum & Beebe *Psychiatric Treatm.* xii. 228/1 Psychodynamics broadly refers to an understanding of the interactions of conscious, unconscious, and reality.

Hence also **psychodyna·mically** *adv.*

1961 in Webster. **1972** L. J. Saul (*title*) Psychodynamically based psychotherapy. **1973** *Pediatric Clinics N. Amer.* XX. 743 As a reflection of our Freudian legacy in medicine, the traditional approach to parents in the pediatric setting has been psychodynamically oriented.

psychodysleptic, -endocrine, -endocrinology, etc.: see *psycho-.

psychogalva·nic, *a.* [ad. G. *psycho-galvanisch*(*er reflex*) (O. Veraguth 1907, in *Monatsschr. f. Psychiatrie und Neurol.* XXI. 387): see Psycho- and Galvanic *a.*] Involving changes in the electrical conductivity of the skin associated with emotional changes; chiefly in *psychogalvanic reflex* or *response* (abbrev. PGR s.v. *P II) = **galvanic skin response*.

1907 *Brain* XXX. 191 It is of interest to ascertain whether the psycho-galvanic reflex runs a parallel course with the complex indices. **1917** C. R. Payne tr. *Pfister's Psycho-Anal. Method* 336 Secretion of tears, sighing, psychogalvanic phenomena, changes in the pulse, etc. **1936** *Discovery* July 201 Some remarkable results have been obtained from 'psycho-galvanic' experiments. **1949** Wimsatt & Beardsley in D. Lodge *20th Cent. Lit. Crit.* (1972) 352 The affective critic is today actually able..to measure the 'psychogalvanic reflex' of persons subjected to a given moving picture. **1960** Koestler *Lotus & Robot* I. iii. 119 The psycho-galvanic reflex (used in lie detectors) reflects changes in the electrical properties of the body surface in response to emotional stimuli. **1973** *Times* 17 Feb. 14/5 Orme-Johnstone, in a study of the psychogalvanic response..showed that meditators had fewer responses than control subjects.

Hence **psy:chogalvano·meter**, a galvanometer used to measure the psychogalvanic response; **psy:chogalvanome·tric** *a.*

1935 H. F. Dunbar *Emotions & Bodily Changes* iii. 87 The usefulness of the psychogalvanometer for the measurement of emotions or for research in the field of psychosomatic relationships. *Ibid.* 592 (Index), Psychogalvanometric apparatus. **1936** K. Dunlap *Elements o[f] Psychol.* iv. 183 For the simplest psychogalvanometric use, a d'Arsonval galvanometer is connected directly to two electrodes. **1956** *Electronic Engin.* XXVIII. 36 The earliest psycho-galvanometer consisted of a Wheatstone bridge connected to the palms of the subject's hands by means of metal electrodes. **1959** *Listener* 17 Sept. 443/1 A horrid array of forceps, scalpels, stethoscopes, electrocardiographs, and psychogalvanometers.

psychogenesis. Add: **2. b.** The psychical origin or cause to which mental illness or behavioural disturbance may be attributed.

1920 S. Freud in *Internat. Jrnl. Psycho-Anal.* I. 125 (*heading*) The psychogenesis of a case of female homosexuality. **1939** C. G. Jung in *Jrnl. Mental Sci.* LXXXV. 1002 Psychogenesis of schizophrenia..in the first place means the question: Can the primary symptom..be considered as an effect of the psychological conflicts and other disorders of an emotional nature or not? **1972** O. L. Zangwill in Cox & Dyson *20th-Cent. Mind* II. vii. 188 There was also greater tolerance for the deviants and eclectics, provided that they subscribed to the general idea of psychogenesis—i.e. the belief that neurosis has a psychological rather than a physical cause. **1979** *N.Y. Rev. Bks.* 25 Oct. 25/1 Schorske's seeming demolition of Freud's Oedipal concept may in fact provide clues..for liberating Freud's heroic Complex from its imprisonment in the depths of individual psychogenesis.

Hence **psychogene·tic** *a.* = *psychogenic *a.*

1896 *Amer. Naturalist* XXX. 443 There is a great series of adaptations secured by conscious agency, which we may throw together as 'psycho-genetic'. **1904** *Jrnl. Philos., Psychol. & Sci. Methods* I. 328 Hume..had quite unwittingly furnished what..should have been regarded as a logical deduction and justification—rather than the mere psychogenetic description, which it purported to be—of the realistic belief. **1915** M. Prince *Psychol. of Kaiser* viii. 67 So long as these so-called psycho-genetic thoughts are there unmodified..he [*sc.* the Kaiser] could not get rid of his fixed fear of the democracy if he would. **1975** *Times Lit. Suppl.* 24 Oct. 1253/4 If he really did regard [Queen] Anne's physical disorders as psychogenetic, eighteenth-century medicine was of precious little use to her case anyway.

psychogene·tics. [f. Psycho- + *genetics.] **1.** The branch of psychology which is concerned with the effects of breeding or inheritance on behaviour.

1951 C. S. Hall in S. S. Stevens *Handbk. Exper. Psychol.* ix. 304/1 This encouraging trend will ultimately give status and stature to an interdisciplinary science of psychogenetics. The psychogeneticist of the future will presumably be trained in the methods and techniques of both genetics and psychology. **1954** *Brit. Jrnl. Psychol.* XLV. 309 This book should be studied..by the psychogeneticists. **1960** P. L. Broadhurst in H. J. Eysenck *Exper. in Personality* I. i. 5 The use of different strains, particularly of rats and mice, has long been a favourite method of psychogenetics. Bagg was a pioneer in the application of 'the methods of genetics to the study of conduct'. **1975** *Nature* 30 Oct. 832/2 The perspectives of ethology, physiological psychology, psychogenetics, social psychology..are all represented to some degree.

2. = Psychogenesis 1.

1964 *Eng. Stud.* XLV. Suppl. 104 Locke's empiricism takes its place in the field of psychogenetics; he asks the question: where does human understanding come from?

psychogenic (-dʒe·nik), *a.* [f. Psycho- + *-genic.] Having a mental or psychological origin or cause.

1902 Baldwin *Dict. Philos. & Psychol.* II. 382/2 The paralyses and anaesthesias of hysteria,..the pains and fears of neurasthenia, the ameliorations following appeals to faith or prayer are instances of psychogenic action. **1926** J. I. Suttie tr. *Ferenczi's Further Contrib. Theory & Technique Psycho-Anal.* viii. 107 A psychogenic disturbance of the voice. **1964** M. Critchley *Developmental Dyslexia* xii. 71 Only too often the backwardness in reading is deemed either an environmental, or a psychogenic problem, or both. **1973** *Sci. Amer.* Sept. 124/1 Whatever the cause of a psychotic disorder, be it biological or psychogenic, the mental content of the psychosis must reflect the input to the mind and how that input is refracted by the mind's history and functional state.

Hence **psychoge·nically** *adv.*

1933 *Jrnl. Nerv. & Mental Dis.* LXXVII. 587 (*heading*) 'Colitis'—psychogenically motivated. **1973** *Lancet* 9 June 1296/1 The psychogenically impotent subject can in fact learn to induce erection at will.

psychogeography, -geographic(al: see *psycho-.

psy:chogeria·tric, *a.* and *sb.* Also with hyphen. [f. Psycho- + *geriatric *a.*] **A.** *adj.* Of or pertaining to mental illness or disturbance in the old. Of a person: old and mentally ill or disturbed.

1961 *Guardian* 19 June 8/4 The number of psycho-geriatric patients is certainly rising with increased longevity. **1965** *Lancet* 18 Sept. 583/2 A psycho-geriatric unit was established at Severalls Hospital in 1961. **1971** *Brit. Med. Jrnl.* 23 Oct. 235/1 Among psychogeriatric patients those most likely to respond appear to be those whose initial degree of memory and intellectual impairment was not of an extreme degree. **1973** *Radio Times* 1 Feb. 41/1 A 69-year-old widow, who lives with her unmarried son and regularly attends a local authority psychogeriatric Day Centre. **1975** *Daily Tel.* 1 Dec. 13/1 There was an acute shortage of psychogeriatric beds in hospitals. **1977** *Lancet* 1 Jan. 27/1 Many doctors and social workers cannot formulate a 'psychogeriatric' problem in any other terms but as the need to get it instantly off their hands.

B. *sb.* An old person who is mentally ill or disturbed.

1971 *Observer* 12 Dec. 8/4 The psycho-geriatrics form only one category of the..mentally ill. **1973** *Listener* 19 Apr. 507/1 They're all referred by the GP... They're all psychogeriatrics.

Hence **psy:chogeria·trics**, the branch of medicine concerned with mental illness and disturbance in old people.

1967 *Brit. Jrnl. Psychiatry* CXIII. 175/1 The subspeciality of geriatric psychiatry is coming to be known as 'psycho-geriatrics', an abbreviation which is not a very happy one, but which seems to be becoming generally accepted. **1972** *Lancet* 8 July 73/2 The working party are not convinced..that there is any need for the recognition of a specialty of psychogeriatrics.

psychogogue, var. Psychagogue in Dict. and Suppl.

psychogram. Add: **2.** *Psychol.* A summary or diagram of someone's personality, esp. one based on his psychological history, responses to tests, etc. [ad. G. *psychogramm* (W. Stern *Differentielle Psychol.* (1911) III. xxii. 327).]

1918 J. Ward *Psychol. Princ.* xviii. 433 It will be possible to construct what has been called a psychogram of the concrete individual. **1924** *Jrnl. Nerv. & Mental Dis.* LX. 227 Neither was it possible to mention so much of the history of the disease as would have been necessary to prove..the correctness of the psychogram. **1935** H. Read in *Social Credit Pamphleteer* XII. 13 The æsthetic criterion is overcome by the force of the creative invention; the picture becomes a 'psychogram'. **1948** L. Spitzer *Linguistics & Lit. Hist.* 15 Linguistic deviations, of which the philologist may take stock in order to build up his 'psychogram' of the individual artist.

psychograph. Substitute for def.: A photographic image attributed to a supernatural or spiritualistic cause. (Further examples.)

1920 *London Mag.* July 443/1 Most puzzling of all forms of super-normal pictures is the psychograph—so-called because it is assumed to be psychic in its origin and production. **1939** H. Price *Fifty Yrs. Psychical Res.* i. 35 If a message in writing or a drawing spontaneously appears on a photographic plate, with or without it being exposed in the camera it is known as a scotograph or a psychograph. **1973** D. A. Spencer *Focal Dict. Photogr. Technologies* 496 All available evidence suggests that these psychographs were fakes or the result of a combination of chemical fog and wishful thinking.

2. = *psychogram 2.

1909 *Q. Rev.* Oct. 500 This is no caricature, but almost a psychograph of the spirit which permeates many if not most of the descriptive reports of cricket matches in popular sporting papers. **1921** *Education* XLI. 513 A character psychograph of the individual is obtained. **1932** C. Landis in K. S. Lashley *Stud. in Dynamics of Behav.* 299 In order to visualize more clearly the results of the tests, three psychographs were drawn to represent the performance of each subject.

3. = *psychobiography a.

1932 *Sunday Times* 6 Mar. 8/2 It was with some anxiety I saw Dame Una Pope-Hennessy was committed to writing a psychograph of Walter Scott. **1961** *Times Lit. Suppl.* 29 Sept. 637/2 Professor Edward Wagenknecht has been driven to compose a 'psychograph', in which he competently balances opinion against opinion in the hope of discovering what Hawthorne was really like. **1967** *Amer. N. & Q.* Sept. 14/2 Forrest, first of the American tragic actors in this assemblage of 'psychographs'. **1974** *Times Lit. Suppl.* 11 Oct. 1130/3 Dickens was the principal exemplar, and Wilson's penetrating psychograph, 'The Two Scrooges', coincided with George Orwell's revaluation in focusing upon a great novelist whose very popularity had caused him to be critically neglected.

psychographer. Add: **2.** = *psychobiographer.

1912 G. Bradford *Lee the American* 269 The prince of all psychographers is incontestably Sainte-Beuve. **1930** *London Mercury* Feb. 378 He does not attempt a new 'life', but only a new character-study from the point of view of the 'psychographer'.

psychography. Add: **1.** Also, = *psychobiography. (Further example.)

1929 G. Bradford in E. C. Wagenknecht *Man C. Dickens* p. xi, Psychography discards chronology, does not concern itself in any way with the sequence of external fact, except in so far as such is absolutely necessary to make clear the background. *Ibid.* i. 13 Before proceeding definitely to the psychography of Dickens, it is interesting and amusing to speculate on just what his own attitude toward this sort of inquiry would be.

3. *Psychol.* The making of a psychogram (sense *2); the systematic experimental examination of an individual's personality. [ad. G. *psychographie* (W. Stern *Differentielle Psychol.* (1911) III. xxii. 327).]

1921 *Education* XLI. 510 Psychography may be defined as the science of making graphic records of mental traits. **1927** A. A. Roback *Psychol. of Character* xxiii. 426 Psychography..records a person's *total* reactions (moral, temperamental, physical and intellectual) under all sorts of conditions. **1938** G. W. Allport *Personality* xv. 404 Psychography has a striking advantage to offset its limitations. It is a method particularly well suited to the *comparative* study of personality, which..demands the use of common traits.

psycho-histo·rical, *a.* Also without hyphen. [f. Psycho- + Historical *a.* (*sb.*).] **1.** (In Dict. s.v. Psycho-.)

2. Of or pertaining to the psychological analysis or interpretation of historical events and characters. Also **psycho-histo·ric** *a.*, in the same sense.

1945 I. Asimov in *Astounding Sci. Fiction* Apr. 10/1 It would be a psycho-historic experiment of my own. *Ibid.* 30/2, I know quite a detailed version of Hari Seldon's psycho-historical claptrap. **1964** E. Erikson *Insight & Responsibility* v. 206 What we may call psycho-historical actuality, that is, the sum of historical facts and forces which are of immediate relevance to the..anticipations and..apprehensions in the individuals involved. **1970** R. J. Lifton *Hist. & Human Survival* 3 This psycho-historical approach..stems from a general uneasiness among practitioners of both psychology and history abou the capacity of their traditional methods to describe and explain man during the latter part of the twentieth century.

Hence **psycho-histo·rically** *adv.*

1957 W. Abell *Collective Dream in Art* 5 Psycho-historically considered, art is one of the cultural symbols into which society projects existent states of underlying tension. **1968** *Partisan Rev.* 27 The principle of 'death and rebirth' is as valid psychohistorically as it is mythologically.

psycho-hi·story. Also without hyphen. [f. Psycho- + History *sb.*] **a.** The analysis and interpretation of historical events with the aid of psychological theory; also = *psychobiography b.

1934 *Reunion* I. 34 Judged by this profound philosophical test, so many of the glibly clear solutions of psycho-history are unsatisfactory. **1942** I. Asimov in *Astounding Sci. Fiction* June 30/2 The terms I use are at best mere approximations, but none of you are qualified to understand the true symbology of Psycho-History. **1957** W. Abell *Collective Dream in Art* 7 The energies involved in such conflicts are neither exclusively material nor exclusively psychological... The further we penetrate into the insights of psycho-history, the more likely we are to discover the means of mastering its disruptive forces. **1972** *Sat. Rev.* (U.S.) 25 Mar. 98/2 The roots of psycho-history may go back to Sigmund Freud's *Leonardo da Vinci: A Study in Psychosexuality*, published in Vienna in 1916. **1976** *Times Lit. Suppl.* 30 Jan. 117/1 Attempts have been made to explain Hitler's personality and ideology, in part at any rate, in terms of his childhood experiences. These works of psychohistory vary in their perceptiveness.

b. A treatise on or study in psycho-history; a psychobiography.

1972 *Sat. Rev.* (U.S.) 25 Mar. 98/3 Another psycho-history to be published..next fall is a study of Hitler by Dr. Walter Langer. **1976** *New Yorker* 17 May 60/1 Erik Erikson, in 'Gandhi's Truth', a biographical exploration that he calls a 'psycho-history', attaches considerable importance to their relationship.

So **psycho-histo·rian,** an expert in or writer of psycho-history.

1934 *Reunion* I. July 34 The psycho-historians have created a new and uneasy fashion; and while we can welcome an *exposé* of some of the lies of history such as Mr. Belloc is making, there are other much-quoted verdicts of ecclesiastical historians which are more epigrammatic than true. **1949** *Astounding Sci. Fiction* Nov. 21/1 It is enough for a Psychohistorian..to know his Biostatistics and his Neurochemical Electromathematics. **1970** *Daily Tel.* 23 Feb. 9/6 A trained psycho-historian..who had witnessed the Franco-Prussian War of 1870, would have realised that the probability of a World War had been raised to higher than 70 per cent. **1975** L. de Mause *Bibliogr. of Psychohist.* p. viii, What the new psycho-historians are creating is a radical empiricism.

psychoid (səi·koid), *sb.* and *a.* [f. Psych(e or Gr. ψυχή + -oid.] **A.** *sb.* A name variously given to vital forces that appear to direct the functions and reflex actions of the living body.

1908 H. Driesch *Sci. & Philos. of Organism* III. iii. iii. 82, I therefore propose the very neutral name of 'Psychoid' for the elemental agent discovered in action. 'Psychoid'—that is, a something which though not a 'psyche' can only be described in terms analogous to those of psychology. **1930** E. Bleuler in *Psychiatric Q.* IV. 43 Bodily functions, too, are integrated to a high degree... Hence we have good grounds for bringing the bodily functions, too, under *one* conception. This summary, the body soul, I have called the psychoid... We cannot do otherwise than regard the psyche as a specialization of the psychoid of the organism. **1931** A. Wolf in W. Rose *Outl. Mod. Knowl.* 575 Just as the development of an animal is directed by an entelechy, so its behaviour is directed by an analogous psychoid, or an inborn intelligent urge to action. **1935** H. F. Dunbar *Emotions & Bodily Change* i. 47 Let us therefore call the body-soul the psychoid. Now the relationship of psyche to soma becomes clear.

B. *adj.* Pertaining to these forces (see also quot. 1960).

1930 *Psychiatric Q.* IV. 43 With human beings we have a number of reactions which are half psychoid and half psychic... When we scratch ourselves..how much of this is reflex action, and how much conscious action? **1934** E. B. Strauss tr. *Kretschmer's Text-bk. Med. Psychol.* iv. 39 The discrete components of an intended act are associated with a minimal degree of conscious participation or are even unconscious; i.e. they are predominantly psychoid. **1944** J. S. Huxley *On Living in Revolution* iv. 50 All the activities of the world-stuff are accompanied by mental as well as material happenings; in most cases, however, the mental happenings are at such a low level of intensity that we cannot detect them; we may perhaps call them 'psychoid' happenings, to emphasize their difference in intensity and quality from our own psychical or mental activities. **1960** R. F. C. Hull tr. *Jung's Structure & Dynamics of Psyche* in *Coll. Wks.* VIII. iii. 177 If I make use of the term 'psychoid' I do so with three reservations: firstly, I use it as an adjective..; secondly no psychic quality in the proper sense of the word is implied, but only a 'quasi-psychic' one such as the reflex-processes possess; and thirdly, it is meant to distinguish a category of events from merely vitalistic phenomena..and from specifically psychic processes.

psychokine·sis. Also with hyphen. [f. Psycho- + *kinesis.] **1.** A psychic power by which some people are held to be able to move objects by other than physical means. Cf. *telekinesis* s.v. Tele-. Abbrev. PK., Pk. s.v. P II.

1914 H. Holt *On Cosmic Relations* (1915) I. xiv. 216 Now assuming Telekinesis to be established, perhaps we are as nearly ready to consider what I shall call Psychokinesis as people were a generation ago to consider Telekinesis. **1943** *Jrnl. Parapsychol.* VII. 22 Psychokinesis seems better for a general term to cover both effects than telekinesis, which leaves out the psychical and emphasizes distance. **1952** *Sci. News* XXIII. 53 In particular the proponents of telepathy, clairvoyance, psychokinesis and the like have reached the conclusion that these phenomena are not affected by distance or orientation. **1957** *Times Lit. Suppl.* 13 Dec. 760/3 Mr. Rose seeks to establish the power of alleged rain-makers by testing their ability to will (by 'psychokinesis') the fall of a die. **1973** *Times* 1 Dec. 16/4 The other group of tests dealt with..psychokinesis, and included Mr Geller's apparent ability to bend spoons, nails and other metal objects.

2. Activity or development within the psyche or spirit. *rare.*

1920 H. L. Eno *Activism* iv. 46 The collective activity of psychons, also, which we shall call 'psychokinesis' differs, with its various combinations upon its own plane, in intensity.

psychokine·tic, *a.* Also with hyphen. [f. Psycho- + Kinetic *a.* (*sb.*).] **1.** Of or pertaining to psychokinesis (sense *2).

1904 G. S. Hall *Adolescence* II. xviii. 724 We are now coming to study and utilize every psycho-kinetic equivalent or analogue between the higher and the lower faith. **1948** J. G. Bennett *Crisis in Human Affairs* vii. 133 This we shall call the 'psycho-static' view, according to which the *psyche*, or essential nature of man, is unchangeable... The alternative we shall call the 'psycho-kinetic' view, which asserts the possibility of movement or transformation within man's psyche.

2. Of or pertaining to psychokinesis (sense *1).

1943 [see PK., Pk. s.v. *P II]. **1950** *Mind* LIX. 453 The random number generator will be subject to the psycho-kinetic powers of the interrogator. **1962** V. Nabokov *Pale Fire* 165 Hazel was involved in some appalling 'psychokinetic' manifestations.

psycholeptic: see *psycho-.

psycholingui·stic, *a.* and *sb.* [f. Psycho- + Linguistic *a.* and *sb.*] **A.** *adj.* Of or pertaining to psycholinguistics (see sense B below).

1936 J. Kantor *Objective Psychol. of Grammar* iv. 55 (*heading*) The psycholinguistic situation analyzed. **1948** *Mind* LVII. 531 The psycholinguistic point of view does not allow one to answer a question such as 'What is a number?' without having ascertained first the purpose of the question. **1953** J. B. Carroll *Study of Lang.* iv. 120 Psycholinguistic analysis might suggest the units of selection in messages and better ways of gauging their semantic content. **1959** Schuell & Jenkins in L. F. Sies *Aphasia Theory & Therapy* (1974) xi. 212 Osgood (1953) has presented a psycholinguistic model for aphasia..; the model was constructed and deficits which should result from interruptions of psycholinguistic rather than neurological processes were deduced. **1967** D. G. Hays *Introd. Computational Linguistics* xi. 187 Psycholinguistic evidence is now regarded as relevant to decisions about syntax. **1970** *Language* XLVI. 87 It was evident that the phonemes as analysed did indeed match the psycholinguistic units that the subject was manipulating in the use of his own language. **1977** P. Strevens *New Orientations Teaching of Eng.* ii. 32 Modern psycholinguistic studies of the way a small child learns his mother tongue.

B. *sb. pl.* (const. as *sing.*). The branch of linguistics which deals with the interrelation between the acquisition, use, and comprehension of language, and the processes of the mind. Cf. *linguistic psychology* s.v. *linguistic a. b.

1936 Allport & Odbert *Psychol. Monogr.* XLVII. i. 25 From the standpoints of the psychology of personality and psycho-linguistics the complete record is of more value. **1948** *Mind* LVII. 530 *Psycho-linguistics*.., a term which is intended to cover the spoken and written word, as well as gestures and such physical actions as are used by human being[s] to influence each other's activities. It also studies the associations which relate such acts with the mental processes which occur in the minds of the parties concerned. **1952** *Language* XXVIII. i. 115 Psycholinguistics is not linguistics plus psychology; it is a resultant of the two. **1959** *Word* XV. 192 Psycholinguistics s a relatively new discipline developing along the border between linguistics and psychology. **1967** S. Saporta *Psycholinguistic Theories & Generative Grammars* 7 Let us understand the central question in psycholinguistics to be the study of whatever psychological processes contribute to the acquisition, production, and comprehension of language. **1974** P. Dickinson *Poison Oracle* ii. 36 My field is psycholinguistics... The study of the effect of language on the mind. **1978** *English Jrnl.* Dec. 63/2 The findings of social linguistics and psycholinguistics are presented in terms meaningful to the younger reader.

Hence **psycholi·nguist,** a student of or specialist in psycholinguistics; **psycholingui·stically** *adv.*; **psy:cholinguisti·cian** *rare* = *psycholinguist.

1953 J. B. Carroll *Study of Lang.* iv. 121 It is also possible that the mass statistics..will provide the psycholinguist with a rewarding set of material for study. **1964** *Language* XL. 226 The problem of isolating the psycholinguistically distinctive units of sequential encoding..has elicited some interesting experiments with pausal phenomena. **1970** *New Scientist* 24 Sept. 615/1 It is a widely held view among psycholinguists..that phonology and syntax are unique to man. **1975** M. Bradbury *History Man* vi. 106 Do you mean am I a structuralist or a Leavisite or a psycho-linguistician or a formalist or a Christian existentialist or a phenomenologist? **1976** *Amer. Speech 1974* XLIX. 80 We can answer it psycholinguistically by claiming that two items are collocates of each other if they belong to a single remembered set. **1977** *Verbatim* Dec. 1/1 The psycholinguists working with infants by and large ignored the obvious; eye contact and touch between mother and child.

psychologese (səikǫlǒdʒī·z). *colloq.* [f. Psycholog(y + -ese.] Language in which technical terms in psychology are used for effect.

1961 R. Hoggart *Auden* iv. 118 On occasions..he wrote 'psychologese' in poor verse. **1974** *Publishers Weekly* 4 Nov. 70/1 The mixture of psychologese and unctuous 'poetry' goes far toward placing this among the sillier books of its sort. **1979** *N.Y. Rev. Bks.* 25 Oct. 11/3 To paraphrase Chris Edwards's psychologese, 'His new ego identity is being built up as the old one is being destroyed.'

psychologic, *a.* Add: Now freq. *poet.* (Further examples.)

1943 I. A. Richards *Basic Eng. & its Uses* i. 17 No one yet knows what the fundamental factors [in changes of population] are. They may be economic, but they may equally well be psychologic. **1948** M. W. Thorner *Psychiatry in Gen. Pract.* vi. 162 For a disease so pronounced in its clinical symptoms and so apparently free of a demonstrated psychologic cause, manic-depressive disease is singularly barren of any distinguishing physical findings. **1951** R. Graves *Poems & Satires* 26 Their hairy bellies warming With buzz of psychologic wit And homosexual swarming. **1954** [see *gerontology]. **1967** *Listener* 19 Jan. 91/3 You've said too much, The lot of you, of psychologic such and such, A more grotesque miscellany I've never read.

psychologics (further examples: still *rare*).

1893 M. Beerbohm *Let.* 15 Aug. (1964) 50 *Me voici* talking self and psychologics. **1943** *Amer. Speech* XVIII. 220 General semantics, as a psychology (or 'psychologics', as Korzybski prefers to call it),..claims to put the Prince of Denmark back into *Hamlet*.

psychologic (səi·kǫlǫdʒik, səikǒlǫ·dʒik), *sb.* Also with hyphen. [f. Psycho(logy + Logic *sb.*] The practice of logical reasoning based on psychological observations and judgements rather than on abstract propositions.

1912 F. C. S. Schiller *Formal Logic* xxiv. 393 Provided he [*sc.* the formal logician] will let us frame a science which will concern itself with the aspects of intellectual functioning which are excluded from the Ideal of Pure Thought, let him restrict 'logic' to what *he* means thereby. We shall merely..adopt another term. Let us call this other study Psychologic... Formal Logic may be left to its own devices henceforth, and Psychologic will study real knowing without impediment. **1931** N. Isaacs in *Proc. Aristotelian Soc.* XXXI. 225 The term 'psycho-logic' has been used by Dr. Schiller in connection with his position. Before I met it in his writings, I happened to be led to the same coinage, some years ago, under the pressure, as I believe it to be, of the same facts... In my view (as in Dr. Schiller's) logic needs to be based on psychology through and through... That is the purport of 'psycho-logic': an expressly intermediate study that starts from clearly psychological facts, but examines these with logical intent, and attempts to show that a logic emerges from them, and of what kind. **1935** *Mind* XLIV. 471 The purely empirical *psychologic* which substitutes judgments for propositions as the subjects of logical discourse, frankly seeks the co-operation of psychology and is willing to be a handmaid of the sciences. **1953** Mays & Whitehead tr. *Piaget's Logic & Psychol.* iii. 25 These three difficulties force us to interpolate between psychology and axiomatic logic a *tertium quid*, a 'psycho-logic',..related to these in the same way as mathematical physics is related to pure mathematics and experimental physics. **1967** *Listener* 19 Oct. 492/2 The most likely resolution is a form of psychologic: the person gradually and unconsciously changes his values and beliefs so as to make them consistent with what he says and does. **1973** I. L. Child *Humanistic Psychol.* vii. 95 An account of 'psycho-logic' that he..worked out several years ago.

psychological, *a.* (*sb.*) Add: **1.** (Earlier and further examples.) Also *fig.* Also *absol.*

a **1688** R. Cudworth in J. H. Muirhead *Platonic Tradition in Anglo-Saxon Philos.* (1931) 64 Wherefore we have proposed another psychological hypothesis: that..there must of necessity be in the soul one common focus or centre in which all these kinds may meet. **1802** *Monthly Mag.* XIV. 35/1 The grand drama of the Leipzig Easter-fair is certainly..very attractive and entertaining, whether one belong to the crowd of busy actors, or be only an idle looker-on—whether one have on his nose a pair of statistical, or psychological spectacles. **1951** M. McLuhan *Mech. Bride* (1967) 143/2 They have a more organic approach than the Germans, whose attitude is closer to the psychological. **1974** R. Assagioli *Act of Will* (1975) v. 48 This knowledge..enables us to make countless..applications of those psychological laws.

2. a. (Further examples.) Also, affecting or pertaining to the mental and emotional state of a person.

1929 *Language* V. 212 Linguistics may thus hope to become something of a guide to the understanding of the 'psychological geography' of culture in the large. **1942** E. Fromm *Fear of Freedom* iii. 49 Significant changes in the psychological atmosphere accompanied the economic development of capitalism. **1958** R. I. Perusse in Daugherty & Janowitz *Psychol. Warfare Casebk.* ii. 34 The expressions..'psychological operations', and 'target' should..be avoided. US observers can vouch for the discomfiture of foreign peoples at being considered by us as a fitting subject for manipulation. **1958** *Times* 15 July 11/5 Mr. Sylvester..said that to have coal stocks lying at the pithead or anywhere else had a psychological effect on the men in the industry. **1962** E. Cleaver in A. Dundes *Mother Wit* (1973) 11/1 The destructive psychological impact of this standard of beauty. **1974** M. Taylor tr. *Metz's Film Lang.* viii. 190 Since the advent of the talking film, we have had the 'psychological comedy' and the 'dramatic comedy'.

c. = *mental *a.*[1] 1c. *colloq.*

1952 M. Allingham *Tiger in Smoke* i. 18 If it's Elginbrodde himself, he's 'psychological'.

3. Special collocations: *psychological hedonism* (Philos.), the theory that the constitution of the human mind is such that men will always choose what is pleasurable; hence *psychological hedonist*; *psychological novel*, a type of novel in which the main interest lies in the mental and emotional aspects of the characters; hence *psychological novelist*; *psychological warfare*, the use of propaganda or other means designed to undermine the morale or allegiance of one's opponents; so *psychological war* (cf. *psy-war* s.v. *psy-); *psychological weapon*, some particular action or reasoning designed to undermine resolution or morale in an opponent.

1884 H. Sidgwick *Methods of Ethics* (ed. 3) i. iv. 40 There is, however, one view of the feelings which prompt to voluntary action... I mean the view that volition is always determined by pleasures or pains actual or prospective. This doctrine—which I may distinguish as Psychological Hedonism—is often connected and not seldom confounded with the method of Ethics which I have called Egoistic Hedonism. **1943** *Mind* LII. 45 Butler is, in fact, right in rejecting Psychological Hedonism. **1961** J. Hospers *Human Conduct* iv. 147 We must distinguish two varieties of psychological hedonism: the variety which says that all that people ever desire is pleasure or satisfaction..and the variety which says that people desire many things but these things are all desired solely for the sake of the pleasure or satisfaction they will bring to the agent. **1903** G. E. Moore *Principia Ethica* iii. 70 It is these two different theories which I suppose the Psychological Hedonists to confuse. **1969** F. Vivian *Thinking Philosophically* v. 114 The psychological hedonist pushes these discoveries about our motives to what seems a logical conclusion. **1855** Geo. Eliot in *Westm. Rev.* July 288 After courses of 'psychological' novels..where life seems made up of talking and journalizing. **1959** *Oxf. Compan. French Lit.* 676/1 He has produced..purely psychological novels, depending for plot and interest on the workings of the characters' minds and their reaction to the outside world. **1960** Beckson & Ganz *Reader's Guide Lit. Terms* (1961) 176 This term *psychological novel* is descriptive of content rather than form or technique and is applied to work as formally conventional as the novels of C. P. Snow and as unconventional as those of James Joyce. **1915** A. D. Gillespie *Let.* 14 Mar. in *Lett. from Flanders* (1916) 49 He can tell a rattling good story, which many of those modern psychological novelists, with their elaborate analysis of character and of sensation, quite fail to do. **1970** G. Jackson *Let.* 22 Mar. in *Soledad Brother* (1971) 187 The truth would aid the convict in the psychological war—con against cop. **1940** *Current Hist.* Jan. 52 (*heading*) Psychological warfare and how to wage it... Psychological warfare is the fight conducted by the state with psychological weapons to strengthen its own prestige..and to weaken that of the enemy. **1946** L. J. Margolin (*title*) Paper bullets: a brief story of psychological warfare in World War II. **1949** *Sun* (Baltimore) 5 Feb. 17 Miss Gillars..is accused..of betraying her country by aiding Hitler's psychological warfare program over a period of more than four years. **1957** G. E. Wright *Bibl. Archaeol.* ix. 161/1 Tiglath-pileser..says that he 'overwhelmed' Menahem (evidently not by actual fighting but by psychological warfare!). **1974** M. Babson *Stalking Lamb* xxi. 157 We have no objection at all to helping in what she calls her 'psychological warfare'. **1940** Psychological weapon [see *psychological warfare* above]. **1944** J. S. Huxley *On Living in Revolution* iv. 44 Incomplete or unsatisfactory peace aims, which will have a..lower efficiency as psychological weapons.

psycholo·gico-, *rare.* [f. by analogy with Logico-.] Taken as comb. form of Psychological *a.*, Psychology in the sense 'psychological and...'.

1869 W. James *Let.* 22 May in R. B. Perry *Tht. & Char. W. James* (1935) I. 296 Charles S. Peirce has been writing some very acute and original psychologico-metaphysical articles in the St. Louis philosophic *Journal*. **1942** *Mind* LI. 178 At this stage of our ignorance we should concentrate on psychologico-mathematical patterns.

psychologism. Add: **b.** The tendency to explain in psychological terms matters which are considered to be more properly explained in other ways.

1905 *Mind* XIV. 530 Psychologism..has been universal in English philosophy from the beginning. **1937** D. Katz in R. B. Cattell et al. *Human Affairs* iii. 36 According to this tendency logic is nothing but the psychology of thinking, mathematics nothing but the psychology of mathematical thinking... This tendency is usually called psychologism. **1945** K. R. Popper *Open Society* II. xiv. 87 The structure..of the social environment, as opposed to the natural environment, is man-made; and therefore it must be explicable in terms of human nature, in accordance with the doctrine of psychologism. **1950** R. Carnap *Logical Found. Probability* ii. 40 One of the important achievements in the development of modern logic has been the gradual elimination of psychologism. **1975** *Times Lit. Suppl.* 9 May 502/2 From Descartes onwards, with some notable exceptions, philosophical semantics was crippled by psychologism.

psychologist. Add: **1. b.** A person who is not an expert in psychology, yet has, or claims to have, insight into the motivation of human behaviour. *colloq.*

1896 W. Cucher *Theatrical World of 1896* 56 In a word (though he would probably not know the meaning of the word), he must be a profound psychologist. **1951** A. P. Herbert *Number Nine* xv. 203 Why on earth had he answered all those perilous questions?... The 'mad Admiral'..must be a pretty subtle psychologist. **1957** P. Lafitte *Person in Psychol.* 1 Psychology has several meanings... It may mean the person's ordinary conduct of his affairs and it does mean this when he has done something ingenious or subtle and so thinks of himself as a bit of a psychologist.

2. Phr. *psychologist's fallacy* (see quots.).

1890 W. James *Princ. Psychol.* I. vii. 196 The great snare of the psychologist is the confusion of his own standpoint with that of the mental fact about which he is making his report. I shall hereafter call this the 'psychologist's fallacy' *par excellence*. **1902** Baldwin *Dict. Philos. & Psychol.* II. 382/2 *Psychologist's fallacy*, the fallacy, to which the psychologist is peculiarly liable, of reading into the mind he is examining what is true of his own; especially of reading into lower minds what is true of higher. **1931** *Brit. Jrnl. Psychol.* XXI. 243 A danger to be avoided known as the 'psychologist's fallacy'. This arises from the fact that the experimenter is apt to suppose that the subject will respond to a stimulus or an order in the same way as he himself would respond in the circumstances.

psychologistic (səi:kǫlǒdʒi·stik), *a.* [f. Psycholog(ism + -istic.] Of, pertaining to, or characterized by psychologism.

1929 *Mind* XXXVIII. 360 His sustained and masterly demonstration of the self-ruinous character of all such 'psychologistic' theories. **1931** W. B. Gibson tr. *Husserl's Ideas* 18 Within this view of things there grows up..a transcendental-phenomenological Idealism in opposition to every form of psychologistic Idealism. **1935** *Amer. Speech* X. 247/1 That the distinction between *langue* and *parole* does not imply any psychologistic assumptions, is shown..by the fact that it has served as a starting point to the phoneme conception of Professor Twaddell, who resolutely protests against all psychologistic definitions of the phoneme. **1957** G. Ryle in M. Black *Importance of Lang.* (1962) 165 Where Frege attacked psychologistic accounts of thinking from the outside, they attacked them from the inside. **1976** *Brit. Jrnl. Sociol.* XXVII. 304 A sociological approach is apt to imply that such groups collude in their subordinate position and then to provide psychologistic interpretations of their behaviour.

Hence **psy:chologi·stically** *adv.*

1964 I. C. Jarvie *Revolution in Anthropol.* iii. 97 The psychological problems of religion..are the only problems of religion that can be tackled psychologistically.

psychologize, *v.* Add: **1.** (Earlier and later examples.)

1830 W. Jacobson *Let.* in J. F. Maurice *Life F. D. Maurice* (1884) I. ix. 111 Neither stay away rusticating and psychologizing, but come here and mind your books. **1967** *Listener* 19 Jan. 91/2 'Let us please psychologize In this wise,' Suggested Freud. **1974** *Sci. Amer.* Aug. 115/1 The texts still like to simplify; generally we psychologize about the local event and regard the suffused experience as secondary.

3. (Examples.)

1940 V. J. McGill in M. Farber *Philos. Ess.* 231 This work was a reaction against his earlier volume in which he [*sc.* Husserl] had attempted to psychologize arithmetic. **1957** *Listener* 9 May 743/2 We have been led to psychologize (to use John Dewey's rather horrid word) and to humanize learning.

4. (Earlier example.)

1877 D. D. Home *Lights & Shadows of Spiritualism* v. 264 Dear old gullible souls who could be readily psychologized into believing that they were eating a piece of the moon in shape of 'green cheese'.

psychologizing *vbl. sb.* (further examples); also *attrib.*; also **psycho·logized** *ppl. a.*; **psychologization** (further examples).

1895 W. James *Coll. Ess. & Rev.* (1920) 393, I never find myself actively taking up the soul, so to speak, and making it to do work in my psychologizing. **1922** C. E. M. Joad *Common-Sense Theol.* v. 233 The so-called 'psychologising' tendency in modern thought. **1956** E. H. Hutten *Lang. Mod. Physics* vi. 224 We are once more in danger of falling victims to the psychologizing attitude of epistemology. **1961** E. Nagel *Structure of Science* iii. 45 Various psychologized versions of the Aristotelian requirement have enjoyed wide currency. **1966** J. J. Katz *Philos. Lang.* v. 261 This argument is a psychologization of an argument of Goodman's in some recent work of his on the concept of confirmation in science. **1970** A. Toffler *Future Shock* (1974) x. 221 The key to the post-service economy lies in the psychologization of all production. **1976** *Spare Rib* Oct. 38/4 A continuous stream of psychologising and explaining..leaves a reader little room to breathe or get curious about the characters.

psychologizer (səikǫ·lǒdʒəizəɹ). [f. prec. + -er[1].] One who psychologizes, in any sense.

1895 'Mark Twain' in *N. Amer. Rev.* CLX. 49 The Observer of Peoples has to be a Classifier, a Grouper, a Deducer, a Generalizer, a Psychologizer. **1931** W. B. Gibson tr. *Husserl's Ideas* iii. iii. 273 The psychologizers everywhere will take offence at this; they are already disinclined to distinguish between judging as an empirical experience and judgment as 'Idea', as essence. **1966** *Listener* 4 Aug. 174/1 The 'spiritualizers' are giving way to the 'psychologizers'. Enter 'psychological man'.

psychologue. For *rare*[−1] read *rare* and add further examples.

1890 W. James *Let.* 8 July (1920) II. 1 A great chance for some future psychologue to make a greater name than Newton's. **1971** K. Millett *Sexual Politics* (1972) III. vi. 330 It is vaguely depressing to see a literary man vending the same trash as those hundreds of psychologues and quacks.

psychology. Add: **1. a.** (Earlier and further examples of general use.) In mod. use no longer concerned with the soul.

1653 tr. *J. de Back's Discourse* in W. Harvey *Anat. Exercises* sig. H7ᵛ, I call the generall doctrine of man *Anthropologie*, the parts of which, I do ordain to be, according to this division, *Psychologie*, *Somatologie*, and *Hæmatologie*, into the doctrine of the soul, bodie, and blood... *Psychologie* is a doctrine which searches out mans Soul, and the effects of it. *Ibid.* sig. H8ᵛ, I do bind up the order of *Psychology* in few words. *a* **1680** R. Cudworth *Treat. Freewill* (1838) 19 The vulgar psychology, or the now generally received way of philosophizing concerning the soul, doth either quite baffle and betray this liberty of will, or else render it absurd, ridiculous, or monstrous. **1892** W. James *Coll. Ess. & Rev.* (1920) xx. 317, I wished, by treating Psychology *like* a natural science, to help her to become one. **1897** C. H. Judd tr. *Wundt's Outl. Psychol.* i. 3 The assignment of this problem to psychology, making it an empirical science coordinate with natural science and supplementary to it, is justified by the method of all the *mental sciences*, for which psychology furnishes the basis. **1910** *Amer. Jrnl. Psychol.* XXI. 72 Although great writers and poets have frequently made the most penetrating generalisations in practical psychology, the world has always been slow to profit by their discoveries. **1930** W. Köhler *Gestalt Psychol.* vi. 167 In psychology too, the influence of *gestalt* has been demonstrated..in very primitive be-

haviour. **1973** C. D. Kernig *Marxism, Communism & Western Society* VII. 98 Starting from the assumption that Soviet psychology is grounded on dialectical materialism, psychological historiography attempts to show how far materialist, and later dialectical and finally Marxist-Leninist thinking moulded the character of psychology.

In mod. usage, the signification of the word has broadened from the narrow concept illustrated in the Dict. to include (*a*) the scientific study of the mind as an entity and in its relationship to the physical body, based on observation of the behaviour and activity aroused by specific stimuli; and (*b*) the study of the behaviour of an individual or of a selected group of individuals when interacting with the environment or in a given social context. So *experimental psychology*, the experimental study of the responses of an individual to stimuli; *social psychology*, the study of the interaction between an individual and the social group to which he belongs.

(*a*) **1895** *Amer. Jrnl. Psychol.* VII. 78 Experimental Psychology was in its origin, and has remained for a considerable extent in its development, a German science. **1927** *Psychol. Rev.* XXXIV. 126 They adopt in regard to them [*sc.* instincts] the attitude common to the stimulus-response school of psychology, which purports to base the development of human behavior, of character and personality, upon the innately determined reactions of the organism to objective stimuli. **1940** Hilgard & Marquis *Conditioning & Learning* i. 2 The conditioned response was called the unit of habit by psychologists to whom habit was the most important concept in psychology. **1953** C. E. Osgood *Method & Theory in Exper. Psychol.* p. v, I have covered the major portion of what is called experimental psychology, including sections on sensory processes, perception, learning, and symbolic processes. **1968** *Internat. Encycl. Social Sci.* XIII. 78/2 Existential psychology is a comprehensive psychology whose aim is an integration of the observations of different psychologies into an explanatory theory about human behavior in its lived international entirety. **1976** H. Brown *Brain & Behavior* i. 5 The study of the relation between brain structure and behavior gives physiological psychology the unique mission of trying to resolve an old and basic puzzle of philosophy and science, often referred to as the 'mind-body problem'.

(*b*) **1896** G. Le Bon *Crowd* ii. 32 What we know of the psychology of crowds shows that treatises of logic need on this point to be rewritten. **1908** W. McDougall *Social Psychol.* 18 Social psychology has to show how, given the native propensities and capacities of the human mind, all the complex mental life of societies is shaped by them and in turn reacts upon the course of their development and operation in the individual. **1922** E. Glover *Roots of Crime* (1960) 4 For the first time in the history of British criminology a meeting of Justices of the Peace has invited a psycho-analyst to lecture on the psychology of crime. **1948** A. L. Kroeber *Anthropol.* (rev. ed.) viii. 323 Clinical psychology was the first recognized branch of psychology that attempted to deal with whole human beings, as distinct from..special aspects of the mind. **1963** Gough & Jenkins in M. H. Marx *Theories Contemp. Psychol.* xxix. 456 The study of verbal learning has not led to the study of verbal behavior in general, to the psychology of language. **1975** W. S. Sahakian *Hist. & Syst. Psychol.* xix. 425 The World War II years found him interested in the psychology of morale and human engineering psychology.

2. a. The attitude or outlook of an individual or a group on a particular matter or on life in general.

1899 G. Le Bon *Psychol. of Socialism* iv. 39 We find in the working classes two well-defined subdivisions, each with a different psychology. **1908** F. M. Ford *Let.* Dec. (1965) 29 Thanks for yr. letter: of course I understand yr. psychology &, God forbid that you shd. restrain yr. irritation before men of good will. **1928** *Daily Tel.* 11 Sept. 10/5 The psychology of the workaday world has infected him with its disquiet. **1931** F. L. Allen *Only Yesterday* ii. 20 War-time psychology was dominant; no halfway measure would serve. **1954** Koestler *Invisible Writing* xxiv. 264 Was not the psychology of the masses an infinitely more complex phenomenon?

b. The nature of an event or phenomenon considered from the point of view of psychology.

1892 C. G. Chaddock tr. *Krafft-Ebing's Psychopathia Sexualis* p. iv, It is not the intention of the author to lay the foundation of a psychology of the sexual life, though without doubt psychopathology would furnish many important sources of knowledge to psychology. **1929** B. Russell *Marriage & Morals* xvi. 182 The psychology of adultery has been falsified by conventional morals. **1932** F. C. Bartlett *Remembering* ii. 16 There is..adequate reason for beginning our detailed study of the psychology of remembering with an investigation into the character and conditions of perceiving and imagining. **1964** B. B. Gilligan tr. I. Lepp (*title*) The psychology of loving.

3. *attrib.*, as *psychology journal, student.*

1971 D. Crystal in E. Ardener *Social Anthropol. & Lang.* 194 This research, largely reported in psychology journals.., is methodologically unsatisfactory in many respects. **1890** W. James *Princ. Psychol.* II. xviii. 56, I have myself for many years collected from each and all of my psychology-students descriptions of their own visual imagination. **1972** G. W. Kisker *Disorganized Personality* (rev. ed.) xv. 494 Psychology students were employed by one investigator as 'companion counselors'.

psychomachy. Delete †*Obs. rare* and add later examples. Also (now more usually) in L. form **psychomachia.**

[**1927** H. Waddell *Wandering Scholars* i. 20 In his most famous and most considerable work, the *Psychomachia*, the Battle of the Soul, he [*sc.* Prudentius] has done more than set the stage for the struggle between the spirit and the flesh.] **1936** C. S. Lewis *Allegory of Love* ii. 55 The favourite theme of the Middle Ages—the battle of the virtues and the vices, the Psychomachia, the *bellum intestinum*, the Holy War. *Ibid.* 73 A good man, even in a panegyric, can now be good only as a result of a successful psychomachy. **1954** —— *Eng. Lit. in Sixteenth Cent.* i. i. 92 'Bewtie and the Prisoneir' is a neat but slightly frigid psychomachy. **1955** D. Davie *Brides of Reason* 22 There were minds Aware of themselves, and figuring this In psychomachia. **1972** *Times Lit. Suppl.* 10 Nov. 1353/1 Ruskin's criticism..becomes a psychomachia dramatizing his own mental state.

psychometric, *a.* Add: (Further examples.)

1879 *Brain* II. 149 (*heading*) Psychometric experiments. **1943** *Amer. Jrnl. Psychiatry* C. 181/2 The various psychometric tests can only touch the surface of this situation. **1970** *Jrnl. Gen. Psychol.* LXXXII. 101 Intelligence measures included two psychometric indices of 'general intelligence'. **1973** R. C. Dennis *Sweat of Fear* vi. 37, I had to find a way of telling her I had psychometric knowledge of her predicament.

B. *sb. pl.* (const. as *sing.*). The science of measuring mental capacities and processes; the application of methods of measurement to the various branches of psychology.

1930 *Proc. & Addr. Amer. Assoc. Stud. Feeble-Minded* XXXV. 94 To most persons who know the term at all, psychometrics is fairly synonymous with the use of intelligence tests. **1934** J. O'Connor (*title*) Psychometrics: a study of psychological measurement. *Ibid.* p. xiv, Every measuring instrument has two major aspects. In the field of psychometrics these are called reliability and validity. **1958** *Times* 27 May 2/2 (Advt.), Applications are invited for the post of Lecturer in Education with special interests in psychometrics, statistics and educational psychology. **1972** *Guardian* 24 Feb. 13/8 Psychometrics is unable to investigate the nature of intelligence. **1973** *Nature* 23 Mar. 279/2 Professor Meredith delivers a blistering attack on traditional psychometrics (i.e., the alleged 'measurement' of IQs, reading ages, and so on).

psychometrical *a.* (further example); **psychometrically** *adv.* (examples corresponding to sense 2 of Psychometry); **psychometrize** *v.* (further examples).

1903 *Nature* 3 Sept. 409/1 It appears doubtful if any other community, European or Polynesian, has been psychometrically investigated under more favourable conditions as regards both absence of disturbing factors and simplification of method. **1922** E. Wallace *Crimson Circle* ii. 14 The police might sneer at Yale's psychometrical powers. **1950** P. Tabori *Harry Price* ix. 183 He proposed..to have the box psychometrized first by the best mental mediums he could obtain. **1975** *Weekend Mag.* (Montreal) 18 Oct. 20/1 Learning of McMullen's psychic abilities, he asked him to psychometrize some artifacts. **1977** *Lancet* 1 Jan. 7/2 These children have been re-examined medically and psychometrically.

psychometrician (səi:kometri·ʃən). [f. prec. + -ian.] An expert in or practitioner of psychometrics.

1950 in Webster Add. **1953** J. B. Carroll *Study of Lang.* i. 4. The psychometrician..tries to make tests of 'verbal intelligence' or tests of achievement in various language arts. **1965** *Language* XLI. 353 Factor analysis has been intensively cultivated by psychologists, particularly psychometricians. **1975** *Sci. Amer.* July 128/1 In seeking to devise 'culture-free' tests of intelligence psychometricians are pursuing a chimera.

psychometrist (səikǫ·mètrist). [f. Psychometer + -ist.] **1.** = Psychometer 1.

1864 T. L. Nichols *40 Yrs. Amer. Life* II. ii. 20 Then came psychometrists, who could tell the lives, characters, fortunes, and diseases of people they had never seen. **1879** *Spiritual Notes* May 148/1 The psychometrist enters a room and is impressed with a vision of events that in some mysterious way have left their traces on its material fabric. **1900, 1903** [in Dict. s.v. Psychometric *a.*] **1966** E. Palmer *Plains of Camdeboo* xvii. 289 Hearing of a psychometrist in Johannesburg, I took my piece of wood to her and asked her what she could tell me of its history. **1976** *Oxford Times* 6 Feb. 15/1 Guest speakers and demonstrators include..a psychometrist—for whom vibrations received from personal belongings create a mental picture of the owner.

2. = *psychometrician.

1932 *Psychol. Exchange* I. 11 The deserving psychometrists are often not discriminated from those who have just taken up mental testing. **1964** M. Critchley *Developmental Dyslexia* xiv. 82 The formal intelligence quotient was given as 65, but this was certainly a serious underestimate due to lack of cultural rapport with the psychometrist. **1974** *Maclean's Mag.* (Toronto) May 12/2 He worked as a psychometrist measuring the abilities of children.

psychometry. Add: **1.** (Later examples.)

1922 E. Wallace *Crimson Circle* vi. 38 'Nothing is absurd,' said the Commissioner quietly. 'The science of psychometry has been practised for years.' **1959** *Times Lit. Suppl.* 1 May 254/4 Such objects..enjoy a semi-consciousness drawn from the human beings who have known them. In this there is an undeniable truth, as the modern practice of 'psychometry' proves. **1966** E. Palmer *Plains of Camdeboo* xvii. 290 Most of us can on occasions sense atmosphere, and psychometry seems to be this power greatly developed. **1975** G. W. Knight *Jackson Knight* i. 44, I know now, from the witness of psychometry, that inanimate objects can indeed become impregnated by qualities gathered during their past.

2. (Earlier and later examples.) Also, psychometrics (see *psychometric *a.* B).

1879 F. Galton in *Brain* II. 149 Psychometry..means the art of imposing measurement and number upon operations of the mind, as in the practice of determining the reaction-time of different persons. **1971** *Brit. Med. Bull.* XXVII. 35/1 Hughes and his colleagues are conducting further studies, including the use of electro-encephalography and psychometry. **1976** H. M. van Praag in *Advances in Drug Therapy of Mental Illness* (World Health Organization) 127 The third field in which psychotropic drugs have served as a pacemaker, that of psychometry... One wants to know (that is, to measure) what it [*sc.* the drug] does, and what it does not do.

psychomimetic, -motility: see *psycho-.

psycho-motor, *a.* Add: Also without hyphen. **a.** (Further examples.)

1879 *Jrnl. Mental Sci.* XXIV. 677 The 'Psychiatrischer Centralblatt' for August.., 1877, gives a *résumé* of the inaugural dissertation..of Dr. Pasternaki, who has studied the question of what he calls the Psychomotor Centres of the Brain. **1961** *Aeroplane & Astronautics* CI. 679/1 It was agreed that the animals selected for this programme should be capable of learning simple psychomotor tasks which could be performed during flight. **1976** Smythies & Corbett *Psychiatry* v. 62 In depressed patients speech tends to be slow, monotonous, and sad-sounding, in keeping with the general psychomotor slowing.

b. *Med.* Applied to a partial seizure or epileptic attack (distinct from grand mal and petit mal) characterized by a state of altered consciousness in which simple or complex automatisms may be performed for which there is subsequently at least partial amnesia.

1938 *Amer. Jrnl. Psychiatry* XXV. 268 The three main manifestations of epilepsy (grand mal, petit mal and psychomotor epilepsy) are each accompanied by a distinct pattern of disrhythmia. **1961** *Listener* 7 Dec. 967/2 An attack of psychomotor epilepsy. **1974** M. C. Gerald *Pharmacol.* xi. 212 Diphenylhydantoin..is also useful against psychomotor seizures.

Hence **psychomoto·ric, -moto·rical** *adjs.*, of or pertaining to psychomotor activity; **psychomoto·rically** *adv.*

1964 L. Kaiser in D. Abercrombie et al. *Daniel Jones* 102 Both ways of coding and decoding show a high degree of similarity, psychomotoric patterns based on language leading to neuromuscular activity, which in its turn leads to spatial coding in both cases. **1969** *Indian Mus. Jrnl.* V. 83 Music is a much directer form of expression (psychomotorically), than painting, sculpture..and ornaments. *Ibid.* 84 These psychomotorical expressions are not in the least outbursts of uncontrolled emotions. The hands of the singers accompany in refined movements the path of the voice.

psychon (səi·kǫn). [f. Psycho- + *-on[1].] A hypothetical unit of nerve impulse or energy. Hence **psycho·nic** *a.*

[**1906** H. W. Armit tr. *A. Forel's Hypnotism* i. 6 It is here that one must seek the transition from the conceived to the unconceived, and not in the strong and repeatedly conceived 'psychomes'. [*Note*] The author apologizes for this term. He has introduced it for brevity['s] sake to express each and every psychical unit.] **1920** H. L. Eno *Activism* iv. 45 Since we are already familiar with one form at any rate of this higher activity in the psychic processes, let us call these units 'psychons'. *Ibid.*, The term 'psychone' was proposed by Forel for the psychic aspect of a hypothetical unit of the nerve process. **1927** P. & W. R. Bousfield *Mind* i. 22 As the basis of the 'immaterial substance' we may postulate a second order of 'ons' which are, like protons and electrons, fashioned out of the ether. Let us call these 'ons' by the name of 'psychons'. **1931** W. M. Marston et al. *Integrative Psychol.* xiii. 314 There is no doubt whatsoever that the psychons in the sensory system differ from those in the motor system. *Ibid.*, We suggest, therefore, that from the point of view of an objective psychology, exterior to its subjects, this psychonic energy is consciousness. **1968** C. L. Burt *Psychol. & Psychical Res.* 45 As unit I would rather start with the 'pure ego', which I envisage as a sort of Leibnizian monad. This, I think, on the plane of natural science we might justifiably treat as an 'elementary particle'—a 'psychon', as I styled it.

psychoneural, -neuroendocrine, -neuroendocrinology, etc., **-neurology,** etc.: see *psycho-.

psychoneuro·sis. *Psychol.* Also with hyphen. [f. Psycho- + Neurosis.] Any of various functional nervous disorders attributed to emotional or psychological causes, often accompanied by manifestations of anxiety, and distinguished from a psychosis by the maintenance of contact with the external world; also, in psychoanalytic theory, a mental disorder attributed to repressed unconscious conflict or fantasy (as distinguished from an 'actual' neurosis, attributed to anxiety caused by present frustration of sexual drive).

1883 [in Dict. s.v. Psycho-]. **1903** W. James in *Proc. Soc. Psychical Res.* XVIII. 32 The parasitic ideas of psycho-neurosis, and the fictitious personations of planchette-writing and mediumship. **1913** E. Jones *Papers on Psycho-Anal.* v. 125 Freud has pointed out that it is necessary to separate the 'actual neuroses' from the 'psychoneuroses'. *Ibid.* 129 We next come to the psychoneuroses proper... The symptoms result from the activity of certain unconscious mental processes..which the patient is unable spontaneously to recall to his memory. **1924** J. Riviere et al. tr. *Freud's Coll. Papers* II. xxi. 253 There always remains as a common feature in the ætiology both of the psychoneuroses and the psychoses the factor of frustration. **1938** *Jrnl. Aviation Med.* IX. 177/2 McFarland and Barach found that patients, on whom a diagnosis of psychoneurosis had been made, appeared to be more severely affected at an atmosphere of 10 per cent oxygen than were the normal controls. **1948** *Sci. News* VIII. 109 The conditions which benefit most from group psychotherapy are the psychoneuroses—particularly anxiety states and reactive depressions. **1959** J. Strachey *Freud's Compl. Wks.* XX. 79 In these cases—in the psychoneuroses—the reason for the accumulation of undischarged excitation was a psychological one: repression. But what followed was the same as in the 'actual' neuroses. **1972** Zax & Cowen *Abnormal Psychol.* viii. 231 The one feature common to all forms of psychoneurosis is the presence of anxiety. *Ibid.* 232 In DSM-I (1952) six types of psychoneuroses were delineated: anxiety reaction; conversion reaction; dissociative reaction; phobic reaction; obsessive-compulsive; and depressive reaction. **1977** Miller & Swift *Words & Women* iv. 67 Hysteria now refers in technical use to a specific psychoneurosis that may affect anyone, male or female.

psychoneuro·tic, *sb.* and *a.* *Psychol.* Also with hyphen. [f. Psycho- + Neurotic *sb.* and *a.*] **A.** *sb.* A person suffering from a psychoneurosis.

1902 *Buck's Handbk. Med. Sci.* (rev. ed.) V. 28/1 In the psychoneurotic..the tendency toward imperfect mental development becomes more and more accentuated. **1924** *Proc. 7th Internat. Congr. Psychol.* 148 Though a genius is frequently a psycho-neurotic, it would be quite untrue to say that the majority of psycho-neurotics tend towards genius. **1936** *Psychoanal. Rev.* XXIII. 1 Thoughts of death and dying occur as every-day symptoms among psycho-neurotics of the anxiety, hysterical and obsessive types. **1972** Zax & Cowen *Abnormal Psychol.* viii. 231 The psychoneurotic maintains good contact with reality.

B. *adj.* Of, pertaining to, or characterized by psychoneurosis.

1909 *Jrnl. Abnormal Psychol.* June–July 144 Every psycho-neurotic symptom is to be regarded as the symbolic expression of a submerged mental complex of the nature of a wish. **1938** *Jrnl. Aviation Med.* IX. 177/2 When 50 per cent oxygen was given, the physiologic symptoms of the psychoneurotic patients were less marked. **1967** *Spectator* 1 Dec. 679/1 When one looks at the incidence of psychoneurotic disorders..it is found that these also were commoner in..flat-dwellers.

psycho-optic(al: see *psycho-.

psychopath. (Further examples; cf. *psychopathy.)

1927 *New Republic* 21 Sept. 128/2 Terms not so long ago confined to specialists are handled familiarly by the laity: moron, inferiority complex, mental age,..paranoid delusions, psychopaths. **1955** D. J. West *Homosexuality* ix. 106 Psychopaths are the last people to try to battle against their instincts; they just obey first impulses regardless of social codes. Being incapable of prolonged or deep personal attachments, they seek only an immediate outlet for their lust. **1967** *Listener* 20 Apr. 529/3 The term psychopath is bandied about in such a way as to make it cover almost any mental disorder... However the psychopath has now achieved legal status in the Mental Health Act of 1959 as having 'a persistent disorder or disability of mind..which results in abnormally aggressive or seriously irresponsible behaviour'. **1967** M. Argyle *Psychol. of Interpersonal Behaviour* i. 21 It is one of the marks of the psychopath that he will engage in social behaviour in so far as it is..profitable to do so, but he has no intrinsic attraction to other people at all. For the psychopath there is no particular difference between people and things. **1972** *Observer* 31 Dec. 23/4 If she's a psychopath I'm a fruit cake. She's just a girl who needs love.

psychopathic, *a.* (*sb.*) Add: **1. a, b.** (Further examples.) For 'mental disease' read 'mental disorder, now *spec.* psychopathy'.

1899 [see *hereditary *a.* 2 a]. **1932** *Sun* (Baltimore) 19 Sept. 2/2 The court..found that Duker is afflicted with a definite mental ailment or disorder known as psychopathic personality, which had reduced his mental and moral responsibility and control but that he is sane according to the legal standard. **1949** *Brit. Jrnl. Psychol.* XL. 12 The Psychopathic Personality (P.P.P.) is one of the major problems of the Prison Commission. **1957** R. F. C. Hull tr. *Jung's Compl. Wks.* I. 111 In many psychopathic illnesses there are persons who think unclearly and are prone to flights of ideas, who are ruthlessly egocentric..but who can hardly be said to be suffering from chronic mania. **1959** *Mental Health Act* 7 & 8 *Eliz. II* c. 72. i. § 4 In this Act 'psychopathic disorder' means a persistent disorder or disability of mind (whether or not including subnormality of intelligence) which results in abnormally aggressive or seriously irresponsible conduct on the part of the patient, and requires or is susceptible to medical treatment. **1968** [see *moral *a.* 7 a]. **1976** *Times* 4 Aug. 5/7 All we can do is protect society from them. Grossly psychopathic people cannot be befriended. **1977** P. Way *Super-Celeste* i. 53 Such men..work to please whatever passions and psychopathic urges drive them personally.

Hence **psychopa·thically** *adv.*

1961 in Webster. **1972** *Lancet* 18 Nov. 1069/2 The psychopathically aggressive, the rigidly authoritarian.

psycho-pathology. (Now usu. without hyphen.) Add: The science of the mental or psychological causation of disorders and abnormalities. (Further examples.)

1951 D. B. Klein *Abnormal Psychol.* 5 The study of mental disorder bulks large in the field of abnormal psychology, so that a good portion of this field might be regarded as coterminous with that of psychopathology. **1964** Gould & Kolb *Dict. Social Sci.* 555/1 This psychoanalytic psychopathology has..been employed to explain..not only mental symptoms but also the physical lesions of so-called psychosomatic diseases; not only disease but also the traits and quirks of normal or deviant personalities. **1968** J. Zubin in Zubin & Shagass *Neurobiol. Aspects of Psychopathology* 289 If we define personality as the systematic aspect of a person's behavior, and psychopathology as those aspects of this systematic behavior attributable to illness, an important question arises regarding the possible connections between premorbid personality and psychopathology. **1969** T. Freeman *Psychopathology of Psychoses* i. 2 Psychopathology has a wider subject-matter than the study of 'conscious psychic events'... To be comprehensive it must take account of psychoanalysis... It must also be based on certain aspects of neurology and internal medicine.

b. A mentally or behaviourally disordered state.

1947 D. Jones in R. Hague *Dai Greatcoat* (1980) ii. 139 *Everything* one does is conditioned by one's psychopathology. **1952** Mettler & Curry in F. A. Mettler *Psychosurgical Problems* i. 16 It does not follow because a patient is discharged that his psychopathology is fundamentally altered. **1970** *Jrnl. Gen. Psychol.* LXXXIII. 61 Classifying states of psychopathology into discrete, mutually exclusive categories..does not seem to be possible. **1973** Sandler & Davidson *Psychopathology* i. 8 Tests designed to reveal the developmental history of the patient's psychopathology.

psychopathological *a.*, **-pathologist** (further examples); also **psy:chopatholo·gic** *a.*, **psy:chopatholo·gically** *adv.*

1891 *Ann. Rep. Board of Regents Smithsonian Inst. 1889–90* 636 One can thus see the links which form the psycho-pathologic chain of human life, at one end of which we may find insanity and at the other criminality. **1919** W. S. Maugham *Moon & Sixpence* i. 7 The mystic sees the ineffable and the psycho-pathologist the unspeakable. **1928** *Guy's Hosp. Rep.* LXXVIII. 458 Psychopathologically, from the content of the hypochondriacal complaints..an anal-erotic basis for some hypochondrias is strongly suggested. **1936** *Jrnl. Nerv. & Mental Dis.* LXXXIV 450 Psychopathologically, we undertake an analysis of the situation. **1971** J. J. Shapiro tr. *Habermas's Toward Rational Society* iii. 42 The use of psychopathological concepts has become necessary for the identification and explanation of a political state of affairs. **1976** *Observer* 11 Jan. 21/4 Interesting little psychopathological tour de force—or, as they say in Soho, kinks and kicks.

psychopathy. Add: **1.** In mod. use, personality disorder that lacks a physiological basis, characterized by markedly impulsive, egocentric, irresponsible, and antisocial behaviour, and an inability to form normal relationships with others, sometimes accompanied by aggressiveness or charm and manifested at all levels of intelligence; the state of such a disorder. (Further examples.)

1923 C. Mackenzie *Parson's Progress* xix. 263 Personally I have found my knowledge of psychopathy of the greatest value in the confessional. **1948** *Amer. Jrnl. Sociol.* LIII. 361/1 Caldwell indicates nomadism, inability to withstand tedium, and irresponsibility as characteristics of psychopathy. **1953** I. Skottowe *Clin. Psychiatry* ii. 32 Three qualities may be discerned biographically which are a common expression of psychopathy. The first is the lack of persistence..; the second is a curious coldness of heart..; the third is an inability to defer the immediate satisfaction of appetites and desires. **1972** G. Sereny *Case of Mary Bell* ii. i. 148 The condition of psychopathy is not generally identified with mental retardation. **1972** Zax & Cowen *Abnormal Psychol.* x. 312 In trying to understand the causes of psychopathy a variety of emphases can be identified, including hereditary, neurological, environmental, and sociocultural. **1976** *Church Times* 12 Mar. 5/2 He turned loving eyes on the tormenting thugs... They, in their psychopathy, could not see him as a man like themselves.

psychopharmaceutical: see *psycho-.

psy:chopharmaco·logy. Also with hyphen. [f. Psycho- + Pharmacology.] The branch of science concerned with the way drugs affect the mind and behaviour.

1920 D. Macht in *Johns Hopkins Hosp. Bull.* XXXI. 167/1 The number of contributions to the domain of what we may be permitted to call 'psychopharmacology' is certainly very meagre. **1935** *Jrnl. Nerv. & Mental Dis.* LXXXI. 161 The psycho-pharmacology of sodium amytal. **1958** A. Huxley *Let.* 2 Feb. (1969) 845, I have been asked..to do a piece on the ethical, religious and social implications of psychopharmacology. **1969** *New Scientist* 10 Apr. 80/1 LSD..may be remembered as a herald of psychopharmacology. **1970** A. Toffler *Future Shock* (1971) viii. 160 New knowledge..in psychopharmacology, made many Freudian therapeutic measures seem quaintly archaic.

Hence **psy:chopharmacolo·gic** (chiefly *U.S.*), **-lo·gical** *adjs.*, of or pertaining to psychopharmacology; (of a drug) psychoactive; **psy:chopharmacolo·gically** *adv.*; **psy:chopharmaco·logist,** an expert or specialist in this subject.

1939 *Jrnl. Mental Sci.* LXXXV. 406 (*heading*) A psychopharmacological study of schizophrenia with particular reference to the mode of action of cardiazol, sodium amytal and alcohol in schizophrenic stupor. **1958** *Science* 10 Jan. 61/2 Depressions not accompanied by anxiety and tension are less apt to be ameliorated by a psychopharmacologic agent. **1960** *Clin. Pharmacol. & Therapeutics* I. 257/2 Studies by psychopharmacologists and comparative psychologists concerning the effects of the new psychotropic drugs on experimentally conditioned emotional responses of animals. **1964** M. Gordon *Psychopharmacol. Agents* I. i. 3 One of the new psychopharmacological agents, reserpine, had a difficult time gaining acceptance in the Western hemisphere. **1967** *Guardian* 1 July 12/3 A psycho-pharmacologist said: 'I wouldn't be dismayed if the use of marijuana were made legal.' **1970** *Nature* 5 Sept. 1008/1 Psychopharmacological experiments are beginning to yield data relevant to urban crowding as a health hazard. **1971** *Amer. Jrnl. Psychiatry* CXXVIII. 695/1 Peyote contains more than ten alkaloids, the most psychopharmacologically significant of which is mescaline. **1977** M. E. Jarvik *Psychopharmacol. in Pract. of Med.* 21 In complicated cases the individual should be referred to a psychopharmacologically oriented psychotherapist. **1979** [see *psychobiologist].

psycho-philosopher, -philosophy: see *psycho-.

psychophone·tics, *sb. pl.* (const. as *sing.*). *Linguistics.* Also with hyphen. [f. Psycho- + Phonetics *sb. pl.* Cf. Pol. *psychofonetych* (J. B. de Courtenay 1894, in *Rozprawy Akad. Umiejebności: Wydział Filol.* 2nd Ser. V. 129).] That branch of phonetics which deals with the mental correlates of speech-sound production. So **psychophone·tic** *a.*, **psychophone·tically** *adv.*

1934 *Maître Phonétique* Jan.–Mar. 3, ɔ:l ðiːz difrənsiz in prənʌnsieiʃn..duː nɔt igzist saikoufənetikəli. **1934** *Ibid.* Apr.–June 44 Doctor Arend..naturally quotes from the later one [*sc.* work by J. B. de Courtenay] published in 1901, though there was a much later work on phonetics and psycho-phonetics published in 1927. **1936, 1950** [see *physiophonetics *sb. pl.*]. **1956** Jakobson & Halle *Fundamentals of Lang.* ii. 11 *The mentalist view*... In the oldest of these approaches,..the phoneme is a sound imagined or intended, opposed to the emitted sound as a 'psychophonetic' phenomenon to the 'physiophonetic' fact. **1966** M. Pei *Gloss. Linguistic Terminol.* 225 *Psychophonetics*, the treatment of a phoneme as the image aimed at in the speaker's mind (Courtenay, Entwistle). **1968** J. W. F. Mulder *Sets & Relations in Phonol.* 20 Articulatory phonetics has achieved results that are more widely accepted and provide a safer guide than the younger acoustic phonetics (let alone the still younger 'perceptive' phonetics, which could well be called 'psycho-phonetics'). **1968** Black & Singh in B. Malmberg *Man. Phonetics* v. 106 For the sake of convenience, psycho-phonetics will be treated here as (a) speech production, (b) speech acoustics, and (c) speech perception.

psycho-physic, *a.* and *sb.* **A.** *adj.* (Earlier example.)

1887 *Amer. Jrnl. Psychol.* I. 140 The conditions are as infinitely complicated as the psycho-physic constitution of man.

B. *sb.* **psycho-physics** (now usu. as one word). (Earlier and further examples.)

1878 *Rep. Brit. Assoc. Adv. Sci. 1877* II. 95 Most of you are aware of the recent progress of what has been termed Psycho-physics, or the science of subjecting mental processes to physical measurements and to physical laws. **1937** *Amer. Speech* XII. 228/1 On the psycho-physics of speech. **1944** *Jrnl. Optical Soc. Amer.* XXXIV. 66/1 Physicists..have surrendered almost the entire field of psychophysics to the psychologist. **1973** C. D. Kernig *Marxism, Communism & Western Society* VII. 91/2 A distinction is made between (a) classical psychophysics in which the physiology of the senses is studied with the help of refined..techniques..; (b) activation (arousal) research, which is concerned with..conditions for the release and course of affective and motivational states. **1976** S. George *Fatal Shadows* 37 The new scientists, men whose discoveries of gifted psychics had shot them to the top of psychophysics.

psycho-physical, *a.* Add: Also without hyphen. (Earlier and further examples.) *Psycho-physical isomorphism* (see quot. 1932).

1847 H. E. Lloyd tr. *Feuchtersleben's Med. Psychol.* iii. 151 Habit, likewise, greatly modifies the psycho-physical character. **1872** *Westm. Rev.* XLII. 177 As a result of experiments according to all these three methods, Fechner arrives at what he calls a general 'psycho-physical law', and also 'Weber's law'. *Ibid.* 188 Fechner's law..embodies and illustrates the law of relativity; but it has a psycho-physical value over and above this. **1892** C. G. Chaddock tr. *Krafft-Ebing's Psychopathia Sexualis* v. 403 Large lips, idiotic expression,..and an awkward attitude complete the picture of psychophysical degeneration. **1894** Creighton & Titchener tr. *Wundt's Human & Animal Psychol.* 448 The principle of psychophysical parallelism..refers always to a parallelism of elementary physical and psychical processes. **1932** B. Petermann *Gestalt Theory* II. iii. 110 He commences with a general leading principle, that of psychophysical isomorphism. This is the assumption that a co-ordination exists between the domain of the experiences and that of the physiological processes..and that it is a co-ordination in the sense of congruence or isomorphism in regard to their systematic properties. 1941, 1956 [see *interactionism]. 1942 W. Koehler *Dynamics in Psychol.* ii. 43 If an experience A may vary in a specific way, its correlate a must be capable of corresponding variations. When consistently applied, this point of view leads to the principle of psychophysical isomorphism. 1972 *Sci. Amer.* May 30/1 Young's celebrated three-color theory of color vision, published in 1802, was formulated entirely on psychophysical evidence. 1972 O. L. Zangwill in Cox & Dyson *20th-Cent. Mind* ii. vii. 174 While still at Chicago, Watson had been much troubled by the psychophysical mould in which instruction in experimental psychology was still cast.

Hence **psychophy·sically** *adv.*, by psychophysical means; as regards psychophysics.

1847 H. E. Lloyd tr. *Feuchtersleben's Med. Psychol.* v. 307 The so-called proximate cause of idiocy can..be no other than a psycho-physically impeded or depressed vital process. 1894 W. James *Coll. Ess. & Rev.* (1920) 360 Do they mean that introspection acquaints them with a part of the emotional excitement which it is psycho-physically impossible that incoming currents should cause? 1948 L. Spitzer *Linguistics & Lit. Hist.* iv. 139 Man, this psycho-physically conditioned being. 1973 *Nature* 21 Sept. 159/2 Gibson and others have shown psychophysically that the gradient of density of visual texture is an important clue to the orientation in depth of textured surfaces.

psycho-physicist. Add: Now usu. without hyphen. (Further examples.)

1951 S. S. Stevens *Handbk. Exper. Psychol.* i. 43/2 The psychophysicists have elaborated subtle measures that try to transcend the inconstancy of the mercurial response. 1975 *Nature* 20 Nov. 201/1 Even though psychophysicists do not necessarily share this confidence, it is worth trying to relate the neurophysiology to measurable psychophysical function.

psychophysicotherapeutics: see *psycho-.

psycho-physiology and derivatives. Add: Now usu. without a hyphen. (Further examples.) Also **psy:chophysiolo·gic** *a.* (chiefly *U.S.*), **psy:chophysiolo·gically** *adv.*

1892 Van Liew & Beyer tr. *Ziehen's Introd. Study of Physiol. Psychol.* vii. 144 The projection and arrangement of our sensations with reference to time, the same as with reference to space, cannot be explained psychophysiologically. 1909 Webster, Psychophysiologic. 1912 L. Bloomfield in C. F. Hockett *Leonard Bloomfield Anthol.* (1970) 37 Sound-change and analogy..are, respectively, psycho-physiologic and psychologic processes. 1927 J. Riviere tr. *Freud's Ego & Id* ii. 31 Psycho-physiology has fully discussed the manner in which the body attains its special position among other objects in the world of perception. 1936 *Psychol. Rev.* XLIII. 411 The loudness scale will probably be utilized extensively by psychophysiologists and acoustical engineers. 1958 A. R. Radcliffe-Brown in M. N. Srinivas *Method Social Anthropol.* i. iii. 45 The determination of what mental differences are correlated with these differences of cerebral structure is a task for the psychologist or psychophysiologist. 1960 *Times Lit. Suppl.* 3 June 355/1 The new sciences of sociology and psychophysiology have both increasingly proposed that whether dealing with the crowd or the creative mind there are new skilled and precise ways of obtaining obedience, consent and, indeed, cooperation. 1972 *Sci. Amer.* June 91/1 The psychophysiological basis of how contrast enables the visual system to distinguish contours. 1976 *Jrnl. Psychiatric Res.* XIII. 7 Three members of each group received a secondary or tertiary diagnosis of 'psychophysiologic gastrointestinal disorder'. 1977 *Nature* 28 Apr. 831/1 The intragroup social pressure determines the gonadal development of the subdominants. Their testes are smaller and show little or no mature testicular tissue. Low ranking males are psychophysiologically castrated.

psycho-political, -politics: see *psycho-.

psychopomp. Add: Also **psy·chopompos.** Also, the spiritual guide of a (living) person's soul; a person who acts as a guide of the soul. (Further examples.)

1920 Webster, Psycho-pomp.., psycho-pompos. 1941 Auden *New Year Let.* i. 27 For though the Janus of a joke The candid psychopompos spoke. 1946 *Antiquity* XX. 168 Hermes psychopompos, the mediator between the upper and the nether world. 1951 K. W. Bash tr. *Jacobi's Psychol. C. G. Jung* (ed. 5) iii. 135 It is therefore 'an important function of the higher..super-personal animus that it guides and accompanies as a true Psychopompos the wanderings and transformations of the soul'. 1958 L. Durrell *Balthazar* vi. 144 If I had been in your shoes and the whole damn thing wasn't just a lie to make yourself more interesting to the psychopomps—I'd..well, I'd bloody well try and sleep with him again. 1965 M. Bradbury *Stepping Westward* vii. 336 He's a psychopomp, that's what he calls himself. You know that word? I think he gets it from Jung. A soul-saver, or something. A man who leads the spirit onward. 1971 *Southerly* XXXI. 12 The concept of the nymphet as psychopomp seems..a grotesque travesty of the Beatrice myth.

Hence **psychopo·mpically** *adv.* (*rare*).

1908 R. Brooke *Lett.* (1968) 121, I, Hermes-like, am coming to fetch you psychopompically to Hell.

psycho-prismatism: see *psycho-.

psy:choprophyla·xis. *Med.* Also with hyphen and (*rare*) anglicized as **-prophylaxy.** [f. Psycho- + Prophylaxis.] A method intended to reduce or eliminate labour pains in which prenatal women are given an understanding of natural childbirth in order to obtain their physical co-operation with the process.

1958 I. Bonstein *Psychoprophylactic Preparation for Painless Childbirth* 5 It will be necessary to use them [sc. drugs] each time various difficulties prevent a confinement, prepared by psychoprophylaxy, from being continued until the end. 1960 D. A. Myshne tr. *Velvovsky's Painless Childbirth through Psychoprophylaxis* viii. 167 By psychoprophylaxis of labour pain we imply a system of measures aimed at preventing the appearance and development of labour pain and effected through influences exerted on the higher divisions of the central nervous system. 1965 *Observer* 1 Aug. 5/6 Dr. Pierre Vellay ..has made Paris a world centre of psycho-prophylaxis. 1972 J.-P. Clerc in N. Morris *Psychosomatic Med. in Obstetr. & Gynæcol.* 76 The first object of psychoprophylaxis is to eliminate the pain which is due to socio-cultural factors and may thus be regarded as psychosomatic. 1978 *Jrnl. R. Soc. Med.* LXXI. 663 Attempts have been made to measure psychological factors.., the benefits of psychoprophylaxis..and attitudinal aspects to antenatal education, but none have the precision of outcome which the obstetrician expects from tests of somatic function.

So **psy:choprophyla·ctic** *a.*, **psy:choprophyla·ctically** *adv.*

1958 I. Bonstein *Psychoprophylactic Preparation for Painless Childbirth* 5 The theory of the superior nervous activity, established by I. P. Pavlov.., is the basis of the psychoprophylactic preparation for painless childbirth. 1960 D. A. Myshne tr. *Velvovsky's Painless Childbirth through Psychoprophylaxis* viii. 199 Most of the time we consider the use of drugs for the purpose of 'enhancing' the pain prevention in psychoprophylactically well prepared women unnecessary. 1964 *New Statesman* 4 Dec. 874/3 Women in this country are most carefully taught how to cooperate with the ordinary maternity unit without abandoning their psychoprophylactic training. 1965 *Observer* 1 Aug. 5/6 The French Government is showing great interest in psycho-prophylactic obstetrics. 1972 E. D. Bing in N. Morris *Psychosomatic Med. in Obstetr. & Gynæcol.* 71 Soon our first psychoprophylactically trained women gave birth, and our teachers had become the *monitrices* in order to assist the young mother in labor and delivery.

psycho-sensory: see *psycho-.

psychose·xual, *a.* Also with hyphen. [f. Psycho- + Sexual *a.*] Involving the mental and emotional aspects of the sexual impulse.

1897 H. Ellis *Stud. Psychol. Sex* I. iii. 73 (*heading*) Psychosexual inversion. *Ibid.* iv. 99 In other cases there is some degree of psycho-sexual hermaphroditism. 1913 A. A. Brill tr. *Freud's Interpretation of Dreams* v. 200 Anxiety in dreams may be of a psychoneurotic nature, or it may originate in psychosexual excitements, in which case the anxiety corresponds to a repressed libido. 1946 *Mind* LV. 347 There are chapters sketching types of psycho-neurosis, psychosis, psycho-sexual disorder and other social disorders. 1963 *New Yorker* 29 June 84 The great unresolved psychosexual dilemma of modern man and woman. 1970 G. Greer *Female Eunuch* 43 Not all the massage in the world will ensure satisfaction, for it is a matter of psycho-sexual release. 1970 *Nature* 17 Oct. 203/2 Experimental studies to determine the psychosexual effects of pornography. 1976 S. Hynes *Auden Generation* v. 147 The [poem's] problem is set entirely in psycho-sexual terms.

Hence **psy:chosexua·lity; psychose·xually** *adv.*

1910 *Jrnl. Abnormal Psychol.* Apr.–May 67 Not only must the physician himself be able to approach the subject without prudishness and lewdness, but he must perforce know something about psychosexuality. 1912 A. A. Brill tr. *Freud's Sel. Papers on Hysteria* (ed. 2) xi. 202 We..prefer to speak of psychosexuality, thus laying stress on the fact that the psychic factor of the sexual life should neither be overlooked nor underestimated. 1925 Psychosexuality [see *fixation 3 b]. 1934 Webster, Psychosexually. 1956 *Nature* 14 Jan. 54/2 Through anthropological investigations, we have some idea how they develop psychosexually in other cultures. 1971 Cauthery & Cole *Fundamentals of Sex* v. 170 They are pathologically promiscuous but rarely obtain much pleasure from sex and are usually underdeveloped psychosexually.

psychosis. Add: **1.** (Further examples.) In mod. use, any mental illness or disorder that is accompanied by hallucinations, delusions, or mental confusion and a loss of contact with external reality, whether attributable to an organic lesion or not.

1924 J. Riviere tr. *Freud's Coll. Papers* I. iv. 72 The ego rejects the unbearable idea..and behaves as if the idea had never occurred to the person at all. But, as soon as this process has been successfully carried through, the person..will have developed a psychosis, and his state can only be described as one of 'hallucinatory confusion'. 1939 E. Glover *Psycho-Analysis* iv. 34 When deep guilt erupts, as in the depressive psychoses, we find that the self-accusations are not justified in reality. They are delusional. 1946 R. B. Cattell *Description & Measurement of Personality* ii. 30 Psychoses are both endogenous and exogenous, organic and psychogenic... Some exogenous psychoses are..organic, in..that the provoking environmental influence—e.g., alcohol or syphilis—is organic. 1957 C. Pfeiffer in H. Abramson *Neuropharmacol.* (1959) 231 We have tried the Akerfeldt test in the model LSD psychosis in human volunteers. 1973 *Sci. Amer.* Sept. 117/3 The severe mental disorders we have learned to deal with more effectively are the psychoses, the most prominent of which are schizophrenia and manic-depressive psychosis. Psychoses are severe disorders characterized by profound and pervasive alterations of mood, disorganization of thought and withdrawal from social interactions into fantasy.

psycho-so·cial, *a.* Also without hyphen. [f. Psycho- + Social *a.*] Pertaining to the influence of social factors on an individual's mind or behaviour, and to the interrelation of behavioural and social factors; also, more widely, pertaining to the interrelation of mind and society in human development.

1899 F. W. Moore tr. *Gumplowicz's Outl. Sociol.* ii. 83 There are also psycho-social phenomena, such as language, customs, rights, religion etc., arising from the action of social elements with or upon the individual mind. 1903 *Amer. Jrnl. Sociol.* VIII. 762 In another quarter it is held that sociology is concerned only with the action of human groups on one another—social phenomena—and the influence of the group on its individual members—psycho-social phenomena. 1927 Ogburn & Goldenweiser *Social Sci.* xxiv. 303 Economics should also profit from such studies as the social psychologists, in the light of their knowledge of psycho-social mechanisms, see fit to make of specific institutional influences and controls. 1953 J. S. Huxley *Evolution in Action* i. 12 We may call these three phases [of evolution] the inorganic or, if you like, cosmological; the organic or biological; and the human or psycho-social. *Ibid.* vi. 134 Psycho-social evolution..operates by cultural transmission. 1954 [see *identity 10 c]. 1958 *Listener* 3 July 12/1 The possibility of conscious purpose arose and for the first time became a factor in man's subsequent evolution, which may be defined by Sir Julian Huxley's term as psycho-social. 1970 *Nature* 31 Oct. 422/1 The original aim of this investigation was to compare the psychosocial characteristics of a group of heroin pushers with those of a group of heroin users who did not sell drugs. 1971 I. G. Gass et al. *Understanding Earth* ix. 125/1 The noösphere..produces idea-systems by processes of psycho-social evolution. 1977 P. B. & J. S. Medawar *Life Science* vi. 53 It is because of the primacy of language as the agency which provides the link between one generation and the next that exosomatic evolution is often referred to as 'cultural' or 'psychosocial' evolution.

Hence **psycho-so·cially** *adv.*

1946 [see *psychosomatically *adv.*]. 1972 *Sci. Amer.* July 81/3 Is it in fact possible to attribute the retarded growth of psychosocially deprived children to sleep patterns that inhibit the secretion of growth hormone?

psycho-sociology, etc.: see *psycho-.

psychosoma·tic, *a.* and *sb.* [f. Psycho- + Somatic *a.* and *sb.*] **A.** *adj.* **a.** Involving or depending on both the mind and the body as mutually dependent entities.

1863 C. Reade *Hard Cash* II. xi. 119 The nocturnal and diurnal attendance of a Psycho-physical physician, who knows the Psychosomatic relation of body and mind. 1930 M. W. Calkins in C. Murchison *Hist. Psychol. in Autobiogr.* I. 44 This biological form of personalistic psychology studies the psychophysical, or better the psychosomatic organism. 1933 H. Devine *Rec. Adv. Psychiatry* (ed. 2) i. 1 The purpose of this chapter is to discuss the modern concept of 'psychosomatic unity', or, as it is sometimes termed, the concept of 'the organism-as-a-whole'. 1956 E. L. Mascall *Christian Theol. & Nat. Sci.* vii. 270 A human being is not just a spirit that is temporarily condemned to inhabit a material garment, but is a highly complicated psycho-somatic unity, in which the body is an essential constituent. 1976 *Verbatim* Sept. 11/2 The biblical view of personality was psychosomatic (and distinct from Platonic dualism—'the body is the prison-house of the soul'). So in both the Hebrew and Christian Scriptures, what we would call psychic or spiritual states are freely ascribed to physical organs and other parts of the human anatomy.

b. Applied to physical disorders caused or aggravated by mental, emotional, or psychological factors, and (less commonly) to mental or emotional disorders caused or aggravated by physical factors.

1938 S. Beckett *Murphy* x. 219 Murphy..did not suffer from this—er—psychosomatic fistula. 1947 J. Steinbeck *Wayward Bus* 170 She called her mother's [headaches] psychosomatic and psychotic. 1950 A. Huxley *Themes & Variations* i. 67 Hypertension, neurosis, psychosis and all the varieties of psycho-somatic disorders. 1957 *New Biol.* XXII. 83 These are the so-called psychosomatic disorders. The distinctive feature of these illnesses is that, although they are of nervous origin, they come to the attention of patient and physician through some malfunctioning of an organ. 1958 H. L. & R. R. Ansbacher *Indiv. Psychol. of Adler* ii. xi. 286 The role of organ inferiority in psychosomatic disturbances proper will be discussed in the next chapter. 1964 *Ann. Reg. 1963* 414 Mental strain..had often led to psychosomatic illness. 1975 B. Wood *Killing Gift* v. iii. 150 We've built up a really solid case, at least statistically, for psychosomatic influence on breast-cancer incidence.

c. Applied to the branch of medicine concerned with the relations between the mind and the body.

1939 (*title of periodical*) Psychosomatic medicine. 1939 *Psychiatry* II. 465/1 Psychosomatic Medicine covers a different and broader field. Its object is to study in their interrelation the psychological and physiological aspects of all normal and abnormal bodily functions and thus to integrate somatic therapy and psychotherapy. 1946 *Jrnl. R. Aeronaut. Soc.* L. 263/2 The American Army had a separate section of the Medical Service dealing with what was called psychosomatic medicine. 1960 *20th Cent.* Mar. 267 Psychosomatic research has revealed that almost every common disease can have an emotional component. 1971 *Country Life* 9 Sept. 646/4 Psychosomatic medicine today appears to be making a full-circle exploration.

B. *sb. pl.* (const. as *sing.*). The field of study concerned with the relationship between mind and body.

1941 *Amer. Jrnl. Psychiatry* XCVII. 781 We will take up the present day research trends in the following order:..psychosomatics (psychophysiology). 1941 *Psychosomatic Med.* III. 332, I surveyed ten years' literature in the field of psychosomatics. 1946 B. Mittelman in P. L. Harriman *Encycl. Psychol.* 678 It is obvious..from its problems and methods that psychosomatics represents a synthesis of several streams of psychological and medical investigation. 1966 *N.Y. State Jrnl. Med.* LXVI. 3157/1 The view of the skin as a major organ of communication is not a new one; it is a basic concept of psychosomatic medicine. But if the ECM proposed here are more fully identified, some of the 'metaphors' of psychosomatics will turn out to be descriptions of real events. 1975 B. Wood *Killing Gift* iv. i. 131 His very next paper ..was on psychosomatics... He chaired a..conference on the psychosomatics of cancer.

Hence **psychosoma·tically** *adv.*, in a psychosomatic manner; *esp.* through the (unconscious) effect of the mind on the body; **psychosoma·ticist**, an expert or specialist in psychosomatic medicine; **psychoso·matism** (*rare*), psychosomatics; **psychoso·matist** = *psychosomaticist.

1854 J. C. Bucknill *Unsoundness of Mind* 15 The psycho-somatists find in the liability of the cerebral instrument to disease, a reasonable basis for the irresponsibility of the insane. 1946 *Lancet* 10 Aug. 190/2 Epidemiology..was founded and formulated mainly upon experience gained by the study of infectious diseases, but its use as a framework of interpretation is equally applicable to many modes of morbid behaviour that are manifestations of disturbances of emotional development—i.e., of 'life' viewed psychologically, psychosomatically, and psychosocially. 1957 *Time* 4 Nov. 56/2 It was the story of a girl who went psychosomatically deaf in emotional flight from her role as the ears of a deaf father, mother and brother. 1960 *20th Cent.* Mar. 268 Psychosomatists believe..that each of us possesses a built-in homeostatic mechanism designed to keep us healthy. *Ibid.* 269 This does not mean that psychosomatists attribute all illnesses, invariably, to states of mind. *Ibid.*, Psychosomatism is not, therefore, a 'craze'. Until less than a century ago every great medical school—Hippocratean and Christian alike—subscribed to what Mr. Jelly sneers at as the 'argot' of treating 'the whole man'. 1960 *Spectator* 14 Oct. 555 The first practising psychosomatist—if the term is allowable—that I had met. 1962 N. E. Whitten in A. Dundes *Mother Wit* (1973) 418/1 The only reason that the effects of a spell can be removed is that they are psychosomatically caused in the first place. 1962 C. L. Buxton *Study of Psychophysical Methods for Relief of Childbirth Pain* i. 2 The very environment thought by psychosomaticists to contribute to difficult, prolonged and painful labor is everywhere present in this episode. 1971 *Time* 7 June 96/2 The family physician bucks the case to a psychosomaticist, who flounders in jargon. 1976 *Times Lit. Suppl.* 10 Sept. 1105/2 She was psychosomatically deaf for a week.

psychosomimetic: see *psychotomimetic *a.* and *sb.* **psychosphere, -stimulant**: see *psycho-.

psychosu·rgery. [f. Psycho- + Surgery.] Brain surgery intended to alter the behaviour of patients with certain kinds of severe mental stress or disorder.

1936 *Q. Cumul. Index Medicus* XIX. 249/1 First attempts at psychosurgery using leukotome; technic and results. 1938 *Jrnl. Nerv. & Mental Dis.* LXXXVIII. 589 (*heading*) Psychosurgery. Effect on certain mental symptoms of surgical interruption of pathways in the frontal lobe. 1952 Mettler & Candis in F. A. Mettler *Psychosurgical Probl.* xvi. 319 As a result of such adverse criticism practitioners of psychosurgery were forced to restrict the operative procedure to more rostral areas. 1973 *Nature* 23 Mar. 222/3 Unlike the classical lobotomy operation, ..psychosurgery now rarely involves actually cutting into the brain... Psychosurgery, which is designed to alter behaviour, is usually distinguished from the treatment of epilepsy and the removal of brain tumours. 1977 M. Edelman *Political Lang.* iv. 58 Some psychiatrists..see political demonstrators or ghetto rioters as sick, calling for drugs or psychosurgery.

Hence **psychosu·rgical** *a.*; also **psy·chosurgeon**, a surgeon specializing in psychosurgery.

1945 Webster Add., Psychosurgeon. 1946 *Canad. Med. Assoc. Jrnl.* LV. 435/1 In 1941 the research unit of the Toronto Psychiatric Hospital embarked upon the psychosurgical treatment of hopelessly mentally ill patients. 1949 M. Clark *Medicine on March* v. 99 Summing up their ten years' work, the two psychosurgeons declared: 'The results are sufficiently good to warrant the use of prefrontal lobotomy..for the relief of the very serious and chronic forms of mental disease.' 1950 [see *leucotomy]. 1952 B. Wolfe *Limbo '90* (1953) xxii. 368 The phony rapist's phony premises become the premises of the psycho-surgeon's science. 1972 *Lancet* 8 July 70/1 The modern psychosurgeon must be versatile enough to employ different operations for different symptom complexes. 1972 M. Crichton *Terminal Man.* vi. 46 Computer scientists and neurobiologists had worked together for several years. From that association had come Form Q, and programs like George and Martha, and new psychosurgical techniques.

psychosyndrome, -synthesis, etc.: see *psycho-.

psycho-technic (-te·knik), *sb.* and *a.* Also without hyphen. [ad. G. *psychotechnik* sb. (H. Münsterberg *Grundzüge der Psychotechnik* (1914) i. 1): see Psycho- and Technic *a.* and *sb.*] **A.** *sb.* = *psychotechnology. Also *pl.*

1926 *Psychol. Rev.* XXXIII. 402 In the applied division [of psychology], sometimes called psychotechnics, the aim is to make psychological facts useful in many different directions. 1927 *Daily Express* 17 June 12/3 The value of psycho-technic..is that it enables the railway company to eliminate at once the employee who will not be reliable. 1928 H. P. Weld *Psychol. as Sci.* xv. 266 Applied psychology is evidently to be classed with the technical sciences. It may be considered as 'psychotechnics. 1952 *Brit. Jrnl. Psychol.* XLIII. 83 He is continental in tending to leave to 'psychotechnics' areas which in England would fall within respectable occupational psychology.

B. *adj.* = *psycho-technical *a.*

1932 *Brit. Jrnl. Psychol.* XXV. 77 These results are of value in..forcing us to realize the importance of attitudes in all judgements..in the ordering of psychotechnic test results. 1957 W. Abell *Collective Dream in Art* 4 The result has been the gradual emergence of a synthesis which we may call the..'psycho-technic' theory of culture. 1973 *Daily Tel.* 27 Nov. 12/6 Peter Riding's programme vividly simulated part of the 'psycho-technic test' to be undertaken as well as a medical check, written questions on motoring and..an examination in actual driving.

psycho-te·chnical, *a.* Also without hyphen. [f. Psycho- + Technical *a.*] Pertaining to or concerned with the application of psychological facts or knowledge to practical problems in industry, employment, education, etc.

1903 H. Münsterberg in *Harvard Psychol. Stud.* I. 654 The science of pedagogy is a psycho-technical discipline which makes education mechanical. 1927 *Daily Express* 17 June 12 Electrical machinery, levers, mechanical puzzles, and complete paraphernalia for psychotechnical tests. 1962 *Guardian* 20 July 5/7 To the applicants for jobs, the 'psychotechnical' examinations were at once ludicrous and frightening.

psy:chotechno·logy. [f. Psycho- + Technology.] The area of study concerned with the practical application of tested knowledge about the human mind or brain. Hence **psy:chotechno·logist**, an expert or specialist in this.

1923 F. A. Kingsbury in *Ann. Amer. Acad. Pol. & Social Sci.* CX. 5/1 Psychotechnology, or applied psychology, is interested in acquiring facts and principles only in so far as they can be turned directly to account in the solution of practical problems, in industry, selling, teaching or other fields of human endeavour. *Ibid.* 8/2 The careful psychotechnologist submits his hypothetical solution of concrete problems to prolonged and severe experimental tests. 1928 H. P. Weld *Psychol. as Sci.* xv. 280 Psychotechnology is a product of this century, and there can be little doubt that it will, in years to come, extend its activities. 1947 *Harvard Univ. Comm. Place of Psychol. in Ideal Univ.* 11 Values of the science and of psychotechnology may be exemplified more readily than enumerated. *Ibid.* 23 Already the demand greatly exceeds the supply of trained psychologists, psychotechnologists, ..etc. 1953 J. B. Carroll *Study of Lang.* vii. 196 In Europe applied psychology is widely identified as psychotechnology. 1973 *Sci. Amer.* Sept. 117/1 The potential power of this developing 'psychotechnology' is..creating concern about unwarranted intrusions into personal privacy and individual rights. 1973 R. L. & R. K. Schwitzgebel (*title*) Psychotechnology: electronic control of mind and behavior.

psycho-therapeu·tic, *a.* and *sb.* Add: Now usu. without hyphen. **A.** *adj.* Also, of, pertaining to, or characterized by psychotherapy. (Further examples.)

1914 A. A. Brill tr. *Freud's Psychopathol. Everyday Life* v. 93 The psychotherapeutic procedure which I employ in the solution and removal of neurotic symptoms. 1957 *Times Lit. Suppl.* 1 Nov. 659/4 That organized religion has been the greatest psychotherapeutic system ever invented is one of Jung's most famous dicta. 1963 A. Heron *Towards Quaker View of Sex* 51 All the large teaching hospitals have psychiatric out-patient departments of varying size and with more or less psychotherapeutic help to hand. 1970 *Jrnl. Gen. Psychol.* LXXXIII. 194 Laffal has applied this approach in studies of..psychotherapeutic interviews. 1976 *Times Lit. Suppl.* 16 July 872/2 The spiritual healers either operate on the margin or they must be certified members or associates of a recognized psychotherapeutic fraternity.

B. *sb.* (Earlier and further examples.) Also = Psycho-therapy in Dict. and Suppl. Also (*rare*) in sing. form **psychotherapeutic.**

1872 D. H. Tuke *Illustrations of Influence of Mind on Body* p. ix, The medical reader.., I hope, may be induced to employ Psycho-therapeutics in a more methodical way than heretofore. 1903 F. W. H. Myers *Hum. Personality* II. 515 Suggestions involving..bodily odour and chemical conditions have thus far been confined to psycho-therapeutics. *Ibid.* 527 That form of psychotherapeutic which consists in a clairvoyant diagnosis.. followed perhaps by advice avowedly based upon a recollection of earthly learning. 1906 *Boston Med. & Surg. Jrnl.* 8 Nov. 542/1 A fundamental difficulty in psychotherapeutics up to this time has been a lack of method. 1933 *Mind* XLII. 114 Abnormal psychology.. has generalised with a freedom astonishing to those who are familiar with the logical austerities of science. This may be allowed to psychotherapeutics, but not to anything that calls itself psychology.

Hence **psychotherapeu·tically** *adv.*, by means of psychotherapy, in a psychotherapeutic manner; **psychotherapeutist** (further examples).

1906 *Amer. Jrnl. Med. Sci.* Oct. 499 The neurasthenic women..rested in bed, were isolated, and treated 'psychotherapeutically' much as were the hysterical. *Ibid.* 520 The psychotherapeutist should be an honest man and an expert clinician. 1909 H. Münsterberg *Psychotherapy* p. ix, Since that time I have never ceased to work psychotherapeutically in the psychological laboratory. 1957 *Listener* 12 Sept. 401/1 The psycho-therapeutist..no longer confines his attention to the classical mental disorders, but,..has set up advisory centres to deal with problems of life or..mental hygiene, arising at almost any phase from infancy to senescence. 1962 J. Money *Reading Disability* i. 17 Reading tuition is individualized, scholastically and psychotherapeutically, as much as economics permit.

psycho-therapy. Add: Now usu. without hyphen. (Earlier and later examples.) In mod. use, the treatment of disorders of the mind or personality by psychological or psychophysiological methods. [Cf. van Renterghem & van Eeden *Clinique de Psycho-thérapie Suggestive* (Brussels, 1889).]

Quot. 1853, an isolated use, represents a different sense.

1853 *Jrnl. Psychol. Med. & Mental Path.* VI. 268 (*heading*) Psychotherapeia, or the remedial influence of mind. 1892 F. W. van Eeden in *Med. Mag.* I. 233 As a general term for our treatment we selected in 1889, 'Suggestive Psycho-therapy'. We called psychotherapy every description of therapeutics that cures by means of the intervention of the psychical functions of the sufferer. This title is borrowed from Hack Tuke... Psycho-therapy..has..had the misfortune to be taken in tow by hypnotism. 1897 T. H. Kellogg *Text-bk. Mental Dis.* xi. 497 By the term psychotherapy is signified..every means and every possible agency which primarily affects the psychical rather than the physical organization of the patient in a curative direction. 1906 *Amer. Jrnl. Med. Sci.* CXXXII. 499 Prof. Dejerine was treating the psychoneuroses, especially hysteria and neurasthenia, by isolation and psychotherapy. 1947 *Nature* 4 Jan. 38/2 Psychotherapy may be useful for the criminal, but it is prolonged, and an impractical treatment with present resources. 1958 *Sunday Times* 15 June 13/3 By psychotherapy is meant the systematic application of psychological principles to the treatment of psychogenic ill-health and maladjustment. 1963 A. Heron *Towards Quaker View of Sex* 50 A number of psychotherapy clinics which give treatment for sexual difficulties. 1976 Smythies & Corbett *Psychiatry* ii. 19 Psychotherapy consists very largely in helping people to grow up, to exchange the egocentric child's role for the mature role of the adult.

Hence **psychothe·rapist**, a specialist in or practitioner of psychotherapy.

1909 A. A. Brill tr. *Freud's Sel. Papers Hysteria* iii. 55, I was not always a psychotherapist but like other neuropathologists I was educated to the use of focal diagnosis and electrical prognosis. 1923 *Daily Mail* 19 Jan. 7 An earnest warning to nervous persons to avoid spiritualism is given by Dr. W. Stekel, the Viennese neurologist and psycho-therapist. 1930 R. S. Woodworth *Psychology* (ed. 8) xiii. 568 Many psychotherapists avoid the use of hypnosis because, as they say, it does not get to the root of the trouble. 1976 Smythies & Corbett *Psychiatry* xvii. 291 Most psychotherapists refuse to give specific advice as to what their patient's conduct should be in cases where ethical problems are concerned.

psychotic, *a.* Add: **a.** (Examples.)

1920 C. S. Read *Milit. Psychiatry in Peace & War* ii. 21 These figures..include pure epilepsy without any psychotic complications. 1949 *Endeavour* VIII. 37/1 The use of electric shock therapy for the treatment of certain types of psychotic patients is sometimes complicated by injuries suffered during the electrically produced convulsions. 1957 *Jrnl. Brit. Interplanetary Soc.* XVI. 8 Among psychologists, 'psychotic break' refers to the breaking loose of feeling from its previously adequate controls. 1965 J. Pollitt *Depression & its Treatment* i. 4 In advanced states the patient lacks insight and becomes psychotic. 1971 *Brit. Med. Bull.* XXVII. 77/1 An equally strong case can be made for the opposite view, that psychotic and neurotic forms of depression lie on the same continuum.

B. *absol.* as *sb.* A person with a psychosis.

1910 *Jrnl. Nervous & Mental Disease* XXXVII. 633 Thus arise the well-known 'explanation-delusions' of the psychotic. 1921 E. J. Kempf *Psychopathology* xiv. 718 Many ask the question.., 'Why do all neurotics and psychotics have sexual difficulties?' 1939 J. Dollard in A. Dundes *Mother Wit* (1973) 278/2 Psychotics in American hospitals display disorders of perception. 1958 M. Argyle *Relig. Behaviour* ix. 109 Many psychotics believe themselves to be religious leaders, prophets or mystics. 1962 *Lancet* 8 Dec. 1212/1 It might be that because some beds had been freed from psychotics, patients with neuroses or personality disorders..could be admitted to hospital. 1975 B. Meggs *Matter of Paradise* vii. i. 189 You're asking..whether..we might hope to identify a deteriorated psychotic *before* he enters government.

Hence **psycho·tically** *adv.*

1961 in Webster. 1977 *Irish Press* 29 Sept. 6/5 The ingredients include a weak and unsuccessful father who died young and an over protective mother whose motives were selfish rather than loving, and who was almost psychotically indifferent to George's progress except to record the inconvenience it caused her.

psychoticism (səikǫ·tisiz'm). [f. Psychotic *a.* + -ism.] The condition or state of being psychotic or of displaying psychotic tendencies; esp. as a factor showing liability to psychosis included in certain types of personality assessment.

1950 H. J. Eysenck in *Jrnl. Personality* XIX. 129 His [sc. Kretschmer's] theory can best be represented in terms of two orthogonal axes, one measuring schizothymia-cyclothymia, the other normality-psychotic abnormality or 'psychoticism'. 1955 *Jrnl. Mental Sci.* CI. 876 The effect of insulin therapy is not dependent upon the initial degree of psychoticism as used in the present sense. 1957 R. B. Cattell *Personality & Motivation* xvi. 719 At present a psychoticism measure would be considerably contaminated and distorted. 1957, etc. [see *neuroticism]. 1960 Payne & Hewlett in H. Eysenck *Exper. in Personality* II. 11 It was demonstrated that only one factor was necessary to account for the differences observed between the three groups. This factor was labelled 'psychoticism'. 1964 [see *introversion 1 b]. 1977 *New Society* 7 Apr. 33/3 This book aims to establish psychoticism..alongside neuroticism, extraversion and intelligence... The attribute it measures is a halfway stage to psychosis, as neuroticism is to neurosis.

psychotogenic (səikǫ:todʒe·nik), *a.* [f. as next + *-genic.] = *psychotomimetic *a.*

1956 *Arch. Neurol. & Psychiatry* LXXV. 122/1 To the neurologist the psychotogenic agents offer a special challenge. 1962 *Lancet* 27 Jan. 200/2 There is impressive evidence of the psychotogenic effects of some adrenaline derivatives. 1971 *Nature* 19 Nov. 152/2 The differential behavioural response of subjects to d and l-amphetamine might provide clues to psychotogenic mechanisms.

Hence (as back-formations) **psycho·togen**, a psychotomimetic substance; **psycho:toge·nesis**, the production of a psychosis or psychosis-like state.

1959 *Neuropharmacology: Trans. 4th Conf., 1957* 223 Relationship between endogenous substances that are found in the brain—epinephrine, norepinephrine, and..serotonin—and the possible relationship to the exogenous psychotogens. 1960 *Clin. Pharmacol. & Therapeutics* I. 251/2 The recovery of a chronically psychotic patient coincidentally with the administration of some drug is as dramatic as..the drug-induced psychosis. Recovery, however, is an even less satisfactory foundation for a psychopharmacologic theory than is psychotogenesis. 1971 *Nature* 19 Nov. 152/2 If noradrenergic mechanisms and behaviourally correlated motor stimulation were critical in psychotogenesis then one might expect d-amphetamine to have ten times the potency of the l form in inducing psychosis as a behavioural effect. 1974 M. C. Gerald *Pharmacol.* xvii. 318 Drugs in this same category are also termed hallucinogens, psychedelics, psychotogens, psychodysleptics, and so forth. 1976 *Nature* 8 Apr. 490/1 For many years the presence of an endogenous psychotogen in schizophrenia has been sought. 1977 *Lancet* 27 Aug. 449/2 Further study of psychotogenesis and of the relation between morphinoids and endogenous neurotransmitters and neuromodulators will probably..clarify the biochemistry both of addiction and of schizophrenia.

psychotomimetic (səikǫ:tomime·tik, -məime·tik), *a.* and *sb.* Also †**psychoso-.** [Orig. formed as *psychosomimetic*, f. Psychos(is + -o + Mimetic *a.*, and later altered to match Psychotic *a.*] **A.** *adj.* Having an effect on the mind orig. likened to that of a psychotic state, with abnormal changes in thought, perception, and mood and a subjective feeling of an expansion of consciousness; of or pertaining to a drug with this effect.

1956 R. W. Gerard in *Neuropharmacology: Trans. 2nd Conf., 1955* 132 Let us at least agree to speak of 'so-called' psychoses when we are dealing with them in animals... Along that same line, I have liked a term which I have been using lately—'psychosomimetic'—for these agents instead of 'schizophrenogenic'. 1957 *Neuropharmacology: Trans 3rd Conf., 1956* 205 (*heading*) Effects of psychosomimetic drugs in animals and man. 1957 H. Osmond in *Ann. N.Y. Acad. Sci.* LXVI. 417 The designation 'psychotomimetic agents' for those drugs that mimic some of the mental aberrations that occur in the psychoses had been suggested by Ralph Gerard and seemed especially appropriate. 1958 M. Jarvik in H. H. Pennes *Psychopharmacology* ix. 204 They have been called psychoso- or psychotomimetic, psychotherapeutic, tranquilizing, ataraxic,..or, as a general class, psychochemical, neurotropic,..and a number of other names. 1962 Henderson & Gillespie *Text-bk. Psychiatry* xi. 258 The hallucinogenic (or psychotomimetic) drugs most often employed have been mescaline and lysergic acid (LSD 25). 1964 D. F. Downing in M. Gordon *Psychopharmacological Agents* I. xiii. 562 Psilocybin and psilocin produce a psychotomimetic effect in man..which is similar to that produced by mescaline or LSD-25. 1967 *Sunday Mail* (Brisbane) 27 Aug. 5/1 In the field of drugs, a medical friend says that if you want another term for psychedelics you could call them..psychotomimetic. 1970 E. Goode *Marijuana Smokers* vii. 172 The general psychotomimetic questions include: 'Is your skin sensitive?' 'Are you happy?' 'Are colors brighter?' 1970 D. H. Efron *Psychotomimetic Drugs* 5 The use of these psychotomimetic substances in ancient times was either for religious, ceremonial or recreational purposes. 1975 Brimblecombe & Pinder *Hallucinogenic Agents* i. 3 It has been pointed out several times..that there are well-defined differences between mental states induced by the majority of pyschotomimetic drugs and those encountered in mental illness.

B. *sb.* A psychotomimetic drug.

1957 *Ann. N.Y. Acad. Sci.* LXVI. 418 We are using Gerard's term 'psychotomimetics' generally for compounds that have been called schizogens, psychotica, psychotogens, phantastica, hallucinogens and elixirs. If one believes that the importance of these compounds lies in their capacity to mimic the mental illnesses called psychoses, psychotomimetics would be the choice. 1970 A. T. Shulgin in D. H. Efron *Psychotomimetic Drugs* 25 Ibogaine..is another example in the family of psychotomimetics, with complex structures and no resemblance to known metabolic materials. 1974 M. C. Gerald *Pharmacol.* xvi. 299 The psychotomimetics (hallucinogens) such as LSD or mescaline produce profound alterations in behavior.

Hence **psycho:tomime·tically** *adv.*

1963 *Tetrahedron* XIX. 2073 Hashish (marihuana), the psychotomimetically active resin of the female flowering tops of *Cannabis sativa* L. 1970 L. Harmon in D. H. Efron *Psychotomimetic Drugs* 349 So for Dan's N.I.M.H. hospitality (T'was psychotomimetically hip) Great thanks for new lights on mentality And a most unforgettable TRIP. 1973 R. Mechoulam *Marijuana* p. xiii, The structure of only one [cannabinoid], the psychotomimetically inactive cannabinol, had been fully elucidated.

psychotropic (səikotrōᵘ·pik, -trǫ·pik), *a.* and *sb.* Also **-trophic** (-trōᵘ·fik). [f. Psycho- + *-tropic, *-trophic.] **A.** *adj.* Affecting a person's mental state; psychoactive; *spec.* = *psychotomimetic *a.*; of or pertaining to a drug of this kind.

1956 M. Rinkel in *Neuropharmacology: Trans. 2nd Conf., 1955* 240, I had considerable conversation on this subject with Dr. Goodman..and Dr. Loewi of Utah University. They made the very good proposal of calling all these drugs which affect the mind 'phrenotropic or psychotropic'. This general term would allow for a number of subdivisions: drugs that are beneficial; those which may cause psychosis; [etc.]. 1962 *Listener* 13 Dec. 1003/1 In recent years extravagant hopes have been centred on the psychotropic drugs, drugs which will relieve agitation and depression. 1968 *Sunday Mail* (Brisbane) 10 Nov. 1/2 Traffic in psychotropic drugs—ranging from LSD to sedatives—had reached 'epidemic proportions', the International Narcotics Control Board warned yesterday. 1968 A. Goldstein et al. *Princ. Drug Action* vi. 474 Despite the nearly universal exposure of the population to the 'legal' psychotropic drugs (alcohol, caffeine, and nicotine), some become habituated and some do not. 1970 *Nature* 7 Feb. 485/1 The essence of the commission's code is a catalogue of drugs of dependence—'psychotrophic drugs' as they are now called. 1972 J. I. M. Stewart *Palace of Art* vii. 66 Art, like a tomtom or a psychotropic drug, can loosen up the mind of an individual exposed to it. 1977 *It* June 2/4 If you enter a state of non-ordinary reality, as you do when you use psychotropic plants, it is only to draw from it what you need in order to see the miraculous character of ordinary reality. 1977 *Lancet* 24–31 Dec. 1326/1 None of the subjects were taking psychotrophic medication at the time of admission. 1978 *Guardian Weekly* 1 Jan. 10/1 Large numbers of prisoners are being given psychotropic drugs as a form of social control rather than medical treatment.

B. *sb.* A psychotropic drug.

1976 *Nature* 29 Apr. p. ix (Advt.), Astra Chemicals is the UK subsidiary of Scandinavia's largest pharmaceutical group which specialises in the research, manufacture and marketing of ethical drugs in the fields of local anaesthetics, bronchodilators, cardiovasculars and psychotropics. 1977 *Proc. R. Soc. Med.* LXX. 766/2 The majority are adults who have deliberately swallowed an overdose of drugs, chiefly in the category of 'psychotropics'.

Hence **psychotro·pically** *adv.*

1962 *Jrnl. Sci. & Industr. Res.* XXIA. 421/2 The psychotropically active hydrazines having a monoamine oxidase inhibiting effect are pronounced antidepressants.

psycho-visual: see *psycho-.

psychro- (səikro), comb. form of Gr. ψυχρός cold (cf. Psychrometer, etc.).

psychrophi·lic *a. Biol.* [*-philic], (of an organism, esp. a bacterium) capable of growing at temperatures close to freezing, or having an optimum temperature that is low; so **psy·chrophil, -phile** *sb.*, a psychrophilic organism; also as *adj.*, = prec.; **psy·chrosphere**, the colder, deeper part of the oceans; hence **psychrosphe·ric** *a.*; **psychroto·lerance** *Biol.* [ad. G. *psychrotoleranz* (Horowitz-Wlassowa & Grinberg 1933, in *Zentralbl. f. Bakteriol., Parasitenkunde und Infektionskrankheiten* (Abt. 2) LXXXIX. 58)], the property of being able to grow at temperatures close to freezing; (introduced, like *psychrotrophic*, because of the ambiguity of *psychrophilic*); so **psychroto·lerant** *a.*; **psychrotro·phic** *a. Biol.* [*-trophic] = *psychrotolerant* adj. above; so **psy·chrotroph**, a psychrotrophic organism.

1928 P. H. Foster in C. M. Hilliard *Textbk. Bacteriol.* viii. 95 Psychrophiles are organisms which develop at or very near the freezing point. 1956 *Nature* 16 June 1106 Such barophilic bacteria..are also psychrophil and very stenothermal. 1959 *New Scientist* 3 Dec. 1111/1 Psychrophils also ferment carbohydrates, decompose proteins,..and generally go about their business like other bacteria, except that their metabolism seems to be somewhat slowed up. 1959 *Bacteriol. Rev.* XXIII. 99/1 The unique property of psychrophilic bacteria is the ability to grow well at 0 C. This was recognized from the very beginning of the study of psychrophiles... The essentially erroneous concept that psychrophiles are distinguished by their ability to grow most rapidly below 20 C did not arise until later. 1969 *Nature* 15 Mar. 1031/1 Obligate psychrophiles—organisms able to grow well at 0°C but incapable of growth at moderate temperatures—provide suitable test organisms in this respect. When the obligate psychrophile *Micrococcus cryophilus* is grown in optimal conditions at 20°C, and then transferred to 30°C—5°C above the maximum for this organism—growth halts very quickly. 1897 Lehmann & Neumann *Atlas & Essentials of Bacteriol.* 98 Psychrophilic bacteria: minimum at 0°, best at 15°–20°, maximum at about 30°. These varieties usually live in water. 1958 W. C. Frazier *Food Microbiol.* xxiv. 304 At refrigerator temperatures, proteolysis by psychrophilic bacteria like *Pseudomonas* is most likely, and molds may follow. 1963 J. L. Stokes in N. E. Gibbons *Rec. Progress Microbiol.* 190 The maximum growth temperatures of many psychrophilic microorganisms can be quite high. 1964 [see *mesophilic* adj. s.v. *meso-]. 1975 R. R. Gillies *Lect. Notes Med. Bacteriol.* iii. 15 Psychrophilic bacteria, i.e. those which grow best at temperatures below 20°C, are non-pathogenic for man but exist in soil and water. 1956 A. T. Bruun in *Nature* 16 June 1106 (*in figure*) Psychrosphere. 1957 —— in *Mem. Geol. Soc. Amer.* LXVII. 641 The division of the hydrosphere into a warm troposphere and a cold stratosphere, or the Warmwassersphäre and Kaltwassersphäre in the terms of Wüst (1950) is the most pronounced division in the oceanic water masses. In the following the terms thermosphere and psychrosphere are used for these divisions to stress this salient ecological factor. *Ibid.*, The limit of the psychrosphere may be at about 100 meters in the eastern parts of the oceans where there is upwelling, whereas in western regions it may be as deep as 700 meters. 1976 *Nature* 8 Apr. 513/2 Three major elements were involved in the evolution of oceanic circulation during the Cainozoic... The third involves the development of the present-day system of bottom waters of the world ocean, the 'psychrosphere'. 1977 *Ibid.* 2 June 399/2 Aside from benthic foraminifera, psychrospheric ostracods also occur in some Early and Middle Miocene cores, indicating water-depths in excess of 1000 m. 1977 *Sci. Amer.* June 50/3 The bacteria's only way to adapt to the environmental conditions of the deep sea appears to be the acquisition of barotolerance and psychrotolerance. 1959 *Bacteriol. Rev.* XXIII. 98/2 The recognition that low temperatures are not optimum for organisms that grow at 0 C led to the introduction of several other names as replacement for psychrophile. These included.. psychro-tolerant. 1970 *Sci. Jrnl.* May 19 Slime is caused by cold resistant, or psychrotolerant, bacteria which are not a health hazard but which produce an objectionable smell. 1979 *Nature* 21 June p. x (Advt.), An excellent survey is thus provided of the importance of psychrotrophic bacteria and of techniques involving psychrotolerant and psychtrotrophic micro-organisms. 1963 M. Ingram in N. E. Gibbons *Rec. Progress Microbiol.* 185 To avoid such misconceptions, the term psychrotroph.. has recently been proposed for all organisms able to grow at temperatures near 0 C. 1968 *New Scientist* 18 Apr. 117/2 Farm contamination from..unclean dairy equipment, will allow psychrotrophs to proliferate if milk is stored at 5°–10°C. 1960 B. P. Eddy in *Jrnl. Appl. Bacteriol.* XXIII. 189 The writer considers that the word psychrotrophic..suggested by Dr. D. A. A. Mossel, should be used for bacteria able to grow at +5° and below, whatever their optimum temperatures. 1975 Campbell & Marshall *Sci. of providing Milk for Man* xxiii. 499 Bulk cooling and storage..have made psychrophilic and psychrotrophic bacteria the primary organisms in raw milk.

psycter. (Examples of the form **psykter.**)

1931 *Times Lit. Suppl.* 16 July 563/1 There is a psykter signed by Duris. 1948 A. Lane *Gk. Pottery* iv. 48 Cooler (psykter) to be filled with ice or water and lowered into a wine-krater. 1973 'D. Halliday' *Dolly & Starry Bird* vi. 75 The tables were crowded with skyphoi and kylikes and psykters.

psylla (si·lă). *Ent.* Also **Psylla.** [mod. L. (E. F. Germar *Systematis Glossatorum Prodromus* (1811) 14), f. Gr. ψύλλα flea.] Any of the jumping plant-lice belonging to the genus so called or closely related genera, which include several pests.

1852 T. W. Harris *Treat. Insects Injurious to Vegetation* (ed. 2) 203 The pear-tree, in Europe, is subject to the attacks of..the pear-tree Psylla. 1891 *Insect Life* IV. 127 We had occasion some time since to abstract Dr. F. Loew's remarks on the Psyllas which inhabit the pear. 1918 W. A. Davis *Study Indigo Soils Bihar* 8 In 1907, two diseases appeared simultaneously—the so-called 'wilt' disease and the less serious insect pest 'psylla'. 1933 *Jrnl. R. Hort. Soc.* LVIII. 283 'Eucalyptus Psylla'... Deals with a Psylla attacking Eucalypti in English greenhouses. 1972 Swan & Papp *Common Insects N. Amer.* xiii. 141 Pear Psylla... A major pest of pears, first introduced into the East in 1832.

psyllid (si·lid). *Ent.* Also **Psyllid.** [ad. mod.L. *Psyllid-æ*, family name, f. prec.: see -id³.] A jumping plant-louse belonging to the family Psyllidæ, which includes several species that damage plants by causing galls or spreading virus diseases. Also *attrib.*

1899 D. Sharp in *Cambr. Nat. Hist.* VI. viii. 580 Sometimes these Psyllid galls are mere changes in form of a limited part, or parts, of a leaf. 1909 H. Maxwell-Lefroy *Indian Insect Life* 743 Kieffer describes Indian gall-making Psyllids. 1922 *Nature* 3 June 714/1, I also find a winged termite, a psyllid,..some small spiders. 1954 S. H. Skaife *Afr. Insect Life* xi. 111 The jumping plant lice, or psyllids.., are small insects about the size of aphids and they look something like tiny cicadas. 1962

[see *lerp]. 1969 *New Scientist* 2 Oct. 19/1 The Australian plantsucking psyllid bug *Cardiaspina albitextura*..lives on eucalyptus leaves. 1972 Swann & Papp *Common Insects N. Amer.* xiii. 141 The psyllids, or jumping plantlice, are tiny insects.

psylly. Restrict *Obs. rare* to this form and add later examples of the form **psyllium.** Add to def.: Also, (a preparation of) the seeds of this plant or of *Psillium ovata* or *P. indica*, used as a laxative. Also *attrib.*

1897 W. T. Fernie *Herbal Simples* (ed. 2) 436 In France these Psyllium seeds..are widely prescribed as a laxative. 1932 R. C. Wren *Potter's Cycl. Bot. Drugs* (ed. 4) 281 Psyllium is used successfully in dysentery in the tropics. 1959 *Times* 30 Dec. 8/6 The mailman dragged in a sack of black Psyllium seeds from France. 1977 *Martindale's Extra Pharmacopoeia* (ed. 27) 929/1 Psyllium, on account of its content of mucilage, is used as a demulcent.

psy-op, -war: see *psy-.

pt-. In English words beginning with *pt-* the initial *p* is no longer pronounced. The Dictionary's convention of giving both (pt-) and (t-) for the pronunc. of each word of this kind has therefore been abandoned.

pteridine (te·ridīn). *Chem.* Also † **-in.** [ad. G. *pteridin* (C. Schöpf et al. 1941, in *Ann. d. Chem.* DXLVIII. 83): see *pterin and *-idine.] A synthetic yellow crystalline solid, $C_6H_4N_4$, which has a bicyclic structure formed from fused pyrazine and pyrimidine rings; any derivative of this, many examples of which occur naturally, esp. as insect pigments and vitamins of the B group.

1943 *Brit. Chem. & Physiol. Abstr.* A. II. 281 The structure of leucopterin..as 2-imino-6:8:9-trihydroxypteridine is confirmed. This and similar CO compounds are named as (enolic) OH-derivatives of pteridine. 1948 *Nature* 2 Oct. 524/2 Reaction of the diaminopyrimidine in aqueous solution with glyoxal bisulphite gives pteridine which crystallizes from alcohol in pale yellow plates, m.p. 140°. 1951 *Jrnl. Chem. Soc.* 474 All natural pteridines have at least one amino- and one hydroxy-substituent. 1962 [see *drosophila]. 1968 L. A. Paquette *Princ. Mod. Heterocyclic Chem.* xi. 384 Another vitamin of the B group, B_2 or riboflavin.., also contains the basic pteridine ring system.

pterido-. Add: **pteridologist** (earlier and later examples); **pteridology** (earlier and later examples); **pteridomania** (later examples); **pteridophyte,** add to etym. after *Pteridophyta*: (E. Haeckel *Generelle Morphologie der Organismen* (1866) II. p. xxxix); also *attrib.*; (earlier and later examples); so **pteridophy·tic** *a.*; **pte·ridosperm** [ad. mod.L. *Pteridospermeæ* (Oliver & Scott (1904) in *Phil. Trans. R. Soc.* B. CXCVII. 239)], a fossil plant belonging to the class Pteridospermeæ or the order Pteridospermales, which include seed-bearing plants resembling ferns.

1845 E. Newman in *Phytologist* I. 273, I am disposed to believe that our Pteridologists have rarely taken that comprehensive view of the characters of ferns which is requisite for their classification in accordance with nature. 1979 *N.Z. Jrnl. Bot.* XVII. 98/1 These two eminent pteridologists do not yet entirely agree on a classification for the Cyathaceae. 1855 G. B. Wollaston in *Phytologist* New Ser. I. 171, I venture with the greatest diffidence,..single-handed, into the battle-field of Pteridology. 1980 *Nature* 7 Feb. 608/1 An overview of the..biochemical, physiological and genetical research which has taken place during the past 40 years in experimental pteridology. 1969 D. E. Allen *Victorian Fern Craze* p. xi, No one..would have thought of filling empty carboys with greenery and building up their present vogue had not Ward himself first prepared the ground... A twentieth-century 'Pteridomania'? 1970 *New Scientist* 7 May 296/1 Pteridomania had many social aspects yet it seems to have been almost forgotten outside botanical circles. 1880 C. E. Bessey *Botany* xx. 437 The epidermis of Angiosperms does not differ in any marked way from that of the Gymnosperms and the Pteridophytes. 1910 Coulter & Chamberlain *Morphol. Gymnosperms* i. 4 (*heading*) Vascular anatomy of pteridophytes. 1938 J. C. Schoute in F. Verdoorn *Man. Pteridol.* i. 3 By these discoveries the range of the Pteridophyte canon of morphology was much enlarged. 1956 B. Cobb *Field Guide to Ferns* 36 Each of the four classes of the Pteridophytes has its own characteristic behavior in producing, bearing, and propagating its spores. 1978 *Fern Gaz.* XI. 349 The pteridophyte flora of Réunion Island is characterized by a high number of species. 1898 *Bot. Gaz.* XXV. 305 In pteridophytic types of embryogeny..it is always possible to distinguish the segment which is the homologue of the originally distal segment. 1977 A. Hallam *Planet Earth* 252 This has been effected by the elimination of the free-living sexual stage (prothallus) of the pteridophytic plants. 1904 Oliver & Scott in *Phil. Trans. R. Soc.* B. CXCVII. 240 The further development of our knowledge of the Pteridosperms will form one of the chief objects of palæo-botanic investigation in the near future. 1931 A. C. Seward *Plant Life through Ages* ix. 147 Evidence..eventually proved that the great majority of the Carboniferous 'ferns' were seed-bearing plants—pteridosperms. 1940 J. Walton *Introd. Study of Fossil Plants* xi. 138 The seed cupules of the Pteridosperms..were borne on fronds or leaves. 1974 G. W. Burns *Plant Kingdom* xix. 449/1 Ovules of the Paleozic pteridosperm *Medullosa* were probably terminally attached on the pinnae.

pterin (te·rin). *Chem.* Also **-ine.** [a.G. *pterin* (Wieland & Schöpf 1925, in *Ber. d. Deut. Chem. Ges.* LVIII. 2178), f. Gr. πτερ-όν wing, feather: see -in¹.] Any of a class of naturally occurring pteridine derivatives found esp. as insect pigments; more generally, any pteridine.

1934 *Chem. Abstr.* XXVIII. 1353 (*heading*) Occurrence of pterins in wasps and butterflies. 1943 A. H. Cook tr. *Mayer's Chem. Natural Coloring Matters* v. 264 The pterins occur as small colored or colorless granules in association with chitin. 1950 *Thorpe's Dict. Appl. Chem.* (ed. 4) X. 264/1 The name 'pterin' denotes a group of heterocyclic compounds, of both natural and synthetic origin, the molecules of which contain..a pyrazine ring fused to a pyrimidine ring. 1954 *Sci. News* XXXIV. 91 The purines and pterines constitute a major source of colour to the wings of butterflies;..the latter as white, red, yellow, and orange pigments, which are also found in the integument of wasps. 1970 A. L. Lehninger *Biochem.* xxiv. 543 Folic acid..contains three characteristic building blocks (1) a substituted pterin, (2) *p*-aminobenzoic acid, and (3) glutamic acid.

ptero-. Add: **pte·robranch** [ad. mod.L. *Pterobranchia*, f. F. *ptérodibranche* (H. de Blainville 1816, in *Bull. Sci. Soc. Philomatique* XXVIII. 122)], a small marine, usually colonial, animal belonging to the class Pterobranchia of the phylum Hemichordata; also *attrib.*; **pte·romorph(a)** [*-morph], in certain mites, a winglike appendage attached to the cephalothorax; **pteropleu·ron** *Ent.*, in Diptera, the section of the thorax from which the wings arise.

1949 *New Biol.* VI. 118 The construction of the tubes in which both Graptolites and Pterobranchs live gives good reason for accepting Kozlowski's thesis. 1962 D. Nichols *Echinoderms* xiv. 175 Grobben has shown the principal transformations necessary to convert a pterobranch hemichordate into an echinoderm. 1968 A. S. Romer *Procession of Life* vii. 141 The pterobranchs are tiny animals, only a fraction of an inch in size, living in colonial fashion within branching tubes, with, superficially, the appearance of tiny stemmed flowers. 1978 *Nature* 23 Nov. 318/3 Examples of arborescent growth occur among sponges, hydroids, graptolites, pterobranchs. 1907 *Ibid.* 12 Dec. 142/2 The species [of mites] are to be called *Oribata bostocki*, distinguished by the pteromorphæ being attached to the anterior margin of the abdomen instead of its lateral margin. 1952 Baker & Wharton *Introd. Acarology* viii. 387 Within the Aptyctima we find several natural groups based on..the possession or lack of pteromorphs or wings. 1959 T. E. Hughes *Mites* viii. 113 When big pteromorphae are present, they can usually be depressed by special muscles. 1962 *New Scientist* 20 Sept. 628/2 One group of mites is unique in possessing hinged outgrowths similar to wings, called pteromorphs. 1972 L. Gozmány tr. *Balogh's Oribatid Genera of World* 22 The pteromorpha is a horizontal or inferiorly deflected chitinous lamella. 1884 C. R. Osten-Sacken in *Trans. Entomol. Soc.* 503 Pteropleura, situated under the insertion of the wing, and behind the mesopleural structure. 1951 Colyer & Hammond *Flies Brit. Isles* 24 The wings arise from the pteropleuron. 1977 Richards & Davies *Imms's Gen. Textbk. Entomol.* (ed. 10) II. iii. 961 The pteropleuron (= dorsal part of mesepimeron) lies below the root of the wing.

pterodactyl. For **-yle** read † **-yle** and add: **2.** *Aeronaut.* Also **Pterodactyl.** An obsolete kind of tailless pusher aeroplane with swept-back wings and a very short fuselage.

1926 G. T. R. Hill in *Jrnl. R. Aeronaut. Soc.* XXX. 528 During the summer of 1925, all outstanding design questions were finally settled, and I therefore propose..to give a more detailed description of the main features of the aeroplane, which I have called the Pterodactyl, on account of its supposed resemblance to the prehistoric lizards of that breed. 1935 *Jrnl. R. Aeronaut. Soc.* XXXIX. 823 A recent proposal for a large tailless flying boat having a pterodactyl wing plus stubs. 1960 C. H. Gibbs-Smith *Aeroplane* xiii. 103 The *Pterodactyls* contributed to a development culminating in the swept-back and delta-wing forms of today.

pteroic (te·ro,ik), *a. Biochem.* [f. *pter(idine + *-oic.] *pteroic acid*: a synthetic crystalline solid from which the pteroylglutamic acids are formally derived by peptidic linkage of glutamic acid residues to the carboxyl group; *p*-(2-amino-4-hydroxypteridin-6-ylmethyl) aminobenzoic acid, $C_{14}H_{12}N_6O_3$.

1946 [see *pteroylglutamic *a.* a]. 1954 A. White et al. *Princ. Biochem.* xii. 263 The linkage coupling the *p*-aminobenzoic acid..and glutamic acid is a peptide bond. Hydrolysis of this bond gives pteroic acid and glutamic acid. 1963 W. Shive in Florkin & Stotz *Comprehensive Biochem.* XI. vii. 90 Among the synthetic derivatives, pteroic acid, because it is effective in replacing folic acid for *S[treptococcus] faecalis* R but not for *L[actobacillus] casei*, and the pteroyldiglutamate..are of biological interest. 1970 T. Shiota in *Ibid.* XXI. 126 The mechanism involving successive additions of L-glutamic acid residues to pteroic acid and folic acid is supported by the work of Griffin and Brown.

pteropod. Substitute for etym. and def.: [ad. mod.L. *Pteropoda*, a. F. *ptéropode* (G. Cuvier 1804, in *Ann. Mus. Hist. Nat.* IV. 232): see Pteropoda.] **a.** A pelagic marine gastropod belonging to the class Pteropoda, which includes molluscs having a modified foot bearing lobes which act as fins; a sea butterfly. (Further examples.)

1877 T. H. Huxley *Man. Anat. Invertebr. Animals* viii. 505 In all Pteropods and Branchiogasteropods, the mantle secretes a cuticular shell. 1890 [see *butterfly snail*]. 1934 W. Beebe *Half Mile Down* ix. 167 At 1825 feet coiled pteropods..appeared by the dozen. 1956 A. Hardy *Open Sea* I. xii. 228 The deep-water pteropods are dark in colour. 1976 *Sci. Amer.* July 95/1 The pteropods, wing-footed marine snails, collect phytoplankton on floating webs of sticky mucus.

b. pteropod ooze, a calcareous marine sediment rich in the remains of the tests of pteropods.

1891 Murray & Renard in *Rep. Sci. Results Voy. H.M.S. Challenger: Deep-Sea Deposits* iii. 223 Pteropod Ooze. This name was employed by Mr. Murray during the cruise of the Challenger to designate those deep-sea deposits in which a very large part of the calcareous organisms consists of the dead shells of Pteropods and Heteropods. 1894 [in Dict.]. 1929 A. Conan Doyle *Maracot Deep* 17 We would bring up..a scoop of pteropod ooze. 1967 *Oceanogr. & Marine Biol.* V. 51 To the South of New Caledonia..there is a small zone of pteropod ooze.

pterosaur. Substitute for first part of etym.: [ad. mod.L. name of order *Pterosauria*, f. generic name *Pterosaurus* (L.F.J.F. Fitzinger *Systema Reptilium* (1843) 15).] (Later examples.)

1907 *Proc. Zool. Soc.* I. 226 It has been frequently assumed that the Pterosaurs enjoyed a bipedal locomotion. 1926 G. Heilmann *Orig. Birds* iv. 140 The Pterosaurs are of interest chiefly because they are flying reptiles. 1933 A. S. Romer *Vertebr. Paleont.* viii. 176 The brain of the pterosaurs..was exceedingly large for a reptile. 1973 J. Updike *Museums & Women* 196 The iguanodon despised these pterosaurs' pretensions. 1975 *Times* 12 Apr. 12/1 The surprise created in the popular mind by Douglas Lawson's discovery of an immense pterosaur—a beast whose wings spanned about 51 feet according to the most reliable computer estimate—is only paralleled by the bewilderment experienced in scientific circles.

pterosaurian *a.* (later example).

1981 *Sci. Amer.* Feb. 95/1 The triangle was supported at the wrist by a uniquely pterosaurian innovation, the pteroid bone.

pteroylglutamic (te:ro,əilglᵉūtæ·mik), *a. Biochem.* [f. *pter(idine + *-oyl + Glutamic *a.* in Dict. and Suppl.] **a.** *pteroylglutamic acid*: any of a series of derivatives of pteroic acid which have a side chain consisting of one or more glutamic acid residues, and include certain members of the vitamin B complex and other animal growth factors; folic acid.

1946 R. B. Angier et al. in *Science* 31 May 669/1 For the compounds formed from *p*-aminobenzoic acid and *p*-aminobenzoyl-*l*(+)-glutamic acid, the names pteroic acid and pteroylglutamic acid are suggested. 1949, 1955 [see *folacin]. 1967 Pike & Brown *Nutrition* iii. 65 At one time both p-aminobenzoic acid and pteroylglutamic acid were considered to be vitamins, but it is now apparent that the species requirement is for one or the other of the two.

b. With inserted prefix indicating the number of glutamic acid residues present, as *pteroylmonoglutamic acid* (cf. *folic acid), etc. Also *pteroylpolyglutamic acid*.

1946 *Ann. N.Y. Acad. Sci.* XLVIII. 287 Whereas vitamin B_c..contains one glutamic acid residue, vitamin B_c conjugate contains seven. In the system of nomenclature suggested by the authors [*sc.* C. W. Waller et al.], vitamin B_c conjugate is therefore pteroylheptaglutamic acid. 1954 A. White et al. *Princ. Biochem.* xlix. 1035 Animal tissues contain an enzyme, 'vitamin B_c conjugase', which hydrolyzes the naturally occurring pteroylpolyglutamic acid compounds to pteroylmonoglutamic acid and free glutamic acid. 1955 *Chem. & Engin. News* 6 June 2433/1 The pure substance hitherto known as folic acid, folacine, or vitamin B_c shall be named pteroylmonoglutamic acid. Compounds analogous to it but containing several glutamic acid residues united by amide linkages may be named pteroyltriglutamic acid, pteroylheptaglutamic acid, etc. 1970 Passmore & Robson *Compan. Med. Stud.* II. vii. 7/1 Folic acid in therapeutics means synthetic pteroylmonoglutamic acid but the term is sometimes used to embrace all naturally occurring substances with folic acid activity.

Hence **pte:roylglu·tamate,** (a compound or anion of) any of these acids. (Also with inserted prefix.)

1950 R. J. Williams et al. *Biochem. of B Vitamins* C. ii. 290 Both pteroic acid and formylpteroic acid are inactive for *L[actobacillus] casei* or humans, while pteroylheptaglutamate is inactive for bacteria. 1958 Fruton & Simmonds *Gen. Biochem.* (ed. 2) xxxix. 999 The pteroyltriglutamate and heptaglutamate both occur in nature and are as active as folic acid in the nutrition of higher animals. 1963 [see *pteroic *a.*]. 1970 R. W. McGilvery *Biochem.* xviii. 411 (*caption*) The structure of the vitamin, pteroylglutamate, more commonly known as folate. 1970 Passmore & Robson *Compan. Med. Stud.* II. vii. 7/1 In foods, folates are found free and in conjugated forms, the latter being polyglutamates such as pteroyltriglutamate and pteroylheptaglutamate.

ptilinum. Add: [ad. F. *ptiline* (J. B. Robineau-Desvoidy *Essai sur les Myodaires* (1830) i. 10).] (Earlier and later examples.) Hence **pti·linal** *a.*

1853 F. WALKER *Insecta Britannica: Diptera* II. 2 The ptilinum is a soft membrane, which in many species, and especially in the newly-hatched flies, appears between the antennæ and the front. **1925** A. D. IMMS *Gen. Textbk. Entomol.* III. 593 The Ptilinum or frontal sac is a characteristic cephalic organ of Cyclorrhapha and its presence is indicated externally by the arched frontal or ptilinal suture. **1962** GORDON & LAVOIPIERRE *Entmol.* xxvii. 171 Having emerged from the puparium the insect pushes its way up through the soil to the surface by the alternate inflation and deflation of the ptilinum. **1969** R. F. CHAPMAN *Insects* xxii. 441 Once the fly has hardened the ptilinum is no longer eversible and the muscles associated with it degenerate. Its position is indicated in the mature fly by the ptilinal suture.

ptosis. Add: Pl. **ptoses** (-īz). **b.** Delete *rare* and add: Also, of the breasts. (Further examples.)

1909 *Amer. Jrnl. Med. Sci.* CXXXVII. 380 Ptoses of the splenic flexure and descending colon are rare. **1934** S. BECKETT *More Pricks than Kicks* 68 Man with weak bladder and tendency to ptosis of viscera. **1953** *Pageant* Aug. 68 About 4,000,000 young American women suffer in some degree from micromastia (immature breasts) and another 10,000,000 from ptosis (or collapse or sagging of the breasts). **1957** J. LAPIDES in J. G. Allen et al. *Surgery* xlviii. 1311/2 A number of operations have been devised and used for the fixation of the highly mobile kidney—renal ptosis. **1965** R. P. G. SANDON in R. J. V. Battle *Plastic Surg.* xiv. 319 Clothing tends to confine and flatten, rather than support correctly, such an enormous bosom, and with the years the relaxation of the skin allows an increasing element of ptosis.

ptotic *a.* (example).

1969 J. H. DEWEERD in Glenn & Boyce *Urologic Surg.* iv. 122/2 The hypermobile (ptotic) kidney.

ptygmatic (tigmæ·tik), *a. Geol.* [ad. Sw. *ptygmatisk* (J. J. Sederholm 1907, in *Bull. de la Commission Géol. de Finlande* XXIII. 89), f. Gr. πτύγμα, πτύγματ- folded matter: see -IC.] Applied to the highly sinuous and often discordant folding exhibited by the veins in some gneisses and migmatites, and to the veins themselves.

1907 J. J. SEDERHOLM in *Bull. de la Commission Géol. de Finlande* XXIII. 110 [*English summary of original Sw. article.*] The primary folding caused by melting, he designates as ptygmatic...These suggestions are made with every reservation. **1926** G. W. TYRRELL *Princ. Petrol.* xxi. 333 (*caption*) Ptygmatic folding of a quartz vein in amphibolite. **1952** *Geol. Mag.* LXXXIX. 1 The term 'ptygmatic' was originally coined by Sederholm in 1907 (p. 110) to describe 'the primary folding caused by melting' in gneisses and migmatites. The word.., as defined, would embrace most of the contortions, many of which are now included in the term 'flow fold', commonly seen in migmatite zones the world over... The term was later restricted by Sederholm (1926) to those tortuous quartzo-felspathic veins, which occur in areas of granitization. **1970** K. C. JACKSON *Textbk. Lithol.* vii. 420 (*caption*) Ptygmatic folding of a pegmatite dike in gneiss.

Hence **ptygma·tically** *adv.*, in a way characteristic of ptygmatic folds; also (as a back-formation) **pty·gma**, a ptygmatic fold.

1928 *Summary of Progr. Geol. Survey Gt. Brit. 1927* II. 72 Other observers of what are certainly ptygmatically folded veins have not accepted Sederholm's explanation. **1944** *Trans. R. Soc. Edin.* LXI. 228 The foliation (gneissic structure) of the host rock appears to conform to the plications of the 'ptygma'. **1960** *Rep. 21st Internat. Geol. Congr.* XIV. 138 It is concluded that ptygmas could have been formed as the result of development in a passive host or possibly, though improbably, as the result of magmatic flowage. **1971** *Scottish Jrnl. Geol.* VII. 316 Pre-D2 quartz veins are..ptygmatically folded when at high angle to S2.

p-type (pī·təip), *a. Physics.* [f. *P* (repr. *positive*) + TYPE *sb.*[1]] Applied to (a region in) a semiconductor in which electrical conduction is due chiefly to the movement of holes (rather than electrons). Opp. *N-TYPE *a.*

1946 [see *N-TYPE]. **1948** TORREY & WHITMAN *Crystal Rectifiers* iii. 49 A semiconductor that conducts principally by holes in the nearly filled band is referred to as a '*p*-type' semiconductor. *Ibid.*, The impurities added to silicon make it *p*-type. **1970** *New Scientist* 15 Oct. (Suppl.) 6/2 In a practical integrated circuit the triangular half of the rectifier is p-type material which is diffused into the silicon chip in a pattern similar to the wall plan for a house.

pua, var. *PUHA.

pub, *sb.* **1.** For *low colloq.* read *colloq.* and add earlier and later examples.

1859 HOTTEN *Dict. Slang* 78 *Pub*, or *Public*, a public house. **1893** K. MACKAY *Out Back* (ed. 2) II. v. 188 It's Molloy's fault...He got tanked at the pub last night. **1922** JOYCE *Ulysses* 70 Waiting outside pubs to bring da home. **1924** *Truth* (Sydney) 27 Apr. 6 *Pub*, hotel. **1936** M. ALLIS *Eng. Prelude* xxiii. 247 First comes the pub, the *Fox and Hounds*. **1946** *R.A.F. Jrnl.* May 175 There are German beer shops turned into typical English 'Pubs'. **1950** 'N. SHUTE' *Town like Alice* vi. 170 She was surprised at the rapidity of [the town's] growth. In 1928 it was about three houses and a pub. **1970** M. GREENER *Penguin Dict. Commerce* 268 A pub offering accommodation to casual customers will probably be a hotel within the definition of the Hotel Proprietors Act 1956 and the various Innkeepers Acts. **1980** 'D. KAVANAGH' *Duffy* ii. 36 They met at a drinkers' pub near Baker Street Station.

2. *attrib.* and *Comb.*, as *pub-door, food, -friend, -goer, -grub, -keeper, -landlord, manager, meal, mirror, parlour, -singer; pub-going, -running, -spieling* vbl. sbs.; *pub-hunting* ppl. adj.; **pub-crawl**: see *CRAWL *sb.*[1] b; hence as *v. intr.*; **pub-crawler; pub-crawling** *vbl. sb.* and *ppl. a.*; **pub-life**, the society of public houses; **pub lunch**, a lunch eaten in a pub; hence **pub-lunch** *v. intr.*, **pub-luncher; pub rock**, rock music of a type played in public houses; **pub-stiff** *N.Z. slang*, a look-out or sentinel acting on behalf of a licensee selling alcoholic drinks after closing-time; **pub theatre**, a public house at which theatrical performances take place; also, theatrical representation performed in a public house; **pub-time**, (*a*) the hour at which a public house opens or closes; (*b*) the time shown by a clock in a public house, with reference to the custom of advancing this slightly to bring forward closing-time.

1915 Pub-crawl [see *CRAWL *sb.*[1] b]. **1937** *Times Lit. Suppl.* 27 Nov. 910/1 Mr. Lyons does not 'pub-crawl' as a writer in search of copy. **1959** [see *CRAWL *sb.*[1] b]. **1972** J. SYMONS *Players & Game* xxiii. 182 He had taken a girl ..on a mild variety of pub crawl. **1974** *Canadian Mag.* (Toronto) 16 Mar. 2/3 Across Canada, kids aren't packing the discothèques; instead, they're pub-crawling. **1910** *Daily Chron.* 28 Jan. 4/4 These 'pub-crawlers' have captured the illiterate and the unthinking. **1976** J. R. L. ANDERSON *Redundancy Pay* ix. 145 You're turning me into quite a pub-crawler. **1919** 'W. N. P. BARBELLION' *Enjoying Life* 75 Drunken Barnabee's Journal..is rhymed Latin verse..describing the author's 'pub crawlings' up and down the country. **1921** F. B. YOUNG *Black Diamond* vii. 73, I bain't goin' to keep you in pub-crawling any longer. **1973** 'H. CARMICHAEL' *Candles for Dead* vi. 74 A pub-crawling reporter. **1980** I. MURDOCH *Nuns & Soldiers* i. 83 This sort of urban life suited Tim, pub-crawling, wandering, looking in shop-windows. **1960** T. HUGHES *Lupercal* 18 The lamp above the pub-door Wept yellow when he went out. **1977** *Times* 11 June 11/7 English people care more about pub food than they used to. **1959** J. CARY *Captive & Free* xviii. 85 His father had been a steady worker, but completely devoid of ambition; a man whose only interests were football, darts, his pub-friends. **1955** T. H. PEAR *Eng. Social Differences* vi. 160 Peter is primarily a pub-goer. *Ibid.* 161 The social differences which come out in pub-going. *Ibid.* 162 The new pub-going habit for girls and women is a genuine problem. **1978** *Country Life* 19 Oct. 1186/4 There are 'lounge bars' and 'singing bars' and many places advertising 'pub grub'. **1922** JOYCE *Ulysses* 599 He commented adversely on the desertion of Stephen by all his pubhunting *confrères* but one. **1925** W. DEEPING *Sorrell & Son* vi. 57 Our pub-keepers rarely visualize the atmosphere of a garden. **1980** D. FRANCIS *Reflex* vi. 67 The pub-keeper from the Sussex village where he lived. **1909** *Daily Chron.* 17 July 4/7 Mr. Lewis Harcourt's reference to 'the ground and the pub-landlord seeking to hold the common fort'. **1944** WYNDHAM LEWIS *Let.* 5 Jan. (1963) 374, I gather from Augustus that the pub-life of London is functioning as of yore. **1970** G. F. NEWMAN *Sir, You Bastard* v. 130 When he reported, Sneed had a pub lunch with the Governor. **1971** 'F. CLIFFORD' *Blind Side* II. iv. 113 He pub-lunched in Richmond. **1975** *Times* 8 Mar. 10/4 Cheapness is often the only virtue of the British pub lunch. **1971** *Times* 2 June 6/1 The sound of pub-lunchers arising merrily from below. **1977** P. COSGRAVE *Cheyney's Law* iv. 39 He waited, no longer the tweedy city stroller, pub luncher, book buyer. **1973** K. GILES *File on Death* v. 118, I got a letter asking if I had perhaps a vacancy for a pub manager. **1975** P. McCUTCHAN *Very Big Bang* x. 96 He would..snatch pub meals as and when he could. **1974** *Selfridges Bk. of Xmas* 80 Victorian style pub mirrors..£18.75 each. **1977** J. WILSON *Making Hate* vii. 85 A reproduction Edwardian pub mirror. **1929** D. H. LAWRENCE *Pansies* 132 Little fleets ..that put to sea and boldly sink Armadas In a pub parlour, in literary London, on certain evenings. **1976** *Star* (Sheffield) 20 Nov., Pub-rock does a dying swan act at the weekend when Sheffield's most celebrated stronghold closes its doors to live entertainment. **1977** *Zigzag* Apr. 39/2 Joe was bored with singing pub rock standards. **1973** K. GILES *File on Death* v. 118, I own the local brewery... Pub-running has problems. **1975** *Radio Times* 3 Apr. 17 George Formby... His songs..have passed.. into the repertoires of every comic, impressionist and pub-singer. **1900** H. LAWSON *Over Sliprails* 38 Jack Drew talked too straight in the paper, and in spite of his proprietors—about pub spieling and such things. **1946** F. SARGESON *That Summer* 63 The pub-stiff that was on the door told us to go upstairs. **1973** *Guardian* 23 Jan. 10/1 Pub theatres are in vogue for the first time since Shakespeare's day. **1976** *Alyn & Deeside Observer* 10 Dec. 11/5 As the pint is pulled downstairs, an audience is held upstairs by one of the best examples of Pub Theatre to be found in London. **1947** DYLAN THOMAS *Let.* 11 June (1966) 314 I'm used to working from after lunch until pub-time. **1968** L. MEYNELL *Death of Philanderer* x. 167 The clock behind the bar would be showing 'pub time', that is..it would be at least five minutes fast.

pub, *v.* Add: **a.** Also without *it.*

1950 *John o' London's* 24 Nov. 614/1 Pubbing through Edinburgh's Old Town and the Leith waterfront. **1960** L. COOPER *Accomplices* III. i. 152 We went pubbing together. **1972** S. CHANCE *Septimus & Minster Ghost* vii. 62 'Can't have you pubbing in your canonicals,' she said, going to the door and looking out into the alley.

b. *intr.* To own or manage a public house.

1936 M. FRANKLIN *All that Swagger* xiv. 130 The profits to be made from fools by pubbing could add to it.

pubarche (piubā·rkī). *Med.* [f. L. *pūb-ēs* pubic hair, groin + Gr. ἀρχή beginning.] The first appearance of pubic hair; chiefly in *premature pubarche*, the premature occurrence of this without other signs of sexual precocity.

1950 L. WILKINS *Diagn. & Treatm. Endocrine Disorders in Childh. & Adolescence* ix. 146/2 We would suggest for this condition the term 'premature pubarche' which does not attempt to define its etiology. **1974** N. D. BARNES et al. in M. M. Grumbach et al. *Control of Onset of Puberty* viii. 223 Premature pubarche or adrenarche, the isolated growth of sexual hair, is another form of precocious sexual development that may be neurogenically determined.

pu·bbish, *a.* [f. PUB *sb.* + -ISH[1].] Of the nature or character of a public house.

1956 D. M. DAVIN *Sullen Bell* 56 You hardly ever use a pub, but like an old puritan you insist it must be as pubbish as possible. **1973** R. LUDLUM *Matlock Paper* ii. 13 The name of the country inn was the Cheshire Cat, and..it was Englishy and pubbish.

pu·bby, *a.* [f. PUB *sb.* + -Y[1].] = prec.

1959 *Good Food Guide* 361 It retains a pleasant pubby atmosphere and there's a good, mildly chaotic restaurant upstairs. **1974** *Times* 5 Oct. 12/8 Balls Brothers wine bar in the Strand had a pubby atmosphere. **1976** *Eastern Daily Press* (Norwich) 16 Dec., In the first, pubs are made to look like anything but pubs, while the 'pubbier than pub' devotees prefer a severe environment..echoing the style of the first urban public bars.

pubertal, *a.* Delete *rare* and add further examples.

1972 *Clin. Endocrinol.* (1973) (B.M.A.) 99 Pubertal development lasts on average about three years in girls and four years in boys. **1976** *Times Lit. Suppl.* 2 Jan. 2/3 A pubertal woman is always potentially taboo [among gypsies] and is actually so after menstruation, childbirth and intercourse. **1979** J. BARNETT *Backfire is Hostile!* v. 49 Some rowdy youths cavorted in the rush and grope of pubertal ritual.

puberty. b. (Further examples.)

1924 A. LIPSCHÜTZ (*title*) The internal secretions of the sex glands: the problem of the 'puberty gland'. **1971** *Canad. Antiques Collector* Sept.–Oct. 4 Arapaho Painted Hide Puberty Robe 67 × 46 inches. **1978** *New York* 3 Apr. 32/2 Masks and helmets used in puberty rites of Sierra Leone and Liberia.

pubic, *a.* Add: **b.** Employed to cover the pubes.

1940 A. UPFIELD *Bushranger of Skies* xi. 130 That Jack Johnson wore only the pubic tassel announced his non-employment by the station. **1959** S. H. COURTIER *Death in Dream Time* ii. 19 The bewildering display of aboriginal weapons and implements..head-dresses and pubic bands.

public, *a.* (*sb.*) Add: **I. 1.** (Earlier and further examples.)

1484 Public administration [see ADMINISTRATION 1]. **1617** Public health [see HEALTH *sb.* 2b]. **1673** J. RAY *Observations Journey Low-Countries* 163 He is entrusted with the management of public monies. **1676** in N. Brent *Sarpi's Hist. Council of Trent* p. xiii, He was in the Publick Employment. **1721** *Mass. House of Representatives Jrnl.* (1922) III. 9 Acts have been Passed..for Striking Bills of Credit, and Issuing out the same, in order to discharge their Publick Debts. **1727** in M. M. Verney *Verney Lett.* (1930) II. xxiv. 101 The main objection against him was his making up of the publick money in the South Sea. **1781** in *Eng. Rep.* (1903) XXVIII. 1028 Treating it as a matter of public policy of the law, and similar to marriage brokage bonds, where, though the parties are private persons, the practice is publicly detrimental, [etc.]. **1785** J. WESLEY *Let.* 7 Apr. (1931) VII. 266, I beseech you..to have no respect of persons.., in disposing of the Yearly Contribution and the Preacher's Fund or any other public money. **1787** M. CUTLER *Jrnl.* 21 July in *Life, Jrnls, & Corr.* (1888) I. iv. 127 Congress would pay more than four millions of the public debt. **1794** Public concern [see CONCERN *sb.* 6]. *c* **1799** *Ess. on Political Society* I. 59 To constitute the desiderated political system, is to constitute a permanent law for regulating the public administration. **1806** in *Documentary Hist. Amer. Industr. Society* (1910) III. 67 The newspaper called the *Aurora*, has teemed with false representations ..to poison the public mind. **1827** J. KENT *Commentaries on Amer. Law* II. IV. 222 Public corporations, are such as exist for public political purposes only, such as counties, cities, towns and villages. They are founded by the government, for public purposes, and the whole interest in them belongs to the public. **1846** *Parl. Papers* I. 257 (*heading*) A bill for providing cemeteries, and promoting public health, in towns and populous districts. **1853** in *Eng. Rep.* (1901) X. 437 Public policy..is that principle of the law which holds that no subject can lawfully do that which has a tendency to be injurious to the public, or against the public good. **1868** TROLLOPE *He knew he was Right* (1869) I. xxxviii. 300 The bias of the public mind. **1883** *Statutes at Large U.S.A.* XXII. 214 Any convict, lunatic, idiot, or any person unable to take care of himself or herself without becoming a public charge. **1887** *Polit. Sci. Q.* II. 212 Public administration is detailed and systematic execution of public law. **1889** G. B. SHAW *Fabian Ess. Socialism* 194 The Manchester

School will urge..the exemption of private enterprise from the competition of public enterprise. *Ibid.* 195 The superior prestige and permanence of public employment. **1894** Public eye [mentioned s.v. EYE *sb.*[1] 8]. **1918** *Current Hist.* Aug. 277/2 The public debt of the..United States.. At Most Recent Date [in millions of dollars] $15,008. **1919** A. M. TODD (*title*) Public ownership of railroads. **1928** *Britain's Industr. Future* (Liberal Industr. Inquiry) 63 In a modern community many services must be run by a Public Concern—meaning by this a form of organisation which..is operated or regulated in the public interest. *Ibid.* 95 We propose..a special class of Company to be designated Public Corporations...The distinction..should depend..mainly on their preponderant position in their own industry or trade. *Ibid.* 243 We stand, not for public ownership, but for popular ownership. **1928** J. BUCHAN *Runagates Club* x. 273 He had been returned to Parliament..but he wasn't much in the public eye. **1937** *Times* 12 Jan. 13/6 Three years ago it was important to use public policy to increase investment. **1943** J. B. PRIESTLEY *Daylight on Saturday* xxxi. 245 A passionate defender of private enterprise (which he was careful always to contrast with..'State ownership' and never with public enterprise). **1943** W. H. CHASE *Sourdough Pot* xix. 123 Some [gold miners] even became public charges with the passing of time. **1943** J. S. HUXLEY *TVA* ix. 69 Much curiosity has been aroused in the public mind. **1955** *Bull. Atomic Sci.* Apr. 112/3 Public employment does bring with it certain obligations beyond those required of citizens in private life. **1955** *Radio Times* 22 Apr. 15/2 It was a gramophone recording that first brought her before the public eye. **1970** E. FLORES in I. L. Horowitz *Masses in Lat. Amer.* ix. 338 It is possible to substitute a cash deposit by Public Debt bonds. **1971** D. HALLIDAY' *Dolly & Doctor Bird* i. 6 Public health is a doctor's concern. **1971** S. A. DE SMITH *Constitutional & Admin. Law* xxiv. 512 Decisions in public administration can be classified in various ways. **1971** P. WORSTHORNE *Socialist Myth* ix. 237 In a democracy..the expenditure of public money must be dependent on popular agreement. **1972** *Guardian* 24 Mar. 14/2 Ministers must..keep open minds about public investment and public enterprise in the regions. **1973** I. M. SINCLAIR *Vienna Convention on Law of Treaties* v. 110 The gradual establishment in common law jurisdictions of the principle that certain types of contract are, by their very nature, injurious to society and therefore contrary to public policy. **1973** *Listener* 26 July 111/1 Mr Maudling said: 'I don't think Michael Foot has the slightest idea what he intends to do in this field of so-called public ownership extension.' **1975** C. STUART *Reith Diaries* 66 The management of public corporations..was his particular field.

2.e. *public service*: service to the community, esp. under the direction of the government or other official agency; consideration of the common good; with *the* spec. = CIVIL SERVICE; also *attrib.*

1570–6 [in Dict., sense 1]. **1645** *Rec. Colony & Plantation New Haven* (1857) 168 The farmers that have butter and cheese were desired to keepe it in their hands, that in case the publicque service require it, they may be furnished. **1706** *House of Commons Jrnl.* 11 Dec. (1742–62) XV. 211/1 Resolved, That this House will receive no Petitions for any Sum of Money, relating to publick Service, but what is recommended from the Crown. **1709** [see SERVICE[1] 11]. **1818** *Ann. Reg. 1817: State Papers* 309/2 They are..not prepared at present to suggest to the House any alteration in *this* mode of conducting *this* important department of the public service. **1857** DICKENS *Dorrit* II. xxviii. 557 It's like a limited game of cricket. A field of outsiders are always going in to bowl at the Public Service, and we block the balls. **1908** I. N. STEVENS *Liberators* 187 The public service corporation had dictated the nomination of the entire Republican State and legislative ticket. **1921** *Daily Colonist* (Victoria, B.C.) 8 Apr. 2/3 John Mitchell, formerly of the public service, Ottawa, and now farming at Landsdowne, Ont., is mentioned as a likely Government candidate. **1921** C. W. TERRY *Pract. Motor Body Building* xxxviii. 254 (*heading*) Public service vehicles. **1926** *Daily Chron.* 13 May 2/5 The Prince of Wales has been paying strictly private visits to public service depots in the London area. **1960** *Encounter* Jan. 42/2 Those public-service advertisements which enjoin us to sneeze into a handkerchief. **1960** *Road Traffic Act* 8 & 9 *Eliz. II* c. 16 § 117 For the purposes of this Act a public service vehicle is a motor vehicle used for carrying passengers for hire or reward. **1972** *Guardian* 21 Nov. 3/2 The Civil Service ('Public Service' is the standard term here [in Canberra]). *a* **1974** R. CROSSMAN *Diaries* (1975) I. 486 The special late-night programmes on the election put on by both commercial and public-service television. **1977** *Times* 2 Sept. 4/3 An independent Public Service Commission consisting of a Chairman and four other members. **1980** *Daily Tel.* 29 Feb. 16 A much broader look ought to be taken at the whole question of public service broadcasting and the way it is financed.

f. *public menace, nuisance*, etc.: anyone or anything obnoxious or annoying to the community. See also sense 9a below.

1638 Public nuisance [see NUISANCE 2B]. **1877** TROLLOPE *Amer. Senator* I. xxvii. 288 'What a very queer bird he is.' 'He is a public nuisance,—and so is the old lady who brought him here.' **1932** KIPLING *Limits & Renewals* 293 She [*sc.* a sow] broke out again and again, till the local body..indicted Mr. Gravell once more as proprietor of a public nuisance. **1952** E. O'NEILL *Moon for Misbegotten* III. 149, I made such a public nuisance of myself that the conductor threatened if I didn't quit, he'd keep me locked in the drawing room. **1955** Public menace [see *CRICKET *sb.*[2] 1c]. **1965** *Listener* 7 Oct. 549/2 Regarded now..as Public Pest No. 1 among vertebrates.., what hell the life of wood-pigeons seems to be. **1977** *Time* 12 Dec. 41/1 Prosecutors saw Barnes as a public menace to put in prison.

g. *public interest*, the common well-being. Also *attrib.* Also, *public welfare.*

1678 BUTLER *Hudibras* III. ii. 102 Both Parties joyn'd to do their best, To Damn the Publick Interest. **1730** BOLINGBROKE *Craftsman* (1731) VII. 22 No Man, who adheres to it, hath the least pretence left him to say that he pursues the publick Interest. **1858** DISRAELI in *Hansard Commons* 27 Apr. 1822 Not..one who proposes a course which will conduce to the advantage of the public interest. **1858** M. ARNOLD *Merope* 119 Let us a union found..Bas'd on pure public welfare. **1901** *Edin. Rev.* Apr. 378 The chief trustee of public welfare is, in this country, Parliament, and with Parliament rests the responsibility of seeing that the interests of the whole community are not subordinated to those of any portions of it. **1934** G. B. SHAW *Too True to be Good* Pref. 15 They voluntarily lived holy lives and devoted themselves to the public welfare in obedience to the impulse of the Holy Ghost within them. **1955** MEYERSON & BANFIELD (*title*) Politics, planning and the public interest. **1971** *Wall St. Jrnl.* 22 July 1/1 To work on coal miners' rights for a Washington public-interest law firm.

h. *public law*: that part of the law pertaining to the state and its relationship with the person subject to it; (see also sense 9a).

1773 J. ERSKINE *Inst. Law Scotl.* I. i. 9 The public law is that which hath more immediately in view the public weal, and the preservation and good order of society. **1923** W. J. BYRNE *Dict. Eng. Law* 519/2 Public law is that part of the law which deals with the State, either by itself or in its relations with individuals. **1973** I. M. SINCLAIR *Vienna Convention on Law of Treaties* v. 110 The *jus publicum* was to be understood in a wide sense as embracing not only public law in the strict sense (that is to say, the law governing relations between individuals and the State) but also rules from which individuals were not permitted to depart by virtue of particular agreements. **1976** J. M. KELLY *Stud. Civil Judicature of Roman Republic* iii. 78 A second class obviously due for exclusion is that which can be broadly labelled 'public law'; legitimate subdivisions of this would be fiscal law, military law, and 'local government' or 'police' law.

i. *public utility*: a service or supply, such as electricity, water, or transport, considered necessary to the community, usu. controlled by a (nationalized or private) monopoly and subject to public regulation. Also (with hyphen) *attrib.*

1903 R. T. ELY *Stud. Evolution Industr. Society* 225 The principal classes of these public utilities are water, light and transportation. **1915** *Political Q.* May 106 Now coal mining is a 'public-utility' industry. **1928** *Daily Chron.* 9 Aug. 7/2 Crops have been destroyed and communications and public utilities have been crippled. **1968** P. A. S. TAYLOR *Dict. Econ. Terms* (ed. 4) 88 *Public utility*, an industry, such as gas, electricity, water and transport facilities, which requires heavy and highly specialised initial investment of capital, on which the return is slow. **1976** H. TRACY *Death in Reserve* xvii. 129 Public utilities worked with the servicemen with an impressive coherence.

j. *public sector*: that part of an economy, industry, etc., which is controlled by the state at any level of government. Usu. with *the.*

1952 [see *PRIVATE *a.* 7 j]. **1969** M. ASH *Who are Progressives Now?* I. v. 122 What we in the public sector miss above all is the sense of involvement of people with progressive ideas with the State system. **1972** *Guardian* 31 Jan. 13/5 Other public sector groups. **1976** F. ZWEIG *New Acquisitive Society* I. ii. 28 The public sector seems to be the most suitable object for pressure groups' claims. **1980** *Jrnl. R. Soc. Arts* Mar. 205/1 The public sector borrowing requirement would be reduced.

3. a. (Further examples.) *public defender* (U.S.), a lawyer employed by the state who represents a defendant who is unable to pay for legal assistance, in criminal cases.

1676 in E. D. Neill *Virginia Carolorum* (1886) 361 For haveing upon specious preferences of publique works raised great unjust taxes [etc.]. **1676** Public servant [see SERVANT *sb.* 2e]. **1741** RICHARDSON *Pamela* IV. xiii. 75 Poor Housekeepers, who..are asham'd to apply for publick Relief. *c* **1810** W. HICKEY *Mem.* (1918) II. xi. 146 The parties complaining were so unreasonable as to refuse any terms, whereby the progress of the public works was impeded. **1845** Public servant [see SERVANT *sb.* 2e]. **1869** *Bradshaw's Railway Manual* XXI. 95 Public Works Loan Commissioners. **1884** B. JERROLD *At Home in Paris* II. xii. 185 That ready kindness of heart and chivalry towards the weak which pervade the 'Public Assistance' of the country. **1891** R. WALLACE *Rural Econ. Austral. & N.Z.* xxxviii. 488 No public-works undertaking can be made economically to suit any purpose whatever when that purpose has only been named without being formulated or its details settled. **1918** *Policeman's Monthly* Oct. 9/2 (*caption*) A public defender needed. **1930** *Economist* 5 Apr. 767/1 The Public Assistance Committees through which the county councils and the town councils of county boroughs are henceforward to administer what will in future be called 'public assistance'. **1931** J. S. HUXLEY *What dare I Think?* iii. 87 A..method for exerting some control over population-growth would be to link it on to public relief. **1932** *N.Y. Times* 23 Mar. 15/1 The steps advocated [at the 1st Public Housing Conference] were..the presenting of questionnaires to candidates for the Legislature seeking to commit them on the question of public housing. **1937** *Statutes at Large U.S.A.* L. I. 887 The term 'public housing authority' means any..public body..authorized to engage in the development or administration of low-rent housing or slum clearance. **1942** E. PAUL *Narrow St.* xxi. 189 She is an ideal public servant, having all the minor ailments possible, a fiendish disposition, short stature and a healthy dislike for mankind. **1961** WEBSTER s.v. Public *adj.* 1 c, Public expenditures. **1961** B. CRUMP *Hang on a Minute* 65 He was working in a Public Works road gang, clearing slips off roads and digging drains and things. **1964** MRS. L. B. JOHNSON *White House Diary* 1 Aug. (1970) 187 He spoke, disheartened, of the enormous quantity of public housing in New York City. **1965** SELDON & PENNANCE *Everyman's Dict. Econ.* 45 One central fund from which all regular public expenditures should be paid. **1965** A. J. P. TAYLOR *Eng. Hist. 1914–1945* vi. 212 Far from welcoming any increase in public spending, let alone advocating it, Labour had inherited the radical view that money spent by the state was likely to be money spent incompetently and corruptly. **1971** *Archivum Linguisticum* II. 50 A working man..on public assistance. **1974** *State* (Columbia, S. Carolina) 15 Feb. 1-B/3 Pickens County Public Defender Joseph Board represented him while Thomas M. Greene..represented the state. **1976** *Times* 21 May 1/5 The combined effects of the Government's public expenditure restraint and pay policies. **1976** *Birmingham Post* 16 Dec. 7/9 Two major hospital schemes in Dudley and Stafford, costing more than £20 million, may be delayed because of the public spending cuts. **1977** *N.Z. Herald* 8 Jan. 2-4/6 Until she breaks the seal and reads what was written in public servant scrawl 24 years ago, she cannot continue her search. **1977** *New Yorker* 27 June 85/3 Queens..has..only eighteen of the city's two hundred and forty-six public-housing projects. **1977** P. JOHNSON *Enemies of Society* v. 71 The new government public-relief system aggravated the evils it was designed to cure. **1977** G. CLARK *World Prehist.* (ed. 3) II. 75 Given the possibility of public works on an adequate scale, it was capable of producing food enough to support society at increasing levels of complexity.

4. a. (Further examples.) *public convenience*: see *CONVENIENCE *sb.* 7 d.

1606 SHAKS. *Ant. & Cl.* II. ii. 234, I saw her once Hop forty Paces through the publicke streete. **1611** CORYAT *Crudities* 290 There are reported to be in Venice..twentie seven publique clocks. *Ibid.* 403 In an open court *sub dio* two publike bathes. **1613** PURCHAS *Pilgrimage* I. v. xvi. 453 They have their publike Meetings and Bankets in their Temples very often. **1699** M. LISTER *Journey to Paris* 150, I never saw in all the Markets once Sprouts.. nor in their publick Gardens any Reserves of old Stalks. **1705** *Boston News-Let.* 24 Sept. 2/1 We know not the certainty of any others besides those mentioned in the Publick Print. **1718** in *Rep. Rec. Commissioners Boston* (1883) VIII. 129 The Projection of an Act for a Publick Market in Boston..Voted disallowed. **1738** W. STEPHENS *Jrnl.* 11 Feb. in *Colonial Rec. Georgia* (1906) IV. 80 Bailiff Parker and Mr. Hugh Anderson..took a Walk first to the publick Garden. **1762** in A. EARLE *Customs & Fashions in Old New England* (1893) 247 At the Public Room of the above Inn will be delivered a series of Moral Dialogues. **1763** J. BELL *Trav. from St. Petersburg* II. 54, I was present at the representation of a kind of farce in the publick street. **1777** Public print [see PRINT *sb.* 11]. **1781** Public street [see NUISANCE 2 c]. **1785** COWPER *Tirocinium* in *Poems* II. 334 And while on public nurs'ries they rely. **1802** C. WILMOT *Let.* 16 May in *Irish Peer* (1920) 69 The most entertaining and pleasant day possible, at 'Bagatelle'... As it is a Publick garden, multitudes of people were parading about. **1804** R. SUTCLIFF *Jrnl.* 31 July in *Trav. N. Amer.* (1811) 42 This morning I was conducted..to one of the Public Baths [in New York City]. **1822** J. C. LOUDON *Encycl. Gardening* 1186 Public Parks, or Equestrian Promenades, are valuable appendages to large cities. **1825** H. WILSON *Mem.* III. 38 We wanted to go to the play..but we had..no private box. I have never in my life, frequented the public boxes. **1832** in *Whig Almanac 1844* 38/1 Within a few years.. restless men have thrown before the public their visionary plans for squandering the public domain. **1848** MRS. GASKELL *Mary Barton* I. i. 1 There are some fields near Manchester..through which runs a public footpath to a little village about two miles distant. **1850** *Eng. Jrnl. Educ.* IV. 434/2 A Public Nursery has recently been established in Nassau Street, Mary-le-Bone, for the purpose of receiving the children of the married industrious poor, during the working-hours of the day. **1880** GEO. ELIOT *Let.* 6 June (1956) VII. 292 Your having learned the news of our marriage by the cable and public prints has always been a vexation to us. **1893** *McClure's Mag.* I. 394/2 There were even days when the Joneses questioned whether they were not running a public telephone, so often did the bell ring. **1898** E. HOWARD *Tomorrow* vii. 73 Their so-called 'public markets'..are by no means public in the same full sense as are our public parks, libraries, water undertakings..which are carried on upon public property, by public officials, at the public expense, and solely with a view to the public advantage. **1903** Public park [see PARK *sb.* 2 b]. **1910** *Bradshaw's Railway Guide* Apr. 1059/2 Private Hotel... Fine public rooms. **1910** W. J. LOCKE *Simon the Jester* xxii. 291, I..went in search of the nearest public telephone office. **1924** J. BUCHAN *Three Hostages* vi. 85, I went into a public telephone-booth. **1926** *Gloss. Terms Electr. Engin.* (Brit. Engin. Stand. Assoc.) 162 *Public call office* (*Pay station*, U.S.A.), a subscriber's station available for the use of the public on payment of a fee, which may be deposited in a coin box or paid to an attendant. **1927** W. B. YEATS *October Blast* 21 They hold their public meetings where Our most renownèd patriots stand. **1927** *Observer* 16 Oct. 11/4 Auto-electric advertising machines are about to be placed in 2,500 public telephone call boxes in London. **1928** D. L. SAYERS *Unpleasantness at Bellona Club* viii. 88 That phone-call..was put through..from a public call-box. *Ibid.* 90 His call came from a public box. **1930** W. S. MAUGHAM *Cakes & Ale* xi. 136 He was rather fond of going down to the Bear and Key..and having a few beers in the public bar. **1932** KIPLING *Limits & Renewals* 386 Improved sanitary appliances and gratuitous public transport. **1933** *Radio Times* 14 Apr. 75/1 The war was newly over..and public clocks had resumed their forgotten chiming. **1933** E. WAUGH *Scoop* I. iii. 50 There was a dense crowd round the public lavatory. **1933** A. G. MACDONELL *England, their England* vii. 100 He covered the twenty yards to the public telephone box. **1943** J. S. HUXLEY *TVA* vii. 50 Over a quarter of the 40,000 square miles of the Southern Highlands is or will shortly be public domain, under either Federal or State ownership. **1952** M. LASKI *Village* xiii. 181 In these public streets, love was easiest spoken of when they talked of the children. **1961** E. WAUGH *Unconditional Surrender* I. i. 23 I'll drop you back at your office. Can't have you using

public transport on your birthday. *Ibid.* III. ii. 230 Guy took to walking every afternoon in the public gardens... There were winding paths, specimen trees, statuary, a bandstand. **1962** J. BRAINE *Life at Top* x. 136 The Warley Council's plan for a new public baths. **1965** *Scotsman* 14 June 8 House contains 2 public rooms, 2 bedrooms, boxroom, scullery and bathroom. **1969** A. CORNELISEN *Torregreca* iv. 139 Our meeting is not entirely private... Our mutual understanding must become public property. **1971** H. CALVIN *Poison Chasers* vii. 83 Two of the security men..came into the public bar, and the rest of the customers..went into the lounge bar. **1971** D. LEES *Rainbow Conspiracy* ix. 134 All I had to look for was a broken stile with a Public Footpath sign. **1971** R. BUSBY *Deadlock* v. 74 A public telephone stood in one corner of the discreetly lit foyer. **1972** P. CLEIFE *Slick & Dead* I. iv. 37 The job of public lavatory attendant. **1973** A. MANN *Tiara* ix. 80 Available on the terrace was a row of public toilets. **1974** M. BIRMINGHAM *You can help Me* iii. 47 There was one caller from a public call-box..and another—not from a public box. **1974** R. C. DENNIS *Conversations with Corpse* xiii. 132 Nothing appeared in the public prints about the missing money. **1975** *Country Life* 2 Jan. 38/3 In 1963 Cypress was given to the City of Charleston as a public park. **1976** P. R. WHITE *Planning for Public Transport* ii. 31 We are concerned with public transport (which is taken to mean modes available for public use rather than any distinction based on ownership). **1977** *Listener* 30 June 861/3 After years of legal wrangles and bankruptcy, Jacques Tati has managed to get his films back into the public domain. **1977** W. MCILVANNEY *Laidlaw* xxii. 94 A pub which from the outside looked as inviting as a public toilet. **1978** 'D. RUTHERFORD' *Collision Course* 121 They paid admission to the Casino and..strolled through the public rooms.

c. (Later examples.)

1869 L. M. ALCOTT *Little Women* II. xi. 162 She excited the suspicions of public librarians by asking for works on prisons. **1972** C. DRUMMOND *Death at Bar* ii. 44 Dubious books submitted to their Members of Parliament by Watch Committees, Purity Leagues and Aldermen who had power over public librarians.

† **e.** *public table* = TABLE D'HÔTE. *Obs.*

1742 M. W. MONTAGU *Let.* 23 May (1966) II. 281 Nothing is cheaper than living in an Inn in a Country Town in France..25 sous for dinner and 30 for supper and lodging of those that eat at the public table. **1842** DICKENS *Let.* 4 Apr. (1974) III. 182 The public table, at this hotel and at the hotel opposite, has just now finished dinner. **1865** TROLLOPE *Can you forgive Her?* II. xxx. 234 At Lucerne they made no acquaintances... They did not even dine at the public table.

f. *to go public*: of a privately-owned company, to seek a quotation on the stock exchange; also in trivial use (passing into sense 5) to reveal oneself, to come out into the open.

1965 H. I. ANSOFF *Corporate Strategy* (1968) iv. 62 Two major alternatives to this end [*sc.* of enhancing the liquidity of the firm's equity] are to 'go public', or to merge the firm with another large one. **1972** *Accountant* 5 Oct. 417/1 It..disregarded the probability that the company would in the near future 'go public'. **1976** 'A. HALL' *Kobra Manifesto* xv. 211 The girl's fever..had either driven or panicked Kobra into the open and in seizing the Boeing they'd gone public. **1977** *Lebende Sprachen* XII. 158/2 This will see the Arabs go public with the new second stage of their economic strategy.

5. a. (Further examples.)

1762 W. SMITH *Discourses Publ. Occasions* (ed. 2) App. 113 This attention to public speaking, which is begun here [in the College of Philadelphia] with the very rudiments of the mother-tongue, is continued down to the end. **1780** Public speaker [see SPEAKER 1b]. **1825** T. MOORE *Mem. Life R. B. Sheridan* x. 322 And, in this great essential of public speaking, must be considered inferior to [etc.]. **1905** G. B. SHAW *Let.* 27 Nov. (1972) 583, I do not know yet exactly how you get your effects, except that it is not in my rather rhetorical, public-speaker kind of way. **1931** *Economist* 28 Mar. 665/2 Mr. Morrison's final conclusion that the proper authority to be set up is a 'business Board' of five members, incorporating what he defined as 'an element of public accountability'. **1940** C. MILBURN *Diary* 25 Dec. (1979) 76 We..listened to our beloved King's speech... How bravely he overcomes the difficulties of public speaking. **1950** B. PYM *Some Tame Gazelle* xxi. 232 She was a confident public speaker and this afternoon's audience of parish women.. held no terrors for her. **1959** *Observer* 18 Oct. 24/7, I am sure she disapproved of the new euphemism: public accountability. **1965** *Mod. Law Rev.* XXVIII. v. 520 Judges..had a near monopoly of the Chairmanships of Royal Commissions, indeed of public inquiries of every sort. **1965** *Times Lit. Suppl.* 25 Nov. 1059/1 The public-speaking Dr. Rosten points a neat moral. **1971** WRAITH & LAMB *Public Inquiries* i. 13 Public inquiries are constituted *ad hoc* to inquire into particular matters, and are for the most part concerned only to establish facts and to make recommendations. **1974** *Times* 17 Nov. 14/2 The important issue is that of public accountability. A body [*sc.* the BBC] which gets all its funds from the public ought to be obliged to answer any question from anyone about how the money is spent. **1975** *Oxf. Compan. Sports & Games* 1112/2 A wrestler at fault is given a warning and if he offends again he is given a public caution. *Ibid.* 1113/1 Should the same wrestler offend again he is given a second public warning. **1976** *Abingdon Herald* 9 Dec., With, it seemed, two men in the ring against him and only the whole crowd for him, Marino gave vent to justifiable retaliation and received the first public warning of the evening. **1979** *Jrnl. R. Soc. Arts* July 511/1 Mr. McWilliam knows all too well that digs at buildings in volumes so important can have a devastating effect at Public Inquiries, and can cause interesting buildings to fall victim to the destroyers.

e. *public address system*: a system comprising microphone, amplifier, and loudspeaker which enables speech or music to be projected to an assembly of people; so *public-address equipment.*

1923 *Electrical Communication* I. IV. 46 Public address systems..developed for the purpose of extending the range of the voice of a speaker addressing an audience. **1950** *Engineering* 24 Nov. 392/2 Public-address equipment using magnetic tapes for announcing and recording..has been installed. **1972** *Police Rev.* 8 Dec. 1597/3 A public address system fitted to each vehicle. **1978** R. V. JONES *Most Secret War* xxi. 176 The effect would be rather like that which occurs in public address systems where the noise from the loudspeakers impinges on the original microphone, and is therefore picked up and relayed back to the loudspeakers again.

f. *Public Lending Right*: the name given by its advocates to authors' (and publishers') entitlement to a fee for books borrowed from public libraries.

1961 *Ann. Reg. 1960* 458 Sir Alan Herbert..celebrated his seventieth birthday by opening a campaign for what came to be known as the Public Lending Right. **1970** *Guardian* 6 Apr. 8/1 His..description on the ballot paper could well be..Ardent Advocate of Public Lending Right... Of the dust-gathering plan to give authors and publishers a bit of the royalty for the books we borrow from the public libraries. **1977** *Time Out* 17–23 June 14/1 Combined with the Public Lending Right, the author would be helped twice over, more sales and more loans.

g. *public-access*: used *attrib.* to designate a form of television in which the general public can produce or contribute to programmes. *U.S.*

1972 *Listener* 6 July 1 Public-Access Television. **1976** *National Observer* (U.S.) 18 Dec. 1/1 All this appeared in recent weeks on a New York City 'public access' cable-TV channel that serves 85,000 families in Manhattan.

6. a. (Further examples.)

1611 CORYAT *Crudities* 205 The Duke sat about the publicke affaires with the other Senators. **1673** J. RAY *Observations Journey Low-Countries* 170 Those who assist the Commonwealth..have liberty granted to them to be present in this Council, and to understand the management of public affairs. **1822** *Sunday Times* 20 Oct. 1/3 (*heading*) Aspect of public affairs. **1937** *Burlington Mag.* Feb. 94/2 The management of public affairs.

7. (Further examples.) Also *transf.*

1932 H. NICOLSON (*title*) Public faces. **1961** Public image [see *IMAGE *sb.* 5 b]. **1962** *Listener* 1 Mar. 366/2 Finally, there is the question of Egypt's public face, the face which she presents to the world. **1967** M. ARGYLE *Psychol. Interpersonal Behaviour* ix. 154 There may be secondary aims [in an assessment interview], such as giving C [*sc.* the candidate] information about the job, or improving the public image of the employing organization. **1976** P. FERRIS *Detective* iii. 46 I'm not really on the board... I was only there today to report on our public image. **1977** C. STORR *Tales Psychiatrist's Couch* 8 Although she was still a difficult woman in her private relationships, she acquired a much easier public face.

II. 9. a. *public enemy* (earlier and later examples); now esp. (passing into sense 2 f above) in *public enemy number one* (orig. U.S.), the first named on a list of wanted criminals; the greatest threat to a community; also *transf.* and in extended use in similar phrases; *public law*: see also sense *2 h.

1756 G. WHITEFIELD *Short Address* 9 We may as lawfully draw our swords, in order to defend ourselves against our common and public Enemy. **1845** Public enemy [see ENEMY *sb.* 1]. **1931** *S. F. Call* (Mag.) 18 June 8 There are two people alive, at least, who love Al Capone—his wife and his kid. Public Enemy No. 1 is to them an idol. **1935** *Daily Mail* 23 Oct. 18/1 (Advt.), We all know who is Public Enemy No. 1 when it's time to get up! The 'Droops'. **1939** LD. CAMROSE in M. Gilbert *Winston S. Churchill* (1976) V. liii. 1081 Well, Winston was Public Enemy No 1 in Berlin, and Eden was the same in Italy. **1940** 'N. BLAKE' *Malice in Wonderland* II. ix. 123 The presence of a public enemy in our midst. **1958** *Listener* 17 July 75/1 Iraq..has been Public Enemy No. 1 to Egypt's propagandists. **1967** M. MURRAY *Ballad of Bonnie & Clyde* (song), 3 Bonnie and Clyde got to be public enemy number one—Running and hiding from ev'ry American lawman's gun. **1978** *Jrnl. R. Soc. Arts* CXXVI. 422/2 In some cities, the car has almost come to qualify for the title of public enemy number one. **1980** *Guardian* 19 Dec. 22/1 The fugitive Irishman..was 'a prominent member of the IRA and a public enemy'.

III. 10. *public-minded* adj. (examples).

a **1706** EVELYN *Diary* an. 1691 (1955) V. 61 This church.., being begged by Dr. Tenison Rector of St. Martines, was set up by that publique minded, charitable & pious Doctor neere my sons dwelling Dover-streete. **1976** *West Lancs. Even. Gaz.* 13 Dec. 6/5 A North Shore woman was particularly public-minded after buying some wrapping paper from a Cleveleys shop recently. She thought the paper might have been a fire hazard and took it round to Blackpool fire station for them to test it.

B. *sb.* **2. b.** With preceding possessive. The particular section of society which is sympathetic to the person or thing indicated.

1921 H. CRANE *Let.* 19 Sept. (1965) 64, I am 'sold out' and will have to rush rhymes and rhythms together to supply my enthusiastic 'public' as fast as I can. **1952** GRANVILLE *Dict. Theatrical Terms* 145 My *public* will hate me in this part.

c. *Sociol.* A collective group regarded as sharing some cultural, social, or political interest but who as individuals do not necessarily have any contact with one another.

1927 J. DEWEY *Public & its Problems* ii. 39 There are associations which are too narrow and restricted in scope to give rise to a public, just as there are associations too isolated from one another to fall within the same public. **1933** F. H. ALLPORT *Institutional Behav.* v. 87 Since the public is no specific group of individuals, but is defined wholly by the range of the common interest in a particular transaction, there may be a separate public for every issue raised. We are compelled, therefore, to think of *various* publics. **1954** G. A. LUNDBERG et al. *Sociol.* xiii. 491 A great source of difficulty has been the varied and confused image of the term 'public'. Clearly one may belong to as many publics as one has interests. **1954** GERTH & MILLS *Character & Social Structure* xv. 435 Publics are composed of people who are not in face to face relation but who nevertheless display similar interests, or are exposed to similar, although more or less distant, stimuli. **1969** G. A. & A. G. THEODORSON *Mod. Dict. Sociol.* 324 Publics are usually large, physically separated, and often quite diverse... Publics have an impact through their voting, buying, noncooperation, financial contributions, letters to the editor, etc.

4. c. Short for 'public bar'. *colloq.*

1957 N. MARSH *Off with his Head* ii. 29 The bar-parlour at the Green Man..lay at right angles to the Public. **1969** M. DUFFY *Wounds* i. 19 The pints of beer she had to pull for the pensioners in the public. **1971** L. LAMB *Worse than Death* vii. 64 They had finished doing the bars, and..were having a cup of tea in the Public.

pu·blically (-ikli), *adv.* [f. PUBLIC *a.* + -AL + -LY[2].] = PUBLICLY *adv.*

1920 E. SITWELL *Bath* i. 20 Goldsmith adds that 'the Masters, struck with such an uncommon instance of good nature, publically thanked him for his benevolence'. **1963** W. SELLARS *Sci., Perception, & Reality* 364 (Index), Publically observable. **1972** D. HOLBROOK *Pseudo-Revol.* viii. 140 Individuals leaping on to the stages of Danish sex clubs and publically copulating. **1974** GAGNON & SIMON *Sexual Conduct* ii. 56 The publically valued institution of marriage. **1977** *Grimsby Even. Tel.* 13 May 2/4 It was sometimes hard for parents to accept publically that their children had mental or physical difficulties.

publication. Add: **2. c.** *attrib.*, as *publication date, day.*

1931 H. CRANE *Let.* 13 June (1965) 373 Will you tell me something about its publication date, etc. **1976** M. HINXMAN *End of Good Woman* xiii. 168 Sometimes I think it's a plot on the part of the publishers to make sure no one can salvage the books from the packing to review them in time for the publication date. **1888** 'MARK TWAIN' *Lett. to Publishers* (1967) 250 No notice should appear before publication-day. **1979** M. RUSSELL *Touchdown* I. 6, I felt an urge to demand how many previous stacks had melted away since publication day.

publicist. Add: **3.** A press or publicity agent.

1930 *Oxford Times* 4 Apr. 7/4 This is the experience of Sir Charles Higham, the famous publicist, who celebrated his 21st anniversary as an advertising agent in Fleet-street on Wednesday. **1942** *Sun* (Baltimore) 4 Nov. 9/4 Most of the testimony..was given by Paul H. De Kruff, bacteriologist and medical publicist. **1969** R. BLYTHE *Akenfield* ii. 60 Well-printed signs of expert publicist talents being employed to disseminate the new *caritas.* **1977** *Times* 12 Feb. 12/6 Mr William Camp's appointment as publicist of the railways... He and his company..will apparently become the overlords of BR publicity.

publicitor (pʊbli·sɪtə̆ɪ). *U.S.* [f. PUBLICIT(Y + -OR.] A press or publicity agent.

1936 *Daily News* (Chicago) 24 Apr. 30/4 But in this solicitous telephonery we detected the fine Italian hand of the publicitor. **1951** *Lincoln* (Nebraska) *Jrnl.* 8 Dec. 6 O—..who rose from the job of sports publicitor..was assailed by the Judge. **1955** *Sun* (Baltimore) 15 Dec. (B ed.) 25/5 'The fancy basketball is out unless we're way on top,' Rod Hundley was saying. The publicitors call him Hot Rod. **1962** [see *COOTIE *sb.*[2]].

publicity. Add: **1.** (Further examples.) Also, public notice; the action or fact of making someone or something publicly known; the business of promotion or advertising; an action or object intended to attract public notice; material issued to publicize.

1826 J. J. AUDUBON *Jrnl.* 16 Dec. (1897) I. 186 Mrs. Rathbone, Senior, refused me the pleasure of naming a bird after her, on account of the publicity, she said. **1842** DICKENS *Let.* 30 Apr. (1974) III. 222, I found the documents of which the inclosed are copies... You will see that they are signed by the first writers in England; and that their object..is *Publicity.* **1847** A. BRONTË *Agnes Grey* xiv. 222 If you add to it by giving publicity to this unfortunate affair, or naming it *at all*, you will find that I too can speak. **1851** *Times* 28 Oct. 5/1 We have reported his proceedings..at greater length than any of the journals devoted to his cause, and we have thus secured to him all the publicity we had it in our power to bestow. **1917** *Electric Railway Jrnl.* 6 Jan. 17/1 Continuous publicity of good work would have softened a public irritation in the day of trouble. **1936** *S. P. E. Tract* XLV. 185 The prejudice that many people feel against advertising is likely to be dispelled if they can be made to believe that it is no more than *publicity.* **1946** E. O'NEILL *Iceman Cometh* I. 38 Why, even at Harvard I discovered my father was well known by reputation, although that was some time before the District Attorney gave him so much unwelcome publicity. **1953** A. HUXLEY *Let.* 21 Dec. (1969) 692 Osmond sent me a copy... I have taken the liberty of forwarding it to Cass Canfield of Harpers, who may like to quote from it in his publicity. **1981** S. RADLEY *Chief Inspector's Daughter* ii. 17 'How come she's featured in a magazine, anyway?' 'Publicity. She's a romantic novelist.'

2. *attrib.* and *Comb.*, as *publicity agent, boy, bureau, drive, expert, film, hand-out, hound, man, manager* (so *-manage* vb. trans.), *-monger, officer, people, ramp, screen, section, stunt, value, woman, worker; publicity-hunting, -seeking* vbl. sbs. and ppl. adjs.; *publicity-loving* ppl. adj.; *publicity-conscious, -minded, -ridden* adjs.; *publicity-wise* adv.; (see also Dict.).

1911 J. C. LINCOLN *Cap'n Warren's Wards* xi. 180 He and his friends needed a representative on the press—a publicity agent, so to speak. **1922** R. G. COLLINGWOOD tr. *Croce's Aesthetic* II. vi. 244 Recognizing the great gifts of Meier as publicity-agent. **1955** W. GADDIS *Recognitions* I. v. 180 The publicity agent looks it over and signs her name to it. **1929** G. ADE *Let.* 8 Dec. (1973) 142 The publicity boys seemed to think it was a great joke to float these wild-eyed stories about my pursuing Dorothy Tennant, [etc.]. **1962** 'M. INNES' *Connoisseur's Case* iii. 33 Tourists sure to come along when the publicity boys do their stuff. **1907** U. SINCLAIR *Industr. Republic* 142 He had an army of experts to help him..skilful lobbyists, newspapers and publicity bureaus. **1959** F. NEWTON *Jazz Scene* v. 74 A publicity-conscious American impresario could advertise a plan to recruit an 'international' orchestra. **1977** H. INNES *Big Footprints* III. i. 217 He had agreed to our taking a camera... Extraordinary how publicity-conscious these men were. **1935** Publicity drive [see *copy-writer]. **1915** *Truth* 4 Aug. 196/1 There are, as one publicity expert puts it, many shirkers doing remarkably well. **1969** *Morning Star* 25 Mar. 4 There has been an increase in short publicity films. **1973** *Screen* Spring/Summer 227 He [*sc.* A. Alexeieff]..made several commercials and publicity films. **1927** S. BENT *Ballyhoo* iii. 86 Washington is this country's premier city of publicity hand-outs. **1979** A. WILLIAMSON *Funeral March for Siegfried* x. 46 York..asked if they had a publicity hand-out on the singers in *The Ring.* **1928** Publicity hound [see *HOUND *sb.*[1] 4 e]. **1936** 'J. TEY' *Shilling for Candles* vi. 62 And they say professional people are publicity hounds! **1971** 'E. LATHEN' *Longer the Thread* (1972) viii. 75 Annie was certainly no publicity hound. **1963** *Times Lit. Suppl.* 17 May 353/2 All the vices from publicity-hunting to jobbery. **1976** C. BERMANT *Coming Home* II. vii. 216 Lord Longford was..regarded as a publicity-hunting fraud. **1938** *Amer. Speech* XIII. 59 A field of education that has suffered much from publicity- and money-loving quacks. **1907** J. C. LINCOLN *Old Home House* 37 You two can be proprietors and treasurers if you want to. But active manager and publicity man—that's yours cheerily, Peter Theodosius Brown! **1932** *Sunday Express* 3 July 9/1 Soon Hammerstein left me with his publicity man, a human talking machine. **1957** *Times Lit. Suppl.* 25 Oct. 638/2 No one, not even a modern film star hounded by his publicity man, has ever courted popular favour more devotedly than Louis Philippe. **1978** I. B. SINGER *Shosha* III. 50 What kind of publicity man will I make, anyway? **1952** L. DURRELL *Let.* 14 Nov. in *Spirit of Place* (1969) 114 My boss Eden whose much advertised tour was publicity-managed by little me. **1908** *Mod. Business* Aug. 86/2 *Publicity Manager.*— Can formulate and carry through an advertising campaign. **1931** 'G. TREVOR' *Murder at School* iii. 49 The rôle of 'writer-up' and general publicity manager. **1974** *Listener* 17 Jan. 66/2 Lady Glencora Palliser..settles down to become..her husband's self-appointed publicity manager and hatchet-woman. **1962** *Observer* 11 Mar. 8/5 I'm a very publicity-minded chap. **1974** *Times* 28 Nov. 17/7 The poor image of universities..[is] emanating in large part from a publicity-minded student minority. **1951** N. MITFORD *Blessing* II. xi. 251 Charles-Edouard's heir was on the way to becoming a publicity-monger. **1962** *Economist* 15 Sept. 1017/1 The publicity-mongers of the various political parties. **1933** *Whitaker's Almanack* 284/2 Colonial Office... Publicity Officer, C. Becker-Platt.. £666. **1966** 'H. MACDIARMID' *Company I've Kept* viii. 186, I..spent a year in Liverpool as Publicity officer for the Liverpool Organization. **1932** *New Yorker* 14 May 44/2, I find..that all those competent publicity people in the shops around town can be counted on to call me up. **1933** DYLAN THOMAS *Let.* Oct. (1966) 31 Patriotism is a publicity ramp organised by holders of excess armament shares. **1961** *Times* 7 Jan. 9/7 A publicity-ridden world. **1911** R. D. SAUNDERS *Col. Todhunter* ix. 128 This amazing projection of himself..on the publicity 'screen' of a newspaper's front page appalled Colonel Todhunter. **1927** F. L. C. FLOUD *Ministry of Agric.* ii. 25 There is also a small publicity section for the issue of information to the Press. **1935** DYLAN THOMAS *Let.* July (1966) 159, I hope to have..a little free love offstage from any publicity-seeking actresses I can find. **1939** 'N. BLAKE' *Smiler with Knife* v. 90 Charges of publicity-seeking. **1966** 'L. LANE' *ABZ of Scouse* Forewd., Be he a dollar-spending tourist or publicity-seeking politician. **1926** 'SAPPER' *Final Count* vii. 195 It was just an advertisement—an elaborate publicity stunt. **1972** P. DICKINSON *Lizard in Cup* xi. 176 The theft of the pictures..being, as Nancy had hinted, a publicity stunt. **1922** FARJEON & HORSNELL *Advertising April* i. 19 The other woman. My successor. What's her name? What's her publicity value? **1978** J. PEARSON *Façades* viii. 149 Beaverbrook..knew the publicity value of controversy. **1958** *Punch* 1 Jan. 50/1 Publicitywise, Sabrina and the Duke of Bedford consolidated their respective resources. **1962** *Listener* 11 Jan. 102/2 Dr. Crick, who publicity-wise appeared the king-pin, was now announced as the colleague (or perhaps subordinate) of Dr. Brenner. **1958** A. HUXLEY *Let.* 22 June (1969) 851 There have been endless contretemps, including, as a last straw, the collapse of the publicity woman with, of all things, chickenpox. **1926** *Amer. Speech* I. 480/1 The public is worked on by a 'publicity worker', assisted sometimes by a person who has the 'technique' of describing 'personal interest' incidents.

publicize (pʊ·blisəiz), *v.* [f. PUBLIC *a.* (*sb.*) + -IZE.] *trans.* To bring to the notice of the public; to make generally known; to advertise.

1928 *Weekly Dispatch* 20 May 14/4 Nowadays the potential star has to be managed and publicised. **1938** M. BRINIG *May Flavin* iv. 363 In my present position it wouldn't do any good to publicize these things. **1943** J. S. HUXLEY *TVA* iv. 25 One of the original aims of the TVA, and one much publicized in its earlier years. **1967** N. FREELING *Strike Out* 78 I'm not in the habit of publicising my private life. **1976** *Guardian* 21 Apr. 6/1 The board does not adequately publicise itself.

Hence **publiciza·tion; pu·blicized** *ppl. a.*, **pu·blicizing** *vbl. sb.*

1932 *New Yorker* 11 June 40/1, I hear that all the.. manufacturers who aren't doing business intend to spend the Summer on that publicized isle [*sc.* Majorca]. **1956** A. H. COMPTON *Atomic Quest* 303 These much-publicized burns have had effects that are no more tragic than those resulting from other forms of modern weapons. **1958** *Jrnl. Amer. Water Works Assoc.* L. 1057/2 Of more value ..will be accurate reporting—even publicizing—of the physical, chemical, and biological difference between the first water, the used supply, and the reclaimed water. **1966** 'H. MACDIARMID' *Company I've Kept* iv. 128 All the disgusting publicisation of our Scottish Queen Mother.

publicly, *adv.* **2.** (Further examples.)

1928 *Britain's Industr. Future* (Liberal Industr. Inquiry) II. vi. 82 We hope that neither thought nor money will be spared to furnish this country with a publicly-owned system of electrical supply. **1969** *Listener* 28 Aug. 267/1 And can capitalism..achieve the rate of growth that planning in a really publicly-owned economy has achieved? *a* **1974** R. CROSSMAN *Diaries* (1976) II. 460 He was hoping to have a properly planned publicly-owned New Town like the others. **1980** *Daily Tel.* 16 Jan. 23/2 The first is a publicly-held corporation chartered by Congress.

public opinion: see OPINION *sb.* in Dict. and Suppl.

public relations. [RELATION *sb.* 6.] **a.** (The establishment or maintenance of) relations, esp. a good relationship, between an organization, firm, etc., and the general public; (also const. *sing.*) the art or practice of establishing or maintaining such good relations; *transf.*, a department or group with responsibility for relations with the public.

1807 T. JEFFERSON *Writings* (1854) III. 89 Questions calling for the notice of Congress, unless indeed they shall be superseded by a change in our public relations now awaiting the determination of others. **1913** *Electric Railway Jrnl.* 16 Oct. 829/1 Effective publicity to deal with questions of public relations and to consider the molding of public opinion by the presentation of real facts. **1917** *Ibid.* 6 Jan. 17/1 This adviser in public relations—for such a man should be far more than a mere publicity agent— should constantly study the temper of the public mind. **1933** *Planning* I. xiv. 5 Public relations may be defined as covering all the contacts between an official organisation and all other bodies with which it may have to deal, directly or indirectly, otherwise than through internal administrative channels. **1943** J. D'ARCY-DAWSON *Tunisian Battle* i. 21 Public Relations were now the possessors of six Simcas. **1945** *Manch. Guardian* 8 Aug. 4/6 The Central Insurance building, now headquarters of Public Relations in Italy. **1958** *Spectator* 8 Aug. 201/2 Some of our more 'enterprising' Cathedral cities have invented a new form of spiritual public relations, which is the Industrial Harvest Festival. **1961** *Observer* 19 Mar. 3/4 He..founded his own firm, first in public relations, then in advertising. *a* **1974** R. CROSSMAN *Diaries* (1976) II. 547 Wedgy is brilliant at public relations.

b. *attrib.*

1923 E. L. BERNAYS *Crystallizing Public Opinion* I. 12 To some the public relations counsel is known by the term 'propagandist'. **1929** [see *HAND-OUT 3]. **1933** *Times* 2 Sept. 15/5 The post of Public Relations Officer in the Post Office is a new one. **1937** *Rep. Proc. 14th Conf. ASLIB* 72, I asked him if he could help me in finding a nice crisp definition of the duties of a Public Relations Officer. He replied, 'His principal function is interpreting his undertaking to the public and the public to his undertaking.' **1944** A. JACOB *Traveller's War* xii. 212 Publicity which verged ever so slightly upon the personal was regarded as improper by the Public Relations department. *Ibid.*, The Public Relations people had asked war correspondents to go slow on stories about Rommel because we had made him into quite a bogey man. **1956** in W. H. Whyte *Organization Man* (1957) I. iii. 26 Every practising public-relations man is an engineer too—a *social engineer.* **1958** M. ARGYLE *Relig. Behaviour* v. 54 There was an elaborate public relations campaign before the meetings, by posters and other publications, by a film, and via the churches. **1961** *Times* 10 Aug. 2/4 (Advt.), Public Relations Officer..experienced in Public Relations work, both with the press and within an organization. **1967** G. F. FIENNES *I tried to run Railway* vi. 67 Before the new timetables I set out to do a public relations job. **1970** G. GREER *Female Eunuch* 151 The public relations experts seek to attract girls to nursing by calling it the most rewarding job in the world. **1976** *Church Times* 30 July 9/2 The public-relations officer improves 'image' but is reluctant to accept quantitative assessments of his own impressions of how much he has done so. **1977** *New Yorker* 19 Sept. 98/2 Mrs. Berner also hired the public-relations firm of Gurtman and Murtha to publicize the contest.

public school. Add: **1. b.** *public school accent, attitude, boy* (examples; hence *-boyish* adj.), *code, English, girl, product, spirit* (further example), *tie, type; public school-bred, -educated* adjs.; also appositive, as *public-school–Oxford* attrib.; also passing into *adj.*

1914 'I. HAY' *Lighter Side School Life* viii. 207 That is the Public School Attitude in a nutshell. *Ibid.* 220 We note a new factor in the composition of the Public School Type—the military factor. **1930** (*title*) The diary of a public school girl. **1930** E. WAUGH *Vile Bodies* x. 174 Knowledgeable young men with..old public-school ties. **1931** D. L. SAYERS *Five Red Herrings* v. 55 Waters.. spoke standard public-school English. **1933** *Granta* 26 Apr. 370/1 It is a movement for unemployed graduates and Public School boys. **1936** J. BUCHAN *Island of Sheep* I. ii. 21 The kind of son I had hoped for..was..the kind of public-school product you read about. *Ibid.* 22 Peter John..didn't care a rush for the public-school spirit. **1938** M. ALLINGHAM *Fashion in Shrouds* x. 162 'He's not in there,' said Jimmy, revealing a stammer and a public-school accent. **1943** M. SCHLAUCH *Gift of Tongues* 264 English writers themselves have sometimes jeered at certain details of 'public school-Oxford' speech as being affected. **1943** F. THOMPSON *Candleford Green* ii. 29 His highly-pitched, public-schoolboyish accent. **1946** P. BOTTOME *Lifeline* i. 10 He was an Eton master, and the Public School code..he believed in. **1952** KOESTLER *Arrow in Blue* xxviii. 263 Public School-bred Foreign Office diplomats like Guy Burgess and Donald MacLean. **1962** *Times* 27 Feb. 13/2 The public-schoolboyish professional Roman soldier. **1962** R. WILLIAMS *Britain in Sixties: Communications* v. 101 In fact 'public-school English'..cannot now become a common speech-form in the country as a whole: both because of the social distinctions now associated with its use, and because of the powerful influence of American speech-forms. **1966** J. CLEARY *High Commissioner* vii. 136 A veneer of public school accent had been laid over the gravel in his voice. **1967** *Listener* 30 Nov. 694/3 'The English public schoolboy', the prototype of unimaginative disciplined conformity. **1971** HALSEY & TROW *Brit. Academics* xv. 421, 41 per cent of the public school educated teachers place themselves on the Left. **1973** 'D. JORDAN' *Nile Green* ii. 12 It's one of the curses of the English public school girl that she never believes in making the bed first thing in the morning. *a* **1974** R. CROSSMAN *Diaries* (1975) I. 131 While she is extremely tough in negotiation she is extremely public school when she's asked..to make a sacrifice in her departmental interests for the good of the nation or for the convenience of the Civil Service. **1975** J. I. M. STEWART *Gaudy* iii. 41 They were entirely English and very public-school. **1975** 'D. JORDAN' *Black Account* xx. 108 One of those public-school ties which carry important messages between all true Englishmen. **1978** CADOGAN & CRAIG *Women & Children First* v. 95 Eighteen-year-old Raleigh..dies still behaving in accordance with the public-school code.

3. (Examples referring to Canada and Australia.) Occas. applied to schools in other countries.

1789 *Nova Scotia Mag.* I. 80/1 This seminary is erected in consequence of a law of this Province, intitled, 'An Act for establishing a public school, in the town of Halifax'. **1872** *Canadian Monthly* June 483/1 Public Schools..are distinguished from those which until recently were entitled Grammar Schools, and were intended to afford instruction in the elements of the classical languages as well as the mother tongue. **1901** 'M. FRANKLIN' *My Brilliant Career* iii. 14 My parents received an intimation from the teacher of the public school..to the effect that the law demanded that they should send their children to school. **1965** *Austral. Encycl.* VIII. 23/1 In Victoria (as in Tasmania) 'public school' generally carries its English meaning... In New South Wales..in 1892..'public school' then meant to most people 'State primary school'. **1968** *Globe & Mail* (Toronto) 17 Feb. 45 (Advt.), Not one street to cross to get to public school. **1972** *Mainichi Daily News* 7 Nov. 5/1 Students in Kanagawa Ken, both Japanese and foreign from public and private schools. **1976** D. HEFFRON *Crusty Crossed* xi. 79 In Big Point, there was nothing to fuss over. Only one school to go to until you passed grade eight. They called that kind of school a public school, which in England was the name for a private school which your parents have to pay quite a lot of money to send you to.

Hence **public-schoo·lish** *a.*, characteristic or suggestive of a public school; **public-schoo·lishness; public-schoo·ly** *a.* = *PUBLIC-SCHOOLISH *a.*

1930 *Observer* 22 June 13 Mr Leslie Mitchell, as the simple Andy, is too public-schoolish in tone and manner. **1930** A. HUXLEY *Let.* 14 June (1969) 337 English literary criticism for the moment is all for being nice and gentlemanly and public-schooly. **1947** 'G. ORWELL' *Eng. People* 39 Many necessary abstract words..are rejected by the working class because they sound public-schoolish, 'tony' and effeminate. *a* **1960** E. M. FORSTER *Maurice* (1971) I. ix. 48 During the previous term he had reached an unusual level mentally, but the vac had pulled him back towards public-schoolishness.

publish, *v.* Add: **I. 2. b.** For *obs.* read '*obs.* exc. *U.S. dial.*' and add later examples with persons as obj.

1886 P. STAPLETON *Major's Christmas* 124 Then say you will marry me, and we will be published to-day. **1975** *Budget* (Sugarcreek, Ohio) 20 Mar. 8/3 Published today in above district were Sam, son of Joe J. Yoders and Mary, daughter of Eli H. Weavers. Their wedding to be Saturday, April 5.

4. a. (Examples with ref. to gramophone records.) Also *absol.* Also by metonymy, with author, etc., as obj.

1918 C. S. LEWIS *Let.* 27 Oct. (1966) 45 He [*sc.* Heinemann] told me that John Galsworthy (who publishes with them) had seen my MS. **1937** J. SQUIRE *Honeysuckle & Bee* 203 Lane..seemed to publish almost all the exciting new authors. **1941** *Sphere* 6 Dec. 361/1 If a publisher chooses to publish you, your reward is almost certain. **1952** SACKVILLE-WEST & SHAWE-TAYLOR *Record Year* 213 It looks like a re-issue of the recording published by Decca during the war and since deleted. **1961** H. M. SILVER in *Webster* s.v., Pressure put on faculty members.. to publish as a condition of appointment or promotion. **1970** J. EARL *Tuners & Amplifiers* vi. 140 Test records are available for channel identification or recognition. Such a record is the excellent HFS69 published by the Haymarket Publishing Group 1971 *Black Scholar* Dec. 23/2 Dr. Ladner is frequently published in professional journals.

b. (Further examples.) *spec.* to make generally available a description or illustration of (an archæological find, a work of art, etc.).

1931 *Oxf. Mag.* 18 June 888/2 H. R. Hall publishes an Egyptian axe in the British Museum. 1968 *Listener* 31 Oct. 580/2 (*caption*) Are you from the BBC? If so it is your duty to 'publish' an open-ended political statement I intend to deliver. 1973 *Oxf. Univ. Gaz.* CIII. Suppl. v. 8 We are grateful to Professor Ashmole, who will publish the head, for kindly consenting to its illustration in this report. 1975 *Times Lit. Suppl.* 10 Oct. 1196/3 The great bronze doors of medieval Europe..are on the whole well known and fully published... Walter Cahn has now published, in *The Romanesque Wooden Doors of Auvergne*, a series of five sets of wooden doors, still extant though considerably damaged. 1976 *Nature* 1 Apr. 415/2 Unfortunately, we cannot know the V_P/V_S ratios in the decreasing stage, because of the scarcity of the data published by the Japan Meteorological Agency before 1950. 1978 *SLR Camera* Sept. 53/3, I wasn't quite sure what I was going to do with the films I shot on the way round but if I wanted to get any published I knew I had to get them back home while the race was still in progress. 1978 *Times Lit. Suppl.* 1 Dec. 1392/3 There are papyri in Greek and Latin (the Egyptian papyri are somewhere else entirely), ostraca (about 4,000 inscribed potsherds, of which only 300 have been published).

c. *intr.* in passive sense. To come into public circulation; to be published.

1928 *Public Opinion* 6 Apr. 325/1 The newspapers do not publish on Good Friday. 1972 *Evening Telegram* (St. John's, Newfoundland) 24 June 1/1 The Evening Telegram will publish Monday, June 26 which is being observed as Discovery Day in Newfoundland.

publishability (pɒbliʃăbi·lĭti). [f. PUBLISHABLE *a.*: see -ILITY.] The quality of being publishable; suitability for publication.

1870 G. H. LEWES *Let.* 8 May in *Geo. Eliot Lett.* (1956) V. 94 When you have read them and decided as to their *publishability* in the Atlantic perhaps you will make me the channel of an offer. 1969 C. DERRICK *Reader's Report* 17 'Success'..means for the most part mere publishability; we are not concerned with the heights. 1974 *Jrnl. Social Psychol.* XCIV. 302 The printed form of the report was rated significantly higher..on the question of publishability than the two mimeographed forms.

publisher. Add: **2. b.** Also *spec.* a newspaper proprietor (*U.S.*).

1911 *Springfield* (Mass.) *Weekly Republican* 6 Apr. 11 Connecticut Publisher Dead. F. R. Swift, owner of the Bridgeport and Waterbury Herald, died last week. 1920 LD. NORTHCLIFFE *Let.* May in W. F. Johnson *George Harvey* (1929) xxxvii. 383 Misunderstandings due to different meanings of the same words are among the basic difficulties of Anglo-American relations. Just now there is a discussion about the price of paper in which the use of the word 'publisher' for 'newspaper owner' confuses our people. 1949 *Manch. Guardian Weekly* 3 Nov. 2/2 Publishers and big advertising agencies in New York..are laying off salesmen, art staff, and layout men. 1974 *Lebende Sprachen* XIX. 39/1 US publisher (of a newspaper)—BE (newspaper) proprietor. Zeitungsverleger (und besitzer).

4. b. publisher's (or **publishers'**) **binding**, a uniform binding provided for an edition of a book before it is offered for sale; **publisher's** (or **publishers'**) **cloth**, a publisher's binding in which cloth is used as the covering material.

1901 D. COCKERELL *Bookbinding* i. 20 For a permanent publisher's binding, something like that recommended for libraries..is suggested. 1924 M. SADLEIR in *Bookman's Jrnl.* Feb. 154/1 It is obvious that no 'publisher's binding', in the accepted sense of lettered durability, was known [in the eighteenth century]. 1928 E. P. GOLDSCHMIDT *Gothic & Renaissance Bookbindings* I. 35 If the bindings with publishers' names or marks are 'original publishers' bindings'..then surely all such bindings must contain books published by the man who signed the binding. 1974 R. MCLEAN *Victorian Publishers' Book-Bindings* 7 This is a picture book showing the richness of publishers' bindings principally in cloth or leather, produced in Britain during the nineteenth century. 1921 T. J. WISE *Bibliogr. Writings J. Conrad* (ed. 2) 1. 44 If only collectors would refrain from purchasing copies of Conrad's books unless they are in the original publishers' cloth,..their position would be perfectly safe. 1935 J. CARTER (*title*) Publisher's cloth: an outline history of publisher's binding in England 1820–1900. 1972 P. GASKELL *New Introd. Bibliogr.* 246 The more fanciful styles of publishers' cloth included gilt blocking over the whole area of the covers.

publishing, *vbl. sb.* **2.** (Further *attrib.* examples.)

1862 R. H. NEWELL *Orpheus C. Kerr Papers* 1st Ser. 380, I'm agent for the great American publishing house of Rushem & Jinks. 1881 'MARK TWAIN' *Lett. to Publishers* (1967) 143 A few days before Canadian publishing-date. 1929 H. CRANE *Let.* 26 Feb. (1965) 339 A marvelous de luxe publishing establishment here. 1937 *Discovery* Mar. 80/1 A book recently published in Moscow by the Biological-Medical Publishing House. 1958 *New Statesman* 1 Feb. 136/3 Within a matter of months every single Yiddish theatre, publishing-house, magazine..etc was closed down. 1978 *Maledicta* I. II. 327 By 'year' we mean the 'publishing year' which is not identical with the 'calendar year'.

pubsy (pʌ·bzi), *a. colloq.* [f. PUB *sb.* + -SY.] Characteristic or suggestive of a public house.

1966 *New Statesman* 30 Sept. 489/1 But the most attractive numbers are in that easy, pubsy English style which suggests Bud Flanagan, Tommy Steele and Julian Slade arm-in-arm, just strolling, lazily flexing their knees and getting nowhere. 1977 *Listener* 7 Apr. 447/3 Return briefly to the pubsy, Fitzrovian atmosphere of Lord Longford's..party.

pucca, var. PUCKA, PAKKA.

Puccinian (putʃi·niăn), *a.* [-AN.] Of, pertaining to, characteristic of, or resembling the works of the Italian operatic composer Giacomo Puccini (1858–1924).

1942 *Scrutiny* XI. 74 We have an exquisitely sensitive line which, the antithesis of Puccinian or Wagnerian hysteria, is almost French and Chausson-like in its delicacy. 1958 *Listener* 20 Nov. 850/3 The cardboard figures of Puccinian melodrama. 1962 *Times* 26 Jan. 16/4 He does not allow them much..Puccinian nobility of soul. 1978 *Gramophone* Jan. 1289/2 On the Italian front, Tebaldi is dominant, giving good, cleanly sung, quite committed versions of several famous Puccinian passages.

So **Puccinie·sque** *a.*, resembling the style of Puccini's works.

1927 *Sunday Express* 19 June 10/4 Yet Tom Burke has sung only once—in 'Rigoletto', which does not give him emotional scope, for he is a tenor of the voluptuously Pucciniesque type. 1961 *Times* 6 Dec. 17/6 As to Pucciniesque lyric passages, tender feelings are unmistakably clear when set to that kind of music.

‖ **puchero** (putʃĕ·ro). Also **puchera.** [a. Sp. *puchero, -a.*] **1.** A glazed earthenware cooking pot. Cf. OLLA[1] 1.

1841 BORROW *Zincali* II. III. App. 125 The puchero, or pan of glazed earth, in which bacon, beef, and garbanzos are stewed. 1846 R. FORD *Gatherings from Spain* x. 113 Most classes are equally satisfied with the Oriental earthenware *ollas, pucheros*, or pipkins.

2. A composite dish of beef or lamb, ham or bacon, and vegetables, cooked as a stew.

1841 [see *GARBANZO]. 1903 CONRAD & HUEFFER *Romance* III. ii. 131 An old woman..cooked his food at an outside fire—his *puchero* and *tortillas*. 1923 *Glasgow Herald* 7 Nov. 6/3 The crop of garbanzo, of which is made the puchero that has superseded the olla of years ago. 1933 H. ALLEN *Anthony Adverse* VIII. lxi. 998 The puchero followed. Brought in separate dishes, it was finally combined into one on the diner's plate.

puck, pook, *sb.*[1] Add: **c. puckfoisted** *a.* (*dial.*), cheated by a demon, bewitched.

1932 H. J. MASSINGHAM *Wold without End* App. 294 Here are a few of the Elizabethan words that were heard at the Globe more than three hundred years ago and are heard to-day in the inns between Chipping Campden and Stow-on-the-Wold..'puckfoisted' for bewitched.

puck, *sb.*[3] **1.** Delete 'in Canada' and add further examples.

1930 *Times* 20 Mar. 7/2 A little later..Benecchi put the puck over his body into the net. 1951 *Sport* 7–13 Jan. 16/2 His old speciality, taking the puck at full speed and boring through to the net, is working overtime. 1971 L. KOPPETT *N.Y. Times Guide Spectator Sports* i. 7 The scoring objective can be stated simply: in football to advance the ball across a goal line; in basketball, hockey, soccer or lacrosse, to get the ball (or puck) into a goal. 1974 *Cleveland* (Ohio) *Plain Dealer* 13 Oct. c. 1/2 Toronto jumped to a 2–0 lead in the first period when Featherstone streaked in for an unassisted goal at 1:35 and Dillon rammed the puck home at 18:48. 1978 *N.Y. Times* 30 Mar. D. 17/6 Wayne Dillon won the draw from Mike Kaszycki but pulled the puck back to an empty spot on the circle.

2. *attrib.* and *Comb.*, as *puck-dribbling, -handling, shot*; **puck carrier**, in ice hockey the player in possession of the puck during play; **puckchaser** *colloq.*, an ice-hockey player; so **puck-chasing** *vbl. sb.*; **puck crown**, ice-hockey championship; **puck pusher** *colloq.* = *puck-chaser*; **puck sense**, natural skill in ice hockey; **puck shy**, of goalkeepers in ice hockey: afraid of being hit by a puck.

1957 *Maclean's Mag.* 28 Sept. 1/2 The top scorers in the League are the best puck carriers. 1921 *Daily Colonist* (Victoria, B.C.) 24 Mar. 11/3 Calgary puckchasers take all the honors. 1979 *Yale Alumni Mag.* Apr. CN 1 The continuing resurgence of puck-chasers and net-stuffers is to be applauded. 1950 *Sport* 24–30 Mar. 13/4 His family later moved to Fort William, in which town he did most of his puck chasing. 1955 *Penticton* (B.C.) *Herald* 17 Mar. 5/3 There is no doubt in my mind—the Vees will bring this puck crown back to Canada. 1974 *Globe & Mail* (Toronto) 28 Jan. S. 2/7 He also gave a few exhibitions of puck dribbling with his skates, the only NHL defenceman who has this unique skill. 1965 *Kingston* (Ontario) *Whig-Standard* 15 May 9/6 Flyers dominated the game with superior skating, checking, passing and puck-handling. 1897 *Medicine Hat* (Alberta) *News* 25 Feb. 1/5 We have a club which can hold its own with the puck pushers from almost anywhere. 1966 *Hockey News* (Montreal) 1 Jan. 13/2 An intangible part of Melnyk's all-round prowess is something called 'puck sense'. 1968 *Globe & Mail* (Toronto) 5 Feb. 17/3 He was lifted with three minutes to play in the second period when struck on the mask over the right eye by a puck shot at close range. 1965 *Ibid.* 29 Dec. 24/2, I think the new rule could cause a goalie to become puck shy.

puck, *sb.*[4] (Further examples.)

1922 JOYCE *Ulysses* 247 One puck in the wind..would knock you into the middle of next week. 1934 J. O'HARA *Appointment in Samarra* (1935) iii. 80 What he should of done was give you a puck in the mouth when you threw the drink at him. 1961 'F. O'BRIEN' *Hard Life* ii. 18 Many a good puck I had myself in the quondam days of my nonage. 1979 N. SMYTHE in E. Berman *Ten of Best Brit. Short Plays* 120 I'll give you a puck in the gob in short order, mate.

puck, *v.* Add: **pu·cking** *vbl. sb.* and *ppl. a.*

1922 JOYCE *Ulysses* 247 Myler Keogh, Dublin's pet lamb, will meet sergeantmajor Bennett, the Portobello bruiser... God, that'd be a good pucking match to see. *Ibid.* 313 The referee twice cautioned Pucking Percy for holding.

pucka, pakka, *a.* (*sb.*) Restrict ‖ to senses a, b, d, e in Dict. and add: **c.** (Further examples.) Now usu. in form **pukka.** Also in general use outside India in various extended senses. Of things: real, not sham; of information: factually correct; of persons: authentic, not pretended; proper or correct in behaviour, socially acceptable. Freq. in *pukka sahib* (cf. *SAHIB 1b), used with allusion to life in the former British Indian Empire.

1894 *Scribner's Mag.* XV. 548/2 The Zinal-Rothhorn or 'Moming' is, to use an Anglo-Indian phrase, a 'puckah' mountain, which means that it is the real thing and not a sham. 1917 W. OWEN *Let.* 17 July (1967) 477 It was better paid than by a pukka Editor's best guineas. 1919 J. BUCHAN *Mr. Standfast* I. v. 113 My boy's at home, convalescing, and if he says you're *pukka*, I'll ask your pardon. 1924 E. M. FORSTER *Passage to India* I. iii. 26 Mrs. Turton..remarked that Mr. Fielding wasn't pukka, and had better marry Miss Quested, for she wasn't pukka. 1929 S. AUMONIER in *Mercury Story Bk.* 389 McLagan and Treadway were pukka soldiers of the old army. 1932 *Daily Express* 27 June 3/3 She wants also to play pukka golf. 1934 'G. ORWELL' *Burmese Days* v. 88 The smell of pukka sahibdom. 1938 N. MARSH *Artists in Crime* xix. 280 Don't be so 'pukka sahib'. 1939 [see *POONAH]. 1942 T. RATTIGAN *Flare Path* I. 14 Pukka gen, sir. 1948 *Observer* 25 Apr. 2/1 Produced for the Government of Southern Rhodesia, that forty minute film..is one of many from British studios that are being specially commissioned to give straightforward information on important subjects—in fact 'pukka gen'. 1955 *Sci. Amer.* Feb. 116/3 The injunctions and other brawling seem now to have subsided, with general agreement that while Buh may not be a pukka sahib, he is a peerless climber. 1966 'G. BLACK' *You want to die, Johnny?* v. 85, I just played pukka sahib greeting faithful old native servant. 1967 SINGHA & MASSEY *Indian Dances* xviii. 157 These barracks once the epitome of pukka British army tradition, for many years echoed all day to the sounds and rhythms of Indian music. 1971 *Daily Tel.* 19 Oct. 19 At one time hotels were classed with brothels in the minds of pukka brewers. 1973 C. MULLARD *Black Britain* III. vii. 75 A list of the reconstituted Board in 1968 showed all the new members, with the exception of two, to be pukka members of the white Establishment. 1973 *Times Lit. Suppl.* 30 Mar. 340/2 A small but pukkah group defending the last tatty remnants of colonial gentility. 1976 *Physics Bull.* Nov. 480/1 What it does show is a pucka trade union doing a proper trade union job. 1977 *Radio Times* 12–18 Nov. 15/1 The two genuinely brown faces in *It Ain't Half Hot Mum* belong to a Pakistani and a Bangla Deshi. He is the only pukka Indian in sight.

puckarow, var. *PUCKEROW.

puckaun (pɒ·kǭn). *Anglo-Ir.* Also **puckawn.** [ad. Ir. *pocán*, a small male goat.] A billy goat.

a 1745 SWIFT *Irish Eloquence* in *Prose Wks.* (1957) IV. 279 His Cows..would hardly give a drop of Milk..For his herd had lost the Puckaun. 1870 P. KENNEDY *Fireside Stories of Ireland* 7 Bring me the giant's puckawn with the golden bells round his neck. 1913 J. STEPHENS *Here are Ladies* 287 Children will dance upon the slightest provocation, so also do lambs and goats; but policemen, and puckauns, and advertisement agents, and fish do not dance at all, and this is because they have hard hearts. 1953 S. BECKETT *Watt* iv. 246 Riley's puckaun again, said Mr Nolan, I can smell him from here.

pucker, *sb.*[1] **2.** (Further U.S. examples.)

1825 J. NEAL *Bro. Jonathan* I. 202 Edith was in tears; Jotham, powerless with amazement;—Miriam, in a 'plaguy pucker'. 1847 J. S. ROBB *Streaks of Squatter Life* i. 15 If I am delayed Gales and Seaton will be very angry, and Blair and Rives get in a pucker.

pucker, *sb.*[2] *rare.* [f. PUCK *v.*] A boxer, a fighter.

1922 JOYCE *Ulysses* 247 The best pucker going for strength was Fitzsimons... But the best pucker for science was Jem Corbett before Fitzsimons knocked the stuffings out of him.

puckeroo (pɒkərū·), *a. N.Z. slang.* Also in various other phonetic spellings and with initial *b.* [ad. Maori *pakaru* broken; also vb., to break.] Useless, broken. Also as *v. trans.* (esp. in pa. pple.), to ruin.

[1844 J. W. BARNICOAT *Jrnl.* (MS.) 160 Gideon [*sc.* a Maori] foreseeing the collision..shouted out Puikero! puikero! puikero! (broken).] 1885 *Short Sk. Life T. Hancock* ii. 19 [Maoris said] 'We will *pukeru* you!' 'Very well', I said, '*pukeru* me'. 1925 FRASER & GIBBONS *Soldier & Sailor Words* 220 *Pakaru*, broken; smashed. A Maori word, in use among the New Zealand Troops. 1941 BAKER *N.Z. Slang* v. 42 [By the 1890's] we had begun to *pukaroo* things, when we broke something, confused an issue, or ruined some plan of action. This is derived with extraordinary simplicity from the Maori *pakaru*, broken 1943 *Amer. Speech* XVIII. 93 *Pukkaroo*, adjective.., (to

make) worthless, useless—it could be used..of an engine that had broken down—..is perhaps from the Maori *pakaru*, to destroy. 1948 R. FINLAYSON *Tidal Creek* vii. 179 The surest way to buckeroo an axe. 1965 S. T. OLLIVIER *Petticoat Farm* i. 14, I come to see if you've got a spare shovel. Mine's puckerooed and I got a cow in the drain. 1970 *N.Z. Listener* 12 Oct. 12/1 Bad show, fighting. I puckerooed things properly last night.

puckerow (pɒ·kərōᵘ), *v. Army* and *Naut. slang.* Also **puckarow, puckero, puckerrow.** [ad. Hind. *pakṛo* imp. of *pakaṛnā* to seize.] *trans.* To seize, lay hold of. Also *intr.* or *absol.* (rare).

1866 G. O. TREVELYAN in *Fraser's Mag.* LXXIII. 390 Fanny, I am cutcha no longer. Surely you will allow a lover who is pucka to puckero. 1876 C. CHAPMAN *Sailor's Life at Sea* iv. 224 Now is the time; let us 'puckerrow' it. 1886 YULE & BURNELL *Hobson-Jobson* p. xix, Hindustani *verbs*..are habitually adopted into the quasi-English by converting the imperative into an infinitive. Thus..to *puckarow*. 1887 *Outing* July 331/1 Charley Wheeler were the lucky man as had 'puckerowd' poor Hans' dry-goods. 1899 F. T. BULLEN *Log of Sea-Waif* xvi. 194 So mechanically did they 'puckarow' those baskets, that often one would pass from the hatch to the gang way empty. 1907 M. ROBERTS *Flying Cloud* iii. 13 What with puckerowing cases, lashing tanks, and frapping stunsail-booms on the deck-house..there was enough to do. 1919 W. H. DOWNING *Digger Dial.* 59 *Puckero*, take; seize. 1931 W. KIRK in *Cabar Feidh* Sept. 389/2 Not all the legislators, robbing poor and rich to-day, can puckarow my talisman—the Badge of Cabar Feidh!

puckery, *a.* **1.** (Earlier example.)

1830 *Massachusetts Spy* 10 Feb. 2/1, I didn't like the set of the shoulders, they were so dreadful puckery.

puckery (pɒ·kəri), *sb. rare.* [PUCK *sb.*[1] + -ERY.] = PUCKISHNESS.

1877 G. MEREDITH *Let.* 24 June (1970) I. 545, I foresee the grin up to the ear tips of exulting Puckery.

puckish, *a.* Add: Hence **pu·ckishly** *adv.*

1972 *Daily Tel.* 20 Nov. 13/5 His feet could not reach the pedals... But when his fingers remained under control..he puckishly turned grace-notes and, in the Gigue, made the music swell imposingly. 1977 *N.Y. Rev. Bks.* 13 Oct. 14/4 Vice Chairman Tower adds puckishly, 'I might say further that the matter of assassinations might be viewed in a broader context of other options that might have been available.'

puckle (pɒ·k'l), *sb.*[2] *Sc.* [Local var. PICKLE *sb.*[2]] An indefinite amount, a few.

1877 G. STEWART *Shetland Fireside Tales* x. 78 A 'puckle o' oo' when da sheep wis rued. 1917 A. S. NEILL *Dominie Dismissed* vi. 86 Aw need hardly say onything aboot the object o' this concert, but it's to get a puckle bawbees to send oot a clean pair o' socks and maybe a clean sark to oor local sojers oot in France. 1930 *Aberdeen Univ. Rev.* XVII. 103 A hinna heard o' im for a gey puckle year an' A doot 'e maun be deid. 1968 E. BUCKLER *Ox Bells & Fireflies* xii. 165 The man with a small nest egg had saved up 'quite a puckle'.

puckster (pɒ·kstəɹ). *N. Amer. colloq.* [f. PUCK *sb.*[3] + -STER.] An ice-hockey player.

1939 *Kansas City Jrnl.* 5 Feb. 23 (*heading*) Greyhounds slaughter Tulsa Pucksters. 1941 *Sun* (Baltimore) 11 Feb. 14/7 (*heading*) Georgetown pucksters schedule five games. 1955 *Toronto Daily Star* 11 Apr. 18/5 Moncton Hawks, as optimistic a band of pucksters as ever came down the Allan Cup hockey trail. 1976 *Bangor* (Maine) *Daily News* 24 Aug. 20/3 U.S. pucksters face long odds.

pud, *sb.*[1] (Further examples.)

1965 R. ERSKINE *Passion Flowers in Business* viii. 108, I saw him clutching your hot little puds. 1968 J. F. STRAKER *SIN & Johnny Inch* 165 How did they get their puds on you? More trickery?

pud (pŭd), *sb.*[2] var. POOD.

1662 in M. Blundell *Cavalier* (1933) vi. 103 Twenty Russe Pud of the tooth of Sea-Horse—each Pud is 40 pound weight. Ten thousand Pud of hemp. 1814 [see POOD]. 1901 A. M. B. MEAKIN *Ribbon of Iron* xvi. 226 From mines discovered in 1866..2,500 puds of gold were extracted during a period of twenty years. 1952 E. H. CARR *Bolshevik Revolution* II. xix. 285 Kalinin estimated the total of relief supplies up to December 1921 at 1,800,000 puds of grain and 600,000 puds of other foodstuffs from home stocks.

pud (pud), *sb.*[3] Also **pudd.** Colloq. abbrev. of PUDDING *sb.* **1. a.** = PUDDING *sb.* 1, 2. Also as second element in *pock-pud* (see POKE-PUDDING). **b.** = PUDDING *sb.* 6 (now the usual sense). Also *transf.* and *fig.*

1706 in J. Watson *Choice Coll. Comic & Serious Scots Poems* I. 61, I leave my Liver, Puds and Tripes. a 1776, 1802, etc. [see POKE-PUDDING 1, 2]. 1828 in P. Buchan *Anc. Ballads & Songs N. Scotl.* I. 261 Whan the puds war sodden. 1914 *Dialect Notes* IV. 164 *Pud*,..pudding. 1943 'R. LLEWELLYN' *None but Lonely Heart* xli. 342 If you lot go to chokey, so do I, for harbouring. So we're all blackbirds in the same old pud. 1951 J. B. PRIESTLEY *Festival at Farbridge* 47 These two have finished mopping up their horrible pink puds. 1955 M. EWER *No Abiding Place* i. 13 Soup, joint, two veg, pud and cheese. 1960 T. COOPER *Winter's Day* III. i. 164, I helped make the pudd. 1976 *Southern Even. Echo* (Southampton) 15 Nov. 10/2 Nostalgic and happy memories of our traditional Christmas 'Pud'.

2. *coarse slang.* = *PUDDING 5b. *to pull one's pud*: see *PULL *v.* 19j.

1939 JOYCE *Finnegans Wake* (1964) 445 There's a lot of lecit pleasure coming bangslanging your way, Miss Pinpernelly satin. For your own good, you understand, for the man who lifts his pud to a woman is saving the way for kindness. 1944 *Publ. Amer. Dial. Soc.* II. 35 *Pud, to pull the (his)*,..to masturbate... Boys and men. Common. 1972 R. A. WILSON *Playboy's Bk. Forbidden Words* 240 *Pud*, the penis; perhaps from *pudding* in *pull the pudding*. 1977 *Amer. Speech* 1975 L. 54 *Pud*, 'penis'.

3. *fig.* = *PUDDING *sb.* 7b; *spec.* an easy college course. Also *attrib.* or as *adj. U.S.*

1938 *Amer. Speech* XIII. 6/2 *Pud*,..an easy job. 1963 *Ibid.* XXXVIII. 167 An easy college course..*pud* adj. 1967 S. B. FLEXNER in Wentworth & Flexner *Dict. Amer. Slang* Suppl. 700/1 *Pud*,..an easy course; a 'snap'... *Adj.* Easy to pass or make a good grade in, as a course or test. 1977 *Amer. Speech* 1975 L. 64 *Pud*,..soft, easy. 'Do you know any pud courses?'

pudder, var. POTHER *sb.* in Dict. and Suppl.

pudding, *sb.* Add: **I. 5. a.** (Earlier and later examples.)

1789 E. BUTLER *Diary* 7 Oct. in G. H. Bell *Hamwood Papers* (1930) 231 A great fat pudding boy brought some. 1903 [see *NON-SIGNIFICANT *a.*]. 1980 A. CORNELISEN *Flight from Torregreca* xi. 267 She is a sallow pudding of a child with a broad flat face.

b. *coarse slang.* The penis.

1719 T. D'URFEY *Wit & Mirth* III. 73, I made a request to prepare again, That I might continue in Love with the strain Of his Pudding. 1961, 1970 [see *PULL *v.* 19j]. 1972 [see *PUD *sb.*[3] 2].

c. *slang.* A fœtus; in phr. *a pudding in the oven* (and similar phrases), a child conceived but not yet born. Cf. *BUN *sb.*[2] 1.

1937 PARTRIDGE *Dict. Slang* 665/1 *With a bellyful of marrow-pudding*,..pregnant. 1965 J. PORTER *Dover Two* vi. 75 'None of us ever suspected that she'd got a pudding in the oven.' 'She was going to have a baby?' asked Dover. 1966 'L. LANE' *ABZ of Scouse* 112 *She's got a pudden in ther uvving*, she is pregnant.

II. 6. a. (Further examples.) Now usu. in British English the course following the main course of a meal, 'afters'.

1909, etc. [see *AFTERS *sb. pl.*]. 1940 S. SPENDER *Backward Son* 12 At lunch there was fruit salad, his favourite pudding. 1954 *Good Housek. Cookery Bk.* (rev. ed.) II. 284 In this section will be found the recipes for suet and sponge puddings, and for some miscellaneous baked puddings. 1968 *New Society* 22 Aug. 266/2 Another course of a meal is called 'sweet' by the non-U. The U word for the course is pudding. 1974 E. AYRTON *Cookery of England* x. 430 Our grandfathers, even our fathers, expected a 'pudding' at least once a day, sometimes twice.

7. b. *U.S. slang.* Something easy to accomplish.

1887 G. W. WALLING *Recoll. N.Y. Chief of Police* xix. 262 It was an 'inside' job from the start... In thieves' slang it was a 'pudding';..the vault, although apparently impregnable, was easy to enter, [etc.]. 1942 BERREY & VAN DEN BARK *Amer. Thes. Slang* § 255/1 Something easy,..*pudding*. 1974 *Guidelines to Volunteer Services* (N.Y. State Dept. Correctional Services) 42 *Puddin*, light action, easy.

8. c. *slang.* A pudding-shaped bomb.

1919 *Athenæum* 25 July 664/1 *Pudding*, i.e. our 60 lb. bomb.

III. 11. a. *pudding course*; *pudding-shaped* adj.; **b.** *pudding-cloth* (examples), *fork, rice, -spoon* (examples), *-stick* (later examples).

1845 E. ACTON *Mod. Cookery* xii. 255 The bird..wrapped in a thin pudding-cloth, closely tied at both ends. 1868 M. JEWRY *Warne's Model Cookery* 482 A pudding-cloth must be kept very clean. 1971 *Country Life* 17 June 1537/2 He tried to do it with oddments of coloured knitting wools on a pudding cloth. 1948 'J. TEY' *Franchise Affair* iv. 40 The gentle monologue went on, all through the pudding course. 1896 *Woman's Life* 15 Aug. 368/1 If the pudding-spoon and fork are grasped from beneath instead of from above, the awkward uplifting of the elbows will be avoided. 1914 JOYCE *Dubliners* 255 Freddy Malins beat time with his pudding-fork. 1974 *Times* 10 Jan. 10/1 Long grain and short or round grain, often called 'pudding' rice. 1895 W. ROBINSON *Eng. Flower Garden* (ed. 4) v. 75 A great many delightful plants..in many cases are jammed into pudding-shaped masses void of form or grace. 1976 *S. Wales Echo* 23 Nov., A pudding-shaped mound in Energlyn near Caerphilly. 1896 Pudding-spoon [see *pudding fork* above]. 1944 A. THIRKELL *Headmistress* iv. 73 Giving a final polish to the pudding spoons with a piece of washleather. 1973 J. WAINWRIGHT *Touch of Malice* 93 Harris..handled the gear-lever like a pudding-spoon. 1852 MRS. STOWE *Uncle Tom's Cabin* I. xviii. 298 Interrupting her meditations to give..a rap on the head to some of the young operators with the pudding-stick that lay by her side. 1878 B. F. TAYLOR *Between Gates* 109 You can get an idea of it by fancying a paddle or a pudding-stick turning into a fiddle.

c. pudding-ball *Austral.* [ad. Aboriginal word], an edible marine fish resembling a mullet, perhaps the sea mullet, *Mugil cephalus*; **pudding basin**, a basin in which puddings are made; *transf.*, applied to a round hat, helmet, or hair-style; also *attrib.*; **pudding chain** *Naut.* (see quot.); **pudding class** = next; **pudding club**: see *CLUB *sb.* 14c; **pudding-face** (later examples); **pudding fender** = sense 4b; **pudding-head** (examples); **pudding-sleeve**, delete † and add later examples; **pudding way** = *pudding club* above.

1847 J. D. LANG *Cooksland* iv. 96 The species of fish that are commonest in the Bay [*sc.* Moreton Bay] are mullet, bream, puddinba (a native word corrupted by the colonists into pudding-ball)... The puddinba is like a mullet in shape, but larger, and very fat; it is esteemed a great delicacy. 1896 *Australasian* 28 Aug. 407/4 'Pudding-ball' is the name for a fish. 1945 BAKER *Austral. Lang.* xii. 21.. Popular fish-names peculiar to the Australian include.. puddingball, corrupted by the law of Hobson-Jobson from the aboriginal puddinba. 1861 MRS. BEETON *Bk. Househ. Managem.* xxvi. 611 (*caption*) Pudding-basin. 1909 *Westm. Gaz.* 3 June 8/3 A grey straw hat of the inverted pudding-basin type. 1925 FRASER & GIBBONS *Soldier & Sailor Words* 231 *Pudding basin*, the British steel shrapnel helmet. (From its shape.) 1951 A. BARON *Rosie Hogarth* I. ii. 19 Each boy's hair close-cropped with a pudding-basin fringe. 1974 *Country Life* 28 Feb. 456/3 A male customer is looking for..shooting and fishing hats, saucy tweed pudding basins and tweed caps. 1977 B. PYM *Quartet in Autumn* i. 1 Now he..had adopted a medieval or pudding-basin style, rather like the American crew-cut of the forties and fifties. 1948 R. DE KERCHOVE *Internat. Maritime Dict.* 561/2 *Pudding chain*, short link chain occasionally used for running rigging. It runs well over sheaves and is easy to belay. It is used for jib halyards and sheets in small trading vessels, but has lately been generally replaced by flexible wire. 1969 E. GÉBLER *Shall I eat you Now?* 88 Girl soon comes..to announce she has a bun in the oven. I'm in the pudding class. 1890 BARRÈRE & LELAND *Dict. Slang* II. 155/1 *Pudding club* (popular), a woman in the family way is said to be in the *pudding club*. 1978 L. DAVIDSON *Chelsea Murders* v. 28 'Was she in the pudding club?'..'Probably. They aren't saying.' 1916 'TAFFRAIL' *Pincher Martin* vii. 116 Orl right, old puddin'-face. Keep yer 'air on! 1950 G. BRENAN *Face of Spain* iv. 84 The Englishman, fresh from the dull hurry of London streets and from their sea of pudding faces. 1883 *Man. Seamanship for Boys' Training Ships R. Navy* (Admiralty) (1886) 186 Pudding fenders are used in the Navy for large boats..and sometimes on lower yards, to take the chafe on the inside part of the quarter yard. 1961 F. H. BURGESS *Dict. Sailing* 164 *Pudding fender*, a fat enclosed bundle of old strands, etc. for use over the side of boats and yachts. 1851 H. MELVILLE *Moby Dick* III. xxii. 152 Pudding-heads should never grant premises. 1893 'MARK TWAIN' in *Century Mag.* Dec. 235/2 Perfect jackass—yes, and it ain't going too far to say he's a pudd'nhead. 1952 S. KAUFFMANN *Tightrope* xiv. 243 Why, you're not doing this at all badly, pudding head. 1978 P. G. WINSLOW *Coppergold* 153, I didn't tell Joss, no matter what that Yorkshire puddenhead thinks. 1910 'MEMBER OF ARISTOCRACY' *Manners & Rules of Good Society* xi. 85 Archbishops, bishops, and clergy should appear in full canonicals, that is black silk full- or pudding-sleeve gowns, cassock and sash bands. 1939 M. B. PICKEN *Lang. Fashion* 136/3 *Pudding sleeve*,..full sleeve held in at wrist, or above. 1960 C. W. CUNNINGTON et al. *Dict. Eng. Costume* 172/1 *Pudding sleeve*,..a large loose sleeve, especially of a clergyman's gown. 1963 'J. PRESCOT' *Case for Hearing* vi. 94 Getting a girl in the pudding way isn't a crime.

pudding-bag. (Earlier and further examples.)

c 1597 T. DELONEY *Jack of Newberie* (1619) iv. sig. G3, The other maide..with the pudding-bagge flapt him about the face. 1858 G. MEREDITH *Let.* 4 Jan. (1970) I. 32 It is a pudding-bag..a quiescent receptacle for Roast Beef, Punch, and mince Pies. 1929 F. BOWEN *Sea Slang* 107 *Pudding bag*, a stocking pennant used as a vane. 1939 F. THOMPSON *Lark Rise* 238 They..yelled Old Hardwick skags! Come..to pick up rags To mend their mothers' pudding-bags. 1943 W. W. GILL in *N. & Q.* 9 Oct. 232/1 *Pudding-bag*, blind alley. 1961 F. H. BURGESS *Dict. Sailing* 164 *Pudding bag*, a stocking or sleeve, used as a weather vane.

pudding-grass. For *Obs.* substitute *Obs.* exc. *Hist.* Add to def.: *Mentha pulegium*. (Later examples.)

1904 G. G. NILES *Bog-trotting for Orchids* x. 132 The Wild Pennyroyal of the ancients..was known in England, during 1500, as Podding-Grasse or Pudding-Grass. 1972 Y. LOVELOCK *Veg. Bk.* III. 336 It [*sc.* pennyroyal] was especially popular as a meat-stuffing, and was thus known as pudding grass or herb.

puddle, *sb.* Add: **1. b.** Also (*colloq.*), a pool of evacuated urine; usu. in phr. *to make a puddle*, with reference to a young child or pet animal (cf. *ACCIDENT *sb.* 1d).

1968 J. LLOYD *Death at Roman Farm* i. 10 Are you sure she hasn't made a puddle? 1972 J. WILSON *Hide & Seek* vi. 107 Can I have a mop to wipe up Mary's puddle? 1977 M. UNDERWOOD *Murder with Malice* iii. 37 Why are you looking at me as if I'd just made a puddle on the floor?

c. Applied *fig.* and humorously to the sea, esp. the Atlantic Ocean; usu. in phr. *this* (etc.) *side of the puddle*. Cf. POND *sb.* 2.

1889 *Ally Sloper's Half Holiday* 6 July 214/2 There seems to be no end to the chaff which the downy dandies across the puddle have to bear. 1902 FARMER & HENLEY *Slang* V. 312 *The Puddle*,..the Atlantic Ocean... In Cornwall, the English Channel. 1978 *SLR Camera* Aug. 21/1 For many years the American company..have made fine enlarging frames (masking frames this side of the puddle) both for retail distribution and for exclusive use by Simmon-Emega.

d. *Rowing.* The circular, rippled, disturbance left in the water after the blade of an oar has been lifted from it at the end of a stroke.

1934 *Times* 17 Mar. 14/1 Holdsworth is rowing better than he has ever done before at No. 2. His puddle is worthy of a man a stone heavier. 1955 R. BANNISTER *First Four Minutes* iii. 39, I could see my oars were making some splendid 'puddles'.

e. A small pool of molten metal, esp. that

formed during welding; a piece of metal solidified from a pool.

1935 C. G. BAINBRIDGE in *Symp. Welding Iron & Steel* (Iron & Steel Inst.) II. 14 A large rod melts slowly and cools the molten puddle, causing rapid solidification. 1942 J. A. MOYER *Welding* v. 43 As the torch flame moves away, the molten metal in the puddle solidifies and joins the two plates into one solid piece. 1958 *Man* LVIII. 64/1 The first flat celts were hammered out of natural copper, and then out of rough casts or puddles of it or of poor bronze, smooth on the lower side, rough and scabbed on the open side. 1975 BRAM & DOWNS *Manuf. Technol.* ii. 47 While one hand manipulates the torch to carry a puddle across the plate, the other adds the correct amount of filler rod.

6. b. puddle-duck (examples); **puddle-jumper** *U.S. slang*, a fast, highly-manœuvrable, means of transport (see quots.), esp. a small light aeroplane; hence **puddle-jumping** *ppl. a.*

1877 *Scribner's Monthly* Nov. 6/1 Presently we heard a shrilly feeble whistle, precisely such as the young puddle-duck of the barn-yard makes in his earliest vocal efforts. 1908 B. POTTER *Tale of Jemima Puddle-duck* 9 Listen to the story of Jemima Puddle-duck, who was annoyed because the farmer's wife would not let her hatch her own eggs. 1975 J. GORES *Hammett* (1976) xix. 132 Puddle ducks..and mud hens..skittered away. 1932 *Daily Progress* (Charlottesville, Va.) 10 Oct. 9 (*caption*) Even the bicycles that these sisters use..seem to be related as they resemble each other almost as much as the owners. They are twins... They were on the way to classes on their 'puddle jumpers' when they halted for this picture. 1941 *Sun* (Baltimore) 23 Aug. 8/8 Only two power-boat tests are slated on today's program..but the 'puddle-jumpers' will take over complete control tomorrow. 1944 *Newsweek* 2 Oct. 31/2 A 'puddle jumper' observation plane with bazookas fixed on the wings dove down and knocked out two of the tanks. 1944 A. M. TAYLOR *Lang. of World War II* 161 *Puddle jumper*,..a nickname for Jeep. 1961 'A. A. FAIR' *Stop at Red Light* (1962) viii. 127, I had to take a puddle-jumper with stops in Chicago, Denver and Salt Lake City. 1971 M. TAK *Truck Talk* 123 *Puddle jumper*, a lightweight truck. 1978 *Detroit Free Press* 16 Apr. (Parade Suppl.) 3/3 Any one..can call his plane an air ambulance even if it's just a 'puddle-jumper' without medical equipment. 1941 *Sun* (Baltimore) 2 Aug. 7/1 They are hoping to receive soon a long-promised consignment of three 'puddle-jumping' Vultee 049 planes in which they will be able to hop up and down from even the smallest corn field.

puddle, *v.* Add: **6.** (Earlier and later examples.) Also in *Opal-mining*.

1853 E. CLACY *Lady's Visit Gold Diggings Austral.* vii. 114 This soil, from being so stiff, would require 'puddling', a work of which he did not seem to relish the anticipation. 1963 *Pix* 13 July 21 Machines are used to 'puddle' (separate and sieve) opal dirt. 1967 S. LLOYD *Lightning Ridge Bk.* (1968) i. 1 Opal dirt can be brought to the surface and examined or puddled.

puddler. Add: **3. a.** Also in *Opal-mining*. (Earlier and later examples.)

1859 *Adelong Mining Jrnl.* 15 July 6/3 A rather unpleasant case occurred on Monday, at Little Bendigo, between an European, and some Chinese puddlers. 1860 *Mining Surveyor's Rep.* (Mining Dept., Victoria) Aug. 214 A valuable piece of ground which could be advantageously worked by the puddlers. 1967 S. LLOYD *Lightning Ridge Bk.* (1968) xiv. 99 Puddlers have ruined the whole Lightning Ridge Field.

b. A puddling machine.

1967 A. KALOKERINOS *In Search of Opal* ii. 19 Modern miners remove the..'pay dirt', in bulk..and spin it in a machine that sifts the dirt out and leaves the nobbies behind... These machines are called 'puddlers' and their variety is almost endless. 1971 J. S. GUNN *Opal Terminol.* 37 There are two dams..at which miners rent sites..where they operate power-driven wet puddlers capable of handling several tons of dirt in one operation.

puddling, *vbl. sb.* Add: **4.** (Earlier and later examples.) Also in *Opal-mining*.

1851 R. TESTER *Wombat Wallaby* 61, I spurred my little mare off, and in doing so she made a plunge, and very nearly bundled me and my mutton into the puddling tub. 1853 E. CLACY *Lady's Visit Gold Diggings Austral.* vii. 117 The great thing is, not to be afraid of over-work, for the better the puddling is, so much the more easy and profitable is the cradling. 1856 *16th Gen. Rep. Emigration Comm.* (Colonial Office) 26 'Puddling machines', which are contrivances for washing the soil by horse-power, appear to be numerous and valuable. 1966 J. HACKSTON *Father clears Out* 62 He was going to do something bigger this time, something better than the tin dish and the cradle—he'd borrow a puddling machine he knew of. 1971 J. S. GUNN *Opal Terminol.* 37 *Puddling tank*, large dam at which wet puddling takes place.

puddy, *a.* (Further example.)

1912 W. DEEPING *Sincerity* ii. 9 Her round, puddy, exquisitely complacent face looked out from between clay-coloured ringlets.

pudendum. Add: **a.** esp. those of a woman. (Further examples.)

1893 T. M. MADDEN *Clin. Gynæcol.* v. 59 The value [for pruritus] of a solution of cocaine freely brushed over the pudendum..is unquestionable. 1922 A. G. MAGIAN *Sex Probl. Women* ii. 31 The Vulva, or Pudendum, includes—(1) The labia majora and minora bounding the pudendal cleft. (2) The mons veneris. (3) The vestibule, [etc.]. 1977 E. J. TRIMMER et al. *Visual Dict. Sex* (1978) v. 58 Sanskrit manuscripts show Indian women with shaved pudenda.

b. *fig.*

1938 S. BECKETT *Murphy* 47 Here are the pudenda of my psyche.

pudent (piū·dĕnt), *a. rare.* [f. L. *pudens, pudentem* pres. pple. of *pudēre* to make ashamed: cf. IMPUDENT.] Having or showing a sense of shame, esp. in regard to matters of a sexual nature; modest; delicate.

1908 G. B. SHAW in W. H. Davies *Autobiogr. Super-Tramp* p. vii, These pudent pages are unstained with the frightful language..of the fictitious proletarians of Mr. Rudyard Kipling and other genteel writers.

‖ **pudeur** (‖ pŭdör). [Fr.; see PUDOR.] A sense of shame or embarrassment, esp. in regard to matters of a sexual nature; bashfulness, modesty, constraint.

1937 WYNDHAM LEWIS *Let.* 21 Nov. (1963) 247 And why this strange *pudeur*? 1959 *Times* 22 Sept. 13/5 The choice of physical type in these figures with their sense of flaunted *pudeur*..acquires a close affinity with Italian Mannerism. 1961 *Spectator* 17 Feb. 221 There was a deep-seated *pudeur*, going back to a finely civilised upbringing in a Victorian working-class home. 1962 I. MURDOCH *Unofficial Rose* xxi. 201 She had in any case, with a sort of *pudeur*, arranged to be out of London. 1963 *Guardian* 15 June 4/7 Spencer's brother, Gilbert..could have given everyone a closer sense of the man than this editing and institutional pudeur. 1968 R. P. WARREN *Incarnations* (1970) 7 The peach has released the bough and at last Makes full confession, its *pudeur* Has departed. 1976 *Listener* 10 June 737/3 It is hard not to be goaded into guessing identities. Pudeur makes the reader bend over backwards to prevent this happening.

pudge[2], var. *PADGE.

pudge, *v.* Var. PODGE *v.*

1904 H. F. DAY *Kin o' Ktaadn* 193 Old Tay..pudges along to the tin box on the mantel. 1932 *Nat. Geogr. Mag.* July 120 Overtaken by darkness on starless nights, the swamp man crawls into a log for safety or sleep. 'We always "pudge" around first to rout out any copperheads.' 1955 E. POUND *Section: Rock-Drill* xci. 75 Farinata pudg'd still there in the cloister.

pudgily (pɒ·dʒĭli), *adv.* [f. PUDGY *a.*[1] + -LY[2].] In a pudgy manner.

1926 *Harper's Mag.* Feb. 351/1 One day she escorted the pudgily tottering six-weeks-old youngsters [*sc.* puppies] on a ramble over the lawn. 1978 R. BARNARD *Unruly Son* iii. 34 He remained pudgily sunk in the easy chair.

pudibund, *a.* Add: (Further examples.) Also ‖ **pudibond.**

a 1922 T. S. ELIOT *Waste Land Drafts* (1971) 103 Pudibund, in the clinging vine. 1923 G. SAINTSBURY *Second Scrap Bk.* iii. 25 My tutor in Scholarship was the late 'Johnny' King, an expert in his subject..and a very good fellow, but rather shy and extremely pudibund. *Ibid.* xxxviii. 269, I understand that Soviet education is not at all pudibund, and that the principles of parenthood are treated and illustrated in it with a fine 'candour'. 1930 D. B. WYNDHAM LEWIS *Stuffed Owl* p. ix, The illiterate, the semi-literate, the Babu,..the hearty but ill-equipped patriot, the pudibond yet urgent Sapphos of endless *Keepsakes* and *Lady's Magazines*.

pudibundery (piū·dibɒ·ndəri). Also ‖ **pudibonderie.** [f. PUDIBUND *a.* + -ERY; cf. F. *pudibonderie*.] Bashfulness, prudery.

a 1913 F. ROLFE *Desire & Pursuit of Whole* (1934) xxii. 249 The pudibundery of Erastian peeping toms. 1915 T. BURKE *Nights in Town* 51 Everyday life is always disgusting to the funny little Bayswaterats, who are compact of timidity and pudibonderie. 1917 E. POUND *Let.* ? Jan. (1971) 107, I have only three quarrels with them: Their idiotic fuss over christianizing all poems they print, their concessions to local pudibundery, [etc.]. 1959 *Listener* 29 Oct. 744/1 This is not an isolated instance of *pudibonderie*.

pudic, *a.* **1.** Delete † *Obs.* and add later examples.

a 1913 F. ROLFE *Desire & Pursuit of Whole* (1934) 247 Water-babies (pudic, though incredibly tattered) wallow in every canal. 1974 *Times Lit. Suppl.* 22 Feb. 169/1 The instructive geographical screen fallen to reveal a pudic Lady Teazle.

pudicity. Delete 'Now *rare*' and add later examples.

1931 C. MACKENZIE *Buttercups & Daisies* xxi. 274 Dodsworth in a turmoil of alarmed pudicity at the prospect of her bedroom being used as a thoroughfare called upon Ralph and Roger to stop their goings on at once. 1958 L. DURRELL *Balthazar* vi. 132 Yet it was accompanied by a delicacy, almost a pudicity, in his dealings with them.

pudor. Delete † *Obs.* and add pronunc. (piū·dǭɹ). (Later examples.)

1922 JOYCE *Ulysses* 508 Woman undoing with sweet pudor her belt of rushrope. 1927 R. FRY *Flemish Art* i. 25 This tinge of sentimental feeling is very discreet. He never abandons himself. He has a certain shy pudor which is very attractive. 1966 *Times Lit. Suppl.* 3 Nov 1012/4 Lawrence's..first head-on collision with the forces of British pudor.

pu-é, var. *PWE.

pueblo. Add: Also with capital initial. **1.** (Earlier and later *attrib.* examples.)

1844 J. GREGG *Commerce of Prairies* I. 132 About two thousand of the insurgent mob, including the Pueblo Indians, pitched their camp in the suburbs of the capital [*sc.* Santa Fé]. 1907 Pueblo Indian [see *CREE *sb.* and *a.*]. 1923 D. H. LAWRENCE *Birds, Beasts & Flowers* 193 Across the pueblo river That dark old demon and I Thus say a few words to each other. 1949 *Nat. Geogr. Mag.* Dec. 783/2 Long-haired Pueblo Indians wrapped in cotton blankets exchange stare for stare with visiting easterners. 1957 T. VEBLEN *Theory of Leisure Class* 6 Some Pueblo communities. 1976 M. & G. GORDON *Ordeal* (1977) viii. 50 The Navajos were..quick to latch onto new ideas. From their next-door neighbours, the Pueblo people, they had borrowed sheep raising and weaving. 1979 *Arizona Daily Star* 5 Aug. (Advt. section) 21/7 This beautiful Territorial..features pueblo fireplace, exposed beams in family room.

2. (Earlier examples.)

1834 A. PIKE *Prose Sk. & Poems* 132 The Pueblos shall mount and prepare to pursue. 1844 J. GREGG *Commerce of Prairies* I. 268 Most of these Pueblos call themselves the descendants of Montezuma.

puer, var. PURE *v.* 1b in Dict. and Suppl.

puerperal, *a.* Add: *puerperal fever, sepsis*, etc., sepsis of the genital tract following parturition, or the fever associated with this.

1768 Puerperal fever [in Dict.]. 1814 J. ARMSTRONG *Facts Rel. Fever called Puerperal* p. vii, Under the common term puerperal fever are comprehended, in the following work, both the ordinary peritoneal inflammation, and the low malignant fever, of lying-in women, and these are considered as modifications of the same disease. 1876 W. S. PLAYFAIR *Sci. & Pract. Midwifery* II. v. v. 302 There is no subject in the whole range of obstetrics which has caused so much discussion and difference of opinion as that to which this chapter is devoted. Under the name of 'Puerperal Fever', the disease we have to consider has given rise to endless controversy. *Ibid.*, If this view be correct, the term 'puerperal fever', conveying the idea of a fever such as typhus or typhoid, must be acknowledged to be in itself misleading, and one that should be discarded. 1935 A. C. BECK *Obstetr. Pract.* xxxvi. 583 Puerperal infection is a wound infection in the birth passages. It often is referred to as puerperal fever, childbed fever, puerperal sepsis, or puerperal septicemia. 1955 *Sci. News Let.* 11 June 373/1 Puerperal sepsis, which is better known as childbed fever. 1974 GREENHILL & FRIEDMAN *Biol. Princ. & Mod. Pract. Obstetr.* lxx. 730/1 The term puerperal infection includes all the inflammatory processes which arise from bacterial invasion of the genital organs during labour or the puerperium. Other terms for this condition are puerperal sepsis, puerperal septicemia, puerperal fever and childbed fever. 1977 J. DONNISON *Midwives & Med. Men* v. 93 The deadly 'childbed' or 'puerperal' fever..regularly closed wards and contributed to maternal death rates as high as 28 per 1,000.

puerperium (pⁱū,əɹpe·riɒm, -pĭ·riɒm). *Med.* L.: see PUERPERY.] The puerperal state or period; *spec.* the few weeks following delivery during which the mother's tissues return to their non-pregnant state.

1890 in BILLINGS *Med. Dict.* II. 410/1. 1894 GRANDIN & JARMAN *Obstetr. Surg.* viii. 164 Many such lacerations heal spontaneously, probably the vast majority if the course of the puerperium is aseptic. 1935 A. C. BECK *Obstetr. Pract.* xv. 232 The term puerperium is applied to the six or eight weeks following labor which are required for the involution of the maternal organism. 1977 *Lancet* Aug. 273/1 Early trials with a Lippes loop inserted in the first week of the puerperium resulted in a high and unacceptable expulsion-rate.

Puerto Rican (pwɜ:ɹto rī·kăn), *sb.* and *a.* Also earlier **Porto Rican** (see note below), **Porto Riquenean.** [f. the name *Puerto* (or *Porto*) *Rico* + -AN.] **A.** *sb.* A native or inhabitant of Puerto Rico, an island in the Greater Antilles group of the West Indies, now a Commonwealth in association with the U.S.A. **B.** *adj.* Of or pertaining to Puerto Rico or its inhabitants.

The name was officially changed in 1932 from *Porto Rico* to *Puerto Rico*. The Commonwealth of Puerto Rico was established on 25 July 1952.

1858 J. T. O'NEIL *Mem. Island Porto Rico* in R. S. Fisher *Spanish West Indies* 152 The Porto Riqueneans..are generally indolent. 1898 R. T. HILL *Cuba & Porto Rico* xv. 146 The Cubans are fired with the spirit of progress and infected with American notions, while the Porto Ricans are plodding along in contentment. *Ibid.* xviii. 166 The Porto Rican Spaniards of the upper class..are the descendants of military men. 1898 *Times* 30 July 7/2 The American troops were received by the entire population.., the piers, balconies, roofs and streets being alive with Puertoricans [*sic*] representing every class. 1899 F. A. OBER *Puerto Rico* xii. 168 Scratch a Puerto Rican and you find a Spaniard underneath, so the language and home customs of Spain prevail here. *Ibid.* 171 The Puerto Rican home life..differs in no important particular from that of Spain and Mexico. 1907 J. H. MOORE *With Speaker Cannon through Tropics* ii. 38 Bright colors are affected by the Porto Rican women. 1926 K. MIXER *Porto Rico* vi. 102 Governor Towner has come to the Island under conditions which have given him an exceptional prestige with Porto Ricans. 1952 S. KAUFFMANN *Tightrope* xvi. 263 Jerry, the building's Puerto Rican shoe-shine boy was just finishing. 1956 'E. MCBAIN' *Cop Hater* (1958) iii. 29 Occasionally, a Puerto Rican wandered into *The Shamrock*. 1962 *Amer. Speech* XXXVII. 18 New York City..contains 80 percent of the Puerto Ricans of the entire United States. 1969 *Listener* 13 Mar. 332/1 On this particular campus, the provocative group was the black and Puerto Rican students' commune, so-called. 1972 D. DELMAN *Sudden Death* (1973) i. 15 It's exactly what that Puerto Rican hot dog would think is funny. 1978 *Langauge* LIV. 424 An example from Romance con-

sonantism would be the uvular *r*, which is at present very widely distributed, from Brazilian Portuguese and Puerto Rican Spanish to French, north and central Italian, and even some speakers of Rumanian. 1979 *Tucson* (Arizona) *Citizen* 20 Sept. 10A/1 They would resort to violence again in the interest of Puerto Rican independence.

puff, *sb.* Add: **1. f.** *Criminals' slang* (orig. *U.S.*). Explosive powder or dynamite used for blowing open a safe.

1904 'No. 1500' *Life in Sing Sing* 251/1 *Puff*, explosive powder. 1926 J. BLACK *You can't Win* ix. 107, I always crush into these powder shacks for my 'puff'.

g. *colloq.* Life, span of existence; usu. in phr. *in (all) one's puff* and varr., in all one's life.

1921 [see *CHEERIO *int.*]. 1922 JOYCE *Ulysses* 338 You never saw the like of it in all your born puff. 1929 WODEHOUSE *Mr. Mulliner Speaking* ix. 301 'Did you ever see a hat like that, Stinker?' 'Never in my puff,' replied his friend. 1938 —— *Code of Woosters* vii. 156 Did you ever in your puff see such a perfect perisher? 1960 K. MARTIN *Matter of Time* 165 That sort of thing's never happened to me in my puff. 1967 A. L. LLOYD *Folk Song in England* iv. 226 Hannah Snell..served for years as a marine..took a public house in Wapping and wore trousers for the rest of her puff. 1972 'A. ARMSTRONG' *One Jump Ahead* i. 9 Here's me actually going to dial nine-nine-nine! Never in all me puff would I've thought it!

2. b. Now usu. with reference to the sleeves of a dress; = *puff sleeve*, sense 9b below.

1884 B. POTTER *Jrnl.* 2 Apr. (1966) 78 Tight long sleeves with puffs to put on over them. 1908 L. M. MONTGOMERY *Anne of Green Gables* xi. 114 They all had puffed sleeves..it was awfully hard there among the others who had really truly puffs. 1968 J. IRONSIDE *Fashion Alphabet* 58 *Puff*, a short sleeve, gathered into the shoulder and into a band above the elbow.

c. A low padded seat or cushion; = *POUF[1] 3.

1877 H. JAMES *American* xii. 195 Valentin was sitting on a puff. *Ibid.* 206 Then she gave a little push to the puff that stood near her, and by a glance at Newman seemed to indicate that she had placed it in position for him.

d. *Cytology.* A short swollen region of a polytene chromosome, active in RNA synthesis. Cf. *PUFFED *ppl. a.* 1d; *PUFFING *vbl. sb.* 3c.

1937 C. B. BRIDGES in *Cytologia* (Fujii Jubilee Vol.) II. 751 Sections 58E and F show another characteristic peculiarity—namely, they often are converted into a much swollen light-staining 'puff' in which the banding is very hard to see. 1957 *Proc. Nat. Acad. Sci.* XLIII. 964 The correlation between the secretory activity in certain cells and the appearance and disappearance of puffs at specific loci in these cells had led many authors to the hypothesis that the genes at the locus of the puff may be actively controlling the secretory process. 1966 *Proc. R. Soc.* B. CLXIV. 284 The two different kinds of 'puff': (1)..the multi-stranded true puffs, which occur only in giant chromosomes: and (2)..the single-stranded loop 'puff' of lampbrush chromosomes. 1974 *Cold Spring Harbor Symp. Quantitative Biol.* XXXVIII. 660/2 Puffs result from the accumulation of RNA and proteins at a band which is being transcribed.

7. (Further examples.)

1774 J. WESLEY *Let.* 8 Jan. (1931) VI. 66, I suppose Mr. Rivington's advertisement is only a puff, as the booksellers call it. 1822 J. ROBISON *Syst. Mechanical Philos.* II. 47 His encomiums..are to a great degree extravagant. resembling more the puff of an advertising tradesman than the patriotic communications of a gentleman. 1916 A. HUXLEY *Let.* 29 Dec. (1969) 118, I lighted in to-day's *Morning Post* on a little puff of myself, apropos of *Oxford Poetry, '16*. 1923 E. WALLACE *Captains of Souls* xlvii. 258 'Ambrose Sault was executed at Wechester Jail..Billet was the executioner.' The hangman always received his puff. 1960 *Punch* 16 Mar. 383/2 Students are advised to omit fine language, puffs for the product, or any form of cosy get-togetherness. 1974 S. CHITTY *Beast & Monk* III. iv. 229 In January 1864 Kingsley reviewed Volumes VII and VIII of Froude's *History of England*.., no doubt with a view to giving his brother-in-law a 'puff'.

8. d. (See also *POOF *sb.*[1]) An effeminate man; a male homosexual.

1902 FARMER & HENLEY *Slang* V. 313/1 *Puff*... 3 (tramps'), a sodomist. 1937 PARTRIDGE *Dict. Slang* 665/2 *Puff*,..a sodomist. 1961 P. WHITE *Riders in Chariot* xi. 414 It was that puf Mortimer would not let me alone. 1967 H. W. SUTHERLAND *Magnie* iv. 63 He'd be a puff boy, this Magnie, and God knows what entertainment he laid on for Arthur. 1974 P. WRIGHT *Lang. Brit. Industry* xi. 95 An infuriated spectator may shout at a plump, sleek referee, 'You nasty little ponce!' (or *puff*).

9. b. (sense 1) *puff train*; (sense 2b) *puff scarf, sleeve*; *puff-sleeved* adj.; (sense 7) *puff merchant, writer*; **puff billiards**, a game resembling billiards, in which a ball is driven about on a table by puffs of air; **puff box** (examples); **puff pipe**, (*a*) a short pipe connected to a trap or valve in a drainage system in order to ventilate it; (*b*) on a vertical take-off aircraft, a pipe out of which compressed air is blown in order to control attitude; **puff port**, on a hovercraft, a vent out of which compressed air is blown in order to control attitude; **puff-shouldered** *a.*, having puffs (sense 2b) on the shoulders.

1897 Puff billiards [see *INDOOR, IN-DOOR *a.* 1]. 1901 *Commercial Advertiser* (N.Y.) 11 May 12/5 Mrs. Hwfa Williams is said to have invented puff-billiards. 1953 P. L. FERMOR *Violins of Saint-Jacques* 74 Usually some newly arrived acquisition from Paris occupied the centre of the room—a magic lantern, a kaleidoscope or..a game of puff-billiards. 1895 *Mongomery Ward Catal.* Spring–Summer 259/1 Puff boxes, made of papier mache... $0.20. 1926–7 *Army & Navy Stores Catal.* 99 Puff Boxes. 1951 R. CHANDLER *Let.* 6 July (1966) 143 Puff merchants ..will go on record over practically anything including the World Almanac, provided they get their names featured. 1894 A. J. WALLIS-TAYLER *Sanitary Arrangement of Dwelling-Houses* ix. 58 A puff-pipe, about one inch in diameter, should be taken from the valve-box through the outer wall, and its free end be also fitted with a brass flap-valve. 1934 *Archit. Rev.* Jan. p. xliv, Puff pipes, always a doubtful practice, although admissable, under certain conditions, are..here abolished and the terminals of the vent pipes being fixed high above all openings to the building ensure a strong current of fresh air throughout the system. 1960 *Aeroplane* XCVIII. 572/1 (*diagram*) Pilot controls..attitude and yaw via 'puff-pipes'. 1965 J. L. NAYLER *Aviation* xiii 188/2 Control in hovering flight was obtained by the 'puff-pipe' system first used in the *Flying Bedstead*. 1972 J. HASTINGS *Plumber's Compan.* 133 In Wiltshire the ornamental end of a puff pipe is a snake's mouth. 1967 *Jane's Surface Skimmer Systems 1967–68* 37/1 Puff ports, to improve in particular low-speed yaw control, and segmented skirts will be incorporated. 1971 R. L. TRILLO *Marine Hovercraft Technol.* v. 93 Puff ports used on the Parkhouse Beckingham Hovercat to assist in directional control also provided rolling moment causing the craft to roll into a turn, thereby enhancing the comfort of the turning manoeuvre. 1880 *Amer. Mail Order Fashions* in *Americana Rev.* (1961) 32 New puff-scarf, satin faced and lined. 1899 A. CONAN DOYLE *Duet* i. 7 A roomful of puff-shouldered young ladies. 1894 B. POTTER *Jrnl.* (1966) 314, I had to take his arm in to dinner, not much encouraged by his scrutiny of my puff-sleeves. 1975 G. HOWELL *In Vogue* 151/2 Little-girl dresses..with..full short skirts, tucks, smocking and puff sleeves. 1883 'MARK TWAIN' *Life on Mississippi* xxxviii. 404 Grandpa and grandma..stiff, old-fashioned, high-collared, puff-sleeved. 1969 *Observer* 21 Dec. 23/4 This smocked, puff-sleeved blouse. 1896 SWINBURNE *Let.* 29 May (1962) VI. 100 When the 'puff-train' did 'anything particularly startling or loud'. 1870 T. A. BROWN *Hist. Amer. Stage* 21/2 In September he quit the business, and soon after obtained the situation of 'puff writer' for the Bowery Amphitheatre.

puff, *v.* Add: **I. d.** Of a fungus: to discharge a cloud of spores suddenly.

1887 H. E. F. GARNSEY tr. *A. de Bary's Compar. Morphol. & Biol. Fungi* iii. 89 As long as the Fungus remains shut up in the damp atmosphere no amount of shaking will cause it to puff. 1953 C. T. INGOLD *Dispersal in Fungi* ii. 27 Once an apothecium has puffed it cannot, as a rule, be induced to do so again for a time.

puff-adder. Add: **1.** (Earlier and later examples.)

1789 W. PATERSON *Narr. Four Journeys Country of Hottentots* 164 The Puff Adder..has its name from blowing itself up to near a foot in circumference. 1915 *Chambers's Jrnl.* July 437/1 Perhaps the most loathsome of the snakes of Natal is the dreaded puff-adder. 1969 *Times* 15 Sept. (Uganda Suppl.) p. vi/8 Puff adders were sacred at Budo, so only Europeans could kill them. 1975 H. B. COTT *Looking at Animals* viii. 160 The difficulty is met by the application of patterns, such as those exhibited by African Rock Python, Gaboon Viper, Puff Adder.

2. *U.S.* The western hog-nosed snake, *Heterodon nasicus*, which belongs to the family Colubridæ but is not dangerous to man.

1882 *Amer. Naturalist* XVI. 566 Twice afterward I noticed this strange habit of the puff adders. 1897–8 'MARK TWAIN' *Autobiogr.* (1924) I. 103 Snakes..liked to lie in it [*sc.* a road] and sun themselves; when they were rattlesnakes or puff-adders, we killed them. 1966 R. C. STEBBINS *Field Guide to Western Reptiles & Amphibians* x. 145 When disturbed it [*sc.* the western hog-nose snake] often spreads its head and neck and strikes with open mouth, hissing, but seldom biting. This behaviour has earned it the names 'puff adder', 'blow viper', and 'hissing adder'.

puff-ball. Add: **3.** *Naut. slang.* (See quot.)

1933 J. MASEFIELD *Bird of Dawning* 263 Bloody Bill China had bonnets on his courses and contrivances that he called puffballs in the roaches of his topsails. *Ibid.* 307 Puff-balls or Save-alls. Extra sails laced to the feet of square sails.

puffed, puft, *ppl. a.* Add: The form **puft** is no longer current. **1. b.** (Examples of *puffed sleeve*.)

1802 C. EDGEWORTH *Let.* 30 Oct. in C. Colvin *M. Edgeworth in France & Switzerland* (1979) 21, I always repeat..that puffed sleives are *si ridicule*..just like our mantua-maker. 1932 'E. M. DELAFIELD' *Thank Heaven Fasting* III. iii. 278, I can remember a lovely pink evening dress you used to wear, with puffed sleeves. 1976 *National Trust* Autumn 25/1 (Advt.), *Ladies overblouses*..with V-cutaway pointed collar, cuffed and gently puffed sleeves.

c. Of cereal grain: expanded by means of high-pressure steam; used esp. in the names of breakfast foods.

1907 *Yesterday's Shopping* (1969) 11/2 *Breakfast Cereals* ..Quaker Puffed Rice—pkt. 0/5½. 1912 *Collier's* 21 Sept. 24/1 Prof. Anderson's process for Puffed Wheat and Puffed Rice requires a terrific heat. 1921 *Daily Colonist* (Victoria, B.C.) 23 Oct. 18/4 At I'Chang I saw the Chinese making puffed rice as for centuries past. 1930 B. S. BRONSON *Nutrition & Food Chem.* xvi. 372 Certain grains are sometimes subjected to a very high steam pressure which is suddenly released, expanding the grains and giving the various 'puffed' cereals. 1944 M. LASKI *Love on Supertax* i. 13 Would it be Puffed Rice or Shredded Wheat this morning? 1957 J. KERR *Please don't eat Daisies* (1958) 104 They all decide to make sandwiches of boiled egg and puffed wheat. 1972 *Sci. Amer.* Jan. 50/3 Most rats limited to exclusive diets of puffed rice, wheat flakes, shredded wheat and macaroni (all 'enriched') were barely able to hold their weanling weight (about 60 grams).

d. *Cytology.* Of part of a chromosome: see *PUFF *sb.* 2d. Cf. *PUFFING *vbl. sb.* 3c.

1938 *Genetics* XXIII. 159 In the giant chromosomes of *Sciarra ocellaris* Comst. certain particular regions appear greatly expanded or 'puffed'. 1965 J. D. EBERT *Interacting Syst. in Development* vi. 110 A given section of a chromosome may appear as a sharp band in most tissues, but as a diffuse 'puff' in one tissue. Or within the same tissue, a given section may be discrete at one time and 'puffed' at another. 1970 *Cold Spring Harbor Symp. Quantitative Biol.* XXXV. 534/1 At any particular stage in development only a subset of these sites is actually puffed and the pattern of puffs changes in a very regular and highly coordinated way as development proceeds.

puffer. Add: **1. a.** Also, *spec.* (chiefly *Sc.*) a small steamboat used for carrying cargo in coastal waters. (Further examples.)

1922 *R. Cruising Club Jrnl. 1921* 98 We got under way half an hour later, having been delayed by the puffer *Anna Bhan*, which had let go almost over our anchor. 1927 [see *CHOO-CHOO]. 1946 J. IRVING *Royal Navalese* 139 *Puffer*, a heavily-built fishing-boat type of vessel, usually fitted with a single cylinder Diesel engine. 1959 *Times* 12 Dec. 9/7 Para Handy and his crew of three run a 'puffer' (a small cargo boat) between the towns and villages on the Firth of Clyde. 1968 'D. HALLIDAY' *Dolly & Singing Bird* x. 112, I saw the anchor light of a big boat, a puffer. 1974 *Times* 7 Dec. 3/2 Mr Alan Pegler bought the majestic old LNER puffer [*sc.* the Flying Scotsman] in 1962. 1975 *Stornoway Gaz.* 5 July 1/9 A call for help was heard, stating that the puffer 'Lady Morven' had broken down and was drifting ashore in Loch Cuan.

c. Substitute for def.: In full, *puffer fish*. A carnivorous globe-fish that can swallow air to inflate itself, belonging to the family Tetraodontidæ, which includes about ninety species found in warm or temperate seas; also, a porcupine-fish belonging to the closely related tropical family Diodontidæ; cf. *puff-fish* s.v. PUFF *sb.* 9b. (Earlier and later examples.)

1814 S. L. MITCHELL *Fishes N.Y.* 473 Puffer... He is called in some places, toad-fish, because his back is mottled with yellow & dark. 1883 *Bull. U.S. Nat. Mus.* XXVII. 428 *Tetrodon nephelus*... Rough Swell-fish; Puffer; Blower; Swell Toad. Gulf of Mexico, abundant. 1884 G. B. GOODE *Fisheries U.S.: Nat. Hist. Aquatic Animals* I. III. 170 The Porcupine Fishes—Diodontidæ. Swell Fishes and Puffers. 1930 *Times Educ. Suppl.* 18 Oct. (Home & Classroom Section) p. ii/2 Puffer fish have the habit of inflating themselves with air, which they swallow. 1941 J. STEINBECK *Sea of Cortez* x. 77 Small spine-covered puffer fish..bloat themselves when they are attacked, erecting the spines. 1947 [see *BLAASOP]. 1962 K. F. LAGLER et al. *Ichthyology* v. 151 A remarkable modification of the stomach exists in the puffers (Tetraodontidae) and porcupinefishes (Diodontidae) which can inflate themselves with water or air to assume often an almost globular shape. 1967–8 *Bahamas Handbk. & Businessmen's Ann.* (ed. 7) 456 Here and there..the wader will find a solitary porcupine fish, also called puffers or blowfish. 1974 M. C. GERALD *Pharmacol.* i. 3 The ovaries of the pufferfish (an excellent source of tetrodotoxin, one of the most powerful poisons known to man).

d. (Earlier and later examples.)

1884 G. B. GOODE *Fisheries U.S.: Nat. Hist. Aquatic Animals* I. I. 14 On the Atlantic coast occurs most abundantly the little Harbor Porpoise *Phocæna brachycion* Cope, known to the fishermen as 'Puffer', 'Snuffer', [etc.]. 1911 *Fisheries U.S. 1908* (U.S. Bur. Census Spec. Rep.) 314/1 Porpoise (*Phocæna communis*).—A cetacean found on the north Atlantic and north Pacific coasts, ascending rivers. It is known as 'harbor porpoise', 'herring-hog', 'puffer', [etc.].

e. A wheel-lock pistol.

1970 G. BOOTHROYD *Handgun* i. 16/2 The French makers equalled the Germans in ingenuity but..were able to combine with that ingenuity an elegance of form that is totally lacking in..the ball-butted Puffer, the German wheellock pistol. 1973 *Country Life* 29 Mar. 881 *Antique Firearms*..Saxon wheel-lock Puffer, dated 1590, length 23 inches. 1973 *Times* 22 May 18/3 A Nuremberg wheel-lock 'puffer' made £7,000.

f. A soft plastic container designed to blow powder on to the skin, etc., when squeezed; freq. *attrib.*, as *puffer bottle, pack*; also *talc puffer*.

1971 *Homes & Gardens* Aug. 89/2 An insect powder in a puffer pack is convenient for this job. 1971 *Petticoat* 24 July 9/3 A must for your handbag is the..*Travel Trio* which contains a puffer talc. 1973 J. WOOD *North Beat* ii. 27 It's fine powder you get in a puffer bottle—puff it on your hair to make it look right grey. 1974 *Harpers & Queen* Sept. 50/1 Talc puffer 55 p. 1978 R. WESTALL *Devil on Road* viii. 52 I'll take..a puffer-bottle for the [cat's] ear-mites.

puffery. Add: **1.** Now chiefly *U.S.* (Further examples.)

1929 D. G. MACKAIL *How Amusing!* 518 The gossip-writers had all contributed their quota of unpaid puffery. 1963 D. OGILVY *Confessions Advertising Man* (1964) vi. 110 The reader finds it easier to believe the endorsement of a fellow consumer than the puffery of an anonymous copy-writer. 1966 *Daily Tel.* 3 Nov. 14/2 Richard Maney, Press agent to some 300 Broadway shows, and..a master of flamboyant puffery. 1970 *Observer* 1 Mar. 13/3 An

American company selling weight reducing pills was prosecuted for misleading advertising. One of its defences was puffery. 1978 J. Carroll Mortal Friends I. ii. 25 If you'll not be subject to the spiritual authority of the Church even on the day of your sacrament, don't blaspheme your martyred countrymen by such puffery.

puffick (pʊ·fik), a. [Repr. colloq. and dial. pronunc. of Perfect.] = Perfect a. (esp. sense 5d). So **pu·ffickly** adv. = Perfectly adv.

1891 Kipling Many Inventions (1893) 3 He knows puffickly well where he is. 1907 E. Nesbit Enchanted Castle iv. 105 You aint allowed to arrest a chap on suspicion, even if you know puffickly well who done the job. 1949 M. Allingham More Work for Undertaker xiv. 176 The chap..was a puffick stranger. 1967 'A. Gilbert' Visitor viii. 139 They'll ask..why you should take a chance like that for a puffick stranger. 1972 C. Drummond Death at Bar ii. 61 She mimicked in Cockney, 'A puffick gentlemen I'm sure, dear.'

puffily (pʊ·fili), adv. [f. Puffy a. + -ly².] In a puffy manner.

1882 Caulfield & Saward Dict. Needlework 415/1 When Petticoats are to be Quilted, the Runnings should be well indented and the satin or silk set up puffily. 1904 H. G. Wells Food of Gods II. ii. 197 He did the rise over by the chalk-pit crest a little puffily. 1963 A. Smith Throw out Two Hands xiv. 150 We could look back at the cloud... It was brooding over us no more, but clearly and puffily to one side. 1975 S. Lauder Killing Time on Corvo i. 7 He looked puffily unhealthy. 1977 C. McCullough Thorn Birds ii. 21 They paused.., the five bright heads haloed against a puffily clouded sky.

puffin¹. Add: **2.** (With capital initial.) The proprietary name of a variety of children's paper-back book or series of books published by Longman Group Limited (see *penguin 2c).

1947 Trade Marks Jrnl. 10 Sept. 535/2 Puffin 648,226. Printed publications, stationery, bookbinding materials, pens and pencils, but not including publications on puffins. Penguin Books Limited..Manufacturers and Publishers.—28th May, 1946. 1960 Penguins Progress 1935–60 54 Each month we publish fifteen to twenty books, varying from Penguin fiction..to Penguin Handbooks and Puffins. 1979 Guardian 29 Oct. 12/6 A list of leading children's writers who are not in Puffin would be a pretty short one.

3. (With capital initial.) The proprietary name of a make of duvet or continental quilt. Also Puffin Downlet.

1959 Trade Marks Jrnl. 18 Feb. 206/2 Puffin... Filled bed coverings in the nature of quilts or eiderdowns. Arthur R. Davis and Company Limited,..Croydon, Surrey; manufacturers and merchants. 1970 'R. Crawford' Kiss Boss Goodbye II. viii. 107 Brenda was lying on her back on top of the Puffin Downlet. 1971 Guardian 29 Sept. 11/2 The cleaning of feather-filled continental quilts is a problem... One reader, taking her Puffin to.. a well-known cleaners..was met with the blankest confusion.

puffing, vbl. sb. Add: **1. b.** The sudden discharge of a cloud of spores by a fungus.

1887 H. E. F. Garnsey tr. A. de Bary's Compar. Morphol. & Biol. Fungi iii. 89 Many of the Discomycetes have the peculiar habit of 'puffing'.., of suddenly discharging a whole cloud of spores. 1953 C. T. Ingold Dispersal in Fungi ii. 27 If a cup-fungus is picked and at once placed near the ear, puffing may occur and is then audible as a hissing sound. 1976 G. C. Ainsworth Introd. Hist. Mycol. vii. 196 Micheli..was also the first to record and illustrate the visible 'puffing' of spores from the ascocarps of discomycetes.

3. c. Cytology. The occurrence or formation of puffs on a chromosome. Also attrib. Cf. *puff sb. 2d; *puffed ppl. a. 1d.

1938 Genetics XXIII. 159 Intermediate degrees of 'puffing' show the bands or discs in various stages of disruption. 1954 Exper. Cell Res. VI. 199 The 'lampbrushes' show a considerable degree of lateral 'puffing' as judged from the fact that the immediately neighbouring branches of the chromosome are only half as thick. 1968 H. Harris Nucleus & Cytoplasm iv. 78 It has been contended that the pattern of puffing shows organ specificity, but the evidence for this does not seem to be at all conclusive. 1970 Cold Spring Harbor Symp. Quantitative Biol. XXXV. 534/1 Periods of greatest puffing activity are seen as the animals molt from one instar to the next and both events, molting and the initiation of a specific sequence of puffing, are primarily controlled by the same hormone,..ecdysone.

5. puffing-hole (see quot.).

1862 J. B. Jukes Student's Man. Geol. (ed. 2) II. x. 220 The sea sometimes gradually forms a passage for itself in the surface above, and if that be not too lofty, forms a 'blow-hole' or 'puffing-hole', through which spouts of foam and spray are occasionally ejected high into the air.

puffing, ppl. a. Add: **1.** Also Puffing Billy, an affectionate name for a steam locomotive or train. Also transf. and attrib.

The original 'Puffing Billy' was built by William Hedley in 1813.

1934 Joyce Let. 25 Apr. (1966) III. 304, I prefer Puffing Billy to Swaggering Bob. 1963 A. Lubbock Austral. Roundabout 172 A 'Puffing Billy' engine, wood-fired, was dredging the silt. 1977 Times 19 Apr. 5/5 (caption) A replica of an early American 'Puffing Billy'.. at the National Railway Museum, York.

puffinry (pʊ·finri). [f. Puffin sb.¹ + -ry.] A place occupied by a breeding colony of puffins.

1954 Fisher & Lockley Sea-Birds iii. 64 There are seven separate puffin-slopes on St. Kilda each of which is larger than..even the largest puffinry in the..Shiant Isles. 1960 Williamson & Boyd St. Kilda Summer xi. 113 It [sc. Stac Lee, St. Kilda] has large guillemot and kittiwake colonies too, and Stac an Armin many razorbills and a big puffinry. 1974 Country Life 14 Feb. 290/2 Much the same is true of the other puffinries on the islands.

puffy, a. Add: **4.** puffy-cheeked, -eyed, -looking adjs.

1922 H. Crane Let. 27 July (1965) 94 The Man Ray photo of J...is really not a good resemblance. The face is not so puffy-looking. 1926 V. Woolf Writer's Diary (1953) 89 In trotted a little puffy-cheeked cheerful old man [sc. Hardy]. 1929 Sunday Dispatch 20 Jan. 16/3 You watch the boy growing puffy-eyed and soft on his fast living. 1957 J. Kerouac On Road (1958) 170 Everybody looked like a broken-down movie extra, a withered starlet; disenchanted stuntmen,..puffy-eyed motel blondes..a lemon lot.

puftaloon (pʊ·ftălūn). Austral. Also **puffta-loon, puftaloony, puffterlooner, puff de loon(ey)**, etc. [Origin obscure; cf. Puff sb. 5.] A small fried cake, spread with jam, sugar, or honey, and usu. eaten hot.

[1853 Mossman & Banister Australia 126 'Leather-jackets'—an Australian bush term for a thin cake made of dough, and put into a pan to bake with some fat... The Americans indulge in this kind of bread, giving them the name of 'Puff balloonera'.] 1871 M. Clarke His Natural Life v. ii, in Austral. Jrnl. VI. 602/2 'Have a puffterlooner, Master Dick,' suggests Derwent Jack, 'or a bit o' sweetcake.' 1908 A. Gunn We of Never Never xiv. 189 The cooking lessons proceeded until the fine art of making 'puff de looneys'..had been mastered. 1935 Bulletin (Sydney) 27 Feb. 20/1 Camped in Tapalin Bend, River Murray, he was frying puftaloons when a light shower caused the hot fat to jump and sting his hands and wrists. 1940 I. L. Idriess Lightning Ridge 82 Puftaloons are tasty though; fry them in fat, then smother them with treacle and swallow 'em while greasy. 1942 E. Langley Pea Pickers 296 She was making puftaloonies all the time we talked and in next to no time had a plate of them in front of us, with a hot cup of tea. 1964 T. Ronan Packhorses & Pearling Boat v. 140 A camp oven full of 'puff de loons' (fried scones to the uninitiated). 1970 P. White Vivisector i. 10 Mumma started telling all she had heard next door, with the kids stuffing on Mrs. Burt's cold puftaloons.

pufter, var. *poofter.

pug, sb.² Add: **II. 9. b.** A net or snood for tying up or holding a bun or knot of hair. Also attrib.

1927 Blackw. Mag. June 747/1 His hair tied in a knot in a little red cloth or pug, on the top of his head. 1967 E. B. Nickerson Kayaks to Arctic x. 92, I had been wearing my hair in a long braid but tonight I coiled and netted it in a pug. 1967 Boston Globe 21 May (Confidential Chat) 17/1 The old fashioned idea of a dark, gloomy building..with an old fashioned old lady with glasses and a pug hair-do for a librarian are far out these days.

III. 13. pug-bitch (sense 7).

1916 E. Pound Lustra 111 Quite plump, with pug-bitch features.

pug, sb.⁴ Add: Also Comb., as pug-mark.

1922 Chambers's Jrnl. Dec. 860/1, I found a good many pug-marks and from them I concluded that the man-eater was a smallish beast. 1946 J. Corbett Man-Eaters of Kumaon 8 Entering the ravine..I found the pug marks of a tiger in some fine earth..; these pug marks showed the animal to be a tigress, a little past her prime. 1974 Country Life 31 Oct. 1302/2 Tigers are elusive... We followed pug marks up hill and down ravine.

pug, sb.⁵ Add: (Further examples.) Hence **pug-glove**, a boxing-glove.

1924 J. Buchan Three Hostages v. 74 The man had been in the ring, and not so very long ago. I wondered at Medina's choice, for a pug is not the kind of servant I would choose myself. 1938 Dylan Thomas Let. 1 June (1966) 198 'Boxed' has the coffin and the pug-glove in it. 1961 Lancet 26 Aug. 447/2 It is well known that boxers, including fair-ground-booth pugs, can tolerate severe direct blows to the head. 1977 Time 19 Dec. 68/2 Hemingway had gone many rounds with pugs, and Journalist Paul Gallico once had his fillings loosened by Jack Dempsey.

pu·gginess². [f. Puggy a.¹ + -ness.] Squatness; stumpiness.

1910 H. G. Wells Hist. Mr. Polly vii. 227 Mr. Hinks.. displayed a freckled fist of extraordinary size and pugginess..to Mr. Polly's close inspection.

pu·ggishness. [f. Puggish a. + -ness.] The nature of or resemblance to a pug.

1924 W. J. Locke Coming of Amos ii. 13 There is a puggishness about her rebellious nose which would disqualify her in a competition of Classical Beauty.

puggle, var. *poggle sb. and a.

puggled (pʊ·g'ld), a. Army slang and Sc. Also Sc. **pagard, pagart**. Var. *poggled a.

1923 G. Watson Roxburghshire Word-bk. 227 Pagart, a. Also pagard... Breathless: 'A was fair pagard; A couldna rin another fitlenth.' 1925 Fraser & Gibbons Soldier & Sailor Words 231 Puggled.., Insane. Mad-drunk. 1968 Sc. Nat. Dict. VII. 274/2 Puggled, -t, at a standstill from exhaustion or frustration, done for, at the end of one's resources. 1973 'J. Patrick' Glasgow Gang Observed xii. 112 One guest in particular..was 'stoatin' aw ower the place—pure puggled she wis'.

puggly, var. *poggle sb. and a.

pugil³ (piū·dʒil). U.S. Mil. [Prob. f. L. stem pugil-, as in Pugilism, etc.] In full, pugil stick. A short pole with padded ends used as a substitute for a rifle and bayonet in military training. Also attrib. and Comb., as pugil bout, training; pugil-armed adj.

1962 Infantry Nov.–Dec. 26/2 Pugil training was first adopted by the Marine Corps... The pugil stick is an oak staff, two inches in diameter, padded on both ends with polyfoam encased in canvas. This stick represents the rifle and is the same length as a rifle with bayonet fixed. Ibid. 26/3 The students engage in pugil bouts, applying all the movements taught in earlier periods with the rifle. Ibid. 28/3 Substitute pugil-armed students for dummies. 1964 A. N. Hardin Amer. Bayonet 1776–1964 188 Modern U.S. Marine Corps practice virtually eliminates fencing bayonets by substituting the 'Pugil Stick' technique. 1967 Britannica Bk. of Year 1966 804/1 Pugil or Pugil stick, a padded club with large rounded ends used in bayonet practice by the military as a substitute for rifle with fixed bayonet. 1976 Billings (Montana) Gaz. 28 June 4–B/1 The first of three sergeants..was accused of violating orders in the conduct of the..pugil stick bouts in which the recruit was pounded into a coma.

pugilant, a. For rare⁻¹ read rare and add further example.

1932 J. Joyce in New Statesman 27 Feb. 260/1 A pugilant gang theirs, per Bantry!

pugilistic, a. (Earlier example.)

1789 Loiterer 27 June 4 A tolerable Proficient in pugilistic Science.

pugilistical, a. (Earlier example.)

a 1790 J. H. Beattie Dial. of Dead III, in J. Beattie Minstrel (1799) II. 195 Some learned innovator..clapping to it [sc. Latin pugil] part of a Greek termination, he made it pugilist; which..gave rise to the adjectives Pugilistic and Pugilistical, as in this example,..'a.. pavilion at Newmarket for Pugilistical exhibitions'.

Puginesque (piū:dʒine·sk), a. [f. the name Pugin (see def.) + -esque.] Of, pertaining to, or characteristic of the English architect A. W. N. Pugin (1812–1852) or his style of architecture; Gothic-revivalist. So **Pu:gine·squery** [-ery] nonce-wd., matters related to Pugin or his architectural style.

1848 C. Kingsley Yeast v, in Fraser's Mag. XXXVIII. 286/1 When they talk Puginesquery, I stick my head on one side attentively, and 'think the more', like the lady's parrot. 1856 F. E. Paget Owlet of Owlstone Edge 210 In her ambition to be Puginesque, she made her husband's chancel look as if it had been decked by a mad haberdasher. 1864 Ecclesiologist XXV. 345 The general idea is Puginesque, the style Middle-Pointed. 1904 A. C. Benson Let. 5 Oct. in Upton Lett. (1905) 234 The roofs and towers of the big house—Puginesque Gothic, I must tell you—came in sight. 1907 E. Gosse Father & Son xii. 339 My Father did not, indeed, forbid me to enter..the stately Puginesque cathedral which Rome had just erected. 1961 E. Waugh Unconditional Surrender II. iii. 72 The Catholic parish church is..a Puginesque structure erected..in the early 1860's. 1975 V. Cunningham Everywhere spoken Against iii. 86 The whole set of Puginesque-Arnoldian assumptions.

pug-mill. (Further examples.)

1902 A. Bennett Anna of Five Towns viii. 169 The press expelled the water, and the pug-mill expelled the air. 1930 Industr. & Engin. Chem. 20 Mar. 14/1 The machinery is..simple in design and operation, including, in addition to the centrifuge, a 'pug mill' in which hot water and steam are used to wash the oil-impregnated sands. 1960 [see *extruder 1]. 1980 M. Forster Bride of Lowther Fell xi. 168, I decided..to ferret out a pug-mill..together with a special plaster sink I would soon be needing.

pugnozzle (pʊ·gnǫz'l), v. nonce-wd. [f. Pug sb.² + Nozzle sb. or nozzle, var. Nuzzle v.; cf. Pug nose.] intr. To move the upper lip and nostrils in the manner of a pug-dog.

1934 S. Beckett More Pricks than Kicks 257 The wretched little wet rag of an upper lip, pugnozzling up and back in what you might call a kind of a duck or a cobra sneer to the nostrils.

Pugwash (pʊ·gwǫʃ). The name of a village in Nova Scotia where in 1957 the first conference was held of scientists concerned to promote international responsibility in the peaceful uses of scientific discoveries, used attrib. of this and subsequent conferences, and also the movement which they generated,

as Pugwash conference, group, etc. Also ellipt.

1957 N.Y. Times 12 July 3/6 Asked if he and his Russian colleague came to the Pugwash meeting as independent scientists, Mr. Topchiev replied [etc.]. 1957 Science 2 Aug. 199/3 These men all signed the 'Pugwash Statement'. 1957 Bull. Atomic Sci. Sept. 244/2 The Pugwash discussion showed clearly that the most important unresolved disagreement among experts concerned not the radiation exposure caused by bomb tests..but the biological consequences likely to be produced by this exposure. 1958 Washington Post 15 Sept. 18/2 Pugwash conferences, so named because the first one took place at Pugwash, Nova Scotia, on the estate of American multimillionaire Cyrus Eaton. 1963 Pugwash Newslet. July 2 Since the Pugwash Newsletter is for private circulation only, it may provide a convenient medium for testing and impromptu..ideas. 1965 New Statesman 16 Apr. 601/1 He [sc. Albert Einstein] died 10 years ago on Sunday, two days after signing the great Russell-Einstein Manifesto which launched the Pugwash Movement of scientists (the 14th meeting took place in Venice this week). 1968 Listener 11 July 36/1 Pugwash..has set up a study group. 1972 Guardian 4 Sept. 10/6 Pugwash is this week spending much of its time discussing problems of developing nations. 1976 Chronicle-Herald (Halifax, Nova Scotia) 19 July 21/1 The banquet meeting of the Canadian and American Pugwash groups.

puha (pu·ha). N.Z. Also **pua, puwha**. [Maori.] A sow-thistle belonging to the genus Sonchus, esp. S. oleraceus; the leaves of this plant used as a vegetable.

1843 E. Dieffenbach Trav. N.Z. II. III. 380 Pua—a sowthistle. 1868 W. Colenso in Trans. N.Z. Inst. I. III. 37 The fresh gum-resin from the Kauri..was commonly chewed as a masticatory..mixed with the inspissated juice of the Puwha, or Sow-thistle. 1905 W. B. Where White Man Treads 177 All was ready for the contents of the haangi—pork, puha, and potatoes. 1947 Coast to Coast 1946 2 Flora Baker gathering puha outside the Bakers' pig-sty. 1963 N.Z. Listener 3 May 5/4 Around in the rank growth, the puha and rariki grew, despised by the Pakeha but loved by the true Maori. 1966 J. K. Baxter Pig Island Lett. 6 Sea-eggs, puha, pork, and kumara. 1978 Islands (N.Z.) Aug. 12 Sometimes they bring food and cook it in my kitchen—mutton-bird with corn and puha—not my favourite dish, so greasy. 1978 P. Grace Mutuwhenua ix. 53 He and his flat-mate had boiled mutton quite often but there wasn't any puha near their place so they had to use cabbage.

pui, var. *poui; **puirt-a-beul**: see *port-a-beul.

puissance. Restrict arch. to senses in Dict. and add: **1. c.** Show-jumping. A competition testing a horse's ability to jump large obstacles. Also attrib.

1951 M. P. Ansell Show Jumping vi. 48 Test (Puissance). This competition is designed to test the horse's ability to jump large obstacles. 1952 R. S. Summerhays Encycl. Horsemen 218/2 'Puissance' Jumping Competition. In this all the straight fences in the course, with the exception of the first, are a minimum height of 4 ft. 7 in. (1 m. 40 cm.). The course usually consists of from six to eight fences, one of which is a double jump counting as one obstacle. 1954 P. Smythe Jump for Joy iv. 68, I lived for the week-ends, and many a lesson was passed in the haze of daydreams about jumping paddocks and over Puissance courses. 1959 Times 10 Aug. 5/6 It was the third triumph of the [Dublin Horse] show for Dundrum, who..shared first place in the puissance with Hollandia. 1974 Country Life 3 Jan. 9/2 Alcatraz, ridden by last year's puissance winner..attempted all three fences. 1975 Oxf. Compan. Sports & Games 281/1 The competition which tests jumping ability alone is known as a Puissance... There is no recourse to the clock, since the object is to test the horse over 'a limited number of large obstacles'.

‖ **pujari** (pūdʒā·ri). Also **poojari, pujaree**. [Hindi, f. Skr. pūjā worship; cf. Poojah, puja.] A Hindu priest.

1813 F. Hamilton Jrnl. Shahabad Survey (1926) 129 The Pujaris who are making a good thing of the ghost have lately been disturbed by a..young Brahman. 1855 H. H. Wilson Gloss. Judicial & Revenue Terms India 396/1 At Benares, the Pandá officiates only on particular occasions, the duties of daily worship being performed by inferior priests or Pujáris in his employ. 1883 Monier Williams Relig. Thought & Life in India ix. 249 Then the Pūjāri, or priest, takes the Bhūta sword and bell in his hands. 1907 B. M. Croker Company's Servant xi. 108 Many baskets of rice were contributed..to the sacrificial pile. On this pile, a drove of buffaloes was killed by the Poojaris. 1967 Singha & Massey Indian Dances i. 33 Offerings of flowers and sweetmeats are placed at the base by the pujarees (worshippers). 1969 Sunday Standard (Bombay) 3 Aug. (Mag. Section) p. iv/3 Next week, Indran left the village, leaving the temple in charge of another poojari. 1969 Enact (Delhi) Nov. 10/3, I will not have any brahmins or pujaris up here, never,..any more than I will go to church for confession or have a mass said to save my soul. 1973 Country Life 11 Jan. 80/1 We..presented our credentials to the pujari (priest).

puka (pu·kă). [Maori.] An evergreen New Zealand shrub or small tree, Griselinia littoralis, or G. lucida, a related, sometimes epiphytic, species, belonging to the family Cornaceæ and bearing leathery ovate leaves and panicles of tiny, greenish flowers followed by purple or black berries.

[1853 J. D. Hooker Bot. Antarctic Voy.: Flora Novæ-Zelandiæ I. 98 Griselinia lucida... Nat[ive] name, 'Poukatea'.] 1889 T. Kirk Forest Flora N.Z. 67 Mr. Colenso informs me that the Native name 'puka' is correctly applied to this species [sc. Griselinia lucida] as well as to Meryta Sinclairii. 1970 M. E. Fisher et al. Gardening with N.Z. Plants I. 59 The puka is usually found perched high in the forks of some forest tree such as the rata.

2. A small evergreen tree, Meryta sinclairii, belonging to the family Araliaceæ, native to Australasia and the Pacific Islands, and bearing large glossy leaves and terminal panicles of greenish-white flowers followed by black berries. Also attrib.

1889 T. Kirk Forest Flora N.Z. 243 The Puka... This noble species is..restricted to a few individuals growing on one or two small islands near the northern extremity of the colony. 1951 Post-Primary School Bull. (Wellington, N.Z.) V. XII. 276 The puka (Meryta sinclairii)..has the largest leaves of any New Zealand tree. 1970 M. E. Fisher et al. Gardening with N.Z. Plants I. 78 The puka is a much branched, small tree from 12 to 20 feet in height... The leaves of this handsome tree are a beautiful glossy green, nine to 20 inches long and half as broad, usually with the broadest part near the apex. They are leathery and strongly veined and the margin is wavy.

pukaki, puka pu, varr. *pukeko, *pakapoo.

pukatea (pukăti·ă). N.Z. [Maori.] A tall forest tree, Laurelia novæ-zelandiæ, of the family Monimiaceæ, native to New Zealand, and distinguished by buttresses at the base of the trunk, pale bark, leathery, obovate leaves, and small clusters of tiny, yellowish flowers; also, the timber obtained from this tree.

1843 E. Dieffenbach Trav. N.Z. I. I. iii. 75 Another tree common in this part of the valley is the pukatea. 1868 W. Colenso in Trans. N.Z. Inst. I. III. 51 From the aromatic leaves and bark of the Pukatea..a valuable essential oil might be extracted. 1882 W. D. Hay Brighter Britain! II. vi. 191 The Pukatea..is a tree of the second largest class. 1949 P. Buck Coming of Maori III. vi. 426 Tree burial was resorted to in the thickly forested Urewera country. Natural hollow trees such as the pukatea were utilized when available. 1950 N.Z. Jrnl. Agric. June 602/3 The rata, pukatea, and titoki from which they [sc. the chairs] are made. 1966 Encycl. N.Z. II. 887/2 Pukatea grows to heights of over 120 ft.

puke, sb.² Add: **1. b.** Matter thrown up from the stomach; vomit. coarse.

1961 in Webster. 1972 D. Lees Zodiac 109, I.. choked back the puke that had rushed to my throat. 1975 New Society 4 Dec. 526/2 At the Black Raven, by Liverpool Street station,.. there is a slight odour of puke and disinfectant.

3. b. (Earlier and later examples.)

1835 A. A. Parker Trip to West & Texas 87 The inhabitants..of Michigan are called wolverines,..of Missouri, pukes. 1843 'R. Carlton' New Purchase II. 47 This Protestant assembly was a gathering of delegates principally from the land of Hoosiers..[with] a small chance of Pukes from beyond the father of floods. 1847 T. Ford Hist. Illinois (1854) ii. 68 The Illinoians.. called the Missourians 'Pukes'... The lower lead mines in Missouri had sent up to the Galena country whole hoards of uncouth ruffians, from which it was inferred that Missouri had taken a 'Puke'. 1908 L. Houck Hist. Missouri III. xxiv. 36 'Hidalgos' the first residents of upper Louisiana and Missouri were called, until in the mouths of the vulgar the name of 'Pukes' was made current. 1944 [see *Missourian sb. and a.].

pukeko (pu·keko). N.Z. Also **pukaki**. [Maori.] The purple gallinule or swamp hen, Porphyrio porphyrio (formerly P. melanotus), belonging to the family Rallidæ and widely distributed in southern Europe, Africa, southern Asia, and Australasia.

1835 W. Yate Acct. N.Z. (ed. 2) ii. 62 Pukeko—A species of water-hen, the size of a well-grown capon. It.. has very long red legs. 1845 E. J. Wakefield Adv. N.Z. I. viii. 228 The pukeko is of a dark blue colour, and about as large as a pheasant. 1853 W. R. Bridges Let. 14 Nov. in Richmond-Atkinson Papers (1960) I. iii. 136, I amused myself with shooting wild duck and pukeko which are very abundant. 1874 A. Bathgate Colonial Experiences vii. 85 In the swamps there is also the pukaki..or swamp-turkey, a bird which rises well and affords good sport. 1884 Lady Martin Our Maoris viii. 114 The place was populous with large black birds, called by the Maoris pu-ke-ko. They have a harsh cry like a corn-crake. 1921 H. Guthrie-Smith Tutira xxii. 209 The Pukeko or Swamp-hen (Porphyrio melanotus) has..proved able to thrive better on dry ground than wet. 1930 L. G. D. Acland Early Canterbury Runs 1st Ser. vii 175 There were thousands and thousands of pukaki in the swamp. 1946 Coast to Coast 1945 110 Poachin' flappers in the swamp... Stringy pukekoes, too, when they're winged. 1957 J. Frame Owls do Cry xviii. 74 Pukekos took long strides through the swamp. 1966 Encycl. N.Z. II. 888/1 Pukekos are not shot in any great numbers during their open season.

pukeru, var. *puckeroo a. and vb. **Pukhto**: see Pushtoo, -tu sb. and a. in Dict. and Suppl.

puking, ppl. a. (Later example.)

1976 'D. Halliday' Dolly & Nanny Bird xv. 200 Order one of your puking friends to go below.

pukka: see Pucka, pakka a. (sb.) in Dict. and Suppl.

puku¹. Now the usual spelling of Pookoo.

1881 [see *lechwe]. 1894 [see Pookoo]. 1900 W. L. Sclater Mammals S. Afr. I. 193 The flesh of the puku is stated by Selous to be even more nauseous and unpalatable than that of the common water-buck. 1946, 1972 [see *lechwe]. 1973 D. Steele Game Sanctuaries S. Afr. 148 The puku is a medium-sized antelope, weighing up to 90 kilograms.

‖ **puku²** (pu·ku). N.Z. colloq. [Maori.] The stomach.

[1905 W. B. Where White Man Treads 96 The Maori is pre-eminently gifted in the selection of suitable nomenclature... He meets a man... He looks him over... If he be massive in girth, what so delicate as reference thereto as 'Puku' (stomach)? 1918 C. H. Weston Three Years with New Zealanders vi. 70 The Medical Officer.. injected the [anti-typhoid] serum in what the Maories call my puku.] 1941 Baker N.Z. Slang. v. 42 By the 1890's..a stomach-ache became, with excellent alliterative effect, a pain in the puku. 1954 Numbers July 5 Let's stick to matters of the puku. 1958 M. K. Joseph I'll soldier no More ii. 50 Eat, Harry boy, eat; get some of this in your puku. 1966 G. W. Turner Eng. Lang. Austral. & N.Z. viii. 170 The [Maori] words potae 'hat' and puku 'stomach' are still current, but North Island words mainly. 1971 N.Z. Listener 19 Apr. 57/1 He was too stonkered and crook in the puku to bother about it at the time. 1978 P. Grace Mutuwhenua xx. 140 Your puku's getting in the way.

puky, a. Restrict rare to sense in Dict. and add: Also **pukey**. **2.** fig. Sick-making, disgusting. colloq.

1965 T. Capote In Cold Blood (1966) i. 37 It was a puky idea. What the hell would they have thought? 1969 W. Garner Us or Them War vi. 53 There'll be all sorts there, most of them pretty pukey and not really my thing.

pul (pūl). Pl. **puls, pooli, puli**. [Pashto, a. Pers. pūl, f. Turk. pul; cf. Gr. φόλλις a small coin.] In Afghanistan, a monetary unit equivalent to one hundredth of an afghani; a coin of this value.

1927, 1934 [see *afghani]. 1941 Whitaker's Almanack 852/1 Afghani (of 100 Puls).

Pul, var. *Peulh sb. and a.

pula (pu·lă). Also †**poola**. [Tswana.] ‖ **1.** Rain. Freq. used as int.: in southern Africa, a traditional salute or expression of good luck.

1827 G. Thompson Trav. & Adv. S. Afr. I. 180 Mattebe ..waved the point [of his assagai] towards the heavens, when all called out 'Poola!' i.e. rain or a blessing. 1842 R. Moffat Missionary Labours & Scenes S. Afr. xxi. 350 The audience shouted, 'Pùla', (rain,) on which he sat down amidst a din of applause. 1934 in C. P. Swart Africanderisms (M.A. thesis, Univ. S Afr.) s.v., When Prince George..uttered the traditional Basuto salute 'Pula' which means rain, a wave of enthusiasm swept the natives in the council chamber. 1974 S. Afr. Panorama Mar. 38/2 'Pula!' shouted the excited crowd of Bantu children.

2. The principal monetary unit of Botswana, consisting of one hundred thebe; a note of this value.

1976 Eastern Province Herald (S. Afr.) 23 Aug. 1 Botswana's new currency, the pula, will be introduced today. 1976 Whitaker's Almanack 1977 980 Botswana... Monetary unit... Pula.

Pulah, var. *Peulh sb. and a.

pulamiting (piū·lămaitiŋ), ppl. a. dial. [Prob. f. unrecorded pulamite vb. + -ing²; cf. Pule v. + Mite² 5.] Of a child or weak person: whining, whimpering. Also **pu·la-mi:ter** [-er¹], one who whines; a sniveller.

1913 D. H. Lawrence Sons & Lovers II. viii. 198 'Me!' exclaimed Morel—'me a good figure! I wor niver much more n'r a skeleton.' 'Man!' cried his wife, 'don't be such a pulamiter!' a 1930 —— in Virginia Q. Rev. (1941) XVII. Suppl. 43/1 You are a base, malingering pulamiting wretch.

pulao, var. Pilau in Dict. and Suppl.

pulaski (pulæ·ski). U.S. Also **Pulaski**. [The name of E. C. Pulaski (1866–1931), Amer. forest ranger, by whom the tool was designed.] A hatchet, of which the head forms an axe blade on one side and an adze on the other. Also attrib.

1924 Frontier Nov. 20, I saw Paul, his back bowed, his Pulaski swingin' like a flail. 1940 [see *hot spot 5 b]. 1946 Trial & Timberline June 91/1 Planting hoes, grub hoes, and Pulaski hoes had been provided by the rangers. 1948 Highway Traveler Aug. 37/1 Besides technical information on fighting fire with chemicals, or with axe and pulaski, smoke-jumpers are taught to bail out from a specially-constructed tower.

pulaskite (pulæ·skait). Petrogr. [See quot. 1891 and -ite¹.] An alkali syenite containing a small proportion of nepheline.

1891 J. F. Williams in Ann. Rep. Arkansas Geol.

Survey 1890 II. iv. 56 Such a rock has not as yet been described and the writer suggests the name pulaskite—that of Pulaski county in which the city of Little Rock and Fourche Mountain are located—as a designation for this type of rock. 1962 W. T. Huang Petrology iv. 129 In alkalic syenites the quartz-bearing type is nordmarkite.. or simply quartz syenite... With a slight deficiency of silica, a little nepheline and sodalite may be present, leading to the formation of pulaskite... Other colored minerals found in pulaskite include aegirite-augite, barkevikite, or arfvedsonite.

pulchritudinous (pʊlkritiū·dinəs), a. orig. U.S. [f. L. pulc(h)ritūdin-, pulc(h)ritūdo beauty + -ous.] Beautiful, graceful, or fine in any way; morally excellent.

1912 L. J. Vance Destroying Angel xv. 217, I love my love with a P because he's Perfectly Pulchritudinous and Possesses the Power of Pleasing. 1914 'High Jinks, Jr.' Choice Slang 17 Pulchritudinous pippen, a pretty damsel. 'A peach.' 1925 Times 13 Dec. 11/6 In an American paper, in which the Yarmouth councillors were described as 'pulchritudinous',..the word actually meant moral excellence. 1949 Chicago Tribune 21 Feb. I. 28/5 By us the hippopotamus..Is never counted pulchritudinous. 1963 Punch 13 Feb. 246/1 Such nice, pulchritudinous girls! 1975 Bookseller 20–27 Dec. 2720/1 An ageing tycoon and a pulchritudinous blonde half his years but nearly equal to him in experience.

Pulfrich (pu·lfrixʸ). The name of Carl Pulfrich (1858–1927), German physicist, used attrib. and in the possessive with reference to an optical illusion first pointed out by him (see Naturwissensch. (1922) X. 553), in which a pendulum that is swinging in a plane perpendicular to the line of sight appears to describe ellipses when one eye is covered with a filter and the other is uncovered.

1925 Brit. Jrnl. Ophthalm. IX. 65 Men, getting on in years, who have been expert in games with a moving ball, find themselves unable to judge its position with their former accuracy... Pulfrich's phenomenon seems to give a probable explanation. 1941 S. H. Bartley Vision vii. 170 Pulfrich's stereoscopic pendulum is another example of a set of conditions which will induce the perception of movement of a kind not corresponding with the real movement of a physical object. 1966 R. L. Gregory Eye & Brain vi. 78 The trading of temporal discrimination for sensitivity with dark adaptation is most elegantly, if somewhat indirectly, observed in a curious and dramatic phenomenon known as the Pulfrich Pendulum Effect. 1974 Vision Res. XIV. 184/2 The only thing that is important in producing a Pulfrich effect is the relative luminance of the objects which move across the retinae whether these be 'target' or 'background' or whether the subject fixates a stationary point or follows a moving object. 1980 Sci. Amer. May 134/2 According to most hypotheses about the Pulfrich illusion, the dark filter delays the perception of the visual signal at the retina of the covered eye.

puli (pu·li). Pl. **pulik**. [Hungarian.] A black, grey, or white sheep-dog belonging to the breed so called, characterized by a long, thick coat having a corded appearance.

1936 Amer. Kennel Gaz. 1 Oct. 62/1 The Puli is used primarily as a sheep herding dog in the hill sections of Hungary. In appearance he has many points of resemblance to the Old English Sheepdog, but is much smaller and more active. 1948 L. Bauvald in B. Vesey-Fitzgerald Bk. Dog II. 616 The Puli is probably one of the aboriginal dog races of the Magyars... Relatively few Pulik find their way to the exhibition bench. 1964 'C. Richards' Gentle Assassin (1965) i. 7 A puli, a Hungarian breed of dog. 1972 Country Life 10 Feb. 329/1 The Hungarian pulik—puli in the singular—are increasing numerically over here. 1973 Times 6 Aug. 26/7 Have a Hungarian Puli puppy..small black sheepdogs with an unusual corded coat. 1978 Times 11 Feb. 2/4 The Hungarian puli, distinguished by its remarkable corded coat, had a class of its own [at Crufts] for the first time.

Pulitzer (piū·litzəɪ; in U.S. also pu·litzəɪ). [Name of J. Pulitzer (1847–1911), Amer. journalist.] Designating any of several annual awards for distinguished work in journalism, letters, and music produced or published in the U.S. Also ellipt.

1918 N.Y. Times 3 June 9/7 The annual Pulitzer Prize of $1,000 for the best play written and produced by an American playwright in 1917..awarded to Jesse Lynch Williams for his comedy, 'Why Marry?' 1934 [see *creative a. 1 b]. 1941 B. Schulberg What makes Sammy Run? i. 20 Sammy was staring..at George Opdyke, the three-time Pulitzer Prize winner. 1955 G. Greene Quiet American I. iii. 38 Why, that account of Road 66..that was worthy of the Pulitzer. 1962 Publishers' Weekly 26 Mar. 16 Pulitzer..pronounced.. pulLitzer"—not, as so many people say, 'pewlitzer'. 1971 R. A. Carter Manhattan Primitive (1972) xiii. 121 The museum's sponsors—a famous concert musician, a film actress, and a Pulitzer Prize-winning novelist. 1972 R. Ludlum Osterman Weekend iii. 52 You were quite successful as an investigative reporter... You were nominated for a Pulitzer, I believe. 1977 Time 31 Oct. 28/1 One..led last year to a Pulitzer-prizewinning photograph of a woman and a little girl plummeting from a collapsed fire escape.

pulka. Add: Also **pulkka**. (Further examples.) Also attrib.

1913 Chambers's Jrnl. Nov. 798/1 The Lapland sledge, or pulk, as it is called, is shaped something like a boat. 1952 Ibid. Jan. 33/2 In a trice he had put on her gay red-and-blue harness and fastened the traces to a pulkha, the canoe-like little sledge of north-east Lapland. 1960 G. Taylor Mortlake III. i. 130 A reindeer drawing a small boat-shaped sledge called a pulkka. 1964 Punch 25 Nov. p. xvii, This store's ski fashion wear includes the Ernst Engel collection, Canadian overboots, patterned Tyrol pullovers, Swedish pulka jackets, furry helmets. 1969 Guardian 31 July 3/4 All our equipment..we pull behind us on pulka sledges..made of wood. 1974 Canad. Consumer Feb. 11/1 A special accessory for family enjoyment is the Norwegian 'pulk' (sled) with a rigid harness for a man or dog to tow even babies along. The pulk..is boatshaped, so that it will not tip on bumpy ground.

pull, sb.² Add: **I. 1. d.** Paired with push: see *push sb.¹ 1d.

2. a. (Earlier example in modern sense.)

1845 Dickens Let. 1 Nov. (1977) IV. 423 The carriage.. is to call for a pull of the first part of the Cricket.

f. (Earlier example.)

1865 Lillywhite's Guide to Cricketers 135 A fast run-getter, little too fond of a pull.

3. c. In fig. phr. to take a pull (at, on oneself), to stop or check (oneself); to pull oneself together. colloq. (chiefly Austral.).

1890 Barrère & Leland Dict. Slang II. 155/2 Pull (society): to take a pull means to stop, check, put an end to. 1916 C. J. Dennis Songs Sentimental Bloke i. 16, I tells meself some day I'll take a pull An' look around fer some good, stiddy job. 1922 Galsworthy Family Man III. ii. 105 Take a pull, old man! Have a hot bath and go to bed. 1942 E. Waugh Put out more Flags iii. 177 Suddenly she found herself weeping in earnest. Then she took a pull at herself. This wouldn't do at all. 1946 T. E. Haughey Railway Reminisc. 21 Look here—, it's about time you took a pull. Just shake yourself up a bit quick. 1953 M. Scott Breakfast at Six xxiv. 202 She may be a wonderful friend, but she'll land you in gaol yet. For heaven's sake, take a pull. 1966 J. Hackston Father clears Out 110 Alf Hodgson talked so much about the Red Range Federal Capital Site Movement that people said he'd be standing for Parliament one of these days if he did not take a pull on himself.

4. b. For U.S. slang read colloq. (chiefly U.S.) and add further examples.

a 1911 [see *island sb. 1 d]. 1937 F. P. Crozier Men I Killed vi. 109 Having been in France for so long and lacking the very necessary 'pull' in influential circles, we were unable to oust the family favourites at the War Office. 1940 [see *bracket sb. 5 c]. 1978 J. Krantz Scruples ii. 57 His future in the giant corporation was assured in the long run through family pull, since he had, on his mother's side, as one said in slang, du piston.

5*. = *lay sb.⁷ 7d. slang.

1969 Fabian & Byrne Groupie xxx. 219 'I'm not going to sleep with you.'..'Why not?' 'Because I'm not an easy pull.' 1973 M. Amis Rachel Papers 33 A mental chant, timor mortis conturbat me, and I began on my clumsiest pull ever. Ibid. 37 It was so obviously me and my pull and Geoffrey and his pull getting together to plan a spotty removal to someone's house.

pull, v. Add: **I. 1. f.** trans. To extract (a tooth). U.S.

1880 'Mark Twain' Tramp Abroad xxiii. 222 A soldier was getting a tooth pulled in a tent. 1915 R. Adair Pract. Oral Hygiene (ed. 2) i. 8 Dr. —— used to pull teeth, but he has got to talking so much about clean mouths that he is losing some of his trade. 1927 M. R. Reidy This Tooth Proposition vi. 110 A long time ago, dentistry consisted mostly of pulling teeth and making plates. 1976 H. MacInnes Agent in Place xix. 202 It was like pulling teeth. But we did learn something important.

3. b. trans. To draw or to be assigned (a task or position); to carry out (a duty). U.S.

1894 Lucky Bag (U.S. Naval Acad. Yearbk.) 67 Pull the sick list,..to get on the sick list when not ill. 1941 Kendall & Viney Dict. Army & Navy Slang 11/1 To pull guard duty,..to do guard duty. For instance, 'I've got to pull K.P.' 1972 Times 13 Apr. 1/8, I feel that my life is more important than having to pull security on this place. 1976 New Yorker 15 Mar. 89/1 How come they got you pulling guard?

6. b. To seize (someone's belongings); to recall or rescind (a document). Also absol.

1967 S. Faessler in Atlantic Monthly Apr. 107/1 One day a month was given over to repossessing merchandise from deadbeats. 'Today I am pulling,' he would say grimly. 1972 H. Kemelman Monday the Rabbi took Off xlv. 263 'They [sc. the police] pulled his passport, didn't they?' 'No... Officially they had just mislaid it.' 1973 R. Hayes Hungarian Game iii. 312 He had moved easily in dip circles until the..State pulled his visa.

II. 7. b. Phr. to pull one's coat: (see quot. 1946). U.S. Blacks.

1946 Mezzrow & Wolfe Really Blues 377/2 Pull somebody's coat, enlighten, tip somebody off. 1971 B. Malamud Tenants 55 The black..said: 'Lesser, I have to pull your coat about a certain matter.' 1972 T. Kochman Rappin' & Stylin' Out 163 If someone is giving you information, he is 'pulling your coat'.

d. In imp. phr. pull the other one, (it's got bells on it) and varr., a statement of disbelief implying suspicion that 'one's leg is being pulled'.

1966 D. Francis Flying Finish v. 63 'They are English mares going to be mated with Italian sires,' explained Conker... 'Pull the other one, it's got bells on,' said the engineer. 1973 'S. Woods' Enter Corpse 112 'Believe it or not, neither Farrell nor I has the slightest interest in the gold...' 'Pull the other one!' said Nelson derisively. 1974 M. Butterworth Man in Sopwith Camel viii. 84 'Pull the other leg, it's got bells on!' she said. 'A bank's a bank, and you've got yourself charge of a bank for no other reason but to dip your fingers into the till.' 1975 D. Bagley Snow Tiger ix. 88 'She doesn't hold the mineral rights.' 'Pull the other one,' scoffed Eric. 1977 J. Bingham Marriage Bureau Murders xii. 146 Pull the other one, it's got bells on it. I saw it all... So don't give me that tripe.

f. To draw or fire (a gun). Const. on. Also absol. U.S.

1854 J. F. Cooper Deerslayer I. iii. 54, I shall not pull upon a human mortal as steadily..as I pull upon a deer. 1883 'Mark Twain' Life on Mississippi p. xxvi, When they happened to meet, they pulled and begun. 1895 Century Mag. June 282/1 He repeated it, and I struck him. He pulled a pistol on me. 1903 S. E. White Forest x. 122 The birds had proved themselves most uncultivated and rude persons by hopping promptly into the trees... I had refused to pull pistol on them. 1926 J. Black You can't Win (1927) xiii. 182 He would have 'pulled' on us. 1952 Sun (Baltimore) 4 July 42/7 Dr. Brady and the would-be bandit were in a middle room..when a second man suddenly appeared and one of them pulled an automatic pistol. 1978 S. Brill Teamsters ii. 70 They couldn't just pull a gun on Hoffa.

g. To stretch and draw (sugar candy, etc.) until it is ready to set. orig. U.S.

1842 W. T. Thompson in Southern Miscellany 10 Dec. 2/6 They's pullin lasses candy in the parlor. 1893 Harper's Mag. Feb. 442 He pulled candy with glee, but also with eager industry, covering platter after platter with his braided sticks. 1948 Good Housek. Cookery Bk. III. 637 Certain toffees..are pulled, which gives them a satiny, silvery look. Attractive effects are achieved by combining pulled and unpulled toffee before cutting it into cushions. The toffee should be pulled immediately it is cool enough to handle.

h. To strain (a muscle or tendon) by abnormal exertion.

1955 M. Allingham Beckoning Lady iii. 35, I pulled a tendon in my foot so I'm stuck at the desk. 1955 R. Bannister First Four Minutes 175 Until then I had never been able to understand how athletes pulled muscles. 1971 Woman's Own 27 Mar. 8/2, I think I pulled a muscle. 1976 P. Harcourt Dance for Diplomats v. 51 'You're still limping.'..'I must have pulled a muscle.'

8. b. to pull in or to pieces: in Bookbinding, simply, to pull.

1901 D. Cockerell Bookbinding I. ii. 34 If the book should prove to be imperfect..the owner should be communicated with, before it is pulled to pieces. This is very important, as imperfect books that have been 'pulled' are not returnable to the bookseller. 1931 A. Esdaile Student's Man. Bibliogr. vi. 194 The book must be 'pulled', i.e. taken to pieces, first.

9. a. (Later fig. example.)

1904 W. N. Harben Georgians 22 So you 'n the old man are still pullin' agin one another?

f. transf. Of the engine of a motor vehicle: to afford (adequate) propulsive force; hence, by metonymy, of the vehicle itself.

1902 C. S. Rolls in A. C. Harmsworth et al. Motors ix. 175 Motor will not 'pull' well or misses fire. 1933 J. Buchan Prince of Captivity III. ii. 282 The driver stopped to examine his engine. 'She pulls badly, mein Herr,' he said. 1974 P. Wright Lang. Brit. Industry i. 24 Another transport term, the bus (or car) won't pull, is from the days of the horse and cart.

11. c. intr. To exert influence or 'root' for (a person, etc.); to sympathize with, favour. Chiefly N. Amer.

1903 C. B. Gilbert in Forum (N.Y.) XXXV. 311 Such committees are exposed to all kinds..of influence..all pulling for this or that applicant. 1922 G. Ade Let. 22 Nov. (1973) 85 Tomorrow I go up to LaFayette to pull for Purdue against Indiana and I hope we may win at least one game. 1949 National Geogr. Mag. Sept. 321/1 I'm usually pulling for the Indians instead of the cowboys. 1968 Globe & Mail (Toronto) 17 Feb. 39 It sure helps to get this evidence that so many people at home are pulling for us. 1970 G. F. Newman Sir, You Bastard ii. 73 The Governor was pulling for him with the Divisional D[etective] C[hief] S[uperintendent].

d. trans. To attract (custom); to secure (patronage). Freq. absol.

1905 Calkins & Holden Mod. Advertising xi. 264 The advertiser likes to know which particular mediums pull best. 1929 L. F. Carr America Challenged 96 Both Republicans and Democrats have tried to pull the farmer vote by favoring legislation which the Populists had demanded. 1938 S. V. Benét Thirteen O'Clock IV. 234 I'd done some advertising copy for the firm that pulled. 1962 R. Stout Gambit (1963) iii. 36 She attracts. She pulls. 1974 S. Marcus Minding Store (1975) xi. 228 The booklets pulled fairly well, both in store response and through the mails. 1976 Record Mirror 3 Apr. 21/1 Brook Benton..can still regularly pull standing-room-only audiences for his live gigs.

e. N.Z. (See quot. 1933.)

1933 L. G. D. Acland in Press (Christchurch, N.Z.) 18 Nov. 15/7 Pull, a dog pulls sheep when he brings them towards his masters. 1935 G. L. Meredith Adventuring in Maoriland v. 47 [The dog] eventually 'pulled' the mob on the slope of a hill. 1938 R. Burdon High Country x. 107 The heading dog is silent and is used to 'pull' or bring sheep back to the shepherd.

f. To earn (a wage or salary). See also *pull down (sense 24 f) and *pull in (sense 25 b). colloq.

1937 'M. Innes' Hamlet, Revenge! II. viii. 197 I'm twenty-two and pulling twelve pounds a week.

g. coarse slang. To pick up a partner (for sexual purposes); spec. to copulate with. So to pull a train, to copulate successively with more than one partner.

1965 C. Brown Manchild in Promised Land i. 15 They

thought that I was one of the guys who had pulled a train on their sister in the park the summer before. Ibid. iv. 112 If you gon pull a bitch, you can'[t] get excited and let her know that you want that pussy so bad you about to go crazy. 1965 Sunday Express 25 July 17/2 As a young man I could never pull (pick up) any birds of my own class. 1973 M. Amis Rachel Papers 23, I could easily pull the village idiotess, who in any case, one windless summer night, had wanked Geoffrey and me off through the school railings, simultaneously. 1973 Boyd & Parkes Dark Number vi. 69 Five years ago you did the big male-menopause bit, didn't you? Skulking off to Paris to prove you could still pull the birds. 1973 P. Cave Speed Freaks viii. 77 'Wanna pull a train for the movie?' Mucky asked Dodo, who was still unclothed. She shrugged resignedly. 1974 H. L. Foster Ribbin' iv. 148 To 'pull a train' is for a female to have consecutive sexual intercourse with numbers of males. The female pulling the train may do so voluntarily for financial remuneration, forcefully, or out of fear. Ibid. 149 Trains are pulled everywhere... Selby..described Tralala pulling endless trains in Brooklyn. 1976 P. Cave High Flying Birds iv. 47 She's certainly worth pulling... But I reckon you can forget it as long as she's got her mother with her.

12. b. Also const. on.

1897 T. De Leon Novelette Trilogy v. 44 He..strode rapidly homeward; pulling hard..on the dead cigar between his lips.

c. trans. To draw (beer) from a keg, etc., by means of a pump or tap.

1969 Sydney Morning Herald 24 May 1/10 During Thursday's strike by hotel staff, the manager..pulled 17½ 18-gallon kegs from two taps between 10 a.m. and 10 p.m. 1975 M. Kenyon Mr Big v. 46 The muscled barmaid pulled pints.

III. 15. b. to pull one's weight: also fig., to perform one's share of work, to take one's share of responsibility. Also to pull weight.

1904 Kipling Traffics & Discoveries 278 They need a lot of working up before they can pull their weight in the boat. 1921 [see Weight sb.¹ 10 c]. 1925 E. F. Norton Fight for Everest: 1924 98 No members of the climbing party pulled more weight in the team than these two by their unostentatious unselfish gruelling work. 1931 Times 27 Feb. 16/5 Referring to people in the administrative grade who did not 'pull their weight', Sir Alfred Woodgate said that assistant principals who had been twice passed over for promotion to principals were a menace to the office and should not be allowed to remain. 1948 M. Laski Tory Heaven x. 138 Lord Starveleigh asked him down to address the electors... We're all expected to pull our weight, you know. 1976 J. B. Hilton Gamekeeper's Gallows xii. 115 How long was he going to put up with me living off the fat of the land in his kitchen, not pulling my weight with his other servants?

16. (Further examples.) Cf. also *pull in (sense 25e).

1907 J. Masefield Tarpaulin Muster 205 The police entered..and 'pulled the joint'—that is, they arrested and fined the proprietor. 1931 'D. Stiff' Milk & Honey Route 189 He's pulled for a vag, his excuses won't do. 'Thirty days,' said the judge. 1950 Wodehouse Nothing Serious 244 Doom had come upon The Cedars... The joint had been pulled. 1970 G. F. Newman Sir, You Bastard 10 They..pulled drunks and bathed tramps, saw children across the road and directed traffic.

17. b. Boxing. In phr. to pull one's punches, to hold back or check one's blows. Also fig., to use less force than one is capable of exerting, to be gentle or lenient, esp. in criticism or punishment.

1934 in Webster. 1937 H. L. Ickes Secret Diary (1954) II. 88 He talked about the judiciary and he didn't pull his punches at any time, although neither was he in any degree personal. 1939 L. Jacobs Rise of Amer. Film 459 Either because Vidor 'pulled his punches' at the revolution or because..he was confused. 1947 People 22 June 5/3 Two of his boys recently fought for two solid hours—and no pulled punches—to provide a minute and a half's action in the actual film. 1955 A. L. Rowse Expansion Eliz. Eng. 133 He charges Ormonde with..pulling his punches in pursuit of the rebels. 1957 D. J. Enright Apothecary's Shop 209 The fact—not a new one—that Eliot doesn't pull his punches. 1960 Times 4 Feb. 11/2 Lady Albemarle's committee have not pulled their punches. 1973 P. O'Donnell Silver Mistress xi. 191 It was a demonstration match. The kicks, chops and punches were pulled at the last instant. 1977 Time Out 28 Jan.–3 Feb. 33/3 The film pulls all its political punches, settling instead for sentimental narrative.

18. (Earlier Cricket example.) Now also used in Baseball (see quot. 1976).

1851 W. Clark in W. Bolland Cricket Notes 143 Never try to pull a straight ball across you. 1943 Amer. Speech XVIII. 105 If he can pull or place the ball he has a better chance of getting a hit. 1976 Webster's Sports Dict. 334/1 Pull,..[in baseball] to hit the ball to the field on the same side of the plate as the batter stands when he takes his normal position in the batter's box.

18*. Oil Industry. To pull up or withdraw (casing, etc.) from a well.

1916 Johnson & Huntley Princ. Oil & Gas Production xiv. 154 The hole in most cases gradually fills up at the bottom with cavings from the walls and sometimes from the roof of the oil sand. This makes it necessary to pull the tubing and clean the well. 1938 L. V. W. Clark in A. E. Dunstan et al. Sci. of Petroleum I. ix. 434/2 In this position the string may be rotated as well as pulled or lowered. 1960 C. Gatlin Petroleum Engin. x. 170/1 A sudden decrease in penetration rate..may mean that the inner barrel is jammed or plugged and the assembly should be pulled for inspection. 1974 P. L. Moore et al. Drilling Practices Man. xii. 308 If the well kicks when pulling the drill string, the formation fluid will enter the entire well bore below the drill string.

IV. 19. b. to pull one's freight, to depart quickly. U.S. colloq.

1895 F. Remington Pony Tracks 252 The wily old fellow..had discreetly 'pulled his freight'. 1905 'O. Henry' in Everybody's Mag. XIII. 814/2 The Kid.. considered it not incompatible with his indisputable gameness to perform that judicious tractional act known as 'pulling his freight'. 1913 J. London Valley of Moon II. xvii. 277, I guess we got a celebration comin', seein' as we're going to pull up stakes an' pull our freight from the old burg. 1926 in J. F. Dobie Rainbow in Morning (1965) 84 He pulled his freight in a hurry.

c. (In Supplement) p. a boner, 19d; p. one's coat, 7b, p. one's freight, 19b; p. a gag, 19e; p. a job, 19f; p. leather 19g; p. the other one, 7d; p. one's pudding, 19j; p. one's punches, 17b; p. rank, 19h; p. a robbery, 19f; p. the rug, 19k; p. stripes, 19h; p. a train, 11g; p. a trick, 19e; p. one's wire, 19j.

d. to pull a boner (etc.), to make a foolish mistake (*boner²). U.S. slang.

1913 [see *boner²]. 1926 Scribner's Mag. Sept. 246/1 The Washington newspaper correspondents are the pick of the land, and their dinners are not the softest spots in the lives of the speakers. It is no place to pull a bloomer. 1929 M. Lief Hangover xv. 234, I pulled an awful boner when I was up in Newport last summer. Ibid., That's nothing... I pulled a dumber thing than that once. 1967 Boston Sunday Globe 23 Apr. B 41/1 Apart from the shabby methods used by construction workers, who have been known to pull some boners, Moscovites are subjected to some truly baffling beauts from bumbling builders.

e. In various colloq. phrases denoting action or speech intended to deceive, shock, or amuse, as to pull a fast one (see *fast a. 11), to pull a gag, to pull a trick, etc.

1914 'High Jinks, Jr.' Choice Slang 17 Pull a punk one (to), to tell a poor joke. 1915 [see *dope sb. 4]. 1922 G. Ade Let. 11 Oct. (1973) 83 The plot of the piece was that George and Frank both loved the same girl and Frank pulled a lot of dirty stuff and deceived the girl for a while but eventually virtue triumphed. 1929 M. Lief Hangover xv. 235, I can tell you about the gag one of our better-known critics pulled in his review of Dillingham's new musical comedy. 1932 E. Wallace When Gangs came to London xxiii. 232 'Fantastical,' suggested Jiggs. 'I'm getting quite used to the word. It's the one you pull when any hard-sense suggestion is made to you.' 1937 G. Heyer They found him Dead xiii. 260 Not that I think anyone would pull the same trick twice. 1940 Wodehouse Eggs, Beans & Crumpets 238 Your aunt..has a right to early information about any rough stuff that is being pulled on the premises. 1957 H. Roosenburg Walls came tumbling Down ix. 208 Just be a little more careful about your company next time you pull a stunt like that. 1976 M. Machlin Pipeline xxviii. 333 For Christ sake, don't think about pulling any movie-type heroics. 1978 Guardian Weekly 2 Apr. 6/5 Many [U.S. coal] miners are now threatening to pull wildcat strikes.

f. to pull a job, robbery (etc.), to commit a crime, usu. theft. colloq. (chiefly U.S.).

1915 Policeman's Monthly Dec. 17/3 He replied that he had often noticed just before they were going to 'pull a job' his partner was happy. 1923 'B. L. Standish' Lego Lamb, Southpaw viii. 58 Yet, by your own confession, you came over here to 'pull a job'. 1937 Research Stud. State Coll. Washington Mar. 19 Some boys think its an honor.. to say..that they pulled jobs with such and such gangsters. 1967 [see *graft v.⁴]. 1972 J. Wambaugh Blue Knight (1973) i. 28 A federal fugitive who..carried a gun and pulled stickups. 1973 E. Bullins Theme is Blackness 160 Tootsie didn't work steady but we still ran together. Even pulled an occasional job. 1973 Philadelphia Inquirer (Today Suppl.) 7 Oct. 14/1, I even pulled three robberies in one night—two drugstores and a haberdashery. 1978 Detroit Free Press 5 Mar. B 1/1, I suggested we pull a robbery.

g. to pull leather, to grasp the saddle horn in order to avoid being thrown from a bucking horse (see also quot. 1933). U.S.

1916 Daily Colonist (Victoria, B.C.) 19 July 5/4 They [sc. bad horses] are still outlaws and..are guaranteed to send almost any rider to pulling leather. 1923 J. H. Cook 50 Yrs. Old Frontier 16 He certainly made me 'pull leather', and I clung to his mane as well in order to keep in close touch with him. 1925 C. E. Mulford Cottonwood Gulch v. 60, I'm pullin' leather, but I'm stickin' to the saddle. 1933 J. V. Allen Cowboy Lore III. 59/2 Pulling leather, holding on to the saddle with the hands while riding a bucking animal, prohibited by the rules of all contests and scorned by all real cowboys.

h. to pull (one's) rank, to employ (one's) superior status in exacting obedience, co-operation, or privilege. Also to pull stripes (etc.). orig. U.S.

1923 Amer. Legion Weekly 23 Feb. 18 Don't pull your rank on him, K.P. You were only a private yourself, once. 1926 Amer. Speech II. 62/2 Give him..officers who do not 'pull rank', and he is well content. 1958 V. Canning Dragon Tree 90 He disliked pulling his rank to claim any personal privileges. 1958 M. K. Joseph I'll soldier no More xiii. 242 Don't you pull your stripes on me, sarge. 1959 N. Mailer Advts. for Myself (1961) 228 Teddy pulled seniority and they gave him his way. 1976 H. MacInnes Agent in Place xiv. 148 'What if he refuses to go with them?' 'They'll be senior men, they'll pull rank.'

i. to pull a (proper name), to imitate, to behave in the manner of (the person named). colloq. (orig. U.S.).

1927 New Republic 9 Mar. 72/1 The following is a partial list of words denoting drunkenness now in common use in the United States... To pull a Daniel Boone. 1931 Technol. Rev. Nov. 67/1 To pull a Lindbergh means to do something heroic, but to go Lindbergh means to get the flying fever in a rather callow manner. 1935 Wodehouse Luck of Bodkins xiii. 133 He'll be much happier in the long run if he gets it into his bean that he can't pull a James Cagney on me every time he's a mite upset.

j. to pull one's pud(ding) or wire: to masturbate. slang.

1944 [see *pud sb.² 2]. 1961 Partridge Dict. Slang Suppl. 1101/2 Pull one's pudding,..may have originated pull one's wire. 1970 J. Osborne Right Prospectus 30 Remember what I said about sex. Keep away from the maids and pretty boys. As for pulling your wire, that's no occupation for a gentleman. 1970 W. Smith Gold Mine xxvi. 61 Jesus... That was ugly. I felt like a peeping tom, watching someone, you know, pulling his pudding.

k. to pull the rug (out) from under (a person or thing), and varr., to weaken or unsettle (something) by an unexpected withdrawal of support or by some other action; to let down or betray (someone). colloq. (orig. U.S.).

1946 Time 23 Dec. 17/3 Strikes, for instance, would pull the rug out from under the best of prospects. 1948 Sun (Baltimore) 15 Dec. 8/3 Although both are reported to feel that United States commitments in western Europe preclude the 'bailing out' of Chiang's regime.. they have chosen to say nothing that would 'pull the rug out' from under Chiang. 1952 Manch. Guardian Weekly 14 Aug. 3/2 What the President [sc. Truman] was talking about was his own part in clinching the nomination of Governor Stevenson. It entailed pulling the rug out from under Mr Harriman. 1966 'W. Haggard' Power House xvi. 179 James Mott had Victor's story... At the worst it could pull the carpet out. 1967 A. Hunter Gently Continental ix. 132 The mat is pulled from under Shelton when he dares to assent to this point. 1973 Physics Bull. Feb. 75/2 Professor Jewkes..proceeds to pull the rug from under many of the assumptions and arguments used to justify government support of 'high technologies'. 1974 W. Garner Big enough Wreath xiii. 196 He did his last job for you... Not too successfully..which is another reason for pulling the rug from under. 1978 Detroit Free Press 2 Apr. C 12/2 When the rug was pulled out from under me in movies and television I went back to the theater.

V. 22. pull away. c. Of a vehicle: to draw away, as from the kerb on starting. Also absol.

1955 H. Kurnitz Invasion of Privacy (1956) iv. 32 The grey convertible..pulled away from the curb. 1974 Sunday Post (Glasgow) 14 Apr. 6/2 As it [sc. a bus] pulled away I was horrified to see a man..pick up the old gents shopping bag and hurry away. 1976 'P. B. Yuill' Hazell & Menacing Jester iv. 44 He was pulling away from the kerb before my reggie reached the upholstery.

23. pull back. f. Sport. To score (a goal) restoring, or serving towards restoring, level terms between two teams.

1976 Northumberland Gaz. 26 Nov. 20/6 Nicholson was on hand to score from the rebound. Annitsford pulled one back but Nicholson was then again on target with a magnificent shot. 1978 Lochaber News 31 Mar. 20/3 In the 60th minute Donald Murchison pulled one back. Five minutes later Mike MacPherson equalised.

24. pull down. f. To earn (money, esp. as a wage or salary). slang.

1917 S. Lewis Job xiii. 192 Good job, too, assistant book-keeper, pulling down his little twenty-seven-fifty regular. 1919 Wodehouse Damsel in Distress xxi. 256 George pulls down in a good year, during the season—around five thousand dollars a week. 1922 [see *berry sb.¹ 1c]. 1933 D. L. Sayers Murder must Advertise v. 78 'So you have become one of the world's workers.'..'Yes; I'm pulling down four solid quid a week.' 1968 New Yorker 9 Nov. 56 How much does your average cornettist pull down per year? 1976 N. Thornburg Cutter & Bone i. 21 He..had been pulling down twenty-five thousand a year.

25. pull in. b. Delete †Obs. rare and add later examples. Cf. prec. sense.

1841 Punch 17 July 6/2 I'm a boy in a school, with a bag of apples, which..I naturally sell at a penny a-piece, and so look forward to pulling in a considerable quantity of browns. 1973 Scotsman 13 Feb. 8/2 The Archbishop of York..pulls in £6000 a year.

e. To arrest (a person); = sense 16.

1893 S. Crane Maggie x. 89 'I'll tump 'im till he can't stand.'..'What's deh use! Yeh'll git pulled in!' 1923 E. Rice Adding Machine vi. 101 You read in the paper all the time about guys gettin' pulled in for annoyin' women. 1933 D. L. Sayers Murder must Advertise ix. 162 We could pull him in any day, but he's not the real big noise. 1956 [see *heeled ppl. a. 2]. 1973 W. M. Duncan Big Timer xxiii. 159 If you hadn't come voluntarily, I'd have pulled you in.

f. intr. Of a locomotive train. to enter a station.

1905 D. G. Phillips Plum Tree 91, I didn't know you till you took out your watch with the monogram on the back, just as we were pulling in. 1929 S. Leacock Iron Man 143 That's your train pulling in now.

g. absol. or intr. Of a driver: to drive a vehicle to the side of a road or off a road (for some specified purpose). Also of the vehicle itself.

1938 G. Greene Brighton Rock III. iii. 122 Notices said: 'Pull in Here', 'Mazawattee tea', 'Genuine Antiques'. 1959 I. Jefferies Thirteen Days i. 13, I was forced to pull off the road on the way back... I would have pulled in thereabouts anyway. 1975 M. Russell Murder by Mile viii. 81 Pulling in for a truck to pass, Hamilton sat tapping the wheel.

26. pull off. c. For slang read colloq. and add further examples.

1918 Policeman's News 25 Feb. 3/2 Criminals can no

longer dispose of loot in Reading without the sleuths having a pretty good insight as to who pulled off the job. 1923 H. G. Wells *Men like Gods* I. i. 6 He was not really clever enough to pull such a thing off. 1968 *Times* 15 Oct. 16/8 Having succeeded in their earlier experiments, there seems no reason why they should not pull off another major 'first'. 1977 *Time* 15 Aug. 13/2 Both looked as if they had just pulled off some master stroke of détente.

d. To steal (something). Cf. sense 2 b. *slang.*
1883 [see *LEATHER *sb.* 2 e (*a*)].

e. Usu. *refl.* To cause (a person) to ejaculate by masturbation. *coarse slang.*
1922 Joyce *Ulysses* 745 How did we finish it off yes O yes I pulled him off into my handkerchief pretending not to be excited. 1961 Partridge *Dict. Slang* Suppl. 1236/1 *Pull oneself off*, (of the male) to masturbate: low: late C. 19–20. 1966 L. Cohen *Beautiful Losers* I. 4 Can an old scholar find love at last and stop having to pull himself off every night so he can get to sleep? 1971 'V. X. Scott' *Surrogate Wife* 139 Spasms shook his entire body as I pulled him off.

f. *intr. Surfing.* To end a ride by bringing one's surfboard out of a wave. Cf. sense 28 g below.
1964 B. Cooper in P. L. Dixon *Men & Waves* (1966) 189, I can't really recall my first wave, but I'm sure I caught an edge and had to pull off. 1967 S. Reid in J. Severson *Great Surfing* 22/2 If someone were sliding faster than you, you had to pull off.

28. pull out. d. For 'Chiefly *U.S.*' read 'orig. *U.S.*' and add earlier and later examples. Also of a ship: to sail out of a harbour or port; of an aeroplane: to emerge from a dive; of a vehicle: to move outwards into another lane of traffic. Also of the driver, etc., or occupants of these means of transport.
1868 *Harper's Mag.* Feb. 293/1 Breakfast over we 'pulled out', for the next station. 1880 'Mark Twain' *Tramp Abroad* xxviii. 287 We got under way..and pulled out for the summit again, with a fresh and vigorous step. 1902 C. J. C. Hyne *Mr. Horrocks, Purser* 105 We pull out from here next Tuesday. 1917 'Contact' *Airman's Outings* 46 We swerved violently, and they pulled out of their dive well away from us. 1938 G. Greene *Brighton Rock* vii. ix. 349 A bus came upon them and pulled out just in time. 1942 T. Rattigan *Flare Path* I. 30, I put the old Wimpey into a dive and..pulled out only a few feet above his head. 1951 *Manch. Guardian Weekly* 15 Mar. 13/2 In the Far West, the San Francisco 'Chronicle' blanketed its whole page with the headline—'Wage board approved increase of 10 per cent—labour members pull out'. 1961 *Listener* 7 Dec. 962/1 Today they must choose—either to stay under an African government or to pull out..to face a new life. 1972 'A. York' *Expurgator* I. iv. 59 The Mercedes was immediately in front of him. He pulled out, into the middle of the road,..and saw the lorry coming at him. 1977 D. Bagley *Enemy* xx. 162 We can pull out and leave Ashton to sink or swim..or we can get him out ourselves.

e. *colloq.* To extend oneself vigorously; to work hard. Cf. *to pull the stops out* s.v. *STOP *sb.*[2]
1866 Trollope *Belton Est.* III. x. 272 There's no getting people really to pull out in this country.

f. Of a drawer, etc.: to be capable of being pulled open.
1943 A. G. Hatcher in *Mod. Lang. Notes* LVIII. 12 Drawers *pull out*..easily.

g. *intr. Surfing.* (See quot. 1963.) Cf. *pull off* (sense 26 f above).
1963 *Surfing Yearbk.* 42/2 *Pull out*, ending a ride by getting your board out of a wave. There are many different ways of pulling out. 1964 J. Severson *Mod. Surfing* xvii. 157 You may also have the opportunity of pulling out before reaching the section. 1968 D. Kahanamoku *World of Surfing* xiii. 127 You can also get wiped out when the wave begins to break in front of you. It's time to pull out of it and avoid an unwanted and unscheduled swim. 1971 *Studies in English* (Univ. Cape Town) Feb. 27 This is called being locked in, because in such a condition it is virtually impossible to pull-out.

28*. pull over. a. See simple senses and Over *adv.*

b. Of a driver: to bring a vehicle to the side of a road or street, or to some other place. Cf. sense 25 g above.
1930 *Morning Post* 12 June 5, I considered that I had not time to pull over to my near side. 1932 *Sun* (Baltimore) 24 Sept. 8/6 Notify him that you are actually about to pass. In most cases he will pull over for you. 1971 *It* 2–16 June 8/1 He's signaling me to pull over! 1972 D. Delman *Sudden Death* (1973) v. 135 The rain so heavy that..the wipers..were unable to cope... 'Can't see too well,' I said. 'I better pull over.'

28. pull round. a.** See simple senses and Round *adv.*

b. *intr.* To recover from sickness or fainting; to come round.
1891 R. Buchanan *Come, live with Me* II. xx. 253 The danger's over..and the little one is pulling round. 1896 *Pall Mall Mag.* Sept. 70 He thinks he's going to pull round again; but I'll bet on his not being alive this day week.

c. *trans.* To restore (a person) to health after sickness, etc.; to put into a healthier or better condition. Also *transf.*
1900 E. Grey *Outrageous Fortune* iv. 37 The attack of meningitis..was fortunately only slight, and the excellent nursing I received..served to quickly pull me round. 1928 *Sunday Express* 29 Apr. 20/1 In the second half Cardiff made a valiant attempt to pull the game round. 1955 A. L. Rowse *Expansion Eliz. Eng.* 230 Smith had pulled the colony round at its lowest point. 1978 *Country Life* 27 July 236/3 'Twas my turn to need help. Denwood pulled me round.

29*. pull to. *trans.* To shut (a door, etc.) by drawing it towards oneself.
1898 [see To *adv.* 4]. 1910 [see *LISTEN *v.* 2 c]. 1922 Joyce *Ulysses* 57 He pulled the halldoor to after him. 1946 W. de la Mare *Three Royal Monkeys* v. 67 He skipped out and pulled-to the door-flap behind him.

30. pull together. a. Also *transf.*
1894 W. Archer in *World* 15 Aug. 25/1 The last act wants a great deal of working-up and pulling together. 1925 J. G. Bruce in E. F. Norton *Fight for Everest: 1924* iii. 63 If the first party of porters could be pulled together again in twenty-four hours' time, they were then to be utilized to keep Camp III supplied from Camp II. 1952 *Listener* 31 Jan. 189/1 He has tried to pull together all that has been said and written about the political struggle between the western allies and Moscow. 1978 *Amer. N. & Q.* XVI. 142/1 A corpus of paintings not pulled together in any previous work.

b. Also, to restore (a person) to a normal condition.
1906 W. S. Maugham *Bishop's Apron* viii. 132 Now come and have tea... I know it'll pull you together.

31. pull up. h. Also *transf.*
1936 N. Streatfeild *Ballet Shoes* xvii. 276 The death of King George in January cut the audiences down to about a quarter..and they never really pulled up again.

pull-. Add: **1.** With advbs., forming adjs.: *pull-along, -down, -on; pull-away* sb. (examples); **pull-apart**, the action or result of being pulled in opposite directions so as to be ruptured; **pull-on**, a garment without fasteners that is pulled on.
1939–40 *Army & Navy Stores Catal.* 829/2 Pull-along engine for child to ride in..25/-. 1967 *Punch* 6 Dec. p. xii, Buy beautifully made wooden pull-along toys for tiny children. 1951 *Special Publ. Soc. Econ. Paleontologists & Mineralogists* No. 2. 89 (*caption*) 'Pull-apart' developed in a laminated bed of clay lying on gravel bed and covered by silts. 1954 *Sun* (Baltimore) (B ed.) 6 Feb. 2/5 As the concrete sets these wires are tightened by special jacks, thus..getting rid of the pull-apart tendency in advance. 1971 *Nature* 2 July 21/1 Fig. 3 shows the initial opening of the proto-North Atlantic Ocean and the pull-apart of North and South America. 1939 T. S. Eliot *Old Possum's Pract. Cats* 17 Abandoning their sampans, and their pullaways and junks, They battened down the hatches on the crew within their bunks. 1950 [see *last-resort* s.v. *LAST *a.* C 2]. 1976 *Drive* May–June 90/2 A fuss-free pullaway from 20mph in top gear and an 11sec time from 30–50mph highlight the car's impressive flexibility. 1907 *Yesterday's Shopping* (1969) 172/1 Electric Indicator... Pull-down Replacement, consisting of 1½ yds. brass chain, pulley and ring. 1971 J. Henderson *Copperhead* (1972) vi. 76 There was a Leitz slide-projector ..and..a pull-down screen. 1978 *Daily Tel.* 24 Oct. 15/2 The pull-down compartment in the front holds six transparent pockets for credit cards. 1919 in C. W. Cunnington *Eng. Women's Clothing* (1952) v. 156 The vogue of the pull-on is still predominant. 1921 *Glasgow Herald* 25 May 4/7 The turban..does not hide all the hair as do the pull-on hat, the cloche, and other popular shapes. 1976 *Billings* (Montana) *Gaz.* 30 June 7–A/1 (Advt.), Debonair styling in delightful prints. Have yours in a V-neck pull-on with roll-up sleeves. 1978 *Detroit Free Press* 16 Apr. (*Detroit Suppl.*) 36 (Advt.), Pull-on pants in petite and average lengths, $15.

2. With sbs. *pull-engine, -toy*; **pull-bell** (later example); **pull-boat**, also *U.S.*, a motorized flat-boat which draws logs over water; also *attrib.*; **pull-bone** *U.S.* = *wish-bone* s.v. Wish *sb.*[1] 4; **pull-cord**, a cord which operates a mechanism when pulled (in quots. *attrib.*); **pull-date** *N. Amer.*, the date stamped on a container of perishable goods indicating when it must be withdrawn as no longer suitable for sale; **pull-hitter** *U.S. Baseball* (see quot. 1955); **pull-stroke**, (*a*) (earlier example); (*b*) in other technical senses, a stroke (of an oar, etc.) effected by pulling towards oneself; **pull-switch**, a switch operated by means of a pull-cord; **pull-tab**, a device, usu. comprising a ring and short tongue of metal, by means of which a tin may be opened; also *attrib.*
1919 R. Macaulay *Three Days* 52 How a pull-bell clangs when it rings! 1903 *Sci. Amer.* 17 Oct. 276/3 In the cypress swamps of Louisiana there are employed what are known as pull-boats, an evolution from the plan of placing a hoisting engine upon a scow and snaking the logs out of the swamp... The endless-rope pull-boat engines have 44-inch winding drums. 1907 'O. Henry' *Trimmed Lamp* 136 In her mind she could hear the girls shrieking over a pullbone. 1963 *Listener* 28 Feb. 399/1 It should be..fitted with either a pull-cord switch or one outside the door. 1978 *Lancashire Life* Nov. 177/1 (Advt.), Mahogany Bracket clock—hour strike and pull-cord repeat. 1973 *Black Panther* 7 Apr. 14/2 What food manufacturer would be so foolhardy as to place his produce bearing a pack date alongside a competitor who uses a pull date. 1977 *Daily Colonist* (Victoria, B.C.) 7 Aug. 2/4 All whipped butter..bearing the pull date of Sept. 12, 1977, or before is being recalled. 1907 *Yesterday's Shopping* (1969) 1072/2 The Pull Engine. With rubber composition tyres... each 16/3. 1937 *Philadelphia Record* 2 Sept. 17/1 Medwick hits to all fields. DiMaggio is mostly a pull hitter. 1955 *Amer. Speech* XXX. 153 The term *pull hitter* denotes a hitter who pulls the ball (i.e., a right-handed hitter who hits to left field or a left-handed hitter who hits to right field). 1972 *N.Y. Times* 4 June v. 2/7 He is primarily a pull hitter to left and some teams stack defensive alignments against him. 1897 K. S. Ranjitsinhji *Jubilee Bk. Cricket* iii. 120 He..applies a marvellous pull-stroke to good-length balls just outside the off-stump. 1968 J. Arnold *Shell Bk. Country Crafts* 74 All the 'work' of a cross-cut is done on the pull-stroke [of the saw], each man alternately pulling and releasing. 1969 *Publ. Amer. Dial. Soc.* LI. 10 *Pull stroke*,..a stroke whereby the kayakist leans far forward, thrusts the paddle into the water and with extreme exertion pulls the kayak forward. 1888 D. Salomons *Managem. Accumulators* (ed. 3) II. ii. 97 The Browett pull switch is fixed near the cornice, and by pulling a cord, the light is put on; also turned off by a similar action. 1971 D. Bagley *Freedom Trap* vi. 130 The light switch in the bathroom was operated, as good building regulations insist, by a ceiling pull-switch from which a strong cord hung to a convenient hand level. 1965 *Economist* 5 June 1150/1 The successful development of pull-tab tops for beer tins, so that they can be opened by hand rather than with an opener. 1978 J. Updike *Coup* vi. 239 Take care..not to drop the pull-tab [on a can of soft drink], once removed, back into the can. 1946 *Sun* (Baltimore) 14 Dec. 14/6 Wagons and other pull-toys should have rope handles. 1958 J. G. MacGregor *North-West of 16* v. 60, I was at the pull-toy stage. I went everywhere dragging behind me a fascinating collection of tin cans, bits of roots..all tied together with string. 1978 *Neiman, Marcus Christmas Bk.* 83 A purple cow pull toy, complete with the N-M brand.

3. Forming phraseological combs. functioning either as sbs. or adjs., as **pull-and-push** *adj. phr.* = *PUSH-AND-PULL *adj. phr.* b.
1950 *Railway Mag.* XCVI. 101 (*caption*) Dolgelly to Barmouth pull-and-push train approaching Barmouth Tunnel. 1959 H. Ellis *Brit. Railway Hist.* II. ii. i. 252 The Great Western and the South Western fought it [*sc.* the electric train] with steam rail motors and pull-and-push trains. 1968 *Railway Mag.* CXIV. 300/1 Pull-and-push trials on the G.N. main line..began in January.

Pullak, var. Polack *sb.* (*a.*) in Dict. and Suppl. **pull-and-push, pull-apart:** see *PULL-.

pull-back. Add: **1. b.** *spec.* An orderly withdrawal of military troops. Also *attrib.* orig. *U.S.*
1951 *Baltimore News-Post* 19 Mar. (Home Final 7th ed.) 1/2 Those who think the Red pullback is leading up to something don't put so much stock in the Red's abandonment of prepared defenses. 1953 *Sun* (Baltimore) 8 May B1/7 (*heading*) Reds continue Laos pullback. 1962 *Listener* 29 Nov. 896/2 *The New York Times* said the Chinese were now 'asking for border talks after a pull-back of troops that would still leave them in full possession of their main territorial objective'. 1971 *E. Afr. Standard* (Nairobi) 10 Apr. 2/7 He was working towards total American withdrawal and would step up the pull-back rate. 1974 *Times* 7 Jan. 4/1 (*heading*) Tel Aviv prepares pull-back offer. 1977 *Time* 10 Oct. 12/3 We had to get very deeply involved in pushing for a pullback and cease-fire.

c. *Cinematogr.* A shot in which the scene is observed to recede.
1957 Manvell & Huntley *Film Music* ii. 34 Long track, mostly in medium or medium-close shot, with one large pull-back during the market-place scene. 1959 W. S. Sharps *Dict. Cinematogr.* 121/2 *Pull back*, the backward movement of a camera from a close to a long shot. 1977 *Listener* 5 May 590/1 Will Frayn appear next time.. on the end of a zoom-in or a pull back?

d. *Tap-dancing.* (See quot. 1957.)
1957 P. Draper in *Dance Mag.* May 60/3 A pull-back is a tap made by the front part of the foot striking the floor while the body is in the air moving backwards. 1975 *New Yorker* 7 July 26/3 The last bit to be recorded..was a wonderful tap dance—a series of time steps, pullbacks, flaps..and pickups done by Michael Bennett himself.

pull-bone to **pull-date:** see *PULL-.

pull-down. Add: **1.** (Later examples.) Also *attrib.*
1938 N. Streatfeild *Circus is Coming* viii. 129 Can't we see the pull-down? Hans says it's more exciting than the build-up. 1962 *Which?* Oct. 305/2 New exercises.. were added... Pull downs (pulling down a bar attached to pulleys and weights). 1965 M. Stewart *Airs above Ground* iv. 58 If this is the last performance, they'll start the pull-down [of the big top] the minute it's over. 1975 Bram & Downs *Manuf. Technol.* vii. 204 (*heading*) Pull-down Broaching. *Ibid.* 205 An advantage of the pull-down method is that where the cutting pressure is evenly distributed around the axis of movement clamping need not be necessary.

2. b. *Cinematogr.* (See quot. 1959.)
1953 [see *ghost-image* s.v. *GHOST *sb.* 14 e]. 1959 W. S. Sharps *Dict. Cinematogr.* 121/2 *Pull down*, the intermittent mechanism used to move film in a camera, projector or printer by engaging the film perforations.

pulled, *ppl. a.* Add: **1. b.** Of wool. orig. *N. Amer.*
1904 J. M. Matthews *Textile Fibres* ii. 24 There is a certain class of wool..known in trade as *pulled wool*; this is obtained from the pelts of slaughtered sheep, and is usually removed from the skin by the action of time, the fibres being pulled out by the roots. 1921 *Daily Colonist* (Victoria, B.C.) 8 Oct. 15/2 The market for pulled wools continues dull. 1934 J. R. Hind *Woollen & Worsted Raw Materials* iii. 19 The removal of the wool is done by plucking or flipping the wool from the skins, which has given these wools the name of 'flipe' or 'slipe'. In the American wool trade, skin wool is known as 'pulled wool'. 1952 H. Haigh *Work of Woolman* vii. 60 The class called 'English Pulled' or 'Scotch Pulled' includes many var-

ieties of wool. 1963 E. M. Pohle in W. von Bergen *Wool Handbk.* (ed. 3) I. ix. 668 The domestic production of pulled wool in 1944 reached an all time high of 74 million pounds.

3. Also, *pulled work* (see quot.).
1967 E. Short *Embroidery & Fabric Collage* ii. 48 *Pulled work.* For this holes are made without withdrawing threads from the ground fabric, but, as the name implies, by pulling and distorting the weave. *Ibid.* iii. 77 (*caption*) Detail of a tablecloth in pulled work, with a border of famous London buildings.

4. (Earlier and later examples.)
1616 W. Browne *Britannia's Pastorals* II. i. 14 In his flesh pull'd downe As hee had liu'd in a beleaguerd towne. 1895 A. W. Pinero *Second Mrs. Tanqueray* 88 You look dreadfully pulled down. We poor women show illness so plainly in our faces.

puller. Add: **1. c.** *Cricket.* A batsman who pulls (sense 18).
1911, 1972 [see *HOOKER[1] 6 b].

5. One who or that which attracts custom; spec. (*N. Amer.*), a person employed to solicit passers-by into a shop. Also *puller-in.*
1894 J. L. Ford *Lit. Shop* ix. 132 The Jewish old-clothing quarter that lies close to the Five Points is near by. The 'pullers-in', as the sidewalk salesmen are termed in the vernacular of the trade, transact business with a ferocity that can be best likened to that of Siberian wolves. 1928 *Sunday Dispatch* 15 July 14/3 Next to the Prince of Wales, Shaw is the best box-office puller in the United States. 1944 *Sun* (Baltimore) 15 Jan. 6/7, I wonder whether radio announcers are descendants of the Gay street and Harrison street pullers-in and the tray men of Holliday and Baltimore streets (circa 1909)... They had to pass a law to stop the pullers-in. 1955 [see *CHEESE-CAKE 2]. 1958 N. Levine *Canada made Me* ii. 57 The cheap clothing stores—with the pullers standing in the doorways like prostitutes. 1970 J. H. Gray *Boy from Winnipeg* 199 Any country family that stopped to look at something in a window was doomed. The 'puller' would come out and sweet-talk them into the store.

pullery[2] (pu·ləri). *Tanning.* [f. Pull *v.* + -ery.] The place in which wool, hair, and bristles are removed from hides.
1903 L. A. Flemming *Pract. Tanning* 1 The relations between the soaking process and the subsequent processes of the beamhouse or pullery, and the tannery are close. 1963 E. M. Pohle in W. von Bergen *Wool Handbk.* (ed. 3) I. ix. 668 The pulling is conducted in the pulleries which..are connected with the large slaughter-houses.

pullet. Add: **1.** (Later *fig.* examples.)
1823 'J. Bee' *Slang* s.v. *Pullet*, in common life, a female barn-door fowl, which has not yet produced eggs. Young women are so denominated, occasionally. 1922 Joyce *Ulysses* 224 Blazes Boylan looked into the cut of her blouse. A young pullet. 1941 J. Smiley *Hash House Lingo* 44 *Pullet*, young woman.

2. For *Tapes pullastra* read *Venerupis pullastra.* Cf. *PALOURDE. (Earlier and later examples.)
1803 G. Montagu *Testacea Britannica* I. 127 This species [*sc. Venus pullastra*]..is frequently eaten by the common people, and in some parts of Devonshire indiscriminately called *Pullers* or *Pullets*. 1974 S. P. Dance *Encycl. Shells* 268/1 Pullet Carpet Shell. Very similar in shape, size and ornament to *V*[*enerupis*] *rhomboides*.

3. pullet disease = *new wheat disease* s.v. *NEW *a.* 10 b.
1941 Jungherr & Levine in *Amer. Jrnl. Vet. Res.* II. 267/2 The majority of the cases occur between the ages of 5 and 7 months..hence the appropriateness of the term 'pullet disease'. 1945 *Vet. Jrnl.* CI. 7 The exact course of the pullet disease syndrome is unknown. 1950 [see *blue comb* (*disease*) s.v. *BLUE *a.* 13]. 1977 R. F. Gordon *Poultry Dis.* vii. 183 It is difficult to obtain an accurate estimate of the incidence of pullet disease today.

pulley, *sb.*[1] Add: **5.** *pulley-wheel* (later examples); **pulley-cone**, a cone grooved and rotating on its axis, forming a set of pulley-wheels of different sizes.
1903 *Harvard Psychol. Stud.* I. 417 A disc..about 50c. in diameter, rotating on a vertical pivot, was driven by a pulley-cone underneath mounted on the same spindle. 1956 G. Taylor *Silver* vii. 154 *Dish Rings*... The earliest type is shaped like a pulley-wheel. 1967 *Antiquaries Jrnl.* XLVII. 227 Globular flagon with moulded base-angle and pulley-wheel rim.

pulley-bone, var. *PULLY-BONE.

pull-hitter: see *PULL-.

pullicate. Delete ?*Obs.* and add: Also 8 pullcat. **b.** (Earlier and later examples.)
1792 P. Freneau *Poems* (1809) I. 31 Hum-hums are here—and muslins—what you please—Bandanas, baftas, pullcats, India teas. 1958 [see *MONTEITH[2]].

pu·ll-in. [f. phr. *pull in*: see Pull *v.* 25.] **1.** The action of pulling anything in or towards one. Also *attrib.* or as *adj.*
1906 [in Dict. s.v. Pull- 1]. 1976 *Offshore Engineer* July 21/1 A flowline pull-in tool has been developed..to meet this problem. 1977 *Ibid.* Apr. 27/2 The port has a step-down diameter which provides a positive stop during pipe pull-in.

2. a. A café or refreshment stand in a lay-by. Also *pull-in café.*
1938 G. Greene *Brighton Rock* III. iii. 128 He didn't speak to her in the bus... The country unwound the other way: Mazawattee tea, antique dealers, pull-ins. *Ibid.* v. ii. 195 A window of Charlie's Pull-in Café. 1959 *Listener* 8 Oct. 593/3 At the pull-in where most of the play was enacted the café owners sometimes struck a slightly false note. 1973 J. Wainwright *Devil you Don't* 18 A blue and white sign warned five miles to the next service area... 'They'll be at the next pull-in.'

b. An entry, recess, or the like where a motor vehicle may pull in; a lay-by.
1954 E. Hyams *Stories & Cream* 163 A sprawling public-house..in front of which was a vast pull-in for motor-coaches. 1972 M. Gilbert *Body of Girl* iv. 43 The site had not been designed as a garage, and..the pull-in in front was not as deep as it should have been. 1976 *Southern Even. Echo* (Southampton) 3 Nov. (Advt.), Required. Workshop,..shop, suitable for antiques restoring. Main road. Good pull-in.

pulling, *vbl. sb.* Add: **3. b.** (Further examples.)
1866 'Mark Twain' *Lett. from Hawaii* (1967) 84 The arraigning of a ship's officers before the courts by the crew to answer for alleged cruelties practiced upon them on the high seas—such as the 'pulling' of captains and mates by the crews of the *Mercury*. 1960 G. A. Glaister *Gloss. Bk.* 333/2 *Pulling*, the removal of the cover, boards, end papers, tapes, and any lining material which, with the softening of old glue and cutting of sewing threads, are necessary stages in the preparation of a book for rebinding. 1975 J. Pidgeon *Flame* ii. 24 Jack Daniels and the D.T.s liked the Jackoranda. They didn't care much for the cramped stage..and the money was always lousy. But it was the best place they knew for pulling.

c. *N.Z.* (See quot.) Cf. *PULL *v.* 11 e.
1947 P. Newton *Wayleggo* (1949) 154 The act of a heading dog bringing sheep back to his master is termed pulling.

5. *pulling boat*; **pulling-bar** = Draw-bar 1; **pulling bone** *U.S.* = *wish bone* s.v. Wish *sb.*[1] 4; **pulling power**, the ability to attract or persuade.
1892 J. G. A. Meyer *Mod. Locomotive Constr.* 528 Fig. 850 shows the wrought-iron pulling-bar which connects the tender to the engine. 1912 A. T. Quiller-Couch *Hocken & Hunken* p. xxiii, The penultimate race (randan pulling-boats) was finishing amid banging of guns and bursts of music. 1975 *Country Life* 2 Jan. 23/2 The RNLI..displays the former Whitby No. 2 lifeboat.. the last pulling boat to have been in the service of the Institution. 1877 Bartlett *Dict. Amer.* (ed. 4) 502 *Pulling-bone*, the common name in Maryland, Virginia, &c, for the yoke-like breast-bone of chickens, by pulling which till it breaks children and young ladies settle which will be the first married. 1942 H. C. Bailey *Dead Man's Shoes* i. 7 Posters..credited by the expert with much more pulling power. 1966 N. Nicolson in H. Nicolson *Diaries & Lett.* 66 Harold Nicolson's importance to Mosley, apart from the increasing pulling-power of his name, was his close connection with Beaverbrook. 1978 P. Bailey *Leisure & Class in Victorian England* vii. 147 Enterprising publicans..abolished the refreshment check ..relying on the pulling power of the entertainment.

Pullman. Add: **a.** (Earlier and later examples.)
1872 W. F. Butler *Great Lone Land* iv. 57 One takes a Pullman..as one takes a Hansom, Pullman and sleeping-car have become synonymous terms. 1951 N. Mitford *Blessing* II. xii. 265 The Bunbury burglar was walking up the Pullman on his way..to the Trianon bar. 1972 *Daily Tel.* (Colour Suppl.) 28 Apr. 49 By the end of this decade, British Rail acknowledges,..the Manchester, the South Wales and the Bristol Pullmans..will have vanished. 1977 *Modern Railways* Dec. 476/2 The other Metro-Cammell Pullmans still mouldering away in sidings have apparently been abandoned so long that rehabilitation is now unacceptably expensive.

b. (Earlier and later examples.)
1869 *Bradshaw's Railway Man.* XXI. 419 Pullman Palace Car Company Stock $72,300. 1873 *Forest & Stream* 28 Aug. 34/1 It was a close pack..the whole scene reminding one forcibly of a 'Pullman' sleeping car. 1954 W. Tucker *Wild Talent* (1955) v. 58 Ray Palmer slept soundly in the topmost Pullman bed. 1955 D. Davie *Brides of Reason* 28 While Pullman sleepers lulled your sleeping head. 1977 *Time* 18 Apr. 22/2 The cabin can accommodate eight passengers on comfortable Pullman seats.

c. *transf.* (Usu. with small initial.) A prefabricated unit of kitchen or bathroom fixtures, compact as in a railway carriage. Usu. *attrib.* Chiefly *U.S.*
[1932 *New Yorker* 23 July 5/2 There are many people who..would be glad to buy a Pullman section to install in their home.] 1967 'L. Egan' *Nameless Ones* iii. 37 A chipped but spotless pullman washstand. 1968 'R. Macdonald' *Instant Enemy* xxv. 155 It was what is called a studio apartment, consisting of one large room with a pullman kitchen. 1973 *Sunday Bull.* (Philadelphia) 7 Oct. (Parade Suppl.) F6 New double sink pullmans with bright colors are available, or, if you want to do-it-yourself, try separate wash basins. 1977 *Chicago Tribune* 2 Oct. I. 33/2 The apartments for the 16 families in the community consisted of two rooms and a Pullman kitchen.

Pullo, var. *Peulh *sb.* and *a.*

pu·ll-off, *sb.* and *a.* [Pull- 1.] **A.** *sb.* **1.** The fact or action of pulling off or of being pulled off, in various special applications.
1859, 1904 [in Dict. s.v. Pull- 1]. 1926 *Gloss. Terms Electr. Engin.* (Brit. Engin. Stand. Assoc.) 133 *Pull-off*, a metal fitting attached to an ear and used on curves for adjusting the position of a trolley-wire in a horizontal plane. 1950 *Richmond* (Va.) *News-Leader* 28 Jan. 1 Detailed on this maneuver chart are the Navy's plans for an all-out 'Operation Pull-Off' to free the U.S.S. Missouri from her Chesapeake Bay mudbank. 1953 R. Knox *St. Paul's Gospel* i. 10 You might get the printer to give you a pull-off of that childish picture all in blue, with the yellows and the reds left out.

2. *spec.* in *Parachuting* (see quot. 1947). Also *attrib.*
1933 *National Geogr. Mag.* May 614 Heels over head at the 'pull off'... The officer climbed out on the lower wing to the outer strut and, holding on with one hand, pulled his rip cord. 1940 *War Illustr.* 26 Jan. 20/2 The 'pull-off' type is used only for training novices in the 'art' [of parachuting]. 1947 M. Newnham *Prelude to Glory* iv. 14 The procedure..was for the pupil..at a signal from the pilot [to] clamber along the lower wing... The pupil let go his hold on the strut, and willy-nilly he became a parachutist. This was known as the 'pull-off' method.

3. = *LAY-BY *sb.* 1 c. Cf. *PULL-IN 2 b.
1969 V. Canning *Queen's Pawn* iv. 63 Gilpin..was waiting in a pull-off down the road with the Land-Rover. 1972 R. K. Smith *Ransom* vi. 264 Just after the city line as you come down the parkway there's a pull-off—a parking area. 1975 V. Canning *Kingsford Mark* x. 163 He..parked the car on a turfed pulloff.

B. *adj.* Designating that which may be pulled off or from which something may be pulled off.
1902 [in Dict. s.v. Pull- 1]. 1962 *Sunday Express* 7 Jan. 13/6 A pull-off calendar still showed the date. 1973 *Daily Tel.* (Colour Suppl.) 11 May 38/1 Much of it comes in supermarket-style cans with pull-off tops.

pull-on: see *PULL-.

pullorum (pulō°·rəm). [mod.L., a. the specific epithet of *Bacterium pullorum* (L. F. Rettger 1909, in *Jrnl. Med. Res.* XXI. 117), f. gen. pl. of L. *pullus* young chick.] The specific epithet of *Bacterium pullorum* (now *Salmonella gallinarum*) used *attrib.* in **pullorum disease** to designate an acute, infectious, often fatal disease of young chicks, which is also known as bacillary white diarrhœa.
1929 *Bull. Mass. Agric. Exper. Station* XLVIII. 2 It was unanimously voted to accept the suggestion of Dr. Leo F. Rattger to change the name of Bacillary white diarrhea to 'pullorum disease'. 1930 M. A. Jull *Poultry Husbandry* xxvi. 411 The pullorum disease, which frequently causes such enormous losses among chicks, has been called 'bacillary white diarrhea'. 1960 *Farmer & Stockbreeder* 23 Feb. 124/1 Among these bacteria are the organisms responsible for respiratory diseases and pullorum disease. 1977 G. A. Cullen et al. in R. F. Gordon *Poultry Dis.* i. 11 Outbreaks of pullorum disease in turkeys are rare in this country.

pu·ll-out, *sb.* and *a.* [Pull- 1.] Also as one word. **A.** *sb.* **1.** The fact or action of pulling out; withdrawal from an undertaking or affair, *esp.* from military involvement or occupation.
1825 [in Dict. s.v. Pull- 1]. 1944 *Daily Progress* (Charlottesville, Va.) 2 Oct. 9/5 A correspondent reported increasing signs of a pull-out..of tens of thousands of German troops. 1968 Mrs. L. B. Johnson *White House Diary* 13 Apr. (1970) 664 Some of the headlines were easing up—'D.C. Curfew Off, Gradual Pull-Out of GI's Starts'. 1976 *Billings* (Montana) *Gaz.* 2 July 3-c/4 A weaker committee motion, specifying a one-year ban for a political pullout, was defeated by the same margin. 1976 P. Henissart *Winter Quarry* v. 60 Most people think a missile pullout is overdue. 1977 P. Theroux *Consul's File* 177, I inherited him [*sc.* a dog] in Saigon... I took him back to the States after the pull-out.

2. In various technical uses. **a.** *Aeronaut.* The transition from a dive or spin to normal flight.
1919 Pippard & Pritchard *Aeroplane Struct.* vi. 54 The combination of terminal velocity with a quick pull out is one which would break practically any aeroplane. 1932 *Discovery* Apr. 114/2 Individual records of 'pull-outs' from a dive have registered high accelerometer readings without the pilots experiencing ill effects. 1943 *Sun* (Baltimore) 20 July 3/2 All of the men were half dazed. They had been flung about the ship in its two dives and pullouts... The plane was under control, but barely so. 1962 F. I. Ordway et al. *Basic Astronautics* xii. 465 During World War II dive bomber pilots found they could minimize the effects of acceleration by.. tightening muscles, and shouting during pullouts from dives.

b. *Surfing.* (See quot. 1967.)
1967 J. Severson *Great Surfing* Gloss., *Pull-out*, steering the board over or through the back of the wave, as to end the ride. 1968 W. Warwick *Surfriding in N.Z.* 13/1 To execute a pullout, guide your board to the top of the wave, then kick it into the wave. 1971 *Studies in English* (Univ. Cape Town) Feb. 27 Like the turn, the pull-out may be forehand or backhand.

3. A self-contained detachable section of a newspaper, magazine, etc. Also, = *fold-out* sb. s.v. *FOLD *v.*[1] 10.
1952 *Conc. Oxf. Dict.* Add. *Pull-out*, page or plate in book that unfolds out from front edge of leaves to facilitate reference. 1955 *Sun* (Baltimore) 28 Oct. (B ed.) 26/4 TV Pull-Out... Have been meaning to write and tell you how wonderful is the new TV pull-out section of *The Sunday Sun*. 1971 S. E. Morison *European Discovery Amer.: Northern Voy.* p. viii, The reproduction of old maps in a book presents typographical problems. Nobody likes

a big pull-out; but if the size is too much reduced, one cannot read the names of places. 1971 *Woman's Own* 27 Mar. 21 Next week..8-page pull-out of dairy dishes. 1977 *Listener* 17 Mar. 332/2 A potential centre-page pull-out for..*Hustler* magazine.

B. *adj.* Designating that which may be pulled out (in various senses).
1881 [in Dict. s.v. Pull- 1]. 1929 'R. Crompton' *William* iv. 86 They're frightened of the big roundabout—an' the pull-out toffee makes them sick. 1950 J. D. Carr *Below Suspicion* xii. 149 Dr. Bierce lowered himself on one of the pull-out seats facing them. 1955 [see sense 3 above]. 1966 *B.B.C. Handbk.* 79 The pull-out map.. shows how..Soviet and Chinese broadcasters have exploited their geographical position. 1979 *Amat. Photographer* Feb. 62/1 The three cameras we're looking at this week..have pull-out or retracting lenses.

pull-over. Add: **4.** (Usu. as one word.) Used *attrib.* or *absol.* to designate articles of clothing that are put on by drawing them over the head; *spec.* (chiefly in *absol.* use) a knitted or woven garment for the upper part of the body; a jumper or jersey.
1907 *Yesterday's Shopping* (1969) 320c/1 The 'Pullover' Storm Coat..is especially designed without any opening when in wear, and, being made without a rubber neck, entirely obviates any discomfort in pulling the garment on or off over the head. 1921 *Daily Colonist* (Victoria, B.C.) 6 Apr. 4/5 (Advt.), Another lot of these smart Wool Pull-Over Sweaters to sell at $2.98. 1925 *Westm. Gaz.* 28 Apr. 3 The vogue of the Pullover has supplanted the waistcoat for golf. 1930 *Daily Tel.* 9 Apr. 15/1 (Advt.), Attractive three-piece suit in tweedknit..designed with.. new tuck-in pullover finely woven to tone with suit. 1940 Graves & Hodge *Long Week-End* iii. 42 Most women in 1919 were wearing jumpers, knitted by themselves as a relief from 'socks for soldiers'; and soon afterwards men, too, began to adopt them under the name of 'pull-over'. 1967 N. Freeling *Strike Out* 82 The young man was.. darning the worn elbow of a pullover. 1977 *New Yorker* 10 Oct. 124/3 (Advt.), Pullover Dress. A most cozy dress in thick pure cotton flannel, brushed inside and out, long known for its wearing qualities.

Hence **pu·llovered** *a.*, wearing a pullover.
1926 *Daily Chron.* 13 May 2/2 'I'll be sorry to leave the old bus tonight,' said the plus foured pull-overed youth at the wheel of the 'General' yesterday afternoon. 1977 *Film & Television Technician* Jan. 10/3 (*caption*) Making whoopee at the fun-packed Animation Social..[were] pullovered Animation Section Chairman, Barry Merritt, Joe Telford and journal editor Roy Lockett.

pull-switch, -tab: see *PULL-.

pull-up. Add: (Now usu. with stress on the first element.) **1. a.** (Earlier *fig.* and later examples.)
1837 Dickens *Pickw.* xxxiii. 344 That's rayther a sudden pull up, ain't it, Sammy? 1842 —— *Let.* 27 Feb. (1974) III. 92, I have so much to say that I could fill quires of paper, which renders this sudden pull-up the more provoking. 1950 A. Huxley *Let.* 6 Aug. (1969) 628 In the plain you and the child can walk abroad without having to take the car and without being fatigued by a pull up. 1980 'M. Innes' *Going it Alone* xvii. 153 If they did a sudden pull up like that, it wouldn't be much good just driving on.

b. A lifting-up, encouragement.
1872 Geo. Eliot *Middlem.* IV. viii. 360 He told Mary that his happiness was half owing to Farebrother, who gave him a strong pull-up at the right moment. 1913 G. de H. Vaizey *College Girl* xxi. 291 Think of all that it means..if we can keep these men from drifting, and give them a pull-up in time!

2. (Further examples.) Phr. *a good pull-up for carmen*, a roadside café; also in various *transf.* and allusive uses.
1899 [in Dict.]. 1925 H. V. Morton *Heart of London* 50 London's tea-shops are of many kinds, from the standardized shop to the good pull-up for millionaires constructed on the Paris plan. 1928 *Sunday Express* 29 Apr. 4/4 It was known in Hollywood as 'The Legs of Carmen', but your censor doubtless attended to that. The censor's office is usually a good pull-up for Carmen. 1935 A. J. Cronin *Stars look Down* II. xx. 442 He went into a workman's coffee-house: *Good pull up for lorrys* was on the sign outside. 1952 M. Allingham *Tiger in Smoke* ix. 153 The pull-up on the corner opens at five... I want..to get a bit o' breakfast. 1965 [see *DINER 2 b]. 1977 *Listener* 24 Mar. 382/2 Our mother..ran..a 'caff'—or what was then known as a good pull-up for car-men.

3. a. The fact or action of pulling something upwards; *spec.* in physical exercise, the action of pulling up the body by means of a bar or beam held by the hands.
1907 M. A. von Arnim *Fräulein Schmidt* lix. 255 'He only wants his wind,' said Vicki... 'It certainly was rather a long pull up,' said I. 1938 *Jrnl. R. Aeronaut. Soc.* XLII. 625 The manœuvres consisted of push-downs and pull-ups from level flight..and push-ups from inverted flight. 1946 J. E. Q. Barford *Climbing in Britain* iv. 63 The main types of holds are as follows:—A Straight Pull Up. This is a hold over which the fingers can curl as over the rung of a ladder. 1960 E. S. & W. J. Higham *High Speed Rugby* 298 Pull-ups on beam, or other horizontal bar. 1971 A. A. Michele *You don't have to Ache* i. 21 Here are some things that you should not be doing:..Do not do push-ups or pull-ups.

b. *attrib.* or as *adj.* Designating that which may be pulled up.
1919 R. Fry *Let.* May (1972) II. 451, I live in a house which has..a real Victorian W.C. with a pull up plug. 1973 *Country Life* 26 July 260/2 The gear lever..is mounted on a floor console with the pull-up handbrake.

‖ **pullus** (pu·lŏs). Pl. pulli. [L., = young chick.] A young bird during the stage before it is fully grown or able to fly.
1774 G. White *Let.* 2 Sept. in *Selborne* (1789) I. xl. 100, I had been..comparing the tails of the male and female swallow, and this ere any young broods appeared; so that there was no danger of confounding the dams with their *pulli*. 1955 R. Spencer in *Brit. Birds* XLVIII. 468 Pull. (pullus)—nestling or chick not yet flying. 1964 A. L. Thomson *New Dict. Birds* 904/2 The bird is technically a 'pullus'..until it is full-grown and flying. *Ibid.*, After the pullus stage a bird is described as 'juvenile' while wearing its first plumage of true feathers.

pully-bone (pu·li,bō°n). *U.S. dial.* Also **pulley-bone.** [f. Pull *v.* + -y + Bone *sb.*] = *wish-bone* s.v. Wish *sb.*[1] 4. Cf. *pull-bone* s.v. *PULL- 2 and *pulling-bone* s.v. *PULLING *vbl. sb.* 5.
1939 B. K. Harris *Purslane* 148 The girls scrambled over the pulley-bone of the turkey. 1947 M. Henry *Misty of Chincoteague* xvi. 152 Somethin' told me to save the pully bone from that marsh hen. 1966 *Publ. Amer. Dial. Soc.* 1964 XLII. 22 *Pully-bone*, the wish-bone or furcula, of a chicken, often pulled after a meal, to determine who is to get married first. 1976 *Amer. Speech* 1973 XLVIII. 180 If we understand that there is a connection between the Southern mountains and coastal plains, the occurrence of *you-all, grea(z)y*, and *pulley bone* 'wishbone' in an area as far away as the Midwest becomes readily explainable.

pulmonic, *a.* Add: **4.** *Phonetics.* Relating to the lungs as the initiator of the air stream used in the articulation of speech sounds.
1942 Bloch & Trager *Outl. Linguistic Analysis* ii. 31 Stops with inner closure at the bottom of the lungs are called *pulmonic*. 1949 J. R. Firth in *Trans. Philol. Soc.* 1948 142 Types of sound which appear to crop up repeatedly in syllabic analysis..are..aitch or the pulmonic onset. 1959 [see *GLOTTALIC *a.*]. 1975 F. R. Palmer in W. F. Bolton *Eng. Lang.* I. 17 Almost without exception the whole of the articulation of sounds in European languages is powered by air expelled from the lungs (it is 'pulmonic egressive').

pulmotor (pv·lmō°təɹ). [f. Pulmo- + L. *mōtor* that which moves.] An apparatus for automatically forcing air or oxygen into and out of the lungs when breathing has ceased or is weak. Also *attrib.* and *fig.*
Formerly a proprietary name in the U.S.
1912 J. W. Paul *Use & Care Mine-Rescue Breathing Apparatus* (U.S. Bureau Mines: Miners' Circular No. 4) (rev. ed.) 25 The pulmotor is intended for use in the resuscitation of persons who have partly or wholly ceased to breathe as a result of inhaling irrespirable gases, of an electric shock, or of drowning. 1913 *Official Gaz.* (U.S. Patent Office) 24 June 1052/1 Drägerwerk..Lubeck, Germany..*Pulmotor*... Mechanical respiratory apparatus and devices for administering oxygen. Claims use since February, 1909. 1928 *Daily Express* 31 Dec. 12/4, I grabbed up my bag and the pulmotor, and was over here in a jiffy. 1940 *Economist* 6 Apr. 618/2 The third view is almost..entirely mechanistic. It belongs in the pulmotor school of economics. 1951 W. Kees in *Furioso* Summer 35 Another fat woman In a dull green bathing suit Dives into the water and dies. The pulmotors glisten. 1974 S. Sheldon *Other Side of Midnight* vii. 180 She debated whether to stay in bed or call a pulmotor squad.

pulp, *sb.* Add: **4. c.** orig. *U.S.* Ephemeral literature, esp. (in derogatory use) that regarded as being of poor quality; popular or sensational writing generally. Freq. *attrib.*, as *pulp artist, fiction, novel, writer*, etc. Also *ellipt.* = *pulp magazine* (sense 5 c below). Also *transf.*
1931 *Frontier* (Missoula, Montana) Nov. 82/1 Even should he fail to publish in the big magazines, and never graduate from the 'pulps', he can rise to as much as ten cents a word. 1945 [see *GLOSSY *sb.* b]. 1945 R. Chandler *Let.* 24 Aug. (1966) 201 Marlowe just grew out of the pulps. He was no one person. 1951 Wodehouse *Old Reliable* ii. 32 Half the best known writers today started on the pulps. 1952 M. Steen *Phoenix Rising* iii. 69 [He] picked up a handful of old pulps. 1966 *New Statesman* 15 July 104/3 There's only one well-known actor... The bulk of the others' experience comes from local rep, TV pulp and Shaftesbury Avenue trivia. 1972 D. E. Westlake *Bank Shot* iv. 24 He'd discovered the pulps..when he was in high school. 1976 *National Observer* (U.S.) 3 July 17/1 When I started..the pulps were gasping their last.
attrib. 1936 Pulp writer [see *ACE 2 c]. 1946 R. Chandler *Let.* 2 Oct. (1966) 24 We have a much better home than an out-of-work pulp writer has any right to expect. 1951 M. McLuhan *Mech. Bride* (1967) 151/1 Why aren't you interested in the private lives of the strippers and pulp artists who upholster our desert landscape? 1955 L. A. Fiedler in D. Lodge *20th Cent. Lit. Crit.* (1972) 464 Wordless narrative: digests, pulp fiction, movies, picture magazines. 1958 *New Statesman* 6 Sept. 294/3 The wretched reader of pulp literature is encouraged to dream of sins and orgies he is forbidden to enact. 1959 *Listener* 30 July 176/3 The pulp novels of Mickey Spillane. 1965 *Ibid.* 27 May 788/1 Feelings should not run too high over sophisticated pulp literature. 1970 G. Greer *Female Eunuch* 164 The bored housewife..intoning the otherwise very forgettable words of some pulp lovesong. 1975 J. McClure *Snake* viii. 102 Constrictors..are certainly not given to crushing anything to a bloody pulp. As pulp fiction would have it! 1976 *Listener* 29 July 122/2 Cody..met up with a pulp novelist..who proceeded to set Cody up as a regular frontier hero in a series of literary adventures. 1977 *Time Out* 17–23 June 35/3 Juicy pulp movie about the organisation's efforts to move in on the truck hi-jacking operation run by Anna and her girls.

5. a. (*a*) *pulp-maker*. **b. pulp paper,** newsprint; paper of similar texture used for books or magazines; **pulp-stone,** (*b*) a stone used like a grindstone for reducing wood to pulp; **pulpwood** (earlier and later examples). **c.** (in sense *4 c) **pulp magazine,** a magazine devoted to popular or sensational literature; also (with hyphen) *attrib.*
1931 *Frontier* (Missoula, Montana) Nov. 83/1 We need some outlets for the work, with pay, of young and enthusiastic writers; something to keep them away from the 'pulp' and 'slick paper' magazines. 1934 *Sun* (Baltimore) 15 Mar. 21/1 He wrote 'Western' fiction for the 'pulp' magazines. 1937 A. Huxley *Ends & Means* xii. 191 Each month the pulp magazines offer to millions of readers their quota of true confessions, film fun, spicy detective stories, hot mysteries. *Ibid.* 207 Pulp-magazine stories are transcriptions of the commonest and easiest day-dreams. 1944 'G. Orwell' in *Horizon* Oct. 239 English imitations of the 'pulp magazine' do now exist. 1954 Koestler *Invisible Writing* III. xv. 186 We churned out a couple of detective stories for pulp magazines. 1968 E. A. McCourt *Saskatchewan* vii. 76 Farwell, as various stories attributed to him suggest, had a pulp-magazine mind. 1975 *Times Lit. Suppl.* 10 Oct. 1174/2 *The Continental Op* is the anonymous narrator of the stories which Dashiell Hammett wrote for the pulp magazine *Black Mask* during the 1920s. 1883 D. A. Wells *Pract. Econ.* (1885) 107 Even the pulp-makers..will find difficulty in marketing their pulp in the immediate future. 1901 *Daily Colonist* (Victoria, B.C.) 31 Oct. 3/2 It is therefore just as desirable that information should be compiled for their [*sc.* fishermen's] use as for the use of miners, lumbermen, pulp-makers or farmers. 1909 Pulp-maker [see *log-lumberer* s.v. *LOG *sb.*[1] 8 b]. 1908 Kipling *Lett. of Travel* (1920) 154 The advertising of Canadian papers,..the brittle pulp-paper, the machine-set type. 1931 *Times Lit. Suppl.* 9 July 542/3 The choice between writing for those pulp-paper magazines that pay by the word and the smooth-paper magazines that pay by the story. 1901 J. H. Pratt in *Mineral Resources of U.S.* 789 Pulpstones differ from grindstones in having a much broader face. 1957 *Encycl. Brit.* XVII. 232/2 Natural pulpstones are 27 to 36 in. wide by 54 in. or more in diameter. Artificial stones are 27 to 54 in. wide and 54 to 72 in. in diameter. 1885 *Rep. New Hampshire Forestry Comm.* 10 Telegraph-poles, pulp-wood, bark, etc. 1928 R. S. Troup *Silvicultural Syst.* xvii. 177 Where coniferous forests are grown solely for pulpwood or mining timber, the clear-cutting system with artificial regeneration is often the only one feasible. 1960 'N. Shute' *Trustee from Toolroom* x. 284 The offcuts were turned into pulpwood for newsprint. 1974 *Globe & Mail* (Toronto) 28 Nov. 45/2 The spruce, which produced its first shoot a few weeks ago, is the first pulpwood tree anywhere to be grown in this way, according to scientists involved in the project.

pulpal (pv·lpăl), *a. Dentistry.* [f. Pulp *sb.* + -al.] Of or pertaining to the pulp of a tooth; *spec.* applied to that surface of a cavity which overlies the pulp.
1908 G. V. Black *Operative Dentistry* II. 6 That wall of a cavity which is to the occlusal of the pulp, and in a plane at right angles to the long axis of the tooth, is called the pulpal wall. 1925 *Dental Rec.* XLV. 627 It has been argued that grinding the enamel in preparation for the crown renders a tooth susceptible to thermal shock, with subsequent destruction of pulpal structures. 1953 J. R. Schwartz *Inlays & Abutments* xiii. 142 The pulpal and axial walls should be flat and at right angles to one another. 1967 *Brit. Dental Jrnl.* CXXIII. 420/1 Pulpal involvement may be unavoidable if preparation for a jacket is undertaken.

pulpectomy (pɒlpe·ktŏmi). *Dentistry.* [f. Pulp *sb.* + *-ectomy.] Surgical removal of the pulp of a tooth (usu. all of it: cf. *PULPOTOMY).
1923 *Dental Items of Interest* XLV. 5 The remaining portion of a vital pulp following excision, or partial pulpectomy, will be destroyed. *Ibid.* 82 Operations upon the pulp are designated as 'Pulpotomy'. In case a portion of the pulp is excised and removed, the operation is termed 'Partial Pulpectomy'. The removal of the entire pulp is 'Pulpectomy'. 1924 F. E. Hogeboom *Pract. Pedodontia* v. 65 (*heading*) Pulpectomy and pulpotomy. *Ibid.*, Pulpectomy is the complete removal of the pulp. 1957 S. B. Finn *Clin. Pedodontics* xiv. 318 Pulpectomy may be complete or partial, depending on whether there is any vital tissue in the canal. 1976 Ettinger & Pinkham in A. J. Nowak *Dentistry for Handicapped Patient* xvi. 287/1 Pulpectomy with appropriate instrumentation and filling is a standard..treatment for pulp pathology.

‖ **pulperia** (pulperī·ă). [Sp. Amer.] In Central and South America and the south-west U.S., a grocery or tavern.
1818 A. Gillespie *Gleanings & Remarks Buenos Ayres* viii. 91 At the intersected corners of almost every street in that capital, pulperias, or dram, and grocery shops, are established, that vend liquors, candles, and other articles. 1840 R. H. Dana *Two Yrs. before Mast* (1841) xxviii. 192 He..came to the Pueblo de los Angelos... Here he went dead to leeward among the pulperias, gambling-rooms, &c. 1859 T. Cochrane *Narr. Services in Liberation of Chili, Peru, & Brazil* I. x. 216 The two months' pay offered the other day could not now effect its purpose, as the whole—and more is due to the Pulperia keepers, to whose benefit, and not that of the seamen, it must have immediately accrued. 1871 H. M. & P. V. N. Myers *Life under Tropics* iii. 21 We were forced by a sudden shower to seek shelter in a way-side pulperia. 1904 Conrad *Nostromo* I. viii. 80 The horseman hammered with the butt of a heavy revolver at the doors of low pulperias. 1905 J. Masefield

Mainsail Haul 14 When Don Alfonso was in the pulperia (that's Spanish for grog-shop), he was a bluin' down that licker. 1936 *Times Lit. Suppl.* 29 Feb. 173/1 'Charlie the Gaucho' opens outside a *pulperia* on the Pampas. 1974 D. Meiring *President Plan* ix. 74 The single *pulperia* which he had visited to buy food.

pulping, *vbl. sb.* Add: *pulping-machine.*
1875 *Encycl. Brit.* I. 327/1 A premium was offered for machines to perform this kind of work [*sc.* pulping of turnips and mangolds for cattle], under the somewhat inappropriate designation of 'pulping-machines'. 1909 *Chambers's Jrnl.* Aug. 518/1 From this pulping machine it passes to the centrifugal pump.

pulpit, *sb.* Add: **4. b.** Also, the harpooner's standing-place on a swordfishing vessel.
1927 G. Bradford *Gloss. Sea Terms* 135/1 *Pulpit*, the harpooning platform on the bowsprit of a sword-fishing vessel. 1959 W. R. Bird *These are Maritimes* v. 132 We noted the 'pulpits' constructed far forward for the use of the man who throws the spear. 1972 E. Staebler *Cape Breton Harbour* i. 16 A bowsprit like a diving board with a metal sort of pulpit on the end of it.

d. A small raised platform or room from which machinery can be observed and controlled.
1880 *Harper's Mag.* Dec. 62 Another shout, and the boy touches another lever in the gallery of levers, irreverently termed the 'pulpit'. 1903 *Electr. World & Engin.* 26 Dec. 1051/2 The operator of the hoisting motor stands in a pulpit above the floor. 1959 *Control* Feb. 97/3 The mill pulpit or control room..is staffed by two rollermen and a member of the metallurgical department who keeps an eye on ingot quality. 1968 'A. Haig' *Sign on for Tokyo* 122 They were sitting in the 'pulpit', three of them, above the bars of the rolling mill.

e. The (pilot's, etc.) cockpit of an aeroplane. *R.A.F. slang.*
1933 D. Grinnell-Milne *Wind in Wires* I. ii. 96 The reason for its unofficial name—'The Pulpit'—was all too obvious. A little three-ply box projected from the front of the machine... The wretched man in this box had.. an unrestricted forward view. 1941 [see *GREENHOUSE 3]. 1942 *Gen* 1 Sept. 14/1 A fighter pilot climbs into the 'pulpit' of his plane.

f. *Yachting.* (See quot. 1961). Also in other water craft.
1961 F. H. Burgess *Dict. Sailing* 164 *Pulpit*, an elevated tubular metal guardrail set up at bow or stern. 1964 *Eng. Stud.* XLV. 23 A pulpit is a raised safety-rail in the bows of a yacht or motor cruiser. 1976 *Yachts & Yachting* 20 Aug. 382/3 (Advt.), Fast Week-ender, excellent condition..two berths, pulpit, full foam buoyancy. 1977 *Mod. Boating* (Austral.) Jan. 110/1 Deck or even pulpit-mounted lights were often hidden from view.

5. d. pulpit-cloth (later U.S. example); **pulpit-rail,** a rail on a ship's pulpit.
1872 *Atlantic Monthly* Mar. 317 Fragments of richly colored altar-pieces, fine pulpit-cloths, and pieces of old carving. 1958 S. A. Grau *Hard Blue Sky* i. 42 Hector walked the full length of the boat and turning settled himself on the pulpit rail. 1974 G. Jenkins *Bridge of Magpies* xv. 226 A light anti-aircraft gun platform.. surrounded by a rusty metal 'pulpit rail'.

pulpitis (pɒlpəi·tis). *Dentistry.* [f. Pulp *sb.* + -itis.] Inflammation of the dental pulp.
1882 *Dental Rec.* II. 444 The tooth became sensitive upon percussion..without signs of pulpitis. 1930 W. H. O. McGehee *Text-bk. Operative Dentistry* xxvi. 776 In the later stages of acute pulpitis, tenderness of the tooth to tapping may arise, from passage of the inflammatory area to the tissues of the apical space. 1977 Curl & Prue in Boundy & Reynolds *Current Concepts Dental Hygiene* vii. 111 Chronic partial pulpitis will tend to be associated with pain.

pulpitum (pu·lpitŏm). *Archit.* [L.: see Pulpit *sb.*] A stone screen separating the choir from the nave and freq. surmounted by a loft housing the organ.
1845 R. Willis tr. Gervase of Canterbury in *Archit. Hist. Canterbury Cathedral* iii. 37 A screen with a loft (*pulpitum*)..separated in a manner the aforesaid tower from the nave. 1908 F. Bond *Screens & Galleries Eng. Churches* iv. 159 The pulpitum differed from the rood screen in many ways. It formed the eastern barrier of the quire, and against the eastern face of it were placed the return stalls. *Ibid.*, The pulpitum always had a spacious loft above it, and carried the organ. 1923 *Trans. Scottish Ecclesiol. Soc.* VII. 101 As originally placed the painting took the shape of a boarded tympanum, occupying the entire space upward from the back of the pulpitum, or loft, to the roof. 1937 *Burlington Mag.* Mar. 128/2 The 'Master of the Naumburg Sculptures'..was responsible.. for the sculptures on the pulpitum in Mainz Cathedral. 1966 J. Fleming et al. *Penguin Dict. Archit.* 179/2 *Pulpitum*, a stone screen in a major church erected to shut off the choir from the nave. It could also be used as a backing for the return choir stalls.

pulpotomy (pɒlpɒ·tŏmi). *Dentistry.* [f. Pulp *sb.* + -o + -tomy.] Surgical removal of the pulp of a tooth (usu. part of it only: cf. *PULPECTOMY).
1923 [see *PULPECTOMY]. 1924 F. E. Hogeboom *Pract. Pedodontia* v. 66 Pulpotomy means the partial removal of the pulp. 1957 S. B. Finn *Clin. Pedodontics* xiv. 318 Pulpotomies can be performed on teeth which have been exposed for as long as 72 hours. *Ibid.* 317 There are four courses of treatment open to the dentist: (1) pulp capping, (2) pulpotomy, (3) pulpectomy with or without apicoectomy, and (4) extraction of the tooth. 1976 Ettinger & Pinkham in A. J. Nowak *Dentistry for Handicapped Patient* xvi. 286/1 The contraindications to pulpotomies in the chronically ill patient are the same as for direct pulp capping.

pulpwood: see Pulp *sb.* 5 b in Dict. and Suppl.

pulpy, *a.* Add: **1.** (Later *fig.* examples in sense *4 c of the sb.)
1939 R. Chandler *Let.* 19 Feb. (1966) 195 *The Big Sleep* is very unequally written. There are scenes that are all right, but there are other scenes still much too pulpy. 1978 *Nature* 27 Apr. 786/1 Even the pulpiest science fiction includes this.

2. *pulpy kidney* (*disease*), a clostridial enterotoxæmia of sheep characterized by rapid post-mortem degeneration of the kidneys.
1927 *N.Z. Jrnl. Agric.* XXXIV. 217 (*heading*) 'Pulpy kidney' disease of lambs. *Ibid.* 227 Out of a mob of 290 lambs twelve died, presumably from 'pulpy kidney'. 1938 tr. F. Hutyra's *Special Path. & Therapeutics Dis. Domestic Animals* (ed. 4) III. 14 In Australia..an epizootic type of severe renal degeneration known as 'pulpy kidney disease' occurs..(the symptoms are paralysis and convulsions, and death occurs in 3 to 7 hours). 1953 *Cape Argus* 21 Mar. 7/3 Pulpy kidney disease is taking a heavy toll among sheep in the Lady Grey district. 1970 'J. Herriot' *If only they could Talk* xxv. 149 The diseases which beset the lambs themselves—swayback, pulpy kidney, dysentery.

pulque. Add: **b.** *pulque alcohol, shop.*
1931 H. Crane *Let.* 21 Sept. (1965) 382 Straight pulque alcohol in each cup. 1836 C. A. Goodrich *Universal Traveller* 139 But, unfortunately, in the lanes near the market are found numbers of pulque-shops (pulquerias). 1910 *N.Y. Even. Post* 16 July (Sat. Suppl.) 3/2 All of the sweet savors of Araby combined could make slight headway against the reek of a pulque shop.

‖ **pulqueria** (pulkerī·ă). [Sp. Amer.] In Mexico, a shop or tavern selling pulque.
1822 J. R. Poinsett *Diary* 28 Oct. in *Notes on Mexico* (1824) v. 49 They go to the pulqueria, and there dance, carouse, and get drunk on pulque and *vino mezcal*. 1847 G. A. F. Ruxton *Adventures Mexico & Rocky Mts.* vii. 43 After leaving the pulqueria, we visited..the dens where these people congregate for the night. 1914 C. J. C. Hyne *Firemen Hot* i. 2 By the time these [dollars] had been passed across the grimy counter of a pulqueria.. they received [etc.]. 1922 *Outward Bound* Nov. 110/2 In the great pulquerias, or saloons..the gramophone is invariably to be found. 1934 S. E. White *Folded Hills* 375 He can play his guitar and entertain the drunkards in the *pulquerias*.

pulsar (pv·lsāɹ). [f. *puls*(*ating st*)*ar*, after *QUASAR.] **1.** *Astr.* A cosmic source of radio signals that pulsates with great regularity at intervals of the order of a second or less, and is believed to be a rapidly rotating neutron star. Also *fig.*
1968 *Daily Tel.* 5 Mar. 21/3 An entirely novel kind of star..came to light on Aug. 6 last year and,..was referred to by astronomers as LGM (Little Green Men). Now..it is thought to be a novel type between a white dwarf and a neutron [*sic*]. The name Pulsar (Pulsating Star) is likely to be given to it... Dr. A. Hewish..told me yesterday: '..I am sure that today every radio telescope is looking at the Pulsars.' 1968 *Time* 26 Apr. 82 Under the careful scrutiny of increasing numbers of scientists around the world, astronomy's newest sensation —the pulsars—continued to beep away last week, confounding observers, with the breathtaking regularity of their signals. 1969 *New Yorker* 19 Apr. 55/1 A tough-looking astronomer, whose ideas stretch beyond the moon to quasars and pulsars. 1972 *Sci. Amer.* July 37/2 Although the pulsar in the Crab Nebula was first discovered at radio frequencies and was later studied visually, most of its energy is emitted in the X-ray range. 1973 L. M. Boston *Memory in House* xi. 132 Oscar is a pulsar, an output of concentrated energy. 1977 *Dædalus* Fall 45 Pulsars, that is, rotating neutron stars, have been detected in the Crab Nebula and in Vela X, a supernova remnant about ten times older than the Crab Nebula.

2. A kind of digital wrist-watch. (A proprietary name in the U.S.)
1970 *Daily Tel.* 14 May 13/7 Production is expected to begin next year, but at first pulsars will be sold in a limited edition at $1,500. 1971 *Official Gaz.* (U.S. Patent Office) 31 Aug. TM240/1 Hamilton Watch Co, Lancaster, Pa... *Pulsar* for watches... First use Apr. 27, 1970. 1977 R. E. Harrington *Quintain* iii. 21 He flipped his wrist and the pulsar wristwatch winked on. 1979 G. Swarthout *Skeletons* 69, I punched my Pulsar. Two minutes past nine p.m.

pulsatance (pɒlsē¹·tăns). *Physics.* [f. Pulsate *v.* + -ance.] The angular frequency of a periodic motion, i.e. 2π times its actual frequency.
1919 A. Campbell in *Proc. Physical Soc.* XXXI. 81 In English a name has not yet been found for $2\pi n$, where n represents frequency... I would suggest that it might be called 'pulsatance'. The termination 'ance' brings it into line with words like inductance and reactance. 1946 *Electronic Engin.* XVIII. 119 The displacement (D) at any point is a periodic time function of the form: $D = \psi \sin \omega t$ ..where ω is the pulsatance and ψ is the amplitude at any point in space. 1957 *Electronic & Radio Engineer* XXXIV. 145/2 Equation (22) represents a phase vibration of natural undamped pulsatance ωn and damping factor ζ.

pulsatility. (Later examples.)
1930 *Amer. Jrnl. Physiol.* XCI. 716 The venous return to the chest must show a pulsatility which is nearly synchronous with the pulsatile outflow. 1977 *Lancet* 3 Sept. 490/2 They observed that the peak venous flow can be increased seven-fold, and its pulsatility thirty fold.

pulsating, *ppl. a.* (s.v. Pulsate *v.*). Add: *spec.* of an electric current or voltage: flowing or acting in one direction but with a periodically varying strength.
1912 Thiess & Joy *Toll Telephone Practice* ii. 20 The pulsating current is used in signaling the exchange and the alternating current for calling the various subscribers on the line by means of the usual code rings. 1960 H. W. Jackson *Introd. Electric Circuits* ix. 202 The tendency of inductance to oppose any change in current can be used to advantage in filtering or smoothing variations in load current that would otherwise occur when the load is fed from a d-c source..which provides a pulsating emf.

pulsational, *a.* Delete *rare* and add further examples.
1969 *Nature* 19 July 279/1 Despite the trend of recent evidence..that radial pulsations may not be the source of variation in pulsars, the pulsational hypothesis has still not been disproved. 1976 *Ibid.* 15 Jan. 88/2 Periods decrease with increase of λ and of polytropic index n..for all different modes of oscillations (pulsational, transverse shear, and toroidal, modes).

Hence **pulsa·tionally** *adv.*
1969 *Nature* 19 July 280/1 Thermonuclear reactions are extremely effective in pulsationally destabilizing a degenerate star.

pulsator. Add: **4.** *Agric.* A device on a milking machine which releases the suction on the teat intermittently so as to simulate the sucking action of a calf.
1907 *Jrnl. R. Agric. Soc.* LXVIII. 133 By another flexible connection the pail communicates with a vacuum pipe, and an air exhaust, by which the pulsators are actuated. 1931 J. B. Davidson *Agric. Machinery* xxiii. 339 The pulsator is an air valve so constructed as to alternate suction and release. 1950 *N.Z. Jrnl. Agric.* LXXX. 568/2 Efficient milking is not possible with leaking pulsators. 1970 R. Jeffries *Dead Man's Bluff* i. 12 The milk began to spurt into the glass jars... The rhythmic clicking of the pulsator, working off the vacuum line, gave him a conscious feeling of well-being.

pulse, *sb.*[1] Add: **2. c.** Phr. *on the pulse* (and variants): through one's own experience (with allusion to Keats's use).
1818 Keats *Let.* 3 May (1931) I. 154 Axioms in philosophy are not axioms until they are proved upon our pulses. We read fine things, but never feel them to the full until we have gone the same steps as the Author. 1970 *Guardian* 23 July 10/3 As I am one of his constituents, the appointment of Sir Robert Grant-Ferris..as Deputy Speaker..has made me feel 'on the pulse'..a frustrating anomaly of our parliamentary system. 1971 R. ap Roberts *Trollope* ii. 42 The problem of *The Warden* is—one might say—proved on our pulses. 1973 *Listener* 6 Dec. 798/3 The committed nationalism of, say, the 19th-century Russian composers, who had felt oppression on the pulse.

4. b. In scientific use now *spec.* (*a*) a train of radio waves, sound waves, or the like, of very short duration; a short burst of radiated energy; (*b*) the more usual term for *IMPULSE *sb.* 4*.
(*a*) 1905 S. R. Bottone *Radium* (ed. 2) iv. 74 A third kind of emanation is also produced by radium... Röntgen rays—ether vibrations—produced as a secondary phenomena by the sudden arrest of velocity of the electrons by the solid matter, producing a series of Stokesian 'pulses' or explosive ether waves, shot into space. 1906 *Nature* 29 Nov. 105/2 The signal produced by a spark discharge consists of a series of violent pulses each consisting of a short train of strongly damped vibrations of definite frequency. 1945 H. D. Smyth *Gen. Acct. Devel. Atomic Energy Mil. Purposes* xii. 131 In this method a neutron source is modulated, i.e., the source is made to emit neutrons in short 'bursts' or 'pulses'. 1947 Crowther & Whiddington *Science at War* 16 Meanwhile the pulse flies on, reaches the aircraft, and is reflected back as an echo. 1969 *Times* 8 Jan. 12/1 Working from the measured length of the successive pulses of energy, it was possible to calculate that the stars concerned would have been as massive as the sun but rather smaller than the earth in size. 1978 *Sci. Amer.* Apr. 38/2 The bats and whales were before us, but now we humans make routine use of pulses of ultrasound (or of microwave) to map the night or the depths.
(*b*) 1932 *Proc. Physical Soc.* XLIV. 77 A transformer.. translated the square-topped current pulsations into voltage pulses, alternately positive and negative, of very short duration. 1949 B. Grob *Basic Television* v. 63 The amplitude of the video signal is divided into two sections, the lower 75 per cent being devoted to the active camera signal while the upper 25 per cent is used for the synchronizing pulses. 1967 *Electronics* 6 Mar. 159/2 A simple change in d-c level cannot be used as a trigger because it locks up the flip-flop against further changes; a pulse is a must. 1975 D. G. Fink *Electronics Engineers' Handbk.* xvi. 6 In the field of radio-frequency interference..the basic response curve of the receiver is defined in terms of its response to regularly repeated pulses... The area under the pulse must be a known constant which is a function of a limited number of circuit parameters.

d. A temporary upward movement of magma through the earth's crust.
1964 *Nature* 13 June 1100/2 Difficulties with this concept have led petrologists..to postulate a pulse mechanism to explain such features as the magnetite layer near the top of the Main Norite Zone of the Bushveld Igneous Complex. 1970 *Ibid.* 25 July 365/1 A more restricted pulse (heave) of magma. 1977 A. Hallam *Planet Earth* 68

Occasionally a new pulse of magma on its way to the surface breaks off fragments that emerge as xenoliths included in lava flows or ash falls.

4*. *Biochem.* A period during which a culture of cells is supplied with an isotopically labelled substrate or substrates. Also *attrib.* Cf. *PULSE *v.* 5.

1960 *Jrnl. Molecular Biol.* II. 308 Phage infection and the subsequent ^{32}P pulse experiment were performed at 28°C. 1961 *Nature* 13 May 580/1 The nascent protein can be labelled by a short pulse of $^{35}SO_4$. 1974 *Ibid.* 25 Jan. 243/1 If a long molecule is labelled at one end (as happens with a short pulse label) and then sheared, the labelled molecules will always appear to have lower molecular weights than the bulk material.

5. Now freq. also in sense 4 b. **a.** *pulse amplitude, height, repetition* (or *recurrence*) *frequency* (or *rate*), *rate* (example of different sense), *train, width.* **b.** *pulse amplifier, analyser, compression, counter* (examples of different sense), *generator, transformer; pulse-amplifying, -counting, -forming, -generating, -shaping* adjs., *pulse-taking* (*lit.* and *fig.*). **c. pulse amplitude modulation** *Telecommunications*, pulse modulation in which variations in the signal are represented by variations in the amplitude of the pulses; **pulse code modulation** *Telecommunications*, pulse modulation in which the actual signal amplitude after each successive interval is approximated by the nearest in value of a set of permitted amplitudes, which is then represented by a short sequence of pulses in accordance with a binary code; **pulse column** = *pulsed column* s.v. *PULSED *ppl. a.* a; **pulse duration modulation** *Telecommunications* = *pulse width modulation* below; **pulse frequency modulation** *Telecommunications*, pulse modulation in which variations in the signal are represented by variations in the frequency of occurrence of the pulses; **pulse jet** *Aeronaut.*, a type of jet engine in which combustion is intermittent, the ignition and expulsion of each charge of mixture causing the intake of a fresh charge; **pulse-la·bel** *v. trans. Biochem.*, to label the metabolites of (cells) by administering a pulse (sense *4*); so **pulse-la·belled** *ppl. a.*, **pulse-la·belling** *vbl. sb.*; **pulse modulation** *Telecommunications*, modulation in which a series of initially identical, regularly recurring pulses is varied in some respect (as amplitude or timing) so as to represent the amplitude of the signal after successive short intervals of time; so **pulse-modulated** *a.*, **pulse modulator**; **pulse position modulation** *Telecommunications*, pulse modulation in which variations in the signal are represented by variations in the time position of the pulses, relative to their unmodulated position; **pulse pressure** *Med.*, the difference between the maximum (systolic) and the minimum (diastolic) pressure of arterial blood; **pulse radar**, radar that transmits pulses rather than a continuous beam of radio energy; **pulse radiolysis**, radiolysis by means of a very short pulse of electrons or other ionizing radiation; **pulse repeater** *Electronics* (see quot. 1971); **pulse time modulation** *Telecommunications*, pulse position or pulse width modulation; so **pulse-time-modulated** *a.*; **pulse width modulation** *Telecommunications*, pulse modulation in which variations in the signal are represented by variations in the width (duration) of the pulses; **pulse-wise** *adv.*, discontinuously; a bit at a time.

1940 *Rev. Sci. Instruments* XI. 44/1 The use of ionization chambers in conjunction with pulse amplifiers permits data to be taken much more rapidly. 1949 [see *pulse height* below]. 1962 SIMPSON & RICHARDS *Physical Princ. Junction Transistors* xv. 371 (*heading*) Video pulse-amplifier equivalent circuits. *Ibid.* viii. 182 This makes the point-contact transistor inherently unstable under certain conditions and makes possible the construction of simple pulse-amplifying or trigger circuits. 1940 *Rev. Sci. Instruments* XI. 45/1 (*caption*) Wiring diagram of the pulse amplitude selector. 1947 [see *KICKSORTER]. 1947 *Bell Syst. Technical Jrnl.* XXVI. 396 When the pulses consist simply of short samples of the speech waves, their varying amplitudes directly represent the speech waves and the system is called pulse amplitude modulation or PAM. 1963 B. FOZARD *Instrumentation Nuclear Reactors* iv. 42 Discrimination against gamma rays is obtained by using a pulse amplitude discriminator in conjunction with the counter. 1972 Pulse amplitude modulation [see *MODULATION 7]. 1947 Pulse analyser [see *KICKSORTER]. 1963 B. FOZARD *Instrumentation Nuclear Reactors* x. 123 In some applications it is required to determine the count rate of pulses of a particular amplitude or..whose amplitudes lie in the band V and $V+\delta V$... Instruments known as pulse analysers are available which give the required result directly. 1967 Pulse code modulation [see PCM s.v. *P II]. 1967 *Times* 7 Feb. 9/3 The Post Office is to start work on the installation..of the world's first pulse code modulation exchange. This technique makes it possible for two ordinary telephone 'pairs' to carry 24 simultaneous conversations. 1976 *B.B.C. Handbk.* 71/2 Pulse Code Modulation is the system developed by BBC engineers for the distribution of high-quality stereophonic audio signals. 1954 R. STEPHENSON *Introd. Nuclear Engin.* ix. 333 The principal advantage of pulse columns is their greater plate efficiency which permits a column of smaller height for a given separation. 1966 *New Scientist* 15 Sept. 609/1 By employing pulse-compression radar techniques the designer can..produce a radar which has long range and yet gives good definition. 1975 D. G. FINK *Electronics Engineers' Handbk.* xxv. 74 Pulse compression is a technique in which a rectangular pulse containing phase modulation is transmitted. When the echo is received, the matched-filter output is a pulse of much shorter duration. 1963 B. FOZARD *Instrumentation Nuclear Reactors* xiii. 166 Two scales..indicate approximately (*a*) the current produced in a reactor instrumentation ionisation chamber..and (*b*) the pulse rate produced by a fission-type pulse counter. *Ibid.* viii. 74 Pulse-counting systems are commonly used to measure radiation intensity in terms of count rate. 1956 S. SEELY *Radio Electronics* xv. 439 In pulse-duration modulation and pulse-position modulation..the signal/noise ratio is proportional to the bandwidth. 1975 Pulse duration modulation [see *pulse position modulation* below]. 1947 LEBACQZ & WHITE in L. N. Ridenour *Radar System Engin.* x. 376 A pulse transformer can be inserted between load and pulse-forming network so that the network can be designed to use the available switching device most efficiently. 1950 Pulse frequency modulation [see *communication(s) engineer* s.v. *COMMUNICATION 12]. 1975 Pulse frequency modulation [see *pulse time modulation* below]. 1975 Pulse-generating [see *pulse transformer* below]. 1931 *Proc. IRE* XXII. 911 (*caption*) Transmitter and pulse generator with cathode ray oscillograph monitor. 1977 *Navy News* June 42 (Advt.), Ideally, applicants should be familiar with oscilloscopes, digital multimeters, pulse generators, frequency counters, etc. 1949 *Atomics* Sept. 57/1 When the gating instrument is in operation, random pulses from a Geiger counter are fed through the pulse amplifier to the pulse height selector. 1952 *Proc. Physical Soc.* B. LXV. 320 An investigation of the pulse heights produced by alpha-particles in various scintillating crystals. 1957 *Economist* 30 Nov. 779/1 Because it sorts out electrical 'kicks' or impulses according to their amplitude—more than 16,000 of them in each of 100 channels..'kick sorter' is the technicians' colloquial name for a Pulse Height Analyser. 1962 F. I. ORDWAY et al. *Basic Astronautics* iv. 137 The pulse height distribution of the incident particles is obtained..by means of a sliding-channel, pulse-height analyzer. 1968 Pulse-height [see *KICKSORTER]. 1946 F. HAMANN *Air Words* 43/1 *Pulse-jet*, a jet plane or motor that..operates in short bursts of power or impulses. 1949 *Aircraft Engin.* Mar. 71/3 No analysis will decide whether ram-jet or pulse-jet is the better—such questions are decided by service experience. 1966 *McGraw-Hill Encycl. Sci. & Technol.* XI. 95/2 In addition to their use on the German V-1 buzz-bomb, pulse jets have been used to propel radio controlled target drones and experimental helicopters. 1961 *Nature* 13 May 581 (*heading*) Unstable ribonucleic acid revealed by pulse labelling of *Escherichia coli*. 1968 H. HARRIS *Nucleus & Cytoplasm* iii. 42 When the pulse-labelled cells were transferred to non-radioactive medium, radioactivity disappeared from the heterogeneous component and appeared in the ribosomal RNA. *Ibid.*, The pulse-labelling revealed a special class of RNA which was not ribosomal RNA or a precursor of ribosomal RNA. 1974 *Nature* 8 Nov. 168/1 Yeast protoplasts were pulse labelled for 30 min with H-adenine, then quickly cooled and lysed by osmotic shock. 1943 *Gloss. Terms Telecommunication* (*B.S.I.*) 65 *Pulse-modulated waves*, recurrent wave-trains in which the duration of the trains is, in general, short compared with the time interval between them. 1962 *Science Survey* III. 279 They also respond to pulse-modulated sounds up to pulse repetition rates of about 800 cycles per sec. 1929 *Proc. IRE* XVII. 1787 It could not be predicted with certainty that the transmitter crystal would provide a suitably constant phase reference for comparison with the echoes, particularly because of the fact that its phase..might be shifted slightly by the pulse modulation of the power amplifiers excited by the crystal circuit. 1945 *Electronics* Jan. 103/3 With pulse modulation, especially at very high carrier frequencies, problems of modulation at the transmitter are greatly simplified. 1975 D. G. FINK *Electronics Engineers' Handbk.* xiv. 28 All pulse modulation schemes require sampling analog signals, and some, such as pulse code modulation.., require the additional quantization of the analog signals. 1965 *Wireless World* July 18 (Advt.), EEV magnetrons, klystrons, pulse modulators,..offer extreme reliability in quality marine electronics. 1945 *Electronic Industries* Dec. 82 (*heading*) Pulse position modulation technic. 1975 D. G. FINK *Electronics Engineers' Handbk.* xiv. 34 In PTM the information is coded into the time parameter instead of, for instance, the amplitude... There are two basic types of PTM: pulse position modulation (PPM) and pulse width modulation (PWM), which is also known as pulse duration (PDM). 1904 J. ERLANGER in *Johns Hopkins Hosp. Rep.* XII. 93 The term 'pulse-pressure' is used in place of the phrase, oscillations of the pressure in the arteries produced by the beat of the heart. It is the difference between the maximum and minimum pressures. 1966 *Lancet* 24 Dec. 1387/1 Fig. 1 shows the mean responses of pulse-rate and pulse-pressure during the insulin-tolerance tests in the six men. 1949 D. G. C. LUCK *Frequency Modulated Radar* ix. 416 Pulse radar is placed at a practical disadvantage, relatively to frequency-modulated radar, by the necessity of operating transmitters at very high peak-power levels. 1966 *McGraw-Hill Encycl. Sci. & Technol.* XI. 211/2 In pulse-radar systems the transmitter and receiver generally share a single antenna. [1960 MCCARTHY & MACLACHLAN in *Trans. Faraday Soc.* LVI. 1187 The technique of pulsed radiolysis has not heretofore been applied to the transient measurement of rapid chemical reactions.] 1961 *Nucleonics* Oct. 54/1 Pulse radiolysis has become feasible within the past few years though the availability of electron accelerators which deliver a very short pulse of high energy electrons of extremely high intensity. 1974 *Nature* 22 Nov. 323/1 On pulse radiolysis of nitrous oxide saturated solutions of either thymidine, cytidine, adenosine or guanosine..transient absorption spectra..attributable to the products of reactions of the hydroxyl radical, were observed. 1963 Pulse rate [see *pulse counter* above]. 1945 *Amer. Speech* XX. 310/1 *PRF*, Pulse Recurrence Frequency. 1953 R. CHISHOLM *Cover of Darkness* xvii. 187 He listened patiently to discussions on megacycles, wave-lengths and pulse-recurrence frequencies. 1949 *Jrnl. R. Aeronaut. Soc.* LIII. 447/2 The Americans now used both words, defining a transponder as a pulse repeater which received and transmitted on different wavelengths. 1971 *Gloss. Electrotechnical, Power Terms* (*B.S.I.*) III. i. 28 *Pulse repeater*, device for receiving pulses from one circuit and transmitting corresponding pulses into another circuit. 1948 L. B. ARGUIMBAU *Vacuum-Tube Circuits* xi. 560 Depending on the pulse height, the multivibrator will synchronize at one fifth, one fourth, or one third of the pulse repetition rate. 1962 *Gloss. Terms Automatic Data Processing* (*B.S.I.*) 57 When the pulse repetition rate is independent of the interval of time over which it is measured it may be called the pulse repetition frequency. 1978 R. V. JONES *Most Secret War* xxiii. 193 If 40 kilometres were its maximum range, its pulse repetition rate should not exceed 3750 per second. 1963 B. FOZARD *Instrumentation Nuclear Reactors* ix. 105 Pulses from the counter are converted by means of an auxiliary pulse-shaping (multivibrator) circuit into rectangular positive pulses of stable amplitude. 1971 J. H. SMITH *Digital Logic* v. 95 Flip flop *A* is used as a pulse shaping circuit and has no logical function to perform. 1950 *N.Y. Times* (City ed.) 20 Apr. 1/1 In the light of today's pulse-taking, it appeared unlikely that the committee would agree on such an approach. 1977 *Proc. R. Soc. Med.* LXX. 425/1 We know that pulse-taking was an important ritual, especially among doctors who did not accept Harvey's discovery of the circulation of blood. 1945 *Electronic Industries* Nov. 91/3 At the transmission end amplitude modulated speech signals are changed into pulse time modulated signals by a tube similar to the cyclophone called the Cyclo-odos. 1944 *Electr. Communication* XXII. 92/1 The merits of another method of transmission applicable to telephony were considered by the Paris Laboratories of the International Telephone and Telegraph Corporation early in 1937... At the time the method was called pulse 'time' modulation. 1975 D. G. FINK *Electronics Engineers' Handbk.* xiv. 28 The control applications [of pulse modulation] are usually confined to the use of pulse time modulation (PTM) and pulse frequency modulation (PFM), where on-off control power can be used to minimize device dissipation. 1951 A. SHEINGOLD *Fund. Radio Communications* xx. 413 If the pulse train is applied to a low-pass filter having an appropriate cutoff value, the signal may be separated from the higher-frequency pulse components. 1975 D. G. FINK *Electronics Engineers' Handbk.* xiv. 28 By interleaving a number of single-channel, low-duty-cycle pulse trains. 1945 *Electronic Industries* Sept. 222 *Pulse transformer*, a special transformer designed to have a frequency response suitable for passing a pulse without materially altering its shape. 1955 *Times* 12 July 2/5 (Advt.), Applicants should preferably be of honours degree standard, with interest in square-loop magnetic devices, ferro-resonant circuits, magnetic amplifiers, or pulse transformers. 1975 D. G. FINK *Electronics Engineers' Handbk.* VII. 18 Lower-power pulse transformers fall into two categories: those used for coupling or impedance matching similar to the high-power pulse transformers, and blocking oscillator transformers used in pulse-generating circuits. 1947 R. LEE *Electronic Transformers & Circuits* ix. 220 Common pulse widths lie between 0·5 and 10 microseconds. 1978 *Nature* 23 Mar. 362/2 The digital nerves of the index finger were stimulated continuously at 3 times the threshold for perception (pulse width 50 μs, 50 shocks s^{-1}) through ring electrodes placed around the finger on either side of the distal interphalangeal joint. 1953 A. T. STARR *Radio & Radar Technique* i. 26 Pulse Width Modulation does not correspond to any normally emitted C. W. system, but corresponds to phase modulations of the frequencies ωr, $2\omega r$,...each multiplied by the original low-frequency signal. 1978 *Gramophone* Apr. 1790/1 Sony offered the first class D (pulse width modulation) power amplifier utilizing VFETs and producing 180 watts per channel. 1909 W. JAMES *Pluralistic Universe* vii. 285 By us it [*sc.* reality] has to be taken pulse-wise, for our span of consciousness is too short to grasp the larger collectivity of things except nominally and abstractly.

pulse, *v.* Add: **4.** (Further examples.)

1954 R. STEPHENSON *Introd. Nuclear Engin.* ix. 332 One of the two liquid phases present is pulsed at the rate of about 60 pulses/min, each pulse causing the liquid in the column to oscillate over a distance of about 0·9 in. 1971 C. J. KING *Separation Processes* xiv. 740 The contents of either a packed column or a plate column can be pulsed by applying intermittent surges of pump pressure to the column. This pulsing promotes mass-transfer rates within the column. 1977 *Design Engin.* July 27/2 Energy in high power radar systems..is pulsed through a magnetron in discrete 'pockets' to allow the returning echo to be related in time with the initial transmission.

5. *trans. Biochem.* To subject (cells in culture) to a pulse of isotopically labelled substrate or substrates.

1960 *Jrnl. Molecular Biol.* II. 320 Uninfected cells were pulsed with ^{32}P under the same conditions and the purified soluble RNA was examined. 1975 *Nature* 14 Aug. 592/1 After various incubation periods the cells were pulsed for 1h with labelled amino acids or nucleosides, collected and macromolecular synthesis measured.

6. *trans.* To apply a pulsed signal to.

1964 *Ann. N.Y. Acad. Sci.* CXV. 665 The..16 digital output lines..can be pulsed individually by a special computer instruction. 1974 *Physics Bull.* June 257/1 The transmitter consists of a small loudspeaker pulsed by a capacitor discharge and the echoes are received by a small conventional microphone.

7. *trans.* To modulate (a wave, beam, etc.) so that it becomes a series of pulses.

1969 *Sci. Jrnl.* Dec. 42/3 Semiconductor lasers whose

output may be modulated up to 1 GHz by pulsing the pump current. 1971 *Nature* 26 Nov. 178/2 High-powered monochromatic beams of laser light..may be modulated or pulsed in times as short as 10^{-11}s. 1975 D. G. FINK *Electronics Engineers' Handbk.* xiv. 28 In usual applications [of pulse modulation], subcarriers are pulsed, time-division-multiplexed, and then used to frequency-modulate a carrier.

pulsed (pɒlst), *ppl. a.* [f. PULSE *v.* + -ED¹.] **a.** Producing or involving pulses.

pulsed column: a tower for solvent extraction in which the natural countercurrent flow of the feed liquid and the solvent has mechanically superimposed on it a rapid reciprocating motion of small amplitude, in order to promote the extraction process.

1946 *Radar: Summary Rep. & Harp Project* (U.S. Nat. Defense Res. Comm., Div. 14) 143/1 Pulsed Doppler shift. 1949 *Jrnl. R. Aeronaut. Soc.* LIII. 439/1 A system employing secondary radar, i.e. a pulsed ground station working in conjunction with a responder, has been designed. 1958 *Engineering* 14 Feb. 205/3 The solvent and water solutions quickly separate after the mixing-contacting process, be this in a packed or pulsed countercurrent contactor column or a mixer-settler unit. 1972 *McGraw-Hill Yearbk. Sci. & Technol.* 268/1 Progress in the development of CW devices has spurred renewed interest in pulsed chemical lasers as..sources of the very-high-energy pulses..needed for plasma heating experiments. 1978 R. V. JONES *Most Secret War* ii. 16 In 1931 W. A. S. Butement and P. E. Pollard..had devised and made a pulsed radio system on a wavelength of about 50 centimetres for detecting ships. 1978 *Nature* 23 Mar. 298/3 Because oxide fuel is highly radioactive, the process is carried out in pulsed columns (rather than mixer settlers) to minimise the contact time between the solvent and the dissolved acid fuel.

b. Consisting of pulses; in the form of pulses.

1949 *Jrnl. R. Aeronaut. Soc.* LIII. 438/1 (*heading*) Measurement of the density of the upper atmosphere by radar—pulsed violet or ultra-violet light. 1955 FRIEDMAN & WEISSKOPF in W. Pauli *Niels Bohr* 155 A pulsed initial neutron beam. 1957 F. HOYLE *Black Cloud* ix. 182 We ought to start sending pulsed messages on the one centimetre [wave-length]. 1971 *Nature* 16 Apr. 426/1 The X-rays from the hitherto unsurprising source Cygnus X-1 are pulsed at a rate of probably about 15 pulses per second.

pulser (pɒ·lsəɹ). [f. PULSE *sb.*¹ or *v.* + -ER¹.] **1.** A device that generates electrical pulses.

1947 LEBACQZ & WHITE in L. N. Ridenour *Radar System Engin.* x. 373 Successful hard-tube pulsers have been made with power outputs up to 3 or 4 Mw. 1973 *Sci. Amer.* June 47/1 The resolution time, or fastest shutter time, of this type of gate is determined by the shortest duration provided by an electrical voltage pulser: about a nanosecond. 1978 *Ibid.* Mar. 58/3 The base houses..a dial pulser to signal the required numbers to the central office and a bell to signal an incoming call to the subscriber.

2. A machine for producing mechanical pulsation in a liquid.

1954 R. STEPHENSON *Introd. Nuclear Engin.* ix. 334 Until satisfactory pulsers are developed..it is doubtful that pulse columns will find any extensive use except for very special solvent extraction operations. 1963 J. H. PERRY *Chem. Engineers' Handbk.* (ed. 4) XXI. 33/1 The pipe connecting column and pulser may be of any length.., but high pressure drop in the transfer pipe contributes to cavitation difficulties.

pulsing, *vbl. sb.* (Further examples.)

1945 *Electronics* Jan. 105/1 The method of pulsing involved frequency-modulating the transmitter ±75 kc by a continuous 170-cycle tone. 1971 [see *PULSE *v.* 4].

pulsive, *a.* Add: **3.** Making a beating or throbbing sound.

1960 'W. HAGGARD' *Closed Circuit* x. 123 He knew a band, quite a small one, quiet and properly pulsive. 1969 G. MACBETH *War Quartet* 43 In wave On gathered wave of pulsive thumping, wings Grazed overhead.

pultrusion (pul-, pɒltrū·ʒən). [f. PUL(LING *vbl. sb.* + EX)TRUSION.] A process for making plastic articles reinforced with glass fibre in which long strands of the reinforcement, encased in liquid resin, are pulled through a heated die that shapes and cures the resin.

1964 OLEESKY & MOHR *Handbk. Reinforced Plastics* v. 324 (*caption*) 'Pultrusion' tank design. 1965 *Mod. Plastics Encycl.* 1966 632/2 Long lengths of reinforced plastics flat strip or sheet..can be produced economically by the pultrusion process. 1968, 1976 [see below].

Hence **pultru·de** *v. trans.*, to make by this process; **pultru·ded** *ppl. a.*

1968 *6th Internat. Reinforced Plastics Conf.* (Brit. Plastics Federation) 6/1/1 The pull-trusion process is the oldest one of the technical processes nowadays used for a continuous production of glassfibre reinforced polyester articles... The percentage of pull-truded products made from glassfibre reinforced polyester has considerably increased. 1971 *Mod. Plastics Encycl.* 1970–1 592/3 Typical volume applications for pultruded products are electrical pole line hardware, ladders, fishing rods and corrosion-resistant structural shapes. 1976 *S9* (N.Y.) May/June 116/2 Plastigage Corporation..has developed and perfected a new method to produce pultruded fiberglass rod ideal for OEM. 1976 *Reinforced Plastics* XX. 295/2 Another advantage of the pultrusion process is the ability to produce a number of similarly dimensioned products at the same time. Currently, the company are pultruding hollow section rods and solid rods at the same time. *Ibid.* 295/3 Early models of the Sky Stunter kite used aluminium tubing for the framework, but this has now been replaced with the pultruded components.

pulu. (Earlier and later examples.)

1833 W. TOLMIE *Jrnl.* 6 Apr. in *Physician & Fur Trader* (1963) 144 Met Madame Boki & retinue, her brows encircled with garlands of pulu. 1917 *Nature* 20 Sept. 58/1 These plants [*sc.* Hawaiian tree ferns] produce at the base of the stipe a great ball of brownish-yellow wool called pulu by the natives, and used by them for stuffing pillows and mattresses.

‖ **pulut** (pū·lut, pūlū). [Mal. (*padi*) *pulut* sticky (rice).] In Malaysia, glutinous rice.

1820 J. CRAWFURD *Hist. Indian Archipel.* I. iv. ii. 360 The most singular variety [of rice] is that called by the Malays *Pulut*.., the *Oryza glutinosa* of Rumphius. This is never used as bread, but commonly prepared as a sweetmeat. 1900 W. W. SKEAT *Malay Magic* 76 A special kind of glutinous rice called *pulut*..is also very generally used for sacrificial banquets. 1972 A. AMIN tr. *Ahmad's No Harvest but Thorn* ii. 10 Our children love *pulut* in the mornings before school. With *pulut* the fullness lasts a long time. *Ibid.* xiv. 148 They would also separate the *Thai* rice from the *pulut* rice.

pulverized, *ppl. a.* (Further examples.)

1926 *Jrnl. Iron & Steel Inst.* CXIII. 507 The author continues his account of practice in the use of pulverised fuel, dealing with the efficiency of the method of powdered coal-firing as compared with other methods. 1950 *Engineering* 28 July 79/1 Where a high proportion of heat recovery was required in the air heater, pulverised-fuel firing was essential. 1976 *Horse & Hound* 10 Dec. 64/2 (Advt.), The latest in animal bedding... Made from pulverised wood—cheaper than straw.

pulverizer. (Further examples.)

1956 [see *OF *prep.* 43 b]. 1967 *Punch* 6 Sept. 360/1 A small East Anglian local authority..would like to dump over 160,000 cubic yards of garbage in the four sidings... If they spend a further £30,000 on a pulveriser (which reduces all forms of household trash to a quarter of its collected volume and renders the result unattractive to flies and rodents) this novel dump should last them for sixteen years. 1971 P. GRESSWELL *Environment* 153 Giant pulverisers can shred a complete car to fist size fragments within seconds.

pulvin, pulvino (pɒ·lvin, pɒlvī·no). *Archit.* [It. *pulvino* pillow.] A cushion cap, impost-block or dosseret.

1907 *Athenæum* 30 Mar. 389/2 The use of the pulvino to enable a thick wall above to be carried on the comparatively slender diameter of the classic column. 1910 G. McN. RUSHFORTH tr. *Rivoira's Lombardic Archit.* I. i. 8 The capitals..supported pulvins ('pulvini') or impost blocks, marked with crosses. *Ibid.* 12 From Ravenna and Naples the pulvin spread over Italy and beyond. *Ibid.* II. vi. 300 The corbel pulvins with rudely curled ends..are derived from the crutch-shaped pulvins, a Lombard creation of the Xth century. 1913 T. G. JACKSON *Byzantine & Romanesque Archit.* I. iv. 52 On the capital they placed a block of stone spreading upwards from the width of the column where it rested on the abacus, to the width of the wall above, and from the top of this stone they sprang their arch, of the full thickness of the wall. This dosseret, pulvino, or impost block is an entirely novel feature. *Ibid.* xi. 171 It is difficult to follow him in claiming the invention of the pulvino for Ravenna on the strength of its use in the church of S. Giov. Evangelista in 425; for he assigns the same date to the much more important Eski Djouma at Salonica where the pulvino is thoroughly developed. 1933 J. A. HAMILTON *Byzantine Archit. & Decoration* ii. 26 Constructive reasons led to the introduction of the impost (pulvino: dosseret) above the capital... The impost was a block, approximately of trapezoidal shape, often carved with a monogram, a cross, or some other device.

pulvinule. Add: **3.** = PULVINUS.

1928 E. HUGHES-GIBB *Life-force in Plant World* vii. 153 Upon scratching or irritating the pulvinule of the terminal leaflet of either of these beans on its under-side, a downward movement very slowly begins. 1975 *Nature* 6 Mar. 69/2 We provide direct experimental evidence for cyclic changes in membrane properties in a circadian system from studies of the pulvinule cells at the base of the leaflets of clover.

pumicate, *v.* (Later example.)

1925 *Chambers's Jrnl.* Nov. 704/2 When it is thoroughly 'pumicated' the coral is rinsed and put into a second bag.

pumice, *sb.* **B. 2. a, b.** (Further examples.)

1891 R. WALLACE *Rural Econ. Austral. & N.Z.* xv. 229 Pumice-topped land..covers unfortunately about thirty per cent. of the area of the North Island. 1950 *N.Z. Jrnl. Agric.* Jan. 17/3 In the north and north-west, where annual rainfall is over 50 in., the soils are classified as yellow brown pumice soils. They are light, fluffy pumice soils formed on volcanic ash. *Ibid.* Feb. 115/2 In its natural state the open pumice country, clothed in a tangled mass of manuka and monoao..looks barren and unattractive. *Ibid.*, The pumice lands of the central plateau area of the North Island consists of soils derived from volcanic-ash showers. 1965 S. T. OLLIVIER *Petticoat Farm* i. 1 Harry stood at the roadside and watched the white pumice dust rising between the bracken at each side of the road.

pumicite (pɒ·misəit). [f. PUMIC(E *sb.* + -ITE¹.] A volcanic ash like pumice in composition but occurring as powder or granules.

1916 E. H. BARBOUR in *Nebraska Geol. Survey* IV. 358 Pumicite has been variously called geyserite, volcanic dust,..and the like... For all of these we have substituted the name pumicite, a self-explanatory term, which we have used for some years. 1949 J. A. BARR in *Industrial Minerals & Rocks* (ed. 2) xxxvi. 752 The principal use of pumicite is for concrete aggregate. 1965 G. J. WILLIAMS *Econ. Geol. N.Z.* xvi. 245/2 Other deposits at Takanini and Hurua [*read* Hunua] have been quoted as a source of diatomaceous material for fibrolite and light-weight concrete but Ritchie pointed out that this is pumicite with no diatoms. 1974 A. C. TENNISSEN *Nature of Earth Materials* vii. 394 Finely ground pumice and unground pumicite are used in general scouring powders... Other uses for unground pumicite and ground pumicite are..in insulation, as filter aids, as poultry litter, [etc.].

pummel, *v.* Add: **a.** Also *transf.* and *fig.*

1927 *New Republic* 12 Oct. 208/2 Once the greater part of the population is pummeled night and morning in underground cattle-cars,..I shall be surprised if there is any energy left. 1972 *Newsweek* 10 Jan. 1/1 For five days U.S. Phantom jets and mammoth B-52 bombers pummeled North Vietnam in the heaviest raids since the 1968 bombing halt. 1977 *Chicago Tribune* 2 Oct. iii. 16/6 The Carthage Redmen..went on to pummel the North Park Vikings 34–8. 1979 *Arizona Daily Star* 19 Apr. 1/1 Typhoon Cecil headed today for Japan after pummeling the Philippines.

pump, *sb.*¹ Add: **I. 1. a.** *spec.* = *petrol pump* (a) s.v. *PETROL 3 b.

1925 F. SCOTT FITZGERALD *Great Gatsby* iv. 81, I had a glimpse of Mrs. Wilson straining at the garage pump. 1972 [see *petrol station* s.v. *PETROL 3 b]. 1974 A. PRICE *Other Paths to Glory* ii. viii. 204 He helps with the odd jobs in the workshop and looks after the pumps.

e. *Physiol.* A mechanism in living cells by which metabolic energy is utilized to cause specific kinds of ion to pass through the cell membrane in the direction opposite to that in which they would pass under ordinary diffusion.

1947 *Arch. Biochem.* XIV. 297 (*heading*) An osmotic diffusion pump. *Ibid.* 298 The essential unit of such a pump consists of the space between two membranes, in which a coupled chemical reaction, utilizing free energy supplied from outside, permits this unit to pump either solvent or dissolved solute into itself through one membrane and out through the other membrane, at a higher chemical potential than that from which it entered. 1964 A. WHITE et al. *Princ. Biochem.* (ed. 3) xxxvii. 727 In the ascending limb of the hairpin-shaped loop of Henle an outwardly oriented sodium pump..operates while the same cells are relatively impermeable to water. 1965 *Nature* 4 Sept. 1099/1 Approximately three sodium ions are expelled for each molecule of ATP split by the pump. 1977 *Sci. Amer.* Aug. 117/3 The transfer of a phosphate group to such a protein could conceivably change the permeability of the membrane to ions,..for example by affecting the activity of an enzyme 'pump' that physically transports ions across the membrane.

III. 6. a. *pump-clip* (examples), *nail, spout.* **pump action** *attrib.* (orig. *U.S.*), designating a type of repeating firearm (see quot. 1964 and cf. *pump gun* s.v. *PUMP *v.* 14); also *absol.*; **b. pump attendant**, a garage hand who serves petrol; **pump hook** (earlier examples); **pump island**, the part of a petrol station on which the pumps stand; **pump-set, pumpset**, a complete pumping installation, comprising a pump, a source of power, and any necessary pipes, valves, filters, etc.; cf. *pumping set* s.v. *PUMPING *vbl. sb.* d; **pump-spear** (earlier example); **pump-turbine** *Engin.*, a machine designed to operate as a pump running in one direction or a turbine running in the other.

1912 *Collier's* 28 Sept. 30/1 (Advt.), The Marlin Pump Action repeating rifle. 1964 H. L. PETERSON *Encycl. Firearms* 249/2 *Pump action*, a popular term describing repeating firearms activated by a horizontally operating slide action. 1973 *Times* 1 Aug. 4/6 The Government decided..not to make self-loading rifles and pump-action shotguns prohibited weapons in its forthcoming Bill. 1977 *Field* 13 Jan. 44/1 (Advt.), A large selection of new and second-hand English weapons... Foreign side by sides, automatics, pump actions and single barrel guns. 1968 A. BINKLEY *What shall I Cry* 10 Harry was pump attendant and not in charge of mechanics. 1972 *Times* 8 Sept. 21/5 The pump attendant's life of opening and shutting filler caps. 1907 *Yesterday's Shopping* (1969) 1060/3 Cooper's patent locking pump clip. 1908 H. G. WELLS *War in Air* ii. 52 Bert stared at these over the card of pump-clips in the pane in the door. 1640 *Archives of Maryland* (1887) IV. 112 For a pump-hook. 1702 in *Essex Inst. Hist. Coll.* (1906) XLII. 161 Inventory of ship... A pumpe Hooke. 1969 *Wall St. Jrnl.* 7 Oct. 19/1 We've seen crop after crop of dolts parade to the pump island. 1974 *Petroleum Rev.* XXVIII. 706/3 Painted in BP or National livery will be the pump islands, canopies, shops, kiosks, [etc.]. 1805 *State Papers & Publick Documents of U.S.* 2 Dec. (1814) I. 455 They robbed the brig of..all her candles, pump nails, locks, and gimblets. c 1889 W. TATE *Princ. Mining adapted to S. Kensington Syllabus* xxi. 158 The thickness of pump sets is calculated by the following formula. 1969 *Capital* (Calcutta) 27 Feb. 354/1 In 1967–68 alone, 250,000 pumpsets, 50,000 private tubewells and 1,000 large State tubewells were installed. 1974 *Petroleum Rev.* XXVIII. 704/1 A sea water injection pumpset. 1702 in *Essex Inst. Hist. Coll.* (1906) XLII. 161 Inventory of ship..Two pump Speares. 1867 'T. LACKLAND' *Homespun* 321 He washes his ruddy face under the pump-spout. 1888 *Harper's Bazaar* 22 Dec. 872/1 When he had filled his pail he took it carefully from the pump spout, and started back to the house. 1934 H. K. BARROWS *Water Power Engin.* (ed. 2) iii. 179 The Baldwin-Southwark Corporation..with the General Electric Company, have recently developed a combined pump-turbine operated by a two-speed motor generator for such plants, model tests of which indicate relatively high efficiencies when acting as either a turbine or a pump. 1977 *Time* 17 Jan. (Advt., verso front cover),

Bill has no idea that the six reversible pump-turbine generator-motors that now supply his area with low-cost electricity were made by Hitachi.

pump, *sb.*² Add: **a.** (Later examples.) Also occas. = *PLIMSOLL 2; in North America, freq. = *court shoe* s.v. *COURT *sb.*¹ 19.

1897 *Sears, Roebuck & Co. Catal.* 203/3 Men's gymnasium shoes... Men's low cut canvas pumps, canvas sole, [etc.]. 1908 *Ibid.* 813/2 A dainty pump of patent coltskin, much in favour with fashionable women. 1928 T. Eaton & Co. *Catal.* Spring & Summer, These smart, attractively-trimmed Pumps can be had in either Black Patent or Honey Beige-shade of leather. 1946 *Sun* (Baltimore) 2 Nov. 3 (Advt.), Two flattering styles to choose from—black suede anklet... and classic black suede sling pump—both mounted on black faille platforms. 1967 *Oxford Mag.* 10 Feb. 205/2 Informed by a girl that she has to wear pumps (court shoes) for her Convocation (degree ceremony) [in Canada]. 1968 J. IRONSIDE *Fashion Alphabet* 132 *Dancing shoes or pumps.* Usually worn by children, they have flat soles and elastic which goes criss-cross round the ankle. Very popular among smart nannies for their charges, especially in bronze leather. 1974 P. WRIGHT *Lang. Brit. Industry* ii. 28 For rubber-soled canvas shoes we have *pumps, plimsolls, gym-shoes* and *squeakers*. 1978 J. KRANTZ *Scruples* vii. 191 Wells Cope, wearing a Dorso sweater, pale beige twill trousers, and black velvet evening pumps embroidered in gold, sat with Harriet.

pump, *v.* Add: **I. 3.** Also, to free from air or other gas by means of a pump or pumps, to evacuate; also with *down* (cf. *pump up* in sense 5) or *out* (cf. *pump out* in sense 2); also *absol.*

1923 *Phil. Mag.* XLVI. 724 The apparatus was pumped out and the residual gas removed as completely as possible. 1935 MILLER & FINK *Neon Signs* viii. 166 The average time required for pumping average lengths of tubing..is shown. 1936 *Physical Rev.* L. 250/1 The tube was opened up for three days, then pumped down. 1952 A. L. REIMANN *Vacuum Technique* ii. 27 Seal-off constrictions are commonly made with a diameter of 6 mm. for pumping water-cooled valves. 1955 KIRK & OTHMER *Encycl. Chem. Technol.* XIV. 547 The tube is sealed onto the manifold of a vacuum system which is capable of pumping down the tube to the order of 10^{-7} mm. of mercury. 1959 N. W. ROBINSON in A. S. D. Barrett *Progress in Vacuum Sci. & Technol.* 25/1 The advocates of oil pumps consider that the ability to pump down to 10^{-6}–10^{-7} mm. Hg without a refrigerant outweighs the disadvantage..of oil contamination. 1971 *Sci. Amer.* Aug. 114/1, I do not add the slurry until the system has been pumped to 10^{-8} torr. 1977 *Ibid.* Jan. 80/2 To measure bearing balls for exo-electron emission would call for..putting the ball in a vacuum chamber, pumping the chamber down and hoping that all this would not interfere with exoelectron emission.

b. *to pump ship* (also *pumpship*), to urinate. *colloq.*

1788 GROSE *Dict. Vulgar T.* (ed. 2) s.v. *Pump*, To pump ship; to make water, and sometimes to vomit. *Sea phrase.* 1886 H. BAUMANN *Londinismen* 147/1 *To pump ship*, sein Wasser abschlagen. 1922 V. WOOLF *Let.* 22 Oct. (1976) II. 572 Its on a par with not pumping shipping before your wife. 1938 J. CARY *Castle Corner* 163 The few passing guests who came now and then to smoke or to pumpship among the stacks. 1939 W. Z. FOSTER *Pages from Worker's Life* iv. 175 He excused himself from the room with the remark that he had 'to go and pump ship'. 1973 'D. RUTHERFORD' *Kick Start* i. 12 A couple of men had come in to pump ship at the stand-up urinals.

II. 6. a. Also freq., to force, inject, or pour (something) into (someone or something).

1899 [in Dict.]. 1940 *War Illustr.* 12 Apr. 367/3 The other six Messerschmitts were circling round him pumping bullets into his 'plane as fast as they could work the guns. 1947 *Sun* (Baltimore) 15 Aug. 12/7 A gunman climbed on the running board of his car and pumped lead into him. 1953 *Times* 31 Oct. 2/7 The atomic energy production division..will be 'generating electrical power which would be pumped into the grid system within the next few years'. 1977 A. THWAITE *Portion for Foxes* 23 Made separate..By actions pumping fear into my blood. 1978 *Guardian Weekly* 29 Jan. 7/2 Moscow started pumping arms into Ethiopia.

b. (Later example.)

1909 *Daily Chron.* 22 Sept. 9/5 My head aches. It pumps and pumps and I can't think.

9. c. *coarse slang.* To copulate. Also *trans.*, to copulate with (a woman).

1730 in Farmer *Merry Songs & Ballads* (1897) II. 204, I work'd at her Pump till the Sucker grew dry, And then I left pumping, a good Reason why. 1937 PARTRIDGE *Dict. Slang* 667/2 *Pump, v.*, to coït with (a woman): low: C. 18–20; ob[solescent]. 1971 R. K. SMITH *Ransom* v. 223 They began to pump on the soft seat... 'We never did it in no Caddy before,' he whispered. 1973 'J. PATRICK' *Glasgow Gang Observed* xii. 108 Skidmarks had come by her name through the boys' practice of kicking her naked behind after they had 'pumped' her. 1976 G. V. HIGGINS *Judgment D. Hunter* xiv. 159 He told me Shanley's pumping Dottie Deininger... Fine-looking woman.

10. a. Also, to shake (a person's hand, or a person by the hand) vigorously.

1912 MULFORD & CLAY *Buck Peters* i. 14 'Tex!..When did you get here? Going to stay?..You look white—sick?' 'City color..,' replied the other, still pumping the hand. 'I'm goin' to stay.' 1938 M. K. RAWLINGS *Yearling* xii. 123 They pumped hands in greeting. a 1951 'J. HACKSTON' in Murdoch & Drake-Brockman *Austral. Short Stories* (1951) 230 He ebbed out looking swamped, with a big man pumping him up and down in a parting, very friendly handshake. 1958 J. COURAGE in *London Mag.* Dec. 26 He pumped my hand pleasantly. 1969 [see *ETIC *a.*]. 1977 *Church Times* 14 Jan. 5/1 Clasping my hand, and pumping it up and down whilst looking intensely into my eyes.

b. (Later examples.)

1928 C. F. S. GAMBLE *Story N. Sea Air Station* xviii. 309 All submarines have a tendency to 'pump' in heavy seas, that is, they tend to move up and down in a vertical plane. 1938 M. K. RAWLINGS *Yearling* xiv. 148 The road under him was a treadmill. His legs pumped up and down, but he seemed to be passing the same trees and bushes again and again. 1976 *N.Y. Times Mag.* 12 Sept. 40/2 He [*sc.* a skateboard rider] pumped from cruise to speed, down the incline, faster until he felt almost weightless like a bird, spinning on wheels that really weren't there.

11. (Earlier examples.)

1791 F. BURNEY *Let.* 7 Nov. (1972) I. 77 She owns she found the greatest difficulty in pumping up decent expressions of concern. 1813 M. EDGEWORTH *Let.* 1 May (1971) 36, I could not pump up any enthusiasm for them.. I have no taste for these hideous old stones.

13*. *intr.* = *HUNT *v.* 7 b.

1901 L. BELL *Electric Power Transmission* (ed. 3) vi. 227 Alternators in parallel are less likely to pump if they have solid poles. 1902 [see *HUNT *v.* 7 b].

13.** *Physics.* To raise (an atom or the like) into or into a higher energy state by irradiation, esp. so as to produce a population inversion and make the substance work as a laser; to excite (a substance or device) in this way. Cf. *optical pumping* s.v. *OPTICAL *a.* 6.

1953 *Rev. Mod. Physics* XXV. 175/1 The vapor is illuminated with circularly polarized light..to pump atoms from the ground state *a*, in which *m* = −½, into state *b*, in which *m* = +½. 1961 *Ann. Reg.* 1960 396 The method of 'pumping' the electrons into their excited state had also to be changed for a continuous method. 1973 *Sci. Amer.* June 52/3 Most substances can be pumped with just the fundamental and second-harmonic pulses emitted by these two lasers. 1973 *Physics Bull.* July 419/3 Perhaps the most important recent development has been the successful operation of the cw rhodamine 6G dye laser, pumped by an argon-ion laser. 1975 *Nature* 28 Aug. 695/2 Laser action over the range 2·5 to 2·9 μm was achieved using the F_A(II) centre in lithium-doped potassium chloride pumped by a krypton ion laser at 647·1 nm.

III. 14. *pump jack*; **pump drill**, a primitive drill in which the shaft is rotated by sliding up and down a cross-piece to which is attached a cord that winds and unwinds about the shaft; **pump gun** orig. *N. Amer.*, a rifle having a tubular magazine and a sliding forearm; so **pump-gunner.**

1865 Pump drill [in Dict.]. 1964 W. L. GOODMAN *Hist. Woodworking Tools* 180 Another primitive method still in use by..natives of New Guinea is the pump drill,..with a flywheel made of stone. 1974 P. W. BLANDFORD *Country Craft Tools* viii. 116 Pump drills were used by many craftsmen. 1906 *Daily Colonist* (Victoria, B.C.) 16 Jan. 10/5 He was using a Winchester pump gun, and in the operation of loading, the gun was fired, the charge striking the left foot. 1970 D. DODGE *Hatchetman* viii. 101 A guard with a pumpgun across his knees sat cross-legged on the floor. 1976 *Shooting Times & Country Mag.* 18–24 Nov., The 16-bore Model 12 is a durable weapon of reasonable weight, very easy to hit with (and to the pumpgunner, at least, having positively classical lines!). 1970 J. BLACKBURN *Land of Promise* v. 74, I had not bought Unzicker's small gasoline engine and pump jack at the sale. 1973 *Times* 1 Dec. 2/3 The Kimmeridge pump jack, familiarly known in the trade as a nodding donkey, seesaws steadily on.

pumpable. Delete *rare*⁻⁰ and add examples; **pumpability** (further examples).

1935 H. MOORE *Liquid Fuels* vi. i. 193 The pour-point test..is only a rough guide in comparing the pumpability of different fuel oils at low temperatures. 1960 *Farmer & Stockbreeder* 12 Jan. 66/2 Agitators make the mass pumpable, through tractor-powered open-vane centrifugal pumps. 1970 *Sci. Jrnl.* Aug. 79/2 Traditional grouts are sand, lime or cement and sufficient water to make the mixture pumpable. 1973 *Nature* 9 Mar. 90/3 For centrifugal pumps with enclosed disk-type impellers there is likely to be a better pumpability with viscoelastic liquids. 1979 *Civil Engineering* Nov. 43/1 The bolt is installed and grouted by means of pumpable resin.

pu·mp-down. Also **pumpdown.** [f. vbl. phr. *to pump down*: see *PUMP *v.* 3.] The action or process of reducing the pressure of air or other gas inside an enclosed volume by pumping. Freq. *attrib.*

1948 *Rev. Sci. Instruments* XIX. 13/1 To obtain a conservative value for the 'pump-down' time, the pump speed..is 0·37 liter per second. 1966 D. G. BRANDON *Mod. Techniques Metallogr.* 184 Preliminary pump-down can be performed with a liquid nitrogen-cooled sorption trap. 1971 *Sci. Amer.* Aug. 114/3 Gas will be liberated from the internal surfaces of the system during pumpdown. 1977 *Design Engin.* July 35/3 This greatly reduces pump-down time.

pumped, *ppl. a.* Add: **3. pumped storage**, the pumping of water to a higher level when demand for electricity is low so that its return to the lower level can be used to generate hydroelectricity when demand is high. Freq. *attrib.*

1927 F. JOHNSTONE-TAYLOR *Water-Power Pract.* x. 163 A total of 450 h.p. being required, pumped storage was resorted to. 1964 *Times Rev. Industry* Jan. 73 A pumped storage scheme..only requires the possibility of constructing two reasonably large storage basins at widely differing levels with a river or adequate rainfall to keep them filled. 1976 *National Observer* (U.S.) 12 June 5/1 In this ancient valley of the New River, American Electric Power Co... wants to build an $845 million pumped-storage project.. that would trap 42,100 acres of water behind two dams.

pumpellyite (pɒmpe·li,əit). *Min.* [f. the name of Raphael *Pumpelly* (1837–1923), U.S. geologist: see -ITE¹.] An iron- and magnesium-bearing hydrous calcium aluminosilicate, crystallographically similar to minerals of the epidote group, which occurs as colourless, greenish, or brown monoclinic crystals and is characteristic of certain low-grade metamorphic rocks.

1925 PALACHE & VASSAR in *Amer. Mineralogist* X. 412 For this mineral we propose the name of Pumpellyite for Raphael Pumpelly, the pioneer student of the detailed paragenesis of the minerals of this region. 1959 W. W. MOORHOUSE *Study of Rocks in Thin Section* vi. 170 The vesicles of basalts are filled with secondary minerals, of which the most typical are zeolites, carbonate, the silica minerals.., feldspar, epidote, pumpellyite, chlorite, and serpentine. 1975 *Lithos* VIII. 72 The textures and crystal forms strongly suggest that the pumpellyite-quartz aggregates are reaction products which developed after equilibrium crystallization of the prehnite and chlorite. 1978 *Nature* 20 July 242/1 Prehnite and pumpellyite have been found in County Cavan, Ireland.

pumper¹. Add: **1. b.** (Examples.)

1874 *Coursing Calendar* Spring 260 All the latter part of a pumper was in favour of Mr Mill's dog. 1879 H. DALZIEL *British Dogs* i. i. 23 Without this [*sc.* a good back] the dog [*sc.* a greyhound] could not endure the exhaustive process of the 'pumpers' he is submitted to.

3. *U.S.* A fire engine that carries the hose and pumps the water.

1915 *Fire & Water Engineering* 14 July 31/3 (*heading*) New Seagrave pumpers tested at Denver. 1919 *Ibid.* 16 July 140/3 Time was, when the motor pumper was still a novelty and many of the departments were still using the horse-drawn steamer to extinguish fires. 1934 W. C. PRYOR *Fire Engine Book* 32 He showed them a big pumper engine... 'Water power alone is not strong enough to throw the water high into the air..so the pumper puts more pressure behind the water.' 1949 J. J. FLOHERTY *Fire Alarm* i. 9 Fire apparatus developed from the man-drawn hand pump to the powerful motorized pumper. 1975 *New Yorker* 10 Mar. 28/3 Pache showed us Aviation's current fire engine ('It's a 1951 Ward LaFrance pumper, and it carries five hundred and fifty gallons').

pump-handle, *sb.* Add: **b.** (Later examples.)

1892 J. E. COX *Five Years in U.S. Army* 83 Performing the 'pump-handle act' with my right arm, I hastened away to obey orders. 1909 R. E. KNOWLES *Attic Guest* xv. 204 They nearly all shook hands in the high pump-handle fashion that was almost unknown in the South.

pumping, *vbl. sb.* Add: **d. pumping set** = *pump-set* s.v. *PUMP *sb.*¹ 6 b.

c 1889 W. TATE *Princ. Mining adapted to Requirements of S. Kensington Syllabus* xlvi. 398 Pumping sets for lifting water to bank vary from 8 to 24 inches, and in extreme cases 36 inches in diameter. 1926 *Power Engineer* XXI. 333 (*heading*) A turbine pumping set. 1957 T. G. HICKS *Pump Selection & Application* vii. 199 (*caption*) Fuel-oil pumping set fitted with 5 gpm 150-psi pumps and heaters. 1962 L. B. ESCRITT *Pumping Station Equipment & Design* iii. 27 Pumping sets are classed into horizontal-spindle and vertical-spindle types.

pumpkin. Add: **2. b.** (Further examples.)

1846 *Spirit of Times* 25 Apr. 97/1 The skins, Indian relics, etc. are 'some punkins' and no mistake. 1849 [see *LAGNIAPPE]. 1859 *Harper's New Monthly Mag.* Sept. 569/2 Gin'ral! you're some punkins! 1903 *McClure's Mag.* XXI. 330/1 He was some pumpkin both in politics and color, and the friend of me and Jones. 1913 J. LONDON *Valley of Moon* (1916) III. vii. 380 Say, friend, you're some punkins at a hundred yards dash, ain't you? 1930 E. POUND *XXX Cantos* xii. 54 Go to hell Apovitch, Chicago aint the whole punkin. 1975 *Publishers Weekly* 21 July 67/3 New England, where the Boston radio team of Eddie Andelman, Jim McCarthy and Mark Witkin is evidently considered some punkins.

4. *pumpkin butter, -eater, patch, pudding, soup, vine* (earlier and later examples); *pumpkin orange, yellow* adjs.; **pumpkin pie** (earlier and later examples); **pumpkin pine** *U.S.*, substitute for def.: a variety of the white pine, *Pinus strobus*; also, the timber from this tree; (later examples).

1893 M. A. OWEN *Voodoo Tales* 6 The place of the vegetables was taken by..little jars of a villainous sweet compound of pumpkin stewed with watermelon-juice and known to all as 'punkin-butter'. 1918 *N. & Q.* 12th Ser. IV. 189/1 Peter, Peter, pumpkin-eater, Had a wife and couldn't keep her. Had another, didn't love her, Causing instantaneous bother. 1962 *Punch* 31 Oct. 648/1 Jake is the pumpkin eater of the title [of a novel]; he tries to put his wife in a pumpkin shell to keep her very well, as the old rhyme says, and it is this that precipitates the crisis. 1974 L. KOENIG *Little Girl* xix. 224 A waitress..wearing a pumpkin-orange uniform. 1935 Z. N. HURSTON *Mules & Men* I. iv. 99 Out dat door John come like a streak of lightning. All across de punkin patch, thru de cotton over de pasture. 1654 E. JOHNSON *Wonder-working Providence of Sions Saviour in New England* 174 This poor Wilderness hath..quince tarts instead of their former Pumpkin Pies. 1784 P. M. FRENEAU *Poems* (1786) 389 Systems they built on pumpkin pies, And prov'd that every thing went round. 1907 *St. Nicholas* May 615/2 Pumpkin pies and strawberry shortcake were also introduced to the French palate and found good. 1851 J. S. SPRINGER *Forest Life &*

Forest Trees 41 The pumpkin Pine is generally found on flat land and in ravines. 1907 *Springfield* (Mass.) *Weekly Republ.* 29 Aug. 15 The virgin white pine has practically disappeared from New England and huge 'pumpkin pines' four and five feet in diameter are now a matter of tradition. 1941 B. A. WILLIAMS *Strange Woman* vii. 576 It was an old pumpkin pine... It was better than six feet through, where I tackled it, about four feet from the ground. 1947 E. H. PAUL *Linden on Saugus Branch* 187 The solid old flooring of pumpkin pine, strewn with sawdust, rumbled and clicked beneath the tread of seamen's boots. 1951 E. M. GRAHAM *My Window looks down East* vii. 65 On the punkin pine bureau, which shines almost as golden as the heart I'd taken from the tree, were two bunches of flowers. 1805 *Indep. Chron.* 26 Dec. 3/1 Clams and oysters, succatouch and pumpkin puddings, turkies, ducks, [etc.]. 1841 A. M. MAXWELL *Run through U.S. during Autumn of 1840* I. 81 Real, genuine, Yankee, new England, pumpkin pudding. 1884 *Cottage Hearth* Apr. 189/1 Pumpkin Soup. Cut the inside and edible part of the pumpkin into large dice. 1955 *Caribbean Q.* IV. ii. 102 After the First Communion, there is a fete for each child, with toasts in vermouth, sheepshead and pumpkin soup. 1810 M. CUTLER *Jrnl.* 9 July (1888) II. 343 Saw the cactus grandiflora, or night-flowering cereus... The plant has a long stem, resembling a pumpkin-vine, but no leaves. 1909 F. B. CALHOUN *Miss Minerva & William Green Hill* xiii. 106 How's he going to sit under a pumpkin vine when he's inside of a whale? 1962 S. WYNTER *Hills of Hebron* iii. 42 Withered pumpkin vines..littered the earth. 1912 J. WEBSTER *Daddy-Long-Legs* 173 Mr. Weaver has painted his barn..a bright pumpkin yellow.

pumpkin-head. Add: **d.** (Examples.)

1876 H. E. SCUDDER *Dwellers in Five-Sisters Court* v. 87 'Pumpkin head!' said the Doctor, more vigorously than politely. 1898 H. FREDERIC *Deserter & Other Stories* 143 You can't raise a plug of [tobacco] in a whole regiment of 'em. Regular pumpkin-heads!

pumpkin-headed, *a.* (Later example.)

1939 WODEHOUSE *Uncle Fred* xvi. 234 You know that pumpkin-headed old man's views on class distinctions.

pump log. *U.S.* A hollowed log used in the construction of a pump or as a water-pipe. Also in *Comb.* as *pump-log borer* (see quot.).

1816 *N. Amer. Rev.* III. 429 He declared also, that the mill for grinding apples, which is an overshot and is fed by a pump log..would often stop during the day. 1858 D. K. BENNETT *Chronol. N. Carolina* 108 He had some men repairing pump-logs, through which water was carried from the mountain side to his hotel. 1879 F. R. STOCKTON *Rudder Grange* xvi. 235 He looked like he'd been drawn through a pump-log. 1965 E. TUNIS *Colonial Craftsmen* iii. 40 Such a one was the pump-log borer—his trade seems to have no generic name—who made wooden pumps and wooden pipes by boring holes lengthwise through logs.

pu·mp-pri:ming, *vbl. sb.* orig. *U.S.* [f. the phr. *to prime a pump* (see PRIME *v.*¹ 4 in Dict. and Suppl.).] The stimulation of commerce or economy by means of investment; also *transf.* and *attrib.* or as *ppl. adj.*

1937 F. D. ROOSEVELT *Public Papers & Addresses* (1941) VI. 520 The things we had done, which at that time were largely a monetary and pump-priming policy.., had brought the expected result. 1938 *Sun* (Baltimore) 5 Jan. 1/7 (*heading*) Eccles urges pump-priming to end slump. *Ibid.* 18 Feb. 15/1 Farm products seemed most likely to benefit from the next 'pump-priming'. 1941 N. ALLEY *I Witness* xxxvi. 300, I spent a pump-priming week at Lisbon as the first step toward getting geared up for war coverage. 1950 *Ann. Reg.* 1949 161 Large development projects whose results would be out of all proportion to the 'pump-priming' required from U.N. 1960 *Guardian* 21 Oct. 24/6 The pump-priming period of the Welfare State. 1961 B. R. WILSON *Sects & Society* 9 The pump-priming activity of the revivalist to generate a distinctive form of religious expression. 1963 *Daily Tel.* 18 Jan. 12/2 To the financial purists, this appears as a bid for straightforward Keynesian pump-priming. Something of the sort is clearly required to stir the American economy out of its present sluggishness. 1978 M. PUZO *Fools Die* xxx. 351 He, Lieverman, would throw in the pump-priming cash, the development money.

Hence **pu·mp-pri:mer,** a financial grant or other action that stimulates economic enterprise.

1953 *Manch. Guardian Weekly* 13 Aug. 9/3 Sir Greville Maginness suggests that the grants will act as 'pump-primers' and encourage the more backward firms. 1962 *Times* 11 May 17/6 Expansion..could be effective as a pump-primer for the economy of the region. 1979 *Nature* 4 Jan. 7/3 Finance has been a constant anxiety, for the initial grants from the Wolfson Foundation and other funds were pump primers.

pumpship: see *PUMP *v.* 3 b.

pun, *sb.*¹ Add: Also **pu·nkin** *rare*, a little pun.

1866 H. JAMES in *Atlantic Monthly* XVII. 197/2 Blunt and I made atrocious puns. I believe, indeed, that Miss Blunt herself made one little punkin, as I called it.

‖ **punalua** (pūnālū·a). *Anthrop.* [Hawaiian.] A relationship term formerly denoting spouses who shared a wife or husband and used by L. H. Morgan in his theory of the evolution of kinship systems for a form of group marriage, assumed by him to have replaced promiscuity and preceded exogamy, in which wives' sisters and husbands' brothers were considered spouses. Hence **punalu·an** *a.*

1860 L. ANDREWS in L. H. Morgan *Anc. Society* (1877) III. iii. 427 The relationship of pūnalūa is rather amphibious [*sic*]. It arose from the fact that two or more brothers with their wives or two or more sisters with their husbands, were inclined to possess each other in common. 1877 L. H. MORGAN *Anc. Society* III. iii. 424 The Punaluan family has existed in Europe, Asia, and America within the historical period. *Ibid.* 428 All the sisters of his wife, own as well as collateral, are also his wives. But the husband of his wife's sister he calls pūnalūa, i.e., his intimate companion. 1889 C. S. WAKE *Marriage & Kinship* ii. 23 The punaluan group can, as will be shown hereafter, be accounted for satisfactorily without assuming the prior existence of the consanguine family. 1915 *Amer. Anthrop.* XVII. 223 He does attribute the change from the older Malayan to the later and more common Turanian form of the system to punaluan marriage as a predecessor of the institution of exogamy and to exogamy itself. 1922 B. MALINOWSKI in *Nature* 22 Apr. 502/2 Starting from promiscuity, mankind went through group marriage, then the so-called consanguineous family or Punalua, then polygamy. 1940 WEST & TORR tr. *Engels's Origin of Family* (1942) ii. 39 The American system of consanguinity..finds, down to the smallest details, its..natural foundation in the punaluan family. *Ibid.* 42 The origin of the matriarchal gens could be derived directly from the punaluan family. 1970 K. MILLETT *Sexual Politics* II. iii. 111 A succession of sexual associations: promiscuity, group marriage, the consanguineous family, the Punalua. *Ibid.* 120 The first course of social change as Engels had charted it was from consanguine group marriage, to the Punaluan consanguine group.

punamu, var. *POUNAMU.

Punan (pūnā·n). Also 9 **Panam.** [Native name.] A group of Dyak peoples inhabiting parts of Borneo, mostly living nomadically in interior jungles; a member of this people.

1838 J. BROOKE *Lett.* (1853) I. 25 The Panams, a race little better than monkeys, who live in trees, eat without cooking, are hunted by the other tribes, and would seem to exist in the lowest conceivable grade of humanity. 1876 *Encycl. Brit.* IV. 58/1 The fifth and lowest [branch of Dyaks] comprises the Manketans and Punans, who are still nomadic and ignorant of agriculture. 1927 *Brit. Weekly* 19 May 154/1 A Punan will never wantonly slay or attack a man of another tribe. 1960 *Spectator* 28 Oct. 662 The..elementary Punan, these..being nomadic forest hunters [of Borneo]. 1964 T. HARRISSON in Wang Gungwu *Malaysia* III. xi. 164 The nearest thing to the Malayan aborigine in Borneo is the nomadic Punan.

punch, *sb.*¹ Add: **2. c.** In mod. use, an instrument for removing a small piece of tissue from a patient.

[1859 S. D. GROSS *Syst. Surg.* II. xi. 631 With a large, sharp saddler's punch the whole of the diseased structures..are then removed.] 1887 *Amer. Jrnl. Med. Sci.* XCIV. 279 (*heading*) The cutaneous punch. 1915 A. MACLENNAN *Surg. Materials & their Uses* vi. 220 It is almost impossible to get punches to cut clean, and a certain amount of tearing out of the uncut tissue must be expected. 1937 C. G. DARLINGTON in S. C. Miller *Oral Diagnosis & Treatment Planning* xxix. 496 Tissue forceps,..scissors, skin or biopsy punch, may be employed, depending on the type and location of the lesion. 1957 G. L. W. BONNEY in Rob & Smith *Operative Surg.* V. 250 The impactor punch is shaped at one end so as to fit on to the convex outer surface of the femur. 1978 *Nature* 9 Mar. 171/2 Skin biopsies were taken from the upper back of 21–39-yr-old male volunteers with an electric biopsy punch 3 mm in diameter.

7. *punch operator*; **punch biopsy** *Med.*, a biopsy in which a punch is used to remove tissue; **punch card** = *punched card* s.v. *PUNCHED *ppl. a.* 2; **punch forceps** *Surg.*, a punch consisting of two hinged parts like a forceps; **punch graft**, a graft of tissue removed (usu. from the scalp) by means of a surgical punch; so **punch grafting** *vbl. sb.*; **punch-marked** *a.*, of a coin: bearing a punch-mark; **punch-press**, a press designed to drive a punch for shaping metal; **punch-ticket** *U.S.*, a railway or other ticket that has been punched.

1941 *Amer. Jrnl. Clin. Path.* XI. 519 We have..performed punch biopsies upon the livers of normal rabbits. 1955 TWISS & OPPENHEIM *Pract. Managem. Disorders of Liver, Pancreas, & Biliary Tract* xiv. 260 Punch biopsy of the liver can be used successfully to make an etiologic distinction between the different forms of cirrhosis. 1976 *Lancet* 11 Dec. 1281/2 Some would not agree that punch biopsies should be routinely used on the face. 1945 J. VON NEUMANN in B. Randell *Origins Digital Computers* (1973) 355 These instructions must be given in some form which the device can sense: Punched into a system of punchcards [etc.]. 1971 K. GOTTSCHALK in B. de Ferranti *Living with Computer* iv. 37 The punch cards embodying the program were deemed a manner of manufacture and not patentable. 1977 *New Yorker* 29 Aug. 41/2 The production by the computer of punch-card decks representing the processed data. 1870 *Brit. Jrnl. Dental Sci.* XIII. 497 A punch forceps for punching holes in the backing for flat teeth, the peculiarity of which lay in the fact, that by a spring interposed between the jaws the plate was liberated from the punch directly the jaws of the forceps were opened. 1958 J. H. OTTY in Rob & Smith *Operative Surg.* VIII. xviii. 93 The antrum is opened by removing a chip of bone wide enough to insert a Hajek's punch forceps, with which the opening is enlarged to the size desired. 1959 *Ann. N.Y. Acad. Sci.* LXXXIII. 465 Of a total of 284 punch grafts, only one of the punch grafts failed to take. 1968 *Plastic & Reconstruction Surg.* XLII. 446 (*heading*) Use of hair-bearing punch-grafts for partial traumatic losses of the scalp. 1976 *Daily Tel.* (Colour Suppl.) 2 July 8/4 Punch grafts require umpteen sessions and are followed by 12-week intervals of frustration before anything grows. *Ibid.* 7/3 Most bald men retain a permanent expanse of hair at the back and sides, and punch grafting has shown that this hair survives even when transplanted into bald areas of the head. 1910 HASTINGS *Encycl. Relig. & Ethics* III. 706/1 On account of this chief characteristic, the term 'punch-marked' is commonly applied to this currency. 1960 H. HAYWARD *Antique Coll.* 230/1 *Punch-marked coins*, flat, square silver coins of India of the last few centuries B.C. Surfaces covered with small punch marks of natural objects, animals and symbols, probably the marks of merchants and states guaranteeing the pieces. 1961 *Evening Standard* 17 July 16/1 (Advt.), Data Processing Department. British Wool Marketing Board—Bradford. Immediate vacancies... Punch operators. 1968 *Brit. Med. Bull.* XXIV. 191/1 The data sheets are handed to a punch operator who types out the information on a keyboard which has a punch attachment. 1911 W. J. KAUP *Machine Shop Practice* xviii. 180 Fig. 162 shows a typical modern punch press. 1935 O. W. BOSTON *Engin. Shop Practice* II. vi. 332 A punch press..is a gap or overhung type of fast, short-stroke press particularly suited for punching dies. 1976 *National Observer* (U.S.) 5 June 7/1 When Nelson Amsdill gets off work as a punch-press operator, he heads for the Veterans of Foreign Wars Post 6691. 1887 C. B. GEORGE *40 Years on Rail* xi. 227 Many cases have been reported where in punch-tickets the bits of pasteboard punched out have been saved and carefully glued in the old places. 1890 *Harper's Mag.* May 908/1 A person..who by many punch-tickets builds up the fortunes of the stockholders.

punch, *sb.*² Add: **1. b.** *transf.* and *fig.* Forceful, vigorous, or effective quality in an activity or in anything spoken or written; vigour, weight, effectiveness. orig. *U.S.*

1911 E. FERBER *Dawn O'Hara* xvii. 254 It lacks that peculiar and convincing quality poetically known as the punch. 1914 'I. HAY' *Knight on Wheels* xvii. 162 The two clerks and the office-boy carried out their duties with what is known in trans-atlantic business circles as 'a punch'. 1919 H. L. WILSON *Ma Pettengill* ii. 64 A gripping drama replete with punch. *Ibid.* 75, I believe he now admits frankly that he wrote most of the play, or at least wrote the punch into it. 1921 D. W. JOHNSON *Battlefields of World War* xii. 535 The attack lost its 'punch'. 1926 *Glasgow Herald* 1 Apr. 5 They lack for the most part the quality of 'punch' which we have come to regard nowadays as one of the principal essentials in a magazine story. 1933 G. ARTHUR *Septuagenarian's Scrap Book* 307 Within a few days French, Americans and British were beginning a forward movement with the necessary 'punch' behind them of which the recent enemy assaults had been devoid. 1947 *E. African Ann.* 1946–7 98/2 (Advt.), Are you satisfied your advertisements have the necessary punch to get their message across? 1955 *Sci. News Let.* 27 Aug. 134/2 Radioiodine loses its punch within weeks. 1968 *Globe & Mail* (Toronto) 13 Jan. 41/8 Oakland Raiders' coach John Rauch said yesterday his team..has the punch to score against the defensively rugged Packers. 1976 *Oadby & Wigston* (Leics.) *Advertiser* 26 Aug. 16/3 Chances were created but there was just no punch up front.

2. Phr. *to beat (someone) to the punch*: of a boxer, to land a blow before his opponent can strike him; also *transf.*, to anticipate or forestall someone in speech or action. *to pull one's punches*: see *PULL *v.* 17 b; *to roll with the punches*: see *ROLL *v.*² 21 f.

1923 H. C. WITWER *Fighting Blood* vii. 226, I beat Hanley to the punch..and he went down on his haunches. 1965 *Listener* 1 July 6/1 The tracking station at Plumeur Bodou is the place that so exultantly beat Britain to the punch in getting the first pictures from America via the satellite Telstar. 1977 *Sunday Times* 3 July 28/3, I feel a batsman uses it as he thinks he will beat a fast bowler to the punch.

3. *attrib.* and *Comb.*, as *punch-packed* adj.; *punch-packing* ppl. a.; **punch-bag**, a stuffed bag suspended at a suitable height on which boxers practise punching; **punch-pull** *v. intr.*, to 'pull one's punches', to refrain from striking as hard as one can, or from expressing oneself forcefully; so **punch-pulling** *vbl. sb.*

1889 A. C. GUNTER *That Frenchman!* 8 Sling from the ceiling a punch-bag such as prize-fighters train with. 1899 *Science Siftings* 25 Mar. 329/2 A fifteen-minute controversy with an active punch-bag. 1927 *Daily Express* 20 July 9/7 His trainer..ordered Dempsey not to box, but to use the punch bag and to shadow box. 1973 D. FRANCIS *Slay-Ride* ii. 25, I..woke at seven feeling like Henry Cooper's punchbag. 1963 *Times* 25 Apr. 15/1 The year's most punch-packed novel. 1936 *Variety* 17 June 26 (Advt.), The punch-packing short stories of the screen. 1961 *New Statesman* 23 June 1010/3 The Bishop taught him how to punch-pull on all outstanding emotional issues. 1959 *Listener* 29 Jan. 224/2 There was some good photography..and a conclusion in which there was no punch-pulling.

punch, *sb.*³ Add: **4.** *punch-pot, -room.*

1960 *Times* 16 Apr. 9/3 A punch-pot measured about 9 in. across its globular body. 1971 *Canadian Antiques Collector* Feb. 16/1 The Museum collection contains..a large and rare punch-pot, the cover with an attractive lemon knop. 1827 DRAKE & MANSFIELD *Cincinnati in 1826* iii. 30 A spacious gallery, with commodious lobbies, punch room, etc. 1841 *Southern Lit. Messenger* VII. 764/1 If you won't go home with me, you can take me down to the punch-room.

Punch, *sb.*⁵ Add: **1. a.** (Earlier *attrib.* examples of *Punch and Judy.*) Also *ellipt.*, = *Punch and Judy show.*

1841 C. Fox *Jrnl.* 18 Feb. (1972) 102 He..teaches us that Punch and Judy men, beggar children and daft old men are also of our species. 1857 C. KINGSLEY *Two Years Ago* I. p. xviii, Those poor idolaters, and their Punch and Judy plays. 1864 [see *FAIR *sb.*[1] 1 a].
b. *as proud as Punch* (earlier examples).
1841 C. BRONTË *Let.* 4 May in C. K. Shorter *C. Brontë & her Circle* (1896) iii. 87 Mrs. White would be as proud as Punch to show it you. 1850 DICKENS *Dav. Copp.* ii. 520, I am as proud as Punch to think that I once had the honor of being connected with your family. 1888 G. B. SHAW *Let.* 20 Sept. (1965) I. 200 Headlam read out about all the gold and silver in the palace,..and the Bishop looked as proud as Punch of owning it all.

punch, *v.*[1] Add: **I. 2. a.** (Further examples.) Also *absol.*
1885 *Nor' Wester* (Calgary, Alta.) 12 Feb. 3/2 It would pay the stockmen to keep men out during the winter to punch the cattle out of the brush during fine weather. 1890 *Stock Grower & Farmer* (Las Vegas, New Mexico) 21 June 4/1 J. O. Phillips..will be initiated into the business of punching cattle. 1906 *McClure's Mag.* May 64/1 About ten year ago I got plumb sick of punchin' cows around my part of the country. 1910 W. M. RAINE *B. O'Connor* 30 We used to punch together on the Hashknife. 1923 'B. M. BOWER' *Parowan Bonanza* xviii. 276 In that case..you'd still be punchin' cows for your dad, most likely. 1946 F. D. DAVISON *Dusty* Forewd., Tailing tame old milkers into the farmyard,..punching stubborn bullocks through the mulga.
3. a. (Later examples, *esp.* in Sport.) Phr. *to punch out*: to knock out, to beat up (*U.S.*).
1920 *Isis* 5 May 9/2 He will get runs, and in the getting of them the ball will be 'punched' very hard. 1929 W. FAULKNER *Sartoris* iv. 301 She put coal on it and punched it to a blaze. 1968 [see *BUNT *sb.*[4] 2]. 1969 *New Yorker* 14 June 44/3 The orthodox way to hit a volley is to punch it, with a backswing so short that it begins in front of the player's body. 1971 *Current Slang* (Univ. S. Dakota) VI. 8 *Punch out*, to beat up; to fight physically. 1976 *National Observer* (U.S.) 12 June 19/2 Young blacks and Puerto Ricans..punched out Moonies who tried to restrain them. 1977 *Detroit Free Press* 11 Dec. 2-D/2 Abdul-Jabbar,..broke his own hand punching out Milwaukee's Kent Benson in the season opener. 1977 RECHIN & PARKER *Crock* 48 One more smart remark about my nose and I'm punching you out.
b. *to punch up*: to assault, beat up (cf. *PUNCH-UP); also *fig.* in *Cinematogr.* (see quots. 1953, 1959).
1953 BERREY & VAN DEN BARK *Amer. Thes. Slang* (1954) §623/8 Punch up one and fade it down, get the picture ready for fading up, as in an opening shot. 1959 W. S. SHARPS *Dict. Cinemat.* 121/2 *Punch up*. In acting, this is to add emphasis to a phrase or action. In filming, the term means to increase picture brightness, and in recording, to bring in a new sound, or to increase the volume or pitch of an existing sound. 1963 *Listener* 31 Jan. 202/1 The folknicks in Washington Square when they punch up the police of a Sunday afternoon.
c. *to punch the* or *a clock*: to clock in or out; so *to punch in*.
1927 *Sunday Express* 8 May 10 Costello flatly refused to 'punch the clock', and had definite ideas about what he would and would not do in connection with his art. 1943 J. B. PRIESTLEY *Daylight on Saturday* i. 2 What happens when you have shown your pass and punched the time-clock? 1944 G. FARWELL in *Coast to Coast 1943* 116 Yesterday I was late punching in. They'll be docking me. 1969 D. CLARK *Nobody's Perfect* iii. 103 We're soft-hearted in the way we treat our staff. Nobody has to punch a clock. 1978 S. BRILL *Teamsters* vii. 292 At the terminal Barkett punched in.
d. To press (a push-button); to operate, switch *on*, or tune *in* (a device) by doing this.
1954 W. TUCKER *Wild Talent* xiv. 211 The man punched the elevator button. 1971 'R. MACDONALD' *Underground Man* x. 61, I punched on the car radio. It was tuned to a local station. 1972 M. KAYE *Lively Game of Death* (1974) vii. 40 Scott punched the intercom and asked to speak with Lasker. 1975 *Gramophone* Sept. 531/1 The user can also punch in any desired FM station frequency and scan for just stereo stations. 1977 *Guardian Weekly* 25 Sept. 19/2 The launch controller punched the destruction button and the rocket with its payload was automatically destroyed.
II. 4. a. (Further examples.) Cf. *KEYPUNCH *v.* 1 a.
1939 J. BERRYMAN in K. Amis *Spectrum* (1961) 167 With a rattle and a whir the calculators punched and sorted the cards. 1964 F. L. WESTWATER *Electronic Computers* vi. 98 The cards have to be punched by hand from information on original documents. 1971 *Daily Tel.* 3 May 2/6 It lights up a wall map, serves a memory bank, punches a tape for computer use and produces a copy in type for control room operators. 1974 J. BANNING *How I fooled World* ii. 13 When I..had to punch tape myself, the computer in London often rejected my copy.
c. Delete *Telegraphy* and add: = *KEY-PUNCH *v.* 1 b. (Earlier and further examples.) Quot. 1876[1] in Dict. belongs to sense 4 b.
1864 C. BABBAGE *Passages from Life of Philosopher* viii. 119 The Tables to be used must..be computed and punched on cards by the machine. 1890 *Jrnl. Franklin Inst.* CXXIX. 301 In order to punch the individual records upon the cards, they are placed one by one in a suitable punching machine. 1946 *N.Y. Times* 15 Feb. 16/3 When the problem is punched on the cards, they are dropped into a slot in a 'reader'. 1952 *Sci. Amer.* Sept. 112/3 The instructions..are punched in the paper tape by a special typewriter keyboard. 1968 *Brit. Med. Bull.* XXIV. 247/1 Use of the program achieves a very considerable saving in effort, which would be further enhanced if..observations of turnover diameter were punched directly on to tape at the time of measurement. 1971 H. LOVE in R. A. Wisbey *Computer in Lit. & Ling. Research* 51, I must not pass over his method of proofreading input, which is to have the text punched-up by two different operators and then use the computer to spot discrepancies.
d. (Later example.)
1977 *Lancet* 26 Nov. 1140/2 Two 3 mm diameter discs are punched out from the filterpaper, one for testing and the other as a control.
6. *to punch out* (Aeronaut.): see quots. *slang.*
1968–70 *Current Slang* (Univ. S. Dakota) III–IV. 98 *Punch out*, *v.* To eject from an aircraft. 1974 *Sunday Times* 16 June 13/2 It never occurred to me to 'punch out' (eject).

pu·nch-ball. [PUNCH *sb.*[2]] **1.** A stuffed or inflated ball suspended at a suitable height for practice punching by boxers. Also *fig.*
1901 *Humane Rev.* II. 218 There would be a large gymnasium with all sorts of appliances, *e.g.*, the punch-ball. 1910 *Cycling* 12 Jan. p. xxi/3 (Advt.), Punch-ball, 10s, cost double. 1927 *Daily Express* 22 June 17/5 He is developing his punch in secret, and..he has broken three punch-balls. 1932 *Pictorial Weekly* 12 Mar. 185/2 In the centre a player is using the punch-ball. 1963 *Times* 13 Mar. 12/5 The Home Office has always been something of a punchball for Mr. Sidney Silverman. 1973 M. AMIS *Rachel Papers* 13 Her buttocks, when she wore stretch-slacks, would dance behind her knees like punch-balls. 1977 *Time* 3 Oct. 50/2 Among the principals: the incomparable Lizzie, a daydreamy beautiful loser, 'punchball' for many lovers, whose flaws prove even more compelling than her easy virtue.
2. *U.S.* A ball-game (see quot. 1932).
1932 *Jrnl. Health & Physical Educ.* May 48/2 *Punch Ball*... The Youngstown, Ohio, playgrounds have promoted a new game... There are fifteen players on a side. A tightly blown rubber ball the size of a basketball is kicked, punched with the fist, or butted with the head as in soccer. 1935 C. F. WARE *Greenwich Village* v. 144 The district abounded in block teams..who played the ubiquitous game of punchball. 1976 *Washington Post* 28 July C6/3 Some of the games we played were called stickball and stoop baseball and punchball.

pu·nch board. *N. Amer.* Also with hyphen and as one word. [PUNCH *sb.*[2]] **a.** A board perforated with holes containing slips of paper which are 'punched' out as a form of gambling, with the object of locating a winning slip.
1912 J. P. QUINN *Gambling & Gambling Devices* 231 (*caption*) Punch board. 1935 J. STEINBECK *Tortilla Flat* xvii. 315 Tito Ralph came in with a box of cigars he had won on a punch board. 1939 *Sun* (Baltimore) 12 Sept. 7/4 The worst form of petty gambling seems to be confined to punch boards. 1949 *Democrat* 15 Dec. 4/1 We would like to amend his paragraph to include punch boards. 1951 *Manch. Guardian Weekly* 15 Mar. 10/4 The sale of punch-boards (gambling machines) was running at about $100 millions a year. 1966 *Times* 28 Feb. (Canada Suppl.) p. x/4 Any dice game, three-card monte, punch board. 1978 *Rugby World* Apr. 50 (Advt.), Now available New Lakeland (tamper proof)..Football cards & Punchboards.
b. *fig.* A promiscuous woman.
1963 *Amer. Speech* XXXVIII. 173 A female who is dated because of her lax sexual habits..*punch*... The word *punch* may well be a shortened form of *punchboard*, which was recorded twice. 1970 G. GREER *Female Eunuch* 267 Girls who pride themselves on their monogamous instincts..speak of the 'campus punchboard'. 1977 J. WAMBAUGH *Black Marble* (1978) iii. 23 There's one woman handler, named Wilma. A punchboard. What the hell, she's a little dumpy, but when you been looking at dogs all day.

pu·nch-drunk, *a.* and *sb.* orig. *U.S.* [f. PUNCH *sb.*[2] + DRUNK *ppl. a.* and *sb.*[2]] **A.** *adj.* Of a boxer or one involved in physical fighting: dazed or stupefied from severe or continual punching; *spec.* exhibiting reduced muscular co-ordination, hesitant speech, slowness of thought, and other signs. Also *fig.*
1918 *Sat. Even. Post* 18 May 12/3 He was in the condition so aptly described as 'punch drunk'. 1927 *Daily Express* 29 July 1, I replied that in my opinion Moore was 'punch-drunk', that is to say that he had taken so much punishment that he could no longer feel it, and his nerves were practically gone. 1934 *Sun* (Baltimore) 2 Mar. 12/7, I am delegated to remind all who may be punch-drunk with winter that the famous Blizzard of Eighty-eight occurred on March 12. 1937 *Daily Mirror* 2 Mar. 12/4 Nowadays the Kid is punch-drunk. His limbs tremble and quiver like a man stricken with ague. His voice is so slurred that one cannot properly understand what he is saying. 1947 *Penguin New Writing* XXX. 126 A film which has been praised by critics who should certainly know better, unless they have themselves become punch-drunk with watching the hallucinated antics of the slap-happy puppets on the screen. 1952 C. DAY LEWIS tr. *Virgil's Aeneid* v. 105 So he called an end to the bout, saving the punch-drunk Dares From further punishment. 1954 J. STEINBECK *Sweet Thursday* 215 Doc's setting over there like he's punch-drunk. 1958 *Times Lit. Suppl.* 5 Sept. 497/3 In this punch-drunk civilization it is perverse to expect me to take seriously the horse itself, or indeed Mr. Kirkup's gracious swan, or Mr. Trypanis's seasick cock. 1959 N. MAILER *Advts. for Myself* (1961) 21, I seem to have turned into a slightly punch-drunk and ugly club fighter. 1974 'A. GARVE' *File on Lester* xxxvi. 129 The papers are terrible this morning... I'm feeling punch drunk and hardly capable of rational thought. 1974 E. BRAWLEY *Rap* (1975) I. i. 31 You and me know he's so..punch-drunk he'd do anything anybody told him.
B. *sb.* One who is punch-drunk (usu. in literal use).
In quot. 1928 used as the name of the condition.
1928 *Jrnl. Amer. Med. Assoc.* 13 Oct. 1103/1 The early symptoms of punch drunk usually appear in the extremities. 1943 *Gen* 16 Jan. 30/2 Your out-and-out punch-drunk is harmless rather than homicidal. 1966 'A. HALL' *Ninth Directive* viii. 78 He spoke with the dulled tone of a punch-drunk. 1969 *Daily Tel.* 13 Nov. 18 'Punch-drunks' are completely unknown in amateur boxing.
Hence **pu·nch-drunkenness**, the condition of being punch-drunk.
1937 *Lit. Digest* 10 Apr. 40/1 The coincidence of boxers developing thickness of speech, unsteadiness of gait,..has prompted London's famed Guy's Hospital to initiate a special study of the cause and cure of punch drunkenness. 1939 J. BERRYMAN in *Astounding Sci. Fiction* May 31 Tiny bloodclots on the lining of the brain, whose pressure on delicate centers often caused them to manifest the symptoms of punch-drunkenness. 1941 *Lancet* 14 June 759/1 There is, however, that specific problem of boxing, 'punchdrunkenness'—the permanent damage due to repeated cerebral injury. 1959 *Daily Tel.* 27 June 7/5 There was 'plenty of evidence to show that punch-drunkenness is a very real hazard to boxers'. 1977 J. PORTER *Who the Heck is Sylvia?* xv. 144 A family which prided itself on never knowing when it was beaten and in which punch-drunkenness was practically an endemic disease.

punched, *ppl. a.* Add: **2.** *punched card*, a card in which a pattern of holes, punched in it in accordance with a prescribed code, represents information; similarly *punched paper*, *punched (paper) tape*. Freq. *attrib.* Cf. *paper tape* s.v. *PAPER *sb.* 12, *perforated tape* s.v. *PERFORATED *ppl. a.* 1.
1876 Punched paper [in Dict.]. 1885 *Electrician* 27 Nov. 57/1 The Wheatstone fast-speed transmitter.., by which one punched tape served for twenty or thirty different wires. 1890 *Jrnl. Franklin Inst.* CXXIX. 301 These punched record cards can easily be read and verified. 1919 A. MACFARLANE *Lectures on Ten British Physicists* 79 To realize the first idea..he had recourse to the device of punched cards similar to those invented by Jacquard for the weaving loom. 1940 W. J. ECKERT *Punched Card Methods in Scientific Computation* 2 Tables of functions are constructed from their differences with great efficiency, either as printed tables or as a file of punched cards. 1948 *Electronics* Aug. 100/1 Approximate positions of the stars, already stored in a punched-card catalog, may be coupled to the servomechanism. 1959 *Engineering* 2 Jan. 5/3 The machine can operate from standard punched tape or can be plugged into a long distance teleprinter circuit. 1962 E. GODFREY *Retail Selling & Organ.* i. 6 Mechanized handling of goods, punched-card accounting systems and advertising direct to the consumer..involve costs which the smaller business cannot afford. 1963 *Listener* 21 Mar. 489/2 A computer..which read a quarter of a million words of Greek prose, translated into its own punched-paper language. 1968 *Brit. Med. Bull.* XXIV. 191/1 The commonest way..of inserting data and program into a computer is via punched paper tape or punched cards. 1975 T. ALLBEURY *Special Collection* i. 5 In Central Intelligence Records they had thousands of simple punched cards..in May 1944. 1980 D. BLOODWORTH *Trapdoor* v. 24 The formidable batteries of punched-card systems and data banks.

puncheon[1]. Add: Also **punchen.** **5. a.** (Earlier and later examples.)
1804 in *Maryland Hist. Mag.* (1909) IV. 9 Houses or cabins..are generally made of heavy timber logs covered with split timbers called 'puncheons' which they pin to the rafters with wooden pins. 1946 C. RICHTER *Fields* 164 The puncheons had holes for seat legs.
b. A piece of timber used in building a railway track or a corduroy road. *N. Amer.*
1843 W. OLIVER *Eight Months Illinois* 236 Trees are split up into what are called puncheons, of three or four inches in thickness, which are laid down on the sleepers. 1955 R. HOBSON *Nothing too Good* xv. 165, I figure that all it will cost you is the axes, shovels, spikes for punchen, crow bars, [etc.].
6. b. For (*U.S.*) read (orig. *U.S.*). (Earlier and later examples.)
1754 J. INNES *Let.* 27 Sept. in *Lett. to Washington* (1898) I. 48, I have erected a punchion Fort. 1784 G. WASHINGTON *Diary* 20 Sept. (1925) II. 294 A Logged dwelling house with a punchion roof. 1843 'R. CARLTON' *New Purchase* xxi. 199 Adjoining the bureau was the puncheon table with its white oak legs. 1940 W. FAULKNER *Hamlet* II. i. 110 The heatless lean-to room was his desert cell, the thin pallet bed on the puncheon floor the couch of stones on which he would lie. 1963 R. SYMONS *Many Trails* xiv. 145 The floor was of the puncheon type—that is, poplar poles laid across stringers and smoothed with an adze, with no attempt at nailing. 1972 E. WIGGINTON *Foxfire Bk.* 33 Green chestnut was split into fence rails, puncheon floors, wide planking, [etc.].

pu·ncheoned, *a.* [f. PUNCHEON[1] + -ED[2].] Covered or laid with puncheons (PUNCHEON[1] 5 a in Dict. and Suppl.).
1843 'R. CARLTON' *New Purchase* xv. 109 And first, the puncheoned area was separated into two grand parts.

puncher. Add: **a.** (Further examples.)
1876 *Jrnl. Soc. Telegr. Engineers* V. 492 When one clerk considers a puncher quite workable, another declares it to be useless. 1915 F. M. HUEFFER *Good Soldier* I. v. 73 Ready to lend you his cigar puncher. 1951 *Sport* 30 Mar.–5 Apr. 10/2 Both have reputations as punchers. *a* 1953 DYLAN THOMAS *Quite Early One Morning* (1954) 33 The clip of the chair-attendant's puncher. 1973 *Irish Times* 2 Mar. 3/3 They [*sc.* a hockey team] depend enormously on John Douglas to initiate their raids and have only one recognised 'puncher' in Tom Jenkinson.

b. For *U.S.* read *N. Amer.* (Earlier and later examples.) Also *attrib.* in *puncher-boy*.
1870 *Daily Territorial Enterprise* (Virginia City, Nevada) 17 Aug. 3/1 All the time the punchers are flying from ox to ox, plying their sticks right and left. 1910 in J. Lomax *Cowboy Songs* 96 But show me a man that sleeps more profound Than the big puncher-boy who stretches himself on the ground. 1912 S. A. WHITE *Wildcatters* 137 Ben had decked him out in puncher's garb. The lariot was correctly coiled at the saddle-horn. 1972 T. A. BULMAN *Kamloops Cattlemen* ii. 16 Tough as these old 'punchers' were, the years gradually took their toll.

punchery. (In Dict. s.v. PUNCH *sb.*[3]) (Later examples.)
1952 *John o' London's* 1 Aug. 724/3 The eighteenth century had a forerunner of our home cocktail bar. This was the Georgian punchery, a magnificent assembly of..bowls, spice dredgers, crystal bottles..and a..punch ladle. 1962 *Times* 20 Jan. 11/3 Well-stocked puncheries magnificently displaying colourful punch bowls.

Punchine (pʊ·nʃəin), *a. rare.* [f. PUNCH *sb.*[6] 2 + -INE[1].] Of or pertaining to the journal *Punch*.
1846 THACKERAY in *Punch* 8 Aug. 59/2 It was this braggart violence of soul that roused the Punchine wrath against Mr. O'Connell.

punching, *vbl. sb.* Add: **a.** Also, a piece of sheet metal cut out by a punch.
1903 *Electr. World & Engin.* 28 Mar. 532/2 The four-pole pieces are made of laminated steel punchings. 1947 R. LEE *Electronic Transformers & Circuits* iii. 75 There is always a certain amount of gap even with punchings stacked alternately in groups of 1. 1951 E. W. WORKMAN in P. Kemp *Electr. Engin.* III. 633/2 In the case of electrical transformers, generators and motors, sheet-steel punchings form a large proportion of the total works cost.
b. *punching machine* (examples); *punching bag* (earlier and later examples); also *fig.*; *punching press* = *punch-press* s.v. *PUNCH *sb.*[1] 7.
1889 *Cent. Dict.*, Punching-bag. 1896 ADE *Artie* i. 4 Say, I like that church, and if they'll put in a punchin'-bag and a plunge they can have my game, I'll tell you those. 1911 *Boxing* IV. 456/2 Once again that old trial hope, Fred Drummond, was dragged from his stall to play the part of punching-bag. 1976 G. SIMS *End of Web* x. 72 Buchanan used him like a punching-bag, hitting him with every combination he knew. 1850 *Rep. U.S. Comm. Patents 1849* I. 185 My improved punching machine. 1878 *Harper's Mag.* Apr. 645/2 The bar then goes to the punching-machine that..bites a bar through the iron. 1962 F. T. DAY *Introd. Paper* viii. 87 Table mats and drip mats..are often produced on a blanking-out or punching machine. *a* 1884 E. H. KNIGHT *Dict. Mech.* Suppl. 730/2 Punching press. 1906 C. H. BENJAMIN *Mod. Amer. Machine Tools* x. 282 (*caption*) 60-inch punching press.

punchless (pʊ·nʃlės), *a.*[1] [f. PUNCH *sb.*[3] + -LESS.] Having no punch to drink.
1903 W. STEVENS *Let.* 26 July (1967) 64, I was looking forward to a cigarless, punchless weary life.

pu·nchless, *a.*[2] [f. PUNCH *sb.*[2] + -LESS.] Lacking a powerful punch; deficient as a boxer.
1950 J. DEMPSEY *Championship Fighting* ii. 11 Punchless performers who can win amateur or professional bouts on points.

pu·nch line. orig. *U.S.* Also with hyphen and as one word. [f. PUNCH *sb.*[2] + LINE *sb.*[2] 23.] Words or a sentence expressing the point of a joke, play, song, etc. Hence **punch-line** *v. intr.*
1921 *Variety* 25 Nov. 8/1 All of their sure-fire punch-lines went over. 1934 S. R. NELSON *All about Jazz* vii. 158 The gentlemen who write lyrics..imagine the public hang on their doggerel—particularly the line known as the 'punch line'. 1937 *Dime Detective Mag.* Nov. 45/2 Go on..give me the punch line. 1944 S. BELLOW *Dangling Man* 158 Yes, things change. *C'est la guerre. C'est la vie.* Good old punch lines. 1957 *Oxford Mail* 17 Oct. 1/2 It was Mr Dulles's punch-line and showed the Russians—and the American people—that President Eisenhower regards the Middle East crisis with great anxiety. 1959 *Time* 14 Sept. 44/2 'I'll kill myself..' said Benny. 'All right,' Truman punch-lined, 'I've got an undertaker friend.' 1961 B. WELLS *Day Earth caught Fire* vii. 107 'Wonder who writes his punchlines?' remarked Reynolds. 1971 *World Archaeology* III. 226 He [*sc.* V. G. Childe] was fond of dramatic punch-lines. 1977 *New Yorker* 27 June 67/1 Reaching the punch line, he erupted in laughter.

pu·nch-up. *slang.* Also without hyphen and as one word. [f. PUNCH *v.*[1] + UP *adv.*[1]] A fight or brawl. Also *fig.*, a fierce or noisy argument.
1958 F. NORMAN *Bang to Rights* 28 The next morning after we had had this little punch up. 1960 H. PINTER *Caretaker* II. 36 Bloke saved me from a punch up. 1963 K. AMIS *One Fat Englishman* iii. 36 The fellow was earning a bigger and better punch-up, oral or physical, with every sentence he spoke. 1966 J. WAINWRIGHT *Evil Intent* 85 He's been responsible for more religious punch-ups than Judas himself. 1967 *New Scientist* 14 Dec. 673/1 Good old-fashioned punch-ups between the holders of rival theories, of the sort that so stimulated 19th century science, are sadly rare today. 1972 J. WILSON *Hide & Seek* viii. 151, I got six months..all because you ruzzers stuck your noses into a private little punch up. 1976 *Daily Mirror* 11 Mar. 9/6 He was fired after an alleged punch-up with another worker. 1978 *Times Lit. Suppl.* 25 Aug. 944/3 Boxing and pub punch-ups were his main amusements.

punchy, *a.*[1] (Earlier example.)
1783 J. WOODFORDE *Diary* 10 Feb. (1926) II. 58 He bought..a short dark Punchy Horse with a Hog main.

punchy (pʊ·nʃi), *a.*[3] [f. PUNCH *sb.*[2] + -Y[1].] Full of punch or vigour.
1926 WHITEMAN & MCBRIDE *Jazz* ii. 41, I would direct a punchy number. 1930 *Observer* 19 Oct. 19 A punchy rhetorical speech on Free Trade. 1937 *Lit. Digest* 4 Dec. 30/3 The English language may some day be as colorful and punchy as it was in Elizabethan times. 1959 *Times Lit. Suppl.* 25 Sept. 545/5 (Advt.), In over a score of punchy entertaining chapters he delights readers. 1971 *Amateur Photographer* 3 Mar. 23/2 The 10-minute playlet with a punchy plot. 1977 *Time* 30 May 55/1 More gregarious than Woodcock, a punchier speaker, a hair more liberal, Fraser signals a change in style rather than substance.

pu·nchy, *a.*[4] *slang* (chiefly *U.S.*). [f. PUNCH *sb.*[2] or *PUNCH-DRUNK *a.* and *sb.* + -Y[1].] = *PUNCH-DRUNK *a.* Hence *transf.*, in a state of nervous tension or extreme fatigue. Also as *sb.* (*rare*).
1937 *Lit. Digest* 10 Apr. 39/2 'Slap-happy' or 'punchy' ex-fighters. 1937 E. HEMINGWAY *To have & have Not* III. xiv. 201 Shut up, slappy... You've got the old rale... You punchies make me sick. 1943 *Gen* 16 Jan. 30/1 He lives in a dream-world..he is, as the boys put it, 'punchy'. 1950 E. B. WHITE *Let.* 12 Nov. (1976) 326 K and I are both pretty well, if a bit punchy. 1958 E. DUNDY *Dud Avocado* II. iii. 209, I am so punchy..that I don't know whether I'm coming or going. 1970 K. PLATT *Pushbutton Butterfly* (1971) xiii. 149 I'm not coming at you because of what a punchy Hell's Angel tea-head told me. 1974 *Summerville* (S. Carolina) *Jrnl.* 24 Apr. 2/3 By the time the serviceman inserted a new tube,..the kids were getting punchy from sitting before a gray screen,..trying to imagine just what it was that the Roadrunner was doing to the wily Coyote. 1977 *Tennis World* Sept. 17/2 A player who breaks up on the court from nervousness is said to be 'punchy', 'gone cuckoo' or simply 'gone'.

punctiliar (pʊŋkti·liǎɪ), *a.* [f. PUNCTILIO + -AR[1].] Of or pertaining to a point of time; = *PUNCTUAL *a.* 5 e.
1906 J. H. MOULTON *Gram. N.T. Greek* I. vi. 109 The Aorist has a 'punctiliar' action, that is, it regards action as a *point*. [*note*] I venture to accept from a correspondent this new-coined word to represent the German *punktuell*, the English of which is preoccupied. 1944 E. A. NIDA *Morphol.* II. ix. 130 If an action is considered as a unit, occurring so to speak, at a 'point' of time, it may be considered 'punctiliar'. In such a case an action which takes place over a considerable extent of time may nevertheless be looked upon as a unitary action of a punctiliar nature. 1964 —— *Toward Sci. Transl.* ix. 199 As a description of the kind of action involved in the verb, aspect serves to differentiate a number of contrasts, of which some of the most common are: (1) complete vs. incomplete, (2) punctiliar vs. continuous, [etc.].

punctilio. B. 5. b. (Later examples.)
1943 *R.A.F. Jrnl.* Aug. 22 All the matchless punctilio of an iron discipline coloured his daily round. 1978 J. UPDIKE *Coup* (1979) iv. 171 These noodly motifs the French had brought, along with military science, the metric system, and punctilio.

punctual, *a.* **II. 2.** Delete † *Obs.* and add: **a.** (Later examples.)
1904 T. HUTCHINSON in Shelley *Wks.* p. iv, Amongst the Editor's Notes at the end of the volume the reader will find lists of the punctual variations in the longer poems. 1930 *Bookman's Jrnl.* XVIII. xiv. (Second Supplement) 15 Both books have been entirely reset for this edition, and in addition to many minor alterations, mainly punctual, there is a new preface. 1931 *Times Lit. Suppl.* 16 Apr. 305/3 A punctual variation (which is also a misprint) between the publisher's imprint in the first two and in the third volume is not recorded.
IV. 5. e. *Gram.* Of action: occurring at a point in time; of aspect or tense: relating to an action or event that occurs at a point in time. Also as *sb.*, the punctual aspect or tense of a verb.
1914 L. BLOOMFIELD *Introd. Study Lang.* v. 145 The Slavic languages distinguish..between..durative and iterative..action..and..punctual and terminative action. 1924 [see *ASPECT *sb.* 9 b]. 1933 L. BLOOMFIELD *Language* xvi. 272 The English categories of *aspect* distinguish between 'punctual' action (some grammarians call it perfective), envisaged as a unit, and 'durative' action..which extends over a segment of time. 1956 J. GONDA *Character Indo-Europ. Moods* iv. 44 The pronounced preference for the aorist may be understood from the predilection for the punctual aspect in formulating prohibitions. 1962 —— *Aspectual Function of Rgvedic Present & Aorist* i. 19 The aorist... is considered 'punctual', whether the initial moment (δακρῦσαι 'burst into tears'), or the end (πεῖσαι 'to persuade, talk over') is indicated. 1971 *Archivum Linguisticum* II. 112 This means that the 'punctual meaning' and the other meanings traditionally thought of as the aorist's are, as far as the sigmatic aorist is concerned, simply another specialization, restricted to a limited dialectal field. *Ibid.* 113 Gonda makes some criticism of the general statement..that VI class presents are 'punctuals', sometimes used to express punctual action, sometimes to describe actions indifferent to duration, but which sometimes can be thought of as punctual.

punctuality. Add: **I. 1. c.** *Gram.* The quality or character of being punctual (sense *5 e); the punctual aspect of a verb.
1962 J. GONDA *Aspectual Function of Rgvedic Present & Aorist* i. 31 The basic notion of the opposition present stem : aorist stem in Greek is that of duration, the present being the marked term, the aorist which is the unmarked term sometimes being indifferent in respect to duration..sometimes expressing punctuality. *Ibid.* 34 The fundamental idea of nondurativeness ('punctuality') in the aorist. 1971 *Archivum Linguisticum* II. 113 Both notions (non-duration and punctuality) became unified at a certain time, that is to say, their meanings became compatible, appearing alternatively according to context.

punctuate, *v.* Add: **3. b.** (*b*). (Earlier and later examples.)
c 1865 E. DICKINSON *Poems* (1955) II. 731 A Flower's unobtrusive Face To punctuate the Wall. 1941 *Penguin New Writing* II. 54 They communicated with each other in a low drone.., punctuated by an occasional deep-throated 'Ah!' 1966 C. M. BOWRA *Memories 1898–1939* v. 98 Roy Harrod..punctuated his speeches with such phrases as 'I dare aver' or 'if you will permit the observation'. 1971 E. MAVOR *Ladies of Llangollen* ix. 159 Great delicacy had to be employed on both sides of a correspondence which was apt to be punctuated with small wounded silences, implied accusations followed by temporary reconciliations. 1977 *Times* 25 Nov. (Christmas Book Suppl.) p. xxxi/4 The novel is one of many nicely turned backward looks that have worthily punctuated the year.

punctuation. Add: **3. c.** *fig.* The repeated occurrence or distribution (*of* something); something that makes repeated or regular interruptions or divisions.
1914 *Sat. Even. Post* 4 Apr. 12/1 The endless punctuation of ties led on and on until even the marshes rose and became level with the tracks. 1933 E. O'NEILL *Ah Wilderness!* I. 20 (*Stage direction*) The bang of firecrackers and torpedoes..continues at intervals..sufficiently emphatic to form a disturbing punctuation to the conversation. 1970 H. BRAUN *Parish Churches* ix. 125 Vertical punctuation disciplines an elevation by sorting it out into orderly compartments and replacing confusion with a pleasant rhythm... Vertical punctuation is of course achieved by the pilasters..later becoming buttresses. Horizontal punctuation is primarily concerned with maintaining a sense of stability. 1977 *Times* 16 Apr. 14/4 It requires a considerable resistance to evidence not to see the Reformation, the Enlightenment, nineteenth-century liberalism and the current decline as stages in a continuous process rather than punctuations in an otherwise stable history.

punctum. Add: **4. b.** *punctum saliens* (earlier and later examples); also *fig.*
1663 R. BOYLE *Consid. Usef. Nat. Philos.* II. i. 18 In Hen-eggs..you may observe the *Punctum saliens*, or Heart, to be ever and anon full of conspicuously red Blood. 1814 M. BIRKBECK *Jrnl.* 18 Sept. in *Notes Journey through France* 83 Paris is the punctum saliens, the organ of political feeling; elsewhere political feeling is absorbed in the love of tranquillity. 1977 *Language* LIII. 56 The 'punctum saliens' is that it was the 'paradigmatic' alternation *u*~*o* which was felt to render the constituent structure opaque.
6. *punctum indifferens* (cf. INDIFFERENT *a.*[1] 8 a), a neutral point.
1923 A. T. QUILLER-COUCH *Shakespeare's Much Ado about Nothing* p. xv, Such a man is the Friar in this play; its steadying sane mind, its *punctum indifferens*. 1932 J. BUCHAN *Sir W. Scott* xiii. 342 This *punctum indifferens* is the peaceful anchorage of good sense from which we are able to watch with a balanced mind the storm outside. 1976 'J. DAVEY' *Treasury Alarm* I. 11 Things may rearrange themselves around me just because I'm there—as a *punctum indifferens*.
7. *Palæogr.* A point used as a (weak) mark of punctuation in medieval manuscripts.
1952, 1975 [see *PUNCTUS]. 1975 *Anglo-Saxon England* IV. 117 The poems in lyric metres are written across the page as prose, but each verse line is separated by at least a punctum (or greater punctuation if the syntax requires it). 1978 *N. & Q.* Oct. 396/1 Each octosyllabic couplet is written as one line, divided by a mid-line punctum.

puncture, *sb.* Add: **4. puncture mark**, a mark made by a needle point, *esp.* in the injection of drugs.
1927 D. L. SAYERS *Unnatural Death* xxiii. 274 Will you go down to the mortuary again and see if you can find any puncture mark on the body. 1935 A. CHRISTIE *Death in Clouds* ii. 23 There was a minute puncture mark on the side of her throat. 1957 D. DU MAURIER *Scapegoat* xxi. 280 The sleeve of her black wool coat fell back, showing the puncture marks between wrist and forearm. 1974 D. RAMSAY *No Cause to Kill* I. 13 She wasn't mainlining, but there's a mark on her arm that could be a puncture mark.

puncture, *v.* Add: **1. d.** (Later examples.)
1927 *Scribner's Mag.* Apr. 450b/2 There is certain value in puncturing the Washington myth which the school histories used to teach. 1974 *Economist* 7 Sept. 16 It is time to puncture the continuing Greek accusation.
e. To punctuate or intersperse.
1899 C. M. M. SHELDON *His Brother's Keeper* xi. 249 The major..made a rattling speech, punctured with frequent amens and hallelujahs from the rest of the army.

3. (Later example.) Now *rare*.
1975 *Country Life* 4 Dec. 1529/2 Many cars suffered... Cowan punctured and spun.

‖ **punctus** (pʊ·ŋktʊ̆s, pu·ŋk-). *Palæogr.* [L., f. *pungere* to prick: cf. PUNCTUM.] A point, a punctuation mark. Freq. in phr. *punctus elevatus* (elēvā·tʊ̆s), a raised point; *punctus interrogativus* (intĕrǫːgātī·vʊ̆s), a question mark; *punctus versus* (vɜ̄·ɪsʊ̆s), a reversed point. See also quots.
1952 P. CLEMOES *Liturgical Influence on Punctuation in Late Old English & Early Middle English MSS* 4 *Punctus elevatus*, a symbol formed by combining *Punctum* and *Podatus*. It denoted the cadence with which a *Colon* ended, generally a gradual lowering of pitch followed by a return to the *Tuba* in one stage. *Ibid.* 5 *Punctus versus*, a symbol consisting of the *Punctum*. It denoted the cadence with which a *Periodus* ended. *Ibid.*, *Punctus interrogativus*, a symbol formed by combining *Punctum* and *Porrectus*. It denoted the cadence with which an interrogative sentence ended. 1954 D. WHITELOCK *Early Eng. MSS. in Facsimile* IV. 18/2 The usual mark of punctuation is the *punctus*, marking minor pauses as well as the end of sentences. *Ibid.*, A sign shaped like a semi-colon, the *punctus versus*, occurs. *Ibid.*, A mark like an inverted semi-colon, the *punctus elevatus*,..never marks the close of a period, but divides a main from a subordinate clause, or one subordinate clause from another. 1957 N. R. KER *Catal. MSS. containing Anglo-Saxon* p. xxxiv, The two marks, the dot and the ;, were reinforced at the end of the tenth century by a third mark, ⁏, which had been used hitherto only in Latin texts. It was known later in the Middle Ages as *punctus elevatus* and is often called now, inaccurately, the 'reversed semicolon'. 1966 P. CLEMOES *Early Eng. MSS. in Facsimile* XIII. 24/1 Four marks are used, namely a simple point placed at about mid-height..a *punctus elevatus*..a *punctus versus*..and a *punctus interrogativus*. 1971 P. J. LUCAS in *Archivum Linguisticum* II. 5 Notably absent are the *punctus elevatus* (or so-called 'inverted semicolon') and the *punctus interrogativus*... His [*sc.* Capgrave's] failure to use both the *punctus elevatus* and the *punctus interrogativus* probably indicates a movement away from liturgical (and formal rhetorical) influence. 1975 *Anglo-Saxon England* IV. 117 This [system of punctuation] consists of a hierarchy of punctum (.), punctus elevatus (⁏), punctus versus (;)..the punctus interrogativus (?) is also used.

pundit. Delete ‖ and add: **a.** (Later examples.)
1891 C. R. DAY *Music of S. India* v. 61 They..were probably composed by some Telegu pandit at the court of Mysore. 1901 [see *KASHMIRI *a.* and *sb.*]. 1953 *Encounter* Oct. 41/2 That this pandit (i.e. 'wise man') has become Prime Minister is one of the caprices of history. 1967 *Guardian* 26 Aug. 7/7 Having received choice jobs during the days of the Maharaja and the British Raj, the Pandits now resent the attempts to redress the balance in favour of..the Moslems. 1971 *Illustr. Weekly India* 4 Apr. 19 After the hair is cut and the *puja* is performed by the *pandit*, the turban with the sehra is placed on the head of the child by the father.
b. Delete (*colloq.* and *humorous*). (Later examples.)
1924 C. E. MONTAGUE *Right Place* xiv. 222 To say things and try to believe them, just because some aesthetic pundit or critical mandarin has said them before. 1938 R. HUGHES *In Hazard* ii. 37 First, this was developing into a true hurricane; and, second, it was not at all where it was thought by the pundits to be. 1941 C. H. WADDINGTON *Scientific Attitude* iv. 51 The architect who wished to build for a scientific and sceptical age had to..find out what was left when scepticism had done its worst. The pundits would say that nothing was left. 1957 *Listener* 5 Sept. 338/1 The British Association..is holding its 119th annual meeting... The pundits have gathered at Dublin. 1976 *Times* 30 Sept. 8/7 Though frowned upon by some pundits as out-of-date and middle-class, *Swallows and Amazons* and its many sequels remain immensely popular with children. 1977 J. I. M. STEWART *Madonna of Astrolabe* iii. 51 Here is what some pundit calls the phantom aesthetic state.

pundit (pʊ·ndit), *v. rare.* [f. prec.] *intr.* To make pronouncements like an expert. Also with quasi-obj.
1959 *Time* (Atlantic ed.) 19 Oct. 61 Huntley..is..inclined to take a panoramic view of the news, more inclined to pundit. 1967 *Punch* 4 Jan. 2/2 Take Alfie Hinds, currently punditing his head off as the BBC's escapological correspondent.

punditry (pʊ·nditri). [f. PUNDIT + -RY.] The characteristics of a pundit; opinions or actions befitting a pundit (sense b).
1926 T. M. HEALY in *Pioneer Ref. Spelling* Apr. 14, I..decry the punditry of Civil Service Commissioners in making so-called orthography a test subject. 1930 *Times Lit. Suppl.* 13 Nov. 932/2 His latest book..blends a good deal of punditry with its collectors' gossip. 1948 J. STEINBECK *Russ. Jrnl.* i. 3 News has become a matter of punditry. A man sitting at a desk in Washington or New York reads the cables and rearranges them to fit his own mental pattern and his by-line. 1958 *Oxford Mag.* 13 Feb. 278/1 All Oxford seemed to be there, with its focus in the ebullient punditry of Sir Isaiah Berlin. 1966 *Listener* 15 Sept. 397/3 It pounded and explored the whole subject of South Africa: past, present, future. Plenty of instant punditry. 1978 *Bull. Amer. Acad. Arts & Sci.* Jan. 22, I have caught him in a moment of punditry and while he was yielding to a weakness common to critics.

pundonor. (Later examples.)
1932 E. HEMINGWAY *Death in Afternoon* xviii. 210 His..body contains enough valour and pundonor to make a dozen bullfighters. 1967 MCCORMICK & MASCAREÑAS *Compl. Aficionado* ii. 45 That outcry was fascinating in its vehemence, for it had the false accents of him who protests too much; one is tempted to think that it could occur only in a land where *pundonor* is on every tongue. 1968 *Medium Ævum* XXXVII. 46 Nor do I think we have a right to assume that the sculptors followed epic pundonor so closely.

pung, *sb.*[2] For *U.S.* read *N. Amer.* and add: **a.** (Later, including Canad., examples.) Also *attrib.*
1886 [see *CLIP *v.*[2] 7]. 1908 L. M. MONTGOMERY *Anne of Green Gables* xix. 208 Her cousins are coming from Newbridge with a big pung sleigh. 1951 E. M. GRAHAM *My Window looks down East* iv. 31, I saw Carl Urlichson bringing home his fishing boat from the shore on sled runners drawn by a big Belgian horse... But I saw more than a boat on a pung. 1952 E. BUCKLER *Mountain & Valley* 171 The pung races on the lake. 1953 *N.Y. Times* 24 July 15/7 Carriages, buckboards and wagons were employed and pungs, sleds with box-like bodies, were popular.

pung (pʊŋ), *v.*[2], *sb.*[3], and *int.* *Mah Jong.* [Chinese.] **A.** *vb. intr.* To take a discarded tile in order to complete a triplet of identical tiles. Also *trans.* **B.** *sb.* A set of three identical tiles; also, the action of the verb. **C.** *int.* The call made by the player performing this action. So pu·nging *vbl. sb.*
1922 R. E. LINDSELL *Ma-Cheuk or Mah-Jongg* 12 A useless domino..is placed face upwards in the middle of the table, and can at once be claimed by any of the other three players who has already a pair or threes of that particular domino. This is called 'parking' or 'punging'. 1923 J. P. BABCOCK *Rules for Mah-Jongg* (ed. 2) ii. 15 Should a tile be discarded and any player have a pair (or three) of this same tile, even though out of his own turn, he may 'Pung', that is he says 'Pung' and takes this discarded tile, placing it with the pair (or three) from his own hand face up in front of him on the table... A Pung which completes a hand takes precedence over any other Pung. 1925 B. TRAVERS *Mischief* v. 86 Louise came in, all fatigued and heated from harbouring red dragons and punging her opponent's wind. 1934 *Neuphilologische Mitteilungen* XXXV. 132 Mah-jongg..*pung* 'set of..three identical tiles'. 1960 R. C. BELL *Board & Table Games* vi. 155 If any of the other three players holds two tiles identical with one just discarded, he may call 'Pung!' and take it out of the pool. Only the last discarded piece may be punged. 1964 E. N. WHITNEY *Mah Jong Handbk.* I. iii. 28 If you have a pair in your hand and *any* player discards an identical tile, you may declare 'pung' and claim the discarded tile instead of drawing from the wall. *Ibid.* 169 *Pung*, claiming a discard that completes a triplet. After punging, the player must meld his completed triplet.

punga, var. *PONGA.

punge, var. *PUNJI.

pungle (pʊ·ŋg'l), *v.* *U.S. colloq.* 9 **pongale, pungale.** [ad. Sp. *póngale* put it down, f. *poner* put, give.] *trans.* and *intr.* To contribute, hand over, or pay. Usu. with *down* or *up*.
1851 *Alta Californian* 19 July 2/3 A singular genius..was 'pongaling down' huge piles of gold at a monte table. 1854 *Pioneer* (San Francisco) Apr. 237 An additional slice of territory and its consequent classical influence upon our language, by the introduction of such precious words..as 'hombre',..'pungle', *et id omne genus*. 1857 *San Francisco Call* 6 Jan. 2/2 'Pungale down, gentlemen; come, pungale', as the vingt-et-un lady used to say. 1884 'MARK TWAIN' *Huck. Finn* v. 33 'I'll ask him; and I'll make him pungle, too, or I'll know the reason why'... Next day he went to Judge Thatcher's and..tried to make him give up the money. 1910 E. S. FIELD *Sapphire Bracelet* xii. 141 I'll have him arrested, and then make him pungle up something handsome before I'll agree not to appear against him. 1959 A. K. LANG in *Alfred Hitchcock's Mystery Mag.* Feb. 71/1 The pusher couldn't pungle up Skreen's three hundred. 1975 J. GORES *Hammett* (1976) xix. 130 Hammett had coffee and pungled up the required fifty cents.

pungo, var. PUNGY in Dict. and Suppl.
1854 W. G. SIMMS *Southward Ho!* iii. 28 Their most innocent name is 'pungo'—a sort of schooner, hailing mostly from Manhattan and Massachusetts. *Ibid.*, For the better oysters..the 'pungos' pay three shillings. 1938 'J. DIGGES' *Bowleg Bill* 24 Next morning the whole harbor is cluttered up with dories, pungoes, and anything down to harness-casks.

pungy. Add: (Earlier and further examples.) Also *attrib.*
1876 T. WESTCOTT *Centennial Portfolio* 32/2 There are models of fish-hatching houses, of fishing-rafts..and also of the oyster-catching material, vessels, pungys, canoes, drags, rafts, etc. 1884 *Forest & Stream* 24 Jan. 526/2 The model is of a round futtock with but little deadrise, with round stern, sharp, or what we call a pungy stern, but a sharp stern like that of a little canoe costs the least. 1939 *Sun* (Baltimore) 4 Apr. 12/7 The pungy was a keel boat with no centreboard. 1941 M. V. BREWINGTON *Chesapeake Bay Bugeyes* 30 Not more than a dozen square stern, or 'pungy bugeyes', seem to have been built. *Ibid.* 34 Up to about 1908 many of the bugeyes were painted 'pungy style' with dark green bends, white rail and 'flesh' colored sides. *Ibid.* 46 The ordinary pungy rig, with sharply raking masts, round mastheads, iron rod cross-trees, upper and lower cap, and with the topmast not fidded but resting in the lower cap. 1942 *Sun* (Baltimore) 28 Dec. 18/5 Several marine railways with tugs..pungies or bateaux. 1950 *Ibid.* 23 Aug. 20/5 The Old Pungy Wave, last of her type, has arrived safely in Detroit, which will be home to her for some while to come. 1967 L. S. TAWES *Coasting Captain* 4 In the fall of 1868 I left the farm and went dredging as cook on a pungy boat.

Punic, *sb.* **3.** (Earlier and later examples.)
1673 J. RAY *Observations Journey Low-Countries* 308 The language of the Natives [of Malta] is a corrupt Arabic or Moresco, introduced by the Saracens, the ancient language before their coming in probably having been Greek, with a mixture of Punick. 1971 S. E. MORISON *European Discovery Amer.: Northern Voy.* i. 11 Phoenician script is so simple that, as with the later Norse runes, it is easy for an overimaginative searcher to read Punic, like Runic, in natural grooves and scratches on rocks.

punily, *adv.* (Later examples.)
1906 BELLOC *Hills & Sea* 56 In Africa, where men build so squat and punily. 1942 W. FAULKNER *Go down, Moses* 193 That doomed wilderness whose edges were being constantly and punily gnawed at by men with plows and axes.

punish, *v.* Add: **B. 3.** (Earlier and further examples.) Also, to abuse (a musical instrument) by playing it badly.
1801 *Sporting Mag.* XIX. 62/2 This desperate contest, comprising sixteen rounds, lasted twenty-one minutes, and we never witnessed a man more *punished* than Burk. 1807 R. SOUTHEY *Lett. from England* III. lxxi. 310 When the [boxing] champion..comes off victor, after suffering much in the contest, he is said to be *much punished*. 1844 DICKENS *Martin Chuzzlewit* xxxvi. 425 Tom, taking up his knife and fork again... 'I shall punish the Boar's Head dreadfully.' 1864 *Sporting Mag.* XLV. 194 When the Eleven come to the wickets, how they punish the ball, and rapidly run-up a long score. 1891 W. G. GRACE *Cricket* xi. 312 It was a treat to watch him punish the bowling. 1930 *Morning Post* 16 July 11/5 Chapman batted remarkably well. He refused to take any risk and yet punished the loose ball. 1934 DYLAN THOMAS in *Listener* 24 Oct. 691/2 Especially when the October wind With frosty fingers punishes my hair. *Ibid.*, Especially when the October wind..With fist of turnips punishes the land. 1942 J. B. PRIESTLEY *Black-Out in Gretley* vii. 149 The bottle of brandy they'd punished was prominent on the little table. 1949 'J. TEY' *Brat Farrar* iv. 34 Ungainly women in unseemly clothes punishing the saddles of broken-spirited horses. 1967 *Observer* 17 Dec. 1/1 An old man punishing a mandolin in Bond Street.

punishable, *a.* Add: **b.** (Later examples.)
1909 *Daily Chron.* 29 Nov. 3/1 If a punishable play is produced, the author and the lessee..should be punished. 1959 *Daily Tel.* 31 Dec. 1/7 Confiscation of the latest issue of Reichsruf, weekly newspaper of the extreme Right-wing German Reich Party, and another party journal,..was ordered by a Bielefeld court today. They were stated to have 'punishable contents'.
c. In sense 3 of the verb.
1910 *Blackw. Mag.* July 106/2 The punishable [ball] escaped scot-free.

punisher. Add: **b.** (Earlier example.) Also *Cricket*.
1812 P. EGAN *Boxiana* 113 The *Lobster* had most powerful *claws*, and was a first-rate *punisher*. 1846 W. DENISON *Cricket: Sketches of Players* 12 If the bowling be at all loose, he is a powerful punisher.

punishing, *ppl. a.* Add: **b.** (Earlier and further examples.)
1811 *Sporting Mag.* XXXVIII. 184/1 The punishing right hand of his adversary. 1846 W. DENISON *Cricket: Sketches of Players* 15 As a batsman he has a good defence, and occasionally is found to be a punishing hitter. 1900 P. F. WARNER *Cricket in Many Climes* v. i. 178 Murray Bisset..is a very sound and stylish bat with a good deal of wrist work, but is perhaps just a little short of punishing power.
c. Excessively severe and exhausting; scarcely tolerable.
1971 N. STACEY *Who Cares?* xiii. 216, I was still involved in a punishing fourteen-hours-a-day programme at Woolwich. 1973 A. ROSS *Dunfermline Affair* 180 He studied his *alter ego* almost every minute of every day for a punishing three years. 1975 *Nature* 6 Nov. 91/2 Barnes had considerable skill as a laboratory bench worker but little opportunity to exercise it in his later years because of the punishing load of advisory and administrative work that he willingly undertook. 1977 *Church Times* 14 Jan. 6/3 The OPEC countries..have been giving real aid to some of the poorer countries which has more than equalled the punishing rise in the price of oil

punishment. Add: **1. b.** *Psychol.* Pain, deprivation, or other unpleasant consequence imposed on or experienced by an organism responding incorrectly under specific conditions so that, through avoidance, the desired learning or behaviour becomes established. Cf. *REWARD *sb.*[1] 4 f.
1907 R. M. YERKES *Dancing Mouse* vi. 99 In general, the method of punishment is more satisfactory than the method of reward, because it can be controlled to a greater extent. 1912 [see *REWARD *sb.*[1] 4 f]. 1949 WOODWORTH & MARQUIS *Psychol.* xvi. 530 Punishment has two important effects. When the child gets a burn from a hot radiator he learns to *avoid* the radiator. When anyone in following a certain lead to his goal meets with punishment..he tends to *shift to another lead*. 1953 B. F. SKINNER *Sci. & Hum. Behav.* xii. 185 In solving the problem of punishment we simply ask: What is the effect of withdrawing a positive reinforcer or presenting a negative. 1956 *Sci. Amer.* Oct 116/3 This finding contradicts the long-held theory that strong excitation in the brain means punishment. 1960 L. M. BAKER *Gen. Experim. Psychol.* xiii. 309 Punishment of undesired behavior soon led to escape activity and satisfactory conditioning in the dog. 1975 FISCHER & GOCHROS *Planned Behav. Change* iv. 54 No positive reinforcement whatsoever is provided for that behavior (as distinct from negative punishment). *Ibid.* 56 When..there is a *decrease* in the probability that the behavior will occur, that stimulus is a negative reinforcer, and the operation is called positive punishment.
2. a. (Earlier and later examples.)
1811 *Sporting Mag.* XXXVIII. 140/1 Silverthorne, with timidity, arising no doubt from punishment in the first round, kept away from his adversary. 1829 P. EGAN *Boxiana* 2nd Ser. II. 97 Burns was not reduced by the punishment he received. 1846 W. DENISON *Cricket: Sketches of Players* 24 The batsman makes up his mind that he shall administer severe punishment. 1929 *Morning Post* 13 July 11/1 Too cautious play, during which loose stuff escaped punishment. 1930 *Daily Express* 6 Oct. 11/5 He took most of the punishment in the first round. 1949 'J. TEY' *Brat Farrar* viii. 69 Now he looked stupid, like a boxer who is taking too much punishment.
b. Of materials, machinery, etc.: excessive use or rough handling.
1930 *Engineering* 11 Apr. 473/3 Steel from which a boiler tube is manufactured should have the following qualities:—It must be capable of withstanding severe punishment during manufacture and also when being rolled into a tube plate and belled. 1955 *Times* 17 May 18/3 Only the finest film-strength oils can withstand the punishment a tractor engine receives.
3. *attrib.* (Further examples.)
1916 W. OWEN *Let.* 14 Mar. (1967) 385 Inspections, punishment parades, & more inspections. 1946 R. CAMPBELL *Talking Bronco* 16 Rather a punishment-parade For friend and enemy alike. 1958 J. TOWNSEND *Young Devils* vi. 54 In one year more than five hundred canings had been officially entered in the punishment book. 1968 L. BERG *Risinghill* 61 Risinghill, like every school, has a Punishment Book.

punitive, *a.* Add: **a.** (Later examples.) Also, in weakened sense: injurious in such a way as to have a deterrent effect.
1959 *Listener* 23 July 154/2 The double, *over* the bid, continues to be punitive. 1959 *Spectator* 14 Aug. 188/1 Punitive box-office taxation. 1973 *Black Panther* 31 Mar. 4/3 Brother Cleophus has been in punitive segregation (lock-up) since the strike. *Ibid.* 20 Oct. 6/1 Punitive damages are assessed only when the judge believes that a defendant has acted deliberately and with malice. 1978 *Lancashire Life* Mar. 107/1 A situation could now develop whereby through punitive financial strategies the independent sector could be so reduced that eventually it could be held to be of no social significance.

Punjab (pʊ·ndʒab, -dʒǭb). Also **Punjaub.** [Hindi *Panjāb*, f. Pers. *panj* five + *āb* water (see below).] The name of an extensive region of the Indian sub-continent, so called from its five rivers, now divided between India and Pakistan, used *attrib.* of its products. **Punjab head**: see quot. 1949.
1907 *Yesterday's Shopping* (1969) 205/2 Punjab Baskets. Nest of 6 baskets, complete—3/0. 1910 *Practitioner* Jan. 18 When a man begins to worry it is time for him to go home, and that applies if he suffers from the so-called Aden or Burma or Punjaub or Madras head. 1949 PARTRIDGE *Dict. Slang* (ed. 3) (Addenda) 1145/2 *Punjab head, have a*, to be forgetful; *Punjab head*, forgetfulness.

Punjabi, Panjabi (pʊndʒā·bi, pan-, -dʒǭ-bi), *sb.* and *a.* Also **Penjabi, Punjabee, Punjaubee.** [f. prec. + *-I.] **A.** *sb.* **a.** The Indo-Aryan language spoken in the Punjab.
1801 *Asiatick Researches* VII. 230, I allude to the *Penjábi* and to the *Brij-bhákhá*. 1838 *Jrnl. Asiatic Soc. Bengal* VII. 711 The Sikhs..carried their hatred..to such an extent as to substitute a vocabulary for their native Punjábi. 1854 L. JANVIER *Dict. Panjábi Lang.* p. iv, The character here adopted, and ordinarily used in writing Panjábi, is that known as the *Gurmukhí*. 1862 R. G. LATHAM *Elem. Compar. Philol.* 219 The following..gives a rough sketch of the grammatical character of the Punjabi. 1921 *Outward Bound* Feb. 74/1 They spoke only Punjabi, of which at that time I knew but three words. 1950 D. JONES *Phoneme* xvi. 84 Such a language as Panjabi which makes use of essential word-tones. 1964 S. K. CHATTERJI in D. Abercrombie et al. *Daniel Jones* 409 In a surrounding group of languages like Panjabi (or Eastern Panjabi), Hindki (Western Panjabi or Lahndi),..and the Himalayan dialects.., we have these new substitutes for aspirates. 1979 *Trans. Philol. Soc.* 192 The very mixed character of the language of the *Ādi Granth*..should not be allowed to obscure the fact that these scriptures provide the best available evidence for Old Panjabi.
b. A native or inhabitant of the Punjab.
1846 *Hist. Punjab* I. ii. 36 In the plains, Patans..are mixed with Jats and Cathis, who compose the bulk of the Punjabis, properly so called. 1878 G. SMITH *Life J. Wilson* xvii. 547 Nanuk, the herd-boy, was the Punjabee or Sikh. 1897 *Daily News* 21 Sept. 5/3 The brunt of the attack fell upon the portion of the camp which was held by the 5th Punjabis. 1969 *Hindu* (Weekly Mag.) 3 Aug. p. iii/1 One of the contributory factors enabling the average Punjabi to drink more milk than his counterpart in other States in the country. 1973 *Guardian* 16 Apr. 5/6 Punjabis are noted for their enterprise—and also for meanness.
B. *adj.* Of or pertaining to the Punjab.
1810 *Asiatick Researches* XI. 277 Nánac, according to *Penjábi* authors, admitted the *Hindú* doctrine of metempsychosis. 1812 W. CAREY *Gram. Punjabee Lang.* p. iv, The Punjabee language is confessedly of mixed origin. 1851 H. B. EDWARDES *Year on Punjab Frontier* I. 30 When all your fat Punjabee dogs are panting in vain after the hare. 1864 *Athenæum* 5 Nov. 597/1 To keep our regular troops..at a strength more than sufficient to render utterly harmless all the turbulent elements of Punjaubee Society. 1886 MRS. EDWARDES *Mem. Sir H. B. Edwardes* II. 315 Before landing at Calcutta, a true Punjabee welcome met him. 1921 *Outward Bound* May 27/1 Ever since I was a boy..these Punjabi lyrics have kept haunting me. 1948 A. TOYNBEE *Civilization on Trial* v. 88 The sonorous Panjabi names of stricken fields in the Anglo-Sikh wars. 1968 *Listener* 26 Dec. 844/3 Surely it's absurd to say..that a child of a..Punjabi family living in a Punjabi community, is just as much an English child as your children or mine. 1971 *Shankar's Weekly* (Delhi) 4 Apr. 4/1 It has been the vain task of the Punjabi elite of West Pakistan to hold down a section of Bengalis as their serfs.

‖ **punji** (pʊ·ndʒi), *sb.* Also **panja, panji(e, punge.** [Origin unknown: prob. from a Tibeto-Burman language.] A sharpened (freq. poisoned) bamboo stake set in a camouflaged hole in the ground as a trap for enemy soldiers (or occas. for animals). Freq. *attrib.* in *punji stake, stick*. So pu·nji *v. trans.*, to fortify with punji stakes; pu·njied *ppl. a.*
1872 E. DALTON *Descriptive Ethnol. Bengal* 11 They [*sc.* the Singphos] are skilled in fortifying naturally difficult positions, using freely the 'panja', a bamboo stake of different lengths sharpened at both ends and stuck in the ground. 1876 R. G. WOODTHORPE in H. L. Thuillier *Gen. Rep. Topogr. Surveys India 1874–5* 60 Two nasty panjied ditches to be crossed. *Ibid.* 61 Steep approaches, very thickly planted with 'panjees'. 1878 G. P. SANDERSON *Thirteen Years among Wild Beasts of India* xvii. 233 Until 1870 this distant abode of the British Lion [*sc.* Tura in the Garo Hills] was defended by a stockade..whilst the neighbourhood was pleasantly *panjied*. The uninitiated may imagine that this *panjieing* is some ornamental arrangement of the grounds, so I must explain that panjies are..a device for..the discouragement, of visitors. They consist of bamboo spikes driven into the ground, almost level with the surface, the earth being scraped away round each so as to form a cup. Hundreds of these are laid in every direction; grass, falling leaves, &c., soon hide them; and if trodden upon they inflict fearful wounds. 1923 *Blackw. Mag.* May 580/2 We had warned the men against 'booby-traps'... Only one man fell into one and got a *panjie*, luckily not poison-tipped, through his leg. 1927 *Ibid.* June 819/1 Others were planting sharp-pointed panjis in the undergrowth round the village to impale the enemy as they rush to the assault. 1950 J. H. WILLIAMS *Elephant Bill* iii. 47 The only effective fence against elephants is what is called the punge... The punge fence, or trap, is made of..sharpened and lightly roasted..bamboo stakes of varying length. 1966 *Time* 4 Feb. 18 Children helped to fashion the village's huts and whittled vicious punji stakes of bamboo. 1969 I. KEMP *Brit. G.I. in Vietnam* iii. 50 Two casualties for evacuation; one had stepped on a *punji* stake, the other had stopped a sniper's bullet. 1973 R. HAYES *Hungarian Game* xxxvi. 216 A workbench directly beneath the window. The thing was loaded with tools; it would be like jumping into a punji pit. 1977 *Time* 21 Nov. 8/1 Three camps in Thailand..which have been set up for some of the thousands of refugees who have run the gauntlet of mines, snipers and *punji* stick booby traps along the frontier to reach freedom.

punk, *sb.*[1] (Further example.)
1928 A. HUXLEY *Point Counter Point* xxix. 478 It amused him to hear the cast-off locutions of duchesses in the mouth of this ageing prostitute... The poor superannuated punk was so gruesome.

punk, *sb.*[2] Add: Also 8 **Punck.**
1. a. (Earlier and further examples.)
1705 R. BEVERLEY *Hist. & Present State Virginia* III. 49 Or else they take Punck, (which is a sort of a soft Touchwood, cut out of the knots of Oak or Hiccory Trees, but the Hiccory affords the best). 1792 J. BELKNAP *Hist. New Hampshire* III. 94 They [*sc.* the Indians] raised a blister by burning punk or touchwood on the skin. 1866 LINDLEY & MOORE *Treas. Bot.* II. 941/2 Punk. Touchwood or vegetable tinder. 1923 J. H. COOK *Fifty Years on Old Frontier* 15 Each man carries a flint and steel, together with a piece of punk. 1924 *Jrnl. Polynesian Soc.* XXXIII. 155 The Maori..carried live fire. To do so he procured..dry material of slow combustion... A kind of punk that grows on trees, [was] used for this purpose. 1936 M. FRANKLIN *All that Swagger* iii. 29 Should the fire die out, a greasy rag, ignited by firing it from the gun, could be applied to punk. 1956 *Te Ao Hou* (Wellington, N.Z.) July 24/2 Little has been recorded of this bracket fungus and its importance in the generation of fire and in carrying fire... In the Thames district..it was known as 'punk' to the early settlers, who learned of its use from the Maoris. 1965 *Austral. Encycl.* VII. 313/2 Such forest pathogens as *Fomes setulosus* and *Polyporus portentosus*..are often called brown punk and white punk respectively. 1972 *Science* 27 Oct. 395/2 In moxibustion only gentle warmth is allowed by the smoldering punk applied on the flesh. 1976 *Yankee* Apr. 107/1 *Wind Bird's* hull looked like a honeycomb with the intervening wood turned to punk.
b. *transf.* Something worthless; foolish or empty talk; nonsense, rubbish. *colloq.*
1869 J. M. HOPPIN *Office & Wk. Chr. Ministry* II. ii. 315 Better have the simplest..thoughts, clearly expressed, than what Carlyle calls 'phosphorescent punk and nothingness'. 1900 ADE *More Fables in Slang* (1902) 212 Well, if they are Right, then I must be Wrong, but to me it is Punk. 1927 D. L. SAYERS *Unnatural Death* xxi. 243 We..men stuff ourselves up with the idea that they're romantic and unemotional. All punk, my son. 1938 'J. BELL' *Port of London Murders* ix. 164, I told him a lot of punk about..a secret process, but..he knew it was all my eye. 1958 'A. GILBERT' *Death against Clock* viii. 109 After all, except for the kids, presents seem to me punk. 1970 'D. HALLIDAY' *Dolly & Cookie Bird* v. 73, I told him what Celeste said about Capricorn and Scorpio, and he said, 'Honestly, Sarah. You don't believe all that punk?' 1973 in *Times* 30 May 5/5, I don't like the family Stein. There is Gert, there is Ep, there is Ein. Gert's writings are punk, Ep's statues are junk, Nor can anyone understand Ein.
c. Bread. *punk and plaster*, bread and butter. *slang.*
1891 *Contemp. Rev.* Aug. 255 Bread is called 'punk' [by tramps]. 1899 'J. FLYNT' *Tramping with Tramps* I. vi. 140 Coffee, a little meat, some potatoes, and 'punk an' plaster' (bread and butter). 1925 [see *MUD *sb.*[1] 2 e]. 1961 R. P. HOBSON *Rancher takes Wife* ii. 45 Jack Lee..took a large slice of my bread, munched on it thoughtfully, and then pronounced it—'Good punk!' 1975 J. GORES *Hammett* (1976) i. 14 'Punk and plaster?' 'You bet.' The waiter picked up his tray... 'What's punk and plaster?' 'Bread and butter. Con talk.'
3. (Earlier and later examples.)
1870 M. S. DE VERE *Americanisms* 157 A Chinese lady of rank in San Francisco was attended by three maids of honor, bearing lighted sticks of punk highly perfumed. 1880 *Harper's Mag.* Dec. 73 Before the ancestral tablets..incense was consumed, punk or joss-sticks. 1953 H. MILLER *Plexus* II. viii. 12 The third night we burned Chinese punk and incense.
3*. *slang.* **a.** [This sense may be influenced by PUNK *sb.*[1]] A passive male homosexual, a catamite; a tramp's young companion or 'gunsel'.
1904 'No. 1500' *Life in Sing Sing* 251/1 *Punk*,..a pervert. 1926 J. BLACK *You can't Win* (1927) x. 129 The 'punks', young bums, were sent for 'mickies', bottles of alcohol. 1927 [see *GUNSEL 1]. 1950 [see *gay-cat* s.v. *GAY *a.* 9]. 1973 B. BROADFOOT *Ten Lost Years* x. 137 They [*sc.* hoboes]'d pick up youngsters as, well—as their playthings. These kids were called punks. 1977 *New Yorker* 24 Oct. 64/3 The involuntary homosexuals tend to be good-looking young men..forced into becoming jailhouse 'punks' by older men serving long sentences.
b. A person of no account, a worthless fellow; a young hooligan or petty criminal. Also *gen.*, as a term of contempt or abuse.
1917 [see *MUTT c]. 1928 M. C. SHARPE *Chicago May* xxxi. 287/1 *Punk*, apprentice thief. 1930 D. HAMMETT *Maltese Falcon* xviii. 216 We've absolutely got to give them a victim... Let's give them the punk... He actually did shoot both of them..didn't he? 1930 *Sat. Even. Post* 26 July 146/2 'Listen to me, you big punk!' he growled ominously. 'What do you think we are—a lot of fools?' 1933 E. HEMINGWAY *Winner take Nothing* 94 This fellow was just a punk..a nobody. 1939 C. R. COOPER *Designs in Scarlet* ii. 18 Punks like him—sixteen, seventeen, eighteen, nineteen years old. *Ibid.* iii. 37 'The punks', as youthful offenders are often called. 1940 *Sun* (Baltimore) 29 Mar. 17/4 This happens to be the Bomber's tenth defense. Most of them bums or punks? 1949 *Chicago Tribune* 10 Dec. 10 This punk must have robbed a bank or got paid off for settin' a forest fire! 1953 W. BURROUGHS *Junkie* iv. 50 Two young punks got off a train carrying a lush between them. 1959 H. NIELSEN *Fifth Caller* xiv. 207, I was a punk then... Fourteen years old and just a little punk... Then I began to fill out. I ain't a punk no more. 1963 T. PYNCHON *V.* vi. 145 There was nothing so special about the gang, punks are punks. 1964 V. S. NAIPAUL *Area of Darkness* ix. 245, I went back to the hotel. The telephone rang. 'Hallo, punk.' 1967 *Boston Sunday Herald* 30 Apr. I. 16/3 Berke has no sympathy for the 'punks' who act up in school, assault teachers or destroy property. 1976 'D. HALLIDAY' *Dolly & Nanny Bird* ix. 113 Punks give their kids a punk childhood which leads to the next generation of punks. 1978 J. UPDIKE *Coup* vi. 246 'Uh——think you've come to the wrong place. Hasn't he?'..'You bet the punk has.'
c. In show business: a youth or novice; a young circus animal. Also *transf.*
1923 *N.Y. Times* 9 Sept. VII. 2/1 *Punk*, an amateur. 1926 *Amer. Speech* I. 282 *Punk*, a baby lion or other young animal. 1926 MAINES & GRANT *Wise-Crack Dict.* 12/1 *Punk*, child in show business. 1942 *Amer. Speech* XVII. 223/2 *Punk*, a boy or any young man not yet professionally dry behind the ears. 1971 *Islander* (Victoria, B.C.) 19 Dec. 6/2 At least 71, cantankerous Lizzie's trunk had been paralyzed as the result of once having overzealously disciplined a young elephant, or 'punk'.
d. Short for (*a*) *PUNK ROCK; (*b*) *PUNK-ROCKER. (Only these senses are widely current outside the United States.)
1974 *New Yorker* 20 May 142/3, I was getting a naïve kick out of watching a woman play rock-and-roll punk. 1976 *New Musical Express* 24 Jan. 20 He's strictly for white dopes high on punks. *Ibid.* 17 Apr. 43/1 Johnny Rotten..has the makings of a good punk. 1976 *Sunday Times* 28 Nov. 37/4 Johnny Rotten and the Sex Pistols are punks. They sing 'Anarchy in the UK.' *Ibid.* 37/5 Punk will fade. Its apologists are ludicrous. 1977 *Evening News* 27 Apr. 11/1 London's growing army of punks have developed a powerful animosity for teds... For the uninitiated, punks..are the ones who match short, ragged hair with short ragged leather jackets. 1978 *Gramophone* May 1954/1 John Cale..could be branded

one of the original punks as a member of Velvet Underground. 1979 *Time* 30 July 76 The music on this record ..is full of brash challenge, like the best punk.

4. a. punk-knot (examples); **punk-oak**, for *Quercus aquatica* substitute *Quercus nigra*; (examples); **punk-wood** (later example).

1920 *Bull. U.S. Dept. Agric.* No. 871. 20 The sporophores, which I find occasionally..on old punk knots from which the original sporophores have fallen and are reviving. 1934 *Forestry* VIII. 155 One of the most characteristic and interesting symptoms of *Trametes pini* rot is the so-called punk knot, a mass of brownish friable substance that develops round embedded branch stubs. 1884 C. S. Sargent *Rep. Forests N. Amer.* 152 *Quercus aquatica*... Possum Oak. Punk Oak... Probably not used except as fuel. 1897 G. B. Sudworth *Nomencl. Arborescent Flora U.S.* 175 Common names [of the water oak] include..Duck Oak, Possum Oak, Punk Oak. 1903 S. E. White *Forest* 180 Sometimes a faint rounded shell..swelled above the level, to crumble to punkwood at the lightest touch of our feet.

b. (In sense 3* above), as *punk band, critic, fan, hater, kid, style; punk-related, -styled* adjs. **punk chic** (after *radical chic* s.v. *RADICAL *a.* 3 f), a fashionable style of design reflecting the unconventional aspects of a punk-rocker's dress or appearance; also in adj. use.

1976 *Melody Maker* 11 Sept. 37/6 Even in Britain there's a punk band called the Suburban Studs. 1977 *Whig-Standard* (Kingston, Ont.) 15 July 11/2 Wherever they go punk bands bring violence. 1977 *Time* 11 July 49/1 Of late, punk chic has even been taken up by a few high-fashion designers. 1977 *Zigzag* Aug. 12/1 ..sickest thing is the Zandra Rhodes 'punk chic' look. 1977 *Zigzag* Aug. 12/1 I dunno about fur coats but anyone who wears them Harpers and Queen punk chic outfits is in order for laying out. 1977 *Rolling Stone* 13 Jan. 20/3 Bangs and Meltzer usually know the difference but most of their followers and fellow punk critics do not. 1977 *Whig-Standard* (Kingston, Ont.) 15 July 11/2 The punk fans fight with other young groups. 1977 *Western Morning News* 1 Sept. 1/3 Mysterious posters..have appeared all over Plymouth in the hope that punk fans will read between the lines. 1977 *Sounds* 1 Jan., Have these Punk haters really thought about what they've heard? 1928 J. M. Sullivan *Criminal Slang* 17 *Punk kid*, a boy who begs and panhandles for yeggmen. 1935 *Amer. Speech* X. 19/2 *Punk kid*, an apprentice who works with only one yegg. 1930 C. R. Cooper *Designs in Scarlet* ii. 22 'We've got to kill 'em,' said a cop bluntly, 'or they'll kill us. Punk kids are dangerous.' 1939 R. Chandler *Big Sleep* xiv. 115 That punk kid?..The kid that works at the store. 1977 *Cosmopolitan* June 87/2 Small time, a punk kid, am I? 1977 *New Yorker* 23 May 102/2 When you were a nobody, a punk kid just starting out, didn't anyone ever lend you a helping hand? 1979 *Ripped & Torn* vi. 3 There's only one restriction: Punk/Punk related stuff only. OK? 1977 *Sounds* 1 Jan., Ten concise memorable songs, in the pop, punk style,..combine to form a dynamite package. 1977 *Ibid.* Certain elements in what we did..created a taste for a certain kind of thing which meant that others could organise punk styled bands.

punk (pɒŋk), *a.* orig. *U.S.* [f. Punk *sb.*²] **1.** Of timber: decayed; rotten, punky.

1902 S. E. White *Blazed Trail* ii. 18 Supplies ran low unexpectedly, trees turned out 'punk'. 1904 —— *Blazed Trail Stories* iii. 49, I cull every log, big or little, punk or sound, that ain't sawed square.

2. *transf.* Devoid of worth or sense; poor in quality; disappointing; nonsensical; 'rotten'. *colloq.*

1896 Ade *Artie* iii. 23 And this crowd up there was purty-y-y punk. *Ibid.* xix 178 They could n't be any punker's there are now. 1916 E. Wallace *Let.* 13 Nov. in M. Gilbert *Winston S. Churchill* (1972) III. Compan. ii. 1582 K. ..I wants you to do a punk interview with. Churchill. 1929 W. Heyliger *Builder of Dam* 4, I call this a punk way to spend an Easter vacation. 1943 *Amer. Speech* XVIII. 89 *Pretty punk* [in New Zealand] describes one of the work performed in America by *lousy*. 1949 *Los Angeles Times* 21 May 6 This jail has about as punk, it was the punkest grub I ever packed away. 1972 *Maclean's Mag.* Mar. 46/2 When my uncle became mayor of Sherbrooke he spoke a pretty punk French.

3. *Comb.*, as **punk-ass** *a. slang* [Ass (= Arse *sb.*)], of a person: worthless, good-for-nothing.

1973 J. Wambaugh *Blue Knight* (1973) vii. 102 A kid, a punk-ass kid, conned me. 1977 *Zigzag* Aug. 12/2 This period of court harassment..went on until July 25th, when I was locked up for good by punk-ass Colombo in Detroit.

punk (pɒŋk), *v. U.S. slang.* [f. *Punk *sb.*³ 3*] *a.*] *intr.* To back out; to withdraw one's support, to quit.

1920 E. Pound *Let.* 11 Sept. (1971) 157 You lay back, you let me have the whole stinking sweat of providing the mechanical means for letting through the new movement. ..Then you punk out, cursing me for not being in two places at once, and for 'seeing no alternative to my own groove'. 1956 'E. McBain' *Cop Hater* (1958) x. 92 We never punk out, but we never go lookin' for trouble, either. 1959 H. Salisbury *Shook-Up Generation* iv. 65 The Chimp, unfortunately, has a tendency to 'punk out' when the fighting gets tough. 1972 C. M. Fuller in W. King *Black Short Story Anthol.* 145 Where was he? She couldn't believe he *punked out* and stayed away from school. 1977 *Zigzag* Aug. 12/1 Holzman punked out, even after he told us that he wouldn't change after all.

punkah, punka. Add: **3.** *punkah-puller*; **punkah-wallah** (earlier example).

1899 Conrad *Lord Jim* (1900) v. 39 The archway from the anteroom was crowded with punkah-pullers, sweepers, police peons, the coxswain and crew of the harbour steam-launch. 1959 *Listener* 13 Aug. 234/2 The Madras Government was still employing 714 'punkah-pullers'. 1857 A. Case *Let.* 6 Aug. in *Day by Day at Lucknow* (1858) v. 133 Last night..a shell burst close to our door, just behind the punkah wallah.

punkie (pɒ·ŋki). *W. Country dial.* [Perh. var. of Punkin, itself a var. of Pumpkin + -ie.] A lantern made by setting a candle in a hollowed-out mangel or similar vegetable. Punkie night (see quots. 1931, 1960).

1931 *N. & Q.* 21 Nov. 372/2 At Hinton St. George, and in the neighbouring village of Lopen, mangolds, scooped out and fitted with candle ends to form lanterns, are known as 'punkies'. During the parade of punkies, the following lines are sung: 'Tis Punkie night to-night, 'Tis Punkie night to-night. Adam and Eve, they won't believe 'Tis Punkie night to-night.' This is said to be a century-old custom..based upon the fact that a party of men from Hinton and Lopen visited Chiselborough Fair..and did not return to their homes as early as promised. Their wives went in search of them, and the attractions of the fair..were so great that by the time they commenced the journey home, their lamps were innocent of oil, and improvised lanterns, made from mangold-wurzels..were utilised. 'Punkie Night' has been once more revived at Hinton St. George. *Ibid.* 26 Dec. 465/2 'Punkie night'..is on Oct. 31, the evening before All Saints' Day, when it is generally celebrated in America as Hallowe'en. 1959 I. & P. Opie *Lore & Lang. Schoolch.* xii. 267 To children in south Somerset a punkie is a home-made mangel-wurzel lantern of more artistic manufacture than those commonly made elsewhere for Hallowe'en. 1960 *Guardian* 10 Nov. 10/2 In Somerset 'Punkie Night', during the third week in October, is a great occasion in the little village of Long Sutton... Punkies..are made from hollowed-out mangels to resemble a face. 1972 *Folklore* LXXXIII. 240 A punkie is a general term for a mangold, turnip or similar vegetable that has been hollowed out, a face marked through the skin, and a candle placed inside. *Ibid.*, Punkie Night was in early November.

punkie, var. Punky *sb.* in Dict. and Suppl.

punkin: see also *pu·nkin.

punk rock (pɒŋk rɒk). [f. *Punk *a.* 2 + *Rock *sb.*⁵] A loud, fast-moving style of rock music characterized by aggressive and deliberately outrageous lyrics and performance. Also *attrib.*

1971 D. Marsh in *Creem* May 43/3 He's [*sc.* Rudi Martinez is] doing the knee-drop, and the splits and every other James Brown move. He's the only one in punk-rock who's still got 'em and he's makin' a comeback. 1974 *Village Voice* 1 June 45/1 (*heading*) When punk rock met the Velvetoo. 1975 *Fusion* Jan. 47 Punk rock top 20. 1975 *New Yorker* 26 May 6/2 On Tuesday, May 27, Manhattan Transfer returns and begins doing its thirties, forties, and fifties routines opposite Canadian punk-rock singer Lewis Furey. 1976 *New Musical Express* 21 Feb. 31/2 A quartet of spiky teenage misfits..playing 60's styled white punk rock. 1976 *Melody Maker* 14 Aug. 25/3 Now we have gone full circle. Critics groan: 'Bring back incompetent, illiterate punk-rock bands, and away with these boring old practising musicians!' 1977 *Gramophone* June 10/2 'Punk rock' groups (or 'new wave', as they are euphemistically called). 1979 *Fortune* 23 Apr. 59/2 At one extreme are hard rock, acid rock, and more recently punk rock—all characterized by the souped-up, violent sound of blaring electric guitars.

Hence **punk-ro·cker**, one who plays or admires punk rock.

1976 *Sunday Times* 28 Nov. 37/3 Punk-rockers hate Mick Jagger (also, Led Zeppelin, Yes and Genesis) as much as they hate politics. 1977 *Sounds* 9 July 36/1, I cannot accept John Peel's suggestion that punk rockers are the only truly socialist representatives we have left. 1978 *Jrnl. R. Soc. Arts* CXXVI. 197/2 What a moral price has been paid in the name of fashion—unhealthy tight corsets, leopard-killing skin coats, the broken promise of false bosoms and the pierced faces of the punk-rockers.

punkster (pɒ·ŋkstəɹ). [f. *Punk *sb.* 3*d + -ster.] = *Punk-rocker.

1977 *New Musical Express* 12 Feb. 20/1 From precocious Nazz punkster to It's-All-My-Own-Work wishful thinker and dabbler. 1977 *Oxford Times* (City ed.) 23 Sept. 6/1 An even more disturbing facet of the punksters—having the nose or cheeks pierced—does not seem to have caught on.

punky, *sb.* Add: The usual spelling now is *punkie*. (Earlier and later examples.)

1769 R. Smith *Tour Four Great Rivers* (1906) 42 We begin to be teazed with Muscetoes and little Gnats called Punkes [*sc.* in New York]. 1903 [see *no-see-em]. 1933 F. H. Chetley *Camping Out* 423 The 'punkeys' and 'midges' can outstrip them [*sc.* mosquitoes] for ferocity and the painful character of the wound which they inflict. 1957 *Biol. Abstr.* XXXI. 1335/1 Punkies of the genus *Culicoides* are an important source of annoyance to personnel engaged in outdoor occupations over much of southern and central Alaska. 1964 Gordon & Lavoipierre *Entomol. for Students of Med.* xxii. 148 Those members of the family Ceratopogonidae which suck blood are known to entomologists as 'biting midges', but in some parts of the world they are given local names, such as 'punkies'.

punky, *a.* Add: **a.** (Earlier and later examples.)

1872 W. S. Huntington *Road-Master's Assistant* 117 A bridge may..have a small knot partially decayed, or 'punky', as it is termed. 1903 [see Pitchpoll, -pole *v.* b]. 1955 *Sun* (Baltimore) 9 Aug. 16/8 All of the punky and weak wood was removed and a glass and resin putty applied to smooth out the sleek lines. 1958 *N.Z. Timber Jrnl.* Feb. 62/2 *Punky*, applied to wood showing signs of decay, or to soft spongy heartwood. 1959 E. Collier *Three against Wilderness* iii. 33 She was scurrying around ..gathering punky chunks of wood for the smudge. 1968 C. Helmericks *Down Wild River North* ii. xxiv. 386 We are seated..around a smudgy, punky fire.

b. *transf.* and *fig.* = *Punk *a.* 2.

1886 *Harper's Mag.* Dec. 105/2 George's mother's folks did have a kind of a punky spot somewhere in their heads. 1904 *N.Y. Times* 5 May 8 Written by another man Mr Austin would doubtless find these verses as amusing as the rest of us do—would appreciate their punky pretentiousness. 1926 F. Rickaby *Ballads* 63 Were you punky, were you hollow, You had been a lucky fellow. 1979 *Maclean's Mag.* 2 July 38/2 He doesn't expect to find his punky pubescents through conventional methods.

punnet². Delete *local* and add further examples. Now also a container of other materials and shapes.

1922 Joyce *Ulysses* 289 Punnets of mushrooms. 1943 H. J. Massingham *Men of Earth* ii. 16 Some gave a pound for a punnet of strawberries, others a pound for a punnet of tomatoes. 1955 [see *chip *sb.*¹ 4 b]. 1971 *Morning Star* 13 July 2 Whole families prepared to descend on the regimented lines of strawberries and fill thousands of punnets. 1975 *Times* 27 Nov. 13/7 A marvellous new filler..(80p a punnet), which claims to fill everything from plaster to concrete.

punny (pɒ·ni), *a.* [f. Pun *sb.*¹ + -y¹.] Consisting of, or characterized by, a pun or puns. Hence **pu·nnily** *adv.*

1961 Webster, Punny. 1974 R. Quirk *Linguist & Eng. Lang.* vii. 116 A confession, punnily derivative from Hannah More. 1976 *Glasgow Herald* 26 Nov. 5/7 Chris McClure..is known nowadays as Christian which provides the punny title of his impending album release, 'The First Christian'. 1977 *Time Out* 28 Jan.–3 Feb. 17/2 Risqué, funny and predictably punny. 1979 *Daily Tel.* 6 Feb. 15/6 (*caption*) 'I love ewe.' This funny punny card is one of a series of Valentine postcards done by cartoonist Haro.

punster. (Further examples.)

1965 W. S. Allen *Vox Latina* 107 In the sixteenth century we find punsters identifying e.g. *habitaculum* with French *habit à cul long*. 1978 *Detroit Free Press* 16 Apr. 14C/1 The latest from the most outrageous living punster, bad jokester and molester of the language.

punt, *sb.*¹ Add: **1. b.** [f. Punt *v.*²] A push with a punt-pole.

1897 *Geogr. Jrnl.* IX. 12 Only practice enables one..to guide the raft by means of timely punts at the surrounding rocks with the pole with which one is armed.

3. punt-gun, -gunner (later examples).

1958 L. Durrell *Mountolive* xvi. 302 Time for the..tuning in of the long punt-guns. 1972 *Shooting Times & Country Mag.* 4 Mar. 21/2 A pair of fine Welney Wash punt guns. 1956 C. Willcock *Death at Flight* xv. 202 'Wire cartridges,' he said. 'Punt-gunners use these to get greater range.' 1971 *Country Life* 28 Oct. 1129/2 Famous alike as punt-gunner, eel-fisher, mole-catcher and skater.

punt, *sb.*³ Add: **1.** (Further examples.) Also in other varieties of football.

1876 *World* (N.Y.) 19 Nov. 3/4 Princeton..now played all together on the ball, the captain himself being instrumental, with a good punt, in securing the second goal. 1887 H. Hall *Tribune Bk. Open-Air Sports* 125 A goal may be won..by kicking the ball..over the cross-bar of the goal of the defence, except by a 'punt'. 1921 [see *fly-kick* s.v. *fly *sb.*² 8]. 1941 *Daily Progress* (Charlottesville, Va.) 14 Jan. 11 A player can elect to run back a punt from scrimmage if the ball is caught in the end zone. 1965 *Sun-Herald* (Sydney) 4 July 51 Denis Aitken won the long-distance kicking competition with a punt of 66½ yards. 1975 *Times* 25 Aug. 9/8 A massive punt downfield from [goal-keeper] Parkes. 1979 *Arizona Daily Star* 5 Aug. 9/1 The Packers overtook the Chiefs with 12:57 remaining in the third quarter after Kansas City's Jimmy Edwards fumbled a Green Bay punt.

3. *attrib.* and *Comb.* in sense 1, as *punt return, returner*; **punt kick** = sense 1; hence **punt-kick** *v. intr.*

1876 *Sun* (N.Y.) 20 Nov. 3/1 A Harvard man redelivers the ball by a fine 'punt' kick. 1960 E. S. & W. J. Higham *High Speed Rugby* ii. 27 Be sure you can punt-kick with either foot *accurately*. 1965 *Advertiser* (Adelaide) 17 July 25 Sturt back pocket player Brenton Adcock follows through with a long punt kick at training. 1961 J. S. Salak *Dict. Amer. Sports* 345 *Punt return* (football), a planned maneuver for running back a punted ball. 1967 *Boston Sunday Herald* 14 May ii. 5/5 A kickoff and punt-return man. 1970 *Globe & Mail* (Toronto) 25 Sept. 31/7 A second injury to punt returner Bryan De-Marchi.

punt, *sb.*⁶ [f. Punt *v.*¹ (perh. infl. by Punt *sb.*³).] A bet, a gamble. Phr. *to take a punt* (Austral.), to take a chance or risk.

1965 J. O'Grady *Aussie English* 71 To 'take a punt at' anything is the equivalent of to 'have a go'. 1969 *Sydney Morning Herald* 7 June 25/9 Melbourne..selectors have 'taken a punt' in naming 20-year-old Russell Collingwood as centre half-forward. 1976 *Daily Tel.* 27 Mar. 2/3 People will still have a punt on Wimbledon. 1978 O. White *Silent Reach* xxiv. 253 Blackness and silence. So take a punt... He..eased the pencil torch out of his bag. 1979 *Ibid.* 29 Jan. 17/8 As a punt, or straightforward gamble with money that can be written off without hardship, there is some appeal in Carr Boyd Minerals.

punt (punt), *sb.*⁷ Also Punt. [Ir. = 'pound'.] The Irish monetary unit, until 1979 equivalent to £1 sterling. Also *Comb.*

1975 *Irish Times* 24 May 13/2 Do we devalue below sterling? Or do we stabilise our punt? Or maybe even attempt to revalue it upwards? 1978 *Observer* 17 Dec. 2/2 The Irish Government's decision to join the European Monetary System and break the link between Ireland's pound (now the punt) and Sterling came at the end of 10 days of hectic negotiations. 1979 *Ibid.* 8 Apr. 4/6 An advertisement in last Friday's *Derry Journal*..told readers: 'Punt holders are welcome in the North.'.. But when I was there last week the small shops in the..Bogside were taking three pence off the Punt, as the Irish pound is called, when one went to buy a packet of cigarettes.

punt, *v.*¹ Add: **c.** *to punt around*, in police slang: to patrol. Also as *sb.* in phr. *to have a punt around.*

1970 P. Laurie *Scotland Yard* 293 *Punt around, to*, to patrol. 1974 G. F. Newman *Price* ii. 58 Thought I'd have a punt around, see who's about. 1977 P. Moyes *To kill a Coconut* vii. 99 To 'punt around' is to patrol.

punt, *v.*³ Add: **1. a.** (Further examples.) Also in other varieties of football.

1905 [see *drop *v.* 24 b]. 1961 *Dallas Morning News* 10 Oct. ii. 1 He..punted once for 39 yards and caught one pass for 13 yards. 1967 *Sun-Herald* (Sydney) 16 Apr. 67 Ryan coolly punted the ball straight through the middle and Geelong had won by a point. 1972 G. Green *Great Moments in Sport: Soccer* xiii. 123 Gregg immediately punted the ball far down-field, well over the half-way line. 1974 *Plain Dealer* (Cleveland, Ohio) 26 Oct. 4-D/6 On their next offensive series, the Falcons were forced to punt.

‖ **punta** (pu·nta). [It., lit. 'point'.] The narrow upper part of straw grown in Tuscany for plaiting. Also *attrib.*

1929 *Daily Express* 26 Jan. 5/2 Rough straws and picture hats are always popular... These are being shown of openwork tuscan or punta straw. 1968 J. Ironside *Fashion Alphabet* 253 *Punta*, straw made of the upper part of the grain stalk.

punt-about. (Later examples.) Also *attrib.*

1917 Kipling *Diversity of Creatures* 255 The 'tump-tump' of the puntabouts before the sides settled to games. 1924 —— *Debits & Credits* (1926) 97 An ancient, but air-tight puntabout-ball. 1963 *Times* 10 May 4/3 A puntabout at Twickenham..afforded an opportunity to bid godspeed..to England's Rugby football team.

Punt e Mes (punt *e* me·s). Also **Punt e mes.** [It. (Piedmontese dial.), lit. 'point and a half'.] An Italian aperitif, made in Piedmont.

1956 C. G. Bode *Wines of Italy* x. 100 There are two kinds, Carpano 'Vermouth'—somewhat closer to the usual line of red vermouth, and Carpano 'Punt e Mes', with the bitter tang still more emphasized. 1959 W. James *Word-bk. Wine* 152 *Punt e mes.* The Italian vermouth, of which one sees much in Italy but little elsewhere, is of the sweet type but with a pronounced bitter after-taste. 1965 *Harper's Bazaar* Jan. 45/1 A Negroni ..equal parts of Campari, gin, and Punt e Mes. 1966 R. Ardrey *Territorial Imperative* (1967) vii. 265 You.. take a table in the islet's little restaurant. You order a *Punt e Mes* or a second-rate omelet. 1974 *Times* 14 Mar. 9/6 (Advt.), *Vermouths & Aperitifs*..Pimms..Punt e Mes..Cinzano.

punter¹. Add: **2.** Also, one who gambles on football pools.

1951 *Sport* 16–22 Mar. 22/2, I know of many punters who have decided to follow one system and then after a short losing spell switched to another system. 1976 *West Lancs. Evening Gaz.* 8 Dec. 1/7 A Great Eccleston punter has scooped £26,082 on the pools.

3. *slang.* A name for a member of various classes of criminal, esp. one who assists in the commission of a crime (see quots.).

1891 *Answers* 4 Apr. 338/1 Having filled the premises with pictures,..the auctioneer engages the assistance of what are known in the business as 'punters'. The 'punter' ..is the auctioneer's confederate, and it is his duty..to make sham bids. *Ibid.* 338/3 In addition to bidding, a 'punter' will often assist the auctioneer in cajoling the public. 1941 Baker *N.Z. Slang* vi. 52 We have also acquired [this century] some underworld slang of our own:.. *punter*, an assistant of a pickpocket who diverts the victim's attention while robbery is committed. 1973 'J. Patrick' *Glasgow Gang Observed* iii. 28 'Punter' was 'a normal man where you live who never gets caught', to whom you took stolen goods and traded them in for guns, 'blades', or money. *Ibid.* 29 They were the people who sold bottles of wine at extortionate prices on Sundays... The word was also generalized to mean a member of a 'team' or gang, as in the much used phrase 'Ya Cumbie punter'.

4. *slang.* The victim of a swindler or confidence trickster.

1934 P. Allingham *Cheapjack* xv. 187 But when the grafter decides that it is time for him to get the punter's money, he leans casually against the stall. *Ibid.* 320 *Punter*, a grafter's customer, client or victim. 1962 [see *kite *sb.* 4 c]. 1974 G. F. Newman *Price* viii. 253 They were three card tricksters. Their patter never changed, but still punters stood for it.

5. *colloq.* A customer or client; a member of an audience or spectator; *spec.*, the client of a prostitute.

In some contexts almost synonymous with *person* (but depreciatory).

1965 *Sunday Times* (Colour Suppl.) 24 Oct. 66/3 There is plenty of irrational judgement about..but like all free-market operators, the traders have to concentrate it on people—on each other and on the 'punters' (dealer buyers). 1968 D. Braithwaite *Fairground Architecture* iii. 60 Described by veteran showmen as 'a good oncer'—that is, a ride the *punters* would normally go on once—and once only—this must have been the least successful of his inventions. 1969 *Jeremy* I. iii. 22/2 *Punter*, client. 1970 *Sunday Times* 15 Mar. 60/5, I [*sc.* a prostitute] always make the punter wear a rubber. 1975 *Ibid.* (Colour Suppl.) 23 Feb. 25/2 There's nuthin' but deid punters walkin' up and doon, wi' their beds under their airms. 1975 *Times* 20 Sept. 2/7 Their clients were known as 'punters' and the youths as 'rent boys'. Some boys, once corrupted, became male prostitutes. 1976 *Sunday Mail* (Glasgow) 28 Nov. 12 (*caption*) Ya eejit! Ah'm talkin' aboot thae punters inra Housa Lords..yir upper-crust. 1977 *Record Mirror* 7 May 17/2 The punters were well pleased. Some people even..said they preferred my sound, as far as I was concerned I played crap that night. 1977 *Drive* Sept.–Oct. 112/1 The more confused you are, the more likely you are to accept his offer. Because you are the punter. 1978 *Observer* 12 Feb. 3/5 Irene, a 19-year-old prostitute, was giving the glad eye to prospecting punters on the side streets of Chapeltown, Leeds. 1980 *Ibid.* 6 Apr. 33/3 Some of the punters [*contextually* pilgrims to Lourdes] were elderly and they were all tired.

pu·nter³. [f. Punt *v.*³ + -er¹.] In various forms of football, one who punts (cf. Punt *v.*³ 1 a).

1890 in Webster. 1910 W. Camp *Bk. of Foot-Ball* viii. 314 Accuracy should be an aim of the punter as much as distance. 1956 V. Jenkins *Lions Rampant* iv. 58 Thus ..., an enormous punter as well as goal-kicker, sent them back..yards at a time. 1970 *Globe & Mail* (Toronto) 25 Sept. 31/7 Hamilton Tiger-Cats have signed University of Toronto punter and halfback Paul McKay. 1977 *New Yorker* 3 Oct. 111/1 Curry, up to now a demon punter, could not average better than twenty-nine yards a kick.

‖ **puntilla** (punti·lʸa). Also (erron.) **puntillo.** [Sp., dim. of *punto* point.] In bull-fighting, a dagger used to give the *coup de grâce* to the bull.

1838 *Q. Rev.* LXI. 419 The butchers..are able infallibly to dart the 'puntilla' into the spine. 1924 E. Hemingway *In Our Time* 23 One of the *cuadrilla* leaned out over his neck and killed him with the *puntillo*. 1932 [see *monosabio]. 1967 McCormick & Mascareñas *Compl. Aficionado* i. 21 The death of the bull at the butcher's *puntilla* was incidental to the central ritual.

Hence **puntillero** (puntilʸĕ·ro), an assistant at a bullring who uses the *puntilla*.

1910 *Encycl. Brit.* IV. 790/2 Should the bull need a *coup de grâce*, it is given by a *chulo*, named *puntillero*, with a dagger which pierces the spinal marrow. 1923 W. J. Locke *Moordius & Co.* x. 144 An indistinguished member of the quadrilla, dignified by the sonorous title of puntillero, knelt down and with a poignard gave him [*sc.* the bull] the *coup de grâce*. 1970 A. Fowles *Dupe Negative* xi. 152 On his faraway face had been the look a bull has the instant the *puntillero* strikes home.

punting, *vbl. sb.*¹ (In Dict. s.v. Punt *v.*¹) Also, *spec.* in Football Pools.

1951 *Sport* 16–22 Mar. 22/2 My advice is to make sure that the system or method of punting you adopt is a good one.

punting, *vbl. sb.*³ (In Dict. s.v. Punt *v.*³) (Further examples.)

1910 W. Camp *Bk. of Foot-Ball* viii. 313 In punting, the ball is kicked with the instep and not with the toe. 1974 *Liverpool Echo* (Football ed.) 31 Aug. 3/3 His timing and immaculate positioning, together with his straight-backed, stiff-legged punting are..faithfully reproduced by Alec. 1979 *Tucson* (Arizona) *Citizen* 20 Sept. 11D/8 UA's Barry Kramer is second in punting with a 42·3 average.

punto¹. Add: **7.** *Lacework.* Used (= Point *sb.*¹ 31) in phrases to denote various kinds of Italian lace and embroidery, as *punto a maglia* (mesh stitch); *punto a rilievo* (erron. *relievo*), (stitch in relief); *punto in aria* (stitch in the air), i.e. needlepoint lace used as a border; *punto tagliato* (cut-work); *punto tirato* (drawn-work).

1865 F. B. Palliser *Hist. Lace* iv. 47 Punto tagliato. —Cut-work. *Ibid.*, Punto in aria.—Worked on a parchment pattern, the flowers connected by brides: in modern parlance, Guipure. *Ibid.*, Punto tagliato a fogliami.—The richest and most complicated of all points, executed like the former [*sc.* *punto in aria*], only with this difference, that all the outlines are in relief, formed by means of cottons placed inside to raise them. *Ibid.* 49 Punto a maglia quadra.—Lacis; square netting, the Modano of the Tuscans. 1865 Punto a rilievo [see *gros point* s.v. *gros *a.*]. 1881 C. C. Harrison *Woman's Handiwork* i. 81 (*caption*) Old punto tirato, or Italian drawn-work. 1881 Punto a maglia [see *darned *ppl. a.*¹ 2]. 1900 E. Jackson *Hist. Hand-Made Lace* 192 *Punto a rilievo*, the Italian name for Venice Raised Point Lace. *Ibid.* 205 *Punto a Maglia.*—Lacis, or darned netting, much used for curtains and bed furniture. 1905 L. A. Tebbs (*title*) New lace embroidery (punto tagliato). 1953 M. Powys *Lace & Lace-Making* iv. 9 *Embroidered linen*, Needlepoint, Italian, 16th century. Drawn thread work, with the pattern left in the linen. Punto Tirato e Tela Lasciato. The design typical of the period. *Ibid.* 11 *Gros point de Venise*, 17th century. Heavy raised Venetian Point. The grand effect is produced by the relief which gives the look of carved ivory and it is sometimes called 'Punto Tagliato a Fogliami', lace resembling cut or carved leaves. 1960 H. Hayward *Antique Coll.* 230/1 *Punto in aria embroidery*,..used to describe a very early form of lace, the cutwork fabric reduced to a strip supporting the needlepoint work. 1974 *Encycl. Brit. Micropædia* X. 386/3 From 1620 Venetian raised lace (in Italian *punto a relievo*, in French *gros point de Venise*) developed distinct from flat Venetian (*point plat de Venise*).

punto banco (pu·nto bæ·ŋko). [In form answers to It. *punto* point, *puntare* to bet and *banco* bank: cf. also Punt *sb.*² and *banco *int.*] A gambling game resembling baccarat (see also quot. 1976¹). Also *attrib.*

1973 'R. MacLeod' *Burial in Portugal* vi. 116 The rows of tables..offering everything from roulette and boule to punto banco and blackjack. 1976 E. L. Figgis *Gamblers Handbk.* 125 Punto-Banco..is played entirely as a casino game at table with bases for 12 players, and spaces for the casino 'chips operator' between bases 6 and 7 and croupier between spaces 1 and 12. The actual banker and punter are both phantom inasmuch as all players must bet with the casino, not with each other. They may either bet 'Banco', that the phantom bank will win the coup, or 'Punto', that the phantom punter will win the coup. 1976 *Daily Record* (Glasgow) 23 Nov. 13/5 It has the biggest layout in Britain with 18 roulette, dice and punto banco tables all going under the plum-and-gold silk shades.

puny, *a.* and *sb.* Add: **A.** *adj.* **4. c.** In bad condition or health; physically weak; ailing. *U.S. dial.*

1838 K. De R. Kennedy in N.E. Eliason *Tarheel Talk* (1956) 289, I found your dear Aunt Catherine in a very puny state, not entirely confined, but obliged to rest herself on the bed more or less every day. 1866 C. H. Smith *Bill Arp* 170 Me and him like to have fit, and perhaps would, if I hadn't been puny. 1904 W. N. Harben *Georgians* xvii. 163 Little Minnie begun to fail; she got so puny she spit up ever'thing she ate. 1943 T. Pratt *Barefoot Mailman* i. 7 Don't you go making fun of sickness. Mister Dewey Durgan here has been puny the last few days and needs the best advice. 1947 *Publ. Amer. Dial. Soc.* viii. 33/1 [S.W. Ohio] *Puny*, in poor health, thin, emaciated. 1979–80 *Verbatim* Winter 14/1 [In Missouri] *puny* was 'confined to bed', *poorly* meant 'chronically ill', and *bad sick* meant it was time to call the undertaker.

B. *sb.* **4. a.** (Later *arch.* example.)

1922 Joyce *Ulysses* 386 Thou chuff, thou puny, thou got in the peasestraw.

pup, *sb.*¹ Add: **2. a.** (Later examples.)

1856 *Porter's Spirit of Times* 15 Nov. 172/3 There were three pups..a parcel of supercilious fellows, who, with a piece of glass stuck in their eye, survey the crowd as if contamination dwelt amongst them. 1870 J. K. Medbery *Men & Myst. Wall St.* 31 Down in the cock-pit the Commodore's 'pups', as the merciless, cacophonic 'street' argot denominates the broker friends of Vanderbilt, are making an ineffective rally.

b. A youthful or inexperienced person, a beginner; a young 'blood'. *colloq.* (chiefly *U.S.*).

1890 *Punch* 7 June 270/2 You ride very nicely indeed for a 'pup'. 1903 G. H. Lorimer *Lett. Self-Made Merchant* ix. 118 Chauncey's father was the whole village, barring the railroad station and the saloon, and, of course, Chauncey thought that he was something of a pup himself. 1903 A. H. Lewis *Boss* vi. 48 'Here's a pup,' cried Big Kennedy, with his hand on my shoulder, 'I want you to look over.' 1938 *Amer. Speech* XIII. 228/1 The youngest intern..may be assigned to routine laboratory tests and called a *pup* or *junior*. 1948 F. Blake *Johnny Christmas* ii. 70 A fresh pup breezin' in, cool as you please, and takin' over Bent's? 1977 *Transatlantic Rev.* lx. 36, I can remember my Daddy brought me down once when I was a young pup.

c. Also, *Sopwith Pup*. A familiar name for the Sopwith Scout Tractor, a small, fast, aeroplane used for combative and instructional purposes in the war of 1914–18.

1917 *Jane's All World's Aircraft* 103b *Sopwith Scout Tractor*. Known as the 'Pup', and one of the fastest machines in the world... The 'Sopwith Pup' on active service has passed the 25,000 feet level with a Naval pilot. *a* 1918 J. T. B. McCudden *Five Years in R. Flying Corps* (1919) 181, I had taken charge of another machine for fighting instruction, this time a Sopwith Scout, vulgarly termed a 'Pup'. 1918 H. G. Wells *Joan & Peter* xiii. 613 They [*sc.* air force officers] made a language for themselves an atrocious slang of facetious misnomers,..the machines were "buses" and 'camels' and 'pups'. 1928 C. F. S. Gamble *Story of North Sea Air Station* xiii. 211 A single-seater machine, the Pup, which was also a product of the Sopwith Aviation Company, passed into the Service during this year [*sc.* 1916]. 1977 J. Cleary *High Road* i. 20 Pups, Camels..even some Spads and Nieuports. *Ibid.* 21 What use was a Sopwith Pup to a couple intent on adding to the postwar baby boom?

d. A four-wheeled trailer drawn by a tractor, lorry, or other road vehicle. *U.S. slang.*

1951 *Amer. Speech* XXVI. 308/2 *Pup*, a narrow four-wheel trailer. They can be 'buttoned up' in tandem and will follow the tractor, just as puppies will follow their mother. 1960 *Newsweek* 20 June 91/1 Compact, 1½-ton 'pup' semi-trailers are hitched behind regularly scheduled intercity passenger buses. 1978 *Detroit Free Press* 14 Apr. 16D/2 On two of three trials earlier, without the modifications, the second tanker or 'pup' of the same truck bounced the wheels of its safety guard sharply against the ground.

3. (Earlier and later examples.) Also, the young of the sea lion or of rats or mice.

1815 *Sydney Gaz.* 22 Apr. 1/1 The *pups* or young seal were also indiscriminately slaughtered. 1824 A. Earle *Jrnl. Residence Tristan d'Acunha* 30 Aug. in *Narr. Residence N.Z.* (1832) 354 We started off early for Elephant

Bay to procure the skin of a [seal] pup, in order to convert it into caps... The pups were nearly all black. 1937 *Discovery* May 140/2 Sea lions..resort to the island [*sc.* Lady Julia Percy Island] to breed, to raise their pups. 1944 E. C. Wood in A. N. Worden *UFAW Handbk. Care & Managem. Lab. Animals* (1947) vi. 122 The individual pups [*sc.* young rats] are of low birth-weight. 1952 L. H. Matthews *Brit. Mammals* ix. 266 The milk teeth of seal pups are shed or absorbed at a very early age. 1972 *Nature* 3 Nov. 21/2 The rats were under close observation... Mortality among offspring was very low..only 1 dead pup out of 138 born. 1975 *Storktalk* (Ches. & N. Wales Branch of Nat. Childbirth Trust) Nov. 16 The female mouse lost 54% more pups than did control mice giving birth in the normal settings for mice. 1979 *Guardian* 14 Sept. 4/4 After failing to persuade the Canadian government to stop the killing of Harp seal pups in Newfoundland, the RSPCA is to campaign for a ban on the importation of Harp seal products.

4. (Earlier and later examples.)

1898 W. B. Haskell *Two Years in Klondike* 253 Every creek has its pups, and if any of them become of considerable importance they may have pups also. 1902 *Pop. Science Monthly* July 232 The principal streams [in the Klondike region] are known as creeks; the short steep tributaries which flow into them as 'gulches'; and the streamlets which feed these as 'pups'. 1916 *Yukon Territory* (Canada Dept. Interior) I. iii. 27 On Hunker creek, below the mouth of Seventy Pup, practically all the gold occurred in a shattered porphyry bedrock. 1973 P. Berton *Drifting Home* v. 69 The little pup creeks gurgling down through the thick forests.

5. a. (Further examples.) Now usu. *fig.* and freq. *pass.*, e.g. *he was sold a pup*. Hence, *to buy a pup*.

1902 Kipling in *Collier's Weekly* 6 Dec. 8/3, I wouldn't have sold old Van Zyl a pup like that..I must hunt him up and explain. 1927 W. Deeping *Kitty* viii. 99 He was not the sort of man to advise a brother officer to buy a pup. 1930 W. S. Maugham *Cakes & Ale* xiv. 165 The public has been sold a pup too often to take unnecessary chances. 1968 *Scottish Daily Mail* 9 Aug. 5/6 The Basset is the aircraft the RAF did not want in the first place. They were sold a pup, in more ways than one. 1978 A. Ryan in Hookway & Pettit *Action & Interpretation* 75 They may insist that..we worry incessantly about what price our virtues will fetch, and about whether we shall be sold a pup by others. 1978 *SLR Camera* Aug. 36/1 Letters have arrived on my desk from petrified owners..wanting to know more, terrified that they had bought a pup.

b. *the night's* (*only*) *a pup*, the night is 'young' or not far advanced; 'it is still early'. Also, occas., *the day's* (*only*) *a pup*. *Austral. colloq.*

1915 H. Lawson *Coll. Prose* (1972) I. 913 The night was not even a pup yet—it was broad daylight, being Northern summer. 1921 K. S. Prichard *Black Opal* xii. 104 You're not taking her away yet, Michael? The night's a pup! 1928 'Brent of Bin Bin' *Up Country* x. 167 The night is only a pup yet. 1934 T. Wood *Cobbers* xi. 138 'What's the worry?' they say; 'the day's a pup.' 1947 K. Tennant *Lost Haven* i. 28 'A man's got to get a bit of sleep.' 'Night's only a pup, hen,' Alec suggested, mildly. 1949 L. Glassop *Lucky Palmer* viii. 73 We'll get him in. The day's only a pup yet. 1968 G. Dutton *Andy* xii. 198 'Are you thinking of driving out to Hangingstone to-night?' 'It's only forty miles and the night is a pup.'

6. *pup-trained* adj.; **pupfish**, a small killifish belonging to the genus *Cyprinodon*, esp. *C. macularius*, found in fresh or saline water in desert regions of California and Nevada; **pup joint** *Oil Industry*, a piece of drill pipe of less than the standard length; **pup-tent** orig. *U.S. Mil.*, a small tent or bivouac, a dog-tent; a shelter-tent, *spec.* one comprising two shelter-halves carried separately (see Shelter *sb.* 3 in Dict. and Suppl.).

1958 *Copeia* 232/1 Ten live pupfish were captured with a fine-mesh seine. 1973 *Daily Colonist* (Victoria, B.C.) 1 Apr. 9/8 The Owens pupfish survived for eons in a changing environment—from a vast inland lake to a string of shallow desert pools. 1976 P. B. Moyle *Inland Fishes California* 251 Even the isolated and harmless pupfishes are being threatened by man's activities. 1937 *Amer. Speech* XII. 153/2 *Pup-joints*, short joint used to 'nipple out'. 1972 L. M. Harris *Introd. Deepwater Floating Drilling Operations* xii. 139 Adequate pup joints are necessary for final space out. Where 40-ft riser joints are selected, one pup joint each of 20-ft, 10-ft and 5-ft are needed. 1863 in *Ohio Archaeol. & Hist. Q.* (1929) XXXVIII. 651 About 10 a.m. we..pitched our pup-tents. 1917 *Collier's Mag.* 21 Apr. 13 When a halt is reached, each pair of men set up their tent poles, stretch the line as a ridge, tie their canvas sheets together, and peg them down as a cover. The result is familiarly called a 'pup' tent. 1929 *T. Eaton & Co. Catal.* Spring & Summer 373/3 'Pup-Tent' For Boys... Just the thing you need during your holidays... Fitted with two uprights, guy ropes and pegs. 1953 K. Tennant *Joyful Condemned* xiv. 123 They had equipped themselves for the camping holiday..: a pup-tent, blankets, a frying-pan. 1977 C. McFadden *Serial* xxv. 56/1 Carol can't pick up any signals if you're a coupla boy scouts in a pup tent. 1928 Kipling in *London Mag.* Dec. 693/2 The janitor's kitten had not been pup-trained and leaped on the table, to make sure.

pupa. Delete ‖ and add: **1.** (Earlier and later examples.)

1773 G. White *Let.* 8 July in *Nat. Hist. Selborne* (1789) ii. xv. 156 The black shining cases or skins of the *pupæ* of these insects [*sc.* forest-flies]. 1898 A. S. Packard *Text-bk. Entomol.* iii. 621 In some cases the obtected pupa remains within the loose envelope formed by the old larval skin. 1928 G. H. Carpenter *Biol. Insects* vii. 172 The pupa..resembles the adult insect much more closely than the larva. 1932 Riley & Johannsen *Med. Entomol.* viii. 116 The pupae of mosquitoes and midges are quite active. 1964 V. Wigglesworth *Life of Insects* vi. 94 Within the pupa a most complex development proceeds.

3. (Earlier and later examples.)

1788 G. White *Let.* in *Nat. Hist. Selborne* (1789) ii. xlvi. 252 All [the crickets] that I have seen at that season [*sc.* March] were in their pupa state. 1898 A. S. Packard *Text-bk. Entomol.* iii. 625 (*heading*) The pupa state.

puparìate (piupĕᵊ·ri,ēⁱt), *v.* [f. Pupari(um + -ate³.] *intr.* (See quot. 1973.)

1973 Fraenkel & Bhaskharan in *Ann. Entomol. Soc. Amer.* LXVI. 419/1 Formation of the puparium can be succinctly expressed by the term pupariation, with to pupariate as the appropriate verb. 1976 *Nature* 8 July 137/1 The posterior, which had pupariated, contained ovaries. 1978 *Ibid.* 20 Apr. 719/1 All the others were apparently full-term third instar larvae which had been produced normally and then had failed to pupariate.

pupariating (piupĕᵊ·ri,ēⁱtiŋ), *vbl. sb.* = next.

1939 Metcalf & Flint *Destructive & Useful Insects* (ed. 2) vi. 154 In the Diptera the formation of the motionless puparium may be called pupariating, to distinguish it from the shedding of the final larval skin, or true pupating.

pupariation (piupĕᵊri,ēⁱ·ʃən). [f. Pupari(um + -ation.] (See quot. 1973.)

1973 Fraenkel & Bhaskharan in *Ann. Entomol. Soc. Amer.* LXVI. 418/2 Formation of the puparium in cyclorrhaphous flies occurs many hours before that of the pupa and should be consistently termed pupariation to distinguish it from the process of pupation. 1977 Richards & Davies *Imms's Gen. Textbk. Entomol.* (ed. 10) I. xix. 367 The third-instar larva [of Cyclorrhaphan Diptera] eventually stops feeding and from then until the onset of puparium formation (pupariation) it may be referred to as the post-feeding larva. 1978 *Nature* 20 Apr. 720/1 In the larviparous tsetse fly, eggs hatch normally and development is not interfered with until the time of pupariation.

pupate, *v.* Add: (Later examples.) Also *fig.*

1939 Duncan & Pickwell *World of Insects* iv. 60 In its preparation to pupate, the swallowtail larva crawls to the underside of a stem. 1977 *Sci. Amer.* May 143/1 One fly larva, armed with antienzymes, feeds on the take and pupates cheerfully in the depths of the pitcher. 1979 *Times* 14 Nov. 12/7 The Parliamentary Labour Party, which has always lived on the edge of a metamorphosis into a cowardly rabble, has already begun to pupate, and will shortly complete the process.

pupation. (Later examples.)

1905 V. L. Kellogg *Amer. Insects* ii. 45 The larva, just before pupation, has spun a protecting silken cocoon about itself. 1932 Riley & Johannsen *Med. Entomol.* viii. 116 Just before pupation the larvae of many insects spin a cocoon. 1973 [see *pupariation].

pupelo. (Earlier example.)

1806 *Salem Register* 7 Apr. 1/2 Do you not deny to the poor labourer..the common refreshment of a little toddy, and stint him with a glass of *pupelo*?

pupil, *sb.*¹ Add: **3. b. pupil power**: see *power *sb.*¹ 4 f; **pupil-room**, also, † the pupils' room in a barrister's chambers; **pupil–teacher** *a.*, designating the relation between pupils and teachers; *esp.* in phr. *pupil–teacher ratio*.

1849 Thackeray *Pendennis* I. xxix. 285 In the pupil-room of Mr. Hodgeman, the special pleader,..six pupils were scribbling declarations. 1958 J. Townsend *Young Devils* vi. 47 How vitally important a good pupil–teacher, teacher–headteacher and teacher–environment relationship had been considered at my training college. 1960 *Where?* Winter 16 *Pupil–teacher ratio*, the number of pupils to a teacher in a school. 1974 *Times* 16 Jan. 13/1 The union knows that pupil teacher ratios are better in London than in some other parts of the country. 1978 C. Hookway in Hookway & Pettit *Action & Interpretation* 40 There will be a network of pupil–teacher relations connecting them.

pupillography (piŭpilǫ·grăfi). [f. Pupil *sb.*²: see -graphy.] The recording and analysis of the movements and size of the pupils of the eyes.

1940 *Arch. Neurol. & Psychiatry* XLIV. 227 (*heading*) Pupillography: its significance in clinical neurology. 1958 *Arch. Ophthalm.* LIX. 358/2 The purpose of clinical pupillography is the detection and localization of pathological processes within the nervous centers and pathways of pupillary control. 1976 *Drive* May–June 20/1 In the late 1960s, Dr Yoss and his colleagues at the Mayo Clinic in Rochester, Minnesota, developed a technique, known as infra-red pupillography, for measuring loss of alertness.

Hence **pupillogra·phic** *a.*, **-gra·phically** *adv.*; also **pu·pillogram**, a record obtained in pupillography; **pu·pillograph**, an apparatus used for pupillography.

1940 *Arch. Neurol. & Psychiatry* XLIV. 227 Lesions.. produce characteristic modifications of the pupillary movements and reflexes which are disclosed by pupillographic examination. *Ibid.* 228 Before the Argyll Robertson stage..there can be detected, pupillographically, a long evolution of seven well defined stages [of degeneration]. *Ibid.* 229 Pupillograms dealing with intraocular conditions, such as lesions of the macula and iris. 1958 *Arch. Ophthalmol.* LIX 355/2 (*heading*) The electronic pupillograph. 1970 *Aerospace Med.* XLI. 1340/2 The pupillographic scanner measured the maximal horizontal diameter of the exposed portion of the pupils. *Ibid.* 1341/2 Three of the 32 well-rested pilots performed, pupillographically, in a less-than-perfect manner. *Ibid.* 1342/2 Data from his pupillogram are shown in Figure 5.

pupillometer. Add: Also (*rare*) **pupilometer.** (Examples.)

1920 *Jrnl. Optical Soc. Amer.* IV. 77 Various pupilometers have been suggested and used for determining the size of the pupil. 1975 L. S. Sasieni *Princ. & Pract. Optical Dispensing & Fitting* (ed. 3) iv. 105 The Pupillometer consists of two parallel tubes each containing a convex lens. 1975 *Sci. Amer.* Nov. 110/2 We have also been using an electronic pupillometer that scans the eye and automatically measures the diameter of the pupil while the experiment is in progress.

Hence **pu:pillome·tric** *a.*; **pu:pillome·trics**, the study of psychological influences on the size of the pupil.

1968 E. H. Hess in F. M. Bass et al. *Appl. Sci. in Marketing Managem.* 431 Pupillometrics..is an area of psychological study that is based on our finding that the pupils..dilate when we see something pleasant or positive. 1968 F. J. Van Bortel in *Ibid.* 440 A new technique of pupillometric measurement. 1977 M. P. Janisse *Pupillometry* iii. 35 The area of pupillometric research that is at once the most popular and the most controversial concerns interest and attitudes. 1980 D. Bloodworth *Trapdoor* xxv. 153 Pupillometrics—which means measuring the dilation of the pupil against ostensible emotions of the subject.

Pupin (piŭpi·n). *Teleph.* The name of Michael I. *Pupin* (1858–1935), U.S. physicist born in Imperial Hungary, used *attrib.* or in the possessive to designate equipment, methods, and principles introduced by him, as **Pupin cable**, a telephone cable provided with loading coils at regular intervals so as to reduce attenuation and distortion of the signal; **Pupin coil** = *loading coil* s.v. *loading *vbl. sb.* 5; **Pupin's law** (see quot. 1911). (No longer current.)

1900 *Amer. Jrnl. Sci.* CLX. 64 Pupin's interrupter may be modified to serve as an alternator in the following way. 1901 *Sci. Abstr.* IV. 1043 An intermittent direct current is produced by means of a Pupin interrupter. 1905 *Ibid.* B. VIII. 352 An adjacent insulator carries a vacuum lightning discharger contained in an ebonite case, to protect the Pupin coil from damage. 1908 *Ibid.* XI. 495 The range of transmission through a Pupin cable as compared with that through an open line having the same damping constant is smaller. 1911 J. A. Fleming *Propagation of Electr. Currents* iv. 122 Pupin reduced the solution of the problem to a verbal statement, which may be called Pupin's Law, as follows: If there be a non-uniform cable line loaded with inductance coils at equal intervals, and if we consider the total inductance and resistance to be smoothly distributed along the line, then these two lines, the non-uniform and uniform lines, having the same total resistance and inductance, will be electrically equivalent for transmission purposes as long as one half of the distance between two adjacent coils expressed as a fraction of 2π taken as the wave length, is an angle so small that its sine has practically the same numerical value as that angle in circular measure. 1934 A. L. Albert *Electr. Communication* xi. 282 It remained for Pupin successfully to solve the difficult problem of loading... His method is known as the series or Pupin system of loading. 1955 P. R. Bardell *Magnetic Materials in Electr. Industry* vii. 180 These coils are often referred to in the literature as 'Pupin' coils but are now usually known as 'Loading Coils'. 1958 [see next].

pupinized piŭ·pĭnaizd), *ppl. a. Teleph.* [f. prec. + -ize + -ed¹.] Of a telephone cable: provided with loading coils at regular intervals, so as to reduce attenuation and distortion of the signal. (No longer current.)

1910 *Electrician* 13 May 178/1 The attenuation coefficient determines the efficiency of any telephone circuit, and thus of a 'pupinised' circuit, so termed because it differs from an ordinary circuit by the introduction of inductance coils at definite points according to the formulæ of Prof. Pupin. 1913 *Sci. Abstr.* B. XVI. 152 (*heading*) Design of pupinised metallic and phantom circuits. 1933 *Jrnl. R. Aeronaut. Soc.* XXXVII. 463 Although great progress has been made in the construction of pupinised telephone cables, submarine telephony is still limited to relatively short distances, since intermediate amplifiers cannot be fitted. 1958 *New Scientist* 9 Oct. 990/1 Born.. 100 years ago on October 4, Pupin was to leave his name perpetuated in a million 'Pupin coils', used in telephone lines known as 'pupinised lines'.

puppet, *sb.* Add: **3. a.** For 'A human figure' read 'A small figure, human or animal' and add: Now also applied to a similar figure moved by rods, or to one in the form of a glove: see *glove puppet* s.v. *glove *sb.* 6, *rod puppet* s.v. *rod *sb.*¹ 12. (Later examples.)

The original sense is now sometimes distinguished as *string puppet* s.v. *string *sb.* 32.

1934 [see *puppeteer]. 1958 *Oxf. Mag.* 6 Feb. 250/2 *The Water Babies* is said to be the first full-length play to have been performed in this country by puppets. 1967 *Oxf. Compan. Theatre* (ed. 3) 776/1 There are many different types of puppets, including the Hand- or Glove-Puppet, the Rod-Puppet, the Marionette, which are all rounded figures, and the flat puppets of the Shadow Show and the toy theatre.

b. Also, a country or state which is ostensibly independent but is actually under the control of some greater power.

1933 A. J. Toynbee *Survey Internat. Affairs* 452 In the wider field of international diplomatic negotiations over

the Sino-Japanese dispute, the Japanese government deliberately gave formal recognition to their puppet in Manchuria. 1945 *Evening Standard* 20 Dec. 3 The rôle she [*sc.* Siam] played as a 'puppet' of the Japs in South-East Asia, made it essential that she should give restitution to the people who were harmed. 1976 *Survey* Summer–Autumn 18, I am not depressed about the large-scale non-fulfilment by the Russians (and their puppets) of the Helsinki agreements.

9. a. *puppet administration, army, government, leader, régime, ruler, state, troops.*

1938 *Ann. Reg. 1937* 267 The threat to Chinese sovereignty constituted by..the puppet administration set up by her in East Hopei. 1968 I. Deutscher *Marxism in Our Time* (1972) 178 The support of the population permits them to make use of the jungle, while the Americans and their South Vietnamese puppet army cannot do this. 1931 *Economist* 14 Nov. 892/2 At Mukden, the puppet Chinese Government set up by the Japanese military authorities is reported to have proclaimed its independence. 1938 *Ann. Reg. 1937* 269 In December the Japanese set up two puppet Governments in North China. 1974 M. B. Brown *Economics of Imperialism* xii. 299 After the tanks came massive economic support for the Soviet puppet government. 1934 C. Lambert *Music Ho!* iii. 182 The gangster film or the comic strip would seem more suitable mediums in which to treat the self-appointed puppet leaders..of the people. 1937 E. Snow *Red Star over China* 33 In 1935..the puppet régime of east Hopei was set up. 1971 *Standard* (Dar es Salaam) 7 Apr. 1/1 Those present [at the conference] were undecided about the relative merits of the Sihanouk Government and the American puppet regime. 1980 *Times* 2 Jan. 9/2 Mr Babrak Karmal, the new puppet ruler of Afghanistan, appears to have been instructed to hold out a conciliatory hand to the rebels. 1935 (*title*) The puppet state of 'Manchukuo'. 1947 *Sun* (Baltimore) 15 Aug. 12/8 We also have to ask ourselves whether strategically we can afford to let the Russians create another puppet state on the Asian mainland, just a few miles away from Japan. 1946 *News Chron.* 25 Feb. 1/1 Remnants of the Japanese forces in Manchuria, assisted by former puppet troops and Chinese reactionaries, were operating against the Red Army.

b. *puppet-like* (example); **puppet-master**, also *fig.*

1965 W. Lamb *Posture & Gesture* iii. 41 That the puppet-like pin-up girl's beauty is only skin deep is often revealed the moment she opens her mouth. 1965 M. Allingham *Mind Readers* ix. 97 Much more worrying was the question of the mind behind their experiments... Who was the puppet master? 1976 H. Wilson *Governance of Britain* 10 None of the prime ministers of my experience..has been either a puppet or a puppet-master.

puppeteer (pɒpĕtiᵊ·ɹ). [f. Puppet *sb.* + -eer.] One who operates puppets; *spec.* one whose occupation is the creation, management, or exhibition of puppet-shows; also *fig.* (cf. Puppet *sb.* 3 b in Dict. and Suppl.).

1930 A. Gerstenberg *Comedies All* 175 (*play-title*) The puppeteer. 1934 *Church Times* 30 Nov. 614/4 Every puppeteer has his own whimsies. Mr. William Simmonds.. makes as well as manipulates his own puppets. 1947 *Sun* (Baltimore) 14 Oct. 5/4 A cartoon in the Communist party newspaper *Pravda* today pictured the United States as a puppeteer manipulating votes of delegates in the United Nations. 1958 *Times* 10 July 4/5 The Rumanian and Polish puppeteers, at least, are set on their own underivative ways towards truly creative expression. 1969 A. R. Philpot *Dict. Puppetry* 8 The compiler of the Dictionary has had around forty years as a professional puppeteer, some twenty years as an instructor and nearly as long as editor of a puppetry magazine. 1972 M. Sheppard *Taman Indera* 69 A puppeteer and a group of musicians were maintained by many northern Malay rajas in the nineteenth century. 1979 *West Lancs. Even. Gaz.* 23 Nov. 25 (Advt.), Puppeteers available—International Cabaret Puppeteers, Scarborough.

puppet-show. Add: (Further examples.) Also *transf.* and *fig.*

1774 J. Harrower *Diary* in *Amer. Hist. Rev.* (1900) VI. 87 This night finishes the Puppet shows, roape dancings &c, which has continowed every night this week in town. 1795 tr. *C. P. Moritz's Travels* 88 Electricity happens at present to be the puppet-show of the English. 1807 *Salmagundi* xi. 262, I have seen that great political puppet-show—an Election. 1836 [see *looker *sb.* 1 b]. 1914 *Amer. Rev. of Reviews* Jan. 102/1 The puppet show does not flourish in our American cities. 1951 Lambert & Marx *Eng. Popular Art* i. 6 Bunyan's contemporaries ..could see their secular prototypes..in pageants and puppet shows.

puppet showman (earlier example).

1715 R. Powel (*title*) Second tale of a tub: or the history of Robert Powel the Puppet-Show-Man.

puppet-valve. Add: The spelling *poppet-valve* is now usual. (Further examples.) Hence **poppet-valved** *a.*

1835 *Amer. Railroad Jrnl.* 25 Apr. 245/3 Let the valve be..of the kind called puppet valves. 1877 *Jrnl. Franklin Inst.* CIII. 13 These dimensions would have been reduced by the use of the poppet in place of the lantern valve. 1912 *Motor Traction* 12 Oct. 328/2 The writer has in his possession a list of some 250 poppet-valved engines, together with such data as valve diameters, [etc.]. 1919 W. H. Berry *New Motoring* xvi. 126 The poppet-valved engine, badly made and of unsuitable materials, can be an atrocious production. 1961 F. A. S. Brown *Nigel Gresley* xvii. 122 With poppet valves there is less restriction of steam flow through the ports and consequently less throttling of the steam supply to the cylinders. 1970 *Railway World* Dec. 532 The poppet valves were actuated by a semi-internal form of Walschaerts valve gear.

puppie-show, var. *poppy-show.

puppy, *sb.* Add: **2. b.** For '= Pup *sb.*¹ 2' read 'cf. Pup *sb.*¹ 3 in Dict. and Suppl.' Also, the young of a shark; so *puppy shark*.

1934 W. Beebe *Half Mile Down* iv. 85, I saw five sharks milling around the foot of the ladder. Two were yard-long puppies. *Ibid.* 310 The uppermost one, about two feet in length, was a puppy shark. 1962 *Amer. Speech* XXXVII. 194 The name *puppy shark* may be applied to any small shark.

6. puppy fat, excessive fat in a child or adolescent causing a condition of plumpness which is freq. outgrown; **puppy foot** *U.S. slang* (see quots.); **puppy-hole** *Eton slang*, a pupil-room; **puppy-love** (earlier example); **puppy-tooth**, a small dog-tooth or houndstooth check.

1937 M. Allingham *Dancers in Mourning* x. 138 A large sulky youth in a black suit..too small for his puppy-fat body. 1940 M. Dickens *Mariana* v. 152 You grew late, now you're getting your puppy-fat late. 1972 A. Christie *Elephants can Remember* xiii. 175 She was beautiful... Not when she was about thirteen or fourteen. She had a lot of puppy fat then. 1907 *Hoyle's Games* 410 *Puppy foot*, the ace of clubs. 1932 *Daily Progress* (Charlottesville, Va.) 26 Feb. 6/6 The ace of clubs is often called the puppyfoot. 1961 Webster, *Puppyfoot*, ..a card of the club suit in a pack of playing cards. 1912 G. Frankau *One of Us* i. 10 Idled in 'puppy-hole'. 1922 S. Leslie *Oppidan* vi. 75 A list of lines due was hung in his *puppy-hole*. 1940 M. Marples *Public School Slang* 144 *Puppy-hole*,..pupil-room, when boys work with their tutors. 1834 W. A. Caruthers *Kentuckian in N.Y.* I. 175 Oh! it is nothing more than puppy love! 1960 *Times* 22 Jan. 14/3 One [coat], particularly good, in black and white puppy tooth silk. 1961 *Sunday Express* 26 Feb. 7/1 A trio of dark puppytooth checks. 1968 J. Ironside *Fashion Alphabet* 218 *Hound's-tooth*, (in smaller versions —dog-tooth or puppy-tooth) is a variety of Broken check.

puppyish, *a.* Add: (Later examples.) Also in sense 2 of Puppy *sb.*

1852 F. E. Smedley *Lewis Arundel* xl. 351 His whole demeanour *blasé* and puppyish in the extreme. 1925 F. Scott Fitzgerald *Great Gatsby* (1926) iii. 61 Girls were putting their heads on men's shoulders in a puppyish, convivial way. 1931 [see *pawing *vbl. sb.* and *ppl. a.*]. 1978 P. Harcourt *Agents of Influence* ii. 45 'My daughter, Sally.'..She was round, puppyish, charming.

Hence **pu·ppyishly** *adv.*; **pu·ppyishness.**

1817 H. C. B. Campbell *Jrnl.* 10 Oct. in *Journey to Florence* (1951) 103 Mr Cornwall was puppieshly vulgar. 1941 *Scrutiny* X. 77 He becomes in short by an inevitable process Frank Churchill,..a suspicion of the original puppyishness and lack of nice feeling still attached to his character. 1949 M. Mead *Male & Female* iii. 67 There is.. much giggling puppyishness among boys.

pupton (pɒ·ptən). [Now the usu. form of Poupeton; ad. F. *poulpeton, poupeton* (Littré 1718).] (See quots. s.v. Poupeton and below.)

1723 J. Nott *Cook's & Confectioner's Dict.* sig. C 2, *To make a Pupton of Apples.* Make the Apples into a Marmalade, with Sugar and Cinnamon; then add..Eggs,.. grated Bread and some Butter; then form it as you please. ..Let it be bak'd in a slow Oven, and then turn it upside down, on a Plate, for a second Course. 1747 H. Glasse *Art of Cookery* ii. 45 A French Pupton of Pigeons. *Ibid.* ix. 83 To make a Pupton of Apples. Do them over a slow Fire... When it is quite thick..let it stand till cool. Beat ..eggs, and stir in..grated Bread, and..Butter; then form it into what shape you please, and bake it. 1944 G. Heyer *Friday's Child* viii. 84 The dinner..consisted of a broiled fowl..followed by a pupton of pears. 1975 *TV Times* 31 May–6 June 54/2 *Pupton of rabbit*... Joint the rabbit... Simmer... Make the stuffing... Add the egg yolks, mushrooms and..asparagus... Cover with.. bacon rashers... Bake.

pupunha (pupu·nyă). Also **pupuña.** [Pg., f. Tupi.] In full, *pupunha palm*. A South American palm tree, *Bactris* (or *Guilielma*) *gasipaes*, which has a spiny stem and yields edible red or yellow fruit about two inches long.

1853 A. R. Wallace *Palms of Amazon* 10 His children are eating the agreeable red and yellow fruit of the Pupunha or peach palm. 1860 [see *murumuru]. 1961 *Times* 13 May 9/7 Bananas, pupuña palms, pineapples and tobacco will grow. 1966 E. J. H. Corner *Life of Palms* vii. 171 The fruits of the American pupunha..are not eaten raw but roasted.

purau (pū·rau). Also **purao, puro(w).** [Tahitian.] A small evergreen tree, *Hibiscus tiliaceus*, belonging to the family Malvaceæ, native to littoral regions of the tropics, and bearing pale yellow flowers fading to deep red; also, the light wood or the fibre produced by this tree; = Mahoe¹ 1 b. Also *attrib.*

1790 W. Bligh *Narr. Mutiny on Bounty* 49 The trees that came within our knowledge were the manchineal and a species of purow. [1865 B. Seemann *Flora Vitiensis* 18 *H*[*ibiscus*]..*tiliaceus*... Nomen vernac... Tahitiense, teste Solander, 'Purau'.] 1892 Stevenson & Osbourne *Wrecker* 2 In the whole length of the single shoreside street, with its grateful shade of palms and green jungle of puraos, no moving figure could be seen. 1894 —— *Ebb-Tide* i. 2 At the far end of the town of Papeete, three such men were seated on the beach under a *purao*-tree. 1907 J. Masefield *Tarpaulin Muster* iv. 68 Some said it was the leaf of the *puro* bush. 1933 *Jrnl. Polynesian Soc.* XLII. 306 The purau appears to be a native of most of the islands of the Pacific. 1952 R. Finlayson *Schooner came to Atia* v. 29 Where was he going there under the dark purau trees. 1959 L. M. Noble in A. H. McLintock *Descr. Atlas N.Z.* 82/2 The traditional rectangular hut with gable roof constructed of purau sticks and palm leaves is still common [in the Cook Islands].

Purcellian (pɒ̄ɹse·liăn), *a.* and *sb.* [f. the name of Henry *Purcell* (*c* 1659–95), English composer + -ian.] **A.** *adj.* Of, pertaining to, or characteristic of Purcell or his style of composition. **B.** *sb.* One who admires or imitates the style of Purcell.

1889 G. B. Shaw in *Star* 21 Feb. 4/3, I daresay many of the Bowegians thought that the unintentional quaintnesses of the amateurs in the orchestra were Purcellian antiquities. 1932 A. K. Holland *Henry Purcell* ii. i. 119 Liszt, whose songs, in their attention to immediate detail, are in the line of the Purcellian 'scena'. 1942 E. Blom *Music in England* vi. 91 Only Turner and Croft, both doctors of music and both Purcellians, counted for a good deal, and may have influenced Handel. 1949 *Scrutiny* XVI. 78 The finest passages in the third Quartet seem to be recovering a more stable rhythmic norm, without any sacrifice of Purcellian and madrigalian intensity. 1959 *Listener* 4 June 972/1 There is some danger of mistaking for Purcellian influence on Handel what is really the influence of Lully on both. 1975 *Gramophone* Jan. 1379/2, I must also especially commend..the direction of that devoted Purcellian, Sir Michael Tippett. 1978 *Early Music* Oct. 577/2 Purcellian alto-clef tenors may have to adopt that term unless they relish some neologism like Rimsky's 'tenore altino'.

purchasabi·lity. [f. Purchasable *a.*: see -ility.] Capability of being bought.

1904 F. Lynde *Grafters* vii. 91. There isn't any doubt about his purchasability.

purchasable, *a.* Add: **b.** (Later examples.) Also as *sb.*

1957 L. MacNeice *Visitations* 46 And from its branches muffled doves Drummed out the purchasable loves. 1966 *Listener* 17 Nov. 734/1 Much attention is given in this book to James Bond's exact social position and..to his use of purchasable objects. 1972 *Village Voice* (N.Y.) 1 June 13/4 Grocery stores tack on a 10-cent charge for the bag you carry your over-priced purchasables home in.

purchase, *sb.* Add: **I. 6. c.** *compulsory purchase*, the enforced purchase of privately-owned land or property usu. by a local authority under statutory powers of compulsion. Freq. *attrib.* in *compulsory purchase order*.

1869 *Bradshaw's Railway Man.* XXI. 40 Extra land, 10 acres; compulsory purchase, 2 years; completion of works, four years. 1932 *Act* 22 & 23 *Geo. V* c. 48 § 25(2) They may..be authorised to purchase that land compulsorily by means of an order (in this Act referred to as a 'compulsory purchase order'). 1962 J. Braine *Life at Top* x. 138 Hewley was the leader of the Labour group on the Council... And Hewley thought that the compulsory purchase wouldn't go through. *a* 1974 R. Crossman *Diaries* (1975) I. 67 Planning permissions and compulsory purchase orders. 1976 *S. Wales Echo* 26 Nov., Plans are in hand for compulsory purchase of 100 houses.

IV. 16. purchase tax, a tax levied (between 1940 and 1973) on goods bought at a rate that was higher on luxuries than on more essential goods.

1940 *Act* 3 & 4 *Geo. VI* c. 48 § 18 A tax, to be called purchase tax, shall be charged,..on the wholesale value of all chargeable goods bought under chargeable purchases. 1940 *Manch. Guardian Weekly* 25 Oct. 293 The purchase tax came into operation on Monday [20 Oct.], amid some confusion and protest... The tax imposes 33⅓ per cent on the wholesale value of luxury goods and 16⅔ per cent on other more essential commodities. 1944 M. Laski *Love on Supertax* iv. 50 Add on the Purchase Tax and the increased Purchase Tax and Special Tax on Luxury Goods. 1947 J. Hayward *Prose Lit. since 1939* 10 The publishers..fought successfully to prevent the imposition of a purchase-tax on books. 1959 *Daily Tel.* 9 Apr. 1/7 The Trades Union Congress yesterday welcomed the Chancellor's cuts in purchase tax. 1972 *Times* 27 Jan. 14/1 Ribena..was held not to be a drug or medicine and therefore not exempt from purchase tax.

purchase, *v.* Add: **6. c.** With money or its equivalent as the subject.

1805 M. G. Lewis *Bravo of Venice* ii. vi. 214 Will ten thousand sequins purchase your departure from the republic? 1904 L. Tracy *King of Diamonds* iii. 35 An establishment where threehalfpence would purchase a cup of coffee and a 'doorstep'. 1916 G. B. Shaw *Androcles & Lion* p. xciv, Such pleasures as money can purchase are suppressed.

d. *absol.*

1850 T. S. Arthur *Golden Grains* 50 He purchased largely and had the goods forwarded before he left the city. 1904 R. M. Williamson *Bits from Bookshop* x. 77 The great public libraries where..books are lent out for hire to those who wish to read but cannot purchase.

purchase-money. (Earlier and later examples.)

1720 *Rec. Early Hist. Boston* (1883) VIII. 146 The which purchase money to be Invested in Some Real Estate for the use of this Town. 1723 Duchess of Buckingham *Let.* 1 Aug. in *Lett. to & from Henrietta, Countess of Suffolk* (1824) I. 115 Half the purchase-money, at least, will go to build him another [house] to his mind. 1892 M. A. Jackson *Life & Lett. Gen. T. J. Jackson* viii. 114 He might be permitted to emancipate himself by a return of the purchase-money. **1961** *New Eng. Bible Acts* v. 2 He kept back part of the purchase-money. **1972** C. Drummond *Death at Bar* ii. 47 Alwyn did not pay over the purchase money.

purchasing, *vbl. sb.* Add: **b.** *purchasing agent, manager, officer*; **purchasing power** (earlier and later examples); hence purchasing power parity (see quots. 1918, 1939).

1921 *Daily Colonist* (Victoria, B.C.) 3 Apr. 1/2 G. W. Wooster, treasurer, and G. L. McNichol, purchasing agent, two of the oldest employees of the company, have retired into private life. **1969** *Times* 2 May 34 (Advt.), Purchasing Manager. A large British Company in the chemical field requires a manager to establish and develop a new central section to be responsible for all research concerning materials, services and sources of supply. **1963** *B.S.I. News* Apr. 24/2 Purchasing officers should make appropriate sample tests of goods received to ensure that they complied with the contract requirements. **1824** J. S. Mill in *Westm. Rev.* II. 42 Those commodities are also the measure of its purchasing power. **1930** *Economist* 11 Jan. 78/2 Good chain-store sales this month indicate a satisfactory volume of purchasing power. **1979** *Bull. Amer. Acad. Arts & Sci.* Mar. 42 Those prices will rise, and in addition, wages and other prices will also tend to rise in order to maintain their original 'purchasing power'. **1918** G. Cassel in *Economic Jrnl.* Dec. 413 At every moment the real parity between two countries is represented by this quotient between the purchasing power of the money in the one country and the other. I propose to call this parity '*the purchasing power parity*'. **1939** J. De Vegh *Pound Sterling* ii. 75 This point of view, known as the purchasing power parity theory, rests fundamentally on the assumption that a change in the internal purchasing power of a currency will, under certain conditions, affect the merchandise balance of the country involved. **1965** Seldon & Pennance *Everyman's Dict. Econ.* 350 Purchasing power parity is…only a partial explanation of exchange rates, although when other circumstances are generally stable it can give a rough guide to them.

purdah, Delete ‖ and add: **1. a.** (Later examples.) (See also quot. 1952.)

1898 C. P. Stetson *Women & Economics* iv. 66 Some air has come through the purdah's folds, some knowledge has filtered to her eager ears from the talk of men. **1927** *New Republic* 21 Sept. 127/2 Miss Mayo speaks as though the seclusion of women behind the purdah were universal throughout India. **1952** S. Selvon *Brighter Sun* i. 4 He didn't feel any sexual excitement… Even when he had looked at her face under the purdah—the white sheet thrown over them.

b. (Later examples.)

1968 *Times* 6 Apr. (Pakistan Suppl.) p. viii/5 In Pakistan today the observance of purdah is, in the broadest terms, in inverse ratio to social status. **1971** R. Russell tr. *Ahmad's Shore & Wave* iv. 39 She very rarely observed purdah and on the day in question was returning from school. **1973** *Language for Life* (Dept. Educ. & Sci.) xx. 293 Mothers may be at work all day, or live in purdah, or speak no English.

c. *transf.* Seclusion; (medical) isolation or quarantine; secrecy. Usu. in phrases *in, into,* (etc.), *purdah*.

1928 J. Galsworthy *Swan Song* ii. v. 143 The diagnosis of Kit's malady (sc. measles) was soon verified, and Fleur went into purdah. **1957** G. B. Stern *Seventy Times Seven* 182 He was supposed to be in purdah with Nicola and deeply occupied with those unspeakable Memoirs. **1958** *Times* 23 Oct. 13/5 The voluminous Cable papers… had been kept in purdah by the family piety of the late Miss Tuckwell. **1963** *Times* 27 Feb. 10/5 Mr. Maudling, Chancellor of the Exchequer, from now on will be in purdah whenever questions touch upon the Budget. **1977** D. Bagley *Enemy* xxxii. 314 When I came out of purdah, but before I was discharged, I went to see her.

3. *purdah girl*; *purdah costume, curtain, glass, party* (example).

1936 Purdah costume [see *burka]. **1937** *Times* 23 Apr. p. xxxviii/1 In the first class compartment there are four special ventilators and purdah glass louvres above the side windows. **1965** *Times* 31 May 7/1 In Peshawar is the university and a college for women, where there are mixed debates with a purdah curtain dividing the sexes. **1971** R. Russell tr. *Ahmad's Shore & Wave* x. 114 Let purdah girls play the men up. **1973** *New Society* 26 Apr. 196/1 A gilded door of purdah glass, which meant that Lady Jersey could see all of Lord Jersey approaching, while he could delight only in outline of her filmy silhouette. **1975** P. Mason in C. Allen *Plain Tales from Raj* 16 One formidable old lady… would sit in a purdah party for Indian ladies before her dinner party to the province.

purdahed, *a.* (Later examples.)

1969 L. Durrell *Spirit of Place* (1969) 103 All the houses in the Turkish quarter have musharabaya trellis windows for purdah-ed girls. **1959** *Encounter* Sept. 33/1 Somewhat secluded (although not purdah-ed) women.

Purdey (pɜ̄ˑɹdi). The proprietary name of firearms and parts manufactured by the firm founded by James Purdey (1816–68).

1884 F. F. R. Burgess *Sporting Fire-Arms for Bush & Jungle* viii. 85 For ordinary shot-guns, the Purdey bolt is a very good fastening. **1931** F. T. Fremantle *Bk. of Rifle* iii. 44 The Purdey was a four-grooved rifle, with an increasing twist. **1937** M. Sharp *Nutmeg Tree* ii. 17 At twelve the boy was to be given his father's old 20-bore, at eighteen the Purdey 12. **1961** C. Willock *Death in Covert* ii. 31 Crumbe-Howard… was standing casually with a Purdey twelve-bore open. **1970** R. A. Steindler *Firearms Dict.* 185/1 *Purdey side lock*, perhaps the most famed & best of all side lock actions… The Purdey design differs in several important points from the standard side lock design… Even a well-used Purdey is considered a 'good buy'. **1974** *Trade Marks Jrnl.* 20 Mar. 468/1 *Purdey*… Guns, rifles and ammunition. James Purdey and Sons Ltd.,… London. **1976** *Shooting Times & Country Mag.* 9–15 Dec. 38/2 (Advt.), These weapons have the usual Purdey engraving with traces of original hardening colour.

pure, *a.* (*sb., adv.*) Add: **I. 1. e.** *pure tone*, a tone composed of a single frequency and represented by a sine wave.

Earlier called a *simple tone*.

1900 *Encycl. Brit.* XXXI. 751/2 Considerable difference of opinion exists as to whether beats can blend so as to give a sensation of tone; but König, by using very pure tones of high pitch, appears to have settled the question. **1963** H. Fletcher *Speech & Hearing* iv. 167 The masking effect of one pure tone by another was determined. **1961** *Lancet* 22 July 197/2 Pure-tone audiometry is only one item in a whole range of tests that can be used to build up a complete picture of the condition of a patient who has a hearing-loss. **1976** L. H. Schlesinger *Sound, Man & Building* i. 30 The beat grows progressively lower and vanishes altogether when $f_1 = f_2$. This gives rise to a method of extremely accurate determining the unknown frequency of a pure tone with the aid of the calibrated output of a sine-wave generator by simply listening in.

f. *Forestry.* Of a wood or plantation; consisting of trees of only one species.

1889 W. Schlich *Man. Forestry* I. ii. iii. 177 Such woods may be composed of one species only, or they may contain a mixture of two or more species; in the former case they are called 'pure woods', and in the latter 'mixed woods'. **1927** *Forestry* I. 11 This century saw the return in an increasing degree to pure crops in normal selection stands. **1948** *Misc. Publ. Univ. Michigan Mus. Zool.* LXVIII. 16 The creosote bush, usually in pure stands, covers great expanses of the broad desert basins. **1976** H. L. Edlin *Nat. Hist. Trees* v. 64 The plantations just described are examples of more or less even-aged stands of a single species, also called a pure even-aged forest.

II. 2. b. *pure line* [tr. G. *reine linie* (W. Johannsen *Erblichkeit in Populationen und in reinen Linien* (1903) 9)], an inbred line of descent; also *attrib.*, an individual belonging to such a line.

1906 R. H. Lock *Rec. Progress Study Variation* iv. 110 If we were to carry on this investigation to the case of bisexual inheritance, we should find that the different pure lines would become crossed and confused together. **1926** J. S. Huxley *Ess. Pop. Sci.* ii. 22 Beans are self-fertilising, so that if instead of treating the sample as a whole, he kept the beans produced from each plant separate, he could be sure of dealing with a hereditarily pure stock, or as it is usually called, a pure line. **1932** *Discovery* Oct. 320/2 In the cotton industry we have the magnificent success of the 'pure lines' particularly in the Egyptian cottons. **1947** *Ann. Rev. Microbiol.* I. 27 Rkt. later showed that an apparently inexhaustible succession of immunologically distinct variants may appear in a single animal. This work…is of especial interest since the strain of *T. brucei* used was a 'pure line' derived from a single trypanosome. **1958** Sea & Owen *Gen. Genetics* xvi. 335 Selection within pure lines…is ineffective because it is biologically meaningless. **1965** Laughmore *Old Tom's Harvest* i. 5 When I get my credit-bank money to buy pure-line seeds them rirefields gwine be planted again.

d. Also, with reference to the arts (chiefly music, painting, and poetry): used of an art in its absolute, essential, or most objective form; freq. in contrast to that which is representational, didactic, or commercial in intent. Also used of an artist whose work is of this sort.

1901 A. C. Bradley *Poetry for Poetry's Sake* 28 Pure poetry is not the decoration of a preconceived and clearly defined matter: it springs from the creative impulse of a vague imaginative mass pressing for development and definition. **1924** [see *Orphic a. (sb.)]. **1924** G. Moore *Anthol. Pure Poetry* 34 If your acquaintance with the poetry of to-day…something that the poet creates outside of his own personality, we there might compile a book that would be a real advancement in the study of poetry—an anthology of pure poetry. **1926** H. Read *Reason & Romanticism* iii. 59 Pure poetry, Mr. Moore holds, is born of admiration of 'the only permanent world, the world of things'. **1927** *New Criterion* V. 10 Whither would the notion of 'pure art' lead us if pushed to its farthest logical extremity? To an art completely isolated from everything but its own chosen 'pure' operations and the object to be created as such. **1929** E. Wilson *I Thought of Daisy* iii. 183, I thought…of those other efforts, those efforts more characteristic of our time, which aimed, also, at an absolute beauty, at an art wholly independent of the appetites and agonies of men—paintings which represented nothing, 'pure poetry' devoid of ideas. **1934** C. Lambert *Music Ho!* iii. 174 The recent invention by certain critics of a hitherto unknown art described as 'pure music' has resulted in the criticism of music becoming more and more detached from any images, and the music of verse together. I then believed in *pure poetry*, as it was then called. **1941** G. Orwell in *Listener* 29 May 768/1 James Joyce, was…about as near to being a 'pure' artist as a writer can be. **1946** A. I. Bachrach *Brit. Music of our Time* viii. 118 The superficial æstheticians who insist on the necessity of music being 'pure', the implication being that any music that has the remotest connection with 'literature' is necessarily impure, and therefore incapable of being successful. **1967** M. Wilson *Eng. Lit. in Sixteenth Cent.* i. i. 92 This mixture of conventions enables the poets to remain recognizable occasional and yet at the same time to become almost 'pure' poetry. We celebrate the royal wedding…; yet equally, we wander in a world of beautiful forms and colours. **1955** *Times* 9 May 5/1 The exhibition would attract considerable attention and must help to break down the barrier which existed between commercial and pure art. **1959** D. Cooke *Lang. Mus.* v. 231 Mozart's Fortieth Symphony and Vaughan Williams's Sixth… would both appear to be 'pure' music, since we have no evidence that their composers ever imagined that they expressed anything at all. *Ibid.* 234 Our own age has retained the romantic period's conception of Mozart as a 'pure' composer, but adopted a different attitude. **1978** P. Griffiths *Concise Hist. Mod. Music* v. 47 His [sc. Debussy's] creative energies were directed…into works of pure music.

g. Philos. and Psychol. *pure ego*: the essential, transcendental 'self' distinguished from the empirical 'self', esp. in phenomenological contexts.

1890 W. James *Princ. Psychol.* I. x. 292 The constituents of the Self may be divided into two classes, those which make up respectively—(a) The material Self; (b) The social Self; (c) The spiritual Self; and (d) The pure Ego. *Ibid.* 296 By the Spiritual Self, so far as it belongs to the Empirical Me, I mean a man's inner or subjective being, his psychic faculties or dispositions, taken concretely; not the bare principle of personal Unity, or 'pure' Ego. **1925** *Mind* XXXIV. 320 In the second edition footnotes the reader of the Logical Studies learns that Husserl has changed his views on this crucial point, and has come to accept definitely…Natorp's Pure Ego. **1931** W. R. B. Gibson tr. *Husserl's Ideas* ii. iv. 173 The pure Ego… **1960** [see *transcendental]. A. Huxley *Let.* 29 June (1969) 635 A deeper layer of 'Original Virtue', which is one of peace, illumination and insight, which seems to be on the fringes of the Pure Ego or Atman. **1961** C. W. Allport *Devel. of Personality* vi. 129 Kant argued that… the knowing self is just there, a transcendental or pure ego. **1974** G. L. Brandon *Murdoch's Idea of Phenomenology* I § 52. 328 The pure ego is therefore the subject of transcendental constitution, the ego pole of intentionality, the centre and point of departure of every intentional function.

h. Ibid. *pure culture*, a culture in which only one species or clone is present; also *attrib.*

1893 *Ann. Bot.* VII. 120 The method adopted by De Bary of cultivating each species separately in pure cultures…that of so-called 'pure cultures'—has been of inestimable value. **1930** *Forestry* IV. 60 It seemed possible…that the fungus were identical in culture… might also yield some information…on the origin of the mycorrhizal habit in trees. **1953** *New Biol.* XIII. 110 Pure-culture technique is a 'stock-in-trade' of the micro-biologist. **1973** G. D. Bowen in Marks & Kozlowski *Ectomycorrhizae* v. 166 Many…of mycorrhizal fungi in pure culture could use nitrate as the sole source of nitrogen.

3. a. *pure and simple* (earlier and later examples).

1860 Geo. Eliot *Let.* 7 June (1954) III. 302 But the most ignorant journalist in England would hardly think of calling me a rival of Miss Mulock—a writer who is read only by novel readers, pure and simple, never by people of high culture. **1954** C. S. Lewis *Eng. Lit. in Sixteenth Cent.* 4 Even when hills are praised for not despising lowly plains we have still hardly reached the realm of metaphor pure and simple. **1977** *Lancs. Life* Nov. 58/3 They attract on a variety of levels: as toys or fun-things pure and simple, or as tactile objects, [etc.].

IV. 8. d. *the pure quill*, the genuine article, the real thing. *N. Amer. dial.* or *colloq.*

1888 E. W. Lewis *Sword of Shatter* 33 There's hairs of six different colors sticking in the splinters, and those blood stains are the pure quill. **1899** *N. & Q.* 23 Sept. 248/1 One of your correspondents…states that the expressions 'the cheese', 'pick of the basket', &c., although now almost obsolete on this side of the Atlantic, are still to be heard in America. The expression 'the pure quill', having a similar meaning, I have often heard used in Canada and in the States. **1917** *Dialect Notes* IV. 327 That tobacco is the pure quill. **1924** H. W. Raine *Hoopla* 188 His brand was the pure quill, he also exhibited…a row of half-gallon fruit-jars, both containing one large live rattler. **1956** N. Algren *Walk on Wild Side* ii. 101 A pint of Bottled-in-the-Barn. They drank it down to the half-pint mark. 'That stuff is so good a feller can't hardly bite it off.' They told Luke. 'It's the pure quill.'

B. *sb.* **5.** Delete † and add earlier and later examples.

1845 G. Dodd *Brit. Manuf.* V. ix. 189 A solution called the 'pure' or the 'pewer' is prepared in a large vessel, and into this the skins are immersed. **1946** *Ward's Dict. Appl. Chem.* (ed. 4) VII. 164/2 Modern artificial bates have replaced almost completely the older 'dung bate' or 'pure', an infusion of dog- or, less often, pigeon-dung.

6. A genuine person. *rare.*

1924 W. M. Raine *Troubled Waters* xix. 201 You-all are losing a better man than Missie ever had. He's a pure, Mac is.

C. *adv.* **1.** (Later U.S. dial. examples.) Also, with verbs: just, simply; really, truly.

1928 J. Peterkin *Scarlet Sister Mary* iii. 27 My jaws pure leak water just to look at em. *Ibid.* 35 What you done put on yo' face? I pure smell you. **1932** W. Faulkner *Light in August* (1933) xv. 332 He was pure crazy by now, standing on the corner and yelling at whoever would pass. **1937** *Frontier & Midland* Autumn 13/2 His was pure accidental and it was with a shotgun he was unloading. **1961** W. Faulkner in *Sat. Even. Post* 18 Mar. 38/1, I would pure cut a throat if it would bring you any ease.

D. a. *pure-blooded* (earlier, later, and *fig.* examples), *-hearted, -natured, -souled, toned* adjs.; *pure-mindedness*; *pure-relational Linguistics* (see quots.); *pure-watered a.*, of unmarred brilliance (cf. Water *sb.* 20).

1850 L. H. Garrard *Wah-To-Yah* vii. 109 The unfair horsetrader might have taken my scalp for presuming to dictate to him, a pure-blooded Cheyenne. **1892** W. James in *Philos. Rev.* I. 149 They have quite as little [aptitude] as the pure-blooded philosophers have for discovering particular facts. **1902** *Rep. Kansas State Board Agric. 1901–02* II. 63 A quarter of a billion acres of grass, nurturing 10,000,000 head of cattle…can be doubled in value to a single decade, if only pure-blooded sires are used in all the cow herds during this time. **1832** J. G. Whittier in S. T. Pickard *Life & Let. J. G. Whittier* (1894) I. iii. 108 Those who…have sent to us pure-hearted. **1860** G. Meredith *One of our Conquerors* III. vii. 135 He might have put a reluctant faith in the pure-mindedness of these aspirations, without reverting to her origin. **1855** Bagehot *Coll. Works* (1965) I. 319 They are emphatically pure-natured and firm-minded. Instinctively casting aside the coarse temptations and crude excitements of a vulgar earth, [etc.]. **1913** J. Masefield *Daffodil Fields* 12 Gentle she seemed, pure-natured, thoughtful, wise. **1921** E. Sapir *Language* v. 107 *Pure Relational Concepts* (purely abstract): normally expressed by affixing nonradical elements to radical elements, or by their inner modification, by independent words, or by position; serve to relate the concrete elements of the proposition to each other, thus giving it definite syntactic form. *Ibid.* vi. 145 Languages that keep the syntactic relations pure and that do not possess the power to modify the significance of their radical elements by means of affixes or internal changes. We may call these *Pure-relational non-deriving languages* or, more tersely, *Simple Pure-relational languages*. These are the languages that cut most to the bone of linguistic expression. *Ibid.*, Languages that keep the syntactic relations pure and that also possess the power to modify the significance of their radical elements by means of affixes or internal changes. These are the *Pure-relational deriving languages* or *Complex Pure-relational languages*. **1944** R. A. Hall *Hungarian Gram.* (Language Monograph No. 21) 22 There are three fundamental types of suffixes which are added to substantives: derivational (the plural suffix), concrete-relational (the personal possessive suffixes, expressing ownership of the object denoted by the noun, on the part of the person indicated by the suffix), and pure-relational (twenty suffixes, including the accusative ending, which indicate the relation between the word to which they are attached and the rest of the sentence). **1945** *Language* XXI. 255 This analysis results in three pure-relational categories in Hungarian: case suffixes, suffixed postpositions, and free postpositions. The former two classes can be grouped together as pure-relational suffixes, to preserve both Hall's (original) Sapir's) term and class. **1963** N. N. Poppe *Tatar Manual* ii. 14 Pure-relational suffixes are added to the concrete-relational (possessive) suffixes… The pure-relational suffixes serve to denote the relations between an object and other objects or between an object and an action. The system of pure-relational forms is what is commonly called 'declension'. **1920** F. M. Ford *Let.* 28 Oct. (1965) 124 When you write unscrupulous villain and I, the pure-souled idealist join forces how that dovecote will flutter! **1923** F. L. Packard *Four Stragglers* ii. iii. 220 A girl, high-minded, pure-souled. **1869** J. G. Whittier *Among Hills* 36 Through her his civic service shows A purer-toned ambition; No double consciousness divides The man and politician. **1851** H. Melville *Moby Dick* III. vii. 36 In the clear air of day…the pure-watered diamond drop will healthful glow. **1930** Blunden *English Poems* (rev. ed.) 64 Here yet its fruit-trees shield love-nooks, Its well's pure-watered diamond.

b. *pure-bred* (further examples); also *absol.* and *fig.*; *pure-breeding*, producing genetically similar progeny.

1903 *Biometrika* II. 171 Pure-bred mice usually are inbred. **1919** W. N. P. Barbellion *Jrnl. Disappointed Man* 179 If only I were pure-bred science or pure-bred art. **1923** D. H. Lawrence *Birds, Beasts & Flowers* 177 Sensitive mother Kangaroo. Her sensitive, long, pure-bred face, Her full antipodal eyes. **1927** Haldane & Huxley *Animal Biol.* ii. 68 The original pure-bred snow-whites could give nothing but rose-combs. **1937** K. Blixen *Out of Africa* iii. 91 A stud-farm of purebreds. **1945** [see *grade *sb.* 7]. **1976** *Sci. Amer.* Sept. 186/5 The object is merely to incorporate new traits into the heterogeneous population, not to create a purebred variety. **1976** *Cumberland News* 3 Dec. 15/3 In the live cattle section the beef purebred bulls made a come-back with a sharp rise in entries from 294 in 1975 to 318 this year. **1927** Haldane & Huxley *Animal Biol.* vi. 140 There is one more point to be considered in a pure-breeding race. **1962** ... the pure-breeding types in the second generation. **1964** D. Michie in G. H. Haggis et al. *Introd. Molecular Biol.* viii. 206 Griffith injected mice with living pneumococci of a pure-breeding strain lacking the polysaccharide capsule characteristic of most members of this species.

c. *pure-cone, -rod adjs.*, having only cones, or rods, as photoreceptors; *pure food attrib.*, of or concerned with the maintenance and promotion of purity in food through the control of additives, avoidance of the use of chemical fertilizers, or the like; *pure-jet Aeronaut.*, *usu. attrib.*, denoting engines, aircraft, etc., in which all thrust is provided directly by reaction to the exhaust jet, without the assistance of fans or propellers; *pure-rod a.*: see *pure-cone* adj. above.

1942 G. L. Walls *Vertebrate Eye* iii. 73 Perfectly familiar to all is the increase of visual acuity with intensity… If we knew very accurately this relationship for pure-rod and pure-cone animals, we would expect to find their curves of acuity-versus-intensity to be kinks less. **1962** *Science Survey* III. 245 One of the American ground squirrels, one of the few mammalian species known to have a pure-cone retina and is so strongly diurnal. **1970** *Jrnl. Marine Biol. Assoc. U.K.* LI. 454 There is no evidence of summation in early development which is to be expected in a pure-cone retina. It is, however, surprising that there is also little evidence of summation in the two pure-rod eyes (leptocephalus and macrurid). **1894** *Jrnl. Franklin Inst.* Apr. 267 Senator Paddock, of Nebraska…after years of futile struggle, succeeded in having the Senate pass what is known as the Pure Food Bill. **1913** *Collier's Weekly* 16 Aug. 24/2 The clubwomen of Idaho are banded together to have their State known as a pure-food State. **1923** Wodehouse *Inimitable Jeeves* xvi. 214, I was feeling more or less like something the Pure Food Committee had rejected. **1966** *Punch* 14 July p. xiii/2 The Pure Swanee is another of the popular pure-food preservation... **1970** *Sci. Amer.* Apr. 120/1 ... **1960** ... in *Mus. with Tony Head* xiv. 161 He was…a pure-food fanatic with a hatred of chemical fertilisers. **1946** *Jrnl. R. Aeronaut. Soc.* L. 360/2 The weight of a pure jet engine, as compared with that of the aeroplane in which it had to be fitted…was small. **1959** *Daily Tel.* 2 Mar. 16/4 A pure-jet plane may evade interception. **1960** C. H. Gibbs-Smith *Aeroplane* xvi. 227 Turboprop engines are ideal for commercial airliners whose operations take them too far from the optimum conditions of altitude and speed necessary for the economic use of the pure jet. **1969** *Jane's Freight Containers 1968–69* 415/1 By…1975, Lockheed will possess unrivalled knowledge of operating large, pure-jet freighters. **1942** Pure-rod [see *pure-cone* above]. **1962** *Science Survey* III. 242 In 'pure-rod' eyes the retinal structure is always the limiting factor for visual acuity and in these eyes it is always poor. **1970** [see *pure-cone* above].

pure, *v.* Add: **1. b.** Also **puer.** (Earlier and later examples.)

1845 G. Dodd *Brit. Manuf.* V. ix. 190 After being 'pured' for some time, the skins are taken out and scraped. **1911** D. J. Law in G. Martin *Industr. & Manuf. Chem., Organic* xix. 580 The goods are then 'puered', which operation consists in paddling in a weak warm infusion of fermented dog-dung.

Hence **pu·ring** *vbl. sb.*

1897 [see *graining *vbl. sb.*]. **1897** *Hide & Leather* 22 May 21/1 After puring, rinse well and work on flesh side. **1898** *Ibid.* 17 Dec. 25/2 The excrement used in puring should be as fresh as possible, when manure being used for hides, pig manure for calfskin and dog's dung for goats. **1964** H. Hodges *Artifacts* xii. 150 This process of plumping, while softening, gives the substantive adverial function. **1974** *Materials & Technol.* V. xii. 401 Puering and bating made it easy to get rid of short hairs, lime soaps, and cementing substances in the skin.

pure blood, *sb.* and *a.* [f. Pure *a.* + Blood *sb.*] **A.** *sb.* **1.** Unmixed inheritance or ancestry. Also *fig.*

1776 [see Blood *sb.* 8]. **1884** J. F. Maurice *F. D. Maurice* II. ii. 62 A number of the political economists of pure blood, who were…the fiercest in opposition to co-operation. **1945** M. F. A. Montagu *Introd. Physical Anthropol.* vii. 201 The term 'blood-relationship'…enshrines the belief that all biological relationships are reflected in, and are to a large extent, determined by the character of the blood. Such terms as 'pure blood', 'mixed blood', and 'good blood' further reflect that meaning. **1963** [see *alsatian]. ... **1971** *Ibid.* 22/2 The pure-blood Chinese...

2. An animal or breed of unmixed inheritance.

1882 *Harper's Mag.* May 895/1 The half and quarter breeds…seem to have…greater powers of resistance than the pure-bloods. **1884** *Rep. Vermont Board Agric.* XIV. 166 Having bred pure bloods for almost thirty years. **1903** *Rep. Kansas State Board Agric. 1901–2* II. 63 Fifty per cent. can be added to the value if pure-bloods only are used in the northern half of this territory.

B. *adj.* Also **pure-blood.** Of unmixed inheritance or ancestry; pure-bred.

1900 *Trans. State Agric. Soc. Michigan* X. 355 The Durham cattle will keep us easy…as the pure blood, elegantly-constructed sprightly Devon. **1888** *Rep. Vermont Board Agric.* X. 49 Why don't you get some pure blood Holsteins? **1963** *English Studies* XLIV. 21 Only he and the other bears and the deer are pure-blood, and he is not so much distinguished from these other animals by his size and his age as he is representative of all pure blood, of all wildness before blood is mixed, of the wilderness itself. *Ibid.* 22 A pure-blood line Old Ben.

pure D, puredee (piūᵊɹ·dī), *a.* *U.S. dial.* [f. Pure *a.* + D. (? = Dee b.)] Thoroughgoing, 'regular'. Also as *adv.*

1941 J. Street *In my Father's House* ix. 148 I'm powerful fond of Woody. He's a pure D man. *Ibid.* xvi. 346 Mama has got pure D fidgets. **1948** *Sat. Even. Post* 6 Dec. 110 It takes a pure D humdinger to hunt birds… If a dog's got pure D hous sense and a fellow's got bad hearts, he can play. **1953** *Time* 15 June 62/2 When I heard it, it goes a bit…, at least. **1953** R. Harwin *Home is Upriver* i. 8 Kip's lip curled at this slovenly practice, one which Pa had always called puredee shif'less. **1964** *Amer. Folk Music Occasional* i. 92 This State you can drink, this State you are pure-dee God-given water.

purée, *sb.*[2] Delete ‖ and add: Now usu. with pronunc. (piuᵊ·ɹēi). (Earlier and further examples.)

1707 J. Mortimer *Whole Art of Husbandry* 593 This small Beveridge, or Cider Kin and Puree… is made for the common drinking of Servants, &c. supplying the place of Small-beer. **1723** J. Nott *Cook's & Confectioner's Dict.* sig. Dr. *Artichokes in Puree*… take them out of the Water, and make them into Puree: then strain them through a Sieve as you do Peas. *Ibid.* sig. F8v Take a Quart of clung Peas, boil them…bruise them to a Mash…and strain the clear Puree. **1929** A. Blackwood *Dudley & Gilderoy* xvi. 183 Of flight and nuts, of hot sunshine, foliage, flowers, of numerous companions of sex, age, nests and eggs—of all these his golden dreams formed a lovely *purée*. **1951** *Good Househ. Home Encycl.* 623/1 *Purée*, fruit, vegetable, meat or fish pounded or sieved into a finely divided pulp. The thickness of the purée depends on the amount of liquid present before sieving: a purée of cooked green peas and potatoes…is very stiff and can be piped for decoration.

purée (piūᵊ·ɹēi), *v.* [f. prec.] *trans.* To make into a purée. Also *fig.* Hence **pu·réed** *ppl. a.*

1934 Webster, *Purée*,…to boil to a pulp and rub through a sieve. **1948** *Good Housek. Cookery Bk.* ii. 55 Pea purée—Boil vegetables and rub through a sieve [etc.]. **1951** *Good Househ. Home Encycl.* 225/1 Sieves are used… for puréeing foods. **1959** J. Thurber *Years with Ross* xiv. 223 He puréed his own peas. **1961** *Listener* 13 Aug. 331/2 A combined grinder-liquidizer…purées fruit and cooked vegetables, in seconds. **1963** Hume & Downes *Penguin Cordon Bleu Cookery* ix. 257 (*heading*) Puréed and Mousseline potatoes. **1973** *Daily Tel.* (Colour Suppl.) 9 Nov. 79/1 Purée the sugar, butter, powdered almonds and most of the cointreau in a blender to obtain a light and frothy cream. **1977** *Time* 21 Feb. 47/3 There are certain pups…and this is one of them—that can be called 'blender drama': puréd bits of other, better works.

purely, *adv.* **2. b.** Delete † *Obs.* and add later U.S. regional examples.

1938 M. K. Rawlings *Yearling* xiv. 140, I purely hate to think the Forresters has trapped 'em. **1952** B. Harwin *Home is Upriver* xx. 198 I'd purely like to see that old woman. She'd be glad. **1970** S. Ellis *Bond* 1vii. 28/1 purely wish you wouldn't point that thing at me…there's all kinds of accidents can happen with a gun. **1975** J. F. Burke *Death Trick* iv. 63 Managers of casbahs [i.e. hotels] like the Castlereagh purely loathe the sight of cops.

4. b. (Earlier example.)

1796 M. Edgeworth *Old Poz* in *Parent's Assistant* (ed. 2) 2nd Ser. II. 55 I'm glad to see your worship look so purely.

‖ **pur et simple** (pür e sæ̃pl), *a.* [Fr.] = *pure and simple* s.v. Pure *a.* 3 a (the more usual form).

1856 *Sat. Rev.* 5 Apr. 451/1 Mr. Disraeli fights for a Bluebook, *pur et simple*. **1864** Bagehot *Coll. Works* (1965) II. 300 Inherent eccentricity, oddity *pur et simple*, is immiscible in the great ocean of universal thought. **1871** Lytton *Coming Race* xvi. 157 The great-grandfather was a magnificent specimen of the Batrachian genus, a Giant Frog, *pur et simple*. **1880** E. W. Hamilton *Diary* 18 Nov. (1972) I. 77 Chamberlain…says that resort to such [extraordinary] powers *pur et simple* must entail his resignation.

Purex (piūᵊ·ɹeks). [App. f. Pur(ification + Ex(traction.] The name of an industrial process for separating the plutonium and uranium from spent uranium fuel by using tri-*n*-butyl phosphate as a solvent.

1956 *Proc. Internat. Conf. Peaceful Uses of Atomic Energy* IX. 471/1 A schematic chemical flowsheet for the Purex process is shown. **1976** *Sci. Amer.* Dec. 31/3 When the separated uranium and plutonium streams emerge from the Purex process, they contain only about a millionth as much radioactivity due to fission products as the feed materia lidid.

‖ **purga** (pu̇ᵊ·ɹgä, pu̇ᵊ·ɹgä). [Russ.] A blizzard of very fine snow in the U.S.S.R.

1880 L. F. Gowing *Five Thousand Miles in Sledge* v. 75 A purga was raging—one of those fierce snowstorms which visit with especial violence the eastern shores of Asiatic Russia. **1977** P. E. Lydolph *Geogr. U.S.S.R.* (ed. 3) xvii. 377/2 Much of the high winds at Baranul are associated with cyclonic storms during winter that may produce strong blizzards, the so-called *buran* or *purga*. **1978** *Soviet Geogr.* XIX. 574 A *purga* is not just any snowstorm; it is a violent storm associated with an invasion of cold air.

purgator. Delete † *Obs. rare*⁻¹ and add later example.

1933 K. Malone *Deor* 15 He conceded the possibility that ll. 31–34 may be a later insertion, made to give the whole a religious turn, but evidently had his doubts about yielding even these lines to the purgator.

purgatory, *sb.* Add: **4. c.** A swamp, esp. one difficult to cross. Also *attrib.* local *U.S.*

1831 J. M. Peck *Guide for Emigrants* iii. 308 In the low prairies near the Wabash, are purgatories, as the inhabitants call them, which are almost impassable in the wet season. **1834** — *Gaz. Illinois* iii. 172 The eastern part, towards the Wabash, contains some wet land and purgatory swamps.

5. (Earlier example.)

1841 C. H. Hartshorne *Salopia Antiqua* 537 *Purgatory*, the pit grate of a kitchen fire place; by falling through which the ashes become purer.

purge, *sb.* Add: **2. a.** In more recent use, the removal (from a political party, army, etc.) of persons regarded as undesirable. Also *transf.* and *fig.*

1933 H. G. Wells *Shape of Things to Come* iii. § 6. 302 The eternal espionage, censorship and 'purges' of the G.P.U. **1939** *Spectator* 1 Dec. 772/2 M. Shakhtin... characterized the 'purges' as a move to stifle every critical voice in the ranks of the *A. P. of L.* **1940** *Ann. Reg. 1939* 204 The Munich bomb…furnished a welcome pretext for a 'purge' of the Jews. **1946** A. Huxley *Let.* 27 Oct. (1969) 553 See the recent accounts of Russian purges of insufficiently patriotic and Marxist writers. **1957** R. N. C. Hunt *Guide to Communist Jargon*

xii. 47 The statements elicited at the purge trials of the middle 'thirties. **1958** *New Statesman* 15 Feb. 186/3 In this he was supported by Ernst Wollweber, the second of the three victims of the new purge, who had been Minister of Security from 1953 until last autumn, when he was accused of 'leniency towards the class enemy' in general and the Harich group in particular. **1964** L. Hellman *Unfinished Woman* vii. 80, I did not even know I was there in the middle of the biggest purge 1937. **1967** G. F. Newman *Sir, You Bastard* viii. 258 They liked nothing better than a sordid purge in an institution. **1974** *Guardian* 24 Jan. 3/1 The word 'purge', with its unhealthy overtones of Stalinism, is naturally frowned on in Yugoslavia… But…almost every party organisation has seen changes in its top leadership… As the purges have reached their climax, party members are being purged for indulging simply in 'factionalism'. **1976** *Survey Summer–Autumn* 137 The Chinese nation faces mounting crises. Deeply-rooted factionalism and a recent history of repeated purges contribute to the grave uncertainties of today. **1977** *New Yorker* 1 Aug. 50/3 To the extent that Coops politics were pro-Soviet politics, there was a falling away after every event like the purge trials or the Hitler-Stalin pact.

c. Removal of one fluid by flushing with another. Freq. *attrib.*

1958 J. B. Gardner in H. W. Cremer *Chem. Engin. Practice* VI. 234 Since traces of acetylene are present in the atmosphere…it is necessary to take steps to prevent a dangerous accumulation occurring during continuous plant operation… Essentially two techniques—purging and adsorption—or a combination of them, are generally employed. In the first, a small bleed of liquid oxygen from the main bath is maintained to keep the acetylene concentration at a suitable low figure… Withdrawal of the liquid product itself in liquid oxygen plants constitutes a large purge and little difficulty is experienced on such units. **1960** V. B. Guthrie *Petroleum Products Handbk.* xi. 38 The sweet natural-gas purge is preferable. **1970** [see *headed* s.v. *head *sb.* 66]. **1976** *Offshore Platforms & Pipelining* 140/2 Oil then can be flushed from the lines by pumping through the surface manifold down the purge line.

purge, *v.*[1] Add: **1. d.** To rid of one fluid by flushing with another.

1960 V. B. Guthrie *Petroleum Products Handbk.* iii. 37 The following precautions should be observed to prevent moisture from entering LP-Gas supplies… Purge new containers being put into service. **1962** W. Schirra in *Into Orbit* 51 In the final stages of descent, a snorkel opens automatically at about 20,000 feet, and brings in fresh, cool air from the outside which purges the hot suit and gives us our first whiff of the briny ocean. **1973** *Daily Tel.* 25 Oct. 1/5 The tank had been 'purged' 18 months ago with nitrogen to force out the gas.

2. a. (Further examples in sense 'to rid of persons regarded as politically undesirable'.)

1936 [see sense 3 a below]. **1942** [see *udenerin a.]. **1945** *Daily Express* 22 May 1 Tito's officials are still purging towns and villages of Italian Fascists and placing local committees in charge.

3. a. (Later examples.) In recent use, to remove (a person regarded as politically undesirable), freq. by drastic methods. Also *transf.*

1936 *Sun* (Baltimore) 11 Mar. 1/3 Reports that the A.A.A. Administrator, who a year ago 'purged' the AAA of a number of its 'left wing' members…is himself being 'purged'. **1938** G. Orwell *Homage to Catalonia* xi. 224 The Russian Consul-General…has since been 'purged'. **1939** Joyce *Finnegans Wake* 72 Purged out of Burke's. **1945** *New Statesman* 8 May 297 The sooner the more extreme elements in it are purged the better; it has throughout been a misfortune that an excellent government here should contain members who would be more at ease collaborating against the U.S.S.R. **1958** *Spectator* 20 June 791/1 Saburov, now said to have been purged with a number of his supporters. **1974** [see *purge *sb.* 2 a]. **1976** *Survey* Winter 161 Peterson was not immediately purged in 1935, but was sent to a military position in the Ukraine.

purgee (pəɹdʒī·). [f. Purge *v.*[1] + -ee¹.] A person who is politically 'purged', expelled from an organization, or excluded from public life.

1938 *Kiplinger Washington Let.* (Kiplinger Washington Agency) 3 Sept. 4 Reelection of purgees will strengthen non-New Dealers in next Congress. **1958** *Economist* 1 Nov. 421/2 Mr. Kish, a former 'purgee' and suspected 'war criminal', became prime minister in February last year. **1963** *Probl. of Communism* July/Aug. p. x/1 Widow of two of Stalin's purgees, Grigori Sokolnikov and L. P. Serebriakov, Serebriakova herself spent a decade in prison.

purger. Add: *spec.* One who carries out a political purge.

1938 *Sun* (Baltimore) 20 June 3/2 (*heading*) Purgers reported purged in Ukraine as Soviet foes. *Ibid.* 23 Aug. 6/5 A number of Senator Tydings' admirers have been wondering who constitute the Maryland purgers.

purging, *vbl. sb.* Add: **1. c.** Removal of political opponents.

1938 *Sun* (Baltimore) 27 Aug. 16/8 Four others marked for purging by the Washington group. **1940** E. Hemingway *For Whom Bell Tolls* xv. 308 Here it reports the purging of many of the famous Russians. **1974** J. White tr. *Poulantzas's Fascism & Dictatorship* vi. 338 The judiciary…suffered least from purging after national socialism came to power.

d. = *purge *sb.* 2 c.

1950 *Sun* (Baltimore) 3 Jan. 9/5 When…natural gas is allowed to flow into the line it will 'push the inert gas or oxygen-short-air ahead of it. This will be allowed to escape until tests indicate pure natural gas has completely filled the line,' an engineer explained. This process is known as purging. **1958** [see *purge *sb.* 2 c]. **1973** J. G. Tweeddale *Materials Technol.* II. ii. 34 Various methods are used to purge the molten metal of gas. Various methods of removal can be used, such as reaction with a suitable flux, or by purging, in which an insoluble carrier-gas is bubbled through the liquid to create large surface areas at which the hydrogen can gather and be carried away. **1980** *New Scientist* 3 July 18/1 The purging of 57 000 curies of krypton from the containment building at TMI is an essential step in the 1,900 million clean-up operation.

‖ **puri**[1] (pū·ri). [Indonesian.] An Indonesian palace (the examples refer to Bali).

1937 M. Covarrubias *Island of Bali* vi. 158 Women of the aristocracy…live restricted and secluded in the palace, the *puri*, usually going out only in groups to festivals. **1961** P. Nasr *Aims for Oblivion* viii. 129, I saw ahead… the carved gateway of the Rajah's *puri*, or palace. **1971** *Country Life* 17 June 1544/1 The *puris* or palaces of the nobility…are entered through elaborately sculptured split doorways.

‖ **puri**[2] (pū·ri). [Hindi.] A small round cake of unleavened wheat-flour deep-fried in ghee or oil.

1954 J. Corbett *My India* v. 64 The wonderful occasion when they had been able to fill their bellies…with halwa and puris. **1960** R. P. Jhabvala *Householder* iii. 73 The puris were very good, piping hot. **1972** ... **1977** *Shankar's Weekly* (Delhi) 11 Apr. 24/3 ... **1971** ... cakes of the coarsest grain, flour wheat-cakes, fried in ghee... **1976** *Punch* 11 Aug. 230/3 All those crowding an Indian confectionery shop and eating sweets, puris [etc.]…is typical Indian style were either Indians or Pakistanis.

purine. Now with pronunc. (piūᵊ·rīn). Add: b. *def.*: Now known to be a bicyclic structure consisting of fused imidazole and pyrimidine rings; *spec.* adenine or guanine, two substituted purines found in nucleic acids, etc. Freq. *attrib.* (Earlier and later examples.)

1898 *Chem. News* 16 Dec. 304/1 (*heading*) Molecular transformation in the group of purines. **1921** *Spectator* 21 May 658/1 Purin bodies, precursors of uric acid,…and purin-free diets have likewise had their day, although as inculcating abstemiousness the latter have played a useful part. **1952** *Sci. News* XXIV. 43 The DNA isolated from different organs of the same animal species seems to be constant in composition with respect to the relative amounts of purine and pyrimidine bases but differs from samples obtained from other species. **1954** *New Biol.* XVI. 15 Nucleic acids are built up of a long chain of purine and pyrimidine nucleotides arranged in a chain. **1966** ...ribose nucleic acid the purine is either adenine or guanine. **1970** *Nature* 25 July 379/2 Naturally occurring cytokinins known so far have a purine nucleus as the essential chemical moiety. **1973** R. G. Krueger et al. *Introd. Microbiol.* x. 307/2 Purines play important roles in metabolism as well as being structural components.

purinergic (piūᵊrinɜ̄·ɹdʒik), *a.* Physiol. [f. prec. + Gr. ἔργ-ον work + -ic.] Of a nerve-fibre: that liberates, and is stimulated by, a purine derivative.

1971 *Courier-Mail* (Brisbane) 16 Jan. 8/1 Professor Burnstock and colleagues…have just published a paper presenting evidence that this third type of autonomic nerve fibre acts by releasing…a purine nucleotide. Because the they tentatively propose to call the nerves 'purinergic'. **1972** G. Burnstock in *Nature* 24 Jan. 483/1 In the early 1960s powerful nerves were found to supply the gut which were neither cholinergic nor adrenergic… Evidence has recently been presented that the transmitter substance released from these nerves may be ATP or some related purine nucleotide. It would therefore seem reasonable…to propose that these nerves be termed 'purinergic'. **1977** *Lancet* 19 Nov. 1065/1 Work on the non-adrenergic non-cholinergic (purinergic) system will probably shed more light on, and possibly lead to more rational pharmacological approaches to the deranged internal sphincter.

puriri. 1. Substitute for def.: A New Zealand forest tree, *Vitex lucens*, belonging to the family Verbenaceæ and bearing compound leaves and axillary clusters of red flowers; also, the hard, durable timber of this tree. Also *attrib.* (Earlier and further examples.)

1835 W. Yate *Acct. of N.Z.* (ed. 2) ii. 43 Puriri (*Vitex littoralis*)—This tree, from its hardness and durability has been denominated the New-Zealand Oak. **1838** J. S. Polack *N.Z.* II. 393 The Puriri…is a wood whose durability equals any of the timbers in the country. **1863** A. S. Atkinson *Jrnl.* 19 Sept. in *Richmond-Atkinson Papers* (1960) II. 64 One [ball from a rifle] pitched…in a puriri log. **1961** ... **1925** L. Cockayne *N.Z. Plants* iii. 39 Birds also fertilise a few New Zealand plants, amongst others the puriri. **1952** *Landfall* VI. 31 The framework of this haystack cover stands on puriri uprights which have long since sunk into the ground are practically everlasting. **1969** *N.Z. Listener* 13 Mar. 5/4 These fascinating little owls had their nest in a clump of astelia in a puriri tree. **1973** Atkinson & Bell in ... N.Z. ... Large number of kohekohe, puriri, karaka and mahoe are also present.

2. *Comb.* **puriri moth,** a large green moth, *Hepialus* (or *Charagia*) *virescens*, of the family Hepialidæ, whose larvæ bore into the wood of the puriri and certain other trees.

1966 *Encycl. N.Z.* II. 590/1 The puriri or ghost moth is the largest native moth of New Zealand. **1971** *N.Z. Listener* 6 Sept. 17/1, I hoped I might be able to find a green puriri moth on that soft and cloudy summer night.

purism. Add: **2.** *Art.* (With capital initial.) An early twentieth-century movement in painting arising out of a rejection of cubism and characterized by a return to the representation of recognizable objects with emphasis on purity of geometric form.

1921 A. Ozenfant *Foundations Mod. Art* p. xi, I have sought to formulate those tropisms which are most clearly apprehended. On them I base the doctrine I have termed 'constants'. I call it 'Purism'. **1959** *Archit. Rev.* CXXV. 356/2 Jeanneret's contribution to Purism was curious. It is the work of a follower, but the pictures have greater presence than those they emulate. **1961** M. Levy *Studio Dict. of Art Terms* 92 *Purism*, a movement in modern painting and sculpture, founded in 1918 by the painters, Amédée Ozenfant, Le Corbusier, and Brancusi. Purism was a reaction against the analytical spirit of Cubism and sought to remake, and thus purify, the world of objects, etc. **1973** *Times* 27 Nov. 12/5 For a time in the early twenties Settanrock worked in a style known as Purism, associated with Leger and Ozenfant, of simplified brightly coloured abstractions of machine forms.

‖ **puris naturalibus** (piūᵊ·ris nætiuræ·libəs), *phr.* [L.: see *in puris naturalibus* s.v. In *Lat. prep.* 2.]

1920 L. F. Hamilton *Vanished Pomps of Yesterday* (rev. ed.) ix. 50 Dick and I spent hours there swimming, and basking *puris naturalibus* on the rocks. **1974** *Listener* 17 Jan. 84/2 In the joy of being idle… With a bun and towel asking Barney…

purist. Add: **3.** *Art.* (With capital initial.) An adherent of Purism (see *purism 2).

1939 in Webster Add. **1959** H. Read *Conc. Hist. Mod. Painting* vi. 215 Between 1919 and 1925 [sic] this Purists had a decisive influence on the development of abstract art throughout Europe and America. **1974** *Encycl. Brit. Micropædia* VIII. 309/3 There were many painters…who, like the Purists, were attracted to a machine-inspired aesthetic.

4. *attrib.* or as *adj.*

1939 in Webster Add. **1945** Koestler *Yogi & Commissar* iii. ii. 155 Not even the most purist critic could expect a sudden jump to total equalitarianism. **1959** *Il. Read Conc. Hist. Mod. Painting* vi. 216 Nicholson began a decorative painter of great charm, and then came under the Purist influence… and his painting developed purist in a way that... **1972** ... *Amer. Art* ... **1962** ... **1963** G. W. Allen *Vox Latina* ii. 35 The purist pronunciation… **1971** *Guardian* ... 15 The purist would therefore be justified in reading the nominative plural forms [fi̯ɛs, fi̯ɛs, fi̯ɛ̃s], de respectively. **1978** *Gramophone* Jan. 1307/3 Disc and cassette reproduce about equally well, though I suspect we would like both recorded with...a more purist recording technique were adopted.

Puritan, *sb.* and *a.* Add: **B.** *adj.* **a.** *Puritan conscience*: a strict individual conscience requiring high standards; *Puritan ethic*: the belief in the redemptive value of work.

1901 Kipling *Let.* in C. Carrington *Rudyard Kipling* (1955) xiii. 318 Her Puritan conscience which she has inherited from the New England forbears still makes her take life too hard seriously. **1932** Q. D. Leavis *Fiction & Reading Public* ii. ii. 97 (*heading*) The puritan conscience. **1972** C. Weston *Poor, Poor Ophelia* (1973) v. 18 He seemed just another…the puritan ethic—work for the night is falling. **1975** *Listener* 16 Oct. 517/2 Hale White… did manage to write 'the Puritan consciousness', also the nastiness...of the puritan conscience trying to beat out a narrow path for itself. **1977** *Ibid.* 7 Apr. 434/1 The Puritan ethic of discipline, work, responsibility.

c. *Puritan spoon* (see quot. 1960).

1960 G. Taylor *Silver* iv. 86 In the reign of Charles I the so-called Puritan spoon…began to appear. **1960** H. Hayward *Antique Coll.* 233/1 *Puritan spoon*, a 1960 with stem with flat stem, straight top edge and nearly oval bowl, the earliest form of English flat-stemmed spoon. **1971** *Country Life* 10 June (Suppl.) 58 (Advt.), Commonwealth 1659 Puritan Spoon.

Purkinje (s.v. Purkinjean *a.*). Add: *Purkinje('s) phenomenon* or *shift* (also *phenomenon of Purkinje*), a decrease in the apparent brightness of light of long wavelength (e.g. red) compared with light of short wavelength (e.g. blue) when the degree of illumination falls; [described by Purkinje in *Mag. für die ges. Heilkunde* (1825) XX. 225].

1900 C. Weiland tr. *M. Tscherning's Physiologic Optics* xvii. 260 A comparison of the brightness of two different colors is not easy…and the result depends besides on the phenomenon of Purkinje. **1910** R. Greenwood *Physiol. of Special Senses* xii. 101 A particular case of adaptation which is of much interest is 'Purkinje's Phenomenon'. **1949** H. C. Weston *Sight, Light & Efficiency* i. 20 The 'Purkinje phenomenon'…is familiar to everyone who has noted…that when green leaves and red flowers are seen in twilight the green appears brighter compared with the red than it would in daylight. **1969** *Sci. Jrnl.* Sept. 79/2 The most familiar effect of the duality of human vision is the so-called 'Purkinje shift'.

purl, *sb.*[1] Add: **1.** (Earlier and later examples.) Also *fig.*

1867 'Ouida' *Under Two Flags* I. iii. 47 In front of that Stand was an artificial bullfinch which promised to treat most of the field to a 'purler'. **1921** H. G. Ponting *Gt. White South* 282 All went well till…on a very slippery surface I came an awful 'purler' on my shoulder. **1931** J. Masefield *Wanderer* ... 209 You seemed to go a fearful purler. **1963** *Spectator* 23 Nov. 630/3 Trevelyan's Indian career ended in a magnificent purler. **1976** *Church Times* 2 July 7/3 This is not just a…catalogue of classic military purlers (the Crimea, Kut, Gallipoli, Singapore, Arnhem, etc.). **1976** *Horse & Hound* 3 Dec. 6/2 Even Up went a real purler at the second last.

2. Something of surpassing excellence. orig. *Austral.* Cf. *pearler *sb.*[3]

1941 Baker *Dict. Austral. Slang* 57 *Purl, purler*, something excellent, outstandingly good. **1966** *New Statesman* 9 Dec. 852/1 Bobby Charlton draws a screen of men and shoots that purler from 30 yards. **1973** *Listener* 8 Mar. 300/3, I hope they are all purlers, but it is absolute purler. **1980** L. Martell *A Murder or Three* iii. 35 Never thought he'd even get round to having any [children], then they produce a purler like that…a real little darling.

‖ **puro** (pū·ro). [Sp., lit. 'pure'.] A cigar (in Spanish-speaking countries and south-west U.S.).

1851 J. L. Stephens *Incidents of Travel* I. 26, I offered for her choice a cheap and a puro. **1845** R. Ford *Hand-bk. Travellers Spain* ii. 194 Ferdinand VII…when meditating a treacherous coup, would dismiss the unconscious victim with a royal *puron* (ed. 1846, *puro*). **1963** W. McGivern *Choice of Assassins* (1964) iii. 51 To talk…with his friends over sherry and puros.

puro, var. *purau.

puromycin (piūᵊrəmai·sin). *Biol.* and *Pharm.* [f. Purine + -o + *mycin.] An antibiotic which is produced by the fungus *Streptomyces albo-niger* and is used esp. to treat sleeping sickness and amœbic dysentery.

1953 C. W. Waller in *Jrnl. Amer. Chem. Soc.* LXXV. 2025/1 The new antibiotic, Puromycin, isolated from the mold *Streptomyces alboniger*, has been found to be active against certain bacteria and Trypanosomes. **1960** M. E. Florey *Clinical Application of Antibiotics* IV. vi. 166 Puromycin…had the unique advantage of not only arresting the growth of trypanosomes but also of being able to kill them. **1968** W. Hayes *Genetics of Bacteria & their Viruses* (ed. 2) xvi. 288 Puromycin has a close structural resemblance to the terminal amino-acyl adenosine of t-RNA and interferes with protein synthesis, both in vivo and in vitro, by inserting itself at ribosomal site No. 2 (etc.). **1978** *Bio Systems* X. 194/1 Puromycin is another commonly used available antibiotic which, because of its toxicity, is not used therapeutically but rather as a research tool.

purow, var. *purau.

purparty. (Later example of form pourparty.)

1920 *Eng. Hist. Rev.* Jan. 30 To each co-heir was allotted, as a permanent pourparty, a definite manor or castle for a chief seat.

purple, *a.* and *sb.* Add: **A.** *adj.* **3.** (Further examples with ref. to literary composition.) *purple patch* (earlier and later examples), *passage* (later examples). See also *purple-patchery*.

1834 Coleridge in *Rev. de Litt. Comparée* (1927) VII. 253 Admirably reasoned on this Essay is, I yet regard it but as one of the rich Purple Patches of the Robe of Casuistry. **1931** *Times Lit. Suppl.* 9 June 464/1 The task lying between the Emperor and his Empress Nourmahal, in *Gowagatte*, is admirable purple comedy. **1936** C. Connolly *Let.* 26 June in *Romantic Friendship* (1975) 145 He realises his epic to be but a collection of purple patches. **1940** *Times* New Year Lit. II. 27 And yet so show complete conviction, Requires the purpler kinds of diction. **1941** H. Hayskraft *Murder for Pleasure* x. 216 The old, whipped-up undergrowing and purple patch. **1962** L. C. Weir *Word & Symbol* (1967) 194 An honoured place in English literature through Milton's 'purple passage' speech. **1975** V. Cunningham *Everywhere Spoken Against* iii. 84 Arnold's famous purple patch about the last enchantments of the Middle Ages. **1975** *Language for Life* (Dept. Educ. & Sci.) xi. 164 Some teachers encourage children to strive for effect, to produce the purple patch, the stock response. **1975** C. N. Parkinson *Fantasy* iii. 28 One [style] is 'purple' and highly emotive. **1977** *Gramophone* Sept. 507/3 One is grateful to be spared one of Wilde's purpler passages.

b. *purple prose* (for 'Ruskin fashion'...

5. b. 1. e. *visual purple*: see *visual *a.* and *sb.*

7. a. = Blueness 4. *rare.*

1930 D. H. Lawrence *Phoenix II* (1968) 489, I should show the public that here is a fine novel, apart from all 'purple' and all 'words'.

b. *the purple*: purple passages; esp. in *to sub the purple* (Journalists' slang), to sub-edit purple passages.

1958 A. R. Robertson *Justice of Heart* iii. 33 A well-known outside contributor from whose copy he had, in his own words, 'subbed the purple'. *Ibid.* vi. 84 The 'subbing of the purple' was a painful business for a journalist with a predilection for the purple passage.

8. *slang.* **a.** = *purple-heart 3. **b.** = *LSD[3].

1968 C. Drummond *Death & Leaping Ladies* v. 112, I heard her on at the Doc…about some Purples to key them up but he let the ceiling. **1971** E. E. Landy *Underground Dict.* 156 *Purple*, LSD.

C. I. 1. a. *purple-crested, -crowned, -flowered* (examples), *-hearted, -leaved* (later examples), *-lidded, -nosed* (example), *-spotted, -tinged, -zoned* adjs.

1756 Pope tr. *Homer's Odyssey* IV. xix. 263 A mantle purple-ting'd, and radiant vest. **1788** J. Woodforde *Diary* 8 July (1927) III. 36 To Mr. Aldridge for 6 Yards of purple spotted Cotton…0.12.0. **1846** D. J. Browne *Trees Amer.* 22 *Magnolia purpurea*, The Purple-flowered Magnolia. **1862** J. G. Whittier in *Atlantic Monthly* Apr. 443/2 The purple-crowned, Bluebird tenanter ... **1870** O. Wilde *Poems* ... **1880** ... **1895** ... purple-hearted rose of the Spring. **1882** ... **1917** ... **1922** Joyce *Ulysses* 6 The lips of the Pipers ... **1931** ... **1953** ... **1958** ... **1979** ... (see page)

sphere is by the photosynthetic green and purple bacteria, of which the purple sulfur bacteria are best known. **1971** Berkeley & Campbell in *Hamfer & Norris Methods in Microbiology* ...

b. *purple-black* (later example), *-blue* (examples), *-brown* (further examples), *-dark, -grey, -pink* adjs. and sbs.

1845 J. R. Lowell in *Harbinger* 2 Aug. 122/3 Far away on Katahdin thou towerest, Purple-blue with the distance. **1956** ... **1881** ... **1961** ... the purple-dark hill. **1926** ... **1928** V. Woolf *Orlando* vi. 243 The wine-blue purple-dark hill. **1930** J. Dos Passos *42nd Parallel* 143 ... **1952** ... **1957** Purple-blue [see *braccofernal]. **1960** ... **1966** ... **1970** M. Hynes *Med. Bacteriol.* (ed. 8) 484 A colony subcultured on to the medium gives a purple-pink colour from NH_3 production in 2–8 hours.

2. a. *purple airway* [*airway 2], a route reserved for an aircraft on which royalty is flying; *purple death slang*, a cheap Italian red wine; *purple haze slang* = *LSD[3]; **purple lake** [Lake *sb.*[6]], a purple pigment; **purple membrane**, a membrane found within the cell membrane of the bacterium *Halobacterium halobium*; **purple zone** = **purple airway*.

1958 *Sunday Times* 30 Mar. 5/1 Now that the Duke of Edinburgh flies his own aircraft, so-called 'purple airways' are more frequently given to commercial pilots. Purple airway is a reserved track for a Royal flight. **1947** D. M. Davin *Gorse Blooms Pale* 199 Everyone goes for the purple death, lady. **1967** J. Hersier in *20 Greatest Songs* (1975) 28 Purple haze is in my brain lately things don't seem the same. **1970** *Times* 14 Mar. 13/3 The American LSD…has been coming in…under such exotic names as 'purple haze', and 'blue cheer'. **1971** *Current Slang* (Univ. S. Dakota) VI. 9 Purple haze, LSD cut with metamphetamine. **1821** *London Mag.* Sept. 290/2 The purple-lake coloured sky. **1886** *Bradshaw's Railway Man.* XXI. 264/2 (Advt.), Reds…Crimson Lake—Scarlet and Purple ditto. **1960** R. L. Hiler *Notes on Technique of Painting* ii. 103 *Crimson lake, Purple lake*, etc., now usually made from alizarin…Also prepared from cochineal. **1968** Stoeckenius & Kunau in *Jrnl. Cell Biol.* XXXVIII. 344/1 The purple band, henceforth designated 'purple membrane'...contained considerably less RNA and slightly less lipid than the orange-red fraction. **1975** *Nature* 11 Dec. 496/1 *Halobacterium halobium* is indigenous to warm saline pools exposed to bright sunlight... Under such conditions the cells synthesise a special region of the cell membrane (the 'purple membrane'). **1970** *Daily Tel.* 17 Aug. 1/7 A 'purple zone' was not in operation…because the Prince was flying only in the immediate area of Tangmere.

b. *purple bacterium*, [ad. G. *purpurbacterium* (T. W. Engelmann 1888, in *Bot. Zeitung* 663)], any of a group of bacteria containing a purple photo-active pigment; *purple finch*, for Latin name substitute *Carpodacus purpureus*; (later examples); **purple gallinule**, substitute for def.: (*a*) the swamp-hen, *Porphyrio porphyrio*, found in parts of Europe, southern Asia, Africa, and Australasia; cf. Porphyrio; (*b*) a similar North American bird, *Porphyrula martinica*, which is smaller than the swamp-hen and has yellow legs instead of red ones; (examples); **purple heron**, a heron with greyish-blue plumage, *Ardea purpurea*, found in central and southern Europe, Africa, and parts of Asia; (later examples); **purple martin**, a large North American swallow, *Progne subis*; (earlier and later examples); **purple sandpiper**, a small wading bird, *Calidris maritima*, found in northern parts of Europe, North America, and Asia; (earlier and later examples).

1900 A. C. Jones tr. *Fischer's Structure & Functions of Bacteria* 19/4 (*Index*), Purple bacteria. **1924** W. H. Lang tr. *Strasburger's Text-bk. Bot.* (ed. 4) ii. i. 237 The Purple Bacteria, which develop in water rich in decomposing organic matter in the absence of oxygen and the presence of light, contain…a green and a red pigment. **1957** G. E. Hutchinson *Treat. Limnol.* I. xiii. 795 The purple bacteria…are photosynthetic green…in water which hydrogen sulfide is oxidized in the hydro-

... **1971** ... **1858** ... purple finch... **1884** ... **1912** ... **1930** ... **1960** ... **1965** ... purple gallinule... **1898** ... **1940** ... **1967** ... purple heron... **1880** ... **1906** ... **1962** ... purple martin ... **1802** ... **1831** ... **1940** ... **1963** ... **1977** ... purple sandpiper ... **1835** ... **1846** ... **1960** ...

c. *purple cone-flower*, a perennial herb belonging to the genus *Echinacea* of the family Compositæ, native to North America, and bearing flowers with a dark central disc and purplish rays; (examples); **purple loosestrife**, a large perennial herb, *Lythrum salicaria*, belonging to the family Lythraceæ, widely distributed in temperate regions, and bearing purple flowers in clusters; (examples); **purple moor-grass**, a perennial grass with purplish panicles, *Molinia cærulea*, native to Europe and Asia, formerly called blue moor-grass; **purple osier**, a large shrub, *Salix purpurea*, belonging to the family Salicaceæ, native to Europe, North Africa, and central Asia, and distinguished by its purplish bark; **purple willow** = **purple osier*.

1848 A. Gray *Man. Bot. Northern U.S.* 223 (*heading*) Purple cone-flower. **1857** [see coneflower]. **1900** L. H. Bailey *Cycl. Amer. Hort.* II. 511/2 Purple Cone-flower. **1939** *Nat. Geogr. Mag.* Aug. 220/2 Striking contrast is provided by the cone-flower and sunflower... **1974** ... the prairie notably the purple coneflower, the butterfly milkweed, and the prickly pear. **1954** J. Y. Ivaskin *Macmillan Wild Flower Bk.* 453 The purple coneflower bears disc-flowers which are purplish in colour. **1974** M. C. Davis *Near Woods* ii. 21 A purple cone-flower for example. **1961** ... purple loosestrife [see Loosestrife 1]. **1861** R. Bentley *Man. Bot.* ii. ix. 518 *Lythrum Salicaria*, Purple Loosestrife, is a common British plant. **1977** *New Yorker* 5 Sept. 98/1 Only the moist thickets of the purple loosestrife and other plants clogs the length of the market. **1859** L. H. Grindon *Manchester Flora* 459 Purple Moor-grass… Everywhere on moorish ground. **1904** M. A. Johnstone *Plant Ecol.* viii. 93 Where water lodges even very plentifully the dominant grass is the purple moor-grass. **1979** R. Grounds *Ornamental Grasses* viii. 147/1 Purple moor grass…Decidedly one of the most popular garden grasses. **1850** J. D. Hooker *Student's Flora Brit. Islands* 342 *Salix purpurea*, Purple Osier… Purple Osier… Purple Osier gets its name from the red or purple bark which clothes the thin but tough twigs. **1866** *Treas. Bot.* ... **1901** J. Rees *Hill Trees & Shrubs* ... **1776** W. Withering *Bot. Arrangement Veg. Gt. Brit.* 602 Purple Willow. Leaves serrated; smooth; spear-shaped… Banks of rivers. **1838** J. C. Loudon *Arboretum & Fruticetum Britannicum* III. 1480 *Salix purpurea*… the purple Willow. **1924** W. J. Bean *Trees & Shrubs Hardy in Brit. Isles* II. 457 Purple Willow. A shrub with thin, graceful branches forming a loose-habited, spreading bush, 10 to 18 ft. high. **1960** ... Purple willow…grows wild…in fens, marshes, and on riverbanks.

pu·rple-heart. Also as one word or two separate words. [f. Purple *a.* + Heart *sb.*]

1. a. A large tree of the genus *Peltogyne*, belonging to the family Leguminosæ and native to areas of tropical rain forest in

Central and South America and the West Indies; also, the dark purplish-brown timber of this tree. Also *attrib.*

1796 [in Dict. s.v. Purple *a.* C. 2 c]. 1825 C. Waterton *Wanderings in S. Amer.* 24 Wallaba, purple-heart,.. and mora, are met with in vast abundance, far and near, towering up in majestic grandeur. 1845 [in Dict. s.v. Purple *a.* C. 2 c]. 1902 G. S. Boulger *Wood* I. v. 99 Hepplewhite and Sheraton employed Mahogany..for small articles such as tea-caddies, whilst in the inlaid work of the period it was used..with other dark woods, such as Rosewood, Laburnum, and Purple-heart. 1924 Record & Mell *Timbers Trop. Amer.* II. 235 There is considerable variation in the size, abundance, and arrangement of the pores in different specimens of purpleheart. 1947 J. C. Rich *Materials & Methods of Sculpture* x. 285 Amaranth wood or Purpleheart is a rich, violet-colored, tropical hardwood imported from the Guianas. 1951 J. C. Fennessy *Sonnet in Bottle* VI. i. 200 There were no tall trees, no soaring palms or smooth-stemmed purplehearts. 1951 *Archit. Rev.* CIX. 288 Floors are in polished purpleheart hardwood. 1956 *Handbk. of Hardwoods* (Forest Prod. Res. Lab.) 193 The different species of purpleheart vary in size at maturity. 1959 P. Capon *Amongst those Missing* 215, I haven't noticed any purpleheart trees about. 1963 *Times* 26 Jan. 4/2 A nineteenth-century kingwood commode.., its sides inlaid with chrysanthemum branches in purpleheart, made £280. 1974 *Country Life* 30 May 1338/1 A second [chest], walnut with panels and banding in purple heart, ebony and ivory.

b. An evergreen tree of the genus *Copaifera*, belonging to the family Leguminosæ and native to tropical America or the West Indies.

1866 [see Purple *a.* C. 2 c]. 1963 Robertson & Gooding *Bot. for Caribbean* xxiii. 186 Purple-heart (*Copaifera pubiflora*).

2. *U.S.* (In form **Purple Heart.**) A decoration bestowed on a member of the armed services wounded in action. Also more fully, *Purple Heart Award*.

A decoration consisting of a heart-shaped piece of purple cloth was instituted by George Washington in 1782, but later fell into disuse. The present bronze enamelled medal was instituted in 1932.

1932 *Army & Navy Jrnl.* 27 Feb. 602/4 Awards of the Purple Heart for acts or service performed prior to Feb. 22, 1932, will be confined to the following persons. 1948 E. E. Cummings *Let.* 27 Aug. (1969) 185 The hyperscientific climax of this hero (a prominent killer, holder of Silver Stars & Clusters & Purple Hearts galore)'s experience. 1974 *Sumter* (S. Carolina) *Daily Item* 23 Apr. 4A/4 He is a recipient of the Purple Heart Award and Army Commendation Medal. 1977 J. Wambaugh *Black Marble* (1978) ii. 15 Mason returned to Pasadena wearing a Bronze Star and a Purple Heart.

3. Usu. *pl.* A familiar term for tablets of the stimulant Drinamyl, so named because of their shape and colour. Also *attrib.*

1961 [see *Drinamyl]. 1962 *Daily Tel.* 31 Aug. 21/1 People involved in the investigation had already had 'purple heart' tablets prescribed for them by their own doctors. 1964 [see *dexamphetamine]. 1968 *Times* 30 Nov. 4/7 A mixture of the stimulant amphetamine and the depressant amylobarbitone... A similar mixture of drugs was popular a few years ago in the guise of 'purple hearts'. 1971 N. Stacey *Who Cares?* xvi. 276 They became more responsible, they took more interest in life, they stopped taking purple hearts and they settled down in their homes, their schools and their jobs. 1973 H. Miller *Open City* xvi. 179 Drugs. Purple Hearts, amphetamines. The bloke was passing the stuff to kids.

purpo·rtedly, *adv.* [f. Purported *ppl. a.* + -ly[2].] Allegedly, ostensibly.

1949 *Scrutiny* XVI. 10 The poem..was..Frere's.. *Prospectus and Specimen of an Intended National Work* purportedly written by the brothers Whistlecraft. 1957 J. Hollander in N. Frye *Sound & Poetry* 76 Orsino's appetite at the start of the play is purportedly for Olivia. 1964 D. F. Galouye *Counterfeit World* xii. 105 The firm I purportedly represented. 1967 R. Stein *Great Cars* 113/1 I've never driven one. But purportedly one of these hotter types would go from 0 to 100 in 17 seconds! 1976 *Daily Tel.* 1 Mar. 2/1 Cases of purportedly sterilised women becoming pregnant had recently become more common.

purpose, *sb.* Add: **I. 3.** (Earlier and later examples in the context of novel-writing.)

1863 *Q. Rev.* Apr. 488 He never sinks so nearly to the level of the ordinary sensation-novelist as when he is writing 'with a purpose'. 1874 *Cornh. Mag.* Aug. 192 His romances are not to be confused with 'the novel with a purpose' as familiar to the English reader. 1932 *Week-end Rev.* 19 Mar. 371/1 This is a pity; for though Isabel is a 'novel with a purpose', it is also, in a high degree, a work of imagination.

III. 14. *purpose-built, -designed, -made* adjs.

1959 *Times* 9 June 11/6 Local authorities have indeed made remarkable progress in..adapted houses and small purpose-built homes. 1962 *Economist* 17 Mar. 980/2 New [bowling alley] centres will mostly be what has come to be known as 'purpose-built'. 1972 *Computers & Humanities* VII. 11 The need for a 'purpose built' command language is described in..'A Command Language for Text Processing'. 1977 *Modern Railways* Dec. 473/2 Rail movement of propylene in two weekly trainloads of purpose-built bogie tanks. 1961 *Economist* 24 June 1347/2 Special trays adapted for fitting on to the arm of the 'purpose-designed' Bingo chair. 1971 J. Howlett in B. de Ferranti *Living with Computer* ii. 10 Purpose-designed experiment. 1975 *Language for Life* (Dept. Educ. & Sci.) xiii. 208 It would be wrong to assume that nothing can be done unless the spaces are purpose-designed. 1930 *Times Educ. Suppl.* 11 Jan. 11/4 In some places there are 'purpose-made' bricks. 1938 *Archit. Rev.* LXXXIV. 208 (*caption*) Ketton stone has been used for the stone dressings, the facing bricks being eleven inches wide and purpose-made. 1974 *Country Life* 21 Mar. 686/1 Wearable outfits, purpose-made for women who..like inconspicuous clothes.

purpose, *v.* **6.** For † *Obs.* read 'Now *rare*' and add later example.

1924 W. J. Locke *Coming of Amos* v. 53 What was the use of a stick purposed to beat neither beast nor man?

purposive, *a.* Add: **2. b.** Relating to conscious or unconscious purpose as reflected in human and animal behaviour or mental activity. Hence **pu·rposivism,** the theory that all human or animal activity is purposive; **pu·rposivist** *sb.* and *a.*

1884 W. C. Coupland tr. *von Hartmann's Philos. of Unconscious* I. B. v. 285 For us, who have already become acquainted with the purposive activity of the Unconscious.., there is here..fresh support for our view. 1912 W. McDougall *Psychol.* i. 29 If we make our notion of purposive activity or behaviour wide enough to include these phenomena of bodily organization in the animal kingdom, it must also include the similar purposes of plant growth. 1932 E. C. Tolman (*title*) Purposive behavior in animals and men. *Ibid.* i. 12 Behavior as behavior, that is, as molar, *is* purposive and *is* cognitive. *Ibid.* xxv. 423 Our psychology is a purposivism; but it is an objective, behavioristic purposivism, not a mentalistic one. 1936 J. Kantor *Objective Psychol. of Gram.* v. 69 The second group of purposivists carry speech farther away from the individual than the first group. For them, speech is primarily an instrument for achieving social purposes. 1940 R. S. Woodworth *Psychol.* (ed. 12) xvii. 583 The purposivist school emphasizes the importance of striving and goal-seeking. 1947 G. Murphy *Personality* vi. 125 (*heading*) Purposivism. 1953 J. Strachey tr. *Freud's Interpretation of Dreams* in *Compl. Wks.* V. 528 It can be shown that all that we can ever get rid of are purposive ideas that are *known* to us; as soon as we have done this, *unknown*—or..'unconscious'—purposive ideas take charge. 1962 H. Cantril in J. Scher *Theories of Mind* 339 It becomes increasingly clear that we must include in our consideration the purposive behavior of the organism of which mind is an aspect.

purposively, *adv.* (In Dict. s.v. Purposive *a.*) (Further examples.)

1927 E. & C. Paul tr. *Ludwig's Bismarck* II. vii. 192 Unless we were more intimately and purposively united with our other fellow countrymen. 1939 P. Gordon *New Archery* II. viii. 89 Never varying except purposively, to correct a mistake. 1949 Wellek & Warren *Theory of Lit.* xx. 298 Literary study within our universities..must become purposively literary. 1965 *New Statesman* 10 Sept. 343/1 Ministers also speak purposively (this, currently, is a vogue adverb along Whitehall) about measures on rating and leasehold reform. 1973 H. Kemelman *Tuesday the Rabbi Saw Red* iii. 32 Dean Hanbury walked toward them purposively.

purposiveness. (In Dict. s.v. Purposive *a.*) (Further examples.)

1932 A. H. Gardiner *Theory of Speech & Lang.* iv. 181 The characteristic feature of the sentence, as opposed to mere unintelligible words, is its purposiveness. 1965 E. E. Harris *Found. of Metaphys. in Sci.* viii. 163 'Purposiveness' is the word that sums up these properties, but it is a word which precipitates controversy both as to its precise meaning and as to its legitimate applicability. 1974 G. Sommerhoff *Logic of Living Brain* ii. 23 The peculiar purposiveness found in living nature.

‖**purpurissum** (pɒ̄ɹpiūri·sɒ̆m). *Obs.* exc. *Hist.* [L.: see Purpurisse.] = Purpurisse.

1611 Coryat *Crudities* 266 Thou maist easily discerne the effects of those famous apothecary drugs.. *stibium, cerussa,* and *purpurissum*. For..the Cortezans..adulterate their faces..with one of these three. 1934 *Discovery* Nov. 323/2 From the artists' materials discovered at Pompeii Professor Pozzi was able to obtain a test-tube full of the mysterious colour *purpurissum*, the actual tincture of the murex used for the Roman purple dye twenty centuries ago.

purpurite (pɒ̄·ɹpiūrəit). *Min.* [f. L. *purpur-a* Purple *a.* and *sb.* + -ite[1].] A phosphate of trivalent manganese and trivalent iron, $(Mn,Fe)PO_4$, occurring as red or purple orthorhombic crystals (sometimes altered to dark brown or black) and differing from heterosite in containing more manganese.

1905 Graton & Schaller in *Amer. Jrnl. Sci.* CLXX. 146 Chemical analysis shows that the material is a new mineral, being a hydrous manganic ferric phosphate—the only manganic phosphate known. The most striking feature of this mineral is its purple or dark red color, and for this reason it has been named purpurite. 1951 [see *heterosite]. 1971 *Mineral. Abstr.* XXII. 226/1 A review of phosphate minerals from Brazilian pegmatites... The minerals described are..roscherite, purpurite, saléeite, [etc.].

purpurogallin (pɒ̄:ɹpiŭrogæ·lin). *Chem.* [ad. F. *purpurogalline* (A. Girard 1869, in *Compt. Rend.* LXIX. 866), f. *purpurine* Purpurin with inserted *-o* and *-gall* (f. *pyrogallique* Pyrogallic *a.*): named after the unrelated purpurin by analogy with the preparation of that substance by oxidation of alizarin.] An orange-red crystalline dye, first prepared by mild oxidation of pyrogallol, which is now known to occur in some oak galls and is a tetrahydric phenol, $C_{11}H_8O_5$, consisting of fused tropolone and trihydroxybenzene rings.

1872 *Jrnl. Chem. Soc.* XXV. 703 Purpurogallin, the substance obtained by Girard..from pyrogallic acid by the action of silver nitrate, or of potassium permanganate and sulphuric acid, is the principal product of the oxidation effected by lead peroxide, hydrogen peroxide, [etc.]. 1919 *Ibid.* CXV. 1329 The investigation of the red colouring matter derived from the 'red pea gall' has..to some extent proved disappointing. It was found that dryophantin..was in no way allied either to the flavones or to the anthocyanins, but..consisted of purpurogallin and two molecules of dextrose. On the other hand, it must be mentioned that purpurogallin has not previously been found in nature. 1968 Kirk & Othmer *Encycl. Chem. Technol.* (ed. 2) XVI. 191 Purpurogallin.., a red-brown to black mordant dye, results from electrolytic and other mild oxidations of pyrogallol.

purr, *sb.*[1] (Further examples of the sound made by a mechanical device.)

1971 G. Ewart *Gavin Ewart Show* I. 12 At the lawn-mower's purr I stop for a moment. 1974 R. Rendell *Face of Trespass* xviii. 168 The powerful purr of a Jaguar sports.

purr, *v.* Add: **2. b.** (Examples of mechanical devices.)

[1916 G. B. Shaw *Androcles & Lion* II. 42 The lion..purrs like a motor car.] 1922 Joyce *Ulysses* 507 His lawn-mower begins to purr. 1962 L. Deighton *Ipcress File* xxx. 190 Jay's Rolls purred along the Cromwell Road. 1974 P. Wright *Lang. Brit. Industry* i. 16 Their engines *purr* or *tick over* sweetly. 1978 *Times* 3 Apr. 12/3 The white Cadillac purred to a halt.

purra, var. *Poro.

purrer. (Earlier and later examples.)

1826 *Blackw. Mag.* XX. 326/1 Invisible to every living soul..except Cyprus, our cat, a perpetual purrer. 1972 *Village Voice* (N.Y.) 1 June 92/3 (Advt.), Adopt handsome male cat: 'Purry' Japanese Harlequin,..short silky fur, great purrer, playful.

pu·rringly, *adv.* [f. Purring *ppl. a.* + -ly[2].] In a purring manner; while purring.

1907 *Westm. Gaz.* 25 May 6/2 Her tail is all unfolded and she walks purringly. 1925 *Glasgow Herald* 18 Aug. 8 Zizi..would purringly allow herself to be stroked and fondled. 1939 Joyce *Finnegans Wake* II. 234 Quite purringly excited. 1946 E. O'Neill *Iceman Cometh* I. 56 (*Purringly*) Come now, Lieutenant, isn't it a fact that you're as guilty as hell? 1964 D. Francis *Nerve* xvii. 217 Buttonhook [*sc.* a horse] was..neighing purringly in her throat when we opened her door.

Purrow, var. *Poro.

‖**pur sang** (pür saṅ). [ad. F. *pur-sang* thoroughbred animal, f. *pur* pure, *sang* blood.] Phr. used adjectivally (freq. following a *sb.*) or adverbially to mean: of the full blood, without admixture, through-and-through, genuine.

1864 G. A. Sala *Quite Alone* I. xii. 194 The Countess was a Frenchwoman, pur sang. 1868 *Sat. Rev.* 14 Mar. 340/2 It is only the old-fashioned sort, not girls of the period *pur sang*, that marry for love. 1911 J. Ward *Realm of Ends* xi. 225 To the speculative mind *pur sang* there is nothing satisfactory about such a view. 1934 C. Lambert *Music Ho!* III. 192 *Pur-sang* exoticism of the fruity Oscar Wilde order is indeed extinct. 1941 *Mind* L. 52 The subjective, *pur sang*, is that which is wholly dependent upon this or the other particular subject. 1947 D. Mahon *Studies in Seicento Art & Theory* I. 16 Many of Guercino's late pictures appear far from classic *pur sang* even by comparison with some contemporary work. 1958 *Listener* 24 July 133/3 A fusion of these two traditions, bearing in mind that they are not always found *pur sang* even at the beginning. 1961 'W. Haggard' *Arena* xviii. 155 He wasn't a Lohmeyer but he was Lohmeyers *pur sang*. He'd despise a Sabin Scott. 1975 *Times Lit. Suppl.* 14 Feb. 162/2 It is in fact possible to be a sociologist *pur sang* and not a black (white, yellow, piebald, Scots, Croat, Methodist, Muslim, etc, etc) sociologist.

purse, *sb.* Add: **I. 3.** (Later examples.)

1967 *Boston Globe* 5 Apr. 51/1 Race horse owners, irked at the New York state legislature for failing to approve the money necessary for increased purses, [etc.]. 1976 *Columbus* (Montana) *News* 27 May 2/5 The four-day tourney..will offer a 72-hole medal play with a $25,000 purse. 1976 *Scotsman* 24 Dec. 15/2 Valsecchi refused a £25,000 purse offered by British match-makers to stage the defence in England as he preferred not to give up the advantage of fighting at home.

II. 6. e. A woman's handbag. *N. Amer.*

1955 W. Gaddis *Recognitions* I. ii. 77 A girl walking alone, swinging her purse. 1957 *New Yorker* 12 Jan. 32/2 Bernadett's purse hung over her arm. 1979 *Kingston* (Ont.) *Whig-Standard* 5 Apr. 24/3 The type of purse and the way you carry it can be enough to make purse snatchers or pickpockets think twice before choosing you as a victim.

III. 10. a. *purse-pocket*; (sense 3) *purse distribution, end, money, offer, winnings.*

1968 *Globe & Mail* (Toronto) 13 Jan. 40/4 There would have to be a reduction of approximately 20 per cent in the aggregate purse distribution. 1928 *Sunday Express* 16 Dec. 21/1 A purse end of £800 is more than Johnny need expect to receive in the States for his first fight. 1898 *Kansas City Star* 19 Dec. 3/1 Bingen's share of the purse money amounted to only $4,650. 1971 *Sunday Express*

(Johannesburg) 28 Mar. 22/2 In Britain, many boxers sell batches of tickets to their followers instead of receiving purse money. 1973 *Times* 16 Mar. 13/6 The long awaited return match between Bobby Arthur and John Stracey for the British welterweight title..is now up for purse offers. 1922 Joyce *Ulysses* 430 Bloom pats..pursepocket. 1970 *Globe & Mail* (Toronto) 28 Sept. 20/4 In each of those starts he was carrying between 119 and 123 pounds because of his purse winnings in South America.

11. purse-bag, a handbag, often having a purse incorporated or attached; **purse-belt** = *money-belt* s.v. *Money *sb.* 8; **purse boat** (further examples); **purse-club** (earlier example); **purse silk** (example).

1907 *Yesterday's Shopping* (1969) 404/3 Roan leather 'Modern' Purse Bag, with inside pocket for gold. 1914 G. B. Shaw *Fanny's First Play* III. 198 Putting down..her purse-bag. 1901 Kipling *Kim* x. 263 A worn old purse-belt embroidered with porcupine quill-patterns. 1943 *R.A.F. Jrnl.* Aug. 25 N.A.A.F.I. are producing those useful purse belts which you may wear under your jacket. 1911 *Oysterman & Fisherman* Sept. 25/2 Conant Brothers Company, Incorporated make a specialty of the construction of Purse boats, used so widely in purse net fishing along the coast. 1950 *Richmond* (Va.) *Times-Dispatch* (Mag. Sect.) 23 July 5/1 Next 'over the side' go powered purse boats, bearing the captain, seine-setters, other crewmen and a purse net. 1790 J. Woodforde *Diary* 25 May (1927) III. 192 The Purse-Club..came to my House this Morning with Cockades in their Hats. 1880 L. Higgin *Handbk. Embroidery* i. 6 (*heading*) Purse silk is sometimes used for diapering, and..where a raised effect is required.

purse, *v.* Add: **4.** Also with *out*.

1896 O. Schreiner *Story Afr. Farm* I. xii. 114 Pursing out his lips, and waving his hand, he solemnly addressed the boy.

pursed, *ppl. a.* Add: **1.** (Later examples with *up*.)

1937 V. Woolf *Years* 289 Maids..with their inscrutable, pursed-up faces. 1955 P. Larkin *Less Deceived* 34 Threading my pursed-up way across the park.

purse-net. Add: **1.** (Further examples.) Also *attrib.*

1911 [see *purse boat* s.v. *purse *sb.* 11]. 1931 *Sun* (Baltimore) 29 Jan. 6/2 Fish that sold for approximately $1,000 was garnered..by ten purse-net fishing boats. *Ibid.* 26 Feb. 5/6 Opponents of purse nets and buck net fishing in the Chesapeake Bay. 1971 *Stornoway Gaz.* 10 July 3/5 Over the past two weeks, a succession of large Norwegian pursers have been visiting the port, some to land purse nets for repairs. 1977 *Young's Sporting Appliances* (S. Young & Sons Ltd.) 4 Purse Net Line.. Purse Net Rings.

Hence **pu·rse-netting.**

1931 *Sun* (Baltimore) 13 Mar. 12/4 (*heading*) Mr. Denmead shows evils of purse netting. 1961 *Listener* 24 Aug. 269/1 The deadly purse-netting ships, with the aid of these depth-recorders, could pin-point the shoals at the exact depth when the fish were on spawning levels.

pursepick. For † *Sc. Obs.* read *rare* and add later example.

1977 J. Wambaugh *Black Marble* (1978) vi. 76 Pigeon droppers, pursepicks, muggers. Don't walk the Boulevard at night.

purse-proud, *a.* (Later examples.)

1930 L. G. D. Acland *Early Canterbury Runs* 1st Ser. v. 113 He was not in the least purse-proud or pompous, but always modest and unassuming. 1967 *Guardian* 27 June 8/3 One of Britain's biggest manufacturers of swimming pools..claims to have captured 40 per cent of a purse-proud market.

purser. Add: **2. b.** Also in the possessive in various combs. and phrases, as *purser's crab* (slang), a naval uniform boot; † *purser's dip* (see quot. 1867) *obs.*; *purser's name*, a false name under which, formerly, a man was entered on the books of a ship in the Royal Navy; *a purser's shirt on a handspike*, used as a type of the ill-fitting.

1810 J. Moore *Post-Captain* v. 23 There is nothing of him left but ribs and trucks. His coat fits him like a purser's shirt upon a handspike. 1821 P. Egan *Life in London* II. v. 308 The greatest anxiety with most individuals..is to appear what they are not; to copy some stylish *hero* for their *model*, but whose *dress* at most they merely *imitate*, and which generally fits them after the manner of a '*purser's shirt upon a handspike*'! 1828 *Night Watch* II. 82 Tom, when he was impressed into his Majesty's service, had taken the 'purser's name' literally '*un nom de guerre*' of Thomas Call, in which his warrant as boatswain was subsequently made out... Mr. Call's name, however, was in reality Thomas Whistle. 1829 Marryat *Naval Officer* I. viii. 231, I was down in one of the wings, reading by the light of a purser's dip—*vulgo*, a farthing candle. 1847 H. Melville *Omoo* 232 Some, to be sure, had for the sake of formality, shipped under a feigned cognomen, or 'Purser's name'. 1867 Smyth *Sailor's Word-bk.* 550 *Purser's dip*, the smallest dip-candle. 1878 *Detroit Free Press* 12 Jan. (Suppl.) 2/5 It fits him like a purser's shirt on a handspike. 1913 T. T. Jeans *John Graham Sub-Lieutenant R.N.* 18 'Now for the purser's "crabs"!' the Model gurgled, when I'd..produced a pair of service pattern boots. 1924 G. H. A. Willis *Royal Navy as I saw It* 83 Modern ships with wire hawsers instead of ropes, and iron decks and ladders, conduce to the wearing of shoes or 'purser's crabs'. 1927 P. Riley *Memories* 89 Each mess being allowed a few small candles, known as 'Purser's dips', to last the week. 1970 P. O'Brian *Master & Commander* x. 264 'Mr Dillon, who have we aboard that speaks Italian? John Baptist is an Italian.' 'And Abram Codpiece, sir—a purser's name.'

4. a. A ship using purse-nets. **b.** A fish caught in a purse-net.

1961 *Listener* 24 Aug. 269/1 The use of these pursers was forbidden in the cod waters of Lofoten. 1971 [see *purse-net 1]. 1973 *Stornoway Gaz.* 27 Jan. 1/1 The quality of the ring-net herring was mixed to very poor and soft, and the pursers very poor and soft.

purserette (pɒ̄:ɹsəre·t). [f. Purser + -ette.] A female purser on a ship or other form of transport.

1959 *Times* 17 Mar. 4/4 Miss Suzette Pienaar, who has joined the Union-Castle Line to become an Afrikaans-speaking 'purserette'. 1960 *Aeroplane* XCVIII. 453/1 Inside of 707 is rather fun... The Purser (we note the big-ship touch) introduces us to the Purserette (we come ashore again rapidly). 1970 *New Scientist* 28 May 431/2 A new technique that British Rail is using to make uniforms for hovercraft purserettes—the low-flying equivalent of the air hostess. 1975 *Times* 19 Feb. 14/7 Of the 90,000 sea staff on British ships..only a thousand or two are women and..practically all are in 'female roles': nurses, stewardesses, purserettes.

purse-seine. Add: (Earlier and later examples.) **purse-seine-net** (later examples); *purse-seine boat, fisherman, fishery* (further examples), *fishing*.

1870 *Amer. Naturalist* IV. 515 Purse-seines are used to the best advantage in capturing [mackerel]. 1884 *Bull. U.S. Nat. Museum* No. 27. 697 Purse-seine boat... This model represents the class of boats exclusively used in the mackerel purse-seine fisheries of New England. 1909 Purse-seine fishing [used s.v. Purse-seiner]. 1935 B. R. Hubbard *Cradle of Storms* vi. 95 Floating canneries had been allowed to invade the False Pass area, and powerful purse-seine boats had come in with them. 1960 R. Kirkbride *Innocent Abroad* x. 74 These purse-seine boats moved like cats' eyes over the water, luring shoals of fish with the beams of their kerosene pressure lamps. 1971 *Stornoway Gaz.* 7 Aug. 3/7 Its provisions apply to all nets, with the exception of purse-seine nets and ring nets, constructed to take fish while being towed or hauled through the sea by or from a fishing boat. 1971 *Country Life* 9 Dec. 1642/2 It was Gardenstown men who took over the first boat built specially for purse-seine fishing. 1975 *New Yorker* 22 Dec. 54/2 When the enormous purse seine (some three hundred fathoms long by eighty fathoms deep) is cast, the boat moves in a circle round a herring shoal, pulling the net into the shape of a pocket in a pool table. 1976 *Times* 14 Feb. 2/3 Huge nets, purse seine nets, that can trap up to 200 tons of fish in a single cast. 1976 *Quoddy Tides* (Eastport, Maine) 13 Aug. 2/1 Herring purse-seine fishermen in the Bay of Fundy have organized themselves with government assistance to change the purse-seine fishery from one based on fish meal to one based on food.

Hence **purse-seiner** (further examples); **pu·rse-seining.**

1919 *Dialect Notes* V. 58 The purse-seiners are reaping a harvest here just now, lots of fish and a good price. Island Co. Times, Coupeville (Washington). 1941 Steinbeck & Ricketts *Sea of Cortez* i. 7 The purse-seiners of Monterey ..are dependable work-boats. 1963 *Times* 25 Feb. (Canada Suppl.) p. ix/5 Present fishing methods are gill-netting, purse-seining and trolling. 1973 *Stornoway Gaz.* 27 Jan. 1/1 Last Wednesday's total of 7,612 crans from purse-seiners and pair trawlers was a record for the port of Stornoway. 1973 *Sunday Times* 10 June (Colour Suppl.) 44/4 The third and most controversial method is purse-seining—where a net is put round a school of tuna, the base is closed and the whole lot hauled on board. 1979 P. Benchley *Island* i. 13 Commercial purse-seiners telling their wives when to expect them home.

pursuant, *sb.* and *a.* **A.** *sb.* Restrict † *Obs.* to sense in Dict. and add: **2.** One who pursues; a pursuer. *rare.*

1924 W. J. Locke *Coming of Amos* xiii. 163 Amos..ran ..followed also at a run by Hamilton, thereby giving.. visitors..the impression of pick-pocket and pursuant. 1978 *Maledicta* 1977 I. 232 Little Bit's longtime pursuant and then casual lover.

pursuit. Add: **I. 2. c.** In track cycling, any of various kinds of competitive race (see quots. 1961 and 1975).

1938 *Encycl. Brit. Bk. of Year* 100/1, 2 miles team pursuit. 1961 F. C. Avis *Sportsman's Gloss.* 146/2 *Australian pursuit*, a track race in which a number of riders, starting from different points on the track and equally spaced out, try to catch the rider in front, whereupon the rider caught drops out of the race. *Ibid.* 149/1 *British pursuit*, a track race between two teams starting at opposite sides of the track, the members of each team standing ready to ride a lap when the previous rider of the team has been round. *Ibid.* 162/1 *Italian pursuit*, a track race between two teams starting at opposite sides of the track, the riders being in file, and who one by one drop out of the race as the track is lapped; thus, the leader of the file rides one lap, the second two laps, and so on. 1975 *Oxf. Compan. Sports & Games* 237/1 *Individual pursuit*.. Two riders take part in each race, starting from stations on opposite sides of the track and attempting to gain on each other. Victory goes to the first rider to reach his home station on completing the distance or, less often, to the rider who overtakes his opponent. *Ibid.*, *Team pursuit*, over 4 km., is between amateur teams of four riders. Each rider leads the team for one lap or half a lap, then swings up on the end banking and drops back to the end of the file. Thus, after another three laps or another lap and a half he finds himself leading the team once more.

III. 11. *attrib.* and *Comb.*, as *pursuit force, party, squadron*; (sense *2 c) *pursuit cyclist, race*; **pursuit aeroplane, aircraft, airplane** (*U.S.*), **biplane, plane** = *Fighter 3; **pursuit-flight** [repr. G. *reihen* sb. (E. Christoleit 1929, in *Beitr. Fortpflanzungsbiol. Vögel* V. 45)], a flight in which one or more male birds follow or attack a female.

1937 *Discovery* Sept. 277/2 The present type of pursuit aeroplane weighs perhaps some 4500 lb. 1940 *Jrnl. R. Aeronaut. Soc.* XLIV. 485 A Curtiss YP-36 type pursuit aeroplane. 1931 *Flight* 25 Dec. 1265/1 The contracts placed with the Detroit Corp...include five Lockheed two-seater pursuit aircraft of the type Y1P-24. 1928 Chatfield & Taylor *Airplane & its Engine* xv. 267 The pursuit airplane is..purely an offensive type. 1920 H. Woodhouse *Textbk. Appl. Aeronaut. Engin.* iii. 93 The D.H.5 Pursuit Biplane. 1970 *Soviet Weekly* 8 Aug. 14 In preparing for the world championship in England the Soviet pursuit cyclists did a lot of road work. 1930 J. S. Huxley *Bird-Watching & Bird Behaviour* iii. 54 Almost immediately he will fly at her, she will fly off, and the two will turn and twist through the air in what may be called the pursuit flight—a regular part of courtship in yellow-hammers and many other small birds. 1940 H. C. Witherby et al. *Handbk. Brit. Birds* III. 234 The pursuit-flights [of mallards]..have been discussed at length by German authors. 1954 D. A. Bannerman *Birds Brit. Isles* III. 376 Mr. Hartley describes tail-fanning by both sexes [of swallows] in pursuit flight. 1968 P. A. Johnsgard *Waterfowl* vi. 51 When the female is involved in incubating..these pursuit flights take a different form. 1945 *Diamond Track* (Army Board, N.Z.) 33/1 The pursuit force opened out into desert formation. 1909 *Daily Chron.* 5 July 1/6 All available attendants were mustered as a pursuit party. 1918 E. S. Farrow *Dict. Milit. Terms* 479 Pursuit plane. 1932 *Flight* 17 Nov. 1099 (*caption*) A new Boeing pursuit plane. 1962 R. B. Fuller *Epic Poem on Industrialization* 1 Zooming aloft In a pursuit plane. 1908 T. A. Cook *Olympic Games* 188 Pursuit Race Three laps (1·807 kilometres) Teams of four to start. First three to count in each heat. 1928 E. Hemingway *Men without Women* 190 In a pursuit race, in bicycle racing, riders start at equal intervals to ride after one another. 1961 J. S. Salak *Dict. Amer. Sports* 346 A pursuit race may have two or more contestants, who are started at equal distances behind and ahead of the nearest contestant or contestants so that the circuit of the track is divided equally by the starting points. A pursuit race may be run to a finish or for a specified distance. In either case the winner is he who has caught and passed all contestants or remaining contestants. 1976 'A. Hall' *Kobra Manifesto* i. 13 He could be chased..by pursuit squadrons of the Yugoslavian air arm.

purty (pɒ̄·ɹti), *a.* and *adv.*, repr. Irish and U.S. local pronunc. of Pretty *a.* and Pretty *adv.* Also (before a nasal consonant) **purt', purt** (*U.S.*).

1829 G. Griffin *Collegians* II. xxiii. 177 Dhrinking away! Wisha, long life to you says I, if that's the way; a purty fruit the tree bears in you, says I. 1844 'J. Slick' *High Life in N.Y.* I. i. 1 Purty much alone. 1860 C. M. Yonge *Hopes & Fears* viii. 312 To be shure, an' it's not such a purty young lady as yourself that need be taking the trouble. 1898 J. D. Brayshaw *Slum Silhouettes* 50, I can hear you purty well. *a* 1911 [see *hon[2]]. 1922 Joyce *Ulysses* 165 Three Purty Maids from School. 1926 J. Black *You can't Win* (1927) vi. 67 I'd take fifty cents of it purty pronto. 1938 R. E. Bass in B. A. Botkin *Treas. S. Folklore* (1949) III. i. 461 Them things was purt' nigh as handy as the eggs from them hens. 1946 *Amer. Speech* XXI. 98 *Purt nigh*, nearly. 1977 *Time Out* 28 Jan.–3 Feb. 17/3 Lawdy lawdy Miss Linda Lewis sings real purty. 1977 *Time* 17 Oct. 54/1 The frog-voiced, razor-witted Daumier of Dogpatch for purt' near 44 years casually told an assistant: 'You can stop cutting the paper. I'm not going to draw any more.'

Purum (pu·rum). *Anthrop.* Pl. **Purum, Purums.** The name of a tribe of mongoloid peoples living near the Indo-Burmese border, whose kinship system is characterized by matrilateral cross-cousin marriage; also *attrib.* or as *adj.*

1912 J. Shakespear *Lushei Kuki Clans* II. iii. 150 It is said that 'Pu rum' means 'hide from tiger', which connects them closely with the Lamgang legend. *Ibid.* 153 Among the..Purum and Lamgang marriages must be made within the clan, but not within the family. 1945 T. Das *Purums* 2 Terms of address used by male and female members of one sib in respect of the members of the remaining sibs also helped to form a correct idea about the ancient laws of Purum marriage. 1958 R. Needham in *Amer. Anthropologist* LX. 83 He also worked these out from Purum statements about who ought to marry whom. 1967 R. Fox *Kinship & Marriage* viii. 217 The picture of the Murngin then is of a 'Purum' type community with clans but with lineages as the operating units marrying asymmetrically. 1971 W. Wilder in R. Needham *Rethinking Kinship & Marriage* ix. 214 Some recent discussions..are about as far removed from Purum facts as it is possible to get and still call the material 'Purum'.

pus. Add: **b.** *pus-yellow* adj.

1922 Joyce *Ulysses* 511 Virag..claps..on the wall a pusyellow flybill.

puschkinia (puʃki·niă). [mod.L. (J. M. F. Adams 1805, in *Nova Acta Acad. Petropolitanæ* XIV. 164), f. the name of Apollos Mussin-*Puschkin* (d. 1805), Russian chemist and plant collector + -ia[1].] A small spring-flowering bulbous plant of the genus so called, belonging to the family Liliaceæ, and bearing spikes of blue or white cup-shaped flowers; also called the striped squill.

1820 *Curtis's Bot. Mag.* XLVIII. 2244 (*heading*) Squill-like Puschkinia. 1914 G. Jekyll *Colour Schemes for Flower Garden* (ed. 3) 6 The colour scheme begins with the pink of *Megasea ligulata*..and later the blue-white of *Puschkinia*. 1925 A. J. MacSelf *Bulb Gardening* xi. 197 The flowers of Puschkinia are blue and white, arranged in a short close-set spike on a stalk only a few inches long. 1959 *Times* 22 Aug. 9/4 Most of the other small bulbs—muscari, chionodoxas, puschkinias—can be grown indoors. 1974 H. G. W. Fogg *Compl. Handbk. Bulbs* vii. 122/2 As long as they are not forced, puschkinias can be grown indoors like crocuses.

Puseyite. b. (Earlier examples.)

1839 J. B. White *Let.* Aug. in *Life* (1845) III. x. 131 That association, called the *Puseyite* party, from which we have those very strange productions entitled *Tracts for the Times*. 1843 J. S. Mill *Let.* 23 Oct. in *Wks.* (1963) XIII. 603 The Puseyite review the British Critic.. almost exhausts language in admiration of me & my book.

push, *sb.*[1] Add: **I. 1. b.** (Examples from *Cricket*.)

1888 R. H. Lyttelton in A. G. Steel et al. *Cricket* ii. 72 There is..a good length ball on the legs to which this push can be usefully applied if the batsman..cannot make use of the sweep to leg. 1898 K. S. Ranjitsinhji *With Stoddart's Team* (ed. 4) xii. 233 [MacLaren].. chiefly obtained his runs by his 'push' in the slips. 1921 G. R. C. Harris *Few Short Runs* iii. 58 [W. G. Grace] introduced what was then a novel stroke,..viz., the push to leg with a straight bat off the straight ball. 1976 *Evening Post* (Nottingham) 14 Dec. 18/4 Both were caught by wicketkeeper Ved Raj off Lal's bowling, Fletcher playing an indeterminate defensive push.

d. Paired with *pull*, esp. to convey the concept of a force.

1878 *Proc. R. Soc. Edin.* IX. 610 The ear does distinguish, as it were, between push and pull on the tympanum. 1932 Andrade & Huxley *Introd. to Science* iii. 63 Electric and magnetic forces act across perfect emptiness, as if with invisible pulls and pushes. 1966 L. Basford *Sci. of Movement* xii. 33/1 We usually think of a force as the push or pull needed to move something.

e. *to give* (a person) *the push*, to eject (a person), to throw out; to dismiss, *esp.* from employment. *colloq.*

1899 C. Rook *Hooligan Nights* ii. 23 He was employed as a chucker-out... His regular business..was 'giving mugs and other barmy sots the push out of pubs'. 1923 T. E. Lawrence *Let.* 23 Mar. (1938) 404 Nothing else showed up, after I got the push from the R.A.F. 1933 D. L. Sayers *Murder must Advertise* ix. 158 He told me to string him along. And afterwards..to give him the push. 1957 W. Camp *Prospects of Love* III. iii. 160 Mummy had her..to work here..but she was quite hopeless..and Mummy gave her the push. 1968 'P. Hobson' *Titty's Dead* xv. 155 His landlady's given him the push. 1976 S. Barstow *Right True End* III. xiv. 209 'Hedley Graham has started a month's notice.' 'You don't mean he's..?' 'Got the push? No. He gave Maurice Kendall his resignation on Friday.'

4. b. Also, *spec.*, a military advance (first widely used in the latter stages of the war of 1914–18). Also *fig.*

1916 *Punch* 7 June 407 (*caption*) The far-reaching effect of the Russian push. 1916 F. M. Ford *Let.* 7 Sept. (1965) 75 The Big Push was too overwhelming for one to notice details; it was like an immense wave full of debris. 1918 J. M. Grider *War Birds* (1927) 260 Henry told us that there is going to be a big push shortly. Push? What's a push to us? That's for the Poor Bloody Infantry to worry over. We push twice a day, seven days in the week. 1929 E. W. Springs *Above Bright Blue Sky* 69 I've shed many a tear over you. I heard that you were killed during the push in front of Amiens. 1935 *Sun* (Baltimore) 15 July 1/8 A marked push toward early completion of the Administration's 'must' program was expected. 1942 *R.A.F. Jrnl.* 30 May 33 The only original officer of the Wing who had been in the first push. 1964 *Wall St. Jrnl.* 5 Feb. 1 We're stepping up our drive on all fronts..and that includes our whole Northern push on housing..and voter registration. 1976 S. Barstow *Right True End* III. xiv. 223 They joined up together in gangs in that war—Pals—and in a big push they sometimes died together. 1978 *Time* 3 July 17/1 The top-priority items are the kind of antitank and antiaircraft weapons that could be used to repulse a Soviet push across the border.

d. The act of selling drugs illicitly (cf. *push *v.* 13 c).

1973 J. Wainwright *High-Class Kill* 58 The push was made in one of the city's public parks. The main pusher was one of those men nobody ever really sees.

II. 8. a. (Further examples.)

1923 T. E. Lawrence *Let.* 21 May (1938) 422, I met your cousin once, at a push in London: had no proper talk of him. 1955 D. W. Maurer in *Publ. Amer. Dial. Soc.* xxiv. 174 A crowd is, to a pickpocket, a *tip*, a *press*, a *crush*, or a *push*... 'Three troupes is up against this push already.'

9. (Further examples.) Also *attrib.*

1903 R. Bedford *True Eyes & Whirlwind* xx. 127 The nightly push club assembled. 1911 [see *nit *sb.*[2] 1]. 1914 *Sat. Even. Post* 4 Apr. 12/2 'The whole push is hungry, Kid,' he said. 'I'm hungry.' 1926 Kipling *Debits & Credits* 307 You're from Sydney, ain't you?..I know how your push talk, well enough. 1927 [see *mob *sb.*[1] 5 b]. 1964 C. Mackenzie *Life & Times* III. 182 Presently there burst into the room half a dozen of the rowing 'push'. 1967 *Sunday Mail Mag.* (Brisbane) 12 Nov. 3/1 Experts on push warfare in Sydney in the early 1870's rated The Rocks Push as the No. 1 team of larrikins in the city. 1973 *Nation Rev.* (Melbourne) 31 Aug. 1436/1 He was portrayed almost as another Keynes—or, at the very least, the intellectual peer of the Friedman-Galbraith-Samuelson push.

push, *v.* Add: **I. 1. b.** *to push up daisies*: see *daisy *sb.* 1 c.

f. *to push the bottle* (earlier example).

1788 J. Woodforde *Diary* 20 Aug. (1927) III. 44 Mr. Atthill being Chairman pushed the Bottle about pretty briskly.

g. Also in *Cricket*.

1893 *Cricket* 26 Oct. 442/1 Box..has a style of getting off his ground when a ball is directed to his legs, with the intention of..'pushing' it to the 'leg'. 1920 D. J. Knight in P. F. Warner *Cricket* 34 If he [*sc.* the batsman] is pushing the ball away to long leg, he must face long leg. 1963 A. Ross *Australia 63* iii. 76 He moved quick enough up the wicket to Titmus, but having got there was content to push.

h. Also, *to push away* i.e. from the shore. *push off*: also, *fig.*, to depart, go away (freq. *imp.*). Also without *off* and *to push along*.

1740 *Proc. Sessions of Peace London & Middlesex* May 164/1 He..heard somebody a cursing and swearing, and a Woman..say, *d——n it, push off, or go off*. 1918 K. E. Harriman *Wine, Women & War* (1926) 39 Grand day to be pushing off for Bordeaux. 1923 Wodehouse *Inimitable Jeeves* xvii. 241 He helped himself absently to a handful of my cigars and pushed off. 1931 A. Christie *Sittaford Mystery* xxiii. 192, I shall be pushing along now. So long. 1947 Wodehouse *Full Moon* vii. 141 I'll be pushing along. 1949 J. B. Priestley *Delight* 231 This is my view, not yours. Push off! 1955 G. Freeman *Liberty Man* I. i. 21 Goodnight, Maur. I'll be pushin'. I've 'ad a day. 1964 R. Jeffries *Embarrassing Death* iii. 25 Bill finished his drink. 'I'd better be pushing.' 1973 E. Page *Fortnight by Sea* viii. 89 She must be quite certain to leave when the girl with the frizzy hair decided to push off. 1976 *National Observer* (U.S.) 26 June 16/4 A man in a small sailboat pushes away from the shore of the Atlantic and never is seen again.

k. *to push* (someone) *around*, to move or cause (someone) to be moved roughly from place to place, to manhandle. Freq. *fig.* (orig. *U.S.*), to browbeat, bully, domineer over. Also, *to push about*.

1923 H. C. Witwer in *Cosmopolitan* Aug. 45/2 Look at the pushing around he's getting because he hauled off and inherited a million. 1930 D. Runyon in *Liberty* 8 Nov. 24/1 After..Johnny gets on the strong-arm squad, he never misses a chance to push Big Jule around. 1942 R. Chandler *High Window* iii. 29 If anybody tries to push Linda around, he'll have to push me around first. 1949 'M. Innes' *Journeying Boy* i. 12 The father doted on the son, the son pushed the father around. 1963 D. Ballantyne in C. K. Stead *N.Z. Short Stories* (1966) 153 The Aussie..has made it bloody clear he *won't* be pushed about. 1964 M. Argyle *Psychol. & Social Probl.* xiv. 177 Resistance to change on the part of industrial workers is reduced if they play some part in making the decision and its augmentation. Not only is the feeling of being pushed about avoided, but those concerned are able to set up the new social system to their satisfaction. 1973 'J. Patrick' *Glasgow Gang Observed* xix. 170 The Glasgow gang boy feels that he is being pushed around, that he has no control over the social conditions which predetermine his future. 1974 N. Freeling *Dressing of Diamond* 93 Thought you could come and push me about. Not the first. But I'm still here. 1976 *National Observer* (U.S.) 26 June 6/2 America has pushed these people around too much, too long, and it's natural that they feel resentment and react violently.

l. Phr. *to push* (someone's) *face in*, to punch (someone) on the nose. *slang.*

1930 'R. Crompton' *William—The Bad* ix. 228 I'll go and find the blighter and push his face in for him. I never heard of such beastly cheek!

m. Fig. phr. *to push the boat out*, to be generous, *esp.* in paying for rounds of drinks. *slang* (orig. *Naut.*).

1937 J. Curtis *You're in Racket, Too* iii. 39 This bloke you're meeting up the Old Jacket and Vest to-night, let him push the boat out, the bastard. Surely he can pester for a tightener if you're hungry. 1946 J. Irving *Royal Navalese* 140 *Push the boat out, to*, a boatwork term used to imply paying for a 'round of drinks'. 1962 'J. Le Carré' *Murder of Quality* i. 10 'Fielding's giving another dinner party tonight.' 'He's pushing the boat out these days.' 1977 B. Pym *Quartet in Autumn* x. 90 'Pushing the boat out, aren't you?' said Norman, with unusual jollity, as Ken topped up his glass.

n. Phr. *when push comes to shove* and varr., when action must back up threats; when the worst comes to the worst. *colloq.* (orig. *N. Amer.*)

1958 Murtagh & Harris *Cast First Stone* vii. 105 Some..judges..talk nice and polite... Then, when push comes to shove, they say 'Six months in the workhouse'. 1970 *Calgary* (Alberta) *Herald* 4 May 57/1 If push comes to shove, make good the threat. 1977 *National Observer* (U.S.) 22 Jan. 12/4 When—to use common parlance—push comes to shove, I have a great deal of faith in American youth. 1981 *Guardian* 10 Jan. 19/8 (*heading*) Push comes to shove.

7. a. Also, *to push along*.

1902 'Mark Twain' in *Harper's Weekly* 6 Dec. 5/1 Push along, cabby, push along—no great lot of time to spare.

II. 8. a. Delete 'Now *rare*' and add later example.

1975 *N.Y. Times* 10 Apr. 29/2 Former Governor Terry Sanford reportedly was one of the men pushing hardest for the primary repeal.

b. *trans.* To approach (a certain age). *colloq.*

1937 S. V. Benét in *Sat. Even. Post* 18 Sept. 42/4 I'd kind of like to beat out Ike Leavis... To hear him talk, you'd think nobody had ever pushed ninety before. 1953 R. Chandler *Long Good-Bye* xxiii. 148 When you're young..you can absorb a lot of punishment. When you are pushing forty you don't snap back the same way. 1959 *Housewife* Oct. 134/2 Maria's a bit old... Pushing seventy, you know. 1962 *Woman's Own* 18 Aug. 16/1 All these women, either pushing 40, or looking back at it without too much regret, have been good box-office for years. 1976 *National Observer* (U.S.) 2 Oct. 12/5 Flicka is pushing 50, but she still wears her frosted hair shoulder length.

10. a. (Further examples.) Also, with *along*.

1911 H. B. Wright *Winning of Barbara Worth* xxix. 411 Give your horse a drink but don't wait to rest. You can push him from now on as hard as you like. 1962 *Which?* Oct. (Car Suppl.) 118/2 It was the back wheels which eventually broke away if the car was pushed too far. 1971 'H. Calvin' *Poison Chasers* vii. 90 Dai was pushing the Land Rover all out, but it was still too slow for me. 1972 'I. Drummond' *Frog in Moonflower* 18 The driver pushed the bus along... It was doing well over sixty now.

c. *Bridge.* To try to force (an opponent) into a higher and more doubtful contract by overcalling him. Also *intr.*

1927 M. Work *Contract Bridge* 149 *Push*, to overbid for the purpose of inducing the opponents to assume a losing contract. 1934 G. F. Hervey *Mod. Contract Bridge* xxii. 247 If you know a player is determined to play every hand, you can 'push' much more successfully against him than against the player who knows when to leave off bidding and when to double. 1959 *Listener* 24 Dec. 1118/2 When East accepted the invitation to game he was pushed beyond game. 1980 *Guardian Weekly* 21 Dec. 23/5 West cunningly bid only 5S[pades] in the hope of being allowed to play in 6S when he was pushed there.

11. a. (Further examples without advb. extension.)

1952 *Sun* (Baltimore) 22 Mar. 6/4 Even if steelworkers push their productivity, a very large share of their production goes..into war materials. 1966 A. Sachs *Jail Diary* iii. 34 He only asked one question all the time, and did not even push that one. 1970 B. Mather *Break in Line* v. 60 'Once is funny, twice is cheeky,' he grunted. 'Don't push things, boy.'

d. *to push one's luck*: see *luck *sb.* 3.

e. Phr. *to push it, things*, to cause (an action) to be rushed; to hurry, cut fine. *colloq.*

1967 H. Dalmas *Fowler Formula* iii. 31 [We] could have her by Christmas... It would be pushing things a little, but they said it could be done. 1971 'F. Clifford' *Blind Side* IV. iii. 165 Fourteen twenty-five?—or is that pushing it a bit?

13. a. (Further examples.) Also (now *obs.*?) with *off*.

1873 *Punch* 26 Apr. 178/2 Why do not the managers imitate another class of persons who push off drugs by means of puffing. 1936 D. Powell *Turn, Magic Wheel* II. 140 He saw a bad month ahead explaining to Dennis why his book was not being pushed. 1949 Wodehouse *Uncle Dynamite* xiv. 237 She was always complaining that her last publishers wouldn't push her books. 1977 *Jrnl. R. Soc. Arts* CXXV. 124/2, I think the improvement grants we have are fairly good. They need to be pushed more.

c. To peddle (drugs) illegally. Also *absol.* *slang* (orig. *U.S.*).

1938 *Amer. Speech* XIII. 190/1 *To push*, to peddle narcotics, especially as a sub-agent or small-time dealer. 1953 W. Burroughs *Junkie* ii. 33, I decided right then I would never push any more tea [*sc.* marijuana]. 1956 'E. McBain' *Pusher* (1959) 37 'How would I know..even if he was supplying himself and others besides?' 'Was he pushing?' 1959 J. Osborne *World of Paul Slickey* II. ix. 71 It will surely bug you when there is..no tea to push. 1968 B. Turner *Sex Trap* xvi. 154 'Are you the man?..You pushing or aren't you?' 1977 'J. Fraser' *Hearts Ease in Death* xv. 171 Was Billy Nesbitt buying amphetamines..and selling them to other kids? Was he, in fact, pushing drugs?

14. (Further examples.)

1863 Trollope *Small House at Allington* in *Cornh. Mag.* XVIII. 272 'They'll be very pushed about money,' said Mr. Boyce. 1946 *R.A.F. Jrnl.* May 170 He is occasionally a little pushed by the constant stream of callers. 1967 P. Moyes *Murder Fantastical* xiv. 209 Sorry we can't invite you to lunch, Tibbett, but what with the funeral and the Fête..Vi's a bit pushed. 1972 K. Benton *Spy in Chancery* viii. 85 We think his boss may be pushing him. 1978 G. A. Sheehan *Running & Being* xii. 173 You frequently read that a runner would have done better if he only had someone to push him during a race.

push-. Add: **a.** (*a*) *push-basket, -boat, -net, -nipple, -plane*; (*b*) *push-pole, -rod* (*attrib.* examples), *-stick*; (*c*) with advbs. forming sbs. and adjs., as *push-along, -in, -on, -out* (also *push-down *sb.* and *a.*, etc.), indicating (*sb.*) the act of pushing in the direction specified; (*adj.*) that pushes or is pushed in the direction specified. **b. push-bicycle** (earlier example); **push-bike** *colloq.*, a push-bicycle; hence as *v. intr.*, to ride a push-bicycle; also **push-biking** *vbl. sb.*; **push-car,** restrict *U.S.* to senses in Dict. and add: (*a*) (example); (*d*) a perambulator; **push-cart,** (*a*) (earlier and

later examples); also *attrib.*; (*b*) a perambulator; **push-chain** *Linguistics*, a sound shift in which one phoneme approaches a second and this in turn shifts so that their differentiation is maintained; also *attrib.*; **push-chair,** a small, wheeled, usu. folding chair in which a child can be pushed along; **push-cycle** (further example); **push-cyclist,** a rider of a push-cycle; **push drive** *Cricket*, a drive (Drive *sb.* 1 d) in which the ball is pushed instead of struck; **push fit,** a fit which enables a part to be pushed into a hole by hand but does not allow free rotation; **push-foot** = *push-pedal*; **push hold** *Mountaineering* = *pressure hold* s.v. *Pressure *sb.* 10; **push-in,** (*a*) *U.S. slang*, a certainty; (*b*) *Hockey*, the act or action of pushing the ball into play from the side-line; (*c*) *Austral. slang* (see quot. 1979); **push money** *U.S. slang* = Spiff *sb.*; **push moraine** *Physical Geogr.*, an arc-shaped moraine formed by an advancing or re-advancing glacier or ice-sheet which pushes material before it into low ridges; **push-out,** (*a*) *sb.*, one who is made to leave, esp. school; *slang*; (*b*) *adj.*, that pushes out; **push pass** *Sport*, a pass effected by pushing rather than hitting or kicking the ball; **push-penny** = *push-halfpenny*; **push-pit** *Naut.* [formed humorously after *pulpit *sb.* 4 f], a raised safety rail in the stern of a boat; **push plate,** a plate attached to a door by which it may be pushed open; **push-process** *v. trans. Photogr. colloq.*, to develop (a film) in such a way as to increase or maximize its effective speed; so **push-processing** *vbl. sb.*; **push-shot** = *push-stroke*; **push-start** *v.*, to start (a motor vehicle or engine) by pushing (the vehicle), usu. after failure of normal procedures; also as *sb.* (*lit.* and *fig.*); **push-through,** (*b*) an instrument for cleaning the bore of a rifle (cf. *pull-through* s.v. Pull- 1); (*c*) used *attrib.* to designate things in which one part is pushed through another; **push-towing** *vbl. sb.*, the propulsion of a line of connected unpowered barges by a powered one at each end; also *loosely* (see quot. 1959); so **push-tow,** a line of such vessels; also *attrib.* and as *v. trans.*

1977 *Grimsby Even. Tel.* 14 May 9/2 (Advt.), Pedigree pushalong fur horse excellent condition, £4.50. 1956 *Harper's Mag.* May 20/2 She threads her pushbasket along the alleys of the super market. 1906 *Bazaar, Exch. & Mart* 16 Nov. (Suppl.) 2042/3 Exchange [motor-cycle].. for good make 25in push bicycle and cash. 1913 'I. Hay' *Happy-go-lucky* xiv. 180 Luckily I had the old push-bike with me, and I managed to find my way down here. 1914 C. Holme *Lonely Plough* xx. 236 Strenuous figures with bare knees and flapping overcoats push-biked past them. 1918 S. P. B. Mais *Schoolmaster's Diary* xvi. 253, I 'push-biked' the eight miles into Lewes. 1920 *Isis* 3 Nov. 3/1 Self-advertisement, or the man who rides a push-bike with both hands in his trouser pockets. 1970 'D. Halliday' *Dolly & Cookie Bird* viii. 123 Derek..thought of a push-bike... He didn't want to be followed. 1926 *Punch* 8 Dec. 643/1 Music, Greek Plays, 'push-biking' tours—All figure in his pages. 1972 *Guardian* 22 Feb. 11/3 If you take to push-biking..you will need some pedal-pushers. 1928 P. C. Chambliss in J. Schoettle *Sailing Craft* 202 The patent stern affords means of fixing davits by which bugeyes may hoist their motor yawls or push boats. 1967 *Guardian* 17 June 9/6 The pushboat picks them [*sc.* barges] up..loaded or unloaded. 1884 E. W. Nye *Baled Hay* 225 A section-crew..riding down that mountain on a push-car. 1922 Joyce *Ulysses* 240 Edy.. was rocking the chubby baby to and fro in the pushcar. 1893 E. King *Joseph Zalmonah* ix. 105 Some hundreds of 'push-carts' like Ben Zion's were ranged within the narrow limits of Hester Street. 1897 F. Moss *Amer. Metropolis* III. ix. 202 The visitor may stand at one point and see without moving..sidewalk merchants and push-cart vendors. 1909 *Daily Chron.* 10 Dec. 5/4 She ran into the..street, and there found the push-cart, and saw the man hurrying away with the baby wrapped up in a travelling rug. 1921 *Daily Colonist* (Victoria, B.C.) 12 Oct. 16/3 (Advt.), Child's wicker push-cart, price $5. 1931 J. T. Adams *Epic of Amer.* xii. 346 Many of the other 'great' bankers..had the souls of pushcart peddlers. 1973 *Amer. Speech* 1969 XLIV. 265 All the level 3 stores operated on the supermarket plan with pushcarts and terminal checkout booths. 1952 A. Martinet in *Word* VIII. 11 It may often be difficult to tell whether we have to do with a B→A→ chain, or drag-chain, or an A→C→ chain, or push-chain. 1969 R. D. King *Hist. Linguistics & Generative Gram.* viii. 194 If one rejects the gradualness of phonological change..and the notion that language abhors merger, push chains are deprived of their major source of plausibility. 1972 M. L. Samuels *Linguistic Evol.* iii. 31 If one phoneme shifts, others will also shift in such a way that the differentiation is preserved ('push-chain mechanism'), while others again will automatically increase their area of possible realisation by moving into the vacated space ('drag-chain mechanism'). 1977 *Language* LIII. 239 Graphemic change provides evidence for a push chain. 1921 *Sunday at Home* Feb. 257/2 Up the hill she struggled... She was throwing her weight against a small push-chair with a carpet seat. 1963 *Times* 25 May 9/5 As the mothers come out of the shops they pop sweets into the mouths of the two-year-olds sitting in pushchairs. 1972 J. Wilson *Hide & Seek* i. 19 She hesitated, wondering whether to pop Jamie in his pushchair and go after them. 1931 D. L. Sayers *Five Red Herrings* ii. 32 He had the body on the floor of the tonneau and on top of it he had a push-cycle, which has left tarry marks on the cushions. 1915 W. H. L. Watson *Adventures Despatch Rider* v. 63 We stopped and questioned a 'civvy' push-cyclist. 1927 *Daily Express* 27 Dec. 3/7 A push-cyclist..writes to protest against being forced to show a red light behind. 1920 D. J. Knight in P. F. Warner et al. *Cricket* (new ed.) 28 If the ball it not struck on the half-volley, but a little later, it [*sc.* the drive] becomes what is known as the push drive, and is in fact the ordinary forward shot. 1918 D. T. Hamilton *Gages, Gaging & Inspection* ii. 38 Push fits..are for shafts that are forced into a hole by hand and that would be free to rotate without seizing, but not free enough to rotate under anything but a very slow speed. 1960 *Practical Wireless* XXXVI. 330/1 A 2½in. length of steel knitting needle ground down to a push fit inside the nylon bearing. 1900 G. D. Hiscox *Horseless Vehicles* ii. 37 The movement..was made by a push-foot connection from a three-throw crank shaft. 1957 R. G. Collomb *Dict. Mountaineering* 122 *Push Hold.* (*American.*) A pressure hold. 1976 D. Clark *Dread & Water* v. 107 Zoom lens showing handholds—push hold, jug-handle, fingers clenched on a small hold. 1948 *Daily Progress* (Charlottesville, Va.) 22 July 11/1 The statement that William and Mary [College] is a push-in for top honors in the Old Dominion is just a lot of wild talk. 1970 *Sunday Tel.* 9 Aug. 24/6 The push-in, the latest addition to the sporting glossary, makes its international debut.. today... The new rule..becomes operative for British clubs at the start of the season. 1976 Read & Walker *Advanced Hockey for Women* v. 119 Occasionally the opportunity may arise to send the ball directly to the middle of the field from a push-in. 1976 *Sunday Mail* (Brisbane) 7 Nov. 47/11 They then walk home—and are followed by the 'push-in' merchants, the teenage savages who push their victims into their apartments from behind, slam the door and then lace into them. 1979 *Courier-Mail* (Brisbane) 20 Jan. 18/6 *Push-ins*, mugging at the door. 1939 C. Morley *Kitty Foyle* (1940) xxx. 296, I was getting twenty-eight a week and my push money extra. 1960 V. Packard *Waste Makers* (1961) xix. 231 The spiff or PM is the 'push money' offered as a reward for each item of the brand sold. 1890 T. C. Chamberlin in *Bull. Geol. Soc. Amer.* I. 28. A glacier deposits material at its margin in three ways: (1) It pushes matter forward mechanically, ridging it at its edge, forming what may be termed push moraines. 1913 *Zeitsch. für Gletscherkunde* VII. 310 Part of the glacier margin was bordered by a push moraine from 5 to 8 feet high. 1960 B. W. Sparks *Geomorphology* xiii. 292 Push moraines are a specialised form of end moraine caused by a readvance of an ice sheet thrusting till, or some similar deposit, up into low ridges. 1979 J. Rabassa et al. in C. Schlüchter *Moraines & Varves* 68/2 In March 1977, the ice front had already advanced over the proximal part of the fluvioglacial plain.., bulldozing its upper sedimentary cover into a set of push-moraines. 1920 W. T. Grenfell *Labrador Doctor* i. 7 The shrimp fishermen..used push-nets in the channels at low tide. 1976 *Weekend Echo* (Liverpool) 4/5 Dec. 9/8 (Advt.), Shrimp push nets for sale, £12.50. 1902 *Engin. Rev.* (N.Y.) May 15/2 The sections [of the boiler] are united by malleable iron push-nipples coated with copper, and fitting accurately reamed holes in the sections. 1926–7 *Army & Navy Stores Catal.* 285/1 Massage Bath Shower and Shampoo Set complete, with large rubber push-on unions. 1974 K. Clark *Another Part of Wood* ii. 69 The old push-on variety [of pianola]..gave the executant much more control than the later one-piece model. 1970 *Britannica Bk. of Year 1969* 798/3 *Pushout*, a student dropped from school for unsatisfactory performance. 1973 *Times* 17 Dec. 2 The growing number of girls who are becoming homeless are not 'drop-outs', as generally thought, but 'push-outs'. 1974 *Florida FL Reporter* XIII. 43/2 The 'push-out' rate of minority students is a national disgrace. 1977 *Design Engin.* July 73/2 They are easily installed by simply squeezing into punched or drilled holes in 1·5mm cold-rolled steel sheets, and resist pushout forces of 26olb. 1963 *Times* 25 Feb. 4/3 Their forwards..used the push-pass far too often on a surface which demanded hard hitting. 1977 *Time Out* 28 Jan.–3 Feb. 6 (Advt.), Push pass... There are at least 26 familiar football terms in this puzzle. 1872 B. Jerrold *London* xviii. 146 Benches where they are playing push-penny. 1975 *Country Life* 11 Dec. 1677/4, I am..looking for examples of the following regional inn sports: aunt sally (Oxfordshire)..push penny (Lincolnshire)..actually played in English pubs today. 1964 *English Studies* XLV. 23 The pulpit is in the bows; a similar device at the stern has become known..as a *push-pit*. 1976–7 *Sea Spray* (N.Z.) Dec./Jan. 90/1 (Advt.), It does not get chipped or rattle against the pushpit. 1928 V. G. Childe *Most Anc. East* iii. 54 A steep-ended scraper or push-plane. 1977 G. Clark *World Prehist.* (ed. 3) v. 214 Wood-working equipment, manifested most notably in heavy bifaces and picks and in high-backed push-planes. 1907 G. A. T. Middleton *Mod. Buildings* VI. xiv. 112/2 The double bolts as supplied for swing doors are the proper pattern to use... They are actuated from the inside by a small push plate. 1963 W. C. Huntington *Building Construction* (ed. 3) xv. 661 Push plates or door pulls are provided on the closing stile as required. 1906 Push-pole [in Dict.]. 1971 Push-pole [see *kilhig]. 1977 *Sat. Rev.* 23 July 3/2 (Advt.), The 200 [*sc.* 200 a.s.a. film] can be 'push-processed' to 400 speed... Dealer can sell you a kit, including directions for 'push-processing'. 1979 *Amat. Photographer* 10 Jan. 90/1 All these fast films can be push-processed to produce even higher speeds. 1934 *Jrnl. R. Aeronaut. Soc.* XXXVIII. 191 Push rod valve mechanism for air-cooled engines..has been almost universally adopted during the last few years. 1973 *Times* 18 Oct. 35/3 The Polski 125P saloon..has the same body as the old Fiat 125 and a 1500cc push rod engine. 1909 P. A. Vaile *Modern Golf* v. 84 The push-shot is a dead straight ball, one of the straightest when well played. 1925 *Country Life* 15 Aug. 244/2 The push shots or placing shots... You can steer and guide these strokes with tolerable accuracy. 1957 S. Moss *In Track of Speed* xiv. 182 Mechanics rushed out and push-started us. 1965 D. Lodge *Brit. Mus. is falling Down* vi. 107 He prepared to push-start his scooter. 1973 *Advocate-News* (Barbados) 29 June 3/3 (Advt.), Maybe you have an idea. And all it needs is a push-start to get it off the ground. 1977 *Daily Tel.* 12 Jan. 10/4 One of my minor objections to automatic transmissions is that they can't be push started. 1979 K. O'Hara *Searchers of Dead* viii. 80 Owen..once gave me a push-start when my battery was flat. 1922 *Woodwork Machinery Reg.* in *Statutory Rules & Orders* (1923) 276 A suitable push-stick shall be kept available for use at the bench of every circular saw which is fed by hand, to enable the work to be carried on without unnecessary risk. 1947 J. Charlesworth *Law of Negligence* xviii. 388 Failure to use a 'push-stick'..may amount to contributory negligence. 1920 G. Burrard *Notes on Sporting Rifles* 68 Greener's 'push through' is an excellent invention for all ultra small bores. 1970 *Which?* Aug. 237/2 Slip-over threading points are better than push-through points. 1979 D. Francis *Whip Hand* xvi. 195 The push-through switch on a table lamp. 1955 *Bull. Soc. Naval Architects & Marine Engineers* Feb. 12/1 Single-screw tugs have been push-towing for many years. 1955 F. Marbury *Push-Towing in Waves* (MS. thesis, Mass. Inst. Technol.) i. 1 The standard river pushtow cannot operate in waves. 1964 *Marine Engineering Log* July 59/1 The economy and flexibility of push-tow operations are gaining favor with Japanese maritime interests. 1970 *1st Internat. Tug Conf. 1969* 272/1 Petroleum barges could be push-towed. *Ibid.* 362/1 In 1957 the first real push-tow..appeared on the Rhine. 1973 *Guardian* 22 Jan. 6/5 Push-tow craft..are, basically, floating boxes which can carry 140 tons and be locked together in a procession of nine, operated by two tugs. 1974 *Encycl. Brit. Macropædia* III. 758/2 These assemblies of unpowered and individually unmanned barges are known, somewhat illogically, as push tows. 1955 F. Marbury *Push-Towing in Waves* (MS. thesis, Mass. Inst. Technol.) vi. 22 The basic conclusion..is that as far as these tests extend pushtowing in waves is feasible. 1959 G. Walker *Traffic & Transport in Nigeria* iii. 47 'Push-towing' has now become the accepted practice. Power craft have two barges lashed to the forequarters, a third being pushed ahead. 1972 *Daily Colonist* (Victoria, B.C.) 29 Mar. 40/5 The tugs are intended for use in..push-towing of such barges in moderate sea conditions.

pu·sh-and-go, *sb. phr.* and *adj. phr.* Also **push and go.** [f. Push-: see Go *sb.* and *v.*] **A.** *sb. phr.* The ability to develop and prosecute a scheme vigorously (see quots.); enterprise, initiative, ambition.

[1915 D. Lloyd George in *Hansard Commons* 9 Mar. 1277 We are on the look out for a good, strong business man with some go in him who will be able to push the thing through and be at the head of a Central Committee.] 1915 *Times* 10 Mar. 14/5 The Government should..get a business man at the head of the organization. The Government were on the look-out for a good, strong business man with some push and go in him, who would be able to put the thing through. 1916 Ld. Fisher *Let.* 26 Jan. in M. Gilbert *Winston S. Churchill* (1972) III. *Compan.* II. 1398, I said what was required was '*Push and Go*'! (NOT '*Wait and See*'!). 1959 F. M. G. Willson in *Polit. Stud.* VII. 224 Recruits from business, industry, and trade unions begin with Sir Eric Geddes, one of Lloyd George's men of 'push and go'..graduating to the Cabinet as the first Minister of Transport in 1919.

B. *adj. phr.* **a.** Ambitious, enterprising, pushing.

1932 Kipling *Limits & Renewals* 80 He is one of the push-and-go type..the flower of the Higher Counter-jumpery.

b. Of a motorized toy etc.: having a mechanism that stores and releases the momentum generated by a preliminary push.

1958 *New Scientist* 9 Jan. 15 This novel type of shunting locomotive is a larger-scale version of a child's 'push-and-go' toy. 1959 *Oxf. Mail* 21 Jan. 6/3 The soft plastic trains and cars had their wheels removed very promptly and the push-and-go 'engines' soon fall out.

pu·sh-and-pull, *adj. phr.* and *sb. phr.* Also **push and pull.** [f. Push-: see Pull *sb.*[2] and *v.* and cf. *push-pull.] **A.** *adj. phr.* Involving pushes and pulls, esp. alternately. **a.** *gen.*

1914 H. Carrington *Probl. Psychical Research* xii. 371 A straight push-and-pull action is easier to accomplish than the more detailed and complicated action of forming words and letters. 1949 Koestler *Insight & Outlook* xiii. 192 The mechanistic push-and-pull physics of the last century. 1960 *Times* 3 Oct. (Advt. Suppl.) p. ii/2 The Skid-Stac attachment..consists of a load carrying plate and push-and-pull rack.

b. Designating (the operation of) a 'reversible' train, which may journey in either direction without having its engine turned about. Also, of the locomotive engine itself. See also **pull-and-push* s.v. Pull- 3.

1939 K. G. Fenelon *Brit. Railways To-day* iii. 61 The most hopeful solution would appear to be the adoption of the push-and-pull type of train, which can be driven from either end, and which can take extra vehicles if required. 1955 C. J. Allen *Gt. Eastern Railway* vi. 66 The Edmonton & Cheshunt line..was..reopened, on March 1st, 1915, to serve some munition factories in the neighbourhood, and was worked by a two-coach 'push-and-pull' unit with 2-4-2 tank No. 1311. 1965 K. Hoole in *Regional Hist. Railways Gt. Brit.* IV. xii. 219 Push-and-pull units, first used between Hartlepool and West Hartlepool in 1905, became a familiar sight throughout the system. 1975 G. Bye in G. W. Knight *Jackson Knight* v. i. 367 A much planned..trip on the 'Tivvy Flyer', the push and pull train on the Tiverton and Bampton branch-line, was, alas, never made.

PUSH-BUTTON

B. *sb. phr.* **a.** *U.S. Mil.* (See quot. 1929.) Also *attrib.*

1920 *Official Hist. 315th Infantry U.S.A.* 28 The greater part of the time was devoted to..the 'push and pull' exercise. **1921** F. T. Floyd *Company F Overseas* 37 These rides remind a soldier of that bit of army exercise popularly known as 'push and pull'. **1929** *Papers Mich. Acad. Sci., Arts & Lett.* X. 317 *Push and pull*,..sighting and aiming drill.

b. *fig.* Tug of war.

1958 *Spectator* 4 July 19/3 The dramatic centre of the book is..the push-and-pull between Yule and his devoted but mutinous daughter.

pu·sh-button, *sb.* and *a.* Also **pushbutton, push button.** [f. Push- + Button *sb.*] **A.** *sb.* A button that is pressed with the finger to effect some operation, usu. by closing or opening an electric circuit.

1878, 1901 [in Dict. s.v. Push-]. **1912** L. Weaver *House & its Equipment* 124 The multiple contact switch, which consists of a little board..on which are arranged a number of push buttons. **1920** C. Sandburg *Smoke & Steel* 218 She used to keep a houseful of girls in kimonos and three pushbuttons on the front door. **1935** *Times* 2 Feb. 9/5 Special signal lights will face pedestrians and push-buttons will be fitted to the posts. **1943** T. Horsley *Find, fix & Strike* 64 The range is point-blank... The pilot's thumb, already on the push-button on top of the throttle bar, jabs hard against the stop. **1957** *Railway Mag.* Mar. 159/2 The pantograph is raised by pressing the push-button in the driving trailer. **1976** *Gramophone* May 1835/1 Below the main controls are the following: sockets for stereo microphone and headphones, push-buttons for low filter, high filter,..and power on-off.

B. *adj.* **1. a.** Operated or effected by pressing a push-button.

1916 *Inland Printer* LVII. 830/2 (*caption*) 'Push-button control'. **1936** *Discovery* Apr. 113/1 The diesel engine is started electrically and the push-button control for this purpose is mounted in the cab adjacent to the driving positions. **1943** *Gramophone* Dec. 107/1 The radio has 3 bands with push-button tuning. **1957** *Observer* 25 Aug. 11/1 It is no doubt true that guided missiles, nuclear warheads, flame-throwers, and push-button apparatus do not in the slightest invalidate the elementary military virtues of *esprit de corps* and self-discipline. **1957** *Amer. Speech* XXXII. 313 The following list of descriptive terms that appear on the aerosol containers of..shaving creams... *push-button lather.* **1965** *Wireless World* July 8/2 (Advt.), Stereo Amplifier..push-button selection. **1978** S. Brill *Teamsters* iv. 127 He took his seat..alongside a table bearing two separate push-button phones.

b. *push-button war(fare)*, warfare conducted by means of (nuclear) missiles launched by the press of a button.

[**1945** *Life* 20 Aug. 17/1 There may be devastating 'push-button' battles.] **1945** *Richmond* (Va.) *Times-Dispatch* 9 Nov. 6/2 (*heading*) 'Pushbutton' war seen by atom men. **1946** *Ibid.* 21 Feb. 12/6 The push-button warfare forecast for the future. **1948** *New Republic* 29 Nov. 15/3 The vision of the clean, fast, economical impersonal push-button war grows dim. **1955** T. H. Pear *Eng. Social Differences* vi. 139 In what historical perspective can we see the changes which radio, television, faster-than-sound travel, push-button warfare..and social medicine are causing in our social life? **1958** *Daily Tel.* 28 June 6/3 If you are thinking in terms of push-button warfare.

2. Characterized by the use of push-buttons, *spec.* implying technological advancement; fully automated or mechanized.

1946 *Birmingham* (Alabama) *News* 3 Feb. 1/3 The Army Air Forces came forth Saturday with a real push-button plane. **1955** G. Freeman *Liberty Man* I. iii. 39 One might have thought he was a ship's boy serving on a graceful old tea clipper, rather than an efficient piece of mechanism in a modern push-button navy. **1960** *Times* 12 Jan. 13/5 The married woman of the future will live in an increasingly push-button home. **1962** *Lancet* 19 May 1081/1 The push-button type of hospital, which already exists in some countries, is not the sort of place in which one would choose to be ill. **1973** *Times* 17 Oct. 14/4 If increased mechanization should be decided on..engineering would ..offer a vastly increased number of 'push-button' jobs.

3. Easily obtainable, as at the press of a button; instant.

1947 *Sun* (Baltimore) 20 Mar. 2/1 Political rulers..have regularly resorted to manufacturing money by one process or another. This measure is the latest of these attempts at 'push-button' money. **1967** *Listener* 22 June 821/3 Some part of the price we are paying for the alleged boon of push-button entertainment..can already be discerned. **1972** [see *Jesus 3 b].

pu·sh-down, *sb.* and *a.* Also **pushdown.** [f. vbl. phr. *to push down*: see Push *v.* 1 b.] **A.** *sb. Aeronaut.* A manœuvre in which an aircraft in level flight loses altitude and resumes level flight.

1938 [see *pull-up 3 a].

B. *adj.* **1.** *Computers* and *Linguistics.* Being or pertaining to a linear store or list that receives and loses items at one end only, the first to be removed on any occasion being always the last to have been added.

1961 *12th Symp. Appl. Math.* 104 These problems seem amenable to solution by the application of techniques based on the use of what some computer people have come to call a 'pushdown' store. **1963** N. Chomsky in R. D. Luce *Handbk. Math. Psychol.* II. 343 Evidently pushdown storage is an appropriate device for accepting (generating) languages..which have..nesting of units (phrases) within other units, that is, the kind of recursive property that.. we called self-embedding. **1963** *IEEE Trans. Electronic Computers* XII. 872 A 'push-down' list is one that is manipulated in a last-in, first-out manner. **1967** D. G. Hays *Introd. Computational Linguistics* ii. 31 The importance of pushdown storage lies in the fact that it has exactly enough power to deal with context-free languages. **1972** R. Quirk et al. *Gram. Contemp. Eng.* xi. 736 The *wh*-element can be fronted from a position in a clause subordinate to the *wh*-clause (a pushdown *wh*-element); for example the informal: I don't remember *which shelf* he told me I was to fetch it from.

2. That may be or is designed to be pushed down.

1977 *Custom Car* Nov. 18/1 The Escort has sprouted chrome push-down bonnet catches.

pushed (s.v. Push *v.* in Dict.), *ppl. a.* Add: **1.** (Later example.)

1969 *Jane's Freight Containers 1968–69* 286/1 The Rhône..will be open to pushed convoys of 3,000 tons.

2. *Comb.*, as *pushed-back, -down, -up* ppl. adjs.

1922 Joyce *Ulysses* 45 They wait, their pushedback chairs..around a board of abandoned platters. **1948** P. White *Aunt's Story* iv. 80 Theodora had gone. There was only the pushed-back furniture. **1971** D. E. Westlake *I gave at the Office* (1972) 20 The pushed-down button for the line in use goes right on. **1878** *Q. Jrnl. Geol. Soc.* XXXIV. 566 Pushed-up mounds or long ridges of gravels ..are a conspicuous feature along the shores of the Polar basin. **1962** *Listener* 5 Apr. 617/2 Mr Thomas has the commanding quality of a real *Heldentenor*, not a pushed-up baritone as are many Wagnerian tenors.

pusher. Add: **1. a.** (Further examples.)

1909 *Daily Chron.* 12 Oct. 4/6 It is a very difficult matter for an agent to canvass in a legitimate manner, as these special 'pushers' have told such glowing yarns of 'increased bonuses and profits'. **1929** [see *passer 3 d]. **1946** Wodehouse *Money in Bank* xix. 159 He was not without his dark suspicions of that big-hearted pusher of oil shares. **1954** *Sun* (Baltimore) 5 June 1/8 Perry had come to this city..to act as a 'pusher' of the stolen cash. **1973** *Amer. Speech 1969* XLIV. 259 *Pusher*, locomotive that helps to start a heavy train or to push a train up a grade.

c. A girl, a young woman; *spec.* a prostitute. *slang.*

1923 J. Manchon *Le Slang* 236 *Pusher*,..girl, une typesse, une gonzesse. **1936** J. Curtis *Gilt Kid* 116 Mr. Bloody Bedbug's up here having a good time with my pusher. **1944** A. Wykes in *Penguin New Writing* xix. 105 A pusher for me. I'm off the beer, but I could use a judy. **1971** B. W. Aldiss *Soldier Erect* 19 Nelson and his pusher took the chance to sneak away, and I managed to manoeuvre Sylvia as far as the kitchen.

d. One who peddles drugs illegally. *slang* (orig. *U.S.*).

1935 J. Hargan *Gloss. Prison Lang.* 6 *Pusher*, one who retails drugs. **1948** H. L. Mencken *Amer. Lang.* Suppl. II. 681 A marihuana smoker is a *viper*..and a peddler is a *pusher*. **1951** *N.Y. Times* 14 June 1/1 Encouraged by 'pushers' of narcotics who sometimes offered free samples to beginners. **1956** 'E. McBain' *Cop Hater* in *87th Precinct* (1959) 60 Junkies are easy to trace. Talk to a few pushers, zing, you're in. **1959** *Guardian* 3 Dec. 9/2 High-powered city detectives..looking for 'junkies' (drug-addicts) and 'pushers' (drug-peddlers). **1967** E. Wymark *As Good as Gold* ix. 140 People like Crane..were called 'pushers' and were usually addicts themselves. **1976** *Howard Jrnl.* XV. i. 46 Western loathing for temptation is vented.. upon the scapegoats of the junkie and the pusher.

2. b. *Naut.* The seventh mast of a seven-masted schooner.

1902 *Boston Even. Transcript* 23 July 20/3 The name of the masts, by the way, are in order, fore, main, mizzen, spanker, jigger, driver, and pusher. **1909** *Shipping Illustr.* 25 Dec. 327/1 As is now well known, the sixth mast was denominated the driver and the seventh the pusher.

c. *Aeronaut.* An aircraft having an airscrew behind the main wings. Freq. *attrib.*

1913 *Flight* 7 June 613/2 The 'pusher', as this machine is familiarly called to distinguish it from a tractor biplane of the same make. **1915** [see *nacelle 2 b]. **1918** Cowley & Levy *Aeronautics* i. 5 The pusher type is much less efficient as a flying machine than a tractor. **1922** *Encycl. Brit.* XXX. 20/2 The first biplanes..were of the 'propeller' type, colloquially 'pushers'; almost all monoplanes were 'tractors'. **1940** *Sun* (Baltimore) 23 Aug. 4/5 The new Nazi types are reported to be a 'pusher' fighter..and a medium bomber. **1955** *Sci. News Let.* 19 Feb. 114/1 Another small 'pusher' propeller mounted between the double tail assembly gives thrust for level flight. **1969** K. Munson *Pioneer Aircraft 1903–14* 104/2 Later in 1906 Blériot converted this machine into the Blériot IV, installing two Antoinettes driving pusher propellers. **1976** *National Observer* (U.S.) 25 Sept. 17/1 Kiceniuk's aerobike looks like some experimental light airplane. It has red wings..and a 6½-foot pusher propeller.

3. An implement, in profile resembling a rake, used by infants to push food on to a spoon or fork; also, a piece of bread used for this purpose.

1926–7 *Army & Navy Stores Catal.* 606/3 Child's silver spoon and food pusher, in case—18/-. **1937** Partridge *Dict. Slang* 671/1 *Pusher*,..a finger of bread used as a feeding-implement. **1939–40** *Army & Navy Stores Catal.* 578/1 Child's silver spoon and food pusher, in case. **1957** J. Kirkup *Only Child* v. 84 Among the cutlery were my own two personal pieces—a spoon and a 'pusher'; the latter was an inelegant little implement which I used to push food on to my spoon. **1959** *Sunday Times* 25 Oct. 20/3 The traditional pusher is..on the way out..dropped ..in favour of the spoon alone. **1963** C. Mackenzie *Life & Times* I. 155 The pusher, a small piece of crust which one was always being adjured to use more carefully to assist the cut up meat on to one's fork.

4. A push-chair. *Austral. colloq.*

1953 A. Upfield *Murder must Wait* vii. 60 Several prams and pushers parked in an alcove. **1966** G. W. Turner *Eng. Lang. Austral. & N.Z.* viii. 180 What he calls a *pushchair*..or a friend from Adelaide [calls] a *pusher*. **1970** K. Giles *Death in Church* vi. 151 With her patent folding, plasticised pusher she intended taking the twins for a walk. **1979** *Verbatim* Summer 8/1 When a headline announces that *pushers* are to be allowed on Adelaide buses, the permission extends not to 'peddlers of drugs' but to 'a child's pushchair'.

5. *attrib.* and *Comb.*, as **pusher set**, a baby's spoon and pusher; **pusher-tug** (see quot. 1970).

1939–40 *Army & Navy Stores Catal.* 543/3 Baby spoon (loop) and pusher set—Gift box 5/-. **1951** *Catal. Exhibits, South Bank Exhib., Festival of Britain* 125/1 Pusher set... Feeding set. **1970** *Guardian* 19 Sept. 18/5 A 'train' of three barges is propelled by a 300 hp twin-screw 'pusher-tug'. **1973** Hrůša & Coxon tr. *Kozák's Ships* 188/1 A new towing system..involves the pusher-tug pushing a group of closely connected barges.

pushful. (Further examples.)

1931 Wodehouse *If I were You* xiv. 163 What a pushful young devil you are. **1938** E. Waugh *Scoop* III. 272 He must be a very pushful fellow, inviting himself here like this. **1970** *Rep. Comm. on University Press* (Univ. Oxford) 71 This more 'pushful' approach. **1974** 'W. Haggard' *Kinsmen* ix. 93 The tiresomely modern bishop ..was pushful and very far to the Left.

pushfulness. Add: (Further examples.) Also *fig.*

1926 R. M. Caven *Gas & Gases* ii. 38 The great characteristic of a gas or vapour is its pushfulness: it is always pushing. *Ibid.* 39 The property of a gas which we have colloquially called its pushfulness..with more propriety we should call the expansive power. **1958** *Economist* 25 Oct. 297/2 Moscow and Peking have divided, by tacit agreement, their zones of interference: China in Asia, the Soviet Union in the Middle East and Africa. Even so, the pushfulness of the two has varied remarkably. **1968** *Listener* 29 Aug. 280/2 The Dick Whittington legend.. with its twin themes of individual pushfulness and the escape from provincial stagnation.

pushiness (pu·ʃinės). **1.** *Philos.* [f. Push *v.* + -y[1] + -ness.] A term used by A. N. Whitehead (1861–1947) for the property inherent in a material object which enables it to be apprehended and identified by touch (see quots.).

1920 A. N. Whitehead *Concept of Nature* ii. 43 We are left with spatio-temporal positions, and what I may term the 'pushiness' of the body. **1927** B. Russell *Outl. Philos.* x. 118 We must give up what Whitehead admirably calls the 'pushiness' of matter. **1944** E. Nagel in P. A. Schilpp *Philos. Bertrand Russell* 339 It seems to me grotesque to say that the 'pushiness' of matter can disappear as a consequence of a new analysis or redefinition of matter.

2. [f. *pushy *a.* + -ness.] = Pushfulness in Dict. and Suppl.

1968 *Economist* 2 Mar. 10/2 Claims in areas of particular union pushiness—say, for engineering draughtsmen—should be looked at very carefully indeed. **1976** *Times Lit. Suppl.* 23 Jan. 79/1 She had a kind of insistent pushiness in the interests of her family that brooked no contradiction.

pushing, *vbl. sb.* Add: **a.** (Further examples, in sense *13c of the vb.)

1962 'K. Orvis' *Damned & Destroyed* ix. 61 My boss don't go for guys that goof like that. So he bounced me fast. I'm through pushing. **1971** B. Malamud *Tenants* 148, I wouldn't want him to go back to numbers, or pushing, or anything like that. **1974** P. McCutchan *Call for Simon Shard* iv. 42 The body had contained no residue of heroin, so pushing was more likely to be the answer.

pushing, *ppl. a.* Add: **b.** (Further examples.)

1765 C. Brietzcke *Diary* 8 Aug. in *N. & Q.* (1964) CCIX. 13/1 Said Nothing.., for fear he should think me pushing. **1966** *Listener* 27 Oct. 613/2 Lesser men might think him pushing or selfish or out for his own ends.

pushmi-pullyu, pushme-pullyou (pu·ʃmī-pu·lyū). [f. phrs. *push me* and *pull you*: see Push *v.* 1 a.] A fabulous creature resembling a llama, but with a head at both ends, invented by Hugh Lofting (1886–1947) in *Doctor Dolittle* (see quot. 1922); hence (with spelling rationalized), applied allusively to incoherent or ambivalent attitudes or policies.

Widely popularized by the film version of *Doctor Dolittle* (1967).

1922 H. Lofting *Doctor Dolittle* x. 92 Pushmi-pullyus are now extinct... They had no tail, but a head at each end, and sharp horns on each head... Only half of them slept at a time. The other head was always awake—and watching. **1964** *Daily Tel.* 5 May 16/2 With one hand it [sc. the Government] may give them incentives to get out of London. With the other, it already gives them incentives to stay where they are... The total effect of these push-me-pull-you policies must be conjectural. **1972** *Guardian* 11 Jan. 12/2 The Push-me Pull-you Bill. **1972** *Times* 28 Nov. 14/6 The [Labour] party's imitation of a Pushme-Pullyou over the European Parliament. **1974** *Economist* 21 Dec. 52/2 Wilsonologists are now trying to work out..whether his pushme-pullyou performance was due to..agnosticism on the common market..or..a shrewd eye on the polls. **1975** W. Percy *Message in Bottle* i. 19 Man's theory about himself doesn't work any more..because its parts are incoherent and go off in different directions like Dr. Doolittle's *pushmi-pullyu.* **1976** *Times Lit. Suppl.* 2 Apr. 399/3 Children likewise seem to need jabberwockies and pushmipullyous [*sic*] to help them learn the boundaries of the natural order.

pushmobile (pu·ʃmŏbīl). *U.S.* [f. Push *v.* + *-mobile.] (See quots.) Also *attrib.*

1911 A. N. Hall *Handicraft for Handy Boys* xxiv. 364 A Pushmobile is a unique form of home-made wagon that has been developed from the simple wagons which the boys used to make for coasting, and for pushing from behind, when the automobile was unknown. It is patterned as nearly as possible after an automobile, and it is pushed by the mechanician, who runs behind, while the driver rides and attends to the steering. **1952** *Milwaukee* (Wisconsin) *Jrnl.* 31 May (Green Sheet) 2/2 (*heading*) Try it! An airplane pushmobile. *Ibid.*, In this push-mobile airplane illustration I have not shown dimensions, for no two boys would work this out alike. **1974** J. Heller *Something Happened* 545 We made push-mobile scooters out of ball-bearing roller skates.

pu·sh-off, *sb.* (*a.*) [f. vbl. phr. *to push off*: see Push *v.* 1 b.] **1. a.** (In Dict. s.v. Push- b.)

b. The, or an, action of pushing down with the foot so as to propel oneself into the air.

1949 Shurr & Yocom *Mod. Dance* v. 165 In the leap, the push-off from back foot onto the forward foot, gives impetus to the leap. **1960** E. S. & W. J. Higham *High Speed Rugby* iii. 38 The take-off is from the *right* foot and consists of a vigorous push-off, so that the *left* foot can take a fairly generous step diagonally to the left.

2. *attrib.* or as *adj.*, designating something that pushes off, *spec.* a powered frame or bar that pushes material from the tines of a buckrake or the like; also *absol.*

1957 C. Culpin *Farm Machinery* (ed. 5) x. 277 A hydraulically operated push-off device can be used in conjunction with a front-mounted buckrake, and this outfit is more suitable than the simpler tipping type for loading most types of vehicles. **1970** *Financial Times* 13 Apr. 8/6 A new twin-ram push off buckrake. *Ibid.*, The push-off assembly is moved forward by two hydraulic rams. **1976** *Billings* (Montana) *Gaz.* 5 July 9-c/4 (Advt.), Used F10 D loader with hay basket, steel teeth, push-off, manure fork & grapple fork.

pushover (pu·ʃōᵘvəɹ). Also **push-over, push over.** [f. vbl. phr. *to push over*: see Push *v.* 1 b.]

1. Something easily accomplished or overcome: an easy task or victory; a 'cinch'. *slang* (orig. *U.S.*).

1906 *Outing* Jan. 461/2 To me it looks like a push-over. **1926** *Amer. Mercury* Dec. 465/2 The combination is a push-over on Loew's or any other time. **1927** *Vanity Fair* (N.Y.) Nov. 67/2 Among some of Conway's more famous expressions are:..'It's a push-over' (a 'cinch'; easy to accomplish); [etc.]. **1931** E. Linklater *Juan in Amer.* II. xiii. 147 Those Princeton guys have been boasting that this game's a pushover for them. **1943** *Amer. Speech* XVIII. 256 Americanisms which have wide currency in Australia:..pushover, [etc.]. **1951** 'J. Wyndham' *Day of Triffids* vii. 133 If Brigham Young could bring it off in the middle of the nineteenth century, this ought to be a push-over. **1973** P. Malloch *Kickback* xxi. 133 About the security van... It's going to be hard to take... Eight years ago they were a push-over.

2. Someone who is easily pushed over or overcome. *slang* (orig. *U.S.*). **a.** *Boxing.* A mediocre fighter.

1926 *Variety* 29 Dec. 7/4 A push-over, which means a fighter with round heels along cauliflower alley, was, by the same token, a dame on rockers in another circle. **1958** C. Williams *Man in Motion* (1959) iii. 27 He was a long way from being a push-over. He was a little heavier than I am, and he could really punch.

b. A woman who makes little resistance to demands for sexual intercourse; an easy 'lay'.

1926 [see sense 2 a above]. **1929** E. Wilson *I thought of Daisy* i. 16 Oh, Myra Busch is a push-over!.. She's got round heels! **1936** [see *cinch *sb.* 2]. **1949** H. Wadman *Life Sentence* II. i. 49 Then you came along with Lawrence —the dark reasons of the blood, and so on. Naturally I was a pushover for you. **1955** D. Barton *Glorious Life* xlvi. 155 She was a pushover, hardly worth the elaborate build-up. **1978** M. Puzo *Fools Die* xlvi. 487 Why the hell shouldn't she be a pushover? Weren't men pushovers for girls who fucked everybody?

c. An easy victim.

1934 *Sun* (Baltimore) 3 May 12/7 The would-be cracksmen have come to regard a policeman as a natural push-over. **1941** W. Stevens *Let.* 13 Jan. (1967) 385, I suppose Denmark was a push-over on account of the pastry they eat there. **1959** F. Richards *Practise to Deceive* vii. 106 You tell me that I'm..such a pushover—that a good-looking man can..wrap me around his little finger? **1975** D. W. S. Hunt *On Spot* v. 83 Since then our overseas suppliers have never been quite sure that we are a push-over at any price they like to ask.

d. Const. *for.* One who is readily influenced by or susceptible to the attraction of something; a 'sucker'.

1944 H. Croome *You've gone Astray* xii. 123 Are you quite advertisement-proof yourself?.. I'm not. I'm a pushover for Vanity. **1946** 'J. Tey' *Miss Pym Disposes* xviii. 184 I'm a push-over for passing plates. It must be the gigolo in me. **1956** S. Ertz *Charmed Circle* 96 He was always trying new tooth pastes and was a 'pushover'.. for all the advertisements he saw. **1975** *New Yorker* 21 Apr. 139/1 This department, always an old pushover for a picture horse, picks Foolish Pleasure.

3. *Rugby Football.* The action whereby one side in a scrum pushes the ball over the opponents' goal line, esp. in *pushover try.*

1958 *Observer* 14 Dec. 24/2 A 'pushover' try by Blackheath..was the only score in a game in which the players could be heard ploughing their way through the mud. **1959** *Ibid.* 15 Mar. 32/8 The Welsh pack wheeled..to try a pushover. **1960** *Times* 7 Mar. 4/7 After 25 minutes came a genuine pushover. **1977** *Western Mail* (Cardiff) 5 Mar. (Rugby Suppl.) 4/3 J. J. Williams's disallowed try in that game, I felt, was only as dubious as the England push-over try, also disallowed.

push-pin. Add: **2.** Chiefly *U.S.* (See quot. 1961.)

1923 *Geyer's Stationer* 5 May 42 (Advt.), Extensive advertising has created big sales for Moore push-pins. Glass heads—steel points. **1926–7** *Army & Navy Stores Catal.* 448/2 Glass push pins... It is a steel point with a glass handle, and is surprisingly strong in wood and plaster..easily inserted, and as easily withdrawn. **1942** *Amer. Cinematographer* Apr. 188/3 A story board is a large 4 × 8 foot piece of wallboard or celotex, on which the story sketches are pinned in rows with aluminium push-pins. **1961** Webster, *Pushpin*, a steel point having a projecting glass or metal head for sticking into a wall or board and used chiefly as a picture hook or as an indicator on a map. **1974** C. C. Woodard *Cable Television* vi. 138 A pushpin is stuck in the map at that location; and that pushpin's number is written in the Work Requested section.

push-pull (puʃ,pu·l), *a.*, *sb.*, and *adv.* [f. Push- + Pull *v.* or *sb.*[2]] **A.** *adj.* **1.** Characterized by, caused by, or being a forced reciprocating motion; responding to or exerting both pushes and pulls. Also *transf.* and *fig.*

Push-pull train (see quot. 1966).

[**1894** *Phil. Mag.* XXXVIII. 301 They..show that the 'push and pull' theory is capable of giving an adequate account of the action of the telephone.] **1929** *Prof. Paper Inst. P.O. Electr. Engineers* No. 124. 34 The frequency characteristics of a Western Electric..'push-pull' carbon transmitter [*sc.* a microphone]. **1934** [see *multipole *a.*]. **1940** *Chambers's Techn. Dict.* 687/2 *Push-pull microphone*, a carbon microphone in which two carbon-granule cells are mounted on either side of a stretched diaphragm, so that amplitude distortion arising in one is largely balanced out by the opposite phase amplitude distortion in the other. **1951** *Engineering* 10 Aug. 178/3 'Push-pull' fatigue tests on welded bridge members were continued. **1959** [see *facia 2]. **1962** R. B. Fuller *Epic Poem on Industrialization* 50 Basic structural stability..by segregated satisfaction of isolated articulating push-pull forces. **1963** *Times Rev. Industry* June 117/1 Fork Truck Attachments... Drum forks, brick handling forks, push-pull device. **1966** K. Möller *Amer. & Brit. Railway English* 31 *Push-pull, reversible train*..a type of locomotive-hauled suburban train fitted with driving control apparatus connected to the engine, at the rear end. **1971** *Engineering* Apr. 110/2 (Advt.), A responsive industrial control system taking care of loads from a few ounces to over 1000 lb through push-pull cables. **1972** *Modern Railways* Sept. 364 Instances continue to occur of Glasgow-Edinburgh push-pull trains being worked by single locomotives. **1972** *Science* 20 Oct. 311/3 Cyclic AMP and cyclic GMP function in opposite directions, that is, in a push-pull fashion to exert long-term control over neuronal excitability in the sympathetic ganglion. **1978** A. Huxley *Illustr. Hist. Gardening* iv. 119 The push-pull weeder hoe—with a flat oblong blade sharpened on both edges.

2. *Electronics.* Having or involving two matched valves or transistors that operate 180 degrees out of phase on identical alternating inputs, so that they conduct for alternate half-cycles and their combined output is the sum of each acting alone, making possible increased power without reduced efficiency.

1924 *Wireless World* 4 June 277/2 With the push-pull amplifier one may employ smaller and therefore less expensive valves. **1925** *Motor* 8 Dec. 980 B/1 The Push-pull Wireless Circuit. **1932** *Oxford Times* 23 Sept. 22/5 Some manufacturers stock 'pairs' of carefully matched valves for push-pull amplification. **1945** *Electronic Engin.* XVII. 431/2 A more satisfactory way of cancelling or reducing cathode self-bias distortion is to use push-pull stages with common self-bias. **1955** *Radio Times* 22 Apr. 30/1 Table radiogram with 6-watt 'push-pull' output. **1965** *Wireless World* July 329/2 Fig. 8 shows a push-pull 55 kc/s oscillator which provides erase and bias signals. **1970** J. Shepherd et al. *Higher Electr. Engin.* (ed. 2) xxiv. 778 The fact that both *p–n–p* and *n–p–n* transistors are available enables push-pull circuits to be designed without transformers... If a *p–n–p* and an *n–p–n* transistor are fed from the same drive, a given input swing will cause one transistor to conduct more while the other conducts less, giving a push-pull operation. **1974** Harvey & Bohlman *Stereo F.M. Radio Handbk.* v. 127 The driver transistor TR_1 provides a common phase signal drive to the bases of the output pair but since they have complementary characteristics, the operation is in effect push-pull.

3. *Cinemat.* (See quot. 1973.)

1934 *Jrnl. Soc. Motion Picture Engin.* July 52 In addition to its inherent freedom from ground noise, the push-pull sound track has other advantages. **1938** *Encycl. Brit. Bk. Year* 498/1 Although not new in 1937, the use of push-pull sound recording increased considerably during the year. **1959** *B.S.I. News* Sept. 25 Sound records and scanning area of 35 mm double width push-pull sound prints (normal and offset centreline types). **1973** D. A. Spencer *Focal Dict. Photogr. Technologies* 498 *Push-pull sound track*, optical sound track on a cine film divided into two equal parts which were exposed to light modulated in opposite phase.

B. *sb.* Chiefly *Electronics.* A push-pull arrangement or state; esp. in adv. phr. *in push-pull.*

1929 *Exper. Wireless & Wireless Engineer* VI. 307/1 A pair of valves, (or banks of valves), working in opposite phase, commonly called 'Push-pull'. **1932** *Oxford Times* 23 Sept. 22/5 Push-pull gives the last stage a much greater power-handling capacity. **1943** *Electronic Engin.* XVI. 55/1 The advantages to be gained by the use of push-pull for deflection are so great that unbalanced time-bases are rarely employed in cathode-ray tube circuits. **1948** A. L. Albert *Radio Fund.* ix. 360 Radio-frequency power amplifiers often are operated in push-pull. **1962** A. Nisbett *Technique Sound Studio* 276 Movement of the stylus produces variations in the magnetic flux, which in turn generates a current in a coil (or in two coils situated on paths which are favoured alternately, and operate in push-pull). **1962** *Listener* 7 June 1006/2 One could not help speculating on the strange symbiosis or state of push-pull—call it what you like—which exists between him and his age. **1975** G. J. King *Audio Handbk.* iv. 83 The output stages of hi-fi amplifiers employ two transistors in push-pull.

C. *adv. Electronics.* In a push-pull manner.

1947 R. Lee *Electronic Transformers & Circuits* v. 109 Operation may sometimes be improved by the use of two tubes connected push-pull. **1978** *Nature* 6 Apr. p. xxxiii/2 Errors existing between the DC reference and the detected signal are amplified and applied push-pull to a transverse field electro-optic light modulator.

push-push (pʊ·ʃpʊʃ). [f. Push *v.*] (See quots.)

1907 *Westm. Gaz.* 13 Dec. 12/1 The only means of conveyance for travellers in this delightful part of India has been the 'push-push',..resembling a bathing-machine, which is impelled by relays of coolies. **1921** *United Free Church Miss. Rec.* June 187/2 All rode wherever they went, or stayed at home, if they did not care to hire the 'push-push', an unwieldy machine like a long bathing-coach on four wheels, drawn and shoved by eight or ten men.

Pushtoo, -tu, *sb.* and *a.* Add: Also **Pushto.** (Earlier and later examples.)

Pashto is now the usual form. Further examples are added here of the spellings *Pakhto, Pukhto.*

1784 H. Vansittart *Let.* 3 Mar. in *Asiatick Researches* (1790) II. 67 A book written in the Pushto language by Husain. *Ibid.* 68, I also submit a specimen of their language, which is called by them Pukhto; but this word is softened in Persian into Pushto. **1790** *Asiatick Researches* II. 76 The Pushto language, of which I have seen a dictionary, has a manifest resemblance to the Chaldaick. **1933** L. Bloomfield *Lang.* 62 *Afghan* (*Pushto*), with some 4 million speakers. **1939** L. H. Gray *Found. Lang.* 320 Besides Persian, the Modern Iranian dialects are Kurdish.., Balōčī and Afghān or Puštū, each with two principal sub-divisions, and..Ossetic. **1955** *Times* 25 May 10/3 The demand for the independence of the Pakhto- (or Pushtu-)speaking people could possibly assume some apparent validity. **1956** J. Whatmough *Language* 29 In the Pamirs dialects of Persian proper and some related dialects as Kurdish, Pashtu (in Afghanistan).., also have maintained their hold. **1962** Chavarria-Aguilar & Penzl in Householder & Saporta *Probl. Lexicogr.* 238 Pashto is taught in the elementary schools of West Pakistan's Pashto-speaking areas. **1964** H. H. Paper tr. *Shafeev's Short Gramm. Outl. Pashto* Introd., Pashto is the language of the people who inhabit Afghanistan and the northwest part of Pakistan. The Afghans themselves call their language paṣ̌tō (in the east paxtō). **1965** *Language* XLI. 529 Only very rarely is there a discrepancy between the Pashto examples and the English translations. **1973** *Times* 22 Mar. (Pakistan Suppl.) p. ii/2 In Karachi, the provincial capital, Urdu-speakers probably account for about 70 per cent of the population, with Punjabis and Pashto-speakers—the latter mainly members of an itinerant Pathan labour force—making up a further 20 per cent. **1974** *Times* 30 Apr. 7/7 It should be ascertained whether the Pushto-speaking people wanted to stay with Pakistan, merge with Afghanistan or have an independent country.

Pushtun, var. *Pakhtun *sb.* and *a.*

pu·sh-up, *sb.* and *a.* Also **pushup, push up.** [f. vbl. phr. *to push up*: see Push *v.* 1 b.] **A.** *sb.* = *press-up; also, an exercise on parallel bars in which the body is supported by the bent forearms and raised by straightening the arms. Also *attrib.* Chiefly *U.S.*

1906 *Amer. Mag.* LXIII. 139/1 First they put him on the parallel bars and beseeched him to do many push-ups, prodding him gently to further exertion when he showed signs of fatigue. **1943** *Sun* (Baltimore) 16 June 8/6 Ten pushups and work details are standard punishments for other minor offences. **1952** J. Steinbeck *East of Eden* xlvi. 516 William C. Bunt died right on the armory floor in the middle of a push-up. His heart couldn't take it. **1958** *Times* 26 Feb. 8/4 Half the boys examined could not do a single push up. **1968** M. Richler *Cocksure* vii. 44 Tomasso..did push-ups on his office carpet every morning. **1973** *Black Panther* 24 Mar. 6/2 Sometimes they make you remain in a push-up position on your knuckles until your knuckles begin to bleed. **1978** S. Brill *Teamsters* iii. 107 Twenty push-ups..was all the exercise a busy union leader should have time for.

2. The act or process of picking a pocket in which the victim's arm is pushed away from his pocket by an accomplice; also *attrib.*, as *push-up man, mob. Austral. slang.*

1919 V. Marshall *World of Living Dead* 69 He acts as chief amongst his 'push-up' and 'break' men, associates skilled in their way, but unpossessed of his dexterity. **1938** F. D. Sharpe *Sharpe of Flying Squad* i. 15 Pick-pockets are known as 'Wizzers' or 'The Push Up Mob'. *Ibid.* 332 *Push Up* (*The*), picking pockets.

3. A muskrat's resting-place, formed by pushing up vegetation through a hole in the ice. *N. Amer.*

1936 K. Conibear *North Land Footprints* 254 There's no danger of catching her [*sc.* a fox] either; she didn't go near the push-up. **1956** H. S. M. Kemp *Northern Trader* (1957) iv. 57 He indicated..the little 'pushup' wherein the muskrat would come to sun himself on the warmer days. **1959** E. Collier *Three against Wilderness* xxviii. 297 The ice should be sound enough for us to get onto it afoot to start staking the muskrat push-ups.

B. *adj.* **1.** That pushes or may be pushed up.

1963 *N.Y. Times* 17 Nov. 12 This slipon coverall..in deftly cut cotton with..push-up sleeves. **1966** A. E. Lindop *I start Counting* iv. 71 The big push-up windows. **1972** D. Lees *Zodiac* 145 A door that..had one of those push-up bars like the emergency door of a cinema. **1977** *Detroit Free Press* 11 Dec. 6-c/2 (Advt.), Seamless push-up bra with underwire, removable padding.

2. *Computers.* (See quots. 1966, 1977.) Cf. *push-down *a.* 1.

1966 C. J. Sippl *Computer Dict.* 149/2 *Push-up list*, a list of items where the first item is entered at the end of the list, and the other items maintain their same relative positions in the list. **1969** P. B. Jordain *Condensed Computer Encycl.* 406 The pushup-list concept is used whenever there is a queue of approximately equal-priority requests that are waiting to be serviced. **1977** P. Quittner *Problems, Programs* 375 *Pushup list*, a list that is constructed and maintained so that the next item to be retrieved and removed is the oldest item still in the list, that is first in, first out (FIFO).

push-wainling (pu:ʃwē̆inliŋ). *nonce-wd.* Also **pushwainling.** [f. Push- + Wain *sb.*[1] + -ling *suffix*[1] 2.] A perambulator.

1878 W. Barnes *Outl. Eng. Speech-Craft* 72 *Perambulator* (the child's carriage), push-wainling. **1908** A. C. Swinburne *Let.* 22 Jan. (1962) VI. 211, I met..a fair friend..who beamed..from the depth of her pushwainling (I hope you never use the barbaric word 'perambulator'?)... The happy term 'pushwainling' for a baby's coach of state is what makes him [*sc.* W. Barnes] immortal in my eyes. **1962** *Listener* 16 Aug. 257/1 He [*sc.* W. Barnes] was also a philologist, the kind that..advocates such coinages as 'two-horned rede-ship' (dilemma) and 'pushwainling' (perambulator).

pushy (pu·ʃi), *a. colloq.* (orig. *U.S.*). [f. Push *sb.*[1] or *v.* + -y[1].] Unpleasantly forward or self-assertive; aggressive.

1936 M. Mitchell *Gone with Wind* viii. 142 It [*sc.* Atlanta] had nothing whatever to recommend it—only its railroads and a bunch of mighty pushy people... Restless, energetic people from the older sections of Georgia. **1959** T. Griffith *Waist-High Culture* xi. 148 The more talented..can be counted on to disqualify themselves further by seeming too pushy. **1963** M. Beadle *These Ruins are Inhabited* xii. 187 A retired-colonel type.. would..turn and glare because you were being pushy. **1969** *New Yorker* 14 June 44/2 His speaking style..sounds pushy. If I'm in a bad mood, it bugs me. **1971** *Nature* 20 Aug. 510/2 Is it..that pushy polytechnics will in future be encouraged to usurp the position of the weaker universities in the academic pecking order? **1972** M. Babson *Murder on Show* vi. 71, I don't mean to be *pushy*... I just thought one had a duty as a citizen. **1979** *N.Y. Rev. Bks.* 25 Oct. 49/1 He faced the rise to autonomous power during the war of pushy new groups—generals, industrial managers, the secret police. **1980** *Times* 29 Feb. 13 The poor dancers gibber earnestly through its minimal dance content, pushy violence and unmotivated antics.

puss, *sb.*[1] Add: **3. c.** int. *puss, puss*: used to imply that the person addressed is a 'cat' (see Cat *sb.*[1] 2 a).

1926 H. Nicolson *Let.* 14 May in J. Lees-Milne *Harold Nicolson* (1980) xi. 235 The man was merely a prig..he would look very foolish..in Gordon Square (Puss, puss, puss). **1936** A. Christie *Murder in Mesopotamia* vi. 47 'We've been so very worried about dear Mrs. Leidner, haven't we, Louise?'.. 'Puss, puss,' I thought to myself. **1948** D. Ballantyne *Cunninghams* xviii. 95 'Stuck-up, if you ask me,' Joy said. 'Puss puss,' Ralph said. **1954** 'M. Cost' *Invitation from Minerva* 75 'Your cinema career was short-lived anyway.' 'Puss-puss,' she warned.

5. Puss in the corner: (earlier and later examples); also called **Puss, Puss.**

1709 W. King *Useful Trans. in Philos.* v. 43 The English Plays have barbarous sounding Names, as..Puss in a Corner..and the like. **1926** 'R. Crompton' *William—the Conqueror* xiii. 240 Now, what shall we play at first?.. Puss in the Corner? **1969** I. & P. Opie *Children's Games* vi. 207 The fun of 'Puss in the Corner' is that the players themselves negotiate when they are going to run; its disadvantage is that it is normally for five players, no more and no less. *Ibid.*, Names: 'Puss in the Corner' and 'Puss, Puss' (both common).

5*. = *pussy *sb.* 5*. *coarse slang.*

Quot. 1664 may not exemplify this meaning, claimed for it by Farmer and Henley.

1664 Cotton *Scarronides* 107 Æneas, here's a Health to thee, To Pusse and to good company. And he that will not do, as I do, Proclaims himself no friend to Dido. **1902** Farmer & Henley *Slang* V. 333/1 *Puss*... The female *pudendum*..also *pussy* and *pussy-cat*. **1935** in A. W. Read *Lexical Evidence from Folk Epigr. in W.N. Amer.* 71 She may (not?) be a cat trader's daughter, but she's got some puss. **1978** I. M. Gaskin *Spiritual Midwifery* (rev. ed.) I. 32 'Vagina' is the medical term, a Latin word, but I prefer to use 'puss' because it sounds friendlier. *Ibid.* 76 A loose mouth makes for a loose puss which makes the baby come out easier.

6. *puss-house, -purr*; **puss boot, shoe** *Jamaica* (see quots. 1961 and 1970).

1942 L. Bennett *Jamaica Dialect Verses* 36 She.. Put awn wan tear-up frack Shove har foot eena wan ole puss boot An go. **1961** F. G. Cassidy *Jamaica Talk* vi. 114 Tennis shoes with rubber soles and canvas tops are widely known in Jamaica as *puss boots* or *puss shoes*. **1970** *Country Life* 26 Feb. 510/3 We [in Jamaica] say 'puss boots' for plimsolls. **1869** J. S. Mill *Let.* 16 Jan. (1910) II. 177 Among the other additions there is a puss-house. **1935** T. S. Eliot *Murder in Cathedral* i. 43 Puss-purr of leopard, footfall of padding bear.

puss (pus), *sb.*[2] *dial.* and *slang* (chiefly *Ir.* and *U.S.*). [a. Ir. *pus* lip, mouth.] A (discontented, pouting) mouth; a sour or ugly face; the mouth or face (considered as the object of a blow).

1890 D. A. Simmons *Words & Phr. Armagh & S. Donegal* in *Eng. Dial. Dict.* (1903) IV. 653/2 He has an ugly puss. **1891** J. Maitland *Amer. Dict. Slang* 213 *Puss* (P[rize] R[ing]), the mouth. **1898** G. Bartram *White-Headed Boy* 40 Say I'm the besht man, or I'll break your puss. **1910** P. W. Joyce *Eng. as we speak it in Ireland* 309 'He had a puss on him', i.e. he looked sour or displeased—with lips contracted. **1911** C. B. Chrysler *White Slavery* viii. 67 She gets 'a slam in the puss' (slugged, struck in the face). **1932** J. T. Farrell *Young Lonigan* iii. 111 He twisted his lips in sneers, screwed up his puss. **1936** 'F. O'Connor' *Bones of Contention* 210 Are you a dummy or what to be standing there with that idiotly bloody smile on your puss? **1953** [see *belt *sb.*[4]]. **1961** C. McCullers *Clock without Hands* iv. 81 When you looked at the picture I didn't like the look on your puss. **1971** A. Burgess *MF* xiii. 149 You can get her to keep quiet about it, threaten her with a sock on the puss and that. **1973** 'J. Patrick' *Glasgow Gang Observed* v. 49 Ah don't fancy the look o' his puss. Go ower an' stab him fur me. **1978** *Guardian* 2 Apr. 18/3 On the air, Frost's pasty puss looked like Nixon's with the air let out of it.

puss, *v. rare.* [f. Puss *sb.*[1]] *intr.* To move or act like a cat, silently and stealthily.

a **1953** Dylan Thomas *Adventures Skin Trade* (1955) 101 They pussed and spied around the room, unaware of their dancing.

puss-cat. (Later examples.)

1915 J. Galsworthy *Bit o' Love* i. 19 Old puss-cat! **1957** [see *hep-cat].

pussens (pu·sėnz), playful elaboration of Puss *sb.*[1]

1922 Joyce *Ulysses* 55 Milk for the pussens, he said.

pusser (pʊ·səɹ), repr. naut. pronunc. of Purser (sense 2 b). Also *attrib.* and in the possessive, as issued by, or characteristic of, a naval purser (cf. *purser 2 b).

1903 [see *matlo(w)]. **1916** 'Taffrail' *Pincher Martin* ii. 13 The articles comprising Martin's kit, even down to his 'pusser's dagger' or seaman's knife. **1925** Fraser & Gibbons *Soldier & Sailor Words* 232 *Pusser's crabs*, seamen's boots. (Navy—lower-deck.) *Ibid.*, *Pusser's dip*, a candle. **1929** F. Bowen *Sea Slang* 107 *Pusser's grins*, sneers. **1943** Baker *Dict. Austral. Slang* (ed. 3) 62 *Pusser*, that which conforms to Naval regulations, e.g., 'pusser's cow', tinned milk; 'pusser's duck', a naval seaplane; 'pusser's waggon', a warship; 'pusser's rig', naval clothes. (R.A.N. slang.) **1944** J. Mallalieu *Very Ordinary Seaman* 90 All the discomfort of a small ship and the pusser routine of a big one. **1948** Partridge *Dict. Forces' Slang* 149 *Pusser's duck*, a Supermarine 'Walrus' flying-boat. *Ibid.*, *Pusser's issue*, clothing, tobacco, food, etc., provided by the Admiralty. **1964** J. Hale *Grudge Fight* iv. 69 'Hot water,' he said, 'plenty of pussers soap—and elbow grease, got it?' *Ibid.* vi. 91 A pair of pusser's long pants. **1973** *Daily Colonist* (Victoria, B.C.) 29 Aug. 2/2 Then, of course, there was Navy pusser rum—not to be confused with any other make of rum. **1977** *Navy News* June 6/3 And dancing was in pusser's shoes on planks of wood laid on the grass. **1977** *Ibid.* Aug. 18/4 Pusser's rum, obtained commercially in Gibraltar, was poured from wicker-work covered jars.

pu·ssful. *Ir. nonce-wd.* [f. *puss *sb.*[2] + -ful.] Something to fill a person's (discontented) mouth.

1922 Joyce *Ulysses* 197 The drouthy clerics do be fainting for a pussful.

pussivanting (pu·sivæ:ntiŋ), *ppl. a.* and *vbl. sb. S.W. dial.* Also **puzzivanting.** [Corruption of Pursuivant *sb.* (*a.*) or *v.*] Causing a disturbance, intruding, meddling, fussing.

1880 Courtney & Couch *Gloss. Words Cornwall* 45/2 *Pussivanting*, part., fussing; meddling. In the latter part of the seventeenth century the *Poursuivants* came into the county to search out all those entitled to bear arms. **1888** 'Q' *Troy Town* xvii. 203 'This 'ere pussivantin' may be relievin' to the mind, but I'm darned ef et can, be good for shoe-leather.' (Note: in the Fifteenth Century, so high was the spirit of the Trojan sea-captains, ..that King Edward IV sent poursuivant after poursuivant to threaten his displeasure. The messengers had their ears slit for their pains; and 'poursuivanting' or 'pussivanting' survives as a term for ineffective bustle.) **1915** Galsworthy *Bit o' Love* i. 17 There's puzzivantin' folk as'll set an' gossip the feathers off an angel.

pussy, *sb.* Add: **2. a.** (Earlier example.) Also (*Austral.*), a rabbit.

1715 T. Cave *Let.* 26 Oct. in M. M. Verney *Verney Lett. of 18th Cent.* (1930) I. xvii. 342 The Dog is very young and has seen but few Pussies, but..I doubt not of his having Appear'd a profess'd enemy to your Hares by this Time. **1941** Baker *Dict. Austral. Slang* 58 *Pussy*, a rabbit.

3. a. (Further examples.) Also, a finicky, old-maidish, or effeminate boy or man; a homosexual.

1925 S. Lewis *Martin Arrowsmith* vi. 65 You ought to hear some of the docs that are the sweetest old pussies with their patients—the way they bawl out the nurses. **1932** A. Christie *Thirteen Problems* xi. 193 '*The dame de compagnie*, you described, I think, as a pussy, Mrs. Bantry?' 'I didn't mean a *cat*, you know,' said Mrs. Bantry. 'It's quite different. Just a big soft white purry person. Always very sweet.' **1941** —— *N or M?* iii. 38 Old boarding-house pussies. Nothing to do but gossip and knit. **1942** Berrey & Van den Bark *Amer. Thes. Slang* § 405/2 *Pussy*, an effeminate boy. **1952** M. Tripp *Faith is Windsock* iv. 73 'Your rear gunner is a hit with the ladies.' 'Jake knows how to make the pussies purr; it's an old Jamaican custom.' *a* **1957** J. Cary *Captive & Free* (1959) x. 50 Some of those old pussies, especially the males, are just longing to put you in a corner. **1958** L. Durrell *Mountolive* viii. 157 'I first met Henry James in a brothel in Algiers. He had a naked houri on each knee.' 'Henry James was a pussy, I think.'

4. b. *Criminals' slang.* A fur garment.

1937 'D. Hume' *Halfway to Horror* 4 Those who steal furs handle them as 'pussies'. **1960** *Observer* 25 Jan. 5/2 If it was tom or pussies (furs) it was probably one of the big buyers. **1972** J. Wainwright *Night is Time to Die* 129 The coat... Ten to one, a fur coat, and there was always somebody ready to lift a pussy. **1973** 'B. Graeme' *Two & Two make Five* vii. 66 From one house they stole every piece of Regency silver..from another..they restricted themselves to jewellery, toms and pussies.

5*. The female pudendum. Hence, sexual intercourse; women considered sexually. *to eat pussy*, of a man: to engage in sexual intercourse. *coarse slang.*

1879–80 *Pearl* (1970) 268 Her legs are wide open showing the red lips and clitoris of her pussey. **1913** L. Strachey *Ermyntrude & Esmeralda* (1969) ii. 12 I'm also sure that it's got something to do with the thing between our legs that I always call my Pussy. **1922** F. Harris *My Life & Loves* I. iii. 61 By thinking of Lucille and her soft, hot, hairy 'pussy', I grew randy again. **1940** C. McCullers *Heart is Lonely Hunter* i. iii. 37 She crossed over to the opposite wall and wrote a very bad word—*pussy*. **1959** N. Mailer *Advts. for Myself* (1961) 98 This is the magical evil of the big city, but he is wary of being taken in: 'I come to see pussy..and I ain't seen pussy yet.' **1962** J. Baldwin *Another Country* i. i. 63 You wouldn't be putting that white prick in no more black pussy. **1965** 'A. Hall' *Berlin Memorandum* xi. 105 You go to town on the tits and pussy, symbolising carnality till it moans. **1967** M. McClure *Freewheelin Frank* i. 8 When we talk about eating pussy we make it sound as dirty and vulgar as possible. **1973** A. Powell *Temporary Kings* v. 258 Louis's stuffed a charming little cushion with hair snipped from the pussies of ladies he's had. **1976** J. O'Connor *Eleventh Commandment* v. 70 He killed about five prostitutes, cut them to pieces and stuffed various objects up their pussies. **1978** J. Krantz *Scruples* ii. 21 There was nothing, he had discovered, like flying a girl away for a weekend to insure as much pussy as you could eat. **1979** *Maclean's Mag.* 12 Mar. 25/3 As one blonde in a black leather coat bluntly replied, 'I sell pussy, not opinions.'

6. a. (Earlier and later examples.) Also *fig.* Cf. also sense 3.

1842 *Amer. Pioneer* I. 182, I walked up very carelessly among the soldiers .and concluded they could never fight with us. They appeared to me to be too pussy. **1930** D. L. Sayers *Strong Poison* xvi. 197 Mrs. Pegler, a very stout, pussy old lady with a *long tongue* (!)

b. pussy bow = *pussy-cat bow*; **pussy four-corners** = *Puss in the corner* s.v. Puss *sb.*[1] 5; **pussy hair** *slang*, a woman's pubic hair; **pussy-hoisting** *slang*, stealing fur garments; **pussy mob** *slang*, a gang of fur thieves; **pussy palm,** = Palm *sb.*[1] 4 and *pussy-willow; **pussy posse** *U.S. slang* (see quot. 1963); **pussy power** (see quots.); **pussy-talk,** feminine gossip; **pussy-whip** *v. trans.* (slang), = Hen-peck *v.*

1972 *Times* 28 July 10/1 His satin faconne shirts tie in a neat pussy bow. **1922** Joyce *Ulysses* 477 He plays pussy fourcorners with ragged boys and girls. **1972** R. D. Abrahams in T. Kochman *Rappin' & Stylin' Out* 231 When the pepper tree begin to bear It burn off all of Jennifer' pussy hair. **1975** R. H. Rimmer *Premar Experiments* i. 68 The wild disarray of your pussy hair beneath your panties. **1962** Parker & Allerton *Courage of his Convictions* i. 82 Then I got three years for pussy-hoisting from a warehouse in the City. **1967** M. Procter *Exercise Hoodwink* xiii. 91 He became the wheel man of a 'pussy' mob... The Flying Squad caught him with a car load of stolen furs. **1936** N. Streatfeild *Ballet Shoes* ix. 134 The catkins and pussy palm showed there would not be much more winter. **1978** *Guardian Weekly* 26 Mar. 19/1 They used to start coming in April like the returning swallows and house martins. Then they arrived for the daffodils and pussy palm. **1963** R. I. McDavid *Mencken's Amer. Lang.* xi. 730 *Pussy posse*, the vice squad. **1973** *Times* 22 Mar. 8/7 The police do their best. They have special teams of detectives (known as pussy posses) who mount drives against the girls. **1970** G. Greer *Female Eunuch* 126 Women in America are reported to be manipulating their menfolk by pussy-power, which is wheedling and caressing, instead of challenging. **1970** *New York* 16 Nov. 48/1 Her specialty at political meetings was the Pussy Power speech. With it Elaine Brown originated the concept that a woman's function is to use her body to entice men into the Panther Party. **1937** Auden & MacNeice *Lett. from Iceland* xii. 161 It looks like a week of pussy-talk. **1963** *Amer. Speech* XXXVIII. 173 One informant noted that a male..is said to be *pussy whipped*, a term one of the authors recalls having heard in the Navy in 1956. **1973** C. & R. Milner *Black Players* vi. 161 White men (and square Blacks) are thought to be 'pussy-whipped' by their wives. **1978** J. Krantz *Scruples* viii. 230 Some men are pussy whipped from the day they are born, some have it happen to them later in life, some never.

pussy (pu·si), *v.* [f. the sb.] *intr.* (With advbs.) To behave or move like a cat (see also quot. 1973).

1943 K. Tennant *Ride on Stranger* xi. 134 Buzz off, Pop. You don't want to be pussying around. **1952** C. Armstrong *Black-Eyed Stranger* ii. 17 He came pussying up. **1973** 'J. Patrick' *Glasgow Gang Observed* 235 *Pussyin' aroun'*, playing about, mostly used in a sexual context.

pussy (pʊ·si), *a.*[2], **pussel** (pʊ·səl), *a.* Also **pussle, puzzle.** Chiefly U.S. dial. corruptions of Pursy *a.*[1] Mainly in **pussy-, pussel-gutted** *adjs.*, corpulent, obese; also **pussy-, pussel-gut,** a corpulent stomach; (*pl.*) a fat person (see also quot. 1976[1]); hence **pussel-gut** *v. trans.* (nonce), to render obese.

1844 'J. Slick' *High Life in N.Y.* II. 89 As..pussy as a turkey-gobbler. *Ibid.* 92 As pussy and pompous as a prize pig jest afore killing time. **1886** F. T. Elworthy *W. Somerset Word-bk.* 598 What a *pussy* old fuller th'old Zaddler White's a-come; I can min' un when he used to go a-courtin, a slim young spark. **1892** S. Hewett *Peasant Speech Devon* 115 'Er's drefful pussy tū-day, an can't walk vast nur var. **1906** *Dialect Notes* III. 152 *Pussy-gutted*, adj., corpulent. 'He's terrible *pussy-gutted*.' **1907** *Ibid.* 197 *Pussy*.., adj., corpulent. 'He didn't use to be so *pussy*.' **1909** *Ibid.* 361 *Pussle-gutted*, adj.. same as *pussy-gutted*. *Ibid.*, *Pussy-gutted, adj.*, corpulent, having a large abdomen. Often used as a term of contempt. 'You low-lifed, *pussy-gutted* scounderl.' *Ibid.* 402 *Pussy guts, n. phr.*, a corpulent man. 'See that old *pussy guts*.' **1933** M. K. Rawlings *South Moon Under* xiii. 133 Sort o' pussle-gutted, eh? **1935** W. Faulkner *As I lay Dying* 10 You pussel-gutted bastard. *Ibid.* 35 He has pussel-gutted himself eating cold greens. **1942** Z. N. Hurston *Dust Tracks on Road* viii. 143 Goat-bellied, puzzle-gutted,..knock-kneed..so-and-so. **1946** *Amer. Speech* XXI. 99 A *body* who has gained weight enough to show signs of obesity is said to be *fleshy* or *pussy* (pursy). **1949** 'J. Nelson' *Backwoods Teacher* ix. 88 A lantern-jawed ol' varmint with a big golden watch chain acrost his ol' pussy-gut. **1959** W. Faulkner *Mansion* 55 Old pussel-gutted Hampton that could be fetched along to look at anything, even a murder, once somebody remembered he was Sheriff. **1976** C. S. Brown *Gloss. Faulkner's South* 157 In northern Florida, the pot-bellied little mosquito-fish, or gambusia, is called the pusselgut. **1976** *N.Y. Times Mag.* 10 Oct. 111/2 All watched over by a savage God, by the dead and by pussel-gutted deputies.

pussy-cat. Substitute for entry:

pu·ssy-cat. 1. A nursery word for a cat.

1805 *Songs for Nursery* 40 Pussy cat, pussy cat, where have you been? I've been to London to see the queen. **1837** [in Dict.]. **1844** 'J. Slick' *High Life in N.Y.* I. 154 As affectionate as a pussy cat. **1871** E. Lear *Nonsense Songs, Stories, Bot. & Alphabets*, The Owl and the Pussy-cat went to sea In a beautiful pea-green boat. **1933** [see *baby-talk* s.v. *baby *sb.* B. 1 d]. **1955** *Sci. News Let.* 26 Mar. 203/1 For better-fed pussycats, add to their diet a good dash of personal attention and a heaping tablespoonful of affection.

2. A willow or hazel catkin: cf. Pussy *sb.* 4.

1850 C. M. Yonge *Henrietta's Wish* xv. 216 The silver 'pussycats' on the withy. **1861** [in Dict.]. **1889** E. Peacock *Gloss. Words Manley & Corringham, Lincolnshire* (ed. 2) 421 *Pussy-cat*.., the catkins of the willow.

3. Applied to a person (cf. Puss *sb.*[1] 3, Pussy *sb.* 3 in Dict. and Suppl.); now esp. one who is attractive, amiable, or submissive.

1859 J. A. Symonds *Let.* Apr. (1967) I. 184 Dalrymple's brother is going to be married to the Lady Edith Dalhousie: I wrote a solemn letter of congratulae [*sic*] to the old pussey cat! **1864, 1881** [in Dict.]. **1955** *Amer. Speech* XXX. 119 *Pussy cat*, a pilot who is overcautious, fearful, or reluctant. **1959** *Times* 17 Aug. 12/7 Ronder, a sly pussy-cat of a man, able to scratch as well as purr. **1964** L. Nkosi *Rhythm of Violence* 25 Jimmy: (*to Mary*) Don't worry pussycat! **1973** E. Jong *Fear of Flying* 89 'Some men claim to be afraid of me.' Adrian laughed. 'You're a sweetheart,' he said, 'a pussy-cat—as you Americans say.' **1975** P. G. Winslow *Death of Angel* iv. 104 He can be a dear, but he's also one of the chief pussycats of the psychic world. **1976** C. Dexter *Last seen Wearing* v. 36 The secret sex-life of a glamorous Hollywood pussycat. **1978** G. Vidal *Kalki* i. 7, I was the one who paid the alimony... Women wrote me ugly letters. I was not apparently, a pussycat.

4. Cattiness, spitefulness. *rare.*

1911 W. J. Locke *Glory of Clementina Wing* xxiv. 361 Let us have a straight talk like sensible women, and put the pussy-cat aside.

5. *Comb.*, as *pussy-cat-like* adj.; **pussy-cat bow,** a soft, floppy bow.

1964 *Sunday Express* 2 Feb. 18/3 For dressy occasions a pussycat bow..under a high, round jacket collar. **1967** [see *gentian 2 b]. **1977** J. Bingham *Marriage Bureau Murders* ii. 21 Her white silk blouse, with the pussy cat bow tied at the neck, lent a touch of femininity. **1881** Pussy-cat-like [in Dict.].

pu·ssyfoot, *sb.* [f. Pussy *sb.* + Foot *sb.*] **1.** One who moves stealthily or warily.

1914 Jackson & Hellyer *Vocab. Criminal Slang* 68 *Pussy foot*... A detective. **1916** *Dialect Notes* IV. 279 *Pussy-foot, v.i.* To be sly, intriguing, or underhand. 'That girl goes pussy-footing around.' Also *n.* 'She's a regular pussy-foot.' **1977** 'E. Crispin' *Glimpses of Moon* xii. 257 Grateful that the creature [*sc.* a cat] was in both senses a pussyfoot, Fen drank some champagne.

2. [f. the nickname 'Pussyfoot' of an American supporter of Prohibition, W. E. Johnson (1862–1945), given to him on account of his stealthy methods when a magistrate.] An advocate or supporter of prohibition; a teetotaller. Also allusively.

1919 *Punch* 23 July 86 *Gloomy Policeman.* 'You've had enough. Better go home.' *Reveller*... 'Shurr-up—Pussyfoot!' **1920** 'Sapper' *Bull-Dog Drummond* vi. 146 We are all confirmed Pussy-foots, and have been consuming non-alcoholic beer. **1921** T. Burke *Outer Circle* 169 The tea arrived, a viscid, leathery fluid of Pussyfoot vintage. **1922** Ld. Riddell *Some Things that Matter* ii. 28 Mrs. A., a 'pussyfoot', with an ardent desire to interfere with other people's habits. **1946** G. Millar *Horned Pigeon* x. 137 There was the heavy drinker... And there was the pussy-foot who said 'poison'.

3. *attrib.* or as *adj.* **a.** Teetotal; without alcohol; non-alcoholic. **b.** Soft; easy.

1923 D. H. Lawrence *Birds, Beasts & Flowers* 15 Even the word Marsala will smack of preciosity Soon in the pussy-foot West. **1940** Dylan Thomas *Portrait of Artist as Young Dog* 217 He'd be knocking back nips without a thought that on the sands at home his friend was alone and pussyfoot at six o'clock. **1973** D. Miller *Chinese Jade Affair* xvii. 156, I was trying to deflect the inevitable course of the evening with a 'Pussy-foot' cocktail. **1974** *Country Life* 17 Oct. 1108 Covering 38 laps of the circuit.. ensured this was no genteel pussyfoot operation.

So **pussy-footed** *a.*, having a light step; elusive; evasive; **pussy-foo·tedness; pu·ssyfootism,** teetotalism, advocacy or enforcement of prohibition.

1893 Pussy-footed [see Pussy *sb.* 6]. **1919** *N.Y. Times* 7 Jan. 4/6 The Republican Party..was evidently in imminent danger of taking a 'pussy-footed' position on the war. **1923** *Daily Mail* 23 July 7 In Tudor England people sang the music they liked, and read the books they liked. They had real freedom, and there was no pussy-footism. **1924** D. S. Barry *Forty Years in Washington* v. 106 Ingalls once said of Senator William B. Allison that he was so pussy-footed he could walk from New York to San Francisco on the keys of a piano and never strike a note. **1926** 'A. Berkeley' *Wychford Poisoning Case* vii. 78 They reached the Man of Kent and ordered the night-caps to which their position as residents entitled them, in defiance of the dictates of a maternal government, pussy-footism and all the other futilities which order our lives for us in these days. **1931** *Times Lit. Suppl.* 10 Sept. 685/1 He was pussyfooted and quick to spring. **1957** *Times* 10 May 13/4 This letter may sound cautious, perhaps pussy-footed, almost priggish... We must tread softly. **1964** *Daily Tel.* 9 Mar. 14/2 There is nothing pussy-footed about this economic strategy... It is a bold mixture of more competition and more responsibility. **1966** *Economist* 30 Apr. 450/1 Politically here is confirmation..of the essential caution, not to say pussy-footedness, of the Wilson Government. **1980** *Jrnl. R. Soc. Arts* Mar. 181/2 Aesthetics is a pussy-footed way of referring to beauty.

pu·ssyfoot, *v.* [f. as prec.] **1.** *intr.* To tread softly or lightly to avoid being noticed; to proceed warily; to conceal one's opinions or plans; to behave evasively or timidly. Also with *it*.

1903 *Atlanta Constitution* 20 Mar. 3 Vice President Charles Warren Fairbanks is pussy-footing it around Washington. **1916** [see *pussyfoot *sb.* 1]. **1918** C. Sandburg *Cornhuskers* 73 Who pussyfoots from desk to desk with a speaking forefinger? **1928** *Observer* 5 Feb. 18/1 While most papers are still 'pussy-footing' on the Presidency they called their editors together and afterwards announced a unanimous decision. **1931** E. Thompson *Farewell to India* 203 Trying to coax a horse to wait while I pussy-footed up to him. **1934** D. L. Sayers *Nine Tailors* III. ii. 286 When I got out through the porch, I had to pussyfoot pretty gently over that beastly creaking gravel. **1949** *Time* 9 May 25/2 The ones who pussy-footed, side-stepped, straddled, carried water on both shoulders and compromised were left at home. **1951** E. Paul *Springtime in Paris* viii. 155, I saw you pussyfooting around the exhibition. **1973** *Times* 16 Oct. 6/6 A Labour Government should not 'pussyfoot around' with reform of the Official Secrets Act but scrap it. **1975** B. Wood *Killing Gift* (1976) iv. i. 129 Why do you pussy-foot, captain?.. Why not just say it—you think Jennifer Gilbert killed him. **1977** *Jrnl. R. Soc. Arts* CXXV. 626/1 We have 'pussy-footed' round this issue of profit for years. **1980** *Brit. Med. Jrnl.* 29 Mar. 937/1 It is time someone was honest enough to stop pussyfooting about.

2. [f. *pussyfoot *sb.* 2.] *trans.* To render teetotal; to impose prohibition on. *rare.*

1921 [implied in *pussyfooting *vbl. sb.*].

So **pu·ssyfooting** *vbl. sb.* and *ppl. a.*

1921 *Q. Rev.* Jan. 100 The tyranny that would ensue from the Pussy-footing of Canada is too horrible to contemplate. **1928** *Collier's* 29 Dec. 38/1 The wrappings which..the pussy-footing politicians impose upon a candidate. **1956** G. P. Kurath in A. Dundes *Mother Wit* (1973) 107/2 Certain qualities seem to predominate... These are whole-bodied movements,.. dynamics from pussy footing to violent acrobatics, rhythmic complexity. **1974** J. Cleary *Peter's Pence* vi. 187 Authority had been given to the pussyfooting amateur..and nothing had gone right. **1976** *Times* 16 Feb. 8/7 In the face of political dogma, 'pussy-footing' and ill-informed decision making, is Mr Laker downhearted? **1977** *Time* 8 Aug. 1/1 To hell with what timid, pussy-footing diplomats think!

pu·ssyfooter. [f. *pussyfoot *v.* and *sb.* + -er[1].] **a.** One who pussyfoots (in any sense of the verb). **b.** An advocate or supporter of prohibition.

1927 *Sat. Even. Post* 24 Dec. 9/1 A good politician is a natural-born pussy-footer. **1928** *Daily Express* 28 Dec. 8/3 The pussyfooters..have given a weary and blasé world a new game to play. **1932** *N.Y. Times* 20 May 10/4 The conditions which are attached to its operation make plain its insincerity. It is, therefore, on that very account beginning to attract the favorable attention of the trimmers and the pussyfooters. **1946** S. H. Holbrook *Lost Men Amer. Hist.* 160 The appeasers and pussyfooters of 1850 also provided that any territories that might come into the Union later could do so with or without slavery.

pussy-willow. orig. *U.S.* Substitute for def.: A popular name for several species of willow or their soft, fluffy catkins, which appear before the leaves; esp., in North America, the glaucous willow, *Salix discolor*, and, in Great Britain, the goat willow, *Salix caprea.* (Earlier and further examples.)

1869 J. G. Fuller *Flower Gatherers* 52 The aments appear before the leaves, and are covered with hairs so soft and silken that children often call them Pussy-Willows. **1878** Mrs. Stowe *Poganuc People* xvii. 182 Then the pussy-willows threw out their soft catkins. **1893** Dartnell & Goddard *Gloss. Words used in Wiltshire* 126 Pussy-willow, *Salix*. **1924** A. D. Sedgwick *Little French Girl* II. i. 103 Sometimes it [*sc.* Alix's skin] was grey, like pussy-willow. **1939** F. Thompson *Lark Rise* i. 1 There were violets under the hedges and pussy-willows out beside the brook. **1949** *Lisle* (Illinois) *Eagle* 31 Mar. 5/4 The spring motif decoration of jonquils and pussy willows.. gave a gay and festive setting. **1958** R. D. Meikle *Brit. Trees & Shrubs* 198 In recent years the childish 'Pussy Willow' has tended to replace these older names [of 'Palm' and 'Goat Willow']. **1969** *Canadian Antiques Collector* Aug. 20/1 Pussywillows are arranged in one of a collection of..sugar bowls. **1976** *Burnham-on-Sea Gaz.* 20 Apr. 12/9 All [the congregation] carried branches of pussy willow which had been cut locally for the occasion [*sc.* Palm Sunday].

puszta (pu·stā). Also **pussta, puzta.** [Hungarian = plain, steppe, waste.] The flat treeless country of Hungary; a plain in Hungary.

1842 F. W. Faber *Styrian Lake* 324 The hailstorms with white oars across the putzas [*sic*] roam. **1852** T. Ross tr. *Humboldt's Trav.* II. xvii. 86 The widely extended pastures, which reach in every direction to the horizon, are called in the country, *Puszta.* **1896** *Daily News* 9 June 7/6 Only a nation of horsemen who have the Pussta to practise upon could turn out such a number of first-class horses. **1927** *Daily Express* 14 Dec. 9/1 They are the Chicos, as the 'cowboys' are called, and the Pusztas, or prairies, are to be found only a few hours' journey from Budapest. **1947** M. R. Shackleton *Europe* v. xxvii. 334 South-east of Kecskemét the soil is impregnated with salts and there is a large area of *puszta* (='waste'), known as the Bugác steppe. **1972** *Guardian* 4 Nov. 14/4 The Great Hungarian Plain... Pleasant to lunch here, serenaded by Hungarian Gipsy bands as you eat your puszta steak. **1973** *Country Life* 11 Jan. 74/1 There can be few areas of Europe that are flatter..than the great plains, the *puszta*, of Hungary.

put, *sb.*[1] Add: **1.** (Later *fig.* example.)

1974 B. Brophy in *New Statesman* 28 June 929/1 The jacket, an unsuccessful but not dishonourable put at the manner of Magritte.

5. *attrib.*, as *put option.*

1881 *Guide Oper. Stocks* 15 A Put Option should be obtained when a decline in the market is expected to take place. **1961** *Daily Mail* 18 Sept. 13/4 In the past three weeks 'put' options (where a fall in the shares is expected) have been an outstanding feature of the option market. **1977** *Private Eye* 4 Mar. 17/1 One suggestion was that some of the shares had come from Jim himself as a result of a 'put' option held on him personally by former lieutenant Herbert Despard.

put, *v.*[1] Add: **I. 8. c.** (Earlier and later examples with reference to a stream.)

1773 P. V. Fithian *Jrnl.* (1900) 56 From his house we see the Potowmack, and a fine River putting from it. **1810** F. Cuming *Sk. Tour Western Country* xiii. 97 The creek..puts in from the Virginia side. **1903** A. Adams *Log of Cowboy* 347 The trail on leaving the river led up Many Berries, one of the tributaries of the Yellowstone putting in from the north side.

II. 10. f. Also const. *ellipt.*

1819 Keats *Let.* 22 Sept. in G. G. Williamson *Keats Mem. Vol.* (1921) 120 Chowder died long ago—Mrs. H. laments that the last time they *put him* (*i.e.* to breed) he didn't take.

III. 20. a. Also in colloq. phr. *to put paid to*: to deal finally or effectually with (a person); to terminate (aspirations, hopes, etc.); to eliminate or put an end to (something).

1919 *Boy's Own Ann.* XLI. 457/2 She [*sc.* a destroyer].. was about to proceed to sea on her mission of 'putting paid' to U-boats. **1931** T. R. G. Lyell *Slang* 606 You can put paid to any friendship that ever existed between him and me. **1951** J. B. Priestley *Festival at Farbridge* II. iii. 344, I thought one time Tanhead might ha' swung 'em, but Commodore put paid to him all right. **1955** 'E. C. R. Lorac' *Ask Policeman* v. 54 He and his premises..were put paid to by a land mine. **1957** J. Braine *Room at Top* xvi. 144, I wanted to put paid to Communism once and for all. **1959** *Listener* 30 July 183/3 The translator's deficiencies put paid to the book altogether. **1971** G. Household *Doom's Caravan* ii. 40 The return journey..put paid to my only pair of formal trousers. **1976** *Economist* 13 Mar. 13/2 [That choice] would also probably put paid to any hopes of fully reintegrating France into the Nato alliance.

25. a. Also in U.S. dial. phr. *to put* (someone) *in the dozen(s)*, to force (someone) to 'play the dozens' (cf. *play *v.* 16 e); *spec.* to insult (a person) by referring to his mother in a derogatory way.

1939 J. Dollard in *American Imago* Nov. 8 Herbert had been put in the Dozens by another boy in the following manner: the boy said, 'Your mama needs a bath.' **1941** W. A. Percy *Lanterns on Levee* xxiii. 301 'Some fool nigger puts you in the dozen.'..'What's putting you in the dozen?' 'That's sho nuff bad talk.' 'Like what?' 'Well,' said Ford, modest and hesitant, 'that's talking about your mommer.' 1973 A. Dundes *Mother Wit* 299 To be 'put in the dozens' is to be put in a bad or losing position. 1974 H. L. Foster *Ribbin'* v. 226 If a teacher is attempting to really stop the dozens playing, just..holding your elbow, could be considered as putting another boy in the dozens.

26. b. Also const. *down.*
1960 R. Williams *Border Country* I. ii. 58 He was able to rent two strips of garden..and these he put down one to gooseberries and currants, the other to potatoes.

27. a. (*c*) *to put* (a person) *through it*: to impose a severe test on (a person); to subject (a person) to an ordeal or trying experience.
1872 G. P. Burnham *Mem. U.S. Secret Service* p. vii, *Put 'em through*, subjecting persons to a thorough searching ordeal. 1922 A. A. Milne *Red House Mystery* vi. 50 Everybody else is bundled off except me, and I get put through it by that inspector as if I knew all about it. 1923 Wodehouse *Inimitable Jeeves* iv. 48 Aunt Agatha ..was putting the last of the bandits through it in the voice she usually reserves for snubbing waiters in restaurants. 1935 *Discovery* Oct. 311/2 The work of the pupils whom he 'put through it'. 1940 H. G. Wells *Babes in Darkling Wood* I. ii. 59, I am afraid we have put you through it, rather. 1959 P. McCutchan *Storm South* xii. 179 Evidently she'd been put through it in the interval, for she was crying bitterly. a 1976 A. Christie *Autobiogr.* (1977) VIII. ii. 380 Mad as a hatter... My goodness, he must have put you through it now and again!

d. Also *to put* (a person) *through* (a school, college, etc.): to pay the cost of educating (a person). Also const. *ellipt.* Chiefly *N. Amer.*
1908 L. M. Montgomery *Anne of Green Gables* xxx. 338 I'd love to be a teacher. But..Mr Andrews says it cost him one hundred and fifty dollars to put Prissy through. 1943 *Deb. House of Commons* (Canada) 4 Feb. 161/2 Voluntary committees should be set up throughout Canada to pick out..boys and girls with a view to seeing that they are put through university. 1949 *Manch. Guardian Weekly* 27 Jan. 13/2 He..put himself through Emory College.

V. 35*. put across.
a. *to put it across* (a person): (*a*) to visit with retribution or punishment; to get even with.
1915 E. Wallace *Man who bought London* iv. 39 He won't half put it across you people. 1918 'D. Valentine' *Man with Clubfoot* xxi. 309 When you..put it across 'der Stelze'..you settled a long outstanding account we had against him. 1923 M. Arlen *These Charming People* 238 There was something—well, indecent, in talking about a man dead nine years or more as though he were alive and still wanting to 'put it across' Antony at every turn. 1928 *Daily Mail* 6 Aug. 14/6 You are a master of mob tactics, but we will put it across you yet. 1929 Wodehouse *Mr. Mulliner Speaking* iv. 129 It was his intention to..confront his erring man-servant and put it across him in no uncertain manner. 1936 —— *Laughing Gas* xvi. 179, I was glad that I had put it across him. My pride was involved. There are some remarks which one does not forgive. 1978 *Rugby World* Apr. 38/2 Meyer was a sports nut who enjoyed nothing more than seeing his pupils put it across the golden youth of Eton and Winchester.
(*b*) to impose upon; to deceive, to delude; to convince by deceit.
1919 E. P. Oppenheim *Strange Case J. Thew* II. vi. 235 'Well,' she exclaimed, 'he does put it across you, doesn't he?' 1923 H. C. Bailey *Mr. Fortune's Practice* i. 25, I say, you have put it across us in the Dean case. 1927 *Observer* 27 Mar. 6/4 It would be difficult for a greedy, hysterical, shameless, half-insane revivalist..to 'put it across' ever-increasing audiences. 1928 *Daily Express* 26 May 13/4 How Mother Cuckoo manages to 'put it across' certain inoffensive countryside birds. 1934 D. L. Sayers *Nine Tailors* 63, I hope our friend doesn't put anything across the good Rector. 1936 A. L. Rowse *Mr. Keynes & Labour Movement* 19 They succeeded in putting it across large sections of the middle classes that Labour's economics meant financial ruin. 1959 D. Eden *Sleeping Bride* xiv. 117 Don't Let Blandina put it across you. She isn't as ill as she pretends to be.

b. To make acceptable or effective; to convey the significance of. Cf. *across *prep.* 2 b.
1922 S. Anderson in R. L. White *S. Anderson/G. Stein* (1972) 15 The author had done a thing we Americans call 'putting something across'—the meaning being that she had, by a strange freakish performance, managed to attract attention to herself. 1923 H. Crane *Let.* 13 Apr. (1965) 131 This 'new consciousness' is something that takes a long while to 'put across'. 1927 M. Diver *But Yesterday* II. xxiii. 263 The Exchange reported, 'No answer.' She was out—naturally; very busy putting it across! 1935 [see *copy-writer* s.v. *copy *sb.* C]. 1938 E. Bowen *Death of Heart* II. iv. 247 Supposing she had a wish to be put across, who could do this for her better than Eddie could? 1943 W. S. Churchill *Second World War* (1951) IV. 839 We must be ready with our plans in the Eastern Mediterranean, and put it hard across Turkey to come in with us. 1943 J. S. Huxley *TVA* 129 The TVA was managing to put across a good deal of its plan. 1945 Mencken *Amer. Lang.* Suppl. I. 449 He [*sc.* C. T. Onions in 1936] noted that *to put it across*, *to get it across*, and *to put it over* were already 'firmly domiciled' in England... They really got their vogue in the United States as baseball terms. 1959 *Times Lit. Suppl.* 9 Jan. 15/3 Many readers, however, dazzled by Mr Graves's gifts as a prose entertainer, by his ability to put across Third Programme material with a Light Programme zing, may not give the poems the attention they deserve. 1970 'D. Halliday' *Dolly & Cookie Bird* ii. 11 If you don't put yourself across, who'll do it for you? 1977 *Wandsworth Borough News* 7 Oct. 5/1 'Help police fight crime by helping yourself'—that is the message the police are trying to put across to the public.

c. *Baseball.* To pitch (a ball) directly over home plate.
1936 Mencken *Amer. Lang.* (ed. 4) 191 The history of baseball terms also deserves to be investigated, for many of them have entered the common speech of the country, e.g...*to put it* (or *one*) *across* (or *over*). 1943 *Amer. Speech* XVIII. 106 If the pitcher throws a straight ball with good control, he is said..to *put it over*, to *put it across*, or to *put it right in there.* If he has speed, he may..*put over a fast one.*

38. put away.
f. (*a*) (Further examples.)
1924 [see *buffy *a.*[1]]. 1958 *Punch* 8 Oct. 469/1 The object of a wine-tasting is not to put the stuff away but to assess the relative values of a varied assortment of bottles. 1969 G. Greene *Trav. with my Aunt* I. viii. 76 Between us we can probably put away half a bottle of vodka. 1976 R. Hill *Another Death in Venice* I. i. 6 You look well enough..but you don't deserve to, not the way you were putting it away.
(*b*) (Earlier and later examples.) Also, to commit to an old people's home; to confine to a mental institution.
1872 G. P. Burnham *Mem. U.S. Secret Service* p. vii, *Put away*, sent to the State Prison, after conviction. 1938 N. Marsh *Death in White Tie* xvi. 179 She became hopelessly insane... He arranged to have her put away. 1952 *Sun* (Baltimore) 2 June 14/3 There is less social pressure on people to make a place for grandparents at home, less feeling that it would be disgraceful to have them put away. 1971 S. Phillips *Death in Sheep's Clothing* v. 48 The mother is nearly frantic now, she is always afraid 'they' are going to 'put him away'. 1973 W. M. Duncan *Big Timer* xxi. 138 He was an inspector then. He put me away. 1974 P. M. Hubbard *Thirsty Evil* iv. 40, I said, 'But can it go on? Won't they have to—?' 'Put him away? I suppose so, if he gets worse.'
(*c*) (Further examples.)
1909 Galsworthy *Silver Box* I. iii. 32 *Mrs. Jones.* We've not got a home, sir. Of course we've been obliged to put away most of our things. *Barthwick.* Put your things away! You mean to—to—er—to pawn them? *Mrs. Jones.* Yes, sir, to put them away. 1926 Maines & Grant *Wise-Crack Dict.* 12/1 *Putting away his ice*, pawning his diamond.
(*d*) Delete *dial.* and add earlier and later examples. Also, to kill.
1588 Greene *Pandosto* f. A4v, Deuising with himself a long time how he might best put away Egistus without suspition of treacherous murder, hee concluded at last to poyson him. 1847 A. Brontë *Agnes Grey* p. xiv, A reward, I should have greatly valued,..were he [*sc.* a dog] not now in danger of being 'put away'. 1920 E. Wallace *Daffodil Mystery* viii. 70 If I could only put her away for it! 1932 E. Waugh *Black Mischief* viii. 311 The dogs had long been rounded up and painlessly put away. 1971 E. Lemarchand *Death on Doomsday* ix. 137 I'd like to see old Peplow put away decently. 1974 M. Butterworth *Man in Sopwith Camel* xiii. 165 What kinda guy puts a buddy away for three lousy dollars?

39. put back.
g. *trans.* With personal object: to cost. *colloq.*
1909 *Dialect Notes* III. 402 'How much did that put you back?' 'Six dollars.' 1958 B. Ruck *Third Love Lucky* iv. 31 It puts you back five shillings for a quarter of an hour.
h. To return or 'plough back' (money, etc.). Cf. *plough *v.* 9 g.
1930 *Economist* 19 July 112/2 In view of Mr. Snowden's recent refusal to consider abatement of income tax on company reserves employed for re-equipment the recent tendency to reduce the proportion of earnings 'put back' is significant. 1931 *Ibid.* 11 July 59/1 The percentage 'put back into the business' during the past twelve months, 15·7 per cent., compares unfavourably with the figure of 18·6 per cent. ascertained for the twelve months ended June 30, 1930.

40. put by.
g. (Earlier example.)
1795 J. Woodforde *Diary* 31 July (1929) IV. 216 We had it [*sc.* the pork] taken up and put by for them against another Day.

41. put down.
a. (Further examples.)
1795 J. Woodforde *Diary* 29 June (1929) IV. 210 We were put down at the White Hart in Stall Street. 1841 Dickens *Let.* 2 May (1969) II. 276 'Mind Coachman' as the old ladies say 'you take me as fur as ever you go, and don't you put me down till you come to the very end of the journey.' 1933 [see *nitwit]. 1981 R. Barnard *Mother's Boys* iv. 48 This ruddy cough. It's the climate... They shouldn't have put people down in this climate.
d. (Further examples.) Now more usually, to disparage, to criticize forcefully, to humiliate.
1923 G. M. Trevelyan *Manin & Venetian Revolution* vi. 112 The principal speaker was Avesani, an eloquent and able lawyer who at once put Palffy down when he tried to speak in a tone of authority. 1958 *Amer. Speech* XXXIII. 225 When somone *puts* you *down* he criticizes you unfavorably, he *fluffs* you. 1961 Rigney & Smith *Real Bohemia* p. xvi, *Put down, to*, to humiliate, or tell off, or part company with someone. 1969 *Down Beat* 20 Mar. 31/3 It became fashionable to put him down as too much of a showman and not enough of a jazzman. 1972 D. Delman *Week to Kill* 86 So why did you put him down that way, in front of me? 1972 W. Labov *Language in Inner City* viii. 350 Sounding is only one of the many ways of putting someone down. 1977 Miller & Swift *Words & Women* p. x, We ourselves had for years been innocently using the words and grammatical forms that put our own sex down.
g. Delete 'Now chiefly *dial.*' and add further examples.
1936 W. Holtby *South Riding* IV. v. 253 Best have him [*sc.* a dog] put down, mercifully. 1942 G. Kersh *Nine Lives Bill Nelson* vii. 41 You could of put Bill down with a Humane Killer. 1958 *Times* 20 Nov. 3/1 An unwanted husband is as easily 'put down' as any other domestic pet. 1971 *Daily Tel.* 19 Aug. 3/6 Kim was ordered to be put down last year after he had bitten two people. 1977 *Guernsey Weekly Press* 21 July 4/7 One of the Jersey police dogs which entertained the large crowd at the recent open day of the local force at Les Vauxbelets has had to be put down.
j. (Earlier example.)
1791 J. Woodforde *Diary* 14 Oct. (1927) III. 306, I dont know that I ever eat a better Hare tho' we had put it down for an old one by skinning it.
m. *Cricket.* (*a*) To hit (a wicket), dislodging a bail. (*b*) With a batsman as subject: to stop or strike (a difficult delivery) without attempting to score. (*c*) With a bowler as subject: to deliver (a ball). (*d*) With a member of the fielding side as subject: to drop a catch.
17.. [In Dict. sense 41 a]. 1816 W. Lambert *Cricketer's Guide* (ed. 6) iii. 39 If the Striker should move of his ground, with an intention to run, he [*sc.* the wicket-keeper] must then do his best endeavour to put down the Wicket, which is called *stumping out.* 1841 in *Cricket Q.* (1967) V. i. 13 *Put down,..to put down a ball* or *the bowling*.. Dean putting the bowling down. 1860 *Baily's Monthly Mag.* Oct. 41 With rare patience did he stop at home and skilfully put down the slows, rarely even attempting to hit them. 1893 R. Daft *Kings of Cricket* xv. 260, I have often seen little men put down with ease a bumping ball which many taller men would let hit their fingers. 1906 A. E. Knight *Compl. Cricketer* iv. 150 When bowlers or wicket-keepers neglect this precaution [of keeping behind the wicket], it may happen that the throw forces them back upon the wicket which they are unable to legitimately put down. 1924 A. C. MacLaren *Cricket Old & New* xiv. 140 On sticky wickets I should doubt if he [*sc.* J. T. Hearne] ever put down a bad ball. 1955 I. Peebles *Ashes* vii. 67 At 26 he had another bit of luck when Hole put him down at first slip.
n. *U.S.* To preserve (food).
1843 *Knickerbocker* XXI. 436 Daniel Gilbert's property..cut up very handsomely, (to borrow the common figure upon such occasions, derived from the putting down of pork for the winter). 1881 S. O. Jewett *Country By-Ways* 40 He's put down a kag of excellent beef. 1889 R. T. Cooke *Steadfast* xxi. 229 Who'll put down my pork and beef as Almiry did?
o. *Aeronaut.* To land (an aircraft or spacecraft). Also *intr.* (with the craft or the pilot as subject).
1933 C. K. Stewart *Speech of Amer. Airman* (Univ. of Akron thesis) 85 *Put down, to*, to land. 1939 *War Pictorial* 6 Oct. 7/3 Orders are to 'put down' the machine on the two-acre landing-deck of a naval aircraft-carrier. 1946 *Sun* (Baltimore) 21 Dec. 17/1 The badly damaged C-47 landed at Phillips Field, while the Eastern Airliner put down at Washington. 1958 'N. Shute' *Rainbow & Rose* i. 14 They put her [*sc.* a freighter] down at Launceston and taxied in. 1972 T. Lilley *K Section* xl. 183 Can you get a chopper..to bring in the District Commissioner? ..It can put down on the padang in front of the police station. 1976 'L. Black' *Healthy Way to Die* ii. 17 The helicopter put down, the engine cut, the rotors gradually slowed. 1976 *New Scientist* 24 June 683/1 If there are no hitches the Viking lander should put him down early on 5 July.
p. *Jazz.* To establish (a rhythm or a style); to play or perform. Chiefly *U.S.*
1944 D. Burley *Orig. Handbk. Harlem Jive* 145 *Put down*, say, perform, describe, do. 1952 B. Ulanov *Hist. Jazz in Amer.* (1958) vi. 67 He put down a good walking beat. 1953 *Down Beat* 11 Feb. 16/3 Those old masters have really put something down, and it'll be a long, long time before those basic sounds change. 1968 *Ibid.* 7 Mar. 19/3 But the *tenor* saxophonists..reasoned that Coleman had been away from the source too long to know the hot licks that Harlem was putting down now.
q. *U.S. slang.* To reject or abandon.
1953 D. Wallop *Night Light* xii. 135 You really ought to put school down and play full-time. 1959 L. Lipton *Holy Barbarians* 102, I put that scene [*sc.* domesticity] down when I got divorced. 1964 *Amer. Folk Music Occasional* i. 62 My mother was the mother of all those kids and my father look like he wanted to put her down, leave her. 1966 E. Liebow *Behavior & Values of Streetcorner Negro Men* (Ph.D. thesis, Catholic Univ.) v. 111 Richard..once 'put down' a woman of thirty or so, foregoing the pleasures of her automobile as well, because 'She's too old'.
r. To replace (the receiver of a telephone), usu. abruptly, in the course of a call; to 'hang up' (the telephone) *on* someone. Freq. in phr. *to put the phone down.*
1966 A. E. Lindop *I start Counting* xxi. 263 He picked up Leonie's telephone..and got through to our flat... When he put down the phone he picked up Leonie. 1970 'M. Carroll' *Bait* v. 67 He put the phone down on me before I could say a word. 1972 T. Lilley *K Section* xl. 187 'You know where I am now; keep in touch.' Carter put the phone down. 1975 C. Fremlin *Long Shadow* iv. 33, I was..so startled, and shocked. I..just put the phone down. 1979 K. M. Peyton *Marion's Angels* v. 77 Geoff put down the receiver and explained gloomily to Marion what was expected of them.
s. To put (a child) to bed.
1968 C. Armstrong *Balloon Man* i. 5 Johnny bounced out of his healthy three-and-a-half-year-old sleep at 6 a.m. The trouble was she had to put him down so early. 1971 D. Devine *Dead Trouble* iv. 33 Sarah Caine was putting Timmy down for his afternoon nap when the

telephone rang. 1978 P. Niesewand *Underground Connection* 152 'Is the baby asleep?' 'I think so. She went off very quickly when I put her down.'

42. put forth.
g. (*b*) (Later example.)
1924 R. Macaulay *Orphan Island* xx. 262 Like some lovely fruit that puts forth, ripens, and tumbles, over-mellow, to the ground.

44. put in.
b. Also *spec.* in *Cricket*, (*a*) to send (a member of one's team) in as batsman.
1823 *Lady's Mag.* July 390/2 David Willis, who, injudiciously put in first..was bowled out, without a stroke, from actual nervousness. 1833 [in Dict.]. 1888 A. G. Steel in Steel & Lyttelton *Cricket* iv. 200 It is as well not to put in two hard-hitters together if possible, as it often tends to make one hit against the other.
(*b*) To cause (a team, usu. the opposing one) to take first innings.
1836 *New Sporting Mag.* Oct. 360 Eton having won the toss, put Winchester in. 1859 *All Year Round* 23 July 305/2 The town won the toss for innings, and put their men in first. 1900 P. F. Warner *Cricket in Many Climes* 212 Lord Hawke, on winning the toss, put the other side in. 1976 J. Snow *Cricket Rebel* 78 It was to be his [*sc.* Mike Denness's] last as captain, after putting the Australians in and then losing the match.
g. Also *to put in the leather* = *to put the boot in* s.v. *boot *sb.*[3] 1 b.
1943 J. Phelan *Lett. from Big House* ii. 30 Almost before he reached the ground the party piled on him. Some punched and cursed, others..'put in the leather'. 1952 M. Allingham *Tiger in Smoke* iii. 57 Someone has been 'putting in the leather'... That was done with a boot.
h. Delete ?*Obs.* and add later U.S. examples.
1855 *Harper's Mag.* Oct. 602/1 The unfortunate victim hollowed out, 'Oh, Moses, if you have any love for your brother, *put in*, and divide this fight!' 1901 W. N. Harben *Westerfelt* 290 You wus tellin' me..'at the lan' an' house wus in yore name an' her'n, an' 'at I had no right to put in.
j. (Further example.)
1972 J. Aiken *Butterfly Picnic* i. 9 The hours I have put in hanging about for her on station platforms.
l. To inform against; to 'frame'; to secure the conviction of (a person); to send to prison. Also *transf.* *slang.*
1922 A. Wright *Colt from Country* 153 'I might have a chance with the girl again.' 'After what you did to put her in?' laughed the detective. 'I like your hide.' 1951 S. Mackenzie *Dead Men Rising* i. 52 Nothing would give me greater pleasure than to put you in, only that's about the one thing I've never done in my life. 1958 D. Niland *Call me when Cross turns Over* vii. 174 Don't put me in. Don't try to hang anything on me. 1966 P. Cowan *Seed* iv. 106, I suppose when they make you a prefect you'll put us in.
m. To let in (the clutch of a motor vehicle).
1928 J. Galsworthy *Swan Song* III. iv. 246 'This is where I put in my clutch,' she said, 'as they say in the "bloods"!' 1943 A. Ransome *Picts & Martyrs* xvii. 167 He put in his clutch and drove off. 1976 'E. McBain' *Guns* (1977) vii. 174 Colley puts in the clutch and manipulates the gear shift.

45. put off.
h. (Later example.) Also, to cause (someone) to be mistaken.
1918 A. Bennett *Pretty Lady* xxii. 146 'That's not *you*, Frankie!' said the Major with a start of recognition... 'Yes, sir,' said Molder... 'It was the red hat put me off,' the Major explained.
o. (Further examples.) Now usu., to offend, to disconcert; to cause (a person) to lose interest in or enthusiasm for something.
1909 F. Barclay *Rosary* ix. 77, I am so afraid of her putting Dal off. He is so fastidious. 1928 *Observer* 19 Feb. 6/3 The prefatory note, with its apparently exaggerated claim, rather put me off. 1932 'E. M. Delafield' *Thank Heaven Fasting* I. ii. 34 A man is very quickly put off, if he thinks that a girl hasn't even taken the trouble to remember what he looks like. 1949 D. Smith *I capture Castle* ix. 134 He'll end by putting them off us. 1973 L. Meynell *Thirteen Trumpeters* v. 80 I'm in grave danger of becoming virtuous. To see those acres of fat Germanic flesh spread out by the pool is enough to put me off for life.

46. put on.
a. (Further theatrical examples.)
1924 A. Huxley *Let.* 29 Apr. (1969) 229 Playfair, who is producing it for the 300 club performance, seems to think that it will make a very good entertainment and has some hopes of getting it put on for a run. 1941 L. A. G. Strong *Bay* 192 A couple of new plays that some amateurs were putting on. 1977 A. Morice *Scared to Death* i. 7 Presumably, if his play is any good, this David Winter would have put it on anyway?
e. (Further examples.) Also, to impose on, to take advantage of; to puzzle or deceive intentionally. *colloq.*
1949 D. Smith *I capture Castle* xiv. 290 'We shall be ashamed of our callousness if father really is going off his head.' 'He isn't—he's putting it on or something.' 1958 *Times* 12 Nov. 3/3 Miss Mollie Sugden's wife has got into the habit of 'putting on' her husband because the husband..rather enjoys being 'put on'. 1958 *Amer. Speech* XXXIII. 225 When a hipster *puts* someone *on* he is pulling his leg (perhaps putting him on a stage to be laughed at). 1960 Willmott & Young *Family & Class in London Suburb* x. 111 'Some of the parents at the school seem to put it on a bit,' said Mr. Prior, a bank manager whose children go to a local preparatory school, 'you do get a bit of the old blue-blooded attitude among them.' 1964 H. E. F. Donohue *Conversations with Nelson Algren* xi. 272 She's putting me on and I'm putting her on, and she marvels at her good fortune in meeting me, I'll marvel at my good fortune in meeting her. 1966 T. Pynchon *Crying of Lot 49* vi. 167 Has it ever occurred to you, Oedipa, that somebody's putting you on? That this is all a hoax? 1967 'G. Bagby' *Corpse Candle* (1968) x. 133 Greg was forever putting people on... He'd do it just for fun. The poetry was his way of putting the English faculty on. 1977 *Sci. Amer.* Dec. 17/3 Persi's brief description of the Rockwell prediction method was so outlandish that I assumed he was putting me on.
f. (*a*) Also, *to put it on.*
1933 E. Hemingway *Winner take Nothing* (1934) 35 It's terrible..the way I put it on. 1967 A. Diment *Dolly Dolly Spy* vii. 98 She had put on a lot of weight... I could see her checking herself off against Veronica—who has definitely been putting it on. 1971 'J. J. Marric' *Gideon's Art* i. 11 'You both take sugar?'..'Not for me,' Slater said, slapping his rounded belly. 'I'm putting it on again.'
(*c*) (Earlier and later examples.)
1868 *Baily's Mag.* Sept. 246 The last wicket fell for 689, six players thus putting on nearly as many hundred runs. 1921 *Glasgow Herald* 17 Oct. 13/7 In the second half P. R. Johnstone scored, and afterwards G. A. Able put on another for Stepps. 1975 [see *outfield *sb.* 3 a]. 1977 *World of Cricket Monthly* June 32/1 Haroon and Imran put on 34.
(*d*) Of a taxi-driver: to join (the end of a rank).
1930 A. Armstrong *Taxi* xii. 164 'Putting on' is the taxi man's expression for coming on at the end of the rank. A driver will say he 'put on sixth cab at the so-and-so', meaning he came on the so-and-so rank when there were only five other cabs there. 1939 H. Hodge *Cab, Sir?* 22, I decide to put on a hotel rank.
l. (*a*) (Earlier examples.)
1836 *New Sporting Mag.* XI. 360 Mr. Paterson's bowling was again very reasonably put on. 1859 *All Year Round* I. 306/1 They put on bowler after bowler,..but they could not get us out.
(*b*) In slang phr. *to put it on* (a person), to charge to (someone else).
1895 *People* 6 Jan. 16/5 Arter all the brass..was nearly all gone, Selby says, 'I'll go round to the Mug agin, and put it on him (make him pay) for another bit.' 1944 L. Glassop *We were Rats* i. 6 I'll have a pint at the Royal tomorrer and put it on the blonde.
(*c*) *colloq.* To draw the attention of or introduce (a person) *to* a particular person or thing.
1895 *N.Y. Dramatic News* 12 Oct. 5/3 Mr. Jack is always a newspaper man's friend, and only too pleased to put one on 'to a good thing' in the shape of news. 1901 O. Wister in *Lippincott's Monthly Mag.* Aug. 199 We're awfully obliged for the way you are putting us on to this. 1902 H. G. Wells *Let.* 2 Sept. in H. Wilson *Arnold Bennett & H. G. Wells* (1960) 83 Accept I pray you my warmest thanks. And also for putting me on to that quite brilliantly done and (as Dr. Robⁿ Nicoll would say) most unpleasant book, *Le Journal d'une Femme de Chambre.* 1924 A. Christie *Poirot Investigates* vii. 165 A friend of mine in the City put me on to a very good thing, and..I have money to burn. 1926 H. J. Laski *Let.* 21 Feb. in *Holmes-Laski Lett.* (1953) II. 833 He also put me on to a new American life of Godwin. 1949 A. Christie *Crooked House* xii. 93, I could put you on to a couple of the tame psychiatrists who do jobs for us. 1977 C. McCarry *Secret Lovers* iii. 33 He put us on to some people who turned out to be..useful.

47. put out.
b. (*b*) Also *fig.*
1937 C. Carmer *Hurricane's Children* 105 He wore waistcoats that would put your eyes out.
(*d*) (Earlier example.)
1780 J. Woodforde *Diary* 15 July (1924) I. 289 John had a fall lately..and put out his shoulder bone, being a little merry.
c. Also in *Baseball*, to cause (a batter or runner) to be 'out'; in *Boxing*, to knock out.
1845 in *Appleton's Ann. Cycl.* XXV. 77/2 A runner can not be put out in making one base, when a balk is made by the pitcher. 1910 J. Driscoll *Ringcraft* iii. 84, I have..not infrequently put opponents 'out' with a blow on the neck. 1912 C. Mathewson *Pitching in a Pinch* 107 Snodgrass was put out trying to get to third base.
d. For 'Now *rare* or Obs.' read: Now only in slang use, to kill (a person). (Perh. a *fig.* use of sense e (*b*).)
1917 W. Owen *Let.* 25 Apr. (1967) 452 For twelve days we lay in holes, where at any moment a shell might put us out. 1935 E. Wallace *Mouthpiece* xvii. 225 That's the offer the gentleman made—five hundred quid to put you out and keep me mouth shut. 1975 'E. Lathen' *By Hook or by Crook* xii. 114 The minute his stomach started acting up, he would've been yelling for the cops. He had to be put out fast.
f. (*e*) (Earlier example.)
1861 T. Hughes *Tom Brown at Oxford* III. xvi. 290 Don't you lose heart because he won't put himself out for you.
i. Delete 'Now *rare.*' and add further examples. Also, to broadcast.
1938 H. Nicolson *Diary* 20 Feb. (1966) 323 On the late news it is put out that Eden has resigned. 1965 G. Melly *Owning Up* xi. 135 His version of 'Rock Island Line'..was put out as a single and rose to be top of the Hit Parade. 1966 *Listener* 13 Jan. 78/1 Earlier this year Midland Region and Anne Owen put out..an unusually direct and perceptive investigation of present-day standards of honesty. 1978 *Times* 26 July 4/2 The BBC says that whatever it films and tapes it is entitled to put out.
k. (*b*) *intr.* Of a river or natural formation: to extend or stretch (in relation to a specified point). *U.S.* See also sense 8 c in Dict.
1755 [in Dict., sense 8 c]. 1840 C. F. Hoffman *Greyslaer* I. 116 A ledge of bald rock to the left yonder..puts out from the ridge. 1878 J. H. Beadle *Western Wilds* 311 Commenced the ascent of the Buckskin, a low range of partially-wooded hills, putting out across the plateau nearly to the Colorado.
m. (*a*) (Earlier example of sense 'to plant out'.) Also to send out (a domestic pet) for exercise, etc.
1851 Mrs. Gaskell *Let.* 7 Apr. (1966) 149 We are sowing very few annuals this year..& relying on putting out the greenhouse things for a summer show. 1917 D. Canfield *Understood Betsy* ii. 46 'Mother, did you put Shep out?' 1925 Wodehouse *Carry On, Jeeves!* ii. 40 When he has put the cat out and locked up the office for the night, he just relapses into a state of coma. 1974 *Listener* 10 Oct. 462/1 The BBC's nightly *Campaign Report*..at an hour when most voters are putting the cat out. 1977 'E. Crispin' *Glimpses of Moon* viii. 128, I was snug in bed... And then..I remembered..that I ought to have put Sal out... She barks rather a lot.
(*c*) (Earlier and further examples.) Also, to place (articles) for collection by tradesmen.
1653 R. Verney *Let.* in M. M. Verney *Mem.* (1894) III. iv. 112, I will keepe but one woeman kind, who must wash my small Linnen (bed & board linnen shall bee put out). 1873 A. J. Munby *Diary* 18 Feb. in D. Hudson *Munby* (1972) 322, I should like very well to clean his boots..and I said to Tarrant 'If you put 'em out I'll clean 'em with pleasure, along with mine.' 1975 'D. Jordan' *Black Account* v. 33 It was late; Sue was in her kimono and putting out milk bottles.
n. *intr.* Of a woman: to offer oneself for sexual intercourse. Also const. *for* (a man). *slang.*
1947 *Horizon* Sept. 202 'Maybe all the whores'll be puttin' out free on New Year's!' Muggleston shouted. 1961 J. Heller *Catch-22* xiii. 131 The beautiful..countess and her beautiful..daughter-in-law, both of whom would put out only for Nately, who was too shy to want them, and for Aarfy, who was too stuffy to take them and tried to dissuade them from ever putting out for anyone but their husbands. 1975 D. Lodge *Changing Places* vi. 232 If she won't put out the men will accuse her of being bourgeois and uptight. 1977 I. Shaw *Beggarman, Thief* III. i. 178 Sometimes those plain-looking little dolls are powerhouses when it comes to putting out. 1978 M. Puzo *Fools Die* vi. 80 He was especially challenged if a girl had a reputation for only putting out for guys she really liked.

48. put out of.
d. *to put out of misery* (examples).
1792 J. Woodforde *Diary* 16 May (1927) III. 351 My poor old Horse, Punch..was shot by Ben this Morning to put him out of his Misery. 1855 C. Kingsley *Westward Ho!* III. xii. 353 Writhing in his great horror, he called to Cary to kill him and put him out of his misery. 1911 *Maclean's Mag.* Oct. 286/1 Get the gun, for God's sake, an' put me out of my misery. 1923 G. Atherton *Black Oxen* xxvi. 145 Tell them all about it... Put them out of their misery. 1957 D. Robins *Noble One* v. 59 Then I'll *have* to stalk him and put him out of his misery. 1975 A. Christie *Curtain* xi. 113 We were talking of euthanasia... 'Does the person most concerned ever wish to "put himself out of his misery", as we say?'

49. put over.
b. Delete † *Obs.* and add later examples.
1871 'Mark Twain' *Lett. to Publishers* (1967) 55 If you can without fail issue the book on the 15th of May—putting the Sketch book over till another time. 1926 J. Black *You can't Win* xxii. 343 We went to court again the next day, but were put over twenty-four hours on the plea of the police that witnesses were on their way from Canada. 1978 H. Kemelman *Thursday the Rabbi walked Out* (1979) xxx. 145 The only thing to do is to put it over for a week.
i. *to put it* (*all*) *over* (*on*), to excel or surpass (in a particular enterprise); to defeat or trounce.
1898 F. P. Dunne *Mr. Dooley in Peace & War* (1899) 172 I've seen..Fitz beat Corbett; an', if I live to cillybrate me goold-watch-an'-chain jubilee, I may see some wan put it all over Fitz. 1905 J. London *Let.* 24 June (1966) 175 If Hillquit..didn't put it all over Bierce—I'll quit thinking at all. 1944 *Living off Land* viii. 155 So far as bushcraft is concerned, he [*sc.* the Aboriginal] could put it all over you. 1973 *Time Out* 2 Mar. 15/2 The teachers..only had time for the Thomas boys; we were treated like shit. So we started throwing our weight around, we put it over on them.
j. To make acceptable or effective; to convey or communicate; to present convincingly; = *to put across* (sense 35* b above).
1912 R. A. Foley in *Mag. Maker* Dec. 8 He saw his opportunity and he 'put it over'. 1914 G. Atherton *Perch of Devil* ii. 298 You don't go into any business..and put it over without running the risk of being shot. 1928 *Daily Express* 18 Apr. 11/2 Is it true that you wanted a star name to put the play over? *Ibid.* 11 July 9/3 On the screen you..are fascinated by the extraordinary way in which he 'puts himself over'. 1929 J. B. Priestley *Good Companions* II. i. 252 He's a find. Works hard, got personality, puts it over all the time. 1931 F. L. Allen *Only Yesterday* viii. 213 The president emeritus of Harvard had had no professional talent to put over his funeral in a big way. 1935 *Motion Picture* Nov. 6/2 Clark Gable plays one of those powerful, he-men rôles in which he glories. And he puts it over with a bang! 1958 *Times* 1 Sept. 3/6 About Mr. Presley's ability to 'put over' a song in his own particular way there can be no two opinions. 1966 *Listener* 17 Mar. 380/2, I did not know how to select what I wanted to do or really put over emotion. 1978

D. Murphy *Place Apart* iii. 59 They blamed 'that Paisley'... They agreed with his anti-ecumenism..but they didn't like the way he put it over.
k. To impose (something false or deceptive) *on* a person; to best or upstage (someone); to achieve by deceit.
1912 J. Sandilands *Western Canad. Dict.*, *Put one over on* him, catching him with the latest puzzling by-word or smart saying... A Winnipeg newspaper recently put up the heading, 'Put one over on Bernard Shaw'. 1914 'High Jinks, Jr.' *Choice Slang* 17 *Put one over, to*: to beat by strategy, 'to hornswoggle'. 1916 H. L. Wilson *Somewhere in Red Gap* i. 19 Funny, the way the little man tried to put it over on us, letting on he was just puzzled—not really bothered, as he plainly was. 1923 R. D. Paine *Comrades of Rolling Ocean* viii. 130 Who calls it a crime to put one over on the Custom House flatties? 1928 A. S. W. Rosenbach *Books & Bidders* 117 One of the greatest hoaxes ever planned was put over by a French forger. 1945 C. Williams *All Hallows' Eve* 35 A fellow who's put it over all America and bits of England is likely to know where he is. 1958 *People* 4 May 8/3, I cannot see her letting any of the Italian or French sexpots put one over on her. 1967 *Listener* 5 Jan. 37/1 Christmas, after all, is essentially an 'old' festival (however much Batman may seem to have put one over on Santa Claus this year). 1972 Wodehouse *Pearls, Girls, & Monty Bodkin* x. 150 It's low. It isn't done. You can't do the dirty on a business competitor just to stop him from putting it over on you in a business deal. 1976 *Church Times* 30 July 7/2 She may have been fleeced in Florence, robbed in Ravenna, grossly overcharged in Ostia..; but *Baedeker* at least has not tried to put one over on her. 1979 *Jrnl. R. Soc. Arts* CXXVII. 650/1 We are not appearing to put something over on the public.
l. *Baseball.* = *put across* (sense 35* c above).
1936, 1943 [see sense 35* c above].

49*. put there.
In imp. phr. *put it* (or *her*, etc.) *there*: shake hands! *colloq.* (orig. *U.S.*).
1898 R. Hughes *Lakerim Athletic Club* i. 3 'Put her there, Punk; you're a white man!' Tug had to exclaim; and the two captains shook hands. 1915 A. Conan Doyle *Valley of Fear* II. i. 154 'Put it there,' he said. A hand-grip passed between the two. 1925 *New Yorker* 20 June 14/1 Well, I'll be damned. Glad ta see ya. Put it there. 1931 O. Nash *Hard Lines* 50 Put it there, Mr. Linthicum, put it there! 1947 Wodehouse *Full Moon* vii. 168 'I'm engaged!'..'Well, I'm dashed,' said Freddie. 'Put it there, pardner.' So beaming was his smile, so cordial his handshake, that Tipton found his last doubts removed. 1970 *Private Eye* 13 Mar. 16 Glad to meetcha! Put it there!

50. put through.
b. Delete 'Chiefly U.S.' and add later examples.
1929 T. H. Burnham *Engineering Econ.* xv. 199 Rush orders are difficult to put through, even in well-organized works. 1966 'J. Hackston' *Father clears Out* 54 Put through a second lot of tailings, but not from the same place.
c. (Examples.)
1891 F. C. Allsop *Telephones* vi. 98 In an exchange system any of the stations wishing to communicate with any other must first ring up the central station, and request to be 'put through' to the other station. 1916 'Boyd Cable' *Action Front* 86 Ask to be put through to the inquiry office. 1928 D. L. Sayers *Unpleasantness at Bellona Club* viii. 59 That phone-call you asked me to trace..was put through..from a public call-box. 1949 A. Christie *Crooked House* xvii. 139 He lifted the receiver—listened and then said: 'Put her through.' 1973 J. M. White *Garden Game* 182, I found the number and dialled Whitehall... I was put through to the Home Office.
d. *Econ.* (See quot. 1959.)
1959 *Economist* 21 Mar. 1099/1 Where the market is narrow, as it can be for example in rubber and tea shares, the jobbing system may not work either smoothly or perfectly. The brokers in these shares then find it convenient to 'marry' the buying and selling orders. The normal practice has been for such a deal to be 'put through' a jobber at a very small turn for him... The stock exchange council..has now proposed a change in the rules governing these 'put through' deals. 1978 *Times* 17 Nov. 21/8 The principle of the put-through deal involves the broker finding a buyer for a large line of stock which one of his clients has on offer. The jobber then puts the shares through his books at a mutually agreed price but does not necessarily make such a good turn on it as he would if he was buying them from the broker and selling them on himself.

51. put to.
f. (Earlier example.)
1791 J. Woodforde *Diary* 8 Aug. (1927) III. 291 We were rather put to for a Dinner in so short a time however we did our best and gave them some Beans and Bacon, mince Veal, Neck of Mutton [etc.].

52*. put under.
a. *trans.* To kill or bury (a person).
1879 R. A. Sterndale *Afghan Knife* II. vii. 75, I wanted to see your bonny face once more, in case these blackguards put me under. 1958 C. Watson *Coffin, scarcely Used* iii. 27 There'll be some pressure to have him put under without any unseemly inquiries.
b. To render unconscious by means of an anaesthetic or by hypnosis.
1962 L. Payne *Too Small for his Shoes* v. 94 Given him something to put him under. Be right as rain. 1963 E. Lanham *Monkey on Chain* xiv. 207 He put Dora under and learned conclusively that she went down to Bleecker Street. 1971 P. O'Donnell *Impossible Virgin* xii. 235 'Is Willie going to give the ether?' 'Yes. I'll put her under myself.'

53. put up.
a. (*c*) (Earlier and later examples.)
1848 Trollope *Kellys & O'Kellys* II. ii. 46 Brien was saddled..and Pat was put up. 1953 E. Coxhead *Midlanders* i. 32 Don't suppose you've yet been on horseback, miss? We'll put you up and see how you like it.
(*d*) (Earlier examples.)
1838 Dickens *Let.* Nov. (1965) I. 465, I don't know what they put up at the Theatre for that night. 1852 *Punch* 11 Dec. 257/1 The entertainments this week have been of a slight and desultory character, the management being..glad to 'put up' anything they could get.
(*e*) In imp. phr. *put them* (or *'em*) *up*: (i) a challenge to raise the fists before a fight; (ii) a command to raise the hands above the head. *colloq.*
1923 E. Wallace *Captains of Souls* xliv. 240 I'm going to give you the damnedest lacing you ever had..put 'em up! 1937 Partridge *Dict. Slang* 672/1 *Put 'em up!*, raise your arms!: from ca. 1860... Put up your fists!..late C. 19–20.
(*f*) To place (a military or other decoration) on one's uniform or other clothes.
1959 M. Gilbert *Blood & Judgement* xiv. 147 He could easily have put up a medal ribbon he wasn't entitled to. 1961 E. Waugh *Unconditional Surrender* 5 He had been trained in the first batch of temporary officers..had twice put up captain's stars and twice removed them; their scars were plainly visible on his shoulder straps.
g. Also *to put up a fight*, to acquit oneself well in a contest. Also *fig.*
1919 H. Crane *Let.* 7 Mar. (1965) 13 Mrs. Brooks is afflicted with consumption against which she is doubtless putting up a strenuous Scientific fight. 1928 —— *Let.* 27 Mar. (1965) 320, I put up quite a fight, but neither of us were in much condition.
i. Delete † *Obs.* and add later examples.
1912 Galsworthy *Justice* II, in *Plays* II. 59 *Judge.* Call the next case. *Clerk of Assize.* (To a warder) Put up John Booley. 1949 F. Sargeson *I saw in my Dream* 75 Anyhow he'd been sacked and put up for it, and he'd only got six months probation. 1960 'M. Underwood' *Cause of Death* xii. 152 The clerk of the court..said in a loud clear voice, 'Put up David Lucas.' 1964 J. Prescott *Case for Court* ix. 175 Mr. Rose asked for the Sorensens to be put up at once so that the Chief Constable might make his application... The two accused were brought up into the dock. 1976 *Howard Jrnl.* XV. i. 42 There are a number of minor errors:..On p. 20 the prisoner is sitting in the dock before he has been put up.
j. (*a*) (Further examples.) Also, to propose for an honour or award.
1840 Lytton *Money* (ed. 4) I. i. 30 Shall I put you up at the clubs? 1967 N. Marsh *Death at Dolphin* vi. 154 We'll put you up for the Police Medal. 1971 J. R. L. Anderson *Reckoning in Ice* vii. 143 He was..a sailor, and I'd put him up for the Mariners. We met at the club occasionally.
(*c*) (Further example.)
1969 'R. Gordon' *Facts of Life* 140, I spend all my time putting up for jobs. In the last six months, I've been to Liverpool, Exeter, Oxford, and York.
m. (*b*) Delete *rare* and add further examples.
1909 P. A. Vaile *Mod. Golf* xvi. 211, I am directing my manufacturers' energies to producing the exact amount of marking required [on a golf ball]... I should not have troubled with it had it not been 'put up to me', as the American would say. 1913 F. H. Burnett *T. Tembarom* xxiv. 306 'Oh, well, I just put it up to them.'..'You mean that you made them feel that they alone were responsible.' 1924 Galsworthy *White Monkey* I. viii. 58 I'll put it up to Mr. Desert; if he speaks for you, perhaps it may move Mr. Danby.
n. (*a*) (Later examples.)
1916 *Daily Colonist* (Victoria, B.C.) 2 July 4/5 Sidney women, under Mrs. Wheeler, have started putting up jam for the boys at the front. 1924 T. S. P. Strangeways *Technique of Tissue Culture* 39 To put up the cultures take the tissues or organs which have been set aside for cultivation and cut up into suitably sized fragments. *Ibid.* 73 If more cultures are desired, put them up in a similar way. 1951 [see *mid *a.*, *sb.*[1], and *adv.* 1 c]. 1954 J. R. R. Tolkien *Fellowship of Ring* I. iv. 107 He produced a large basket from under the seat... 'Mrs. Maggot put this up for Mr. Baggins, with her compliments.' 1970 *Nature* 19 Dec. 1139/2 When either bone marrow or circulating blood cells from humans, mice or rats are put up in culture in a freshly made medium containing calf serum, few if any colonies of haematopoietic cells grow. 1971 R. Thomas *Backup Men* v. 34 He's helping me put up some marmalade.
(*e*) (Earlier example.)
1865 'Mark Twain' in *N.Y. Sat. Press* 18 Nov. 249/2 And so the feller..put up his forty dollars along with Smiley's.
(*f*) In imp. phr. *put up or shut up*: defend yourself or be silent. *colloq.* (chiefly *U.S.*).
1878 F. H. Hart *Sazerac Lying Club* 167 'P.U. or S.U.' means put up or shut up, doesn't it? 1884 *National Police Gaz.* (U.S.) 26 Apr. 1 (*caption*) Put up, shut up, or get! 1889 'Mark Twain' *Connecticut Yankee* xl. 512 This was a plain case of 'put up, or shut up'. 1952 *Manch. Guardian Weekly* 1 May 3/4 The old alternatives will be revived: put up or shut up—get out or get on to the Yalu and beyond. 1976 *Billings* (Montana) *Gaz.* 17 June 6-C/1 It wasn't a case of put up or shut up because the money was voted as a sincere effort to clean up the mess.
q. (*c*) Sense (*b*) used without following *to* and adjunct: to annoy, to vex (a person).
1930 H. G. Wells *Autocracy of Mr. Parham* IV. i. 266 This cheap Mussolini at Westminster is putting us up some! 1960 T. McLean *Kings of Rugby* xi. 160 Hill's protest was more likely to restore the true spirit of the game than..some other method of retaliation by the Canterbury men who believed that they were being put up.
r. (Earlier example.)
1699 M. Lister *Journey to Paris* 25 There are an infinite number of Busto's of the Grand Monarque every where put up by the Common People.
u. To judge, regard, or assess (a person, situation, etc.) in a particular way. *U.S.*
1877 'Mark Twain' in *Atlantic Monthly* Nov. 590/1 Would you like to have me explain that thing to you?.. Now, this is the way I put it up. 1880 —— *Tramp Abroad* xx. 192 'Didn't I put you up right?' 'Oh, yes.' 'Sho! I spotted you for *my* kind the minute I heard your clack.' 1895 *Century Mag.* Sept. 674/2 And Jack says to himself, 'Well,..I done what I could! What is to be will be.' That's about the way I put it up.

VI. 54. a. *to put the boot in*: see *boot *sb.*[3] 1 b; *to put the fear of God into*: see *fear *sb.* 3 e; *to put next to*: see *next *a.* 13 c; *not to put it past someone*: see *past *prep.* 3 b; *to put* (a person's) *pot on*: see *pot *sb.*[1] 13 k; *to put the wind up* (a person): see Wind *sb.*[1] 10 b; *to put* (one) *wise* (*to*): see Wise *a.* 3 b (*b*).

put, *ppl. a.* Add: **put-together** (52: in quots. 52 d); **put-upon** (23 f (*b*)); also *absol.* as sb. *to stay put*: see Stay *v.*[1] 6 b.
1950 in E. C. Richards *Diary of E. R. Chudleigh* 23 Such a 'put-together' mob of wild cattle required at least six to eight experienced stockmen. 1957 J. Kerouac *On Road* (1958) 21 Country boys in a put-together jalopy. 1970 *Times* 2 June 8/2 The essential of the put together look which stays put is a belt. 1920 *Quill* (N.Y.) Nov. 12 Lulu is the ideal poor relation of fiction, the put-upon slavey. 1966 M. Kelly *Dead Corse* iv. 53 Those who follow unquestioning, docile, simple... The put-upon. 1976 *Listener* 6 May 586/2 Juliet Mills was very good as Cady's put-upon wife. 1980 G. Mitchell *Uncoffin'd Clay* iii. 32 Having to cook..a sensitive charwoman would regard as victimisation or, in her parlance, a put-upon.

put-. Add: **put-away** *Lawn Tennis* and *Rackets* = *kill *sb.* 2 c; also *attrib.*; **put-back** (later example); **put-down,** (*a*) an act of putting (a person) down, a snub; also *attrib.*; (*b*) *attrib.*, with reference to the act of alighting from a vehicle; **put-in,** (*a*) *U.S. colloq.*, one's turn to speak, one's affair; (*b*) the act of putting the ball into a scrum in rugby football; **put-out,** (*a*) (earlier and later examples); *U.S.*, an annoyance or inconvenience (?*obs.*); **put-through,** (*a*) a measure of the number of persons or objects which have been put through a process; (*b*) *Econ.*, a financial transaction in which a broker arranges the sale and the purchase of shares simultaneously; also *attrib.* Cf. *put-on *sb.*
1969 *New Yorker* 14 June 75/1 He intercepts, and sends a light and graceful putaway past Graebner, down the line. 1977 *Ibid.* 25 July 70/2 Connors..also carried off the next three games on the strength of some fine, deep approaches and remarkable put-away volleys. 1913 D. H. Lawrence *Love Poems* p. lviii, An' mind... Ye slip not on the slippery ridge Of the thawin' snow, or it'll be A long put-back to your gran' marridge. 1962 J. Baldwin *Another Country* II. iv. 335 Flattery will get you nowhere, son. Or is that a subtle put-down? 1968 *Punch* 23 Oct. 593/2 Michael Denison sustains an appropriately truculent pout and Dulcie Gray delivers tart and catty put-downs with relish. 1972 G. Lyall *Blame the Dead* xiv. 100 He'd picked me up at the put-down place for Euston station... It's a one-way underground street. 1973 *N.Y. Times* 18 Feb. I. 24/1 He [*sc.* Trudeau] doesn't rise to bait—with choice epithets and that put-down Gallic shrug of his. 1974 S. Alsop *Stay of Execution* ii. 160 He [*sc.* Dean Acheson] detested silliness, and he was justly famous for his put-downs—when he put down a fool, the fool was left in no doubt that he was a fool. 1976 *National Observer* (U.S.) 20 Mar. 14/2 The woman whom former Attorney General John Mitchell immortalized with his famous putdown, 'Katie Graham's gonna get her tit caught in a big fat wringer', is hardly the 'bitch bringing down Presidents'. 1977 Miller & Swift *Words & Women* vi. 100 Some speakers and writers use Ms. only as a put-down. 1853 'Mark Twain' in *Hannibal* (Missouri) *Jrnl.* 25 May 3/1 Never speak when it's not your 'put-in'. 1903 W. N. Harben *Abner Daniel* xxxv. 301 This ain't no putin o' mine, gracious knows. I hain't got nothin', an' I don't expect to lose or gain by what is done. 1962 *Times* 11 Jan. 4/3 The Navy came out better in the matter of put-ins against the head. 1975 *Sunday Tel.* 2 Mar. 30/1 He may have lost confidence as the game developed after being penalised four times for a crooked put-in. 1833 J. Neal *Down-Easters* I. vi. 83, I shouldn't think twould be any put-out to you to take somebody else. 1843 A. S. Stephens *High Life N.Y.* II. 32 Don't be uneasy about the trouble, it won't be no put out to Captain Doolittle. 1885 *California Athlete* 19 Dec. 5/1 He assisted yesterday in fourteen put outs. 1891 N. Crane *Baseball* vi. 44 An 'assist' is given to every player who handles the ball in assisting a put-out or other play of the kind. 1904 R. H. Barbour *School & College Sports* 200 *Put-out*, a play by which a batsman or a base-runner is retired. 1972 *Evening Telegram* (St. John's, Newfoundland) 24 June 21/8 He..led the league's shortstops in fielding percentage, putouts, assists and double plays. 1958 *Punch* 8 Jan. 84/1 He..gave me the acreage, cost, cubic capacity and passenger put-through. 1959 [see *put through* s.v. *put *v.*[1] 50 d]. 1968 *Economist* 4 May 64/1 Even in the leaders trading is often very narrow, and the resulting prices (on the basis of which an increasing amount of shunting and 'put throughs' now go on) are not struck on the total volume of trading. 1973 *Daily Tel.* 7 June 21 Trading on the Paris Bourse traditionally has consisted either of very big 'put-throughs' or of small deals for private individuals.

‖ **puta** (pu·tă). *slang.* [Sp.] A whore, a slut.
1967 McCormick & Mascareñas *Compl. Aficionado* iii. 72 You must be like a young priest—do not go near the *putas*, and keep away from all women as long as you can. 1968–70 *Current Slang* (Univ. S. Dakota) III–IV. 99 *Puta*, a promiscuous girl. 1969 E. Bishop *Compl. Poems* 207 Under the false-almond tree's Leathery leaves, a childish *puta* Dances. 1971 L. Gribble *Alias the Victim* vii. 121 You tricky bitch of a *puta*.

pu:t-and-ca·ll. *Econ.* (See quot. 1905.) Also *attrib.*
1892 *Congress. Rec.* 6 June App. 448/1 Members of exchanges do not deal in '*puts and calls*'. 1893 [see Put *sb.*[1] 4.] 1905 F. Bower *Dict. Econ. Terms* 133 *Put and call*, on the Stock Exchange it is a common practice to arrange to buy or sell a certain number of shares at option at a fixed price within a specified time. The arrangement may be (1) to buy shares, known as the 'Call', (2) to sell shares, the 'Put', or (3) to buy or sell at option, called the 'Put and Call'. 1929 *Sun* (Baltimore) 26 June 1/6 She started business for herself, dealing in options, or in Wall Street slang the 'put and call' or 'put and take' brokerage business. 1962 S. Strand *Marketing Dict.* 599 Put and call contracts are written for 30, 60 or 90 days, or longer. 1973 *N.Y. Law Jrnl.* 4 Sept. 7/5 (Advt.), Lawyers, their clients and those involved in the put and call trading process.. need to have up-to-date knowledge on the nature of these investments. 1975 G. V. Higgins *City on Hill* ix. 216 He ..bought silver futures, puts and calls, I think.

pu:t-and-ta·ke. 1. A gambling game played with a six-sided top. Also *transf.*, the top with which the game is played. Also *attrib.* and *fig.*
1922 *Daily Mail* 5 Jan. 5/4 (*heading*) Put and take in court. *Ibid.* For playing Put and Take in a recreation ground a youth was charged under the Gaming Act at Hull yesterday. 1922 *Vet. Jrnl.* XXVIII. 105 A rough-haired fox-terrier dog, 'Jazz',..was brought to me with the history that a brass 'Put and Take' top, 'put' on the hearthrug, had been 'taken' up and swallowed by him. 1940 Graves & Hodge *Long Week-End* viii. 132 In 1922 the craze was for a simple gambling device known as 'Put and Take'..a small six-sided top which players..spun in turn. 1960 R. C. Bell *Board & Table Games* v. 146 Put and Take. Each player..spins a six sided teetotum and obeys the instructions on the face falling uppermost. 1970 *Guardian* 9 Dec. 1/5 The deadly game of put-and-take brought on by the electricity generating workers' work-to-rule. 1972 *Observer* 3 Sept. 32/1 Each juror was issued with two bronze tokens with a shaft through the middle, rather like a put-and-take.
2. *Econ.* = put-and-call.
1929 [see *put-and-call].
3. The stocking of streams and lakes with fish for anglers to catch. Usu. *attrib.*
1943 *Sun* (Baltimore) 26 Jan. 6/3 (*heading*) Put-and-take fishing planned by W. Virginia. *Ibid.* The revision would take the form of stocking a much larger number of legal size trout on a 'put-and-take' basis in rivers and creeks which are not year-around trout streams. 1973 *Country Life* 21 June 1804/1 The rainbow trout is..the ideal stock fish for enclosed waters..where fishing is increasingly on a 'put and take' basis. 1974 *Ibid.* 26 Sept. 831/3 The cultivation of sizeable trout which are stocked makes Traws a sort of put-and-take fishery unlike most put-and-take places.

putanism. For † *Obs. rare*—0 read *rare*, and add example.
1922 P. Nielsen *Black Man's Place in S. Afr.* 57 Immorality is rife amongst Natives.., but neither can putanism amongst the whites be denied.

putcheon, var. Putcher in Dict. and Suppl.

putcher. Examples of the form *putcheon.*
1898 *Birmingham Daily Post* 16 Dec. 8/6 Heavy catches of eels in 'putcheons'. 1945 J. Moore *Portrait of Elmbury* iv. 147 Jim also earned £4 by selling eel-putcheons which he'd made out of withies. 1968 J. Arnold *Shell Bk. Country Crafts* 258 Salmon traps, putchers or putcheons, are quite different in structure. *Ibid.* 260 The putcheons are arranged in a stout, permanent framework, forming a 'barrage' extending for scores of yards across the Severn grounds.

puthery: see *pothery *a.*

pu·t-on, *sb.* Chiefly *N. Amer.* [Put-: cf. Put *v.* 46 e in Dict. and Suppl.]. A deception, a ruse, a hoax.
1937 Partridge *Dict. Slang* 672/2 *Put-on*, a deception, subterfuge, excuse..from ca. 1860. 1949 H. Hornsby *Lonesome Valley* xxiv. 316 He knew there was no put-on; that she was not talking just to make him feel better. 1967 *New Yorker* 24 June 34/3 What was once an occasional surprise tactic—called 'joshing' around the turn of the century and 'kidding' since the twenties—has been refined into the very basis of a new mode of communication. In all its permutations, this phenomenon is known as the 'put-on'. It occupies a fuzzy territory between simple leg-pulling and elaborate practical joke. 1968 'E. McBain' *Fuzz* ix. 140 Meyer thought the call was a put-on, nobody had a name like Carlyle Butterford. 1970 *Globe & Mail* (Toronto) 28 Sept. 8/7 (Advt.), Wool carpet excellence is assured by the Woolmark label. Protecting you from a put-on. 1973 *Sat. Rev. Society* (U.S.) May 76/3 A wild mishmash of put-on, fantasy, and cultivated lunacy. 1975 *New Yorker* 17 Nov. 125/1 There is no hint of satire here—or, to be fair, of put-on. 1977 *Time* 4 Apr. 44/2 Much of the tone of such writing is personal, confessional, full of macho bellicosity and show-biz put-on.

‖ **putonghua** (pŭtuŋhwā). Also pu-, p'u-t'ung-hua. [Chinese *pŭtōnghuà*, f. *pŭtōng* common + *huà* spoken language.] The standard spoken language now in general use throughout the People's Republic of China, based on the northern dialects, esp. that of Peking. Cf. *Kuo-yü.
1950 J. De Francis *Nationalism & Lang. Reform in China* v. 94 He expressed his views..in the following statement: '..China now has a *p'u-t'ung hua* (common language) which can serve as a general standard.' 1968 P. Kratochvíl *Chinese Lang. Today* v. 164 This concept of the so-called *pŭtōnghuà* 'Common Language' is largely based on the latter stage of *guóyŭ* (the two terms *guóyŭ* and *pŭtōnghuà* are now almost synonymous). 1971 R. Newnham *About Chinese* 75 The Communists took up the ['spoken language'] of the older 'literary revolutionists', renaming it *pŭtōnghuà* or 'common speech'. The inspiration became firmly social. Pŭtōnghuà is now what the West generally thinks of when it speaks of 'Mandarin'... It is adopted by all the communications media in China (the phrase 'national language' is not heard), while in print it is virtually standard. 1973 T. R. Tregear *Chinese* vi. 124 The *pai hua* has had a profound influence since today it forms the basis of *p'u t'ung hua*, 'usual words' now in universal use throughout the land. 1976 W. H. Canaway *Willow-Pattern War* xv. 156 He spoke enough *pu-t'ung hua*, the Common Chinese, for Shao to give him instructions before we left. 1978 *Nagel's Encycl.-Guide: China* 68 A common spoken language, now known as *putonghua* (Common Language) was imposed throughout China.

put-put (pʊ·tpʊt), *sb.* orig. *N. Amer.* Also putt-putt. [Echoic.] A muffled explosive sound characteristic of an internal-combustion engine. Also applied to objects which make such a sound, as a machine-gun, a motorized boat or bicycle, etc. Also *attrib.*
1905 *Rudder* Feb. 61/2 Already the class of small launches, of which the converted [St. Lawrence] skiffs form the majority, has achieved the distinction of a special title, of unknown origin, but of universal use; 'put-put', or more briefly, 'put'. The words themselves are now almost as common to the ear as the familiar voice of the two-cycle motor which called them into being. 1929 M. A. Gill *Underworld Slang* 9/2 *Put-put*, machine gun. 1930 J. P. Burke in *Amer. Mercury* Dec. 457/1 *Putt-putt*, an out-board motorboat used in liquor running. 'A sneaker's no good. Water's too shallow. Got to use a putt-putt.' 1959 P. Capon *Amongst Those Missing* 165 The others could hear it now, a quick-fire 'put-put-put' like a distant motor-bike. 1964 E. A. Nida *Toward Sci. Transl.* iii. 31 An onomatopoeic expression, such as *putt-putt* or *choo-choo*, is presumably an imitation of the very sound made by the object in question. 1965 S. T. Ollivier *Petticoat Farm* iv. 46 A roar of laughter drowned the put-put of the engine. 1967 *Guardian* 28 Dec. 5/3 The little put-put boat which carries passengers ashore. 1968 *Sunday Mail Mag.* (Brisbane) 15 Sept. 4/2 Anchored right in the middle was a small, old putt-putt type launch. 1974 J. Mitchell *Death & Bright Water* xii. 134 [He] listened to the putt putt of the two-stroke as he rode back to Kronis. 1977 *Time* 19 Dec. 13/2 Does he really think the new cruise missile is no better than Hitler's high-flying, inaccurate put-put? 1978 M. Z. Lewin *Silent Salesman* xv. 80 The cop..got on his putt-putt, and went away.

put-put (pʊ·tpʊt), *v.* orig. *N. Amer.* Also putt-putt. [Echoic.] *intr.* To make an intermittent explosive sound characteristic of an internal-combustion engine. To move, making such a noise. Also quasi-*trans.* Also *fig.* Hence put(t)-putting *vbl. sb.*
1905 *Outing* July 389/1 In and out between them trim little launches go *put-putting*. 1939 A. Keith *Land below Wind* vii. 113 The motor put-puts back a horrid petrol odour. 1955 'D. Cory' *Phoenix Sings* v. 82 A peculiar *put-put-put-put*... The put-putting noise was increasing... I made out the shape of a great sailing-barge..its tiny donkey-engine chugging away. 1958 *Spectator* 4 July 12/2 The diesel engine put-put-puts its warm gargle. 1961 J. C. Lilly *Man & Dolphin* xi. 153 A dolphin..naturally uses other sounds to convey and receive 'meaning':..putt-putting and whistles for exchanges with other dolphins. *Ibid.* 158 Lizzie [*sc.* a dolphin], near the hydrophone, putt-putted. 1973 'D. Jordan' *Nile Green* xliii. 220 The helicopter..came in fast over the desert, the put-putting loud overhead. 1974 R. Jeffries *Mistakenly in Mallorca* viii. 78 Old Morley keeps boasting about how he putt-putts all his money out of England.

putrefied, *ppl. a.* Add: ¶ **2.** *U.S. dial.* A malapropism for 'petrified'.
1848 G. F. Ruxton in *Blackw. Mag.* June 714/2 'I've seen a putrefied forest.' 'La, Mister Harris, a what?' 'A putrefied forest, marm.' *Ibid.* 715/1 'I shows him the piece I chipped out of the tree, and he called it a putrefaction too; and so, marm, if that wasn't a putrefied peraira, what was it?' 1896 'Mark Twain' in *Harper's Mag.* Sept. 536/2 Jubiter..was just fairly putrefied with astonishment.

putrid, *a.* Add: **3.** Used as a mere intensive: dreadful, awful, appalling. *colloq.*
1883 'Mark Twain' *Life on Mississippi* ii. 37 La Salle drew from these simple children of the forest acknowledgements of fealty to Louis the Putrid, over the water. 1898 *Windsor Mag.* Dec. 40/1 You're an ass—a putrid ass. 1902 S. J. Cotes *Those Delightful Americans* 104 Last night at billiards you first said your luck was 'rotten', and then you got excited and declared it was 'putrid'. 1913 'I. Hay' *Right Stuff* p. vi, He seems to have perfectly putrid notions about some things. 1931 D. L. Sayers *Five Red Herrings* iv. 45 Some putrid fool sliced a ball..and got me slap-bang in the eye.

‖ **putsch** (putʃ). [Swiss G., orig. knock, thrust, blow.]
a. A revolutionary attempt.
1920 *Times* 3 June 15/5 The possibility of a *Putsch* continues to exercise the minds of all parties. 1922 *Q. Rev.* Jan. 125 King Charles has made his second attempt to ascend the Hungarian Throne. In the circumstances outlined above it was doomed to failure. So was Louis Napoleon's second *coup d'état*—*Putsch* is the modern word—at Boulogne. 1930 *Economist* 4 Oct. 612/2 The officers were charged with conspiring..to secure the neutrality of the Reichwehr in the event of another 'Putsch' by the revolutionaries of the Right. 1945 A. J. P. Taylor *Course of German Hist.* xi. 192 No one who took part in the *putsch* was punished. 1950 [see *blanquism]. 1968 A. Coates *Myself a Mandarin* i. 5 Since the end of the Second World War the population [of Hong Kong] had topped the million mark, and the place was thus technically already overcrowded when the communist *putsch* began. 1975 *N.Y. Times* 29 Nov. 27/2 Allende, of course, is gone—a suicide in September 1973 when the current President, Gen. Augusto Pinochet Ugarte, seized power in a bloody putsch. 1981 *Listener* 1 Jan. 12/2 The greatest achievement had been to keep the [Ghanaian] army putsch a secret.
b. *colloq.* Any sudden vigorous effort or campaign.
1938 A. Campbell *Flying Blind* xi. 89 He grasped it firmly, and flexing his muscles prepared a putsch. 1940 C. T. Carr in *Mod. Lang. Rev.* XXXV. 71 *Putsch*, a coup d'état... The word was apparently borrowed round about 1920 and is now quite common in English newspapers.. spreading to English slang in the non-political significance of a 'push forward'. 1953 M. McCarthy *Groves of Academe* x. 215 Criticism..has been reduced to a minimum... No poet of any real merit has been excluded... You..are too impatient. You want to make a *putsch* for the sake of tighter control, more daring methods of promotion, but violence is unnecessary. 1970 *New Scientist* 30 July 221/2 The present step-by-step attack on brucellosis is much more likely to succeed than a premature *putsch*. 1973 *Observer* 14 Jan. 29/6 Apart from a much-needed putsch on our chimneys and exhaust pipes there is little more that the public health departments can do to increase the level of our good health.
Hence **pu·tsching** *vbl. sb.*; **pu·tschism**, the advocacy of a *putsch* or of the violence associated with a *putsch*; **pu·tschist**, an advocate of or participant in a *putsch*; also *attrib.* or as *adj.*
1898 A. P. Atterbury tr. *Sombart's Socialism 19th Cent.* iv. 73 Putschism..is the fanatical tendency towards street struggle, faith in the barricade. *Ibid.* v. 113 The Putschists, Clubists, and Blanquists were utopists, who through conspiracies and street riots would through all time control economic development. 1923 *Glasgow Herald* 26 Oct. 9 The Separatist 'Putschists' have succeeded in maintaining their position. 1937 E. Snow *Red Star over China* IV. v. 167 P'eng Pai..formed a Soviet, which, following a policy of putschism, was soon destroyed. 1940 K. Mannheim *Ideology & Utopia* 125 The ideology of 'putschist' groups led by intellectuals. 1954 P. Toynbee *Friends Apart* v. 65 The Communist Party.. would have regarded it as a piece of futile 'diversionism', 'putschist' and disorganised. 1966 'Han Suyin' *Mortal Flower* I. iv. 128 The practice of shooting deserters and of inflicting corporal punishment, both of which smack of putschism. 1968 *Economist* 1 June 44/1 Mr Cecil King has gone the same way that he came in—by a boardroom putsch. Seventeen years ago it was Mr King who did the putsching. 1974 J. White tr. *Poulantzas's Fascism & Dictatorship* IV. ii. 169, 1921. A series of 'putschist' attempts in Prussia by the KPD. *Ibid.*, The Comintern, at its Third Congress, passed a severe judgment on this 'putschism'. 1975 *New Left Rev.* Nov.–Dec. 66 When the putschists struck at Nicosia in July 1974, only a few EDEK members were armed and ready to resist. 1979 *China Now* Mar./Apr. 24/2 The essentially putschist character of both Lin Biao and the Four.

putteed, *a.* (Further example.)
1929 E. Bowen *Last September* xii. 141 The last they saw of him was a putteed leg.

putter, *sb.*[1] Add: **8.** *putter down* = putter off (*b*); putter up (*b*) (later examples).
1906 *P.T.O.* 16 June 16/2 Three men as a rule take an active part in a forgery—the 'putter-up', the capitalist who finds the necessary funds; the 'blacksmith', the actual forger; and the 'putter-down', who actually presents the forged document and obtains the money. 1926 *Clues* Nov. 162/1 *Putter-down*, the party who passes forged checks for the real forger. 1929 C. Humphreys *Great Pearl Robbery* i. 12 The police knew that Grizard was the 'putter-up'. 1975 M. Crichton *Great Train Robbery* ii. 18 Edward Pierce..accumulated sufficient capital to finance large-scale criminal operations, thus becoming what was called 'a putter-up'.

putter (pʊ·təɹ), *sb.*[3] [Echoic. Cf. *put-put *sb.*] A muffled explosive sound characteristic of an internal-combustion engine, as an outboard motor, etc. Also applied to an engine or vehicle which makes such a sound.
1942 'N. Shute' *Pied Piper* 224 There was a fishing-boat..coming in from the sea; faintly they heard the putter of an engine. 1948 G. Greene *Heart of Matter* II. I. ii. 118 Across the river the tinkering in the launch went on: the sharp crack of a chisel, the clank of metal, and then again the spasmodic putter. 1964 J. Masters *Trial at Monomoy* ii. 56 The putter of the marine diesels and the slap and sigh of the sea. 1969 *Listener* 12 June 814/1 We heard the first putter of outboard motors that, by mid-morning, become the background noise of the region. 1975 *Islander* (Victoria, B.C.) 27 July 7/1 We had a small open boat with an inboard engine. This 'putter' would

PUTTER

provide slow but reliable transportation. 1979 R. Laidlaw *Lion is Rampant* xiv. 111, I could hear the putter of farm machinery.

putter, *v.*² Delete '*dial.* and *U.S.*' and add earlier and later examples.

1878 L. M. Alcott *Under Lilacs* xii. 130 Ben infinitely preferred to watch ants and bugs..rather than 'putter' over plants with long names. 1878 L. C. Bell in *Wide Awake* Jan. 24/1 Every morning in the midst of his chores, Max found time for a long, hovering, puttering visit. *Ibid.*, Max likes to 'putter' with the housework, too. 1907 J. M. Synge *Let.* 3 June (1971) 153 Yesterday we puttered about, and today we are going for another long expedition. 1925 R. Frost *Let.* 20 June (1964) 174, I am free to putter my days out without even writing any more. 1925 J. G. Macleod in *Oxf. Poetry* 26 His still moving body Like a strange motor-boat propelled by nothing puttered round The headland 1931 D. L. Sayers *Five Red Herrings* xxiv. 278 He would be the one person who might habitually see Campbell having breakfast and puttering about the house. 1952 *Arena* (N.Z.) XXXI. 5 Real man of mystery he was these days. Puttering round the whare at all hours. 1960 M. K. Joseph *I'll Soldier no More* 150 Tired, they putter slowly back to billets. 1977 G. Durrell *Golden Bats & Pink Pigeons* v. 121 The Box fish puttered to and fro like some weird, orange boat.

putter (pʊ·təɹ), *v.*³ [Echoic. Cf. *put-put *v.*] *intr.* To make an intermittent explosive sound characteristic of an internal-combustion engine; to move, making such a sound. Hence **pu·ttering** *vbl. sb.* and *ppl. a.*

1937 M. Lane *At Last Island* ix. 270 The boat puttered and back-fired out of the harbour. 1947 J. Steinbeck *Wayward Bus* 197 The rain had diminished so that there was only a faint puttering on the roof. 1956 J. Masters *Bugles & Tiger* i. 31 A groaning truck..backfired and puttered steadily down the road to the plains. 1958 *Times* 7 July 9/4 The mower must depend on human exertion and not be of the petrol-puttering kind. 1971 P. Crampton tr. *Heyerdahl's Ra Expeditions* vii. 163 Our first hesitant moves were now being followed by excited journalists and experienced old salts on board the puttering vessels which circled about us. 1975 *New Yorker* 28 Apr. 98/3 They [*sc.* Hanoians] rent rowing shells or go for a ride in the motorboats that putter back and forth between its islands.

putting, *vbl. sb.*¹ Add: **9.** *putting on, out, up*: further examples.

1930 A. Armstrong *Taxi* xii. 163 'Putting on' is the taxi man's expression for coming on at the end of the rank. 1968 *Listener* 31 Oct. 566/1 They acknowledged their debt to McLuhan and paraded his definition of modern myths—the putting on of an audience and its environment. 1947 S. C. Adams in A. Dundes *Mother Wit* (1973) 519 The younger generation are largely indifferent either as to the necessity of joining the church, or, if they are already members, as to the 'putting out' of the church. *c* 1806 D. Wordsworth *Jrnl.* (1941) I. 258 She did not much encourage us to go, because..it was a long way, 'and there was no putting-up for the like of us'. 1907 J. G. Millais *Newfoundland* iv. 76 During the month of September the big stags keep to themselves in various 'putting up' spots..near the lakes and rivers. 1909 *Daily Chron.* 16 June 1/2 It was the biggest fight he had ever undertaken, but he was going to win it..or if he did not win he was going to give the other side a rare 'putting up'. 1914 *Chambers's Jrnl.* Aug. 536/1 Herrings cause similar bubbles, which fishermen call 'putting up'.

putting, *vbl. sb.*² Add: **1. b.** **putting course** *poet.* = *putting green*; **putting-green**: also, a miniature golf course.

1945 J. Betjeman *Coll. Poems* (1958) 116 Over the putting-course rashes were seen Of pink and of yellow among the burnt green. 1966 —— *High & Low* 62, I will not go to Finsbury Park The putting course to see. 1977 *Evening Post* (Nottingham) 27 Jan. 14/1 (Advt.), Local amenities bowls, tennis, putting green, paddling pool.

putting, *ppl. a.* Add: **putting-off,** disconcerting, off-putting, repellent; cf. *off-putting* ppl. adj. s.v. Off-put in Dict. and Suppl.

1928 M. Arlen *Lily Christine* vi. 86 The idea of anyone living..her life 'bravely'..is, to tell the truth, slightly embarrassing—'putting-off', the phrase is. 1932 S. Gibbons *Cold Comfort Farm* i. 9 'Would it impress them with my efficiency?' 'No... It would be *too* putting-off.' 1945 C. Williams *All Hallows' Eve* 184 She was very putting-off, and only said: 'Pray, nurse, do not interfere.' 1959 *Sunday Times* 22 Mar. 24/5 The first act was so putting-off that I should not have been much surprised if many viewers had accepted the B.B.C.'s invitation to 'Follow The Fleet', with Fred Astaire and Ginger Rogers, on the other channel.

putto. (Examples in *sing.*)

1914 C. F. Bell *Drawings by Old Masters in Christ Church* 44 Fiammingo, François du Quesnoy, *attributed to*... A nude putto playing with a goat. 1931 B. Berenson *Ital. Pictures of Renaissance* 109 Bramantino (*Bartolommeo Suardi*)... Milan... 16. Fresco: Putto under Vine. 1968 *Listener* 22 Aug. 247/1 The omission in Poussin's painting of anything corresponding to the *putto* in the Bordone leads to a strange placing of the left leg. 1973 *Daily Tel.* 13 Feb. 13/6 On one side are two coats of arms in Baroque mantling held together by ribbons in the hand of a flying putto.

‖ **puttony** (pu·tonʸ). Pl. **puttonys.** [Hungarian; cf. Hung. *puttonyos* holding as much as goes into one *puttony* (e.g. *öt puttonyos tokaji* five-basket Tokay).] In Hungary, a basket or dosser made of wooden staves or wickerwork used to transfer grapes from the vineyard to the wine-press.

1958 A. L. Simon *Dict. Wines* 156/1 They [*sc.* over-ripe grapes] are gathered in wooden vessels known as *puttony*, holding about 25 quarts... When the label on the bottle records '1 puttony', it means that about 10 per cent of the grapes used were *trockenbeeren*..; if '3 puttony', the proportion was 30 per cent. 1959 W. James *Word-Bk. of Wine* 190 Pickers carry a small container called a puttony, into which they put selected overripe berries..; the number of full puttonyos mixed with the ripe but not overripe grapes in the fermentation cask determines the richness of the wine, the range being from one to five (and occasionally six) puttonyos. 1967 A. Lichine *Encycl. Wines & Spirits* 426/2 The collar label on every bottle of Tokaji Aszu will state: 3 Puttonos, 4 Puttonos, etc.—the export agency simplifies the spelling by removing the 'y'. *Ibid.* 522/2 The buckets or *puttonys* of raisin-dry, concentrated grapes added to the fermentation..are shown on each bottle of the vatting. *Ibid.* 523/1 The overripe, dried, shrivelled berries, picked separately and put into the little *puttony* pails, are worked in a trough... Alcoholic content and other characteristics of course vary with the *puttonys* content. 1972 *Guardian* 26 Jan. 9/5 In theory there can be a Tokay of 6 puttonyos; but in practice only the five turns up.

putt-putt, var. *put-put *sb.* and *v.*

putty, *sb.* Add: **3. b.** Phr. *up to putty*, worthless, useless. *Austral. colloq.*

1916 *Anzac Bk.* 32/1 A man's got a chance to hit back there, but down 'ere it's up to putty. 1953 D. Stivens *Gambling Ghost* 24 'I don't hear anything, Cabbage-tree,' said Thunderclap. 'Your hearing's up to putty,' said Cabbage-tree. 1965 *Telegraph* (Brisbane) 5 July 8/5 *Up to putty*, no good.

c. Used *fig.* to designate one who is easily influenced or malleable. Freq. in colloq. phr. *to be* (*like*) *putty in* (someone's) *hands*.

1924 H. Crane *Let.* 3 Feb. (1965) 173, I was quite exemplary of both sides of my family in not being made of any putty—knowing what I want to do, and sticking it out. 1946 W. S. Maugham *Then & Now* ii. 3 You are infatuated with the man. You're like putty in his hands. 1979 D. Kyle *Green River High* vii. 90, I was putty in her hands... The arguments *were* very attractive.

5. c. (Earlier and later examples.)

1883 G. C. Davies *Norfolk Broads & Rivers* i. 5 All the other Broads have bottoms of black mud..so soft that a yacht's anchor will not hold in it, so that large blocks of iron ballast are used instead, which will not drag through the 'putty,' as the mud is locally called. 1961 P. Moyes *Sunken Sailor* ii. 33 My adorable wife has put us on the putty. On a falling tide.

d. (Further examples.)

1915 *T. Eaton & Co. Catal.* Spring & Summer Suppl. 1/2 Natty tailored Suit... In light Putty (Tan) shade only. 1926–7 *Army & Navy Stores Catal.* 698/1 *Hose*..Botany Wool, in Black, Grey, Tan, Nigger, Putty, Beige, Nude.

e. A former type of golf ball made of some material other than gutta-percha.

1891 R. Forgan *Golfer's Handbk.* 39 The 'putty' being the popular name for the 'Eclipse'. 1900 A. E. T. Watson *Young Sportsman* 293 Several kinds of composition balls, known generically as 'putties' in contradistinction to the 'gutties' or gutta-percha balls..have failed to take the place of those made of the raw material.

f. *Naut.* (See quots.)

1946 J. Irving *Royal Navalese* 141 *Putty*, the ship's painter. 1961 F. H. Burgess *Dict. Sailing* 165 *Putty*,..a ship's painter.

6. *putty beige, colour* (examples), *face, grey, shade* (example), *white*; *putty-cool* adj.; **putty-blower** (earlier example); **putty-head** *U.S.*, a stupid person; **putty-headed** *a.*, stupid, soft-hearted; **putty-hearted** *a.*, lacking in courage, cowardly; **putty medal** *jocular*, a worthless reward for insignificant service or achievement (cf. *medal *sb.* 2b); **putty-shooter** = *putty-blower*.

1969 *Sears Catal.* Spring/Summer 10 Putty beige. 1862 R. H. Newell *Orpheus C. Kerr Papers* 1st Ser. 156 [The muskets] are inferior to the putty-blowers of our innocent childhood. 1933 *Burlington Mag.* Sept. 122/1 The olive-grey celadon glaze has the same peculiar tint, with a slight suggestion of putty colour. 1979 'G. Black' *Night Run from Java* iii. 32 Banana palms..bleached to a kind of putty colour. 1970 R. Lowell *Notebk.* 69 Our bedroom, putty-gray and putty-cool. 1927 M. Sinclair *Hist. A. Waring* xvii. 88 Charlie, in spite of his putty face, was handsome in a heavy way. 1931 W. Faulkner *Sanctuary* vii. 68 'Yes, putty-face!' the woman cried. 1969 Putty-grey [see *putty-cool* above]. 1856 M. J. Holmes *L. Rivers* 370 He got so engaged about the darned 'liquor law', and the putty-heads that made it, that he'd no idee 'twas so late. 1873 'Mark Twain' *Gilded Age* xliii. 393 In a word, the great putty-headed public loves to 'gush'. 1885 R. L. Stevenson *Prince Otto* ii. i, A springless, putty-hearted, cowering coward. 1898 R. E. F. Cohen in W. A. Morgan '*House' on Sport* 378 Not even the proverbial putty medal or a memento of any sort was awarded to the man who upheld the honour of his University by beating his rival in the water. 1958 M. Kelly *Christmas Egg* (1965) iii. 195 'You know what you'll be given for all this?'..'A putty medal. Sooner have a cheque.' 1972 *Guardian* 12 Sept. 15/7 Putty medal blues... The Americans are busy mounting a major public inquest on their showing at Munich and on the Olympic Games. 1930 *Daily Express* 6 Oct. 11/6 (Advt.), The Barry 'militaire' (in fashionable putty-shade). 1896 D. C. Beard *Amer. Boy's Bk. of Sport* xxxiii. 395 When people depend for their dinner or personal safety upon a 'putty-shooter' you may be sure that they learn to shoot with great accuracy. 1971 *Guardian* 20 July 9/3 Caramel..and putty-white make the season's signature combination.

put-up, *ppl. a.* Add: Now usu. with stress on first syllable. **1.** (Further examples.) Also *absol.* as *sb.*

1903 *Outing* XLII. 660/2 'Why, man,' he exclaimed, 'it's a graft—a dirty put-up game. Can't you see it?' 1923 H. G. Wells *Men like Gods* i. ii. 19 The whole of this business is, as they say nowadays, a put-up thing. 1936 J. Curtis *Gilt Kid* xv. 148 He would believe that it was a put-up. 1941 [see *line *sb.*² 13 g]. 1974 N. Freeling *Dressing of Diamond* 213 There's going to be a lot saying it's a put-up job.

2. a. (Further examples.)

1846 Dickens *Pictures from Italy* 124 An English lady ..who always carries..a put-up parasol. 1897 G. B. Shaw *Let.* 26 Mar. (1965) 738 He sends Felix to bid for fashionable put-up plays.

put-u-p-able-with, *a. rare.* [f. phr. *to put up with*: see Put *v.*¹ 53 p (*b*) + -able.] That may be put up with; tolerable.

1812 M. Edgeworth *Tales of Fashionable Life* VI. 37 The accommodations, and everything of that nature, now is vastly put-up-able with!

pu·t-you-up. Also **Put-u-up.** [f. phr. *to put up*: see Put *v.*¹ 53 o (*a*)]. A sofa or settee which can be converted into a bed. Also *attrib.*

The form *Put-u-up* is a proprietary term.

1924 *Trade Marks Jrnl.* 20 Aug. 1898 Put-u-up... Settees convertible into bedsteads. Greaves & Thomas,.. London,..manufacturers. 1948 G. V. Galwey *Lift & Drop* viii. 206 He did have a 'Put-u-up' in the office for firewatching. 1966 M. Cronin *Jump Gun* iv. 45, I found some blankets, wrestled with the put-u-up and coaxed it into some semblance to a couch. 1966 A. E. Lindop *I start Counting* i. 14 The sofa was one of those big double-bed Put-U-Ups. 1973 *Country Gentlemen's Mag.* Mar. 180/2 (*heading*) Fitted cabin trunk..ideal service cadet or world rover, £12. Or exchange settee Put-u-Up bed. 1973 *Country Life* 22 Mar. 753/1 A sofa bedstead (an early form of put-you-up). 1978 *Morecambe Guardian* 14 Mar. 9/2 Besides traditional beds, there are the convertible put-you-up types which are essential when space is short.

putz (puts, pʊts). *U.S.* [a. G. *putz* decorations, ornaments.] **1.** *dial.* In Pennsylvanian Dutch homes, a representation of the Nativity scene traditionally placed under a Christmas tree.

1902 *N.Y. Times Mag.* 14 Dec. 15/2 Only the chosen few can afford to have a really impressive 'putz' which fills half a room, and represents a landscape in miniature... This more elaborate '*putz*' requires not only money for its erection, but artistic handiwork. 1926 *Ladies' Home Jrnl.* Dec. 82/2 The putz is simply the pictured story of the Nativity, built near or at the base of the Christmas tree. 1938 A. Hark *Hex Marks Spot* 186 Everybody's curious to see what kind of putz everybody else has this year, so they go around visiting. 1970 L. M. Feinsilver *Taste of Yiddish* i. 44 In Pennsylvania, Jews who know Yiddish are often startled during the Christmas season by ads inviting the public to some company's 'putz'... This German word for decoration means, in Pennsylvania Dutch, a Nativity scene.

2. *slang.* [Yiddish.] **a.** The penis.

1934 H. Miller *Tropic of Cancer* (1935) 34 [She] ought to have better sense than be tripped up by every guy with a big putz who happens to come along. 1968 L. Rosten *Joys of Yiddish* 298 *Putz*, rhymes with 'nuts'. Literally, *putz* is vulgar slang for 'penis'... *Putz* is not to be used lightly, or when women or children are around. It is more offensive than *shmuck*;..*putz* has a pejorative ambience. 1969 P. Roth *Portnoy's Complaint* 101 He simply cannot —*will* not—control the fires in his putz, the fevers in his brain.

b. A fool, a simpleton; an objectionable person.

1964 W. Markfield *To Early Grave* (1965) vii. 127 What I think is—you're a *putz*. *P, U, T, Z!* 1966 'E. V. Cunningham' *Helen* (1967) v. 66 'What are you telling me? That you fell for her—love at first sight?' 'Don't be a putz. I run a gambling house. I don't fall in love.' 1975 A. Bergman *Hollywood & Le Vine* (1976) v. 56 He understood life's mysteries and tragedies, this gold-plated putz. 1975 *Publishers Weekly* 21 July 60/1 Leaving their 'putz' of a son Harry home to nurse his ulcers. 1978 J. Krantz *Scruples* i. 6 'You,' she said, enunciating clearly, 'are a putz, a schmekel, a schmuck, a schlong, and a shvantz. And a WASP putz, at that.'

‖ **Putzfrau** (pu·tsfrau). [Ger.] A charwoman.

1927 J. Joyce *Let.* 25 July (1966) III. 162 Mrs Purefoy is not a Putzfrau. 1977 *Time* 8 Aug. 22/3 Says one Greek *Putzfrau* (charwoman): 'I know the West Germans wish us to hell.'

puukko (pū·ko). Also **puuko.** Pl. **puukot.** [Finn.] A type of knife used in Finland.

1952 *Chambers's Jrnl.* Jan. 37/1 He fumbled desperately for his sharp-bladed 'puukko', expecting to be set upon forthwith by a snarling fury. 1959 A. Glyn *I can take it All* ii. 36 His knife, his puukko, a large weapon with a plain handle and a double-edge blade in an embossed leather scabbard. 1964 C. Gavin *Fortress* xi. 188 A red braided belt from which swung..the Finnish knife, the *puuko*, in its heavy leather sheath. 1964 G. Lyall *Most Dangerous Game* iii. 24, I caught the glint of *puukot*, those nasty little hook-ended Finnish knives. *Ibid.*, The first *puukko* merchant dodged.

puwang, puwha, varr. *pawang, *puha.

puya (pū·yā). Also **puza.** [mod.L. (G. I. Molina *Saggio della Storia Naturale del Chile* (1782) 160), a. Amer. Sp., f. Sp. *puya* goad.] A herbaceous or woody plant of the genus so called, sometimes as large as a small tree, belonging to the family Bromeliaceæ, native to dry regions of the Andes, and distinguished by rosettes of spiny leaves and blue or yellow flowers borne singly or in large panicles or racemes.

1809 tr. *J. I. Molina's Geogr., Nat. & Civil Hist. Chili* I. iii. 130 The trunk of the *puya*..is used for cork throughout Chili. 1847 *Curtis's Bot. Mag.* LXXIII. 4309 (*heading*) Altenstein's Puya; gigantic variety. 1885 *Pall Mall Gaz.* 11 Mar. 11/1 We mounted over rocks and more dust for some 2,000 feet, among puzas and succulent and prickly plants. 1902 *Westm. Gaz.* 23 May 12/1 The blue Puya is known to frequenters of Kew by the beautiful picture of it painted by Miss North in Chili. 1920 *Nature* 8 Apr. 160/2 On a sandstone plateau [in Bolivia]..was growing the gigantic 'Puya'. 1963 W. Blunt *Of Flowers & Village* 240 The Chilean puyas, if it is true that they are fertilized by hummingbirds, are exceptions. 1974 T. Morrison *Land above Clouds* 149 The tallest flower spikes in the world grow from solitary stands of a giant bromeliad, the Puya.

puzzivanting, var. *pussivanting *ppl. a.* and *vbl. sb.*

puzzle, *sb.* Add: **4.** *puzzle-card, -jug* (examples), *-map, -picture, -ring* (later example), *-thing*; *puzzlewise* adj. and adv.; **puzzle-box,** a puzzle in the form of a box; *spec.* in *Psychol.*, a box with no obvious connection between its door and the opening device, designed to test the learning abilities of an animal in trying to release itself; also *attrib.*

1866 Trollope *Claverings* in *Cornh. Mag.* XIII. 396 Another girl..was engaged with a puzzle-box. 1908 M. F. Washburn *Animal Mind* x. 232 The dropping off of useless movements is further illustrated in those experiments where animals are required to work some kind of mechanism. This may be called briefly the puzzle-box method. 1921 R. S. Woodworth *Psychol.* (1922) xiii. 308 (*caption*) A puzzle box. The animal must here reach his paw out between the bars and raise the latch, *L.* 1966 H. C. Lindgren et al. *Psychol.* iv. 105/1 Guthrie and Horton.. have used the behavior of a cat trying to escape from a puzzle box as a basis for demonstrating this concept of learning. 1970 E. R. Guthrie in W. S. Sahakian *Psychol. of Learning* iii. 58 An account of the behavior of cats in a puzzle-box is here offered, with the hope that he [*sc.* the reader] will begin to see himself..in a multitude of puzzle-box situations. 1853 Puzzle-card [see *conversation card]. 1878 Puzzle-jug [see *Caughley]. 1960 R. G. Haggar *Conc. Encycl. Continental Pott. & Porc.* 375/1 The potter, Cornelis Hendricksz (born 1566), is stated to have made surprise jugs or puzzle jugs at Haarlem. 1980 R. Rubens *Cosway Miniature* ii. 13 Bonnie's greatest love was old china..Regency-striped saucers and Swansea lustre frogs and puzzle jugs. 1870 *Food Jrnl.* Nov. 533 The arrangement of the different collections and classes reminding one of a puzzle-map well shaken in a bag. 1886 Kipling *Lispeth* in *Plain Tales from Hills* (1888) 5 There was an old puzzle-map of the World... She used to put it together. 1906 B. L. Taylor *Extra Dry* 73 (*caption*) Puzzle picture. Find the man who is paying for the drinks. 1929 W. Faulkner *Sartoris* ii. vi. 151 Pieces of a patient puzzle-picture. 1978 *Country Life* 21 Sept. 850/4 Roy Strong..treated Yeames's *And when did you last see your father?* as a puzzle picture which 'leaves us to fill in what has gone before..in a highly tantalizing way'. 1974 J. Gardner *Corner Men* v. 36 Long slim fingers and a Greek puzzle ring. 1781 J. Woodforde *Diary* 16 Mar. (1924) I. 304 To 7 pieces of wood, a Puzzle thing, pd o. o. 6. 1914 W. J. Locke *Fortunate Youth* xii. 163 Paul stood ruminating puzzlewise on the audacious behest. 1950 *Mind* LIX. 174 No doubt all this sounds stale and naïve to puzzle-wise professional philosophers.

puzzledly, *adv.* (Later examples.)

1951 D. Knight *Turning On* (1967) 144 He peered at Mazurin puzzledly. 'Is that what you're for?' 1964 *Economist* 10 Oct. 114/1 They puzzledly ask what the issues..really are. 1975 J. Grady *Shadow of Condor* (1976) xvi. 251 Captain Roe looked at his executive officer puzzledly.

puzzledness. (Later example.)

1935 *Theology* XXXI. 152 The first thing which strikes a simple reader is the apparent puzzledness of the accounts in the three Gospels.

puzzlist (pʊ·zlist). *U.S.* [f. Puzzle *sb.* + -ist.] One who devises puzzles.

1961 *Times Lit. Suppl.* 24 Feb 127/1 Readers in this country are unlikely to accept 'ticktack toe' for 'noughts and crosses' or speak of their favourite composer as a 'puzzlist'. 1970 *Sci. Amer.* Feb. 112/3 This cryptarithm (or alphametic, as many puzzlists prefer to call them) is an old one of unknown origin. 1971 *Ibid.* Oct. 106/3 Let us combine the rules of the two rival puzzlists by allowing both steps and hops, as in Halma.

puzzolan, puzzuolana, etc., varr. Pozzolana in Dict. and Suppl.

‖ **pwe** (pwe). [Burmese.] Also **poi, pooay, pu-é.** A Burmese festival which includes drama, dancing, sports, or other entertainments.

1861 *Chambers's Encycl.* II. 443/1 A *pooay*, or theatrical representation, is a very favourite amusement. 1876 *Encycl. Brit.* IV. 556/1 The historical books are then read, as well as the *Pu-es* or dramatic productions. 1878 A. Fytche *Burma* II. i. 21 It is a strange and curious sight to see the large crowds of Burmese assembled for the night to witness the performance of a *pooay*, or play. 1884 T. H. Lewin *Fly on Wheel* vii. 213 The night after my arrival at Cox's Bazaar, I was invited to attend a Burmese 'poi', or dramatic representation. 1905 *Statesman* (Calcutta) 23 Aug. 5/3 What the Chief Judge had to decide was whether a foot race fell within the definition of a 'pwe'. A 'pwe' ordinarily means a puppet show or other theatrical or dramatical performance, or a native cart, pony, boat or water race held for the public entertainment. 1908 *Athenæum* 29 July 254/3 A story with a Burmese Pwe dancer for heroine. 1929 F. T. Jesse *Lacquer Lady* i. xii. 88 Each of the Princes..has his own pandal erected and has pwès acted for seven days. 1934 'G. Orwell' *Burmese Days* viii. 128 No one with eyes in his head could resist a pwe-dance. *Ibid.* 129 They're having a pwe—that's a kind of Burmese play; a cross between a historical drama and a revue. *Ibid.* 134 The pwe girl began dancing again. 1936 F. Richards *Old-Soldier Sahib* xix. 323 The Burmese were having a *pooay*, a festival which lasted seven days and was entirely devoted to gambling and enjoyment. 1971 *Nat. Geographic* Mar. 349/1 Burmese still obey, as seen by their enthusiasm for the *pwe*, a marathon of drama, singing, dancing, and joke telling. 1974 P. Gore-Booth *With Great Truth & Respect* 205 It was one of the famous Burmese *Pwes* in the open and it went on all night.

‖ **pya** (pī·a). [Burmese.] A Burmese monetary unit, the hundredth part of a *kyat*; a coin of this value.

1952 [see *kyat]. 1962 B. Fergusson *Return to Burma* x. 201, I grudge the fare!.. I'd rather have the money! Twenty *pyas*!.. I could do with twenty *pyas*. 1971 *Whitaker's Almanac* 1972 984 Burma .. Coins.. *Pyas* 50, 25, 10, 5, 1. 1975 P. Theroux *Great Railway Bazaar* xvii. 181 A Burmese with a telescope urged me to have a look. I paid my fee of 25 pyas (five cents).

Pybuthrin (pəibū·þrin). Also **pybuthrin.** [Blend of *pyrethrin and *butoxide* (f. But(yl + Oxide *sb.*).] A proprietary name for an insecticide compounded of pyrethrins and piperonyl butoxide.

1951 *Trade Marks Jrnl.* 3 Jan. 5/2 *Pybuthrin*... Chemical substances used for veterinary and sanitary purposes, insecticides, fungicides and preparations for destroying vermin. Cooper, McDougall & Robertson Limited.. Manufacturers and Merchants. 1958 *Times* 7 July 2/7 The simplest way of dealing with them [*sc.* red mites], and with lice at the same time, is to spray the birds while at roost.. with a fine sprayer containing pybuthrin. 1971 *Homes & Gardens* Aug. 89/1 A spray containing pybuthrin for rapid effect and chlordane for persistence will control ants in house or garden for a period of two to three months outside and up to a year indoors.

pycnic, var. *pyknic.

pycnidial, *a.* (Examples.)

1923 *Nature* 21 Apr. 553/1 The hyphomycete stage [of *Polythrincium Trifolii*] is followed by a pycnidial stage. 1971 P. H. B. Talbot *Princ. Fungal Taxon.* x. 146 Most of the pycnidial Deuteromycotina have slimy spores. 1977 *Lancet* 26 Mar. 672/2 Various pycnidial fungi related to the *Phoma* sp. isolated produce the mycotoxin responsible for lupinosis in sheep and cattle.

pycnidiospore. (Earlier example.)

1880 C. E. Bessey *Bot.* xvii. 294 The cavities are called pycnidia, and the small bodies pycnidiospores.

pycnidium. (Later examples.)

1938 G. M. Smith *Cryptogamic Bot.* I. xii. 416 If the fertile layer lies in a cup- or flask-shaped cavity that is open from the beginning, the cavity and the surrounding tissue constitute a pycnidium. 1966 K. Tubaki in Ainsworth & Sussmann *Fungi* II. iv. 127 Simple or branched sporophores may line a hollow flask-shaped fruit body, the pycnidium.

pycnium (pi·kni,ɒ̆m). *Bot.* Pl. **pycnia.** [mod.L., f. Gr. πυκνός thick.] In rust fungi of the order Uredinales, a fruit-body resembling a pycnidium. So **py·cnial** *a.*, of or pertaining to a pycnium; **py·cniospore,** a spore from a pycnium.

1905 J. C. Arthur in *Bot. Gaz.* XXXIX. 221 For the sorus of the initial stage [of uredineal fungi], usually.. called spermogonium, pycnidium, etc., I propose pycnium..; derivatives pycnial, pycniospores, etc. 1926 *Mycologia* XVIII. 90 The inefficient sori (pycnia) are present or absent in both macrocyclic and microcyclic rusts. 1929 J. C. Arthur et al. *Plant Rusts* i. 6 The pycnia produce pycniospores. 1937 Pycnial [see *æcium]. 1937 *Nature* 8 May 800/2 Pycniospores..were present in the nectar. 1946 K. S. Chester *Nature & Prevention of Cereal Rusts* v. 49 The pycnia occur on both leaf surfaces. 1976 G. C. Ainsworth *Introd. Hist. Mycol.* v. 132 Isolated monosporidial infections gave rise to pycnia which produced pycniospores and nectar but no aecia developed as they frequently did when two or more pycnial pustules were adjacent to one another.

pycno-. Add: **pycnochlo·rite** *Min.* [ad. G. *pyknochlorit* (J. Fromme 1903, in *Min. und Petrogr. Mitt.* XXII. 70)], a chlorite, (Mg, Fe^{2+},Al)$_6$(Si,Al)$_4O_{10}(OH)_8$, having the same silicon content as clinochlore (2·8–3·1 atoms per formula unit) but more iron (1·5–3 atoms); **py·cnocline** *Physical Geogr.*, a thin layer separating water of different densities; **py·cnogon** = *pycnogonid*; **pycnogonid,** add to etym.: [f. mod.L. class name *Pycnogonida*, f. generic name *Pycnogonum* (M. T. Brünnich *Entomologia* (1764) 84)]; also *attrib.*; (later examples); **pycnome·tric** *a.*, involving or employing a pycnometer; hence **pycnome·trically** *adv.*; † **pycnomo·rphous** *a. Cytology* [ad. G. *pyknomorph* (F. Nissl in *Neurol. Centralblatt* (1894) XIII. 683, (1895) XIV. 70), f. Gr. μορφή form, shape], characterized by much darkly staining matter; **pycnospore**: also = *pycniospore* s.v. *pycnium; (earlier and later examples).

1903 *Mineral. Mag.* XIII. 375 Pyknochlorite... A greyish-green, compact chlorite occurring in a quartz and calcite vein in the gabbro of the Radauthal, Harz. It has the same general formula..as clinochlore, but differs from this in containing much more ferrous iron and in its compact (πυκνός) texture. 1960 *Amer. Mineralogist* XLV. 797 The co-existing chlorite occurs in fairly large pale green crystals and shows the typical anomalous interference colors. Its analysis shows it to be fairly rich in Mg and Al, and following the classification of Hey (1954) it may be termed a pycnochlorite, with Fe (total):(Fe + Mg) = 0·273 and Si 2·83, on the basis of 14 oxygens (anhydrous). 1973 *Nature* 2 Mar. 28/1 Microscopic studies reveal that the metamorphic boundary involves the replacement by quartz and a chlorite mineral of fixed composition (pycnochlorite). 1978 *Ibid.* 20 July 243/1 Chlorites occurring as matrix in greywacke and amygdule fillings and groundmass replacement in spilite are either pycnochlorite or diabantite. 1957 G. E. Hutchinson *Treat. Limnol.* I. v. 282 When a wind blows over a thick layer of water lying over a second layer of greater density, not only will the surface level be raised at the lee end but the pycnocline, or plane separating the two layers of different density, will be tilted in the opposite direction. 1967 *Oceanogr. & Marine Biol.* V. 278 Changes in the sinking rate..are well substantiated... In pycnoclines a retardation of passive organisms is frequent, sometimes associated with a synthesis of pigments. 1976 *Nature* 2 Sept. 8/1 Over large areas of the present-day ocean, a permanent density discontinuity (pycnocline) arises as a consequence of the latitudinal variation in the intensity of incident radiation from the Sun. 1927 *Proc. Imper. Acad. Japan* III. 610 (*title*) Notes on some pycnogons living semi-parasitic on holothurians. 1935 T. H. Savory *Arachnida* xvi. 172 Ever since the first pycnogon was described..the problem of their affinities has been debated. 1959 A. C. Hardy *Open Sea* II. v. 100 Sea-slugs, ascidians, sea-spiders (pycnogons) and spider-crabs, starfish and brittlestars—all these, and more, may be in just one haul. 1935 T. H. Savory *Arachnida* xvi. 172 The Pycnogonid crawls about, extremely slowly. 1973 P. E. King *Pycnogonids* i. 7 The pycnogonid body is considerably reduced. *Ibid.* 8 The pycnogonids have a wide geographical and bathymetric range. 1925 *Arch. Internal Med.* XXXV. 133 Specific gravity determinations were made by the pyknometric method. 1938 *Trans. Faraday Soc.* XXXIV. 1214 (*heading*) Pyknometric studies on chemical equilibrium. 1934 *Jrnl. Chem. Soc.* 498 The samples of water obtained by combustion were carefully distilled..and their densities were measured pyknometrically. 1976 *Nature* 3 June 438/3 There is a reasonable agreement between X-ray and pycnometrically determined densities in the minerals of the oldest rocks. 1899 L. F. Barker *New Syst. Constituent Neurones* xi. 123 Nissl consequently designates the extremely darkly stained cells as pyknomorphous cells, or cells in which the stainable portions are arranged relatively most closely. 1903 *Med. Chron.* XXXIX. 19 The stained, chromophile, or tigroid substance of nerve cells is regarded as nutritional substance. When it is abundant the cell is described as being in a pyknomorphous condition. 1887 H. E. F. Garnsey tr. *Ade Bary's Compar. Morphol. & Biol. Fungi* v. 246 Pycnidia: receptacles.. producing gonidia which are known as pycnospores. 1938 G. M. Smith *Cryptogamic Bot.* I. xii. 416 In addition to forming conidia or pycnospores, a mycelium may also form large thick-walled spores.

pycnosis, var. *pyknosis.

pycnotic, *a.* Add: The form **pyk-** is now usual. **2.** *Cytology.* Displaying pyknosis.

1910 in *Lippincott's New Med. Dict.* 798/1. 1926 *Arch. Neurol. & Psychiatry* XVI. 134 This change..was characterized by pyknotic shrinking of the nuclei and an increase in cytoplasm. 1936 J. Krafka *Textbk. Histol.* i. 3 In old senescent cells a pycnotic nucleus is produced by a condensation of chromatin to the extent that no ground substance shows. 1957 C. P. Swanson *Cytol. & Cytogenetics* ii. 34 The pycnotic state..persists into interphase to form what were formerly called pro-chromosomes. 1974 *Nature* 11 Oct. 509/1 Within 3 h of furosemide administration..single cell necrosis with pyknotic hepatocytes showing eosinophilic degeneration was..occasionally present.

pye-dog, pie-dog. For '*Anglo-Ind.*' read 'orig. *Anglo-Ind.*' and add further examples.

1884 Kipling *Departmental Ditties* (1886) (ed. 2) 52 Glare down old Hecate..And bid the pie-dog yell. 1924 *Blackw. Mag.* Sept. 355/1 The men of Bokkos and their dogs—the sorriest-looking pie-dogs in all Africa..go forth to get what they can. In a poor lot these Bokkos pies are the poorest. 1940 F. Stark *Winter in Arabia* 107 The Saint who is buried in the tomb below has pye-dogs who slink in to him at night. 1954 M. K. Wilson tr. *Lorenz's Man Meets Dog* i. 14 There are lots of localities in the near East where Pie dogs and golden jackals abound, yet never intermingle. 1977 *Times* 25 June 15/5, I tied red, white and blue ribbons round the neck of my pye-dog.

pyeenock (pəi,ī·nɒk), dial. var. of Peony.

1911 D. H. Lawrence *White Peacock* ii. ix. 354 There's a fine show of pyeenocks this year.

pyelo-. Add: **py·elogram,** an X-ray photograph showing the pelvis of the kidney; † **py·elograph,** a pyelogram; hence **pyelo·graphy** [ad. G. *pyelographie* (Voelcker & Lichtenberg 1906, in *Münchener med. Wochenschr.* 16 Jan. 105)], **py·eloplasty** *Surg.*, a plastic operation on the pelvis of the kidneys.

1923 R. Knox *Radiogr. & Radio-Therapeutics* (ed. 4) I. facing p. 388 (*captions*) Pyelogram—dilated pelvis with kinking of the ureter... Pyelogram of a normal kidney. 1952 M. E. Florey *Clin. Appl. Antibiotics* xvii. 507, 25 days after the operation a pyelogram revealed no abnormalities. 1980 *Brit. Med. Jrnl.* 29 Mar. 930/1 We in Britain do not feel that intravenous pyelograms are necessary before every hysterectomy. 1913 *Jrnl. Amer. Med. Assoc.* 18 Jan. 184/2 A pyelograph is taken while the fluid is being injected and the pelvis or the ureter is kept as full as possible at the time the exposure is being made. 1914 *N.Y. Med. Jrnl.* XCIX. 1057/2 Doctor Furniss, in making pyelographs, had until recently been injecting argyrol or collargol with a syringe. 1906 *Jrnl. Amer. Med. Assoc.* 14 Apr. 1149/2 Pyelography.—Voelcker and Lichtenberg have coined this term for radiography of the kidney and ureter after these structures have been filled with a solution of a silver salt. 1975 *Daily Colonist* (Victoria, B.C.) 23 Dec. 2/1 In pyelography, a contrast medium, the 'dye', is injected intravenously. 1913 C. H. Chetwood *Practice of Urol.* xxxi. 587 (*heading*) Pyeloplasty (Fenger's operation). 1976 *Lancet* 20 Nov. 1109/2 The boy with a horseshoe kidney had a pyeloplasty and 1 of the boys with obstructed congenital megaureter had a reimplantation.

Pygmalion (pigmē̆i·liŏn). The name of a play by George Bernard Shaw (1856–1950), used quasi-advb. in *not Pygmalion likely*, a joc. euphemism for the phrase 'not bloody likely' which occurs in Act III of the play (see *bloody *adv.* 2, quot. 1914) and was the occasion of a public sensation at the time of the first London production in 1914. Also used *attrib.* of utterances regarded as mildly shocking.

1949 Partridge *Dict. Slang* (ed. 3) Add. 1121/1 *Not Pygmalion likely!* Not at all likely; certainly not! 1960 *Guardian* 8 Mar. 7/2 (*heading*) Not Pygmalion likely. 1960 *Times* 28 Apr. 14/5 Mr. S. M. Nutley..said: 'The trouble really began when Alderman Mrs. K. Sheridan was speaking about the council fleecing tenants and used a pygmalion word.' 1964 N. Squire *Bidding at Bridge* 185 So we pass? Not pygmalion likely! 1967 G. Fallon *Rendezvous in Rio* xiii. 106 'Are you thinking of joining in?' 'Not Pygmalion likely,' Bland returned brusquely. 1967 A. Wilson *No Laughing Matter* ii. 96 You bloody bird! No, no, Mouse, Mr Polly and I were just talking Pygmalion talk! 1976 *Times* 18 Mar. 11/5 My immediate reaction was to say, 'Not Pygmalion likely'.

Pygmalionism (pigmē̆i·liŏniz'm). *Psychol.* Also **pygmalionism.** [f. *Pygmalion* a character in Greek mythology + -ism. According to Ovid (*Metam.* x. 243–97), Pygmalion was a King of Cyprus who made a statue of a beautiful woman and loved it so deeply that Aphrodite gave life to it.] The condition of loving a statue, image, or inanimate object; love for an object of one's own making.

1905 H. Ellis *Stud. Psychol. Sex* IV. 188 Pygmalionism, or falling in love with statues, is a rare form of erotomania founded on the sense of vision and closely related to the allurement of beauty. 1923 —— *Dance of Life* vii. 328 We find records of Pygmalionism and allied perversities in Lucian. 1940 Hinsie & Shatzky *Psychiatric Dict.* 453/1 *Pygmalionism*,..the condition of falling in love with a creation of one's own. 1946 'M. Innes' *From London Far* iii. iv. 201 'Did you ever happen to hear of something called Pygmalionism?' '..It's a fancy name for iconolagnia'. 1954 H. T. F. Rhodes *Satanic Mass* vi. 52 After the kiss, accounts agree that the priestess offered herself to the God by an act of pygmalionism. 1966 J. Cohen *Human Robots* iv. 66 We may infer that the Greeks, who had a highly developed visual sense, were inclined to *Pygmalionism*. *Ibid.*, We may regard *Pygmalionism* as a manifestation of a more general tendency to excitement induced by a partner's passivity.

pygmoid (pi·gmoid), *a.* [f. Pygmy *sb.* + -oid.] Resembling a pygmy; having (some of) the characteristics of a pygmy. Also as *sb.*

1933 R. G. Austin tr. O. Menghin in *Antiquity* VII. 242 Mr Clark is perfectly correct in stating (p. 12) that I connect the Mughem men with the Grimaldi and Bushman types, treating them as pygmoid (not as pygmies). 1958 *Listener* 2 Oct. 507/1 The majority of these little people whom you see outside the forests in the north-east [*sc.* of the Congo] are not pygmies. They are pygmoids, the offspring of a liaison between a pygmy and a normal-sized Negro. 1965 E. E. Evans-Pritchard *Theories Primitive Relig.* v. 102 The Pygmies and Pygmoids of Africa and Asia. 1976 Eveleth & Tanner *Worldwide Variation in Human Growth* vii. 190 In New Guinea where one encounters numerous short-statured populations there can be no clear separation of pygmoid groups. 1977 P. Johnson *Enemies of Society* xvii. 226 The Veddas, a pygmoid people of primeval hunters living in the interior of Ceylon.

pygmy, pigmy, *sb.* and *a.* Add: **C.** *pygmy-cup, -folk*; **pygmy-flint** *Archæol.*, a type of microlith.

1936 *Proc. Prehist. Soc.* II. 223 The urns comprise two food-vessels, a pigmy-cup and an encrusted urn. 1963 H. N. Savory in Foster & Alcock *Culture & Environment* iii. 43 The Breach Farm barrow, with its dry-stone wall kerb and its fine biconical Pygmy Cup. 1907 T. R. Holmes *Anc. Britain* 82 Of all stone implements the most curious are the tiny objects which are known as 'pygmy flints'. 1930 F. Elgee *Early Man in N.E. Yorkshire* v. 31 The pygmy-flint men lived by hunting and fishing. 1963 *Field Archaeol.* (Ordnance Survey) (ed. 4) 8 Various palæolithic objects like hand-axes and choppers, microliths ('pygmy' flints), arrow and lance heads. 1788 W. Collins *Ode on Pop. Superstitions Highlands of Scotl.* 18 In..small vaults a pigmy-folk is found.

pygo-. Add: **pygo·pagus** = *pygopage* [a. F. *pygopage* (I. G. St.-Hilaire 1830, in *Ann. des Sci. nat.* XX. 338)]; so **pygo·pagous** *a.*

1895 *Teratologia* II. 274 Several of the pygopagous twins of whom there are scientific records, survived birth and lived for a number of years. 1866 *Trans. Med. Soc. State of N.Y.* XXIV. 224 The symmetrical pygopagus is exceedingly rare. 1903 J. W. Williams *Obstetrics* xxxix. 680 Ischiopagi and pygopagi, as a rule, call for complicated and difficult manœuvres before delivery can be effected. 1959 *Jrnl. Chronic Dis.* X. 84 A wooden carving from the Solomon Islands suggests conjoined twins of the pygopagus type with the union of the bodies and heads and the extremities shortened by achondroplasia.

pyinkado (pyi-ŋkādo, pī,i-ŋk-). Also † **pingadoo, pyengadu, py(i)ngado, pynkado,** [Burmese.] The heavy timber of the tree *Xylia xylocarpa* (formerly *X. dolabriformis*), which belongs to the family Leguminosæ and is native to Burma and parts of India; also, the tree itself. Also *attrib.*

1832 W. Roxburgh et al. *Flora Indica* II. 543 It [*sc.* *Minosa xylocarpa*] is called Pingadoo in Pegu, where it is used for knees, crooked timbers, &c. in ship building. 1875 T. Laslett *Timber & Timber Trees* xxi. 129 The Pyengadu, or Iron-wood tree,..is a species of Acacia, of straight growth, found in the Burmese forests. 1885 W. T. Oldreave in Rattray & Mill *Forestry & Forest Products* xii. 381 Pynkado..is said to be a species of acacia. 1896 W. R. Fisher in W. Schlich *Man. Forestry* V. i. ii. 117 In London..doubtless Pyngado..and other heavy Indian woods might be used with advantage [for wooden paving]. 1902 G. S. Boulger *Wood* i. v. 92 Pynkado or Pyengadu..is the Ironwood of Pegu. 1934 'G. Orwell' *Burmese Days* 69 At the edge of the stream there was a huge dead pyinkado tree festooned with spidery orchids. 1940 *Archit. Rev.* LXXXVII. 47 For the remainder of the building sound-proofing floors are used finished with 3 in. pyinkado strips. 1951 *Dict. Gardening* (R. Hort. Soc.) IV. 2295/1 Pyingado..is extremely hard, heavy, strong, and durable. 1956 *Handbk. of Hardwoods* (Forest Prod. Res. Lab.) 194 Pyinkado grows to a height of 100–120 ft. *Ibid.* 195 Pyinkado is unsuitable for plywood manufacture because of its weight. 1971 F. H. Titmuss *Commerc. Timbers of World* (ed. 4) 263 Pyinkado is a difficult timber to work.

pyjamas, pajamas, *sb. pl.* Add: *pyjamas*, pronounced (pădʒā·măz), is now standard in the U.K., *pajamas* in the U.S. The second part of def. is now the prevailing sense. In extended use, applied to a similar day-time or evening garment worn by women (see also *beach-pyjamas* s.v. *beach *sb.* 4, *palazzo pyjamas* s.v. *palazzo 3). Also (*occas.*) *sing.*, as *pyjama*. **a.** (Further examples.)

1878 E. S. Bridges *Diary* 6 Sept. in *Round World in Six Months* (1879) iii. 37, I relinquished my English *chemise de nuit* and took to pyjamas—bedclothes are not used at this time of year [in Japan]. 1886 *Girl's Own Paper* 23 Oct. 59/1 The pattern for this month..is a combination nightgown, or lady's 'pyjama'. 1897 [see *sleeping-suit* s.v. Sleeping *vbl. sb.* 2 b]. 1932 *Barker's Spring Catal.* This ideal pyjama is made of a very soft washing cotton. 1932 *Boston Even. Transcript* 6 Aug. 1 Clad in pajamas and admitting to police that she was returning home from a party, Mary Callahan, twenty four ..was arrested at seven o'clock this morning. 1936 A. Christie *ABC Murders* xvi. 120 Girls passed him..in summery frocks and pyjamas and shorts. 1968 J. Ironside *Fashion Alphabet* 62 *Pyjamas*, blouse or shirt and wide-legged trousers worn for lounging or for beach wear —introduced by Chanel in the late 1920s. 1976 *Washington Post* 19 Apr. A12/4 (Advt.), Pre-school boys' pajamas reduced. 1978 *Neiman-Marcus Christmas Bk.* 32 The ambient glow of soft panne velvet for party pajamas.

b. *pyjama-clothes, coat, cord, dress, jacket, leg, pants, suit* (earlier and later examples), *top, trousers* (later examples); *pyjama-clad* (later examples), *-legged, -like* adjs.; **pyjama bottom,** the bottom half of a suit of pyjamas, pyjama trousers; usu. *pl.*; **pyjama case,** a bag or other container in which pyjamas can be kept when not being worn; **pyjama party,** a party at which those present are dressed in pyjamas; also **pyjama-and-bottle-party.**

1928 *Sunday Dispatch* 5 Aug. 15/2 Mention was made of the splendid work of Mrs. X—Y—for her pyjama-and-bottle party. 1959 R. Condon *Manchurian Candidate* (1960) i. 10 The..movie actor..had opened the door of the hotel suite wearing only pyjama bottoms. 1972 J. Wainwright *Requiem for Loser* iii. 48 He..stepped out of his pyjama bottoms and began dressing himself. 1973 *Black World* Oct. 55/1 None of the kids had on a complete outfit of clothes: some were in a pajama top—or a bottom. 1925 'R. Crompton' *Still—William* ix. 164 Thrusting his..paper fleet into his pyjama case. 1976 W. J. Burley *Wycliffe & Schoolgirls* viii. 151 Lying on it [*sc.* the bed] was a pyjama case in the shape of a dog with 'Jane' embroidered across it. 1921 R. Macaulay *Dangerous Ages* i. 2 Her slight, straight, pyjama-clad body. 1976 'L. Black' *Healthy Way to Die* iv. 38 The pyjama-clad legs dangling inside the silken dressing-gown. 1939 Auden & Isherwood *Journey to War* 43 The Cantonese, in their light pyjama-clothes. 1916 M. Diver *Desmond's Daughter* ii. v. 71 A Punjab Cavalryman in a turban and silk pyjama coat. 1978 C. Storr *Winter's End* v. 68 She wore a blue and white striped pyjama coat. 1917 E. Fenwick *Diary* 18 Feb. in *Elsie Fenwick in Flanders* (1981) 143 He tried to hang himself with his pyjama cord. 1972 *Times* 28 July 10/1 Top coats with pyjama cord belts. 1967 *Guardian* 2 Nov. 7/6 For luscious evening attire..a pyjama dress in hot pink and orange. 1891 E. Dowson *Let.* 1 July (1967) 206, I am more elaborately vested, in a Pyjama jacket. 1976 C. Dexter *Last seen Wearing* xxxviii. 261 The top button of the pyjama jacket already undone. 1933 A. Thirkell *High Rising* ii. 48 Tony, by now in what he called his pyjama-legs, executed a dance of joy. 1977 *Transatlantic Rev.* lx. 145 The water isn't as cold as I figured, but when the bottom of my pajama-leg gets wet, I get a little nauseous. 1960 Pyjama-legged [see *culotte 2]. 1960 Koestler *Lotus & Robot* i. i. 41 Another table in the Mascot's dining-room was occupied by an Egyptian gentleman in a pyjama-like attire. 1956 'E. McBain' *Cop Hater* (1958) iv. 38 He was wearing pajama pants and nothing else. 1980 G. Lord *Fortress* i. 7 She pulled down her pyjama pants. 1910 *Westm. Gaz.* 13 Apr. 5/3 A pyjama party held a couple of days ago at the residence of Mrs. Edwin Avon, a well-known member of Chicago society. 1928 A. Waugh *Nor Many Waters* ii. 36 They'd thought of making a dressing-gown and pyjama party of it, so you can guess what it'll be like from that. 1933 Dylan Thomas *Let.* (1966) 67 It sounds as though you'd invited me to a pyjama party. 1978 S. Sheldon *Bloodline* xii. 157 They had often invited Elizabeth to their pajama parties. 1883 C. Bell tr. *Haeckel's Visit to Ceylon* xx. 329 The rest of our attire consisted of that particularly light and airy white flannel garment, known throughout India as a pajama suit. 1973 'G. Black' *Bitter Tea* iii. 41 She was..wearing the kind of pyjama suit some women go shopping in [in Malaysia]. 1949 N. Mitford *Love in Cold Climate* i. vi. 66 Lady Montdore..in bed..wearing what appeared to be a man's striped pyjama top under a feathered wrap. 1976 C. Dexter *Last seen Wearing* xxxiii. 225 She..fastened all but the top button of her pyjama top. 1932 D. C. Minter *Mod. Needlecraft* 146/2 For pyjama trousers cut straight down. 1975 W. J. Burley *Wycliffe & Pea-Green Boat* i. i. 10 A tall, skinny young man in pyjama trousers.

pyjamaed, *a.* Add: (Later examples.) Also **pyjama'd.**

1922 F. Hamilton *P.J. the Secret Service Boy* i. 47 Mr. Davenant sleepily extended a pyjama'd arm. 1929 D. Hammett *Dain Curse* (1930) xvi. 182, I..let in Jack Santos, pajamaed, bathrobed, and slippered. 1959 D. Campbell *Evening under Lamplight* 47 Pyjama'd figures were clambering..into the shadowy bedroom. 1959 P. McCutchan *Storm South* ix. 124 Her pyjama-ed legs. 1974 'D. Meiring' *President Plan* xvii. 158 She knew where he slept... A light went on and..he was there, pyjamaed.

pyjams, pyjies (pəi--), colloq. abbrevs. of Pyjamas, pajamas *sb. pl.* Also **pygies** and (redupl.) **pyjimjams.**

1926 D. L. Sayers *Clouds of Witness* iv. 99 Why do girls wear such mimsy little pyjimjams in this damn cold climate? 1929 P. Sturges *Strictly Dishonorable* ii. 139 Now go and get the pygies and things. 1937 Partridge *Dict. Slang* 674/2 *Pyjams*, abbr[eviation of] *pyjamas*. 1960 J. Betjeman *Summoned by Bells* vii. 66 House-slippers, sponge-bag, pyjams. 1962 J. Braine *Life at Top* i. 8 But Daddy has to earn pennies..for pyjies and frocks.

pyknic (pi·knik), *a.* Also **pycnic.** [f. Gr. πυκνός thick, close-packed + -ic.] In Kretschmer's theory of human physical and corresponding temperamental types, designating a stocky physique with a rounded body and head, thickset trunk, and a tendency to fat, usu. accompanied by a cycloid temperament; also *absol.*, a person belonging to this type. Cf. *asthenic *a.* b; *athletic *a.* 3; *leptosomic *a.* Phr. *pyknic practical joke* (see quot. 1964).

1925 W. J. H. Sprott tr. *Kretschmer's Physique & Character* i. ii. 29 The pyknic type..is characterised by the pronounced peripheral development of the body cavities (head, breast, and stomach), and a tendency to..fat about the trunk. *Ibid.*, The pyknics tend emphatically to a covering of fat. 1940 W. H. Sheldon *Varieties of Human Physique* (1963) iii. 32 Much of the confusion associated with Kretschmer's terminology arises from the fact that his term 'pyknic' actually applies to a physique combining endomorphy and mesomorphy. 1942 —— *Varieties of Temperament* iv. 109 The mother was a PPJ (pyknic practical joke). She was slim-waisted and active..,but after the first pregnancy she came into her full endomorphic blossom. 1958 A. R. Radcliffe-Brown *Method in Social Anthropol.* i. iv. 103 Psychiatry affords an example of a 'special psychology', as do attempts to define psychological 'types'—..pycnic, asthenic. 1960 J. Comas *Man. Physical Anthropol.* vi. 340 A well-developed thorax predominates over the shoulders in the pyknic type. 1964 L. J. Bischof *Interpreting Personality Theories* (1970) xi. 431 *Pyknic practical joke* (PPJ).., the PPJ refers to a person who has a muscular mesomorphic body in adolescence but in later life balloons out into obesity to become an endomorph. 1971 J. Z. Young *Introd. Study Man.* xxxix. 576 Kretschmer..found that his pyknics tended to be what Jung called extroverted. 1975 A. Ferraro in S. Arieti *Amer. Handbk. Psychiatry* (ed. 2) IV. 103/1 Badia found that the megalosplanchnic type of Viola, or pycnic type of Kretschmer, discloses a tendency to chronic changes in the blood vessels of the heart.

pykno-: for words beginning thus see also Pycno- in Dict. and Suppl.

pyknolepsy (pi·knolepsi). *Med.* [ad. G. *pyknolepsie* (Schröder: see *Monatsschr. für Psychiatrie und Neurol.* (1916) XL. 281), f. Gr. πυκνός thick, crowded, after *narkolepsie* Narcolepsy.] An epileptic condition in which brief attacks similar to petit mal occur many times in a day. Hence **pyknole·ptic** *a.*

1922 *Q. Cumulative Index Current Med. Lit.* 1921 533/2 (*heading*) Pyknolepsy. 1924 *Brain* XLVII. 98 Pyknolepsy, in spite of its long duration and the great frequency of the attacks, does not impede mental development nor give rise to psychical defects. *Ibid.*, Of the many that have been used the name pyknolepsy is recommended for use by English writers... It allows us to coin a handy adjective, 'pyknoleptic', by analogy with epileptic. 1952 F. A. Elliott et al. *Clin. Neurol.* vii. 133 In pyknolepsy, the attacks cease with puberty and may not recur. 1972 P. H. Hoch *Differential Diagnosis in Clin. Psychiatry* iii. xiii. 395 Grand mal, petit mal,.. or other subgroup forms..such as the narcoleptic, pyknoleptic and so forth—have a special metabolic formula of their nervous system. 1975 S. Arieti *Amer. Handbk. Psychiatry* (ed. 2) IV. xiii. 320/2 The incidence of absence attacks varies from very few, often in the morning, to a great many, up to 100 or more per day ('pyknolepsy').

pyknosis (piknōᵘ·sis). *Cytology.* Also **pycnosis.** [f. Gr. πυκν-ός close, compact + -osis.] The contraction of a dying cell, or of its nuclear material, into a densely staining mass or masses.

1900 Dorland *Med. Dict.* 552/1 *Pyknosis*, a thickening; especially degeneration of a cell in which the protoplasmic substance becomes more dense and the size of the cell smaller. 1926 *Arch. Neurol. & Psychiatry* XVI. 135 In general there is a progressive shrinking and pyknosis of the nucleus. 1946 [see *hyperchromatosis 2]. 1950 A. W. Ham *Histol.* v. 60 The changes that occur in nuclei as, or after, individual cells die in the living body are of three sorts. The commonest one is called pycnosis; this consists of a shrinkage of the nuclear material into a homogeneous hyperchromatic mass. 1972 *Physics Bull.* Mar. 147/1 The biological end points that will be studied include glycogen accumulation, nerve cell pyknosis, nerve cell injury or loss and glial reaction. 1978 *Nature* 25 May 306/2 In minced muscle grafts sarcoplasmic structure was rapidly lost, and most muscle nuclei seemed to undergo pyknosis, fragmentation and lysis.

pyknotic, var. Pycnotic *a.* in Dict. and Suppl.

py korry (pəi kǫ·ri), *int. N.Z. slang.* [Maori corruption of *by golly*.] = *by golly* s.v. Golly *int.*

1938 R. D. Finlayson *Brown Man's Burden* 32 'Py korry, that right!' Wi admitted to himself. 1941 Baker *N.Z. Slang* ix. 71 There is..not much to distinguish the authenticity of an expression like *py korry*! (by God) from one like *rekureihana* (regulation) except that the former is colloquial. 1943 J. A. W. Bennett in *Amer. Speech* XVIII. 94 The Maori treatment of certain English words is conventionally indicated by such spellings as *plurry* and *py korry* for 'bloody' and 'by golly'. 1961 J. Reid *Kiwi Laughs* 12, I have steered clear in this selection of the 'Py korry, Hori' type of alleged Maori humour. 1966 G. W. Turner *Eng. Lang. Austral. & N.Z.* x. 200 Maori English interlarded with *plurry* and sentences like 'Py korry, that the nice baby, eh?' belongs to the language of journalists rather than the language of Maoris.

pykrete (pəi·krīt). Also **Pykrete.** [f. the name of G. N. *Pyke* (1894–1948), an Englishman involved in Combined Operations (where pykrete was invented) during the war of 1939–45 + Conc)rete *a.* and *sb.*] A frozen mixture of ice and wood pulp or sawdust.

1948 *Jrnl. Glaciol.* I. 96 In February 1943 the..outlook was suddenly transformed by the discovery that the inclusion of a small percentage of wood pulp improved the mechanical properties of ice in a spectacular manner. The discovery was made by Mark and Hohenstein, working at the Brooklyn Polytechnic. In view of the similarity to concrete and in honour of the originator of the bergship project, the frozen wood pulp was given the code name of pykrete (Pyke's concrete). *Ibid.* Pykrete ..was ductile and could even be machined on a lathe. *Ibid.* 104 As a protection against explosives pykrete is weight for weight as good as concrete. 1960 *New Scientist* 28 Apr. 1081/3 The aircraft carrier project showed that the engineering properties of ice are greatly improved by 'alloying' it with sawdust (to make 'pykrete'). 1966 *Ibid.* 3 Feb. 284/3 Just as the Eskimo had learned to stiffen and toughen ice by freezing moss into it, so Pykrete owed its strength to fibres of cellulose that blocked the spread of cracks.

Pylian (pī·liăn, pəi--), *sb.* and *a.* [f. Gr. πύλος, L. *Pylos* Pylos: see -ian.] **A.** *sb.* A native or inhabitant of the Homeric town of Pylos in the southern Peloponnese, traditionally regarded as the birthplace of Nestor and the name of his dynasty, and usually identified with Messenian Pylos at the northern end of Navarino Bay. Hence, by extension, a native or inhabitant of the territory ruled by Nestor or his dynasty. **B.** *adj.* Of or pertaining to Pylos or its inhabitants.

1611 Chapman *Homer's Iliads* ii. 28 The Pylians and their townes. ? 1614 —— tr. *Homer's Odysses* iii. 32 Soone they reacht the Pylian throngs and seates, Where Nestor with his sonnes sate. *Ibid.*, When the Pylians saw These strangers come: in thrust did all men draw About their entrie. 1725 Pope in *Homer's Odyssey* I. 142 This was a very solemn sacrifice of the Pylians. 1846 G. Grote *Hist. Greece* II. i. xviii. 16 The Pylians, together with the great heroic family of Nêleus and his son Nestôr, who preside over them, give place to the Dorian establishment of Messênia, and retire to Athens, where their leader Melanthus becomes king. 1934 A. Toynbee *Study of Hist.* I. 403 In the Homeric epic, Pylos is not called 'Minyan' as Orchomenos is, nor are the Pylians called 'Minyae'. *Ibid.*, The Greek inhabitants of..the *ci-devant* Pylian domain. 1965 *Language* XLI. 315 Scribes who use different orthographies may have come from different localities within the Pylian territory.

pylon. Delete ‖ and add: **1. a.** (Later *transf.* examples.)

1930 *Morning Post* 9 Aug. 11, 200 men have been employed excavating granite for the facing of the bridge piers and pylons. 1974 *Sci. Amer.* Nov. 145/1 The Bayonne bridge lacks the huge pylons of Sydney Harbor, which contain the thrust visually as well as in Newtonian fact.

2. *Aeronaut.* Also † **pylone** [F. *pylône*]. **a.** A tall structure used to mark out the course round which aeroplanes fly (or, formerly, in launching them); also, by extension, a structure round which cars drive on a race-track.

1909 *Flight* 13 Mar. 143/1 The machine is brought to earth conveniently close to the pylone. 1909 *Westm. Gaz.* 16 Oct. 9/3 After a successful round of the course his aeroplane came to earth near the second pylon on the south side. 1913 A. E. Berriman *Aviation* Pl. facing p. 38 (*caption*) The lower picture illustrates a similar machine banking while turning about one of the pylones at the Hendon Aerodrome. 1970 *Pop. Mechanics* Oct. 106/1, I still had the third and best run to make. A pylon was placed in the centre of the pad. Instructions were to hit the brakes as before and *steer around the pylon*, brakes full on! 1977 *Sci. Amer.* Oct. 74/3 The craft had to take off unassisted from level ground in a wind of 10 miles per hour or less, fly in a figure-eight pattern around two pylons half a mile apart and pass over a 10-foot hurdle at the start and finish.

b. A post on some early aircraft to which wires for supporting or warping the wing were attached; also, in modern aircraft, a pillar that projects from a wing or fuselage to support an engine, rotor, weapon, or the like.

1912 *Aero* Aug. 236/1 The machine bears..a resemblance to a torpedo boat on account of its squat 'funnels', which are..the..pylons carrying the wing bracing wires. 1919 Pippard & Pritchard *Aeroplane Structures* xi. 142 The pylon bracing..comes into operation (1) In high speed flight. (2) In landing. *Ibid.*, The vertical components of the loads in the pylon wires AD, CD throw an extra load in the interplane strut BE. 1955 Liptrot & Woods *Rotorcraft* iii. 20 The rear-end ring [of the fuselage] carries the pylon on which is mounted the tail rotor. 1959 *Times* 26 Feb. 10/6 On the Boeing 707-120..the engines are mounted separately on pylons beneath the wings. 1969 K. Munson *Pioneer Aircraft 1903–14* 106/1 As flown for the first time at Issy on 23 January 1909, it had a 30 h.p. R.E.P. engine.., and a small kite-shaped fin was fixed above the wing-warping pylons. 1979 *Daily Tel.* 29 May 1/4 The airline said it believes the attachments of the engine pylon to the wings of its aircraft are sound.

3. *Surg.* A temporary, unjointed, artificial leg.

1920 *Lancet* 14 Feb. 373/2, I will endeavour to illustrate the most important details in the manufacture of a thigh pylon. 1945 Thomas & Haddan *Amputation Prosthesis* ii. 49 It is the opinion of many that the most effective and rapid shrinkage and adaptation of the stump takes place with the use of a pylon or a temporary prosthesis. 1971 P. J. R. Nichols *Rehabil. Severely Disabled* II. iii. 107 Many elderly patients fitted with a satisfactory pylon are reluctant to exchange it for a definitive limb, which is heavier and 'more difficult' to use.

4. A tall structure erected as a support; *spec.* a lattice-work metal tower for overhead electricity lines.

1923 E. Shanks *Richest Man* iii. 52 Half a mile up the mountain, a cable, a thin black line, traversed the crystal air, borne up on pylons. 1930 Auden *Poems* 67 Pylons fallen or subsiding, trailing dead high-tension wires. 1942 J. Lees-Milne *Ancestral Voices* (1975) 51 This unconfined, Thames estuary is rather exciting, sprinkled as it is with drifting pylons, factory chimneys and distant gasometers. 1966 J. Betjeman *High & Low* 67 Encase your legs in nylons, Bestride your hills with pylons O age without a soul. 1971 *Nature* 12 Nov. 62/3 A commercial application of the hovertrain would operate on pylons spaced up to 150 foot apart and 25 to 30 foot off the ground. 1972 R. Adams *Watership Down* xviii. 104 They had heard the unnatural humming of a pylon in the summer air. 1977 *Times* 19 Jan. 14/2 The North-Western Electricity Board were understandably forbidden to string wires on overhead pylons up the valley.

b. Used *attrib.* to designate those poets of the nineteen-thirties (chiefly Auden, Day Lewis, MacNeice, and Spender) who used industria scenes and imagery as themes of their poetry.

Spender's poem 'The Pylons' was published in 1933.

[1935 H. A. Mason in *Scrutiny* III. 405 In *Vienna* Spender appears very clumsily dressed in the robes of Eliot (chiefly *Ash Wednesday*) the 'pylon' imagery and possible other borrowed garments.] 1951 H. Sergeant *Tradition in Making of Mod. Poetry* I. iii. 44 His [*sc.* Wilfrid Gibson's] method of recording factual details of the industrial background to many of his poems furnishes a parallel with that of the 'pylon' school of the thirties. 1957 R. Hoggart *W. H. Auden* 14 His first links were made with others who were to become writers and publicists in what has variously been called the Thirties Group, the Pylon School and the Auden Group. 1958 *Listener* 4 Dec. 924/2 The trouble with most of the 'Pylon Poets'—with the honourable exception of W. H. Auden—is that to them industry was still too much of a new thing. 1961 *Ibid.* 24 Aug. 284/1 After Eliot..there appeared Auden and Spender and the 'pylon' school of the nineteen-thirties. 1973 *Commentary* Dec. 53/2 After the withering of 30's illusions it became fashionable to laugh at 'Pylon' poetry.

5. *U.S.* A small pillar or column, used to accommodate a sign or signal.

1934 *Sun* (Baltimore) 10 Oct. 7/1 A proposal to replace the safety pylons with an overhead signal light, with pedestrians waiting on the sidewalk until ready to board a street car, was made yesterday. 1977 *Washington Post* 24 Mar. D.C. 5 Officials have recommended changes in the station that include an end to total dependence upon station names lettered sideways on upright pylons located along the station platforms, requiring passengers to crane their necks to read them.

pyloric, *a.* (*sb.*) (Examples of *pyloric stenosis.*)

1900 S. & W. S. Fenwick *Ulcer of Stomach & Duodenum* i. ii. 41 Pyloric stenosis is a frequent result of gastric ulcer. 1970 H. M. Spiro *Clin. Gastroenterol.* xvi. 272/1 The characteristic physical finding of pyloric stenosis is the succession splash. Shaking the patient's abdomen or grasping the stomach through the abdomen and shaking it will elicit a loud gurgling sound.

pyloro-. Add: **pylo·rospasm,** spasm of the pylorus.

1898 J. C. Hemmeter *Dis. Stomach* iii. ix. 643 (*heading*) Pyloric spasm (pylorospasm, cramp, convulsion, spasm of the pylorus). *Ibid.* 644 Under narcosis the pylorospasm relaxes. 1960 Jones & Gummer *Clin. Gastroenterol.* xix. 564 Pylorospasm, so frequently invoked as a cause of symptoms in peptic ulcer, gall-bladder disease, and chronic appendicitis, has been considered as a possible cause [of hypertrophic pyloric stenosis] but without any convincing evidence in its support.

pyo-. **pyocyanin,** delete formula and add: now known to be 5-methyl-9-oxo-5,9-dihydrophenazine, $C_{13}H_{10}N_2O$; now usu. **-ine** (-īn); (further examples); **pyode·rma,** pyodermia; **pyorrhœa** (also, *U.S.*, -rrhea), also, *spec.* (in full pyorrhœa alveolaris) a purulent inflammation of the tissues surrounding the teeth that results in shrinkage of the gums and loosening of the teeth; (further examples).

1947 *Sci. News* V. 90 Many bacteria in presence of certain organic substances, which they activate, reduce a molecule such as pyocyanine to its colourless leuco form. 1949 H. W. Florey et al. *Antibiotics* I. xii. 549 Pyocyanine, a substance which is now recognized to be bactericidal and to which pyocyanase probably owes some of its activity... This is the blue pigment to which 'blue pus', characteristic of infection by *Ps*[*eudomonas*] *pyocyanea*, owes its name. 1957 G. A. Swan in Swan & Felton *Phenazines* x. 176 Pyocyanine, the first phenazine compound discovered in nature. 1976 *Ann. Rev. Microbiol.* XXX. 247 The purified enzyme contains FAD, which functions when pyocyanine is the electron donor. 1930 *Arch. Dermatol. & Syphilol.* XXI. 151 The case was presented simply as pyoderma. 1930 *Ibid.* XXII. 655 The term 'pyoderma' denotes a purulent infection of the skin due to pyogenic organisms, ordinarily staphylococci. 1936 *Ibid.* XXXIII. 811 Pyodermas and ulcerations of the skin have been described under various names. 1974 Passmore & Robson *Compan. Med. Stud.* III. xix. 102/2 A rare skin lesion which is almost specific for ulcerative colitis and Crohn's disease is pyoderma gangrenosum; intra-epidermal bulla form and contain clear fluid which soon becomes milky and frankly purulent, but is sterile. 1875 *Dental Cosmos* XVII. 278 Your correspondent.. while not very definite in his descriptions, is sufficiently so to indicate the disease as 'pyorrhoea alveolaris' of the French writers. 1921 *Daily Colonist* (Victoria, B.C.) 25 Mar. 7/6 (Advt.), Be suspicious of any tenderness or bleeding of the gums. This is usually the first stage of Pyorrhea—an insidious disease of the gums that destroys the teeth. 1975 J. Symons *Three Pipe Problem* xii. 93 The brick and mortar shaking loose like teeth with pyorrhoea.

pyocyanase (pəi,osəi-ănēⁱz). *Med.* [a. G. *pyocyanase* (Emmerich & Löw 1899, in *Zeitschr. f. Hygiene u. Infektionskrankheiten* XXXI. 10), f. mod.L. *pyocyan-eus* (f. Gr. πύο-ν pus + κυάνεος dark blue), former specific epithet of the source bacterium + *-ase* *-ase.] An antibiotic preparation, orig. thought to be an enzyme, which was obtained from cultures of the bacterium *Pseudomonas aeruginosus* and was formerly used to treat a number of infections, esp. diphtheria.

1900 *Jrnl. Chem. Soc.* LXXVIII. II. 159 Thus pyocyanase, the enzyme of *Bacillus pyocyaneus*, destroys the deadly effect of the diphtheria toxin. 1908 *Lancet* 21 Mar. 899/1 If pyocyanase came in contact with leucocytes, their plasma was dissolved so that the granules and the nuclei only remained and these were..immobilised. 1949 H. W. FLOREY et al. *Antibiotics* I. i. 24 After 1914 the mention of pyocyanase for clinical use almost entirely disappeared from the literature. 1969 *Listener* 5 June 781/1 The experiments of Florey and Chain on pyocyanase..went to show that pyocyanase was a complicated mixture of substances, all equally poisonous to microbes and to mice.

pyramid, *sb.* Add: **1.** Also *Great Pyramid*, the pyramid of the fourth-dynasty pharaoh Cheops at Giza; freq. used (usu. *attrib.*) with reference to its supposed mystical powers. (*Great*) *Pyramid prophecy*, the prediction of events of worldwide importance, based on a belief in the occult significance of the internal measurements of the Great Pyramid; pyramidology.

1802 E. A. KENDAL tr. *Denon's Trav. in Upper & Lower Egypt* I. 102 Herodotus relates that he was informed the great pyramid was the tomb of Chæops. 1859 J. TAYLOR *Great Pyramid* p. vi, I have confined my observations to the Great Pyramid alone. 1877 [in Dict.]. 1937 E. GILL *Let.* 7 July (1947) 389, I did go and see the great Pyramid! and went up & into its middle! Nought but exclamation marks will convey to you its amazing & marvellous mad grandeur! 1948 A. CHRISTIE *Taken at Flood* I. iv. 36 Did you read the book on the Pyramid prophecies I sent you?.. Really explains everything. 1958 L. DURRELL *Balthazar* vi. 150 It gave me the respite I needed to have a go at his heart. It was silent as the Great Pyramid. 1960 M. BOUISSON *Magic* 288 The case of the Great Pyramid prophecy for the date of 20 August 1953 seems to us..inexplicable. 1961 E. WAUGH *Unconditional Surrender* II. v. 145 His objections..were..occult, being in someway based on the dimensions of the Great Pyramid. 1972 *Guardian* 5 Oct. 17/6 Innumerable errors of the Shakespeare cypher and Great Pyramid Prophecy variety. 1976 *Listener* 19 Feb. 199/1 Books on ESP, UFOs, the mystic powers of the Great Pyramid..are..strong runners in the publishing stakes.

5. b. *Finance.* A structure of financial control achieved by a small initial investment; *spec.* in *Stock Exchange*, (*a*) a series of increases in stock acquired from the increased value of stocks already held; (*b*) a system by which a controlling interest in a holding company leads to control of a series of companies and their subsidiaries. orig. *U.S.*

1911 in WEBSTER. 1932 *New Yorker* 14 May 22/1 The bankers who were setting up the biggest financial pyramids of yesterday are replaced by other steel-nerved bankers today. 1971 *Financial Mail* (Johannesburg) 26 Feb. 701/1 A further development came in 1969 when, at the height of the boom, an investment pyramid, Bivec, was floated. It had a 50 per cent interest in BBH and also controlled the properties of both Berzack and Illman. *Ibid.* 703/3 For the cautious investor seeking soundness, the yields are tempting, with the pyramid the more attractive and accessible share.

c. *U.S.* A form of lottery in which each participant recruits two or more further participants. Also *attrib.*

1949 *Washington Post* 22 Mar. 1/6 All night long people would call me up to ask how the pyramids work. *Ibid.* 11/1 He personally believes pyramid clubs are illegal and violate the State lottery laws. 1955 *Britannica Bk. of Year* 489/2 The gullibility of some members of the public gave notoriety to the *Pyramid-Party*, a new version of the old chain-letter game, in which the individual, by paying an initial subscription and by recruiting two new members for the scheme, hoped eventually to reach the top of the 'pyramid', a position which would (theoretically) involve a considerable monetary profit.

d. *attrib.* Used to designate: (*a*) a system of profit involving extensive subcontracting of work; (*b*) a sales market in which each buyer secures the participation of further buyers. See also sense 10 below.

1964 *Daily Tel.* 1 Apr. 24/5 Douglas sub-contracted the work to another firm, made only a plastic cover itself and then charged on the basis of the total cost. The Senate report..was on public hearings in 1962 into 'pyramid' profits of this type. 1970 *Toronto Daily Star* 24 Sept. 12/1 Pyramid sales is a system whereby goods are sold, often at an inflated price, but a reduction in price is offered to purchasers who supply the names of others who buy the product. 1972 *Observer* 5 Nov. 13/2 Pyramid distributors..can and may make more money by recruiting other people to sell products. 1973 *Daily Tel.* 1 Feb. 3 Scotland Yard detectives have obtained a warrant for the arrest of..an American businessman who controls Koscot Interplanetary (U.K.), a pyramid firm selling cosmetics.

8. a. (Later examples.) *spec.* Formations of men or pieces in sports and games.

1899 A. H. QUINN *Pennsylvania Stories* 25 It was Penn's ball. The pyramid started with the cheers of ten thousand back of it. 1948 C. DAY LEWIS *Otterbury Incident* iv. 39 Peter..who is super at gym,..began a routine of tumbling, pyramids, etc. 1969 R. C. BELL *Board & Table Games* II. iii. 58 (*caption*) Initial position of pieces in 'Pyramid'. 1973 *Guardian* 28 Mar. 15/2 There was nothing new about one line [bingo] games or games such as the 'pyramid' and the 'sandwich'.

10. *pyramid-building* (also *fig.*); *pyramid-shaped* adj.; (*spec.* with ref. to the supposed mystical powers of pyramids: cf. sense 1 above) *pyramid energy, freak, power*; **pyramid-selling** *vbl. sb.*, the selling of goods by a pyramid system (see sense 5d (*b*) above); also *pyramid-sell* vb. trans.

1961 L. MUMFORD *City in Hist.* v. 152 All this..was pyramid-building, both in the Egyptian and later Keynesian sense of the words. 1973 C. SAGAN *Cosmic Connection* ix. 67 Harold Urey has perceptively referred to the space program as a kind of contemporary pyramid-building. 1976 *National Observer* (U.S.) 30 Oct. 17/1 (Advt.), Discover pyramid energy... Send $7.50..for this 3"×5" Pyramid Energy Generator. 1977 *Undercurrents* June–July 19/3 The Book of Revelation has remained a happy hunting ground for Jehovah's Witnesses, UFO and Pyramid freaks, and amateur apocalyptics of all denominations. 1976 *Globe & Mail* (Toronto) 24 Apr. 1/1 A book entitled Psychic Discoveries Behind the Iron Curtain says that Russia has been into pyramid power for a long time. *Ibid.* 1/2 Mrs Kelly isn't prepared to say whether it was pyramid power or whether the pyramid created a psychological effect that led to Casey's release from headaches. 1975 D. BLOODWORTH *Clients of Omega* xiv. 135 Why have you been pyramid-selling confidential information on the side to all and sundry? 1972 *Daily Tel.* 30 Mar. 3/1 A company whose cosmetic business is said to involve 'pyramid selling'—a system whereby a franchiser sells to others the right to market goods—was banned in the High Court yesterday from operating its bank account. 1973 D. FRANCIS *Slay-Ride* iv. 51 Always full of get-rich-quick schemes..I even heard him on about pyramid selling once. 1942 PARSONS & STALLARD *Dis. Eye* (ed. 10) xxxii. 657 A pyramid-shaped gauze dressing, with its apex against the wound is firmly applied. 1976 *Billings* (Montana) *Gaz.* 7 July 1-B/1 Green-uniformed troops..patrolled the pyramid-shaped twin buildings where Olympic teams are living.

py·ramid, *v.* [f. the sb.] **1.** (In Dict. s.v. PYRAMID *sb.*)

2. *trans. Finance.* **a.** To accumulate (assets); *spec.* in *Stock Exchange*, to build up (stock) from the proceeds of a series of advantageous sales. Also *absol.*

1901 G. H. LORIMER *Lett. Self-Made Merchant* (1903) v. 64 He'd invent a system for speculating in wheat and go on pyramiding his purchases till he'd made the best that Cheops did look like a five-cent plate of ice cream. 1927 P. MARKS *Lord of Himself* ii. 23 He pyramided his winnings and piled gold on gold..and finally saw himself a millionaire three times over. 1961 'E. LATHEN' *Banking on Death* (1962) xiv. 113 He started pyramiding; put up twenty dollars and got the banks to lend him eighty to a hundred dollars. 1976 *National Observer* (U.S.) 13 Mar. 3/4 (Advt.), And if you are older than 30, it is true that you do not have as long a period of time to pyramid your savings.

b. To set up (a company) as part of a pyramid (see sense *5 b (*b*) of the sb.).

1942 E. PAUL *Narrow St.* xxiv. 212 With the money Stavisky borrowed he floated several companies and sold stock, pyramiding one concern upon the other until he had a finger in practically every financial pie in France. 1955 A. S. LINK *Amer. Epoch* II. xiv. 312 The promoter might pyramid one holding company on top of another almost indefinitely.

3. To distribute (assets or costs), esp. to pass on (costs) by means of a pyramid (sense *5 d (*a*)) of subcontracted work.

1933 *Sun* (Baltimore) 19 Aug. 1/6 Cotton manufacturers are attempting to make abnormal profits by pyramiding their labor costs and the processing tax. 1973 *Time* 25 June 86/2 Southwestern pyramids its commissions to reward the chain of students and executives above the salesman for each sale.

4. *fig.* To arrange in the form of a pyramid; *gen.*, to pile up.

1945 L. MUMFORD in *Archit. Rev.* XCVII. 6/1 Centres like New York, which continue to pyramid their mistakes, will descend with Gadarene swiftness into the abyss. 1948 J. STEINBECK *Russ. Jrnl.* iii. 41 The canned goods are piled in mountains, the champagne and wine from Georgia are pyramided. 1964 GOULD & KOLB *Dict. Social Sci.* 287/1 Power may be pyramided as in the army or relatively evenly divided as in fellowships. 1976 *National Observer* (U.S.) 10 Apr. 21/1 He will pinch powdery tobacco between his thumb and forefinger, pyramid it on the back of his opposite hand, bring it to his nostrils, and sniff.

5. *intr.* To become rich; to acquire greatly increased value or wealth. Also with *up*.

1960 I. JEFFERIES *Dignity & Purity* vii. 134 There is something about the spectacle of..Gobbo pyramiding up on property—houses, flats and so on, that the ordinary person needs must have—that I don't quite like. 1962 K. ORVIS *Damned & Destroyed* ii. 21 The same ounce of heroin..has pyramided in black-market value.

Hence **py·ramided** *ppl. a.*, **py·ramiding** *vbl. sb.* and *ppl. a.*

1930 J. R. AIKEN *Eng. Present & Past* IV. xi. 226 In the words *uppermost, furthermost, innermost, hindermost*, and several others like them, we have the comparative degree combined with the doubly superlative *most*,..causing a triple pyramiding of inflections. 1933 *Sun* (Baltimore) 19 Aug. 1/1 Couzens said he 'knew of no other city in the whole world where there was such an orgy of pyramiding of corporations and the fixing of fictitious values and earnings'. 1941 *Ibid.* 25 Mar. 14/3 Because the projects are in the jurisdictions of differing metal trades unions, the unhappy welders have had to pay tribute not to just one union, not to just one local of one union, but to many metal trades unions and to many locals of metal trades unions. It is this pyramiding of initiation, membership and 'permit' fees on the welders that Mr. Hillman tried to stop. 1951 M. MCLUHAN *Mech. Bride* (1967) 128/1 Production for use? Yes. But for the briefest possible use consistent with the rigging of the market for the pyramiding of profits. 1957 D. L. BOLINGER in *Publ. Amer. Dial. Soc.* XXVIII. 18 Imputations..share the characteristic of the larger class of pyramided Qs to which they belong and may be inverted with little or no change of meaning. *Ibid.* 21 How—why Qs with *that* and their pyramiding are discussed. 1958 *Times* 22 Nov. 7/7 The pyramiding of prosperity, American style, poised more and more on the expanding leisure of a consumers' State. 1967 *Economist* 17 June 1248/4 There is strong feeling against companies..which retain control of an empire with the minimum of capital through 'pyramiding'. 1976 G. W. MCKENZIE *Econ. Euro-Currency System* vi. 78 It is possible that euro-banks may place their dollar assets with other euro-banks. This raises the possibility of a 'pyramiding' chain of inter-bank deposits being created. 1977 *Time* Feb. 33/1 Winter stress can be aggravated by the thought of pyramiding fuel bills.

pyramidal, *a.* (*sb.*). Add: **3. c.** *pyramidal orchid* or *orchis*, an orchid, *Anacamptis pyramidalis*, which is native to Europe and North Africa, and bears dense spikes of deep pink flowers. (Earlier and later examples.)

1778 W. HUDSON *Flora Anglica* (ed. 2) II. 383 Pyramidal Orchis. 1858 A. IRVINE *Illustr. Handbk. Brit. Plants* 316 Pyramidal Orchis... Flowers in a very dense, short, ovate spike, of a beautiful rose colour. 1951 V. S. SUMMERHAYES *Wild Orchids Brit.* iii. 51 The pyramidal orchid is a very beautiful example of perfect adjustment to pollination by butterflies and moths. 1977 M. ALLAN *Darwin & his Flowers* xi. 198 The Pyramidal Orchid..has its parts arranged very differently from *Orchis mascula*.

py:ramido·logy. [f. PYRAMIDO- + -LOGY.] The study of or theories about the mathematical or occult significance of the measurements of the Great Pyramid. Hence **py:ramidolo·gical**; **pyramido·logist.**

1924 DAVIDSON & ALDERSMITH *Great Pyramid* I. p. xi, The reader..will probably have realised that Pyramidology, for over sixty years, has consisted of intuitions, and theories based on these intuitions. 1948 'N. SHUTE' *No Highway* i. 4 'Call yourself a scientist, and you don't know pyramidology!'.. 'Well, I don't. What is it?' 'It's all about the Great Pyramid, in Egypt. Prophecies, and all that sort of thing.' 1954 A. HUXLEY *Let.* 12 Dec. (1969) 719 What may be called the Baconian-pyramidological-cryptographic-spiritualist-theosophical syndrome afflicts a large percentage of the human race. 1964 *Listener* 23 July 117/2 Taylor..believed that he had found various mathematical truths in its [*sc.* the Great Pyramid's] measurements which showed him that the Egyptian priests knew most..of the secrets of the universe... Taylor was, therefore, the founder [1859] of the cult of pyramidology. 1972 *Guardian* 5 Oct. 17/7 Some comfort for would-be biological pyramidologists—cycles are real. 1974 *Nature* 2 Aug. 448/1 Somewhere about here one crosses the transitional zone between statistics and pyramidology. *Ibid.*, Like the pyramidologists he clearly believes that man should be very alert to the hidden meanings buried in the depths of what appear to be fairly straight forward objects or events.

Pyramidon[2] (piræ·midǫn). *Pharm.* Also **pyr-, -one** (-ōᵘn). [a. G. *pyramidon* (W. Filehne 1896, in *Berliner klin. Wochenschr.* XXXIII. 1061), f. *pyrazolon* *PYRAZOLONE with inserted *amid-* (see *AMIDO-).] A white crystalline solid used as an anti-pyretic and analgesic; 4-dimethylamino-1, 5-dimethyl-1-phenylpyrazolin-3-one, $C_{13}H_{17}N_3O$.

Formerly a proprietary term in Britain, and still registered as one in the U.S.

1898 *Jrnl. Chem. Soc.* LXXIV. II. 656 Dimethylamidophenyldimethylpyrazolone, or pyramidone, prepared by Filehne and Spiro, and recommended by them as a substitute for antipyrin..yields bluish-violet colours when oxidised by ferric chloride, nitric and nitrous acids, and the halogens. 1898 *Official Gaz.* (U.S. Patent Office) 6 Dec. 1582/2 Remedy for certain named disease. Farberwerke, vormals Meister, Lucius & Brüning, Höchst-on-the-Main, Germany... *Pyramidon*... Used since December, 1896. 1903 *Brit. Med. Jrnl. Suppl.: Weekly Epitome of Current Med. Lit.* 21 Nov. 79/3 Having heard a good account of pyramidon he proceeded to employ it in a case which he and the nurse were able to observe very closely. 1908 A. BENNETT *Jrnl.* 14 Dec. (1932) I. 300, I..saw a chemist make me a cachet of pyramidon. 1925 W. GERHARDIE *Polyglots* xlix. 358 She took pyramidon for her head, and aspirin for her cold. 1942 *R.A.F. Jrnl.* 30 May 35 A German industrialist in 1942 testified that in the Rhineland it was almost impossible to buy soothing drugs, like pyramidon, any more.

pyran (pəiᵊ·ræn). *Chem.* Also **-ane** (-ēin). [f. *PYR(ONE + -AN, -ANE.] **a.** A heterocyclic compound, C_5H_6O, having a doubly unsaturated six-membered ring consisting of five carbon atoms and one oxygen atom; two isomers (differing in the positions of the double bonds) are possible, only one of which, CH₂CH:CH·CH:CH·O (*γ-pyran*), has been isolated, as a colourless, unstable oil. **b.** Any derivative of either isomer containing a pyran ring. Freq. *attrib.*, as **pyran ring**, a ring (which may be saturated) of five carbon atoms and one oxygen atom.

[1901 *Jrnl. Chem. Soc.* LXXX. I. 559 7-Hydroxy-2-phenyl-4-methylbenzopyran.] 1904 *Ibid.* LXXXVI. I. 816 (*heading*) Properties of oxygen in the pyran ring. 1927 *Ibid.* 3139 The normal sugars are thus seen to be representatives of the parent form indicated by pyran and the labile or γ-sugars have as their parent substance furan. 1952 K. VENKATARAMAN *Chem. Synthetic Dyes* II. xxiv. 742 The γ-pyran ring..occurs in the anthocyanins, the red, violet and blue coloring matters of flowers and fruits. 1953 FRUTON & SIMMONDS *Gen. Biochem.* xvii. 374 The ring in the cyclic form of glucose is related to the heterocyclic compound pyrane. 1962 *Jrnl. Amer. Chem. Soc.* LXXXIV. 2453/1 γ-Pyran is extremely unstable at room temperature, particularly when exposed to air. 1965 KICE & MARVELL *Mod. Princ. Org. Chem.* xix. 360 The two double bonds can be conjugated or unconjugated, and two types of pyran rings are known. *Ibid.*, Like enol ethers in general these rings are unstable, and α-pyran itself is unknown.

pyranometer (pəirănǫ·mĭtəɹ). [See quot. 1916 and -METER.] An instrument for measuring the amount of radiation incident from the entire sky on a horizontal surface.

1916 ABBOT & ALDRICH in *Smithsonian Misc. Coll.* LXVI. No. 7. 2 The name Pyranometer, selected for the instrument we have devised, is taken from Greek words (πῦρ, fire; ἀνά, up; μέτρον, a measure) signifying that which measures heat above. The name was chosen with reference to the fact that the instrument is designed to measure the energy of radiation to or from a complete hemisphere lying above the measuring surface. 1967 *Jrnl. Appl. Meteorol.* VI. 688 (*heading*) An integrating pyranometer for climatological observer stations and mesoscale networks. 1973 *Nature* 16 Feb. 448/2 These data were obtained by subtracting the solar radiation measured by a temperature compensated Eppley pyranometer from the mean hemispherical all-wave radiation obtained from three Fritschen-type net radiometers converted to hemispherical operation.

pyranose (pəiᵊ·r-, pi·rănōᵘz). *Chem.* [f. *PYRAN + -OSE².] A structure containing a saturated pyran ring, frequently assumed by sugars; a sugar having this structure. Freq. *attrib.*

1927 GOODYEAR & HAWORTH in *Jrnl. Chem. Soc.* 3141 On the basis of the conclusions now reached, the formula (A) must be accepted as a general representation of a normal sugar, which might be described as a pyranose. 1953 FRUTON & SIMMONDS *Gen. Biochem.* xvii. 380 Another important hexose found in nature is..fructose... The open-chain form..is in equilibrium with the corresponding pyranose and furanose forms; in solutions of the free sugar, the pyranose form predominates. 1957, 1963 [see *FURANOSE]. 1971 *Nature* 26 Nov. 220/1 The anomers α- and β-D-mannose differ structurally only in that H and OH at the C_1 atom of the pyranose ring are interchanged.

pyranoside (pəir-, piræ·nǒsəid). *Chem.* [f. prec. + *-IDE.] Any glycoside in the pyranose form.

1932 *Jrnl. Chem. Soc.* 2254 Usually the product is also contaminated with small amounts of the corresponding pyranosides. 1934, 1966 [see *FURANOSIDE]. 1970 R. W. MCGILVERY *Biochem.* xxviii. 715 Furanosides and pyranosides can be distinguished by the number of moles of periodate consumed in their oxidation.

pyrazinamide (pəir-, pirăzi-nǎməid). *Pharm.* [f. next + AMIDE.] A white crystalline powder, pyrazine-2-carboxamide, CH:N·CH:CH·N:C·CONH₂, which is used in the treatment of tuberculosis, usu. in conjunction with other drugs.

1952 *Amer. Rev. Tuberculosis* LXV. 515 The antituberculous activity of pyrazinamide..in mice lies between that of streptomycin and that of PAS. 1961 *Lancet* 2 Sept. 533/2 With 1 g. of streptomycin daily on six days a week..and 1–1·5 g. of pyrazinamide in a single daily dose by mouth, 32..of the 57 patients..attained bacteriological quiescence by the end of the year. 1974 R. M. KIRK et al. *Surgery* ii. 29/2 Ethionamide, pyrazinamide, viomyocin and capreomycin may be used in the treatment of tuberculosis when it proves resistant to the usual combination of streptomycin, INAH, and PAS.

pyrazine (pəiᵊ·r-, pi·răzīn). *Chem.* [ad. G. *pyrazin* (A. T. Mason 1887, at suggestion of V. Merz, in *Ber. d. Deut. Chem. Ges.* XX. 267). f. *pyridin* PYRIDINE with inserted *az-* (see AZO-).] A weakly basic white crystalline solid, CH:N·CH:CH·N:CH; any substituted derivative of this.

[1887 *Jrnl. Chem. Soc.* LII. I. 493 Xenylenedihydropyrazine.] 1888 *Proc. Chem. Soc.* IV. 107 The author [*sc.* Mason] adopts Widman's nomenclature..: the 'Ketines'.., 'Pyrazines'.., 'Aldines'.., and all compounds containing a ring of four carbon-atoms and two nitrogen-atoms in para-position are now termed Paradiazines, or Piazines. 1926 H. G. RULE tr. *J. Schmidt's Text-bk. Org. Chem.* III. viii. 699 Pyrazine itself is produced by the condensation of amino-acetaldehyde or amino-acetal. It..has an odour of heliotrope. 1967 M. H. PALMER *Struct. & Reactions Heterocyclic Compounds* iii. 89 Pyrazines are readily prepared by the self-condensation of α-aminoketones. 1970 *Acta Crystallographica* B. XXVI. 979/1 Pyrazine..and $Cu(NO_3)_2$ form an anhydrous crystalline complex with a 1:1 pyrazine to $Cu(NO_3)_2$ ratio.

pyrazole (pəiᵊ·r-, pi·răzǒul). *Chem.* [ad. G. *pyrazol* (L. Knorr 1885, in *Ber. d. Deut. Chem. Ges.* XVIII. 311). f. *pyrrol* PYRROL with inserted *az-* (see AZO-).] A weakly basic white crystalline solid, CH:CH·CH:N·NH; any substituted derivative of this.

1887 *Jrnl. Chem. Soc.* LII. II. 665 In order to examine whether the formation of pyrazoline-derivatives from phenylhydrazine and acids of the acrylic series is analogous to the formation of pyrazole-derivatives from unsaturated ketones, cinnamyl hydrazine was prepared and its products of decomposition investigated. 1926 H. G. RULE tr. *J. Schmidt's Text-bk. Org. Chem.* III. iii. 566 Pyrazole differs strongly from pyrrole in its remarkable stability and more definitely basic character. 1938 C. D. HURD in H. Gilman *Org. Chem.* I. vii. 679 β-Diketones and β-keto esters give rise to..pyrazoles with hydrazines. 1968 [see *PYRAZOLONE]. 1970 *New Scientist* 5 Mar. 447/1 By dosing rats heavily with pyrazole, a compound that inhibits alcohol dehydrogenase, Krebs and his group have shown that quite high levels of alcohol can build up in the bloodstream if not removed.

pyrazoline (pəir-, piræ·zǒlīn). *Chem.* [ad. G. *pyrazolin* (L. Knorr 1887, in *Ann. d. Chem.* CCXXXVIII. 144): see prec. and -INE⁵.] Any of three isomeric compounds, $C_3H_6N_2$, which are dihydro derivatives of pyrazole; *spec.* CH₂·CH₂·CH:N·NH (*2-pyrazoline*), a colourless basic liquid, the only one of the three so far prepared; also, any substituted derivative of any of these compounds.

1887 [see *PYRAZOLE]. 1938 R. C. FUSON in H. Gilman *Org. Chem.* I. i. 30 Pyrazolines are unstable toward heat and decompose to give cyclopropane derivatives. 1961 G. M. BADGER *Chem. Heterocyclic Compounds* v. 223 Pyrazolines are much less stable than the corresponding pyrazoles and are attacked by oxidizing agents. 1967 C. H. JARBOE in R. H. Wiley *Pyrazoles* vi. 177 Examples of each of the three tautomeric pyrazoline structures..are well known, 2-pyrazoline being by far the most common. 1968 [see *PYRAZOLONE].

pyrazolone (pəir-, piræ·zǒlōᵘn). *Chem.* [ad. G. *pyrazolon* (L. Knorr 1887, in *Ann. d. Chem.* CCXXXVIII. 145): see *PYRAZOLE and -ONE.] Any keto derivative of a pyrazoline; *spec.* NH·N:CH·CH₂·CO (*5-pyrazolone*), a weakly acidic white crystalline solid; also, any of their substituted derivatives, several of which on coupling with diazo compounds give rise to commercially important dyestuffs.

1887 *Jrnl. Chem. Soc.* LII. 601 Many of the compounds obtained by the action of ethyl acetoacetate and its derivatives on the primary aromatic hydrazines have already been described by the author as quinizines... The author [*sc.* Knorr] now regards these substances as pyrazolones. 1926 H. G. RULE tr. *J. Schmidt's Text-bk. Org. Chem.* III. iii. 566 The oxygen of pyrazolones may also be removed by heating..with phosphorus oxychloride, when chloro-derivatives are formed. 1961 G. M. BADGER *Chem. Heterocyclic Compounds* v. 226 Theoretically, there are three types of..pyrazolone, but only derivatives of 3-pyrazolone and of 5-pyrazolone are known. 1968 *Kirk-Othmer Encycl. Chem. Technol.* (ed. 2) XVI. 763 Neither pyrazoles nor pyrazolines have found extensive uses, but pyrazolones have been widely utilized as fabric dyes and pigments, as food coloring agents, in color photography, as photographic developing agents, and as pharmaceuticals. 1977 *Lancet* 23 Apr. 905/1 Reversible agranulocytosis has also been described with pyrazolone derivatives (such as phenylbutazone).

Pyrenean, -aean, *a.* and *sb.* Add: The form Pyrenean is now standard. Also **Pyreneean.**

A. *adj.* **a.** (Further examples.)

1865 [see *hair-net* s.v. *HAIR *sb.* 10]. 1892 C. M. YONGE *Old Woman's Outlook* 50 The gorgeous Pyrenean anemone, brilliant scarlet with a black or purple centre. 1895 *Army & Navy Co-op. Soc. Price List* 1089/1 (*caption*) Dressing Jacket, Pyrenean Wool, with girdle. 1925 G. JEKYLL *Colour Schemes for Flower Garden* (ed. 6) i. 2 Some groups of the pale early Pyrenean Daffodil gleam level on the ground. 1931 *Times Lit. Suppl.* 21 May 399/3 General Beatson..is already known to the reading public as the author of three monographs on Wellington's operations in the Pyreneean area. 1936 A. W. CLAPHAM *Romanesque Archit.* vi. 132 It is curious that the XP monogram maintained its place in the Pyrenean countries well into the twelfth century. 1966 M. R. D. FOOT *SOE in France* vii. 155 This donnish figure made some highly unacademic contacts with the Pyrenean smugglers.

b. Pyrenean (mountain or **† guard) dog,** a large, heavily built, white dog of the breed so called, often with grey or brown markings on the head, distinguished by a thick, shaggy, double coat; **Pyrenean sheepdog,** a small, fawn or grey, long-coated sheepdog, often with white markings, belonging to the French breed so called; also = prec.; **Pyrenean wolf-hound** = *Pyrenean mountain dog* above.

1851 H. MAYHEW *London Labour* I. 358/2 The collars most in demand are brass. One man pointed out to me the merits of his stock, which he retailed from 6d each..to 3s—for collars seemingly big enough for Pyrenean sheep dogs. 1865 M. EYRE *Lady's Walks in S. France* xiv. 168 The Pyrenean sheep-dogs equal the largest Newfoundland in size. They are very sagacious and tractable, and usually snow-white. 1871 Pyrenean wolf-hound [see WOLF-HOUND]. 1885 C. M. YONGE *Two Sides of Shield* I. iv. 40 There's Basto, the big Pyrenean dog. 1894 R. B. LEE *Hist. & Descr. Mod. Dogs Gt. Brit. & Ireland* (*Non-Sporting Division*) iv. 107 The so-called Pyrenean guard dog we often see on our show benches, is a dog some one hundred and twenty pounds weight and more. 1922 R. LEIGHTON *Compl. Bk. Dog* v. 62 The beautiful white-coated Pyrenean Dog is also essentially a Mastiff. 1927 E. C. ASH *Dogs* II. IX. ii. 589 There is..the remarkable similarity of the St. Bernard of about that time to the Pyrenean sheep-dog 'Cabbas', the property of her Majesty Queen Victoria. 1931 A. C. SMITH *About our Dogs* xxiii. 364 So great was my admiration of the Pyrenean Mountain Dogs..that it is rather difficult for me to write with becoming restraint. 1945 C. L. B. HUBBARD *Observer's Bk. Dogs* 184 Pyrenean Sheepdog... This is quite a distinct race from the Pyrenean Mountain Dog,..being much smaller and seldom used for guard work. 1964 *Vogue* 15 Apr. 90 A great heap of Pyrenean mountain dog called Addo de Fontenoy. 1978 *Times* 11 Feb. 2/4 Those shaggy dogs masquerading as overcoats called Pyrenean mountain dogs.

B. *sb.* **b.** (Later example.)

1779 H. SWINBURNE *Trav. through Spain* viii. 58 The prospect is..grand..bounded by the mountains of Roussillon. The true Pyreneans appear only through some breaks in that chain.

c. A Pyrenean mountain dog or sheepdog.

1922 R. LEIGHTON *Compl. Bk. Dog* v. 62 The Pyrenean has the same massive body. 1931 A. C. SMITH *About our Dogs* xxiii. 364 The biggest of the Pyreneans are taller than Newfoundlands. 1950 A. C. SMITH *Dogs since 1900* xii. 272 The Pyreneans..have temperaments that make them most pleasing companions. 1971 F. HAMILTON *World Encycl. Dogs* 73 The Pyrenean is the smallest [French sheepdog]. 1976 *Drive* July–Aug. 31/2 Mischka ..is the Wilsons' fifth Pyrenean.

pyrethrin (pəiri·þrin). [a. G. *pyrethrin* (Staudinger & Ruzicka 1924, in *Helvetica Chimica Acta* VII. 181): see PYRETHRUM and -IN¹.] Any of a class of insecticidal terpenoid esters which are obtained from flower heads of *Chrysanthemum cinerariæfolium* and related species, or have been synthesized; *spec.* either of two such compounds (*pyrethrins* I and II) which are the major active principles of pyrethrum powder.

1924 *Chem. Abstr.* XVIII. 1819 A mixt. of the semicarbazones..of pyrethrin I..and pyrethrin II. *Ibid.* 2135 Myrcene and N_2CHCO_2Et give 31% of an acid, $C_{11}H_{16}O_2$, which yields a very slightly active pyrethrin. 1934 *Discovery* Sept. 251/2 Although the difference between pyrethrin I and II is of chemical and biological importance, their action is sufficiently similar for it to be disregarded in practical work. 1951 H. H. SHEPARD *Chem. & Action Insecticides* viii. 146 The pyrethrin content is higher in flowers produced in cool mountain valleys than where the mean temperature is high. *Ibid.* 150 The pyrethrins and cinerins, together referred to generally as pyrethrins, are chemically unstable. 1971 *Inside Kenya Today* Mar. 43/2 Pyrethrin-based insecticides have proved patently safe. 1971 *Nature* 15 Oct. 441/1 Six esters are found in the pyrethrum of which the most important and plentiful are known as pyrethrins I and II.

pyrethroid (pəiri·þroid). *Chem.* [f. as prec. + -OID.] Any substance possessing the terpenoid structure and insecticidal properties characteristic of the pyrethrins.

1954 *Analytical Chem.* XXVI. 604/1 In recent attempts to estimate small quantities of pyrethrins in flour which had been stored in pyrethrum-treated cotton bags..the extraction..was found to be difficult and there were apparent limitations in the accuracy of the analysis of pyrethroids. 1956 *Nature* 25 Feb. 357/2 It has now also been established that the pyrethroids undergo detoxification in the adult housefly. 1971 *Ibid.* 15 Oct. 441/2 Organochlorine and organophosphorus insecticides, many of which tend to break down less quickly than the pyrethroids. 1977 *Protecting World's Crops* (Shell Internat. Petroleum Co.) 2 Insecticides fall into four main categories: chlorinated hydrocarbons, organo-phosphorus, carbamates and synthetic pyrethroids.

pyrethrum. Add: **2.** [Adopted as a generic name in A. Haller *Enumeratio Methodica Stirpium Helvetiæ* (1742) II. 720.] Substitute for def.: A composite plant of the genus formerly so called, now included in the genus *Chrysanthemum* or the subgenus *Tanacetum*. (Later examples.)

1939 A. CUMMING *Hardy Chrysanthemums* ii. 27 Efforts of the writer to cross this species with the garden chrysanthemum have been unsuccessful, which may indicate..that it is a true pyrethrum. 1951 *Dict. Gardening* (R. Hort. Soc.) I. 469/1 The forms [of chrysanthemum] most widely grown are varieties of *C. morifolium*..and of *C. roseum*, generally known in gardens as Pyrethrums. 1964 G. B. SCHALLER *Year of Gorilla* (1965) iv. 86 Many Europeans have settled in the rift mountains. There they grow tea and white-flowered pyrethrum, used in making insect powder. 1976 *Hortus Third* (L. H. Bailey Hortorium) 267/1 Feverfews. These are forms of C[*hrysanthemum*] *Parthenium*, sometimes known as pyrethrum. *Ibid.*, Pyrethrums. These are derived from *C. coccineum*, and bloom in late spring and summer.

b. Substitute for the names of the plants listed in Dict.: *Chrysanthemum* (or *Tanacetum*) *cinerariifolium* or *C. coccineum*. (Further examples.)

1888 *Insect Life* I. 145 Pyrethrum powder was freely used. 1902 *Chambers's Jrnl.* 22 Feb. 191/1 A house where a case of fever had occurred would be fumigated with burning pyrethrum, contiguous dwellings being treated in the same manner. 1955 *Sci. News Let.* 10 Sept. 169/2 Pyrethrum is harmless to human beings and animals, but fatal to flies. 1964 *Which?* Apr. 115/1 The risk is reduced by using..short-lived insecticides, such as derris or pyrethrum. 1978 C. JEFFREY in V. H. Heywood *Flowering Plants of World* 268/3 *Tanacetum cinerariifolium* is the main commercial source of natural pyrethrum, used as an insecticide.

Pyrex (pəiᵊ·reks). Also **pyrex.** [Invented word.

Cf. the following quot.: 1957 *Amer. Speech* XXXII. 290 The assistant secretary of the [Corning Glass] company wrote me as follows: The word *pyrex* is a purely arbitrary word which was devised in 1915 as a trade-mark for products manufactured and sold by Corning Glass Works... We had a number of prior trade-marks ending in the letters *ex*. One of the first commercial products to be sold under the new mark was a pie plate and in the interests of euphonism the letter *r* was inserted between *pie* and *ex* and the whole thing condensed to *pyrex*.]

The proprietary name of a hard, heat-resistant, borosilicate glass. Freq. *attrib.*

1915 *Amer. Cookery* Aug.–Sept. 159 (Advt.), Pyrex ('fire-glass') Glass Dishes for Baking. *Ibid.*, Pyrex is a new-process glass, fire-proofed to withstand the heat of the hottest oven. 1916 *Official Gaz.* (U.S. Patent Office) 1 Aug. 245/1 Corning Glass Works, Corning, N.Y. Filed July 10, 1915 *Pyrex*... Glass. Claims use since May 20, 1915. 1917 *Trade Marks Jrnl.* 10 Jan. 30 *Pyrex*... All goods included in Class 15. Corning Glass Works..New York, U.S.A.... 11th October 1916. 1927 *Glasgow Herald* 1 July 10 Housewives no longer use iron pots and pans. Their kitchenettes are bright with aluminium and pyrex ware. 1932 AUDEN in *Rev. Eng. Stud.* (1978) Aug. 295 Tea was served, Poured by the secretary from a pyrex teapot. 1932 *Discovery* June 199/2 The door itself is fitted with a Pyrex glass window for inspection of the flame. 1958 *Times Lit. Suppl.* 7 Feb. 70/2 Fashionable art critics' jargon which attributes organic qualities to Mr Moore's bronzes or Mr Frank Lloyd Wright's pillars of Pyrex glass. 1961 R. M. DASHWOOD *Provincial Daughter* 107 Odd pieces of Cornish Ware, Pyrex, Willow pattern. 1976 'Z. STONE' *Modigliani Scandal* IV. iv. 178 Moore took out his false teeth..and dropped them in a Pyrex beaker.

Pyribenzamine (piribe·nzămīn). *Pharm.* [f. PYRI(DINE + BENZ(O- + AMINE.] A proprietary name in the U.S. for the antihistamine tripelennamine hydrochloride.

1946 *Proc. Soc. Exper. Biol. & Med.* LXII. 65/1 B-dimethylaminoethyl benzhydryl ether (Benadryl) and pyridil-N¹-benzyl-N-dimethylethylenediamine (Pyribenzamine)..have proven effective in histamine, and anaphylactic shock and in the management of some allergic conditions in man. 1946 *Official Gaz.* (U.S. Patent Office) 27 Aug. 511/1 Ciba Pharmaceutical Products, Inc., Summit, N.J.... *Pyribenzamine* for preparation indicated for use in allergic conditions. Claims use since July 12, 1945. 1962 F. J. FERGUSON *Drug Therapy* xliv. 346 Recommended agents are promethazine, or Phenergan (if marked sedation is desired), tripelennamine, or Pyribenzamine (for lesser degree of sedation), and phenindamine (Thephorin). 1974 M. C. GERALD *Pharmacol.* ii. 30 Antihistamines such as tripelennamine (Pyribenzamine) do not cure allergies, but they do prevent the agonist, histamine, from initiating an allergic response.

pyridazine (piri·dăzīn). *Chem.* [f. PYR(O- + -ID⁴ + *AZINE.] The weakly basic colourless liquid $C_4H_4N_2$, 1,2-diazine; also, any derivative of this.

1895 *Jrnl. Chem. Soc.* LXVIII. 301 When it is heated..with 5 per cent. hydrochloric acid at 200°, it yields pyridazine. 1926 H. G. RULE tr. *Schmidt's Text-bk. Org. Chem.* VIII. i. 697 Pyridazines can often be prepared by the oxidation of their dihydro-derivatives. 1975 R. F. BROWN *Org. Chem.* xxviii. 902 The three isomeric diazines are 1,2-diazabenzene (pyridazine), 1,3-diazabenzene (pyrimidine), and 1,4-diazabenzene (pyrazine).

pyridine. Add: Now with pronunc. (pi·ridīn). **b. pyridine nucleotide,** either of the two oxidizing co-enzymes di- and triphosphopyridine nucleotide (co-enzymes I and II); (sometimes with added *di-* or *triphosphate* respectively).

[1936 *Chem. Abstr.* XXX. 8262 Cozymase is an adenine-pyridine-nucleotide, which is a H-transporter because the pyridine changes to dihydropyridine.] 1937 *Proc. R. Soc.* B. CXXII. 355 The question then arose whether or not the action of pyridine nucleotide triphosphate was specific and if it could be replaced by pyridine nucleotide diphosphate. *Ibid.* 359 The two pyridine nucleotides were active as 'V' factor, the limit of the activity of pyridine nucleotide triphosphate being about 1/600,000,000. 1951 WHITBY & HYNES *Med. Bacteriol.* (ed. 5) xvi. 282 The V factor is the di- or tri-phosphate of pyridine nucleotide, coenzymes that act as intermediate hydrogen receptors in cytochrome and other respiratory mechanisms. 1966 S. P. COLOWICK et al. in Florkin & Stotz *Comprehensive Biochem.* XIV. i. 4 The pyridine nucleotides are thus concerned in virtually all biosynthetic and degradative processes involving oxidation-reduction steps.

pyridostigmine (pi:ridosti·gmīn). *Pharm.* [ad. G. *pyridostigmin*: see PYRIDINE and *neostigmine* s.v. *NEO- 1 b.] The ion $(CH_3)_2N·CO·O·C_5H_4N^+CH_3$ or its bromide derivative, a whitish crystalline powder similar in action to neostigmine but weaker and longer-acting and giving rise to fewer side-effects.

1953 *Jrnl. Amer. Med. Assoc.* 12 Sept. 175/1 Since 1948, a pyridine homologue of neostigmine (Pyridostigmine), was given a therapeutic trial in 23 patients with myasthenia gravis associated with pseudoparalysis. Results of animal experiments..showed that Pyridostigmine is five times less toxic than neostigmine. 1961 D. DUNLOP et al. *Textbk. Med. Treatm.* (ed. 8) 875 Myasthenia Gravis... The drug which is most generally useful is pyridostigmine bromide, B.P.C. (Mestinon)... Pyridostigmine gives a very satisfactory smooth control with few parasympathomimetic side-effects, and the long action without sudden withdrawal effects makes it particularly suitable for treatment during the night. 1980 *Sci. Amer.* Apr. 38/2 Reversible inhibitors of acetylcholinesterase such as pyridostigmine (administered by means of pills swallowed a short time before nerve-gas exposure) may improve the prognosis.

pyridoxal (piridǫ·ksæl). *Biochem.* [f. as next + *-AL².] One of the forms of vitamin B_6 (derived from pyridoxine by oxidation of the 4-hydroxymethyl group to aldehyde), which usu. occurs in mammals as the phosphate ester and is a co-enzyme in a number of metabolic processes, notably transamination. Cf. *PYRIDOXINE.

1944 E. E. SNELL in *Jrnl. Biol. Chem.* CLIV. 313 (*caption*) Pyridoxal. *Ibid.*, When pyridoxamine or pyridoxal is used as a standard of comparison with *S*[*treptococcus*] *lactis*, values for the 'B_6' content of natural materials are obtained similar to those indicated by yeast assay, instead of the absurdly high values obtained against a pyridoxine standard. 1955 *Sci. News Let.* 1 Oct. 211/1 The cancer cells..need these seven vitamins: choline, folic acid, nicotinamide, pantothenate, pyridoxal, riboflavin and thiamine. 1966 E. R. M. KAY *Biochem.* xxiii. 302 Transamination reactions appear to involve a transfer of the amino group to the pyridoxal, forming pyridoxamine and a keto acid. 1970 [see *PYRIDOXAMINE]. 1974 *Nature* 9 Aug. 502/2 Pyridoxine, the major dietary form of vitamin B_6, is rapidly converted in the body to pyridoxal phosphate.., the coenzyme form.

pyridoxamine (piridǫ·ksămīn). *Biochem.* [f. as next + AMINE.] One of the active forms of vitamin B_6 (related to pyridoxine by replacement of the 4-hydroxymethyl group by an aminomethyl group), which is usu. present in mammals as the phosphate ester and is a co-enzyme in protein metabolism. Cf. *PYRIDOXINE.

1944 E. E. SNELL in *Jrnl. Biol. Chem.* CLIV. 315 (*caption*) Pyridoxamine. 1944 [see *PYRIDOXAL]. 1950 *Thorpe's Dict. Appl. Chem.* (ed. 4) X. 316/2 Schlenk and Fisher..working with the glutamic-aspartic transaminase system from pig-heart, have suggested that the reversible interconversion of pyridoxal and pyridoxamine in the prosthetic group forms the basis of the enzyme system. 1961 [see *PYRIDOXINE]. 1966 [see *PYRIDOXAL]. 1968 J. MARKS *Vitamins in Health & Dis.* 94/2 The B_6 group are rapidly converted in the body into the co-enzymes pyridoxal phosphate and pyridoxamine phosphate... These co-enzymes play an essential role in protein metabolism. 1970 R. W. MCGILVERY *Biochem.* vi. 120 The substituted pyridine ring of pyridoxal phosphate..cannot be synthesized by vertebrates, and its dietary precursors are lumped together as vitamin B_6, which includes pyridoxal, pyridoxamine, and the corresponding alcohol, pyridoxine.

pyridoxic (piridǫ·ksik), *a.* *Biochem.* [f. as next + -IC.] *pyridoxic acid* (more explicitly *4-pyridoxic acid*), an inactive oxidized derivative of pyridoxine (the 4-hydroxymethyl group having been oxidized to carboxyl), which is the form in which excess vitamin B_6 is usu. excreted.

1944 HUFF & PERLZWEIG in *Jrnl. Biol. Chem.* CLV. 355 A fluorescent compound appearing in urine after the ingestion of pyridoxine was isolated and identified as 2-methyl-3-hydroxy-4-carboxy-5-hydroxymethylpyridine (4-pyridoxic acid). 1950 W. SHIVE in R. J. Williams et al. *Biochem. of B Vitamins* D. viii. 657 Pyridoxic acid is the chief metabolic product of either pyridoxine, pyridoxal or pyridoxamine. 1968 BAKER & FRANK *Clinical Vitaminol.* vii. 79 In subjects on a normal diet, pyridoxic acid excretion accounted for only about half the intake of vitamin B_6.

pyridoxine (piridǫ·ksīn, -in). *Biochem.* Also **-in.** [f. PYRIDINE with inserted OX- 1.] One of the three common forms of vitamin B_6, a colourless, weakly basic, crystalline solid which occurs esp. in cereals, liver oils, and yeast, is also manufactured commercially, and is readily interconverted inside the body to the other forms of the vitamin, pyridoxal and pyridoxamine, with which it usu. occurs; 3-hydroxy-4,5-di(hydroxymethyl)-2-methylpyridine, $C_8H_{11}NO_3$; = *ADERMIN.

1939 [see *ADERMIN]. 1941 *Science* 5 Dec. 545/2 *In vitro* experiments with pyridoxine..failed to demonstrate the presence of the unknown substance. 1948 *New Biol.* IV. 24 Roots of a hybrid [tomato] are apparently able to synthesise pyridoxine and nicotinamide better than the roots of its parents. 1961 *Lancet* 16 Sept. 623/1 Pyridoxine deficiency induced by these drugs is an alternative mechanism invoked to explain the liver damage, and Coursin..found low levels of circulating pyridoxal and pyridoxamine in a patient showing iproniazid toxicity. 1970 L. J. HARRIS in J. Needham *Chem. of Life* vi. 164 In 1934, a second component of the vitamin B_2 complex, at first called vitamin B_6..and now generally known as pyridoxin, was identified. 1972 *Materials & Technol.* V. xix 679 Pyridoxine hydrochloride is produced commercially by synthesis... It is used in multi-vitamin preparations and in larger quantities to counteract drug effects. 1974 M. C. GERALD *Pharmacol.* x. 182 Deficiencies of pyridoxine (vitamin B6) and niacin cause convulsions in infants and pellagra, respectively.

Now also called **pyrido·xol** [-OL].

1959 *Recueil des Travaux Chimiques des Pays-Bas* LXXVIII. 226 (*heading*) An improved synthesis of vitamin B_6 (pyridoxol). [*Note*] The name pyridoxol was proposed at the 17th IUPAC congress at Stockholm in 1953. 1966 MAHLER & CORDES *Biol. Chem.* viii. 341 Numerous nutritional studies..have established that a deficiency of pyridoxol (pyridoxine)..results in many lesions in protein metabolism. 1973 ZEFFREN & HALL *Stud. Enzyme Mechanisms* viii. 157 The nutritional factor vitamin B_6 was known in 1934 to be involved in protein metabolism. Chemical studies showed that its structure was 3-hydroxy-4,5-di(hydroxymethyl)-2-methylpyridine.., called pyridoxol.

pyriform, *a.* Delete '(Only in scientific or technical use.)' and add further examples.

1913 *Cunningham's Text-bk. Anat.* (ed. 4) 164 The apertura piriformis [*ed.* 3 (1909): pyriformis]..which lies below and in part between the orbits, is of variable shape and size—usually piriform [*ed.* 3: pyriform], it tends to be long and narrow in Europeans. 1928 V. G. CHILDE *Most Anc. East* v. 117 The piriform mace has a very long history in Babylonia. 1932 *Times Lit. Suppl.* 20 Oct. 743/1 His performances on the putting green with an archaic piriform wooden putter were marked by a deadly accuracy. 1948 A. BRODAL *Neurol. Anat.* x. 327 The olfactory fibres ultimately transmitting olfactory impulses to the cerebral cortex are those passing to the piriform lobe. 1948, 1962 [see *PALÆOPALLIUM]. 1971 *Country Life* 15 July 184/1 Shapes of fashionable cruet glass followed those of silver. Castors were at first pyriform. 1973 M. CROWELL *Greener Pastures* 167 Birds like the murre..lay triangular, or pyriform, eggs that roll in tight circles. 1973 *Nature* 3 Aug. 314/2 A small snout with a small piriform aperture projected from a broad and very flat face.

pyrimethamine (pirime·þămīn). *Pharm.* [f. *PYRIM(IDINE + ETH(YL + AMINE.] A white crystalline solid, 2,4-diamino-5-*p*-chlorophenyl-6-ethylpyrimidine, $C_{12}H_{13}N_4Cl$, which is given orally for the prophylaxis and suppression of malaria.

1953 *Brit. Med. Jrnl.* 31 Jan. 253/1 Pyrimethamine..is the latest of the new synthetic anti-malarial drugs to be tested by us... This compound is produced by Burroughs Wellcome & Co. under the trade name 'daraprim'. 1961 *New Scientist* 20 Apr. 120/1 Pyrimethamine resistance has appeared in some districts in East Africa. 1966 *Lancet* 24 Dec. 1382/1 The combination of dapsone to shorten the red-blood-cell life-span and pyrimethamine to inhibit erythropoiesis has now been used successfully on five patients with secondary polycythæmia. 1976 *Ibid.* 4 Dec. 1257/1 When a dose of 25 mg weekly is taken for malarial prophylaxis, pyrimethamine is unlikely to produce adverse hæmatological effects except in those actually or potentially folate deficient.

pyrimidine (piri·midīn, pəir-). *Chem.* Also **† -in.** [ad. G. *pyrimidin* (A. Pinner 1885, in *Ber. d. Deut. Chem. Ges.* XVIII. 760) f. *pyridin* PYRIDINE with inserted *-im* (f. *imid* IMIDE in Dict. and Suppl.).] A colourless, crystalline basic solid, CH:N·CH:N·CH:CH; any substituted derivative containing this ring structure, *spec.* cytosine, thymine, or uracil, pyrimidines found in nucleic acids, etc. Freq. *attrib.*

1885 *Jrnl. Chem. Soc.* XLVIII. II. 751 The author substitutes the formula R·C⟨N·C(OH)⟩CH / ⟨N:CMe⟩ for that previously assigned, terming the nucleus pyrimidine. 1886 *Ibid.* L. 45 The author has tried to prepare the corresponding pyrimidines from various amidines. 1899 *Ibid.* LXXVI. I. 639 Pyrimidine..melts at 20–22°..; it has a penetrating, stupefying odour. 1924 *Nature* 12 Apr. 524/1 In these bodies [*sc.* the nucleic acids] a carbohydrate group is associated with a phosphoric acid group and also with a purin or pyrimidin base. 1947 *Sci. News* IV. 109 The I.C.I. men started on something quite different: the pyrimidine ring, an assemblage of atoms found in uric acid, in the chromosomes of the nucleus of living cells, and widely in all living things. 1952, 1954 [see *PURINE]. 1960 *New Biol.* XXXI. 40 The two pyrimidines, which are smaller molecules than the purines, are named thymine and cytosine. 1962 D. J. BROWN *Pyrimidines* iv. 116 Until recently, pyrimidine was an exceedingly rare substance, but several good methods of preparation have now made it readily available in quantity. 1976 *Ann. Rev. Microbiol.* XXX. 92 Other auxotrophs such as those requiring amino acids, purines, and pyrimidines do not assume a colonial or semicolonial growth habit if the biochemical supplement in the medium is limiting.

pyritic, *a.* Add: *spec.* applied to a process for smelting sulphide copper ores with pyrites so that oxidation of the latter produces all the necessary heat.

1897 HUNTINGTON & MCMILLAN *Metals* (new ed.) 331 This is the process known as pyritic smelting, and there is some prospect that it may take a prominent position in the metallurgy of copper, although at present it is mainly employed in the extraction of gold and silver from complex ores. 1926 D. M. LIDDELL *Handbk. Nonferrous Metallurgy* II. xxvii. 944 There are three distinct processes in blast-furnace smelting: (1) The reduction process... The blast oxidizes the carbon of the coke and but little of the sulphur in the ore... (2) The pyritic process, in which raw massive sulphides are smelted in a highly oxidizing atmosphere without the addition of carbonaceous fuel... (3) The partial- or semipyritic process. *Ibid.* 945 The essential requirements for pyritic smelting are siliceous material which is high in free silica, and heavy pyrite ore... The pyrite ore not only furnishes the heat for the operation, but also the sulphur for the matte required. 1974 D. AVERY *Not on Queen Victoria's Birthday* xiv. 280 'Pyritic smelting', developed by Lawrence Austin and Robert Sticht at [*sic*] Montana between 1887 and 1896 and introduced at Mount Lyell in Tasmania in the latter year,..provided a means of producing much cheaper copper.

pyro-. Add: **1. pyroce·llulose,** a form of nitrocellulose containing slightly less nitrogen than gun-cotton (see quots.); **py·roclast** *Geol.*, a pyroclastic rock fragment; **pyroclastic** *a.* (earlier and later examples); also as *sb.*, a pyroclastic rock or rock fragment; **pyromagnetic** *a.*, more widely, pertaining to or exhibiting pyromagnetism; (further examples); **pyroma·gnetism**; [ad. G. *pyromagnetismus* (W. Voigt 1901, in *Nachrichten v.d. K. Ges. d. Wissensch. zu Göttingen* (*Math.-phys. Klasse*) I. 1)], magnetism that is dependent on the temperature of the material; **pyromania** (later examples); **pyromaniac** *sb.* (later examples), *a.* (example); **pyromaniacal** *a.* (example); **pyroma·nic** *a.*, of or relating to pyromania; **pyrome·tallurgy,** metallurgy in which high temperatures are employed for the extraction of metals; hence **py:rometallu·rgical** *a.*, **pyrome·tallurgist**; **py·rosphere** *Geol.* = *BARYSPHERE; **pyrosy·nthesis,** synthesis by the action of heat; hence **pyrosynthe·tic** *a.*

1906 E. M. WEAVER *Notes Military Explosives* iv. 123 *Pyrocellulose*, a soluble nitrocellulose of so called definite percentage of N(12·4), corresponding to the molecular formula, $C_{30}H_{38}(NO_2)_{12}O_{25}$, claimed to have been produced by Mendeléef; it possesses just sufficient content of O to burn all of the C to CO, the H to H_2O. 1920 O. W. WILLCOX in A. Rogers *Industr. Chem.* (ed. 3) xlvi. 1076 Nitrocellulose of from 12·50 to 12·70 per cent of nitrogen is called pyrocellulose, or simply pyro, and is the material from which smokeless powder for cannon is made. 1951 KIRK & OTHMER *Encycl. Chem. Technol.* VI. 36 Various types or grades of nitrocellulose are characterized by their nitrogen contents and the following names are used: pyroxylin, 8–12% nitrogen; pyrocellulose, 12·6±0·1% nitrogen; guncotton, 13·3±0·1% nitrogen. 1974 *Encycl. Brit. Macropædia* VII. 87/1 It was..the most important type of smokeless powder used by the Allies in World War I. It was made from a nitrocotton of relatively low nitrogen content, called pyrocellulose, because that type is quite soluble in ether-alcohol. 1920 A. HOLMES *Nomencl. Petrol.* 193 *Pyroclasts*, a general term for fragmental deposits of volcanic ejectamenta. 1934 *Bull. Amer. Assoc. Petroleum Geologists* XVIII. 1573 The bentonite beds in the basal McLure shale represent purer beds of originally vitric pyroclasts. 1944 A. HOLMES *Princ. Physical Geol.* XX. 443 The great clouds of gases, vapours, and pyroclasts that are the most conspicuous feature of explosive eruptions may be luminous or dark. 1972 *Nature* 21 Jan. 157/1 This eruption was extremely violent: an estimated 1·4 km³ of pyroclast flow and fall was emitted. 1887 J. J. H. TEALL in *Geol. Mag.* Decade III. IV. 493, I venture to suggest that..we should distinguish between the three types of clastic rocks at present recognized by using the terms epiclastic, cataclastic, and pyroclastic... *Pyroclastic*—Fragmental rocks of volcanic origin. The same terms may be applied to the structures which characterize the rocks in question. 1903 *Bull. U.S. Geol. Survey* No. 213. 73 The gravels of Slate Creek contain..a certain proportion of material derived from the older quartzites, pyroclastics, and granite intrusives occurring on the south side of its lower valley. 1939 W. H. TWENHOFEL *Princ. Sedimentation* viii. 291 The coarse-grained pyroclastics fall near the places of expulsion. 1976 P. FRANCIS *Volcanoes* iv. 127 All volcanic rocks..may turn up either as lavas or pyroclastics. *Ibid.* v. 158 Pumice is the best-known of all pyroclastic..rocks. 1901 Pyromagnetic [see *piezomagnetic* adj. s.v. *PIEZO-]. 1931 S. R. WILLIAMS *Magn. Phenomena* v. 164 A pyromagnetic crystal must show a magnetic moment at room temperature. 1975 *Jrnl. Appl. Physics* XLVI. 2250/1 The low-frequency pyromagnetic effect has been used to study the behavior of a ferromagnetic material both in the low-temperature region as well as near its Curie temperature. 1901 Pyromagnetism [see *piezomagnetism* s.v. *PIEZO-]. 1956 *Soviet Physics: JETP* III. 436/2 Recently the opinion has been expressed that pyromagnetism, piezomagnetism, etc., are impossible. 1973 *Jrnl. Appl. Physics* XLIV. 424/1 Another interesting application of pyromagnetism revealed by the present study is the possibility of verifying, or determining, the relationships between crystalline anisotropy constants and the magnetization near the critical point. 1895 Pyromania [see *ONOMATOMANIA b]. 1937 *Times* 7 Oct. 11/2 Mr. A. Lawson-Walton ..said that there was no evidence of spite.., and it seemed that the accused had a kind of pyromania and delighted in making fires. 1929 W. S. SADLER *Mind at Mischief* x. 141 We have the same condition in the case of certain types of pyromaniacs. 1967 *Listener* 6 Apr. 466/2 Jeanne Moreau lends to the role..more credibility than was apparent in her full-length portrayal of the pyromaniac school-mistress. 1972 G. W. KISKER *Disorganized Personality* (ed. 2) viii. 260/1 The defiance is usually aimed at the police in their role of father image. Pyromaniacs of this type go about setting fires indiscriminately. 1873 G. H. LEWES *Probl. Life & Mind* 1st Ser. I. 234 Phases which manifest homicidal, kleptomaniacal, and pyromaniacal instincts. 1926 J. I. SUTTIE tr. *Ferenczi's Further Contrib. Theory & Technique Psycho-Anal.* xxxi. 258 There were quite a number [of cases] in which incendiaries set fire to their *beds*, as though to indicate the..enuristic primitive source of their pyromanic character trait. 1968 G. JONES *Hist. Vikings* I. ii. 52 According to *Ynglinga Saga*, this pyromanic imbecility cost the Ynglings their realm of Uppsala. 1917 E. OBERG *Machinery's Encycl.* II. 204/2 There are three methods by means of which copper may be obtained from its ores: 1. By the pyro-metallurgical or dry method. 1971 *Daily Tel.* 29 Apr. 25 (Advt.), Applicants should have..a strong pyrometallurgical background and a minimum of 3 years' smelting or related development experience. 1960 *Times* 6 Apr. 3/1 (Advt.), Applications are invited..for appointment to the posts of Pyrometallurgist, [etc.]. 1974 *Daily Tel.* 2 May 23 (Advt.), The Pyrometallurgist will work in the Smelter, which has an annual production capacity of approximately 84,000 tonnes of copper. 1909 WEBSTER, Pyrometallurgy. 1957 *New Scientist* 26 Sept. 20/1 Pyrometallurgy, the study of metals in the molten state, may find an application in the treatment of metals which are 'hot' in the sense of being highly radioactive. 1973 R. D. PEHLKE *Unit Processes of Extractive Metallurgy* i. 5 Following the mining and concentration of minerals, their extraction is accomplished by application of chemical metallurgy in one of the three areas of extractive metallurgy: pyrometallurgy, hydrometallurgy, or electrometallurgy. 1900 *Geogr. Jrnl.* XV. 88 A coloured diagram showing an ideal section of the Earth on the hypothesis that within the solid lithosphere lies a pyrosphere of intensely high temperature. 1963 D. W. & E. E. HUMPHRIES tr. *Termier's Erosion & Sedimentation* i. 1 Glyptogenesis is the process of sculpturing of the lithosphere through the agency of the atmosphere, hydrosphere, biosphere and pyrosphere. 1947 *Bull. Geol. Soc. Amer.* LVIII. 1232 (*heading*) Pyrosyntheses of telluride minerals. 1955 *Jrnl. Amer. Chem. Soc.* LXXVII. 1048/2 (*heading*) Pyrosynthesis of aspartic acid and alanine from citric acid cycle intermediates. 1961 *Amer. Mineralogist* XLVI. 823 Differential thermal pyrosynthesis may be considered a modification of differential thermal analysis which allows investigation under closed system conditions. A record is obtained of thermal reactions which occur during synthesis by heating elemental constituents to the fusion point. 1956 *Amer. Scientist* XLIV. 357 Pyrosynthetic experiments.

2. pyrobe·lonite *Min.* [ad. G. *pyrobelonit* (G. Flink 1919, in *Geol. Föreningens i Stockholm Förhandl.* XLI. 436), f. Gr. βελόνη needle], a basic vanadate of manganese and lead, $MnPb(VO_4)(OH)$, occurring as red, transparent, needle-shaped crystals; **pyrobi·tumen,** any of a class of native hydrocarbons that differ from the bitumens proper in being relatively hard, infusible, and insoluble in organic solvents; hence **py:robitu·minous** *a.*; **pyrochlore** (further examples); also, any of a group of minerals that includes pyrochlore, microlite, betafite, and obruchevite, the members of which have the general formula $A_2B_2O_6(O,OH,F)$, where A may be sodium, potassium, calcium, cerium, or certain other elements, and B may be niobium, tantalum, titanium, or certain other elements; freq. *attrib.* in *pyrochlore group*; **py·rolite** [-LITE], † (*a*) an artificial rock (see quot. 1848) (*obs.*); (*b*) a mixture proposed as the primitive material of the earth's upper mantle (see quots. 1962, 1975); **pyrophillite,** var. *pyrophyllite*.

1920 *Chem. Abstr.* XIV. 1097 (*heading*) Pyrobelonite, a new lead-manganese vanadate from Långbanshyttan. 1969 *Canad. Mineralogist* X. 117 The specimen..was Harvard 94831 from the type locality, Långban, Sweden. It consists largely of massive to well-crystallized hausmannite in contact with, and cut by, calcite. The pyrobelonite occurs as very fine grains with a few small crystals (commonly < 100 μ in largest dimension) primarily in the hausmannite. 1903 C. RICHARDSON in *Science* 13 Mar. 420/1 The evidence thus obtained has been carefully analyzed, and the following classification of the native bitumens deduced: Gas... Petroleum... Maltha. Solid Bitumens... Pyrobitumens: Practically insoluble in chloroform or heavy petroleum hydrocarbons. 1951 K. K. LANDES *Petroleum Geol.* iv. 127 The solid hydrocarbons may be subdivided into four main groups: petroleum bitumens, pyrobitumens, disseminated bitumens, and oxygen-bearing bitumens. 1965 Pyrobitumen [see *IMPSONITE]. 1918 H. ABRAHAM *Asphalts & Allied Substances* xi. 149 They [*sc.* pyrobitumens] are grouped into five classes, viz.: elaterite, wurtzilite, albertite, impsonite, and asphaltic pyrobituminous shales. 1937 *Bull. Amer. Assoc. Petroleum Geologists* XXI. 122 Regardless of the possible economic value of the Brazilian algal deposits and other pyrobituminous sediments, those now forming in fresh-water ponds and the related geologically young deposits have great scientific interest. 1906 J. P. IDDINGS *Rock Minerals* II. 464 Pyrochlore group. Pyrochlore, RNb_2O_6. $R(Ti,Th)O_3$.NaF. Koppite, $R_2Nb_2O_7$.½NaF. Microlite, $Ca_2Ta_2O_7$.pt. 1941 *Amer. Mineralogist* XXVI. 504 A study of the available analyses of pyrochlore shows that both cerium and titanium are invariably present in appreciable amounts and must therefore be regarded as essential constituents. *Ibid.* 505 Koppite... Winchell (1933) describes this mineral as 'a pyrochlore containing K' while Brandenberger (1931) states that koppite should be regarded as an iron-columbium pyrochlore. 1959 [see *PANDAITE]. 1977 *Amer. Mineralogist* LXII. 404/1 The pyrochlore group comprises those multiple cubic oxides having the following characteristics: (*a*) essential amounts of niobium, tantalum, and titanium, either individually or in combination (*b*) the space group *Fd3m*, (*c*) the pyrochlore structure as defined by Gaertner (1930) and Brandenberger (1931), and (*d*) the general formula $A_{2-m}B_2O_6(O,OH,F)_{1-n}·pH_2O$... The recommended subgroups are: Pyrochlore Subgroup in which Nb+Ta>2 Ti and Nb>Ta. [Etc.] 1848 *Mining Jrnl.* 4 Nov. 521/1 Mr. Twining's object is to form, by chemical means, a comprehensive series of petreous substances which he proposes to designate..pyrolite or artificial lava, as..being of igneous origin. 1962 A. E. RINGWOOD in *Jrnl. Geophysical Res.* LXVII. 860/1 Immediately below the M discontinuity, the mantle consists dominantly of dunite and peridotite... This zone passes downward..into the primitive 'pyrolite'. [*Note*] Peridotite is an unsatisfactory name for the hypothetical primitive mantle material, chemically equivalent to 1 part basalt plus 4 parts of dunite. Since a rock of this composition would crystallize dominantly as a mixture of olivine and pyroxene, the name 'pyrolite' is suggested. 1975 *Sci. Amer.* Mar. 56/1 Assigning an appropriate chemistry to the residual peridotite, one arrives at the hypothetical composition of the upper mantle. Pyrolite (pyroxene-olivine rock) is the name given to one of these hypothetical peridotites. 1946 J. R. PARTINGTON *Gen. & Inorganic Chem.* xviii. 506 Montmorillonite shows the same X-ray pattern as pyrophillite, which occurs crystalline in slates. 1975 TINDALL & THORNHILL *Blandford Rock & Mineral Guide* II. 96 This structure can extend indefinitely in a two-dimensional network or 'sheet'; it is found, for example, in the mineral pyrophillite, $Al_2Si_4O_{10}(OH)_2$.

3. a. pyropho·sphate, a salt or ester, or the anion, of pyrophosphoric acid; a group or linkage formed from two condensed phosphate groups.

1836–41, 1869 [in Dict.]. 1912 E. FEILMANN tr. *Molinari's Inorg. Chem.* 348 The pyrophosphates..give a precipitate with copper salts, which is soluble in excess of pyrophosphate. 1950 N. V. SIDGWICK *Chem. Elements* I. 746 Ethyl pyrophosphate $Et_4P_2O_7$ can be made from the silver salt and ethyl iodide. 1957 G. E. HUTCHINSON *Treat. Limnol.* I. xii. 728 Though pyrophosphate plays an important role inside the organism, it is easily hydrolyzed and only orthophosphate is likely to be of importance in the environment. 1970 AMBROSE & EASTY *Cell Biol.* vii. 248 There is another type of reaction, catalysed by enzymes known as phosphorylases, in which a sugar phosphate reacts with another sugar to form a disaccharide and inorganic pyrophosphate.

b. pyroca·techol = *CATECHOL, *pyrocatechin* s.v. PYRO- 3 b.

1890 *Proc. Chem. Soc.* VI. 90 The very high price of pyrocatechol renders it desirable to discover improved methods of preparing it. 1932 I. D. GARARD *Introd. Org. Chem.* xiv. 199 Pyrocatechol is used as a photographic developer. 1956 *Nature* 28 Jan. 184/2 Copper cyanide, though it accelerates considerably the rate of autoxidation of pyrocatechol.., is not very superior to cupric ions alone as regards catalytic activity on pigment formation from pyrocatechol.

pyrobelonite to **-cellulose:** see *PYRO- 1, 2, 3 b.

Pyroceram (pəiᵊ·roseræ:m). Also **pyroceram.** [f. PYRO- + CERAM(IC *a.* (*sb.*).] A proprietary term in the U.S. for a type of strong, heat-resistant glass which has been heat-treated so that it consists entirely of microscopic crystalline domains.

1957 *New Scientist* 23 May 27/1 The name of this fabulous stuff, which Dr. Stookey invented, is pyroceram. It is harder than flint, light as aluminium, stronger (in ratio to its weight) than stainless steel. *Ibid.* 28/2 The Corning Glass Works has at least a thousand different formulæ for pyrocerams. 1957 *Amer. Ceramic Soc. Bull.* XXXVI. 279/1 Pyroceram is melted and formed like glass, but with a formula containing one or more nucleating agents. 1958 *Official Gaz.* (U.S. Patent Office) 20 May TM 75/1 Corning Glass Works... *Pyroceram*... First use Feb. 7, 1957. 1965 *New Scientist* 4 Nov. 341/1 Housewives who can afford to pay for..coffee pots made of pyroceram are by now accustomed to being told that 'this is the material used in American rocket noses'. 1968 *McGraw-Hill Yearbk. Sci. & Technol.* 38/2 Controlled devitrification to give glass-ceramics or Pyrocerams depends upon the availability of adequate nuclei.

pyrochlore, -clast(ic: see *PYRO- 2, 1.

pyro-electric, *a.* Also **pyroelectric.** Add: Also applied to the effect exhibited by such crystals and to devices employing it. (Further examples.)

1902 H. A. MIERS *Mineralogy* 480 The pyro-electric property [of tourmaline], which was first observed at the beginning of the eighteenth century,..can be very easily shown by means of Kundt's dusting method. 1922 GLAZEBROOK *Dict. Appl. Physics* II. 598/1 The existence of a true pyro-electric effect has been questioned by several investigators. 1973 *Physics Bull.* Mar. 161/2 The flame monitor uses two telescopes, each containing either a photoelectric or pyroelectric cell, to pinpoint and monitor a particular flame. 1979 'R. CASSILIS' *Arrow of God* III. v. 60 Beneath a layer of Wordsworth we packed ..half a dozen pyroelectric-vidicon cameras.

pyrogen. Restrict *rare* to senses a and b and add further examples of sense c.

1955 *Times* 30 Aug. 4/3 We have now reached the stage where bacterial pyrogens in pure form can, with advantage, replace the older materials and methods for producing a general stimulation of the defence mechanisms of the body. 1957 *New Scientist* 12 Dec. 25/1 Rabbits, too, played their part..in pyrogen tests, to ensure the safety of injectable solutions. 1961 M. HYNES *Med. Bacteriol.* (ed. 7) iii. 28 Fluids for parenteral use must be pyrogen-free as well as sterile. 1973 *Nature* 16 Nov. 162/2 It is not known whether the malarial parasite produces a pyrogen, like bacteria, or whether the malarial fever results from destruction of red blood cells.

pyrogenetic, *a.* Add: **2.** *Petrol.* Of a mineral: crystallizing from a magma at a high temperature.

1920 A. Holmes *Nomencl. Petrol.* 193 *Pyrogenetic minerals*, a term applied to the primary magmatic minerals of igneous rocks, excluding those due to pneumatolytic, hydrothermal, and thermodynamic processes. . .The solidification of a magma may constitute a continuous process beginning with indubitable pyrogenetic minerals, and yet finishing with a well-defined hydrothermal series of minerals. **1923** *Mineral. Mag.* XX. 146 In the granites, tourmaline, muscovite, and topaz. . behave as pyrogenetic minerals and commence to crystallize at an early stage but. . their crystallization continued to a late stage in the consolidation of the rock. **1950** F. H. Hatch et al. *Petrol. Igneous Rocks* (ed. 10) iii. 163 The separation of these pyrogenetic minerals leaves the liquid relatively enriched in H_2O and various other components of low atomic and molecular weights. **1954** H. Williams et al. *Petrogr.* i. 9 The first minerals to form from magma are usually anhydrous. . . Such minerals are called pyrogenetic.

pyrogenic, *a.* Add: **1.** (Further example.)

1904 A. W. Grabau in *Amer. Geologist* XXXIII. 230 Returning now to the. . chemically deposited rocks, we may readily distinguish four groups. . . The first. . includes the well recognized Igneous rocks, to which the term *pyrogenic* is applicable.

4. *Chem.* Caused by the application of heat.

1887 *Jrnl. Chem. Soc.* LII. I. 572 (*heading*) Pyrogenic reactions. **1912** *Ibid.* CI. II. 1453 One of the authors was engaged in examining the pyrogenic decomposition of American turpentine with the object of obtaining isoprene in quantity. **1920** *Ibid.* CXVIII. I. 589 (*heading*) Pyrogenic acetylene condensations.

pyrogenicity (pəiːrodʒɪ̆ni·siti). [f. prec. + -ity.] The property of producing fever; freq. *attrib.*

1956 *Nature* 17 Mar. 497/1 The procedure. . to be used in toxicity, pyrogenicity and sterility tests. **1973** *Ibid.* 16 Nov. 162/2 We took advantage of this differential sensitivity to test the pyrogenicity and nature of malarial parasites. **1977** *Lancet* 2 July 47/2 Intravenous pyrogenicity tests in rabbits were negative.

pyrolite: see *pyro- 2.

pyrolyse (pəiᵊ·rolaiz), *v.* Also **-lyze, -lize** (both chiefly *U.S.*), **-lise.** [f. *pyrolysis, after *hydrolysis, hydrolyse*.] **1.** *intr.* To undergo pyrolysis. Const. *to.*

1929 C. D. Hurd *Pyrolysis of Carbon Compounds* ii. 12 Who would predict that the groupings ≡C—CH_2OH and =N—CH_2OH pyrolyze differently? **1938** *Jrnl. Amer. Chem. Soc.* LX. 2420/1 Phenyl acetate pyrolyzed smoothly into ketone and phenol. **1970** *Sci. Jrnl.* May 68 (Advt.), The Pye range of liquid chromatographs is capable of detecting all organic compounds which vaporise or pyrolyse at temperatures up to 700°C. **1974** *Physics Bull.* Feb. 56/2 The fuel, in approaching the flame in the absence of oxygen, tends to pyrolyse to soot and other products of incomplete combustion. **1977** *Engin. Materials & Design* Aug. 26/2 On exposure to high temperatures the resin binder may undergo some further cross-linking. . . Thereafter, the surface layer of resin begins to pyrolise.

2. *trans.* To decompose by heating; *loosely*, to cause to undergo any chemical change by heating.

1932 *Jrnl. Amer. Chem. Soc.* LIV. 3632 Hydroanisamide was pyrolyzed and found to yield no compound which corresponded in composition to trimethoxylophine. **1959** *Times Rev. Industry* Dec. 18/1 Chlorodifluoromethane. This substance is a gas. . which, on pyrolizing at about 800 deg. C., forms tetrafluoroethylene. **1973** *Nature* 23 Mar. 232/2 The results of pyrolysing 50 mg portions of the lunar samples indicated that carbon was present in the gases. **1976** *Amer. Scientist* LXIV. 625/1 The second sample was pyrolized first at 350°C to attempt to drive off most of the water and increase sensitivity to organics.

Hence **py·rolysed** *ppl. a.*; also **py·rolysable** *a.*, capable of being pyrolysed; **pyro·lysate** [after *distillate, filtrate*, etc.], a product of pyrolysis.

1934 Nash & Howes *Princ. Motor Fuel Prep. & Appl.* x. 496 T. S. Wheeler and I.C.I. Ltd., have suggested the removal of hydrogen from pyrolyzed gas before it is passed to a second stage. **1953** D. H. R. Barton in E. H. Rodd *Chem. Carbon Compounds* IIB. xvi. 755 The acid fraction of the pyrolysate. . on dehydrogenation. . gave 1:2-dimethylnaphthalene. . and the anhydride. **1961** *Flight* LXXIX. 836/2 Modern finishing schemes for the internal and external surfaces of military and civil aircraft must supply protection against moisture condensation, heat, pyrolised ester lubricants, hydraulic fluids, etc. **1972** *Sci. Amer.* Oct. 83/2 The lunar samples contained less than one p.p.m. of volatile and pyrolyzable organic matter. **1976** *Ibid.* Mar. 41/3 Clusters of benzene rings, as in pyrene, benzopyrene and perylene, are. . commonly found in pyrolysates. **1977** *Nature* 10 Feb. 493/3 Molecular analysis of the surface dust by pyrolysis-gas chromatography-mass spectrometry. . show that it is virtually free of all pyrolysable carbon compounds.

pyrolysis (pəirǫ·lisis). [f. Pyro- + *-lysis.] Decomposition of a substance by the action of heat; *loosely*, any chemical change produced by heating.

1890 Billings *Med. Dict.* II. 420/1 *Pyrolysis*, dry distillation, decomposing by heat. **1928** W. A. Gruse *Petroleum & its Products* ix. 164 The growth of the petroleum industry, and of the gas-making processes using petroleum, prompted a number of interesting theoretical researches on low-temperature, and particularly on high-temperature, pyrolysis. [*Note*] The use of this word was suggested in 1918 by W. A. Hamor. **1928** *Fuel in Sci. & Pract.* VII. 539/2 Benzene has been found to be an important product of pyrolysis of methane between 875. . and 1,100 deg. Cent. **1943** *Endeavour* Jan. 27/2 In 1924–25 Kennaway succeeded in producing carcinogenic tars by the pyrolysis of petroleum, skin, hair, yeast and cholesterol. **1954** *Chem. & Industry* 13 Nov. 1418/1 It is possible that the presence of the polycyclic hydrocarbons. . is due to the pyrolysis of acetylene which is known to occur in cigarette smoke. **1968** A. A. Baker *Unsaturation in Org. Chem.* ii. 20 Modern investigations on the pyrolysis of carbon compounds have shown that when ethyl alcohol is passed through a glass tube at 610°–630°C, the yield of ethylene is only about 9 per cent. **1975** *New Scientist* 7 Aug. 315/2 Gas evolved by pyrolysis from any organic material will pass to a detector to check for the presence of carbon-14. **1977** *Nat. Westminster Bank Q. Rev.* Aug. 68 Another promising possibility is the use of Pyrolysis which involves the heating of the refuse in the absence of air in order to produce gas, liquid and char which can all be used as fuel.

pyrolytic (pəiroli·tik), *a.* Also **-litic.** [f. *pyrolysis: see *-lytic.] Of, involving, or produced by pyrolysis; *pyrolytic carbon* or *graphite*, a strong, heat-resistant, highly-ordered form of graphite deposited as a vapour from products of hydrocarbon pyrolysis and used esp. in rocket-engine nozzles and as a coating in missile nose-cones, etc.

1909 in *Cent. Dict.* Suppl. **1922** B. T. Brooks *Chem. Non-Benzenoid Hydrocarbons* i. 36 It is. . readily understood that small differences of operating temperature may cause very great difference in the character of the pyrolytic products. **1936** *Chem. Abstr.* XXX. 6176 Some general laws of pyrolytic reactions. . are proved thermodynamically. **1946** *Electronic Engin.* XVIII. 66 A new component. . for use in electrical communications equipment. . is the pyrolytic or cracked carbon resistor. **1961** *Aeroplane* C. 70/1 The big change in the graphite situation of recent years has been the development of a new type of product deposited from the vapour and termed pyrolytic graphite. This is distinguished from the so-called conventional graphite by having a highly oriented structure, and by a higher density. **1966** *Economist* 20 Aug. 754/2 New fuels, pellets rolled and coated in pyrolytic carbon. . have overcome a good deal of the radio-active contamination problem. **1970** *Daily Tel.* 22 Jan. 18 Makers of the coloured pipes said the bowls were made of 'pyrolitic graphite', a very hard substance used in the making of nose-cones for missiles. **1972** DePuy & Chapman *Molec. Reactions & Photochem.* ii. Enol ethers are readily converted to ketones under pyrolytic conditions. **1976** *National Observer* (U.S.) 4 Sept. 3/3 Another test with the pyrolitic release experiment, which looks for evidence that carbon dioxide is being taken up by something in the soil (perhaps life), confirmed the instrument's earlier findings.

Hence **pyroly·tically** *adv.*, by pyrolysis.

1956 *Amer. Scientist* XLIV. 356 This reaction was not highly reproducible pyrolytically. **1975** *Nature* 5 June 474/1 We have studied the influence of high temperature on the tensile fracture strengths of pyrolytically deposited silicon carbide fibres.

pyromagnetic to **-metallurgy:** see *pyro- 1.

pyrometric, *a.* (Examples of *pyrometric cone.*)

Cf. quots. 1800, 1839 in Dict.

1947 J. C. Rich *Materials & Methods of Sculpture* ii. 46 Pyrometric cones. . are used to determine and thereby control the firing temperatures of the kiln. **1964** H. Hodges *Artifacts* i. 40 Today potters use small cones of clay—pyrometric cones—which melt below the maturing point of the wares being fired. **1977** *Western Living* (Vancouver) Apr. 25/3 The way you tell the temperature. . is with pyrometric cones which are little triangular objects made of different combinations of ceramic materials.

pyrone (pəiᵊ·rōᵘn). *Chem.* [ad. G. *pyron* (Haitinger & Lieben 1885, in *Sitzungsber. d. österreichischen Akad. d. Wissensch. in Wien* XCI. (Abt. II.) 923): see *pyro- and -one.] Either of two unsaturated heterocyclic compounds, $C_5H_4O_2$, which are mono-keto derivatives of the pyrans; *spec.* CH:CH·O·CH:CH·CO (γ- or 1, 4-*pyrone*), a colourless, basic, crystalline solid; also, any heterocyclic ketone or lactone containing the ring structure characteristic of either isomer. Freq. *attrib.*, as *pyrone ring.*

1891 *Jrnl. Chem. Soc.* LX. I. 458 The pyrone is almost insoluble in water or alkalis. *Ibid.* II. 939 The conversion of α-pyrone into pyridine derivatives. **1913** T. H. Pope tr. *Molinari's Org. Chem.* 626 Chelidonic acid, . . which is found in celandine, loses CO_2 giving comanic acid and pyrone. **1938** G. H. Richter *Textbk. Org. Chem.* xxxi. 659 The pyrones are interesting also because the hetero oxygen atom is basic forming oxonium salts with strong acids. **1962** K. Venkataraman in T. A. Geissman *Chem. Flavonoid Compounds* iv. 94 Isoflavones undergo hydrolysis with opening of the pyrone ring under mild conditions of alkali treatment. **1963** L. F. & M. Fieser *Topics in Org. Chem.* ii. 103 α-Pyrone has the properties expected of a doubly unsaturated δ-lactone. *Ibid.* 104 Representative natural γ-pyrones are kojic acid, formed by bacterial fermentation of carbohydrates; maltol, isolated from the bark of the larch tree; and yangonin, from the roots of the kava shrub. **1972** J. M. Tedder et al. *Basic Org. Chem.* IV. iii. 139 The reactivity of the carbonyl group is reduced so much by the conjugation. . that the α-pyrone ring is much more stable towards alkali than the unsubstituted coumarin.

pyronin (pəiᵊ·rōnin). *Chem.* and *Biol.* Also **-ine** (-īn), and with capital initial. [a. G. *pyronin* (trade name), prob. f. *pyro-* Pyro- + -*in* -in¹, -ine⁵ with inserted *n*.] Any of a class of synthetic red xanthene dyes employed chiefly as microscopic stains; now *esp.* either of two such dyes, called *pyronin G* or *Y* and *pyronin B.*

1895 *Jrnl. Chem. Soc.* LXVIII. I. 47 Formaldehydetetramethylamidofluorimum chloride, the zinc double salt of which is known commercially as pyronine. **1906** *Practitioner* Nov. 666 Stained with pyronin and methyl green, the nuclei were large, rounded, pale, and contained chromatin in 'lumps'. **1952** K. Venkataraman *Chem. Synthetic Dyes* II. xxiv. 745 Pyronines are obtained by condensing *m*-dialkylaminophenols with formaldehyde in presence of concentrated sulfuric acid, and oxidizing the xanthene derivative. **1957** *Nature* 31 Aug. 440/1 Both the nucleolus and the puffs stain bright red with pyronin. **1960** E. Gurr *Encycl. Microscopic Stains* 340 Pyronin B. . is in fairly frequent use in bacteriology and in animal and plant histology. *Ibid.* 341 Pyronin Y finds its chief application. . in conjunction with methyl green for demonstrating nucleic acids. **1960** *New Biol.* XXXI. 99 Their cytoplasm is unusually strongly stained by dyes, such as pyronine, which specifically combine with ribonucleic acid. **1974** H. C. Cook *Man. Histol. Demonstration Techniques* ii. 43 For better results the pyronin should be of 'Y' or 'G' type in preference to the 'B' variety.

pyrope. 1. Delete † *Obs.* and add later *attrib.* example.

1948 R. Graves *Coll. Poems, 1914–1947* 237 He carved his law on tables of sapphirus, Jerusalem shines with his pyrope gates.

2. Also *pyrope garnet.* (Further examples.)

1959 *Times* 31 Oct. 9/5 Four. . types of garnets interest the collector. . the red varieties pyrope and almondine being the most popular. . . Carbuncles are the large oval red pyrope garnets which were cut *en cabochon.* **1976** *Nature* 8 Apr. 517/2 In the Elie Ness vent, high pressure pyroxenes and pyrope garnets indicate rapid elevation from depths in excess of 60 km. **1977** A. Hallam *Planet Earth* 34/3 It is likely that the mantle consists dominantly of magnesium olivine,. . together with smaller amounts of enstatite. ., diopside. ., pyrope garnet ($Mg_3Al_2Si_3O_{12}$) and perhaps some phlogopite mica.

pyrophoric, *a.* Add: *pyrophoric alloy*, an alloy (usu. of iron and cerium) that emits sparks when scratched or struck with a file or the like.

1906 C. A. von Welsbach *U.S. Patent 837,017* What I claim is—1. A pyrophoric alloy, containing cerium alloyed with iron; substantially as and for the purposes described. **1950** *Thorpe's Dict. Appl. Chem.* (ed. 4) X. 328/2 The pyrophoric alloy, in the form of a small rod, is generally pressed firmly against a hard steel wheel with a roughened surface. The shower of sparks produced on rapidly rotating the wheel is projected on to a cotton wick impregnated with. . lighter fluid.

pyrophosphatase (pəirofǫ·sfătēᵢz, -s). *Biochem.* [f. *pyrophosphate* s.v. *pyro- 3a + *-ase.] Any enzyme which hydrolyses pyrophosphate esters or ions, or pyrophosphoric acid; *inorganic pyrophosphatase*, an enzyme capable of hydrolysing the free acid or its ions, liberating orthophosphate.

1928 H. D. Kay in *Biochem. Jrnl.* XXII. 1446 Both kidney extract and takadiastase. . contain, therefore, a pyrophosphatase. *Ibid.* 1448 Pyrophosphatase is widely distributed in mammalian tissues. **1951** *Jrnl. Biol. Chem.* CXCII. 87 Bauer. . found that autolysates of slowly dried bottom yeast were a rich source of inorganic pyrophosphatase. **1970** R. W. McGilvery *Biochem.* xi. 209 The inorganic pyrophosphate. . is in turn hydrolyzed by the ubiquitous inorganic pyrophosphatase with an additional liberation of free energy. **1976** *Soil Biol. & Biochem.* VIII. 391/1 An assay procedure for soil pyrophosphatase activity.

pyrosphere to **-synthetic:** see *pyro- 1.

pyrotechnic, *a.* and *sb.* Add: **A.** *adj.* **2. a.** Also more widely, capable of being ignited for technical or military purposes. (Further examples.)

1919 H. B. Faber *Military Pyrotechnics* I. I. v. 45 In the latter part of the eighteenth century and the beginning of the nineteenth, there was great activity on the part of those enthusiastic over pyrotechnic devices as war instruments. **1922** A. St. H. Brock *Pyrotechnics* iii. 13 Pyrotechnic compositions and gunpowder are inextricably mixed together in early European records. **1953** Kirk & Othmer *Encycl. Chem. Technol.* XI. 324 In addition to illuminating and signal compositions there are many other types of pyrotechnic compositions such as smoke, incendiary, whistle, dark-fire, tracer, and igniter compositions. **1970** *Guardian* 18 Apr. 1/1 The astronauts fired the pyro-technic bolts that connect the command ship and the service module. **1974** *Encycl. Brit. Micropædia* VII. 186/2 Napalm is also employed in formulating a pyrotechnic gel containing gasoline. ., powdered magnesium, and sodium nitrate.

B. *sb.* **2. c.** (Later examples.)

1921 F. Scott Fitzgerald *Let.* a 12 Dec. (1964) 150 The cruel Hebrew God, against whom such writers as

even Mark Twain. . have delivered violent pyrotechnics from time to time. **1970** S. Schoenbaum *Shakespeare's Lives* VII. x. 686 Such pyrotechnics of rationalization must be a Baconian's envy. **1977** *Rolling Stone* 7 Apr. 44/2 No emotional pyrotechnics for him, just calm, deathly calm.

3. A device or material which can be ignited to produce light, smoke, or noise, e.g. for purposes of display or illumination.

1919 H. B. Faber *Military Pyrotechnics* I. 7 The art of manufacturing military pyrotechnics. **1948** W. Haynes *Amer. Chem. Industry* IV. ix. 130 When demand shrank to peacetime uses, chiefly in matchheads and pyrotechnics. ., European manufacturers. . began cutting prices. **1953** Kirk & Othmer *Encycl. Chem. Technol.* XI. 332 Commercial pyrotechnics mainly having to do with *sound* include cannon crackers. . and trick cigars. **1957** *Spaceflight* I. 51/2 A Fifth-of-November rocket or catherine wheel is a pyrotechnic, and modified forms of these have been used to ignite liquid propellant rockets. **1964** F. G. W. & M. G. Jones *Pests of Field Crops* xvi. 355 Smokes are made by combusting a mixture of pesticide and a suitable pyrotechnic. **1972** *Materials & Technol.* IV. xix. 732 The chemical reactivity of the ingredients and their particle size have very significant effects on the ignition characteristics and burning rates of pyrotechnics.

pyrotechnician. Delete † *Obs.* and add later example.

1979 F. Morton *Nervous Splendour* (1980) xxx. 316 Pyrotechnicians labored. . for Easter Sunday: a symphonic fireworks.

Pyrotenax (pəirote·næks). *Electr.* Also **pyrotenax.** A proprietary name for a make of robust, heat-resistant copper-sheathed cable with magnesia insulation.

1937 *Nature* 20 Nov. 887/2 'Pyrotenax' cable has a copper conductor, magnesia insulation, and copper sheath. **1949** *Trade Marks Jrnl.* 2 Feb. 95/2 Pyrotenax. . . Terminal plates and boxes for electric cables. Pyrotenax Limited, Hedgeley Road, Hebburn-on-Tyne. ., Manufacturers. **1958** Molloy & Say *Electr. Engineer's Ref. Bk.* (ed. 9) xxx. 84 Tailor-made lengths of Pyrotenax cables, tested at 2,000 V. after water immersion. **1958** *Oxf. Univ. Gaz.* 7 Mar. 694/1 At the same time this part of the Library was rewired with pyrotenax. **1960** *Official Gaz.* (U.S. Patent Office) 17 May TM115/2 Societe Alsacienne de Constructions Mecaniques [*sic*], Paris. . . *Pyrotenax.* Owner of French Reg. No. 2,373, dated June 28, 1952. . . For electrical conductors, [etc.].

pyroxene. Add: In mod. use, any of a group of silicates characterized by the presence of single chains of SiO_4 tetrahedra and prismatic cleavages at nearly 90 degrees, the general formula of which is approximately $XY(SiO_3)_2$. (Further examples.)

1888 J. P. Iddings tr. *Rosenbusch's Microsc. Physiogr. Rock-Making Minerals* 202 The orthorhombic pyroxenes become transparent in various colors, according to the position of the section and to the iron percentage. **1906** J. P. Iddings *Rock Minerals* II. 286 (*heading*) Pyroxene group. *Ibid.*, The triclinic pyroxenes are less closely related to the other forms, and are less frequently met with as rock-making minerals. **1959** *Dana's Man. Mineral.* (ed. 17) 434 The pyroxenes crystallize at higher temperatures than their amphibole analogues and hence are generally formed earlier in a cooling igneous magma. **1966** W. A. Deer et al. *Introd. Rock-Forming Minerals* II. 99 The pyroxene group includes both orthorhombic and monoclinic minerals. The orthorhombic sub-group consists essentially of the compositional series $MgSiO_3$-$FeSiO_3$ while the monoclinic sub-group includes members having a wide range of chemical composition. *Ibid.*, The general formula of the pyroxene group may be expressed $X_{1-p}Y_{1+p}Z_2O_6$ where X = Ca,Na; Y = Mg,Fe^{+2},Mn,Li,Ni,Al,Fe^{+3},Cr,Ti; Z = Si,Al. **1975** D. Shelley *Man. Optical Mineral.* vii. 143 Most pyroxenes form rather stumpy prismatic crystals, though occasionally, as in the case of aegirine, the crystals are more elongate. **1978** *Sci. Amer.* Apr. 125/2 Peridotite. . is composed mainly of olivine and another silicate mineral, pyroxene.

pyroxeni·tic *a.*, of or pertaining to pyroxenite.

1933 R. A. Daly *Igneous Rocks & Depths of Earth* ix. 189 The speeds of seismic waves cannot be said to demonstrate a peridotitic or pyroxenitic layer between the 40-kilometer and 60-kilometer levels. **1979** *Nature* 5 Apr. 545/1 Gorgona is one of the rare places in the world where young pyroxenitic komatiites exhibiting typical quenched spinifex textures. . occur.

pyroxenoid (pəirǫ·ksĕnoid). *Min.* [f. prec. + -oid.] Any of a small group of triclinic silicates formerly classed as pyroxenes but now differentiated from them on structural grounds.

1937 H. Berman in *Amer. Mineralogist* XXII. 389 The so-called 'triclinic pyroxenes' are not included here in the pyroxenes because the writer believes they are more properly considered as a separate group, with no isomorphous relations to any of the pyroxene minerals, and with physical and chemical properties clearly differing from those of the pyroxenes. To these pyroxene-like minerals we here give the name pyroxenoids. **1942** Rogers & Kerr *Optical Mineral.* (ed. 2) II. 275 Pyroxenoids include rhodonite, bustamite, pectolite, and wollastonite. **1966** J. Sinkankas *Mineral.* II. 487 The pyroxenoids are species whose chemical compositions resemble the pyroxenes, but [whose] crystal structures differ slightly but importantly in the way the chains are linked and arranged. **1977** A. Hallam *Planet Earth* 136 Rhodonite is closely similar in structure to the pyroxene group (for this reason it is sometimes called a pyroxenoid).

pyroxferroite (pəirǫksfe·ro,əit). *Min.* [f. Ferro- + -ite¹, after next.] A yellow pyroxenoid, $Fe_6Ca(SiO_3)_7$, that has been found on the moon and is an iron-rich analogue of pyroxmangite.

1970 E. C. T. Chao et al. in *Proc. Apollo 11 Lunar Sci. Conf.* I. 65 Pyroxferroite was first recognized by the LSPET (1969) as an unidentified yellow mineral that seemed to be concentrated in vuggy areas of the Type B rock. *Ibid.* 75 Lindsley (1967) recently synthesized pyroxferroite of composition $Ca_{0.15}Fe_{0.85}SiO_3$ at pressures from 10 to 17·5 kbar and temperatures from 1130 to 1250°C. **1973** Sorrell & Sandstrom *Rocks & Minerals of World* 60 The major lunar minerals are calcic plagioclase. . and pyroxene. . . Common are olivine. . and pyroxferroite, $CaFe_6(SiO_3)_7$, a new mineral similar to the pyroxenes.

pyroxmangite (pəirǫksmæ·ŋgəit). *Min.* [f. Pyrox(ene + Mang(anese + -ite¹.] A manganese- and iron-containing pyroxenoid, $(Mn,Fe)SiO_3$.

1913 Ford & Bradley in *Amer. Jrnl. Sci.* CLXXXVI. 169 (*heading*) Pyroxmangite, a new member of the pyroxene group and its alteration product, skemmatite. **1937** *Amer. Mineralogist* XXII. 720 The discovery of a further occurrence of pyroxmangite among the Lewisian rocks of Scotland has provided additional data on this interesting mineral. *Ibid.*, The pyroxmangite forms an important constituent of a manganiferous schist interbedded with a series of para-gneisses. . . Pyroxmangite in this rock occurs in pink grains of ¼–½ mm. average grain size. Exceptionally grains up to 5 mm. diameter may appear as porphyroblasts. **1963** W. A. Deer et al. *Rock-Forming Minerals* II. 201 Pyroxmangite is a mineral of metamorphic or metasomatic rocks, being found typically in manganese-rich assemblages in association with spessartine garnet, tephroite,. . or rhodochrosite. **1970** *New Scientist* 15 Jan. 94/2 Some 68 elements forming minerals familiar to geologists were sorted out of the lunar soil and rock, along with three new minerals. They. . were identified as a titanium-chromium spinel, a ferro-pseudobrookite and a pyroxmangite.

pyrrol. The spelling pyrrole is now the usual form. **a.** Amend formula to read C_4H_5N and add: Also, any derivative of this containing a pyrrole ring. (Further examples.)

1902 *Jrnl. Chem. Soc.* LXXXII. I. 54 Better yields of pyrroline are obtained by reducing pyrroles with zinc and hydrochloric acid than by using zinc and acetic acid. **1926** H. G. Rule tr. *J. Schmidt's Text-bk. Org. Chem.* 521 Pyrroles are aromatic in character and possess points in common with both phenols and aromatic amines. **1954** *New Biol.* XVI. 33 The tendency of the related pyrroles to combine with metal atoms to form highly coloured reactive compounds, such as hydrogenase and cytochrome, and probably later chlorophyll and haemin, points the way to the evolution of enzyme systems and photosynthesis. **1972** J. M. Tedder et al. *Basic Org. Chem.* IV. ix. 454 On reduction with hydriodic acid haem yields eight comparatively simple pyrroles.

b. *attrib.* and *Comb.*, as **pyrrole base,** any of a series of bases containing a pyrrole ring; **pyrrole nucleus, ring,** a doubly unsaturated ring of four carbon atoms and one nitrogen atom.

1851, 1875 Pyrrol base [in Dict.]. **1913** Bloxam & Lewis *Bloxam's Chem.* (ed. 10) 782 The metals are not present as bases, but as integral parts of the complex molecules, probably exercising their subsidiary valencies, and uniting a number of pyrrol nuclei. **1926** H. G. Rule tr. *J. Schmidt's Text-bk. Org. Chem.* 523 This oxidation has recently been recognised as a valuable means of determining the orientation of substituents in the pyrrole nucleus, and also for detecting the presence of a pyrrole ring in substances of unknown constitution. **1970** R. W. McGilvery *Biochem.* xxi. 494 The basic unit of porphyrins is the pyrrole ring, with four of these linked to form the large porphyrin ring.

Hence **pyrro·lic** *a.*

1909 in *Cent. Dict.* Suppl. **1912** *Chem. Abstr.* VI. 2749 Pyrrolic α-, β- and γ-diketones. **1955** *Endeavour* July 135/2 (*caption*) Porphobilinogen, the simplest pyrrolic substance known to be a precursor of haem and porphyrins. **1972** J. M. Tedder et al. *Basic Org. Chem.* IV. ix. 464 The blood-red, tripyrrole microbial pigment, prodigiosin, provides an example of the biogenesis of pyrrolic compounds.

pyrrolidine (pirǫ·lidīn). *Chem.* [ad. G. *pyrrolidin* (Ciamician & Magnaghi 1885, in *Ber. d. Deut. Chem. Ges.* XVIII. 2080): see Pyrrol and *-idine.] A saturated heterocyclic compound, $(CH_2)_4NH$, which is a colourless, pungent, strongly basic liquid obtained esp. by catalytic reduction of pyrrole; any substituted derivative of this. Freq. *attrib.*, as **pyrrolidine nucleus, ring,** a saturated ring of four carbon atoms and one nitrogen.

1885 *Jrnl. Chem. Soc.* XLVIII. II. 1243 Pyrrolidine. . C_4NH_9, is obtained with other products from hydropyrroline in the above reaction; it shows great similarity in properties to piperidine, hence the name given to it. **1926** H. G. Rule tr. *J. Schmidt's Text-bk. Org. Chem.* 532 Certain important degradation products of the *coca* and *atropa* alkaloids have been identified as carboxylic acids of pyrrolidine. **1956** I. L. Finar *Org. Chem.* II. xiv. 531 The presence of this pyrrolidine nucleus also accounts for the formation of pyrrole when nicotine zincichloride is distilled. **1963** T. Robinson *Org. Constituents of Higher Plants* xii. 256 Leete. . using 1, 4-labelled putrescine fed to tobacco found that it was an efficient precursor of the pyrrolidine ring in nicotine. **1976** Streitwieser & Heathcock *Introd. Org. Chem.* xxxv. 1065 Pyrrolidine and piperidine are typical secondary amines.

pyrrolidone (pirǫ·lidōᵘn). *Chem.* [ad. G. *pyrrolidon* (J. Tafel 1889, in *Ber. d. Deut. Chem. Ges.* XXII. 1861): see prec. and -one.] Either of two isomeric mono-keto derivatives of pyrrolidine; *esp.* NH·$(CH_2)_3$·CO (2-*pyrrolidone*), a colourless crystalline solid with weakly basic properties; any substituted derivative of either isomer.

[**1889** *Jrnl. Chem. Soc.* LVI. 961 Methylpyrrolidone.] *Ibid.* 1211 The first product. . is very unstable, condensation immediately taking place, with formation of a derivative of pyrrolidone. **1926** H. G. Rule tr. *J. Schmidt's Text-bk. Org. Chem.* 517 The 2-keto-derivative of pyrrolidine is commonly known as 'pyrrolidone'. *Ibid.* 520 Since succinimides are readily prepared in quantity, this process also renders the pyrrolidones easy of access. **1951** I. L. Finar *Org. Chem.* I. xvi. 309 When reduced with sodium and ethanol, succinimide forms pyrrolidine,. . and when reduced electrolytically, it forms pyrrolidone. *Ibid.* xxx. 617, 2-Pyrrolidone is the lactam. . of γ-aminobutyric acid. **1975** Gutsche & Pasto *Fund. Org. Chem.* xxvii. 817 The reaction between pyrrolidone and acetylene in the presence of potassium hydroxide to yield N-vinylpyrrolidone.

pyrroline. Now pronounced (pi·rǫlīn). Mark sense in Dict. † *Obs.* and add: **b.** formerly also **-in.** A partially reduced derivative of pyrrole having the formula C_4H_7N, of which three isomers are possible; *esp.* HN·CH_2·CH:CH·CH_2 (3-*pyrroline*), a colourless, basic liquid obtained by reduction of pyrrole with nascent hydrogen; also a substituted derivative of any of these compounds. [Named in Ger. as *pyrrolin* (Ciamician & Dennstedt 1883, in *Ber. d. Deut. Chem. Ges.* XVI. 1539).]

1884 Roscoe & Schorlemmer *Treat. Chem.* III. II. 610 Pyrrol unites with nascent hydrogen, forming pyrrolin, C_4H_6(NH), an oily liquid that boils at 90°–91°, and acts as a strong base. **1902** [see *pyrrol a]. **1926** H. G. Rule tr. *J. Schmidt's Text-bk. Org. Chem.* 517 Dihydro-pyrroles are known as pyrrolines, and the completely reduced tetrahydro-pyrroles as pyrrolidines. **1961** G. M. Badger *Chem. Heterocyclic Compounds* ii. 17, 3-Pyrrolines have been isomerized to the corresponding 1-pyrrolines by heating with Raney nickel. *Ibid.* 18, 2-Pyrrolines have been poorly characterized. **1972** J. M. Tedder et al. *Basic Org. Chem.* IV. ix. 463 Another group of tetrapyrrolic macrocyclic compounds found in nature are the corrins which are made up of four partially reduced pyrrole (pyrroline and pyrrolidine) rings.

pyrrolizidine (piːrǫli·zidīn). *Chem.* [ad. G. *pyrrolizidin* (G. Menschikoff 1936, in *Ber. d. Deut. Chem. Ges.* LXIX. 1802), f. *pyrrolidin* *pyrrolidine with inserted *-iz* (ult. f. *azo-* Azo-) after *chinolizin* quinolizine, an analogous compound containing six-membered rings.] A colourless basic liquid, $C_7H_{13}N$, which has a structure consisting of two fused pyrrolidine rings sharing a carbon and a nitrogen atom; any derivative of this. Freq. *attrib.*, as **pyrrolizidine alkaloid,** any of a large class of toxic alkaloids based on this structure which occur widely in plants, esp. in the genera *Senecio, Crotalaria,* and *Heliotropium.*

[**1936** *Chem. Abstr.* XXX. 6379 A satd. tertiary base $C_8H_{15}N$. . for which is proposed the name 2-methylpyrrolizidine.] **1939** *Ibid.* XXXIII. 5850 P[relog] and H[eimbach] have found a method of preparing the hitherto unknown pyrrolizidine (I). . . The free I is a water-sol. oil of basic odor, b.p. 148°. **1950** *Thorpe's Dict. Appl. Chem.* (ed. 4) X. 710/2 All Senecio and related alkaloids are esters of a cyclic amino-alcohol of pyrrolizidine type with aliphatic or alicyclic acids. **1968** *New Scientist* 20 June 619/2 Many pyrrolizidine alkaloids. . not only exert their effects as liver poisons but may also produce lung damage. **1975** K. Nakanishi et al. *Natural Products Chem.* II. x. 299 Recent studies on the metabolism of the alkaloids have shown that it is not the pyrrolizidine alkaloids themselves, but their pyrrole derivatives, formed as metabolites, which exhibit toxicity.

pyrus. Add to etym.: Adopted by Linnæus (*Hortus Cliffortianus* (1737) 189) as a generic name. Substitute for def.: A small tree of the genus so called, belonging to the family Rosaceæ and widely cultivated for the sake of its blossom or its fruit, the pear; also, a tree or shrub once included in this genus and now in a separate one, esp. the japonica (*Chænomeles* species) or the rowan (*Sorbus* species). (Earlier and later examples.)

1849 Thoreau *Week Concord Riv.* 92 The shad make their appearance early in May, at the same time with the blossoms of the pyrus, one of the most conspicuous early flowers. **1914** W. J. Bean *Trees & Shrubs Hardy in Brit. Isles* II. 269 All the Pyruses like a loamy soil. **1930** F. K. Ward *Plant Hunting on Edge of World* vi. 120 As for the Pyrus, its numerous clusters of reddened berries presently turned snow white.

pyruvate (pəirū·vēᵢt). *Biochem.* [f. Pyruv(ic *a.* + -ate⁴.] **1.** A salt or ester, or the anion, of pyruvic acid; *loosely*, denoting either anions or the acid itself.

1855 H. Watts tr. *Gmelin's Hand-bk. Chem.* IX. 419 The Pyruvates are prepared by saturating the dilute acid with the base. **1877** [in Dict. s.v. Pyruvic *a.*]. **1905** *Jrnl. Chem. Soc.* LXXXVIII. I. 572 Butyroin. . yields a pyruvate which boils at 134–138° under 12 mm. pressure. **1946** *Nature* 7 Sept. 350/1 Pyruvate accumulates in the blood of thiamin-deficient animals. **1955** *Sci. Amer.* Jan. 76/2 They [*sc.* Rickettsiæ] also proved capable of oxidizing slowly two other substances, pyruvate and succinate, which, like glutamic acid, are oxidized by most animal and plant tissues. **1970** Ambrose & Easty *Cell Biol.* vi. 183 Two molecules of pyruvate are finally produced. *Ibid.*, Normally, muscle respires aerobically, oxidizing pyruvate via the Krebs cycle. ., but during violent exercise oxygen cannot reach the tissues fast enough. In this case muscles obtain extra energy by reduction of pyruvic acid to lactic acid. **1970** *New Scientist* 23 Apr. 168/1 Glycolysis is the process which oxidizes carbohydrates to pyruvate, ready to enter the citric acid cycle and be burnt up completely. **1976** *Ann. Rev. Microbiol.* XXX. 157 Pyruvate in polysaccharides of *R. trifolii, R. meliloti, R. radicicolum*, and Pn 27 functions as a major determinant in serological specificity.

2. Special Comb.: **pyruvate kinase,** any enzyme which catalyses the transfer of a phosphate group between adenosine triphosphate and pyruvic acid; **pyruvate oxidase** = **pyruvic oxidase.*

1951 S. P. Colowick in Sumner & Myrbäck *Enzymes* II. I. xlvi. 123 'Creatine kinase', 'pyruvate kinase' and '3-phosphoglycerate kinase' will refer to the respective enzymes which catalyze the reversible reaction of ATP with these substrates. **1970** *New Scientist* 23 Apr. 168/2 The enzyme pyruvate kinase. . is slightly different in liver and kidney from that in other tissues. **1959** *Jrnl. Vitaminol.* V. 94 Recently, further attempts were made to elucidate whether vitamin B_{12} activates the pyruvate oxidase system in B_{12}-deficient rats. **1970** Ambrose & Easty *Cell Biol.* vi. 188 Pyruvate. . is first converted to acetyl-coenzyme A by combination with coenzyme A. (Catalysed by pyruvate oxidase.)

pyruvic, *a.* Add to etym.: [ad. mod.L. *pyruvicus*, in *acidum pyruvicum* pyruvic acid (J. J. Berzelius 1835, in *Ann. der Physik u. Chem.* XXXVI. 5)]. *pyruvic acid*, add: The acid occurs widely in living organisms as an intermediate in many metabolic processes, notably glycolysis; (further examples); *pyruvic kinase* = **pyruvate kinase*; *pyruvic oxidase*, any enzyme or enzyme complex which catalyses the oxidation of pyruvic acid.

1927 M. Bodansky *Introd. Physiol. Chem.* ix. 226 There is. . no convincing evidence that physiologically lactic acid is converted into pyruvic acid, although the change is known to occur *in vitro* under the influence of hydrogen peroxide. **1941** *Adv. Enzymol.* I. 147 Even the pyruvic acid oxidase of *Bacterium Delbrueckii*. . might be called a composite thiamin enzyme as it contains flavin adenine nucleotide besides thiamin pyrophosphate, both of which are possibly bound to the same protein. **1945** *Jrnl. Biol. Chem.* CLIX. 543 The pyruvic oxidases of *Proteus vulgaris* and *Escherichia coli* do not appear to require inorganic phosphate for activity. **1951** *Sci. News* XXII. 80 An illuminated preparation of spinach chloroplasts can reduce carbon dioxide plus pyruvic acid to malic acid. **1951** Sumner & Myrbäck *Enzymes* II. II. 1433/1 (Index), Pyruvic kinase. **1962** S. G. Waley in A. Pirie *Lens Metabolism Rel. Cataract* 357 Thiamine (as the pyrophosphate) is the coenzyme in the decarboxylation of the α-keto acids, pyruvic acid and α-ketogluric acid. **1963** Conn & Stumpf *Outl. Biochem.* xi. 201 The conversion of pyruvate to acetyl-CoA is catalyzed by the enzyme complex known as pyruvic oxidase. **1967** Batsakis & Briere *Interpretive Enzymol.* xii. 242 Pyruvic kinase deficiency leads directly to an impairment of ATP synthesis. **1970** [see *pyruvate 1]. **1971** C. J. Gray *Enzyme-Catalysed Reactions* vii. 291 Porcine pyruvic oxidase catalysed the synthesis of α-acetolactate from pyruvate and the hydroxyethylthiamine pyrophosphate.

Pytchley (pəi·tʃli). The name of a village in Northamptonshire, used *attrib.* and *ellipt.* to denote a famous hunt centred there. *Pytchley (riding) coat* (see quots. 1907, 1963).

1866 J. Blackwood *Let.* 26 Apr. in *Geo. Eliot Lett.* (1956) IV. 245, I enjoyed her. . laughing at his French accent. . . It is like the House of Commons. . laughing at Bright for talking of the *Pittchley*. **1867** 'Ouida' *Under Two Flags* I. iv. 74 It lay in the Melton country, and was equally well placed for Pytchley, Quorn, and Belvoir. **1907** *Yesterday's Shopping* (1969) 323/1 Pytchley Riding Coat. Full Skirt to cover saddle. **1919** J. Buchan *Mr Standfast* xiv. 257 He used to hunt with the Pytchley. **1935** *Encycl. Sports* 294/1 About 1750, the modern system of hunting was introduced in the Quorn country by Meynell and by Lord Spencer in the Pytchley Hunt. **1955** M. Allingham *Beckoning Lady* ii. 16 Two seasons with the Pytchley foxhounds. **1963** Bloodgood & Santini *Horseman's Dict.* 157 *Pytchley coat* (More usually called *Shadbelly, Swallow tail* or *Cutaway*), tight-fitting, Regency, double-breasted hunting coat (either scarlet or black) worn with a double-breasted buff waistcoat and named after its originators, the thrusters, or hard riding members of the Pytchley Hunt, England. **1969** A. Horsbrugh-Porter in A. S. C. Ross *What are U?* 51 It would be tedious to delve into the date which constituted the name of 'the shires', denoting the Pytchley, Quorn, Belvoir, Fernie and Cottesmore Hunts.

python¹. Add: **3.** *python-steak, -stretch.*

1953 R. Campbell *Mamba's Precipice* ii. 26 He and Nyali had had python-steak for supper. **1923** D. H. Lawrence *Birds, Beasts & Flowers* 177 The great muscular python-stretch of her tail.

python² (pəi·þǒn). *Mil.* [A code name.] Leave granted at the end of the 1939–45 war to members of the British forces who had served a long period overseas. Also *attrib.*, as *python leave.*

1945 L. Durrell *Spirit of Place* (1969) 82, I took down a pomegranate. . and tried to send it to her with a friend on python. **1945** *Punch* 22 Aug. 166/1 Naturally they are all either due for Python or their Age and Service Groups, and the last few days we have suffered from a constant round of farewell parties. **1949** D. E. Stevenson *Vittoria Cottage* xiv. 95 When men come home from FARELF after doing their three years they say they are *coming home on Python*. . . FARELF. . means Far Eastern Land Forces. **1959** I. Jefferies *Thirteen Days* iii. 37 Your python must be coming up soon. **1969** A. G. Thomas in L. Durrell *Spirit of Place* 82 Under Python leave any soldier who had been in the Middle East for more than four years was granted one month at home with his family and then three months in some unit in Britain. **1976** R. Lewin *Slim* xv. 252 *Python*, the scheme for repatriation of men who had served a minimum of three years and four months in the Far East.

pyxie (pi·ksi). *U.S.* Also **pixy.** [Abbrev. of mod.L. *Pyxidanthera* (A. Michaux *Flora Boreali-Americana* (1803) I. 152), f. Gr. πυξίδ-ιον box + ἀνθηρά, fem. of ἀνθηρός flowery.] In full, *pyxie moss.* A small, prostrate, evergreen shrub, *Pyxidanthera barbulata*, belonging to the family Diapensiaceæ, native to limited areas of New Jersey and the Carolinas, and bearing tiny white flowers; also called the pine-barren beauty (see Pine-barren b).

1882 *Harper's Mag.* June 65 Among her [*sc.* Nature's] treasures is the delicate pyxie. . a little prostrate trailing evergreen. **1892** *Amer. Folk-Lore* V. 100 *Pyxidanthera barbulata*, pyxie moss. **1916** J. W. Harshberger *Vegetation New Jersey Pine-Barrens* xvi. 240 The flowering-moss, or pyxie, is usually a prostrate or creeping plant. **1925** *Scribner's Mag.* July 35/1 Innumerable clusters of oval-leaved *Diapensia lapponica* in rounded clumps like red pin-cushions (closely resembling what is called pixy-moss). **1951** E. W. Teale *North with Spring* xxviii. 276 If we had been a few weeks earlier, we would have found. . the pyxie of the pine barrens, the matted, mosslike flowering plant. *Ibid.*, One year, on the 31st of March,. . I came upon a dense mass of pyxie moss,. . across which a host of tiny waxy-white flowers were scattered. **1973** Robichaud & Bell *Vegetation New Jersey* xii. 217 The most commonly known of these [pine-barren plants] include the turkey-beard, pyxie moss, goat's-rue, [etc.].

Q

Q. Add: **I. 1.** *attrib.* Used *spec.* to designate one of the two main groups of languages which developed from Common Celtic, so called because its distinctive phonological features include the retention of IE. **qʷ*, as *Q-Celt*, a speaker of *Q-Celtic*. Cf. *P-Celtic* s.v. *P I. 1.

1891, 1913 [see *P I. 1]. **1944** J. Whatmough *KEATIKA* 49 The possibility that traces of *q*-Keltic may lurk hidden in the magico-medical formularies of Marcellus of Bordeaux. **1962** T. C. Lethbridge *Witches* vi. 72 This adds to the evidence which suggests that the Iceni were 'Q' Celts, speaking a form of Gaelic. **1972** [see *P I. 1].

2. *Q-boat, Q-ship*, an armed and camouflaged merchantman used as a decoy or to destroy submarines; also *ellipt.*; cf. *decoy *sb.*² 6, *mystery ship* s.v. *mystery¹ 13. Hence *Q car*, a disguised police car.

1920 *Blackw. Mag.* Mar. 325/1 They had complied with the regulations that dictated that no uniform must be shown abroad sailing 'Q's'. **1918** *Army & Navy Gaz.* 10 Aug. 501/1 Among the anti-submarine measures initiated and encouraged by Mr. Churchill and Lord Fisher were the 'Q' boats, the mystery attaching to which has now been dispelled by Sir Eric Geddes. . . The 'Q' boat may be briefly defined as a decoy. **1976** R. Moore *Dubai* iv. 51 We're talking about making your dhow into a high-speed Q-boat. **1937** *Times* 13 Apr. p. xxvii/2 Among the cars used by the London police are a number to which the name 'Q' is applied. . . If inside a small and unimpressive body there is an engine that will develop the highest speed attainable by the most powerful vehicle the 'Q' car is complete. **1961** *Guardian* 29 Sept. 2/3 Three men jumped out of a badly-damaged car, which crashed into the side of a lorry. . and in turn was rammed by a police 'Q' car. **1976** L. Henderson *Major Enquiry* xiv. 88 She. . doubled back. . to the waiting Q car. The watch was being kept by Sheehan and Milton. **1919** *Boy's Own Paper* July 458/1 One of the finest examples of coolness, discipline, and good organisation in the history of Q-ships. **1946** *Daily Tel.* 15 May 5/4 After his experience with two British 'Q' ships, the 10,000 ton Kolchak. . and the motor vessel Alfred Jones,. . which nearly led to his destruction in 1941, he thought U-boat captains perfectly justified in not attempting rescue work after torpedoing ships. **1972** J. Broome *Convoy is to Scatter* i. 25 The Q-ship's lure-power lay in her half-sunken appearance appealing to the U-boat captain for his coup-de-grâce.

4. Used with reference to its shape, *spec.* in *Skating*. Also *attrib.*

1852 G. Anderson *Art of Skating* vi. 73 The Q Figure. Start with a curve on the outside forwards, then change the edge to inside forwards, and finish with a circle outside backwards, all on the one foot, without setting down the other. **1935** *Encycl. Sports* 559/2 A difficult but beautiful figure called the Q. . . The figure bears a pretty distinct resemblance to the letter.

5. Repr. clipped pronunc. of 'thank you'; = *kew.

1925 Wodehouse *Sam the Sudden* ii. 13 The conductor presented himself, punch in hand. 'Fez, pliz.' 'Valley Fields,' said Kay. 'Q,' said the conductor. **1956** J. Latimer *Sinners & Shrouds* xxiii. 181 'Son of a bitch!' 'I beg your pardon?' 'I beg yours.' 'Q.'

II. 1. c. *q.d.s.* = *quater in die sumendus* 'to be taken four times a day'; *q.i.d.* = *quater in die* 'four times a day'; *q.s.* (earlier example).

d. Q.E.D. (earlier and later examples). Also as *sb. phr.*

a **1662** J. Symcotts in *Publ. Beds. Hist. Rec. Soc.* (1951) XXXI. 101 For the shaking of the hands: Take rosemary bruised q.s. and apply it to the wrists. **1760** L. Sterne *Tr. Shandy* II. xix. 168 If. . people can walk about and do their business without brains,—then certes the soul does not inhabit there. Q.E.D. **1932** *Times Lit. Suppl.* 7 Jan. 1/3 Matisse, with his frugal presentation of purely aesthetic values to a purely aesthetic appetite, appears to be the Q.E.D. of French painting as looked at in perspective. **1955** R. J. Schwartz *Compl. Dict. Abbrev.* 149/1 *q i d*, four times a day (*quater in die*—Latin). **1960** Laurence & Moulton *Clin. Pharmacol.* 454 q.d.s., quater in die sumendus, four times a day (q.i.d. and q.q.h are sometimes used). **1975** J. Mitchell *Smear Job* xviii. 159 He hates himself. . . He drinks. Q.E.D. **1977** *Lancet* 20 Aug. 376/1 Two subjects who inhaled 400 μg salbutamol q.i.d. from the start. **1978** *Jrnl. R. Soc. Med.* LXXI. 464/2 Abbreviations such as bds and qds will not be understood by foreign readers.

2. a. Q, Quartermaster, Quartermaster-General or -Sergeant; also *attrib.* or as *adj.*, and *ellipt.* for the Quartermaster's or Quartermaster-General's Department; = Quarto 2. **b.** Q and A, question and answer (esp. *attrib.*); Q.B.I. (*R.A.F. slang*), quite bloody impossible, (applied to flying conditions); also *ellipt.*; Q.C. (earlier examples); QCD, quantum chromodynamics; QED, quantum electrodynamics; Q.F., q.f., quick-firing; also *ellipt.*, a quick-firing gun; Q.I., quartz-iodine; Q.M. (examples); also *attrib.*; Q.M.G. (examples); Q.M.S., Quartermaster-Sergeant; QS [?f. *quadraphonic-stereophonic*: cf. SQ s.v. *S 4a], a designation (proprietary in the U.S.) of audio equipment used with reference to a system of quadraphonic recording and reproduction; QSO, quasi-stellar object (i.e. a quasar); QSS, quasi-stellar source (of radio waves); Q.T., q.t. (later examples).

1916 G. Frankau *Poetical Wks.*, (1923) I. 223 And the Boche shells; and 'Q.' still issues bromo. *Ibid.* 227 No more I'll turn the mordant line till 'Q' clerks blush incarnadine. **1918** *Punch* 2 Jan. 15/2 Military experts will tell you that this is a 'Q.' war, meaning thereby that the Quartermaster-General's department is the one which matters. **1919** W. S. Churchill in M. Gilbert *Winston S. Churchill* (1977) IV. Compan. I. 456 Another inroad on 'Q' should it seems to me be made by transferring all discipline to the Adjutant General. **1930** H. Belloc *Wolsey* v. 126 It was certainly he who did all the 'Q' work, to him all the letters were addressed; he gave the orders, bought provisions, organised transport, [etc.]. **1942** W. S. Churchill *Second World War* (1951) IV. I. xx. 311 The arrangements for bringing off the wounded would alone open up a vista of Q problems. **1976** D. Clark *Dread & Water* v. 119 A well-run army Q store. **1871** H. H. Furness in *New Variorum Ed. Shakespeare* I. p. ix, I have very seldom noted the *variæ lectiones* of the First Quarto. . . When referred to in the textual notes it is designated as (Q_1). **1936** *Times Lit. Suppl.* 23 May 440/2 The stage directions of the stolen Q1 and of the authoritative Q2 were not contradictory. **1964** F. Bowers *Bibliogr. & Textual Crit.* v. iv. 157 If Q is a memorially reconstructed 'bad quarto'. ., its reading derives ultimately from the prompt-book. **1954** W. R. & F. K. Simpson *Hockshop* v. 127 We stalled until we could get the police into the Q. and A. contest. **1976** B. Bova *Multiple Man* (1977) i. 14 McMurtie wanted. . to know if I'd planned a Q and A session after the speech. **1938** *Times* 3 Mar. 7/3 Instructions. . as to height and position to be kept when flying in controlled areas during 'Q.B.I.' conditions. **1942** *Tee Emm* (Air Ministry) II. 69 So now you fly in Q.B.I. *Ibid.* 143 He waited for some Q.B.I. And rushed aloft, the beam to try. **1870** A. J. Munby *Diary* 14 May (1972) 284 Came Vernon Lushington Q.C., and I did greet him friendly. **1887** L. George *Let.* 25 Feb. (1973) 18 A Q.C. of high standing. **1976** QCD [see *quantum chromodynamics* s.v. *quantum 7a]. **1979** *Nature* 1 Feb. 349/3 The one essential difference between QED and QCD is that whereas there is but one type of electrical charge in QED, the colour charge has three independent varieties. **1969** *Physics Bull.* June 223/2 The energy splitting between the $2S_{\frac{1}{2}}$ and $2P_{\frac{1}{2}}$ states of the hydrogen atom—the Lamb shift—. . arises entirely from higher order effects in qed. **1975** *McGraw-Hill Yearbk. Sci. & Technol.* 115/1 The detection of positrons from overcritical electric fields would constitute an important test of QED. **1890** G. S. Clarke *Fortification* Pl. xxviii, Balance pillar mounting for 4·7-inch Q.F. gun. **1902** *Encycl. Brit.* XXXI. 347/2 Endeavouring. . to produce a more powerful gun than the then existing Q.F. **1915** Kipling *Fringes of Fleet* 1 They gave her Government coal to burn And a Q.F. gun at bow and stern. **1972** D. Dakin *Unification of Greece* xi. 157 The Greek government ordered 144 7·5 mm Q.F. Schneider-Canet mountain guns. **1976** *Yorkshire Evening Press* 9 Dec. 21/1 (Advt.), Escort Mexico, 'K' reg. 60,000 miles. Q.I. headlamps, inertia belts. **1977** J. Hedgecoe *Photographer's Handbk.* 34 Q.I. lamps generate considerable heat, and must be ventilated. **1916** *Wipers Times* 12 Feb. (1918) 11/1 Obtainable from all Q.M. stores. **1933** M. Lowry *Ultramarine* i. 49 It's good of you to ask me in, Q.M., thanks. **1907** *Field Service Pocket Bk.* viii. 160 Q.M.G. (Maj.-Gen.). **1918** in M. Gilbert *Winston S. Churchill* (1977) IV. Compan. I. 367 As regards the latter the General Staff will inform QMG of our requirements and he will take up the matter with you. **1977** 'D. MacNeil' *Wolf in Fold* xvi. 165 I'm not leaving all those tents. . . Just think, the trouble there'd be with the QMG's department! **1916** *Anzac Bk.* 65, I am a Q.M.S. . . We have a Quartermaster, but of course, I do all the work. **1969** V. de S. Pinto *City that Shone* ix. 207 A genial horsy character called Bob Duffield, the Q.M.S. and myself. **1971** S. Hill *Strange Meeting* ii. 157 The day I went to the village to see the Q.M.S. **1972** *Wireless World* Feb. 55/2 A way of avoiding the mislocalization. . in the simple matrix technique has been adopted by Sansui in their QS system. **1975** *Official Gaz.* (U.S. Patent Office) 8 Apr. TM123/1 Sansui Electric Company Limited,. . Tokyo, Japan. Filed Dec. 13, 1972. *QS*. . . For disc-type music recordings. . . First use Oct. 23, 1970; in commerce Sept. 19, 1972. **1975** G. J. King *Audio Handbk.* vii. 167 Image error results when a basic QS decoder is used to play an SQ record. *Ibid.* 168 SQ records also yield good stereo. ., but QS records are less objectively accommodating in this respect. **1964** *New Scientist* 13 Aug. 393/3 The objects known variously as superstars, quasars, quasi-stellar objects or (for short) QSO's continue to cause intense interest among astronomers. **1973** *Nature* 23 Nov. 205/1 Although it is the majority view that QSO redshifts are cosmological in origin and related to distance by Hubble's law, several workers have proposed that QSOs may be more local objects. **1977** J. Narlikar *Struct. Universe* iii. 87 In looking for new QSOs, the astronomer picks upon star-like objects showing a marked ultraviolet radiation excess. **1965** Sandage & Wyndham in *Astrophysical Jrnl.* CXLI. 328 To the present time there have been positive identifications of nine quasi-stellar radio sources (hereinafter called 'QSS'). **1973** QSS [see *quasi-stellar a.]. **1910** A. Bennett *Clayhanger* II. xxi. 315 Mind you this is strictly q.t.! Nobody knows a word about it, nobody! **1922** Joyce *Ulysses* 610 Sailing under false colours after having boxed the compass on the strict q.t. somewhere. **1972** *New Yorker* 17 June 24/1 (*caption*) This is strictly on the q.t., Senator.

III. As a symbol. **1.** *Q* or *q* in *Physics* represents electric charge. [f. the initial letter of *quantity*.]

1846 W. Thomson in *Cambr. & Dublin Math. Jrnl.* I. 91 Denoting by Q, Q' the quantities of electricity constituting the charges before, and q, q' after contact, we shall have [etc.]. **1879** *Encycl. Brit.* VIII. 22/1 The law of electric force between two quantities q and q' now becomes Force = qq'/d^2. **1938** G. P. Harnwell *Princ. Electr. & Magn.* i. 11 **F** is the force in dynes exerted by the charge q_1 on the charge q_2. **1973** L. J. Tassie *Physics of Elem. Particles* xix. 40 The antiparticle of a particle of charge Q and baryon number B, has charge $-Q$ and baryon number $-B$.

2. *Theol.* [Prob. abbrev. of G. *quelle* source.] The symbol used to denote the hypothetical source of the passages shared by the gospels of Matthew and Luke, and not found in that of Mark.

1901 J. Moffatt *Historical New Testament* 266 It is still hotly disputed. . whether Matthew had access to any sources besides Q and Mark. **1920** *Jrnl. Theol. Stud.* XXI. 286 'Real Aramaism may be allowed ungrudgingly in those parts of the New Testament which are virtually translated from Aramaic oral or written sources', i.e. Mark and Q. **1935** R. H. Lightfoot *Hist. & Interpretation in Gospels* ii. 27 Dr. Armitage Robinson. . maintained. . that he himself was the first to use the symbol. . . In the 'nineties of the last century, he was in the habit. . of alluding to St. Mark's gospel as P (reminiscences of St. Peter), and to the presumed sayings-document as Q, simply because Q was the next letter after P in the alphabet. **1955** A. M. Farrer in D. E. Nineham *Stud. in Gospels* 56 We can conceive well enough how St. Luke could have both read St. Matthew's book as it stands, and written the gospel he has left us. Then at one stroke the question is erased to which the Q hypothesis supplied an answer. **1965** J. H. Roberts *Q Document* i. 33 The Q document is a hypothetical document invented by German biblical historians in the 1800s to explain a gap in our knowledge of the early Christian Era. . . They called this document the *quelle* or 'source' document. Later this was shortened to 'Q'. **1978** F. Neirynck in *Ephemerides Theologicae Lovanienses* LIV. 123 It seems to be a fair conclusion that he [*sc.* J. Weiss] substituted Q (= *Quelle*) for Λ (= Λόγια).

3. Also *Q factor.* The ratio of the reactance of an inductor or capacitor to its electrical resistance; more widely, a parameter of any oscillatory system representing the degree to which it is undamped, equal to 2π times the ratio of the mean total energy of the system to the energy that must be supplied each cycle to sustain the oscillations. So **Q-meter,** an instrument for measuring the *Q* of a component.

1931 *Proc. IRE* XIX. 874 Let $Q = \omega_r L/R$. **1932** F. E. Terman *Radio Engin.* ii. 39 Tubing. . has. . a better current distribution. . than does either flat- or edgewise-wound strip and hence has a better *Q* in proportion to the amount of conductor material employed. **1933** K. Henney *Radio Engin. Handbk.* vi. 109 For a coil, $Q = \omega L/R$. For a condenser, $Q = 1/\omega RC$. **1938** *Admiralty Handbk. Wireless Telegr.* II. § F19 Good coils often have a Q of the order of 100. **1943** F. E. Terman *Radio Engineers' Handbk.* XIII. 916 *Q* meters are frequently used to measure reactance and resistance (or conductance) of choke coils, dielectrics, etc., by the substitution method. **1943** *Electronic Engin.* XVI. 33/3 The two crystals were operated at 'Q' values of 20,000 and 5,000. **1948** P. M. Morse *Vibration & Sound* (ed. 2) ii. 25 Another method of expressing this is in terms of the '*Q* of the system', where $Q = (\omega_0 m/R_m)$ is the number of cycles required for the amplitude of motion to reduce to $(1/e^{\pi})$ of its original value. **1965** *Wireless World* July 338/1 A technique. . which had resulted in inductors with good Q factors of 50 to 80. *Ibid.* Aug. 413/1 A Q-meter can be used to establish the effective series resistance. **1971** [see *quality factor* s.v. *quality *sb.* 13]. **1975** D. G. Fink *Electronics Engineers' Handbk.* XI. 13 By lowering the *Q* of the optical cavity, the laser cannot oscillate, and a large inverted population builds up. When the *Q* is restored, a single 'giant pulse'. . is generated.

4. *Psychol.* Used in factor analysis to designate personality testing methods used to obtain correlations between the persons tested, by requiring each subject to rate in order those personality traits that seem most applicable to himself. Usu. as *Q-sort, -technique.*

1935 G. H. Thomson in *Brit. Jrnl. Psychol.* July 75 Then we have $Y'Y = Q$. . where Q is a p-square matrix of q-correlations, each correlation being between *two persons*, not between two tests. **1936** W. Stephenson in *Ibid.* Apr. 345 Following Prof. G. H. Thomson's suggestion, I shall use *Q* as the sign for correlations between persons, so distinguishing them from correlations such as r_{12} between two tests. . . It is convenient to designate all previous factor analysis as *r* technique, and this new inverted form as *Q* technique. **1952** R. B. Cattell *Factor Analysis* vii. 93 He [*sc.* Stephenson] has particularly urged a method in which each subject writes down a set of traits or questionnaire-like statements about himself in order of their *significance* for his own personality (*Q-sort*). **1954** A. Anastasi *Psychol. Testing* xx. 543 This approach..bears a certain resemblance to the procedure proposed by Stephenson in his 'Q-sort' technique. **1967** M. Argyle *Psychol. Interpersonal Behaviour* vii. 118 The so-called 'Q-sort' in which subjects are asked to place a series of statements on cards in order, with the cards which apply most to themselves at the top. **1972** *Jrnl. Social Psychol.* LXXXVIII. 84 The Q-sort variant known as the own-categories technique was used.

5. A unit of energy equal to 10^{18} British thermal units (very nearly 10^{21} joules).

1952 *Resources for Freedom* (President's Materials Policy Commission, U.S.) IV. xv. 213/1 In the first 18½ centuries of this era, the total input to the energy system of the world was about 6Q, equivalent to some 225 billion short tons of bituminous coal. [*Note*] 1·0Q = $1·0 \times 10^{18}$ B.t.u. **1971** *Nature* 29 Oct. 593/1 The present annual energy consumption rate of the world is 0·2Q. **1978** *Jrnl. R. Soc. Arts* CXXVI. 605/2 The earth and its atmosphere intercepts some 5200 Q of solar energy each year, one Q representing one million, million, million British Thermal Units.

qabab, var. *Kebab.

‖ **qadi, qadhi, qazi,** varr. Cadi.

1885 T. P. Hughes *Dict. Islam* 255/1 It becomes a Muslim not to covet the appointment of Qāẓī. **1899** *Folk-Lore* X. 409 So strict a Musalman as one must presume a Qazi to be. **1906** F. A. Klein *Relig. of Islam* iv. 201 He appoints the Qâdi or judge, whose office and duty is to examine law-suits. **1918** G. Bell *Let.* 6 Mar. (1927) II. xvii. 448, I sat in a row with the Qadhi, the Mudir of Church Lands.., the Judge of Appeal and so on and so on. **1955** G. E. von Grunebaum *Islam* i. 11 The canon-law judge, the qāḍī, will find himself unable to take care effectively of all contingencies. **1959** *Listener* 19 Nov. 888/1 Subsequent legislation provides for a similar *qadi's* court at Kombo St. Mary. **1971** *Illustr. Weekly India* 4 Apr. 47/1 The imperative that the law must be enforced by a Muslim *qazi* became meaningless in British India. **1977** *Times* 21 Nov. 7/3 At the service, a *qadi* read a sermon in which he said abandoning Jerusalem was like abandoning Mecca.

‖ **qaimaqam,** var. Kaimakam.

1961 *Times* 17 July 11/2 The then Shaikh of Kuwait, who in the following year was invested with the rank of qaimaqam. **1970** H. Trevelyan *Middle East in Revolution* 182 When in 1899 Mubarak as Sabah of Kuwait murdered his brothers, he sought protection with the Turkish Government which gave him the honorary title of Qaimaqam or Sub-Governor of a district.

‖ **qanat** (kanā·t), var. *Kanat.

1944 G. C. Thompson *Tombs & Moon Temple of Hureidha* 10 The Persian *qanat*, for instance, is seen on the coast, and in the Yemen. **1953** A. Smith *Blind White Fish in Persia* i. 19 Shortly after the collection of facts had begun, we learnt of the qanats. It was written that there were 100,000 miles of them in Persia, that they were artificial underground water channels and that the deepest was over 1,000 feet. *Ibid.* iii. 55 As with birds before an island, the qanat wells warned you of the approach of a town. **1958** A. J. Toynbee *East to West* lix. 176 One sees line upon line of qanat-made molehills. **1966** P. English *City & Village in Iran* i. 19 By the sixth century B.C.,..qanat technology was known on the Central Plateau. *Ibid.* iii. 50 The qanat-watered towns and villages on the slopes of the Kuhi Jupar. **1976** *Apollo* Apr. 302/3 Seen from the air, the ventilating shafts of these *qanats* punctuate the buff-coloured plains like perforations in paper. **1976** *Times* 17 Aug. 11/6 If the underground water resources of a semi-arid region are so immense that they can keep the 'qanats' running for the whole year then the unexploited ground-water resources of the British Isles must be incalculable.

‖ **qanon, qanun,** varr. Kanoon.

1874 C. Engel *Descr. Catal. Musical Instruments S. Kensington Museum* 208 The *kanoon*, or *qânon*, an instrument especially appertaining to the Arabs and Persians, is, like the *santir*, a kind of dulcimer evidently of high antiquity in the East. **1931** C. S. Hurgronje *Mekka in Latter Part of 19th Cent.* 44 Much worse is their habitual accompaniment of song with musical instruments: especially the *gabûs*, a four-stringed instrument which much resembles the *kemênjeh*, only that its strings are of gut instead of horsehair, and also the well known *qânûn* (guitar). **1957** H. G. Farmer in E. Wellesz *Anc. & Oriental Mus.* xi. 444 The psaltery was attributed to Al-Fārābi,..but the instrument is not mentioned by him under its millennium-old name of *qānūn*... As the *qānūn* it was known in Muslim Spain in the eleventh century, and in the fourteenth century it was mounted with sixty-four strings, tuned tricordally, in Persia. **1976** D. Munrow *Instruments Middle Ages & Renaissance* 21/4 Little metal flaps fitted on the modern Arab qānūn.. enable the players to alter the pitch of a course of strings with an adroit flick of the fingers. *Ibid.* 23/2 The Arabic qānūn today is a large psaltery played with great virtuosity in Middle Eastern orchestras, and is a direct descendant of the forerunner of the European psaltery.

‖ **qasida** (kasī·da). Also 9 **kaszyde; kasida(h, quasida.** [Arab. *ḳaṣīda*.] An Arabic or Persian panegyric or elegiac poem or ode, usu. having a tripartite structure.

1819 J. L. Burckhardt *Trav. Nubia* 354 Like the eastern Arabs, they celebrate the praises of their warriors in the Kaszyde. **1842** McG. de Slane in Ibn Khallikan *Biogr. Dict.* I. p. xxxiii, The opinion held in the schools that the ancient *kasîdas* were masterpieces of art contributed also to the perversion of good taste; their plan and ideas were servilely copied, and it was by refinement of expression alone that writers could display their talent. **1885** *Encycl. Brit.* XVIII. 656/1 Those principal forms of poetry now used in common by all Mohammedan nations—the forms of the *ḳaṣīda* (the encomiastic, elegiac, or satirical poem), the *ghazal* or ode [etc.]. **1903** C. Huart *Hist. Arabic Lit.* ii. 10 According to the ancient rules..the author of a *qaṣīda* must begin by a reference to the forsaken camping-grounds. Next he must lament, and pray his comrades to halt, while he calls up the memory of the dwellers who had departed. **1907** R. A. Nicholson *Lit. Hist. Arabs* iii. 76 This fashion centres in the *Qaṣída*, or Ode, the only form, or rather the only finished type of poetry that existed in what..may be called the classical period of Arabic literature. **1927** F. Krenkow in T. Houtsma et al. *Encycl. of Islam* II. 796/1 An Arabic (or Persian etc.) ḳaṣīda is a very artificial composition; the same rhyme has to run through the whole of the verses, however long the poem may be. *Ibid.* 796/2 The form of the ḳaṣīda has survived to modern times and I have specimens by poets still living where we find..a description of a desert-ride by persons who live in Cairo and travel by railway. **1934** *Times Lit. Suppl.* 27 Sept. 641/2 The four main types of Persian poetical composition are still..the *qasida*, the *ghazal*, the *masnavi* and the *ruba'iy*. **1940** F. Stark *Winter in Arabia* 45 Iuslim last night brought a singer of Qasidas. **1958** L. Durrell *Balthazar* iv. 82 He was delighted to hear some music and listened with emotion to the wild *quasidas* that the old man sang. **1964** *Listener* 25 June 1036/1 This translation..from the Spanish of an Arabic-Andalusian *qasida* fragment from eleventh-century Toledo epitomizes one aspect of this breath of fresh air from the east which began to revivify the poetry of southern Europe. **1971** G. Household *Doom's Caravan* iii. 158 My translation of Shakespeare's sonnets into kasidahs.

‖ **qat** (kat). Also **quatt.** [Arabic: var. Kat; *khat.] = Kat; also, the narcotic drug obtained from the leaves of this plant. Also *attrib.*

1958 *Times* 16 May 7/2 Lifting the ban on the importation into Aden Colony of qat, the narcotic leaf, was recommended by the qat commission of enquiry. **1963** *Times* 12 Mar. 12/7 Qat-chewing is not designed for the promotion of conversation. **1966** 'S. Harvester' *Treacherous Road* xi. 104 Ledgers and records..gave details of his exports of coffee, *qat*, fruit. **1968** *Daily Colonist* (Victoria, B.C.) 17 Dec. 5/2 A new product..is chewing gum containing..'Quatt', a mild narcotic produced from the tender leaves of a tea-like bush growing in profusion throughout the Middle East. **1971** [see *gob-stopper* s.v. *gob *sb.*[3]]. **1978** *Guardian Weekly* 17 Sept. 9/1 Normally, the day's qat session begins in the afternoon but during Ramadan it is chewed through most of the night.

Qatabanian (kæ:tăbē^i^·niăn), *a.* and *sb.* Also **Catabanian, Kata-, Qatha-.** [f. *Qatabān* (see below) + -ian.] **A.** *adj.* Of or pertaining to the kingdom of Qatabān in south Arabia, or its ancient Semitic dialect or language. **B.** *sb.* A native or inhabitant of Qatabān; the Qatabanian language. Also **Qataba·nic** *a.*

1926 A. Musil *Northern Heǵâz* 310 It is named after the Catabanian, Esbonitan, and Scenitan Arabs. **1934**, **1936** [see *Minæan *sb.* and *a.*]. **1939** L. H. Gray *Foundations of Lang.* 364 *South Arabic* is known only from inscriptions (Minaean, Sabaean, Qathabānian, and Ḥadramautian) ranging, perhaps, from the eighth century B.C. to the sixth A.D., and by..modern dialects. **1948** D. Diringer *Alphabet* II. ii. 225 The Qatabanian kingdom, with its capital at Tamna'. *Ibid.* 226 The South Arabian inscriptions..are generally divided into five groups: The Minaean..the Sabaean..the Himyaritic..the Qatabanic and the Hadhramautic. **1951** W. F. Albright in H. H. Rowley *Old Testament & Mod. Study* (1961) 9 The most archaic South-Arabic dialect, Qatabanian, shows uses of enclitic *m* more closely parallel to Proto-Sinaitic as deciphered by the writer than does any other Semitic tongue. **1959** A. F. L. Beeston *Qahtan* I. 8 *N* is the twelfth letter of the Qatabanian alphabet. *Ibid.* 12 And when the overseer of šmr announces that he desires Qatabanians to make (trading) journeys among the tribes,..then Qatabanians may trade on their own account with the tribes. **1971** B. Doe *Southern Arabia* i. 22 It is possible to note that the languages of the Minaeans, Sabaeans, Qatabanians and Ḥadramant were similar but with differing dialects. *Ibid.* ii. 70 The Qatabanian kingdom was for centuries a neighbour south-west of Saba', and the capital city was Timna', now also known as Hajar Quḥlan (Koḥlān) in the Wadi Baihān. **1973** A. K. Irvine in D. J. Wiseman *Peoples Old Testament Times* xii. 299 The apparent relationships of the South Arabian languages may suggest that while Minaean, Qatabanian, and Ḥaḍrami could have a north-eastern origin, Sabaean came rather from Central or North Arabia.

Qatari (kæ̆tā·ri, ‖ gatā·ri), *sb.* and *a.* [f. *Qatar* (see below) + *-i.] **A.** *sb.* A native or inhabitant of the state of Qatar in the Persian Gulf; also *Comb.*, as *Qatari-born* adj. **B.** *adj.* Of or pertaining to Qatar.

1959 R. Hay *Persian Gulf States* 110 The Ruler's Courts exercise jurisdiction over Qataris. **1960** *Geogr. Jrnl.* CXXVI. 447 The transcription used in this article is a conventional one, with a few exceptions which show the Qatari pronunciation of certain consonants, vowels and diphthongs. **1964** *Ann. Reg. 1963* 316 New regulations controlling foreign investment laid down that any foreign firms engaging in commerce or industry..must have Qatari partners holding at least 51 per cent of the capital. **1970** *Guardian* 10 Apr. 11/7 Qatar is regarded..as a protégé of Saudi Arabia, a point some Qataris indignantly deny. **1976** *Times* 3 Sept. (Qatar Suppl.) p. i/1 Qatari-born civil servants occupy many of the top-level posts. *Ibid.* p. i/7 Western expatriates as well as Qataris indisputably live well, usually employing a domestic servant or nanny... All educated Qataris speak English as their second language. **1978** *Financial Times* 22 Feb. 20/2 A further licence has been granted to a group of Qatari nationals to start a thirteenth bank, but it is not known when, or whether, this bank will commence business. **1979** R. S. Zahlan *Creation of Qatar* 118 For the next thirty years, no indication of the number of Qataris is available.

Qazaq, var. *Kazakh.

qazi: see *qadi.

‖ **qere, Q^e^re,** varr. Keri.

1941 R. H. Pfeiffer *Introd. Old Testament* I. v. 93 The *kethib* (written) is the consonantal text; the *qere* (read) is an accepted reading differing from the consonantal text. *Ibid.* 95 Long after the introduction of the *qere* 'Lord' for *YHWH*..vulgar expressions in the text..were removed by substituting a euphemism in the reading (*qere*). **1958** F. Kenyon *Our Bible & Anc. MSS.* (ed. 5) iv. 78 Such variations were known by the name of Q^e^re ('read') and k^e^thib ('written'), the latter being the reading of the text, and the former that of the margin, which was to be substituted for the other when the passage was read. **1968** J. Barr *Compar. Philol. & Text of Old Testament* x. 246 Here, however, there is a rather peculiar *Qere*. *Ibid.* 247 The translation at 1 Sam. 25. 14 makes it extremely probable that the LXX is a translation not of the present *Kethibh* at 14. 32 but of the present *Qere*. **1969** R. K. Harrison *Introd. Old Testament* (1970) iv. ii. 213 The preferred reading was known as *Q^e^re*, meaning 'that which is to be read', as contrasted with the sacred consonantal text, the *K^e^thibh* or 'written' Scripture. The *Q^e^re* on occasions may have constituted a genuine variant reading.

Q fever (kiū· fī:vəɹ). [f. the initial letter of Query *sb.*[1] + Fever *sb.*[1]] A disease caused by the rickettsia *Coxiella burnetii* that is variable in symptoms and often resembles influenza.

1937 E. H. Derrick in *Med. Jrnl. Australia* 21 Aug. 282/1 The suspicion arose and gradually grew into a conviction that we were here dealing with a type of fever which had not been previously described. It became necessary to give it a name, and 'Q' fever was chosen to denote it until fuller knowledge should allow a better name. **1947** *Ann. Rev. Microbiol.* I. 342 The serological diagnosis of Q fever is at present the only satisfactory one for routine work, since the clinical picture is such that Q fever may be confused with a number of other diseases. **1964** E. H. Derrick in *Queensland's Health* Dec. 11/2 'X' is a recognised term for an unknown quantity. But Australia already had an 'X disease', now known as Murray valley encephalitis. However, the rest of the alphabet was open. Query also signified the unknown. 'Q (for query) fever' it became. *Ibid.*, Many have wrongly assumed that the 'Q' stands for Queensland. **1978** *Jrnl. R. Soc. Med.* LXXI. 765 There appear to have been no further military outbreaks of Q fever until the Cyprus epidemic of 1974.

‖ **qi** (tʃi). Also **ch'i, Qi,** etc. [Chin. *qì* air, breath.] The physical life-force postulated by certain Chinese philosophers; the material principle.

1850 *Chinese Repository* XIX. 370 The following short expression of the doctrine of the Yih King is that in which probably all the literati would agree. It is from the 49th section of Chú-fútsz's entire works. 'All things..come only from the Great Extreme (*t'ái kih*). The Great Extreme is the primordial substance (*k'í*) which, moving along, divided and made two *k'í*; that which in itself has motion is the *Yang*, and that which had rest,..is the *Yin*.' **1917** S. Couling *Encycl. Sinica* 436/2 The nature of man consists of this *Li*, or the Ethical Principle... In its essence it is absolutely pure and good, but seeing that it is inseparable from the material element *Ch'i*..it is from Man's birth to a greater or less extent impeded and tainted. **1958** W. Willetts *Chinese Art* II. vii. 586 Corresponding to this formal cause of each existence was its material cause, *ch'i*. **1964** K. K. S. Ch'en *Buddhism in China* xiv. 395 Chang Tsai (1020–1077) put forward a metaphysical system based on the theory that ch'i, ether or matter, existed at the beginning of the world. He held that ch'i consolidated itself into things at the beginning, and that things dissolved into ch'i in the end. **1971** F. Mann *Acupuncture* vi. 57 To the ancients the cornerstone of the theory of acupuncture, the concept whereby they explained its effects and action, was Qi, the energy of life. **1972** *Which?* Feb. 49/2 The energy of life (called Ch'i) flows along various 'meridians' in the body, and acupuncturists believe that if needles are inserted at..points along these meridians..the flow of energy in the body can be corrected. **1973** *Lancet* 14 July 58/1 They are not connected to internal organs, and Qi or anything else cannot flow along them. **1978** *Nature* 26 Oct 697/1 Arguably the most original Letter so far is a report of some experimental results on the physical basis of the traditional *yunqi* therapy—the curing of disorders through the passing of *qi* (pneuma) from doctor to patient without bodily contact.

Qiana (ki,ā·nă). Chiefly *U.S.* Also **qiana.** [Invented word.] A proprietary name in the U.S. for nylon.

1968 *Official Gaz.* (U.S. Patent Office) 3 Sept. TM25/1 Qiana. For Yarns of Man Made Fibers..First use May 15, 1968... E. I. du Pont de Nemours and Company. **1969** *Science Year* 274/1 Qiana is said to have color, clarity, and luster equal to or better than most luxurious silks. **1971** *New Yorker* 10 July 69 The commendable and reasonable sedate bathing suits produced by Edith Lances of sea-blue Qiana, which is a man-made silk. **1975** *Times* 13 May 12/4 Washable qiana jersey dress... Approx. £39.50. **1977** *Monitor* (McAllen, Texas) 26 June 1c/2 The mother of the bride wore a pale blue gown of

Qiana with a fitted long-sleeved jacket. **1979** *Farmington* (New Mexico) *Daily Times* 27 May 3A/5 (Advt.), Give him his favorite Qiana shirt.

qibla(h, var. Kiblah in Dict. and Suppl.

qibli, var. *gibli.

‖ **qiviut** (ki·vi,ŭt). [Eskimo.] The underwool of the arctic musk-ox; fibre made from this.

1965 *Sci. News Let.* 12 June 370/1 Many woolen manufacturers are enthusiastic about the principal product of the musk ox, its underwool, which the Eskimos call 'qiviut'. The fiber is similar to that of cashmere but about twice as long and half as thick. **1968** *Beaver* Winter 37/2 Eskimos call him [*sc.* the musk-ox] 'umingmak', the bearded one, and have long known his wonderful underwool as qiviut, a fibre far warmer than the silken down of the northern grey goose. **1972** *Guardian* 26 June 12/7 Collection of qiviut begins at the end of April. Bulls yield 6lb, and cows 5lb. **1979** R. Fiennes *Hell on Ice* viii. 125 The early explorers shot many musk-oxen. They took the soft wool that lies beneath the long brown hair. This *qiviut* was valuable to the Eskimo.

Q-spoi·ling, *vbl. sb.* *Physics.* [f. *Q III. 3 + Spoiling *vbl. sb.*[1]] = *Q-switching *vbl. sb.* So **Q-spoi·l** *v. trans.*, **Q-spoi·led** *ppl. a.*

1963 *New Scientist* 24 Oct. 201/2 A technique known as 'Q-spoiling' is used to store the laser's energy and liberate it in brief but tremendous bursts. **1966** *Appl. Physics Lett.* IX. 285 A technique involving the use of a magnetic field in 'Q'-spoiling a ruby laser cooled to 77°K is discussed. **1970** *Physics Bull.* Mar. 116/2 By temporarily 'Q-spoiling' the laser resonator by inserting a Kerr cell shutter between the high reflectivity mirror and the ruby laser rod, the energy storage capacity of the ruby was increased above the level at which relaxation oscillation would normally begin. *Ibid.*, By 1965 powers of 10^9 W in pulses of duration ~ 10 ns were produced by following the Q-spoiled oscillator with a series of ruby amplifiers. **1975** D. G. Fink *Electronics Engineers' Handbk.* xi. 13 The typical output of an optical laser consists of a series of spikes occurring during the major portion of the time that the laser is pumped... Q switching (Q spoiling) is a means of obtaining all the energy in a single spike of very high peak power.

Q-switch (kiū·,switʃ), *sb.* *Physics.* Also **q-switch.** [f. *Q III. 3 + Switch *sb.*] A means of suddenly increasing or decreasing the *Q* of a laser by effectively unblocking or blocking the optical path to one of the mirrors.

1963 *Jrnl. Appl. Physics* XXXIV. 1000/2 (*heading*) Faraday effect as Q-switch for ruby laser. *Ibid.*, The Q-switch..was placed between the ruby and the separated end mirror. **1966** *New Scientist* 20 Oct. 93 Q-switches are employed to obtain a very powerful pulsed output from a laser by allowing the laser to store up energy; when it reaches a maximum the blockage is quickly removed, and an intense pulse of radiation is emitted. **1973** *Jrnl. Appl. Physics* XLIV. 4067/1 The oscillation of the laser with the cavity closed is sensed by a fast photodiode whose output signal opens the electro-optical *Q* switch. **1979** *Nature* 5 July p. vii/2 A compact pulsed ruby laser, that incorporates a q-switch.

So **Q-swi·tching** *vbl. sb.*, the process of pumping a laser that has a low *Q*, and so cannot oscillate, and then suddenly increasing the *Q* so that the stored energy is released in a single pulse of very high power; **Q-switch** *v. trans.*, to subject (a laser) to this process; **Q-swi·tched** *ppl. a.*

1963 *Jrnl. Appl. Physics* XXXIV. 1000/2 We wish to report the successful application of the Faraday effect (magneto-optic shutter) as a Q-switching technique for a ruby laser. *Ibid.* 3407/1 (*heading*) Q-switched $CaWO_4$: Nd^{3+} laser. *Ibid.*, A $CaWO_4$:Nd^{3+} laser was successfully Q-switched. **1965** *Wireless World* July 351/3 The ruby laser is Q-switched by a rotating prism. **1968** *McGraw-Hill Yearbk. Sci. & Technol.* 223 Saturable absorbers have recently been used very successfully to Q-switch ruby and neodymium-doped lasers. **1968** *New Scientist* 24 Oct. 205/1 Q-switching is a technique for producing giant laser pulses by preventing lasing action until a large amount of energy has been pumped into the atoms responsible. **1970** *Sci. Amer.* Mar. 41 They used the 104-inch telescope to transmit and detect pulses of 50-nanosecond duration produced by a 'Q-switched' (short-pulse) ruby laser. **1974** *Physics Bull.* Jan. 13/1 A four stage Q switched ruby laser is used to fire a pulse every three seconds, with a pulse length of a few tens of nanoseconds and an output energy of about 3 J.

qua, *adv.* Add: The pronunc. (kwā) is also common.

Qua, var. *Kwa.

Quaalude (kwā·lūd). *Pharm.* A proprietary name for methaqualone; also, a tablet of this.

1966 *Official Gaz.* (U.S. Patent Office) 12 Apr. TM72/1 William H. Rorer, Inc., Fort Washington, Pa... *Quaalude.* For Sedative-hypnotic tablets. First use Aug. 5, 1965. **1967** H. Beckman *Dilemmas in Drug Therapy* 186/2 As 'newer' hypnotics I list the following:..methaqualone (Quaalude), 150 mg. at bedtime. **1968** *Trade Marks Jrnl.* 4 Dec. 2107/2 *Quāālude*... Sedative-hypnotic pharmaceutical preparations. William H. Rorer, Inc... Fort Washington. **1974** *Saturday Night* (Toronto) July 21/2 A quick trip to an interior bedroom, a little Quaalude to relax, a little coke to get the performing ego's motor humming and you were ready to face down King Kong. **1977** *Rolling Stone* 13 Jan. 31/1 A doctor had prescribed the chalky white Quaaludes to help her sleep at night. **1979** *Guardian* 9 Jan. 5/8 Quaalude (Mandrax in Britain)..has a reputation as an aphrodisiac and is one of the fastest rising drugs of abuse in the US.

quack, *sb.*[1] Add: **1. b.** *slang* (orig. *Austral.* and *N.Z.*). A doctor (with no implication that he is unqualified); also in *Mil.* use, a medical officer.

1919 W. H. Downing *Digger Dial.* 40 *Quack*, a medical officer. *c* **1926** 'Mixer' *Transport Workers' Song Bk.* 43 And ask me if I want a 'sub'. For to take me to the 'quack'. **1943** *Coast to Coast 1942* 29 Might be he lose his leg if we don't get him across right away to the quack. **1945** C. H. Ward-Jackson *Piece of Cake* (ed. 2) 51 *Quack*, medical officer. **1960** J. Iggulden *Storms of Summer* 169 I'll get the quack at the Bush Hospital to have a look at it in the morning. **1962** Granville *Dict. Sailors' Slang* 93/1 *Quack, the*, medical officer. Jocular. **1976** D. Ireland *Glass Canoe* 136, I go along to this quack and he says Get back to the surf and get some green vegetables into you.

quack (kwæk), *sb.*[4] *U.S.* [var. of Quick *sb.*[2]] Couch-grass, *Agropyron repens*, a European grass with creeping roots, widely naturalized elsewhere. Cf. Quick *sb.*[2]

1833 L. C. Beck *Bot. N. & Middle States* 416 A troublesome weed. Couch Grass. Quack. **1872** *Rep. Vermont Board Agric.* I. 289 He who sets out to subdue a piece of quack must resolve on no half-way measures. **1909** *N.Y. Even. Post* (semi-weekly ed.) 11 Mar. 5 In conquering the quack he did the one thing that could have enabled him to get a crop from that unfertilized soil. **1930** *Times Educ. Suppl.* 31 May 248/1 It [*sc.* couchgrass] has a good many names: squitch, scutch, quack..are all in use. **1948** H. A. Jacobs *We chose Country* 189 The big garden across the road, where we fought quack instead of weeds, really established us.

2. *Comb.* **quack-grass** = prec.

1822 A. Eaton *Man. Bot.* (ed. 3) 404 *Triticum repens*, wheat-grass, couch-grass, quack-grass... Very troublesome in fertile soil, and useful in barren sand. **1839** J. Buel *Farmer's Compan.* xiv. 151 One of our neighbours has been enabled completely to eradicate quack-grass in his Indian corn. **1884** G. Vasey *Agric. Grasses U.S.* 108 Quack grass... There has been a good deal of discussion relative to this grass, some pronouncing it one of the vilest of weeds. **1949** *This Week Mag.* 17 Sept. 2/2 The quackgrass and the sassafras is getting the best of him. **1970** *Daily Progress* (Charlottesville, Va.) 24 May 4/2 Burning robs the topsoil of its fertility..and actually increases the growth of quackgrass, weeds, and other unwanted, troublesome, perennial plants.

quack, *v.*[1] **3.** (Later example.)

1925 *Scribner's Mag.* Oct. 385/1 Time..has not obliterated the love of being quacked.

quacker (kwæ·kəɹ). *colloq.* [f. Quack *v.*[2] + -er.] One that quacks; a duck. Also *fig.*

1846 *Swell's Night Guide* 75 Jest pipe her—she turns her ogles up like a croaking quacker (dying duck). **1965** *New Statesman* 18 June 980/2 The noisiest quackers have now felt a wind of deflation beginning to stir about the world and have fallen silent. **1978** *Times* 23 Feb. 16/4 Though ducks are plentiful along the Thames..two dozen quackers were brought along.

quacky, *a.*[1] (Earlier example.)

1836 *Southern Lit. Messenger* II. 327 The critical department of this work..is in our opinion decidedly quacky.

quad, *sb.*[3] Add: Also in other senses of Quadruplet. (Further examples.)

1951 M. Abercrombie et al. *Dict. Biol.* 182 An armadillo (*Dasypodus*) always has identical quads from a single ovum. **1974** *Publishers Weekly* 26 Aug. 302/3 The good sports books seem to come in quads or pairs: four Babe Ruth books, two on [etc.]. **1975** *Nature* 3 Apr. 379/2 If the same pairs, triples, quads.., of references are observed in a number of articles, there is an implied consensus of opinion. **1978** *Daily Tel.* 4 Dec. 2/6 A five-year-old Friesian cow has given birth to quads. **1979** *Daily Tel.* 16 Nov. 3/4 Quads were born by Caesarean operation at St. Mary's Hospital, Paddington.

quad, *sb.*[5]: see *quad *a.*[1]

quad (kwǫd), *sb.*[6] *Teleph.* [Abbrev. of Quadruplex *a.* and *sb.*] A group of four insulated conductors twisted together, the conductors usu. forming two circuits. Also *attrib.* or as *adj.*, as *quad cable.*

1922 *Bell Syst. Techn. Jrnl.* I. 72 The cable is of quadded construction, that is, the wires are first wrapped with dry paper for insulation and twisted into pairs and then two pairs are twisted into what is called a quad. **1940** *Chambers's Techn. Dict.* 691/1 *Quad* (or *star-quad*) *cable*, lead-covered cable in which the unit group is four paper-insulated conductors twisted together, opposite conductors forming the go and return circuits of a four-wire channel respectively. **1941** A. E. Knowlton *Stand. Handbk. Electr. Engineers* (ed. 7) xxii. 2075 A quad is formed by twisting together two individual pairs,..in the same manner that each pair is formed. **1958** J. R. G. Smith *Elem. Telecommunications Pract.* v. 72 The quads are then made up in the number of layers required for the size of the cable.

quad (kwǫd), *sb.*[7] [Abbrev. of Quadrilateral *a.* and *sb.*] **1.** *Naut.* A four-sided jib used on racing yachts. Also *attrib.*

1937 *Sun* (Baltimore) 4 Aug. 14/1 Endeavour's board of strategy had ordered a new quadrilateral Genoa jib made of synthetic silk and similar to the big 'quad' which has proved so effective in Ranger's windward work. **1937** *Yachting Monthly* LXIII. 376/1 Both yachts broke out their Genoa jibs, but whereas Ranger's was a double-clewed sail, or 'quad', that aboard the challenger was somewhat smaller. *Ibid.* 378/1 Vanderbilt sprang his own choice in head rigs and set a wonderful quad Genoa. **1938** *Britannica Bk. of Year* 712/1 A feature of 'Ranger's' rig was the efficient use of quadrilateral jibs and particularly of a large 'quad' Genoa jib made of rayon.

2. *Radio.* An aerial in the form of a square or rectangle broken in the middle of one side. Freq. *attrib.*

1961 *Amateur Radio Handbk.* (ed. 3) xiii. 385/2 A quad aerial is practicable for 14 Mc/s but it is, of course, a one-band aerial; however, 21 and 28 Mc/s quads can be nested instead inside a 14 Mc/s structure without serious interference. **1962** *Flight Internat.* LXXXI. 596/1 A quad-helix aerial could be seen... It is one of 52 Cubic installations furnished for acquisition, telemetry and communications for manned orbital flights. **1976** Perkowski & Stral *Joy of CB* xi. 125 Even greater directionality is achieved with a stacked quad consisting of four elements.

quad, *sb.*[8]: see *quad *a.*[2]

quad (kwǫd), *sb.*[9] [Abbrev. of Quadrillion.] A unit of energy equal to 10^{15} British thermal units (very nearly 10^{18} joules).

1974 *Newsweek* 7 Oct. 84/2 The current standard energy forecast is that U.S. demand will double in the next ten years to nearly 145 'quads' (or quadrillion BTU's). **1976** *Sci. Amer.* Jan. 21/2 Without any new initiatives the need for imported oil will rise steadily from about 12 quads at present to more than 60 in the year 2000. **1976** *National Observer* (U.S.) 17 Apr. 8/3 The nation's energy consumption in 1974 was about 73 quads. **1980** *Nature* 10 Apr. 501/2 The first assumes a level of energy consumption in the US of 71 quads, equal to estimates of consumption in 1975 when the study was started.

quad, *a.*[1] Add: Also in other contexts in the sense of Quadruple *a.* (Further examples.)

1961 *Times* 30 Oct. (Brit. Posters Suppl.) p. iii/6 Posters ranging from quad-crown to crown-folio. **1967** *Electronics* 6 Mar. 158/1 One manufacturer..offers a 'quad 2-input OR gate' that actually has three 2-input AND gates and one 3-input AND gate feeding into a NOR gate. **1972** G. V. Higgins *Friends E. Coyle* xviii. 106 Eddie Coyle drove the old Sedan de Ville cautiously, the quad headlights on high beam. **1972** *Country Life* 10 Feb. 341/2 Asked what brought her to the forecourt apart from the imminent need for petrol, she replied unhesitatingly: 'Quad stamps and promotion offers.' **1973** D. Robinson *Rotten with Honour* 176 How much should I ask him for it?.. I'll try two-fifty and maybe throw in quad stamps. **1974** *Sunday Times* 16 June 29/6 Their time of 4 min 49·80 sec...was only 20 seconds slower than Eton's record in the eights, clearly indicating the potential speed of quad scullers.

B. *sb.*[5] Also **Quad.** A vehicle with four-wheel drive; *spec.* (see quot. 1948). orig. *U.S.*

1919 'I. Hay' *Last Million* xiv. 224 Smaller vehicles of American design, known as 'Quads'. These possess the unusual feature of a drive upon either axle. **1941** *Illustr. London News* 24 May 657/2 The third drawing shows 'Quads', powerfully mechanised units, which haul and guide the new guns into position and out. **1948** Partridge *Dict. Forces' Slang* 150 *Quad*, a four-wheel drive, gun-towing vehicle. **1961** W. Vaughan-Thomas *Anzio* vi. 122 The Recce Regiment cars were trying to tuck themselves behind the walls as shelter for the night, a Quad, towing an anti-tank gun, was vainly trying to do the same thing.

quad (kwǫd), *a.*[2] and *sb.*[8] **A.** *adj.* Abbrev. of *quadraphonic *a.*

1970 *Sat. Rev.* (U.S.) 27 June 56/1 Their new Quadraphonic Processor is an add-on bringing true (and quite well simulated) quad sound to two-channel systems. **1972** *Observer* (Colour Suppl.) 22 Oct. 56/4 The manufacturers have been to great pains to make their quad equipment 'compatible'. **1976** *Washington Post* 19 Apr. c8/4 (Advt.), 'Jesus Christ Superstar' 7:30, 9:30. Shown in quad sound and Big Screen Projection. **1977** *Listener* 20 Oct. 507/3 With quad transmissions now a regular feature of Radio 3, the argument that the BBC cannot afford to extend stereo coverage seems a little thin. **1980** *Broadcast* 7 July 8/3 A 17-year-old 2" quad VTR machine.

B. *sb.*[8] Abbrev. of *quadraphony.

1971 *Esquire* Nov. 228/2 The necessity of making quad compatible with mono and stereo recordings and broadcasts is probably going to necessitate the use of some system that matrixes the four channels down to one. **1974** *Radio Times* 27 June 5/1 Quad adds a depth and perspective to radio that is missing in stereo. **1975** *Time Out* 15 Aug. 49/3 They give a live electronic concert (in quad) of music by little-known younger composers.

quad, *v.*[1] Add: (Further example.) Hence **qua·dded** *ppl. a.*, **qua·dding** *vbl. sb.*

1914 H. Pender *Amer. Handbk. Electr. Engineers* 1355 *Quadded* or *phantomed cable*,..a cable adapted for the use of phantom circuits. **1922** [see *quad *sb.*[6]]. **1962** *Engineering* 16 Feb. 245/1 Once insulated, the conductor wires are either twisted into pairs or 'quadded' in groups of four. *Ibid.* 245/2 After twisting or quadding each conductor wire is given a 'ring through' test with a bell and battery.

quadrangle, *sb.* Add: **1. b.** *Palmistry.* (See quot. 1883.)

1883 Frith & Heron-Allen *Chiromancy* 138 The Quadrangle is that part of the human hand comprised between the line of the Head and the line of the Heart

and between the line of Fate and the line of Apollo. **1895** H. Frith *Pract. Palmistry* III. i. 121 The Quadrangle is an extremely important space, for upon its width and general appearance the mind and the disposition of the man or woman may be estimated and 'reckoned up'. **1934** C. de Saint-Germain *Study of Palmistry* IV. 313/1 A cross in the Quadrangle touching the Line of Heart—Influence of the opposite sex on the subject... A cross in the Quadrangle touching the Line of Head—The subject will exert in the matters of love or friendship more influence on the other person than the said person will exert on him. **1952** J. Malcolm *Frith's Pract. Palmistry* (rev. ed.) xviii. 112 The Quadrangle should be regular and wide in the centre, and it should expand at both ends. This indicates good health, honesty and trustworthiness. **1971** *Cheiro Bk. of Fate & Fortune* xli. 122 When the quadrangle is abnormally wide in its entire length, it denotes want of order in the brain, carelessness of thought and ideas, an unconventional nature, and one imprudent in every way.

quadrangular, *a.* (Later *Comb.* example.)

1854 *Poultry Chron.* I. 431 It is a spacious, quadrangular-shaped house, built of a greyish stone.

quadrant, *sb.*[1] Add: **4. b.** *spec.* (i) *Naut.* A metal frame, shaped as the quadrant or sector of a circle, that is fixed to the rudder head or stock and to which the steering ropes or chains are attached.

1885 H. Paasch *From Keel to Truck* 32/1 *Tiller*..*quadrant*, barre de gouvernail en quadrant. Quadrant als Ruderpinne. **1894** *Ibid.* (ed. 2) 223 Steering-quadrant. **1923** *Glasgow Herald* 3 Feb. 8/7 The modern helm, or its equivalent, the quadrant, is placed out-board. **1961** *Lloyd's Register of Shipping: Rules & Regs. Construction Steel Ships* 57/1 Tillers and quadrants are to be shrunk on or bolted to the rudder head. **1976** *Oxf. Compan. Ships & Sea* 731/2 It is by means of the quadrant, with the assistance of a steering engine, that force is applied to turn the rudder.

(ii) (See quot. 1940.)

1885 H. Paasch *From Keel to Truck* 80/1 *Quadrant, reversing*, secteur de changement de marche. Umsteuerungs-Quadrant. **1940** *Chambers's Techn. Dict.* 691/1 *Quadrant*, a slotted segmental guide through which an adjusting lever (e.g. a reversing lever) works. It is provided with means for locating the lever in any desired angular position. **1959** *Weekly Times* (Melbourne) 30 Sept. 1 (Advt.), At your fingertips is the 'automatic brain' of your Ferguson [tractor]—the amazingly simple control quadrant that gives you complete 4-way work control. Without moving your hand more than six inches, you can raise and lower implements or hold them rigidly in any position. **1971** 'D. Rutherford' *Clear the Fast Lane* 46 He took the sharp curve..snicking the gearbox quadrant into second. **1977** *R.A.F. News* 8–21 June 11/2 In such foggy conditions..quadrant operators would be unable to sight and pinpoint the bomb and rocket hit points.

d. (Freq. with capital initial.) A street or part of a street curved in a quarter-circle; *spec.* the eastern end of Regent Street adjoining Piccadilly Circus in London.

1822 S. Leigh *New Picture of London* (new ed.) vi. 289 Quadrant, extending from Piccadilly to Glasshouse-street, ornamented by handsome colonnades supported by about 140 cast-iron pillars. **1847** Thackeray *Van. Fair* (1848) xxxviii. 348 The bearded savages..who..scowl at you..n the Quadrant arcades. **1875** A. E. Housman *Let.* 9 Jan. (1971) 6 The Quadrant, Regent Street, and Pall Mall are the finest streets. **1885** *List of Subscribers, Brighton* (South of Eng. Telephone Co.) 3 Farringdon B.—5, North-street-quadrant. **1974** F. Selwyn *Cracksman on Velvet* II. 131 The wagon rumbled the length of Nash's elegant quadrant. **1974** J. Gardner *Return of Moriarty* 131 They were in a house near Regent Street, near the Quadrant.

6. quadrant electrometer (earlier example); **quadrant method** *Archæol.*, a method of dividing up a site to be excavated (see quot. 1954).

1777 T. Cavallo *Compl. Treat. Electr.* III. iii. 161 Fig 7th. represents Mr. Henly's quadrant electrometer. [**1930** A. E. van Giffen *Die Bauart der Einzelgräber* 7 Bei den Hügeluntersuchungen wird von mir mit Vorliebe nach der sog. Quadrantenmethode..vorgegangen. Ich verwendete sie zuerst..im Jahre 1916.] **1939** G. Clark *Archæol. & Society* iv. 97 For dealing with round barrows with internal structures of timber Dr A. E. van Giffen of Groningen has evolved what he terms the 'quadrant method'. **1954** M. Wheeler *Archaeol. from Earth* viii. 95 It is known as *quartering* or the *quadrant method*. The mound is marked out into four quarters by two strings, laid preferably to the cardinal points of the compass and over the approximate centre. Opposite quarters are then excavated in turn, a balk 1½–3 feet wide being left between each quadrant in such a fashion as to give a complete transverse section across the mound in both directions. **1967** L. de Paor *Archæol.* ii. 49 The 'quadrant method'..is much used in the excavation of circular barrows and other small mounds... It will be observed that one of the objects of the method is to preserve to the last moment two full cross-sections.

quadrantal, *a.*[1] (Examples in the sense of *quadrantic *a.*)

1914 *Trans. Ophthalm. Soc.* XXXIV. 209 Uncomplicated quadrantal unrecognised hemianopsia. **1918** *Arch. Ophthalm.* XLVII. 126 Early homonymous upper quadrantal defects in Stage I.

quadrantanopia (kwǫ:drǣntănō^u^·piă). *Ophthalm.* [mod.L., f. Quadrant *sb.*[1] + Gr. ἀν- priv. + -ωπια sight.] A quadrantic loss of perception in an eye.

1942 I. S. Tassman *Eye Manifestations of Internal Dis.* iii. 75 The quadrantanopias affecting both fields may be homonymous, bitemporal, or binasal, inferior, superior, or crossed. **1964** S. Duke-Elder *Parsons' Dis. Eye* (ed. 14) xxiv. 358 Rare cases of homonymous quadrantanopia have been reported, in which corresponding quadrants of each field—the upper or lower half of one temporal, and the upper or lower half of the other nasal—have been lost. **1979** *Internat. Rehabilit. Med.* I. 53/1 The presence of a hemianopia or quadrantanopia was assessed by a confrontation method and confirmed by peripheral perimetry.

quadrantanopsia (kwǫ:drǣntănǫ·psiă). *Ophthalm.* [mod.L., f. as prec. + Gr. ὄψις sight.] = prec.

1910 *Lippincott's New Med. Dict.* 802/1 *Quadrantanopsia*, a sector defect of the eye with blindness limited to a portion of the visual field and due to cortical disease. **1947** F. H. Adler *Gifford's Textbk. Ophthalm.* (ed. 4) iv. 46 The result of a lesion in this region [*sc.* Meyer's loop] is to produce an isolated quadrantanopsia. **1964** D. O. Harrington *Visual Fields* (ed. 2) vii. 134 The quadrantanopsia [ed. 1: -opia] is truly congruous..in a patient with a cortical lesion. **1977** Tate & Lynn *Princ. Quantitative Perimetry* i. 44 Testing of both eyes assures that the pattern is not a pair of right and left homonymous quadrantanopsias.

quadrantic (kwǫdræ·ntik), *a.* *Ophthalm.* [f. Quadrant *sb.*[1] + -ic.] Involving the upper or lower part of one side of the visual field.

1914 *Trans. Ophthalm. Soc.* XXXIV. 212 Lower quadrantic right hemianopsia of four years standing. **1938** W. S. Duke-Elder *Textbk. Ophthalm.* II. xxix. 1228 If the sector defect is bounded by vertical and horizontal radii, the defect is called quadrantic anopia. **1976** *Lancet* 6 Nov. 1007/1 Neurological signs are absent—with the occasional exception of an isolated sixth-nerve palsy or an inferior-nasal quadrantic visual-field defect.

quadraphonic (kwǫdrăfǫ·nik), *a.* and *sb.* Also **quadro-, quadri-.** [f. Quadra-, Quadro-, Quadri- + Phonic *a.* (*sb.*).] **A.** *adj.* Produced by or pertaining to a system of sound recording and reproduction that employs four signal sources, two or more channels, and four loudspeakers, these being placed so that the original front-to-back sound distribution may be reproduced as well as the side-to-side one of stereophony.

1969 *High Fidelity* Sept. 63/1 The four channels might be used for double ping-pong effects—perhaps a quadriphonic version of 'Switched on Bach'. **1970** [see *quad *a.*[2]]. **1971** *Sci. Amer.* Aug. 13/3 One of my main interests ..is music—from rock to Bach—preferably in live concert or quadraphonic sound. **1972** *Sat. Rev.* (U.S.) 25 Mar. 34/2 The quadrophonic sound proves very satisfactory on stereo equipment. **1974** *Radio Times* 27 June 55/3 Raymond Raikes talks about the quadraphonic experimental broadcast which can be heard at 12.5. **1975** D. G. Fink *Electronic Engineers' Handbk.* xix. 81 Figure 19-147 summarizes the track and playback head locations for open-reel, eight-track cartridge and cassette tapes for both stereo and quadriphonic operation. **1975** *Jrnl. Audio Engin. Soc.* XXIII. 3/1 These presentations were made through each of the three basic types of quadraphonic system under test, namely, 4–4–4, 4–3–4, and 4–2–4. **1975** G. J. King *Audio Handbk.* vii. 168 All 'quadraphonic' records will play through a stereo system with varying degrees of accuracy. **1977** *Times* 13 Apr. 1/8 Quadraphonic broadcasts—radio in the round—are to be transmitted for an experimental period of 12 months by the BBC. **1977** *Time* 19 Sept. 56/2 At the starting gun in Rhode Island Sound this week, thousands of people will be watching in everything from little outboards to palatial cruisers with bars and quadraphonic sound systems. **1978** *Daily Tel.* 22 Feb. 16/6 (Advt.), BMW 528... Aluminium sports wheels. Quadrophonic radio and cassette player.

B. *sb.* = *quadraphony.

1972 *Esquire* June 55/3 (Advt.), Should you wait until quadraphonic is perfected? **1977** *Homes & Gardens* Nov. 161 Quadraphonic was boosted..when the BBC developed Matrix H, a new method of mixing and encoding quadraphonic signals.

Hence **quadrapho·nically** *adv.*, in a quadraphonic way; by means of quadraphony.

1970 *High Fidelity Mag.* Nov. 76/2 In classical music recorded quadriphonically the two front channels serve the same use as the left and right channels of conventional stereo. **1973** [see *matrix 6 e]. **1976** *Gramophone* Mar. 1441/2 The hall sound and orchestral layout were beautifully conveyed quadraphonically.

quadraphonics (kwǫdrăfǫ·niks), *sb. pl.* (usu. const. as sing.). Also **quadro-, quadri-.** [f. prec.: see -ic 2.] = *quadraphony.

1970 *High Fidelity Mag.* Nov. 76/2 (*heading*) The quadriphonics sweepstakes. **1972** *Observer* (Colour Suppl.) 22 Oct. 55/2 How could one have been taken in for so long by stereo which merely made you a member of a concert audience when quadraphonics elevate you to a member of the orchestra? **1974** H. W. Hellyer *Stereo Sound* ix. 182 Quadrophonics purports to give us back that lost illusion of reality by supplying the lost ambience. **1976** *Which?* May 99/1 One purpose of quadrophonics is to recreate this ambience properly.

quadraphony (kwǫdrǫ·fŏni). Also **quadro-, quadri-.** [f. as prec.: see -y[3].] Quadraphonic reproduction; the use of quadraphonic techniques.

1969 *High Fidelity* Sept. 3 Has quadriphony finally arrived? **1970** *Ibid.* Jan. 38/2 With musical material emanating from all four corners of the room, we found that visitors stood every which way to listen. Perhaps the most descriptive terminology for these two techniques, therefore, would be polarized and unpolarized quadriphony, respectively. **1972** *Daily Tel.* 20 Jan. 4 (Advt.), Where stereophony ends quadraphony begins. **1973** *B.B.C. Handbk. 1974* 74/1 Quadrophony involves four sound channels leading to four loudspeakers, normally placed in the four corners of a room. **1974** *Radio Times* 27 June 55/3 With your front two speakers bringing you Radio 2 on VHF Stereo.., and another Stereo Receiver driving your two back speakers and tuned to Radio 3 on VHF Stereo.., you will be able to hear this programme in Quadraphony. **1975** G. J. King *Audio Handbk.* ix. 211 For quadraphony or four-channel stereo four microphones are required. **1976** *Sci. Amer.* Sept. 144/1 (Advt.), Lowest effective stylus mass (0·39 mg) in quadriphony.

quadraplegic, var. *quadriplegic *a.* and *sb.*

quadrasonic (kwǫdrăsǫ·nik), *a.* Also **quadro-, quadri-.** [f. Quadra-, Quadro-, Quadri- + *sonic *a.*] = *quadraphonic *a.*

1970 *Time* 28 Sept. 73/2 Davis is looking forward to Columbia's further development of quadrisonic sound, a kind of double-stereo system that was introduced on tapes last year by Vanguard. **1970** *Rolling Stone* 12 Nov. 40/5 Phillips of Holland..has shown some very promising four-channel tapes and cassette machines, called a 'quadrosonic' system. **1971** *Hi-Fi Sound* Feb. 66/2 A compact system with good bookshelf speakers..can be quadrasonic if you like. **1973** *New Yorker* 24 Sept. 113/1 Be a part of New York's newest disco... Appreciate the elite atmosphere of total quadrasonic sound.

quadrat. Add: **3.** *Ecology.* Each of a set of small measured plots of land, formerly usu. one metre squares, used in studying the local distribution of plants and animals. Also *transf.* (see quot. 1960).

1905 F. E. Clements *Res. Methods Ecol.* iv. 161 Vegetation exhibits both development and structure, and is, in consequence, open to exact methods of inquiry. In the search for feasible methods, it was quickly seen that the quadrat, first used for determining the abundance of species, furnished the key to the problem. *Ibid.* 164 The unit size of quadrat is the meter, and when the term is used without qualification, it refers to the meter quadrat. **1922** *Ecology* III. 158 He does not say whether the quadrats as counted were contiguous or scattered. **1939** Clements & Shelford *Bio-Ecol.* v. 196 Cross checking rendered possible by his use of four methods, viz: trapping, censuses, hare transects, and quadrats of droppings. **1950** *Jrnl. Ecol.* XXXVIII. 108 There has thus grown up in botanical ecology a study of the distribution of species in 'quadrats' or small samples of fixed area (usually 1 sq. m. or less) in which only the presence or absence of particular species is recorded and not its numbers. **1960** *New Phytologist* LIX. 1 In the point quadrat method of vegetational analysis..thin needles are passed vertically through grassland..and the number of contacts between needles and foliage is recorded. **1974** *Nature* 25 Oct. 713/2 It is difficult to estimate the number of snails in natural populations since they move around too slowly for ordinary capture-recapture methods, and their patchy distribution makes quadrat or transect counts unreliable.

quadratic, *a.* and *sb.* Add: **A.** *adj.* **2.** Also *quadratic form* (see *form *sb.* 5 d); *quadratic programming*, a technique analogous to linear programming but dealing with a quadratic rather than a linear objective function.

1859 [see *form *sb.* 5 d]. **1896** *Bull. Amer. Math. Soc.* III. 100 The general theory of quadratic forms is not taken up. **1965** C. H. Springer et al. *Adv. Methods & Models* viii. 232 *Quadratic Programing*—a name given to a problem which looks much like a linear program, except the objective function is of second degree, i.e., contains squared terms in it. **1968** E. T. Copson *Metric Spaces* ix. 141 This space was first studied by David Hilbert in his work on quadratic forms in infinitely many variables. **1974** Adby & Dempster *Introd. Optimization Methods* v. 153 Although this is an effective strategy for quadratic programming, it can multiply unnecessary function evaluations for general *f*.

quadratically, *adv.* (Examples.)

1928 *Physical Rev.* XXXI. 74 Such an equation has no quadratically integrable solution. **1955** *Rev. Sci. Instruments* XXVI. 116/1 The mode separation varies quadratically as the limit is approached. **1974** Adby & Dempster *Introd. Optimization Methods* iv. 90 A quadratically terminating method..which makes use of conjugate gradient vectors..has been utilized in procedures for constrained optimization because of its modest memory requirements. **1979** *Sci. Amer.* May 32/3 This implies that in the past the storehouse has grown quadratically with time, as Engels said, not linearly, as Rescher argues.

quadrature. Add: **3. a.** More widely, the calculation of the area bounded by, or lying under, a curve. (Further examples.)

1911 E. B. Wilson *Adv. Calculus* xi. 313 It is therefore customary to restrict the application of the term 'area' to such simple closed curves as have $l_u = 0$, and to say that the quadrature of such curves is possible, but that the quadrature of curves for which $l_u \neq 0$ is impossible. **1942** H. M. Bacon *Differential & Integral Calculus* i. 3 The desire was to find a *square* equal in area to the area bounded by the given curve (in this case, a circle). For this reason the problem has been called the problem of quadrature. *Ibid.* xiv. 410 The definite integral, which solves the problem of quadrature mentioned in Chap. I, suggests the notation for integrals, the ∫ being a conventionalized form of 'S' for 'sum'. **1968** Fox & Mayers *Computing Methods* ix. 170 Apart from the rare possibility of being able to perform the quadrature by expressing the indefinite integral in closed form.., we have various general methods and some special methods.

4. e. *Electr.* A phase difference of 90 degrees. Usu. *attrib.* or in phr. *in quadrature* (*with*).

1889 T. H. Blakesley *Alternating Currents of Electr.* (ed. 2) xiii. 117 The only induction in the secondary coil is derived from the core, and is, therefore, as regards phase, in quadrature with the magnetization. **1892** S. P. Thompson *Dynamo-Electric Machinery* (ed. 4) xxii. 628 The waves of self-induced electromotive-force will lag exactly a quarter-period behind those of the current, or will be 'in quadrature' with them. **1940** *Chambers's Techn. Dict.* 691/2 *Quadrature transformer*, a transformer designed so that the secondary e.m.f. is 90° displaced from the primary e.m.f. *Ibid.* 705/1 *Reactive component*, the term now preferred for the component of the vector representing an alternating quantity which is in quadrature (at 90°) with some reference vector... Also called quadrature component. **1944** *Electronic Engin.* XVII. 58/2 The only unusual section is the reactance modulator or quadrature tube, so called because the anode circuit is back-coupled to the grid..in such a way as to cause a 90-degree phase difference between grid and anode voltages. **1967** *Electronics* 6 Mar. 120/2 The current in these resistors is in quadrature with the current in A_2's input resistor. **1974** Harvey & Bohlman *Stereo F.M. Radio Handbk.* iv. 65 The normal signal is again divided by 2 to give a 19kHz quadrature signal for feeding back along the phase-lock loop.

quadrennially, *adv.* (Later examples.)

1932 H. G. Wells *Work, Wealth & Happiness of Mankind* xii. 599 It could go to the country triennially or quadrennially for new blood and the elimination of persons who had become unpopular. **1972** *Publishers' Weekly* 24 Jan. 62/3 Campaign biographies are quadrennially the cream cheese of publishing.

quadri-. Add: **I. a. quadricentennial** (example); **quadrilingual** (later examples); **quadripa·schal,** including four passovers (cf. *bipaschal, *tripaschal* s.v. Tri- 1).

1977 *Times* 19 Apr. 14/6 The three-year quadricentennial celebrations [in California] of Sir Francis Drake's voyage round the world. **1962** Quadrilingual [see *francophone *sb.* and *a.*]. **1969** *Internat. Herald Tribune* 6 Nov. 14/2 (Advt.), Young American, excellent education U.S.A.-Europe, Ph.D. quadri-lingual, well traveled,..seeks unparalleled + challenging position. **1883** Quadripaschal [see *tripaschal* s.v. Tri- I. 1]. **1908** J. Hastings *Dict. Christ* II. 185/1 The *long period* theory..holds that there were four Passovers in the ministry, and is hence called the *quadripaschal* theory.

b. quadrinucleate (examples); **quadrivo·ltine** [It. *volta* time, turn], (of a silkworm moth) producing four broods in a year; also as *sb.*

1956 *New Biol.* XXI. 93 At last the little quadrinucleate amoeba breaks free as a whole. **1973** *Nature* 3 Aug. 293/1 Of these multinucleate compartments 88% were binucleate, 9% were trinucleate, and 3% were quadrinucleate. **1888** Quadrivoltine [see *bivoltine *a.*]. **1969** R. F. Chapman *Insects* xxxv. 719 In *Bombyx mori*..univoltine, bivoltine and quadrivoltine strains are known.

II. a. quadricycle (later examples); also **quadracycle; quadripa·resis** *Med.*, paresis of both arms and both legs; hence **quadripare·tic** *a.*; **quadriva·lency** = *quadrivalence.*

1963 Bird & Hutton-Stott *Veteran Motor Car* 90 Between about 1898 and 1908 more than 100 different makes of bicycle, tricycle, quadricycle, tricar, fore-car and light car proper..were powered by De Dion Bouton engines. **1972** *Sci. Amer.* May 104/3 By the second half of 1891..the Peugeot quadricycle..had covered the 1,200 kilometers (745 miles) from Paris to Brest and back to Paris in 10 days. **1979** *Time* 8 Jan. 80/3 Accordingly, Wilson is proposing a two-seat lunar vehicle or quadracycle, made of lightweight metals. **1956** *Jrnl. Neurol., Neurosurg. & Psychiatry* XIX. 163/1 The patients remained quadriparetic and almost totally unresponsive from the time of the accident. *Ibid.* 170/2 The patient had a quadriparesis, complete in the left leg and more marked in the left arm than the right. **1977** *Lancet* 27 Aug. 461/1 By discharge on day 32 he was fully mobile with only a residual spastic quadriparesis. **1932** *Nature* 19 Nov. 756/2 The quadrivalency of carbon. **1937** A. Findlay *Hundred Yrs. of Chem.* ii. 40 The theory of molecular constitution..rested on two main postulates, the quadrivalency of carbon..and the capacity of the carbon atoms for mutual linking.

quadridentate (kwǫdride·ntē^i^t), *a.* [f. Quadri- + L. *dentātus* (see Dentate *a.*).]

1. *Bot.* and *Zool.* (In Dict. s.v. Quadri- I. b.)

2. *Chem.* Of a ligand: forming four separate bonds. Of a complex: formed by such a ligand.

1925 Morgan & Smith in *Jrnl. Chem. Soc.* CXXVII. 2031 We have now succeeded in identifying a group capable of quadruple attachment to metallic atoms, this being the first known case of a quadridentate group in co-ordination complexes. **1937** *Chem. Rev.* XXI. 98 Probably the first synthetic quadridentate compounds.. were the copper and bivalent platinum compounds of ethylene bisthioglycolic acid. **1956** R. W. Parry in J. C. Bailar *Chem. Coordination Compounds* v. 239 The base is a quadridentate molecule in which the four nitrogen atoms can be expected to occupy the corners of a tetrahedron. **1972** M. L. Tobe *Inorg. Reaction Mechanisms* vi. 70 (*caption*) Trigonal bipyramidal complex of a 'tripod' quadridentate ligand showing how the trigonal symmetry of the ligand forces the configuration in the complex.

quadrilateral, *a.* and *sb.* Add: **B.** *sb.* **c.** *Eccl.* The four essential principles of Anglicanism, orig. enunciated in 1870 and approved by the Lambeth Conference of 1888 as a basis for the reunion of the Christian Church. Freq. *Lambeth Quadrilateral.* (Transf. use of sense b in Dict.)

1870 W. R. Huntington *Church-Idea* vii. 157 1st. The Holy Scriptures as the Word of God. 2d. The Primitive Creeds as the Rule of Faith. 3d. The two Sacraments ordained by Christ Himself. 4th. The Episcopate as the key-stone of Governmental Unity. These four points, like the four famous fortresses of Lombardy, make the 'Quadrilateral' of pure Anglicanism. **1902** [see *Lambeth 1]. **1925** J. W. Suter *Life & Lett. William Reed Huntington* vii. 162 It might, with some reason, be maintained that Dr. Huntington's chief claim to lasting fame was his invention and promulgation of the Quadrilateral... It was in a sermon preached at Worcester [Mass.], January 30, 1870, that he for the first time set forth the term. **1944** W. Temple *Church Looks Forward* ii. 14 The Lambeth Conference has repeatedly offered as a basis of negotiation the famous Quadrilateral. **1954** Rouse & Neill *Hist. Ecumenical Movement* v. 250 When Huntington had boiled Anglicanism down to its irreducible minimum, there remained his basis for a united Church—a platform later to be known as the 'Chicago-Lambeth Quadrilateral'... The House of Bishops at the General Convention of 1886 [at Chicago] finally adopted his 'Quadrilateral' and it was reaffirmed in slightly modified form by the Lambeth Conference of 1888. **1957** *Oxf. Dict. Chr. Ch.* 781/2 Lambeth Quadrilateral... The text of the Articles is as follows: A. The Holy Scriptures of the Old and New Testaments,..as being the rule and ultimate standard of faith. B. The Apostles' Creed..and the Nicene Creed... C. The two Sacraments ordained by Christ Himself—Baptism and the Supper of the Lord... D. The Historic Episcopate, locally adapted.

quadrilla, var. *cuadrilla.

1921 E. E. Cummings *Let.* 22 Apr. (1969) 75 Gentlemen in cockades..come out again heading a procession of quadrillas, bandilleros, and espadas, all of whom bow. The bull is admitted: the quadrillas—each matador has his own group—try him with cloaks. **1923** W. J. Locke *Moordius & Co.* x. 140 The proud procession of the quadrillas, matadors, banderilleros, picadores.

quadrille, *v.*[2] Add: (Later examples.) Also *transf.*

1903 Ld. R. Gower *Rec. & Reminisc.* 59 Teaching us how to quadrille and how to valse. **1905** W. H. Hunt *Pre-Raphaelitism* I. ii. 24, I..rejoiced with the happy birds quadrilling around the sentinel trees.

quadrillé, *a.* Add: Also **quadrille.** (Later examples.) Also *ellipt.* as *sb.*

1895 *Montgomery Ward Catal.* Spring & Summer 38/3 Red leather vest pocket memorandum... 40 leaves... quadrille ruling. **1907** *Yesterday's Shopping* (1969) 354/2 *Manuscript and Account Books.* Ruled either with Feint lines, Cash Column, or Quadrille. **1926** *Paper Terminol.* (Spalding & Hodge, Ltd.) ii. 22 *Quadrille*, paper ruled on the surface or in watermark with a multitude of small squares. Used by draughtsmen, diary makers, and for foreign correspondence. **1960** R. G. Haggar *Conc. Encycl. Continental Pott. & Porc.* 375/1 *Quadrillé*, a diapered ground pattern with quatrefoils and squares much favoured at Chantilly, painted in blue third quarter of the eighteenth century. **1969** R. Mayer *Dict. Art Terms & Techniques* 320/1 *Quadrille paper*, paper faintly ruled with small squares. Also, paper so patterned with a watermark or a plate finish. **1978** *E & A Office Equipment Catal.* 122/1 Memo Pads... Quadrille 5 mm.

quadrillion. Add: **b.** (Further examples.)

1975 *Offshore* Sept. 246/2 Southern areas of the North Sea off the northeast coast of England produced a record 129 quadrillion BTU's last year. **1976** *Sci. Amer.* Jan. 21/1 The ERDA projections are expressed in terms of quads, or quadrillions (10^{15}) of British thermal units (B.t.u.). **1977** *Ibid.* Apr. 71/1 When *n* is 31, the total number of possible binary trees is 14,544,636,039,226,909, and each of these 14 quadrillion trees will be optimum for some set of assumed frequencies for the 31 words.

quadrillionth *sb.* (example).

1976 *New Yorker* 3 May 30/3 Further advances.. should make it possible to record even faster chemical reactions—reactions that take place in femtoseconds, or quadrillionths of a second.

quadriphonic, -phony, -plane, varr. *quadraphonic *a.*, *quadraphony, *quadruplane.

quadriplegia (kwǫdriplī·dʒ^i^ă). *Path.* [mod.L., f. Quadri- + Para)plegia.] Paralysis of both arms and both legs.

1921 *Brain* XLIV. 428 To elicit nociceptive reactions of the upper or lower limbs in our cases of hemiplegia and quadriplegia potentially painful stimuli were usually necessary. **1948** *Brit. Med. Jrnl.* 14 Feb. 289/2 A man aged 56 suffered damage to his cervical cord..following a blow. The resultant spastic quadriplegia was thought to be due to a haematomyelia. **1974** Luckmann & Sorensen *Med.-Surg. Nursing* xxxviii. 444/1 Quadriplegia is the same as paraplegia except that the level of injury to the spinal cord is cervical and thus the upper extremities are affected as well as the lower extremities.

quadriplegic (kwǫdriplī·dʒik), *a.* and *sb.* *Med.* Also **quadra-, quadru-.** [f. prec. + -ic.] **A.** *adj.* Suffering from quadriplegia. **B.** *sb.* A quadriplegic person.

1921 *Brain* XLIV. 438 Three of the quadriplegic patients ..showed extensive reflex reactions involving the trunk and limbs. **1961** Webster, Quadriplegic, *n.* **1962** *Lancet* 6 Jan. 3/1 The patient emerging from stupor was at first mute and quadriplegic. **1965** *Listener* 27 May 772/2 Young girls.. both of whom, as a result of their injuries, became quadriplegics, that is they became permanently paralysed from the neck downwards. **1969** *Age* (Melbourne) 24 May 5/6 A Geelong man who became a quadraplegic following a car accident in October 1967, was awarded damages of $100,000 yesterday. **1971** *Daily Tel.* 25 May 3/7 He was left a quadruplegic and will spend the rest of his days in a wheelchair. **1971** *Rand Daily Mail* 4 Sept. 13/4 The youth injured his spine in the accident and was totally paralysed from the neck down. Like all quadriplegics, he is tied to a wheelchair existence for life. **1974** Passmore & Robson *Compan. Med. Stud.* III. liv. 4/1 The quadriplegic patient presents greater problems than those of the paraplegic with regard to rehabilitation and resettlement. **1975** 'M. Fonteyn' *Autobiogr.* II. x. 241 It was a long time before I realized what everyone else knew perfectly well—that Tito was incapacitated and probably a quadruplegic for life. **1980** *Times* 8 Feb. 4/7 He is a quadriplegic after having sustained multiple injuries in a fall.

quadriplex, var. *quadruplex *a.* 3.

quadripole (kwǫ·dripō^u^l). [ad. F. *quadripôle* (L. J. Collet 1926, in *Ann. des Postes, Télégraphes et Téléphones* XV. 939): see Quadri- and Pole *sb.*[2]] **1.** *Electr.* A network or device having one pair of input terminals and one pair of output terminals; *esp.* one that is passive.

1928 *Sci. Abstr.* B. XXXI. 103 The author considers the quadripole, its..velocity of propagation, characteristic impedance and reflection properties. The amplifying valve is considered as an example of a quadripole. **1935** G. P. Harnwell *Princ. Electr. & Magn.* iv. 115 In order to effect such a transfer [of power] a device with two pairs of terminals must be used. The transfer circuit itself generally contains no source of power and hence is known as a passive quadripole. **1952** *Electronic Engin.* XXIV. 264/1 The method of dealing with the steady state analysis of passive quadripoles in terms of the matrix of the linear transformation expressed by the network equations was introduced by Strecker and Feldtkeller in 1929. **1955** [see *eigen-]. **1962** *Newnes Conc. Encycl. Electr. Engin.* 629/2 A quadripole may be designed with dissipative elements as an attenuator, or with reactive elements as a filter.

2. Var. *quadrupole.

quadrisonic, var. *quadrasonic *a.*

quadrivalent (see below), *a.* and *sb.* [f. Quadri- + L. *valēnt-em*, pres. pple. of *valēre* to be worth.] **A.** *adj.* **1.** *Chem.* (kwǫdrivē^i^·lĕnt). Having a valency of four; capable of combining with four univalent atoms.

1865 A. W. Hofmann in *Chem. News* 13 Oct. 176/1 The nitrogen and carbon atoms, respectively trivalent and quadrivalent, are provided with three and four arms, indicating the three and four combining units respectively distinguishing these atoms. **1869**, **1880** [in Dict. s.v. Quadri- I]. **1922** A. D. Udden tr. *Bohr's Theory of Spectra* III. iii. 109 While an element like titanium in the fourth period already shows a marked tendency to occur with various valencies, on the other hand an element like zirconium is still quadri-valent like carbon in the second period and silicon in the third. **1964** J. W. Linnett *Electronic Struct. Molecules* ii. 30 The carbon atom..is, of course, quadrivalent. **1978** *Sci. Amer.* Feb. 58/2 Cerium, alone among the trivalent rare earths, can be oxidized to the relatively insoluble quadrivalent state.

2. *Cytology.* (kwǫdri·vălĕnt). Applied to a meiotic structure composed of four wholly or partly homologous chromosomes joined together.

Less common than the corresp. sb. (see *B. 2).

1898 *Zool. Jahrb.* (Abt. für Anat.) XII. 44 Frequently also in the loose spirem are found quadrivalent chromosomes with a constriction about the middle. **1929** *Jrnl. Genetics* XXI. 18 The hyacinths offered an opportunity for the comparative study in the same cell of the structure and behaviour of bivalent, trivalent and quadrivalent chromosomes of distinct types.

B. *sb.* **1.** *Chem.* A quadrivalent element. *rare*[−1].

1880 [in Dict. s.v. Quadri- II].

2. *Cytology.* (kwǫdri·vălĕnt). A quadrivalent group of chromosomes.

1923 *Proc. Nat. Acad. Sci.* IX. 109 In one quarter and more of the pollen-mother-cells this regular distribution does not occur with regard to all of the 12 quadrivalents. **1946** [see *microsporogenesis]. **1952** [see *bivalent *a.* 2]. **1975** *Nature* 17 Apr. 595/2 A sister plant which was a monosomic ($2n = 41$)..formed some trivalents, quadrivalents.., pentavalents and hexavalents, in addition to bivalents.

quadrivial, *a.* and *sb.* **A.** *adj.* **2.** Delete † *Obs.* and add later examples. Also *transf.*

1886 S. S. Laurie *Rise & Constit. Universities* 61 Practically under the name of dialectic, logic was a quadrivial study. **1912** *Encycl. Relig. & Ethics* V. 172/2 The 'trivial' arts were Grammar, Rhetoric, and Dialectics... The 'quadrivial' arts were Geometry, Arithmetic, Astronomy, and Music. **1949** *Author* Winter 40/2 A quadrivial passion for theology and literature, horses and international peace. **1964** C. S. LEWIS *Discarded Image* vii. 196 The four Quadrivial Arts must here be summarily dismissed.

quadro (kwǫ·dro), *a.* and *sb.*[2] Colloq. abbrev. of *QUADRAPHONY, *QUADRAPHONIC *a.* Cf. *QUAD *a.*[2] and *sb.*[8]

1972 *Guardian* 28 July 13 The critics say that the real reason for quadro is pure commercial pressure to sell more records. **1976** *Which?* May 99/1 It's possible to use quadro to create a totally new listening experience by getting the sounds to surround you completely. **1977** *Gramophone* Aug. 357/1 This will necessitate the introduction of decks equipped with two heads to accommodate both stereo and quadro-tapes.

quadroon. Add: **2.** *quadroon ball.*

1805 J. F. WATSON in *Amer. Pioneer* (1843) II. 236 These colored women have..their weekly balls, (called quartroon balls) at which none but white gentlemen attend. **1880** G. W. CABLE *Grandissimes* iii. 19, I saw the same old man, at a quadroon ball a few years ago. **1948** *Chicago Tribune* (Grafic Mag.) 8 Feb. 18/3 Most notorious of the carnival affairs, was the Quadroon ball, given by the young men of the town for their mistresses and friends.

quadrophonic, -phony, -sonic, varr. *QUADRAPHONIC *a.*, *QUADRAPHONY, *QUADRASONIC *a.*

quadrumvirate. Delete † *Obs. rare* and add pronunc. (kwǫdrʊ·mvirět). (Later examples.)

1923 *Contemp. Rev.* Feb. 151 He [*sc.* Mussolini] formed a quadrumvirate..to whom he entrusted full powers. **1955** *Times* 3 May 11/3 He is the third to go of the quadrumvirate who for much of the post-war period held the leadership of the Trades Union Congress largely in their hands. **1958** P. KEMP *No Colours or Crest* viii. 155 Colonel Fiqri Dine,.. one of the original quadrumvirate of chiefs that had helped Zog in his early days of power. **1974** *Times Lit. Suppl.* 29 Mar. 315/3 Osler..one of the remarkable quadrumvirate who established the Johns Hopkins Medical School.

quadrupedal, *a.* and *sb.* Add: **A.** *adj.* **1.** (Further examples.) Also, using all four feet for walking or running; *transf.*, of a person: on hands and knees.

1854 [see *BIPEDAL *a.* 2]. **1897** *Proc. R. Soc.* LX. 412 The posture assumed suggests the taking of a forward step in quadrupedal progression. **1914** CHESTERTON *Wisdom of Father Brown* x. 249 Seeing him thus quadrupedal in the grass, the priest raised his eyebrows rather sadly. **1971** *Nature* 12 Mar. 86/1 The four gaits [of the kangaroo] identified were a slow progression, a walk, a quadrupedal bound and a bipedal hop.

2. (Later example.)

1971 *Nature* 30 Apr. 577/1 Those early or middle Miocene dryopithecines..have been dubbed 'dental apes'—primates which apparently combined a hominoid dentition with limbs which retained a more primitive quadrupedal monkey-like morphology.

Hence **quadrupe·dally** *adv.*

1847 W. J. BRODERIP *Zool. Recreations* II. 179 Ask the zoologists, and one will tell you that the jackal..is the impure source of all that is quadrupedally good and amiable. **1952** RIESEN & KINDER *Postural Devel. Infant Chimpanzees* vi. 46 (*table*) Creeps (quadrupedally for chimpanzees). **1976** *Nature* 29 Jan. 305/2 At these speeds the mice either 'trotted' with the hind legs moving independently or 'galloped' quadrupedally with the hind legs moving together.

quadruplane (kwǫ·drʊplěⁱn). Also **quadri-**. [f. QUADRU- + PLANE *sb.*[3]] An aeroplane having four sets of wings, one above another.

1909 A. BERGET *Conquest of Air* 141 Naturally we can make triplanes or quadriplanes, but one must not proceed too far in this direction. **1909** *Times* 17 Aug. 10/5 Major Baden-Powell will attempt a flight with his quadruplane. **1919** *Jane's All World's Aircraft* 60 a (*caption*) Experimental Armstrong-Whitworths.—Two of the F.K. 10 Type Quadruplanes. **1937** *Times* 5 Oct. 9/3 He [*sc.* Major Baden-Powell] made two aeroplanes, one with swivelling propellers to obtain direct lift, and the other a kind of quadruplane.

quadruple, *a.* Add: **b.** *quadruple expansion*, the use of four stages of expansion in a compound steam engine, the same steam expanding successively in four cylinders; usu. *attrib.*

1885 R. SENNETT *Marine Steam Engine* (ed. 2) i. 23 In some few cases, in which steam of 150 to 180 lbs. has been used, quadruple expansion engines have been fitted,..but there is not sufficient evidence yet to show that the additional complication thus introduced is compensated for by any marked gain in economy. **1894** J. J. ASTOR *Journey in Other Worlds* iv. 48 The electricity generated by..slow-moving quadruple-expansion steam engines, provides the power required to run our electric ships. **1919** *Jane's Fighting Ships* 260/2 Kashima..machinery: 2 sets 4 cylinder vertical quadruple expansion. **1952** FOX & MCBIRNIE *Marine Steam Engines & Turbines* vii. 137 The quadruple-expansion engine is not now so common as formerly, and modern practice appears to favour the triple-expansion engine with fairly high initial superheat.

quadruplegic, var. *QUADRIPLEGIC *a.* and *sb.*

quadrupler (kwǫdrū·plər). [f. QUADRUPLE *v.* + -ER[1].] A device that makes something four times as great.

1941 MILLMAN & SEELY *Electronics* xii. 416 The circuit.. can be extended from a doubler to a quadrupler by adding two tubes and two condensers. **1946** *Nature* 5 Oct. 477/1 During the years 1929–32 they developed together the voltage quadrupler steady potential generator of 600 kilovolts. **1947** L. B. YOUNG in C. G. Montgomery *Technique Microwave Measurem.* vi. 368 The crystal-oscillator frequency is multiplied to 5 Mc/sec by means of a push-push quadrupler.

quadruplet. Add: **2. b.** *Mus.* A group of four notes to be played in the time of three of the same value.

1873 H. C. BANISTER *Music* 13 Other irregularities.. such as four notes for three, termed a Quadruplet. **1938** *Oxf. Compan. Mus.* p. xlviii (*heading*) Irregular rhythmic groupings (duplets, triplets, quadruplets, etc.). **1946** P. HINDEMITH *Elem. Training for Musicians* ix. 117 The names of these newly established values are: duplets, triplets, quadruplets, and so forth up to decuplets... Some of these terms are so awkward (linguistically) that they are hardly ever used.

quadruplex, *a.* and *sb.* Add: **A.** *adj.* **3.** *Genetics.* Also **quadri-**. Of a tetraploid individual: having the dominant allele at some particular locus represented four times.

1923 [see *DUPLEX *a.* 1e]. **1932** SANSOME & PHILP *Rec. Adv. Plant Genetics* v. 182 There are five possible types of zygote in the tetraploid, quadruplex SSSS, triplex SSSs, duplex SSss, simplex Ssss and nulliplex ssss. **1963** [see *NULLIPLEX *a.*].

quadrupole (kwǫ·drʊpōᵘl), *sb.* and *a.* *Physics.* Also **quadri-**. [ad. Du. *quadrupool*: see QUADRU-, QUADRI-, and POLE *sb.*[2]] **A.** *sb.* **a.** A multipole of order $l = 2$ (cf. *MULTIPOLE *sb.*). Usu. *attrib.*

1922 *Proc. Sect. Sci. K. Akad. Wetensch. Amsterdam* XXIII. 939 Assuming the molecules to act on each other as electric quadrupoles with constant quadrupole moment ..Burgers has calculated the quadrupole moment of the hydrogen molecule. **1927** *Proc. Cambr. Philos. Soc.* XXIII. 930 A method of calculating the probability of switches due to radiation by the quadripole moment seems to be supplied by Dirac's recent theory of the interaction between matter and radiation. **1932** *Proc. R. Soc.* A. CXXXVIII. 666 Assuming for the field of the nucleus the field of a quadripole, instead of that of a dipole. **1957** [see *DIPOLE 1]. **1957** *Sci. News* XLIII. 87 There is an electrical property of nuclei..called the quadrupole moment, which is a measure of the departure of the nucleus from a spherical to an ellipoidal shape. **1970** G. K. WOODGATE *Elem. Atomic Struct.* iii. 50 Pure quadrupole radiation arises when two parts of the charge distribution are oscillating like dipoles out of phase so that the dipole contribution vanishes.

b. An arrangement of four magnetic (or electric) poles, of alternate polarity, pointing at the same volume of space (used to focus beams of sub-atomic particles).

1954 *Rev. Sci. Instruments* XXV. 289/2 The introduction of the quadrupoles Q_1 and Q_2 introduces both vertical and horizontal focusing effects for a given energy. **1961** D. LUCKEY in D. M. Ritson *Techniques High Energy Physics* ix. 429 A single quadrupole can be used to obtain vertical focusing in a broad-range uniform-field spectrometer. **1969** *IEEE Trans. Nucl. Sci.* XVI. 728/2 Work on the dipole magnetic circuits as well as on cryogenic temperature iron core quadrupoles and sextupoles has continued. **1976** *McGraw-Hill Yearbk. Sci. & Technol.* 387/2 Even when dipoles or quadrupoles are to be operated at flux densities considerably above 2 T,..addition of iron shields around the windings has important advantages.

B. *adj.* Having or pertaining to two pairs of magnetic (or electric) poles.

1955 *Rev. Sci. Instruments* XXVI. 220/1 A second possibility..is to use a succession of electric or magnetic quadrupole lenses. **1960** *Ann. Rev. Nucl. Sci.* X. 163 A single quadrupole magnet has a focusing action for motion in one plane (either horizontal or vertical) and a defocusing action in the other plane. However, two or more quadrupole magnets may be used to give a net focusing action in both planes. **1976** *McGraw-Hill Yearbk. Sci. & Technol.* 387/1 Usually, uniform or quadrupole fields are required over considerable distances along the beam path. **1979** *Sci. Amer.* May 64/3 The ZGS lacks the quadrupole and sextupole magnets that are employed for focusing in many other accelerators.

Hence **quadrupo·lar** *a.*

1950 *Physical Rev.* LXXIX. 698/1 A situation in which any quadrupolar splitting of the nuclear resonance in a magnetic field is small compared to the magnetic resonance frequency itself. **1959** G. TROUP *Masers* ii. 28 This statement can be generalized, because the dipole moment may be electric or magnetic; or the moment may be of higher order, quadrupolar, for example. **1972** *Science* 26 May 903/3 The author directly proceeds to describe the effects due to quadrupolar and hyperfine interactions between electrons and nuclei.

quæstor. Add: **2.** The chief financial officer, the Treasurer, of the University of St. Andrews and, formerly, of other universities, esp. in Scotland.

1673 J. RAY *Observations Journey Low-Countries* 85 This Senate chuses..a Quaestor, who gathers up the University Revenue and Rents [at Heidelberg]. **1754** *Session Papers* 5 Mar. 12 In effect the Quaestor of the Library, Quaestor of the University, Aerarii universitatis quaestor, falls to be deemed as one and the same Thing. **1920** H. SCOTT *Fasti Ecclesiæ Scoticanæ* III. 133/1 He was Quaestor [*sc.* collector of revenues] of the Univ. of Glasgow in 1577. **1946** R. G. CANT *Univ. St. Andrews* 109 [The Senatus] authorised the conferring of degrees, it administered the university finances, controlled the common Library, and appointed the common officers, such as the Librarian, the Quaestor, and the Archbeadle. **1966** *Times* 17 June 3/2 (Advt.), Applications are invited for the post of an additional Assistant Quaestor in the University, in the Quaestor & Factor's department, St. Andrews.

quæstorian, *a.* Delete † *Obs. rare*⁻¹ and add later examples.

1879 LEWIS & SHORT *Latin Dict.* 1503/1 *Quaestōrius*,..of or belonging to a quæstor, quæstorian. **1976** *Classical Q.* XXVI. 99 Under the Republic the *calles* of Italy were..a quaestorian *provincia*. *Ibid.* 106 At some point in the second century the occasional quaestorian *provinciae* of the *calles* and the *aquae* were devised.

quag, *sb.* Add: Also with pronunc. (kwǫg). (Earlier *fig.* example.) Also *transf.*

1842 I. TAYLOR *Anc. Christianity* II. VIII. 480 Thoughtless thousands of the people are thus beguiled into the filthiest quags of 'abominable idolatry'. **1904** *Daily Chron.* 18 May 3/4 Her clothes were a quag of blood.

quagga. Add: Now also with pronunc. (kwǫ·gă). Also **quagger.** (Further examples.)

1815 *Times* 25 July 1/4 To be Sold..Two beautiful Animals of the Zebra species, called Quaggers; they are perfectly docile..being two of the handsomest ever imported to this country. **1899** *Pall Mall Gaz.* 21 Nov. 2/1 It [*sc.* Cape Colony] was the great home of the brown quagga. ('Kwokka'..is the proper pronunciation of the name of the old friend of my childhood's natural history.) **1937** *Nature* 25 Dec. 1079/2 The blaauwbok..and the quagga have already vanished. **1966** E. PALMER *Plains of Camdeboo* viii. 141 Of all our vanished creatures we mourn these quaggas most of all. **1974** *Nature* 11 Oct. 468/2 The last quagga died in the Amsterdam Zoo in 1883, but it is thought that this one had outlived by several years the wild quagga in South Africa.

quagginess. (In Dict. s.v. QUAGGY *a.*) (Later example.)

1940 *Chambers's Techn. Dict.* 692/1 *Quagginess*, a term used to indicate the defective condition of timber having shakes at the heart of the log.

quaggy, *a.* Add: Also with pronunc. (kwǫ·gi).

1. (Later examples.)

1956 PETERSON & FISHER *Wild Amer.* xxxiii. 356 The banks..were aproned by mud—quaggy and adhesive. **1969** P. DICKINSON *Pride of Heroes* 98 Putting his foot into a quaggy area, which sent..stinking inky ooze between shoe and sock.

2. (Later example.)

1968 G. JONES *Hist. Vikings* II. iv. 139 Alas, Einar's later-acquired nickname [*sc.* Thambarskelfir] has nothing to do with..a bowstring; it refers to his pendulous and quaggy belly.

quagmire. Add: Also with pronunc. (kwǫ·g-maiᵊɹ).

quahaug, quahog. Add: (Earlier and later examples.)

1753 *Southampton* (N.Y.) *Rec.* (1878) III. 6 The Trustees shall have the care of the Fishery of Quogue. **1781** S. PETERS *Gen. Hist. Connecticut* 262 The oysters, clams, quauhogs, lobsters, crabs, and fish, are innumerable. **1815** *Topogr. & Hist. Wareham* in *Coll. Mass. Hist. Soc.* (1846) 2nd Ser. IV. 289 The quahaug clam is common. **1870** *Amer. Naturalist* III. 354 Fragments of Quahaug valves ..are quite abundant. **1934** E. REYNARD *Narrow Land* v. 249 He handed over his Old Woman's recipe for quahaug fritters. **1949** R. J. SIM *Pages from Past* 66 The quahog, or hardshell clam, is deservedly the most famous of all. **1960** S. PLATH *Colossus* 24 The gritted wave leaps The seawall and drops onto a bier Of quahog chips. **1967** *Boston Sunday Herald Mag.* 26 Mar. 11/1 Anybody who uses potatoes in quahog pie is no Cape Codder. **1977** *Time* 4 July 37/1 On the Fourth, New Englanders will be flocking to Clam Shacks for rolls stuffed with batter-fried whole quahogs or steamers.

Hence as *v. intr.*, to dig or collect quahaugs; **quahau·ging** *vbl. sb.*

1905 J. C. LINCOLN *Partners of Tide* iv. 76 How's the quahaugin' nowadays? **1949** K. KNIGHT *Bass Derby Murder* 122, I was down to the pond quahoggin', all afternoon.

quai (kěⁱ, ‖ ke). [Fr.: see QUAY *sb.*] **1.** A public way constructed on the quay or embankment of a stretch of navigable water, usu. having buildings along the land side; *spec.* such a street on either bank of the Seine in Paris.

1870 [see KIOSK 2]. **1873** BROWNING *Red Cotton Night-Cap Country* II. 89 One whose father's house upon the Quai Neighboured the very house where that Voltaire Died mad and raving. **1927** C. CONNOLLY *Let.* 11 Feb. in *Romantic Friendship* (1975) 251 The solidarity of this town [*sc.* Bordeaux] with its respectable houses and cobbled quais. **1949** E. POUND *Pisan Cantos* lxxxi. 110 And at first disappointed with shoddy The bare ram-shackle quais. **1954** I. MURDOCH *Under Net* xiv. 191 The cloudless light drew a wash of colour along the grey façades of the *quais*. **1963** V. GIELGUD *Goggle-Box Affair* xi. 103 The *quais* and the cafés of the Left Bank in spring. **1977** *Time* 28 Nov. 31/3 The canal that once passed along the quai has been replaced by a Métro station.

2. *ellipt.* for *QUAI D'ORSAY.

1960 N. MITFORD *Don't tell Alfred* iv. 47 They know absolutely everything you do..what impression Alfred makes at the Quai..and so on. **1973** H. TREVELYAN *Diplomatic Channels* i. 20 During General de Gaulle's later years, French diplomats who received their instructions from the Quai were handicapped, since the Quai often did not know what policy the General was pursuing through his inner circle.

quaich, quaigh. Add: Also in extended use: a drinking vessel or trophy of similar design.

1971 *Timber Trades Jrnl.* 14 Aug. 54/3 Play in the morning will be for the Brownlee trophy against bogey under handicap and for the Granton quaich for the best net score. **1975** *Listener* 5 June 728/2, I drink a ceremonial draught from an immense, lacquered quaich [at Koriyama, Japan]. **1976** D. MARLOWE *Nightshade* iii. 44 Lapotre had arrived accompanied by a tall negro... He was a *houngan*, a voodoo priest... Lapotre stopped before a woman..carrying a souvenir (a mahogany quaich).

Quai d'Orsay (ke dorsę). The name of a *quai* (see *QUAI) in Paris, used by metonymy for the French Ministry of Foreign Affairs, which is situated there. Also *attrib.*

1922 W. S. MAUGHAM *On Chinese Screen* xix. 74 He represented certain important French interests in China and was said to have more power at the Quai d'Orsay than the minister himself. **1927** N. WAINWRIGHT tr. *Dekobra's Madonna of Sleeping Cars* xii. 167 The Quai d'Orsay would register a formal protest. **1933** G. ARTHUR *Septuagenarian's Scrap Bk.* 33 The Foreign Office and the Quai d'Orsay must have put their cards on the table with quite amazing confidence in one another's goodwill. **1940** H. G. WELLS *New World Order* § 11. 151 The Germans..have to get on with collectivisation..and they cannot give themselves to that if they are artificially divided up and disorganised by some old-fashioned Quai d'Orsay scheme. **1958** L. DURRELL *Mountolive* viii. 158 Your Quai d'Orsay people shock me. **1969** B. WEIL *Dossier IX* viii. 63 Asher chuckled. 'I suppose the Quai d'Orsay felt it was going to be caught with its knickers down again!' **1975** C. MOTT-RADCLYFFE *Foreign Body in Eye* vii. 138 Preliminary negotiations had taken place in Paris for several weeks between Sir Maurice Peterson from the Foreign Office, and St Quentin from the Quai d'Orsay.

quaies kateah, var. *QUAISS KITIR *int.*

quai hai, var. QUI-HY in Dict. and Suppl.

quail, *sb.* Add: **4*.** *U.S. slang.* A girl, a young woman.

1859 *Yale Lit. Mag.* XXIV. 291 (Th.), [The Freshman] heareth of 'Quails', he dresseth himself in fine linen, he seeketh to flirt with ye 'quails'. **1901** *Dialect Notes* II. 146 *Quail*, a young lady student at co-educational institution. Wesleyan Univ. **1904** *Hartford* (Connecticut) *Courant* 4 Oct. 1 Because she was hazed by the young women students at Wesleyan, one 'quail', as the boys call them, who was a freshman here last year did not return to Wesleyan this fall. **1935** J. HARGAN *Gloss. Prison Lang.* 6 *Quail*, a girl. **1935** A. J. POLLOCK *Underworld Speaks* 93/1 *Quail*, an attractive girl, not of age. **1947** *Time* 6 Oct. 68/1 A less active sport is 'piping the flock', when Cal males watch Cal 'quails' preening in the sun on the steps of Wheeler Hall. **1970** *Women Speaking* Apr. 5/1 For any woman..man has a strange conglomeration of terms:.. quail, squab, [etc.].

5. *quail-bagger, -bagging, -shot, -time, -track, -trap.*

1879 *Harper's Mag.* Oct. 703 The..advice offered by a circle of quail-baggers and other by-standers. *Ibid.*, The conclusion that a quail-bagging expedition was regarded as an event of considerable importance. **1865** 'MARK TWAIN' in *N.Y. Saturday Press* 18 Nov. 249/2 He got the frog out.. and filled him full of quail-shot. **1897** *Outing* XXX. 94/2 Ever since last quail-time I have been casting rather dubious glances at a certain old gun. **1842** *Yale Lit. Mag.* VIII. 96, I can't always decipher quail tracks. **1855** *Trans. Mich. Agric. Soc.* VI. 495 One acre of quail track corn planted on muck land. **1807** *Salmagundi* 1 Oct. 312 He was particularly adroit in making our quail-traps. **1845** Quail-trap [see *fishing-light* s.v. *FISHING *vbl. sb.*[1] 5 a].

quain (kwěⁱn), *sb.* [Prob. var. of QUOIN *sb.*] In the poetical terminology of G. M. Hopkins: an angle, a wedge-like corner; angularity. Hence **quain** *v.*[2], **quai·ning** *vbl. sb.*[2]

1868 G. M. HOPKINS *Jrnls. & Papers* (1959) 170 Swiss trees are, like English, well inscaped—in quains. *Ibid.*, Before sunrise..saw a noble scape of stars—..Cassiopeia on end with her bright quains pointing to the right. *Ibid.* 171 The straight quains and planing of the Alps were only too clear. *Ibid.* 176 Sycomores grew on the slopes of the valley, scantily leaved, sharply quained and accidented by perhaps the valley winds. **1871** *Ibid.* 205 And if you look well at big pack-clouds overhead you will soon find a strong large quaining and squaring in them which makes each pack impressive and whole. *Ibid.* 206 Below it [*sc.* the bud]..is a half-moon-shaped sill as if once chipped from the wood and this gives the twig its quaining in the outline. *a* **1889** *Ibid.* 290 The figure may be repeated runningly, continuously, as in rhythm (ABABAB) or intermittently, as in alliteration and rhyme (ABCDABEFABGH). The former gives more tone, *candorem*, style, chasteness; the latter more brilliancy, starriness, quain, margaretting. **1953** W. H. GARDNER in G. M. Hopkins *Poetry & Prose* 114 In July the principal stars of this constellation form a sort of flattened W on end —its two base angles ('quains') pointing to the right.

quaint, *sb.*[2] [f. the adj.] An odd, unusual, or strange person.

1939 J. CARY *Mr. Johnson* 112 'He's a comic, isn't he?' 'A perfect quaint.' **1959** B. ALDISS *Canopy of Time* 164 What's it matter what a broken-down quaint like Stayker said or didn't say?

quaint, *a.* (*adv.*) **A.** *adj.* **2.** For † *Obs.* read 'Now *arch.*' and add later examples.

1742 W. SHENSTONE *Schoolmistress* xii, With quaint arts the giddy crowd she sways. **1889** 'MARK TWAIN' *Yankee* iv. 37 This quaint lie was most simply and beautifully told. **1970** C. HAMPTON *Philanthropist* i. 13 *John* puts the revolver into his mouth and presses the trigger. Loud explosion. By some quaint device, gobs of brain and bright blood appear on the whitewashed wall.

8. b. Of furniture: designed in the style of *art nouveau*.

1897 *Furnit. & Decoration* XXXIV. 197/1 That new style called 'Quaint', which seems to be carcase without the spirit of the new style promulgated by the Arts and Crafts and other societies. **1952** J. GLOAG *Short Dict. Furnit.* 377 A fashion in furniture design, corresponding with the New Art movement at the end of the 19th and the opening of the present century, was known as the quaint style. **1975** *Country Life* 2 Oct. 852/3 The spindly chairs and tables of the 'quaint' vogue.

C. *quaint-carved, -looking, -mouthed, -sounding* adjs.

1838 J. R. LOWELL *Class Poem* IX. 11 What quaint-mouthed sentences! and how profound! **1859** J. G. WHITTIER *On Prayer Bk.* in *Independent* (N.Y.) 15 Sept. 1/1 The quaint-carved, Gothic door. **1922** R. LEIGHTON *Compl. Bk. Dog* xii. 178 Most people are well acquainted with the personal appearance of this quaint-looking dog. **1957** A. N. PRIOR *Time & Modality* 55 'The True' and 'The False' are certainly quaint-sounding objects to be named by phrases like 'The conquest of Gaul by Caesar'.

‖ **quaiss kitir** (kwəis kitᵊ·ɹ), *int. Mil. slang.* Also **kwais ketir, quash kateer, quies kiteer,** etc. [ad. Egyptian Arab. *kway·yis*, dim. cl. Arab. *kayyis* fine + *ktīr*, f. cl. Arab. *katīran* very.] Very good! Very well! O.K.! Also **quies, quash, quois** *a.*, good, nice, satisfactory.

1898 G. W. STEEVENS *With Kitchener to Khartoum* ii. 13 When the recruit made a bull..the white sergeant, standing behind him with a paper, cried '*Quaiss kitir*'—'Very good'. **1919** W. H. DOWNING *Digger Dial.* 40 *Quies* (Arab.), good. *Quies-kiteer* (Arab.), very good. **1925** FRASER & GIBBONS *Soldier & Sailor Words* 233 *Quash*, good; nice. An Arabic word (khwush) in use colloquially on Eastern Fronts. *Quash kateer*, very good. **1947** D. M. DAVIN *Gorse blooms Pale* 199 Our outfit was parked in a cornfield off the track and very quois it was too. We used to go swimming in the Liri every day. **1965** BROPHY & PARTRIDGE *Long Trail* 168 *Quaies kateah*, very good. Arabic. Very common with troops in Egypt or with Regulars that had been there. **1967** W. H. CANAWAY *Mules of Borgo San Marco* ix. 101 'They'll take us off to Germany and make us have nowt but sausages and beer.' Sergeant Entwistle said, 'Sausages and beer, kwais ketir, I wish I had some now instead of this muck.' **1967** *Sunday Times* (Colour Suppl.) 10 Sept. 46/4 *Quiess*, good!, capital! 'Quiess kateer' = very well, the answer to the question 'enta quiess?', you well?

quaite (kwěⁱt), a representation of an affected pronunc. of QUITE *adv.* Also **quate**.

1929 [see *CULTURED *ppl.a.* 2]. **1933** W. S. MAUGHAM *Sheppey* I. 28, I always say I quaite understand. Noblesse oblige if you know what I mean. **1962** WODEHOUSE *Service with Smile* v. 77 'Do you mean no *beer*?' 'Quate. I shall be keeping an eye on you.' **1965** K. GILES *Some Beasts no More* i. 6 'So you're one of those funny little men who slide about big offices as if they didn't quaite belong there.' Hell, yes, she'd said *quaite*. **1979** G. PETRIE *Hand of Glory* iii. 37 She is quaite raight... Ay would not wish to seem *grotesque*.

quake, *sb.* Add: **1. a.** (Further examples of the sense 'an earthquake'.)

Now apprehended as a colloq. abbrev. of *earthquake*, and occas. written *'quake*.

1905 *Westm. Gaz.* 14 Nov. 2/1 Even the most violent quakes in the vicinity of Mount Etna are rarely felt with any force across the straits. **1956** R. ST. B. BAKER *Dance of Trees* vii. 96 It was a serious 'quake though few people lost their lives. **1973** *Express* (Trinidad & Tobago) 1 Feb. 5/1 The quake shook Mexico City and nearby provinces. **1973** 'D. SHANNON' *No Holiday for Crime* (1974) v. 83 Glasser's Ford had been demolished by the quake last August. **1976** *Nigerian Chron.* 18 Aug. 1/2 The epicentre of the quake appeared to be in the Celebes Sea, South-West of Mindanao. **1977** *Time* 3 Jan. 27 (*caption*) In May, a quake centered in the Northeastern Italian region of Friuli killed nearly 1,000.

b. *attrib.* and *Comb.*

1931 *Daily Express* 21 Sept. 2/4 (*heading*) More 'quake shocks in Baluchistan. **1937** *Discovery* Feb. 63/1 Quake-proof reservoirs of water. **1960** *Daily Tel.* 27 May 14 Apparently the depth of the Pacific Ocean makes it specially liable to serve as a vehicle for these 'quake waves'. **1973** *Daily Colonist* (Victoria, B.C.) 21 Oct. 5/5 A 36-year-old business executive built a concrete quake-proof shelter in his yard.

quake, *v.*[1] Add: **4. quake grass** = QUAKING-GRASS.

1814 O. RICH *Synopsis Genera Amer. Plants* 10 Briza.. Quake Grass. **1909** *Daily Chron.* 25 June 7/2 We used to call 'em 'quake grass', and 'cats' tail'.

Quaker. 2. a. Substitute for def.: A member of the Religious Society of Friends, founded by George Fox in 1648–50, distinguished by its stress on the 'Inner Light' and rejection of sacraments, ordained ministry and set forms of worship; noted also for pacifist principles and simplicity of life, formerly in particular for plainness of dress and speech. (Earlier and later examples.)

1651 T. HALL *Pulpit Guarded* 15 We have many Sects now abroad; Ranters, Seekers, Shakers, Quakers, and now Creepers. *Ibid.* 29 A Bastard-brood of Arrians, Arminians,.. Quakers, Ranters. **1930** G. B. SHAW *You never can Tell* in *Wks.* VIII. 208 She is too militant an Agnostic to care to be mistaken for a Quaker. She therefore dresses in as business-like a way as she can. **1924** —— *Saint Joan* p. xlvi, In war, for instance, we suppress the gospels and put Quakers in prison, muzzle the newspapers, [etc.]. **1941** A. HUXLEY *Let.* 17 Nov. (1969) 470 England and America owe an incalculable debt to the Quakers for the way in which they have educated successive generations of rulers to realize that a theocentric opposition is a thing of enormous value to the society containing it. **1972** J. G. DAVIES *Dict. Liturgy & Worship* 329/1 For Quakers the difference between cleric and layman is irrelevant. **1978** J. A. MICHENER *Chesapeake* 380 In the name of God and Jesus Christ they must be set free, and no man dare call himself a Quaker and a slave-holder, too.

b. (*d*) Substitute for def.: One of several European, grey or brown, noctuid moths belonging to the genera *Orthosia* or *Agrochola*, esp. *O. stabilis*, the common quaker, or *O. cruda*, the small quaker; (examples).

1775 M. HARRIS *Eng. Lepidoptera* 41 Quaker... Of a plain brown colour, having a small ring in the middle, and a whitish line near the edge. **1907** R. SOUTH *Moths Brit. Isles* 1st Ser. 328 The Small Quaker... Most specimens of this species have the fore wings pale greyish ochreous. *Ibid.*, (*heading*) The Common Quaker. **1948** W. J. STOKOE *Caterpillars Brit. Moths* I. 312 The Small Quaker..appears to be common throughout England and Wales. *Ibid.* 313 The Common Quaker..is on the wing during March and April. **1968** *Oxf. Bk. Insects* 74/2 Small Quaker... Dingy, undistinguished little moths, usually with a dusky spot on each forewing. *Ibid.*, Common Quakers visit sallow blossoms. *Ibid.* 78/2 Red-line Quaker (*Agrochola lota*). All four wings of this moth are a dingy blackish-grey colour.

c. Also *ellipt.* for *quaker-colour*, etc.

1923 *Daily Mail* 21 Feb. 14 (Advt.), In Black, Nigger, Putty, Fawn, Quaker.

3. a. *quaker cap*; also designating various subdued colours, as *quaker-brown, -green, -grey.* **c. Quaker City,** Philadelphia, U.S.A.; **Quaker collar** (see quot. 1957); **quaker-ladies** (examples); **quaker-meeting** (earlier and later examples); **Quaker Oats,** a proprietary brand of oats used esp. for making porridge as a breakfast food; **Quaker state,** Pennsylvania.

1851 MRS. GASKELL *Let. c* 28 Mar. (1966) 147, I have got a new silk gown, quaker-brown coloured. **1822** M. EDGEWORTH *Let.* 16 Mar. (1971) 373 Enter Mrs. Fry in drab colored silk cloak and plain borderless quaker cap. **1851** MRS. STOWE *Uncle Tom's Cabin* (1852) I. xiii. 198 'Nicely,' said Ruth, taking off her little drab bonnet, and..displaying..a round little head, on which the Quaker cap sat with a sort of jaunty air. **1856** M. J. HOLMES *Homestead* iv. viii. 220 Grandma, in rich black silk and plain Quaker cap, was hovering near her favorite child. **1836** T. POWER *Impressions Amer.* I. 51 It was night before we gained the Quaker city. **1903** *Critic* Aug. 190 Sketches of Philadelphia life and society by a New York woman who..does not find the Quaker city so 'slow' as is generally represented. **1975** *Country Life* 2 Jan. 44/1 The First Troop, Philadelphia City Cavalry,..celebrated its 200th anniversary on November 15, 1974, the night we flew into the 'Quaker City'. **1957** M. B. PICKEN *Fashion Dict.* 267/2 *Quaker collar*, broad flat collar, similar to Puritan collar. **1974** M. HIGGINS *Changeling* ii. 8 Dark dress with wide Quaker collar. **1869** *Bradshaw's Railway Man.* 460/3 (Advt.), Greens. Brunswick Greens, all shades. Quaker ditto. Emerald Green. **1880** *Harper's Mag.* Nov. 906/1 The powders are most deceptive in color;..black appears a purplish-gray; Vandyck brown, Quaker gray. **1922** JOYCE *Ulysses* 498 In quakergrey kneebreeches and broadbrimmed hat. **1953** 'N. BLAKE' *Dreadful Hollow* 27 The modest Quaker-grey of the house. **1871** *Scribner's Monthly* II. 102 In yonder woods, where hepatica, and May-blossoms, and Quaker-ladies twinkle into life. **1946** D. C. PEATTIE *Road of Naturalist* v. 58 There are bluets around Stonybrook in Jersey..called also 'innocence' and 'Quaker ladies'. **1954** Quaker-ladies [see *INNOCENCE 6]. **1659** in *Compact with Charter & Laws of Colony of New Plymouth* (1836) II. 125 Others thinke it meet to p[er]mitt some p[er]sons to frequent the Quaker meetings to endeavor to reduce them form [*sic*] the error of theire wayes. **1704** S. SEWALL *Diary* 23 May (1879) II. 102 Convers'd with Mr. Noyes, told him of the Quaker Meeting at Sam. Sawyers. **1848** J. F. COOPER *Oak Openings* II. i. 9 The silence resembled that of a Quaker meeting. **1974** *Encycl. Brit. Macropædia* VII. 743/2 Friends were hounded by penal laws for not swearing oaths,..for going to Quaker meetings, and for refusing tithes. **1894** *Trade Marks Jrnl.* 5 Dec. 984 Pure Quaker Oats... Rolled white oats for use as food. The American Cereal Company,..Chicago, Illinois. **1901** B. S. ROWNTREE *Poverty* viii. 285/2, 2 lbs. Quaker oats, 5½d. **1921** R. MACAULAY *Dangerous Ages* i. 11 The annoyances and disappointments..such as quaker oats because the grape-nuts had come to an end. **1980** G. GREENE *Dr. Fischer* iv. 28 Do you happen to know anything about porridge? Real porridge I mean. Not Quaker Oats. **1896** *Peterson Mag.* Mar. 309/2 It [*sc.* Pennsylvania] has been long and favorably known as 'The Quaker State', in honor of the Society of Friends. **1934** G. E. SHANKLE *State Names* ii. 142 Five nicknames are given to the State of Pennsylvania; namely, the *Coal State*, the *Keystone State*, the *Oil State*, the *Quaker State*, and the *Steel State*. **1948** MENCKEN *Amer. Lang.* Suppl. II. 598 State nicknames..of Pennsylvania... *Quaker State*.

Quakerdom (earlier example).

1824 R. SOUTHEY *Let.* 3 Apr. in *N. & Q.* (1975) Sept. 403/2 My designs upon George Fox have, as you may suppose, excited a stir throughout all Quakerdom.

Quakeress. (Earlier example.)

1721 *New-England Hist. & Geneal. Reg.* (1876) XXX. 61 [Baptism of] John Rennolds, the little child of John Rennolds, his wife a Quakeress, not consenting.

Quakerish, *a.* Add: (Earlier example.)

1743 in F. Chase *Hist. Dartmouth Coll.* (1891) i. 5 [He] made a great show of sanctity, by means whereof he was under advantage to propagate his Quakerish notions.

Hence **qua·kerishness**.

1785 G. A. BELLAMY *Apology for Life* (ed. 2) I. xiv. 80 Deceived..by the Quakerishness of my dress, (excuse the new coined word).

Quakerly, *a.* (Later examples.)

1842 DICKENS *Amer. Notes* I. vii. 235 Philadelphia,..is a handsome city, but distractingly regular. After walking about it for an hour or two, I felt that I would have given the world for a crooked street. The collar of my coat appeared to stiffen, and the brim of my hat to expand, beneath its quakerly influence. **1879** *Church Times* 21 Mar. 187/1 The Quakerly 'simplicity' which the Presentation Company is seeking to force upon so comparatively insignificant a matter as the Worship of the King of Kings. **1958** J. SYKES *Quakers* I. ii. 49 This is the basis of Quakerly action.

quaking, *ppl.a.* Add: **1. b.** *quaking pudding* (see quot. 1971), *tart*.

1648 Quaking tart [see CUSTARD 1]. **1709** W. KING tr. J. H. van Slonenbergh in *Useful Trans. Philos.* III. 52 White Bread and Butter and Quaking-pudding. **1747** H. GLASSE *Art of Cookery* ix. 112 Quaking Pudding. Cream.. Eggs..Flour..boil it. **1971** R. HOWE *Mrs Groundes-Peace's Old Cookery Notebk.* 129 Quaking pudding, a pudding made of breadcrumbs, cream, eggs and spices.

quaking asp(en. *U.S.* Also **quakenasp**. The North American aspen, *Populus tremuloides*, a tall tree belonging to the family Salicaceæ; also, the soft white wood of this tree.

1822 J. FOWLER *Jrnl.* 1 June (1898) 143 The timber on the mountains Heare is Pitch Pine Spruce Pine Hemlock and quakingasp. **1825** W. H. ASHLEY in H. C. Dale *Ashley-Smith Explorations* (1918) 152 This range of mountains is.. closely timbered with pine, cedar, quaking-asp. **1845** J. PALMER *Jrnl.* 30 July in *Jrnl. Trav. Rocky Mts.* (1847) 36 Occasionally there is a grove of quaking aspen. **1848** See Ast[?] 13. **1878** J. H. BEADLE *Western Wilds* xii. 186 The town is in a grove of quaking asp. **1902** O. WISTER *Virginian* iv. 55 They took us..through a thicket of quaking asps. **1905** N.Y. *Even. Post* 2 Sept. (Sat. Suppl.) 1/6 Here seen quakenasp groves on the summer range. **1929** E. HOUGH *Sagebrusher* 4 A few quaking asps standing near the cabin door likewise gave motion and brightness to the scene. **1947** J. J. ROWLANDS *Cache Lake Country* 19 Up on the sandy ridge the quaking aspen grows. **1963** S. A. GRAHAM et al. *Aspens* 1 Two species of aspen trees are indigenous to..the so-called Lake States. These are the trembling (quaking) aspen..and the bigtooth (large-toothed) aspen.

qualification. 6. b. (Later example.)

1785 J. WOODFORDE *Diary* 26 Sept. (1927) III. 143 Ben returned by Dinner, brought a Qualification for my sporting this year for which I am to pay 2 Guineas and 1 Shilling.

qualified, *ppl.a.* Add: **5. a.** *qualified privilege.*

1972 *Times* 16 Mar. 9/7 The defence was a denial of the words in the statement of claim, a plea of qualified privilege, a plea of absolute privilege, and also a plea of justification. **1973** *Scotsman* 21 Feb. 10/3 The occasion on which the alleged remarks were made was in his view clearly one where qualified privilege applied.

b. *euphem.* for 'bloody', 'damned', etc. *slang.*

1886 KIPLING *Plain Tales from Hills* (1888) 121 He was.. told not to make a [qualified] fool of himself. **1938** D. L. SAYERS *Have his Carcase* xxvi. 353 'I wish we'd never come up against this qualified case,' added the Superintendent bitterly. **1949** 'E. C. R. LORAC' *Still Waters* iii. 39, I..knocked my head on those qualified hooks.

B. *ellipt.* as *sb.* One who is or those who are eligible for a position, military service, etc.; one who possesses a professional qualification.

1920 *Westm. Gaz.* 22 Apr. 14/1 In 1916 of 443,385 persons fully qualified for service (the German) State took only 221,853; and it is estimated that in 1911 the State will take only about 39 per cent. of the qualified. **1972** *Accountant* 12 Oct. 12/1 (Advt.), Newly qualifieds—Birmingham.

qualifier. Add: **1.** Also, one who makes himself eligible for a tournament, or for the final rounds of a tournament, as in golf or lawn tennis. Also *transf.*, a preliminary round of a competition. Also *attrib.*

1903 *Daily Chron.* 7 May 8/4 Out in 36, he came home in a good 73, and..made certain of a place among the qualifiers. **1920** *Glasgow Herald* 15 July 8 [Rife shooting] Along with the Prince of Wales's tie were decided the ties in the Qualifier competition. **1951** *Sport* 16–22 Mar. 20/3 Although there are still quite a few more qualifying races due between now and April 7th it looks doubtful that we shall get a better qualifier. **1976** *Evening Today* Aug. 8/1 Everyone's distinctively was a total of sixteen players including the sole British qualifier Corinne Molesworth. **1976** *Western Mail* (Cardiff) 21 Nov. 16/2 Irish jumping star Rannow Rambler will be sent to Chepstow for an Embassy 'Chase qualifier next Saturday week. **1977** *Daily Express* 29 Mar. 32/4 England put nine goals past Luxemburg 16 years ago in the away leg of a World Cup qualifier.

2. (Further examples.) Also, a qualifying phrase or subordinate clause.

The specialized applications of quots. 1892, 1965, and 1972 are not in general use.

1795 J. ELPHINSTON *Princ. Eng. Lang. Digested* II. viii. 183 Instead of a definition or picture, a name or noun was invented, instead of its specification, the qualifier or adjective. *Ibid.* 186 A sentence may also be complex and compound. Complex it becomes, when the subject or object has a qualifier joined. **1892** H. SWEET *New Eng. Gram.* I. § 34. 13 When we distinguish between *many men*, *all men*, and *some men* or *few men*, we call *many*, *all*, *some*, *few* are attribute-words; they are only qualifiers. **1925** GRATTAN & GURREY *Our Living Lang.* xvii. 104 A Qualifier may be a Substantive or a Case-Phrase, or even a Sentence. **1933** O. JESPERSEN *Essent. Eng. Gram.* vii. 67 *Little* is sometimes a qualifier (*a little girl*), sometimes a quantifier (*a little bread*). **1965** N. C. STAGEBERG *Introd. Eng. Gram.* xv. 226 The fourth structure-class contains the qualifiers. The qualifier position is the one just before an adjectival or an adverbial. *Ibid.*, It is evident that certain inflected words like *very, quite*, and *rather* can be called qualifiers; and when an inflected word like *pretty* and *mighty* appears in the same position, consider it a qualifier by position. **1972** J. MUIR *Mod. Approach Eng. Gram.* i. 5 An element of structure may precede the head element; this is called the modifier; and an element of structure may follow the head element; this is called the qualifier. Thus.. *the big boy with red hair* is a nominal group with three elements of structure..a modifier (*the big*), an element (*boy*) and a *q* element (*with red hair*) giving the structure *mhq*.

qualify, *v.* Add: **I. 1. b.** (Further examples.) Also used of attributive nouns, qualifying phrases, or subordinate clauses.

1892 H. SWEET *New Eng. Gram.* I. § 34. 14 Thus *very* in *very many* men qualifies the attribute-word *many*. **1924** O. JESPERSEN *Philos. Gram.* vii. 96 In any composite denomination of a thing or person... we always find that there is one word of supreme importance to which the others are joined as subordinates. This chief word is defined (qualified, modified) by another word, which in its turn may be defined (qualified, modified) by a third word, etc. **1929** G. H. MCKNIGHT et al. *Gram. Living Eng.* ii. 202 By far the largest class of adjectives is that used in qualifying nouns. **1947** A. C. BAUGH *Short Hist. Eng. Lit.* 81 Frequently..noun-adjectives have been placed after the nouns they qualify by phrases:—*bird-cage, book-review*, etc. **1959** H. BURTON *Hardlie Amer. Proal.* II. 127 Add to each of the following sentences one phrase to qualify the subject word, one phrase to qualify the object word, and one phrase to qualify the verb. **1973** R. S. SANCHEZ *Linguistic Evol.* v. 88 *Sow* is usually either modified by *my/his/her*, etc. or qualified by an *of*-group, whereas *own* is normally preceded by the definite article. **1975** [see *QUALIFYING *ppl. a.*].

3. b. Also, to become eligible for an old-age pension.

1911 *Rep. Labour & Social Conditions in Germany* (Tariff Reform League) III. 92 The man cannot draw his pension until he is 70 years of age, except through invalidity; he qualifies after one year's payment. **1937** W. E. COLLINSON *Contemp. Eng.* 83 The Old Age Pensions Act (1908) supplied the language with the verb to qualify (*for* the pension).

6. (Later example.)

1940 A. D. GRANT *Students' Gloss. Scottish Legal Terms* 71 *Qualify*, to make out or establish, as in the expression, *to qualify a title*.

qualifying, *ppl. a.* (Examples in *Gram.*: see QUALIFY *v.* I b in Dict. and Suppl.)

1892 H. SWEET *New Eng. Gram.* I. § 36. 15 When attribute-words are used in this way [*sc.* as in *give me that red book, not the blue one*], we call them *qualifying* attribute-words. **1924** O. JESPERSEN *Philos. Gram.* vii. 106 The less specified degrees of hardness are required for the production of good types, which are produced by *qualifying*.. **1933** L. BLOOMFIELD *Lang.* 12 ... **1949** GRATTAN & GURREY *Our Living Lang.* xvii. 103 But it is not only variety of gender which distinguishes these qualifying nouns from Substantives. In many languages some of them have special distinguishing suffixes. *Ibid.* xii. 261 *Expansion in the Subject:*—(b) Qualifying Phrases. **1929** G. H. MCKNIGHT et al. *Gram. Living Eng.* x. 202 By the use of qualifying words, the pronoun *man*.. might be so narrowed in meaning as to refer to one definite individual. **1975** R. A. CLOSE *Ref. Gram.* ii. 66 *Qualifying* nouns, the..

† qualimeter (kwǫli·mītəɹ). *Radiology. Obs.* [f. QUALI(TY *sb.* + -METER.] An apparatus for measuring the penetrating power of a beam of X-rays.

1903 H. BAUER in *Arch. Roentgen Ray* XV. 308 Whether this Roentgen Qualimeter, as we may call it, has all the requisites for the standard measure of hardness so much sought for by every Roentgenologist is a question which I must leave for others. **1913** R. KNOX *Radiography* 58 It is here that the qualimeter is particularly useful, since different degrees of hardness are required for the production of good types of various parts of the body. **1926** P. K. BOWES *X-Ray Apparatus* x. 98 This qualimeter can only be used with any degree of success on the coil discharge, using the same X-ray tube all the time, and never altering the rate of interruptions.

qualisign (kwǫ·lisain). *Philos.* [f. QUALI(TY *sb.* + SIGN *sb.*] A term originally used by C. S. Peirce in his theory of signs (see quot. *c*1903).

c **1903** C. S. PEIRCE *Coll. Papers* (1932) II. 142 Signs are divisible by three trichotomies; first, according as the sign in itself is a mere quality, is an actual existent, or is a general law... A sign may be termed a *Qualisign*, a *Sinsign*, or a *Legisign*... A Qualisign is a quality which is a Sign. It cannot actually act as a sign until it is embodied; but the embodiment has nothing to do with its character as a sign. **1934** *Mind* XLIII. 496 The specific quality of voice (a qualisign) by which I recognize a friend (a dicent sign)... **1970** *Jrnl. Philos.* 17 Dec. 902 It is pretty plain that Miss Stebbing means by 'type-word' what we call a 'word' and she does not recognize our category of 'word-type' or Peirce's 'qualisign'. **1973** J. FITZGERALD *Peirce's Theory of Signs* iii. 65 The division into Qualisign, Sinsign, and Legisign, is based on the mode of existence of the sign vehicle, not on any relationship to its object.

qualitated, *pa. pple.* For *Obs. rare*⁻¹ read *rare* and add later example.

1949 *Mind* LVIII. 54 The succession of somatic fields which go to constitute a single somatic sense-history may be..uniformly qualitated.

qualitative, *a.* Add: **b.** *Chem.* *qualitative analysis*, identification of the constituents (as elements, ions, or functional groups, etc.) present in a substance. Cf. *quantitative analysis* s.v. *QUANTITATIVE *a.* 3 b.

1842 [in Dict.]. **1923** R. M. CAVEN *Quantitative Chem. Anal.* i. 19 Often the methods of qualitative analysis are available for the quantitative separation of metals in solution. **1955** E. C. PIGOTT *Ferrous Anal.* (ed. 2) 13 The spectrograph is reliable, not merely for the rapid identification of materials, but also for a preliminary qualitative analysis of unknown samples. **1966** SHRINER & SOLTERS *Introd. Mod. Org. Anal.* iii. 20 In the classical scheme of organic qualitative analysis, the derivatives prepared were characterized by their melting points and the identification was then made from a table where melting point was obtained with the derivative of a known material. **1964** CHERONIS & MA *Org. Functional Group Anal.* i. 3 Qualitative organic analysis aims to identify one or more organic compounds present in an unknown.

qualitiedness. For *rare*⁻¹ read *rare* and add later example.

1960 *Mind* XLIX. 321 My present treatment is thus concerned rather with 'qualitiedness' than with 'quality'.

quality, *sb.* Add: **I. 4. a.** (Further examples.)

1669 M. LISTER *Journey to Paris* 180 A Lady of Quality, Madam M...; asked me, What I had seen in Paris that most pleased me. **1712** [see PEOPLE *sb.* 6 b]. **1852** M. ARLEN *Piracy* II. i. 60 I'd forgotten that such a phrase was ever made by the men for fine women—a woman of quality!

b. For 'Now *arch.* or *vulgar* and *dial.*' read 'Now *dial.* or rather *arch.*' (Further examples.)

1850 P. METCALF *Life St. Winefride* (1917) 82 He and his Son..received the Holy Sacrament of Baptism, at which the greatest Quality of that County were pleased to stand Patrines. **1853** MRS. GASKELL *Cranford* xiv. 270 He's dazed at being call'd one of the quality. **1894** 'MARK TWAIN' in *Century Mag.* Feb. 550 He was the highest quality in this county—one Virginia stock. First families, he was. **1904** M. CORELLI *God's Good Man* (ed. 2) xxi. 475 The quality don't seem to care for no one 'cept theirselves. **1961** *Times* 25 Apr. 20/1 In the period of the second marriage Opie..never lacked sitters among the 'quality'. **1978** M. KENYON *Deep Pocket* xiv. 164 A grouse-shoot, led,.. it's simply loaded with quality.

II. 8. a. *spec.* in *phr. the quality of life.*

1943 J. B. PRIESTLEY *Daylight on Saturday* xxxi. 253 The plans already maturing that would give all our citizens better security, better opportunities, and a nobler quality of life. **1955** E. SEVAREID *Newsmakers* CBS Radio broadcast script) No. 1 Nov. 1.. In the foundations of Library of Congress) 6 He [*sc.* Adlai Stevenson] seems disturbed about the quality of American life, when most politicians measure it only in quantity. **1965** A. STEVENSON *Let.* (1 Sep.) (Severaid MS. Collection, Library of Congress), I have.. read Severaid Newsmakers CBS Radio, November 29, 1955.. Your summarization of my anxieties about America is still quite my own.. I have reached the quality of..more utterances along that line. **1966** A. SCHLESINGER in *N.Y. Times Mag.* 4 Mar. 60/3 The [liberal's] belief in this way.. A Government which says it concerns itself with the quality of life..cannot be without a broadcasting policy. **1972** J. MANN *Mrs Knoz's Profession* ii. i 'Vic was going to make it a good thing, and the quality of life, which he had.. world would be closer to the hearts of local constituents.

1977 M. EDELMAN *Polit. Lang.* viii. 151 The consequence is a decline in the quality of life, springing from a lowering of real income. **1979** *Nature* 24 May 311/2 Monitoring of trace constituents of the atmosphere is becoming increasingly more important as their influence on the quality of life of growing concentrations of several compounds which, in sufficient concentrations, are toxic.

c. *ellipt.* for *quality newspaper* (see sense 13 below).

1970 *Guardian Weekly* 25 July 11 The 'qualities'..need to earn a greater percentage of their income from advertising than the 'populars'. **1972** *Times* 7 Dec. 18/2 The 'qualities'.. **1978** *Times* 18 Mar. 4/1 The 'qualities' are the *Daily* and *Sunday Telegraph*, *Financial Times*, *Guardian*, *Observer*, *The Times* and *Sunday Times*. **1977** *Vole* I. 26/2 Many of the 'quality' journalists are good writers.

d. *Engin.* The proportion by weight of vapour in a mixture of vapour and the parent liquid.

1958 H. A. GOLDING *Theta-Phi Diagram* iv. 52 (*in figure*) Dryness fraction or quality curve. **1937** CREPT & PURDY *Steam Boilers* (ed. 2) iii. 24 The quality of steam in average practice is..from about 97 to 99 per cent. **1977** G. F. HEWITT in Butterworth & Hewitt *Two-Phase Flow & Heat Transfer* ii. 24 Data on flow patterns for a particular geometry and fluid pair can be plotted directly in terms of the velocities, flow rates, etc. of the phases. Alternatively, it is often convenient for a single-component fluid to plot the results in terms of mass flux *G* and quality *x*.

e. *Radiology.* The penetrating power of a beam of X-rays.

1903 PUSEY & CALDWELL *Pract. Applic. Röntgen Rays* II. v. 309 The quality of the rays, and their intensity vary greatly. **1948** B. J. LEGGETT *Theory & Pract. Radiol.* II. vi. 162 The more or less exact measurement of the quality of radiation is of importance in all branches of radiotherapy. **1968** M. B. HOLLANDER *Ultrasoft X-Rays* i. 3 Most articles published since 1945 have described the quality in terms of half-value layer. **1973** HARNER & RIES *Conc. Textbk. Radiation Therapy* vii. 154 The choice of the radiation quality is determined by the site and the size of the lesion.

f. The degree to which reproduced sound resembles the original; fidelity.

1913 G. F. ROWELL *Hints about Gramophone* 9 He revels in the loudest records he can buy, and so long as the noise is satisfactory the musical quality does not trouble him in the least. **1928** A. E. GREENLEE *Amplification & Distribution of Sound* xvi. 230 The sales literature of manufacturers will provide much of useful information as to quality.. what may be expected in the way of quality of reproduction. **1971** EARL *Pickups & Loudspeakers* i. 9 The tape element involved in the production of a disc record detracts very little from the overall quality these days.

IV. 13. a. (sense 4) *quality gentleman, -woman*; (sense 8) *quality mark*; *quality-tested* adj.; (sense 9c) *quality audience, food, note, producer*; *spec.* = of a high cultural standard, or quality magazine, newspaper, as *quality magazine, newspaper, paper, press, programme, publisher, Sunday*, etc.; *quality control*, the maintenance of the desired quality in a manufactured product, esp. by means of critical examination of a proportion of the output and its comparison with the specification; also *transf.*; hence *quality controller*, one whose responsibility this is; *quality factor* = Q III. 3.

1903 *Times* 10 Oct. 43/1 It has a fairly large and very vociferous 'quality' audience. **1935** E. S. PEARSON (*title*) The application of statistical methods to industrial standardisation and quality control. **1943** R. E. WAREHAM in J. P. Young *Materials & Processes* xviii. 569 Quality control methods are based on the laws of probability and statistics. **1968** *Brit. Med. Bull.* XXIV. 220/1 Examination of the bacteria sources has proved to be one of the most valuable ideal ways in which a man may train himself to look for a fact. **1960** *Physics Bull.* July 250/1 Special attention has been given to quality control. **1980** *Proc.* ... some of the conference and the paper has given to the laboratories that have the 'quality control' of the sheep. **1977** P. JOHNSON *Enemies of Society* xi. 158 This forced academic growth leads to an inevitable collapse of quality-control. **1978** *Jrnl. R. Soc. Arts* CXXVI. 200/2 The result was anarchism, disorder, inadequate production planning, cost and quality control and all the rest. **1963** *Listener* 28 Nov. 891/2 What is the point of a quality controller? **1955** *Electronic Engin.* XXVII. 11 The ratio of the series reactance to the series resistance of a reactor is defined as Q, its quality factor. **1967** CONDON & ODISHAW *Handbk. Physics* (ed. 2) iv. vii. 108/1 Frequently..the dielectric, serves as the figure of merit, called its 'quality factor' Q of the coil. **1971** *Nature* 24 Dec. 461/1 Recent measurements of the quality factor, Q, for mechanical vibrations generated by dropping parts of the Apollo lunar module and by impacting objects indicate that the Q of the Moon is larger than the Q on the Earth by a factor of 10 to 100. **1927** *Sunday Express* 27 Mar. Quality food..the Quality gentleman. **1961** *Wine & Food* Winter 240 Shoppers should..be well aware of the quality food that comes from their own farmers and growers. **1908** J. M. SULLIVAN *Criminal Slang* 1 *A quality gentleman*.. **1941** *Times* 22 May (Advt.), Issued by The National Magazine Company Limited..who publish such quality magazines as 'Good Housekeeping', 'Harper's Bazaar', 'Connoisseur', etc. **1954** W. FAULKNER *Fable* 151 The quality-mark and warrant of man's immortality: his deathless folly. **1959** *New Statesman* 14 Mar. 362/3 Independent consumer guidance—i.e., journals..which establish quality-marks. **1967** M. MCLUHAN in... Quality newspapers, however.. **1960** A. WILSON *Middle Age Mrs. Eliot* ii. 86 The quality of the world of the quality newspapers. **1971** *Glass. Terms Materials Handling* (B.S.I.) v. 32 *Quality mark*, a mark which appears on some kinds of timber or steel. **1956** *English Summer* 4 Quality newspapers decline alarmingly, serious journals go bankrupt. **1961** *Punch* 11 Jan. 74 .. Whitbreads, for example, is the beer most favoured by forward-looking Pops [i.e. Pop People]. The advertising strikes a 'quality' note. **1962** *Listener* 11 Oct. 569/2 You can see such reporting even in the so-called 'quality' papers. **1974** *Spectator*... in the quality press. **1956** *Economist* 7 Feb. 514/1 Firms..in country towns—whose recent progress has been based on lower wage scales than those paid by the 'quality' producers in the leading centres. **1969** *Guardian* 5 Dec. 12/6 Ideally..the union would like to see..'quality' programmes analogous to the quality papers. **1961** *Ibid.* 20 Oct. 12/4 The struggle between the commercial and the quality publisher is fierce. **1964** *Punch* 11 Jan. 68/3 When may a young man have been fluent?... **1974** *Times* 22 May 20/6 The quality dailies have done rather better than the popular journals. **1934** *Sun* 30 Oct. 9 In that case, *there* are both 'quality-tested rayon fabrics'. **1974** *Times* 15 Aug. (India Suppl.) v/1 (Advt.), India's cottons..are quality tested. **1966** K. L. MORGAN in A. Dundes *Mother Wit* (1973) 598 I never questioned the implication that..my white ancestors were 'quality' whites.

b. Passing into *adj.*

1962 E. WILLIAMS *Brit. in Sixties: Communications* iii. 29 The division of space between advertising and editorial material is not, then, governed by whether a paper is 'quality' or 'popular'. **1972** *Britain 1972* (Central Office of Information) xvii. 429 The national newspapers..fall into two categories: popular and quality.

qualming, *ppl. a.* Delete † *Obs.* and add later *poet.* example.

1923 AUDEN *Nones* 40 How will you answer when from their qualming spring The immortal nymphs fly shrieking.

qualmless (kwā·mlės, kwǫ·m-), *a.* [f. QUALM *sb.*[1] + -LESS.] Having or feeling no qualms. So **qua·lmlessness.**

1849 T. ARNOLD *Let.* 22 Nov. in *N.Z. Lett.* (1966) 158 The beautiful scenery of the Sound did not appear to advantage in this tempestuous weather, nor to tell the truth, was I in that state of quietude and qualmlessness as to my internals, which would allow me to enjoy it. **1905** *Westm. Gaz.* 4 Mar. 5/2 Picture of Ronald absolutely qualmless facing charging rhinoceros. **1927** W. DEEPING *Kitty* xix. 244 Any qualms that he had about Kitty's capacity for happiness and for making him happy had evaporated. That he had suddenly become qualmless was not surprising in the beginning disappeared... By the end of January she was qualmless.

qualup (kwǫ·lʊp). The name of an estate in Western Australia, used *attrib.* in **qualup bell** to designate a local shrub, *Pimelea physodes* (cf. *PIMELEA), which has greyish-green leaves and terminal yellow-green bracts forming bells round the flowers.

1921 E. H. PELLOE *Wildflowers W. Austral* 50 'Qualup Bell'.. An erect shrub of about 3 ft., glabrous except the flowers. **1944** *Ibid.* 37 'Qualup bell'... It is a native of Qualup—and carries bell flowers which are tinged with yellow, green and purple. **1962** *Times* 11 Nov. (W. Austral. Suppl.) p. iv/2 The lovely qualup and mountain bells; the hoveas, the myrtles.

quandary, *v.* (Later example.)

1794 W. B. STEVENS *Let.* 19 Nov. in *Jrnl.* (1965) 206 Your Grandfather's sentiments are so far *coram* (read) that it seems to be quandaried (that's not a dictionary word I believe) at whose cost it should be.

‖ **quand même** (kan mẽm). [Fr., lit. 'when the same'.] All the same, even so, nevertheless.

1818 H. WILSON *Mem.* III. 179 She was however dreadfully agitated, quand même. **1854** GEO. ELIOT *Let.* 21 Apr. (1954) II. 151, I shall always love her *quand même*. **1908** W. JAMES *Will to Believe* (1897) 163 Or accepted my faith *quand même*, a systematic and infatuated optimism like that ridiculed by Voltaire in his Candide, is one of the possible ideal ways in which a man may train himself to look on life. **1966** *Pinnacles* (Greene) ii. 77 First we hear Mr. Bradley convicting things of absurdity; next, calling on the absolute to vouch for them *quand même*. **1952** M. SPENCER *Society* xi. The *quand même* to be...its ideals. **1960** J. DE JONE (1967) 756 We have no plans..for a holiday. We shall have a good time *quand même*.

quandong. 1. Substitute for def.: **a.** A small Australian tree, *Santalum acuminatum*, belonging to the family Santalaceæ, and bearing racemes of small greenish-white flowers; also, the globular red fruit of this tree. **b.** A forest tree found in northeastern Australia, *Elæocarpus grandis*, belonging to the family Elæocarpaceæ, and distinguished by grey bark and axillary racemes of bell-shaped, greenish-white flowers; also, the blue berries of this tree. Also *attrib.* (Further and later examples.)

1862 R. HENNING *Let.* 2 Nov. (1966) 114 He [*sc.* an emu] ate ours quandongs, a sort of wild plum that grows in the bush. They look very like the common black plums you preserve, but they are quite inside. **1918** E. J. BANFIELD *Confessions of Beachcomber* I. ii. 22 The shiny blue quandong (*Elæocarpus grandis*), misleading and insipid. **1939** J. L. FINLAYSON *Red Centre* viii. 84 Of the sweet fruits, the quandong and quince are the most popular. **1956** F. CLUNE *Roaming round Darling* xvii. 162 Then there is the quandong-tree, which has a small fruit..red in colour..grows in mallee country. **1945** *Coast to Coast* 1944 88 Give me your quandong stones... **1969** L. SMITH... Quandongs are the.. fruit. **1956** *Blue bowls*... **1937** *The Open Countries*... Quandong lemonade. **1961** G. DUTTON *Flock* ... the blue fig, and sometimes, but incorrectly, as quandongs. **1963** [see *NATIVE *a.* 13 d]. **1976** *Observer* (Colour Suppl.) 24 Jan. 4/1 Yatungka..had gone east a great distance to collect the fruit of the quandong tree. *Ibid.*, Yatungka arrived at the camp about midday, carrying a dish full of quandongs.

2. *Austral. slang.* A disreputable person who lives by his wits (see also quot. 1977).

1919 K. TENNANT *Foveaux* 11 In this crowd of low heels, quandongs and rippety men, she looked at her ease and yet not of them. **1973** F. HUELIN *Keep Moving* 176 *Quandong*, woman who sponges on men without 'paying off'. **1977** J. RAMSAY *Cop it Sweet* 75 *Quandong*, female who makes a practice of remaining virtuous after being wined and dined.

quango (kwæ·ŋgo). Also with capital initial and in form **QUANGO**. Pl. **quangos**. [Acronym f. the initial letters of *quasi non-governmental organization*: see note below.] A semi-public administrative body outside the civil service but financed by the exchequer and having members appointed by the government. Also *attrib.*

The expansion given above appears in the evidence set out below from 1967. The expansion 'quasi-autonomous national government[al] organization' and its variants (now common) are first attested in 1976.

1967 A. PIFER in *Ann. Rep. Carnegie Corp. N.Y.* 3 Is the private quasi-nongovernmental organization] is now becoming a new genus of organization which represents a noteworthy experiment in the art of government... We may call it the *quasi nongovernmental organization*.] **1973** C. HOOD in *New Society* 16 Aug. 386/1 It was the Americans who first drew attention to the importance of what they have labelled the 'grants economy', the 'contract state' and the 'quasi non-governmental organisation' (Quango). **1975** *Listener* 2 Oct. 433/1 What our American cousins describe as 'quangos', which..are the quasi non-governmental organisations..from the University Grants Committee to the British Tourist Authority. **1976** *Observer* 2 May 1/2 A new species of animal is multiplying in the undergrowth of Britain—the QUANGO, or Quasi Autonomous National Government Organisation... The quango..is defined as one which is financed by public funds, and..as part of a State.. While millions of workers have their pay limits rigidly fixed, their union bosses are able to increase their incomes by becoming members of 'quangos'. The word, newly-coined, stands for Quasi Autonomous National Government Organisations. **1977** *New Society* 17 Mar. 531/1 New sits in the House of Lords and has an assured place among the 'quangos' on which leading trade unionists serve. **1978** *Economist* 5 Aug. 20 A quango concern [*sic*] about everything from the Price Commission to the Police Complaints Board and the British Waterways Board. **1978** *Daily Tel.* 25 Nov. 16 Baroness Young will be declaring a personal interest when she opens a timely debate in the Lords today on the growth of Quangos, Quasi-Autonomous Non-Governmental Organisations. **1979** *Daily Tel.* 8 Aug. 14 Anthony Barker of Essex University, describes the gathering at his university... *Ibid.*, ...for having, he claims, invented the word quango. **1979** *Times* 10 Sept. ... **1980** *Times* 15 Feb. 15/3 It seems impossible to write about the quango or the..quango... Government, however intent upon abolishing 'Quangos' (Quasi-Autonomous Non-Governmental Organisations), would kill off the Advisory Council on the Penal System. **1980** T. SHARPE *Ancestral Vices* ix. 70 He's some sort of Quango... A Quasi Autonomous Non-Governmental Organization, as you very well know.

Hence **quango-cracy**, quangos regarded collectively; the power or influence attributed to quangos; **quangocrat**, a member of a quango regarded (usu. unfavourably) in terms of his authority.

1979 *Daily Tel.* 17 Apr. 18/1 Mr Callaghan's quangocrats were chosen to perform socialist functions, so were most of their institutions. **1979** *Observer* 23 Sept. 5/2 The sacked quangocrats are almost all unpaid 'part-timers'. **1979** *Daily Tel.* 29 Oct. 16/3 The original attack of the Conservatives concerned cosy bogies..of 'union hierarchies which interlock with quangocracy'. **1980** *Times* 12 Jan. 3/2 Those great beneficiaries of quangocracy, the trade unions and the academics of the left, are beginning to fight back.

quank (kwǫŋk), *sb.* [Echoic.] A representation of sounds made by animals and birds. Hence as *v. intr.*, to utter such a sound. Also **quan·king** *vbl. sb.*

1931 *Chambers's Jrnl.* Mar. 178/1 He could even hear the nasal laugh of the zebra, the resonant 'Quank' of the gnus, the rattle of bones and horns as they fought together. **1934** E. RICHARDSON *Living Island* 215 Offener we hear quanking among the tidal pools..as males beat their way to the enticing sounds—for it is the female [duck] which quanks.

quanset, var. *QUONSET, QUONSET.

quant, *sb.*[1] Add: For def. read: A pole for propelling a barge, esp. one with a cap at the top and a prong at the bottom to prevent it sinking in mud. Also *attrib.*, as *quant-pole*.

1901 *Academy* 26 Oct. 393/1 He..was a large family boat immovable. A quant-pole stood rigidly upright beside it. **1904** *Oxf. Junr. Encycl.* (rev. ed.) IV. 189/1 On the Norfolk Broads, boatmen often propel their 'wherries' [sailing barges] for short distances by 'quanting' with very long, heavy poles called 'quant-poles'.

2. (Later examples in senses 1936 and 1945).

1924 *Trans. Newcomen Soc.* III. 42 All the framing and gearing of these mills are of wood, the only important parts of iron being the wrought iron gudgeons upon which the shafts revolve, and perhaps the 'quants' or spindles which drive the runner stones. **1937** R. HEWSON *Windmills in Sussex* ii. The drive from above is called 'quant-drive' and is the most usual drive in a windmill. *Ibid.*, This chute vibrates against the lower part of the stone-shaft, which is called the 'quant' and which is not circular, but ribbed. **1945** *Archit. Rev.* XCVIII. 78/1 When the stones are over-driven the nuts are mounted on vertical spindles or 'quants' which drive the 'runner stones' from above. **1957** FREESE *Windmills & Millwrighting* iii. 48 The feed shoe is agitated by a..crank-pole or 'quant', because it oscillates a quant-pole when freed from its upper bearing (the glut-box) in order to disengage the stone-nut.

† quant (kwǫnt), *sb.*[3] Used for *QUANTUM 5. *Obs.*

1918 *Phil. Mag.* XXXV. 204 Sommerfeld gave a generalization of the quant-conditions which enabled him to explain the fine structure. **1920** *Ibid.* 307 The assumption that the recombining electron comes from one orbit with a higher quant number leads to an equation of the right types. **1926** *Nature* 18 Dec. 874/1 It would seem inappropriate to speak of one of these hypothetical entities as a corpuscle of light, a light quantum or a light quant, if we are to assume that it spends only a minute fraction of its existence as a carrier of radiant energy. **1932** STILES & WALSH tr. *Castelfranchi's Rec. Adv. Atomic Physics* II. iv. 460 The number of 'light quants', or grains, the magnitude of which depends solely on the colour.

quanta, pl. of QUANTUM in Dict. and Suppl.

quantal, *a.* Restrict † *Obs.* to sense in Dict. and add: **2. a.** Composed of discrete units; varying in steps, not continuously.

1917 F. H. PRATT in *Amer. Jrnl. Physiol.* XLIV. 518 Regarding the effective energy content of a physiological system discharging always to full capacity as a physiological quantum, I purpose to use the derivative *quantal* to express ..the conception of structural quanta from which integers of energy is effects discontinuously graded. *Ibid.*, A series of responsive values would be quantal when composed of discontinuous steps. **1919** *Ibid.* XLIX. 38 That tetanic contractions are quantal is shown even more clearly in figure 15. **1958** *Oxf. Univ. Gaz.* 2 Oct. 84/1 Three problems arising in the analysis of quantal data. **1964** *Language* XL. 206 The most salient characteristic of these linguistic primes..is their discreteness, their discontinuity... It is obviously their quantal attributes which distinguish linguistic primes from the species units of other behavioural sciences, e.g. the 'culture trait' of anthropology.

b. *Biol.* and *Med.* Of an effect or response (see quots.).

1933 J. H. Gaddum in *Med. Res. Council Special Rep. Ser.* No. 183. 5 The term 'quantal response' is used in this paper for any 'all-or-none' biological reaction, i.e. a reaction of such a type, or observed under such conditions, that only the bare fact of its presence or absence in each animal is recorded. **1954** *Proc. R. Soc. Med.* XLVII. 203 The third kind of method used..to assay drugs depends upon quantal effects. These effects are not measured but recorded as positive or absent and the number of positive effects is counted. **1974** M. C. Gerald *Pharmacol.* iii. 58 Does the rat die at a given dose of poison or not? Let us assume that this quantal (all-or-none) response follows a normal distribution.

c. *Physics.* Of, pertaining to, or being a quantum or the quantum theory.

1936 C. G. Darwin in *Nature* 28 Nov. 909/1 One may not infrequently see in learned journals such a phrase as 'This may be proved by quantum theoretical methods.'.. The proper English form would be *quantum theory methods*, though even that is very clumsy, and *quantum methods* is quite good enough... The right procedure is to coin the adjective *quantal*. To justify its adequacy it is only necessary to notice the impossibility of finding anything that would be *quantally* right, but *quantum-mechanically* wrong. **1951** E. M. Corson *Perturbation Methods* iii. 36 If **q**, **p** are to be canonically conjugate in the quantal sense. **1954** *Jrnl. Physiol.* CXXIV. 571 The number of quantal units responding to a nerve impulse fluctuates in a random manner. **1975** *Physics Bull.* Dec. 545/3 Anybody who wishes to learn the mathematics of coherence, classical and quantal, of light fields and their interaction with atoms can do no better than turn to this book. **1978** *Nature* 12 Jan. 191/1 The transmitter, synthesised and packaged on the spot, is released in quantal, countable packets.

Hence **qua·ntally** *adv.*

1936 [see 2 c above]. **1957** *Psychol. Rev.* LXIV. 137/1 Such preactivation can be produced by gradual learning or quantally by instruction. **1970** *Nature* 5 Sept. 1006/2 Vesicles slotted between dense projections are thought to release transmitter quantally by exocytosis through the membrane into the cleft. **1978** *Ibid.* 13 July 136/1 The small standard error in each case shows clearly that the V_1 change is quantal and, hence, that the junctional conductance itself changes quantally.

quantasome (kwǫ·ntăsōᵘm). *Bot.* [f. *quanta*, pl. of Quantum + *-some[4]. (The quantasome was believed to be the fundamental body capable of photosynthesis.)] One of numerous small proteinaceous particles found in chloroplasts.

Quot. 1962 is only the earliest of several such seeming coinages, all in papers by Professor R. B. Park and his colleagues at the Lawrence Radiation Laboratory and Botany Department, Berkeley, California.

1962 M. Calvin in *Science* 16 Mar. 889/1 Here we can see the lamellae on its flat side showing a granular structure, made up of fairly uniform oblate spheroids which we have called quantasomes. **1964** *Science* 22 May 1009/1 The quantasome as seen in a two-dimensional crystalline array is 185 Å long, 155 Å wide, and 100 Å thick. The surface of the quantasome appears to contain four or more subunits. The molecular weight..is 2×10^6. This..corresponds to a chlorophyll content of 230 chlorophyll molecules per quantasome. **1968** R. Rieger et al. *Gloss Genetics & Cytogenetics* 370 According to recent evidence, the quantasomes do not participate in photoreduction reactions but show Ca^{++}-dependent ATPase activities. **1976** Coombs & Greenwood in J. Barber *Intact Chloroplast* i. 12 The concept of a quantasome as the structural counterpart of a functional photosynthetic unit in the full classical sense of this term is in doubt.

quantifiable, *a.* (Later examples.)

1953 D. Riesman in *Amer. Scholar* XXIII. 24 The quantifiable measure of longevity. **1966** J. Ellis in C. E. Bazell *In Memory of J. R. Firth* 80 Both phonological and formal meaning correspond to the information of information theory in being dependent on, and quantifiable in terms of, the number of oppositions in the given system. **1967** *Times Rev. Industry* July 29/1 A first-class service with quantifiable savings in the company's own accounts department s the best possible selling point. **1972** *Daily Tel.* 6 Apr. 21/4 What we wanted were not quantifiable results but actual feelings. **1974** *Sci. Amer.* May 43/1 Reconstruction, redevelopment and the redesign of approach roads and nternal streets are destroying the evidence of the city's past at a quantifiable rate.

qua·ntifiable, *sb.* *Linguistics.* [f. the adj.] = *mass noun* s.v. *Mass *sb.*[2] 10 d.

1957 *College English* XVIII. 351/1 These are the quantifiables, such as *furniture* and *milk* and *news*. **1961** R. B. Long *Sentence & its Parts* ii. 39 Quantifiables such as *courage, fun, pneumonia, milk, spaghetti, machinery,* and *furniture* are not made plural, though it is true that some quantifiables have pluralizer status also.

quantification. Add: (Later examples.) Esp. in *Logic*, as *quantification theory*: theory concerned with quantifiers or with giving formal expression to the scope of variables in general propositions. Cf. *Quantifier.

1918 C. I. Lewis *Survey of Symbolic Logic* i. 24 He [sc. Lambert] reconstructs the whole of Aristotelian logic by the quantification of the predicate. **1940** *Brit. Jrnl. Psychol.* Jan. 233 Quantification pertains to a statistical population of persons, assessed or measured for amounts of that particular quality. **1940** W. V. Quine *Math. Logic* p. vi, Quantification theory, like the preceding part of logic, is expounded within the medium of metamathematics; its presentation in any other medium appears disadvantageous, indeed, because of subtleties having to do with the so-called bound and free occurrences of variables. **1949** R. K. Merton *Social Theory* iii. 109 What appears as a tendency in research for quantification (through the development of scales) can thus be seen as a special case of attempting to clarify concepts sufficiently to permit the conduct of empirical investigation. **1950** W. V. Quine *Methods of Logic* (1952) § 28. 166 Quantification theory was founded by Frege in 1879. **1962** E. W. Beth *Formal Methods* iii. 50 If a semantic tableau for the sequent Ø/*Z* is closed, then the formula *Z* will be called a tautology (of quantification theory). **1966** A. D. Biderman in R. Bauer *Social Indicators* ii. 75 Quantification was not until fairly recently an essential element of the definition of statistics. **1969** N. I. Styazhkin *Hist. Math. Logic Leibniz to Peano* iii. 123 It would not be an exaggeration to say that Lambert's theory of quantification actually contains all the basic results of the studies in quantification made by W. Hamilton in the nineteenth century. **1971** *World Archaeol.* III. 122 Groups of granaries ..do not permit quantification of men by compound. **1973** J. Hintikka *Logic, Language-Games & Information* iii. 53 Quantification theory may be characterized from this point of view as the study of the phrases 'there is', and 'for every' over and above the study of the words 'not', and 'or', which are already studied in propositional logic plus whatever terms are required to express predication. **1974** H. Wang *From Math. to Philos.* 143 The many attractive properties of the first order or restricted predicate calculus (quantification theory) have suggested the convenient identification of it with first order logic, pure logic, or just logic. **1979** *Amer. Speech* LIV. 9 Scholars have criticized, sometimes effectively, the validity of the prescriptivists' quantifications. **1979** *Dædalus* Summer 77 A quantification of love that would be comic were it not that its effects are so awful.

quantificational (kwǫntifikē̇ⁱ·ʃənăl), *a.* [f. Quantification + -al.] Of, pertaining to, or relating to quantification. Hence **quantifica·tionally** *adv.*

1940 W. V. Quine *Math. Logic.* ii. 89 We avoid the labor of writing out specific sequences of quantificational axioms and ponentials. **1951** —— *Ibid.* (rev. ed.) ii. 81 The advent of quantification opens up a wider class of logical truths: statements which are true by virtue of their structure in terms of joint denial and quantification. These may be *quantificationally* true. **1955** *Jrnl. Philos.* LII. 753 Sententially and quantificationally valid schemata are presented. **1957** N. Chomsky in Saporta & Bastian *Psycholinguistics* (1961) 266/2 A 'quantificational' sentence such as 'everyone in the room knows at least two languages'. **1966** *Jrnl. Philos.* LXIII. 699 The first-order quantificational calculus with identity. **1975** *Times Lit. Suppl.* 5 Dec. 1466/3 He is equally curt with some of the most familiar arguments in favour of God's existence: Aquinas's third way is guilty of an elementary quantificational fallacy.

quantified, *ppl. a.* Add to def.: Resulting from quantification or the use of quantifiers. (Further examples.)

1951 J. Łukasiewicz *Aristotle's Syllogistic* iv. 84 Every quantified expression..consists of three parts. **1972** P. T. Geach *Logic Matters* ii. 69 The wrong idea that a universally quantified subject-term stands for the whole class of *Ss* can be put across. **1974** H. Wang *From Math. to Philos.* iii. 113 The idea that truth functions governing quantified expressions must be explained in terms of propositions in which truth functions do not govern quantified expressions.

quantifier (kwǫ·ntifəiəɹ). [f. Quantify *v.* + -er[1].] **1. a.** *Logic.* Something which quantifies, esp. an expression (such as 'all' or 'some') that indicates the scope of a term to which it is attached; *existential quantifier*, a quantifier that asserts that there exists something for which the proposition following it is true or valid. Also *attrib.*

In quot. 1876 'a person who quantifies'.

1876 *Mind* I. 213 Quantifiers of the predicate insert the word *some*, and Boole uses a special symbol V, to mark the partial character of the identity. **1885** C. S. Peirce *Coll. Papers* (1933) III. xiii. 232 If the quantifying part, or Quantifier, contains Σ_x, and we wish to replace the *x* by a new index *i*, not already in the Quantifier, and such that every *x* is an *i*, we can do so at once by simply multiplying every letter of the Boolian having *x* as an index by x_i. **1896** —— in *Monist* VII. 32 My general algebra of logic.. consists in simply attaching indices to the letters of an expression in the Boolian algebra, making what I term a Boolian, and prefixing to this a series of 'quantifiers', which are the letters *Π* and *Σ*, each with an index attached to it. **1940** W. V. Quine *Math. Logic* ii. 106 Distribution of a universal quantifier through an alternation is not valid, nor is distribution of an existential quantifier through a conjunction. *Ibid.* 110 The last occurrence of '*y*' in (5) refers back to the last quantifier occurrence in (5). **1951** J. Łukasiewicz *Aristotle's Syllogistic* iv. 84, I denote quantifiers by Greek capitals, the universal quantifier by *Π*, and the particular or existential quantifier by *Σ*. *Π* may be read 'for all', and *Σ* 'for some' or 'there exists'. **1956** A. Church *Introd. Math. Logic* iv. 288 The notion of propositional function and the use of quantifiers, originated with Frege in his Begriffsschrift of 1879... The terms 'quantifier' and 'quantification' are Peirce's. **1963** O. Wojtasiewicz tr. *Łukasiewicz's Elem. Math. Logic* 95 Detachment..is the same as in the quantifier-free sentential calculus. **1965** B. Mates *Elem. Logic* x. 159 A string of universal quantifiers containing *n* quantifier-occurrences. **1971** *Sci. Amer.* Aug. 96/3 Perhaps the only detail in our plan..that requires special comment is the treatment of the quantifiers ∀ (meaning 'for all') and ∃ ('there exists'). **1976** J. S. Gruber *Lexical Struct. Syntax & Semantics* i. iii. 63 This kind of relationship with *not*, interestingly, is the same that occurs relating the universal and existential quantifiers.

b. *Linguistics.* A word or phrase indicative of quantity. Also *attrib.* and *Comb.*

1924 O. Jespersen *Philos. Gram.* vi. 85 All these quantifiers, as they might be called, differ from ordinary qualifying adjectives in being capable of standing alone..as when we say 'some (many, all, both, two) were absent'. **1933** —— *Essent. Eng. Gram.* vii. 67 *Little* is sometimes a qualifier (*a little girl*), sometimes a quantifier (*a little bread*). **1951** S. F. Nadel *Found. Social Anthropol.* 43 Thus I am not considering the 'formal' parts of language (conjunctions, prepositions, 'quantifiers', etc.). **1969** W. A. Cook *Introd. Tagmemic Analysis* iv. 107 *Quantifier tagmeme*, filled by numerals (num) such as: *one, two, first, second*; or quantitative adjectives. **1970** G. Carden in *Linguistic Inquiry* I. 287 The higher-S analysis requires an as-yet unspecified rule of Quantifier Lowering (QL) to move the quantifier from the high-S down to its surface-structure position. **1972** J. J. Lamberts *Short Introd. Eng. Usage* vii. 137 These are generally taken as plural unless a 'quantifier' like *a pair of* precedes the noun. **1977** *Word 1972* XXVIII. 88 If compared adjectives are to be analyzed as containing underlying quantifier-NPs in sentences such as 1 f and 1 g, then, ideally, we would expect them to be so analyzed elsewhere as well. **1978** *Language* LIV. 83 Certain quantifier constructions in Japanese were shown above to be exceptions.

2. *gen.* One who quantifies.

1963 *Economist* 9 Feb. 513/2 All that can be asked of the quantifiers is that they explore the sources thoroughly. **1970** *Computers & Humanities* V. 1 'Quantifiers' assemble in conferences, workshops and symposia to thrash out common problems and share research methods and findings.

quantify, *v.* Add: **2.** (Further examples.)

1949 *Cape Argus* 5 Nov. 5/4 It pleased them immensely.. to hear him [sc. Winston Churchill] fall on Sir Stafford Cripps for having recently used the word 'quantify'. **1962** *Times Lit. Suppl.* 16 Nov. 875/1 Though we can list all these factors, we cannot quantify them. **1971** I. G. Gass et al. *Understanding Earth* 40 Studies using the law of superposition and fossil faunas and floras could only produce a *relative* time scale, and efforts were made during the 19th and early 20th centuries to quantify it. **1971** *Country Life* 3 June 1350/1 What is so often missing from the survey equation is..an adequate technique for quantifying the results. **1977** *Modern Railways* Dec. 472/3 Railfreight's performance in chemicals transport is hard to quantify in terms of market share. **1979** *Dædalus* Summer 76 Lear is quantifying love, confusing it with other things that can be measured.

quantifying, *ppl. a.* (Later examples.)

1955 A. N. Prior *Formal Logic* 210 The presence..of..a quantifying element. **1976** *Times Lit. Suppl.* 30 Jan. 104/4 Those who would expect the author..to have little regard for received wisdoms or sacred cows (or even quantifying historians) will not be disappointed.

quantile (kwǫ·ntəil). *Statistics.* [f. L. *quant-us* how much, how great: see *-il, -ile 2.] Each of any set of values of a variate which divide a frequency distribution into equal groups, each containing the same fraction of the total population; also, any one of the groups so produced, e.g. a quartile, decile, or percentile.

1940 *Suppl. to Jrnl. R. Statistical Soc.* VII. 83 It is not shown that the distribution of a quantile tends to normality in large samples. **1961** Kendall & Stuart *Adv. Theory Statistics* (ed. 2) II. xxxii. 513 X_p is the *p*-quantile of the distribution, i.e. the value below which 100*p* per cent of the distribution lies. **1961** L. G. Parratt *Probability & Exper. Errors in Sci.* ii. 79 If there are *M* intervals, each interval is called a quantile or an *M*-tile, or sometimes a fractile.

† quantimeter (kwǫnti·mītəɹ). *Radiology. Obs.* [f. Quanti(ty + -meter.] An apparatus for measuring the quantity of X-rays administered. Hence **quantime·tric** *a.*

1906 R. Kienböck in *Arch. Roentgen Ray* XI. 17/2 In 1905, at the Roentgen Congress at Berlin, I introduced the new method of direct dosimetry—the 'quantimetric method'. *Ibid.*, My quantimeter consists essentially of..a strip of photographic paper..and a normal scale of graduated tints, with which it is to be compared. **1915** R. Knox *Radiography* 282 The Kienböck Quantimeter.—This method is based on the discoloration of bromide of silver under the influence of X-rays. **1926** P. K. Bowes *X-Ray Apparatus* x. 95 (*heading*) Kienböck's quantimeter.

quantitate (kwǫ·ntitē̇ⁱt), *v.* *Med.* [f. Quantit(y + -ate[3].] *trans.* To ascertain the quantity or extent of.

1960 *Anat. Rec.* CXXXVIII. 395/1 Because their appearance is sudden the time required for development can be measured and used to quantitate the rate of metabolic change produced by glucose. **1962** Burchfield & Storrs *Biochem. Appl. Gas Chromatogr.* i. 117 All the components of the mixture are converted to the same substance and can be quantitated from a single curve relating detector response to amount of CO_2 or methane. **1969** *Adv. Appl. Microbiol.* XI. 51 The susceptibility of a penicillin to β-lactamases is usually quantitated by measuring its rate of hydrolysis to penicilloic acid. **1976** *Nature* 22 Jan. 236/1 Vitamin E content of the filtrate was quantitated by absorbancy at 292 nm. **1978** *Dædalus* Spring 27 If there is a danger, quantitate it.

So **quantita·tion**, the action or process of quantitating.

1959 *Jrnl. Lipid Res.* I. 76/2 The separation and quantitation of the shorter acids could have been improved by use of a lower temperature and longer column. **1964** *Analytical Biochem.* VII. 295 Quantitation of each peak was made by triangulating the peak, cutting out, and weighing on an analytical balance. **1975** *Nature* 30 Oct. 828/1 Quantitation of the spectra was aided by the use of a Nicolet 1020A signal averager.

quantitative, *a.* and *sb.* Add: **A.** *adj.* **3. b.** Chem. *quantitative analysis*, measurement of the amounts of constituents present in a substance. Cf. **qualitative analysis*.

1849 [in Dict., sense 3]. **1913** Cumming & Kay *Quantitative Chem. Analysis* 109 One of the most difficult problems met with in quantitative analysis is the selection of good methods of separation. **1961** D. & B. A. Ambrose *Gas Chromatogr.* x. 161 With the integral detectors described.. quantitative analysis is simple: the detector response is directly proportional to the mass of material, and the step height permits the analysis to be calculated in accordance with the property being determined (e.g. titre or volume).

4. (Later examples.)

1933 C. D. Buck *Compar. Gram. Greek & Latin* 93 Long vowels are shortened before other vowels in various dialects. .. When the second vowel is short it may be lengthened, resulting in what is known as 'quantitative metathesis'... Homer often shows the older forms.., but also in many cases the shortening and quantitative metathesis. **1973** A. H. Sommerstein *Sound Pattern Anc. Greek* ii. 69 Quantitative Metathesis. The need for this rule arises chiefly from the vocalism and accentuation of certain third-declension genitive case forms. **1978** *Language* LIV. 441 Thus the order of the rules required by this analysis is Pre-German Accentuation followed by Quantitative Ablaut followed by Germanic Accentuation.

5. *Chem.* Of a procedure or a reaction: acting on the whole quantity of a particular substance or species; having an efficiency or a yield of 100 per cent. Hence also used of the yield or product of such a process.

1905 *Proc. R. Soc.* A. LXXVI. 116 Its [sc. a possible new element] quantitative extraction from thorium salts has not yet been investigated. **1907** *Chem. Abstr.* I. 1539 The yield is almost quantitative and the product very pure. **1923** [see *Qualitative *a.* b]. **1930** W. T. Hall *Textbk. Quantitative Analysis* xi. 140 For practical purposes, a reaction is complete or quantitative, as we often say, when less than 0·1 mg. remains in solution. **1962** Cotton & Wilkinson *Adv. Inorg. Chem.* x. 194 Diborane..is obtained in essentially quantitative yields by reaction of metal hydrides with boron trifluoride. **1964** N. G. Clark *Mod. Org. Chem.* xi. 204 On careful combustion there remains a quantitative residue of metallic silver.

B. *sb.* **c.** = *Quantifier 1 b.

1924 H. E. Palmer *Gram. Spoken Eng.* 45 Quantitatives and Numericals (matʃ), (meni), (faiv).., etc.

quantitatively, *adv.* Add: **b.** *Chem.* Completely, entirely; with a yield of 100 per cent.

1911 F. Soddy *Chem. Radio-Elements* 1. 26 The radium may be precipitated quantitatively by sulphuric acid. **1950** *Thorpe's Dict. Appl. Chem.* (ed. 4) X. 447/1 An equal quantity of barium chloride is added to ensure that the radium is carried quantitatively with the insoluble residue. **1974** *Nature* 1 Feb. 291/1 Prephenate is quantitatively converted to phenylpyruvate in 10 min at acid *p*H (0·1 N HCl).

quantitativist (kwǫntitē̇ⁱ·tivist), *sb.* and *a.* [f. Quantitativ(e *a.* + -ist.] **a.** *sb.* A person for whom quantity is a criterion of value. **b.** *adj.* Resulting from an evaluation of quantity.

1957 R. K. Merton *Social Theory* (rev. ed.) x. 396 The local influential is typically concerned with knowing *as many* people as possible. He is a quantitativist in the sphere of social contacts. **1972** *Human World* Feb. 21 The.. quantitativist illusion that the level of enlightenment must correspond to the amount of time spent between the school walls. **1974** B. Pearce tr. *Amin's Accumulation on World Scale* II. iii. 452 It is by means of concrete historical explanation that we must account for each period of price increase in the nineteenth century, and not by means of a general quantitativist explanation.

‖ **quantité négligeable** (kaňtite negliʒāb'l). [Fr., lit. 'negligible quantity'.] A factor of no account, something insignificant.

1886 T. P. White *Ordnance Survey U.K.* vi. 98 It is certain that among the details which would not be a *quantité négligeable*, would figure the trees of any particular district or country. **1913** S. Shaw *William of Germany* viii. 151 The resolve that as Emperor he would not allow Germany to be overlooked, to be treated as a *quantité négligeable*, in the discussion or decision of international affairs. **1921** Baron von Margutti *Emp. Francis Joseph & his Times* viii. 205 The old Sovereign apparently still regarded the Slavs as a *quantité négligeable*, as they had been at the beginning of his reign. **1973** E. Osers tr. *Waldheim's Austrian Example* 7 The geopolitical position of Austria within the contact zone or testing ground of the great ideological power-groups of our time has given this small country a meaning that is far from that of a *quantité négligeable*.

quantitive, *a.* (Later examples.)

1958 T. G. E. Powell *Celts* 182 It cannot be argued here whether it was prestige, or quantitive representation, that won for it eventually an exclusive position. **1959** *Times* 9 Sept. 16/7 The Four No Trump bid is regarded more often than not as quantitive.

quantity. Add: **1. d.** In surveying, *bill of quantity* (or *quantities*) (see quot. 1964).

1877 B. Fletcher *Quantities* i. 5 The operations necessary to produce the schedule, or bills of quantities, from which builders make up their tenders, are: [etc.]. **1964** J. S. Scott *Dict. Building* 32 *Bill of quantities*, a list of numbered items, each of which describes the work to be done in a civil engineering or building contract. Each item shows the quantity of work involved... Those contractors who wish to do the work return the bill, with an extended price opposite each item. **1972** *Guardian* 20 June 10/6 When the architect and engineer have produced drawings, the quantity surveyor can begin 'taking off' (which really means reading the drawings) and 'working up' (which means determining the total quantities of the materials and labour requirements)... He can then produce his 'Bills of Quantity'.

13. (sense 8) *quantity output, production*; **quantity surveyor**: for def. read: a surveyor who estimates the quantities of labour and materials required for building and engineering work; (later examples); **quantity theory (of money)** the hypothesis that prices correspond to changes in the monetary supply; so *quantity theorist*.

1888 M. Frewen *Econ. Crisis* i. 5 More emphatic still is John Stuart Mill's statement of the 'quantity theory'. 'That an increase of the quantity of money', wrote Mill, 'raises prices, and a diminution lowers them, is the most elementary proposition in the theory of currency.' **1895** *Econ. Jrnl.* V. 103 So far as concerns the possible causes on the side of *money* for the fall in prices, Lex denies that the 'quantity theory' affords any ground for speaking of an appreciation of gold. **1903** J. L. Laughlin *Princ. Money* vii. 225 (*heading*) History and literature of the quantity theory of money. **1908** *Westm. Gaz.* 8 May 8/3 Bills which had been through the hands of the quantity surveyor and architect. **1912** I. Fisher *Purchasing Power of Money* p. vii, The main contentions of this book are at bottom simply a restatement and amplification of the old 'quantity theory' of money. **1919** *Brit. Manufacturer* Nov. 42/1 Quantity output may mean cheap production, but the manufacture of more modest quantities need not be much inferior in this aspect. *Ibid.*, An immense home market.. has encouraged him to undertake big quantity production. **1928** E. O'Neill *Strange Interlude* v. 159 The room is a typical sitting room of the quantity-production bungalow type. **1931** *Times Lit. Suppl.* 19 Feb. 124/2 The quantity theorists have always been baffled by variations in the public's habits and other factors. **1968** *Internat. Encycl. Social Sci.* X. 433 In its most rigid and unqualified form the quantity theory asserts strict proportionality between the quantity of what is regarded as money and the level of prices. **1972** Quantity surveyor [see 1 d above]. **1979** *Jrnl. R. Soc. Arts* CXXVII. 445/1 The quantity surveyor's functions in the construction process may be described as cost planning, cost control, and the attainment of value for money expended.

quantizable (kwǫ·ntəizăb'l), *a.* *Physics.* [f. as next + -able.] Capable of being quantized.

1935 J. Dougall tr. *Born's Atomic Physics* v. 101 Certain quantities can only take values which are whole numbers—they are called quantisable quantities. **1946** *Nature* 31 Aug. 309/1 Our solar system is quantizable according to the equation: $n \times 137^{\frac{1}{2}}$ = orbital impulse/(2 × planetary spin).

quantization (kwǫntəizē̇ⁱ·ʃən). [f. next + -ation.] The action of quantizing; the fact or state of being quantized. **a.** *Physics.* Cf. *Quantize *v.* 1.

1922 *Rep. Brit. Assoc. Adv. Sci. 1921* 473 In some cases.. the quantisation can be done in several ways and each leads to a different set of permissible orbits, but the energy of each has the same set of values and so they give the same spectrum. **1922** *Proc. Cambr. Philos. Soc.* XXI. 80 (*heading*) A general condition for the quantisation of the conditionally periodic motions with an application for the Bohr atom. **1925** *Nature* 5 Dec. 849/1 It seemed possible to formulate certain general laws, the so-called rules of 'quantisation', by means of which the stationary states were to be chosen from the continuous manifold of such motions. **1940** Glasstone *Text-bk. Physical Chem.* i. 63 The existence of 2*j* + 1 magnetic quantum numbers for each value of *j* implies that the electrons can take up 2*j* + 1 different orientations in a magnetic field..: in other words, there should be a quantization of electron orbits. **1975** *Physics Bull.* July 311/1 To see the departure of the specific heat of solids from the classical Dulong Petit law due to quantization of lattice vibration energies we often need to go to temperatures of the order of 100 K.

b. *Telecommunications.* Cf. *Quantize *v.* 2. Usu. *attrib.*, as *quantization distortion, level, noise*.

1947 *Bell Syst. Techn. Jrnl.* XXVI. 395 PCM involves the application of two basic concepts. These concepts are..the time-division principle and the amplitude quantization principle. **1948** *Proc. IRE* XXXVI. 1324/2 Representing the signal by certain discrete allowed levels only is called quantizing. It inherently introduces an initial error in the magnitude of the samples, giving rise to quantization noise. **1951** *Ibid.* XXXIX. 44 (*heading*) Quantization distortion in pulse-count modulation with nonuniform spacing of levels. **1953** A. T. Starr *Radio & Radar Technique* i. 32 Quantization could be used with any known system of modulation, e.g. the signal wave could be quantized and then used to modulate the amplitude or frequency of a carrier. **1975** D. G. Fink *Electronics Engineers' Handbk.* iv. 44 A quantization noise..is inevitably associated with all quantized signals. This noise can be made as small as desired by choosing enough quantization levels or, equivalently, making each quantization level small enough.

quantize (kwǫ·ntəiz), *v.* [f. Quantum + -ize.] **1.** *trans.* *Physics.* To apply quantum theory to; *esp.* to restrict the number of possible values of (a quantity) or states of (a physical entity or system) so that certain variables can assume only certain discrete magnitudes that are integral multiples of a common factor.

1922 *Rep. Brit. Assoc. Adv. Sci. 1921* 473 For the specific heats of gases..it is necessary to 'quantise' rotations instead of vibrations. **1958** *New Scientist* 27 Feb. 29/1 Yukawa had been very impressed by the success of quantum mechanics..in explaining phenomena connected with the ordinary electromagnetic field. Such a field can be 'quantised' into photons—discrete packets of light waves. **1973** *Physics Bull.* Nov. 656/2 This amplification of gravity produces a theory which cannot be quantized by standard methods, however. **1975** *Nature* 17 Apr. 560/2 The problem of quantising gravity is a pressing one if a unified theory of all the forces of nature is to be obtained.

2. *trans.* *Telecommunications.* To approximate (a signal varying continuously in amplitude) by one whose amplitude is restricted to a prescribed set of discrete values.

1947 *Bell Syst. Techn. Jrnl.* XXVI. 409 Over and above these effects,..the background noise which is present to a greater or lesser extent in all communication circuits, is quantized by the PCM system. **1953** *Electronic Engin.* XXV. 148/1 If the signal is sampled before being quantized, harmonics (including the first) of the pulse repetition frequency will be present. **1953** [see *Quantization b]. **1972** *Physics Bull.* Jan. 44/1 In the new technique the signals from the eddy current probes are quantized by digital means. **1977** F. G. Stremler *Introd. Communication Syst.* vii. 356 Suppose..we wish to quantize one-half a cycle of a one-volt (peak) sinusoid using eight discrete levels.

3. *transf.* and *fig.*

1956 *Kenyon Rev.* XVIII. 412 The sounds..are then quantized into their phonemes. **1965** *Revue Internat. de la Documentation* XXXII. 21/2 To do this it is necessary to quantize the texts by dividing them into convenient units. **1969** *Language* XLIV. 14 As Hoenigswald..remarked, 'The doctrine of gradual phonetic change may turn out to be a remnant from pre-phonemic days,' when the multi-dimensional continua of speech had not yet been successfully quantized. **1974** *Sci. Amer.* Dec. 132/2 Because curves can be coded to any desired precision by numbers, a symphony, like a painting or a poem, can be quantized and expressed by a number chain.

Hence **qua·ntizing** *vbl. sb.* and *ppl. a.*

1923 E. N. da C. Andrade *Structure of Atom* ix. 147 The principle laid down for the quantising of angular momentum gives $mvr = nh/2\pi$, where *n* is a whole number. **1948** *Bell Syst. Techn. Jrnl.* XXVII. 456 Since quantizing noise is uniformly distributed throughout the signal band, its interfering effect..is probably similar to that of thermal noise with the same mean power. **1973** *Appl. Physics Lett.* XXIII. 41/1 Because of the quantizing longitudinal magnetic field, the conduction band is split into Landau levels. **1975** D. G. Fink *Electronics Engineers' Handbk.* iv. 44 The process of quantizing is irreversible since regardless of how small the quantization level *Q* is taken to be, an unresolvable uncertainty of ±*Q*/2 is associated after quantizing with each amplitude value.

quantized (kwǫ·ntəizd), *ppl. a.* [f. prec. + -ed[1].] Subject to the restrictions imposed by quantization; able to occur with certain discrete values only.

1923 H. L. Brose tr. *Sommerfeld's Atomic Struct. & Spectral Lines* iv. 212 Both [conditions] together demand that the electron move only in certain 'quantised' circles. **1937** G. Gamow *Struct. Atomic Nuclei* iv. 67 The atomic nucleus, being a quantized system, is in general capable of existence in any one of a number of states of different energy. **1939** V. Rojansky *Introd. Quantum Mech.* iv. 145 The energy of a free particle is not quantized. **1966** C. Kittel *Introd. Solid State Physics* (ed. 3) v. 134 The lattice contribution to the heat capacity of solids..approaches zero as the temperature approaches zero; this can be explained only if the lattice vibrations are quantized. **1973** *Physics Bull.* Dec. 715/3 The flux which is trapped in a superconducting loop is quantized in units of *h*/2*e*.

quantizer (kwǫ·ntəizəɹ). *Electronics.* [f. as prec. + -er[1].] A device that quantizes a signal applied to it.

1948 *Bell Syst. Techn. Jrnl.* XXVII. 450 An actual quantizer (staircase transducer) has a finite overload value which must not be exceeded and hence can have only a finite number of steps. **1953** *Electronic Engin.* XXV. 146/2 The channel modulators of a pulse-length modulated T.D.M. system feed a common quantizer and coder. **1967** *Electronics* 6 Mar. 127/2 Sierra also offers a 'quantizer' that puts thermal contours..on the scope. **1975** C. L. & J. W. S. Liu *Linear Systems Analysis* ii. 48 Consider a discrete quantizer whose output y(*n*) at any instant *n* is equal to the integral part of the input *x*(*n*) if the fractional part of *x*(*n*) is less than 0·5.

Quantometer (kwǫntǫ·mītəɹ). Also **quantometer.** [f. Quant(ity + -ometer.] A type of automatic spectrograph, used esp. for the analysis of alloys.

Formerly a proprietary term in the U.S.

1927 *Official Gaz.* (U.S. Patent Office) 3 May 12/1 Francis Cutler Ellis..Chicago, Ill...*Quantometer*... Instrument for diagnosis and food and remedy testing by analysis of radiant energies. Claims use since Sept. 1, 1926. **1947** *Jrnl. Iron & Steel Inst.* CLVI. 78 A commercial model, the Quantometer, consists of source, dispersing, and recording units operating as one instrument. **1958** *Engineering* 21 Feb. 61/2 (Advt.), To meet the need for still greater

speed there is now in use at Corby an automatic direct-reading spectrographic analyser known as a Quantometer. **1974** J. H. Dixon tr. *Torasov's Spectroscope* i. 6 The quantometer is used for the simultaneous quantitative emission analysis of several elements in steels or alloys.

quantophrenia (kwǫntofrī·niă). [f. Quant(itative *a.* + -o + *-phrenia* as in *hebephrenia*.] A term used for an obsession with and exaggerated reliance upon mathematical methods or results, esp. in research connected with the social sciences. So **quantophre·nic** *a.*

1956 P. A. Sorokin *Fads & Foibles in Mod. Sociol.* (1958) vii. 103 When the true quantitative method is replaced by pseudomathematical imitations; when the method is misused and abused in various ways; when it is applied to phenomena which, so far, do not lend themselves to quantification..then the approach misfires. Under these conditions, use of mathematical method becomes a mere quantophrenic preoccupation having nothing in common with mathematics and giving no cognition of the psychosocial world... The tidal wave is at present so high that the contemporary stage of the psychosocial sciences can be properly called *the age of quantophrenia and numerology*. **1964** *Encounter* Sept. 72 There is quantophrenia—an obsession with statistics as the sole ground of certitude in a changing world. **1975** *Times Lit. Suppl.* 14 Feb. 162/5 Lundberg remains a sociologist honest and reflective enough to have tried to give 'quantophrenia' and testability ..a firm intellectual base.

quantum. Delete ‖. For 'Pl. **quanta** (*rare*)' read: Pl. **quanta** (*rare* except in senses 5 and 6), † **quantums** (sense 5 only). **1. a.** *spec.* in *Law*, an amount, a sum (*of* money payable in damages, etc.).

1898 in *Southern Reporter* XXIII. 718/2 The quantum of damages as fixed by the lower court is, we think, too low. **1912** *Law Rep.* (House of Lords Appeal Cases) 688 The quantum of damage is a question of fact, and the only guidance the law can give is to lay down general principles. **1945** *Tulane Law Rev.* XIX. 626 In a large majority of jurisdictions the pecuniary condition of the defendant has no bearing on the *quantum* of compensatory damages awarded the plaintiff in an action for personal injuries. **1951** *Scots Law Times* 11 Aug. 181/1 There can never be any binding precedents on quantum of damages because a sum reasonable in the circumstances of case A might be grossly unreasonable in case B. *Ibid.*, Quantum must be considered afresh in every case having regard to the particular circumstances of that case. **1970** *Internat. & Compar. Law Q.* XIX. 126 Strict liability with an unbreakable limit should confine litigation to questions of quantum. **1974** *Times* 6 Feb. 7/1 The Court of Appeal dismissed an appeal on quantum of damages by Horizon Holidays Ltd.

5. *Physics.* A minimum amount of a physical quantity which can exist and by multiples of which changes in the quantity occur.

This use of *quantum* originated in Ger. in two classic papers by Planck and by Einstein. Planck introduced the concept of a quantum in *Verh. d. Deutsch. Physik. Ges.* (1900) II. 237ff. In that paper he assumed that the energy of an oscillator is always an integral multiple of an 'energy element' (G. *energieelement*, p. 242), i.e. a quantum (sense *5 a), but he did not call it a quantum; however he did use the word in a passing reference to the electronic charge ('das Elementarquantum der Elektricität', p. 245: = sense *5 b).

Einstein, in *Ann. d. Physik* (1905) XVII. 132ff., assumed that light is radiated in the form of what he called 'energy quanta' (G. *energiequanta*, p. 133: = sense *5 a).

The affinities of the following isolated use are not clear; it seems not to be related to Planck's use, and may derive rather from the senses in Dict.

1902 Ld. Kelvin in *Phil. Mag.* III. 257 According to the well-known doctrine of Aepinus,..positive and negative electrifications consist in excess above, and deficiency below, a natural quantum of a fluid, called the electric fluid, permeating among the atoms of ponderable matter. *Ibid.* 259 The neutralizing quantum of electrons [= 'atoms of resinous electricity'] for any atom or group of atoms has exactly the same quantity of electricity of one kind as the atom or group of atoms has of electricity of the opposite kind. The quantum for any single atom may be one or two or three or any integral number, and need not be the same for all atoms... The differences of quality of the atoms of different substances may be partially due to the quantum-numbers of their electrons being different.

a. A discrete quantity of electromagnetic energy proportional in magnitude to the frequency of the radiation it represents.

1910 *Sci. Abstr.* A. XIII. 556 The absorption of the corresponding light-quantum. **1913** *Rep. Brit. Assoc. Adv. Sci. 1912* 407 Assuming that an oscillator can only emit definite, discontinuous quantums of energy, Planck showed that their magnitude is proportional to the frequency. **1913** *Phil. Mag.* XXVI. 19 These calculations strongly suggest that an electron of great velocity in passing through an atom and colliding with the electrons bound will loose energy in distinct finite quanta. **1929** D. H. Lawrence *Pansies* 28 Look then Where the father of all things swims in a mist of atoms Electrons and energies, quantums and relativities. **1934** A. J. Mee *Physical Chem.* xix. 721 We can, in a few cases, induce fluorescence of shorter wavelength than that of the absorbed light, since the energy emitted is not only that of the absorbed quantum, but also that inherent in the system before the absorption. **1965** *Physical Rev. Lett.* XV. 1013/1 He^+ ions decaying spontaneously from the 3*S* state do so via the 2*P* levels in ~ 1×10^{-9} seconds with the emission of a 1640Å quantum followed by a 303Å quantum. **1978** *Sci. Amer.* June 69/2 Conversely, when the atom or molecule is de-excited, it drops back one full step or more and the energy difference is either radiated away as a quantum of electromagnetic energy or transferred directly, through a collision, to another atom or molecule.

b. An analogous discrete amount of any other physical quantity (as momentum, electric charge).

1914 *Chem. Abstr.* VIII. 1050 (*heading*) Existence of quantities of electricity which are smaller than the charge of the elementary quantum or the electron. **1923** H. L. Brose tr. *Sommerfeld's Atomic Struct. & Spectral Lines* iv. 199 We see that the rotator is to be quantised not in energy quanta but in quanta of moment of momentum... The moment of momentum must be a whole multiple of $h/2\pi$. **1931** H. P. Robertson tr. *Weyl's Theory of Groups* i. 43 The constant of proportionality was equal to the quotient of the *h* obtained by Planck from black body radiation and the elementary quantum of electric charge *e*. **1958** *Nature* 31 May 1524/1 (*heading*) Detection of single quanta of circulation in rotating helium II. **1969** *Sci. Jrnl.* Jan. 87/2 There is a possibility that one day a minimum absolute quantum of length may also be found in the universe. **1973** *Sci. Amer.* Jan. 88/3 Waves of elastic crystal vibrations generate quanta of sound called phonons.

c. More fully *quantum of action*. = **Planck's constant*.

1913 *Phil. Mag.* XXVI. 2 Whatever the alteration in the laws of motion of the electrons may be, it seems necessary to introduce..a quantity foreign to the classical electrodynamics, *i.e.* Planck's constant, or as it often is called the elementary quantum of action. **1922** *Rep. Brit. Assoc. Adv. Sci. 1921* 473 The essential feature of the [quantum] theory is the existence of a universal constant, the quantum $h = 6{\cdot}55 \times 10^{-27}$ erg sec., which in some way..controls exchanges of energy. **1923** B. Russell *ABC of Atoms* vi. 80 The quantity *h*, Planck's quantum, has been found to be involved in all the very minute phenomena that can be adequately studied. **1933** *Nature* 25 Mar. 422/2 A causal description in the classical sense is possible only in such cases where the action involved is large compared with the quantum of action. **1956** E. H. Hutten *Lang. Mod. Physics* v. 179 We can measure simultaneous values of both parameters only in such a way that the numerical product of their inaccuracies is, at best, equal to the quantum of action h.

d. *fig.*

1960 R. W. Marks *Dymaxion World of B. Fuller* 10/1 Fuller regards all human experiences as energy events finite in extent. All experiments performed, books written, thoughts expressed, and structures completed, are finite energy events. Together they form a totality, a cornucopia of patterned quanta. **1962** P. Strevens *Papers in Lang.* (1965) v. 67 Teaching takes place by *quanta*. Whether the teacher realizes it or not, he can teach only in steps, though these vary in size. **1969** *Daily Tel.* (Colour Suppl.) 10 Jan. 32/2 Generalisations serve a purpose, but true understanding is made up of many discrete *quanta*. I can describe the Atlantic littoral thus and so; but I *know* it—thus: [etc.]. **1977** *Times* 4 Aug. 8/6 A fine quantum of derring-do ranging from icy Sweden to a storm-threatened Scottish islet.

6. *Physiol.* Orig. a small voltage of which integral multiples go to make up the end-plate potential measured at a neuromuscular junction; hence, the unit quantity of acetylcholine corresponding to this, multiples of which are released to transmit a nerve impulse across the junction.

1952 Fatt & Katz in *Jrnl. Physiol.* CXVII. 120 The experiment throws some new light on the action of calcium at the nerve-muscle junction: lack of calcium apparently reduces the e.p.p. in definite 'quanta'. **1954** del Castillo & Katz in *Ibid.* CXXIV. 560 It has been suggested that the end-plate potential (e.p.p.) at a single nerve-muscle junction is built up statistically of small all-or-none units... A convenient picture of how hundreds of such quanta..can build up an e.p.p. of, say, 70–80 mV is provided by the hypothesis that separate parcels of acetylcholine (ACh), released from discrete spots of the nerve endings, short-circuit the muscle membrane. *Ibid.* 574 Recent evidence indicating that ACh release occurs in discrete quanta. **1970** J. W. Phillis *Pharmacol. of Synapses* ii. 17 In addition to being released by stimulation, individual quanta are released from the terminal spontaneously... At the neuromuscular junction, the number of quanta available for immediate release is probably of the order of 1000. **1978** *Nature* 9 Feb. 561/1 Acetylcholine..is released from stimulated nerve terminals in packets or quanta, each containing roughly 10,000 molecules.

7. a. *attrib.* and *Comb.* (in sense *5), as *quantum energy, hypothesis, law, physics* (hence *quantum physicist*), *property*; **quantum advance** = **quantum leap*; **quantum chemistry,** the branch of physical chemistry concerned with the explanation of chemical phenomena in terms of quantum mechanics; so **quantum chemist,** an expert or specialist in this; **quantum-chemical** *a.*; **quantum chromodynamics** [Chromo-, after **quantum electrodynamics*], a quantum field theory in which the strong interaction is described in terms of an interaction between quarks that is mediated by gluons, both kinds of particle being assigned a quantum number called 'colour'; abbrev. *QCD; **quantum condition,** a condition resulting from, or forming part of, the application of quantum theory to a system; a condition that selects from the states allowed by classical physics those that are consistent with quantum theory; **quantum defect,** a number representing the degree to which an energy level of an atom with a single valence electron is displaced from the corresponding level of the hydrogen atom, being the amount by which the true principal quantum number of the level exceeds the effective value of the number; **quantum dynamics** = *Quantum mechanics; hence **quantum-dynamical** *a.*; **quantum effect,** a physical effect attributed to the existence of quanta; **quantum efficiency,** the proportion of incident photons that are effective in causing the decomposition of a molecule, the emission of a photoelectron, or similar photo-effect; **quantum electrodynamics,** the part of quantum field theory concerned with the electromagnetic field and its interaction with electrically charged particles; so **quantum-electrodynamic, -dynamical** *adjs.*; abbrev. *QED; **quantum electronics,** the branch of physics concerned with the practical consequences of the quantization of energy states and their interaction with electromagnetic radiation; so **quantum-electronic** *a.*; **quantum field theory,** a field theory that incorporates quantum mechanics and the principles of the theory of relativity; **quantum increase,** a sudden large increase; cf. **quantum jump*; **quantum jump,** an abrupt transition between one stationary state of a quantized system and another, with the absorption or emission of a quantum; also *transf.*, a sudden large increase or advance; **quantum leap,** a sudden large advance; cf. **quantum jump*; **quantum level,** an energy level in a quantized system; **quantum liquid,** a liquid that exhibits quantum effects on the macroscopic scale; **quantum number,** a number which enters into the expression for the value of some quantized property of a system (usu. a particle, atom, or molecule) and can assume only certain integral and sometimes half-integral values; also *transf.*, the property so characterized; **quantum orbit,** an orbit (of an electron in an atom) defined by a set of quantum numbers; **quantum solid,** a solid that exhibits quantum effects on the macroscopic scale; **quantum state,** a state of a physical (esp. atomic) system that is defined by a set of quantum numbers; a quantized state; **quantum statistics,** the statistics of the energy distribution of particles when the quantization of energy is taken into account; cf. **Bose–Einstein statistics*, **Fermi–Dirac statistics*; hence **quantum-statistical** *a.*; **quantum transition** = **quantum jump* (*lit.* sense); **quantum yield** = **quantum efficiency*. Also *Quantum mechanics, *Quantum theory.

1974 *Sci. Amer.* June 105/3 When these inferences are taken together with the differences between the two material cultures..one is led to conclude that the Upper Paleolithic represents a quantum advance in human cultural evolution. **1960** *McGraw-Hill Encycl. Sci. & Technol.* XI. 145/2 The most useful of these methods, the variation method, has produced most of the important quantum-chemical concepts. *Ibid.* 146/1 The success of this method depends..on the ability of the quantum chemist to guess at trial functions which are good approximations and at the same time contain parameters in such a form that *W* can be minimized without undue labor. **1970** *Sci. Amer.* Apr. 54/1 Since the introduction of the fundamental wave equation of quantum mechanics by Erwin Schrödinger in 1926, much of the work of quantum chemists has been focused on its solution for specific chemical systems. **1944** H. Eyring et al. (*title*) Quantum chemistry. **1963** *New Scientist* 14 Mar. 582/3 After a rather long incubation period, the new subject of 'quantum chemistry' has got into its stride and is gaining rapidly in strength. [**1976** *Physical Rev. Lett.* XXXVI. 1521 In an exact color SU(3) gauge theory (chromodynamics), the dynamics of confinement may be similar to the dynamics of the string model.] **1976** *Nature* 12 Aug. 538/1 The 'gauge field theories' are underlying not only weak-electromagnetic but perhaps also strong interactions (the new jargon here being quantum chromodynamics, or QCD, analogous to quantum electrodynamics). **1979** *Sci. Amer.* Aug. 157/2 No one has yet been able to derive the confinement of quarks from the underlying theory of quantum chromodynamics. **1923** *Physical Rev.* XXII. 547 This variation principle includes formally in a single equation the results of classical dynamics and the Sommerfeld quantum conditions. **1955** O. Klein in W. Pauli *Niels Bohr* 99 The importance of transformation groups for the formulation of quantum conditions in field theories..has been strongly emphasized. **1974** G. Reece tr. *Hund's Hist. Quantum Theory* vi. 89 The 'stationary states' were selected from the possible classical motions by 'quantum conditions' $I_k = hn_k$. **1930** Ruark & Urey *Atoms, Molecules & Quanta* vii. 194 The quantum defect is a measure of the departure of the spectral term from the hydrogenic term having the same total quantum number. **1970** G. K. Woodgate *Elem. Atomic Struct.* vi. 103 We consider the sequence iso-

electronic with sodium... The ionization potentials for this sequence,..together with the quantum defect δ(*s*) ..are given below. **1932** *Physical Rev.* XL. 406 In solving the wave mechanical perturbation problem the distance between the interacting structures has been treated as a fixed parameter. Then quantum dynamical reasoning has been abandoned, and the remainder of the problem has been solved by the method of classical statistics. **1927** *Proc. R. Soc.* A. CXIII. 621 (*heading*) The physical interpretation of the quantum dynamics. **1967** Condon & Odishaw *Handbk. Physics* (ed. 2) vii. i. 3/1 These laws of quantum dynamics must involve the universal Planck constant..in an essential way; and the quantum laws must go over asymptotically into the classical laws, not involving *h*, as the scale of the phenomena is increased. **1914** *Chem. Abstr.* VIII. 3141 E. detd. the at. ht. of highly compressed He at temps. from 18° to 32° abs... A small quantum effect is apparent. **1946** *Physical Rev.* LXIX. 195 (*heading*) Quantum effects in the interaction of electrons with high frequency fields and the transition to classical theory. **1975** *McGraw-Hill Yearbk. Sci. & Technol.* 114/2 The interaction between two charged particles, classically treated, is given ..by Maxwell's electrodynamic equations, if the particles are in relative motion. The inclusion of quantum effects has led to a more general theory, called quantum electrodynamics. **1926** *Trans. Faraday Soc.* XXI. 453 In the photochemical isomeric change of maleic and fumaric acid, the quantum efficiency was found to be much smaller than unity. **1940** Glasstone *Textbk. Physical Chem.* xiii. 1135 According to the law of the photochemical equivalent one mol of absorbing substance should decompose for every $2{\cdot}854 \times 10^8/\lambda$ kg.-cal. of radiation absorbed... This relationship..permits the law to be tested experimentally. The results are expressed in terms of the quantum efficiency. **1978** *Nature* 16 Mar. p. xiv/3 Two versions of the new EMI photomultiplier tube..are available with..typical peak quantum efficiency of about 22%. **1965** *Physical Rev. Lett.* XV. 1013 (*heading*) Measurement of the quantum-electrodynamic level shift in the *n* = 3 state of (He^4). **1956** *Physical Rev.* CI. 1410 The quantum-electrodynamical fourth-order corrections for the intervals of the triplet fine structure of helium are calculated. **1927** *Proc. R. Soc.* A. CXIV. 243 Hardly anything has been done up to the present on quantum electrodynamics. **1955** L. Rosenfeld in W. Pauli *Niels Bohr* 70 They questioned the logical consistency of quantum electrodynamics by contending that the very concept of electromagnetic field is not susceptible, in quantum theory, to any physical determination by means of measurements. The measurement of a field component requires determinations of the momentum of a charged test-body; and the reaction of the field radiated by the test-body in the course of these operations would..lead to a limitation of the accuracy of the field measurement, entirely at variance with the premises of the theory. **1971** *Sci. Amer.* June 64/3 The laws of electricity and magnetism as they are now embodied in the equations of quantum electrodynamics represent the one and only area in physics where a single quantitative description has proved valid over the entire range of experiments for which it has been tested, from cosmic dimensions down to 10^{-18} centimeter. **1976** B. Bova *Multiple Man* (1977) i. 11 The President..was protected by an invisible laser-activated shield... Fool-proof quantum-electronic security. **1959** *Jrnl. Appl. Physics* XXX. 956/1 An international conference on Quantum Electronics—Resonance Phenomena will be held..on September 14–16, 1959. The conference will consider basic problems in physics and electronics which are important to the increasing use of molecular and atomic resonance in masers, atomic clocks, and related devices. **1965** *Wireless World* Aug. 386/2 The film..covers a wide range of applications including quantum electronics in transistors, lasers and masers. **1972** *Physics Bull.* Sept. 562/1 Quantum electronics as a field of study dates from the use of stimulated emission for microwave amplification in 1954. **1921** *Discovery* Sept. 227/2 When the quantum energy of the exciting radiation exceeds this amount the whole K series [of X-rays] is excited. **1948** *Physical Rev.* LXXIV. 224/1 (*heading*) On infinite field reactions in quantum field theory. **1956** H. Umezaw *Quantum Field Theory* i. 11 The present quantum field theory was formulated (Heisenberg and Pauli (1929)) by extending quantum mechanics so as to satisfy the relativity requirements and to treat the various transmutations [of particles into one another]. **1978** *Sci. Amer.* Feb. 128/1 (*caption*) The creation and annihilation of particles and antiparticles is the characteristic process that distinguishes quantum field theories from 'classical' field theories such as Maxwell's or Einstein's. **1914** *Rep. Brit. Assoc. Adv. Sci. 1913* 378 The quite definite result is obtained that..the exchange of energy between matter and ether must take place by finite jumps of amount..$\epsilon = h\nu$. This is, of course, the hypothesis, spoken of briefly as the quantum-hypothesis, which was first suggested by Planck. **1968** F. L. Pilar *Elem. Quantum Chem.* i. 8 Planck's own feeling, which persisted for many years, was that the quantum hypothesis itself could have no basic significance but rather was an artificiality which would eventually be replaced with a more reasonable alternative. **1973** *Sci. Amer.* May 8/2 Intellectual and other exchanges with China appear to be on the verge of a quantum increase. **1974** *Daily Tel.* (Colour Suppl.) 16 Aug. 17 Throughout this enormous area..people have been dying of diseases aggravated by malnutrition, just as they have died every year since man first inhabited the Sahel, but so far there has been no *quantum* increase in human mortality as a result of the lengthy drought. **1927** N. V. Sidgwick *Electronic Theory of Valency* ii. 18 Bohr's theory is based on two fundamental postulates... The second..is that the electron radiates energy..only when it passes in a 'quantum jump' from one of these stationary states to another of smaller energy. **1937** E. C. Kemble *Fund. Princ. Quantum Mech.* viii. 290 The variation in the distribution function thus defined..is commonly interpreted as due to 'quantum jumps' from one energy level to the other caused by the radiation. **1955** *Sci. News Let.* 19 Feb. 116/2 Radioactive fall-out is the 'third quantum jump' in the history of modern weapons. The first quantum jump, Dr. Lapp explained, was the A-bomb that shattered Hiroshima. **1961** *Flight* LXXX. 907/1 On the subject of launch operations, Mr Debus claimed that a 'quantum jump' was necessary to meet the challenge of the lunar programme. **1966** *Guardian* 1 Jan. 8/1 This new escalation (or 'quantum jump', to use the latest addition to the bewildering mixture of metaphors) would look better if it were preceded by another bombing pause. **1974** G. Reece tr. *Hund's Hist. Quantum Theory* x. 129 While the emission and absorption of light seemed to be connected with a quantum jump, its dispersion did not. **1975** *Chinese Econ. Stud.* VIII. iv. 52 In old China, the issue of legal tender reached astronomical figures, leading to galloping inflation and quantum jumps in prices. **1916** *Chem. Abstr.* X. 1722 (*heading*) The quantum law and the structure of the hydrogen atom. **1967** Quantum law [see **quantum dynamics*]. **1970** *New Scientist* 3 Dec. 372/1 The ability of marine technology to take 'quantum' leaps in innovation means that a laissez-faire approach to the ocean mineral resources can no longer be tolerated. **1973** *Daily Tel.* (Colour Suppl.) 2 Nov. 27/2 Hovercraft, like many inventions of modern technology, are supposed to progress in quantum leaps. **1977** *New Yorker* 13 June 108/2 The imperial Presidency did not begin with Richard Nixon although under him abuses of the office took a quantum leap. **1931** G. Gamow *Constitution of Atomic Nuclei* iii. 63 Some elements have rather complicated γ-ray spectra, while others have only a few lines... These facts are evidently connected with the strength of the initial excitation and the relative position of the quantum-levels in the different nuclei. **1960** Chalmers & Quarrell *Physical Examination of Metals* (ed. 2) xvi. 751 Above the true quantum levels there are further, 'empty' levels, representing states in which nuclei may exist momentarily before disintegrating. **1950** F. London *Superfluids* I. 1 The 'superfluids' or 'quantum liquids' probably exhibit the most conspicuous phenomena of macroscopic physics which have not yet been integrated into molecular theory. **1967** [see *quantum solid* below]. **1975** *Physics Bull.* July 311/3 The electrons in a metal constitute a quantum liquid in this sense at all temperatures up to the melting temperature. [**1902** Quantum-number: see sense 5 above]. **1920, 1922** [see *N I. 4 b]. **1926,** etc. [see *L 6* b]. **1939** G. Herzberg *Molecular Spectra & Molecular Struct.* I. i. 15 The azimuthal quantum number *l* gives us therefore the orbital angular momentum of the electron in units $h/2\pi$. *Ibid.* 18 In a magnetic or electric field, a precession of the angular momentum of an atom takes place... While classically the precession could take place at any angle to the field direction, according to quantum theory, only those angles are possible for which the components of the angular momentum in the direction of the field have the discrete values $m_l(h/2\pi)$, where $m_l = l, (l-1), (l-2), .., -l$... m_l is called the magnetic quantum number of the electron. **1967** W. R. Hindmarsh *Atomic Spectra* vi. 70 If the magnetic field is so strong that its interaction with the magnetic moment of the atom is much stronger than the spin-orbit interaction, the orbital and spin angular momenta are no longer even approximately conserved, so that *L* and *S* are not well-defined quantum numbers. **1968** J. Bernstein *Elementary Particles & their Currents* xiii. 211 The strong interactions of the strange particles are characterized by a new conserved quantum number in addition to the isotopic spin. This quantum number is the 'strangeness', *S*, or, equivalently, the 'hypercharge' *Y*. **1976** *Sci. Amer.* Jan. 45/1 Some quantum numbers, such as spin angular momentum and electric charge, are invariably conserved. **1923** H. L. Brose tr. *Sommerfeld's Atomic Struct. & Spectral Lines* ii. 67 The quantum theory asserts that all these quantum orbits are stationary states of motion, that is they are traversed without radiation being emitted. **1928** Quantum orbit [see *Packet *sb.* 1 h]. **1946** *Mind* LV. 161 The philosopher finds support in the quantum physicist's principle of complementarity. **1964** M. McLuhan *Understanding Media* (1967) vii. 73 Werner Heisenberg..is an example of the new quantum physicist whose over-all awareness of forms suggests to him that we would do well to stand aside from most of them. **1931** H. Johnston tr. *Planck's Universe in Light of Mod. Physics* 22 The Principle of Relativity..has proved itself a reliable and eloquent guide in the new regions of Quantum Physics. **1971** *Sci. Amer.* Mar. 75 Quantum physics normally deals with natural phenomena on a submicroscopic scale. **1978** *Ibid.* Dec. 128/3 Quantum physics predicts that captured electrons can have only discrete energies, corresponding to the atomic energy levels described above. **1927** A. S. Eddington *Stars & Atoms* 68 The property here referred to (the quantum property) is the deepest mystery of light. **1978** *Nature* 16 Mar. 291/3 The quantum properties of electromagnetic radiation. **1967** *Sci. Amer.* Aug. 85/2 Solid helium..is the only known example of a 'quantum solid', just as liquid helium is the only known example of a 'quantum liquid'. **1921** *Chem. Abstr.* XV. 1843 Schottky discusses..the Nernst heat theorem from the point of view of the internal 'quantum state' of the mols. composing the system under discussion. **1946** [see *Polaron]. **1972** *Sci. Amer.* Oct. 101/1 The individual nucleons move in discrete quantum states just as the electrons in the atom exist in discrete quantum states. **1958** *Physical Rev.* CXI. 1460 (*heading*) Quantum statistical theory of electron correlation. **1932** *Jrnl. Chem. Soc.* 373 We shall.. discuss the important conclusions which can be drawn from the interpretation of molecular spectra regarding the so-called quantum statistics in their relation to the nuclei. **1935** Pauling & Wilson *Introd. Statistical Mech.* viii. 219 The quantum statistics resulting from the acceptance of only antisymmetric wave functions is considerably different. **1972** *Physics Bull.* Dec. 709/1 It is our great good fortune that the only two substances that remain liquid down to the absolute zero obey different statistics (³He: Fermi–Dirac, ⁴He: Bose–Einstein) and thus allow us to study the differing effects of quantum statistics on condensed matter. **1924** *Physical Rev.* XXIV. 330 The term to be retained is..the combination overtone asymptotically connected to the particular quantum transition under consideration. **1927** Fisher & Hartree tr. *Born's Mech. of Atom* ii. 54 Quantum transitions can be caused by light and by molecular impacts. **1927** *Jrnl. Amer. Chem. Soc.* XLIX. 2451 These new facts make possible a better conception of the mechanism of the photochemical decomposition of ammonia... Warburg measured the quantum yield and found it to be 4 quanta per molecule for light of wave length 2025–2140 Å. **1971** *Jrnl. Appl. Physics* XLII. 567/2 The quantum yield of the Ag–O–Cs cathode for visible light is less than 5×10^{-4} electron per photon.

b. Passing into adj. (cf. *Quantal *a.*).

1922 *Rep. Brit. Assoc. Adv. Sci. 1921* 474 This suggests that many phenomena which at present are thought to be satisfactorily explained by dynamics are really quantum phenomena. **1924** *Physical Rev.* XXIV. 340 This connection of the classical and quantum differential absorption we shall term the correspondence principle for absorption. **1951** *Ibid.* LXXXII. 116/2 The quantum nature of the exchange of energy between free electrons and electromagnetic fields. **1967** *Sci. Amer.* Aug. 85/2 Perhaps the most perplexing characteristic of solid helium..is the fact that in spite of its unique quantum nature solid helium behaves in so many respects as a purely classical, or nonquantum, solid. **1975** *Nature* 20 Mar. 223/3 The book is concerned mainly with electrodynamics (both classical and quantum).

quantum mecha·nics. *Physics.* [f. Quantum + Mechanics.] A mathematical theory of the motion and interaction of (esp. subatomic) particles that was developed from the old quantum theory and incorporates the concept of wave–particle duality, the uncertainty principle, and the correspondence principle; cf. **matrix mechanics*, **wave mechanics*.

This is not quite the sense in quot. 1922.

1922 *Rep. Brit. Assoc. Adv. Sci. 1921* 473 The spectrum theory is far the most important branch of the quantum theory, as it has led and is still leading to extensions of quantum mechanics. **1925** *Proc. R. Soc.* A. CIX. 642 (*heading*) The fundamental equations of quantum mechanics. **1935** B. Russell *Relig. & Sci.* vi. 151 The challenge has come through the study of the atom by the new methods of quantum mechanics. **1956** E. H. Hutten *Lang. Mod. Physics* v. 179 The mathematical formulae of quantum mechanics permit us to predict future events only within a certain margin of error. **1956** H. Umezawa *Quantum Field Theory* i. 1 Quantum mechanics and relativity are respectively characterised by the constants *h* and *C* and, when *h* and (1/*C*) are taken as vanishingly small, the results of these theories are identical with those of Newtonian mechanics. **1974** G. Reece tr. *Hund's Hist. Quantum Theory* x. 132 The first strictly valid version of quantum mechanics that was both logical and capable of generalization was provided by Heisenberg's paper of July 1925. **1978** *Sci. Amer.* Feb. 126/3 There is no workable theory of gravitation that is consistent with the principles of quantum mechanics. *Ibid.* 131/1 The present understanding of the fundamental laws of nature arose from three principles: special relativity, general relativity and quantum mechanics.

Hence **quantum-mechanical** *a.*, **-mechanically** *adv.*

1927 *Proc. R. Soc.* A. CXIV. 715 One can obtain the result..by using a quantum-mechanical argument. **1936** Quantum-mechanically [see *Quantal *a.* 2 c]. **1949** Koestler *Insight & Outlook* xi. 156 Quantum-mechanical concepts slowly permeate all spheres of physics, chemistry, and..biology. **1963** G. Troup *Masers & Lasers* (ed. 2) iv. 58 It can be shown quantum-mechanically..that there is a definite probability that A will make the transition 1 → 2 while B makes the transition 3 → 1. **1968** G. Ludwig *Wave Mech.* I. ii. 9 These equations..are the basis even today for the construction of modern accelerators, since quantum-mechanical effects are not involved in the motion of the particles in these devices. **1971** *Nature* 15 Jan. 158/2 The existence of such a low-lying σ transition would be a surprise to theoreticians, but quantum-mechanically it is not entirely out of the question.

‖ **quantum meruit** (kwǫ·ntŏm me·rŭ,it). *Law.* [L., 'as much as he has deserved'.] A reasonable sum of money to be paid for services rendered or work done, when the amount due is not determined by any provision constituting, or forming part of, a legally enforceable contract (see also quot. 1959). Also *attrib.* and as quasi-*adv.*

1657 H. Grimston tr. *G. Croke's Rep.* (Charles I) 77 It is the usuall way to lay down in certainty, *viz.* That he should pay for it *tantum quantum meruit, &c.*, and then to averre what it is reasonably worth, which being the common course and alwaies allowed, Judgment was therefore affirmed. **1659** in E. Bulstrode *Rep.* III. 86 How shall a Taylor be paid upon a *quantum meruit*? **1685** J. Keble *Rep.* I. 422 In action by executors on *indebitatus* for wares sold to pay *quantum meruit*, not shewing what the particulars were. **1718** W. Salkeld *Rep.* II. 557 *Quantum meruit* for meat, drink, &c. **1729** G. Jacob *New Law-Dict.* sig. H h h h 2, *Quantum meruit*, i.e. how much he has deserved, is a Man's Action of the Case,.. grounded upon the Promise of another, to pay him for doing any Thing so much as he should deserve or *merit*. If a man retains any person to do work or other thing for him..without any certain agreement; in such case the law implies that he shall pay for the same, as much as they are worth, and shall be reasonably demanded; for which *Quantum meruit* may be brought. **1832** in P. Bingham *Rep.* VIII. 16, I agree that, when a special contract is in existence and open, the plaintiff cannot sue on a *quantum meruit*. **1904** *Law Rep. King's Bench Div.* II. 329 Remuneration upon the basis of a quantum meruit in respect of work and labour done for the company at its request. **1919** *Brit. Manufacturer* Nov. 36/2 The owners could, therefore, sue under the old contract (the charter party) for 23s., the agreed freight, on each of the 1208 tons of steel billets; and, in addition, could sue on the new implied contract for a *Quantum Meruit* in respect of the 987 tons of general cargo. So the only question is the rate per ton at which their *Quantum*

Meruit should be calculated. **1959** Jowitt *Dict. Eng. Law* II. 1452/2 Where the failure to complete performance of the contract is due to the fault of the other party, the party not in default has the right to sue on a *quantum meruit* for the services which he has done under it. In its early history the action for *quantum meruit* was, no doubt, a genuine action in contract... In many cases the action is now founded on what is known as 'quasi-contract'. **1962** A. Turner *Law of Trade Secrets* iv. iv. 346 Other expressions, like 'unjust enrichment', sometimes used by the courts are recovery in *quantum meruit* and recovery under 'quasi contracts', 'constructive contracts', or, contracts 'implied-in-law'. **1964** *Mod. Law Rev.* XXVIII. 353 The carrier deserves remuneration under the *quantum meruit* rule. **1973** *N.Y. Law Jrnl.* 1 Aug. 13/3 However, since the former attorneys have the right to elect whether they will accept their compensation on the basis of a presently fixed dollar amount quantum meruit or whether, still on the basis of quantum meruit, they will accept a contingent percentage instead [etc.]. **1975** S. J. Stoljar *Hist. Contract at Common Law* ix. 109 If, as often happened, the work was undertaken at no specified price, the plaintiff could turn to another count in assumpsit becoming known as the *quantum meruit*, a claim not for a fixed but for a reasonable amount: what the work merited or was worth.

quantum sufficit. Add: **a.** (Further examples of *quant. suff.*)

1840 Barham *Patty Morgan* in *Ingol. Leg.* 1st Ser. 60 One glance was enough, Completely '*Quant. suff.*' As the doctors write down when they send you their 'stuff'. **1907** G. B. Shaw *John Bull's Other Island* p. xxv, It was hardly reasonable to ask Parnell to shed blood *quant. suff.* in Egypt..and then to expect him to become a Tolstoyan or an O'Connellite in regard to his own country. **1964** C. S. Lewis *Discarded Image* i. 9 Popular iconography.. wishing to summon up the idea of the Medieval, draws a knight errant with castles, distressed damsels, and dragons *quant. suff.* in the background.

b. (Later example of *quant. suff.*)

1863 *Fraser's Mag.* Feb. 156/2 A *quant. suff.* is beaten up with water, which is strained off after standing half an hour.

quantum theory. *Physics.* [f. Quantum + Theory¹.] A theory of matter and energy based on the concept of quanta (sense *5); *spec.* the branch of physics that was developed from the ideas in Planck's paper of 1900 and Einstein's of 1905 (see *quantum 5), was extended by Bohr (1913) in relation to atomic structure, and later evolved into quantum mechanics and quantum field theory; *old quantum theory*, the early form of the theory, based on classical mechanics, prior to the development of wave mechanics and matrix mechanics in the mid-1920s.

[**1911** *Sci. Abstr.* A. XIV. 1702 The quanta theory of Planck and Einstein must be modified considerably to give a quantitative interpretation of the results obtained.] **1912** *Monthly Notices R. Astron. Soc.* LXXII. 677 The constant of nature in terms of which these spectra can be expressed appears to be that of Planck in his recent quantum theory of energy. **1926** *Times Lit. Suppl.* 19 Aug. 544/3 Relativity theory and quantum theory have not yet been properly assimilated. **1927** *Proc. R. Soc.* A. CXIV. 181 This equation..was obtained originally by Sommerfeld from relativistic considerations with the old quantum theory. *Ibid.* 243 The new quantum theory, based on the assumption that the dynamical variables do not obey the commutative law of multiplication, has by now been developed sufficiently to form a fairly complete theory of dynamics. **1958** W. Heisenberg *Physics & Philos.* vi. 106 Quantum theory does not allow a completely objective description of nature. **1970** G. K. Woodgate *Elem. Atomic Struct.* vi. 103 For small *l* the electron orbit is highly eccentric (to use the language of the old quantum theory). **1972** *Physics Bull.* Sept. 548/1 This book consists of five short chapters, on classical mechanics and the old quantum theory, the new quantum theory, many particle systems, valence theory and quantum theory of chemical reactivity. **1978** *Sci. Amer.* Feb. 132/3 Quantum theories of the gravitational force still have serious difficulties with infinities.

Hence **quantum-theoretical** *a.*, **-theoretically** *adv.*; **quantum theorist**, an expert or specialist in quantum theory.

1920 *Chem. Abstr.* XIV. 1637 (*heading*) Quantum theoretical principles of photochemistry. **1931** *Physical Rev.* XXXVIII. 1787/1 No quantum theoretical calculations have been made for particles of very high energy. **1935** *Amer. Jrnl. Math.* LVII. 429 The algebra II is known to the quantum theorist from the process of 'superquantizing'. **1939** G. Herzberg *Molecular Spectra & Molecular Struct.* I. iii. 84 Quantum theoretically, emission of radiation takes place as a result of a transition of the oscillator from a higher to a lower state. **1959** K. R. Popper *Logic of Sci. Discovery* ix. 222 In his derivation of the uncertainty relations, Heisenberg follows Bohr in making use of the idea that atomic processes can be just as well represented by the 'quantum-theoretical image of a particle' as by the 'quantum-theoretical image of a wave'. **1968** G. Ludwig *Wave Mech.* i. ii. 19 These classical results were transformed by Bohr into quantum-theoretical results. **1976** *Nature* 1 July 17/1 General Electric advertised for a quantum theorist to look into the fundamental quantum processes involved in fluorescence.

quarantinable, *a.* Delete *rare* and add earlier and later examples.

1863 *Laws N.Y. State* ccclviii. 576 With existing quarantinable disease on board,..merchandise of the first class shall be landed at the quarantine warehouse. **1906** *Daily Colonist* (Victoria, B.C.) 25 Jan. 8/1 Foreign coasting vessels touching at Victoria to the number of 947 required inspection. No quarantinable disease occurred. **1961** *Times* 25 Aug. 12/2 There are six quarantinable diseases.

quarantine, *sb.* Add: **2. a.** Also, a period of seclusion or isolation after exposure to infection from a contagious disease; *transf.*, (a period of) isolation imposed in a similar way on an animal or thing. Freq. in phr. *in quarantine*. Also *fig.*

1855 [see *ask *v.* 4 c]. **1879** *Investigation of Diseases of Swine* (Special Rep. No. 12, U.S. Dept. Agric.) 151 All strange hogs must be kept in quarantine for fourteen days before being allowed to run with healthy herds. **1891** *Boston Jrnl.* 7 Jan. 2/3 A rigid quarantine against firearms and firewater on the reservations of the Northwest is one of the prime requirements of the Indian problem. **1913–14** *Wellcome's Nurse's Diary* 209 Isolation required after exposure to: Asiatic Cholera..12 days' quarantine. **1922** *Encycl. Brit.* XXX. 925/2 Formerly great stress was laid on the value of quarantine; all plant imports were grown in a quarantine ground under the supervision of a Government botanist until it was certain that they had no disease. **1952** *Oxf. Jun. Encycl.* X. 357/2 All dogs.. have to be isolated in quarantine for 6 months in case they may be carrying rabies. **1971** *Sci. Amer.* Oct. 49/2 To guard against the possibility..of introducing pathogenic organisms from the moon, the lunar samples were placed in quarantine for seven weeks. **1978** W. Garner *Möbius Trip* (1979) ii. 60 Putting him in emotional quarantine.

b. *spec.* in international politics, a blockade, boycott, or severance of diplomatic relations intended to isolate a nation, or the isolation caused by such action.

This use arose from a speech by F. D. Roosevelt, President of the U.S. (see quots. 1937).

1937 *N.Y. Herald Tribune* 6 Oct. 1/5 (*heading*) President calls for 'quarantine' of aggressors. *Ibid.* 1/8 President Roosevelt today challenged the effectiveness of a policy of neutrality in keeping the United States at peace and advocated instead a collective 'quarantine' of aggressor nations. **1938** *Sun* (Baltimore) 16 Nov. 1/8 Ambassador Wilson will not return soon to his post... It may even imply a 'quarantine' or an effort to quarantine Germany. **1945** *Richmond* (Va.) *News-Leader* 4 Oct. 2/7 (*heading*) Argentina faces diplomatic 'quarantine' by Pan-America. **1962** *Daily Tel.* 23 Oct. 1/2 Mr. Kennedy announced the following actions in response to the military build-up in Cuba. The blockade against delivery of offensive weapons. The 'quarantine' would be extended if necessary, to other types of cargo and carriers. **1975** *Ibid.* 1 Oct. 1/7 A call by the International Transport Workers' Federation..for a 48-hour quarantine of services to and from Spain.

4. *quarantine-flag* (earlier example), *-ground*, *kennel*; *quarantine-breaking* adj.

1808 *Deb. Congress U.S.* 9 Mar. (1852) 1753 The ship arrived at the quarantine ground, near the harbor of Boston. **1835** J. E. Alexander *Sk. Portugal* xi. 265 After some delay before we could get our yellow quarantine-flag struck, we were allowed to land. **1852** G. Coggeshall *Second Series of Voyages* xiii. 82 We were requested to proceed immediately to the quarantine ground. **1867** 'Mark Twain' in *Daily Alta California* (San Francisco) 18 Oct. 1/4 This kind of conversation did no good, further than to give a sort of dismal interest to our quarantine-breaking expedition, and so we dropped it. **1942** E. E. Dale *Cow Country* 203 Wide strips were left for trails across the Outlet and lands were also set aside for quarantine grounds. **1976** T. Heald *Let Sleeping Dogs Die* i. 11 To prevent it [*sc.* rabies] being imported all dogs coming in to Britain had to spend six months in quarantine kennels. **1977** *Hongkong Standard* 14 Apr. 8/5 A friend visited the Government Quarantine Kennels at Shatin recently and was distressed and appalled at the neglect of the poor animals awaiting their death, particularly the puppies.

quarantine, *v.* Add: **1. a.** Also *transf.* and *fig.*; *spec.* in sense 2 b of the *sb.*

1860 *Harper's Mag.* June 137/2 Duelling had become epidemic among the midshipmen at the Gosport navy-yard. A determined effort was made..to suppress the practice. The entire body of reefers were 'quarantined', i.e. confined to the limits of the yard. **1870** W. M. Baker *New Timothy* i. 13 The business of these [ministers] is with human nature, and from exactly that are they quarantined for years. **1937** *N.Y. Times* 6 Oct. 1/8 President Roosevelt today pledged his Administration to a 'concerted effort' with other peace-loving nations to 'quarantine' aggressor nations. **1938** [see *nose count]. **1938** [see *quarantine *sb.* 2 b]. **1945** *Sun* (Baltimore) 28 Sept. 11/2 At school, they find themselves 'quarantined' and they are the butt of jibes and social ostracism. **1953** P. C. Berg *Dict. New Words* 132/1 *Quarantine, v.t.*, to isolate (a nation). **1962** *Listener* 12 Apr. 623/2 They think that, given the Communist Powers' publicly proclaimed hostility to the West, the whole lot of them should be quarantined, as the U.S.A. quarantines China. **1976** *O.E.D. Suppl.* s.v. *L* 7, LRL, Lunar Receiving Laboratory (building where astronauts and lunar samples are quarantined for a period after returning from the moon). **1980** *Early Music* Apr. 255/2 In this setting the melody is not quarantined in the tenor register.

c. To isolate (an area) by the imposition of quarantine.

1890 *Stock Grower & Farmer* 24 May 7/3 The state [*sc.* of Nebraska] is strictly quarantined against all cattle from Texas. **1955** *Sci. Amer.* June 82/2 An outbreak of disease may be localized by quarantining the infected area.

3. *intr.* To go into quarantine.

1928 *Daily News* 7 Aug. 7/3 The Mauretania..is expected to 'quarantine' at New York at 10 a.m. to-morrow.

‖ **Quarant' Ore** (kwarant ōᵊ·re). Also **Quarantore.** [It., contraction of *quaranta* forty + *ore*, pl. of *ora* hour.] = *forty hours* s.v. *forty *a* e.

1623 *Ven. Eng. College, Rome MS. Scritt.* 29.5.1a. f.4, By the church doore, which by chaunce was open by reason that the Quarante Hore were celebrated there at the same tyme. **1839** N. Wiseman *Let.* in P. Devine *Life Fr. Ignatius of St. Paul* (1866) iii. viii. 255 My idea was borrowed from my..friend, Charles Weld, and consisted in *Quarant' Ore*,..making the circuit of all England, so that by day and night the Adorable Sacrament might be worshipped through the year. **1859** A. D. Hope *St. Philip Neri* v. 16 It was also the means of introducing into Rome, A.D. 1548, the devotion of the Quarant' Ore, which had been first practised in Milan A.D. 1534. **1890** Gasquet & Bishop *Edward VI & Bk. Common Prayer* iv. 54 The devotions known as the benediction of the Blessed Sacrament and the *Quarantore*. **1923** C. Mackenzie *Parson's Progress* xi. 135 If the authorities remonstrated with him for holding such a service as Creeping to the Cross on Good Friday, he remonstrated with them for allowing such vulgar innovations as the Stations of the Cross or the Devotion of the Quarant'ore. **1974** *Oxf. Dict. Chr. Ch.* (ed. 2) 524/2 *Forty hours' devotion* (also known as the *Quarant' Ore* or *Quarantore*), a modern Catholic devotion in which the Blessed Sacrament is exposed..for a period of *c.* forty hours, and the faithful pray before it by turns throughout this time.

quare (kwēᵊɹ), *a. dial.* Also **quair.** [repr. dial. pronunc. of Queer *a.*¹] = Queer *a.*¹ Also, in Ulster English, used as a general intensifier, esp. in phr. *quare and —*, very, extremely. Hence **qua·rely** *adv.*

quare fellow = *queer fellow* s.v. *queer *a.*¹ 1 a.

1805 E. Cavanagh *Let.* 20 Aug. in *Russ. Jrnls. M. & C. Wilmot* (1934) ii. 179 Tis *quair* things I have been seeing! **1805** —— *Let.* 4 Oct. in *Ibid.* ii. 185 It was *quairly* made. **1871** E. Eggleston *Hoosier Schoolmaster* iii. 32 'What a quare boy Shocky is!' remarked Betsey Short, with a giggle. 'He just likes to wander 'round alone.' **1880** W. H. Patterson *Gloss. Words Antrim & Down* 81 Quare, Queer, *adj.* very '*quare* an' nice' = very nice. **1896** J. Barlow *Mrs. Martin's Company* 13 Sure I know the roof's quare and bad. **1896** M. Hamilton *Across Ulster Bog* xi. 92 You're mended quarely this last while. **1900** M. O'Neill *Songs of Glens of Antrim* 17 Now we're quarely betther fixed. **1938** M. K. Rawlings *Yearling* vi. 55 Hit's mighty quare you toted a dog along wouldn't be no good to you. **1941** [see *jist]. **1949** C. Graves *Ireland Revisited* i. 21 One aged groom said: '..There's a fine, mettlesome lot of gerrls here today, but they look very quare in thon harness.' **1956** B. Behan *Quare Fellow* i. 4 What was the commotion last night round in D. Wing? Did the quare fellow get a reprieve?.. Now which quare fellow do you mean? The fellow beat his wife to death..was reprieved. *a* **1966** 'M. na Gopaleen' *Best of Myles* (1977) 49 I've a quare bit of news for you. The brother's nose is out of order. **1977** G. B. Adams in D. Ó Muirithe *Eng. Lang. in Ireland* 67 Mine ye, A was quare an gled tae get anntae ma settle bed thaat naght.

quarenden. Add: Now usually with spelling **quarrenden.** (Further and later examples.)

1851 R. Hogg *Brit. Pomol.* 67 Devonshire Quarrenden. .. A very valuable and first-rate dessert apple. **1870** Trollope *Vicar of Bullhampton* vii. 40 The quarantines are rare this year. **1905** *Westm. Gaz.* 11 Aug. 10/1 One grower in the West of England obtained 20s. a bushel for his Devonshire Quarrendens. **1907** *Ibid.* 31 Aug. 7/2 English apples..are a poor crop, except Worcesters and Quarantines—the latter an early cheap fruit. **1921** *Contemp. Rev.* Oct. 559 The Quarrendens are gone. September saw them out. **1945** H. J. Massingham *Wisdom of Fields* vii. 133 Red and sweet Quarrendons on the orchard trees. **1969** *Oxf. Bk. Food Plants* 48/1 'Devonshire Quarrendon'. Known before 1650, it was possibly originally French. It has a deep crimson fruit with white juicy flesh.

quaresimal (kware·simăl), *a. rare*⁻¹. [ad. It. *quaresimale* Lenten.] Of a meal: having the qualities of Lenten fare; meagre, austere.

1923 Joyce *Let.* 17 Dec. (1966) III. 84 Can we not have a quaresimal dinner somewhere together?

quark (kwǭɹk, kwāɹk), *sb. Physics.* [Invented word, associated with 'Three quarks for Muster Mark!' (Joyce *Finnegans Wake* (1939) ii. iv).

'I employed the sound "quork" for several weeks in 1963 before noticing "quark" in "Finnegans Wake", which I had perused from time to time since it appeared in 1939... The allusion to three quarks seemed perfect... I needed an excuse for retaining the pronunciation quork despite the occurrence of "Mark", "bark", "mark", and so forth in Finnegans Wake. I found that excuse by supposing that one ingredient of the line "Three quarks for Muster Mark" was a cry of "Three quarts for Mister..." heard in H. C. Earwicker's pub.'—M. Gell-Mann, private let. to Ed., 27 June 1978.]

Any of a group of sub-atomic particles (orig. three in number) conceived of as having a fractional electric charge and making up in different combinations the hadrons, but not detected in the free state.

1964 M. Gell-Mann in *Physics Lett.* VIII. 214/2 A simpler and more elegant scheme can be constructed if we allow non-integral values for the charges. We can dispense entirely with the basic baryon b if we assign to the triplet t the following properties: spin ½, z = −⅓, and baryon number ⅓. We then refer to the members u⅔, d⁻⅓, and s⁻⅓ of the triplet as quarks q and the members

of the anti-triplet as anti-quarks q̄. [*Note*] James Joyce, Finnegan's [*sic*] Wake (Viking Press, New York, 1939) p. 383. **1965** *New Scientist* 4 Mar. 575/2 Just as atoms are composed of particles (protons, neutrons and electrons) so may the heavy particles themselves be made up of combinations of simpler entities, called 'quarks'. **1967** *Observer* 23 Apr. 2/6 If quarks exist, they would represent a more fundamental building brick of matter than any yet known. **1972** *Daily Colonist* (Victoria, B.C.) 24 Feb. 5/2 The physicists hope to make the first observation of 'quarks', which many theorists believe are the fundamental building blocks,..by studying the activity of a rare and elusive sub-atomic particle called the omega-minus. **1973** L. J. Tassie *Physics of Elementary Particles* xi. 146 Mesons have *B* = 0 and are made up of a quark and an antiquark. **1973**, **1974** [see *parton]. **1976** *Sci. Amer.* Nov. 50/1 The *u* quark has a charge of +2/3, and the *d* quark and the *s* quark each have a charge of −1/3. **1977** *Nature* 21 July 201/1 Quarks..have not been found free in nature; and if present theories are correct they never will be: they are thought to be permanently confined to the interior of the particles they compose. *Ibid.* 204/1 Recently hadrons containing a new, heavier quark—the c-quark (for 'charmed')—have been discovered. The mass of the c-quark is thought to be roughly 1,500 MeV. The old quarks are much lighter: the u- and d-quarks may even be massless while the s-quark's mass is about 300 MeV. **1978** [see *psi 1 b]. **1978** *Sci. Amer.* Oct. 67/1 The upsilon resonances present physics with an embarrassment of riches: an unexpected family of new particles composed of an unexpected fifth quark.

quark, var. *quawk *sb.*

quarrel, *v.* Add: **2.** Also with *over*.

1868 Mayne Reid *White Squaw* xxviii. 133 Ere long they [*sc.* wolves] could be seen skulking through the enclosure and quarrelling over the corpses. **1883** G. Moore *Mod. Lover* I. xii. 244 Here a group of Cupids quarrelled over some masks and arrows. **1939** G. B. Shaw *In Good King Charles's Golden Days* 24 She has put us to shame for quarrelling over a matter of which we know nothing. **1961** *Middle East Jrnl.* XV. 3 The Istiqlal quarreled over foreign policy, labor politics and economic development.

quarrenden, var. Quarenden in Dict. and Suppl.

quarried, *ppl. a.*² Add: **b.** *Physical Geogr.* Eroded or broken off by glacial quarrying; = *plucked *ppl. a.* 4.

1909 *Jrnl. & Proc. R. Soc. N. S. Wales* XLIII. 265 Moutonnées..if large..appeared to be abraded on the up slope, and heavily quarried on the downslope. **1930** *Prof. Papers U.S. Geol. Survey* No. 160. 90/2 Muir..described long trains of glacially quarried blocks which he had observed in the vicinity of Tenaya Lake.

qua·rried, *a.* [f. Quarry *sb.*³ + -ed².] Of flooring: paved with quarries. Of a window: decorated with quarries.

1842 G. Francis *Dict. Arts* s.v. *Quarry*, Quarried pavements are by no means uncommon in old village churches. **1856** Geo. Eliot in *Westm. Rev.* X. 56 In those days, the quarried parlour was innocent of a carpet. **1954** M. Rickert *Painting in Brit.: Middle Ages* 231 *Quarried glass*, window panels divided into squares or diamonds, each containing an ornamental or heraldic motif.

quarrion (kwǫ·riən). Also **quar(r)ian.** [Prob. Aboriginal name.] An Australian parrot, *Leptolophus hollandicus*, which has grey plumage with white and yellow patches; = Cockatiel. Also *attrib.*

1901 A. J. Campbell *Nests & Eggs Austral. Birds* 622 The Grey and Yellow Top-knotted Parrot ('Quarrion', native name among bushmen) flies round about water-holes. **1934** *Bulletin* (Sydney) 26 Sept. 21/3 Quarians caught by broken wings on telephone-wires and emus held by the leg in fences are other casualties I've come across. **1938** N. W. Cayley *Austral. Parrots* 112 The Cockatiel (also called Cockatoo Parrot and Quarrian) was met with during Cook's voyage. **1943** W. Hatfield *I find Australia* v. 87 Quarrion parrots and ring-necks, rosellas and parakeets,..and magpies and butcher-birds (singing shrikes) added their morning warbles to the screeching and trilling. **1964** *People* (Austral.) 16 Dec. 38 The quarrians, sometimes known as cockaties or cockatoo parrots, are far from home. **1966** Eastman & Hunt *Parrots Austral.* 176 Call-note in flight is distinctive, and is a field mark in indicating the quarrion's presence long before it is sighted.

quarromes, quarron. For † *Obs.* read *Obs.* or *arch.* and add: Also **quarroms, quarrons,** (Later examples.)

1846 *Swell's Night Guide* 128/2 *Quarroms*, a body. **1922** Joyce *Ulysses* 48 White thy fambles, red thy gan And thy quarrons dainty is. **1932** Auden *Orators* iii. 105 Salmon draws Its lovely quarrons through the pool.

quarry, *sb.*¹ **4.** Delete † *Obs.* and add later example.

1884 T. Speedy *Sport* xix. 360 We have not above half-a-dozen times seen the peregrine in the act of making a quarry.

quarry, *sb.*² Add: **4.** *quarry-face, -pit* (later example), *-wagon*; **quarry-water** (earlier example).

1893–4 R. O. Heslop *Northumb. Words* II. 557 *Quarry-fyess*, the quarry face; its perpendicular side. **1936** *Discovery* Oct. 317/1 The skull..is thought to be still buried in the quarry-face. **1974** *Environmental Conservation* I. 38/1 Quarry-face risks are by no means confined to high country where population is sparse. **1911** J. Masefield *Everlasting Mercy* 4 In the old quarry-pit they say Head-keeper Pike was made away. **1937** Blunden *Elegy* 15 Above the square With plodding quarry-waggons filled. **1838** C. Lyell *Elements of Geol.* i. iv. 74 It is desirable to shape the stones which are to be used in architecture while they are yet soft and wet, and while they contain their 'quarry-water'.

quarry, *sb.*³ Add: **2.** (Later examples.) Also occas. round in shape.

1913 F. S. Eden *Anc. Glass* iv. 82 Round quarries, set close together in rows, are..formed in lieu of rectangular quarries. **1970** H. Braun *Parish Churches* viii. 111 The glass of medieval days was..set as a mosaic of diamond-shaped 'quarries' fixed together with a network of delicate tooled strips.

attrib.

1899 R. Glazier *Man. Hist. Ornament* 98 'Quarry glass', square or diamond in shape, with brown enamel details, was frequently used, where simple masses were desired. **1971** *Country Life* 20 May 1248/1, I have had an estimate made..for filling all the nave and one chancel window with quarry glass of a very pleasing though simple kind.

4. *Comb.*, as **quarry-tile** (see quot. 1940); also *attrib.*; hence **quarry-tiled** *a.*

1940 *Chambers's Techn. Dict.* 692/2 *Quarry tile*, the common unglazed, machine-made paving tile not less than ¾ in. in thickness. **1953** [see *chip-board* s.v. *chip *sb.*¹ 9 a]. **1966** *Listener* 28 July 128/2 Rough concrete and quarry-tile floors like a farmhouse kitchen. **1970** G. F. Newman *Sir, You Bastard* 258 He rapped his knuckles against the brown quarry-tiles in frustration. **1976** *Outdoor Living* (N.Z.) I. ii. 9/2 You might choose to have concrete, bricks or quarry tiles or it might suit the house more to have a timber surface for your sunny area. **1979** *Arizona Daily Star* 5 Aug. (Advt. Section) 18/9 See this family oriented 3 bedroom home with its quarry tile floors. **1960** *Farmer & Stockbreeder* 22 Mar. 66/3 The covered bullock yard at Drayton has..a flat quarry-tiled feeding floor edged with a 6 in kerb for silage.

quarry, *v.*² Add: **1. b.** (Later examples.)

1936 R. Campbell *Mithraic Emblems* 57 The gypsies quarried from the gloom, For their carouse a silver hall. **1958** L. Durrell *Balthazar* vi. 140 Were these words of Pursewarden's quarried from his own experience? **1975** *New Yorker* 29 Apr. 6/1 (Advt.), Dick Wellstood, a subtle and inventive pianist, reproduced the raw materials.. that the old-master pianists of the thirties and forties quarried their styles out of.

2. b. *Physical Geogr.* = Pluck *v.* 1 b in Dict. and Suppl.

1874 *Overland Monthly* Aug. 179/1 The size of the blocks, their abundance along the line of dispersal, and the probable rate of motion of the glacier which quarried and transported them, form data by which..the rate of block denudation may be reached. *Ibid.* 180/1 They had been quarried from the base of the ridge. **1909** *Jrnl. & Proc. R. Soc. N. S. Wales* XLIII. 264 Frequently the ice impact had been of such nature that a rock block had been quarried across the dominant joint structure. **1955** Longwell & Flint *Introd. Physical Geol.* xii. 191 The bottom of the glacier breaks off blocks of bedrock and quarries them out, especially from surfaces unsupported on their downstream sides. **1976** J. E. Sanders et al. *Physical. Geogr.* x. 346 Typically, the remaining mountain rim towers high above the bottom of a cirque because centuries of frost wedging enables the glacier to quarry deeply into the rock.

quarrying *vbl. sb.* (further examples).

1904 *Jrnl. Geol.* XII. 574 The glacier will be efficient as the agent for débris removal; the result, therefore, must be quarrying and excavation, and basal sapping. **1969** D. J. Easterbrook *Princ. Geomorphol.* xvi. 314 Storm waves are especially effective where rocks along the shore are highly jointed or bedded, and are thus vulnerable to quarrying.

quart, *sb.*² (Later *attrib.* examples.)

1976 D. Munrow *Instruments of Middle Ages & Renaissance* 44/3 Praetorius gives the following sizes of curtals: 1. Quint Bass; 2. Quart Bass. **1977** *Early Music* Oct. 570 (Advt.), [Recorders] The standard range is from Garklein to Bass, the Michael Praetorius range extends to a Quart Bass in C.

quartal (kwǭ·ɹtăl), *a. Mus.* [f. L. *quartus* fourth + -al.] Of a harmony: based on the interval of the fourth.

1937 *Musical Q.* XXIII. 178 Once we make our choice in favour of quartal harmony, we must be ready to accept all the logical consequences. **1938** J. Yasser *Mediæval Quartal Harmony* iii. 87 Unlimited freedom in unison and octave parallelism turned out to be one of the greatest obstacles to completing the rationalization of the quartal harmonic system. **1944** W. Apel *Harvard Dict. Mus.* 619/1 Quartal harmonies have been recommended to replace tertian harmonies in harmonizations of Gregorian chant. **1955** A. Hughes in *New Oxf. Hist. Mus.* (rev. ed.) II. 329 The two-part version of Brit. Mus. Harl. 524 seems to have been written in 'quartal' harmony, i.e. harmony based on the fourth as the most important interval. **1970** W. Apel *Harvard Dict. Mus.* (ed. 2) 6/2 Attempts have been made to replace this style with..archaic idioms such as quartal harmonies or parallel organum.

quartel, var. *cuartel.

quarter, *sb.* Add: **I. 1. b.** (Earlier and later examples.) Cf. *mauvais quart d'heure.

[*a* **1851**: see Music *sb.* 11.] **1875** Trollope *Way we live Now* II. lxii. 70 He was prepared..to console himself when the bad quarter of an hour should come with the remembrance that he had garnered up a store. **1909** *Daily Chron.* 30 Aug. 4/7 The 'bad quarter of an hour' we all know was first given a name by the heartless Louis XIII., who, looking at his watch on the day of the execution of Cinq-Mars, supposed that the poor young fellow 'passait alors un mauvais quart d'heure'. **1922** C. Mackenzie *Altar Steps* xxi. 233 Mark fancied that it would be the prelate who would have the unpleasant quarter of an hour. **1937** M. Sharp *Nutmeg Tree* xviii. 232 Susan was in for a bad quarter of an hour.

II. 6. a. Delete † *Obs.* and add later examples.

1959 I. & P. Opie *Lore & Lang. Schoolch.* ix. 167 A one-man High-Street confectioner..was found to be offering..*Bassett's Liquorice Allsorts 7d.* per quarter. **1977** *Jackson's of Piccadilly Price List* 2/2 [Ox tongue] £0.48 per qtr [i.e. quarter].

7. d. *U.S.* The fourth part of a mile.

1827 J. F. Cooper *Prairie* I. iv. 56, I can make myself heard a mile in these open fields, and his camp is but a short quarter from us. **1868** H. W. Woodruff *Trotting Horse* vii. 84 What's the use of a horse going a quarter fast? Now, they must go a quarter fast before they can go a mile fast. **1878** J. H. Beadle *Western Wilds* ii. 31 It was weeks before I could walk a quarter.

8. c. (Further examples.) Also without article, as *quarter of an hour*, etc. Also (*Sc.* and *N. Amer.*), 'a quarter *of* (or *till*) (a certain hour)', a quarter to (the hour specified).

1871 *Sci. Amer.* 11 Feb. 102/2 When everything was tightened..and the propellor arranged to cause elevation, it was just quarter of one o'clock. **1894** A. Robertson *Nuggets* 165 His Excellency the Governor wants to see you, detective, at a quarter to eleven sharp. **1912**, etc. [see *of *prep.* 4 c]. **1913** C. Mackenzie *Sinister St.* I. ii. i. 141 In the 'quarter' (as the break was now called) Michael would stand on..the step that led down..into the schoolground. **1920** J. S. Clouston *Carrington's Cases* ix. 135, I found myself sitting in a first-class smoking carriage with nearly quarter of an hour to spare. *Ibid.* xiii. 237 It was then quarter-past eleven. **1933** P. Godfrey *Back-Stage* i. 14 Once more the call-boy appears. 'Shall I call 'the quarter', sir?' **1949** H. Kurath *Word Geogr. Eastern U.S.* ii. 30/2 In the greater part of the Midland..*quarter til eleven* is current. **1952** M. Laski *Village* vii. 119 If I'm not there by quarter to, you'll know I couldn't make it. **1963** R. I. McDavid *Mencken's Amer. Lang.* 298 Americans may say *quarter to*, *quarter of* or *quarter till*, the last being characteristic of Pennsylvania and its dependencies, including the upland South. **1966** H. Kemelman *Saturday the Rabbi went Hungry* (1969) ii. 21 He said..that traffic would be heaviest between a quarter of and a quarter past seven. **1969** A. Glyn *Dragon Variation* v. 142 He checked the time on his Omega Seamaster. It would be just a quarter of three in New York.

e. *Sport.* One of four equal periods of play in a match; also *gen.*, the fourth part of the time taken to play a match.

1911 P. H. Davis *Football* viii. 115 The periods of the game, the halves,..were replaced by quarters. **1922** P. D. Haughton *Football & how to watch It* ix. 191 In contrast to the preceding period this quarter was marked by excellent play. **1954** *New Yorker* 6 Nov. 87/1 The play of the afternoon came in the middle of the final quarter. **1963** *Times* 29 Apr. 4/6 Mellor soon recovered their balance to take a 3–0 lead in the first quarter. **1969** *Sun-Herald* (Sydney) 13 July 48/2 Footscray made a great fight of it in the final quarter. **1972** J. Mosedale *Football* ix. 130 Playing on an 80-yard field the teams were deadlocked near the end of the fourth quarter. **1976** *Eastern Even. News* (Norwich) 9 Dec. 19/8 In the last quarter, Reading gave UEA and their supporters a scare by scoring a well-worked try which was converted. **1979** *Tucson* (Arizona) *Citizen* 20 Sept. 1D/2 The young Warriors battled Sahuaro to a 7–7 draw through three quarters before falling, 16–13, in the season opener.

III. 14. a. Also *spec.*, the Latin Quarter of Paris (see *Latin *sb.* 5).

1919 W. S. Maugham *Moon & Sixpence* xxvii. 117 Lots of fellows in the Quarter share a studio. **1926** E. Hemingway *Sun also Rises* I. v. 37 'What do you do nights, Jake?'.. 'Oh, I'm over in the Quarter.' *a* **1967** A. Ransome *Autobiogr.* (1976) xii. 120 In those days the Quarter did its best for hard-up students, and I was able to furnish my studio for next to nothing.

15. c. (Earlier and later examples.)

1724 H. Jones *Present State of Virginia* iv. 36 The Negroes live in small Cottages called Quarters. **1760** G. Washington *Diary* 26 Feb. (1925) I. 131 Began Plowing the Field by the Stable and Quarter for Oats and Clover. **1799** I. Weld *Trav. N. Amer.* xi. 84 Their quarters, the name whereby their habitations are called, are usually situated one or two hundred yards from the dwelling house. **1909** 'O. Henry' *Roads of Destiny* xvii. 282 Almost the entire population of the quarters volunteered their aid. **1916** J. B. Thoburn *Stand. Hist. Oklahoma* I. 261 'The quarters'..formed a picturesque feature of the old time plantation life. **1935** Z. N. Hurston *Mules & Men* i. iv. 85 It sauntered on down the bark-covered road and into the quarters just as if it had really wanted to come. **1949** B. A. Botkin *Treas. S. Folklore* iv. i. 551 The 'South's tradition of good cooking'..belonged originally to the 'big house' rather than to the 'quarters' and the cabin.

26*. In various colloquial shortened and abbreviated forms. **a.** *U.S. Football.* = *quarterback *sb.* 2 a.

1893 W. C. Camp *Bk. College Sports* 120 The criss-cross or double pass is another excellent example of a disguised play, the ball being passed by the quarter to one of the backs. **1907** *St. Nicholas* (N.Y.) Sept. 1013/2 A line man could..take the ball from the quarter. **1914** P. Withington *Bk. Athletics* 58 In handling the team the quarter must have absolute command.

b. (*a*) = *quartermaster-sergeant* s.v. Quartermaster 2 c; (*b*) = Quartermaster 1 a.

1917 A. G. Empey *Over Top* 304 Quartermaster-Sergeant, or 'Quarter' as he is called. A non-commissioned officer in a company who..takes charge of the company stores. **1963** M. Lowry *Ultramarine* ii. 60 Well, it's your business to get me up, quarter.

c. *pl.* = *quarter-finals* (see sense 30 below).

1978 *Guardian Weekly* 5 Feb. 24/2 The other semi-final was disappointing. Roscoe Tanner..had upset Bjorn Borg in the quarters. **1978** *Times* 4 July 19/3, I had never won a match on grass at Wimbledon and here I am in the quarters.

V. 27. a. *quarter-bottle, century, hour, litre, truth* (after Half-truth); *quarter-armed, -hourly, -striking, -witted* (after Half-witted *a.*) adjs.

1881 F. Day *Fishes Gt. Brit. & Ireland* I. 239 *Gasterosteus gymnurus*... The quarter-armed or smooth-tailed stickleback. **1907** *Yesterday's Shopping* (1969) 99/2 *Saumur, sparkling*..In original hampers of 1 Dozen ¼ bots. **1915** H. G. Wells *Boon* ix. 333 One of those quarter-bottles of Perrier Jouet on a tray. **1977** J. R. L. Anderson *Death in City* v. 81, I ordered a quarter bottle of cognac. **1902** *Westm. Gaz.* 21 July 4/1 To put the result in quarter-century periods. **1920** H. G. Wells *Outl. Hist.* xxix. 265/2 The opening quarter century of the Christian era was troubled by a usurper. **1979** *Bookseller* 23 June 2818/1 The Warsaw Bookfair continues towards its quarter century. **1883** 'Mark Twain' *Life on Mississippi* xxxvi. 392 My uneasy spirit kept dragging me back at quarter-hour intervals. **1977** *Detroit Free Press* 11 Dec. 24-A/1 The head of the department [should] have at least ..90 quarter hours of criminal justice courses completed. **1929** J. Owen *Shepherd & Child* iv. 46 The church clock.. had a quarter-hourly chime. **1978** J. Sherwood *Limericks of Lachasse* iv. 48 [He] had drunk only a quarter-litre of light carafe wine. **1959** *Times* 6 Mar. 12/5 A Breguet gold and enamelled quarter-striking, quarter-repeating clock watch. **1977** *Gay News* 7–20 Apr. 22/2 It was such dangerously oversimplified quarter-truths that led to the vilification of ——. **1979** *Daily Tel.* 12 Dec. 16 Mr Timothy Raison's article is a distressing collection of quarter truths and specious arguments. **1864** A. Wallace *Scottish Tales* iii. 38 A quarter-witted individual from Muthil. **1972** P. Green *Shadow of Parthenon* 128 They vaguely assume their young readers to be either quarter-witted miniature adults or innocent prelapsarian angels.

29. *quarter-knee* (Knee *sb.* 7 a).

1941 C. O'Brien *Sea-Boats, Oars, & Sails* ii. 22 Breast-hook and quarter-knees..connect the gunwales with the stem and transom respectively.

30. quarter-ball *Billiards*, a ball that strikes another so that a quarter of the one overlaps a quarter of the other; **quarter-binding** (examples); now also, this style of binding using materials other than leather; **quarter-bloke** *Mil. slang*, a quartermaster(-sergeant); **quarter-blood** *U.S.*, one whose descent is only one fourth derived from the blood of a particular race (esp. Amer. Indian) or breed; also as quasi-*adj.*; **quarter-boat** *U.S.*, a boat containing living quarters for river workmen; **quarter-bound** *a.* (examples); **quarter-bred,** (*b*) *N.Z.* = *quarterback *sb.* 1; also *attrib.* or as *adj.*; **quarter-breed** *U.S.*, the offspring of a half-breed and a white; a quarter-blood; also *attrib.*; **quarter-calf** = *quarter-binding* (in calf); **quarter-caste** *Austral.* and *N.Z.*, a person of mixed breed, having one-quarter Aboriginal or Maori and three-quarters white descent; also *attrib.*; **quarter-chord** *Aeronaut.*, a quarter of the length of a chord of an aerofoil, *spec.* such a distance measured backwards from the leading edge; freq. *attrib.*, as *quarter-chord line, point*; **quarter-elliptic(al)** *adjs.*, applied to a leaf spring having the profile of a quarter of an ellipse; **quarter-finals** *pl.*, the four matches constituting the round before the semi-finals in a tournament; also *sing.*, one of these four matches; also *attrib.*; **quarter girth measure,** *Hoppus measure; **quarter-in-the-slot** *a. U.S.*, actuated by the fall of a quarter inserted through a slot (in a machine) (cf. *penny-in-the-slot* s.v. Penny 12 b); **quarter leather** = *quarter-binding* (in leather); **quarter-light,** (*b*) a small triangular side-window on a motor vehicle for ventilation and the admission of light; also *attrib.*; **quarter-moon** (*a*) delete † and add later examples; also *attrib.*; **quarter-note,** substitute for def.: (*a*) = *quarter-tone*; (*b*) *U.S.* a crochet; **quarter peal** *Bell-ringing*, a peal comprising one quarter of the number of changes in a full peal; **quarter pole,** add: *U.S.* (earlier and later examples); **quarter-racing** *U.S.*, the holding of quarter-races; also *attrib.*; **quarter-round** (further examples); **quarter-saw** *v. trans.*, to saw (a log) radially into quarters; to produce (a board) by quarter sawing; **quarter sawing** *vbl. sb.*, the method or action of producing boards by sawing a log radially into quarters and then sawing each quarter into boards so that the growth rings make angles of greater than 45° with the faces of the boards; **quarter-sawn** *ppl. a.*; **quarter-section** (earlier example); **quarter stretch** *U.S.*, (a part of) a racecourse that is a quarter of a mile long; **quarter-tonal** *a.*, based on quarter tones; so **quarter-tonality**; **quarter-tone** (further *attrib.* examples); **quarter-track,** (*b*) a track recorded on magnetic tape so that four such tracks can be accommodated side by side; usu. *attrib.*; also as *adv.*, using this width of tape; **quarter-turn,** (*b*) (example); also as *v. trans.* and *intr.*; also **quarter-turning** *vbl. sb.* (see quot. 1901); **quarter-wave** *a. Physics*, having a thickness or a length equal to a quarter of the wavelength transmitted or received; *quarter-wave plate*, a plate of a birefringent substance cut parallel to the optic axis and of such a thickness that it introduces a time difference of a quarter of a period between ordinary and extraordinary rays passing normally through it.

1873 J. Bennett *Billiards* 34 If the half of one overlaps the half of the other, it is a half ball; and so on for a quarter ball. Anything less than a quarter ball is called a fine ball. **1912** Monk & Lawrence *Text Bk. of Stationery Binding* 140 Quarter binding. 55. **1932** A. F. Collins *Book Crafts for Schools* iii. 27 'Quarter-binding'..has the stronger material..on the back, and the weaker material ..on the sides. **1978** A. W. Johnson *Thames & Hudson Man. Bookbinding* 216 *Quarter binding*, an economical covering method in which the spine and part of the sides are covered in one material and a cheaper one is used on the remainder. **1919** *Athenæum* 1 Aug. 695/2 The Q.M.S. (the colour-sergeant or 'Flag' of the Old Army) is always called the 'Quarter Bloke' or 'The Bloke'. **1920** *Punch* 18 Aug. 137/2 It's great..To eat a daintier kind of grub than quarter-blokes provide. **1944** *Gen* 30 Dec. 55/2 Nickly overstepped the mark when he suggested to the quarter-bloke..that he was flogging the rations. **1950** C. MacInnes *To Victors Spoils* i. 21 I'll drop back there and talk to his quarter-bloke. **1845** *Knickerbocker* Mar. 236 Of this description was a quarter-blood [*sc.* Indian], of great beauty. **1873** J. H. Beadle *Undevel. West* xix. 355 He had four children, only quarter-blood, but differing very much in shade. **1878** —— *Western Wilds* ii. 26 The straight black hair, and nose just aquiline enough to give piquancy to the countenance, indicated the quarter-blood. **1943** *Sun* (Baltimore) 11 Dec. 11/5 Medium shorn domestic fleeces have had a further small sale, mostly quarter-blood. **1929** *Ibid.* 23 Oct. 1/3 The President and his immediate party left Cincinnati..aboard the Greenbrier..and three other light craft—quarter boats, they are called—for the remaining members of the party. **1962** A. Davison *In Wake of Gemini* 228 A quarterboat is an operational center for the maintenance and repair department of the U.S. Army Corps of Engineers, Mississippi River Commission. **1888** C. T. Jacobi *Printer's Vocab.* 108 *Quarter bound*, books bound with back only in leather. **1929** A. J. Vaughan *Mod. Bookbinding* ii. 121 (*caption*) A Quarter Bound Book. *Ibid.* iv. 217 *Quarter Bound*, where the back and some portion of the sides only of the binding consist of one material, and the remainder of the sides of another. **1960** G. A. Glaister *Gloss. Bk.* 338/1 Quarter-bound: a book bound with either leather spine and cloth sides, or cloth spine and paper sides. **1979** *London Rev. Bks.* 25 Oct. 11/2 (Advt.), Quarter-bound in leather. **1891** R. Wallace *Rural Econ. Austral. & N.Z.* xviii. 259 In 1890, the better portion of the greasy quarter-bred wool fetched 1s. 2½d. *a* **1948** L. G. D. Acland *Early Canterbury Runs* (1951) 370 *Comeback*, a sheep three-quarters merino and one quarter long wool..but in New Zealand I think these sheep are often called quarterbreds or quarter-backs. **1826** T. L. McKenney *Sk. Tour to Lakes* (1827) 387 Three were full blood, the remainder half breeds, and quarter breeds. **1880** *Harper's Mag.* Dec. 31 All four were of mixed blood their mother having been a beautiful French quarter-breed. **1909** *Lady's Realm* Feb. 465/1 El-Soo was a full-blooded Indian, yet she exceeded all the half-breed and quarter-breed girls. She was a treasure. **1956** *Library* XI. 81 The binding has no corner-pieces; and so is properly called 'quarter-calf'. **1948** D. Ballantyne *Cunninghams* i. i. 6 Being up the duff to a young quarter-caste..was no joke. **1952** R. Finlayson *Schooner came to Atia* x. 51 The crew with the exception of the quartercaste mate were all Maori. **1966** *Times* 28 Mar. (Austral. Suppl.) p. ix/4 Aboriginal people..vary from full-bloods to quarter-castes. **1976** *Times Lit. Suppl.* 9 Apr. 417/1 Prindy, an eight-year-old quarter-caste who sees himself as purely aboriginal, in spite of his fair hair and grey eyes. **1979** *Church Times* 9 Mar. 4/3 Bishop Reeves is a 'quarter-caste' Maori and a graduate of St. Peter's College, Oxford. **1946** *Jrnl. R. Aeronaut. Soc.* L. 436/2 The centre of pressure is at the half-chord instead of approximately at the quarter-chord point. **1947** C. F. Toms *Introd. Aeronaut.* i. 26 The quarter-point of the mean chords lies very nearly on the quarter-chord line of the wing. **1957** L. L. Beckford *A.B.C. of Aeronaut.* 98/2 This angle, known as Sweep Back, is measured between the lateral axis and a line drawn a quarter-chord back from the leading edges. **1959** J. L. Nayler *Dict. Aeronaut. Engin.* 207 The quarter-chord line is the line joining the quarter points of the chords across the span of the wing. **1967** *Jane's Surface Skimmer Syst.* 1967–68 12/1 Aerofoil section NACA 0009. Sweepback at quarter-chord 30°. **1926** C. T. B. Donkin *Elem. Motor Vehicle Design* xviii. 253 The bending moment in the chassis frame set up by the reactions of a quarter-elliptic spring is very considerable, and the type is seldom used for heavy cars except in the form of a short subsidiary spring. **1963** Bird & Hutton-Stott *Veteran Motor Car* 53 The characteristic reversed quarter-elliptic rear springs appear on the 5-litre model. **1909** R. W. A. Brewer *Motor Car* xiv. 140 When quarter-elliptical springs are used they may be either shackled to the side springs or pinned to them direct. **1927** *Daily Express* 28 July 1/5 Miss Helen Jacobs..scored a signal success in the quarter-finals of the Essex County Invitation Tournament. **1932** *News Chron.* 23 Sept. 2/4 The quarter-finals of the Canadian Women's Open Golf Championship. **1941** G. Marx *Let.* 23 June (1967) 27 Art plays Frank Parker in the quarterfinals of the 55th Annual Los Angeles Tournament. **1976** *Lancs. Even. Post* 7 Dec. 15/5 Morecambe's quarter final game at Runcorn is a vital one. **1977** *Whitaker's Almanack 1978* 583/2 Newcastle United were fined £4000..for fielding a weakened team in the quarter-final of the Anglo-Scottish Cup. [**1894** W. Stevenson *Wood* 194 The Hoppus measure by string, quarter girth.. on round timber, is an overmeasure in favour of the buyer.] **1924** [see *Hoppus]. **1954** W. E. Hiley *Woodland Managem.* ix. 127 Quarter girth or 'hoppus' measure. The volume of a felled log is determined by its length and the area of the cross section half way along it. The length is measured by a long tape,..and the middle girth by a specially marked short tape, known as a quarter-girth tape. **1903** R. L. McCardell *Conversat. Chorus Girl* 80 Mama de Branscombe had a quarter-in-the slot gas meter put in. **1938** L. M. Harrod *Librarians' Gloss.* 124 Quarter leather. A term used to describe a book with a leather spine and cloth sides. **1963** B. C. Middleton *Hist. Eng. Craft Bookbinding Technique* xi. 160 It was not uncommon on the Continent in the Middle Ages for books to be bound in quarter-leather. **1938** *Times* 13 Oct. 8/4 The quarter lights are taken as far aft as possible. **1972** *Police Rev.* 8 Dec. 1588/2 The campaign includes 200,000 posters, 1½ million car quarter-light stickers, and 250,000 shop-window stickers. **1976** 'Z. Stone' *Modigliani Scandal* iii. i. 113 He..[was] driving with the quarter-light open and enjoying the fresh air. **1947** L. P. Hartley *Eustace & Hilda* v. 91. The quarter-moon was resting on the roofs. **1977** A. Desai in P. Collenette *Winter's Tales 23* 19 He..went about dividing the melon into quarter-moon portions. **1958** Quarter note [see *eighth *a.* 3]. **1959** Westrup & Harrison *Collins Music Encycl.* 524/1 *Quarter-note*, American for 'crotchet'. **1888** A. P. Heywood *Treat. on 'Duffield'* vii. 53 (*heading*) (1296) Quarter peal. **1931** E. Morris *Hist. & Art Change-Ringing* iii. 53 The above twice repeated will come round at the quarter peal end. **1980** *Times* 7 Apr. 3/1 Bellringers at the parish church rang a quarter peal. **1857** Quarter pole [see *peanut boy* s.v. *peanut 1 b]. **1868** H. Woodruff *Trotting Horse* xxxi. 259 At the quarter-pole she had recovered her stroke. **1975** *New Yorker* 3 Mar. 73/1 Susan's Girl..led past the quarter pole. **1784** J. F. D. Smyth *Tour in U.S.* I. ii. 22 In the southern part of the colony and in North Carolina, they are much attached to *Quarter racing*, which is always a match between two horses, to run one quarter of a mile streight [*sic*] out. **1889** *Harper's Mag.* Sept. 554/1 Foot-racing for the men and quarter-racing for the horses. **1974** *New Yorker* 29 Apr. 83/2 His journey into the center of the quarter-racing world. **1963** C. R. Cowell et al. *Inlays, Crowns, & Bridges* iii. 23 A quarter-round bur for contra-angle handpieces is used to cut 2-mm. pinholes. **1975** *New Yorker* 24 Feb. 56/3 He..addressed the axe to the quarter-round log. **1898** S. B. Green *Forestry in Minnesota* 299 Quarter-sawing... The log is first quartered and then sawed into boards, cutting then alternately from each face of the quarter of the log. **1907** Webster, Quarter-saw, *v.t.*; -sawed or -sawn. **1931** *Harper's Mag.* June 62/1 No one rived the beams for his ceiling, or quarter-sawed oak for his chairs. **1934** *Archit. Rev.* LXXVI. 64/1 When logs are riven or quarter-sawn the large rays which form the silver grain are revealed to the fullest extent. **1966** A. W. Lewis *Gloss. Woodworking Terms* 74 Quarter-sawn boards shrink less and are less liable to warp then other boards. **1968** Quarter sawing [see *rift sawing* s.v. *rift *sb.*² 4]. **1974** *Islander* (Victoria, B.C.) 29 Dec. 15/2 When the logs have reached the proper degree of dryness staves will be quarter sawn from them. That is, the flat boards will radiate out from the centre of the log like segments of an orange. **1806** *Deb. Congress U.S.* (1852) 9th Congress 2 Sess., App. 1032 The public lands are now sold in sections, half sections, and quarter sections. **1834** *Southern Lit. Messenger* I. 182/1, I pulled and pulled till I got out of sight, and turned down the quarter stretch. **1883** H. Watterson *Oddities Southern Life* 439 He ran a quarter stretch down the low grounds of the base. **1934** S. R. Nelson *All about Jazz* i. 13 Maddening is that persistent beat of the tom-toms and these quarter-tonal intervals gradually rising to an overwhelming climax. **1930** *Proc. Musical Assoc.* Apr. 92 We are dogged by such words as tonality,..quartertonality, modality. **1947** *Penguin Music Mag.* Sept. 11 The romantic composers of the nineteenth century..needed a new instrument—neo-modality, atonality, polytonality, quarter-tonality. **1891** C. R. Day *Music & Mus. Instruments S. India & Deccan* ii. 20 The quarter-tone system used in Syria. **1934** C. Lambert *Music Ho!* v. 311 The quarter-tone quartets of Aloys Haba..differ from the quartets of Brahms only through being written in the quarter-tone scale. **1959** Westrup & Harrison *Collins Music Encycl.* 524/1 Alois Hába has written a number of compositions..for quarter-tone piano, and for quarter-tone harmonium. **1978** P. Griffiths *Conc. Hist. Mod. Music* v. 53 Ives had explored atonality, free rhythm, quarter-tone harmony. **1962** R. E. B. Hickman *Magnetic Recording Handbk.* (ed. 3) iii. 29 Present day domestic magnetic tape recorders normally employ dual track (half-track) recording on ¼ in. tape, although four track (quarter track) recording, with alternate tracks recorded in the same direction, is also common practice. **1962** A. Nisbett *Technique Sound Studio* iv. 82 Correction can be made by..replaying only part of the full recorded track: e.g. replaying..a half-track recording quarter-track. **1967** P. Spring *Tape Recorders* iv. 41 A correct head alignment is very much more important than in the case of a quarter-track tape recorder. **1975** G. J. King *Audio Handbk.* x. 226 (*caption*) A mono recorder lays one track on each half of the tape, the track then having approximately the width of two quarter tracks. **1934** D. L. Sayers *Nine Tailors* ii. i. 73 Rector was saying the other day as she [*sc.* a bell] did soon ought ter be quarter-turned. **1954** W. Faulkner *Fable* 303 It made a rigid quarter-turn. *Ibid.* 384 Some twenty men with a sergeant, who halted and quarter-turned and stood them at ease.

1964 G. C. Kunzle *Parallel Bars* ix. 403 Quarter turn into handstand on one bar, squat off with straight legs dismount. **1901** H. E. Bulwer *Gloss. Technical Terms Bells* 3 *Quarter-turning*, re-attaching a bell to its 'stock' at right angles (or less) to its former position with reference to the latter, in order that the 'clapper' may strike on a fresh segment of the 'sound-bow'. **1979** *Church Times* 5 Oct. 5/1 The bells [were] unsafe as they were. The bell-founders had recommended quarter-turning. **1882** L. Wright *Light* xiv. 298 This then is our 'quarter-wave' plate, which should be at once mounted between two glasses in balsam, and its working planes marked on the edges by scratching with a diamond, or quartz crystal. **1937** Jenkins & White *Fund. Physical Optics* xvi. 357 When a quarter-wave plate is oriented at an angle of 45° with the plane of the incident polarized light the emerging light is circularly polarized. **1943** C. L. Boltz *Basic Radio* xvii. 265 The ½λ aerial is in two halves... From the centre ends hang two wires ¼λ long. The feeder is then tapped to the quarter-wave line by moving the contacts up or down until the correct impedance is found. **1970** D. W. Tenquist et al. *University Optics* II. iii. 109 Quarter-wave plates are usually made of quartz or of mica (sandwiched between glass plates). **1977** S. W. Amos *Radio, TV & Audio Techn. Ref. Bk.* xxi. 28 It is sometimes necessary to be able to change the input impedance of a λ/2 dipole from its normal value of 73Ω without inserting a separate quarter-wave transformer.

quarter, *v.* Add: **10. a.** Also, of birds of prey flying over their hunting grounds.

1919 T. A. Coward *Birds Brit. Isles* I. 309 Sharing with other Harriers the habit of closely and diligently quartering the ground with buoyant easy flight, the Hen-Harrier more frequently interrupts its progress by hovering. **1946** D. C. Peattie *Road of Naturalist* iii. 39 The hawks and the falcons quartered the dazzling playa. **1976** L. Brown *Brit. Birds of Prey* x. 121 When returning to the roost in the evening they [*sc.* Montagu's harriers] often travel straight and quarter the ground much less than they would when hunting.

d. (Earlier and later examples.)

1883 'Mark Twain' *Life on Mississippi* iii. 51, I see a black something floating on the water away off to stabboard [*sic*] and quartering behind us. **1938** M. K. Rawlings *Yearling* xx. 260 His bear was quartering from him, but he was able to draw a bead on the left cheek from the rear.

11. a. Also *fig.*

1824 C. A. Bowles *Let.* 24 Jan. in E. Dowden *Corr. Southey* (1881) 48, I keep quartering, or trying to quarter, for a yard or so, and then down goes the wheel into the old groove. I cannot keep out of blank verse.

qua·rterback, *sb.* **1.** *Austral.* and *N.Z.* (See quots.)

1891 [see *come-back *sb.*² 4]. **1940** E. C. Studholme *Te Waimate* 96 Using Merino rams on the cross-bred ewes, the progeny being known as Quarter Backs.

2. *U.S. Football.* **a.** (In Dict. s.v. Quarter *sb.* 30.) Add to def.: by whom the team's play is usu. co-ordinated and directed. Usu. in modern practice, a player lining up behind the centre of an offensive team, who calls the signal to initiate a play and receives the ball when it is snapped back by the centre. (Further examples.)

1947 *Sun* (Baltimore) 15 Aug. 12/6 He is a stocky man who..has the build of a quarterback. **1979** *Arizona Daily Star* 5 Aug. c. 9/1 Oakland quarterback Ken Stabler.. played the first quarter and part of the second.

b. *transf.* A supporter or critic of a football team. Also *downtown quarterback*, an interested supporter of the home team; *grandstand quarterback* (in quot. *fig.*); *Monday morning quarterback*, one who engages in reductive 'post-mortem' criticism of a game; also *fig.*

1932 B. Wood *What Price Football* vi. 100 A kind of sportswriter known to football players and coaches as a 'Monday morning quarterback'... Not content with reporting the game..the writer must analyze it. **1947** *Collier's* 15 Nov. 17 (*heading*) Penn's downtown quarterbacks were a gloomy lot when George Munger became grid coach. **1949** *Ibid.* 19 Nov. 18/3 The quarterbacks in the stands consider Pop Warner's system..obsolete. **1976** President Ford in *Sunday Sun* (Baltimore) 10 Oct. K. 4/4 Somebody who sits in Washington D.C., 18 months after the Mayaguez incident, can be a very good grandstand quarterback.

c. *fig.* One who directs or masterminds an operation; a leader.

1961 in Webster. **1968** *Globe & Mail* (Toronto) 10 July 27/5 Van Burkleo is an original-thinking quarterback who can run his own show. **1971** *Sci. Amer.* Aug. 38/1 It was my privilege to be working in Colombia as 'quarterback' of a team of investigators. **1978** *Guardian Weekly* 1 Jan. 18/1 King Juan Carlos I, Franco's 39-year-old successor, and Premier Adolfo Suarez, the monarch's 44-year-old political quarterback.

3. *attrib.* and *Comb.* (sense *2), as **quarterback club**, an association of supporters actively interested in promoting their team's success; **quarterback sack** = *sack *sb.*¹ 1 j; **quarterback sneak** (see quot. 1966).

1948 *Life* 1 Nov. 110/2 Factory workers as well as merchant princes belong to the Quarterback Club, an organization of last-ditch rooters who gather..to put questions to the Ohio State head coach..and to see movies of the team's latest games. **1951** *Sun* (Baltimore) 9 Oct. 21/1 Special funds contributed by..downtown quarterback clubs. **1974** *Plain Dealer* (Cleveland, Ohio) 26 Oct. 6-D/2 Alzado..leads the defense in quarterback sacks with four. **1923** J. Wilce *Football* viii. 105 *Quarterback sneak*,..R. takes the ball from centre directly, pauses a moment as the line charges forward, then sneaks through any opening that may appear between the guards. **1947** *Time* 3 Nov. 74/3 The story Bob liked best was Dad's quarterback sneak. **1966** J. R. Winter *Lang. Pro Football* 132 *Quarterback sneak*, play where quarterback moves straight ahead into line behind charge of his center and guards immediately after taking snap.

qua·rterback, *v. U.S. Football.* [f. the sb.] *trans.* **a.** To play quarterback for (a team); to direct as quarterback.

1948 *Sun* (Baltimore) 13 Jan. 13/5 McCann..has 'quarterbacked' the LaSalle team to nine consecutive victories. **1950** *Life* 9 Oct. 66/2 Bob calmly quarterbacked the team to its first victory. **1972** J. Mosedale *Football* v. 62 Detroit, coached by Buddy Parker and quarterbacked by Bobby Layne, exhibited a strange mastery. **1979** *Tucson* (Arizona) *Citizen* 20 Sept. 11 D/1 He used to..quarterback the Sun Devils from 1974 to 1977.

b. *fig.* To direct or co-ordinate (an operation).

1952 *Britannica Bk. of Year* 666/2 *Quarterback*,.. (extension of *n.* in U.S. football terminology). To direct (1945). **1971** *Daily Colonist* (Victoria, B.C.) 26 Oct. 1/7 Bush..quarterbacked what UN observers regarded as Washington's worst diplomatic defeat. **1976** *Guardian Weekly* 7 Mar. 18/3 Relations were particularly strained after the 1972 election success, which Wehner felt he had quarterbacked without credit.

So **qua·rterbacking** *vbl. sb.*, the action of playing or directing as quarterback; also *fig.*, esp. *Monday morning quarterbacking* (see *quarterback *sb.* 2 b).

1947 *Time* 17 Nov. 75/2 (*heading*) Quarterbacking by telephone. **1948** *Sat. Even. Post* 18 Sept. 24/3 There also should be less of the Western Union quarterbacking from the bench, which frequently backfired in the face of the masterminds. **1950** *Sport Life* Feb. 9/1 Royal, nominated by the coach to take over for Mitchell at the quarterbacking post. **1960** *Guardian* 9 May 6/6 The realities of sheer, hard political quarter-backing. **1966** 'H. B. Taylor' *Triumvirate* iv. 30 There was the usual Thursday morning quarterbacking over what I had put in the paper and how I'd put it in. **1972** *Science* 9 June 1083/2 But Monday-morning quarterbacking is an easy sport. **1977** *N.Y. Rev. Bks.* 13 Oct. 14/3 Despite President Ford's admonition against 'Monday morning quarterbacking' about presidential responsibility for assassination plots, Rockefeller has said..that there was 'White House knowledge and/or approval of all major undertakings'.

quarter-deck. Add: **b.** (Further *attrib.* examples.) Also *transf.*

1850 H. Melville *White-Jacket* xxiii. 117 See, White Jacket, all round they have *shipped their quarter-deck faces again*... I afterward learned that this was an old man-of-war's-man's phrase, expressive of the facility with which a sea-officer falls back upon all the severity of his dignity, after a temporary suspension of it. **1893** W. C. Russell *Emigrant Ship* I. iv. 53, I saluted him with a quarter-deck flourish. **1927** H. D. Capper *Aft — from Hawsehole* p. xii, There are close upon three thousand naval officers of 'hawsehole origin'..who are, or have been, performing fine quarter-deck service for their county. **1976** *National Trust* Autumn 17/3 Still unchanged are the 'quarter-deck', a path laid so that the children could reach the river without getting muddy [etc.].

qua·rter-deck, *v.* [f. the sb.] *intr.* To walk up and down as on a quarter-deck. So **qua·rter-decking** *vbl. sb.*

1901 E. F. Benson *Luck of Vails* xviii. 207 He continued quarter-decking about the room for a few times in silence. **1913** Mrs. H. Ward *Coryston Family* viii. 164 The quarter-decking began again; and Lester waited patiently on a slowly subsiding frenzy. **1924** Kipling *Debits & Credits* 87 There was Potiphar..quarter-decking serenely below the Pebble-ridge. **1954** 'M. Coles' *Not for Export* i. 20 'This it is', said Spelmann, swinging round at each turn of his quarter-decking..'to work hard and be successful.'

quarterer. Add: Also, a quartering bird (sense 10 d of Quarter *v.*).

1892 W. W. Greener *Breech-Loader* vii. 209 When the shooter facing No. 2 trap gets a quarterer to the left from No. 5.

quarter-horse. *U.S.* A small stocky horse belonging to a variety recognized as a breed in 1941 and noted for agility and speed over short distances. Also *attrib.*

1839 *Spirit of Times* 11 May 114/3 *Martha Malone*.. was evidently short of work, not fine enough drawn, and too much on the quarter-horse order for a hard race under a hot sun and over a dusty track. **1845** T. J. Green *Jrnl. Texian Exped.* x. 136 Nausea which caused me to break for the door like a quarter-horse to relieve my distress. **1887** *Outing* May 115/1 You would not think him a quarter-horse for he looks like a clumsy sleepy old plug. **1907** S. E. White *Arizona Nights* i. i. 11 A quarter-hoss couldn't have beat me to that shack. **1944** *Sun* (Baltimore) 16 May 15/3 Cross-breeding..has developed a new type of quarter horse with a thoroughbred strain. **1954** *Ibid.* 4 Jan. 16/4 An honest to goodness cowboy who runs with a football like a quarterhorse chasing a steer. **1970** *Telegraph* (Brisbane) 17 Dec. 12/3 In America.. quarter horse racing is a popular sport. **1973** 'I. Drummond' *Jaws of Watchdog* xvi. 211 Holman Walker was a quarter-horse man and bred Appaloosa ponies..on his ranch. **1979** *Sunset* Apr. 6/2 The horse that helped tame the West—the quarterhorse—is still popular for its versatility and beauty.

quarter-jack. Add: **1.** (Further example.)

1971 *Country Life* 10 June 1444/3, I was fortunate in being on the spot to take this photograph when the Quarter Jack was brought down..for repairs.

3. [Jack *sb.*¹] *Mil. slang.* = Quartermaster 2 a.

1930 G. McMunn *Behind Scenes in Many Wars* xiv. 300 Fresh caviare..annoyed our men when they got a ration of it and complained of 'that black jam, what the quarter-jack had said was fish'.

quarterly, *a.* and *sb.* Add: **A.** *adj.* **3.** Special combinations, as *quarterly-meeting*, (*a*) in the Society of Friends (Quakers): a general meeting of all the local monthly meetings of a district; (*b*) in the Methodist Church: an administrative meeting of society officials within a circuit.

(*a*) **1675** in *Extracts Minutes Yearly Meeting of Friends, London* (1783) 63 Advised that the church's testimonies and judgments against disorderly and scandalous walkers ..be recorded in the respective monthly and quarterly meetings. **1675–7** G. Fox *Jrnl.* (1911) I. 267 About this time [*sc.* 1656] I was moved to sett uppe ye mens Quarterly meetinges throughout ye nation though in ye north they was setled before. **1837** J. J. Gurney *Autobiogr.* in J. B. Braithwaite *Mem.* (1854) I. v. 85, I well remember insisting in our Quarterly Meeting, on the reading of the advice..respecting what ought to be the character of representatives. **1869** Beck & Ball *London Friends' Meetings* vii. 69 In most parts of the country, the origin of Quarterly Meetings may be traced to those periodical gatherings of Friends from different but adjoining counties, known as *General Meetings*, and held at intervals with more or less regularity in different places. **1912** W. C. Braithwaite *Beginnings of Quakerism* xiii. 336 By Quarterly Meetings we must, I think, understand General business meetings for a district. **1965** *Friend* 19 Nov. 1393/2 We welcome the proposals of the Revision Committee that..the name Quarterly Meeting be changed to General Meeting [etc.]. **1974** G. Hubbard *Quaker by Convincement* iv. i. 181 General Meetings were until recently Quarterly Meetings, but they now occur as necessary. They..offer an opportunity for Friends from a wide area to meet and discuss matters of common interest.

(*b*) **1750** J. Wesley *Jrnl.* 22 Aug. (1756) 61 We had a Quarterly Meeting, at which were present the Stewards of all the Cornish Societies. **1807** J. Nightingale *Portraiture of Methodism* xxix. 303 The *Quarterly-meetings* are composed of all the travelling-preachers in the circuit where such meetings are held; of the leaders and stewards of the society; and of such of the local-preachers and members as may be invited by any of the travelling-preachers or stewards. **1898** B. Gregory *Side Lights on Conflicts of Methodism* x. 501 Take..the essential part of our economy which, next to the Class meeting, the Conference, and the Circuit, is the oldest and most central part of our economy—the Quarterly Meeting. **1963** R. E. Davies *Methodism* App. i. 173 The Circuit consists of a group of Societies in a neighbourhood (some contain as many as fifty, others as few as five or even less); its affairs are directed by the Quarterly Meeting, consisting of the ministers, deaconesses, and local preachers of the Circuit, and the leaders and trustees of each Society. **1972** *Methodist Recorder* 6 July 16/1 A number of familiar names in the life of the churches will be changed if the proposals are accepted: circuit quarterly meeting becomes *circuit meeting*.

B. *sb.* (Earlier examples.) Also, lesson notes for Sunday schools, issued every three months.

1830 W. Sewall *Diary* 15 May (1930) 131/1 Methodist quarterly commenced. **1857** [see *Christmas number* s.v. *Christmas *sb.* 4]. **1908** L. M. Montgomery *Anne of Green Gables* xi. 110, I got a quarterly from Mr. Bell for you and you'll go to Sunday School to-morrow.

quartermaster, *sb.* Add: **2. d. quartermaster captain**, an officer in the U.S. army with the rank of captain having duties similar to those of a quartermaster.

1907 *N.Y. Even. Post* (Semi-Weekly ed.) 13 May 6 The person enjoying the title of quartermaster captain (a rank that causes our British cousins to smile).

quartermaster *v.* (example).

1936 W. James *Gangways & Corridors* i. 10 He had quarter-mastered in Cuba during the Spanish American war.

qua·rtermistress. [f. after Quartermaster.] An officer in the W.A.A.C. having the duties of a quartermaster.

1917 *Times* 13 Aug. 3/1 The W.A.A.C. will be controlled by a Chief Controller, and the following appointments are authorized:—..Quarter mistress Class I. Attached to Depôt. 2 roses.

quarter-sessions. Add: Now usu. as two words without hyphen. **1.** (Further examples.) Also, *court of quarter sessions*, a court of record held quarterly before two or more justices of the peace in counties and before a recorder in certain boroughs.

Quarter sessions were abolished on 1 Jan. 1972 by the Courts Act, 1971, and their duties largely passed to the newly-instituted Crown Court.

1889 *Act* 52 & 53 *Vict.* c. 63. sect. 13 § 14 The expression 'court of quarter sessions' shall mean the justices of any county, riding, parts, division, or liberty of a county, or of any county of a city, or county of a town, in general or quarter sessions assembled, and shall include the court of the recorder of a municipal borough having a separate court of quarter sessions. **1916** *Act* 6 & 7 *Geo. V* c. 50 § 38 A justice of the peace in England when committing trial a person charged with burglary shall commit him for trial before a court of assize unless,..he thinks it expedient in the interest of justice to commit him for trial before a court of quarter sessions. 1967 *Act* 15 & 16 *Eliz. II* c. 58 § 8 A court of quarter sessions shall not have jurisdiction to try an indictment for any offence for which a person may be sentenced to death. 1972 *Daily Tel.* 1 Jan. 2 The ancient legal structure of assizes and quarter sessions disappears today.

3. *Comb.* **quarter-sessions rose**, var. of Quatre Saisons, the name of a variety of perpetual rose, *Rosa damascena* var. *bifera*, bearing pink or white flowers.

1892 C. M. Yonge *Old Woman's Outlook* vi. 137 The *Quatre Saisons* rose might very fairly become the Quarter Sessions rose. 1937 Partridge *Dict. Slang* 676/1 Quarter-sessions rose. A 'perpetual' rose.

quartet, quartette. Add: **1. a.** (Earlier example.)

1773 C. Burney *Present State of Music in Germany* I. 313 [Hasse] added, that..his trios, quartets and concertos for instruments, were so numerous that he should not know many of them again if he was either to see or hear them.

b. *transf.* (See quot. 1965.)

1943 T. S. Eliot (*title*) Four quartets. 1965 *New Statesman* 21 May 802/3, I asked what the four of his Quartets were four of, and he [sc. T. S. Eliot] answered that..'quartet' suggested Chamber Music composed for close attention by a small audience who had a fair understanding of what they were listening to.

4. *Comb.*, as **quartette table** *Obs.* exc. *Hist.*, = *quartetto table* s.v. *QUARTETTO 3.

1856 F. S. Cozzens *Sparrowgrass Papers* vii. 89 In one door-way stood a tray of delicate confections, upon two slender quartette tables. 1857 *Harper's Mag.* Mar. 453/1 On this quartette-table we will lay the portfolio. 1925 Penderel-Brodhurst & Layton *Gloss. Eng. Furnit.* 133 *Quartette Tables*, nests of four small tables of light make and diminishing size enclosed within each other and made to draw out. Sheraton, who may have been the first to devise them, called them 'Quartetto tables'.

quartetto. Restrict ? *Obs.* to senses in Dict. and add: **3.** *Comb.*, as **quartetto table** *Obs.* exc. *Hist.*, one of a nest of four small tables.

1802 T. Sheraton *Cabinet-Maker's & Upholsterer's Drawing-Bk.* (ed. 3) in J. M. Bell *Chippendale, Sheraton & Hepplewhite Furnit. Designs* (1900) 148 (*caption*) Cabinet and Quartetto Table. 1803 —— *Cabinet Dict.* 293 *Quartetto table*, a kind of small work table made to draw out of each other, and may be used separately. 1842 [see sense 2 in Dict.]. 1944 C. W. Drepperd *Primer Amer. Antiques* 248/2 Quartetto Tables. Sets of tables..resting in and under largest one. 1975 [see *NEST *sb.* 6].

quartic, *a.* and *sb.* For **a, b** read **A, B** and add: **A.** *adj.* **1.** (Example.)

1905 R. W. H. T. Hudson (*title*) Kummer's quartic surface.

2. Applied to a steering wheel shaped like a rectangle with rounded corners.

1973 *Country Life* 17 May 1417/2 A 'quartic' steering-wheel..is a two-spoke design shaped rather like a television screen. 1974 *Daily Tel.* 17 Sept. 6/5 With two other sporting 1750cc versions also announced today, it becomes the first Allegro to abandon the controversial squared-off 'quartic' steering wheel in favour of a conventional round wheel.

‖ **quartier** (kärtye). [Fr.: cf. Quarter *sb.* 14 a.] **1.** In France: a neighbourhood, district. Also *transf.* Also *ellipt.* = *Quartier Latin* below.

1828 Lytton *Pelham* I. xxiii. 192, I love that *quartier*—if ever I went to Paris again I should reside there. 1850 Thackeray *Pendennis* II. vii. 73 I'm thinking of becoming a moral man..with a good reputation in my *quartier*. 1864 Mrs Gaskell *French Life* in *Fraser's Mag.* Apr.–June 435/1 The new Boulevard..making a clear passage..through the densely populated *quartier*. 1896 E. Dowson *Let.* 19 Mar. (1967) 346 It is a refreshing change after Paris and the Quartier. 1967 J. Porter *Chinks in Curtain* 190 The nuns have a mission in Montmartre. The Princess will work there among the—er—ladies of the quartier. 1976 *Times* 31 July 12/5 A collection of *quartiers* each expressing its own set of neighbourhood loyalties and prejudices.

2. Quartier Latin (kärtye latæ̃), the Latin Quarter (of Paris). Also *attrib.*

1857 C. Kingsley *Two Yrs. Ago* I. i. 17 Tom Thurnall went to Paris, and became the best pistol-shot..in the Quartier Latin. 1861 in D. du Maurier *Young George du Maurier* (1951) 92 Physical prowess & muscular development—the natural antidotes to morbid Quartier-latin Romance. 1903 Chesterton *Robert Browning* v. 111 The thieves' kitchens and the studios of the Quartier Latin. 1936 C. Day Lewis *Friendly Tree* viii. 112 If everyone who comes out of the King's Head is a Bohemian, Kings Ampnett will soon rival the Quartier Latin. 1979 *Guardian* 30 Apr. 10/6 The authentic atmosphere of Paris's Quartier Latin in the late 1950s.

‖ **quartiere** (kwā:rtiẹ·rẹ). [It.; cf. prec.] In Italy: a district or area (of a city). (In quot. 1888 *transf.*)

1888 H. James *Aspern Papers* I. iii. 52 We walked together along the sala... I was afraid it would not form part of my *quartiere*. 1967 P. E. H. Durston *Mortissimo* (1968) xii. 95 Ran into them in a little antique shop. Down in your *quartiere*. 1979 'C. Brand' *Rose in Darkness* xi. 140 This is the quartiere of the thieves.

quartile, *a.* and *sb.* Restrict *Astr.* and *Astrol.* to senses in Dict. and add: **B.** *sb.* **2.** *Statistics* (kwǭ·ɹtail). The first and third of the three values of a variate which divide a frequency distribution into four groups, each containing one quarter of the total population (the second value of the three, the mean, is sometimes also included); also, any of the four groups so produced.

1879 D. McAlister in *Proc. R. Soc.* XXIX. 374 As these two measures, with the mean, divide the curve of facility into four equal parts, I propose to call them the 'higher quartile' and the 'lower quartile' respectively. It will be seen that they correspond to the ill-named 'probable errors' of the ordinary theory. 1931 *Biometrika* XXIII. 360 The distribution of the quartile is not so closely normal as that of the median. 1957 W. Martin in R. K. Merton *Student-Physician* III. 203 Among students in the top quartile, 28 per cent have high estimations of their ability to perform as a doctor should. 1964 Bruner & Potter in J. S. Bruner *Beyond Information Given* (1974) v. 85 For each picture, the point at which it was first recognized by any subject was obtained.., and likewise the point at which a quarter of the subjects recognized the objects (first quartile). 1977 *N.Y. Times* 4 Apr. 60 (Advt.), The top income half of the nation (those households with incomes above the national median) controls about three-fourths of total income... The top quartile always controls about 48%.

b. quartile deviation, the semi-interquartile range.

1911, etc. [see *semi-interquartile range* s.v. *SEMI- 5 a]. 1972 *Jrnl. Social Psychol.* LXXXII. 209 Table 1 shows the medians, quartile deviations, and composite ranks.

quart-pot. Add: **a.** Also *Austral.*, a billy-can of this measure for boiling tea-water, etc.

1838 H. Capper *South Australia* (ed. 2) 71 List of other articles provided for the passengers.—One wooden mess bowl, one ditto platter, one mess bread basket, one tin quart-pot, two three-gallon hawse buckets.—For each mess of six passengers. 1863 [see sense b below]. 1881 A. C. Grant *Bush Life in Queensland* I. v. 43 One of the quart-pots..was boiling madly. 1901 M. Franklin *My Brilliant Career* i. 4, I refilled the quart-pot in which we had boiled our tea with water from the creek. 1936 A. Russell *Gone Nomad* iii. 20 We had boiled the quartpots at a waterhole a mile or so back and were continuing on our way. 1941 *Coast to Coast* 155 He was a swaggie all right, with his roll of old blue blanket across his shoulders and his quart pot dangling from it.

b. (Earlier example.)

1863 R. Henning *Let.* 29 June (1966) 131 We..then made a fire, boiled some 'quart-pot tea', and sat down..to enjoy our dinner.

quartz. Add: **2. a.** *quartz crystal* (further simple and *attrib.* examples; cf. *quartz clock, watch* in sense *c), *dolerite*, *-fret* (poet.) [Fret *sb.*[1]], *lead* [Lead *sb.*[2] 6a], *leader* [Leader[1] 13c]; **quartz glass**, a glass consisting almost entirely of silica; = **silica glass*.

1933 K. Henney *Radio Engin. Handbk.* vii. 167 Piezo-electric quartz crystals provide standards of frequency, when permanently connected into suitable vacuum-tube circuits and allowed to oscillate continuously. 1959 R. H. Baker *Astron.* (ed. 7) iii. 77 In the quartz-crystal clock now in use in observatories, a quartz crystal controls the frequency of an electronic oscillator, imposing its natural frequency of vibration on the current. 1970 *Observer* (Colour Suppl.) 3 May 27/1 The quartz crystal clock; most accurate domestic timepiece of all. 1977 T. Allbeury *Man with President's Mind* vii. 66 It was a small quartz-crystal watch. 1911 G. W. Grabham in C. T. Clough et al. *Geol. Glasgow District* iv. 146 A very interesting feature of the quartz dolerites is their instability at temperatures slightly above the consolidation point. 1970 R. J. Small *Study of Landforms* iii. 73 Among the more important of these are..quartz-dolerite sills of the areas inland from Dippin Head. 1883 G. M. Hopkins *Poems* (1967) 96 All The thick stars round him roll Flashing like flecks of coal, Quartz-fret, or sparks of salt, In grimy vasty vault. 1903 *Amer. Jrnl. Sci.* CLXVI. 469 Quartz glass.—A very full account of the behaviour of this material has been given by H. Heraeus. 1912 J. W. Mellor *Mod. Inorg. Chem.* xl. 773 Quartz glass is used for the manufacture of elastic threads to suspend the delicate parts of electrical instruments. 1941 *Thorpe's Dict. Appl. Chem.* (ed. 4) V. 606/1 Quartz glass has an extremely small coefficient of linear thermal expansion,..and articles made from it can be heated to redness and quenched in cold water without fracture. 1965 Phillips & Williams *Inorg. Chem.* I. xiv. 553 Soda-glass is conveniently worked at a lower temperature than borosilicate glass, and the latter at a lower temperature than quartz glass. 1866 'Mark Twain' *Lett. from Hawaii* (1967) 272 There are hundreds of men in California who are sitting on their quartz leads..and hoping for the day when they will pay. 1884 R. C. Reid *Rambles on Golden Coast N.Z.* x. 105 A slate face in which there are numerous quartz-leaders.

b. *quartz mill* (examples).

1860 H. J. Hawley *Jrnl.* 22 May in *Wisconsin Mag. of Hist.* (1936) XIX. 341 There are a Quarts Mills runing which crush the quarts that are raised from the Lodes. 1908 *Chambers's Jrnl.* Sept. 640/2 Natthey..had started a little quartz-mill on his claim.

c. Used *attrib.* to designate devices in which use is made of quartz, usu. for its optical or its piezo-electric properties; as *quartz oscillator, spectrograph, temperature sensor*; **quartz clock**, a clock in which great accuracy is achieved by employing a vibrating quartz crystal in an oscillatory electric circuit to determine the frequency of the current; **quartz-halogen** *a.*, applied to electric light bulbs which have a quartz envelope and contain the vapour of a halogen (usu. iodine) in order to prevent deposition on the envelope of tungsten from the filament; so **quartz-iodine** *a.*; **quartz lamp**, a lamp in which the envelope is of quartz rather than glass, so that ultraviolet light will pass through it; **quartz-locked** *a.*, maintained at a constant speed of rotation by means of a quartz crystal, employed as in a quartz clock; **quartz watch**, a watch in which a quartz crystal is employed as in a quartz clock.

1937 *Ann. Reg. 1936* 63 Scheibe and Adelsberger (*Phys. Zeit.*) described the comparison of three 'quartz' clocks with the mean astronomical time obtained from three observatories during the 30 months ending June, 1935. 1976 R. Condon *Whisper of Axe* i. xxii. 144 A collection was taken up to present him with a quartz clock. 1968 *Listener* 21 Mar. 391/3 Headlights..should make use of the extra brightness of the new quartz-halogen lamps. 1970 *AA Bk. of Car* 163/4 Quartz-halogen bulbs do not suffer from blackening as ordinary bulbs do. 1979 *Arizona Daily Star* 5 Aug. c5/3 (Advt.), Quartz-halogen fog light kit. Two fog lights and wiring kit. 1965 *Economist* 27 Feb. 929/2 The new quartz-iodine bulbs are being imported to Britain in growing though limited numbers... In France, quartz-iodine headlamps are now standard equipment for the Citroen Pallas. 1973 D. Kyle *Raft of Swords* (1974) xii. 122 'Lights, please.' He switched them on, pivoting the 1,000-watt quartz-iodine beams downward. 1922 *Brit. Med. Jrnl.* 4 Nov. 852/1 The recognition of the antirachitic action of sunlight..is the outcome of Huldschinsky's observations of the curative effect on rickets exerted by the ultra-violet rays emitted by the mercury-vapour quartz lamp. 1933 *Burlington Mag.* Sept. 139/1 If it [sc. the painting] could be closely studied with all the resources of the quartz-lamp and the microscope it seems hardly possible that it should not betray its later date. 1977 McGuinness & Stein *Building Technol.* xiv. 406/1 The quartz lamp has about three to four times the life of a normal incandescent. 1977 *Time* 14 Nov. 7 (Advt.), First they designed an all-direct-drive-motor-system. Using one motor for each reel and another quartz-locked motor for the tape capstan. 1938 *Proc. Physical Soc.* L. 413 The quartz oscillator not only is widely used for stabilizing the frequency of transmitting stations but also constitutes the most accurate means of recording the passage of time. 1905 *Astrophysical Jrnl.* XXII. 129 From negative by Professor R. W. Wood with quartz spectrograph of the Johns Hopkins University. 1966 *McGraw-Hill Encycl. Sci. & Technol.* XII. 590/1 Two types of quartz spectrographs are in common use for recording ultraviolet and visible spectra. 1978 *Sci. Amer.* Mar. 144/1 (Advt.), The HP 2804 digital thermometer achieves exceptional accuracy because HP calibrates each quartz temperature sensor. 1974 *Country Life* 5 Dec. 1763 (Advt.), Seiko sold the first quartz watch. 1976 *Times* 5 Feb. 17/5 The quartz watches now being produced fall into two groups..the traditional watch face..and digital watches.

qua:rtzofeldspa·thic, *a.* *Petrol.* Also **-fels-**. [f. Quartz + -o + Feldspathic, felspathic *a.*] Containing a high proportion of quartz and feldspar.

1839 H. T. de la Beche *Rep. Geol. Cornwall, Devon & W. Somerset* vi. 175 It has chiefly a quartzo-felspathic base with crystals of felspar. 1973 *Nature* 11 May 62/2 The oldest known rocks in the world are quartzofeldspathic gneisses (the Amîtsoq gneisses) from the Godthaab area of West Greenland.

quasar (kwē[1]·sāɹ, -zāɹ). *Astr.* [f. *quas(i-stell)ar *a.*] Any of a class of celestial objects that give a star-like (i.e. unresolved) image on a photograph and have a spectrum showing a large red shift, usu. taken to indicate great remoteness and immense power.

1964 H.-Y. Chiu in *Physics Today* May 21/2 So far, the clumsily long name 'quasi-stellar radio sources' is used to describe these objects. Because the nature of these objects is entirely unknown, it is hard to prepare a short, appropriate nomenclature for them so that their essential properties are obvious from their name. For convenience, the abbreviated form 'quasar' will be used throughout this paper. 1964 *Observer* 14 June 1/1 For the past twelve months, astronomers have been obsessed with these 'quasars' or 'quasi-stars' as they are now called. 1968 L. Durrell *Tunc* i. 18 Here was Vibart persuading poor Felix to quit quasars and debouch into memoirs. 1968 *Times* 13 Nov. 16/2 One view is that the red-shift comes about because quasars are remarkably distant at the boundaries of the known universe. The other..is that quasars are extremely massive objects, and that the light from them is reddened by gravitational effects. 1971 *Sci. Amer.* May 55/2 The term quasar..was originally applied only to the starlike counterparts of certain strong radio sources whose optical spectra exhibit red shifts much larger than those of galaxies. Before long, however, a class of quasi-stellar objects was discovered with large red shifts that have little or no emission at radio wavelengths. 'Quasar' is now commonly applied to starlike objects with large red shifts regardless of their radio emissivity. *Ibid.* 56/2 Assuming that they are at cosmological distances, one can easily show that many quasars are from 50 to 100 times brighter than entire galaxies containing hundreds of billions of stars. 1973 *Nature* 6 July 19/2 Two neutral-coloured stellar objects..have been shown to be QSOs with redshifts of 3·40 and 3·53 respectively. This has stimulated interest in other radio sources identified with stellar objects which lack the

ultraviolet or blue colour excesses normally associated with quasars. 1978 *Daily Tel.* 8 June 2/5 The nearest known quasar in space, an object the size of the solar system but emitting the energy of 100 galaxies, has been discovered by an X-ray experiment aboard an American satellite. 1978 Pasachoff & Kutner *University Astron.* xxviii. 713 We now use the term quasar to include both quasi-stellar radio sources and quasi-stellar objects.

Quashee, quashie. Add: Also 8 **Quashy**, 9 **Quashi.** (Earlier and later examples.)

1778 P. Thicknesse *Year's Journey* (ed. 2) II. xlv. 104 When Quashy, found the physicians had given his master over, he stole his breeches..and went off with them... The indignant master..never recovered his *faithful* black, nor his *departed* breeches. 1825 C. Waterton *Wanderings* iv. 279 Quashi's fiddle was silent. 1960 *Tamarack Rev.* xiv. 7 'Can't catch Quashie, catch his shirt' is a West Indian proverb, so the occupiers turned to gold of another colour: brown gold, sugar.

quasi, *adv.* and *pref.* Add: Also with pronunc. (kwē[1]·zəi, kwā·zi). **I. 2.** As a prefix *quasi-* has remained a common formative element. The following is a selection from some of the more frequently occurring modern formations.

a. With sbs.: *quasi-belief, -continuum, -copula, -definition, -dereliction, -equilibrium, -existence, -implication, -jazz, -marriage, -miracle, -modal, -molecule, -monopoly, -neutrality, -object, -partner, -quotation, -quote, -religion, -science, -semi, -sensation, -statement, -substance, -totality, -universal, -verb.*

1925 C. D. Broad *Mind & its Place in Nature* iv. 217 The quasi-belief which is an essential factor in all perceptual situations. 1942 *Mind* LI. 245 When I read about Captain Costigan or about Mr. Micawber I knew perfectly well and all the time that there were no such persons. There is no temporary quasi-belief or make-believe..as there is in the case of the mirage. 1966 D. G. Brandon *Mod. Techniques Metallogr.* 179 The conduction electrons in a metal occupy a quasi-continuum of energy levels. 1979 *Sci. Amer.* May 108/1 In this region, known as the quasi-continuum, the additional rotational and translational states provide all the 'fine tuning' necessary to match the photon frequency with the quantum gap between vibrational levels. 1934 Webster, Quasi copula. 1963 F. T. Visser *Hist. Syntax Eng. Lang.* I. ii. 153 We have a similar complete change of status when verbs of this kind function as quasi-copulas, as in 'this meat eats tough'. 1966 *Eng. Stud.* XLVII. 51 An interesting..problem arises with quasi-copulas i.e. border-line cases where the verb may stand between 'copula' and 'full verb'. 1927 C. R. S. Harris *Duns Scotus* ii. vi. 188 We are..able to arrive at a quasi-definition in which we can describe it [sc. the divine essence] more perfectly than by means of any of its other attributes. 1939 *Mind* XLVIII. 541 And this he does by discerning (between the lines) the frequent interpolations of new quasi-definitions demarcating new distinctions of meaning of the terms involved. 1978 *Dædalus* Summer 28 Let us note three elements. First, a quasidefinition of the marginal man. 1950 D. Gascoyne *Vagrant* 8, I stand still in my quasi-dereliction. 1905 *Jrnl. Geol.* XIII. 393 The surface ever wearing down, the waste..continually exported by the winds, a nearly level rock-floor,..everywhere slowly lowering at the rate of sand and dust exportation, is developed over a larger and larger area; and such is the condition of quasi-equilibrium for old age. 1964 *Amer. Jrnl. Sci.* CCLXII. 793 As the landscape slowly degrades ..rivers have a tendency to remain in quasi-equilibrium. 1978 *Dædalus* Spring 25 Most states of nature are quasiequilibria, the outcome of competing forces. 1909 W. M. Urban *Valuation* v. 127 An aspect..is given a quasi-existence. 1944 M. Black in P. A. Schilpp *Philos. B. Russell* 241 Hamlet and the Snark, the philosopher's stone and the round square, being all characterised by predicates, must all, in some versions of this position, have their being in a multiplicity of distinct limbos, realms of *Sosein, Aussersein* and *Quasisein* in which to enjoy their ambiguous status of partial or quasi-existence. 1951 *Mind* LX. 355 The truth-table for 'quasi-implication' in Professor Reichenbach's three-valued logic. 1973 J. J. Zeman *Modal Logic* ii. 21 We may refer to such formulas as 'quasi-implications'. 1947 R. de Toledano *Frontiers of Jazz* 70 Hundreds of musicians, playing in all the jazz and quasi-jazz styles. 1977 *Time Out* 28 Jan. 47 (*caption*) The already expansive Runt canvas—everything from Philly soul harmonies through wittily timeless psychedelia and inspiring quasi-jazz—has added two more excellent albums. 1926 W. J. Locke *Stories Near & Far* 166 Quasi-marriage bond. 1976 *National Observer* (U.S.) 4 Sept. 13/3 'What does it matter what we think?' says a friend of mine, three of whose sons have opted for quasi marriage. 1893 *Mind* II. 210 That seems to me but an excessively clumsy way of stating in terms of a *quasi*-miracle the very truth which Stumpf and I express by saying that likeness is an immediately ascertained relation. 1971 J. Anderson in A. J. Aitken et al. *Edin. Stud. Eng. & Scots* 71 Certain 'quasi-modals', which satisfy some but not all of the criteria. 1972 W. Labov *Lang. in Inner City* ix. 376 The quasimodals produce many problems which are not fully resolved. 1968 C. G. Kuper *Introd. Theory Superconductivity* xi. 181 A pair of electrons in the immediate vicinity of the Fermi surface can form a bound quasimolecule. 1975 *McGraw-Hill Yearbk. Sci. & Technol.* 115/2 Electron ejection from an atom or a quasimolecule is due to the time-varying electric field acting on the electron as the collision partners pass each other. 1934 *Planning* II. xl. 6 Where monopoly or quasi-monopoly powers are taken there shall be an independent chairman and other independent members of the Industry Board administering the scheme. 1980 *Jrnl. R. Soc. Arts* Mar. 207/1 Once a housebuilder has bid successfully for what little land is available..a quasi-monopoly situation is created. 1934 Webster, Quasi neutrality. 1962 W. B. Thompson *Introd. Plasma Physics* ii. 9 Because of their inertia, the electrons will oscillate about the initially charged region but with a very high frequency, so that quasi-neutrality is preserved in the mean. 1963 F. T. Visser *Hist. Syntax Eng. Lang.* I. iv. 453 With verbs that do not usually take a direct object..the construction with indefinite *it* as quasi-object is also frequently met with. 1967 Quasi object [see *incomplete *a.* 2]. 1848 Bouvier *Law Dict.* (ed. 3) II. 401 *Quasi partners*, partners of lands, goods, or chattels, who are not actual partners, are sometimes so called. 1930 M. Clark *Home Trade* 3 Quasi-partners are those who have played, but no longer play, an active part. 1867 G. M. Hopkins *Let.* 15 Aug. (1956) 41 There are quotations or quasi-quotations fr. the Bible in it. 1943 *Mind* LII. 267 It may be doubted..whether the additional typographical complexity of the device of 'quasi-quotation' is worth the bother. 1937 W. V. Quine in *Jrnl. Symbolic Logic* II. 146 An expression beginning and ending in corners is to denote the expression which we obtain, from the expression between the corners, by replacing all Greek letters by the expressions which those Greek letters are intended to denote. The corners may thus be viewed as 'quasi-quotes'; but they must not be confused with ordinary quotation marks. 1949 *Mind* LVIII. 524 The quasi-quotes would not be needed if '⊃' and '~' were being used autonymously. 1934 Webster, Quasi religion. 1952 C. P. Blacker *Eugenics* 112 Once the importance of eugenics was grasped, once its principles had 'been accepted as a quasi-religion, the result will be manifested in sundry and very effective modes of action which are as yet untried, and many of them unforeseen'. 1977 P. Johnson *Enemies of Society* xiv. 193 Pseudo-sciences, irrational quasi-religions, and phantasmagoric utopias. 1874 W. Wallace tr. *Hegel's Logic* 21 The quasi-sciences..are founded on an act of arbitrary will alone, such as Heraldry. 1924 W. B. Selbie *Psychol. Relig.* ii. 33 Frazer also regards it [sc. magic] as a quasi-science, in fundamental opposition to religion. 1976 *Word 1971* XXVII. 77 To me this is quasiscience, if not pseudoscience. 1974 P. Wright *Lang. Brit. Industry* xvii. 162 *Quasi-semis*, joined by their garages, certainly act as *semis*, though their Latin *quasi* sounds so foreign to English speech. 1979 *W. Lancs. Even. Gaz.* 23 Feb. 17 (Advt.), Quasi semi conveniently situated to all schools. 1922 Joyce *Ulysses* 674 What were Stephen's and Bloom's quasisimultaneous volitional quasisensations of concealed identities? 1948 *Mind* LVII. 194 Necessary statements, then, might be called 'quasi-statements', to indicate that they neither *mention* the expressions of which they are composed, nor *use* them to talk about the non-linguistic world. 1972 *Jrnl. Symbolic Logic* XXXVII. 421 A quasi-statement is a statement or a question. 1925 C. D. Broad *Mind & its Place* iii. 99 Even so extreme a dualist about Mind and Matter as Descartes occasionally suggests that a mind and its body together form a quasi-substance. 1943 *Mind* LII. 336 'Space' is the name of an entity, a quasi-substance, though according to Kant a mind-dependent one. 1941 *Mind* L. 389 A mechanism in the modern sense of the term, *viz.*, as a quasi-totality of serial and reciprocating temporal causes. 1977 *Dædalus* Fall 141 History is also connected more generally, more largely to the quasi-totality of the social sciences. 1890 W. James *Princ. Psychol.* I. xii. 475 The nominalists, on their side, admit a *quasi*-universal, something which we think *as if it were* universal, though it is not. 1957 *Publ. Amer. Dial. Soc.* xxviii. 31 The affirmative quasi-verb *better*. 1972 *Language* XLVIII. 466 Quasi-verbs..constitute an inflectional category which is rare, or possibly unique, among Indo-European languages.

b. With adjs. (and advbs.): *quasi-æsthetic, -arithmetical, -automatic* (hence *-automatically*), *-classic, -continuous, -crystalline, -divine, -elastic, -eternal, -ethical* (hence *-ethically*), *-Fascist, -feudal, -general, -governmental, -grammatical, -hallucinatory, -historical, -horizontal, -independent, -instantaneous, -legal, -logical, -marital, -material, -mathematical, -mechanical, -mechanistic, -metallic, -metaphysical, -military, -miraculous, -molecular, -monastic, -mythical, -neutral, -normal, -official, -optical, -ossianic, -periodic, -permanent, -personal, -philosophical, -physical, -purposive, -religious, -scientific, -simultaneous* (hence *-simultaneously*), *-stationary, -technical, -thermodynamic, -totalitarian, -transitive, -universal.*

1909 W. M. Urban *Valuation* vi. 152 In certain quasi-aesthetic combinations of utilities—as in a festal meal ..—a detail..may acquire an extraordinary value. 1963 R. M. Hare *Freedom & Reason* ix. 174 The universal, quasi-aesthetic ideal of not having addicts about the place. 1944 *Mind* LIII. 242 One could of course use as M a quasi-arithmetical calculus whose integers are construed as particulars. 1965 *Language* XLI. 490 A quasi-arithmetical notation for syntactic description. 1890 W. James *Princ. Psychol.* II. xxvi. 523 The intermediary terms of an habitual series of acts leading to an end are apt to be of this *quasi*-automatic sort. 1963 F. T. Visser *Hist. Syntax* I. ii. 152 Intransitive verbs used to represent the action—as quasi-automatic, or self-originated. *Ibid.* 159 An interesting sequence of intransitive verbs and 'transitive' verbs used quasi-automatically. 1905 O. Jespersen *Growth & Struct. Eng. Lang.* vi. 123 Authors sometimes coin quasi-classic words without finding anybody to pass them on, as when Milton writes 'our *inquisiturient* Bishops' (Areop. 13). 1890 W. James *Princ. Psychol.* I. xiii. 531 The jingling of the bells on the horses of a horse-car passing the door..and the rumbling of the vehicle itself,..to our ordinary hearing merge together very readily into a *quasi*-continuous body of sound. 1973 *Jrnl. Amer. Chem. Soc.* XCV. 8301/1 The Raman spectrum of titanium tetraiodide in solution has been obtained previously..by use of 694·3 nm excitation of a quasicontinuous ruby laser. 1946 *Nature* 31 Aug. 297/2 Such scattering [of sound waves] has been ascribed either to the quasi-crystalline structure of a liquid. 1964 G. H. Haggis et al. *Introd. Molecular Biol.* ix. 220 Wilkins, Franklin and others.., by drawing out threads from concentrated DNA solutions with careful control of humidity, have been able to get the molecules into a highly-orientated quasi-crystalline form and to obtain in this way more detailed diffraction patterns. 1941 Wyndham Lewis *Let.* 21 Oct. (1963) 302 The 'Royal Academy' seems to them a quasi-divine institution. 1955 R. Graves *Crowning Privilege* 113 The living poet hero is a modernism; I think I am right in saying that Petrarch was the first poet to receive quasi-divine honours during his lifetime. 1899 *Phil. Mag.* XLVIII. 80 The destructive effects of earthquakes are..due to the propagation of quasi-elastic disturbances. 1972 *Sci. Amer.* Oct. 103/3 The observation of pairs of alpha particles with this unique signature emerging from heavy nuclei was taken to indicate the presence of essentially free alpha particles at the nuclear surface. Such scattering is called quasi-elastic because the particles are not totally free but are somewhat bound to the target nucleus. 1976 *Nature* 15 July 177/1 In these plots, the vertical lines around $A = 32$ correspond to the incident [32]S ions that have undergone peripheral elastic or quasielastic interactions with the loss or gain of a few nucleons. 1895 *Psychol. Rev.* II. 124 [Parts of mental contents do not] have an eternal or quasi-eternal individual existence, like the parts of objects. 1932 H. H. Price *Perception* viii. 223 We must not..think that nearness is always an 'advantage' (if we may go on using this quasi-ethical language). 1949 *Mind* LVIII. 203 The deductions which welfare economists make are quasi-ethical. 1909 W. M. Urban *Valuation* vii. 214 The feeling..is quasi-ethically qualified. 1938 *Political Q.* IX. 1. 133 The new Germany, imperialist Japan and quasi-Fascist Poland must all, at the least, be democratised. 1954 Koestler *Invis. Writing* iv. xxxv. 376 Greece, too, lived under a quasi-Fascist dictatorship. 1960 —— *Lotus & Robot* ii. x. 228 Japan was able to..build a quasi-capitalistic state on a quasi-feudal foundation. 1977 *Dædalus* Summer 50 The straitjackets of public regulations, quasi-feudal traditions, financial dependence, and intellectual routine which have so often paralyzed the universities of postwar Europe. 1896 W. Caldwell *Schopenhauer's Syst.* ix. 47 This fact..warrants us in calling his whole philosophy a *quasi* general overturning of the philosophy of the idea. 1948 *Mind* LVII. 50 If 'If Nero had been Seneca' means anything at all, it is a quasi-general proposition which can be analysed either as 'If Nero had had the character of Seneca' or 'If Seneca had been emperor' or in some similar fashion. 1961 *Ethical Outlook* May–June 93/2 If corporations are not to run away with us, they must become quasi-governmental institutions. 1977 P. Johnson *Enemies of Society* xii. 167 The University Grants Committee, the quasi-governmental body which distributes state cash. 1953 H. A. Hatzfeld *Crit. Bibliogr. New Stylistics* 1 A theoretical, communicable, analytical, quasi-grammatical language of the critic. 1955 J. L. Austin *How to do Things with Words* (1962) vi. 68 'I state that' seems to conform to our grammatical or quasi-grammatical requirements. 1890 W. James *Princ. Psychol.* ii. xx. 220 *Optical objects not actually present*..being imagined now with a quasi-hallucinatory strength. 1979 C. E. Schorske *Fin-de-Siècle Vienna* vii. 330 A quasi-hallucinatory, erotic dream experience. 1880 A. H. Sayce *Introd. Sci. Lang.* ii. vii. 98 Many of the ballads are quasi-historical. 1933 *Mind* XLII. 396 Objective Spirits..*regulate* individual personality, making personality, by participation, a quasi-historical thing, or, as I should perhaps have said, a sui-thing. 1956 *Nature* 21 Jan. 113/2 The salient place in theory taken by horizontal wind divergence..was stressed by showing its relation..with the development of circulations, cyclonic and anti-cyclonic, through the vorticity equation applied to quasi-horizontal motions on a rotating Earth. 1967 *Oceanogr. & Marine Biol.* V. 82 The analysis of the water masses of the southwest Pacific show them to be very clearly stratified,..with the different water masses lying in quasi-horizontal layers. 1933 *Mind* XLII. 385 There is..no point in '*reifying*' the system, and then treating it as something quasi-independent of its parts, because, in these cases its behaviour is inferable from a knowledge of the parts. 1966 S. Beer *Decision & Control* xv. 381 Cybernetic insights show, in particular, that the totality of the organization ought to be made up of building-blocks that will be called quasi-independent domains. 1958 *Bull. Geol. Soc. Amer.* LXIX. 111/2 In geologic writing..the standards for contemporaneity and instantaneousness are more flexible. Perhaps, H. and G. Termier's (1956) expression quasi-instantaneous (*quasi-instantané*) could be recommended as a convenient term to indicate instantaneousness in the geologic sense. 1960 C. S. Lewis *Studies in Words* 214 The words in a great poet's phrase ..strike the mind as a quasi-instantaneous chord, yet, strictly speaking, each word must be read or heard before the next. 1911 Webster, Quasi legal. 1951 E. E. Evans-Pritchard *Social Anthropol.* i. 14 The sample includes studies of..quasi-legal institutions..and of the entire social organization..of one or other people. 1965 H. Kahn *On Escalation* ix. 172 The declaration of war is just one of a series of legal and quasi-legal measures..which have a role in escalation situations. 1902 W. James *Var. Relig. Exper.* iv. 106 Philosophers usually profess to give a quasi-logical explanation of the existence of evil. 1960 *New Biol.* XXXI. 135 They form part of an entrancing quasi-logical world, but it has little in common with the pedestrian, irrational world of the empiric. 1959 G. D. Mitchell *Sociol.* 64 His wife's sisters are, therefore, in a relationship to him which Radcliffe-Brown has described as quasi-marital. 1978 *Times Lit. Suppl.* 1 Dec. 1391/4 The eighty-six men assigned to the 'dysfunctional' group also lacked any quasi-marital partnership and often expressed regret about their homosexuality, but they were sexually more active and promiscuous. 1876 *Nation* 8 June 369/1 It [sc. liberty] and necessity being alike indemonstrable by any quasi-material process, must be *postulated* if taken at all. 1924 W. B. Selbie *Psychol. Relig.* xiv. 269 This 'soul' was regarded as similar to the body in form and nature, and as having a quasi-material existence of its own. 1870 W. S. Jevons in *Phil. Trans. R. Soc.* CLX. 516 In Boole's system the same groups are indicated by certain quasi-mathematical symbols as follows. 1957 C. Vereker *Devel. Polit. Theory* iv. 147

Men..pursued their own happiness..by a comparison of their experienced and expected pleasures and pains in this quasi-mathematical manner. 1920 W. R. Sorley *Hist. Eng. Philos.* xii. 265 He had started in his thinking with the quasi-mechanical view of a fixed norm of belief existing in the past. 1942 *Mind* LI. 166 The 'quasi-mechanical' reproductive and associative processes. 1923 T. P. Nunn *Education* xi. 136 Quasi-mechanistic theory, leading us to think of a man's self as built up of instincts much as a machine is built up of wheels. 1961 *Chicago Rev.* XV. 1. 94 Sub-human and quasi-mechanistic powers. 1858 Carlyle *Fredk. Gt.* I. iv. iv. 417 The voice..was of clangorous and penetrating, quasi-metallic nature. 1968 C. G. Kuper *Introd. Theory Superconductivity* xv. 260 One electron per spine atom is in a π-state, and exhibits quasimetallic properties in one dimension. 1890 W. James *Princ. Psychol.* I. v. 138 To urge the automaton-theory upon us..on purely *a priori* and *quasi*-metaphysical grounds, is an *unwarrantable impertinence in the present state of psychology*. 1959 *New Biol.* XXX. 59 So impressive is this 'organic unity' that its existence has been claimed to set living matter apart from non-living and has more than once been elevated into a quasi-metaphysical postulate. 1895 M. Pemberton *Impregnable City* xvii. 126 Men in quasi-military uniforms..contributed to the impression of the scene. 1964 Dentler & Cutright in I. L. Horowitz *New Sociol.* 425 Undemocratic leadership by a quasi-military elite. 1974 tr. *Wertheim's Evolution & Revolution* 127 It is clear that the term 'mutiny' mostly refers to an insurrection within a military or quasi-military apparatus. 1890 W. James *Princ. Psychol.* I. xii. 474 They invent..as the vehicle of the knowledge of universals..an Ego, whose function is treated as quasi-miraculous. 1904 *Jrnl. Philos.* I. 541 An experience that knows another can figure as its *representative*, not in any quasi-miraculous 'epistemological' sense, but in the definite practical sense of being its *substitute* in various operations. 1968 C. G. Kuper *Introd. Theory Superconductivity* i. 2 In 1951 Onsager conjectured that conduction electrons might form quasi-molecular pairs bound by the Fröhlich interaction. 1974 *Physics. Bull.* Oct. 467/1 Fernandez..gave a review of the radiative and other properties of resonant states observed in heavy ion collisions. Usually known as quasimolecular states, these resonances represent states at approximately 30 MeV excitation in the compound system. 1979 *Sci. Amer.* Dec. 125/3 The resulting structure is called a quasi-molecular state. 1938 *Ann. Reg. 1937* 241 President Azaña emerged from his quasi-monastic retirement in a villa near Gerona to appoint a new Government. 1964 P. F. Anson *Bishops at Large* iv. 95 The quasi-monastic character of the life. 1890 W. James *Princ. Psychol.* II. xx. 211 Difficulties arise which have made psychologists appeal to new and *quasi*-mythical mental powers. 1963 M. H. Abrams in N. Frye *Romanticism Reconsidered* 53 Quasi-mythical agents tend to recur. 1905 *Daily Chron.* 11 May 5/1 The Admiral asserts that he is not the only one who has provisioned in neutral or quasi-neutral waters. 1962 W. B. Thompson *Introd. Plasma Physics* iv. 45 Also, from the extremely small value..it is clear that a conductor resembles a plasma in remaining quasi-neutral. 1909 O. Lodge *Survival of Man* i. 4 Actually more inventive sometimes of other and quasi-normal methods of explaining inexplicable facts. 1965 *Math. in Biol. & Med.* (*Med. Res. Council*) v. 228 The form of this distribution (Figure 2) is negatively exponential, high on the left and tapering away to the right quite unlike the normal and quasi-normal distributions with which we deal intuitively in most everyday situations. 1882 *Mind* VII. 186 Hegel's philosophy..must, now that it has become quasi-official, make ready to defend itself as well as to attack others. 1965 H. Kahn *On Escalation* ii. 44 A quasi-official move ordered by a government. 1940 *Chambers's Techn. Dict.* 693/2 *Quasi-optical waves* (Radio), electromagnetic waves of such short wavelength that their laws of propagation are similar to those of visible light. 1965 *B.B.C. Handbk.* 115 The signals which carry domestic broadcasting programmes are usually designed to be received by ground-wave on medium and long waves and within a quasi-optical range for television and VHF sound broadcasting. 1956 Auden in *Listener* 26 Jan. 137/1 The Paid Announcer..with his quasi-ossianic prose Cuts in upon the lovers, halts the band, To name a sponsor or to praise a brand. 1895 *Funk's Stand. Dict.*, *Quasi-periodic*, noting a function in which the increase of the variable by a fixed amount is equivalent to the multiplication of the whole function by another function. 1923 *Proc. R. Soc.* A. CIII. 97 These quasi-periodic atmospherics showed considerable diversity of form, the most frequently recurring form was..similar to a uniformly damped sinusoid. 1962 W. B. Thompson *Introd. Plasma Physics* vii. 164 In such situations, the orbit of particles along the magnetic field lines is also quasi-periodic, and again an adiabatic invariant may be formed. 1927 B. Russell *Outl. Philos.* iv. 50 The act of writing produces quasi-permanent material structures. 1967 *Oceanogr. & Marine Biol.* V. 77 The Solomon Divergence is a quasi-permanent phenomenon of the Coral Sea. 1909 W. M. Urban *Valuation* xii. 354 Quasi-personal constructions of the group or the nation. 1934 Webster, Quasi-philosophical. 1943 *Mind* LII. 100 If 'practical' means 'relevant to human happiness and misery' there is hardly anything more practically important than have been certain philosophical or quasi-philosophical ideas. 1960 *Spectator* 22 July 137 Vast quasi-philosophical works of great power. 1881 W. James *Let.* 18 Dec. in R. B. Perry *Tht. & Char. W. James* (1935) I. 620 One gains an aesthetic pleasure in what you write that is of a *quasi* physical order. 1953 G. E. M. Anscombe tr. *Wittgenstein's Philos. Investigations* i. 121 You interpret a grammatical movement made by yourself as a quasi-physical phenomenon which you are observing. 1964 E. Palmer tr. *Martinet's Elem. Gen. Linguistics* iv. 131 French has a class of 'adjectives' characterized by uses which are quasi-predicative. 1907 H. Rashdall *Theory of Good & Evil* II. iii. iv. 373 Biology now finds that it cannot get on without the idea of 'quasi-purposive' behaviour in accounting for the growth of the individual organisms. 1943 *Mind* LII. 346 A rationalist may be one who, admitting a hierarchy of being and value, seeks to explain the universe as exemplifying the highest degree of quasi-purposive significance, or he may be one who, like the Positivists [etc.]. 1838 J. W. Semple *Kant's Relig. within Boundary of Pure Reason* p. ix, That it concerns us islanders *to know* the religious or quasi-religious opinions entertained by our next-door neighbours on the Continent, no sane man, I apprehend, can doubt. 1906 W. James *Coll. Ess. & Rev.* (1920) 465 In the writings of this youthful Italian [sc. Papini]..I find..a tone of feeling well fitted to rally devotees and to make of pragmatism a new militant form of religious or quasi-religious philosophy. 1964 P. F. Anson *Bishops at Large* xi. 505 A quasi-religious community known as 'The Clerks Secular of St. Basil'. 1977 P. Johnson *Enemies of Society* xii. 169 Marxism is a system of quasi-religious prophecy. 1979 *Dædalus* Summer 156 The Ford scheme aroused Dewey's intense and quasi-religious sense of his mission to the democracy. *a* 1873 Mill *Three Ess. Relig.* (1874) 204 It is sometimes..wrapt up in a quasi-scientific language. 1970 G. E. Evans *Where Beards wag All* xx. 230 Many archaeologists appear purposefully to avoid the kind of folk-life material discussed here because they see in it some threat to their lately acquired quasi-scientific respectability. 1977 *Listener* 15 Dec. 782/3 Meanwhile, the mis-spelling of the quasi-scientific term *minuscule*, as *miniscule* is now so common it is close to becoming accepted English. 1922 Quasi-simultaneous [see *quasi-sensation*]. 1946 F. E. Zeuner *Dating Past* xii. 353 Another instance of quasi-simultaneous evolution of classes and orders is provided by primitive vertebrates. 1956 *Nature* 28 Jan. 178/1 It [sc. the radio telescope] may be used to scan through a small angle in the meridian plane, recording five separate declinations quasi-simultaneously. 1930 *Physical Rev.* XXXV. 944 For the occurrence of raditionless transitions it is essential that the material system..be in a quasistationary state of an energy equal to the energy of some aperiodic motion of the system. 1967 *Oceanogr. & Marine Biol.* V. 38 The method first determines the wind-induced elevations of sea-level at a particular place assuming that the changing wind conditions over the North Sea and the English Channel are quasi-stationary. 1906 *Spectator* 7 Apr. 545/1 When metaphysics begin we flounder among quasi-technical platitudes. 1961 R. B. Long *Sentence & its Parts* ix. 224 This kind of thing is commoner in quasi-technical use than in general use. 1956 *Nature* 25 Feb. 369/1 Thermodynamic, quasi-thermodynamic and non-thermodynamic methods of investigating the electrochemistry of clays. 1965 Phillips & Williams *Inorg. Chem.* I. vii. 235 We shall show how by investigation of the mechanisms of chemical reactions it is possible to break them down into a series of elementary reactions, each of which may be treated by a quasi-thermodynamic approach. 1946 W. S. Churchill *Victory* 80 Controls under the pretext of war or its aftermath which are in fact designed to favour the accomplishment of quasi-totalitarian systems..are a fraud which should be mercilessly exposed to the British public. 1953 M. Lowry *Sel. Lett.* (1967) 337 B.C. at the moment has no government at all, though both of them are totalitarian... Or quasi-totalitarian. 1927 B. Russell *Analysis of Matter* xii. 118 In the case of similarity, we have a relation which is capable of degrees, and may be called 'quasi-transitive'—*i.e.* if *A* is very like *B*, and *B* is very like *C*, then *A* must be rather like *C*. 1963 F. T. Visser *Hist. Syntax* I. ii. 135 (*heading*) Quasi-transitive verbs. 1974 tr. *Wertheim's Evolution & Revolution* 177 Brinton's unsuccessful attempt to introduce a quasi-universal, cyclical model of revolution..is symptomatic of the danger of any attempt to deal with the genesis and process of revolutions in general. 1977 *Jrnl. R. Soc. Arts* CXXV. 579/1 Most significant of all is the emergence and quasi-universal adoption of the full-fledged *nagara śikhara*.

c. [through It.] With pronunc. (kwā·zi). Used in musical directions: as if, almost, as *quasi-parlando, parlato* [It., 'speaking', 'spoken'] sb. and advb.

1908 R. Dunstan *Cyclopaedic Dict. Music* 325/1 *Quasi parlato*, as if spoken. 1945 Britten & Slater *Peter Grimes* ii. i. 183 Ellen *semplice* (*quasi parlato*) Nothing to tell me, nothing to say. 1959 *Listener* 5 Mar. 432/3 The *quasi-parlando* which Puccini uses so effectively as a stage language seems on the screen so desperately sluggish in effect..that [etc.]. 1972 V. C. Clinton-Baddeley *To study Long Silence* i. 45 A fine girl..twirling a malacca cane, and singing, *quasi parlando*, in a fruity baritone. 1975 *Country Life* 13 Nov. 1312/2 Singers do not *have* to be provided with tunes—in *Tristan*..they get by with almost continuous *quasi parlando*.

qua·si-co:ntract. *Law.* [Quasi *adv.* and *pref.*] (See quot. 1959[1].) Hence **qua:si-contra·ctual** *a.*

[1618 in *Eng. Reports* (1907) LXXX. 353 For though there was no actual contract, yet there was a kind of contract in law, so it is ex quasi contractu.] 1727–41 [see Quasi *adv.* and *pref.* 2 a]. 1806 W. D. Evans tr. *Pothier's Treat. Law of Obligations, or Contracts* I. i. i. 69 A Quasi contract is the act of a person permitted by the law which obliges him in favour of another, without any agreement intervening between them. 1847 *Louisiana Supreme Court Ann. Rep. 1846* I. 380 The act of a party in taking as security for a loan of money made to an agent for his private use, a pledge of claim against a third person, known by the lender to belong to the principal of that agent, forms a quasi-contract. 1893 W. A. Keener *Treat. Law of Quasi-Contracts* i. 16 A statutory obligation which does not rest upon the consent of the parties, is clearly quasi-contractual in its nature. 1936 R. M. Jackson *Hist. Quasi-Contract in Eng. Law* II. 127 The name 'quasi-contract' has become the general jurisprudential term for obligations which do not fall into the categories of contract or tort. 1959 Jowitt *Dict. Eng. Law* II. 1456/2 *Quasi-contract*, liability, not exclusively referable to any other head of law, imposed upon a particular person to pay money to another particular person on the ground that non-payment of it would confer on the former an unjust benefit. *Ibid.*, The main heads of quasi-contractual liability are: (1) money paid by the plaintiff at the request of the defendant, including payment by sureties and contribution between joint contractors; [etc.]. 1962 A. Turner *Law of Trade Secrets* iv. iv. 348 We may consider relief under quasi-contract and relief for unjust enrichment to be for practical purposes the same.

qua·sifi:ssion. *Nuclear Physics.* [f. Quasi *adv.* and *pref.* + Fission.] A type of interaction between heavy ions in which the reaction products have kinetic energies similar to those expected for fission of a compound nucleus but an angular distribution that is quite different.

1974 F. Hanappe et al. in *Physical Rev. Lett.* XXXII. 738/1 A new kind of reaction was observed, which seems to occur instead of complete fusion; therefore we called it 'incomplete fusion', but the name 'quasifission' is perhaps better. 1976 *McGraw-Hill Yearbk. Sci. & Technol.* 297/1 It is now believed that for these quasifission heavy-ion projectile reactions, nuclear encounters which achieve an initial distance of approach or sufficient overlap..lead to the situation where the projectile's kinetic energy is essentially dissipated, and the nuclei remain in contact for a brief period of time, exchanging large numbers of nucleons, and then begin to separate.

‖ **quasi in rem** (kwē[1]·səi in rem), *adv. phr.* *U.S. Law.* [L., 'as if against a thing'.] Used of proceedings in a court with limited jurisdiction, having no power over a defendant's person (see quots.). Also *attrib.*

1887 *U.S. Reports* CXIX. 187 There is, however, a large class of cases which are not strictly actions *in rem*, but are frequently spoken of as actions *quasi in rem*, because, though brought against persons, they only seek to subject certain property of those persons to the discharge of the claims asserted. 1904 *Atlantic Reporter* LVII. 555/2 In suits quasi in rem, that is, where the suit is against the person in respect of the res—where, for example, it has for its object partition or the sale or other disposition of defendant's property within the jurisdiction to satisfy plaintiff's demand by enforcing a lien upon it—personal service within the jurisdiction or appearance is not necessary. 1910 H. C. Black *Law Dict.* (ed. 2) 608/1 *Quasi in rem.* A term applied to proceedings which are not strictly and purely *in rem*, but are brought against the defendant personally, though the real object is to deal with particular property or subject property to the discharge of claims asserted. 1963 C. A. Wright *Handbk. Law of Federal Courts* x. 236/2 State statutes which commonly permit so-called 'quasi-in-rem' actions, which are commenced by attachment or garnishment of property within the state, with notice to the person outside the state against whom the claim is asserted. 1970 *Internat. & Compar. Law Q.* XIX. 181 One finds an interesting description..[of] the American *quasi in rem* jurisdiction. 1972 *N.Y. Law Jrnl.* 22 Aug. 12/5 The court is mindful of the quasi in rem judgment granting replevin to plaintiff by a Florida court of competent jurisdiction.

Quasimodo (kwā:zimō[u]·do). Also (*erron.*) Quasimoto, and with small initial. [f. the name of the hunchback in Victor Hugo's novel *Notre-Dame de Paris*.] A surfing feat performed in a crouching position (see quots. 1962 and 1963).

1962 *Austral. Women's Weekly* 24 Oct. (Suppl.) 3/3 *Quasimodo*, trick riding, with body bent nearly double, with one hand stretched out in front and the other behind. 1963 *Surfing Yearbk.* 43/1 *Quasimoto*, riding on the nose of the board in a crouched position, with one arm forward and one arm back. 1965 J. Pollard *Surfrider* ii. 18 If you are a 'hot dogger', you're an expert, you can perform quick turns, head dips, quasimodos. 1970 *Studies in English* (Univ. Cape Town) I. 31 Another form of crouch is the *Quasimodo*.

qua·sipa:rticle. *Physics.* [f. Quasi *adv.* and *pref.* + Particle *sb.*] An excitation of a many-body system that has some of the properties of a free particle, such as momentum and position.

1957 R. T. Beyer tr. L. D. Landau in *Soviet Physics: JETP* III. 921/1 The role of the gas particles in this classification is assumed by the 'elementary excitations' (quasi-particles), each of which possesses a definite momentum. 1968 C. G. Kuper *Introd. Theory Superconductivity* xii. 193 As the idea of quasiparticles has developed, it has come to permeate the whole of solid-state theory. Thus, for example, the Debye theory of specific heat can be regarded as a quasiparticle theory in which the quasiparticles are phonons. 1968 [see *exciton]. 1974 D. M. Adams *Inorg. Solids* ix. 283 Phonons are the quasiparticles of energy associated with excitation of one of the modes of vibration of the perfect crystal.

qua:si-ste·llar, *a.* *Astr.* [f. Quasi *adv.* and *pref.* + Stellar *a.*] Giving a star-like (i.e. unresolved) image on a photograph but believed not to be a star; chiefly in *quasi-stellar object* (= *quasar), *(radio) source*.

1963 *Sci. Amer.* Dec. 54/1 In recognition of their small size, and for lack of a better name, they are called quasi-stellar radio sources. 1964 *New Scientist* 28 May 532/1 The highly enigmatic 'quasi-stellar objects' whose discovery over the past year or so has been of considerable excitement to astronomers. 1964 *Listener* 20 Aug. 266/1 Radio astronomy has already revealed to us an entirely new universe... But the record for excitement and unexpectedness must surely go to the discovery within the last year of the quasi-stellar radio sources. 1973 *Nature* 23 Nov. 185/1 The physical nature of quasistellar objects is as puzzling now as it was 10 years ago when their large redshifts were discovered. 1973 Smith & Jacobs *Introd. Astron. & Astrophysics* xix. 493 Some QSOs are strong radio sources; indeed, optical identification of such quasi-stellar sources (QSS) first drew attention to the QSOs. The majority of QSOs, however, are weak radio sources or altogether radio-quiet. 1977 J. Narlikar *Struct. Universe* vii. 215 When quasi-stellar objects were first discovered, some with red-shifts close to $z = 2$, it was hoped that they would settle the cosmological issue once for all... Unfortunately QSOs have made the issue more complicated than before. 1978 Pasachoff & Kutner *University Astron.* xxviii. 713 Sources in this new class resembled the quasi-stellar radio sources in being quasi-stellar and in having large redshifts but did not emit any detectable radio radiation. These were called quasi-stellar objects (QSO's).

quassia. Add: **3.** *quassia chips* (examples), *extract*; **quassia cup** (example).

1884 A. J. E. Wilson *Vashti* xxiv. 237, I have been forced to drink out of quassia-cups until my whole being has imbibed the bitter. 1900 G. S. Saunders in E. T. Cook *Century Bk. Gardening* 508 Quassia chips should be soaked..and then boiled. *Ibid.* 509 Quassia extract may be bought... This is a very useful insecticide. 1926 *Army & Navy Stores Catal.* 479/2 Quassia Chips –lb. –/5. 1965 M. Thomas *Grannies' Remedies* 14 A ½ oz. of quassia chips boiled in a pint of water and mixed with 4 oz. of treacle draws flies and kills them. 1969 E. H. Pinto *Treen* 39 Quassia chips were formerly used for washing lousy heads.

quate, var. *Quaite.

quater-centenary. Delete *rare* and add: (kwæ:təɹsentī·nări, kwē[1]:təɹ-). (Later examples.) Also as *adj.*

1904 J. Stalker *John Knox* p. v, In 1905 not only Scotland..will be celebrating..the Quatercentenary of the birth of the greatest of Scotsmen. 1906 *Daily Chron.* 25 Sept. 4/4 That is why the Quatercentenary of the University [of Aberdeen] has created an unparalleled amount of interest in the North. 1955 *Times* 13 July 9/4 The recent quater-centenary celebrations of Queen Mary's School, Walsall. 1971 *Oxford Times* 4 July 28/5 The Jesus College quatercentenary celebrations included a concert of rarities. 1977 *Times* 31 Mar. 14/3 The Jews fled to nearby Cochin... There, in 1567, they built Jewtown and in the following year completed their Synagogue... Its quater-centenary was celebrated..in 1968.

quatercentennial (kwæ:təɹsente·niăl, kwē[1]:təɹ-), *a.* [f. L. *quater* four times + Centennial *a.* (*sb.*).] Pertaining to a four-hundredth anniversary or celebration.

1964 *Listener* 30 Apr. 730/3 Scholars don't regard their function this quater-centennial year as involving..scratching pocky boils.

quaternary, *a.* and *sb.* **A.** *adj.* **1. a.** Add: This sense is now *Obs.* in *Chem.*

b. *Chem.* Of ammonium and phosphonium ions and salts: in which the central atom forms four bonds to organic radicals; also applied to the central atom, and extended to analogous compounds of other elements. Of a carbon atom: bonded to four other carbon atoms.

1871 *Jrnl. Chem. Soc.* XXIV. 570 Hofmann considers that Drechsel and Finkelstein had tertiary and quaternary phosphonium salts under examination, and not a salt of the primary phosphine. 1871 *Chem. News* 9 June 275/1 Quaternary substitution. 1903 A. J. Walker tr. *Holleman's Text-bk. Org. Chem.* I. 46 A carbon atom which is only linked to one other carbon atom is called primary. .. If it is linked..to four, quaternary. 1910 N. V. Sidgwick *Org. Chem. Nitrogen* ii. 22 The quaternary hydroxides are strong bases. 1951 I. L. Finar *Org. Chem.* I. xxx. 622 Pyridine forms quaternary salts when heated with alkyl halides, *e.g.*, pyridine methiodide or *N*-methylpyridinium iodide, $C_5H_5N^+{\cdot}CH_3)I^-$. 1955 J. G. Davis *Dict. Dairying* (ed. 2) 879 Quaternary ammonium germicides have received considerable attention in the past 10 years in relation to their use as sterilising compounds. 1972 Cotton & Wilkinson *Adv. Inorg. Chem.* (ed. 3) xiii. 390 The stibonium compounds are the most difficult to prepare and are the least common. These quaternary salts, excepting the hydroxides,..are white crystalline compounds. 1972 *Materials & Technol.* V. x. 304 Among the more complex quaternary surfactants are some which include two quaternary nitrogen atoms. 1975 Gutsche & Pasto *Fund. Org. Chem.* iii. 61 If it carries no hydrogen but is attached to four carbons it is called a 'quaternary carbon'.

2. Also *absol.*

1910 *Encycl. Brit.* II. 344/2 The beginning of archaeology..may be broadly held to follow on the last of the geological periods, viz., the Quaternary. 1946 L. D. Stamp *Britain's Struct. & Scenery* ii. 16 The Quaternary is not really comparable in duration or importance with the other great eras. 1977 A. Hallam *Understanding Earth* 234 The Pleistocene is a subdivision embracing the period from the beginning of the Quaternary until 10,000 years ago.

3. a. Of or belonging to the fourth order or rank; fourth in a series. Also *spec.* (see quot. 1961).

1874 H. W. Beecher *Plymouth Pulpit* II. 486 The first comprehensive determination breaks itself up into subsidiary determinations, so that the primary will becomes secondary, the secondary becomes tertiary, and the tertiary quaternary. 1924 O. Jespersen *Philos. Gram.* vii. 96 A tertiary word may be further defined by a (quaternary) word, and this again by a (quinary) word. *Ibid.* 97 Quaternary words..may be termed *sub-subjuncts*. 1961 J. Gottman *Megalopolis* xi. 576 One wonders whether a new distinction should not be introduced in all the mass of nonproduction employment: a differentiation between *tertiary* services—transportation, trade in the simpler sense of direct sales, maintenance, and personal services—and a new and distinct *quaternary* family of economic activities—services that involve transactions, analysis, research, or decision-making, and also education and government. Such quaternary types require more intellectual training and responsibility. 1973 *New Society* 15 Nov. 386/3 The 'tertiary' sector contains at least three divisions—'tertiary' proper..; 'quaternary' (information exchange and decision-making); and 'quinary' (research, development and education). 1975 J. B. Goddard *Office Location in Urban & Regional Devel.* ii. 11 The intra-urban location of head office functions and of independent firms in the quaternary sector.

b. *Biochem.* *quaternary structure*, the relative configuration of polypeptide sub-units in a protein molecule, being structure of an order higher than the tertiary.

1958 J. D. Bernal in *Discussions Faraday Soc.* XXV. 14 (*caption*) Hierarchy of polymer complexes:.. (*d*) quaternary structure (homogeneous type), linked groups of tertiary molecules, haemoglobin structure..; (*e*) quaternary structure (heterogeneous type)—linking of different types of ternary protein and primary ribonuclease, tobacco mosaic virus. 1964 G. H. Haggis et al. *Introd. Molecular Biol.* iv. 81 The presence of the nucleic acid chain apparently increases the cohesion between the sub-units in successive turns of the helical quaternary structure. 1977 *Lancet* 26 Nov. 1116/1 The four glycoprotein hormones..possess a common quaternary structure characterised by two dissimilar polypeptide chains called the α and β subunits.

B. *sb.* **2.** *Gram.* In Jespersen's terminology: a word or phrase that belongs to the fourth order or rank.

1937 O. Jespersen *Analytic Syntax* 121 The possibility of having quaternaries, quinaries, and so forth. 1946 —— *Mod. Eng. Gram.* V. i. 3 In other combinations we may have quaternaries or quinaries, e.g. *a not* (5) *particularly* (4) *well* (3) *constructed* (2) *plot* (1).

3. *Chem.* A quaternary ammonium compound.

1947 *Ann. Rev. Microbiol.* I. 173 Jacobs and associates ..studied the relation between structure and bactericidal effects in the hexamethylene tetramine groups. This was followed in 1928 by Hartmann & Kägi's work..with other quaternaries. 1955 J. G. Davis *Dict. Dairying* (ed. 2) 880 The degree to which complexes of this kind will be found depends upon the particular quaternary used. 1968 Kirk & Othmer *Encycl. Chem. Technol.* (ed. 2) XVI. 860 Some quaternaries form hydrates or other solvates.

quaternate, *a.* (Later example.)

1908 *Scott's Autumn List*, *Lady Beauclerc and Socialism* is the title of the last book of the Rev. H. T. Perfect's quaternate work on Lady Beauclerc's life.

quaternion. Add: **1. a.** (Further examples.) *spec.* a set of four poems.

1815 Scott *Guy M.* III. iii. 42 A species of florid elocution, which often became ridiculous from his mis-arranging the triads and quaternions with which he loaded his sentences. 1964 C. S. Lewis *Discarded Image* iv. 68 He accepts the classical quaternion of virtues, Prudence, Temperance, Fortitude and Justice. 1967 J. Hensley *Wks. Anne Bradstreet* p. xxiv, The Quaternions follow the structure of Thomas Dudley's own 'On the Four Parts of the World', now lost. 1976 *Times Lit. Suppl.* 27 Aug. 1049/1 The formal elegies and quaternions she [sc. Anne Bradstreet] laboured over in imitation of Du Bartas.

4. c. quaternion group, the group which is generated by multiplication of the unit quaternions, *i*, *j*, and *k*.

1911 W. Burnside *Theory of Groups of Finite Order* (ed. 2) viii. 132 The group defined by these relations is known as the quaternion-group. 1949 H. Zassenhaus *Theory of Groups* iv. 116 We wish to find non-abelian groups of order p^n which contain only one subgroup of order p. An example is the quaternion group. By the theorem of Hölder it is defined by the relations $A^4 = 1$, $BAB^{-1} = A^{-1}$, $B^2 = A^2$. 1972 F. J. Budden *Fascination of Groups* xv. 245 The simplest group in this class is Q_4 of order 8 ($n = 4$), and this is usually known as the quaternion group, though in fact all the dicyclic groups may be realised as groups of quaternions.

quaternize (kwǫ·tə̆naiz), *v.* *Chem.* [f. Quatern(ary *a.* + -ize.] **a.** *trans.* To convert (a tertiary compound, esp. an amine, or an atom) into a quaternary form. **b.** *intr.* To undergo this process.

1951 *Jrnl. Polymer Sci.* VI. 513 This observation suggests that some of the reluctance of 2-pyridyl ethers to quaternize may also be due to steric effects. *Ibid.*, We may consider the butyl salt to be a chain polymer with approximately every other pyridine quaternized. 1965 *Jrnl. Chem. Soc.* 6831 A specimen obtained on quaternising (−)-6-methoxytropan-3α-ol. 1969 *Encycl. Polymer Sci. & Technol.* XI. 333 By copolymerizing vinylpyridine with styrene, quaternizing the nitrogen atoms in the pyridine rings of the resulting copolymer, and..reacting with Li^+TCNQ^- and TCNQ a product is obtained that can be cast from solution into a film with a conductivity of 10^{-2} ohm^{-1}-cm^{-1}. 1970 J. Williamson in H. W. Mulligan *Afr. Trypanosomiases* vii. 167 The dimethylated diamino dye, Acridinium Yellow, in which the ring N atom was quaternized. 1973 *Nature* 27 Apr. 605/2 A similar reaction of nicotinamides quaternized with alkyl groups was also..investigated.

So **qua·ternized** *ppl. a.* Also **quaterniza·tion**, the process of quaternizing a compound or atom.

1949 *Jrnl. Polymer Sci.* IV. 103 An excess of butyl bromide over pyridine content was always used, so that the irreversible quaternization reaction would prevail. 1951 *Ibid.* VI. 523 Quaternized copolymers of styrene and 4-vinylpyridine. 1965 Phillips & Williams *Inorg. Chem.* I. xv. 556 The quaternization reaction $EtI + Et_3N \rightarrow [Et_4N]^+I^-$ is, at 100°, 80 times faster in benzene..than in hexane. 1966 *Trans. Amer. Soc. Artific. Internal Organs* XII. 151 Simple contact of such a quaternized surface with aqueous sodium heparinate is sufficient to yield a heparinized surface. 1976 *Nature* 15 July 221/2 Quaternisation of the nitrogen atom generally caused a marked decrease in ixodicidal activity.

‖ **Quatorze Juillet** (katǫrz ʒwiyẹ). Also **quatorze juillet** and *ellipt.* **quatorze.** [Fr., lit. fourteenth of July.] In France, the anniversary of the fall of the Bastille (see Bastille *sb.* 3) on 14 July 1789, observed as a national holiday. Also *attrib.*

1934 Webster, Quatorze juillet or (le) quatorze. 1951 R. Senhouse tr. *Colette's Last of Chéri* 196 One day, not long before the Quatorze Juillet, Charlotte Peloux was lunching with them. 1955 *Times* 18 July 6/1 *Le Quatorze Juillet*—never referred to by French people under any other name, although its official title is *fête nationale* and its most common English name, more explicitly, 'Bastille Day'—has come and gone, as in other years, in a flurry of military parades, undisciplined queues for free matinées at State theatres, firework displays, and ubiquitous *bals populaires* in the streets. 1966 H. Yoxall *Fashion of Life* xxiv. 233 A firework display..far more lavish than anything I'd seen in Paris on the *Quatorze Juillet*. 1971 *Guardian* 15 July 13/4 The 'quatorze' is..the type of public jamboree which carries no obligations. 1977 E. Ambler *Send no more Roses* viii. 175 It's the Quatorze today. The servants want to..go to the local fête. *Ibid.* x. 214 We had eaten simply..so that the servants could get off early to their local Quatorze juillet fête.

‖ **quatre-couleur** (katr,kulǿr), *a.* [Fr., f. *quatre* four + *couleur* colour.] Of objets d'art: made of or decorated with carved gold of several (esp. four) different colours. Hence **quatrecouleurs** *sb. pl.*

1959 *Times* 19 Dec. 9/4 The one [thimble] on the left has fruit and flowers in *quatre-couleur* work round the band. 1960 H. Hayward *Antique Coll.* 231/2 *Quatre-couleurs*, the art of combining various colours of carved gold in a decorative scheme. The colour of gold is determined by the nature of the alloy. 1967 *Times* 7 Mar. 21/4 (Advt.), A quatre-couleur gold box..by Barrière, 1768. 1975 *Catal. Important Gold Boxes & Objects of Vertu* (Sotheby, Monaco) 66 A rectangular quatre-couleur gold and enamel snuff-box. 1978 *Times* 5 Aug. 9/1 Snuff-boxes ..in imitation of gold boxes with..'quatre-couleurs' decoration.

quatro (kwa·tro). *W. Indies.* Also **cuatro.** [ad. Sp. *cuatro*, lit. four.] A small four-stringed guitar, of a kind originating in Latin America.

1955 *Caribbean Q.* IV. ii. 101 The band includes a home-made banjo, a cuatro (small 4-stringed Spanish guitar)..and shac-shacs. 1958 E. Borneman in P. Gammond *Decca Bk. Jazz* xxi. 261 The instruments used were..guitar, quatro, conga drum, bongos, maracas, cencerro. 1965 'Lauchmonen' *Old Thom's Harvest* v. 59 His guitar-pickney quatro hung over his shoulder and across his back. 1968 E. Lovelace *Schoolmaster* i. 8 There is the orchestra..with the fiddle, and the quatros and flute and tambourines. 1974 *Sunday Advocate-News* (Barbados) 3 Feb. 13/7 The programme is dedicated to the composers of the early tent brigade; men who made music with 'Cuatro, bottle and spoon'. 1975 S. Marcuse *Dict. Mus. Instruments* 135/2 *Cuatro*.., 1. guitar of Puerto Rico, with 5 courses of strings, 4 pairs and a single chanterelle, played with a plectrum; 2. a small guitar of Venezuela, with 4 strings.

‖ **quatsch** (kvatʃ). Also **quatch.** [Ger.] Nonsense, rubbish. Freq. as *int.*

1907 M. A. von Arnim *Fräulein Schmidt* liv. 218 '*Quatsch*,' said Onkel Heinrich, with sudden and explosive bitterness. *Ibid.* 219 *Quatsch* is German for silly, or nonsense, and..is more rude than either. 1915 Wyndham Lewis *Lett.* (1963) 72, I never did nor ever shall, as you probably divine, despite 'quatch' about malevolence. 1939 Joyce *Finnegans Wake* (1964) 520 Quatsch! What hill ar yu fluking about, ye lamelookond fyats! 1947 D. M. Davin *Gorse blooms Pale* 165, I notice he commits himself to nothing about Zionism. A lot of quatsch and schmaltz, if you ask me. 1962 K. O'Hara *Double Cross Purposes* iii. 31 'I'm rusty in my law after two years..' '*Quatsch*. You're worse than rusty.' 1976 *Times Lit. Suppl.* 4 June 671/2 Fantasy masquerading as history, pure *quatsch* dressed in purple kitsch. 1979 N. Freeling *Widow* vii. 42 'Oh Quatsch,' she said. 'I..know how to look after myself.'

quattie (kwǫ·ti). *W. Indies.* [Corruption of Quarter *sb.*] A penny halfpenny; money or a coin of the value of 1½d.

1859 Trollope *West Indies & Spanish Main* ii. 20 'And now de two quatties,' he said. I knew nothing of quatties then, but I gave him the sixpence. 1873 C. J. G. Rampini *Lett. from Jamaica* ix. 94 'Quattie', a penny-half-penny—the 'quarter' of sixpence. 1893 R. Bithell *Counting-House Dict.* (rev. ed.) 254 *Quattie*, a small silver coin used in the West Indies, worth about 1½d. English. 1961 F. G. Cassidy *Jamaica Talk* ix. 209 The *tup*, of course, is an abbreviation of *twopence*, but nowadays means the same as *quattie*: 1½d. 1971 *Jamaican Weekly*

Gleaner 3 Nov. 25/2 Pound' wort' a fret nebber pay quattie wort' a debt. **1975** *New Yorker* 12 May 37/1 He put every penny, every quattie of what we had into a small herd and a prize black bull.

quattroce·ntism. [f. Quattrocent(o + -ism.] The fifteenth-century style in Italian art.
1905 W. H. Hunt *Pre-Raphaelitism* II. xiii. 367 It was pointed out to them that our pictures had never attempted quattrocentism.

quattrocento. Add: Also *attrib.* or as *adj.*
1921 A. Huxley *Let.* 31 May (1969) 197 For my taste, at least, Florence is too tre- and quattrocento. **1955** *Times* 20 May 3/7 The settings [in a film] were purely *quattrocento*, very scholarly, and very pretty. **1965** F. Raphael *Darling* xxvi. 131 The pictures started with quattrocento pieces. **1977** *N.Y. Rev. Bks.* 24 Nov. 36/4 The charming *cassone* fronts of a minor painter called Apollonio di Giovannie, which look to us like decorative charades in *quattrocento* costume. **1979** *Jrnl. R. Soc. Arts* CXXVII. 627/2 We know virtually everything we could hope to know about the decoration of a quattrocento chapel.

quaver, *sb.* Add: **1.** Also *Comb.*
1728 Chambers *Cycl.* s.v. *Rest*, The Quaver-Rest of common time.

quaver, *v.* **1. a.** Delete 'Now *rare*' and add: Also, with *adv.*, to go with a tremulous or quivering movement.
1943 A. Ransome *Picts & Martyrs* xv. 144 The three-cornered white flag..quavered up to the masthead. **1953** C. Mackenzie *Passionate Elopement* xxx. 270 Old Tabrum would quaver in from time to time to survey the comfort of his guests, regaling them with some particularly choice floral anecdote.
3. b. *trans.* To utter with a quaver or in a quavering tone.
1872 A. C. Steele *Broken Toys* I. vii. 102 'Oh, yes, I was the upper-housemaid,' the old woman quavered. **1897** W. W. Jacobs *Skipper's Wooing* iii. 36 'I'd rather you stayed,' he quavered. 'I would indeed.' **1912** *Red Mag.* 1 Mar. 513/2 'Gus!' she quavers. 'Oh, Gus!' **1947** *Punch* 5 Mar. 206/2 'Thank heavens you're here!' he quavered as I went up to him. **1972** 'J. Herriot' *It shouldn't happen to Vet* xv. 105 'Have you got a drop o' whisky handy, Jim?' he quavered.

quaverous (kwē[1]·vərəs), *a.* [f. Quaver *v.* + -ous.] Tremulous, quavering.
1918 J. F. Bridge *Westm. Pilgrim* xi. 146, I can still see two of these old gentlemen..with hardly a quaverous note to mark their years, valiantly voicing 'I saw lovely Phyllis'.

quavery, *a.* Delete *rare* and add later examples.
1965 House & Storey *Lett. C. Dickens* I. 557 A quavery line is drawn under 'Ill! *He* ill!', and another under 'Any time while the—'. The writing of the last three lines gets progressively more shaky. **1975** *Daily Tel.* 10 Nov. 2/6 A woman's voice in the crowd soared above the others, quavery but sweet.

quawk (kwǭk), *sb.* *U.S.* Also **quark, quauk, quock.** [Imitative; cf. Quawk *v.* and Squawk *sb.*] **1.** The black-crowned night-heron, *Nycticorax nycticorax*, which is widely distributed in temperate and tropical regions; = Qua-bird.
1844 J. E. DeKay *Zool. N.Y.* II. 227 The Black-crowned Night Heron, or Quawk,..derives its popular name from the deep guttural cry. **1867** *Amer. Naturalist* I. 344 Many..were all agog to cover themselves with glory by shooting a quawk. **1877** W. Whitman *Specimen Days* (1882–3) 100, I find [in New Jersey].. Cheewinks, Quawks, Ground robins. **1895** F. M. Chapman *Handbk. Birds Eastern N. Amer.* 136 Black-crowned Night Heron; Quawk. **1926** A. C. Bent *Life Hist. N. Amer. Marsh Birds* 197 The familiar night heron or 'quawk' is one of the best known and most widely distributed of our herons. **1962** R. S. Palmer *Handbk. N. Amer. Birds* I. 475 Characteristic *quock* of this heron has been widely used as a vernacular name: 'squawk', 'quock', etc. **1968** *Times* 28 Feb. 11/7 During a recent expedition to the Falkland Islands I was lucky enough to see and hear..the Quark. They are birds of great beauty, heron-like in appearance.
2. The cry of a duck or night-heron; = Quack *sb.*[2]
1863 'G. Hamilton' *Gala-Days* 73 For the heavy booming of cannon rose the 'quauk!' of ducks. **1895** F. M. Chapman *Handbk. Birds Eastern N. Amer.* 137 Occasionally they [sc. black-crowned night herons] utter a loud, hoarse *quawk*, the origin of their common name. **1962** R. S. Palmer *Handbk. N. Amer. Birds* I. 476 The Black-crown's note can be expressed as *quock*.

quay, *sb.* Add: **b.** *quay-rail*, *-side* (examples), *-stone* (later example); **quay crane** = **wharf crane*; **quay-punt** (in full, **Falmouth quay-punt**), a small fore-and-aft-rigged half-decked two-masted sailing boat, orig. used on the river Fal for transporting stores between ship and shore.
1969 *Jane's Freight Containers 1968–69* 286/3 Stevedoring companies who already have modern mechanised equipment at their disposal (quay-cranes, pontoon-cranes, trucks and elevators). **1977** *Hongkong Standard* 12 Apr. (Business Suppl.) 4/5 The group has a total of seven quay cranes. **1886** D. Kemp *Man. Yacht & Boat Sailing* (ed. 5) 341 Table of offsets (Falmouth quay punt). **1925** *Yachting Monthly* XXXIX. 39/2 A quay punt before the war cost about £120 to build. **1971** *Country Life* 20 May 1224/1 They were the bum boats of the western world and ranged from such rugged deep-keel craft as the Quay Punts of Falmouth to the graceful Deal galleys. **1936** Dylan Thomas in *Contemp. Poetry & Prose* 53 Let the first Peter from a rainbow's quayrail Ask the tall fish. **1903** *Westm. Gaz.* 31 Dec. 5/3 He saw another man climbing up the quayside ladder. **1928** *Daily Tel.* 7 Feb. 14/1 The foundations..rested in the rock found 70 ft. under the two quaysides. **1974** *Times* 12 Nov. 3/1 Quayside fish merchants at Hull. **1979** *Jrnl. R. Soc. Arts* CXXVII. 663/1 Local fishermen..used to sell their hake, cod and herring on the quayside. **1938** Dylan Thomas in *20th Cent. Verse* Jan./Feb. 3, I make this in a warring absence when Each ancient, stone-necked minute of love's season Harbours my anchored tongue, slips the quaystone.

quean. Add: **1.** (Later examples.) Now *arch.*
1924 E. Sitwell *Sleeping Beauty* xiv. 47 My eyes are dim,—I yet can see You, lazy quean! Go work! **1969** *Listener* 10 Apr. 503/3 Nora (an old quean who thinks she's an old queen).
3. *slang.* A male homosexual of effeminate appearance. Cf. *queen *sb.* 11*.
1935 D. Lamson *We who are about to Die* xv. 294 We did hear startling tales..of 'family' life, of marriage ceremonies, of fights with knives for the favor of some 'quean', as the perverts are called in prison. **1937** Partridge *Dict. Slang* 676/1 *Quean*; incorrectly *queen*, a homosexual, esp. one with girlish manners and carriage; low: late C. 19–20; ob. except in Australia. **1968** J. R. Ackerley *My Father & Myself* xii. 127, I did not want him to think me 'queer' and himself a part of homosexuality, a term I disliked since it included prostitutes, pansies, pouffs and queans.

queasily, *adv.* (Later examples.)
1973 *Times Lit. Suppl.* 2 Feb. 118/2 We are growing somewhat queasily disillusioned with the achievements of our post-Renaissance civilization. **1976** *New Society* 4 Mar. 502/2 *Super Natural Cookery*, a typically camp title, written in the queasily flirtatious style which seems to afflict all Aquarian cookery books.

queasy, *a.* Add: (Later examples of spelling queazy.)
1906 C. M. Doughty *Dawn in Brit.* VI. 64 Proud queazy, faltering-kneed, Men, fat of the lean people's gifts. **1912** Galsworthy *Inn of Tranquility* 187, I would think, Sirs, that you should rather blame the queazy state of Pranza's stomach.
6. queasy-stomached *a.* (later example).
1802 W. Gifford tr. *Juvenal's Satires* vi. 292 Why waste the wine and cakes The queasy-stomach'd guest, at parting, takes?

Quebec (kwĕbe·k, kĕ-). The name of the city and province in eastern Canada, used *attrib.* in *Quebec heater*, a kind of solid-fuel, domestic heating stove with a tall, cylindrical fire-box.
1918–19 *T. Eaton & Co. Catal.* Fall & Winter 461 The new Quebec heaters. Built on the popular 'Quebec Idea' pattern..burn coal, coke and any kind of wood fuel. **1927** *Toronto Daily Star* 8 Oct. 2 Dominion Circulation 'Quebec' Heater..gets the heat into every corner of the house. **1948** H. MacLennan *Precipice* (1949) i. 20 In the winters there had been a Quebec heater in the pavilion. **1968** E. A. McCourt *Saskatchewan* iii. 42 The long glorious winter evenings we spent huddled behind the old Quebec heater.

Quebecker (kwĕbe·kəɹ, kĕ-). Also **Quebecer.** [f. prec. + -er[1].] = *Québecois *sb.* a.
1836 *Bytown* (Ottawa) *Independent* 24 Feb. 3/2 We think the Quebecers are not such flats as to pay for measuring timber, merely because they receive it. **1837** *Montreal Transcript* 14 Jan. 2/2 The match..was played on Tuesday last, at Three Rivers, when the Quebeckers proved victorious. **1890** [see *new *a.* 5 d]. **1946** H. Croome *Faithless Mirror* xiv. 153 Other Canadians go overseas..the Quebeckers stay at home. **1957** *Listener* 17 Oct. 593/1 The typical Quebecer is still a Royalist, a Conservative, even if he votes Liberal. **1963** *Times* 25 Feb. (Canada Suppl.) p. vii/3 We've got to keep control of new industries in Quebec for Quebecers. **1973** *Times Lit. Suppl.* 26 Oct. 1295/2 The backward, Church-controlled, inferiority-complex-ridden Quebecker of yesteryear. **1977** *Time* 7 Feb. 21/1 The man who wants to lead 6 million Quebeckers out of Canadian confederation. **1979** *Amer. N. & Q.* Mar. 119/2 Radwanski believes that Trudeau is the man best equipped to meet Lévesque's challenge. Most Quebecers appear to think that, too.

‖ **Québecois** (kebękwa), *sb.* and *a.* Also **Québecquois.** [Fr., f. *Quebec.] **A.** *sb.* **a.** A native or inhabitant of the city or province of Quebec, esp. one who is a French-Canadian. **b.** The French spoken in Quebec. **B.** *adj.* Of or pertaining to Quebec or its inhabitants.
1873 J. M. LeMoine *Maple Leaves* 171 The County and town of Joliette preserve the name of another distinguished Canadian, a Québecquois, Louis Joliette. **1954** T. H. Raddall *Muster of Arms* 135 She was speaking in English but the mind behind the tongue was Quebecois. **1960** *Times* 21 Nov. (Canada Suppl.) p. ii/7 Mr. Dorion is the most typically Quebecois of the French-Canadian ministers. **1962** *Maclean's Mag.* 17 Nov. 4/1 Editors of French language newspapers who had wailed about the 'stupidity' of rural Québecois in the last federal election. **1963** *Ibid.* 6 Apr. 23/1 Her French sounds more Parisian than what we who don't speak French well call Quebecois. **1970** *Globe Mag.* (Toronto) 26 Sept. 17/1 The manuscripts of Pauline Archange is a first-person story of a Quebecois slum child. **1974** *Times Lit. Suppl.* 14 June 634/2 Her little book assembles..comments by prominent Québécois. *Ibid.* 634/3 M Étheir-Blais is a Franco-Ontarian, not a Québécois. **1977** H. Giles *Lang., Ethnicity & Intergroup Relations* 2 Québécois air controllers engaged in industrial action for a refusal at their not being allowed to use French as a medium of air communications. **1977** *Listener* 23 June 808/2 The beleaguered English-speaking Québécois. **1978** *Maledicta* II. 61 For the most part, Québecois imprecations flourished on the spot in response to the needs of a people living in conditions of moral and material misery.

quebracho. Add: Also **quiebrahacha.** (Further examples.) Also, the timber of these trees.
1889 G. S. Boulger *Uses of Plants* v. 159 Red Quebracho..is a hard wood from the River Plate... Other Quebrachos..are also used in South America. **190[illegible]** W. R. Fisher *Schlich's Man. Forestry* (ed. 2) V. i. vii. 572 Experimental use is also made of pitch pine and quebracho wood [for railway sleepers]. **1913** *Chambers's Jrnl.* 28 June 480/2 Quebracho ('break-axe'), one of the hardest, heaviest, and most durable woods known. **192[illegible]** *Nature* 8 Sept. 376/2 The principal sources of this [tannin] are oak galls and bark, mangrove bark, Myrobalans, Quebracho, and *Acacia decurrens*. **1924** Record & Mell *Timbers Trop. Amer.* ii. 508 White quebracho is the name most commonly applied to *Aspidosperma quebracho-blanco*..of Argentina, Paraguay, and southwestern Brazil. **1933** H. Allen *Anthony Adverse* v. xxi. 409 It was full of a thousand lush and exotic odours from..juniper and lantana; the fragrant quiebrahacha, tamarind and rotting mastic leaves. **1955** *Times* 30 June 18/2 The production of Quebracho extract during the year amounted to 81,644 tons. **1969** T. C. Thorstensen *Pract. Leather Technol.* ix. 140 Quebracho extract is used in the manufacture of heavy leathers and in the vegetable re-tanning of chrome tanned upper leathers. **1974** [illegible] Meiring *President Plan* ix. 74 A big fire of *quebracho* glowed and flickered.

‖ **quebrada** (kebrā·dā). [Sp., fem. of *quebrado*, pa. pple. of *quebrar* to break.] **1.** A mountain stream in S. America.
1833 *Blackw. Mag.* Aug. 144/2 Next morning I rode out on my mule, to take my last dip in the *Quebrada* of the *Loseria*, which was a rapid in a beautiful little rivulet distant from Panama about three miles. **1920** *Chambers's Jrnl.* Dec. 979 The many quebradas which flow down the mountain ravines. **1973** K. Benton *Craig & Jaguar* v. 45 The Chasco [river] was fed by the *quebradas* that brought down meltwater from the high ranges.
2. A ravine in S. America.
1845 *Encycl. Metrop.* XIV. 565/1 Abrupt precipices..occur in every part of the parent chain of the Andes near the equator, and diversify its appearance with the most horrid chasms, or rents, here called Quebradas, varying from 100 feet to 4 or 5,000 feet in depth. **1860** Mayne Reid *Odd People* 456 The stupendous ravines termed 'barrancas' & 'quebradas', which intersect the Cordilleras of the Andes in other parts of South America. **1927** *Blackw. Mag.* Aug. 229/2 We will build another house in the quebrada at Palalle Grande. **1949** *Américas* Sept. 17 The men wore rags and undoubtedly lived in one-room shanties in the quebradas. **1974** *Encycl. Brit. Macropædia* I. 1135/1 To the west are the Salto-Jujeña, cut by canyons called *quebradas* through which run small rivers.

Quechua (ke·tʃwā), *sb.* and *a.* Also **Quichua, Kechua,** and other varr. [Sp., ad. Quechua *k'echua, k'eshua*, plunderer, despoiler.] **A.** *sb.* **a.** An Indian people of Peru and neighbouring parts of Bolivia, Chile, Colombia, and Ecuador; a member of this people. **b.** The language (actually a group of related languages) spoken by this people. **B.** *adj.* Of or pertaining to this people or their language.
1840 *Penny Cycl.* XVIII. 6/2 They [sc. the Peruvian Indians] speak the Quichua language, which is generally called the language of the Incas, and which is used by all the natives of South America, from Quito near the equator, to Tucuman in La Plata, 27° S. lat. *Ibid.* 9/1 As the aborigines who inhabit this extensive country speak one language, the Quichua, it must be supposed that they belong to one race. **1843** J. C. Prichard *Nat. Hist. Man* xlv. 431 Among the Peruvian nations, the dominant race were the Quichuas, or Incas, distinguished by their language, which is the Quichuan. **1860** [see *Aymara]. **1877** [see *Chibcha]. **1891** D. G. Brinton *Amer. Race* ii. i. 203 The Kechua..was spoken by an unbroken chain of tribes for nearly two thousand miles from north to south. **1908** [see *Cholo, cholo]. **1921** [see *Aymara]. **1927** *South Amer.* Sept. 71/2 We take with us a native helper who can speak Quechua. **1937** R. H. Lowie *Hist. Ethnol. Theory* xiv. 258 The absence of hoes from all of South America would be of real interest were it not that their undoubted use among the Quichua, past and present, eliminates the problems otherwise raised. **1948** A. L. Kroeber *Anthropol.* (ed. 2) x. 431 The Aztec, Otomi, Maya, Quechua, or Aymará Indians were economically oppressed by their conquest. **1950** J. A. Mason in J. H. Steward *Handbk. S. Amer. Indians* VI. 197 Quechua (Kechua, Quichua, Keshwa, etc.) is the South American analogue of Aztec. *Ibid.*, Today probably several millions of Indians in Perú, southwestern Ecuador, western Bolivia, and northwestern Argentina speak Quechua, and most of them nothing else. **1952** *Language* XXVIII. 366 (*title*) Semantic components in Kechua person morphemes. **1966** *Chambers's Encycl.* VIII. 366/2 Quechumaran comprises two branches..the Quechuan branch, spoken from Colombia to Chile, has as its main member Quechua, of Peru, Ecuador and Bolivia. This is by far the most important native language of South America. **1973** [see *pick *v.*[1] 20 f]. **1974** *Times* 6 Dec. 15/7 The Quechua Indian women of Southern Peru. *Ibid.* 15/8 The Quechua hitherto have lived in inconveniently

scattered farmholdings. **1977** C. F. & F. M. Voegelin *Classification & Index of World's Languages* 21 Quechua and Aymara were said to be genetically related as early as Steinthal (1890).

Quechuan (ke·tʃwăn), *a.* [f. prec. + -an.] Also **Quichuan, Kechuan.** = prec. B. Also as *sb.*
1843 [see prec.]. **1853** F. L. Hawks tr. *Rivero & Tschudi's Peruvian Antiquities* v. 93 We here present a short review..of the character of the Quichuan or Peruvian tongue. **1862** R. G. Latham *Compar. Philol.* 482 The capital Cuzco, Quichuan as it is in many respects, is a town upon Aymara ground. **1901** A. H. Keane *Central & S. Amer.* 207 The Incas were originally a tribe of Quichuan stock. **1931** W. A. Read *Louisiana-French* 142 A French adaptation of Spanish *ñapa*, which is taken in turn from Kechuan *yapa*, 'a present made to a customer'. **1933** L. Bloomfield *Language* iv. 73 *Kechuan*, the language of the Inca civilization. **1966** [see prec.]. **1972** W. B. Lockwood *Panorama Indo-Europ. Lang.* iii. 35 A large majority of Aymarans and Quechuans are monoglots. **1973** K. Benton *Craig & Jaguar* xii. 190 The men, under Huaman's directions, spiced with obscene Quechuan oaths, lowered it.

Queckenstedt (kwe·kĕnstet). *Med.* The name of Hans Heinrich *Queckenstedt* (1876–1918), German physician, used *attrib.* and in the possessive to denote the procedure proposed by him (*Deutsch. Zeitschr. f. Nervenheilk.* (1916) LV. 326) of compressing a patient's jugular veins while observing the pressure of the cerebrospinal fluid (the absence of a marked increase in this indicating obstruction of the subarachnoid space).
1928 R. J. E. Scott *Gould's Med. Dict.* (ed. 2) 1175/2 *Queckenstedt's test*, in healthy persons pressure on the veins of the neck causes a prompt rise in the pressure of the cerebrospinal fluid, which disappears on release of the pressure on the veins. **1934** D. Munro in *Practitioners Library of Med. & Surg.* V. xi. 803 If..the Queckenstedt test is positive, that is, if flow of cerebrospinal fluid is partly or completely interfered with by a block in the subarachnoid space, the rise on jugular compression will be totally absent if the block is complete, or greatly slowed and diminished in height if the block is incomplete. **1968** Passmore & Robson *Compan. Med. Stud.* I. xxiv. 76/2 In performing Queckenstedt's test, it is as well not to squeeze or even seem to squeeze the trachea. **1970** R. G. Feldman in Keefer & Wilkins *Medicine* xliii. 957 Compression of the jugular vein (Queckenstedt maneuver) has been used to observe the response of intracranial pressure to a change in venous pressure.

queen, *sb.* Add: **2. c. Queen Mum,** colloq. alteration of Queen-mother, used with affectionate reference to Queen Elizabeth, the Queen Mother (b. 1900).
1960 L. R. Banks *L-Shaped Room* ix. 135, I kept it a treat. I could've had the Queen Mum to tea there and not been ashamed. **1965** J. Potter *Death in Office* x. 101 Mrs Barber..extending a gloved hand from the sitting position: her imitation of the Queen Mum no doubt. **1974** J. Gardner *Corner Men* xiii. 185 What do you think I do all day..? Play canasta with the Queen Mum and help feed the royal corgis? **1980** *Times* 4 Aug. 4/1 The dear old Queen Mum is..'the best loved lady in the land'.
d. *Queen and country*: the objects of allegiance for a patriot whose head of state is a queen.
1706 Farquhar *Recruiting Officer* v. 72, I endeavour by the Example of this worthy Gentleman to serve my Queen and Country at home. **1861** C. M. Yonge *Young Step-Mother* xxix. 443 His son got his death fighting for his queen and his country. **1900** *Times* 2 Apr. 7/1 It was a keen joy..to be allowed to fight for his Queen and country. **1906** W. S. Churchill *Lord Randolph Churchill* (1907) xxi. 792 Night after night he had risen in his place to discharge..his duty—as he would have phrased it—to 'Queen and country'. **1948** F. Thompson *Still glides Stream* ii. 41 Thank God he died for his Queen and country! **1966** J. Gardner *Amber Nine* i. 28 Anyway, got to go. Queen-and-Country as my lovable boss would say. **1977** *Sounds* 9 July 20/6 Step up the beaches and annihilate somebody for Queen and country, get a medal pinned on your chest.
e. *ellipt.* as *the Queen*: the national anthem 'God save the Queen'.
1898 J. D. Brayshaw *Slum Silhouettes* 37 The curtain fell at last, and the band struck up the 'Queen'. **1916** M. Diver *Desmond's Daughter* iv. iv. 341 They're playing 'The Queen'. I *must* be on the spot to say good-bye to people. **1965** 'W. Haggard' *Powder Barrel* ix. 86 The police band..crashed into The Queen in time in a formal way. **1970** *Daily Tel.* 18 Aug. 13/7 Wherever the Prince was present at a function organised by the association three anthems were played—the Queen, 'Land of My Fathers' and 'God Bless the Prince of Wales'.
3. b. *Queen Mary.* (*a*) Used *attrib.* in *Queen Mary hat*, a variety of toque popularized by Queen Mary (1867–1953), wife of King George V, who favoured it because it enabled the public to have a clear view of her face; so *Queen Mary toque*. (*b*) A type of long low-loading road trailer (in allusion to the Cunard passenger liner: see below).
(*a*) **1938** L. Bemelmans *Life Class* ii. iii. 140 She is dressed in a trotteur, and wears a hat with a thousand cloth violets, a Queen Mary toque. **1947** *New Yorker* 22 Nov. 38/1 A Queen Mary hat protected her from the autumnal sunlight. **1950** J. D. MacDonald *Brass Cupcake* (1955) xv. 157 She sold power shovels, sang baritone and wore Queen Mary hats. **1967** L. J. Braun *Cat who ate Danish Modern* ix. 76 There was enough blue blood to float a ship... You never saw so many pince-nez and Queen Mary hats. **1977** G. Markstein *Chance Awakening* x. 25 Golly sat down in the deck chair..with the grey-haired gentleman..and the old ladies with Queen Mary toques.
(*b*) **1943** C. H. Ward-Jackson *It's a Piece of Cake* 50 *Queen Mary*, a type of long, low-loading, articulated vehicle specially designed for the road transportation of airframes. **1949** *Jrnl. R. Aeronaut. Soc.* LIII. 822 The Council wish to acknowledge their appreciation of the 'Queen Mary' which was kindly provided by Hawker Aircraft Ltd. for the transport of some of the aircraft of the Shuttleworth Collection. **1960** *Picture Show Ann. 1961* (Austral.) 151/2 *Queen Mary*, a 60ft. trailer used for moving scenery. **1968** I. Lambot *Queen dies First* v. 34 There's a signal in, from the salvage boys. They'll be sending a Queen Mary on Monday to shift it.
c. *the Queens*: the Cunard passenger liners, 'Queen Mary' and 'Queen Elizabeth'.
1949 P. Duff *Brit. Ships & Shipping* i. 28 The two *Queens* of the North Atlantic did invaluable service in every theatre of war. **1956** H. Grattidge *Captain of Queens* 291 Both *Queens* had the same 118 foot breadth, but at 1,031 feet the *Elizabeth* eclipsed the *Mary*'s length by ten clear feet. **1959** *Daily Tel.* 9 Apr. 1/4 Plans to replace the 'Queens' must be modern and far-reaching. **1968** O. Wynd *Sumatra Seven Zero* vi. 85 The first clang of metal sounded like a mid-Atlantic collision between the two Queens. **1970** W. G. Roberts *Quest for Oil* xi. 116 This may seem slow compared with the 30 knots of the 'Queens' or even with the 20 to 25 knots of the majority of other passenger liners.
4. *Queen of Spain fritillary*, an orange butterfly with black markings, *Argynnis lathonia*, belonging to the family Nymphalidæ and widely distributed in Europe, N. Africa, and parts of Asia.
1775 M. Harris *Eng. Lepidoptera* 3 Fritillaria, Queen of Spain... Orange brown spotted with black. **1866** [see Fritillary 2]. **1906** R. South *Butterflies Brit. Isles* 91 The Queen of Spain Fritillary... This butterfly is not unlike a small example of the Silver-washed Fritillary. **1976** *Country Life* 18 Mar. 680/2 The remaining nine fritillaries include three occasional migrants, the Queen of Spain fritillary, the cardinal and the weaver's fritillary.
5. d. (Further examples.) Also, a woman who has pre-eminence in an unspecified sphere. See also *beauty queen* s.v. *beauty *sb.* III b, etc.
1847 C. Brontë *Jane Eyre* II. i. 14 Most of them.. looked handsome; but Miss Ingram was certainly the queen. **1958** *Spectator* 22 Aug. 247/1 A robust, jolly-looking person, more like a hockey queen than a film star. **1962** E. Lucia *Klondike Kate* 9 Rare instances of chivalry and devotion were exhibited by the miners toward this frontier queen. **1979** C. MacLeod *Family Vault* (1980) xxiv. 213 She decided to become a society queen and married a man who had the cash but not the inclination.
e. *slang.* An attractive woman; a girl-friend, female partner.
1900 *Dialect Notes* II. 53 *Queen*,..an attractive girl. **1914** 'High Jinks, Jr.' *Choice Slang* 17 *Queen*, a pretty girl. 'A Beauty'. **1937** J. T. Farrell *Fellow Countrymen* 181 Wouldn't it be luck if a ritzy queen fell for him! **1944** C. Himes *Black on Black* (1973) 196 My queen 'gan bouncin' out her twelve-dollar dress. **1952** S. Selvon *Brighter Sun* x. 207 Same ting happen wen my old queen was sick. **1955** P. Sillitoe *Cloak without Dagger* xiv. 128 Both gangs used hatchets, swords, and sharpened bicycle chains..and these were conveyed to the scenes of their battles by their 'queens'. **1975** *Globe & Mail* (Toronto) 11 June 3/7 Since some of the members have no respect for the law, they refuse to enter into a legal marriage. They view it as an unnecessary burden and responsibility. Instead, some Rastafarians have many 'Queens'.
6. b. *queen of puddings*, a pudding made of breadcrumbs, milk, and other ingredients, freq. with a layer of meringue on top; *queen-of-the-meadow(s*, (*a*) substitute for def.: the meadow-sweet, *Filipendula ulmaria*, native to Europe and Asia and naturalized in eastern North America; (further examples); (*b*) *U.S.* = *joe-pye weed; *queen-of-the-night*, a variety of night-blooming cereus (see *night *sb.* 14), esp. *Selenicereus grandiflorus*, which is native to the West Indies and bears fragrant white flowers; *queen-of-the-prairie*, substitute for def.: a perennial North American herb, *Filipendula rubra*, found in meadows and prairies and bearing clusters of small pink flowers; (examples); *Queen of the West*: Cincinnati, Ohio (cf. **Queen city* in sense 13). Similarly with an *of* phrase to designate other cities.
1784 *Mem. Amer. Acad.* I. 451 Queen of the Meadows. Blossoms red or purple. In moist pastures. **1835** C. F. Hoffman *Winter in West* I. 130 It is in vain for thriving Pittsburg or flourishing Louisville..to dispute with Cincinnati her title of 'Queen of the West'. **1838** H. Martineau *Retrospect* II. 254, I should prefer Cincinnati as a residence... The 'Queen of the West' is enthroned in a region of wonderful and inexhaustible beauty. **1840** *Knickerbocker* XVI. 157 In this way we glided in our broad-horn past Cincinnati, the 'Queen of the West' as she is now called. **1851** *San Francisco Picayune* 19 Sept. 2/4 Some person, gifted with a sufficient amount of patience, may undertake to compile the history of San Francisco..the Queen of the Pacific. **1852** H. R. Noll *Bot. Class Bk. & Flora Pennsylvania* 100 S[piræa] lobata, Murr. Queen of the Prairie. **1892** *Amer. Folk-Lore* V. 98 *Eupatorium purpureum*, Queen of the meadow. **1898** C. A. Creevey *Flowers of Field, Hill & Swamp* 146 Queen-of-the-prairie... A stately, beautiful plant adorning the meadows and prairies south and west of Pennsylvania. *Ibid.* 484 Meadow-sweet. Queen-of-the-meadows... A slender, reddish-stemmed shrub, 2 to 6 feet high. **1911** Queen-of-the-meadow [see *Indian turnip* s.v. *Indian *a.* 4 b]. **1917** M. Byron *Pudding Book* iii. 72 Queen of Puddings... Soak a pint of breadcrumbs in boiling milk, and the yolks of four eggs well beaten. **1920** Britton & Millspaugh *Bahama Flora* 294 *Selenicereus grandiflorus*... Queen-of-the-Night. Often cultivated. **1949** H. Hornsby *Lonesome Valley* xxii. 291 The wind was working among the alder bushes and the willows and queen of the meadow. **1963** M. Patten *Puddings & Desserts* (recipe no. 389) Queen of puddings. **1968** Peterson & McKenny *Field Guide to Wildflowers N. Amer.* 284 Queen-of-the-prairie... Flowers deep pink. **1971** *Fashion Panorama* (Ceylon) Apr.–June 21 The Queen of the night was in full bloom outside and its heavy and overpowering scent reached her from the garden. **1972** E. Wigginton *Foxfire Bk.* 242 Take one root from a queen-of-the-meadow plant. **1977** H. Fast *Immigrants* i. 30 For almost nine weeks, the shattered city [sc. San Francisco], known not only as the 'Queen of the Pacific' but as the 'queen of larceny' as well, entered into a period of benign brotherhood.
10. a. Substitute for def.: A small scallop, *Chlamys opercularis*, found off several parts of the coast of north-western Europe; = Quin; (later examples).
1901 E. Step *Shell Life* 84 The Quin or Queen..is more nearly circular in shape, thin and smooth. **1928** Russell & Yonge *Seas* iii. 74 Another animal which can move about is the scallop, especially the smaller 'queen'. **1959** A. C. Hardy *Open Sea* II. vi. 143 The smaller and delicious 'queens'..may occasionally be brought in by trawlers..in sufficient quantities to be marketed. **1971** *Country Life* 21 Oct. 1040/1 Last year nearly 5,000 tons of queens..were brought into Scottish ports.
11. (Later examples.)
1934 P. Wade *Siamese Cat* iv. 45 Not only should the queen herself be excellent, but her pedigree must be above suspicion. **1954** D. Hartley *Food in England* 660 You cannot keep a cat on milk only... Nursing queens should be given water to drink and solid food. **1960** *Amer. Speech* XXXV. 300 Cat fanciers use the name queen in speaking of their litter-bearing female cats. **1977** *Proc. R. Soc. Med.* LXX. 3/1 This calcium deficient diet produced..fractures of vertebrae and limb bones in growing kittens and young zoo felids. Calcium deficiency also occurred in lactating queens and their young litters.
11*. A male homosexual, esp. the effeminate partner in a homosexual relationship. *slang.* Cf. *quean 3.
1924 *Truth* (Sydney) 27 Apr. 6 *Queen*, effeminate person. **1929** M. Lief *Hangover* vi. 100 'What's those?' 'You know—all those queens.' **1930** E. Waugh *Vile Bodies* 61 'Now what may *you* want, my Italian queen?' said Lottie as the waiter came in with a tray. **1938** N. Marsh *Artists in Crime* ix. 127 We met the chap that runs the place. One of those die-away queens. **1952** A. Wilson *Hemlock & After* i. v. 88 Anyone would think he was just another routine, harmless old queen. **1962** [see *faggot, fagot 6 b]. **1971** F. Forsyth *Day of Jackal* xx. 333 He must be..how marvellous! A handsome young butch looking for an old queen to take him home. **1977** *New Yorker* 24 Oct. 64/2 There are only a handful of 'queens' at Green Haven at any one time—men with feminine characteristics they do their best to enhance. The queens are usually given women's nicknames.
12. a. *queen-woman.*
1904 W. B. Yeats *Stories of Red Hanrahan* 20, I heard under a ragged hollow wood, A queen-woman dressed out in silver, cry.
13. Queen At, A.T. *Mil. slang* (see quot. 1943); **queen bee,** for 'also *transf.*' read 'also *transf.* and *fig.*' and add earlier and later examples; *spec.* (*Mil.*) an automatically-controlled aeroplane used as a target in firing practice; **queen-cake** (later examples); **queen cat** (later examples); **Queen City** *N. Amer.*, an epithet applied to the chief or pre-eminent city (*of* a region) (cf. sense *6 b); **queen conch** (later examples); **queen-excluder** (examples); **queen-fish,** (*a*) *U.S.* (examples); (*b*) a large Australian marine fish, *Scomberoides sanctipetri*, of the family Carangidæ; **queen olive,** the particularly large fruit of certain varieties of olive; **queen-pin** *colloq.*, a woman who controls the (successful) organization of a specified institution or event; (see King-pin in Dict. and Suppl.); **queen pudding** = **queen's pudding*; **queen scallop** = *queen *sb.* 10 a; **queen-side** *a.* *Chess*, of or pertaining to the side of the board in which both queens start the game; **queen-size** *a.*, of an extra large size, though occas. in a series (as of beds), smaller than **king-size*; also *queen-sized* adj.; **Queen staysail,** a triangular main topmast staysail in a schooner yacht (see quot. 1948); **queen-stitch** (earlier examples); **queen substance,** a pheromone produced by a queen bee and given to the colony's workers to prevent the production of more queens; **queen trigger-fish,** a deep-bodied, blue and yellow marine fish, *Balistes vetula*, found in the Indian and Atlantic

Oceans; **queenwood,** substitute for def.: an Australian evergreen tree, *Daviesia arborea*, of the family Leguminosæ, or its wood; (examples).
1943 Hunt & Pringle *Service Slang* 54 *Queen At*, a Chief Commander of the A[uxiliary] T[erritorial] S[ervice]. **1947** N. Streatfeild *Grass in Piccadilly* 33 That queen A.T. of yours must have been a holy terror. **1807** R. Southey *Lett. from England* II. xxx. 41 Wherever one of the queen bees of fashion alights, a whole swarm follows her. **1935** *Sun* (Baltimore) 18 July 2/6 King George today saw the British fleet repel an attack by robot planes. The feature of the war game was the fight between the new aircraft guns on the battleship and the radio-controlled 'queen bee' flying machines. **1938** *Times* 26 Aug. 11/1 'Queen Bees'—pilotless, wireless controlled target machines—were only available at two of the..camps. **1943** C. H. Ward-Jackson *It's a Piece of Cake* 50 *The Queen Bee*, the Director of the Women's Auxiliary Air Force; or the senior W.A.A.F. officer on a station. **1951** A. Christie *They came to Baghdad* xv. 139, I thought it was just some female who was coming out to boss things. A kind of Queen Bee. **1956** N. Streatfeild *Judith* i. 44 Beatrice became a queen bee in London's civil defence force. **1960** P. Stanton *Village of Stars* 63 She walked into the W.R.A.F. sitting-room... There was a little radio in the corner... The Queen Bee had no doubt been wangling the Comforts Fund. **1973** G. Bromley *Chance to Poison* v. 79 She's a very dominating character..Queen Bee of the Women's Institute—without her it would collapse. **1978** R. V. Jones *Most Secret War* xxxix. 356 We had in fact evolved our own 'Queen Bee' remote controlled aeroplane for use as an anti-aircraft target in the years before the war. **1894** W. B. Yeats *Land of Heart's Desire* 32, I will have queen cakes when you come to me! **1977** *Radio Times* 12–18 Mar. 16/4 They added a domestic touch by selling their own home produce, little queen cakes and jam. **1893** J. Jennings *Domestic or Fancy Cats* iv. 31 At what age should the queen cat breed? **1960** *Amer. Speech* XXXV. 300 Has this name [sc. queen] arisen from the often-observed imperious bearing of queen cats? **1838** B. Drake (*title*) Tales and sketches from the Queen City [= Cincinnati]. **1844** in C. Cist *Cincinnati Misc.* (1845) I. 9/1 [Cincinnati] is now familiarly called the Queen City of the West. **1870** *Colorado Gazetteer* 40 Denver, the principal city and capital of Colorado—the Queen City of the Plains—is the county seat of Arapahoe county. **1879** Whitman *Specimen Days* (1882) 147 So much for my feeling toward the Queen City of the plains and peaks [= Denver]. **1880** *Harper's Mag.* Dec. 70 Local prejudice..and proverbial procrastination..unite to keep 'Chinatown' practically a sealed book to the better-class denizens of the Queen City of the Pacific [= San Francisco]. **1943** *Colorado Mag.* Jan. 15 The Queen City of the Plains [= Denver] started in 1878. **1949** *Bull. Hist. & Philos. Soc. Ohio* Apr. 99 That enthusiastic booster for the 'Queen City' [= Cincinnati], Dr. Daniel Drake. **1979** M. G. Eberhart *Bayou Road* v. 47 How *could* the Yankees have injured..New Orleans, the Queen City, so completely. **1918** *Chambers's Jrnl.* Aug. 541/2 It is the Queen conch my friend has come to buy. **1975** M. Humfrey *Sea Shells W. Indies* i. 29 The powerful Queen Conch..may weigh more than five pounds. **1881** T. W. Cowan *Brit. Bee-keeper's Guide Bk.* [vii. 33 One of the features of this hive is the possibility of preventing swarming, by confining the queen..by placing a zinc excluder..near the front of the hive.] *Ibid.* 134/1 Queen-excluder. **1887** F. R. Cheshire *Bees & Bee-keeping* II. iii. 74 This [sc. the restriction of the queen] is now accomplished by what is called 'excluder-zinc', or 'queen-excluder'. **1930** W. Herrod-Hempsall *Bee-keeping* I. ix. 447 The first queen excluder, made from wood, was invented and used in Scotland in 1849. **1976** T. Hooper *Guide to Bees & Honey* iv. 76 A queen-excluder will be necessary for each [hive]. **1883** J. J. Lalor *Cycl. Political Sci.* II. 217/2 The queen-fish, the bagre and the roncador are..well known in California. **1905** D. S. Jordan *Guide to Study of Fishes* II. xx. 354 The queenfish, *Seriphus politus*, of the California coast, is much like the others of this series... It is a very choice fish. **1937** Z. Grey *Amer. Angler in Austral.* x. 111 The queen fish, a beautiful silvery dolphin-like leaper, is one of the greatest fish I have caught. **1951** T. C. Roughley *Fish & Fisheries Austral.* i. 60 The queenfish has a wide distribution from northern New South Wales to the north-west coast of Western Australia, and..is considered to be one of the fastest and most spectacular game-fish. **1965** A. J. McClane *Stand. Fishing Encycl.* 223/1 The queenfish, *Seriphus politus*, is elongate... Its body is bluish and the fins are yellowish. Growing to about a foot, it occurs from central California to Baja California in shallow water. **1969** *Northern Territory News* (Darwin) *Focus '69* 63/1 The next biggest catch is the threadfin..followed by the mackerel..and queenfish. **1911** *Daily Colonist* (Victoria, B.C.) 15 Apr. 10/1 (Advt.), Spanish Queen Olives, bottle, 50c or $1.00.. Rowat's Selected Queen Olives, bottle 50c. **1974** Queen olive [see *manzanilla 2]. **1961** *Guardian* 16 Jan. 4/2 A break for the 'queen-pin'..is utterly essential if you are to keep going. **1972** *Daily Tel.* 21 Jan. 13/1 Welcome to Elaine May..not just as a voice but as the queen-pin—director, author and actress. **1891** T. F. Garrett *Encycl. Pract. Cookery* II. 267/2 Queen pudding. **1971** *Jean Bowring Cookbook* 227 Queen pudding. [Recipe follows]. **1959** A. C. Hardy *Open Sea* II. vi. 143 (*caption*) The queen scallop..showing the swimming action. **1972** *Aquaculture* I. 280 The fish were fed a mixed diet of fresh or frozen chopped herring and queen scallop (*Chlamys opercularis*) meat, at a rate of 750 g twice weekly. **1941** F. Reinfeld *Keres' Best Games of Chess* 70/1 Richter prefers to retain the Queen-side Pawns, even at the cost of exchanging Rooks. **1966** J. R. Capablanca *Last Chess Lectures* (1967) i. 37 His Queen-side majority of Pawns could be converted into a passed Pawn. **1959** *Punch* 28 Oct. 371/1 A motel in Los Angeles advertises Queen-size beds. **1967** *Boston Sunday Herald* 30 Apr. (Bedding Suppl.) 1/5 Queen size is the answer if a king-size bed doesn't fit your plans... Its 60-by-80-inch innerspring mattress is six inches wider and five inches longer than the old double size. **1973** *Publishers Weekly* 23 July 66/3 An appealing and handsomely produced queen-size book. **1976** *Washington Post* 19 Apr. A9/4 (Advt.), Traditionally styled Queen Size Sleep Sofa and matching love-seat combination. **1979** *Arizona Daily Star* 5 Aug. (Advt. Section) 8/8 It is beautifully designed, complete with queen-size bed. **1955** *Sun* (Baltimore) 19 Mar. 9/4 Mrs Daniel J. Flood, wife of a Democratic Congressman from Pennsylvania, is introducing a new fad here—'queen-sized' colored cigarettes to match her costume. **1975** A. Bergman *Hollywood & Le Vine* (1976) ix. 123 A queen-sized mattress. **1978** *New York* 3 Apr. 74 (Advt.), And for just $50 more, we'll transform the Sofa into a queen-sized sleeper convertible! **1944** H. A. Callahan *Rigging* 130 The late J. Rogers Maxwell introduced a funny little staysail on his famous schooner *Queen* and it has always been known as the queen staysail. **1948** L. F. Herreshoff in *Rudder* Aug. 58 Because previous staysails had to be lowered away in tacking, when my father designed the schooner Queen he did away with the triatic stay and in its place ran a stay called a 'fresh water stay' between the topmast heads. This staysail with which a schooner can tack is called a 'Queen staysail', as it was first used on the schooner Queen. **1631** J. Taylor *Needles Excellency* (1634) sig. A2, col. 2, Bred-stitch, Fisher-stitch, Irish-stitch, and Queen-stitch. **1841** Lady Wilton *Art of Needle-work* (ed. 3) xx. 317 There are..ferne- and queen-stitches. **1954** C. G. Butler in *Trans. R. Entomol. Soc.* CV. 14 It is necessary for the bees to have physical contact with their queen in order to obtain this 'queen substance'. **1972** *Sci. Amer.* Sept. 56/3 The 'queen substances' are outstanding in the complexity and pervasiveness of their role in social organization. **1924** J. T. Nichols in J. O. La Gorce *Bk. Fishes* 166/2 The gaudy colors of the Queen Trigger-fish..are an exception among such forms. **1971** 'D. Halliday' *Dolly & Doctor Bird* xii. 166 Bahamian waters are full of extraordinary fish, from Striped Grunts to Queen Triggerfish. **1889** J. H. Maiden *Useful Native Plants Austral.* 415 *Daviesia arborea*..'Queen-wood'. This wood is hard, close-grained, with beautiful pink streaked lines. **1902** G. S. Boulger *Wood* ii. 300 Queen-wood.. North-eastern Australia... Streaked with pink, hard, close-grained, susceptible of a fine polish.
14. a. *Queen's speech*: see *speech *sb.*[1] 8 d.
b. queen's chair, a makeshift seat (cf. *queen's cushion*); **queen's cloth** (later example); **queen's conch** = *queen conch* (sense 13 in Dict. and Suppl.); **queen's cushion,** for 'Jamieson, 1808' in Dict. read 'Jamieson, 1825'; (examples); **Queen's English,** the English language regarded as under the guardianship of the Queen; hence, standard or correct English; **Queen's Guide,** a holder of the highest award in Guiding, *spec.* during the reign of a queen; **queen's head** (earlier and later examples); **queen's pattern** (see quots. 1910 and 1957); **queen's pudding,** a steamed suet pudding (occas. used for *queen of puddings*, sense 6 b above); **Queen's Scout,** a holder of the highest award in Scouting, *spec.* during the reign of a queen; **Queen's shilling:** also *transf.* and fig.; (earlier and later examples); **Queen's staysail** = **Queen staysail*; **queen's stuff** (later example); **queen's taste:** in phr. *to the* or *a queen's taste*, to perfection; **queen's ware,** (*a*) (earlier and later examples); (*b*) (earlier and later examples); **queen's weather** (earlier and later examples); **queen's woman** *slang* (now *Hist.*), a prostitute who received medical attention under the terms of the Contagious Diseases Acts of the 1860s.
1965 S. T. Ollivier *Petticoat Farm* vi. 77 Henry's buggy reins, tied to a cream-can lid, had formed a queen's chair for Harry. **1818** M. Edgeworth *Let.* 29 Oct. (1971) 130 Tell me which you prefer the Merino or the Queens cloth... The queens cloth comes to a guinea the dress cheaper. **1812** E. Weeton *Let.* June in *Jrnl. of Governess* (1969) II. 93, I have inquired the price of shells... Yours are conch shells; these are called Queen's conches. **1963** *Times* 27 Apr. 11/2 The pink queen's conch shell from the Caribbean. **1825** J. Jamieson *Suppl. to Etym. Dict. of Scottish Lang.* II. 253/1 *Queen's*, also *king's cushion*, a mode of carriage, whether in sport, or from necessity... Two persons, each of whom grasps his right wrist with his left hand, with the other lays hold of his neighbour's wrist, so as to form a seat of four hands and wrists conjoined. On these the person, who is to be carried, seats himself, or is seated by others, putting both his arms, for greater security, round the necks of the bearers. **1873** C. M. Yonge *Pillars of House* IV. xl. 161 The hands were clasped, queen's-cushion fashion, beneath her, the necks were bent for the arms to be thrown round them. **1592** T. Nashe *Strange Newes* sig. B1ᵛ, He must be running on the letter, and abusing the Queenes English without pittie or mercie. **1848** [see *on *prep.* 6 a]. **1864** H. Alford *Plea for Queen's Eng.* (ed. 2) 2 The *Queen's English* is not an unmeaning phrase, but one which may serve to teach us some profitable lessons with regard to our language. **1867** F. S. Cozzens *Sayings of Dr. Bushwhacker* 82 In fact, that arbitrary style of speaking which is commonly known as the Queen's English. **1885** *Punch* 4 July 5/2 (*heading*) The Premier's Primer; or Queen's English as she is wrote. **1902** F. Hume *Fever of Life* 146, I! Oh, how can you? I speak the Queen's English. **1975** *Verbatim* Dec. 15/2 One irate caller said, according to the network [NKH], 'How dare you deprive us of our one chance to hear the Queen's English?' **1968** M. E. Brimelow *Guide Handbk.* iv. 70 If a Guide has..taken a full and active part including earning badges, in all the Eight Points of the Programme..she can qualify as a Queen's Guide. **1976** *Milton Keynes Express* 9 July 5/7 Three North Bucks girls received Queen's Guides awards at Milton Keynes College of Education, Wolverton, on Tuesday evening. **1840** *Chambers's Edin. Jrnl.* 11 July 193/2 The perplexed purchaser immediately devotes a queen's head, as he most irreverently calls it, to the purpose of asking the editors what he is to do. **1879** Trollope *John Caldigate* III. x. 132 That stamp, that effigy, that two-penny queen's-head. **1915** *Chambers's Jrnl.* Sept. 599/1 When the new stamp was introduced it was invariably called the 'queen's head'. **1769** *Catal. Worcester Porcelaine* in J. E. Nightingale *Contrib. towards Hist. of Eng. Porcelain* (1881) 95 Twelve fluted handle cups and saucers, 6 coffee cups, and two tea pots plain Queen's pattern 2l. **1910** R. L. Hobson *Worcester Porcelain* vii. 58 The catalogue of a sale of Worcester porcelain at Christie's, in 1769, includes several references to a 'Queen's pattern', which was no doubt the same as the traditional 'Queen Charlotte's pattern' of today. **1928** W. B. Honey *Old Eng. Porcelain* viii. 167 A design in Oriental style long popular at Worcester..consists of vertical or spirally curved panels alternately red on white and white on blue, with gilding... It was variously known as the 'whorl', 'spiral', 'catherine-wheel' and 'Queen's pattern'. **1957** Mankowitz & Haggar *Encycl. Eng. Pottery & Porcelain* 186/1 *Queen's pattern*, a counter-changed pattern consisting of alternate radiating whirling bands of red-on-white and white-on-blue ornament with gilded embellishments used at Worcester from c. 1770 onwards. **1974** K. Royce *Trap Spider* i. 12 The cutlery was mid-Georgian Queen's pattern. **1884** *Cassell's Dict. Cookery* 675/2 *Queen's Pudding*. [Recipe follows.] **1917** N. Soyer *Standard Cookery* 271 Queen's Pudding. Ingredients.—Eight ounces of finely-chopped suet [etc.]. **1935** G. Greene *Basement Room* 9 It was a pudding he liked, Queen's pudding with a perfect meringue. **1955** *Radio Times* 22 Apr. 14 St. George's Parade of Queen's Scouts. **1962** L. Deighton *Ipcress File* iii. 24 He picked the limp Raven off the..table like a Queen's scout with a rucksack. **1975** *Scout Handbk.* xxix. 274 Beyond the Membership Badge you'll aim for the Venture Award and the Queen's Scout Award. **1877** G. W. Godfrey (*title*) The Queen's Shilling. **1975** N. Luard *Travelling Horseman* vi. 161 If you've had enough of the Queen's shilling, try Pat Foley. **1976** *Listener* 1 Apr. 403/2 He took the Queen's shilling and joined the Royal West Kents..as a private soldier. **1926** *Yachting Monthly* XLI. 244/1 Above the mainstaysail was another triangular sail, commonly known as a 'Queen's' staysail. **1845** S. Judd *Margaret* ii. xi. 358 Rose had on..a queens-stuff habit of the same colour. **1902** W. N. Harben *Abner Daniel* xxxiii. 279 You worked 'im to a queen's taste—as fine as split milk. **1911** R. D. Saunders *Col. Todhunter* ix. 126 It's the best and truest thing I ever saw in my life! They've got you finished off to the Queen's taste. **1767** J. Wedgwood *Sel. Lett.* (1965) 58 The demand for this said *Creamcolour*, Alias *Queens Ware*, Alias *Ivory* still increases. **1863** W. Chaffers *Marks & Monograms on Pott. & Porc.* 120 The principal inventions of Wedgwood were, 1, the cream-coloured table ware, afterwards called *Queen's ware*; [etc.]. **1872** 'Mark Twain' *Roughing It* lix. 432 By-and-by he went home to his lodgings—an empty queensware hogshead—and employed himself till night trying to make up his mind what to buy with it. **1897** [see *merchant *sb.* 1 d]. **1900** F. Litchfield *Pottery & Porcelain* iii. 32 [Thomas Whieldon's] celebrated cream ware, called 'Queen's ware'. **1906** *Dialect Notes* III. 152 *Queensware*.., ordinary crockery. 'You can get queensware at Hansard's grocery or the ten-cent store.' **1961** *Connoisseur New Guide to Antique Eng. Pottery, Porcelain & Glass* 54 [Wedgwood's] Queensware was copied by most of the potters of his time. **1851** *Illustr. London News* 7 June 512/2 The 'Queen's weather', as it has been styled, did not hold good to-day, for..the rain began to come down. **1893** S. Grand *Heavenly Twins* I. ii. iv. 234 'Queen's weather!' he remarked. 'Yes,' she answered, looking out at the sparkling water. **1937** M. V. Hughes *London Home in Nineties* x. 167 The 'Queen's weather' of glorious sunshine began to work in the early part of the year [1897]. **1979** G. St. Aubyn *Edward VII* v. 251 There could hardly have been a more convincing demonstration of 'Queen's Weather' than on..the day of the Diamond Jubilee... The sun burst through the clouds. **1871** *Rep. R. Comm. Admin. Contag. Dis. Acts* I. 14 in *Parl. Papers* (C. 408) XIX. 1 Some of them are called 'Queen's women'; some exhibit the printed order to attend the periodical examination as a certificate of health. **1981** F. K. Prochaska *Women & Philanthropy* vi. 205 One effect of the [Contagious Diseases] Acts was the creation of an outcast and more professional class of prostitute, 'Queen's women' as they were sometimes called.
c. queen's lace *U.S.*, the wild carrot or Queen Anne's lace, *Daucus carota*.
1906 M. E. W. Freeman *By Light of Soul* 52 The fields ..were white and gold with queen's lace and golden rod. **1947** L. M. Beebe *Mixed Train Daily* 88 This freight train, carrying its passengers in the caboose and wading pleasantly through springtime Arkansas meadows brave with daisies and queen's lace, is the Graysonia, Nashville and Ashdown's morning redball.

queen, *v.* Add: **4. b.** (Earlier example.)
1842 *Chess Exemplified* 78 If the pawn have the move—it will queen.

Queen Anne. Add: **b.** (Earlier and later examples.)
1863 A. J. Munby *Diary* 12 May (1972) 160 The house.. is noble, having a fine Queen Anne front and elaborate scrollwork gates. **1878** C. L. Eastlake *Hints on Househ. Taste* (ed. 4) i. 25 The recently revived taste for the so-called 'Queen Anne' style..for that domestic type of brick architecture which prevailed..from the Caroline to the Georgian period. **1879** M. E. Braddon *Vixen* I. xiii. 253 The Queen Anne kettle was hissing merrily over its spirit-lamp. **1913** W. J. Locke *Stella Maris* ii. 13 A Queen Anne gem of a tiny house in Kensington. **1936** *Archit. Rev.* LXXIX. 213 (*caption*) This house claims to be the first brick-built house of the 'Queen Anne' Revival; No. 1 Palace Green, Kensington, by Philip Webb. **1964** S. Nowell Smith *Edwardian England* iv. 152 Tudor, Jacobean, Queen Anne, and Chippendale pieces..often jostled each other in the same room. **1977** *Times* 28 May 22/7 The audio and radiogram..designs are in Jacobean, Regency, Queen Anne and Scandinavian contemporary.

c. In the possessive in the names of plants: **Queen Anne's (double) daffodil,** a daffodil, *Narcissus capax plenus*, with pale yellow double flowers; **Queen Anne's (double) jonquil,** a small variety of *Narcissus jonquilla* with clusters of double, yellow, fragrant flowers; **Queen Anne's lace,** a popular name for various umbelliferous plants bearing clusters of small white flowers, esp., in North America, the wild carrot, *Daucus carota*, and in Britain, cow parsley, *Anthriscus sylvestris*.
1806 *Curtis's Bot. Mag.* XXIV. 934 Varies with very double flowers, and is then called by some Gardeners 'Queen Anne's Jonquil'. **1886** G. Nicholson *Illustr. Dict. Gardening* II. 413/1 A double form of this species [sc. *Narcissus jonquilla*] is known as Queen Anne's Jonquil. **1889** *Jrnl. R. Hort. Soc.* XI. 90 Its name of 'Queen Anne's Daffodil' was no doubt originally given in honour of Queen Anne of Austria. **1894** *Amer. Folk-Lore* VII. 89 *Daucus Carota*, Queen Anne's lace, somewhat general. **1913** W. P. Eaton *Barn Doors & Byways* 273 Wild carrot bears a dainty, flat-topped white bloom sometimes as large as a saucer, and a long bed of them will often appear like a strip of delicate embroidery along the wayside, making their more aristocratic title of Queen Anne's lace entirely applicable. **1926** W. de la Mare *Connoisseur* 318 Only the rotting sleepers remained, matted with weeds and bordered with Queen Ann's [*sic*] lace, golden rod and Michaelmas daisy. **1930** F. A. Pottle *Stretchers* 204 The fields..are rioting with the delicate blooms of Queen Anne's lace. **1934** E. A. Bowles *Handbk. Narcissus* viii. 89 *N*[*arcissus*] *eystettensis*,.. otherwise known as capax plenus or Queen Anne's Double Daffodil, is another plant of mysterious origin. *Ibid.* 90 Queen Anne's Daffodil may be a garden hybrid of the sixteenth century. *Ibid.* xiv. 137 Its double form is known as 'Queen Anne's Jonquil'. **1946** D. C. Peattie *Road of Naturalist* i. 21 In Maryland the Queen Anne's lace was dancing under the apple trees that made wide pools of shadow. **1951** M. Jefferson-Brown *Daffodil* x. 117 Queen Anne's Double Daffodil is another flower with which this monarch's name is associated. *Ibid.*, Queen Anne's Double Jonquil..is but six inches high. **1955** W. Gaddis *Recognitions* iii. ii. 760, I remember..tall weeds including Queen Anne's laces. **1961** M. Fish *Cottage Garden Flowers* iii. 29 *N*[*arcissus*] *capax plenus* (Syn. *N. eystettensis*) is the flower that was known as 'Queen Anne's Double Daffodil'. **1961** P. Synge *Collins Guide to Bulbs* 236 *N*[*arcissus*] *jonquilla floreplena* is the double form often known as Queen Anne's Double Jonquil. **1971** P. Larkin in *Listener* 29 July 144/1 Young-leafed June,.. With hedges snowlike strewn,.. Lost lanes of Queen Anne's lace. **1974** M. Ingate *Sound of Weir* xix. 164 A dragonfly..settled for a moment among the litter scent of Queen Anne's lace. **1978** *Church Times* 23 June 14/5, I waded into waist-high cow-parsley (or, if you prefer it, Queen Anne's lace, hemlock or bogy's gruel).
Hence also **Queen A·nn(e)ish** *a.*, suggestive of, or designed in, a Queen Anne style; **Queen A·nnery,** a style (of utterance, etc.) characteristic or reminiscent of the reign of Queen Anne.
1881 G. M. Hopkins *Let.* 29 Oct. in Hopkins & Dixon *Corr.* (1955) 83 The language is a quaint medley of Middle-Ages and 'QueenAnnery'. **1926** *Spectator* 24 July 154/2 Gradually Queen-Annish cornices began to creep in. **1943** J. Lees-Milne *Ancestral Voices* (1975) 269 It is a low, rambling, attractive house, Queen Anne-ish but not of architectural distinction. **1975** *Country Life* 9 Oct. 909/1 The Queen Anne-ish windows fill the room with light.

queenie (kwī·ni). Also **queeny.** [f. Queen *sb.* + -ie, -y[6].] **1. a.** An effeminate male, a homosexual; used esp. as a term of address. **b.** *Comb.*, as *queenie-fashion* advb.
[*c* **1736** Pope *Poems* (1954) VI. 367 The Motley Race of Hervey queenies [*var. reading* Harvequini's] And Courtly Vices, Beastly Venyes.] **1933** N. Ersine *Underworld & Prison Slang* 60 *Queenie*.., a homosexual man. **1936** J. Curtis *Gilt Kid* viii. 79 'You're not a man. You're a pouf.'.. 'I'll show you who's a pouf.' 'Call yourself a man do you this morning, Queenie? Well you wasn't one last night, see. You gets into bed and goes straight off to kip.' **1948** D. Ballantyne *Cunninghams* xx. 261 The dark, good-looking kid..who always had his shirt collar turned up queeny-fashion. **1966** 'L. Lane' *ABZ of Scouse* II. 87 *Queenie*, an effeminate male. **1974** R. Bly tr. *Hamsun's Hunger* i. 41, I felt a hand on my shoulder, and turned: 'Queeny' was saying good morning... The sight of this aimless, painted-up creature somehow enraged me.
2. (With capital initial.) An affectionate informal term for Queen Elizabeth II.
1976 C. Storr *Unnatural Fathers* xiii. 131 'The Queen absolutely refuses.'.. 'So Queenie's turned up trumps, has she? Good for her!' **1979** *Time* 13 Aug. 6/1 When Britain's Queen Elizabeth II arrived in Lusaka,..she was cheered by Zambians everywhere she went as 'Queenie! Queenie!'.

queening, *vbl. sb.* (In Dict. s.v. Queen *v.*) Add: Also *attrib.* in *queening square*, the square on a chess board on which a particular pawn may be queened.
1922 A. Emery *Elements of Chess* iv. 32 Under no circumstances whatever can a solitary pawn on either rook's file be advanced to the queening square, if it is possible for the adverse king to take up a position in front of it on the same file. **1937** M. Euwe *Strategy & Tactics in Chess* viii. 168 In choosing the queening square Black takes care that, at all cost, the White Rook is standing on the adjacent file. **1966** *New Statesman* 9 Sept. 371/3 The draw is secured either by stalemate or by perpetuating the Black Bishop's command of the queening square. **1969** A. Glyn *Dragon Variation* ii. 58 If he takes your Knight now, you play Pawn on and then Queen it, and his own Pawn protects the queening square from his Bishop.

queenly, *a.* Add: **3.** Characteristic or reminiscent of a male homosexual (see *queen *sb.* 11*).
1968 *Listener* 11 July 59/3 Paul Scofield..is using his queenliest *Staircase* manner to play a bisexual scriptwriter. **1976** *Times* 5 Apr. 6/8 A..rumour went around that the play was really about homosexuals. Certainly some of the dialogue does have a queenly tinge.

queenright (kwī·nrəit), *a.* [f. Queen *sb.* + Right *a.*] Of a colony of social insects, esp. bees: provided with a queen. (In quot. 1932, 'relating to the presence of a queen'.)
1932 E. B. Wedmore *Man. Beekeeping* ii. 38 In the writer's opinion the bees in the portion of the hive in question conclude that the queen is failing, and he calls the impulse [to rear a second queen] 'queenright supersedure'. **1966** Laidlaw & Eckert *Queen Rearing* ii. 33 A strong queenright colony can care for 45 grafted cells a day. **1971** E. O. Wilson *Insect Societies* xvii. 333/2 Marchal..also reported the presence of laying workers in queenright wasp colonies. **1974** *Country Life* 2 May 1041/1 The owner of the colony who hasn't an ear for the contented hum of what apiarists call a 'queenright-hive' waits until his slow wits tell him that tragedy has struck.

Queensberry (kwī·nzbəri). Also (*erron.*) **-bury.** The name of Sir John Sholto Douglas (1844–1900), eighth Marquis of Queensberry, used *attrib.* in *Queensberry rules*, a code of rules drawn up in 1867 under his supervision to govern the sport of boxing in Great Britain; also, the name by which the present rules of boxing are known; also *transf.*
1895 G. B. Shaw in *Sat. Rev.* 28 Sept. 410/1 There the contest was in the presence of a court, with measured ground and due formality—under Queensberry rules, so to speak. **1899** A. Conan Doyle in *Strand Mag.* XVIII. 368/1 It's twenty rounds, two-ounce gloves, Queensberry rules. **1931** *Times Lit. Suppl.* 11 June 472/3 Ever since 1866, when the Queensberry rules were introduced, pugilism has lost much of its romantic interest. **1960** *Observer* 24 Jan. 7/2 You would have thought that the average citizen would be only too anxious to see property being protected and would not worry so much about the Queensberry Rules being observed. **1975** *Oxf. Compan. Sports* 110 Broughton's rules were so sensible that they formed the basis of the London Prize-ring Rules drawn up in 1838..and even of the Queensberry Rules introduced towards the end of the nineteenth century in London. **1977** *New Yorker* 13 June 121/1 An at last complete love affair with a married woman..was in the Queensberry rules; she could not object.

Queensland (kwī·nzlănd). [The name of a state in north-eastern Australia.] **a.** In the names of local trees or their woods, esp. **Queensland kauri,** a tall evergreen tree, *Agathis robusta*, of the family Pinaceæ; **Queensland maple,** either of two evergreen trees, *Flindersia brayleyana* or *F. pimenteliana*, of the family Rutaceæ; **Queensland nut** = *macadamia; also, the fruit of this tree; **Queensland walnut,** an evergreen tree, *Endiandra palmerstonii*, of the family Lauraceæ. **b. Queensland blue** = **blue heeler* s.v. *blue *a.* 12 a; **Queensland sore,** a sore which does not heal readily because of scurvy.
1956 G. Casey in *Coast to Coast 1955–56* 80 It was a noble animal, a true Queensland-blue, with jaws like an alligator, sagacity in its eye, and limbs like a well-muscled leopard. **1977** C. McCullough *Thorn Birds* iv. 80 The big Queensland blue fawning and cringing at Father Ralph's feet. **1889** J. H. Maiden *Useful Native Plants Austral.* 414 'Queensland Kauri', or 'Dundathu Pine'. Wood of a light yellow colour..largely used by joiners and cabinet-makers. **1932** R. S. Troup *Exotic Forest Trees in Brit. Empire* 32 *Agathis robusta*..Queensland kauri. A very large tree attaining 160 ft. in height. **1919** R. T. Baker *Hardwoods Austral.* ii. 33 The soft woods are those most generally found on the market, such as Red Cedar, Red Bean, Queensland Maple, Beech, and Pines. **1928** E. H. F. Swain *Timber & Forest Products Queensland* 147 Placed on the Southern markets later under the name of Queensland Maple, the wood has risen to fame under this unsuitable trade appellation. **1967** A. Rule *Forests Austral.* xii. 140 Australia's cabinet timbers—Queensland walnut, Queensland maple, red cedar, silky oak, black bean and so forth—were found to make fine sliced veneers. **1881** *Official Rec. Sydney Internat. Exhib. 1879* 750 Queensland Nut... A small-sized tree with a very dense foliage. **1928** E. H. F. Swain *Timbers & Forest Products Queensland* 472 The Queensland Nut is one of the finest edible nuts. **1929** [see *macadamia]. **1950** *N.Z. Jrnl. Agric.* Dec. 527/1 A number of Queensland nut trees growing in Northland are producing good crops annually. **1963** A. Lubbock *Austral. Roundabout* 93 A great, dark tree, rather like an ilex,..bears the Queensland nut. **1892** G. L. James *Shall I try Australia?* xxiv. 242 'Queensland Sores'..are, I believe, generally attributed to excessive thinness and poverty of the blood, caused by the great heat and an absence of vegetable diet. **1945** Baker *Austral. Lang.* iii. 62 *Barcoo rot*, *Kennedy rot* or *Queensland sore*, a festering sore difficult to cure under inland conditions—it rapidly disappears when the sufferer eats plenty of fruit or green vegetables. **1919** R. T. Baker *Hardwoods Austral.* ii. 339 'Queensland Walnut'..is a good commercial timber. **1967** Queensland walnut [see *Queensland maple* above].

Quee·nslander. [f. prec. + -er[1].] A native or inhabitant of the state of Queensland in Australia.
1861 J. D. Lang *Queensland, Australia* viii. 238 A letter addressed to the editor of the 'Moreton Bay Courier',..is signed 'A Queenslander'. **1883** G. W. Rusden *Hist. Australia* III. xvii. 235 The gallant conduct of the 'Queenslander', a well-managed newspaper in Brisbane, enables me to use the words of witnesses or neighbours. **1899** E. W. Hornung *Amateur Cracksman* 176 The Queenslander handed it to me. **1924** Kipling *Debits & Credits* (1926) 316 'An' what about your Queenslander?' the Australian asked. **1963** *Times* 12 Mar. p. vi/3 The wise detective could deduce from such a sign that Queenslanders are easy-going to an incredible degree, that they have a loathing for formality and they live in a warm climate. **1978** R. V. Jones *Most Secret War* xvi. 130 A buccaneering Queenslander, Sidney Cotton, who had been a pilot in the Royal Naval Air Service.

queeny (kwī·ni), *a.* Also **queenie.** [f. *queen *sb.* 11* + -y[1].] Effeminate; also as *adv.*, in an effeminate manner.
1936 N. Marsh *Death in Ecstasy* xxiv. 299 Alleyn may talk queeny, but he's doped that out. **1937** —— *Vintage Murder* vii. 72 He talks with a corker sort of voice. Not queeny, but just corker. **1952** *Arena* (N.Z.) xxxi. 2 Up in front would be Slick, the leader [sc. a sheepdog], stepping along in a queenie way. **1965** G. McInnes *Road to Gundagai* xv. 265 Thereafter he said he'd rather play football with the other fellows: reading aloud was a bit 'queeny'. **1977** 'J. Le Carré' *Honourable Schoolboy* i. 19 'So what's happened, Lukie?' whined the dwarf, in his queeny Greenwich Village drawl.

queer, *a.*[1] Add: **1. a.** *queer fellow*, an eccentric person. Also used, esp. in Ireland and in nautical contexts, with varying contextual connotations (see quots.). Cf. *quare *a.*
Possibly some examples illustrate Queer *a.*[2]
1712 [in Dict.]. **1883** J. F. Keane *On Blue Water* 212 Remembering that incident, the 'queer-fellow's' disappearance didn't alarm me very much. **1910** D. W. Bone *Brassbounder* 64 D'ye think th' queer-fella' is goin' t' pay them prices for 'is kit? **1922** [see *middle leg* s.v. *middle *a.* 6]. **1932** J. W. Harris *Days of Endeavour* 17 No matter what ship you serve your time in, you'll find there'll be a queer-feller. **1936** J. Curtis *Gilt Kid* vi. 60 He'd a good mind to tear over and spoil her lark with the queer fellow. **1939** J. Brophy *Queer Fellow* 10 When I am 'making up' a story,.. I am never my normal self, the man that other people know. Nor dare my normal self return for a moment in the hope of catching the other one, The Queer Fellow, as they say in Ireland, at work. **1958** M. Procter *Man in Ambush* xii. 134 Mobsters, queer fellows, bar flies and layabouts. **1961** Partridge *Dict. Slang* Suppl. 1240/2 *The queer fella*, the person that happens to be in command: Regular Army: late c. 19–20. **1962** Granville *Dict. Sailors' Slang* 93/2 *Queer fella*, any merchant seaman who does not conform to the average type. A nautical eccentric. **1966** 'L. Lane' *ABZ of Scouse* ii. 87 *Whur's ther queer feller?* Where is the boss or foreman whose name I don't know?
b. Of a person (usu. a man): homosexual. Also in phr. *as queer as a coot* (cf. Coot *sb.*[1] 2 b). Hence, of things: pertaining to homosexuals or homosexuality. orig. *U.S.*
1922 *Pract. Value of Scientific Study of Juvenile Delinquents* (Children's Bureau, U.S. Dept. of Labor) 8 A young man, easily ascertainable to be unusually fine in other characteristics, is probably 'queer' in sex tendency. **1931** G. Irwin *Amer. Tramp & Underworld Slang* 153 *Queer*, crooked; criminal. Also applied to effeminate or degenerate men or boys. **1936** J. G. Cozzens *Men & Brethren* i. 24 'He's not queer, or something, is he?' 'Lord, no! Worse than that. He's a convert.' **1937** *Listener* 10 Mar. (Suppl.) p. vii/2 'Queer', in a specifically sexual sense—a word imported from America. **1939** C. Isherwood *Goodbye to Berlin* 296 Men dressed as women?.. Do you mean they're *queer*? **1952** [see *camp *a.* (and *sb.**)]. **1958** P. Mortimer *Daddy's gone a-Hunting* xxx. 169, I suppose they're queer as coots. **1960** [see *bent *ppl. a.* 5 c]. **1963** [see *gay *a.* 2 c]. **1974** *Amer. Speech 1971* XLVI. 81 *Female homosexual*, lesbian, screwball, queer, lady-lover, minty. **1975** *Times* 9 July 14/2 *Bombus fragrans*..sprinkles himself..with attar of roses, and..is as queer as a coot. **1976** A. White *Long Silence* i. 10 'I say, Peter, you're not turning *queer* by any chance, are you?' The thought that I might be queer had haunted me.
c. In U.S. colloq. phr. *to be queer for* (someone or something): to be fond of or 'keen on'; to be in love with.
1953 W. Burroughs *Junkie* ii. 28 She began talking about Jack. 'I'm queer for Jack,' she said. 'He works at being a thief just like any job.' **1956** J. Baldwin *Giovanni's Room* i. ii. 45 Actually, I'm sort of queer for girls myself. **1957** M. Shulman *Rally round Flag, Boys!* iv. 51 When..the cellars were finally snug and dry, Waldo promptly persuaded the homesteaders to fill them with.. a huge, gleaming variety of tools. This took no great persuasion, for..the average commuter was queer for tools. **1977** *Time* 28 Mar. 54/2 The sister (Carol Potter) is crazy about him and Francis is queer for her brother (Reed Birney), or so he fears.
2. (Earlier and further examples.)
1781 S. Crisp *Let.* 1 Mar. in W. H. Hutton *Burford Papers* (1905) 60, I have been very queer for some time, sleepless and indigestion. **1837** [see *earthquaky *a.*]. **1889** J. K. Jerome *Three Men in Boat* i. 14 So I set my face against the sea trip. Not, as I explained, upon my own account. I was never queer. But I was afraid for

George. 1922, 1938 [see *COME v. 66 f]. 1952 A. CHRISTIE *Mrs McGinty's Dead* iv. 28 Either the husband's taken queer, or the old mother... With old McGinty, at least it was only she herself who came over queer. 1978 'F. PARRISH' *Sting of Honeybee* iv. 43 Jake's off queer, wi' a rumblin' stummick.
3. (Earlier and later examples.)
1811 *Lex. Balatron.*, *Queer Street*, wrong, improper, contrary to one's wish. It is queer street, a cant phrase, to signify that it is wrong or different to our wish. 1821 P. EGAN *Real Life in London* I. xi. 186 Limping Billy was also evidently in *queer-street*. 1829 —— *Boxiana* 2nd Ser. II. 503 Gas let fly right and left, give Pope a tremendous blow over his left *ogle*, putting him a little into Queer-street. 1952 A. WILSON *Hemlock & After* III. i. 208 He enjoys a little flutter..and if he finds himself in Queer Street now and again, I'm sure no one would grudge him his bit of fun. 1963 *Times* 8 May 9/2 He felt that the levy should not be applied so rigidly as to force companies into Queer Street if their costs rose faster than their incomes. 1980 J. WAINWRIGHT *Man of Law* xlvii. 222 If Patsold talks, Webb's in queer street.
4. *queer-shaped* (earlier example).
1876 H. SIDGWICK in A. & E. M. Sidgwick *Henry Sidgwick* (1906) v. 323 Stone hovels that a generation ago were the ordinary houses here: things with a hole in the roof, low, queer-shaped.

queer, *a.*² and *sb.*¹ For **a, b.** read **A, B.** and add: **A.** *adj.* **b.** Of coins or banknotes: counterfeit, forged.
1740 *Ordinary of Newgate, his Account* III. 15/1 Instead of returning the good Guinea again, they used to give a *Queer One*. 1812, etc. [see SCREEN *sb.* 2]. 1848 *Ladies' Repository* Oct. 316/2 *Queer*, counterfeit. *Queer screen*, counterfeit paper money. *Queer wedge*, counterfeited silver money. *Queer ridge*, counterfeited *gold money*. 1877 J. HABBERTON *Jericho Road* xvi. 151 'Let's give him fifty [dollars] to send her.' 'Fifty queer?' asked Mr. Lodge. 'No, fifty straight,' said the little man. 1890 *Buckskin Mose* ii. 34 At the same time he pulled out of his pocket a lot of 'queer' or counterfeit bills. 1941 R. CHANDLER in *Detective Story Mag.* Sept. 52 If it was discovered to be queer money, as you say, it would be very difficult to trace the source of it.
B. *sb.*¹ **a.** (Later examples.) Also (*U.S.*), forged paper currency or bonds. Phr. *to shove (the) queer*: see SHOVE *v.* 10a.
1821 P. EGAN *Life in London* II. i. 154 That admired sort of Life in London, all jostling against each other in the Park... The Duke and the 'Dealer in Queer'—the Lady and her Scullion [etc.]. 1847 *National Police Gaz.* (U.S.) 9 Jan. 137/1 'Bogus' is base coin, 'queer' is counterfeit paper. 1859 [see SHOVE *v.* 10]. 1889 G. GISSING *Nether World* II. xi. 233 'Got any *queer* to put round?'..'You know what he meant, Bob?' Bob nodded and became reflective. *Ibid.* III. ii. 38 He opened it, and showed about a dozen pieces of money—in appearance half-crowns and florins... 'The snyde' or the 'queer' is the technical name by which such products are known. 1898 A. M. BINSTEAD *Pink 'Un & Pelican* xi. 240 He hardly ever uttered the spurious coins himself..and, consequently, seldom had any 'queer' about his person. 1926 *Flynn's* 16 Jan. 640/2 After I coughed up an' promised to quit the queer he give me th' gate. 1949 E. L. IREY *Tax Dodgers* v. 112 An alcoholic engraver..turned out the best 'queer' that ever competed with the Bureau of Engraving's product and Lustig took over the distribution of the counterfeit money. 1954 W. R. & F. K. SIMPSON *Hockshop* ix. 232 Eagle-eyed concessionaires..always on the lookout for shovers of the queer. 1981 'E. LATHEN' *Going for Gold* iii. 37 Nobody's laying off any queer on the Sloan [Bank].
b. *on the queer*: living dishonestly; *spec.* engaged in the forging of currency.
1905 C. H. DAY *Actress & Clerk* ii. 22 Only just feeling of you to see if you was on the queer. 1909 R. A. WASON *Happy Hawkins* 277 Dick may have been on the queer all right, but he was smooth enough to hide it. 1910 C. E. B. RUSSELL *Young Gaol-birds* x. 150 Convinced that he could get along as well 'on the queer', *i.e.*, by thieving, as he could by keeping straight. 1935 *Amer. Speech* X. 11/1 Boys who are *on the queer* are handsomely equipped to print anything from twenty dollar bills to fake government bonds. 1942 BERREY & VAN DEN BARK *Amer. Thes. Slang* § 494/2 Counterfeit; forge,..*be on the green goods,—the queer* or *the spud(s)*.

queer (kwiəɹ), *sb.*² *slang.* [f. *QUEER *a.*¹ 1b.] A (usu. male) homosexual. Also in *Comb.*, as *queer-bashing* vbl. sb., the attacking of homosexuals; hence *queer-basher*.
[1932 AUDEN in *Rev. Eng. Stud.* (1978) Aug. 294 An underground cottage frequented by the queer.] 1935 *Amer. Speech* X. 19/2 *Queer*... A male homosexual. 1936 L. DUNCAN *Over Wall* xx. 277 There was even a little room..where the 'fairies', 'pansies', and 'queers' conducted their lewd practices. 1946 E. WAUGH *Jrnl.* 21 Nov. (1976) VI. 663 The headmaster, an old queer. 1952 A. WILSON *Hemlock & After* I. iii. 58, I quite like queers if it comes to that, so long as they're not on the make. 1959 ANON. *Streetwalker* I. i. 18 Jackie, one of the commercial queers..is a tall, gangling boy with long hair..combed into carefully casual curls. 1965 *Spectator* 19 Feb. 239/1 Smith pursues Lulu in Valletta only to discover that she is a queer. 1966 [see *FAGGOT, FAGOT *sb.* 6b]. 1970 *Times* 5 Feb. 2/3 Four of 12 youths said to have taken part in a 'queer bashing' expedition on Wimbledon Common on September 25 were found Guilty of murder. *Ibid.* 25 Nov. 9/5 (*heading*) 'Queer-bashers' lose appeal. 1975 [see *PANSY 3a]. 1977 *New Wave Mag.* No. 7. 6 To fight the National Front, the queer-bashers and any other diseases.
Hence **quee·rdom**.
1965 *New Statesman* 9 July 58/1 Its climactic evocation of high Hapsburg queerdom at its annual drag ball. 1977 *Daily Express* 29 Jan. 7/2 This is a groin-directed compound of mime, ballet and freak show which, as a mere heterosexual, I take to be a celebration of the erotic imagery of queerdom.

queer, *v.* Add: **1.** (Later example.)
1854 W. HARCOURT *Let.* in A. G. Gardiner *Life W. Harcourt* (1923) I. iv. 76 The American Minister..spat on the floor all dinner-time. I hear he does this to queer the Britishers, and does not practise those manners at home.
2. a. (Later examples.) Also, with a person as object: to spoil the reputation of, to put (a person) in bad odour (with someone); to spoil (a person's) undertaking, chances, etc.
1895 J. L. FORD *Bohemia Invaded* 91 Without having you come in and queer me right in the middle of it [*sc.* a story]. 1895 E. W. TOWNSEND *Chimmie Fadden* 38 De Duchess gives me de orders, an' I wasn't goin' to queer meself wid 'er any more. 1904 *N.Y. Tribune* 12 Jan. 2/1 Van Wyck will queer the whole thing. His appearance before the National Committee will recall..things that knocked Tammany out in 1901. 1913 *Dialect Notes* IV. 11 *Queer*, *v.* To spoil the reputation (or good impression) a person has made or is trying to make. 'That *queered* me with the teacher.' 1919 E. O'NEILL *Moon of Caribbees* 17 Bella... Don't talk so loud... Think I wants the ole captain to put me off the ship, do you? *Yank.* Yes, nix on hollerin', you! D'yuh wanta queer all of us? 1927 [see *HORN *v.* 3c]. 1941 *Sun* (Baltimore) 25 Aug. 10/1 (*heading*) Queering the oil conservation drive.
b. *to queer the pitch*: to interfere with or spoil the business (of a tradesman or showman) (cf. PITCH *sb.*² 11a); now freq. *to queer one's pitch* (in more general use). Hence in similar phrases, as *to queer the game, the job*, etc.
1846 *Swell's Night Guide* 47 Rule iv... Nanty coming it on a pall, or wid cracking to queer a pitch. 1866 M. MACKINTOSH *Stage Reminisc.* vii. 93 The smoke and fumes of 'blue fire' which had been used to illuminate the fight came up through the chinks of the stage, fit to choke a dozen Macbeths, and—pardon the little bit of professional slang—poor Jamie's 'pitch' was 'queered' with a vengeance. 1875 T. FROST *Circus Life* xvi. 278 The spot they select for their performance is their 'pitch', and any interruption of their feats, such as an accident, or the interference of a policeman, is said to 'queer the pitch'. 1889 E. SAMPSON *Tales of Fancy* 38 They could not understand it when their pitch was queered, and one or two of the gang arrested. 1890 *Punch* 16 Aug. 74/3 Wy, they'd queer the best pitches in life, if they kiboshed the Power of the Quid! 1901 *Windsor Mag.* Dec. 204/1, I think you and I between us have queered the game. 1911 L. MERRICK *Position of Peggy Harper* III. i. 287 'You leave the contract to me..' 'I can do all that's wanted... You'd go asking too much and queering the job for me.' 1919 H. JENKINS *John Dene of Toronto* (1920) i. 17 'Suppose the Germans were able to sink a ship without even showing their periscopes?'.. 'Oh, shucks!' cried John Dene in disgust. It would queer the whole outfit... It would mighty soon finish the war.' 1927 *Observer* 4 Dec. 19/4 It may conceivably queer the pitch of Mr. de Valera, who..is about to approach the American public for a substantial sum. 1934 J. E. NEALE *Queen Eliz. I* xix. 334 Elizabeth..tried to break off... He went on, determined to queer the pitch for Coke and his supporter Burghley. 1973 E. LEMARCHAND *Let or Hindrance* iv. 31 He's a decent lad... He would never have risked queering Wendy's pitch with Eddy. 1977 *Rolling Stone* 5 May 50/1 Since trying to crash a closed Stones party the first night would likely queer the whole deal, I decided to check out El Mocambo with a local reporter.

queerie (kwiə·ri). *slang.* [f. *QUEER *a.*¹ 1b + -IE.] One who is queer; esp. one who is soft, effeminate, or homosexual.
1938 *Amer. Mercury* Oct. 181/1 Any boy showing interest in [classes] is correctly considered a 'queerie'. 1940 L. L. STANLEY *Men at their Worst* xx. 201 In every prison are many sex perverts. We know them the minute they step through the front gates on their way into prison. They are known as the 'Queens', the 'Fairies', the 'Queeries', and by other names. 1942 BERREY & VAN DEN BARK *Amer. Thes. Slang* § 405/1 Effeminate man or masculine woman; homosexual,..*queerie.* 1951 *N.Y. Jrnl.-American* 14 Aug. 3/1 Why then does he want to get his hands on these millions? To pay the wages of a lot of queeries in the State Department? 1972 B. RODGERS *Queen's Vernacular* 165 Queer,..*queerie* ('that little "queerie" is the only one I know who shoots Sal Hepatica').

quee·ringness. *rare*⁻¹. [f. QUEER *v.* + -ING¹ + -NESS.] Aptness to distort or to falsify.
1955 E. BOWEN *World of Love* ii. 49 She mistrusted the past's activity and its queeringness..who did not speak of it either with falsifying piety or with bitterness.

queerness. Add: **2.** Homosexuality. Cf. *QUEER *a.*¹ 1b.
1956 L. MCINTOSH *Oxford Folly* 104 Some of the most brilliant Oxford figures are queers, and the others simply ape their mannerisms without having the courage to imitate their queerness. 1958 *Observer* 4 May 15/5 A play which says more about the simple, non-tragic aspects of queerness than anything our theatre has so far permitted. 1965 *New Statesman* 9 July 58/1 Osborne has taken queerness as his subject precisely as Nabokov took the love of pubescent girls.

quelea (kwī·liă). [mod.L., the specific name of *Emberiza quelea* (Linnæus *Systema Naturæ* (ed. 10, 1758) I. 177), perh. f. med.L. *qualea* QUAIL *sb.*; adopted as a generic name by H. G. L. Reichenbach (*Avium Systema Naturale* (1850) pl. lxxvi).] An African weaver-bird of the genus so called, belonging to the family Ploceidæ, esp. the red-billed dioch, *Quelea quelea*, which is an important pest of grain crops; = *DIOCH. Also *attrib.*
1930 C. F. BELCHER *Birds Nyasaland* 316 (*heading*) Black-fronted Quelea. 1936 E. L. GILL *First Guide S. Afr. Birds* 26 Southern Pink-billed Weaver, Quelea Finch... The adult male in breeding plumage, with his black face surrounded by a halo of pink, is readily recognizable. 1957 BENSON & WHITE *Check List Birds N. Rhodesia* 123 Red-billed Quelea... Only numerous in drier, more open areas. 1964 [see *DIOCH]. 1969 N. W. PIRIE *Food Resources* ii. 67 The weaver bird, or quelea, infests 2m square miles of Africa. 1969 *Sci. Jrnl.* Apr. 15/1 Quelea birds..raid cereal crops in swarms numbered in millions. 1973 *Bokmakierie* XXV. 46/2 Colleagues in Kenya handed Mr. Skead over to the Quelea Control officers. 1975 *Times* 19 Feb. 14/6 Other exhibits [at the Centre for Overseas Pest Research] relate to army-worms, quelea birds, and termites.

‖**Quellenforschung** (kve·lənfo:rʃuŋ). [Ger., f. *quelle* source + *forschung* research.] The study of the sources of, or influences upon, a literary work.
1958 C. S. LEWIS in *Times Lit. Suppl.* 28 Nov. 689/3, I wonder how much *Quellenforschung* in our studies of older literature seems solid only because those who knew the facts are dead and cannot contradict it? 1966 J. WAIN in *Punch* 27 Apr. 616/2, I hadn't read the book when I wrote *Hurry On Down*, but I had read Orwell's literary and social criticism... Conclusion of *Quellenforschung.* 1975 *Times Lit. Suppl.* 29 Aug. 974/4 Intertextuality would appear to be so universally applicable to medieval literature as to imply, at best a statement of the obvious, at worst *Quellenforschung* in sheep's clothing.

‖**quemadero** (kemaðe·ro). [Sp., f. *quemar* to burn.] In Spain and former Spanish territories, a place where convicted heretics were executed by burning. Also *transf.*
1855 W. H. PRESCOTT *Hist. Reign Philip II* I. II. iii. 353 The place of execution—the *quemadero*, the burning-place, as it was called—was a spot selected for the purpose without the walls of the city. 1874 W. H. RULE *Hist. Inquisition* I. xiv. 208 Outside the city..was a hearth, or place of burning. As our own language is too poor to provide a name for such a thing, we consent to borrow from Spanish its peculiar and exclusive designation, and call it the *quemadero*. The quemadero was a piece of pavement devoted to the single use of burning human bodies. 1908 H. C. LEA *Inquisition in Spanish Dependencies* vi. 206 It was not until 1596 that the municipality [*sc.* Mexico], at a cost of four hundred pesos, constructed a *quemadero* or burning place, where concremation could be performed decently and in order. 1932 C. ROTH *Hist. Marranos* ii. 43 A *quemadero*, or burning place, was constructed in the Campo de Tablada. 1934 A. HUXLEY *Beyond Mexique Bay* 189 On each mound were..hearths of broken potsherds, blackened with smoke—the *quemaderos*, or burning places of the Indians. 1960 S. BECKER tr. *Schwarz-Bart's Last of Just* (1961) I. 11 He died very old..on the vast white slab of the *Quemadero* in Seville. Around him, scattered among the fagots, was the daily ovenful of three hundred Jews.

quench, *sb.* Delete *rare* and add: **1.** (Further example.)
1972 A. D. FRANKLIN in Crawford & Slifkin *Point Defects in Solids* I. i. 33 The special property of ductility possessed by many metals allows thin wires to be drawn, which may be very rapidly quenched, at maximum cooling rates of 10⁸ deg/sec or higher. With such rapid quenches, one may hope to retain the equilibrium defects present at the high quench temperature.
2. *Electronics.* The process of stopping an oscillation, esp. in a superregenerative receiver; a signal used for this. Freq. *attrib.*, as **quench frequency**, the frequency with which oscillations are stopped.
1938 *Proc. IRE* XXVI. 94 The use of a rectangular wave quench voltage would not be practicable in most applications of superregenerative receivers. *Ibid.* 96 In a given design of a separately quenched superregenerative receiver there is a particular quench frequency which gives maximum sensitivity. 1948 *Electronics* Sept. 98/3 This action..is eliminated by restricting the frequency content of the quench. 1950 J. R. WHITEHEAD *Super-Regenerative Receivers* vii. 125 A super-regenerative receiver with grid quench and a.g.s. controlling the oscillator grid bias. 1959 G. TROUP *Masers* vii. 118 A 600 c/s quench frequency was used. 1965 *Wireless World* July 336/2 Quench oscillators in super-regenerative receivers..have..set their own problems. 1975 D. G. FINK *Electronics Engineers' Handbk.* IX. 56 Electron current flow is initiated by an rf input signal and is terminated at the end of the rf input signal either by a voltage pulse or a dc bias voltage applied to a quench electrode.

quench, *v.* Add: **I. 1. a.** (Further example with *out.*)
1863 E. WETHERELL *Old Helmet* (1864) I. xi. 230 In Africa they sit in the darkness of centuries, till almost the spark of humanity is quenched out.
d. *Radio.* To cause (the spark in a spark transmitter) to cease by mechanical means, so that the secondary (aerial) circuit is no longer coupled to the primary; hence, to stop (oscillation).

1910 G. W. PIERCE *Princ. Wireless Telegr.* xxiii. 267 The spark is quenched when the energy in the primary attains its first minimum. 1913 *Chambers's Jrnl.* Mar. 232/2 The oscillatory current in the aerial, and therefore the wave-train radiated, continue long after the spark has been quenched. 1927 O. F. BROWN *Elements of Radio Communication* iv. 53 The spark is produced between projecting studs on a rapidly revolving metal disc and two fixed electrodes... The rotation of the disc will rapidly increase the distance between the studs and the electrodes, so that the spark is quenched and the oscillation in the primary circuit ceases. 1938 *Proc. IRE* XXVI. 76 In a typical superregenerative receiver the regenerative coupling between the plate and grid circuits of the detector tube is great enough so that self-sustained oscillations are produced, and these oscillations are periodically quenched, by applying..an alternating voltage having a frequency much lower than that of the oscillations. 1959 G. TROUP *Masers* vii. 117 These authors measured the noise figure of an ammonia maser amplifier operated superregeneratively: that is, oscillations were allowed to build up and then quenched. 1966 *McGraw-Hill Encycl. Sci. & Technol.* I. 362/1 A regenerative detector in which the oscillations are periodically stopped or quenched is called a superregenerative detector.
3. d. *Physics.* To suppress (luminescence); hence, to de-excite (an atom that would otherwise give rise to this effect).
1928 *Proc. Nat. Acad. Sci.* XIV. 851 The results show that hydrogen quenches the resonance radiation of cadmium as effectively as it does that of mercury. 1932 *Jrnl. Amer. Chem. Soc.* LIV. 572 The apparent decrease in quenching at high pressures or temperatures does not at all preclude the possibility that some fraction of the mercury atoms are being quenched to the normal state. 1954 C. ZWIKKER *Physical Properties of Solid Materials* xiii. 230 Fluorescence may be quenched by radiation, *e.g.* infra-red of too long a wavelength to excite fluorescence. The quenching photons raise electrons from the crystal lattice..to the copper ions..and thus inhibit the recombination effect. 1976 *Sci. Amer.* June 47/2 (Advt.), While the list of molecules which will react with $^1O_2$ is growing rapidly, the list of molecules which will quench $^1O_2$ back to O_2 is much smaller.
e. *Physics* and *Chem.* To suppress (the orbital angular momentum of an electron and the associated magnetic moment).
1932 J. H. VAN VLECK *Theory of Electric & Magn. Susceptibilities* xi. 282 Solids or solutions in which interatomic forces quench the orbital angular momentum but leave the spin free. 1955 TOWNES & SCHAWLOW *Microwave Spectrosc.* vii. 175 In nonlinear molecules, the orbital motion of electrons is almost completely 'quenched' or suppressed, and a spin momentum is the only angular momentum in the molecule of distinctly electronic origin. 1962 COTTON & WILKINSON *Adv. Inorg. Chem.* xxiv. 508 The electric fields of other atoms, ions, and molecules surrounding the metal ion in its compounds interfere with the orbital motion of the electrons so that the orbital angular momentum and hence the orbital moment are wholly or partially 'quenched'. 1971 J. D. PATTERSON *Introd. Theory of Solid State Physics* iv. 240 The cubic field acts to 'quench' the orbital angular momentum.
f. To prevent (the discharge in a Geiger counter) from continuing too long and reducing the possible counting rate; also with the counter as obj.
1940 *Physical Rev.* LVII. 1036/1 If we merely assure ourselves that the counter wire is falling somewhat below the starting potential with each discharge, then we can be sure that the discharge is quenched after the first stage and we will have a clean, fast pulse. 1942 POLLARD & DAVIDSON *Applied Nucl. Physics* iii. 30 A very common device to quench a counter is to employ a vacuum tube. 1958 O. R. FRISCH *Nucl. Handbk.* xv. 14 The discriminator circuit used with Geiger counters..should provide facilities for quenching the counter for a period of several hundred microseconds after each pulse. 1963 W. E. BURCHAM *Nuclear Physics* vi. 218 It is the function of the alcohol in the gas filling to 'quench' the discharge. 1975 K. H. GOULDING in Williams & Wilson *Biologist's Guide to Princ. & Techniques Pract. Biochem.* vi. 178 To overcome this, the tube is quenched by the addition of a suitable gas, which reduces the energy of the ions.
II. 7. *Physics.* To change from the superconducting state to the non-superconducting state.
1969 *Sci. Jrnl.* Apr. 42/2 Increasing current is passed through the superconductor until the material 'quenches' (goes normal). 1975 *Physics Bull.* May 214/1 The normal metal (copper or combinations of copper and cupronickel) is still required to protect the conductor when it reaches the limit of its current carrying capacity and 'quenches' (ie undergoes a transition to the normal state).
III. 8. Combs. (from sense 2b): **quench-ageing,** changes in the properties of steel, notably hardening, which occur after the metal has been quenched from a high temperature (see quot. 1968); **quench-cracking,** fracture of a metal caused by thermal stresses during rapid cooling; **quench-hardening,** hardening of steel by heating it above a critical temperature for some time, quenching rapidly, and then allowing further slow cooling; also = *quench-ageing* above; so **quench-harden** *v. trans.*
1935 *Trans. Amer. Soc. Metals* XXIII. 1049 To one of the three most important examples of aging, found in practically all soft steels, the designation 'Carbonizing' has been given for purposes of this discussion. It has also been called 'sub-critical quench-aging'. 1938 *Jrnl. Iron & Steel Inst.* CXXXVIII. 247P The usual theory put forward to explain the process of quench-ageing, whether in steel or in any other age-hardening alloy, is that it is caused by the precipitation from super-saturated solid solution of particles of the solute in a highly dispersed form on the lattice of the solvent,..preliminary to precipitation. 1961 G. E. DIETER *Mech. Metallurgy* v. 137 Quench aging is a type of true precipitation hardening that occurs on quenching from the temperature of maximum solubility of carbon and nitrogen in ferrite. 1968 E. R. PETTY *Physical Metall. of Engin. Materials* v. 92 These changes involve an increase in hardness, elastic limit and tensile strength, accompanied by a fall in ductility and impact resistance, and may occur in low carbon steels finished by rapid cooling from a softening temperature above 600°C or by cold working. In the former case the phenomenon is referred to as quench ageing while the latter is known as strain ageing. 1971 *Engineering* Apr. 20/1 The absence of the defects of material or liquation on the surface of the pins and journals is of particular importance..where these areas are to be hardened by flame or induction, as the risk of quench cracking is almost entirely eliminated. 1973 J. G. TWEEDDALE *Materials Technol.* I. vi. 172 There is usually a limiting rate of cooling from the outside for any given steel, beyond which it is impractical to go because too-rapid contraction from the outside may cause quench-cracking. 1934 H. O'NEILL *Hardness of Metals & its Measurement* vi. 201 Mehl..has reported that quench-hardening a pearlitic steel does not alter its compressibility. *Ibid.* 202 Ordinary quench-hardening practice by continuous rapid cooling to room temperatures will produce martensite if the rate is sufficient to preserve austenite down to Ar″. 1961 G. E. DIETER *Mech. Metallurgy* v. 146 Quench hardening results in an increase in yield stress and a decrease in the rate of strain hardening. 1969 D. K. ALLEN *Metallurgy Theory & Pract.* vii. 194/2 Most all carbon steels can be quench-hardened but the hardness does not become appreciable until the carbon content..reaches about 0·35 percent. *Ibid.* 196/2 The second requirement for quench hardening is that the steel be heated to the recommended hardening temperature and held for a sufficient length of time to allow the steel to become fully austenitized.

quenched, *ppl. a.* (s.v. QUENCH *v.*). Add: **a.** Also with *out* and in other senses of the vb. (Further examples.)
1881 O. WILDE *Poems* 211 The quenched-out torch, the lonely cypress-gloom. 1938 [see *QUENCH *sb.* 2]. 1946 [see *quenching rate* s.v. *QUENCHING *vbl. sb.* 2]. 1958 [see *AGGREGATE *sb.* 6]. 1963 B. FOZARD *Instrumentation & Control Nucl. Reactors* v. 50 Organically quenched counters are characterised by high starting and operating voltages.
b. Radio. *quenched spark*, a spark in a spark transmitter that is extinguished mechanically soon after it begins (see *QUENCH *v.* 1d); so *quenched gap*, a spark-gap designed to bring this about.
1910 G. W. PIERCE *Princ. Wireless Telegr.* xxiii. 269 The quenched spark is..economical in transmitting energy, and is favorable to sharp tuning. 1927 O. F. BROWN *Elements of Radio Communication* iv. 53 The two methods most frequently employed for quenching are either the use of a rotating spark gap or a specially designed spark gap known as the 'quenched gap'. 1962 J. H. & P. J. REYNER *Radio Communication* vii. 294 Owing to the rapid cooling a very high spark frequency may be used, and quenched spark sets operated with a spark frequency of 1,500 per second or more.

quencher. a. (Further examples.)
1950 H. W. LEVERENZ *Introd. Luminescence of Solids* iv. 132 A phosphor center may function as..a poison (or killer, or quencher), by having the excited-state equilibrium level sufficiently near or above *f* so that radiationless transitions predominate. 1961 G. R. CHOPPIN *Exper. Nuclear Chem.* v. 61 The effect of multiple discharges due to failure of the quencher is included..but will be negligible for a good tube operating at the proper plateau voltage. 1971 *Nature* 13 Aug. 444/3 It is well known that paramagnetic ions are efficient quenchers of electronically excited states. 1976 *Sci. Amer.* June 47/2 (Advt.), This is because the excitation energy of $^1O_2$ is unusually low; a quencher molecule to relieve $^1O_2$ of this energy must have an even lower excited state.

quenching, *vbl. sb.* Add: **1.** (Further examples.)
1908 J. A. FLEMING *Elem. Man. Radiotelegr. & Radiotelegh.* 338 (Index), Quenching noise of an electric spark. 1928 *Proc. Nat. Acad. Sci.* XIV. 849 (*heading*) The quenching of cadmium resonance radiation. 1943 B. F. WELLER *Radio-Technol.* iv. 114 Quenching may be effected by a separate valve,..or the reacting detector valve may be arranged to oscillate at the quenching frequency, as well as the radio-frequency. 1963 B. FOZARD *Instrumentation Nucl. Reactors* ii. 25 Because of the need for 'quenching' in a Geiger–Mueller counter for example, its detailed design may be quite different from that of an ionisation chamber. 1972 DE PUY & CHAPMAN *Molec. Reactions & Photochem.* iii. 37 Sensitization and quenching are important methods for determining the spin multiplicity of excited states responsible for photochemical reactions.
2. *quenching crack, medium, rate, trough.*
1926 A. SAUVEUR *Metallogr. & Heat Treatment of Iron & Steel* (ed. 3) xv. 220 Water quenching is to be preferred to oil quenching if it can be performed without producing quenching cracks. 1966 C. R. TOTTLE *Sci. Engin. Materials* x. 224 The strain in the transformed martensite is tensile, in the circumferential direction, and so radial cracks form in the martensite to relieve the stress; these are known as quenching cracks. 1932 E. GREGORY *Metallurgy* iv. 112 Water is obviously the cheapest quenching medium and is invaluable for tools and purposes where an extremely hard surface is desired. 1946 *Nature* 31 Aug. 308/1 Experiments with various iron-carbon alloys quenched in various ways tend to show that the amounts of ferrite, martensite and retained austenite obtained in the quenched specimen are independent of the quenching-rate so long as a certain critical rate..is not exceeded. 1896 F. S. MEYER *Handbk. Art Smithing* ii. 19 In the front part of the forge are found, as a rule, a quenching trough, hollows and receivers for fuel and slack. 1973 *Canad. Antiques Collector* May–June 7 (*caption*) The stone quenching trough from the oldest smithy in eastern Ontario.
quenching *ppl. a.* (later examples).
1954 [see *QUENCH *v.* 3d]. 1958 W. K. MANSFIELD *Elem. Nucl. Physics* vi. 50 Positive ions arriving at the cathode are sometimes able to eject an electron. If this were to occur..a continuous series of pulses might be observed. This is prevented in a Geiger counter by the inclusion of a quenching agent. 1966 D. G. BRANDON *Mod. Techniques Metallogr.* iii. 154 The addition of a small amount of a second, 'quenching', gas..serves to prevent secondary electron emission by the positive ion bombardment of the cathode.

quenchless, *a.* (Further examples.)
1838 W. HOWITT *Rural Life Eng.* II. I. ii. 35 The Romances of Scott..have..piled quenchless fuel on this social flame. 1895 YEATS *Poems* 12 And with quenchless eyes and fluttering hair A beautiful young man followed behind. 1952 C. DAY LEWIS tr. *Virgil's Aeneid* IV. 78 And consecrated their quenchless flames. 1976 *New Yorker* 15 Nov. 59/1 Vaccaro and Jack Smith, the underground filmmaker, shared a quenchless passion for 'Siren of Atlantis', 'White Savage', 'Cobra Woman', and other nineteen-forties epics starring Miss Montez.

quenelle. (Earlier and later examples.) Also **quenelle de volaille,** a ball made with chicken or other fowl meat.
1845 E. ACTON *Mod. Cookery* vi. 180 *French Forcemeat called Quenelles.* This is a peculiarly light and delicate kind of forcemeat. 1846 [see *CROÛTON]. 1861 MRS. BEETON *Bk. Househ. Managem.* 202 *Veal Quenelles*... If the quenelles are not firm enough, add the yolk of another egg. 1889 J. WHITEHEAD *Steward's Handbk.* IV. 420/1 *Richelieu garnish*, quenelles of chicken, cockscombs and slices of fat livers in brown onion sauce. 1936 LUCAS & HUME *Au Petit Cordon Bleu* 53 Decorate the tops of each *paupiette* with small fillets of anchovy and the *quenelles* with strips of anchovy. 1976 *Punch* 27 Oct. 737/1 Is it also prejudice..to prefer *quenelles* to fish cakes, to hate Coca-Cola and adore wine? 1977 C. MCFADDEN *Serial* (1978) vi. 18/1 She could really dig *quenelles* about now.

quenselite (kwe·nsĕləit). *Min.* [ad. G. *quenselit* (G. Flink 1925, in *Geol. Föreningens i Stockholm Förhandl.* XLVII. 377), f. the name of P. D. *Quensel* (b. 1881), Swedish mineralogist: see -ITE¹.] An oxide of lead and manganese, $PbMnO_2(OH)$, found as black, tabular, monoclinic crystals.
1926 *Mineral. Abstr.* III. 110 Quenselite, another new mineral from Långban, Sweden, occurs as small (1 mm.) pitch-black crystals with calcite and baryte in crevices of the granular haematite ore. 1958 *Proc. Nat. Inst. Sci. India* A. XXIV. 95 This is probably the first reported occurrence of quenselite in manganese ores of metamorphic origin. 1971 *Zeitschr. für Kryst.* CXXXIV. 331 The significance of the quenselite structure lies in its role as a connecting link between certain of the Pb oxides and the lithiophorite-chalcophanite group... In addition to red and yellow PbO, quenselite has structural similarities to Pb_3O_4.

quercetin. Add to def.: 3,3′,4′,5,7-Pentahydroxyflavone, $C_{15}H_{10}O_7$. (Further examples.)
1949 *Thorpe's Dict. Appl. Chem.* (ed. 4) IX. 300/2 Quercetin gives red-brown, brown-orange, bright orange and olive-black shades on wool mordanted with chromium, aluminium, tin, and iron, respectively. 1962 T. R. SESHADRI in T. A. Geissman *Chem. Flavonoid Compounds* ii. 9 Quercetin and its glycosides can be conveniently extracted by borax and can be liberated by acidification.

quercitron. Add: **2.** Special Combs. **quercitron lake, yellow,** the yellow pigment obtained from quercitron bark, yielding quercetin and rhamnose on hydrolysis; quercetrin.
1886 H. C. STANDAGE *Artists' Man. Pigments* iv. 43 Yellow Lakes (Madder Yellow,..Quercitron Yellow or Lake). 1918 PERKIN & EVEREST *Natural Org. Colouring Matters* xix. 628 Quercitron-yellow lake, Flavin-lake, or Dutch pink can be made..by precipitating a decoction of quercitron bark containing alum with chalk. 1934 Quercitron lake [see *Italian pink]. 1947 L. S. PRATT *Chem. & Physics Org. Pigments* vii. 65 Quercitron lake is a yellow coloring matter made from the inner bark of a species of oak, *Quercus tinctoria*, that is indigenous to North America.

‖**querencia** (kere·nþia). [a. Sp. *querencia* lair, haunt, home ground, f. *querer* to seek, desire, f. L. *quærere* to seek.] **1.** *Bullfighting.* The part of the arena where the bull takes his stand; stamping ground.
1932 E. HEMINGWAY *Death in Afternoon* xiii. 150 A querencia is a place the bull naturally wants to go to in the ring; a preferred locality. That is a natural querencia and such are well known and fixed, but an accidental querencia is more than that. It is a place which develops in the course of the fight where the bull makes his home. 1957 R. CAMPBELL *Portugal* vi. 115 He [*sc.* the bull] may choose his *querencia* for some strategical advantage—near the body of a dead horse, for instance. 1964 *Listener* 27 Aug. 317/1 It is when the bull leaves his *querencia*—the place where he feels safe—that he falls a victim to delusion. 1974 F. NOLAN *Oshawa Project* i. 3 Some taunted fighting bull seeking its *querencia*.

2. *fig.* A (person's) favourite place; home ground, refuge.
1952 R. CAMPBELL *Lorca* i. 8 Andalusia is Lorca's *querencia*. 1977 A. SCHOLEFIELD *Venom* III. 98 Returning always to the centre of the gold carpet for there, like a bull in a ring, he had instinctively made his *querencia*, his territory.

Queres, var. *KERES.

‖**Querflöte** (kvɛ̄·ɹflötə). *Mus.* [a. G. *querflöte* cross-flute, f. *quer* transverse + *flöte* flute.] **1.** A transverse flute, blown through an opening at the side; = *cross-flute* s.v. *CROSS B.
1876 STAINER & BARRETT *Dict. Mus. Terms* 373/1 *Querflöte* (*Ger.*),..The flute played sideways, as opposed to the flute which was blown at one end, and held straight in front of the performer. 1914 H. M. FITZGIBBON *Story of Flute* iii. 30 (*caption*) Prætorius' Bass Querflote, 1620. 1959 WESTRUP & HARRISON *Collins Mus. Encycl.* 525/1 *Querflöte*,..'Cross' or 'transverse flute', i.e. the modern flute as distinct from the recorder or *Blockflöte*. 1976 D. MUNROW *Instruments Middle Ages & Renaissance* 53/3 Back in medieval times the different playing positions of the two instruments had provided a means of distinction..: hence the use of..*Querflöte*, or *Querpfeife* (German, cross flute) for the transverse flute.
2. An organ stop that emits a sound resembling that of a flute.
1921 G. A. AUDSLEY *Organ-Stops* 217 *Querflöte*... The name..has been frequently used by German organ-builders to designate the stop which, in its voice, imitates, as closely as practicable in organ-pipes, the tone of the Flute of the orchestra. 1966 P. WILLIAMS *European Organ 1450–1850* 286 *Querflöte* (Ger. 'cross flute'), properly, an open cylindrical metal or wood flue stop (usually 4′), over-blowing to the first or second overtone due to the pipe's narrow scale, large foot-holes and fairly low cut-up.

‖**querida** (keri·ða). [Sp. *querida*, pa. pple. *querer* to seek, desire, f. L. *quærĕre* to seek.] A sweetheart, darling: freq. used as a term of address. Also **querido** (-ðo), the male equivalent.
1846 R. FORD *Gatherings from Spain* xx. 274 His short-petticoated *querida*. 1926 W. N. BURNS *Saga of Billy the Kid* xiv. 185 In every *placeta* in the Pecos some little señorita was proud to be known as his *querida*. 1963 E. LININGTON *Death of Busybody* i. 9 Be careful now, *querida*. Lock both doors on your way home. 1970 KOENIG & DIXON *Children are Watching* iii. 23 Did her *querido* have to go back to work at the restaurant? 1976 'S. WOODS' *My Life is Done* 40 Everything will be well, *querido*.

querl, *sb.* (Earlier and later examples in form *quirl*.)
1854 B. F. TAYLOR *Jan. & June* 23 [The grape vine's] aspirations were manifested in the display of divers mermaidish-looking ringlets, with two or three dainty 'quirls' therein. 1871 L. M. ALCOTT *Little Men* v. 78 Sally, loading her pie with quirls and flourishes. 1889 R. T. COOKE *Steadfast* xv. 162 A hundred resolute little quirls above the low forehead. 1950 *Publ. Amer. Dial. Soc.* XIV. 55 *Quirl*, a *curl*, as on a watermelon vine. A melon is supposed to be nice when the quirl is dead.

querl, *v.* Add: Also **quirl.** (Examples.) So **querled** *ppl. a.*, **que·rling** *vbl. sb.* and *ppl. a.*
1787 *Amer. Museum* II. 571/1 She thought there was something alive in her side, for, to use her own expression, she plainly perceived a tickling and quirling in it. *Ibid.* 574/1 She next complained of a quirling pain, that would last three or four hours with the utmost violence. 1830 *Northern Watchman* (Troy, N.Y.) 30 Nov. 3/5 We..come out of the plagid lock, wrong eend foremost, all quirled up in a h—l of a twist. 1840 J. F. COOPER *Pathfinder* I. xiii. 206 One of his hands coiled a rope against the Sun, and he called it *querling* a rope, too, when I asked him what he was about. 1890 *Dialect Notes* I. 75 'Quirled way up'.. 'Quirl, both noun and verb, is familiar to me.' 1893 H. A. SHANDS *Some Peculiarities of Speech in Mississippi* 52 *Quirl*,..this word is largely used by negroes, and to some extent by white people, for *curl*. It is also thus used in New England. In Mississippi a snake is nearly always said to be *quirled* or *quoiled up*, instead of *curled* or *coiled up*. 1944 *Publ. Amer. Dial. Soc.* II. 30 *Quirl*, to curl. 'Does hit *quirl* like a pig's tail?'.. Common.

querulist. (Later example.)
1922 C. E. MONTAGUE *Disenchantment* iv. 52 The querulist of the book took it hard..that more kind words did not come to the men.

query, *sb.*¹ Add: **2. b.** *spec.* in the Society of Friends, an item in a formal list of questions issued for the guidance of Friends; now freq. in phr. *Advices and Queries* (see quot. 1954).
1654 BURROUGH & HOWGILL (*title*) Answers to severall queries put forth to the despised people called Quakers. 1701 G. KEITH *Answer to 17 Queries Quarterly Meeting Quakers, Oxford* 3 Those seventeen Queries ye sent me being only Queries, contain little or nothing Affirmatively or Negatively, by way of position. 1768 in *Extracts Minutes Yearly Meeting Friends, London* (1783) 269 This meeting directs, that the 11th query remain as it now stands. 1797 *Encycl. Brit.* XV. 737/2 At the quarterly-meeting are produced written answers from the monthly-meetings, to certain queries respecting the conduct of their members, and the meeting's care over them. 1898 *Friends' Q. Examiner* 483 For about a century it was uncertain whether the interrogations addressed to the meetings of Friends should be termed 'questions' or 'queries'. The former term was chiefly employed up to 1762, but 'queries' has held the field since 1783. 1921 R. M. JONES *Later Periods of Quakerism* I. iv. 134 At first the Queries were formal questions asked for the sake of securing information in reference to the number of members suffering under persecution. *Ibid.* 135 As fresh moral issues arose,..the list of Queries enlarged. They grew in number and in importance until they embodied almost all the essential aspects of the Quaker moral ideal, and they furnished a kind of silent confessional for each individual member, as well as a moral measuring rule to guide the Overseers in their work of looking after the flock. 1928 *Advices & Queries* (Society of Friends) 5 The Queries being directed in recent years to arouse the thought and conscience..rather than to obtain specific information. 1954 H. LOUKES *Friends face Reality* viii. 107 Quakers have performed this process of moral illumination of each other, while at the same time guarding against the danger of setting up an external moral code, by a system of 'Advices and Queries', moral and spiritual reflections couched in the form most calculated to set the individual searching his own heart.
3. Also written out as a quasi-*adj.* or -*adv.*, preceding the word(s) to which it refers.
1942 *Jrnl. R. Naval Med. Service* XXVIII. 21 Admitted with a diagnosis of 'query' septicæmia. 1953 R. LEHMANN *Echoing Grove* 61 'You reminded me of someone I once knew...' (Query her sister?) 1967 G. F. FIENNES *I tried to run Railway* iii. 21 It would have had to be an unwary Hun that let me get near him with my pike—design Circa 1500; origin query Birmingham Small Arms Company. 1977 J. MCCLURE *Sunday Hangman* x. 109 '*Murders*...,' snapped Colonel Muller. 'Query murders,' corrected Kramer, recognising an urgent need..to treat the situation as routinely as possible.

query, *v.* Add: **1. b.** (Later examples with direct speech as obj.)
1905 *Smart Set* Sept. 119/2 'Been here long?' I queried. 1976 B. FREEMANTLE *November Man* iii. 38 'The elections?' queried Hollis.
2. For *rare* read 'Now chiefly *U.S.*' and add further examples.
1943 *Sun* (Baltimore) 30 Nov. 10/7 He queries me concerning the passage. 1952 *Time* 14 Apr. 17/1 Before an issue of *Time* goes to press..a twin-bed position may be dummied, a stringer queried for a checking point. 1974 *Spartanburg* (S. Carolina) *Herald* 18 Apr. A6/2 College Students..who were queried at Iowa State University, have some curious ideas about what 'academic freedom' means. 1977 *Daily Times* (Lagos) 11 Jan. 20/2 When these officers were queried, they felt unhappy.
3. a. (Later example.)
1961 [see *EAR *sb.*¹ 5].

‖**quesadilla** (kesadī·lʸa). [Sp.] A variety of turnover, usu. with a cheese filling.
1944 E. ZELAYETA *Elena's Famous Mexican & Sp. Recipes* iv. 35 Quesadilla. Tortilla Stuffed with Cheese. Take fresh tortillas..place generous piece of Monterey cream cheese..in the center, and fold it over as you would a turnover... Cook lightly, turning often until cheese is melted. 1957 *House Beautiful* Sept. 176/4 Quesadillas. Small-size tortillas made into turnovers and filled with cheese or squash flowers or sometimes chicken *molé*. They are baked or deep-fat fried. 1963 *Sunset* Jan. 70/1 Quesadillas (Fruit and Cheese Turnovers). 1965 *House & Garden* Sept. 223/3 Quesadillas. These turnovers are made with unbaked tortillas stuffed with a variety of fillings. 1978 *Chicago* June 249/1 Start with nachos..and quesadillas (miniature tortillas filled with guacamole or thick white cheese and onions—60c).

quest, *v.*¹ **1. a.** (Later examples.)
1954 J. I. LLOYD *Beagling* 142 *Quest*, draw for a hare. 1969 M. PUGH *Last Place Left* x. 60 We heard the shot then saw the spaniel questing. 1976 *Shooting Times & Country Mag.* 16–22 Dec. 25/2 Two short toots on the horn..had hounds questing among the tufts of coarse grass and gorse bushes.

questing, *vbl. sb.* (Later example.)
1923 M. SADLEIR *Desolate Splendour* iv. 72 From externals only need the weary questing of a stranger mind seek teaching or enlightenment.

que·stingly, *adv.* [-LY².] In a questing manner.
1926 R. CLEMENTS *Stately Southerner* 89, I..looked questingly right and left.

question, *sb.* Add: **II. 3. b.** (Earlier and further examples.) From the 18th cent., *spec.* a question put in Parliament by a Member to the Government or to a Minister.
1549 *House of Commons Jrnl.* 28 Mar. 21/1 In the Question, it is agreed, That the Number, which said No to the Bill, be the greater Number by One Person. 1559 *Ibid.* 17 Apr. 60/1 Carnesew declared to the House, that Thrower, Servant to the Master of the Rolls, did say against the State of the House, that if a Bill were brought in for Womens Wyers in their Pastes, they would dispute it, and go to the Question. 1614 *Ibid.* 13 Apr. 464/1 Mr. Hackwill:—That nothing to pass, by Order of the House, without a Question; and that no Order, without a Question Affirmative and Negative: And that ordered to be, upon the Question. 1778 *Parl. Reg.* 16 Dec. 181 Lord Newhaven put a variety of questions to the gentlemen belonging to the board of ordnance. 1844 T. E. MAY *Treat. Law, Privileges, Proceedings & Usage of Parl.* viii. 166 Any member may propose a question, which is called 'moving the house', or, more commonly, 'making a motion'. *Ibid.* 171 In the commons, when the motion has been seconded, it merges in the question, which is then proposed by the speaker to the house, and read by him. 1908 A. E. STEINTHAL tr. *Redlich's Procedures House of Commons* II. VII. vi. 241 Requests for information, 'Questions' are regularly addressed by members of the House to the Government, and at times to the Speaker or to private members. 1929 G. F. M. CAMPION *Introd. Procedures House of Commons* iv. 124 Oral Questions are by far the most numerous. To them is allotted the whole of 'Question time' proper, *i.e.* from not later than three o'clock to not later than a quarter to four. 1956 P. HOWARTH *Questions in House* i. 17 Apart from the procedural reasons, there were also reasons of a political or a constitutional nature why the custom of asking parliamentary questions developed slowly. 1958 S. HYLAND *Who goes Hang?* xvi. 72 As the only office-holder present, he knew about the Colonial Secretary's statement due at the end of Questions. 1971 P. D. G. THOMAS *House of Commons in 18th Cent.* ii. 30 The eighteenth century saw the evolution of the Parliamentary question. *Ibid.* 32 Questions in the House proper may well have been established practice long before the first instance found of a question put to and answered by a minister. 1976 *Ann. Rep., Howard League for Penal Reform 1975/76* 6 On the Bail Bill..the two organizations suggested a number of amendments... We are grateful to several MP's..who asked Questions, some at our suggestion.
d. (Later examples.)
1930 G. B. SHAW *Apple Cart* I. 17 You cant. You mustnt. Of course not. Out of the question. 1977 A. ECCLESTONE *Staircase for Silence* iv. 76 Anything like an attempt to impose a parochial structure, new or old, was out of the question.
5. c. *no questions asked*: with no need to give an account of oneself or one's conduct.
1948 M. LASKI *Tory Heaven* ix. 121 Under the old system, I could have had the pair of them and no questions asked. 1962 WODEHOUSE *Service with Smile* x. 160 Give him a skipper and a little daughter..and he could have made straight for the reef of Norman's Woe, and no questions asked. 1968 *Listener* 4 July 30/2 They can..sign a contract with a sports or tobacco firm and will then be..able to play lawn tennis for profit 365 days in the year—and no questions asked.
6. a. *a good question*: see *GOOD *a.* 14c.
d. In various proverbial phrases and expressions, as *ask me no questions and I'll tell you no lies* and varr.; *a civil question deserves a civil answer* and varr. Also *ask a silly question and you get a silly answer*: see *SILLY *a.*
1773 GOLDSMITH *She stoops to Conq.* III. 51 Ask me no questions and I'll tell you no fibs. 1818 SCOTT *Ht. Midl.* I. ix. 247 If ye'll ask nae questions, I'll tell ye nae lees. 1844 T. C. HALIBURTON *Attaché* 2nd Ser. II. iv. 62 Let me give you a piece of advice;—Ax me no questions, and I'll tell you no lies. 1853 —— *Sam Slick's Wise Saws* II. ii. 48 Give a civil answer to a civil question. 1858 S. A. HAMMETT *Piney Woods Tavern* xxvii. 285 The Squire there asked me a civil question, and that deserves a civil answer,—at least that's manners where I come from. 1900 H. LAWSON *Over Sliprails* 135 'Where did you buy the steer, father?' she asked. 'Ask no questions and hear no lies.' 1955 W. C. MACDONALD *Destination Danger* xii. 140 Quist smiled. 'Ask me no questions, I'll tell you no lies.' 1970 V. CANNING *Great Affair* xii. 221 'What has happened to Sarah?'..'Ask no questions hear no lies.'
III. 7. b. *question-answerer, -answering* (examples), *-asking* (further examples), *-beggar, -begging* (earlier and later examples), *-putting* (example), *raising*. **d. question-master,** the chairman of a discussion panel (*PANEL *sb.*¹ 5b), by whom the questions are put; **question pitch,** the rising intonation of an interrogative sentence; **question time** (earlier example); *spec.* a time set apart in Parliament for Members to question Ministers; **question word,** an interrogative pronoun, etc., used to introduce a question.
1900 *Westm. Gaz.* 20 Jan. 9/2 (Advt.), It is also a great question-answerer, or work of reference. 1972 *Computers & Humanities* VII. 9 Simmons is particularly good in describing the extensive progress made in second-generation question-answering systems. 1977 *Dædalus* Fall 123 The various attempts to process natural language by machine—analysis and synthesis of speech, automatic translation, question-answering,..and the like. 1938 *Ann. Reg. 1937* CLXXIX. 303 To the same medley of generations in England belong Mr. Richard Aldington's *The Crystal World*..and Mr. W. H. Auden's question-asking *Spain*. 1972 *Jrnl. Social Psychol.* LXXXVII. 9 English praise delivered to small groups of these children by a familiar Anglo adult exerted some degree of reinforcing effects on their question-asking behavior. 1935 A. P. HERBERT *What a Word!* viii. 229 The chief kind is the Question-beggar, the epithet or phrase which assumes or imputes that the question under discussion has been conclusively answered already. 1824 J. BENTHAM *Book of Fallacies* IV. i. 213 (*heading*) Fallacies of Confusion, the object of which is, to perplex, when Discussion can no longer be avoided... Question-begging appellatives. 1863 GEO. ELIOT *Romola* III. xiii. 130 There was no argument more widely convincing than question-begging phrases in large type. 1910 A. SIDGWICK *Application of Logic* vii. 217 Question-begging in the extended sense..occurs just so far as any attempt is made on the part of either disputant to prevent question-*raising*. 1911 H. G. WELLS *New Machiavelli* I. iv. 113, I scoffed at that pompous question-begging word 'Evolution'. 1957 *Times Lit. Suppl.* 27 Dec. 782/2 He is not above question-begging in the most ingenuous way. 1979 C. MOULE in M. Goulder *Incarnation & Myth* v. 138 Evidence of this kind in no way depends..upon question-begging theories of scriptural authority. 1946 L. MACNEICE *Dark Tower* (1947) 165 Listeners will have the privilege of hearing a number of experts on Truth, all of them equally infallible; our question-master is the March Hare. 1952 *Gloucestershire Echo* 3 Oct. 6/2

It has become traditional for the Festival to end..with a Brains Trust. With Gilbert Harding as question-master and..a varied team. 1977 'E. CRISPIN' *Glimpses of Moon* ii. 36 The bright, uncommitted fashion of a television question-master..in a quiz. 1933 L. BLOOMFIELD in Saporta & Bastian *Psycholinguistics* (1961) 244/2 *Yeah?* and *Is that so?* with a peculiar modification of the question-pitch, have been used as facetious vulgarisms expressing disbelief. 1964 C. C. FRIES in D. Abercrombie et al. *Daniel Jones* 244 Formal yes-or-no questions, along with question-pitch. 1884 E. W. HAMILTON *Diary* 30 July (1972) II. 663 My main points are:..2. Confinement of question-putting to Private Members' nights [etc.]. 1910 Question-raising [see *question-begging* above]. 1959 *Times* 25 Sept. 8/4 Curious and question-raising as they are, the megapodes are worth a more serious..programme of research. 1852 MRS. GASKELL *Let.* 4 Sept. (1966) 197 (*heading*) Saty schoolroom, Question-time. 1891 W. FRASER *Disraeli & his Day* 381 Colonel Makins, the..Member for Essex..said, 'They have got it hot this afternoon about a Dissolution.' I replied, 'Oh, nonsense!' This was during 'Question-time'. 1936 H. NICOLSON *Diary* 3 Dec. (1966) 281 Members crowd in as question-time draws to its end. 1976 H. WILSON *Governance of Britain* vii. 132 Harold Macmillan, a highly successful performer at Question time. 1924 H. E. PALMER *Gram. Spoken Eng.* 263 In Direct Questions, the question-words are said to be interrogative; in Indirect Questions, they are said to be conjunctive. 1964 E. ULDALL in D. Abercrombie et al. *Daniel Jones* 274 Question-word question: 'What did he think they were doing?' 1978 *Language* LIV. 86 In English, questions are typically initiated by question words or verbs, so as to distinguish them from declarative sentences.

question, *v.* **5. b.** Delete † *Obs.* and add later example.
1879 G. MEREDITH *Egoist* III. xiv. 291 At the game of Chess it is the dishonour of our adversary when we are stale-mated: but in life..such a winning of the game questions our sentiments.

questionability. (Later examples.)
1966 *Listener* 1 Sept. 317/3 It is in the fact that only *one* of this grossly neglected composer's works has appeared that the questionability lies. 1969 R. HARPER *World of Thriller* ii. 51 Only occasionally for most men is life reduced to total questionability by any particular situation.

question and answer. A dialogue consisting of alternate question and answer. Also (with hyphens) *attrib.*; occas. also **question-answer.**
1817 KEATS *Let.* 10 Sept. (1931) 39 My dear Fanny, Let us now begin a regular question and answer — a little pro and con. 1839 [see QUESTION *sb.* 7c]. 1908 MRS. H. WARD *Diana Mallory* II. xii. 237 The trivial question-and-answer of the tea-making. 1940 N. MARSH *Surfeit of Lampreys* (1941) xiii. 187 She maintained a question-and-answer attitude, replying in the most meagre phrases. 1941 L. MACNEICE *Poetry of W. B. Yeats* i. 14 He may be answering quite different questions from mine but the question-answers which he evolves are the same kind of organism, and result from the same kind of activity as my own question-answers. 1945 C. S. LEWIS *Great Divorce* 41 That question-and-answer conception of thought only applies to matter of fact. 1957 E. BOTT *Family & Social Network* ii. 42 The question-and-answer pattern of fact-collecting. 1960 *Guardian* 9 June 9/1 Police interrogators..now hold a daily..question-and-answer session with the former Nazi. 1965 *Language* XLI. 387 The question-answer pair *What does he do? He draws cartoons* can be analyzed in the same way. 1977 *Oxford Diocesan Mag.* Oct. 20/2 It was decided to organise..a question-and-answer programme on an electronic screen. 1980 *English World-Wide* I. i. 28 It is not easy to elicit syntactic information by using the short direct question-answer technique.

questionary, *sb.*² Delete *rare* and add later examples. Also *attrib.*
Now largely superseded by *QUESTIONNAIRE, exc. in *Med.* use.
1951 *Lancet* 7 July 23/1 The questionary method used in this particular study has certain limitations. 1957 *Brit. Med. Jrnl.* 7 Sept. 550/2 The clinical concept of the disappointed undergraduate is therefore given some support by the answers to a general questionary. 1959 *Times* 5 Sept. 10/2 The proposed Welsh dialect atlas, information for which was gathered by means of a questionary of about 1,000 items. 1970 *Jrnl. Gen. Psychol.* Jan. 97 How did you answer when the item was difficult? And why did you answer in such a manner? (verbalization questionary). 1977 *Lancet* 27 Aug. 417/2 After 21 days, the patient was interviewed by one of us..using a standard questionary.

questionee. Delete *rare*⁻¹ and add earlier and later examples.
1838 CARLYLE *Lett. to Mill &c.* (1923) lix. 164 Your answer is according to your question, and your questionee,—'as the fool thinks the bell clinks'. 1905 *Grand Mag.* Feb. 131 The..questionee does not overlook the fact that [etc.]. 1953 *Rep. Sel. Comm. Delegated Legislation* 27/2 in *Parl. Papers 1952–3* IV. 115, I am not sure that the questioner could not answer that rather better than the questionee. 1971 J. WAINWRIGHT *Last Buccaneer* II. 183 He..murmured his questions in a very low voice. This forced the questionee consciously to *listen*.

question mark. (In Dict. s.v. QUESTION *sb.* 7 d.) **1.** *Gram.* A mark of interrogation, represented by the sign ?
1905 T. F. & M. F. A. HUSBAND *Punctuation* II. vi. 74 A question-mark is sometimes placed in the middle of a sentence. In such a position it concentrates attention on certain elements of the thought. 1930 M. A. PINK *Dict. Correct Eng.* 147 To complete the list of stops we may mention here that the Question Mark (?) is used at the end of direct questions. 1960 KIERZEK & GIBSON *Macmillan Handbk. Eng.* II. 278 A question mark is used after a direct question but not after an indirect question. 1971 N. STACEY *Who Cares?* xii. 207 We agreed that the article should be called 'A Mission's Failure'. But I thought it ought to have a question mark after it.
2. *fig.* A point about which there is uncertainty or doubt; an unresolved problem, an enigma. Also, a person whose character is unknown or unfathomable (usu. in some particular respect). Freq. in phr. *a question mark hangs* (etc.) *over* (something), there is doubt about (that thing).
1869 [see *question-mark* s.v. QUESTION *sb.* 7d]. 1924 R. MACAULAY *Orphan Island* xxv. 322 Across the future of Orphan Island..is scrawled a question mark. 1945 A. HUXLEY *Time must have Stop* xxx. 286 Contemporary science..is engaged in destroying, not only things and lives, but entire patterns of civilization. So we find ourselves faced with yet another set of question marks. 1952 M. ALLINGHAM *Tiger in Smoke* xv. 215 Luke was destined to become one of the great policemen... The man was a living question-mark. 1957 J. S. HUXLEY *Relig. without Revelation* (rev. ed.) iii. 62 The great question-mark of our continuance after death. 1958 *Daily Express* 23 July 4/3 It is a question mark which has been ringing in my mind. 1960 *Economist* 15 Oct. 215/2 Question marks now loom hugely over the future of the white-dominated states of southern Africa. 1963 *Listener* 24 Jan. 152/1 His death will take its place as one of the great question marks in English political history. 1971 A. PRICE *Alamut Ambush* viii. 95 Razzak, the unknown quantity,..wasn't quite such a question mark since he'd turned up at the Ryle reception. 1973 *Times* 15 Oct. 22/1 The project..is near to completion of its first stage. But a question mark hangs over the second stage. 1974 *Plain Dealer* (Cleveland, Ohio) 19 Oct. 6-D/2 Caffery..could become a question mark if his knee does not respond to a couple of operations. 1978 G. GREENE *Human Factor* II. iii. 85 A question mark kept him awake for a long while: had there always been a taxi rank so close to Davis's flat? 1979 *Nature* 15 Feb. 506/1 The question mark hanging over the whole meeting was precisely what principle was at stake.
3. (With hyphens.) *attrib.*
1962 *Listener* 15 Nov. 835/2 The effect of changing Byron's tragic question-mark ending to a kind of 'redemption'..is quite unconvincing. 1966 J. S. COX *Illustr. Dict. Hairdressing* 124/1 *Question-mark curl*, a stand up curl with a very long stem, like a quaver in music. 1973 M. AMIS *Rachel Papers* 130 Shaving-cream bubbled inside the nearer of his question-mark ears.
Hence **que·stion-marked** *a.*, accompanied by a question mark.
1950 G. BARKER *True Confession* VI. 33, I will not care who Or what you are, save palliation Of the question marked heart. 1975 C. WESTON *Susannah Screaming* (1976) x. 52 The guff was already sorted, checked where possible, question-marked where not.

questionnaire (kwestyənē³·ɹ, ‖ke-). [a. F. *questionnaire*, f. *questionner* to ask questions.]
a. A list of questions by which information is sought from a selected group, usu. for statistical analysis; a questionary.
The word was resisted by purists (see Fowler *Mod. Eng. Usage* (1926) 479/1) for many years after its first use in English. Some retained a Fr. pronunc. (kestionę̄r), whilst others preferred the Eng. word QUESTIONARY *sb.*² The anglicized pronunc. is now dominant.
1901 E. B. TITCHENER *Exper. Psychol.* I. I. xii. 197 The questionary or 'questionnaire' is a series of questions bearing upon the matter to be investigated, and submitted to a large number of persons for introspective answer. 1920 *Glasgow Herald* 20 Aug. 7/1 Valuable information, never previously collected, is being obtained through a questionnaire by the Federation of British Industries concerning the fuel requirements of the great industrial centres. 1924 W. B. SELBIE *Psychol. Relig.* i. 4 A careful study of the phenomena of religious experience derived mainly from biographies, introspection, and a systematic use of the questionnaire. 1931 *Times Lit. Suppl.* 19 Feb. 126/2 There was issued recently the report of a study (by *questionnaire*) of two hundred marriages. 1952 *Shell Aviation News* June 9/2 The moment for the distribution of a printed questionnaire is obviously in flight, when passengers are frequently bored and glad of any distraction. 1975 *New Yorker* 21 Apr. 45/2 The reports involve the verification of routine information that has already been supplied by citizens in response to questionnaires included in application forms for insurance, for employment, for mortgages or apartment leases. 1978 S. BRILL *Teamsters* viii. 312 Drivers..responded to questionnaires placed as advertisements in trucking magazines.
b. *attrib.*
1924 R. M. OGDEN tr. *Koffka's Growth of Mind* ii. 45 With the aid of Hall's questionnaire-method, one of his students has collected a large mass of material concerning children's play. 1941 J. S. HUXLEY *Uniqueness of Man* xi. 231 The questionnaire method is widely used. 1949 M. MEAD *Male & Female* 457, I have had access to enormous varieties of..original questionnaire blanks. 1964 I. L. HOROWITZ *New Sociology* 6 Specialized techniques of questionnaire design [etc.]..make the interviewing process into the end of research rather than merely its instrument. 1978 *Regional Lang. Stud.—Newfoundland* VIII. 31 The purpose of the project was..to give experience in questionnaire design and administration.

questionous (kwe·styənəs), *a.* *rare.* [f. QUESTION *sb.* + -OUS.] Given to asking questions; inquisitive.
1893 R. BRIDGES *Humours of Court* II. i. 914 Of late you are grown questionous and prying.

questor. 3. (Further example.)
1977 *Times Lit. Suppl.* 23 Dec. 1498/5 The prosing of that coruscating bore Dr Emily Brightman, a notator and questor of the first water.

‖**Questore** (kwestō·re). [It. *questore*, f. L. *quæstor*, contraction of *quæsitor* investigator, f. *quærere* to seek, enquire.] = QUESTOR 2 b.
1943 I. ORIGO *Diary* 10 Dec. in *War in Val d'Orcia* (1947) 120 His family succeeded in speaking to the Questore, and inquired what charge there was against him. 1969 G. GREENE *Travels with my Aunt* II. vii. 296 If the *Questore* had described me as a rat, I would have had no objection. 1972 K. BENTON *Spy in Chancery* v. 36 The Questore of Rome had done his polished best to make the luncheon for the Interpol delegates a success.

‖**Questura** (kwestū·rā). [It. *questura*, f. L. *quæstura* the office of a quæstor: see prec.] In Italy: the police station or headquarters; the police.
1907 *Daily Chron.* 22 Aug. 4/4 He had been kicked out of the Central Questura, whither he had gone to give information, because the Neapolitan city police were in the pay of Camorrist assassins. 1950 E. HEMINGWAY *Across River & into Trees* viii. 65, I fill out a slip there for the Questura. 1965 'W. HAGGARD' *Hard Sell* ii. 15 Charles Russell took a taxi from the *questura* back to his hotel. 1975 'D. RUTHERFORD' *Mystery Tour* vii. 149 No one leaves before the inspector from the Questura in Varese gets here.

‖**quête** (kɛ̄t). [Fr., quest.] The traditional act of begging for food or alms to the accompaniment of folk-song; *spec.* as part of a folk-play. Also *attrib.*, as *quête song*. Cf. QUEST *sb.*¹ 7.
1903 E. K. CHAMBERS *Mediaeval Stage* I. vi. 119 Hardly a rural merry-making..is without its procession; if it is only in the simple form of the *quête* which the children consider themselves entitled to make. *Ibid.* viii. 168 The rest..have either become..mere *quête* songs, or..have taken on a Christian colouring. 1933 —— *Eng. Folk-Play* 13 Structurally, the piece falls into three parts: the Presentation.., the Drama.., the *Quête*. *Ibid.* 21 To his normal lines Father Christmas may add others which..are also found as independent *quête*-songs of the Christmas season. 1967 A. L. LLOYD *Folk Song in England* ii. 102 Here is another set of wassailing verses.— The begging motive..is important... We call such pieces *quête* songs. 1979 R. PALMER *Eng. Country Songs* 15 A number of such songs..provided the opportunity for a *quête*.

quetsch (kvetʃ, kwetʃ), *sb.* Also **quetsche,** † **quitch.** [a. G. *quetsche*, dial. form of *zwetsche* plum.] A variety of plum with oval, dark-skinned fruit; also, the liqueur made from plums of this kind. Also *attrib.*
1839 C. MCINTOSH *Orchard & Fruit Garden* 327 The German Quitch Plum is dried and preserved in immense quantities. 1842 J. C. LOUDON *Suburban Horticulturist* III. iv. 559 Quetsche... A good bearer, and well adapted for drying. 1860 R. HOGG *Fruit Man.* 251 Quetsche... Fruit medium sized, oval... Skin dark purple... A culinary plum. 1936 BENTLEY & ALLEN *Trent's Last Case* xi. 130 His wife was a Lorrainer and responsible for the Quetsch, the liqueur made from her father's plums. 1940 [see *MIRABELLE]. 1961 *Sunday Times* 16 July 36/6 Of plums and damsons there will not be a single one, and worst of all there will be no quetsch for jam making. Quetsch jam is one of the very best. 1966 P. V. PRICE *France: Food & Wine Guide* 51 Two [plums] that may be met with in open tarts are *mirabelles*..and *quetsches*. 1969 *Listener* 2 Jan. 31/1 The [Romanian] national drink..is a plum brandy like quetsch or slivovitch. 1975 WOOD & CROSBY *Grow it & cook It* vi. 234 Quetsche... October. Long oval, black. When stewed has the flavour of Carlsbad plums. 1977 M. JANCATH *Seatag* II. v. 99 A heavy lunch of Quenelles with sauerkraut and..a Quetsch tart.

quetzal. 1. For Latin name substitute the spelling *Pharomachrus mocino*. (Later examples.)
1930 R. MACAULAY *Staying with Relations* ii. 26 Above their heads a quetzal, bright emblem of his country, his lovely tail caught in a liquorice vine. 1950 *Caribbean Q.* II. II. 24 The gorgeous plumage of the Macaw, the Quetzal and the Wild Turkey were sewn or gummed, feather by feather, onto cotton cloth to form resplendent cloaks. 1961 *Guardian* 22 May 5/4 The quetzal is a bird of rainbow plumage which symbolises Central America. 1978 *Washington Post* 7 July B2/3 The normal heart rejoices to think of wolves and quetzals flourishing in the great world.
2. (Pl. **quetzales.**) The name of a silver Guatemalan coin, initially equivalent to one U.S. dollar, and comprising 100 centavos.
1928 *Whitaker's Almanack* 778 Revenue (Budget, 1927–28) *Quetzals* 11,031,102. 1962 R. A. G. CARSON *Coins* 433 The coinage reform [in Guatemala] of 1924 created a new unit the quetzal in silver with subdivisions and with multiples in gold..on obverse the quetzal, a Central American bird of the parrot family. 1974 *Nat. Geographic* Nov. 673 For five quetzals (five dollars, U.S.) I savored a grilled filet mignon, [etc.]. 1977 *Westworld* (Vancouver, B.C.) May–June 20/2 The average income of a [Guatemalan peasant] family ranges from 200 to 300 quetzales a year (a quetzal equals one American dollar).

Quetzalcoatl (ketsa:lko,ā·t'l). Forms: 6 Quecalcoatl, 7 Quetzaalcoatl, Quezalcoatl, 8 Quatzalcoatl, Quezalcoatl, 8- Quetzalcoatl. [a. Nahuatl *quetzalli* (see QUETZAL) + *coatl* snake.] The Plumed Serpent of the Toltec and Aztec civilizations, traditionally known as the god of the morning and evening star, later as the patron of priests, inventor of books and of the calendar, and as the symbol of death and resurrection. Hence **Que:tzalcoa·tlian** *a.* and **Que:tzalcoa·tlism.**

1578 T. NICHOLAS tr. *L. de Gomara's Pleasant Hist. Conqu. Weast India* 203 There was one rounde temple dedicated to the God of the ayre called Quecalcouatl. **1604** E. GRIMSTONE tr. *Acosta's Naturall & Morall Hist. E. & W. Indies* v. ix. 354 In *Cholula* which is a commonwealth of *Mexico*, they worshipt a famous idoll which was the god of marchandise, being to this day greatly given to trafficke. They called it *Quetzaalcoalt.* **1613** PURCHAS *Pilgrimage* I. VIII. ix. 656 They had sacrificed ten children ..to *Quezalcouatl* their god. *Ibid.* 657 Their chiefe god was *Quezalcouatl*, god of the Aire. **1725** J. STEVENS tr. *de Herrera's Gen. Hist. Amer.* II. II. v. vi. 375 There were forty or more great or small, and other lesser Temples.. which being all of different Sizes, and each of them dedicated to a several God, there was one among them round, consecrated to the God of the Air, call'd Quezalcoatl. **1726** —— *Ibid.* III. II. x. iii. 206 At Chulula, a City near Mexico, they ador'd a famous Idol that was the God of Commerce... His Name was Quatzalcoatl, he stood in a very lofty Temple, in a spacious Square, with Gold, Silver, Feathers, and costly Cloaths about him, bearing the Figure of a Man, his Face like a Bird... His name signify'd Snake of rich Feathers. **1787** C. CULLEN tr. *Clavigero's Hist. Mexico* I. II. 88 The Toltecas..built in honour of their beloved god Quetzalcoatl, the highest pyramid of Cholula. **1843** W. H. PRESCOTT *Hist. Conqu. Mexico* I. I. iii. 53 A far more interesting personage in their mythology was Quetzalcoatl, god of the air, a divinity who, during his residence on earth, instructed the natives in the use of metals, in agriculture, and in the arts of government. **1907** L. SPENCE *Mythol. Anc. Mexico & Peru* ii. 20 The worship of Quetzalcoatl was antipathetic if not directly opposed to that of the other deities of Anahuac. **1924** D. H. LAWRENCE *Let.* 15 Nov. (1962) 820 Well, I shall try and finish my *Quetzalcoatl* novel [sc. *The Plumed Serpent*] this winter. **1926** —— *Plumed Serpent* xxvii. 459 If you want to be so—so abstract and Quetzalcoatlian, then bury your head sometimes, like an ostrich in the sand, and forget. **1934** A. HUXLEY *Beyond Mexique Bay* 300 The Indians..can..practise whatever queer blend of catholicism and Quetzalcoatlism pleases them best. **1955** W. GADDIS *Recognitions* I. i. 56 True, many stirred with indignant discomfort..to find they had been attending, not Christ, but..Balder, Attis, Amphion, or Quetzalcoatl. **1973** *Guardian* 23 Mar. 12/4 Quetzalcoatl, the ancient god of the Toltecs.

queue, *sb.* Add: **2.** (Further examples.)

1904 L. HEARN *Japan: an Attempt at Interpretation* xii. 257 All classes excepting the nobility, samurai, Shinto priests, and doctors, shaved the greater part of the head, and wore queues. **1947** R. BENEDICT *Chrysanthemum & Sword* iv. 77 Insignia and distinctive dress of caste were outlawed—even queues had to be cut. **1959** E. TUNIS *Indians* 117/1 The Hopi had brown skins and straight black hair. Men wore it either in a queue bound up in the back or in the long bob they inherited from the Basketmakers. **1976** 'D. FLETCHER' *Don't whistle 'Macbeth'* 22 One of her habitual wigs..that..ended in a pert *queue* at the back.

3. (Further examples.) Also *transf.* and *fig.* *to jump the queue*: see *JUMP *v.* 10c.

1903 E. CHILDERS *Riddle of Sands* xxvi. 298, I joined a *queue* of three or four persons who were waiting their turn, flattened myself between them and the partition till I heard him walk out. **1943** E. M. ALMEDINGEN *Frossia* ii. 64 Paulina had a mind above bread queues and unlit streets. **1951** *Jrnl. R. Statistical Soc.* B. XIII. 152 My own interest in the subject arose from a correspondence.. about queues of taxis in station yards and of customers in retail shops. **1953** *Times* 5 Nov. 4/2 It would be for the Commons to discuss whether the claim of the judges on salaries in the queue of claims should be met before others. **1956** *Newsweek* 9 Jan. 43/1 In Leningrad, Gershwin's music and Heyward's 'Porgy' were anticipated by a two-day queue for tickets priced up to $15 apiece in rubles. **1958** *Listener* 20 Nov. 839/3 After the war the railways had to take their place in the queue after housing and housing repairs. **1966** *Rep. Comm. Inquiry Univ. Oxf.* II. 279 In arts and social studies, most of those with a college post before a university post were tutorial fellows in the 'queue' for a CUF lectureship. **1968** *Sci. Amer.* Aug. 96/1 Airplanes stacked over an airport, shoppers,.. freight cars lined up for unloading at a railroad terminal and messages seeking a free path through a telegraph network all have one thing in common: they are members of a queue, or a line waiting for a service. **1969** *Listener* 28 Aug. 267/3 Are we going to wait until Marxism and socialism have conquered the world, and then stand there last in the queue, waiting for its return to us? **1977** *Spare Rib* May 19/4 Women in poor areas are always at the end of the queue for anything.

6. (Perh. a different word.) A barrel or cask capable of holding approximately one and a half hogsheads of liquid, usu. wine.

1777 P. THICKNESSE *Year's Journey* I. vi. 47 The carriage of a queue of wine from Dijon to Dunkirk..costs an hundred livres..but if sent in the bottle, the carriage will be just double. **1851** C. REDDING *Hist. & Descr. Mod. Wines* v. 91 The names applied in various wine districts of France to the casks which they use, differ without reference to the measure; in the department of the Marne, the *tonneau* is called the *queue*. **1931** W. E. MEAD *Eng. Medieval Feast* iii. 81 In 1385–6 Jean de Neele declared that his household used in one year between six and seven 'queues' of verjuice or between 2,346 and 2,737 litres. **1956** *Atlantic Monthly* June 94/2 In Burgundy the barrel is called *pièce* and contains from 226 to 228 liters, in the Mâconnais 215 liters, in the Beaujolais 216 liters, in Alsace 114 liters. In the Champagne it's called a *queue* and contains 216 liters.

7. *attrib.* and *Comb.*, as *queue day, discipline, driving, form, number, system, theory* (hence *queue-theoretic* adj.); **queue-barging** *vbl. sb.*, = *QUEUE-JUMPING.

1977 *Time Out* 30 Sept.–6 Oct. 15/1 The elaborate queue system is an attempt to eliminate queue barging. **1908** *Daily Chron.* 4 Aug. 3/4 It was queue day at the Franco-British Exhibition yesterday. At 6 o'clock..a line of people a quarter of a mile long extended on either side of the Flip Flap. **1951** *Jrnl. R. Statistical Soc.* B. XIII. 152 The queue-discipline is the rule or moral code determining the manner in which the customers form up into a queue and the manner in which they behave while waiting. **1972** *Guardian* 29 Aug. 2/1 The high standard of British queue discipline. **1970** *Sunday Tel.* 20 Dec. 7/5 Yet another factor contributing to fast 'queue' driving in fog on motorways..is that drivers with their families as passengers tend to drive quickly for fear that a car behind might ram them. **1902** *Westm. Gaz.* 14 Nov. 10/1 From the pens to the steps of the car the intending passengers will go in queue form, as now adopted with so much success at most of the theatres. **1956** R. BRADDON *Nancy Wake* I. i. 9 Each day they received queue numbers so that they could take up their correct positions next morning. **1941** *New Statesman* 27 Dec. 523/2 The argument that the queue system is fair to everybody. **1966** S. BEER *Decision & Control* ix. 176 This thoroughly basic situation is so important in operational research as applied to dynamic systems that a whole branch of mathematical statistics, known as queue theory, has been developed round it. *Ibid.* 178 Some of the earliest queue-theoretic notions were developed around the problem of the doctor's waiting room.

queue, *v.* Add: **2. b.** *trans.* To cause to form a queue; to arrange (persons or things) in or as in a queue or queues.

1928 *Daily Express* 8 Oct. 1/1 The foot and mounted police..had queued the concourse into twisting lines of people. **1973** P. C. SANDERSON *Interactive Computing in BASIC* ii. 23 Multiplexors..checking for transmission errors, and storing and queuing the messages received.

c. *intr.* To stand in a queue; to form *up* in a queue; to take one's place at the end of a queue; also *fig.*

1933 *Observer* 5 Mar. 23/4 There were stuffs at the White City which made French buyers queue up. **1938** E. BOWEN *Death of Heart* I. iv. 71 They hung their hats and coats in the annexe cloakroom, and queued up for the mirror. **1945** 'TACKLINE' *Holiday Sailor* i. 9 Whilst we queued-up before him to have our cap-tallies—*not* cap-ribbons, we now discovered—secured. **1949** E. TAYLOR *Wreath of Roses* i. 11 They have to do all the wretched jobs not even a paid servant will do—queue for tomatoes, etc. **1955** *Times* 1 Aug. 8/7 Everywhere people are queueing—even at the bureau de change and of course at the cafeteria. **1964** L. DEIGHTON *Funeral in Berlin* xxiii. 128 Do you think that the whole of Germany was queueing up to fight Bolshevism? **1976** C. DEXTER *Last seen Wearing* xx. 155 The suspects are beginning to queue up, aren't they, Lewis? **1978** D. FRANCIS *Trial Run* i. 17 We are damned lucky to have been given the few weeks' option. They've got other buyers practically queueing for it.

3. *trans.* To follow or track (a person's steps, etc.).

1906 HARDY *Dynasts* II. v. i. 254 Perhaps within this very house and hour, Under an innocent mask of Love or Hope, Some enemy queues my ways to coffin me.

Hence **queue·ing** *ppl. a.*

1949 N. MITFORD *Love in Cold Climate* I. ix. 91 The large crowd in Park Lane was rewarded by good long stares into the queueing motor cars. **1976** M. RUSSELL *Double Deal* xi. 88, I don't happen to be the queueing type.

queueing, queuing (kiū·iŋ), *vbl. sb.* [f. prec.]

a. The action of waiting in a queue. Also const. *up*.

1927 *Brit. Weekly* 21 Apr. 55/3 When the public-houses opened their doors in the evening there was no queuing-up. **1946** G. MIKES *How to be an Alien* I. 44 Queueing is the national pastime of an otherwise dispassionate race. The English are rather shy about it, and deny that they adore it. **1948** M. LASKI *Tory Heaven* ii. 28 James was delighted to see a row of taxis... There was none of that queuing he had been led to expect. **1951** *Jrnl. R. Statist. Soc.* B. XIII. 180, I assumed that a newly arriving vehicle could always find somewhere in the station yard to unload, so that the problem was in effect not one of queueing. **1956** L. H. C. TIPPETT *Statistics* (ed. 2) vii. 125 This is typical of a number of congestion problems that arise in telephony, in road and rail traffic, in the queuing of patients at a hospital, and so on. **1958** *Manch. Guardian Weekly* 22 May 15/4 Even with a Ponselle a day or two of queueing was all you needed. **1967** *Times Rev. Industry* Aug. 28/2 These techniques include stock control, linear programming and queueing. **1979** *Bookseller* 23 June 2816/3 To avoid queueing..we did not ask visitors to register attendance.

b. *attrib.*, as **queueing theory**, the mathematical study of the structure and behaviour of queues of people or articles.

1951 *Jrnl. R. Statist. Soc.* B. XIII. 168 The different people associated with a queueing system will assess its merits and demerits in different ways. *Ibid.* 181 The congestion should be measured at the peak, but this would need the non-steady solution of a complex queueing problem with non-steady traffic conditions. **1954** *Science News* XXXIV. 112 One particular application [of operational research] is that of queueing theory. This was employed during the design of London Airport and has also been used for such diverse subjects as omnibus routing, toll gate staffing, and determining the number of grinding wheels required by a toolroom. **1966** *Listener* 3 Feb. 162/2 Queueing theory has in fact been used in this calculation in an attempt to relate the time on the waiting list to the number of beds made available, and the demand for these beds. **1974** GROSS & HARRIS *Fundamentals Queueing Theory* i. 1 'How long must a customer wait?' and 'How many people will form in the line?' Queueing theory attempts (and in many cases succeeds) to answer these questions through detailed mathematical analysis.

queue-jumping. [f. QUEUE *sb.* + JUMPING *vbl. sb.*] Pushing forward out of one's turn in a queue; also *fig.* Cf. *JUMP *v.* 10c. Hence (as a back-formation) **queue·-jump** *v.*; also **queue·-jumper**, one who jumps a queue.

1959 *Guardian* 22 Oct. 1/1 Mrs. Braddock..complained of Tory queue-jumping. **1960** *Ibid.* 17 June 2/3 There are three types [of private patients]—the snobs..the queue-jumpers..and the business executive. **1965** M. DRABBLE *Millstone* 67 Afraid that I would be accused of queue-jumping, I rose to my feet and went in search of authority. **1968** *Daily Tel.* 13 Nov. 14/7 A major hindrance to smooth traffic-flow is the queue-jumper, who invariably gets stuck. **1972** *Ibid.* 30 Mar. 16 Private patients in National Health hospitals..are widely regarded as rich queue-jumpers. **1973** *Listener* 6 Sept. 298/3 News reports of Asians who were trying to 'queue-jump' into Britain. **1975** J. PIDGEON *Flame* i. 7 Daniels, having queue-jumped up the crowded stairs..gazed out above the lowered, pebbled window. **1976** *Daily Tel.* 20 Dec. 3/8 The row started when shop stewards complained that the women had queue jumped a union waiting list of people wanting to become ferry drivers.

queuer (kiū·əɹ). [f. QUEUE *v.* + -ER[1].] One who waits in a queue.

1948 'J. TEY' *Franchise Affair* xxii. 260 This was fare that not even the most optimistic queuer outside the court had anticipated. **1952** *Time* 6 Oct. 64/1 The queuers were hoping for standing room. Reserved seats had been gone since July. **1958** *Times Lit. Suppl.* 21 Nov. 669/4 They become refugees somewhere across the border, queuers for charitable soup, squatters on alien schoolroom floors. **1960** V. JENKINS *Lions down Under* xv. 230 Some queuers who had been waiting from 3.30 a.m. **1976** *Times* 10 June 10/3 He watched a senior army officer walk straight to the head of a long taxi queue. The tired queuers grumbled curses.

Queuetopia (kiutōᵘ·piă). [Blend of QUEUE *sb.* + UTOPIA; cf. *SUBTOPIA.] A humorous designation of Great Britain under Labour or Socialist rule, supposedly characterized by universal queueing. Also *transf.*

Said to have been coined by Winston Churchill.

1950 *Manch. Guardian Weekly* 2 Mar. 9/1 (*heading*) 'Queuetopia'. Few of our national disorders have made better campaign material than..the continual queues, the swelling bureaucracy.., and the general mechanisation of the British Way of Life. **1975** S. POTTER *Changing Eng.* 82 London..has far too many *queuetopias* at its bus stops and supermarket checkouts.

‖ **que voulez-vous** (kə vule vu). [Fr., lit. 'what do you want?'] An expression denoting mild exasperation or resignation; 'what do you expect?', 'what can one expect?'.

[**1830** C. CLAIRMONT *Let.* 28 Mar. in J. Marshall *Life & Lett. Mary Shelley* (1889) II. xxi. 202 He [*sc.* Trelawny] receives all his impressions through his heart, I through my head. *Que voulez vous? Le moyen de se recontrer* when one is bound for the North Pole and the other for the South?] **1841** W. M. THACKERAY in *Britannia* 5 June 363/2 No doubt she was dancing away last night..and finished the morning at the Courtille. *Que voulez vous?* it is her nature. **1878** H. JAMES *Europeans* I. vi. 239 The Baroness gave a little philosophic shrug. 'Que voulez-vous? They are princes.' **1880** G. GISSING *Let.* 21 Dec. in J. Korg *George Gissing* (1965) iii. 90, I fear they put me down for a prig, an upstart, an abominable aristocrat, but *que voulez-vous?* **1923** W. J. LOCKE *Moordius & Co.* xxi. 288 *Que voulez-vous*, mademoiselle? A train-omnibus stopping at every station is bound to be late. **1935** LADY FORTESCUE *Perfume from Provence* 35 Are there not floods..all over the world? *Que voulez-vous?* **1945** 'O. MALET' *My Bird Sings* I. v. 37 The poor Comte is hopelessly in love..but *que voulez-vous?* She will not have him.

‖ **queyu** (kwē̆·u). Also **keweyu, kuyu, kway, queyou,** etc. [Guyana Creole, app. from a Cariban language.] In Guyana and neighbouring regions, a small apron-like garment worn by the women of certain Amerindian tribes, consisting of a panel of coloured beads set in intricate geometrical patterns and surrounded by a fringe of cotton.

1796 J. G. STEDMAN *Narr. Five Years' Exped.* I. xv. 386 The women wear an apron of cotton, with party-coloured glass beads strung upon it, which they call *queiou*. This covering is no great size, being only one foot in breadth by eight inches in length..but being heavy..it answers all the purposes for which it was intended. **1806** G. PINCKARD *Notes on West Indies* II. 444 Sometimes, instead of the band, the women use a small apron about three or four inches square, which being tied around the waste [*sic*], and left to hang loose before, serves by way of a fig-leaf. These aprons they call *kways*. **1866** R. DUFF *Brit. Guiana* xi. 261 The only covering which the females wore was the quieyoo, an article of dress, worked out of seeds of trees, about ten inches long, and six or eight broad, hung in front of the person by a string fastened round the loins. **1867** W. T. VENESS *El Dorado* 141 '*Cuyus*', or '*Queyus*', the entire dress of Indian women, of the Accawai tribe. **1895** *Timehri* June 144 The queyus too were remarkable owing to their small bead surface, the greater extent

being taken up by wide cotton fringes. **1904** W. H. HUDSON *Green Mansions* v. 71 Oalava herself would be ready to bestow her person—queyou, worn fig-leaf-wise, necklace of accouri teeth, and all—on so worthy a suitor as myself. **1912** J. RODWAY *Guiana* 216 Geometrical patterns of most intricate lines are found on basket-work, old pottery, *queyus* or aprons, and on their [*sc.* the Indians'] painted faces. **1923** W. E. ROTH tr. *R. Schomburgk's Trav. Brit. Guiana, 1840–1844* II. xii. 379 The queyu of the woman was made out of seed pips. **1924** *38th Ann. Rep. U.S. Bureau Amer. Ethnol.* xxi. 446 The Creole terms kway.., queyu, kuyu, etc., applied to the glass-bead apron, is [*sic*] apparently identical with that of the original cotton loin-cloth guayuco of the Orinoco Indians. **1964** V. G. C. NORWOOD *Jungle Life in Guiana* v. 99 The commonest form of covering adopted by Indian women generally is a small apron made from..coloured glass beads strung on cotton strands, this latter form originally Acawoian, called a 'quayo'. **1965** J. YDE *Material Culture of Waiwái* 199 The bead apron, *keweyu*.., is as indispensable a garment to the women as is the *kamisa* to the men.

‖ **quia timet** (kwī·ă ti·met). *Law.* [L., lit. 'because he fears'.] An action brought to prevent a possible future injury. Also as *attrib.* or *advb. phr.*

1628 E. COKE *First Part of Institutes of Lawes of Eng.* II. vi. 100 There be 6. Writs in Law that may be maintained quia timet, before any molestation, distresse, or impeading. **1697** *Cases Argued & Decreed in High Court of Chancery* 223 It was objected, that the Daughter is not of Age, and so this Bill is *quia timet* only;..and the Court would be vexed with vain Suits if any one might be admitted to sue only *quia timet*, to prevent a remote Possibility. But the Court answered, that Suits *quia timet* only were proper in Law and Equity. **1815** H. MADDOCK *Treat. on Princ. & Pract. High Court of Chancery* I. ii. 178 The denomination of Bills *Quia Timet* was borrowed, probably, from the Title of some ancient Writs at the Common Law. **1860** J. J. S. WHARTON *Law-Lexicon* (ed. 2) 619/1 *Quia Timet Bill*, it is filed for the purpose of quieting a present apprehension of a probable or possible future injury to property. **1927** P. G. OSBORN *Conc. Law Dict.* 41 *Bill Quia Timet*, a proceeding in the old Court of Chancery for providing against an apprehended injury. Now replaced by the injunction. **1961** *Times* 14 Feb. 5/1 Dismissing with costs their *quia timet* claim for an injunction. **1971** [see *LIE *v.*[1] 12].

quibblingly, *adv.* (Later example.)

1901 W. J. CRAIG *King Lear* 117 *note*, Perhaps 'roarer' in *Tempest* I. i. 18, quibblingly applied to the raging waters.

quiche (kīʃ). [Fr., ad. Alsatian *küchen* (G. *kuchen*).] An open flan or tart with a savoury filling. Also *attrib.* *quiche Lorraine*: properly, a quiche containing a savoury custard with bacon or ham; also used of other types of quiche.

1949 A. L. SIMON *Dict. Gastron.* 199/1 *Quiche*, a savoury custard in an open tart, a Lorraine *spécialité*. **1951** E. DAVID *French Country Cookery* 92 *Quiche Lorraine*.. Make a pastry... Onto the pastry spread the bacon ..beat the 2 eggs into the cream..pour on to the pastry.., bake for about 30 minutes. **1960** —— *French Provincial Cooking* 206 There have been various evolutions in the composition of a *quiche*... A *quiche* is a flat open tart... The fillings, of course, vary enormously. **1966** *Daily Tel.* 18 Oct. 13/5 After a soup..we had a smoked salmon *quiche*. **1967** *Woman's Day* (Austral.) 5 June 55 Most widely known is Quiche Lorraine, which traditionally contains only eggs, cream or milk, and bacon or ham. **1969** S. BURNFORD *Without Reserve* v. 180 So we sat in the sun on the dock, eating the quiche Lorraine that Mary had providently packed for just such an occasion. **1970** *Islander* (Victoria, B.C.) 16 Aug. 8/1 Some writers tell us this savory custard tart belongs to German cookery... Designed as a main dish, the quiche originally was baked in a pie pan..and was basically a custard, sometimes flavored with bacon. **1972** K. STEWART *Times Cookery Bk.* viii. 107 Roll the pastry out..slightly larger all round than an 8-inch flan or quiche tin. **1979** P. SIMMONDS *Mrs Weber's Diary* 27 Quiches are marvellous! They're all out of my freezer. Now, the vegetarian ones are at the front.

Quiché (kītʃē̆ⁱ·), *sb.* and *a.* [Native name.] **A.** *sb.* **a.** Name of a people inhabiting the western highlands of Guatemala; a member of this people. **b.** One of the principal languages of the Mayan family, spoken by this people. **B.** *adj.* Of or pertaining to this people.

1823 J. BAILY tr. *Juarros' Statistical & Commercial Hist. Guatemala* II. ii. 168 In all these places the Quiché language is spoken. *Ibid.*, It may be inferred..that the greater part of the province of Sapatitlan,..was a colony of the Quichées. **1823** [see *KEKCHI *sb.* and *a.*]. **1883** *Encycl. Brit.* XVI. 208/1 The *Popol-Vuh* or national book of the Quiché kingdom of Guatemala. *Ibid.*, After this comes the creation of the four men and their wives who are the ancestors of the Quichés. **1933** A. HUXLEY *Let.* 24 Mar. (1969) 368 From there to Chichicastenango, which is the centre of the native life of the Quiché Indians, who thickly inhabit the plateaus. **1934** [see *LADINO[1] 2]. **1948** A. L. KROEBER *Anthropol.* (rev. ed.) xviii. 794 Highland Maya, of upland Guatemala, speaking languages like Quiché, Cakchiquel, and Mam. **1950** I. VELIKOVSKY *Worlds in Collision* vi. 131 In the years of this gloom, when the world was covered with clouds and shrouded in mist, the Quiché tribe migrated to Mexico. **1963** *Times* 7 May 14/6 Judas is dressed in the costume of a Solola tribe—it is told that they were jealous of the Quiche and appealed for help to the Spaniards, who sent de Alvarado to liquidate them. **1968** *Language* XLIV. 191 It is in Quiché that the fascinating legends and history of the *Popol vuh*, 'the sacred book of the ancient Quichés', were recorded. It is also in Quiché and its near relatives of the 'Quichean' group that we have the largest accumulation of other written documents. **1974** *Encycl. Brit. Micropædia* VIII. 353/1 The Quiché Maya had an advanced civilization in pre-Columbian times. *Ibid.*, Written records of Quiché history and mythology are preserved in the *Popol Vuh*, written down in the Quiché language.. shortly after conquest by the Spaniards in 1524. **1977** *Language* LIII. 261/2 In a final appendix, a 100-word vocabulary list is given in..Quiché (also Mayan), and in Proto-Mayan as reconstructed by Kaufman.

Hence **Quiche·an** *a.*, applied to the subgroup of the Mayan family of languages to which Quiché belongs.

1956 N. A. MCQUOWN in *Internat. Jrnl. Amer. Linguistics* XXII. 195/2 It differs from Kroeber..in suggesting a closer link between Quichean and Kekchian than between Quichean and Mamean. **1968** [see above]. **1978** *Language* LIV. 496 Kaufman's paper, 'New Mayan languages in Guatemala', summarizes identifications of 'new languages' made since 1969: Sacapultec and Sipacapa in the Quichean subgroup; [etc.].

Quichua, Quichuan: see *QUECHUA *sb.* and *a.*, *QUECHUAN *a.*

quick, *a.*, *sb.*[1], and *adv.* **A.** *adj.* **I. 2. c.** Delete † *Obs.* and add later examples of *quick flesh*; now also *quickflesh*.

1926 T. E. LAWRENCE *Seven Pillars* (1935) xliii. 251 If such animals [*sc.* camels] were taken suddenly inland for long marches over flints or other heat-retaining ground, their soles would burn, and at last crack in a blister; leaving quick flesh..in the centre of the pad. **1928** H. A. MANHOOD *Nightseed* 56 Men worn away to quickflesh, their eyes staring, reddened at the rims, men who coughed and coughed with a dry, torn-paper sound, mud to the waist. **1956** D. LESSING in *New Statesman* 30 June 768/3 With quickflesh contest if you need: There is no argument with bone.

II. 9. *quick water* (further N. Amer. examples). Also *transf.*

1857 THOREAU *Jrnl.* 30 July in *Maine Woods* (1864) 276 The Indian navigator naturally distinguishes by a name those parts of a stream where he has encountered quick water and forks. **1894** *Harper's Mag.* Apr. 782/1 That quick water's the Mahkin Rapids. **1905** L. MOTT *Jules of Great Heart* xxi. 260 Nearer and nearer sounded the quick water of the thoroughfare between Lac des Rochers and the dead-water of Rivière du Renard. **1951** H. E. GILES *Harbin's Ridge* xxiii. 201, I couldn't say a word for the knot in my throat, and my eyes stung with quick water. **1958** *Montreal Star* 22 Oct. 14/3 On the subject of water also there is the definition of quickwater. .. It used to be used in the Maritimes, a term designating water running rapidly but not broken by rapids.

10. Delete 'Now *rare*'. (Further examples.) Cf. *quick-clay* in sense *D.

1895 *Trans. Australasian Inst. Mining Engin.* III. 141 *Quick*. Veins are said to be quick when productive, and dead when non-productive. *Quick-ground*, ground in a loose incoherent state; soft watery strata, e.g., running sand. **1901** *Norges Geol. Undersøgelse* No. 32. 221 All kinds of soft clay are often called 'quick' clay; in a more restricted sense it means clay which has the property of being comparatively stiff when it lies in its original bed, but becomes fluid when it is set in motion. **1963** MEANS & PARCHER *Physical Prop. Soils* xi. 333 The velocity of the upward flowing water required to cause the soil to become quick. **1967** A. R. JUMIKIS *Introd. Soil. Mech.* iv. 32 Quicksand is not a special type of soil, but a condition. Any granular material through which an upward flow of water takes place may become 'quick' under proper hydraulic conditions. **1978** *Sci. Amer.* Nov. 143/2 Sand does not become quick without an influx of water, because any extra water separates out on top of a bed of closely packed sand, creating a situation similar to the ones encountered on the beach and in the demonstration with a bottle.

III. 19. d. *Cricket.* Of a bowler.

1967 [private letter from Mr. R. Bowen]. **1976** J. SNOW *Cricket Rebel* 36, I was not fast enough to be classed as a genuine quick bowler.

25. b. *quick one*: an alcoholic drink intended to be taken rapidly. *colloq.*

1928 D. L. SAYERS *Unpleasantness at Bellona Club* ix. 105 They had a quick one together. **1936** A. HUXLEY *Eyeless in Gaza* xlii. 503 After the second 'quick one' in the bar of the theatre. **1948** R. A. KNOX *Mass in Slow Motion* viii. 80 The conductor slipping in to the Corner house to have a quick one. **1959** B. COBB *Don't lie to Police* (1960) xii. 194 We go in a bunch at half-past eleven and have a quick one, or rather one or two quick ones. **1968** *Listener* 19 Dec. 811/3 We've time for a quick one and then we'll go and do our bit of business. **1976** G. MOFFAT *Over Sea to Death* vii. 79 Ken Maynard came into the cocktail lounge.. 'Just in time for a quick one... Two lagers, please.'

28. c. with *about*.

1937 W. H. SAUMAREZ SMITH *Let.* 29 July in *Young Man's Country* (1977) ii. 85, I shall have to be very quick about writing this letter as the Air Mail goes from the Club in half an hour.

IV. 29. *quick-eared* (later example), *-footed, -worded* adjs.

1920 D. H. LAWRENCE *Lost Girl* viii. 182 He turned like a quick-eared animal. **1839** W. C. BRYANT in *U.S. Democratic Rev.* Apr. 406 Here the quick-footed wolf.. crushed the flower Of Sanguinaria. **1938** *Times Lit. Suppl.* 5 Mar. 158/1 A very quick-footed batsman. **1954** J. R. R. TOLKIEN *Two Towers* III. iv. 81 It [*sc.* a song] is..quick-worded, and soon over.

B. (*sb.*[1]) **4.** Also, the sensitive part of a horse's foot, above the hoof. Also *attrib.* (Later U.S. examples.)

1925 W. G. L. TAYLOR *Saddle Horse* i. 81 The hoof is pared down to the quick in streaks, leaving only enough for the animal to stand on. **1940** W. FAULKNER *Hamlet* iii. 60 The newcomer darting between Houston and the raised hoof and clapping the shoe onto it and touching the animal's quick with the second blow of the hammer. **1949** D. F. MONTGOMERY *Essentials of Horsemanship* iv. 37 The sensitive sole or 'quick' inside the foot follows the shape of the hoof. **1954** W. FAULKNER *Fable* 196 They was trying..to pull the quick shoe. **1963** M. C. SELF *Compl. Bk. Horses & Ponies* iii. 56 As with your finger nail, we must be careful not to trim too close to the 'quick', or sensitive part of the horse's foot.

8. *Cricket.* A fast bowler.

1960 I. PEEBLES *Bowler's Turn* 63 He was a fine player of every type of bowling, fast of foot against spin, and strong and resolute against the quicks. **1977** *World of Cricket Monthly* June 66/3 He..still considers Lindwall the greatest of quicks he faced.

C. *adv.* **1. a.** (Later examples.)

This use is now usually avoided in educated speech and writing, though found in some standard colloq. constructions.

1901 M. FRANKLIN *My Brilliant Career* xxxii. 272 Lizer, shut the winder quick. **1922** JOYCE *Ulysses* 47 He [*sc.* a dog]..pissed quick short at an unsmelt rock. **1936** C. SANDBURG *People, Yes* 83 Some men dress quick, others take as much time as a woman. **1968** *Listener* 11 July 38/3 I've never known a journey go so quick. **1979** *Times* 23 Nov. 5/4 The brash and selfish values of a 'get rich quick' society.

3. a. *quick-acting* (later examples), *-drying, -firing* (later examples), *-growing* (later examples), *-loading, -seeing, -setting* (later examples), *-speeding, -springing, -surprising, -talking* adjs.

1931 A. HUXLEY *Music at Night* I. 12 Chemically pure pornography..is a quick-acting emotional drug. **1960** *Farmer & Stockbreeder* 8 Mar. 88 'Nitro-Chalk' 21, a granular fertilizer containing 21%N, is free-flowing, quick-acting. **1869** *Bradshaw's Railway Man.* XXI. 460/3 (Advt.), Varnishes..Qck. Dryg. Copal..Qck. Drying Oak. **1913** V. B. LEWES *Oil Fuel* 91 Explosions..from leakages of volatile spirit used in making up anti-fouling and quick-drying paint..led to extended investigations being made. **1969** *R. & E. Coordinator* Apr. 8/1 A meter and a modified IGT Printability Tester are used to measure the drying time and penetration of quick drying inks into paper. **1890** G. S. CLARKE *Fortification* xiv. 207 Quick-firing guns require only two or three men..to work them. **1940** *Chambers's Techn. Dict.* 691/1 *Q.F. guns*, quick-firing guns. These may be guns, or howitzers, that are loaded with ammunition having brass cartridge cases, either attached to the shell or separate. **1979** A. FOX *Threat Warning Red* i. 3 That twin 4.5" turret-radar-controlled, quick-firing automatic. **1941** J. S. HUXLEY *Uniqueness of Man* III. vi. 100 The quick-growing beast..suffers. **1968** *Trees* Spring 15 The algaroba..is a quick-growing tree. **1874** J. W. LONG *Wild-Fowl Shooting* 37 A quick-loading [powder] flask, i.e., one having a large feed-hole to the charger, should also be used. **1925** J. GREGORY *Maid of Mountain* xxxvi. 327 Very keen, quick-seeing eyes withal. **1962** H. C. WESTON *Sight, Light & Work* (ed. 2) viii. 229 It is desirable to select workers who are quick-seeing for objects of the apparent size with which they will have to deal. **1887** J. NEWMAN *Notes on Concrete* iv. 25 The hardening of slow-setting cements is generally considered more trustworthy than that of quick setting cements. **1963** C. R. COWELL et al. *Inlays, Crowns, & Bridges* ii. 7 Very deep parts of a cavity..should have a sedative sub-lining of quick-setting zinc oxide and eugenol. **1973** J. WAINRIGHT *Touch of Malice* 188 She stiffened—as if quick-setting concrete was suddenly working in her veins. **1919** V. WOOLF *Night & Day* xxviii. 416 The quick-speeding silver moon. **1663** R. HEAD *Hic et Ubique* sig. A2ᵛ, Your sublime dignity, quick-springing wit. **1911** E. M. CLOWES *On Wallaby* xi. 308 To grapple with all the quick-springing mass of undergrowth which leaps to life, almost in a night. **1937** BLUNDEN *Elegy* 42 And though you marked my last arising, My next shall be as quick-surprising. **1963** *Punch* 4 Sept. 358/2 Gerry is a show-off, a quick-talking egotist. **1980** P. LIVELY *Judgement Day* v. 51 Those long-haired quick-talking women.

b. *quick-spread* adj.

1895 KIPLING *Second Jungle Book* 8 The quick-spread ears of the deer caught the last sentence.

D. Special combs.: **quick-action,** *attrib.* of apparatus having a quick action; **quick bread,** a bread or cake that can be prepared quickly, usu. one made with a leavening agent that permits immediate baking; **quick-break** *a.* *Electr. Engin.*, applied to (the action of) a switch designed to break a circuit and stop a current quickly regardless of the speed with which it is operated; **quick buck** chiefly *U.S.* [*BUCK *sb.*[8]] = *fast buck* s.v. *FAST *a.* 11; **quick-cake,** a cake that can be prepared and baked in a short time; **quick-change** *attrib.* (later examples); also *transf.* and *ellipt.*; **quick-change** *v. intr.*, to perform a 'quick change'; *trans.*, to change (clothes) quickly; **quick-clay** [tr. Norw. *kvikkleire*, formerly *kvikler*], clay that is quick (sense 10); **quick death** *U.S.* = *sudden death* s.v. SUDDEN *a.* 3b in Dict. and Suppl.; **Quick Dick** *Mil. slang*, a quick-firing gun; **quick-disconnect,** *attrib.* of couplings and the like that can be

quickly disconnected; **quick-fire:** also *fig.*; **quick-firer** (earlier and later examples); also *fig.*; **quick-fix** *a.*, that can be quickly fixed into place; also *fig.*; **quickgold** *fig.* [prob. modelled on QUICKSILVER *sb.*], living or liquid gold; **quick-heel** *v. intr.*, in Rugby Football, to heel rapidly from a scrum; **quick kill,** a sudden or rapid victory (cf. KILL *sb.*[1] 2 in Dict. and Suppl.); also *attrib.*; **quick-knit** *a.*, used (*a*) of very thick wool with which a garment can be knitted in a short time; (*b*) of a garment made with such wool; also *ellipt.*; **quick-look** *Astronautics*, used *attrib.* with reference to the rapid provision of information; **quick-lunch(eon,** *attrib.* of a person or establishment selling lunches that can be served and eaten quickly; also *fig.*; **quick-minded** *a.*, having a quick or ready mind; quick-witted; **quick reference** *attrib.*, giving quick and easy access to information; **quick-release,** *attrib.* of any device designed for rapid release; also *ellipt.*; **quick-return** (examples); more widely, applied to any reciprocating motion or mechanism in which the speed in one direction is greater than the speed in the other; also *ellipt.*; **quick-seller,** an article, esp. a book, that sells quickly; **quick-service** *attrib.*, that is characterized by quick service; **quick-spot** (see quot.); **quick-start** *a.*, pertaining to or characterized by rapid starting; **quick-stick(s)** (earlier and later examples); also as *v. intr.* (see quot. 1935); **quick succession,** a change in ownership of property twice within a limited period; used *attrib.* of remission of part of capital transfer tax (formerly estate duty) in such an eventuality; **quickthorn,** delete † and add: esp. a hawthorn; (earlier and later examples); **quick trick,** in Bridge, a card or combination of cards which should furnish a trick in the first or second round of the suit; a trick won 'on top'; *attrib.*, of a system valuing the hand according to the holding of such cards; **quick turnover,** *attrib.* of a person concerned with selling goods as rapidly as possible after they have been bought or produced; **quick worker** *colloq.*, one who rapidly achieves intimacy with persons of the opposite sex.

1909 *Cent. Dict.* Suppl., Quick-action *a.* **1960** *Farmer & Stockbreeder* 1 Mar. 72/1 Four quick-action jacks adjust the tilt to vary the throughput. **1920** M. WILSON *Cook Book* 36 Quick breads include griddle cakes, waffles, muffins, Sally Lunn's, shortcakes and biscuits. **1940** *Quantity Food Service Recipes* 20 (*heading*) Quick breads. Baking powder biscuits..muffins..griddle cakes. **1960** A. E. BENDER *Dict. Nutrition* 106/1 *Quick breads* include biscuits, muffins, popovers, waffles and griddle cakes. **1970** *Islander* (Victoria, B.C.) 29 Nov. 8/2 Use it [*sc.* cranberry sauce] as an ingredient in pies, quick bread and basting sauce. **1891** J. B. VERITY *Electr. up to Date* v. 62 (*caption*) 'Quick-make and quick-break' switch. **1900** *Electrician* 21 Dec. 325/2 Mr. Duddell..pointed out a danger with concentric cables and metal break, quick-break switches. **1930** MOYER & WOSTREL *Industr. Electr. & Wiring* xvii. 382 Circuits carrying large currents.. should be provided with either a quick-break switch or a circuit breaker. **1962** *Newnes Conc. Encycl. Electr. Engin.* 728/1 Shunt limit switches range from the very simple, in which speed of break is entirely dependent on the speed of operation, to considerably more elaborate designs in which some form of quick-break action is incorporated. **1960** *Christian Herald* July 12/2 This is most discouraging to those who stoop to make a 'quick buck' through propagating immorality. **1972** *National Observer* (U.S.) 27 May 20/2 Richard Chamberlain is no dripping-behind-the-ears graduate of the school of 'charm acting' (his phrase) out for a quick buck at the expense of the classics. **1980** R. BARNARD *Death in Cold Climate* xvii. 193 Dreaming of luxury, of the quick buck dubiously acquired. **1925** J. GREGORY *Maid of Mountain* ii. 15 I'll stir up a quick-cake for him. **1896** Quick change [see *trick-change* s.v. TRICK *sb.* 13]. **1905** *Daily Chron.* 13 Mar. 5/7 Mr. Balfour's first..Ministry may reasonably be dubbed the 'Quick-change Ministry'. **1906** *Ibid.* 12 Nov. 6/4 He quick-changed with the deftness and speed of a Fregoli. **1928** *Collier's* 18 Aug. 18/3 We had to quick-change our hats, put on badges [etc.]. **1939** T. S. ELIOT *Old Possum's Bk. Pract. Cats* 22 As knockabout clowns, quick-change Comedians..They had an extensive reputation. **1973** C. EGLETON *Seven Days to Killing* i. 14 They stripped off their uniforms and changed into civilian clothes..in a time which would have been a credit to a troupe of quick-change artists. **1901** Quick clay [see sense A. 10 above]. **1950** *Géotechnique* II. 58 A very soft and extremely sensitive clay, known in Norway as *kvikkleire* (quick-clay). **1968** R. W. FAIRBRIDGE *Encycl. Geomorphol.* 640/2 A special class of mudflows are those developing in quick clays which spontaneously liquefy and may flow readily on very gentle slopes, rafting houses, roads and trees appreciable distances. **1972** Quickclay [see *ILLITE]. **1942** Z. N. HURSTON in A. Dundes *Mother Wit* (1973) 225/1 I'm quick death and easy judgment. **1958** *Washington Post* 20 June A 16/4 Miss McKeever advanced to the final..by beating medalist Mrs. Thomas Konopa.. on the 19th green with a birdie in a quick death playoff. **1918** *Sat. Even. Post* 31 Aug. 34 A fifty-millimeter gun which they dubbed Quick Dick played on them with direct fire. **1936** J. G. HARBORD *Amer. Army in France 1917–1919* xvii. 283 The time of warning usually varied from practically nothing with the 'Quick Dicks' as the boys called the Austrian 88's..to as many as five seconds with the heavier calibers. **1962** F. I. ORDWAY et al. *Basic Astronautics* xii. 468 Quick-disconnect pins allow the astronaut to detach himself from a conventional aircraft type seat..once the period of acceleration is over. **1969** *Jane's Freight Containers 1968–69* 577/3 Two hydraulic pressure lines to trailer, with quick-disconnect couplings. **1954** *Encounter* Mar. 70/2 Such chiefly subjective variables are difficult to elicit in fairly short, single, quick-fire interviews. **1960** I. PEEBLES *Bowler's Turn* 44 He was an even timer with a beautiful economical action, and a sure quick-fire return. **1977** *Cleethorpes News* 27 May 5/5 They have a polished act which includes skilful harmony, impressions and quickfire jokes. **1891** MARQUIS OF SALISBURY *Let.* 20 Sept. in G. Cecil *Life Salisbury* (1932) IV. 314 They must carry Maxims and quickfirers. **1933** F. RICHARDS *Old Soldiers never Die* xii. 162 When we had no time to write letters we sent field-service post cards which we called 'quick-firers'. We simply wrote the address on them and signed our names and dates of sending on the backs. **1956** Quick-firer [see *MORALE 3]. **1959** *Archit. Rev.* CXXV. p. xcv (Advt.), Quick-fix reflectors and diffusers, heavy duty bi-pin lampholders. **1972** *Times* 30 Nov. 18/5 Quick-fix, switch-operated adaptor fittings. **1976** *National Observer* (U.S.) 25 Dec. 5/4 On the one hand, he is urged to make quick-fix tax cuts and get the economy moving. **1877** G. M. HOPKINS *Poems* (1967) 66 The grey lawns cold where gold, where quickgold lies! **1954** I. MURDOCH *Under Net* xx. 276, I took two gulps of the whiskey; it ran through me like quickgold. **1936** *Times* 30 Nov. 5/4 A pack who shoved to the last man and quick-heeled from the tight and loose. **1969** *Listener* 14 Aug. 221/3 Wilson..lost interest in the 'quick kill' approach and tried to make a deal with Smith. **a 1974** R. CROSSMAN *Diaries* (1975) I. 382 James Callaghan, for example, said he thought that..we should try to get a quick kill... In that case the preparations for the quick kill should have taken place *before* U.D.I. **1935** *Home Notes 2nd Knitting Bk.* p. iii (Advt.), W. B. Kwiknit The original quick-to-knit wool. *Ibid.*, W. B. Kwiknit is a thick 2 ply, ideal for outdoor sportswear. **1957** *Vogue Knitting Bk.* 16 (Advt.), Two such outstanding wools as Patons Quickerknit Botany and Quickerknit Baby Wool. **1960** *Ibid.* 9 Lister Lavenda quick-knits are the.. chunkiest, quickest-to-knit-with wools in knitting today! **1962** J. WADE *Running Sand* vi. 64 Some Tyrolean quick-knit jumper. **1966** G. N. LEECH *Eng. in Advertising* xv. 141 More or less common types of compound structure: 'quick-knits'; 'speedy-knit'; 'flip-tops'. **1972** C. FREMLIN *Appointment with Yesterday* xi. 84 Milly turned down the third remnant of tattered quick-knit cardigan. **1964** *Proc. Joint Computer Conf.* (*Spring*) 125/1 The requirement of the programming system for OGO was to provide quick-look analysis and control of the status of the spacecraft and selective experiments on board the satellite. **1966** *Electronics* 3 Oct. 134 Displaying quick-look performance data for evaluation by the astronaut. **1975** *Geos* (Dept. Energy, Mines, & Resources, Canada) Spring 8/1 A 'Quick Look' facility attached to the satellite station at Prince Albert, Sask., provides black and white photography of Arctic sea-ice within 20 minutes of the satellites pass over Canada. **1979** *Nature* 3 May 47/1 We concluded that to within the accuracy of the SAS 3 quick-look data timing (± 1 s), the onset of the optical burst was coincident with that in X rays. **1903** *Everybody's Mag.* Aug. 191/1 He figured them out with the stub of a blue pencil..sometimes on the slippery edge of reeking quick-lunch counters. **1903** *N.Y. Even. Post* 24 Sept. 8 The quick lunch man a few blocks away from the grocery store. **1909** CHESTERTON *Tremendous Trifles* 242 One of those quick-lunch restaurants in the City. **1911** E. M. CLOWES *On Wallaby* iii. 51 The haste of the Fisher Government to do things lately moved Alfred Deakin to describe its proceedings as 'quick-lunch legislation'. **1930** [see *DRIVE-IN]. **1975** *New Yorker* 22 Sept. 96/3 Men whose money derives from, and whose deepest loyalties adhere to, insurance companies,..quick-lunch chains, and the like. **1928** J. BUCHAN *Runagates Club* viii. 220 At a quick-luncheon counter he got into talk with a man. **1852** W. BAGEHOT *Coll. Works* (1965) I. 346 But he was a quick-minded..man of the world. **1908** *Daily Chron.* 24 Aug. 1/3 They say they never met such a quick-minded man. **1938** L. M. HARROD *Librarian's Gloss.* 124 Quick-reference books, Books which are essentially of a reference character, such as directories, dictionaries, and gazetteers. **1978** *Early Music* Oct. 599/2 Part II is a quick-reference chart summarizing the advice given on French in Part I. **1905** *Internat. Libr. Technol.* LXII. XLI. 48 Instead of moving the valve handle to this [slow-release] position..it is moved to the extreme left to quick-release posititiion... The brake-cylinder air rushes out, allowing the release springs to release the brakes suddenly. **1916** G. FRANKAU *Guns* 11 Now the foul clay cakes on britching strap and clogs the quick-release. **1933** *Gloss. Aeronaut. Terms* (*B.S.I.*) 73 *Quick release*, a device enabling the user to clear himself from the parachute and/or the harness. **1942** *Tee Emm* (Air Ministry) Sept. 135/2 Immediately you are grounded, turn on your back and unlock the quick release mechanism. **1961** *AWA Techn. Rev.* XI. Fig. 8 (*caption*) The sub-chassis can be removed by means of quick-release fasteners. **1972** [see *HARNESS *sb.* 4c]. **1976** J. WAINWRIGHT *Walther P.38* 58, I saw him bend to work the quick-release mechanism. The car gave a gentle heave as the weight of the caravan left its rear bumper. **1864** Quick return [see *RAM *sb.*[1] 5e]. **1894** W. J. LINEHAM *Text-Bk. Mech. Engin.* v. 169 The tool cuts in one direction only, and the back stroke is wasted. To minimise this loss, and at the same time reverse the stroke without changing the continuous rotation of main shaft, ingenious motions called quick returns have been devised. **1915** [see *KINEMATICALLY *adv.*]. **1930** *Engineering* 14 Mar. 341/3 Starting, stopping and the quick-return motion are all effected through a plate clutch. **1964** S. CRAWFORD *Basic Engin. Processes* viii. 218 Shaping machines are fitted with a quick-return mechanism. **1926** *Ironmonger* Suppl. 16 Jan. 50 (Advt.), Dealers who stocked early are now enjoying the profits from this quick-seller. **1934** *Archit. Rev.* LXXV. 11 True, academicians like Herkomer ..deigned to use photographic labour-saving devices for quick-service portraiture. **1976** H. MACINNES *Agent in Place* vii. 63 A hamburger at a Madison Avenue quick-service counter. **1873–4** G. M. HOPKINS *Note-Books & Papers* (1937) 223 Every visible palpable body has..a centre of illumination or *highspot* or *quickspot*. **1950** *Archit. Rev.* CVIII. 424 Quickstart or starter switch control gear can be supplied and the four-lamp fittings can be arranged for two-circuit control. **1962** A. NISBETT *Technique Sound Studio* viii. 147 Quick-start techniques fall into two categories, depending on the type of drive employed by the turntable. **1977** *Gramophone* Oct. 716/2 Technics have applied the quartz control direct-drive principle of their high-torque (quick-start) SP10 Mk. 2 turntable..to a lower priced deck. **1835** *Dublin Univ. Mag.* Apr. 391/2 'All's right,' said Denis, putting the musket to his shoulder; 'I see them. Now stand clear, boy, and hand along fresh cartridges cleverly. I'll give them a blaze in quick sticks—nothing like a long range; stand clear!' **1860** HOTTEN *Dict. Slang* (ed. 2) 196 *Quick sticks*, in a hurry, rapidly; 'to cut *quick sticks*', to be in a great hurry. **1935** A. J. POLLOCK *Underworld Speaks* 94/1 *Quicksticks*, to escape from the law. **1936** J. B. PRIESTLEY *They walk in City* xvi. 483 She can pop into that kitchen an' dish yer up something nice in quicksticks. **1966** *Listener* 29 Sept. 461/1 Then, with Mr Buchanan safely making up, it was quick-sticks for the Hippodrome and a performance of *Mercenary Mary*. **1914** *Act* 4 & 5 *Geo. V.* c. 10 § 15 Relief in respect of quick succession where property consists of land or a business. **1936** G. M. GREEN *Death Duties* v. 123 If any such other allowance is available on the second death, the 'quick succession' allowance is computed first and the other allowance is made against the reduced duty. **1967** E. RUDINGER *Wills & Probate* 14 Estate duty might have to be paid twice on the same property... There is a reduction in duty, known as quick succession relief, which helps a bit in this situation. **1973** *Times* 6 Oct. 19/2 Quick succession relief will reduce the double burden to some extent. **1611** A. STANDISH *Commons Complaint* 44 Have a good ditch double or treble set with quicke Thornes,..which hedge will last well three yeares. **1838** J. C. LOUDON *Arboretum et Fruticetum Britannicum* II. 836 Three rows of quickthorns shall be set in each ridge. **1971** *Country Life* 21 Oct. 1066/1 A variety of hedge plants—elm, ash, quickthorn, dogwood—..show the boundary hedge line is of great age. **1927** M. C. WORK *Contract Bridge* iii. 58 Two quick tricks..is the minimum strength with which a Contract denial should be made. **1955** I. FLEMING *Moonraker* viii. 81 It's a famous Culbertson hand... He used it to spoof his own quick-trick conventions. **1958** *Listener* 23 Oct. 669/2 East's King might fill West's club suit to produce nine quick tricks at No Trumps. **1977** *Homes & Gardens* Feb. 17 Your five quick tricks ought to be enough to see him [*sc.* your partner] home in Five Diamonds. **1951** M. MCLUHAN *Mech. Bride* (1967) 129/2 If there's anything this type of quick-turnover gent can't see..it's cold facts. **1956** Quick-turnover [see *EASY *a.* 13]. **1938** E. WAUGH *Scoop* II. ii. 157, I will say you're a quick worker. Sorry to barge in on the tender scene. **1969** O. HESKY *Sequin Syndicate* v. 48 'But there's something going on.' 'Well,' the old man said cheerfully, 'that's all right, isn't it? I didn't think Tarni was such a quick worker, though.'

quick, *v.*[2] Add: **2.** *trans.* To coat with mercury by immersion. Cf. QUICKEN *v.* 4b, *QUICKING *vbl. sb.*[2] 2.

1873 E. SPON *Workshop Receipts* I. 308/1 A little of this solution is poured into a basin, and with a brush dipped therein they stroke over the surface of the metal to be gilt, which immediately becomes quicked. **1891** G. E. BONNEY *Electro-Platers' Handbk.* v. 112 Brass and silver are best quicked in a solution of the double cyanide of mercury and potassium. **1923** W. R. COOPER *W. G. McMillan's Treat. Electro-Metallurgy* (ed. 4) vi. 116 Many articles are 'quicked' before being subjected to the operation of depositing other metals, especially silver and gold, upon their surfaces.

quicken, *v.* Add: **5. b.** Also, to make (a slope) steeper.

1838 *Civil Engin. & Archit. Jrnl.* I. 376/2 Retaining walls, or quickening the slopes, might perhaps get over the difficulty.

quickening, *vbl. sb.* (Later examples in sense 6b of the verb.)

1890 BILLINGS *Med. Dict.* II. 424/2 *Quickening*.., first sensation of movement of the foetus in a pregnant woman, occurring generally in the first or second week of the fifth month. **1922** JOYCE *Ulysses* 377 Send us, bright one, light one, Horhorn, quickening and wombfruit. **1975** *Church Times* 27 June 20/5 A foetus that had not reached the time of quickening (twelve to fourteen weeks old).

quick-freeze (kwi·k,frīz), *v.* Also written as one word. [f. QUICK *adv.* + FREEZE *v.*] *trans.* To freeze (perishable material) rapidly so that it can be stored at a low temperature for a long time. Also *absol.*

1930 *Popular Science* Sept. 27/2 Obviously, it would take longer to quick-freeze a six-pound cut of beef than a half-pound fillet. **1940** *Daily Progress* (Charlottesville, Va.) 20 Mar. 1/5 The range of food which can be quick-frozen has..brought a variety of business combinations into the field. **1957** *Daily Mail* 5 Sept. 11/5 In Florida.. they are quick-freezing water-melon concentrate..and making it available to any hospital anywhere in the country. **1959** *Times* 4 Nov. 23/11 (*heading*) Crosse & Blackwell to quick-freeze. **1967** *Heretaunga Plains* (School Publications Branch, N.Z.) 48/1 Fifteen tons of raspberries were quickfrozen and two more tons processed.

Hence **qui·ck-freeze** *attrib.*, that consists of or is used for quick-freezing; also *absol.*; **qui·ck-freezing** *vbl. sb.* and *ppl. a.*; **qui·ck-frozen** *ppl. a.*

1930 *Popular Science* Sept. 26/1 Clarence B. Birdseye.. succeeded in placing quick-frozen fish on the market. *Ibid.* (*caption*) Oysters, sealed in a package, are turned solid by quick-freeze process. **1932** *Sun* (Baltimore) 26 May 29/2 The comparatively new process known as quick-freezing. **1940** *Daily Progress* (Charlottesville, Va.) 20 Mar. 1/5 Quick-frozen foods..are rapidly becoming an important part of the American food production and distribution scene. **1943** J. S. HUXLEY *TVA* 103 TVA research has also led to the marketing of new types of quick-freezing machinery. **1945** NELSON & WRIGHT *Tomorrow's House* vi. 74/2 A kitchen..will almost inevitably have a quick-freeze unit. **1950** *Times* 27 Feb. 4/5 No points are now needed for canned pork and some other imported ready-cooked and quick-frozen meats. **1951** *Good Housek. Home Encycl.* 514/2 Set the dial at 'maximum' or 'quick freeze' about an hour before the mixture is ready. **1959** *Times* 4 Nov. 23/11 The plant.. would provide some quick freezing facilities..to enable the company to enter the quick freezing field on a limited scale. **1965** *Supermarket & Self-Service* (Johannesburg) June/July 13/3 Quick-frozen polywrapped broilers. **1973** *Press & Jrnl.* (Aberdeen) 3 Aug. 3/2 The processing and quick-freezing of prawns. **1976** *Woman's Day* (U.S.) Nov. 95/1 Another argument against performing a biopsy..is that it depends on quick-frozen sections of the suspicious tissue.

quick-grass (in Dict. s.v. QUICK *sb.*[2]). Add: **1.** (Later example.)

1878 *Golden Hours* X. 200/2 She tripped lightly past a knot of Quick Grasses.

2. *S. Afr.* = *KWEEK.

1931 E. P. PHILLIPS *Introd. Study S. Afr. Grasses* vi. 79 Quick grass. *Cynodon Dactylon*; *C. incompletus*; *Stenotaphrum secundatum*. **1972** *Stand. Encycl. S. Afr.* V. 320/2 The quick-grasses (*Cynodon* spp.)..are amongst those grasses that are commonly planted as lawns.

quickie (kwi·ki). *colloq.* Also **quickey, quicky.** [f. QUICK *a.* + -Y[6], -IE.] **1. a.** A cinematographic film that is made quickly and cheaply. See also *quota quickie* s.v. *QUOTA *sb.* 4.

1926 *Amer. Mercury* Dec. 465/1 Motion pictures which are ground out wholesale by the studios at the rate of one a week are called *quickies*. **1937** *Times* 13 Nov. 8/1 It is not handicapped, as the quickie is, by the expenditure of £1 a foot. **1946** *Sun* (Baltimore) 4 Oct. 12/1 The possibilities of the subject are barely touched on in 'Down Missouri Way', for this is a quickie, made..on a limited budget. **1958** *Observer* 16 Mar. 15/2 Those early wartime quickies. **1961** *John o' London's* 14 Sept. 307/4 An equally pleasant semi-amateur quicky, *One More River*. **1977** *Time* 22 Aug. 43/3 Producer Charles Band plans to return by Christmas with another quickie titled *Laser Blast*.

b. In various extended and *transf.* senses: anything produced or carried out quickly.

1940 *Washington Post* 6 July 5/7 The publishers have their firecrackers, too. They call them 'quickies'. They are books pushed through the presses to meet the headlines of the day, to go off with a bang, even if they are but rubbish in the grass when the holiday is over. **1941** B. SCHULBERG *What makes Sammy Run?* x. 187, I may not have time to see you again. This trip is one of those quickies. **1942** *Gen* 1 Sept. 14/1 Then he [*sc.* a fighter pilot] 'screams downhill' and sends the German into the 'drink' with a 'quickie'. **1943** *Newsweek* 8 Feb. 56/1 Some observers interpreted the growing trend toward 'quickies'—undeclared strikes in the form of work stoppages for a few hours or a day or two—as the rank and file's way of spurring their leaders to a crackdown on the WLB for more money. **1944** *Sun* (Baltimore) 19 July 22/8 It [*sc.* a meeting] looked like 'one of those quickies which the Mayor held in his office'. **1948** *Variety* 25 Aug. 54/1 Deitz, due back by air from a quickie to Paris last week. **1950** W. HAMMOND *Cricketers' School* xvi. 151 Close was bowling right-arm off-spinners, and 'quickies' with the new ball. **1952** *Word Study* Feb. 8/1 In a publicity release Alfred A. Knopf, Inc., describes one of its Fall books, Richard G. Baumhoff's *That Dammed Missouri*, as follows: 'Though the appearance of this book could not be better timed, it is not what in publishing circles is called a 'quicky' (or, 'quickie')—that is, a book written and rolled off the presses hot on the heels of a national news event.' **1957** *Spectator* 15 Mar. 341/1, I think we've just got time for a quickie, and it's a real tickler from Mr. Bumple, of Bedford. **1958** [see *LIBRARY[1] 2*]. **1969** A. GLYN *Dragon Variation* v. 142 Debbie had been keen on a Mexican divorce, a quickie in Mexico City followed by a honeymoon in Acapulco. **1970** *Times* 13 Jan. 8/3 The usual mishmash of buzzers, bonus points and quickies to confuse the looker-on. **1971** *Petticoat* 17 July 38/3 Here's a quicky for you when you get a blank spot about what to make for that summertime snack or light supper. **1975** R. H. RIMMER *Premar Experiments* II. 210 Yesterday they were asking some of the girls if they were hookers or 'hos'. Kathy told Mohammed that a tough Irish kid offered her ten dollars for a 'quickie'.

c. A rapidly-taken alcoholic drink; a short drink.

1941 BAKER *Dict. Austral. Slang* 58 *Quickie*, a drink taken quickly. **1942** BERRY & VAN DEN BARK *Amer. Thes. Slang* § 101/4 Cocktail names of slang and facetious origin..quickie. **1943** S. LEWIS *Gideon Planish* xv. 143, I guess that calls for a drink. Let's make it a quickie, because I got to go out shopping with George. **1947** B. MARSHALL *Red Danube* iii. 25 'I could do with a quicky'... Lined along the bar the other high ups..were having quickies too. **1959** H. HOBSON *Mission Ho. Murder* iii. 21 Have you finished with the bottle?.. Give it to the musicians, there's just about three quickies. **1970** H. MCLEAVE *Question of Negligence* xxiv. 220 'Drink?' Conway-Smith asked. 'Just a quickie,' she replied. **1974** P. HAINES *Tea at Gunter's* xv. 156 Are you in the bar, dear?.. Ronald and I just dropped in for a quickie—we're on our way back from a party.

2. *Cricket.* A fast bowler.

1934 *Evening News* 21 June 1/1 Whatever chance England may or may not have, it will be a better one if there are two or three 'quickeys' in the side in any of the remaining Tests. **1963** R. GILCHRIST *Hit me for Six* i. 15 It just happened that the first team wanted a 'quickie'. **1966** B. JOHNSTON *Armchair Cricket 1966* 108 *Quickie*, a slang term for a fast bowler. **1977** *News of World* 17 Apr. 20/8 Their other unknown quickie, Len Pascoe.., isn't as fast as Lillee or Thomson.

3. *attrib.* or as *adj.*

1927 *Daily Express* 12 Dec. 13 The most eminent 'quickie' producer is Phil Goldstone, who can make a full-length film..in eight days. **1936** *Sun* (Baltimore) 21 Dec. 8/2 In recent months there have been scores of 'sit-down' or 'quickie' strikes in the automobile factories. **1940** *Common Sense* Mar. 20/1 [Cordell Hull] found his way to a quicky law school in the county. **1944** *Sun* (Baltimore) 15 Nov. 11/2 Another quickie bridge and the town was ours. **1959** M. DOLINSKY *There is no Silence* iv. 57 Virtually every model in the country lived on these quickie love affairs. **1960** *Guardian* 27 Oct. 9/6 He made as many as eight 'quickie' Westerns..a year. **1969** A. GLYN *Dragon Variation* v. 142 Joann's rumour that there now were quickie divorces in Nevada too was apparently unfounded. **1976** H. NIELSEN *Brink of Murder* ii. 19 He's kept a packed travel-bag in his office for these quickie trips. **1979** *Yale Alumni Mag.* Apr. (Suppl.) cn14/3 The deadline conflicts with a planned quickie vacation in Puerto Rico.

quicking, *vbl. sb.*[2] Restrict *rare* to senses in Dict. and add: **2.** Coating with mercury by immersion; also *concr.*

1863 *Brit. Patent 1512* 1 This Invention consists in preserving the silvering or quicking applied on glass and objects made of silvered or quicked glass. **1873** E. SPON *Workshop Receipts* (Ser. 1) 307/2 When a sufficient quantity of mercury is dissolved, the articles to be gilt are put into the solution, and stirred about with a brush till they become white. This is called quicking. **1923** W. R. COOPER *W. G. McMillan's Treat. Electrometallurgy* (ed. 4) vi. 116 Quicking is..often resorted to in order to increase the adhesiveness of deposited metals on objects which would have no action on the bath; for the mercury..retains a bright surface when exposed to the air for a period which would suffice to provide a film of oxide upon an unquicked surface. *Ibid.* 117 The quicking-solutions more commonly used are: the per-nitrate or proto-nitrate of mercury..or the cyanide of mercury. **1930** FIELD & WEILL *Electro-Plating* 104 A thin, bright film of mercury is applied by simple immersion in a suitable solution. This process is called quicking. **1952** H. SILMAN *Chem. & Electro-Plated Finishes* (ed. 2) 478/2 (Index), 'Quicking' solutions.

qui·ckish, *a.* [f. QUICK *a.* + -ISH[1].] Somewhat quick; in *Cricket*, of a bowler, fast-medium.

1900 O. AGNUS *Jan Oxber* iii. 194 Be quickish downstairs, mind 'ee. **1955** *Times* 9 July 4/7 Neame and Parker, for Harrow, made steady progress against Sinclair and Douglas Pennant, a quickish left-hander. **1963** *Times* 4 June 3/6 He is tall, well built, and quickish, and Hampshire think that they may have a real find in him. **1977** *Sunday Times* 3 July 28/6 Nottinghamshire's quick and quickish bowlers had two wickets..in the first four overs. **1978** *Gramophone* Apr. 1713/1 The quickish pulse, less accommodatingly shaped than Karajan's.., does occasionally suggest a nature ramble being hastened along by a warden who senses rain in the offing.

quicklike, *adv.* Chiefly *U.S.* [f. QUICK *a.* + -LIKE.] In a quick manner; quickly.

1913 G. STRATTON-PORTER *Laddie* viii. 235 He.. looked down the hole I showed him and he cried out quicklike. **1951** H. E. GILES *Harbin's Ridge* xxi. 186 He slewed his eyes at me quick-like. **1978** A. PRICE *'44 Vintage* vi. 70 We've to nip in quick-like.

quickness. Add: **6.** Mobility or plasticity (of soil).

1969 *Engin. Geol.* III. 135 A series of experiments.. was carried out with the purpose to investigate if quickness could be produced by leaching of a clay deposited in salt water. **1972** *Nature* 28 Jan. 220/2 It may be the hitherto neglected non-clay mineral fraction which is responsible for quickness.

quickset, *sb.*[1] and *a.* Add: **A.** *sb.* **1. b.** (Later example.)

1938 M. HADFIELD *Everyman's Wild Flowers & Trees* 63 Common Hawthorn... A valuable hedge plant, as such, called quick-set.

2. (Later example.)

1973 R. ADAMS *Watership Down* xlv. 375 Hazel halted among the quickset on top of the nearer bank.

B. *adj.* **b.** Also *fig.*

1938 L. MACNEICE *Earth Compels* 56 Columns of ads, the quickset road to riches. **1948** [see *HAIR-TRIGGER b].

quick-set (kw·ikset), *a.*[2] [f. QUICK *adv.* + SET *v.*] **1.** Also **Quickset.** Applied to a type of surveyor's level in which the foot-screws in the levelling head are replaced by a ball and socket joint to facilitate quick setting.

1930 S. W. PERROTT *Surveying for Schools* xiv. 114 In the case of the Quickset level, there are no foot-screws, the telescope and bubble being set approximately level by a ball and socket joint. **1948** B. G. MANTON *Highway Surveying & Setting Out* iii. 58 Many instruments of the 'quick-set' type are fitted with a prismatic device which enables the bubble position to be seen from the eye-piece end of the telescope. **1971** R. J. P. WILSON *Land Surveying* x. 215 The essential difference between the quickset level and other types is that it is a tilting level without footscrews in the levelling head.

2. That hardens or dries quickly.

1967 KARCH & BUBER *Offset Processes* vii. 266 Quickset inks are also used when printing on coated paper stock. **1970** R. JOHNSTON *Black Camels of Qashran* xi. 178 The quickset cement was dry and hard.

quick step, quickstep. Add: **1.** (Earlier U.S. example of quasi-*adv.* use.)

1864 *Sunset Stories No. 1* 10 Stir round, can't you, Jem? take your fingers out of your mouth, and get some kindlins, quick step.

3. A quick dance; *spec.* a fast modern ballroom dance in 4/4 time.

1880 [see *LASSU]. **1927** V. SILVESTER *Mod. Ballroom Dancing* 25 For Charleston, Quickstep, and Tango the right hand should be held further round. *Ibid.* 64 Owing to the speed at which foxtrots are played, it is not possible to do the slow foxtrot unless the orchestra plays slowly, hence the evolution of the Quickstep. This dance is now done mixed in with the Charleston. **1937** E. PORTER *Music through Dance* viii. 142 The dances of the post-war period were still the Quickstep, and the Foxtrot in its quick and slow variations. **1955** *Radio Times* 22 Apr. 7/2 The Gold Star Trophy..will be awarded for the best all-round performance in the four basic ballroom dances—the waltz, foxtrot, tango, and quickstep. **1976** *Times* 11 June 14/6 The bandleader..changed the tempo to what the programme said was a foxtrot but which..was more of a quickstep.

qui·ck-step, *v.* [f. the sb.] **1.** *intr.* To march in quick time; also quasi-*trans.* and *fig.*

1906 *Daily Chron.* 27 Aug. 4/4 They quick-step it up and down the asphalted front at Hythe. **1961** *Time* 28 Apr. 22/2 The G.O.P. majority quick-stepped behind his program. **1964** G. MCDONALD *Running Scared* vi. 78 The streets were filled with people..all slim and quick-stepping. **1975** *New Yorker* 17 Feb. 101/1 Lightning Mandate, who won a division of the recent Malibu, was right behind him, and these two probably quickstepped themselves out of the money.

2. To dance the quick-step.

1935 C. DAY LEWIS *Time to Dance* 32 For no silver posh plane was their pigeon, no dandy dancer quick-stepping through heaven.

Hence **qui·ck-stepping** *ppl. a.*

1908 *Daily Chron.* 1 Oct. 7/3 The quick-stepping figure in white flannels glanced around at the click of the latch. **1936** C. DAY LEWIS *Friendly Tree* i. 14 Her gait was delicate, quick-stepping.

quick time. b. (Later Jamaican examples.)

1956 in Cassidy & Le Page *Dict. Jamaican Eng.* (1967) 372/1 You better come here quick time!.. Him run quick-time an' tell him mumma. **1971** *Jamaican Weekly Gleaner* 3 Nov. 5/1 The real tourist types..did not miss the chance to dress quick time in tee-shirt and so forth emblazoned with 'Miami, Florida'.

‖ **Quicunque vult** (kwīku·ŋkwe vult). Also **Quicumque vult.** [L.] The Athanasian Creed, so called from its opening words *quicumque vult* (*salvus esse*) 'whosoever will (be saved)'. Also *ellipt.*, as *Quicunque*, and *fig.*

c **1400** *Mandeville's Travels* (1967) xvi. 106 Seynt Athanasie, that was bisshopp of Alisandre, that made the psalm *Quicumque vult*. *a* **1530** *Myroure of oure Ladye* (1873) II. 139 This psalme *Quicumque vult*, enformeth vs fyrste in faythe of the godhed and after in faythe of the manhode of cryste. *Ibid.* III. 312 And the thyrde crede that ys. *Quicunque vult*, was made by a holy bysshop. called Athanasius. **1567** J. JEWEL *Def. Apol. Ch. Eng.* II. 83 The Creede called, *Quicunque vult*, written, as some thinke, by Athanasius. **1724** D. WATERLAND *Crit. Hist. Athanasian Creed* ii. 25 Robertus Paululus, Presbyter of Amiens, in the Diocess of Rheims, speaking of the Offices recited at the Prime, observes that the Piety of good Christians had thereunto added the *Quicunque vult*, that the Articles necessary to salvation might never be forgotten any hour of the Day. **1855** F. PROCTER *Hist. Book Common Prayer* II. i. 214 Another title of this Creed was 'Psalmus *Quicunque vult*': hence the custom of reciting it antiphonally. **1877** *Encycl. Brit.* VI. 562/2 Nothing definite as to the authorship of the Quicunque can be rested on such resemblances. **1910** *Jrnl. Theol. Studies* XI. 401 (*heading*) A critical text of the *Quicumque Vult*. *Ibid.* 402 The first three are the earliest known MSS of the *Quicunque*. **1921** G. SAMPSON *English for the English* ii. 32 Let *Quicunque vult* and threatened excommunications of all kinds remain in the realm of theology and outside the realm of education. **1963** AUDEN *Dyer's Hand* 54 Herewith, then, what I might describe as..a kind of private *Quicunque vult*. **1964** J. KELLY *Athanasian Creed* vii. 126 For these reasons the Quicunque deserves to retain its place among the normative formularies of Christendom. **1967** H. CHADWICK *Early Church* xv. 235 It was a theologian living in Southern Gaul, or perhaps in Spain, who produced the catechetical compendium.. *Quicunque Vult*, which soon (if not from the start) passed under the august title of the Creed of St. Athanasius. **1980** *Times* 20 Nov. 17/7 The very bases of belief..the Catechism, the Quicunque Vult, the Thirty-nine Articles.

quid, *sb.*[2] Add: **1. a.** (Further examples.) Also, one pound sterling. *slang.*

1907 G. B. SHAW *Major Barbara* II. 241, I ad two quid saved agen the frost; an Ive a pahnd of it left. **1917** A. G. EMPEY *Over Top* 304 *Quid*, Tommy's term for a pound or twenty shillings... He is not on very good terms with this amount as you never see the two together. **1929** W. P. RIDGE *Affectionate Regards* 71 Milton received only ten quid for the first edition of 'Paradise Lost'. **1951** *People* 3 June 2/2 It took less than a couple of quid on the down trip. **1959** I. JEFFERIES *Thirteen Days* xi. 183 You buy a car, it costs you a thousand quid; but you get a gir like that free. **1968** K. Weatherly *Roo Shooter* 74, I was thinking of moving on a bit but there are still enough here to make a few quid. **1971** *Venerabile* XXV. III. 191 It is surprising what difficulties the good old English quid can cause. **1977** C. McCullough *Thorn Birds* vii. 160 Do you want to go after Auntie Mary's thirteen million quid?

b. Phr. *quids in*: in luck or profit; well off for money. *slang.*

1919 *Athenæum* 1 Aug. 695/2 *Quid's in*, for a stroke of good fortune. **1939** W. Allen *Blind Man's Ditch* 236 We'll be quids in to-morrow. **1960** O. Manning *Great Fortune* xix. 226 Anyone who financed the trip would be quids in. **1969** J. N. Chance *Abel Coincidence* x. 187 If you know about people and they know nothing about you, you're quids in at the starting grid. **1976** *News of World* 14 Mar. 19/4 And to make sure you are quids in anyway, we'll give you as well the starting price odds to £10 each way on whichever horse does win.

c. Phr. *the full quid*: (see quot. 1959). *Austral.* and *N.Z. slang.*

1946 *Coast to Coast 1945* 106 'There's some say Lizzie's not the full quid either,' he said. **1959** Baker *Drum* II. 111 *Full quid*, in full possession of one's faculties. A person who is said to be *ten bob in the quid* or any smaller sum down to *tuppence in the quid*, is held to be stupid. **1960** N. Hilliard *Maori Girl* III. vi. 213 Not that she was simple in the sense that she was short of the full quid. **1972** I. Moffitt *U-Jack Society* xiv. 227 We avoid individuality as firmly as we suspect joy ('You're not the full quid!'). **1975** *Sydney Morning Herald* 5 July 9 It's perfectly clear that not all members of our community are the full quid.

quid, *sb.*[3] Add: **2.** = Cast *sb.* 19 and *pellet *sb.*[1] 2 c.

1834 [see *pellet *sb.*[1] 2 c]. **1879–81** G. F. Jackson *Shropshire Word-bk.* 315 Them owls..sin a mouze..an' ketchen 'im..an' chawen 'im..'an crushen 'im, an' sooken 'im till theer inna nuthin' left on 'im, an' then they droppen the quid.

quid, *v.*[2] **1.** (Further example.)

1902 J. Masefield *Salt-Water Ballads* 66 Quiddin' bonded Jacky out a-lee.

‖ **quidditas** (kwi·ditas). [L.: see Quiddity.] = Quiddity 1.

1878 *Encycl. Brit.* VIII. 758/1 This matter is differentiated into particular things..through the addition of an individualizing principle (*hæcceitas*) to the universal (*quidditas*). **1911** *Ibid.* XXIV. 354/2 The additional determinations are as truly 'form' as the universal essence. If the latter be spoken of as *quidditas*, the former may be called *haecceitas*. **1934** 'H. MacDiarmid' *Stony Limits* 16 Lyin' in wait in vain for a single grey drop To quicken into a perfect quidditas. **1976** 'M. Innes' *Gay Phoenix* iii. 41 A man's identity—his *quidditas*, as the learned might say.

quidlet (kwi·dlėt). *slang.* [f. Quid *sb.*[2] + -let.] A sovereign; one pound sterling; (see also quot. 1912).

1911 L. Tracy *Sylvia's Chauffeur* v. 96 [He] handed Dale a fiver—five golden quidlets, if you please! **1912** J. W. Horsley *I Remember* xi. 254 'Quidlet', for half a sovereign, has recently been coined from the older 'quid'. **1940** A. W. Upfield *Bushranger of Skies* xvi. 183 It cost four thousand quidlets—Australian.

quid pro quo, *sb.* Add: **2.** *attrib.* (Earlier example.)

1838 J. S. Mill in *Westm. Rev.* Aug. 489 We did not expect that the *petite morale* almost alone would have been treated, and that with the most pedantic minuteness, and upon the *quid pro quo* principles which regulate *trade*.

3. With substantial elements considered discretely.

1939 S. de Madariaga *Christopher Columbus* xii. 136 The contractual sense, that attitude which sees every event of life as a transaction and expects and demands a definite *quid* for every *quo*. **1961** *Daily Tel.* 1 Sept. 12 She could well take all and give nothing in return, pocket the quos as well as the quids. **1979** M. McCarthy *Cannibals & Missionaries* xi. 304 Conditions for the committee's release..had never been 'aired'... Not a *quid* or a *quo* vouchsafed.

quidsworth (kwi·dzwɒɹþ). *slang.* [f. Quid *sb.*[2] + Worth *sb.*[1]] The amount of anything which may be bought for one pound.

1966 P. O'Donnell *Sabre-Tooth* v. 83 Modesty was after that ten million quidsworth of diamonds. **1968** 'O. Mills' *Sundry Fell Designs* xix. 192 There's thousands of quidsworth of equipment in there. **1977** F. Branston *Up & Coming Man* v. 51, I went to a Chinese takeaway and bought a couple of quidsworth of indigestion.

‖ **quien sabe** (kie·n sa·be). [Sp.] 'Who knows?', 'who can tell?' Also *attrib.*

1836 in *Papers of M. Buonaparte Lamar* (1921) I. 436 Austin..will be elected and will do well provided he selects a *good* Cabinet—and an honest one—quen Sabe—. **1846** J. W. Abert *Jrnl.* 17 Oct. in *Rep. Exam. of New Mexico* (1848) 51 To all our other questions with regard to this ancient town, we received the usual Mexican reply of 'quien sabe'. **1849** T. Arnold *Let.* 7 July (1966) 124, I wonder what you are doing now. Whether the 'daily possibility of falling in love'..has ripened into a certainty..? Quien sabe? as Matt used to say. **1864** *Weekly New Mexican* 23 Dec. 2/4 We cannot trust an answer in the common vernacular to which we are accustomed, and must reply in all the Spanish we are master of, *quien sabe*. **1925** D. H. Lawrence *Reflections on Death of Porcupine* 110 What makes the difference? *Quien sabe!* **1933** A. Huxley *Let.* 13 Aug. (1969) 372 Do you think I am kind and unpossessive? Quien sabe? **1947** M. Lowry *Under Volcano* i. 38 But why had all this happened? he asked himself now. Quién sabe? **1949** *Southwestern Rev.* Summer 235/1 One yarn thrown in as a sort of *quien sabe?* item suggests an even more unpalatable morsel. **1965** L. Meynell *Double Fault* I. i. 14 'Does this mean that he is..too wealthy to work, or what?' '*Quien sabe?* Maybe he's just on holiday.' **1976** A. White *Long Silence* iii. 28 It's one thing for us all to make our decision here..but there—quien sabe?

quiescence. Add: Also, the action of making quiet or calm.

1859 Trollope *Bertrams* viii. 71 He had been useful as a great oil-jar, from whence oil for the quiescence of troubled waters might ever and anon be forthcoming.

quiescent, *a.* and *sb.* Add: **A.** *adj.* **3.** *Electronics.* Corresponding to or characterized by an absence of an input to a device ready to receive one.

1923 E. W. Marchant *Radio Telegr. & Teleph.* vi. 84 Attempts have been made to arrange the transmitter in such a way that the speech-current will act as a switch for starting up the continuous waves at the transmitting end. This arrangement of circuit has been called the 'Quiescent aerial' system. **1952** E. Armitage *Wireless Fund.* ix. 167 The advantage of Class B amplification is that the steady anode current flowing through the circuit when the valve is quiescent is very much smaller than under Class A conditions. **1965** *Wireless World* July 325/2 This imposes a problem on the restricted signal handling capacity of Tr1 due to its very low quiescent current. **1975** D. G. Fink *Electronics Engineer's Handbk.* ix. 56 In the absence of an rf input signal, these amplifiers remain quiescent even with full operational voltage applied.

quies kiteer, quiess kateer, varr. *quaiss kitir *int.*

quiet, *a.* Add: **I. 1. b.** Also phr. *quiet American*: freq. used ironically, as of an undercover agent or spy.

1955 G. Greene (*title*) The quiet American. **1963** *Listener* 10 Jan. 96/3 She has much data on these delightful grasshopper people, though a certain 'Quiet American' ingenuousness is difficult to digest. **1963** *Times* 23 Feb. 4/5 Mr. H. F. Johnson..is a model quiet American. **1973** *Times* 11 Jan 10/6 There never was much 'reality' about Washington's presence in Vietnam from the moment when the first quiet Americans moved in. **1980** J. McNeil *Spy Game* ix. 99 I've heard of you.. The Quiet American, no less.

2. b. (Further examples.)

1853 C. Brontë *Villette* I. xiv. 257 Her dress was almost as quiet as mine. **1895** G. B. Shaw in *Sat. Rev.* 28 Sept. 409/1 He associates low tones ('quiet colors' they call them in Marshall & Snellgrove's) with dignity and decency. **1957** *Observer* 28 July 5/6 Both Hardy Amies and Victor Stiebel are masters of the art of inserting contrasting lines which define the figure in the quietest way. **1977** *Spare Rib* May 15/1 Quiet shades of blue, brown and grey were almost *de rigueur*.

d. Of the sun: marked by an absence of all transient and localized emission of radio waves such as accompanies sunspots. Of other celestial objects: = *radio-quiet* adj. s.v. *radio *sb.* 7. Also, in *Geophysics*, marked by no local fluctuations of magnetism.

1946 *Nature* 2 Nov. 632/2 Edlén's recent work..shows that the coronal matter is normally at a temperature approaching 10^6 degrees. We should therefore expect to find black-body radiation of about 1 metre wave-length having a normal (quiet sun) intensity corresponding nearly to $T = 10^6$. **1961** *I.U.G.G. Chron.* No. 34. 6 To December 1963 (i.e. approximately up to the commencement of the proposed International Year of the Quiet Sun). **1962** *Nature* 24 Mar. 1145/1 During magnetically quiet days. **1963** *Daily Tel.* 18 Mar. 19 (*heading*) 60 nations will seek 'quiet sun' secrets. **1966** *Sci. Amer.* Nov. 54/1 At the beginning of each [sunspot] cycle the surface of the sun is quiet, disturbed only by the 'granulation' effect. **1974** *Nature* 13 Sept. 129/1 The seamount possesses a strong magnetic signature..in marked contrast to the main part of the Rockall Trough which is magnetically quiet. **1977** *Sci. Amer.* Aug. 32/1 A number of giant elliptical galaxies radiate prodigiously at radio wavelengths, where stars and normal galaxies are quiet.

II. 3. a. Phr. *anything for a quiet life*: see *life *sb.* 3 c.

d. *quiet number*, an easy job (cf. *number *sb.* 6 f (iii)). *Naut. slang.*

1948 Partridge *Dict. Forces' Slang 1939–1945* 129 *Number, quiet*, an easy job at sea or ashore. **1977** *Navy News* July 18 (*caption*) 'Got a nice quiet number for you after the Review,' he says...'Just scoop up any odd little bit of gash,' he says.

4. a. *quiet-room*, (*a*) a room set aside for quiet activities; (*b*) a room especially designed so as not to transmit any noise made within it, usu. in a mental institution.

1938 [see *pad *sb.*[3] 1 c]. **1968** M. Torrie *Your Secret Friend* i. 15 The Sixth have the boudoir as their 'quiet' room and the bedroom is now the staff common-room. **1976** [see *padded *ppl. a.*[1]]. **1977** C. H. Jaques *Dragon Century 1877–1977* xxi. 241 A complete re-organisation..with..the Doctor's surgery and consulting room taking over the Quiet Room, while the old Dining area is divided up into new Quiet Room, Headmaster's study and Secretary's office. **1977** *Spare Rib* Jan. 15/2 The second half of the day passes so quickly and imperceptibly that at first I don't grasp why it's suddenly got so noisy in our 'quiet' room.

c. (Further examples.)

1891 [see Think *sb.* 1]. **1905** Chesterton *Heretics* xiv. 181 A place where a man can have what is somewhat fantastically called a quiet chop. **1953** R. Usborne *Clubland Heroes* II. 99 An off-day for a good Buchan type was: a cold bath or plunge, a big breakfast, a quiet pipe and then off to the hills in filthy tweeds. **1963** E. Summers *Where No Roads Go* v. 72 I'll go out on the verandah for a quiet smoke. **1977** *Private Eye* 1 Apr. 23/3 (Advt.), A quiet word about hair transplants. **1977** *Zigzag* Apr. 28/2 I can just see the nice young couples out for a quiet Sunday drink.

d. Of a period of time: spent in seclusion for the purposes of prayer, meditation, etc.; *quiet time*: a daily session of private Bible study or prayer.

1884 *Lichfield Diocesan Mag.* Jan. 10 A *Quiet* day, to which all the clergy and lay readers of the archdeaconry were invited, was held at Stoke, on Tuesday Dec. 18th. **1896** C. T. Studd *Let.* in N. P. Grubb *C. T. Studd* (1933) xi. 106, I have had such a good day to-day, early up and a quiet time for most of the day and the Lord has been opening up the Word. **1934** R. Macaulay *Going Abroad* xxx. 263 That must have been about the same time I was having *my* quiet time. **1935** *Methodist Recorder* 1 Aug. 5/4 A large number of ministers assembled..for the 'Quiet Day'. **1945** [see *sharing *vbl. sb.*[2] 2]. **1957** J. R. W. Stott *Being a Christian* 23 What many people call the daily 'quiet times', first thing in the morning and last thing at night. **1960** I. Kuhn *In Arena* iii. 35 Some students were trying to let classwork reading [of the Bible] do for personal quiet time. *Ibid.* 36 Letting the day's business occupy the central place and trying to fix quiet time with the Lord somewhere shoved into the odd corner. **1967** M. Griffiths *Take my Life* iii. 61 Maintaining a regular 'Quiet Time' unhurriedly in the Lord's presence, reading the Bible, hearing His word and responding in prayer from the heart, becomes harder. **1973** *Franciscan* XV. 169 A quiet afternoon during the Pentecost season was conducted by the Dean of Worcester.

III. 8. For **1873** *Slang Dict.* read **1863** Hotten *Dict. Slang* (ed. 2) and add further examples.

1862 *Otago: Goldfields & Resources* 35 Unless men can work [the gold] on 'the quiet', they are not likely to make 'piles' so rapidly as Messrs. Hartley and Riley. **1873** 'Mark Twain' *Gilded Age* xi. 112 The other day he let me into a little secret, strictly on the quiet. **1903** A. H. Lewis *Boss* 59 They've put out a lot of money on the quiet among my own people. **1967** N. Freeling *Strike out where not Applicable* 36 She has a good act of letting Francis rule the roost, but on the quiet I think she makes the decisions.

9. *quiet-eyed* (examples), *-footed, -tinted, -toned, -voiced, -walled*; *quiet-smiling* adjs.

1895 W. B. Yeats *Poems* 197 The Druid, gray, wood-nurtured, quiet-eyed. **1956** R. Finlayson in *Landfall* (N.Z.) X. 12 A handsome..Jersey, sleek and quiet eyed. **1954** J. R. R. Tolkien *Fellowship of Ring* I. ii. 62 Long after, but still very long ago, there lived by the banks of the Great River on the edge of Wilderland a clever-handed and quiet-footed little people. **1952** R. Campbell tr. *Baudelaire's Poems* 84 An huntress born, sure-eyed and quiet-smiling. **1909** *Westm. Gaz.* 14 Sept. 1/2 It included the following: Shirts, 72;..socks of quiet-tinted silk, 2 dozen; hats, evening suits, smoking coats. **1960** *Times Lit. Suppl.* 25 Nov. 1049/4 A quiet-toned, carefully tender book. **1940** T. S. Eliot *East Coker* ii. 9 Had they deceived us Or deceived themselves, the quiet-voiced elders. **1974** P. Gore-Booth *With Great Truth & Respect* 394 What disturbed me immensely was their quiet-voiced extremism. **1865** G. M. Hopkins *Poems* (1967) 34 Those charms accepted of my inmost thought The towers musical, quiet-walled grove.

quiet, *v.* Add: **2. d.** *Electronics.* To reduce automatically the gain of (a radio receiver) in the absence of a usable signal; = *squelch *v.*

1950 J. K. Henney *Radio Engin. Handbk.* (ed. 4) xvii. 821 The purpose of these circuits is to squelch or quiet the receiver when sufficient signal for satisfactory reception is not present. **1960** Cooke & Markus *Electronics & Nucleonics Dict.* 453/2 *Squelch*, to quiet a receiver automatically by reducing its gain in response to a specified characteristic of the input.

4. (Further examples of *to quiet down*.) Now chiefly *N. Amer.*

1897 A. Beardsley *Let.* 17 Mar. (1970) 279 There has been no return of haemorrhage. The lung too is quieting down. **1914** G. B. Shaw *Misalliance* 85 You would then be charged and imprisoned until things quieted down. *Ibid.* 93 Let us postpone the discussion. Wait until Monday: we shall have Sunday to quiet down in. **1916** *Daily Colonist* (Victoria, B.C.) 30 July 7/4 The trade had quieted down somewhat lately and some mills have lacked an adequate supply of logs. **1944** *New Yorker* 25 Mar. 84/2 Moved back into London when things quieted down. **1974** *Sci. Amer.* July 47/2 The effect of the drugs is often dramatic, with the children quieting down, paying attention to their schoolwork and in some cases doing better in school.

‖ **quieta non movere** (kwiē·tā nōᵘn movēᵊ·ri). [L., lit. 'not to move settled things'.] A maxim expressing preference for the *status quo*; 'let sleeping dogs lie'.

1771 H. Walpole *Let.* 26 Mar. in *Corr.* (1937) VII. 289 My father's maxim, *Quieta non movere*, was very well in those ignorant days. **1854** W. Bagehot in *Prospective Review* X. 526 It lived on the fat of the land; *quieta non movere*, was its motto. **1887** *Athenæum* 27 Aug. 276/2 But was the book quite worth publishing? 'Quieta non movere' holds good even of dormant articles. **1905** D. M. Wallace *Russia* I. xix. 373 *Quieta non movere* is her

[*sc.* the Russian Church's] fundamental principle of conduct. **1960** *Encounter* XIV. III. 88 *Quieta non movere* is the motto of many once aggressive..radicals.

quieten, *v.* Add: **1.** Also const. *down.*

1902 C. Hyne *Mr. Horrocks, Purser* 37 Mr Horrocks had given the wink to the chief steward to go and quieten down the Second-Class passengers. **1908** G. A. Birmingham *Spanish Gold* xxi. 296 We got them quietened down after a bit.

2. (Further example.)

1897 *Daily Tel.* 28 Aug. 6/4 It [*sc.* Afghanistan] is beginning to quieten down now, in my opinion.

quietener. Add: Also 9 **quietner.** (Earlier example.)

1856 *Punch* 19 July 22/1 The conjugal powders are called in the town of Bolton 'Quietners'... These quietners are sold at one penny each.

qui·etening, *ppl. a.* [-ing[2].] That quietens or becomes quiet.

1905 *Daily Chron.* 25 Mar. 7/3 The presence of a large addition of police has had a quietening effect on the..operatives on strike. **1907** *Smart Set* Mar. 26/1 She lay there on her pillow, grateful..for the sheltering, homely realities, that enmuffled her and gave tangibility to her quietening thoughts.

quieting, *vbl. sb.* Add: **2.** *Radio.* **a.** The automatic reduction of the gain and therefore the noise of a receiver when there is no usable signal.

1937 F. E. Terman *Radio Engin.* (ed. 2) xiii. 561 In tuning a sensitive receiver provided with automatic volume control, the noise output between stations is high because, when no signal is being received, the A.V.C. system increases the sensitivity of the receiver to the maximum possible value. Arrangements for eliminating this interstation noise are variously known as *Q* circuits, quieting systems, squelch circuits, etc. **1959** R. L. Shrader *Electronic Communication* xvii. 554 (*caption*) A squelch interstation quieting, or Q, circuit.

b. The reduction of the noise level in a receiver caused by the presence of an input at the frequency to which it is tuned.

1949 *Proc. IRE* XXXVII. 1373/2 The quieting-signal-sensitivity test input is the least unmodulated signal input which, when applied to the receiver through the standard dummy antenna, reduces the receiver noise by a factor of 20 decibels. **1960** *IRE Trans. Vehicular Communications* Dec. 32/2 Two methods of measuring sensitivity are recognized... The oldest and probably the most used is the quieting method. This procedure measures the on-frequency signal level required to reduce the noise output of a receiver by 20 db. **1974** Harvey & Bohlman *Stereo F.M. Radio Handbk.* iii. 41 Receiver sensitivity must be quoted together with the corresponding noise figure; a typical figure for a good circuit is 2μV for 30 dB quieting.

quietish (kwəi·ĕtiʃ), *a.* [f. Quiet *a.* + -ish[1].] Somewhat quiet.

1913 R. Brooke *Let.* May (1968) 464, I wasn't *sick*: just quietish, and I had a bleedin' headache. **1925** G. S. Gordon *Let.* 13 May (1943) 177 The house seems quietish without you! **1939** N. Marsh *Overture to Death* xiii. 125 Though lately it's been quietish—hasn't it, Mr. Alleyn? **1977** *Hot Car* Oct. 49/1 It's obviously got to be neat, safe and quiet-ish.

quietism. 2. (Later example.)

1976 *Gramophone* Aug. 266/1 In its place there is..almost a sense of quietism.

quietist. Add: **1.** (Further examples.) Also *fig.*

1923 W. Deeping *Secret Sanctuary* xxiii. 241 In love he had become a Quietist. **1924** —— *Three Rooms* xxix. 268 She sat like a quietist, hands folded, her brown eyes benignly equivocal.

quietistic, *a.* Add: (Later examples.) Also *transf.*

1909 *Quarterly Rev.* July 117 Altogether, quietistic analysis breaks down while leaving the reality and value of the experience untouched. **1973** B. R. Wilson *Magic & Millennium* xii. 387 Among less-developed peoples the autonomous introversionist movement would appear to rely on the quietistic prophet. **1978** *Gramophone* July 231/1 He does not overpoint the ostinato of No. 4, which remains essentially quietistic.

quietlike, *a.* and *adv.* For *Sc.* read 'orig. *Sc.*' and add later examples of advb.

In some or all of these examples the formation may have been influenced by the parenthetic use of *like* (see Like *adv.* 7 in Dict. and Suppl.).

1909 J. Masefield *Tragedy of Nan* II. 31 He was fiddlin' quiet-like, all the time 'e were a-singing. **1913** W. de la Mare *Peacock Pie* 98 Calling me, 'Sam!'—quietlike. **1976** *New Yorker* 8 Mar. 102/2 Someone said, 'It wasn't no mortar round,' real quietlike. **1977** I. Shaw *Beggarman, Thief* I. iv. 49 Wesley turned to him and said, quiet-like, 'Shut your big trap about Americans, limey.'

quietly, *adv.* Add: **a.** Also, surreptitiously, without attracting public notice.

1961 *Minnesota Rev.* I. III. 349 In the prison camp's Black Market civilian clothes were quietly bought..for him. **1976** *N.Y. Rev. Bks.* 15 Apr. 22/4 When the recent coal rush got underway, companies would quietly obtain mineral leases for as little as twenty-five cents an acre from the Interior Department.

b. *just quietly* advb. phr., confidentially, between ourselves. *Austral.* and *N.Z. colloq.*

1937 Partridge *Dict. Slang* 448/2 *Just quietly* was, in the G[reat] W[ar], a tag-c[atch] p[hrase] among New Zealanders. It had virtually no meaning. **1941** Baker *Dict. Austral. Slang* 40 *Just quietly*, between you and I. **1952** E. Lambert *Twenty Thousand Thieves* II. 123 'That Chips Prentice is a soldier and a half. Just quietly, he's up for a decoration.' Dick found no cause for surprise at this. Chips with a decoration seemed natural. **1966** G. W. Turner *Eng. Lang. in Austral. & N.Z.* viii. 177 Much New Zealand colloquialism is shared with Australia, e.g. *just quietly* 'between you and me', [etc.].

quiff (kwif), *sb.*[1] *U.S.* and *dial.* [Var. of Whiff *sb.*[1]] Also **quift. 1.** A puff or whiff of tobacco smoke. Also *fig.*

1831 J. M. Galloway *Poems* 27 Thou'st warm'd my nose at mony a speil; Ae quiff o' thee [*sc.* a pipe] Has made me play wi' care and skill. **1840** *Southern Lit. Messenger* VI. 447/2 A quiff would now and again ascend and hang like a tropical cloud over the hemisphere of his cranium. **1866** J. E. Brogden *Provincial Words* 159 *Quiff*, a puff. *Ex.* Should you like a quiff? **1876** F. K. Robinson *Gloss. Whitby* 149/2 *Quiff*, a whiff, a puff of smoke, an exhalation. 'I got a *quiff* on 't,' caught the scent. **1889** *Brighouse News* 14 Sept. 3/7 Hah nivver heeard a quift on't.

2. A puff or blast of wind.

1912 J. Masefield *Dauber* v. 268 She came within two shakes of turning top, Or stripping all her shroud-screws, that first quiff.

quiff, *sb.*[2] *dial.* and *slang* (esp. *naut.*). [Origin unknown.] A clever trick or dodge; a hint.

1881 *Advertiser Notes & Queries* I. 77/2 *Quiff*. What is the origin of this word, so often used in the sentence, 'I'll teach thee a quiff', meaning something clever. It is often heard in Cheshire. **1890** Barrère & Leland *Dict. Slang* II. 164 *Quiff*..(Tailors), a word used in expressing an idea that a satisfactory result may be obtained by other than strictly recognised rules or principles. **1925** Fraser & Gibbons *Soldier & Sailor Words* 223 *Quiff, a*, any specially ingenious smart, tricky, or novel or improvised way of doing anything. (Navy). In the Army used of any drill method peculiar to a battalion, and not usually done in others. Where the wording of the Drill Book is vague, units often read different meanings into the phraseology and invent their own 'Quiffs'. **1925** N. Lucas *Autobiogr. of Crook* v. 72 I'll give you one quiff, right now, because I like your face and your nerve. Never touch the dope, it's hell—and worse than that. **1928** *Weekly Dispatch* 13 May 10/4 Suddenly a faint grey blur on the horizon in the expected direction. The seaman blinks his eyes—an old quiff which prevents many a false alarm—and then makes his report. **1933** J. Masefield *Bird of Dawning* 107 It was young Mr. Abbott worked that quiff on you, sir. **1961** F. H. Burgess *Dict. Sailing* 166 *Quiff*, a trick or artifice that makes a job easier.

quiff, *sb.*[3] Also **quif.** [Origin obscure: cf. *quiff *sb.*[1]] A curl or lock of hair plastered down on the forehead, worn orig. by soldiers; more recently, a tuft of hair brushed upwards over the forehead. Also *attrib.*

1890 Barrère & Leland *Dict. Slang* II. 164 *Quiff* (military), the small curl on a soldier's temple just showing under his glengarry or forage cap. **1908** *Daily Chron.* 19 Mar. 4/4 He wears a quif of hair soaped down on his forehead in a slimy arc that nearly touches his eyebrows. **1919** H. G. Jenkins *John Dene of Toronto* XVI. 255 He's quite a nice youth, with black hair greased into what I think he would call a 'quiff'. **1925** H. G. Wells *Christina Alberta's Father* II. iii. 224 He had..highly oiled and entirely subjugated sandy hair with an army 'quif' on the forehead. **1929** P. Gibbs *Hidden City* xlii. 206 Revealing his well-plastered hair curled into a quiff over his forehead. **1953** N. Jacob *Morning will Come* iv. 72 A man with a short clipped beard and his hair parted in what was called a 'militiaman's quiff'. **1965** M. Bradbury *Stepping Westward* ii. 108 He could see the quiffs of his three cabin-mates as they moved round getting dressed. **1968** J. Ironside *Fashion Alphabet* 197 *Quiff*, where the hair falls forward over the forehead before being brushed back. Made fashionable by Elvis Presley and Cliff Richard, so adapted for girls. **1969** H. E. Bates *Vanished World* iii. 38 There were..plenty of men who were great dandies in that generation [his grandfather's]—the moustache-waxers, the quiff-plasterers, [etc.]. **1976** *Times* 8 Mar. 6/3 Mr Reagan..is certainly the best turned out candidate. The 1950s quiff is immaculate.

quiff, *sb.*[4] *dial.* and *slang.* [Origin obscure: cf. *quiff *v.*[1]] A young woman; *spec.* a prostitute, a 'tart'.

1923 G. Watson *Roxburghshire Word-bk.* 244 *Queef*,..an engaging girl. **1931** *Amer. Speech* VI. 440 *Quiff*, a cheap prostitute. **1966** R. H. Rimmer *Harrad Experiment* 107 'This looks like a rich quiff,' King Arthur said. 'I think we'll look into her suitcase.' **1973** L. Snelling *Heresy* I. i. 6 If only there was some other quiff about I might be able to deal with her indifference.

† **quiff,** *v.*[1] *Obs. coarse slang.* [Origin obscure.] *intr.* To copulate *with*. As **qui·ffing** *vbl. sb.* in quots.

1719 T. Durfey *Wit & Mirth* V. 243 By quiffing with Cullies three Pound she had got. **1796** Grose *Dict. Vulgar T.* (rev. ed.), *Quiffing*, rogering.

quiff, *v.*[2] [f. *quiff *sb.*[3]] *trans.* To arrange hair into a quiff. Also with *up*.

1940 R. Lehmann in *Folios of New Writing* Spring 101 There was one [sailor] in particular, large, with a genial, knobby raw-beef face and a flaxen curl quiffed up in the forefront of his sailor cap. **1972** *Daily Tel.* 24 Jan. 11/5 Fringes can be quiffed up too. **1977** *West Briton* 25 Aug. 5/8 They turned up in three-quarter length jackets, drainpipe trousers and shoelace ties. Their hair was quiffed and oiled.

qui-hy. Add: Also 9 **qui-hye; quoi hai, quai hai,** etc. (Further examples.) See also *koi hai.

1848 J. H. Stocqueler *Oriental Interpreter* 195/1 *Qui-hye! qui-hi!* or *koee-hye!* 'Who is there?' or 'Who waits?' In domestic establishments in Bengal..a servant..is summoned to the presence by the foregoing exclamations. Hence, the Europeans who reside in Bengal are called *Qui-hyes*, to distinguish them from the residents of Bombay, Madras, or Ceylon. **1858** G. F. Atkinson *Curry & Rice* Pref., The 'Qui Hye' of Bengal, the 'Mull' of Madras, and the 'Duck' of Bombay. **1939** 'E. Bell' *Memory be Good* i. 15, I must have been a bit of a nuisance to the old *quoi hais* who wanted to read quietly in their deck-chairs [on an India-bound Anchor Line boat]. **1965** B. Sweet-Escott *Baker Street Irregular* viii. 240 Most of Gavin's principal assistants were drawn from British business houses operating in the Far East... There was an inevitable tendency for some of them to regard themselves as old qua'hais. **1977** *Times* 25 June 15/4 An Old Quai Hai, as the diminishing band of servants of the Raj still living in this country [*sc.* India] are affectionately known.

quill, *sb.*[1] Add: **1. c.** pl. *spec.* = Pan-pipe. *U.S.*

1883 J. C. Harris *Nights with Uncle Remus* xiii. 69 Uncle Remus declared that Brother Rabbit could perform upon the quills, an accomplishment to which none of the other animals could lay claim. **1886** *Century Mag.* Feb. 521/2 But to show how far the art of playing the 'quills' could be carried..see this 'quill tune'..from a gentleman who heard it in Alabama. **1952** B. Ulanov *Hist. Jazz in Amer.* (1958) iii. 20 The homemade instruments of the Negro are described in some detail, the tambo, bones, quills, fife, triangle. **1970** *Western Folklore* XXIX. 231 Blues singer Big Joe Williams..recalls..a two-stringed cigar box guitar, a cane fife, a set of pan pipes or 'quills', and an upturned bucket, which served as a drum.

e. The whistle of a steam locomotive. *U.S.*

1945 F. H. Hubbard *Railroad Avenue* ii. 9 With its interpretive tone the ballast scorcher could make that quill say its prayers or scream like a banshee. **1961** *Listener* 24 Aug. 270/2 The fabled Casey Jones..was a 'quill artist' of note, who always carried with him his own quill (that is what they used to call a chime in the deep South).

f. An improvised straw or channel through which narcotics may be sniffed or smoked; the narcotic itself. *U.S.*

1935 A. J. Pollock *Underworld Speaks* 94/2 *Quill*, choicest grade opium. **1970** C. Major *Dict. Afro-Amer. Slang* 95 *Quill*, folded matchbook cover in which a narcotic is held and smoked or sniffed. **1971** *Black Scholar* Sept. 36/1 He..rolled a ten dollar bill up into a quill and gave the coke and quill to Christine, who snorted up half of the line on the card.

2. d. A hollow sleeve rotating in bearings which is used to transmit the drive from a motor to a concentrically-mounted axle.

1910 *Engineering* 12 Aug. 246/3 A gearless concentric motor for each driving-axle is mounted on a quill flexibly connected to the driving-wheels. **1930** *Ibid.* 6 June 722/1 Two new types of drive had been developed... The first consisted of a geared quill surrounding the driving axle and carrying two crankpins, the latter being connected by a flexible linkage to two crankpins on the driving wheels. **1968** D. W. & M. Hinde *Electr. Traction Systems & Equipment* ii. 32/2 A certain amount of experimental work has been carried out with the motor armature shaft of the hollow or quill pattern. **1975** Bram & Downs *Manuf. Technol.* vii. 208 The spindle rotates in the quill to provide the rotary motion for cutting tools.

3. g. Phr. *the pure quill*: see *pure *a.* 8 d.

8. a. (sense 1 b) *quill-machine*; (sense 3 b) *quill-case, -cleaner*.

1795 J. Woodforde *Diary* 28 Mar. (1929) IV. 186 Mr. Thorne..applied a Caustic to it just touching the part with it with a small kind of very fine hair pencil in a Quill-Case. **1968** *Canadian Antiques Collector* Nov. 25/2 It would seem that the ink bottle was usually on the right of the inkstand, the pounce on the left, with the quill standing in the central bottle, which was the quill cleaner. **1971** *Country Life* 1 July 23/1 This shelf carried writing equipment: inkpot, quill cleaner and sand box. **1846** G. Dodd *Textile Manuf. Gt. Brit.* VI. 182 Sail-making. The quill machines..have a considerable number of quills arranged in a row, and made to rotate rapidly.

b. quill drive, (the apparatus for) the transmission of power from a motor by means of a quill (sense *2 d); **quill shaft** = sense 2 d above; **quillwork,** a type of embroidery, using the quills of a porcupine, done by North American Indians; hence **quillworker.**

1912 Sheldon & Hausmann *Electr. Traction* 306/1 (Index), Quill drive. **1927** R. E. Dickinson *Electr. Trains* vi. 111 There are several other forms of suspension; e.g. the quill drive in which the motor-armature is on a hollow 'quill' inside which is the axle of the wheel. **1970** Lightband & Bicknell *Direct Current Traction Motor* iii. 47 The great majority of direct-current traction motors in service are either axle-suspended or fitted with some form of quill drive. **1934** *Jrnl. R. Aeronaut. Soc.* XXXVIII. 738 The turbine itself is mounted on a quill shaft which telescopes the pinion shaft, the latter being attached to the quill shaft at the low pressure end of the turbine. **1949** *Ibid.* LIII. 143/1 As originally designed the gear was a compound epicyclic gear, the sun gear of which was driven by a quill shaft from the front end of the com-

pressor. **1843** *Knickerbocker* XXII. 164 The Indians prepare it in bark, curiously ornamented with quill work and beads. **1908** *Encycl. Relig. & Ethics* I. 827/2 Closely akin to beadwork is quillwork, especially among the Plains Indians (now done in its purity by few except the Eskimos, the tribes of the north-west coast, and the northern Athapascans). **1966** L. Cohen *Beautiful Losers* I. 97 With a bowed head she received the compliments which the quillwork on her deerskin gown evoked. **1976** *San Antonio* (Texas) *Express* 8 Dec. 5-B/2 Quillworkers' tools have not changed. Today they still consist of some awls, strands of sinew and a knife.

quill, *v.* Add: **5.** *trans.* To write (with a quill), to pen.

1890 J. Coghill *Poems, Songs, & Sonnets* 67 This screed whilk he's juist new dune quillin'. **1945** J. Dickson in *Sc. Nat. Dict.* (1968) VII. 309/3 For each and a' the cheque's been quilled Wi' nae successors. **1977** *Even. Standard* 18 July 13/2 In 1677..Henry Vaughan quilled the immortal lines [etc.].

quill-driving, *ppl. a.* (*pres. pple.*). (Further example.)

1877 O. Wilde *Let.* Aug. (1962) 47, I had two jolly letters yesterday, one from Bouncer who is quill-driving or going to.

quiller (kwi·ləɹ), *sb.*[2] [f. Quill *v.* + -er[1].] One who quills material, esp. into the form of a ruff.

1853 Mrs. Gaskell *Ruth* II. vii. 172, I shall quill up a ruff for you. You know I am a famous quiller of net.

quillet, *sb.*[3] (Earlier example.)

1872 C. M. Yonge *P's & Q's* ix. 95 Rolling up her papers into little quillets.

quilling, *vbl. sb.* Add: **2.** (Earlier example.)

1790 A. M. Woodforde *Let.* 3 Sept. in *Parson Woodforde Soc. Jrnl.* (1972) V. iii. 56 Lady Bacon was dress'd in a striped muslin Gown and Coat..and a quilling of Black Lace at the edge.

3. *U.S.* The art of blowing distinctive sounds on the whistle of a steam locomotive. Cf. *quill *sb.*[1] 1 e.

1945 F. H. Hubbard *Railroad Avenue* ii. 8 The engineman put on a whistle of his own with a tone that suited him and then practised a technique of blowing it that would be distinctive. This was called 'quilling' and was a highly developed art. **1947** *Richmond* (Va.) *News Leader* 13 May 13/5 But the art of 'quilling', or 'making her talk', went out with electric and diesel locomotives, with their shrill horns and pneumatic whistles. **1966** *Listener* 14 Apr. 542/1 The variety of weird sounds he was able to extract from the six-tone engine whistle of his own property—an art known in railroad vernacular as 'quilling'.

quillon. Add: **1.** (Earlier and later examples.)

1884 R. F. Burton *Book of Sword* 125 The quillons may be either straight—that is disposed at right angles—or curved. **1978** N. K. Sandars *Sea Peoples* 158 The sword has a tapered blade, flanged tang and quillons.

2. *Comb.*, as **quillon-dagger** (see quot. 1960).

1950 *Proc. Prehist. Soc.* XVI. 24 The 'short iron sword' from Woodcuts, Dorset,..is actually a quillon-dagger of the 13th or 14th century A.D. **1960** H. Hayward *Antique Coll.* 232/1 *Quillon dagger*, a type of dagger with a simple cross-guard.

quilly, *a.* Delete *rare* and add: **2.** Consisting of or covered by quills.

1935 E. R. Eddison *Mistress of Mistresses* xviii. 355 A porcupine's quilly rump. **1938** T. H. White *Sword in Stone* xx. 294 Archimedes [*sc.* a hedgehog] got married, and brought up several handsome families of quilly youngsters.

quilt, *sb.*[1] Add: **1. d.** A layer of warm, thick material placed over the frames of a bee-hive to prevent draughts and contain the bees.

1870 *Amer. Bee Jrnl.* June 258/2 We finally had some little quilts (or whatever you choose to call them) made, and they answer admirably. **1873** *Brit. Bee Jrnl.* Nov. 100/1 His quilts, for so they are called by him, are laid close to the tops of the frames; they are not sufficiently heavy to crush the bees, even if laid directly on them, and they mould themselves to any possible condition. **1904** J. R. G. Digges *Irish Bee Guide* 50 Sheets and Quilts are required upon the frames or supers to preserve heat; to prevent draught; and to keep the bees from ascending into the roof... The quilts should be of felt, carpet, or other warm material. **1927** *Chambers's Jrnl.* XVII. 91/2 Place between the tops of the frames containing the combs and the bottom 'quilt' or cover two strips of wood an inch or two apart. **1952** H. Mace *Bee-Keeper's Handbk.* i. 15 Although many bee-keepers still use quilts, an increasing number have returned to an improved form of Crown board. **1962** A. S. C. Deans *Bees & Beekeeping* x. 66 November. Carry out a periodic inspection of all hives to make sure that mice have not settled under the roofs and on top of the quilts if crown boards are not in use.

3*. The interior of a cricket ball (see quot. 1921).

1882 *Baily's Mag.* Nov. 391, I took up the inside of a [cricket] ball just newly finished..and laid it on a bench, and hammered it with a wooden mallet, which rebounded without making the slightest impression on the substance which is called 'the quilt'. **1921** *Dict. Occup. Terms* (1927) §688 *Quilter*.., wraps worsted thread, by hand, round a cork to make quilt, or core, of cricket ball.

quilt, *v.*[1] Add: **1. a.** *transf.* and *fig.* (Further examples. in sense 'to cover as with a quilt'.)

1924 R. Campbell *Flaming Terrapin* v. 75 The lilies..quilted the land with snow. **1930** —— *Adamastor* 79 The gorgeous Ram..whose great pelt is rolled To quilt a thousand hills with fire.

b. *spec.* with a ball as object.

1886 W. H. Long *Dict. Isle of Wight Dial.* 53 *Quilt*,..to cover a ball with a network of twine.

quilt, *v.*[3] Add: **a.** Also *Austral.*

1945 Baker *Austral. Lang.* 120 One of the inevitable consequences..has been the development of an extensive vocabulary of fighting terms. Here are some of the best..*roll into, vacuum, quilt* and *stoush* a person. **1973** D. Stuart *Morning Star Evening Star* 111 More than one bloke I've seen Joe quilt good and proper for trying to make a joke of it.

b. *Cricket.* To hit (the ball, bowling, etc.) about the field with great force, usu. for a sustained period of time.

1866 *Baily's Mag.* Feb. 92 Mr Lyttleton had an early taste of the lobs; these he quilted awfully. **1867** *J. Lillywhite's Cricketers' Compan.* 69 That punishing bats-man, Mr. Lucas, 'quilted' the Colts' bowling tremendously. **1897** K. S. Ranjitsinhji *Jubilee Bk. Cricket* ii. 56 A batsman may get bowled first ball, a bowler may be quilted all over the field without getting a wicket, but both can redeem themselves by good fielding.

quilted, *ppl. a.* **2.** Delete † and add later examples.

1925 E. Sitwell *Troy Park* 50 One candle spills out thick gold coins Where quilted dark with tree shade joins. **1968** *National Observer* (U.S.) 3 June 15/1 This quilted personality has regenerated a sorrowful team. **1971** A. Sampson *New Anat. Brit.* xxxi. 561 The English pattern of hedgerows and quilted landscape.

quilter. Add: (Examples in sense of an apparatus.)

1895 *Montgomery Ward Catal.* Spring & Summer 264 Each..Sewing Machine will be supplied..with..1 braider, 1 binder, 1 quilter, 4 hemmers. **1908** *Sears, Roebuck Catal.* 41/2 We furnish with every [sewing] machine a complete set of accessories, consisting of one quilter, six bobbins..and one instruction book. **1964** *McCall's Sewing* v. 74/2 The quilter has a short open foot and an adjustable or removable space guide that may be used to the right or left of the needle.

quilting, *vbl. sb.*[1] Add: **3.** (Earlier and further N. Amer. examples.)

1768 in *Essex Inst. Hist. Coll.* (1879) XVI. 260 Quilting at my house. **1770** J. Parker *Diary* 22 Feb. in *New-England Hist. & Geneal. Reg.* (1915) LXIX. 10 Naby went to Mr Wildes to Quilting. **1825** J. Neal *Bro. Jonathan* I. 58 She returned however to the quilting and Peters..to the study of Mr. Harwood. **1913** *Atlantic Monthly* Dec. 826/2 Zobbie often met Pauline at the quiltings and other gatherings at the homes of non-partisans. **1939** L. M. Montgomery *Anne of Ingleside* xxxii. 228 The Ladies' Aid is going to have their quilting at Ingleside. **1975** *Budget* (Sugarcreek, Ohio) 20 Mar. 15/6 Mrs. Eli Y. Byler had 2 quiltings last week and will have another one this week.

4. quilting-bee (earlier and later examples); also *Canad.*; **quilting day** *N. Amer.*, a day devoted to a quilting-party; **quilting-feast** (example); **quilting frame** (examples); **quilting match** *U.S.* = *quilting-party*; **quilting-party** (earlier and later examples); also outside the U.S.

1832 S. G. Goodrich *System of Univ. Geogr.* vii. 167 The females also have similar meetings called 'quilting bees', when many assemble to work for one, in padding or *quilting* bed coverings or *comforters*. **1921** *Daily Colonist* (Victoria, B.C.) 5 Oct. 6/5 It was decided to hold a quilting bee at the home of Mrs. Scott. **1976** 'D. Halliday' *Dolly & Nanny Bird* x. 126 He looked..like a lush from a quilting bee. **1939** L. M. Montgomery *Anne of Ingleside* xxxii. 230 The quilting day was more like June than October. **1968** Quilting feast [see *quilting-party* below]. **1739** *Pennsylvania Gaz.* 15 Nov. 4/1 Just imported and to be sold by John Brientnall, in Chestnut-street Tenter-hooks of several sizes, fit for Butchers, Skinners, Fullers, and for Quilting-Frames. **1854** M. J. Holmes *Tempest & Sunshine* xx. 274 Said she, 'Mighty good opinion Mr. Quilting-frames has of me (alluding to Mr. Miller's height); glad I know his mind.' **1908** L. M. Montgomery *Anne of Green Gables* xxi. 235 If Marilla..was actuated by any motive save her avowed one of returning the quilting-frames. **1813** M. L. Weems *Drunkard's Looking Glass* (ed. 2) 5 He does not trouble his head about asking the Fool where he has been, whether at a Funeral or a Wedding..or a Quilting-match. **1881** A. B. Alcott *New Connecticut* 108 The Wolcott Dialect..Quilting match. **1833** S. Smith *Life & Writings J. Downing* 139 A few others..wouldn't invite poor Mrs. No-tea to their husking and quilting parties. **1835** *Knickerbocker* VI. 180 It so happened that there was a great quilting-party invited to Tecumseh-Place, which assembled all the principal young people of the county. **1907** *St. Nicholas* (N.Y.) 1044/2 She has gone, with her grandmother whom she was visiting, to a quilting-party. **1948** *Minneapolis Star* 17 Sept. 31/1 Nobody sees Nellie home when the women of Halvarson Bowers post 187, VFW auxiliary, have a quilting party. **1968** J. Arnold *Shell Bk. Country Crafts* xxix. 310 It was..possible for as many as six people..to be sewing at the same time. Such gatherings became known as quilting parties or feasts.

quilting, *vbl. sb.*[2] (Earlier example.)

1829 P. Egan *Boxiana* 2nd Ser. II. 242 The *quilting* Bob had previously received, rendered him in a great measure incapable of taking advantage of his adversary's distress.

quim (kwim). *coarse slang.* [Origin obscure: perh. rel. to Queme *a.* or *v.*; cf. Quaint *sb.*] **1.** The female external genital organs; the vagina.

An example of 1613 cited in Farmer & Henley's *Dict. Slang* has not been traced.

c **1735** *Harlot Un-mask'd* (Ballad), Tho' her Hands they are red, and her Bubbies are coarse, Her Quim, for all that, may be never the worse. **1796** Grose *Dict. Vulgar Tongue* (rev. ed.), *Quim*, the private parts of a woman: perhaps from the Spanish *quemar*, to burn. **1846** *Swell's Night Guide* (rev. ed.) 57 Vell, rattle my dice box for a musty quim, but that Sall is a splitter! c **1863** 'Philo Cunnus' *Festival of Passions* II. 7 Gently pulling up my shift his hand touched my thigh, and instantly I sighed as it was laid on my quim. **1882** in 'P. Fraxi' *Catena Librorum Tacendorum* (1885) 267 My imagination fills the empty galligaskins with cosy bottoms and hirsute quims. **1922** Joyce *Ulysses* 550 Were you brushing the cobwebs off a few quims? **1936** H. Miller *Black Spring* 253 'Now,' he says, 'I'm going to pay you as usual,' and taking a bill out of his pocket he crumples it and then shoves it up her quim. **1951** N. Coghill tr. *Chaucer's Cant. Tales* 113 He made a grab and caught her by the quim And said, 'O God, I love you!' **1966** P. Willmott *Adolescent Boys* iii. 50, I got my hand on her tit and I thought well, that's all right. So I thought I'd try for her quim. **1974** H. R. F. Keating *Underside* ii. 25 Is it worse to have it on me belly than to have it in me quim?

2. A woman; women collectively. *N. Amer. slang.*

1935 A. J. Pollock *Underworld Speaks* 94/2 *Quim*, a female. **1974** *Saturday Night* (Toronto) Jan. 35/2 The key to success in this contest is a flashy car; and if the car is both expensive and impressive 'you have to beat the quim off with a hockey stick'.

quin, *sb.*[2] [Shortened f. Quintuplet.] One of five children born at one birth; such a child in later life. Also *attrib.*

1935 *Dionne Quintuplets growing Up* (caption), My, what big girls the 'Quins' are getting to be. **1936** W. Thornton *Country Doctor* 127 In Portugal in 1866, one quin baby attained an age of 50 days. **1937** R. Macaulay *I would be Private* I. i. 16 Another little boy! Win's got quins! Can you beat it?... It took me and her poor father ten years to get five, and here's you and Win done it in a year. **1951** L. Barker *Truth about Dionne Quins* i. 11 Dr. Dafoe..held Press conferences..in addition to his quin column, a newspaper feature, written by various ghost writers. *Ibid.* xxiii. 178 Emilie, the happy-go-lucky tomboyish quin,..was not inclined to worry. **1968** *Economist* 15 June 49/1 Despite the title [of the book], Dr Scheinfeld is less good on supertwins, on triplets, quads and quins. **1976** *Liverpool Echo* 6 Dec. 1/1 The condition of Fiona, the fourth quin, has deteriorated and she has had difficulty in breathing since birth.

quinacridone (kwinæ·kridōᵘn). *Chem.* [f. Quin(oline + Acrid(ine + -one.] Any of four synthetic isomeric compounds, $C_{20}H_{12}N_2O_2$, or their substituted derivatives, which have a heteroaromatic structure consisting of a string of five fused rings (three benzene and two 4-pyridone arranged alternately) and which include a class of usu. red or violet pigments. Also *attrib.* and in *Comb.*, as *quinacridone red, violet*, etc.

[**1896** *Jrnl. Chem. Soc.* LXX. I. 261 Hydroxyquinacridone.] **1906** *Ibid.* XC. I. 459 Quinacridone,..crystallises in yellow needles..and dissolves in concentrated sulphuric acid to a yellow solution with a greenish-blue fluorescence. **1958** *Chem. Abstr.* LII. 10215 Dihydroquinacridones..can be oxidized to quinacridones, pigments of good light fastness. **1963** *Jrnl. Oil & Colour Chemists' Assoc.* XLVI. 29 Quinacridone red yellowish, quinacridone red bluish and quinacridone violet which were issued by Du Pont in 1958,..were the first of the quinacridones. Chemically, all three pigments are one and the same unsubstituted linear *trans*-quinacridone. **1969** R. Mayer *Dict. Art Terms & Techniques* 321/2 Besides this relatively yellowish or scarlet shade, bluish, magenta, and violet shades are also made from the same dyestuff (linear quinacridone) that yields quinacridone red. **1972** *Materials & Technol.* V. xi. 358 Quinacridone reds, maroons and violets are a comparatively recent introduction. These very expensive pigments have excellent light-fastness in full colours and in pastel shades. **1973** Tilak & Ayyangar in R. M. Acheson *Acridines* (ed. 2) viii. 603 Diketoquinolinoacridines are designated by the trivial name 'quinacridones'. They can be linear *trans* (71) and *cis*..or angular... The compound 71, which is usually referred to as quinacridone, and its derivatives are..valuable as high quality pigments.

quinacrine (kwi·năkrīn). *Pharm.* [f. Quin(ine + Acr(id)ine.] A name for mepacrine (hydrochloride). Also *attrib.* and in *Comb.*, esp. with reference to the use of quinacrine or quinacrine mustard to stain chromosomes.

Formerly a proprietary term in the U.S.

1934 *Official Gaz.* (U.S. Patent Office) 23 Oct. 771/1 Société des Usines Chimiques Rhone-Poulenc, Paris... *Quinacrine*... Pharmaceutical product to be used in the treatment of malaria. Claims use since Feb. 20, 1933. **1934** *Chem. Abstr.* XXVIII. 3792 Quinacrine is chloro-2-diethylaminopentylamino-5-methoxy-7-acridine dihydrochloride. **1936** *Trade Marks Jrnl.* 4 Mar. 267/1 *Quinacrine*... Chemical substances used in medicine and pharmacy. Société des Usines Chimiques Rhone-Poulenc, ..Paris..; manufacturers. **1960** C. Achebe *No Longer at Ease* xi. 105 He said the poor man must be suffering from malaria, and the next day he bought him a tube of quinacrine. **1970** *Nature* 6 June 897/1 The quinacrine

staining test shows up the number of *Y* chromosomes in a human interphase (non-dividing) cell. *Ibid.* 961/2 (*caption*) A group of quinacrine-stained spermatozoa showing fluorescence. **1971** *Oxford Times* 26 Mar. 2/5 Samples of fluid surrounding the foetus were treated with quinacrine dihydrochloride, an anti-malaria drug. The effect was to make the 'Y' chromosomes..take on a fluorescent glow when viewed through an ultra-violet microscope. **1975** Fraser & Nora *Genetics of Man* ii. 10/1 Quinacrine binds preferentially to certain regions of metaphase chromosomes to produce characteristic banding patterns ('Q-bands').

2. Special Comb.: **quinacrine mustard,** a nitrogen mustard derived from quinacrine and used as a fluorescent stain for chromosomes.

1957 R. Jones et al. in *Jrnl. Org. Chem.* XXII. 783/2 We here wish to report on the conversion..of 2-methoxy-6,9-dichloroacridine to 2-methoxy-6-chloro-9-[4-bis(β-chloroethyl)-amino-1-methylbutylamino]acridine (quinacrine mustard). **1970** [see *mustard *sb.* 2* b]. **1970** *Nature* 4 July 101/1 The trivial name 'quinacrine mustard' has been used..to describe at least two chemical species. *Ibid.* 101/2 May I..suggest that if authors wish to continue to use the expression 'quinacrine mustard' this be restricted to the true quinacrine derivative [*sc.* 2-methoxy-6-chloro-9-[4-*bis*(2-chloroethyl)amino-1-methyl-butylamino] acridine]. **1971** *New Scientist* 18 Mar. 606/1 When hamster and bean chromosomes were stained with..quinacrine mustard, and viewed with long wavelength ultraviolet light, the chromosomes fluoresced differently along their length.

quinalbarbitone (kwinælbā·ɹbitōᵘn). *Pharm.* [f. L. *quin-que* five + Al(lyl + *barbitone.] The compound 5-allyl-5-(1-methylbutyl)-barbituric acid, which is used as a sedative-hypnotic, esp. for pre-operative sedation, usu. in the form of its sodium salt, $C_{12}H_{17}N_2O_3Na$, a white powder often known by the proprietary name *Seconal. Cf. *secobarbital.

1951 *Addendum to Brit. Pharmacopœia 1948* 46 Quinalbarbitone Sodium should be kept in a well-closed container. **1960** *Brit. Med. Jrnl.* 19 Mar. 872/2 For older children quinalbarbitone ('seconal') may be given. **1971** 'D. Halliday' *Dolly & Doctor Bird* xiv. 199, I..stayed behind to administer a mild dose of quinalbarbitone. **1977** *Proc. R. Soc. Med.* LXX. 773/2 Barbiturates took pride of place as the commonest lethal drug, and quinalbarbitone sodium (Tuinal) headed the list.

quinarius (kwinā·riɒs). Pl. **quinarii** (-i,ī). [L.: see Quinary *a.* and *sb.*] = Quinary *sb.* 2 a.

1601 P. Holland tr. *Pliny's Nat. Hist.* II. xxxiii. iii. 463 At what time ordained it was, that the Denarius or Denier should go for tenne Asses or pounds of brasse money; the halfe Denier, Quinarius, should be currant for five; and the Sesterce reckoned worth two and a halfe. a **1666** Evelyn *Diary* an. 1645 (1955) II. 398 Now other observations I made in Rome are these amongst other, As to Coynes & Medails, that 10 Asses make the Roman Denarius, 5 the quinarius. **1708** Kersey *Dict. Anglo-Britannicum, Quinarius*.., a Roman Coin of the Value of Five Asses, equal to 3 Pence 3 Farthings English. **1771** [see Sesterce]. **1840** [see Denarius 1]. **1962** R. A. G. Carson *Coins* 110 The new system comprised three denominations in silver—the denarius worth ten asses, the quinarius of five asses and the sestertius of two and a half.

quinary, *a.* and *sb.* Add: **A.** *adj.* **2.** Of or belonging to the fifth order or rank; fifth in a series.

1924 [see *quaternary *a.* 3]. **1953** *Amer. Econ. Rev.* May *Papers & Proceedings* 365 Logically and empirically, quinary industries as we shall define them are not once more a residual category. These industries comprise medical care, education, research, and recreation... The principle that guides this grouping is that they all have to do with the refinement and extension of human capacities. **1973** *New Society* 15 Nov. 386/3 The 'tertiary' sector contains at least three divisions—'tertiary' proper (pure service provision..); 'quarternary' [*sic*] (information exchange and decision-making); and 'quinary' (research, development and education).

B. *sb.* **3.** Something that belongs to the fifth order or rank.

1937, 1946 [see *quaternary *sb.* 2].

quinazoline (kwinæ·zŏlīn). *Chem.* [ad. G. *chinazolin* (A. Weddige 1887, in *Jrnl. f. prakt. Chem.* XXXVI. 142), f. *chinolin* Quinoline with inserted *az-* (see Azo-).] A yellow, basic, crystalline solid, $C_8H_6N_2$, which has a bicyclic structure formed from fused benzene and pyrimidine rings; any substituted derivative of this.

1887 *Jrnl. Chem. Soc.* LII. II. 1044 Anhydroacetylorthamidobenzamide is regarded as a derivative of a hypothetical base, quinazoline,..isomeric with cinnoline and quinoxaline. **1903** *Ibid.* LXXXIV. I. 446 By oxidising this base with potassium ferricyanide, quinazoline..was at last successfully prepared; it..crystallises from light petroleum in glistening flakes resembling naphthalene, has a bitter, burning taste, and..is odourless at the ordinary temperature. **1926** H. G. Rule tr. *J. Schmidt's Text-bk. Org. Chem.* 700 Quinazolines..are strong bases, which are readily reduced to their dihydro-compounds. **1950** *Thorpe's Dict. Appl. Chem.* (ed. 4) X. 346/2 A variety of substituted quinazolines have been prepared recently as part of the search for effective antimalarials. **1968** L. A. Paquette *Princ. Mod. Heterocyclic Chem.* ix. 326 Quinazoline 3-oxide..shows the same reactivity as quinazoline toward nucleophilic reagents, but the addition products frequently eliminate water with the net effect that 4-substituted quinazoline results.

‖ **quincaillerie** (kæ̃nkāyəri). [Fr.] **a.** (See quot. 1883). **b.** A hardware or ironmonger's shop. Also *attrib.*

1883 J. W. Mollett *Illustr. Dict. Art & Archæol.* 272 *Quincaillerie*, a general term for all kinds of metallurgical work in copper, brass, iron, etc. **1951** W. Sansom *Face of Innocence* xi. 164 Past the paper shop... Past the dark leathern hole of the shoe-menders, past the sharp quincaillerie. **1966** P. V. Price *France: Food & Wine Guide* 83 Spend an hour in the hardware department of a good store or a *quincaillerie*. **1968** *Guardian* 3 May 9/1 Red-faced *quincaillerie* reps sit in the dark brown bar totting up their accounts over a Ricard.

quince. Add: **1. b.** Phr. *to get on* (*a person's*) *quince*, to irritate or exasperate. *Austral. slang.*

1941 Baker *Dict. Austral. Slang* 58 *Get on one's quince*, to annoy or aggravate deeply. **1948** *Sydney Morning Herald* 3 July 9/1 Aw, can it boss! You're gettin' on me quince. **1963** A. E. Farrell *Vengeance* ii. 19 These bloody trees are getting on me quince! **1974** D. O'Grady *Deschooling Kevin Carew* 95 In an unguarded moment, he told Bill Moynihan 'This joint is getting on my quince.'

3. *quince cream* (examples), *jam, jelly, marmalade* (earlier and later examples), *pudding*; *quince-coloured, -flavoured* adjs.

1907 Quince-coloured [see *low-sized* s.v. *low *a.* 21]. **1723** J. Nott *Cook's & Confectioner's Dict.* sig. Ee 4 (*heading*) To make quince cream. **1974** J. Grigson *Eng. Food* 216 (*heading*) Gooseberry, pear, apple or quince cream. **1950** D. Gascoyne *Vagrant* 55 And whilom most becomingly strums On his poignantly Quince-flavoured lute! **1936** *Farmhouse Fare* 113 Quince jam... Put the pulp through a sieve, or mash very finely with wooden spoon. **1978** R. V. Jones *Most Secret War* xxxvii. 330 My moment came when I asked him one day whether he had ever made quince jam. **1861** Mrs. Beeton *Bk. Househ. Managem.* 795 *Quince jelly*... To every pint of juice allow 1 lb. of loaf sugar. **1971** M. McCarthy *Birds Amer.* 35 They stole quinces..and she put up quince jelly. **1636** P. Massinger *Great Duke of Florence* iv. ii. sig. H3, This Quince-Marmalade Was of my owne making. **1728** E. Smith *Compleat Housewife* (ed. 2) 190 (*heading*) To make white Quince Marmalade. **1832** L. M. Child *Amer. Frugal Housewife* 118 Quince Marmalade.—To two pounds of quince put three quarters of a pound of nice sugar. **1963** A. L. Simon *Guide Good Food & Wines* (rev. ed.) iv. 292/1 Its name in Portuguese—*Marmelo*—is the origin of the name *Marmalade*, which was at first applied to no other but *Quince Marmalade*. **1723** J. Nott *Cook's & Confectioner's Dict.* sig. Ee 6 (*heading*) To make a quince pudding.

Quincke (kvi·ŋkə). *Med.* The name of Heinrich I. *Quincke* (1842–1922), German physician, used in the possessive and with *of* to designate an acute form of urticaria, described by him in 1882 (*Monatshefte für prakt. Dermatol.* I. 129–31).

1894 *Jrnl. Nerv. & Mental Dis.* XXI. 627 The recognition of angioneurotic œdema as a special form of disease is generally credited to Quincke who described it..as acute circumscribed œdema, and it has by some authors been designated as Quincke's disease. **1933** E. A. Cockayne *Inherited Abnormalities of Skin* xiv. 370 (*heading*) Angioneurotic oedema. Acute circumscribed oedema. Quincke's oedema. **1934** Dore & Franklin *Dis. of Skin* iv. 46 Giant urticaria sometimes called acute circumscribed œdema of Quincke.., is characterized by the rapid development on the cutaneous or mucous surfaces of large patches of localized œdema, varying in size from a hazel-nut to an ordinary orange. **1974** Warin & Champion *Urticaria* i. 5 Bannister..protested at the use of the term Quincke's disease.

qui·ncunx, *v. rare.* [f. the sb.] *trans.* To put in quincunx arrangement.

1847 *Simmonds' Colonial Mag.* June 165 Some [say] that the bushes are not near enough together, and that I ought to quincunx them.

Quinean (kwəi·niăn), *a.* [f. the name of the U.S. logician Willard Van Orman *Quine* (b. 1908) + -an.] Of, pertaining to, or characteristic of Quine or his theories.

1966 S. Beer *Decision & Control* viii. 170 For the record, this is what the three propositions look like in formal Quinean terms. **1972** J. J. Katz *Semantic Theory* vi. 243 Quinean arguments do not apply to the analytic-synthetic distinction..such as the one developed here. **1978** C. Hookway in Hookway & Pettit *Action & Interpretation* 31 Once we recognise that relativism of this kind is at least an option in translation theory, we can see how to construct a Quinean response to the charge that the argument involves arbitrary discrimination in favour of the theory of nature and against translation theory.

quinella (kwine·lă). orig. *U.S.* Also **quinela, quiniela.** [ad. Amer. Sp. *quiniela* in same sense.] A form of (usu. totalizator) betting in which the punter is required to select the first two place-getters in a race or other contest, not necessarily in the correct order.

1942 Berrey & Van den Bark *Amer. Thes. Slang* § 734/6 *Quinella*, a separate betting pool in which the bettor attempts to pick the first and second horses of a given race. **1944** *Amer. Speech* XIX. 231/2 *Quinella*, Saturday and Sunday off together. The term is used in dog races where the better must pick two winners on a single ticket. **1949** *Rocky Mountain News* 2 July 9/1 A parlay or quiniela is a combination wager coupling the first two animals to finish in one race... Odds on parlay and quiniela are heavy and, when hit, the return is great. **1956** *Sun* (Baltimore) 6 Oct. (B. ed.) 9/1 A petition for quinella wagering at Maryland thoroughbred tracks. **1964** A. Wykes *Gambling* viii. 193 Or there is the 'quiniela', in which the bettor tries to pick the first and second horse of a race. **1967** *Punch* 5 July 20/2 Incomprehensible to the non-punter are notices which attest the management's determination to accept, or not to accept, Castellas, Couplets, Duellas, Eliminators, Jackpots, Plums, Quinellas and Yankees. **1969** *Australian* 7 June 36/1 She picked..first and second in two races—including the final quinella which paid $7.60. **1971** [see *nap *sb.*[6] 2 c]. **1974** *Dominion* (Wellington, N.Z.) 11 July 9/9 Two relative outsiders..returned a quinella of $473.60 when they were first and second in the Seatoun Handicap. **1977** *Listener* 30 June 847/3 Didn't put number five in the Quinela.

quingenary (kwindʒī·nări), *sb.* and *a.* [ad. L. *quingēnārius*, f. *quingēnī*, distrib. of *quingentī* five hundred.] **A.** *sb.* A five-hundredth anniversary or the celebration of this; = Quincentenary *sb.*

1926 Fowler *Mod. Eng. Usage* 72/2 Quingē'nary, sescē'nary, septingē'nary. **1965** E. Gowers *Fowler's Mod. Eng. Usage* (ed. 2) 83/1 *Quatercentenary* and *quincentenary*..are unlikely to be ousted by *quadringenary* and *quingenary*.

B. *adj.* Of an ancient Roman military unit: consisting of five hundred men.

1969 G. Webster *Roman Imperial Army* iii. 148 There are examples of the six barrack blocks of a quingenary fort at the Welsh fort of Gellygaer. *Ibid.* 149 According to Hyginus, the quingenary unit was composed of 380 infantry and 120 cavalry. **1976** E. N. Luttwak *Grand Strategy of Roman Empire* ii. 122 The new formations were clearly useful in bridging the gap between the legions and the quingenary *auxilia*.

quinhydrone. For 'green crystalline substance' read 'brown crystalline substance with a green lustre' and add: Also, any analogous molecular compound formed by a quinone and another aromatic compound. (Earlier and later examples.)

1857 H. Watts tr. *Gmelin's Hand-bk. Chem.* XI. 164 Quinhydrone. [*Ibid.*, When kinhydrone is heated, it sublimes partly undecomposed, partly converted into yellow kinone.] **1908** *Chem. Abstr.* II. 104 In the reaction of α-hydronaphthoquinone with quinone, the former is oxidized to α-naphthoquinone and the latter reduced to hydroquinone, and these react to give the mixed quinhydrone. **1963** I. L. Finar *Org. Chem.* (ed. 4) I. xxvii. 670 Quinhydrones are a group of coloured substances formed from quinones and other aromatic compounds. Some are believed to be charge-transfer complexes. *Ibid.* 671 Another type of quinhydrone is the one formed between one molecule of benzoquinone and one molecule of quinol... It is believed that the rings of the two molecules are parallel and held together by hydrogen bonds at both ends.

2. *attrib.* **quinhydrone electrode,** an electrode (usu. of platinum) immersed in a test solution to which quinhydrone has been added, which has a potential dependent upon the pH of the solution and can thus be used to measure the pH potentiometrically.

1921 E. Biilmann in *Jrnl. Soc. Leather Trades' Chemists* V. 27 In the quinhydrone electrode two hydrions disappear, and at the same time a mol. of quinone is hydrogenated to one of hydroquinone. **1930** G. D. Elsdon in C. A. Mitchell *Recent Advances in Analytical Chem.* I. vii. 250 Certain difficulties are encountered in the determination of the *p*H value of milk. These have been overcome to a considerable extent..by the use of the quinhydrone electrode. **1969** H. Rossotti *Chem. Applications Potentiometry* viii. 92 The quinhydrone electrode can..be used as a probe for hydrogen ions in the acidity range $1 \leqslant \text{pH} \leqslant 8$.

qui·nine, *v.* [f. the sb.] *trans.* To dose or treat with quinine.

1927 *Bulletin* (Glasgow) 18 Mar. 5/5 The 'choleric colonel from India'..is apparently more sinned against than sinning. His medical man has 'quinined' him.

quinion (kwi·niɒn), abbrev. of Quinternion.

1897 R. Garnett in *Bibliographica* III. 37 Their [*sc.* Sweynheym and Pannartz's] spacious premises are choked with unbound sheets in quinions. **1927** E. K. Rand in W. M. Lindsay *Palaeographia Latina* v. 53 A quinion may consist of a quaternion ruled four leaves at a time plus an extra leaf. **1959** P. H. Blair in *Moore Bede* 11/2 It contains thirteen gatherings of which II, IV, V and VIII–XII inclusive are regular quinions, that is, each of five bifolia making ten leaves after folding.

quinnat. For the specific epithet of the Latin name substitute *tschawytscha*; also *attrib.* (Later examples.)

1874 J. S. Hittell *Resources Calif.* 407 The most important fish of California is the quinnat salmon. **1881** [see *king salmon* s.v. *king *sb.* 13 a]. **1907** T. W. Lambert *Fishing in Brit. Columbia* 73 They are the king salmon or quinnat, a large fish running up to over 80 lb. **1925** J. T. Jenkins *Fishes Brit. Isles* 216 The Chinook, Quinnat or King Salmon.., the most valuable species [of Pacific salmon], attaining a length of 4 to 5 feet. **1948** *Pacific Discovery* July 26/2 When referring to 'salmon' we will mean the true salmon (also known as quinnat, chinook or

QUINO

king) which invariably dies after spawning. **1962** L. WEDLICK *Fishing in Austral.* III. 87 It is unfortunate that the king or chinook salmon—known as quinnat salmon in Australia—has never been successfully acclimatized in our streams, for this fish is the largest of all the Pacific salmon.

quino, var. KENO in Dict. and Suppl.

quinoid (kwi·noid), *a.* and *sb.* *Chem.* [f. QUIN(ONE + -OID.] **A.** *adj.* = *QUINONOID *a.* **B.** *sb.* A quinonoid compound.
1900 E. F. SMITH tr. *V. von Richter's Org. Chem.* (ed. 3) II. 364 The colored alkali salts are quinoid derivatives of phenolphthalein. **1907** *Chem. Abstr.* I. 300 (*heading*) Oxidation of benzidine. (VIII Communication on quinoids.) **1908** *Ibid.* II. 827 The quinoid theory of colour. **1938** G. H. RICHTER *Textbk. Org. Chem.* xxx. 627 The *ortho* and *para* quinoid structures show strong chromophoric properties. **1949** ENGLISH & CASSIDY *Princ. Org. Chem.* xx. 396 It is thought that some metabolic processes involve a reduction-oxidation system involving quinoid and hydroquinoid molecules. **1961** B. J. STOKES *Org. Chem.* xxiv. 437 Addition of alkali to phenolphthalein causes a change in structure and a molecule containing a quinoid ring is made. **1975** LEWIS & PETERS *Facts & Theories Aromaticity* v. 76 The quinones and quinoids generally are usually thought to be non-aromatic yet they apparently contain the magic 6π-electrons in a ring, an electron arrangement which is often regarded as characteristic of aromaticity.
Hence **quinoi·dal** *a.*
1907 *Chem. Abstr.* I. 302 The presence of methylene destroys the ability of the two phenyl groups to assume the quinoidal state. **1975** *Nature* 20 Feb. 625/1 Its infrared spectrum in $CHCl_3$ further substantiates a quinoidal structure, showing a strong carbonyl absorption at 1,654 cm^{-1} and an olefinic absorption at 1,584 cm^{-1}.

quinol. (Earlier and later examples.)
1879 *Jrnl. Chem. Soc.* XXXVI. 464 (*heading*) Derivatives of quinol (hydroquinone). **1964** [see *CATECHOL]. **1976** *Nature* 17 June 621/3 The central biological roles of quinones, quinols and chromanols had become as acceptable as they were surprising.

quinonoid (kwi·nonoid), *a.* *Chem.* [f. QUINON(E + -OID.] Being, resembling, or characteristic of a quinone; being or exhibiting a molecular structure typified by that of quinone, viz. a doubly unsaturated ring with double bonds to two substituents (usu. in *para* positions).
1878 *Chem. News* 5 July 9/1 (*heading*) Quinonoid body found in a species of agaric. **1892** *Proc. Chem. Soc.* VIII. 101 It was maintained that in the case of azo-dyes, the rosanilines, methylene-blue, &c., colour was conditioned by a quinonoid structure. **1908** *Jrnl. Chem. Soc.* XCIV. I. 806 The meaning of the term quinonoid must be widened to include substances..in which the quinonoid oxygen atom is displaced by a bivalent imine or hydrocarbon group. **1951** I. L. FINAR *Org. Chem.* I. xxvii. 545 The yellow colour of *p*-benzoquinone is due to the presence of the quinonoid structure =⟨=⟩=. **1956** *Nature* 14 Jan. 79/2 The quinonoid product of the catecholase reaction..is an active oxidase in the cresolase reaction. **1974** *Ibid.* 20 Dec. 710/2 The sclerotisation and tanning of insect cuticles is generally thought to result from a crosslinking of the cuticular proteins by quinonoid derivatives of tyrosine.

quinoxaline (kwinǫ·ksălīn). *Chem.* [ad. G. *chinoxalin* (O. Hinsberg 1884, in *Ber. d. Deut. Chem. Ges.* XVII. 319), f. *chin-olin* QUINOLINE + *-oxal* (f. *glyoxal* GLYOXAL) + *-in* -INE⁵. So named on account of its structural similarity to quinoline and its preparation from glyoxal.] A weakly basic colourless crystalline solid, $C_8H_6N_2$, which was first prepared by reaction of glyoxal and *o*-phenylenediamine, and has a bicyclic structure formed from fused benzene and pyrazine rings; any substituted derivative of this compound.
1884 *Jrnl. Chem. Soc.* XLVI. 1052 He [*sc.* Hinsberg]..proposes to call this series of compounds quinoxalines. The formula of quinoxaline (the lowest homologue) is undoubtedly $C_6H_4<^{N:CH}_{N:CH}>$. **1887** [see *AZINE]. **1926** H. G. RULE tr. *J. Schmidt's Text-bk. Org. Chem.* 700 Quinoxalines..are weakly basic compounds, which may be reduced to hydro-quinoxalines, but are stable towards oxidising agents. **1951** I. L. FINAR *Org. Chem.* x. 189 It [*sc.* glyoxal]..combines with *o*-phenylenediamines to form quinoxalines..; *e.g.*, with *o*-phenylenediamine it forms quinoxaline itself. **1974** *Nature* 20 Dec. 654/1 The quinoxaline chromophores of echinomycin are similar in size to the quinoline chromophore of chloroquine, which is known to bind to DNA by intercalation.

quinquagint (kwi·nkwădʒint). *nonce-wd.* [ad. L. *quinquāgintā* fifty.] A set of fifty persons or things.
1843 THACKERAY *Irish Sk.-Bk.* II. xiv. 264 There are 220 voters, it appears;..but as parties are pretty equally balanced, the votes of the quinquagint..carry an immense weight.

quinquennial, *a.* and *sb.* Add: **A.** *adj.* **1. b.** (Earlier example.)
1847 J. S. MILL *Let.* 19 Nov. in *Wks.* (1963) XIII. 725 The number of deaths..is *less* in each quinquennial period.
2. (Later examples.)
1903 *Westm. Gaz.* 14 Jan. 5/2 The Board of the London Hospital..has been impelled to issue its quinquennial appeal two or three weeks in advance of the appointed time. **1955** *Times* 3 May 15/2 A quinquennial valuation of the 'Royal' life and annuity business was made at December 31.
B. *sb.* **3.** A fifth anniversary.
1903 *Westm. Gaz.* 14 Jan. 5/2 The hospital only begs widely every five years, and this year is our quinquennial.

quinquennially, *adv.* (Later examples.)
1972 *Nature* 4 Feb. 246/3 Is this fraction to be reviewed annually or quinquennially, and, if so, by whom? **1979** *Daily Tel.* 3 Jan. 2/1 There is no excuse for the use of phrases like reversionary bonus compounded quinquennially, when what is meant is a five-yearly pay-out with interest added.

quinta. Add to def.: Also, a country estate; in Portugal, a wine-growing estate; in S. America, a house or estate on the outskirts of a town. Also *attrib.* (Earlier and further examples.)
1754 FIELDING *Let.* Aug.–Sept. in A. Dobson *At Prior Park* (1912) 143 He hath a little Kintor..or Villa at a Place called Jonkera. *a* **1770** A. HERVEY *Jrnl.* an. 1748 (1953) 75, I used, too, to go out to my friend Mr. Mayne and Barn's quinta, where I was very much at ease. **1818** C. A. RODNEY *Let.* in *Amer. State Papers: Foreign Relations* (1834) IV. 219 The small farms, or quintas, in the neighborhood of cities, are in fine order. Those around Buenos Ayres..are, by irrigation, in the highest state of culture. **1870** *Weekly Standard* (Buenos Aires) 7 Sept. 2/5 The people in the quintas and suburbs were all flocking into Montevideo. **1878** E. CLARK *Visit S. Amer.* x. 128 The elegant quintas and country houses that surround the city. **1931** *Discovery* Oct. 310/2 The dipleidoscope and the gravity escapement are both due to a gifted inventor who worked at my Quinta [in Madeira]. **1960** *Spectator* 15 Jan. 91/3 'Noval' is the name of the *quinta*, or estate, from which the port comes. **1969** J. MANDER *Static Soc.* v. 132 Bolivar's *quinta*, a modest eighteenth-century lodge on the slopes overlooking the city, may still be inspected. **1976** F. GREENLAND *Misericordia Drop* I. i. 14 The property..belonged to..the venerable hospital that employed him... Periodic inspection of this *quinta* was one..of Dr Sá's duties. **1976** *Times* 6 Nov. 13/2 Quinta wines—ports from one particular vineyard.

quintal, kintal, kentle. a. (Further eastern N. Amer. and later examples.)
1645 in *Deeds Suffolk Co., Mass.* (1880) I. 65 One thousand Kintall of dry Cod fish. **1712** N. TREVANION in D. W. Prowse *Hist. Newfoundland* (1895) x. 272/2 And planters very backward in paying, he got only one hundred quintals of fish this season. **1897** KIPLING *Captains Courageous* vii. 144 Then pray back my son to me! Pray back a nine-thousand-dollar boat an' a thousand quintal of fish. **1912** N. DUNCAN *Best of Bad Job* xxv. 169 A quintal here an' a quintal there. **1918** A. HUXLEY *Let.* 12 Aug. (1969) 160 He drains..twelve hogsheads of rich milk, to say nothing of the nineteen Imperial Kilderkins of cream and the thirty two quintals of sugar. **1964** *Newfoundland Q.* Spring 27/1 A prior pre-quintal charge is reserved from the entire voyage's value for this labour called 'fish-making'. **1967** *Times* 23 Nov. 5/7 An agreement for the sale of 200 tons of Sudan cotton c.i.f. Bombay, at a price of Rs. 393.68 per quintal. **1974** D. AVERY *Not on Queen Victoria's Birthday* vi. 102 By February 1815 the price of imported south American copper had risen to 20 dollars a *quintal* (£8 per 100 lb).

quinte·ssenced, *ppl. a. rare.* [f. QUINTESSENCE *v.* + -ED¹.] Reduced to its quintessence; quintessential.
1898 G. E. B. SAINTSBURY *Short Hist. Eng. Lit.* VII. vii. 467 Its charms..consist in extreme strangeness, in quintessenced or preternatural art.

quintessential, *a.* Add: **B.** *sb.* The most essential part of a thing; a quintessential element.
1899 'MARK TWAIN' in *Harper's Mag.* Sept. 529/1 These are the very quintessentials of good citizenship. **1916** J. G. HUNEKER *Ivory, Apes, & Peacocks* 37 He, too, dreamed of quintessentials, of the sheer power of golden vocables and the secret alchemy of art.

quintessentially. (Later examples.)
1936 *Delineator* CXXIX. 49/2 All of those early musicians in New Orleans were quintessentially swing players. **1958** [see *BALDING *a.*]. **1971** *Country Life* 24 June 1584/3 Montgeoffroy is everything that is quintessentially French. **1978** *Language* LIV. 275 That master negotiator and persuader, the quintessentially pragmatic Odysseus.

quintile, *a.* and *sb.*¹ Restrict *Astrol.* to sense in Dict. and add: **2.** *Statistics.* Any of the four values of a variate which divide a frequency distribution into five groups, each containing one fifth of the total population; also, any of the five groups so produced.
1951 *Brit. Jrnl. Nutrition* V. 199 For the purposes of classification of the subjects into categories of fatness, we have used the twentieth, fortieth, sixtieth and eightieth percentile. These limit values are sometimes called 'quintiles'. **1973** *Jrnl. Genetic Psychol.* CXXIII. 79 Each class's ranked students were divided into quintiles: high, medium-high, middle, medium-low, and low status. **1974** *Nature* 22 Nov. 294/2 These are quintiles for temperature and terciles for rainfall. **1976** *Lancet* 6 Nov. 979/2 Apart from smoking, where no such division into quintiles is practicable, the five measures used are systolic and diastolic blood-pressure, overweight, [etc.].

‖ **quinton** (kæ̃ntoñ). [Fr.] A musical instrument of the viol or violin families, having five strings (see also quots. 1941, 1954).
[**1870** C. ENGEL *Descr. Catal. Mus. Instr. in S. Kensington Mus.* 77 Five-stringed viol. Called by the French *Quinton*.] **1889** [see *PARDESSUS 2]. **1916** STANFORD & FORSYTH *Hist. Mus.* ix. 195 True viols have no fewer than six strings apiece... But the makers are beginning to see the necessity of reducing the number... They are just introducing a modified type of instrument with five strings. They call these the *quintons*. **1941** [see *PARDESSUS 2]. **1954** *Grove's Dict. Mus.* (ed. 5) VI. 1037/1 The rather ugly word quinton seems to have originated as a 19th-century dealers' name to describe a curious hybrid sort of viol-violin... The instruments appear to belong to the late 18th century... The name drifted into reference books where it was misapplied to the perfectly normal treble and *par-dessus* French viols of the mid-16th century, which often have only five strings. **1961** T. DART in A. Baines *Mus. Instr. through Ages* 189 During the eighteenth century certain hybrids between the two families of viols and violins made a brief appearance on the musical scene—for instance, the quinton, the baryton, and the arpeggione—but few musicians regretted their equally abrupt departure. **1976** *Early Music* July 361/2 The quinton is built basically like a violin, with arched back, shallow ribs and pointed bouts, but with sloping viol shoulders.

quintuple, *sb.* (Later example.)
1975 N. CHOMSKY *Logical Struct. Linguistic Theory* ix. 327 A set of ordered quintuples ((*Z*, *K*, *Z*¹, *Pr*⁽¹⁾, *Pr*⁽ᴵ⁾)).

quintuplicate, *sb.* Add: **a.** Also, *in quintuplicate*.
1941 F. SCOTT FITZGERALD *Last Tycoon* iii. 41 Miss Doolan's notes would be typed in quintuplicate. **1968** H. EDWARDS *Thirty Years Spiritual Healer* v. 51 In Holland, no one can be admitted to a healing meeting unless he becomes a member of the society, and even then records in quintuplicate have to be made for each person attending. **1978** B. NORMAN *To nick a Good Body* vii. 55 Statements..typed out in quintuplicate.

quintu·pling, *vbl. sb.* [f. QUINTUPL(E *v.* + -ING¹.] Fivefold increase, multiplication by five.
1975 *N.Y. Times* 25 Sept. 65/7 The quintupling of oil prices in the last two years, most economists agree, has accelerated world inflation and intensified the recent slump in world business. **1976** *Times Lit. Suppl.* 20 Feb. 193/3 The burden on the British balance of payments of a quintupling of the price of oil.

‖ **quintus.** *Mus.* [a. L. *quintus* fifth.] (See quot. 1883.)
1883 GROVE *Dict. Mus.* III. 61/1 *Quintus*.., the Fifth Part in a composition for five Voices: called also Pars quinta and Quincuplum. **1954** *Ibid.* (ed. 5) VI. 570/1 When a *quintus* is needed half of it is written on the left-hand page below the *tenor* and the remainder (*reliquium*) below the *bassus*, on the right-hand page. **1977** *Early Music* July 419/3 Those intending to sing through the motet volume should note that the quintus is not in the same vocal range throughout and six voices are needed.

quinze. Add: *Comb.* Also, *quinze table*.
1782 J. HARE *Let.* 11 Feb. in *15th Rep. R. Comm. Hist. Manuscripts* App. VI. 576 in *Parl. Papers 1897* (C. 8551) LI. I. 1 A deep Quinze Table has taken away from Richard and Charles almost the whole profits of the Bank. **1803** *Lett. Miss Riversdale* II. 190 'I never before knew Mrs. Blandford guilty of affectation,' said Lord Lessingham, from the quinz-table.

quip, *v.* Delete 'Now *rare*' and add: **1. b.** With speech as direct object: to say or reply as a quip.
1950 *New Yorker* 25 Mar. 26/3 'Looks like somebody lost their head!' quipped Detective Garnet P. Quail. **1965** D. LODGE *Brit. Mus. is falling Down* viii. 132 'Someone taking my name in vain?'.. 'In bane,' Adam quipped, and laughed immoderately. **1974** *Times* 11 Oct. 10/1 'I became a tremendous bargain. I went from a million to nothing,' she quipped to reporters. **1978** *Woman's Own* 16 Sept. 12/1 My daughter was rather amused and quipped: 'Most parents have problem teenagers but we have a problem mother.'
2. (Later examples.) Also, const. *with*.
1908 *Smart Set* June 50 Audrey in her blithesome way Would quip and jest with roguish glee. **1942** Z. N. HURSTON *Dust Tracks on Road* xii. 225 The educated Negro..is fighting entirely out of his class when he tries to quip with the underprivileged.

quipper. (Later examples.)
1951 O. NASH *Family Reunion* 36 She'll nogg his eggs and she'll toast his kippers, And disparage the quips of the current quippers. **1960** *Guardian* 19 Nov. 3/6 His intervention has at least provided the quippers with a not very good quip.

quippery (kwi·pəri). *rare.* [f. QUIP *sb.* or *v.* + -ERY.] The uttering or bandying of quips; quips collectively.
1960 *Guardian* 22 July 10/6 This is campaign quippery, and we shall have a bellyful of it before the winter.

quippish, *a.* (In Dict. s.v. QUIP *sb.*) Add: (Example.) Hence **qui·ppishly** *adv.*, **qui·ppishness**.
a **1834** S. T. COLERIDGE in *N. & Q.* (1888) 7th Ser. VI. 501/2, I prefer Fuller's [version] as more quippish and

adagy. **1909** A. BENNETT *Literary Taste* vi. 31 Charles Lamb's essay on *Dream Children*..enlivened by a certain quippishness concerning the children. **1976** *New Scientist* 22 July 190/3 Another speaker quippishly suggested that we should treat them [*sc.* 'core' journals] like apples.

quipster. (In Dict. s.v. QUIP *sb.*) (Later examples.)
1962 *Punch* 15 Aug. 224/2 The one who provides the old Russian sayings is his nearest equivalent to [Bob] Hope's tarmac quipster. **1976** *National Observer* (U.S.) 14 Aug. 16/2 He doesn't mind being called Gooey Lombardo or Guy Lumbago; he smiles when quipsters dub him the Pied Piper of the geriatric set. **1977** *Time* 6 June 39/2 A quipster with no audience for his one-liners?

quire, *sb.*¹ Add: **3.** *quire signature* [SIGNATURE *sb.* 6a]; *quire-folded* adj.; **quire-wise** *adv.* (examples).
1882 J. SOUTHWARD *Pract. Printing* xiii. 117 Folio sheets are sometimes required to be..folded within each other, or *quirewise*. **1888** C. T. JACOBI *Printers' Vocab.* 108 *Quire folded*.., folded in quires—not sent in 'flat'. *Ibid.* 109 *Quirewise*, jobs of single leaves printed on both sides of the paper. **1922** JOYCE *Ulysses* 120 The nethermost deck of the first machine jogged forward its flyboard with sllt [*sic*] the first batch of quirefolded papers. **1957** N. R. KER *Catal. MSS. containing Anglo-Saxon* p. xl (*heading*) Quire-Signatures and Leaf-Numbers. **1978** *Anglo-Saxon England* VII. 232 It may have its own series of quire signatures.

Quirinal (kwi·rināl). [ad. It. *Quirinale* (also used), f. L. *collis Quirīnālis*.] The name of the royal (now presidential) palace on the Quirinal hill in Rome, and hence used to designate the Italian monarchy or government, *esp.* as distinct from the Vatican. Also *attrib.*
1851 W. E. GLADSTONE tr. *Farini's Roman State* II. i. 3 The Civic Guards and the commonalty gave the wonted expression of their thanks in the Piazza of the Quirinal. **1881** E. W. HAMILTON *Diary* 7 Dec. (1972) I. 195 Errington..whom suspicious people regard as a probable diplomatic agent at the Quirinal, seems..to be going rather too fast ahead. **1905** J. WEBSTER *Wheat Princess* vii. 78 Being interested in the domestic arrangements of kings, she was insistent that they visit the Quirinal. **1917** N. DOUGLAS *South Wind* xxxiv. 408 A feeling of joyous elation at the prospect..of a battle between the Vatican and the Quirinal. **1922** *Contemp. Rev.* Nov. 583 The complete accord which exists in practice between the Vatican and the Quirinal. **1927** E. M. FORSTER *Aspects of Novel* viii. 193 He..sets him out to explore [Roman] society..café, studio, Vatican and Quirinal purlieus are all reached. **1949** N. MITFORD *Love in Cold Climate* I. x. 107 Friendship with royal personages only ever began for her when their days of glory were finished. Tsarkoe-Selo, Schönbrunn, the Quirinal,..and the island of Corfu had never known her, unless among an enormous crowd in the state apartments. **1979** N. SLATER *Falcon* v. 91 He looked more like a country squire than Her Majesty's Ambassador to the Quirinale.

quiring (kwaiə·riŋ), *vbl. sb.* [f. QUIRE *v.*¹ + -ING¹.] **a.** A series of signatures indicating the order of quires. **b.** The arrangement of a series of quires.
1922 *J. & J. Leighton's Catal. Old & Interesting Bks.* N.S. III. 209 Without printed quiring, &c., but original MS. signatures remain. **1957** N. R. KER *Catal. MSS. containing Anglo-Saxon* p. xxv, A book in which the sheets are arranged like this shows two hair sides where one quire ends and the next begins... The collation of such books is simple and the original quiring can be made out even if the leaves have been mounted separately. **1968** G. POLLARD in *Bodl. Libr. Rec.* VIII. 82 Having thus established the arrangement of the first twenty-eight leaves, we must now consider the quiring at the opposite end of the book. **1975** N. BARKER in R. W. Hunt et al. *Studies in Bk. Trade in Honour of G. Pollard* 11 The quiring may have been put in and later trimmed away.

quiring: see CHOIRING *vbl. sb.* and *ppl. a.*

quirk, *sb.*¹ Add: **4. b.** A peculiar feature or result (*of* an event); a peculiarity, an anomaly, a freak.
1961 *Yale Rev.* LI. 190 It is one of the ironic quirks of history that the viability and usefulness of nationalism and the territorial state are rapidly dissipating at precisely the time that the nation-state attained its highest number. **1973** A. H. SOMMERSTEIN *Sound Pattern Anc. Gr.* ii. 74 The other two rules..are mere quirks, survivals that play no active role in the system of the language, but merely go on existing. **1976** *Scotsman* 20 Nov. (Weekend Suppl.), In the San Blas Islands, off the coast of Panama, they found that through a quirk of history they had come a step closer to home. **1976** *Star* (Sheffield) 29 Nov. 12/7 Perhaps..it was something of an unfortunate quirk of the draw that they had to meet in the first round of the Roebuck Cup.
5. Delete † '*Obs. rare*⁻¹' and add later example.
1928 GALSWORTHY *Swan Song* II. x. 188 That indefinable look of a damned soul..awakened within Soames..the queerest little quirk of sympathy.
7. e. (Earlier and further examples.) Also *attrib.* or as *adj.*
1799 A. YOUNG *View Agric. Lincoln* 29 Window shutters quirk, ogee and astragal with ½ and 2 heights and 4 panels, at 12 d. **1842** GWILT *Encycl. Archit.* § 2367 A two-panel door, square on one side, with quirk ovolo and bead upon the other. **1882** NODAL & MILNER *Gloss. Lancs. Dial.* 220 *Querk, sb.*, a moulding in joinery. **1966** J. FLEMING et al. *Penguin Dict. Archit.* 181/2 *Quirk*, a sharp v-shaped incision in a moulding and between mouldings.

tinker out of Bedford, A vagrant oft in quod. **1933** *Sun* (Baltimore) 11 July 1/5 According to the representations of the other four Mr. Fullerton was in no way responsible for the incident which put them in quod. **1968** *Listener* 18 July 72/3 Now, one of this chap's maternal uncles.. has got to pay a 50 quid debt or go to quod.

quod, *v.* (Later examples.)
1888 J. Runciman *Chequers* 80 A woman answered, 'You've struck me, you swine; and if I've got a black eye I'll quod you, sure as I'm yere. Ain't I lushed you, and fed you, and found your clobber long enough?' **1923** D. L. Sayers *Whose Body?* iii. 60 That's her story. Sugg's delighted,..and quodded Thipps on the strength of it. **1930** R. H. Mottram *Europa's Beast* v. 136 In England the police would have 'quodded' her.

quodlibetal, *a.* (Later example.)
1975 Alluntis & Wolter tr. Duns Scotus (*title*) God and creatures: the quodlibetal questions.

quodlibetarian. b. (Later examples.)
1943 Beerbohm *Lytton Strachey* 22 That agile and mellifluous quodlibetarian, Dr. Joad. **1966** *Duckett's Reg.* Feb. 14/2 The clever quodlibetarians called it an obvious strategic move.

quoi hai, var. Qui-hy in Dict. and Suppl.

quoin, *sb.* Add: **4. quoin-wedge** = sense 2.
1923 D. H. Lawrence tr. *Verga's Mastro-don Gesualdo* I. iv. 71 We want more man-power—a crane!—or tie a pulley-wheel up there to the beam of the roof—then a quoin-wedge underneath.

quoit, *sb.* Add: **3. c.** The backside, the buttocks. Phr. *to go for one's quoit*, to hurry. *Austral. slang.*
1941 Baker *Dict. Austral. Slang* 58 *Quoit*, the buttocks. *Ibid.*, *Go for one's quoits*, to travel quickly, go for one's life. **1951** E. Lambert *Twenty Thousand Thieves* x. 165 See those jokers sitting on their quoits over there? **1952** J. Cleary *Sundowners* i. 42 Going for the lick of his coit up the street. **1954** T. A. G. Hungerford *Sowers of Wind* xiv. 176 Gawd, he blew the tripes outa me for nothing at all, and then he kicks a Nip in the coit. **1972** J. Bailey *Wire Classroom* x. 82 'I think he needs a good kick up the coit,' says Cromwell.

quoit, *v.* **1.** (Later example.)
1871 L. W. M. Lockhart *Fair to See* II. xi. 15 The quoiters quoited.

quoiter. Add: **b.** A curler.
1833 J. Cairnie *Essay on Curling & Artificial Pond Making* 93 He was a grand quoiter, he never missed a shot. **1899** J. Kennedy *Compl. Sc. & Amer. Poems* (ed. 3) 128 May quoiters' joys be mair an' mair.

quoiting, *vbl. sb.* Add: Also, = Curling *vbl. sb.*^{1} 2. (Further examples.)
1811 J. Ramsay *Acct. Game Curling* 20 From one end of Scotland to the other, it was always named *kuting*, to curl, meaning nothing more than to slide upon the ice. In some parts of Ayrshire..it is pronounced *coiting*. **1884** J. Taylor *Curling* 74 He had seen Bryan o' the Sun Inn and the deil *quitin'* (curling) on the Auld Water.
Comb. **1853** W. Watson *Poems* 71 The lee-side was cheer'd by the quoitin'-stane roar. **1969** R. Welsh *Beginner's Guide Curling* ii. 15 Kuting stones, channel stanes or loofies are the oldest curling implements known to us.

quokka (kwǫ·kă). [Aboriginal name.] A small short-tailed wallaby, *Setonix brachyurus*, found in restricted coastal areas of south-western Australia.
1863 J. Gould *Mammals Austral.* II. 38 At Augusta.. its [the short-tailed wallaby's] native name, Quāk-a, is the same as at King George's Sound. **1943** C. Barrett *Austral. Animal Bk.* xi. 96 The quokka or short-tailed pademelon..inhabits the coastal districts of South-western Australia. **1963** *Spectator* 25 Jan. 102/2 The quokka, indigenous to Western Australia, is a 'species of rat', albeit rather friendly. **1968** *New Scientist* 29 Feb. 455/2 At night the place [*sc.* Rottnest Island] quivers with wallabies about the size of hares. They are what the Aborigines call quokkas. **1976** *Nature* 1 Jan. 42/1 Female quokkas mate soon after giving birth.

quoll. (Later examples.)
1924 *Truth* (Sydney) 27 Apr. 6 *Quoll*, aboriginal name of native cat. **1970** *Courier-Mail* (Brisbane) 15 June 12/5 In 1955, during intensive searches for taipans in scrubs and huge lantana thickets at Chatsworth near Gympie, I was astonished when I captured a savage little quoll (native cat) almost a thousand miles south of what I thought was its home country. **1978** *Ibid.* 18 Feb. 18/8 And talking of this native cat or Quoll, it is a marsupial and a carnivore and it is a clever and persistent hunter.

quondam, *adv. sb.*, and *a.* **B.** *sb.* Delete † *Obs.* and add later examples in *spec.* sense (see quot. 1962).
1888 A. Blomfield *Let.* in H. H. Henson *Retrospect* (1942) I. ii. 30 My dear Henson (If I may as a quondam thus familiarly address you). **1962** A. Sampson *Anatomy of Britain* xiii. 214 The most worldly college has been All Souls... The fellows and ex-fellows (called Quondams) are supposed to be the cream of Oxford intellectuals... Lord Curzon, the Viceroy of India, 'an enthusiastic quondam'.

Quonset, quonset (kwǫ·nsèt). orig. and chiefly *U.S.* Also (erron.) **quanset.** [f. the name of *Quonset* Point, Rhode Island, where the article was first made.] Used *attrib.* and *ellipt.*, esp. in *Quonset hut*, of a kind of prefabricated building consisting of a semi-cylindrical corrugated metal roof on a bolted steel foundation.
A proprietary name in the U.S.
1942 *Collier's* 19 Sept. 21/1 The boys practiced erecting on deck their 'Quonsett huts', queer little igloos, the roofs of which are designed to catch rainwater to be saved for a sunny day. **1943** *Pop. Mechanics* Apr. 61/1 'Quonset Huts', those portable barracks, begin replacing the tent city. **1946** *New Yorker* 16 Mar. 22/1 The first American-made Nissens, or Quonsets, were sent to England under lend-lease in June, 1941. **1946** *Official Gaz.* (U.S. Patent Office) 19 Nov. 287/2 Great Lakes Steel Corporation, Wilmington, Del... Filed Mar. 15, 1946. *Quonset.* For readily erectable buildings, knock-down buildings, portable buildings, and prefabricated buildings. Claims use since September 1941. **1949** 'P. Michaels' *This Perverse Generation* viii. 67 From sturdy, roomy architectural loveliness to quonset hut—civilization marches on! **1957** J. Kerouac *On Road* xiii. 90 A tremendous aluminum Quonset warehouse. **1959** E. Tunis *Indians* 32/2 There seems to have been a quonset-shaped house, too, in Virginia and North Carolina. **1966** T. Pynchon *Crying of Lot 49* iv. 82 She parked in an enormous lot next to a quonset building painted pink and about a hundred yards long. **1972** E. Staebler *Cape Breton Harbour* ii. 24 A long scalloped row of traps made of raw wood slats, shaped like a quanset hut. **1973** 'B. Mather' *Snowline* iii. 30 Nissen and Quanset huts hastily erected in World War II. **1974** *Spartanburg* (S. Carolina) *Herald-Jrnl.* 21 Apr. A6/5 The family's quonset hut home on Oahu Island's north shore was washed away by a swollen river. **1977** *Time* 27 June 48/2 The final dozen hippies..last week evacuated their grimy quonset hut on the edge of town.

Quoom, var. *Qum.

quorate (kwǭ^{ə}·rĕt), *a.* [f. Quor(um + -ate^2.] Of a meeting: attended by a quorum (and thereby constitutional). Hence **inquo·rate** *a.*, not attended by a quorum.
1969 *PI* (University College London, Students' Union) 20 Oct. 3/2 Even barely quorate General Meetings could be dominated by an organised minority. **1971** J. Henderson *Copperhead* xv. 192 This meeting is now closed. We're quorate without you. **1973** *Times Higher Educ. Suppl.* 11 May 14/1 In a tiny department of three, what happens if the head is wed to one of the other two? The department meeting becomes quorate during intercourse. **1974** *Times* 13 May 17/6 The meeting..was an inquorate one and therefore had no validity and was entirely unofficial. **1976** *Cherwell* 30 Jan. 2/3 A barely quorate JCR passed a motion..that freshmen at the college should 'fag' for second and third year men. **1978** *Evening News* 22 Apr. 4/3 The motion was passed but there were several people who were in strong disagreement and walked out making the meeting in-quorate.

Quorn (kwǭɹn). The name of a village in Leicestershire (now Quorndon), used *attrib.* and *ellipt.* to denote a famous hunt centred there.
1867 'Ouida' *Under Two Flags* I. iv. 74 It lay in the Melton country, and was equally well placed for Pytchley, Quorn, and Belvoir. **1904** A. E. W. Mason *Truants* ii. 16 He hunted with the Quorn that winter. **1933** Kipling in *Strand* Feb. 126, I..was a Gentleman in Red When all the Quorn wore woad, Sir. **1966** J. Betjeman *High & Low* 23 The rumble of the railway drowned The thunder of the Quorn. **1971** *Guardian* 30 Sept. 16/8 Melton was frequented by many of the nobility and gentry who came for the Quorn hunting season. **1972** *Country Life* (Suppl.) 26 Oct. 11/1 *Hotel*..in the heart of the Quorn countryside.

quota, *sb.* Add: **1.** Also *Eccl.*, the proportion of the funds of a parish that is contributed to the finances of the diocese; in full *diocesan* (formerly *parochial*) *quota*. (Later examples.)
1911 *Archbishops' Committee on Church Finance Rep.* iv. 45 We recommend... That a system of parochial assessment..should be adopted... That the amount of this apportionment or contribution (which may conveniently be called the Parochial Quota) should from time to time be fixed by the Diocesan Board of Finance. **1919** W. B. Incledon *Vicar Reconstructs* vi. 27 So the Diocesan Finance Committee has been worrying you about the diocesan quota. My advice is to let them worry until you have your parochial finance scheme in proper working order. **1936** W. K. L. Clarke *Almsgiving* ii. i. 82 The diocesan income in its turn is raised by assessing the parishes, each of which is expected to pay its 'quota'. *Ibid.* vi. 115 In many parishes the diocesan quota is paid reluctantly. **1976** *Church Times* 2 July 12/1 Any extra money raised by a parish for some specific occasion..is engulfed by the diocese by an increase in quota. **1978** *Ibid.* 29 Sept. 12/5 Parishes are already struggling to meet vastly increased quotas.
2. b. In a system of proportional representation, the minimum number of votes required to elect a candidate.
1857 T. Hare *Machinery of Representation* 17 No person shall be returned as a member to serve in parliament for whom there shall not be recorded the full quota or number of votes specified. **1930** C. F. Strong *Mod. Polit. Constitutions* viii. 177 Instead of having to gain an absolute majority, the candidate needs only to reach the *quota*, i.e. the number of votes cast divided by the number of seats to be filled. **1943** J. F. S. Ross *Parliamentary Representation* xxiii. 219 The *quota* has next to be determined: this is done by dividing the grand total of votes by a number which is one more than the number of members to be elected. The quota is the whole number next above the result of this division. **1955** Lakeman & Lambert *Voting in Democracies* v. 91 If the number of envelopes containing..the Socialist list amounts to one d'Hondt quota, the Socialist candidate whose name appears first on the list is declared elected. **1973** *Irish Times* 2 Mar. 8/1 Carlow-Kilkenny. Seats, 5... Valid vote, 46,717;.. Quota, 7,787. **1975** H. Berrington in S. E. Finer *Adversary Politics & Electoral Reform* 281 Neither Labour nor the Conservatives would have 'wasted' many votes—the Tories would have polled two quotas, with little to spare, and Labour would have accumulated three.
3. The maximum number (of immigrants or imports) allowed to enter a country within a set period; a regulation that imposes such a restriction on entry to a country. Also *transf.*
1921 *Washington Post* 21 May 4/5 North European countries may not furnish the full 20 per cent of their entire quotas each month. **1921** *U.S. Congr. House Report* No. 169. 1 The law provides that not in excess of 20 per cent of the yearly quota allotted to any country may enter in one month. **1930** W. K. Hancock *Australia* xii. 232 In recent years we have extended its operation by instituting 'quotas', which limit the immigration from Southern Europe. **1931** *Foreign Affairs* Apr. 401 On the score of its domestic economic consequences alone, the wisdom of an embargo or fixed import quota is disputable. **1931** *Contemporary Review* Aug. 221 It has not yet been realised, even by some Free Traders, how insidious and dangerous a form of protection is that latest expedient of the tariff-mongers, known as the 'Quota'. **1932** E. Bowen *To North* xxi. 224 'I have decided to go to America.' 'You couldn't stay more than six months, though, because of the quota.' **1949** [see *immigration]. **1973** *Nation Rev.* (Melbourne) 31 Aug. 1448/2 The board doesn't call for quotas on imports. **1976** *Survey* Winter 128 Quotas upon Jews in higher education had, to be sure, been a staple of the Soviet scene since the postwar period.
4. *attrib.* and *Comb.*, as *quota act, film, immigrant, immigration, law, limit, period, plan, restriction*; **quota method,** the statistical method of using quota samples (usu. for opinion polls); **quota quickie** [*quickie 1a], a cheap cinematographic film, rapidly made outside the United States to offset American films shown in other countries; **quota sample,** a sample that is chosen so that various categories of individual (when classified by age, sex, social class, and the like) are represented in the same proportions as in the general population; **quota sampler,** one who devises or uses quota samples; **quota sampling,** the use of the quota method; **quota system,** a law or custom restricting the number or proportion of persons or goods that may be admitted to a country or an institution; also, one prescribing the minimum number of persons to be admitted.
1938 *Times* 22 Feb. 19/3 British film production has made considerable progress during the past ten years as a result of the first Quota Act. **1939** 'N. Blake' *Smiler with Knife* ii. 29 Those extras, dressed up to look like rustic ancients for a British quota film. **1974** R. Harrison *Rex* ii. 29 The British film industry had only just begun to stir [in 1927], with Gaumont Films, and then with American quota films—'quota quickies' made over here by American companies, on a budget of £1 per foot of film and never mind the quality, so that they could then unload their Hollywood products on England. **1924** *U.S. Laws & Statutes* (1925) XLIII. 155 When used in this Act the term 'quota immigrant' means any immigrant who is not a non-quota immigrant. **1965** *Listener* 2 Sept. 340/2 Some [Australian] commentators advocate quota immigration for Asians. **1922** *Proc. Conf. Social Work* XLIX. 460 Just what effect the quota law has had on immigration.. cannot be accurately determined. *Ibid.* 465 Under the quota law..we have exclusion for the first time on the basis of numbers alone. **1975** J. Cleary *Safe House* 10 Every country..had a quota limit when..asked to extend its sympathy into acceptance of the Jews as immigrants. **1946** Hauser & Leonard *Govt. Statistics for Business Use* xii. 356 A commonly used type of sample design is that usually referred to as the 'quota' or 'in-ratio' method of sampling. **1953** W. G. Cochran *Sampling Techniques* v. 105 Sampling theory cannot be applied to quota methods which contain no element of probability sampling. **1971** *Guardian* 9 June 13/4 Teams of interviewers were sent out to a random sample of areas to find certain numbers of consumers determined by sex, age, and social class. This is known as the quota method of sampling. Random sampling—where lists are drawn from an 'unbiased' source such as the electoral register—is more rigorous, but much more expensive and is rarely used for commercial work. **1958** *Economist* 18 Oct. 280/1 For the first two quota periods (6 months) of the current year, the permitted output averaged 118 tons per month. **1977** *Grimsby Even. Tel.* 13 May 1/5 It showed we have no confidence in him and his quota plans. **1936** *Economist* 5 Dec. 456/2 The primary objective of the Board of Trade..was the complete extirpation of the 'quota quickie'. **1948** H. Wilson in *Hansard Commons* 21 Jan. 226 To protect this section of the industry against the so-called 'quota quicky' shorts. **1976** *Oxf. Compan. Film* 574/1 'Quota quickies'—cheap programme-fillers made with local personnel and technical resources but financed from the US—kept standards down while fulfilling legal requirements. **1924** J. S. Huxley in *Spectator* 20 Dec. 981/1 The 1924 Immigration Law takes the quota idea as its basis... The quota restrictions do not apply to Canadians. **1938** *Ann. Reg. 1937* 17 In Great Britain there were no exchange controls and no quota restrictions save in the case of iron and steel imports. **1944** *Jrnl. Marketing* July 26/2 The current method, perhaps most-widely employed in the selection of respondents in market surveys and in polls of opinion, is that of 'in ratio' or 'quota sampling'... The essentials of this method consist in:..(3) the fixing of quotas for each enumerator in such a way that the re-

spondents..will include the specified proportion of each class of the population agreed upon. **1952** *Economist* 20 Sept. 689 (Advt.), Checking the quota sample against a random sample. **1966** *Guardian* 25 Apr. 8/5 Under the quota method..people are interviewed in the street... The random sample, which involves contacting named individuals..is much more expensive than the quota system. **1974** *Times* 15 Feb. 5/7 The Gallop Poll..plans to do a mixture of random and quota samples. **1952** *Jrnl. R. Statistical Soc.* A. CXV. 412 Quota samplers invariably attempt, as one of their controls, an economic or social breakdown of the sample. **1958** C. A. Moser *Survey Methods Social Investigation* vi. 105 Quota samplers generally claim that instructions to, and constraints on, interviewers are sufficient to guard against the main dangers of selection bias. **1944** *Jrnl. Marketing* July 29/1 One of the most important advantages of area sampling over..quota sampling is that which reduces the dependence of the investigator on knowledge of the characteristics of the population. **1950** M. Parten *Surveys, Polls, & Samples* i. 31 Up to 1946, the nationwide tabulations were usually based on about 3,000 cases selected according to a stratified quota sampling procedure. **1973** *Guardian* 27 Oct. 13/5 Quota sampling sets the interviewer the.. task of finding people who are socially representative... Random means what it says, and should..be based on the picking of names at random from the electoral register. **1924** *N.Y. Times* 8 Dec. 17/1 A bill amending the immigration law and abolishing the quota system is being drafted for presentation to Congress. **1934** A. G. Macdonell *How like an Angel* ii. v. 86 It was the business of the All-British Company to produce seventeen exceptionally bad and cheap films every year in England in order to allow two hundred and forty-six exceptionally bad and expensive films to be imported every year into England from Hollywood. This is called the Quota System. **1936** *Discovery* Dec. 374/1 Under the quota system..2,894 Japanese were entitled to admission [to Brazil] during 1935. **1969** *Listener* 14 Aug. 201/1 *The quota system.* Thirty per cent of first features shown in this country have to be made here. **1976** *National Observer* (U.S.) 14 Feb. 11/5 Demanding what amounts to a quota system for college and university faculty hiring.

quotation. Add: **3. c.** A short passage or tune taken from one piece of music to another or quoted elsewhere.
1906 E. Newman *Elgar* v. 147 The clarinet softly gives out a quotation..from Mendelssohn's 'Calm Sea and Prosperous Voyage' overture. **1942** G. Abraham *Beethoven's Second-Period Quartets* 42 A note-for-note quotation of a figure much used in the first movement of Mozart's great Quartet in C major, K.465. **1960** *New Oxf. Hist. Music* III. v. 156 The following quotation will.. serve to show the extent to which the late Gothic composers intentionally subordinated the natural verbal rhythm to the rhythmic compulsion of the music. **1972** *Jazz & Blues* Oct. 26/3 Flashes of humour in the form of oblique quotations.

quotatious (kwotē^{i}·ʃəs), *a.* [f. Quotati(on + -ous.] Fond of using quotations; inclined to quote.
1903 *Daily Chron.* 10 Apr. 3/2 The myriad quotations of that most 'quotatious' of authors are identified every one—no easy task with such a perverter of phrases as Hazlitt. **1910** *Ibid.* 29 Jan. 6/1 A certain person endeavoured to quench my flaming optimism by a stream of frigid quotation... I sought knowledge from the quotatious person.

quotative, *a.* Add: **2.** Of a construction or expression: indicating that the speaker is quoting a word, phrase, etc., attributed to another person.
1927 L. Bloomfield in *Amer. Speech* II. 438/2 Whatever is hearsay and not the speaker's own experience has the predicate verb or particle in a special *quotative* form. **1957** D. L. Bolinger in *Publ. Amer. Dial. Soc.* xxviii. 26 A reflex..may be re-worded to make it less quotative. **1975** *Language* LI. 804 Normally a sentence containing the quotative clitic translates with 'one says'.

quotative (kwō^{u}·tătiv), *sb.* [f. the adj.] A quotative word or expression (see prec.).
1902 *Amer. Anthropol.* IV. 401 The quotative *wAnsū'ga*, 'they say', is used extensively in the Skidegate dialect. **1957** D. L. Bolinger in *Publ. Amer. Dial. Soc.* xxviii. 95 With other verbs, notably the outright quotatives that connote the adoption or assertion of a view..concord gives a different meaning. **1965** *Language* XLI. 68 A few further elements—pseudo-constitutes.., multiple constitutes.., and quotatives. **1975** *Ibid.* LI. 804 The first example is provided by quotatives in Luiseño.

quo·tativeness. [f. Quotative *a.* + -ness.] **a.** Fondness for the use of quotations. **b.** The quality of being quotative (sense *2).
1886 G. B. Shaw in *Pall Mall Gazette* 26 July 5/1 A certain facetiousness and quotativeness. **1957** D. L. Bolinger in *Publ. Amer. Dial. Soc.* xxviii. 27 The quotativeness of the echo suggests that it be classed as a frag Q.

quote, *sb.*^{2} Add: **2.** (Later examples.)
1922 T. S. Eliot *Let.* ? Jan. in E. Pound *Lett.* (1951) 236 Do you mean not use the Conrad quote or simply not put Conrad's name to it? **1950** G. B. Stern *Ten Days of Christmas* i. 27 The title must be a quote. **1959** *Times Lit. Suppl.* 23 Jan. 45/4 The blurb..and 'quotes' selected from the American Press inaccurately suggest brashness. **1968** *Listener* 25 July 108/2 Don't ask me questions, since I have no wish to figure as the father of all the quotes in your stories. **1978** *Guardian Weekly* 15 Oct. 7/3 A quote from Dayan is painted on one of the twisted gun doors: 'The Bar-Lev line looks like a piece of cheese with a hole in it.'
b. (Later examples.)
1920 Wodehouse *Coming of Bill* i. ii. 24 Below the signature, in what printers call 'quotes', a line..'Bear the torch and do not falter.' **1937** *Daily Express* 4 Feb. 6/3 New use for 'quotes' ('inverted commas') came to light at the Scottish Literary luncheon in London yesterday. **1955** T. H. Pear *Eng. Social Differences* iii. 90 The upper class fashion of speaking in 'quotes'—'I don't mind if I do' (in a pronounced Cockney accent). **1969** 'J. Morris' *Fever Grass* ii. 24 He'd have witnesses if I put him in the dock. Witnesses in quotes, I mean. **1976** *New Yorker* 16 Feb. 37/1 Freezes over close quote, paragraph.
3. = Quotation 6.
1959 *Daily Mail* 8 June 2/3 'Quotes' for readers. The following list of prices is a selection from readers' requests for quotation of some of their shares which do not fluctuate sufficiently to be quoted daily. **1965** E. Gundrey *Foot in Door* xii. 91 She was shown a long list of things that needed doing..and was given a quote for 'about £28'. **1970** *Globe & Mail* (Toronto) 25 Sept. 4/4 The Duke price was consistently lower than other quotes. **1976** J. Drummond *Funeral Urn* iv. 14 Do the work, will I?.. I'll give you a quote before I get stuck in. **1980** *Daily Tel.* 3 Jan. 15/6 Yesterday..he was appointed chairman of construction company Tebbitt, which has a market value of £1·3 million. 'It is a small start, but I needed a vehicle with a quote and Tebbitt is exactly right.'

quote, *v.* Add: **3. b.** Also, to repeat a statement by (someone); to give (a person's name) as the authority for a statement. Freq. in phr. *don't quote me*.
1953 A. Christie *Pocket Full of Rye* ii. 12 Of course, I *may* be wrong—don't quote me, for Heaven's sake. **1965** *Word Study* Apr. 6/2 We might happen to have such a quotation from Public Enemy No. 1, and we wouldn't hesitate to quote him. **1973** *Times* 15 Feb. 18/5 They won't be here at all in three years time. And you can quote me on that too. **1976** 'R. Macdonald' *Blue Hammer* xviii. 150 'Who had reason to kill him?' 'I don't know. His wife, perhaps... Don't quote me, but I wouldn't put it past her.'
4. a. Also *transf.*, of a composer or musical composition: to reproduce or repeat (a passage or tune that forms part of another piece of music).
1946 E. Blom *Everyman's Dict. Mus.* 138/1 Dies irae. The 2nd section of the Requiem Mass, orig[inally] associated with a distinct plainsong theme which has been frequently used or quoted by var[ious] comp[oser]s. **1954** *Grove's Dict. Mus.* (ed. 5) II. 955/1 Var. XIII [of the 'Enigma Variations'] quotes from Mendelssohn's 'Calm Sea and Prosperous Voyage' Overture. **1975** R. S. Gold *Jazz Talk* 213 *Quote*,..of a vocalist or soloing instrumentalist to insert a phrase from another tune into the one being played.
c. *quote..unquote*: a formula used in dictation to introduce and terminate a quotation. Freq. *transf.*, in speech or writing, introducing and terminating words quoted (or ironically imagined to be quoted) from the speech or writing of another.
1935 E. E. Cummings *Let.* 3 Oct. (1969) 145 The Isful ubiquitous wasless&-shallbeless quote scrotumtightening unquote omnivorously eternal thalassa pelagas or Ocean. **1950** 'S. Ransome' *Deadly Miss Ashley* xvii. 198 She says, quote, 'What girl wouldn't?' unquote. **1956** *Times* 5 Dec. 1/5 (Advt.), Today, America, you sure are quote in the Big Time unquote. **1958** B. Hamilton *Too Much of Water* xi. 245 But he did have, quote, a jolly good reason for bumping off one special person, unquote. **1961** P. Ustinov *Loser* viii. 140 He expressed the personal opinion that the picture was quote great for America unquote. **1973** D. Robinson *Rotten with Honour* 8 The British... see too many people like you in London... East Germans, Bulgarians, and Rumanians, all of them quote diplomats unquote.
7. Also, to name a racehorse *at* specified odds; *absol.* or with a person as *obj.*, to give (someone) a quotation for goods or services. (Earlier and later examples.)
1865 *Atlantic Monthly* May 575/1 The artist is like the stock which is to be quoted at the board and thrown upon the market. **1888** *Economist* (Chicago) 3 Nov. 4/3 The effect of quoting Gas Trust upon the ticker..has been to stimulate trade in this stock here. **1934** *Collier's* 11 Aug. 48/2 Black Gold was quoted at 200 to 1 for the Coffroth event. **1938** *Times* 29 Sept. 19/1 War Loan..was being quoted 95 middle in inter-office dealings. **1971** *Timber Trades Jrnl.* 14 Aug. 71 (Advt.), Your machines..reconditioned, re-installed... Let us quote you.

quotee. (In Dict. s.v. Quote *v.*) (Earlier and later examples.)
1813 W. S. Walker *Let. to Sir Walter Scott* in *N. & Q.* (1961) Jan. 19/2 You see I am addicted to comparisons, like your quotee Fluellen. **1926** *Eng. Jrnl.* May 395 The quotee protests. Professor Baker has asked that attention be called to a quotation by Professor Campbell..which he feels does not present his meaning fairly. **1973** *Publishers' Weekly* 7 May 19/2 Ruland was not a manufactured quotee.

quotha, *interj.* (Later examples.)
1917 W. Owen *Let.* 4 Feb. (1967) 432 Distaste? Distaste, Quotha? **1958** L. Durrell *Mountolive* v. 111 A fellow-romantic quotha!

‖ **quot homines tot sententiae** (kwǫt hǫ·minēiz tǫt sente·nti,əi). [L.] An observation on the diversity of opinions, deriving from Terence *Phormio* ii. iv. 14 *quot homines tot sententiae: suus cuique mos* 'there are as many opinions as there are men: to each his own way'.
1539 R. Taverner tr. *Erasmus's Proverbes or Adagies* f. xiii, Quot homines, tot sentenciæ. So many heades, so many iudgementes. **1575** G. Gascoigne *Certayne Notes of Instruction concerning Making of Verse or Ryme in English*, And therwithall I pray you consider that *Quot homines, tot Sententiæ*, especially in Poetrie. **1602** W. Watson *Quodlibeticall Questions concerning Relig. & State* 343 They follow each one of them their owne priuate foule spirits of deceit and error, & so *quot homines tot sententiæ*, So many men so many minds. **1869** *Fraser's Mag.* LXXX. 68/1 Here all is to be pleasure. The opinions as to what is pleasure vary as a matter of course. *Quot homines tot sententiæ.* **1969** *Listener* 13 Nov. 680/2 A visitor from another planet might well have marvelled at the fertility of the human race in generating opinion—*quot homines tot sententiae* with a vengeance. **1975** *Times* 13 Nov. 17/6 No one has ever agreed entirely about Kipling; *Quot homines, tot sententiae*.

quotidian, *a.* and *sb.* Add: **A.** *adj.* **3.** (Later example.)
1978 *Studies in Eng. Lit.: Eng. Number* (Tokyo) 121 Malory..omits many of the 'quotidian' actions of chivalric life.
Hence **quoti·dianism.**
1920 A. Huxley *Limbo* 261 'It is our cheap press. The ephemeral overwhelms the permanent, the classical.' 'This journalism,' I agreed, 'or call it rather this piddling quotidianism, is the curse of our age.'

quotient. Add: **1. b.** *quotient ring*; **quotient group** = *factor group* s.v. *Factor *sb.* 9.
1893 *Bull. N.Y. Math. Soc.* III. 74 The quotient-group of any two consecutive groups in the series of composition of any group is a simple group. **1911** W. Burnside *Theory of Groups of Finite Order* (ed. 2) iii. 39 Herr Hölder has introduced the symbol *G/H* to represent this group; he calls it the quotient of *G* by *H*, and a factor-group of *G*. **1958** R. V. Andree *Mod. Abstract Algebra* iv. 101 The order of the quotient group is the order of *G* divided by the order of *H*, providing *G* is finite. *Ibid.* viii. 188 The term *quotient ring* is also used because of the similarity to the concept of quotient groups. **1965** Patterson & Rutherford *Elem. Abstract Algebra* iii. 101 If *S* is an ideal in the ring *R*... The set of all cosets of *S* in *R* is a ring:..this is called the..quotient ring of *R* by *S*. **1972** F. J. Budden *Fascination of Groups* xxi. 409 A group which resolves into a succession of *cyclic* quotient groups in this way is called a 'soluble' group.

Quran, Qur'an, Qur(')anic, varr. Koran^1, Koranic *a.*
1876 T. P. Hughes in A. Qádir *Qurán, Transl. into Urdu Language* p. iii, There is no *authorized* translation of the Qurán in any language. **1885** —— *Dict. Islam* 483/2 *Qur'ān*.., the sacred book of the Muhammadans. **1905** W. St. Clair Tisdall *Orig. Sources of Qur'ân* iii. 63 The *Source* of the rest of the Qur'ânic account of the murder is the legend in the *Pirqêy Rabbî Elî'ezer*. **1919** H. U. W. Stanton *Teaching of Qur'ân* 5 The best studies on quranic theology in English are the pamphlets by Rev. W. R. W. Gardner. **1931** *Times Lit. Suppl.* 11 June 459/2 The tales ..vary from Qur'anic legends to popular stories of the most ribald and grotesque description. **1932** *Ibid.* 17 Mar. 185/1 He had forgotten the Quran and could not recite one of the suras. **1939** L. H. Gray *Foundations of Lang.* 363 *Arabic*, famous as the language of the Qur'ān. **1954** *Scott. Jrnl. Theol.* VII. 334 The non-expert will find this an eminently readable and absorbing book, and one that might well stimulate to a lasting interest in the Qur'an and the Islamic world. *Ibid.*, Again, the book is the fruit of a lifetime's devotion to Qur'anic studies. **1971** *Nigerian Jrnl. Islam* II. 45 Thus, the emphasis in the Quranic School is on the moral development of the child. **1972** *Computers & Humanities* VI. 195 Arabic, the native language of 100 million people, is also used by many more millions as the language of the Quran and Islamic Law. **1976** *Daily Tel.* 6 Apr. 11/3 Their loans have been supplemented by Qur'ans from the British Library's own collection. **1980** *Oxf. Diocesan Mag.* Feb. 9/1 The total restriction of women to their homes..has been a matter of social custom, not Quranic law.

QWERT, QWERTY, qwerty (kwǝ̄ɹt, kwǝ̄ɹti). Part of the series of letters that label the first row of letter keys on typewriters in English-speaking countries; also **qwert yuiop,** the full series in that row. Also (in form *qwerty*) used *attrib.* or as *adj.* to designate a keyboard or machine that incorporates this type of non-alphabetical lay-out.
1929 *Times Lit. Suppl.* 11 July 552/2 The 'qwerty' keyboard appears first on the Yost in 1887. **1961** *Courier-Mail* (Brisbane) 5 June (*heading*) 'QWERTS' girls are in demand. **1962** *Which?* Dec. 356/2 The keyboards of all the machines were laid out in the traditional—and irrational—pattern, sometimes called 'qwert yuiop', which gives the left hand a lot of work to do, and its little finger too big a share of that. **1967** *Crescendo* Dec. 15/1 As soon as I had the virgin sheet of paper threaded into my type-writer I discovered that I was at peace with the world. Not a single hostile thought came to mind. I wrote QWERT a couple of times and gazed at my brainchild. **1975** *Nature* 16 Oct. 556/1 Input is usually by Qwerty keyboard, either direct entry or off-line, using punched tape. **1976** *Times* 9 Nov. 16/7 Mutterings..are to be heard among non-French secretaries employed by the European Commission in Brussels over plans to introduce a standardized typewriter based on the French AZERTY keyboard... The Commission..points out that if English QWERTY machines had been chosen..this decision would have been just as open to accusations of discrimination. Germans..operate QWERTZ machines, while Italians..prefer QZERTY.

Qy. (Earlier example.)
1819 M. Edgeworth *Let.* 17 Apr. (1971) 195 We had been presented to the (Qy.) Duchess of Sussex.

R

R. Add: **I. 1. c.** Phonetics. *r-less* adj.; *r-colour*, the modification of a vowel sound caused by a following *r*, as in the U.S. pronunciation of *bird*, etc.; hence *r-coloured* adj., *r-colouring*. Also *intrusive r* (see introductory note in Dict.); *linking r*: see *linking *ppl. a.* d.
1887 *Trans. Philol. Soc. 1885–6* 3 The intrusive *r* has actually produced an additional syllable in modern English. **1902** H. L. Wilson *Spenders* xxiv. 277 Her speech bore just a hint of the soft *r*-less drawl of the South. **1909** O. Jespersen *Mod. Eng. Gram.* I. 372 In literature the intrusive *r* is frequently indicated as a characteristic mark of vulgarity; the oldest example, perhaps, is in Smollett. **1928** I. C. Ward *Phonetics of Eng.* xiii. 130 There is no doubt that the intrusive r is spreading; even in districts where it has not been known, the younger generation is using it. **1935** J. S. Kenyon *Amer. Pronunc.* (ed. 6) 158 In Southern American speech, instead of accented ɜ, an '*r*-colored' vowel varying to ɚ is often heard. *Ibid.* 191 The retroflexion is slight, or replaced by raising and retraction of the tongue, but..the vowel is still '*r*-colored', giving the impression of an *r* sound. *Ibid.* 193 In South England..the '*r* color' itself disappeared, leaving the sound ɜ. **1940** *Maître Phonétique* Oct.–Dec. 63 ðə nouteiʃnz..witʃ dinout prisaisli vauəlz wiθ r-kʌlariŋ. **1941** *Language* XVII. 240 This occurs frequently in the mixed dialect of those who have both '*r*-pronouncing' and '*r*-less' forms in their speech. **1950** D. Jones *Phoneme* xvi. 82, r-colouring, when vowels are said with simultaneous lowering of the soft palate. *Ibid.*, r-coloured vowels are found with significant function in various types of American and British English. **1965** *Canad. Jrnl. Linguistics* XI. i. 65 Nine free vowels occur under stress in all dialects..; a tenth occurs only in *r*-less dialects. **1977** P. Strevens *New Orientations Teaching of English* xii. 151 In American English, in all words spelled with *r* there is an *r* sound which occurs simultaneously with the vowel before it. (..The vowels in such cases are said to be *r*-coloured.)
II. 1. b. (Examples.) Also (occas.) as *v. intr.*
1816 *Catholicon* II. 264 Obituary.. On the 24th inst. Mr. Cornelius Peter Murphy..possessed of a heart glowing with the most generous sentiments, he contracted his illness by the devotedness of his friendship to a deserving youth, from whom, during the course of his long and fatal malady, he could not be separated. R.I.P. **1917** A. G. Empey *Over Top* 306 'R.I.P.' In monk's highbrow, 'Requiescat in pace', put on little wooden crosses over soldier's graves... Tommy says like as not it means 'Rest in pieces', especially if the man under the cross has been sent West by a bomb. **1962** *Punch* 5 Sept. 334/1 We had a field mouse RIP-ing under the cupboard. **1976** *Liverpool Echo* 22 Nov. 4/1 Fortified by rites of Holy Church (R.I.P.). Requiem Mass Thursday, November 25.
2. a. R = *rand *sb.*^{2} 2; ®, registered (of a trademark: incorporated in *Statutes at Large U.S.A. 1946* (1947) LX. i. 436); R, restricted (rating) (*U.S.*); R, reverse (as on the selector mechanism in a vehicle with automatic transmission); R, r, right; also *spec.* of a stage (examples); R (*Bacteriol.*) = *rough *a.* 1 e; R, rupee (examples); R.A., Royal Academy or Academician (further examples); R.A., Royal Artillery; R.A.A.F., Royal Australian Air Force; R.A.C., Royal Armoured Corps; R.A.C., Royal Automobile Club; R.A.E., Royal Aircraft Establishment; R.A.F., Royal Air Force (see as main entry in Suppl.); R.A.F. [G. *Rote Armee Fraktion*], Red Army Faction (in West Germany); R.A.F.V.R., Royal Air Force Volunteer Reserve; RAM (*Computers*), random-access memory; R.A.M., Royal Academy of Music; R.A.M.C., Royal Army Medical Corps; R and B, R & B, R'n B, r'n'b = **rhythm and blues*; R and D, R & D, research and development (orig. *U.S.*); R and R, R & R, rest and recreation (leave) (orig. *U.S.*); R. and R., R.'n' R., r'n'r = *rock and roll; R.A.O.C., Royal Army Ordnance Corps; R.A.P., Regimental Aid Post; R.A.S.C., Royal Army Service Corps; R. Aux. A.F., Royal Auxiliary Air Force; R.B.C., red blood cell or corpuscle; red blood (cell) count; R.B.E., relative biological effectiveness (of radiation); R.B.I. (*Baseball*), run batted in; RC (*Electronics*), resistance/capacitance (or resistor/capacitor); R.C., r.c., reinforced concrete; R.C., Roman Catholic (examples); R.C.A., Radio Corporation of America; R.C.A.F., Royal Canadian Air Force; R.C.M., radio (or radar) counter-measures; R.C.M.P., Royal Canadian Mounted Police; R.D., refer (also loosely understood as return) to drawer (of cheque); R.D.C., Rural District Council; R.D.F., radio direction-finding, -finder (in quots., referring to radar); also as *v. trans.*, to employ R.D.F. against; RDV, rdv = Rendezvous *sb.* (orig. *U.S.*); R.E., religious education; R.E., Royal Engineers; r.f., R.F., radio-frequency; R.F., representative fraction; usu. *attrib.*; R.F.A., Royal Field Artillery; R.F.A., Royal Fleet Auxiliary; R.F.C., Royal Flying Corps; R.F.D., rural free delivery (of letters) (*U.S.*); R.G.A., Royal Garrison Artillery; R.G.N., Registered General Nurse; Rh, rhesus (blood group); usu. *attrib.*; R.H.A., Royal Horse Artillery; R.I., religious instruction; RIAA, Record (since 1970, Recording) Industry Association of America; R.I.A.F., Royal Indian Air Force; R.I.B.A., Royal Institute of British Architects; R.I.C., Royal Irish Constabulary; R.I.N., Royal Indian Navy; R.K., religious knowledge; R.M., Reichsmark; R.M., Resident Magistrate (examples); R.M., Royal Marines; R.M.A., Royal Marine Artillery; R.M.C., Royal Military College (at Sandhurst); R.M.L.I., Royal Marine Light Infantry; r.m.s., R.M.S. (chiefly *Electr.*), root mean square; usu. *attrib.*; R.M.S., Royal Mail Steamer (also Ship); R.N., Registered Nurse; R.N., Royal Navy (examples); R.N.A.S., Royal Naval Air Service; R.N.D., Royal Naval Division; R.N.L.I., Royal National Life-boat Institution; R.N.R., Royal Naval Reserve; R.N.V.R., Royal Naval Volunteer Reserve; R.N.Z.A.F., Royal New Zealand Air Force; ROA [Russ. *Rússkaya osvobodítel'naya ármiya*], the Russian Liberation Army; R.O.C., Royal Observer Corps; R.O.K., Rok (rǫk), Republic (also Relief) of Korea; also *pl.*, soldiers of the Republic of Korea; ROM (*Computers*), read-only memory; R.O.P., rop, run of paper (as of advertisements not booked for a specific position in a newspaper); also *fig.*; also in colour printing (see quot. 1967); ROSLA (also with pronunc. rǫ·zlă), raising of the school-leaving age; RoSPA (rǫ·spă), Royal Society for the Prevention of Accidents; RP, rp = *received pronunciation* s.v. *received *ppl. a.* 1 b; R.P.M., r.p.m., resale price maintenance; r.p.m., R.P.M., revolution(s) per minute; RPV, remotely piloted vehicle (orig. *U.S.*); R.Q. (*Med.*), respiratory quotient; rRNA, ribosomal RNA; R.S., rs, received standard; formerly, received speech; R.S.A., Royal Society of Arts; also *pl.*, R.S.A. examinations; R.S.F.S.R. [Russ. *Rossíiskaya Sovétskaya Federatívnaya Sotsialistícheskaya Respúblika*], the Russian Soviet Federative Socialist Republic; RSJ, rolled steel joist; RSLA = *ROSLA* above; R.S.M., Regimental Sergeant Major; R.S.P.B., Royal Society for the Protection of Birds; R.S.P.C.A., Royal Society for the Prevention of Cruelty to Animals; R.S.V., Revised Standard Version (of the Bible); RSV (*Biol.* and *Med.*), Rous sarcoma virus; R/T, R.T., radio-telegraph or -telephone; usu. *attrib.*; RTE, Radio Telefis Eireann, the official broadcasting authority of the Republic of Ireland; R.T.O., Railway Transport(ation) Officer, Railroad Transportation Officer; R.T.U. (*Mil.*), returned to unit; R.U.C., Royal Ulster Constabulary; RV, rateable value; RV (earlier RecV), recreational vehicle, as a motorized caravan (*U.S.*); hence *RVer*; *RVing* ppl. a.; R.V., r.v. = Rendezvous *sb.* and *v. intr.*
See also *R.A.D.A., *RDX, *REM *sb.*^{3}, *R.E.M.E., *RNA, *R.O.T.C. (as main entries).
1961 *Times* 27 Jan. 19/4 Offers of 'one-ninetyfour' and 'one-ninetyfive'..were chalked up as 'R1.94' and 'R1.95'. **1961** *Africana Notes & News* Mar. (recto rear cover), Subscription R2 per annum..Holt, B. Place-Names of the Transkeian Territories, 1959, R0·75. **1971** J. McClure *Steam Pig* iv. 40 She kept her money in the post office, just over R200. **1925** *Registration of Trade-Marks* (U.S. Congress Senate, Comm. on Patents) 20–1 Jan. 8 It shall be the duty of the registrant to accompany a registered trade-mark with the words 'Registered in U.S. Patent Office',..or by letter 'R' in a circle, thus ®. **1977** *Gloss. Terms Unfair Competition* (U.S. Trademark Assoc.), ®, one of several notices prescribed by law to indicate that a mark is registered in the U.S. Patent and Trademark Office. **1966** *Acronyms & Initialisms Dict.* (Gale Research Co.) 58 R... Restricted (Military document classification). **1973** *Daily Colonist* (Victoria, B.C.) 6 Feb. 2/3 The Strawberry Statement, the MGM version of a campus rebellion..was rated R (no one under 17 admitted without parent or guardian). **1976** *New Yorker* 12 Jan. 70/2 Peckinpah was forced to trim 'The Killer Elite' to change its R rating to a PG. *Ibid.*, Many of these theatres wouldn't have taken it if it had an R and the kids couldn't go by themselves. **1959** R = reverse [see L = low s.v. *L 7]. **1846** J. R. Planché *Bee & Orange Tree* ii. 7 On (R.) a Cavern. Tempest. A Vessel is seen in distress. When it is out of sight, enter (R.) from Cavern, Princess Amy. **1893** G. B. Shaw *Let.* 27 Apr. (1965) I. 392 The old style—the Princess & the audience grouped R, and Adrienne beginning L in profile. **1976** M. S. Hoque *Hunger* i. i. 1, Moina and Latif appear—R. They are just visible by the door. **1977** *Rolling Stone* 24 Mar. (*caption*) (Opposite, l to r): John, Mick, Christine McVie, Lindsey Buckingham and Stevie Nicks. **1920** J. A. Arkwright in *Jrnl. Path. & Bacteriol.* XXIII. 359 The R form grows in colonies which have a more or less jagged outline, are flatter and often have an irregular, rough, or dull surface and are slightly opaque. **1973** Klainer & Geis *Agents of Bacterial Dis.* i. 23 Rough (R) colonies have a dry, flat, irregular, wrinkled appearance and are generally formed by cells that lack a capsule. **1885** Kipling *Let.* 31 July in C. Carrington *Rudyard Kipling* (1955) iv. 67 One Proprietor offered My Mother Rs 1,000 for an Anglo-Indian story. **1971** *Shankar's Weekly* (Delhi) 4 Apr. 9/3 'It won't cost much.' 'No, about Rs. 10,000.' **1829** J. Constable *Let.* 5 Apr. (1965) III. 21, I beg my best regards to Mrs Leslie—I am always dear Leslie. / your obliged friend / John Constable R.A. **1890** Lloyd George *Let.* 10 June (1973) 28 He had numerous R.A.'s & in fact I should fancy his picture gallery alone must have aggregated £10,000 in value. **1970** *Oxf. Compan. Art* 547/2 He was trained as a chorister in the Chapel Royal, and later received an allowance..to study at the R.A. Schools. **1815** J. Kane *List Officers R. Regiment Artillery* 65 List of Subaltern Officers of the Corps of R.A. Drivers. **1955** *Times* 16 June 4/3 Both achieved a creditable rate of bangs per minute, the R.E. with various demolitions and a set piece assault by flail tanks, the R.A. with gunfire. **1936** *Age* (Melbourne) 5 May 13 (*caption*) Aircraftsmen making adjustments to fuselage and bomb racks on the R.A.A.F. Hawker Demon at the Exhibition. **1955** *Times* 21 June 9/5 Melbourne, June 20... Four hundred soldiers, police and bushwalkers, helped by R.A.A.F. Dakota aircraft are searching 5,000ft. Mount Baw Baw for Mihran Haig. **1977** *Parade* (Melbourne) Sept. 22/1 The RAAF Lockheed Hudson, carrying a VIP load, was about to land at Canberra from Melbourne. **1942** Partridge *Dict. Abbrev.* 81/1 *R.A.C.*, Royal Armoured Corps; armoured fighting vehicles and tanks. **1950** *Jrnl. R. United Service Inst.* XCV. 289 The Royal Armoured Corps, as such, did not come into being until April 1939 (A.O. 58/1939)... In the same Army Order it was also stated that on transfer to the R.A.C. the R.T.C. would be re-designated Royal Tank Regiment. **1908** *Autocar Handbk.* (ed. 2) xxvi. 201 Members of most of the best clubs require only one proposer when joining the R.A.C. **1934** *Glasgow Herald* 11 Apr. 13/3 The R.A.C. will continue to press for a regulation that all pedal cyclists should be compelled to carry red rear lamps. **1977** J. Bingham *Marriage Bureau Murders* vi. 61 A large, respectable hotel.. mentioned in the A.A. and R.A.C. handbooks. **1926** *Encycl. Brit.* I. 20/2 (*heading*) The R.A.E. Bubble Sextant. **1977** *R.A.F. News* 11–24 May 11/2 Over at the R.A.E.'s Air Transport Flight. *Ibid.* 11/3 The Experimental Flying Squadron..is widely referred to as the sharp end of RAE flying. **1977** *Time* 19 Sept. 8/3 It was signed 'Kommando Siegfried Hausner, R.A.F.'—referring to a terrorist who died after a 1975 attack on the West German embassy in Stockholm. **1980** C. Moorehead *Fortune's Hostages* viii. 155 The freeing of six jailed 'RAF' prisoners. **1938** *Times* 2 Feb. 18/6 (*heading*) New branch of R.A.F.V.R. **1951** *Sunday Pictorial* 21 Jan. 13/6 (Advt.), They must undertake to fly with the R.Aux. A.F. or R.A.F.V.R. during their subsequent reserve service. **1957** R. K. Richards *Digital Computer Components & Circuits* vii. 347 'Random access storage' (or RAM, for 'random access memory'). **1977** *Design Engin.* July 15/2 The MM5799..contains 1,536 8-bit instructions in its ROM, and its RAM can store 96 BCD digits of 4 bits each. **1891** G. B. Shaw in *World* 23 Dec. 15/2, I am not in the habit of advising novices to lay the foundations of their vocal methods in the R.A.M. **1954** *Grove's Dict. Mus.* (ed. 5) 271/1 The R.A.M. continues its own separate examination in London..of music teachers and performers. **1900** *Morning Post* 25 July 5/6 Surgeon-Captain Rupert Fawssett, R.A.M.C. *a* **1944** K. Douglas *Alamein to Zem Zem* (1946) 128 Presently an R.A.M.C. warrant officer came. **1971** S. Hill *Strange Meeting* i. 67 Dick's in the R.A.M.C. but he's gone out to Egypt. [**1949** *Billboard* 12 Nov. 110/2 Could score in pop as well as b & r mart.] *Ibid.* 31 Dec. 83, I Ain't Fattening Frogs for Snakes..with its catchy tag, could have a pop as well as r and b future. **1955** L. Feather *Encycl. Jazz* (1956) 70 Today's top R & B artists, contemporary equivalents of the Bessie Smiths and other 'race record' favourites of the 1920s. **1967** L. Deighton *London Dossier* i. 30 Deafening foreground music is provided by a giant jukebox. This is where the working City mod goes for his mid-day transfusion of R'n B. **1973** *Publishers Weekly* 14 May 48/3 Devoted to r'n'b groups, blues and r'n'b vocalists. **1977** *Rolling Stone* 13 Jan. 55/1 'Autobiography'..and 'Shakey Ground' have no doubt been formulated by producer David Rubinson to bring Snow into an R and B-flavored pop mainstream. **1952** Pohl & Kornbluth in *Galaxy* Aug. 129/1 'How's Research and Development doing on the Venus question?'..'R. and D. is in there punching.' **1966** *Inland* (Inland Steel Co., Chicago) Autumn 5/1 The American iron

and steel industry alone will spend $200 million on R & D this year. **1978** *Nature* 2 Mar. 2/2 A reprocessing plant for throughput of about 1,000 tons per annum will cost in excess of £500 million, including radioactive waste storage, some R & D expenditure (say 5% of the total) and a decade of operating costs (perhaps 25% of the total). **1953** *Britannica Bk. of Year* 639/2 A similar idea is contained in the expression R. and R., the initials standing for 'Rest and Recreation'. **1966** *New Statesman* 14 Oct. 549/2 R & R = Rest and Recreation. Local leave for American troops, mostly in Hong Kong, Bangkok and Singapore. **1968** *Telegraph* (Brisbane) 3 May 1/7 American troops visiting Australia on R and R leave are to be briefed on the exploits of Diggers in Vietnam. **1977** J. Gardner *Werewolf Trace* xv. 129 Tonight was for R & R, tomorrow he would have another go at her to get the facts straight. **1955** *Life* 18 Apr. 168 On a list of 10 top juke box best-selling records last week, six were r'n r. **1958** *Listener* 9 Oct. 572/1 The musical basis of R.'n' R...is the blues. **1973** *Publishers Weekly* 10 Dec. 39/1 This fat collection is devoted solely to writing about music (though not restricted to r'n'r). **1977** *Sounds* 1 Jan. 4/3 From head-bashing r&r, through Sixties pop..to an exquisite kick-in-the-teeth for Definitive Punk Rock. **1918** *Times* 28 Nov. 9/3 (*heading*) R.A.S.C., R.A.V.C., and R.A.O.C. **1930** *War Office Regulations for Equipment of Army (Provisional)* ii. i. 3 Scabbards, bayonet, No. 1... R.A.O.C. on mobilization *if especially ordered*. **1971** S. Milligan *Adolf Hitler* iii. 75 Thirty Signallers drove to the R.A.O.C. Depot at Reigate in a three-ton truck. **1942** E. Waugh *Put out More Flags* ii. 154 'I don't think you mentioned the R.A.P., did you, Smallwood?' 'R.A.P. sir. No, sir, I'm afraid I don't know where it is.' **1948** E. H. Smith *Guns against Tanks* 29 The 26th Battalion's Medical Officer..who earned the admiration of the gunners by bringing his RAP truck to within fifty yards of the forward positions. **1954** J. Masters *Bhowani Junction* xix. 174, I came to the R.A.P., swallowed, and went down to report to Captain Chaney. **1918** R.A.S.C. [see R.A.O.C. above]. **1955** *Times* 12 Aug. 4/7 In the permanent hutment camp by the wartime airfield..the Army Emergency Reserve transport columns of the R.A.S.C. are carrying out in succession their annual 15 days' training. **1976** *Daily Mail* (Hull) 16 Dec., Kenneth William Tully Bodfield, T.D., Major late R.A.S.C. **1948** *R.A.F. Rev.* Jan. 19/1 The short title of the Royal Auxiliary [Air Force] will be R. Aux. A.F. **1972** L. Hunt *Twenty-One Squadrons* 12 It was on 16th December 1947 that King George VI gave permission for the 'Royal' prefix and from that date until the squadrons disbanded in March 1957 they were R Aux AF units. **1922** *Indian Med. Gaz.* LVII. 126/2 The results of blood examination before treatment of soamin was R.B.C. 3,000,000, W.B.C. 3,500. **1968** Passmore & Robson *Compan. Med. Stud.* I. xxvi. 2/2 The red cell count (RBC) ranges from 4·5–6·5 million/mm^3 in men and 3·9–5·6 million/mm^3 in women. **1978** *Nature* 16 Feb. 674/2 Ageing of circulating RBC. **1954** P. E. Smith in A. J. Fleming et al. *Mod. Occupational Med.* xiii. 179 When irradiation with two different types of radiation requires different doses to achieve the same biological effect, the radiations are said to have a relative biological effectiveness (R.B.E.) different from one. The R.B.E. is the inverse of the dosage ratio required for equal effect. **1961** G. R. Choppin *Exper. Nuclear Chem.* ii. 10 The RBE (relative biological effectiveness) is defined as the ratio of the absorbed dose in rads of gamma radiation to the absorbed dose of the given radiation which is required to give the same biological effect. **1975** S. C. Bushong *Radiologic Sci. for Technologists* xvi. 286 Diagnostic x-rays have an RBE of approximately 1·0. **1951** *Sun* (Baltimore) 6 Oct. B-1/7 (*heading*) Yankee hurler says own R.B.I. gave him biggest thrill. **1976** *Billings* (Montana) *Gaz.* 1 July 4-E/1 Rusty Staub..had three RBI's, including a two-run homer on Monday. **1979** *Honolulu Advertiser* 8 Jan. C-2/5 Ahu got his six R.B.I.s on three hits, including two doubles and a grand-slam homer. **1932** C. L. Boltz *Everyman's Wireless* x. 190 The stage gain from an R.C. stage is E/*e*, where E is the voltage (A.C.) developed across the anode resistance and *e* is the voltage (A.C.) applied between grid and filament. *Ibid.*, A 2-valve circuit incorporating detector and one stage of L.F. amplification with R.C. coupling. **1965** *Wireless World* July 326/1 RC coupling has been employed throughout, in preference to direct coupling, to minimize drift and facilitate the setting up procedure while still preserving the lower audio frequencies. **1932** *Civil Engin.* Sept. 17 Each column is supported by two vibro piles capped with a small R.C. slab. The whole area between the pile caps is covered with a 4 in. R.C. slab. **1953** *Archit. Rev.* CXIV. 305/1 Floors are of r.c. slab with a lightweight screed containing embedded heating coils. *a* **1762** Lady M. W. Montagu in *Lett.* (1967) III. 92 If the S[cripture] is true the R.C. Religion is false because contrary. **1820** J. Milner *Suppl. Mem. Eng. Catholics* II. 176 If it be true, that the Lords Grenville and Grey had any arrangements..inconsistent with the integrity and safety of the R.C. Religion, we declare, that we consider such arrangements as foreign to the obvious meaning of the Resolution we have signed. **1860** Queen Victoria *Let.* 7 Jan. in R. Fulford *Dearest Child* (1964) 227 Dear good Leopold Hohenzollern arrived... Oh! if only he were not a R.C.!! **1977** *Belfast Telegraph* 24 Jan. 10/1 (Advt.), R.C. Gent, of good family background, with means, wishes to meet teacher, nurse, or respectable farmer's daughter. **1922** *Radio enters Home* (Radio Corp. of Amer.) 7/1 The distributors of R.C.A. sets and apparatus have been selected carefully. **1938** *Rev. Sci. Instruments* IX. 219/1 (*caption*) The tube is an RCA 32. **1967** Cox & Grose *Organization & Handling Bibl. Rec. by Computer* ii. 48 Both Linotron and RCA Digiset have this facility for checking enlarged characters. **1924** *Pay & Allowance Regulations R. Canad. Air Force* IV. 14 Officers..not carried on the General List, R.C.A.F., but..attached to the Royal Canadian Air Force for flying duty. **1943** *Times* 13 Dec. 2/2 A Royal Canadian Air Force Sunderland of Coastal Command with a mixed R.C.A.F. and R.A.F. crew. **1958** 'Castle' & 'Hailey' *Flight into Danger* ii. 27 'I know it looks a bit like RCAF,' Dun was saying, fingering his great bush of a moustache. **1945** *Electronic Industries* Sept. 222 *RCM*, radar counter measures. **1947** *Jrnl. R. Aeronaut. Soc.* LI. 432/2 The job of the R.C.M. aircraft of the new Group was to impair these systems... Consequently we developed a jamming screen for the purpose of blinding the enemy's early warning system. **1978** R. V. Jones *Most Secret War* xxxiii. 289, I was amazed to sit through meetings of the Radio Counter Measures Board right up to the end of 1942 discussing whether it was advisable for us to start an 'R.C.M. War'. **1920** *Globe* (Toronto) 9 Feb. 3/4 The great increase in the membership of the force consequent of the extensions of its duties to Eastern Canada and the absorption of the Dominion Police within the organization of the R.C.M.P. **1967** *Canad. Ann. Rev. 1966* 21 Prime Minister Pearson had asked for any information in the RCMP files which indicated impropriety or wrongdoing. **1977** *Financial Times* 11 Nov. 5/5 Prime Minister Pierre Trudeau ordered a Royal Commission to investigate the RCMP's activities. **1913** W. T. Rogers *Dict. Abbrev.* 163/2 *R.D.* (bank.), refer to drawer (of cheque). **1938** N. Marsh *Death in White Tie* iii. 41 One of my cheques has been returned R.D. **1974** M. Butterworth *Man in Sopwith Camel* ii. 24 No..girl was going to rest a wet RD stamp on one of Stryvers' cheques by accident! **1904** *Local Govt. Jrnl.* 2 Jan. 3/1 (*heading*) Contracts open. 3/2 Granite.—The Clerk, R.D.C., Midhurst. *a* **1974** R. Crossman *Diaries* (1975) I. 114 At the time, I wasn't too alarmed when Alderman Reeves, the chairman of the R.D.C., called the decision the worst he had ever seen. **1938** F. A. Lindemann 8 Mar. in R. V. Jones *Most Secret War* (1978) iv. 40 Lest too much reliance be placed upon the R.D.F. methods, it is perhaps worth pointing out that certain difficulties may easily be encountered in actual use. **1942** E. Waugh *Put out More Flags* i. 28 We've got a most valuable invention called R.D.F. *Ibid.*, The German air-bases are too far away for them to be able to attack us. If they do, we'll R.D.F. them out of the skies. **1963** L. Deighton *Horse under Water* xxiii. 101 Just behind me on the bridge Singleton was admiring the R.F.D. and the electronic depth-gauge. **1975** S. Johnson *Urbane Guerilla* iii. 147 We didn't RDF him... The trouble is RDF-ing, radar direction finding, requires a cross-bearing. **1976** *Oxf. Compan. Ships & Sea* 685/2 In its earliest days, radar in Britain was known as RDF, the initial letters of radio direction finding, but the name was changed early in the Second World War to avoid confusion with H/F D/F, high frequency direction finding. **1955** R. J. Schwartz *Compl. Dict. Abbrev.* 152/3 *RDV*, rendezvous (US Army). **1973** 'A. Hall' *Tango Briefing* iv. 45 I'm speaking from base. We shall need a little more time to set up the radio, so the next rdv is for 15.00 hours tomorrow. **1963** R. Acland *We teach them Wrong* 9, I told the deputy head master that I was a Lay Reader in the Church of England and could perhaps help with R.E. lessons. **1972** *Guardian* 29 Aug. 13/7 RE has been crippled by dogmatism and an arid biblicism. **1978** *Times* 3 Oct. 15/7 Some LEA's have been bold enough to recast their RE curriculum. .. The RE Council keeps a monitoring eye on emerging methodologies for RE. **1848** *Brit. Army Despatch* 24 Nov. 307/3 Captain Chapman, R.E., has returned to Zante from Cephalonia. **1877** *Army Circular* Dec. App. 18 Each Officer of R.E. below the rank of Major..is allowed 2 public horses. **1921** V. E. Inglefield *Hist. 20th (Light) Div.* iii. 37 Rations, R.E. material, etc., had to be brought up a long way from the dumps under very difficult conditions. **1959** I. Jefferies *Thirteen Days* xi. 175 If that's all you have to say you can get out. This is the R.E. office. **1918** W. H. Eccles *Wireless Telegr. & Teleph.* (ed. 2) 471 The r.f. generator 1 and microphone 2 induce currents in circuit 5. **1930** *Proc. IRE* XVIII. 1339 A relatively simple outfit is used comprising an r-f oscillator of variable frequency. **1931** *Electronics* July 17/1 A new impregnation compound for r.f. transformers affords better protection in humid atmosphere. **1956** *Proc. CERN Symposium* I. 64/2 Synchrotron oscillations (if an rf is present) are damped. **1967** *Electronics* 6 Mar. 2 (Advt.), Sweep Oscillators with RF and marker plug-ins meet virtually all of your swept frequency testing requirements. **1886** H. D. Hutchinson *Military Sketching made Easy* i. 12 A French map may show a scale of mètres, but to be generally useful to Englishmen it would be necessary to add to it a scale of yards with the same R.F. **1969** G. C. Dickinson *Maps & Air Photographs* vii. 102 In countries using the metric system scales of so many centimetres to a kilometre usually also give R.F.s with convenient round numbers. **1900** W. S. Churchill in *Morning Post* 27 Mar. 7/7, 1 battery Corps artillery (R.F.A.). **1974** A. Price *Other Paths to Glory* vi. 71 See how those casualties in the first two years [of the First World War] came from the..8th Hussars and the RFA—I'll bet they were all regulars. **1911** *Shipping World* 15 Mar. 276/1 (*caption*) The R.F.A. 'Burma'. **1931** *Jane's Fighting Ships* 84/2 Oilers (R.F.A.). **1913** *Flight* 12 Apr. 404/1 (*heading*) Motto for the R.F.C. **1933** V. Brittain *Testament of Youth* ii. vii. 293 As we left the harbour a transport of the R.F.C. cheered us. **1977** *R.A.F. News* 27 Apr.–10 May 2/2 The final meeting of the North West Royal Flying Corps Association..was attended by Mr. Walter Sumner, a member who is now the sole RFC veteran to be a pensioner at the Royal Hospital, Chelsea. **1903** (*title*) R.F.D. news; devoted to the interests of the Rural Free Delivery Carriers of the U.S. **1903** *R.F.D. News* Feb. 21/2 Pa doesn't go to town for mail; we've got the R.F.D. **1974** M. Hoyt *Thirty Miles for Ice Cream* xv. 187 Our mail used to be delivered by an R.F.D. mailman riding behind a horse in a sort of outhouse on wheels, painted red, white and blue, and lettered U.S. Mail. **1907** *Field Service Pocket Bk* vi. 120 R.G.A. **1909** *Army & Navy Gaz.* 6 Feb. 140/1 Sir F. D. Blake, Bt., late Northumberland R.G.A. (Mil.). **1924** *Nursing Mirror Pocket Encycl. & Diary 1924* 85 *R.G.N.*, Registered General Nurse (Scotland). **1975** *Irish Independent* 27 May 16/5 (Advt.), Resident R.G.N., or S.E.N., required for day duty in nursing home in Dublin south east. **1940** Landsteiner & Wiener in *Proc. Soc. Exper. Biol. & Med.* XLIII. 223 The capacity possessed by some rabbit immune sera produced with blood of Rhesus monkeys, of reacting with human bloods that contain the agglutinogen M has been reported previously. Subsequently it has been found that another individual property of human blood (which may be designated as Rh) can be detected by certain of these sera. **1954** A. E. Mourant *Distribution Human Blood Groups* iii. 12 The *Rh* blood groups depend upon three very closely adjacent loci on each of a pair of chromosomes, which may be called the *C*, *D* and *E* loci. **1958** L. P. Strean *Birth of Normal Babies* iv. 60 If the fetus is Rh-positive and the mother is Rh-negative, then the mother is actually immunized by the developing child's blood and she develops antibodies against the Rh-positive factor. **1968** Passmore & Robson *Compan. Med. Stud.* I. xxvi. 19/2 Only the ABO and Rh blood group systems contribute significantly to isoantibody incompatibility in the giving of blood transfusions. **1854** *Times* 20 Oct. 7/1 Captain Maude's Battery, R.H.A., was most useful. **1916** Ld. E. Hamilton *First Seven Divisions* 27 The 119th Battery R.F.A. was at this time just south-west of Eloges, and L Battery R.H.A. just north-east of Andregnies. **1962** M. Carver *El Alamein* iv. 51 While F Battery R.H.A. put down a 'stonk' the carriers counter-attacked and drove the enemy infantry back. **1961** *Where?* III. 16/2 *Religious instruction* (RI): The only subject which state schools are obliged to teach by law: in county and voluntary controlled schools. **1976** *Evening Post* (Nottingham) 15 Dec. 13/3 (Advt.), A teacher required in January and April for Independent School.. (3) R.I. with an interest in general counselling for girls 11 and over. **1952** *Billboard* 16 Feb. 18/1 Selection of an executive secretary for RIAA..was put off until early next week. **1971** J. Earl *How to choose & use Pickups & Loudspeakers* iii. 90 It is often necessary to attenuate the signal from a piezo cartridge before applying it to the RIAA amplifier input. **1975** *Gramophone* Jan. 1421/3 Also the RIAA correction curve for discs shows only the slightest deviation from the ideal playback curve. **1946** *Civil & Milit. Gaz.* 2 June 11/3 Disciplinary standards in the R.A.F. and R.I.A.F. **1971** R. Russell tr. *A. Ahmad's Shore & Wave* xv. 161 On one side stood Hasan, looking very smart in his R.I.A.F. uniform. **1913** W. T. Rogers *Dict. Abbrev.* 167/1 *R.I.B.A.* (Soc.), Royal Institute of British Architects. **1938** *Times* 7 Feb. 9/2 The article consists of a memorandum by the Public Relations Committee of the R.I.B.A. and correspondence therein between the Minister of Health and the R.I.B.A. Council. *a* **1974** R. Crossman *Diaries* (1975) I. 86, I had a difficult task when I had to go to the R.I.B.A. and give the certificates and prizes for the best-designed houses of the year. **1903** Somerville & 'Ross' *All on Irish Shore* 196 Two tall constables of the R.I.C. stood at the door of the cottage. **1972** A. Hezlet *'B' Specials* i. 4 The I.R.A. issued a proclamation referring to the R.I.C. as spies and traitors. **1978** J. Carroll *Mortal Friends* i. iv. 47 He saw a private soldier, not an RIC Auxiliary, but an honest to God Tommy. **1946** *Civil & Milit. Gaz.* 2 June 11/3 R.I.N. ratings. **1959** R. K. [see *physical torture* s.v. *physical *a.* 7]. **1968** G. Mitchell *Three Quick & Five Dead* i. 24 'Edward teaches history and something he calls R.K.' 'Religious Knowledge,' said Laura. 'They used to call it Scripture in my young day.' **1875** *Anglo-Brazilian Times* 23 July 6/1 On Hamburg the rates for Bank paper at 90 days have been 442 Reis per R.M. **1963** L. Deighton *Horse under Water* 251 Peterson had a *Reisepass*..and 200 RM. **1888** V. Martin *Diary* 24 June in M. Collis *Somerville & Ross* (1968) iv. 56 Jostled as to our seat by Brady R. M. **1899** Somerville & 'Ross' (*title*) Some experiences of an Irish R.M. **1947** *Daily Gleaner* (Kingston, Jamaica) 3 Nov. 12/5 Some cases before the R.M. Court... His Hon. Mr. S. T. B Sanguinetti, R.M. observed that he had left over the case from last week. **1827** J. Wright *Let.* 13 Nov. in P. H. Nicolas *Hist. Record of Royal Marine Forces* (1845) II. viii. 313 (*heading*) Deputy Adjutant-general's office, R.M. **1923** *Admiralty Fleet Order No. 1643* 22 June The R.M. Artillery at Eastney, and the R.M.L.I. at Forton will be combined into the Portsmouth Division, R. M. *Ibid.*, *Shoulder Letters.*—Letters 'R.M.' to be worn by all ranks. **1962** S. Bassett *Royal Marine* iii. 69 If I got high marks, I'd be appointed Superintendent RM Signals, a Corps appointment. **1883** A. Cooper Key *Let.* in L. Conway-Gordon *Case of Lt. Lewis Conway-Gordon* (1884) 17, I might be able to give you a final answer to your request as to the removal of the names of the two Officers from..the list of Lieutenants, R.M.A. **1931** *Times Lit. Suppl.* 26 Mar. 246/3 After 1894 the R.M.A. were no longer sent to sea in small craft, but assigned only to first-class ships and flagships. **1893** W. S. Churchill *Let.* 10 Sept. in R. S. Churchill *W. S. Churchill* (1967) I. Compan. i. vii. 412 Tonight I go back to RMC. **1956** J. Masters *Bugles & Tiger* ii. 36 Bert King..led me up the steps and into the low grey pile before me—the Royal Military College, Sandhurst. From that moment..I was Gentleman Cadet John Masters, of No. 5 Company of the R.M.C., as I soon learned to call it. **1968** A. Powell *Military Philosophers* i. 55 Chu enjoyed the RMC so much he wants to go to Eton. **1889** W. Richards *Her Majesty's Army* II. 195 Lieutenant H. Earle, R.M.L.I., was sent as an envoy to the King of Dekra. **1916** 'Taffrail' *Pincher Martin* ix. 166 He..hastily told the marine corporal of the watch to turn out twenty marines..and then to inform Captain Hannibal Chance, R.M.L.I., that a Zeppelin was in sight. **1930** *Times Lit. Suppl.* 3 July 542/3 It is the evacuation of the Y Beach at Gallipoli, in which the 1st Battalion was concerned, with one company of the 2nd South Wales Borderers and the Plymouth Battalion R.M.L.I. (consisting entirely of men specially enlisted for the War). **1980** *Globe & Laurel* July/Aug. 241/1 The original Forton RMLI Cadet Band was disbanded on amalgamation of the RMLI and RMA. **1897** A. Hay *Princ. Alternate-Current Working* vii. 93 When we speak of so many volts of an alternating P.D., or so many amperes of an alternating current, we thereby mean their R.M.S. values. **1940** *Jrnl. Acoustical Soc. Amer.* XI. 278/1 Peak and r.m.s. pressures in one-eighth-second intervals..from the voices of six men and five women. **1945** *Electronic Engin.* XVII. 737/3 The oscillator output voltage was 20V R.M.S. **1977** S. A. Book *Statistics* ii. 40 The standard deviation is the square root of the mean of the squared deviations. (In some applied contexts, it is referred to as the 'root-mean-square deviation' or the 'rms deviation'.) [**1852** *Nautical Mag.* XXI. 91 To the Secretary R.M.S.P. Company.] **1870** *Weekly Standard* (Buenos Aires) 12 Jan. 1/3 The R.M.S. La Plata arrived in port..with the following passengers. **1936** *Discovery* June 181/2 R.M.S. '*Queen Mary*', the greatest achievement of British shipbuilding. **1976** *Oxf. Compan. Ships & Sea* 684/1 *R.M.S.*, the prefix, short for Royal Mail Ship, placed before the name of a British merchant ship with a licence to carry the Royal Mails. **1938** *Amer. Speech* XIII. 228/2 She is an 'R.N.', which means *registered nurse*—one who has graduated from an accredited school of nursing. **1974** *Publishers Weekly* 11 Feb. 62/2 Irish Katie, R.N. in a nursing home. **1846** H. D. Chads *Let.* in *Madras Artillery Record* VI. 323 [signed] H. D. Chads, Captain R.N. **1946** 'Tackline' *You met such Nice Girls in Wrens* xiii. 144 It is very strange..why R.N. officers are so much meaner with their clothing-coupons than R.N.V.R. officers. **1975** 'J. Bell' *Victim* iv. 52 Wing-Commander Redfern, Late R.A.F., his wife, Amanda, and Commander Pilcher, R.N., retired. **1918** W. S. Churchill *Let.* 10 Sept. in M. Gilbert *W. S. Churchill* (1975) IV. vii. 144 How much flying, for instance, is done by the RNAS for the 45,000 first-rate fighting men and skilled men they employ? **1977** *Air Mail* Spring 21/1

R

So this RAF phrase, like 'port' and 'starboard', came with the RNAS when the flying sailors merged with the RFC in 1918 to become the Royal Air Force. **1914** R. Brooke *Let.* 24 Sept. (1968) 619 The nucleus of the R.N.D. is marines, Naval Reserve, etc.—more or less trained men. **1916** W. S. Churchill *Let.* 29 Nov. in M. Gilbert *W. S. Churchill* (1977) IV. Compan. I. 34, I rejoiced to read of the glorious achievements of the RND. **1924** *Life-Boat: Centenary Number* 33 Small wonder that its sister-organization in little Holland has kept its eye upon the largest and oldest organization, the R.N.L.I. **1948** *Life-Boat* Mar. 157 The R.N.L.I. Roadless Tractor..can take the boat over all types of beach. **1977** *Navy News* June 1/2 The Merchant Navy and fishing fleets will be represented, as well as organizations like the R.N.L.I., Trinity House and H.M. Coastguard. **1868** *Times* 1 July 2/2 (Advt.), John Gray, R.N.R., Commander. **1902** C. J. C. Hyne *Mr. Horrocks, Purser* 85 What's the use of being R.N.R. if you don't let people know it? **1977** *Stornoway Gaz.* 27 Aug. 3/1 As were many other Lewismen, Alick was a member of the R.N.R. **1905** *Text Book for Royal Naval Volunteer Reserve* 3 The smartness, efficiency, and the credit of the R.N.V.R. depend upon the following principles. **1934** *Brassey's Naval & Shipping Ann.* 25 New rules were notified in Fleet Orders dated April 6, 1933, respecting the rank of Commodore in the R.N.R. and R.N.V.R. **1977** *Listener* 25 Aug. 241/2 Lambert was given a commission in the RNVR. *c* **1944** *Mod. Jun. Dict.* (Whitcombe & Tombs) 454/1 R.N.Z.A.F., Royal New Zealand Air Force. **1947** *Air Force News* (Cairo) 14 Jan. 1/2 The first two senior RNZAF officers nominated for participation in the exchange scheme with the RAF. **1959** *Listener* 23 July 146/2 We follow the fortunes of Vlasov's Army of Liberation, the ROA, during the last months of the war. **1974** T. P. Whitney tr. *Solzhenitsyn's Gulag Archipelago* I. I. vi. 253 As for the leaflets reporting the creation of the ROA, the 'Russian Liberation Army', not only were they written in bad Russian, but they were imbued with an alien spirit that was clearly German. **1942** Partridge *Dict. Abbrev.* 84/2 *R.O.C.*, Royal Observer Corps..A civilian body that scans the skies. Modest and selfless. **1947** *R.A.F. Rev.* Sept. 20/2 We hope to live up to its claim, as did the R.O.C. **1950** *N.Y. Times* 2 July IV. 1/6 The man who took over direction of operation ROK—Relief of Korea—in these dire circumstances was General MacArthur. **1950** *Life* 11 Sept. 51/2 (*heading*) The Durable Roks. *Ibid.*, But the Rok (for Republic of Korea) army was more durable than anybody thought. **1972** P. M. Bartz *South Korea* 2/2 South Korea's correct title is the Republic of Korea, usually abbreviated R.O.K. **1966** *IEEE Trans. Electronic Computers* XV. 502/1 Transformer memories are read-only memories (ROM) which make use of magnetic coupling..between a set of interrogation lines and a set of sense lines. **1975** *Sci. Amer.* May 37/3 ROM's are nonvolatile: their contents cannot be altered during the operation of the computer, and the retention of stored data does not depend on a supply of power. **1979** *Personal Computer World* Nov. 82/1 Monitors are held in ROM so that they are available and running as soon as the computer is started. **1947** K. M. Wells *Owl Pen Reader* (1969) I. 97 Our goats..were neither registered, accredited, pure-bred, or R.O.P. **1955** R. J. Schwartz *Compl. Dict. Abbrev.* 155/3 *rop*, run of paper (advertising). **1961** *Penrose Ann.* LV. 110 Only in recent years was the reproduction of colour transparencies taken up by means of ROP printing, where the full range of colours is produced by superimposing the three basic colours. **1967** E. Chambers *Photolitho-Offset* 276 *R.O.P.*, an American term (run of paper) applying when colour halftones are printed at the same time as the type matter. **1972** J. R. Bridge in *Mathematics in School* Sept. 8 (*title*) ROSLA and after. *Ibid.* 9/2, I believe ROSLA will be rewarding for groups 2 and 3. **1977** *Jrnl. R. Soc. Arts* CXXV. 300/1 When the additional ROSLA year operated, the prospect of a further year of standard education seems to have been unpalatable to some of the potential beneficiaries. **1948** *Fire Protection & Accident Prevention Year Book 1948–49* III. 139 The Royal Society for the Prevention of Accidents (RoSPA)... The aim of the Society is the prevention of accidents of all kinds. **1955** *Proc. National Industrial Safety Conf.* 7 Attract new members to your local groups, encourage them to become members of RoSPA. **1965** *Autocar* 24 Sept. 609/1 The leaflet recently published by RoSPA in conjunction with the Ministry of Transport. **1889** A. J. Ellis *Early Eng. Pronunc.* v. 6 Other Abbreviations in Frequent Use... rp., received pronunciation, or that of pronouncing dictionaries and educated people. **1964** R. H. Robins *Gen. Linguistics* ii. 57 The pronunciation characteristic of this type of English has been called 'received pronunciation' or 'R.P.'. **1964** *English Studies* XLV. 26 There is no Standard Accent of English, and almost the only people who think there *is* are a small number of RP speakers who feel that their accent is..in some way superior. **1977** P. Strevens *New Orientations Teaching of English* xi. 138 In Britain, the non-regional accent is RP. **1960** *Times Rev. Industry* Mar. 3/2 In the absence of price competition, distributors may compete by offering extra services. R.p.m. does not guarantee that they do. **1964** *Financial Times* 31 Jan. 6/4 However, the pending abolition of R.P.M. has confused some M.P.s in deciding their attitudes. **1966** J. F. Pickering *Resale Price Maintenance* i. 15 In a number of trades r.p.m. was enforced privately through the mechanism of the relevant trade association. **1978** *Bookseller* 17 June 3183/1 In Australia and in Sweden the abolition of r.p.m. has had these very effects. **1906** *Trans. Inst. Engineers & Shipbuilders in Scotland* XLIX. 30 Especially should this be done in cases where..the weight increases more rapidly in inverse proportion to the R.P.M. and the diameter than it does with other types. **1931** *Discovery* Nov. 344/2 Running the engine at a higher r.p.m. **1950** *Down Beat* 5 May 16 (*heading*) Here's LP, 45 RPM jazz list. **1966** R. Thomas *Spy in Vodka* (1967) xix. 218 The engine was responding nicely..and I was estimating the rpm's needed for the next bend. **1970** *Air Force Mag.* Oct. 40/2 This study of the Remotely Piloted Vehicle (RPV) potential consisted of detailed examinations of presently available technologies. **1972** *Observer* 13 Aug. 1/6 Remotely Piloted Vehicles (RPVs) that promise a bizarre kind of 'bloodless' warfare. **1977** *Time* 23 May 33/3 That will make being a pilot a cushy job: he sits at a TV console 200 miles away and gets the RPV to provide surveillance or relay radio messages or pinpoint targets for precision bombing. **1905** W. H. Howell *Textbk. Physiol.* xxxviii. 632 In connection with other data,..the R.Q. may be used to throw light upon the character of the nutrition. **1968** Passmore & Robson *Compan. Med. Stud.* I. xlii. 4/2 Ingestion of glucose within minutes restores the blood glucose concentration and the working ability, without changes in the RQ. **1967** *Jrnl. Molecular Biol.* XXIII. 149 Abbreviations used: rRNA, ribosomal RNA. **1977** *Nature* 8 Dec. 473/3 It is probable that no rRNA gene is integrated into the chromosomal DNA of the macronucleus [of *Tetrahymena*]. **1889** A. J. Ellis *Early Eng. Pronunc.* v. 3 They all spoke 'received speech' (abbreviated to rs.) in 'received pronunciation' (abbreviated to rp.) **1934** H. C. Wyld *S.P.E. Tract* xxxix. 605 With this type [of spoken English]..I contrast what, for want of a better name, I call *Received Standard* (henceforward referred to in this paper by the initials R.S.). **1964** C. Barber *Ling. Change Present-Day Eng.* ii. 22, I shall use the expression Received Standard (R.S.) to refer to all aspects of the language. **1920** *R. Soc. Arts Syllabus Examinations, 1921* p. xviii (Advt.), A selection from many books suitable for R.S.A. examinations. **1973** M. Amis *Rachel Papers* 64 Not that age was a helpful grouping criterion, the elder lot ranging as they did from fifteen (a delinquent ghoul studying for RSAs) to nineteen (myself). **1980** *Jrnl. R. Soc. Arts* Feb. 145/1 Nor is the RSA alone. **1923** *Nation* 31 Jan. 130/1 The Russian Socialist Federated Soviet Republic (RSFSR)..and the Transcaucasian Socialist Federated Soviet Republic (Georgia, Azerbaijan, and Armenia) are concluding the present treaty on their consolidation into one united state. **1975** *Whitaker's Almanac 1976* 956/1, By the 1947 Peace Treaty with Finland, the district of *Petsamo*..was added to the territory of the R.S.F.S.R. **1940** *Chambers's Techn. Dict.* 733/1 *R.S.J.*, rolled steel joist. **1954** *Archit. Rev.* CXV. 334 The extension has a reinforced concrete frame with the exception of the top floor of the main block, which has a RSJ portal frame and a flat roof of asbestos cement decking. **1978** *Private Eye* 17 Mar. 15 We had..three RSJs put across the ceiling to stop the upstairs coming downstairs. **1971** *Circular* (Dept. Educ. & Sci.) No. 8. 1 The effect of raising the school leaving age (RSLA) is to substitute 16 for 15 in all the provisions of the Education Acts which deal with the upper limits of compulsory school age. **1979** *Rep. on Educ.* (Dept. Educ. & Sci.) No. 95. 1/1 The raising of the school leaving age to 16 (RSLA)—on 1 September 1972 with effect from 1 September 1973—marked the achievement of a long held objective. **1913** W. T. Rogers *Dict. Abbrev.* 170 *R.S.M.* (tit.), Regimental Sergeant-Major. **1955** E. Waugh *Officers & Gentlemen* I. ix. 107 The RSM sent up the rocket which announced the start of the exercise. **1978** R. V. Jones *Most Secret War* x. 90 The Grenadiers had sent an R.S.M. to Dover to pick up any Grenadiers and drill them. **1907** *Bird Notes & News* II. 52/2 In our parish nearly all belong to the R.S.P.B. **1937** *Discovery* Jan. 21/2 We are not planning to have our own bird sanctuaries with watchers to protect them, as this work is already being carried out with great success by the R.S.P.B. **1979** *Birds* Summer 34/1 The RSPB has seven reserves with major reedbeds. **1870** *Animal World* (Suppl.) 1 July 177/1 (*heading*) Annual report of the R.S.P.C.A. **1924** Fairholme & Pain *Century of Work for Animals* iii. 53 At the present time the R.S.P.C.A., working in England and Wales, alone employs two hundred and three full-time inspectors. **1978** *Daily Tel.* 18 July 1/8 The RSPCA is to launch an investigation today after about 2,800 of a cargo of 4,000 mink were found dead at Manchester airport. **1961** F. F. Bruce *English Bible* xiv. 186 The R.S.V. dispenses with one of the most distinctive features of what English-speaking people have come to regard as 'Bible language'. **1974** *Oxf. Dict. Chr. Ch.* (ed. 2) 171/2 The RSV is widely used not only in America but also in Britain..and other English-speaking countries. **1956** *Jrnl. Nat. Cancer Inst.* XVI. 365 Lesions of the central nervous system produced by a duck variant of the Rous sarcoma virus (RSV). **1974** J. D. Acton et al. *Fund. Med. Virol.* xxiii. 301 Some strains of RSV will induce transformation of mammalian (mouse, rat, hamster, primate, human) fibroblast cells. **1942** *Tee Emm* (Air Ministry) II. 64 If the R/T transmission is a bit distorted, 'Say again' is a set expression. **1971** 'D. Halliday' *Dolly & Doctor Bird* xiii. 180 The R/T isn't functioning, and neither are the radar or echo sounder. **1975** J. Wyllie *Butterfly Flood* (1977) xxxii. 156 Keep within R.T. range of us. **1969** W. B. Emery *Nat. & Internat. Syst. Broadcasting* vi. 117 RTE now has a repertory company of twenty-eight actors. **1979** J. J. Lee *Ireland 1945–70* 173 Over 50 per cent of RTE television broadcasting featured imported programmes at the end of the 1960s. **1917** *B.E.F. Times* 8 Sept. (1918) 4/2 There is also a rumour that the R.T.O. of Bath has not got a decoration in the new list. **1919** G. S. Gordon *Let.* 20 June (1943) 95 We then drove over to the Gare de Lyon, where we hung about the R.T.O.'s office till 11, getting tickets made out, and our luggage registered. **1930** *Amer. Speech* V. 385 *R.T.O.*, Railroad transportation officer. **1955** E. Waugh *Officers & Gentlemen* I. v. 45 Guy sought aid of the RTO and was rebuffed. **1978** Ld. Lovat *March Past* I. viii. 132 His last appointment was that of Railway Transport Officer (RTO) at Euston Station. **1917** Kipling *Book of Words* (1928) 147 I'm not defending ragging—I've known cases where everyone who took part in it ought to have been R.T.U. **1976** K. Bonfiglioli *Something Nasty in Woodshed* ix. 103 Mortdecai [*sic*] would never wear the coveted red tabs on his khaki. 'RTU' (Returned to Unit) would follow his name for ever. **1922** in M. Gilbert *Winston S. Churchill* (1977) IV. Compan. III. 1948 The Secretary of State for the Colonies..stated that the RUC were not up to strength. **1941** T. J. Campbell *Fifty Years of Ulster* xxiii 322 The R.U.C. are an armed force, provided with revolvers and rifles. **1977** *Irish Times* 8 June 11/8 Last night the injured RUC man was stated to be seriously ill in hospital. **1975** *Ibid.* 9 May 24/1 (Advt.), Long Lease. G.R. £12·50. R.V. £25. **1976** *Dumfries & Galloway Standard* 25 Dec. 17/7 (Advt.), Modernised ground floor flat... R.V. to be reassessed. **1967** *Wheels Afield* June 42 (*heading*) Handling your RV in traffic. **1968** *Trailer Life* June 52/2 For early Rec V construction the terms cracker box or bullets served as type designation. **1973** *Ibid.* Apr. 134/2, I sure wish there was some way to allow RVers to utilize those great areas. *Ibid.* 194/2 An RVing woman's luck in being able to take along a wardrobe for every occasion. **1978** *Sunday Sun-Times* (Chicago) 1 Jan. 120/2 The RV industry..has regained its feet. **1942** Partridge *Dict. Abbrev.* 86/1 *r.v.* or *R.V.*, (place of) rendezvous. (Common to all three services.) **1973** C. Egleton *Seven Days to Killing* xiv. 163 He had no way of knowing whether Tarrant would have set up a fail-safe RV or not. **1975** N. Luard *Travelling Horseman* vi. 162 We'll r.v...in the boozer, and I'll take him over to the garage.

b. Substitute 1825 for date of first example in Dict. and add further examples.

1828 *New Jersey Eagle* (Newark, N.J.) 23 May 3/3 The three R's—honest 'Rithmetic, Reading & 'Riting I think I can say, I'm no fool in. **1976** *Times* 26 Mar. 6/3 In some primary and secondary schools the three Rs have been neglected or devalued. **1979** *Jrnl. R. Soc. Arts* CXXVII. 483/1 Writing is, after all, one of the three R's on which education is based.

3. Also as *v. intr.*

1969 R. V. Beste *Next Time I'll pay my Own Fare* viii. 111 The Duchess de Santine Miorna requests the pleasure of Detective-Inspector John Gage's company to dinner tonight... R.S.V.P... Gage R.S.V.Ped in Spanish. **1978** *Observer* 19 Feb. 7/1 The printed invitations to 27 journalists were delivered... We all RSVP'd.

4. R or r (*Physics*): abbrev. of *roentgen* (*unit*) s.v. *ROENTGEN 2. Also *r-unit*.

1922 *Physics Abstr.* A. XXV. 508 The instrument is calibrated in terms of a unit R (the Röntgen), i.e. the ionisation produced in 1 sec. by 1 gm. of Ra at a distance of 20 mm. after filtration by 0·5 mm. of Pt. **1938** R. W. Lawson tr. *Hevesy & Paneth's Man. Radioactivity* (ed. 2) iv. 58 The γ-radiation from 1 mg. Ra..supplies 8·6 *r* per hour at a distance of 1 cm. *Ibid.* xxiv. 260 The maximum permissible daily dosage of neutrons is estimated to be one-tenth that for γ- or X-rays, or 0·01 *r* unit. **1962** *Newnes' Conc. Encycl. Nuclear Energy* 205/1 Thus for X- and γ-radiation of energy between 0·3 and 3 MeV, 1 *r* is nearly equivalent to 1 rad in water or soft tissue. **1973** Kaye & Laby *Tables of Physical & Chem. Constants* (ed. 14) iii. 285 The unit of exposure is the roentgen (R) which is equal to $2{\cdot}58 \times 10^{-4}$ coulomb per kilogram of air.

III. As a symbol. **1.** *Chem.* [Initial letter of *radical*.] R is used in chemical formulæ to represent an unspecified radical or group of radicals (usu. organic).

1866 Watts *Dict. Chem.* IV. 219 In these formulæ, R+ represents a positive organic radicle, and R− a negative radicle: no negative organic radicle as such, has yet been introduced into these compounds. *Ibid.* 227 Stannic compounds of the form $Sn^{iv}R^{4}+$ are colourless mobile liquids. **1872** [see *polymerization]. **1909** C. A. Keane *Mod. Org. Chem.* viii. 130 In the formula for ketones, the radicals R and R′ may be either like or unlike. **1932** I. D. Garard *Introd. Org. Chem.* vi. 74 The aldehydes have one of these valences attached to carbon and one to hydrogen, so that for any hydrocarbon radical, R the aldehyde is R—C=O (with H below the C). **1950** N. V. Sidgwick *Chem. Elements* I. 30 It can be shown that the cyanamides $H_2N{\cdot}CN$ and HRN·CN are highly associated in benzene, while $R_2N{\cdot}CN$ is monomeric. **1966** Williams & Fleming *Spectrosc. Methods in Org. Chem.* ii. 38 The absorption maximum of the nitroaniline XXXVI (R=Me) is at 385 mμ..showing a red shift and marked reduction in intensity from that of the parent compound XXXVI (R=H) at 375 mμ.

2. Physical Chem. *R* is used to denote the *gas constant*, i.e. the constant of proportionality in the equation of state for an ideal gas, $PV = nRT$, where P = pressure, V = volume, n = the number of moles of gas, and T = absolute temperature; now usu. taken to be 8·314 joule kelvin^{-1} mole^{-1}. [Arbitrary: first used by J. D. Van der Waals 1873, *Over de Continuïtet van den Gas- en Vloiestoestand* (Leiden) 58.]

1880 *Phil. Mag.* IX. 393 In relation to pressure, volume, and temperature, gases follow..with a certain degree of approximation, the laws of Mariotte and Gay-Lussac, which can in common be expressed by the following equation—pv = RT..wherein..R is a constant dependent on the nature of the gas. **1895** C. S. Palmer tr. *W. Nernst's Theoret. Chem.* I. i. 31 The factor R is only conditioned by the unit of measure chosen, but is independent of the chemical composition of the gases in question. **1940** Glasstone *Textbk. Physical Chem.* ix. 654 The value of *R* for solutions is thus very close to that of *R* usually accepted for a gas, namely 0·0821 liter-atm.., so that it is possible to write $\Pi V = RT$, where *R* may be taken as the gas constant [and Π = osmotic pressure]. **1978** P. W. Atkins *Physical Chem.* 13 In this expression *R*, the gas constant, is another fundamental constant with the value 8·3 J K^{-1} mol^{-1}.

3. Biol. **a.** *r* is used in formulæ and elsewhere to denote the rate of increase of a population, usually representing the factor by which its size is multiplied in each generation; or the value which this factor would have if resources (food etc.) were unlimited.

1918 *Q. Publ. Amer. Statistical Assoc.* XVI. 123 Where *b* = birth rate per head per annum, *r* = natural rate of increase per head per annum... The formulae thus obtained give the relationship between *b* and *r*. **1954** Andrewartha & Birch *Distribution & Abundance of Animals* iii. 33 In nature, one or several components may predominate to determine the *actual* rate of increase, which we shall call *r*. **1967** MacArthur & Wilson *Theory of Island Biogeogr.* vii. 150 Clearly..*r* will be increased by a habitat or food shift which increases the density of available food. **1971** A. S. Boughey *Fundamental Ecology* iii. 97 The biotic potential is usually expressed by a factor representing the intrinsic rate of natural increase *r*. **1977** J. L. Harper *Population Biol. of Plants* i. 3 There is a growing tendency to talk and write about *r*-phases in the life of a population—phases when it explodes with near-exponential growth after a disaster or after a new colonization into an unexploited environment.

b. *r selection*, the form of natural selection which acts on populations having ample resources and little or no competition.

1967 MacArthur & Wilson *Theory of Island Biogeogr.* vii. 149 In an environment with no crowding (*r* selection), genotypes which harvest the most food (even if wastefully) will rear the largest families and be most fit. **1973** P. A. Colinvaux *Introd. Ecology* 618/2 *r-Selection*, selection for an opportunist strategy leading to an opportunist species. Traits are preserved which use energy to obtain a high natural rate of increase 'r'. **1976** *Nature* 12 Feb. 478/2 MacArthur and Wilson coined the terms *r* selection and *K* selection to describe two general kinds of selection they believed could be functioning in nature (*K* refers to carrying capacity and *r* to maximal intrinsic rate of natural increase, r_m). **1979** *Sci. Amer.* Jan. 26/3 The poles of evolutionary adaptation, *r* selection increasing the number of offspring (the prodigal mackerel roe with half a million eggs) and *K* selection ensuring survival (the proverbial royal litter, only one cub born at a time, but that one a lion).

4. Spectroscopy. *R* is used to denote the *Rydberg constant* s.v. *RYDBERG I.

1920 *Phil. Mag.* XXXIX. 47 If the nucleus be a simple point-charge, the negatived total energy *W* belonging to any one of the stationary orbits is given by $W/ch = \kappa^2 R/n^2$,.. where *n* is an integer, *c* the light velocity in vacuo, *h* Planck's constant, and *R* the Bohr expression of Rydberg's constant. **1926** G. Birtwistle *Quantum Theory of Atom* ii. 21 In 1913, Niels Bohr..gave a theory of the hydrogen spectrum which ..led..to an expression for the Rydberg constant *R* in terms of known physical constants. **1965** R. N. Dixon *Spectroscopy & Struct.* ii. 31 Since the lowest energy level has *n* = 1, and the highest *n* = ∞, the Rydberg constant R_H corresponds to the minimum energy required for the removal of the electron. **1978** P. W. Watkins *Physical Chem.* xiv. 427 Balmer..pointed out that the wavelength of the light in the visible region..fitted the expression $1/\lambda = R(\frac{1}{4} - 1/n^2)$, n = 3, 4..where *R* is a constant now called the Rydberg constant, and having the value 109 677 cm^{-1}, or $3{\cdot}289\,8 \times 10^{15}$ Hz.

5. Psychol. *R*, or *r*, is used in some methods of factor analysis in which traits or abilities are the variables (see quots.); also *attrib.*, as *R, r technique*. Cf. *Q III. 4.

1925 *Brit. Jrnl. Psychol.* July 75 Where *R* is the matrix of the *r*-correlation coefficients. **1927** C. Spearman *Abilities of Man* vi. 73 The form recently preferred is given below. In it, as usual, the letter *r* stands for any correlation, whilst its two subscripts indicate the two abilities (tests, school marks, etc.) that are correlated. **1936** *Brit. Jrnl. Psychol.* Apr. 345 It is convenient to designate all previous factor analysis as *r* technique, and this new inverted form as *Q* technique. **1950** R. B. Cattell *Personality* ii. 30 In *R* technique one would point to a particular trait-indicator test and say, 'This is the test measure of the surface trait or factor in question'. **1974** W. B. Arndt *Theories of Personality* viii. 99 In an actual application of the procedure just described, fifty or more measurements would be made on one hundred or more subjects. This technique, based on normative data, is called the *R-technique*.

6. *Chem.* [Abbrev. of L. *rectus* right.] *R* is used to designate (compounds having) a configuration about an asymmetric carbon atom in which the substituents, placed in order according to certain rules, form a clockwise sequence when viewed from a particular direction. Opp. *S (see S 13).

1956 R. S. Cahn et al. in *Experientia* XII. 83/2 The suggested indications for asymmetry leading, under the sequence and conversion rules, to a right- and left-handed pattern, are capital italic *R* and *S* respectively, where *R* derives from the Latin *rectus*, meaning 'right', and *S* from *sinister*... For the description of pseudo-asymmetric atoms, it is proposed to employ the corresponding lower-case italic symbols, *r* and *s*. *Ibid.* 85/2 The simplest case may be exemplified by bromochloroacetic acid (XV). Priority of the groups (a, b, c, d) is here determined simply by the atomic numbers as Br, Cl, C (of CO_2H), H. Hence, by the conversion rule..formula (XV) represents an (*R*)-form. **1966** Rakoff & Rose *Org. Chem.* xiv. 477 Going from OH to CHO to CH_2OH traces a clockwise path, so the configuration about the asymmetric center in D-glyceraldehyde is rectus, and the compound would be designated (R)-glyceraldehyde. **1971** *Sci. Amer.* Aug. 46/3 One enantiomeric form of carvone, *R*-carvone, has a strong odor of spearmint; the other form, *S*-carvone (which is geometrically a mirror image..) has the odor of caraway. **1973** *Times* 20 Jan. 16/6 Most of the tests were therefore carried out using the R analogue which seemed to be free from side effects.

7. *Astr.* [Initial letter of *rapid*.] *r-process*: a process believed to occur in stars in circumstances of high neutron flux (e.g. in supernova explosions), in which heavy atomic nuclei are formed from lighter ones by a combination of rapid neutron captures and slower beta decays.

1956 F. Hoyle et al. in *Science* 5 Oct. 612/2 We have distinguished two conditions under which the Neutron capture can take place, a slow (*s*) process and a rapid (*r*) process. *Ibid.*, The *r*-process we associate with the explosion of supernovae, the time scale being as small as 10 to 100 seconds. **1975** *Nature* 2 Oct. 362/1 The measured charge distribution in the vicinity of the Earth at ∼0·5 GeV per nucleon strongly favours synthesis of these cosmic rays by the r-process (rapid neutron capture) within the last 10^7 yr. **1977** J. Narlikar *Struct. Universe* ii. 49 Whereas the s-process produces proton-rich nuclei, the r-process produces neutron-rich nuclei.

8. *Microbiology.* [Initial letter of *resistance*.] *R* is used to denote certain plasmids which confer drug-resistance on bacteria and can be transferred to other bacteria by conjugation.

1961 *Gunma Jrnl. Med. Sci.* X. 59 It is strongly suggested that there is a transmissible drug-resistance factor (R) which makes R^- cells drug resistant when infected with R^+ cells following cell-to-cell contact. **1962** *Jrnl. Bacteriol.* LXXXIV. 902/2 *R* or *R factor* is a general term for the infectious drug-resistance factors (Mitsuhashi, 1960; at the Meeting of Microbial Genetics at Mishima, Japan, it was agreed by investigators in this field to use the term 'R' for the multiple drug-resistance factor). **1967** *Jrnl. Gen. Microbiol.* XLIX. 97 Genetic alterations resulting in the loss or acquisition of F or R pili are strictly correlated with loss or gain of ability to transfer the chromosome by conjugation. **1969** A. M. Campbell *Episomes* iii. 41 Cells harboring R can transfer it to other cells. **1975** *Sci. Amer.* July 28/3 Antibiotic-resistant *E*[*scherichia*] *coli* isolated in many parts of the world..were found to contain plasmids, designated *R* factors..carrying the genetic information for products that in one way or another could interfere with the action of specific antibiotics.

‖ **ra²** (rā). *Physical Geogr.* Pl. **ras,** ‖ **raer.** [Norw.] One of the terminal moraines near the coast in southern Norway and Sweden that are in the form of long ridges of gravel and clay with a covering of large stones.

1902 *Geol. Mag.* Decade IV. IX. 319 Outside the terminal ra, that is, between the moraine and the coastline,..there is a widely spread deposit of clay. **1957** J. K. Charlesworth *Quaternary Era* II. xxx. 627 The radially dispersed erratics and diverging striae transverse to the morainic *Raer*..finally overcame both prejudice and honest conviction. *Ibid.* xlii. 1172 The *Ras*, traceable as far as Stavanger, have been correlated with the double moraines observed in many fjords and fjord-valleys.

‖ **raad** (rāt), *sb.*¹ *S. Afr.* [Du., = council; freq. as shortening of *Heemrad, *Volksraad, etc.] A council, an assembly; *spec.* (usu. with capital initial) the legislative assembly of one of the former Boer republics (*Hist.*).

1850 R. Gray *Jrnl.* 8 May in *Jrnl. Bishop's Visitation Tour in 1850* (1851) 25 These men have formed themselves into a Republic, and have their 'Raad' (Council). **1856** C. J. Andersson *Lake Ngami* v. 57 He laid his complaint before the chief of the tribe; and a 'raad', or counsel, was held. **1873** F. Boyle *To Cape for Diamonds* xii. 153 The gentlemen..obtained their grant under solemn seal and bond of the Transvaal Parliament, or *raad*. **1930** *Times Lit. Suppl.* 7 Aug. 635/2 The tribal council has become a 'raad' consisting of the chief, ex-officio, and of members elected at the mass meeting. **1940** F. B. Young *City of Gold* 261 So the Raad sat..voting Burgers down. **1963** S. Cloete *Rags of Glory* vi. 53 Field Kornet Adriaan de la Rey, the member of the Raad who had been against the war. **1973** *Hansard S. Afr.* 8 Feb. 297 The Minister of Labour..says that in no circumstances will he sit in the same 'raad' as a non-white.

raad (ra,ad), *sb.*² [Arabic.] The electric catfish, *Malapterurus electricus*, belonging to the family Malapteruridæ, found in the Nile and other rivers of central and western tropical Africa, and distinguished by electric cells in the fatty tissue just beneath the skin.

In quot. 1878 = Torpedo *sb.* 1 a.

1869 A. Günther in J. & B. H. Petherick *Trav. Cent. Afr.* II. 240 (*Raad* or *Raasch*.) The electric Silurus is spread over the whole of tropical Africa. **1878** *Encycl. Brit.* VIII. 8/2 The Arabians had long before given this fish [*sc.* the torpedo] the name of *raad* or lightning. **1931** J. R. Norman *Hist. Fishes* viii. 149 This species is used by the Arabs for food, and they refer to it as the Raad or Thunder-fish. **1957** E. Le Danois *Fishes of World* iv. 130 The electric cat-fish.., the raad or thunder-fish, lives in African rivers.

rab³. Chiefly *Cornish* (orig. *dial.*). [Shortened f. Cornish *rabman, -men* in the same sense. Ultimate origin obscure. Also recorded from Wales (quot. 1910).

There is no apparent connection with Scottish and Northumberland dial. *raab, rab* (the fall of a cliff, a mass of broken rock, etc.), which is of Scandinavian origin (see e.g. S.N.D.).]

Rough or stony subsoil; rubble, gravel.

[**1769** W. Borlase *Antiquities Cornwall* 452/1 *Rabman*, rubble; that mixture of clay and stone which has not been moved since the flood, and generally lies over the karn.] **1868** *Proc. Soc. Antiquaries London* IV. 164 The floor..is composed of the hard subsoil of the country, called by the Cornish 'rabman'. **1880** Courtney & Couch *Gloss. Words Cornwall* 46/1 *Rab*, decomposed granite used for mending roads. **1910** W. M. Morris *Gloss. Demetian Dial.*, *Rab*, rough, stony soil; sub-soil. **1928** *Jrnl. Brit. Archaeol. Assoc.* XXXIV. 153 The paving stones had been laid on the rab or subsoil. **1912** *Antiquity* XXVI. 90 The site of the hut was cleared and levelled by cutting back into the hillside and spreading the excavated rab over the lower part of the floor. **1961** E. Clark *Cornish Fogous* ix. 68 A drain passes from the fogou to the outer wall, the floor of which is composed of the natural rab... The floor of the main structure is of stoneless rab. **1962** *Punch* 30 May 836/3 Clutching for dear life at the poor rab (the local name for the pink subsoil) this tree has grown.

rabab, rababa, varr. *rebab.

rababoo, var. *rubbaboo.

rabanna (rābæ·nă). [Malagasy.] A fabric woven from raffia.

1883 *Encycl. Brit.* XV. 172/2 The chief articles of export being..rôfia palm cloths (*rabànnas*) and fibre. **1969** R. T. Wilcox *Dict. Costume* (1970) 281/1 *Rabanna*, a textile imported from Madagascar. Used for hats and bags.

rabap, var. *rebab.

rabat². Delete *Sc.*, for † *Obs.* read '*Obs.* exc. *Hist.*', and add: **a.** (Later examples).

1865 F. B. Palliser *Hist. Lace* viii. 126 Suddenly, at the court of Henry [III of France], the fraise gave way to the rabat, or turn-down collar. **1953** M. Powys *Lace & Lace-Making* vii. 77 With the coming of the wig or natural long curled hair the ruff was given up and men wore their lace in front falling under the chin in the form first of a cravat or Rabat and later in ruffles.

b. A type of turned-down clerical collar. Also, = Stock *sb.*¹ 44 b. Also, a similar garment worn by a layman.

1889 in *Cent. Dict.* **1931** D. Attwater *Catholic Encycl. Dict.* 441/1 *Rabat*... Part of the dress of the French secular clergy. A white *rabat* is worn by the Brothers of Christian Doctrine, university professors, magistrates, barristers and attorneys. **1936** J. G. Cozzens *Men & Brethren* 122 You can go courting, if you like; but you have to go in a rabat, so to say. **1966** H. Kemelman *Saturday the Rabbi went Hungry* (1967) ii. 18 In spite of the clerical collar and black silk rabat, he looked more like a football player than an Anglican minister. **1975** *New Yorker* 3 Feb. 23/1 He is dressed in an ultra-attractive robin's-egg-blue rabat and brown velvet suit.

rabat (rābæ·t), *v. Geom.* Also **rebat(e.** [ad. F. *rabattre* to lower, fold back, rabat.] *trans.* To rotate (a plane) about its line of intersection with another plane, *spec.* the horizontal plane, until the two coincide. Hence **raba·tted** *ppl. a.*, **raba·tting** *vbl. sb.* Also **raba·tment,** the process of rabatting.

1885 *Encycl. Brit.* XIX. 801/1 In rabatting the plane π_1 its trace OB with the plane π_2 will come to the position OD. **1908** L. N. G. Filon *Introd. Projective Geom.* i. 7 For practical purposes..it is convenient to rotate one of the two planes about *x* until it coincides with the other plane. A figure in the former plane rotates with it, but is fixed in it. Such a process is termed rabatting. *Ibid.*, If..we rabat the figure 2 upon the plane a_1 we obtain a new figure 3 in the plane a_1. *Ibid.* 10 Locus of vertex of projection during rabatment. **1931** A. H. Jameson *Contour Geom.* i. 13 A..way to find the true angle between two lines or the true shape of any figure drawn on a plane, is to 'rebate' the plane in which the two lines or the figure lies, i.e. to rotate the plane into a horizontal position about one of the contour lines as axis. Draw a vertical section across the plane at right angles to the contours and draw circular arcs.., giving the rebated contours. *Ibid.* (*heading*) Rebatement. **1949** N. L. Reece *Newnes Building Geom.* 206 A rebattement is the name given to the construction when a figure lying in an inclined plane is turned or 'rebatted' so as to lie in the horizontal plane. **1950** G. A. Hanby *Geometry* viii. 92 Find the true shape of the surface *abcdef* by rabatment. *Ibid.* 93 The true shape of the section is rabatted to the left to keep it clear of the plan, so with *T* as centre, rabat *b′c′* to B′C′.

rabbi, *sb.*¹ **2. b.** Delete † *Obs.* and add later examples.

1855 Macaulay *Hist. Eng.* III. xi. 98 One of the great Presbyterian Rabbies..might well doubt whether..he should be benefited by a comprehension. *a* **1894** Stevenson *Weir of Hermiston* (1896) i. 24 'I can't see it,' said the little Rabbi... 'No, I cannae see it.' **1932** *Amer. Speech* VII. 278 [New York Post Office] *Rabbi*, one who aids in the securing of a special privilege or favor. **1969** *New Yorker* 3 May 64/2, I asked him if he had done any thinking lately about..violence... 'Mitch Ginsberg has been my rabbi in that,' the Mayor answered. **1972** B. F. Conners *Don't embarrass Bureau* (1973) II. 192 You're damn lucky. You came out of the thing pretty clean. You got a rabbi down at the Bureau?

rabbi (ræ·bi, ræ·bəi), *sb.*² [Corruption of Rabat².] = Stock *sb.*¹ 44 b. Cf. *rabat² b.

1909 J. A. Nainfa *Costume of Prelates* iii. 50 Our Roman Collar..consists of two parts, a starched circle of white linen—the collar, and a piece of cloth or silk to which the collar itself is fastened by means of buttons, hooks, etc., and has been given the..name of 'rabbi'. **1948** H. J. McCloud *Clerical Dress & Insignia of R. C. Ch.* viii. 74 The rabbi is a loose breast piece of silk or woolen material. **1978** J. Carroll *Mortal Friends* IV. ii. 386 Father McShane..was carefully decked out, too, but clerically and with such fastidiousness as to seem almost the dandy. He wore his black suit and rabbi, of course, with the spotless high collar which was too tight.

rabbinic, *sb.* Add: **2.** *pl.* The study of the writings or doctrines of the rabbins.

1905 *Jewish Encycl.* XI. 93/2 In 1892 Schechter was elected reader in rabbinics [at Cambridge]. **1973** *Jewish Chron.* 2 Feb. 16/5 Dr Nicholas de Lange, lecturer in rabbinics at Cambridge University. **1976** *N.Y. Times* 30 Dec. 26 A Russian-born professor of rabbinics.

rabbit, *sb.*¹ **1. a.** For *Lepus* substitute *Oryctolagus*. Add: Also, one of several North American animals of the same family, esp. the varying hare, *Lepus americanus*. (N. Amer. examples.)

1634 W. Wood *New Englands Prospect* vi. 22 The Rabbets be much like ours in England. **1743** M. Catesby *Nat. Hist. Carolina* II. p. xxviii, The Monax. This animal is about the bigness of a wild rabbet. **1831** J. J. Audubon *Ornithol. Biogr.* I. 268 Small hares, or, as we usually call them, Rabbits, are also frequently caught. **1842** [see *Orkneyman]. **1872** R. L. Dashwood *Chiploquorgan* 88 There is a species of hare..mis-called a rabbit, which is numerous but hardly eatable. *c* **1897** 'Mark Twain' *Autobiogr.* (1924) I. 97 The sumptuous meals—well, it makes me cry to think of them. Fried chicken,..squirrels, rabbits, pheasants. **1907** *St. Nicholas* July 835/1 Sometimes rabbits and prairie-dogs scampered among the bushes. **1958** J. G. MacGregor *North-West of 16* v. 69 Many an Alberta farmer and business man is alive today because of rabbits. **1969** M. M. Firestone in Halpert & Story *Christmas Mumming in Newfoundland* 64 Many 'rabbits' (varying hares) are caught in snares.

b. *to buy the rabbit* (slang), to conclude a transaction unfavourably, to fare badly.

1825 J. Neal *Bro. Jonathan* II. xviii. 156 If that air invoice aint ready soon, thee'll buy the rabbit, I guess! **1935** A. J. Pollock *Underworld Speaks* 52/2 *He bought the rabbit*, a criminal case in court poorly handled by attorney; got the worst of it in a business deal.

c. Used (freq. *fig.* or allusively) with reference to the conjuring trick of producing a rabbit from a hat (cf. *hat *sb.* 5 c).

1877 E. Sachs *Sleight of Hand* xviii. 183 The production of..rabbits from a hat is always very startling. **1906** Kipling *Puck of Pook's Hill* 99 I've seen a man take rabbits out of a hat, and he told us we could see how he did it, if we watched hard. **1932** L. Golding *Magnolia Street* II. iv. 322 His wife..gave the impression of having emerged like a rabbit from the Conjurer's top-hat. **1938** M. Allingham *Fashion in Shrouds* vii. 108 There you are... Once more the veteran conjuror staggers out with the rabbit. **1940** A. Christie *Sad Cypress* II. ii. 121 You want me..to be the conjuror. To take out of the empty hat rabbit after rabbit. **1965** 'D. Shannon' *Death by Inches* (1967) xx. 245 Well, you pulled the rabbit out of the hat. **1967** *Guardian* 21 Sept. 7/1 Will man..control or stop the ever increasing flow of white rabbits..out of our technological top hats? **1975** *Times* 20 Sept. 6/5 Almost any of the Poirots of the 1930s..produce the authentic rabbit-from-the-hat shock that is the whole aim of their sort of book.

d. *pl.* Also *white rabbits*. Repeated as a good-luck charm, esp. on the first day of a month (see quots.).

1920 'D. Yates' *Courts of Idleness* II. ii. 195 On the first day of the month you have to say 'Rabbits'. If you say it to me first, I have to give you a present, and if I say it to you first, you have to give me a present. **1949** H. Nicolson *Diary* 31 Dec. (1968) 178, I hear the clock strike midnight and say 'rabbits'... That is the end of 1949. **1959** I. & P. Opie *Lore & Lang. Schoolch.* xiii. 299 'On the first morning of the month,' notes a typical informant, 'before speaking to anyone else, one must say 'White rabbits, white rabbits, white rabbits' for luck.' Subject to minor modifications the utterance of this spell appears to be the accepted routine throughout Britain. Some children feel it is enough just to cry 'Rabbits'. **1972** Evans & Thomson *Leaping Hare* xvi. 233 A Claydon (Suffolk) woman told us she used to say *Hares, Hares* before going to bed on the last day of the month, and *Rabbits, Rabbits*, when she got up in the morning. **1977** *Times* 20 Aug. 12/2, I took the opportunity of asking them what was the magic word which they..should have said, first thing that morning... They replied 'Rabbits' except for a few..who said 'White Rabbits'.

2. a. (Later examples.) *spec.* (slang) a poor performer at any game; a novice; also *attrib.*

1904 *Daily Mail* 29 Jun. 4/6 Terms now used in describing the game of cricket... 'Googlies, rabbits'. **1906** *Westm. Gaz.* 8 May 1/3 Nearly every eleven has a 'rabbit' or two at the end. **1908** A. W. Myers *Compl. Lawn Tennis Player* 184 There was no draw at all, the manager..merely selecting the four semi-finalists and filling in the gaps with the other players, most of them 'rabbits'. **1924** *Punch* 4 June 620 (*caption*) *Nervous beginner* (*to caddy*). 'I-er-suppose you get an occasional 'rabbit' here?' **1927** *Observer* 17 Apr. 17/2 Fencing is no more considered to be a feeble pastime for 'Rabbits', for those boys who cannot play the more vigorous games of youth. **1930** A. E. M. Foster *Contract Bridge for All* 7 Many people of the 'rabbit' class, and even average Auction players, are deterred from giving it a trial at all. **1932** A. J. Worrall *Eng. Idioms* 12 That girl is a rabbit. She's afraid to say 'Boo' to a goose. **1932** *Sun* (Baltimore) 24 Sept. 20/2 He found the appearance of the young women improved by abandonment of 'rabbit rules'... Freshman girls were required..to refrain from using cosmetics... But now the 'rabbits' can appear in all their glory of pink lips and powdered noses. **1947** *People* 22 June 7/4 Engines roar and the four 'rabbits' get away as best they can, but definitely not in the style of champions. **1952** E. O'Neill *Moon for Misbegotten* I. 10 I'll bet you're a mile away by now, you rabbit! **1957** *Sun* (Baltimore) 9 Apr. 20-B/3 'I think we owe it to the sponsors,' Ford explained after his remarkable closing 66... 'They help us when we're rabbits (a term for novices) and we should help them when we become winners.' **1974** J. I. M. Stewart *Gaudy* v. 95, I must have been accustomed to think of wee Dreichie as what we called a rabbit, meaning a timid boy wholly without aptitude either for games or for ragging around. **1976** *0–10 Cricket Scene* (Austral.) 41/2 Dennis Amiss..will have a special desire to prove to the Australian public that he is no longer Dennis Lillee's 'rabbit'. *a* **1976** A. Christie *Autobiogr.* (1977) VI. iv. 320 He could get no fun out of playing [golf] with a rabbit like me. **1979** *Daily Tel.* 14 Apr. 13/2 In '*Rabbits Review*' B. P. Floyd aims..to cater with a light touch for the poorer player.

d. Also in *gen.* use of a horse. *slang.*

1900 F. P. Dunne *Mr. Dooley's Philos.* 170 'Well,' says th' horse rayporther, 'they's a couple iv rabbits goin' to sprint around th' thrack at th' fair groun's,' he says. **1940** W. Faulkner *Hamlet* ii. 47 He lifted his own reins. 'Come up, rabbits,' he said. 'Let's hit for town.'

e. Liquor; a bottle of beer. *to run the rabbit*: see quots. 1916, 1941. *Austral. slang.*

1895 E. Gibb *Thrilling Incidents Convict System in Australasia* 46 'Ikeing the rabbit for a fake for his Bingy, and making a coil of a conkey myrnionger'... Convict slang ..it may be freely translated as having surreptitiously concealed some liquor under the excuse that one was ill and it was required for medicine, and ('making a coil') complaining loudly of some fancied grievance on the part of a ('conkey myrnionger') contemptible or ignorant newly-arrived convict. **1916** C. J. Dennis *Songs of Sentimental Bloke* Gloss. 124 *Rabbit, to run the*, to convey liquor from a public-house. **1941** Baker *Dict. Austral. Slang* 58 *Rabbit*, a bottle of beer. *Ibid.* 62 *Run the rabbit*, to secure liquor, often illegally, e.g. after hours.

f. A smuggled or stolen article (see also quot. 1945). *Naut.* and *Austral. slang.*

1929 F. Bowen *Sea Slang* 109 *Rabbit*, property stolen from the Royal Dockyards, most frequently used in Devonport. **1945** Baker *Austral. Lang.* viii. 163 *Rabbit*, an article made by a sailor at sea as a gift to a friend or girl. As verb, to scrounge. **1955** G. Freeman *Liberty Man* I. i. 11 All at once he remembered his presents for them. 'Rabbits' they called them in the navy. **1958** *Times* 10 Feb. 11/6 'Making rabbits' is a collective term for seamen's 'hobbies'. **1961** F. H. Burgess *Dict. Sailing* 167 *Rabbit*, an article unlawfully obtained and smuggled ashore.

g. [Shortening of *rabbit-and-pork.] A conversation, a talk. Also, a lingo. *slang.*

1941 [see *rabbit-and-pork* below]. **1950** P. Tempest *Lag's Lexicon* 173 To have a 'rabbit' = to have a pow-wow. **1958** F. Norman *Bang to Rights* III. 155 We still had quite a heated rabbit about it. **1962** R. Cook *Crust on its Uppers* i. 20 Moody rabbits in Spanish bars. **1976** E. Ward *Hanged Man* xxvii. 171 Touchy old place, Glasgow...you can't understand that Scotch rabbit they talk.

h. = *pig *sb.*¹ 8*.

1949 *Amer. Speech* XXIV. 33 The piece of steel or iron dropped or pushed through racked pipe to remove obstructions is known as a rabbit. **1975** G. Anderson *Coring* v. 95 The core is not completely out of the barrel until a metal slug, called 'the rabbit', appears.

i. A pneumatically or hydraulically propelled container used to convey material into a nuclear reactor or other place where it is to be irradiated.

1950 S. Glasstone *Sourcebk. Atomic Energy* xiii. 356/2 The study of the delayed-neutron emitters of short life is facilitated by the use of a device referred to colloquially as a 'rabbit'; by this means a sample of fissionable material, after exposure to a high density of neutrons,..is rapidly transferred to a counter where the emission of the delayed neutrons is registered automatically. **1954** R. Stephenson *Introd. Nuclear Engin.* x. 367 The rabbit (holder for the material to be irradiated) is rapidly shot into and out of the reactor by about 100 lb pressure of filtered air or carbon dioxide. **1967** J. G. Wills *Nuclear Power Plant Technol.* 318 'Rabbits' often consist of small cylinders of aluminium or plastic, moved by air pressure through a long pipe.

3. a. *rabbit-blood, farm, fence, fur, -hutch* (earlier and later *lit.* and *fig.* examples), *netting, snare, soup, trap, -warren* (earlier and later *lit.* and *fig.* examples), *wire*.

1923 D. H. Lawrence *Birds, Beasts & Flowers* 207 Eagle of the Rockies..Lifting the rabbit-blood of the myriads up into something splendid. **1964** M. Hynes *Med. Bacteriol.* (ed. 8) xxviii. 438 Culture is effected on Novy and McNeal's medium or in the water of condensation at the bottom of a rabbit-blood agar slope. **1900** J. K. Jerome *Three Men on Bummel* v. 106 A man starting a rabbit farm with twelve selected rabbits..must, at the end of three years, be in receipt of an income of two thousand a year. **1939–40** *Army & Navy Stores Catal.* 982/1 1¼″ Mesh for Rabbit fences, proof against the smallest Rabbits. **1944** F. Clune *Red Heart* 53 Colson travelled on to Birdsville, crossing the old rabbit fence (built in 1886) to keep vermin out of Queensland. **1873** *Practical Mag.* I. 282 (*heading*) Rabbit fur as a substitute for wool and cotton. **1973** 'D. Halliday' *Dolly & Starry Bird* iii. 37 A long coat of gray glacé snakeskin, edged..with lime green rabbit fur. **1743** W. Ellis *Mod. Husb.* June xviii. 141 An old Rabbit hutch, that had several Rooms in it. **1859** [see *penny plain* adj. s.v. *penny 12 c]. **1905** *Pall Mall Mag.* July 28/2 The *concierge* in his 'rabbit hutch' down below smiles and even sometimes whistles in tune. **1965** G. McInnes *Road to Gundagai* ix. 134 Hurstbridge station was a low gravel platform with a small rabbit hutch of a booking office. **1977** P. G. Winslow *Witch Hill Murder* II. 135 The collection of rabbit hutches that was now the Manor. **1907** *Yesterday's Shopping* (1969) 664/2 Rabbit and Hare Nettings..4 ft. wide, for half grown rabbits. **1915** Kipling *Let.* 22 Aug. in C. Carrington *Rudyard Kipling* (1955) xvii. 436 Don't forget the beauty of rabbit netting overhead against hand-grenades. **1973** *Country Life* 1 Mar. 511/1 The only solution is a deer fence..using large mesh, light-gauge wire above the ordinary rabbit netting. **1907** *Yesterday's Shopping* (1969) 664/3 Rabbit snares..A few made up ready for use generally in stock. **1978** P. O'Donnell *Dragon's Claw* viii. 159 There was a noose of wire round his neck which..tightened like a rabbit snare. **1845** E. Acton *Mod. Cookery* I. 37 (*heading*) Rabbit soup à la reine. *Ibid.* 38 (*heading*) Brown rabbit soup. **1960** *Good Housek. Cookery Bk.* (rev. ed.) 65/2 (*heading*) Rabbit soup. **1824** *Cobbett's Weekly Register* 27 Mar. col. 797 It is the invariable practice of the farmers to have a number of rabbit-traps constantly set on the farms. **1856** C. Patmore *Angel in Ho.* II. Prol. 3 But she turn'd pale, for now the beast Found stock-still in the rabbit-trap,..Unglobed himself. **1976** J. B. Hilton *Gamekeeper's Gallows* vii. 61 Brunt ..came upon a rabbit trap. **1766** Smollett *Trav.* I. 32 Open downs, where there is a rabbit warren. **1821** M. Wilmot *Let.* 10 Apr. (1935) 104 This Rabbit Warren in the air. **1905** *Birmingham Institute Mag.* Oct. 187 The Institute has been graphically..described as a rabbit-warren. **1973** 'S. Harvester' *Corner of Playground* III. vii. 219 Their dull dusty sunless offices in the chaotic rabbit-warren of officialdom. **1906** Kipling *Puck of Pook's Hill* 259 A quick movement of the hand as though he were pegging down a rabbit-wire. **1945** N. Collins *London belongs to Me* III. xxxix. 301 The special visitors' room where they saw their loved ones through a screen of rabbit-wire. **1973** 'M. Campbell' *Halfbreed* iv. 30 We took Daddy's rabbit wire and strung it across two small green trees on either side of the footpath.

b. *rabbit-catcher* (earlier example; also *transf.*), *-courser, farmer, -farming, -hunting, -inspector, -shooting* (later example), *-trapping*; *rabbit-hunt* vb.

1724 Swift *Wks.* (1941) X. 146 Rabbet Catcher. I'll Ferret him. **1955** *Sun* (Baltimore) 23 Feb. 5-B/2 Engineers ..have stretched a nylon tennis net across a miniature landing field and are 'serving' Model F86 Sabre Jets into it. The purpose of the experiment is to test a new device, called a 'rabbit catcher'. **1875** Rabbit-courser [see *dog-racer* s.v. *dog *sb.* 17 a]. **1900** J. K. Jerome *Three Men on Bummel* v. 106, I have never met a rabbit farmer myself worth two thousand a year. **1889** G. F. Morant (*title*) Hutch rabbit-farming in the open [in K. W. Knight *The book of the rabbit*]. **1943** J. Stuart *Taps for Private Tussie* xvii. 172 Uncle Mott cut wood for the fireplace in the mornins and rabbit-hunted in the afternoons. **1873** *Trans. Illinois Dept. Agric.* X. 65 They prevailed on him to suspend his rabbit-hunting, and 'show them 'round'. **1953** N. Tinbergen *Herring Gull's World* iv. 27 Rabbit-hunting is a regular feature in the Dutch North Sea sand dunes. **1936** F. Clune *Roaming round Darling* xiii. 114 They were collected by the rabbit-inspector a number of years ago. **1978** H. Wouk *War & Remembrance* xxxix. 400 The slaughter that ensued was mere rabbit-shooting by our aircraft and submarines. **1880** W. Carnegie *Practical Trapping* 20 The same sort of gins, the use of which I advocated for rabbit-trapping, will do. **1935** *Discovery* June 168/1 Gamekeepers in winter in Ross-shire often come across wild-cat tracks in the snow when rabbit-trapping.

c. *rabbit-coloured, -eared* (later examples), *-faced, -hearted, -mouthed, -scared, -toothed* adjs.

1953 R. Graves *Poems* 18 Such gross-headed, rabbit-coloured litters As soon they shall be happy to desert. **1939** *Times Lit. Suppl.* 30 Dec. 756/1 Once there was a hasty glimpse of the rabbit-eared bandicoot. **1977** *Jersey Even. Post* 26 July 4/1 And the rabbit-eared bandicoots, small marsupials, only come out to feed at night. **1905** E. F. Benson *Image in Sand* i. 12 He was a rabbit-faced little man. **1920** *Chambers's Jrnl.* Dec. 812/1 'Rabbit-hearted' is an expression commonly used not only by white races, but also by red and brown people. **1909** *Cent. Dict.* Suppl., Rabbit-mouthed,..having a rabbit-mouth; harelipped. **1956** H. Gold *Man who was not with It* (1965) xxix. 272 His soft and rabbit-mouthed touchiness. **1936** *Partisan Rev.* III. 21/1 Standing there, his big gun smoking, Rabbit-scared, alone. **1971** *Maclean's Mag.* Oct. 3/2 We've never seemed so rabbit-scared as a nation as we did in August. **1800** D. Wordsworth *Jrnl.* 14 May (1941) I. 37 The grassy-leaved rabbit-toothed white flower. **1963** D. Lessing *Golden Notebk.* II. 246 His mouth is a rabbit-toothed hole, and his eyes are sunk in scar tissue. **1980** 'T. Hinde' *Sir Henry & Sons* iv. 33, I am round-faced and stocky... I became rabbit-toothed.

d. instrumental, as *rabbit-browsed, -haunted, -nibbled* adjs.

1923 Kipling *Land & Sea Tales* 81 A little rabbit-browsed clearing of turf. **1921** F. B. Young *Black Diamond* iv. 38 Evening visits to rabbit-haunted banks. **1947** W. de la Mare *Coll. Stories for Children* 30 The rabbit-nibbled turf.

4. rabbit-and-pork *Rhyming slang* = Talk *sb.*, *v.* (usu. *ellipt.*: see sense 2 g above and *rabbit *v.* 6); **rabbit ball** *U.S.*, a baseball that is springy in construction and lively in action; also *fig.*; **rabbit-bandicoot,** substitute for def.: a small Australian marsupial belonging to the genus *Macrotis* of the family Peramelidæ, living in a burrow and having rabbit-like ears; cf. *rabbit-eared* in sense 3 c above; (later examples); **rabbit-beagle,** a beagle used for the hunting of rabbits; so **rabbit-beagling** *vbl. sb.*; **rabbit berry,** substitute for def.: a deciduous North American shrub, *Shepherdia argentea*, of the family Elæagnaceæ, or its red berries; (earlier and later examples); **rabbit brush, bush,** substitute for def.: a western North American shrub of the genus *Chrysothamnus*, esp. *C. nauseosus*, belonging to the family Compositæ and bearing clusters of yellow flowers; (earlier and later examples); **rabbit drive** *U.S.*, a driving together of jack rabbits for slaughtering; **rabbit fever,** a vernacular name for tularæmia; **rabbit-fish,** (*a*) (earlier and later examples); (*b*) substitute for def.: *U.S.*, the smooth puffer, *Lagocephalus lævigatus*, or the spotted balloon fish, *Cyclichthys schoepfi*; (earlier and later examples); (*c*) a small herbivorous fish belonging to the family Siganidæ, found in tropical Indo-Pacific seas and bearing venom glands on the fins; **rabbit flea,** one of several fleas which infest rabbits, esp. the European rabbit flea, *Spilopsyllus cuniculi*, or the North American rabbit flea, *Cediopsylla simplex*; **rabbit food** (also **rabbit's food**), food such as is eaten by rabbits; hence (*slang*), lettuce; green salad; **rabbit-foot:** see also *rabbit's foot; **rabbit-foot (clover)** (earlier and later examples); **rabbitfoot grass,** an annual grass, *Polypogon monspeliensis*, distinguished by soft hairs on its flower-head, native to Europe and widely naturalized elsewhere; **rabbit-hawk** *U.S.*, the red-tailed hawk, *Buteo jamaicensis*, or the hen harrier, *Circus cyaneus*; **rabbit('s) pea** *U.S.*, a perennial herb, *Tephrosia virginiana*, belonging to the family Leguminosæ and bearing

white, pink, or yellow flowers; also called goat's rue and wild sweet pea; **rabbit-proof** *a.*, proof against rabbits; esp. of a fence, that excludes rabbits (in Austral. *spec.* such a fence marking a border between States); *ellipt.*, such a fence; **rabbit-rat,** substitute for def.: an Australian rodent belonging to either of the genera *Mesembriomys* and *Conilurus*, distinguished by long ears and a bushy tail, esp. the white-footed tree rat, *C. albipes*, the only one not restricted to northern parts of the country; (examples); **rabbit test,** a pregnancy test in which rabbits are used; **rabbit tobacco** *U.S.*, the sweet everlasting, *Gnaphalium obtusifolium*, belonging to the family Compositæ, and bearing clusters of fragrant white flowers; also, the dried flowers of this plant, used as a substitute for tobacco; **rabbit tooth** *slang* = Buck-tooth (usu. *pl.*).

1941 G. Kersh *They die with their Boots Clean* I. 27 He uses slang... Talk is Rabbit, or Rabbit-an'-Pork. **1960** *Spectator* 4 Mar. 326 We only allow ourselves a second to remember that rabbit-and-pork is talk. **1971** *National Times* (Austral.) 13 Dec. 20/1 (*heading*) Cockneys lay claim to their rabbit and pork. **1922** *N.Y. Times* 5 June 10/3 The officials who control the destiny of the big leagues let it be known at the opening of the current season that the 'rabbit' ball had seen its day. **1937** *Sun* (Baltimore) 18 Aug. 8/3 There does not seem to be any question of the changes that have been worked in baseball by the lively rabbit ball. **1973** *Times* 15 Aug. 7/3 The rabbit-ball..jumps like a rabbit. **1977** *Time* 11 Apr. 17/1 Rawlings Co. now makes the official major league baseball.., and the scuttlebutt is that Rawlings is turning out a rabbit ball. **1923** F. W. Jones *Mammals S. Austral.* I. 154 The animal..is usually termed the Common Rabbit-Bandicoot, but it would be most misleading to apply the term 'common' to it to-day. **1941** E. Troughton *Furred Animals Austral.* 69 The very descriptive name of rabbit-bandicoot was provided by the early colonists who regarded them with a certain amount of tolerance because of their extremely useful share in the destruction of mice and insects. **1970** W. D. L. Ride *Guide Native Mammals Austral.* vii. 104 There are generally thought to be only two species, the Common Rabbit Bandicoot and the small Central Australian Yallara or Lesser Rabbit Bandicoot. **1824** *Sporting Mag.* XIV. 312/2 There is no prettier sport for youth than rabbit beagling... Rabbit beagles should never be permitted to run hare. **1888** H. Dalziel *Brit. Dogs* (ed. 2) I. xvi. 226 Beagles may be fairly classed as Hare-Beagles and Rabbit-Beagles, other distinction than size being minor. **1804** J. Whitehouse *Jrnl.* 24 Aug. in *Orig. Jrnls. Lewis & Clark Exped.* (1905) VII. ii. 52 We found some red berreys which they call Rabbit berrys. **1891** [see *mountain tea* s.v. *mountain 9d]. **1952** A. G. L. Hellyer *Sanders' Encycl. Gardening* (ed. 22) 454 *S*[*heperdia*] *argentea*, 'Rabbit Berry', 'Buffalo Berry',..scarlet fruits. **1914** E. Stewart *Lett. Woman Homesteader* 18 Our horse was midside deep in rabbit brush, a shrub just covered with flowers that look and smell like goldenrod. **1927** W. Cather *Death comes for Archbishop* 95 The sandy soil of the plain.. was splotched with masses of blooming rabbit brush. **1946** D. C. Peattie *Road of Naturalist* i. 17 The burro bush and rabbit brush are the natural sons of the desert. **1955** *Daily Progress* (Charlottesville, Va.) 2 Nov. 11/5 That's rabbit-brush, a hearty range perennial that moves in quickly when other vegetation is killed. **1980** *Blair & Ketchum's Country Jrnl.* Oct. 46/3 The first mule I ever owned was rescued from a prairie dog town in southern Colorado, where she had lived for who knows how long on fresh air and rabbit-brush. **1852** H. Stansbury *Exbl. & Survey Valley of Gt. Salt Lake* 235 The only vegetation today has been a little dwarf artemisia, grease-bush, rabbit-bush. **1972** R. & R. Wright *Cariboo Mileposts* 16 Grass and trees are scarce, with only sage and rabbit bush covering the ground. **1979** *Tucson* (Arizona) *Citizen* 28 Apr. 3A/9 Pollen count (yesterday)..Rabbit bush 4. **1887** *Lisbon* (N. Dakota) *Star* 23 Dec. 7/1 Several hundred people..assembled to engage in the rabbit drive. **1963** R. Symons *Many Trails* xii. 119 A hunt in the manner of the California rabbit drives. **1977** *New Yorker* 11 July 43/3 We have rabbit drives there. Drive the rabbits from one end of the island to the other and kill them. **1925** *Jrnl. Amer. Med. Assoc.* 25 Apr. 1244/1 A man..working in the Washington, D.C., market went to his physician..in 1921, for treatment for what he informed the physician was 'rabbit fever', adding that 'rabbit fever' was well known among market men... This was the first case of tularemia to be reported for the eastern United States. **1955** *Sci. News Let.* 16 July 43/1 Some species of ticks are occasional spreaders of Rocky Mountain spotted fever and tularemia (rabbit fever). **1973** *Daily Colonist* (Victoria, B.C.) 29 July 2/2 Then he mentioned that he had been doing some squirrel hunting and the doctor immediately ordered a blood test which showed he had tularemia, or 'rabbit fever'. **1828** J. Fleming *Hist. Brit. Animals* 173 The specimen [of *Chimæra*], from which the preceding description was taken, was sent from Unst, where it is termed the Rabbit-fish. **1842** J. E. DeKay *Zool. N.Y.* IV. 330 The Lineated Puffer..is called Rabbit-fish, according to Schoepfi, on account of the whiteness of its flesh. **1884** *Bull. U.S. Nat. Museum* No. 27. 428 Spiny Box-fish; Rabbit-fish; Swell Toad. East coast of the United States. **1897** *Rep. N.Y. Forest, Fish, & Game Comm.* II. 224 Rabbit-fish; Smooth Puffer.—Occasionally taken in the fall in Gravesend Bay. **1905** D. S. Jordan *Guide to Study of Fishes* II. xxiv. 423 In the [American] rabbit-fishes..the body is box-shaped. **1925** J. T. Jenkins *Fishes Brit. Isles* 349 From its great cutting teeth it [*sc.* *Chimæra monstrosa*] is known to the Shetlanders as the Rabbit-fish. **1941** R. Faherty *Big Old Sun* xi. 255 'I'm a blow-puffing rabbitfish,' he sputtered. **1953** J. L. B. Smith *Sea Fishes S. Afr.* 328 Rabbitfishes. Compressed ovate body with slippery skin and minute concealed scales. ..Curious small herbivorous fishes of reefs and weeds of the tropical Indo-Pacific. **1962** K. F. Lagler et al. *Ichthyol.* iv. 32 Non-fatal but nevertheless painful to man..are stings of the venomous sharks,..rabbitfishes, and dragonets. **1973** *Aquaculture* I. 361 Rabbitfish are widely spread throughout the Indian and the western and central Pacific oceans. **1975** *Times* 5 Dec. 12/3 The deep-sea species..can be portrayed as menacing horrors. The specific name for one of them, rabbitfish, is Chimaera Monstrosa. **1904** *Proc. U.S. Nat. Museum* XXVII. 368 In the United States the cat, dog, and rabbit fleas..will readily attack the human being. **1925** A. D. Imms *Gen. Textbk. Ent.* III. 663 The rabbit flea.. commonly affects the ears of hares and rabbits. **1963** O. Breland *Animal Life & Lore* i. 26 In some regions, it [*sc.* myxomatosis] is transmitted by rabbit fleas. **1967** J. M. Brownjohn tr. *Grzimek's Four-Legged Australians* xii. 250 For some years, European rabbit-fleas refused to propagate themselves in Australian research centres. **1975** *Times* 23 June 4/2 The rabbit flea..spreads myxomatosis. **1907** *Yesterday's Shopping* (1969) 58/1 Rabbit Food..Pigeon Food..Foal Food. **1936** *Amer. Speech* XI. 44/2 *Rabbit's food*, lettuce. **1941** J. Smiley *Hash House Lingo* 45 *Rabbit food*, lettuce. **1959** I. & P. Opie *Lore & Lang. Schoolch.* ix. 163 Tomatoes are generally 'squashers', and 'rabbit's food' is any green salad. **1972** A. Price *Colonel Butler's Wolf* ix. 98 You can both come back with me and eat pounds of rabbit food. **1817** A. Eaton *Man. Bot.* 84 *Trifolium..arvense* (field clover, or rabbit-foot) heads cylindric. **1850** S. F. Cooper *Rural Hours* 125 The downy 'rabbit-foot', or 'stone-clover', the common red variety..[is] introduced. **1889** *Cent. Dict.* s.v. Clover, Other species, mostly weeds of little value, are the yellow or hop clover,..the stone, hare's-foot, or rabbit-foot clover. **1954** *Amer. Speech* XXIX. 15 *Trifolium arvense*, a cosmopolitan woolly-headed clover,..is known in America both as the rabbitfoot (clover) and hare's-foot (clover)... Rabbitfoot clover is the commonest designation in the United States, whereas hare's-foot is the usual British term. **1935** A. S. Hitchcock *Man. Grasses U.S.* 350 *Polypogon monspeliensis*..Rabbitfoot Grass. **1954** C. E. Hubbard *Grasses* 285 Annual Beard-grass. *Polypogon monspeliensis*... Known in N. America as 'Rabbitfoot Grass'. **1968** F. W. Gould *Grass Systematics* v. 172 Most common and widespread is..rabbitfoot grass, present in moist areas along streams. **1851** *De Bow's Rev.* XI. 54 1st, Rabbit hawk. **1880** G. W. Cable *Grandissimes* vii. 43 A great rabbit-hawk sat alone in the top of a lofty pecan-tree. **1904** 'O. Henry' *Heart of West* 64 The other eye noticed a rabbit-hawk sitting on a dead limb in a water-elm. **1964** *Publ. Amer. Dial. Soc.* XLII. 22 Rabbit-hawk. The marsh hawk (*Circus cyaneus*), so called because of its flying low over the pastures in search of rodents. **1938** M. K. Rawlings *Yearling* xvii. 199 A rabbit-pea vine was in blossom. **1976** *Hortus Third* (L. H. Bailey Hortorium) 1101/1 *Tephrosia..virginiana*..Goat's rue, catgut, rabbit's pea. **1832** *Useful & Ornamental* in *Brit. Husbandry* (1840) III. III. 26 The fence of a forest-tree nursery requires to be rabbit-proof. **1894** W. Robinson *Wild Garden* (ed. 4) xv. 209 Periwinkle, which is named amongst rabbit-proof plants, is generally eaten to the ground in severe weather. *a* **1902** H. Morant in *Penguin Bk. Austral. Ballads* (1964) 212 But once we're through the rabbit-proof..it's 'West-by-North' again. **1957** R. Stow *Bystander* 29 That's the coldest little bitch this side of the rabbit-proof fence. **1961** *Times* 19 Apr. 14/7 Her rabbit-proof fence is not high enough. **1976** D. Hewett *Bonbons & Roses for Dolly* 28 Best little ticket takers this side of the rabbit-proof. **1837** G. Bennett *Catal. Specimens Nat. Hist. Austral. Mus.* 6 The Rabbit Rat of the Colonists. Habitat, Interior of Australia. **1863** J. Gould *Mammals Austral.* III. 1 White-footed Hapalotis... The Rabbit Rat of the Colonists. **1879** A. R. Wallace *Australasia* iii. 55 Bandicoots and rabbit-rats, are small animals with sharp nose and long claws, allied to the kangaroos. **1941** E. Troughton *Furred Animals Austral.* 305 The various species..are sometimes called 'rabbit-rats' in reference to the rather large ears. **1970** W. D. L. Ride *Guide Native Mammals Austral.* ix. 142 Little is known of the White-footed Tree-rat of eastern Australia; early settlers called this the Rabbit Rat because of its rounded form and long ears. It has not been seen alive in this century. **1949** S. T. De Lee *Safeguarding Motherhood* 133 (Index), Rabbit test. **1958** H. Speert *Obstetr. & Gynecol. Milestones* xxviii. 244 The urine of pregnant women contains a gonadotropic substance simulating the secretion of the anterior pituitary in its effect on the mouse ovary. Applying this observation to the rabbit, Friedman proceeded to develop the pregnancy test known by his name, popularly as the 'rabbit test'. **1977** E. Leonard *Unknown Man No. 89* i. 7 The guy was a gynecologist. So he went in with Rita for her rabbit test, the concerned hubby. **1880** J. C. Harris *Uncle Remus* xiii. 66 'Den he drawd de rockin'-cheer in front er de fier, he did, en tuck a big chaw terbarker.' 'Tobacco, Uncle Remus?' asked the little boy, incredulously. 'Rabbit terbarker, honey.' **1909** 'O. Henry' *Options* 200, I don't give a pipeful of rabbit tobacco whether Queen Sophia Christina or Charlie Culberson rules these fairy isles. **1936** M. Mitchell *Gone with Wind* xxix. 488 'You all got any chewing tobacco, Scarlett?' 'Nothing but rabbit tobacco. Pa smokes it in a corn cob.' **1937** *Amer. Speech* XII. 235/1 On all the poor land in the middle and far West there is a weed known as..rabbit tobacco. **1964** *Publ. Amer. Dial. Soc.* XLII. 22 Rabbit tobacco. Life everlasting.., used in many folk remedies for catarrh; also chewed and smoked by boys. **1800** Rabbit tooth [implied in *rabbit-toothed* above]. **1915** W. Owen *Let.* 4 Apr. (1967) 330 Will he be 13 or 14 next birthday?.. Are his rabbit-teeth humanising? **1980** E. Leather *Duveen Let.* viii. 98 He was tall, thin, with large rabbit teeth.

rabbit, *v.*¹ Add: **3.** *intr.* To go; to move quickly; to run away. *dial.* and *colloq.*

1887 *Rep. & Trans. Devonshire Assoc.* XIX. 77 Miss—du rabbut 'bout en awl wethurz. **1937** D. Jones *In Parenthesis* IV. 71 You can't find the lane—the one way—you rabbit to and fro. **1960** Wentworth & Flexner *Dict. Amer. Slang* 416/1 *Rabbit*..*v.i.*, to move quickly, to run; specif., to flee, to escape. **1962** J. F. Straker *Coil of Rope* iii. 22 Susan.. kept skipping from one side of David to the other... David said irritably, 'For heaven's sake stop rabbiting around!' **1972** J. Wambaugh *Blue Knight* (1973) ii. 33, I noticed another junkie watching me. He was trying to decide whether to rabbit or freeze.

4. To copulate. *rare.*

1919 J. Masefield *Reynard the Fox* 16 I'll learn 'ee rabbit in my shed.

5. *trans.* To borrow or steal. *Austral. Naut. slang.* Cf. *rabbit *sb.*¹ 2 f.

1943 Baker *Dict. Austral. Slang* (ed. 3) 63 *To rabbit*, to borrow, 'scrounge'. (R.A.N. slang.) **1953** K. Tennant *Joyful Condemned* xxi. 198 Why were Australian Navy men better at 'rabbiting' little valuable articles than Americans?

6. [See *rabbit-and-pork* s.v. *rabbit *sb.*¹ 4.] *intr.* To talk, to discourse volubly; to gabble. Freq. const. *on*. *colloq.*

1950 P. Tempest *Lag's Lexicon* 173 One who 'rabbits' all the time is one who never stops talking. **1959** *Encounter* Mar. 63/1 The next thing I knew, I was rabbiting away to a geezer. **1960** *News Chron.* 16 Feb. 6/6 She don't want to stand rabbiting away about colourful denizens. **1963** 'A. Garve' *Sea Monks* iii. 108 Then stop rabbitin' an' get that wall cleaned. **1967** A. Diment *Dolly Dolly Spy* iv. 39, I let him rabbit on about the twilight hours of the Third Reich. **1976** J. Bingham *God's Defector* vii. 99 You go into a pub with a short-back-and-sides and people stop rabbiting and stare at you. **1977** *Guardian Weekly* 9 Oct. 20/3 A girl reporter from Rolling Stone rabbits on idiotically about the Maharishi.

rabbitish, *a.* Delete *rare* and add later examples.

1941 E. R. Eddison *Fish Dinner* xiii. 225 Young man a bit rabbitish by the look of him: doesn't seem to know quite what to do. **1978** J. Symons *Blackheath Poisonings* I. 24 Bertie, who was fair and pale, with pinkish eyes, had a slightly rabbitish appearance.

ra·bbit-o. *Austral. slang.* Also **Rabbit-O, rabbit-oh.** [f. Rabbit *sb.*¹ + *-o².] An itinerant seller of rabbits as food. Also *attrib.*

[**1908** T. E. Spencer *Budgeree Ballads* 84 When I ought to think of bizness, I can only think of you, And instead of 'Rabbit-oh!' I sings out 'Liza!'] **1911** A. Wright *Gambler's Gold* 75 Engaged in the hopeless task of trying to win the Rabbit-O man's money. **1945** K. Tennant *Time enough Later* x. 150 Mrs. Drew knew all about her neighbours from the butcher and the grocer and the rabbit-o. **1975** *Sydney Morning Herald* 9 Apr. 1 Now 65, he is probably the last rabbit-oh in Sydney.

ra·bbit punch. Also **rabbit's punch.** [f. Rabbit *sb.*¹ + Punch *sb.*²] A sharp, chopping blow to the back of the neck delivered with the side of the hand. Also *fig.*

1915 E. Corri *30 Years Boxing Referee* 175 The occipital punch is well described by its other name, the 'rabbit-punch', derived from the way in which a gamekeeper puts a rabbit out of pain. **1936** *Daily Tel.* 30 Jan. 20/7 The only incident to which any exception could be taken was Petersen's habit of using the 'rabbit punch'—a cuff downwards on the back of the neck. **1958** *Daily Mail* 7 June 5/7 Simpson admitted to the police that he had killed the dog. He said he..gave it a 'rabbit punch'. **1959** I. & P. Opie *Lore & Lang. Schoolch.* x. 202 A 'Rabbit's Punch' is delivered by pulling a child's head forward, usually by his hair, and slicing the back of his neck with the side of the hand. **1968** A. Diment *Bang Bang Birds* iii. 27 Another gust of warm wind caught me in a rabbit punch. **1973** G. Sims *Hunters Point* xiii. 123 As he toppled forward Buchanan delivered a murderous rabbit punch.

So **ra·bbit-punch** *v. trans.*; **ra·bbit-punching** *ppl. a.* and *vbl. sb.*

1936 R. Chandler in *Black Mask* Jan. 18/2 Somebody rabbit-punched him from behind. **1940** Dylan Thomas *Portrait of Artist as Young Dog* 72 The strange boy rabbit-punched me twice. **1968** T. Stoppard *Real Inspector Hound* (1970) 11, I dream of champions chopped down by rabbit-punching sparring partners. **1971** *Times* 16 Feb. 8/5 A controversial disqualification for rabbit punching. **1972** T. Coe *Don't lie to Me* (1974) xx. 161 He rabbit-punched me in the back of the neck.

rabbitry. Add: **1.** Also, rabbit-breeding.

1968 *Punch* 12 June 858/3 Anyone with distressful childhood memories of pet does unconcernedly making away with their blind and naked young may like to know that commercial rabbitry has just about eradicated such uneconomic vice.

2. *slang.* In sport, poor performers (*collect.*). Also, poor play or performance in any game. Cf. *rabbit *sb.*¹ 2 a.

1930 *Observer* 25 May 16 The rabbitry..is faithful to the definition of the game [*sc.* golf] as 'pedestrianism round the margin of the links'. **1932** A. Marshall *Mrs. Jim* v. 44 Her two younger girls..were made welcome at these [tennis] parties, so that Mrs. Gurdon's rabbitry should not be put to shame.

rabbit's ear. Also **rabbit ear.** [f. Rabbit *sb.*¹ + Ear *sb.*¹] **1.** A perennial herb, *Stachys olympica* (formerly *S. lanata*), belonging to the family Labiatæ, distinguished by greyish foliage and dense white tomentum covering the whole plant; usually called lamb's ears.

1928 V. Woolf *Let.* 15 Oct. (1977) III. 545 Might I beg some Saviour's flannel or rabbit ear? *Ibid.* 22 Oct. 547 A thousand thanks for the rabbits ears. **1949** H. Nicolson *Let.* 15 June (1968) 171, I think of it [*sc.* a garden] as *cineraria* in masses, Rabbit's Ears in masses,..predominantly grey.

2. An indoor television aerial consisting of a base supporting two stiff wires that form a V. *U.S.*

1967 *Boston Sunday Herald Mag.* 16 Apr. 32/2 (Advt.), Top back lifts for rabbit ears—doors fold flat against sides for televiewing. 1972 T. Kenrick *Tough One to Lose* ii. 30 There wasn't much... A TV set in the corner that needed rabbit's ears. 1978 *Tucson Mag.* Dec. 37/1, I use rabbit ears with my set and get a perfect picture.

ra·bbit's foot. [Rabbit *sb.*[1] 4.] **1.** Also **rabbit-foot.** The foot of a rabbit carried to bring luck; also *transf.* Phr. *to work the rabbit('s)-foot* (U.S.), to cheat, to trick.

1879 L. Hearn *Amer. Miscell.* (1924) I. 185 After the girl told that [ghost] story, Banjo Jim seldom passed along the row at night without a rabbit's foot in the breast pocket of his woolen shirt. 1902 W. N. Harben *Abner Daniel* 309 Pole worked the rabbit-foot on them back there. 1922 *Sunset* Dec. 10/2 Presently the word went round that I was a 'rabbit's foot'—a bringer of good luck—and the gamblers began to give me money to place for them. 1948 *Salt Lake Tribune* 17 Dec. 34/6 A dimestore rabbit's foot paid off with one of 1948's biggest football surprises and landed his team in the Delta bowl. 1972 C. Weston *Poor, Poor Ophelia* (1973) vii. 45 Don't you know a charm when you see one? This is my rabbit's foot.

2. Special combs.: **rabbit's foot (clover)** = *rabbit-foot clover* s.v. Rabbit *sb.*[1] 4 in Dict. and Suppl.; **rabbit's foot (fern)** *U.S.*, an epiphytic fern, *Polypodium aureum*, native to tropical America and cultivated elsewhere as a house plant; **rabbit's foot grass** = *rabbitfoot grass* s.v. *Rabbit *sb.*[1] 4.

1878 H. M. Jackson *Bits of Travel at Home* 186 [There grows] Rabbit's foot, May-weed, shepherd's purse. 1968 Peterson & McKenny *Field Guide to Wildflowers N. Amer.* 246 Rabbit's-foot Clover... The soft silky foliage and the fuzzy, grayish-pink heads of bloom are unmistakable. 1972 G. Beine *Land of Coyote* 90 These are rattle boxes, and there, some rabbitsfoot clover. 1951 E. Graham *My Window looks down East* vii. 63 The rabbit's foot fern and all the other green things looked so radiant and fresh. 1964 F. G. Foster *Gardener's Fern Bk.* 182 Golden Polypody... Commonly called 'rabbit's-foot' fern. Actually, the 'rabbit's foot' is a colorful orange-brown or white, scaly rhizome. 1973 J. L. Faust *N.Y. Times Bk. House Plants* 107 Ferns..best for indoors are..coarse-leaved polypody..and rabbit's-foot. 1973 *Times* 20 Oct. 16/7 Upright 'rabbits' [sic] foot', cock-spur or barnyard grasses have been common.

rabbity, *a.* Add: (Further examples.) Also, suggestive or characteristic of a rabbit. Also *Comb.*, as *rabbity-faced, -looking* adjs.

a 1930 D. H. Lawrence *Last Poems* (1932) 98 There are too many people on earth Insipid, unsalted, rabbity, endlessly hopping. 1931 D. L. Sayers *Five Red Herrings* xvi. 177 The rabbity-faced fellow in the train. 1937 C. Day Lewis *Starting Point* iii. iv. 284 Darling, you're not going to turn me into a rabbity little wage-earner, are you? 1963 Auden *Dyer's Hand* 120 A leporello who, in real life, is a rabbity-looking..professor. 1965 M. Shadbolt *Among Cinders* xi. 96 I'd been starving on rabbity salads. 1974 T. P. Whitney tr. *Solzhenitsyn's Gulag Archipelago* I. ii. ii. 544 But somehow this is hard for our rabbity brains to grasp. 1976 *Gramophone* Nov. 878/3 My own family's verdict on the book [sc. *Watership Down*] was that it was 'extraordinarily rabbitty'. 1978 G. Greene *Human Factor* iii. v. 144 He thought he recognised all the faces except for one woman in a shabby rabbity fur.

rabble, *sb.*[1] (and *a.*). **2. e.** Delete entry.
The word printed 'rabble' in the 1766 ed. of *Swift's Lett.* is 'babble' in the author's MS.

rabble, *sb.*[2] Add: **4.** *Comb.*, as **rabble-arm** = sense 3.
1905 *Electrochem. & Metall. Industry* May 194/1 The upper four hearths are provided with two rabble-arms each, the sixth and seventh with three rabble-arms, in order to increase the stirring and discharge rather on one side of the furnace.

rabble, *v.*[2] Add: **2.** Also, to behave as a rabble; to go *off* as a rabble.
a 1861 T. Winthrop *John Brent* (1883) ix. 76 He dismissed his audience with an..injunction to keep closer to the train..and not be 'rabbling off to catch grasshoppers'. 1970 O. Chadwick *Victorian Church* II. vi. 355 The agitators began to disturb and rabble at services.

rabble, *v.*[3] Add: (Earlier example.) Cf. *Ravel *v.*[4]
1860 C. Tomlinson *Useful Arts & Manuf.* 2nd Ser. ii. 17 The door of the furnace is removed, and the liquid mass well *rabbled* or stirred.

rabble-rouser (ræ·b'lˌraʊzəɹ). [f. Rabble *sb.*[1] + Rouser.] **a.** One who practises rabble-rousing; a demagogue.
1843 'R. Carlton' *New Purchase* I. xxii. 211 Nothing surpasses the munificent promises..of a genuine rabble rouser, just before an election. 1926 *Even. Standard* 12 July 3/2 You need not be a Tammany politician to know the meaning of boodle, gerrymander, carpet-bagger, wirepuller, rabble-rouser. 1959 'M. M. Kaye' *House of Shade* xiv. 190 The murder of a rabble-rouser would probably be considered as of little account. 1972 *Listener* 28 Dec. 899/1 Thomas Paine..was never a rabble-rouser. 1978 J. A. Michener *Chesapeake* 543 Paul..astounded the more conservative Steeds by prevailing upon an itinerant Methodist rabble-rouser to spend five days at Devon.

b. Something, esp. music, that excites an audience.
1958 *New Statesman* 1 Feb. 133/3 The symphony ends with a noisy finale which seemed no more than an unusually prolonged rabble-rouser. 1965 G. Melly *Owning Up* iv. 38 They chose their fastest, loudest numbers, 'rabble rousers' was the trade name, in the hope of making some impact on the audience.

rabble-rousing (ræ·b'lˌraʊzɪŋ), *a.* and *sb.* [f. Rabble *sb.*[1] + Rousing *ppl. a.* and *vbl. sb.*[1]] **A.** *adj.* Tending to arouse the emotions of a rabble or disorderly crowd, esp. for political ends; demagogic, inflammatory, excitatory.
1802 [in Dict. s.v. Rabble *sb.*[1] 5]. 1933 *Sun* (Baltimore) 8 Aug. 8/2 They are as old as rabble-rousing nationalism itself. 1951 T. Sterling *House without Door* xi. 126 Mouthing the rabble-rousing opinions of Communist filth. 1969 *N.Y. Rev. Bks.* 30 Jan. 12/4 Convinced that this was a rabble-rousing exaggeration, Booth set out to compile his own statistics. 1976 *Eastern Even. News* (Norwich) 9 Dec. 2/1 Mussolini made a rabble-rousing appearance in Milan after the Allies had landed in the south.

B. *sb.* The act or process of arousing the emotions of a crowd; demagoguery, trouble-stirring.
1933 *Sun* (Baltimore) 18 Sept. 8/2 It is not necessary to pay a great deal of attention to this form of economic rabble rousing. 1962 D. H. Lawrence in G. B. Shaw *Platform & Pulpit* p. xii, He was not concerned with rabble-rousing or spell-binding. 1974 A. Ross *Bradford Business* 9 Heated rabble-rousing in our seats of learning.

Hence (as a back-formation) **ra·bble-rouse** *v. intr.*, to arouse the emotions of a crowd by a demagogical harangue.
1959 *Time* 15 June 36/3 He rabble-rouses more fluently in English than in Chinese. 1971 *Guardian* 1 July 11/6 Did he [sc. Oswald Mosley] make speeches in such a way as to rabble-rouse?

Rabelaisian, *a.* (and *sb.*) Add: **A.** *adj.* (Earlier example.)
1817 Lady Morgan *France* ii. viii. 167 He talked of recovery,..and still emitted some of those sparks of Rabelaisian humour, attributed to him by contemporary wits.

B. *sb.* (Earlier example.)
1882 Hardy *Let.* 17 May (1978) I. 106 We Rabelaisians have not as yet reached the state of enlightenment which distinguishes you Noviomagians, & do not include ladies in our company.

Rabelaisianism. Add: (Further example.) Also, a Rabelaisian feature or characteristic.
1908 A. Bennett *Old Wives' Tale* ii. ii. 166 The robust Rabelaisianism of his more private conversation. 1922 *Times Lit. Suppl.* 5 Jan. 9/1 His [sc. Balzac's] mere Rabelaisianisms sometimes bring a later accent into what should be a very simple atmosphere.

‖ **Rabfak** (ræ·bfæk). Also **rabfac.** [a. Russ. *rabfák*, f. *rab(óchiĭ) fak(ul'tét)* workers' school.] A workers' school, established after the Russian Revolution, to prepare workers and peasants for higher education. Also *attrib.*
1928 *Observer* 19 Feb. 12 Among the candidates for admission to the local 'rabfac', or workers' high school, appeared a young woman..who announced herself as 'Sheem Sheem', daughter of Sun-Yat-sen. 1943 E. M. Almedingen *Frossia* ix. 316 'Rabfaks?' she asked. 'Oh yes, I have heard of them. Workers' Faculties, are they not?' *Ibid.* 337 A few [teachers] deplored the Rabfak system—always in cautious undertones. 1960 *Twentieth Cent.* June 573 In 1922 he [sc. Kruschschev] was sent by the Party for a three-year adult education course at a 'Rabfak' school.

rabies. Now usu. with pronunc. (rēi·biz, -iz). Add to def.: A contagious virus disease of dogs and other warm-blooded animals, which produces paralysis or a vicious excitability and in man causes a fatal encephalitis with throat spasm upon swallowing and convulsions. (Earlier and later examples.)
1598 J. Florio *Worlde of Wordes* 307/2 *Rabbino, Rabi, Rabis*, the Rabbies. 1967 Swain & Dodds *Clinical Virol.* xiii. 184 Eradication of rabies can be achieved only when the total elimination of the reservoir of animal infection is possible. It has been achieved in Great Britain by rigid quarantine laws which govern the importation of all livestock. 1976 *Daily Tel.* 20 July 3/3 A 32-year-old teacher, fined £300 at Uxbridge for contravening the rabies regulations had the penalty reduced on appeal to £100 yesterday. 1976 T. Heald *Let Sleeping Dogs Die* ii. 30 'But surely he got an injection?' 'Not even anti-tetanus. Let alone a rabies jab. It was only a little bite.' 1977 D. A. Warrell in C. Kaplan *Rabies: the Facts* iii. 32 In man the disease called rabies is a severe inflammation of the brain and spinal cord..associated with invasion of these tissues by rabies virus. 1980 *Sci. Amer.* Jan. 109/1 Fox rabies is particularly serious in Europe, where the disease has spread steadily at the rate of about 30 kilometers per year from east to west since World War II.

‖ **Rabkrin** (ræ·bkrin). [a. Russ. *rabkrin* f. *rab(óche)-kr(est'yánskaya) in(spéktsiya)* worker–peasant inspectorate.] An organization established in 1920 by Lenin to examine the conformity of state organizations to official policy.
1928 *Observer* 18 Mar. 19/5 The Rabkrin (the Russian abbreviation for the Commissariat of Workers' and Peasants' Inspection) is a supreme controlling and auditing department, which is supposed to expose deficiencies in the work of State and industrial institutions. 1949 I. Deutscher *Stalin* vii. 230 The Rabkrin, as the Commissariat was called, was set up to control every branch of the administration.

Racah (ra·kā). *Physics* and *Chem.* The name of Giulio *Racah* (1909–65), Italian-born Israeli physicist, used *attrib.* with reference to his work in quantum mechanics, as *Racah coefficient* or *parameter*, either of two coefficients representing electrostatic interactions within a system of equivalent charged particles, esp. electrons within an atom.
1952 *Physical Rev.* LXXXVIII. 581/2 The coefficients of the transformation have been given by Racah..in terms of his *W* function and are called the Racah coefficients. 1959 *Astrophysical Jrnl.* CXXIX. 441 By means of a few simple formulae, all multiplet strengths..can be expressed in terms of only two basic quantities, viz., the Racah coefficients and the coefficients of fractional parentage. 1962 Cotton & Wilkinson *Adv. Inorg. Chem.* xxvi. 595 The Racah parameters are measures of the energy separations of the various Russell-Saunders states of an atom. 1966 Phillips & Williams *Inorg. Chem.* II. xxiii. 162 The total electron-electron repulsion energy of each *LS* state arising from a d^n configuration can be calculated and expressed in terms of so-called Racah parameters. 1968 A. B. P. Lever *Inorg. Electronic Spectroscopy* vii. 207 The Racah inter-electronic parameter *B* (and to a lesser extent *C*) is a function of ligand, central ion and stereochemistry. 1975 *Physics Bull.* Apr. 169/3 The classification of particle and nuclear properties uses techniques such as spinors and Racah algebra which are not for the first degree student.

race, *sb.*[1] **III. 8. c.** Delete 'Now chiefly *U.S.*' and add further examples.
1901 M. Franklin *My Brilliant Career* xiv. 117 They have cut races between the two creeks. 1912 B. E. Baughan *Brown Bread* 99 Little runnels and 'races' of water led through the plain from the mountain rivers. 1941 I. L. Idriess *Great Boomerang* xxxi. 243 We would take the water from a creek on one side of a mountain and by means of a race (channel) take it completely around the mountain. 1976 *Jrnl. Lakeland Dial. Soc.* 35 Ah thowt Ah wud ga up t' race an' then cross t' beck on t' steppin steans.

e. (Earlier example.)
1833 J. C. Loudon *Encycl. Archit.* 470 The back wall of the barn is to be sunk sufficiently deep for the wheel of the threshing-mill and the race (horse-course) from it.

f. (Earlier and later examples.) Also *Austral.*
1865 M. A. Barker *Station Life in N.Z.* (1870) v. 34 The newly-shorn [sheep]..have passed thro' a narrow passage, called a 'race'. 1878 E. S. Elwell *Boy Colonists* 214 They made a 'lead' in the stockyard for branding the cattle. This was something like a 'race' for drafting sheep, with a swing gate. 1934 T. Wood *Cobbers* iv. 41 'Bullicks come aboard along a *race*. This is a race,' and he pointed to narrow gangways, railed in on both sides, which sloped from the main deck down to the cattle deck. 1950 *N.Z. Jrnl. Agric.* Apr. 373/3 The units [of the pig house] are usually placed side by side under one continuous roof, a service race being provided along the front. 1963 A. Lubbock *Austral. Roundabout* 180 The cattle were in the yards and the drovers and dogs were putting them through a 'race'—two rows of wooden fencing with a swing gate at the end. 1977 *N.Z. Herald* 8 Jan. 4–7/9 (Advt.), At present dairy and beef. Good race and fencing, tidal boundary, ample hay storage.

g. (Example.)
1825 J. Nicholson *Oper. Mech.* 104 The stones of the race are hewn to a mould, and laid in their places with great care.

h. Each of the two grooved rings of a ball or roller bearing.
1903 *Sci. Amer. Suppl.* 2 Feb. 22689/1 The rollers are made to fit the inner and outer treads of the roller race. 1907, 1908 [see *ball-race* s.v. *Ball *sb.*[1] 22]. 1930 *Engineering* 4 Apr. 462/1 There are two rings of rollers running side by side between hardened and ground inner and outer races. 1960 [see *Coin *v.*[1] 3 b]. 1968 *Autocar* 25 Jan. 49/2, I drove the 2-litre car at Monte Carlo and we had transmission trouble there which was bad luck because it was a ball race that broke. 1971 B. Scharf *Engin. & its Language* xii. 135 Ball bearings..consist of..an inner race, which is a grooved ring firmly attached to the shaft, and an outer race in the stationary housing. The balls which are free to rotate between the races are kept apart by means of a cage. 1980 *Dirt Bike* Oct. 33/1 You may even need to replace the balls and races if they're dented or worn.

IV. 10. c. *fig.* An electoral contest for public office. Phr. *to make the race*, to run for public office (see also quot. 1881). *U.S.*
1855 I. C. Pray *Mem. J. G. Bennett* 288 He had been the first to start many of them upon the ground for a successful political race. 1881 H. W. Pierson *In Brush* 132 To 'make the race' was to secure an election. *Ibid.* 133 This pursuit of office was always spoken of as a 'race'. 1903 *N.Y. Even. Post* 17 Sept. 1 Mr. Cutting ran up stairs to tell Dr. Gould..that Mr. Gront would make the race. 1949 *Dallas Morning News* 1 May 1/7 He might perhaps consider making the Senate race. 1976 *National Observer* (U.S.) 6 Nov. 24/3 New totals for..U.S. Senate races and gubernatorial races will churn out every five minutes.

d. As the second element in Comb. with a defining sb., as *armament race, arms race, moon race, space race*, etc.: see the sbs. in Suppl.

e. In colloq. phr. *to be in the race*, to have a chance. Usu. in neg. contexts. *Austral.*
1945 M. Trist *Now that we're Laughing* x. 73 With you and Daffy dressed up, none of us others will be in the race. 1953 T. A. G. Hungerford *Riverslake* x. 227 'See that bloke?' He pointed down the road after the vanished car. 'A few years ago he wouldn't have been in the race to own a car like that.' 1956 J. T. Lang *I Remember* vi. 34 The trade unions realised that if the Chinese could get away with long hours and low pay they would not be in the race to get better conditions.

f. *Electronics.* In a switching circuit, a condition in which the time a component secondary circuit or device takes to operate has to be taken into account (as when two are required to operate simultaneously, though in practice one will operate before the other). Freq. *attrib.*
1954 *Jrnl. Franklin Inst.* CCLVII. 170 In a composite transition matrix the presence of two or more of the digits 'I' indicates that at least two of the secondary relays are simultaneously unstable, and that a race condition exists. 1958 S. H. Caldwell *Switching Circuits* xii. 469 The race condition in this instance can lead to false operation and we designate this as a critical race. 1969 J. J. Sparkes *Transistor Switching* vii. 161 It is customary to arrange the logic so that all races are removed or rendered non-critical. 1975 J. C. Boyce *Digital Logic* ix. 267 Races are characterized by arrows that skip rows on flowtables, since more than one gate must change at the same time to allow the operating point to follow the arrow.

V. 11. a. (sense 10) *race-boat* (earlier and later examples), *-colt, driver, -ground* (earlier examples), *mare, record, report, -rider* (later examples), *-time*; (sense 8 f) *race gate, shed.* **b. race-ball** (later example); **race card** (later examples); **racecaster** orig. *U.S.*, a radio or television broadcaster who reports on horse-racing; **race game,** a board game simulating a horse-race in which rival counters proceed at the throw of a dice; also *transf.*; **race gang,** a group of petty criminals who frequent race-meetings; **race-glass:** now usu. in *pl.* (examples); **race-goer,** a frequenter of race-meetings; also **race-going** *a.* and *sb.*; **race-mark,** a mark attached to pigeons before a particular race; hence **race-mark** *v. trans.*, to supply (pigeons) with race-marks; **race-path,** (*a*) a race-track; (*b*) the channel along which water flows to a mill-wheel; **race-reader,** (*a*) one who forecasts the performance of horses in a given race; (*b*) (see quot. 1953); also, a race commentator; hence **race-reading; race stand** (earlier example); **race-track** orig. *U.S.* = Race-course 1 a; also *transf.* and *attrib.*; **race train,** a special train which runs to and from a race-meeting; **race-trough,** a plank with raised edges along which goods are passed in loading or unloading ships or wagons; **race walking,** the act or practice of competing in a walking race; hence **race walker; race walk** *v. intr.*

c 1838 W. H. Murray in M. R. Booth *Eng. Plays of 19th Cent.* (1973) IV. 162, I saw Charlotte at the race-ball, and fell over head and ears in love with her. 1839 *Spirit of Times* 15 June 177/1 It is most probable we would still have continued to get our race boats from Philadelphia. 1972 C. Mudie *Motor Boats & Boating* 144 The race boat hull form is not of great value for cruising boats. 1967 *Listener* 14 Sept. 325/3, I got hold of a race-card and I said to him: 'This one's got a marvellous chance.' 1979 D. Francis *Whip Hand* ii. 23 It took a course in the country to..run out of racecards. 1938 *Amer. Speech* XIII. 239/2 *Newscaster* and *sportscaster* are now common terms in *Variety*. *Racecaster* is also found. 1969 *Australian* 24 May 35/1 The caller will be 3DB's race-caster, Bill Collins. 1850 'M. Tensas' *Odd Leaves Life Louisiana 'Swamp Doctor'* 47 She couldn't 'tend races, and have a race-colt of her own to comfort her 'clinin' years. 1935 A. G. Kennedy *Current Eng.* xiv. 613 In a recent issue of a widely read city paper..the following headings appeared..'Race Driver Pinned under Flaming Car.' 1972 *N.Y. Times* 3 Nov. 45/2 Look at that guy jumping lanes... You have race drivers the same way, but they never amount to nothin'. 1895 *Montgomery Ward Catal.* 235/2 The Grand Race Game, consists of a substantially cloth bound, folding board..printed in colors to represent a race track... Has six colored cardboard horses and riders..two wooden dice cups and two dice. 1903 J. M. Falkner *Nebuly Coat* xiii. 187 A 'race-game' where the little leaden horse is moved steadily forward. 1930 A. P. Herbert *Water Gipsies* xiv. 196 In that light the hare seemed tinier and the greyhounds toys... The place might have been some monstrous nursery 'race-game'. 1973 *Daily Tel.* (Colour Suppl.) 29 June 40/2 Backgammon is a race game..and the precursor of every modern board-pieces-and-dice game. 1931 M. Allingham *Look to Lady* xxiv. 250 Gipsies and race gangs always hate each other. 1937 E. Rickman *On & off Racecourse* xii. 271 The term 'race gang' is still a very favoured one by newspaper sub-editors. 1930 L. G. D. Acland *Early Canterbury Runs* 1st Ser. x. 251 He once asked Moore why he didn't put race gates into his yards. 1865 *Let.* 11 June in Ld. W. Lennox *My Recoll.* (1874) II. 153 General Fleury almost forced a race-glass into the hands of the Emperor. 1938 F. D. Sharpe *Sharpe of Flying Squad* xix. 204 From the coach descended a number of the boys one of whom was carrying a pair of race-glasses. 1974 D. Francis *Knock Down* (1976) xii. 137 The runners went down the far side and we lifted race-glasses to watch. 1880 *Baily's Monthly Mag.* Feb. 71 Race-goers are not, as a rule, early drinkers. 1948 *Sunday Pictorial* 18 July 13/4 It has now been in operation for some time, and many race-goers are fed up with it. 1975 D. Francis *High Stakes* i. 9 Racegoers were hurrying towards the stands to watch the imminent steeplechase. 1848 *Sporting Life* 2 Sept. 324/2 A great favourite with the race-going public. 1929 S. Ertz *Galaxy* xv. 332 He had always been so busy with his..shooting, his race-going, and latterly his horses. 1963 *Times* 23 Jan. 3/5 Waiting for any crumbs that may be going are the racecourses, the owners and the racegoing public. 1977 D. Francis *Risk* ii. 12 Near the course the crawling racegoing jams would mean half an hour for the last mile. 1698 J. Collier tr. *Tertullian* in *Profaneness & Immorality Eng. Stage* vi. 253 We have nothing to do with the Frensies of the Race-Ground..or the Barbarities of the Bear-Garden. 1727 in *Maryland Hist. Mag.* (1912) VII. 400 This Vestry Resolve to meet on Thursday..at the race Ground near Mr Bensons. 1853 *Southern Lit. Messenger* XIX. 70/1 He brought with him a small race mare which excited the acquisitiveness of his father. 1976 *Times Lit. Suppl.* 15 Oct. 1296/2 Park Top..began to emerge as the outstanding race-mare of her decade. 1890 *Homing News* 3 Jan. 14/3 (Advt.), He is not certain about the race marks. 1928 *Sunday Dispatch* 8 July 22/3 Birds competing in the..race from Marennes,..will be racemarked at No. 5 platform. 1837 *Knickerbocker* X. 413 The only race-path known in this new settlement was that on which the husband and wife contended for the prize of domestic comfort. 1853 F. W. Thomas *J. Randolph* 84 Along the devious narrow race-path to the mill-dam. 1951 E. Rickman *Come racing with Me* iii. 23 Some practised race-readers tend to become ridiculously self-opinionated. 1953 P. G. Berg *Dict. New Words* 133/1 *Race reader*,..an expert attached to a radio commentator who helps in giving a broadcast of a horse-race. 1955 *Radio Times* 22 Apr. 29/3 Racing at Newmarket... Commentary by Raymond Glendenning, assisted by Tom E. Webster as race-reader. 1968 'J. Welcome' *Hell is where you find It* ii. 33 'Mountpatrick still well clear,' came the race-reader's voice. 'Then Blue Soldier, Mark Twain, Kitchener..all in a group together.' 1963 'J. Prescot' *Case for Hearing* iv. 60, I think you can rule out..the gift of race-reading in advance by looking into a crystal ball. 1976 *Horse & Hound* 10 Dec. 6/3 Michael O'Hehir, whose Telefis Eireann race-reading they [sc. the BBC] have taken in previously. 1893 *Outing* XXII. 101/1 Goldsmith Maid left the turf with a race record never equalled. 1977 *N.Z. Herald* 5 Jan. 1–11/1 He set a race record for the track when he went 1 m 58·5 s. 1934 T. S. Eliot *Rock* i. 29 Many read nothing but the race reports. 1827 J. F. Cooper *Prairie* p. xxiii, She is no great race-rider. 1973 *Country Life* 22 Feb. 457/2 Lester Piggott came from a family of race-riders and trainers. 1950 *N.Z. Jrnl. Agric.* Apr. 375/2 The race shed was..popular many years ago. 1829 P. Egan *Boxiana* 2nd Ser. II. 30 The Race-stand, at ten bob per nob, was opened for their reception. 1702 Lady Verney *Let.* 25 Aug. in M. M. Verney *Verney Lett.* (1930) I. vii. 113 We shall have Company at Claydon, it being our Race-time. 1893 W. B. Yeats *Celtic Twilight* The race-time came round. 1862 *N.Y. Tribune* 6 Mar. 6/6 We don't think that Tennessee is likely to be much of a battle-ground hereafter. There's more probability of her being a race-track. 1927 *New Republic* 21 Sept. 120/2 No American town is more completely absorbed in race-track gambling. 1945 *Transit News* (Capital Transit Co., Washington, D.C.) 15 June, I picked an early straight on the race track... Translated into plain English, the operator said: 'I signed up for a day's work on 16th street.' 1958 *New Scientist* 30 Jan. 18/3 We..began the construction of an aluminium torus of 12-in. bore diameter,..and in it were two straight sections—making the torus into a race-track. 1963 *Times* 24 May 16/6 The problem is best presented in terms of the dimensions of the race-tracks round which nuclear particles..are caused to circle repeatedly while gaining energy at each circuit. 1973 D. MacKenzie *Postscript to Dead Letter* 15 A set of Dufy racetrack prints on the walls. 1938 F. D. Sharpe *Sharpe of Flying Squad* xx. 223 The gymer gets out of the race train and boards the Underground railway. 1977 *Times* 18 June 12/8 The race train from Waterloo to Sandown Park. 1842 Carlyle in *Cornh. Mag.* (1922) Oct. 496 A huge high pier of wood..lowered down upon us a long race-trough of wood, by the side of which at due distances some four men stationing themselves [etc.]. 1973 F. Wakefield et al. *Track & Field Fundamentals for Girls & Women* (ed. 3) 253 Because some people do not want to race walk, the field of competition is small. 1962 *Sport of Race Walking* (Race Walking Assoc.) 34 Race-Walking can only be as strong as its Judges. ..To this end they have striven..to select men of integrity, mainly former race-walkers themselves. 1972 Pickering & Harris *Olympics* 74/2 The difference between the fast, fair race walker and the ordinary person at 3½ miles per hour is the straight leg action. 1954 *Times* 20 Sept. 3/3 Delegates [to the annual meeting of the Road Walking Assoc.] decided to change the name to Race Walking Association following the Amateur Athletic Association's authority to develop track walking. 1976 *Cumberland News* 26 Nov., The Olympic Games authorities seem hell bent on pushing race walking into the distant background.

race, *sb.*[2] Add: **I. 2. d.** (Further examples.)
The term is often used imprecisely; even among anthropologists there is no generally accepted classification or terminology.
1936 *Nature* 18 Apr. 636/2 The races or types into which the anthropologist groups the varieties of *Homo sapiens* are ideal types. 1959 *New Biol.* XXIX. 69 From the U.N.E.S.C.O. statement we can define 'race' as 'a division of man, the members of which, though individually varying, are characterized as a group by certain inherited physical features as having a common origin'. 1971 R. M. & F. M. Keesing *New Perspectives in Cultural Anthropol.* 51 It is at this point that the term 'race' becomes relevant. Though in popular usage it is emotionally charged and imprecise, it has a straightforward and important meaning in evolutionary biology. A race is a geographically separated, hence genetically somewhat distinctive, population within a species.

II. 11. a. Now found in almost unlimited *attrib.* and *Comb.* uses: caused by, based on, of or pertaining to race, as *race-aversion, -blood, -conflict, culture, discrimination, division, equality, -experience, -feeling, -hatred* (further examples), *-heritage, -history, -improvement, -inheritance, instinct, law, line, -mixture, -name, prejudice, pride, problem, quarrel, -question, relationship, solidarity, superiority, -survival, tension, -type, war; race-conscious, -hating, -perpetuating, -proud* adjs.
1897 'Mark Twain' *Following Equat.* xxi. 207 It must have been race-aversion that put upon them a good deal of the low-rate intellectual reputation which they bear. 1906 W. H. Fleming *Slavery* 37 The one is based on a supposed duty to God; the other on a supposed duty to one's race-blood. 1880 A. W. Tourgée *Invis. Empire* xii. 513 Any one who asked the support of colored men as against a Democratic nominee was precipitating a race-conflict. 1949 *Caribbean Q.* I. ii. 28 Countless little stories..about..present life, in country and town..in race-conflict, and class-conflict. 1927 *Observer* 5 June 5/3 Frenchmen are not so race-conscious as either Englishmen or Americans. 1977 P. Johnson *Enemies of Society* viii. 106 *Nigger*..is now frequently employed by the more race-conscious blacks, but only among themselves. 1909 C. W. Saleeby (*title*) Parenthood and race culture. An outline of eugenics. 1917 *Cases Argued U.S. Supreme Court: Lawyers' Ed.* (1918) 155/2 Plaintiff is not in a position to raise the issue of race discrimination, not being himself a negro. 1906 *Westm. Gaz.* 21 Feb. 2/3 That simple principle [of One Vote One Value]..at once supplies a strong motive for those who once had everything to gain from the race-division to talk about 'bringing both races together'. 1974 *Race* XV. 462 The present race divisions are projected into the past as though they were always a feature of South African society. 1911 G. Spiller *Papers on Inter-Racial Problems* i. 31 It becomes a vital matter to grapple with the problem of race equality. 1890 O. Wilde in *19th Cent.* Sept. 443 The imagination is the result of heredity. It is simply concentrated race-experience. 1888 Kipling *City of Dreadful Night* (1891) 18 A casual reference to Hindus and Mahometans... There is race-feeling, to be explained away. 1944 J. S. Huxley *On Living in Revolution* 169 The actual physical kinship, which is frequently claimed as 'race feeling', must be fictitious. 1941 Auden *New Year Let.* iii. 68 Self-respect drives negroes from The one-crop and race-hating delta. 1901 *Times* 5 Aug. 7/2 The object of these documents has usually been..to fan the race-hatred of the Dutch in South Africa. 1935 *Economist* 27 July 175/2 The new excesses are confined to the special domains of class hatred, race hatred and hatred of religion. 1976 *Birmingham Post* 16 Dec. 5/2 Intent to stir up race hatred. 1911 W. James *Some Probl. Philos.* i. 4 Philosophy, thus become a race-heritage, forms in its totality a monstrously unwieldy mass of learning. 1894 *Psychol. Rev.* Nov. 651 The one criticism which I would venture to make upon this paper..is that it neglects the phylogenetic point of view, the considerations from race-history. 1907 W. James *Pragmatism* v. 169 The most primitive ways of thinking..may remain as indelible tokens of events in our race-history. 1903 *Daily Chron.* 29 July 4/5 We have a great deal yet to learn on matters bearing upon race-improvement. 1909 W. James *Meaning of Truth* viii. 214 Dr. Schiller has shown that all our truths, even the most elemental, are affected by race-inheritance with a human co-efficient. 1901 —— *Let.* 3 Mar. (1920) II. 141 Empire anyhow is half crime by necessity of Nature, and to see a country like the United States..perversely rushing to wallow in the mire of it, shows how strong these ancient race instincts be. 1942 'G. Orwell' *War-time Diary* 22 Mar. in *Coll. Ess.* (1968) II. 412 German propaganda is..offering..emancipation to the Kaffirs and stricter race laws to the Boers. 1960 *Twentieth Century* Nov. 407 Race-laws make camps almost impossible within the Union. 1978 G. Greene *Human Factor* ii. i. 62 'I fell in love.' 'Yes. So I see. With an African girl... You broke their race laws.' 1883 G. W. Williams *Hist. Negro Race* II. xxviii. 543 Race lines must be obliterated. 1891 *Congress. Rec.* App. 17 Jan. 101/1 At Marion, Ind.,..when the Democrats were attempting to have a rally,..they were attacked by the colored people, the race line being distinctly drawn by that race. 1905 O. Jespersen *Growth & Structure Eng. Lang.* iii. 47 There we had a real race-mixture, where people speaking two different languages were living in actual contact in the same country. 1935 Huxley & Haddon *We Europeans* ix. 278 From what has been said, it will be clear that 'race-mixture' has in the past been beneficial. 1924 Race-name [see *Atestine *a.* and *sb.*]. 1950 Partridge *Here, There & Everywhere* 17 The other self-confident Asiatic race-names are fully qualified. 1915 R. Lankester *Diversions of Naturalist* xxi. 194 Natural automatically-growing mechanisms of life-saving or race-perpetuating importance. 1890 O. Wilde in *19th Cent.* Sept. 457 Criticism will annihilate race-prejudices, by insisting upon the unity of the human mind in the variety of its forms. 1913 J. London *Let.* 25 Aug. (1966) 395 First of all..by stopping the stupid newspaper from fomenting race prejudice. 1920 H. Crane *Let.* 6 Mar. (1965) 35, I am as anti-Semitic as they make 'em, but Frank's comments cannot afford to be ignored merely because of race prejudice. 1942 E. Paul *Narrow St.* xii. 91 Guy delivered a concise impassioned talk against race prejudice. 1956 L. Kuper *Passive Resistance in S. Afr.* 18 Then Dr. Naicker commented on..the United Party's pandering to race-prejudice to catch votes. 1905 W. Baucke *Where White Man Treads* 276 On our side race prejudice, race pride, preaching honesty, yet unblushingly swindling him and each other. 1973 A. Dundes *Mother Wit* 2/1 The relationship between folklore and race pride..corresponds to the relationship between folklore and nationalism in the nineteenth century. 1890 A. W. Tourgée *Pactolus Prime* xi. 141 If every one could do as much, the race-problem would soon be solved. 1923 O. Schreiner *Thoughts on S. Afr.* vii. 296 To..attempt to comprehend or deal rationally with race-problems. 1980 *Bananas* Aug. 7/1 Talking about Korea, Chicago, war, the race problem. 1937 E. Muir *Coll. Poems* (1960) 72 Now I am shackled to a Grecian dolt, Pragmatic, race-proud as a pampered colt. 1931 F. L. Allen *Only Yesterday* iii. 68 If a white man stood up for a Negro in a race quarrel, he might be kidnapped and beaten up. 1889 *Boston Jrnl.* 26 Dec. 2/4 Time only can solve the race-question in the South. 1920 L. Stoddard *Rising Tide of Color* xi. 293 She [sc. Japan] should not allow her immigration to be treated as a race-question. 1908 R. S. Baker *Following Colour Line* x. 217, I have found a sharper feeling and a bitterer discussion of race relationships among the Negroes of the North than among those of the South. 1942 Z. N. Hurston in A. Dundes *Mother Wit* (1973) 25/1 'Race Solidarity' looked like something solid in my childhood, but like all other mirages, it faded as I came close enough to look. As soon as I could think, I saw that there is no such thing as Race Solidarity in America with any group. 1901 E. A. Ross in *Ann. Amer. Acad. Pol. Sci.* XVIII. 67 (*title*) The causes of race superiority. 1951 J. Masters *Nightrunners of Bengal* v. 58 She was goading herself to wipe out a sense of race superiority she presumed him to have... She wanted..him to acknowledge beauty in an Indian woman. 1933 A. N. Whitehead

Adventures of Ideas vi. 97 We can observe insects performing elaborate routine actions..which yet are essential either for their own individual survival or for race-survival. 1954 P. Mason *Ess. Racial Tension* iii. 45 One would expect race tensions to be most acute..in the country where there is a temperate climate. 1974 *Allendale* (S. Carolina) *County Citizen* 24 Apr. 6/3 We found ourselves discussion-slanted toward race tension and struggles. 1864 W. D. Whitney in *Ann. Rep. Board of Regents Smithsonian Inst. 1863* 113 The kind and amount of modification which external circumstances can introduce into a race-type is as yet undetermined. 1892 Kipling *Lett. of Travel* (1920) 30 Seven million negroes..their race-type unevolved. 1927 Peake & Fleure *Priests & Kings* 181 'Race-type' in a general sense is a very difficult matter to define. 1897 *Chicago Tribune* 28 July 3/7 This gave the negroes an excellent chance to start a 'race' war. 1977 P. Johnson *Enemies of Society* xix. 247 He realized he was taking part in a race-war, as well as a class-war.

b. Of, pertaining to, or designating a style of music, originating among Blacks of the Southern U.S. (cf. Race *sb.*[2] 6 b), freq. in a twelve-bar sequence (see also quot. 1938).
1926 H. Niles in W. C. Handy *Blues* 31 Listen to the 'race records', for this craft is *sui generis*. 1927 *Jrnl. Abnormal & Social Psychol.* Apr.–June 12 'Race blues'..are not always what they seem. 1935 *Vanity Fair* (N.Y.) Nov. 71/3 Negro bands play 'race music' (a curious euphemism spread by phonograph companies). 1938 *Collier's* 30 Apr. 24/4 We were afraid to advertise Negro records. So I listed them in the catalogue as 'race' records and they are still known as that. 1942 Partridge *Usage & Abusage* 208/2 'Race (phonograph) recordings' for recordings made by Negroes. 1946 R. Blesh *Shining Trumpets* (1949) vi. 145 It was considered authentic enough for the uncritical Victor Company to issue in its race catalogue. 1946 Mezzrow & Wolfe *Really Blues* (1957) ix. 161 Preaching blues was strictly race music. 1952 B. Ulanov *Hist. Jazz in Amer.* (1958) iv. 32 Their masterpieces appeared on the so-called 'race' labels of the record companies. 1968 P. Oliver *Screening Blues* 5 In the ensuing months more stores carried Race records, specially pressed for the Negro market... Race records from jazz to vaudeville to rural blues reached the remotest districts. 1976 A. Murray *Stomping Blues* iv. 50 The period of the race catalogs was also the decade of the so-called revolution in race consciousness known as the Harlem Renaissance. 1977 *Times* 17 Aug. 14/4 Negro styles traditionally stigmatized as 'race' music.

c. Special combs., as **race consciousness,** emotionally based awareness of those differences between people or social groups that can be ascribed to racial factors; the supposed intuitive awareness of a common heritage shared by members of a race or culture; **race-gap,** a difference between racial groups; **race man** *U.S. colloq.*, a Black, esp. one who advocates the rights of Blacks; **race memory,** (*a*) subconscious memory of events in the history of one's race or of the human race which, it is suggested, is transmitted genetically; **race relations,** a term for such social contacts between racial groups living within a particular area as arise from or are affected by differences in cultural origin or skin colour; freq. *attrib.* or as *adj.*; **race riot,** a riot that results from racial hostility; hence **race rioting; race suicide,** the self-extinction of a racial group through failure to reproduce itself sufficiently, esp. of one with high cultural standards and a low birth-rate in competition with a racial group having lower standards and a high birth-rate; the self-destruction of a race; also *attrib.*; **race theory,** a hypothetical assertion that some racial groups are endowed with specific 'superior' qualities; hence **race theorist,** an advocate of a race theory; **race-thinking** (see quot. 1937); also **race-thinker.**

1905 Race consciousness [see *Amalgamation 2]. 1926 G. Callaway *Native Probl. in S. Afr.* 2 It is conceivable that the Native people of South Africa might have lived alongside of the Europeans without developing a strong race consciousness. 1968 *Internat. Encycl. Social Sci.* XIII. 269 Relationships which are capable of producing race conflict and race consciousness. 1890 W. James *Will to Believe* (1897) 260 We here..catch the only glimpse it is allotted to us to attain of the working units themselves, of whose differentiating action the race-gaps form but the stagnant sum. 1936 R. L. Abbot in *Chicago Defender* 13 June 16/5 One Race man, finding out this outrage, fired on the officers. 1942 Z. N. Hurston in A. Dundes *Mother Wit* (1973) 25/1 A 'Race Man' was somebody who always kept the glory and honor of his race before him... It was a mark of shame if somebody accused: 'Why, you are not a Race Man (or woman).' People made whole careers of being 'Race' men and women. They were champions of the race. 1969 *Publ. Amer. Dial. Soc.* li. 29 Names used..by both Negroes and Whites [for Negroes who demand equal status with whites]..civil rights man, mau mau, race man. 1974 Yi-Fu Tuan *Topophilia* xiii. 209 The upper shadies can identify emotionally with the ghetto poor; they are recognized by the poor as Race Men, that is, supporters of black causes. 1904 *Folk-Lore* XV. 349, I have heard this belief referred to a 'race-memory' of antediluvian reptiles. 1912 A. Conan Doyle *Lost World* i. 10 That race-memory which we call instinct. 1934 R. Knox *Still Dead* xi. 138 A cave has, for all of us, an atmosphere of..terrifying mystery. The anthropologists would tell us..that it is due to race-memory. 1950 [see *Oldest *a.* 3]. 1972 C. Fremlin *Appointment with Yesterday* xiv. 110 A race-memory of the days when servants weren't quite real, and so it didn't matter what they heard. 1911 *Pol. Sci. Q.* XXVI. 193 (*title*) Race relations in the Eastern Piedmont region of Georgia. 1925 *Scribner's Mag.* July 12/2 On two occasions great intercollegiate conventions of students have dealt with race-relations,..and war itself. 1934 *Race Relations* I. 32/1 We have to deal in this country not only with relations between English and Dutch but also between Jews and Gentiles, and between Whites and Coloured, Whites and Indians, as well as between Whites and Bantu... Hence, we decided to invite certain men..to give us their views on how race relations problems strike them. 1965 *Act 13 & 14 Eliz. II* c. 73 (*heading*) Race Relations Act 1965... An Act to prohibit discrimination on racial grounds in places of public resort; to prevent the enforcement or imposition on racial grounds of restrictions on the transfer of tenancies; to penalize incitement to racial hatred. *Ibid.* § 2 For the purposes of securing compliance with the provisions of..this Act..there shall be constituted a board to be known as the Race Relations Board, consisting of a chairman and two other members appointed by the Secretary of State. 1970 *Oxf. Univ. Gaz.* 30 Apr. Suppl. 14 During the course of the year two visiting Fellows in Race Relations were appointed in collaboration with St. Antony's College. 1977 *Whitaker's Almanack 1978* 348/2 A Lords amendment to the Race Relations Bill..was reversed in the Commons on Oct. 27. 1890 *Our Day* May 406 Race Riots in the South. 1921 *Palestine Weekly* 2 Dec. 779/2 With regard to the actual question as to which side initiated the race riot, the Commission speaks with definiteness and precision. 1928 F. Hurst *President is Born* xxiv. 250 Race-riots out in Chicago. 1958 *Daily Mail* 3 Sept. 6/6 After three nights of race-riots in their streets the people who live in Notting Hill have been asked to put themselves under a voluntary curfew. 1979 *Dædalus* Spring 103 Race riots broke out in Marseilles in 1973 that left six Algerians dead. 1968 *Economist* 20 July 43/1 The second problem is the emergence of race rioting as a regular, not to say an annual, occurrence. 1901 E. A. Ross in *Ann. Amer. Acad. Pol. & Social Sci.* July 88 The American farm hand, mechanic and operative might wither away before the heavy influx of a prolific race from the Orient... For a case like this I can find no words so apt as 'race suicide'. 1904 *Daily Chron.* 9 June 3/2 I'm with the President on this race-suicide question. 1936 M. Plowman *Faith called Pacifism* 14 If war has become race suicide by a perfectly natural process of evolution, why should we continue to call it 'war'? 1945 C. F. McCleary (*title*) Race suicide. 1921 *Times Lit. Suppl.* 25 Aug. 543/1 In defiance of the German race-theorists, and similar superficial or prejudiced observers, Dr. Curtius insists that French culture..cannot be dismissed with the formulae 'esprit' and 'décadence'. 1949 Koestler *Promise & Fulfilment* 334 With the exception of the 'race-theorists' nearly all modern authorities hold that Jewish characteristics are a product of sustained environmental pressure. 1895 W. D. Babington (*title*) Fallacies of race theories. 1945 Koestler *Yogi & Commissar* ii. ii. 192 Within a century or two..race-theory and Jew-baiting would have shrunk to episodes of the past. 1937 J. Barzun *Race: Study in Mod. Superstition* x. 263 Then came the 'biological revolution' and race-thinkers pinned their hopes on anatomy. *Ibid.* i. 17 We must..see what men who have thought and written about race think it is. Their ideas form, not a definition of race, for they all disagree among themselves, but a type of thinking, which I shall call race-thinking. 1965 *Listener* 11 Nov. 740/2 This kind of thinking involves what are, in fact, vague figures. It has been described as 'race thinking'. People who think this way..are becoming racists.

race, *v.*[1] Add: **2. c.** Also of an electric motor, car engine, etc. (Further examples.)
1893 S. R. Bottone *How to manage Dynamo* ii. 29 The dynamo..if shunt-wound..will race and go much faster than usual. 1907 C. W. Brown *Petrol Engine* i. 3 Neglect of this matter will cause the engine to develop a knocking sound, especially..when 'racing' with the governor out of action. 1980 *Sci. Amer.* Jan. 118/3 If the load on the turbine was suddenly removed, then unless the turbine was shut down promptly it would pick up speed and race, conceivably until it flew to pieces.

3. a. Also *refl.*
1963 *Daily Tel.* 17 July 1/4 He said he got the impression that America was 'racing herself' in this quest [to the Moon].

4. a. (Further examples.)
1906 E. Dyson *Fact'ry 'Ands* ii. 17 She raced her work. 1945 *ABC of Cookery* (Ministry of Food) vii. 22 Take care not to race the boiling or the eggs may crack.

d. To cause (an engine) to race; to 'rev'.
1932 *New Yorker* 23 July 14/3 The cop got on his machine, raced his engine,..and throttled down. 1973 W. McCarthy *Detail* ii. 117 He started the car and raced the engine noisily.

e. *to race off*, to seduce (a woman); to hurry (a woman) away in order to seduce her. *Austral. slang.*
1965 W. Dick *Bunch of Ratbags* xii. 185 Three of Knuckles's boys had raced Sharon off to the park to see if they could do any good for themselves. 1967 M. Wilding in *Coast to Coast 1965–6* 250 Perhaps Peter thought he would try to race her..off. He relished the phrase, race off. He had not heard it in England. 1969 W. Moxham *Apprentice* vii. 87 'That's one bird you won't race off,' Rufe said. 'I know her and the bloke she's with.' 1978 *TV Week* (Austral.) 24 June 46/4 Luckinbill blabs to his wife Cannon that he raced off most of her best friends.

raceable (rēi·săb'l), *a.* [f. Race *v.*[1] + -able.] **a.** Of a racehorse: capable of being raced.
1965 D. Francis *For Kicks* iii. 40, I might well be given a raceable horse to look after.

b. Of a racecourse: fit for horse-racing.
1976 *Daily Record* (Glasgow) 4 Dec. 27/7 It is raceable at present. 1976 *Horse & Hound* 10 Dec. 6/1 He got to Chepstow [racecourse] only to find..that although the ground there was raceable thick local fog meant that proceedings could never get off the ground.

ra·ceabout. *U.S. Naut.* [f. Race *v.*[1] + About *adv.*] A sloop-rigged racing yacht with a smaller keel and larger sailyards than those of a knockabout (see *Knock-about, Knockabout *a.* 3 b). Also *attrib.*
1897 *Forest & Stream* 6 Mar. 194/2 The next size, the 'raceabout', or the *fin de siècle* perversion of the knockabout, promises to be more popular and numerous. 1899 *Ibid.* 25 Feb. 157/1 The yacht will have two rigs, knockabout and raceabout, which can be shifted one for the other in half an hour. 1905 *St. Nicholas* Aug. 865 There were other prizes, of course: the much-coveted silver cup,..pennants for the raceabouts and halfraters, and a first money prize of twenty-five dollars for the fishermen's sloops. 1907 *Forest & Stream* 30 Nov. 863/1 The old jib and mainsail sand-bag rig has been replaced by the modern raceabout. 1927 E. P. Morris *Fore & Aft Rig in Amer.* 75 It [sc. the knockabout rig] has been somewhat diverted from its original purpose by being used on small racing boats, which are not knockabouts in the strict sense and have been ironically called 'race-abouts'.

race-horse. Add: Now usu. without hyphen.
1. a. (Earlier and later *attrib.* examples.)
1839 Dickens *Let.* 25 Nov. (1965) I. 605 Barnaby moves—not at racehorse speed. 1868 H. Woodruff *Trotting Horse* ii. 47, I do not undertake to disparage the method pursued by the race-horse men. 1962 D. Francis *Dead Cert* iii. 27 Pete Gregory, racehorse trainer. 1973 *N.Y. Law Jrnl.* 1 Aug. 11/6 An arrest of other persons..who were involved in 'racehorse policy' gambling activities.

3. *transf.* and *fig.* Anything racy, sleek, or speedy. Also *attrib.*
1845 W. G. Simms *Wigwam & Cabin* 1st Ser. 32 He had the sanguine, the race-horse temperament. 1903 [see *monthly ship* s.v. *Monthly *a.* 4]. 1953 K. Tennant *Joyful Condemned* xviii. 164 He sat rolling a very thin cigarette, known as a 'racehorse'. 1974 *State* (Columbia, S. Carolina) 15 Feb. 5-B/2 Freshman flash Adrian Dantley triggered a racehorse offense for Notre Dame with 27 points. 1978 *English Jrnl.* Dec. 72/1 What Charles Cooper refers to as 'race-horse studies'..often compared tactics for forcing students through the hoops handed down from the textbook writers.

4. racehorse lizard, a small Australian lizard, *Amphibolurus caudicinctus*, the ring-tailed dragon.
1937 *Discovery* May 137/1 The Racehorse or Cycling Lizard..runs at an incredible speed. *Ibid.*, Place a Racehorse Lizard on a pink handkerchief and it will quickly assume that colour. 1978 O. White *Silent Reach* iv. 50 The little racehorse lizards skittering away through the spinifex with their tails up.

race-knife. Add: (Examples.) Also *ellipt.*
1923 G. Sturt *Wheelwright's Shop* vi. 30 The various points for cross-cutting the tree were scratched with a 'race'—a sort of knife with point turned back and sharpened at the bend for this especial purpose. 1964 H. Hodges *Artifacts* vii. 109 A row of wedge-pits or triangular slots was cut with an axe-shaped tool (race, gad or jad). 1969 E. H. Pinto *Treen* 401 The saddler's race or race knife,..used for marking leather, is the same tool that is also used by coopers, lumbermen, and carpenters to mark or scribe logs or timber sections, or to register tallies.

racemate. Add: **a.** (Earlier example.)
1835 *Rec. Gen. Sci.* II. 101, 3.20 grs. of neutral racemate of lead.

b. A racemic (sense *b) form of a compound.
1907 *Jrnl. Chem. Soc.* XCI. i. 906 In the case of substances known only in the liquid state, no general method has as yet been obtained to establish in a sufficiently convincing manner the existence of liquid racemates. 1936 *Discovery* Nov. 341/2 A solution of the racemic form (or racemate as it is sometimes called) is treated with the active form of a Base..obtained from Nature. 1953 *Jrnl. Amer. Chem. Soc.* LXXV. 4587/2 The possibility of strong intramolecular hydrogen bonding in the case of α-phenethyl phthalate would offer an alternative explanation for the similarity in the spectra of the enantiomorphs and racemate. 1973 C. H. Snyder *Introd. Mod. Org. Chem.* viii. 129 Since attack by bromide ion is equally likely at either carbon, the product is a racemate. 1973 D. Whittaker *Stereochem. & Mechanism* i. 8 Racemates..often have different melting points from the optical isomers.

racemic, *a.* Add: **a.** (Earlier example.) [ad. F. (*acide*) *racémique* (printed *racenique*) (J. L. Gay-Lussac *Cours de Chimie* (1828) xxiv. 23).]
1835 *Rec. Gen. Sci.* II. 97 The term racemic acid given by the French is preferable.

b. Composed of dextro- and lævorotatory isomers of a compound in equal molecular proportions, and therefore optically inactive.
1892 *Jrnl. Chem. Soc.* LXII. ii. 822 The fact that the racemic modification of gulolactone cannot be obtained by crystallisation from water is of general interest, but not without analogy. 1894 G. M'Gowan tr. *Bernthsen's Text-bk. Org. Chem.* (ed. 2) 39 Optically inactive modifications, which result like racemic acid from the conjunction of two active components, are termed 'racemic' modifications. 1926 H. G. Rule tr. *J. Schmidt's Text-bk. Org. Chem.* 37 In a few instances it is possible, by allowing a solution of the racemic mixture to crystallise under certain conditions, to obtain the two enantiomorphs depositing individually. 1947 *Endeavour* VI. 97/2 Pasteur not only showed that the effect on polarized light was due to asymmetry of molecular architecture, but devised methods of resolving racemic mixtures. 1964 N. G. Clark *Mod. Org. Chem.* iii. 34 The racemic modification of a substance is sometimes said to be 'externally compensated' because its optical inactivity is due to the balancing of two equal and opposite rotations which do not occur within the same molecule. 1971 *Sci. Amer.* May 6/1 L-Leucine has a bitter taste, D-leucine has a sweet taste and the racemic mixture (both) is tasteless.

Hence †**ra·cemism,** the state of existing in a racemic form; †**ra·cemoid** *a.* = Racemic *a.* b.
1896 *Proc. Chem. Soc.* 23 Apr. 97 The ethereal salts..were examined for racemoid compounds, but the quantity found

was not sufficient to account for the low activity. 1897 *Jrnl. Chem. Soc.* LXXI. 990 Notwithstanding these marked differences, which may be accepted as conclusive evidence of racemism, this modification of the inactive acid has the same melting point as the active compound. 1902 *Proc. Chem. Soc.* XVIII. 198 Amygdalic acid is racemoid with respect to its asymmetric carbon atom. 1904 *Science* 5 Aug. 178/2 In the cases of some double nuclei (as the camphor group) racemism appears to be impossible, owing to the peculiar molecular structure.

racemization (ræːsĕməizēi·ʃən). *Chem.* [f. prec. + -ization.] Conversion of an optically active substance into a racemic form.
1895 A. Eiloart in *Sci. Progress* III. 503 If the temperature of racemisation[1] lies below the ordinary temperature, the difficulty of separating the active isomers is accounted for. [*Note*] [1] This term is used to indicate the formation of the inactive aggregate from two isomers of opposite activity. 1927 Arnall & Hodges *Theoret. Org. Chem.* II. vi. 59 Pinene and limonene undergo racemisation on simple heating. 1950 R. C. Fuson *Adv. Org. Chem.* vi. 111 When L-2 iodoöctane is treated with sodium iodide in acetone solution..racemization occurs. 1966 Phillips & Williams *Inorg. Chem.* II. xxix. 455 The complexes $[Cr(C_2O_4)_3]^{3-}$ and $[Co(C_2O_4)_3]^{3-}$ will not exchange oxalate with radioactive oxalate in aqueous solution under conditions in which complete racemization takes place. 1975 *Kingston* (Ontario) *Whig-Standard* 5 Sept. 7/5 The ages were obtained by the so-called racemization method. This records the extent to which molecules of aspartic acid in a specimen have altered their configuration from the form that occurs in living bones to its mirror image.

racemize (ræ·sĕməiz), *v. Chem.* [f. as prec. + -ize.] **a.** *trans.* To convert (an optically active substance) into a racemic form. **b.** *intr.* To undergo conversion to a racemic form.
1896 [implied in *racemized *ppl. a.* below]. 1902 *Proc. Chem. Soc.* XVIII. 198 The author concludes that in alkaline solution the glucoside is racemised by the catalytic action of the hydroxyl ions. 1913 *Jrnl. Chem. Soc.* CIII. 607 The hydrogen *d.* camphorate..racemises within twenty hours of making up the aqueous solution. 1938 G. H. Richter *Textbk. Org. Chem.* xvii. 335 Many natural products which occur only as one optical antipode may be racemized and resolved into both optical antipodes. 1965 G. Hallas *Org. Stereochem.* ii. 35 Several compounds are known to racemize via non-ionic planar intermediates. 1971 *Nature* 12 Mar. 107/1 In fossil material..the free or bound amino-acids slowly racemize. 1976 Lowry & Richardson *Mechanism & Theory Org. Chem.* v. 218 The excess of rate of loss of optical activity..over rate of product formation..means that some process racemizes the substrate more rapidly than the substrate can form products.

Hence **ra·cemized** *ppl. a.*, **ra·cemizing** *vbl. sb.*
1896 *Proc. Chem. Soc.* 23 Apr. 97 Examining..for the presence of racemised salt by converting it into zinc lactate. 1904 *Jrnl. Chem. Soc.* LXXXV. 1253 The racemising effect of the alkali during hydrolysis of the esters. 1940 *Biochem. Jrnl.* XXXIV. 300 Small amounts of partially racemized aspartic acid, as well as of glutamic acid, can be isolated from both normal and malignant tissue protein material. 1951 C. R. Noller *Textbk. Org. Chem.* xvii. 273 The racemizing action of halide ion was known to Kekulé in 1864. 1972 S. J. Weininger *Contemp. Org. Chem.* vii. 158, 2-Octyl bromide in aqueous ethanol forms, among other products, 2-octanol that is partly racemized.

racemous, *a.* (Later example.)
1821 W. P. C. Barton *Flora N. Amer.* I. 79 In this case, the floriferous ramuli must be considered as racemous branches.

racer[1]. **2. b.** For 'the name of several species of American snake' substitute: a North American snake belonging to the genera *Coluber* or *Masticophis*, esp. a variety of *C. constrictor*. (Earlier and later examples.)
1823 E. James *Acct. Exploring Exped. Pittsburgh to Rocky Mts.* I. 267 *Coluber constrictor*—Racer. 1866 J. C. Gregg *Life in Army* 18 This species, were called racers on account of their great speed. 1900 *Ann. Rep. Bd. Regents Smithsonian Inst. 1898* 794 The *Zamenis constrictor* is the 'black snake' of the East and the 'blue' and 'green racer' of the West. 1926 J. K. Strecker in J. F. Dobie *Rainbow in Morning* 76 W. S. Blatchley..speaker of many 'lies' about the blue racer, or blacksnake. 1946 J. Stuart *Tales from Plum Grove Hills* 85 They slid to the foot of the mountain like racer snakes before a new-ground fire. 1954 R. C. Stebbins *Amphibians & Reptiles Western N. Amer.* 373/2 When foraging, like most racers, the head and neck are held well above the ground. 1969 A. Bellairs *Life of Reptiles* I. iii. 103 Even the fastest snakes such as the North American racers and whip-snakes..cannot exceed more than about four miles an hour.

3. a. (Further examples.)
1903 *Sci. Amer.* 8 Aug. 96/1 Barney Oldfield, on the Ford-Cooper racer..has been steadily lowering track records. 1923 [see *Chirpily *adv.*]. 1935 A. J. Cronin *Stars look Down* iii. viii. 553 They swooped and soared and dived from giddy heights on the Giant Racer until the whole glittering Fun Fair spun around them in one glorious daze. 1974 D. Ramsay *No Cause to Kill* ii. 158, I wanted a bicycle, and she ran right out and bought me an English racer. 1978 *Dumfries Courier* 20 Oct. 11/1 The newcomers range from fairly basic family saloons to the exotic Mazda RX7 and BMW's M1 sports racer.

5. An article of clothing designed in a racing style; *spec.* (*pl.*) a pair of swimming trunks (*Austral.* and *N.Z.*).
1969 *New Yorker* 20 Sept. 149/2 (Advt.), Baby him with our soft, zip-front creepers... Racer in medium blue. 1971 *Telegraph* (Brisbane) 21 Sept. 18/1 (Advt.), Ash boys racers in bright florals..and plain colours. 1977 *N.Z. Herald* 8 Jan. 2–4/8 (Advt.), Speedo togs in quick-drying Bri-nylon. Racers, boys..$4.75.

raceway. For *U.S.* in Dict. read 'Chiefly *U.S.*' and add: **1. a.** (Earlier examples.)
1828 in S. Jenkins *Story of Bronx* (1912) ix. 199 Fourteen mill sites, each fifty by one hundred feet, were mapped out along the raceways. 1837 *Knickerbocker* IX. 254, I was jerked out with great spite, and, with an imprecation, thrown into the raceway.

b. An artificial channel of running water for the rearing of fish.
1897 *Man. Fish Culture* (U.S. Comm. Fish & Fisheries) 74 The object of these boards is to form four avenues leading to the raceway, so that one or two pugnacious fish can not command the approach and keep back spawning fish inclined to enter. 1913 W. E. Meehan *Fish Culture* vii. 104 There has been a tendency in recent years for fish-culturists to abandon raceways. 1953 H. S. Davis *Culture & Dis. Game Fish* ii. 21 Raceways for holding small fingerlings are usually constructed of concrete. 1972 *Aquaculture* I. 229 Water requirements for 'raceway' production of turbot would be very low compared with the requirements in current trout farming practice. 1976 *San Antonio* (Texas) *Express* 23 Sept. 8-G/1 The facility includes a laboratory, office building, 20 ponds, eight concrete raceways and two wells.

2. Also, a groove in a type-setting machine along which types are moved; also, a groove in which ball-bearings run.
1898 *Inland Printer* Nov. 178/1 This machine will select the type, place them in a raceway and move them along until a line is set up. 1946 *Richmond* (Va.) *Times-Dispatch* 8 Dec. iv. 6-D/1 The balls are held in place by two concentric steel rings. They run in grooves or 'raceways' cut in the rings. A retainer or separator is usually inserted to keep them from rubbing against each other.

3. A metal pipe or plastic tube enclosing electrical wires; piping or tubing so used.
1897 F. C. Moore *How to Build* iv. 58 Conduits or raceways for carrying wires through the house should be of iron or other metal. 1964 R. F. Ficchi *Electrical Interference* ix. 168 Metal boxes, cabinets and fittings, or noncurrent-carrying metal parts of other fixed equipment, if metallically connected to grounded cable armor or metal raceway, are considered grounded by such connections. 1976 Lieberman & Rhodes *Compl. CB Handbk.* v. 105 Stick the snake through the 'raceway' (the space in which the tail-light wires run through the trunk).

4. A track or circuit on which harness races, etc., take place; a racecourse. Also *attrib.*
1936 *Sun* (Baltimore) 19 Oct. 12/6 Topping..had purchased the Maserati car which Philippe Etancelin, of France, drove in the Roosevelt raceway test. 1942 *Ibid.* 24 Aug. 11/4 Saratoga Raceway's new harness racing venture. 1958 *Washington Post* 30 Aug. A12/5 Raceway officials told the commission the plant would be ready for night racing by next March 1. 1968 *Globe & Mail* (Toronto) 17 Feb. 43 (*heading*) Greenwood Raceway. Saturday, Feb. 17. 1979 *Beautiful Brit. Columbia* Winter 33 The fans..turning to the racing form for the next race in to-night's harness racing card at Cloverdale Raceway.

rachel (răʃe·l). Also **Rachel.** [f. *Rachel*, the stage-name of Elisa Félix (1820–58), French actress.] A light, tannish colour (used orig. and chiefly of face-powder). Also *attrib.* or as *adj.*
1887 *Illustr. London News* 6 Aug. 163/1 (Advt.), Toilet powder... In three tints: Blanche, for fair skins; Naturelle, for darker complexions; and Rachel, for use by artificial light. 1907 [see *Papier]. 1907, 1927–8 [see *Naturelle *a.*]. 1936 M. Kennedy *Together & Apart* iv. 320 'If you must use powder at your age, do at least find a more becoming shade.' ..'It's not dead white,' muttered Elisa furiously. 'It's Rachel.' 1970 'D. Halliday' *Dolly & Cookie Bird* xii. 195 She has facial plates like a rachel armadillo.

rachi-, rachio-. Add: **rachi·schisis** (-skisis) [Gr. σχίσις cleavage] = *Myelocele 1 b; **rachitome,** (*b*) *Palæont.*, a labyrinthodont belonging to the suborder Rhachitomi; usu. written *rh-*; **rachitomous** *a.* (earlier and further examples); usu. written *rh-*.
1890 Billings *Med. Dict.* 427/2 *Rachischisis*, congenital posterior fissure of spinal column, a form of spina bifida. 1900 *Boston Med. & Surg. Jrnl.* CXLIII. 458/2 (*heading*) A case of rhachischisis. 1901 T. M. Rotch *Pediatrics* (ed. 3) v. 301 Rhachischisis is one of the principal forms of congenital defects of the spine. It is characterized by a deficiency of the vertebral arches either complete or partial. 1963 K. M. Laurence in A. P. Norman *Congenital Abnormalities in Infancy* ii. 26 Myelomeningoceles, myelocystoceles, hydromyelocoeles, and syringomyelocoeles, localized rachischisis and myelocoeles are all essentially the same lesion, and best regarded as myelocoeles. 1966 Wright & Symmers *Systemic Path.* II. xxxiv. 1234/2 Spina Bifida... The severest and rarest form is rachischisis, in which the spinal canal is open to the exterior, either for a short distance or over its whole length. 1947 *Bull. Museum Compar. Zoöl. Harvard Coll.* XCIX. 103 In the skull roof [of *Edops*], a primitive character is the presence of a distinct intertemporal element, lost in characteristic rhachitomes. 1964 *Jrnl. Animal Morphol. & Physiol.* XI. 7 The temnospondyls, forming the 'main line' of labyrinthodont evolution, began with Carboniferous types of primitive structure but with rhachitomous vertebrae, evolved in the later Carboniferous and Permian into typical rhachitomes, and eventually gave rise to stereospondylous forms. 1971 E. C. Olson *Vertebr. Paleozool.* iv. iv. 591 In the Lower Carboniferous limestone of Scotland..is the Gilmerton ironstone from which a number of labyrinthodonts have come:..*Loxomma*, a rhachitome; and..an anthracosaur, *Crassigyrinus*. 1882 E. D. Cope in *Amer. Naturalist* XVI. 334 (*heading*) The rhachitomous Stegocephali. 1884 —— in *Ibid.* XVIII. 30 Rachitomous vertebræ from the same locality are of larger size and resemble those of Eryops. 1947 *Bull. Museum Compar. Zoöl. Harvard Coll.* XCIX. 102 The skull pattern is typically rhachitomous in key features, such as the firm fusion of cheeks and table. 1964 [see *rachitome* above]. 1971 E. C. Olson *Vertebr. Paleozool.* iv. iv. 591 Although rhachitomous amphibians occur early, in the upper Mississippian, the limbs from this age have not been worked out.

rachill, var. Ratchel in Dict. and Suppl.

rachitogenic (rækitodʒe·nik), *a.* [f. Rachit(is + -o + *-genic.] Tending to cause rickets.
1932 *Biochem. Jrnl.* XXVI. 202 The rachitogenic property of Steenbock's diet is due to its high value for Ca/P together with its lack of vitamin D. 1976 *Lancet* 20 Nov. 1132/2 The rachitogenic activity of oatmeal.

Rachmanism (ræ·kmăniz'm). [f. the name of Peter *Rachman* (1919–62), a London landlord + -ism.] Exploitation of slum tenants by unscrupulous landlords. Hence **Ra·chman,** any such unscrupulous landlord. Also **Ra·chmanite** *a.*, of or resembling a Rachman, and other nonce-ish derivatives.
1963 H. Wilson in *Guardian* 23 July 2/3 The disease of Rachmanism is to buy controlled properties at low prices, and to use every means..to bring about evictions which..have the effect of decontrolling the property. 1963 *Daily Tel.* 23 July 10/2 (*Editorial*) Emotion appeared to get the better of precision in Mr. Harold Wilson's opening contribution to the Rachman debate yesterday... More powers..will not remove the basic conditions in which the Rachmans of this world can thrive. That can only be done by increasing the supply of housing space and bringing down rents. 1963 *Guardian* 7 Aug. 8/3 (*heading*) Rachmanship lives on. *Ibid.* 8/5 The sprawling, sordid acres of what has become known as Rachmanland. 1963 *Daily Tel.* 30 Aug. 19/3 Bringing Rachman-type racketeers to heel. *Ibid.* 22 Nov. 28/4 'Rachman-like' landlords illegally withheld deposits totalling more than £1,400. 1965 *Economist* 6 Mar. 976/2 The poorer tenants who have suffered most from Rachmanite and neo-Rachmanite intimidation. 1968 *Guardian* 13 Sept. 3/3 Rachmanism—harassment of unwanted tenants—has simply grown more subtle since the 1965 Rent Act. 1969 D. Widgery in Cockburn & Blackburn *Student Power* 137 The same crisis which forces council rents up and allows the domination of the Rachmans of Islington, Moss Side and Liverpool 9. 1973 C. Mullard *Black Britain* ii. iv. 46 Rachman-type landlords offered as little as they could for as much as they could get. 1973 *Times* 29 Dec. 11/4 Rachmanite landlords..make millions out of office blocks and luxury flats. 1975 *Times* 8 Jan. 15/3 On a fair rents basis..the transaction..would not be attractive to the speculative builder or the Rachman. 1977 *Jrnl. R. Soc. Arts* CXXV. 116/2 Recently, we have added to the problem by extending the area of control to include furnished tenancies. Mr. Tilbe of *Shelter* will no doubt tell you in two weeks time that this was necessary to prevent Rachmanism. 1981 *Times* 9 Feb. 17/5 The long-term tenant's legitimate need for protection against the Rachmanite landlord.

Rachmaninovian (ræːχmænindou·viăn), *a.* and *sb.* [f. the name of Sergei Vasilyevich *Rachmaninov* (1873–1943), Russian pianist and composer.] **A.** *adj.* Of or resembling the style or the works of Rachmaninov. **B.** *sb.* An admirer of Rachmaninov.
1958 *Times* 3 Nov. 14/4 In their recital..there was only one original composition, a Toccata by Murgatroyd Farrar, which made suitably Rachmaninovian noises. 1962 *Times* 5 July 15/1 A verbose Rachmaninovian Scherzo by John White. 1973 *Times* 27 July 15/4 Mr Previn drew the right pliable phrasing from the LSO without overdoing the succulence (like some older Rachmaninovians one could name). 1976 *Gramophone* May 1816/3 The Rachmaninovian flavour of 'Whitechapel'..making a delightful treasure of sound. 1977 *Ibid.* Nov. 874/1 What Rachmaninovians ought to be shouting for now, however, is a recording of the *Liturgy*, Op. 31.

racial, *a.* **a.** (Further examples.)
A word of considerable frequency in the 20th century. The examples that follow illustrate some of the more usual modern collocations.
1889 [see *Nurtural *a.*]. 1892 F. W. Gage *Negro Problem* iii. 56 If it be demonstrated that individual members of the race under favourable circumstances are capable of mature mental development, then the question of racial development is settled. 1899 A. Nutt in *Folk-Lore* June 146 In determining the relative importance of either element for racial discrimination in folklore, I was guided by observation of man in the civilized stage. 1899 C. Waldstein *Expansion of Western Ideals* 141 An historical basis for German unity was not enough; an ethnological, racial unity had to be established. 1914 G. K. Chesterton *Wisdom of Father Brown* ix. 266 An attitude we must always remember when we talk of racial prejudices. 1929 H. Miles tr. *P. Morand's Black Magic* i. 63 He supported..racial equality. 1935 Huxley & Haddon *We Europeans* ix. 286 Ethnic intercrossing and culture-contacts have proceeded so far that 'racial purity', like complete isolationism or self-sufficiency, is impossible of attainment. 1942 H. A. Wallace *Century of Common Man* (1944) vi. 32 In June of 1941 he [sc. Roosevelt] issued an executive order prohibiting racial discrimination in the employing of workers by national defence industries. 1942 Z. N. Hurston in A. Dundes *Mother Wit* (1973) 32/2, I did not have to consider any racial group as a whole. 1943 E. H. Brooks *Bantu in S. Afr. Life* ii. 3 When I say 'a similar point of view', I mean the doctrine of racial domination: there are Africans who still think that the Europeans can be driven into the sea. 1947 'G. Orwell' in *Tribune* 7 Feb. 12/2, I should

like to think that the position of the racial minorities could be safeguarded. **1954** H. GIBBS *Background to Bitterness* 7 Racial conflict between the groups has not been witnessed on a major scale for many years. **1954** *Harvard Law Rev.* XXIV. 80 Judge Edgerton wrote that as the Supreme Court had..recognized that enforced racial segregation in housing was unconstitutional, it followed that enforced racial segregation in schooling was even more so. **1955** B. SCHWARTZ *Amer. Constitutional Law* ix. 224 There has been a profound change in recent years in the attitude of that tribunal [sc. the Supreme Court] toward racial discrimination. **1958** *Spectator* 22 Aug. 239/1 The Little Rock High School must resume racial integration when the new term began. **1960** J. RAE *Custard Boys* II. xix. 210 You think this is a case of racial prejudice and you're probably right. **1960** 'I. ROSS' *Murder out of School* i. 7 There's none of what the papers like to call 'racial tension' at Mark Hopkins [School]. **1967** *Boston Sunday Herald* 7 May III. 4/1 'Racial imbalance in Parochial Schools' is the topic for a panel discussion. **1971** R. BENDIX in A. Bullock *Twentieth Cent.* xv. 357/1 Racial minorities..constitute a lower class as women obviously do not. **1971** *Publishers' Weekly* 2 Aug. 46 Mr. Fuller finds that the anthology's one story by a black author—an Eldridge Cleaver story first published in *Playboy*—is racial tokenism. **1976** *CRC Jrnl.* July 3/1 All of the recent immigration debates..have connected the general anxiety about immigration with the current racial tension. **1977** *Whitaker's Almanack 1978* 595 South African Government declared that where feasible there should be an end to racial segregation on buses. **1979** MILES & PHIZACKLEA *Racism* i. 17 A reaction by blacks in Britain to racial discrimination and violence.

racialism (rēɪ·ʃăliz'm). [f. RACIAL *a.* + -ISM.] Belief in the superiority of a particular race leading to prejudice and antagonism towards people of other races, esp. those in close proximity who may be felt as a threat to one's cultural and racial integrity or economic well-being.

1907 *Daily Chron.* 2 Jan. 6/5 The two principal planks in the party platform are opposition to all racialism and co-operation with the Government. **1910** *Westm. Gaz.* 11 Apr. 10/3 What appears to me to be the greatest results of the Botha–Smuts Government is the abolition of Racialism and the construction of roads. **1925** E. S. JONES *Christ of Indian Road* ii. 67 Amid the racial clash and bitterness there stands one who is the Son of man. Racialism withers at his touch. **1934** R. MACAULAY *Going Abroad* xv. 127 A Cape Afrikander ..had renounced Dutch racialism and the detestation of the English. **1938** *Sun* (Baltimore) 2 Sept. 10/1 The Italian Jews are thus to be added to the victims of Hitler's imbecile 'racialism', now adopted by Mussolini as a sop to superior force. **1940** R. BENEDICT *Race: Science & Politics* vii. 215 Racialism has become involved in scientific absurdities under the Third Reich. **1955** *Times* 20 Aug. 5/5 The Prime Minister spoke on race relations, commenting that in the last session of Parliament there had been less racialism in debate and more moderation. **1960** *Spectator* 6 May 650/1 The two main convictions of racialism are, firstly..'The highest aim of human existence is the conservation of the race..the maintenance of the racial stock unmixed'..and secondly, that once a man's mind is made up about this, he can never think of changing it. **1971** S. ABBOTT *Prevention of Racial Discrimination* i. 16 Britain's long history of colonialism overseas does not sufficiently explain the present racialism in this country. **1975** *Daily Tel.* 13 Nov. 16/3 Racism, or racialism, or racial discrimination,..covers everything from a vile form of monomania to the innocent preference of human beings for association with their own kind. **1977** P. JOHNSON *Enemies of Soc.* ii. 25 Racialism was linked to wishful-thinking, and almost deliberate self-deception.

racialist (rēɪ·ʃălist), *sb.* and *a.* [f. RACIAL *a.* + -IST.] **A.** *sb.* A partisan of racialism; an advocate of a racial theory.

1917 *Deb. House of Commons Canada* 5870/2 We all become nationalists in the true sense of the word, as distinguished from provincialists and racialists. **1930** *Observer* 22 June 13/4 Some of its characters said things that were calculated to make the blood of headstrong racialists boil. **1937** *Discovery* July 224/2 Curiously enough.., the 'nigger' is much more likely to be treated with contempt by the half-educated in England than among the politically-organised racialists of Germany. **1939** A. TOYNBEE *Study Hist.* IV. 19 We can even drive the racialists out of their one remaining Italian stronghold by finding an alternative explanation for the rise of the Roman Republic. **1940** R. BENEDICT *Race: Science & Politics* i. 6 The racialists have rewritten history to provide the scion of such a race with a long and glamorous group ancestry. **1958** *Times Lit. Suppl.* 28 Mar. 164/3 It is easy today for Britain to see Hertzog as a bitter, anti-British racialist, who deprived the remaining Cape Africans of their vote. **1960** *Spectator* 6 May 650/1 A racialist..lives according to what most people think is a fantasy. **1977** M. WALKER *National Front* iv. 85 He [sc. A. K. Chesterton] went on to warn of the perils of racialist extremism, while wholeheartedly agreeing with the racialists' arguments about '..mongrelization'.

B. *adj.* Of, pertaining to, or characterized by racialism.

1946 W. S. KNICKERBOCKER *20th Cent. Eng.* 81 It would be, however, an error to consider this Nazi literary history simply as racialist. **1952** B. DAVIDSON *Rep. S. Afr.* I. vii. 75 Members even of the highly racialist Electors' Union of Kenya..have expressed to me their horror at the explosive possibilities induced by white policy in South Africa. **1960** [see *Africanistic* s.v. *AFRICANIST *sb.* (and *a.*)]. **1971** E. POWELL *Let.* in *Observer* 14 Mar. 8/6 The adjective 'racialist' has gained a strange sort of currency in recent years and seems to wear all sorts of meanings. I have even once or twice heard it applied to myself.

Hence **raciali·stic** *a.*

1960 *Guardian* 14 Dec. 16/2 The extreme racialistic African leaders. **1969** *Daily Tel.* 18 Jan. 18/4 Coomaraswamy was more than a little influenced by the sort of racialistic sentiment applied to art that has become one of the curses of the 20th century. **1977** P. JOHNSON *Enemies of Society* xix. 248 The statement is purely racialistic.

racialization (rēɪ·ʃăləizēɪ·ʃən). [f. RACIAL *a.* + -IZATION.] The process of making or becoming racialist in outlook or sympathies. Hence **ra·cialize** *v. trans.*

1918 *Encycl. Relig. & Ethics* X. 557/2 Why should the most progressive Muslim populations be affected most powerfully by 'racialization', which is clearly a retrogressive tendency? **1930** *Month* Dec. 485 A Catholic, following St. Paul, will repudiate this attempt to racialize the universal genius of Christianity. **1977** M. BANTON *Idea of Race* 18 There was a social process, which can be called racialization, whereby a mode of categorization was developed.

racially, *adv.* Add: (Further examples.) Freq. linked with a ppl. adj. to form adjs., as *racially-blended, -integrated, -selected.*

1914 'SAKI' *When William Came* x. 170 The record of your racially-blended supper-party. **1921** J. BRYCE *Mod. Democracies* I. xiv. 163 Where a racially distinct body of unwilling subjects is included within a State..are they to be reckoned as part of the people? **1962** Racially-integrated [see *COLOUR-BLINDNESS b]. **1976** *Drum* (E. Afr. ed.) Sept. 13/2 The team pulled out of the Olympics..in protest against New Zealand sending a rugby team to play in South Africa against racially-selected sides. **1976** E. K. FRANCIS *Interethnic Relations* xxii. 280 Particular populations that are racially distinct from their social environment occasionally show typical mental and cultural differences.

raciation (rēɪsiˌēɪ·ʃən). [f. RACE *sb.*² + -ATION; cf. *SPECIATION.] The evolutionary development of distinct biological races.

1952 *Sat. Rev.* (U.S.) 5 July 16/3 The methods he [sc. Edgar Anderson] describes are very like those used by the anthropologist interested in tracing the wanderings of peoples, their mixture, and 'raciation'. **1971** *Nature* 28 May 250/1 This approach to microsystematics has been of particular value in the study of raciation..in commercially important marine fishes.

racing, *vbl. sb.*¹ Add: **1.** (Further examples.)

1901 *Chambers's Jrnl.* Apr. 221/2 The *Turbinia* has been run..in almost all states of the sea, and on no occasion has the slightest symptom of racing occurred. **1912** G. GREENHILL *Dynamics of Mech. Flight* v. 106 Racing of the screw is due chiefly to variation of axial flow. **1980** G. M. FRASER *Mr American* xviii. 336 Pip's method of travel was..constant racing of the engine.

2. (Some functionally indistinguishable from *racing* used as ppl. adj.) *racing bicycle, canoe, car, change* (*CHANGE *sb.* 1 g), *correspondent, cycle, cyclist, driver, establishment* (earlier example), *guide, motor-car, motorist, page, result, rig, stable* (later example), *stud* (earlier example), *-track*; **racing colours**, the colours (COLOUR, COLOR *sb.* 6 a) by which an owner's racehorses are identified; **racing demon** (see *DEMON 2 f) (examples); **racing dope** *U.S.*, information about races contained in a *dope-sheet* (see *DOPE *sb.* 5); **racing flag** *Naut.* (see quot. 1961); **racing game** = *race-game* s.v. *RACE *sb.*¹ 11 b; **racing pigeon**, a homing pigeon taking part in competitions to complete a specific journey as quickly as possible; **racing-plate** (see quot.).

1910 *Encycl. Brit.* III. 915/1 Wood rims are used on racing bicycles. **1959** I. & P. OPIE *Lore & Lang. of Schoolchildren* x. 191 Who's teacher's pet boy and was given a racing bicycle? **1876** Racing-canoe [see *long-spooned* s.v. *LONG *a.*¹ 16]. **1932** *Man* May 106 The coffin is placed in a large racing canoe. **1909** Racing car [see *power producer* s.v. *POWER *sb.*¹ 18]. **1977** M. KENYON *Rapist* iv. 44 He said he..had been an important racing car driver. **1959** I. FLEMING *Goldfinger* vii. 86 James Bond flung the DB III through the last mile of straight and did a racing change down into third. **1907** *Yesterday's Shopping* (1969) 302/2 About seven days required to execute orders for Racing Colours. **1955** W. GADDIS *Recognitions* I. v. 196 She handed a folded twenty-dollar bill to a boy wearing her racing colors. **1961** A. CLARKE *Later Poems* 87 At Maynooth, instead of skulls, His racing colours were displayed. **1961** E. WAUGH *Unconditional Surrender* III. ii. 240 His brief experience as a racing correspondent seemed irrelevant to the zeitgeist. **1976** 'J. WELCOME' *Grand National* i. 11 That is Andrew Mostyn, our chief racing correspondent. **1976** *Eastern Even. News* (Norwich) 22 Dec. 11/3 (Advt.), Racing cycles at discount prices. **1974** *Times* 29 Oct. 17/4 Scores of club racing cyclists. **1945** N. MITFORD *Pursuit of Love* ix. 74 The Kroesigs obviously longed for bridge, and did not seem to care so much for racing demon when it was offered as a substitute. **1977** *Times* 24 Dec. 10/2 Try racing demon for the party card game if you have a large table and fast-playing, shouting screaming players. **1931** F. L. ALLEN *Only Yesterday* iv. 81 Workmen forgot to be class-conscious as they..studied the racing dope about Morvich. **1961** *Daily Tel.* 21 May 10, I used to be a racing driver. **1977** 'D. CORY' *Bennett* iii. 88 Shop girls identify themselves with film stars, bank clerks with racing drivers. **1811** J. STEELE *Let.* 29 Jan. in *Papers* (1924) II. 649 There is nothing..which wd. afford me greater pleasure than to see a respectable racing establishment at this place. **1860** 'VANDERDECKEN' *Yarns* 135 Cut the racing flag clear, and send a hand aloft to lash it to the stump as a signal that you'll fight to the last. **1961** F. H. BURGESS *Dict. Sailing* 167 *Racing flag*, a private flag hoisted when racing, instead of the burgee; it is hauled down only on retiring, or when a race is completed. **1860** C. M. YONGE *Hopes & Fears* I. I. v. 123 She beheld his sister..at the racing game... Honor waited, however, till the little white horseman had reached the goal. **1890** —— *More Bywords* 154 The 'racing game'..which was now spread on the dining-table, with all the young people playing in high glee. **1909** A. L. BRUCE *Bridge-Fiend* 12 A peppery, red-faced old gentleman, who was reading a racing-guide..was then appealed to. **1909** *Westm. Gaz.* 21 Oct. 5/1 Fewer accidents have happened to aviators in proportion to their numbers than to racing-motor-car drivers in the same period. **1906** *Chambers's Jrnl.* Apr. 347/1 The trophy which is the prize for which racing-motorists compete. **1948** 'J. TEY' *Franchise Affair* xiv. 156, I went to rest every afternoon with.. the racing page of the daily paper. **1978** *Islands* (N.Z.) Aug. 78, I got the paper of course. For the racing page really. **1910** A. H. OSMAN *Pigeon Bk.* xiii. 148 It is impossible to say what breeds have and have not been used to 'make' the racing pigeon. **1933** *Discovery* Nov. 344/1 Racing and homing pigeons are often captured. **1960** *Farmer & Stockbreeder* 26 Jan. 4/2 Demand is expected to broaden as the racing-pigeon season approaches. **1977** *Wandsworth Borough News* 16 Sept. 15/5 Literally rescued from the teeth of a predatory cat, a blue racing pigeon now awaits a claimant at the home of Mr. I. A. McWilliam.' **1958** J. HISLOP *From Start to Finish* 174 *Racing-plate*, a light shoe (usually made of some form of aluminium) with which horses are shod when they race. **1926** Racing result [see *bedtime story* s.v. *BEDTIME]. **1976** 'J. WELCOME' *Grand National* ii. 27 He turned on the radio to get the racing results. **1906** CONRAD *Mirror of Sea* viii. 39 Of those three varieties of fore-and-aft rig, the cutter—the racing rig *par excellence*—is of an appearance the most imposing. **1981** E. WARD *Baltic Emerald* xxiv. 189 A racing stable with high pasturelands for gallops. **1828** T. CREEVEY *Let.* in *Creevey Papers* (1963) xii. 241 We started about 3 for Petworth..Sefton's object being to see Lord Egremont's Racing Stud before dinner. **1913** *O.E.D.* s.v. *Track* sb. 6 b, Racing-track. **1917** [see *MORNING *sb.* 7 ¶]. **1929** W. E. COLLINSON *Spoken Eng.* 72 They've built a greyhound racing-track just near the house.

Racinian (rasi·niăn), *a.* and *sb.* Also **Racinean**. [f. the name of Jean *Racine* (1639–99), French dramatic poet.] **A.** *adj.* Of, pertaining to, characteristic of, or resembling Racine or his writings. **B.** *sb.* An admirer or imitator of Racine.

1927 *Sunday Times* 13 Mar. 8/3 He [sc. Otway] still remains the most Racinian of all our poets. **1931** *Times Lit. Suppl.* 1 Jan. 1/3 Shakespearian, Racinian or Sheridanesque convention. **1946** *Month* May–June 229 His brilliant and enthusiastic appreciation of the great cycle of tragedies, above all *Athalie*, will please the most ardent Racinian. **1950** M. MCCARTHY in *Reporter* 18 July 37/2 This Racinean world, where stepmother Phèdre and grandmother Athalie queened it. **1948** L. SPITZER *Linguistics & Lit. Hist.* 178 The Racinian, the Vergilian power of poetic alchemy whereby brute reality is transmuted. **1962** *Listener* 30 Aug. 315/2 This Racinian reading of modern life which is dramatized in terms of passion versus reason, will versus duty. **1974** *Ibid.* 8 Aug. 185/2 *A Month in the Country*..harks back to Racine. There is the Racinian web of emotional incompatibilities.

raciology (rēɪsiǫ·lŏdʒi). [f. RACE *sb.*² + -OLOGY; cf. F. *raciologie.*] The study of the races of man; racial composition. Hence **raciolo·gical** *a.*; **racio·logist**.

1924 *Glasgow Herald* 7 Feb 6/5 Societies were formed for the study of their language and raciology, just when the authentic gipsies themselves had begun to disappear. **1926** *Ibid.* 27 Jan. 10/4 A new and searching process of selection.. will result in a revisal of our preconceived notions of African raciology. **1939** C. S. COON *Races of Europe* viii. 286 Von Eickstedt, the most articulate of the modern German raciologists. **1950** E. W. COUNT *This is Race* 703 For a commentary on Buffon's raciology, see Scheidt. *Ibid.* 734 The Russian raciologists were very actively engaged..in combating..'bourgeois' racism in general. *Ibid.* 735 Some other works of Fleure have a raciological bearing.

racism (rēɪ·siz'm). [f. RACE *sb.*² + -ISM; cf. F. *racisme* (Robert 1935).] **a.** The theory that distinctive human characteristics and abilities are determined by race. **b.** = *RACIALISM.

1936 L. DENNIS *Coming Amer. Fascism* 109 If..it be assumed that one of our values should be a type of racism which excludes certain races from citizenship, then the plan of execution should provide for the annihilation, deportation, or sterilization of the excluded races. **1938** E. & C. PAUL tr. *Hirschfeld's Racism* xx. 260 The apostles and energumens of racism can in all good faith give free rein to impulses of which they would be ashamed did they realise their true nature. **1940** R. BENEDICT *Race: Science & Politics* i. 7 Racism is an *ism* to which everyone in the world today is exposed. **1952** M. BERGER *Equality by Statute* 236 Racism, tension in industrial, urban areas. **1952** *Theology* LV. 283 The idolatry of our time—its setting up of nationalism, racism, vulgar materialism. **1960** *New Left Rev.* Jan./Feb. 21/2 George Rogers saw fit to kow-tow to the incipient racism of his electorate by including a line about getting rid of 'undesirable elements'. **1964** GOULD & KOLB *Dict. Social Sci.* 571/2 *Racism* is a newer term for the word *racialism*... There is virtual agreement that it refers to a doctrine of racial supremacy. **1971** *Ceylon Daily News* (Colombo) 18 Sept. 8/5 Mr. Seneviratne is welcome to his ideal of inter-racial marriages as panacea for Racism. **1972** J. L. DILLARD *Black English* iii. 90 In the British sailors' reactions to the slaves.., the very early existence of racism is as well documented as the difference in language. **1974** M. FIDO *R. Kipling* 50/2 In *The Story of Muhammad Din* he wrote one of the most economical and bitter attacks on British racism ever penned. **1976** *Plain Dealer* (Cleveland, Ohio) 4 Mar. A2/4 The Vatican radio said,..'Racism might have different faces but it will always be reprehensible.' **1977** M. WALKER *National Front* vi. 155 A strike of the Asian workers against racism in the factory.

racist (rēɪ·sist), *sb.* and *a.* [f. RACE *sb.*² + -IST.] **A.** *sb.* = *RACIALIST *sb.*

1932 M. EASTMAN tr. *Trotsky's Hist. Russ. Revol.* i. 27 This brief comment completely finishes off not only the old philosophy of the Slavophiles, but also the latest revelations of the 'Racists'. **1934** H. G. WELLS *Exper. Autobiogr.* I. iii. 107 So much for the Hitlerite stage of my development, when I was a sentimentalist, a moralist, a patriot, a racist. **1940** R. BENEDICT *Race: Science & Politics* vii. 214 Classic German racists..ascribed all achievements beyond the Alps to infiltrations of northern blood. **1959** *New Statesman* 30 May 754/2 They see nothing to be gained..if they are dismissed and replaced by fanatical racists. **1965** *San Francisco Examiner* 15 Apr. 34/5, I recently heard a man denounced as a racist for having observed that the rate of illegitimacy in New York is 14 times as high among the Negro population as among the white. **1973** A. DUNDES *Mother Wit* p. xii, Folklore has been used as the tool of racists.

B. *adj.* = *RACIALIST *a.*

1938 E. & C. PAUL tr. *Hirschfeld's Racism* xv. 201 Elective affinity laughs at the maxims and prohibitions of racist wiseacres. **1938** *Mag. Digest* Aug. 22 The racist revue, *Archiv für Biologie und Rassengesellschaft*, one of the organs of the National Socialist Party, published an article..on 'The utility of aerial bombardments from the point of view of racial selection and social hygiene'. **1938** *Sun* (Baltimore) 14 Nov. 6/2 On Thursday..Rome approved new decrees increasing the severity of Italian Fascism's new 'racist' principles. **1940** R. BENEDICT *Race: Science & Politics* vii. 188 The racist traditions..of the fair, blue-eyed narrow-heads. **1957** P. WORSLEY *Trumpet shall Sound* App. 268 Racist doctrines and rule by force 'worked' to a degree in the short run of Nazidom, they failed in the (not very) long run. **1960** *Guardian* 23 Mar. 8/2 The President is trying to knock out the racist props from under the present immigration law. **1970** E. BULLINS *Theme is Blackness* (1973) 167 I'm too mature and sophisticated to get sucked in by racist arguments. **1979** *Globe & Mail* (Toronto) 28 Nov. 8/4 [Mr. Levesque] can't bear any suggestion that he or his party could be racist, could treat non-francophones as second-class citizens.

Hence **raci·stic** *a.* (*rare*).

1950 E. W. COUNT *This is Race* 734 Combating racistic theories. **1963** *Observer* 7 Apr. 22/2 This society is dedicated to pleasure and not over-concerned with the big racistic abstractions.

rack, *sb.*¹ Add: **4. a.** (Later examples.)

1899 H. T. TIMMINS *Nooks & Corners of Shropshire* 65 We go down a rough footpath, or 'rack', as they call it hereabouts. **1904** G. A. B. DEWAR *Glamour of Earth* v. 81, I came down the rack—the narrow path which is cut through ripe underwood fifteen years old, and marks the end of one lot and the beginning of another. **1919** T. WRIGHT *Romance of Lace Pillow* xii. 110 What a relief..to be absolutely free for a few hours; to be able to..roam the ridings, racks, and glades of Yardley Chase. **1957** *Brit. Commonw. Forest Terminol.* II. 149 *Rack*, (*a*) A narrow woodland track maintained for inspection and communication and for extraction of poles, etc. by hand or animal haulage.

e. (Earlier and later U.S. examples.)

1832 J. F. WATSON *Historic Tales of Olden Time N.-Y. City* 27 The 'Racks' so called, along the [Hudson] river, were Dutch names for Reaches. **1930** *Amer. Speech.* V. 164 The Dutch navigators divided the Hudson into *racks* or *reaches*. The former word remains in Claverack.

rack, *sb.*² Add: **3. e.** *to stand* (or *come*) *up to the rack*: to face or bear the consequences of what one has undertaken; to take one's share of hard work or responsibility. *U.S.*

1834 D. CROCKETT *Narrative of Life* iv. 61, I was determined to stand up to my rack, fodder or no fodder. **1835** —— *Col. Crockett's Tour* 69 It was a hard row to hoe; but I stood up to the rack. **1837** R. M. BIRD *Nick of Woods* II. xiv. 183 But, you see, captain, there's a bargain first to be struck between us, afore I comes up to the rack. **1848** J. F. COOPER *Oak Open.* II. iii. 43 The English used to boast that the Americans wouldn't 'stand up to the rack', if the baggonet was set to work. **1890** *Stock Grower & Farmer* 12 July 4/2 For several years cattlemen have been severe losers but most of them have stood pluckily to the rack.

4. b. *spec.* One on which items of clothing are transported and displayed for sale. Phr. *off the* (or *a*) *rack* = *off the peg* adv. phr. s.v. *PEG *sb.*¹ 1 e.

1948 H. MCCLENNAN *Precipice* (1949) ii. 189 Shipping clerks pushing racks of women's dresses. **1962** W. SCHIRRA in *Into Orbit* 47, I acted as a kind of consultant tailor on the pressure suit. It is not possible just to walk in and buy one off the rack. **1976** 'R. BOYLE' *Cry Rape* xx. 91, I chose a simple navy shirtmaker dress from the $20-and-under rack. **1976** *Times* 2 Nov. 12/2 In the women's outfitting department, there was..a scramble around the racks of camel coats. **1978** R. LUDLUM *Holcroft Covenant* xxxiii. 385 His suit was off a rack.

5. g. A large, vertical, metal framework, usu. of standardized dimensions, for supporting items of telephonic or electronic equipment and allowing ready access to them.

1893 PREECE & STUBBS *Man. Telephony* xix. 311 An even more effective contrivance for cable racks..is shown in fig. 240. **1906** J. POOLE *Pract. Telephone Handbk.* (ed. 3) xxi. 299 Condenser Rack.—This frame is for the accommodation of the 2½-microfarad condensers used in connection with the incoming junction lines... The frame is 7 feet 3½ inches wide and 7 feet 10 inches high. **1930** *Proc. IRE* XVIII. 1320 The amplifiers are mounted on relay racks and connected by twin lead wire pulled in rigid conduit. **1951** *Short Wave Mag.* May 179/1 The left-hand rack, No. 1, starting at the bottom, contains the filament supplies for all transmitters; the 1000 v. HT supply for the 430 and 144 mc exciters; [etc.]. **1977** *Gramophone* June 118/2 In the professional world it is common practice for tuners, preamplifiers, power amplifiers, equalizers, etc. to be mounted on slotted panels, which are mounted vertically into racks. Now several domestic manufacturers are also mounting their units in neat vertical racks, but usually they are less than the professional 48cm (19-inch) width.

h. *U.S.* (See quots.)

1903 *Nation* (N.Y.) 6 Aug. 115/2 Another Americanism we miss under Racks, the technical name for the side plankings or buffers of our ferry slips. **1905** *N.Y. Even. Post* 20 Dec. 1 Three of the Lackawanna 'racks', as the arrangement of piles to fit the ferryboats are called, were left intact.

i. *N. Amer.* A set of antlers. Also *attrib.*

1945 *Richmond* (Va.) *Times-Dispatch* 27 July 14/1 There is no real means of comparing a rack of antlers killed in Bath County and one in New Kent, unless they are placed side by side. **1958** *Outdoor Life* Sept. 34/1 I'd shot moose in British Columbia but never a really big one. This trip I was determined to get a trophy rack. **1971** D. C. BROWN *Yukon Trophy Trails* i. 22 'Wow, he's sure got a big rack,' someone else yelled. **1976** *Listener* 15 Apr. 466/2 The moose ..had a rack of five points, which meant that it was five years old and almost fully grown. **1978** L. L. RUE *Deer N. Amer.* iv. 66 A deer with more than four points is called a rack buck... Some racks are large but have few points, some are small but have more points.

j. *U.S. Naut. slang.* (See quot. 1962.)

1955 C. KENTFIELD *Alchemist's Voyage* I. iii. 68 'Where's D'Alessio?' 'In his rack.' **1962** *Amer. Speech* XXXVII. 288 A Marine's bed is not a *sack*, but a *rack*. He *hits the rack* or *puts in rack time*. **1963** *Ibid.* XXXVIII. 78 The term *rack* was borrowed by the Marines from the Navy, and it began to supersede *sack* as the popular term in Marine speech during the early 1950s.

6. b. (Later example.)

1965 G. MCINNES *Road to Gundagai* ix. 134 Up again, straining on the rack-and-pinion of the Rigi.

c. (Further examples.)

1903 *Baedeker's Northern Italy* 13 From Capolago to the Monte Generoso, rack-and-pinion railway in 56 minutes. **1958** R. LIDDELL *Morea* II. ii. 55, I took the rack and pinion railway up to Calávryta. **1969** *Observer* (Colour Suppl.) 23 Mar. 29 Rack-and-pinion steering 'can be twirled from lock to lock with the flat palm of one hand'. **1972** *Modern Railways* Sept. 334 This was overcome on the BOB [sc. Berner Oberland Bahn] by the use of rack-and-pinion operation with gradients as steep as 1 in 8. **1973** *Country Life* 18 Oct. 1172/1 The Haflinger, a forward-control platform truck..seems to wind on inexorably, rather like a rack and pinion train climbing a mountain. **1978** *Daily Tel.* 16 Aug. 10/6 The ride is on the firm side with the handling being safe and predictable from the rack and pinion steering.

8. b. = *rack-rail* in sense 9.

1909 *Westm. Gaz.* 7 Aug. 7/2 The greater part of the line would traverse exceedingly difficult country, necessitating.. possibly a few short lengths of rack.

9. rack car, (*b*) *U.S. Logging*: see quot.; **rack chain** (later examples); **rack chase** *Printing*, a chase having racked sides into which fit two adjustable bars; **rack-meat**, fodder placed in racks for horses; **rack mounting** *vbl. sb.*, the use of standardized racks for supporting telephonic or electronic equipment; so **rack mount** *sb.* and *v. trans.*; **rack-rail, railway** (later examples); **rack saw**, (*b*) see quot.; **rack-way**, (*b*) a path through a wood, *esp.* one used for timber extraction.

1958 W. F. MCCULLOCH *Woods Words* 145 *Rack car*, a railroad car specially equipped with stakes or racks to handle pulpwood. **1958** J. HISLOP *From Start to Finish* iv. 20 Do not leave your horse tied up by the rack-chain, in your hurry to get away. **1963** E. H. EDWARDS *Saddlery* xxii. 167 Usually a rope..to the rear of a head collar is best for tying up unless one has rack chains. **1882** J. SOUTHWARD *Practical Printing* vi. 72 Rack chases for fixing small formes on presses are made the size of a press table, and obviate the use of furniture. **1898** —— *Mod. Printing* I. ix. 66 Rack chases..are made to fit the carriage of a press and the bed of a machine. **1743** W. ELLIS *Mod. Husbandman* Dec. vii. 46 To..further their Fattening, by enough of dry, hearty Trough and Rack-meat in Time. **1849** G. A. DEAN *Essays on Construction of Farm Buildings & Labourers' Cottages* 23 Many persons consider that the racks are best placed by the sides of the mangers..others, that horses who work hard should have no rack-meat given to them, considering that they satisfy their hunger much quicker..from the manger. **1965** *Wireless World* July 2 (Advt.), Series 'Y' instruments are housed in strong metal cases and, in some instances, can be rack-mounted. **1976** *Physics Bull.* Jan. 9 Available in either a rack-mount or a cabinet configuration, it is designed to be used by persons with little or no previous experience with signal averagers. **1978** *Chicago* June 106/1 Rack mounts, for instance, are 'in'... These racks will hold pre-amps, amps, equalizers, tuners and tape decks. Some can even accommodate a turntable. **1940** *Chambers's Techn. Dict.* 697/1 *Rack mounting*, the use of standard racks..for mounting panels carrying apparatus..with a uniform scheme of wiring. **1977** *Gramophone* June 118/1 A Sony rack mounting amplifier using pulse width modulation. **1979** *Sci. Amer.* June 8/2 (Advt.), The 5315B is essentially the same instrument housed in a metal case for rack mounting or stacking. **1918** *Chambers's Jrnl.* Jan. 13/1 See hazardous bridges being built, and the rack-rail employed to surmount steep gradients. **1931** *Times Educ. Suppl.* 21 Feb. p. iii, An engraving showing a Blenkinsop rack-rail engine and train. **1913** *Chambers's Jrnl.* Jan. 128/2 This railway introduces a cheaper means of ascending rugged mountains than the rack-railway laid upon the ground. **1931** *Times Educ. Suppl.* 21 Feb. p. iii, John Blenkinsop, the inventor of the rack railway, died 100 years ago. **1973** C. BONINGTON *Next Horizon* xii. 183, I..plunged through the deep powder snow..down to the rack-railway track that led up to Kleine Scheidegg. **1971** F. C. FORD-ROBERTSON *Terminol. Forest Science* 209/2 *Rack saw*, a head saw (circular or band) with a travelling table operated by rack-and-pinion. **1727** D. EATON *Let.* 16 May (1971) 120 The rack ways in Priors Haw are all brush'd up regularly... I was yesterday in Priors Haw whilst Mr. Goods servant was tything the brush-wood that was cut out of the rack ways. **1969** *Gloss. for Landscape Work* (*B.S.I.*) v. 40 *Rackway*, a narrow unpaved pathway left or cut through a tree crop to give access and to facilitate the extraction of timber to a wider ride or road.

rack, *sb.*⁴ **1. a.** Delete 'Now only *dial.*' and add later examples.

1964 J. MASTERS *Trial at Monomoy* iii. 101 Mary Tolley began to serve the main course, a rack of lamb. **1972** *New York* 12 June 63 Purée of cold carrot soup, rack of lamb, cauliflower provençale. **1974** *Observer* (Colour Suppl.) 22 Sept. 60/3 (Advt.), Fresh river trout followed by rack of highland lamb. **1977** *Time* 21 Nov. 44/3 Recently he ordered a hotel restaurant billboard repainted after noticing that the rack-of-lamb dinner on it 'looked raw'.

3. c. *U.S. colloq.* **rack of bones**, a skeleton; an emaciated person or animal. Also *rackabone, rackerbone, rack-o'-bones.*

1804 J. ORDWAY in Lewis & Ordway *Jrnls. Western Explor.* (1916) 128 We saw the rack of Bones of a verry large fish. **1854** M. J. HOLMES *Tempest & Sunshine* iv. 58 Turn that old rackerbone of yourn straight round, and turn down that ar street. **1856** G. D. BREWERTON *War in Kansas* xxxi. 314 Indeed she was to all appearances but a mere rack of bones, over whose unpicturesque outline nature had condescended to draw an angular wrinkling of skin. **1877** J. HABBERTON *Jericho Road* xvi. 146 Ain't it bad enough to be a good-for-nothin' rack of bones that's no comfort to myself? **1900** *Congress. Rec.* XXXIII. App. 6 Mar. 117/2 A Western farmer had a college-bred son who went off preaching... He came back with an old rackabone. **1911** J. C. LINCOLN *Cap'n Warren's Wards* ix. 140 If she fell on that poor rack-o'-bones,..'twould be a final smash. **1949** *Sat. Even. Post* 2 Apr. 97/2 Mount that rack o' bones you call a horse and ride in front o' me.

rack, *sb.*⁵ Add: **2. c.** *U.S.* **rack-heap**, (*a*) a heap of wreckage; (*b*) (see quot. 1958).

1883 'MARK TWAIN' *Life on Miss.* xxiii. 258 There was only one boat advertised..a Grand Tower packet... She was a venerable rack-heap, and a fraud to boot. **1889** P. BUTLER *Personal Recollections* vii. 72 There were in the river heaps of drift-wood, called 'rack-heaps', dangerous places into which the water rushed with great violence. **1892** 'MARK TWAIN' in *Sun* (N.Y.) 13 Mar. 18/2 Fridolin entered..with a tall skeleton stalking in his rear... The testimony of this wandering rackheap of unidentified bones. **1909** —— *Is Shakes. Dead?* i. 18 When the *Pennsylvania* blew up and became a drifting rack-heap freighted with wounded and dying. **1958** W. F. MCCULLOCH *Woods Words* 145 *Rackheap*. *a.* A piled-up drift or heap of logs and trees in a river. *b.* Sometimes applied to a heap of logs piled up ready to be splashed down a river.

rack, *v.*² Add: **3. b.** (Earlier and later examples.)

1743 W. ELLIS *Mod. Husbandman* Dec. iv. 29 When the Landlord came to rack up the Horse for all Night. **1960** G. E. EVANS *Horse in Furrow* ii. 43 The baiters' mates.. were expected..to *rack the horses up* for the night—that is, to fill their racks with fodder.

c. (Earlier example.)

1856 'STONEHENGE' *Brit. Rural Sports* 330/2 The lad first racks up his horse, so that he cannot lie down, but can reach his manger.

d. *fig.* To chalk up, to notch up; to achieve, to score. *N. Amer.*

1961 in WEBSTER. **1970** *Globe & Mail* (Toronto) 28 Sept. 18/3 The winners won the statistical battle by a wide margin, racking up 22 first downs to 16 for the losers. **1974** E. MCGIRR *Murderous Journey* 82 I've got some leave of absence piled up... I would have racked up close on a month. **1976** *Billings* (Montana) *Gaz.* 16 June 3-c/1 Billings began racking up runs in the fourth and fifth innings, while Missoula managed only one more run. **1977** *Time* 14 Nov. 25/1 Over the short run the U.N. vote may even have played into the hands of South African Prime Minister John Vorster, who is anxious to rack up a big majority in the country's Nov. 30 elections. **1978** G. VIDAL *Kalki* vi. 139 CBS had racked up a Nielsen rating of 36.3, the highest ever in that particular time slot. **1979** *Sci. Amer.* Dec. 30/3 She professionally ferries light aircraft (very often Beeches), the number of engines your choice, out of the U.S. to anywhere at all, having racked up almost 350 ocean crossings in 17 years of action.

4. a. (Later examples.) *spec.* in the *Oil Industry*, to place (lengths of drill pipe) in a pipe rack or derrick.

1949 *Our Industry* (Anglo-Iranian Oil Co. Ltd.) (ed. 2) ii. 39 When the drill pipe is being withdrawn, it is uncoupled in 'stands' of three 30-ft. lengths, these 90-ft. 'stands' being racked upright in the derrick. **1970** W. A. SMITH *Gold Mine* xxix. 81 Big King..wiped down his glossy shoes and racked them. **1971** C. BONINGTON *Annapurna South Face* xiii. 163 Mick got ready for the next pitch,..racking his pitons on karabiners slung to one side so that he could free them easily. **1973** J. W. JENNER in Hobson & Pohl *Mod. Petroleum Technol.* (ed. 4) iv. 120 The floormen..swing the bottom end of the stand away from the table and it is lowered on to the rig floor, at which time the derrickman..pulls the top of the stand over and racks it against the side of the derrick. **1974** *BP Shield Internat.* Oct. 18/2 The second noise was..the drilling pipe being racked in the derrick.

5. a. (Later example.)

1906 *Westm. Gaz.* 18 Aug. 14/2 If the image is too big, rack out the camera a little and bring the board nearer. If too small, rack in and push the board away.

rack, *v.*⁴ Add: **a.** (Later examples.) Also *trans.* with a distance as object.

1887 E. CUSTER *Tenting on Plains* vi. 187 He [sc. a horse] is very affectionate, and he racks a mile inside of three minutes. **1935** H. DAVIS *Honey in Horn* xi. 175 He saddled and bridled the mare..and racked out on the road. *Ibid.* xv. 254 When the wagon went out of sight..he spurred up and racked after it.

b. Used *transf.* of vehicles or of persons. *to rack off* (Austral.), to go missing, 'get lost'.

1935 Z. N. HURSTON *Mules & Men* I. iv. 95 Pretty soon the log-train came racking along. **1975** *Sun-Herald* (Sydney) 29 June, [*title of record*] Rak Off Normie. **1980** *Courier-Mail* (Brisbane) 10 Apr. 36 (*caption*) 'Gimme ya money mate or I'll shoot ya!' 'No... Now rack off!'

rack (ræk), *v.*⁷ *Building.* [var. RAKE *v.*³] *trans.* To build (a brick wall) by stopping each course a little short of the one below, so that the end slopes (usu. temporarily until the work is completed). Usu. with *back*. Cf. RAKING *vbl. sb.*³ in Dict. and Suppl.

1873 F. ROBERTSON *Engin. Notes* ii. 35 In repairing masonry where there is a crack or junction, or where new work is to be connected with old, the adjoining ends should be racked back from each other, as it were in ascending steps, and the resulting wedge-shaped void subsequently built in. **1904** C. F. & G. A. MITCHELL *Brickwork & Masonry* ii. 77 (*caption*) Angles of walls racked preparatory to building. *Ibid.* 78 The base of the corner is extended along the wall, and is racked back as the work is carried up. **1945** E. L. BRALEY *Brickwork* iii. 58 Usually five or seven courses are built at each corner, the work being racked back, e.g. first of all three stretchers, then four headers and one closer, then two stretchers, two headers and a closer, one stretcher, and finally the heading face of the top brick. **1972** S. SMITH *Brickwork* iv. 17 When building a wall, it is usual to raise the 'quoins' (corners) first, 'racking back' the work as necessary.

rackan hook. Also **reckon hook** (the usual form). [f. RACKAN + HOOK *sb.*] = RACKAN-CROOK.

1645 *Essex County* (Mass.) *Probate Rec.* (1916) I. 50 Estate of William Goog of Lynn... One gridiron & recke hookes [etc.]. **1647** *Ibid.* 99 Estate of John Jarrat of Rowley ..Reckon hooks & some small things, 4s. **1867** B. BRIERLEY *Marlocks of Merriton* 41 His eyes still intent upon the 'rackan'-hook' hanging in the kitchen. **1961** M. W. BARLEY *Eng. Farmhouse & Cottage* III. v. 175 In such houses as these, cooking was usually done in the house body. There, along with the iron range and the reckon hook, were the bakestone and the wooden boards..with which oatcakes were made.

rack-bone. (Later example.)

1831 W. YOUATT *Horse* ix. 153 The other neck, or rack-bones, as they are denominated by the farrier,..are of a strangely irregular shape.

rackensak (ræ·kənsæk). [Prob. altered form of *Arkansan.*] *U.S. colloq.* A native of Arkansas. ? *Obs.*

1845 [see *CORN-CRACKER 1]. **1854** *Putnam's Mag.* III. 665/2 Artillerists and dragoons, suckers and rackensacks, were all mixed up in confusion.

racker². (Later examples.)

1891 *Harper's Mag.* Aug. 366/1, I have seen more than one racker of true Norman blood. **1903** A. D. MCFAUL *Ike Glidden* xiv. 108 Lickety got ter puffin' up his ole hoss, soze you'd a thought it was the Millbridge Racker.

racket, *sb.*² Add: **3. b.** (Earlier example.)

1846 P. J. DE SMET *Oregon Missions* (1847) xiv. 193 The savages travel over these marshy places in *Rackets*.

racket, *sb.*³ Add: **2. c.** A dance: see quots.

1882 L. O. CARPENTER *J. W. Pepper's Universal Dancing Master* 33 *Racquette*..Make three galop steps or slides to the left, throwing the foot out in second position... Slide to right [etc.]. **1882** P. V. CARTIER *Practical Illustrated Waltz Instructor* 45 *The Racquet*..Take two long galop slides with left foot on accent, and as right foot is brought up to left foot for second time, rest, and hold left foot in air. Repeat by sliding with right foot, etc. **1885** A. DODWORTH *Dancing* vii. 51 Racket Waltz (One-Slide Racket in Waltz Time). *Ibid.* 52 Changes are made..by alternating the one-slide racket with the three-slide. **1935** D. N. CROPPER *Dance Dict.* 54 *Racket*, popular 6/8 number of the 'nineties'. Basic step: a waltz form with leap (1) slide (&) change-cut (2).

3. a. Now usually, any scheme or procedure which aims at obtaining money or effecting other objects by unusual, illegal, and often violent means; a distinctive form of organized crime.

1928 *Daily Express* 14 Sept. 1/1 The 'racket' has for years been distinctively a Chicago institution; and it has been found to be such a profitable form of crime there that it is spreading to the other large cities of the Middle West. **1931** *Sun* (Baltimore) 4 Apr. 1/4 One racket in New York State alone—that of fake securities—is known to total approximately $100,000,000 a year. **1940** E. GILL *Autobiogr.* vii. 259 It [sc. politics] is all a confused business of ramps and rackets—pretended quarrels and dishonest commercial schemings, having no relation to the real interests of peoples, neither to their spiritual nor their material welfare. **1944** M. LASKI *Love on Supertax* xii. 117 You organised all this Black Market racket, didn't you? **1950** G. BRENAN *Face of Spain* ii. 54 Of all the rackets recorded in history, the Spanish Inquisition, during the first hundred years of its career, was perhaps the most mean and repulsive. **1956** 'C. BLACKSTOCK' *Dewey Death* iv. 83 Mr. Wilson is now telling everybody that I.L.D.A. is the secret headquarters of the drug racket. **1974** J. GARDNER *Return of Moriarty* 31 All our family is affected if we start to lose in any racket, any lay. **1977** *Times* 29 Nov. 14/2 Ulster by the middle of 1974 was suffering from rackets and violent crime on a scale equal to some of Europe's most notorious cities.

b. In more weakened senses: an activity, a way of life; a line of business.

1891 [in Dict.]. **1907** R. DUNN *Shameless Diary of Explorer* xviii. 251 The Professor is working his faith-in-God-and-self, and line-of-least resistance racket, a mite too strong. **1916** J. BUCHAN *Greenmantle* i. 4, I thrive on the racket and eat and sleep like a schoolboy. **1927** *Vanity Fair* (N.Y.) XXIX. 132/3 'What's your racket?' meaning 'What do you do for a living?' **1930** *Sun* (Baltimore) 12 Feb. 10/7 My satisfaction would be complete if there were a 100 per cent rush for the doors that would entirely eliminate the encore racket. **1931** F. L. ALLEN *Only Yesterday* vii. 172 At the beginning of the decade advertising had been considered a business..by the end of the decade many of its practitioners..were beginning to refer to it—among themselves—as a racket. **1936** *Amer. Speech* XI. 274/2 Nowadays a *racket* may be a legitimate business... A man may say..'I rather like the racket I'm in', referring to his business. **1938** *Downside Review* LVI. 100 It is true that the phrase 'muscling in on the culture racket' reflects a development of English word-usage from which the present reviewer had perhaps been preserved by monastic seclusion. **1942** *R.A.F. Jrnl.* 3 Oct. 11, I sold insurance, and..that's the racket to develop your wits. **1944** J. S. HUXLEY *On living in Revol.* 23 What with football, racing, the cinema, the theatre, popular literature, and holiday resorts, recreation is today one of the most profitable commercial rackets. **1978** J. UPDIKE *Coup* (1979) vi. 239, I am in the insurance racket. I am a claims adjuster.

4. *to stand the racket* (further examples). Also, (*c*) to pay.

1846 *Swell's Night Guide* 132/2 *Stand the racket*, treat, pay for all. **1904** G. K. CHESTERTON *Napoleon of N.H.* III. iii. 168 'Can we do fifteen hundred pounds?' 'I'll stand the racket.' **1905** *Pall Mall Mag.* Dec. 678/2 If there is trouble, it will be for Great Britain to stand the racket. **1930** *Punch* 19 Feb. 204/3 If her..friend had been a sportsman he'd have stood the racket himself.

6. *attrib.* and *Comb.*, as *racket-buster, -busting* (sb. and adj.), *-ridden* (adj.), *ring*.

1940 *Sun* (Baltimore) 21 Nov. 1/2 Sol Gelb..had been assigned by the New York 'racket-buster' to watch the hearing. **1959** *Times Lit. Suppl.* 30 Jan. 55/4 Mr Danforth was senior investigator..from 1935–1951, when former Governor Thomas E. Dewey was the courageous D.A...and his famous racket-busting took place. **1972** 'H. HOWARD' *Nice Day for Funeral* iv. 58 Until the motive is established beyond doubt this case remains part of the DA's racket-busting programme. **1978** *Time* 3 July 55/2 Died. Luther W. Youngdahl, 82, unflappable federal judge who..was appointed to the bench in 1951 after five years as a racket-busting Republican Governor of Minnesota. **1931** F. D. PASLEY *Muscling In* v. 138 New York stood revealed as the most racket-ridden city in the country. **1973** *Black Panther* 5 May 2/2 It is widely known that Inman is himself a king pin in the city's organized crime and racket rings.

racket, var. *RACKETT.

racket, *v.*¹ **1. b.** (Later *transf.* example.)

1861 F. W. ROBINSON *No Church* I. iv. 95 An improvident young man, who..would racket away all the money he might be able to leave her.

racket, *v.*² Add: **2. a.** (Later examples.) Also const. *about, along, around.*

1885 B. POTTER *Jrnl.* 6 May (1966) 141 How is it these high-heeled ladies who dine out..can racket about all day long, while I..am so tired toward the end of the afternoon that I can scarcely keep my feet? **1914** W. OWEN *Let.* 21 Dec. (1967) 309, I racketed about all Saturday making luggage out of lumber. **1916** 'BOYD CABLE' *Action Front* 197 A dozen paces away two of the machine-guns were clattering and racketing in rapid gusts of fire. **1929** M. DE LA ROCHE *Whiteoaks* xiv. 188 Aha..that's what I like to hear! Young lads racketing about! **1936** A. RANSOME *Pigeon Post* vi. 69 But do you think we'll hear it?' said Mrs. Blackett, 'when we're racketing about and busy with other things.' **1967** J. C. HOLMES *Nothing More to Declare* i. 20 We read it in an empty subway car racketing along under the deserted streets. **1970** G. GREER *Female Eunuch* 331 The first significant discovery we shall make as we racket along our female road to freedom is that men are not free. **1977** 'L. EGAN' *Blind Search* i. 12 That girl racketing around heaven knows where or with what sort of characters. **1977** W. M. SPACKMAN *Armful of Warm Girl* 34 Guests..racketing up into the bedroom.

racketeer (rækĕtīə·ɹ), *sb.* orig. *U.S.* [f. RACKET *sb.*³ + -EER.] A member of a gang or association of criminals practising extortion, intimidation, violence, and other illegal acts on a large scale; any person making easy money by such means. Also *transf.*, one who achieves an easy result by illegitimate means.

1928 *Time* 9 July 14 In the old days it was a mark of distinction to be seen at gangster funerals, but during the Loesch prosecutions, probably not even U.S. Senator Deneen of Illinois would care to be seen near the bier of a 'racketeer'. **1928** *Daily Express* 14 Sept. 1/4 'Racketeers'..now control 150 lines of business in Chicago, and collect an enormous tribute for immunity from their violence. **1929** *Sun* (Baltimore) 15 Nov. 1/6 'Spike' along with six police captains and a dozen politicians and racketeers, is accused of participation in the profits of gambling machines placed in speakeasies. **1931** *Times* 24 Sept. 11/2 The campaign against gangsters and 'racketeers' in New York City has resulted to-day in the arrest of..one of the most powerful 'labour racketeers' in the city. **1935** J. T. FARRELL *Judgment Day* iv. 76 We got to get a strong man in the White House..to kick out the bankers and grafting politicians and racketeers. **1939** *Scrutiny* VII. 439 The older generation of middlebrow propagandists, whom *Scrutiny* used to refer to as literary racketeers. **1948** *Sunday Pictorial* 18 July 7/1 The public are completely at the mercy of these racketeers. **1956** 'C. BLACKSTOCK' *Dewey Death* iii. 52 You romantic writers are as much a menace to the community as drug racketeers. **1967** *Wall Street Jrnl.* 24 Apr. 32/2 Rosenberg, according to Illinois authorities, was secretly associated with..an important Chicago racketeer. **1978** *Cornish Guardian* 27 Apr. 3/1 Metrication will be an open invitation for every spiv and racketeer to cheat the British public.

racketeer (rækĕtīə·ɹ), *v.* *U.S.* [f. the sb.] **a.** *trans.* To subject to racketeering. **b.** *intr.* To engage in fraudulent business.

1928 *Time* 30 Jan. 11/2 In 36 years in Chicago I have never been held up, robbed, or racketeered. **1933** G. B. SHAW *Polit. Madhouse in Amer.* 56 What is the use of paying you money to racketeer with? **1934** *Words* Nov. 5/2 *To press-agent*,..*to service*,..*to gesture*, *to racketeer*..are new, and most of them are obviously American.

racketee·ring, *vbl. sb.* *U.S.* [f. *RACKETEER *sb.* + -ING¹.] The business of racketeers; a system of organized crime directed chiefly to extorting money from business firms by intimidation, violence, or other illegal methods. Also *attrib.*

1928 *N.Y. Times* 18 Aug. 15 Two gang murders within the last week prompted Judge Edwin O. Lewis..to order the August Grand Jury to delve to the bottom of 'racketeering' in Philadelphia. **1928** *Daily Express* 14 Sept. 1/4 'Racketeering' is the new word that has been coined in America to describe the big business of organised crime. **1929** *Sun* (Baltimore) 10 Jan. 1/6 The defendants are charged with compelling..manufacturers to pay tribute to them by threatening to call strikes. The indictment was one of the first to be returned here in connection with a Federal investigation of racketeering. **1930** *Observer* 19 Oct. 17 He had hoped..perhaps to introduce and organise 'racketeering' processes. *Ibid.*, The only new detail is the paid protection of blackmail which now exists in many cities, described by the term 'racketeering'. **1931** *Sun* (Baltimore) 29 Jan. 1/6 Another effort is to be made to prohibit congressional nepotism now commonly recognized as 'pay-roll racketeering' on the part of members of both the House and Senate. **1941** L. B. NAMIER *Conflicts* (1942) 163 Look at this Jew! What did he do in the war? Some racketeering? **1959** *Ann. Reg. 1958* 187 Only on three major Bills was the President defeated..the attempt to regulate the affairs of trade unions to cut out racketeering [etc.]. **1978** S. BRILL *Teamsters* ii. 41 It was at the time when he was being charged with racketeering that Bufalino joined Hoffa's legal team.

racketee·ring, *ppl. a.* [f. *RACKETEER *v.* + -ING².] Characterized by or engaging in rackets.

1931 *Times* 30 July 11/3 Sometimes employers trying to operate an 'open shop' hire strike breakers... In other cases 'racketeering' gangs take the initiative in intimidating employers with 'open shops'. **1967** *Sunday Times* 8 Oct. 24/8 Kim, at this period, 'gave the impression of being a complaisant passenger in a racketeering upper-class world'.

racketiness (ræ·kĕtinĕs). [f. RACKETY + -NESS.] The quality of being rackety; fondness for noise, excitement, etc.

1939 C. DAY LEWIS *Child of Misfortune* III. ii. 270 No doubt racketiness was just part of the fashionable sexual lure then. **1979** *Listener* 1 Nov. 508/2 Her racketiness and smart friends arouse less appreciation.

rackett (ræ·kĕt). Also **racket, ranket(t.** [a. G. *rackett, rankett.*] **1.** A Renaissance musical instrument of the oboe family, consisting of a squat cylinder containing nine parallel channels joined alternately at top and bottom to form a continuous tube nine times the length of the cylinder.

1876 STAINER & BARRETT *Dict. Mus. Terms* 374/1 *Rackett, Rankett*, (1) an obsolete wind-instrument of the double bassoon kind. **1891** C. R. DAY *Catal. Mus. Instr. R. Milit. Exhib. 1890* 100 Racket. This beautiful instrument is constructed in the form of an ivory cylinder, and it is played by means of a rather large double reed. **1910** F. W. GALPIN *Old English Instr. of Music* ix. 167 A yet shorter instrument of bass pitch with a cylindrical-shaped tube.. was called the Racket. **1939** A. CARSE *Musical Wind Instr.* xiv. 206 In the racket or sausage-bassoon the air-passage is doubled and redoubled to such an extent that the sounding-length of the tube is quite nine times as long as the body of the instrument. **1961** A. BAINES *Mus. Instr. through Ages* ix. 232 The deep soft buzz of one racket among recorders, cornetts, etc., made a better effect than a whole consort of them [sc. mixed instruments]. **1966** —— *Europ. & Amer. Mus. Instr.* 98 The *rackett* contains a number of short parallel bores connected in series to make up a total windway of a metre or more. **1968** *Radio Times* 26 Sept. 48 A unique collection of medieval instruments—including rackett, rebec, crumhorn. **1970** *Daily Colonist* (Victoria, B.C.) 26 Feb. 28/1 They have chosen to play medieval music on a number of rare and little-used instruments, including the krummhorn, rauschfeiffer, ranket, and baroque oboes. **1976** D. MUNROW *Instr. Middle Ages & Renaissance* 46/1 The rackett's narrow cylindrical bore consists of no less than nine parallel channels drilled in a wooden or ivory cylinder and connected alternately top and bottom.

† **2.** An organ stop. *Obs.* exc. *Hist.*

1876 STAINER & BARRETT *Dict. Mus. Terms* 374/1 *Rackett, Rankett*..(2) An organ stop of 16 ft. or 8 ft. pitch now obsolete. **1897** H. RIEMANN *Dict. Mus.* 629/1 *Rackett* (Ranket),..in the organ an obsolete reed-stop almost entirely covered, of quiet tone (16 and 8 feet). **1962** S. IRWIN *Dict. Pipe Organ Stops* 160 *Rankett*, a very old form of short-resonatored Reed stop, at 16' or 8' on both manual and pedals.

rackety. Add: **1.** (Later examples.)

This and sense 2 are tending to merge.

1975 I. MURDOCH *Word Child* 257 It was raining, and a rackety wind was sweeping the rain in little wild gusts across the windows. **1976** A. POWELL *Infants of Spring* v. 80 In the middle age-group of most houses there inclined to occur a cluster of fairly rackety boys, from whom the housetutor might expect trouble. **1977** *Daily Tel.* 20 Jan. 12/1 Crosby did not much like Harvard, but he seems to have been a fairly conventional undergraduate there, even if wilful and rackety.

2. (Later examples.)

1927 C. CONNOLLY *Let.* 11 Feb. in *Romantic Friendship* (1975) 250 One misses the thrilling rackety journey to the wagon restaurant. **1961** A. RITNER *Seize Nettle* 158 The big basket of clothes to be coaxed through the rackety old washer. **1974** C. MILNE *Enchanted Places* xix. 129 A room designed—as a nursery should be—for doing things in, messy things, rackety things, rough-and-tumble things. **1975** J. SYMONS *Three Pipe Problem* xviii. 201 He unlocked the door, switched on the engine, and listened to its rackety coughing.

Rackhamesque (rækăme·sk), *a.* [See -ESQUE.] Characteristic of or resembling the drawings of Arthur Rackham (1867–1939), book illustrator.

1935 *Forestry* IX. 15 There was also strong feeling about the way in which Rackhamesque trees of the forest were rapidly being replaced by pines. **1936** 'G. ORWELL' *Keep Aspidistra Flying* iv. 88 The trees..twisted themselves into whimsy Rackhamesque attitudes. **1961** S. GILRUTH *Drown her Remembrance* iii. 30 A few isolated olive trees, gnarled and twisted into weird Rackhamesque shapes.

racking, *vbl. sb.*² Add: **2.** Shelving designed to be functional and inexpensive rather than decorative.

1937 G. FRANKAU *More of Us* viii. 91 While Art Department hummed like dynamo As frenzied hands tore pictures from their racking. **1976** *Gloss. Documentation Terms* (*B.S.I.*) 52 *Racking*, shelving, usually of a cheaper quality, used for storage purposes in non-public stacks and areas of a library.

racking, *vbl. sb.*³ Add: **2.** *spec.* Distortion of a structure under shear. (Further examples.)

1957 *Brit. Commonw. Forest Terminol.* II. 149 *Racking*, in timber testing, the application of loads to an assembly, tending to deform it in shear. **1976** W. J. PATTON *Construction Materials* 386 *Racking*, tendency of a rectangular frame to distort from its rectangular shape due to lack of stiffness against shear forces. **1977** *Engin. Materials & Design* Aug. 17/1 A batch of fifty radiators made in this way have been subjected to tests against thermal shock cycling, pulsating pressure, vibration and racking.

racking, *vbl. sb.*⁴ (Later example.) Also *attrib.*, as *racking event, horse.*

1974 *Marlboro Herald-Advocate* (Bennettsville, S. Carolina) 18 Apr. 10/3 In ladies racking, Sherry Jean Nolan..rode King to a first-place win. *Ibid.*, Larry Griggs rode King to a first place victory in the junior racking event. **1974** *Greenville* (S. Carolina) *News* 23 Apr. 11/2 Friday performances, beginning at 1 p.m. and 7 p.m., have the pleasure horse classes,..along with three racking horse classes.

racking, var. RAKING *vbl. sb.*³ in Dict. and Suppl.

ra·ck-jobbing, *vbl. sb.* [f. RACK *sb.*² + JOBBING *vbl. sb.*² 2.] The supplying of goods to a retailer for display on racks on condition that the supplier undertakes to accept unsold stock after an agreed period. Hence (as a back-formation) **ra·ck-job** *v. intr.* and *sb.* (used *attrib.*). Also **ra·ck-jobber.**

1959 *Economist* 12 Dec. 1090/1 A complete 'rack jobbing' service. Here the supplier takes responsibility for the stock and display from week to week; the retailer simply provides the space, and accepts a lower margin on the goods that are sold. **1964** *Credit Trends* Sept. 5 Some supermarket operators are moving these lines, and the profession of rack-jobbing has grown up. The retailer leases out shelf space to these specialist concessionaires in return for an agreed profit margin related to normal earning for the space used. **1967** *Economist* 15 July 238/3 Smith and other wholesalers can help by teaching them [sc. newsagents] and stocking for them—as some paperback publishers already rack-job for the small newsagent or supermarkets. **1968** *Times* 29 Nov. p. iv/5 E.M.I., Decca and Pye have recently set up a joint company called Record Merchandisers to exploit what is referred to in the trade as rack jobbing. This is a system where the servicing company supplies the records for display on racks in non-conventional outlets (such as stationers and supermarkets) taking full responsibility for what is put on display and taking back unsold stock. **1969** *JEMF Quarterly* V. III. 101 In recent years, the introduction of rack-job merchandising of LPs in supermarket and other retail outlets has obviated the need for printed graphics to complement record distributions. **1977** *Rolling Stone* 19 May 14/4 It further states that Klein instructed 'another person' to sell the records at a profit to wholesalers, rack jobbers and distributors.

ra·ckman. *U.S.* [f. RACK *sb.*² + MAN *sb.*¹] A man who distributes newspapers from the publishing office to local newspaper racks.

1943 *Sun* (Baltimore) 17 Sept. 20/2 The..Court of Appeals upheld today a lower court decision that rackmen distributing papers..for the publishing company of the Baltimore Sun were not engaging in interstate commerce within the meaning of the Fair Labor Standards Act. **1944** *Ibid.* 18 Jan. 17/7 (*heading*) Rackmen decision refused review.

rack-out, *a.* [f. RACK *v.*² 5.] Designed to rack out.

1893 *Photogr. Ann.* 333 This camera is well known. It has double extension leather bellows... The extension is rack-out, but by an ingenious arrangement instantly extended as required.

rack-pin. 1. (Earlier example.)

1832 *Blackw. Magazine* XXXII. 471 Friend, if thou be'st not nautical, thou knowest what a *rack-pin*, something of the stoutest, is.

rack-renting, *vbl. sb.* (s.v. RACK-RENT *v.* in Dict.) (Earlier and later examples.)

1663 E. BUTTERFIELD *Let.* 1 Feb. in M. M. Verney *Memoirs of Verney Family* (1899) IV. ii. 43, I hate this rack-renting 'tis worse than usury. **1840** J. S. MILL in *Edin. Rev.* LXXII. 46 Much alteration may be requisite in the system of rack-renting and tenancy at will. **1963** *Economist* 3 Aug. 421/1 The rack-renting of London's tenantry. **1969** *Listener* 12 June 815/3 Communism appeals to hundreds of thousands of peasants who hate corruption, rack-renting and foreign intervention.

racks (ræks). *Television slang.* (See quots.)

1960 O. SKILBECK *ABC of Film & TV* 104 *Racks*, colloquial term for the vision control department (T.V.) between cameras and vision mixer. **1974** *Some Technical Terms & Slang* (Granada Television), *Racks*, the television control area between studio camera and control box.

‖ **raclette** (raklet). [Fr., = scraper.] **1.** *Archæol.* [A. Cheynier 1930, in *Bull. Soc. Préhist. Française* XXVII. 488] An end-, or side-scraper, of a type discovered in the valley of the Vézère, dating from the Early Magdalenian age. Also *attrib.*

1931 *Proc. Prehist. Soc. E. Anglia* VI. 322 Dr. André Cheynier,..working at Badegoule, has obtained from a special layer several hundreds of..tools, to which he has given the name of Raclettes. **1936** *Nature* 11 July 79/2 (*heading*) An Early Magdalenian 'raclette' industry in the Lower Thames valley.

2. A fondue-like dish consisting of cheese melted before an open fire, scraped on to the plate, and served with potatoes. Also *attrib.*

1949 A. L. SIMON *Dict. Gastron.* 200/2 *Raclette*, the name given in the Valais Canton of Switzerland to the local *Fondue*. **1958** *Times* 15 Nov. 11/6 There is dried meat of the Valais..and there are *raclettes* and *fondues*. **1961** *Times* 23 Mar. 16/7 The Seiler family arranged an enormous *raclette* party on the slopes of the Riffelalp. **1971** *Vogue* 15 Sept. 118/1 *Raclette*..is..a speciality of the Valais... A whole side of cheese is grilled in front of a brazier..the sizzling bits scraped on to your plate..served with potatoes boiled in their skins and gherkins. **1974** *Times* 4 Feb. 17/5 Six helpings of raclette cheese with potatoes.

racloir (raklwā·ɹ). Chiefly *Archæol.* [Fr., = scraper.] A scraper, esp. of a type discovered amongst the remains of the Mousterian period of the Middle Palæolithic period.

1892 P. L. SIMMONDS *Commercial Dict. Trade Products* 311/2 *Racloir*, a scraper; a grater; an instrument to strike off the heaped corn in a measure. **1923** *Nature* CXII. 118/2 The latest group which is found upon the Stoke Newington 'floor' is a clearly-defined Mousterian industry, with fine examples of both racloirs and of the equally characteristic trimmed-flake points. **1935** *Antiquity* IX. 118 [Flint] blades of Upper Palaeolithic facies, racloirs, tranchets, and carinated fragments. **1956** A. L. ARMSTRONG in D. L. Linton *Sheffield* vi. 94 Zones II and III yielded quartzite hand-axes, racloirs, and scrapers displaying great skill and a refined technique.

racon (rēɪ·kǫn). orig. *U.S.* [f. *RA(DAR + BEA)CON *sb.*] = *radar beacon* s.v. *RADAR².

1945 *Army & Navy Jrnl.* 18 Aug. 1534/4 *Racon*, radar beacons. Stations which serve as the radar equivalents of lighthouses. **1947** L. A. TURNER in L. N. Ridenour *Radar System Engin.* viii. 246 Beacons of the synchronous sort just described have been variously called 'radar beacons', 'responder beacons', 'racons', and 'transponders'. **1958** *Proc. Inst. Electr. Engineers* CV. B. Suppl. No. 8. 351/1 A racon has recently been developed at the Admiralty Signals and Radar Establishment for use on lightvessels, the intention..being that a number of such racons shall be fitted around the shores of Great Britain on both lightvessels and lighthouses. **1967** B. KNOX *Blacklight* vii. 142 'The screen was registerin' one o' those blacklight beacons.'..'The deep-water racon,' nodded Carrick. **1977** *Globe & Mail* (Toronto) 24 Feb. B16 Lakes shipping..can also obtain guidance from racons (shore-based radar reflectors).

raconteur. (Earlier and later examples.)

1828 J. C. YOUNG *Jrnl.* 3 July in *Memoir Charles Mayne Young* (1871) I. v. 169 Sir Charles is a handsome, thorough-bred gentleman, and a capital *raconteur*. **1922** JOYCE *Ulysses* 604 A gifted man, Mr Bloom said of Mr Dedalus senior, in more respects than one and a born *raconteur* if ever there was one. **1937** *Discovery* Oct. 326/1 Mrs. Johnson says little about herself, indulges in no purple passages, and without the conscious effort of the *raconteur* she manages to introduce many good stories and telling anecdotes. **1958** L. DURRELL *Mountolive* xv. 296 The inevitable anecdote of a famous *raconteur* to round off the letter. **1972** J. MOSEDALE *Football* iii. 35 (*caption*) Jimmy Conzelman functioned as quarterback, coach, raconteur, songwriter..and promoter.

racoon, raccoon, *sb.* Add: **a*.** The skin or fur of the racoon.

1815 C. WILT *Let.* in J. C. Luttig *Jrnl. Expedition Upper Missouri* (1920) 130 Raccoon from your country will not bring 62½c in Kentucky. **1901–2** *T. Eaton & Co. Catal.* Fall & Winter 42/1 Alaska Sable... Black Persian Lamb... Raccoon. **1976** 'D. HALLIDAY' *Dolly & Nanny Bird* ii. 28 Hefty young men clad in Timberwolf, Raccoon, Scimmia, Tibetan Yak and Natural Unplucked Nutria.

b. racoon-cap *U.S.*, a cap made from the dressed skin of the racoon; **racoon dog**, a mammal about the size of a fox, *Nyctereutes procvonoides*, belonging to the family Canidæ, native to eastern Asia, and distinguished by thick greyish-brown fur and black, racoon-like markings on its head; so **racoon-like dog** (in same sense); **racoon oyster**, substitute for def.: a small, brown-shelled oyster, *Ostrea frons*, found in clusters off the shores of south-eastern North America; (earlier and later examples).

1840 *Knickerbocker* XVI. 163 He then made me a rakish raccoon-cap, with a flaunting tail to it. **1848** in H. Howe *Hist. Coll. Ohio* 151 For..several years after the war, raccoon-caps, with fur outside..were almost universally worn. [**1833** J. E. GRAY *Illustr. Indian Zool.* II. plate 1 (*caption*) Racoon-faced Dog.] **1868** *Proc. Zool. Soc.* 522 Raccoon Dog. Tail short, bushy. **1876** A. R. WALLACE *Geogr. Distrib. Anim.* I. x. 226 The quadruped figured is the curious racoon dog. **1959** *Times* 23 Feb. 10/5 Two of the strangest members of the dog family arrived recently at the Regent's Park Zoo. They are the maned wolf..and the raccoon-dog from Siberia. **1974** L. E. BUELER *Wild Dogs of World* 217 In Japan..the raccoon dog was once common to all the principal islands. **1890** ST. G. MIVART *Dogs, Jackals, Wolves, & Foxes: Monogr. Canidæ* 135 The Raccoon-like Dog is an inhabitant of Japan, the valley of the Amoor, and China. **1931** *Proc. Zool. Soc.* 174 A female Raccoon-like Dog..lived..5 years. **1964** L. S. CRANDALL *Managem. Wild Mammals in Captivity* 280 The raccoon-like dog..is a small grayish animal,..with a black facial mask which is the basis for its name. **1834** J. J. AUDUBON *Ornith. Biogr.* II. 504 Shrimps..have been detained at low water on the banks of racoon oysters, a kind of shell-fish so named under the idea that they are eaten by that quadruped. **1854** W. G. SIMMS *Southward Ho!* iii. 28 They procure the ordinary 'racoon oyster'—the meanest of the tribe. **1885** *Harper's Mag.* Jan. 219/1 When the mangrove grows on the outer edge of the water-line, and drops its aerial roots,..the spat of the raccoon oyster finds a lodgement.

racquetball (ræ·kĕtbǫl). orig. *U.S.* Also **racquet ball.** [f. RACQUET + BALL *sb.*¹] A game resembling paddle ball played with a light ball and a racket in a four-walled handball court. Also *attrib.* Hence **ra·cquetba:ller**, one who plays racquetball.

1972 WICKSTROM & LARSON (*title*) Racquetball-paddleball fundamentals. **1974** *Wall Street Jrnl.* 12 June 1 Mr. Kendler split with the International Racquetball Association last year to form a rival organization, the National Racquetball Club, which sponsors a professional tour for 16 top-ranked racquetballers. **1976** *Milton Keynes Express* 23 July 17/4 Over the five week period they will include trampoline, squash, volley-ball, racquet ball, gymnastics, football, table tennis, five a side football,..chess and draughts. **1978** *Monitor* (McAllen, Texas) 29 May 3B/7 The newest indoor sports craze for physical fitness nuts is racquetball. The game is a cross between tennis and handball. **1979** *Tucson Mag.* Apr. 55/1 Racquetball courts and numerous other recreation centers are spotted throughout the city.

racy, *a.* **3.** (Further examples.)

1901 'P. BEE' *Vagaries of Men* 107 Women who tell racy stories..can rouse a great deal of enthusiasm in a room full of men. **1955** *Times* 19 Aug. 8/2 Lieutenant-Colonel R. J. T. Hills..contributes to the summer number of the *Household Brigade Magazine* some racy memories of Combermere, the Household Cavalry barracks at Windsor. **1971** D. POTTER *Brit. Eliz. Stamps* xv. 179 Every new Great Britain stamp appears on the front cover in full colour, and there is hardly a week without a racy, but informative piece, on some aspect of Great Britain collecting. **1973** *Nature* 27 July 241/2 The introductory passages are autobiographical in content and colourful, frank and uninhibited in style—racy is the only word.

rad, *sb.*¹ Add: Also **Rad.** (Earlier and later examples.)

1820 LADY J. CAMPBELL *Let.* 18 Nov. in Duke of Argyll *Intimate Society Lett.* (1910) II. 654 We shut all our shutters for fear our lights shd seem *Rads* too. **1898** H. N. PAGE *Red Rock* xxxiv. 411 He..was abusing Leech and Still and pretty much all the Rads. **1912** J. GALSWORTHY *Eldest Son* I. ii. 27 Plenty of time to work up the constituency before we kick out these infernal Rads. **1973** R. HAYES *Hungarian Game* viii. 61 A clumsy bribe and a gambit about student rads.

rad (ræd), *sb.*³ Abbrev. of RADIAN.

1913 GODFREY & SIDDONS *Four-Figure Tables* 38 Radians and degrees... Rad. **1960** C. L. MCCLURE *Theory of Inertial Guidance* i. 11 The total inertial angular velocity of the terrestrial geoid: $\omega_{EI} = 7{\cdot}2921 \cdot 10^{-5}$ rad/sec...measured in radians per mean solar second. **1973** *Nature* 6 Apr. 372/3 The Mars 2 and 3 experiments were intended to give a continuous scan of the planet's surface using a field of vision of 0·01 rad (15 km from a distance of 1,500 km).

† **rad** (ræd), *sb.*⁴ *Obs.* [f. RAD(IATION.] A unit of X-ray dose (see quots.).

1918 S. RUSS in *Arch. Radiol. & Electrotherapy* XXIII. 226 [I venture to put forward a suggestion for a new unit, by which the dose of X rays administered to a patient may be measured; the new unit is primarily intended for use in what is known as 'deep therapy'.] *Ibid.* 231 The radium capsule, when applied to malignant cells, causes complete inhibition of their proliferation after an exposure thereto for one hour; this latter quantity is the suggested unit, viz., the Rad. **1922** S. RUSS et al. in *Lancet* 4 Feb. 213/1 As yet there is no international unit by which X rays are measured, and during the course of this work we have employed the 'rad' as our working unit—i.e., the minimum dose of X rays which..is needed to prevent the growth of Jensen's rat sarcoma when inoculated into normal rats.

rad (ræd), *sb.*[5] Abbrev. of RADIATOR. Also *attrib.*

1935 F. BRETT YOUNG *White Ladies* v. vi. 558 The rads are stone-cold. **1967** L. DEIGHTON *Expensive Place* xxvii. 169 He..opened up the rad cap. **1975** J. SYMONS *Three Pipe Problem* iii. 28 It was an air bubble, we've bled the rads. **1977** *Hot Car* Oct. 19/2 The fan should be left off and an electric one fitted in front of the rad.

rad (ræd), *sb.*[6] Pl. **rad, rads.** [f. initial letters of *radiation absorbed dose.*] A unit of absorbed dose of ionizing radiation, corresponding to the absorption of 100 ergs of energy per gramme of absorbing material (0·01 joule per kilogramme).

In the International System of Units replaced by the gray, equal to 100 rads.

1954 *Brit. Jrnl. Radiol.* XXVII. 243/1 The rad is the unit of absorbed dose and is 100 ergs per gramme. **1957** *Financial Times Ann. Rev. Brit. Industry* 87/2 For example, a 1,000 curie cobalt source has a total power output of 15 watts and gives a radiation intensity of the order of 100 rads per second of high penetration at a convenient operating distance. **1968** *Times* 16 Dec. 7/4 During the flight the three astronauts are expected to receive radiation doses of less than one rad a man, less than from chest X-ray series. **1971** J. Z. YOUNG *Introd. Study Man* xxviii. 393 The minimum dose of radiation that will produce a doubling of the spontaneous mutation rate is likely to be in the region of 10 rad per 30 years. **1972** 'J. LANGE' *Binary* 64 Two rad cartridges... Bars of plutonium-238 oxide. That's a radioactive isotope. **1976** *Lancet* 6 Nov. 993/1 The intended tumour dose was 4000 rads in four weeks. **1976** *Path. Ann.* XI. 368 (*caption*) Follicular thyroid carcinoma excised 26 years after approximately 700 rad were applied for enlarged tonsils and adenoids.

rad, *a.*[2] Add: Also **radd.** (Further examples.)

1861 R. QUINN *Heather Lintie* 56 We'd na be rad o' scath frae wather, Though snow was wreathin'. *a* **1870** H. S. RIDDELL *Poet Wks.* (1871) II. 250 That ye might saints and angels meet And no be rad. **1930** in *Sc. Nat. Dict.* (1968) VII. 319/3 I'm radd ye ken mair aboot youres and sick like than yer buiks.

R.A.D.A., RADA (rā·dă). Also **Rada.** [Acronym f. the initials of *Royal Academy of Dramatic Art*, founded in 1904 and granted a royal charter in 1920.] One of the leading acting schools in Britain, at present situated in Gower Street, London. Also *attrib.*

1921 *Times* 19 July 8 (*heading*) R.A.D.A. Ex-Students First Production... The author of the play was Mr. Kenneth Barnes, Director of R.A.D.A., and his maiden effort was a worthy one. **1937** W. S. MAUGHAM *Theatre* xx. 191, I always say you're the greatest actress on the stage. I've learnt more from you than I did all the years I was at the R.A.D.A. **1949** *Listener* 21 Apr. 676/1 They were, in fact, R.A.D.A. students at play. **1962** J. BRAINE *Life at Top* iv. 68 A different voice, the voice of the red-haired girl from RADA whom now I dimly remember talking to earlier. **1965** *Observer* 19 Sept. 1/5 The new man for Rada? **1968** *Listener* 26 Dec. 873/3 The melodies are dim, the lyrics wanly snobbish, the costumes are rainwear ads and the Cockney accents (except Breeze's) RADA-provincial. **1977** L. MEYNELL *Hooky gets Wooden Spoon* xiv. 179 Sergeant Fitt could certainly have won a medal at RADA for..complete impassiveness.

Rada (rā·dă). [App. ad. *Allada*, the name of a former principality of Dahomey (now Benin).] The name of a group of deities of West African derivation venerated in Haiti; the worship of these deities. Also *attrib.*

1929 W. B. SEABROOK *Magic Island* I. i. 11, I knelt at last before the great *Rada* drums. *Ibid.* iii. 43 Some..danced, not the mad *Rada* of the night, but boisterous, gay Congo dances. **1937** M. J. HERSKOVITS *Life in Haitian Valley* viii. 149 When *vodun* deities are discussed in Mirebalais, most often two 'companies' of them are mentioned, the Rada and the Pétro 'squads'. **1941** J. G. LEYBURN *Haitian People* vii. 145 To the spirits..not drawn from Christianity the name 'loa' is applied. There seem in general to be two classes of these: those who have an African origin, and those with a Haitian background only. The former group is called 'Rada', the latter 'Petro'. **1957** *Amer. Anthropologist* LIX. 821 Members of two African-derived religions, Yoruba 'Shango'..and Dahomean 'Rada'..consider themselves also to be Roman Catholics.

radappertization (rădæ:pəɹtaizē[1]·ʃən). [f. L. *rad-iāre* to furnish with rays, shine + F. *appertization* kind of heat treatment of food (f. the name of Nicolas-François *Appert* (d. 1841), French inventor: see -IZATION).] The treatment of food with ionizing radiation so as to reduce the number of micro-organisms sufficiently to prevent future spoilage in the absence of recontamination (see quot. 1964). Cf. *RADICIDATION, *RADURIZATION.

1964 H. E. GORESLINE et al. in *Nature* 17 Oct. 237/2 Type 1 is the application to food of doses of ionizing radiation sufficient to reduce the number and/or activity of viable organisms to such an extent that very few, if any, are detectable in the treated food by any recognized method (viruses being excepted) while no spoilage or toxicity of microbial origin is detectable no matter how long or under what conditions the food is stored in the absence of recontamination... The following are names we suggest... Type I, radappertization. **1966** *Proc. Internat. Symp. Food Irradiation* 352 Salmonellae could be said to be moderately radiation-sensitive when bacteria are generally considered.. and therefore the dose requirement for their inactivation in food will be very much lower than that needed for radappertization (4·5–5·0 Mrad). **1975** *Appl. Microbiol.* XXX. 811 (*heading*) Low-temperature irradiation of beef and methods for evaluation of a radappertization process.

Hence **rada·ppertize** *v. trans.*, **rada·ppertized** *ppl. a.*

1974 *Jrnl. Food Sci.* XXXIX. 806 The accepted procedure for estimating the 12D dose of radappertized food has shortcomings that ought not to be ignored. **1975** *Appl. Microbiol.* XXX. 811/1 The wholesomeness of beef radappertized..with ^{60}Co gamma rays and with 10 MV electrons. **1977** *Jrnl. Food Sci.* XLII. 338 (*heading*) Variables affecting the acceptability of radappertized ground beef products.

radar (rē[1]·dāɹ). orig. *U.S.* [f. *radio detection and ranging.*] **1. a.** A system for detecting the presence of objects at a distance, or ascertaining their position or motion, by transmitting short radio waves and detecting or measuring their return after being reflected; also (*secondary radar*), a similar system in which the return signal consists of radio waves that a suitably equipped target automatically transmits when it receives the outgoing waves. Cf. *RADIOLOCATION.

1941 *N.Y. Times* 18 Nov. 8/4 The Navy undertook a special enlistment campaign today to recruit men for training in maintenance of the radio device known as 'Radar', which is used to locate ships and aircraft that are hidden by fog or darkness. **1943** *News Chron.* 9 Feb. 4/6 He described Radar as 'probably the most dramatic new weapon to come out of this war'. **1943** *Times* 24 June 4/6 It is expected also to improve 'radar', the device for detecting enemy aircraft and ships. **1946** *Electronics* Apr. 130/3 Frequency-modulation radar determines the distance to a reflecting surface..by measuring the frequency shift between transmitted and reflected waves. **1957** *Economist* 7 Sept. 831 (Advt.), The performance of modern aircraft must be matched by the radio, radar and Doppler navigational aids necessary for their safe and efficient operation. **1959** K. HENNEY *Radio Engin. Handbk.* (ed. 5) xxv. 34 Secondary radar, or the ATC radar-beacon system, solves the problem of identification of the individual aircraft in air-traffic control. **1960** J. D. HAIGH *Radiolocation Techniques* xiv. 213 Another advantage of secondary radar is that if the 'responding' transmitter is made to radiate on a frequency different from that of the 'interrogating' transmitter, the received picture will be completely free from all permanent echoes and no targets other than those with responding transmitters will be seen. **1962** R. M. PAGE *Origin of Radar* i. 15 The name 'radar' was coined from the words *Radio Detection And Ranging* by two U.S. Naval officers, F. R. Furth and S. M. Tucker. **1971** D. W. SCIAMA *Mod. Cosmol.* i. 2 The distance of the Sun is determined most accurately by radar. **1977** C. MCCULLOUGH *Thorn Birds* xvii. 451 The field which had fascinated him since he first got acquainted with radar: electronics.

b. (An) apparatus or an installation used for this system.

1945 *Electronics* Apr. 92/1 The free-space radar equation relates the power radiated from the radar..to the power delivered to the terminals of the receiver. **1945** *Proc. IRE* XXXIII. 740/2 An army radar, the SCR-270, gave warning of the impending attack on Pearl Harbor. **1959** K. HENNEY *Radio Engin. Handbk.* (ed. 5) xxv. 34 Present-day precision approach radars look only 20 deg in azimuth and 6 deg in elevation. **1967** *Listener* 9 Feb. 185/3 A 'thin area defence' for the whole United States, consisting of a dozen Spartan batteries and the relevant radars, would cost about $4,000 million. **1971** *Sci. Amer.* Oct. 73 (Advt.), APQ-140 is a digitally controlled phased array radar system that does the job formerly requiring several radars. **1977** *Navy News* Aug. 34 (Advt.), We're working on a wide variety of radars for applications throughout the world.

c. *fig.* An intuitive perception or sense of awareness attributed to or regarded as a characteristic of a person.

1950 'D. DIVINE' *King of Fassarai* xxix. 262 Thirty-kid power radar. Nothing ever gets by 'em. **1959** *Listener* 1 Jan. 31/3 With that political radar with which all good Levantine rulers are endowed, he succeeded for years. **1975** D. M. DAVIN *Closing Times* vi. 142 With that radar sense of quality that governed his taste in poetry and other people's clothes, he chose her own favourite scarf. **1976** P. ALEXANDER *Death of Thin-Skinned Animal* xviii. 183 That's what Joan thinks..intuition, female radar, or maybe she read it in the tea-leaves.

d. *transf.*

1963 [see *LIDAR]. **1974** *Physics Bull.* Jan. 11/2 Like ordinary radar, laser radar is based on the detection of a signal scattered from some object. If a laser beam is emitted into the atmosphere, some of the radiation will be absorbed by atmospheric molecules, some will be scattered by them.. and some will be scattered by aerosols. **1974** *Sci. Amer.* Mar. 83/2 One of the more promising instruments is an acoustic-radar detector, in which a noise signal is transmitted from the ground and is deflected by interaction with the vortex.

2. *attrib.* and *Comb.* *radar altimeter, astronomy, beam, dish* [*DISH *sb.* 4 b], *echo, equation, equipment, eye, operator, reconnaissance, set, signal, station, system*; *radar-controlled, -directed* ppl. adjs.; *radar-ranging* vbl. sb.; **Radar Alley** (see quot. 1971); **radar beacon,** a radio transmitter that automatically transmits a return signal when it receives a signal from a radar transmitter; *esp.* one that transmits a coded signal enabling it to be identified; **radar fence,** a line of radar stations for giving warning of intrusions into the air space behind it; **radar man, radarman,** a man trained to operate radar equipment; **radar map,** a map compiled from radar observations; hence **radar-map** *v. trans.*, **-mapping** *vbl. sb.*; **radar net,** a network of radar stations, esp. a radar fence; **radar picket,** a picket-ship specially equipped with radar; **radar plotter,** one who plots the direction and course of objects from radar observations; **radar scanner,** a rotatable aerial for transmitting and receiving radar signals; **radar screen,** the screen of a radarscope; also *fig.*; **radar-sonde,** a sonde which can be tracked by radar so that information on the wind may be obtained as well as the usual meteorological information; **radar speed detector, trap,** etc. = **radar trap*; **radar-track** *v. trans.*, to track by radar; also *fig.*; **radar trap,** a speed trap in which speed is measured using radar and the Doppler effect.

1971 M. TAK *Truck Talk* 127 *Radar Alley*, Interstate 90, between Cleveland and the New York line; so named because of the numerous radar traps set on the road. **1976** PERKOWSKI & STRAL *Joy of CB* 174 The name of Radar Alley..aptly applies to the Ohio Turnpike..that is heavily patrolled by bears. **1946** *Electronics* Apr. 130/2 A detailed description of an f-m radar altimeter may now be given, following the declassification of the model AN/APN-1 altimeter. **1951** *Jrnl. Brit. Interplanetary Soc.* X. 101 This means a vertical descent using reverse rocket braking in conjunction with a radar-altimeter and landing legs. **1978** *Nature* 10 Aug. 540/1 Balloon-borne radar altimeters have also been used to map surface topography in Antarctica. **1959** DAVIES & PALMER *Radio Studies of Universe* i. 3 This discovery laid the foundation for the second branch of the science of radio astronomy, that of radar astronomy which probes the nearest inhabitants in space—the meteors, moon and planets—using the radio-echo technique. **1969** *Sci. News Yearbk.* 83 Radar astronomy had done particularly well with its studies of nearer planets, especially Venus. **1945** *Electronic Engin.* XVII. 685/2 By this means our airborne forces were enabled to direct themselves to any given point in hostile country where an advance party had already installed a 'Radar beacon'. **1958** *Listener* 13 Nov. 779/1 Navigational aids, such as radio stations and radar beacons, are few. **1959** [see sense 1 a above]. **1965** D. K. BARTON in R. S. Berkowitz *Mod. Radar* ii. 12 A modern radar beacon weighing a few pounds can receive signals 100 db below 1 watt and retransmit them as 100-watt pulses, providing tremendous range extension. **1958** *Times* 9 Oct. 10/2 When the rockets burn out,..the missile coasts to the target, riding along the centre of a radar beam which is directed by the pilot of the launching aircraft. **1959** N. MAILER *Advts. for Myself* (1961) 183 The other-directed man is flexible... His movement is 'radar-controlled',..his 'taste' rather than his work the primary concern... He obeys 'the process of paying close attention to the signals from others'. **1978** R. V. JONES *Most Secret War* xxxi. 265 He also concluded that some of the lights were radar-controlled. **1947** CROWTHER & WHIDDINGTON *Science at War* 75 H.M.S. *Warspite* fired her first radar-directed broadside. **1952** *Electronic Engin.* XXIV. 126/1 A 3 centimetre radar installation with a V.H.F., D.F. receiver, the radar dish being mounted on a common rotatable shaft with the D.F. aerial. **1969** *New Yorker* 11 Oct. 51/1 In the gaunt shade of oil rigs and radar dishes. **1946** *Electronic Engin.* XVIII. 149/2 The technique of setting the calibration pip against the radar echo is one which gives surprisingly accurate results. **1962** F. I. ORDWAY et al. *Basic Astronautics* iii. 65 The first radar echoes were returned from the Moon in 1946. **1945** D. G. FINK in *Electronics* Apr. 92/1 The basic factors concerned are the formation and propagation of radio beams, and the reflection of radio waves when they encounter a sudden change in the electrical properties of the transmission medium... The equation which links these factors may be termed appropriately the radar equation. **1966** *McGraw-Hill Encycl. Sci. & Technol.* XI. 200/2 The strength of echo signals is related to the parameters of the radar system by the radar equation. **1944** *Jrnl. R. Aeronaut. Soc.* XLVIII. 103 Tests of various wireless and Radar equipment. **1976** *Daily Mirror* 16 July 2/2 The cases are believed to contain aircraft parts and radar equipment. **1957** R. WATSON-WATT *Three Steps to Victory* xliii. 249 British fighters could entrust themselves to the..far-seeing radar-eye. **1977** *R.A.F. News* 5–18 Jan. 9/1 The Shackleton airborne early warning aircraft of Lossiemouth's 8 Squadron..act as airborne radar eyes searching for low-flying aircraft. **1948** *Sat. Even. Post* 21 Aug. 27/1 There are wide gaps in the Soviet radar fence. Russia is too huge to be rimmed completely with twirling antennas. **1957** R. WATSON-WATT *Three Steps to Victory* (*caption to plate facing p.* 320) The 350 ft. steel lattice masts on 'T' site. Combined with ultra-short-wave-radiation these ensured that the outer edge of the radar fence could be set far enough from the British coasts. **1942** *Radio-Craft* Jan.–Feb. 332/1 'Radar' men will operate the newly perfected radio device which locates planes in flight. **1946** P. CARTER in Aldiss & Harrison *Decade 1940's* (1975) 109 He was at once geologist, radarman, vibration expert and navigator. **1966** M. WOODHOUSE *Tree Frog* x. 75 Pzenica, the radar man from Poland, was standing over me. **1977** *R.A.F. News* 5–18 Jan. 8/2 The Buchan radarmen help to look after the safety of these fliers. **1960** *Sci. Amer.* Aug. 59/1 The radar mapping experiment marks the first trial of this technique. *Ibid.* 59/2 (*caption*) Radar map of the moon. **1962** F. I. ORDWAY et al. *Basic Astronautics* iii. 65 In January, 1960, a preliminary radar map of the Moon was made from 440 mc observations performed with the 84-ft antenna at Millstone Hill, in Massachusetts. **1970** *New Scientist* 19 Feb. 361/1 Each Victor can radar-map the

entire Mediterranean in one seven-hour sortie. **1948** *Sun* (Baltimore) 3 June 2/3 Representative Vinson (D., Ga.) introduced a bill to authorize construction of a radar net along the coasts of the United States and Alaska. **1964** G. LYALL *Most Dangerous Game* vi. 48 The border could be flown all right... The Finns didn't have a complete radar net along it, and most pilots knew just where the Finnish radar stations were. **1945** *Yank* 5 Oct. 16/3 Radar operators and observers were known as 'radio operators'. **1967** L. B. ARCHER in Wills & Yearsley *Handbk. Management Technol.* vii. 129 Some of the early experiments [in ergonomics] were to determine the best spacing of markings on instrument dials so that the pilot or gunner or radar operator could read them rapidly and accurately. **1952** *Jane's Fighting Ships 1952–53* p. vii, Among the recent U.S. photographs added to this issue are those of the..radar picket *William T. Powell*. **1966** *McGraw-Hill Encycl. Sci. & Technol.* XII. 264/2 Some U.S. Navy destroyers are specially fitted as radar pickets, and some British ones as aircraft fighter directors. **1957** *Technology* Dec. 358/3 This new principle of design may be contrasted with the old by considering the radar plotter tracking targets on his radar tube. **1978** R. V. JONES *Most Secret War* xxiii. 196 She..had been taken on with two other girls at Bawdsey by Watson-Watt to see whether girls would make good radar-plotters. **1976** *Sci. Amer.* Feb. 51/1 The second astronomical experiment consists of precise radar-ranging measurements of the distance between the earth and the inner planets Mercury and Venus. **1959** *Daily Tel.* 7 Mar. 7/3 An area equal to the whole of the Mediterranean could be covered in a single radar reconnaissance by one aircraft. **1946** *Electronics* Oct. 35 (Advt.), Aircraft radar scanner. **1947** *Daily Progress* (Charlottesville, Va.) 15 Dec. 2/3 New radar scanner to watch the skies for airplanes approaching New York airport is now in experimental operation. **1976** 'A. YORK' *Dark Passage* xiii. 164 Her radio aerial and the radar scanner on her wheelhouse roof remained visible. **1945** *Instruments* XVIII. 745/1 Its beam-splitter..allows operator to view radar reflection in normal manner while image on radar screen is reflected up into camera. **1958** *New Statesman* 18 Jan. 64/1 What is needed is 'patience': the West is to sit behind the radar screens and rocket bases waiting for a change of heart, and perhaps a change of regime, in Russia. **1977** J. BINGHAM *Marriage Bureau Murders* v. 58 He was reluctant to switch from thinking about her... He kept her within the radar screen of his mind. **1946** *Electronic Engin.* XVIII. 2 The operation of a Radar set as a position finder. **1947** *Daily Mail* 22 May 3/6 Radar sets which will 'see' through fog are to be fitted in long-distance passenger trains as soon as materials are available. **1969** *Times* 25 Mar. 12/6 By bouncing radar signals off Venus, scientists from the United States have counted the numbers of electrons in the space between the planets. **1949** *Jrnl. R. Aeronaut. Soc.* LIII. 441/1 At T.R.E. the wind-finding responder is being adapted for use as a 'radarsonde', meteorological instruments to measure pressure, temperature and humidity being mounted towards the top of the responder. **1956** *Nature* 17 Mar. 520/1 Progress was made in the measurement of solar radiation and in the development of the new radar-sonde. **1957** *Times* 4 Oct. 6/4 A London electronics engineer described to Leyland magistrates to-day the effects of laboratory and outside tests he had carried out on a radar speed detector to assess the effects of extraneous sources on the accuracy of the instrument. **1962** *Daily Tel.* 22 May 24/6 [He] complained through his solicitor that police radar speed traps were 'un-British' when he was fined £10..for speeding... Police said Cadbury went through a radar speed check at 58 mph. **1971** *Observer* 26 Sept. 9/5 Radar speed traps are being planned for a 60-mile stretch of the River Nene..to stop boats from exceeding the seven m.p.h. limit. **1945** *Times* 15 Aug. 5/6 By December 1935, the experimental work was sufficiently advanced for the Air Ministry to decide on establishing a chain of five radar stations on the east coast of England... This was the first operational radar system installed anywhere in the world. **1958** *New Statesman* 22 Feb. 223/2 Radar stations at sea picked up 'blips' suggesting that an air armada was flying at 2,000 mph towards the interior of the United States. **1978** R. V. JONES *Most Secret War* xxiii. 196 It happened that this sortie was also plotted by a German radar station north-west of Brest. **1945** Radar system [see *radar station* above]. **1956** *Tuscaloosa* (Alabama) *News* 31 July 11/6 The Italian Line said the Doria radar-tracked the oncoming Stockholm before the fatal crash. **1973** M. AMIS *Rachel Papers* 24, I felt a hand traverse the lower areas of my back. In seconds—radar-tracked by my whisker-sensitive pubic hairs—it was treading air above my groin. **1962** *Punch* 18 July 74/3 A motorist has been acquitted of a speeding charge because the radar trap that provided the evidence against him could not be put on oath. **1975** *Globe & Mail* (Toronto) 4 June 1/6 A check with the radar trap involved showed the limousine was doing a mere 71 miles an hour in a 70-mile-an-hour zone.

radarscope (rē[1]·dāɹskō[u]p). Also **radar scope.** [f. *RADAR + -SCOPE.] A cathode-ray oscilloscope on the screen of which radar echoes are represented for observation; also, the screen itself.

1948 *Nat. Geographic Mag.* Feb. 255/1 Then an operator 'talks them down', watching each plane on the radarscope. **1958** *Observer* 22 June 12/6 The future of mankind might lie at the mercy of an officer reading a radarscope in the Canadian Arctic, who might conceivably mistake an electronic storm..for the trajectory of a Soviet I.C.B.M. **1959** K. HENNEY *Radio Engin. Handbk.* (ed. 5) xxv. 34 Precision Approach Radar... Distance, bearing, and elevation above ground are presented continuously on the face of the radar scope. **1966** M. WOODHOUSE *Tree Frog* xxi. 154, I can never stand by a radarscope..without recalling the Rubaiyat; the moving Finger writes. **1972** *Sci. Amer.* Jan. 52/3 The image..is a direct representation of the scene ahead of the driver rather than the usual maplike plan view of the radarscope.

Radcliffian (rædkli·fiăn), *a.* Also **Radcliffean.** [f. the name of (Mrs.) Ann *Radcliffe* (1764–1823), English 'Gothic' novelist.] Of or characteristic of Mrs. Radcliffe or her works.

1801 M. EDGEWORTH *Belinda* II. xx. 265 Here is a Radcliffean tour along the picturesque coasts of Dorset and Devonshire. **1884** *Spectator* 19 Jan. 91/1 There is, too, a Radcliffian eeriness about his castle, with its mysterious noises, secret passages, and buried rooms. **1931** *Times Lit. Suppl.* 26 Feb. 151/3 The 'Radcliffian' novel, or *roman noir*, as the French call it. **1966** *English Studies* XLVII. 287 But symbol, action, and the perceiving sensibility. add to the Radcliffean question, 'what is the mystery of the veil?'

raddle, *v.*[2] (Later example.)

1966 J. F. H. THOMAS *Sheep Farming Today* v. 57 Recently there has been introduced a breast harness device which holds a colour marking crayon on rams in lieu of raddling their breast wool with colour.

raddled, *ppl. a.*[2] Add: (Further examples.) Also *fig.*, worn, worn out.

1901 W. BARRY *Wizard's Knot* vii. 74 Raddled like scabby sheep with red paint. **1922** GALSWORTHY *In Chancery* II. vii. 621 An 'old Johnny' in a gown and long wig, looking awfully like a funny raddled woman, came through a door. **1922** JOYCE *Ulysses* 96 Outside them and through them ran raddled sheep bleating their fear. **1969** *Listener* 30 Jan. 156/2 Ruth Gordon's Minnie might be a typical New York eccentric, a raddled Carmen Miranda. **1975** *Nature* 28 Aug. 738/2 They were all kept in identical conditions of housing and management, and oestrus was detected by raddled vasectomised rams at 0900 daily. **1978** *Time* 3 July 10/3 By day she rests, and her face, without makeup, softens but still retains the raddled quality of hard living.

raddling (ræ·dliŋ), *vbl. sb.* [f. RADDLE *v.*[2] + -ING[1].] The action of marking sheep with raddle, or some other red substance.

c **1873** J. ALBERY *Fortune* III, in *Dramatic Wks.* (1939) I. 574 Mark 'em [*sc.* sheep] all with two red letters, Johnny; they always do that;—raddling they call it, I think.

radiæsthesia, etc., varr. *RADIESTHESIA, etc.

radial, *a.* and *sb.* Add: **A.** *adj.* **2. d.** *Bot.* and *Timber.* Applied to a longitudinal section or cut along a radius or diameter, and to the surface so exposed. Also as *adv.*, as *radial-sawn* adj.

1881 J. S. GAMBLE *Man. Indian Timbers* 175 Medullary rays fine, equidistant, prominent on a radial section. **1924** HOLMAN & ROBBINS *Textbk. Gen. Bot.* iv. 104 The annual rings appear in radial section as more or less parallel bands which impart to the wood its characteristic grain. **1938** H. E. DESCH *Timber* ii. 19 The rays are usually just visible to the naked eye on radial surfaces, where they appear as narrow, horizontal ribbons. **1958** *N.Z. Timber Jrnl.* Apr. 61/2 Radial sawn. **1969** H. L. EDLIN *What Wood is That?* i. 20 (*caption*) The radial cut through a birch-log, which splits it into two, becomes convex after seasoning. **1972** H. J. DITTMER *Mod. Plant Biol.* viii. 151 Radial or quarter-sawed wood is taken through the radius of the log.

e. Of a road, route, etc.: running directly from the centre of a town or city to an outlying district (usu. as part of a system of such roads). Hence also used of transport services using such routes.

[**1909** H. I. TRIGGS *Town Planning* ii. 88 In cities laid out upon the radial system, the radiating streets form as it were the skeleton of the system.] **1937** [see *parkway* s.v. *PARK *sb.* 7]. **1942** *Policy on Rotary Intersections* (Amer. Assoc. State Highway Officials) 1 Each road approaching a rotary hereafter is called 'radial road'. **1948** T. SHARP *Oxford Replanned* 7 Radials. Oxford has a highly developed pattern of radial roads which all centre on Carfax. **1957** *Encycl. Brit.* XXII. 335 Wren suggested the use of radial streets integrated with the rectangular plotting of intervening areas, broad streets and location of industry outside the city limits. **1976** P. R. WHITE *Planning for Public Transport* v. 93 The major growth during the twentieth century was in improvement of principal radial roads beyond the limit of the then built-up areas, and construction of bypasses or ring roads. *Ibid.* 106 Even on the radial Green Line Services, the heaviest loads are often between inner and outer suburban centres.

3. c. *radial engine*, a type of internal-combustion engine (used chiefly in aircraft) having its cylinders fixed radially around a rotating crankshaft. Hence *radial-engined* adj.

1909 R. KENNEDY *Flying Machines* vii. 115 Miesse aeromotor: 8-cylinder radial engine, having the cylinders arranged in pairs and in an horizontal position. **1934** *Discovery* Dec. 353/1 The tendency in Great Britain is to develop both types, the large in-line engine being composed of four banks of cylinders forming an H, and the corresponding radial engines with two circles of cylinders one immediately behind the other. **1942** *R.A.F. Jrnl.* 16 May 14/1 Single-seat, low-wing, radial-engined monoplane fighters..are standardised by the Army. **1963** BIRD & HUTTON-STOTT *Veteran Motor Car* 21 The interesting radial-engined Enfield-Allday. **1971** L. J. K. SETRIGHT *Power to Fly* vi. 162 Each of the 9 cylinder banks had its own overhead camshaft, a most uncommon luxury in radial engines.

d. Also *radial-ply*. Denoting a tyre in which the layers of fabric are laid with the cords running at right angles to the circumference of the tyre, and the tread is strengthened by further layers of fabric running round the circumference. Also *ellipt.* as *sb.*

1964 *Economist* 26 Sept. 1191/2 British manufacturers have shied away..from buying components at more than minimal cost. They are changing their attitudes..with the advent of the radial-ply tyre, whose greater expense is.. compensated for by better wear. *Ibid.* 1254/1 Firestone has been experimenting with fibreglass cords for radials. **1966** *Ibid.* 22 Oct. 388/2 Michelin, pioneers of radial-plies, have yet another new tyre. **1967** *Autocar* 5 Oct. 19/2 Radial tyres..add appreciably to the car's safety margin, especially in the wet. **1971** *Guardian* 23 Mar. 18/4 Patents for radial-ply tyre design were first registered in about 1912. It was not until 36 years later that Michelin..introduced the first radial for commercial sale. **1972** *Practical Motorist* Oct. 82/1 Since July 1 it has been illegal to fit a cross-ply tyre and a radial on the same axle. **1972** C. WESTON *Poor, Poor Ophelia* (1973) viii. 42 His car whirled up, radial tires singing on the glossy pavement. **1976** *Field* 18 Nov. 1021 (Advt.), The..radial ply tyres produce a very peaceful ride.

4. c. *radial-flow* adj.: being or employing a turbine, pump, or the like in which fluid is forced to move at right angles to an axis of rotation.

1881 *Encycl. Brit.* XII. 527/1 For radial flow turbines the wheel may have the form A or B, fig. 191, A being most usual with inward, and B with outward flow turbines. **1914** W. M. WALLACE *Hydraulics* vii. 101 It is usual to designate turbines as radial flow or parallel flow according as the water flows in a radial or axial direction through the wheel. **1958** *Technology* Mar. 29/3 The basic principles of the Derwent, radial-flow, jet engine are described. **1972** J. M. K. DAKE *Essentials of Engin. Hydraulics* vi. 160 The centrifugal pump and the Francis turbine are examples of radial flow reaction machines.

d. Astr. *radial velocity*, the velocity of a star or other body along the line of sight of an observer.

1895 *Nature* 13 June 155/2 (*heading*) Measurement of radial velocities. **1930** R. H. BAKER *Astron.* viii. 315 Radial velocities of the stars up to 30 km./sec.. are usual. **1966** J. D. KRAUS *Radio Astron.* viii. 353 The wisps of nebulosity show a large radial velocity of over 5,000 km sec^{-1}, indicating a rapid expansion. **1977** F. N. BASH *Astron.* iii. 67 The Doppler effect..gives astronomers a way of measuring radial velocity.

e. *Physical Geogr.* Of (a pattern of) drainage: being or involving a pattern of streams flowing outwards from a central dome or other elevated region.

[**1910** LAKE & RASTALL *Text-bk. Geol.* iv. 86 The arrangement of the principal valleys, in most of which are considerable lakes, is distinctly radial.] **1932** W. H. EMMONS et al. *Geol.* vi. 106 (*caption*) A radial stream pattern developed on the surface of the dome-like uplift of the Black Hills in South Dakota. **1939** A. K. LOBECK *Geomorphol.* xv. 513 A radial drainage pattern is characteristic of very young and undissected domes. **1954** W. D. THORNBURY *Princ. Geomorphol.* v. 126 In radial patterns,..the individual stream patterns may well be dendritic or pinnate, and radial designates more their arrangement with respect to each other than it does the stream pattern. **1969** *Geography* LIV. 199 The almost perfect radial drainage developed on the andesite volcano of Mount Egmont.

6. *radial energy*: in the writings of Teilhard de Chardin, a form of energy postulated to be independent of the conventional laws of thermodynamics and to tend to produce increasing organization and complexity in both the physical and spiritual worlds; it was held to be manifest, for example, in the evolution of living organisms and in the development of ideas. Cf. **tangential energy*. [Introduced in Fr. *c* 1938 in *Le Phénomène Humain* (1956) 62.]

1959 B. WALL et al. tr. *Teilhard de Chardin's Phenomenon of Man* I. ii. 65 In each particular element this fundamental energy is divided into two distinct components: a tangential energy which links the element with all others of the same order (that is to say, of the same complexity and the same centricity) as itself in the universe; and a radial energy which draws it towards ever greater complexity and centricity—in other words forwards. *Ibid.* iii. 72 Spiritual energy, by its very nature, increases in 'radial' value, positively, absolutely, and without determinable limits, in step with the increasing chemical complexity of the elements of which it represents the inner lining. **1965** *Listener* 15 Apr. 558/2 We really need different kinds of language to deal with the concepts proper to biology and those proper to thermodynamics. One thinks at once of Teilhard's brilliant pioneering neologisms, such as 'radial' and 'tangential' energy, phrases devised precisely to meet some at least of the difficulties. **1969** A. RICHARDSON *Dict. Christian Theol.* 333/1 It is through increase in radial energy that decisive 'critical points' are reached, whether, for example, at the molecular level or at the leap from instinct to thought.

B. *sb.* Restrict subject-label *Anat.* to senses in Dict. and add: **4.** *Aeronaut.* A radial engine.

1920 G. C. BAILEY *Complete Airman* xii. 93 The rotary is very similar in appearance to the radial. **1944** P. H. WILKINSON *Aircraft Engines of World* 44 The M-11 engine is a 100 h.p. 5-cylinder air-cooled radial which is used in secondary training planes. **1960** C. H. GIBBS-SMITH *Aeroplane* xiv. 110 The 247..was powered by two 550-h.p. Pratt and Whitney *Wasp* radials.

5. a. A radial road or route. **b.** *Canad.* A train or tram running on a radial route.

1948 [see sense A. 2 e above]. **1966** *Globe & Mail* (Toronto) 12 Sept. 16 (*caption*) The radials: commuter dream of the auto killed. **1972** J. MOSHER *Adultery* III. xiv. 134 Father had already gone off on the half-past six radial.

6. (See quot. 1956.)

1956 W. A. HEFLIN *U.S.A.F. Dict.* 417/2 *Radial*, any one of a number of lines of position radiating from an azimuthal radio-navigation facility,..identified in terms of the bearing of all points along that line from the facility. **1971** *Flying*

Apr. 42/2 Then via the Coyle 270 radial until intersecting the 113 radial of Modena. **1972** R. L. TAYLOR *Instrument Flying* viii. 114 The number that now appears under the OBS index is your course to the station (inbound it will always be the reciprocal of the radial you are on).

7. A radial-ply tyre (see sense A. 3 d above).

radiance. Add: **3.** The radiant flux emitted by unit area of a source into unit solid angle.

1917 H. E. IVES in *Astrophysical Jrnl.* XLV. 43 If now we divide this radiant intensity in any direction by the projected area of the opening in that direction.., we obtain the specific radiant intensity or radiance in that direction. *Ibid.* 48 'Radiance', as a time-saving word for 'specific radiant intensity', is frankly taken from the dictionary, as the appropriate synonym for 'brightness'. **1944** *Jrnl. Optical Soc. Amer.* XXXIV. 253/2 The luminance of any surface is the product of the radiance of that surface by the absolute luminosity *K* of the energy radiated by the surface. **1951** [see **luminosity curve*]. **1972** *Science* 22 Sept. 1100/1 Radiance gradients determined from data gathered by the infrared spectrometers aboard the Nimbus III and Nimbus IV satellites.

radiant, *a.* and *sb.* Add: **A.** *adj.* **1. d.** Designed to send out radiant heat.

1936 FABER & KELLS *Heating & Air-Conditioning* iv. 81 The warming effect of convectors is slower than with the radiant type of heater. **1936** *Economist* 28 Mar. 738/1 Our High Beam fires have proved to be the most efficient radiant gas fires in the world. **1951** *Good Housek. Home Encycl.* 35/2 More recently the use of 'radiant radiators'..has become more common. **1957** *Heating, Ventilating, Air Conditioning Guide* XXXV. 399 Radiant heaters..have a refractory directly above the burners which is heated to incandescence. **1962** *Listener* 13 Sept. 411/3 It [*sc.* an electric cooker] has two radiant plates. **1954** J. J. BARTON *Heating & Ventilating* xii. 318 The principal heat requirements are provided by convection, with a small radiant unit providing sufficient beamed heat for topping-up. **1977** *Direct Electric Heating* (Electricity Council) v. 21/1 Direct electric heating appliances and equipment... They may be divided into four basic groups, namely—(i) Low intensity radiant systems incorporated in the building fabric. (ii) Individual radiant-convector units. (iii) Forced warm air systems. (iv) Hot water radiator systems.

3. Also *radiant heating*; *radiant-heated* adj.

1912 *Cassier's Mag.* XLI. 569/1 (*heading*) Radiant heating. *Ibid.* 570/1 The demonstrations were continued..at the experimental works of the Radiant Heating Syndicate. **1962** *Punch* 24 Jan. 178/2 The 'tropical radiant-heated indoor pool'. **1966** *McGraw-Hill Encycl. Sci. & Technol.* XI. 213/1 Any radiant-heating system using a fluid heat conveyor may be employed as a cooling system by substituting cold water or other cold fluid.

b. *radiant efficiency*, the ratio of the radiant flux emitted to the power consumed; *radiant energy*, energy in the form of (usu. electromagnetic) radiation; *radiant flux*, the rate of flow of radiant energy; *radiant power* = *radiant flux* above.

1914 *Rep. Brit. Assoc. Adv. Sci. 1913* 435 The true remedy must be sought for in increasing the 'radiant efficiency' of the [gas] fire. *Ibid.* 436 Part of the radiant energy is directly determined..by using a radiation calorimeter. **1917** *Astrophysical Jrnl.* XLV. 44 Through this space radiant energy is passing at a certain rate. This rate constitutes the radiant flux through the space. **1923** *Sci. Papers U.S. Bureau of Standards* No. 475. 132 The ratio of luminosity to radiant power—the luminous efficiency. **1932** HARDY & PERRIN *Princ. Optics* i. 19 If the human eye were uniformly sensitive to radiation of all wave lengths, the radiant power expressed in watts would provide an adequate method of evaluating the flux. **1936** FABER & KELL *Heating & Air-Conditioning of Buildings* iv. 74 Their [*sc.* gas fires'] radiant efficiency is about 50 per cent. (with an additional 10 per cent. convection). **1944** *Jrnl. Optical Soc. Amer.* XXXIV. 252/2 When radiant flux is evaluated with respect to its capacity to evoke the brightness attribute of visual sensation it is called luminous flux. **1970** M. V. KLEIN *Optics* iv. 121 The radiant energy density..is the radiant energy contained in a unit volume of space. Radiant power or radiant flux..is the time rate of change, or rate of transfer, of radiant energy.

B. *sb.* **1. b.** Each of the units in a radiant fire that are designed to radiate heat.

1914 *Rep. Brit. Assoc. Adv. Sci. 1913* 436 A little later a more marked step in the evolution of the new radiating fire consisted in joining the two or three fireclay pieces into one, and thus making the firefront consist of a series of hollow fireclay columns (now known as radiants). **1936** *Economist* 28 Mar. 738/1 Our 'Thermo-XX Beam' radiant..was the first radiant designed to emit an increased proportion of short infra-red rays, and its introduction marked a new stage in the development of the gas fire. **1953** 'N. SHUTE' *In Wet* 173 They sat down in arm chairs before the radiants of the electric fire. **1958** *Woman's Journal* Mar. 20/2, I..sat down by my hissing gas fire. One of the radiants was missing.

radiantly, *adv.* Add: **3.** By means of radiant energy.

1948 R. W. SHOEMAKER *Radiant Heating* i. 11 A radiantly heated room.

radiata. Add: **2.** [a. the specific epithet of *Pinus radiata* (D. Don 1837, in *Trans. Linn. Soc.* XVII. 442).] In full, *radiata pine*: = *INSIGNIS, *Monterey pine* s.v. MONTEREY in Dict. and Suppl.

1953 *N.Z. Jrnl. Forestry* V. 374 The development of major industries, dependent wholly or substantially on radiata pine as a basic raw material, is well advanced. **1959** A. MCLINTOCK *Descr. Atlas N.Z.* p. xiii, Plantation and shelter belts of introduced (exotic) trees, mainly radiata pine and macrocarpa, break up the farmlands. **1963** *Times* 11 Feb. 9/1 At Kaingaroa..273,000 acres of radiata pines were planted during the depression. **1972** *Daily Colonist* (Victoria, B.C.) 24 May 5/4 It was found, however, that the radiata pine (Monterey pine) would flourish in this otherwise 'dead' [volcanic] soil. **1973** T. E. SIMPSON *Kauri to Radiata* vii. 310 By 1959 radiata would have achieved prominence in sawn output over that of our indigenous timbers.

radiate, *sb.* Add: **3.** A classical coin with rays issuing from the device.

1939 *Oxoniensia* IV. 61 The coins discovered during the exploration of the Cemetery are listed below... 1 Radiate *Antoninianus*. **1974** *Sci. Amer.* Dec. 122/3 The radiate was by then [*sc.* A.D. 301] reckoned to be equal to two of the seldom-seen *denarii*.

radiate, *v.* Add: **1. b.** More widely, to emit energy of any kind in the form of rays or waves.

1927 I. B. CRANDALL *Theory of Vibrating Systems & Sound* ii. 55 Dissipation due to radiation is usually of much greater importance than dissipation due to friction in the neck of the resonator, provided the resonator is so situated that it can radiate. **1960** *Practical Wireless* XXXVI. 414/2 The VHF transmitter also radiates during daylight but continues throughout the hours of darkness. **1962** A. NISBETT *Technique Sound Studio* iii. 64 Remembering that it is the soundboard that is radiating, an alternative is to move the piano well away from the wall and stand the microphone on a chair at the back. **1978** *Nature* 2 Mar. 37/1 The ability of degenerate dwarfs..to radiate at X-ray wavelengths has created much theoretical interest.

3. d. *Biol.* Of an animal or plant group, to spread from its area of greatest concentration into new habitats.

1923 F. W. JONES *Mammals S. Austral.* I. 24 A stock will become progressively altered by adaptation to its environment as it radiates from its centre of domicile. **1949** W. C. ALLEE et al. *Princ. Animal Ecol.* xxxiv. 661/1 The Australian marsupials..radiated into a great variety of habitats. **1957** P. J. DARLINGTON *Zoogeogr.* ii. 58 Over a longer period of time, the whole of the Ostariophysi may have radiated from the Old World tropics. **1978** *Sci. Amer.* Sept. 111/1 Many of the early amphibian lineages developed rather large body sizes and radiated into the available habitats, becoming herbivores and predators on many food items in aquatic, semiaquatic and terrestrial settings.

4. a. More widely, to emit (energy of any kind) in the form of rays or waves. Also with *away*.

1937 *Discovery* Nov. 331/1 It is possible that the Baird video transmitter..will be removed to Birmingham, since only small alterations would be required to make it radiate the signals generated by Marconi-E.M.I. cameras. **1971** *Nature* 29 Jan. 304/1 The compactness of the hot plasma cloud suggests that it is probably gravitationally confined and, as the thermal energy is radiated away in a fraction of a second, continuously replenished in some way. **1978** *Ibid.* 27 Apr. 784/2 Stars radiate their heat away, and must burn nuclear fuel to keep going.

c. To transmit (radio waves); to broadcast.

1923 *Radio Times* 28 Sept. 12/3 A ninety-minute excerpt ..will be simultaneously radiated from all other stations. **1951** *Times* 4 Jan. 7/6 Items, mainly music, are selected from one of the three home services. They are radiated simultaneously by a 25-kw. frequency modulated (f.m.) transmitter..and a standard amplitude modulated transmitter. **1956** *B.B.C. Handbk.* 1957 59 Thirty-nine high-power, short-wave transmitters..radiate the BBC's External Services programmes. **1967** *Listener* 30 Mar. 423/3 All three programmes in the United Kingdom will eventually be radiated in colour on 625 lines. **1972** *Radio Times* 28 Sept. 42/1 World News is also radiated on medium waves.

radiation. Add: **1. b.** In mod. use (usu. in *sing.*), energy transmitted in the form of rays, waves, or sub-atomic particles; in non-*techn.* use *spec.* ionizing radiation. (Further examples.)

1896 *Strand Mag.* July 108/1 If a solid object is placed in the path of this [negative] stream..it may become the seat of the production of that..which is variously known as Röntgen radiation or X-radiation. At the solid object the new radiation springs into being, and then travels away from it in all directions, in very much the same way that ordinary light would do. **1899** *Phil. Mag.* XLVII. 109 The remarkable radiation emitted by uranium and its compounds has been studied by its discoverer, Becquerel. **1934** H. M. VERNON *Princ. Heating & Ventilation* iii. 50 The intensity of the radiation from a coal fire varies enormously. **1958** *New Statesman* 25 Oct. 545/3 This week's inquest on the second serviceman to die after the Christmas Island tests reveals once more the astonishing medical and scientific ignorance of the effects of radiation. **1962** A. NISBETT *Technique Sound Studio* iii. 52 With stringed instruments, very little of the sound we hear is directly from the strings; what we are actually listening to is the radiation from a sounding board to which the strings are coupled. **1966** *McGraw-Hill Encycl. Sci. & Technol.* XIV. 265/2 The trapped radiation consists of protons and electrons constrained to bound orbits by the geomagnetic field. **1972** *Daily Tel.* (Colour Suppl.) 3 Mar. 17 Our planet is constantly being bombarded by all kinds of radiation—from natural radio waves to infra-red, visible light, ultraviolet, X-rays and gamma radiation. **1978** *Sci. Amer.* July 82/1 Although bees are most sensitive to ultraviolet radiation, it is not the color that they learn best. To the bee ultraviolet radiation is sky radiation, and it seems not to be expected as a pure color at food sources.

2. In mod. use, the emission of energy of any kind in the form of rays or waves, esp. electromagnetic waves. (Further examples.)

1908 C. C. F. MONCKTON *Radio-Telegr.* iv. 69 There will be a certain amount of radiation possibly from the tree back towards the radiator, and a consequent distortion of the field. **1934** *Discovery* Dec. 347/2 The noise in aeroplanes can be reduced..; little can be done for reducing the external radiation of noise. **1956** *B.B.C. Handbk.* 1957 9 The radiation of signals from the transmitting aerial.

3. c. *Biol.* The spread of an animal or plant group from an area in which its greatest concentration is or was found. Cf. *adaptive radiation* s.v. *ADAPTIVE *a.*

1949 W. C. ALLEE et al. *Princ. Animal Ecol.* xxxiv. 662/2 The early radiation of the orders of insects is even more remarkable than that of the orders of mammals. **1957** P. J. DARLINGTON *Zoogeography* vii. 414 The most profound patterns of geographical radiation, reflecting spread of great, dominant groups from definite centers, are found among the animals which disperse most slowly over land. **1977** *Sci. Amer.* May 28/3 In the course of their wide and long-lasting radiation these apes seem to have encountered increasingly cooler environments. **1978** *Nature* 17 Aug. 662/1 The *Dicroidium* complex, a diverse group of seed ferns characteristic of the Triassic, became nearly entirely extinct and was replaced by a Jurassic radiation of conifers.

4. In mod. use freq. *attrib.* with the sense 'ionizing radiation', as *radiation dosage, dose, dosimetry, level, meter, monitor*; *radiation-induced, -proof* adjs.; in the names of bodily disorders caused by ionizing radiation, as *radiation cachexia, carcinogenesis, dermatitis, osteitis, ulcer*; **radiation accident,** an accident involving potentially hazardous exposure to ionizing radiation; **radiation badge,** a badge that changes its appearance when a prescribed dose of ionizing radiation has been received; cf. *film badge* s.v. *FILM *sb.* 7 b; **radiation belt** *Astr.*, a region surrounding a planet where charged particles accumulate under the influence of the planet's magnetic field; **radiation burn,** a burn caused by over-exposure to ionizing radiation; **radiation chemistry,** the study of chemical changes arising from the impact of ionizing radiation; (cf. *RADIOCHEMISTRY); hence *radiation-chemical* adj.; *radiation chemist*; **radiation counter** = *COUNTER *sb.*[3] 3 b; **radiation damage,** damage caused by ionizing radiation; **radiation efficiency** *Telecommunications* (see quot. 1977); **radiation field,** an extent of space in which there is radiation; *spec.* the space around an aerial in which there is a continuous outward flow of energy, separated from the aerial by the induction field; **radiation fog,** substitute for def.: fog formed when the ground loses heat by radiation and cools overlying moist air; (further examples); **radiation frost,** frost which occurs when the ground loses heat by radiation; **radiation genetics,** the branch of biology concerned with the genetic effects of ionizing radiation; **radiation hazard,** a risk to health owing to the presence of ionizing radiation; **radiation injury,** an injury caused by over-exposure to ionizing radiation; **radiation pattern,** the way in which the intensity of the radiation from an aerial or other source varies in different directions from it; **radiation pressure,** mechanical pressure exerted by electromagnetic radiation or by sound waves; **radiation pyrometer,** a pyrometer which functions by measuring radiant energy; hence *radiation pyrometry*; **radiation resistance,** the part of the electrical resistance of an aerial that is due to its radiating properties, being the ratio of the radiated power to the mean square current in the aerial; an analogous property of a sound radiator; **radiation sickness,** disease caused by exposure of the body to ionizing radiation; **radiation therapy, treatment,** medical treatment by means of radiation, such as X-rays or ultraviolet light.

1954 A. HOLLAENDER *Radiation Biology* I. 1255/2 (Index), Radiation accidents. **1970** PASSMORE & ROBSON *Compan. Med. Stud.* II. xxxiii. 1/1 Further knowledge of the acute effects of high doses of radiation in man has been provided by the outcome of radiation accidents. **1964** C. HODDER-WILLIAMS *Main Experiment* I. i. 18 Some of you are not wearing radiation badges. **1959** *Times* 14 Feb. 8/4 By a 'radiation belt' is meant a region in which incoming charged particles are trapped and held captive by the magnetic field of the earth. **1959** *Sci. Amer.* Mar. 47/2 The radiation belts obviously present an obstacle to space flight. **1962** *Daily Tel.* 2 May 1/8 American scientists have said it may take 10 or even 100 years before the radiation belts return to their present conditions [after the explosion of a hydrogen bomb]. **1977** F. N. BASH *Astron.* xiii. 398 Jupiter's radiation belts are not only more intense, but also much larger than the similar Van Allen belts that girdle the earth. **1949** *Surg., Gynecol., & Obstetr.* LXXXVIII. 609 (*heading*) Surgical treatment of radiation burns. **1960** *Gloss. Atomic Terms* (Atomic Energy Authority) 42 Patients subject to intense irradiation, e.g. in

cancer therapy by X or gamma rays sometimes get surface or skin radiation burns. **1961** R. D. BAKER *Essent. Path.* vii. 127 Heavy dosages of x-ray, especially to the upper abdomen, may lead to radiation sickness, nausea and vomiting shortly after the radiation; radiation cachexia, weakness, anemia, and leukopenia several weeks after exposures. **1947** *Radiology* XLIX. 359/2 The question of radiation carcinogenesis in the lungs cannot be fully dismissed. **1951** *Jrnl. Chem. Educ.* XXVIII. 407/2 Not all instruments shown in Table 1 are equally useful for the radiation chemist. *Ibid.* 419/2 From the more theoretical point of view radiation chemical reactions should be studied to determine how internal conversion and energy transfer affect yields. **1940** *Chem. Abstr.* XXXVI. 229 (*heading*) The significance of radiation chemistry and its procedures for the science and technology of glass. **1951** *Jrnl. Chem. Educ.* XXVIII. 416/2 We may note finally in the radiation chemistry of water the anisotropic distribution of H_3O^+ and OH^- ions, particularly in cases of heavy particle bombardment. **1961** G. R. CHOPPIN *Exper. Nuclear Chem.* xii. 192 Most of the early research in radiation chemistry used natural nuclear emissions. **1974** J. E. WILSON *Radiation Chem. of Monomers, Polymers, & Plastics* iv. 185 A study of the radiation chemistry of small molecules is helpful in understanding the radiation chemistry of polymers. **1979** *Nature* 15 Feb. 583/3 Work was soon under way in the β and γ-ray spectrometry of (mainly) naturally radioactive species, and in radiation chemistry. **1947** *Nucleonics* Sept. 23/1 (*heading*) Crystal and Cerenkov radiation counters. **1941** *Jrnl. Appl. Physics* XII. 293/1 Cases of severe radiation damage to the hands of radiologists and technicians. **1957** BENEDICT & PIGFORD *Nuclear Chem. Engin.* iii. 66 The extent to which fuel elements of a power reactor can be irradiated may be limited..by physical changes in the fuel caused by radiation damage. **1970** PASSMORE & ROBSON *Compan. Med. Stud.* II. xxxiii. 7/2 That harmful mutations may result from radiation seems almost certain... The demonstration of microscopic radiation damage to human chromosomes makes the occurrence of finer damage almost beyond question. **1961** R. D. BAKER *Essent. Path.* xx. 542 Radiation dermatitis, usually from overexposure to X-rays and less commonly to radium, results in similar atrophic changes in the skin..which, after a number of years, become squamous cell carcinoma. **1934** *Radiology* XXIII. 738 (*heading*) Some mathematical aspects of radiation dosage. **1922** tr. Friedrich & Glasser in Kroenig & Friedrich *Princ. Physics & Biol. of Radiation Therapy* 241 (*heading*) The distribution of the radiation dose in intracorporeal radium and mesothorium therapy. **1970** PASSMORE & ROBSON *Compan. Med. Stud.* II. xxxiii. 1/2 Other important aspects of radiation dose are the rate of delivery, whether it is fractionated in time and the duration of the radiation-free periods between doses. **1937** *Brit. Jrnl. Radiol.* X. 600 (*heading*) Radiation dosimetry. **1968** *Courier-Mail* (Brisbane) 16 Nov. 15 (Advt.), The Australian Atomic Energy Commission is in the process of establishing a programme of research and development in the field of radiation dosimetry and radiation standards. **1913** J. ERSKINE-MURRAY *Handbk. Wireless Telegr.* (ed. 4) xix. 367 (*heading*) A direct experimental method for the determination of the radiation efficiency, earth resistance and other constants of a transmitter. **1977** S. W. AMOS *Radio, TV & Audio Technical Ref. Bk.* xxi. 5 Antenna radiation efficiency = power radiated by antenna/input power to antenna. **1924** *Physical Rev.* XXIV. 339 The energy which is removed from a radiation field..by the positive absorption of quanta by *N* oscillators all in $(n-1)$ quantum states. **1928** STERLING & KRUSE *Radio Man.* 663/1 (Index), Radiation field of antenna. **1948** A. L. ALBERT *Radio Fund.* xiii. 494 At a distance greater than, say, a few wavelengths from the antenna, the induction field approaches zero, and the radiation field exists... The energy in the radiation field does not return to the antenna when the antenna voltage and current die out. **1971** *Nature* 20 Aug. 572/1 The ultrasonic field was monitored by a small, sensitized, thermistor probe which could be positioned at any point in the radiation field by means of a micromanipulator attached to the irradiation chamber. **1937** G. T. TREWARTHA *Introd. Weather & Climate* iii. 109 Radiation fog is at its worst in the vicinity of large cities where the air is rich in hygroscopic smoke particles. **1971** W. HILLEN *Blackwater River* xiii. 121 Morning radiation fog shrouded the lake..and filled the lower valley. **1889** *Q. Jrnl. R. Meteorol. Soc.* XV. 23 It was not an ordinary radiation frost, but one in which the entire valley was filled with an almost uniformly cold atmosphere. **1906** *Daily Chron.* 8 Mar. 7/4 It will surprise many people.. to know that a slight radiation frost actually occurred on the grass early yesterday morning. **1945** R. BUSH *Frost & Fruitgrower* iii. 13 Cloud coming up when a radiation frost is beginning sends the temperature up surprisingly quickly. **1959** *Chambers's Encycl.* VI. 93/2 The occurrence of radiation frost at night is associated with the growth of an inversion in the surface layers of air. **1934** *Lancet* 28 July 214/1 A new subject to many members is 'radiation genetics'. **1956** C. AUERBACH *Genetics in Atomic Age* viii. 69 In the twenty-eight years since its beginning, radiation genetics has developed into a large and flourishing branch of research. **1945** H. D. SMYTH *Gen. Acct. Devel. Atomic Energy Mil. Purposes* vii. 73 Two types of radiation hazard were anticipated—neutrons generated by the pile, and alpha-particles, beta-particles, and gamma rays emitted by products of the pile. **1964** C. HODDER-WILLIAMS *Main Experiment* I. iii. 34 She's batty on the subject of radiation hazards. She used to flirt with the CND. **1946** U.S. Atomic Energy Comm. Rep. MDDC-700 (*title*) Radiation-induced changes in ultraviolet absorption spectra of urine. **1962** *Times* 12 July 3/1 The effects of radiation-induced mutations on mouse populations. **1971** *Brit. Med. Bull.* XXVII. 66/1 The Survey is occasionally referred to as a source of information about mongols and leukaemia, but never in connexion with radiation-induced cancers. **1942** *Radiology* XXXIX. 663 (*heading*) Influence of the medium on radiation injury of sperm. **1970** PASSMORE & ROBSON *Compan. Med. Stud.* II. xxxiii. 7/2 In other respects the histological appearance of the tissue are those of a nonspecific inflammatory response although some features suggest a radiation injury. **1955** Radiation level [see *RADIOBIOLOGY]. **1947** *Nucleonics* Sept. 61/1 (*heading*) Beta and gamma radiation meter. **1961** *Economist* 23 Dec. 1220/2 'Survival biscuits'.., water, first-aid kits and radiation meters. **1951** GRAY & MARTENS *Radiation Monitoring in Atomic Defense* xii. 91 Emergency workers, particularly radiation monitors, should be equipped with dose meters for their own protection. **1955** [see *film badge* s.v. *FILM *sb.* 7 b]. **1964** C. HODDER-WILLIAMS *Main Experiment* I. iii. 29 Recessed radiation monitors placed at fifty-foot intervals. **1926** *Acta Radiol.* VI. 399 (*heading*) Radiation osteitis. **1948** A. L. ALBERT *Radio Fund.* xiii. 496 (*caption*) The radiation patterns of a half-wave antenna. **1978** *Nature* 5 Jan. 48/2 The radiation pattern of the antenna is similar to that of a half-wave dipole. **1978** *Sci. Amer.* Aug. 120 (*caption*) Flask arrangement for measuring the radiation pattern from a conventional grate. **1901** *Physical Rev.* XIII. 308 Radiation pressure, from its nature, must reach its maximum value instantly. **1905** [see *RADIOMETER 2]. **1926** H. C. MACPHERSON *Mod. Astron.* vi. 96 It is now generally admitted that the chief factor in producing comets' tails is the radiation pressure. **1966** *McGraw-Hill Encycl. Sci. & Technol.* XI. 317/1 Since the radiation pressure is $p = 2E$, where E is the energy of the acoustic wave per unit area, a direct measurement of this pressure will determine the energy of a plane wave. **1976** *Nature* 2 Sept. 15/2 Theoreticians.. remained frustrated in their attempts to explain motions in the great straight tails [of comets] by means of solar radiation pressure alone. **1952** B. WOLFE *Limbo* '90 (1953) IV. xiv. 220 Heat-proof and radiation-proof name plates. **1904** *Physical Rev.* XIX. 422 The best types of radiation pyrometers that are at present available to the experimentalist. **1959** *Techn. Survey Dorman Long (Steel) Ltd.* (Iron & Coal Trades Rev.) 59/2 Each furnace also has two radiation pyrometers focussed on the crown of the roof ten feet on either side of the tap-hole. **1972** G. F. WARNKE in H. H. Plumb *Temperature* IV. I. ii. 503/2 The instrument engineer seeking to employ infrared radiation pyrometers for process temperature measurement and control. **1904** *Physical Rev.* XIX. 422 Under the term radiation pyrometry may be grouped all those methods in which the temperature of bodies is estimated from the radiant energy emitted, either in the form of visible light radiation or of the longer infra red waves. **1922** GLAZEBROOK *Dict. Appl. Physics* I. 643/1 In radiation pyrometry generally the term 'full radiator' or 'black body' denotes one that will absorb all the radiation it receives. **1913** J. ERSKINE-MURRAY *Handbk. Wireless Telegr.* (ed. 4) xiv. 223 The fourth numeric of the antenna, namely *r*, the radiation coefficient or 'radiation resistance', is that on which transmission actually depends. **1938** F. E. TERMAN *Fund. of Radio* xiv. 389 The magnitude of the radiation resistance depends..upon the point in the antenna system at which the resistance is considered as being inserted. *Ibid.* xvii. 435 The presence of the air in contact with the vibrating diaphragm produces a mechanical radiation resistance..which varies with frequency. **1975** D. G. FINK *Electronics Engineers' Handbk.* xix. 55 A loudspeaker designed to resonate at a low frequency combines decreasing velocity with increasing radiation resistance to yield a uniform response within the frequency range where the assumptions hold. **1924** *Lancet* 23 Aug. 365/1 Dodds and Webster have recently summarised the literature of radiation sickness in *The Lancet*. **1948** *Sci. News* VII. 8 Radiation sickness, its prevention, its treatment, is..a problem for the medical services. **1961** R. D. BAKER *Essent. Path.* vii. 127 Heavy dosages of x-ray, especially to the upper abdomen, may lead to radiation sickness, nausea and vomiting shortly after the radiation. **1978** *Sci. Amer.* May 49/3, 10 percent of the people exposed to 150 rads will die from radiation sickness. **1922** tr. Krönig & Friedrich (*title*) The principles of physics and biology of radiation therapy. **1961** R. D. BAKER *Essent. Path.* vii. 127 Radiation therapy of the cervix under abnormal circumstances may result in damage to the mucosa of the rectum or of the urinary bladder. **1980** *Daily Tel.* 4 Dec. 6/8 It can implant radioactive seeds by needle when..patients can no longer sustain external radiation therapy. **1924** *Lancet* 4 Oct. 725/1 (*heading*) Chloride metabolism in radiation treatment. **1968** *Brit. Med. Bull.* XXIV. 190/1 Someone without training using the computer procedure can produce a competent radiation-treatment plan. **1970** PASSMORE & ROBSON *Compan. Med. Stud.* II. xxxiii. 7/2 A radiation ulcer of the skin often fails to heal.

radiational (rē[1]di,ē[1]·ʃənăl), *a.* [f. prec. + -AL.] Of, pertaining to, or involving radiation.

1902 *Encycl. Brit.* XXXIII. 213/1 This wider theory [of the electric telegraph] is founded upon Maxwell's theory of electromagnetic radiation, and may be accordingly termed the radiational theory. **1949** KOESTLER *Insight & Outlook* xi. 156 The 'particles' in the atom..are capable of persisting as functional wholes in the teeth of mechanical, thermal and radiational disturbances. **1964** M. MCLUHAN *Understanding Media* (1967) xix. 198 The wheel and the road expressed and advanced this explosion by a radiational or center-margin pattern. **1969** *Sci. Jrnl.* Jan. 64/2 Analysis of the tides at other localities suggests that the ratio of radiational: gravitational tide of O·2 is fairly typical.

Hence **radia·tionally** *adv.*

1969 R. BUCKMINSTER FULLER *Operating Man. Spaceship Earth* vi. 76, I would like to inventory rapidly the system variables which I find to be by far the most powerful in the consideration of our present life-regenerating evolution aboard our spaceship as it is continually refueled radiationally by the Sun and other cosmic radiation.

radiationless (rē[1]di,ē[1]·ʃənlės), *a.* *Physics* and *Chem.* [f. RADIATION + -LESS.] Not involving the emission of electromagnetic radiation.

1919 *Proc. Nat. Acad. Sci.* V. 590 For any real motion, where the velocity is less than that of light,..radiationless orbits, other than straight lines uniformly described, are impossible. **1931** *Proc. R. Soc.* A. CXXXII. 668 It is uncertain whether these γ-rays are actually emitted or whether the β-ray lines from which their existence is inferred are due to a radiationless transition. **1939** *Trans. Faraday Soc.* XXXV. 142 The radiationless transfer of energy from A to A' is caused by the close electrical coupling (resonance) between the electrons in atoms close together as in crystals. **1974** *Nature* 15 Mar. 196/1 The excited molecule loses its energy, first, by radiationless conversion to the lower excited singlet state and finally by fluorescence from this state at 28,600 cm^{-1}.

Hence **radia·tionlessly** *adv.*

1974 GILL & WILLIS *Pericyclic Reactions* v. 137 However, the BO 'hole' provides a mechanism for the system to relax radiationlessly into the ground state.

radiative, *a.* **a.** Add to def.: Occurring by means of radiation; involving or accompanied by the emission of radiation. (Further examples.)

1894 *Mem. R. Astron. Soc.* LI. 145 There are two causes affecting the distribution of density, which respectively cease at different known points, namely, convection at a uniform distribution, and unequal radiative cooling at the distribution given on p. 144. **1930** *Proc. R. Soc.* A. CXXIX. 2 Radiative transitions in the nucleus resulting in the emission of a γ-ray. **1934** *Nature* 10 Feb. 211/1 The energy of these rays corresponds roughly to that which would be emitted in the radiative combination of a neutron and a proton. **1958** *New Scientist* 16 Jan. 14/1 A radiative heat loss due to this process (technically known as *bremsstrahlung* radiation) is thus an unavoidable feature of a thermonuclear reactor. **1973** W. K. ROSE *Astrophysics* ii. 63 The transport of energy in the interior of a star can be caused by radiative transport, convection, or electron conduction. **1977** J. NARLIKAR *Struct. Universe* ii. 29 The star therefore is no longer completely convective, and it develops a 'radiative core'. That is, the energy transport in the core takes place by radiation rather than convection.

b. *radiative capture*, capture of a particle by an atomic nucleus with accompanying emission of one or more gamma rays; *radiative equilibrium*, a state of equilibrium in which the total energy flux emitted is equal to that absorbed; a state of equilibrium in which radiation is the predominant energy transport mechanism.

1926 A. S. EDDINGTON *Internal Constitution of Stars* v. 99 Radiative equilibrium has a natural precedence over convective equilibrium, since in radiative equilibrium convection ceases, whereas in convective equilibrium radiation remains and tends to destroy it. **1934** D. BRUNT *Physical & Dynamical Meteorol.* vii. 138 It is impossible for the upper region of the atmosphere to be kept in an isothermal condition by radiative equilibrium when it is only irradiated from below by long-wave radiation. **1935** *Physical Rev.* XLVII. 508/1 (*heading*) Radiative capture of protons by carbon. **1953** B. STRÖMGREN in G. P. Kuiper *Sun* ii. 40 In regions in radiative equilibrium, i.e., where the transport of energy..is entirely by radiation, the transport by convection and conduction being negligible by comparison, the temperature gradient is given by equation (4*a*). **1962** *Newnes Conc. Encycl. Nuclear Energy* 287/1 In radiative capture the compound nucleus de-excites through intermediate states to the ground state, each step yielding a γ-ray.

Hence **ra·diatively** *adv.*, by means of (esp. electromagnetic) radiation; with emission of radiation.

1957 F. HOYLE *Black Cloud* ix. 177 The volume of information that can be transmitted radiatively is enormously greater than the amount that we can communicate by ordinary sound. **1969** *Nature* 16 Aug. 727/1 The permitted lines are probably radiatively excited from the lowest levels. **1973** *Physics Bull.* Apr. 243/1 The introduction of thallium into the crystal allows the exciton to decay radiatively by luminescence.

radiator. Add: Also, anything that emits sound waves.

1927 *Physics Abstr.* A. XXX. 415 (*heading*) Performance and design of sound radiator consisting of an acoustic transformer and horn. **1936** P. M. MORSE *Vibration & Sound* vii. 240 Few radiators of sound are so much longer than they are wide that they behave like long cylinders, but many radiators behave like spherical sources. **1962** A. NISBETT *Technique Sound Studio* iii. 60 'Cellos and basses differ from violins and violas in that they are much more efficient as radiators in the lower register. **1977** G. PORGES *Appl. Acoustics* xii. 140 A dipole is a less efficient radiator of sound than a monopole, and a quadrupole is even less efficient.

b. Add to def.: Now usu. a tank or compact array of pipes, having a large surface area, which is heated usu. by circulating hot water and gives off warmth also by convection. Also applied to various kinds of electric space heater. (Earlier and additional examples.)

1851 C. CIST *Cincinnati* 213 The introduction of evaporatory radiators and registers. **1891** W. P. BUCHAN *Ventilation* xv. 166 The so-called steam or hot-water 'radiators' heat principally by conduction, as may be easily proved by suspending one thermometer, say, a foot high right above the 'radiator' and another right in front of it and six inches from it. *a* **1910** 'MARK TWAIN' *Europe & Elsewhere* (1923) 178 The German stove..is lovely,..compared with any 'radiator' that has yet been intruded upon the world. **1959** N. MAILER *Advts. for Myself* (1961) 153 She has shut the window and neglected to turn off the radiator. The room is stifling. **1966** N. CHAPMAN *Heating* 37 A storage radiator..provides the most comfortable conditions in a room early in the morning because the air is warmer from the heat given off during the charging period. **1975** R. H. WARING *All about Home Heating* ii. 14 Modern radiators are usually either of panel type or radiant panel, of relatively slim construction and with inlet and outlet connection at the bottom on each end. Panel radiators have a ribbed surface. Radiant panel radiators consist of a radiator tube attached to the back of a metal sheet..or sandwiched between two metal sheets for complete enclosure.

c. Part of the cooling system of a liquid-cooled internal-combustion engine, consisting of a large bank of thin tubes in which the circulating fluid (freq. water) is cooled by the surrounding air after passing round the engine.

1900 *Sci. Amer. Suppl.* 25 Aug. 20617/3 The present water circulating plan..has..the defect of complicating the mechanism by the addition of tanks, radiators and pumps, causing multiplied trouble. **1905** *Daily Chron.* 6 July 5/2 Lancia had negotiated half of the round when his radiator suddenly burst, and left him stranded. **1932** KIPLING *Limits & Renewals* 80 Private Gillock..was stage-whispering me for leave to 'put a shot into his radiator'. **1935** C. G. BURGE *Compl. Bk. Aviation* 271/2 In a single-engined aircraft the radiator may be mounted below the fuselage. **1970** K. BALL *Fiat 600, 600D Autobook* iv. 40/2 Maintenance of the cooling system is confined to periodical flushing of the radiator, refilling with antifreeze solutions for cold weather or clean water for warm, [etc.].

d. An aerial for transmitting (and often also receiving) radio waves.

1903 *Sci. Amer. Suppl.* 4 July 23000/3 The 'aerial', or radiator; that was, the tall wire which..sent electric waves off into the ether. **1947** D. G. FINK *Radar Engin.* v. 303 The radiator forms the radiated beam and presents the absorption area for reception of the echoes. **1952** E. A. LAPORT *Radio Antenna Engin.* ii. 94 The type of radiator that is generally used for medium-frequency broadcasting is the straight uniform vertical with its lower end near ground. **1973** W. E. KOCK *Radar, Sonar, & Holography* ii. 42 In radars and sonars, this ability to steer the beam electrically, that is, without moving the radiator, permits much larger arrays to be used.

2. *attrib.* and *Comb.*, (chiefly sense *1 c), as *radiator cap, fluid, grill(e*.

1913 *Collier's* 11 Jan. ii. 9/1 If it has a new radiator cap or a new form of rear spring suspension..it is immediately surrounded by a crowd of motor sharps. **1948** M. LASKI *Tory Heaven* ix. 130 James managed to undo the radiator cap and poured the water in. **1979** *Arizona Daily Star* 5 Aug. c 5/3 (Advt.), Includes pressure-checks of radiator cap, hoses and fittings for leaks. **1972** D. E. WESTLAKE *Bank Shot* ii. 16 Glass was broken, chrome was bent, and..radiator fluid was making a green puddle on the blacktop. **1938** *Decorative Art* 53/2 The floral designs over the bookcases and the radiator grille are hand-painted on wood. **1959** *Motor* 4 Mar. 168/1 Wider than hitherto, the radiator grille is flanked by combined sidelamps and turn-signal flashers. **1973** D. MILLER *Chinese Jade Affair* xxi. 212 Potter was still in the Mini... I kicked the radiator grill in passing.

radical, *a.* and *sb.* Add: **A.** *adj.* **1. c.** Philos. *radical empiricism*: a name given by W. James (1842–1910) to a philosophical position according to which even underlying postulates are regarded as hypotheses to be verified (see quot. 1897); more generally, a rigorous or sceptical empiricism; hence *radical empiricist*, one who adopts this position. Similarly *radical pluralism, pluralist*.

1897 W. JAMES *Will to Believe* p. vii, Were I obliged to give a short name to the attitude in question, I should call it that of *radical empiricism*... I say 'empiricism', because it is contented to regard its most assured conclusions concerning matters of fact as hypotheses liable to modification in the course of future experience; and I say 'radical', because it treats the doctrine of monism itself as an hypothesis. *Ibid.* p. ix, This is pluralism, somewhat rhapsodically expressed. He who takes for his hypothesis the notion that it is the permanent form of the world is what I call a radical empiricist. **1904** —— *Ess. Radical Empiricism* (1912) ii. 90 These are the main features of a philosophy of pure experience... In my own mind such a philosophy harmonizes best with a radical pluralism. **1911** —— *Some Prob. Philos.* x. 164 He concluded that such realities as present beings, past events and causes, steps of change and parts of matter, must needs exist in limited amount. This made of him a radical pluralist. **1949** *Mind* LVIII. 369, I remember well the dismay in an undergraduate philosophical society when one of the members..announced that he had adopted the 'radical pluralism' of this author [*sc.* Russell]. **1965** P. A. BERTOCCI in B. B. Wolman *Scientific Psychol.* xvi. 295 Radical empiricism, synoptic examination as well as analysis—this is the methodology that alone can provide hypotheses which..will not lose their anchorage in what is phenomenally given. **1973** K. B. MADSEN in D. Berlyne *Pleasure, Reward, Preference* xi. 286 We expect a radical empiricist to be a materialist rather than a dualist.

3. c. Also *radical reformation*.

1781 C. WYVILL *Polit. Papers* (1794) I. vi. viii. 341 While independent men, supported by large bodies of their fellow citizens, have the virtue thus to resist corruption..the hope of a radical reformation cannot be ill-founded.

d. *Politics.* Advocating 'radical reform' (see sense 3 b in Dict.) or any thorough political and social change; representing or supporting the extreme section of a political party; hence, in more recent use (orig. *U.S.*) left-wing, revolutionary. Also in *Comb.* with sense 'radical and —'.

This sense overlaps and merges with sense B. 5 c in Dict. Because of political changes the examples given are of course not restricted to those of 'the extreme section of the Liberal party' in Britain as stated in the Dict.; on the continent of Europe in the present century, parties bearing the title of 'Radical' have in fact freq. tended towards the centre or even a conservative standpoint.

1832 J. S. MILL *Let.* 17 Sept. in *Wks.* (1963) XII. 117 Several friends of mine, radical-utilitarians of a better than the ordinary sort. **1835** *Knickerbocker* VI. 92 The tendency of Americans, instead of being aristocratic, is decidedly radical. **1844** C. M. YONGE *Abbeychurch* xiv. 303 The window ..was adorned with all the worst caricatures which had found their way to Abbeychurch, the portraits of sundry radical leaders. **1847** *Semi-Weekly News* (Fredericksburgh, Va.) 21 Oct. 2/2 The Barnburners with their 'Radical democracy' can never long govern this great State. **1865** *Atlanta Daily Intelligencer* 1 Oct. 3/1 The radical Republicans are now proposing a compromise on the negro-suffrage question. **1884** E. W. HAMILTON *Diary* 6 Oct. (1972) II. 699 He is sure the Opposition will insist on our strengthening our Navy and he believes that even the 'Radical Economists' will advocate this. **1905** *Daily Chron.* 30 June 6/2 It is reported that..a Radical-Protectionist Government will take office. **1908** *Ibid.* 6 Apr. 7/4 Under Mr. Asquith, the balance of power in the Government shifts from the Radical-Socialist to the more moderate Radical side. **1912** W. E. WALLING *Socialism as it Is* iii. 166 While one element is growing more radical another is growing more conservative and the breach between the Independents and the other Labourites is widening. **1927** W. IRWIN *How Red is America?* ii. 47 The half-century when the radical parties were getting their foothold in Europe and America. **1940** B. MIALL tr. *Salvatorelli's Conc. Hist. Italy* xx. 616 Since there was an anti-ministerial movement even in the Radical party, Giolitti resigned. **1950** THEIMER & CAMPBELL *Encycl. World Politics* 171/2 The radicals..are associated with..the *Union démocratique socialiste et radicale*, in the *Rassemblement des Gauches*..; they still use the old name of radical-socialists, but are not socialist. **1953** *Manch. Guardian Weekly* 1 Jan. 2 He deplored the 'pinks', 'radical cliques' and other victims of the 'Communist line'. **1965** E. NOLTE in Rogger & Weber *European Right* 297 The antigovernmental Right..often attacked Hitler for being a Catholic or insufficiently radical. **1968** *Ann. Reg. 1967* 5 Mr Thorpe represented the radical wing [of the Liberal Party], while Mr Lubbock epitomized the suburban vote. **1969** *Ann. Reg. 1968* 491 British students (with the exception of a radical handful at the London School of Economics in 1967) had been quiet and conservative. **1969** [see *MARXIST *a.*[1]]. **1969** G. S. JONES in Cockburn & Blackburn *Student Power* 38 A country [*sc.* the U.S.] where the hysteria of the Cold War had previously smothered all radical politics. **1974** A. J. GREGOR *Fascist Persuasion in Radical Politics* 16 The Fascist emphasis..on the 'primacy of politics', on a 'collectivistic' rather than a 'liberal' social order..are all emphases that have become more and more characteristic of *all* radical political ideologies.

e. Characterized by independence of, or departure from, what is usual or traditional; progressive, unorthodox, or revolutionary (in outlook, conception, design, etc.).

1921 *Daily Colonist* (Victoria) 20 Mar. 24/1 In appearance the Coats car is attractive, but not at all radical in design. **1928** E. O'NEILL *Strange Interlude* I. 15 A natural tendency toward a prim provincialism where practical present-day considerations are concerned (though he is most liberal—even radical—in his tolerant understanding of the manners and morals of Greece and Imperial Rome!). **1938** H. A. MURRAY *Explor. in Personality* iii. 148 Radical sentiments: the origination, promulgation or defence of sentiments, theories or ideologies that are novel, questionable or opposed to tradition. **1958** *Listener* 23 Oct. 648/1, I would describe as radical paintings that make a difference to our ideas about art. The American contribution to radical art in this sense is particularly to be seen in the big picture. **1962** *New Statesman* 25 May 768/2 A true modernist, a radical functionist, would have rejected this basic proposition. **1971** C. HAMPDEN-TURNER *Radical Man* i. 17 While Conservative Man is caused to behave, Radical Man imagines and reasons autonomously. **1977** *Western Mail* (Cardiff) 5 Mar. (Rugby Suppl.) 2/7 The 'radical rugby thinker' Jika Travers from Australia and Oxford University, capped as a war veteran at 27.

f. Special collocations in senses 3 d and e, as *radical chic*, the fashionable affectation of radical left-wing views or of dress, style of life, etc., associated with such views; also *transf.*, those who embody such an affectation; *radical feminism*, advocacy of radical left-wing views designed to counter the traditional dominance of men over women; hence *radical feminist* adj. and sb.; *radical left* = *NEW LEFT; *radical right*, extremist conservative or fascist views favouring group action to protect or re-instate certain social traditions.

1970 T. WOLFE in *New York* 8 June 40/2 Radical Chic invariably favors radicals who seem primitive, exotic, and romantic, such as the grape workers..the Panthers..and the Red Indians. *Ibid.* 56/1 Radical Chic..is only radical in style; in its heart it is part of Society and its traditions. Politics, like Rock, Pop, and Camp, has its uses; but [etc.]. **1973** *Guardian* 5 Mar. 20/4 (*heading*) For a taste of radical chic—try a plum jam buttie. **1977** *Rolling Stone* 24 Mar., Right now bisexuality is the big radical chic on campuses. **1977** *Daily Tel.* 16 July 12 The environmental lobby, a cause largely invented by the Manhattan radical chic. **1980** *Church Observer* Apr. 18/2 By translating the actions of ancient Judea into modern radical chic, the whole story [*sc.* *Monty Python's Life of Brian*] comes dangerously close to life today and threatens to upset the safe, neat and tidy arrangement we favour in Sunday morning religion. **1923** A. R. WADIA *Ethics of Feminism* i. 19 These would also have to be studied so as to enable us to judge how far they afford a stable basis for the advocates of radical feminism. **1971** S. FIRESTONE *Dialectic of Sex* ii. 16 In the radical feminist view, the new feminism is not just the revival of a serious political movement for social equality. **1976** *Papers on Patriarchy Conference London 76* 3 Radical feminists/separatists place importance on the subjective, personal perceptions of oppression to provide the guidelines for 'political' action. **1977** *Rolling Stone* 24 Mar., I've read some radical feminist reviews of your book which gripe about your excessive closeness to men. **1969** *N.Y. Times Mag.* 9 Feb. 34/1 They are members of a newly assertive radical left group called Women's Liberation. **1970** *New Yorker* 26 Sept. 136 The Wallace voters and the radical right—often ideologically distinct, though not always—and the radical left. **1977** *Times* 21 Sept. 1/6 Although the 'radical left' are still only a minority in the institutes of higher education, they deserved serious..rebuttal. **1954** T. TAYLOR *Grand Inquest* (1955) p. xvi, The Communist *Weltanschauung* and conception of the individual's relation to the State have much more in common with the nationalism of the radical right than with the liberal internationalism of the left. **1965** E. WEBER in Rogger & Weber *European Right* 20 The violence and brutality of the radical Right became acceptable to conservatives who viewed them as defenders against the threat of a radical Left. **1970** C. HAMPDEN-TURNER *Radical Man* (1971) ix. 267 This explains the relative efficiency and speed of organized groups on the Radical Right and the Communist Left. **1977** M. WALKER *National Front* v. 117 The [Monday] Club accepts..the name 'Radical Right'.

5. *radical word* (earlier and later examples).

1605 BACON *Adv. Learn.* II. f. 59[v], They [*sc.* the Chinese] haue a vast multitude of Characters, as many (I suppose), as Radicall words. **1921** E. SAPIR *Language* ii. 29 Radical-words may and do occur in languages of all varieties.

B. *sb.* **1. c.** One of the set of basic Chinese characters (usu. reckoned to number 214) which, occas. in modified form, constitute elements, freq. with semantic significance, in the composition of other characters and are a means by which characters can be classified and arranged in dictionaries. Cf. *PHONETIC *sb.*

[**1736** R. BROOKES tr. *Du Halde's Hist. China* II. 393 As in Hebrew there are radical letters..so likewise the Chinese have their radical characters; for instance the characters of mountains, of trees,..under which must be sought all that belongs to mountains, trees.] **1824** J. JOHNSON *Typographia* xii. 368 (*heading*) Table of the Chinese Radicals. *Ibid.* 369 Each Radical is placed at the head of a new family of characters. **1834** *Chinese Repository* May 31 The most conspicuous portions of characters were adopted as 'heads of tribes', which in Europe have been called keys and radicals... The radicals..rarely have any relation to the sounds of the characters of which they form component parts. **1874** [see *PHONETIC *sb.*]. **1907** W. HILLIER *Chinese Lang.* i. 4 We..divide Chinese characters into two parts—one, the *sound* indicator, to which the name 'phonetic' is generally given; the other, the *idea* indicator, which is commonly called the 'radical'... The radicals are limited in number, there being only 214 of them altogether. **1921** *New China Review* III. 390 The sign..does not figure..as a phonetic but rather as the genuine hieroglyph, to which a classifying radical has been added. **1948** R. A. D. FORREST *Chinese Lang.* ii. 38 After we have deducted the 'sheep' device from each of the three examples, the part remaining in each case is known as the 'radical', or better, as the 'signific'. **1973** *Sci. Amer.* Feb. 54/1 A dictionary published in 1971..has merged some radicals, reducing the number to 189. **1978** *Nagel's Encycl.-Guide: China* 88 Character 14 was described as consisting of a phonetic and a radical; the phonetic showed the reader that the word was pronounced in much the same way as 'door', and the radical showed that the meaning was linked with 'ear'.

5. a. Also *gen.*, an advocate of any thorough political or social change; one who belongs to the extreme section of a political party; a member or supporter of a radical movement (cf. sense *A. 3 d), a left-winger or revolutionary. Also *transf.*

1822 J. Q. ADAMS *Diary* 16 June in *Memoirs J. Q. Adams* (1875) VI. iii. 22 General Scott..said Archer was a Radical and inclined to be Jacobinical. **1829** *Western Monthly Rev.* II. 593 The schism in the Methodist Church..exists between the sticklers for the ancient structure of episcopalianism.. and the *radicals*, who seem to be contending for a more democratic form of church government. **1847** *Semi-Weekly News* (Fredericksburgh, Va.) 21 Oct. 2/2 The Barnburners are the progressives, the radicals. **1862** O. W. NORTON *Army Lett.* (1903) 129, I suppose the radicals have got enough of Burnside now. **1876** *San Francisco Daily Examiner* 10 Oct. 2/1 The real issue with the Radicals is to divert public attention from the unexemplified corruption of the Radical party during the past eight years. **1888** W. H. DAWSON *German Socialism* i. 22 The association was joined by South German Radicals, North German students... All that came of it was an outbreak..at Frankfort on April 3rd, 1833. **1912** W. E. WALLING *Socialism as it Is* iii. 35 Whether the radical of to-day, the 'State Socialist' favors political democracy or not, depends on whether these 'passive beneficiaries' of the new 'altruistic' system are in a majority. **1921** *N. Amer. Rev.* Aug. 316 Liberals are merely terror-stricken Radicals. **1938** N. M. BUTLER *Family of Nations* vi. 86 There are very few radicals who are liberals; radicals are almost without exception advocates of compulsion in some one of its forms. **1942** E. PAUL *Narrow St.* iv. 35 Some radical that year [*sc.* 1923] (they called them anarchists, not communists, then) had taken a pot shot at Léon Daudet. **1946** E. O'NEILL *Iceman Cometh* (1947) I. 10 There is a foreign atmosphere about him, the stamp of an alien radical, a strong resemblance to the type Anarchist as portrayed, bomb in hand, in newspaper cartoons. **1950** [see sense A. 3 d above]. **1962** S. E. FINER *Man on Horseback* iv. 59 The Egyptian army officers were..right-wing radicals..with motivations much more akin to the Nazi storm-troopers. **1976** *Guardian* 8 Nov. 2/6 Hundreds of police sealed off the red brick social club where the Right-wing radicals gathered. **1977** *New Yorker* 1 Aug. 49/1 A radical—someone who continued to defend the Soviet revolution after other Socialists had turned against it.

c. *radical-minded* adj.

1890 W. JAMES *Princ. Psychol.* I. 24 The more radical-minded reader can always read 'ideational process' for 'idea'.

ra·dicalish, *a.* [f. RADICAL *a.* and *sb.* + -ISH[1].] Characterized by, or suggestive of, political radicalism.

1837 R. H. BARHAM *Let.* 29 Apr. (1870) II. 24 There is a sort of Radicalish tone about *Oliver Twist* which I don't altogether like.

radicalism. Add: **1. c.** Views or principles favouring radical social or political change and reform (see *RADICAL *a.* 3 d).

1899 C. B. R. KENT *Eng. Radicals* i. 4 When contentment reigns, and there is satisfaction with the present, then radicalism languishes. **1925** tr. *Trotsky's Lenin* 45 What a contrast Vera Ivanovna, with her indefinite radicalism, her subjectivity, and her confusion presented to Vladimir Ilyich. *Ibid.* 85 The shell of intellectual radicalism. **1938** H. A. MURRAY *Explor. in Personality* iii. 226 The S [*sc.* subject] favours modern art, the rejection of sex taboos, socialism,..progressive schools, the humane treatment of criminals, etc. Radicalism is usually opposed to authority, to any force that restrains liberty. **1960** A. B. ULAM *Unfinished Revol.* iii. 78 The element of ethnic hostility..may become subdued..if that radicalism is assimilated into socialism. Or it may become expanded if the radicalism is absorbed into a revolutionary movement with nationalistic or fascist characteristics. **1968** *Internat. Encycl. Social Sci.* XIII. 299/1 Radicalism is a part of the general theme of the growth of rationalistic ethics. **1974** J. WHITE tr. *Poulantzas's Fascism & Dictatorship* vi. ii. 282 Like the urban petty bourgeoisie, the small-holders are at once drawn towards 'democratic' radicalism, *and*..especially likely..to give massive support to Bonapartist forms of State.

d. Psychol. *radicalism–conservatism*: an axis of attitude measurement sometimes used for personality testing.

[**1930** T. F. LENTZ in *Jrnl. Social Psychol.* I. 537 Since the original purpose was to measure radicalism and conservatism, statements were chosen which would involve this characteristic. The statements used are drawn from various fields of interest: ethics, education, feminism,..sex and marriage, politics, race relations.] **1944** G. MYRDAL *Amer. Dilemma* II. App. ii. 1038 The place of the individual scientist in the scale of radicalism-conservatism has always had strong influences upon both the selection of research problems and the conclusions drawn from research. In a sense it is the master scale of biases in the social sciences. **1953** H. J. EYSENCK *Structure of Human Personality* vii. 224 His [*sc.* Carlson's] second factor, radicalism-conservatism, opposed belief in God to an attitude favourable to pacifism and communism. **1965** R. BROWN *Social Psychol.* x. 535 Our problem..is that communism and fascism are opposite extremes on a left to right or radicalism-conservatism dimension.

radicality. 1. (Later example.)

1979 *Dædalus* Winter 30 The radicality of these changes.. had lent credence to the set of beliefs described above.

radicalization. (In Dict. s.v. RADICALIZE *v.*) (Later examples in sense *3 d of the adj.)

1935 *Communist* XIV. 884 (*heading*) The radicalization of the masses. *Ibid.* 885 The bourgeoisie..are alarmed at the rapid radicalization of the workers. **1953** R. BENDIX in Bendix & Lipsett *Class, Status & Power* 605 The radicalization of people who had not participated actively in party politics..gave a major impetus to the rise of fascism. **1960** *New Left Rev.* Nov.–Dec. 7/2 This is a question of the radicalisation of Labour politics. **1969** G. S. JONES in Cockburn & Blackburn *Student Power* 53 British students..are for the first time showing a certain radicalization. **1971** *Nature* 19 Feb. 513/1 What might have been a valuable public discussion on the ethical problems of modern biology..seems to have tailed off into discussion of the process of radicalization of science and even some of the familiar problems of Vietnam. **1977** *Time* 27 June 21/1 As in so many other places, the suppression of legitimate, moderate opposition leads to radicalization.

radicalize, *v.* Add: **1.** (Later examples in sense *3 d of the adj.) Hence **ra·dicalized** *ppl. a.*

1966 *New Statesman* 11 Feb. 185/2 Volunteers returning from two years living with poor people..find that their notions about the world have been radicalised. **1969** *N.Y. Rev. Books* 30 Jan. 17/2 The strongest force for 'radicalizing the campus' may not, after all, be the Cossacks' whips, as many radical tacticians—including those at Columbia—have thought. **1972** *Maclean's Mag.* Sept. 6/2 Being in prison for eight days..helped to radicalize me, as it did so many others. **1976** *National Observer* (U.S.) 24 Apr. 17/1 After all, isn't economic freedom one very important kind of human freedom? Obviously it is, and yet many college students, most particularly the radicalized students of the late 1960s, have been unable to make the connection. **1979** *Dædalus* Summer 19 It disrupts the delicate balance of pluralistic societies and radicalizes everyone.

radicalizing, *ppl. a.* (Later examples.)

1905 *Daily Chron.* 1 May 4/5 The united French Socialist party from now henceforward..will draw to itself..all the radicalising elements in the country. **1973** *Bulletin* (Sydney) 25 Aug. 26/2 It was an amazing revelation and a radicalising one. **1978** *New York* 3 Apr. 34/3 Those themes also spoke to a significant section of the radicalizing, college-educated young, the members of Students for a Democratic Society.

radically, *adv.* Add: **3.** *Comb.* *radically-minded* adj.

1978 J. ANDERSON *Angel of Death* x. 112 He's..radically-minded, and doesn't go much for that old-fashioned religion.

radicel. (Earlier and later examples.)

1819 J. LINDLEY tr. *Richard's Observ. Fruits & Seeds* 68 A slight rim round the base of the radicel. **1944** S. PUTNAM tr. *da Cunha's Rebellion in Backlands* i. 31 Their principal roots are atrophied..and in their place is a wide expanse of secondary radicels, clustered in sap-swollen tubercles.

radicidation (ræːdisəidē[i]·ʃən). [f. L. *radi-āre* to furnish with rays, shine + *oc-cīd-ere* to strike down, kill + -ATION.] The treatment of food with ionizing radiation so as to reduce the number of micro-organisms in it to an undetectable level (see quot. 1964). Cf. *RADAPPERTIZATION, *RADURIZATION.

1964 H. E. GORESLINE et al. in *Nature* 17 Oct. 237/2 Type II is the application to foods of doses of ionizing radiation sufficient to reduce the number of viable specific non-spore-forming pathogenic micro-organisms (other than viruses) so that none is detectable in the treated food by any standard method... The following are the names we suggest for these types of treatment:..Type II, radicidation. **1972** *Poultry Sci.* LI. 277 (*heading*) Poultry feed radicidation. **1973** N. F. LEWIS et al. in *Radiation Preservation of Food* (Internat. Atomic Energy Agency, Vienna) 203 'Radicidation' is synonymous with a 'bactericidal' effect of radiation where a specific pathogenic species is eliminated. Thus, when the process is used solely for destroying enteropathogenic and enterotoxinogenic organisms belonging to the genus *Salmonella*, it is referred to as '*Salmonella* radicidation'.

radiculitis (rædikiŭləi·tis). *Path.* [ad. F. *radiculite*, f. *radicule* radicle (f. L. *rādicula*, dim. of *rādix* root): see -ITIS.] Inflammation of the root of a spinal nerve.

1907 *Jrnl. Amer. Med. Assoc.* 5 Oct. 1227/2 (*heading*) Radiculitis. **1940** S. A. K. WILSON *Neurol.* xiii. 260 For self-evident anatomical reasons, radiculitis can scarcely occur by itself. *Ibid.* 261 Radiculitis and 'spinal neuritis' will not here be distinguished. **1973** *N.Y. Law Jrnl.* 19 July 12/2 Her physicians testified that she had suffered from severe cervical sprain and cervical radiculitis since the accident.

radiesthesia (rē[i]:di,ĭspī·ziă). Also **radiæsthesia.** [ad. F. *radiesthésie*: see RADIO- and *ÆSTHESIA.] The detection of radiation by, or by means of, the body: a process believed by some to be responsible for the operation of dowsing rods, pendulums, and the like as means of locating buried substances, diagnosing illness, etc.

1935 *Jrnl. Brit. Soc. Dowsers* II. 337 The field of radiesthesia is beset with many pitfalls for the unwary. **1950** V. D. WETHERED *Radiesthetic Approach to Health & Homeopathy* iv. 30 The term radiesthesia has been given to the detection by human reactions of ultra-fine radiations or influences such as are given off by the body. **1958** B. NICHOLS *Sweet & Twenties* xi. 146 A few doctors who are making the first, hesitant steps into the uncharted science of radiaesthesia. **1960** *Times* 19 July 18/4 Radiaesthesia and radionics were studied and believed in by many persons. **1960** *Spectator* 28 Oct. 653/2 Radiesthesia may be described as the utilising of psychic powers for the diagnosis of disease, with the help of some mechanical apparatus. **1962** M. ASH *Health, Radiation & Healing* iv. 81 The practice of radiesthesia is based on two fundamental facts: first the muscular response of human beings to fields of radiating energy, and secondly the radiating energy emitted by the human body itself. **1975** *Homes & Gardens* Nov. 63/1 More than 20 doctors in this country practise radiesthesia and some belong to the Medical Society for the Study of Radiesthesia.

So **radiesthe·sic** *a.*, radiesthetic; **radie·sthesist**, a radiesthetist; **radiesthe·tic** *a.*, of or pertaining to radiesthesia; **radiesthe·tically** *adv.*; **radie·sthetist**, one skilled in the techniques of radiesthesia.

1934 *Jrnl. Brit. Soc. Dowsers* I. 142 *Study of radiesthetic fields*, by Maurice Alby, describes in rather complicated detail the nature and position of fields surrounding underground water. **1936** *Discovery* Dec. 395/2 Radiesthesists have not hesitated to borrow, often irrationally and incautiously, from the terminology of physics. **1939** MABY & FRANKLIN *Physics of Divining Rod* 436 Many investigators have thought that dowsers work radiesthetically. **1950** F. A. ARCHDALE *Elem. Radiesthesia & Use of Pendulum* i. 9 Radiesthesists have different methods, although they achieve the same results or at least hope to, that is to develop and intercept the reflexes resulting from their sensitiveness to radiations emanating from other bodies, either animal, vegetable or mineral. **1957** V. D. WETHERED *Introd. Med. Radiesthesia & Radionics* xiii. 130 A foreign radiesthetist travelling in a train on the Continent was able to diagnose accurately what a fellow traveller was suffering from. **1959** B. COPEN *What Radiesthesia is & What it can Do* 15 All Radiesthesic research, diagnosis or therapy does not involve vivisection or cruelty to animals in any way. **1959** M. CLEMENT tr. *A. Mermet's Princ. & Pract. Radiesthesia* iii. 36 No insulating substance is known which can affect radiesthetic radiations. *Ibid.* 37 Some reveal themselves to be radiesthetists at the first attempt, sometimes while undertaking an experiment for fun. **1962** M. ASH *Health, Radiation & Healing* iv. 88 He showed me how he took a radiesthetic reading by allowing a pendulum held in his hand to swing radially away from a patient's body and moving the pendulum gradually outwards until it began to swing in a different direction. **1977** D. V. TANSLEY *Dimensions of Radionics* p. xvii, Out of the investigation we made, there developed in due course his remarkable discovery that it was possible, radiesthetically, to find the archetypal pattern of any substance. **1978** COPEN & KOWA *Pendulum* 7 There are radiesthetists in every country of the world, many who use it [*sc.* radiesthesia] as a hobby, others as a full-time profession.

†**radiferous** (rădi·fĕrəs), *a. Obs.* [f. *RADI(UM + -FEROUS.] Containing or yielding radium.

1903 W. J. HAMMER *Radium* 18 They used two small bulbs..., one containing one gramme of radiferous barium chloride. **1913** J. COX *Beyond Atom* vi. 89 Several investigators made a thorough search of all the known radiferous minerals.

radio (rē[i]·dio), *sb.* orig. *U.S.* [Independent use of the initial element of *RADIO-TELEGRAM, *RADIO-TELEGRAPHY, etc.] † **I. 1.** A message sent by wireless telegraphy or telephony; a radio-telegram. *Obs.*

[**1906** *Internat. Radiotelegraphic Convention: Regulations* (Internat. Radiotelegr. Conf., Berlin) 34 Radiotelegrams bear the service instruction 'Radio' in the preamble.] **1915** R. H. DAVIS *With Allies* 2 For any exhibition they gave of excitement or concern, the news the radio brought them might have been the result of a by-election. **1920** *Glasgow Herald* 10 Aug. 7 In reply the Polish Government sent the following radio. **1923** R. D. PAINE *Comrades of Rolling Ocean* xiv. 245, I shall have to get a radio off to my wife to come on from Ohio and meet me. **1924** R. KEABLE *Recompence* (1926) i. 18 There's a radio in. The *Balmoral* sailed a fortnight after we did.

II. 2. a. The transmission and reception of radio-frequency electromagnetic waves, esp. as a means of communication that does not need a connecting wire; wireless telephony or telegraphy.

Orig. in *attrib.* use only. *Radio-receiver* in quot. 1903 is prob. f. *RADIO- 4 and not necessarily to be taken as evidence of a word *radio*.

1903 Radio-receiver [see sense *5 a]. **1907** L. DE FOREST in *Electr. World* 22 June 1270/1 This factor, damping, is of far more vital import than any regulation of wave-lengths... Radio chaos will certainly be the result until..regulation is enforced. **1911** Radio-communication [see sense *5 a]. **1912** Radio station [see sense *7]. **1913** Radio operator [see sense *5 a]. **1914** Radio transmitter [see sense *5 a]. **1917** *Electr. Experimenter* Jan. 650 (*heading*) Election returns flashed by radio to 7,000 amateurs. **1919** *Pop. Sci. Monthly* Mar. 116/3 Instead of taking bearings by known landmarks, the bearings are determined from known wireless stations by means of radio. **1921** *Sci. Amer.* 2 July 5/1 Armstrong became interested in radio and erected a radio station at his home. **1922** C. W. TAUSSIG (*title*) The book of radio: a complete, simple explanation of radio reception and transmission. **1924** *Glasgow Herald* 26 Jan. 11/5 At the time when radio is in its infancy, experimentalists midway in the United States summoned their friends to hear the Atlantic waves and Pacific surf simultaneously. **1948** A. L. ALBERT *Radio Fund.* x. 380 In radio, the feed-back coil of an oscillator is sometimes called a tickler. **1960** C. H. GIBBS-SMITH *Aeroplane* xi. 76 In August [1910], radio was used for the first time between the ground and an aeroplane in flight. **1964** R. H. BAKER *Astron.* (ed. 8) xvii. 505 (*heading*) Tracing of spiral arms by radio. **1975** FINK & MCKENZIE *Electronics Engineers' Handbk.* XVIII. 62 In radio, polarization usually refers to the electric vector. **1976** PERKOWSKI & STRAL *Joy of CB* i. 6 It's the initial onslaught that is difficult to take, and that is somewhat the condition that we find ourselves in today with CB radio.

b. Organized wireless broadcasting in sound; the sound broadcasting network or service as a whole; sound broadcasting considered as a medium of communication or as an art form.

1922 *Sci. Amer.* June 376/2 Radio today is a continuous performance. You purchase your ticket in the form of a receiving set..and then listen in..to the music of today.. the news of the minute, stock quotations, and so on. **1944** W. C. GREET *World Words* p. v, For effective radio.. pronunciation is not an opportunity to be elegant but an everyday problem of what to do with..words. **1946** *B.B.C. Year Bk.* 29 Plays were a popular form of radio before the beginning of the war. **1951** *Ann. Reg. 1950* 415 Not only did these forums..make good radio, but they were also excellent publicity for the B.B.C. **1958** *Listener* 25 Sept. 482/1 It [*sc.* a play] was also made into some very good radio by the adaptation of the prologue spoken by Luxury and her daughter Poverty. **1960** *B.B.C. Handbk.* 33 In addition to the series of Party Political Broadcasts, the Budget broadcasts, and the Ministerial broadcasts (on sound radio and television), there were reports on Parliamentary topics..in the Home Service. **1966** *Listener* 2 June 816/3 Going back over reviews of the past three months I cannot find a dozen productions which were unequivocally radio and nothing else. **1967** *Ibid.* 18 May 653/1 In the evening they will have radio to listen to, television to watch, and darts and billiards as well in a light and attractive recreation room. **1972** G. GREEN *Great Moments in Sport: Soccer* xxii. 186 They invited me..to examine the organisational set-up they were planning—for spectators, press, radio, television and the rest. **1977** *Daily Tel.* 9 Feb. 11/2 The programme itself was not an outstanding piece of radio. **1978** *Times* 12 June 3/2 Mainly because of economies, radio had become very run down... Some equipment had not been replaced, studios were becoming less suitable. *Ibid.* 3/3 He welcomed the competition of commercial radio.

c. (Preceded by a proper name, esp. of a place.) A particular radio station or network.

1943 C. J. ROLO *Radio goes to War* xxii. 207 The Berlin radio continued to rely most heavily on divisive propaganda. **1958** *Whitaker's Almanack 1959* 582/1 Moscow radio announced that Russia had launched an earth satellite. **1967** *Listener* 12 Jan. 58/1, I do recall that Cairo radio—as well as many Western sources—had interpreted the Soviet warning to Britain and France requiring them to cease operations as implying a threat to bomb London and Paris by missiles. **1968** *Ibid.* 27 June 824/1 Ask Goose Bay Radio if they have any other traffic in this area. **1975** *Whitaker's Almanack* 593/1 The Ethiopian army took control of the national radio station in Addis Ababa and of the independent Voice of the Gospel Radio owned by the World Lutheran Federation. **1978** *Oxford Mail* 20 Feb. 1, Twelve Egyptian soldiers died and 19 were injured in the commando raid at Larnaca airport, Cyprus radio said today.

d. (With capital initial.) Forming the first part of the proper names of particular radio stations or services (the second part freq. being a place-name); *Radio 1, 2, 3, 4*, (also *Radio One*, etc.), the four national radio networks of the BBC (inaugurated on 30 Sept. 1967 in place of the programme services that had existed previously).

1920 *Wireless World* Jan. 587/2 A new Dutch wireless company, called the Nederlandsche Telegraaf Maatchappij 'Radio-Holland' has been formed... 'Radio-Holland' has been formed... 'Radio-Holland' acquires the rights of wireless installations on Dutch mercantile vessels..and the contracts relating thereto. **1926** *Encycl. Brit.* I. 455/1 In 1924–25 the Cie. Française de Radiophonie set up 'Radio Paris'..and provincial stations at Toulouse and Lyons. **1938** *Ann. Reg. 1937* 161 On New Year's Eve, a message from General Smuts, one of the founders of the League, was transmitted from 'Radio-Nations', the League wireless station at Geneva. **1958** *Economist* 25 Oct. 331/1 Radio Free Europe..concentrates on Poland, Czechoslovakia, Hungary, Rumania and Bulgaria, while Radio Liberation broadcasts to the Soviet Union, in no less than seventeen languages. **1964** *Daily Tel.* 13 May 1/8 Radio Atlanta, Britain's second floating commercial 'pirate' radio station, went on the air with programmes for the first time yesterday. **1967** *Listener* 17 Aug. 194/1 The new 247 metres network will be known as Radio 1. The 1500 metres and VHF network will be Radio 2, and..the present Third Network will become Radio 3 and the Home-Service network Radio 4. **1968** *B.B.C. Handbk.* 29 Our first local station, Radio Leicester, began broadcasting on 8 November 1967, followed shortly afterwards by Radio Sheffield and Radio Merseyside. **1973** P. DICKINSON *Gift* ix. 139 Penny was listening to Radio One. **1976** *Daily Tel.* 30 June 1/4 Radio Uganda, monitored in Nairobi, gave no immediate indication of the 'penalties' involved. **1976** *Times* 29 Sept. 16/6 That all this may be entering the field of immodesty is redeemed by a quick smile and the admission that perhaps Radio 3 takes itself too seriously. **1978** *Bookseller* 1 July 54/1 Mike Stevenson, author of Ward Lock's biography of cricketer *Ray Illingworth*..has been interviewed on Radios Leicester, Leeds, Cleveland, Pennine, Piccadilly, Hallam and City.

3. Radio equipment; *spec.* a receiving set.

[**1913**: cf. *radio operator* in sense *5 a]. **1917** *Electr. Experimenter* May 3/1 When the German spies..found that it was not very healthy to operate their outfits in attics or in house chimneys..they simply put their radios in touring cars, cleverly concealing the aerial wires inside of the car bodies. **1925** H. L. FOSTER *Trop. Tramp with Tourists* 97 It fairly shrieked with the blare of jazz—of jazz from radios, jazz from mechanical pianos. **1936** KING EDWARD VIII in *Manch. Guardian Weekly* 6 Mar. 185/1 Science has made it possible for me..to speak to you all over the radio. **1941** AUDEN *New Year Let.* II. 36 He moves on tiptoe round the room, Turns on the radio to mark Isolde's Sehnsucht for the dark. **1968** *New Society* 22 Aug. 265/2 Non-U *radio*/U *wireless* is no longer true; the U call it a radio too. **1973** J. PATTINSON *Search Warrant* ii. 28 There was a load of noise... It sounded like a radio going full belt on a pop-music channel.

4. = *radio spectrum* in sense 7 below; radio wavelengths.

1968 *Physical Rev. Lett.* XXI. 1540/1 NGC1275 and 3C120..are a hundred times more luminous in the radio than most of the Seyferts. **1975** *Nature* 3 Jan. 7/1 It [*sc.* the Crab nebula] is unique in that it has been detected over the entire electromagnetic spectrum from radio through infra-red and the visual to X rays and γ rays.

5. *attrib.* **a.** In general uses, as *radio aerial, antenna, apparatus, beacon* (= *BEACON *sb.* 6 d), *beam, bearing, black-out, cabinet, communication, countermeasure, detector, fade-out, fix, intercept, link* (= *LINK *sb.*[2] 3 f), *marker, mast, message, operator, receiver* (in quot. 1903 prob. f. *RADIO- 4), *relay, room, set, shop, traffic, transmission, transmitter, valve*.

1949 E. B. MOULLIN (*title*) Radio aerials. **1968** *Times* 16 Oct. 8/8 It also seems to have a large radio aerial more suitable for a trip to the planets. **1927** B. F. DASHIELL *Popular Guide to Radio* v. 71 (*heading*) The use of radio antennas and grounds. **1972** K. BONFIGLIOLI *Don't point that Thing at Me* xi. 89 It was the same car..but overnight it had..acquired a suit of whitewall tyres and another radio antenna. **1912** *Statutes U.S.A.* XXXVII. I. 303 The President..may cause the closing of any station for radio communication and the removal therefrom of all radio apparatus. **1916** *Lit. Digest* (N.Y.) 1 Jan 13/1 It is conceivable that this small body of men might have neither sending or receiving radio-apparatus. **1919**, etc. Radio beacon [see *BEACON *sb.* 6 d]. **1966** D. FRANCIS *Flying Finish* ii. 27, I flew contentedly along..checking my direction by the radio beacons over which I passed. **1923** E. W. MARCHANT *Radio Telegr.* i. 12 A ship coming into harbour will pick up the radio beam at the instant at which its direction is towards the ship. **1966** *McGraw-Hill Encycl. Sci. & Technol.* IX. 20/1 Radio beams that are transmitted from moving aircraft to the ground will have an apparent change of frequency. **1977** Radio beam [see *radio detector* below]. **1978** R. V. JONES *Most Secret War* xvii. 139 Milch, the Head of the Luftwaffe, was advising Goering that the current policy of night attacks was useless without special radio-beam devices, like the new X-Gerät. **1935** Radio bearing [see *FIX *sb.* 3]. **1957** *Encycl. Brit.* XVI. 174/1 The accuracy of radio bearings ..[is] dependent upon properly functioning equipment and skilful operation. **1958** *New Scientist* 6 Mar. 8/3 The storm and its associated radio blackout. **1925** *Scribner's Mag.* Sept. 19/1 Have you seen a radio cabinet which..actually does not look like one? **1911** *Statutes U.S.A.* XXXVI. I. 629 (*heading*) An act to require apparatus and operators for radio-communication on certain ocean steamers. **1912** *Ibid.* XXXVII. I. 308 The expression 'radio communication' as used in this Act means any system of electrical communication by telegraphy or telephony without the aid of any wire connecting the points from and at which the radiograms, signals, or other communications are sent or received. **1942** *Electronic Engin.* XV. 116/1 Planck's constant *h*, is a universal one which is applicable to all radiations, including.. the wavelengths used in radio-communication. **1947** *Jrnl. R. Aeronaut. Soc.* LI. 425/2 The use of radio countermeasures, introduced for the first time with such telling effect in the last war, will be a prominent feature in all future large-scale conflicts. **1978** R. V. JONES *Most Secret War* xxii. 180 With our nightfighters and guns powerless, radio countermeasures were our only means of defence. **1936** *Discovery* Mar. 70/1 Each of these observatories is equipped..with..a radio detector, which enables the specialist to learn at once which station is transmitting. **1977** C. FORBES *Avalanche Express* x. 106 It takes two radio-detector vans..five minutes to take a fix on a secret transmitter—to plot from two locations the cross-point of the radio beams indicating where the transmission is coming from. **1937** Radio fade-out [see *FADE-OUT 3]. **1942** Radio fix [see *FIX *sb.* 3]. **1977** C. FORBES *Avalanche Express* xxi. 218 The operator in the radio-detector van..reached for the radio-telephone... 'We have a radio fix, sir... Positive.' **1974** G. MARKSTEIN *Cooler* liv. 192 It was a radio intercept by security monitoring. The message, decoded, read: 'Stand by 24-hourly.' **1928**, etc. Radio link [see *LINK *sb.*[2] 3 f]. **1971** A. DIMENT *Think Inc.* xii. 208 'I love you,' I said and cut the radio link. **1978** *Dumfries Courier* 20 Oct. 10/2 A special Police control room on site at the NEC will have radio links with Police helicopter, car, motor cycle and foot patrols. **1933** *Nat. Geogr. Mag.* May 618/2 Radio-range and radio-marker beacons. **1942** B. A. SHIELDS *Air Pilot Training* xxx. 515 Low-powered radio stations, called radio markers, ..are placed along the airways to serve as radio fixes. **1950** 'D. DIVINE' *King of Fassarai* v. 38 The metereological station was completed... The radio masts went up. **1976** *Cumberland & Westmorland Herald* 27 Nov., The committee approved the Corporation's application to build a radio mast almost 50 ft. high and a modular equipment building. **1916** *Lit. Digest* (N.Y.) 1 Jan 13/2 The radio-message containing this intelligence is flashed over the hills. **1974** M. HASTINGS *Dragon Island* xiii. 112, I have to..send a radio message to Djakarta. **1913** *Year-bk. Wireless Telegr.* 96 The radio operator..must furnish to the inspector evidence that he is 'skilled in the use of the apparatus'. **1925** *Scribner's Mag.* July 44/2 Hank Quiller was rated as chief radio operator on board the S.S. *Omega*. **1974** G. MARKSTEIN *Cooler* lxii. 218 If we find your killer we find your mysterious radio operator. **1903** C. H. SEWALL *Wireless Telegr.* iv. 154 The first radio-receiver in which cause and effect were observed and recognized was devised by Hertz in 1886. **1929** J. H. MORECROFT *Elem. Radio Communication* vii. 220 With coils having a power factor of about 1 per cent, as is the case with the average radio receiver, one tuned circuit will not give sufficient selectivity to eliminate interference. **1976** B. JACKSON *Flameout* (1977) I. 22 They all wore Bellboy radio receivers in their shirt pockets, in case they could not be reached by telephone. **1926** *Wireless World* 1 Sept. 307/1 The wireless signals received in this Radio Relay Office are relayed to the Central Radio Office in the same building. **1927** B. F. DASHIELL *Popular Guide to Radio* xi. 195 The radio relay scheme, whereby a program from a central station is broadcast, received and rebroadcast by other stations, has been successfully tried out. **1966** *McGraw-Hill Encycl. Sci. & Technol.* V. 520/2 Radio relays are used for simultaneous transmission of up to hundreds of telephone conversations over a trunk route. **1921** R. D. PAINE *Comrades of Rolling Ocean* xiv. 244 A message from the radio-room, sir. **1976** 'J. FRASER' *Who steals my Name?* ii. 15 Later..the radio room springs into life, dispatching police cars..over the face of the city. **1913** *Proc. IRE* I. 43 The purpose of this paper is to describe some recent radio sets designed for the Marconi Wireless Telegraph Company of America to meet the new specifications of the United States Navy. **1926** S. LEWIS *Mantrap* ii. 22 How necessary for hardy camp-life are the portable radio set, the pneumatic cushion. **1971** *Daily Tel.* 13 Sept. 1/8 The radio ham..heard two men planning the raid over short-wave radio sets. **1974** E. JONES *Barlow comes to Judgement* 127 He works in a radio shop in Bayswater. **1927** *Jrnl. Franklin Inst.* CCIV. 240 Beam stations are carrying regular radio traffic between England and Canada. **1973** D. KYLE *Raft of Swords* (1974) I. ix. 88 There had been remarkably little radio traffic. So he would sit in the radio room. **1935** C. G. BURGE *Compl. Bk. Aviation* 506/1 The number of channels available for radio transmission is limited. **1974** G. MARKSTEIN *Cooler* li. 181 The interception stations..noted every illicit radio transmission. **1914** R. STANLEY *Text Bk. Wireless Telegr.* xix. 300 The range of a given size of radio transmitter has been greatly increased since the time when coherer detectors were used in conjunction with a Morse tape machine or siphon recorder. **1970** V. CANNING *Great Affair* x. 167 He probably had a secret radio transmitter and receiver somewhere. **1929** *Radio Times* 8 Nov. 434/2 (Advt.), The Radio Valves—with the only filament that has stood the test of time. **1970** P. DICKINSON *Seals* ix. 178 It provided a short-cut in the mass-production of radio valves.

b. Connected with, participating in, or transmitted as part of organized sound broadcasting, as *radio acting, actor, actress, adaptation, announcer, audience, ballad, broadcast, broadcasting, bulletin, celebrity, comedian, commentator, commercial, company, critic, criticism, drama, dramatist, interview, journalism, journalist, listener, news, organization, personality, play, producer, production, programme, reporter, revue, script, serial, series, spot, star, talk, writer*.

1940 *Radio Times* 23 Aug. 6/4 Frederick Allen..had also done a considerable amount of radio acting and singing before becoming a BBC announcer. **1968** *Daily Tel.* (Colour Suppl.) 15 Nov. 32/3 He returned to a contract with Mr Beaumont, radio-acting, poetry recitals, bit parts in films. **1938** *Encycl. Brit. Bk. of Year* 123/1 The widespread unionization of radio actors. **1975** *Times* 20 Sept. 8/4 That uncommon breed, the exceptional radio actor. **1972** P. BLACK *Biggest Aspidistra in World* I. iv. 39 Jenny was played by Lilian Harrison, the leading radio actress of her day. **1931** T. H. PEAR *Voice & Personality* viii. 94 Holt Marvell's radio-adaptation of..*Carnival*. **1927** *Scribner's Mag.* Apr. 437/2 We must listen to radio announcers who insist that the instant programme is most colorful. **1970** 'T. COE' *Wax Apple* (1973) iv. 29 His voice was deep and resonant, like that of a radio announcer. **1932** *Radio Times* 29 July 251/3 Christopher Stone keeps the radio audience amused with a selection from the new discs. **1972** P. BLACK *Biggest Aspidistra in World* I. vii. 56 Mabel Constanduros, who in 1925 introduced the Buggins family..aiming..at the entire radio audience listening in its own family groups. **1960** *Times* 16 Aug. 5/2 Singing the Fishing; radio ballad by Ewan MacColl and Charles Parker. **1956** H. KURNITZ *Invasion of Privacy* xviii. 117 The late radio broadcasts..had given the Morley case a big spread. *a* **1974** R. CROSSMAN *Diaries* (1976) II. 549 A B.B.C. experiment for live radio broadcasts of extracts from the House of Commons debate. **1922** Radio broadcasting [see *RADIOGRAPHY 2]. **1975** *Listener* 25 Dec. 853/3 The poem, with its five voices, is suited to radio broadcasting. **1965** *B.B.C. Handbk.* 65 During the day well over twenty million people listen to at least one radio bulletin. **1948** E. E. CUMMINGS *Let.* 20 Aug. (1969) 184 A charming & handsome & tall & sweet youth named Billy; think he later became a radio-celebrity. **1930** H. NICOLSON *Diary* 18 Oct. (1966) 57, I have become 'famous' as a radio comedian, and shall never be able to live down the impression thus acquired. **1980** S. BRETT *Dead Side of Mike* vi. 60 The programme was merely a showcase for the talents of a once-loved radio comedian. **1938** Radio commentator [see *COMMENTATOR 2 c]. **1953** *Manch. Guardian* 21 May 1/4 Raymond Gram Swing was the most powerful and admired radio commentator working between the two major allies during the war. **1978** *Listener* 27 July 111/3 The skilled radio commentator..was there to tell you what is happening. **1951** M. MCLUHAN *Mech. Bride* (1967) 117/2 The big hefty heartiness of this is very familiar in the radio commercials. **1980** *Broadcast* 7 July 17/2 (Advt.), Production of radio commercials, voice-overs, stereo programmes. **1920** *Sci. Amer.* 24 July 79/1 A leading radio company is about to begin construction of a super-powered radio station. **1938** JOYCE *Let.* 6 June (1966) III. 424 His (my son's) experience of broadcasting in the U.S.A...is that these Radio Companies are all in watertight compartments and that the director who has charge of the singing almost resents even a friendly introduction from the director, say, who controls the sports programme. **1974** *IBA Evidence to Annan Comm.* 46 The local radio companies should retain their creative initiative. **1929** *Vox* 9 Nov. 3/1 Heaven forbid that I should try to emulate the new Radio critics of the *Daily Express* who have apparently been told..to sit down immediately after tea and listen right through the evening, after which they are expected to discuss their indigestion attractively next morning. **1966** *B.B.C. Handbk.* 14 Every critic (and who is not his own television and radio critic?) would do well to temper his occasional rage with the thought that there is much to admire about the BBC. **1976** *Listener* 15 Jan. 42/1, I stopped being radio critic of the *Guardian*. **1940** R. S. LAMBERT *Ariel & all his Quality* vii. 173 It seemed desirable for the BBC to try and build up a..solid school of radio criticism. **1978** *Listener* 27 July 111/2, I was writing radio criticisms for *The Listener*. **1925** *Glasgow Herald* 1 Sept. 8 One of the many limitations of radio-drama will be the impossibility of introducing any but audible actions of a simple kind. **1951** M. MCLUHAN *Mech. Bride* (1967) 157/2 Soap operas are written and acted quite as well as the ordinary evening radio drama. **1929** *Radio Times* 8 Nov. 388/2 Conrad has, curiously, attracted the radio-dramatist. **1944** Radio dramatist [see *ACTUALITY 4 b]. **1926** G. FRANKAU *My Unsentimental Journey* ii. 37 A new form of torment, the 'radio interview'. **1974** *Guardian* 21 Mar. 1/3 Mr Edward Short, Leader of the House, said in a radio interview. **1968** *Listener* 21 Mar. 380/2 When the war was over Ed Murrow went back home, the 'first authentic original' of radio journalism. **1977** *Times* 14 May 10/5 Mary Goldring is a radio journalist... She has now concluded three reports on contemporary India. **1926** *Public Opinion* 2 July 17/2, 27,000,000 persons are now radio listeners in the United States. **1974** *Listener* 24 Jan. 123/3 Faithful radio listeners were entitled to their regular programmes. **1940** J. FLANNER in *New Yorker* 7 Dec. 60/2 To Parisians, the most trusted radio news is what they get from the American short-wave broadcasts. **1973** A. PRICE *October Men* iii. 46 Not a word in the morning paper..or on the radio news. **1962** A. NISBETT *Technique Sound Studio* x. 168 At a radio organization such as the BBC it is easy to feel spoilt for choice. **1941** J. W. WELCH in *Listener* (1978) 18 May 626/1 The BBC is now building up Joad as a radio personality. **1972** P. BLACK *Biggest Aspidistra in World* I. vii. 56 Tommy Handley..in 1929..was already a radio personality. **1924** *Variety* 24 Dec. 35/4 Gene Rouse, announcer for WOAW, has written a 'radio play'. **1973** M. AMIS *Rachel Papers* 75, I slammed the door, so that the sounds of the radio-play on the kitchen wireless were reduced to an underground rumble. **1955** T. H. PEAR *Eng. Social Differences* ix. 195 'Living by the clock' is a virtue in a radio-producer. **1974** *Listener* 14 Feb. 219/2 The radio producer hears the play over and over again. The lives and all the action become crystal-clear to him. **1959** D. COOKE *Lang. Mus.* iv. 200 A private tape-recording of a radio-production of *King Lear*. **1978** *Listener* 2 Feb. 152/1 *The Beggar's Opera*..is a splendid new radio production. **1922** *Variety* 10 Mar. 7/2 Among the theatres which will provide acts exclusively for the 'Star's' radio programs are the Shubert, Orpheum,..Royal and 12th streets. **1925** A. H. MORSE *Radio* v. 78 There need be no limitation of the public enjoyment of the radio programme. **1972** *Sat. Rev.* (U.S.) 27 May 18/2 In Israel I was invited to go on a radio program to discuss the problem. **1926** *Wireless World* 18 Aug. 231/1 The radio reporters are kept busy all the summer, carrying their microphones to the important sporting events, [etc.]. **1975** *Times* 6 Jan. 12/8 Why did Sir Keith tell a radio reporter..that it [*sc.* a speech] was meant to be 'light-hearted'? **1929** *Radio Times* 8 Nov. 395/3 The brilliant little skit..which enlivened a recent radio revue. **1941** B. SCHULBERG *What makes Sammy Run?* ii. 30 He had written a radio script. **1962** A. NISBETT *Technique Sound Studio* ix. 156 Such a 'radio script' will probably have to be reinterpreted. **1958** M. KENNEDY *Outlaws on Parnassus* ii. 28 A Dickens novel had to end sometime, whereas a radio serial can go on for ever. **1977** S. BRETT *Star Trap* iv. 40, I can't do Friday... Doing a pilot of a radio series. **1973** *Black Panther* 28 Apr. 11/2 Bobby has layed out his program..in countless radio spots, interviews and discussions. **1932** *Radio Times* 29 July 279/3 (Advt.), A permanent..record of your favourite radio stars. **1980** P. ABLEMAN *Shoestring's Finest Hour* i. 11 Lonely women who ring up famous radio stars..in the hope of getting to sleep with them. **1940** R. S. LAMBERT *Ariel & all his Quality* vii. 173 Experts are able..to express themselves freely..about current radio drama, music, talks, and television programmes. **1943** D. POWELL *Time to be Born* i. 8 My days are filled with my war committees and my refugee children and my radio talks. **1977** *Listener* 16 June 800/3 BBC Television now appears to be giving radio talks. **1944** L. MACNEICE

Christopher Columbus 8 The radio writer has to think of words in the mouths of actors.

c. Designating devices controlled or operated by radio, as *radio bomb*; vehicles equipped with radio for receiving information, directions, etc., as *radio cab, car, taxi, van*.

1974 D. SEAMAN *Bomb that could Lip-Read* xxiv. 243 Once they discover it was a radio bomb, they will take this hamlet apart. **1977** *Times* 18 July 1/4 (*heading*) Provisional IRA widen use of radio bombs. **1955** J. B. PRIESTLEY in Priestley & Hawkes *Journey down Rainbow* xii. 173 Most of the taxis down here are like our radio cabs in London, but the voice of the distant operator, calling cabs and giving addresses, is always left turned on. **1977** F. WELDON in *Winter's Tales 23* 190 Maureen's on the phone, calling radio cabs. **1925** *Sci. Amer.* Nov. 308/1 The Yard has seven radio-equipped motor cars attached to the Criminal Investigation flying squad... These radio cars not only aid in detecting crime but also perform a helpful service in regulating heavy traffic along the highways. **1967** *Listener* 19 Jan. 95/2 WINS reporters were there with their radio cars and tape recorders inching the story along every few minutes or so with eye-witness reports, [etc.]. **1973** 'E. MCBAIN' *Hail to Chief* i. 3 Two radio-car cops, on routine patrol. **1962** *Spectator* 13 Apr. 486 Lots of tourist information, though we do have radio-taxis here too. **1977** E. AMBLER *Send no more Roses* x. 215, I had used the time..to check out the local radio-taxi services. **1950** J. FLANNER in *New Yorker* 8 Apr. 79/1 On the street, crowds had collected and a radio van had arrived. **1974** N. FREELING *Dressing of Diamond* 10 They [*sc.* the police] say they've two radio vans.

d. Chiefly *Astr.* Connected with the natural emission of radio waves (freq. denoting objects or entities which emit radio waves in unusually large quantities or are being considered as sources of radio waves), as *radio brightness, emission, emitter, flux, galaxy, noise, observatory, sky, source, sun, universe*.

1960 RODMAN & VARSAVSKY tr. *I. S. Shklovsky's Cosmic Radio Waves* iii. 174 The radio brightness of the hypothetical objects, averaged over time, would thus exceed that of the sun by some ten orders of magnitude. **1974** *Sci. Amer.* Aug. 26/3 In the direction of this cloud of ionized gas, designated 30 Doradus, there is a decrease in the radio brightness of the sky. **1949** *Nature* 12 Nov. 816/1 Only a proportion of the flares have associated radio bursts, the bigger flares being the more likely to produce strong radio emissions. **1958** *Listener* 27 Nov. 870/1 One of the earliest of the post-war surprises was the discovery by Appleton and Hey that the sun spots and flares which occasionally appear on the solar surface are associated with large and irregular increases in the solar radio emissions. **1978** *Nature* 14 Sept. 111/1 One of the most important features of the jovian decametric radioemission comes from the geometry of the observed radiation. **1954** *Ann. Reg. 1953* 373 Future accurate measurements of the positions of cosmic radio emitters. **1951** *Monthly Notices R. Astron. Soc.* CXI. 366 The intensity of the radio flux from M31 observed on the Earth at a wavelength of 1·89 metres is 10^{-24} watts/square metre/c.p.s. **1960** RODMAN & VARSAVSKY tr. *I. S. Shklovsky's Cosmic Radio Waves* vi. 356 The relative radio emitting power of radio galaxies is 10^3 and even 10^5 times as great as for normal galaxies. **1973** *Sci. Amer.* Sept. 72/3 Many 'classic' radio galaxies consist of an optically bright galaxy situated halfway between two radio-emitting regions. **1979** *Jrnl. R. Soc. Arts* CXXVII. 582/1 The so-called 'radio galaxies', whose power output in radio waves exceeds the total galactic luminosity of all the stars. **1933** *Gen. Electr. Rev.* XXXVI. 201/1 The radio-noise meter..detects radio noise, measures its intensity, and locates its source. **1946** *Nature* 17 Aug. 234/1 The solar radio noise from sunspots is also characterized by strong fluctuations. **1977** *New Yorker* 19 Sept. 137/1 In 1964, two radio astronomers..made the unexpected discovery that there was cosmic radio noise entering their system which they could do nothing to get rid of. **1958** *Ann. Reg. 1957* 472 The principal radio-observatories developed during the previous few years. **1969** *Times* 18 Apr. 12/6 The Parkes radio observatory in New South Wales. **1959** DAVIES & PALMER *Radio Studies of Universe* iv. 46 The early maps of the radio sky were made with small aerials which could not readily distinguish radio sources from the background of radio emission. **1950** *Monthly Notices R. Astron. Soc.* CX. 519 The five major extra-galactic nebulae in the selected area are listed.. together with the radio sources which appear to be associated with them. **1961** *New Scientist* 5 Jan. 50/1 About 30 per cent of the accurately measured and suitably placed radio sources can be identified with visible galaxies. **1971** *Sci. Amer.* May 56/3 Before 1960 radio astronomers had identified and catalogued hundreds of radio sources. **1961** WEBSTER, Radio sun. **1965** M. R. KUNDU *Solar Radio Astron.* i. 1 Decimeter-wave observations of the radio sun during a solar eclipse. **1974** G. L. VERSCHUUR *Invisible Universe* iii. 32 (*heading*) The radio sun and planets. **1960** RODMAN & VARSAVSKY tr. *I. S. Shklovsky's Cosmic Radio Waves* vi. 355 Further investigations of the variations in the cosmic radio-wave background at high galactic latitudes will undoubtedly reveal new peculiarities of the 'radio universe'.

6. *Comb.* (cf. *RADIO- 4): *radio-controlled* (so *radio-control* vb. trans.), *-emitting, -equipped, -linked, -minded, -receiving, -steered, -transmitting* adjs.

1959 K. VONNEGUT *Sirens of Titan* vii. 121 Without Boaz, their real commander, to radio-control them, they fought bitterly. **1979** *Amat. Photographer* 30 May 162/3 The Post Office refusing him a license to radio-control a camera. **1936** *Punch* 4 Mar. 273/1 Then possibly another sequel in which the generations are estranged over the question of small rubber electric radio-controlled fish. **1958** I. ASIMOV *Naked Sun* i. 9 The radio-controlled flight would be smooth; there would scarcely be any sensation of motion once the plane was airborne. **1976** L. ST. CLAIR *Fortune in Death* xi. 109 The gates swung wide—radio-controlled. **1960** Radio-emitting [see *radio galaxy* in sense 5 d above]. **1971** *Sci. Amer.* May 56/3 Object 3C 48 was thought to be a unique kind of radio-emitting star in our own galaxy until 1963. **1935** C. G. BURGE *Compl. Bk. Aviation* 499/1 By means of radio and land-line the pilot of a radio-equipped 'plane is to-day in closer touch with the ground than the driver of a normal road vehicle. **1963** A. LUBBOCK *Austral. Roundabout* 36, I wouldn't like to say what my radio-equipped bikes and plane save me in time and labour. **1974** H. R. F. KEATING *Bats fly up for Inspector Ghote* iii. 33 Ghote, in yet a third radio-linked car, would be a useful addition to the team. **1930** *Wireless World* 10 Dec. 655/3 To the housewife anxious to please her radio-minded family I would say fill the Christmas pudding this year with a fair sprinkling of miniature fuse lamps instead of with sixpences. **1922** *Glasgow Herald* 21 Apr. 10 Already the number of radio receiving outfits installed in private houses runs into seven figures. **1936** *Discovery* Mar. 69/2 Some 70 radio-receiving observatories all round the earth. **1917** *Nature* 2 Aug. 442/2 Attempts to develop a radio-steered torpedo. **1935** C. G. BURGE *Compl. Bk. Aviation* 511/1 The purpose of the radio compass is to determine whether or not one is flying directly towards a radio-transmitting station. **1959** *Observer* 3 May 17/5 The Authority has now developed radiotransmitting tide gauges which will continuously relay to Gravesend the height of the water at all points of navigational importance. **1978** R. V. JONES *Most Secret War* xliv. 422 We had not known beforehand of these radio-transmitting samples.

7. Special Combs. (cf. *RADIO- 4): **radio altimeter**, an altimeter which functions by emitting a radio signal and measuring the time it takes to be reflected back from the ground; **radio amateur**, one who makes a hobby of picking up, and often also transmitting, radio messages; **radio compass**, a radio direction-finder used for the purpose of navigation; **radio contact**, the state or an instance of being in communication by radio; **radio dial** = *DIAL *sb.*[1] 6 e; **radio direction-finder** = **direction-finder*; so **radio direction-finding**; **radio dish** = *DISH *sb.* 4 b; **radio echo** = *ECHO *sb.* 1 d; **radio energy**, energy transmitted in the form of radio waves; **radio engineering**, the branch of engineering concerned with the design, construction, and operation of radio equipment; so **radio engineer**; **radio ham** *colloq.* [*HAM *sb.*[1] 6], = *radio amateur* above; **radio industry**, the radio engineering or sound broadcasting industries; **radio licence**, a licence certificate that owners of radios are required to have; (such licences for radio only were abolished in the U.K. in 1971); **radio-loud** *a. Astr.*, emitting significant quantities of radio waves; **radio man**, (*a*) a man who operates, repairs, or otherwise deals with radios; (*b*) a man employed in sound broadcasting; **radio map** *Astr.*, a diagram showing the strength of the radio emission from different parts of the sky; **radio microphone** (see quot. 1962); also (*colloq.*) **radio mike**; **radio navigation**, navigation by means of radio signals; so **ra:dio-naviga·tional** *a.*; **radio net**, a system of intercommunicating radio sets, operated esp. by a police force or similar body; **radio network**, a system of radio stations for navigation, communication, or broadcasting; a sound broadcasting organization or channel; **radio pager** = *PAGER *sb.*[3]; so **radio paging** *vbl. sb.*; **radio-pho·nograph** *U.S.* = *RADIO-GRAMOPHONE; **radio pill** *colloq.* = *endoradiosonde* s.v. *ENDO-; **radio pirate** = *PIRATE *sb.* 4 b; **radio-qui·et** *a.*, emitting a negligible amount of radio waves; hence **radio-qui·etness**; **radio-radar**, used *attrib.* of devices, systems, etc., combining radio and radar; **radio range**, (*a*) = *radio spectrum* below; (*b*) a radio beacon transmitting directional radio signals which can be used by aircraft possessing appropriate receiving apparatus to determine the bearing of the transmitter; **radio shack**, a small building housing radio equipment; (esp. *Naut.*) a radio room; **radio show**, (*a*) an exhibition of radio equipment, etc.; (*b*) a radio programme, usu. featuring light entertainment; **radio signal**, a radio message; a group of radio waves transmitted or emitted by any source; **radio silence**, deliberate abstention from radio transmission; failure to communicate by radio; **radio-silent** *a.*, (*a*) maintaining radio silence; (*b*) = *radio-quiet* adj. above; **radio sounding**, the use of radiosondes or radar techniques for investigating the atmosphere, sea bed, or the like; so **radio sounder**; **radio spectrum**, the radio-frequency part of the spectrum of electromagnetic radiation; also, the spectrum of any particular source at these frequencies; **radio star**, any discrete source of radio waves outside the solar system (rarely a star in the usual sense); **radio station**, a radio-transmitting installation or establishment; a sound broadcasting establishment or organization; **radio telescope** *Astr.*, an apparatus or installation for detecting and recording radio waves from the sky with great sensitivity and a high degree of resolution, consisting essentially of a large sensitive directional aerial together with a receiver and recording equipment; (in quot. 1929, a fictitious apparatus in which a 'radio contrivance' is attached to an optical telescope); **radio wave**, an electromagnetic wave having a frequency within the range used for telecommunication; (cf. *RADIO-FREQUENCY); usu. *pl.*; so **radio wavelength.**

1940 *Chambers's Techn. Dict.* 698/2 Radio altimeter. **1953** R. CHISHOLM *Cover of Darkness* I. xii. 123 My Mosquito had a radio altimeter, a device which gave absolute readings of height. **1968** *Times* 15 Nov. 8/6 The radio-altimeter, which made only one measurement of the probe's altitude, must have been wrong. **1916** *Lit. Digest* (N.Y.) 1 Jan. 13/1 But there will be a lone radio amateur on the alert who has seen the approaching fleet. **1977** *N.Z. Herald* 5 Jan. I. 1/8 Mr Kilpatrick said radio amateurs—there are almost 4000 in New Zealand—were anxious to have aerial installation defined because of the growing public interest in town planning and the environment. **1918** *Flying* 14 Aug. 150/3 The radio-compass and wireless log signals will doubtless be pressed into service in the age of commercial aviation. **1946** *Happy Landings* July 9/1 A severe thunderstorm can be detected by intelligent use of the Radio Compass. **1966** *McGraw-Hill Encycl. Sci. & Technol.* III. 332/2 The modern radio compass uses a nondirectional antenna in combination with a bidirectional loop antenna to provide a unidirectional bearing indication. **1958** *Times* 18 Jan. 8/7 It will enable the masters or pilots of vessels coming in and out of Southampton to coordinate their movements with other shipping by direct radio contact with the radio information centre. **1962** V. GRISSOM in *Into Orbit* 130, I was in radio contact with..the helicopters which were on their way to pick me up. **1975** T. ALLBEURY *Special Collection* v. 34 He had a long radio contact with London and..gave full details. **1976** H. MACINNES *Agent in Place* xvii. 185 'Emil is sleeping on board.' 'You've radio contact with him?' 'Of course.' **1934** WEBSTER, Radio dial. **1974** *Listener* 7 Nov. 593/3 Twiddling his radio dial to hear what is top of the pops. **1922** *Sci. & Invention* May 10/1 (*caption*) The radio direction finder aboard this vessel can locate the transmitting station within one half a degree accuracy. **1966** *McGraw-Hill Encycl. Sci. & Technol.* IV. 232/2 This ground-based radio direction finder, operating at frequencies of 2 to 20 megacycles, is used mainly for navigational assistance in the long-distance en-route zone. **1920** *Radio Rev.* Oct. 644 Radio direction finding..has become a practical possibility owing to the use of powerful amplifiers. **1974** *Encycl. Brit. Macropædia* XII. 903/2 Radio direction finding..developed in two ways. First, radio transmitters, or 'beacons', were sited..to enable ships or aircraft to fix their positions. Second, ground DF stations that could pick up radio signals sent out by a ship or an aircraft were built. **1960** *Aeroplane* XCVIII. 366/2 The largest steerable radio 'dish' in the World, at Jodrell Bank in England. **1977** *Nature* 9 June 478/1 Australia leapt into the big league of astronomical nations with the building of the giant 64-metre radio dish at Parkes, New South Wales. **1928** Radio echo [see *ECHO *sb.* 1 d]. **1947** *Sci. News* V. 36 Radio echoes do not come from the meteors themselves but from the lengthy filament of highly conducting gas which forms their trail or streak. **1975** *Nature* 30 Oct. 780/2 A radio-echo technique gave the surface velocity relative to a layer that reflected electromagnetic waves—evidently from a level close to the base of the ice. **1946** *Proc. IRE* XXXIV. 558/2 Presumably, radio energy could..be focused by means of lenses made of a material such as plastic or glass which is transparent at the transmitter frequency. **1955** *Sci. Amer.* Mar. 36/1 The radio energy given forth by..Cygnus A, startles even astronomers. **1974** F. W. COLE *Fund. Astron. Solar Syst. & Beyond* xiv. 379/1 Certain peculiar galaxies emit thousands of times more radio energy than does an average galaxy. **1912** *Electrician & Mechanic* Aug. 140/1 The Institute of Radio-Engineers comprises the bodies formerly known as the Society of Wireless Telegraph Engineers and the Institute of Wireless Engineers. **1937** *Discovery* Apr. 111/1 Radio engineers have recorded an increasing number of sudden and complete fadings affecting reception on short-wave wireless transmission. **1974** *Encycl. Brit. Macropædia* XV. 429/2 Marconi's faith in the successful commercial operation of the system was more than justified, and radio engineers elsewhere were quick to change from skepticism to enthusiasm. **1917** *Wireless World* Apr. 10 The name 'wireless system' in radio engineering seems now to have no scientific meaning. **1942** P. C. SANDRETTO *Princ. Aeronaut. Radio Engin.* p. v, It is necessary to explain how I determined the point where ordinary radio engineering ends and aeronautical radio engineering begins. **1975** D. G. FINK *Electronics Engineers' Handbk.* p. xiii, This new Handbook is the first to be devoted to the field of electronics engineering at large. Earlier important handbooks..treated the field primarily from the point of view of the first important application in the field—radio engineering. **1928** Radio ham [see *HAM *sb.*[1] 6]. **1951** H. M. WATSON et al. *Understanding Radio* (ed. 2) xxvi. 642 The exploits of the radio hams in times of disaster..are well known. **1971** *Daily Tel.* 13 Sept. 1/8 The radio ham..heard two men planning the raid over short-wave radio sets. **1926** *Wireless World* 18 Aug. 229/1 American business men engaged in the radio industry. **1933** *Radio Times* 14 Apr. 94 (Advt.), Mullard valves have always taken the radio industry ahead. **1951** M. MCLUHAN *Mech. Bride* (1967) 22/2 This is not a situation peculiar, for example, to the radio, movie or book industries. **1928–9** *T. Eaton & Co. Catal.* Fall & Winter 245/3 Radio License... The law requires that every radio set be licensed. **1969** *Morning Star* 9 Aug. 5 Heavier fines for TV and radio licence dodgers have been called for by the Postmaster General. **1975** *Times* 26 Sept. 15/4 His father was one of the original radio licence holders. **1978** *Nature* 14 Sept. 91/3 Less than 10% of these [QSOs] are turning out to be 'radio-loud'. 1921 R. D. Paine *Comrades of Rolling Ocean* iv. 73 The radio man reports storm signals hoisted all the way from Key West to Norfolk. 1928–9 *T. Eaton & Co. Catal.* Fall & Winter 245/3 Our radio goods..are purchased and inspected by qualified radio men. 1945 M. Lowry *Let.* 6 June (1967) 46 A great friend of mine who was at college with me—a well-known radioman in Canada. 1977 *Time* 3 Jan. 35/2 He lied about his age to get into the Navy and served as a radioman in the Pacific during World War II. 1977 *New Statesman* 2 Sept. 298/1 This is not because Bush has a particularly high quota of veteran radio men. 1978 Pasachoff & Kutner *University Astron.* viii. 221 (*caption*) A radio map of the sun made at a wavelength of 2·8 cm with the 100-meter dish of the Max Planck Institute for Radio Astronomy. 1962 A. Nisbett *Technique Sound Studio* 267 *Radio microphone*, microphone and small transmitter sending a signal which can be picked up at a distance of up to perhaps several hundred yards. 1978 *Broadcast* 6 Feb. 12/3 Licences which authorize..mobile radiotelephone, radiomicrophones, radiopaging devices. 1980 J. Ball *Then came Violence* (1981) xvi. 144 He picked up the radio microphone. 1974 *Listener* 14 Mar. 330/3 Rix..had me fitted with a radio mike, which is..a small and highly sensitive transmitter, enabling me to record impressions unobtrusively. 1931 B. Jones *Avigation* xv. 274 (*heading*) Radio navigation. 1951 *Sci. News* XXII. 110 Positions were fixed by radio navigation. 1974 *Encycl. Brit. Macropædia* XII. 908/2 Loran..is a radio-navigation system that permits a ship to locate its position accurately by timing the arrival of pulses from synchronized shore transmitters. 1921 *Brit. Pat.* 161,448 (*title*) Improvements in or relating to radio-navigational systems. 1958 *Times* 18 Aug. 8/3 The Nantucket field had no instrument landing system, no high intensity approach lights, and only a minimum of radio navigational aids. 1978 R. V. Jones *Most Secret War* xxiv. 210 The drive at last started for us to emulate the Germans in their radio navigational techniques. 1941 *Sun* (Baltimore) 15 Oct. 5/5 The Twenty-ninth Division has its full complement of radio nets, and is maintaining its communications of this nature with considerable success. 1976 C. Egleton *State Visit* ix. 84 A re-broadcast system had been installed which allowed them to monitor the police radio net. 1935 C. G. Burge *Compl. Bk. Aviation* 503/1 Air transport is essentially international in character, and the organization and control of the radio networks, if they are to benefit air transport, must also be applied on an international basis. 1966 N. Wymer *From Marconi to Telstar* viii. 85 Since the war all the great nations have steadily expanded their overseas services until today the radio network covers the entire world. 1972 J. Mosedale *Football* iv. 52 New York..is the communications capital of the world—home to national magazines and the television and radio networks. 1968 Radio pager [see *PAGER *sb.*^{2}]. 1960 Radio paging [see *PAGE *v.*^{1} c]. 1978 *Times* 3 Nov. 27/4 The Post Office itself has listed the main telecommunications services ..envisaged for the years 1985 and 2000... By 1985 there will be..radiopaging, confravision (conference television), viewphone. 1925 *Scribner's Mag.* Sept. 80 (Advt.), Model 50 Radio-Phonograph Combination. Price $325. 1952 Auden *Nones* 18 According to the gospel Of the radio-phonograph. 1979 T. Gifford *Hollywood Gothic* (1980) xxi. 216 An ancient Philco radio-phonograph. 1957 *Nature* 4 May 898/1 This 'radio pill', as it is termed, was designed by Dr. V. K. Zworykin and developed by engineers of the Radio Corporation of America at Camden, New Jersey. 1962 *New Scientist* 10 May 288/3 A radio pill contains an electronic circuit that generates a radio frequency signal. 1970 *Sci. Jrnl.* June 84/1 Pressure changes within the vagina and uterus have also been measured, by using a tiny device known as the radio-pill. 1933 Radio pirate [see *PIRATE *sb.* 4 b]. 1964 *Daily Tel.* 11 May 20 (*heading*) Radio 'pirates' problem for Cabinet. 1965 *Astrophysical Jrnl.* CXLI 1560 Members of the class called here quasi-stellar galaxies..resemble the quasi-stellar radio sources..in many optical properties, but they are radio-quiet. 1977 *Sci. Amer.* Aug. 38/3 Many quasars have no detectable radio emissions. In fact, the great majority of quasars may be radio quiet. 1971 D. W. Sciama *Mod. Cosmol.* v. 73 It must be emphasised that the radio-quietness of ethse new objects is only relative. 1949 *Sun* (Baltimore) 26 July 18/3 The wing tip 'radomes', as the compact radio-radar installation is called, were developed for the Air Force. 1966 M. Woodhouse *Tree Frog* viii. 62 A radio-radar control system with a range of seven hundred miles. 1976 G. H. Morrison in L.-H. Lee *Characterization of Metal & Polymer Surfaces* I. 362 Our laboratory has been involved in an examination of steel strands in the cables suspending a 525-ton feed platform in the world's largest radio-radar telescope at Arecibo, Puerto Rico. 1926 *Physical Rev.* XXVII. 202 The ionic term will be negligibly small compared to the electronic term, except for very long waves outside of the usual radio range. 1929 *Proc. IRE* XVII. 2147 The so-called 'aural' type of directive radiobeacon, or 'radio range' as it is now called, was finally considered to be most applicable to the Airways Service. 1942 P. C. Sandretto *Princ. Aeronaut. Radio Engin.* i. 7 By the late fall of 1929, there was installed in the United States a line of radio ranges extending from New Jersey to Iowa. 1949 *Nature* 12 Nov. 816/2 The radio range from about 1 cm. to 20 m. 1951 *Oxf. Jun. Encycl.* IV. 291/1 The Radio Range, a long-range navigational aid..transmits the letters A..and N..in Morse code simultaneously in different directions. If the pilot is flying directly on his proper course, he will be midway between the paths of the two signals. 1960 Rodman & Varsavsky tr. *I. S. Shklovsky's Cosmic Radio Waves* ii. 84 Thus if the sun appears to be the dominating source of radiation at optical frequencies, in the radio range it plays a much more modest role. 1966 *McGraw-Hill Encycl. Sci. & Technol.* XI. 255/2 *Consol*... This radio range navigation aid provides a number of characteristic signal zones that rotate in a time sequence. 1946 R. E. Higginbotham *Wine for my Brothers* v. 97 He paused a moment before the door of the radio shack. 1973 D. Kyle *Raft of Swords* (1974) xiv. 152 Bill Harrison was sitting in his radio shack at Bella Bella. 1976 'M. Nelson' *Crusoe Test* iii. 42 He had veered off into the captain's room..taken the radio shack key... He had only needed seconds in the radio room. 1922 *Moving Picture Stories* 4 Aug. 22/2 When the Women's Radio League of America some months ago asked me to join them and appear at their exhibition at a radio show, I thought it a unique invitation. 1932 *Radio Times* 29 July 239/2 Peter Creswell will produce *Ball and Dance*, a German radio show built up from scenes at famous balls of history. 1940 G. Marx *Let.* 10 Oct. (1967) 26 I'm..discussing a radio show that I might do... A kind of human interest story with a slightly wacky father, who, of course, would be me. 1971 D. Nathan *Laughtermakers* ii. 50 When the series was over Milligan went to Australia, where he did thirteen radio shows similar to the Goons. 1976 B. Took *Laughter in Air* i. 11 Theatre magnates still held the whip hand, making offers they couldn't refuse to Arthur Askey and Jack Warner for stage versions of their hit radio shows. 1923 E. W. Marchant *Radio Telegr.* ix. 100 The problem of finding the direction from which a radio signal is coming has been referred to already, and its practical importance..is obvious. 1937 *Discovery* Jan. 3/2 Radio signals can travel round the world and not be lost in space as would be the case if the ionosphere did not exist. 1962 F. I. Ordway et al. *Basic Astronautics* iii. 46 The planet [*sc.* Mercury] emits natural radio signals. 1969 *Times* 16 Jan. 4/7 Measurements of the radio signals from sulphur hydride may be a valuable check of estimates of the amount of sulphur in interstellar space. 1974 L. Deighton *Spy Story* xix. 208 That radio signal obliges us to continue with the pick-up. 1946 Radio silence [see *RADIOLOCATOR]. 1959 R. Collier *City that wouldn't Die* vii. 106 [He] exultantly broke radio-silence: 'I've got two dirty great Huns in my sights!' 1970 A. Dekker *Divers Diamonds* ii. 15 *Toledo* [*sc.* a submarine] has subsequently maintained total radio silence and has gone without trace. 1977 *Observer* 3 Apr. 1/6 The control tower said sharply, 'Radio silence please, I will continue to call up KLM.' 1978 *Peace News* 25 Aug. 8/2 This 'strike' lasted some three days, and rumour had it that GCHQ feared that the Russians were maintaining 'radio silence', the traditional prelude to offensive action! 1976 B. Lecomber *Dead Weight* xiii. 154 Filing incomplete flight plans and going radio-silent for long periods is bloody silly. 1977 *Sci. Amer.* Aug. 38/3 Only later, when the sky was searched at optical wavelengths for bright blue and ultraviolet objects, were the radio-silent quasars discovered. 1931 *Flight* XXIII. 278/1 The trials proceeding in America with a radio-sounder have been successful and appear to promise good results. 1969 *Times* 20 Jan. 8/1 Alouette 1, the first of a series of satellites built in Canada..is equipped with a radio sounder which probes the atmosphere beneath the satellite. 1929 *Bull. Amer. Meteorol. Soc.* X. 220 The radio sounding balloons to be released from the *Graf Zeppelin* will employ a radio sending device developed under the direction of P. A. Moltchanoff. 1936 *Meteorol. Mag.* LXXI. 5 (*heading*) Radio-sounding of the atmosphere. 1958 Miller & Parry *Everyday Meteorol.* i. 29 The radio-sounding balloon..has the advantage that it need not be visible. 1963 *Times* 31 May 16/2 A form of radio sounding, similar to radar, may provide a new means of charting the depth of rock surfaces covered by snow and ice, as in Greenland and Antarctica. 1929 *Bell System Techn. Jrnl.* VIII. 313 Fortunately this frequency was so located in the radio spectrum that a band of the desired width..could be obtained. 1932 *Proc. IRE* XX. 96 Ultra-short waves in point-to-point propagation resemble light waves rather than the longer and more conventional waves of the radio spectrum. 1964 R. H. Baker *Astron.* (ed. 8) iv. 116 The only known emission line in the radio spectrum was first observed..in 1951. 1978 *Nature* 8 June 431/2 Samples of QSOs with flat radio spectra were chosen. 1949 *Sci. Amer.* Sept. 38/1 The small spots are tiny enough to be considered 'radio stars'. 1957 *New Scientist* 27 June 32/1 Ryle concludes that his weak radio stars are evidence of more crowded days when the universe was young. 1963 *Times* 20 Apr. 8/4 The first true radio stars—stars (in the ordinary sense) that emit radio waves at sufficient intensity to be detected and identified by radio telescopes—have been discovered by Sir Bernard Lovell. 1969 *Times* 18 Apr. 12/5 Pulsars, the radio stars whose clock-like regularity has so far eluded explanation. 1912 *Statutes U.S.A.* XXXVII. 1. 303 Every Government station on land or sea shall have special call letters designated and published in the list of radio stations of the United States. 1934 Joyce *Let.* 20 Nov. (1966) III. 328 You don't say what those radio stations pay. 1968 A. Diment *Bang Bang Birds* iii. 41 One of the local radio stations gave me the news. 1977 C. McCullough *Thorn Birds* xv. 340 Progress had finally come to Gillanbone in the shape of an Australian Broadcasting Commission radio station. 1929 *Amazing Stories* June 202/1 Well, what do you think of it?.. How do you like my radio-telescope?' 1948 *Newsweek* 18 Nov. 98/2 The newer radio telescope..is designed to gather radio static in the microwave region. 1953 *N.Y. Times* 19 Apr. E9/5 The foundations of Britain's million-dollar radio telescope.. are now being built at Jodrell Bank, Cheshire. 1969 *Times* 25 Mar. 12/6 The experiment was carried out with the giant radio telescope carved out of a natural bowl in the hills of Puerto Rico. 1976 L. Deighton *Twinkle, twinkle, Little Spy* xxiii. 229 Two huge radio telescopes, the dishes about sixty feet across. 1916 *Electr. Experimenter* IV. 486/3 If the radio waves were powerful enough to travel from Mars to the moon, they..could travel from the moon to Mars. 1946 *Nature* 3 Aug. 150/1 As radio-waves are reflected by obstacles of any kind, they can be used in darkness as well as daylight, in thick fog or other obscuring atmospheric conditions, as light to show whether the way is open or not. 1969 G. Lyall *Venus with Pistol* xxxiii. 212 The dark air between us slowly started to hum like radio waves. 1977 *Times* 20 June 5/2 Through vibrations, the radio wave can transmit conversation and noise back from the room it is aimed at. 1937 *Proc. Nat. Acad. Sci.* XXIII. 178 The low opacity in the radio wavelengths, arising from the small relative size of the particles, will selectively permit the escape of these longer wavelengths. 1972 *Sci. Amer.* Aug. 51/3 Neutral atomic hydrogen emits and absorbs radiation at the radio wavelength of 21 centimeters and can be readily observed by radio telescopes.

radio (rēɪ·dio), *v.* [f. the sb.] **a.** *trans.* To transmit or send (a message or information) by radio. **b.** *intr.* To send a message, etc., by radio; to give information or make a request by radio (with dependent clause). (In both senses freq. with advbs.)

1919 *Pop. Sci. Monthly* Sept. 116/2 He radios the information to the ship. 1926 H. T. Wilkins *Marvels Mod. Mech.* 213 As soon as the observer spots a shoal of fish, he marks a square on the chart,..and at once radios to the port. 1926 *Glasgow Herald* 20 Dec. 9 The British ship Defender has radioed that it has saved two members of the crew of the schooner Lincoln. 1937 G. Frankau *More of Us* xiii. 136 Let Pink flay Anti-Pink, or vice versa, 'Delicious weather', radio'd still our purser. 1958 *Industr. & Engin. Chem.* Mar. 22A/2 Explorer has radioed back information that the temperature inside is between 20° and 50 °C., tolerable enough for a human passenger. 1958 'Castle' & 'Hailey' *Flight into Danger* ii. 29 Let me know if she gets any worse and I'll radio ahead. 1969 *New Yorker* 12 Apr. 68/2 A satellite..radioed information about the fields of low-energy particles far above the earth. 1970 *Daily Tel.* 14 Oct. 1/3 The lifeboat later radioed back that the dead man and the others were being taken by the trawler to Boulogne. 1972 *Oxford Times* 25 Feb. 1/8 Our beat policemen radio through if congestion is building up anywhere. 1973 J. Rossiter *Manipulators* xxvi. 244 He had to get away before Jackson found Bradley's body and radio'd back. 1977 *Daily Tel.* 18 Mar. 1/7 The police radioed for assistance and a detachment of Irish troops arrived. 1978 J. Irving *World according to Garp* xv. 313 Go radio our position.

Hence **ra·dioed** *ppl. a.*, transmitted or reported by radio.

1943 J. Flanner in *New Yorker* 29 May 42/3 A radioed appeal from General de Gaulle in London. 1953 A. Huxley *Let.* 9 Aug. (1969) 682 Robots responsive to the radioed will of their masters. 1973 'A. Hall' *Tango Briefing* xii. 149 London wanted photographs and a full radioed report of the freighter's cargo. 1977 'W. Wingate' *Fireplay* i. 12 *Doneska* could have foundered anywhere from her last radioed position off Los Angeles to close by Valparaiso.

radio-. Add: **2.** [Now apprehended as a comb. form of Radiate *v.* or Radiation.] Esp. connected with ionizing radiation. **radiode·nsity,** the degree to which a material will absorb ionizing radiation; radio-opacity; **ra:diodermati·tis,** dermatitis caused by X-rays or other ionizing radiation; **ra:diodiagno·sis,** the diagnosis of disease by means of X-rays or other ionizing radiation; hence **ra:diodiagno·stic** *a.*; **radio-eco·logy,** the study of the ecological effects of radioactive materials and ionizing radiation; hence **ra:dio-ecolo·gical** *a.*; **radio-eco·logist**; **radiogene·tics,** the study of the genetic effects of ionizing radiation; hence **radiogene·tic, -ical** *adjs.*; **ra:diolumine·scence,** luminescence caused by ionizing radiation; hence **ra:diolumine·scent** *a.*; **ra:dio-micro·meter,** an instrument for measuring minute degrees of infra-red or microwave radiation; **radiomime·tic** *a.*, of (the action or properties of) a substance: producing effects upon living cells resembling those produced by ionizing radiation; **ra:dionecro·sis** *Med.*, necrosis caused by excessive exposure to ionizing radiation; **ra:dio-pasteuriza·tion,** pasteurization of food by exposure to ionizing radiation; so **radio-pa·steurized** *a.*; **ra:diopharmaceu·tical** *a.* and *sb.*, (being or pertaining to) any radioactive compound or preparation which is administered to a patient for the purpose of radiotherapy or diagnosis; **ra:diopharmaco·logy,** the use of drugs in radiology; also, the study of physiology and the metabolism of drugs by means of radiopharmaceuticals; so **ra:diopharmacolo·gical** *a.*, **-pharmaco·logist**; **radiopha·rmacy,** the preparation and use of radiopharmaceuticals; a laboratory for this work; so **radiopha·rmacist**; **ra:dio-prote·ction,** the prevention or countering by chemical means of the harmful effects produced in living tissues by ionizing radiation; so **ra:dio-prote·ctive** *a.*, being or possessing this property; **ra:dio-prote·ctor,** a substance possessing this property; **radio-resi·stant** *a.*, resistant to the action of ionizing radiation; so **radio-resi·stance**; **radiose·nsitive** *a.*, sensitive to the action of ionizing radiation; so †**radiose·nsitiveness, radiosensiti·vity**; **radiose·nsitize** *v. trans.*, to make (more) radiosensitive; so **radiose·nsitizing** *vbl. sb.*; also **ra:diosensitiza·tion**; **radiose·nsitizer,** a substance which is used to increase the sensitivity of particular organisms or tissues to ionizing radiation; **ra:diosteriliza·tion,** (*a*) the process of rendering sterile by means of ionizing radiation; (*b*) the process of rendering free from micro-organisms by means of ionizing radiation; also (in either sense) **radioste·rilized** *a.*; **radiosu·rgery,** the use of beams of ionizing radiation in surgery; so **radiosu·rgical** *a.*; **ra:diotranslu·cent** *a.* = *RADIOLUCENT *a.*; hence **ra:diotranslu·cency.**

1936 B. J. M. Harrison *Textbk. Roentgenol.* iii. 62 Of the media of greater radiodensity than the tissues the most commonly used is sulphate of barium. 1977 *Proc. R. Soc. Med.* LXX. 518/2 If the stone is still *in situ* then the chemical composition may be assessed by the radiodensity. 1903 *Progressive Med.* III. 161 Heidingsfeld's case was a bullous radiodermatitis. 1930 *Times Educ. Suppl.* 2 Aug. p. iv/3 The Cross of the Legion of Honour has been conferred on Dr. Jean Chabry, whose experimental work in radiology has resulted in an attack of radiodermatitis, necessitating amputation of his right arm. 1968 A. Rook et al. *Textbk. Dermatol.* xv. 359/2 Chronic radiodermatitis is not an inflammatory process and should strictly be termed roentgen atrophy or perhaps roentgen poikiloderma. 1904 F. P. Foster *Appleton's Med. Dict.* 1676/2 *Radiodiagnosis*, diagnosis by means of Röntgen ray examination. 1910 A. Abrams *Diagnostic Therapeutics* iv. 627 Radio-diagnosis is more accurate than percussion in defining the dimensions of the organ. 1978 *Lancet* 25 Feb. 434/1 The current practice of treating radiodiagnosis as cost-free, risk-free, and done in a flash has seriously affected medical standards over the past three decades. 1907 *Jrnl. Amer. Med. Assoc.* 17 Nov. 1392/2 (*heading*) The correlation of clinical and radiodiagnostic findings. 1961 *Lancet* 29 July 257/1 Dr. C. Pickard..was..critical of past and future planning for radiodiagnostic departments. 1956 E. P. Odum in *Conf. Radioactive Isotopes in Agric.* (U.S. Atomic Energy Comm.) 102/2 Radio-ecological research at AEC installations..has been handicapped by (1) lack of prior knowledge of the environment, and (2) uncontrolled experimental conditions. 1975 *Nature* 3 Jan. p. xiii (Advt.), The successful candidate should have knowledge and some experience in marine radioecological research and related techniques. 1959 E. P. & H. T. Odum *Fund. Ecol.* (ed. 2) xiv. 477 This rather surprising finding, repeatedly documented by radioecologists working at the Nevada Test Site.., is apparently to be explained by the fact that the smaller particles which fall at a distance stick to the leaves of plants and dissolve more readily. 1956 E. P. Odum in *Conf. Radioactive Isotopes in Agric.* (U.S. Atomic Energy Comm.) 100/1 Only now that we have some familiarity with the functional aspects of our ecosystem are we ready to begin controlled experiments in radio-ecology. 1964 *Oceanogr. & Marine Biol.* II. 256 Radioecology or radiation ecology is the branch of ecology which concerns itself with the dispersion and interaction of radionuclides in and with the physical, chemical, and biological environment. 1974 *Nature* 13 Dec. 618/2 The two fundamental problems in radioecology are to determine how radionuclides migrate within biogeological systems and how ionising radiations affect microorganisms, plants and animals. 1971 *Radiation Bot.* XI. 119 (*heading*) Radiogenetic effects of gamma- and fast neutron irradiation on different ontogenetic stages of the tomato. *Ibid.*, Pollen, which has advantages for radiogenetical studies, seems to hold little promise for mutation breeding purposes. 1950 *Genetics* XXXV. 56 (*heading*) On the interpretation of the dose-frequency in radiogenetics. 1955 [see *RADIOBIOLOGY]. 1963 *Biol. Abstr.* XLI. 642/2 Valuable data have been obtained in the field of radiogenetics. 1911 *Chem. Abstr.* V. 3059 The intensity of the radioluminescence is proportional to the distance in mm. 1946 *Thorpe's Dict. Appl. Chem.* (ed. 4) VII. 405/2 Radioluminescence of solids induced by *X*-rays finds considerable technical application in industrial and medical fields. 1966 G. F. J. Garlick in P. Goldberg *Luminescence of Inorg. Solids* xii. 689 It was the existence of the particle-excited radioluminescence of uranyl salts that led Becquerel to the discovery of radioactivity in 1896. 1919 *Chem. Abstr.* XIII. 2806 A long discourse on the underlying principles and the preparation of radioluminescent paints. 1887, 1888 Radio-micrometer [in Dict. s.v. Radiometer]. 1908 *Rep. Brit. Assoc. Adv. Sci. 1907* 621 The first thing is to tune up the receiver accurately. This can be done by a Duddell radio-micrometer, which measures the received [radio] energy satisfactorily although it is very small. 1966 *McGraw-Hill Encycl. Sci. & Technol.* XI. 319/1 The radiomicrometer was invented by C. V. Boys in 1887 to avoid the limitations of a separate thermocouple and galvanometer. However, because it is delicate and inconvenient, it has virtually gone out of use. 1947 P. Dustin in *Nature* 14 June 796/2 These effects are remarkably similar to those of ionizing radiations: the mitotic poisons of the trypaflavine type are radiomimetic. 1965 *New Scientist* 25 Nov. 586/2 The still unidentified agent causing the haemorrhage is a radiomimetic compound—that is, it poisons the bone-marrow very much as radiation does. 1974 R. M. Kirk et al. *Surgery* ii. 9/2 Irradiation and radiomimetic drugs..administered in the treatment of neoplasms, cause delay in wound healing by their damaging effects on dividing cells. 1933 Ward & Smith *Rec. Adv. Radium* viii. 85 In this way radio-necrosis may result in the more radio-resistant tumours without destruction of the tumour. 1963 *New Scientist* 9 May 334/3 Early treatment of certain irradiation accident cases with 'vasodilators' may prevent radionecrosis altogether. 1977 *Lancet* 27 Aug. 460/1 Radionecrosis of the brain may follow therapeutic irradiation of the pituitary. 1959 *Internat. Jrnl. Appl. Radiation & Isotopes* VI. 128/1 Extensive investigations..will be necessary before radio-pasteurization without refrigeration can be recommended for meats. 1968 *Biol. Abstr.* XLIX. 2446/2 Freshly killed pre-rigor fish respond better toward the radio-pasteurization process than do post-rigor fish. 1966 E. R. Killam et al. in *Proc. Interant. Symposium Food Irradiation* 839 A petition was submitted to the FDA on April 29, 1966 for the approval of radio-pasteurized strawberries for public consumption. 1971 *Jrnl. Food Technol.* VI. 82 Blackening in radio-pasteurized shrimps can be effectively controlled by pre-blanching treatment. 1960 *Jrnl. Amer. Med. Assoc.* 10 Sept. 162/1 The production of radiopharmaceutical compounds that concentrate in organs, such as colloidal gold in the liver and chlormerodrin..in the kidneys. 1963 P. F. Belcastro in H. M. Burlage et al. *Physical & Technical Pharmacy* xvii. 701 Radiopharmaceuticals can be used as therapeutic agents for treating specific diseases more efficiently than by traditional methods. 1966 G. V. LeRoy in G. A. Andrews et al. *Radioactive Pharmaceuticals* xxxvii. 669 There is a continuing disagreement about the appropriate dose of almost all the radiopharmaceutical agents currently in use. 1977 *Lancet* 23 Apr. 907/2 Adverse reactions to radiopharmaceuticals are rare, and are generally due to the carrier portion rather than to the isotope itself. 1973 Radiopharmacist [see *RADIOCHEMIST]. 1963 *Biol. Abstr.* XLIV. 479/2 (*heading*) Radio-pharmacological investigations of the mechanism of action of sympathetic alpha and beta receptors in the region of the cardia of the rabbit. 1960 *Jrnl. Amer. Med. Assoc.* 10 Sept. 165/2 The clinician, biochemist, physiologist, or radiopharmacologist will eventually discover substances that will concentrate in the pancreas and adrenal and parathyroid glands. 1968 *Australasian Radiol.* XII. 239/1 Pancoast, in 1914, appears to have initiated radiopharmacology when he used morphine to stimulate gastric peristalsis. 1976 M. Tubis in Tubis & Wolf *Radiopharmacy* xv. 406 Radiopharmacology..is concerned with the use of labeled compounds, the 'radiophores' carriers of radioactivity, to demonstrate the distribution, deposition, kinetics of metabolism, turnover, and the excretion. *Ibid.*, Radiopharmacy..is the science and art of preparing and dispensing the labeled compounds of pharmaceutical quality that are used in nuclear medicine for diagnosis and therapy. 1977 K. Kristensen et al. *Quality Control in Nuclear Med.* xxxiii. 271/1 The radiopharmacy should be designed so that dispensing and radioactive waste handling do not interfere with or contaminate each other. 1957 *Brit. Jrnl. Radiol.* XXX. 97/1 This..underlines the importance of the liver in radio-protection. 1975 *Internat. Jrnl. Radiation Biol.* XXVIII. 41 It appears that cystamine may be limiting the availability of reducing equivalents and thus providing radioprotection to lipogenesis. 1956 *Brit. Jrnl. Radiol.* XXIX. 623/1 Scientists..have studied the radio-protective activity of cysteamine and cystamine. 1958 *Ibid.* XXXI. 339/2 All the radio-protective substances effective in mice have the common action of lowering body temperature. 1975 *Biochem. & Biophys. Res. Communications* LXVII. 1170 The increase..in the amount of exogenous superoxide dismutase associated with the eluted bone marrow stem cells was also accompanied by an enhancement in the radioprotective effect of the enzyme on the proliferative capacity of the cells. 1960 *Internat. Jrnl. Radiation Biol.* II. 231 (*heading*) Sub[s]trates as radioprotectors of hexokinase. 1977 *Nature* 3 Nov. 15/3 A significant contribution to radiobiology was the demonstration that SOD is an effective radioprotector exerting a protective role even when administered after radiation doses have been delivered. 1927 *Cancer Rev.* II. 397 The degree of radio-resistance seems to be more marked in proportion as the treatment has been unwisely prolonged. 1957 *Brit. Jrnl. Radiol.* XXX. 97/1 A higher concentration of cysteamine in the liver increases radio-resistance. 1929 *Radiology* XIII. 316/2 Squamous carcinoma, malignancy of varying grades, radioresistant. 1959 *Internat. Jrnl. Appl. Radiation & Isotopes* VI. 157/1 The alkaline phosphatase of milk is extremely radio-resistant. 1977 *Lancet* 27 Aug. 460/1 The endocrine-active adenomas are characteristically more radio-resistant than the endocrine-inactive variety. 1920 *Amer. Jrnl. Roentgenol.* VII. 53/1 Remarkable examples of radio-sensitive tumors are ectodermal and basal-celled epitheliomata derived from the basal-celled layers of the epidermis. 1956 C. Auerbach *Genetics in Atomic Age* viii. 80 Tissues in which division is going on are so much more radiosensitive than tissues in which cell division has ceased. 1976 *Nature* 17 June 588/1 If these cells are less radiosensitive, ionising radiation could favour their overgrowth. 1921 *Arch. Radiol. & Electrotherapy* XXV. 348 A further factor in tissues is the blood or lymph content; the more this is, the more is the radio-sensitiveness. 1924 *Brit. Jrnl. Radiol.* XXIX. 270 The radio-sensitivity was found to vary greatly according to the stage of cellular division. 1971 G. G. Luce *Body Time* v. 167 Perhaps the rhythm of radiosensitivity may be traced to cycles of activity in the bone marrow and spleen where blood is formed. 1951 *Jrnl. Chem. Educ.* XXVIII. 414/2 Radiosensitization may be significant in radiation chemistry. 1976 *Radiology* CXIX. 221 This..would ostensibly lead to some degree of tumor reoxygenation ('radiosensitization') as well as direct chemotherapeutic effects. *Ibid.* 725/1 Cetylpyridinium chloride did not radiosensitize bacteria suspended in nutrient broth. 1953 *Brit. Jrnl. Cancer* VII. 316 This compound (Synkavit) has a small but useful effect as a clinical radiosensitiser. 1972 *Lancet* 22 Sept. 638/2 A true radiosensitiser is a chemical which increases the cell-killing effect of a given dose of radiation. 1936 *Biol. Abstr.* X. 1905/1 His researches concern the radio-sensitizing effect of metabolic exchanges and the degree to which substances modifying the metabolism also modify the radiosensitivity. 1953 *Brit. Jrnl. Cancer* VII. 314 A radiosensitising chemical agent..in combination with radiotherapy should produce a mean survival time after treatment double that after radiotherapy only. 1978 *Jrnl. R. Soc. Med.* LXXI. 672 The radiosensitizing properties of the group of nitroimidazoles. 1964 *Jrnl. Econ. Entomol.* LVII. 756/1 For mosquito sterilization..the desired goal of highly competitive, yet permanently sterile, males is more readily attainable with chemosterilization than radiosterilization. 1966 E. R. Killam et al. in *Proc. Internat. Symposium Food Irradiation* VIII. 842 The radiosterilization of certain cuts of beef may require irradiation at this temperature [*sc.* −80 °C]. *Ibid.*, Future developments in cryogenics should lower the..cost of radiosterilized meat products. 1967 *Jrnl. Econ. Entomol.* LX. 696 (*heading*) Mating competitiveness in radiosterilized males. 1975 *Ibid.* LXVIII. 595/2, 300 radiosterilized ♀ and 4200 radiosterilized ♂ were released each evening. 1933 *O.E.D. Suppl.* s.v. *Radio-* 2, Radio-surgery. 1963 *New England Jrnl. Med.* 19 Sept. 597/2 In 1954 we set out upon a study in clinical medicine in which these high-energy protons and alpha particles have been used in various forms of so-called 'bloodless' surgery, or radiosurgery. 1973 *Internat. Jrnl. Radiation Biol.* XXIV. 229 (*heading*) Split-brain cats prepared for radiosurgery. 1929 *Daily Express* 16 Jan. 9/6 Even if the growth should have extended to the glands, radio-surgical methods would offer a good prospect of eradication. 1959 *Probl. Oncol.* V. 98 (*heading*) Radio-surgical treatment of skin cancer involving a free skin graft. 1973 *Internat. Jrnl. Radiation Biol.* XXIV. 239 A 3 mm beam of protons is too wide for safe radiosurgical use in the rat brain. 1974 A. Henry in R. M. Kirk et al. *Surgery* xv. 295/2 X-rays show an area of radiotranslucency in the metaphysis, which may cross the growth plate to involve the epiphysis. 1959 *New Biol.* XXX. 25 If the process is observed radiologically, the solid lungs become radio-translucent rapidly, as if a light has been turned on. 1964 L. Martin *Clin. Endocrinol.* (ed. 4) iv. 149 Destructive bone lesions..cause sharply demarcated radiotranslucent areas in affected bones.

3. a. Connected with radioactivity, as **ra:dio-allergoso·rbent** *a.* [*ALLERG(Y + -O + *SORBENT *sb.* (*a.*)], in *radioallergosorbent test*, a form of radioimmunoassay for measuring antibodies to an allergen (see quot. 1967); **radioa·ssay** *sb.*, an assay performed by measuring radioactivity from a radioisotope; also as *v. trans.*; **ra·dio-atom,** an atom of a radioactive substance; **radioau·tograph** = *AUTORADIOGRAPH; also **ra:dioautogra·phic** *a.*; **radioauto·graphy**; **ra:diochroma·togram,** a chromatogram of a radioactively labelled preparation which is recorded or measured by means of a radiological technique; **radioco·lloid,** a radioactive substance in colloidal form; hence **radiocolloi·dal** *a.*; **radio-da·ting** *vbl. sb.*, isotopic dating; **ra·dio-element,** a radioactive element; **radioha·lo** = *pleochroic halo*; **radio-i·odinate** *v. trans.*, to label (a substance) with radio-iodine; so **radio-i·odinated** *ppl. a.*; **ra:dio-iodina·tion**; **ra·dioligand,** a radio-labelled compound that has a strong chemical affinity for a particular receptor; **radio-nu·clide,** a radioactive nuclide; **radiopu·rity** = *radiochemical purity*; **ra·dioscan,** a determination of the distribution of radioactive material (esp. a tracer) in a sample, an organ, etc.; **ra:dioste:reoa·ssay,** any biological assay technique in which the test substance is determined by allowing it to bind to a suitable protein or antibody in competition with a known quantity of radioactively labelled material, the extent of reaction being measured radiologically; (usu. applied to non-immunological methods: cf. *radioimmunoassay* s.v. *RADIOIMMUNO-); **radiotoxi·city,** the property of a radioactive substance of being injurious to a living organism when present in its tissue; hence **radioto·xic** *a.*; **radiotra·cer,** a radioactive tracer.

1967 L. Wide et al. in *Lancet* 25 Nov. 1105/2 An in-vitro method, called the radioallergosorbent test, has been developed for the detection of allergen-specific antibodies of a new immunoglobulin class, provisionally called IgND. *Ibid.* 1106/1 The principle of the method, the radioallergosorbent test (R.A.S.T.) is as follows: an allergen coupled to an insoluble polymer is added to the serum to be investigated, if antibodies to the allergen are present they should react with the conjugate; after the removal of all unbound serum components ^{125}I-labelled anti-IgND antibodies are then added, they will bind to the antibodies of the IgND class which have reacted with the polymer-coupled allergen; the uptake of labelled antibodies, measured in terms of radioactivity, on the particles is essentially proportional to the IgND allergen antibodies. 1971 *Internat. Arch. Allergy & Appl. Immunol.* XLI. 443 The radioallergosorbent test (RAST) was applied for quantitative estimation of IgE antibodies to various common allergens. 1977 *Lancet* 22 Oct. 847/2 Radioallergosorbent tests..for specific IgE antibody were also positive. 1951 *Nucleonics* Nov. 60 (*heading*) A simple inexpensive sample changer for the radioassay of..simple samples. 1963 *Analytical Biochem.* V. 89 The radioassay of cholesterol-C^{14} digitonide by gas flow technique..using methanol as the solvent for solution and plating. 1970 *Steroids* XV. 470 The method..requires only a simple extraction and an alumina thin layer chromatographic separation prior to radioassay. 1972 *Nature* 22 Dec. 463/2 Out of a series of a hundred [silver] bars, we have radio-assayed seven bars picked at random. 1905 *Phil. Trans. R. Soc.* A. CCIV. 209 It may..be supposed that occasionally one of the outlying revolving electrons, comprising the radio-atom, lapses into a position which results in a slow loss of energy from the atom in the form of radiation. 1947 *Instruments* XX. 712/1 The qualities which make radio-atoms ('hot' atoms) useful to science and industry make them hazardous to handle. 1941 *Jrnl. Appl. Physics* XII. 328 Slides of radio-autographs showing the distribution of phosphorus in various plant tissues will be shown. 1956 *Sci. Amer.* Nov. 144/2 (*caption*) Radioautographs indicate how Isoniazid is concentrated in brain tissue. 1974 K. N. Prasad *Human Radiation Biol.* xxi. 426 For the radioautographs of soluble materials, the tissue sections must be cut in a frozen state and then dried in a vacuum. 1947 *Radiology* XLIX. 327/2 Extensive radioautographic studies were made of these organs in which a high degree of selective localization took place. 1967 F. O. Schmitt in G. C. Quarton et al. *Neurosciences* 211/2 The radioautographic method using tritiated precursors lends itself well to the determination of the fate of axoplasm moving cellulofugally, both down the axon and possibly out along the dendrites. 1978 *Bull. Amer. Acad. Arts & Sci.* Feb. 17 These radioautographic experiments have revealed that in normal female somatic cells, one member of each pair of X chromosomes always replicates much later than all the other chromosomes including the second X chromosome. 1941 *Jrnl. Appl. Physics* XII. 446/1 Stout and his co-workers have followed the metabolism of inorganic phosphate in the leaves and fruit of the tomato plant by..the technique of radio-autography. 1973 *Nature* 20 Apr. 523/2 Whole-body radioautography is a useful method for studying the distribution of radio-labelled compounds among all the organs and tissues of an experimental animal. 1952 F. P. W. Winteringham et al. in *Nucleonics* Mar. 56/1 When the radiochromatogram is plotted as net rate of count against distance along the strip, *w* [*sc.* the total weight of labelled component] is proportional to the area enclosed by the relevant part of the curve. 1972 *Physics Bull.* May 298/1 At the 1971 Physics Exhibition Panax showed a unique rapid imaging system which promised to shorten the time to locate the radioactive regions on thin layer radiochromatograms. 1930 *Chem. Abstr.* XXIV. 1279 Recent expts. on solns. of the 'radiocolloid' Th C (Bi). 1956 *Nature* 28 Jan. 184/1 It is more likely..that the high concentration of chromate ions in the resin promotes the formation of a radio-colloid, which is then adsorbed on the resin. 1977 *Proc. R. Soc. Med.* LXX. 522/2 In the normal subject the radio-colloid is taken up avidly by the mononuclear phagocytes of the liver and only a small amount is taken up by other organs. 1936 O. Hahn *Appl. Radiochem.* 275/1 (Index), Radiocolloidal particles, size of... Radiocolloidal thorium X. 1950 *Thorpe's Dict. Appl. Chem.* (ed. 4) X. 433/1 Some portion..of the radiocolloidal phenomena is due to adsorption of the radioactive ions on particles of dust and solid impurities fortuitously present in the solutions. 1969 *New Scientist* 25 Sept. 632/1 Radiodating has revealed..that the lunar samples are at least 3100 million years old. 1975 K. H. Goulding in Williams & Wilson *Biologist's Guide to Princ. & Techniques Pract. Biochem.* vi. 196 The assumptions made in radiodating are sweeping and hence palaeontologists and anthropologists who use this technique can only give very approximate dates to their samples. 1903 *Contemp. Rev.* May 709 In the Periodic table of elements arranged in the ascending order of their atomic weights the three radio-elements are therefore at the extreme end. 1937 *Discovery* Mar. 65/2 A search for new radio elements of very short life whose existence is suspected. 1967 *New Scientist* 15 June 675/1 The main problem here is to ensure that dangerous long-lived radio-elements are not inadvertently produced. 1971 *Science* 20 Aug. 728/1 A new type of composite radiohalo has been found with rings attributable both to the ^{210}Po decay sequence and to ^{213}Po and possibly ^{213}Bi. 1974 *Nature* 13 Dec. 564/1 Polonium radiohaloes occur widely and not infrequently (total about 10^{11}–10^{20}) in Precambrian rocks. 1971 *Ibid.* 4 June 322/1 This protein was radio-iodinated to a specific activity of 0·6 μCi/μg with ^{125}I-iodide. 1955 *Ibid.* 26 Mar. 536/1 (*heading*) Radioiodinated human serum albumin. 1970 *Ibid.* 3 Oct. 58/1 Radioiodinated polypeptide hormones are widely used as tracers for both radioimmunoassay and *in vivo* metabolic studies. 1977 *Lancet* 13 Aug. 355/1 Dr Chait and his colleagues have studied the metabolism of very-low-density lipoprotein (V.L.D.L.) in a patient with type III hyperlipoproteinæmia..by injection of radioiodinated V.L.D.L. from a donor with endogenous hypertriglyceridæmia (type IV). 1957 *Jrnl. Laboratory & Clin. Med.* XLIX. 128 (*heading*) A method for radioiodination of antibody protein. 1974 *Nature* 25 Jan. 175/2 Several groups of investigators have subjected lymphocytes to enzymatic radioiodination of their cell surface proteins. 1972 *Jrnl. Clin. Endocrinol. & Metabolism* XXXIV. 130/1 Radioligand assays enable the biological activity of hormones at the target cell to be evaluated without the additional and variable effect of metabolism *in vivo*. 1978 *Nature* 8 June 472/1 This radioligand, which has been used in identifying and quantifying β-adrenoreceptors in a variety of intact and disrupted cell preparations, has enabled us to study cellular cyclic AMP accumulation and binding to β-adrenoreceptors in similar experimental conditions. 1947 T. P. Kohman in *Amer. Jrnl. Physics* XV. 356/2 *Radionuclide* should replace *radioelement* and *radioisotope* in most applications. 1963 *Engineering* 20 Sept. 378/3 The use of short-lived radio-nuclides makes it imperative to transfer the sample from the irradiation area to the activity-measuring location as speedily as possible. 1976 J. Follett *Doomsday Ultimatum* 73 What is the exact nature of the radionuclides stored.. and what will happen if they are released? 1977 I. M. Campbell *Energy & Atmosphere* vii. 187 The radionuclide ^{14}C is produced naturally in the atmosphere by cosmic radiation but the level was almost doubled by the atmospheric nuclear-weapon testing of the 1960's. 1956 *Internat. Jrnl. Appl. Radiation & Isotopes* I. 227/2 The only disadvantage of the method of measurement is its extreme sensitivity to gamma-emitting impurities, although this can be turned into an asset in the determination of radio-purity. 1973 *Weed Res.* XIII. 340 The radiopurity of the isolated metabolites was checked by thin-layer chromatography..and autoradiography. 1965 *Amer. Rev. Respiratory Dis.* XCII. 959/2 (*caption*) The per cent of total pulmonary blood flow calculated from the radioscan for the diseased lung. 1966 *Amer. Jrnl. Cardiol.* XVIII. 819/2 The radioscan accurately reflected the pattern of arteriolar-capillary blood flow. 1974 *Nature* 1 Nov. 68/1 The radio-purity of these compounds was determined by radioscan of thin layer chromatograms. 1967 B. E. P. Murphy in *Jrnl. Clin. Endocrinol.* XXVII. 973/2 Such methods have been termed by the author 'competitive protein binding (CPB) analysis' and by R. P. Ekins 'saturation analysis', but because of their basic similarity to radioimmunoassays..the name 'radiostereoassay' has also been suggested [by the writer] as an analogous term which could be applied to both types of assay. 1974 *Nature* 22 Feb. 563/2 Before it can act, vitamin D must first be converted to its 25-hydroxy derivative..in the liver, and this chief circulating metabolite can be measured by radio-stereo-assay, thus providing a precise index of vitamin D nutritional status. 1950 *Nuclear Sci. Abstr.* IV. 398/1 Rat erythrocytes were irradiated with 6,030r and then injected subcutaneously..in order to determine the severity and nature of radiotoxic effects. 1975 *Nature* 27 Mar. 278/3 The dose limit for bone is based on uniformly deposited ^{226}Ra: but Pu, collecting at the bone surface, is considered ~5 times as radiotoxic. 1946 D. Antony et al. *Radiotoxicity of Injected Sr^{89} for Rats, Mice & Rabbits* (U.S. Atomic Energy Comm. Rep. MDDC 1540) 6 Radium, hitherto the only substance of which the radiotoxicity had been extensively studied. 1961 G. R. Choppin *Exper. Nuclear Chem.* ii. 12 Some of the commonly used nuclides are listed..according to their relative radiotoxicities. 1977 S. L. Barker in B. A. Rhodes *Quality Control in Nuclear Med.* xxviii. 243/1 Prepared formulations of long-lived nuclides present no special problems with the exception of possible radiotoxicity. 1950 *Mining Engin.* Mar. 364/2 Radiotracers were demonstrated to be of considerable value in the study of a typical mineral-collector system, dithiophosphate-galena. 1977 *Lancet* 19 Nov. 1072/2 No abnormal localisation of the radiotracer was observed in muscles of any of the controls.

b. Prefixed to the names of chemical elements and compounds: (i) Designating a radioactive isotope (usu. one prepared artificially) of the named element († occas. of an element other than that named), as *radio-caesium, -chlorine, -cobalt, -gold, -iron, -potassium, -silver, -sodium*, etc.; † **radio-acti·nium,** a thorium isotope (mass number 227) which is produced by beta decay of actinium and is an alpha emitter of half-life 18·2 days; **radio-i·odine,** esp. iodine 131, an artificial isotope with a half-life for beta decay of about 8 days, which is widely used as a tracer (esp. in investigating thyroid function), and for radiotherapy of the thyroid gland; **radio-lea·d,** a mixture of isotopes of lead, together with some bismuth and polonium, which constitutes the longer-lived product of the decay of radon; sometimes *spec.* lead 210, the major radioactive component of this mixture, which has a half-life for beta decay of 21 years; **radio-pho·sphorus,** *spec.* phosphorus 32, an artificial isotope which is widely used as a tracer and decays by beta emission with a half-life of 14·3 days; **radio-stro·ntium,** *spec.* strontium 90, a beta emitter with a half-life of roughly 28 years which occurs among uranium fission products; (in quot. 1941 referring to strontium 85); † **radio-tellu·rium,** a former name for a polonium isotope of mass number 210 (originally thought to be tellurium), which occurs in the radium decay series and decays by alpha emission with a half-life of 138 days; † **radio-tho·rium,** thorium 228, an alpha emitter of half-life 1·91 years which occurs in small amounts in natural thorium and is formed by the beta decay of mesothorium II; (see also quot. 1950). Also *RADIO-CARBON.

1906 O. Hahn in *Nature* 12 Apr. 560/1, I have found that a new product is present in actinium which is intermediate between actinium and actinium X, and..will be called.. 'radio-actinium'. 1926 R. W. Lawson tr. *Hevesy & Paneth's Man. Radioactivity* xxiii. 164 The β-radiation of actinium, the existence of which it is necessary to assume in order to explain its transformation into radioactinium, is too weak to be detected. 1955 *Bull. Atomic Scientists* Oct. 287/3 The long-lived useful products radiocaesium and radiostrontium will be separated from the rest of the fission products and concentrated into radioactive sources. 1935 *Physical Rev.* XLVIII. 571/1 During a study of the beta-radiation from a sample of silver radiochloride..it was found possible to follow the decay of the radiochlorine. 1949 *Atomics* Oct. 75/2 Radio-cobalt and radio-strontium have been used to dissipate charge from textile machinery. 1959 *Listener* 22 Oct. 675/2 The new knowledge does not only apply to atomic bombs and radio-cobalt but to the mechanisms of biology as well. 1955 *Sci. News Let.* 19 Mar. 184/3 A gram of radiogold, costing about $25, can irradiate tissues with the power that would be obtained from a $20,000 chunk of radium. 1938 R. W. Lawson tr. *Hevesy & Paneth's Man. Radioactivity* (ed. 2) 303/1 (Index), Radio-iodine. 1940 *Amer. Jrnl. Physiol.* CXXXI. 135 The radio-iodine was prepared by the Berkeley cyclotron and converted to sodium iodide. 1961 *Lancet* 2 Sept. 551/1, 8 mC of radioiodine was administered..to achieve a permanent remission of the hyperthyroidism. 1970 Passmore & Robson *Compan. Med. Stud.* II. vi. 8/2 The rapidity of uptake and the speed with which the plasma is cleared of radioiodine provide measures of the activity of the thyroid gland. 1960 *Proc. Soc. Exper. Biol. & Med.* CIV. 442/2 This calculation is based on the assumption that loss of radioiron from liver between 0–3 hrs is negligible. 1903 *Phil. Mag.* V. 585 Thus the radio-lead described by Hoffmann and Strauss and by Giesel cannot be regarded as a new element until it is shown that it has permanent activity of a distinctive character. 1910 *Nature* 24 Feb. 492/1 The most natural source of polonium is radium D (radio-lead), which grows polonium and has a period of half-transformation of about twenty years. 1910 A. T. Cameron *Radiochem.* v. 52 The lead obtained from pitchblende was strongly and permanently radioactive—hence the name radio-lead. 1926 R. W. Lawson tr. *Hevesy & Paneth's Man. Radioactivity* xiv. 118 Polonium..can be prepared..from solutions of the salts of radio-lead, which.. contains an isotope of bismuth (RaE) and three isotopes of lead (RaD, RaG, Pb). 1941 *Jrnl. Appl. Physics* XII. 440/2 The detection of radio-lead by its radioactivity is more than a million times more sensitive than the ordinary chemical and physical methods. 1938 R. W. Lawson tr. *Hevesy & Paneth's Man. Radioactivity* (ed. 2) x. 122 For every 10^7 α-particles of high energy value..only 6 are effective in the activation of aluminium, i.e. in the production of radio-phosphorus. 1951 *New Biol.* X. 36 Mosquito larvae grown in water containing phosphorus-32 (radiophosphorus) produce radioactive adults whose presence in a swarm can be picked out immediately. 1963 Ball & Hooper in Schultz & Klement *Radioecology* 227/1 The movement of radiophosphorus through the ecosystem of a cold water stream was studied by adding spikes of approximately 23 millicuries of phosphorus-32 to the water during the summers of 1958, 1959, and 1960. 1973 P. A. Colinvaux *Introd. Ecol.* xiv. 207 Radiophosphorus has a half-life of 14 days. 1948 *Sci. News* VII. 38 The release of heat from radio-potassium must have been 200 times greater than from uranium and thorium combined. 1938 R. W. Lawson tr. *Hevesy & Paneth's Man. Radioactivity* (ed. 2) v. 118 A 40-fold yield in the preparation of radio-silver. 1971 *Nature* 10 Dec. 347/1 The reaction ^{109}Ag (*n*, 2*n*)^{108m}Ag does seem to be important in thermonuclear bomb production of radiosilver. 1935 *Physical Rev.* XLVII. 17 Doubtless radio-sodium will find many uses in the physical and biological sciences. 1951 *New Biol.* X. 39 In animals, it has also been shown that radiosodium is actively transferred inwards across a frogskin membrane into a solution of higher sodium concentration than that of the external medium. 1941 *Jrnl. Appl. Physics* XII. 456/2 Radio-strontium has a half-life of 55 days, can be readily prepared in relatively large amounts, and emits very energetic beta-particles. 1946 *Chemistry* Jan. 20/1 The use of radio-strontium in the treatment of metastatic carcinoma in bone and for other diseases can now be tried in a much larger field than was ever possible before. 1957 *New Scientist* 9 May 30/1 He..argues that..the level of radio-strontium in human bones will eventually rise to between 5 and 20 'sunshine units' if test firings of bombs continue indefinitely at their present rate. 1958 *Times* 12 Nov. 4/2 A further report will be published shortly by the Atomic Energy Research Establishment, Harwell, giving the 1957 figures for radiostrontium in soil, herbage, animal bone, and milk samples from the United Kingdom. 1972 Goldman & Bustad (*title*) Biomedical implications of radiostrontium exposure. 1904 *Technics* II. 173/1 There is at present a good deal of evidence that the radio-active substance, separated from pitchblende by Marckwald, and called by him radio-tellurium, is in reality the fifth product of the disintegration of the radium atom. 1906 *Phil. Mag.* XII. 361 A bismuth rod coated with radiotellurium was used as a source of α rays. 1962 O. Hahn in *Coll. Papers Ld. Rutherford* I. 168 Eventually it turned out that polonium and radiotellurium were identical, and the latter name had to be dropped, although polonium is in fact a higher homologue of tellurium. 1905 *Jrnl. Chem. Soc.* LXXXVIII. II. 789 The author [*sc.* Ramsay] hence considers it very probable that the radioactive power of thorium is to be attributed to small quantities of this element, radiothorium. 1921 *Phil. Mag.* XLI. 572 A much weaker source of α rays was obtained by dipping a nickel plate for a few seconds in a more dilute solution of radio-thorium. 1950 *Thorpe's Dict. Appl. Chem.* (ed. 4) X. 435/1 The term radiothorium is loosely applied to the mixture of isotopes that can be separated from natural thorium and which contain the bulk of its activity. Such radiothorium consists of a mixture of meso-ThI, meso-ThII, radium, and actinium isotopes, respectively, Th-*X*, a radium isotope, and their decay products, including true radiothorium or ^{228}Th.

(ii) Designating a compound containing a radioactive label.

1935 Radiochloride [see *radiochlorine* in (i) above]. 1951 *New Biol.* X. 40 A sample of blood is withdrawn and mixed with radiophosphate, in the form of sodium or chromium phosphate. 1952 *Ibid.* XIII. 64 The addition of anti-thyroid substances to the incubating fluid resulting in a diminished formation of radiothyroxine and radiodiiodotyrosine. 1963 *Amer. Jrnl. Obstetr. & Gynecol.* LXXXVII. 208/1 One hundred microcuries of sodium radiochromate is added and the mixture allowed to incubate at room temperature for 25 minutes. 1977 *Lancet* 5 Feb. 303/2 We infer that the clots were formed after radio-fibrinogen had been administered.

4. Connected with radio (cf. *RADIO *sb.* 6, 7), as † **ra:dio-condu·ctor,** a former term for a coherer; **radiohe·liograph** *Astr.*, an interferometric radio telescope system designed to record instantaneous high-resolution pictures of the sun as observed at radio wavelengths; **radiome·teorograph** [Meteorograph] = *RADIOSONDE; so **ra:diometeoro·graphy**; **radiophy·sics,** the branch of physics concerned with the properties and applications of radio waves; hence **radiophy·sical** *a.*; **radiophy·sicist**; **ra:dio-teleme·tering** *vbl. sb.*, **ra:dio-tele·metry,** telemetry by means of radio; hence **ra:dio-teleme·tric** *a.*; **radio-te·letype,** a teletype which transmits and receives information by radio.

1898 Radio-conductor [in Dict. s.v. Radio- 2]. 1906 S. R. Bottone tr. *Mazzotto's Wireless Telegr. & Telephony* vii. 166 Branly gave the name of 'radio-conductors' to the tubes with filings, a name which some prefer, as it only points to the fact, leaving the true nature of the phenomenon unexplained. 1962 *Nature* 18 Aug. 649/1 The Ford Foundation has announced the grant of 550,000 dollars to the Radio-physics Division of the Australian Commonwealth Scientific and Industrial Research Organization for the construction of a radioheliograph for photographing the sun in its natural radio emission. 1966 *New Scientist* 7 Apr. 27/1 The main circle of the radioheliograph will observe events in the Sun's atmosphere with all the detail and definition to be expected from a dish aerial of the same diameter. 1973 *Sci. Amer.* Oct. 72/3 Spatial structure in the corona is studied on a second-by-second basis by a radioheliograph composed of 96 linked antennas each 45 feet in diameter at Culgoora in Australia. 1932 *Nature* 31 Dec. 1006/1 What is believed to be the first kind of such radio-meteorograph was devised a few years ago by Prof. Moltchanoff, of Leningrad, records of temperature and pressure being obtained in several test ascents in January, 1930. 1944 C. P. Lent *Rocket Res.* 74/2 Improved balloons and radiometeorographs..are now being developed by meteorologists. 1934 *Monthly Weather Rev.* LXII. 221 (*heading*) Radiometeorography as applied to unmanned balloons. 1974 R. Rodman tr. *Al'pert's Radio Wave Propagation & Ionosphere* (ed. 2) II. p. ix, Various radiophysical and radio-engineering problems. 1961 *Flight* LXXX. 531/1 From the point of view of Soviet radiophysicists and astronomers the West Ford project, if carried out, may have consequences dangerous to artificial satellites, and especially to those with a man on board. 1929 *Compt. Rend. des Séances des Commissions* (Union Radio Scientifique Internationale) II. 29/2 The Commission of Radiophysics held in Brussels in 1928. 1947 *Nature* 18 Jan. 103/1 The Division of Radiophysics [of the Australian Council for Scientific and Industrial Research] was formed early in the War. 1960 Rodman & Varsavsky tr. *I. S. Shklovsky's Cosmic Radio Waves* i. 11 It is shown in radiophysics that if a receiving antenna is used as a transmitting antenna the power radiated by it in each direction will have the same angular dependence $A(\theta,\phi)$. 1976 *Nature* 24 June 663/1 *The Australian Journal of Physics* has had a reputation..for specialising in radiophysics and high energy physics. 1946 *Trans. Amer. Inst. Electr. Engineers* LXV. 865/1 The accuracy and stability of the airborne radio telemetering equipment depend upon a high degree of stability of the plate supply. 1949 *Electronic Engin.* XXI. 209/1 The data.. is transmitted to the ground by radio telemetering methods. 1974 *Physics Bull.* Oct. 443/2 Launch and subsequent collection of data by radiotelemetric methods demand access to major ground facilities. 1976 L. Brown *Birds of Prey* 60 Advanced radiotelemetric techniques may be the only method of learning much detail about many forest raptors. 1951 *Rev. Sci. Instruments* XXII. 2/1 Almost every application of radio telemetry involves some form of multiplexing— i.e., the transmission of several channels of information by the same radio carrier. 1967 E. L. Gruenberg *Handbk. Telemetry & Remote Control* iv. 2 The development of radio telemetry has been principally centered around the drone and missile programs of the armed forces. 1974 *Country Life* 13 June 1572/2 Roding behaviour and its purpose are still not fully understood, and until radiotelemetry or a similar technique is used to discover what different individuals are doing, we have to watch and guess. 1939 H. K. Morgan *Aircraft Radio & Electr. Equipment* xii. 325 It may be that by 1945 it will be considered profitable to equip transports with a tape or page radio teletype. 1949 Koestler *Promise & Fulfilment* II. iii. 236 Communications..are rather precarious, depending as they do on the vagaries of the American consul's radio teletype. 1976 *S9* (N.Y.) May/June 137 (*caption*) Bart also monitors RTTY (radioteletype) and amateur radio SST (slow-scan television) signals.

radio-actinium: see *RADIO- 3 b (i).

radioa·ctivate, *v.* Also with hyphen. [f. as next + -ATE^3.] *trans.* To make radioactive. Also *fig.* So **radioa·ctivated** *ppl. a.*; **radioa·ctivating** *vbl. sb.*

1903 *Electr. World & Engin.* 28 Mar. 529/2 An account of the investigation of the radio-activating process at the foot of waterfalls. 1949 M. Muggeridge *Affairs of Heart* iii. 47 Even the atomic bomb..may prove a great deception, only serving to radio-activate yet another attempt to re-define human rights. 1971 P. Kruger *Princ. Activation Analysis* vi. 223 The energies of most gamma-ray transitions from radioactivated nuclides are less than 2 MeV. 1978 *N.Y. Times* 30 Mar. c 19/2 Both Russia and America are looking for an excuse to try out on each other a new variation of the neutron bomb that will radioactivate the earth for 1,000 years.

ra:dioactiva·tion. *Chem.* [f. next + -ATION.] The process of making radioactive; freq. *attrib.*; **radioactivation analysis,** chemical analysis in which a sample is made radioactive by exposure to radiation and its components are then identified, and their concentrations measured, by radiochemical methods; also called *activation analysis*.

1938 *Brit. Chem. Abstr.* A. I. 112/2 (*heading*) Artificial disintegration and radioactivation. *a* 1947 Clark & Overman in *Determination Trace Elem. by Radioactivation Anal.* (U.S. Atomic Energy Comm. Rep. MDDC–1329) (1947) 1 The potential uses of radioactivation analysis for the determination of trace amounts of substances have been recognized for some time. 1950 *Ann. Rep. Progr. Chem.* XLVI. 285 Although radioactivation analysis was first used in 1936, it is still in the early stages of development. 1960 *Times* 19 July (Royal Society Suppl.) p. xv/1 The geochemist is particularly trying..new techniques of analysis such as the isotope dilution and radioactivation methods. 1971 P. Kruger *Princ. Activation Analysis* iii. 62 The most copious and controllable sources of neutrons for radioactivation are found in nuclear reactors. 1971 *Nature* 15 Oct. 506/2 Radioactivation analysis..is now well established in its own right as an important tool in science and technology.

radioactive (rēɪdio,æ·ktiv), *a.* Also with hyphen. [f. Radio- 2 + Active *a.*] **1.** Of an atomic nucleus, a substance, etc.: (capable of) undergoing spontaneous nuclear decay involving emission of ionizing radiation in the form of particles or gamma rays; *spec.* of an element: consisting of a radioactive isotope.

1898 *Nature* 28 July 312/1 On a new radio-active substance contained in pitchblende. 1904 E. Rutherford *Radio-Activity* v. 161 This increase of activity is due to the continuous production by the radium of the radio-active emanation or gas, which is occluded in the radium compound. 1913 *Q. Rev.* July 117 The disintegration of atom after atom of the radio-active element. 1926 R. W. Lawson tr. *Hevesy & Paneth's Man. Radioactivity* i. 4 We now define a substance as being radioactive when the atoms of which it is composed disintegrate spontaneously, and regardless of whether or not the emission of rays can readily be detected in the process. 1935 *Nature* 9 Nov. 754/1 The radio-active phosphorus in the urine..was then investigated. 1952 *New Biol.* XIII. 63 The introduction of radio-active iodine (I^{131}) has provided a tool..of the greatest value..to the comparative physiologist. 1955 *Times* 18 Aug. 8/6 Radioactive caesium recovered from waste fission products has been made available to the Royal Marsden Hospital.. for treatment of deep-seated cancer. 1957 *Times Lit. Suppl.* 15 Nov. p. iv/3 Death waits for us on the zebra crossing; and tomorrow morning's milk may be radioactive. 1959 *Daily Tel.* 10 Mar. 16/3 He also asked the secretary..to advise us of the possibility of any appreciable rise in radioactive content in the river which may result. *Ibid.* 23 Mar. 18/3 The rate of descent from the stratosphere of radio-active debris, including strontium 90, from nuclear explosions. 1961 G. R. Choppin *Exper. Nuclear Chem.* vi. 77 The average time of existence τ of a radioactive atom before decay is the reciprocal of the decay constant. 1966 C. R. Tottle *Sci. Engin. Materials* i. 23 Almost all elements of atomic number greater than $Z = 83$ (bismuth) are naturally radioactive. 1969 N. W. Pirie *Food Resources* v. 127 Marine organisms concentrate several of the radioactive isotopes enormously. 1977 *Nature* 14 Apr. 585/1 Any radioactive chemical which enters one organism may be transported a considerable distance as it travels through the food chain.

2. Of a process, phenomenon, etc.: of, pertaining to, involving, or produced by radioactivity.

1903 *Phil. Mag.* V. 580 These rays have not yet been sufficiently examined to make any discussion possible of the part they play in radioactive processes. 1906 *Nature* 25 Oct. 634/1 The escape of the emanation causes a radio-active contamination of the laboratory which renders delicate experiments on radio-activity or ionisation very difficult. 1919 *Phil. Mag.* XXXVII. 537 Results showed..that..the H atoms from a glass α-ray tube were a product of radioactive disintegration. 1926 R. W. Lawson tr. *Hevesy & Paneth's Man. Radioactivity* vii. 78 Ions are produced along the tracks of the radioactive rays. 1938 *Ibid.* (ed. 2) iv. 55 (*heading*) Absorption measurements with γ-rays as applied to radioactive analysis. 1956 A. H. Compton *Atomic Quest* v. 303 At the instant of the flash there is a burst of radioactive rays of enormous magnitude. 1961 G. R. Choppin *Exper. Nuclear Chem.* i. 3 The formation of a more stable nucleus as a result of radioactive decay is accompanied by a release of energy. 1969 Bennison & Wright *Geol. Hist. Brit. Isles* i. 6 Only the limitations of radioactive dating.. are here dealt with. 1971 *Nature* 6 Aug. 367/2 The dating of the earlier part of the Pleistocene period still presents considerable problems, but further exploitation of the various 'radioactive clocks' should eventually overcome these difficulties. 1977 *Ibid.* 10 Mar. 106/1 Even the most resolute proclaimers of radioactive doom seem to have convinced themselves that fusion power will be gentle and on a homely scale, and that radioactive pollution will be a thing of the past when dirty fission gives way to clean fusion.

3. *fig.* (Possibly also influenced by *RADIO- 4.)

1905 S. Macnaughtan *Lame Dog's Diary* x. 128 Eliza has found that London is radio-active, hence enjoyable. And Eliza had only been once to the Royal Institution when she said it! 1909 A. Bennett *Glimpse* x. 71 She did not begin to live, socially, till her body was at rest... Then her individuality would be radioactive whereas the individuality of Inez spent itself mildly without ceasing in a persuasive appeal to the sight. 1919 D. H. Lawrence in *Eng. Rev.* June 477 Our plasmic psyche is radio-active, connecting with all things, and having first-knowledge of all things. 1923 L. P. Smith *S.P.E. Tract* XII. 57 This radio-active quality of popular idiom, this power to give out life and never lose it. 1934 *Punch* 5 Sept. 280/2 The invalid son, who ultimately comes out at least all even in the contest with the radio-active Audrey. 1955 Koestler *Trail of Dinosaur* 12 The logic of expediency leads to the atomic disintegration of morality, a kind of radioactive decay of all values. 1974 *Times* 27 May 6/1 His [*sc.* Byron's] was a radio-active personality that had shattering effects on all who came in contact with him.

4. Special collocations: **radioactive constant,** the average proportion of nuclei of a given radioactive nuclide which will decay in a given time; now usu. called *decay constant*; = **dis-integration constant*; **radioactive equilibrium,** a condition in which the quantities of radioactive daughter nuclides in a material remain constant because each is decaying at the same rate as that at which it is being formed; **radioactive indicator** = *INDICATOR 2 b (ii); **radioactive series,** a series of radioactive nuclides each member of which decays into the next, together with a non-radioactive end-product; the series of transformations relating such a set of nuclides; (four such series exist among the nuclides heavier than lead: see quot. 1974); **radioactive tracer** (see *TRACER^1); **radioactive waste,** waste material that is radioactive, esp. spent nuclear fuel.

1903 Rutherford & Soddy in *Phil. Mag.* V. 581 The law of radioactive change may therefore be expressed in the one statement—the proportional amount of radioactive matter that changes in unit time is a constant... λ may therefore be suitably called the 'radioactive constant'. 1923 Glazebrook *Dict. Appl. Physics* IV. 574/2 λ, the radioactive constant, represents the average fraction of the number of atoms which break up per unit time. 1942 J. D. Stranathan *'Particles' of Mod. Physics* viii. 325 (*heading*) Relationships among the several radioactive constants. 1904 E. Rutherford *Radio-Activity* vii. 189 In uranium and thorium compounds there is a continuous production of active matter which keeps the compound in radio-active equilibrium. 1923 Glazebrook *Dict. Appl. Physics* IV. 576/1 If a radioactive mineral is sealed up so that the products of transformation are allowed to accumulate, a stage is ultimately reached in which the amount of each product formed is equal to the amount transformed per unit time throughout the series... Material which has reached this state is said to be in radioactive equilibrium. 1946 *Physical Rev.* LXIX. 672/2 Since the age of the earth is much greater than the life of C^{14} a radioactive equilibrium must exist in which the rate of disintegration of C^{14} is equal to the rate of production. 1968 Musset & Lloret *Conc. Encycl. Atom* 95/2 Radioactive equilibrium is also referred to when a radioactive body is produced by nuclear reaction, such as bombarding a target with a beam of particles having a constant intensity. After a certain time, the body formed in the nuclear reaction also acquires a constant radioactivity through the balance of its rates of formation and decay. 1923 *Biochem. Jrnl.* XVII. 439 (*heading*) A contribution to the application of the method of radioactive indicators in the investigation of the change of substance in plants. 1943 Radioactive indicator [see *INDICATOR 2 b (ii)]. 1923 *Phil. Mag.* XLVI. 647 There are only four complete radio-active series of the type that is known. 1926 R. W. Lawson tr. *Hevesy & Paneth's Man. Radioactivity* xxiv. 184 The element thorium is the parent element of a radioactive series. 1956 I. Asimov *Inside Atom* v. 85 The entire set of changes is an example of a radioactive series. This particular one we have been talking about is the uranium-238 series. 1974 *Encycl. Brit. Micropædia* VIII. 378/2 *Radioactive series*, any of four independent sets of unstable atomic nuclei that decay through a sequence of nuclear transformations until a stable nucleus is achieved... Three of the sets, the thorium series, uranium series, and actinium series..are headed by naturally occurring species of unstable nuclei whose half-lives are comparable to the age of the Earth... The fourth set, neptunium series.., came to light after the discovery of induced radioactivity (1934). 1947 C. D. Coryell in C. Goodman *Sci. & Engin. of Nucl. Power* I. vii. 249 In general, radioactive wastes from separation processes must be stored indefinitely in systems free from leaks. 1958 *Engineering* 21 Feb. 236/2 The problem of radioactive waste disposal is finally solved only by the complete natural decay of the constituent fission product activity—a process which may take hundreds of years. 1969 N. W. Pirie *Food Resources* v. 127 Nuclear power stations discharge slightly radioactive waste into the sea near the shore. 1977 *Nature* 10 Mar. 109/2 One of the major obstacles to have arisen to the expanded use of nuclear power has focused on the reprocessing and storage of radioactive waste.

Hence **radioa·ctively** *adv.*, by radioactive decay; with radioactive material; by means of a technique dependent upon radioactivity.

1925 J. Joly *Surface-Hist. of Earth* ix. 147 This method would not be reliable..if the lead was in either case unstable—i.e. ultimately changed radioactively into something else. 1936 O. Hahn *Appl. Radiochem.* vi. 171 In the Kjeldahl determination of nitrogen, the ammonia formed is led into a solution of lead nitrate that has been radioactively activated. 1948 *Physical Rev.* LXXIV. 279/1 Individual atoms of this isotope transform radioactively. 1959 *Oxf. Univ. Gaz.* 16 Mar. 795/2 In the event of nuclear warfare, farm products, particularly milk, obtained from a radio-actively contaminated soil might be positively harmful. 1968 R. A. Lyttleton *Mysteries Solar Syst.* ii. 65 The release of radioactively produced energy means that the internal temperatures of the planets will have risen from their initial values. 1976 *Nuclear Sci. Internat.* Nov. 36/2 Direct measurements of bronchial clearance rates (by inhalation of radioactively labelled dust followed by measurements of radiation from the chest) show little difference between smokers and non-smokers.

ra:dioacti·vity. Also with hyphen. [f. Radio- 2 + Activity.] **1. a.** The property or condition of being radioactive; (the field of study concerned with) the phenomena displayed by radioactive materials. Hence, the radiation emitted by a radioactive material, or such material itself in a dispersed form.

1899 *Nature* Nov. 71/1 On the radio-activity induced by the Becquerel rays. 1900 [see Radio- 2]. 1902 *Harper's Mag.* Aug. 364/1 For days Professor Curie was unable to approach his electrometers..owing to his acquired radio-activity. 1920 *Discovery* Apr. 122/1 This was the first discovery in the science which later became known as radio-activity. 1947 *Sci. News* V. 55 Between them, these three series of radioactive elements include the whole of the known natural sources of radioactivity. 1955 *Sci. Amer.* Aug. 35/1 Radioactivity is measured in curies: one curie is equal to the radioactivity from one gram of radium (37 billion atoms disintegrating per second). 1955 *Bull. Atomic Scientists* Sept. 253/1 Radioactivity can travel in the air over large distances, and its very nature and action are unknown and unfamiliar to most people. 1962 S. G. Waley in A. Pirie *Lens Metabolism Rel. Cataract* 359 When the enzyme is treated with isotopically labelled iodoacetate.. and the enzyme is boiled, no radio-activity is released. 1969 *Daily Tel.* 23 Jan. 1/4 A small Swiss nuclear research and training reactor at Lucens was closed yesterday after developing a leak of radio-activity. 1978 L. Deighton *SS-GB* xvi. 135 Do you know what radio-activity is?.. It's the emission of radiation from unstable atomic nuclei—alpha particles, nucleons, gamma rays, electrons and so on.

b. *fig.*

1922 C. E. Montague *Disenchantment* (1924) xv. 210 In men and women of high mental vitality, in places where any of the radio-activity of gifted teaching breaks out for a while..the mind is easily delighted. 1928 *Manch. Guardian Weekly* 27 July 74/1 That radio-activity and that consummate technique shine out from poem after poem.

2. The property of emitting radio waves.

1959 *Sci. Amer.* Feb. 66/1 (Advt.), This is Messier 87—one of many galaxies and nebulae radiating radio energy throughout time and space. What primeval force motivates this celestial radioactivity? 1972 *Nature* 20 Oct. 440/2 It seems unlikely that Cygnus X-3 is unique with respect to its radioactivity and it may therefore represent only one member of a large number of objects that exhibit highly variable radio emission, and which may have been overlooked in classical radio surveys. 1978 *Ibid.* 11 May 131/1 We show here that these non-Io-related..decametric radio activities are affected by solar wind conditions around Jupiter.

radioallergosorbent, -assay: see *RADIO- 3 a.

radio astronomy. Also with hyphen and as **radioastronomy.** [f. *RADIO *sb.* + Astronomy.] The branch of astronomy concerned with the study and interpretation of radio waves reaching the earth from space, and with the astronomical use of radio-echo techniques.

1948 *Sci. News Let.* 1 May 279/1 Radioastronomy is a new branch of astronomy only recently announced, Dr. Shapley stated. By use of high-frequency radio waves meteors are tracked in their flight. 1951 *Sci. News* XXI. 40 There are two basic methods of observing in radio astronomy. In the first, we use the familiar processes of radar to explore space around us... The second method is the one in general use. Radiation of radio wavelengths, emitted..by extra-terrestrial bodies, is collected. 1951 *Nature* 7 July 17/1 Dr. Alfred Charles Bernard Lovell has been appointed to a newly created chair of radio-astronomy in the University of Manchester. **1958** *Listener* 27 Nov. 869/2 Out of the cataclysm of a world war have emerged two technical developments which are creating a revolution in astronomical observations—radio astronomy and the earth satellite. **1966** *McGraw-Hill Encycl. Sci. & Technol.* XI. 247/1 Large radio antennas designed for radio astronomy are now used for radar, radio communications, and satellite telemetering. **1973** 'D. Halliday' *Dolly & Starry Bird* xiii. 188 Innes is a Steady-State man, having done a sabbatical on radio astronomy at Cambridge. **1977** J. Narlikar *Struct. Universe* iii. 66 Radio-astronomy has played an important part in galactic explorations. For example, 21-cm radio wavelength observations are useful for detecting neutral hydrogen. **1978** R. V. Jones *Most Secret War* xlix. 486 We supplied German radar components..to both Bernard Lovell and Martin Ryle, to help them in their start on radio-astronomy.

Hence **radio astronomer,** a person engaged in radio astronomy; **ra:dio-astrono·mical** *a.*, of or pertaining to radio astronomy; obtained by means of radio telescopes.

1949 *Nature* 12 Nov. 815/2 Three British research organizations which have played a major part in radio-astronomical research. **1952** *Ibid.* 1 Mar. 350/1 It appears likely that..experimental radio astronomers will concentrate their attention on devices for achieving increased angular accuracy. **1959** *Listener* 31 Dec. 1152/2 Interest in peculiar galaxies has been stimulated recently by radio-astronomical observations. **1962** *New Scientist* 5 Apr. 827/3 The primary gamma-ray picture of the Universe might resemble the radioastronomical one. **1968** *Times* 26 Oct. 4/4 A new pulsar has been discovered by radio astronomers at Jodrell Bank. **1973** C. Sagan *Cosmic Connection* vi. 47 We see..in radio-astronomical studies of the interstellar medium, a profusion of simple and complex organic molecules. **1976** *Time* 27 Dec. 37/2 Radioastronomers..have beamed coded signals toward the stars to let any other civilization know that intelligent life exists on earth.

radio-atom, -autograph (etc.): see *Radio- 3 a.

†**radiobe** (rēˡ·di,ōᵘb). *Obs.* [f. *Radi(o- 3 + -obe, after Microbe.] A cell-like body observed to form in large numbers in gelatin solutions in the presence of radium salts, which was formerly claimed to be a living organism owing its existence to radioactivity.

1905 J. B. Burke in *Nature* 25 May 79/2 As these bodies cannot be identified with microbes, on the one hand, nor with crystals on the other, I have ventured..to give them a new name, *Radiobes*, which might..be more appropriate as indicating their resemblance to microbes, as well as their distinct nature and origin. **1905** *Daily Chron.* 20 June 4/4 Tubes of bouillon containing radium and tubes without radium were stopped up with cotton-wool, subjected to a temperature far above the boiling point of water, under pressure, for half an hour. The control tubes which contained no radium were then watched, and 'nothing happened'... But the surface of the beef-gelatine in the other tubes began to show a peculiar 'growth'. This 'growth'..was examined by a very high power of a microscope and found to consist of minute rounded objects which looked like bacteria... Like living cells they contain nuclei and these have been photographed through the microscope... They exhibited a property possessed by no crystals; a property possessed by living things alone... When they reach the maximum size already named, they subdivide... Mr. Burke calls them radiobes. **1908** *Encycl. Relig. & Ethics* I. 26/1 Mr. Butler Burke inclines to the conclusion that they are organisms on the border lines between microbes and crystals, and, provisionally, he names them 'radiobes'. **1920** *Punch* 7 Jan. 7/2 Let scientists on various fronts Indulge in their atomic stunts, Or harness to our prams and punts The puissant radiobe.

radiobio·logy. Also with hyphen. [f. Radio- 2, *3 + Biology.] The branch of biology concerned with the effects on living organisms of radiation and radioactivity, and with the application in biology of radiological techniques.

1919 *Med. Sci. Abstr. & Rev.* I. 358 In radio-biology, when we wish to show the selective action of X-rays we usually choose, as an example, one of the glands of external secretion. **1935** *Discovery* Aug. 225/1 A Congress of Radiobiology was held..last year in Venice. **1955** *Times* 18 Aug. 6/1 The organization that he thought necessary would consist of experts in radiophysics and radiobiology, including radio-genetics, and would be empowered to promote essential research in its field and to organize continuous general supervision of the radiation level of the world. **1962** *Lancet* 26 May 1106/2 The subjects cover almost every aspect of radiobiology, from radiation chemistry to straightforward radiotherapy. **1972** *Physics Bull.* Mar. 147/1 Typical of the radiobiology experiments is one carried out by a team at the Ames Laboratory of NASA in which some 56 pocket mice..were irradiated by the 250 MeV/nucleon nitrogen ion beam.

Hence **ra:diobiolo·gic** (chiefly *U.S.*), **-lo·gical** *adjs.*, of or pertaining to radiobiology; **ra:diobiolo·gically** *adv.*; **radiobio·logist.**

1929 *Radiology* XII. 454/2 (*heading*) Radio-biologic investigations on eggs of Ascaris. **1931** *Gen. Electr. Rev.* XXXIV. 98 (*heading*) Instruments for radiometric and radio-biological investigations at the Desert Sanatorium and Institute of Research, Tucson, Arizona. **1945** C. W. Wilson *Radium Therapy* iii. 81 Innumerable radio-biological experiments and clinical studies have shown that..a number of purely physical factors influence the biological effects produced by high-voltage radiation. **1946** *Nature* 2 Nov. 601/2 Radiobiologists should be grateful for accurate data such as these. **1955** *Times* 18 Aug. 6/1 The problem of extending the present international standards of radiobiological protection from the occupational workers..was discussed. **1956** *Proc. Internat. Conf. Peaceful Uses Atomic Energy* XI. 3/1 It is our belief that, if we take sufficient care radiobiologically to look after mankind..the rest of nature will take care of itself. **1961** *Ann. N.Y. Acad. Sci.* XCV. 828 (*heading*) Radiobiologic observations on human hemic cells *in vivo* and *in vitro*. **1971** *New Scientist* 8 Apr. 108/2 Even if radiobiologists could quantify the deaths and deformations caused by radiation, planners and government consultants feel that these numbers then need to be translated into economic terms. **1976** *Nature* 22 Jan. 209/1 Among the many important radiobiological findings recently summarised are the following: the mutation rate depends on sex, on the type of germ cell irradiated, on radiation quality (X rays, neutrons), [etc.]. **1977** *Lancet* 20 Aug. 411/1 Complete recovery of thyroid function after prolonged ^{131}I-induced hypothyroidism can be explained radiobiologically.

radio-cæsium: see *Radio- 3 b (i).

radio-ca·rbon. Also **radiocarbon.** [f. *Radio- 3 b + Carbon.] **1.** A radioactive isotope of carbon; *spec.* = *carbon 14*, which is formed in trace amounts by the effect of cosmic rays on atmospheric nitrogen. Also *ellipt.* for *radio-carbon dating*.

1940 *Physical Rev.* LVII. 549/2 Large quantities of nitrogenous material have been exposed to neutrons for several months and will be shortly worked up for radio-carbon. **1946** *Ibid.* LXIX. 672/1 The purpose of this letter is to..suggest that radiocarbon might be found in living matter especially in connection with the concentration of C^{14} for tracer uses. **1956** W. F. Albright *Archaeol. of Palestine* (rev. ed.) i. 22 Radiocarbon has a 'half life' of some 5,600 years, and the count loses any significance beyond 25,000–30,000 years ago. **1957** *Times* 11 Sept. 6/2 Recent studies using radiocarbon indicated that the yield of photosynthesis by the plankton of the oceans was at least equal to that of the land flora, and might be several times greater. **1963** G. M. B. Dobson *Exploring Atmosphere* i. 12 The radio-carbon formed in the upper atmosphere becomes radio-carbon dioxide and is gradually mixed throughout the whole atmosphere. **1970** *Nature* 4 Apr. 45/1 The chronology of several glacial stages has been recognized in the microfossils and dated by radiocarbon. **1976** P. L. Brown *Planet Earth* iii. 77 Radiocarbon then combines with oxygen to form $C^{14}O_2$ which is diffused through the atmosphere and then is absorbed by plants via photosynthesis and ultimately by all living things.

2. *attrib.* and *Comb.* (usu. with reference to radio-carbon dating), as *radio-carbon age, content, method, year*, etc.; **radio-carbon dating,** a method of isotopic dating which is applicable to dead organic matter and in which the proportion of carbon 14 (which has decreased at a known rate since the death of the sample material), is measured and compared with the known natural abundance of the isotope; hence *radio-carbon date*; *radio-carbon-dated* ppl. adj.

1949 *Antiquity* XXIII. 113 A method of dating dead pieces of formerly living substances (such as wood and bone) by means of their radiocarbon content. **1949** *Science* 23 Dec. 679/2 These results indicate that the two basic assumptions of the radiocarbon age determination method—namely, the constancy of the cosmic radiation intensity and the possibility of obtaining unaltered samples—are probably justified for wood up to 4600 years. **1950** Arnold & Libby (*title*) Radiocarbon dates. **1951** *Amer. Jrnl. Sci.* CCXLIX. 257 (*heading*) Radiocarbon dating of Late-Pleistocene events. *Ibid.*, All of the dates are of the right order of magnitude, with a few exceptions where it seems likely that the stratigraphic position of the sample, and not the radio-carbon age, has been incorrectly given. *Ibid.* 268 The radio-carbon date, 12,148 ± 700 years. **1956** M. Wheeler in A. Pryce-Jones *New Outl. Mod. Knowl.* 399 A radio-carbon dating gives 7538 B.C. (with a possible error ± of 350 years) for the settlement. **1957** G. E. Hutchinson *Treat. Limnol.* I. i. 8 The dating of the events, based primarily on the varve chronology, is in fair accord with the radio-carbon chronology. **1963** D. W. & E. E. Humphries tr. *Termier's Erosion & Sedimentation* iii. 55 (*caption*) Old peat deposits (radio-carbon-dated to the Early Holocene) dot the area 'drowned' by the Flandrian transgression. **1966** *Radiocarbon* VIII. 534 The result of a radiocarbon determination is commonly expressed as an age given in radiocarbon years. *Ibid.*, The conversion of a radiocarbon age..to a true calendar year makes necessary certain assumptions with respect to: (1) the half-life of C^{14}, (2) the production rate of C^{14} by cosmic rays, (3) the size of reservoirs into which C^{14} is distributed and the exchange rate of this distribution. **1973** *Nature* 1 June 266/1 Three radiocarbon laboratories, at La Jolla, Philadelphia and Tucson, have obtained radiocarbon dates over the past decade for specimens of bristlecone pine already dated dendrochronologically, thereby allowing the 'correction' of the radiocarbon scale. **1973** *Nation Rev.* (Melbourne) 31 Aug. (Suppl.) 2/3 Until 1972 only two radio-carbon-dated remains of Chinese origin were known. **1978** *New Scientist* 2 Mar. 599/2 The dating of carbon in glacier varves would extend the calibration of the radiocarbon dating curve.

radiocast (rēˡ·diokɑst), *sb.* and *v.* *U.S.* [f. *Radio *sb.* + -cast, after Broadcast.] **A.** *sb.* A radio broadcast.

1931 *Daily Progress* (Charlottesville, Va.) 10 Feb. 5/3 J. B. Priestley..is going to Tahiti, via New York, to write his next novel. He said in a radiocast he'd need a small island to recover from the fright he expects to receive on the gigantic island of Manhattan.

B. *v.* To broadcast by radio; so **ra·diocasting** *vbl. sb.*

1931 *Amer. Speech* VI. 253 Where the writer wishes to leave no doubt that he means *to radio-broadcast*, he is taking now to *to radiocast*. **1940** *Christian Sci. Monitor* 16 Mar. (Mag. section) 3/3 (*caption*) From this radiocasting structure, 400 feet high, music without static is being sent to listeners. **1947** Partridge *Usage & Abusage* 260/1 Both 'to *radio*' and 'to *broadcast*' are infinitely preferable to either *radiocast* or *radio-broadcast*.

radioche·mical, *a.* (*sb.*) Also with hyphen. [f. *Radio- 3 + Chemical *a.*] **1.** Of, pertaining to, or considered in terms of radiochemistry; *radio-chemical purity*, the state of being free from radioactive impurities. Also as *sb.*, a radioactive chemical.

1915 F. Soddy *Chem. Radio-Elements* (ed. 2) I. 46 In a condition of complete radio-chemical purity. **1946** *Chem. & Engin. News* 10 Dec. 3168/1 Laboratory facilities for radio-chemical research are classified according to the level of radioactivity involved in the operations. **1961** G. R. Choppin *Exper. Nucl. Chem.* ix. 129 Radiochemical purity is frequently of greater importance than chemical purity; it may be better to have a milligram of inert impurity in the final sample than 10^{-9} gm of radioactive contaminant. **1971** *New Scientist* 1 Apr. 26/2 The US market is estimated to be worth over $80 million a year with some $25 million spent on radiochemicals, $40 on radiopharmaceuticals, and about $20 million on basic radioisotopes and sealed sources. **1973** D. L. Horrocks in Moghissi & Carter *Tritium* i. 34 The total fissions are determined by radiochemical assay of a fission product (i.e., ^{140}Ba) of known yield. **1978** *Nature* 11 May p. xii/1 (Advt.), Our supply of radiochemicals of the highest quality and technical specifications.

2. (With hyphen.) Of or pertaining to chemical changes caused by radiant energy. *rare*.

1921 D. H. Lawrence *Let.* 2 Mar. (1962) II. 512 The sun is dangerous these months—it has a radio-chemical action on the blood which simply does for me. **1935** *Mind* XLIV. 545 In general, chemical reactions are coming to be conceived in the light of the 'radio-chemical' hypothesis, *viz.* that no structural modification can take place except as the result of quantities of radiant energy imparted in rhythmic pulses.

Hence (sense 1) **radioche·mically** *adv.*, by a radiochemical method or process; in terms of radiochemistry.

1915 F. Soddy *Chem. Radio-Elements* (ed. 2) I. 44 Another and very important sense [of the term 'pure'], which may conveniently be termed 'radio-chemically' pure. **1923** Glazebrook *Dict. Appl. Physics* IV. 590/1 A product may be prepared 'radio-chemically' pure, in which the radio-element may be mixed with a certain amount of inactive matter, but is free from substances chemically analogous to itself. **1957** *Jrnl. Biol. Chem.* CCXXIX. 443 A substance subsequently shown to be radiochemically pure was treated in a similar manner. **1975** *Nature* 1 May 77/1 The plutonium in each right testis was measured radiochemically.

radioche·mistry. Also with hyphen. [f. *Radio- 3 + Chemistry.] The chemistry of radioactive substances; sometimes held to include also *radiation chemistry* s.v. *Radiation 4. Hence **radioche·mist.**

In quot. 1904 perh. = *radiation chemistry*.

1904 *Jrnl. Physical Chem.* VIII. 506 Under chemical energetics we find the mass law relations for equilibrium and reaction velocity,..electrochemistry, photochemistry and radiochemistry. **1910** A. T. Cameron (*title*) Radiochemistry. **1911** F. Soddy *Chem. Radio-Elements* 11 A special branch of chemistry, which may be appropriately termed 'Radio-Chemistry' has come into existence. **1938** R. W. Lawson tr. *Hevesy & Paneth's Man. Radioactivity* (ed. 2) xxiv. 247 The study of the chemical effects of the rays from radium is sometimes called 'radio-chemistry' in the narrower sense... However, we shall confine the use of the term 'radio-chemistry' to the chemistry of the radio-elements, or the study of their chemical properties and reactions. **1952** F. E. Zeuner *Dating Past* (ed. 3) iv. x. 346 One can confidently expect the radio-chemist to be aware of the difficulties. **1953** E. Simon *Past Masters* III. 141 A radio-chemist [should] be called in to carry out tests on the stone fabric of this and the other huts in order to determine age. **1971** I. G. Gass et al. *Understanding Earth* ii. 42/1 Direct measurements of long half-lives..have been carried out in many physics and radiochemistry laboratories. **1973** *Nature* 3 Aug. p. xiii. (Advt.), South African Atomic Energy Board. Radiochemist or Radiopharmacist. Applications are invited.

radio-chlorine to **-element:** see *Radio- 2, 3, 4.

ra·dio-frequency. Also radio frequency, radiofrequency. [f. *Radio *sb.* + Frequency.] **1.** A frequency in the range used for telecommunication; greater than that of the highest audio-frequency and less than that of the shortest infra-red waves (i.e. between about 10^4 and 10^{11} or 10^{12} Hz).

1915 *Electrician* 2 July 463/1 Mercury arc rectifiers can be operated at a good efficiency even at radio frequencies. **1937** *Discovery* Mar. p. xxiii/1 Loud Speaker output at radio frequencies. **1976** A. White *Long Silence* xi. 95 You can often get close enough..without them being aware of it, especially if they're listening to a radio frequency.

2. *attrib.* Pertaining to (electromagnetic radiation having) such a frequency; operating at or having such a frequency; employing alternating current having such a frequency.

1915 *Engin. Mag.* XLIX. 253/2 The usual radio frequency transformers. **1919** *Wireless World* May 75/1 The arc transforms..the 1,500-volt direct current power into radio frequency energy. **1922** *Nature* 20 May 650/2 A six-tube

amplifier having three stages of radio-frequency amplification. **1943** *Electronic Engin.* XV. 344/2 When the bridge goes off balance a radio-frequency current flows through the meter. **1946** *Nature* 10 Aug. 194/1 Radio-frequency heating is not an economic proposition for heating stable liquids, but may prevent serious losses of activity in heat-sensitive ones. **1955** J. G. Davis *Dict. Dairying* (ed. 2) 828 The use of radio-frequency heating for the pasteurization of milk. **1957** *Endeavour* XVI. 187/1 Electronic equipment to generate the radio-frequency radiation and to measure its absorption by the sample. **1965** *B.B.C. Handbk.* 116 For best results on short waves, a receiver should incorporate a tuned radio-frequency amplifier preceding the frequency-changer stage. **1967** M. Chandler *Ceramics in Mod. World* ii. 76 Other methods of drying, of more limited and special application, include infrared and radio-frequency drying. **1970** G. K. Woodgate *Elem. Atomic Struct.* viii. 148 Many precise measurements of g_J..have now been made by the methods of radio-frequency spectroscopy.

radiogenetic(al, -ics: see *Radio- 2.

radiogenic (rēˡdiodʒe·nik), *a.* [f. *Radio- 3, 4 + *-genic (in sense 1 after *photogenic*).] **1.** Well suited for broadcasting by radio; providing an attractive subject for a radio broadcast.

1928 *Radio Times* 24 Aug. 342/2 Their object is to discover..a form (or forms) of drama which shall be truly 'radiogenic'. **1931** T. H. Pear *Voice & Personality* xii. 149 England has greater artistic variety of 'radiogenic' material. **1943** S. Lewis *Gideon Planish* xxxii. 418 The..Great Leaders: Governor Blizzard and..the dazzle-sounding, radiogenic Winifred Marduc Homeward. **1946** *Electronic Engin.* XVIII. 207/3 Transatlantic Quiz, even if radiogenic, is not pictorial and there seems no good reason for transplanting it into television. **1959** *Listener* 27 Aug. 332/2 Radiogenic in the extreme, Miss Jacob led her interlocutors more of a dance than anyone else in this series since Thurber. **1975** *Encounter* Sept. 43 This short play..has appeared in book form; but so totally radiogenic is its very nature that the printed page cannot represent it.

2. Produced by or resulting from radioactive decay or ionizing radiation.

1935 *Jrnl. Amer. Chem. Soc.* LVII. 470/2 There seems to be no definite trend in the relation of the atomic weight of uranium lead to the period during which the radiogenic lead has been forming. **1947** *Endeavour* VI. 104/1 The present rates of production of radiogenic lead are known with a remarkable degree of accuracy. **1960** *New Scientist* 5 May 1114/3 No evidence of radiogenic mutations has been discovered so far. **1970** *Nature* 5 Dec. 906/1 It could be..that the Moon is expanding slowly as a result of radiogenic heating. **1971** *Brit. Med. Bull.* XXVII. 68/1 There were about 15 times as many spontaneous cases as radiogenic cases.

Hence **radioge·nically** *adv.*, by means of radioactive decay.

1956 *Sci. News* XXXIX. 14 Any helium nuclei produced radiogenically would all be of mass 4. **1970** *Nature* 23 May 692/1 These dykes have been radiogenically aged at 2,420 million years.

radio-gold: see *Radio- 3 b (i).

ra:diogonio·meter. Also with hyphen. [f. *Radio- 4 + Goniometer.] = *Goniometer 2. Hence **ra:diogoniome·tric, -me·trical** *adjs.*, of, pertaining to, or by means of a radiogoniometer; **ra:diogonio·metry,** direction-finding by means of a radiogoniometer.

1908 L. H. Walter tr. Bellini & Tosi in *Electr. Engin.* 5 Mar. 348/1 In the present article it is proposed to treat in a more detailed manner the theory and construction of the instruments above referred to, and to which the name of radio-goniometers has been given by the authors. **1913** *Year-bk. Wireless Telegr.* 310 The radiogoniometer..consists of two coils wound over and at right angles to each other, each coil being connected to one of the directive aerials. **1921** *Nature* 23 June 542/2 Radiogoniometry and atmospheric influences. **1927** *Daily Express* 13 Oct. 11/5 A radiogoniometric aerial, an ingenious piece of wireless apparatus which enables the exact position of an air liner in flight to be located. **1932** *Times Lit. Suppl.* 9 June 429/4 He had several narrow escapes..before a cleverly-hidden radiogoniometer ran the Whisperer to earth. **1935** C. G. Burge *Compl. Bk. Aviation* 480/1 Visual radio-goniometry, or the actual wireless compass, will no doubt come into being when a solution is found to a number of the present wireless problems. **1936** *Discovery* Apr. 125/2 Radiogoniometrical control of such a projectile..is quite beyond accomplishment at present. **1961** *Engineering* 16 June 825/1 A new radiotelephone transmitter-receiver..may be used with a radiogoniometer unit. **1971** Watson & Wright *Radio Direction Finding* iv. 46 The radiogoniometer is designed to give maximum accuracy and sensitivity..to the direction finding equipment. **1978** D. Bloodworth *Crosstalk* xvi. 136 Interception and radiogoniometrical checks show irregular traffic on one high-frequency waveband.

radiogram¹. Add: **1.** (Further examples.) Cf. *Radiograph *sb.* 2.

1921 A. V. Knox in A. V. & R. Knox *Gen. Pract. & X-Rays* I. i. 21 A radiogram may be defined as a shadow-picture of structures lying in different planes reproduced on a flat surface. **1962** G. Cremer-Bartels in A. Pirie *Lens Metabolism Rel. Cataract* 444 (*caption*) Radiogram after digestion with phosphatase. **1975** B. Wood *Killing Gift* (1976) I. 13 She could be crippled if that hip is broken and I don't set it. And if I have to set it without a radiogram, I can't be sure of doing it properly.

2. = *Radio-telegram.

1904 *Prelim. Conf. Wireless Telegr. Berlin 1903* 10 It seems to us inadvisable..to insist on the interchange of radiograms between ships on the high sea. **1925** H. L. Foster *Trop. Tramp Tourists* 172 Radiogram just came in. The railway can only furnish us with six cars. **1929** *Star* 21 Aug. 12/4 The wireless message was in reply to the following radiogram. **1938** E. Waugh *Scoop* I. v. 90 William hastened to consult him about a radiogram which had arrived that morning. **1949** *Radio Times* 15 July 6/1 I'm from the radio room. Here's a radiogram for you. **1966** *McGraw-Hill Encycl. Sci. & Technol.* XIII. 423/2 In overseas communication, telegraph messages usually are referred to as cablegrams or radiograms, depending on the overseas transmission medium. **1980** L. St. Clair *Obsessions* ii. 58 Helen had replied with her own radiogram: 'What's wrong? No shipboard romance?'

radiogram² (rēˡ·diogræm). Abbrev. of next.

1932 G. Wilson *Gramophones* viii. 43 If you have no electricity in your house you should avoid the radio-gram. **1933** *Sunday Referee* 2 July 16/2 In the living-room..is a ply-wood built-in book-case, at one end of which is..a built-in electric radiogram and loud-speaker. **1935** M. Egan *Dominant Sex* III i. 86 She switches on the radiogram and dances gaily for her own amusement. **1938** Auden & Isherwood *On Frontier* II. ii. 77 Valerian and Stahl..are listening to the Leader's speech on the radiogram. **1945** J. Betjeman *New Bats in Old Belfries* 31 Softly croons the radiogram, loudly hoot the owls. **1970** J. Earl *Tuners & Amplifiers* i. 21 The old-style radiogram with inbuilt speakers..rarely yields good stereo reproduction. **1977** *Gramophone* Jan. 1217/1 The widespread development of the music centre as successor to the radiogram.

radio-gra·mophone. [*Radio *sb.* + Gramophone in Dict. and Suppl.] A radio and gramophone combined in a single cabinet (with a speaker).

1927 *Wireless World* 19 Oct. 539/1 (*heading*) A combined radio-gramophone installation. **1930** *Times Educ. Suppl.* 15 Feb. p. iv/4 We may reasonably expect a radio-gramophone to work without an outside aerial. **1935** *Economist* 23 Nov. 1042/1 Radio receiving sets and radio-gramophones had formed the largest part of the sales in the home markets. **1976** *Broadcast* 29 Nov. 19/1 Granny and her pre-war radiogramophone.

radiograph, *sb.* Add: **1.** (Earlier example.)

1880 D. Winstanley in *Chem. News* 30 Apr. 205/1, I will now ask your attention to the description of another and much more perfect apparatus, one which continuously records the intensity of thermal radiation in which it is exposed. This instrument I have called the 'Radiograph'.

2. Add to def.: Now also made using other forms of ionizing radiation. (Further examples.)

1923 Glazebrook *Dict. Appl. Physics* IV. 618/1 Tool-marks and fine mould-marks often show up in a radiograph. **1948** *Sci. News* VII. 104 A new type X-ray tube permits radiographs to be made with exposures of 1/500,000th second. **1966** *McGraw-Hill Encycl. Sci. & Technol.* XI. 304a/1 Radiographs made with γ-rays have high resolving power because of the absence of scattering. **1971** *Sci. Amer.* Oct. 16/3 Radiographs of the chest in persons on a starvation diet have indicated that the heart shrinks in size. **1972** *Nature* 15 Sept. 157/2 It has been the practice among nuclear physicists to take 'radiographs' with beams of accelerated light nuclei to determine the location of detector targets..relative to the position of the beam.

†**3.** = *Radio-telegraph. *Obs.*

1904 *Prelim. Conf. Wireless Telegr. Berlin 1903* 5 It is to him [*sc.* Popoff] that we owe the first radiograph apparatus.

radiograph (rēˡ·diograf), *v.* [f. the sb.] *trans.* To make a radiograph of; to study by radiography. Also *fig.* Hence **ra·diographing** *vbl. sb.*

1896 [in Dict. s.v. Radiograph *sb.*]. **1897** *Treatment* I. 43/2 It is almost routine practice..to radiograph fractures. **1908** *Sci. Abstr.* A. XI. 105, 1M of No. 5 rays will suffice for the radiographing of a hand. **1924** *Observer* 6 Apr. 12/3 He [*sc.* Byron] has been radiographed to the bone. **1940** J. A. Ross *Handbk. Radiography* xii. 112 Various methods of examination have been devised in an attempt to radiograph movement. **1951** L. P. Dudley *Stereoptics* vi. 106 The tube-film distance adopted in radiographing the subject must be the same as that adopted in radiographing the wire model. **1977** *Lancet* 19 Nov. 1059/2 Each section was photographed in colour, radiographed, drawn in black-and-white and compared with the scanner image at the corresponding level.

radiographer (rēˡdiǫ·grăfəɹ). [f. prec. + -er¹.] One who practises radiography; a person qualified to operate radiographic equipment.

1896 [in Dict. s.v. Radiograph *sb.*] **1907** *Oxf. Univ. Gaz.* 19 Feb. 395/2 The Committee for appointment of Honorary Medical Officers will shortly proceed to the Election of a Radiographer. **1917** *Med. Jrnl. Australia* 5 May 386/1 The radiographer, who had made a screen examination, reported the presence of a large aneurysm of the aortic arch. **1958** *Times* 4 Aug. 9/3 Recent figures have revealed a serious shortage of more than 500 radiographers in the National Health Service. **1971** [see *Radiologist]. **1975** E. Lawson *Seeing through You* i. 16 The radiographer was turned towards her patient, one hand outstretched to guide him safely against the cassette containing the X-ray film.

radiographic (rēˡdiogræ·fik), *a.* [f. as prec. + -ic.] **1.** Of, pertaining to, or carried out by means of radiography.

1896 [in Dict. s.v. Radiograph *sb.*]. **1921** *Lancet* 22 Jan. 175/2 (*heading*) Radiographic appearances of pyorrhœa. **1933** *Jrnl. Franklin Inst.* CCXXVI. 183 The radiographic method is useless..in detecting cracks in which the two conjugate surfaces are pressed closely together, leaving no open space. **1976** *Offshore Platforms & Pipelining* 167/1 The most effective way to determine weld quality is by radiographic inspection.

†**2.** = *Radio-telegraphic *a.* *Obs.*

1904 *Prelim. Conf. Wireless Telegr. Berlin 1903* 5 It was Hughes..who laid, in 1877, the first stone of radiographic practice by his detailed experiments. **1907** *Liverpool Post* 10 Sept. 7 On Wednesday night..the Lusitania will..get into radiographic touch with the American coast.

Also **radiogra·phical** *a.* (*rare*) = sense 1; also *fig.*; **radiogra·phically** *adv.*, by means of radiography; as regards radiography.

1898 Radiographic, -ically [in Dict. s.v. Radiograph *sb.*]. **1925** *Jrnl. Anat.* LIX. 149 Dubreuil employed lead pellets, periodically measuring the intervals between the pellets, radiographically. **1931** S. Beckett *Proust* 63 He describes the radiographical quality of his observation. The copiable he does not see. **1977** *Lancet* 14 May 1053/1 Although similar clinically and radiographically, the two syndromes show marked differences when pulmonary surfactant is examined.

radiography (rēˡdiǫ·grăfi). [f. Radio- 2 + -graphy.] **1.** The science or process of making radiographs.

1896, 1898 [in Dict. s.v. Radiograph *sb.*]. **1922** G. W. C. Kaye *Pract. Application of X-Rays* vi. 85 When the art of radiography had sufficiently advanced in medicine, it extended its scope to industry. **1948** *Sci. News* VII. 104 Radiography of rapidly-moving enclosed machine parts, such as pistons or the impeller blades of turbines, has become practical. **1948,** etc. [see *neutron radiography* s.v. *Neutron 2]. **1966** C. R. Tottle *Sci. Engin. Materials* vi. 147 With γ-rays, down to 10^{-3} Å, the penetration through metals is even greater; hence the use of X- and γ-rays for radiography. **1971** *World Archaeol.* III. 240 My colleague, Miss Theya Molleson, assisted in the radiography of the skeleton.

†**2.** = *Radio-telegraphy. *Obs.*

1904 *Prelim. Conf. Wireless Telegr. Berlin 1903* 5 It is due to radiography that communication has been created between parts of the globe which had previously been deprived of it. **1922** *Hotel World* 15 Apr. 6/2 Mr. Eastman, in charge of the radio broad-casting station in Chicago,..said: 'When I took charge of this work I knew very little about radiography.'

radiohalo, -heliograph: see *Radio- 3 a, 4.

radioimmuno- (rēˡ:dio,iːmiuno, -imiūːno). [f. Radio- 2 + *Immuno-.] Formative element in terms pertaining to analytical techniques combining immunological and radioisotopic methods. (In the following words secondary stresses vary as indicated above, and are not in general marked in each word.) **radioimmunoa·ssay,** an immunological assay in which the test sample is determined by allowing it to react with a prepared antiserum in competition with a known quantity of radioisotopically labelled antigen, the extent of reaction being measured from the amount of radiation emitted (see quot. 1974); hence **radioimmunoa·ssayable** *a.*, capable of determination by radioimmunoassay; **radioimmunoche·mical** *a.*, deriving both from immunology and from radiochemistry; employing radioisotopically labelled antigens and antibodies as reagents for chemical analysis; hence **radioimmunoche·mically** *adv.*; **radioimmunoe:lectrophore·sis,** immuno-electrophoresis carried out using radioisotopically labelled samples, usu. as a means of studying the formation or binding of proteins; so **radioimmunoele:ctrophore·tic** *a.*; **radioimmunoprecipita·tion,** the use of radioisotopically labelled antigen or antibody in a precipitin test, the radioactivity in any precipitated complex being measured.

1961 *Jrnl. Clin. Investigation* XL. 1086/1 A specific radioimmunoassay of human growth hormone..has been devised. **1966** *Lancet* 24 Dec. 1389/2 Glucose was measured by the potassium-ferricyanide method..and insulin by radio-immunoassay. **1974** R. M. Kirk et al. *Surgery* ii. 36 To carry out radioimmunoassay, an antiserum is first raised by injecting the substance repeatedly into laboratory animals, and later withdrawing serum containing antibody. A known concentration of antigen, labelled with a radioactive marker, is placed in competition with the test serum for combination with the antibody. **1977** *Time* 24 Oct. 52/1 Yalow and her late collaborator, Dr Solomon Berson, devised a sensitive new biological analytic technique called the radioimmunoassay (RIA). **1973** *Nature* 27 July 230/1 The serum level of radioimmunoassayable (RIA) growth hormone. **1978** *Ibid.* 12 Jan. 178/2 The radioimmunoassayable enkephalin content of the basal ganglia does not change after cortical ablation. **1968** *Gastroenterology* LV. 317/2 The supernatant solutions were analyzed in triplicate for gastrin by a radioimmunochemical technique *Ibid.* 326/2 The gastrin contents..were measured radioimmunochemically. **1977** *Lancet* 26 Mar. 666/2 Blood obtained by antecubital-vein puncture was assayed radioimmunochemically for insulin, glucagon, gastrin, and pancreatic polypeptide. **1962** Morse & Heremans in *Jrnl. Laboratory & Clin. Med.* LIX. 893 This method of immunoelectrophoresis, followed by autography, will be referred to as radioimmunoelectrophoresis. **1970** *Nature* 12 Dec. 1086/2 Radioimmunoelectrophoresis was also performed on dialysed supernatant from cell cultures incubated with either amino-acids labelled

with ^{14}C or with reconstituted protein hydrolysate similarly labelled. **1975** *Ibid.* 11 Dec. 547/2 Radioimmunoelectrophoresis of serum from snakes injected with HEA revealed that the gamma precipitin line specifically bound HEA^{131}I. **1962** *Jrnl. Immunol.* LXXXIX. 744/2 Similar radio immunoelectrophoretic technique was applied in insulin-binding antibodies from human patients. **1962** *Jrnl. Clin. Investigation* XLI. 260/1 Our observations with radioimmunoprecipitation confirm the information obtained with more orthodox immunologic procedures. **1971** *Jrnl. Immunol.* CVI. 1167/1 The radioimmunoprecipitation test (RIP) was performed by the microtiter method as modified by Sever..using V-bottom thin plastic plates. **1974** *Nature* 25 Jan. 176/3 In addition the antigen was detected by radioimmunoprecipitation in the supernatant fluid of a few cultures which had been serially passaged.

radioimmunology (rēˡdio,imiuno·lǫdʒi). [f. Radio- 2 + *Immunology.] The application of radiological techniques in immunology.

1971 *Biol. Abstr.* LI. Ann. Cum. Index 4200/1 (*heading*) Radio immunology. **1976** *Scand. Jrnl. Immunol.* V. 609 (*heading*) Unified mass-action theory for virus neutralization and radioimmunology.

So **ra:dioimmunolo·gic** (chiefly *U.S.*), **-lo·gical** *adjs.*, combining radiological and immunological methods; of or pertaining to radioimmunology; **ra:dioimmunolo·gically** *adv.*

1965 *Jrnl. Clin. Endocrinol.* XXV. 1043 (*heading*) A radioimmunological assay method for insulin using insulin-^{125}I and gel filtration. *Ibid.* 1457 (*heading*) Radioimmunologic measurement of human placental lactogen in plasma by a double antibody method. **1970** *Ibid.* XXXI. 679/2 HGH antibodies were determined radioimmunologically. **1976** *Science* 24 Dec. 1428/3 Radioimmunologic techniques have markedly increased the sensitivity with which viruses can be detected. **1977** *Lancet* 7 May 1006/2, 194 rural men, aged 55–74 and leading the same agricultural life, were screened for milk antibodies by a radioimmunological method.

radio-iodinate(d, -iodination: see *Radio- 3 a. **radio-iodine, -iron:** *Radio- 3 b (i).

radioisotope (rēˡdio,ai·sǫtōᵘp). Also with hyphen. [f. *Radio- 3 + *Isotope.] A radioactive isotope.

1946 *Chem. & Engin. News* 10 Dec. 3168/1 The availability of radioisotopes of nearly all elements in quantities hitherto unachievable. **1950** *Times* 8 May 4/4 Also being shown are the machines and methods used for the extraction and synthesis of C-14, a radio-isotope of carbon, which has many uses in industrial, medical, and biological research work. **1958** *Economist* 8 Feb. 496/2 Radio-isotope departments are being set up in the Royal Hospital in Baghdad and the University of Shiraz for the diagnosis and treatment of disease. **1976** *Daily Colonist* (Victoria, B.C.) 15 Apr. 5/5 A freshly fallen meteorite..contains radioisotopes which decay in a matter of days or weeks. **1980** *Brit. Med. Jrnl.* 29 Mar. 931/1 The clinical use of radioisotopes has developed over the last 30 years from a technical science into a recognisable clinical specialty.

Hence **ra:dioisoto·pic** *a.*, **-isoto·pically** *adv.*

1956 *Nature* 7 Apr. 639/1 The kinetics of biological systems as studied by radioisotopic methods. **1970** *Ibid.* 13 June 1025/1 Detection of regulatory proteins present in such small amounts would probably require radioisotopically labelled proteins of very high specific activity. *Ibid.* 24 Oct. 383/1 Radioisotopic tracer experiments have established that carbonate from seawater is incorporated into the skeleton by many corals. **1978** *Ibid.* 19 Oct. 667/2 We used a radioisotopically labelled complementary DNA probe..generated from an *in vitro* reverse transcriptase reaction.

radioize (rēˡ·dio,aiz), *v.* *U.S.* [f. *Radio *sb.* + -ize.] *trans.* To equip with radio.

1922 *Sci. & Invention* May (Advt., rear cover), Radioize your phonograph with a guaranteed adapter. **1950** *Sun* (Baltimore) 19 July 13/1 Russia is in the middle of an all-out campaign to 'radioize' the entire population of its sprawling Soviet Socialist Republics.

ra·dio-label, *v.* and *sb.* *Biol.* and *Chem.* Also **radiolabel.** [f. *Radio- 3 + Label *v.* in Dict. and Suppl.] **A.** *v. trans.* To label with a radioactive isotope (see *Label *v.* 2). So **ra·dio-labelled** *ppl. a.*, **radio-la·belling** *vbl. sb.*

1953 *Adv. Biol. & Med. Physics* III. 149 It may be possible to demonstrate the existence of such autoantibodies by the use of radio-labeled sera. **1962** *Jrnl. Immunol.* LXXXIX. 559/1 In the present study with poliovirus, this hindrance was overcome by radiolabeling the virus. **1970** *Nature* 16 May 649/1 In radiolabelling experiments, synchronized cultures were incubated with radioactive amino-acids for the required time and radiolabelled γG_{2a} was assayed. **1972** *Science* 16 June 1226/3 The tube is then washed, radiolabeled HBAb is added, and the new mix incubated. **1976** *Chem. in Brit.* XII. 375 (*heading*) Radiolabelling water's courses. *Ibid.* 379/1 Radiolabelling has become the preferred method for studying the metabolic fate of foreign compounds in biological systems.

B. *sb.* A radioactive label (*Label *sb.*¹ 7 d).

1972 *Science* 16 June 1226/3 The use of a radiolabel makes the technique expensive. **1978** *Nature* 12 Jan. 111/3 That these phenomena are not due to breakdown and reincorporation of the radiolabel is shown by the negligible label in juice if the radiolabelled protein presented is albumin.

radiolarian, *a.* and *sb.* Add: Also *Geol.* **A.** *adj.* *radiolarian chert*, a cryptocrystalline type of radiolarite; *loosely* = *Radiolarite; *radiolarian earth*, unconsolidated siliceous rock formed from the remains of radiolaria; *radiolarian ooze*, a siliceous marine sediment rich in the remains of the tests of radiolarians.

1876 [see *Ooze *sb.*² 2 b]. **1890** *Geol. Mag.* VII. 144 (*heading*) Radiolarian chert in the Ballantrae series. **1911** [see *Ophiolitic *a.*]. **1913** Hatch & Rastall *Petrology of Sedimentary Rocks* I. iv. 143 One of the most notable instances of the abundant presence of this division of the Protozoa is afforded by the so-called Radiolarian earths of Barbados. *Ibid.* II. iii. 233 An interesting example of the metamorphism of a nearly pure siliceous rock is found in the Galloway district, where certain radiolarian cherts of Arenig age are altered by the Loch Doon granite. **1926** W. H. Twenhofel *Treat. Sedimentation* v. 377 Radiolarian ooze is confined to the Pacific and Indian oceans. **1944** A. Holmes *Princ. Physical Geol.* xv. 317 The siliceous remains persist to greater depths, some of them down to 5,000 fathoms, the average for radiolarian ooze being about 3,000. **1953** *Caribbean Q.* III. iii. 179 Barbados..has a foundation of rocks... On top of this is a layer of deep sea deposit (radiolarian earth) and on top of this a slab of white limestone. **1971** B. W. Sparks *Rocks & Relief* ix. 306 Certain beds of radiolarian chert..are sufficiently resistant to form series of hog-back ridges across both outcrops. **1971** [see *Ooze *sb.*² 2 b]. **1975** Fox & Heezen in Nairn & Stehli *Ocean Basins & Margins* III. x. 430 Piston cores obtained from the base of one escarpment recovered Early Eocene radiolarian ooze.

radiolarite (rēˡdiolǣˡ·rait). *Geol.* [a. F. *radiolarite* (F. Jaccard 1904, in *Bull. des Laboratoires de Géol.*, etc. (Univ. of Lausanne) II. viii. 79): see Radiolaria *sb. pl.* and -ite¹.] A type of homogeneous fine-grained siliceous sedimentary rock formed mainly from skeletal remains of radiolarians; also applied loosely to other sedimentary rocks similarly formed.

1910 *Proc. Sect. Sci. K. Akad. Wetensch. te Amsterdam* XII. 141 In the year 1894 I [*sc.* G. A. F. Molengraaff] discovered in the basin of the Upper Kapoewas in Western Borneo cherts and hornstones, consisting almost entirely of tests of Radiolaria, which I described as deep-sea deposits. Such rocks are also known as Radiolarite. **1924** J. G. A. Skerl tr. *Wegener's Orig. Continents & Oceans* ii. 20 The practically non-calcareous 'radiolarites' of the Alps. **1938** Hatch & Rastall *Petrol. of Sedimentary Rocks* (ed. 3) x. 203 In Great Britain, two principal types of radiolarite may be distinguished, one represented by the jaspers and jasper-like cherts, and the other by the lydite type of chert. **1945** M. F. Glaessner *Princ. Micropalaeont.* ii. 11 Radiolarite- and serpentine-zones in close association form important structural elements in many folded zones of different age. **1972** *Gloss. Geol.* (Amer. Geol. Inst.) 587/1 *Radiolarite*,..(a) the comparatively hard, very fine-grained, chert-like, homogeneous, consolidated equivalent of radiolarian earth. (b) Indurated radiolarian ooze. (c) A term that is often applied as a syn. of radiolarian earth. **1977** A. Hallam *Planet Earth* 254/1 Ancient radiolarian-rich sediments are known as radiolarites, and many of these are believed to be deep-water deposits.

radiole (rēˡ·diōᵘl). *Zool.* [f. L. *radiolus*, dim. of *radius* Radius.] One of the spines or prickles of a sea-urchin. Also *radiole spine*.

1929 *Encycl. Brit.* VII. 900/2 Primarily radioles serve for protection, but the larger radioles may be used like stilts for locomotion or for digging. **1979** *Nature* 8 Nov. 135/3 Differing feeding processes in echinoids were correlated with differing fine structure of their radiole spines.

radio-lead: see *Radio- 3 b (i).

radioless (rēˡ·diolės), *a.* [f. *Radio *sb.* + -less.] Without radio broadcasting or radio equipment.

1937 *Daily Express* 3 Mar. 8/3 'Wilky' was an R.A.F. war pilot, and after the armistice was one of the heroes to work the radioless, single-engined air-mail service for the Rhineland Army. **1938** *Times Lit. Suppl.* 29 Jan. 69/1 Sir John [Reith] would have liked to impose a radioless Sabbath on the listening public. **1973** H. Gruppe *Truxton Cipher* xviii. 189 Cutter had been radioless, his ship last reported as destroyed.

radioligand: see *Radio- 3 a.

ra:dioloca·tion. Also with hyphen. [f. *Radio- 4 + Location.] The term orig. used in Britain for *Radar; the determination of the position and course of ships, aircraft, etc., by means of radar.

1941 *Flight* 26 June 430/1 In the Battle of Britain the advantages of radiolocation were even more apparent. **1941** *Hutchinson's Pict. Hist. War* 14 May–8 July 180 Radiolocation has been described by Air Chief Marshall Sir Philip Joubert as Britain's secret weapon against the German bomber. Rays are sent out and any aircraft or ship in their path immediately reflects a signal. **1942** *R.A.F. Jrnl.* 30 May 30 By what, then, is the R.A.F. kept in the air?..Its radio-location girls,..its operations staff? **1947** J. Hayward *Prose Literature since 1939* vi. 43 The public should be disabused of the notion that science..is merely another word for technology; and that its purpose is..to develop and supply 'modern conveniences'..in the form of..radio-location for the General Staff. **1976** M. Gilbert *Winston S. Churchill* V. xxiii. 659 The existence of 'radar' was not publicly acknowledged until 1941, when it was referred to as 'Radiolocation'.

Hence **ra:dioloca·te** *v. trans.*, to locate (an aircraft, etc.) by means of radar; **ra:dioloca·ted** *ppl. a.*; **ra:dioloca·tor,** an apparatus for radiolocation.

1941 *Flight* 26 June 430/1 They could rely on the vast radiolocator system to tell them in plenty of time when the enemy were coming and from what direction. **1942** *Times* 20 Jan. 2/2 Members of the A.T.S. have taken over the working of the radiolocators, predictors, and height-finders. **1943** *Times* 25 Oct. 3/4 Many experts are convinced that the attack was..ordered from a German patrol boat that had radio-located the aircraft. **1945** *Electronic Engin.* XVII. 679/3 It is no exaggeration..to say that the first object to be radiolocated was the Heaviside layer. **1945** *Jrnl. Inst. Electr. Engineers* XCII. I. 342/1 In this method of determining the distance of a radiolocated object, short radio-frequency pulses are sent out at regular intervals. **1946** P. Carter in Aldiss & Harrison *Decade the 1940s* (1975) 108 He passed by the radio locator and the radioman. Their jobs would come later, meantime radio silence was enforced. **1947** Crowther & Whiddington *Science at War* 18 The radiolocator requires as strong an echo as possible.

radiologic (rēˡdiolǫ·dʒik), *a.* Chiefly *U.S.* [f. as next + -ic.] = next.

1909 in *Cent. Dict.* Suppl. **1914** *Jrnl. Amer. Med. Assoc.* 28 Mar. 980/2 Increased gastric peristalsis has been commonly listed among the radiologic signs of duodenal ulcer. **1945** *Radiology* XLIV. 82/2 His interest in radiology has made him an energetic participant in local and national radiologic organizations. **1975** S. C. Bushong (*title*) Radiologic science for technologists. **1978** *Detroit Free Press* 5 Mar. 9/1 Career training offered at Madonna for deaf students includes..radiologic technology, business administration, [etc.].

radiological (rēˡdiolǫ·dʒikăl), *a.* [f. next + -ical.] **a.** Of, pertaining to, or concerned with radiology.

1909 *Proc. R. Soc. Med.* II. I. (Electro-Therapeutical Section) 5 Two most interesting papers were those by Béclère, of Paris, and Alban Köhler, of Wiesbaden, on the 'Radiological Exploration of the Liver'. **1923** *Daily Mail* 7 Aug. 5/6 In the radiological department of the hospital there he was shown his hand with the X rays. **1947** *Radiology* XLIX. 345/2 One of the old ideas running through the entire radiological literature is that small doses of X-rays exert a stimulating effect. **1955** *Bull. Atomic Scientists* May 171/1 The composition and possible radiological effects of the super-bomb tested in the Pacific last spring. **1969** P. Jacobs in D. Sutton *Textbk. Radiol.* I. ii. 31/2 Since the surgeon is likely to explore the lesion as soon as it is suspected clinically, one will seldom nowadays observe the radiological phases corresponding to the pathological changes mentioned above. **1976** *Globe Democrat* (St. Louis, Missouri) 18 Sept. 8f/3 Six sections of radiological shield manufactured here..are being shipped by rail to..a nuclear power plant.

b. Of warfare, weapons, etc.: involving the deliberate release of ionizing radiation or radioactive material in harmful quantities.

1951 *Britannica Bk. of Year* 686/2 *Radiological warfare*, warfare with radioactive material. **1952** B. Wolfe *Limbo '90* (1953) vii. 79 H-bombs..supplemented with radiological-warfare dust. **1958** *Ann. Reg. 1957* 478 Radiological warfare was now a distinct possibility through the use of low altitude explosions of high fission energy yield which could contaminate large areas beyond the range of physical damage. **1963** *New Scientist* 28 Mar. 679 (*heading*) A radiological arms race in the Middle East? *Ibid.* 679/3 The radiological bomb would, presumably, come into the same military category as a persistent gas.

Hence **radiolo·gically** *adv.*

1924 *Brit. Jrnl. Dermatol.* XXXVI. 516 When studying radiologically the osseous lesions..I had been struck by the clear delimitation of the pathological loci. **1955** *Sci. News* XXXV. 87 It was shown radiologically that the sphincter is capable of shutting off the venous return. **1979** *Guardian* 1 May 3/3 The Windscale site is known to be radiologically 'dirty'.

radiology (rēˡdiǫ·lŏdʒi). [f. Radio- 2 + -logy.] **a.** The medical use of X-rays, esp. in diagnosis; also extended to include the diagnostic use of other forms of radiation. †**b.** (See quot. 1905.) Cf. *roentgenology* s.v. *Roentgen-, Roentgeno-.

1900 *Pop. Sci. Monthly* May 110/1 An International Congress of Medical Electrology and Radiology has been connected with the International Congress system of the Paris Exposition, 1900. **1905** A. M. Clerke *Syst. Stars* (ed. 2) vi. 80 The many suggestions of 'radiology' (as the new science of radioactivity might be designated) cannot be inconsiderately set aside. **1928** *Daily Express* 6 Dec. 7/4 A specialist in radiology and electrical treatment. **1938** S. C. Shanks et al. *Text-bk. X-Ray Diagnosis* I. p. vii, Diagnostic radiology is becoming an increasingly complex specialty, and it is difficult for one person to be equally expert in all its branches. **1959** W. T. Moss *Therapeutic Radiol.* i. 23 The selective destruction of tissues forms the basis of therapeutic radiology. **1979** Bartrum & Crow *Case Stud. in Ultrasound* p. v, Even more than in conventional radiology, the satisfactory application of ultrasound requires that the practitioner know the clinical history. **1980** D. Sutton *Textbk. Radiol. & Imaging* p. vii/1 When the last edition of this book was published in 1975 Radio-isotope Scanning was well established and Ultrasound had already made a major impact on radiology... Nuclear Medicine has also steadily progressed. Imaging, as these new disciplines are usually called, is now a major force which has profoundly influenced the practice of radiology.

Hence **radio·logist,** a person employing ionizing radiation or radioactive material in any field, esp. a medically qualified practitioner of the diagnostic use of X-rays.

1906 *Arch. Roentgen Ray* XI. 20/1 Hitherto the majority of radiologists have been accustomed to work without any measurement of the Roentgen light. **1926** *Encycl. Brit.* XXXII. 284 This region of the intestinal tract is becoming an open book to the radiologist. **1955** *Sci. News Let.* 11 June 372/2 Because the field of radiation itself is so well-defined, City of Hope radiologists and physicists found it possible to dispense with the customary heavy, lead-lined

door. **1958** *Times Lit. Suppl.* 2 May 233/3 The radiologists in their helmets of lead, the scientists tracking down the atom, cast their strange shadows on the Pyramids. **1971** *Lancet* 29 May 1124/1 A radiologist to the N.H.S. is medical and a radiographer non-medical.

radiolucent (rēˡdiol¹ū·sěnt), *a.* [f. Radio- 2 + Trans)lucent *a.*] Transparent to X-rays. So **radiolu·cency,** the state or property of being radiolucent.

1917 K. Thoma *Oral Roentgenol.* iv. iv. 197 The Roentgen evidence of alveoloclasia is due to the dissolution of bone and replacement by radiolucent pathological tissue. **1936** B. J. M. Harrison *Textbk. Roentgenol.* iii. 62 Of the relatively radiolucent substances those used are air, carbon dioxide, and oxygen. **1940** *Bull. Johns Hopkins Hosp.* LXVI. 91 The cutaneous region of the foetus is usually represented by a dark line, due to radiolucency. **1961** *Dental Progr.* I. 177/2 Most of the difficulty in diagnosis was caused by general graininess of the film combined with poor contrast between areas of varying radiolucency. **1973** *Daily Tel.* 22 Oct. 13/6 The plastics used in the modern motor car..are radiolucent and do not reveal themselves on an X-ray film.

radioluminescence, -ent: see *Radio- 2.

radiolysis (rē'di,ǫ·lisis). *Chem.* [f. Radio- 2 + *-lysis.] Decomposition of a compound by the action of ionizing radiation. Hence **radioly·tic** *a.*, of, pertaining to, or formed as a result of, radiolysis.

1948 *Jrnl. Physical & Colloid Chem.* LII. 516 The kinetics of the radiolysis of pure water and of indirect action on solutes are discussed. **1950** *Ann. Rev. Physical Chem.* I. 120 Negative ions make important contributions to the general radiolytic process only in special cases. **1951** *Jrnl. Amer. Chem. Soc.* LXXIII. 532/2 H_2O_2 produced in the radiolysis may affect the pH. **1963** Dawson & Sowden *Chem. Aspects of Nuclear Reactors* II. iii. 96 The discovery..that cupric ion would act as an effective homogeneous catalyst to promote the recombination of radiolytic gas within the reactor core..was of the highest importance to homogeneous aqueous reactor technology. **1977** J. Weisman *Elem. Nuclear Reactor Design* ii. 35 Ions which participate in electron-transfer reactions can combine with H or OH ions thus preventing recombination and leading to an increased radiolysis.

Ra·diometal. Also **radiometal.** An alloy consisting largely of nickel and iron in approximately equal proportions and having a high magnetic permeability.

1940 *Chambers's Techn. Dict.* 699/1 *Radiometal*, an alloy of permalloy type. Contains iron 50%, nickel 45%, and copper 5%. **1941** *Jrnl. Brit. Inst. Radio Engineers* II. 101 A group of metallurgists and electrical technicians concentrated on the problem of high permeability materials and carried out intensive research on alloys with 80%, 50% and about 36% nickel content. The outcome of the work was the commercial introduction of the well known alloys Mumetal (78% Ni), Radiometal (48% Ni) and Rhometal (36% Ni). **1946** *Nature* 13 July 54/2 The stator core consists of radiometal laminations 0·010 in. thick. **1951** *Electronic Engin.* XXIII. 330/1 The smoothing choke is wound on a radiometal core. **1957** W. J. John *Mod. Electr. Engin.* I. iv. 116/2 Radiometal, or Permalloy B, has a saturation of 16,000 gauss and a maximum permeability of 10,000 to 30,000 at H = 0·3 to 0·4 oersted... Composition: nickel 50 per cent, iron 50 per cent.

radiometeorograph(y: see *Radio- 4.

radiometer. Add: **2.** Also, more generally, any device used to detect, or measure the intensity of, electromagnetic radiation (freq. *spec.* infra-red). Also extended to instruments (in the first cases adapted from Crookes's device) used to measure the intensity of sound by means of its radiation pressure. (Further examples.)

1905 R. W. Wood in *Physical Rev.* XX. 113 It occurred to me that a mill-wheel or radiometer driven by these [sound] waves would be useful for purposes of demonstration in treating of radiation pressure. **1920** *Q. Jrnl. R. Meteorol. Soc.* XLVI. 399 (*heading*) The ether differential radiometer. **1927** I. B. Crandall *Theory Vibrating Syst. & Sound* iv. 180 The radiometer is a simple torsion balance; a thin hollow metal box filled with air, on being submerged serves as a very good totally reflecting vane. **1930** R. H. Baker *Astron.* x. 396 Abbot,..working with the 100-inch reflector, made use of delicate radiometers, having vanes of blackened bits of fly-wings, to measure the energy in different parts of the spectrum. **1963** G. L. Pickard *Descriptive Physical Oceanogr.* vi. 92 The downward directed component of the long-wave radiation term..is determined by means of a radiometer. **1966** *McGraw-Hill Encycl. Sci. & Technol.* XI. 317/1 Other radiometers use the pressure of sound waves to deflect a spherical body. *Ibid.* 317/2 The Crookes radiometer survives in jewelers' windows as a 'perpetual motion' device. **1969** *Times* 19 Feb. 13/6 An infra-red radiometer will map the temperature across the surface of Mars. **1973** *Nature* 20 Apr. 506/2 The radiometer was a load-switched superheterodyne microwave receiver. **1977** *R.A.F. News* 11–24 May 11/1 Radiometers to measure infra-red and solar radiation.

3. An instrument for determining the amount of X-radiation administered to a patient.

[**1904** *Sci. Abstr.* A. VII. 362 A 'chromo-radiometer' has been invented by Holzknecht, which consists of two parts—a measurer and a comparative scale.] **1912** *Med. Ann.* 73 The..method of using Holzknecht's new radiometer for the more exact measurement of the x-ray dose. **1918** R. Knox *Radiogr.* (ed. 2) II. 437 The radiometer is so sensitive that it is possible to measure with accuracy the pastille tint when it has not become nearly such a dark colour as in the case of the Sabouraud method. **1934** H. Davies *Pract. X-Ray Therapy* iii. 31 (*caption*) Holzknecht radiometer, showing both halves of the pastille in position.

radiometric (rēˡdiome·trik), *a.* [f. Radio- 2 + *-metric.] **1.** Of or pertaining to the radiometer (sense 2) or its use.

1877 *Chem. News* 12 Jan. 21/2 (*heading*) New radiometric experiments. **1883** [in Dict. s.v. Radiometer]. **1904** *Electrician* 28 Oct. 58/2 Prof. Nichols also suggests a new radiometric receiver, whose action is based on the electrostatic and electrodynamic action between equal resonators. **1927** H. N. Russell et al. *Astron.* II. xxi. 735 With a suitable thermoelectric device the heat radiation of the stars may now easily be measured and the results expressed as 'radiometric magnitudes'. **1964** *Yearbk. Astron. 1965* 111 For Mars, information about ground temperatures may be obtained from radiometric observations, and the amount of insolation is easily calculable. **1977** A. Hallam *Planet Earth* 111 Magnetic, electromagnetic and radiometric techniques are commonly adapted to airborne surveys.

2. Of, pertaining to, or involving the measurement of radioactivity or ionizing radiation; *radiometric dating*, isotopic dating.

1906 *Sci. Abstr.* A. IX. 50 Sabouraud and Noiré..point out that their radiometric value is not in any way diminished. **1938** R. W. Lawson tr. *Hevesy & Paneth's Man. Radioactivity* (ed. 2) xviii. 169 This method of 'Radiometric Microanalysis' has rendered it possible..to carry out a determination of the nitrogen content of organic substances to the order of magnitude of some ten-thousandths of a milligram. **1951** *Engineering* 2 Feb. 148/1 Radiometric methods are frequently used in the investigation of uranium-bearing minerals. **1962** F. I. Ordway et al. *Basic Astronautics* v. 191 Gamma ray scintillation spectrometers make radiometric assays. **1968** *Sci. Jrnl.* Oct. 122/3 *Geologic Time* is concerned with how the present knowledge of the Earth's history has been built up;..palaeomagnetism and continental drift; biostratigraphy and radiometric dating. **1969** *Beaver* Summer 34/2 The coal deposits there are overlaid by volcanic ash determined as 72–73 million years old by radiometric dating processes. **1972** *Science* 2 June 977/1 Radiometric ages obtained for the Apollo 14 examples..cluster around a value of $3{\cdot}9 \times 10^9$ years.

Hence (in either sense) **radiome·trically** *adv.*, by a radiometric method.

1920 *Jrnl. Franklin Inst.* CLXXXIX. 27 The energy was determined in the spectrophotometric laboratory, visually on a König-Martens Spectrophotometer, by a substitution method of comparison with a radiometrically calibrated Mazda lamp. **1962** F. P. W. Winteringham in *Radioisotopes & Radiation in Entomol.* (Internat. Atomic Energy Agency) 117 P^{32} and S^{35} are readily differentiated radiometrically. **1968** *Palaeogeogr., Palaeoclimatol., Palaeoecol.* V. 69 The paleobotanical event has been radiometrically dated at about 12,000,000 years B.P. **1972** *Science* 2 June 978/1 The times of formation of lunar craters have been determined radiometrically. **1973** *Physics Bull.* Apr. 239/3 The centre is equipped with radiometrically and photometrically calibrated sources and a range of calibrated detectors for continuous wave and pulsed radiation.

radiometry (rēˡdiǫ·mėtri). [f. Radio- 2 + -metry.] The use of a radiometer; *spec.* the detection and measurement of infra-red radiation.

c **1890** [in Dict. s.v. Radiometer]. **1906** *Sci. Abstr.* A. IX. 50 They have used this method of radiometry in 5,000 cases of tinea tonsurous without accident. **1923** Glazebrook *Dict. Appl. Physics* III. 718/2 The second method of attack is by means of thermal radiometry with non-selective receivers. **1934** *Discovery* Oct. 282/2 The combination of visual observation at the telescope with the employment at some of the great observatories of the resources of photography, spectroscopy, radiometry, and polarimetry. **1977** I. M. Campbell *Energy & Atmosphere* iii. 44 One of the methods which has been used to measure the temperature profile of the atmosphere as a function of altitude, namely infrared radiometry from satellites.

radio-micrometer, -mimetic, -necrosis: see *Radio- 2.

radionic (rēˡdiǫ·nik), *sb.* and *a.* [f. Radi(o- 2 + *-onic*, after *Electronic *a.*] **A.** *sb. pl.* (const. as *sing.*) † **1.** *U.S.* Electronics, esp. those aspects of electronics connected with radio. *Obs.*

1943 *Radio News* May 4/2 In its simplest form, radionics would be understood by we Americans as being an all-inclusive term of any equipment or science where the use of vacuum tubes is employed [*sic*]. **1943** *Proc. IRE* XXXI. 192 He [*sc.* Hitler] was stopped because the RAF had gone him one better on the new weapon which had paced his early victories, the only new weapon this war has produced: Radionics.

2. The study and interpretation of radiation believed by some to be emitted by and to characterize living and other substances, and to be detectable by skilled use of various complicated electrical instruments.

1954 *Brit. Jrnl. Radiesthesia* I. iii. 18 Radiesthesia..applied to the use of the pendulum, Radionics..applied to instrumental detection and use, and Dowsing, the use of the Divining Rod. **1960** *Times* 21 June 5/4 The plaintiff alleged that the defendant is an exponent of and practitioner in the pseudo-science of radionics, and that..he fraudulently represented that there were associated with substances distinctive waves, vibrations or radiations capable of affecting a device..called a Delawarr Diagnostic Instrument. **1960** *Spectator* 28 Oct. 653 The founder of what today is known as radionics was Dr. Albert Abrams (1863–1924), a..physician who came to believe that the basis of disease was atomic or electronic..and that disease could therefore be treated by giving healthy radiations to neutralise the unhealthy ones. **1969** B. Copen *Radionic Computer Handbk.* 8 Radionics is the science of radiation detection which uses the extra sensory perception of the operator..and automatic instruments, of which this computer is but one. Modern Radionics is a combination of the older Radiesthesia (detection with pendulum) and the more modern semi-automatic instruments (Radionics). **1976** T. Graves *Dowsing* xii. 124 *Radionics*, the specific form of medical dowsing which uses as its instrument a 'box' containing a number of dials in a particular sequence or pattern, seems to be a compound-word formed from 'radiesthetic electronics'—so the word is another product of the tangle over the assumed physical basis of dowsing.

B. *adj.* **1.** Of or pertaining to radionics (sense 1); electronic. orig. *U.S.*

Quot. 1963 represents an independent use not connected with the orig. use of radionics (sense A. 1 above).

1943 *Radio News* May 75/1 Radio News will use 'radionic' in preference to 'electronic' wherever it is more descriptive. **1943** *Proc. IRE* XXXI. 193 This one [*sc.* war], at the front, is run by radio and radionic devices. **1963** *Spectator* 15 Nov. 616/3 Even in this radionic age, a Prime Minister cannot hope to make his way in the country without establishing his mastery over the House of Commons.

2. Of, pertaining to, or practising radionics (sense 2).

1947 *Radiesthesia* III. 58 The development of a new technique in Radionic Diagnosis. **1969** B. Copen *Radionic Computer Handbk.* 54 Any potency of the lower order may be chemically analysed, but the higher potencies are non-analytic in nature, by the orthodox system, but are by the Radionic system. **1972** D. V. Tansley *Radionics & Subtle Anat. of Man* 7 Today, most if not all radionic practitioners would agree that it is their belief that man does have what is referred to as an etheric body. **1975** *Homes & Gardens* Nov. 63/2 Sometimes a diagnostic box helps to make the diagnoses and may also broadcast the treatment. The instrument is not electrically powered but gains its energy from the patient's witness which is put into the box... The radiesthetist determines the diagnosis and then uses the radionic box to broadcast the selected rate of the remedy.

radionuclide: see *Radio- 3 a.

ra:dio-opa·que, *a.* Also **radiopaque.** [f. Radio- 2 + Opaque *a.*] Impervious to X-rays. So **ra:dio-opa·city, radiopa·city,** the state or property of being radio-opaque.

1917 K. Thoma *Oral Roentgenol.* iv. iv. 197 Unsanitary concretions, such as salivary..calculi, can be seen on account of their radiopacity. **1926** *Jrnl. Amer. Med. Assoc.* 19 June 1883/1 In the stenotic larynx, the extent of the destruction may be visualized by the introduction of radiopaque substances. **1940** *Brit. Jrnl. Radiol.* XIII. 261/2 To determine the depth of the foreign body, the observer places a radio-opaque guide-mark on the skin..; then he moves the X-ray tube horizontally. **1961** *Dental Progr.* I. 178/2 Radio-opacity of the palatal vault blotted out the apices of the maxillary teeth. **1970** H. McLeave *Question of Negligence* xxvii. 228 They had trundled in a portable X-ray machine..injected a radio-opaque substance into a neck artery. They ran off six plates. **1971** G. H. Bourne *Ape People* vii. 186 A series of cinemagraphs were taken with radiopaque material being injected into different chambers of the heart. **1971** *Brit. Dental Jrnl.* CXXX. 430/1 The localised radiopacity caused by the superimposition of shadows can be very misleading, therefore patients should be instructed to remove their ear-rings before the picture is taken.

radio-pasteurization, -ized: see *Radio- 2.

radiophare (rēˡ·diofɛ̄ˡɹ). [f. *Radio- 4 + Phare.] A navigational radio beacon.

1915 W. H. Eccles *Wireless Telegr. & Telephony* 362 The French Government is installing wireless lighthouses or radiophares round the French coast. **1922** *Nature* 20 May 650/1 Until two or three years ago the radiophares—or radio-beacons as they are called in America—were purely stations for giving ships their positions. **1966** *McGraw-Hill Encycl. Sci. & Technol.* IV. 231/2 Radio stations erected specifically for use with the ADF are known as NDB (non-directional beacons or radiophares).

radiopharmaceutical, -pharmacist, -pharmacology (etc.), **-pharmacy:** see *Radio- 2.

radiophone. Add: **2.** Also **radio-phone, radio phone.** = *Radio-telephone.

1919 *Wireless World* May 105/2 (*heading*) Radiophones over London. **1922** *Sci. Amer.* Sept. 160/1 The receiver can be worked on very short waves, well below 200 meters, thus opening up a new field of wave lengths for radio-phone broadcasting. **1926** *Popular Radio* IX. 91 (*caption*) The first radiophone booth on an ocean liner. **1940** N. Monks *Squadrons Up!* ii. 39 The boys [*sc.* pilots] wisecracked to each other into their radio-'phones. **1971** J. Brunner *Honky in Woodpile* xii. 90 A luxuriously-equipped Mercedes convertible—it even boasted a radiophone. **1978** R. Ludlum *Holcroft Covenant* xxviii. 322 We use a radiophone off Cap Camarat.

radiophonic, *a.* Add: **2.** Pertaining to or designating synthetic sound produced by electronic means and the use of tape recorders, usually for use in broadcasting in conjunction with conventional material.

1958 *Times* 24 May 10/2 B.B.C.'s Radiophonic Workshop. A 'workshop' for producing synthetic sounds, partly by electronic oscillators and partly by trickery with conventional sounds recorded on tape has been set up by the

B.B.C. at their Maida Vale studios. **1958** *Observer* 22 June 14/3 He [*sc.* Patrick Magee] must be the only actor who can sound as if he's talking through a radiophonic filter. *Ibid.* 6 July 14/2 Those radiophonic plays where an old woman's memories or a young man's nightmares are made the slender excuse for the latest in ghost-train noises. **1960** *BBC Handbk.* 67 Special effects..created..for the occasion by 'radiophonic' devices. **1961** *Listener* 16 Nov. 834/1 Michael Bakewell employed radiophonic effects to communicate the variant senses of time experienced by the normal and the mentally ill characters in this double bill. **1972** *Ibid.* 21 Dec. 872/3 Not a *Goon Show* script, but..one of Michael Mason's radiophonic workshop larks.

Hence **radiopho·nicist,** an exponent of radiophonic sound; **radiopho·nics** *sb. pl.*, the production and use of radiophonic sound; the sounds themselves; **radio·phonist** = *radiophonicist* above.

1958 *Listener* 28 Aug. 319/3 Our local radiophonicists and 'concrete' music men have too often seized upon noise in the higher octaves to drive home dramatic nails that would have seemed excruciating even to Jael. **1962** A. Nisbett *Technique Sound Studio* xii. 203 Radiophonics does not in general attempt to assert itself as an art form in its own right; it is always an element in a larger picture. **1963** F. C. Brooker (*title*) Radiophonics in the B.B.C. **1976** *Listener* 21 Oct. 511/3 The tinkling celeste tune—now, alas, abandoned in favour of radiophonics—that used to introduce *Listen with Mother.* **1977** *Listener* 18 Aug. 217/2 What was crushing to the fond memory was the way the poverty had affected the radiophonics, too. **1977** *Times* 3 Sept. 10/3 Isaac Asimov's *Foundation Trilogy*..on Radio 4 certainly provided a field day for the radiophonists.

radio-phosphorus: see *radio- 3 b (i).

radiophoto (rē^i·diofoto). [f. *Radio- 4 + Photo.] A photograph transmitted by means of radio.

1942 *Wireless World* Sept 217/2 The first..radiophoto service between America and Egypt was recently opened. **1961** *New Scientist* 13 July 91/2 If the radiophotos show recognizable evidence of hurricane breeding in the clouds, hurricane-hunting planes will be sent out. **1971** *Encycl. Americana* XXIII. 121n/1 The channel capacity of an HF system is limited to four voice or radiophoto channels.

radiophoto·graphy. Also with hyphen. [f. *Radio- 2 + Photography.] **1.** = *Radiography 1 (see also quot. 1939).

1897 *Chem. News* 26 Feb. 102/2 (*heading*) Radiophotography of the soft parts of man and the lower animals. **1934** *Jrnl. R. Anthrop. Inst.* LXIV. 70 It may be demonstrated by radiophotography that bronze, on exposure to the atmosphere, immediately begins to oxidize. **1939** *Jrnl. Amer. Med. Assoc.* 6 May 1844/1 Radiophotography, or indirect radiography, consists in simultaneously photographing the thoracopulmonary images as they are observed at radioscopic examination.

2. The transmission of photographs by means of radio.

1915 *Wireless World* Apr. 60/1 The number of interruptions per second required is very high, as in radio-photography. *Ibid.* 60/2 A system of radio-photography.. would be of great military use.

Hence **radiopho·tograph** = *Radiograph *sb.* 2; (in either sense) **ra:diophotogra·phic** *a.*

1897 *Chem. News* 26 Feb. 102/2 We have been able to place the muscles, the ligaments, and the tendons in such a state that they have yielded radio-photographic images. **1915** *Wireless World* Apr. 60/2 Even were they possessed of radio-photographic apparatus, the received message would be unintelligible unless they knew the exact speed. **1924** *Times* 13 Sept. 12/5 In this section are some radio-photographs showing considerable advance on any hitherto.

radiophysics (etc.): see *radio- 4. **radio-potassium:** *radio- 3 b (i). **radio-protection, -protective, -protector:** *radio- 2. **radio-purity:** *radio- 3 a.

ra·dioreceptor. *Physiol.* [f. Radio- 2 + Receptor.] †**1.** Written **radio-receptor.** A sensory receptor which responds to electromagnetic radiation of any kind. *Obs.*

1927 [see *mechanoreceptor]. **1935** *Brit. Jrnl. Psychol.* XXV. 266 As Parsons describes, the receptors differentiated along three main lines, chemo-, radio-, and mechano-receptors.

2. *radioreceptor assay*, a radiological assay for hormones in which the test sample, together with a known quantity of radioactively labelled hormone, is allowed to bind to a standard preparation containing receptor sites.

1973 *Science* 1 June 968 A radioreceptor assay with a sensitivity of 5 nanograms per milliliter has been developed for mammalian and avian pituitary prolactin, placental lactogenic hormones, and human growth hormones, using a membrane receptor preparation isolated from rabbit mammary glands. **1974** *Nature* 29 Mar. 436/2 At this time the PL concentrations measured by radioreceptor assay were almost entirely due to the cross reaction of pituitary prolactin in the receptor assay. **1978** *Ibid.* 20 Apr. 730/2 The fact that affinity to rat opiate receptors was demonstrated in the plasma using a radioreceptor assay..indicates that a loss of the postulated biological activity by metabolic or other processes can be excluded.

radio-resistance, -ant: see *radio- 2. **radioscan:** *radio- 3 a.

radioscopy. Add: *spec.* = *fluoroscopy (earlier and later examples); **radioscopic** *a.* (earlier and later examples); also †**ra·dioscope** = *fluoroscope.

1897 [see *radiotherapeutic *sb. pl.* and *a.*]. **1897** *Treatment* I. 43/2 It makes his own heart beat faster as the observer sees for the first time thrown upon the radioscopic screen a living heart in action. **1898** *Amer. Jrnl. Med. Sci.* CXV. 464 The capsules are readily seen (in thin persons) with the radioscope. **1908** *Chem. Abstr.* II. 229 (*heading*) The radioscope and the radiograph applied to the inspection of tubercular meats. **1915** R. Knox *Radiogr.* 188 Radioscopy, or the examination of a patient with the fluorescent screen, is a method of great value, as a diagnosis can often be made from it alone, to be subsequently confirmed by radiographic exposures. **1928** B. J. Leggett *Theory & Pract. Radiol.* III. x. 459 (*heading*) Radiographic and radioscopic rooms. **1937** M. Campbell in *Brit. Encycl. Med. Pract.* VI. 352 Radioscopy is of great help; the presence of a large left ventricle or of a dilated left auricle in mitral stenosis,..may clinch a doubtful diagnosis. **1946** *Surg. Clinics N. Amer.* Oct. 1286 Such long radioscopic exposures are only possible without danger of dermatitis when a very small radioscopic field is used. **1979** *SLR Camera* Sept. 5/1 Now the holiday season is on us perhaps a word, or two, about the precautions needed to safeguard films from damage by radioscopic screening is pertinent.

radiosensitive(ness, etc.: see *radio- 2. **radio-silver, -sodium:** *radio- 3 b (i).

radiosonde (rē^i·diosǫnd). *Meteorol.* Also **radio-sonde, radio sonde.** [a. G. *radiosonde* (P. Moltchanoff 1931, in *Beiträge z. Geophysik* XXXIV. 36), f. *radio-* Radio- + *sonde* probe, sounding-line.] A small package of meteorological instruments which is carried through the atmosphere by balloon or other means and automatically transmits measurements of conditions at various heights by radio. Freq. *attrib.* So **radiosondage** (-sǫ·ndēdʒ), sounding of the atmosphere by radiosonde.

1937 *Geogr. Jrnl.* XC. 381 The use of the radio-sonde, from which automatically transmitted W/T signals can be transformed into data of temperature and pressure. **1939** *Meteorol. Gloss.* (Meteorol. Office) (ed. 3) 153 Radio-sondages (Radio-soundings). **1940** *Manch. Guardian* 30 Jan. 6/6 The staff of our Meteorological Office..receives great help from an instrument called the radio sonde, which is sent up attached to a small balloon and automatically reports by wireless the air conditions up to a great height. **1946** *Electronics* May 123/3 Several methods of radiosonde tracking have been used in order to determine the speed and direction of the wind at various altitudes. **1948** *Times* 23 Feb. 6/7 The radio-sonde equipment..was insufficiently sensitive to detect the rapid temperature changes within the first few hundred feet of the sea surface. **1951** T. F. Malone *Compendium Meteorol.* 1215/2 The parachute radiosonde was designed to be launched from a weather-reconnaissance plane. **1955** W. Girvan *Flying Saucers & Common Sense* ix. 110 The weather-balloon..crossed the coast near Eastbourne and then burst, the radio-sonde falling into the Channel at approximately 3.30 p.m. after descending slowly by parachute. **1959** H. R. Byers *Gen. Meteorol.* (ed. 3) vi. 101 Several types of rawinsonde systems, combining radiosondage with tracking of the radiosonde to get wind drift of the balloon, have been developed. **1960** *Times* 19 July (Royal Society suppl.) p. x. 6 Preparations are being made to launch a hydrogen-filled balloon carrying a radiosonde transmitter. **1976** B. Jackson *Flameout* (1977) xii. 205 The radiosonde balloons from the unnamed Air Force base in Nevada.

radiostereoassay: see *radio- 3 a. **radio-sterilization, -ized:** *radio- 2.

Radiostol (rē^i·diostǫl). Also **radiostol.** [f. Radio- 2 + *-stol.* (f. *ergosterol from which calciferol can be prepared by irradiation with light).] A proprietary name for a preparation of calciferol (vitamin D_2).

1927 *Trade Marks Jrnl.* 27 Apr. 737/1 *Radiostol*... Chemical substances prepared for use in medicine and pharmacy. The British Drug Houses, Limited. **1928** *Daily Express* 27 Jan. 3/3 It is possible to give margarine the rich food value of the best butter by treating it with radiostol. **1934** *Nature* 2 June 821/2 Modifications [in the diet]..were additions of oatmeal, olive oil, cod liver oil or radiostol (irradiated ergosterol), and milk, [etc.]. **1967** *Martindale's Extra Pharmacopoeia* (ed. 25) 258/2 Radiostol (*British Drug Houses*). Calciferol, available as Capsules..; as a Solution [etc.].

radio-strontium: see *radio- 3 b (i). **radio-surgery, -surgical:** *radio- 2.

radio-te·legram. [f. *Radio- 4 + Telegram.] A telegraphic message sent by radio.

1902 *Sci. Amer. Suppl.* 15 Nov. 22474/1 The radio-telegrams arrived from Poldhu constantly and surely at the receiver on board the 'Carlo Alberto'. **1904** *Prelim. Conf. Wireless Telegr. Berlin 1903* 16 We desire to grant to existing systems a fair share of the charge to be collected for radio-telegrams. **1913** *Wireless World* Aug. p. xxx (Advt.), Hand your message in at any Telegraph Office, where full particulars concerning radio-telegrams can be obtained. **1920** *Glasgow Herald* 2 Apr. 5/7 The charge for such radio-telegrams is 10½d. per word.

radio-te·legraph. Also **radiotelegraph.** [f. *Radio- 4 + Telegraph *sb.*] A means of sending telegraphic messages by radio rather than along a wire. Usu. *attrib.* So **ra:dio-telegra·phic** *a.*, **-telegra·phically** *adv.*, **-tele·graphist** (all *rare*); **radio-tele·graphy,** wireless telegraphy.

1898 J. Munro in *Electrician* 21 Jan. 428/2 'Wireless telegraphy' is not a bad technical term; but if a more scientific name be desirable would not Radiotelegraphy or Ray Telegraphy be preferable to 'Space Telegraphy'? which Dr. Lodge employs. **1902** *Nature* 25 Sept. 538/2 The creation of a radio-telegraphic station communicating with the stations established..by the Marconi companies in London and in America. **1903** *Ibid.* 23 Apr. 590/1 The establishment on the coast and on the islands off the Italian coast of a system of twelve Marconi radio-telegraph stations of an average range of 300 kilometres. **1906** *Westm. Gaz.* 19 Dec. 2/2 Wireless telegraphy, or 'radio-telegraphy', as it is more technically called. **1907** *Athenæum* 14 Sept. 308/1 The Report and Evidence of the Radiotelegraphic Convention Committee. **1907** *Daily Chron.* 11 Nov. 6/6 The Amalgamated Radio-Telegraph Company, Limited. **1908** J. A. Fleming *Radiotelegr. & Radiotelephony* p. vi, Expositions of electrical phenomena which are..unnecessary to the practical radiotelegraphist. **1918** *Wireless World* June 192 (*heading*) Senatore Marconi radiotelegraphically expresses the national sentiment of Italy. **1921** *Jrnl. R. Soc. Arts* 9 Dec. 68/2 Marconi..was able to dispatch wireless messages across the Atlantic which made long-distance radio-telegraphy a demonstrated achievement. **1945** *Sun* (Baltimore) 23 Oct. 1/3 (*heading*) Radio telegraph plan announced. **1955** *Times* 12 Aug. 5/3 A feature of the year has been the development in leased channel operations, that is, in the provision of direct customer-to-customer private radio-telegraph circuits. **1966** *McGraw-Hill Encycl. Sci. & Technol.* I. 364/2 Much of the radiotelegraph traffic of the world uses AM telegraphy, although there had been extensive conversion to frequency-shift (frequency-modulation) telegraphy since 1944. *Ibid.*, Most aviation and marine radiotelegraphy uses AM manual methods.

radio-telemetering, -telemetric, -y: see *radio- 4.

radio-te·lephone. Also **radiotelephone, radio telephone.** [f. *Radio- 4 + Telephone *sb.*] A transmitting and receiving set for radio-telephony.

1909 *Daily Chron.* 15 Feb. 1/2 Mr. Lee De Forest,..whose radio-telephone system has been adopted by the American navy. **1922** *Encycl. Brit.* XXXII. 1027/2 Compact lightweight sets of radio-telephone transmitters and receivers. **1935** H. G. Wells *Shape of Things to Come* 15 There will be a radio telephone arrangement on his chest no more obtrusive than a modern breast pocket. **1957** *Practical Wireless* XXXIII. 517/1 Cable and Wireless Ltd. announce that a radiotelephone service has been opened between Saudi Arabia and Bahrein. **1966** P. O'Donnell *Sabre-Tooth* xiv. 187, I will arrange for you to speak to her briefly over the radio-telephone. **1973** *Sat. Rev. Society* (U.S.) Mar. 52/2 An immaculately uncluttered teakwood desk, flanked by..two telephones, a radio-telephone and an intercom.

So **ra:dio-telepho·nic** *a.*; **ra:dio-telepho·nically** *adv.* (*rare*); **radio-tele·phony,** telephony in which the signal is transmitted by radio over part of the route; wireless telephony.

1908 J. A. Fleming (*title*) Radiotelegraphy and radio-telephony. *Ibid.* 325 We have..in the combined radio-telephonic transmitter and receiver, a wonderful transformation of energy. *Ibid.* 329 Transmitting speech radio-telephonically from Paris..to Dieppe. **1923** *Radio Times* 28 Sept. 12/1 We are most enthusiastic admirers of this remarkable radio-telephonic invention. **1930** *Daily Express* 8 Sept. 3/7 The principal means of communication; both in the brigade and battalions, will be radio-telephony. **1935** *Times* 12 Nov. 20/5 It is a strange thing that in radio-telephony across the Atlantic..some wave-length, which has been behaving admirably, will rather swiftly fade and fail. **1966** *McGraw-Hill Encycl. Sci. & Technol.* I. 364/1 Since radio is the only way to communicate with ships and aircraft, hf AM radiotelephony has remained essential to these operations. **1974** *Guardian* 22 Mar. 13/7 Will you see to it that the cars used by public figures are fitted with..radio-telephonic methods of communication?

radio-teletype, -tellurium: see *radio- 4, 3 b (i).

ra:diotherapeu·tic, *sb. pl.* and *a.* Also with hyphen. [f. Radio- 2 + Therapeutic *a.*] **A.** *sb. pl.* = *radiotherapy. **B.** *adj.* Of, pertaining to, or employing radiotherapy. Hence **ra:diotherapeu·tically** *adv.*; †**ra:diotherapeu·tist** = *radiotherapist.

1897 *Treatment* I. 43/1 (*heading*) Radiography, radioscopy, and radiotherapeutics. **1904** *Bristol Medico-Chirurg. Jrnl.* XXII. 43 What this variation is is a question for the radio-therapeutists to solve. **1906** *Arch. Roentgen Ray* XI. 20/1 It is to be hoped that in radiotherapeutic practice the use of the dosimeter may become universal. *Ibid.* 20/2 Radio-therapeutists are divided into..those who believe in the large (so-called) measured dose, and those who administer small unmeasured doses at intervals. **1932** *Discovery* Aug. 255/2 The report urges the establishment of radio-therapeutic centres associated with large general hospitals. **1937** *Amer. Jrnl. Obstetr. & Gynecol.* XXXIV. 50 In the surgically or radiotherapeutically induced menopause, much can be done by prophylactic psychotherapy. **1975** *Nature* 13 Mar. 97/1 The Medical Research Council is to set up a new unit for research in clinical oncology and radio-therapeutics at Cambridge. **1977** *Proc. R. Soc. Med.* LXX. 591/2 This is a very comprehensive review of the present state of knowledge of the investigation and treatment, medical, surgical and radiotherapeutic, of these lesions.

radiothe·rapy. [f. Radio- 2 + Therapy.] The treatment of disease by means of X-rays or other forms of ionizing radiation.

1903 *Boston Med. & Surg. Jrnl.* CXLIX. 325/1 He had been interested in comparing the effects of phototherapy and radiotherapy. **1904** *Westm. Gaz.* 29 Dec. 2/1 A working knowledge of the technique of radio-therapy. **1955** *Bull. Atomic Sci.* Oct. 287/3 These will become available in source strength varying from kilocurie sources of radio-caesium for use in radiotherapy to megacurie sources for industrial applications. **1970** *Sci. Jrnl.* Mar. 62/3 Where a cancer has become disseminated to other areas, treatment with drugs or radiotherapy is used. **1974** 'H. Carmichael' *Motive* iv. 47 Dr. Egan will see you. He's in the Radio-Therapy department.

Hence **radiothe·rapist,** one who practises radiotherapy.

1918 R. Knox *Radiogr.* (ed. 2) II. 388 It is easy..to understand the complexity of the problem which confronts the radiotherapist in dealing with morbid growths. **1934** *Lancet* 29 Sept. 697/1 The patient should..be examined in consultation with an expert radiotherapist before any operation on the primary tumour or glands. **1976** *Ibid.* 6 Nov. 992/1 The introduction of improved techniques has enabled radiotherapists to treat relatively large volumes of the body with comparative safety.

radiothon (rē^i·dioþǫn). *U.S.* [f. *Radio *sb.* + *-a)thon.] A prolonged radio broadcast by a person or group, usu. as a fund-raising event.

1964 *Richmond* (Va.) *Times-Dispatch* 26 Jan. 6B/1 With only an hour to run, the radiothon had netted the March of Dimes here more than $1,200. **1974** *State* (Columbia, S. Carolina) 15 Feb. 20-A/3 A 15-day try for a world record radiothon is being made by WCOS-AM disc jockeys to benefit the Heart Fund. **1976** *Long Island Traveler-Watchman* 8 July 1/3 (*caption*), The radiothon, to raise funds for Kiwanis youth programs, will be broadcast WRIV.

radiothorium: see *radio- 3 b (i). **radio-toxic(ity, -tracer:** *radio- 3 a. **radiotranslucency, -ent:** *radio- 2.

ra·diovision. Also with hyphen. [f. *Radio- 4 + Vision *sb.*] **1.** The combination of a radio programme with a specially prepared film strip or series of slides, esp. as an educational aid. Freq. *attrib.*

1964 *Guardian* 14 Feb. 7/2 The technique..is to issue film strips to accompany selected broadcasts; it is becoming known as 'radiovision'. **1965** *B.B.C. Handbk.* 70 The use of radio-vision in the teaching of languages was the subject of an experiment in fifty schools in the autumn of 1963. **1966** *Listener* 24 Sept. 786/2 In a radiovision talk of great virtuosity..Mr Morris explained the nature of an atmospheric depression. **1971** *Daily Tel.* 4 May 2/2 The radio-vision programmes were more easily taped and film-strips could be stored for parent evenings.

2. Radio broadcasting accompanied by the simultaneous showing of the broadcasters on television.

1980 *Daily Tel.* 6 Nov. 19/4 Miss Monica Sims, controller of Radio 4 and a former head of children's television, has been asked to chair an immediate feasibility study on the possibilities of 'radiovision' and report by next spring to Sir Ian Trethowan, B B C director-general.

radish. Add: **2. radish communist,** one who professes communism but is not sincerely devoted to it. Also *ellipt.*

1920 *Times* 31 Oct. 11/1 A 'radish' is a man who fervently professes devotion to the Communist cause while harbouring a secret longing for its overthrow. Red outside, but white.. inside. The epithet was invented by Trotsky. **1966** *Listener* 29 Sept. 445/1 Stalin would speak disparagingly of Mao's men as being 'not real communists', mere 'margarine communists', 'radish communists'—red on the outside and white on the inside.

radishy (ræ·diʃi), *a.* [f. Radish + -y^1.] Resembling or suggestive of a radish.

1861 H. Mayhew *London Labour* III. 64/2 The matches were tied..to his [*sc.* the guy's] radishy and gouty fingers.

radium (rē^i·diǝm). [f. L. *radius* ray, Radius: see -ium.] **1.** *Chem.* **a.** [a. F. *radium* (P. Curie et al. 1898, in *Compt. Rend.* CXXVII. 1217).] A radioactive element, chemically a member of the alkaline earth metals, which occurs in small amounts in uranium ores, notably pitchblende; atomic number 88, symbol Ra.

1899 *Chem. News* 6 Jan. 1/2 These different reasons lead us to believe that the new radio-active substance contains a new element, to which we propose to give the name of radium. **1903** *Daily Mail* 11 Sept. 3/1 All the speakers recognised that the discovery of radium, with its apparent power of emitting heat for ever without diminution, has opened the door to something like a new world of science. **1904** *Daily Chron.* 7 Jan. 5/1 It is quite as good as any other assertion to say that an ounce of radium is worth the British Empire; no more having yet been obtained than about the weight of a lump of sugar. **1909** 'O. Henry' *Roads of Destiny* xxi. 358 'Change the treatment,' says I... 'Call a consultation or use radium or smuggle me in some saws or something.' **1933** Bowing & Fricke in O. Glasser *Sci. of Radiol.* xv. 281 The paramount advantage of the use of radium in medicine..consists in the proved fact that the rays of radium have a selective action on cancer cells. **1950** *Thorpe's Dict. Appl. Chem.* (ed. 4) X. 446/1 The history of radium refining during the 50 years, 1898–1948,..covers the rise of radium from a scientific curiosity to a commodity of almost fabulous value and wide importance, and its subsequent relegation to a minor role following the development of the atomic pile. **1958** *Daily Express* 11 Mar. 7/1 An escape of radium at a hospital led to the dumping of tons of material down a disused pit shaft. **1974** *Encycl. Brit. Micropædia* VIII. 382/1 Metallic radium has high chemical reactivity. It dissolves in water with vigorous evolution of hydrogen.

b. (Followed by capital letter.) Designating substances (mostly radioactive) subsequently identified as isotopes of other elements, which are formed successively in the radioactive series of radium: *radium A*, polonium 218; *radium B*, lead 214; *radium C*, bismuth 214 together with some polonium 214 (*radium C_1* or *C′*) and thallium 210 (*radium C_2* or *C″*); *radium D*, lead 210; *radium E*, bismuth 210; *radium F*, polonium 210; *radium G*, lead 206, the non-radioactive end-product of the series.

The substances now designated *radium E* and *F* were in the first instances named *radium D_1* and *E* respectively.

1904 E. Rutherford in *Phil. Mag.* VIII. 636 For convenience, the products in the active deposit will be termed Radium A, Radium B and Radium C, respectively. *Ibid.* 641 Following the nomenclature suggested, radium C gives rise to the β ray product, which will be called Radium D, while radium D changes into the α ray product, which will be called Radium E. **1905** —— in *Nature* 9 Feb. 342 In order to avoid confusion, I have called the new radium product 'radium D_1'. If no further intermediate products of radium are brought to light, it would be simpler to call it radium E and to call the α ray product (polonium) radium F. **1905** —— in *Phil. Mag.* X. 293 This rayless product will be called radium D. The β ray product which arises from it will be called radium E. The α ray product (previously termed radium E) will be called Radium F. **1910** *Westm. Gaz.* 16 Feb. 4/1 Sir William Ramsay's description..of 'Radium D', one of the mysterious products of radium, as 'rather dull-looking, like lead'. **1911** *Phil. Mag.* XXII. 628 Fajans has suggested that the name radium C_2 should be given to the new product of period 1·4 minutes. **1913** *Nature* 28 Aug. 659/2 That radium-G and lead are identical is supported by much indirect evidence, though no direct proof has been advanced. **1933** Failla & Quimby in O. Glasser *Sci. of Radiol.* xiii. 249 In about a month the equilibrium amounts of radon and radium A, B, and C will have accumulated, and the preparation has a maximum beta and gamma ray activity. **1936** *Discovery* July 218/2 The radium E was produced..through the bombardment of bismuth, with deuterons at an energy of 5,500,000 electron volts. **1966** R. Oliver *Radiation Physics in Radiol.* vi. 71 Radium then disintegrates to form radon which is a gas at normal temperature and pressure. There follows a long series of disintegrations through nuclides referred to as radium A (RaA), B, C, C′, C″, D, E, F and G. The last nuclide (radium G) is a stable isotope of lead. **1974** *Encycl. Brit. Micropædia* VIII. 382/1 A sample of radium, sealed to prevent the escape of gaseous radon, reaches radioactive equilibrium in about a month and becomes a powerful source of gamma radiation, due especially to radium C (bismuth-214).

2. A smooth, plain fabric with the sheen of silk (see quots.).

1904 *Daily Chron.* 12 Mar. 8/5 The newest ribbon is radium, and what a future lies before it! **1930** *Daily Express* 6 Oct. 5/2 (Advt.), Radium velvet. Rich quality panne velvet giving a fashionable brilliance for afternoon and evening wear. **1957** M. B. Picken *Fashion Dict.* 269/1 *Radium*, smooth, pliable lingerie fabric of synthetic yarn in plain weave, with dull finish and natural sheen of silk. **1970** R. T. Wilcox *Dict. Costume* 281/2 *Radium*, a lustrous, plain, smooth silk or rayon, which has crispness, yet supple, draping quality.

3. a. *attrib.* and *Comb.*, as (sense *1) *radium atom, bromide, chloride, salt, sulphate, treatment*; (sense *2) *radium poplin, silk, velvet*; *radium-bearing, -coloured* adjs.

1903 *Daily Mail* 11 Sept. 3/1 Illustrating the amazing properties of a radium atom. **1926** R. W. Lawson tr. *Hevesy & Paneth's Man. Radioactivity* xxiii. 161 St. Joachimstal in the Erzgebirge is the most important source of radium-bearing ores in Europe. **1904** *Chem. News* 24 June 301/1 As radium bromide yields electrolytic gas, containing an excess of hydrogen, the pressure gradually rose. **1913** *Med. Ann.* 647 Each capsule contains ·0002 mgram radium bromide. **1966** P. Astbury tr. *G. Amaldi's Nature of Matter* iv. 125 Pierre and Marie Curie..had to examine about seven tons of pitch-blende..in order to prepare a single gram of radium bromide. **1902** *Harper's Mag.* Aug. 360/2 M. Curie possesses about two to three hundredths of a gram of chemically pure radium chloride. **1904** *Daily Chron.* 2 Sept. 8/3 A magnificent mantle fully trained and made of radium-coloured taffetas mousseline, the blue, pink, and moonlight shades of which mingle with one another. **1952** C. W. Cunnington *Eng. Women's Clothing* 296 *Radium poplin*, a silk and wool textile looking like a silk poplin. **1905** Kipling *Actions & Reactions* (1909) 121 An expense of one hundred and seventy-odd pounds..for radium salts and such trifles. **1956** J. K. Robertson *Radiology Physics* xii. 228 When radium needles are used, the dose delivered to the region treated depends on..the amount of absorption by the materials enclosing the radium salt. **1972** Barnes & Rees *Conc. Textbk. Radiotherapy* viii. 178 Radium is prepared as the salt, radium sulphate, and mixed with a suitable filler, it is sealed into thin-walled metal capsules. **1936** *Discovery* Nov. 351/2 Apparatus for radium treatment recently installed at the Hampstead annexe of Westminster Hospital. **1930** Radium velvet [see sense 2 above].

b. Special Combs.: **radium beam,** a beam of gamma radiation from a radium source, used in radiotherapy; **radium bomb,** a container holding a large quantity of radium and used in radiotherapy as a source of a gamma ray beam; **radium burn,** a burn caused by over-exposure to radiation from radium; **radium clock,** a device utilizing the β-rays of radium to charge two electroscopic leaves, which discharge at regular intervals when they diverge sufficiently to touch two earthed metal plates; **radium emanation,** the radon isotope of atomic weight 222, which is the first product of the radioactive decay of radium; cf. *emanation 2 c; **radium needle,** a needle containing radium which can be inserted into tissue for radiotherapy; **radium plaque** = *plaque 3; **radium therapy,** radiotherapy using radiation from radium.

1933 *Jrnl. Amer. Med. Assoc.* 12 Aug. 533/2 The work before the radium beam therapy research will be to..discover how far the present limited field of operation for a mass radium unit..may be extended. **1940** *Ibid.* 16 Mar. 999/1 Treatment of carcinoma of the mouth and throat by the radium beam is at least as satisfactory as by surgery or interstitial radium. **1956** J. K. Robertson *Radiology Physics* xii. 227 In radium beam therapy, the source contains several grams of this radioactive element. **1930** *Brit. Med. Jrnl.* 19 July 98/1 An attempt to measure accurately the gamma radiation field, in air, of the radium 'bomb' in use at Westminster Hospital annexe. **1952** W. M. Levitt *Handbk. Radiotherapy* iii. 42 It is probable that the radium bomb can do nothing that super-voltage x-ray therapy cannot do at least as well and little that deep x-ray therapy cannot do. **1908** *Jrnl. Amer. Med. Assoc.* 30 May 1871/1 (*heading*) Radium burns of the skin. **1926** R. W. Lawson tr. *Hevesy & Paneth's Man. Radioactivity* xxv. 211 Radium burns are especially troublesome, because..even if they mend there remains a supersensitiveness of the skin. **1940** S. Cade *Malignant Dis. & Treatm. by Radium* 246 Radium burns of the skin or mucous membrane may take 8 to 10 months to heal. **1905** W. Hampson *Radium Explained* 69 We are now able to understand the action of the radium clock. **1905** E. Rutherford *Radio-Activity* (ed. 2) iv. 123 This 'radium clock' should work at a sensibly uniform rate for many years, but..the number of β particles emitted would decrease exponentially with the time, falling to half value in about 1200 years. **1926** R. W. Lawson tr. *Hevesy & Paneth's Man. Radioactivity* xxv. 211 The so-called 'radium clock'..illustrates how the rays from radium may be utilized indirectly to give rise to motion. **1901** *Nature* 13 June 157/2 The radium emanation..preserved its radiating power for several weeks. **1910** *Daily Chron.* 17 Feb. 3/3 Take the next product, the gas which it is continually giving off, and which is called radium emanation. That can be worked with because it is only half gone in four days. **1946** F. E. Zeuner *Dating Past* x. 318. As radium decays, a gas called radium-emanation is formed which, in turn, emits another atom of helium and thereupon changes into a solid substance, called radium A. **1921** *Pennsylvania Med. Jrnl.* XXIV. 218/1 Radium needles should be inserted directly into the glands. **1959** C. L. & J. A. Martin *Low Intensity Radium Therapy* iii. 33 Low intensity radium needles still seem to provide the best available medium for interstitial therapy. **1960** A. Huxley *Let.* 20 May (1969) 890 For cases like mine, radium needles are now standard procedure at the University of California. **1919** *Amer. Jrnl. Roentgenology* VI. 134/1 It is..a very simple matter to treat a small lesion with a radium plaque. **1962** J. Thewlis *Encycl. Dict. Physics* VI. 171/1 Formerly, radium plaques were used as β-emitters, but these have now been almost completely replaced by applicators containing strontium-90. **1904** *Med. Electrol. & Radiol.* V. 336 (*heading*) Radiumtherapy. **1905** *N. Amer. Jrnl. Homœopathy* LIII. 720 It would seem wise to postpone judgment as to the worth of radium therapy in dermatology. **1931** G. B. Shaw *Doctors' Delusions* 28 Nobody would dream of excluding radium therapy from the medical curriculum merely because more nonsense has been written about radium than about the philosopher's stone. **1964** L. Deighton *Funeral in Berlin* xxix. 152 Began work as a representative [for] radium therapy machinery 1948. Assigned to Northern Spain as radium therapy equipment salesman 1949.

radiumize (rē^i·diǝmaiz), *v.* [f. *Radium + -ize.] *trans.* To subject to the action of radium. Hence **ra·diumized** *ppl. a.*

1906 *Jrnl. Amer. Med. Assoc.* 21 July 184/2 Two fine examples of spindle-celled sarcoma..have been thoroughly radiumized. **1914** *Chambers's Jrnl.* Oct. 688/2 The process of radiumising the soil. **1920** *Amer. Jrnl. Roentgenology* VII. 54/1 Such radiumized tissue will not grow when inoculated in mice. **1928** *Daily Tel.* 10 July 9/5 Work..is being pursued upon the immunity conferred upon rats and mice by X-rayed or radiumised tumour tissues. **1939** *Jrnl. Amer. Med. Assoc.* 1 Apr. 1280/1 (*heading*) Radiumized health pad.

radius. Add: **3. d.** *radius of action*: in *Aeronautics*, the distance that an aircraft can cover so as to leave sufficient fuel for its return to base.

1908 *Aeronaut. Jrnl.* Apr. 44/2 In a 10 hours' run a distance of 300 kilm. (186 miles) might be traversed, the radius [of] action being 150 kilm. **1918** *Flying* 20 Mar. 188/3 The problem of any extension of the radius of action is almost entirely one of increased efficiency..of modern aircraft.

5. radius rod (further examples).

1907 [see *chain case* s.v. *chain *sb.* 19]. **1946** *Happy Landings* July 2/1 The inner radius rod of the starboard undercarriage was pulled away from the undercarriage. **1970** K. Ball *Fiat 600, 600D Autobk.* 165/1 *Radius rod*, pivoted arm confining movement of a part to an arc of fixed radius.

radius (rē^i·diǝs), *v.* [f. the sb.] *trans.* To round off, make (a corner or end) curved.

1938 J. Healey *Metal Aircraft* ii. 14 Tubular rivets are a reamer fit, so having reamered the hole to size, radius the edge of the metal slightly. **1954** *Electronic Engin.* XXVI. 538/1 All corners and bends should be 'radiused', i.e. finished with as large a radius as possible. **1962** *Engineering* 6 July 10/1 The effect of radiusing the corners is also discussed. **1972** Greer & Howell *Mech. Engin. Craft Stud.* II. III. 157 (*table*) Millsaw. For sharpening circular saws, radiusing slots etc.

Hence **ra·diused** *ppl. a.*

1954 *Archit. Rev.* CXV. 144/1 Radiused blocks are made for use at corners. **1959** *New Scientist* 31 Dec. 1339/2 Sharply radiused members are more expensive than straight members. **1975** *Mariner's Mirror* LXI. 406 A slightly more radiused stem rabbet where it meets the keel.

radix. For 7–9 read 7– and add: Also with pronunc. (ræ·diks). **1. b.** (Further *attrib.* examples.)

1950 W. W. Stifler *High-Speed Computing Devices* vi. 80 For any radix arithmetic the basic tables corresponding to the addition and the multiplication tables of decimal arithmetic can be written. *Ibid.* 87 Corresponding representations of the same numbers for radices 2, 3, 4, 5, 8, and 10 are tabulated. **1960** N. R. Scott *Analog & Digital Computer Technol.* vii. 224 To represent the *r* digits in a radix *r* system by *r* binary digits is highly inefficient, and this figure of merit unduly penalizes radices not close to *e*. *Ibid.* 227 If numbers can have *n* digits to the left of the radix point, the radix complement of a negative number is formed by adding the radix raised to the *n*th power to the negative number. **1969** P. B. Jourdain *Condensed Computer Encycl.* 412 If a number is added to its radix complement, the result is a 1 followed by a 0 for every position in the original number. Radix complement is used in some computers..and desk calculators for representing negative numbers. **1970** O. Dopping *Computers & Data Processing* xvii. 280 In radix sort, the records usually pass through the sorting device—a computer or a card sorter, as many times as there are digits in the sorting key.

radome (rē^i·dōᵘm). [Blend of *Radar and Dome *sb.*] A dome or other structure, transparent to radio waves, protecting a radar aerial.

1945 in *Amer. Speech* XX. 310/2 *Radome*, housing enclosing a radar scanner. **1949** *Sun* (Baltimore) 29 Dec. 5/1 Supported by air pressure.., the balloon-like buildings, called radomes, are ideal for the housing of large radar antennae. **1951** *Electronics* Aug. 89/2 The radar antenna is enclosed in a streamlined radome aft of the big bomb bay. **1962** *Guardian* 3 Oct. 3/7 The Air Ministry should..mitigate the 'nuisance' of a station in the National Park by keeping buildings..away from the main road..and by making the radomes a pale blue to tone with the sky. **1968** *New Scientist* 21 Mar. 631/2 The *Vladimir Komarov* is distinguished by two massive radomes of some 50ft diameter and a smaller radome amidships. **1973** C. Mason *Hostage* x. 136 Radar picket aircraft..with grotesque radomes projecting above and below the fuselages. **1977** *Time* 4 Apr. 13/1 A mushroom-shaped 'radome' 30 ft. in diameter and 6 ft. thick sprouts from the rear of the grey fuselage on two large struts.

radon (rē^i·dǫn). *Chem.* [a. G. *radon* (C. Schmidt 1918, in *Zeitschr. f. Anorg. Chem.* CIII. 114): see *Rad(ium and *-on^2.] **1.** A short-lived radioactive element which belongs to the group of noble gases and occurs naturally in trace amounts as a result of the decay of radium and other radioactive elements; orig. *spec.* the longest-lived isotope, radon 222, having a half-life of 3·82 days. Atomic number 86, symbol Rn (orig. Ro). Cf. *radium emanation* s.v. *radium 3 b.

1918 *Jrnl. Chem. Soc.* CXIV. ii. 306 Radium emanation is given the name Radon, Ro, which at once indicates its origin and its relationship to the argon group. **1927** *Observer* 3 Apr. 20/2 The Radium Institute sends radium, or rather radon, its active principle, to hospitals all over the country. **1938** R. W. Lawson tr. *Hevesy & Paneth's Man. Radioactivity* (ed. 2) xxiii. 227 The first five disintegration products of the gas radon are isotopes of the metals polonium, lead, bismuth, or thallium. **1942** S. Tolansky *Introd. Atomic Physics* xiii. 218 A body exposed for a short time to radon coats with an active deposit which emits α-, β-, and γ-radiation and exhibits a regular decay. **1974** *Environmental Conservation* I. 24/1 Uranium miners are known to suffer from an increased risk of lung cancer from inhaled radon. **1977** *Time* 22 Aug. 8/2 The radon in these waters is supposed to be good for everything from paralysis to curvature of the spine.

2. Special comb.: **radon seed,** a short tube containing radon that is used in radiotherapy as a source of alpha radiation.

1925 A. E. H. Pinch *Clin. Index Radium Therapy* 61 Treatment by the burying..of numerous unscreened radon 'seeds'..will often prove effective. **1930** *Sunday Times* 12 Oct. 24/2 Medical evidence showed that the child was placed under an anæsthetic and radon seeds..were placed in the growths. **1966** Henschke & Hilaris in G. H. Fletcher *Textbk. Radiotherapy* i. 43/1 Ninety radon seeds each 0·75 cm. were permanently implanted through 17 needles and the uterus was sutured over the implant.

radula. 2. (Later examples.)

1901 E. Step *Shell Life* iii. 42 The number of these teeth to one tongue or radula varies to a remarkable extent. **1928** Russell & Yonge *Seas* ix. 202 In common with many other members of the snail family they [*sc.* limpets] possess a very characteristic feeding apparatus consisting of a long horny ribbon, made up of many rows of fine teeth, and known as the 'radula'. **1959** A. C. Hardy *Open Sea* II. vi. 128 A radula is a remarkable structure found in the mouths of all typical gastropods; it is a long ribbon, bearing a vast number of transverse rows of sharp horny teeth. **1975** *Sci. Amer.* Feb. 106/3 The snail combines the functions of teeth and tongue in a single organ: the radula, a toothed, filelike muscle inside the mouth.

radurization (rædiūᵊraizē^i·ʃǝn). [f. L. *radiāre* to furnish with rays, shine + *dur-āre* to make hard, preserve + -ization.] The treatment of food with ionizing radiation so as to enhance its keeping qualities by killing many of the micro-organisms in it (see quot. 1964). Cf. *radappertization, *radicidation.

1964 H. E. Goresline in *Nature* 17 Oct. 237/2 Type III is the application to foods of doses of ionizing radiation sufficient to enhance keeping quality by causing substantial reduction in the numbers of viable specific spoilage micro-organisms... The following are the names we suggest for these types of treatment... Type III, radurization. **1973** N. F. Lewis et al. *Radiation Preservation of Food* (Internat. Atomic Energy Agency) 201 'Radurization' is essentially a pasteurization treatment that results in prolonging shelf-life of foods by a selective control of spoilage microflora. **1977** *Biol. Abstr.* LXIII. 5003/1 Irradiation preservation of Korean fish: I. Radurization of croaker, yellow corvenia and roundnose flounder.

radwaste (ræ·dwē^ist). orig. *U.S.* Also **rad-waste.** [Short for *radioactive waste.*] = *radioactive waste* s.v. *radioactive *a.* 4.

1973 *Trans. Amer. Nucl. Soc.* XVI. 176/1 (*heading*) A cyrogenic approach to fuel reprocessing gaseous radwaste treatment. **1975** *Proc. Symp. on Reliability of Nuclear Power Plants* (Internat. Atomic Energy Agency) 373 A computerized reliability-risk model has been developed to simulate the rad-waste system. **1978** *Times* 28 July 1 Principal components of typical radwaste calcin[ation]. **1979** *Nature* 15 Mar. 219/1 The most popular procedure advocated by the nuclear power establishment during the past 25 yr has been to incorporate the radwaste into a borosilicate glass.

Rætian, Rætic: see *Rhætian, *Rhætic.

R.A.F. Also (*colloq.*) **raf, raff** (ræf). [f. initial letters of *Royal Air Force*, founded in 1918 on the amalgamation of the Royal Flying Corps with the Royal Naval Air Service.] The British Air Force or (*collect.*) members of this organization.

1920 M. Baring *R.F.C., H.Q.* xxi. 276 On the 20th of May we started on a long expedition to the R.A.F. Headquarters. **1924** G. Bell *Let.* 2 July (1927) II. xxiv. 701 The most interesting thing which happened during this week was a performance by the R.A.F., a bombing demonstration. **1941** W. S. Churchill *Into Battle* 310 Operating from new Greek bases, the R.A.F. attack Bari and Brindisi, and bomb military objectives in Naples. **1946** 'Tackline' *You met such Nice Girls* vii. 73 And it is a peculiar thing, but the Raff and the Wavy Navy do not mix at parties, and in fact the only place the Wavy Navy like the Raff to be is in the air. **1950** C. MacInnes *To Victors the Spoils* ii. 227 They're Raf bods, escaped prisoners. **1954** 'E. C. R. Lorac' *Shroud of Darkness* xvi. 173 I'd fly the plane for you if the Raf'd let me. **1957** M. Swan *Brit. Guiana* iv. 76 He was a big man, in his late twenties, with an R.A.F. moustache, wearing a bush-hat and a bush-shirt whose breast pockets bulged with papers. **1965** J. Porter *Dover Two* v. 60 'A decent lad like our Rex..in the Raf.' 'R.A.F., Dad... I've told you before not to call it Raf!' **1974** S. Milligan *Rommel* 186, I never dreamed, one day he, I, and a lone RAF erk called Sellers.. would make a sort of comic history. **1980** J. Ditton *Copley's Hunch* I. i. 11 For a Raff bloke, that's good going. You're not trained to make full use of ground cover, are you?

Hence as *v. trans.* (see quot. 1940) and *intr.* (*rare*).

1930 T. E. Lawrence *Let.* 8 Jan. (1938) v. 675, I spend innocent days R.A.F.ing. **1940** *Daily Mail* 28 Aug. 3/1 Yesterday I heard: 'He'll get R.A.F.'d if he doesn't mind.' There's surely a rousing neologism in this—to 'raff' the Nazis instead of the old 'strafing Fritz'. Why not say 'Berlin has been raffed to blazes?'

‖**rafale** (rafal). [Fr., lit. a gust of wind.] A series of bursts of gun-fire; a roll of drums. Also *fig.*

1903 P. de B. Radcliffe tr. *G. Rouquerol's Tactical Employment Quick-Firing Field Artillery* II. i. 33 To obtain the instantaneous effect, to produce that which he [*sc.* Gen. Langlois] vividly termed the *rafale*, or shell-storm, he conceived a special device which he called '*échelon* fire'. **1914** *Sphere* 3 Oct. 8/1 The second diagram shows a 'rafale', or 'shell-storm'. This is the method practised by batteries of French artillery to prevent the advance of infantry. **1916** *Chambers's Jrnl.* Sept. 604/2 The ·75, by rafale and curtain tactics, is able to isolate an attacking force by keeping the supports at bay. **1922** *Public Opinion* 28 July 85/1 If I had a few private batteries I should fire a private rafale in honour of the best book of the year. **1928** *Blackw. Mag.* Jan. 69/1 This was delivered with a slobbering roll of 'r's' like a rafale of water-logged kettle-drums. **1931** E. Linklater *Juan in Amer.* II. xii. 137 Now the staccato ear-splitting *rafale* of cheering rowels them afresh.

raff, *sb.*^1 Add: **3. b.** *spec.* Ore which requires re-crushing; *raff-wheel*, a wheel for lifting such ore.

1867 *Ure's Dict. Arts* (ed. 6) II. 72 The hopper is continuously charged, and that portion which is not reduced sufficiently fine is returned by the raff wheel to be recrushed. **1902** *Trans. Inst. Mining & Metall.* X. 459 The stuff rejected by..[a cylindrical trommel] is brought back by means of a Raff wheel and re-crushed.

raffe. Substitute for entry:

raffee (ræfi·). Also **raffeé.** [Of obscure origin.] (See quots. 1880 and 1891.) Also *attrib.*

1880 D. Kemp *Man. Yacht & Boat Sailing* (ed. 2) 547 *Raffee*, the square topsail set flying on the foretopmast of schooners, and formerly often set on cutters and ketches above the squaresail. Sometimes this topsail is triangular in shape, like a scraper. **1891** H. Patterson *Illustr. Naut. Dict.* 144 *Raffeé Rail*, a sail in the shape of an equilateral triangle △ which is sometimes set over the highest yard... This sail is common to English schooner yachts rigged to carry a squaresail, as the *raffeé* is set over the yard. **1922** *Field* 8 July 59/1 A square sail and a raffee, or the topsail set over it, are such old-fashioned sails that many modern yachtsmen have never seen them. **1942** C. Crockett *House in Rain Forest* i. 18 The southeast trades filling our square-sail and raffee. **1976** *Oxf. Compan. Ships & Sea* 687/1 *Raffee*, another name for the sail in a square-rigged ship known as a moonraker, set only in light weather.

Rafferty (ræ·fǝti). *Austral.* and *N.Z. slang.* [Eng. dial. corruption of *refractory* (Eng. Dial. Dict.).] Used *attrib.* or in the possessive, as *Rafferty('s) rules*, no rules at all, esp. in boxing.

The customary initial capital suggests that the word is felt by many to be the Irish surname *Rafferty*.

1928 *Bulletin* (Sydney) 5 Jan. 37/4 M.Q. (and Rafferty) Rules. **1935** *Sydney Morning Herald* 28 Dec. 11 Rafferty rules may suit Mr Keenan and the Communist party, but they are repugnant to the trade union movement. **1941** Baker *Dict. Austral. Slang* 58 *Rafferty rules*, no rules at all, applied to any system, organisation or contest run in slip-shod fashion. **1958** A. Wall *Queen's English* xxxii. 112, I do not know that the Queensberry Rules ever acquired any figurative usage; but the 'Rafferty Rules' certainly did. This term means no rules at all; in Australian, and hence New Zealand slang, it means any free and easy way of running things; 'Rafferty' here is thought to be an English dialect corruption of 'refractory'. **1964** H. P. Tritton *Time means Tucker* (ed. 2) 34 The Show adjourned at noon for the races. They seemed to be run on the 'Rafferty Rules' principle, but I heard no complaints. **1974** *Bulletin* (Sydney) 18 May 63 Rafferty's rules predominate. **1977** *Financial Times* 17 May 37/8 Because of the nature of the town and its 'Rafferty's rules' violence is a way of life and it is a well known haunt for criminals and tribal outcasts.

raffia. Add: **1.** (Further examples.)

1906 *Westm. Gaz.* 26 Sept. 8/1 Mr. William H. Hunt.. announced the discovery, in the leaves of the rafia palm, of a product which..may be classed between wax and gum. **1958** C. Achebe *Things fall Apart* I. viii. 57 Obierika was sitting outside under the shade of an orange tree making thatches from leaves of the raffia-palm.

2. Also, extensively employed in the making of baskets, lamp-shades, mats, and similar articles. (Further examples.)

1901 M. White *How to make Baskets* ii. 11 It is a rare thing to find a material at once so soft and so strong as raffia. **1912** *Educ. Handwork* Nov. 201/1 The materials most suitable for weaving are, wool, bast or raffia, and cane. **1937** A. H. Crampton *Raffia Work & Basketry* 7 Raffia work although allied to the ancient craft of Basket Making, may be termed a modern craft. **1951** 'R. Brinley' *Raffia Work* i. 11 One of the chief advantages of working with raffia is the very low cost of the material. **1978** P. Van Greenaway *Man called Scavener* vi. 84 He noticed a confusion of raffia, macramé and pieces of knitting.

3. *attrib.* and *Comb.*, as (sense 2) *raffia bag, basket, cloth, fibre, grass, lace, mat, needle, tape, work, workbag*; *raffia-embroidered* adj.

1932 S. Gibbons *Cold Comfort Farm* viii. 121 Raffia bags and linen bags embroidered with hollyhocks. **1960** G. Durrell *Zoo in my Luggage* iii. 77, I bent down, picked up a raffia bag and held it aloft. **1914** S. G. Fitzgerald *Priscilla Juniors' Basketry Bk.* 19 (*heading*) Handle for raffia basket. **1977** G. Scott *Hot Pursuit* iii. 25 Dried fish, piled in raffia baskets, on the pavement. **1932** D. C. Minter *Mod. Needlecraft* 59/1 Raffia work is really another form of embroidery... The materials for working on are.. hessian, raffia cloth, and woven straw. **1967** E. Short *Embroidery & Fabric Collage* iii. 84 Fabrics with unusual textures, raffia cloth, for instance, can be decorated with simple embroidery such as bands of drawn threads. **1904** *Daily Chron.* 3 May 8/3 A pretty and attractive novelty..is the raffia embroidered cushion. **1906** *Westm. Gaz.* 26 Sept. 8/1 The natives gather the rafia fibre. **1910** M. T. Priestman *Handicrafts in Home* 207 Delicate strands of raffia fibre should be secured for this purpose. **1904** *Daily Chron.* 3 May 8/3 The embroidery is worked with raffia grass dyed in various colours. **1906** *Queen* 5 May 757/1 Raffia lace hats are the choicest things in headgear that ingenuity has ever devised out of vegetable fibre. **1914** S. G. Fitzgerald *Priscilla Juniors' Basketry Bk.* 6 (*heading*) Woven raffia mat. **1953** E. Simon *Past Masters* IV. 263 Raffia mats, cutlery and glasses defined the full number of places. **1976** *Daily Times* (Lagos) 22 Sept. 16/4 To build a crib to store maize harvested from a one-hectare plot requires only six 12-ft (360-cm)-long bamboo poles; another six of such poles of 180 cm length each; 45 raffia mats; [etc.]. **1914** H. C. Walker *Rafia Work* 7 The reasons for urging the claims of rafia work are many... It requires no tools beyond rafia needles, wool needles..and knitting needles. **1932** D. C. Minter *Mod. Needlecraft* 226/2 Use a small packing needle, or a raffia needle, and a backing of Helvellyn canvas with Persian, or Straight, or Shetland rug wool. **1979** *Dryad Catal.* 89/3 Raffia needles..For coiled raffia basketry. **1907** *Daily Chron.* 5 Jan. 9/1 Sometimes the flowers require staking, and this should be done..with stiff, straight wires or sticks, to which the stems should be fastened with West's Raffia Tape. **1939–40** *Army & Navy Stores Catal.* 956/1 Raffia Tape..balls in coloured string nets. **1908** M. E. Morgan *How to dress Doll* vii. 65 Little girls who know how to do raffia work can easily make such a hat. **1974** C. Fremlin *By Horror Haunted* 88 He attended his own classes in Braille and raffia-work. **1928** *Chambers's Jrnl.* 24 Mar. 261/1 From a corner of one of the baskets she unpacked her raffia workbag.

raffinate (ræ·finē^it). [ad. G. or F. *raffinat*, f. G. *raffinieren* (F. *raffiner*) to refine + *-at* -ate^1 (as in *distillate, filtrate*, etc.).] The refined fraction which results after removal of impurities by solvent extraction, *spec.* in oil refining. Also *attrib.*

The term was first used in oil refining, in connection with the solvent extraction process invented by the Romanian chemical engineer L. Edeleanu and introduced on a commercial scale at Rouen *c* 1911.

1928 L. Edeleanu *U.S. Pat.* 1,661,565 2/1 The finished hot raffinate is taken from the last evaporator 21 by the pump 23 and passed..to the storage tank. **1932** *Jrnl. Inst. Petroleum Technologists* XVIII. 919 Dr. Edeleanu could lay claim to a further distinction, in having added two words to the English language—the words 'edeleanize' and 'raffinate' —words which, if not already in the dictionary, soon would be. **1941** W. L. Nelson *Petroleum Refinery Engin.* (ed. 2) xxvii. 617 Elaborate equipment is required to distill the solvent (or oil) from the extract and raffinate solutions. **1950** *Jrnl. Amer. Chem. Soc.* LXXII. 12/2 Now the flask in the 'R' position contained penicillin which was emptied into a larger container as raffinate pool. **1958** *Engineering* 14 Feb. 205/2 The presence of this dissolved salt in the fission-product raffinate stream limits the degree of concentration which may be achieved by evaporation while still keeping all the material in solution. **1970** W. G. Roberts *Quest for Oil* ix. 95 The raffinate is given a final sweetening and is ready for use as premium kerosene.

‖**raffiné** (rafine), *a.* (*sb.*) [Fr.] Of manners or judgement: refined. Also as *sb.*, a person distinguished by the possession of refinement in manners, action, or feeling.

1876 [see *méfiance]. **1883** *Atlantic Monthly* Aug. 179/1 The ingenious Catherine—she was a *raffinée*. **1920** D. H. Lawrence *Lost Girl* i. 10 No French marquis..could have been more elegant and *raffiné*. **1943** *Scrutiny* XI. 317 He is an older and wiser Tonio Kröger who has broken away from the precious and the *raffiné*. **1966** *Punch* 2 Feb. 173/1 The waiter, the actor, the communist, the painter all have their says [in a play] but it is through the analyst and the raffiné aristocrat..that we learn of hate-objects. **1970** *New Yorker* 23 May 126/3 He has a tendency..to favor his vigorous vulgarians at the expense of his effete *raffinés*.

raffinose. Substitute for entry:

raffinose (ræ·finōᵘz, -s). *Chem.* [a. F. *raffinose* (D. Loiseau 1876, in *Compt. Rend.* LXXXII. 1058), f. *raffiner* to refine: see -ose^2.] A non-reducing trisaccharide sugar found in sugar-beet, cotton seed, and many cereals.

1876 *Jrnl. Chem. Soc.* XXX. 398 At 20° water dissolves one-seventh of its weight of raffinose. **1881, 1894** [in Dict.]. **1934** *Industr. & Engin. Chem.* Apr. 462/1 The cottonseed meal has been the raw material from which most of the small supply of pure raffinose has been obtained. **1950** *Thorpe's Dict. Appl. Chem.* (ed. 4) X. 468/1 The arrangement of the hexoses in raffinose is galactose–glucose–fructose. **1970** A. L. Lehninger *Biochem.* xi. 227 Raffinose (fructose, glucose, galactose) is found in abundance in sugar beets and many other higher plants.

raffle, *sb.*^1 Add: **3.** *attrib.*, as *raffle prize, ticket.*

1976 *Milton Keynes Express* 16 July 9 The raffle prize of a 10 foot canoe went to Mr Sheldrick of Tandra, Bean Hill. **1976** *Jrnl.* (Newcastle) 26 Nov., Mr. Large produced a bundle of official raffle tickets offering Michael's models, which include a gypsy caravan, as prizes.

raffle, *sb.*^2 **1. a.** Delete † and add later example.

1921 G. C. Shedd *Lady of Mystery House* xix. 171 Probably the drunken raffle were seeking far and near to take me.

2. a. (Further examples.)

1895 Kipling *Day's Work* (1898) 343 He..was pushed and prodded through the slack back-waters of the Lower Fourth, where the raffle of a school generally accumulates. **1906** *Macmillan's Mag.* Aug. 755 A heavy cattle-boat limping past us..with its raffle of pens and its sour sweet reek. **1977** A. Hunter *Gently Instrumental* iv. 50 It was a pleasant-enough spot up there, in spite of the raffle of the yard below.

raffle, *v.*^1 Add: **2.** Also const. *off.*

1889 'Mark Twain' *Connecticut Yankee* xv. 175, I shan't know what to do with them: unless I raffle them off. **1976** *Washington Post* 7 Nov. K2/3 We'll raffle off a 'possum and award a prize to the wearer of the biggest beehive hairdo.

Raffles (ræ·f'lz). The name of A. J. *Raffles*, hero of *The Amateur Cracksman* (1899) and other books by E. W. Hornung (1866–1921), used allusively of a gentleman who engages in crime, esp. burglary. Also *attrib.*

1908 'O. Henry' *Gentle Grafter* 142 It's part of my business..to play up to the ruffles when I want to make a riffle as Raffles. **1930** G. Smithson *Raffles in Real Life* i. 16 What an ignominious ending to the ambitious ideas that had filled my mind when I set out on the perilous career of a modern Raffles! *Ibid.* 18 By steady, if somewhat painful, steps, I had graduated from a middle-class cracksman to a Raffles of the gentleman type. **1953** J. Trench *Docken Dead* viii. 113 It [is] a good thing to look to one's way of retreat. Have you thought of that, my good Raffles? **1960** *John o' London's* 31 Mar. 384/3 An educated renegade..in the classic *Raffles* tradition. **1974** *Listener* 21 Feb. 242/3 Damiaen van Doorninck, a Dutch Naval officer, made a reputation..as a cracksman—'the Raffles of Colditz'.

rafflesia. Substitute for def.: A stemless, leafless, parasitic plant of the genus so called, of the family Rafflesiaceæ, native to Java and Sumatra, and remarkable for the size of its flowers. (Later examples.)

Quot. 1820 in Dict. should be dated 1821.

1933 L. Ainsworth *Confessions Planter in Malaya* 111 This plant..is known as the 'Rafflesia', and is one of the largest parasites in the world. **1954** R. E. Holttum *Plant Life in Malaya* xiv. 192 Rafflesia plants are only found in deeply shaded forest.

‖**rafik** (rafi·k). Also **raffik, rafiq.** [ad. Arab. *rafīk.*] In Arabia: a companion or escort (see quot. 1920).

1856 R. F. Burton *Pers. Narr. Pilgrimage to El-Medinah* III. 84 The payment of a small sum secures..a 'Rafik', and this 'friend', after once engaging in the task, will be faithful. **1888** C. M. Doughty *Trav. Arabia Deserta* I. viii. 235 The Arabian rafik, often an enemy, is a paid brother-of-the-road, that for a modest fee takes upon him to quit the convoy from all hostile question and encounter of his own tribesmen. **1911** T. E. Lawrence *Let.* 18 June (1954) 170 'Am I not your friend, your raffik?' said the zaptieh. **1913** G. Bell *Let.* 20 Dec. (1927) I. xiii. 314 We have a Rafiq, a comrade of the Ghiyatah with us—we fetched him from Dumeir to stand surety for us if we met his tribe. **1920** *Handbk. of Arabia* (Admiralty) I. 21 The one thing needful is a *rafiq*, i.e. a companion derived from the tribe through whose range one must pass, or from some tribe allied with it or authorized to share its range.

raft, *sb.*^1 Add: **4.** esp. ducks; also, a group of other aquatic animals. (Further examples.)

1848 W. E. Burton *Waggeries & Vagaries* 70 We've shoals of shad, hull rafts of canvass-back ducks, and no eeend of terrapins. **1872** *Fur, Fin & Feather* 26 The great collections [of ducks] are termed rafts. **1949** Sprunt & Chamberlain *S. Carolina Bird Life* 135 The Greater Scaup ..congregates there in large flocks or 'rafts', as they are called. **1952** *Chambers's Jrnl.* Jan. 19/2 That black mass far away to starboard, what is it—brent geese? And the bigger one over there? It looks like a good raft of wigeon. **1959** E. Collier *Three against Wilderness* xxi. 210 A raft of newly hatched geese had been perched on the beaver house. **1961** *Guardian* 4 Mar. 4/1 Rafts of widgeon cry *wee-too* from the steamer lines. **1972** S. Burnford *One Woman's Arctic* ii. 48 There were long rafts of murres out in the middle. **1975** *Country Life* 16 Jan. 132/3 A tiny band of sea otters.. had grown to a raft of 130. **1975** P. A. Johnsgard *Waterfowl N. Amer.* II. 347 Scaup in such rafts do not all forage at the same time.

4*. *Building.* A layer of reinforced concrete forming the foundation of a building.

1903 *Engineering* 17 Apr. 517/1 Finally, the Co-operative Wholesale Society resolved to adopt their architect's recommendation to have recourse to a raft of Hennebique ferro-concrete over the whole area of the ground. **1936** *Concrete & Constructional Engin.* XXXI. 423 There are a great many districts where rafts are the best and often the only solution to foundation problems. **1970** R. Jeffries *Dead Man's Bluff* iii. 21 A concrete raft on which was a feeding trough and about fifty Kow Kennels. **1978** *Daily Tel.* 12 May 19/3 All the floors in the houses are dropping out because they were built on a concrete raft. The raft is now sinking into the peat below.

5. *raft-man* (further examples), *-voyage*; *raftwise* adv.; **raft-deck,** an under-water protecting deck formerly used to cover the unarmoured parts of some warships; **raft-duck,** add after 'scaup or blackhead duck': (*Athya marila*); (later examples); **raft foundation** = sense 4* above; **raft-fowl** (for 1718 read 1709 in def. and quot.).

1886 *Encycl. Brit.* XXI. 822/2 But the sailors of 1854–1860 did not take the view that buoyancy and stability.. were the vital parts, needing defence by armour or by a raft-deck. **1872** E. Coues *Key to N. Amer. Birds* 289 Greater Scaup Duck... Raft Duck. Flocking Fowl. **1975** P. A. Johnsgard *Waterfowl N. Amer.* II. 347 The 'rafting' behaviour of migrant and wintering scaup is well known and indicated by their vernacular names—'raft duck', 'flock duck', and 'troop duck'. **1904** C. F. Marsh *Reinforced Concrete* 541/1 (Index), Raft foundations. **1910** F. Rings *Reinforced Concrete* iv. 61 For ordinary level or raft foundations wire meshing or expanded metal are extremely useful. **1974** *Encycl. Brit. Macropædia* III. 458/1 Of mat, or raft, foundation, there are two types—the beam-and-slab type and the flat-slab type. **1828** A. Sherburne *Mem.* vi. 234 The raft men had the privilenge [*sic*] of cooking, and sleeping under cover, in the boat. **1847** C. Lanman *Summer in Wilderness* xviii. 111 The principal anglers for this fish are steamboat hands and raftmen. **1903** *Blackw. Mag.* Feb. 228/1 The figures of the raftmen seem to walk magically on the water. **1963** A. Smith *Throw out Two Hands* xv. 153 We wished..to be a raft in the air, and to pay as little attention to our conveyance as raft-men do. **1891** 'Mark Twain' *Let.* 1 Oct. (1917) II. 558 A pedestrian tour in Europe doesn't begin with a raft-voyage for hilarity and mild adventure. **1905** 'Q' *Shining Ferry* I. vi. 61 A hatch opened in her bows, through which the long balks of timber were thrust..to be laid raftwise and lashed together with chains.

raft, *sb.*^2 Delete '*dial.* and *U.S.*', and '(Used disparagingly)' and substitute 'orig. *dial.* and *U.S.*' (Earlier and later examples.) Freq. in phr. *a whole raft of* (persons or things).

1830 W. A. Ferris *Life in Rocky Mts.* (1940) vi. 29 We.. would have fought a whole raft of them. **1922** S. Lewis *Babbitt* viii. 116 They say there's a whole raft of stuff being smuggled across at Detroit. **1936** P. Bottome *Level Crossing* xxiv. 298 You've got to stand on what results there are, and not take on a whole raft of things that have nothing to do with it. **1947** 'N. Shute' *Chequer Board* x. 298 There's a whole raft o' things here that I never seen before. **1959** *Sunday Times* 7 June 21/5 Only very rich companies with a raft of employees build them. **1972** *Guardian* 19 Oct. 14/4 A year or two of keeping house, loving and looking after a raft of kids. **1977** *Time* 3 Jan. 36/1 There were a whole raft of programs in the '60s followed by eight years when there was no attempt to work with any degree of compassion. **1977** *Guardian Weekly* 16 Oct. 20/2 It has a raft of actors with the energy and skill to bring humour, depth and resonance to even the smallest parts. **1978** P. Theroux *Picture Palace* ix. 71 There was a whole raft of photographers in New York at the time. **1979** *Amer. Speech* LIV. 44 He reserved what he called 'a great raft' of *Beowulf* materials for a more leisurely time.

raft, *v.*^1 Add: **1. a.** (Further examples.) Also *transf.*

1896 Kipling *Seven Seas* 26 With cedars out of Lebanon Which Hiram rafted down. **1921** *Discovery* Feb. 48/1 The water hyacinth..causes an annual loss of one-fourth of the value of the logs rafted down the river. **1961** *Nature* 3 June 856/2 Sial blocks..being rafted to down-welling sites. **1972** *Sci. Amer.* May 62 (*caption*) In this schematic view the lithosphere is thicker under the continent it is rafting toward a subduction zone.

b. (Earlier and later examples.)

1689 H. Kelsey *Jrnl.* 22 July in *Kelsey Papers* (1929) 30 This morning tryed to gett over ye mouth of it but could not so..went up ye river to Raft our selves over. **1847** *Knickerbocker* XXIX. 314 We crossed it on the following day, rafting over our horses and equipage with much difficulty. **1972** *Nat. Geographic* Oct. 469/2 Wielding huge sweeps, Peruvians raft bananas down the Ucayali River.

2. (Further examples.) Also with *up*, and in sense 4 of Raft *sb.*^1

1883 J. Fraser *Shanty, Forest & River Life* xxix. 340 The timber is floated in single pieces down all the numerous tributaries of the Ottawa, and then 'rafted-up' at 'the mouth' of each. **1924** R. Campbell *Flaming Terrapin* ii. 38 Stacked with flaming spears Old Ocean shone, as swaying through the Night He rafted up his monstrous chandeliers. **1950** E. Hemingway *Across River* xli. 242 They [*sc.* ducks] must really be rafted up out there, the Colonel thought. **1976** *Yachts & Yachting* 20 Aug. 353/1 In St. Peter Port I have seen yachts rafted-up almost wall-to-wall on the buoys, like the trots used to be at Cowes. **1978** J. A. Michener *Chesapeake* 665 There must have been three thousand ducks rafted there beneath a frozen late-rising moon.

3. (Later examples.)

1844 J. C. Frémont *Rep. Exploring Expedition* (1845) 251 We had expected to raft the river. **1955** E. Pound *Classic Anthol.* I. 17 Ready to raft the deep, Wade shallow or dive for gain. **1974** *Marlboro Herald-Advocate* (Bennettsville, S. Carolina) 22 Apr. 6/2 During the four day exercise, members of the 541st engineer company rafted the Rhine river and secured tactical positions.

5. *intr.* Of an ice floe: to be driven on top of or underneath another floe. Also *trans.*, to drive (ice) in this way.

1883 Hatton & Harvey *Newfoundland* III. iii. 301 Or, under pressure of the storm, it frequently happens that the ice is 'rafted', as the sealers call it; that is, the fragments are piled in layers one over the other to the height of thirty or forty feet, being lifted by the swell and hurled forward as if from huge catapults. **1919** E. Shackleton *South* i. 11 In obedience to renewed pressure this young ice 'rafts', so forming double thicknesses of a toffee-like consistency. **1939** *Beaver* Mar. 13/2 On the sea and large lakes ice seldom forms smoothly. Early storms break it and pack it in confusion, and pressure causes it to 'raft'.

raft (ræft), *v.*^3 *Southern dial.* [Origin unknown.] *trans.* To rouse; to disturb, disquiet, or unsettle. Hence **ra·fted** *ppl. a.*^2, **ra·fting** *ppl. a.*^2

1851 *Gloss. Provincial Words Dorset* 6 *Raft*, to irritate. **1895** Hardy *Jude* IV. iv. 290 'I think you are rafted, and not yourself,' he continued. 'Do go back and make up your mind to put up with a few whims.' **1896** —— *Under Greenwood Tree* (rev. ed.) I. iv. 32 They should ha' stuck to strings. Your brass-man is a rafting dog—well and good; your reed-man is a dab at stirring ye—well and good; your drum-man is a rare bowel-shaker—good again. **1904** in *Eng. Dial. Dict.* V. 11/1 When a sick person is at the point of death, old nurses think it wrong to disturb the spirit of the dying by speaking to him, lest by doing so his spirit may be 'rafted', i.e. disturbed by earthly thoughts, and so bereft of the consolation of religion [Hants.]. **1920** Hardy *Coll. Poems* 377 My rafted spirit would not rest.

rafted (rɑ·fted), *ppl. a.*^1 [f. Raft *v.*^1 + -ed^1.] **1.** Of floating ice: piled up as a result of one floe having been driven on top of another.

1897 J. W. Tyrrell *Across Sub-Arctics of Canada* vii. 92 Toward the north end of the lake we passed great piles of rafted ice on the shore. **1924** R. J. Flaherty *My Eskimo Friends* III. ii. 93 On the tenth of April we came face to face with a gigantic pile of rafted ice. **1960** G. Blanchet *Search in North* viii. 109 Sunlight created strange effects from rafted ice. **1974** L. Deighton *Spy Story* xix. 203 We'll go south until we find the end of the rafted ice.

2. Of swimming birds: gathered into a dense flock.

1962 R. Haig-Brown *Whale People* x. 89 They covered the canoe with brush and allowed it to drift in on the rafted ducks. **1978** J. A. Michener *Chesapeake* 844 With a gentle ..push he launched the skiff toward the rafted geese.

rafter, *sb.*[2] Add: **1.** (Further examples.)

1891 C. Roberts *Adrift in Amer.* 206 The rafters were engaged in making the rafts up. **1905** 'Q' *Shining Ferry* vi. 70 In fifty strokes he brought her alongside the barque where the rafters—twenty-five or thirty—were at work. **1936** [see *cross-cutter]. **1954** A. M. Bezanson *Sodbusters invade Peace* xxii. 160 Rafters kept coming quite a while. They all finally got tired waiting for God to freeze the rivers again, and came down on rafts.

2. One who travels on a raft.

1978 *TV Bk.* (Detroit Free Press) 16–22 Apr. 21/2 Adventures of a group of white water rafters on the Chatooga River in South Carolina. **1979** *Sunset* Apr. 38 (*caption*) Jagged, glacier-dotted Mount Moran hobnobs with the clouds as rafters laze along Jackson Lake towards shore for Teton camping.

rafter, *v.* Add: **1.** Also *fig.*

1935 C. Day Lewis *Time to Dance* 64 A hungry soul Urged them to try new air-routes, and their skill Raftered the sky with steel.

4. *N. Amer.* Of ice: = *raft *v.*[1] 5.

1792 G. Cartwright *Jrnl.* II. p. vii, Raftering of ice. Ice is said to rafter, when, by being stopped in its passage, one piece is forced under another, until the uppermost ones rise to a great height. **1861** L. De Boilieu *Recoll. Labrador Life* viii. 100 It is a sad sight to see a ship on the weather edge of ice not enabled to work off; for when the ice begins to rafter she is thrown up, falls over, and becomes like corn between two millstones, and is literally ground up. **1908** N. Duncan *Every Man for Himself* ii. 60 The ice begun t' drive an' grind an' rafter. **1924** R. J. Flaherty *My Eskimo Friends* iii. iii. 99 Miles and miles of ice, raftering and rearing and overriding us it fought its way to the sea. **1964** *Newfoundland Q.* Spring 16/3 Evidently, just like frozen masses of ice raftered, one layer rising above the other by pressure, the crust of the earth broke and travelled southward.

raftered, *ppl. a.* Add: Also *transf.*, and in sense 4 of *rafter *v.*

1916 N. Duncan *Billy Topsail, M.D.* xvii. 130 It was six miles from the edge of the raftered ice to the first island. **1924** R. Campbell *Flaming Terrapin* ii. 32 She skimmed along—Till, raftered by the forest,..She saw the monsters that the jungle breeds.

rafterless (rɑ·ftəɹlės), *a.* [f. Rafter *sb.*[1] + -less.] Having no rafters.

1854 'G. Greenwood' *Haps & Mishaps* vii. 128 This is a picturesque, roofless, rafterless edifice, in a good state of preservation. **1943** L. B. Lyon *Evening in Stepney* 20 Who tremble, yet dare not call On the crushed bones to bear witness And rafterless heaven to fall.

rafting, *vbl. sb.* Add: **a.** (Earlier and later examples.)

1697 H. Kelsey *Jrnl.* 16 Aug. in *Kelsey Papers* (1929) 95 Our 3 boats went to ten shilling creek to rafting. **1905** 'Q' *Shining Ferry* i. vi. 69 'Have they begun the rafting?' 'Bless your life, they've been working all night. There's one raft finished.' **1920** W. T. Grenfell *Labrador Doctor* ix. 180 On then swept the floe, crashed into the fixed ice, shattered its edge, rose up out of water over it, which is called 'rafting', forced itself on the unfortunate ship. **1967** *Vogue* Jan. 76/1 Jamaica has..skin-diving, rafting, cheap rum. **1975** B. L. Fairbank *Cruising Guide to Lake Ontario* 19 Rafting alongside is accepted practice on crowded weekends, although it is good manners to ask permission of the boat you are coming alongside.

b. rafting chain, a chain used to bind logs together into a raft; **rafting distance,** the distance that can be traversed in a raft; **rafting works** (see quot. 1969).

1842 A. Langton *Jrnl.* 4 Nov. in *Langton Records* (1904) 319 His errand..was to get a rafting chain. **1904** *Daily Chron.* 29 Jan. 3/3 Unlike Crusoe he has no ship within rafting distance filled with everything he might want. **1931** *Sun* (Baltimore) 20 Mar. 10/6 The Mississippi Logging Company..maintained a boarding house..on the way to the rafting works at West Newton. **1969** L. G. Sorden *Lumberjack Lingo* 94 *Rafting works*, booming grounds. A place where logs are held.

rafting (rɑ·ftiŋ), *ppl. a.* [f. Raft *v.*[1] + -ing[2].] Of ice: that rafts (see *raft *v.*[1] 5).

1883 Hatton & Harvey *Newfoundland* iii. iv. 311 When they are in danger from 'rafting' ice, or fragments of floes.. the self-sacrificing affection of the mothers leads them to brave all dangers. **1935** *Discovery* Mar. 77/1 A suitable block-and-tackle is essential in order that the boat may be hauled far enough up the shore to be safe from 'rafting' ice. **1966** *Weekend Mag.* (Montreal) 19 Mar. 34/1 Each year.. seals congregate on the rafting ice pressing in around the shores of Canada's Magdalen Islands.

rafwire (ræ·fwəiᵊɹ). *Aeronaut.* Also **Rafwire, RAFwire, raf-wire.** [f. the initial letters of the *R*oyal *A*ircraft *F*actory + Wire *sb.*] A kind of wire having a flattened, semi-streamlined cross-section formerly used as bracing wire on aircraft.

1918 W. E. Dommett *Dict. Aircraft* 38 *Rafwire*, wires of flattened section, approaching to streamline shape, designed originally at R.A.F. **1920** *Techn. Rep. Advisory Comm. Aeronaut. 1915–16* 15 Recent Laboratory tests have shown that raf-wires of the type developed at the Factory show little or no aerodynamic disadvantage as compared with wires of stream-line section. **1922** *Encycl. Brit.* XXX. 34/2 The elliptical-section wires were called 'Rafwires', to distinguish them when they were standardized. **1933** *Flight* 4 May 412/2 External bracing is by RAFwire. **1933** *Jrnl. R. Aeronaut. Soc.* XXXVII. 788 The 'singing' of a rafwire when yawed in the wind at an angle greater than the stalling angle is an example of this type of oscillation.

rag, *sb.*[1] Add: **I. 1. f.** *pl.* Personal clothing or garments of any kind. Also in *sing.*, a garment, esp. a dress or coat. Cf. *glad rags* s.v. *glad *a.* 4 f. *colloq.* (orig. *U.S.*).

1855 *Knickerbocker* XLV. 502 Oh! the robe was of *moire antique*, (a very expensive 'rag'). **1883** 'Mark Twain' *Life on Mississippi* iii. 43, I stood up and shook my rags off and jumped into the river. **1903** *N. & Q.* Dec. 513/1 'Raggie' is of course diminutive or fond for 'rag', *i.e.* coat, tunic. I remember my uncle, writing to congratulate me on passing into the R.M. Academy, Woolwich, many years ago, asking me if I was 'going to sport the blue rag or the red one'—R.A. or R.E. **1906** E. Dyson *Fact'ry 'Ands* x. 126 In their secon' best baggin', they're sort iv subdood... Look at ther difference when they get inter ther rags. **1966** 'L. Lane' *ABZ of Scouse* II. 88 *Rags*, any form of clothing. **1974** H. L. Foster *Ribbin'* iv. 171 *Rags*, clothing.

g. (*from*) *rags to riches*: used variously to describe a 'fairy-tale' rise from poverty to wealth; esp. as *attrib. phr.*

1947 R. de Toledano *Frontiers of Jazz* 148 Goodman was the first real rags-to-riches success in the swing-jazz field. **1953** *Gramophone* Dec. 256/2 The Irish flavour is readily apparent in *Begorrah* as sung by Ray Burns... This is great fun, and infinitely preferable to the more commonplace *Rags to Riches* verso. **1954** M. Ewer *Heart Untouched* ix. 156 Isn't this a Cinderella story—a rags to riches? **1959** *Times Lit. Suppl.* 13 Nov. 658/2 The story he has to tell is..a classic American rags-to-riches story with nothing lacking. **1965** 'H. Carmichael' *Post Mortem* x. 120 One of those spectacular companies that came up from nowhere. You've heard of them—from rags to riches in five years. **1972** D. Lees *Zodiac* 34 It stands up as a rags to riches yarn. **1977** *Cornish Times* 19 Aug. 9/2 Last week's *Cornish Times* spelt out a success story with the rare theme of rags to riches by sheer hard labour.

2. b. (Further examples.)

1922 E. Sitwell *Façade* 14 Limp in bright crackling rags of laughter. **1924** R. Campbell *Flaming Terrapin* iii. 45 Their spirits shed their gross Rags of despair. **1963** S. Cloete (*title*) Rags of glory.

c. (Later examples.)

1811 *Lexicon Balatronicum*, s.v. *Rag*, ..Money in general. The cove has no rag; the fellow has no money. **1846** *Swell's Night Guide* 14 The pleasure-seeker may gain admission, if his appearance proclaim that he is in possession of the *rag*—the *tin* to defray the unavoidable demands upon his purse.

d. A familiar name applied to the Army and Navy Club in London. In full *the Rag and Famish* (see quot. 1908[1]). *slang.*

1858 Trollope *Three Clerks* II. i. 5 He delighted in the Rag and Famish, and there spent the most of his time. **1908** Nevill & Jerningham *Piccadilly to Pall Mall* vi. 235 The familiar name of the 'Rag', by which it is generally known, was invented by Captain William Duff, of the 23rd Fusiliers. .. Coming in to supper late one night, the refreshment obtainable appeared so meagre that he nicknamed the club the 'Rag and Famish'. **1908** 'One of Old Brigade' *London in Sixties* xvii. 224 These touts and store-keepers and bonnet-shop keepers will make the Rag a den of thieves, by Gad. **1941** E. Nash *I liked Life I Lived* iv. 33 Cairnes, who was a most hospitable man, invited me to dine with him once a week at the Rag, while the sales of the book were in full swing... No member of the Army and Navy Club was aware that Cairnes had written the book. **1974** R. McDouall *Clubland Cooking* 12 Unlike the clubs on the south side of Pall Mall..the Junior Carlton and the Rag owned their freeholds.

e. Phr. *to knock all to rags*: to knock senseless. *U.S.*

1889 'Mark Twain' *Connecticut Yankee* xxxiii. 432 The blow came crashing down and knocked him all to rags.

3. a. (Further examples.) Also, a napkin worn during menstruation, a sanitary towel; esp. in phr. *to have the rag(s) on.*

1782 J. Trumbull *M'Fingal* iv. 97 O'er heaps of rags, he waves his wand, All turn to gold at his command. **1811** *Lexicon Balatronicum*, s.v. *Rag*, bank notes. **1816** *Deb. Congress U.S.* 29 Jan. (1854) 775, I say cash, sir, for we, there, have nothing of that circulating medium which the gentleman from Virginia..denominates rags. **1846** *Swell's Night Guide* 129/1, *Rag*, money—I've no rag; meaning, I've no notes. **1859** J. W. Cole *Life & Times C. Kean* I. i. 8 Our old friends of the Dublin gallery, who, in days of yore, never failed to cry, 'Up with the rag!' even before the act-drop, so classically designated, had time to reach the ground. **1906** E. Dyson *Fact'ry 'Ands* xvii. 233 Ther revolvin' arm [of a machine] was bent out, 'n' it got home a left lead 'n' er right cross, 'n' ther rag went in from ther Pelican's corner. **1920** J. Ferguson *Northern Numbers* 101 The lights are lowered and the 'rag' divides. **1929** A. Conan Doyle *Maracot Deep* 200 Has your rag commissioned you to obtain an interview? **1948** *Amer. Speech* XXIII. 249/2 *Riding the rag*, menstruating. **1955** D. W. Maurer in *Publ. Amer. Dial. Soc.* xxiv. 115 That working stiff had over two C's in rag on him. **1961** Partridge *Dict. Slang* Suppl. 1242/2 *Rags (on), have the*, to be having one's period. **1970** G. Greer *Female Eunuch* 51 Male disgust [for menstruation] expressed in terms like *having the rags on.* **1974** *National Skat & Sheepshead Q.* Mar./Apr. 7 The bills [*sc.* paper money] wear out down here very fast due to the humidity. They all become 'rags' in a short period. **1977** J. I. M. Stewart *Madonna of Astrolabe* i. 27 A fugitive rag put out by one of our junior members. **1978** *Maledicta* II. 50 There were several references to menstruous conditions or activities, found equally commonly in both male and female rest rooms ('Sue Ellen's on the rag' [etc.]).

c. Colloq. phrases. (*a*) Miscellaneous, as *to drop the rag* (U.S.): to give the signal, to give notice; *to take the rag off* (*the bush* or *hedge*) (chiefly U.S.): to excel, to surpass everything or everyone; to take the palm. Also *to chew the rag*: see *chew *v.* 3 g. (*b*) Expressing anger, as *to get one's rag out* and varr.: to become or make angry; *to lose one's rag*: to lose one's temper; *on the rag* (U.S.): angry, irritable.

1810 *Norfolk* (Va.) *Gaz.* 19 Sept. 2/3 This 'takes the rag off the bush' so completely, that we suppose we shall hear no more..about the Chesapeake business. **1837** *Davy Crockett's Almanack Wild Sports of West* I. iii. 40, I can take the rag off—frighten the old folks—astonish the natives—and beat the Dutch all to smash. **1868** *Accrington Times* 16 May 5/3 These three elegant flags ull teck th' rag off th' edge, un ull be mich admir'd bi all them ut's i' th' love o' fine arts. **1880** 'Mark Twain' *Tramp Abroad* xx. 194 I've got to stay here, till the old man drops the rag and gives the word,—yes, *sir*, right here in this—country I've got to linger till the old man says *Come!* **1897** *Halifax Courier* 12 June, He's getten his rag drawn. **1901** W. N. Harben *Westerfelt* 3 That gal certainly takes the rag off'n the bush. **1902** —— *Abner Daniel* 264 You are a jim-dandy, young man... That's all there is about it. You take the rag off the bush. **1914** D. H. Lawrence *Prussian Officer* 185 An' that got your rag out, did it? **1927** W. E. Collinson *Contemp. Eng.* 116 Anger is expressed by such phrases as.. he got shirty or hairy, he got his rag out, [etc.]. **1938** G. Greene *Brighton Rock* vii. vii. 329 'I've told you before how I won't stand...' 'You needn't get your rag out,' the Boy said. **1959** I. & P. Opie *Lore & Lang. Schoolch.* x. 178 They taunt the person [who is easily provoked]:..'Don't lose your bait' ('rag', 'rise', or 'wool'). **1960** L. Cooper *Accomplices* i. vi. 60 Roger was definitely shirty about that. .. He really got his rag out. **1969** *Current Slang* (Univ. S. Dakota) I–II. 65 *On the rag*, in a bad mood.—College males, Arizona. **1975** Hill & Thomas *Give Little Whistle* x. 95 Allison lost his rag with me over two goals by Leicester's Mike Stringfellow, both of which he considered were offside. **1977** *Rolling Stone* 16 June 31/1 *Time* has Joan Baez on the rag.

6. b. The fibrous pithy part of an orange, lemon, or other citrus fruit.

1895 *U.S. Dept. Agric. Yearbk. 1894* 196 The fruit resulting is usually of poor quality, inclined to be large and rough, with a thick rind and abundant rag.

II. 8. a. *rag-baby* (examples), *-carpet* (examples), *-carpeting*, *doll* (earlier and later examples), *-rug*, *-torch*, *-wick*; *rag-carpeted* adj.

1809 *Deb. Congress U.S.* 20 Jan. (1853) 1165 If they insist upon dressing up, in their own ways, their rag-babies,..it is not for me to interfere. **1900** J. de F. Shelton *Salt-Box House* xvii. 143 Dolls were almost as mythical as fairies, but a 'rag-baby' was loved. **1837** *Southern Lit. Messenger* III. 333 There was a snug little bed room..and a comfortable good-sized one for Charlotte, with a neat rag carpet on it. **1904** M. E. Waller *Wood-Carver of 'Lympus* 72, I have begged Aunt Lize to take up the rag-carpet. **1845** C. M. Kirkland *Western Clearings* 185, I led the young gentleman through the shop into the rag-carpeted sitting-room. **1813** *Niles' Weekly Reg.* III. 329/1, 24 yards rag carpeting. **1853** J. Ruskin *Let.* 17 Nov. in *Wks.* (1904) XII. p. xxxiv, She thought me so wise that anybody might make an idol of me ..but when she got to talk to me, I turned out only a *rag doll* after all. **1972** M. Woodhouse *Mama Doll* xiii. 194 She slumped like a rag doll. **1923** E. Sitwell *Bucolic Comedies* 42 The witch's rag-rug takes its flight. **1937** [see *kewpie]. **1969** M. Harris *Kind of Magic* 30 By the fire stretched a lovely rag rug. **1973** J. Thomson *Death Cap* xiii. 177 Finch was reminded of his grandmother's bedroom... There had been the same kind of rag rug on the floor; the same marble-topped wash-hand stand. **1894** Kipling in *To-day* 6 Jan. 5/1 And the doolie-bearers lit the noisome, dripping rag-torches. *a* **1918** G. Stuart *40 Yrs. on Frontier* (1925) I. 31 A tin lamp holding about a quart of lard with a rag wick in its spout which, when lighted, would cast a strong light for several yards.

b. *rag-gatherer* (earlier example), *-picking.*

1704 *Visits from Shades* iii. 21 Rag-gatherers, Cynderwomen, and Oyster Wenches wou'd disclaim her Acquaintance. **1931** 'D. Stiff' *Milk & Honey Route* ix. 101 It is not surprising to find the moper turning to rag-picking if the notion comes to him to earn a living. **1966** *National Observer* (U.S.) 5 Dec. 15/3 There are ways to hold rag-picking down. One can, for example, specialize in $1 bills.

c. *rag-and-bone gatherer, man* (examples), *merchant, shop, warehouse*; *rag-and-bottle man, merchant, warehouse.*

1901 B. S. Rowntree *Poverty* iii. 35 Rag and bone gatherer. Married. One room. One child. **1904** E. Nesbit *Phoenix & Carpet* xii. 236 An insane millionaire who amused himself by playing at being a rag-and-bone man. **1960** 'H. Carmichael' *Seeds of Hate* xix. 157 Someone had sold them to a rag-and-bone man. **1963** *Times* 6 Mar. 9/3 Four Soviet rag and bone merchants have been sentenced to death in Azerbaijan for heading a gang which robbed the state of hundreds of thousands of roubles. **1895** C. M. Yonge *Long Vacation* xix. 188 Transforming the draperies from the aspect of a rag-and-bone shop to a wonderful quaint and pretty fairy bower. **1939** W. B. Yeats *Last Poems* 31 In the foul rag and bone shop of the heart. **1848** Mrs. Gaskell *Mary Barton* I. v. 74 Public-houses, pawnbrokers' shops, rag and bone warehouses, and dirty provision shops. **1904** E. Nesbit *Phœnix & Carpet* xii. 229 It's the rag-and-bottle man's day to-morrow... He will take it away. *a* **1902** S. Butler *Way of All Flesh* (1903) lv. 254 A rag and bottle merchant in..the last stage of dropsy. **1852** Dickens *Bleak Ho.* (1853) v. 35 A shop, over which was written, *Krook, Rag and Bottle Warehouse.*

9. rag-book, a book for children of which the pages are made of untearable cloth; **rag-box,** (*a*) a box in which rags are contained; (*b*) *slang*, the mouth; **rag-chawing, -chewing** *vbl. sbs.*, protracted discussion or argument (cf. *to chew the rag* s.v. *chew *v.* 3 g); also *attrib.*; **rag content,** the proportion of rag in paper; freq. *attrib.*; **rag end,** the extreme and untidy end of something; cf. Fag-end 2; **rag frame** *Mining* (see quot. 1964); **rag-front,** in a carnival or circus: a façade or banner made of painted canvas; **rag-head** *N. Amer. slang*, one who wears a turban or cloth about the hair; **rag-lamp** *U.S.*, a lamp in which a rag serves as a wick; **rag running,** whippet-racing; **ragsack-man** *nonce-wd.*, a ragman bearing a sack; **rag-shop** (earlier example); also *fig.*; **rag-store** *U.S.* = *rag-shop*; **ragtop** *U.S. slang*, a convertible car with a soft hood (see also quot. 1971); also *attrib.*; **rag trade** (further examples); now usu. applied to the manufacture and sale of women's garments; freq. in humorous or ironical use; also *attrib.*

1905 *Athenæum* 16 Dec. 833/1 The improvements recently made in the productions called rag-books are strikingly exemplified in *Dog Toby*. **1974** P. Dickinson *Poison Oracle* iv. 111 He had packed..rag books, fruit, favourite toys. **1801** D. Wordsworth *Jrnl.* 12 Nov. (1941) I. 79, I put the rag-boxes into order. **1890** Kipling in *Scots Observer* 28 June 149/1 Now all you recruities what's drafted to-day, You shut up your rag-box an' 'ark to my lay. **1885** *Santa Fé Weekly New Mexican* 1 Oct. 1/3 After a few minutes rag-chawing a verdict of 'came to his death from unknown causes', is promptly rendered. **1904** 'H. McHugh' *I'm from Missouri* v. 66 The news of the proposed joint debate spread like wildfire, and it soon became patent that whoever won the rag-chewing contest would also win the election. **1937** G. Frankau *More of Us* xii. 130 Great work Lord Bubbles put in presently Over their teas and pastries and rag-chewing. **1976** Perkowski & Stral *Joy of CB* viii. 86 In the evening you can set up your rig at home for extended rag-chewing sessions. **1930** *Official Gaz.* (U.S. Patent Office) 7 Oct. 25/1 Rag Content Paper Manufacturers. **1957** J. B. Calkin *Witham's Mod. Pulp & Paper Making* (ed. 3) ii. 20 Rag papers cover a considerable spread of products from 100 per cent rag to the so-called 'rag content' papers which are made from various percentages of wood and rag fibers. **1967** Karch & Buber *Offset Processes* xi. 479 The finer, longer-lasting paper, made from cotton and cloth clippings, is called 'rag content', or lately, 'cotton content' paper... Rag content is usually 25 percent, 50 percent, 75 percent, or 100 percent. **1976** *Cincinnati Enquirer* 16 Sept. a-19/3 The documents and letters are as legible as the handwriting of their authors. The high rag content of the paper has preserved them. **1917** E. Pound *Lustra* 192 And the booths Were scattered align, the rag ends of the fair. **1904** *Eng. Dial. Dict.*, Rag frame. **1920** *Conquest* Nov. 17/1 The stream is dammed and the sludge or slime settles, and is allowed to flow through launders which feed automatically-tilting tables of the most ingenious structure... These tables are called rag frames. **1964** A. Nelson *Dict. Mining* 358 *Rag frame*, a broad, slightly inclined wooden frame for the rough concentration of slimes. **1926** *Variety* 29 Dec. 7/4 The outdoor show game with its 'rag front', 'silver men', [etc.]. **1927** K. Nicholson *Barker* 150 *Rag front*, painted canvas banners. **1921** *Dialect Notes* V. iii *Raghead*, a Hindu; any Asiatic. From the turbanned Asiatics who are common on the campus [of the University of California]. **1970** C. Major *Dict. Afro-Amer. Slang* 96 *Raghead*, black male who wears a scarf tied around his head to protect an expensive hairdo. **1975** *Canadian Mag.* 8 Mar. 6/1 East Indians are called 'rag-heads' if they continue to wear the traditional turban of the Sikh religion. **1889** 'Mark Twain' *Connecticut Yankee* xlii. 531 He had re-instituted the ancient rag-lamp. **1893** —— in *Cosmopolitan* Nov. 59/2 The house was shut up tight and the rag lamps lighted. **1927** *Daily Express* 25 May 12 A little more foresight and push..might have made 'rag running' a very popular entertainment. **1922** Joyce *Ulysses* 429 A sackshouldered ragman bars his path. He steps left, ragsackman left. **1829** P. Egan *Boxiana* 2nd Ser. II. 643 'It's the Bank of Ireland,' said an Irish swell, 'to a rag shop.' **1894** G. B. Shaw *Let.* 23 Apr. (1965) I. 427 You have a perfect rag shop of old ideas in your head. **1903** —— *Man & Superman* p. xiii, His profligacy and his dare-devil airs have gone the way of his sword and mandoline into the rag shop of anachronisms and superstitions. **1869** 'Mark Twain' *Innoc. Abroad* xvi. 157 Filthy dens on first floors, with rag stores in them (the heaviest business in the Faubourg is the chiffonier's). **1882** Rag store [see *junk store* s.v. *junk *sb.*[1] 5]. **1955** *Sun* (Baltimore) 27 Aug. b11/1 Every American manufacturer with a 'ragtop' in his line will be represented. **1971** M. Tak *Truck Talk* 127 *Rag top*, 1. a low-sided trailer with metal bows over the top to support a tarpaulin... 2. an open top van with a tarpaulin covering over the top. **1974** D. Westheimer *Olmec Head* xvii. 235 Get a ragtop trailer. That's one with a fabric cover instead of a solid top. **1976** *Springfield* (Mass.) *Daily News* 22 Apr. 4/3 (*caption*) The last U.S. built convertible, a Cadillac Eldorado, rolls along the assembly at the General Motors' plant in Detroit Wednesday. It ended an era for ragtops that began 74 years ago. **1979** T. Gifford *Hollywood Gothic* (1980) ix. 100 Eddie's ragtop had a small tear that let the rain draw a bead on the back of the seat. **1907** *Daily Chron.* 31 Dec. 8/4 They do an enormous business with the 'rag trade'—that is to say, the wholesale drapers, silk mercers, hosiers, and so on. **1938** *Times Lit. Suppl.* 11 June 405/2 The few years which she spent as fashion-goods buyer in what is known to its members as 'the rag trade'. **1957** J. Coates *Ship of Glass* 241, I know that line. It's going to be fashionable... Forgive the digression but I'm in the rag trade. **1967** *Listener* 2 Feb. 168/3 These delicate and puzzling effects..were seized upon with glee by rag-trade designers and window dressers. **1975** R. Butler *Where All Girls are Sweeter* i. 2 I'd sold her to a man in the rag-trade.

rag, *sb.*[3] Add: (Earlier and later examples.) Now usu. a programme of satirical revues, frivolous stunts, etc., organized by students to raise money for charities. Also *attrib.*

1864 H. Sidgwick in A. & E. M. Sidgwick *Henry Sidgwick* (1906) ii. 111 They enjoy beer, tobacco and students' 'rags'. **1885** *Punch* 5 Dec. 273/1 We had a good rag when he was away. **1905** *Westm. Gaz.* 25 Apr. 3/3 It [*sc.* Sheridan's 'Critic'] has been left alone of late except for an occasional 'rag' performance at a charity matinée. **1924** *Glasgow Herald* 26 Feb. 9/7 Liberals played up skilfully in their interrogative zest, and the P.M.G...found the 'rag' more embarrassing than any miners' indiscipline. **1930** J. Buchan *Castle Gay* iv. 60, I do not wish to have my name associated with an undergraduate—'rag', I think is the word. **1946** L. P. Hartley *Sixth Heaven* iii. 76 He organised one or two rags..of the more painful kind. **1958** *Oxford Mail* 15 Feb. 1/1 A 1902 James and Browne vintage car removed from the Imperial College, South Kensington, London, by students of Southampton University for their 'rag' day. **1974** *Times* 4 Nov. 14/6 The university's Rag Week..[with] the 24-hour piano-playing marathon..and the joke kidnapping of the president of the athletic union.

rag (ræg), *sb.*[5] orig. *U.S.* [Of obscure origin: perh. f. Ragged *a.*[1] 3; see *ragtime.] **1.** A dance or ball; esp. a variety of dance performed to ragtime music. ? *Obs.*

1896 *Dialect Notes* I. 423 *Rag*, dance, ball. 'We can go to *rags*.' **1899** *Musical Rec.* (Boston) 1 Apr. 158/1 The negroes call their clog-dancing, 'ragging', and the dance, a 'rag'. **1914** 'High Jinks, Jr.' *Choice Slang* 17 *Rag*, one of the newer gyrations now included under the category of dancing. **1923** *Dialect Notes* V. 218 *Rag*,..dance.

2. A musical composition written in ragtime, a ragtime tune.

1897 W. H. Krell (*song-title*) The Mississippi rag two-step: the first rag-time two-step ever written. **1897** T. Turpin (*song-title*) Harlem rag: two step. **1916** *Variety* 25 Aug. 8 Ash..is seen daily on the streets playing rag dance numbers. **1922** T. S. Eliot *Waste Land* ii. 21 But O O O O that Shakespeherian Rag—It's so elegant So intelligent. **1947** G. Sklar *Two Worlds J. Truro* iii. 24 They listened to rags and stomps, to fox trots and marches. **1957** G. Lascelles in S. Traill *Concerning Jazz* 77 Few of the original rags were written, and those which were, had often no bass part added beyond the conventional harmonies. **1977** *New Yorker* 19 Sept. 96/2 She would play some Menotti, Barber, and Gershwin, a piece by Paul Tufts, a Seattle composer, and some Scott Joplin rags.

3. *attrib.* and *Comb.* in sense 'ragtime', as *rag music, musician, rhythm*; *rag-flavoured* adj.

1959 'F. Newton' *Jazz Scene* vi. 103 Rag-flavoured numbers also became part of the staple repertoire of New Orleans jazz. **1934** C. Lambert *Music Ho!* ii. 95 The *Ragtime*, like the piano *Rag Music*, is an abstract pattern created out of the raw material of certain syncopated devices. **1955** R. Davies in J. McCarthy *Jazzbk. 1955* 37 They left the Crescent City..intending to disseminate through the dance halls of Chicago the rag music they had created. **1976** R. Sanders in D. Villiers *Next Year in Jerusalem* 198 One of the great Negro rag musicians, Ben Horney, a composer and singer. **1923** R. H. Myers *Mod. Music* 65 Darius Milhaud has sought, by the use of rag-rhythms, to evoke the exotic yet intensely human atmosphere of the Bar and its inhabitants.

rag, var. *raga.

rag, *v.*[2] Add: **1. a.** (Earlier example.) Also, to examine or question.

1739 *Proc. Sessions of Peace* June 107/2 On Monday Night Bird and Clark came to their House to ragg (scold) her Grandfather for what he had talk'd of, concerning them. **1908** A. S. M. Hutchinson *Once aboard Lugger* i. iv. 47 Not one had ever worked. Each had been 'ragged' on a subject of which he knew absolutely nothing.

b. (Further examples.)

1891 *Spectator* 3 Jan. 3/2 The revellers went round and 'ragged' several men in their rooms. **1956** 'C. Blackstock' *Dewey Death* vii. 156 You're always ragging me, and I know you think I'm an ass. **1975** *Times* 30 Dec. 4/2 The President is now ragged mercilessly on national television, by talk show hosts, by comics, and in cartoons.

rag (ræg), *v.*[5] orig. *U.S.* [f. *rag *sb.*[5]] **1.** *intr.* To play, sing, or dance in ragtime. Also const. *it.*

1905 *Dialect Notes* III. 152 *Rag*,..to dance. 'Everybody *rag* as pooty (puti) as you can.' **1923** R. D. Paine *Comrades of Rolling Ocean* viii. 137 They were dancing on the pavement of the public market or ragging it on the smooth white streets. **1928** F. Scott Fitzgerald in *Sat. Even. Post* 29 Sept. 118/3 Oh, listen!.. Do you know how to rag? **1946** B. Treadwell *Big Bk. Swing* 125/1 *Rag*, to play the blues and jazz. **1971** C. C. Adams *Boontling* 237 *Rag*, to dance in 'ragtime' form, considered indecent in valley dance halls in the Boontling era.

2. *trans.* To convert (a melody, etc.) into ragtime; to play ragtime music on (an instrument).

1917 *Lit. Digest* 25 Aug. 28/2 The jazz bands take popular tunes and rag them to death to make jazz. **1922** H. L. Foster *Adventures Trop. Tramp* v. 47 The camp victrola was broken and..I was the only man in camp that could rag the piano. **1949** R. Blesh *Shining Trumpets* viii. 181 The violin played the melody straight while Bolden ragged it. **1956** G. P. Kurath in A. Dundes *Mother Wit* (1973) 108/1 The slaves ragged and syncopated their clog dances. **1960** [see *anti-[1] 2 c].

raga (rā·gă). Also **raag, rag** (rāg). [a. Skr. *rāga*, colour, passion, melody; Hindi *rāg*, a mode in music, music.] **1.** In Indian music, a melodic type which provides a framework for improvised melodies; such an improvised melody. Also *fig.*

1788 W. Jones in *Asiatick Researches* I. 264 The beautiful allegories of the Hindus in their system of musical modes, which they call Rágás, or Passions, and suppose to be Genii or Demigods. **1807** *Asiatick Researches* IX. 447 The Indian Rágas and Ráginís are fixed respectively to particular seasons of the year and times of the night or day. **1891** C. R. Day *Music & Mus. Instr. S. India & Deccan* ii. 23 Mode and râga are..perfectly distinct from each other. *Ibid.* 24 In almost all these works a somewhat similar classification of the râgs and râginis has been adopted. **1924** E. M. Forster *Passage to India* vii. 80 The song is composed in a raga appropriate to the present hour, which is the evening. **1934** C. Lambert *Music Ho!* v. 289 Although the scales of folk music may vary from the simple pentatonic scales of the Hebridean to the complicated ragas of the Hindu, the same outlook on tonality is implied. **1944** W. Apel *Harvard Dict. Mus.* 332/2 The ragas fall under the classification of melody-types... A raga is a much more specialized tonal frame than a mode; it prescribes..not only a scale and a center-tone (*amsa*), but also the avoidance of certain tones. **1958** *Times* 30 May 16/5 The Indian *raga* provides the structure for an improvisatory kind of music which.. approaches the articulation of speech. **1962** I. Holst *Tune* iv. 53 It is impossible to memorize fragments of a raga during a performance. **1967** 'La Meri' *Spanish Dancing* (ed. 2) v. 67 The *Rag* (roughly, melody) has a definite mood throughout. **1969** *Sunday Standard* (Bombay) 3 Aug. (Magazine Section) p. iv/6 Instead of outlining the form of his 'raag' in sinuous, unbroken contours he seems to have preferred segment sketching. **1970** W. Burroughs *Speed* vii. 151 On stage, a couple of young guys played a mediocre raga. **1971** J. Mandelkau *Buttons* vii. 81 'The Truth is the Only Law!' It's sewn on the inside shoulder of William's jacket and has become my daily prayer. My holy raga. **1972** P. Holroyde *Indian Music* 277 *Ragas* developed as an arrangement of intervals so that seven notes are in a certain relationship with each other and thus define a melody. **1975** *Guardian Weekly* 25 Jan. 21/3 The score by Alan Lloyd can best be described as Schubertian raga, a non-stop prelude (vamp till ready) to the verbal fugues on the stage. **1977** Y. Menuhin *Unfinished Journey* xii. 258 A *raga* is a scale-cum-melody, a given sequence of notes whose interrelationships are already determined.

2. *attrib.* and *Comb.*, as *raga form, quality, system*; **raga rock,** rock music characterized by improvisation, etc., in the style of a raga.

1968 *Jrnl. Mus. Acad. Madras* XXXIX. 7 The classical music of India has thus for its aim the delineation of the Raga-forms. **1972** *Last Whole Earth Catalog* (Portola Inst.) 3/1 Fuller's lectures have a raga quality of rich nonlinear endless improvisation full of convergent surprises. **1960** *Melody Maker* 30 Apr. 12/1 America's..Byrds come up with a new formula! Raga-rock, based on Eastern musical forms. **1967** P. Welles *Babyhip* (1968) xv. 108 'Leave me alone,' she said, 'you're some kind of sex fiend. And anyway, I'm looking for raga rock.' She turned the station to another. **1957** A. Bake in *New Oxf. Hist. Mus.* I. iv. 213 In the classical *râga* system such a new creation usually bears the name of its parents, as, for instance, *Megh-Malhâr.*

‖ **ragazzo** (raga·tso). Pl. **ragazzi**; fem. **ragazza.** [a. It. *ragazzo* boy, *ragazza* girl.] In Italy: a youngster; a lad or a young girl.

1862 Borrow *Wild Wales* I. xxiv. 283 When I was a ragazzo I knew many from the Lake of Como, who dressed much like yourself. **1897** A. T. Ritchie *Let.* Mar. (1924) xi. 239 Two nice little ragazzi showed us the way. **1906** W. De Morgan *Joseph Vance* xliv. 404 In a day or two Beppino was bored, and..I noticed that he was keeping at a respectful distance from every ragazza. **1957** 'N. Culotta' *They're Weird Mob* (1958) ii. 24 'What please is a sheila?' 'A sheila? A bint. A ragazza.' **1975** *Publishers Weekly* 27 Jan. 230/1 A novel concerning a *ragazza* of the streets grown to movie star.

rag-bag, *sb.* Add: **a.** (Earlier and later examples.)

1820 M. Wilmot *Let.* 7 Aug. (1935) 76 Well, and well and well, what have I got to say in my *ragbag* of a brain? I have a hundred odds and ends. **1850** Dickens *Dav. Copp.* xlviii. 490 Sheets in the rag-bag. **1853** C. Brontë *Villette* I. ix. 174 Your mind..seems..chaotic as a rag-bag. **1917** E. Pound *Lustra* 181 And that the 'modern world' Needs such a rag-bag to stuff all its thought in. **1964** A. Sexton *Sel. Poems* 57, I hid in the kitchen under the ragbag.

b. (Further examples.) Also *spec.* a sloppily-dressed woman, a slattern. *slang.*

1888 Kipling *Under Deodars* 70 If I were a man I would perish sooner than be seen with that rag-bag. **1937** Partridge *Dict. Slang* 683/2 *Rag-bag*,..a slattern. **1973** 'W. Haggard' *Old Masters* vi. 75 Lord Tokenhouse was no sort of economist, indeed he despised the whole ragbag wholeheartedly. **1976** P. Cave *High Flying Birds* iii. 31 She was neither attractive nor plain; not a raver or a ragbag. **1980** *Jrnl. R. Soc. Arts* Feb. 149/2 The design courses are in danger by their very existence of joining Art, Music and Technical Drawing in the ragbag of the system.

2. *attrib.*

1907 *Daily Chron.* 13 Mar. 8/4 The association of wealth and rag-bag poverty in London is one of her most remarkable features. **1948** *Life* 13 Sept. 152/2 In the old days.. some of them little 'rag-bag' shows used to carry a lot of grift. **1953** S. Kauffmann *Philanderer* vi. 104 The living-room, with its rag-bag summer cottage furnishings. **1968** *Listener* 31 Oct. 583/2 There is certain to be some inherent weakness in the direction of generalisation and 'rag-bag' proliferation of studies that I deplore. **1980** *Jrnl. R. Soc. Arts* Feb. 154/1 That is what I call the ragbag attitude.

rage, *sb.* Add: **7. a.** Also const. *after.*

1790 *Loiterer* 2 Jan. 4 This prevailing rage after knowledge.

b. (Further examples.)

1802 *Monthly Mag.* 1 Oct. 253/1 The *rage* for the dotting style of engraving..is on the decline. **1811** Byron *Let.* 15 Dec. (1973) II. 149 Tomorrow, I dine with Rogers & am to hear Colridge, who is a kind of rage at present. **1834** Lytton *Last Days of Pompeii* I. i. 173 Sylla is said to have transported to Italy the worship of the Egyptian Isis. It soon became 'the rage'—and was peculiarly in vogue with the Roman ladies. **1861** K. Stone *Jrnl.* 28 Aug. in *Brokenburn* (1955) 48 Plaiting palmetto for baskets has been the rage for several days. **1881** [see Colloquialism 2]. **1940** Graves & Hodge *Long Week-End* iii. 38 After the war the new fantastic development of Jazz music and the steps that went with it, became, in the contemporary phrase, 'all the rage'. **1951** [see *dead *a.* 30].

rageless, *a.* For *Obs. rare*[−1] read *rare* and add later example.

1948 R. Graves *Coll. Poems* 218 My self reversed, my rage-less part, a slimy yellowish cone.

Hence **ra·gelessness,** absence of rage or rages.

1904 E. F. Benson *Challoners* v. 101 London, tired with its spinster ragelessness, rose at them as trout rise in the days of May fly.

rager. a. (Later example.)

1925 G. Murray tr. *Aeschylus' Eumenides* 4 The ragers sleep: the Virgins without love.

ragesome (rēᶦ·dʒsəm), *a. U.S. rare.* [f. Rage *sb.* + -some.] Rageful, angry.

1913 G. Stratton-Porter *Laddie* xvii. 580 He can be awful ragesome when he's excited.

‖ **raggare** (ra·garə). Pl. **-(s).** [Sw., f. *ragga* to pick up (girls).] In Sweden: a member of a gang of youths who cruise about in cars; a street-tough, a teddy-boy.

1964 *Pix* 25 Jan. 45/3 By midnight even most of the 'raggare'—Sweden's beatniks—have disappeared from the streets. **1971** *Daily Tel.* 2 Aug. 4/8 About 300 'raggares'—Sweden's motorised teddy boys—clashed with 90 police in..Stockholm yesterday. **1977** *Time* 25 July 10/1 To their bitterest enemies, a group of restless young toughs known as the Raggare, the Assyrians are despised 'black-skulls', to be attacked with chains and clubs.

ragged, *a.*[1] Add: **II. 6.** Also of mood or condition: tired, run-down. Colloq. phr. *to run* (one) *ragged* (orig. *U.S.*): to exhaust or debilitate (a person).

1925 *New Yorker* 5 Sept. 10/3 This eighteen-year-old youngster ran Bill Johnston, the Californian, ragged. **1951** W. Stevens *Let.* 26 Mar. (1967) 712 This is simply typical of the sort of thing that runs one ragged. **1969** M. Pugh *Last Place Left* xxix. 213 Sorry, sir, I'm pretty ragged. Is Miss Drummond okay? **1970** A. Draper *Swansong for Rare Bird* v. 35 We really ran the teachers ragged. **1977** P. Hill *Liars* (1978) iii. 36 All four of them were now feeling mentally ragged.

7. *ragged-edged, -looking* (earlier example) adjs.; **ragged edge** *U.S. slang*: in phr. *on* (also *in*) *the ragged edge*, on the extreme edge or verge; also *transf.*, in a state of distress or resourcelessness; **ragged r** (see quots.).

1885 *N.Y. Mercury* 10 Jan. 4/7 It seems fair to assume that father, daughter and her child sailed yesterday for Paris, leaving poor Tom on the ragged edge. **1889** 'Mark Twain' *Connecticut Yankee* xvi. 196 He was always on the ragged edge of apprehension. **1892** —— *Amer. Claimant* ii. 28 It was away out in the ragged edge of Washington and had once been somebody's country place. **1935** A. J. Pollock *Underworld Speaks* 84/1 *On the ragged edge*, slight chance to make good; down and out. **1916** 'Boyd Cable' *Action Front* 164 The face of one house was marked by a huge splash, with solid centre and a ragged-edged outline of radiating jerky rays. **1833** F. Witts *Diary* 8 Apr. (1978) 91 One tract of common over which we passed is called Coalpit-heath, a ragged-looking spot. **1755** J. Smith *Printer's Gram.* iv. 117 Black letter..has two different r's, one of which is called the ragged r [ꝛ], and is particularly used after letters that round off behind. **1969** H. Carter *View of Early Typogr.* iii. 62 The ragged r—the r that followed letters that had once been round.

ragged (rægd), *ppl. a.*[2] [f. Rag *v.*[2]] That has suffered ragging, teasing, or annoyance.

1903 *Westm. Gaz.* 11 May 6/2 The 'ragged' officer was allowed leave of absence and has not yet returned to duty.

ragged (rægd), *ppl. a.*[3] [f. *rag *v.*[5]] Of music: that has been converted to ragtime.

1956 G. P. Kurath in A. F. C. Wallace *Men & Cultures* (1960) 155 They represent the following steps:..Two-step (ragged). **1958** C. Wilford in P. Gammond *Decca Bk. Jazz* ii. 40 Ragtime lives on in jazz, for all jazz is based on ragtime, though it is true some parts are more ragged than others.

raggedy, *a.* Substitute for entry (s.v. Ragged *a.*[1]):

raggedy (ræ·gėdi), *a.* Chiefly *U.S.* and *dial.* Also 9– **raggety; raggity.** [f. Ragged *a.*[1] + -y[1].] **1.** Of a ragged form or appearance.

a. Of persons. Cf. Ragged *a.*[1] 6.

1890 J. W. Riley in *Century Mag.* Dec. 318/2 Oh, the Raggedy Man! He works fer Pa; An' he's the goodest man ever you saw! **1894** [in Dict.]. **1939** J. Steinbeck *Grapes of Wrath* xxiv. 462 A raggedy kid with no shoes. **1959** *Encounter* Dec. 17/2 The covey of raggety small boys. **1960** C. Day Lewis *Buried Day* iii 38 The bridge..was lined..with raggedy men. **1975** *New Yorker* 1 Dec. 48/2 A lady's.. raggedy kids lined up by the track where they'd put pennies.

b. Of clothing. Cf. Ragged *a.*[1] 5 a.

1893 *Trans. Devonshire Assoc. Adv. Sci., Lit., & Art* XXV. 201 My smock's a-got cruel raggety, I sim. **1929** D. Runyon in *Hearst's Internat.* Oct. 62/2 Mostly she is wearing raggedy clothes and busted shoes. **1967** H. Porter in *Coast to Coast 1965–6* 171 Underwear got more raggedy. **1978** J. Carroll *Mortal Friends* iii. v. 322 It serves me right for asking what raggedy clothing you'd flung off.

2. Of a rough, irregular, or straggling form. Cf. Ragged *a.*[1] 2. **a.** *gen.*

1896 'Mark Twain' in *Harper's Mag.* Aug. 358/1 Raggedy white patch between the shoulders..looked like somebody had hit him with a snow-ball. **1899** [in Dict.]. **1938** L. Mumford *Whither Honolulu?* v. 16 Cities throughout America have been..over-extending themselves in raggedy fragments. **1955** J. Masters *Coromandel!* i. 23 Voy sucked his raggedy moustache.

b. Of branches, plants, etc.

1912 J. Stephens *Crock of Gold* v. 40 There was a raggedy blackberry hedge all round the field. **1927** D. H. Lawrence *Mornings in Mexico* 12 Like rather raggedy green buds climbing to the sun.

3. Of music or rhythm: irregular, uneven, broken. Cf. Ragged *a.*[1] 3 b.

1949 R. Blesh *Shining Trumpets* viii. 186 The tempo is buoyant, the beat is alive, and the rhythm is very *raggedy*. **1975** *New Yorker* 19 May 113/2 A casual encounter between a boy and a girl who compete to the piled-up, off-center rhythms of many raggedy dances.

4. *Comb.*, as **Raggedy Andy** *U.S.*, a rag-doll, the male counterpart of Raggedy Ann (see next); **Raggedy Ann** orig. *U.S.*, a rag-doll with short, mop-like, red hair; also *attrib.*; **raggedy-ass(ed)** *a. U.S. slang* (orig. *Mil.*) [cf. Ass, vulgar and dial. var. Arse in Dict. and Suppl.], of persons: inexperienced, raw; also *transf.*

1920 J. Gruelle *Raggedy Andy Stories* 2 Gran'ma had told Daddy..that at the time Raggedy Ann was made, a neighbor lady had made a boy doll, Raggedy Andy. **1974** *News & Reporter* (Chester, S. Carolina) 22 Apr. 5-a/1 He was honored with a family party at his home with guests enjoying Raggedy Andy Cake. **1918** J. Gruelle *Raggedy Ann Stories* Pref., To the millions of children and grown-ups who have loved a Rag Doll, I dedicate these stories of Raggedy Ann. **1957** M. B. Picken *Fashion Dict.* 124/2 *Raggedy Ann costume*, costume taken from child's story book about a stuffed doll. Consists of bright colored patched skirt, patched apron, simple white blouse, white socks and black shoes. Usually worn with ragged woolen wig. **1966** J. S. Cox *Illustr. Dict. Hairdressing* 124/2 *Raggedy Ann Bob*, a short hair style for women similar to the wind-blown bob. **1967** *Southerly* XXVII. 209 Katrina..brushed the short, raggedy-ann hair until it shone. **1970** *New Yorker* 14 Nov. 50/3 She removed the clothes from a Raggedy Ann doll and stuck a pin..into the center. **1976** *Times* 8 July 12/1 (*caption*) Raggity Ann patched up trousers. **1977** *Redbook* Mar. 81/2 Kim wanted a Raggedy Ann cake. **1930** T. Fredenburgh *Soldiers March* vii. 50 The Raggedy Ass Cadets are out today. **1956** B. Holiday *Lady sings Blues* (1973) vi. 57 I'd have to travel five hundred to six hundred miles on a hot or cold raggedy-ass Blue Goose bus. **1967** E. Liebow in T. Kochman *Rappin' & Stylin' Out* (1972) 406 A nice woman does not dress in 'raggedy-ass clothes'. **1969** J. A. McPherson *Hue & Cry* 58 Who taught you the moves when you were just a raggedy-ass waiter? **1929** in J. J. Niles et al. *Songs My Mother never taught Me* 182 The raggedy assed cadets are on parade. **1971** R. Flanagan *Maggot* 252 Respect what man, you raggedy-assed little fuck?

ragger[2]. [f. Rag *v.*[2] + -er[1].] One who rags or teases; *spec.* a participant in a student rag.

1903 *Speaker* 7 Feb. 451/1 There is much to be said in favour of the 'raggers'. Mere 'ragging' as distinguished from ..persistent and brutal bullying..never yet did a youngster any harm. **1905** *Westm. Gaz.* 15 May 7/3 One of the raggers received a bullet in the mouth, and is seriously injured. **1909** H. G. Wells *Ann Veronica* xi. 220 Ann Veronica decided that 'hoydenish ragger' was the only phrase to express her. She was always breaking rules. **1930** *Daily Express* 6 Nov. 1/2 Guys were then thrown on the blazing piles amid the wild shrieks of the 'raggers'.

raggie (ræ·gi). *Mil.* Also **raggy.** [f. Rag *sb.*[1] + -ie.] † **1.** A mess jacket. *Obs.*

1843 F. J. Bellew *Mem. Griffin* I. ix. 126 As for myself, in my scarlet raggie, brimstone facings,..and regulation sword, in my own opinion, I looked quite the god of war. **1903** *N. & Q.* 26 Dec. 512/2 In India, in the early part of last century,..the scarlet 'shell' jacket, or mess jacket—almost the only uniform then worn in that country—was invariably called a 'raggie', and this not jocularly or as slang, but seriously and as a matter of course.

2. *Naval slang.* (See quot. 1912.)

1912 'Aurora' *Jock Scott* xiv. 170 A 'raggy' is a friend whom you know so intimately that you feel you could with confidence keep your brass-rags in the same bag as his. **1914** 'Bartimeus' *Naval Occasions* xiv. 111 'If I don't get no letter this mail—so 'elp me I stops me 'arf pay,' he confided grimly to a 'Raggie'. **1916** 'Taffrail' *Carry On!* 27 Men who are friendly with each other are 'raggies', because they have the free run of each others' polishing paste and rags; but if their friendship terminates they are said to have 'parted brass-rags'. **1946** J. Irving *Royal Navalese* 143 *Raggie*, a very close friend.

ragging, *vbl. sb.*[2] (Further examples.) Also *attrib.*

1888 E. Dowson *Let.* Nov. (1967) 19 After a good deal of ragging with Chitty J. two days ago, the affair was settled. **1920** *Chambers's Jrnl.* 1 May 337/1 An ugly ragging mood was astir. **1932** *Daily Tel.* 8 Oct. 12/4 Ragging in the army, such as we have at home (this was said just after one of the so-called 'ragging scandals' in the Guards) would be impossible here. **1942** *R.A.F. Jrnl.* 2 May 31, I miss the comradeship, the ragging, the talks in Mess.

ragging, *vbl. sb.*[3] Add: **2.** (Earlier example.)

1860 C. Tomlinson *Useful Arts & Manuf.* 2nd Ser. *Dressing ores* 10 That portion [of a dredging] occupying the bottom of the sieve, called ragging, which is also in a marketable state.

3. = Straggling *vbl. sb.*[2]

1850 [see Straggling *vbl. sb.*[2]].

ra·gging, *vbl. sb.*[4] orig. *U.S.* [f. *rag *v.*[5]] The act or practice of playing, singing, or dancing in ragtime.

1899 [see *rag *sb.*[5] 1]. **1913** *Collier's Mag.* 15 Feb. 6/2 The worst of these dance halls..are habitually frequented by people of the fashionable and so-called decent class, who go..for the purpose of joining in the 'ragging'. **1914** 'High Jinks Jr.' *Choice Slang* 17 *Ragging*, the act of doing the 'Rag'. **1936** *Harper's Mag.* 3 Apr. 570/2 'Jamming', 'cat-time', 'swing', 'riffing', 'getting off', 'going to town', 'ragging', 'gut-bucketing',..are names for the *hot* performance, which is the heart and soul of jazz. **1958** *Life* (Internat. ed.) 13 Oct. 96/2 The rhythm was called ragtime (after a Negro clog dance sometimes called 'ragging').

raggle, *sb.*[1] (Earlier and later examples.)

1835 in *Sc. Nat. Dict.* (1968) s.v. *Raggle* v.[1], n.[1] **1956** *Scotsman* 22 Sept. 9/1, I am..taken aback by the nonchalance with which these..men stroll about among the chimney pots..uttering deep thoughts about raggles and sarking, flashing and skews and Rafit verges.

raggle-taggle (ræ·g'l₁tæ:g'l), *a.* and *sb.* Also **wraggle-taggle.** [App. fanciful var. Rag-tag.]

A. *adj.* Of a group of persons: ragged, rambling, straggling; disreputable, unorganized (freq. used in allusion to the song referred to in quot. 1904). Hence of a person and in extended use of appearance, etc.

1904 in Sharp & Marson *Folk Songs from Somerset* 1st Ser. 19 What care I for my house and my land? What care I for my money, O? What care I for my new wedded lord? I'm off with the wraggle taggle gipsies, O! **1913** C. Mackenzie *Sinister St.* I. i. ii. 29 He made up his mind..that it was better to be a raggle-taggle wanderer than anything else. **1923** J. Buchan *Midwinter* i. 34 My companions are the moor-men and..the raggle-taggle gypsies. **1933** W. Starkie *Raggle-Taggle* ii. 7 It was no easy task to convince my family of my sanity in wishing to follow the raggle-taggle Gypsies. **1936** C. Day Lewis *Friendly Tree* 132 Like the lady who ran away with the wraggle-taggle gipsies. **1942** C. Barrett *On Wallaby* vi. 122 It was a Paddy's market day and a raggle-taggle crowd had gathered to bargain shrewdly and noisily. **1970** N. Bawden *Birds on Trees* v. 83 It's not easy for someone of my generation to accept this raggle-taggle look, we had a bit more self-respect. **1977** O. Jacks *Autumn Heroes* viii. 118 How would you rate your raggle-taggle band of poorly armed warriors?

B. *sb.* **a.** A wanderer, a rover. **b.** A straggling, unorganized, or disreputable collection of persons.

1933 W. Starkie (*title*) Raggle-taggle. **1958** *Times* 19 Mar. 7/4 President Nasser's permanent Afro-Asian guests—that Cairo raggle-taggle ranging from the permanent representatives of the Algerian National Liberation Front on the one hand to the office of the Deputy Imam of Oman on the other. **1972** 'H. Calvin' *Take Two Popes* ii. 18 The Curia..considered *all* non-Italians as an unfortunate raggle-taggle.

raggling. Add: Also 5 **raggalyne,** 7– **raglin(e, ragling. a.** *Build.* Substitute for def. = Raggle *sb.*[1]

1500 in *Sc. Nat. Dict.* (1968) s.v. *Raggle* v.[1], n.[1] **1683** [in Dict. s.v. Raggle *v.*]. **1704–1898** in *Sc. Nat. Dict.* (1968) s.v. *Raggle* v.[1], n.[1] **1929** H. Marwick *Orkney Norn* 137/1 *Raglins*, the top of the side walls of a house, the space between top of wall and the slates. **1964** J. S. Scott *Dict. Building* 255 *Raglet* or *raggle* or *raglin*, a thin groove, in stone often dovetailed, cut in stone or in a mortar joint of brickwork to receive the end of a lead flashing, which is fixed by burning in or wedging.

b. *Mining.* (See quot. 1886.)

1839 Ure *Dict. Arts* 985 There is a simple mode of conducting air from the pit bottom to the forehead of the mine, by cutting a ragglin, or trumpeting, as it is termed, in the side of the gallery. **1886** J. Barrowman *Gloss. Scotch Mining Terms* 53 *Raggling*, a channel cut in the side of a mine and covered with boarding to serve as an airway.

raggy (ræ·gi), *a.*[2] *slang.* [f. Rag *v.*[2] + -y[1].] Annoyed; irritated.

1900 G. Swift *Somerley* 21 He was jolly raggy about us taking his old gee.

raggy (ræ·gi), *a.*[3] orig. *U.S.* [f. *rag *sb.*[5] + -y[1].] Of music: pertaining to or resembling ragtime; characterized by ragtime.

1933 *Fortune* Aug. 92/1 At sixteen he began to play raggy music for Washington society with Louis Thomas' orchestra. **1944** W. Russell in M. T. Williams *Art of Jazz* (1960) iv. 36 His feeling for a joyful, raggy, and stompy rhythm. **1952** B. Ulanov *Hist. Jazz in Amer.* (1958) xv. 181 Duke wove tricky, raggy, endlessly inventive variations around the Miley theme. **1958** in P. Gammond *Decca Bk. Jazz* xv. 185 Arthur Schutt, whose raggy piano sounded on many a Mole or Nichols session. **1972** *Jazz & Blues* Dec. 30/2 Billie's very raggy piano.

raggy, var. *raggie.

raging, *ppl. a.* Add: **1. d.** Highly successful, tremendous; also as a mere intensifier. *raging favourite*, 'hot' favourite. *colloq.*

1886 H. Baumann *Londinismen* 151/2 A raging favourite. **1889** 'Mark Twain' *Connecticut Yankee* xxxi. 398 He..was doing a raging business. **1894** —— in *St. Nicholas* Mar. 400/2 A raging lot of sand. **1977** *Hongkong Standard* 12 Apr. 12/2 Raging favourite Orange Peel was pushed to the limit by Glynn Parry.

‖ **ragini** (rā·gini). [a. Skr. *rāgiṇī*, lit. coloured, impassioned.] In Indian music, a modification of a raga.

1788 W. Jones in *Asiatick Res.* I. 264 The Nymphs of Musick are the thirty Ráginis or Female Passions. **1807**, **1891** [see *raga]. **1944** W. Apel *Harvard Dict. Mus.* 333/1 The theoretical system..knows 6 (main) ragas and 30 *raginis*, each of which are duplicated according to the sa-grama or the ma-grama scheme (perfect or augmented fourths). **1948** B. Bose *Acre of Green Grass* 41 In classical Indian music..words are no more than pleasing containers of the *ragini*, or the melodic pattern. **1954** *Grove's Dict. Mus.* (ed. 5) IV. 457/1 Seven were chosen in the end..to serve as basic modes or pure *jâtis*. From these seven *jâtis* the whole elaborate structure of *râgas* and *râginis* was eventually developed. **1968** *Indian Mus. Jrnl.* V. 33 He minutely studied the nature of various *rāga-s* and *rāginī-s*, to find out their suitability for expressing different emotions.

Raglan. Add: Now usu. with small initial.

a. (Earlier and later examples.)

1863 'G. Hamilton' *Gala-Days* 27 A thousand considerations, in the shape of raglans..induce you to modify your view. **1926** *Daily Colonist* (Victoria, B.C.) 5 Jan. 2/6 (Advt.), Overcoats... Some with half belts, full belters and raglan-back styles. **1948** M. Laski *Tory Heaven* xii. 166 A thick raglan overcoat over his dinner jacket. **1976** 'J. Ross' *I know what it's like to Die* ii. 11 His thick raglan overcoat.

b. *attrib.* Applied to the shoulder or sleeve of any (esp. knitted) garment designed after the style of a raglan. Also *absol.*, a garment with such sleeves.

1906 *Daily Chron.* 4 Oct. 3/4 As for the Raglan shoulders they are only to be worn with sporting clothes. **1930** M. Story *Individuality & Clothes* ii. ii. 204 Raglan and kimono sleeves tend to increase the breadth of the figure. **1957** M. B. Picken *Fashion Dict.* 269/2 *Raglan sleeve*, sleeve with long armhole line extending to neckline. **1966** *Illustr. London News* 26 Feb. 5 (Advt.), The knitwear is made of Shetland wool and is fully fashioned with raglan sleeves. **1969** *Sears Catal.* Spring/Summer 26 *Cardigan sweater* interlock knit of Orlon acrylic... Rib knit cuffs on the long sleeves, raglan shoulders. **1978** *Detroit Free Press* 2 Apr. 5D/1 One reason for the great popularity of 'knit-from-the-neck-down' raglans is the fact that there are no sleeves to set in. **1980** *Times* 11 Nov. 10/5, I liked..the reversible coats with raglan shoulders.

c. *Comb.*, as *raglan-coated, -sleeved* adjs.

1975 J. Symons *Three Pipe Problem* xviii. 207 Twenty yards ahead..stalked the raglan-coated figure of Sheridan Haynes. **1974** S. Marcus *Minding Store* (1975) iv. 81 The first raglan-sleeved fur coat.

ragman[1]. Add: **3. a.** (Later examples.)

1966 F. Shaw et al. *Lern Yerself Scouse* 24 *De ragman*, the old-clothes man. **1976** *New Yorker* 23 Feb. 39/2 The street down which will sometimes come, on his rattling wagon, a ragman.

4. [See *rag *sb.*[5]] A musician who plays ragtime music.

1938 J. R. Morton in *Downbeat* Sept. 4/1 Blues players who could play nothing else... What we call 'ragmen' in New Orleans. **1950** Blesh & Janis *They all played Ragtime* (1958) vi. 108 Following 1907–8 there comes a second generation of ragmen. **1970** C. Major *Dict. Afro-Amer. Slang* 96 *Ragmen*, jazzmen who play that type of music.

Ragnarök (ræ·gnārök). Also 8 **Ragnarockur; Ragnarok, Ragnarökr.** [ON. *ragna rök*, f. *ragna* gen. of *regin* the gods + *rök* destined end (a conventional usage in this phrase of *rök* the course of events from origin to end, their causes and consequences); properly as above, but the form *Ragna rökr* (*rökr* twilight, a more readily comprehensible substitute for the archaic *rök*) occurs in the prose Edda, and the mod. sense is freq. founded upon this (see Cleasby-Vigfusson ed. 2).] In Scandinavian mythology, the destruction of the gods or the twilight of the gods; *spec.* the last battle of this world, in which gods and men will be defeated by monsters and the sun will grow dark. Also *fig.*

1770 T. Percy tr. *Mallet's Northern Antiquities* I. vi. 119 The end of the world, that is,..the time of that universal desolation of nature which was to be followed by a new creation, and what they called *Ragnarockur*, or the Twilight of the Gods. **1866** *Chambers's Encycl.* 525/2 This terrible age of destruction, the Ragnarök, or twilight of the gods, will be marked by a three years' winter of hard frost. **1887** A. Lang in *Contemp. Rev.* Nov. 693 If the battle between the crocodile of Realism and the catawampus of Romance is to be fought out to the bitter end—why, in that Ragnarök, I am on the side of the catawampus. **1899** A. H. Yapp *Cuckoo* iii. 178 The ragnarok of the cuckoos—the last band of the migrating males. **1933** A. Thirkell *High Rising* xi. 199 Zeus dethrones Chronos..Ragnarok swallows up the Gods. **1953** G. Turville-Petre *Origins Icelandic Lit.* i. 15 The giant..predicts the *Ragnarøk* (Fall of the Gods) and the manner of Óðinn's death. *Ibid.* ii. 57 The treacherous Loki was bound in chains, but he will break them when the gods' Doomsday (*Ragnarøk*) draws near. **1965** H. M. Smyser in Bessinger & Creed *Medieval & Linguistic Stud.* 110 It is not the place where Odin's *einherjar* fight daily as they await Ragnarøkr. **1978** *Islands* (N.Z.) Aug. 86 Ragnarök and Loki's frames Lie curing in the dust.

rag-tag. Add: **B.** Passing into *adj.* Of form or appearance: ragged, raggle-taggle; disreputable; disorderly, unorganized, straggling.

a. Of persons, etc.

1883, 1884 [see sense 1 a in Dict.]. **1969** *Telegraph* (Brisbane) 19 Aug. 19/4 In unison a ragtag band of motorcyclists drank a toast. **1978** S. Brill *Teamsters* ix. 322 These were by and large a ragtag group of overaged ne'er-do-wells.

b. Of things.

1922 Blunden *Shepherd* (ed. 2) 30 When on the green the rag-tag game had stopd. **1969** *Jrnl. Amer. Chem. Soc.* XCI. facing p. 4946 (Advt.), Such rag-tag, sometimes illegible copies probably stay in your files..permanently. **1977** *Rolling Stone* 5 May 11/1 The audience was evenly divided between under-18s in ragtag street wear and those first-generation rock fans now moving into Sisley jeans affluence.

rag-time. Substitute for entry:

ragtime (ræ·gtəim). orig. *U.S.* Also **rag-time.** [Prob. f. *rag *sb.*[5] + Time *sb.*] **1. a.** A musical rhythm characterized by a syncopated melodic line and regularly-accented accompaniment, evolved among American Negro musicians in the 1890s; hence, music (esp. for the banjo and piano) of this character, the immediate precursor of jazz.

1897 W. H. Krell *Mississippi Rag* (title-page of sheet-music), The first rag-time two-step ever written. **1898** *Étude* Oct. 285/3 'Rag time' is a term applied to the peculiar, broken, rhythmic features of the popular 'coon song'... Unfortunately, the words to which it is allied are usually decidedly vulgar, so that its present great favor is somewhat to be deplored. **1899** *Musical Rec.* (Boston) 158/1, I feel safe in predicting that rag-time has come to stay. **1900** *Musical Courier* 23 May 20/2 'Rag-time' is a rhythm which is the most characteristic feature of what may be called American negro music. **1906** 'O. Henry' *Four Million* 238 They sing 'Home Sweet Home' in ragtime. **1916** A. Huxley *Let.* 7 Aug. (1969) 109, I have been sleeping out on the roof..spending most of the night in conversation or in singing folk-songs and rag-time to the stars. **1934** C. Lambert *Music Ho!* iii. 206 Jazz, or to be pedantically accurate, 'ragtime',..has suddenly achieved the status of a 'school'. **1938** *Sun* (Baltimore) 6 Sept. 8/7 Under their ministrations, the simplest ragtime becomes jazz. **1956** B. Edwards in S. Traill *Play that Music* 59 Ragtime, to give it its contemporary title [in 1920], was absolute anathema to both my parents. **1957** G. Lascelles in S. Traill *Concerning Jazz* 76 In the late nineteenth century there started a music which was a rather modern development of the old dance rhythms such as the schottische and polka... This development was later to be known as 'ragtime', and was essentially piano music. **1968** *Blues Unlimited* Nov. 23 Ragtime one probably loves or loathes. **1976** *New Yorker* 8 Mar. 32/3 He started to play, and before long he was lost in some wildly effervescent ragtime.

† **b.** A piece of music in rag-time; = *rag *sb.*[5] 2. *Obs.*

1914 G. B. Shaw *Misalliance* p. cviii, If they [*sc.* our young people] had learnt what can be done with syncopation from Beethoven's third Leonora overture, they would enjoy the ragtimes all the more. **1916** *Oxford Song Bk.* p. iii, An authority has just informed me that 'rag-times are "back numbers" now'.

2. a. *attrib.* and *Comb.*, as *ragtime accompaniment, band, melody, music, party, rage, record, saloon, singer, song, sound, tune, wedding.*

1901 Ragtime accompaniment [in Dict.]. **1911** I. Berlin (*song-title*) Alexander's Ragtime Band. **1949** R. Blesh *Shining Trumpets* vii. 156 Buddy Bolden's Ragtime Band of 1893, generally considered the first jazz band. **1906** *Westm. Gaz.* 23 Apr. 7/2 He is amusing the crowd with ragtime melodies. **1921** R. D. Paine *Comrades of Rolling Ocean* v. 75 Through an open hatch rose the rag-time melodies of a piano. **1897** Ade *Pink Marsh* 159 He told of his belief that the angels in heaven played 'rag-time' music. **1977** *Monitor* (McAllen, Texas) 9 Jan. 5c/3 Tichenor..says *ragtime music* peaked in the 1920s and died in the Depression. **1960** *Times* 29 Sept. 16/6 She pushes him downstairs during a ragtime-party. **1900** *Musical Courier* 23 May 20/1 (*heading*) The rag-time rage. **1906** Taylor & Gibson *Extra Dry* 71 Putting a ragtime record in the graphophone he pulls the throttle wide open. **1903** *Outing* Aug. 552/1 He has as many eyes for the ragtime saloon. **1917** E. Wallace *Kate plus Ten* (1930) v. 74 A peer of the realm and a ragtime singer. **1914** Ragtime song [see *high kick* s.v. *high *a.* 21]. **1927** *Jrnl. Abnormal & Social Psychol.* XXII. 19 The earliest ragtime songs, like Topsy, 'jes' grew'. **1974** *Times* 27 Apr. 9/8 We've ended up with a ragtime sound you might even have heard from the piano accompanying the silent movies. **1913** Kipling *Diversity of Creatures* (1917) 282 The boys and girls at the piano played the rag-time tunes of their own land. **1922** F. Scott Fitzgerald *Beautiful & Damned* ii. i. 131 Kept telling me she wished this was a ragtime wedding.

b. *attrib.*, passing into *adj.* Ragged; irregular, inferior, disorderly; disreputable, mean. *slang.*

1919 H. Crane *Let.* 17 June (1965) 20 Your remarks 'about the ladies' really hurt me with a kind of ragtime vulgarity. **1926** F. M. Ford *Man could stand Up* ii. 119 A Hun up against a Tommie looked like a Holbein *lansknecht* fighting a music-hall turn. It made you feel that you were indeed a rag-time army. *c* **1926** 'Mixer' *Transport Workers' Song Bk.* 47 Note the constant drop in wages..Endorsed by every rag-time press That the master-class command. **1929** *Papers Mich. Acad. Sci., Arts & Lett.* X. 317/2 *Ragtime Army*, the Australasian forces. **1940** Graves & Hodge *Long Week-End* ii. 26 The topic of 'this rag-time f—g peace' succeeded that of 'this bloody f—g war'. *Ibid.* iii. 38 The more extravagant 'rag-time' dances had not been socially approved. 'Rag-time' was an adjective of reproach; a ragtime regiment was a disorderly and untrustworthy one. **1948** V. Palmer *Golconda* ii. 14 But what if some really big concern swallowed up the three ragtime companies and planned to open up the whole mountain? **1974** P. Wright *Lang. Brit. Industry* xiv. 142 Hated shifts in that [electrical engineering] industry are the *rag-time* or *spare-shift*, one necessary to make up the required number of hours for the week.

Hence **ra·gtimer,** one who plays ragtime; **ra·gtim(e)y** *a.*, suggestive of ragtime; **ra·gtiming** *a.*

1912 G. Frankau *One of Us* x. 92 Where the Rat Mort's rag-timing Ethiope Dins in one's ears. **1915** D. O. Barnett *Let.* 7 Jan. 37 Been making out forms of times. Feel rather ragtimy. **1927** *Daily Tel.* 1 Nov. 9/3 Most nimble of ragtimers at the piano. **1950** Blesh & Janis *They all played Ragtime* (1958) i. 26 Under Joplin he quickly became an adept ragtimer and was soon playing at dances in and around Sedalia. **1952** B. Ulanov *Hist. Jazz in Amer.* (1958) xxii. 306 They played their ragtimey work..with a finish and polish. **1974** *Country Life* 7 Nov. 1360/1 The ragtime music chosen..was the least..heard of Scott Joplin, supplemented by less well-known ragtimers such as..Joseph Lamb.

Ragusan (rāgū·zăn), *sb.* and *a.* Also † **Ragusian. A.** *sb.* **a.** An inhabitant of Ragusa, now Dubrovnik, on the Adriatic coast of Yugoslavia. **b.** An extinct dialect of Dalmatian, spoken formerly by inhabitants of Ragusa.

1652 Heylin *Cosmogr.* ii. 197 Languste, environed about with very high Mountains, in which are the Ragusians Farms. **1905** *Westm. Gaz.* 29 Apr. 4/1 The advent of Napoleon I. early in the nineteenth century deprived Ragusans for ever of their freedom. **1933** [see *Dalmatian *a.* and *sb.* 2]. **1967** D. S. Parlett *Short Dict. Lang.* 38 Ragusan extinct since beginning C17.

B. *adj.* Of or pertaining to Ragusa.

1788 Gibbon *Decl. & F.* V. vi. 613 The Apulian and Ragusian vessels fled to the shore. **1799** R. Smelt *Let.* 6 Dec. in B. Ward *Dawn Catholic Revival* (1909) II. xxxi. 220 We sailed from Palermo on Saturday night, November ye 2nd: altogether seventeen sail, Imperial, Sicilian, Ragusian and American. **1932** *Times Lit. Suppl.* 14 Jan. 27/1 If all the Ragusan men knew Italian, most of the ladies to whom their love poems were dedicated spoke only Slovinski. **1971** *Textile Hist.* II. 10 In 1420 the Ragusan government arranged for a colony of traders from the Italian textile town of Prato to live in Ragusa.

rag-wheel. 1. (Earlier and later examples.)

1812 *Niles' Reg.* II. 393/2 A rag wheel [of a mill machine] of three feet diameter is kept in complete motion. **1848** *Rep. Comm. Patents 1847* (U.S.) 79 The feeding is usually effected by a band taking into a ratchet, or rag wheel. **1873** J. M. Bailey *Life in Danbury* 9 He employed a carpenter from a place twenty miles distant to make a new rag-wheel.

ragworm. Substitute for def.: a polychæte worm belonging to the family Nereidæ, esp. *Nereis diversicolor*, found in sand or under stones and often used as bait for fish. (Earlier and later examples.)

1865 J. C. Wilcocks *Sea-Fisherman* 99 The larger Rag Worms, are found by digging in stoney ground. **1906** *Daily Chron.* 6 Jan. 8/6 An enormous specimen of the king ragworm has been found by a bait-digger at Southend-on-Sea. **1928** Russell & Yonge *Seas* 34 It [*sc. Nereis diversicolor*] is frequently used as a bait, being known as 'rag-worm' in many parts. **1967** *Daily Tel.* 21 Oct. 14/6 There were other ways of catching bass than with ragworm or sandeels. **1978** B. Gledhill *Cod Fishing* ii. 58 The ragworm likes hard ground usually.

rah, *int.* and *sb.* Add: Also **ra.** (Earlier and later examples.)

1870 D. J. Kirwan *Palace & Hovel* xxiv. 372 The 'Rah, 'Rah, 'Rah, of Harvard pierces the air... Oxford has just got into her careless, easy swing. **1887** *Harper's Mag.* Feb. 395/1 The junior class filed into the green enclosure amidst the 'rahs of their friends. **1889** 'Mark Twain' *Connecticut Yankee* xxxiii. 421 'Rah for protection—to Sheol with free-trade! **1905** *N.Y. Even. Post* 29 June, Harvard almost immediately increased her stroke, and the way their cut-water slid along called forth the nine long 'rahs again and again. **1917** R. Frost *Let.* 3 Dec. (1972) 20 Rah rah rah for some other college than Wellesley. **1924** H. T. Lowe-Porter tr. *Mann's Buddenbrooks* II. vii. iv. 24 A voice.. shouts suddenly: 'Heine Seehas is 'lected—'rah for Heine Seehas!' **1942** Ade *Let.* 1 Feb. (1973) 227 We didnt play basket-ball or foot ball [at school in the 'seventies and eighties'] and we never learned to stand up on our hind legs and let out a rah-rah. **1972** 'E. Lathen' *Murder without Icing* (1973) iv. 41 'Way to go, Billy!' 'Rah! Rah! Billy Siragusa!' **1977** *Lancashire Life* Mar. 56/1 Ra-ra-ra! Give a cheer from the sidelines for Accrington, the town that is instilling new life into American baseball.

‖ **rahat lokum** (rāhat lǫku·m). Also **rahat lakoum, lahkoum, lakuhm,** etc. [a. Turk. *rahat lokum*, ad. Arab. *rāḥat al-ḥulqūm* throat's ease.] Turkish delight. Also (occas.) *ellipt.* as *rahat*. Cf. *locoum.

1856 R. F. Burton *Personal Narr. Pilgrimage to El-Medinah* III. 362 Squares of Rahah, a comfiture highly prized in these regions, because it comes from Constantinople. **1861** *Punch* 12 Jan. 12/1 Rahat lahkoum, or lumps of delight! **1894** [see *delight *sb.* 4]. **1900** *Confectioners'*

Union Handbk. 167 Butter-Scotch, nougat, rahat lakuhm. *Ibid.* 169 Hawes, J., & Son...Rahat Lakoum and water mould fancies. **1907** *Yesterdays' Shopping* (1969) 32/1 Rahat La Koum or Turkish Delight. **1931** A. J. Cronin *Hatter's Castle* I. vii. 126 They come off to the ship in boats at Port Said and sell very good rahat-lakoum which is an excellent sweet. **1931** *Discovery* Nov. 359/2 Today the Turks here [on the island of Ada Kaleh] live by some gardening and fruit cultivation, a little cigar manufacture, the preparation of rahat ('Turkish Delight') and mild catering for the few individuals who visit the island. **1935** M. Morphy *Recipes of All Nations* 767 The Turks are extremely fond of sweetmeats..and among the most popular is *rahat el hulkum*. **1945** C. S. Forester *Commodore* xviii. 198 He had eaten Westphalian ham and Italian beccaficoes and Turkish rahat lakoum. **1960** *Times* 24 Oct. (Financial Rev.) p. xiv/6 The sweetmeat was called Rahatlokum, but today it is better known as Turkish delight. **1963** *Punch* 20 Nov. 748/2, I..went for coffee and rahat lakoum. **1968** C. Roden *Bk. Middle Eastern Food* xv. 295 *Rahat Lokum*, Turkish Delight. This little sweet epitomizes luxury, pleasure and leisure. **1970** [see *Locoum].

rah·ing, *vbl. sb.* = Hurrahing *vbl. sb.*

1904 *Daily Chron.* 25 July 7/7 There was not nearly so much 'rahing' and flag-waving as in 1899.

rah rah (rā rā), *sb.* and *a. slang* (orig. *U.S.*). Also **rah rah rah, ra ra.** [Reduplication of Rah *int.*] **A.** *sb.* A shout of support or encouragement, as for a college team: see Rah *int.* and *sb.* in Dict. and Suppl. **B.** *adj.* with hyphens. Of or pertaining to college, collegiate; (of behaviour, etc.) characteristic of college students; marked by the generation of enthusiasm or excitement, as in cheer-leading, etc. *U.S.*

separately and, owing to these unavoidable termini, will always be running off the totalitarian rails. **1953** K. Amis *Lucky Jim* xxii. 228 He resolved not to run off the rails again. He cleared his throat, found his place, and went on in a clipped tone. **1954** T. S. Eliot *Confidential Clerk* II. 63, I make decisions on the spur of the moment, But you'd never take a leap in the dark; You'd keep me on the rails. **1955** G. Greene *Quiet American* II. ii. 132 All my conversations with Pyle seemed to take grotesque directions. Was it because of his sincerity that they so ran off the customary rails? **1971** *Daily Tel.* 20 Oct. 12/7 They are proud, stubborn and steady—even if dad appears to have gone off the rails. **1975** M. Babson *There must be Some Mistake* xvi. 128 Would John have gone off the rails like this if she had been paying enough attention?

c. *N. Amer.* A railwayman.

1938 L. M. Beebe *High Iron* 223/2 *Rail*, railroad man. **1960** [see *biscuit-toss* s.v. *biscuit 3 b]. **1974** *Maclean's Mag.* Jan. 16/2 She spent too much time..listening to a bunch of young 'rails' repeat lies handed down over the years.

d. *Electronics.* A conductor which is maintained at a fixed potential and to which other parts of a circuit are connected.

1960 H. Carter *Dict. Electronics* 255. **1965** *Wireless World* Aug. 399 The common rail for input signals is the positive line, while the common connection to the thyristor is the negative line. **1977** *Gramophone* Feb. 1344/1 Gramophone inputs are to feedback pairs on a 25-volt rail giving a reasonable overload margin. **1979** *Personal Computer World* Nov. 3 (Advt.), 30 Amp, 8v power supply, 5 Amps on ±16v rails (all rails are separately fused).

5. a. (Further examples.) *spec.* **British Rail,** the name of the national railway of Britain.

1843 *Ainsworth's Mag.* III. 361 My lord and lady start, per rail, To London for the Season. **1848** J. J. Ruskin *Let.* 17 Mar. in M. Lutyens *Ruskins & Grays* (1972) xi. 98 For God-sake be done with Rails and Shares—or you will not have a Business, for who will confide in Railway people I am not clear. **1858** Queen Victoria *Let.* 8 May in R. Fulford *Dearest Child* (1964) 103 We went by rail, nice, quick! **1867** in 'Mark Twain' *Innoc. Abr.* (1869) i. 21 They can by rail go on to Florence. **1884** E. W. Hamilton *Diary* 5 Aug. (1972) II. 666 His solution of the financial difficulty is that we should guarantee 'the Rails', the slang phrase for the Preference Debt to which the Railroads are assigned. **1892** B. Potter *Jrnl.* 30 July (1966) 243 The Volunteers broke up and several companies went off by rail. **1926** *Daily Colonist* (Victoria, B.C.) 15 July 17/1 The advance in rails has slowed up temporarily, although a strong investment demand is reported for several of the high grade issues. **1935** *Economist* 27 July 191/1 Rails and utilities..have hardly participated at all in this week's upward movement. **1965** *Evening Standard* 4 Jan. 9/3 (*heading*) Now it's British Rail—Beeching's last new look for trains, stations, men. *Ibid.*, A new image for British Railways—'British Rail'—is being launched tonight when Dr. Beeching opens an exhibition showing the new symbols, train liveries and uniforms at the Design Centre in the Haymarket. 'There's been a lot of criticism about the new name,..' Mr. George Williams, the railways Director of Industrial Design, said at a preview of the exhibition today. 'Personally I think that passengers..will soon find themselves saying British Rail.' **1965** *Observer* 31 Jan. 12/8 British Rail officials admit there appears to be an injustice. **1976** *Illustr. London News* Nov. 52/4 Fruit and vegetables now tend to go increasingly by road..where a few years ago they travelled by rail. *Ibid.* 53/3 The productivity a head has more than quadrupled since 1945, whereas..for British Rail [it has] only doubled. **1981** *Daily Tel.* 3 Mar. 18/2 Lucky evening travellers tonight..will each be presented with a miniature of British Rail sherry.

c. Usu. *pl.* A railway station.

1854 *Poultry Chron.* I. 117/2 The middle and humble classes..are oft times virtually prohibited from attending if bad weather; sets in more especially if out of the way of rails. **1939** H. Hodge *Cab, Sir?* xv. 222 The 'Rails' are railway stations, as distinct from the Underground.

6. a. *rail-under* (poet. adv.); **rail-bird** *U.S.*, (*b*) one who watches from the rails or sidelines (*lit.* and *fig.*); **rail-cut,** a length of timber cut off for a rail; **rail-fence**: see as main entry in Suppl.

1931 *Daily Progress* (Charlottesville, Va.) 26 Mar. 13/7 Louisville's railbirds are..watching and clocking the work-outs of Derby eligibles. **1947** *Sun* (Baltimore) 4 Nov. 16/1 With the grandstand crowded when regular railbirds pushed their way into the shelters, betting dropped away. **1957** [see *cookie 4]. **1959** *Washington Post* 8 Oct. A21 Political railbirds out here suspect that Brown is beginning to think of himself as something more than a favorite son. **1971** *Daily Colonist* (Victoria, B.C.) 27 May 54/3 He gave them a three-week course of instruction in lacrosse fundamentals, banned railbirds who might make the boys feel self-conscious then turned them loose in lacrosse games. **1836** D. B. Edward *Hist. Texas* IV. 69 The farmers often get it measuring two rail cuts in length. **1881** *Scribner's Monthly* Feb. 503/2 The pole fence was laid after the same fashion of a rail fence only the poles were longer than rail-cuts. **1930** J. Masefield *Wanderer of Liverpool* 23 The ship..Beaten rail-under by tempest and deluged by billows.

b. *rail-bed*; **rail bond,** an electrical connection between consecutive lengths of rail in a railway or tramway.

1880 'Mark Twain' *Tramp Abroad* xxix. 306 There was no level ground at the Kaltbad station; the railbed was as steep as a roof. **1969** E. W. Morse *Fur Trade Canoe Routes* II. vi. 78 The portage is rough, and at its western end leads into an abandoned rail-bed once used for logging. **1893** in K. Hedges *Amer. Electr. Street Railways* (1894) iii. 22 Each joint of the rails is supplied with two rail bonds of No. 0000 copper wire, each only 12 inches long. **1907** Wilson & Lydall *Electr. Traction* I. vi. 107 The 'Protected' rail bond is made by fusing terminals of solid copper upon a loop of flattened copper wire.

c. *rail-bridge, -end, -fare, -service, -side, tanker*; *rail-borne, -minded, -mounted* adjs.; **rail-bus,** (*a*) a vehicle resembling a bus but running on a railway track; (*b*) in Denmark, etc.: a tramcar running on tram-lines set in the road; **rail-car,** (*a*) = Car *sb.* 2 (earlier and later examples); (*b*) (see quot. 1949); **railcard,** a pass entitling the holder to reduced fares on the railway; **rail-cutting,** the destruction of railway communications; **rail-head,** (*a*) the furthest point reached by a railway; (*b*) the point on a railway from which branch-line or road transport of supplies begins; hence *railhead facilities*; also *fig.*; **rail-line,** a railway line; **rail link,** a railway service joining two established transport systems; **rail-motor,** a passenger train which consists of a single coach attached to a small locomotive or having its own engine; a rail-car; also *attrib.*; **railplane** (see quots.); **railsickness,** sickness caused in a passenger by the motion of a train.

1928 *Britain's Industr. Future* (Liberal Industr. Inquiry) IV. xxiii. 313 In Germany the tonnage of canal and river traffic is equivalent to one-fifth or one-sixth of rail-borne traffic. **1976** *Illustr. London News* Nov. 52/4 The trend has been for rail-borne freight to lose ground to passenger traffic and to road transport. **1963** *Times* 8 June 14/3 Two-day talks between English and French Government officials on whether there should be a Channel rail bridge or road rail tunnel ended in London yesterday. **1978** H. R. F. Keating *Long Walk to Wimbledon* IV. 59 The massive yellow-brick rail-bridge. **1933** *Morning Post* 30 Aug. 10/4 The London and North-Eastern Railway Company will put the new..stream-lined Diesel-electric 'railbus' into regular service on the suburban and outlying railway systems round Newcastle, within the next two weeks. **1956** *Railway Mag.* Mar. 195/1 The 'railbus' advocated for branch-line use by a correspondent in your January issue may have disadvantages. **1968** *Drive* Spring 37/2 British Rail could save many of their rural routes by introducing rail buses—a sort of single-decker diesel tramcar, operated by a driver-conductor not as a train but as if the vehicle were on the open road. **1976** J. Tate tr. *Bodelsen's Operation Cobra* viii. 42 The access road..is to be blocked..where the rail-bus cuts across it. *Ibid.* x. 54 The empty road along which the rail-bus ran. *Ibid.* xi. 56 Frederik cycled across the rail-bus tracks. **1977** *Modern Railways* Dec. 485/3 An early example was the German MAN railbus built in 1932, which remained in service for 30 years. **1834** *Knickerbocker* III. 112 After two hours past in this fair presence on rail-cars, I returned with my head running most uncomfortably upon this new acquaintance. **1860** J. S. C. Abbott *South & North* ix. 206 Thence, in rail-cars..through the heart of Alabama. **1934** *Discovery* Nov. 314/1 The term railcar is a convenient one to apply to the fast self-contained passenger units now running on many of the world's railways. **1949** *Richmond* (Virginia) *Times-Dispatch* 27 Oct. 4/2 This new-fangled transport is called a 'rail-car'.., principally for the reason that it is built compactly into a single unit... It operates much on the same principle as a streetcar, with controls at each end so that it can travel in either direction. The car, with a seating capacity of 90 persons, is especially designed for local passenger traffic. **1959** A. McLintock *Descr. Atlas N.Z.* 63 New Zealand Railways: Some Facts..Wagons 18,650..Railcars 23. **1963** *Times* 27 Feb. 5/1 Fiat was the first to start mass production of railcars. Its products are used by the railways not only of Italy but of a number of other countries. Fiat railcars are in service in Spain, Portugal, Egypt, Greece, Yugoslavia, Argentina, Brazil, Mexico, Venezuela, India and elsewhere. **1971** *Railway World* Mar. 116 There are also several diesel locomotives and a couple of railcars. **1976** *Sci. Amer.* Jan. 27/2 After the cooling period the fuel will in the future be shipped in specially protected trucks or railcars to a chemical-reprocessing plant. **1977** *Times* 16 Mar. 6/4 The senior citizens' Railcards will become available from April 1 for a full year regardless of the date of purchase. **1978** *Oxford Consumer* Mar. 18/1 Railcards for the 14–17 yr olds will be able to be purchased at most local stations from the above mentioned date. **1899** *Westm. Gaz.* 9 Dec. 5/3 We shall hear a good deal more of rail-cutting operations on the part of the enemy. **1944** Rail-cutting [see *interdiction 4]. **1869** W. Barnes *Early England* 106 When the railway was taken into the hands of more learned men, we had..the *terminus* instead of the rail-end. **1955** R. W. & M. L. Settle *Saddles & Spurs* xii. 205 The first rail was laid in Sacramento October 26, 1863. Two years later rail-end had reached Colfax, fifty-five miles away. **1976** S. Hynes *Auden Generation* vii. 229 Details of landscape—the mountains, the pass, the rail-end—take on symbolic meanings. **1974** *Times* 22 Oct. 14/4 I'll pay rail fares, of course. Second class. **1976** B. Williams *Making of Manchester Jewry* vi. 157 If the synagogue was prepared to pay the rail fare of a Jewish pauper as far as Hull, the society undertook to see him across the North Sea. **1905** *Daily Chron.* 14 June 4/2 The political rail-head..has not got beyond Balfour Junction, and there are no definite lines of policy laid down beyond that point. **1905** *Athenæum* 24 June 781/2 When Lord Kitchener, during the operations of the Soudanese war, sternly relegated the war correspondents to the railhead, he earned the hostility of those who regard the distribution of news as of more importance. **1915** A. D. Gillespie *Lett. from Flanders* (1916) 243 There are some hills not far away, beyond the rail-head from which I marched in February. **1941** I. L. Idriess *Great Boomerang* xvii. 119 Now mineral wealth comes in—copper at the Duchess, with a railhead at Dajarra. **1955** *Times* 22 July 9/7 The Indian Government pays for the long, expensive haul from Pathankot, the railhead on the plains. **1961** *Times* 30 June 9/4 Railwaymen have come to talk of such goods stations as 'railheads'. **1972** *Oxford Times* 5 May 4/1 Culham, Clifton Hampden, Stadhampton and Little Milton are all on the route from the Didcot railhead—and lorries are due to start rolling in three weeks' time... No-one..expected Didcot to be the railhead for the materials. **1973** *Times* 29 Nov. 16/7 Every factory and warehouse and all the rest—is provided with railhead facilities. **1974** *Times* 8 Jan. 2/5 We have said we would go ahead provided all those who wanted a railhead were prepared to make good the financial shortfall. **1979** *Jrnl. R. Soc. Arts* CXXVII. 412/2 This was in general only temporary until the pipeline, gathering stations and railhead installations..had all been fully run in. **1961** E. F. McKinney *Educ. in Violence* 365 Garrard's Covington raid and Rousseau's Opelika raid cut two-thirds of the rail lines he had to break. **1976** *Jrnl.* (Newcastle) 26 Nov., Holly Avenue, a quiet street sandwiched endways between Osborne Road and the rail-line. **1978** *Amer. Poetry Rev.* Nov./Dec. 6/3 The linear travel of the rail-line has become three-dimensional. **1975** *Guardian* 21 Jan. 12/1 The rail link from Folkestone to London. **1976** *Illustr. London News* Nov. 52/1 The first rail link with Britain by train and boat had been opened with the Calais Docks station in 1849. **1963** *Times* 23 May 13/7 Switzerland is the most rail-minded country in the world. **1906** *Westm. Gaz.* 5 June 5/3 An excursion train on the Great Western line colliding..with an empty rail-motor coach. **1927** *Observer* 13 Nov. 13/3 'Rail motors' or 'motor trains', may either take the form of self-contained vehicles having a steam or petrol engine built into the coach, or of trains hauled by very small engines and arranged to be driven from either end. **1962** *Coast to Coast 1961–62* 202 Rattling along on a rail-motor somewhere south-west of Bundaberg, recollection nagged busily and painfully. **1967** G. F. Fiennes *I tried to run a Railway* iii. 24 They allocated a railmounted gun..to Norfolk. **1969** *Jane's Freight Containers 1968–69* 241/3 All these cranes are rail-mounted, pneumatic tyred cranes will not be used. **1933** *Sun* (Baltimore) 25 Sept. 6/8 A railplane car, built along the lines of airplane architecture and designed to carry passengers over railroad tracks at ninety miles an hour, was announced today by the Pullman Car and Manufacturing Corporation. **1968** S. E. Ellacott *Everyday Things in Eng. 1914–68* xii. 185 A gallant pioneer effort..to revolutionize rail travel by suspending a carriage on an overhead rail..was the invention of a Scot, George Bennie, who built his first railplane in 1929... Ironically, a monorail service was running with apparent success in Tokyo in 1957. **1976** *Illustr. London News* Nov. 29/2 The most effective way of providing a 'rapid transit' is to improve the rail service. **1892** Swinburne *Lett.* (1962) VI. 30, I have got over the unnerving effects of railsickness. **1928** *Daily Tel.* 17 July 4/5 Freehold railside factory. **1959** *Listener* 8 Jan. 50/1 Iron ore is brought down to rail-side by country carts from the nearby mountains. **1958** *Times Rev. Industry* Feb. 74/1 Railtankers standing in readiness then loaded up and took the crude oil the remaining 250 miles to the sea at Philippeville for storage. **1979** *Jrnl. R. Soc. Arts* CXXVII. 406/1 The oil would be transmitted by pipeline to a rail terminal for transmission by rail tanker.

rail (rēⁱl), *sb.*⁶ [Origin uncertain: cf. prec.] A hot-rod or dragster.

1962 *Punch* 17 Oct. 560/2 A dragster, or rail, is the most skeletal vehicle of all. **1965** *Daily Mail* 2 Oct. 5/5 There is no lonelier place on earth than the cockpit of a rail... A rail? That is race-jargon for a dragster. **1977** *Hot Car* Oct. 42/2 A reasonable crowd showed to watch rails, gassers, comp altered, and street saloons race together.

rail, *v.*² **5.** (Later examples.)

1916 E. W. Hamilton *First Seven Divisions* 142 Four Army Corps were railed up from the eastern frontier. **1936** R. C. K. Ensor *England, 1870–1914* ix. 299 It cost as much at that time to rail coal from the Rhondda to North Dorset as to ship it 3,000 miles to Alexandria. **1973** *Sunday Times* 7 Oct. 46 Forty-four-thousand gallons of sterile milk are daily railed from Anand to Bombay. **1975** *Times* 27 Dec. 9/7 Next year's Motorail brochure has just come... For many years I railed my car to Scotland. Not again, at £100 a time.

rail, *v.*⁴ Add: **1. b.** (Later examples with *on.*)

1611 Bible *Mark* xv. 29 And they that passed by, railed on him. *Ibid. Luke* xxiii. 39 And one of the malefactors..railed on him, saying, If thou be Christ, save thy selfe and us. **1855** Macaulay *Hist. Eng.* III. xii. 213 His very soldiers railed on him in the streets of Dublin.

railage (s.v. Rail *sb.*² in Dict.). Add: (Later examples.) Also *attrib.*

1903 *Daily Chron.* 19 June 5/2 Food and forage..are continually coming forward from the coast at high cost for railage. **1907** *Westm. Gaz.* 19 Jan. 7/1 Welsh smokeless coal is now 19s. per ton at the pit's mouth, and to that has to be added 8s. 4d. per ton for railage to London. **1955** *Times* 3 June 10/6 Further increases in the cost of commodities and stores, the latter resulting largely from the higher railage rates introduced in recent years. **1972** P. Newton *Sheep Thief* i. 14, I would require two horses..and I would like to take my own. This would involve the cost of railage.

railed, *ppl. a.* Add: **1.** (Later examples with *off.*)

1930 *Times Educ. Suppl.* 23 Aug. 364/4 A railed-off enclosure. **1973** 'D. Halliday' *Dolly & Starry Bird* ix. 119 The man we were following..gazed..at the railed-off pieces of terrazzo on the pavement.

railer². Add: **2.** One who travels by rail.

1874 J. Albery *Two Roses* I. 12 Wherever you go there's Stone before you... Stone's a railer. **1889** F. E. Gretton *Memory's Harkback* 103 Your constant 'railers' are blindly ignorant of the localities they scud over.

3. *Racing.* A runner that stays close to the rail.

1958 *Times* 29 Nov. 7/7 A 'railer' will always stick to the rails and a slow starter will always be such.

rail-fence. orig. *U.S.* [f. Rail *sb.*² + Fence *sb.*] **1.** A fence made of wooden posts and rails. Hence *rail fencing*.

1649 *Charlestown* (Mass.) *Land Rec.* (1878) 110, I doe sell..five Akers of planting Land,..Bounded on the..

TROLLOPE He knew he was Right (1869) I. xxxvii. 292 The real disgrace of England is the railway sandwich. 1915 J. BUCHAN Thirty-Nine Steps vii. 177, I never ate a meal with greater relish, for I had had nothing all day but railway sandwiches. 1972 Railway sandwich [see railway tea below]. 1876 'L. CARROLL' Hunting of Snark III. 30 You may seek it with thimbles—and seek it with care; You may hunt it with forks and hope; You may threaten its life with a railway-share; You may charm it with smiles and soap. 1942 C. MILBURN Diary 26 Feb. (1979) 130 The King and Queen slept in our railway siding on Tuesday night. 1976 A. WHITE Long Silence vi. 43 The railway sidings became one of the principal marshalling yards. 1866 GEO. ELIOT Let. 11 Sept. (1956) IV. 309 My 6/- editions are never on the railway stalls. 1849 DE QUINCEY in Blackw. Mag. Oct. 492/1 Interesting personal communications,..revelations of impressive faces..could not have offered themselves amongst the hurried and fluctuating groups of a railway station. 1892 B. POTTER Jrnl. 25 Aug. (1966) 252 A large proportion of the seven thousand spectators poking away through the small Railway Station. 1964 G. L. COHEN What's Wrong with Hospitals? vi. 115 Their baronial board-room..dignified by a staircase for consultants only, in Railway Station Gothic. 1852 Jewish Chron. 10 Dec. 78/2 The poor..are induced to go..from town to town; each congregation..assisting to pay a portion of the dividends on railway-stock, under the name of..charity. 1838 Civil Engin. & Archit. Jrnl. Oct. 358 Railway Switch Signal. 1872 GEO. ELIOT Middlem. III. lvi. 227 The infant struggles of the railway system. 1981 Times 10 Feb. 6/6 The railway system and the port of Mombasa have.. benefited from the growing volume of tea exports. 1972 'G. NORTH' Sgt. Cluff rings True iv. 31 Travellers..reviving themselves with railway tea and railway sandwiches. 1845 Punch VIII. 101 An applicant for information at a Railway Terminus. 1942 E. WAUGH Put out More Flags iii. 228 They reached the classic columns of the railway terminus. 1976 G. SEYMOUR Glory Boys iv. 42 The big railway termini of North London. 1839 F. WITTS Diary 4 May (1978) 158 Meantime the passengers receive a railway ticket to London which purports to be worth 5s 6d. 1895 E. LEAR in Nonsense Songs & Stories 87 On his nose there was a Cricket,—In his hat a Railway-Ticket. 1912 W. OWEN Let. 4 June (1967) 140 Does the Railway Ticket Problem assume a different aspect now? 1977 Lancashire Life Dec. 60/2 In those wartime days, holiday-period railway tickets were at a premium. 1847 Railway timetable [see railway sandwich above]. 1932 D. L. SAYERS Have his Carcase xxvi. 352 Do you happen to have a railway time-table on you? 1980 Times 12 Aug. 10/1 He wants the centre of the British railway system moved from London to Birmingham..to rewrite the entire railway timetable around Birmingham. 1881 A. BEGG Gt. Canadian N.W. 106 The railway is constructed to a point on the eastern side of the river opposite Selkirk, so that there is every chance of its becoming a railway town. 1888 Lippincott's Monthly Mag. XLII. 783 There stood close to the opposite wall a large piano of the class known as the 'grand', rare enough among the railway towns west of the Mississippi States. 1943 Sun (Baltimore) 13 Dec. 5/2 The Peiping Hankow railway town of Sinyang. 1976 A. WHITE Long Silence vi. 44 She propelled the creaking vehicle..up and down the railway tracks. 1891 Murray's Handbk. India & Ceylon p. xv, In Bombay, the Indian A.B.C. Guide and the Indian Railway Travellers' Guide.. give..the railway routes for all India. 1980 G. M. FRASER Mr American xxvi. 552 The vaguely hostile silence of the British railway traveller. 1838 Civil Engineer I. 390/1 A new Railway Truck, the invention of Mr. Robert Grant, of Maine. 1839 F. WITTS Diary 4 May (1978) 158 There the carriages are placed each on a railway truck. 1841 DICKENS Let. 13 June (1969) II. 302 One is hoisted bodily, carriage and all, on a Railway Truck. 1867 QUEEN VICTORIA Jrnl. 22 Aug. in D. Duff Victoria in Highlands (1968) 223 We went by the side of Eildon Hills, past an immense railway viaduct. 1976 C. DEXTER Last seen Wearing xxi. 162 He..slept beneath a railway viaduct. 1976 A. WHITE Long Silence xiv. 121 They might hide themselves in a railway wagon. 1943 J. FLANNER in New Yorker 29 May 45/1 The Nazis..had decided to kill the French railroad men's resistance movement..by inviting twelve thousand French railway workers to go to Germany. 1979 P. ALEXANDER Show me Hero xv. 161 Crumbling terraced cottages, once occupied by railway workers. 1869 Bradshaw's Railway Manual XXI. 371, 17,500,000 thalers were to be applied to railway works. 1976 Milton Keynes Express 16 July 7/7 The part time station at the Wolverton is still dependent on men from the railway works.

4. railway beetle = railroad worm s.v. *RAILROAD sb. 3 c; **railway bull** = railroad bull s.v. *RAILROAD sb. 3 c; **railway crossing** = level crossing s.v. LEVEL a. 3 b; **railway edition**, a cheap edition of a book suitable for reading on a railway journey; **railway guide**, a train timetable; **railway hotel**, an hotel sited near to a railway station for the convenience of travellers; **Railway Institute**, a (social) club building for railway workers, esp. in India; **railway label**, an address or destination label stuck on a passenger's luggage; **railway letter** (see quot. 1933); **railway novel** (earlier and later examples); **railway pass**, a ticket authorizing the holder to travel by rail; **railway rug** (earlier example); **railway sickness** = railsickness s.v. *RAIL sb.^{1} 6 c; **railway spine** (later examples); **railway time**: see as main entry; **railway volume** = *railway edition; **railway warrant** = *railway pass; **railway whistle**, a whistle blown by the guard as a signal to the driver to start the train; **railway wrapper**, a travelling-cloak.

1915 E. R. LANKESTER Diversions of Naturalist 234 A peculiar grub-like female glow-worm, three inches long, is found in South America, which produces a red light at each end of the body and numerous points of green light on each side of it. It is called the 'railway-beetle' in Paraguay. 1973 B. BROADFOOT Ten Lost Years ii. 27 [To] the town cop or the railway bull..you automatically became a criminal. 1907 Yesterday's Shopping (1969) 1040/2 Railway Crossing, finely japanned, to be opened and shut..8 in. long. 1937 Discovery May 144/2 A railway crossing west of Scarborough. 1859 G. H. LEWES in Blackw. Edin. Mag. July 101/1 Twice or thrice have the railway editions been out of print. c 1838 W. H. MURRAY in M. R. Booth Eng. Plays of 19th Cent. (1973) IV. 161 That woman..[is] as difficult to understand as an Act of Parliament, or a Railway Guide, or a Weather Almanack. 1855 MRS. GASKELL North & South II. vii. 87 One of the very common mistakes in the 'Railway Guide' as to the times when trains arrive at the smaller stations. 1932 D. H. LAWRENCE Etruscan Places i. 13 The inestimable big Italian railway-guide says the station is Palo. 1847 F. A. KEMBLE Rec. Later Life (1882) III. 265 To this Hull Railway Hotel is attached a magnificent Railway Station (or rather vice versâ). 1871 GEO. ELIOT Let. 20 Aug. (1956) V. 179 There is a palatial bedroom here, but if you reject that, there is a Railway Hotel, just opposite the station. 1911 BEERBOHM Zuleika Dobson ii. 13 She took a night's sanctuary in some railway-hotel. 1972 Country Life 5 Oct. 799/1 The Great Northern, designed by Lewis Cubbitt..was far more restrained than any of the railway hotels of the next decade. 1937 K. BLIXEN Out of Afr. iv. 299 The High Court was set in Nakuru, in the Railway Institute. 1954 J. MASTERS Bhowani Junction I. v. 42 There were two Railway Institutes..one for Europeans and Anglo-Indians, and one for Indians... Ours was a fine big building with a dance floor and card rooms and a bar. 1979 Times 26 Nov. 14/7 Giving song recitals at the Railway Institute at Lahore. 1909 BEERBOHM Yet Again 125 Railway-labels are..crudely coloured, crudely printed. 1917 Science Progr. XI. 685 The railway labels jealously preserved on travellers' portmanteaux. 1933 Post Office Guide Jan. 42 The following Railway Companies..under agreement with the Postmaster-General, accept and convey letters, both on week-days and on Sundays, by the next available train or ship, either to be called for at the Station of address or to be transferred there to the nearest Post Office letter box. Such letters are called Railway Letters. 1971 D. POTTER Brit. Eliz. Stamps xii. 132 In 1969 two other railway lines decided to avail themselves of the privilege of accepting and carrying railway letters. 1857 C. M. YONGE Dynevor Terrace I. ii. 12 'Reading... See here', as he held up maliciously a railway novel. 1896 B. MATTHEWS Bookbindings IV. i. 239 In England the railway novel is incased in boards sheathed in paper. 1981 V. POWELL Flora Annie Steel i. 6 The yellow-backed novels.. known as 'railway novels' and bought to beguile the long haul from Harrow to the West of Argyll. 1901 KIPLING Kim vii. 178 Colonel Creighton's railway pass lay in his hand, and Kim..was still lord of two rupees seven annas. 1915 W. B. YEATS Reveries 5 An uncle called me out of bed one night, to ride..to Rosses Point to borrow a railway-pass from a cousin. 1853 Illustr. London News 12 Nov. 409/2 Blankets and railway rugs have formed the chief protection against the low temperature. 1895 SWINBURNE Lett. (1962) VI. 85, I am now sufficiently recovered from railwaysickness. 1895 G. B. SHAW in Liberty 27 July 6/2 The connection between degeneration and 'railway spine'. 1954 W. MAYER-GROSS et al. Clin. Psychiatry iv. 135 The days of neurasthenia and railway-spine at the time of Beard. 1869 D. G. ROSSETTI Let. Aug. (1965) II. 710 In the above year [1862] Messrs. Routledge reprinted it as a shilling railway volume. 1919 J. BUCHAN Mr. Standfast ix. 183, I..emerged in the uniform of a British private... I had a railway warrant made out in my name for London. 1978 T. ALLBEURY Lantern Network iii. 30 He had been released..with a railway warrant to Southampton. 1854 DICKENS Seven Poor Travellers in Househ. Words VIII. Extra Christmas No. 4/1 A..young man connected with the Fly department, and well accustomed to the sound of a railway whistle which Ben always carries in his pocket. 1847 Railway wrapper [see railway sandwich, sense *3 a]. 1860 DICKENS Uncomm. Trav. (1861) i. 1 No hotel-room tapestried with great-coats and railway wrappers is set apart for me. 1866 G. M. HOPKINS Let. 14 Dec. (1956) 34 Tell Anne to pack my railway-wrapper (which will help to keep the books fr. jolting).

railway, v. Add: **3.** To provide with railways.

1917 H. MACFALL Germany at Bay xii. 243 When Russia was gunned and munitioned and well railwayed, she was more than a match for the Germans.

railwayana (rēⁱ·lwēⁱ,ā·nā). [f. RAILWAY sb. + ANA suffix.] Material pertaining to railways; railway relics.

1970 Sunday Mail Mag. (Brisbane) 7 June 9/3 In the wake of steam's demise, the collection of railway relics has become almost a national obsession. Enthusiasts have invented a new word—Railwayana... On May 21, Sotheby's held its 'first sale of Railwayana'. 1970 Railway Mag. Oct. 541/2, I discovered, in one of the many stalls selling 'railwayana', a line-drawing of Highland Railway No. 103. 1972 Times 30 Dec. 14/1 When it proved possible to buy a section, preservationists realized there was more to 'railwayana' than static museums. 1975 Times 15 Jan. 15/5 British Rail might do better to invest their pension funds in their own surplus railwayana. Steam locomotive nameplates.. now change hands at £200–300 each.

railwaydom (rēⁱ·lwēⁱdəm). [f. RAILWAY sb. + -DOM.] Railways considered collectively; the railway world.

1881 Punch 17 Sept. 132/1 The public ought to devise some means of putting the screw on railwaydom. 1885 Ibid. 10 Oct. 170/2 For one of the ways By which Railwaydom pays Is to keep us at work day & night, my dear!

railway time. [RAILWAY sb.] A standard time adopted throughout a railway system to supersede local time for railway operations (in Great Britain, London time before the adoption of Greenwich Mean Time).

1847 DICKENS Dombey (1848) xv. 155 There was even railway time observed in clocks, as if the sun itself had given in. 1851 London at Table i. 19 A private note,..has been sent out, naming eight, railway time, and at that hour to a minute the guests are seated. 1898 Murray's Handbk. India (ed. 3) p. xvi, Railway time throughout India is Madras time... Karachi time is 52 min. behind railway time... Allahabad [time is] 7 min. before railway time. 1950 Chambers's Encycl. XIII. 641/1 Till the end of the first half of the 19th century local time was largely used in Great Britain, but the development of railways caused a need for a common system of time-keeping throughout Great Britain; thus the use of 'railway' i.e. Greenwich time became fairly general. 1968 Guardian 17 Feb. 8/4 A 24-hour electric clock was built into a wall at Greenwich, and this master clock controlled, by 'galvanism', another clock at London Bridge station, and all the other stations followed London Bridge. Greenwich Mean Time, or Railway Time, as it was called, prevailed. 1975 Times 21 June 12/2 From 1840 Greenwich or 'London' time began to be used as 'railway time' throughout Great Britain.

rain, sb.^{1} Add: **1. b.** Also in fig. phr. to know enough to come in out of the rain, and varr., to be sensible enough to act prudently in a given situation. Cf. *RAIN v. 1.

1848 F. A. DURIVAGE Stray Subjects 95 Ham was one of 'em—he was! He 'knew sufficient to get out of the rain'. 1884 'MARK TWAIN' Huck. Finn xiv. 122 De man dat think he kin settle a 'spute 'bout a whole chile wid a half a chile doan' know enough to come in out'n de rain. 1894 STEVENSON & OSBOURNE Ebb-Tide i. vi. 102 You seem to think underwriters haven't got enough sense to come in out of the rain. 1920 'SAPPER' Bull-Dog Drummond ii. 47 Either, James, I am a congenital idiot, and don't know enough to come in out of the rain; or we've hit the goods. 1932 'A. GILBERT' Body on Beam i. 23 A girl who's lived..in Menzies Street knows when to come in out of the rain. 1941 B. SCHULBERG What makes Sammy Run? i. 9 He didn't know enough to come in out of the rain and he died of..dumbness. 1973 J. WAINWRIGHT Devil you Don't 47 Come on in, out of the rain, Sugden—your brain is growing moss.

c. rain or shine: see *SHINE sb.^{1} 3 b.

3. b. (Further example.)

1963 C. D. SIMAK They walked like Men xxiv. 142 It was raining. Not much of a rain, just the beginning of a rain, cold and miserable.

4. a. (Further examples.)

1935 T. S. ELIOT Murder in Cathedral ii. 73 A rain of blood has blinded my eyes. 1942 R.A.F. Jrnl. 2 May 4 The continuous rain of pamphlets in all languages told of the enemy's defeats. 1975 J. G. EVANS Environment Early Man Brit. Isles iv. 81 It is difficult to work out vegetational structure from the pollen record due to the wide area from which the pollen 'rain' derives.

5. a. rain-blast, -blur, -cult, -curtain, -dew, -ditch, -drift, -dust, -gem, -land, -light, -mist, -pearl, -pipe (further examples), -shine, -song, -spout, -squall, -stain, -storm (further examples), -tear, -vapour, -washing, -world.

1863 G. M. HOPKINS Poems (1967) 12 But if the rain-blasts be unbound And from dank feathers wring the drops. 1930 E. POUND XXX Cantos vii. 27 Passion to breed a form in shimmer of rain-blur. 1923 L. SPENCE Gods of Mexico i. 11 We shall..attempt to descry..an incipient rain-cult. 1926 M. LEINSTER Dew on Leaf ii. iii. 159 The rain-curtain that swayed like a grey chiffon veil before Rhona's window. 1954 J. R. R. TOLKIEN Fellowship of Ring i. viii. 146 Frodo heard a sweet singing running in his mind: a song that seemed to come like a pale light behind a grey rain-curtain. 1922 JOYCE Ulysses 416 The air is impregnated with raindew moisture. 1949 E. POUND Pisan Cantos lxxiv. 12 Tovarish blessed without aim wept in the rainditch at evening. 1910 W. DE MORGAN Affair of Dishonour xiii. 207 The rain-drift..strengthened ever from the seaward. 1912 GALSWORTHY Inn of Tranquility 48 The mist had thickened to a white, infinitesimal rain-dust. 1931 BLUNDEN To Themis 27 The impulses of April, the rain-gems, the rose-cloud. 1930 T. S. ELIOT Ash-Wednesday 18 Not On the mainland, in the desert or the rain land. 1957 LD. HAILEY Afr. Survey xiv. 1011 Subject to a fixed annual payment to the former 'rainland landholders'. 1904 W. H. HUDSON Green Mansions x. 143 This subdued rain-light did not last long. 1893 KIPLING Seven Seas (1896) 76 Then softly as a rain-mist on the sward, Came to the Rose the Answer of the Lord. 1936 Discovery Aug. 242/2 A leaf which is slightly damp with dew or rain-mist. 1968 'HAN SUYIN' Birdless Summer vi. 136 The peaks were shrouded in rain-mist. 1879 O. WILDE in Time Apr. 30 Brush the rain-pearls from the eucharis. 1926 M. LEINSTER Dew on Leaf ii. iv. 191 Eastern music..was to him..as soft as whispered words of love, delicate as tumbling rain-pearls. 1913 J. MASEFIELD Daffodil Fields 92 The water..gurgled through the rain-pipe to the butt. 1969 Sears, Roebuck & Co. Catal. Spring/Summer 924/2 All guttering and rainpipes have double-locked seams and rolled edges for added strength. 1960 C. DAY LEWIS Buried Day i. 15, I am still haunted by the rainshine of orchards in the vale of Evesham. 1907 N. CURTIS Indians' Bk. 365 The Rain-Youth made the rain-songs and gave them to the Navajos. 1954 J. R. R. TOLKIEN Fellowship of Ring i. vii. 140 It seemed plain to them that the song was a rain-song, as sweet as showers on dry hills, that told the tale of a river from the spring in the highlands to the Sea far below. 1922 JOYCE Ulysses 470 Boys from high school are perched on the..rainspouts, whistling and cheering. 1962 Publ. Amer. Dial. Soc. xxxviii. 38 Rain spouts and water spouting predominate in Midland territory. 1978 J. A. MICHENER Chesapeake 488 The two Steeds, tumbling from their widow's walk, had caught momentarily on rainspouts edging the roof, and then fallen heavily into flowerbeds. 1849 N. KINGSLEY Diary (1914) 80 A few rain squalls headed off this forenoon to the NE. 1902 CONRAD Youth 40 Before sunset a thick rain-squall passed over the two boats..and that was the last I saw of them for a time. 1930 Times Educ. Suppl. 24 May 4/2 Dense rain-squalls forced her to deviate from her course. 1923 W. DE LA MARE Riddle 183 The mosses and rain-stains and

frost-flowerings of centuries of autumns and winters. 1952 DYLAN THOMAS Let. 6 Nov. (1966) 381 This tumbling house whose every rain-stain..I know in my sleep. 1905 Macmillan's Mag. Nov. 42 Everything European was washed off, as is paint off a woman's face in a rain-storm. 1979 'A. BLAISDELL' No Villain need Be ix. 147 Sunday morning broke bright and clear..which was normal for Southern California after a rainstorm. 1909 E. POUND Personae 13 Cloud and rain-tears pass they fleet! 1922 JOYCE Ulysses 416 Those burgeoning stars overhead, rutilant in thin rainvapour. 1886 HARDY Mayor Casterbr. II. xxii. 306 The walls..had been worn by years of rain-washings to a lumpy crumbling surface. 1970 T. HUGHES Crow 43, I am the uncrowned Of the rainworld.

b. rain-affected, -beaten (later examples), -bedraggled, -blown, -blurred, -born, -bruised, -burdened, -cold, -darkened, -dishevelled, -drenched (further examples), -fed, -filled, -flawed, -fragrant, -gorged, -heavy, -laden, -laid, -logged, -loud, -molested, -murmured, -pitted, -pocked, -rusted, -shimmery, -sleeked, -slicked, -soaked (further examples), -sodden, -stained, -starred, -streaked, -sunken, -sweet, -swept, -varnished, -washed (further example), -weathered, -wet, -worn adjs.

1905 Daily Chron. 26 July 1/7 A brilliant innings by Darling was the redeeming feature of Australia's batting on a rain-affected wicket at Manchester. 1962 Times 30 Aug. 3/3 Essex won the toss, decided to bat on a rain-affected wicket. 1976 J. SNOW Cricket Rebel 26 The rain affected wicket was a little suspect all through. 1914 W. B. YEATS in Poetry (Chicago) May 60 The pale unsatisfied ones Appear and disappear..With all their ancient faces like rain-beaten stones. 1932 D. GASCOYNE Roman Balcony 7 From the rain-beaten roses under the balcony. 1909 Daily Chron. 21 Aug. 1/2 (heading) Rain-bedraggled Suffragettes removed by Police. 1922 JOYCE Ulysses 434 Her dark den furtive, rainbedraggled. 1917 R. GRAVES Fairies & Fusiliers 24 I'm away to the rain-blown hill. 1901 'L. MALET' Hist. Sir R. Calmady v. i. 384 Actuality of rain-blurred, wind-scourged town without, and anger-begetting memories of Brockhurst within. 1965 M. MORRIS in Overland XXXI. III. 11 A gateway that still bore, rain-blurred and tattered, the printed notices. 1862 G. M. HOPKINS Poems (1967) 114 Then while the rain-born arc glows higher Westward on his sinking sire. 1916 D. H. LAWRENCE Amores 130 The rain-bruised leaves are suddenly shaken. 1932 D. GASCOYNE Roman Balcony 15 Clouds rear, Dark and ominous, rain-burdened. 1916 E. POUND Lustra 22 Grey olive leaves beneath a rain-cold sky. 1928 V. WOOLF Orlando v. 224 Her eyes slowly lowered themselves down and down till they came to the rain-darkened earth. 1962 I. MURDOCH Unofficial Rose i. 13 Her bright hair..hidden now except where a few rain-darkened ends clung to her neck. 1917 D. H. LAWRENCE Look! We have come Through! 56, I listen For the sluicing of their rain-dishevelled petals. 1901 Contemp. Rev. Mar. 437 This process of disafforesting is ruinous in a rain-drenched country. 1932 D. GASCOYNE Roman Balcony 75 To wait in the weary, rain-drenched queues. 1979 Arizona Daily Star 5 Aug. c1/4 The drive found the rain-drenched fairway. 1892 KIPLING Barrack-Room Ballads 192 Till he heard as the roar of a rain-fed ford the roar of the Milky Way. 1979 Nature 27 Sept. 251/3 Some good progress has certainly been made in research on dry-land farming, largely aimed at reducing risks in cultivation in rainfed areas. 1942 W. FAULKNER Go down, Moses 356 The tent-flap falling on the same out-waft of faint and rain-filled light. 1963 A. LUBBOCK Austral. Roundabout 4 Across the paddock the sheen of water glittered from the rain-filled dams. a 1918 W. OWEN Poems (1963) 95 Or be you in the gutter where you stand, Pale rain-flawed phantom of the place. 1916 JOYCE Portrait of Artist (1969) 228 The brother's face was bent upon her fair rainfragrant hair. 1917 KIPLING Diversity of Creatures 404 Rain on rain-gorged channels raised the water-levels round them. 1943 D. GASCOYNE Poems 1937–1942 59 Beside the stolid opaque flow Of rain-gorged Thames. 1942 W. FAULKNER Go down, Moses 240 Now the woods ahead of them and the rain-heavy air were one uproar. 1916 JOYCE Portrait of Artist (1969) v. 176 The rainladen trees of the avenue evoked in him, as always, memories of the girls and women in the plays of Gerhart Hauptmann. 1921 W. DE LA MARE Mem. Midget viii. 47 We..mounted into a four-wheeled cab, and once more were in motion in the rain-laid dust. 1970 G. E. EVANS Where Beards wag All viii. 90 On some Suffolk farms it was used in the 'fifties, though admittedly for special jobs like..the saving of wind- or rain-laid crops. 1960 T. HUGHES Lupercal 27 Rain-logged, wind-unroofed, The manor farm hulked its last use As landmark. 1926 A. HUXLEY Two or Three Graces 173 Peddley broke the rain-loud silence. 1845 LONGFELLOW To Old Danish Songbook in Poems (1846) 377 Yellow are thy time-worn pages, As the russet, rain-molested Leaves of autumn. 1942 W. FAULKNER in Story Mag. May–June 51/2 The tent, the rain-murmured canvas globe, was filled with it once more. 1916 H. G. WELLS Mr. Britling II. iii. 285 Men shouted and women sobbed and cowered, and flares played upon the rain-pitted black waves. 1931 V. WOOLF Waves 227 One bone lay rain-pocked and sun-bleached. 1947 AUDEN Age of Anxiety (1948) ii. 49 O the drains are clogged, Rain-rusted, the roofs of the privies Have fallen in. 1905 Academy 21 Oct. 1103/1 The rabbit that scuttled before us, the league-footing hare That shot from her form with tawny and rain-sleeked coat. 1964 Listener 9 Jan. 67/1 In darkness outside Foxes and rain-sleeked stones and the dead. 1970 Globe & Mail (Toronto) 28 Sept. 12/4 She braked and skidded on rain-slicked pavement but had no real chance to stop. 1890 KIPLING Departmental Ditties (ed. 4) 94 The rotten, rain-soaked khud. 1975 N. LUARD Robespierre Serial iii. 11 They'd gathered in his suite at the Vendôme as soon as they got back from the Loti, without even pausing to change their rain-soaked clothes. 1904 Westm. Gaz. 10 May 2/1 The Cossacks were drawn up on a large, rain-sodden, muddy field. 1978 Cornish Guardian 27 Apr. 4/8 In the second half it was the clay club who played the game to suit the rain-sodden conditions. 1895 M. PEMBERTON Impregnable City II. xiii. 270, I..watched for a while the sleeping island through the rain-stained glass. 1958 L. DURRELL Balthazar iii. 62 The gonfalons bellied like sails in the rain-starred afternoon. 1956 H. GOLD Man who was not with It (1965) xii. 100 The carnival smelled the same, and The rain-streaked sign swung there. 1980 Jrnl. R. Soc. Arts July 530/1 She made an artificial stone, but this was nothing like the dismal, rain-streaked concrete of today. 1916 BLUNDEN Harbingers 20 Rainsunken roof, grown green and thin For sparrows' nests and starlings' nests. 1913 W. DE LA MARE Peacock Pie 127 Feathered birds in the rain-sweet sky. 1932 BLUNDEN Face of England 177 Those blackening rain-swept fields. 1978 R. LUDLUM Holcroft Covenant xxxi. 366 His last assignment had been at Kennedy Airport during a rainswept night when a cordon of police surrounded the glistening fuselage of a British Airways 747. 1867 J. R. LOWELL in Atlantic Monthly July 99 The candle she held in the door, From rain-varnished tree-trunk Flashed fainter. 1965 New Statesman 19 Nov. 803/1 We shared a rainwashed picnic with some shy, boyish chaps. 1940 J. BETJEMAN Old Lights for New Chancels 39 The rain-weathered streets of adjacent Rumsaa. 1916 D. H. LAWRENCE Amores 49 All round the yard it is cluck, my brown hen, Cluck, and the rain-wet wings. 1928 E. SITWELL Five Poems 16 The airs like rain-wet shrinking petals curl. 1904 W. DE LA MARE Henry Brocken vi. 51 She trod with cautious foot and peering eye the green, rainworn paths. 1968 T. KINSELLA Nightwalker 11 A rain-worn, delicate Stone shape.

c. rain-giver, -maker (see as main entry in Suppl.); rain-bringing, -repellant, -repelling, -resistant, -resisting adjs.

1922 W. G. KENDREW Climates of Continents v. 25 But in the south, south-west winds begin to make their appearance and the southern limit of the [Sahara] desert is found about 18° N. lat. where these rain-bringing winds become predominant in summer. 1946 F. E. ZEUNER Dating Past vii. 203 The secondary effects of the glaciation..affected the Mediterranean, through the deviation of many rain-bringing depressions. 1922 W. G. KENDREW Climates of Continents xx. 110 The south-west monsoon is essentially the rain-giver of India. 1927 P. M. LARKENS in Sudan Notes & Rec. IX. 46 But in spite of this it is Liwa's spirit, and not Mbali, that is supposed to be the rain-giver here. 1968 Daily Tel. 4 Nov. 11/5 Best outfits..included a gleaming parchment kid coat, rain-repellant. 1892 C. M. YONGE Old Woman's Outlook 178 Sheaves..built up in the rain-repelling arrangements. 1958 Times 6 Oct. 13/1 And all the tweeds are mothproofed as well as rain-resistant. 1963 Economist 20 July 256/2 Novelties (such as 'rain-resistant' petunias..from Japan). 1952 R. LEIGHTON Compl. Bk. Dog (rev. ed.) viii. 126 The outer-coat [of the Alsatian] is also close,..so that it is rain-resisting.

6. rain-belt, a stretch of land much subject to rain; also fig.; **rain bonnet** chiefly U.S., a plastic fold-up bonnet worn as a protection against the rain; **rain boot** U.S. (see quot. 1975); **rain-cape**, a waterproof cape (CAPE sb.^{2} 3) furnishing protection from rain; hence **rai·ncaped** a., wearing a rain-cape; **rain-charm**, an object, action, or incantation used by a rainmaker to summon rain; **rain check, cheque** chiefly U.S., (a) a ticket given to a spectator at an outdoor event providing for a refund of his entrance money or admission at a later date, should the event be interrupted by rain; transf., a ticket allowing one to order an article before it is available, and to collect it when it becomes so; (b) fig. (see quot. 1930); also, esp. in phr. to take a rain check on, to reserve the right not to take up a specified offer until such time as it should prove convenient; **raincoat** (examples); used attrib. of someone or something thought to be lewd or unseemly, esp. in phr. raincoat brigade, disreputable, raincoat-clad, frequenters of cinemas that show pornographic films (also in extended use); hence **rai·ncoated** a., wearing a raincoat; **rain crow** U.S., the yellow-billed or black-billed cuckoo, Coccyzus americanus or C. erythrophthalmus; cf. RAINBIRD 2; **rain dance**, a dance performed by a tribal group in the hope of summoning rain; **rain date** U.S., an alternative date upon which an outdoor activity can be held if rain should cause the intended date to be unsuitable; **rain-day** Meteorol., a day, commencing for statistical purposes at 9 a.m. G.M.T., on which the recorded rainfall is not less than 0·01 inch or 0·2 mm.; **rain dog** (see DOG sb. 10 a and quot.); **rain-fly**, a blood-sucking, greyish fly, Hæmatopota pluvialis, of the family Tabanidæ; **rain frog**, a name used in North and Central America for a small tree frog or spring peeper belonging to the genus Hyla; **rain-god** (earlier and later examples); also **rain-goddess**; **rain-hat, hood**, a head-covering designed spec. to afford protection against the rain; **rain jacket**, a short raincoat designed in the shape of a jacket; also, a small protective covering worn by a dog; **rain-jungle** = *RAIN FOREST; **rain load**, the weight of rain on an airship; **rain-pitting** (example); **rain-shadow**, an area of small annual rainfall, brought about because it is sheltered from prevailing rain-bearing winds by a range of hills (see also quot. 1955); hence **rain-shadowing**, the action of producing a rain-shadow; **rain-stone**, a stone believed to possess certain qualities and used in primitive rain-making rituals; **rainsuit**, a jacket and leggings designed to protect one against the rain; **rain-temple**, a temple in which supposedly rain-provoking rituals are enacted; **rain-tree**, substitute for def.: (a) an evergreen shrub, Brunfelsia undulata, of the family Solanaceæ, native to Jamaica and bearing white, bell-shaped flowers; (later example); (b) = *GUANGO, *SAMAN^2; **rain-wash** (further examples); **rainwear** (see quot. 1953). See also *RAIN FOREST.

1878 R. J. HINTON Handbk. Arizona vii. 201 The Santa Cruz [Valley], up to Tubac, marks the western limit of a notable rain-belt. 1948 E. WAUGH Loved One 69 Water played everywhere from a buried network of pipes, making a glittering rain-belt waist-high. 1968 Punch 19 June 892/3 Weinsoff can have his name on his key tag..hatband and sponge (and Mrs. Weinsoff on her pot holder, thimble and rain bonnet). 1975 S. LAUDER Killing Time on Corvo vi. 53 She grinned chummily, tucking strands beneath the rain bonnet. 1951 Sun (Baltimore) (B ed.) 23 May 7/6 (Advt.), The only rainboot made of Norlon, the wonder plastic!.. Folds neatly and compactly in a fit-in-your-purse plastic pouch. 1975 Listener 25 Dec. 845/1 American rain boots.. are a kind of mid-calf galoshes which pull over your shoes. 1921 Daily Colonist (Victoria, B.C.) 1 Oct. 3/5 (Advt.), Children's good quality tan colored raincapes to fit girls 4 to 12 years; made with lined hood. 1922 D. H. LAWRENCE in Dial Jan. 54 Fat cab-men, whose rubber rain-capes flapped like wings in the wind. 1976 Southern Even. Echo (Southampton) 1 Nov. (Advt. section) 8/6 Cindico Pushchair Raincape, royal blue, new, unwanted gift. 1977 Observer (Colour Suppl.) 7 Aug. 41/2 Dress properly for riding a bike... Avoid rain capes that flap about the wheels. 1922 JOYCE Ulysses 406 They hear the heavy tread of the watch as two raincaped shadows pass the new royal university. 1973 A. PRICE October Men iii. 41 The rain-caped policeman materialised out of a gap in the hedge to stop the car. 1890 J. G. FRAZER Golden Bough I. iii. 403 The story that the body of Osiris enclosed in a coffer was thrown by Typhon into the Nile perhaps points to a custom of throwing the body of the victim, or at least a portion of it, into the Nile as a rain-charm. 1936 E. E. EVANS-PRITCHARD in Essays Soc. Anthrop. (1962) viii. 188 On one occasion, at Tambura, a rainmaker buried his rain charm with the body of his son, in revenge for the latter's death, which he suspected to have been caused by magic. 1884 St. Louis (Missouri) Post-Dispatch 26 May 8/2 The heavy rain yesterday threw a damper over local operations. At each of the parks the audience had to be content with three innings and rain checks. 1919 Nat. Geogr. Mag. Aug. 103 Even the sport-loving Britons are said to have admired and wondered at the American dough-boy..issuing occasional rain-checks in mid-inning when the downpour of bursting shells became too distracting. 1930 J. LAIT Big House 6 A parole is a 'rain check'. 1939 R. CHANDLER Big Sleep xi. 83 The Sternwoods have money. All it has bought them is a rain cheque. 1959 P. H. JOHNSON Humbler Creation xviii. 120 Westlake said, 'I'll take a rain-check. Be back.' Maurice, not certain of this idiom which he vaguely knew to be American, watched him go. 'Has he had enough?' he asked Kate... 'I should make a guess that he has to have a drink.' 1970 Washington Post 30 Sept. D3/5 (Advt.), If our stores cannot perform this work within the time indicated we will give you a raincheck enabling you to have work done within 30 days at the advertised price. 1976 L. DEIGHTON Twinkle, Twinkle Little Spy xiv. 141 'Let me take a rain-check.' 'On a love affair?' I said. 1976 New Yorker 26 Apr. 31/1 Levin's project is to array miniature open-air versions of a hundred New York restaurants side by side along the Central Park Mall on Saturday, May 22nd (rain check for Sunday). 1977 Time 14 Nov. 41/3 Unless he can cash his rain check by early 1978, the President will run smack into France's March elections. 1830 J. F. WATSON Ann. Philadelphia App. 52 As a defence from rain, the men wore 'rain coats', and the women 'camblets'. 1871 A. B. MITFORD Tales Old Japan II. 3 The farmers, dressed in their grass rain-coats. 1897 Sears, Roebuck Catal. 187/1 Men's Double Texture Mackintosh Rain Coats..Made from good quality diagonal cloth..suitable to wear in place of an overcoat. 1925 Scribner's Mag. Sept. 238/2 His raincoat was split up the back, under his belt. 1976 Vogue Jan. 22/4 Raincoats in rust, khaki and navy blue..£74. 1976 Times Lit. Suppl. 9 July 842/2 The serious press in England is still being penalized for having so long tolerated less high-minded efforts at disclosure by the raincoat press. 1976 Observer 12 Dec. 24/4 Mindful of its duty to the raincoat brigade, however, the film abandons investigation at regular intervals to provide the expected, and inordinately protracted, bouts of titillation. 1977 New Yorker 23 May 33/2 O.K. for the raincoat brigade, but rather bland for the longhairs. 1930 W. FAULKNER As I lay Dying 74 Slowly he strokes his hands on his raincoated thighs. 1974 D. FRANCIS Knock Down i. 10 The rain-coated assembly looked.. miserable. 1806 M. LEWIS Jrnl. 16 July in Orig. Jrnls. Lewis & Clark Exped. (1905) V. 205, I saw both yesterday and today the Cookkoo or as it is sometimes called the rain craw. 1831 J. J. AUDUBON Ornith. Biogr. I. 19 The Dutch farmers of Pennsylvania know it [sc. the yellow-billed cuckoo] better by the name of Rain Crow. 1872 E. COUES Key to N. Amer. Birds 190 American cuckoos..are..noted for their loud jerky cries, which they are supposed to utter most frequently in falling weather, whence their popular name, 'rain crow'. 1880 G. W. CABLE Grandissimes xxviii. 237 The dismal ventriloquous note of the rain-crow. 1899 B. TARKINGTON Gentleman from Indiana viii. 118 The rain-crow that sat on the fence. 1917 T. G. PEARSON Birds Amer. II. 130/1 Both species are known as the 'Rain Crow' because of the belief—especially among farmers—that their guttural cry predicts rain. 1935 H. DAVIS Honey in Horn xv. 237 A rain-crow..sang like clanking a little copper bell. 1946 G. STIMPSON Bk. about Thousand Things 55 The

American cuckoo is known to country people as the rain or storm crow because its plaintive note..is regarded as a sign of rain or storm. 1930 J. M. BUTTREE Rhythm of Redman 89 It is the privilege of all to improvise the new song for each rain dance. 1968 E. MCCOURT Saskatchewan 6 Old Chief Sheepskin, nominally a Christian, summoned his braves to perform a rain dance. 1977 Time 18 Apr. 25/1 When the wet season in Northern California turned up bone dry, about 2,000 San Franciscans staged a modern rain dance in the Hyatt Regency Hotel. 1972 Village Voice (N.Y.) 1 June 34/5 The Blue Mountain Paper Parade will present choreographer Barbara Roan's latest work, 'Waystation/Truckers Only,' on Saturday, June 3, at 3 p.m. on the Brooklyn Heights Promenade... The rain date is June 4. 1975 New Yorker 7 July 11/3 Wednesday, July 2, at 8:30 (rain date, July 3), 'La Traviata'. 1978 Chicago June 201/2 It's open house at the Durant's on June third..Stop in between ten a.m. and five p.m. $1.50 admission. Rain date is June fourth. 1906 Brit. Rainfall 1905 123 This improvement may be traced by the steady increase in the number of rain days reported. 1928 Nature 14 Apr. 591/2 The variations of the number of rain-days over the British Isles are much less than the corresponding variations of rainfall. 1976 Southern Even. Echo (Southampton) 2 Nov. 13/6 According to meteorological office statistics, rainfall in the city was more than double the average and the number of 'rain-days' ranked as the highest in 50 years. 1866 'MARK TWAIN' in Sacramento Union 24 Aug. 3/2 What the sailors call 'rain dogs'—little patches of rainbow—are often seen drifting about the heavens in these latitudes. 1921 E. STEP Brit. Insect Life 231 This is the Clegg or Rain-fly.., the grey, dusty-looking Fly that you have no suspicion is sitting on your hand or neck until its sharp lancet is thrust into your flesh. 1952 E. F. DAGLISH Name this Insect 277 Rain-fly or Clegg... Common in fields, meadows, woods, and about hedgerows. 1958 J. CAREW Black Midas v. 79 A money-spider was struggling with a rain fly. 1971 Country Life 21 Oct. 1035/1, I am not partial to the bite of the rainfly. 1827 T. L. MCKENNEY Sk. Tour to Lakes 158 We found the few people who live near its mouth..[with] rain frogs on the logs of their huts to sing them to repose. 1938 M. K. RAWLINGS Yearling i. 6 A rain frog sang a moment and then was still. 1958 J. CAREW Wild Coast iv. 59 Outside, rain frogs were complaining to the stars. 1962 S. WYNTER Hills of Hebron v. 70 From the dark hollows rain-frogs cracked sharp sad notes. 1864 Rain-god [see *MGANGA]. 1968 New Larousse Encycl. Mythol. 438 (caption) Tlaloc, the Aztec rain god enjoyed a similar popularity and was also one of the chief gods of the pantheon. 1904 Rain-goddess [see Earth-Mother s.v. *EARTH sb.^{1} II.]. 1967 J. R. CRAWFORD Witchcraft & Sorcery in Rhodesia xii. 188 We go to Hosana, that is the rain goddess, and appeal to Hosana for rain. 1921 Daily Colonist (Victoria, B.C.) 28 Oct. 7/1 (Advt.), Women's and children's rain hats at $1.50 each. 1955 E. POUND Classic Anthol. IV. 208 With a crowd of rain-hats And clicking hoes Out goes the weed To mulch and rot. 1967 Punch 21 June 907/1 Prizes for the best set of answers include restrung celebrity-used rackets, foldaway rain hats and transistor radios for getting the cricket commentaries. 1977 Chicago Tribune 2 Oct. XIII. 4/1 Besides, you have only a mile and a half to go, and you have boots, raincoat, and rain hat. 1964 'E. PETERS' Flight of Witch ii. 34 Did she go off wearing her rain-hood, and her heavy shoes, a night like this? 1977 P. KEMP in P. Collenette Winter's Tales 23 49 She was busy.. untying a transparent plastic rain-hood that she was wearing over a rayon head-square. 1975 Country Life 29 May 1424/2 The newest thing in rainwear is the suit... Next to the suit, the rain jacket is very much around. 1976 Evening Times (Glasgow) 1 Dec. 26/6 Rain jackets, royal, sky, red and yellow, all sizes from £8·99. 1960 Spectator 30 Sept. 493/2 The Brookes..stayed a hundred years as landgraves of an impoverished swatch of rain-jungle. 1964 R. PERRY World of Tiger i. 3 There are..less than four thousand scattered over India,..from Goa to the rain-jungles of Assam. 1928 C. F. S. GAMBLE Story N. Sea Air Station xv. 249 Rain load also had to be reckoned with, which may go as far as 3,000 kg. 1879 Encycl. Brit. X. 294/1 Sun-cracks, Rain-pittings, &c.—Proofs may not infrequently be found that during deposition aqueous strata have been laid bare to air and sun. 1902 H. J. MACKINDER Britain & Brit. Seas 165 This dryness to leeward of the heights has been termed their rain-shadow. 1955 Sci. News Let. 2 Apr. 212/1 Prevailing winds..smack against the mountains, and are then forced down the eastern slopes. These winds are so strong they keep warm, moist air from the Gulf of Mexico from penetrating deep inland. The air is sinking rather than rising. Result: no moisture falls. Weathermen call this effect the rain shadow. 1974 M. PEISSEL Great Himalayan Passage vi. 112 This entire area is within what is known as the 'rain-shadow' of the Himalayas, a sheltered spot where clouds never break. 1936 Geogr. Jrnl. LXXXVII. 11 Valley routes have certain advantages as well as disadvantages for air travel. Among their advantages I would put..the finer weather, owing to local rain-shadowing. 1897 W. E. ROTH Ethnol. Stud. N.-W.-Central Queensland Aborigines xii. 167 The rain-stick, koo-roo-mun-do, is formed of a thin piece of a species of 'white' wood, about 20 inches long, on to the top of which is fixed a mass of the ordinary cementing-substance: into this the three 'rain-stones', pieces of white quartz-crystal, are stuck. 1932 Times Lit. Suppl. 1 Dec. 915/2 Such rain-stones may be of quartz, of which the transparency is suggestive of water. 1975 Islander (Victoria, B.C.) 2 Feb. 3/1 When the medicine man..wishes to produce rain he plunges his rain-stone in water, takes a split-top peeled cane in hand and beckons with it to the clouds. 1965 Harper's Bazaar Feb. 39 News now, the rainsuit, newer still if it's trousers. 1966 Daily Tel. 14 Nov. 10/7 Not certain of the weather? Then for town or country a rainsuit could be the answer. 1974 M. HOYT 30 Miles for Ice Cream x. 114 So I, for one, wear the rubber pants of a rain suit when I'm shanty-fishing. 1977 New Yorker 27 June 74/2 Six miles downstream, I had added..a rain suit, hood to heel. 1907 R.S. RATTRAY Some Folklore Stories & Songs in Chinyanja II. ix. 118 Next morning at dawn everyone comes together and they go to the rain temple. 1911 J. G. FRAZER Golden Bough: Magic Art (ed. 3) I. v. 250 When the rains do not come..the people of Central Angoniland repair to what is called the rain-temple. 1911 Chambers's Jrnl. Feb. 206/1 The rain-tree..can withstand extreme climatic fluctuations, needs but little care in its cultivation, and grows rapidly. 1924 RECORD & MELL Timbers Trop. Amer. 204 The 'samán' or rain tree..belongs to this genus. 1939 R. C. MARSHALL Silviculture Trinidad & Tobago 119 Rain Tree..is a short-boled tree with an enormous wide spreading crown. 1961 20th Cent. Jan. 64 Rain-trees, whose leaves close up at night, enfolding moisture which in the morning the unfurling foliage discharges. 1963 [see *GUANGO]. 1969 [see monkey-pod (tree) s.v. *MONKEY sb. 17 b]. 1975 Bangladesh Observer 25 July 5/6 Rain-tree..is found in abundance all over Bangladesh. 1976 Hortus Third (L. H. Bailey Hortorium) 185/2 Brunfelsia..undulata..Rain tree. Slender, evergreen shrub. 1942 Rain-wash [see *BAD LANDS]. 1965 G. J. WILLIAMS Econ. Geol. N.Z. vii. 80/1 In his view all the components of the schist except the quartz and gold and such resistant heavy minerals as zircon were reduced to clay and generally removed by rain-wash. 1977 A. HALLAM Planet Earth 81/1 Because of the dominance of rainwash over other erosional forces, these features are most common in warmer climates. 1953 P. C. BERG Dict. New Words 134/1 Rainwear, garments suited for wearing in rain; e.g. 'Britain's Best Rainwear'. 1958 People 4 May 15/7 (Advt.), Have your suit for £1 down and add whatever else you need—rainwear, shoes, shirts, sportswear, etc. to your account. 1975 Country Life 6 Feb. 347/3 Ismat also sell..T-shirts and rainwear. 1977 J. AIKEN Last Movement i. 33 It was still pouring, so..Gina..put herself back into her red plastic rainwear.

rain, v. Add: **I. 1.** Phr. to go (or come) in when it rains: to take measures for one's own safety; to exercise ordinary prudence; to shift for oneself. U.S. Cf. *RAIN sb.^{1} 1 b.

a 1852 F. M. WHITCHER Widow Bedott Papers (1856) iii. 36 He was a safily feller—dident scarcely know enough to go in when it rained. 1866 'MARK TWAIN' Lett. from Sandwich Islands (1937) 84 A..majority..that knew just about enough to come in when it rained. 1873 —— Choice Humorous Wks. III. 524, I perceive that thou art none of them that know not to come in when it doth rain. 1906 Springfield (Mass.) Weekly Republ. 12 July 2 Every citizen of Vermont who is capable of going in when it rains ought to understand [etc.]. 1923 E. F. WYATT Invis. Gods I. ii. 19 Hetherington Marshfield, who doesn't know enough to go in when it rains!

b. Also const. on.

1925 R. TORRENCE Hesperides 60, I was weak as a rained-on bee. 1937 Burlington Mag. June 262/1 Rained-on, as it were, by Nature. 1972 J. GORES Dead Skip xvi. 113 Two big recently rained-on cardboard boxes of trash.

II. 6. a. (Further example.)

1871 W. D. HOWELLS in Atlantic Monthly Dec. 722/2 It was raining one of those cold rains.

b. Phr. if it should rain porridge, he would want his dish and varr., denoting a person's recurrent bad luck or management.

1670 J. RAY Coll. Eng. Proverbs 191 If it should rain pottage, he would want his dish. 1732 T. FULLER Gnomologia 112 If it should rain Porridge, he'd want a Dish. 1889 C. H. SPURGEON Salt-Cellars 257 If it rained porridge, the lazy man would have no basin. 1950 K. S. PRICHARD Winged Seeds ii. 29 Unluckiest man ever I knew. If it was raining pea soup, he'd only have a fork. 1970 R. BEILBY No Medals for Aphrodite (1971) v. 169 Gawd, we're an unlucky battalion, we are. If it was rainin' virgins we'd be washed away with a poofta, dinkum!

7. a. (Further examples.)

1939 Ann. Reg. 1938 260 Air-raids rained bombs on Barcelona, Tarragona, and a number of peaceful seaside towns where there was no trace of any military objective. 1977 Time 14 Mar. 53/1 Fans rained bottles and cans on to the ground.

8. b. pass. and intr. Of particulate matter: to be removed from the atmosphere as a result of being incorporated into raindrops as they form. Cf. *RAIN-OUT.

1975 Nature 13 Nov. 134/2 The importance of this observation is that it makes nonsense of the assumption..that practically everything that can..be rained out as air ascends into the stratosphere, actually is rained out and thus removed. 1979 McGraw-Hill Yearbk. Sci. & Technol. 153/1 The ^{3}H generated was largely injected into the upper atmosphere, from which it 'rained out' into the oceanic-hydrologic system.

9. a. (Later examples.)

1924 C. MACKENZIE Old Men of Sea ii. 17 The sky had rained itself out. 1944 T. D. CLARK Pills, Petticoats & Plows 88 Behind him at home was a cotton crop which had been rained out. 1976 G. MOFFAT Short Time to Live ix. 85 She was a walker rained off the hill.

b. to be rained out (U.S.) or off, of an outdoor event (esp. a match), an airline flight, etc.: to be terminated or cancelled because of rain. So **rained-off** a.

1928 Chicago Tribune 18 June 27/7 (heading) Sox, Boston series final is rained out. 1937 Sun (Baltimore) 18 May 17/8 Today's [baseball] game was rained out. 1955 Times 24 May 10/2 An open-air meeting was rained off, but pickets patrolled the dock entrances carrying sandwich boards. 1960 C. DAY LEWIS Buried Day i. 20 Their tiny tragedies—a rained-off picnic, a broken toy. 1964 Observer 12 Jan. 32/1 Rain check..is the receipt or counterfoil of a ticket taken for a baseball game which entitles you to see another match if the one you wanted to see is 'rained off'. 1969 'E. LATHEN' When in Greece ix. 98 Unfortunately the planes are rained out, but the train should get her here before midnight. 1974 Union (S. Carolina) Daily Times 19 Apr. 6/1 In the American League..Minnesota clipped California 3–2 and Baltimore's game at Detroit was rained out. 1977 'J. LE CARRÉ' Hon. Schoolboy xvi. 309 Watching a rained-off cricket match where the contestants wanted only to go back to the pavilion. Ibid. xxii. 525 The match was rained off... Another date would be fixed. Stand by, they said.

rain-bird. 1. For Gecinus substitute Picus. (Later examples.)

1913 H. K. SWANN Dict. Eng. & Folk-Names Brit. Birds 187 Rain-bird, Rain-fowl, or Rain-pye: The Green Woodpecker... It is still a country belief that when the cry of this bird is much heard rain will follow. 1979 Country Life 20 Sept. 820/3 Now I hardly ever hear the rainbird, as the green woodpecker is called.

rainbow. Add: **1. a.** Phr. the end of the rainbow, the rainbow's end: with allusion to the proverbial belief in the existence of a crock of gold (or something else of great value) at the end of a rainbow. Cf. rainbow-chase at sense 4 d below.

1836 W. D. COOPER Gloss. Provincialisms Sussex 16 Go to the end of the rainbow, and you'll find a crock of money. 1916 R. E. PEACH (title) Rainbow's end. 1971 A. DIMENT Think Inc. ii. 21 Every wornout hack of an agent retires there... It's the end of the rainbow for every British spy. 1973 R. LEWIS Blood Money i. 12 He'd found the jackpot again, the end of the rainbow. 1976 Times 9 Aug. 10/5 Until recently West Germany has [been]..relegating reunification to the dream world at the rainbow's end. 1977 Times 27 Apr. 11/1 Holyhead could do with some oil at the end of the rainbow... It has seen a substantial decline in trade.

3. b. (Earlier and later examples.)

1779 P. FRENEAU in U.S. Mag. Feb. 85 The rainbow cuts the deep of varied green. 1909 Westm. Gaz. 13 Feb. 16/2 If there is a river in which rainbows should grow large and wax fat it is the Thames. 1940 R. PERTWEE Master of None v. 27, I caught a sizeable rainbow with it [sc. a rod] last season. 1963 Times 8 June 13/2 The rainbow is a fish of American origin. 1977 F. PARRISH Fire in Barley iii. 33 The Mullett family had eaten a good many of his fat rainbows. 1979 Fisherman's Weekly 21 June 5/4 Top fish was a rainbow of 7lb. 7oz. taken by Eddy Ouslie of Exeter.

3*. A capsule containing the barbiturates Amytal and Seconal, one end of which is red and the other blue. slang (orig. U.S.).

1968–70 Current Slang (Univ. S. Dakota) III–IV, 100 Rainbows, n. A type of barbiturate in red and blue capsules. (Drug users' jargon.) 1972 Sunday Sun (Brisbane) 2 July 14/3 The barbiturate addict takes red devils..rainbows.. all 1972 junkie names for various drugs. 1972 J. WAMBAUGH Blue Knight (1973) xvi. 303 One lousy time I dropped a red devil and a rainbow with some guys at school, and that's all the dope I ever took. 1976 M. MILLAR Ask for me Tomorrow xiv. 115 Getting their kicks by mixing drinks and drugs, like..the high school kid carrying a flask of vodka to wash down the rainbows.

4. a. rainbow flower (later example), light (later examples), -space.

1928 BLUNDEN Retreat 67 What angel dropped her rainbow-flowers In that horizon blue of ours? 1854 THOREAU Walden 218 It was a lake of rainbow light, in which, for a short while, I lived like a dolphin. 1924 R. GRAVES Mock Beggar Hall 78 Then since laws move in rainbow-light Let faith be therefore strong. 1918 E. SITWELL Clowns' Houses 14 A glassy ball that clowns have hurled Through the rainbow-space of laughter.

c. rainbow-gay, -happy, -sweet, -tailed, -tinted (earlier and later examples) adjs.

1893 KIPLING Seven Seas (1896) 69 O rainbow-gay the red pools lay that swilled and spilled and spread. 1940 BLUNDEN Poems 1930–1940 190 How comes the rainbow-happy shower! 1942 L. HUGHES Shakespeare in Harlem 20 So if you want to know beauty's Rainbow-sweet thrill, Stroll down luscious, Delicious, fine Sugar Hill. 1929 Oxf. Poetry 32 Conjure..rich waterlilies lightly to be embraced by rainbowtailed delirious dragonflies. 1827 DISRAELI Viv. Grey IV. vi. vi. 196 The sun..lent additional brilliancy to the rainbow-tinted birds of paradise. 1897 E. L. VOYNICH Gadfly viii. 352 The surplices of the choristers gleamed, rainbow-tinted, beneath the coloured windows.

d. rainbow-bird Austral., the bee-eater, Merops ornatus, a small, brightly coloured bird belonging to the family Meropidæ and native to northern Australia; **rainbow boa**, a large iridescent snake, Epicrates cenchris, of the family Boidæ, found in forest areas of northern South America; **rainbow cactus**, a small, cylindrical cactus, Echinocereus pectinatus, native to southwestern North America and bearing red flowers and spines in bands of various colours; **rainbow-chase** fig., a quest which is rendered pointless by the illusory nature of its object; hence **rainbow-chaser, -chasing**; **rainbow-fish**, for New Zealand substitute Australia; also, a brightly coloured wrasse belonging to the family Labridæ; (examples); **rainbow-serpent**, in Australian aboriginal mythology, a large snake associated with water; **rainbow wrasse** (earlier and later examples).

1911 J. A. LEACH Austral. Bird Bk. 107 Rainbow-bird, Aust. Bee-eater,..Spinetail, Pintail, Merops ornatus. 1933 Bulletin (Sydney) 11 Oct. 21/3 One of the loveliest birds of N. Australia..is the rainbow-bird, which, with its green coat, black tail, and bright orange throat, well deserves its name. 1944 A. RUSSELL Bush Ways v. 28 The mutton-birds make underground nesting burrows, and so, too, do the rainbow-bird..and the white-backed swallow. 1963 A. LUBBOCK Austral. Roundabout 167 A rainbow-bird darted across the road, a brilliant flash of emerald, blue and orange. 1975 I. ROWLEY Bird Life 253 Hole-nesting birds such as pardalotes, kingfishers, rainbow-birds, parrots, mutton-birds and penguins can all be captured while they are visiting the nest. 1910 R. L. DITMARS Reptiles of World IV. 231 The Rainbow Boa..derives its name from a gorgeous iridescence playing over the scales of a healthy example. 1937 A. H. VERRILL Strange Reptiles ix. 120 We found two of the lovely iridescent rainbow boas, one barely

five feet in length, the other a trifle over six feet. 1958 J. CAREW Black Midas vi. 119 On our way back I saw a rainbow boa coiled around a branch. 1965 R. & D. MORRIS Men & Snakes viii. 198 But what is the true life span of a snake?.. An anaconda managed 28 years, a rainbow boa 27. 1972 M. RICHARDSON Fascination of Reptiles vii. 79 A relation [of the Cuban boa] is the rainbow boa..found between the south of Mexico and northern Argentina. 1892 J. G. LEMMON in G. W. James In & Around Grand Canyon (1900) xxxi. 325 Rainbow cactus, with bright-coloured zones. 1930 J. M. BREAZEALE Color Schemes of Cacti 14 (caption) A flower of the rainbow cactus..about three-fourths natural size. 1976 Express-News (San Antonio, Texas) 27 Nov. 10-c/1 We..saw a spiny little cactus growing upright out of the ground like a fat ear of corn. Its bands of colors were the colors of the rainbow—and that's its name—the rainbow cactus. 1886 St. James's Gaz. 2 June 10 A fact which had led Mr. Rylands off a rainbow-chase after a visionary Chancellorship. 1892 Courier-Jrnl. (Louisville, Kentucky) 1 Oct. 1/8 The rainbow chasers of the Administration are not idle these days. 1925 D. SENIOR Rainbow Chasers iii. 33 'We are all Rainbow-Chasers in youth,' I retorted. 'And even in old age the hope of finding the Crock of Gold is hard to kill.' 1904 N.Y. Even. Post 1 Sept. 7 Early in the campaign he had told his associates that it was of no use to go rainbow chasing after Massachusetts, Wisconsin, or Illinois. 1908 Hampton's Broadway Mag. Nov. 599/1 We had no business whatever to go rainbow chasing. 1888 G. B. GOODE Amer. Fishes 205 In this limpid pool were many gorgeously-colored species, the angel-fish, the parrot-fish, the rainbow-fish. 1908 E. J. BANFIELD Confessions of Beachcomber I. iv. 156 In the rainbow and parrot fishes they [sc. pharyngeal teeth] are highly specialised. 1927 [see *GUPPY^1]. 1955 I. S. R. MUNRO Marine & Fresh Water Fishes Ceylon 183 Rainbow Fish, Wrasses... Mostly brilliant coloured species. 1962 D. W. TUCKER tr. Sterba's Freshwater Fishes of World 807 Dwarf Rainbowfish, Black-lined Rainbowfish. Northern Australia. 1973 Stand. Encycl. S. Afr. IX. 237/2 Rainbow-fishes... Tropical fishes, known also as wrasses, mostly living about coral reefs in warm seas. 1977 J. M. THOMSON Field Guide Sea & Estuary Fishes Austral. 92 The female Rainbow Fish is less colourful than her mate. 1926 A. R. RADCLIFFE-BROWN in Jrnl. R. Anthrop. Inst. XXIX. 19 There is found in widely separated parts of Australia a belief in a huge serpent which lives in certain pools or water-holes. This serpent is associated, and sometimes identified, with the rainbow... Hence the rainbow-serpent may come to occupy an important place in the beliefs and customs relating to medicine-men and the practice of magic. 1930 Oceania I. 270 It [sc. the snake represented] is both the rainbow-serpent who brings spirit babies into water-holes, and also a 'quiet' python which is eaten. 1950 ELKIN & BERNDT Art in Arnhem Land i. 17 A diffusion from the south of Rainbow Serpent mythology. 1965 R. & D. MORRIS Men & Snakes i. 19 By far the most spectacular snakes in Australian aboriginal art are the mythical rainbow serpents. These usually live deep in waterholes during the dry season, but take to the thunder clouds when the rains come, sometimes appearing in the sky as rainbows. 1977 Bulletin (Sydney) 22 Jan. 65/1 At the foot of the monolith is an extensive Aboriginal art gallery including an ochre rock painting of the sacred rainbow serpent. 1836 W. YARRELL Hist. Brit. Fishes I. 291 (heading) The Rainbow Wrasse. 1972 Oxf. Bk. Vertebrates 62/1 Coris julis (Rainbow Wrasse), common in the Mediterranean and adjacent Atlantic, appears only rarely in British waters.

rain forest. Also with hyphen. [tr. G. regenwald (A. F. W. Schimper Pflanzengeographie (1898) III. iii. 281): see RAIN sb.^{1} and FOREST sb.] A dense forest in an area of high rainfall with little seasonal variation, esp. a tropical forest characterized by a rich variety of plant species. Also attrib.

1903 W. R. FISHER tr. Schimper's Plant-Geogr. I. iii. 260 The Rain-forest is evergreen, hygrophilous in character, at least thirty meters high, but usually much taller, rich in thick-stemmed lianes, and in woody as well as herbaceous epiphytes. 1922 W. G. KENDREW Climates of Continents 327 The air is always moist, and the forests are very luxuriant. Dense rain-forest, with rubber, vanilla, and cacao, flourishes up to about 4,000 feet. 1926 T. F. CHIPP in Tansley & Chipp Aims & Methods in Study of Vegetation x. 207 The tropical rain forest is a type developed under abundant water supply, with high temperature of little variation, and but a short, if any, dry season. 1937 ALLEE & SCHMIDT Hesse's Ecol. Animal Geogr. xxi. 428 This rain-forest reaches its largest continuous extent in South America. 1952 P. W. RICHARDS Tropical Rain Forest i. 1 The name 'Rain forest' is commonly given, not only to the evergreen forest of moist tropical lowlands..but also to the somewhat less luxuriant evergreen forest found at low and moderate altitudes on tropical mountains, and to the evergreen forests of oceanic subtropical climates. 1956 Nature 25 Feb. 367/2 A detailed entomological survey..has been commenced in and around Ilobi, a typical rain-forest belt village fifty miles from Lagos. 1960 N. POLUNIN Introd. Plant Geogr. xiv. 430 In lowland rain forest any luxuriant herbaceous ground-vegetation is found chiefly in clearings..where illumination is above the average. 1973 Sci. Amer. Dec. 59/1 Sizable areas of rain forest still stand in Amazonia, Africa, Borneo and New Guinea, but..the rain forest is retreating. 1974 Country Life 9 Oct. 894/3 Apes, monkeys, rhinoceroses, okapis, bongoes, tapirs and antelopes are just some of the other rain forest animals dependent on this habitat for their survival. 1978 Vole Dec. 25/1 Tropical rain forests are one of the world's main remaining wild places.

rai·nmaker, rain-maker. [f. RAIN sb.^{1} + MAKER.] **a.** A member of a tribal community believed or claiming to be able to procure rain by the use of magic. **b.** One who attempts to cause rainfall by a technique such as seeding. Hence **rai·n-making** sb. and a.

1775 [see RAIN sb.^{1} 5 c]. 1775 J. ADAIR Hist. Amer. Indians 87 Rain-making, in the Cheerake mountains, is not so dangerous an office, as in the rich level lands of the Chikkasah country, near the Mississippi. 1856, 1889 [see RAIN sb.^{1} 5 c]. 1890 J. FRAZER Golden Bough I. i. 13 The third, who was called 'the rain-maker', had a bunch of twigs with which he sprinkled water from a vessel. 1903 Folk-lore Sept. 252 The sorcerers..are capable of rain-making, sun-making, and wind-making. 1930 E. R. B. GRIBBLE 40 Yrs. with Aborigines ix. 87 One old fellow was a noted rain-maker. 1934 V. G. CHILDE New Light Most Anc. East i. 10 These are ruled by rain-maker magicians or by divine kings who were until recently ritually slain. 1960 Times 22 Apr. 9/2 The rainmaker must try to induce artificially the formation of ice crystals. 1971 Islander (Victoria, B.C.) 18 July 12/3 Here in Canada, where lawmakers are now eyeing rainmakers suspiciously, I might not have been able to try at all. 1976 G. A. BROWNE Slide (1977) 14 The Air Force had been conducting rain-making manoeuvres above the Mojave and Death Valley. 1978 D. BATES in C. Allen Tales from Dark Continent vi. 88 They took chickens or pots of beer to a rain-maker in order to pray for rain.

rai·n-out. Also rainout. [f. RAIN v. + OUT adv.] **1.** U.S. The termination or cancellation of an outdoor event because of rain. Cf. *RAIN v. 9 b.

1947 Richmond (Va.) Times-Dispatch 1 June 12-C/3 Barring professional uncertainties such as rainouts, wrecks, engine trouble, etc., a racer has a chance for good money. 1967 Boston Sunday Herald 14 May II. 3/5 National League figures show the senior circuit was hit hardest with 20 rainouts compared with 11 for the same period last year. 1977 New Yorker 15 Aug. 24/3 The day after the rainout, the sun emerged, and that evening Jimmy Buffett and his band, the Coral Reefers, finally made it to Wollman Memorial Rink, in Central Park.

2. [After *FALL-OUT.] Incorporation into raindrops of radioactive debris from a nuclear explosion and its localized deposition on the Earth's surface (see quot. 1974).

1954 Science 7 May 619/1 The extent of the rainout was much greater than that detected in the area from any previous nuclear detonation. Ibid. 620/1 Although there were several days of rain immediately after the rainout, the activity was firmly adsorbed on the pavement and disappeared at a rate about equal to that for decay alone. 1955 Sci. News Let. 25 June 406/3 'Rain-out' might take place instead of fallout, Dr. Lapp suggested, thus producing localized areas of contamination 'hotter' than the surrounding region by a factor of ten or more. 1974 Population Dose Evaluation 492 Several mechanisms have been conceived for deposition from the atmosphere to the earth's surface. In addition to dry deposition, there is washout by rain falling through a cloud of activity in the atmosphere, or rainout in which the activity is incorporated into rain drops at the time of their formation.

rai·nproof, rain-proof, a. (and sb.). [f. RAIN sb.^{1} + PROOF sb.] Impervious to rain. Hence as sb., a rainproof garment, esp. a raincoat. Also **rai·n-proofed** (a) rendered impervious to rain; (b) wearing a rainproof; **rai·nproofer**, a manufacturer of rain-proof fabrics.

1831, 1870 [in Dict., s.v. RAIN sb.^{1} 5 c]. 1902 Daily Chron. 7 Jan. 6/3 The greatcoat is to be made of rain-proofed drab-mixture cloth. 1908 Ladies' Field 25 July p. iii/3 (Advt.), J. W. Elvery & Co., waterproofers and rain-proofers,..London, W. 1923 W. DEEPING Secret Sanct. xiii. 136, I was a wiser virgin than you. I did take a rainproof with me. 1960 News Chron. 6 May 8/4 His identically rainproofed escorts prepare to shoot him. 1965 D. FRANCIS For Kicks ii. 29 Everything from under-clothes to washing things, jodhpur boots to rainproof, jeans to pyjamas. 1967 N. FREELING Strike Out 34 The figure wears a black track-suit now gone greenish, with khaki rainproof trousers and an English suède windcheater. 1977 M. KENYON Rapist xii. 165 George in rainproof hat.

rain-shower. (Later examples.)

1910 W. OWEN Let. 29 Dec. (1967) 66 We have been prevented from going this morning by the first rain-shower of the week. 1981 L. DEIGHTON XPD xxvi. 214 London..The chilly climate with frequent rain showers.

rain-water. Add: **b. rain-water goods**, exterior pipework, guttering, etc., designed to conduct rain-water from a building; **rain-water head**, a collecting piece, freq. ornamental in design, at the top of a drainpipe.

1949 J. F. L. D'ESTÉ in A. C. Martin Mod. Pract. Plumber (ed. 3) III. xiv. 321 (caption) Surveyor's dimensions for quantities of rainwater goods. 1955 Times 8 July 3/3 The Government would have to look at the restrictive practices and rings in the building industry, particularly in the rain-water goods section. 1981 London (North) Telephone Directory: Yellow Pages Jan. 401/1 Bond & White Ltd. Suppliers—sanitary ware—plumbing—rainwater goods. 1876 W. P. BUCHAN Plumbing xiii. 86 Fig. 149 shows conductor with rain-water head, carrying off water from gutter. 1936 P. E. THOMAS Mod. Building Pract. IV. 225 The lower end of a gutter..can be formed so as to discharge direct into a rain-water head. 1963 Times 8 May 6/2 Mr. Lipton (Brixton, Lab.) asked if the Minister would give a list of the persons whose initials were inscribed on the rainwater heads in the new buildings in Downing Street.

rainy, a. Add: **2. c.** rainy season: in certain, esp. tropical, regions, an annually recurring season of heavy rain (in Meteorol., of at least one month's duration).

1720 DEFOE Capt. Singleton 135 We could not expect to reach it till an other rainy Season would be upon us. 1817 S. R. BROWN Western Gazetteer 13 The rainy season.. commences after midsummer. 1872 R. G. MCCLELLAN Golden State xxii. 294 December..and the succeeding months until May are termed winter, or the 'rainy season', in California. 1910, 1922 [see *BAI-U]. 1977 'J. LE CARRÉ' Hon. Schoolboy xvii. 410 'This [tarmac road is] where he lands?' 'Only in the rainy season.'

5. rainy-shimmery, -sounding, -wet adjs.

1930 J. DOS PASSOS 42nd Parallel 27 Through the rail of the bridge we can look way down into the cold rainy shimmery water. 1896 A. E. HOUSMAN Shropshire Lad xxvi. 37 Overhead the aspen heaves Its rainy-sounding silver leaves. 1952 R. CAMPBELL tr. Baudelaire's Poems 70 There the suns, rainy-wet, Through clouds rise and set.

raion, var. *RAYON^2.

raise, sb.^{1} Add: **3.** spec. in Mining, a sloping shaft excavated from the lower end. Cf. RISE sb. 10 b. (Further examples.)

1877 [in Dict.] 1898 S. J. TRUSCOTT Witwatersrand Goldfields xiii. 293 It being usual in that mine for the man who is driving the levels with machines to come back and put up the raises. 1930 Economist 26 Apr. 951/2 The work done by means of drives, winzes, raises and incline shafts to open up new ground. 1973 L. J. THOMAS Introd. Mining vi. 167 Most raises are in the orebody and follow the footwall in grade in narrower stopes.

4. (Further examples.)

1845 J. J. HOOPER Some Adventures Simon Suggs iv. 48 The chances were altogether favourable for making a 'raise'. 1878 J. H. BEADLE Western Wilds II. 41 At last I made a little raise..and concluded to come home. 1900 S. HANDSAKER Pioneer Life (1908) 35 The two brothers 'made quite a raise' in the California mines soon after their discovery. 1914 'HIGH JINKS, JR.' Choice Slang 15 Make a raise, to secure a loan.

5. a. (Earlier examples.) Also, an increase in the price, rate, or value of a thing. Cf. RISE sb. 16 a.

1728 Maryland Hist. Mag. (1923) XVIII. 335 You wil certainly find Crops short this year..which I hope may contribute to the Raise of that on hand. 1883 'MARK TWAIN' Life on Mississippi xxxix. 366 France and Italy.. cracked on such a rattling impost that cotton-seed olive-oil couldn't stand the raise. 1904 [see *JACK v.^{1} 1 b]. 1931 W. G. MCADOO Crowded Yrs. xxx. 469 A gigantic raise in [freight] rates would have added materially to the inflation.

b. An increase of a stake or bet at poker; in Bridge, a higher bid in the same denomination as a previous bid by one's partner.

1821 Hoyle's Games Improved 164 The player who last goes the double, raise, or brag, has the right, in his turn, of increasing either. 1887 'S. CUMBERLAND' Queen's Highway vi. 277 You feel certain that every 'raise' he makes will be his last. 1887 J. W. KELLER Draw Poker 11 Limit, a condition made at the beginning of the game limiting the amount of any single bet or raise. 1921 C. E. MULFORD Bar-20 Three vii. 86 He had a reputation to maintain, and he saw the raise and returned it. 1923 [see *PRE-EMPTIVE a. 2]. 1929 [see *GOOD a. 22 j]. 1959 [see *LIMIT sb. 2 g (b)]. 1964 Official Encycl. Bridge 192/2 A raise to two spades would be appropriate when one spade has been overcalled by two hearts. 1976 SCOTT & KOSKI Walk-In (1977) xxxii. 236 They were making another raise in that poker game, they were threatening to break off diplomatic relations.

c. An increase in wages or salary. Cf. RISE sb. 15 b. Chiefly U.S.

1898 Scribner's Mag. Oct. 489/1 A. J. Packer..had begun to ponder doubts of his wisdom in agreeing to the second 'raise'. 1902 G. H. LORIMER Lett. Merchant xiii. 187, I ear-marked Charlie for a raise and a better job right there. 1921 H. L. MENCKEN Amer. Lang. (1922) iv. 131 When her wages are increased she does not get a raise, but a rise. 1934 T. WILDER Heaven's my Destination 28, I keep getting raises all the time. 1956 S. ERTZ Charmed Circle xiii. 217 She could go on working. She had lately had a raise. 1968 Globe & Mail (Toronto) 3 Feb. 37/1 'Stay in shape and I'll give you a raise next season,' advised coach George Imlach. 1971 C. FICK Danziger Transcript (1973) 25 My bureau.. were delighted that I got to Cambodia... I got a raise.. when I went back to Cuba. 1977 Time 10 Jan. 46/2 Workers strike like clockwork to protest high prices, and nearly always win raises from management.

raise, v.^{1} Add: **I. 4. b.** (Later examples.)

1874 J. W. LONG Amer. Wild-Fowl Shooting ix. 157 Watch this old fool of a duck coming, and see me 'raise her'. 1976 Globe & Mail (Toronto) 21 Jan. 36/1 A jack rabbit was loping his way. It was only 250 yards from where he first raised it.

II. 8. a. (Earlier and later Amer. examples.)

1657 in Essex Inst. Hist. Coll. (1865) VII. 40/1 The said John norman is..to be paid in corne & cattell the one halfe att or before the house be raised. 1712 S. SEWALL Diary 15 July (1879) II. 355, I, and Mr. Gerrish went to Hog-Island and saw the Barn Rais'd. 1846 Knickerbocker XXVIII. 338 After the usual amount of eating, drinking, swearing, and joking, the house..was raised and covered in. 1879 Harper's Mag. June 142/1 If a man raised a house or barn, the rum flowed freely. 1943 W. FAULKNER in Sat. Even. Post 13 Feb. 70/3, I told you we would meet here tomorrow to roof a church... We'll meet here in the morning to raise one.

10. a. (Further examples.)

1824 A. HODGSON Lett. from N. Amer. II. 208 One of my young Canadian female companions..was raised, as they say here, in Portsmouth. 1846 J. HALL Wilderness & War Path 160 'I can't back out,' said he, 'I never was raised to it, no how.' 1882 G. C. EGGLESTON Wreck of Red Bird 3 Maum Sally was born and 'raised', as she would have said, in 'Ole Firginny'. 1929 D. RUNYON in Hearst's International Oct. 63/1 She slips this baby off to her sister in a little town in Spain to raise up. 1953 Manch. Guardian Weekly 15 Jan. 13/3 It [sc. Wenatchee, Wash.] is a pleasant town of sixteen thousand home-loving people, mostly engaged in raising nice children and very good eating apples. 1977 'J. LE CARRÉ' Hon. Schoolboy xv. 354 The American wife asked Jerry where he was raised and..where his home was.

11. c. To establish contact with (a person, etc.) by radio or telephone.

1929 *Amer. Speech* V. 49 *Raise*, to secure [radio] communication with. 1969 'J. Morris' *Fever Grass* xxii. 208 Raise McKay on that [radio] set of yours. 1974 'M. Hebden' *Pride of Dolphins* III. i. 210 She's gone off the air... We can't raise her. 1976 G. Seymour *Glory Boys* vii. 87 She raised Jimmy, still waiting beside the receiver. 1979 *Daily Tel.* 3 Jan. 1/1 A British Airways plane which tried to get into Teheran had to turn back to Kuwait when it could not raise air traffic control.

14. e. (Further examples.)

1890 Kipling *Barrack-Room Ballads* (1892) 53 Ship me somewheres east of Suez, where the best is like the worst, Where there aren't no Ten Commandments an' a man can raise a thirst. 1930 'Sapper' *Finger of Fate* 79 He grinned and said, 'We're not all savages, Mrs. Dankerton. Even though there aren't no Ten Commandments, and a man can raise a thirst.'

III. 17. e. Fig. phr., *to raise its (ugly) head*, to make an (unwelcome) appearance; to present itself as a (troublesome) subject for attention. Cf. *rear *v.*¹ 10 b.

1822 Scott *Peveril* II. i. 27 The ancient superstition..is raising its head. 1930 Wodehouse *Very Good, Jeeves!* ix. 230, I am starving on my feet. Well, when I tell you that it's weeks since a beefsteak pudding raised its head in the house, you'll understand what I mean. 1966 *Listener* 28 July 141/3 The subject of money for the arts raised its head again when *New Release*..investigated the facts behind the Authors' Society recent publication about the stipends of professional writers. 1971 Wodehouse *Much Obliged, Jeeves* xiii. 133 The snag which had raised its ugly head was one of formidable—you might say king-size—dimensions.

f. Other phrases. *to raise one's eyebrow(s)*: see *eyebrow 1 c; *to raise the roof*: see *roof *sb.* 1 e (*a*).

19. d. *spec.* in relation to consciousness: to heighten (sensitivity or awareness).

1970 K. Millett *Sexual Politics* (1971) I. ii. 38 The hope of seeking liberating radical solutions of their own seems too remote for the majority to dare contemplate and remains so until consciousness on the subject is raised. 1976 *Spare Rib* Dec. 22/2 We're raising consciousness, affecting some concrete issues like age discrimination, putting lousy pay on the agenda too. 1977 *Rolling Stone* 7 Apr. 53/3 My efforts to raise the consciousness of whites who are so against Indians in the States were bound to be stopped by the FBI sooner or later.

19*. *Phonetics.* To articulate (a vowel) with the tongue closer to the roof of the mouth. Cf. *raising *vbl. sb.* 1 d.

1874 H. Sweet *Hist. Eng. Sounds* in *Trans. Philol. Soc.* *1874* 506 To assume that the low-narrow [ė] was first widened, and then raised to the mid position, would be to ignore the fundamental laws of short vowel change. 1914 H. C. Wyld *Short Hist. Eng.* vii. 136 Old tense *ē* was raised to [ī] at least by the end of the first third of the sixteenth century. 1934 C. Davies *Eng. Pronunc.* 5 ME *ā*. Was early fronted and raised to [ɛ̄]. 1957 E. J. Dobson *Eng. Pronunc. 1500–1700* II. 612 The view that ME *ę̄* tended to be raised to ME *ẹ̄* is strongly supported by the parallel case of ME *ǭ*, which is shown by seventeenth-century evidence and by that of the modern dialects to have been raised to ME *ọ̄*, ModE [uː]. 1959 A. Campbell *Old Eng. Gram.* 122 By the tenth century *ǣ* of whatever origin had been raised to *ē* in Kt. [*sc.* Kentish]. 1968 Chomsky & Halle *Sound Pattern Eng.* III. 255 The environments where /ī/ and /ū/ are lowered to [ē] and [ō] are distinct from those where /ē/ and /ō/ are raised to [ī] and [ū].

20. b. Also, *to raise Ned* (U.S. slang). *to raise Cain*: see also *Cain² 1 b; *to raise hell*: see *hell *sb.* 10 n, q; *to raise hob*: see *hob *sb.*¹ 2 b.

1848 J. R. Lowell *Biglow Papers* 1st Ser. 69 Your fact'ry gals..'ll go to work raisin' promiscoous Ned. 1904 J. C. Lincoln *Cap'n Eri* ii. 28 The boy sort of run loose, as yer might say. Went to school when he had to, and raised Ned when he didn't, near's I can find out.

21. Also *fig.*

1921 H. Crane *Let.* 1 Oct. (1965) 65 It will be time for me to raise my voice in praise of Anderson soon, as his new book..is on the market.

22. See also *sand *sb.*² 7 e.

23. a. (Further examples.)

1851 H. Melville *Moby Dick* I. xxxv. 259 Whosoever of ye raises me a white-headed whale..he shall have this gold ounce. 1928 Belloc *Chanty of Nona* 1 Before it was morning he raised Lundy Light.

24*. To reach the crest or summit of (a hill, ridge, etc.). *U.S.*

1804 J. Ordway in Lewis & Ordway *Jrnls. Western Explor.* (1916) 168 We raised a Steep bank back of this bottom. 1866 'Mark Twain' *Lett. from Hawaii* (1967) 291 We 'raised' the summit of the mountain and began to canter along the edge of the crater. 1872 —— *Roughing It* xli. 287, I 'raised the hill' overlooking the town. 1934 J. R. Barrows *Ubet* 280 Every time I would raise a ridge, I expected to see him; for the signs were fresh.

25. a. (Earlier and later examples in *gen.* sense.)

1838 J. W. Carlyle *Let.* 28 Nov. (1903) I. 71 We have.. *raised* (as dear Mary used to say) a capital easy chair. 1948 'N. Shute' *No Highway* vi. 171, I..told Miss Learoyd to see if she could raise two cups of tea. 1973 M. Woodhouse *Blue Bone* xv. 162 Would you please see if you can raise us all a drink? 1976 *Daily Mirror* 16 July 3/4 He..was living on what his wife Susan could raise by selling her furniture and jewellery.

29. (Later N. Amer. examples.)

1824 P. Ogden *Jrnl.* 20 Dec. in *Publ. Hudson's Bay Rec. Soc.* (1950) XIII. 5 The Kootonnies & Flat Heads are likewise here waiting our arrival intending to raise Camp together. 1837 W. Irving *Rocky Mts.* I. vi. 78 On the following morning, just as they were raising their camp, they observed a long line of people pouring down a defile of the mountains. 1855 A. Ross *Fur Hunters of Far West* II. xi. 61 The next morning on raising camp, I ordered Martin's horses to be loaded and we set off.

IV. 32. a. Also *Cards* (orig. *U.S.*), to lay a higher stake than (one's opponent); to increase (a stake or bid); freq. *absol.* and with partner as obj. *to raise out* (earlier example). Also *fig.* Cf. *raise *sb.*¹ 5 b.

1821 *Hoyle's Games Improved* 163 After the first three cards are dealt, but before taking in, the eldest hand after seeing his cards, may raise the ante. 1864 W. B. Dick *Amer. Hoyle* 165 When *any* player makes a bet, it is the privilege of the *next player to the left* to *raise him*, or..to deposit in the pool the amount already bet by his adversary, and make a still higher bet. 1872 'Mark Twain' *Roughing It* 332 'I have to pass, I judge.' 'How?' 'You've raised me out, pard.' 1890 Champlin & Bostwick *Young Folks' Cycl. Games & Sports* 269/1 The third player may stay out, see, or raise the second player. 1901 R. F. Foster *Poker* 40 Twice the amount of the blind is the amount of the ante, unless some player has raised it. 1951 E. Culbertson *Bidding & Play in Duplicate Contract Bridge* v. 55 In rubber bridge he might take a chance and raise partner to three of the minor suit. 1959 T. Reese *Bridge Player's Dict.* 142 North opens one spade and South..raises to two spades. 1965 *Listener* 4 Nov. 735/2 The only course open to him therefore is to raise to Four Clubs. 1981 P. Van Greenaway *'Cassandra' Bell* xv. 183 I'll raise you. Any odds you care to name I walk out of here a free man.

33. Also *absol.*

1869 'Mark Twain' *Lett. to Publishers* (1967) 25, I had a bargain about concluded for the purchase of an interest in a daily paper and when everything seemed to be going smoothly, the owner raised on me.

V. 35. For † *Obs.* read '*Obs.* exc. *U.S.*' and add further examples.

1770 C. Carroll *Let.* 5 Sept. in *Maryland Hist. Mag.* (1918) XIII. 61, I am quite indifferent whether Stephenson takes or Refuses the tob[acc]o, as I think the Price will raise again. 1785 G. Washington *Diary* 22 Sept. (1925) II. 415 The Water having raised,..I could form no accurate judgment of the progress. 1808 in J. H. Beadle *Undevel. West* (1873) xxi. 410 Should the accused person or persons raise up with arms in his or their hands. 1819 T. Forsyth in *Minnesota Hist. Coll.* (1880) III. 143 As the Mississippi was raising, the current was very strong. 1861 *Trans. Illinois Agric. Soc.* IV. 102 The milk sours before the cream all raises. 1911 H. P. Fairchild *Greek Immigration to U.S.* 70 About 1,000 houses are vacant in Athens, and yet the prices of rent have raised 15 to 20 per cent.

37. *intr. Mining.* To drive a raise (Raise *sb.*¹ 3 in Dict. and Suppl.).

1898 S. J. Truscott *Witwatersrand Goldfields* xiii. 294 With machine drills it costs slightly more to sink a winze than to raise. 1973 L. J. Thomas *Introd. Mining* i. 9 The connections may be made by driving upwards, known as rising or raising, or working downwards, known as winzing.

raised, *ppl. a.*¹ Add: **3. a.** Also *raised eyebrows*, eyebrows raised in censure or query (see *eyebrow 1 c).

1968 P. Durst *Badge of Infamy* v. 39 'I'll bet that causes some raised eyebrows,' Gina said. 1974 E. Ambler *Dr. Frigo* II. 82, I expected another violent reaction. None came. He merely glanced with raised eyebrows at Delvert, who nodded.

b. *raised beach* (earlier and later examples); *raised bed*, a flower-bed, at a higher level than the adjacent garden; *raised bog*, an area of acid, peaty soil, esp. that developed from moss, in which growth is most rapid at the centre, giving rise to a domed shape.

1834 *Proc. Geol. Soc.* II. 102 These are the only instances which the author could discover of a raised beach on this part of the coast. 1959 J. D. Clark *Prehist. S. Afr.* vi. 164 The earliest Middle Stone Age is probably that associated with the 20-foot raised beach at Blind River, East London. 1969 Bennison & Wright *Geol. Hist. Brit. Isles* xvi. 366 The evidence of uplift is found in the form of marine-cut erosion platforms, or raised beaches where these erosion surfaces have a cover of beach material. 1910 L. B. Meredith *Rock Gardens* ii. 25 The best and most usual form [of small rock garden] is a raised bed. 1959 C. Spry *Favourite Flowers* xxiii. 166, I wanted a raised bed of well-drained agreeable soil. 1974 *Country Life* 21 Mar. 650/1 Raised beds housing all manner of plants became characteristic features. 1938 *New Phytol.* XXXVII. 425 Since the primary lakes are frequently of glacial origin, raised bogs have generally proved to be extremely satisfactory sources of long continuous profiles illustrative of post-glacial forest and climatic history. 1946 *Proc. Prehist. Soc.* XII. 4 Raised-bogs very commonly occur in peri-glacial regions, often upon the sites of lakes created by the laying down of terminal moraines across glaciated valleys. 1966 F. H. Brightman *Oxf. Bk. Flowerless Plants* 88 Sometimes fen becomes choked with vegetation, and acid conditions develop at the top, so that species of Sphagnum can grow on the surface, forming 'raised bog'. 1976 *Nature* 23 Sept. 281/1 Raised bogs represent a special type of peat bog... They..have their own water regime, with all of the moisture supplied from the atmosphere.

4. a. (Further examples.)

1873 *Young Englishwoman* Aug. 406/2 The medallions are worked on brown American cloth..in raised embroidery. 1900 E. Jackson *Hist. Hand-made Lace* 216 *Raised work*, in bobbin lace this term denotes the raised edge worked down one side of leaves and flowers. Honiton and Duchess each have occasionally raised work, which heightens the effect of the lace considerably. 1960 B. Snook *Eng. Hist. Embroidery* 86 Stump work originally was known as 'raised work', its present name probably not being used before 1894.

c. *raised bands* (see quots.).

1835 J. Arnett *Bibliopegia* i. 23 The old mode of sewing on raised bands combined many advantages. 1846 G. Dodd *Brit. Manuf.* 6th Ser. iv. 96 'Raised bands'..are sometimes used for ornament in the better kinds of books; they consist of little strips of leather or cord pasted across the back of the book before it is covered. 1875 [see Panel *sb.*¹ 8 c]. 1901 D. Cockerell *Bookbinding* i. 26 The public having become accustomed to raised bands on the backs of books, and the real bands being sunk in the back, the binders put false ones over the 'hollow'. 1952 J. Carter *ABC for Bk.-Collectors* 147 When a book is bound..the gathered sections are sewn on to horizontal cords or bands... When the boards are covered, these cords (unless sunk in grooves to make a flat spine) will stand out in the form of ridges. These are known as raised bands. 1972 P. Gaskell *New Introd. Bibliogr.* 148 The cords themselves could be placed either outside the backs of the folded sheets, where they would show as raised bands across the spine of the book, or in slots sawn into the folds.

d. Of a cake, biscuit (see U.S. sense s.v. Biscuit 1), etc.: made with baking-powder or other raising agent. *U.S.*

1889 R. T. Cooke *Steadfast* xvii. 189 Then it [*sc.* the election cake] wore only the style of 'raised cake'. 1890 *Harper's Mag.* Oct. 707/1 I've got raised biscuit for supper. 1907 *N.Y. Even. Post* (semi-weekly ed.) 18 July 5 The everlasting repetition of salt meats, potatoes, and raised biscuit in their bill of fare. 1914 G. Atherton *Perch of Devil* i. 28 I've got fried chicken..and raised biscuit. 1937 E. K. Haines *Cook Bk.* xii. 360 *Raised muffins*..A good old-fashioned change from the modern quick muffin.

6. *Phonetics.* Articulated with the tongue in a higher position.

[1888 H. Sweet *Hist. Eng. Sounds* 2 Intermediate positions are: *retracted*..and *advanced*, *raised*..and *lowered*.] 1942 J. S. Hall *Phonetics Gt. Smoky Mountain Speech* (Amer. Speech Repr. & Monogr. No. 4) 15 There is a tendency in some speakers to use a tense, slightly raised [ɪ]. 1957 E. J. Dobson *Eng. Pronunc., 1500–1700* II. 614 The struggle of the raised ME *ę̄* to replace the normal ME *ę̄* was in StE [*sc.* Standard English] a long-drawn-out process. *Ibid.* 642 The raised pronunciation was not accepted in any but careless speech in the seventeenth century. 1972 M. L. Samuels *Linguistic Evol.* iii. 44 There is often a process of systemic regulation..by the selection of raised variants.

raiser. Add: **1. a.** Also, a nurseryman who breeds or cultivates new varieties of plants.

1707 J. Mortimer *Whole Art of Husbandry* xvi. 315 A planter or raiser of trees ought to consider the under soil, as well as the superficies of the earth. 1845 *Florist's Jrnl.* VI. 198 For the beginner, whose chance of competing with the more established raiser is so remote, this suggestion holds forth the most flattering hopes. 1921 *Sat. Westm. Gaz.* 10 Sept. 16/2 We owe this new race [of freesias] largely to Continental raisers. 1970 *Daily Tel.* 17 Oct. 9/1 New roses from the same raiser..are Golden Times and Rosy Mantle.

2. (Further example.)

1908 *Animal Managem.* 121 The utility of boiled foods in the service is limited to their occasional use as condition raisers for horses which require fattening.

4. *Bridge.* **a.** A player who increases his partner's bid. **b.** A card or combination of cards which justifies the increase by a player of his partner's bid. Cf. *raise *v.*¹ 32 a.

1912 F. Irwin *Fine Pts. Auction Bridge* 89 No guarded queens nor guarded jacks in side-suits, may be counted as raisers. 1929 M. C. Work *Compl. Contract Bridge* iv. 34 In the latter case the raiser must have Hearts stopped at least once. *Ibid.* v. 100 It is impossible to translate each individual element directly into raisers. 1964 *Official Encycl. Bridge* 447/1 *Raiser*, the player who bids for a greater number of tricks in a suit first bid by his partner.

raisin. Add: **2. d.** The dark purplish-brown colour of raisins.

1909 *Westm. Gaz.* 4 Sept. 3/2 The long-suffering mole is to find serious rivals in coal-dust grey, elephant's breath, and a purplish black called raisin. 1927 *Daily Express* 7 Mar. 6 (Advt.), Light mulberry, new blue, raisin, rosewood. 1971 *Guardian* 7 Sept. 9/1 A choice of colours: greengage,.. fig, raisin, grape, black.

3. *raisin bread, brew, cake, pudding, -wine* (earlier and later examples); *raisin-coloured* adj.

1902 A. Bennett *Anna of Five Towns* xii. 316 The delicacies which differentiate high tea from tea..hot toast, sardines with tomatoes, raisin-bread, currant bread [etc.]. 1965 W. R. Harding *Days of Henry Thoreau* x. 183 Thoreau experimented frequently with his bread making and soon learned that an unleavened variety was the simplest. When he added raisins to the dough, it was said that he became the inventor of raisin bread. 1980 A. Auswaks *Trick of Diamonds* iii. 80 Toasted raisin bread and butter. 1919 H. Crane *Let.* 27 Dec. (1965) 28, I got dreadfully drunk on dreadful raisin brew. *Ibid.* 29 This fellow of the raisin brew is another poor soul like myself, in Akron exile from N.Y. 1907 *Yesterday's Shopping* (1969) p. iii/1 Raisin cakes. 1973 M. Amis *Rachel Papers* 83 Two kinds of sandwiches, raisin cake, sliced ham, unlimited tea. 1939 Spender & Gili tr. *Lorca's Poems* 47 Raisin-coloured shoes. 1767 J. Woodforde *Diary* 24 July (1924) I. 64, I gave them a fine ham..and a good rich raisin pudding. 1861 Mrs. Beeton *Bk. Househ. Managem.* 671 (*heading*) Baked raisin pudding (plain and economical). 1723 J. Nott *Cook's & Confectioner's Dict.* sig. Ff1v (*heading*) To make raisin wine. 1976 'Trevanian' *Main* (1977) viii. 165 The proprietor of the Greek restaurant..keeps refilling her glass with raisin wine.

raising, *vbl. sb.* Add: **1. a.** (Further examples.)

1842 Dickens *Amer. Notes* II. ii. 58 Down Easters, and men of Boston raising. 1929 M. C. Work *Compl. Contract*

Bridge p. xv, Any advice given for bidding, raising, etc., applies when the score is 'love-all'. 1967 *Gloss. Mining Terms* (*B.S.I.*) ix. 11 *Raising*, the process of excavating a shaft from the bottom upwards. 1978 *Sci. Amer.* July 112/3 Raising (making a new bet by putting more money into the pot than is required for calling).

b. (Examples in sense 10 a of vb.)

1929 D. Runyon in *Hearst's International* Oct. 63/1 Madame La Gimp figures a baby is not apt to get much raising-up off of her as long as she is on Broadway. 1972 J. S. Hall *Sayings from Old Smoky* 113 In my raisin' up [illegible] you would set up with sick people.

c. (Earlier Amer. examples.)

1651 *Rec. Watertown, Mass.* (1894) I. 25 [For] raising of the meeting house. 1711 J. Green *Jrnl.* 6 June in *Essex Inst. Hist. Coll.* (1869) X. i. 93, I went to ye raising ye New Meeting House at Col. Gardner's.

d. *Phonetics.* Articulation (of a vowel) with the tongue closer to the roof of the mouth; an instance of this. Cf. *raise *v.*¹ 19*.

1874 H. Sweet *Hist. Eng. Sounds* in *Trans. Philol. Soc.* *1874* 533 The short vowels do not seem to have changed much in the last few generations. The most noticeable fact is the [illegible] in the tongue [illegible] by raising the tongue into the mid-front-wide, resulting in the familiar *ceb* for *cab*. [illegible] gradually spreading among the upper classes. 1909 O. Jespersen *Mod. Eng. Gram.* I. viii. 231 The great vowel-shift consists in a general raising of all long vowels with the exception of the two high vowels. 1934 C. Davies *Eng. Pronunc.* 7 From the fourteenth century until the present day ME. *ǭ* underwent a gradual raising and rounding. 1957 E. J. Dobson *Eng. Pronunc. 1500–1700* II. 633 There are parallels enough for isolative raisings occurring in spite of, but hardly because of, a combinative raising because of a following r. 1959 A. Campbell *Old Eng. Gram.* 72 Just as all back vowels undergo raising to fronting by i-umlaut, so certain front vowels are subject to raising. 1972 M. L. Samuels *Linguistic Evol.* iii. 44 In ME. we find wide-spread new raisings to [i]. 1973 *Language* LI. 997 The asymmetry between front and back vowels is due to the raising rule.

3. a. (Earlier example.)

1857 *Hunt's Merchants' Mag.* XXXVI. 755 Mr. Pease claimed it [*sc.* the tobacco] as his own raising and pointed to his mark to corroborate his statement.

4. *raising-gig* (example).

1952 *Chambers's Jrnl.* Aug. 455/2 For really fine-quality face cloths..nothing has yet been found to equal the slower and less rigorous action of the traditional west of England teazle raising-gig.

raisiny (rēɪ·zini), *a.* [f. Raisin + -y¹.] Like or suggestive of (the taste of) raisins. Also *Comb.*

1864 J. A. Grant *Walk across Afr.* 157 Plantain-wine..is a sweet raisiny-tasting wine; if aerated, nearly equal to sparkling hock in richness of flavour. 1975 *Times* 5 Apr. 11/5 Fortified dessert wines—rather more raisiny than port.

‖**raison d'état** (rɛzõ deta). [Fr.] = *reason of state* s.v. Reason *sb.*¹ 5 b.

1869 Mill *Subj. Women* iii. 92 The *raison d'état*, meaning the convenience of the government,..was deemed a sufficient explanation and excuse for the most flagitious crimes. 1939 A. Toynbee *Study of Hist.* VI. 38 The god who was worshipped by the Imperial Dynasty was not..a product of *raison d'état*. 1958 E. H. Carr *Socialism in One Country* I. 6 *Raison d'état* is tough enough to emerge unscathed from the revolutionary turmoil. 1964 *New Statesman* 20 Nov. 648/2 The Third Reich, whose ideologists preferred the declamatory pathos of Fichte to Hegel's dispassionate concern with raison d'état. 1976 *Times Lit. Suppl.* 10 Nov. 1483/2 [illegible] for *raisons d'état* by a post-war Labour Lord Chancellor. 1977 *Times* 14 Jan. 17/4 The Elysée's short-sighted conception of *raison d'état*.

raison d'être. (Earlier and later examples.)

1864 J. S. Mill *Let.* 18 Mar. (1910) II. 3 Modes of speech which have a real *raison d'être*. 1898 A. James *Diary* 2 Dec. (1964) 69 The French ladies of the 18th century..whose whole *raison d'être* was [illegible]. 1927 C. Connolly *Let.* 1 May in *Romantic Friendship* (1975) 301 With Desmond gone I have no *raison d'être*. 1959 J. Braine *Vodi* vii. 102 [illegible] *raison d'être* [illegible]. 1975 J. G. Evans *Environment Early Man Brit.* 104 A possible *raison d'être* for these sites was [illegible].

‖**raisonneur** (rɛzɔnœr). [Fr., lit. 'one who reasons or argues'.] A character in a play, etc., who gives expression to the author's message, standpoint, or philosophy. Also *transf.* and (*nonce-wd.*) as *v. intr.*

1903 M. Beerbohm in *Sat. Rev.* 5 Dec. 700/2 There is an old man [in Gorki's *Lower Depths*], in whom we dimly descry a 'raisonneur'. 1913 G. B. Shaw *Quintessence of Ibsenism* 177 Poins, who was originally meant to be the raisonneur of the piece, and the chief figure among the prince's dissolute associates. 1950 E. H. Carr *Bolshevik Revol.* I. ii. 38 The Mensheviks..were primarily men of theory; in Bolshevik terminology they were *raisonneurs*, 'dry-as-dust archivists', the 'party intelligentsia'. 1955 *Times* 24 May 3/7 The *raisonneur*, for example, the family friend who announces the thesis of the play, draws the moral, and preaches his little sermons to the parties concerned, is bound to strike us as something of a nuisance. 1959 *Listener* 29 Oct. 749/1 He is clearly introduced as the play's *raisonneur*, the doughty exponent of a radical viewpoint. 1963 Wodehouse *Stiff Upper Lip, Jeeves* xii. 97, I saw that the time had come to be a *raisonneur*... 'Are you sure,' I said, raisonneuring like nobody's business, 'that you were altogether wise in confining him to spinach and what not?' 1969 M. R. Booth *Eng. Plays of 19th Cent.* II. 345 It seems rather unfair that..the worldly *raisonneurs*..are responsible for breaking up love affairs between young people and..forcing unhappy wives to stay with unpleasant and incompatible husbands, yet are widows of means. 1980 *Times Lit. Suppl.* 23 May 581/1 [illegible] is a scatterbrained raisonneur, on stage throughout, directing her two lovers..to their respective concealment in kitchen or spare bedroom.

raj. Add: **b.** *spec.* the British dominion or rule in the Indian sub-continent (before 1947). In full, *British raj*. Also *transf.*

1859 M. Thomson *Story of Cawnpore* xvi. 229 [illegible] 1908 *Daily Chron.* 21 June 4/4 The Indian agitators who represent the British raj as the enemy of the Indian people [illegible]. 1955 *Times* 25 Aug. 9/7 It was effective against the British raj in India, and the conclusion drawn here is that the British knew that they were wrong. 1969 R. Millar *Kut* xv. 288 Sir Stanley Maude had taken command in Mesopotamia, displacing the raj of antique Indian Army commanders. 1971 *Illustr. Weekly India* 4 Apr. 4/1 Though it appears paradoxical, in the last days of the Raj, the British were the most assiduous wooers of Gandhi. 1976 *Times* 6 Nov. [illegible] 1977 [illegible] 1978 I. Isaacs in H. M. Patel et al. *Say not the Struggle Nought Availeth* 251 The post-independence régime is still in its early stages [illegible].

Rajasthani (rādʒăstā·ni), *sb.* and *a.* [f. the name *Rajasthan* (see below) + -i.] **A.** *sb.* A collective term for the dialects spoken in Rajasthan, a state in north-western India; also, a member of the people of Rajasthan. **B.** *adj.* Of or pertaining to this language or people; *spec.* used of a style of dancing.

1901 [see *Jaipuri]. 1933 L. Bloomfield *Lang.* iv. 63 A number of dialect areas..cover the larger part of India and include such great languages as..Panjabi (16 millions), Rajasthani (13 millions), Bihari (36 millions). 1957 [see *Jaipuri]. 1958 [see *Mewati sb. and a.]. 1967 D. C. Cooke *Color*..Embassy (1966) v. 40 The Rajasthani women labourers. 1969 [see *Kathak]. 1972 W. B. Lockwood *Panorama Indo-Europ. Lang.* xii. 216 The language Rajasthani..is essentially a linguist's term to denote a theoretical concept. In the past there was a tendency to [illegible] through Hindi/Urdu. 1975 A. B. Shah in H. M. Patel et al. *Say not the Struggle Nought Availeth* 100 Rajasthanis have settled down in all parts of the country without coming into conflict with the local people. 1976 H. R. F. Keating *Filmi, Filmi, Inspector Ghote* vi. 56 Some simple Rajasthani peasant girl in *ghagra* and *choli*. *Ibid.* viii. 64 Tiny mosaic fragments of purple and gold Rajasthani enamel work.

raja yoga (rī·dʒă yōᵘ·gă). Also with capital initials. [Skr., f. *rājan* king + *yoga* Yoga.] A form of Yoga by which the practitioner attains control over his mind and emotions. Also *attrib.* Hence **raja-yogin**, one who practises raja yoga.

1885 M. N. Dvivedi *Rāja Yoga* i. 44 The Vedāntic process of attaining to this state of Brahma generally described as *Rājayoga* is purely mental, and deals entirely with rules for fostering [illegible]. 1913 E. F. Benson *Thorley Weir* v. 107 For three years he lived a life of meditation that would have done credit to a student of Raja Yoga [illegible]. 1966 D. Wood *Yoga Dict.* 127/1 *Rāja-yoga*, the system of yoga in which the man takes himself as king (the *rāja*) of all his mental and bodily possessions and powers. 1969 J. Hewitt *Yoga* 14 Others of the Raja Yoga school consider only a little Hatha necessary. *Ibid.* 15 Raja Yoga is designed to..put our mental house in order and concentrate our scattered energies. 1960 R. Rolland in *Ibid.* 27 If [illegible] raja yogin. 1977 C. McFadden *Serial* (1978) xviii. 42/1 She had like mutated over the years through..hatha and raja yoga.

Rajmahali (rādʒmahā·li), *sb.* and *a.* Also **Rajmahal.** [f. the name of the *Rajmahal* hills of northern India + -i.] = *Maler *sb.* and *a.*

1848 *Jrnl. Asiatic Soc. Bengal* XVII. 11. 553 Comparative Vocabulary of the Aboriginal languages of Central India: 5. Rājmahal. 6. Rajmahali. 7. Gondi. 1853 J. S. Taylor [illegible] Rajmahal [illegible]. 1873 E. T. Dalton *Descr. Ethnol. Bengal* viii. [illegible] 1903 G. A. Grierson *Linguistic Survey of India* IV. 446 The Maler also call themselves Sauria, and their language is also known under the name of Rājmahālī, *i.e.*, the language of the Rajmahal Hills.

Rajpoot, rajput. Add: **Rajput** is now the accepted form. (Earlier examples of *Rajpoot* and later example of *Rajput.*)

1829 J. Tod *Rajast'han* I. 137 The poorest Rajpoot of this day retains all the pride of ancestry, often his sole inheritance. 1836 N. Wiseman *Lect. Doctr. Cath. Ch.* I. vi. 174, I consider the soul of the meanest and poorest in the lowest caste, equal, in the estimation of God, to that of the Rajpoot. 1971 *Illustr. Weekly India* 4 Apr. 8/3 The non-Vedic and non-Aryan Kshatriyas..had been admitted to the military class by Shankaracharya. These (along with certain other families who claimed descent from Raja Ram Chandra) became the masters of Rajputana. Their royal families acquired the title of Rajputra or *Rajanya*, which ultimately became 'Rajput'.

attrib. (later examples.)

1931 *Times Lit. Suppl.* 18 June 482/3 They [*sc.* hill paintings] differ not only from the metropolitan art of the Mughal Court, but from the Rajput paintings of other parts of India. 1937 [see *candy pink]. 1974 *Encycl. Brit. Micropædia* VIII. 396/3 Rajput painting, the art of the independent Hindu feudal states in India, as distinguished from the Mughal art of the emperors.

‖**rajpramukh** (rādʒprāmuk). [Hindi, f. *rājya* state + *pramukh* chief.] In the Republic of India between 1948 and 1956, a governor of a state which was formerly a princely state or which resulted from the unification of several princely states.

1949 *Britannica Bk. of Year* 339/1 The remainder [of the Indian states] were grouped under the aegis of a *Rajpramukh* or presiding prince into so homogeneous groups, [illegible]. 1950 G. N. Joshi *Constitution of India* vi. 231 The head of a Part B State is called the Rajpramukh. *Ibid.* xix. 324 As the executive heads of the Union and the State, the President and the Governor or Rajpramukh are given immunity from proceedings in court. 1957 *Listener* 5 Dec. 925/1 [illegible] Rajpramukh of Rajasthan. 1958 C. Mines *East is East* 171 He [*sc.* the Nizam of Hyderabad] became *rajpramukh*—appointed democratic governor—of his own state [illegible]. 1962 *Times* 26 Jan. (India Suppl.) p. xx/4 The office of *Rajpramukh* was abolished.

‖**Rajya Sabha** (rā·dʒyă să·bā). Also **Raj Sabha.** [Hindi, f. *rājya* state + *sabhā* assembly, council.] The upper house of the central Indian parliament.

1948 *Whitaker's Almanack* 791/1 The Union Constitution Committee of the Constituent Assembly recommended the bicameral union of the legislature on the British model with the two Houses to be called the 'Lok Sabha'..and 'Raj Sabha' [illegible] (Index), Rajya Sabha (Council of States). 1955 M. P. Sharma *Govt. Indian Republic* (ed. 2) vi. 144 The Parliament consists of the President and the two Houses—the Council of States and the House of the People now formally styled as Rajya Sabha and Lok Sabha respectively. 1969 *National Herald* (New Delhi) 29 July 4/3 The two Rajya Sabha seats from UP for which by-elections are due on August 13. 1972 *Times of India* 28 Nov. 8/1 (*heading*) Rajya Sabha to debate world situation. 1979 *Daily Tel.* 27 Aug. 1/5 Those in the alignment were willing to accommodate Mr. Mahadev Prasad Varma of the Charan Singh faction as deputy leader of the party in the Rajya Sabha, but not Mr. Amin.

Rakah (rā·kă, ‖ra·kay). [mod. Heb., acronym f. *Reshimah Komunistit Hadashah*, New Communist List (of candidates).] One of the two communist parties in Israel, formed in 1965.

1966 W. Laqueur *Struggle for Middle East* ix. 166 From July 1967 RAKAH was recognized as the official Israeli Communist party by the Soviet bloc. *Ibid.* 167 RAKAH demanded the unconditional surrender of all the Israeli-occupied territories. 1970 A. Dershowitz in *Commentary* Dec. 76/2 They [*sc.* some Israeli Arab citizens] may..belong to the Maoist-oriented Rakah party [illegible] members in the Knesset. *Ibid.*, Sharing office space with an older Arab lawyer active in Rakah. 1974 N. Chomsky *Peace in Middle East?* v. 176 One of the communist parties (Rakah) is 'pro-Arab', a testimonial to Israeli democracy. *Ibid.*, Sabri Jiryis..describes Rakah as 'the Communist sector of the Israeli establishment'. 1977 *Hongkong Standard* 14 Apr. 9/4 Until now, discontent has been expressed through voting for Rakah, the pro-Moscow Israeli Communist Party whose membership is overwhelmingly Arab.

rake, *sb.*¹ Add: **2. a.** *spec.* (*a*) an implement with a blade instead of teeth for gathering money or chips staked in a game of chance; (*b*) (see quot. 1966).

1865 Trollope *Can you forgive Her?* II. xxv. 180 The money..was all drawn back by the croupier's unconscioned rake. 1884 [see Croupier 2]. 1907 *Yesterday's Shopping* (1969) 1009/3 Aluminium Combs—Treasure Rakes, each 1/6... Rakes, Vulcanite. 1/3. 1937 [see *passe]. 1966 J. S. Cox *Illustr. Dict. Hairdressing & Wigmaking* 125/1 *Rake*, a strong comb with large even-sized teeth. Used for removing tangles in long hair. 1972 D. Lees *Zodiac* 47 If you don't want Françoise to blue in all her winnings..you'd better get her away from the table. The way she's going the croupier's going to have to send out for a bigger rake. 1973 'R. Macleod' *Burial in Portugal* vi. 117 A blonde..was the only player to avoid the croupier's rake on the first couple of spins of the wheel.

3*. An act of raking.

1869 'Mark Twain' *Innoc. Abr.* xii. 114 The first rake of his razor loosened the very hide from my face. 1961 Y. Olsson *Syntax Eng. Verb* vii. 107 [From garden talk:] Let me have a rake!

4. *rake-head* (earlier example), *-stem*; **rake-comb** = *rake *sb.*¹ 2 a (*b*); **rake-up,** something concocted; a fabrication.

1790 Rake-comb [see *dressing-comb* s.v. *dressing *vbl. sb.* 5 a]. 1969 E. H. Pinto *Treen* 364 The boxwood H comb is an 18th-century type barber's comb, used on wigs after the coarse or 'rake' comb. 1644 *Essex County, Mass. Probate Rec.* (1916) I. 39 Rakes and rake hedds, 7s. 8d. 1880 Hardy *Trumpet-Major* II. xxiii. 160 For use at home

as rake-stems, benefit-club staves, and pick-handles. 1957 M. Spark *Comforters* v. 95 On the front of the [cigarette] case was a tiny raised crest... 'It's the Hogarth crest. Only a Victorian rake-up, I imagine.'

rake, *sb.*³ Add: **4.** (Earlier and later examples.) Also, the amount carried in a single journey by a railway train.

a 1779 D. Graham *Writings* (1883) II. 59 I'll gar haf-a-crown and haf-a-mutchkin or a rake o' coals do it a'. 1930 *Times Educ. Suppl.* 20 Sept. 397/3 When the boys of the Clachan had finished dinner, their customary task was to fetch the day's 'rake' of water to their homes. 1934 T. Wood *Cobbers* v. 69 Before we could board, however, they said we must see the Rake for the Day go by: all the trees we had seen felled. 1976 *Indian Express* 23 Jan. 5/4 The first rake of 1000 tonnes for Iran was despatched from the Bhilai Steel Plant yesterday. 1979 *Times of India* 17 Aug. 10/2 Actual daily arrivals, according to food department sources, has recently been two rakes or one and half rakes less than what it should be.

8. *spec.* a series of wagons or carriages on a railway or of wagons or trucks in a mine or a factory (in this sense, no longer restricted to *Sc.* and *north. dial.*).

1901 [in Dict.]. 1909 *Daily Chron.* 11 Dec. 4/5 In reaching the shaft some of the men got on to the top of a loaded rake of hutches. 1921 H. Foston *At Front* viii. 60 The ballast engine, with her 'rake' of empty trucks. 1940 L. A. G. Strong *Sun on Water* 224 The train I join is made up locally. The rake of carriages does be waiting at the station, and a tank engine comes in. 1949 D. M. Davin *Roads from Home* II. vii. 175 Geordie Smith on his shunter had just given a rake of meat waggons a bit of a nudge. 1961 *Trains Illustrated* Dec. 762/2 A special seven-car rake of B.R. standard stock provided the final services, hauled alternatively by Class E1 4-4-0 No. 31739 and Class Q1 0-6-0 No. 33029. 1962 *Times* 26 Oct. (Spencer Steelworks Suppl.) p. xviii/3 The operator uses a Beetle to bring forward the first rake of 15 wagons. 1969 *Sunday Standard* (Bombay) 6 July 1/7 Only 27 rakes were available out of the normal complement of 31 rakes for running the scheduled number of trains. 1973 C. D. Garratt *Masterpieces in Steam* 108 The engine was bringing its last rake of wagons through the colliery yard. 1977 *Modern Railways* Dec. 492/1 Positioning two cars at one end of the train would lead to excessive buckling forces when propelling, hence the central position which was also claimed to provide two relatively-short virtually-identical rakes of trailer cars for operational flexibility.

rake, *sb.*⁴ Add: **2. a.** (Later examples.)

1932 *Pictorial Weekly* 19 Mar. 223/2 Most stages have to rise in floor level towards the back in order to make the action visible to the audience... This is known as the 'rake'. 1939 Joyce *Finnegans Wake* (1964) 560 Spotlight working wall cloths. Spill playing rake and bridges. 1951 [see *Cossack 2 e]. 1955 *Times* 10 May 7/7 The front seat is immediately adjustable over a range of 5 in., and further adjustments for rake and height are provided. 1967 *Oxf. Compan. Theatre* (ed. 3) 40/2 The Chicago Opera House.. was a tamed Bayreuth, with a flattened rake and a wedge auditorium. 1973 *Country Life* 22 Feb. 468/3 Front seats have ample adjustment for reach and rake. 1974 B. Forbes *Notes for Life* xii. 92 The rake on the stage of the Theatre Royal, Brighton, is a violent one.

b. The inclination of an edge or face of a cutting tool (or other tool) with respect to some line or plane (freq. to a line perpendicular to the surface of the work). Also *angle of rake, rake angle*.

1888 *Lockwood's Dict. Terms Mech. Engin.* 276 *Rake*. (1) A term usually applied to signify the angles of metal turning tools, as side rake, front rake, &c. (2) The amount of forward angle, or pitch of saw teeth. 1901 *Shop & Foundry Practice* (Colliery Engineer Co., Scranton, Pa.) I. v. 8 This tendency to spring is greatly increased when the tool has insufficient side rake or clearance. 1903 W. H. Van Dervoort *Mod. Machine Shop Tools* xv. 198 The cutting edge of the lathe tool..has what we term an angle of clearance A and an angle of rake B. *Ibid.*, A tool may have front rake..or side rake... A tool without rake requires greater force to drive it through the cut as it tears rather than cuts the metal. *Ibid.* 550/1 (Index), Rake angles. 1923 T. R. Shaw *Mechanisms of Machine Tools* i. 71/1 With a flat top face either so much metal has to be removed that the tool is weakened or the rake angle must be so slight that there would be no proper cutting action. 1938 R. T. Kent *Kent's Mech. Engineers' Handbk.* (ed. 11) III. xxi. 41 Rake angle of a milling cutter..is defined as the angle by which the face of the tooth is displaced back of the radial line drawn from the center of rotation to the cutting edge. 1964 S. Crawford *Basic Engin. Processes* v. 119 Rake angles influence chip formation, tool wear, cutting force, surface finish, and permissible cutting speed. 1975 *Drilling Technol. & Collet Chuck* (Bristol Erickson Ltd.) 3 Figures 1 & 2 illustrate the variation in normal rake angle across the cutting edge of a standard ½″ diameter drill. *Ibid.* 11 Helic angle is the angle between the outer edge of the drill land and the drill axis, and is equivalent to the top rake of a flat cutting tool.

rake, *sb.*⁵ Add: **c.** Phr. *rake's progress* [the title of a series of engravings (1735) by William Hogarth: see Hogarthian *a.*]: a course of dissipation; a progressive degeneration or decline.

1849 Thackeray *Pendennis* I. xx. 183 (*heading*) Rake's progress. 1925 T. E. Lawrence *Let.* 13 June (1938) 476 If you want to trouble yourself still with the rake's progress of this deplorable work. 1937 H. G. Wells *Brynhild* x. 202 I'd have been ashamed not to have been her lover. Another score in the rake's progress. 1950 A. Huxley *Themes & Variations* 250 If the Western Powers had a positive instead of a mainly negative international policy, they would come forward with a plan to check this rake's progress towards human and planetary bankruptcy. 1959 *Listener* 17 Sept. 451/3 The later rake's progress toward corruption, aggression, humiliating subjection to Hitler, pathological megalomania and final catastrophe. 1961 *Times* 9 Feb. 15/2 A literary rake's progress from false materialist to fake romantic. 1976 *Times* 20 Feb. 19/4 Mr Healey has described the course of Britain's recent financial history as a 'rake's progress'.

rake, *v.*¹ Add: **I. 3. b.** (Further examples.)

1865 Trollope *Can you forgive Her?* II. xxviii. 224 The croupier raked it [*sc.* money] all up, and carried it all away. 1913 W. Owen *Let.* 9 Apr. (1967) 182 Unless I raked up matter from the past,..I have had nothing to deliver myself of. 1939 'N. Blake' *Smiler with Knife* ii. 33 Had tea? No? Good, I'll see what Mrs. Raikes can rake up.

c. (Further examples with *in*.) *To rake in the shekels*: see *shekel 2.

1851 *Oquawka* (Illinois) *Spectator* 22 Oct. 1/4 Then, of course, they 'dropped the gate' upon them and 'raked in the pile'. 1893 'Mark Twain' *Lett. to Publishers* (1967) 343 We are at a heavy expense, now, in breaking up housekeeping and raking-in old bills. 1926 *Punch* 22 Dec. 682/2 His having been raked in to complete the officers' team at the last minute in place of an absentee. 1959 *Economist* 11 Apr. 153/2 The Bank of France has 'raked in' part of the working balances of foreign currency held by the French commercial banks. 1969 *Observer* 21 Dec. 28/3 He's raking it in already. Writes 'think pieces' for *Honey* magazine.

d. *to rake down*: (esp.) to win (money) at cards, etc. *U.S. slang.*

1839 *Spirit of Times* 13 July 223/3 If he has anything like as good a horse as the balance, he is certain to *rake down the corn*. 1843 *Ibid.* 18 Nov. 431 She [*sc.* a horse] is a perfect wax figure and all believed that she would rake down the socks. 1846 S. F. Smith *Theatr. Apprenticeship* 151 With one hand he gracefully turned over *four Kings* and a Jack, and with the other tremblingly 'raked down' the pile of bank notes, gold and silver. 1853 J. G. Baldwin *Flush Times Alabama* 8 What lots of 'Ethiopian captives' and other plunder he *raked down* vexed Arithmetic to count. 1882 B. Harte *Flip & Found at Blazing Star* 164 You kin rake down the pile now.

III. 9. a. (Further examples.)

1939 Joyce *Finnegans Wake* (1964) 349 Spraygun rakes and splits them from a double focus: grenadite, damnymite, alextronite, nichilite. 1959 'B. Mather' *Achilles Affair* I. ii. 22 The Nazi patrol..could have raked us with their Schmeissers. 1967 *Boston Sunday Globe* 23 Apr. 17/2 Artillery and air strikes raked the Communist positions in the battle. 1973 *Times* 13 Feb. 7/1 Mr Rashad al-Shawa, the former mayor of Gaza town, narrowly escaped today when his car was raked by automatic weapon fire.

d. (Earlier and later examples.)

1763 J. Bell *Travels from St. Petersburg* II. 78 The hawks generally raked in the pheasants while flying. 1896 A. Austin *England's Darling* II. iii. 42 Until the unseamed falcon learned to wing its way..And, binding, rake its quarry to the ground.

rake, *v.*³ **2.** (Earlier and later examples.)

1842 J. Gwilt *Encycl. Archit.* II. iii. 635 If dwarf wainscoting be framed with two panels in height, add ·016 to the rate... When raked to stairs, ·023 extra. 1930 W. Faulkner *As I lay Dying* 100 The broken hat raked at a swaggering angle. 1964 *Listener* 23 Apr. 664/2 Two of the best modern theatres, merely for the fact that the auditorium is steeply raked. 1977 *Transatlantic Rev.* LX. 83 He spoke with the microphone..his TWA captain's hat raked across one eyebrow.

raked (rēɪkt), *ppl. a.*² [f. Rake *v.*³ + -ed¹.] Inclined from the perpendicular or from the horizontal; also, swept-*back*. (For earlier examples of pa. pple. see Rake *v.*³ 2 in Dict. and Suppl.)

1948 H. Innes *Blue Ice* iv. 96 A single raked-back funnel. 1955 *Times* 5 July 5/6 The well-raked steering wheel gives a sense of precise control at speed without being heavy in manoeuvring. 1972 *Soviet Weekly* 22 Apr. 13 Lecturers use portable TV cameras, and students have screens over their heavily raked desks. 1976 B. Jackson *Flameout* xi. 185 The clean aerodynamic form of the airliner:..the raked-back wings..supporting streamlined engine pods. 1977 *Time* 21 Feb. 41/2 The action is carried out on a raked stage.

ra·ke-off. *slang* (orig. *U.S.*). Also **rakeoff, rake off.** [Cf. *rake *sb.*¹ 2 a (*a*).] A share (of the winnings in gambling, of profits, etc.), a 'cut'; commission. (Freq. with derogatory overtones.)

1888 *Texas Siftings* 28 Jan. 16/1 We always give him a rake-off, so he makes a good enough thing of it. 1890 J. P. Quinn *Fools of Fortune* 188 This percentage is technically known as the 'rake-off', and insures the proprietors of the establishment a handsome royalty on all winnings. 1899 B. Tarkington *Gentleman from Indiana* vii. 95 In oil it's the farmer that gets the rake-off. 1903 *Sun* (N.Y.) 2 Nov. 3 This is a day of rake-offs. The boss..gets his rake-off from every service he renders to his party. The laborer gets his rake-off for selling his vote. 1905 D. G. Phillips *Plum Tree* 61 It means a big rake-off for Dunkirk. Politics is on a money basis nowadays. 1914 'J. H. Keate' *Destruction of Mephisto's Greatest Web* ii. 60 The slot that receives the 'rake off', or percentage. *Ibid.* iii. 91, I would do my best to prepare him a meal. This I was willing to do, as my 'rake off' for the night had only been ten cents. 1926 E. Wallace *Yellow Snake* xxvii. 224 This was something more profitable than the smuggling of cocaine,..a quicker way to fortune than the rake-off of coppers from a forbidden game of fantan. 1929 M. A. Gill *Underworld Slang* 10/1 *Rake-off man*, nickel taker at dice game. 1934 J. T. Farrell *Young Manhood* viii. 121 'Hey, Hugo, what undertaker's giving you a rakeoff?' interrupted Arnold Sheehan. 1934 *Punch* 26 Dec. 722/2 Why don't you make the cost of The Club Dinner inclusive of wines and spirits? I suppose it is because the hotel where you are having the meal won't give you a rake-off on all liquor served? 1936 [see *football pool]. 1951 *Proc. Prehistoric Soc.* XVII. 174 The Burmese ones are reputed better, and always fetch a higher price—in fact they must do so or they would never reach the western villages, as the rake-off from passing them on would be inadequate. 1959 J. Cary *Captive & Free* lxix. 315, I didn't say fifty to you. Sorry. But the agency would give me a rake off on reprints. 1968 *Globe & Mail* (Toronto) 3 Feb. 11/1 Police and politicians in Vancouver may be getting a rake-off from the ring. 1973 'D. Rutherford' *Kick Start* viii. 174 The vendors were waiting... Obviously the courier would receive a handsome rake-off. 1977 *Time* 31 Oct. 23/3 The President again assailed the big oil companies, charging that deregulation of natural gas prices would produce 'big profit rake-offs and huge cost increases to the American consumers'.

raker¹. Add: **5.** = *raking-coal* s.v. Raking *vbl. sb.*¹ 3. *dial.* Cf. Rake *v.*¹ 5.

1858 Geo. Eliot *Scenes Clerical Life* II. 250 The kitchen fire..was kept in under a huge 'raker'—a possibility by which the coal of the midland counties atones for all its slowness and white ashes. *c* 1909 D. H. Lawrence *Collier's Friday Night* (1934) iii. 87 Ernest has come from the cellar with a large lump of coal, which he pushes down in the fireplace so that it shall not lodge and go out... He puts the candle on the table, and puts some coal on the fire, round the 'raker'.

6. An inclined beam or strut.

1882 C. H. Stock *Treat. Shoring & Underpinning* ii. 5 The outer shore is called the top raker, the middle shore the middle raker, and the lowest is called the bottom shore. 1887 G. H. Blagrove *Shoring* ii. 32 The foot of the middle raker thrusts against a cleat bolted to the foot of the bottom shore. 1956 *Archit. Rev.* CXX. 327/2 The existing balcony rakers, the main roof and the back stage parts were retained, but the stalls floor..was rebuilt. 1963 M. J. Tomlinson *Foundation Design & Construction* vii. 404 A large pile cap is provided to counteract the uplift on the backward raker.

raker². Add: **1. a.** (Earlier example.)

1876 *Coursing Calendar* 38 Poacher, going a raker from Cannobie lea, never let the latter next the hare in a well-run course of good length.

b. A good stroke at golf.

1899 *Golf Illustr.* 15 Sept. 393/2 Vardon drove a 'raker' from the first tee, nearly hole high.

raki. Add: Now also used of a liquor made from other ingredients (see quot. 1959) in various countries of eastern Europe and the Middle East. Also, a drink or glass of this. (Further examples.)

1775 R. Chandler *Trav. Asia Minor* lxxv. 255 Not far from us were booths of the Turcomans... Some of them joined us, and one or two wanted raki or brandy. 1845 E. Warburton *Crescent & Cross* I. xxvi. 295 Their dram is distilled from rice, and called *Raki*. 1919 E. H. Jones *Road to En-Dor* iii. 34, I..poured myself out a tot of Raki from Alec's bottle. 1933 'G. Orwell' *Down & out in London & Paris* xvi. 122 Raki, the Arab drink, was very cheap. 1941 'R. West' *Black Lamb* (1942) I. 410 Raki, the colourless brandy loved by Slavs. 1956 R. Macaulay *Towers of Trebizond* xii. 125 We had supper at the khan, and sat on there smoking and sipping raki..while Turks played trictrac at little tables beneath the trees. 1959 W. James *Word-bk. Wine* 154 *Raki*, a fairly general name for spirits in Balkan countries; it may be made from wine, grain, molasses, potatoes, plums, and so on; in Turkey it is a spirit resembling Pernod, which goes milky when diluted. 1969 J. Mavor *Voyage to Atlantis* ix. 206 The old woman with two eye-teeth brought us raki or tsipuro, a 40-proof sort of grappa. 1980 M. Bar-Zohar *Deadly Document* ix. 158 He sat down and ordered a *raki*.

rakia (rā·kia, raki·ya). Also **rakija.** [a. Bulg. *rakia*, Serbo-Croat *rakija*: cf. Raki.] In the Balkan countries: brandy, liquor; = prec.

1845 *Encycl. Metrop.* XXV. 1290/2 *Rakia* is a Dalmatian spirit, drawn from the murk of the wine-press, mingled with aromatics. 1926 *Blackw. Mag.* Apr. 526/1 It was a large scent-bottle filled with *rakija*. 1932 G. Greene *Stamboul Train* iv. i. 179 Ninitch sipped his glass of *rakia*; the heavy plum wine brought tears to his eyes. 1950 V. Canning *Forest of Eyes* iii. 62 He took the *rakia* bottle from his desk drawer and set two glasses. 1959 F. Maclean *Back to Bokhara* ii. 103 Rakia, the plum brandy of Bosnia. 1966 *New Statesman* 11 Nov. 705/2 A Montenegrin town, in 1943, is garrisoned by a tattered Venetian legion... The garrison stifles in rock heat, bloats on rakija. 1980 J. Hone *Flowers of Forest* i. 21 Playing chess over a bottle of rakia somewhere in Yugoslavia.

rakily (rēɪ·kili), *adv. rare.* [f. (*rakey* f.) Rake *sb.*⁵ + -ly².] In a rakish manner; rakishly.

1904 C. Hamilton *Passing of Arthur* xx. 167 His newly-ironed tall hat was rakily cut, and he wore an immaculate light grey frock coat and waistcoat.

raking, *vbl. sb.*¹ Add: **1. a.** (Further example.)

1942 *R.A.F. Jrnl.* 18 Apr. 8 There should be full scale raking down of land and sowing.

c. (Later examples.) Also *fig.*

1868 in *Trans. Illinois State Agric. Soc.* (1870) VII. 434 The supply of hogs..appeared to be made up of the rakings and sweepings of the country. 1977 *Times* 24 Dec. 14/5 A series of ovens had been cut into the back of the rampart, their rakings being deposited in the abandoned ditch of the superseded marching camp.

d. A rebuke or scolding. Also with *down*, = a 'dressing down'. Chiefly *U.S.*

1854 *La Crosse* (Wisconsin) *Democrat* 17 Jan. 2/4 Mr. Wright..gave Smith a small raking down. 1883 J. D. Shields *Life & Times S. S. Prentiss* 125 He cheerfully paid it, vowing that the 'raking down' which Prentiss had given his prosecutor was worth that. 1897 G. Bartram *People of Clopton* iii. 80 He got such a terrific raking-down from Aunt that his long black shadow never darkened our doors again. 1907 *Black Cat* Jan. 7 I'll bet somebody has got a raking for losing it.

raking, *vbl. sb.*³ Add: *spec.* in *Building*, the action of *rack *v.*⁷; also, the resulting arrangement of bricks or stones. Usu. written **racking,** in phr. *racking back*. (Further examples.)

1873 A. M. Lang *J. G. Medley's Roorkee Treat. Civil Engin. in India* (ed. 3) I. xvii. 345 (*heading*) Racking back. 1894 C. F. Mitchell *Building Construction* iii. 132 Racking is the term applied to the method of arranging the edge of a brick wall, part of which is unavoidably delayed while the remainder is carried up. 1902 *Encycl. Brit.* XXVI. 437/1 The foundations must be spread below the column bases... This is accomplished by rackings of stone or brickwork. 1945 E. L. Braley *Brickwork* iii. 60 Racking back is the best method of executing this particular job, as by this means the bonding bricks can be perfectly bedded. 1946 Holgate & McDougall *Bricklaying* vi. 88 Corbling and raking are also largely used in the building of chimney flues. Since a chimney flue always runs upwards at an angle, raking is necessary on one side and corbling on the other. 1964 J. S. Scott *Dict. Building* 254 The normal way of building a *brick* wall consists of first building the corners or ends..in steps rising one course at a time from the middle part of the wall. The gradual increases of height to the corner are called racking back.

raking, *ppl. a.*³ Add: **b.** *raking bond* (see quot. 1876); also, *raking stretcher bond*; *raking-piece* (*b*) (earlier example).

1876 P. G. L. Smith *Notes Building Construction* II. xiii. 221 Raking Bond is of two kinds, *Diagonal* and *Herring-bone*. In both the bricks in the interior of the wall are placed in directions oblique to the face. A course or two of raking bond is sometimes introduced at intervals in thick walls built in English bond. The proportion of stretchers in a brick wall diminishes according to its thickness... The raking courses are therefore useful in giving longitudinal strength to thick walls which are deficient in stretchers. In both kinds of raking bond alternate courses rake in opposite directions. 1937 P. E. Thomas *Mod. Building Pract.* III. 294 The direction of the diagonal or raking bond is changed in each course to further strengthen the bond of the wall. 1884 Raking-piece [see *cut cloth* s.v. *cut *ppl. a.* 12]. 1974 *Bricks: Their Prop. & Use* (Brick Devel. Assoc.) I. 29 (*caption*) *Raking Stretcher bond*. Economical and more interesting than normal Stretcher Bond. Joints tend to become very prominent unless mortar colour is chosen with care.

rakish, *a.*² **1.** (Earlier *Comb.* example.)

1838 Poe *Narr. A. G. Pym of Nantucket* in *Wks.* (1902) III. 146 She was a long, low, and rakish-looking schooner.

‖**rakshasa** (rā·kʃăsă). Also **raksasa, rakshas**; fem. **rakshasi.** [a. Skr. *rākshasa* demon, f. *rakshas* something to be guarded against or warded off.] In Hindu mythology, a malignant demon, esp. one of a band at war with Rama and Hanuman. Also, an artistic representation of such a demon.

1866 *Chambers's Encycl.* VIII. 101/1 *Rakshas*, or *rākshasa*, is, in Hindu Mythology, the name of a class of evil spirits or demons,..imagined..more frequently as mischievous, cruel, and hideous monsters, haunting cemeteries, [etc.]. 1899 F. H. Groome *Gypsy Folk-Tales* p. lxxiv, In the folk-tales of India..a *rakshasi* makes nothing of polishing off the entire population of a city. 1917, etc. [see *pishachi]. 1937 M. Covarrubias *Island of Bali* vii. 178 The most complete statue is that of a wild *raksasa* crowned with skulls. *Ibid.* viii. 239 The giant Rawana, the *raksasa* king, a monster of wickedness and lechery. 1967 Singha & Massey *Indian Dances* x. 106 A pair of enormous discs are often held by an unseen helper on either side of the face of a rakshasa or demonic character [in Kathakali dancing]. 1972 *Daily Tel.* (Colour Suppl.) 12 May 58/3 The most feared Indian vampire, the Rakshasa, was all powerful; it animated dead bodies, snatched babies,..could change into any form and lengthen its arms to 80 miles.

‖**rakshi** (rā·kʃi). Also **raksi.** [a. Nepali *raksi*; cf. Tibetan *rag-ši*.] In Nepal and Tibet, a liquor distilled from rice or grain.

1877 D. Wright *Hist. Nepal* ii. 30 The Newārs, and most of the lower castes consume a considerable quantity of a coarse spirit called Rakshī, which is distilled from rice and wheat. 1954 W. Noyce *South Col.* v. 72 Hospitably entertained with tea..and a not very good *rakshi* or rice spirit. 1959 *Times* 23 May 7/7 We spent half a day at the famous monastery of Thyangboche, where we were entertained to dinner by the head lama and sampled his excellent *rakshi*, the local distillation. 1966 C. Bonington *I chose to Climb* x. 124 The Sherpas had settled down to a celebratory binge, consuming huge quantities of rakshi, a potent spirit distilled from barley or rice. 1971 —— *Annapurna South Face* iv. 50 And then supper—chicken and rice washed down by *raksi*, the local spirit made from distilled millet. *Ibid.* vii. 81 Meanwhile, Pasang had taken the unwise step of making a *raksi* issue to some of them.

‖**raku** (ra·ku). Also **Raku.** [Jap., lit. ease, relaxed state, enjoyment: see quot. 1882.] A kind of lead-glazed Japanese pottery, often used as tea-bowls and similar utensils. Also *attrib.*

1875 Audsley & Bowes *Keramic Art Japan* I. 52 Raku is occasionally covered with lacquer, and it is made in other places than in Kioto. 1882 C. Dresser *Japan* II. iv. 371 Shōgun Taikosama..honoured this particular manufacture with a golden seal, on which the character 'raku' (meaning enjoyment) was engraved... The competition for objects specially valued (as some of these black raku cups) was such that wars were often waged between Daimios with the sole view of possessing certain coveted goods. 1890 B. H. Chamberlain *Things Japanese* 278 The *Raku-yaki* of Kyōto is the parent of all the rest... The *Raku* faience owed much of its popularity to the patronage of the 'tea clubs'. 1960 B. Leach *Potter in Japan* ii. 51 One bowl was old Corean and the other, used alternately, was a 'Doniu' black raku. 1970 *Oxf. Compan. Art* 610 The famous *Raku* is a very soft glazed ware widely used for deliberately mis-shapen vessels. 1979 C. McCarry *Better Angels* II. i. 105 The elegant old man pouring tea into Raku bowls.

râle. Add: Also **rale.** (Earlier and later examples.)

1828 *Glasgow Med. Jrnl.* I. 72 The respiratory murmur is often rendered fainter, and it is accompanied and obscured by certain *râles*, or unnatural sounds. 1963 *Lancet* 12 Jan. 76/1 There were a few dry rales at the right lung base. 1977 *Ibid.* 17 Dec. 1265/2 At this stage there are usually rales in the chest but no definite signs of lung consolidation.

rale (rēᵊl), *a. U.S.* and *dial.* var. of Real *a.*²

1835 D. Crockett *Acct. Col. Crockett's Tour* 60 Folks need not go out of Boston to find rale hospitality. 1873 C. M. Yonge *Pillars of House* II. xvi. 108 A fellow that..makes verses—rale, superior, iligant articles. 1901 M. Franklin *My Brilliant Career* xxviii. 237 'Wot's she like?' 'Oh, a rale little bit of a thing.' 1922 E. O'Neill in *Hearst's Internat.* Mar. 47/1 It was..aisy for a rale man with guts to him, the like of me. 1940 *Amer. Speech* XV. 46 Characteristic of those encountered in the Cumberlands and Ozarks are.. *rale*, real.

rall (ræl). *Mus.* Abbrev. of Rallentando.

1876 Stainer & Barrett *Dict. Mus. Terms* 374/1 *Rall.*, abb. for rallentando. 1886 [see *a tempo]. 1959 *Collins Music Encycl.* 530/1 *Rallentando*,..often abbreviated *rall.*

rallentando. Add: (rælĕntæ·ndo). (Examples.) Also *transf.*

1861 Geo. Eliot *Let.* 6 Oct. (1954) III. 456 Our violoncello..is..a man who is all assent and perpetual rallentando. 1926 *Brit. Weekly* 27 May 161/2 The beauty of the theme was heightened by a perfect rallentando. 1946 A. Christie *Come, tell me how you Live* i. 24 Its [*sc.* the Orient Express's] tempo..gradually slows down in a *rallantando* [*sic*] as it proceeds eastwards. 1965 A. T. Hatto *Nibelungenlied* 349 The strophe of the *Nibelungenlied* is normally a balanced and self-contained unit with a marked *rallentando* in the last half-line owing to its extra filled bar. 1978 R. Donington in J. M. Thomson *Future of Early Music in Britain* 13 What sort of rallentandos and how many of them.

ralliance. (Earlier example.)

1826 T. Jefferson *Writings* (1854) IX. xlviii. 510 The good Old Dominion..will then..become a centre of ralliance to the States whose youth she has instructed.

ra·llied, *ppl. a.*² *rare.* [f. Rally *v.*² + -ed¹.] Subjected to raillery or banter.

a 1704 Locke *Educ.* (1705) 255 The rallied Person also finds his account, and takes part in the diversion.

rallier, *sb.*¹ (Later example.)

1904 F. Lynde *Grafters* i. 11 They..presently found themselves in the thick of the crowd of debarking ralliers.

rally, *sb.*¹ Add: **1. a.** (Further example.)

1928 *Observer* 19 Feb. 27/1 In a belated rally Ivor Jones scored and converted his own try for Llanelly.

2. b. *spec.* a rapid rise in share prices after a fall.

1930 *Economist* 22 Nov. 965/1 Despite all the bad news, a rally in prices made some headway after the liquidation of Monday. 1979 *Daily Tel.* 6 Oct. 23/1 The Index fell..to be 478.8 (down 2.6 at 1 p.m.), but some new account demand after the official close of the market brought an encouraging rally to 480.4.

3. b. For *U.S. colloq.* read 'orig. *U.S.*'. (Earlier and later examples.) Also, a mass-meeting of the supporters of any specified cause.

1840 *Niles' Reg.* 12 Sept. 20/3 Rally of the democracy of Niagara. 1921 *Proc. 3rd Nat. Country Life Conf.* 48 Funds for the school are obtained through rallies held in the churches on the fifth Sunday. 1930 A. P. Herbert *Water Gipsies* iii. 27 At an annual rally of the Boy Scouts..the band was playing the National Anthem. 1960 M. Laurence in *Tamarack Rev.* Autumn 8 'Hey, you Sabina!' Mammi Ama shouted. 'Were you at the rally?' 1973 D. Aarons *Unwritten War* III. viii. 129 His passions were literary, not political. He disliked speeches, rallies, meetings.

c. *dial.* A crowd of persons, a large group.

1837 M. Palmer *Dialogue in Devonshire Dialect* 11. 16 There was a whole rally of us at the Pigeons. 1892 S. Hewett *Peasant Speech Devon* 116 There's a turrabul rally aw'm down there. 1908 *Daily Chron.* 16 Jan. 8/5 There's a whole rally of us driving over in Peter's wagonette.

d. Also **rallye** [Fr.]. A competitive event for motor vehicles, usu. over a long distance on public roads.

1932 *Radio Times* 29 July 267/1 Some of the big motoring events of the year—the Ulster Motor Rally, [etc.]. 1949 C. A. N. May *'Wheelspin' Abroad* v. 79 There can be no guarantee of a 1949 Rallye actually taking place. 1963 P. Drackett *Motor Rallying* i. 9 The true progenitor of the rally was the reliability trial. 1973 E. Lemarchand *Let or Hindrance* viii. 90 He seemed more interested in cars than in girls. Went to rallies, and so on.

4. a. Also, a sustained flurry of blows in a boxing match.

1829 P. Egan *Boxiana* 2nd Ser. II. 20 A terrible rally was the finishing stroke of the round. 1885 G. B. Shaw in *Mag. of Music* Nov. 178/3 Exciting pantomime music to what is called by stage managers and prize-fighters a rally.

b. Also used in similar games.

1955 *Times* 9 May 15/3 Kershaw was now increasing his own pace of stroke which led to many long rallies [Real Tennis]. 1977 'S. Woods' *Law's Delay* III. 125 [The bell] chimes in the hall... We might not have heard it if we were in the middle of a rally [Table Tennis].

5. *attrib.* and *Comb.*, as (sense 3 b) *rally-goer*; (sense *3 d) *rally atmosphere, car, coat, driver, -driving, motorist, plate, seat, wheel, winner*; *rally-bred, -prepared, -proved, -sickened, -toughened* adjs.

1952 W. Leonard *Rallies & Races* i. 19 We touched the real Rally atmosphere. 1977 *Belfast Telegraph* 17 Jan. 10 (Advt.), Safe rally-bred handling, failsafe disc braking all round, safety cage construction. 1976 *Morecambe Guardian* 7 Dec. 18/7 A young Morecambe man had been drinking and was well over the limit for driving when his high-powered rally car smashed into a tree, an inquest at Preston was told on Tuesday. 1973 *Perthshire Advertiser* 17 Feb. 10/1 (Advt.), Gent's rally coats. Concealed hood. 2 in-side pockets. 1952 W. Leonard *Rallies & Races* 14 They are rather nice chaps, these Rally-drivers. 1976 *Cumberland News* 3 Dec. 21/3 Even the hardened rally drivers..said it was the worst spectator behaviour they had ever seen. 1954 C. Meisl tr. *J. A. Grégoire's Best Wheel Forward* 121 The future racing driver passes through three well-defined phases: rally driving, hill climbs and endurance tests. 1973 *Guardian* 26 Jan. 13/1 Monte Carlo Rally drivers..should have been able to check in..on time without having to resort to full-blooded rally driving techniques. 1960 *Ibid.* 11 July 6/7 There were few of the familiar, regular rally-goers. 1955 *Times* 9 Aug. 7/7 In making their cars comfortable living places for several days on end, rally motorists have 'invented' several ideas which could add to the comfort and convenience of the ordinary motorist. 1949 C. A. N. May *'Wheelspin' Abroad* iii. 40 One car.. carrying the official Rallye plates. 1968 *Autocar* 25 Jan. 30/1 Like all the rally-prepared cars we have tried, the Triumph handled beautifully. 1961 *Times* 5 Oct. 11/6 You get the rally-proved, highly-praised Anglia engine. 1976 *Liverpool Echo* 22 Nov. 15/4 (Advt.), Mini 1000, blue, Rostyle wheels, rally seats etc. 1961 *Times* 4 Aug. 11/6 To visit any of Britain's several hundred rally-sickened villages during the sport's 'close' season is an edification that regular rally teams should not miss. 1963 P. Drackett *Motor Rallying* iv. 52 From a rag-bag of memories one recalls the blush brought to the rally-toughened countenance of Gregor Grant. 1966 *Motor Trends* Dec. 52/1 (*caption*) Our great Rally II wheels... Wider white-walls are available. 1951 S. C. H. Davis *Rallies & Trials* viii. 108 The Rally winner is rarely decided by the acceleration and brake test.

rally, *v.*¹ Add: **I. 2*.** To drive (a vehicle) in a motor rally. *colloq.*

1969 *Guardian* 18 Aug. 10/2 He drove it around, occasionally attending vintage meetings, but never rallying the car. 1976 L. Deighton *Twinkle, Twinkle Little Spy* xxii. 220 You want to buy a car?.. One owner. Never raced or rallied. 1977 *Drive* Sept.–Oct. 113/1 The punter-hunter fills his advert with euphemisms. Such as: *1974 model* made in 1973;..*never raced nor rallied*.

II. 6. a. (Further examples.)

1866 Geo. Eliot *Felix Holt* I. iii. 87 People were told they must 'rally' at the coming election. 1905 R. Fry *Let.* 21 Jan. (1972) I. 233, I find that these people will rally and do all they can. 1915 J. Webster *Dear Enemy* 201 His friends rallied about the babies, sold..the studio fittings..paid off the debts. 1940 'N. Blake' *Malice in Wonderland* II. ix. 123 Those who were loudest..are now..the first to rally round. 1963 P. Willmott *Evolution of Community* vii. 75 When one of the neighbour's children married a little while ago we all rallied round. 1978 J. Barnett *Head of Force* iii. 19 The demonstrators marched..to Trafalgar Square where they planned to rally, sing the National Anthem and dismiss.

c. *to rally round the flag*: of a group, to demonstrate loyalty to a cause, as at a moment of impending danger. Also as *attrib. phr. U.S.* (orig. in Civil War use).

1862 G. F. Root *Battle Cry of Freedom* (song) 3 Yes we'll rally round the flag, boys, we'll rally once again, Shouting the battle-cry of Freedom. 1957 M. Shulman (*title*) Rally round the flag, boys! 1964 *New Yorker* 4 Jan. 77 Much of the book is written in a tough, choppy, emotional style, but this approach disguises a spurious rally-round-the-flag vision. 1968 *Listener* 8 Feb. 164/2 The immediate reaction of the American people is to rally round the flag and demand that its honour be upheld.

7. a. (Earlier examples.)

1744 Ellis *Mod. Husbandman* Jan. vi. 60 After the first Mowing..they do not *rally*, as we call it, *i.e.* they do not grow again to much Profit. 1792 H. Newdigate *Let.* Feb. in A. E. Newdigate-Newdegate *Cheverels* (1898) viii. 111 Sally Rally'd last night and sang Charmingly.

rally, *v.*³ Add: Also 8 **ralley.** (Earlier example.)

1728 *Calendar Virginia State Papers* (1875) I. 215 We were like a sow that had lost her pigs, would ralley for a little time and then have done.

rallycross (ræ·likrɒs). [f. *rally *sb.*¹ 3 d + *auto)cross.] A form of motor racing combining elements of rallying and autocross. Also *attrib.*

1967 *Daily Mirror* 26 Aug. 9/5 What is the sport that scales down motor rallying to a seven-minutes sensation?.. Rallycross, which combines the thrills of autocross where old 'bangers' tackle mud-covered hills—with the ordeals of the Monte Carlo rally. 1968 *Observer* 21 Apr. 7/4 (Advt.), 'It's got these ingredients that cut the corrosion cars suffer from when they're left hanging about.' Roger Clark, leading Ford rally driver and rallycross man. 1978 *Morecambe Guardian* 14 Mar. 11/1 Motor sport is having trouble finding suitable space for short track speedway racing and rallycross.

rallying, *vbl. sb.*[1] Add: **a.** Also, the action or practice of participating in a rally (in senses of *sb.*[1]).

1960 S. Turner *Rallying* i. 12 There are four levels of rallying: Closed, Restricted, National and International. 1971 'D. Rutherford' *Clear Fast Lane* 39 The usual changes needed before a production car is ready for rallying. 1978 *Times* 4 July 19/2 He [*sc.* Nastase] mixed the pace and pattern of his rallying with..nonchalant grace.

b. *rallying-round* (later example).

1930 Wodehouse *Very Good, Jeeves!* iii. 69 Cold and haughty. No symp. None of the rallying-round spirit which one likes to see.

rallyist (ræ·li,ist). [f. Rally *sb.*[1] + -ist.] One who competes in a motor rally or rallies.

1961 *Times* 4 Aug. 11/7 With some common sense.. rallyists could avoid strangling their own lusty child. 1963 P. Drackett *Motor Rallying* ii. 28 Another outstanding rallyist, John Sprinzel, writes a regular column for the monthly *Motor Racing*. 1972 *Sci. Amer.* Jan. 11/1 Since becoming involved in this task I've met a fascinating variety of nonacademic types—automobile wreckers, automobile racers, rallyists, manufacturers of technical equipment.

ram, *sb.*[1] Add: **1. c.** *transf.* A sexually aggressive man; a lecher. *colloq.*

1935 *N. & Q.* 23 Nov. 366/2 *Ram*, a male, sexual enthusiast. 1946 *Penguin New Writing* XXVIII. 185 'Yes, it's the Chalk all right,' Willie said. 'The old ram!' he added, happily. 1977 J. Wainwright *Do Nothin' till You hear from Me* viii. 128 One day, May will wise up on the ram she has for a husband and pan him in the chops.

3. c. (Further U.S. example.)

1868 C. A. Dana *Recoll. Civil War* iii. 37 First came seven ironclad turtles and one heavy armed ram.

e. *U.S. Naut.* (See quot. 1961.)

1909 *Sun* (Baltimore) 1 Aug. 14/2 Capt Andrew Hubbard, who later, with all hands, was lost at sea, saw the queer craft coming down stream. He shouted at Captain Insley 'That's certainly a Nanticoke ram.' 1953 *Sun Mag.* (Baltimore) 25 Oct. 30/5 Rams which once hauled lumber are now summer cruise ships. 1961 J. E. Marvil *Sailing Rams* 9 *Ram*, 3 masted bald headed schooner, flat bottom, straight sides without jib boom built and sailed mainly on Chesapeake Bay.

f. *Eton slang.* (See quot. 1977.)

1922 S. Leslie *Oppidan* xvii. 200 A strange procession... There was no variation in the *ram*, as it was called by a football metaphor. 1930 *Daily Tel.* 1 Dec. 21/2 On the last occasion Lori-Phillip touched a rouge, but the ram failed. 1942 J. Lees-Milne *Ancestral Voices* (1975) 30 Then to evensong in [Eton] College Chapel where the traditional ceremonial is invariable. The 'ram' marches in the same deliberate, self-conscious manner. 1977 A. J. Ayer *Part of My Life* ii. 44 After scoring a rouge the attackers could gain an extra point by charging in column and bundling their opponents and the ball into the goal. The column was known as a ram, which was also the name given to the twin columns of Colleger and Oppidan sixth-formers, as they processed into Chapel.

g. An underwater projection from an iceberg or other body of ice.

1952 *Functional Gloss. Ice Terminol.* (U.S. Navy Hydrogr. Office) 22 More rapid melting at the water line than above and below causes a notch to be formed at the water line below which is the ram. 1974 *Encycl. Brit. Macropædia* IX. 160/2 In the Labrador current, sailing at one or two knots toward the North Atlantic shipping routes, this iceberg had a ram (underwater projection), as most icebergs do by the time they enter the warmer waters near the Gulf Stream. 1976 *Jrnl. R. Soc. Arts* CXXIV. 644/1 Ice headlands are elevated because the underwater 'ram' at the seaward side gives more buoyancy.

5. e. The reciprocating arm on which the tool is mounted in a shaping or slotting machine.

1864 D. K. Clark *Exhibited Machinery of 1862* iii. §1. ii. 133 Machines..for shaping levers, cranks, connecting rods... The ram is moved by means of a peculiar crank-motion, with a quick return. 1935 *Buck & Hickman Ltd. Gen. Catal. Tools & Supplies* 126 Hand shaping machine... There are nine different working positions of handle, the ram having three holes and the handle three holes. 1964 S. Crawford *Basic Engin. Processes* viii. 216 Most shaping machines are of the crank type... The ram is located in the top slideway of the body and is reciprocated by the crank mechanism. 1977 *Buck & Hickman Catal. 1977–1979* High speed shaping machine... The ram, carriage and table are all mounted in dovetail guides.

6. a. (sense 1) *ram-faced* adj.

1921 R. Graves *Pier-Glass* 49 Ram-faced lecher, the blood on his own beast head!

b. (sense *3 e) *ram schooner*; *ram type* adj.

1904 *Naut. Gaz.* 14 Apr. 211/2 Geo. K. Phillips & Co., Bethel, Del., have on the stocks a three-masted ram schooner 140 ft. long. 1956 *Sun* (Baltimore) 19 Apr. 38/8 Mr. Katz said he would prove the captain of the ill-fated 'ram type' schooner ran for shelter when hurricane winds were predicted.

7. ram-coupler, a form of coupler used between closely-set organ manuals.

1881 W. E. Dickson *Pract. Organ-Building* xii. 156 The ram-coupler can be used between manuals arranged too closely to admit of tumblers.

ram, *sb.*[3] Add: Also, the centre plank of a coble. Also *attrib.*

1933 *Yachting Monthly* LVI. 108/1 The centre plank [of a coble] is called the 'ram'. 1970 E. J. March *Inshore Craft Gt. Brit.* I. iv. 137 The true coble is built up on a 'ram plank' not a keel. 1973 W. Elmer *Terminol. Fishing* iv. 113 The *ram* denotes the broad central bottom plank in the cobles, which have no keel.

ram, *sb.*[4] Add: **2. a.** The compressive effect experienced by air which is constrained to enter a moving aperture or restricted space (*spec.* the intake of a jet engine); (cf. *ramjet). orig. and freq. *attrib.*, as *ram compression, effect, pressure*.

1944 *Jrnl. R. Aeronaut. Soc.* XLVIII. 445 Air is led from the intake A, under full ram due to the forward speed of the aircraft, to the compressor B. 1945 P. H. Wilkinson *Aircraft Engines of World* 343 Ram effect in flight compresses air to more than atmospheric pressure. 1948 C. E. Chapel *Aircraft Power Plants* iii. 30/2 Ram pressure is that developed in the carburetor air scoop by the forward speed of the airplane. 1953 J. Liston *Power Plants for Aircraft* ii. 65 Compression is obtained by utilizing the forward motion of the aircraft to produce a dynamic pressure or 'ram' in the diverging inlet section of the engine. 1960 D. G. Shepherd *Introd. Gas Turbine* (ed. 2) iii. 84 This increase of temperature and pressure due to aircraft speed is called the ram effect, or simply ram. 1969 W. Thomson *Thrust for Flight* 43 At the high forward speeds made possible by jet propulsion the pressure in the turbine compressor intake can be raised by ram effects to such an extent that the turbines of fast aircraft become very efficient. *Ibid.* 53 Increase of pressure by ram is not a free gift. 1971 P. J. McMahon *Aircraft Propulsion* iii. 111 All the compression needed for it [*sc.* the ramjet] to operate as a heat engine comes from ram compression in the intake due to forward motion.

b. Special Combs.: **ram air,** air which is constrained to enter a moving aperture; freq. *attrib.*; **ram-wing,** a wing-like structure on an air-cushion vehicle which generates lift by means of a ram effect, compressing air between itself and the ground or water surface as it moves.

1953 Jennings & Rogers *Gas Turbine Analysis & Practice* i. 9 Ram air slightly compressed by the forward progress of the airplane enters the impeller, where the pressure of this air is increased by some 10 to 20 in. Hg. 1962 *Engineering* 31 Aug. 258/2 If..all the engines were out, two ram-air turbines provide electric and hydraulic power for essential flying control and aircraft services. 1978 A. Welch *Bk. of Airsports* vi. 92/2 The Para-Foil, or ram-air 'chute, was developed in the late 1960s by Domina Jalbert. 1962 *Air-Cushion Vehicles* Oct. 70/2 Of particular interest.. is a Kawasaki ram-wing craft now being built. 1968 Elsley & Devereux *Hovercraft Design & Construction* i. 13 The ram wing is another type of aerodynamic craft... This is essentially a low-aspect-ratio wing with its trailing edge virtually touching the surface and with endplates sealing the tips to the surface. Suction pressures are developed on the upper surface together with ram or dynamic pressure underneath.

ram, *v.*[1] Add: **6. b.** Of a ship: to force *a way* by ramming.

1914 *Times* (Weekly ed.) 10 Apr. 293/1 The Bellaventure was nine hours yesterday in ramming her way through four miles of ice.

-rama: see *-orama.

ramada (rămā·dă). *U.S.* [Sp.] In the Western U.S.: an (orig. temporary) arbor or similar structure; a porch.

1869 B. I. Hayes *Pioneer Notes from Diaries* (1929) viii. 289, I paid them a dollar for my bath, at the rustic bathing establishment they have constructed, consisting of two goods' boxes sunk in the ground, sheltered by a *ramada*. 1911 H. B. Wright *Winning of Barbara Worth* 201 Every evening under the ramada Barbara sat with her father, often alone. 1949 *Desert Mag.* Apr. 24/1 In a brush ramada the Navajo women weave their blankets while the older children tend the sheep. 1957 G. Shirreffs *Rio Bravo* (1972) i. 8 Someone walked across the boardwalk beneath the *ramada* in front of the building. 1976 *Arizona Republic* (Phoenix, Arizona) 27 May B-15/3 Desert Foothills Scenic Drive..a 17-mile desert drive with ramadas and rest rooms. 1979 *Arizona Daily Star* 5 Aug. (Advt. Section) 17/1 Common areas with playgrounds and ramada.

Ramadan, ramazan. Add: The form Ramadan is now usual. (Later examples.)

1927 C. Connolly *Let.* Mar. in *Romantic Friendship* (1975) 279 It is Ramadan at present and nobody eats till the evening. 1935 H. Edib *Clown & his Daughter* xli. 67 He has been back for some time, and is getting ready to open his shop for Ramazan. 1971 *Sun* (Colombo) 17 Sept. 3 November 8... Ramazan Festival Day. 1976 *Hanna* (Alberta) *Herald* 16 June 10/5 During one month each year the Muslims observe Ramadan.

ramage, *sb.*[2] Restrict *arch.* to senses in Dict. and add: **1. b.** *Anthrop.* A corporate descent group which includes members of both maternal and paternal lineages.

1936 R. Firth *We, the Tikopia* x. 371 One term that might be employed to characterize such kinship groups is 'ramage', for which there is literary authority, though it has fallen out of use. *Ibid.* xvi. 586 The patrilineal principle of descent in the ramage ('joint family' is the translation given of the native term *hoaɔ*) is modified. 1957 —— in *Man* LVII. 6/1 In former publications I have used *ramage* to include the Tikopia unilineal descent group. This, I think, is better described functionally as a lineage. *Ibid.* 6/2 Ramage would then be defined as a corporate descent group of a non-unilinear (ambilineal) character, membership being obtained ambilaterally..according to circumstances. 1963 *Brit. Jrnl. Sociol.* XIV. 24 The word *ramage* has been used to mean a corporate group in which membership may be acquired through either parent. 1976 Hunter & Whitten *Encycl. Anthropol.* 330/1 *Ramage*, nonunilineal descent group composed of individuals who are descended from one ancestor through any combination of male and female links.

Ramage (ræ·mėdʒ), *sb.*[3] The name of Adam *Ramage* (1770–1850), a printer of Philadelphia, used *attrib.* to denote a (usu. wooden) printing-press, or part of one, designed by him.

1827 *Hallowell* (Maine) *Gaz.* 20 June 4/3 For Sale, a small font of Brevier, nearly new; also a Printing Press with a new Ramage Screw. 1874 B. F. Taylor *World on Wheels* I. iii. 24 The cargoes of those boats..was something wonderful,..plows, axes and Bibles, teachers, preachers and Ramage presses. 1923 L. E. Young *Founding of Utah* xxxiii. 349 The old Ramage press had a lever which the printer pulled in the printing of each page. 1949 *Mississippi Valley Hist. Rev.* Mar. 634 It was..printed on a small wrought-iron Ramage press.

Raman (rā·măn). *Chem.* and *Physics.* Also (*rare*) **raman.** The name of Sir Chandrasekhara Venkata *Raman* (1888–1970), Indian physicist, used *attrib.* and in *Comb.* with reference to the Raman effect he discovered, as *Raman band, line, shift* (so *Raman-shifted* adj.), *spectrometer, spectroscopy*, etc.; **Raman-active** *a.*, capable of giving rise to Raman scattering; **Raman effect, scattering,** the scattering of light by a substance with a change in the frequency of the light by an amount which is characteristic of the scattering substance and represents a change in the vibrational, rotational, or electronic energy of the substance; occas. with ellipsis of *effect*; **Raman spectrum,** a spectrum of scattered light showing additional bands produced by the Raman effect.

1928 *Nature* 29 Sept. 477/2 (*heading*) The Raman effect in crystals. 1929 *Ibid.* 24 Aug. 301/1 The Raman spectra for various liquids. 1937 *Jrnl. Amer. Chem. Soc.* LIX. 1139/1 It is generally admitted that the 3654 cm^{-1} Raman band of water vapor corresponds to the symmetric..vibration. 1945 G. Herzberg *Infrared & Raman Spectra of Polyatomic Molecules* iii. 243 All three vibrations of non-linear symmetric..XY_2 are Raman active as well as infrared active. 1947 *Nature* 11 Jan. 60/2 The principal feature noticeable in the spectra is the appearance of a whole series of sharply defined Raman lines. 1947 *Thorpe's Dict. Appl. Chem.* (ed. 4) VIII. 190/1 Fundamental frequencies will be active in Raman scattering only when the mode of vibration of the molecule associated with that fundamental frequency causes an equiperiodic variation in the polarisibility of the molecule. 1950 W. J. Moore *Physical Chem.* xi. 332 Since the Raman spectra are studied with light sources in the visible or ultraviolet, they provide a convenient means of obtaining the same sort of information about molecular structure as is given by the infrared spectra. In many cases, the two methods supplement each other, since vibrations and rotations that are not observable in the infrared..may be active in the Raman. 1958 *Oxf. Univ. Gaz.* 23 Apr. 881 Application of raman spectra to chemical problems. 1962 *Times* 9 Mar. 2/6 (Advt.), The successful applicant will work in one of the following fields:..ultra-violet, infra-red and Raman spectroscopy of aromatic, *N*-heterocyclic and tautomerizable substances. 1966 D. H. Whiffen *Spectroscopy* ix. 110 The Raman lines are always extremely weak with respect to the incident light intensity. 1967 L. A. Woodward in H. A. Szymanski *Raman Spectroscopy* I. i. 16 In practice, the pure rotational Raman shifts are small.. and it is necessary to use a spectrograph of high dispersion and resolving power. 1970 Gilson & Hendra *Laser Raman Spectroscopy* ii. 21 In the basic Raman spectrometer, the sample under examination is subjected to irradiation from a suitable monochromatic source, and the Raman spectrum is observed by use of a system comprising a monochromator, detector, and recorder. 1974 *Nature* 8 Mar. 124/1 Raman-shifted backscatter of laser radar radiation from atmospheric constituents in the troposphere has been reported. 1976 *Jrnl. Chem. Soc. Dalton Trans.* 1148/2 Five of the six Raman-active vibrations of the two octahedral anions.. were clearly observed. 1978 P. W. Atkins *Physical Chem.* xvii. 563 The molecule is active in the Raman because the end-over-end rotation modulates its polarizability as viewed by a stationary observer.

ramanas (ræ·mănăs). [perh. f. Jap. *ranman* full bloom.] *ramanas rose* = *rugosa.

1876 *Garden* 3 May 452/2 (*heading*) The Ramanas Rose of Japan. 1955 N. D. G. James *Forester's Compan.* xxii. 246 Shrubs of value for food and cover..Ramanas rose (*Rosa rugosa*). 1956 [see *Japanese rose* s.v. *Japanese *a.* b].

Ramapithecus (rā:măpi·þikǒs). [mod.L. (G. E. Lewis 1934, in *Amer. Jrnl. Sci.* CCXXVII. 162), f. *Rāma*, the name of an Indian prince of Ayodhyā + Gr. πίθηκος ape.] A fossil anthropoid ape, sometimes considered a hominid, belonging to the genus so called and known from remains found in northern India and East Africa. So **ramapi·thecine** *sb.* and *a.*, a fossil anthropoid closely

related to *Ramapithecus*; pertaining to or resembling an anthropoid ape of this kind.

1934 *Discovery* July 197/2 The other [ape], *Ramapithecus*, is represented by an upper jaw in which the teeth are set in a rounded V, approaching the human form, instead of the U-shape characteristic of the anthropoids. 1965 *Folia Primatologica* III. 90 No finds of *Ramapithecus* or any other dryopithecine or hominid have ever been made in the Tatrot or Pinjor zones. 1968 D. Pilbeam in *Nature* 28 Sept. 1337/1, I would argue..that *Ramapithecus* is probably a hominid,..even if fossil evidence were produced which showed that it was not yet a habitual biped. 1969 J. E. Pfeiffer *Emergence of Man* ii. 47 Imagine an adventurous hominid or prehominid, a primitive *Ramapithecus* perhaps, looking for food in the savanna. 1971 [see *Kenyapithecus]. 1973 *Listener* 10 May 604/3 Some anthropologists would boldly put *Ramapithecus* among the hominids. 1976 R. Ardrey *Hunting Hypothesis* 35 Were the ramapithecines close to the common ancestor of chimp and man? 1977 *Sci. Amer.* May 31/1 Kretzoi now recognizes his fossils as being 'ramapithecine' even if they are not generically *Ramapithecus*. 1978 *Nature* 5 Jan. 12/1 It is of interest and of uncertain significance that not all ramapithecines have been found in woodland deposits.

ramarama (ra·măra:mă). Also **ramiram, rummyrum.** [Maori.] An evergreen New Zealand shrub or small tree, *Myrtus* (or *Lophomyrtus*) *bullata*, belonging to the family Myrtaceæ and bearing small white flowers followed by dark red berries.

1843 E. Dieffenbach *Travels in N.Z.* II. iii. 382 Rama rama—name of a tree (*Myrtus bullata*). 1851 H. W. Richmond *Let.* 15 June in *Richmond–Atkinson Papers* (1960) I. ii. 97 A letter might be written about the tapitap so useful for hammer handles..and the ramiram for ramming in the posts in fencing. 1882 W. D. Hay *Brighter Britain!* II. vi. 196 The Rama-rama..has a good hard wood. 1889 T. Kirk *Forest Flora N.Z.* 273 The ramarama is the largest and most attractive of the New Zealand myrtles. *Ibid.*, The wood of the ramarama is red, straight, and compact. 1928 H. H. Allan *N.Z. Trees & Shrubs* 131 Ramarama..Bushy shrub up to 15 ft; branchlets tomentose. 1946 *Jrnl. Polynesian Soc.* LV. 157 Ramarama, a tree.., often corrupted to rummyrum. 1970 M. E. Fisher et al. *Gardening with N.Z. Plants* i. 75 The ramarama is easily distinguished from all other native plants by its reddish-brown leaves, the spaces between the veins of which are inflated.

Ramayana (rā·māyanā). Also † **Ramayuna, -unu.** [a. Skr. *Rāmāyaṇa*, f. *Rāmā* Rama, the seventh incarnation of Vishnu + *ayana* a going.] An ancient Hindu epic, ascribed to the poet Valmiki. Also *attrib.*

1788 *Asiatick Researches* I. 351 The first Indian Poet was Válmíci, author of the Rámáyana, a complete Epick Poem on one continued, interesting, and heroick, action. 1806 Carey & Marshman tr. *Ramayuna* I. p. ii, The Committee.. have made choice of the Ramayuna of Valmeeki to be the first in the series of translations from the Sungskrit. 1811 W. Ward *Acct. Hindoos* I. p. viii, Next follow Translations of the Contents of the Mŭhabharŭtŭ, the Ramayŭnŭ, and the Shrēē-Bhagŭvŭtŭ. 1841 *Penny Cycl.* XX. 401/1 There are three pieces of his, one in which a domestic subject is treated..and two others taken from the cycle of traditions of the Râmâyana. 1870 F. Richardson *Iliad of East* p. vii, The Râmâyana comprises in all some twenty-four thousand verses, or *slokas*. 1937 M. Covarrubias *Island of Bali* viii. 235 The great epics of the Hindus, the famous *Ramayana* and *Mahabharata*..came by way of Java as propaganda for the ancient Hindus and as part of their religious teachings. 1974 *Times* 5 Nov. 11/4 The play is a stirring episode from the Hindu *Ramayana* epic: Rama's courtship of the lustrous Siti Dewi.

rambai (ra·mbai). Also † **rambé, rambeh.** [Malay.] An evergreen tree, *Baccaurea motleyana*, or a closely related species, belonging to the family Euphorbiaceæ, native to Malaysia, and bearing large dark green leaves and racemes of tiny yellowish-green flowers; also, the fruit of this tree, which is an oval about two inches long with white flesh and pale brown seeds in a smooth brownish-yellow skin. Also *attrib.*

1811 W. Marsden *Hist. Sumatra* (ed. 3) 101 The *chupak*, *ayer-ayer*, and *rambé* are species or varieties of the same fruit [*sc.* lanseh]. 1820 J. Crawfurd *Hist. Indian Archipelago* I. iv. 432 The *Langseh*, *Rambeh*, and *Dukuh* are indigenous fruits. 1839 T. J. Newbold *Straits of Malacca* I. ii. 53 In the valley grow various fruit-trees, such as..the rambai. 1911 [see *Langsat]. 1924 H. N. Ridley *Flora of Malay Peninsula* III. 250 *B*[*accaurea*] *Motleyana*... The Rambeh, cultivated all over the peninsula. 1940 E. J. H. Corner *Wayside Trees of Malaya* I. 241 In habit, trunk, bark and strings of buff-coloured fruits with green seeds and white pulp, the *Rambai* and *Langsat*..are very much alike. 1975 B. M. Allen *Common Malaysian Fruits* 11 Rambai is native to Malaysia and is commonly cultivated in the lowlands. *Ibid.*, Rambai fruits are easily confused with Duku and Langsat.

‖**rambla** (ra·mblă). [Sp., ad. Arab. *ramla*, lit. 'sandy ground'.] **1.** A Spanish ravine, usu. waterless; the dry bed of an ephemeral stream.

1829 W. Irving *Conq. Granada* I. xii. 97 Sometimes their road was a mere rambla, or dry bed of a torrent. 1845 R. Ford *Hand-bk. Spain* I. 398/2 Three long L[eagues], by a *rambla* of red rocks, lead to Berja. 1923 *Blackw. Mag.* Oct. 509/2 Before we reached this *rambla*, coming around a series of sharp bends down a hillside to a lower level, we were tempted by a combination of sun, dust, and the fortuitous appearance of a wayside venta to halt the Colonel. 1960 *Geogr. Rev.* L. 60 Long stretches of drainage channels have been filled up..with erosion debris of rock fragments, gravel, and sand. Such debris-choked channels are called *ramblas*. 1977 *New Yorker* 16 May 34/3 On the perimeter of the terrace were a four-foot-high chain-link fence, then a narrow *rambla*, and then another..fence.

2. A broad street in a city of eastern Spain, built on a shallow watercourse; *spec.* (now usu. in *pl.*) a broad avenue in Barcelona.

1829 A. S. Mackenzie *Year in Spain* i. 37 Our fonda was situated..upon the Rambla, an immense highway through the city, the chief thoroughfare and promenade of Barcelona. 1873 *Amer. Cycl.* II. 304/1 Foremost among its numerous promenades is the Rambla (so called from the Arabic *raml*, sand, applied to a dry river bed, used as a road). 1893 *Johnson's Universal Cycl.* I. 497/1 The city is divided into the old and the new town by a beautiful promenade called La Rambla. 1923 *Blackw. Mag.* Aug. 197/1 The rambla turned into a narrow valley. 1968 *Encycl. Brit.* III. 154/2 Cutting through the old city to the west is the street called the Ramblas leading from the Puerta de la Paz to the Plaza de Cataluña, the centre of the modern Barcelona where are numerous banking houses. 1975 D. Beaty *Electric Train* 34 She had seen the boys and girls promenading up and down in the evening on the *ramblas*.

ramblage (ræ·mblėdʒ). *Manx law.* [f. Ramble *v.* + -age.] Chiefly in phr. *right of ramblage*, the right to ramble over land in addition to passing across it.

1887 *Peel City Guardian* 17 Sept. 5/2 The public will not only be able to claim right of way but also right of ramblage over the whole of the headlands down to the shores. 1898 *Ibid.* 19 Feb. 3/5 The defendant claimed right of ramblage for the people of Rushen. *Ibid.* 3/6 His Honour said that was ramblage. 1930 *Times* 11 Nov. 10/5 Surely the word 'ramblage' is as worthy of preservation as 'the right of ramblage' itself.—The Rev. T. G. Phillips, Wesley Manse, Peel, Isle of Man.

ramble, *v.* Add: **3.** (Earlier and later examples.)

1810 E. Weeton *Let.* 4 June (1969) I. 265 If my time were my own, [I] would ramble the country over. 1930 V. Woolf in *Death of Moth* (1942) 19 The greatest pleasure of town life in winter—rambling the streets of London.

rambler. Add: **a.** In later use, one who walks through the countryside on a specified route, freq. in company with others.

1888 C. F. Wardley (*title*) The rambler's guide to Buxton and neighbourhood, specially compiled for pedestrians. 1932 *Ramblers' Federation Handbk.* III. 15 Many ramblers have made acquaintance with the Batham Gate on visits to the Peak Forest district. 1955 *Times* 11 July 7/1 Such acts of vandalism, said an official of the Peak Park Board, turned the farmers and villagers against genuine ramblers. 1977 *Times* 5 Oct. 14/5 The animal is..in a very unsuitable field and frightening the wits out of innocent ramblers.

c. A type of house, usu. a single-storey suburban building; a ranch house. *U.S.*

1958 *Washington Post* 16 Aug. B1/4 These 3-bedroom, full-basement ramblers and split-foyer ramblers will be sold on terms of no money down to qualifying G.I. buyers. 1967 T. Baird *Finding Out* x. 88 The houses in lots..Cape Cod cottages, ramblers, colonial split-levels, random provincials. 1975 P. Moyes *Black Widower* ii. 20 She would not have traded their tiny frame house..for a modern split-level rambler in any smart suburban development. 1976 *Washington Post* 19 Apr. C17/2 (Advt.), Forestville—Rambler, 3-br. $325.

rambling, *vbl. sb.* Add: Also *Comb.*, as *rambling club*.

1902 *Encycl. Brit.* XXXII. 682/2 Sketching clubs and rambling clubs are formed among young people. 1974 *Country Life* 7 Mar. 471/3 Many rambling clubs..[are] printing on their club programmes that dogs must be kept on a lead on farm land.

rambling, *ppl. a.* **4.** (Earlier examples.)

c 1702 C. Fiennes *Journeys* (1947) 152 There are no good houses but what are old rambling ones [in Bury St. Edmunds]. 1790 *Loiterer* 2 Jan. 10 C—— Castle is a wretched, irregular, heavy, and rambling pile of building. 1816 Jane Austen *Emma* III. vi. 93 The house was..rambling and irregular. 1836 F. Witts *Diary* 12 July (1978) 115 A large rambling mansion first built in the close of the 17th century.

rambo (ræ·mbo). *U.S.* [See Rambure.] A variety of eating or cooking apple which ripens late in autumn and has yellowish, red-streaked skin. Also *attrib.*

1804 J. Mease in A. F. M. Willich *Domestic Encycl.* (Amer. ed.) III. 113/1 Rambo. From Delaware; a fine fall apple. 1817 W. Coxe *View of Cultivation of Fruit Trees* 116 Rambo, or Romanite. This apple is much cultivated in Delaware, Pennsylvania, and New Jersey. 1867 *Hours at Home* V. 318/1 There, too, was the 'rambo-row'—alas! how it is thinned out. 1880 *Harper's Mag.* Aug. 355/1 Outside there were great orchards..with old fashioned Baldwins and Rambos. 1898 M. Deland *Old Chester Tales* 178 Its apples ..were poor enough—hard, gnarly russets, or small, bitter rambos. 1906 *N.Y. Even. Post* (Saturday Suppl.) 5 May 3/2 There were still the meetings of an evening..beneath the rambo apple tree. 1913 G. Stratton-Porter *Laddie* xi. 329 There was a teasing fragrance in the spiced vinegar for pickles, a reminder of winesap and rambo in the boiling cider. 1942 C. Weygandt *Plenty of Pennsylvania* 64 The Rambo apple is not generally raised nowadays.

rambunctious (ræmbʊ·ŋkʃəs, ræ·mbʊ:ŋkʃəs), *a. colloq.* (orig. and chiefly *U.S.*). Also **rambunkshus, rumbunctious,** etc. [Origin unknown: cf. Rumbustious *a.*] Of a person: rumbustious, exuberant; boisterous, unruly; flamboyant; of an animal: wild, high-spirited. Also *transf.* Hence **rambu·nctiously** *adv.*, **rambu·nctiousness.**

1830 *Boston Transcript* 1 Sept. 3/1 If they are 'rumbunctious' at the prospect, they will be 'riprorious' when they get a taste, for a 'copious acquaintance' with Vinegar. 1859 'Dow, Jr.' *New Patent Sermons* 120 Some [men] are mild and peaceable as lambs, while others are as uproarious and rambunctious as tigers. 1866 C. H. Smith *Bill Arp* 54 A plan was set on foot to procure a fierce and rambunkshus animal from the mountains of Hepsidam. 1886 *Galaxy* 1 Oct. 275 'Rampunctious' is belligerent. 1899 *Century Mag.* Aug. 623/1 Och, Misther McGeever, now..I niver heerd no man accuse you of bein' anyway rambunkshus about yer nabor's house. 1902 Kipling *Traffics & Discoveries* (1904) 181 'The petrol will light up and the boiler may blow up.' 'How rambunkshus!' 1904 E. Robins *Magnetic North* I. 123 And it hasn't thought of sleetin'..or anything else rambunksious. 1914 *Blackw. Mag.* July 123/1 Our western bronco retains much of his primitive rambunctiousness. 1922 Joyce *Ulysses* 146 Those slightly rambunctious females. 1926 'J. J. Connington' *Death at Swaythling Court* xvi. 284, I was a bit worried and not feeling particularly rambunctious. 1928 W. A. White *Masks in Pageant* 442 Outside of Vermont, in these expansive, more rambunctious United States, economy is a low virtue ordinarily. 1962 J. E. Hartshorn *Politics & World Oil Econ.* xiii. 193 In the United States these companies, and a host of smaller, even more rambunctiously 'free enterprisers' in the domestic oil business there, get on very comfortably thank you with state agencies. 1962 *Guardian* 4 May 16/5 It then seemed that Mr Fulbright's opponent would be..Dale Alford, a weak but rambunctious man who had promising connections with the Walkerites, the Indignationists, the Birchites. 1971 G. H. Bourne *Ape People* xi. 247 This animal was a rough, tough, rambunctious creature. 1976 *Time* 27 Dec. 13/2 Brezhnev inherited many problems from his rambunctious, buccaneering predecessor, Nikita Khrushchev 1977 *New Yorker* 8 Aug. 47/3 The club that initiated the championship was Prestwick, which occupies a rambunctious stretch of duneland on the east shore of the Firth of Clyde. 1980 *Daily Tel.* 19 Feb. 17/4 J. Edgar Hoover, the then-director of the FBI and the man in charge of all counter-intelligence operations in the United States, personally ordered the probe into Flynn's off-screen activities after the rambunctious star was accused in California of sexually assaulting two under-age girls.

rambutan, -bootan. Add: Also **rambotan.** Also, the tree bearing this fruit, which belongs to the family Sapindaceæ. Also *attrib.* (Further examples.)

1795 tr. *C. P. Thunberg's Travels* (ed. 2) II. 277 The Rambutan (*Nephelium lappaceum*) grows in large clusters and is very generally eaten. 1839 T. J. Newbold *Straits of Malacca* I. ii. 53 In the valley grow various fruit-trees, such as the mangostin,..the rambotan, etc. 1895 *Natural Sci.* VI. 25 He came upon a brisk fight between a large number of Kras and about half a dozen Lotongs (*Semnopithecus*) for the possession of a Rambutan tree..in fruit. 1920 W. Popenoe *Man. Tropical Fruits* x. 328 The rambutan is eaten fresh. 1940 E. J. H. Corner *Wayside Trees of Malaya* I. 592 Mangosteen, Durian and Rambutan generally flower together. 1962 B. Harrisson *Orang-Utan* iii. 106 The only alternatives were some rambutan trees in the back of the garden... We were little surprised that the new trees, though they had only a small supply of ripe rambutan, stimulated a bout of nest-building. 1966 [see *Jack *sb.*[4] b]. 1969 *Oxf. Bk. Food Plants* 100/2 The rambutan is limited to tropical lowlands and requires a high rainfall. 1976 'G. Black' *Moon for Killers* ix. 125, I had two helpings and followed them with a couple of rambutans. 1977 *Borneo Bull.* 7 May 2/2 The corn potential is shown in the centre's annual report for 1976, which also underlines pepper and rambutans as two other valuable income-earning crops.

ramdohrite (ræ·mdōᵊrəit). *Min.* [ad. G. *ramdohrit* (F. Ahlfeld 1930, in *Centralbl. f. Min.*, etc. A. 367), f. the name of Paul *Ramdohr* (b. 1890), Ger. mineralogist: see -ite[1].] A sulphide of lead, silver, and antimony found as long, grey-black prisms.

1931 *Chem. Abstr.* XXV. 2940 Ramdohrite is found as bluish, grey-black prisms in quartz with accompanying pyrite, stannite, sphalerite, jamesonite and pyrargyrite at the Chocaya la Vieja mine, Potosí [Bolivia]. 1954 *Amer. Mineralogist* XXXIX. 161 Most specimens of andorite are syntaxic intergrowths of two distinct species, which have also been observed separately... We propose to call them andorite IV and andorite VI... Ramdohrite = andorite VI. 1972 *Mineral. Abstr.* XXIII. 128/2 The dark phase is thus andorite whereas the light-coloured phase gives a new formula; the formula previously assumed for ramdohrite ($Ag_2Pb_3Sb_6S_{12}$) occupies an exactly intermediate position... The name 'ramdohrite' must be transferred to the Pb-rich phase with formula $AgPb_2SbS_7$. 1975 *Amer. Mineralogist* LX. 621 Another solid solution includes the compositions of andorite ($PbAgSb_3S_6$), ramdohrite ($Pb_6Ag_3Sb_{11}S_{24}$), and fizelyite ($Pb_8Ag_5Sb_{13}S_{30}$).

ramekin, ramequin. Add: (The latter form is now *Obs.*) **b.** A dish in which ramekins or other portions of food are baked and served.

1895 in *Funk's Stand. Dict.* 1946 *Farmhouse Fare* 19 Flake fish... Mix well and pile into buttered ramekins. 1957 *Housewife* Sept. 89/2 Scandinavian saucepan in ovenproof pottery. There is a ramekin..in the same design. 1974 M. Babson *Stalking Lamb* xviii. 129 Sybilla was standing beside the oven, having just transferred the ramekins into it.

1976 *Field* 18 Nov. 1028 (Advt.), These superb oven proof ramekins each depict a different game bird of Great Britain.

ramenas (ramənа·s). *S. Afr.* Also **ramnas.** [Afrikaans, f. Du. *ram*(*m*)*enas* black radish.] The wild radish, *Raphanus raphanistrum*, belonging to the family Cruciferæ; also, formerly, the wild mustard, *Sinapis arvensis*, a similar plant also belonging to the family Cruciferæ.

[1896 R. Wallace *Farming Industries of Cape Colony* vi. 117 Charlock, wild mustard, or 'rominas', *Sinapis arvensis*, L., is a widely prevalent weed of the corn-fields of Cape Colony.] 1913 C. Pettman *Africanderisms* 391 Ramenas..*Raphanus raphanistrum*—Wild mustard is known by this name in the Western Province. 1932 Watt & Breyer-Brandwijk *Medicinal & Poisonous Plants S. Afr.* 56 *Raphanus raphanistrum* L., an introduced species known as Charlock (jointed), Ramenas, Ramnas, and Knopherik, is used by Europeans in the treatment for gravel. 1950 *Cape Times* 8 Aug. 9/6 Weeds such as 'wilde ertjies', ramenas and sorrel,..are regarded by farmers as their biggest enemies. 1953 *Ibid.* 20 May 8 The ramnas, the wild radish of the Cape..is a weed of cultivation also introduced from Europe.

ramet (rēˡ·mĕt). [f. L. *rāmus* branch + -et.] An individual plant belonging to a clone.

1929 A. B. Stout in *Jrnl. N.Y. Bot. Garden* XXX. 33 It is here suggested that the word 'ramet'..be used for a member of the clon. 1938 *Jrnl. Ecology* XXVI. 379 *Silene maritima* was remarkable for the quick completion of first flowering by all ramets on sand and chalky clay as contrasted with the slow completion by ramets on calcareous sand and Potterne soil. 1963 Davis & Heywood *Princ. Angiosperm Taxonomy* x. 345 Cloning has obvious advantages in that the vegetative divisions or ramets are genetically identical. 1977 J. L. Harper *Population Biol. Plants* i. 24 The 'ramet' is the unit of clonal growth.

ramie. Add: Also rami (rami). **a.** and **b.** (Further examples.)

1832 W. Roxburgh et al. *Flora Indica* III. 590 Rami, the Malay name in the Island of Pulo Pinang. A native of the Island of Sumatra, where it is cultivated for its bark, which abounds in fibres of very great strength and fineness. 1839 T. J. Newbold *Straits of Malacca* I. vii. 444 The Rámi Rámi..., the fibres of which the Malays twist into fishing lines, cordage, etc., flourishes on the peninsula. 1868 A. Gray *Field, Forest & Garden Bot.* 299 Ramie, or the Grass-Cloth Plant of China,..is recently planted S.W. for its very valuable textile fibres. 1874 *Rep. Comm. Agric. 1873* (U.S. Dept. Agric.) 262 A treatise has been prepared by Emile Lefranc, of the 'Southern Ramie-Planting Association'. 1895 [see *filasse]. 1906 *Westm. Gaz.* 21 July 5/3 The ramie plant —a tall, bushy member of the nettle family. *Ibid.*, The ramie undergarments are so light and..occupy so little space that they make one covetous on a hot July day. 1909 *Chambers's Jrnl.* Nov. 699/2 The future of the mantle lies in the discovery of some material which can take the place of cotton or ramie. 1909 *Public Ledger* (Philadelphia) 24 June 5/6 (Advt.), Ramie-cloth and natural linen. 1949 *Sun* (Baltimore) 11 Mar. 7/2 The pioneer ramie producers and manufacturers. 1955 *Sci. News Let.* 2 Apr. 213/2 Ramie.. produces a strong silky fiber used to make upholstery material. 1963 R. R. A. Higham *Handbk. Papermaking* xiii. 264 Ramie..and hemp..were used prior to the third century A.D. 1965 'Han Suyin' *Crippled Tree* ii. 22 The weavers of ramie cloth along the water courses are Hakkas ('guest people'). 1976 *Times* 29 Apr. 8/8 A wall hanging made in..herdwick, swaledale, Welsh kemp, raw and treated ramie, flax, alpaca, mohair. 1978 *Nagel's Encycl.-Guide: China* 277 China produces tobacco (400,000 tons in 1956) and fibre-bearing plants—jute, ramie, hemp.

c. A garment woven of ramie fibre.

1922 *Chambers's Jrnl.* Mar. 145/1 Mrs. Godfrey, insufficiently but comfortably clothed in a *rami*, sat under the shade of her veranda. 1928 *Funk's Stand. Dict.*, *Ramie*,..3. (Papua.) A skirt, waist-cloth, or kilt of ramie.

ramin (rami·n). [Malay.] A tree of the genus *Gonystylus*, esp. *G. bancanus*, belonging to the family Thymelæaceæ and native to fresh-water swamps of Malaysia, Sarawak, and the Philippines; also, the light-coloured hardwood obtained from this tree.

1955 *World Timbers* (Timber Development Assoc.) III. 94 There are very large stands of ramin in Sarawak. The timber is comparatively new to the European market, the first shipment being in 1949. 1955 F. G. Browne *Forest Trees of Sarawak & Brunei* ii. 340 *Ramin* can be sawn without difficulty. 1959 *Archit. Rev.* CXXVI. 13 The wall is pale yellow ramin, the floor maple. 1970 *Timber Trades Jrnl.* 21 Mar. 54/1 Ramin, too, is becoming more expensive and shipments increasingly extended. 1977 *Ibid.* 17 Dec. 30/1 The slight improvement in demand has helped prices in the Far East. Ramin remains rather weak, but kerning has started to firm up.

ramisection (ræmise·kʃən). *Surg.* [f. L. *rami-*, comb. form of *rāmus* (see Ramus) + *sectio*, *section-* cutting.] Section of some of the *rami communicantes* so as to prevent sympathetic nervous impulses from reaching some region of the body.

1924 *Med. Jrnl. Australia* 14 June 589/1 After ramisection of the second, third and fourth ganglia has been effected, the cord is divided below the fourth ganglion and the operation completed. 1930 *Lancet* 19 July 127/2 The operation of ramisection, or division of the sympathetic rami to the limbs, was devised by J. I. Hunter and N. D. Royle in the course of some experimental observations upon muscle tone in animals. 1954 E. L. Farquharson *Textbk. Operative Surgery* iv. 104 The operation of pre-ganglionic rami-section came to be recommended in place of ganglionectomy for denervation of the upper limb.

ramisectomy (ræmise·ktŏmi). *Surg.* [f. as prec., blended with *-ectomy.] = prec.

1924 N. D. Royle in *Med. Jrnl. Australia* 26 Jan. 85/2 Sympathetic ramisectomy is a procedure which deals with efferent nerve fibres and prevents discharge from the central nervous system into the affected muscles. 1935 H. P. Jenkins *Terminol. of Operations* vi. 35 Section of nerve roots includes..Ramisectomy.

Ramist, *sb.* (and *a.*) Add: **A.** *sb.* (Further examples.)

1837 K. H. Digby *Mores Catholici* viii. vi. 194 Petrus Ramus, to whom adhered Francis Fabricius, the poet Milton, and others, who obtained the title of Ramists, applied himself [etc.]. 1976 *Times Lit. Suppl.* 2 July 812/4 The same idea had already been popularized on the Continent long before Bacon wrote. As McRae has pointed out, for example, the Ramists had already stressed that knowledge ought to be 'for use'.

B. *attrib.* or as *adj.* (Earlier and later examples.)

1852 H. L. Mansel *Artis Logicæ Rudimenta* (ed. 2) p. xl, The mixed school represented by Keckermann, Aristotelian in matter, Ramist in method. 1955 D. Davie *Brides of Reason* 38 But Wesley's sermons could be methodized According to a Ramist paradigm.

Rami·stic, *a.* = Ramistical *a.*

1962 *Listener* 5 July 19/1 Behind Bacon stands not only the Ramistic Cambridge of his youth, but also the body of London craftsmen and scientists. 1965 K. Charlton *Educ. in Renaissance Eng.* v. 153 Ramistic texts..poured from the presses.

ramjet (ræ·mdʒet). Also **ram-jet, ram jet.** [f. Ram *v.*[1] + Jet *sb.*[3]; see also *ram *sb.*[4] 2.] A simple form of jet engine in which the air used for combustion is compressed solely by the forward motion of the engine. Also *attrib.* and *Comb.*

1942 in *29th Ann. Rep. U.S. Nat. Advisory Comm. Aeronaut. 1943* (1943) 407/2 An experimental investigation of an idealized ram-jet propulsion system was conducted in the.. wind tunnel at the Langley Memorial Aeronautical Laboratory in March 1941. 1945 *Sci. News Let.* 17 Nov. 317/3 A ramjet motor of new and unusual design. 1947 *Sun* (Baltimore) 3 Sept. 20/7 He had been working on the development of the supersonic ram jet engine. 1948 *Times* 2 Feb. 3/2 An aerial missile driven by a 'ram-jet' engine is reported, on its first flight over the desert at Inyokern, California, to have 'far exceeded' the speed of sound. 1955 *Sci. Amer.* Jan. 38/2 Probably the simplest helicopter built so far is the Hiller Hornet. It is powered by ramjets, mounted in the tips of the rotor blades. 1957 *Jane's Fighting Ships 1957–58* 439/3 The ramjet-propelled 'Talos' is capable of engaging both supersonic and subsonic targets. 1967 M. Chandler *Ceramics in Mod. World* vi. 177 The ramjet engine..is little more than a plain tube with a very high-temperature fire inside it. 1974 *Encycl. Brit. Micropædia* VIII. 407/2 Ramjets work best at speeds of Mach 2..and higher. 1977 *Times* 5 Nov. 4/5 (*caption*) Five turbojet engines would power it from take-off until it reached 600 mph when five ramjet engines using liquid hydrogen fuel would take over.

‖**ramkie** (ra·mki). *S. Afr.* Also **raamakie, ramkee,** etc. [Afrikaans, ad. Nama *ramgi-b*, prob. ad. Pg. *rabequinha* dim. of *rabeca* fiddle.] A stringed instrument, roughly resembling a guitar, played by the Hottentots and Bushmen of Southern Africa.

[1790 E. Helme tr. *Le Vaillant's Trav. Afr.* II. vii. 128 The *rabouquin*..is, indeed, a kind of guittar [*sic*].] 1805 *Gleanings in Afr.* xxvii. 232 Others were busily employed in dancing to the music of the ramky, (as they call it,) and seemed highly delighted with their exertions. 1827 G. Thompson *Trav. & Adv. Southern Afr.* I. 391 In the evening we were entertained by a Bushwoman..playing on the *Ramakie*—an instrument about forty inches long by five broad, and having the half of a calabash affixed to one end, with strings somewhat resembling those of a violin. 1835 J. W. D. Moodie *Ten Yrs. S. Afr.* I. 224, I have often listened with great pleasure to the wild and melancholy notes of the 'gorah' and 'ramkee'. *Ibid.* 226 The 'ramkee' is constructed on the same principle as a guitar, by stretching six strings along a flat piece of thin board, with the half of a gourd or 'calabash' at one end, over which a piece of dried skin is strained, on which the bridge is placed. It is played after the manner of a guitar; and in the hands of a skilful performer, makes no contemptible music. 1934 P. R. Kirby *Mus. Instr. of Native Races of S. Afr.* x. 249 Another stringed instrument, the name of which is familiar to most South Africans, is the *ramkie*. 1969 J. M. White *Land God made in Anger* 41 There is everything here..from the two-note whining of the Bushman *ramkie* to Schönberg. 1970 P. Oliver *Savannah Syncopators* 109 *Ramkie, remkie*, three- or four-stringed guitar related to the Portuguese *rabequinha* brought from Malabar to South Africa and developed by the Cape Hottentots. Also *rabekin, ramakienjo, raamakie, ramki*. 1972 *Stand. Encycl. S. Afr.* V. 609/1 Besides stringed instruments, such as the *!goura* (gorah), the *!guba* and the *ramgyb* (ramkie), they [*sc.* the Hottentots] used a set of reed-pipes.

Ramlila, Ram Lila (rā·m,lī:lă). *India.* [Hindi, f. *Rām* the god Rama (cf. *Ramayana) + *līlā* sport, deeds.] A drama, representing episodes from the *Ramayana* and commemorating the victory of Rama over Ravana, performed during the festival of Dusserah. Also *attrib.*

1880 *Encycl. Brit.* XI. 474/2 The principal religious fairs are the following:— At Bilgrám, in September, on the occasion of the *Rám Lila* festival, lasting ten days and attended by about 40,000 persons. 1894 J. C. Oman *Great Indian Epics* I. iii. 71 A large space for the performance of the Ram Lila was kept clear by sepoys. 1935 L. S. S. O'Malley *Popular Hinduism* ii. 59 In North India the *Ramlila* thrills multitudes of villagers every year with the moving story of Rama and Sita. 1953 L. Iremonger in *Caribbean Anthol. Short Stories* 20 There was a *Ram Lila*, and they went. They did not often leave their fields. But one must go to a *Ram Lila*. Dharam and Dharup dressed up..in European shirts and Indian *dhotis*. 1968 *Indian Mus. Jrnl.* V. 32 Omkar..also performed as an actor..in a Rāmlīla party. 1969 *Cultural News from India* Nov. 28 Both shows were largely attended and are now as much a part of the Delhi scene as the yearly 'Ramlila'.

rammelsbergite (ræ·mĕlzbəɪgəit). *Min.* [ad. G. *rammelsbergit* (W. Haidinger *Handb. der bestimmenden Mineral.* (1845) 560), f. the name of K. F. *Rammelsberg* (1813–99), German mineralogist: see -ite[1].] An arsenide of nickel, $NiAs_2$, that occurs as white granular or fibrous masses.

1854 J. D. Dana *Syst. Min.* (ed. 4) II. 61 Rammelsbergite... Trimetric... Slightly ductile. 1913 [see *maucherite]. 1939 [see *pararammelsbergite* s.v. *para-[1] 2 c]. 1963 [see *maucherite]. 1973 Sorrel & Sandstrom *Rocks & Minerals of World* 106 Rammelsbergite..is found with loellingite, skutterudite minerals, sulfides, and other arsenides in medium-temperature veins.

rammies (ræ·miz), *sb. pl.* *Austral.* and *S. Afr. slang.* [Origin unknown: cf. *ramie c.] Trousers.

1919 W. H. Downing *Digger Dial.* 41 *Rammies*, breeches. 1933 *Bulletin* (Sydney) 26 Apr. 20/1 Old Bill watched the youngest jackeroo disrobing... 'If I was you, young feller,' he said, 'I'd leave them rammies on.' 1953 T. A. G. Hungerford *Riverslake* ii. 42 Elastic for the old girl's rammies. 1961 *Personality* 16 May 27 Narrow trousers have ceased to be 'drain-pipes' and are now identified..as 'rammies'.

rammy (ræ·mi), *sb.* *Sc. slang.* [? f. Sc. *rammle* row, uproar, var. Ramble *sb.*[1]: see *S.N.D.* s.v. *rammle*.] A brawl, a fight (esp. between gangs); a quarrel.

1935 McArthur & Long *No Mean City* iv. 45 Evidence about a 'rammy' is always conflicting, never reliable and frequently perjured. 1938 *Evening Standard* (Glasgow) 1 Apr. 17 Gallaher had the body, he was Irish, he laid out two slops in the last rammy. 1944 *Scots Mag.* Oct. 46 Not so long ago, in a dance hall rammy, a fella had laid hands on Hardy. 1967 'H. Calvin' *DNA Business* iv. 49, I enjoy a good rammy... I think I must have an adrenalin hunger. 1973 J. Wood *North Beat* vi. 82 He'd had a rammy with his missus. 1977 *Time Out* 28 Jan.–3 Feb. 10/3 Still, the Villiers at Charing Cross plans a rammy where the only authentic ingredients missing will be the blood and hair on the walls.

ramnas, var. *ramenas.

‖**Ramon Allones** (ramo·n alʸo·nes). The proprietary name of a brand of cigar. Also *attrib.*

1907 *Yesterday's Shopping* (1969) 65/2 *Havana cigars*.. 'Ramon Allones.' ('Flor fina.') 1913 *Trade Marks Jrnl.* 22 Jan. 122 Ramon Allones... Havana cigars and Havana cigarettes. Allones, Limited,..London,..cigar manufacturers. 1965 V. Canning *Whip Hand* xv. 173 He pulled out a cigar... It was certainly Havana, and probably Ramon Allones. 1973 'M. Innes' *Appleby's Answer* i. 11 His droopy moustache held the particular tinge of brown..known to proceed only from the smoking of *Ramon Allones* (or would it be *Romeo y Julieta*?) cigars.

ramonda (rămǫ·ndă). Also **ra(y)mondia.** [mod.L. (A. Richard in C. H. Persoon *Synopsis Plantarum* (1805) I. 216), f. the name of L. F. *Ramond* (d. 1827), French botanist and traveller + -a or -ia[1].] A small perennial herb of the genus so called, belonging to the family Gesneriaceæ, native to mountainous regions of Europe, and having hairy leaves in a basal rosette and single stems of white, pink, or violet flowers.

c 1828 B. Maund *Bot. Garden* I. 83 (*heading*) Borage-leaved Ramonda. 1865 M. Eyre *Lady's Walks in South of France* xxiv. 267 Mrs. Nash took a small *terrine* with a *Raymondia* in it with her to Switzerland. 1907 R. Farrer *My Rock-Garden* xv. 243 Much more beautiful..to my mind, is the Queen of all Ramondias, Queen Natalie's Ramondia. 1931 *Discovery* Nov. 354/1 We stopped..at the gorge of the Treska river to collect ramondia. 1942 T. G. Mansfield *Alpines* i. 80 Ramonda... The name is frequently mis-spelt Ramondia. 1955 L. D. Hills *Alpine Gardening* vii. 248 All the Ramondas like a soil that is both leafy and limy. 1971 D. Bartrum *Rock Gardens* iii. 113 Grow Ramondas in low crevices of a rock wall.

ramp, *sb.*[2] Add: **3. b.** *U.S.* A small, wild onion, *Allium tricoccum*. Cf. *ramps 2.

1923 A. Price *Dreams* 102 A gorge of trout and hot corn pone, Mountain dew that would float a stone, All topped off by a mess of ramps. 1939 *Sun* (Baltimore) 11 Feb. 5/6 Ramps are tiny, green, onion-like bulbs that dot West Virginia's slopes each spring, taste like the food of the gods and smell..a dozen times worse than limburger cheese and burning rubber. 1952 R. Bissell *Monongahela* iii. 40 We

boys would go out on what we called the Hogback near the fort to hunt ramps. 1967 *National Observer* (U.S.) 10 Apr. 6/3 Back in the hills of West Virginia it's ramp season again. 1976 *Daily Colonist* (Victoria, B.C.) 9 Mar. 3/7 The ramps are sprouting and requests are coming in from ramp lovers in many states.

ramp, *sb.*[4] Add: **I. 1. a.** (Further examples.) Also, *spec.* a movable slope or passageway which may be positioned to admit access to another level, as on to a boat or aeroplane.

1893 Kipling *Day's Work* (1898) 2 At either end rose towers of red brick, loopholed for musketry and pierced for big guns, and the ramp of the road was being pushed forward to their haunches. 1901 'Linesman' *Words by Eyewitness* xii. 230 They [*sc.* animals] are then led out..down the slippery ramp, stepping gingerly, much afraid of man's extraordinary devices. 1908 *Animal Managem.* 267 The only difficulty which occurs is from the nervousness of some horses to step across or up the ramp. 1909 *Cent. Dict. Suppl.*, *Ramp*, an inclined traveling platform or carrier for transferring freight from a boat to a dock or warehouse. 1938 *New Yorker* 24 Sept. 29/2 Meek-eyed parents hasten down the ramps To greet their offspring, terrible from camps. 1961 Webster, *Ramp*,..the stairway by which passengers enter the main door of an airplane. 1976 *Lancs. Evening Post* 7 Dec. 1/7 They..were driving off the ramp of the landing craft type vessel when the accident happened.

b. *Railways.* (*a*) The tapering end of a conductor rail, provided to guide the collector shoe on to or off the rail. (*b*) An apparatus used to replace derailed rolling stock on the track.

1922 F. W. Carter *Railway Electr. Traction* v. 218 Ramps are provided at the ends of each length of conductor rail, in order that the shoes may be brought to the contact surface without shock. 1926 *Chambers's Jrnl.* Aug. 539/1 The turned up edge of the ramp guides the wheels of the vehicle back to the rail... Four ramps are used, one for each wheel. 1927 R. E. Dickinson *Electr. Trains* vii. 161 Sudden re-application of motor current occurring at ramps caused pressure surges. 1956 R. A. Hamnett *Brit. Rail Track* iv. 191 At the ends of sections of conductor rail, ramps are provided to pick up the shoe. 1960 *Chemins de Fer* (Bureau Internat. de Documentation des Chemins de Fer) 212/3 Rerailing ramp.

c. An inclined slip road leading on to or off a main highway. Cf. *off-ramp. *N. Amer.*

1952 [see *expressway]. 1965 *Tamarack Rev.* Winter 10 The town must be five or six miles off the highway and one of the county roads connects. One of these days I'll take the ramp, turn off north. 1979 *Upper Valley Progress* (Mission, Texas) 9 Aug. 1/4 Frontage road traffic shall yield the right of way to traffic..leaving an off-ramp on controlled access highways.

d. (See quots.)

1940 *Highway Engin. Terms (B.S.I.)* 12 *Ramp*, a short slope formed to overcome differences in level or for some other special purpose. 1961 Webster, *Ramp*,..a contrivance (as of blocks or wedges of wood) laid parallel in a roadway for passing traffic over lines of hose.

e. A low platform from which competitors leave successively at timed intervals at the start of a motor rally.

1963 P. Drackett *Motor Rallying* iv. 57 The start was in Blackpool, from a Mille Miglia type ramp. 1971 *E. Afr. Standard* (Nairobi) 13 Apr. 1/3 There were 107 starters from the ramp outside City Hall, Nairobi, on Thursday.

4. orig. *U.S.* The point or area at an airport where unloading and loading of aircraft takes place; the 'apron'. Freq. *attrib.*

1947 M. B. Baker *Airline Traffic & Operations* viii. 205 We're not too bothered with other aircraft parked on the ramp. 1961 *New Scientist* 13 Apr. 15/2 On stages up to 1,000 miles in length the turboprop is equal to the jet in ramp to ramp speeds. 1971 *Daily Colonist* (Victoria, B.C.) 24 Aug. 7/1 We are flying at 33,000 feet, the outside temperature is 24 degrees and our expected ramp time in Los Angeles is..oh..about 3.30. 1974 P. S. Smith *Air Freight* x. 399 Ramp handling is the operation of loading and unloading freight on and off an aircraft and moving it across the apron.

II. 5. *Electronics.* An electrical waveform in which the voltage increases linearly with time. Freq. *attrib.*, as *ramp function*, a mathematical expression of the shape of a ramp.

1957 J. D. Ryder *Engin. Electronics* xviii. 618 The response of the first-order system to a so-called ramp input θ_i..is also of interest. 1959 J. Markus *Handbk. Electronic Control Circuits* 281/1 An output pulse begins when the ramp voltage starts increasing and terminates when the ramp and signal voltages are equal. *Ibid.* 281/2 The flip-flop opens a transistor switch, allowing the ramp generator to begin generating a linearly rising voltage. 1962 Huskey & Korn *Computer Handbk.* xviii. 31 The stability is a function of..the stability of the ramp function itself. 1965 *Wireless World* Aug. 399 By the introduction of an integral of the ramp voltage as a feedback term it should be possible to generate an ultra-linear ramp. 1976 *Austral. Jrnl. Physics* XXIX. 187 A second phase-sensitive detector yielding a ramp-function output is used to indicate, by the positive or negative gradient of the ramp, the sense of the Doppler frequency shift, i.e. decreasing or increasing. 1980 *Sci. Amer.* Mar. 73/1 To compensate for echo decay over longer distances, receiver sensitivity is enhanced by ramp gain and a Q filter.

III. 6. *Comb.*, as **rampway** *N. Amer.*, a sloping passageway formed by a ramp between different levels.

1970 I. Petite *Meander to Alaska* ii. xi. 105 Then we walked up the long, cleated rampway to the sidewalk above. 1976 B. Bova *Multiple Man* (1977) vi. 68 We stepped through one of these rampways into a different building.

ramp, *sb.*[5] Add: **a.** (Further examples.) Also, † (see quot. 1812); *spec.* the act or practice of obtaining profit or benefit fraudulently, as by the unwarranted increase of the price of a commodity.

1812 J. H. Vaux *Vocab. Flash Lang.* in *Mem.* (1964) 261 *Ramp*, to rob any person or place by open violence or suddenly snatching at something and running off with it... A man convicted of this offence, is said to have been done for a *ramp*. This audacious *game*, is called by *prigs*, *the ramp*. 1895 J. Caminada *Twenty-Five Years of Detective Life* 161 Watching them perform the 'ramp'—a sudden rush and bustle in which robberies are committed. 1915 *Truth* 2 June 890/2 The ramp in connection with the shares of the East Rand Amalgamated Gold Estates. 1922 *Daily Mail* 20 Nov. 8 (Advt.), Is there a coal 'ramp'? Miners, middlemen and merchants in the coal trade all blame one another for the high price of coal. 1934 R. Macaulay *Going Abroad* xxiii. 195 If I had my way, you would sign a paper..confessing that your whole business is a ramp and a fraud. 1956 *People* 13 May 8/3 From Manchester, Glasgow and Bristol, examples of this growing ramp have reached 'The People'. a 1966 'M. na Gopaleen' *Best of Myles* (1968) 228 You know the limited edition ramp. 1976 W. G. Kerr *Scottish Capital on Amer. Credit Frontier* iii. 77 On their arrival in Dallas, Wellesley and Renshaw discovered that some serious 'ramps', or swindles, had been going on there.

b. bankers' ramp (see quot. 1932). Also *transf.*

1931 J. R. MacDonald in *Times* 26 Aug. 12/3 We are told that this is a bankers' 'ramp', or a conspiracy, or something of the kind, against a Labour Government. 1932 *Ann. Reg. 1931* 68 Mr. Hayday..elaborated the theory of what was known in Labour circles as 'the bankers' ramp'—a financial crisis deliberately engineered for the purpose of forcing Britain to curtail its expenditure on social services, and so remove one of the chief barriers to a reduction of the wage level. 1958 *Times* 10 July 13/2 The orthodox Labour version is that the second Labour Government was broken by a 'bankers' ramp'. 1976 Ld. Robbins *Against Inflation* (1979) xx. 98 What I hope your Lordships will agree is a pure fallacy; namely, the suggestion that what has happened recently is all a pure conspiracy, a banker's ramp and so on.

ramp, *v.*[1] Add: **2. c.** Delete 'Now *dial.*' and add later examples.

1921 [see *Fletcherian *a.*]. 1959 *Listener* 29 Oct. 754/3 The schizanthus..is ramping away and will need potting.

3. a. (Further examples.)

1922 E. R. Eddison *Worm* xiv. 203 It stamped with its silver hoofs, flapping its wings, ramping like a lioness. 1924 R. Campbell *Flaming Terrapin* v. 79 Panthers..And tigers..And lions..They ramped in the morning light. 1974 R. Adams *Shardik* xv. 100 Kelderek..remained constantly near the bear, observing all that it did, attentive to its moods and ways—its frightening habit of ramping from side to side in excitement or anger; [etc.].

5. b. (Later example.)

1951 Dylan Thomas *Poems* (1971) 206 Wherever I ramped in the clover quilts.

6. b. Also with *along*. (Further examples.) Also *transf.*

1933 P. A. Eaddy *Hull Down* 283 *Ramp along*, sailing with all sails drawing to the wind. 1941 J. Cary *House of Children* xxxv. 153 It won't waste your time because you'll learn more too—it's a place for teaching stupid young men how to pass into the army, so you'll simply ramp along.

ramp, *v.*[2] Add: **2. b.** To search (a prisoner) in jail. Also of a prison cell. *Austral.*

1950 *Austral. Police Jrnl.* Apr. 117 *Ramp*, search a prisoner in gaol, as distinct from a search anywhere else. 1979 *Courier-Mail* (Brisbane) 2 Aug. 9/3 He heard noises from Gage's cell, but presumed the cell was being 'ramped' (searched).

rampa·ciously, *adv. rare.* [f. Rampacious *a.* + -ly[2].] In a rampacious or unruly manner.

1915 V. O'Connor *Mary's Meadow Papers* x. 127 During such a wet summer the grass had grown rampaciously.

rampage, *sb.* Add: (Further examples.) Also in U.S. colloq. phr. *to ride* (*a*) *rampage*. Also *fig.*

1906 'Mark Twain' in *Harper's Mag.* Aug. 335 The raven ..sets on her shoulder often when she rides her breakneck rampages. 1927 H. Crane *Let.* 19 Dec. (1965) 313 Her drunken and exclamatory rampage through *Edificios Blancos* [*sc.* H. Crane's *White Buildings*]. 1955 *Times* 29 Aug. 10/5 Mr. Jack Warner, representing Scotland Yard, is indeed a comfort to have at hand when Things are on the rampage. 1967 *Boston Sunday Herald* 2 Apr. (T.V. Mag.) 57/1 A gang on a rampage through Matt Dillon's territory.

rampage, *v.* Add: **3.** *trans.* To rampage about or over (a place).

1905 E. M. Albanesi *Brown Eyes of Mary* i. 7 Where is the beast now? Is she rampaging the premises?

rampart, *sb.* Add: **1*.** *Canad.* A steep bank of a river or gorge. Usu. *pl.*

a 1853 R. Campbell *Two Journals* (retyped from MS, Vancouver Publ. Libr.) 112 Yesterday and today we have been passing through what they call the ramparts—rocks and steep banks along the river. 1921 W. A. Fraser *Red Meekins* I. iii. 32 Cast high on a rampart by a thrust of the waters lay the stern half of their canoe. 1940 *Beaver* June 29/1 After forty miles more came to the ramparts, a large canyon or gorge where for seven miles the river flows between perpendicular cliffs of limestone from one hundred to two hundred feet high. The channel is very deep here—three hundred feet in places. 1973 D. Andersen *Ways Harsh & Wild* i. 48 Here in the upper ramparts there were steep cliffs and mountains rising on each side of the river.

2. *rampart-base, -walk*.

1915 G. Frankau *Tid'apa* v. 25 Green-dark to the rampart-bases, save where, like a wild beast's eye One red light glowered and glimmered in the shadow-tracery, stretched jungle. 1923 R. G. Collingwood *Roman Britain* ii. 30 Along the top [of Hadrian's wall] was a rampart-walk, patrolled by the sentries, and reached by stairs either at a fort, a milecastle, or a turret.

rampike. For '*dial.* and *U.S.*' read '*dial.* and *N. Amer.*' and add: (Earlier and later examples.) Also *attrib.*

1853 S. Strickland *Twenty-Seven Years in Canada West* II. 198 The recently burnt fallow, with its blackened stumps and rampikes did not contribute much to improve the landscape. 1908 C. Mair *Through Mackenzie Basin* 146 The 'rampike' country would be..converted from a burnt-wood region to a bare one. 1936 J. Masefield *Letter from Pontus* 19 With blackened rampikes from old forest fires. 1955 *Jrnl. Canad. Ling. Assoc.* Oct. 6 Apparent survivals from various Scottish and English dialects; such as *bultow, drake, glitter, knap, rampike*. 1961 R. M. Patterson *Buffalo Head* vi. 220 There they stood—three gaunt, upstanding rampikes of charcoal with the humus burnt away from their roots.

ramping, *vbl. sb.*[2] (Earlier example.)

1830 Lytton *Paul Clifford* I. viii. 150 Before this initiatory process, technically termed 'ramping'..had reduced the bones of Paul..to the state of magnesia, [etc.].

ramps. Add: **2.** *U.S.* = *ramp *sb.*[2] 3 b.

1939 *Jrnl. Tennessee Acad. Sci.* XIV. 280 One kind [of wild onion] is much sought after by mountain folk, who call it 'ramps', and gather and eat its bulb with great relish. 1960 *Washington Post* 29 Apr. 12/1 It [*sc.* a smell] will emanate from the cooking of ramps—a wild plant somewhat like an onion. 1976 *Amer. Speech 1974* XLIX. 21 'There's a wild onion, smells like ramps, you know.'

rampsman (ræ·mpsmæ̆n). *slang.* [f. Ramp *sb.*[5] + Man *sb.*[1]: cf. *cracksman*, etc.] One who commits robbery with violence.

1859 Hotten *Dict. Slang* 80 *Rampsman*, a highway robber who uses violence when necessary. 1862 H. Mayhew *Lond. Lab.* Extra vol. 31/2 The '*Rampsman*' or '*Cracksman*' plunders by force; as the burglar, footpad, etc. 1882 *Sydney Slang Dict.* 7/2 *Rampsmen*, burglars, highway robbers. 1932 H. Walpole *Fortress* iii. iv. 467 The perils of London—the cracksmen, the rampsmen, the snorrers and thimble-screwers. 1975 M. Crichton *Great Train Robbery* xiii. 72 Barlow was a reformed buzzer turned rampsman—a pickpocket who had degenerated to plain mugging.

ramrod. Add: **2.** *transf.* † **a.** *Cricket.* (See quots.) *Obs.*

1870 R. B. Mansfield *School-Life at Winchester College* 228 *Ramrod, Raymonder*, names given to a ball bowled all along the ground. 1880 *Baily's Monthly Mag.* Feb. 93 It [*sc.* a quick underhand ball] must be very straight so as to ensure a L.B.W., and provided it pitches half-way—for 'ramrods' are *not* cricket proper—the more irregular the pitch the better.

b. A foreman or manager. *N. Amer.*

1881 E. W. Nye *Bill Nye & Boomerang* 60 John Humpfner, the ram-rod of the New York House, feared that the explosion might break the large French plate glass windows of his palatial hotel. 1905 *Dialect Notes* III. 92 *Ramrod*, mainstay, manager, superintendent. 1942 *Amer. Speech* XVII. 75/1 The man in charge of a herd on the trail [in Nebraska] is the *trailboss* or *ramrod*. 1973 R. Symons *Where Wagon Led* I. i. 10 Jake..acted in between times as a sort of foreman, or in western parlance, 'ramrod' of the outfit.

c. With sexual connotations.

1951 Dylan Thomas *Poems* (1971) 206 Sighed the old ram rod, dying of women. 1974 H. S. Kaplan *New Sex Therapy* i. 17 The glans also becomes enlarged. It does not become hard, however, thus guarding against the possibility that the female will be hurt by the ramrod of the phallus.

3. a. *attrib.* and *Comb.*, as *ramrod-maker, spring*; *ramrod-backed, -like, -rigid, -stiff, -straight* adjs.; **ramrod roll** *U.S.* (see quot.).

1939 'N. Blake' *Smiler with Knife* ii. 33 The major or his ramrod-backed housekeeper. 1861 Ramrod-like [in Dict.]. 1860 Ramrod maker [in Dict.]. 1961 *Countryman* LVIII. 515 We lay ramrod rigid. 1936 M. Mitchell *Gone with Wind* xxviii. 477 The men added as dessert some 'ramrod rolls'..and this was the first time Scarlett had ever seen this Confederate article of diet... The soldiers mixed their ration of corn meal with water, and salt too when they could get it, wrapped the thick paste about their ramrods and roasted the mess over camp fires. 1841 *Ordnance Man. for Use of Officers of U.S. Army* (U.S. Ordnance Dept.) 120 In the musket of the model of 1840, the trigger screw and trigger are taken off after the guard, then the pin for the ramrod spring and the ramrod spring. 1961 *Times* 6 Dec. 17/3 A ramrod-stiff naval officer. 1979 G. Hammond *Dead Game* xii. 150 A slim man, ramrod-straight.

b. *attrib.*, passing into adj. Rigid, inflexible; solemn, formal.

1905 *Pall Mall Gaz.* 18 Dec. 2 Under the pretence that, apart from such ramrod rule, the nation would tumble to pieces. 1920 D. H. Lawrence *Lost Girl* vii. 137 He was a tall Swiss with..a flattish face and a rather stiff, ramrod figure. 1932 *Times Lit. Suppl.* 19 May 366/3 Mrs. Stainlit, an elderly lady of the ramrod breed. 1978 J. A. Michener *Chesapeake* 74 He saw him as..a man of petty foibles and ramrod rectitude.

Hence **ra·mrod** *v. trans.*; to force or drive (something), as with a ramrod; *spec.* (*U.S.*), to manage, direct (a ranch, event, etc.); also, to beat or thrust in the form of a ramrod.

1948 *Popular Western* June 15/2 This is Tex Grant, the star-toter who ramrods this burg. **1955** R. Hobson *Nothing too Good* vi. 50 You're Rich Hobson, ramrodding the Batnuni end of Frontier Company. **1966** S. Heaney *Death of Naturalist* 39 A volley of cold blood ramrodding the current. **1973** R. Symons *Where Wagon Led* I. vii. 118, I ramrodded the Circle Diamond for quite a few years after that. **1976** *Publishers Weekly* 1 Mar. 84/1 The skittish livestock Kingman must ramrod to a distant army post. *Ibid.* 8 Mar. 57/1 Railroad builders par excellence who ramrod tracks across the virgin countryside. **1979** *Tucson* (Arizona) *Citizen* 20 Sept. 11A/3 Scores of volunteers rallied around this charitable event which was ramrodded by the city of Tucson.

ram-sammy (ræ:msæ·mi). *slang* (orig. *dial.*). [Origin unknown.] **a.** A family quarrel; a noisy gathering. **b.** A fight; a scrap.

1891 *N. & Q.* 12 Sept. 206/1 'A ram sammy' = [in the W. country] a family quarrel or, in variation, a noisy gathering. **1902** *Western Morning News* 22 Apr. 5/5 A 'ram sammy' was a family quarrel. **1967** H. W. Sutherland *Magnie* v. 70 She'd sent for Big Hughie in case there was a ram sammy. Magnie was no match for Big Hughie. **1976** J. Wainwright *Walther P. 38* 98, I was there, at this jazz club..in a cramped basement... Some people might have called it a 'jam session'. Some people might have called it a 'ram-sammy'.

Ramsauer (ræ·mzɑuəɹ). *Physics.* The name of Carl Wilhelm *Ramsauer* (1879–1955), Ger. physicist, used *attrib.* with reference to the Ramsauer effect, as *Ramsauer cross-section, free path*; **Ramsauer effect,** the sharp decrease, almost to zero, of the scattering cross-section of atoms of inert gases for electrons with energies below a critical value (first described by Ramsauer in 1921 and independently by Townsend in 1922; now usu. called the Ramsauer–Townsend effect: see next); **Ramsauer minimum,** the minimum in the scattering cross-section for electrons exhibiting the Ramsauer effect.

1930 *Chem. Abstr.* XXIV. 5218 An historical and critical review of the Ramsauer effect. **1938** L. B. Loeb *Atomic Structure* xix. 338 The Ramsauer cross sections..vary in some cases over wide ranges with electron energy. *Ibid.*, Ramsauer's pioneer measurements and the later ones of Ramsauer and Kollath have been carried to such perfection that we today speak of the electron free paths as Ramsauer free paths. **1960** *Physical Rev.* CXVII. 1416/2 For argon, krypton, and xenon $\sigma_t(\epsilon)$ is approximately proportional to ϵ except below the Ramsauer minimum. **1963** *Ibid.* CXXX. 1021/1 This method of analysis might be characterized as a short circuiting of the procedure used by Holtzmark in his remarkable analysis of the Ramsauer cross sections for argon and krypton. **1972** A. Gilardini *Low Energy Electron Collisions in Gases* iv. 338 Argon has a large Ramsauer minimum at ~0·3 eV, so that electrons experience far fewer collisions with molecules at these energies than at lower and higher energies. **1974** G. Reece tr. *Hund's Hist. Quantum Theory* xi. 146 It was possible to explain the Ramsauer effect qualitatively in terms of the diffraction of long waves by small spheres.

Ramsauer–Townsend (ræ·mzɑuəɹ tɑu·nzend). *Physics.* The names of C. W. *Ramsauer* (see prec.) and John Sealy Edward *Townsend* (1868–1957), Irish physicist, used *attrib.* with reference to the Ramsauer effect (see prec.).

1952 Massey & Burhop *Electronic & Ionic Impact Phenomena* i. 9 The remarkable transparency of the heavier rare gases atoms towards electrons of energy 1 eV or so.. will be referred to as the Ramsauer–Townsend effect. *Ibid.* iii. 115 It is now of interest to consider the contribution from higher-order cross-sections. These are very small at the Ramsauer–Townsend minimum. **1963** *Physical Rev.* CXXX. 1022/2 It might be guessed that the Ramsauer–Townsend minimum should occur at approximately the same place in both scattering and momentum transfer cross sections. **1977** P. G. Burke *Potential Scattering in Atomic Physics* vii. 73 The balance of this repulsion with the long range r^{-4} attraction caused by the polarization of the atom in the field of the incident electron is the cause of the Ramsauer–Townsend effect.

ramsayite (ræ·mze,ɪəit). *Min.* [ad. Russ. *ramzait* (E. E. Kostyleva 1923, in *Compt. Rend. de l'Acad. des Sci. de Russie* A. 55), f. the name of Wilhelm *Ramsay* (1865–1928), Finnish geologist: see -ite[1].] A silicate of sodium and titanium, $Na_2Ti_2Si_2O_9$, occurring as orthorhombic crystals.

1924 *Mineral. Abstr.* II. 250 (*heading*) On a new mineral—ramsayite from the Khibinsky and Lovozersky tundras. **1967** *Norsk Geol. Tidsskr.* XLVII. 249 Ramsayite is an important constituent of the mosandrite pseudomorphs... The crystals range in length from about 0·1 mm to 1·5 mm. They are colourless to greyish.

ramsdellite (ræ·mzdělәit). *Min.* [f. the name of Lewis S. *Ramsdell* (1895–1975), U.S. mineralogist + -ite[1].] An oxide of manganese, MnO_2, similar to pyrolusite, occurring as orthorhombic crystals and platy masses of a grey to black colour.

1943 Fleischer & Richmond in *Econ. Geol.* XXXVIII. 278 Ramsdellite. [*Note*] This is the first use of this name in print. The mineral was described in 1932 by Ramsdell.. as a possible dimorph of pyrolusite. **1965** G. J. Williams *Econ. Geol. N.Z.* xiii. 190/2 Coombs (in Reed, 1959) mentioned the existence of ramsdellite (a polymorph of pyrolusite) at Taieri Mouth in ore which consists of pyrolusite with minor manganite. **1969** *Econ. Geol.* LXIV. 221/2 Ramsdellite can very easily be mistaken for pyrolusite. **1972** *Mineral. Rec.* III. 210 Ramsdellite, a relatively common mineral in the Chihuahua geodes, is found as fibrous dendritic masses on quartz but is more commonly found growing as well developed crystals on goethite blades in two different habits.

Ramsden (ræ·mzdĕn). The name of Jesse *Ramsden* (1735–1800), English instrument-maker, used *attrib.* and in the possessive to designate an eyepiece commonly used in astronomical telescopes (see quot. 1847) which he described in 1783 (*Phil. Trans. R. Soc.* LXXIII. 94).

1787 *Phil. Trans. R. Soc.* LXXVII. 55 The telescope.. was an achromatic object-glass of Dollond, of 16 inches focal length, and 2 inches aperture; with a Ramsden's eye-glass, magnifying about 25 times. **1847** R. Potter *Elem. Treat. Optics* I. vii. 117 Ramsden's eyepiece is formed of a plano-convex and a convexo-plane lens of equal focal lengths, set at a distance of about two-thirds of the focal length of either. **1900** R. A. Herman *Treat Geom. Optics* vi. 137 Both Huyghens' and Ramsden's eye-pieces have convergent powers, and the image appears inverted. **1923** Glazebrook *Dict. Appl. Physics* IV. 77/1 When a Ramsden eyepiece is employed even pronounced apparent distortion is of no importance for exact measurements. **1971** Smith & Thomson *Optics* viii. 133 The focal plane of eyepieces of the Ramsden type is close to the first lens.

ramshack (ræ·mʃæk), U.S. Black English var. Ransack *v.*

1901 W. Churchill *Crisis* III. vi. 413 Supper, Miss Jinny. Lawsy, if I ain't ramshacked de premises fo' you bof. **1935** in Z. N. Hurston *Mules & Men* (1970) 323 Me and my buddy and two three more, Going to ramshack Georgy everywhere we go. **1962** W. Faulkner *Reivers* i. 13 Now you got to ramshack the country to locate another back window you can crawl in.

ramshackledom (ræ·mʃæk'ldəm). [f. Ramshackle *a.* + -dom.] = next.

1962 L. G. Pine (*title*) Ramshackledom: a critical appraisement of the Establishment. **1962** *Listener* 15 Nov. 835/2 Originality he [*sc.* Berlioz] possessed..but it does not consist in hopeful random ramshackledom (almost a principle of composition nowadays). **1963** *Guardian* 1 Feb. 9/1 The condition of British ramshackledom exceeds anything hitherto known.

ramshackleness (ræ·mʃæk'lnės). [f. Ramshackle *a.* + -ness.] Ramshackle character or state.

1922 'R. West' in *Public Opinion* 17 Nov. 486/2 These Houses of Parliament are the symbol of a real miracle, a real mixture of ramshackleness and nobility.

ramshackling, *a.* (Further example.)

1951 Dylan Thomas *Poems* (1971) 211 And the tusked, ramshackling sea exults.

ramshackly, *a.* Add: **1.** Substitute for def.: = Ramshackle *a.* 1. (Further examples.)

1897 E. A. Bartlett *Battlefields of Thessaly* xii. 262 There was also the araba with our baggage, a very ramshackly weather-beaten old carriage. **1924** E. M. Forster *Passage to India* xxiv. 217 The chuprassies..filed into the ramshackly room with a condescending air, as if it was a booth at a fair. **1939** in B. A. Botkin *Treas. S. Folklore* (1949) I. iii. 59 The cabin was ramshackly.

2. = Ramshackle *a.* 2

1910 E. M. Forster *Howards End* v. 38 Helen, you must not be so ramshackly. You took this gentleman's umbrella away from Queen's Hall, and he has had the trouble of coming round for it. **1927** —— *Aspects of Novel* i. 28 In the rather ramshackly course that lies ahead of us, we cannot consider fiction by periods. **1976** *Daily Tel.* 8 July 16 The accelerating shift to the Left which has rendered that great ramshackly confederation [*sc.* the Labour party] increasingly vulnerable to the subversive activities of Marxists.

ramshandry (ræmʃæ·ndri), *a.* *nonce-wd.* [Fanciful combination of Ram-stam *a.* and Shandy *a.*] Thoughtless; light-headed; precipitate.

1907 W. De Morgan *Alice-for-Short* xxviii. 280 Don't you go making a runaway match with a ramshandry sort o' half-French girl.

ram's head. Add: **2. b.** (See quots.)

1944 L. T. C. Rolt *Narrow Boat* 206 *Ram's head*, the boatman's name for the wooden rudder post of a narrow boat; usually it is bound with pipe-clayed Turk's-head knots, and occasionally decorated with a horse's tail. **1947** S. Woolfitt *Idle Women* ii. 26 The tiller is called the 'ram's head' and is curved like a swan's neck towards the steerer. It is metal, on the motor, and is painted in sections with.. red, white and blue. *Ibid.* Gloss. 222 *Ram's Head*, in the butty, the post at the top of the rudder, usually highly decorated with Turk's heads and/or horse-hair; in the motor, the steering column. **1976** A. Hill *Summer's End* i. 12 The ram's head was part of the tiller, and by putting our hands to it we got the 'feel' of boat-steering.

ram's horn. Add: **1. a.** Also, in proverbial phr. *as* † *right* (*straight, crooked*, etc.) *as a ram's horn*. (Earlier and later examples). Also *transf.* and *fig.*

c **1320** in Wright & Halliwell *Reliquiæ Antiquæ* (1843) II. 19 As ryt as ramis orn. **1658** Mennis & Smith *Wit Restor'd* 102 Straight as a rams horne is thy nose. **1834** C. A. Davis *Lett. J. Downing* xxxvi. 324 As crooked as a ram's horn. **1842** [see Sheep's eye(s 2]. **1878** Hardy *Return of Native* II. iii. iii. 107 I'm as stiff as a ram's horn, stooping so long. **1914** C. Mackenzie *Sinister Street* II. iii. v. 588 A ramshorn of snuff and glasses of mead waiting for casual callers. **1947** *S. Folklore Q.* Dec. 263 More conventionally..a seasoned reprobate is characterized as 'old as the hills and crooked as a ram's horn'.

attrib. and *Comb.*

1897 *Sears, Roebuck Catal.* 612/2 *Handle bars*,..up or down or ram's horn as desired. Our ram's horn bar is the neatest and most comfortable made. **1909** *Daily Chron.* 17 Aug. 7/4 A great silver-mounted ram's-horn snuffbox. **1934** *Archit. Rev.* LXXVI. 64/3 Material [*sc.* timber] possessing 'ram's horn' figure, is also met with, and 'stripe' or 'ribbon-grain' is relatively common. **1964** J. S. Scott *Dict. Building* 256 *Ram's horn figure*.., a ripple like *fiddleback*.

6. In full, *ram's-horn* (*pond*) *snail*. A herbivorous aquatic snail belonging to the family Planorbidæ, esp. *Planorbis planorbis* or *P. corneus*.

1901, 1926 [see *flat-coil* s.v. *Flat *adj.* 15]. **1950** *Sci. News* XV. 101 These absorption bands have not got precisely the same wave-lengths in the blood of man and in that of the ramshorn pond snail. **1952** J. Clegg *Freshwater Life Brit. Isles* xvi. 254 The larger species of Ram's Horn Snails..live at least two or three years. **1960** *Times* 4 Aug. 1/3 (Advt.), Pond Naturalist would gratefully acknowledge information where to collect Ramshorn Snails. **1971** *Oxf. Bk. Invertebrates* 68 The ramshorns can possibly live in more stagnant conditions than other water-snails.

ram-stam, *a.*, *sb.*, and *adv.* Add: **A.** *adj.* (Later examples.)

1934 M. Watt *Visitors at Birkenbrae* 12 John's that ram-stam..he wad be sure tae come oot wi' something. **1976** *Scottish Rev.* Spring 19 In 'The Steeple Bar, Perth'..his [*sc.* Sydney Goodsir Smith's] ram-stam bacchanalian voice is heard, as naturally as if he had been attuned to spoken Scots from childhood. **1976** *Ibid.* Summer 32 This is a rather ramstam, happy-go-lucky, whimsical procedure.

C. *adv.* (Further examples.)

1910 *Blackw. Mag.* Jan. 31/1 He proposed, ramstam, by telegraph; was accepted, and the lady came home..that he might marry her. **1922** T. S. Cairncross *Scot at Hame* 54, I left my gless a meenit, ran ram-stam. **1967** 'H. Calvin' *Nice Friendly Town* ix. 136 The way young folk nowadays go ram-stam at things without a by-your-leave.

ramus. Add: **1. b.** A major branch of a nerve; *ramus communicans* (pl. *rami communicantes*), a branch of one nerve that joins another; *esp.* one of those joining a sympathetic ganglion with a spinal nerve.

1811 J. Bell *Anat. Human Body* (ed. 3) II. 515 Whilst it [*sc.* the tibial nerve] is yet in the hollow behind the joint formed by the hamstring tendons, it gives off a nerve which comes out from the ham, and descends superficially on the back of the leg. This has been called ramus communicans tibialis. *Ibid.* 517 When it [*sc.* the fibular nerve] arrives at the anular ligament, it is much diminished. Here it divides into the ramus dorsalis pedis profundus, and superficialis. **1893** H. St. J. Brooks in H. Morris *Treat. Human Anat.* v. 864 Each of the thoracic ganglia is connected to the corresponding spinal nerve by two rami communicantes, a white and a grey. The white ramus communicans consists of medullated fibres... The grey ramus communicans or revehent nerve is formed of non-medullated fibres. *Ibid.* 865 Occasionally the number of rami is increased to five or six. **1927** W. Keiller *Nerve Tracts* i. 147 The spinal nerves distal to the root ganglia divide almost immediately into anterior and posterior rami. **1946** *Gray's Anat.* (ed. 29) 1078 The sensory root [of the ciliary ganglion] is formed by a ramus communicans from the nasociliary nerve. **1964** A. J. Berger *Elem. Human Anat.* v. 84 The definitive spinal nerve is very short because it splits into two main branches (or rami) almost as soon as it is formed. The smaller dorsal primary ramus innervates the deep back muscles. **1967** G. M. Wyburn et al. *Conc. Anat.* iii. 93/1 Incise one or two intercostal spaces to expose the intercostal nerves and find their slender connections with the sympathetic trunk, i.e. the grey and white rami communicantes. **1972** C. R. & T. S. Leeson *Human Structure* iii. 11/1 Immediately after emerging from an intervertebral foramen, the spinal nerve divides into a posterior primary ramus and an anterior primary ramus.

3. *Zool.* In the mastax of rotifers, each of the two articulated parts of the incus that are attached to the fulcrum.

1856 *Phil. Trans. R. Soc.* CXLVI. 426 The incus..consists of distinct articulated portions. The principal are two stout rami..resting on what appears to be a slender pedicel (fulcrum, *h*). **1886**, etc. [see *Fulcrum 2 c]. **1972** M. S. Gardiner *Biol. Invertebrates* vii. 236/2 The wall [of the mastax]..is extended into seven firm, sclerotized pieces, the trophi, moved by small individual muscles. These pieces are a single median fulcrum and three pairs of lateral ones, two rami, two unci, and two manubria.

Rana (rā·nă). [Nepali and Hindi *rānā* prince, f. Skr. *rājana* royal: cf. Raja, rajah.] The title used by members of the family which virtually ruled Nepal from 1846 to 1951. Also *attrib.*

1951 *Britannica Bk. of Year* 462/2 Although the relations between the exiled members of the ruling Rana family with the hereditary prime minister..had been unfriendly [etc.]. **1959** *Manch. Guardian* 18 Aug. 5/1 The Ranas married their daughters to Rajput princes. **1971** C. Bonington *Annapurna South Face* iii. 32, I could pick out Rana palaces with spacious courtyards and rusty, corrugated-iron roofs. *Ibid.* 37 The Ranas have lost their power but many of them still hold high office, particularly in the armed forces. However, most of the old Rana palaces..have now been turned into offices or hotels. **1971** K. Kent in *Ibid.* 311 In early 1951 a democratic form of government [in Nepal] was established by proclamation. It kept the existing Rana, Sir Mohan Chamsher, as Prime Minister, but in November 1951 a revision replaced this last Prime Minister with Mr M. P. Koirala of the Congress Party. **1977** *Times* 15 Nov. 15/1 The overthrow of the Rana régime in 1951.

ranai, var. *Lanai.

Ranal (rē[i]·năl), *a.* and *sb.* *Bot.* [f. mod.L. name of order *Ranales* (J. Lindley *Nixus Plantarum* (1833) 9), f. Ran(unculus + L. *-āles*, pl. of *-ālis* -al.] **A.** *adj.* = *Ranalian *a.* **B.** *sb.* A ranalian plant.

1839 T. Baskerville *Affinities of Plants* 73 We shall find there a collection of orders constituting the Ranal and Anonal alliances. **1846** J. Lindley *Veget. Kingdom* 416 In appearance Ranals are singularly different in the same Order.

ranalian (rănē[i]·li,ăn), *a.* *Bot.* Also **Ran-, -ean.** [f. as prec. + -ian.] Of or pertaining to a plant of the order Ranales or the group as a whole.

1915 C. E. Bessey in *Ann. Missouri Bot. Garden* II. 116 The opposite-leaved class..is the first to emerge from the cycadean phylum, appearing as the ranalean complex. *Ibid.* 127 Ranalean evolution has..been one of more and more simplification of flower structure. **1926** *Jrnl. Bot.* LXIV. 83 In following the lead of those who assume a common Ranalian origin for Angiosperms, Mr. Hutchinson has burnt his boats. **1931** A. M. Johnson *Taxonomy of Flowering Plants* i. 11 We have thus a phylogenetic tree of three main branches radiating from the polycarpellate, apocarpous, hypogynous, Ranalian base. **1959** Foster & Gifford *Compar. Morphol. Vascular Plants* xviii. 458 It is clear that these ranalian plants are primitively vesselless. *Ibid.* 480 It is only by such laborious procedures that we shall be able to test the importance of recent investigations on ranalean families. **1972** *Nature* 14 Apr. 353/2 The Magnoliidae association corresponds with the 'Ranalian' and the Hamamelidae-Dilleniidae associations with the 'Rosalian' cluster as conceived above.

ranch, *sb.*[2] For '*U.S.*' read 'orig. *U.S.*'. Add: **1. b.** A single-storey or split-level house.

1960 'E. McBain' *Killer's Payoff* xi. 111 Some real estate agent had decided to give the title 'ranch' to any house that had all of its living space on one floor. **1965** H. Hood in *Tamarack Review* Winter 12 Our house was the first California-redwood ranch in the town of Mount Royal. **1965** 'L. Egan' *Detective's Due* (1966) 10 Big newish expensive homes. The one they wanted turned out to be a split-level ranch with a lot of synthetic stone on its front. **1974** R. B. Parker *God save Child* (1975) x. 73 A one-story house..a low ranch built on a slab.

2. a. (Earlier and later examples.) Also *transf.* Phr. *meanwhile, back at the ranch*: orig. used in Western cowboy stories and films, introducing a subsidiary plot. Also *attrib.*

Quot. 1831 may belong to sense 1.

1831 J. O. Pattie *Personal Narr.* 221 [At] a ranch..I procured a horse for three dollars. **1847** B. Lundy *Life* x. 58 We set off at day-break, and went twenty-one miles to a *ranche*. **1870** J. C. Duval *Adventures Big-Foot Wallace* xl. 247 When I have got through with..my 'inheritance', I shall come back to my ranch here, put on my old buckskins, and run after stock and fight Indians for a livelihood the balance of my life. **1888** R. W. Buchanan *Heir of Linne* II. xx. 121 I've got a cattle ranche in the wild west. **1933** J. D. Higinbotham *When West was Young* 201 A ranch frequently takes the name of the horse or cattle brand of its owner... Other ranches took the name of a locality. **1950** *Sun* (Baltimore) 2 Mar. 8/5 A New Zealander has swopped his cow pastures for a 'whale ranch'... The 'ranch' has been set up in a bay of Arapawa Island off Wellington by Gill Perano, who carries out his 'round-ups' by motor boat. **1958** L. van der Post *Lost World of Kalahari* iv. 72 He went out to East Africa to one of its remote frontier areas and started a ranch of his own. **1963** J. Crist in *N.Y. Herald Trib. Sunday Mag.* 24 Nov. 40/2 Kramer keeps any number of old 'meanwhile, back at the ranch' side adventures going. **1972** *Buenos Aires Herald* 2 Feb. 7/6 He..seduced several women after promising he would marry them later in Venezuela where he said he owned several ranches. **1978** *Observer* (Colour Suppl.) 29 Jan. front cover (*caption*) Meanwhile, back at the ranch...

b. A farm (arable, fruit-growing, etc.); spec., one on which foxes or mink are bred and raised for their fur.

1865 'Mark Twain' in *Californian* 28 Oct. 5/3, I have a ranch of quite unknown extent, Its turnips great, its oats without compare. **1890** *Stock Grower & Farmer* 12 July 5/1 They are pasturing on the Alfalfa ranches. **1900** *Sci. Amer.* 21 Apr. 242/3 There are..no less than thirty-five [Alaskan] islands occupied by proprietors of fox ranches. **1948** [see *mink ranch* s.v. *Mink 4]. **1953** A. R. M. Lower *Unconventional Voy.* 146 The Okanagan with its pleasant little towns and its fruit ranches—everything is a ranch in B[ritish] C[olumbia]. **1958** H. B. Allen in *Publ. Amer. Dialect Soc.* xxx. 6 *Ranch*, denoting an establishment for cattle-raising..although the somewhat prestigious nature of the word has spread it eastward through its adoption by farmers who have large wheat-farms. **1976** *Billings* (Montana) *Gaz.* 4 July 8-D/2 (Advt.), Experienced worker..for dryland grain & livestock ranch South of Billings.

3. a. *ranch-boarding, -building, country* (earlier example), *dog, experience, girl, guitar, hand, hide, -house* (earlier and later examples), *-hut, job, -land, -life, -mark, -overseer*; *ranch-owning* adj.; **ranch egg,** a fresh egg; **ranch mink,** mink bred on a ranch, or its fur; also *ellipt.*, a coat made of ranch mink fur; **ranch wagon,** (*a*) a horse-drawn wagon used on a ranch; (*b*) = *estate car* s.v. *Estate *sb.* 14.

1976 *Rhyl Jrnl. & Advertiser* 9 Dec. 22/3 (Advt.), Reclaimed and new timber for sale..Perspex corrugated sheets, Target board, plywoods, ranch boarding, chipboard, melamine etc. **1926** D. H. Lawrence *Plumed Serpent* vi. 109 The telephone was in the old ranch-building. **1885** *Weekly New Mexican Rev.* 22 Jan. 2/5 Texas..is still as good, if not a better ranch country, than New Mexico. **1897** E. Hough *Story of Cowboy* 227 In the Indian Nations the writer has been out with a pack of greyhounds, which included a good pack of ranch dogs. **1923** D. H. Lawrence *Birds, Beasts & Flowers* 156 Now you've come sex-alive, and the great ranch dogs are all after you. **1908** *Sunset* Dec. 792/2 If you were working with ranch eggs, store eggs or yard eggs, it might be different. **1966** '.. Holton' *Out of Depths* xv. 149 Two ranch eggs with ham. **1976** *Billings* (Montana) *Gaz.* 4 July 8-D/2 (Advt.), Must have Animal Science or Animal Husbandry degree. Previous sales, ranch or feed lot experience preferred. *c* **1894** C. Hoyt *Texas Steer* i, in M. J. Moses *Representative Amer. Dramas* (1925) 17 Oh, I do so want to be something besides a Texas ranch girl. **1951** M. McLuhan *Mech. Bride* (1967) 148/1 There is a story about a little ranch girl. **1947** R. Taylor *Bar Nothing Ranch* xvi. 190 She'd cuddle the ranch guitar and croon heartbreakingly. **1951** Gillis & Myles *North Pole Boarding House* 6 A minister to the spiritual needs of the husky ranch-hands in Alberta's godless foothills. **1977** *Herald* (Melbourne) 17 Jan. 18/7 (Advt.), Sheepskin car seat covers and rugs, ranch hides, kangaroo and calf skin. **1978** *Lancashire Life* Mar. 70 (Advt.), The seating area is in hard wearing ranch hide. **1862** *Harper's Mag.* June 14/1 Having awakened my mule I rode on about five miles further, where I reached a small ranch-house. **1933** J. Steinbeck in *North Amer. Rev.* Nov. 423/2 At the ranch house, he found his mother sitting on the porch. **1930** R. Macaulay *Staying with Relations* xix. 273 At El Refugio,..there's a tiny ranch-hut where I tried for a drink of Mexicali. **1976** *Billings* (Montana) *Gaz.* 4 July 7-D/7 (Advt.), Competent college girl for summer ranch job. **1933** *Times Lit. Suppl.* 9 Feb. 94/2 The arrival of a stranger on the ranchlands of Clinton Prescott coincided with a fierce snowstorm. **1979** *Time* 8 Jan. 40/2 In Texas, Prince Franz Joseph..bought 16,000 acres of ranch land. **1899** 'Mark Twain' in *Harper's Monthly Mag.* (1914) Dec. 4/2 It told me where to begin to talk ranch-life in Carson Valley. **1903** A. M. Binstead *Pitcher in Paradise* x. 110 Sleep, someone, whose name and ranchmarks I have forgotten, once observed, is much overrated. **1952** S. Ellin *Key to Nicholas Street* i. i. 10 A ranch mink that is worth laying your life down for. **1960** *Bull. Retail Trading-Standards Assoc.* Nov. 2 Ranch Mink should..be so described unless it is a mutation mink. **1964** *Harper's Bazaar* Nov. 83 Ranch mink hat. **1976** *Daily Record* (Glasgow) 23 Nov. 18/2 Ranch mink brushes expensive mohair and the whole atmosphere is narcotic and totally unreal. **1926** D. H. Lawrence *Plumed Serpent* vi. 112 He was not really an hotel manager, but a ranch-overseer. **1977** *New Yorker* 29 Aug. 46/3 Harlow comes from a ranch-owning family whose roots were established in northern California over several generations. **1886** T. Roosevelt in *Outing* July 387, I took along the ranch wagon, drawn by four shaggy horses. **1970** *Globe & Mail* (Toronto) 26 Sept. 48/1 (Advt.), 69 Ford 6-passenger ranch wagon, finished in a rich dark blue. **1975** *Country Life* 2 Jan. 32/1 One feature of these New World ranch wagons..is the tailgate.

b. Of a modern building: built in the style associated with a ranch; single-storey; as *ranch bungalow, dormitory, home, house*; also *ranch-style, -type* adjs.

1966 *Globe & Mail* (Toronto) 30 June 30/5 (Advt.), Rustic 7 room ranch bungalow on ½ acre lot, trees galore! **1973** *Irish Times* 2 Mar. 22/6 (Advt.), Delightful American ranch bungalow..in exceptional position and close to all amenities. **1974** *Country Life* 21 Feb. Suppl. 36/1 American Ranch Bungalow in Surrey..£60,000. **1976** *Columbus* (Montana) *News* (Joliet Suppl.) 27 May 2/3 Housing [malefactors] in ranch and dairy dormitories. **1973** *N.Y. Law Jrnl.* 26 July 16/8 (Advt.), Mini-farm, 19 acres, neat 5 room ranch home, almost borders Summit Lake. **1978** *Detroit Free Press* 16 Apr. F9/1 (Advt.), Bar with 2 bdrm ranch home & soft ice cream business. **1952** H. Innes *Campbell's Kingdom* I. ii. 33 The Fergus home was a low, sprawling ranch-house building. **1959** N. Mailer *Advts. for Myself* (1961) 378 Eitel's rented ranch-house in the desert. **1975** *Woman's Jrnl.* Sept. 62/1 Today..she lives in style in a palatial ranch-house in Beverly Hills. **1961** J. Mitford in *Life* 4 Aug. 8/2 Motels..may be Tudor, Queen Anne, Colonial, Knotty Pine, Ranch Style, or Futuristic in décor. **1962** E. Snow *Other Side of River: Red China Today* lx. 461 To people who live in American ranch-style bungalows, all the new multiple dwellings in China would seem primitive. **1970** *Cape Times* 28 Oct. 19/9 (Advt.), Outstanding architect designed, owner built ranch style house on high foundation. **1978** *Detroit Free Press* 16 Apr. (Parade) 20/2 Just before Christmas, Dan Scarborough, the senior state senator from northeast Florida, was entertaining some friends in his family's sprawling ranch-style home. **1956** W. H. Whyte *Organization Man* 11 It is difficult to see the three-button suit as more of a strait-jacket than overalls, or the ranch-type house than old law tenements. **1976** J. Philips *Backlash* (1977) iii. 118 [She] lived in a modern ranch-type house about a mile out of town.

ranch, *v.*[1] Add: **1. a.** (Earlier and later examples.)

1866 B. Harte in *Californian* 26 May 1/1 Ranchin' out this way? **1872** 'Mark Twain' *Roughing It* xxxiv. 242 He had been farming (or ranching as the more customary term is) in the Washoe District. **1976** *Billings* (Montana) *Gaz.* 28 June 5-A/2 The bridegroom graduated..with a degree in agriculture. The newlyweds will ranch near Boyes. **1976** *Laurel* (Montana) *Outlook* 30 June 14/1 He farmed and ranched at Acton until 1974.

b. To let land for grazing.

1910 *Blackw. Mag.* July 126/2 He is 'ranching', that is, letting grazing on the eleven months' system.

2. *trans.* **a.** To put (an animal) on a ranch.

1873 J. H. Beadle *Undevel. West* xxx. 663 Six hundred miles..had worn out my horse, and on the 16th instant I 'ranched him' twenty miles south of Beaver. **1916** *Yukon Territory* (Canada Dept. Interior) x. 177 Foxes should be ranched in woodland areas. **1935** *Discovery* Feb. 50/2 Animals so purchased will be ranched and tended by the vendor. **1980** *Times* 18 Mar. 14/8 Mink being ranched, humanely killed and kept in a life of luxury.

b. To use (land) as a ranch.

1901 B. Harte *Under Redwoods* 74 It caused her to remove to Santa Ana, where her old father had feebly ranched a 'quarter section' in the valley. **1927** *Daily Tel.* 23 Aug. 11/6 The large arable farms..must..be grassed down, and ranched at a nominal rent. **1965** F. Symington *Tuktu* 13 If the central Arctic were being systematically 'ranched' with caribou herds.

Hence **ranched** *ppl. a.*, of an animal: confined to or bred on a ranch. Also *transf.*, of the fur of a ranched animal.

1954 *Economist* 20 Feb. 567/1 Ranched Russian mink. *Ibid.*, Wild furs and ranched furs. **1970** *Daily Tel.* 10 Oct. 14 Tourists who visit emergent countries do so to see the animals in their natural environment—not confined and 'ranched'. **1977** *Harpers & Queen* Sept. 65 Very dark natural ranched mink coats from £2,750.

rancher. (In Dict. s.v. Ranch *v.*[1]) Add: **1.** (Earlier and later examples.)

1836 *Papers of M. Buonaparte Lamar* (1921) I. 337 Capt King went..to Lopes..for the purpose of chastising some ranchers. **1866** *Rep. Indian Affairs* (U.S.) 189 Teamsters, packers, herders, ranchers and miners all over the country have become exasperated. **1953** *Manch. Guardian Weekly* 18 June 3 The surviving farmers and ranchers have bought up the land from the men who went away. **1958** *Publ. Amer. Dialect Soc.* xxx. 6 A man with such a dual [*sc.* wheat and cattle] establishment in central Nebraska..is more likely to identify himself as a rancher rather than as a farmer. **1961** L. van der Post *Heart of Hunter* i. vii. 112 One of the foremost ranchers near Gemsbok Pan was in the Union and coming back by truck across the desert. **1973** *Sunday Advocate-News* (Barbados) 21 Jan. 14/1 There are a few ranchers..on the Brazilian border, a few miners, a few lumberjacks. **1976** *Billings* (Montana) *Gaz.* 4 July 1-A/5 Z— C— was born into a family of ranchers who had come to Montana from Canada a few years before the first Yegen arrived.

2. A modern single-storey house.

1965 *Sun* (Vancouver) 8 Aug. 47/7 (Advt.), Rancher.. close to Burnaby Park and swimming pool. **1968** H. Waugh *Con Game* ii. 23 The Demarest home was a rancher in the $25,000 class. **1970** *Globe & Mail* (Toronto) 28 Sept. 24/3 (Advt.), Lovely buff brick rancher on wide 80 ft. lot. **1978** *N.Y. Times* 29 Mar. B8/6 (Advt.), Beautiful rambling Rancher. 2 yrs old, 7 spac rms, [etc.].

rancheral (rɑnʃēə·răl), *a.* *rare.* [f. Ranchero + -al.] Of or pertaining to rancheros.

1847 G. F. A. Ruxton *Adventures in Mexico* xiii. 94 Of these eight mozos, he who bore away the palm of rancheral superiority..was the third son.

rancheria. Add: (Further examples.) Also **rancheree, rancherie** (rɑ·nʃĕri).

1854 C. E. De Long in *Calif. Hist. Soc. Q.* (1929) VIII. 200 Anderson & I fell in with an Indian Rancheria. **1872** R. B. Johnson *Very Far West Indeed* 197 I'd take a dozen Injuns straight out of the rancherie, an' make a better government out of 'em than they've got up thar. **1901** *Canad. Mag.* XVIII. 179/1 Jim is the policeman at the Indian rancheree. **1901** *Daily Colonist* (Victoria, B.C.) 15 Oct. 3/2 A heavy cargo, including everything from shipments of lumber for Bella Coola and Kitimaat, to new gravestones for various Indians at Northern rancheries. **1914** A. H. Fitch *Junipero Serra* iii. 222 There were many rancherias on the banks of the Colorado. **1955** *Sci. News Let.* 16 Apr. 245/3 These [South American] Indians..live in very small communities called 'rancherias', consisting of from two to 15 individuals. **1963** R. Symons *Many Trails* vii. 71 Although they own small reserves—which they call rancheries—most of them prefer a nomadic life. **1976** *Islander* (Victoria, B.C.) 27 June 2/3 Sent their logs downstream to a booming ground near the Indian rancherie.

ranchero. Add: (Earlier and later examples.) Also *attrib.*

1826 J. E. B. Austin *Let.* 31 Oct. in C. Barker *Austin Papers* (1924) I. ii. 1482 The old Ranchero was much surprised, but overjoyed to see me. **1836** C. J. Latrobe *Rambler in Mexico* ii. 27 You have here..every degree from the substantial *Ranchero*, or proprietor, bespurred,..embroidered vest, and gaudy *serape*,..or the trusty *arriero*, with his long string of mules. **1927** W. Cather *Death comes for Archbishop* iii. 78 He and the *rancheros* had run their church to suit themselves, making a very gay affair of it. **1947** M. Lowry *Under Volcano* iii. 90 The orange juice and ranchero eggs. **1963** *Punch* 20 Feb. 285/2 Women have gone overboard for their ranchero hats. **1968** *Esquire* July 99/3 John Weitz.. designed this yellow-and-black-striped blazer suit. With it, Ranchero tie over a neckband shirt. **1973** *Country Life* 6 Dec. 1970/1 The Mexican heavy knit woollen jacket.. worn by *rancheros*... The coat looks good on either sex. **1977** *Times Lit. Suppl.* 21 Jan. 88/5 In the ranks of the Cristeros [in Mexico 1926–29] there figure small landowners, rancheros, Indians.

ranchette (rɑnʃe·t). *U.S.* [See -ette.] A small, modern, single-storey or split-level house.

1956 R. A. Heinlein in *Mag. of Fantasy & Sci. Fict.* Oct. 23/2, I wanted a ranchette near the plant; she favored a flat in Town. **1966** *Punch* 9 Nov. 692/2 Their split-level Graeco-Moorish ranchette, set on teetering stilts above the green Los Angeles smog. **1979** *Tucson* (Arizona) *Citizen* 3 Oct. 18c/3 (Advt.), 1½ to 2 commercial acre Ranchettes overlooking the city.

ranching, *vbl. sb.* (In Dict. s.v. Ranch *v.*[1]) Add: (Earlier and later examples.) Also, the raising of game and other animals. Also *attrib.*

1863 E. R. Meredith *Let.* May in *Frontier & Midland* (1937) XVII. 288/1 A ranch is properly a grazing farm and the term 'ranching' sometimes means farming but is generally applied in this country [*sc.* Idaho] to taking care of stock. **1870** J. C. Duval *Adventures Big-Foot Wallace* xlvi. 302 We would try our hands at 'ranching'. **1873** J. H. Beadle *Undevel. West* xv. 267 'Ranching' came next, and all this industry is not lost. **1916** [see *mink ranching* s.v. *Mink 4]. **1950** *Manch. Guardian Weekly* 5 Jan. 10 Teamsters carried it through the ranching country. **1971** *Inside Kenya Today* Mar. 52/2 Arrangements have also been finalized for the establishment of a ranching scheme at Jaldesa. **1974** *Listener* 2 May 572/3 [The land] is all earmarked for ranching development. **1975** *Nature* 5 June 449/2 Interest in game ranching of natural animal populations as a food source goes back a long way. **1980** *Daily Tel.* 30 Dec. 7/2 Salmon ranching, inshore fishing and tourism are the immediate projects for economic development.

ranchito (rɑntʃi·to). Also **ranchita.** [Sp., dim. of Rancho.] In the Western U.S., a small ranch or farm.

1850 L. H. Garrard *Wah-to-Yah* xvii. 231 To the Ranchita is something less than a mile. **1906** A. Adams *Cattle Brands* 92 He had sent to a nearby ranchito for a man who had at least the reputation of being quite a hunter. **1976** *New Yorker* 26 Apr. 122/2 Many descendants of the *pobladores* still live in villages that seem dominated by cracked adobe or on tiny *ranchitos* with a garden patch and a few head of cattle.

ranchman. (In Dict. s.v. Ranch *sb.*[2]) (Earlier and later examples.)

1856 *Spirit of Times* 4 Oct. 75/1 The dusty, rusty, rough-clad, huge-pawed creatures, known as ranch-men. **1926** J. F. Dobie *Rainbow in Morning* (1965) 97 A barbecue required the coöperation of the substantial citizens of the country, for they furnished the beef to barbecue. A ranchman might contribute a calf, a yearling, or steer; or he might contribute money. **1949** *Daily Ardmoreite* (Ardmore, Oklahoma) 4 Dec. 10/1 The ranchmen are piling up hay in strategic places to feed stock if the severe cold comes again. **1959** T. D. Clark *Frontier America* xxvii. 640 A ranchman could homestead 160 acres in his own name... A ranchman could purchase public lands outright.

rancho. Add: **1. a.** (Later example.)

1931 *Times Lit. Suppl.* 19 Mar. 214/2 To live in a mud 'rancho', eat frugally..was no hardship for this class of immigrant.

b. (Further examples.) Hence, in extended use, a roadhouse or inn.

1849 J. T. Brooks *Four Months Among Gold-Finders in Alta California* ii. 8 Bradley urged us to proceed a few miles farther, where we could take up our quarters at a rancho belonging to a friend of his. **1854** C. E. De Long in *Calif. Hist. Soc. Q.* (1929) VIII. 199 Took Dinner at the Empire Rancho.

c. (See quots.)

1957 P. Kemp *Mine were of Trouble* iii. 45 *Rancho*, usually some form of stew, was at half past twelve... And, at six-thirty, the evening *rancho*. **1976** *National Observer* (U.S.) 17 July 10/1 A 23-inch TV is in the dining room, and most of the population are able to eat the *rancho*—the Mexican native meals—because it is very good.

ranchy (rɑ·nʃi), *a.* *U.S. slang.* [Perh. var. *Raunchy *a.*] Dirty, disgusting, indecent.

1903 A. M. Binstead *Pitcher in Paradise* xii. 283 Then they brought the monkey in—the sad-faced, bare-based, flea-ranchy old monk. **1959** Ld. Kinross *Innocents at Home* xxii. 196 The bridegroom, an Englishman, declared his intention of having the English as opposed to the American marriage service. This included..the worshipping of her with his body. There was an embarrassed pause at this; and then one of the bridesmaids remarked, 'A bit ranchy, that.'

rancid, *a.* Add: **2.** (Earlier and later examples.) Also as *sb.*

1833 J. Constable *Let.* 3 Apr. (1965) III. 98 He is too fond, of *rancid old art*. **1892** 'Mark Twain' *Amer. Claimant* xv. 156 He couldn't get his breath at first. When he did get it, it came rancid with sarcasm. **1912** B. Pain *Locris of Tower* in *Stories in Grey* 195 Black kid gloves, the most rancid form of gloves. **1924** E. F. Benson *David of King's* v. 79 'When Milton talked of a pansy, he called it "freaked with jet".' 'O Lord, did he really?' said David. 'How frightfully rancid!' *a* **1930** D. H. Lawrence *Old Men* in *Last Poems* 251 The rancid old men that don't die because the gods don't want them..Old people fixed in a rancid resistance to life, fixed to the letter of the law. **1941** E. P. O'Donnell *Great Big Doorstep* ix. 125 If she ain't a rancid! She's sweet as sugar to Evvie jiss to get Evvie in a good humor, then she turn around and talk sarcastic. **1973** *Times Lit. Suppl.* 16 Mar. 304/3 The movement was sufficiently vague for any rancid provincial mediocrity to come to believe that he was himself at the centre.

ranciéite (rænsi,ē[i]·əit). *Min.* Also **rancieite,** † **rancierite.** [ad. F. *rancierite* (A. Leymerie *Cours de la Minéralogie* (1857) II. 329), f. Rancié (formerly also Rancier), name of the mountain near Vicdessos, Ariège, France, where it was first found: see -ite[1].] A hydrated oxide of calcium and manganese, $(Ca,Mn^{II})\,Mn^{IV}_4O_9\cdot3H_2O$, occurring as soft flakes and as compact or friable masses.

1861 H. W. Bristow *Gloss. Mineral.* 313/1 *Rancierite*... Occurs in earthy masses of a deep brown colour. Soils the fingers. **1907** *Mineral. Mag.* XIV. 408 *Rancieite*. (Collection de Minéralogie du Muséum d'Histoire Naturelle, Paris, Guide du Visiteur, 2nd edit., 1900, p. 29 (Ranciéite).) The correct spelling of rancierite (A. Leymerie), a variety of wad. **1923** *Mineral. Abstr.* II. 144 A mineral found by G. Friedel in the haematite of Villerouge, Aude, was identified as rancieite of Lacroix. **1943** *Econ. Geol.* XXXVIII. 594 The manganese is in the form of ranciéite, a finely flaky, soft, brownish-black, hydrous calcium-manganese oxide, mixed with calcium carbonate in the form of calcareous tuffs. **1969** *Mineral. Abstr.* XX. 225/2 Ranciéite, from old workings at Rancié, forms soft brown scales with metallic lustre;.. pleochroic yellow to deep brown. **1979** *Nature* 12 July 137/1 That rancieite belongs to the birnessite group (family) has been established by Bardossy and Brindley.

rand, *sb.*[1] Add: **1.** (Further example.)

1958 *New Biol.* XXVI. 92 The underlying fen peat acts as a reservoir of wetness while the margins of the peat-filled basin insulate the raised centre from the soil water which is draining into the basin... Such a bog is called a raised bog, the sloping slides being called the rand, and the insulating zone round the edge the lagg.

3. c. *Basketry.* (See quots. 1910 and 1912.)

1903 R. M. Jacot *Useful Cane Work* I. p. ix/2 *Trade or workshop terms*,..randing, or a 'rand'. **1910** *Encycl. Brit.* III. 482/2 The chief strokes used in constructing an ordinary basket are:— the 'slew'—two or more rods woven together; the 'rand', rods woven in singly. **1912** T. Okey *Introd. Art of Basket-Making* v. 20 The next section is formed by a *Rand*—one single rod worked alternately in front of and behind each Stake. **1959** D. Wright *Baskets & Basketry* vi. 136 *Rand*: a single rod worked in front of one stake and behind the next.

rand (rænd, ‖ rɑnt), *sb.*[2] *S. Afr.* Also **randt, rant.** Pl. **rands,** ‖ **rande.** [Afrikaans, a. Du. *rand(t)* edge, margin: rel. to Rand *sb.*] **1. a.** In South Africa: a rocky ridge or area of high sloping ground, esp. overlooking a river-valley. **b.** *spec.* *The Rand*, the Witwatersrand, a notable gold-mining area of the Transvaal.

1839 J. Collett *Diary* 27 May in *Voorloper* (1976) 663 Finished making New Kralls to day on Willow fountain rant. **1856** F. Fleming *Southern Africa* v. 109 The country, lying between the Rand and the Fish River, is thickly populated with Fingoes. **1890** *Digger's Doggerel* 28 The best Crushing Spec..on the Rand. **1891** B. Mitford *Romance of Cape Frontier* iv. 23 He stood on the top of the *randt* for a brief blow after his exertions. **1899** G. B. Shaw *Let.* *c* 26–30 Dec. (1972) II. 124 The conflict that was inevitable from the moment that gold was discovered in the Rand..had to come. **1900** A. H. Keane *Boer States* iii. 22 We can here speak of 'rands', that is, ridges of moderate elevation, which, however, are sometimes high enough to form water-partings. **1928** E. Walker *Hist. S. Afr.* xii. 413 Mining areas were proclaimed on the Rand. **1947** H. C. Bosman *Mafeking Road* 3 'I don't think they [*sc.* the stars] would be good for growing mealies on, though,' I answered, 'they look too high up, like the rante of the Sneeuberge, in the Cape.' **1953** D. Lessing *Five* iii. 129 He thought of the old prospectors..panning gold.., washing the grit for those tiny grains that might proclaim a new Rand. **1972** D. Francis *Smokescreen* ix. 123 Most of Johannesburg sank about three feet..after all the reef was out... The Rand gold fields are shallower.

2. (freq. with capital initial.) Pl. *rand* or *rands*. [f. with ref. to sense 1 b.] A unit of decimal currency, orig. equivalent to ten shillings sterling, and containing 100 cents, adopted by the Republic of South Africa in 1961, and subsequently by certain other southern African countries. Also *attrib.*

1961 *Times* 27 Jan. 19/4 There was a boom on the Johannesburg Stock Exchange last night... The occasion was the second dress rehearsal for trading in rand and cents when decimalization overtakes South Africa on the second Tuesday of next month. **1961** *Guardian* 26 May 3/3 A two-rand (formerly one pound) postal order bought in a Cape-town post office. **1970** *Daily Nation* (Nairobi) 16 Jan. 14/5 Lesotho's currency is the South African Rand. **1972** P. Driscoll *Wilby Conspiracy* ii. 37 You want wine? It'll cost you two rands a bottle. **1978** J. Paxton *Dict. European Econ. Community* (rev. ed.) 21 In March 1976 the E.E.C. agreed to provide Botswana with about 90m. Rand in aid over four years.

rand, *v.*[1] Add: **2. c.** *trans.* and *intr.* To weave by randing (sense *1 b).

1959 D. Wright *Baskets & Basketry* ii. 45 After the initial pairing the base may be randed. **1962** *Punch* 1 Aug. 170/3 The basket workers..still keep their rhymed boast: I can rand At your command..Wale all right And keep my stakes in order.

ran-dan, var. *Ran-tan *v.*

randed, *a.* *Basketry.* [f. Rand *sb.*] Woven by randing (sense *1 b).

1907 *Yesterday's Shopping* (1969) 153/1 Basket, fine randed buff wicker. **1912** T. Okey *Introd. Art of Basket-Making* ix. 91 The chief difference to be noted in the methods of Randed work are: that fine White or Buff is normally used. **1959** D. Wright *Baskets & Basketry* ii. 42 For a Randed base..an extra short stick will be needed to make an odd number.

randing, *sb.* Add: **1. b.** *Basketry.* The action or process of weaving rands (sense *3 c); randed work.

1903 R. M. Jacot *Useful Cane Work* I. p. x/2 The basket maker..would speak of driving down the stakes..into the 'randing' or 'pairing' of a bottom. **1912** T. Okey *Introd. Art of Basket-Making* ix. 91 Randing is sub-divided into Coarse, Slight, Light, Fine and Close. **1946** N. Wymer *Eng. Country Crafts* vii. 73 He now fills in the sides by weaving single willows..a stage known as 'randing'. **1964** H. Hodges *Artifacts* x. 146 The simplest weave, *randing*, was done with a single rod passed behind one stake and in front of the next.

‖ **randjie** (rɑ·ntʸi). *S. Afr.* Also **randje, rantjie.** [Afrikaans, ad. Du. *rand(t)je, rantje*, dim. of *rand(t)* *Rand *sb.*[2]] A narrow ridge of rocky ground, not as high as a kopje.

1889 '*Argus*' *Ann. & S. Afr. Directory* 849/1 Stony hills and knolls—known locally as *randjes* and *spitskops*. **1914** L. H. Brinkman *Breath of Karroo* i. 17 Here and there a few kopjes relieved the monotony of the view, and every few miles, randjes, or low stony hills, stretched across the plains. **1939** S. Cloete *Watch for Dawn* 31 The small path that runs along the randjie. **1944** V. Pohl *Adventures Boer Family* i. 11 There were roebuck in the *rantjies*. **1949** M. Leigh *Cross of Fire* v. 99 The track climbed the randjie on which we stood and petered down again into the darkness and the gloom towards Matzana's kraal.

‖ **randkluft** (ra·ntkluft). [Ger., lit. 'edge crevice'.] A crevasse between the head of a glacier and a surrounding rock wall.

1934 C. F. Meade in S. Spencer *Mountaineering* vi. 122 Another sort of chasm closely related to the bergschrund is the *randkluft*. This is the name given to the gap between the glacier and the bare rock-wall of a peak. **1958** *Polar Record* IX. 91 *Bergschrund*, the crevasse which occurs at the head of a cirque or valley glacier and which separates the moving glacier ice from the rock wall and the ice apron attached to it. When the ice apron is absent the gap is known as a Randkluft. **1970** R. J. Small *Study of Landforms* xi. 380 Temperature changes such as these seem insufficient to cause powerful shattering of the rock, and in any case they are presumably still less pronounced in the crack ('randkluft') between the névé and the headwall.

Randlord (ræ·ndlǭɹd). *slang.* Also with small initial. [f. *Rand *sb.*[2] + Lord *sb.*, after Landlord *sb.*] The owner or manager of a gold-field on the Rand in South Africa.

1904 *Daily Chron.* 21 Mar. 5/5 The Randlords' proposal really drives the British workman out of the Transvaal. *Ibid.* 14 May 4/7 Recent newspaper writers, in coining the word Randlords—have they not all done what they could to popularise a wrong pronunciation of Rand? **1936** R. J. M. Goold-Adams *S. Afr. To-day & To-morrow* ii. 54 The black man may..die a premature death, but that—so the Rand-lords say—is not the fault of the mines in which he worked. **1938** J. Cary *Castle Corner* v. 255 There's no religion any more—no honesty either—it's all money—these randlords, as they call 'em. **1955** L. Hotz in Saron & Hotz *Jews in S.Afr.* xix. 357 Among the leading figures from Kimberley who quickly assumed a commanding position on the gold-fields, the so-called Randlords, were Barnato, the Joels, Alfred Beit, Albu, and Lionel Phillips.

random, *sb.*, *a.*, and *adv.* Add: **A.** *sb.* **I. 1. a.** (Later U.S. example.)

1889 'Mark Twain' *Conn. Yankee* xv. 180 Two knights came together with great random.

3. a. (Further examples. Cf. sense *B. 1 b.)

1898 W. A. Whitworth *Expectation of Parts* 7 If a magnitude *s* be divided at random into *n* parts, the expectation of each part is *s/n*. **1921** *Biometrika* XIII. 309 An event happens at random once in a period *m*, therefore its chance of occurring in an interval of time or space δ*t* is δ*t*/*m*. **1931** H. Jeffreys *Scientific Inference* iii. 24 We select at random *m* of the objects... We need a definition of what we mean by *at random*. We mean that every possible selection of *m* objects from the original *n* is equally probable. **1951** *Jrnl. Ecol.* XXXIX. 172 The principle of contagion..is that the groups are distributed at random and that the number of individuals in each group is also random.

4. Also, that which is random; random state, randomness.

1929 R. Bridges *Testament of Beauty* ii. 44 As when a high moon thru' the rifted wrack gleameth upon the random of the windswept night. **1969** *Listener* 13 Nov. 678/3 'There's a lot of random in our songs,' says Paul [McCartney].

8. *Printing.* (see quots.)

1888 C. T. Jacobi *Printers' Vocab.* 110 Random, a special frame used by compositors in making-up. **1898** J. Southward *Mod. Printing* I. iv. 29 Making-up Frames.. consist of an ordinary whole frame, fitted with a 'random'—that is, a sloping board, corresponding to a case, with ledges running along it transversely. **1910** A. Bennett *Clayhanger* I. xii. 101 Under the furniture rack was the 'random', full of galleys. **1922** W. H. Sclater *What Compositor should Know* I. 82 Randoms, on which new composition is placed for the purpose of being 'made up' into column form or page form. **1960** G. A. Glaister *Gloss. Bk.* 341/2 *Random*, the sloping work-top of a composing frame.

B. *adj.* **1. b.** *Statistics.* Governed by or involving equal chances for each of the actual or hypothetical members of a population; also, produced or obtained by a random process (and therefore completely unpredictable in detail); *random distribution*, a probability distribution, esp. the Poisson distribution; *random error*: see *Error 4 d; *random noise* (see quot. 1954); *random number*, a number selected from a given set of numbers in such a way that all the numbers in the set have the same chance of selection; also, a pseudo-random number; *random process*, (a process characterized by) a sequence of random variables (see also quot. 1937); *random sample*, a sample drawn at random from a population, each member of it having an equal or other specified chance of inclusion (sometimes contrasted with *quota sample* s.v. *Quota *sb.* 4); so *random sampling*; *random selection*, a random sample; random sampling; *random variable, variate*, a variable whose values are distributed in accordance with a probability distribution; *random walk*, the movement of something in successive steps, the direction, length, or other property of each step being governed by chance independently of preceding steps.

1898 *Phil. Trans. R. Soc.* CXCI. 230 Every artificial or even random selection of a group out of a community changes not only the amount of variation, but the amount of correlation of the organs of its members as compared with those of the primitive group. **1900** *Phil. Mag.* L. 157 (*heading*) On the criterion that a given system of deviations from the probable in the case of a correlated system of variables is such that it can be reasonably supposed to have arisen from random sampling. *Ibid.* 164 The question we wish to determine is whether the sample may be reasonably considered to represent a random system of deviations from the theoretical frequency distribution of the general population. **1905** K. Pearson in *Nature* 27 July 294/2 (*heading*) The problem of the random walk. **1924** *Bell Syst. Technical Jrnl.* III. 88 (*heading*) Deviation of random samples from average conditions. **1925** F. C. Mills *Statistical Meth.* xvi. 552 Great care is generally needed in securing a purely random selection. The obvious procedure of picking the most readily available cases would by no means meet the condition of random selection. **1927** *Tracts for Computers* xv. p. iii, In order to form this table of random numbers 40,000 digits were taken at random from census reports and combined by fours to give 10,000 numbers. *Ibid.*, These numbers, if truly random, could be used in a very great variety of ways for artificial sampling. **1933** *Forestry* VII. 149 As far as possible a random arrangement of replicated treatments and controls was adopted. **1936** *Jrnl. Ecol.* XXIV. 232 The simplest assumption, and the one most frequently made, concerning the distribution of the individuals of a plant species, is that it is random, i.e. that the chance that an individual shall occur in a given spot is the same for all spots... The chances that 0, 1, 2, 3,..individuals shall occur in a sample area large enough to contain very many individuals, are given by the terms of the Poisson series. *Ibid.* 240 Fig. 5 illustrates the divergence from random distribution. **1937** H. Cramér (*title*) Random variables and probability distributions. *Ibid.* viii. 90 The set of variables Z_τ will be said to define a homogeneous random process if, for $\tau_1 \geq 0, \tau_2 > 0$, the difference $U_{\tau_1\tau_2} = Z_{\tau_1+\tau_2} - Z_{\tau_1}$ is a random variable which is independent of the variable Z_{τ_1} and has a d[istribution] f[unction] which is independent of τ_1. **1938** *Jrnl. R. Statist. Soc.* CI. 147 In colloquial speech the word 'random' is applied to any method of choice which lacks aim or purpose; and this usage is also found in certain sciences. In statistics, however, the word has a somewhat different and more definite significance, closely related to probability. **1939** *Jrnl. Franklin Inst.* CCVII. 747 Corresponding to the constrictions in the vocal passages from which are made the unvoiced sounds the Voder contains an electrical 'random noise' source which by itself produces a continuous hissing sound. **1946** C. E. Weatherburn *First Course in Math. Statistics* ii. 21 The method thus uses the similarity of the marbles to ensure that the selection is random. **1948** *Tracts for Computers* xxv. (*title*) Random normal deviates. *Ibid.* p. iii, They may be regarded as fair random samples from a normal universe having a zero mean and a unit standard deviation. **1949**, etc. [see *Pseudorandom *a.*]. **1951** *Jrnl. Ecol.* XXXIX. 172 In a random distribution the variance is equal to the mean, and the variance divided by the mean..can be used to test departures from a random distribution. **1952**, etc. [see **quota sample*]. **1953** J. B. Carroll *Study of Language* vii. 204 Communication theory is forced to regard messages as random processes. *Ibid.* 245 We use the term *random process* in the statistical sense: a random process is the sampling at random from a population of *potential* events. **1954** T. W. Anderson in P. F. Lazarsfeld *Math. Thinking in Social Sci.* i. 35 Since each individual's sequence of opinions is a random variable (i.e., there is a probability attached to each possible sequence), the total number of individuals holding a given sequence is a random variable. **1954** L. L. Beranek *Acoustics* xiii. 393 Random noise is an acoustical quantity (*e.g.*, sound pressure) or an electrical quantity (*e.g.*, voltage) whose instantaneous amplitudes occur, as a function of time, according to a normal (Gaussian) distribution curve. A common random noise is that resulting from the random motion of molecules of the air... Random noise need not have a flat (uniform) frequency spectrum. *Ibid.*, White noise need not be random. **1959** *Oxf. Univ. Gaz.* 5 Mar. 678/1 Random variates, probability, and likelihood. **1963** B. Fozard *Instrumentation & Control of Nuclear Reactors* vii. 71 Radioactive disintegrations are truly random, *i.e.* the probability of a disintegration is independent of the occurrence of other events. **1967** C. Berners-Lee in Wills & Yearsley *Handbk. Managem. Technol.* 5 In an agricultural experiment to determine the effect of two different fertilizers, *N* and *P* for instance, the classical approach would be to compare unfertilized plots with suitably chosen random selections of plots fertilized with *N*; also to compare with the unfertilized plots a number of plots fertilized only with *P*. **1967** G. Wills in *Ibid.* 186 Random samples..ensure that if we are drawing a sample from the 27,000 paint retailers in the United Kingdom each retailer would have an equal chance of being selected. This is *not* what is generally meant in common parlance by 'random', but it is what the statistician means. **1968** P. A. P. Moran *Introd. Probability Theory* i. 6 A random variable is also sometimes called a 'variate'. *Ibid.* x. 459 The study of random walks is the study of the sums of random variables, these variables varying in complexity from simple independent distributions in one dimension to random variables of a much more complicated character. **1970** W. B. Davenport *Random Processes* ix. 299 Such an infinite family of random variables is commonly called a random sequence or random process (or stochastic sequence or stochastic process). **1971** [see **quota method*]. **1973** Lord & Robinson tr. *Kuttruff's Room Acoustics* viii. 211 The room under investigation..is excited by stationary random noise..with a large frequency bandwidth. **1973** F. E. Fischer *Fund. Statistical Concepts* vii. 139 We can think of most discrete random variables as counts (how many heads, children, spades, or accidents?) and most continuous random variables as measures (how tall, long, heavy, or intelligent?). **1975** R. B. Ellis *Statistical Inference* ii. 17 What counts is making sure that the sample really is random, that at any time any item in the population has as much chance of being chosen as any other item in the population. **1975** *Sci. Amer.* May 48/3 A series of numbers is random if the smallest algorithm capable of specifying it to a computer has about the same number of bits of information as the series itself. **1977** *Private Eye* 1 Apr. 5/2 The naive random walk theory..rules out the application of any device based upon the movement of past prices in the market. **1978** *Sci. Amer.* Apr. 71/2 One way to test for the role of chance in such a situation is to devise a Monte Carlo computer program, which generates random numbers to determine the distances and angles between consecutive directional changes of simulated tracks.

c. *Psychol.* Of activity, movements, etc.: seeming to be without purpose or direct relationship to a stimulus, sometimes thought of as an organism's initial reaction to unfamiliar stimuli, and giving way to directed action as learning takes place.

1905 *Psychol. Bull.* II. 251 Fuhrmann noted prevalence of predicative association and of the egocentric factors; moreover, especially in the beginning of tests, random association. **1911** E. L. Thorndike *Animal Intelligence* vi. 242 If the movements are really random, they occur by virtue of some force that works at random. **1927** J. Adams *Errors in School* x. 304 The knowledge the psycho-analyst acquires by random-answering is knowledge-by-the-way so far as the instructor is concerned. **1927** M. K. Thomson *Springs of Human Action* v. 74 *Undifferentiated activity* (random movement), among the native tendencies none is more primitive and fundamental than activity. **1934** Crozier & Hoagland in C. Murchison *Handbk. Gen. Exper. Psychol.* 20 It is to be noted that there is obtainable in this way a *measure* of 'random' movements and a key to their interpretation. **1935** K. Koffka *Princ. Gestalt Psychol.* xiii. 629 Trial and error may then mean that he gets a 'hunch' from the data... This would no longer be random activity, but activity determined by the nature of the task. **1935** E. L. Thorndike *Psychol. of Wants, Interests & Attitudes* ii. 13 The fact of multiple response or varied reaction..has led to the error of assuming that man at least had a tendency to make responses that were random. *Ibid.*, There doubtless is a residuum of behavior that may be called random. **1948** E. R. Hilgard *Theories of Learning* v. 116 Although it was often convenient to talk about 'random' or 'spontaneous' responses, it was not doubted that stimuli were present to elicit them.

5. *random access* (Computers): used to designate a memory or file all parts of which are directly accessible, so that it need not be read sequentially; esp. one in which the access time for any item is effectively independent of the location and the access time of the item last accessed.

1953 *Proc. IRE* XLI. 1264/2 The random-access property also makes it easier to operate input, output, and external storage devices out of synchronism with the central computer. **1967** *New Scientist* 5 Oct. 13/1 This type of memory, in which all addresses are directly accessible, is known as random-access, to distinguish it from the slower but cheaper type which is scanned sequentially, used for backing stores. **1969** P. B. Jordain *Condensed Computer Encycl.* 414 The true random-access devices are static memories: core, thin film, electrostatic cathode-ray tubes, electronic. Most of the so-called random-access devices—drum, disk, tape loop, magnetic card file..—are really cyclic access devices. **1971** *Publishers' Weekly* 9 Aug. 24/3 The advantage of random access disc storage is that all required files for a specific application will be on-line to the computer when that application is being processed. **1975** T. W. Pratt *Programming Languages* iii. 87 A random access file is organized as a set of unordered records. Access is through an address that indicates the position of the record on the external device.

C. *adv.* **2. a.** *random-blown, -fashioned, -rubbed* adjs.

1871 Tennyson in *Contemp. Rev.* XIX. 12 Tristram.. sank Down on a drift of foliage random-blown. **1906** Hardy *Dynasts* II. v. viii. 287 Ephemeral at the best all honours be,..So random-fashioned, swift, perturbable! **1862** *Illustr. Catal. Internat. Exhib., Industr. Dept., Brit. Div.* II. No. 2253 Castellated circular turret, random rubbed; white quartz.

randomly *adv.* (further examples); **randomness** (earlier and further examples); also **randomi·city** = prec.

1866 J. Venn *Logic of Chance* ii. 30 We must also idealize the 'randomness' of the throwing of the penny. **1921** J. M. Keynes *Treat. Probability* xxiv. 281 Many important differences of opinion in the treatment of probability have been due to confusion or vagueness as to what is meant by Randomness. **1936** *Rev. Sci. Instruments* VII. 459/2 Fluctuations to be expected due to the randomicity of the counting. **1938** *Nature* 14 May 881/2 Four tests for randomness..were applied to these numbers, with satisfactory results in each case. **1957** P. Greig-Smith *Quantitative Plant Ecol.* iii. 51 Departure from randomness of distribution of a species indicates that one or few factors are determining the performance or survival of the species. **1959** *New Scientist* Dec. 1144/3 An interesting point of this arrangement is that the lengths of the delays can be chosen randomly. **1963** H. M. Morris *Twilight of Evolution* ii. 44 The natural tendency of all change is to create a greater degree of disorder and randomness. **1963** T. & P. Morris *Pentonville* iv. 75 Half the men in each group were randomly selected on the basis of throwing dice. **1972** *Science* 27 Oct. 392/3 Environmental randomicity of various kinds may have important implications in triggering evolutionary sequences that would be impossible or unlikely in non-stochastic environments. **1980** *Jrnl. R. Soc. Arts* Feb. 138/1 The cosmos will end in heat death with all matter randomly diffused.

random (ræ·ndəm), *v.* *rare.* [f. the sb.] *intr.* To do something at random, to occur at random.

1889 'Mark Twain' *Conn. Yankee* xxvii. 349 A thought came randoming overthwart this majestic dream. **1921** R. Frost *Let.* 15 Apr. (1964) 127 She wasn't experimenting, poor thing. She was randoming, as Alisande hath it.

randomization (ræ:ndəməizē[i]·ʃən). [f. next + -ation.] The action, process, or result of randomizing.

1926 *Jrnl. Min. Agric.* XXXIII. 510 A process of randomization by which one is selected at random out of the total number of Latin Squares possible. **1934** *Math. Gaz.* XVIII. 294 The process of randomisation has in recent years come to play such a central part in experimental design that it is of some interest to find that it affords a means of resolving one of the oldest paradoxes which arose in discussions of gaming. **1946** G. W. Snedecor *Statistical Methods* (ed. 4) 253 The degree of randomization is no longer evident, but the weekly heights and the block differences are brought into prominence. **1960** *Daily Tel.* 23 Aug. 16/5 Because the experiments were made over five years, using a 'randomisation' method, the Australian results were considered by other experts at the conference to be the first to be absolutely conclusive. 'Randomisation' means that clouds are selected in a completely random manner. **1966** *Lancet* 24 Dec. 1371/2 We tested the randomness of these four groups; except for age-distribution, randomisation was satisfactory. **1980** *Times Lit. Suppl.* 15 Feb. 160 A 'genuinely open society' is, presumably, one in which relative mobility chances have been equalized. But how?.. By the randomization of access to the labour market?

randomize (ræ·ndəməiz), *v.* [f. Random *sb.*, *a.*, and *adv.* + -ize.] *trans.* To render unpredictable, unsystematic, or random in order or arrangement; to employ random selection or sampling in (an experiment or procedure).

1926 *Jrnl. Min. Agric.* XXXIII. 509 The distinction between errors eliminated in the field, and the errors which are to be carefully randomized in order to provide a valid estimate of the errors which cannot be eliminated, may be made most clear by one of the most useful and flexible types of arrangement, namely, the arrangement in 'randomized blocks'. **1939** H. Jeffreys *Theory of Probability* iv. 192 Most physicists, of course, will envy workers in subjects where uninteresting systematic effects can be randomized. **1958** *New Scientist* 21 Aug. 659/3 Having produced a beam of the required energy, the Oak Ridge group then try to randomise it by firing it into a magnetic bottle produced by two mirror coils. **1959** *Test House Rep.* No. OTL/M1/5/59 (Inspectorate of Fighting Vehicles & Mech. Equipm. Test House. Oil Test Lab.) 4 A programme of two tests on each of the seven oils was arranged and the tests randomised to eliminate bias. **1967** *Listener* 26 Jan. 134/2 A number of years ago Sir Ronald Fisher made a statistical analysis of the problem and came to the unexpected conclusion that the best rule in each case was not to have any fixed rule at all, but to randomize one's actions by an appeal to chance. **1971** *Jrnl. Gen. Psychol.* LXXXV. 172 The order of these pronouns was randomized. **1977** *Sci. Amer.* Sept. 42/2 Assume that the deck of cards is finite, so that an umpire can indeed randomize them.

Hence **ra·ndomized** *ppl. a.*, chosen at random; deliberately made random or unpredictable; **randomizing** *ppl. a.*, generating random output; **randomizing** *vbl. sb.*, the act of rendering random.

1926 Randomized [see *Randomize *v.*] **1936** *Nature* 15 Feb. 253/1 It is with applications of the principle of testing null hypotheses by means of randomised and replicated experiments that the first nine chapters of this book deal. **1936** *Jrnl. R. Statist. Soc.* Suppl. No. 3. 118 The tendency of deliberate randomizing is to increase the error. **1938** *Nature* 14 May 881/1 M. G. Kendall and B. B. Smith have designed a randomizing machine. **1955** *Sci. Amer.* Feb. 81/2 The idea is to let chance play a role in the choice of strategy, that is, to use a randomized or 'mixed' strategy. **1962** *Times Lit. Suppl.* 2 Feb. 73/3 To randomized numbers where no fancy boggles The Great Boondogglers go. **1968** P. A. P. Moran *Introd. Probability Theory* i. 44 Decimal digits which were obtained by a 'randomizing machine'. **1970** *Nature* 12 Dec. 1113/1 Each block of twenty trials contained all possible combinations of test stimuli presented in randomized sequence. **1978** *Lancet* 14 Jan. 72/2 The claim that an upright maternal posture during labour improves the efficiency of the uterus to the benefit of both mother and fetus has been investigated in a randomised prospective study.

randomizer (ræ·ndəməizəɹ). [f. prec. + -er[1].] A device which generates random output.

1974 *Sci. Amer.* Apr. 112/2 It is very easy to use a penny as a randomizer for deciding between two alternatives with probabilities expressed by rational fractions. **1976** *Times*

Lit. Suppl. 13 Feb. 172/4 Chickens..which appear capable of influencing mentally an electronic randomizer controlling the switching mechanism of a lamp.

|| **randori** (randō·ri). [Jap., lit. 'informal practice'.] A (session of) informal practice in Judo.
1913 E. J. Harrison *Fighting Spirit of Japan* iv. 65 The non-esoteric branches of judo are called *randori*, in which the pupil freely applies his knowledge in open practice.. with others. **1932** —— *Art of Ju-Jitsu* iii. 30 It is not absolutely necessary for the pupil to master all these details ..before beginning '*randori*' practice. **1954** E. Dominy *Teach Yourself Judo* vi. 63 In 'randori' which is free practice you never find an opponent who is willing to lie passively on his back. **1972** *Oxf. Mail* 1 Aug. 10/3 This initial practice very soon develops into 'randori', the form of training in which most players spend most of their judo time. This involves moving about with a partner.

randy, *a.* and *sb.* Add: **A.** *adj.* **2. b.** Delete *dial.* and add further examples.
c **1888–94** *My Secret Life* III. 280 She'll be randy directly her belly is filled. **1922** [see **pussy sb.* 5*]. **1939** J. Steinbeck *Grapes of Wrath* vi. 69 First time I ever laid with a girl..snortin' like a buck-deer, randy as a billygoat. **1957** W. Camp *Prospects of Love* II. v. 62 Suffers from too much sex, if anything—he's a randy old man. **1965** F. Sargeson *Mem. Peon* iv. 87, I was randy myself at your age. But be careful. These native girls can put you right into hospital if you don't take care. **1978** K. J. Dover *Greek Homosexuality* ii. 58 The gangs or clubs of randy and combative young men.

3. *Comb.*, as (sense 2 b) *randy-arsed* adj.; also with sbs. forming attrib. compounds, as *randy-dog*.
1968 Randy-arsed [see **length sb.* 11 c]. **1963** *Times Lit. Suppl.* 18 Jan. 37/4 Harold Barlow is an Amis character ..with that special randy-dog flavour. **1973** M. Amis *Rachel Papers* 173 Tom, Geoffrey's analogue of my own Sebastian; sixteen, wealthy in pustules, randy-dog smells, sebum-moist hairline, and other adolescentiana.
Hence as *v. trans.*, to render (a person) lascivious (*nonce-use*); **ra·ndiness**, the quality or condition of being randy (sense 2 b); lustfulness.
1911 *Conc. Oxf. Dict.*, Randiness. **1953** W. Cooper *Ever-Interesting Topic* 145 Attending De Foy's series of lectures on sex was indicating in the boys a distinctly higher-than-usual state of—there is no other word for it—randiness. **1967** A. Wilson *No Laughing Matter* iii. 303 This bloody randiness always threatened to suck him down into the vast, empty emotional gulf of Ted's shapeless life. **1976** W. Greatorex *Crossover* 103 He didn't like randy old sods. He thought randiness should be forbidden after the age of, say, forty. **1961** A. Wilson *Old Men at Zoo* v. 278 You've randied him into the looney bin now..the highest-minded little whore that ever almost gave herself out of chastity.

randy, *sb.*² Add: (Further example.) Also *on the* (or *a*) *randy*, 'on the spree'.
1877 E. Peacock *Gloss. Words Manley & Corringham, Lincolnshire* 202/1 Bill's on the randy to-day. **1934** L. MacNeice *Poems* (1935) 18 Over the randy of the theatre and cinema I hear songs. **1940** Dylan Thomas *Portrait of Artist as Young Dog* 67 'Hush! hush! your mother'll be waiting. You must come home.' 'No she won't.'–She's gone on a randy with Mr Robert.'

ra·ndy-da·ndy. Redupl. form of Randy *sb.*²
1927 J. M. Barrie *Old Lady shows her Medals* 34, I have a theatre tonight, followed by a randy-dandy.

Raney (rēⁱ·ni). *Chem.* The name of Murray Raney (1885–1966), U.S. engineer, used *attrib.* in **Raney nickel**, a form of nickel catalyst first prepared by him (see quot. 1932) which has a high surface area and is used in organic hydrogenation reactions.
First described by Raney in U.S. Patent 1,628,190 (1927).
1932 *Jrnl. Amer. Chem. Soc.* LIV. 4116 The Raney catalyst is prepared by alloying equal parts of nickel and aluminum and dissolving out the latter with aqueous sodium hydroxide. **1940** *Ibid.* LXII. 1687/1 The catalytic hydrogenation of azo compounds at normal temperature and pressure over Raney nickel is not only of scientific interest, but also of considerable practical value. **1964** G. H. Haggis et al. *Introd. Molecular Biol.* xii. 325 The cystine was then converted to alanine with the aid of a catalyst called Raney nickel. **1967** L. F. & M. Fieser *Reagents for Org. Synthesis* I. 54 The corresponding 1,4-diketones can be obtained by hydrolysis catalyzed by mercuric chloride or by treatment with Raney nickel.

|| **rangatira** (raŋātī·rā). *N.Z.* Also *erron.* **rangitira.** [Maori.] A Maori chief (male or female), a noble.
1820 *Gram. & Vocab. Lang. N.Z.* (Church Missionary Soc.) 200 *Rănga tira*, a gentleman or lady. **1844** H. Williams *Let.* 27 Jan. in H. Carleton *Life of Henry Williams* (1874) I. 35 We told him that *rangatiras* [gentlemen] do not steal. **1843** E. Dieffenbach *Travels in N.Z.* II. I. vii. 112 The principal person in a tribe is the Ariki; but as he is *per se* a Rangatira, he is rarely called by the former name. **1855** V. Lush *Jrnl.* 23 Sept. (1971) 165 This addition to the Native Ministry is a fine, tall and highly intelligent man, and being by birth a Rangatira (a Chief) he has far, far more influence among his countrymen than Rota has, who formerly was but a slave. **1863** F. E. Maning *Old N.Z.* i. 6 The chief..having made some inquiries..such as, whether I was a *rangatira*. **1882** W. D. Hay *Brighter Britain!* I. 253 The caste styled tana, or chieftains, a degree above that of rangatira, or simple gentlemen-warriors. **1903** *Daily Chron.* 21 Sept. 5/1 This lad, Victor Huia, was born in New Zealand thirteen years ago, and by the Maoris was formally created a native chief, or rangatira, being duly decorated with the sacred feather of the huia. **1936** R. Hyde *Check to your King* 146 Maori rangatira..promise that in future jury trial shall be observed. **1937** N. Marsh *Vintage Murder* xxii. 242 My grandfather was a deeply-instructed rangitira. **1943** — *Colour Scheme* ii. 37, I am a *rangitira*. My father attended an ancient school of learning. He was a *tohunga*. **1962** M. K. Joseph *Pound of Saffron* iv. 65 I'd like to see old Blennerhassett in a flax mat—some rangatira he'd make. **1967** A. & R. Reed *Paddle Wheels on Wanganui* 69 According to local history a rangatira had a beautiful greenstone mere..buried in a certain upriver area. **1978** B. Mason in *Islands* Aug. 18 In the case of our people, he would have been a splendid rangatira, a glory to his iwi.

range, *sb.*¹ Add: **I. 2. c.** (*Austral.* examples in *sing.* and *pl.* of sense 'hill' or 'mountain'.)
1839 J. Mossnett in J. Stephens *Land of Promise* ii. 17 We passed the range at the point where the shingle-splitters have their settlement. **1846** F. Dutton *S. Australia & its Mines* xii. 297 The Ranges, immediately at the back of Adelaide, are at present the principal locality where this ore has been met with in great abundance. **1864** J. Rogers *New Rush* 31 How merrily passes the digger's life..To live in the ranges free from strife. **1901** M. Franklin *My Brilliant Career* v. 31 The further he got into the ranges the more ... **1946** K. Tennant *Lost Haven* (1947) 6 Alec's father..set up a farm for himself far enough off in the ranges to make the ... **1966** 'J. Hackston' *Father clears Out* 11 There's a big fire burning off in the ranges. *Ibid.*, It's over the other side.

d. For *U.S.* read *N. Amer.* (Earlier and further examples.)
Ranges were established by the U.S. Congress on 20 May 1785.
1785 *Jrnls. Continental Congress U.S.* (1933) XXVIII. 376 The geographer shall designate the townships..by numbers progressively from south to north; always beginning each range with number one. **1790** *Deb. Congress U.S.* 27 Dec. (1834) 1834 Mr. Clymer wished to know how much land these seven ranges included. **1811** R. Sutcliff *Trav.* 28 Nov. in *Trav. N. Amer.* (1811) ... Ranges ... **1817** J. M. Peck *Gazetteer Illinois* (ed. 2) i. 76 In numbering the townships east or west from a principal meridian they are called ranges, meaning a range of townships. **1960** Davies & Vaughan *Beyond Old Bone Trail* iii. 23 The land had been split up into townships, ranges, sections, and quarter-sections. Townships and ranges were six miles square. **1977** *Chicago Tribune* 2 Oct. II. 14/6 (Advt.), Section 23 in Chapel Hills Garden Farm area of the North West quarter of Section 21, Township 37 North, Range 13, East of the Third Principal Meridian.

II. 6. a. For '*U.S.*' read 'orig. *U.S.*'. (Further examples.)
1640 *Essex Inst. Hist. Coll.* (1863) V. 170/1 The range of the cattle at the forest river head. **1851** S. Stephens *Jrnl.* 18 Apr. (typescript) II. 50 There is plenty of grass for [stock] here. [*sc.* Waitaia, N.Z.] and an extensive range, more suited for growing young stock..than the more limited boundary of the farm. **1921** *Daily Colonist* (Victoria, B.C.) 6 Apr. 8/3 Here they were able to learn something about a country where the sheep growers are able to raise their herds on expansive ranges. **1946** *Richmond* (Va.) *Times-Dispatch* 24 Nov. ... Turkeys in the Valley of Virginia generally are raised 'on the range', which means that after they're nine weeks..old, in the Summer they're put out in fenced fields with a few shelters or roosts and plenty of room to wander about. **1949** E. Nelson *Backwoods Teacher* iii. 29 He..steals maggie off the range for winter meat. **1963** R. Symons *Many Trails* v. 52 When a rancher spoke in his 'ranges', he meant the natural grazing area around his buildings.

b. For '*U.S.*' read 'orig. *U.S.*' (Earlier and later examples.)
1626 in *Virginia Mag. Hist. & Biogr.* (1894) II. 52, 300,000 acres of land, which will feed such numbers of people, with plentiful range for Cattle. **1959** *Times* 18 May 10/4 There are plenty of range rearers who annually take their pullets off range into winter laying houses without trouble. **1961** *Guardian* 17 May 3/1 There was insufficient land in Britain to keep sufficient birds on range. **1970** 19 *Farmhouse Gaz.* ... There are even 1,000 pullets of ... 1971 ... 13/1 After eighteen years of battery production it was continue to lay for a further two or three years.

7. (Further examples.)
1966 A. Murray *Geogr. Distrib. Animals* ii. 29 Even birds are subject to the same law, although it cannot be expected that rivers should often limit their range. **1908** D. B. Jackson *Ocean Biol. Terms* 221/1 *Range*, the region over which a given species ... grows spontaneously. **1951** *Jrnl. Ecol.* XXXIX. 205 By 'range' is intended the geographical area within which, at a given time, a taxon or a plant community is to be found. **1965** Davies & Heywood *Princ. Angiosperm Taxonomy* xiii. 425 Species tend to occur in more restricted, and often more extreme, habitats at the edge of their range than in the centre of it. **1975** P. A. Johnsgard *Waterfowl N. Amer.* ii. 346 No specific information on home ranges of the greater scaup is available. **1980** *Nature* 3 Jan. 15/2 The recurring theme was the problem of identifying the stresses and stratagems of animals exposed in different parts of their range to widely differing environmental conditions.

8. a. (Further examples.)
1942 [see **level sb.* 4 c]. **1966** *Oxf. Univ. Gaz.* 23 Dec. 430/1 Important general questions of constitutional arrangements..down to those individual points with which you are familiar... That is the range we have at present.

10. b. (Further examples.)
1892 *Photogr. Ann.* II. 549 Three sets..of apparatus which will prove..to give a range for demonstrating purposes that will embrace most experiments that are required. **1921** *Glasgow Herald* 7 Nov. 11 Manufacturers were called upon to make far too many patterns. In preparing their ranges for the particular season, manufacturers are guided largely by the experience of the seasons which have just gone. **1967** E. Short *Embroidery & Fabric Collage* ii. 45 Today there is a wide range of beads and sequins on the market.

c. *Statistics.* The difference observed in any sample between the largest and the smallest values of a given variate.
1911 G. U. Yule *Introd. Theory Statistics* viii. 133 The simplest possible measure of the dispersion of a series of values of a variable is the actual range, i.e. the difference between the greatest and least values observed... The range is..subject to meaningless fluctuations. **1948** G. Herdan *Quality Control* ii. 23 The range being the difference between the greatest and smallest dimensions in one sample of *n* specimens. **1974** A. K. S. Jardine et al. *Statistical Methods for Quality Control* ii. 15, I. Range = class mark of highest class – class mark of lowest class; or 2. Range = upper class limit of highest class – lower class limit of lowest class... It can be seen that range is not uniquely defined in such cases. **1978** *Nature* 3 Aug. 490/2 Seven normal volunteers (four males, three females; mean age 25 yr, range 21–30) were studied.

d. *Math.* The set of values that the dependent variable of a function can take; the set comprising all the second elements of the ordered pairs constituting some given set.
1914 [see **domain sb.* 4 c]. **1959** J. G. Kemeny et al. *Finite Math. Structures* ii. 70 Let *f*(*x*) be the set of *x*, expressed to the nearest year. The range of *f* consists of a set of whole numbers, starting with 0, apparently including all integers up to 100, and even having a few integers above 100 in the set. **1968** E. T. Copson *Metric Spaces* i. 19 The set of all points..which are images of points of *E* is called the range of the mapping. **1977** C. R. Allendoerfer et al. *Elem. Functions* iii. 49 The values of the dependent variable constitute a subset of the range called the range of the function. *Ibid.* 50 It is often difficult to determine the range.

11. a. More widely, the distance anything can travel, as (*a*) *Nucl. Physics*, the maximum distance which an ionizing particle of a given energy can travel in a given medium; (*b*) the maximum distance at which a radio transmission may reliably be received; (*c*) the distance which an aircraft can travel without refuelling, normally under stated assumptions regarding factors such as speed, air speed, and altitude; (*d*) the distance on the earth's surface which a rocket or missile can traverse from launch to landing.
1904 *Phil. Mag.* VIII. 725 The first breakdown of the radium atom is responsible for the *a* particle of..the least range. **1906** G. Eichhorn *Wireless Telegr.* vi. 37 Bearing in mind..the enfeebling influence of obstructions and the curvature of the earth, the range of normal installations is reduced to about 300 miles. **1924** *Harmsworth's Wireless Encycl.* 1633/1 It is a common experience that sets of only one and a half kilowatts to transmit over a range of 1,000 miles. **1926** R. W. Lawson tr. *Hevesy's Man. Radio-activity* vii. 78 Fig. 25..shows the tracks of individual *a*-particles. Almost without exception they are rectilinear, and the ionization produced by the rays ceases quite suddenly, which indicates that they have a definite range. **1938** *N.Y. Herald Aircraft* xix. 818 *Range at economical speed*, the maximum distance a given aircraft can cover while cruising at the most economical speed and altitude at all stages of the flight. *Range at full speed*, the maximum distance a given aircraft can cover at full speed at sea level. **1947** C. F. Toms *Introd. Aeronautics* iii. 141 The range.. depends on the air-speed, on the initial all-up weight.., and ..on the bomb-load. **1947** *Jrnl. Inst. Elect. Engineers* XCIV. i. 176/1 On the Plan Position Indicator these objects will appear in their correct relative position provided we correct for the fact that radar measures slant range and not plan range. **1949** G. P. Sutton *Rocket Propulsion Elements* viii. 246 Altitudes above 250 miles and ranges over 300 miles can generally be attained only with single-step missiles having small payloads or multiple-step missiles. **1963** [see sense 11 c below]. **1963** Stuhlinger & Messmer *Space Sci. & Engin.* xvi. 191 The range of a particle can be obtained directly by integrating the reciprocal of the stopping power over the appropriate energy range. **1967** Ketcham & Martin *Propulsion* iv. 197 It is possible to make an approximation for the maximum range of a ballistic missile from the geometry of the elliptical flight path and the intersections with the spherical earth surface. **1968** Sands & Tellet *VHF-FM Marine Radio* i. 21 The range of VHF transmissions is limited to a little more than line-of-sight distance. **1970** N. Smith *Aviation Fuels* xxxvii. 256 When boiling occurs, the loss of fuel can be serious both from the viewpoint of lost range and from danger to the aircraft structure from excessive tank pressures.

c. (Earlier and later examples.) Also, a strip of land or sea used for testing rockets or missiles in flight between their launch and return to earth.
1862 *St. Andrews Gaz.* 3 Oct. 3/4 At the rifle range, the corps was divided into two squads, the one party firing against the other. **1900** Kipling *Let.* 24 July in C. Carrington R. *Kipling* (1955) xii. 315 We've started a rifle-club in the village... We've got a 1,000 yards range among the downs. **1947** *Jrnl. Brit. Interplanetary Soc.* VI. 191 Some details have appeared of an American rocket-testing range comparable to the projected Anglo-Australian one. **1955** *Times* 19 Aug. 4/6 The crofters last night unanimously agreed to a six-point resolution protesting that the range, which is expected to absorb crofting land in Benbecula and North and South Uist, represented a threat to the Hebridean way of life. **1965** L. E. Foster *Telemetry Systems* vii. 284 The range of these vehicles requires an accurate real-time 'picture' of the vehicle position to assure that it does not go beyond the safety corridor of the range and impact on some populated territory. **1971** Green & Lomask *Vanguard* v. 133 In the fall of 1955 the 15,000-acre missile firing range on the snake-infested and palmetto-covered sand dunes of

Florida flatlands was completing its sixth year as the..Proving Ground for American guided-missile development.

III. 12. a. Also, a gas or electric cooker, typically with a grill, ring burners or plates, and one or more ovens. Now chiefly *U.S.*
1895 *Montgomery Ward Catal.* Spring & Summer 413/2 Gas ranges... This range..has four top burners..and is fitted with movable ovens. **1908** *Sears, Roebuck Catal.* 529/2 Ranges..gas..oil. **1929** *T. Eaton & Co. Catal.* Spring & Summer 416/1 Acme Electric Range... One of the most efficient cooking stoves ever made—with a..aluminum cooking burners; ..fuel saving oven; sanitary high burner tray [porcelain]; choice of closed or open top cooking surface; Astogril broiler in rollout drawer. **1959** *Sears, Roebuck Catal.* Spring & Summer 931/2 Our best Kenmore kerosene-burning range. **1970** *Washington Post* 30 Sept. B13/4 (Advt.), We have everything you'll need to transform your kitchen..including the fantastic Modern Maid flameless electric range with all the newest, most wanted features. **1973** *Sunday Express* (Trinidad) (Suppl.) 1 Apr. 2/5 Today's squared off ranges are designed for a close fit. **1977** *Sci. Amer.* Dec. 51/2 Between them, in ascending order, are a washing machine, a dishwasher, a color television set, a freestanding electric range, a gas clothes dryer, a freestanding gas range, an electric clothes dryer and a refrigerator.

14. e. (Earlier example.)
1798 *Maryland Hist. Mag.* (1931) XVIII. 216, 20 tables Crown glass cutt into Ranges 7 inches high.

h. (See quot.)
1923 *Daily Mail* 28 Apr. 8 Eighty ranges, the young wood of ten acres—a range consists of all but the grown timber of twenty-one years' growth.

IV. 16. a. *range boss, -land, management, rider*; **range egg**, an egg laid by a hen which has ranged outdoors for its food; **range war** *U.S.*, a struggle for the control and use of a cattle or sheep range.
1893 W. L. Cottenden *Ranch Verses* 94 The range boss's outfit rides in through the herd. **1912** *Short Stories* Feb. (early issue) 70/2 He dominates everybody but Ben Whitman and.. dad's range-boss. **1963** *Punch* 19 June 891/1 We..keep hens and if they pack up buy 'range' eggs. **1931** *Sun* (Baltimore) 23 Dec. 15/1 Ranchers in the district are unable to care for them [*sc.* straying horses] and the rangeland is covered with snow. **1949** *Daily Progress* (Charlottesville, Va.) 26 Jan. 2/3 The winter snows have..laid the foundation for a good growth of rangeland grasses next summer. **1958** *Yearbk. Agric.* 1957 (U.S. Dept. Agric.) 765/2 *Range* (or *rangeland*), land that produces primarily native forage plants suitable for grazing by livestock, including land that has some forest trees. **1961** *Times* 7 Feb. 3/1, 40 million acres of rangeland north of Brisbane. **1969** *Science & Technology* Jan. 49/2 Rangelands is an American term now used throughout the world to describe areas where rainfall is too low or too unreliable for crops or sown pastures. **1976** *Billings* (Montana) *Gaz.* 11 July 1-e/5 The fires apparently were started by lightning and began on the rangeland some miles south of Interstate 90 on the Big Horn–Yellowstone County line. **1979** *Arizona Daily Star* 5 Aug. 44/5 Four fires..destroyed more than 3,700 acres of timber and range land in western and central Oregon. **1972** *Agric. Handbk.* (U.S. Dept. Agric.) No. 435. 1/1 Values for each of the range management practices were determined from published reports and from experience of USDA Forest Service experts in this field. **1979** *Arizona Daily Star* 1 Apr. (Advt. Section) 3/5 Natural resource management; a years experience, preferably in range management. **1890** *Stock Grower & Farmer* 28 June 4/4 A few years more will see all the last of the range-rider. **1912** C. Mulford *Buck Peters* (1914) xvi. 167 Sheep [?] ... **1966** 'O. Henry' *Roads of Destiny* xvii. 260 ... **1972** G. Burton *Stone* xiv. 119 ... range war in the Wild West.

b. **range-indicator**; *spec.* in radar, as *range* (*-amplitude*) *display*, *gate*, *mark*, *-marker*, *measurement*, *resolution*, *ring*, *step*.
1946 *Jrnl. Inst. Electr. Engin.* XCIII. IIIa. 1559/1 The Range or Type A display is one of the basic forms of presentation of radar information. **1948** K. Ulleytt *How Radar Works* vii. 101 Range-amplitude, or type A, display on the CRT of a radar receiver would be of very little use with many modern systems, as the information it gives is not sufficient by inspection. **1962** G. J. Wheeler *Radar Fund.* v. 55 A range gate, a sweep of short duration..is generated..and is initiated by the range-finding circuits... **1973** Meyer & Mayer *Radar Target Detection* ii. 29/2 The majority of operational radar detection devices have some collapsing loss. The loss can be caused by..improper range gate width. **1977** *Electronics Lett.* XIII. 416/1 The range gate first resets the tunnel diode D, and any subsequent transition of D which is within the range gate produces a 1 output of the AND gate. **1916** 'Boyd Cable' *Action Front* 131 When the range-indicator told that it was within reach of their shells the first gun opened with a ranging shell. **1949** J. L. Hornung *Radar Primer* vii. 69 A simple kind of range-indicator consists of a cathode-ray tube with a time-base voltage applied to the left and right deflection plates and echo-pulse voltages connected to the top and bottom plates. **1945** *Electronic Industries* Sept. 222/3 *Range mark*, a mark on the CRT screen which indicates distances from the radar set of the various echoes appearing on the screen of the CRT. **1949** H. E. Penrose *Princ. & Pract. Radar* xv. 281 In the simplest calibrating system a generator (range mark generator)..produces at equal intervals of time, short sharp signals or pips..., and these are imposed upon the trace just as though they were incoming signals. **1944** *Princ. Radar* (Mass. Inst. Technol. Radar School) ii. 3 A very common method [of range determination] for search-type sets is the use of range markers generated in the timer. **1977** J. French *Small Craft Radar* ii. 69 Two types of range markers are in common use, the variable range marker giving only one ring which can be set by the operator, and fixed rings which appear at fixed distances usually preset by the Range switch. **1949** H. E. Penrose *Princ. & Pract. Radar* i. 1 When velocity is a known constant and time can be measured, distance can be calculated = velocity × time. This is the basis of range measurement. **1978** R. V. Jones *Most Secret War* xxi. 177 Only later did I find that there was no foundation to our original reasoning that the Y system was based on a range measurement. **1959** R. L. Shrader *Electronic Communication* xxix. 833 Range pulses are applied to the P.P.I. to indicate two or more targets in the same direction but at different distances. **1970** G. Kennedy *Electronic Communication Systems* xv. 625 Another important requirement of short pulses is that they improve the range resolution, which is the ability to separate targets whose distance from the transmitter differs only slightly. **1965** D. G. Lane *Marine Radar* vii. 80 Range rings can be applied to the P.P.I. [*sc.* plan position indicator], but as the P.P.I. is intensity modulated and the trace rotates around the screen's centre, the pulses appear as rings encircling the centre of the screen... Fixed rings are known as calibration or range rings, and a variable ring is known as a variable-range marker. **1977** J. French *Small Craft Radar* ii. 71 To adjust the range.. this circuit is added to the video amplifier, often via a Brightness control or on/off switch so that the operator can check for the possible presence of echoes under the range rings. **1946** *Princ. Radar* (Mass. Inst. Technol. Radar School) (ed. 2) i. 23 Higher precision with a Type A indicator may be attained by the addition of a range step. The horizontal sweep is interrupted very rapidly, producing the effect of a step in the sweep.

d. **range beacon** = *radio range* (*b*) s.v. **radio sb.* 7; **range-finder**, a device used for the estimation of the distance separating the observer from an object; *spec.* (*a*) *Mil.*, usu. attached to a weapon, to estimate the range of a target; (*b*) *Photogr.*, as an aid to focusing a camera, freq. coupled to the lens (see quot. 1958); so **range-finding** *vbl. sb.* (also *transf.*); **range-plate**, a ring-burner inset on the surface of a gas or electric range; **range-proof** *a.* (*U.S.*), unbreakable by heat from a ring burner on a gas or electric range; **range safety crew**, a team of persons responsible for ensuring the safety of people and structures should a rocket deviate from its intended path after launching by destroying it in flight if necessary; **range safety officer**, the principal member of a range safety crew; **range-zone** *Palæont.* (see quot.).
1931 P. V. H. Weems *Air Navigation* xiv. 271 Range beacons..with a Type A Indicator..which operate an ordinary aural receiving set with head phones, or of the visual type which operate vibrating reeds in a special visual indicator. **1935** C. G. Burge *Compl. Bk. Aviation* 503/1 The system.. employs a number of beacons whereby the pilot approaching the aerodrome along the route marked by range beacons is advised of the locality of the 'drome by first an approach marker beacon and secondly by a boundary marker beacon. **1940** A. Black *Story of Flying* xvi. 183 Each range-beacon signal is interrupted every 30 seconds by an identification signal. **1872** [Range-finder] [in Dict., sense 16 b]. **1916** 'Boyd Cable' *Action Front* 38 He..was..followed by his trumpeter and a man with the six-foot tube of a range-finder strapped to the saddle. **1930** *Pop. Sci. Monthly* Dec. 53/3 A new 'gravity range finder' that clips to the camera and eliminates guesswork in judging the distance. **1934** [see **coupled a.* 2 e]. **1958** *Oxf. Mail* 19 May 7/2 *Coupled rangefinder*. This means a rangefinder which you work usually by making two images coincide. It is coupled to the lens, so that when you have focussed the image on the rangefinder the lens is automatically focussed. **1973** *Times* 5 Oct. 16/3 Exactly what the cost of the new tank will be is anyone's guess. One estimate is that it might cost as much as £500,000 by the time one has installed the £50,000 fire control system with its laser range-finder and computer. **1977** J. Hedgecoe *Photographer's Handbk.* 13/1 The range-finder focusing system uses two windows,..one of the actual viewfinder and another placed further along the camera body. *Ibid.* 15/1 The Leica CL..is a sophisticated direct vision 35 mm camera with coupled rangefinder. **1890** *Range-finding* [in Dict., sense 16 b]. **1946** *Jrnl. Inst. Electr. Engineers* XCIII. I. 378/2 The corresponding advance in precision of range-finding was achieved by Pollard in his ..anti-aircraft artillery. **1971** *Nature* 16 Apr. 460/2 Range-finding experiments have shown that the maximal tolerated doses of propylene imine and propane sultone, in distilled water,..were 10 mg/kg and 56 mg/kg, respectively. **1954** *Archit. Rev.* CXVI. 270 72a 'Main' gas rangeplate in cast iron and pressed steel, finished in cream vitreous enamel. **1969** *New Yorker* 29 Nov. 109/1 The ware is ovenproof but not rangeproof. **1966** *Electronics* 17 Oct. 37/1 Range safety crews must monitor the handling and launch of a rocket on the ground, and then watch such flight parameters as critical velocity, position and impact prediction for signs of danger. **1965** L. E. Foster *Telemetry Systems* vii. 284 During the transition from aerodynamic to ballistic missiles, the range safety officer turned to the development of an impact predictor. **1971** Green & Lomask *Vanguard* ix. 162 The displays in the Central Control room gave the range safety officer a second-by-second picture of the path the vehicle was following. **1908** *Daily Chron.* 16 Apr. 4/6 Some 200 of the London Scottish will be quartered at Aldershot for range work and field exercises, their sergeants sharing the mess of the Gordon Highlanders. **1957** E. D. McKee et al. in *Bull. Amer. Assoc. Petroleum Geologists* XLI. 1880 Vertical and horizontal limitations in the absolute three dimensional distribution of individual taxonomic entities in the rocks of the earth's crust provide the basis for biostratigraphic subdivision of strata into range-zones or zones comprising the total body of strata through which specimens of a particular taxonomic entity..range or occur. **1976** H. D. Hedberg *Internat. Stratigr. Guide* vi. 53 A biostratigraphic range-zone may represent the stratigraphic range of some one taxonomic unit.., or of a grouping of taxons, or of a lineage or segment of a lineage, or of any particular paleontological feature whatsoever.

range, *v.*¹ Add: **I. 4.** (Further example.)
1903 M. Beerbohm in *Sat. Rev.* 16 May 615/2 Suppose that, when he disembarked at S. Helena, Napoleon so 'ranged himself' as to become a gentle, agreeable,..old gentleman.

II. 6. a. (Further example.)
1877 G. H. Lewes *Let.* 27 Feb. in *Geo. Eliot Lett.* (1956) VI. 345 When do you think we ought to issue the 7/6 edition? It would of course be made to range with the edition I speak of.

b. Also *trans.*
1841 R. H. Dana *Seaman's Man.* 79 She [*sc.* a ship] may be ranged a little ahead, or deadened, by filling or backing the cross-jack yards.

IV. 11. c. To cast (one's eyes) over a series of objects.
1862 Mrs. H. Wood *Channings* I. i. 11 The master ranged his eyes round the circle.

rangé, *a.* Add: **1.** (later example.)
1894 Gough & Parker *Gloss. Heraldry* 489 *Rangé*, (fr.): arranged in a line.

2. Domesticated, orderly, regular, settled. Also in fem. form **rangée**.
1893 A. Atkinson *Let.* 6 Feb. in F. Stark *Traveller's Prelude* (1950) ii. 29 He is stolider than ever and getting fat—looked terribly rangé and bourgeois. **1906** W. De Morgan *Joseph Vance* xviii. 149 It [*sc.* a public house] is still the George the Fourth, but the gas-jets no longer sow wild oats of lamp-black—they are *rangés*. **1947** *Observer* 17 July 14/3 Beckmesser (Bruce Anderson) displayed a fine voice,..and was amazingly quick on his legs for a tolerably *rangé* mastersinger. **1931** *Times Lit. Suppl.* 29 Oct. 838/2 Experiences that left her mature, to enter possibly, on the next stage of her life definitely *rangée*. **1934** 'A. Bridge' *Ginger Griffin* iv. 46 It's possible for girls to be too *rangées*. **1935** *Essays & Studies* XX. 117 The best printers, whose spelling was far more *rangé* than that of even the most scholarly and learned writers. **1941** 'R. West' *Black Lamb & Grey Falcon* (1942) I. 195 That bias would make it very difficult for Slavs ever to settle down under a government, and lead a *rangé* political life. **1951** R. F. Harrod *Life of J. M. Keynes* iii. 131 By comparison with Strachey he was *rangé*. **1952** 'J. Tey' *Singing Sands* v. 78 He listened..to the talk of the two fellow sheep-farmers... Nothing hounded these large *rangé* creatures. **1977** A. Wilson *Strange Ride* vii. 334 Kipling was a..respected writer in France, particularly in *rangé* circles.

ranged, *ppl. a.*¹ Add: **2.** *transf.* Systematically arranged; ordered.
1869 Tennyson *Pelleas & Ettarre* 152 in *Idylls of King* 313 Whose lightest whisper moved him more Than all the ranged reasons of the world.

3. = **rangé a.* 2.
1899 G. Kegan Paul *Memories* v. 142, I have loved less many more ranged and orderly men.

ranger, *sb.*¹ Add: **1. a.** Also *spec.* = Bush-ranger.
1840 *Sydney Herald* 9 Sept. 2/3 It seems as though the constabulary there, are either too weak-handed or hearted—for the rangers are still there, and at large. **1862** *Mudgee* (N.S.W.) *Liberal* 28 Nov. 2/6 Down on his knees pops our repentant ranger, and earnestly pleads for mercy for the sake of his wife and babes. **1918** C. Featherstonhaugh *After Many Days* 364 The rangers then went on to the store... 'We have come to bail you up, young man.'

2. a. (Further examples.) Now esp., a warden of a park or resort. Also *attrib.*
1938 R. D. Finlayson *Brown Man's Burden* 40 The rangers had spotted them spearing the trout. **1943** *Amer. Speech* XVIII. 242 As the years went on, the rangers and other forest officers began to help with the good work [of map revision]... One name..was 'Pomas Creek', submitted by Ranger Jim McKenzie. **1966** *Weekly News* (N.Z.) 1 Aug. 7/4 Fiordland's chief ranger..emphasises that the authority's huts are 'spartan'. **1970** *Cape Times* 28 Oct. 3/2 (*heading*) Rangers wanted. *Ibid.*, the Simonstown Town Council is inviting applications from people to act as beach rangers during the summer season at R20 a week. **1973** *Sun-Herald* (Sydney) 26 Aug. 85/3 It is no easy job being a national park ranger. **1976** J. van de Wetering *Tumbleweed* xiii. 132 He became a ranger on a nature reserve. **1978** *Nature* 26 Jan. 343/1 According to accounts of the rangers at Volcan Poas National Park, bushes at the high plateau level, 300–500 m, above the lake, were set on fire.

b. orig. *U.S.* An officer who rounds up straying domestic animals or livestock.
1744 *Pennsylvania Gaz.* 15 Nov. 3/3 Any Person or Persons, who have lost one or more of the following Strays, by applying to William Hartley, of Charles Town, Chief Ranger for Chester County,..may be informed where to find them. **1796** in G. Imlay *Topogr. Descr. W. Territory N. Amer.* (1797) 535 [The county court] shall also have power to appoint one register and ranger for the county, who shall hold their offices during good behaviour. **1828** *Cherokee Phoenix* (New Echota, Georgia) 27 Mar. 1/2 The ranger shall be entitled to one dollar for every horse so posted. **1886** *Buck's Handbk. Med. Sci.* II. 304/1 In Mississippi the coroner is also the county ranger, and performs the duties of that office. **1886** [see Pound-keeper]. **1926** A. Webb *Mis' Peters' Special* 50 The ranger's got the cow.

3. a. (Earlier and further examples.) Also in *sing.*
1670 *Massachusetts Hist. Soc. Coll.* (1800) VI. 211, [I] saw one of captain Willet's rangers coming on post on horseback. **1692** *Calendar Virginia State Papers* (1875) I. 38 [Petition of Let^t David Straughan and] 'eight Rangers' [for pay for services]. **1713** *Colonial Rec. Carolina* (1886) II. 32, I have ordered all our Rangers..to march that way. **1733** *Colonial Rec. Georgia* (1905) III. 90 Captain Macpherson with fifteen of the Rangers..cover'd and protected the new Settlers. **1789** in O. Browning *Despatches from Paris* (1910) II. 169 The strongest proof is given of His disinterestedness by His proposing to resign the Office of Ranger (Capitainerie) of different Districts. **1906** *Westm. Gaz.* 2 June 8/3 Governor Ysabel, of the Mexican province of Senora [*sic*]..will be met there by a force of American rangers from Bisbee... It is thought that the arrival of the Rangers from Bisbee will restore order. **1909** 'O. Henry' *Roads of Destiny* xvi. 257 Standifer himself had served the commonwealth as Indian figther, soldier, ranger, and legislator.

b. Chiefly *U.S.* A member of an élite American military unit established in 1941 for close combat and raiding; = **commando* 3 a. Also *attrib.*
1942 *N.Y. Times* 20 Aug. 1/5 The first American troops to receive a baptism of fire in Europe in this war were the men of the United States Ranger Battalion who fought in the Dieppe raid today. It was the first time the name Rangers had appeared in a war communiqué anywhere. **1942** *Newsweek* 31 Aug. 21/3 Mention in last week's communiqués of a detachment of a 'United States Ranger battalion' that had taken part in the Dieppe raid was the first disclosure of the existence of these Commando-type American troops... All Rangers are volunteers, they reported. *Ibid.* 22/1 The Rangers were named after Rogers's Rangers, the rough and crafty Indian fighters of colonial days who battled near the Canadian border. **1961** B. Fergusson *Watery Maze* vii. 180 The Commandos had a number of Rangers, their American counterparts, attached to them for experience. **1976** R. Moore *Dubai* ii. 22 A newly organized unit.., the U.S. Army Special Forces..made up of soldiers who were paratroopers, rangers, and combat men from World War II.

5. In full now *Ranger Guide.* A member of the Girl Guides Association who is in the section for older girls, aged between 14 and (usually) 18. *Ranger Guider*, a leader of a unit of Ranger Guides. Cf. **guide sb.* 2 d.
1921 G. I. J. Potts *Girl Guide Badges* 2 The Service Star for Guiders and Rangers is worn on a red cloth ground. **1929** *Second Bk. of Ranger Games* 72 This is a favourite game with Rangers. **1944, 1969** [see *Guider* s.v. **guide sb.* 2 d]. **1969** [see **guide sb.* 2 d]. **1976** *Ulverston* (Cumbria) *News* 3 Dec. 1/4 She has kept up her membership of the Ranger Guides and still finds time for some needlework. **1977** *National Trust* Spring 21/3 The Ranger Guides of S.E. England were invited to steward, usher and sell programmes. *Ibid.* 21/4 The smartness, efficiency and good humour of the Rangers.. evoked most favourable comment. **1977** *Daily Tel.* 2 June 18 Early photographs showing the Queen, Princess Margaret and other members of the Royal family as Brownies, Guides or Rangers will be among the exhibits. *Ibid.*, The exhibition..also marks the diamond jubilee of the Ranger Guide section. **1977** *Guider* July 319/2 An enthusiastic Ranger Guider will invite older Guides to some of the interesting Ranger activities as 'tasters'.

|| **ranger** (raŋʒe), *v.*¹ [a. F. *se ranger* (also used) to settle down.] *refl.* To settle down. Cf. **rangé a.* 2.
Used only in the infinitive form.
1854 Thackeray *Newcomes* I. xxxii. 320 It is high time that Kew should *ranger* himself... I am sure he will make the best husband..in England. **1883** W. James *Let.* 2 May in R. B. Perry *Tht. & Char. W. James* (1935) I. 753 The time had come that I should *ranger* myself. **1924** J. Buchan *Three Hostages* xiii. 182, I heard somewhere you were goin' to be married... What do you call it—*ranger* yourself? **1979** A. Buchan *Scrap Screen* iii. 40 He desired to *se ranger* himself, having sown his wild oats.

ranger (rēⁱ·ndʒəɹ), *v.*² [f. Ranger *sb.*¹] *intr.* To be a ranger.
1909 'O. Henry' *Roads of Destiny* xvi. 263 We fought Kiowas, drove cattle, and rangered side by side nearly all over Texas. **1979** P. L. Sandberg *Stubb's Run* ii. 9 He had rangered in the Sawcut for seventeen years.

rangette (rēⁱndʒe·t). *N. Amer.* and *N.Z.* [f. Range *sb.*¹ 12 a + -ette.] A small gas or electric cooker.
1959 *Sears, Roebuck Catal.* Spring & Summer 934/1 Kenmore 20-inch Gas Rangettes. **1968** *Globe & Mail* (Toronto) 5 Feb. 25/9 (Advt.), Wanted to buy. Household furniture, fridge, stove, washer, rangette, urgent. **1977** *N.Z. Herald* 5 Jan. 1–20/4 She cooks on an electric rangette.

rangey, rangily: see **rangy a.*

ranginess (rēⁱ·ndʒinės). [f. Rangy *a.* + -ness.] **1.** Capacity for ranging.
1872 *Rep. Vermont Board Agric.* I. 213 A 1100 or 1200 lb horse, with bone, ranginess and endurance.
2. The state of being tall and slender.
1965 G. McInnes *Road to Gundagai* xiv. 257 He was..of a Gary Cooper ranginess.

ranging, *vbl. sb.* Add: **1. a.** *spec.* The action of measuring the distance to an object by radar or other means.
1919 *Sci. Amer.* 17 May 511/3 Sound-ranging is a vast improvement [for]..locating hostile points of fire. **1946** *Jrnl. Inst. Electr. Engineers* XCIII. 378/2 The foundation for precision radar ranging in fire control was firmly laid in 1938. **1965** Filipowsky & Muehldorf *Space Communications Techniques* ii. 136 The reference frequency of 29⅓ or 32 Mc is transmitted..over the microwave link to the receiver where the ranging equipment is situated. **1970** *Nature* 12 Dec. 1024/2 The Smithsonian Astrophysical Observatory has..smaller instruments for satellite ranging.

rangiora (raŋiō²·rä). [Maori.] An evergreen New Zealand shrub or small tree, *Brachyglottis repanda*, belonging to the family Compositæ, and bearing large ovate leaves with white tomentum on the under side and terminal panicles of small greenish-white flowers. Also *attrib.*
1868 W. Colenso in *Trans. N.Z. Inst.* I. III. 38 For wounds..they used the large leaves of the Pukapuka, or Rangiora. **1882** *Trans. N.Z. Inst.* XIV. 357 The Maoris also distinguish the two plants by different names, the present plant being known as Rangiora. **1921** H. Guthrie-Smith *Tutira* vii. 51 Later, appeared slender matapo [*sic*]..and rangiora. **1933** *Bulletin* (Sydney) 9 Aug. 21/2 The large leaves of the rangiora bushes growing near tracks. **1966** *Encycl. N.Z.* II. 785/2 Gums and resins..were obtained from kauri,..rangiora..and wharangi.

ra·ngled, var. Wrangled *ppl. a.*¹
1924 A. J. Small *Frozen Gold* x. 220 The trail led through four miles of rugged, rangled bushland to the Yukon water edge. **1929** R. Bridges *Testament of Beauty* III. 84 Nor the rangled shroud that she wove for his sire.

Rangoon (ræŋgū·n). The name of the capital of Burma, used *attrib.* in **Rangoon bean**, the Lima bean, *Phaseolus lunatus* var. *limensis* (see Lima in Dict. and Suppl.); **Rangoon creeper**, a tropical climbing shrub, *Quisqualis indica*, belonging to the family Combretaceæ, native to south-east Asia, and bearing spikes of white, pink, or red flowers having an elongated calyx tube.
1877 G. M. Woodrow *Hints on Gardening in India* (ed. 2) 110 *Quisqualis Indica* (Rangoon Creeper). A straggling shrub, which may be trained to cover a wall..; flowers vary in colour, from white to rose. **1901** L. H. Bailey *Cycl. Amer. Horticulture* III. 1486/2 This [genus] includes the Rangoon Creeper, a tender woody plant with 5-petaled red flowers. **1922** Joyce *Ulysses* 289 The extremely large wains bring foison of the fields, flaskets of cauliflowers, floats of spinach, pineapple chunks, Rangoon beans, strikes of tomatoes. **1928** K. Gough *Garden Bk. for Malaya* viii. 111 Kopsia and Rangoon creeper are respectively a shrub and a creeper with flowers that shade from pale-pink to deep red-pink. **1971** C. Flieger in E. L. Wardman *Bermuda Jubilee Garden* viii. 163/2 Rangoon creeper. A deciduous clambering shrub with very attractive drooping spikes of elongated pink and red flowers, and a perfume that suggests a mixture of ripe fruits. **1972** Y. Lovelock *Veg. Bk.* I. 55 A variety known as Lima bean.., also Tonga, Burma, Rangoon, Java and Madagascar bean, bears large white seeds which are dried, canned and marketed under the names wax or butter beans.

Rangri (rā·ŋgrī). Also **Rangree.** A form of the Malvi dialect of Rajasthani.
1823 J. Malcolm *Mem. Central India* II. xiv. 166 In Central India they learn the Rangree or Hindui dialect, in which business is commonly transacted. **1878** R. N. Cust *Mod. Languages of E. Indies* ii. 49 Sir J. Malcolm..alludes to Rangri as the Dialect of Hindi taught in the schools of Central India... By Rangri he meant at that time Dialects of Hindi, spoken all over Central India, and now resolved into separate Tribal and Political subdivisions. **1908** G. A. Grierson *Ling. Survey of India* IX. II. 52 The form of Mālvī spoken by Rajputs of Malwa proper is called Rāŋgṛī. It is distinguished by its preference for Mārwāṛī forms. *Ibid.* 53 As stated above, Mālvī, in the Malwa country, has two forms, *viz.*:—Rāŋgṛī..or Rāj-wāṛī, spoken by Rajputs, and Mālvī.., spoken by the rest of the population. **1920** —— *Index of Language-Names* 172 Rāŋgṛī... A form of the Mālvī Dialect.

rangy, *a.* Delete 'Chiefly *U.S.*' and add: Also **rangey. 1.** Also of persons. **a.** (Earlier and further examples.)
1868 H. Woodruff *Trotting Horse* xlvii. 381 The latter was a fine, rangy gelding. **1885** T. Roosevelt *Hunting Trips of Ranchman* 21 The ponies..used for the circle riding in the morning have need rather to be strong and rangey. **1936** F. Clune *Roaming round Darling* xxiv. 252, I counted over four score of rangy mangy looking racehounds all capable of running a kangaroo to earth. **1955** *Times* 20 May 10/5 Some bulls—black and white beasts of the Holmogorky breed, somewhat like British Friesians, but rather bigger and rangier. **1977** J. Bingham *Marriage Bureau Murders* iii. 30 She had been out cubbing one autumn on a great rangy beast. **1977** *Time* 19 Sept. 61/1 The royal government provided no housing materials and no food beyond a few rangy cattle.

b. (Earlier and further examples.)
1876 *Rep. Vermont Board Agric.* III. 215 They were.. light colored, rather rangy sheep. **1899** G. Ade *Doc' Horne* 42 He was considerably over six feet tall, raw boned and rangy. **1910** *Daily Chron.* 18 Apr. 8/3 Truxton King is a tall, raw-boned, 'rangy' American—notice particularly the 'rangy', as it seems to be his chief characteristic—though we don't quite know what it means. **1932** L. C. Douglas *Forgive us our Trespasses* (1937) x. 197 You're rangy enough. Tall as Craig, himself. **1956** C. Willock *Death at Flight* iv. 51 The Americans..would describe such a man as 'rangy'. **1968** J. Lock *Lady Policeman* viii. 76 He's tall, rangy, with skin-tight pants, long dark hair and plum-coloured lipstick. **1971** *Guardian* 28 Aug. 9/1 A rangy 26-year old Australian who walks and talks with equal stamina. **1977** *Time* 3 Jan. 36/3 A freewheeling, gregarious politician, the rangy (6ft. 2in.) Bergland is married to a farmer's daughter and is the father of six children. **1977** J. F. Fixx *Compl. Bk. Running* i. 4 Carver is tall, rangy and sturdily built, but at thirty-three he was out of shape.
Hence **ra·ngily** *adv.*, in a rangy fashion.
1976 'S. Woods' *My Life is Done* 62 He was..very tall and rangily built.

ranid (rēⁱ·nid), *sb.* and *a.* [ad. mod.L. family name *Ranidæ*, f. L. *rāna* frog + -id³.] **A.** *sb.* A frog of the family Ranidæ, which includes typical amphibious frogs. **B.** *adj.* Of, pertaining to, or designating a frog of this family.
1888 *Proc. Zool. Soc.* 509 Our attention was early arrested by the general similarity between the proximal syndesmoses in the Hylids and Ranids and the knee-joint in the higher Vertebrata. **1901** *Ibid.* I. 58 (*heading*) On abnormal ranid larvae from North-Eastern India. **1946** *Nature* 30 Nov. 749/2 In *L*[*eptodactylus*] *pentadactylus*..both the ureter and the Wolffian ducts run separately throughout their course as in ranid frogs. **1957** P. J. Darlington *Zoogeogr.* iii. 170 Ranids have evolved in the main part of the Old World tropics. **1974** H. R. Heusser in B. Grzimek *Animal Life Encycl.* V. vi. 397 In true frogs or ranids..the presacral vertebrae are typically amphicoelous.

ranikaboo: see **rannygazoo.*

|| **ranjau** (ra·ndʒau). Also 8–9 **ranjow.** [Malay.] A stake or caltrop of bamboo or iron placed in the ground to pierce the feet or body of an enemy.
1783 W. Marsden *Hist. Sumatra* 278 *Ranjows* are sharp pointed stakes of *bamboo*, of different lengths, stuck into the ground, in order to penetrate the naked feet, or body, of an enemy. **1839** T. J. Newbold *Straits of Malacca* II. xii. 215 A Malay considers himself completely armed with..the tombak (spear), and a quiver of ranjows, or caltrops, at his back. **1936** G. B. Gardner *Keris* x. 118 While the chief might have used horses for purposes of display, on account of the *ranjau* or pointed stakes planted in the passes they could not have been employed effectively.

rank, *sb.*¹ Add: **1. a.** (Examples in *Teleph.*)
1924 *Brit. Stand. List Terms Telegraphs & Telephones* 13 *Rank of switches*,..the switches which provide for any one stage of call selection. **1929** *P. O. Engin. Dept. Techn. Instr.* XXV. II. 6 The number of ranks of selectors is one less than the required number of digits to call a number on the exchange. **1969** S. F. Smith *Telephony & Telegraphy A* v. 110 If more than 1000 numbers are required, another rank of group selectors can be used.

c. A row of public vehicles waiting to be hired, or the place where these stand; a taxi-rank.
c **1843** J. R. Planché *Extravaganzas* (1879) II. 240 My Minstrel Boy for a cab is gone, In the ranks no doubt he'll find one. **1851** [see sense 1 a in Dict.]. **1903** *Daily Chron.* 29 Sept. 3/1 These proposals include the use of such large ranks as that in Berkeley-square as feeders for smaller ones in the vicinity. **1922** *Daily Mail Year Bk.* 1923 74/2 One London firm having an air-garage with machines waiting for hire always on the 'rank'. **1930** D. L. Sayers *Strong Poison* i. 21 The taxi-driver Burke, who was standing on the rank in Guilford Street, was approached by Philip Boyes. **1962** S. Beckett in *Evergreen Rev.* Jan.–Feb. 16, I learnt there were still some cabmen who spent their day snug and warm inside their cabs on the rank. **1974** 'J. Ross' *Burning of Billy Toober* ix. 88 She returned..by taxi... The driver..picked her up at the rank in the town square. **1978** *Taxi* 16 Feb. 2/2 On Alf's return to the forecourt the dozen cabmen who had gathered decided that a protest boycott was necessary and they began picketing the rank.

3. a. *to close ranks*: see Close *v.* 10 b. Also *fig.*
1796–7, 1873 [see Close *v.* 10 b]. **1941** 'G. Orwell' *Lion & Unicorn* I. iii. 35 England..is a family... It has its private language and its common memories, and at the approach of an enemy it closes its ranks. **1948** W. Churchill *Gathering Storm* I. xxi. 382 The tide of events brought with it a closing of the ranks between England and France, and also at home. **1974** G. Hubbard *Quaker by Convincement* i. 68 Some Friends occasionally suggest that a creed might help to clarify our thoughts... At this suggestion the majority close their ranks, and hold firm to their belief which are not to be contained in the strait-jacket of a creed. **1977** *Oxford Mission Quarterly Paper* Jan.–Mar. 26 When surrounded by astronomical numbers of non-Christians, Christians tend to show their solidarity by closing their ranks.

b. (Earlier *transf.* example.) *to rise from the ranks*, of a private or non-commissioned officer: to receive a commission, to become a commissioned officer; freq. *transf.*, to rise from a lowly social, etc., position. Cf. **other a.* 5 f.
1845 *Punch* VIII. 127 I've flattered Peel; he smiles back thanks... But still he keeps me in 'the ranks'. **1853** Ruskin *Let.* 6 Nov. in M. Lutyens *Millais & Ruskins* (1967) 106 Mr Beveridge [has]..been effecting singular cures..and rose from the ranks—as Jephson did. **1936** B. Kellermann's *Tunnel* II. iii. 84 There was no doubt about the fact that Woolf had risen from the ranks. **1958** J. Wain *Contenders* ii. 27 Baxter, a harmless bore of about thirty-five, had risen from the ranks, so to speak. **1977** *Times* 2 June 15/4 St. George..is acclaimed as a soldier who rose from the ranks to become a tribune.

5. b. (Further *transf.* examples.) Now freq. hyphened. Cf. **rank-and-file a.*
1927 *New Republic* 12 Oct. 205/2 The rank and file have grown tired of the persistent effort of the Communists to subvert trade-union discipline. **1939** H. Nicolson *Diary* 24 Mar. (1966) 392 The ignorance of the Tory rank-and-file in

regard to foreign policy is as terrifying as the prospect of a gardener suddenly driving a Rolls Royce. **1972** *History Workshop Pamphlet No. 6* 37 In the depression of the later 1870s the demand from the rank-and-file for a policy of restriction became very strong. **1976** E. Maclaren *Nature of Belief* ix. 91 Professional theologians might refine beyond recognition the bald credal outlines demanded of the rank-and-file. **1976** tr. Wang Chin-fu et al. in *Yenan Seeds & other Stories* 113 You'd better go to the rank-and-file to find out what they think.

8. b. Also in professional, military, and other walks of life. Phr. *to pull rank*: see **pull v.* 19 h.
1964 M. Banton *Policeman in Community* iv. 116 One Carolina City officer had served in a United States army unit which was stationed alongside the Black Watch..and he commented: 'Why, a man with one stripe in that outfit had more rank than a master-sergeant in ours.' **1968** J. Lock *Lady Policeman* xv. 129 Most days we wouldn't see any rank. **1972** J. R. T. Pollard in G. W. Knight *Jackson Knight* 10 One who stood above rank,..and took just as much interest in the problems and activities of the..College domestic staff as he did in those of the students and his fellow academics. **1977** *Transatlantic Rev.* LX. 21 Why doesn't he wear his rank on his coat, he had once asked his mother.

9. a. *spec.* in *Statistics*, position in a numerically ordered series; the number specifying the position. Cf. **rank v.*¹ 3 b.
1883 F. Galton *Inquiries into Human Faculty* 53 We are often called upon to define the position of an individual in his own series... In reckoning this, a confusion ought to be avoided between 'graduation' and 'rank', though it leads to no sensible error in practice. *Ibid.* 54 All..ranks stand half a degree short of the graduation bearing the same number. *Ibid.*, His rank of No. 5 will correspond to the graduation 4°.5. **1904** *Amer. Jrnl. Psychol.* XV. 81 Rank has..the useful property of allowing any two series to be easily and fairly combined into a third composite one. **1907** *Drapers' Company Res. Mem.* (Biometric Ser.) IV. 10, I term rank the actual position in order of an individual with regard to any variate in a given series obtained by measurement or observation. If v_1 be the 'rank' of an individual for a given character this signifies that in the observed population there are $v_1 - \frac{1}{2}$ individuals with character greater than x. **1936** *Ann. Math. Statistics* VII. 32 Continuous variates expressing these qualities are likely not to be normally distributed... We may therefore resort to the ranks, ignoring any exact values that have been assigned. **1943** M. G. Kendall *Adv. Theory Statistics* xvi. 390 The ranks are ordinal numbers and cannot without justification be operated on by the laws of cardinal arithmetic. **1976** T. D. V. Swinscow *Statistics at Square One* x. 60 The ranks for the two samples are..added separately and the smaller total is used. *Ibid.* xii. 74 Boys K, L, and M are tied at rank 12.

b. *Linguistics.* The position of a unit in a grammatical or phonological hierarchy.
1961 *Word* XVII. 251 The units of grammar form a hierarchy... The relation among the units, then, is that.. each 'consists of' one, or of more than one, of the unit next below... The scale on which the units are in fact ranged.. may be called 'rank'. **1964** M. A. K. Halliday et al. *Linguistic Sci.* 27 The term used to name the hierarchical relation among the units is *rank*; they can be arranged on a scale, and this is known as the 'rank scale'. **1965** J. C. Catford *Linguistic Theory of Translation* i. 6 The units of grammar or of phonology operate in hierarchies—'larger' or more inclusive units being made up of 'smaller' or less inclusive units. They form a scale of units at different ranks... The sentence is a unit of higher rank than the clause. *Ibid.* 9 The concept of rank..is an important one both in theoretical linguistics and..translation-theory. **1971** R. A. Hudson *Eng. Complex Sentences* ii. 70 What a grammar will contain..is not a number of different system-networks, each for a different 'rank' (clause, phrase, etc.) or a different environment (subject, main verb, etc.), but a single network which includes all the grammatical systems needed for the language. **1977** *Language* LIII. 192 Although admitting that a sentence could be regarded as a special type of clause, she in fact treats them as distinct ranks.

9*. *Math.* Used variously, at the discretion of the author, to denote some integer that characterizes the entity being discussed.
1835 *Rep. Brit. Assoc. Adv. Sci.* 1834 528 *i* in the denominator..names, according to the author's nomenclature, the 'order' of the logarithm, and *i*, in the numerator, its 'rank' in that order. **1913** C. E. Cullis *Matrices & Determinoids* ix. 265 The rank of the matrix *A* is the greatest order which a non-vanishing derived determinant can have. **1914** H. Hilton *Homogeneous Linear Substitutions* ii. 50 Suppose that..the determinant itself and all the 1st, 2nd,.., $(m-r-1)$-th minors vanish, but that not all the $(m-r)$-th minors vanish. Then the determinant is said to be of rank *r*. If the determinant does not vanish, it is of rank *m*; if the determinant vanishes but not every first minor vanishes, the determinant is of rank $m-1$, and so on. **1941** [see **nullity* 6]. **1953** N. Jacobson *Lect. Abstr. Algebra* II. i. 22 Any two maximal linearly independent subsets of a set *S* have the same cardinal number. We call this number the rank of the set *S*. **1965** J. J. Rotman *Theory of Groups* xi. 241 If *F* is a free group, the rank of *F*, *r*(*F*), is the number of elements in a free set of generators. **1972** R. J. Wilson *Introd. Graph Theory* viii. 121 The rank $\rho(A)$ of a subset *A* of *E* is defined as the number of elements in the largest partial transversal of $\mathscr{S}$ contained in *A*. **1979** *Proc. London Math. Soc.* XXXVIII. 532 Let *G* be a reduced torsion-free abelian group of finite rank.

9.** *Petrol.* The degree of metamorphic maturity or hardness, esp. of coal.
1914 *Bull. U.S. Bureau of Mines* No. 38. 4 The higher rank ('grade') of coal differs from the respective lower rank of the same genetic type by the effects of the greater metamorphism and devolatilization to which the higher rank of coal has been subjected. **1928** *Jrnl. Chem. Soc.* 2971 What we have termed the 'degree of coalification' of a coal is sometimes..termed its 'rank', a coal of lowest rank being one in which the processes of coalification..have had least result. **1948** *Mem. Geol. Soc. Amer.* No. 30. 55 Metamorphic rank and grade are synonymous terms denoting the stage of metamorphism reached. **1964** A. Nelson *Dict. Mining* 360 Lignite is a low-rank coal whilst anthracite is a high-rank coal. *Ibid.* 89 Coal rank indicates the stage of coalification which any particular coal deposit has reached between peat at one end of the scale and anthracite at the other. *Ibid.* 218 Hitt..came to the conclusion that in a vertical succession at any point in the coalfield the rank of the coals increased with the depth. **1976** *Nature* 1 July 48/1 Coals of various ranks have been treated in electrical arcs. **1979** *Sci. Amer.* Jan. 29/3 When adjustments are made for the inferior heating value of the lower ranks, the recoverable tonnage comes to about 600 billion tons of hard-coal equivalent, enough for more than 200 years' consumption at current rates.

10. (sense 3) *rank-closing*; (sense 8) *rank badge, -class, -holder, -mark, tab*; (sense *9 b) *rank-scale, -unit*; *rank-based, -bound* adjs.; **rank correlation** *Statistics*, the correlation between two ways of assigning ranks to the members of a set; **rank difference** *Statistics*, the difference between two ranks assigned to the same thing; freq. *attrib.* Also **rank order*.
1961 *Men Only* June 37/2 His eyes flicked contemptuously to the rank badges on his right sleeve. **1975** T. Allbeury *Special Collection* v. 32 Wehrmacht men..with their insignia and rank badges torn off. **1966** G. N. Leech *Eng. in Advertising* ii. 11 A rank-based description avoids these confusions, because of the insistence that each sentence should be fully described at all ranks. **1965** J. C. Catford *Linguistic Theory of Translation* ii. 25 The cruder attempts at Machine Translation are rank-bound in this sense..; that is, they set up word-to-word or morpheme-to-morpheme equivalences, but not equivalences between high-rank units such as the group, clause or sentence. **1968** *Meta* XIII. 7 *Replacement*..may be *rank-bound* (as when only word-to-word equivalences are sought) or unbounded (as when equivalences occur between higher rank units such as sentences). **1930** C. G. Seligman *Races of Africa* ix. 221 The Akamba have age-grades, and within them rank-classes. **1966** *New Society* 12 May 5/3 Rank-closing had generally made the seamen as solid as a filled-in ditch... But there was more to the solidarity than just the rank-closing. **1907** *Drapers' Company Res. Mem.* (Biometric Ser.) IV. 25 No two rank correlations are in the least reliable or comparable unless we assume that the frequency distributions are of the same general character..provided by the hypothesis of normal distribution. *Ibid.* 3 Dr. Spearman has suggested that rank in a series should be the character correlated, but he has not taken this rank correlation as merely the stepping stone..to reach the true correlation. **1943** M. G. Kendall *Adv. Theory of Statistics* xvi. 391 A second coefficient of rank correlation which has certain advantages may be obtained as follows. *Ibid.* 408 Up to this point we have considered the problem of rank correlation without reference to any variate system which might underlie the rankings. **1977** S. A. Book *Statistics* xi. 427 The technique, called the Spearman test of 'rank correlation', requires us to rank each set of data and then compute the correlation coefficient of the ranks. **1904** *Amer. Jrnl. Psychol.* XV. 86 (*heading*) Method of rank differences. **1907** *Drapers' Company Res. Mem.* (Biometric Ser.) IV. 39 Mean square of rank differences will be more accurate than mean positive rank difference. **1972** Kagan & Havemann *Psychol.* xiii. 485 In some cases it is convenient to use the rank-difference method, which produces a different coefficient of correlation called ρ. **1951** S. F. Nadel *Found. Social Anthropol.* i. 18 As a rank-holder..features of social life on lower levels may be closed to you. **1928** C. F. S. Gamble *Story N. Sea Air Station* iv. 76 They were entitled to wear the 'curl' on their gold lace rank-marks. **1964** Rank scale [see sense 9 b above]. **1970** G. C. Lepschy *Survey of Structural Linguistics* vii. 124 Along the rank scale five grammatical units (sentence, clause, phrase, word, morpheme) and four phonological units (tone group, foot, syllable, phoneme) are used in English. **1974** 'G. Black' *Golden Cockatrice* xii. 212 Janey, in her neat uniform without rank tabs. **1968** Rank unit [see *rank-bound* above].

rank, *a.*, (*sb.*²), and *adv.* Add: **D.** *Comb.* **c.** *rank-smelling, -tasting.* **d.** *rank-old* adj.
a **1889** G. M. Hopkins *Poems* (1967) 179 What being in rank-old nature should earlier have that breath been. **1904** W. de la Mare *Henry Brocken* vii. 71 Near by me grew some rank-smelling waterside plant. **1919** 'K. Mansfield' *Let.* 21 Nov. in *Lett. to J. M. Murry* (1951) 401, I remember standing in a rank-smelling field. **1921** D. H. Lawrence *Sea & Sardinia* ii. 63 A massive yellow omelette..cooked in the usual rank-tasting olive oil.

rank, *v.*¹ Add: **3. b.** *Statistics.* To assign a rank to (**rank sb.*¹ 9 a).
1907 *Drapers' Company Res. Mem.* (Biometric Ser.) IV. 25 It is easier to rank individuals than to measure their attributes accurately. **1944** *Jrnl. Anat.* LXXVIII. 185/1 The severity of the inflammation can easily be ranked, i.e. given an ordinal number. **1951** Brookes & Dick *Introd. Statistical Meth.* ix. 221 It is possible to rank depths of colour without requiring some form of graduated scale, and to rank dimensions of objects..by comparing one with another and arranging them in sequence. **1977** [see *rank correlation* s.v. **rank sb.*¹ 10].

5. a. (Earlier and later examples.)
1841 *Southern Lit. Messenger* VII. 766/1, I have Mr. Sanford under my command—I *rank* him,..and then I have charge of the *whole* ship. **1860** *Congress. Globe* 10 Dec. 27/3, I shall..submit a few reasons for this opinion..but not until other Senators are heard who rank me in age, experience, and wisdom. **1904** *Delineator* Dec. 933 The Secretary of State ranks all the other members of the Cabinet. **1907** [see **importance* 2 b]. **1976** J. Wainwright *Who goes Next?* 177 Bear..ranked Sullivan, neck-and-neck—they were both deputy chief constables.

b. *U.S. Mil.* To deprive or turn (someone) *out* of quarters, etc., by virtue of superior rank.
1872 F. M. A. Roe *Army Lett.* (1909) 66 Faye has been turned out of quarters—'ranked out', as it is spoken of in the Army. **1891** C. King *Trials of Staff Officer* 184 We were 'ranked' out of those quarters presently. **1932** L. H. Nason *Among Trumpets* 13 What's the good of havin' three stripes if you can't rank somebody out of a bunk or horse or something?

c. *U.S. Blacks.* (See quots.) Cf. **ranking vbl. sb.*¹ c.
1971 C. Mitchell-Kernan in A. Dundes *Mother Wit* (1973) 316/2 'Barbara was trying to *rank* Mary', to put her down by typing her. **1974** H. L. Foster *Ribbin', Jivin', & Playin' Dozens* iv. 171 *Rank*, to insult someone. **1976** *English Jrnl.* Dec. 56/1 'We're ranking people out.'..'What does that mean?' I asked... 'We're saying things about other people to put them down,' answered one helpful student.

6. b. (Further examples.) Also in phr. *to rank high*.
1824 Macaulay in *Knight's Q. Mag.* II. 357 Ovid, Catullus, Tibullus, Horace, and Propertius, in spite of all their faults must be allowed to rank high in this department of the art. **1828** Carlyle *Crit. & Misc. Ess.* (1840) I. 153 This play should rank high among that class of works. **1885** *Nature* 8 Jan. 223/2 The man who ranks 5th from the bottom of a class of 100 males would rank 10th from the top in a class of 100 females. **1932** *N.Y. Times Book Rev.* 3 Jan. 9/5 Nevertheless, as an imaginative humorist he ranks very high, and his omnibus can be warmly recommended to those who have hitherto been unfamiliar with his work. **1973** *Daily Tel.* (Colour Suppl.) 13 July 9/1 Diplomats caught have ranked up to ambassador.

c. (Further examples.) Also, to qualify *for*.
1899 I. Pitman *Commercial Corresp.* xix. 197 Statement of Affairs: Cross Liabilities; Liabilities as Estimated by Debtor; Expected to Rank; [etc.]. **1928** *Daily Mail* 3 Aug. 18/2 The new shares did not rank for the interim dividend. **1930** *Daily Express* 22 May 7/4 Mr. Bottomley's amended statement of affairs showed gross liabilities £116,769, ranking at £115,899. **1976** *Milton Keynes Express* 11 June 13/1 There may well be very considerable sums to be spent on essential repairs which will not rank for subsidy aid.

8. *trans. U.S. slang.* **a.** To betray (a person), to give away (in quot. *refl.*). Also, to apprehend in the act of committing a crime.
1929 [see **fuzz sb.*²]. **1955** D. W. Maurer in *Publ. Amer. Dial. Soc.* XXIV. 175 *Whiz dicks* are on the lookout..to *rank* a pickpocket.

b. To spoil or thwart (an action), esp. in phr. *to rank the play*.
1937 C. Himes *Nigger* in *Black on Black* (1973) 125 The landlady..had sent Mr. Shelton on up to catch him there in the hopes of ranking Fay's play. **1968–70** *Current Slang* (Univ. S. Dakota) III–IV. 101 *Rank*, to spoil.

ranking *ppl. a.* (earlier and later examples; chiefly in sense 5 of vb.).
1863 *Yale Lit. Mag.* XXIX. 80 His two ranking officers both gone. **1885** E. Custer *Boots & Saddles* xiii. 137 The ranking lady had a sabre which her chief had received as a present, and this she waved over the others in command. **1906** *Westm. Gaz.* 4 Apr. 7/2 It was estimated that his [*sc.* the bankrupt's] ranking indebtedness would be about £2,280. **1925** T. Dreiser *Amer. Tragedy* (1926) II. II. xlvii. 73 These tall, close, ranking pines. **1931** *Publishers' Weekly* 20 June 2849/1 The publishing industry of this country..now takes a ranking place in the economic structure. **1962** Plano & Greenberg *Amer. Polit. Dict.* vii. 141 *Ranking member*, that member of the majority party on a legislative committee who ranks first after the chairman in number of years of continuous service (seniority) on the committee. **1970** E. R. Johnson *God Keepers* xii. 132 There was a lot of merit in having the ranking man right where the heat was going to be. **1974** P. Gore-Booth *With Great Truth & Respect* 315, I learned it on the afternoon of a Sunday on which we had arranged for a dinner in honour of Sir James Cassels, the new British Chief of the General Staff. The ranking guest was to be General Muchu Chaudhuri, the Indian Commander-in-Chief. **1976** *Washington Post* 19 Apr. B6/5 Six ranking generals carried the coffin to the simple military grave. **1976** H. Wilson *Governance of Britain* viii. 150, I approached Sir John Arbuthnot, then, as the Americans would have put it, 'ranking' Conservative Member on the Committee. **1978** R. Ludlum *Holcroft Covenant* xv. 179 The offspring of a ranking member of the Third Reich's High Command.

ra·nk-and-file, *a.* [f. *rank and file*: Rank *sb.*¹ 5 b.] **a.** *Mil.* Of or belonging to the rank and file; private; ordinary.
1885 [see **chew v.* 3 g]. **1904** *Westm. Gaz.* 27 Feb. 2/3 Those who know the rank-and-file men of the Russian army will agree. **1907** *Daily Chron.* 19 Dec. 3/3 The memoirs of a rank-and-file man like this are very touching and painful reading. **1966** *Times* 9 July 9/7 Command Orders say..Rank and File Mess is altered to 'soldiers' mess'.

b. *transf.* Of, pertaining to, or designating an ordinary member of a group (as a political party, union, etc.) as opp. to a leader or principal.
1887 [see Rank *sb.*¹ 5 a]. **1931** *Times Lit. Suppl.* 4 June 437/2 He always assumed that the rank-and-file politician was actuated by his own high motives. **1945** Koestler *Yogi & Commissar* III. i. 131 The absence of any rank-and-file influence on the party-line. **1955** *Times* 2 May 17/4 He came in for much criticism for not being in touch with the rank-and-file members of his union. **1974** *Times* 19 Sept. 1/1 No party leader or rank-and-file MP can exclude..the possibility..that..the electorate may once again create a parliamentary deadlock. **1976** *Church Times* 15 Oct. 13/1 More than twenty years ago a refreshing, highly competent and, for rank-and-file Christians in the parishes, most

stimulating and encouraging publishing event took place. 1977 *Times* 7 Sept. 4/3 That might be because many rank-and-file union members are unenthusiastic about the so-called war on Grunwick.

ra:nk-and-fi·ler. orig. *U.S.* [f. as prec. + -ER[1].] A member of the rank and file; an ordinary member (of a group, society, etc.).

1940 *Sat. Even. Post* 27 Jan. 82/1 Others assert he never in his life toiled as an active rank-and-filer. 1943 *Sun* (Baltimore) 1 June 10/2 Many of the AFL transit rank and filers—perhaps even many of their union leaders—agree at heart with Messrs. Green and Katz. 1950 E. H. CARR *Bolshevik Rev.* I. ii. 33 Lenin's followers were rank-and-filers with scarcely a known name among them. 1959 *Encounter* Aug. 74/2 No 'ordinary people', no John Doe, Tommy Atkins, rank-and-filer, man-in-the-street. 1966 *New Statesman* 13 May 680/1 Some 30 members of the union's executive are sea-going ex-rank-and-filers. *a* 1974 R. CROSSMAN *Diaries* (1976) II. 63 Three of the rank-and-filers were ultra-left-wingers and my part of the programme developed into a futile shouting match. 1978 S. BRILL *Teamsters* i. 25 Barkett was more involved in the union than most rank-and-filers.

ranked, *ppl. a.*[1] Add: **2.** *Statistics.* Assigned a position in a series.

1907 *Drapers' Company Res. Mem.* (Biometric Ser.) IV. 36 The actual size of organ corresponding to a bracket rank may differ widely from the size really belonging to the ranked organ. 1931 *Biometrika* XXIII. 396 The mean interval between any two individuals in a ranked series. 1937 YULE & KENDALL *Introd. Theory Statistics* (ed. 11) xiii. 247 We arrange the students *in order* of their ability... We then allot to each student a number which indicates his position... The students are then said to be *ranked*, and the number of a particular individual is his rank.

ranker. **3.** (Further examples.)

1905 *Daily Chron.* 11 May 4/7 The 'ranker' officer..has been explaining how his bankruptcy was due to expenses connected with his promotion. 1926 A. BENNETT *Lord Raingo* I. xxxi. 143 He regretted his humorous remarks to her about the war..for they took no account of the daily torture of millions of young men, including Delphine's own promoted ranker. 1943 J. W. DAY *Farming Adventure* (1949) xviii. 210 The village and the Army discussed ranker-officers with great candour. There had been two of them at the Hall with a previous unit—an ex-farmer and an ex-sergeant-major.

rankest (ræ·ŋkest), *a. colloq.* (chiefly *U.S.*). Now *rare.* [Irreg. f. RANK *sb.*[1] + -EST, as adj. of superlative degree; cf. RANK *v.*[1] 5 and RANKING *ppl. a.*] Of the highest rank; most senior, 'rankingest'.

1907 B. M. CROKER *Company's Servant* xxiv. 257 All the 'Rankest' or senior ladies, had received their due meed of attention. 1926 *Sat. Even. Post* 6 Mar. 154 As he's the rankest man he'll have to take charge while I'm gone. 1930 T. FREDENBURGH *Soldiers March!* 86 It's very hard on a self-respecting Corporal to be seen publicly with the rankest soldier in the outfit.

ranket(t, var. *RACKETT.

Rankine (ræ·ŋkin). [The name of William John Macquorn *Rankine* (1820–72), Scottish physicist and engineer.] **1.** Used *attrib.* or in the possessive to designate concepts propounded by Rankine or arising out of his work, as **Rankine cycle,** a thermodynamic cycle which describes the operation of an ideal composite engine worked by steam or another condensable vapour, and is used as a standard of efficiency; **Rankine efficiency,** the efficiency of an engine relative to that of an ideal engine following the Rankine cycle; **Rankine's** (occas. **Rankine**) **formula,** any of a number of formulæ derived by Rankine in his work in various fields; *spec.* (see quot. 1940).

1868 J. BOURNE *Treat. Steam-Engine* (ed. 8) i. 86/1 (*heading*) Rankine's formula. 1891 S. ANGLIN *Design of Structures* xi. 209 Apply Rankine's formula to determine the breaking weight of a wrought-iron hollow cylindrical column, its length being 10 feet, [etc.]. 1896 R. H. THURSTON in *Jrnl. Franklin Inst.* CXLII. 444 Comparing the efficiency of the best modern steam-engine employing saturated steam, about 0·20, with the Rankine cycle, which is that ideal cycle which constitutes the closest approximation to its method of steam distribution, the ideal case giving an absolute efficiency, 0·25, it is found to have, measured by this latter standard, a relative efficiency of 80 per cent. 1897 J. A. EWING *Steam-Engine* (ed. 2) iv. 120 Rankine's formula ought to be applicable when the amount of superheating is very great. 1907 W. H. P. CREIGHTON *Steam Engine* viii. 196 This engine is one which follows the Rankine cycle, where steam at a constant pressure is admitted into the cylinder with no clearance, and after the point of cut-off is expanded adiabatically to the back pressure. 1913 J. DUNCAN *Appl. Mech. for Engineers* x. 237 At present, most designers rely on the Rankine formula coupled with a liberal factor of safety. 1930 *Engineering* 6 June 739/3 The new Barton turbine is 8¼ per cent. more efficient relative to the Rankine cycle than the older units. 1933 T. H. TAFT *Elem. Engin. Thermodynamics* ix. 130 (*heading*) Effect on the Rankine efficiency of changing conditions. 1940 *Chambers's Techn. Dict.* 702/2 *Rankine's formula*, an empirical formula giving the collapsing load for a given column. 1954 L. PILBOROUGH *Appl. Heat & Heat Engines* xviii. 313 The Rankine efficiency may be used when determining the efficiency ratio of a steam plant. 1967 *Trans. Inst. Engineers & Shipbuilders in Scotland* CX. 21 As turbine efficiency increases with superheat and decreases with pressure the Rankine efficiency gains are modified accordingly. 1976 T. J. REYNOLDS et al. *Struct. Steelwork* (ed. 14) xi. 237 Using Rankine's formula, find the safe axial load for a 308 × 305 × 97 kg universal column 3·3 m high —to be regarded as having ends fixed—using the usual constraints for mild steel and adopting a factor of safety of 4. 1978 *Jrnl. R. Soc. Arts* CXXVI. 608/2 We have not, however, been complacent about more advanced thermodynamic cycles. We have studied a number of possibilities of achieving higher thermal efficiency than the Rankine cycle used in most of our existing plant.

2. Used *attrib.* to designate a temperature scale in which the zero is identified with absolute zero and the degrees are equal in size to those on the Fahrenheit scale. Also *degree Rankine* (or *Rankine degree*), a degree of this scale; *Rankine temperature*, a temperature expressed in terms of this scale.

1941 H. T. WENSEL in *Temperature* (Amer. Inst. Physics) i. 10 The Rankine Scale is essentially the same scale [as the Kelvin]. Temperatures on the Rankine Scale are simply 9/5 of temperatures on the Kelvin Scale. 1941 L. S. MARKS *Mech. Engineers' Handbk.* (ed. 4) 295 Degrees Rankine (R) = degrees Fahrenheit + 459·69. 1962 J. THEWLIS *Encycl. Dict. Physics* VII. 207/1 On the Rankine scale, absolute zero is zero degree Rankine, the freezing point of water is 491·7° and the boiling point of water 671·7°. 1963 EASTOP & MCCONKEY *Appl. Thermodynamics* i. 8 Note that 1 Kelvin degree is equivalent to 1·8 Rankine degree. 1974 P. L. MOORE et al. *Drilling Practices Manual* v. 131 T_s = Temperature at surface, degrees Rankine.

ranking, *vbl. sb.*[1] Add: **a.** (Later examples.) Also, the outcome of such action; ordering, classification. Also *attrib.*

1903 *Daily Chron.* 29 Sept. 3/1 The provision of additional ranking accommodation [for cabs] in suitable places. 1909 *N.Y. Even. Post* 31 Dec. 4/5 William A. Larned..and Hackett and Alexander..lead the tennis players, according to the official ranking. 1926 W. S. BRUCE *Salt & Sense* viii. 64 The men of money are supposed to be above the men of mind. That ranking is entirely wrong. 1939 A. E. TRELOAR *Elem. Statistical Reasoning* iii. 38 Arrangement of a series of values in order of magnitude is known as ranking. 1948 *Sporting Mirror* 21 May 14/1 According to American world ranking he places one higher than our own Alan Paterson. 1962 C. L. BARBER in F. Behre *Contrib. Eng. Syntax* 28 The complete ranking-order of the ten tenses in my material is as follows: 1. Present Simple Active (64%). 2. Present Simple Passive (25%) [etc.]. 1969 C. DAVIDSON in Cockburn & Blackburn *Student Power* 345 The question of ranking and university complicity with the Selective Service System needs to be tied to a general anti-draft and 'No Draft for Vietnam' movement. 1978 *Taxi* 16 Feb. 17/2 All the Airport hotels refused..to allow ranking facilities on their premises.

c. *U.S. Blacks.* (See quot. 1958.) Cf. *RANK *v.* 5 c.

1958 W. B. MILLER in *Jrnl. Social Issues* XIV. III. 16 The term 'ranking', used to refer to the pattern of intra-group repartee, indicates awareness of the fact that this is one device for establishing the intra-group hierarchy. 1974 H. L. FOSTER *Ribbin', Jivin', & Playin' Dozens* v. 183 In a Brooklyn, New York, secondary school the terms ranking and sounding are still used.

rankinite (ræ·ŋkinəit). *Min.* [f. the name of G. A. *Rankin* (1884–1963), U.S. physical chemist + -ITE[1].] A white, monoclinic polymorph of a calcium silicate, $Ca_3Si_2O_7$, of which kilchoanite is another polymorph.

1942 C. E. TILLEY in *Mineral. Mag.* XXVI. 194 For the new mineral the name rankinite is now proposed in honour of Mr. G. A. Rankin, who has contributed so much to our knowledge of the binary and ternary systems in which it was first discovered as a constituent phase. 1961 [see *KILCHOANITE]. 1966 W. A. DEER et al. *Introd. Rock-forming Minerals* I. 77 Rankinite occurs at contacts between limestone and basic igneous rocks, as at Scawt Hill, Northern Ireland, and Ardnamurchan. 1974 *Mineral. Abstr.* XXV. 322/1 Rankinite occurs in the metasomatic formations of the Anakitsk trap massif, in the Lower Tunguska area [of the U.S.S.R.], where it is associated with other high-temperature phases.

rankle, *sb.*[2] Delete *rare*[−1] and add later examples. Also without article, rankling, bitterness.

1913 H. SUTCLIFFE *Open Road* xviii. 275 She won't get the rankle out of her mind. 1922 *Blackw. Mag.* June 710/2, I had come prepared to find they loved us as much as we loved the Germans—or perhaps the rankle might be unkinder, for we did not lose the war. 1941 B. WEBB *Why does God permit Evil?* 117 Suffering ill borne causes rankle in the soul.

rankle, *v.* Add: **II. 5. b.** (Later *N. Amer.* examples.)

1962 E. LUCIA *Klondike Kate* vii. 142 It probably rankled a great many of them just to patronize his theatres. 1971 *Daily Colonist* (Victoria, B.C.) 31 Oct. 4/1 The demonstrations of glee..which understandably rankled President Nixon. 1976 *National Observer* (U.S.) 19 June 7/2 Rome's unbending attitude towards mixed marriages of Roman Catholics and non-Catholics rankled Anglicans. 1978 *National Geographic* Nov. 627/1 The low sum [of money] rankles many Yakimas, who complain of poor management.

rank order, *sb. Statistics.* Also **rank-order.** [f. RANK *sb.*[1] + ORDER *sb.*] An arrangement of the members of a set in order, with consecutive integers assigned to them. Also *attrib.*

1920 *Amer. Jrnl. Psychol.* XXXI. 32 We are left..with the rank-orders of our psychological quantities, given by reference to a fixed but arbitrary extra-psychological scale. 1959 SCHUELL & JENKINS in Saporta & Bastian *Psycholinguistics* (1961) 441/1 Table 3 presents the rank-order correlation coefficients between the order of difficulty of the tests for the entire group, and for each of the subgroups. *Ibid.*, Table 4 presents the tests on which rank-orders of the three subgroups differed by three or more ranks. 1973 *Jrnl. Genetic Psychol.* CXXIII. 79 Rank-order correlations were obtained by correlating the rank of each student on the original test with his rank on the retest. 1975 Z. GOVINDARAJULU *Sequential Statistical Procedures* iii. 259 One can derive an explicit expression for the probability of a rank order.

Hence **rank-order** *v. trans.*, to arrange in such a way.

1972 *Jrnl. Social Psychol.* LXXXVIII. 169 The concepts are rank-ordered in terms of the pre-GSRI attitude change correlations.

ra·nkshift, *sb. Linguistics.* [f. RANK *sb.*[1] + SHIFT *sb.*] A downward shift in the rank of a grammatical unit (see quot. 1966).

1961 M. A. K. HALLIDAY in *Word* XVII. 251 The theory allows for downward 'rank shift': the transfer of a (formal realization of a) given unit to a lower rank... It does not allow for upward rank shift. 1966 G. N. LEECH *Eng. in Advertising* ii. 20 *Embedding* is a shift in rank, whereby a group acts as a word, or a word acts as a morpheme, etc... The nominal genitive is another case of double rank shift: (((Mary)'s aged grandmother)'s faithful servant). The nominal groups 'Mary' and 'Mary's aged grandmother' act as morphemes. 1969 —— *Towards Semantic Descr. Eng.* iii. 52 Rank-shift or downgrading, as forms of subordination, introduce an extra factor of order into semantic structure, so that by reversing the relation of dependence between two predications, one may account for a difference of meaning. 1972 M. L. SAMUELS *Linguistic Evolution* iv. 59 Many later additions to the inventory of English prepositions resulted from a similar process of historical rankshift, e.g. *concerning, regarding, according to, owing to,* and (more recently) *due to.* 1977 *Language* LIII. 192 Although *the men's halls and the women's halls* is a group consisting of two smaller groups, B does not regard this as an instance of 'rankshift': this is restricted to cases where an element of a multivariate structure is filled by an expression of higher or equal rank.

Hence as *v. trans.*, to assign an inferior rank or function to a unit in a grammatical structure. Also **ra·nkshifted** *ppl. a.*; **ra·nkshifting** *vbl. sb.*

1964 M. A. K. HALLIDAY et al. *Linguistic Sci.* 27 In English some clauses are rankshifted and work inside the pattern of a group. If I say 'where I live it always rains', this sentence consists of two clauses, which together make up the structure of the sentence. But if I say 'the house where I live is very damp', the sentence consists, *in its structure*, of only one clause; the clause 'where I live' is rankshifted and operates in the structure of the *group* 'the house where I live'. 1969 *Eng. Studies* L. 36 We shall not go into the question of whether examples here quoted under the heading of headlines with FB structure do not really belong in the section rankshifted clauses. 1969 G. N. LEECH *Towards Semantic Descr. Eng.* ii. 26 There is no defined limit of tolerance to the depth of constituent structure produced by rank-shifting one predication inside another. 1972 *Language* XLVIII. 451 Limitations of space permit both L [*sc.* G. N. Leech] and myself only to raise, but not to answer, the question whether one can express within the notation of logic the operations of rank-shifting and downgrading. 1973 G. W. TURNER *Stylistics* iii. 82 'If you expect to get something good, for nothing, you are likely to get something good for nothing'..in the second clause 'good for nothing' is rank-shifted to behave as one word and become a single qualifier. 1977 *Language* LIII. 192 It follows that *from Glasgow* is rankshifted in *the man from Glasgow*, where it functions as qualifier in a multivariate nominal group.

ra·nnel-balk. *north. dial.* = RANNEL-TREE.

1790 GROSE *Provincial Gloss.* (ed. 2) sig. K4, *Rannel-tree,* cross-beam in a chimney, on which the crook hangs; sometimes called *Rannebauk*; North. 1817 *Edin. Monthly Mag.* June 241 The rusticity of their benisons amused me.—One wished them, 'thumpin luck and fat weans'; another, 'a bien rannle-bauks, and tight thack and rape o'er their heads'. 1859 A. WHITEHEAD *Legends of Westmorland* (1896) 11 Fair shack'd the rannel bawk et swang The keayle pot ower the grate. 1906 H. D. RAWNSLEY *Months at Lakes* 236 A great cauldron of spiced ale..hung on the 'rannel bowk', and was ladled out from time to time into basins and presented to the guests. 1910 W. G. COLLINGWOOD *Dutch Agnes* 47 In the chimney at this time of the year muttonhams hanging from the rannelbalk. 1931 H. S. WALPOLE *Judith Paris* II. i. 218 He was aware of..sides of bacon hanging, the oak settle screened by the 'heck', the 'rannel-balk' or great wooden beam across the chimney.

rannygazoo (ræ:nigăzū·). Chiefly *U.S. dial.* or *slang.* Also **ranikaboo, reinikaboo, renicky-boo.** [Origin unknown.] A prank, trick; horseplay, 'nonsense'. (See also quots. 1901, 1940.)

1901 *Dial. Notes* II. 146 *Reinikaboo*.., a newspaper story which is midway between a fake and a statement of fact; a statement of news out of all proportion and almost out of relation to the facts, yet having a certain origin and shadowy foundation. 1907 S. E. WHITE *Arizona Nights* iii. 255 'You —bluffer!' shouted a voice, 'don't you think you can run any such ranikaboo here!' 1917 *Dial. Notes* IV. 328 *Renicky*..or *renicky-boo*. 'He wants to run some sort of bluff or *renicky* on us.' 1924 WODEHOUSE *Bill the Conqueror* xi. 204 I'll hang around for a while just in case friend Pilbeam starts any rannygazoo. 1940 *Time* 14 Oct. 28 Wilkie went to N.Y.C. to tour Democratic Brooklyn... Still he refused to make a..speech, still turned down..pleas ..to let loose with a ring-tailed, rabble-rousing rannygazoo.

1947 *Sun* (Baltimore) 20 Jan. 1/2 A ranikaboo in Arizona would be known as a prank in other states. 1974 WODEHOUSE *Aunts aren't Gentlemen* vii. 59 Her lips were tightly glued together, her chin protruding, her whole lay-out that of a girl who intended to stand no rannygazoo.

ransacking, *vbl. sb.* (Later example.)

1955 H. ROTH *Sleeper* i. 10 What..happened before the ransacking?

ransom, *sb.* Add: **2. d.** (Earlier and later examples.)

1816 SCOTT *Antiquary* I. iii. 59 Could a copy [of Caxton's 'Game of Chess'] now occur..Lord only knows what would be its ransom. 1875 W. ALEXANDER *Sk. Life among my Ain Folk* viii. 133 Some said Sandy Mutch had taken the farm 'at a ransom'. 1932 A. J. CRONIN *Three Loves* II. xviii. 352 But the price of things... It's shameful. Everything a ransom now.

5. a. *ransom demand, money* (later examples), *package.*

1976 R. L. BOYER *Giant Rat of Sumatra* (1977) x. 150 None of the staff..were aware of the ransom demand. 1969 B. MALAMUD *Pictures of Fidelman* iii. 69 The insurance company..would at once kick in with the ransom money. 1976 T. HEALD *Let Sleeping Dogs Die* vii. 130 These particular villains were interested in ransom money, not selling. 1974 *Aiken* (S. Carolina) *Standard* 24 Apr. 5-A/3 Police said a dummy ransom package with a note asking the alleged abductors for more time was delivered according to the instructions Dantzler said his kidnappers had given him.

b. ransom note, a letter sent by a kidnapper or kidnappers to interested parties demanding ransom money or other satisfaction, and specifying the consequences should this not be forthcoming.

1935 M. M. ATWATER *Murder in Midsummer* xiv. 135 Maybe it's a kidnapping, but there's no ransom notes. 1975 D. PITTS *This City is Ours* viii. 28 That ship might explode if it's tampered with. That's what the ransom note says, and I believe it.

ransomite (ræ·nsəməit). *Min.* [f. the name of F. L. *Ransome* (1868–1935), U.S. mining geologist + -ITE[1].] A hydrated sulphate of copper, ferric iron, and aluminium, $Cu(Fe,Al)_2(SO_4)_4 \cdot 7H_2O$, formed as blue monoclinic crystals in a mine fire.

1928 C. LAUSEN in *Amer. Mineralogist* XIII. 221 Ransomite occurs as crusts lining cavities in the crushed rocks and also as small tufts of radiating crystals. 1970 *Ibid.* LV. 729 Ransomite..occurs only in the United Verde mine at Jerome, Arizona where it formed as the result of a mine fire.

rant, var. *RAND *sb.*[2]

ran-tan, *sb.* **2.** (Further examples.)

1936 I. L. IDRIESS *Cattle King* v. 43 When sober, he worked miracles with his limited materials. When 'on the ran-tan', which was every three months, Charlie did the cooking. 1959 G. SLATTER *Gun in my Hand* iv. 42, I remember you on that trip to Nelson for the footie. You really got on the rantan. I bet you never been back to that pub! 1963 A. PRIOR *Z Cars Again* (1964) ii. 22 Wilson had been out on the ran-tan, came home late.

ran-tan (ræn,tæn), *v. north. dial.* Now *rare.* Also **ran-dan.** [f. the sb.] *intr.* To make a noise with unruly singing, and the beating of pots and kettles, at the house of a man who has beaten his wife. Also *trans.*, with husband as obj. Hence **ra:nta·nning** *vbl. sb.*

1866 J. E. BROGDEN *Provincial Words & Expressions Lincs.* 163 *Ran-dan*, to ride the stang (or pole) connected with agricultural lynch-law, usually applied to husbands who have beaten their wives. 1886 *Folk-Lore Jrnl.* IV. 262 He'd ought to be *ran-dan'd* out o' the town. 1891 *Lincolnshire N. & Q.* II. vi. 186 As the news spreads, 'So-and-so threshed his wife yisterday mornin',' it is accompanied by the comment, 'We must 'ran-tan' him to-night.' 1928 *Observer* 26 Feb. 17/2 Seventeen villagers of Lincolnshire have been fined for 'rantanning'... Rantanning is the 'rough music' of kettle and pan, in which the rustic moralist conveys his sense of outraged propriety.

ranter, *v.* Add: **2.** To join (two edges of cloth) with fine stitching (see also quot. 1902).

1902 M. PRINCE BROWNE *Pract. Work of Dressmaking & Tailoring* II. iii. 87 *Rantering.*—This is a stitch used for joining cloth, and the two edges to be joined are put together—level..with the right side of each piece facing; they must then be neatly 'back-stitched' together by hand, as *near* to the edge of the cloth as possible... A cloth which is too thin to join by 'fine drawing', can be 'rantered' together. 1933 J. E. LIBERTY *Pract. Tailoring* iii. 24 (*caption*) Seaming and rantering. An alternative to stoting for use on loose materials (tweeds, etc.).

ranterpike (ræ·ntəɹpəik). Now *Hist.* Also **rantipike.** [Origin unknown.] (See quot. 1948.) Also *attrib.*

1892 L. ANDERSON *Among Typhoons & Pirate Craft* 38 Like the *Ranterpikes*, that used to ply between Liverpool and Glasgow, before the day of steamers, there was no taking in sail here as long as she would stand up to it. 1929 *Mariner's Mirror* XV. 316/1 They were two of the famous Ranterpikes employed carrying pig-iron and heavy castings to Liverpool. *Ibid.*, The correct name is Ranterpike, not Rantipike. 1948 R. DE KERCHOVE *Internat. Maritime Dict.* 579/2 *Ranterpike*, three-masted topsail schooner or brigantine of the River Clyde of about 250 tons... These vessels were generally engaged in the transportation of pig iron from doubled rope. 1959 S. Clark *Puma's Claw* xii. 143 In 1956 the two of us had used the technique of 'abseil' or 'rappel' for seven hours continuously on a descent in the Alps. 1965 A. Blackshaw *Mountaineering* viii. 240 Roping down... This involves:..pulling the rope round the anchor and down (the '*rappel*' from which the manœuvre takes its name). 1971 D. Haston in C. Bonington *Annapurna South Face* xvii. 209 The descent took only twenty minutes as I was so enraged with myself. I caught Nick up on the last rappel. *Ibid.* 218 Don was already fixing a rappel peg.

rappel, *v.* Restrict † *Obs. rare* to sense in Dict. and add: **2.** (ræpe·l). *intr. Mountaineering.* To make a steep descent on a doubled rope; to rope down. Hence **rappe·lling** *vbl. sb.*
1957 P. Mansfield *Final Exposure* xvi. 239 He *rappelled* downwards, with the feel of good rock beneath his feet and the bite of the rope at his back. 1969 *Time* 22 Aug. 8/2 It is a stirring demonstration ranging from scuba diving..to archery and rappelling (descending a cliff on a double rope). 1974 *Telegraph* (Brisbane) 7 Aug. 23/3 There goes Harry Garner Haskell Jr. rappelling down the side of Abercrombie and Fitch.

‖ **rappen** (rɑ·pən). Also (anglicized) **rap(p).** [Ger., pl. as sing. f. *rappe* raven.] In the German-speaking cantons of Switzerland: the Swiss centime.
1838 *Murray's Hand-bk. Travellers Switzerland* p. viii, 1 batz contains 10 rappen, and = 1½*d.* (nearly) English. The Swiss coins most frequently met with are pieces of 5 batzen, or ½ a Swiss franc; 1 batz, ½ batz, and rappen. *Ibid.* p. ix, 1 French franc = (commonly) 7 batzen or exactly 6 batzen 8 rapps. 1864 *Baedeker's Switzerland* ii. p. xix, The Swiss monetary system has since 1854 been assimilated to that of France. Coins of 5, 2, 1 and ½ fr. in silver; 20, 10 and 5 Rappen (centimes) in plated copper; 2 and 1 Rappen in copper. 1 fr. = 100 Rappen = ..9/½d. 1911 *Encycl. Brit.* XIX. 907/2 Like Belgium, Switzerland had before her adhesion to the Latin Monetary Union adopted the French system, with the franc of 100 centimes or rappen as the unit of value. 1938 *Baedeker's Switzerland* iii. p. xvii, The monetary unit is the *Swiss franc* (fr.) of 100 *Centimes* (c.) or *Rappen* (in the German-speaking cantons). 1960 H. Hayward *Antique Coll.* 233/2 *Rappen*, small Swiss copper coin of late 18th and early 19th cent. with types of shield in wreath and value and date. 1962 R. A. G. Carson *Coins* 326 In 1798 Switzerland was invaded by the revolutionary armies of France and a Helvetic republic was established and a new unified and decimal coinage was instituted with the franc as the unit, divided into 10 batzen, the batz divided in turn into 10 rappen. The system included pieces of 32 and 16 francs in gold; 40, 20, 10 and 5 batzen in silver and in billon the batz, its half and the rap. 1973 *Daily Tel.* 4 Aug. 15/2 In Zurich..the dollar lost 1 rappen to 2·84½.

rapper. Add: **1. a.** (Later example.)
1857 P. Cartwright *Autobiogr.* xix. 276 There is a dark, motley crowd of..spiritual rappers, so called.

† **b.** *slang.* One who tells a downright lie; a professional perjurer. Cf. sense 3 a and Rapping *vbl. sb.*[1] 2 b. *Obs.*
1840 in Fielding *Jonathan Wild* (new ed.) p. lxii, The rapper, I think (as the cant phrase has it), is the most necessary man for your purpose.

c. *U.S. slang.* A complainant, plaintiff; a prosecutor. Cf. *Rap *v.*[1] 1 c.
1904 'No. 1500' *Life in Sing Sing* 252/1 *Rapper*, prosecutor; complainant. 1910 *New England Mag.* July 587 A complaint or charge of crime is a 'rap' and the complainant is the 'rapper'. 1926 *Clues* Nov. 158/2 *Lam* up to the *pagey* and see the *rapper*. 1935 *Jrnl. Abnormal Psychol.* XXX. 364 *Rapper*, the complainant in the case. 1955 D. W. Maurer in *Publ. Amer. Dial. Soc.* xxiv. 108 A *rapper* is a mark who prefers charges against a pickpocket; sometimes he can be bought off with a return of his money. 'If the law won't take, the next thing is to try the rapper.'

d. An itinerant purchaser of antiques; esp. one who buys valuable objects cheaply from credulous householders. Cf. *Knocker 2 e.
1914 H. A. Vachell *Quinneys'* x. 146 Gossip had it that he had begun life as a 'rapper'. 1928 *Daily Express* 29 Feb. 14 The rapper frequently adopts the Aladdin method of offering new lamps for old, undertaking, for example, to give the innocent owner of a little old Queen Anne bureau a brand-new chest of drawers and a cash sum in exchange.

e. *U.S. slang.* A talker; a chatterer. Cf. *Rap *v.*[1] 3 d.
1971 *Time* 22 Feb. 38 Boulez clearly hopes there will be as many rappers as listeners. 1973 C. Milner *Black Players* i. 8 He is recognized as among the best talkers or 'rappers' in the hustling world. 1972 C. Weston *Poor, Poor Ophelia* iii. 18 Let's you and me talk like crazy. You look like a rapper, and I feel like rapping.

rapper: for 'obs. Sc.' in Dict. read 'dial.'

rapping, *vbl. sb.*[1] Add: **1.** Also, = Spirit-rapping (earlier and later examples). Also *transf.*
1848 M. Fox in A. Conan Doyle *Hist. Spiritualism* (1926) I. iv. 62 The children, who slept in the other bed in the room, heard the rapping, and tried to make similar sounds by snapping their fingers. 1894 A. Lang *Cock Lane* 29 The facts of rappings, ghosts, clairvoyance..are very doubtful facts after all. 1923 *Times Lit. Suppl.* 6 Dec. 853/3 He writes in short snappy sentences, the persistent rapping of which makes it very difficult for the reader to keep up his attention for long. 1954 J. F. Rinn *Searchlight on Psychical Res.* vi. 56 'What are the points of your coming exposé?' asked Creelman. 'First the rappings,' Mrs. Kane smiled. 1974 *Encycl. Brit. Macropædia* XVII. 513/1 The spirits,..it is alleged, use different methods [of communication]..., such as rappings, table tippings, [etc.].

2. c. *colloq.* (orig. *U.S.*). The action or practice of talking or chatting; conversation, gossip. spec. *U.S. Blacks*, repartee, banter. Cf. *Rap *v.*[1] 3 d.
1969 *Observer* 16 Feb. 40/6 Pendennis boogaloos, falls by cats into numbers, and lays down heavy outasight rapping. Translation: Pendennis arranges events, calls on people in the money, and produces a lot of fantastic gossip. 1969 T. Kochman in *Trans-Action* Feb. 27/1 While often used to mean ordinary conversation, rapping is distinctively a fluent and lively way of talking, always characterized by a high degree of personal style. *Ibid.* 27/2 Rapping between men and women often is competitive and leads to a lively repartee. 1972 M. J. Bosse *Incident at Naha* i. 60 After this failure of communication, our rapping dragged on like that of two old ladies. 1973 *Black World* Mar. 85 Jiving, bopping, rapping, signifying, sounding—all modes of Afro-American expression—seek to affirm the vitality of the Black American experience. 1977 *Rolling Stone* 13 Jan. 60/2 (Advt.), Multi-racial magazine features workable self-help articles, rapping, fiction.

3. *attrib.*, as **rapping bar**, a pointed iron bar used in founding for loosening patterns from moulds; **rapping iron**, an implement used in basketry to tap the rows of weaving into the desired position; **rapping plate** *Founding*, a metal plate attached to a pattern in order to prevent damage to the pattern when it is loosened from the mould.
1888 *Lockwood's Dict. Mech. Engin.* 277 Rapping bar. 1948 H. W. Baker *Mod. Workshop Technol.* i. iii. 60 For a wood pattern the rapping bar may be pointed and driven into the wood itself; or special rapping plates with suitable holes may be fixed to it. 1960 R. Lister *Decorative Cast Ironwork in Gt. Brit.* ii. 31 The rapping bar is tapped with a mallet to loosen the pattern. 1924 C. Crampton *Cane Work* 12 Rapping iron to tap the rows of weaving to make the work quite level. 1979 *Dryad Catal.* 87/1 Canework tools..Rapping iron for levelling (or regulating) cane weaving. 1885 *Pattern Making* xxii. 158 (*heading*) Rapping plates. 1894 W. J. Lineham *Textbk. Mech. Engin.* ii. 67 Rapping plates have become necessary in order to prevent injury to the pattern by the moulder. 1948 Rapping plate [see *rapping bar* above]. 1960 R. Lister *Decorative Cast Ironwork in Gt. Brit.* ii. 25 To further facilitate the withdrawal, metal rapping and lifting plates are available.., ready drilled and countersunk for screwing to the pattern, and tapped to receive the lifting screw.

rapping, *ppl. a.* **1.** (Further example.)
1933 N. Fodor *Encycl. Psychic Sci.* 321/2 *Rudolf of Fulda*, a chronicle dating from 858 A.D. speaks of communications with a rapping intelligence.

rapport, *sb.* Add: Now usu. with pronunc. (rɑpō^ə·ɪ, ræ-). **2. a.** (Further examples.) Also, harmonious accord, co-ordination. Now freq. used of relations between persons.
1915 'W. N. P. Barbellion' *Jrnl.* Apr. in *Enjoying Life* (1919) 67 It wounds my self-esteem not to be..in direct telepathic rapport with the universe and its beauty. 1919 *Lancet* 8 Feb. 206/2 A lack of complete *rapport* between the muscles and the brain nerve centres. 1934 C. Lambert *Music Ho!* iii. 175 This lack of rapport between the tune and harmony is particularly noticeable in some of the later works of Bartók. 1941 A. White *Let.* 7 Feb. in *Hound & Falcon* (1969) 114 It was one of our good 'rapports' that I should have hit on Huvelin's admirable sayings just as you had written to me quoting some of them. 1957 J. Kerouac *On Road* (1958) iii. xi. 247 Dean's second baby, the result of a few nights' rapport. 1969 *Morning Star* 18 Nov. 2 The animal kingdom is in rapport with its environment, but man is not, unless he makes it so by taming his planet and remaking it to suit his needs. 1976 J. I. M. Stewart *Memorial Service* vii. 104 Jiffy temperately signalled the gratification proper to be felt on establishing this sort of *rapport* with a stranger.

b. (Earlier and later examples.) Also, a feeling of sympathy and co-operation between therapist and patient or tester and subject that is considered necessary for successful therapy or psychological testing. Also *attrib.* and *Comb.*
1845 Poe in *Amer. Rev.* Dec. 564/2, I endeavored to place each member of the company in mesmeric *rapport* with him. 1894 Creighton & Titchener tr. *Wundt's Lect. Human & Animal Psychol.* xxii. 331 There then arises what the animal-magnetism school term the *rapport* of the medium with the magnetiser. 1923 J. T. MacCurdy *Probl. Dynamic Psychol.* xi. 121 Manipulation of *rapport* is thus made the core of psychoanalytic treatment. 1924 J. Riviere tr. *Freud's Coll. Papers* I. 293 In his treatment of her case Brewer could make use of a very intense suggestible rapport on the part of the patient. 1954 A. Anastasi *Psychol. Testing* iii. 50 It is apparent that the establishment of rapport, prior to the administration of the test, is generally an important part of the testing procedure. 1965 F. M. Lopez *Personnel Interviewing* i. 12 Then there are the *rapport* behaviors that help the interviewer to establish a bond of communication with the interviewer. 1970 A. Magonet *Psychotherapy by Hypnosis* i. 3 Mesmer discovered that it was important that there should exist between physician and patient a close interest in and sympathy for each other. He described this as rapport. 1971 D. B. Peskin *Human Behavior & Employment Interviewing* vii. 170 If the response seems neutral, the interviewer applies more of the same rapport-producing small talk and humor. 1976 A. Anastasi *Psychol. Testing* (ed. 4) ii. 34 The training of examiners covers techniques for the establishment of rapport.

3. Substitute for def.: in a state of rapport or close and harmonious relation; in sympathy, in harmony. Usu. of persons. (Further examples.)
1846 Geo. Eliot *Let.* Apr. in J. W. Cross *George Eliot's Life* (1885) I. ii. 116 See what it is to have a person *en rapport* with you, that knows all your thoughts. 1884 *Proc. Soc. Psychical Res.* II. 127 A mesmerised 'subject' who is sufficiently *en rapport* with his mesmeriser. 1933 Wodehouse *Mulliner Nights* v. 151 'I fear we were not exactly *en rapport*,' sighed Sacheverell.

‖ **rapportage** (rapɔ̄rtāʒ). [Fr., 'tale-telling'; Eng. usage is influenced by Reportage.] The reporting or describing of events in writing; mere description, uncreative accounting. Also *transf.*
1903 *Independent* 22 Jan. 210/2 It has nearly disabled prose fiction..by making of it a trade or handicraft, an affair of *rapportage*. 1935 S. Spender *Destructive Element* i. 26 James's earlier books are much fuller of descriptive writing and *rapportage* than the later books. 1957 *Times Lit. Suppl.* 8 Nov. 667/2 In describing his visits to the three North African territories in 1956 he has produced the modern counterpart to those brilliant pieces of pre-war *rapportage*. 1960 K. Clark *Looking at Pictures* 145 The *Snowstorm* is very far from rapportage. It is the essence of all that Turner had discovered about himself and his art during forty years of practice. 1966 C. Mackenzie *My Life & Times* V. 141, I decided it was too soon to write a novel about the war. Experience which had not been 'cooled a long age' might produce *rapportage* instead of genuine creative work.

rapporteur. Restrict † *Obs. rare*[−1] to sense in Dict. and add: ‖ **2.** A person who prepares an account of the proceedings of a committee, etc., for a higher body. Cf. Reporter 1 c.
1791 Ld. Palmerston *Diary* 13 July in O. Browning *Despatches Earl Gower* (1885) 290 The Rapporteur said that the Committees did not consider the king's flight as a constitutional crime in him. 1927 *Daily Tel.* 8 Mar. 11/5 The representative of Holland, as the rapporteur, submitted a report of the Permanent Mandates Commission which was adopted by the Council. 1937 *Nature* 17 Apr. 683/1 Two *rapporteurs* were appointed to summarize the papers. 1949 I. Deutscher *Stalin* v. 143 At the conference he was the *rapporteur* on the problem of nationalities. 1955 *Times* 1 July 8/5 The congress..has appointed five *rapporteurs* on economic and financial policy, agriculture, foreign affairs, and the French oversea territories. 1964 *Ann. Reg. 1963* 175 The recommendation submitted by the Defence Committee's rapporteur..urged full nuclear co-operation between Britain and France. 1977 *Daily Tel.* 24 May 16 Normanton is to be the rapporteur of a study into the industrial and economic aspects of buying arms through a Common Market agency.

rapprochement. (Further examples.)
1905 *Daily Chron.* 19 Jan. 4/3 The retiring Premier is fully justified in claiming as a distinctive mark of his Government that it aimed at a 'rapprochement between peoples'. 1934 C. Lambert *Music Ho!* iii. 207 The same rapprochement between highbrow and lowbrow..can be seen in literature. 1940 [see *Artify *v.*]. 1955 *Times* 17 May 9/5 They should not assume that a Soviet–Yugoslav *rapprochement* was outside the framework of current conciliatory developments. 1958 L. Durrell *Mountolive* xii. 228 With some half-formulated idea of offering the silent figure another chance to open up a discussion with him or seek a *rapprochement*, he rode his horse into the courtyard. 1974 J. White tr. *Poulantzas's Fascism & Dictatorship* iv. iii. 211 It was only with the split in the Socialist Party, in October 1922, that rapprochement with the Maximalists was attempted. 1979 *Dædalus* Winter 78 The Community's structures were designed..to accommodate the Franco-German rapprochement.

rapsca·llionism. *rare.* [f. Rapscallion + -ism.] Rapscallions collectively; the conduct or condition of rapscallions.
1920 Galsworthy *In Chancery* iii. ix. 288 Soho seemed more than ever the disenchanted home of rapscallionism.

rapscallionly, *a.* (Further example.)
1899 G. B. Shaw *Let.* 29 Sept. (1972) II. 106. There being between Maggie [*sc.* Margaret Hobhouse] & myself a sort of rapscallionly freemasonry.

rapt, *pa. pple.* (and *pa. t.*). Add: **I. 3.** Also const. *away.*
1924 A. D. Sedgwick *Little French Girl* ii. xii. 194 The heir..stood with his little shoulders screwed up, his elbows in his hands, rapt away from shyness and self-consciousness by his sincere delight.

raptly, *adv.* Add: **c.** Intently, concentratedly, absorbedly.
1924 E. Power *Medieval People* ii. 52 The crowd of Venetian prisoners and Genoese gentlemen, raptly drinking in all the wonders of Kublai Khan. 1971 *Daily Tel.* 13 Feb. 7/1 There was a delightful sequence of a 7 year-old girl 'cellist playing raptly and well for her judges.

raptor. 1. For † *Obs.* in Dict. read *rare* and add: (Later example.) Also, an abductor.
1884 Addis & Arnold *Cath. Dict.* 436/1 The councils..prohibit subsequent marriage between the raptor and his victim. 1975 *Daily Tel.* 12 Aug. 12 What I had in mind was the ruling that there was no rape provided the raptor believed..that the woman consented.

3. (Later examples.) Also *attrib.* and *fig.*
1933 *Condor* XXXV. 19 (*heading*) Food habits of Southern Wisconsin raptors. *Ibid.*, The ex post facto recording of raptor kills encountered in the field should not be regarded as a proper source of quantitative data. 1963 D. P. Mannix *All Creatures Great & Small* vi. 86 Raptors (birds of prey) do not have the nervous, active minds of the Corvidæ. 1974 *Sci. Amer.* Dec. 156/3 A cruel human raptor..boasted of killing golden eagles with a sawed-off shotgun as he flew beside them. 1976 *Nature* 23 Sept. 321/1 Other relevant predators such as large raptors or felids do not occur on these islands. 1980 *Observer* (Colour Suppl.) 6 Jan. 57/1 Hundreds of thousands of raptors—birds of prey such as stork, buzzard, kite and eagle—commute across this corner of north-western Turkey.

raptorial, *a.* (and *sb.*) Add: **1. a.** *raptorial bird* (further examples).
1919 T. A. Coward *Birds Brit. Isles* I. 330 Most raptorial birds are variable in plumage. 1931 —— *Life of Birds* vi. 46 The flesh-eating or raptorial birds kill birds which we call useful as well as those which are troublesome. 1968 *Nature* 14 Dec. 1098/1 Declining populations of raptorial and fish-eating birds. 1976 *Field* 30 Dec. 1281/1 Goats shot this year have helped to nourish a foundation stock of ten white-tailed eagles, brought over from Norway by special arrangement with the government of that country in an attempt to reintroduce a raptorial bird once native to Britain.

2. (Later examples.) Also *fig.*
1916 *Sci. Progress* XI. 245 'Canto di cigno'..—a droll metaphor having regard to Spallanzani's raptorial countenance. 1933 *Condor* XXXV. 19 Erroneous impressions obscure the true proportions of one prey species to another in raptorial diets. 1955 W. Gaddis *Recognitions* ii. vii. 620 At that instant the room was pierced by a raptorial cry like that of the bird descending. 1973 *Nature* 20 July 179/2 Planktonic crustacea exhibit selective feeding behaviour ranging from passive size selection in *Daphnia* to raptorial feeding in cyclopoid copepods.

rapture, *sb.* Add: **5. e.** *rapture(s) of the deep* or *depths*, nitrogen narcosis.
1953 J. Y. Cousteau *Silent World* ii. 14 (*heading*) Rapture of the deep. *Ibid.* 21 We called the seizure *l'ivresse des grandes profondeurs* (rapture, or 'intoxication', of the great depths). 1955 R. & B. Carrier *Dive* iii. 77 This nitrogen narcosis or 'rapture of the depths'..seems to have varying effects on different types of people. 1955 J. Sweeney *Skin Diving & Exploring Underwater* vii. 89 It has not been established with absolute certainty how nitrogen causes a narcotic effect ('raptures of the deep') when breathed under high pressure. 1962 [see *nitrogen narcosis]. 1971 J. F. Bernard tr. *Cousteau's Life & Death in Coral Sea* 261 The diver's threshold of susceptibility to rapture of the depths can be pushed back..by replacing the nitrogen in one's breathing mixture by a lighter gas, such as helium. 1974 *Petroleum Rev.* XXVIII. 672/1 Nitrogen narcosis, popularly called 'raptures of the deep' but perhaps more accurately described as 'the uglies', is the malady caused by nitrogen under pressure, interfering with the normal function of the nervous system.

rapture, *v.* Add: **b.** *intr.* To express oneself in raptures; to take rhapsodic delight *in* or display ecstatic excitement *over* something.
1908 A. Lipdegraff in *Smart Set* June 133, I rapture in some lonely night-bird's cries. 1965 E. O'Brien *August is Wicked Month* i. 12 She went out and raptured over the tent and said what a genius he was.

raptus. Add: **2.** (Further examples.) Also, an instance of this.
1902 W. James *Var. Relig. Exper.* xvi. 412 In the condition called *raptus* or ravishment by theologians, breathing and circulation are so depressed that it is a question among the doctors whether the soul be or be not temporarily dissevered from the body. 1964 L. Woolf *Beginning Again* i. 32 Beethoven, every now and again, used to have what his faithful disciple called 'a raptus', a kind of volcanic creative outburst... The raptus or inspiration is clearly only a rare and wonderful form of a well-known everyday mental process. 1977 A. Sheridan tr. *J. Lacan's Écrits* vi. 207 The subject had his first attack of anxious confusion with suicidal raptus.

raquette. (Earlier and later examples in sense 3 a of Racket *sb.*[2])
c 1665 P. E. Radisson *Voyages* (1885) 66 We found snowes in few places, saving where the trees made a shaddow, wch hindred the snow to thaw, wch made us carry the raquetts. 1760 T. Jefferys *Nat. & Civil Hist. French Dominions* I. 57 The texture of the raquette or snow-shoe, consists of straps of leather about two lines in breadth, bordered with some light wood hardened in the fire. 1849 J. E. Alexander *L'Acadie* II. 19 It was ludicrous to witness the mishaps of those who figured on the broad racquettes for the first time. 1965 *Canad. Geogr. Jrnl.* Feb. 62/2 The 'raquettes' of today are strung, usually, with especially prepared cowhide, rather than the original deerhide.

ra-ra, var. *Rah-rah *a.*

‖ **rara avis** (rē^ə·rā ē^ɪ·vis, rā·ra a·vis). Pl. **rara avises,** ‖ **rarae aves.** [L., 'rare bird' (Juvenal *Sat.* vi. 165; cf. also Persius *Sat.* i. 46).] **1.** A person of a type seldom encountered; an exceptional person, a paragon. Occas. without article.
1607 G. Wilkins *Miseries Inforst Mariage* sig. A3^v, And by that, thou hast beene married but three weekes, tho thou shouldst wed a *Cynthia rara avis*, thou wouldest be a man monstrous: A cuckold, a cuckold. [1654 E. Gayton *Pleasant Notes Don Quixot* iv. xv. 251 But all to Donna Clara The judges daughter yield, shee's *Avis rara*.] 1748 Smollett *R. Random* II. li. 166 Calling my Lord Strutwell by the appellations of Jewel, Phœnix, *Rara avis*. 1749 Fielding *Tom Jones* III. viii. i. 150 A single instance..is not sufficient to justify us, while we are writing to thousands who never heard of the person, nor of any thing like him. Such Raræ Aves should be remitted to the epitaph-writer. 1813 Southey *Let.* 30 Nov. (1856) II. 338, I sent them, thinking that a char in London must be like a tortoise-shell Tom cat, a *rara avis*. 1825 Mill in *Jrnl. Adult Educ.* (1929) IV. 54 A good doubter may as yet be truly pronounced to be *rara avis*, etc. 1852 'G. Greenwood' *Haps & Mishaps* (1854) iv. 91 A pretty Irish peasant girl we found the rarest of *rara avises*. 1892 D. H. Tuke *Dict. Psychol. Med.* 854 A good nurse for neurotic patients is a *rara avis* indeed. 1919 V. Woolf *Night & Day* xii. 154 'But I do read De Quincey.., more than Belloc and Chesterton, anyhow.' 'Indeed!' exclaimed Mrs Cosham...'You are, then, a *rara avis* in your generation.' 1931 'G. Trevor' *Murder at School* i. 13 That *rara avis*, the headmaster who was also a man of the world. 1955 *Times* 23 May 3/3 The harpist is a *rara avis* among recitalists. 1975 *Listener* 31 July 140/1 Peter Jay is the *rara avis* of broadcasting. He is the high-flyer who descends from the thunderous crags of the *Times* to perch, with no hint of condescension, on the television aerials of the nation.

2. That which is seldom found; an unusual occurrence, etc.; something very remarkable.
1884 J. J. Hissey *Old-Fashioned Journey* vii. 99 A perfect day with us is somewhat of a *rara avis*. 1906 [see *Redouble *sb.*[2]]. 1942 *Burlington Mag.* Feb. 50/2 Truly a *rara avis* among books of memoirs! 1979 *Guardian* 15 Mar. 11/1 It is a moving document, not to say a *rara avis*.

3. *lit.* A rare bird.
1891 H. M. Smith in *Proc. U.S. Nat. Museum* XIII. 171 The Dick Cissel [*sc.* a bird]..is now a veritable *rara avis*.

rare, *a.*[1] (*adv.*[1] and *sb.*) Add: **5. c.** *rare earth* (Chem.), any naturally occurring oxide of an element of the lanthanide series (usu. including lanthanum and freq. also scandium and yttrium); also (*loosely*), any of these elements themselves; a lanthanide. Hence *rare-earth element, metal.*
In earliest examples used in a less restricted sense, including oxides of other metals.
1875 *Jrnl. Chem. Soc.* XXVIII. 1001 (*heading*) The quantivalence of the metals of the rarer [*sic*] earths [= tr. G. seltenen Erdmetalle]. 1877 *Ibid.* XXXI. 49 (*heading*) The quantivalence of the rare earth-metals. 1878 *Chem. News* 13 Sept. 136/2, I found nearly one-half of the known elements represented, and separated a group of oxides belonging to the rare earths. 1902 *Encycl. Brit.* XXVI. 710/1 Helium has since been extracted from a variety of minerals consisting of salts of uranium, yttrium, thorium and other rare earths. *Ibid.* 710/2 In the case of the rare-earth metals, the elements from air and the radio-active elements, the discovery of new elements has frequently been consequent on the introduction of novel methods. 1933 *Discovery* Jan. 1/1 Samarium has an atomic number 62 and a mass of about 150. Chemically it is a 'rare earth'. 1937, etc. [see *Lanthanide]. 1958 *Optima* Mar. 22b/2 Thorium occurs chiefly in association with the rare-earth elements in monazite ore. 1959 *Nomencl. Inorg. Chem.* (I.U.P.A.C.) 6 The name rare-earth metals may be used for the elements Sc, Y, and La to Lu inclusive. 1965 D. Abbott *Inorg. Chem.* i. 36 The third transition series is interrupted after lanthanum by a set of elements known as the Rare Earths or Lanthanons. 1968 *Times* 18 Oct. 16/8 Gadolinium belongs to the little known family of minerals called the rare earths... The name rare earth is something of a misnomer, since many of the minerals in the family are widely distributed throughout the earth's crust. 1972 *Nature* 31 Mar. 197/1 Also concentrated in this liquid are a group of trace elements, including barium, yttrium, rare earths, zirconium, hafnium, phosphorus and niobium, about which are centred many of the geochemical arguments concerning the differences between terrestrial and lunar rocks.

d. *rare bird* = *Rara avis 1, 2.
1890 G. B. Shaw in *Star* 21 Feb. 2/5 She [*sc.* the perfect dancer] is the rarest of rare birds. 1912 R. Lynd *Rambles in Ireland* i. 45 In Ireland..one drunk man is as conspicuous as a thousand sober ones... But he is, comparatively speaking, a rare bird and an exception for all the show he makes. 1934 C. Day Lewis *Hope for Poetry* x. 66 The true lyric poet is a very rare bird indeed. 1950 W. Saroyan *Assyrian* 21 The writer who is only a writer is a rare bird these days, most writers having taken posts at universities, [etc.]. 1962 *Which? Car Suppl.* Oct. 141/1 The foreign cars are still somewhat rare birds. 1977 *Times* 20 Oct. 16/3 That rare bird, a historian who was also a history-maker.

e. *rare book*, a book which is in demand and made valuable by its actual or prospective rarity. Also *attrib.*
[1862 J. H. Burton *Book-Hunter* ii. 210 David Clement..lays it down with authority, that 'a book which is difficult to find in the country where it is sought ought to be called simply rare'.] 1895 W. Roberts *Rare Bks.* 7 This is a very unusual contingency even in the history of rare books. 1910 A. W. Pollard in *Encycl. Brit.* IV. 223/1 The Boccaccio ..went to Earl Spencer (d. 1834) for £750, to pass with the rest of his rare books to Mrs Rylands in 1892. 1930 R. Curie *Collecting Amer. First Editions* vi. 168 (*heading*) Rare books still obtainable. 1941 B. Schulberg *What makes Sammy Run?* v. 97 One of the finest collections of rare books in the country. 1948 J. Carter *Taste & Technique in Book-Collecting* p. ix, I am the first member of the rare book trade to have been appointed Sandars Reader. 1952 —— *ABC for Book-Collectors* 148 The First Folio Shakespeare and the Gutenberg Bible are certainly 'rare books' as the term is generally understood. 1967 E. Grierson *Crime of one's Own* i. 12 The 'rare book' trade, which he was trying to build up. 1976 'O. Bleeck' *No Questions Asked* ii. 22 The Library of Congress is an interesting place... The reading room of the Rare Book Division turned out to be a peaceful place. 1980 *Times* 29 Jan. 14/6 It was a characteristic out-of-season sale, containing faulty copies of rare books.

f. *rare gas* (Chem.) = *inert gas* (*b*) s.v. *Inert *a.* 1 c. (Cf. sense 1 a in Dict.)
1901 M. W. Travers *Exper. Study Gases* xi. 116 Mercury, which somewhat resembles the rare gases with regard to its low boiling-point and monatomic character, would be chemically inactive at 1000° C. 1937 *Discovery* Aug. 227/2 The separation of the 'rare gases' from the atmosphere. 1963 J. H. Pomeroy in H. H. Hyman *Noble-Gas Compounds* iii. 125, I..suspect that the discovery of the rare-gas compounds has been greeted with particular enthusiasm by the producers of textbooks in chemistry, since it gives them..a sort of windfall of unplanned obsolescence. 1972 *Nature* 8 Dec. 345/2 Radon (^{222}Rn)—a radioactive rare gas emanating essentially from large continental land areas.

9. Also used advb. (mainly *poet.*) with adjs. or ppl. adjs. *rare-bred, -coming, -composed, -dear, -feathered, -felt, -grown, -seen, -spoken, -veined.*
1877 *Coursing Calendar Autumn 1876* 327 Westeria..was put out in a bad trial. She is a rare-bred one, being by Contango out of Joan-of-Arc. 1937 Blunden *Elegy* 90 Where the dogs..regard The rare-coming stranger in the yard An excitement not to be missed. 1601 Marston *Jack Drums Entertainment* sig. A2^v, If he could..distill the quintessence of heauen In rare composed Sceanes. 1876 G. M. Hopkins *Wreck of Deutschland* xxxv, in *Poems* (1967) 63 Let him [*sc.* our King]..be a crimson-cresseted east, More brightening her, rare-dear Britain, as his reign rolls. 1904 *Westm. Gaz.* 19 Nov. 9/2 The class for pied, albino, or rare-feathered British birds contains a pure yellow, pink-eyed, yellow-hammer. 1785 T. Dwight *Conquest of Canäan* iv. 97 A rare-felt joy inspir'd the friendless band. 1922 Blunden *Shepherd* 85 And rare-grown daisy in the meadow. 1794 T. Dwight *Greenfield Hill* I. 18 The rare-seen felon startles every mind And fills each mouth with news. 1915 D. H. Lawrence *Rainbow* i. 8 A rare-spoken, almost surly man. 1879 G. M. Hopkins *Duns Scotus's Oxford* in *Poems* (1967) 79 Of realty the rarest-veinèd unraveller.

rare, *a.*[2] Add: **a.** (Later U.S. examples.)
1836 *Public Ledger* (Philadelphia) 19 Apr. 1/3 [Certain persons] in calling for boiled eggs, instead of ordering them to be done rare, order them to be '*boiled soft*'. 1856 *Knickerbocker* XLVII. 249 'Do you like your eggs done rare?' asked the good landlady.

b. Now usu. of grilled beef-steak. (Now also quite customary outside American English.)
1904 *N.Y. Sun* 6 Aug. 5 The waiter took his order for a sirloin rare. 1911 E. Ferber *Dawn O'Hara* ii. 20 I've devoured rare porterhouse and roast beef day after day for weeks. 1940 R. Chandler *Farewell, my Lovely* xxv. 186 Would you like your steak rare or medium, sir? 1977 *Times* 24 Aug. 14/8 A reader ordered a steak, rare, at a Yorkshire roadside café.

rare, *v.* orig. *U.S.* and *dial.* [Var. of Rear *v.*[1]] **1. a.** *intr.* = Rear *v.*[1] 15.
1833 *Sketches & Eccentricities D. Crockett* vii. 92 He just *rared* up upon his hind legs. 1898 H. S. Canfield *Maid of Frontier* 100 Break 'em with a curb an' they rare an' fall back on you. 1938 [see *Jes, jes'].

b. colloq. *to be raring to* (*go*, etc.), to be extremely eager or fully ready to (do something). Also *transf.*
1909 E. Banks *Mystery F. Farrington* iv. 13/2 They make me raring, tearing mad to look at 'em. 1927 F. N. Hart *Bellamy Trial* i. 10 Both sides are rarin' to go, and they are not liable to touch their peremptory challenges [of jurymen]. 1935 Wodehouse *Luck of Bodkins* xv. 167 Keep it crisp, because I'm raring to go. 1957 A. MacNab *Bulls of Iberia* viii. 79 The bull was a *toro de bandera*, the bravest of the brave,..and was 'rarin' to fight'. 1971 J. Tyndall *Death in Lebanon* vii. 114 He's laid it on that the preacher makes some inflammatory remarks..so that the congregation..will be rarin' to go. 1979 *Church Times* 9 Feb. 9/1 We were at the starting-gate and raring to go.

2. *trans.* = Rear *v.*[1] 9.
1901 M. Franklin *My Brilliant Career* v. 24 It was my duty to 'rare the poddies'. 1961 'F. O'Brien' *Hard Life* ix. 67 Well, there's no doubt about it, we rare up strange characters in this country.

raree-show. Add: **2. a.** (Further examples.)
1931 Blunden *To Themis* 22 Colours flying, drums drubbing, boys run miles for the raree-show. 1955 W. Gaddis *Recognitions* ii. vii. 634 He'll show you... He'll put up a real maudlin raree-show for you. 1971 *Daily Tel.* 8 Nov. 9/1 Religious people today protest against 'Jesus Christ Superstar' as a vulgar raree-show cashing in on an adolescent fad.

3. *raree-show box* (earlier example), *-performance.*
1737 *London Mag.* June 324/2, I presume that he [*sc.* Punch] will not be tolerated, either upon the Stage, or even in a Raree-shew Box. 1812 S. Jones *Baker's Biographia Dramatica* (rev. ed.) III. 306 A frivolous raree-show performance.

rarefactional (rē^ərifæ·kʃənăl), *a.* [f. Rarefaction + -al.] Characterized by rarefaction.
1909 in *Cent. Dict.* Suppl. 1971 I. G. Gass et al. *Understanding Earth* xix. 272/1 On either side of the fault, the first waves to leave the source are compressions in the direction of motion and rarefactions away from it. This gives the symmetrical distribution of compressional and rarefactional first motions shown. 1972 *Sci. Amer.* May 58/3 After an earthquake one finds that the seismological stations that have received the first waves can be assigned to one of four geographic quadrants. In two of the quadrants, lying opposite each other, the first waves are compressional; in the other two quadrants the first waves are rarefactional.

rarefied, *ppl. a.* Add: Also *transf.* and *fig.*
1961 *Blackw. Mag.* Oct. 290/1 From the light of common day into the rarified atmosphere of the late eighteenth century. 1977 G. Michanowsky *Once & Future Star* iv. 33 In the rarefied world of cuneiform scholarship, it is known as BM—86378. 1978 N. Moss *What's the Difference?* (ed. 2) 93 *Professor, n*—a less rarefied post than at a British university, since there are usually several professors to a department.

rarefy, *v.* Add: **2. d.** *intr.* To discourse exaltedly. *nonce-use.*
1928 Blunden *Undertones of War* iv. 44, I remember how Limbery-Buse and myself chirped and rarefied over some crayfish and a great cake.

Rarey (rē^ə·ri). The name of the horse-breaker J. S. *Rarey*, used *attrib.* and in the possessive to denote methods or equipment employed by him for the taming of horses. Hence **Ra·reying**, the action or fact of breaking in a horse by Rarey's methods. Cf. Rareyfy *v.*
[1856 J. S. Rarey (*title*) The modern art of taming wild horses]. 1875 S. Sidney *Bk. of Horse* xxvi. 562 The Rarey principle consists in teaching the colt as much as possible without putting him in any pain, and without frightening him by any strange sight or sound. *Ibid.* 565 (*caption*) Horse, with Rarey fittings. *Ibid.* 567 The application of the Rarey straps in the following manner affords a better chance of success than the ordinary exhausting plans of old-fashioned colt-breakers and of circus-riders. 1896 M. H. Hayes *Illustr. Horse-Breaking* (ed. 2) iv. 124 Having 'picked up' the foot, we may secure it..by Rarey's leg-strap, which is about 3 ft. long, and is furnished at one end with a buckle, below which a leather 'keeper' is placed on both sides. *Ibid.* 175 Mr. Norton Smith adopted..a modification of Rarey's system. 1905 S. Galvayne *20th Century Bk. on Horse* 100 (*heading*) The Rarey system. *Ibid.*, It may not be uninteresting to the reader to briefly explain the method of 'Rareying' a horse. 1911 *Encycl. Brit.* XIII. 725/1 The method of subduing a colt by 'galvayning' is as good as any. It is a more humane system than 'rareying', which overcame by exhaustion under circumstances which were not fruitful of permanent results. 1942 Rareying [see *Galvayne]. 1979 *Jrnl. R. Soc. Arts* Oct. 724/2 The 'Rarey' method of throwing a horse is explained.

raring: see *Rare *v.* 1 b.

‖ **rariora** (rē^əriō^ə·rā, rari-), *sb. pl.* [L., neut. pl. comparative of *rārus* rare.] Rare books. Cf. *Rare *a.*[1] 5 e.
1863 *Macm. Mag.* VIII. 36 (*heading*) Rariora of old poetry. 1908 *Daily Report* 26 Aug. 8/3 Such books fetch far better prices in London and Paris than in New York, where the demand for such *rariora* is small. 1932 J. Buchan *Gap in Curtain* iv. 189 There was a fine set of Donne, two of the Shakespeare folios,..besides a quantity of devotional and political *rariora*. 1964 D. Cox in D. Daiches *Idea of New University* ix. 162 Where sufficient copies of not only European but of American *rariora* have been unavailable, microfacsimile has been called in.

rarish, *a.* Add: Also **rare-ish.** (Further examples.)
1875 Browning *Inn Album* iii. 92 Would..I winged were..And so could straightway soar..Back to my nest where broods whom I love best—The parson o'er his parish —garish—rarish—. 1959 N. Marsh *False Scent* (1960) i. 19 It's rare-ish. The frame's contemporary. I'm afraid it's twelve guineas.

‖ **rarissima** (rē^əri·simă, rari·-), *sb. pl.* [L., neut. pl. superlative of *rārus* rare.] Extremely rare books. Also **rari·ssime** *a.* [lit. 'very rarely (*sc.* found)'], extremely rare.
1903 A. Bennett *Truth about Author* iv. 56, I possessed a *rarissime* illustrated copy of *Manon Lescaut*. 1952 *Times Lit. Suppl.* 14 Nov. 752/3 The..books..of which Mr. Wing has been able to locate only a single copy..are not as *rarissime* as one would infer. 1972 *Ibid.* 29 Sept. 1173/3 A few of the important *rarissima*, like Fust and Schoeffer's *Canon Missae* of 1458.

rarity. Add: **6.** *Comb.*, as *rarity value.*
1962 *Listener* 30 Aug. 329/1 Shots..came across with a breezy freshness that is only partly explained by their rarity value on television. 1978 P. McCutchan *Blackmail North* xiii. 138 The cave is *under* a layer of schist... Because of the rarity value the geologists play it down and hope to keep the tourists away.

Rarotongan (ræro̅tǫ·ŋăn), *sb.* and *a.* [f. *Rarotonga*, the name of the largest of the Cook Islands in the South Pacific + -an.] **A.** *sb.* **a.** A native or inhabitant of Rarotonga. **b.** The language of Rarotonga: Cook Islands Maori, a member of the Polynesian group. **B.** *adj.* Of or pertaining to Rarotonga or its language.
1842 M. Russell *Polynesia* vi. 226 The Rarotongans were a most ferocious class of men. 1851 G. Browne *Let.* 8 July in A. Buzacott *Mission Life Islands of Pacific* (1866) xiii. 188 It was resolved to request Miss Buzacott to accept ..a copy of the Bible in Rarotongan and English. 1854 A. Buzacott *Rarotongan & Eng. Gram.* 2/2 The Rarotongan alphabet has five soft letters. 1866 —— *Mission Life Islands of Pacific* xiv. 198 It is scarcely credible how strange it was to Mr. Buzacott to preach..in the English language..his mind being prone to think according to the habits and ideas of the Rarotongans. 1897 *Jrnl. Polynesian Soc.* VI. 10 The word *akono*..appears from its use in the Rarotongan scriptures, to be..akin to 'appointed'. 1910 F. W. Christian *Eastern Pacific Lands* 193 The history of the Christianizing..and progress of these Rarotongan Maori is the brightest chapter in the history of the South Seas. 1947 *Jrnl. Polynesian Soc.* LVI. 197, I wish to acknowledge assistance given to me by..other Rarotongans. *Ibid.* 215 The manuscript..requires translation into modern Rarotongan. 1966 J. E. Buse in C. E. Bazell *In Memory of J. R. Firth* 52 Rarotongan Maori is spoken on the island of Rarotonga in the Southern Cooks. 1977 *Times* 2 Dec. 19/2 Rarotongans are almost wholly dependent on inflation-prone imported tinned goods.

‖ **ras** (rās). [a. Amharic *rās* head, chief, from Arab.: cf. Reis[2], rais.] **1.** The title of a leading citizen. **a.** An Ethiopian king, prince, or feudal lord.
1682 tr. *H. Ludolf's New Hist. Ethiopia* ii. xii. 213 To these succeeded another Chief Officer whom the Ethiopians call Ras, from the Arabic word which signifies a Head. 1710 B. Tellez *Trav. Jesuits in Ethiopia* x. 54 They constituted another call'd Raz, which signifies Head; because he who has that Employment is next the Emperor, Head of all the great Men in the Empire. 1735 Johnson tr. *Lobo & Le Grand's Voy. to Abyssinia* 262 There is now a Generalissimo established under the title of Ras, or Chief. [1759 —— *Rasselas* i. 2 Rasselas was the fourth son of the mighty emperour, in whose dominions the Father of waters begins his course.] 1833 *Penny Cycl.* I. 58/1 Mr. Salt saw..the Ras's wife. 1904 S. Walpole *Hist. Twenty-Five Yrs.* II. xi. 269 Practically, the power was in the hands of several ras, or chiefs—of whom Ras Ali was the foremost—who carried on a turbulent warfare among themselves. 1921 *Glasgow Herald* 14 Sept. 6/3 Most of the power lies in the hands of the important 'Rases', who correspond with the Dukes of mediæval Europe, and although nominally members of a vague State Council, are to a large extent independent rulers, with governors and chiefs in practically feudal subordination to them. 1926 *Ibid.* 27 Apr. 7 Mr Rey said that when he arrived in the capital of Abyssinia, Adis Ababa,..he was most kindly received by the Empress and by the Regent, Ras Tafari, who was much interested in the expedition. The Ras not only gave the necessary permits, without which no one could move in Abyssinia, but supplied also some rifles for Mr Rey's men. *Ibid.*, He had an extraordinary welcome from the Governor, Ras Hailu, who was the last of the hereditary Provincial Governors in Abyssinia. The Ras exercised a feudal despotism in his government, and he maintained the old time hospitality. 1936 E. Waugh *Waugh in Abyssinia* i. 38 The rases and officials copied the Emperor.

b. = Reis[2], rais 2.
1935 *Words* May 7/1 Legionaire Aurelius Philinus of ancient Palmyra set up in 251 A.D. an honorary statue to Septimius Hairan, illustrious senator and head man (*ras*) of Tadmor (Palmyra) accompanied by a bilingual inscription in Semitic and Greek.

2. *transf.* An Italian Fascist leader; a petty despot.
1923 *Glasgow Herald* 16 Oct. 6/4 The strength of the materialist group is to be found in the 'rases' or 'bosses' who have found in the revolution an excellent opportunity to establish themselves as the petty tyrants of their town or district. 1924 *Ibid.* 27 June 8/4 It [*sc.* a declaration] will compel the 'rases' and the physical force party generally to abandon their methods or to resist Mussolini's authority in the open. 1967 C. Seton-Watson *Italy from Liberalism to Fascism* xiii. 594 The local fascist bosses were commonly known as *ras*, a title borrowed from the feudal nobility of Ethiopia. *Ibid.*, Mussolini hit back.., condemning the parochialism of the *ras* and the degeneration of Emilian fascism into the defence of sordid private interests. 1973 P. A. Allum *Politics & Society in Post-War Naples* ix. 301 The leadership of the party has been taken over by a solid conservative block..that has as solid supporters that poor class of bosses (*ras*) of the various provincial centres. 1975 *Times Lit. Suppl.* 11 July 783/3 A biography of Roberto Farinacci, arguably the toughest and unquestionably the most uncouth of the Fascist 'ras'.

ras, var. *Rasa[1,2].

‖ **rasa**[1] (ra·sa). Also **ras.** [a. Skr. *rasa* juice; essence, character; sentiment.] Essence, character, sentiment.
1799 F. Wilford in *Asiatick Researches* VI. 503 The Greeks supposed that mount Parnassus was the favorite abode of the Muses. The Hindus have not limited their residence to any particular spot... They are called Rasa in Sanscrit, in which language this word signifies juice in general, but is more particularly understood as the honied juice of flowers: it implies also any thing which we particularly delight in. *Ibid.* 505 The nine Rasas are represented as beautiful damsels. 1828 H. H. Wilson in *Ibid.* XVI. 118 The Bhakti of the followers of this division of the Hindu faith is supposed to comprehend five Rasas or Ratis, tastes or passions. 1926 in P. Hartog *On Relation of Poetry to Verse* 6 With them [*sc.* Sanskrit authors] what is essential in poetry is something incomprehensible, something which cannot be directly expressed, but only suggested—the aesthetic delectableness which they call *rasa*, a subjective condition of the reader's mind. 1967 [see *Nritya]. 1968 *Indian Music Jrnl.* V. 34 His life became lonely. Perhaps this accounts to some extent for his successful evocation of *Karunā* rasa. 1968 *Jrnl. Musical Acad. Madras* XXXIX. 8 In one sense there is only one Rasa, *Bhakti*, that runs through the songs of Tyagaraja... Thus *Bhakti Rasa* can wear different complexions... In these songs the melody is itself suggestive of the Rasa of the song. 1977 S. Pandit *Approach to Indian Theory of Art & Aesthetics* ii. 32 An object which does not contain *rasa* cannot be classified in the category of an art work, and no experience without it can be called aesthetic.

‖ **rasa**[2] (rā·sa). Also **ras.** [a. Skr. *rāsa* dance, sport.] A rustic Indian dance (commemorating that) performed by Krishna and the Gopis; a festival celebrating this.
1828 H. H. Wilson in *Asiatick Researches* XVI. 92 Another [festival] is the *Rás Yátra*, or annual commemoration of the dance of the frolicksome deity. *Ibid.* 93 The *Rás Yátra* is celebrated at the village of Sivapur. 1887 W. J. Wilkins *Mod. Hinduism* iv. iii. 233 *Rāsa Jātra.* This festival is held to commemorate the sports of Krishna with the milkmaids of Vrindāvana. 1912 *Encycl. Relig. & Ethics* V. 889/2 The *rāsa*, or sportive dance, performed by Kṛṣṇa's cowherds and cowherdesses. 1921 *Glasgow Herald* 29 Dec. 7 Graceful dancing women from Manipur..went through their most famous diversion, the 'Ras' dance, in which 'Krishna', here played by a boy, is teased by his feminine playmates. 1933 S. Sitwell *Canons of Giant Art* 122 And now he pipes for them the pastoral rasa, The dance, the shepherds' dance, the Indian cordax. 1951 L. R. Dayal *Manipuri Dances* p. ix, The main forms of dance are the Ras, which is performed in the months of Vaishak (May), Kartik (November), and Phalgoon (February). 1953 F. Bowers *Dance in India* 130 Manipur's two most celebrated distinctions are polo and Ras Lila... Ras Lila is played according to season, and usually on mountain sides or in far-off temples. 1969 *Femina* (Bombay) 26 Dec. 27/1 Of all the seasonal delights of Winter that I've known, those of Gujarat remain the most clearly defined in my memory, especially the beautiful garba and ras dances, with their unforgettable rhythm and sheer vitality. 1972 N. Hein *Miracle Plays of Mathurā* vi. 129 We have seen that the deeds of Vishnu are *līlās*, 'sports'. Ras is the name of a particular one of these deeds, which was done by Vishnu in his Krishna incarnation. *Ibid.*, Krishna favored the gopis by dancing with them a circular dance, which the Vishnu Purana calls the rasa. *Ibid.* 130 The rāslīlā is a drama of which Braj claims the sole guardianship.

rasagoola, rasagulla, varr. *Rasgulla.

rasant, *a.* (Later example.)
1944 S. Putnam tr. *E. da Cunha's Rebellion in Backlands* v. 256 Eight hundred rifles blazing, eight hundred rifles aimed in a rasant line down the drop of the hill.

‖ **rascasse** (raskas). [Fr.] A small Mediterranean scorpion-fish, *Scorpæna scrofa*, which has reddish skin and spiny fins which can cause painful wounds, and which is used esp. as an ingredient of bouillabaisse.
1921 W. J. Locke *Mountebank* xvi. 206 The wondrous dish [*sc.* bouillabaisse] was set before them..*Rascasse, loup de mer*, mostelle, *langouste*—a studied helping of each. 1940 A. Simon *Conc. Encycl. Gastron.* ii. 81/1 Rascasse..is caught in the Mediterranean only. 1950 E. David *Bk. Mediterranean Food* 54 *Rascasse*..is a red spiny fish. 1957 R. Campbell *Portugal* iv. 67 The *rascasse*..is the chief component of..bouillabaisse. 1961 E. McLeod tr. *Colette's Break of Day* 36 A southern luncheon..salads, stuffed *rascasse* and aubergine fritters. 1975 *Sat. Rev.* (U.S.) 29 Nov. 55/1 The ugly rascasse, without which, they say in Marseilles, a proper bouillabaisse cannot be made.

raschel (ræ·ʃĕl). Also **Raschel.** [ad. G. *Raschelmaschine*, f. the name of the French actress *Rachel* (1820–58): cf. *Rachel.] **a.** A kind of knitting-machine (see quots. 1940, 1968). **b.** The coarse warp knitting produced by such a machine. Also *attrib.* Hence **ra·schel** *v. trans.*, to knit with a raschel machine (in quot. pa. pple.).
1940 *Chambers's Techn. Dict.* 703/1 *Raschel* (Hosiery), the name for the two-bar warp loom, fitted with latch needles. 1957 M. B. Picken *Fashion Dict.* 195/1 *Raschel knitting*,..type of warp knitting resembling tricot, but coarser. Done by machine in plain and Jacquard patterns, often with lacy effect. Used for underwear. 1968 J. Ironside *Fashion Alphabet* 246 *Raschel*, a type of knitting-machine producing ribbed fabric. 1970 *Times* 12 May 11/6 The designer demonstrated the uses..of the acrylic yarn in all its many shapes, in fabrics blended, bonded, Neospun, raschelled, knitted, woven, [etc.]. 1972 *Daily Tel.* 30 Oct. 12/6 Frank Usher will have this look, appropriately braid-edged, in a beautiful raschel knit of shaded brown and beige scallops on cream. 1974 *Ibid.* 7 Jan. 9/5 Acrylics, cotton bouclés and raschels, cotton mixed with linen or Vincel: all these give a crisper look to the stripes in your spring sweater than we've had for some time now. 1974 *Encycl. Brit. Macropædia* XVIII. 182/2 The two types of warp knitting are raschel, made with latch needles, and tricot, using bearded needles. *Ibid.*, In the Raschel machine, the needles move in a ground steel plate, called the trick plate.

Raschig (rɑ·ʃig). Also (*rare*) **raschig.** *Chem. Engin.* The name of Friedrich *Raschig* (1863–1928), German chemist, used *attrib.* († or in the possessive), as **Raschig process**, a process developed by workers in Raschig's chemicals company in which phenol is produced by heating benzene vapour with hydrogen chloride and air over a copper-containing catalyst to yield chlorobenzene, which is then hydrolysed to form phenol; **Raschig ring**, a small cylindrical ring, introduced by Raschig, made of ceramic or other suitable material and used in bulk as a packing material in towers and columns for fractionation, solvent extraction, etc.
1920 *Chem. Abstr.* XIV. 2054 Attention is directed to the advantages attained by the use of Raschig's rings as a filling material for absorption towers in the manuf. of H_2SO_4, HNO_3, HCl, etc. 1926 *Ibid.* XX. 2214 Raschig rings have been used advantageously for clearing of liquids, the turbid liquor being run in at the bottom of a 80-cm. layer of rings, 15×15 mm. 1937 *Discovery* Sept. 282/2 Another form of acetifier contains raschig rings. 1940 *Chem. & Engin. News* 10 Nov. 921/1 The unit, which has been in continuous operation for four months.., employs the Raschig process, a catalytic vapor-phase system for the chlorination of benzene and the hydrolysis of the chlorobenzene produced. The process, invented by Prahl and Mathes of the Raschig organization, Ludwigshafen, Germany, produces a phenol purer than U.S.P. and does so with less than 0·1 per cent of by-product. 1947 Kirk & Othmer *Encycl. Chem. Technol.* I. 16 Raschig rings have the greatest number of applications because they provide a large effective surface of contact between the gas and the liquid phases and they have a large free volume. They are available in sizes from ¼ to 6 in., and the usual design has a height equal to the diameter. 1959 R. J. Hengstebeck *Petroleum Processing* xi. 265 Modern furfural extraction columns contain about 40 to 50 ft of Raschig rings, with redistribution equipment at about 5-ft intervals. 1963 *Economist* 14 Dec. 1193/3 The ICI plant uses the Raschig process. 1972 *Materials & Technol.* IV. viii. 309 A modification of the Dow process is the Raschig process, in which benzene is chlorinated by Cl_2 generated by the catalytic oxidation of HCl with oxygen at 230° C, and HCl is recovered in the hydrolysis at 425° C, promoted by Cu catalyst without alkali.

raser[1]. Add: (Later examples.) Also **razer.**
1684 Ld. Strathmore *Jrnl.* 6 Aug. in *Publ. Scottish Hist. Soc.* (1890) IX. 65, 19 Razers wheat sold at 8 lib 15 sh. s. 1915 C. A. Mercier *Leper Houses* 40 During Lent each had a razer of wheat..and two razers of beans.

‖ **rasgado, rasgueado** (rasgā·do, rasgeā·do). *Mus.* [a. Sp. *rasgado, rasgueado*, pa. ppl. of *rasgar, rasguear* to strum, to make a flourish.] The act of sweeping the strings of a guitar with the fingertips. Also, an arpeggio so produced.
1876 Stainer & Barrett *Dict. Mus. Terms* 374/2 *Rasgado*.., to sweep the strings of a guitar with the thumb, for the purpose of producing a full chord, *arpeggio*. 1944 W. Apel *Harvard Dict. Music* 628/1 *Rasgado*,..in guitar playing, sweeping the strings with the thumb to produce an arpeggio. 1974 *Early Music* July 185/2 Guitar arrangements in the popular *rasgado* or strumming style. 1979 *Guardian* 29 Oct. 13/2 What he [*sc.* Segovia] didn't like was the *rasgeado*..—the boastful chatter, the flourishes of strokes with the tips of the fingers.

‖ **rasgulla** (rasgū·la). Also **rasgoola, rasgula,** etc. [Hindi, f. *ras* juice + *gullā* ball.] An Indian sweet; balls of soft milk cheese soaked in syrup.
1936 E. P. Veerasawmy *Indian Cookery* 207 Rasgollah... Drop these into a rose-flavoured syrup. 1944 M. R. Anand *Barber's Trade Union* 72 A sweetmeat seller hawked: 'Gulab-jaman, rasgula, burfi, jalebi.' 1954 S. Chowdhary *Indian Cooking* 150 Rasgullas are served warm or cold. 1960 *Harper's Bazaar* July 76/2 Tables laden with curries, chutneys and rasgolas. 1961 B. Singh *Indian Cookery* vii. 149 Rasgulla is a delicious Bengali sweet. The land of Bengal has long been famous for the preparation of exquisite sweets and in making rasgullas they have excelled themselves. 1962 Hosain & Pasricha *Cooking Indian Way* 225 Rasagulla... They should be double their original size and floating on the surface of the syrup. 1968 P. Lal *Indian Recipes* 218 Rasogoolas... Drop the rasogoolas in the clean boiling syrup. 1969 *Femina* (Bombay) 26 Dec. 27/2 Our dinner however, in the midst of all this rusticity, was certainly no rustic fare;..and to top it all, delicious ice-cream and *rasagoolas!* 1975 R. H. Christie *Twenty-Two Authentic Banquets from India* xviii. 123 *Rasgullah*, (Milk Croquettes), six tumblers of milk, lemons sufficient in number. 1976 *Times* 18 Aug. 12/7 Try rasgoulas—balls of curd in syrup at 7p each.

rash, *sb.*[3] Add: **2.** *transf.* and *fig.* A proliferation or spate; a sudden outbreak *of* something.
1820 J. Hogg *Winter Evening Tales* I. 312, I was workin' at the loom, wi' my leather apron on, an' a rash o' loom needles in my cuff. 1907 W. De Morgan *Alice-for-Short* xl. 427 There too appear more bells than Poe ever wrote about ..a rash of bells that makes you think before you ring. 1930 R. Campbell *Adamastor* 67 A rash of stars upon the sky, A pox of flowers on the earth. 1933 *Times Lit. Suppl.* 16 Mar. 186/1 Her prose is marked by a rash of exclamation marks. 1956 J. M. Mogey *Family & Neighbourhood* i. 8 By 1870 the pleasant meadow land..had disappeared beneath a rash of bricks and pavements. 1968 B. Hines *Kestrel for Knave* 30 Great rashes of buttercups spread across the fields. 1980 J. McNeil *Spy Game* xix. 189 A piece of open ground which has miraculously escaped the rash of building covering the rest of the site.

rash (ræʃ), *sb.*[5] *Coal Mining.* orig. *dial.* (esp. *S. Wales*) and *U.S.* [Prob. f. Rash *a.*] Usu. *pl.* = *Rashing *vbl. sb.*[2]
1903 A. Strahan *Geol. S. Wales Coal-Field* IV. iii. 26 Black rashes, fireclay, and clift. *Ibid.* 27 Clift, rashes and rock. 1912 *Federal Reporter* (U.S.) CXCIII. 125 The appearance of rash in the eighth and ninth west entries and at the face of the slope justified the belief that the coal in that vicinity of the mine was inferior and unmerchantable. 1917 Gibson & Cantrill *Geol. S. Wales Coalfield* (ed. 2) IV. v. 70 Rashes and clod 1 ft. 3 in. 1964 [see *Rashing *vbl. sb.*[2]]. 1964 Woodland & Evans *Geol. S. Wales Coalfield* (ed. 3) IV. v. 126 At Western the section (from top): coal 7 in, rashes 2½ ft, coal 7 in, fireclay and rashes 5 ft.

rash, *a.* **2. b.** Delete † *Obs.* and add later example.
1876 G. M. Hopkins *Wreck of Deutschland* xix, in *Poems* (1967) 57 The rash smart sloggering brine Blinds her.

rashing (ræ·ʃiŋ), *vbl. sb.*[2] *Coal Mining.* orig. *dial.* (esp. *S. Wales*) and *U.S.* [Prob. f. Rash *a.* + -ing[1].] Usu. *pl.* A loose brittle deposit of shale or poor coal (see quot. 1964). Cf. *Rash *sb.*[5]
1883 W. S. Gresley *Gloss. Terms Coal Mining* 200 *Rashings* (S[outh] W[ales]), loose dirt or shaley beds of rock. 1903 A. Strahan *Geol. S. Wales Coal-Field* IV. iii. 10 Rashings 0 ft. 9 in. 1905 *Maryland Geol. Survey* V. 534 Immediately overlying the 'Big Vein' is a brittle slate interstratified with thin seams of coal known as the 'Wild Coal' or 'Rashings'. Upon exposure to the air the wild coal crumbles and falls. 1917 Gibson & Cantrill *Geol. S. Wales Coalfield* (ed. 2) IV. iv. 47 The Wet-and-Dry Coal is said to lie 14 yds. below the Lantern, and to consist of top coal, 2 ft. 8 ins. thick, bottom coal, 1 ft. 4 ins. to 1 ft. 6 ins. thick, with 6 ins. of soft rashings between. 1954 A. Trueman *Coalfields Gt. Brit.* ii. 23 Black carbonaceous shales, containing abundant plant remains (the 'rashings' of some coalfields). 1964 Woodland & Evans *Geol. S. Wales Coalfield* (ed. 3) IV. iv. 75 *Rashes* or *rashings*, *either* soft carbonaceous shale with streaks of coal *or* highly disturbed, slickensided, comminated shale or mudstone formed by movement parallel to the bedding and usually associated with the roof or dirt bands in coal seams. Normally 'rashes' should be retained for the former definition and 'rashings' for the latter. 1967 I. A. Williamson *Coal Mining Geol.* vii. 55 Batt and rashings are approximately synonymous for shale containing thin coal streaks.

rashleighite (ræ·ʃliˌəit). *Min.* [See quot. 1948 and -ite[1].] A hydrated basic phosphate of copper, aluminium, and iron, $Cu(Al,Fe)_6(PO_4)_4(OH)_8 \cdot 5H_2O$, found as crusts and friable masses of triclinic crystals.
1948 A. Russell in *Mineral. Mag.* XXVIII. 353 The name rashleighite commemorates Philip Rashleigh, F.R.S., F.G.S., F.S.A., 1729–1811,..one of the earliest Cornish mineralogists and famous for having amassed the finest collection of Cornish minerals ever made. 1971 *Country Life* 3 June 1382/2 Turquoise in the form of rashleighite occurs on St. Austell Moor, in Cornwall. 1972 *Amer. Mineralogist* LVII. 1681 Rashleighite..can be described as a substitutional solid solution of turquois and chalcosiderite.

rashling. For '† *Obs. rare*[−1]' read '*rare*' and add later example.
1922 Baroness Orczy *Triumph of Scarlet Pimpernel* v. 57 The young Hotspur and his crowd of rashlings would ere now have been torn from their seats.

‖ **raskol** (ræskǫ·l). Also **rascol.** [Russ.: see Raskolnik.] **1. a.** The schism in the Russian Church which resulted from the reforms of Patriarch Nikon, who excommunicated dissenters in 1667.
1887 A. F. Heard *Russ. Church & Russ. Dissent* ix. 179 Both German Protestantism and Russian Raskol preserve the stamp of their similar religious origin, as issuing each from an established State Church. 1900 'Odysseus' *Turkey in Europe* vi. 288 This energetic and ambitious prelate [*sc.* Nikon]..provoked by his reforms the great schism (or *raskol*) from which spring the various sects of Old Believers. 1908 W. F. Adeney *Greek & Eastern Churches* II. iii. vii. 444 The Raskol obtained new vigour from another source—popular resistance to Peter the Great's Western innovations. 1963 N. V. Riasanovsky *Hist. Russia* xix. 221 The raskol constituted the only major schism in the history of the Orthodox Church in Russia. 1969 K. Minogue in Ionescu & Gellner *Populism* 203 The Russian *raskol* after 1654, although a religious phenomenon, has been taken as a peasant reaction to urban culture.

b. Dissent from an established orthodoxy.
1947 *Partisan Rev.* XIV. 396 Russian revisionism was a heterodoxy, a fanatic schism, a *raskol*.

2. *collect.* A body of dissenters under the raskol (sense *1 a).
1888 'Stepniak' *Russ. Peasantry* II. 441 The *Rascol* proper, the 'Old Believers'.

Raskolnik. (Earlier example.)
1723 in tr. *F. C. Weber's Present State of Russia* I. 82 He was reported to be of the Sect of the Roskolniks, who have entirely separated from the Russian Church, a few Ceremonies excepted.

rasogoola, var. *Rasgulla.

rasophore (ra·zofō^əɪ). Also **rasophor, rhasophore.** [ad. med. Gr. ῥασοφόρος, f. ῥάσον cassock + -φορος bearer.] The lowest grade of monk in the Greek Orthodox Church. Also ‖ **rhasophoria**, the grade of a rasophore.
1887 A. Riley *Athos* v. 68 The monks are divided into two classes, the *dokimos*.., or novice and the *caloyer*.., or professed monk. The caloyers, again, are divided into three grades—*rhasophoria*.., *the little habit*..and *the great habit*. 1934 Webster, *Rhasophore.* 1960 P. Sherrard *Athos* 63 The monk of the lowest grade is called a Rasophore, after the *rason* or tunic which he wears as part of his habit... He is..more than a novice in the western sense. 1964 P. F. Anson *Bishops at Large* i. 37 Almost any alien, provided that he..calls himself an abuna,..pappas, rasophor, starets, synkellos,..vartapet..invariably finds a warm welcome in England.

rasour: see also *Réseau.

rasp, *sb.*[1] Add: **2. c.** A ribbed band or organ in some insects.
1826 Kirby & Spence *Introd. Entomol.* III. xxx. 143 This animal..has on it a double series of rasps. 1871 Darwin *Descent of Man* I. x. 378 The rasp generally consists of a narrow, slightly-raised surface, crossed by very fine, parallel ribs.

4. (Further examples.)
1976 *National Observer* (U.S.) 25 Dec. 4/1 'Christ, can you believe that?' he cries in his staccato, Brooklyn-accented rasp that has been honed just enough so that his 'thats' don't come out 'dats'. 1977 *Rolling Stone* 30 June 113/1 Like many such groups, Detective centers around a guitarist (Michael Monarch, whose aggressive rasp distinguished the earliest Steppenwolf sides).

rasp, *v.*[1] Add: **2. c.** Also with *on.*
1898 F. P. Dunne *Mr. Dooley in Peace & War* 232 But wan day it happened that that whole fam'ly begun to rasp on wan another. 1905 *Pall Mall Mag.* Dec. 674/2 Any reference to the Philippine campaign rasped on his nerves.

d. To grate (the hard crust) off (a roll); also *intr.* for *pass.*
1889 R. Wells *Pastrycook & Confectioner's Guide* ii. 11 French rolls must always be rasped. 1892 —— *Mod. Pract. Bread Baker* 57 They must be well baked, or they will not rasp as all French rolls should. 1908 J. Kirkland *Mod. Baker* II. xxvi. 162 These rolls are occasionally baked with a very hard crust, which is afterwards rasped off.

4. Delete *rare*[−1] and add further examples. Also *absol.* and *transf.* Also with *out.*
1877 *Harper's Mag.* Oct. 664/1 A somewhat harsh clock rasped out the seconds. 1905 *Pall Mall Mag.* Sept. 276/1 Commander McTurk stiffened. 'Ah,' he rasped, 'that's news to me.' *Ibid.* Nov. 543/1 'Really, Bridget!' her brother rasped out, 'I wish you wouldn't interfere.' 1922 Joyce *Ulysses* 442 A bunch of loiterers listen to a tale which their broken snouted gaffer rasps out. 1937 C. S. Forester *Happy Return* x. 124 'Hard-a-starboard,' he rasped at the quartermaster. 1962 [see *Hell *sb.* 10 h]. 1976 W. Greatorex *Crossover* 142 'I'm not thirsty,' Calder said. 'I've had enough for one night.' 'You're telling *me*,' the inspector rasped. 1977 *Time* 28 Nov. 50/3 Rasped an Agriculture official: 'That was a technically accurate statement. But it also was a god-damn lie.'

5. a. (Earlier example.)
1808 S. W. Ryley *Itinerant* I. iv. 91 A blind fiddler, mounted on a three footed stool, rasped away very *seriously* the black *Joke*.

raspberry. Add: **3*. a.** [App. an ellipt. use of *raspberry tart* (*b*) below.] A derisive sound; = *Bronx cheer* s.v. *Bronx 2.
1890 in Barrère & Leland *Dict. Slang* (1890) II. 171/1 The tongue is inserted in the left cheek and forced through the lips, producing a peculiarly squashy noise that is extremely irritating. It is termed, I believe, a *raspberry*, and when not employed for the purpose of testing horseflesh, is regarded rather as an expression of contempt than of admiration. 1899 A. M. Binstead *Gal's Gossip* 144 A loud and offensive noise, like the rending of glazed calico, made by obtruding the wet tongue between the closed lips, and by low cabmen and persons of that class, called a 'raspberry', came from the gallery. 1912 *Confessions of Dancing Girl* iv. 69 The custom of 'guying' a performer and giving him what is called a 'raspberry'. Not a few artistes have had 'raspberries' from the audiences in the Glasgow variety halls. 1932 A. J. Worrall *Eng. Idioms* v. 33 As soon as the speaker rose the crowd gave him the raspberry. 1940 R. Chandler *Farewell, my Lovely* xxix. 211 The kind of bossy knock that makes you want to..emit the succulent raspberry. 1955 [see *Bronx 2]. 1960 G. Durrell *Zoo in my Luggage* vi. 157 To my complete astonishment Minnie responded by..giving a prolonged raspberry of the juiciest variety. 1975 *South Wales Echo* 30 Oct. 8/4 The only answer to that kind of nonsense is a long-drawn-out vintage raspberry.

b. *fig.* A refusal; a reprimand, disapproval; dismissal.
1920 Wodehouse *Damsel in Distress* vi. 71 Convict son totters up the steps of the old home and punches the bell! What awaits him beyond? Forgiveness? Or the raspberry? 1923 —— *Inimitable Jeeves* ix. 89 He was given the respectful raspberry by Jeeves, and told to try again about three hours later. 1927 *Punch* 14 Dec. 649/2, I have embodied the above suggestions in a memo, and they are now on their way to the Army Council... They may even be on their way back, with a raspberry from Somebody Very Senior written across the top left-hand corner. 1942 *New Statesman* 1 Aug. 75/1 An ordinary reprimand is a *raspberry* (often referred to by other ranks as a *razzer*), which has been adopted from that form of civilian disapproval which one hears in the gallery of a theatre. 1947 [see *All *adv.* 2 c]. 1960 O. Manning *Great Fortune* i. 67 Miller made it! Nice scoop for Miller! And a raspberry for the rest of us. 1973 M. Spark *Hothouse by East River* iv. 69 The security officer mutters all the way to the compound about what a raspberry the police are going to get because of this, a raspberry in these days being already an outdated expression meaning a reprimand. A man less set in his limited ways.. would call it a rocket in this English spring of 1944. 1977 *Western Morning News* 1 Sept. 1/2 Controversial punk rock band Sex Pistols blew a raucous raspberry at the Establishment by sneaking into Plymouth to play an unannounced concert at Woods.

3.** The colour of a raspberry, varying from pink to scarlet. Also *attrib.* or as *adj.*
1923 *Daily Mail* 19 Feb. 15 New tailor-mades of black or navy serge have shawl collars of mattelassé silk in a contrasting colour—raspberry, green, or kingfisher blue. 1935 *Times* 2 Oct. 17/4 A raspberry cellophane wrap has been shown over a white and gold lamé gown. 1941 [see *Hackmanite]. 1951 E. Paul *Springtime in Paris* ii. 17 Two neon signs,..one..in a faint raspberry shade, and another..in a luminous tone of white. 1969 'H. Pentecost' *Girl Watcher's Funeral* (1970), ii. i. 88 The raspberry dress that clung to her lush figure. 1972 [see *Lime *sb.*[4] 1 c]. 1978 J. Krantz *Scruples* ii. 24 The floor of the air-conditioned room was covered from one wall to another in thick raspberry carpet.

4. *raspberry bitters, drop, jam* (earlier and later examples), *jelly, noyau, pie, roll, seed, syrup, tart* (earlier examples), *vinegar* (later example); **raspberry beetle**, a beetle of the genus *Byturus*, esp. *B. tomentosus* (formerly *B. urbanus*), the larva of which attacks the fruit of raspberries and related plants;

raspberry fruitworm = prec.; **raspberry jam (tree)** (later examples); also, the wood itself; **raspberry tart** *rhyming slang*, (*a*) the heart; (*b*) a breaking of wind or 'fart'.

1884 E. A. Ormerod *Observations Injurious Insects 1883* 65 A small brown Beetle..proved to be the 'Raspberry Beetle',..which was causing fearful havoc and entirely devouring the Raspberries. 1909 F. V. Theobald *Insects & Other Pests of Orchard, Bush & Hothouse Fruits* 420 There is no insect enemy so serious to raspberries as the Raspberry Beetle. 1959 E. F. Linssen *Beetles Brit. Isles* II. 23 This genus [sc. *Byturus*] includes a well-known pest of raspberries, loganberries and blackberries, commonly referred to as the Raspberry Beetle. 1849 Thackeray *Pendennis* I. xv. 131 He..drank a glass of raspberry bitters at the Clavering Arms. 1897 'S. Grand' *Beth Book* iii. 19 Give me a ha'porth ..of raspberry-drops. 1939–40 *Army & Navy Stores Catal.* 52/2 *Boiled sweets*..Raspberry drops. 1924 *Ann. Rep. Connecticut Agric. Exper. Station* 91 The Raspberry Fruit Worm or Raspberry Beetle has long been known as a pest of red raspberries in the United States. 1945 L. Pyenson *Pest Control in Home Garden* vii. 122 The formula used in these two sprays controls the raspberry fruit worm. 1972 Swan & Papp *Common Insects N. Amer.* 396 The Western Raspberry Fruitworm..is a serious pest of raspberries and loganberries. 1747 Raspberry jam [see Jam *sb.*[2] a]. 1846 [see *gooseberry jam* s.v. *Gooseberry 7 a]. 1865 J. E. Tenison-Woods *Hist. Discovery & Exploration Austral.* II. 68 The other trees beside the palm were known to the men by colonial appellations, such as the bloodwood and the raspberry-jam. 1896 *Australasian* 15 Feb. 313 The raspberry-jam-tree is so called on account of the strong aroma of raspberries given out when a portion is broken. 1948 F. H. Titmuss *Conc. Encycl. World Timbers* 113 Raspberry Jam is rather a difficult timber to work. 1965 *Austral. Encycl.* VII. 389/1 Raspberry-jam, a popular name standardized in the timber trade for *Acacia acuminata* of Western Australia; the dark reddish-brown, durable, close-grained wood of this wattle has a distinct raspberry-like scent. 1861 Mrs. Beeton *Bk. Househ. Managem.* 797 Raspberry Jelly... To each pint of juice allow ¾ lb. of loaf sugar. 1913, 1963 Raspberry noyau [see *Noyau 1 b]. 1828 E. Leslie *Seventy-Five Receipts* 25 Raspberry and apple-pies are much improved by..pouring in a little cream. 1864 *Harper's Mag.* Nov. 735/1 A green cheese, a dish of cucumbers, and two raspberry pies supplied a finish to the entertainment. 1974 J. Grigson *Eng. Food* 206 Raspberry Pie..Filling 1 lb raspberries About 4 oz sugar [etc.]. 1868 A. D. Whitney *Patience Strong's Outings* (1869) xii. 138, I..made her give up..the special raspberry-roll for dinner. 1841 M. Edgeworth *Let.* 23 Mar. (1971) 590 Dr. Lindley..saw that they were raspberry seeds and he put them in ground and they..have grown to real raspberry bushes. 1728 E. Smith *Compl. Housewife* (ed. 2) 192 Two Ounces of Syrup of red Poppies, and as much of Raspberry Syrup. 1965 *Savoy Cocktail Bk.* (ed. 3) 177 *Raspberry Lemonade Cocktail.* Put into tumbler juice of 1 Lemon..sugar.. Raspberry Syrup..ice..water. 1723 J. Nott *Cook's & Confectioner's Dict.* sig. Ff3 (*heading*) To make a Raspberry Tart. 1848 [see Hard-bake]. 1892 *Sporting Times* 29 Oct. 1/2 Then I sallied forth with a careless air, And contented raspberry tart. 1959 I. & P. Opie *Lore & Lang. Schoolch.* i. 9 Breaking wind was, at one time, by the process of rhyming slang, known as a 'raspberry tart'. 1960 J. J. Rowlands *Spindrift* 55 That was the era of raspberry vinegar, lemonade, and homemade root beer.

rasper. Add: **2.** (Later example.)

1929 H. A. Vachell *Virgin* iii. 53 In front was a big solid fence, a rasper.

3. (Earlier and later examples in sense 'anything remarkable or extraordinary in its own way'.)

1844 *Spirit of Times* 19 Oct. 403/2 She promises to be a perfect 'rasper', and will have some 'tall' chronicling in the 'Spirit' before all of her yarn is spun. 1977 Scollins & Titford *Ey up, mi Duck!* II. 52 *Rasper*, another expression of excellence; often describes a good goal in football.

rasping, *vbl. sb.* Add: **2. a.** (Further examples.) In mod. usage, *spec.* breadcrumbs made from baked or stale bread.

1846 *Jewish Manual Cookery* 47 Butter a mold, sprinkle it with raspings. 1945 *ABC of Cookery* (Ministry of Food) iii. 14 *Raspings*, very fine crumbs obtained by grating the crust of stale bread on a fine grater. Browned breadcrumbs are sometimes called raspings. 1951 *Good Housek. Home Encycl.* 629/1 Sprinkle raspings over the fat of cooked ham. 1976 M. Patten *Barbecue* 31/2 Cut the crusts from the loaf; these need not be wasted, but can be turned into crumbs (raspings) as described below.

raspite (ræ·spəit). *Min.* [ad. G. *raspit* (C. Hlawatsch 1897, in *Ann. des K. K. Naturhist. Hofmuseums* XII. 33), f. the name of Mr. *Rasp*, discoverer of the Broken Hill mines, New South Wales, where the first specimen was found: see -ite[1].] A lead tungstate, $PbWO_4$, that is dimorphous with stolzite and occurs as brownish, yellow, or grey prismatic monoclinic crystals.

1898 *Mineral Mag.* XII. 47 Raspite, a new dimorphous form of lead tungstate, is found on some of the stolzite specimens as brownish or yellow monoclinic crystals with a strong adamantine lustre. 1931 J. W. Mellor *Compreh. Treat. Inorg. & Theoret. Chem.* XI. lxii. 793 Stolzite occurs in tetragonal crystals, raspite in monoclinic crystals. 1956 *Mineral Abstr.* XIII. 30 Natural raspite transforms irreversibly to stolzite at about 400° C. 1968 I. Kostov *Mineralogy* ii. viii. 486 Stolzite is uniaxial negative..; raspite biaxial positive.

Rasputin (ræspiū·tin, ræ·spiutin). The acquired name (lit. 'debauchee') of Grigory Yefimovich Novykh (*c* 1872–1916), mystic and favourite at the court of the Russian Emperor Nicholas II, used allusively of one who resembles Rasputin in exercising an insidious or corrupting influence over another or (esp.) over members of the governing class. Also *attrib.*

1937 H. G. Wells *Star Begotten* viii. 152 My professional gifts give me a kind of Rasputin hold on one or two exalted families. 1975 *Economist* 8 Feb. 35/2 Referred to in private ..as a Rasputin or a Svengali, her private secretary, Sr José López Rega, has enlarged his power. 1975 *Times* 23 May 2/2 Mr Wilson..drove home the policy divisions on the Tory front bench by noting that Sir Keith Joseph—'Mrs Thatcher's..mentor and Svengali'..the right hon lady's Rasputin—had been kennelled up for the debate. 1978 G. Mitchell *Wraiths & Changelings* xv. 149 He was becoming a sort of Rasputin where Mrs Crieff-Tweedle was concerned... I'm sure she was coming under..a sinister influence.

So **Raspu·tinism**, the principles and practices held to be characteristic of Rasputin, chiefly with reference to his libertinism and his corrupting influence over government.

1918 *Pall Mall Gaz.* 26 Jan. 3/1 Some exceedingly fine shades of difference were explained, and some striking affinities pointed out. For instance, Bolshevism, in Mr. Wilton's opinion, is nothing but Rasputinism under another name. 1919 W. le Queux *Rasputinism in London* xii. 160 London is surely vicious enough without the evil practices of Rasputinism! 1933 H. G. Wells *Shape of Things to Come* ii. iv. 169 It had a touch of Rasputinism, this revival of the ancient heresy that one must sin *thoroughly* before one can be saved. 1975 *Economist* 8 Feb. 35/3 But a new decree, signed by Sra Perón herself, has made it clear that he [*sc.* her private secretary] will have the last word about whom she sees and what she is told. This latest step towards the legitimisation of rasputinism (the latest word to be coined here [*sc.* in Argentina]) has aroused the armed forces.

raspy, *a.*[1] Add: **2.** (Earlier example.)

1869 L. M. Alcott *Little Women* II. i. 20, I don't wish to get raspy, so let's change the subject.

3. *Comb.*, as *raspy-gaspy*.

1903 Kipling in *Windsor Mag.* Sept. 363/2 She said it in a raspy-gaspy whisper that would have frightened a steam-cow.

rass (rās), *sb.* (*a.*) and *v.* *Jamaica. coarse slang.* [f. Arse *sb.* by metathesis and perh. partly also by metanalysis (of *your arse*).] **A.** *sb.* The buttocks, the arse. Also *transf.* as a term of contempt, and *attrib.* or as *adj.* **B.** *vb. trans.* = *Bugger *v.* 2; also *ellipt.* for 'shove it up your arse' used as an insult.

1790 J. B. Moreton *Manners & Customs in W. India Islands* 154 Then missess fum me wid long switch, And say him da for massa; My massa curse her, 'lying bitch!' And tell her 'buss my rassa!' *c* 1918 in Cassidy & Le Page *Dict. Jamaican Eng.* (1967) 372/2 /raas/ as in 'Raas to you!'—common among schoolboys. 1952 *Caribbean Q.* II. iv. 27 Fred dropped his hoe and hit him hard. Big Joe was up roaring:— 'Rass yah today!' 1959 A. Salkey *Quality of Violence* vii. 101 You rass clothes going rip off... You, class-war rass hole, you! *Ibid.*, They believe they is superior and all that rass! 1965 I. Fleming *Man with Golden Gun* xiii. 173 'Rass, man! Ah doan talk wid buckra.' The expression 'rass' is Jamaican for 'shove it'. 1976 Boot & Thomas *Jamaica* 88 If he gave them any *rass* they'd hit from all sides with a gale of maniacal rhetoric that would reduce the poor man to blubber.

rasse[1]. Substitute *Viverricula indica* for *V. malaccensis* in def. (Later examples.)

1896 R. Lydekker *Handbk. Carnivora* 216 While all the other civets are non-arboreal animals, the Rasse is said to be an expert climber. 1971 L. H. Matthews *Life of Mammals* II. ix. 273 The rasse, *Viverricula indica*, widespread in India and southeast Asia, is smaller but similar [to the African civet] in diet and habits.

‖ **rassenschander** (ra·sənʃa:ndər). *rare.* [erron. f. G. *rassenschande*, f. *rasse* race + *schande* violation.] The violation of the purity of the ('Aryan') race by marriage to one of a different race.

1937 Auden in Auden & MacNeice *Lett. from Iceland* xiii. 201, I ought to be the prize, the living wonder, The really pure from any Rassenschander... The Nordic type, the too too truly Aryan.

‖ **rasta** (rasta), *sb.*[1] [Fr., abbrev. of *Rastaquouère.] = *rastaquouère.

1905 *Truth* 18 May 1267/2 The *rasta*—for the word undergoes contraction—was twenty years ago apt to be unpleasant on account of his airs. 1929 J. Buchan *Courts of Morning* 25 One of his South American colleagues had taken him to dine at a restaurant much in vogue among the *rastas*. 1937 G. Frankau *More of Us* vii. 77 Doyen de Bouche, that celebrated rasta, Inventor of the coup de limonade.

Rasta (ræ·stă), *sb.*[2] Also **rasta.** [Shortened form of *Rastafari or *Rastafarian(ism).] **a.** = *Rastafarian *sb.* **b.** = *Rastafarianism. Freq. *attrib.* and *Comb.* Hence **Ra·staman**, a (male) Rastafarian.

1955 G. E. Simpson in *Social Forces* XXXIV. ii. 167/1 The 'Rasta' people consider Marcus Garvey..as the forerunner of their movement. 1960 M. G. Smith et al. *Ras Tafari Movement in Kingston, Jamaica* 11 Those people who worshipped the Emperor and were locally known as 'Ras Tafaris' or 'Rastamen' came to describe themselves as 'Niyamen'. 1962 *Listener* 1 Feb. 209/1 There are other Rastas, fanatic and militant, whose ideology is a blend of myth, religion, anarchistic politics, and black nationalism. 1965 I. Fleming *Man with Golden Gun* vi. 84 'You carry a gun?' 'Of course. You don't go after the Rastas without one.' 1966 [see *Locksman]. 1976 Boot & Thomas *Jamaica* 79/1 Rasta is not just some half-witted heretic sect selling space in the hereafter. *Ibid.* 93/1 Why should a Rastaman lift a finger to participate in a society that thinks he's just some poor dope-fiend with fried spinach for brains? 1976 *New Musical Express* 27 Mar. 19/6 Doris Day's son Terry Melcher and ex-Beach Boy Bruce Johnston scraping the bottom of the barrel with this well known primary school rasta chant. 1977 McKnight & Tobler *Bob Marley* ii. 31 Remember the words of the song *Jah Live*, a Rasta-minded admonishment to all those who suggest that Selassie's physical and merely temporal death might represent the death of the living God. 1977 *Westindian World* 3–9 June 13/3 They are also reflections of Tosh's world view; a combination of Pan African nationalism and rasta mysticism.

Rastafari, Ras Tafari (ræ:stăfā·ri; *locally also* ra:stafarəi·). Also **Rastafaria.** [f. the name *Ras Tafari* (cf. *Ras), by which Emperor Haile Selassie of Ethiopia (1892–1975) was known from 1916 until his accession in 1930.] A Jamaican sect which believes that Blacks are the chosen people, that the late Emperor Haile Selassie is God Incarnate, and that he will secure their repatriation to their homeland in Africa. Also *pl.*, the members of this sect, and *attrib.* So **Rastafa·rinism, Rastafa·rism, Ras Tafa·rism**; **Rastafa·rite**, a member of this sect.

1953 F. Henriques *Family & Colour in Jamaica* iii. 62 The current expositors of black consciousness in Jamaica are a group of people who call themselves Ras Tafarites. 1955 G. E. Simpson in *Social & Econ. Stud.* (Kingston, Jamaica) IV. ii. 133 The Ras Tafari movement began to take shape about 1930. *Ibid.* 137 Among the favourite chapters of the Ras Tafaris are: Isaiah 43. *Ibid.* 146 Ras Tafarism provides explanations of their plight to economically disadvantaged people. 1960 *Guardian* 12 Apr. 9/3 The 'Rastafari' cult, who have long, matted hair and beards and believe that the drug marihuana..is ordained by the Bible for man's use. 1962 *Listener* 22 Feb. 345/1 The 'uppressors' ..go off into weirdness such as Rastafarinism. 1962 *Times Lit. Suppl.* 10 Aug. 578/4 The surcharge of an equally false nationalistic fantasy, at its most absurd and dangerous in the 'Rastafarism' of Jamaica with its worship of Haile Selassie and its slogan 'Death to the Whites'. 1965 I. Fleming *Man with Golden Gun* iii. 41 He has groups of admirers (e.g. the Rastafari in Jamaica). 1973 *Caribbean Contact* Feb. 15/1 The Rastafaria brethren, too, who venerate Haile Selassie I and who keenly look forward to repatriation to their lost homeland of Africa, remain today faithful Garveyites. 1977 *Times Lit. Suppl.* 7 Jan. 9/5 Inadequate housing and insufficient jobs [in Kingston, Jamaica]..also contributed to the rise of the Ras Tafari with their dreams of escape to Africa.

Rastafarian (ræ:stăfā·riăn, -fē·riăn), *a.* and *sb.* Also **Ras Tafarian.** [f. prec. + -an.] **A.** *adj.* Of or pertaining to the Rastafari sect. **B.** *sb.* A member of this sect. Hence **Ra:stafa·rianism.**

1955 G. E. Simpson in *Social & Econ. Stud.* (Kingston, Jamaica) IV. ii. 134 Emphasis..is placed on love and kindliness to fellow Ras Tafarians. 1960 *Guardian* 28 June 9/4 The bearded Marijuana-smoking Rastafarian sect. *Ibid.*, The Rastafarians run a campaign to get Jamaican negroes to return to Africa. 1963 *Times* 19 Apr. 9/1 Eight people were killed and others injured after members of the ganja-smoking Rastafarian cult went wild with guns and machetes. 1968 F. Henriques *Family & Colour in Jamaica* (ed. 2) xiv. 181 The response of the University was to send a team of experts to investigate and report on Rastafarianism. 1976 *Peace News* 25 June 3/1 A group of Rastafarian West Indians wearing brightly-coloured woolly hats walking along the pavements alongside us, shaking their maracas. 1976 G. Sims *End of Web* x. 69 A couple of Rastafarians whose thick matted tassels of hair hung down over their shoulders. 1977 *Vole* No. 4. 41/2 The cult of Rastafarianism, involving the deification of Haile Selassie, the promise of Ethiopia as their spiritual home, and above all, the rejection of..white society.

Rastaman: see *Rasta *sb.*[2]

‖ **rastaquouère** (rastakwę̄r). Also **rastacouaire, rastaqouère,** etc. [F. *rastaquouère, rastaquère*, ad. S. Amer. Sp. *rastacuero* upstart.] A social intruder or upstart of exaggerated manners or dress, esp. from a Mediterranean or S. Amer. country; a dashing but untrustworthy foreigner. Also *attrib.*

1883 M. E. Braddon *Phantom Fortune* III. v. 110 He was the typical *rastaquouere*, a man of finished manners, and unknown antecedents. 1904 J. T. Grein *Dramatic Crit.* 1902–1903 56 Even *rastaquouères*..do not err in this direction. 1913 M. Larisch *My Past* ix. 104, I rather liked the Baroness, although Count Larisch rudely termed the whole family *rastaquouères*. 1924 J. Buchan *Three Hostages* xiii. 199 The usual *rastaquoère* crowd of men and women drinking liqueurs and champagne. 1930 Belloc *New Cautionary Tales* 28 Ambassadors and Papal Counts, And Rastaquouères from Palamerez. 1930 A. Huxley *Vulgarity in Literature* vi. 34 The *rastacouaire* might display the twin cabochon emeralds at his shirt cuffs and the platinum wrist watch.

1940 J. Buchan *Memory Hold-the-Door* vi. 128 There was a vulgar display of wealth, and a *rastaquouère* craze for luxury. 1975 *Times Lit. Suppl.* 19 Dec. 1511/2 The worst book ever written by an Old Etonian, even by a rather *rastaquouère* Old Etonian.

raster (ræ·stəɹ), *sb.*[2] [a. G. *raster* screen, frame, f. L. *rastrum* rake, f. *rāsum*, supine of *rādĕre* to scrape.] **a.** A usu. rectangular pattern of parallel scanning lines forming or corresponding to the display on a cathode-ray tube; also more widely, with reference to other instruments and techniques involving systematic scanning movements or patterns without the use of a cathode-ray tube. Also *raster pattern, scan*; *raster-scan* vb. trans., *raster-scanning* vbl. sb. and ppl. adj.

1934 Bedford & Puckle in *Jrnl. Inst. Electr. Engineers* LXXV. 64 The path of the spot must, so to speak, be mapped out beforehand into a suitable line *raster*, which is of such a size and shape as to allow the real image of the spot to explore the whole of one picture. [*Note*] This word, imported from the German, is used to mean a scanning field or grating. 1939 *Television & Short-Wave World Pract. Handbk.* No. 1. vi. 54/1 The production of the series of lines, or 'raster' as it is generally termed, is..purely a local function of the receiver and quite independent of any reception of signals. 1940 *Jrnl. R. Aeronaut. Soc.* XLIV. 103 It is claimed that the electron Raster microscope overcomes these difficulties. The principle of the microscope is as follows:— A thin electron beam is made to scan the object in a 'Raster' as in television. 1946 *Jrnl. Inst. Electr. Engineers* XCIII. iiia. 1560/2 A raster approximately an inch square was then substituted for the noise pattern, obtained by applying the output from two saw-tooth oscillators, one of 50 c/s and the other of 10 kc/s, to the two pairs of plates of the tube. 1952 *Electronic Engin.* XXIV. 166/1 If a raster composed of horizontal scanning lines is further divided into the appropriate number of vertical lines, each line will become broken up into dots. 1966 [see *Interlace *v.* 6]. 1968 P. R. Thornton *Scanning Electron Microscopy* i. 8 Three years later [in 1938] v. Ardenne..built the first scanning electron microscope which used two magnetic lenses to provide a small electron spot at the specimen. Two sets of magnetic coils were used to scan the beam across the specimen in a television-like raster. 1969 Barton & Ward *Handbk. Radar Measurement* viii. 227 Raster-scanning pencil beams. These radars can be analyzed as search radars or sequential-processing trackers in both angular coordinates. 1970 *New Scientist* 4 June (Suppl.) 7/2 The screen can be 'raster-scanned' as on a TV screen (i.e. the light is deflected across the screen in a series of lines gradually moving to the bottom). 1973 *Sci. Amer.* Oct. 73/1 The spectroheliograms were made by holding the diffraction grating at one angle, so that only a single wavelength fell on the photomultiplier. The solar image was then scanned in a raster pattern to build up a picture of the sun in that one wavelength. 1973 *Physics Bull.* May 275/1 A flying spot performs a raster scan of the whole picture and with the help of a photodetector transforms the entire optical information—spots and all—into the memory of a large online computer. 1977 *Sci. Amer.* Oct. 84/2 In a circular-scan radar system the raster of the cathode-ray tube rotates synchronously with the antenna.

b. *Cinemat.* and *Photogr.* A fine grid, comprising wires, slits, or lenticular elements, placed in front of the projection screen in some stereoscopic cinematography systems, notably that invented by F. Savoye in 1942. Also *raster screen*.

1952 E. F. Linssen *Stereo-Photogr. in Practice* xxi. 291 In 1945 B. T. Ivanov wrote on 'raster-stereoscopy in the cinema'. *Ibid.* 292 Savoye's Cyclostereoscopic system... The cone allows the two beams to pass through many slits.. with which it is provided, and these same slits (constituting a revolving grid or raster) also act as selectors for the spectators. 1957 K. C. M. Symons *Stereo Photogr.* 205 Another method depends on the provision of a grid or raster screen in front of the projection screen. This method, which has a certain affinity with the parallax stereogram..forms the basis of two methods of projection, one Russian, the other French. *Ibid.* 213 *Raster*, a term used to describe certain autostereoscopic methods which depend on the use of a screen for multiplying and selecting the images. 1958 *Newnes Compl. Amat. Photogr.* 235 An alternative method of projection is the Cyclosterscopic system... This system consists of a metal cone of fine grids or rasters which revolves around the screen and is not noticeable in motion. 1965 *Focal Encycl. Photogr.* (rev. ed.) II. 1207/1 Raster screens consist of an arrangement of vertical wires interposed between the screen surface and the audience... There are also patented screens in which the projection surface is composed of vertical lenticular prisms, or of spherical lenticular elements graduated in size.

Hence as *v. trans.*, to scan (an area) with a beam that goes over it in a raster pattern; **ra·stered** *ppl. a.*, (of a beam) made to scan an area thus.

1975 *Nature* 9 Oct. 521/1 Methods by which the properties of inorganic materials may be measured quantitatively using the interaction of a rastered kilovolt electron beam with a solid. 1978 *Ibid.* 3 Aug. 457/2 Fig. 1 shows..two X-ray maps obtained (using electron microprobe X-ray fluorescence) by rastering the same region with a 30-kV electron beam and collecting in sequence the K Kα and Ca Kα rays.

Rastus (ræ·stəs). *U.S.* [Prob. shortened form of the personal name *Erastus*.] A name applied joc. in a number of songs and moving picture films to a 'typical' Negro, subsequently used as an offensive term for a Black person.

1895 K. Mills *Rastus on Parade* (*song*) 5 No use in talking he's hot stuff, Is Rastus when on Parade. 1909 C. W. Hayes in *Lippincott's Mag.* May 636 (*title*) Rastus's baby. 1910 *Moving Picture World* 7 May 749/2 Rastus in Zululand.—Rastus is an odd-jobs man, that is he does odd jobs when he has to. 1932 F. DuBose *Episodes in Black & White* 1 Two little pickaninnies were sitting by the fire when in came Uncle Rastus with a pumpkin pie. 1932 E. V. White *Chocolate Drops from South* 13 Mandy: 'Rastus, yo' makes me think yo' got the equator on yo'.' Rastus: 'How cum, Mandy?' Mandy: 'Yo' got sech a hot line!' 1944 H. L. Mencken in *Amer. Speech* XIX. 172 In my boyhood *Cuffy* had disappeared and *Sambo* was being supplanted by *Rastus*. 1955 W. Moore *Bring Jubilee* vi. 59, I call him Sambo because it sounds nicer than Rastus. 1965 W. McCord *Mississippi: Long Hot Summer* vi. 163 Leaders like Medgar Evers..discovered in the armed forces a new sense of personal dignity... They returned to their state determined to play roles other than 'Sambo' and 'Rastus'. 1965 'Malcolm X' *Autobiogr.* (1966) ii. 104 Wherever I showed my face.. the audiences 'niggered' and 'cooned' me to death. Or called me 'Rastus'. 1978 J. Blackburn *Dead Man's Handle* vi. 73 On your way, Rastus. Me and Massa's buddies.

rat, *sb.*[1] Add: **1. a.** For *Mu:* substitute *Rattus.* (Later *fig.* examples.)

1888 Kipling *Letters of Marque* (1891) xv. 111 Ram Baksh..headed his two thirteen-hand rats straight towards the morning sun. 1900 R. Barr *Unchanging East* 258 The Turkish Government has a little rat of a boat..which dare not venture out in a storm. 1907 J. Masefield *Tarpaulin Muster* 186 'I've been looking for truth,' he says; 'looking for truth in all these books... There's not a rat of truth in one of them. Not a solid rat, there isn't.' 1977 *Best of Austral. Angler* 12/2 Earlier, Col Noakes and myself had landed some 10 kg kings—comparative 'rats' that chased bait past all day.

b. esp. the North American musk-rat, *Ondatra zibethica*, of the family Cricetidæ, an aquatic rodent hunted for its thick brown fur; also, the pelt of this animal or its flesh used as food. (Examples.)

1584 R. Hakluyt *Discourse concerning Westerne Planting* in *Maine Hist. Soc. Coll.* (1877) II. 27 There is greate store of..bevers, squirrells, badgers, and ratts excedinge greate. 1800 A. N. McLeod *Jrnl.* 18 Nov. in C. M. Gates *Five Fur Traders* (1933) 130 The first paid his Debt, the next gave 40 Ratts en present. 1824 S. Black *Jrnl. Voy. from Rocky Mountain Portage* (1955) 153 Saw no appearance of the Otter, Rat or Mink. 1882 *Edmonton Bull.* 18 Feb. 3/2 They are living principally on rats and jackfish from Buffalo Lake. 1944 J. Martin *Canad. Wilderness Trapping* 48 It is the food which makes the pelt..and in the northwest we get the best rats. 1946 *Sun* (Baltimore) 15 Nov. 18/3 The 'rats' referred to are, of course, the musk-rats of the rich Dorchester marshes. 1953 *Jessen's Weekly* 19 Feb. 5/2. The trapping of Beaver and muskrat in the Huslia area does not show much promise as the ice is from three to six feet thick in the rivers and lakes... Many of the rats have been frozen in.

2. e. (Earlier and later examples.) Also as a general expression of disgust, annoyance, etc.

1886 *Lantern* (New Orleans) 20 Oct. 5/2 What a rotten game. Rats! 1888 *Texas Siftings* 7 Jan. 5/2 Smaller Boy—'Let me shine 'em up, Sir; for I have to support a poor little sick brother at home who is lame and can't see.' Bigger Boy—'Rats! I'm that poor little sick brother myself.' 1901 S. R. Crockett *Cinderella* xxvii. 188 'My cousin has lessons along with the younger children.' 'Rats!' declared Vic, smiling broadly; 'she sees that they do theirs—that's more like it.' 1914 G. B. Shaw *Misalliance* 23 *Mrs Tarleton.* Dont boast, John. Dont tempt Providence. *Tarleton.* Rats! You dont understand Providence. Providence likes to be tempted. 1951 J. Cornish *Provincials* i. ii. 21 'I don't kiss girls,' I said hurriedly. 'I never kiss girls. Never.' 'Oh, rats!' 1976 *National Observer* (U.S.) 21 Feb. 9/2 About a day later another letter from the company turned up in my mailbox. Rats, I thought, they have discovered their mistake and are going to take all the fun out of my life. 1977 *New Yorker* 27 June 30/3 Rats, you sound like a sorority pledge at Sophie Newcomb College.

f. *to give* (a person) *rats*: to give (him) a hard time; to berate, rebuke. orig. *U.S.*

1863 *Sunday Herald* (Boston) 15 Feb. 4/1 Hooker is doing something in the way of giving the rebels 'rats'. 1869 'Mark Twain' *Sk. New & Old* (1875) 48 You may write a blistering article on the police—give the Chief Inspector rats. 1940 F. D. Davison *Woman at Mill* III. 245 She was now going to give me rats, treat me as if I were personally responsible for the short-comings of the land of my birth.

g. *to get* (or *have*) *a rat* (or *rats*): to be eccentric or insane. *Austral.* and *N.Z. slang.*

[1890 Barrère & Leland *Dict. Slang* II. 171/2 (American), 'to have *rats*', to have wild or eccentric fancies.] 1906 E. Dyson *Fact'ry 'Ands* vii. 84 The factory flat loudly asserted that Spats had 'got a rat'. 1908 H. Fletcher *Dads & Dan* 65 In a town a whole population gets rats together, an' though they's all clean daft at times, yet, 'cause they all thinks alike, they don't doubt they's sane. *c* 1926 'Mixer' *Transport Workers' Song Bk.* 12 'Lend us a quid!' 'Lend you a what! Blime, have you got a rat?'

h. *rats and mice*: rhyming slang for 'dice'; a game of dice.

1932 P. P. *Rhyming Slang* 23 *Rats and mice*, dice. 1938 F. D. Sharpe *Sharpe of Flying Squad* xv. 170 We used to play dice with them... Rats and Mice the game was called.

3. a. (Later examples.)

1901 G. B. Shaw *Caesar & Cleopatra* iv. 188 Now, by great Jove, you filthy little Egyptian rat, that is the very word to make him walk out alone into the city. 1927, 1928 [see *double-crosser* s.v. *double-cross, double cross 1]. 1929 [see *Fink *sb.*[2]]. 1945 S. Lewis *Cass Timberlane* xliii. 324 The sort of male once described with relish as 'an agreeable scoundrel'..could now be referred to..as..a louse, a stinker, a rat, a twirp, a crumb, or a goon. 1959 [see *Fink *sb.*[2]]. 1976 *Western Mail* (Cardiff) 27 Nov., He turned to a group of policemen and said, 'I hope you are satisfied, you rats.' 1977 *Rolling Stone* 16 June 43/3 This is the terrible part of me, it's awful and I'm really a rat.

b. Preceded by a specifying sb., applied to one who is associated with or frequents the place specified (originally esp. a dock or riverside: in this context occas. without defining word); see also *rink-rat* (*Rink *sb.*[2] 5), *river rat* (River *sb.*[1] 4 e), *wharf-rat* (*b*) (Wharf *sb.*[1] 3). Chiefly *U.S.*

1864 *Harper's Mag.* Feb. 341/1 At our swimming-place we were often much molested by the river-border citizens of the town, variously known as 'dock-rats' and 'townies'. 1870 *Scribner's Monthly* I. 41 Many of the inmates.., as 'dock-rats', house-thieves, peace-breakers, and horse-stealers, have grown preternaturally quick-witted. 1872 *Harper's Mag.* Oct. 673/2 Our business is with those smaller, but terribly annoying vermin, the 'dock rats', with the river thieves, and with the junk-shops. 1883 J. Greenwood *Tag, Rag, & Co.* 33 Then, again, there's the regler 'rats'. How many of them, sneaking about craft at anchor,.. make a slip and get drowned? *Ibid.* 35 He was drowned, and carried away with the tide, and it wasn't till a week after that the 'rats'..fell in with the body and robbed it. 1890 [see *Extra C. *sb.* b]. 1928 H. Asbury *Gangs of New York* xi. 240 The police found him in company with a gang of notorious little dock rats. 1962 N. Maxwell *Witch-Doctor's Apprentice* iii. 25, I..gathered a lot of information on jungle medicine. I spent hours in bars buying *aguardiente* (local rum) for old jungle rats. 1967 *Boston Sunday Herald* 14 May (This Week Mag.) 2/2 As a kid, I was an airport rat. I rode my bike to the airport, 15 miles each way, and I would hang around and help. 1978 K. Bonfiglioli *All Tea in China* iii. vi. 80 A gaggle of waterfront rats, wasters, the scum of the sea.

4. d. (Earlier and later U.S. examples.)

1824 [see *rat-printer*, sense 7 e below]. 1836 *Proc. Nat. Typogr. Convention, Washington* 4 Martin H. Andrews..has been recognised as being one of the individuals published on the *Rat List* of the Columbia Typographical Society. *Ibid.* 11 Men pronounced *Rats* by one society, shall be considered such by all others. 1851 *Proc. 1st Nat. Convention Journeymen Printers U.S., 1850* 11 The present system is prolific of 'rats'. Our trade should be purged of this vermin. 1868 *Oregon State Jrnl.* 17 Oct. 1/6 The President of the National Typographical Union has pronounced a general amnesty, by virtue of which all expelled members, 'rats', etc., will be admitted to the Unions. 1896 *Typographical Jrnl.* IX. 100 A force of rats were doubled up from the Evening Ledger to get out the paper. 1902 *Encycl. Brit.* XXXIII. 411/1 A strike occurred in Mr Weed's office in 1821 on account of the employment of a non-union man, who was then designated a 'rat'.

e. *U.S.* A new student or freshman, esp. a newly-recruited cadet.

1850 'M. Tensas' *Odd Leaves Life Louisiana 'Swamp Doctor'* 113 There were four or five brother 'Rats' besides myself residing in the hospital, all candidates for graduation, and..all desirous of obtaining sufficient medical lore. 1896 *Bomb* (Virginia Military Institution) 109 An unfortunate 'Rat' whose face was glum, As he often to himself did hum—Guard Duty. 1900 *Dialect Notes* I. 54 *Rat*, a new student. 1930 *Amer. Speech* VI. 129 The freshman class at this institution [*sc.* Alabama Polytechnic Institute] is known as 'The Rats', and any given member is a 'rat'. 1937 W. Couper in J. E. Johnston *Echoes of V.M.I.* 43 One of the recollections of every Rat is the comforter who goes through the barracks purring—'Mister, do you see that tree? Well, all the leaves have got to go, and all the leaves have got to come back, before you..' and you know the rest. 1939 W. Faulkner *Wild Palms* 35 'This is Rat,' she said. 'He is the senior living ex-freshman of the University of Alabama. That's why we still call him Rat.' 1951 *Time* 28 May 50/2 Of all the cadets, the 'rats' of the entering class have the roughest time.

f. A police informer; an informer in a prison. *slang.*

1902 Farmer & Henley *Slang* V. 376/1 *Rat*... 6. (thieves').—A police spy. 1917 *New Republic* 13 Jan. 294/1 In most cases they were 'rats', and the best tools the keepers had. 1929 [see *Heel *sb.*[1]]. 1970 G. Jackson *Let.* 22 Mar. in *Soledad Brother* (1971) 186 You see every time a rat does get put away, the prison authorities always release a different reason for the attack, never that he was an informer. 1977 *New Yorker* 24 Oct. 72/3 Like all prisons, Green Haven is run with the help of informers—'rats'... One way..of rewarding rats is with jobs.

7. a. *rat-cage, -fur, -horde, -land, -plague, -season*; (sense *4 e) *rat rule*.

1936 J. Steinbeck *In Dubious Battle* ii. 21, I did hate being in the rat-cage. 1977 P. Dickinson *Walking Dead* i. i. 17 The neat rank of rat-cages in the animal room. 1907 *Daily Chron.* 24 Aug. 4/7 The hair was gathered up, chignon-fashion, and tied behind with strings made of rat-fur. *a* 1930 D. H. Lawrence *Last Poems* (1932) 189 In the moment of choice, the soul..utterly fails to recognize any more the grey rat-hordes of classes and masses. 1955 J. R. R. Tolkien *Return of King* v. x. 166 Dwarf-coat, elf-cloak, blade of the downfallen West, and spy from the little rat-land of the Shire. 1961 S. Price *Just for Record* vi. 47 He..hung up. The phone went burrup and I knew I'd crossed the border into literary rat-land. 1936 I. L. Idriess *Cattle King* (*caption facing p.* 50) These birds, in immense numbers, harry the 'rat plagues' which occur occasionally in portions of the interior. 1933 *Sun* (Baltimore) 8 Nov. 20/3 The sophomore victory..meant that the freshmen would continue to wear little red caps and obey 'rat' rules. 1939 *Amer. Speech* XIV. 29 *Rat*, new cadet, recruit... *Rat rules*,..see recruit regulations. 1942 Berrey & Van den Bark *Amer. Thes. Slang* § 882/1 *Rat rules*, cadet recruit regulations. 1921 *Beaver* May 14 The rat season closes today. All the hunters are now in.

b. *rat-hunting* (earlier and later examples).

1810 E. Weeton *Let.* 18 Jan. in *Jrnl. of Governess* (1969) I. 223, I set out on my rat-hunting expedition. *a* 1817 Jane Austen *Persuasion* (1818) IV. x. 229 We had a famous set-to at rat-hunting all the morning. 1966 *Beaver* Winter 54/1 All the different Indians started back to their own countries for rat [*sc.* muskrat] hunting.

c. *rat-borne, -eaten, -infested* (later examples) adjs.

1928 *Moderna Språk* Sept. 187 Terms like..*mother-craft, rat-borne* diseases, *wholemeal* bread. 1938 *Sun* (Baltimore) 1 Nov. 22/2 To protect the public health and to prevent the spread of rat-borne disease. 1901 'L. Malet' *Hist. R. Calmady* i. viii. 69 The fusty atmosphere of a cottage garret, right up under the rat-eaten thatch. 1951 P. Abrahams *Wild Conquest* 49 He had a funny, rat-eaten beard. 1916 E. & O. Sitwell *20th Cent. Harlequinade* 23 On to that rat-infested maze. 1974 *Country Life* 14 Mar. 588/1 The pond was a rat-infested rubbish tip.

d. *rat-brained, -fat, -grey, -poor, -shrewd, -souled, -swift, -toothed* adjs.

1971 B. Malamud *Tenants* 75 Those rat-brained Jews. 1930 E. Sitwell *Coll. Poems* 256 To show the shame Of the rat-fat soul to the grinning day. 1937 E. Muir *Coll. Poems* (1960) 71 The light was rat-grey. 1952 J. Steinbeck *East of Eden* ii. 9 A man who might have been well-to-do on ten acres in Europe was rat-poor on two thousand in California. 1960 S. Plath *Colossus* (1967) 63 Rat-shrewd go her squint eyes. 1921 R. Graves *Pier-Glass* 49 Was Sisera then more ripe for the knife or nail Than rat-soul'd Becker? 1969 G. Macbeth *War Quartet* 61 Air gushed in..Rat-swift. 1930 S. Spender *Twenty Poems* 14 And older whores Skuttle rat-toothed into the dark outdoors.

e. **rat-bat** *W. Indies* = Bat *sb.*[1]; **rat-bite fever,** either of two similar fevers of which the bacteria causing it are carried by rodents; **rat cheese** *U.S. colloq.* = *mousetrap *sb.* 2; **rat-fish,** substitute for def.: a fish of the family Chimæridæ, characterized by a long tail, esp. *Hydrolagus colliei*, which is found off the Pacific coast of North America; (later examples); **rat flea,** a flea infesting rats, esp. *Nosopsyllus fasciatus* or the tropical *Xenopsylla cheopis*, which are vectors of the bacillus causing plague; **rat-fucker** *coarse slang*, a base, despicable person (see also quot. 1967[1]); **rat-house,** (*b*) *Austral.* and *N.Z. slang*, a lunatic asylum; **rat-hunt,** a hunt for rats; also *fig.*; **rat-kangaroo,** add to def.: a very small kangaroo belonging to the subfamily Potoroinæ; (earlier and later examples); **rat pack** *slang* (orig. *U.S.*), a gang of disorderly young people; **rat-printer** = sense 4 d; **rat-proof** *a.*, able to keep out rats; hence **rat-proofing** *vbl. sb.*; **rat-run,** (one of) a maze-like series of small passages by which rats move about their territory; freq. *transf.* and *fig.* (usu. in derogatory sense); **rat-snake,** substitute for second part of def.: esp. a colubrid snake of the South Asian genus *Ptyas*, esp. the Indian *P. mucosus*; = *dhaman 1; (later examples); **rat-tight** *a.* = *rat-proof* adj.

1851 P. H. Gosse *Naturalist's Sojourn in Jamaica* 163 All Bats are called by the negroes Rat-bats, probably to distinguish them from Butterflies, to which they give the name of Bats. 1956 J. Hearne *Stranger at Gate* xviii. 142 A cave full of rat-bat droppings. 1910 *Q. Jrnl. Med.* III. 125 To the pathogeny of rat-bite fever I am at present unable to offer any clue. 1910 *Lancet* 11 June 1618/1 The January issue of the Quarterly Journal of Medicine contains an interesting paper by Dr. T. J. Horder on three cases of irregularly periodic fever associated with the bite of a rat, and so alike in their other features as to cause him to group them together under the name of 'rat-bite fever'. 1917 *Jrnl. Exper. Med.* XXV. 42 The clinical symptoms of rat-bite fever are inflammation of the bitten parts, paroxysms of fever of the relapsing type, swelling of the lymph glands, and eruption of the skin. 1924 *Ann. Trop. Med. & Parasitol.* XVIII. 171 The correct name for the causal organism of rat-bite fever is *Spirillum minus*, Carter 1887. 1949 H. W. Florey et al. *Antibiotics* II. xxxi. 1030 In man the results of treating rat-bite fever due to Actino. muris [*sc. Streptobacillus moniliformis*] were from the first exceedingly good on very moderate doses of penicillin. 1970 *New Scientist* 27 Aug. 407/2 The mouse has been incriminated in the transmission of rat-bite fever (or soduku). 1939 *Sun* (Baltimore) 4 May 8/5 Guilty of saying, in regard to macaroni, that 'it is merely associated with a dish whose other component part is rat cheese'. 1952 J. Steinbeck *East of Eden* xvii. 159 Their lunch of..bread and rat cheese. 1976 *Washington Post* 7 Nov. K1/5 We will try to recreate the atmosphere of a country store. Sardines, pickled pig's feet..rat cheese. 1905 D. S. Jordan *Guide to Study of Fishes* I. xxxi. 564 The existing Chimæras are known also as spookfishes, ratfishes, and elephant-fishes. 1936 P. S. Barnhart *Marine Fishes California* 14 *Hydrolagus colliei*..Ratfish..San Diego to Alaska, in cold bottom waters. 1955 *Sci. News Let.* 19 Feb. 121/3 Liver oil from ratfish is the richest source of batyl alcohol. 1965 [see *holocephalian* sb. s.v. *Holo-]. 1974 *Daily Colonist* (Victoria, B.C.) 28 July 17/2 Kevin's prizes..were awarded for a two pound, 15 ounce ratfish, the heaviest landed. 1871 *Hardwicke's Science Gossip* May 99/2 The rat has two kinds of fleas, that is, the banded Rat Flea..and the common Rat Flea. 1907 *Daily Mail* 19 Aug. 7/1 The Plague Commission has decided..that the vehicle of contagion is the rat-flea. 1929 H. E. Ewing *Man. External Parasites* v. 170 The Common Rat Flea, *Ceratophyllus fasciatus* Bosc., is the flea most commonly found on rats in Europe and North America. 1953 L. F. Hirst *Conquest of Plague* xii. 335 Hygiene improvements must have greatly reduced the rat and rat-flea population of the old-fashioned cities. 1953 *New Biol.* XIV. 111 The species from which we have gained most of our knowledge are the tropical rat flea, *Xenopsylla cheopis*, and..the rat flea of temperate countries, *Nosopsyllus fasciatus*. 1977 Richards & Davies *Imms's Gen. Textbk. Entomol.* (ed. 10) II. iii. 941 In India the species mostly implicated [in the transmission of plague] is the rat flea. 1967 *Amer. Speech* XLII. 229 *Rat-fucker*, a tool, usually made from a straight piece of metal coat hanger, approximately six to ten inches long, with a ninety degree bend two inches from each end, in such a manner that it ultimately has the shape of an old car crank handle. 1967 P. Welles *Babyhip* vi. 61 'Scum,' John mumbled... 'Ratfucker, prick,' George said. 1913 A. J. Rees *Merry Marauders* ii. 24 It was a rat-house—an asylum, to be perlite. 1922 A. Wright *Colt from Country* 83 He'll be the long lost boy, instead of the guy that's missed and landed in the rat-house. 1946 F. Sargeson *That Summer* 108 Maybe they'd have to take me away to the rat-house. 1948 V. Palmer *Golconda* xxxiii. 278 Hadn't it been plain all along that there was a streak of madness in the old boy?.. He had done a spell in the rat-house and was only out on sufferance. 1957 I. Cross *God Boy* (1958) ii. 17 You're heading for the rat house, the way you talk. Imagining things. 1825 G. Simpson *Jrnl.* 2 May in F. Merk *Fur Trade & Empire* (1931) 150 The Rat hunts have likewise failed in consequence of the lowness of the Waters, but the returns in Beaver are very fair about 3,000. 1843 *Ainsworth's Mag.* III. 78 If my father's keepers invited me to a private rat-hunt, Tickle was sure to smell a rat. 1961 *Guardian* 1 Dec. 13/1 It is also to be doubted whether the OAS leaders, for all their deliberate use of murder and plastic bombs, want the 'rat hunts'. 1841 Rat-kangaroo [see Kangaroo *sb.* 2]. 1926 Le Souef & Burrell *Wild Animals of Australasia* 232 The rat-kangaroos for the most part live on the surface of the ground. 1944 F. Clune *Red Heart* 21 The two aboriginal guides ran tirelessly.., supplementing the larder by killing lizards, rat-kangaroos, and other small creatures. 1965 D. Morris *Mammals* 66 Rat-kangaroos differ from other members of the family in having well-developed canine teeth. 1978 D. Ovington *Austral. Endangered Species* 76/1 Also known as ..the burrowing rat kangaroo, the boodie is the only burrowing member of the kangaroo family. 1951 R. S. Prather *Bodies in Bedlam* ix. 65 It looked like three or four of L.A.'s juvenile moron gangs, sometimes called rat packs, had taken turns going over the place. 1973 *Observer* 8 Apr. 48/4 Constance throws aside worldly success..and the ratpack to immerse herself in her first love—history. 1974 H. L. Foster *Ribbin'* iii. 91 More recently, gangs, fraternities, cliques, organizations, and 'rat packs' have again begun to be reported. 1977 *Time* 25 July 10/1 This summer Sweden has been hit by a small but ugly wave of racial incidents, three of which have been violent encounters between a rat pack of young Swedish ruffians and a community of Assyrian immigrants from the Middle East. 1824 *Microscope* (Albany, N.Y.) 6 Mar. 191/2 Loren..Webster, chief inkdauber in a rat-printing office at the west. Ralph Walby, nothing at all but a rat-printer. 1931 T. S. Stribling *Forge* xviii. 159 The Lacefield barns were rat proof. 1960 *Farmer & Stockbreeder* 26 Jan. 51/3 The virus is handled inside areas surrounded with a rat-proof fence. 1929 *Times* 2 Nov. 9/5 Surely it would be an economy to employ a man permanently for rat proofing and rat catching at £150 per annum. 1870 *Gentl. Mag.* Sept. 497 The barracks are a mouldy rat-run now. 1893 *Baily's Mag.* Oct. 253/1 The rat-runs had been stopped up, and he killed nearly..a hundred rats before he paused. 1924 Galsworthy *White Monkey* iii. i. 223 Hurrying along the rat-runs of the Tube, she slipped her hand into his pocket. 1940 Blunden *Poems 1930–1940* 202 In roofless barns, in rat-run saps. 1953 *Spectator* 13 Feb. 194/2 She will be able, through her own will and capacity, to escape from the rat-run of her environment. 1974 *Country Life* 24 Jan. 148/4 During weekday rush hours..a normally peaceful residential street can become a rat run. 1882 C. C. Hopley *Snakes* iv. 85 The rat snake..and the *Clothonia* of India are 'said' to suck the teats of cows. 1907 *Country Life* July 328/3 The yellow rat snake or chicken snake is one of the most useful and is entirely harmless. 1927, etc. [see *Dhaman 1]. 1958 R. Conant *Field Guide Reptiles & Amphibians U.S.* 154 (*heading*) Rat Snakes: Genus *Elaphe*. *Ibid.* 155 All the Rat Snakes vibrate their tails rapidly when alarmed. 1969 A. Bellairs *Life Reptiles* I. vi. 242 In some genera such as the rat snakes (*Ptyas*), the viper *Echis* and most sea snakes it [*sc.* the right lung] stretches almost back to the cloaca. 1893 *Jrnl. Soc. Arts* 5 May 623/1 What is wanted is a mode of running the wires..that shall not only be electric-tight, but shall also be water-tight, air-tight, oil-tight, fire-tight, and rat-tight. 1908 *Installation News* II. 33/1 The union between two screw threads does not make a perfectly 'watertight, airtight, gastight, and rat-tight' joint, as the saying is.

rat, *v.*[1] Add: **2. a.** (Earlier and later examples.) Also, to turn traitor; in *Criminals' slang*, to inform.

1812 Southey *Let.* 18 May in *Life & Corr.* (1850) III. 341 W— and C—, I doubt not, ratted upon the Catholic question because they expected the Prince upon that ground would eject Perceval. 1814 Ld. Brougham *Let.* 29 June in *Creevey Papers* (1903) I. ix. 195 The Whigs have just discovered old Sherry to be 'an old and valued friend and an ancient adherent of Fox'. They therefore support him. To be sure, he has ratted and left them—he kept them out of office twice—and he now openly stands on Yarmouth's influence and C[arlton] House. 1910 *Blackw. Mag.* Aug. 256/2 Those who, in the slang of politics, are said to 'rat'. 1934 *Sun* (Baltimore) 20 Aug. 5/1 Misunas..has 'turned State's evidence'—'ratted' in gangland parlance. 1938 [see *Hooligan]. 1938 E. Bowen *Death of Heart* iii. iii. 371 The girl at the switchboard must have ratted. 1969 *Listener* 24 July 102/2 One's feeling for the Chamberlain government was one of such utter contempt that one felt they might very well rat once again. 1974 S. E. Morison *European Discovery of Amer.: Southern Voyages* xx. 480 The captain of *San Gabriel* ratted..and sailed for Spain.

c. With *on.* To default on; to let (someone) down; to behave disloyally towards; (orig. *Criminals' slang*), to inform on.

1932 A. J. Worrall *Eng. Idioms* 12 Of course I won't do that. Do you think I'd rat on a pal. 1938 E. Ambler *Cause for Alarm* xviii. 311 The Italians may rat on that contract. 1938 E. Bowen *Death of Heart* iii. ii. 351 In a small way I have just ratted on Anna. 1938 *Sun* (Baltimore) 1 Sept. 4/2 This pair..have admitted the fatal attack on another prisoner for 'ratting on us'. 1948 *Richmond* (Va.) *Times-Dispatch* 13 Feb. 30/1 But of all persons to rat on the set up McKeever, the planner and perpetrator..should have been the last. 1957 C. MacInnes *City of Spades* ii. xiv. 199 Why isn't Muriel here, anyway? She's ratted on him. 1973 *Times* 1 Dec. 9/6 If they were to rat on these policies he would become one of their strongest opponents and critics. 1974 *Socialist Worker* 2 Nov. 1/2 The Labour government has ratted on these men. 1977 *New Yorker* 24 Oct. 128/2 Finando and the two men he had ratted on..were all transferred to the same prison.

4. *trans.* To search (a person, his belongings, etc.) for things to steal; to pilfer. Chiefly *Austral.* and *N.Z.*

1919 Downing *Digger Dial.* 41 *Rat* (vb.), (1) Search a prisoner or dead body. (2) Pick a pocket. 1925 Fraser & Gibbons *Soldier & Sailor Words* 236 *To rat*, to steal. To search a dead body. 1931 V. Palmer *Separate Lives* 267 'Look here, you slinking cur!' he began. 'You've been ratting other people's property for months.' 1937 J. A. Lee *Civilian into Soldier* 194 There must be a lot of dead Huns to rat. 1941 K. Tennant *Battlers* i. 9 Some thieving (*adjective*) robber was 'ratting' his tucker-box. 1971 J. S. Gunn *Opal Terminol.* 38 *Rat*, to pilfer opal from a miner's hiding place or enter someone's mine and take out opal rock.

rata. Also 8 **ratta(h).** Substitute for def.: An evergreen tree or woody climber belonging to one of several species of *Metrosideros*, of the family Myrtaceae, esp. a New Zealand species, the small *M. lucida* or the much larger *M. robusta*, both bearing terminal clusters of red flowers with long stamens; also, the fruit or the heavy reddish timber of a tree of this kind. (Earlier and further examples.)

[*a* 1771 S. Parkinson *Jrnl. Voy. to South Seas* (1773) 40 E ratta, or e pooratta... This tree, or shrub, grows upon the Tooaroa, or Lower-hills... The flowers are full of beautiful scarlet stamina.] 1792 W. Bligh *Voy. to South Sea* xi. 139 The *rattah*, not much unlike a chestnut,..grows on a large tree. 1829 W. Ellis *Polynesian Researches* I. xiii. 376 The wood of the rata has a fine straight grain. 1847 [see *Nikau]. 1853 A. S. Atkinson *Jrnl.* 7 Sept. in *Richmond-Atkinson Papers* (1960) I. 130 We found a beautiful little scorpion spider—it was under a rata log. 1896 [see *Pakihi]. 1935 [see *Nikau]. 1949 F. Sargeson *I saw in my Dream* 127 Mr Anderson and Dave sat with their backs against the twisted barrel of a fallen rata. 1968 [see *Inanga 2]. 1974 *Nat. Geographic* Aug. 195 The nightmare tree called rata, which begins as an innocent-seeming vine, and, in the end, strangles the tree to which it attaches itself.

ratafia. Add: **1.** (Later examples.) Now applied esp. to a type of aperitif made from grape-juice and brandy.

1907 *Yesterday's Shopping* (1969) 18/1 Essences, for flavouring..ratafia. 1946 A. L. Simon *Conc. Encycl. Gastron.* VIII. 139/2 *Ratafia*, a generic name for a number of Cordials, usually home-made, always sweet and often of very highly alcoholic strength. Ratafia may be made with new wine or grape juice and sufficient spirit to stop its fermentation; being further flavoured with various fruits, herbs and spices; or else by the infusion of the same ingredients in brandy. 1959 W. James *Word-bk. Wine* 155 *Ratafias* were infusions of fruit or herbs in brandy, made by housewives in happier days when brandy was cheap... In the champagne country, ratafia is an aperitif, somewhat stronger than sherry or vermouth, made by mixing some unfermented champagne grape juice with brandy. 1964 *Harper's Bazaar* Nov. 146/2 *Ratafia*..bears no resemblance to our ratafia, which is a home-made liqueur, usually made with almonds. *Ratafia* is an aperitif made with local white wine and brandy. 1973 *Daily Tel.* 12 July 16, I was interested to learn..that Ratafia is an aperitif made in Champagne by mixing brandy with unfermented champagne grape juice. I have 'essence of Ratafia'..which..is made from bitter almonds.

attrib. (Earlier examples.)

a 1711 in E. Hamilton *Mordaunts* (1965) vi. 134 Rattefea biscakes, sugar puffs, chips. 1728 E. Smith *Compl. Housewife* (ed. 2) 100 *A Ratafia Pudding*..Cream..Naples-bisket ..Butter, some Sack, Nutmeg, and Salt..Almonds..Eggs ..bake... Scrape Sugar on it. *Ibid.* 142 (*heading*) To make Ratafia Cream. 1755 H. Glasse *Art of Cookery* (ed. 5) xvi. 285 To make a Trifle. Cover the Bottom of your Dish..with ..Ratafia Cakes.

ratama, var. *Retama.

rat-a-tat. Add: (Further examples.) Also *attrib.* in **rat-a-tat ginger** (see quot. 1959).

1916 N. Douglas *London Street Games* 64 Rat a tat tat, who is that? Only grandma's pussy-cat. What do you want? A pint of milk. 1959 I. & P. Opie *Lore & Lang. Schoolch.* xviii. 381 Illegally knocking at doors... *Rat Tat Tat.* Abertillery and Newbridge in Monmouthshire, Barry in Glamorgan, Coventry. A variant in Lydney is *Rat-a-Tat Ginger. Rat-a-Tat-Tat.* Solihull, near Birmingham, and Grenoside, near Sheffield. 1961 *Times* 17 June 3/2 Any passing rat-a-tat of close volleying.

‖ **ratatouille** (ratatū·y). [Fr.: the final element is app. f. *touiller* to stir up.] † **a.** A ragout. **b.** In full, *ratatouille niçoise*: a dish, originating in Nice, consisting of aubergines, tomatoes, onions, peppers, and other ingredients stewed in olive oil.

1877 *Cassell's Dict. Cookery* 717/2 *Ratatouille* [*sic*]. This is a popular French method of making a savoury dish out of the remains of cold meat. 1946 F. M. Farmer *Boston Cooking-School Cook Bk.* (ed. 8) 472 Ratatouille Niçoise... Place in shallow baking dish, sprinkle with cheese and brown under the broiler. 1950 E. David *Bk. Mediterranean Food* 125 Ratatouille is a Provençal ragoût of vegetables, usually pimentos, onions, tomatoes and aubergines, stewed very slowly in oil. 1956 L. McIntosh *Oxford Folly* 36 She prepared a *ratatouille* of aubergines, peppers, tomatoes, onions, and, of course, garlic. It was the only foreign dish she knew. 1960 E. David *French Provincial Cooking* 242 Ratatouille Niçoise. Aubergines, tomatoes, onions and peppers stewed in oil. 1969 *Times Lit. Suppl.* 25 Dec. 1465/3 On the Riviera, the liberator's cause is seen as a distant grey dream, so easily dulled by ratatouille and red wine. 1973 R. Littell *Defection A. J. Lewinter* iv. 26, I have some cold *ratatouille* in the fridge.

ratbag (ræ·tbæg). *Austral.* and *N.Z. slang.* Also **rat-bag.** [f. Rat *sb.*[1] + Bag *sb.*] A stupid or eccentric person, a fool; an unpleasant person, a trouble-maker. Also *attrib.*, stupid, idiotic, uncouth.

1937 'W. Hatfield' *I find Australia* 138 'You brought one rat-bag *in*,' said Evans to me, 'so now do me a favour by taking one off my hands.' 1948 V. Palmer *Golconda* xiv. 107 Why the hell, Donovan, are you backing that old ratbag, Christy Bangham? 1954 *Coast to Coast 1953–1954* 172 Or a ratbag watchman who'd run amuck if you breathed on a window. 1955 D. Niland *Shiralee* 24 And what a ratbag situation, what a story. 1956 D. M. Davin *Sullen Bell* i. iii. 19 How did you ratbags get here, a highbrow joint like this? 1961 B. Crump *Hang on a Minute, Mate* 164 This'd be the best scrapper among you bunch of ratbags, wouldn't it? 1963 *Australasian Post* 14 Mar. 51/2 'He's not one of your ratbag mob, though, he—.' That cough again... 'He's a gentleman, you see.' 1965 W. Dick (*title*) A bunch of ratbags. 1967 H. Hunter *Case for Punishment* ii. 37 Reading all about our little ratbags?.. A lot of kids from this school are involved. 1970 *N.Z. Listener* 12 Oct. 13/2 A bit late to find that out, you snobbish ratbag wowser. 1973 [see *Lead *sb.*[1] 1 f]. 1976 I. Murdoch *Henry & Cato* i. 42 Well, it's just a game, you know me, scare a few ratbags, does no harm.

rat-catcher. Add: **2.** Unconventional hunting dress. Also *transf.* and *attrib.*

1910 Kipling *Diversity of Creatures* (1917) 310 He came back to the bar, after he'd changed into those rat-catcher clothes. 1928 J. B. Thomas *Hounds & Hunting* 254 *Ratcatcher*—referring to one informally dressed when hunting. 1930 *Field* 29 Nov. 764/1 The self-respecting beginner will want to be turned out properly, in the right 'rat-catcher style'. 1933 A. Powell *From View to Death* viii. 200 Both sons were in ratcatcher and Torquil wore a canary-coloured waistcoat. 1963 M. Malim *Pagoda Tree* xvi. 101 While one could exhibit oneself and did to the Cricket Club in ratcatcher, so to speak, one had to dress for presentation on the Thursday committee night at the Tulyar. 1976 *Horse & Hound* 3 Dec. 32/2 (Advt.), Moleskin ratcatcher trousers. Warm, tough, supple. Shirt-grip self-supporting waist... Slim leg style. 1977 'E. Crispin' *Glimpses of Moon* xi. 205 Nor were they a very distinguished Hunt: the men mostly turned up in ratcatcher (Fen had that morning noted one.. who was wearing a hoicked-up caftan and prayer beads above his shining riding boots).

ratch, *v.*[1] **3.** (Later example.)

1904 Hardy *Dynasts* I. i. iii. 31 The thousands called.. will ratch the lines Of English regiments..To glorious length.

ratch (rætʃ), *v.*[2] *north. dial.* and *Sc.* [f. Rache, ratch *sb.*[1]] **a.** *intr.* To forage for food, to ferret *around*; to ramble or wander *about*. **b.** *trans.* To search thoroughly, ransack. Hence **ra·tching** *vbl. sb.*

1801 'Berwickshire Sandie' *Poems* 73 Hens ratch'd through the house wi' greed. 1859 W. Dickinson *Gloss. Words & Phrases Cumberland* (foreword), Yan wad ratch ivry neukk ov oald Cummerlan. *Ibid.* 91 *Ratch*, to ramble, to search vigorously. 'Ratchan about like a hungry hound.' 1869 A. C. Gibson *Folk-Speech Cumberland* 96 Cook's house was ratch't through an' through. 1971 *Country Life* 9 Sept. 630/1 There's oalas an odd yan or two that'll leave their lambs an' ratch aboot. *Ibid.* 7 Oct. 900/1 Hill sheep can be very active, and when they feel like ratching it takes a very formidable fence to prevent them. 1973 *Guardian* 26 Feb. 10/2 The grass is poor, as yet, so some of the fell-sheep go foraging in and out of the woods..and into any fell-side garden with an open gate... 'Ratching' is part of their nature. 1976 *Jrnl. Lakeland Dial. Soc.* 21 Yan day t' auld Friesian bull gat oot, a' caved an' ratched aroond, Neabody durst gan near him as he rwoared an' scratted t' groond.

ratchell. **1.** (Later example of form rachill.)

1884 R. Hunt *Brit. Mining* 912 *Rachill*, small loose stones that are usually found on the top of the rock forming as the depth increased into the nature of beds.

ratchet, *sb.* Add: **3*.** = *ratchet-knife.

1975 *Globe & Mail* (Toronto) 11 June 3/1 Walking with the road with a ratchet (knife) in your waist, Johnny you're too bad. 1976 *Daily Mirror* 2 Apr. 20/2 *Ratchet*, knife.

4. *ratchet screwdriver*; **ratchet effect** (see quots. 1977); **ratchet knife,** a type of knife popular in Jamaica.

1970 *Times* 13 Apr. 20/7 It appears that there has been a ratchet effect in employment in the service industries. 1977 *New Society* 31 Mar. 643/2 One of the curiosities of political life is what you might call the 'ratchet effect'. This is the process by which one party makes the running over an issue, and gradually winds the other (or others) along after it. 1977 *Listener* 31 Mar. 397/2 It was Sir Keith Joseph who drew attention to the 'ratchet effect' in politics, whereby the right seems to have acquiesced in the changes the left brings about. Curiously, in broadcasting matters, the ratchet effect has worked the other way: the Conservatives broke the BBC's monopoly in the Fifties.., and it has been the left that has acquiesced. 1979 *Dædalus* Spring 122 These efforts are illustrated by the rediscovery of the 'ratchet effect' theory, which simply begins with the common wisdom that prices go up more freely than they go down. 1971 *Jrnl. Commonwealth Lit.* Dec. 140 The DJE..passed by such terms as..*ratchet-knife, trouble.* 1976 Boot & Thomas *Jamaica* 40/1 Just the sort of deft digital flourish you need to whip out the blade of a ratchet knife—which is a particular kind of nicely curved blade in a tapered handle, made in Germany for gutting fish. 1897 *Sears, Roebuck Catal.* 80/1 The old and well-known Gay's Double-Action Ratchet Screw Driver. 1979 S. Brett in *Winter's Crimes 11* 12 A ratchet screwdriver... Just the job for putting up shelves.

ratchet, *v.* Add: **a.** Also *transf.* and *fig.*

1977 *Time* 3 Jan. 44/3 The signal, according to some radio operators who have heard it ratcheting over their headsets, sounds like a 'buzzsaw' or 'the whirring of helicopter blades'. 1977 *Rolling Stone* 16 June 36/1 The movie director, age 34, spirals, ratchets, thrusts his chin like Mussolini.

b. *trans.* To move (something) *up* as by a ratchet. Cf. *ratchet effect* s.v. *Ratchet *sb.* 4.

1977 R. Jenkins *Europe's Present Challenge & Future Opportunity* (Jean Monnet Lect.) 8 Floating exchange rates transmit violent and sudden inflationary impulses... Each new impulse ratchets up the inflationary process. 1979 *Daily Tel.* 9 Aug. 2/7 We are quite clear that the union movement has not been responsible for ratcheting up inflation.

rate, *sb.*[1] Add: **II. 7. a.** Also in phr. *at the rate of knots*: see *Knot *sb.*[1] 3 c.

b. (Further examples.)

1912 in C. B. Smith *Testing Time* (1961) ii. 32 The rate of rising loaded as above has been tested up to 600 feet & found to be at the rate of 155 feet per minute. 1930 C. Dixon *Parachuting* ii. 21 Hampton descended very slowly.. landing after a most pleasant experience, such as the modern parachute gives, in thirteen minutes. This rate of descent worked out at 8 feet per second. The modern rate is about 21 feet per second. 1943 T. Horsley *Find, Fix & Strike* 21 Both pilots throttled back..and used just sufficient engine to give them a rate of sink of 250 feet a minute. 1946 *Happy Landings* July 9/1 Your rate of climb indicator may show as much as 3,500 feet per minute up. 1963 B. Fozard *Instrumentation Nuclear Reactors* xiii. 163 Reactivity must be released slowly by withdrawal of the control rods at a rate which is known to keep the rate of divergence within safe bounds.

8. c. Of a spring: a quantity relating the applied load to the compression or extension produced (see quots. 1959, 1961).

1957 E. B. Jones *Instrument Technol.* III. i. 6 For each value of the measured variable there will be a definite position of the feed-back bellows, and as the spring rate of the bellows and spring is fixed, this must mean a definite value of the transmitted pressure. 1959 *Motor Manual* (ed. 36) v. 123 This ratio can be expressed in terms of pounds weight required to produce a deflection of one inch..and this figure is known as the spring rate. 1961 W. R. Berry *Spring Design* i. 12 The rate *S* of the spring is equal to the load per unit of deflection. 1964 H. A. Rothbart *Mech. Design & Systems Handbk.* xxxiii. 12 Some springs are such that their rate is constant over the entire usable range of deflection.

III. 9. c. (Later examples.) Also, = Rating *vbl. sb.*[1] 2 b.

1963 *Amer. Speech* XXXVIII. 76 *Crow* (the embroidered eagle on the rating badge) was used to designate the insignia of rate for the petty officers of the Navy. 1977 *Navy News* Feb. 18/3 H.M.S. Tartar has no fewer than eight 'stripeys' among the junior rates on board. 1978 *Ibid.* Oct. 4/3 These numbers are calculated from the requirement, which is the number of ratings needed in each rate and branch to fill all sea and shore billets.

IV. 16. (Later examples.)

1921 G. B. Shaw *Back to Methuselah* p. lx, But how, at this rate, did Darwin succeed with the capitalists too? 1930 —— *Apple Cart* i. 20 Oh for Heaven's sake dont contradict her, Joe. We shall never get anywhere at this rate.

17. a. *rate-determining, -limiting* adjs., *-making, -payer* (later examples), *rebate, -setting* adj., *-support, -supported* adj.; **rate-buster** *slang*, a piece-worker whose high productivity causes or threatens to cause a reduction in piece-work rates; hence **rate-busting** *vbl. sb.*; **rate-card,** a list of charges for advertising; **rate constant** *Physical Chem.*, a coefficient of proportionality relating the rate of a chemical reaction at a given temperature to the concentration of reactant (in a unimolecular reaction), or to the product of the concentrations of reactants (in a reaction of higher order); **rate-cutting,** a lowering of charges or of rates of pay; † **rate factor** *Biol.*, a biological factor which influences or determines the speed of a developmental process, and so may affect the phenotype of the mature organism; **rate-fixer,** one who fixes the rates at which piece-workers are paid; so **rate-fixing** *vbl. sb.*; † **rate gene** *Biol.*, a gene which acts as a rate factor; **ratemeter,** an instrument which displays or records the counting rate, usu. averaged over a time interval, of pulses in an electronic counter, esp. those resulting from incidence of ionizing radiation.

1939 Roethlisberger & Dickson *Managem. & Worker* xxii. 522 You should not turn out too much work. If you do, you are a 'rate-buster'. 1948 *Appl. Anthropol.* VII. 5/1 In every work group there is nearly always a very small minority of individuals who refuse to be held back and insist on making as much bonus as they like... In current American industrial literature such workers are referred to as 'rate-busters'. 1972 M. Argyle *Social Psychol. of Work* vi. 113 A rate-buster may be given a raised eyebrow, a look of disapproval, a blow on the arm, or may simply be avoided. 1967 C. Margerison in Wills & Yearsley *Handbk. Managem. Technol.* 31 However, such practices as the prevention of rate-busting are extremely logical and rational from the workers' point of view. 1905 Calkins & Holden *Art of Mod. Advertising* 352 A rate-card is a card or printed sheet giving the advertising rates in a given publication. 1962 *Rep. Comm. Broadcasting 1960* 69 in *Parl. Papers 1961–2* (Cmnd. 1753) IX. 259 These figures represent the cost, as shown in the programme companies' rate-cards, of booking the time taken for advertisements during the year. 1977 *Listener* 31 Mar. 399/2 The NBC and CBS affiliate stations.. are getting worried. Can they hold their advertising rate-cards when audiences are falling? 1927 *Jrnl. Amer. Chem. Soc.* XLIX. 1620 Let K_∞ be the fraction decomposed per second at high pressures, or the unimolecular rate constant. 1950 W. J. Moore *Physical Chem.* xvii. 516 The units of the rate constant depend on the order of the reaction. 1956 *Nature* 21 Jan. 127/1 The results allowed determination of the approximate values of the pseudo first-order hydrolytic rate-constants of the amide groups in these substances. 1972 R. A. Jackson *Mechanism* iii. 28 If the equilibrium constant is known.., and the kinetic order and reaction rate constant is determined for the reverse reaction, the order and rate constant for the forward reaction can be calculated. 1888 *Scribner's Mag.* Oct. 485/2 Its percentage being fixed there is no motive for rate-cutting. 1956 J. A. C. Brown *Social Psychol. Industry* v. 145 The most frequent reasons given by the unorganized worker for such practices were rate-cutting, fear of unemployment, [etc.]. 1978 *Detroit Free Press* 16 Apr. 2A/1 Capitol Cartage and its owners deny that they have used shakedowns, payoffs, bribery, rate-cutting, intimidation and terror to eliminate competition. 1935 *Jrnl. Chem. Physics* III. 113 When the rate determining step shifts to the collision process..we again use well-known statistical methods. 1968 R. O. C. Norman *Princ. Org. Synthesis* xiv. 436 The rate-determining step in the reaction of neopentyl bromide is the formation of the high-energy primary carbonium ion. 1927 *Brit. Jrnl. Exper. Biol.* V. 121 (*heading*) Summary of families carrying accessory rate-factors. 1931 E. B. Ford *Mendelism & Evolution* ii. ii. 29 It is probable that such rate-factors controlling the speed and time of onset of processes in the body, are of very general occurrence. 1932 J. S. Huxley *Prob. Relative Growth* i. 4 Any genes controlling relative size of parts will have to exert their action by influencing the rates of processes, and so fall in line with the numerous other rate-factors whose importance has been summarized by Goldschmidt (1927) and by Ford and Huxley (1929). 1930 *Engineering* 30 May 696/2 In this office, the operations are made out on a master card and forwarded to the rate fixers. The latter carefully estimate the cost of the work, and add the workmen's allowance of 33⅓ per cent. 1907 *Daily Chron.* 19 Mar. 2/6 The operations of a specially-constituted rate-fixing branch as regards piece work in the Royal Carriage Department. 1959 *Times* 14 Jan. 2/5 They [*sc.* candidates] must also be experienced in the running and organising of process planning, ratefixing. 1932 J. S. Huxley *Probl. Relative Growth* v. 230 (*caption*) Diagram to show effects of rate-genes on eye pigmentation. 1938 R. Goldschmidt *Physiol. Genetics* ii. iii. 52 Goldschmidt..first drew attention to rate genes and their importance for an understanding of gene action. 1946 *Nature* 28 Sept. 448/2 A slow heterolysis of nitric acid cannot depend only on proton transfers, and therefore the rate-limiting fission.. must occur in an NO-bond. 1974 M. C. Gerald *Pharmacol.* xii. 219 The rate at which this reaction occurs determines the speed at which alcohol is removed from the body, and hence this reaction is termed 'the rate-limiting step'. 1969 *Jane's Freight Containers 1968–69* 126/1 The formation of a Market Research Department to develop new methods of rate-making. 1949 *Nucleonics* Feb. 74/1 After the charge in the tank circuit of the rate meter has reached equilibrium, the rate meter indicates the counting rate at a glance. 1962 G. A. T. Burdett *Automatic Control Handbk.* ix. 33 The electronic system usually associated with nucleonic switches and gauges incorporating Geiger counters is known as a 'ratemeter'. This unit provides an indication of the rate at which the counter is detecting incident radiation and is usually provided with a variable integrating time control. 1974 *Physics Bull.* Aug. 349/1 The intensity of the x ray reflection is indicated by the frequency of clicks in a loud speaker, by the flashes of a GaP lamp and by the reading of a ratemeter. 1898 G. B. Shaw *Let.* 12 Mar. (1972) II. 15 Ratepayers' Association in the evening. 1955 *Times* 27 June 9/3 No ratepayer can properly gauge the effect of revaluation until he knows what rate his local authority intends to levy in the new dispensation. 1976 *Daily Tel.* 30 June 2/6 Honest citizens and rate-payers alike. 1965 *Economist* 4 Dec. 1050/1 (*heading*) Rate rebates. Helping the needy. *a* 1974 R. Crossman *Diaries* (1975) I. 303 The main issue was whether I should be allowed a special short Bill introducing rate rebates *before* the long-term reform of rating. 1963 *Economist* 7 Sept. 828/1 The foreign shipowners who dominate the rate-setting consortiums. 1966 *Times* 21 Dec. 12/6 Mr. Greenwood, Minister of Housing and Local Government.., moved that the Rate Support Grant Order, 1966, be approved. 1976 *Scotsman* 25 Nov. 9/4 They condemned the reduction in rate support for Scotland as mere transfer of existing expenditure from central government to the ratepayer. 1905 *Daily Chron.* 1 Feb. 6/1 A remission in rates should be made to companies which were the chief ratepayers in districts in which rate-supported tramways run. 1922 Joyce *Ulysses* 710 Ratesupported moribund lunatic pauper. 1961 T. Landau *Encycl. Librarianship* (ed. 2) 296/1 There is only one local authority area in the whole of the U.K. which has not a rate-supported public library.

b. *pl.* used *attrib.* in sense 6 d, as *rates aid, man, rebate, reduction, tribunal.*

1966 *Times* 11 May 1/4 (*heading*) Rates aid to start next April. 1953 Dylan Thomas *Let.* 17 Mar. (1966) 397 Friendly Brown's can wait. These tradesmen and rates-men can't. 1966 *Times* 17 Aug. 8/6 Rates rebates in the first six months' operation of the Rating Act, 1966, are expected to total £6,850,000. 1971 *Reader's Digest Family Guide to Law* 90 (*heading*), How to apply for a rates reduction. 1935 *Economist* 23 Feb. 429/1 The task of the Railway Rates Tribunal will be unenviable when, next November, they are due to fix the rebates for the year 1935–36.

rate, *sb.*[2] (Later example.)
1976 *Shooting Times & Country Mag.* 18–24 Nov., Never fail to give a severe rate and a flick of your whip if any young hound gives a challenging voice to a bullock, cow, or a strange dog.

rate, *v.*[1] Add: **3. a.** Also, to set a high value on, to think much of.
1973 *Times* 10 Feb. 7/7 You can never be sure of Brazil, of course, but I don't rate the South Americans next time. I believe 1974 will be dominated by the Europeans. 1973 *New Society* 12 Apr. 64/2 He would like to play cricket for Surrey, but he doesn't rate his chances. 1976 E. Dunphy *Only a Game?* iv. 104 He's a good honest pro, but somehow Benny doesn't rate him. 1977 *World of Cricket Monthly* June 85/1, I must say we rated our chances going up to Headingley.

4. (Later examples.)
1932 *Sun* (Baltimore) 27 Apr. 15/2 Major General, avoiding interference going to the first turn, was rated in front until reaching the homestretch. 1950 *Amer. Jrnl. Psychol.* LXIII. 521 The *Ss.*..rated each pair on a scale of similarity. 1956 C. Willock *Death at Flight* vi. 76, I still rate your type of scatter-gun old-fashioned. 1962 *Observer* 25 Feb. 21/4, I still rate him the tops. 1976 *Billings* (Montana) *Gaz.* 27 June 4-A/2 Gen. Antonio Ramalho Eanes, a dour disciplinarian pledged to restore law and order, was rated an overwhelming favorite Saturday to win the Portuguese presidency.

6. b. (Later examples.) Also, to be accorded a certain position; to be considered *as*, to count *as*.
1949 *Sun* (Baltimore) 23 May 16/6 She will rate near the top of her class, which means that in racing she will have to give time to most of her competitors. 1959 A. Fullerton *Yellow Ford* xiv. 186 I'm important to him, too, but I don't rate that high. 1961 *Observer* 12 Mar. 29 (Advt.), You can be an important person all the same. You'll certainly rate as one if you own a 'Retinette', the Kodak precision camera. 1965 *Listener* 16 Sept. 425/2 Although Madame de Beauvoir rates as an 'intellectual', her book is a narrative of doings and feelings rather than of ideas and reflections. 1966 J. Porter *Sour Cream* viii. 104 The disappearance of Melkin didn't even rate as a nine-days' wonder. 1976 *National Observer* (U.S.) 18 Dec. 17/1 A show may rate as expected, may languish or may exceed expectations. 1977 *Ibid.* 1 Jan. 1/4 Toy poodles rate dumbest in a test you can give your dog too.

c. *trans.* To merit, to deserve; to be worthy to attain or obtain; to be treated as worthy of, be accorded.
1921 *Collier's* 3 July 22/1 Where does he rate that stuff?.. Where does he fit to grab off that Jane? 1928 *Amer. Speech* III. 220 *Rate, v.*, to obtain, or to be entitled to. 'Did you rate a bid to the Kappa party?' 1940 A. W. Fearn *My Days of Strength* xiv. 153 Fond as the Chinese are of weddings and funerals, only virgins in China rate weddings with all the trimmings. 1957 E. Hyams *Into Dream* II. i. 93 His power of command barely rated his two stripes. 1959 *Listener* 4 June 992/3 Nor is one certain..whether C. W. Brodribb or Harold Child, nice men though they both were, really rate the *Dictionary of National Biography*. 1962 *New Statesman* 28 Dec. 935/1 Christmas cards don't even rate a mention in the *Encyclopaedia Britannica*. 1974 'A. Gilbert' *Nice Little Killing* vi. 82 He got out his old second-hand car—the village bobby didn't rate a panda. 1974 *Howard Jrnl.* XIV. 95 Mannheim, Radzinowicz [etc.]..rate between them nine titles in the bibliography. 1976 *Times Lit. Suppl.* 16 Apr. 460/5 The work of A. Z. Steinberg, for instance, whose contribution to Dostoevsky studies is sadly undervalued, rates no more than a mention in a half-page summary. 1978 G. Greene *Human Factor* v. i. 236, I rate a lifetime in jail. 1979 'A. Hailey' *Overload* III. xii. 257 The statement by the Governor of California in support of the project rated a brief paragraph near the end of the..report.

d. *intr.* To have some standing, to be of importance; to matter, 'count'; to be highly esteemed.
1928 *Amer. Speech* III. 220 Sometimes the word is used in another sense; 'Price is rating pretty high with Betty these days' means that he stands very high in her estimation. 1938 *Chatelaine* Feb. 44/1 Nowadays to really 'rate' you must have more than an attractive face and figure. 1951 M. McLuhan *Mech. Bride* (1967) 112/1 The eye is anxiously turned on the neighbor or friend with a 'How do I measure up?' 'Do I rate?' 1967 *Listener* 14 Sept. 350/3 As a rock group, then, the Kinks don't rate musically.

7. d. To cause to proceed at a moderate pace, to regulate the pace of; *spec.* in horse-racing, to ride (a horse) at a moderate pace, conserving his energy for the finish. Also *intr.* for passive, of a horse: to be ridden in this way. orig. *U.S.*
1920 H. C. Witwer in *Collier's* 3 July 9/2 Ring generalship, that's what you're minus, and the only way you can get it is by experience. You gotta be rated along, not rushed... Many a promisin' kid has been ruined at the start by bein' overmatched. 1946 *Sun* (Baltimore) 28 June 9/2 Villa Nova, escaping the jam at the half-mile pole, was rated to head of stretch, then closed well through last eighth of a mile. 1961 J. S. Salak *Dict. Amer. Sports* 354 *Rating* (harness racing), maintaining an even rate of speed and timing finishing rush. Harness horses are rated to a fraction of a second in miles. 1977 *N.Z. Herald* 5 Jan. 1–12/5 He rated Red Vesta perfectly in front, kicked clear in the straight and won comfortably. 1977 *Time* 20 June 51/2 But Turner's gentle methods have made Slew, a natural front runner, into a sound horse who 'rates kindly', or can tolerate another horse in front of him—at least for a while.

8. To assign a rating (sense *3) to (a piece of equipment, etc.). Const. *at* the value concerned. Usu. in *pass.*
1893 *Trans. Amer. Inst. Electr. Engineers* X. 255 Manufacturers cannot accurately rate their fuse-wire unless the length of the specimen to be used is specified. 1905 *Jrnl. Inst. Electr. Engineers* XXXV. 388 Fuses rated to blow with an excess current of 50 per cent of their normal carrying capacity get far too hot. 1940 *Amat. Radio Handbk.* (ed. 2) ii. 39/1 This valve..is rated at 25 watts dissipation at 500 volts. 1953 J. Liston *Power Plants for Aircraft* i. 7 Nearly all aircraft power plants are capable of developing much more power than that at which they are rated. *Ibid.* iii. 104 By definition, iso-octane was given an octane number of 100 and normal heptane was rated at zero octane number. 1975 D. G. Fink *Electronics Engineers' Handbk.* vii. 6 Such resistors are rated from 5 to 20 kV, have a resistance range of 2,000 Ω to 1,000 MΩ, and are rated from 5 up to 20 W. 1977 *Offshore Engineer* Apr. 75/2 Average life of the new type thallium iodide bulb is rated by UMEL at approximately 500 hours.

rate, *v.*[2] **2.** (Later example.)
1872 A. C. Steele *Broken Toys* II. xxv. 151 Ben Alymer..took up the butt-end of his gun and rated the pointer back.

‖ **raté** (rate), *a.* and *sb.* [Fr.] **A.** *adj.* Ineffective, miscarried; of a person, consistently unsuccessful. **B.** *sb.* A person who has failed in his vocation.
1905 R. Fry *Let.* 21 Jan. (1972) I. 233 His scheme being *raté* is a blow to his prestige. 1910 B. W. Wells *Mod. French Lit.* xiii. 477 The mean spirit of D'Argenton, the poet, who, with his attendant group of *ratés*, the failures of literature and art, forms a sort of mutual admiration club. 1949 E. Hyams *Not in our Stars* v. 47 A raté sensualist, he would willingly have enjoyed an 'affair' with Miriam. 1966 *New Statesman* 9 Sept. 369/1 Yves Montand..is too plump and groomed to make a certified *raté*. *a* 1970 E. Starkie in J. Richardson *Enid Starkie* (1973) iii. 58 The *raté*, the typical Paris figure. 1977 *Times* 7 Sept. 10/7 They are failures... They will all be *ratés* and hopeless when they are 30.

rateable, *a.* Add: **2.** *rateable value*, the value ascribed to a property for the purpose of assessing the rates to be levied on it.
1836 *Act* 6 & 7 *Will. IV* c. 96 Gross estimated Rental. Rateable Value. Rate at 6*d.* in the Pound. 1874 *Act* 37 & 38 *Vict.* c. 54 §4 The gross and rateable value of any land used for a plantation or a wood..shall be estimated as follows. 1909 *Daily Chron.* 13 Aug. 4/3 Mr. Lloyd George is understood..to have promised favourable consideration to the plea of the London Members for some alteration in the 'rateable-value' basis. 1925 *Act* 15 & 16 *Geo. V* c. 90 §2 Every general rate shall be a rate at a uniform amount per pound on the rateable value of each hereditament. 1962 L. Golding *Dict. Local Govt.* 408 The net annual value is in most cases the rateable value. 1971 *Reader's Digest Family Guide to Law* 88/1 Ratepayers do not pay an amount equal to the full rateable value of their property each year. They pay a proportion of it—the rate in the £. 1976 *Evening Post* (Nottingham) 14 Dec. 4/7 The rateable values did not take into account the disadvantages of living in the city centre.

rateably, *adv.* (Later U.S. example of spelling *ratably*.)
1973 *N.Y. Law Jrnl.* 4 Sept. 17/1 The widow's elective share must be paid ratably by all beneficiaries under the will.

rated, *ppl. a.*[1] Add: **2.** Of a numerical characteristic or property: having the value that a device, apparatus, etc., is designed to operate at or attain under normal conditions, or at which other characteristics are evaluated.
1893 *Trans. Amer. Inst. Electr. Engineers* X. 260 Promptness of action requires that the temperature of fusion should not be too far removed from that attained when the wire is being worked at its rated capacity. 1916 [see *rating *vbl. sb.* 3]. 1931 R. N. Liptrot in *Handbk. Aeronautics* (R. Aeronaut. Soc.) ii. 121 From ground level to the rated height the r.p.m. and horse-power progressively increase until at the rated height normal r.p.m. and horse-power are reached. 1942 I. Gleed *Arise to Conquer* ii. 16 One of my jobs was to test them [*sc.* Spitfires] at rated altitude, about 18,000 feet. 1945 G. V. Welbourne *Flight & Engines* xi. 135 The use of two separate jets is but one way of ensuring a sufficiently rich mixture at full rated power and take-off. 1975 D. G. Fink *Electronics Engineers' Handbk.* vii. 9 The surge voltage applicable to electrolytic capacitors is a voltage in excess of the rated voltage which the capacitor will withstand for a specified limited period at any temperature. 1978 *Nature* 6 Apr. 520/1 The wind speed data.. were fed into simulated wind generators having the following characteristics: rotor diameter, 60 m; rated wind speed, 14 m s^{-1}; furling wind speed, 27 m s^{-1}.

ratel[1]. Substitute for def.: The honey badger, *Mellivora capensis*, belonging to the family Mustelidæ, native to Africa and southern Asia, and distinguished by a coat that is light grey on the back and black elsewhere, powerful claws on the front feet, and a diet that includes honey, insects, reptiles, birds, and small mammals. (Later examples.)
1902 *Chambers's Jrnl.* Oct. 667/2 The camel and llama, the badger and ratel, all of which may be seen and compared at the Zoo. 1947 J. Stevenson-Hamilton *Wild Life S. Afr.* xxix. 242 The ratel does not hesitate to attack the most venomous species of snakes. 1961 L. van der Post *Heart of Hunter* iv. 73 He gave me such a vivid picture of the ratel eating snakes that I saw it gobbling up tangles of serpents like spaghetti. 1969 B. Seshadri *Twilight of India's Wild Life* v. 139 The ratel, or honey badger..has lived its life adapting it to many types of forest and plain. 1975 H. B. Cott *Looking at Animals* iii. 73 Ratels are said to trot unhurriedly with a long, swinging stride.

ratelier. Restrict † *Obs. rare*[−1] to sense in Dict. and add: **2.** A set of teeth; a set of false teeth.
1839 Thackeray in *Fraser's Mag.* June 746/1 In the large picture, everybody grins, and shows his whole *ratelier*. 1863 Crown Princess of Prussia *Let.* 1 Sept. in R. Fulford *Dearest Mama* (1968) 261 We are left to the care of a very funny old gentleman of the bedchamber who wears a 'ratelier' and is very tiresome besides. 1916 R. Fry *Let.* 8 Sept. (1972) II. 402, I dread the flashing *râtelier* of that Chicago girl. 1923 R. Nevill *World of Fashion 1837–1922* iii. 62 When alone in his study he [*sc.* Labouchère] would take out his *ratelier* and lay it on the ground.

‖ **ratemahatmaya** (ra:temaha·tmayā). *Sri Lanka.* [Sinhalese, f. *rate* of the district + *mahatmaya* gentleman.] A chief headman of a Kandyan district.
1821 J. Davy *Acct. Interior of Ceylon* v. 147 The duties of Rate-mahatmeyas were similar to those of Dissaves; but their official rank was inferior. 1841 J. Forbes *Eleven Years in Ceylon* iv. 67 The authorities under the Kandian dynasty were thus arranged... The Rate Mahatmeās, chiefs of inferior districts. 1913 L. Woolf *Village in Jungle* ii. 37 Disa Mahatmaya is the title used by villagers in referring to chief headmen or Ratemahatmayas. 1956 R. Pieris *Sinhalese Social Organization* I. iv. 24 The *ratēmahatmayās*, as they were colloquially styled, bore the title *ratērāla* at court.

rater[1]. **1.** Delete 'Now *rare*' and add later examples.
1957 D. T. Herman et al. in Saporta & Bastian *Psycholinguistics* (1961) 540/1 Two raters independently rated each of the reproductions. 1967 M. Argyle *Psychol. Interpersonal Behaviour* x. 182 The use of 'blind' ratings, where the raters do not know which of the individuals they are rating are being trained or belong to a control group. 1973 *Jrnl. Genetic Psychol.* Mar. 40 Raters had no knowledge of whether a given interview protocol was from the psychopathic or nonpsychopathic sample. 1977 Ryan & Carranza in H. Giles *Lang., Ethnicity & Intergroup Relations* ii. 70 Ethnicity of rater did not affect evaluation of the standard speakers. 1979 *Nature* 22 Mar. 357/2 The assessment on HRS was made by two independent raters who showed a satisfactory concordance.

rat fink. *slang* (chiefly *U.S.*). Also **rat-fink, ratfink.** [f. Rat *sb.*[1] + *fink *sb.*[2]] One who is obnoxious or contemptible, esp. (*a*) an odiously pretentious person; (*b*) an informer, a traitor. Also *attrib.* or as *adj.* Hence as *v. trans.*, to inform *on*.
1964 *Guardian* 8 July 7/6 That's the hitcheroo, baby,.. this time when Cliff and Shirley dance on top of Everest it's a send up. That keeps the intellectual rat-finks happy, see. *Ibid.*, This is going to bring them all in: the dads and mums, the squares, the rat-fink intellectuals, and the teenagers. 1965 D. Boroff *State of Nation* 212 In fitting American youth for its destiny in the free world of tomorrow, our schools may be virtually compelling them to become a bunch of ratfinks. 1965 P. De Vries *Let me count Ways* xx. 273 So cool and rat fink. What college did you go to? That made you so cultured and rat fink. 1966 *Listener* 9 June 838/1 The cool cats and the rat finks and the camp hips [in America]. 1969 C. Burke *God is Beautiful, Man* 88 His name was Judas and he was a rat fink. So this dirty rat fink he says to the pres of the gang, Caiaphas, 'What's in it for me if I put the finger on him?' 1973 M. & G. Gordon *Informant* x. 46, I may be a rat fink but I'm not going to be a paid rat fink. 1973 *Houston Chron.* 21 Oct. 10/2 As for informing on an honor violator, the colonel said you didn't exactly rat fink on such a man, you usually confront that man with his conduct and expect him to do the honorable thing. 1975 A. Price *Our Man in Camelot* iii. 52 Gildas.. was..denouncing the rulers of Britain as a bunch of ratfinks who were letting the country go to the dogs. 1976 L. Deighton *Twinkle, twinkle, Little Spy* v. 30 It was Tony Nowak's rat-fink cousin Stefan who put the spaghetti in the piano. 1976 *Courier-Mail* (Brisbane) 19 Feb. 5/2, I still think you were a rat fink. 1977 *New Yorker* 26 Sept. 127 The hairy little hipster Go Go, a ratfink wearing a cross and a yarmulke.

rath (rɑþ), *sb.*[2] A factitious word introduced by 'Lewis Carroll' (see quot. 1855[2]).
Quot. 1855[1] also occurs in the first verse of 'Jabberwocky' in *Through the Looking-Glass* (1871) i. 21.
1855 'L. Carroll' *Rectory Umbrella & Mischmasch* (1932) 139 All mimsy were the borogoves; And the mome raths outgrabe. *Ibid.* 140 *Rath*, a species of land turtle. Head erect: mouth like a shark: the fore legs curved out so that the animal walked on its knees: smooth green body: lived on swallows and oysters.

‖ **Rathaus** (rā·thaus). Also **Rath-haus, Rathhaus.** [Ger., lit. council-house.] A German town hall.
1611 Coryat *Crudities* 619, I..will make mention of their *Prætorium* or Senate house, which they commonly call the Rathausz [in Cologne]. 1855 Geo. Eliot in *Fraser's Mag.* June 703/2 The Markt, a cheerful square [in Weimar], made

smart by a new Rath-haus. 1864 C. M. Yonge *Bk. Golden Deeds* 208 The grim effigy..grins over the door of the Rathhaus [in Freiburg]. 1962 K. O'Hara *Double Cross Purposes* vii. 95, I did see two cars parked outside..the local Rathaus. 1975 *Times* 26 Apr. 10/5 The [Hamburg] Rathaus..has a cellar restaurant of quite Gothic splendour.

rathe, *sb.*[2] **2.** (Further example.)
1875 *Ure's Dict. Arts* (ed. 7) III. 1111 A comb or raithe ..guides the threads with precision on to any length of beam.

rathe, rath, *a.*[1] Add: **1.** (Later example.)
1949 E. Pound *Pisan Cantos* lxxxi. 113 Pull down thy vanity, Rathe to destroy, niggard in charity.
3. (Later example.)
1914 C. Mackenzie *Sinister Street* II. iii. xiv. 787 Spring on these rathe mornings of wind and scudded blue sky was forward with her traceries.

rather, *adv.* **II. 7.** For (*vulgar*) read (orig. *vulgar*) and add: freq. with emphatic pronunc. (rā:ðə·ɹ). Now, or until recently, common also in upper-class or affected speech. (Later examples.)
1885 A. Edwardes *Girton Girl* I. xiv. 280 Rather proud of my own accent... But Arbuthnot puts me in the shade, ra-*ther*. 1905 H. Vachell *Hill* i. 2 'You'll enjoy it—as I did —amazingly.' 'Ra-ther,' said John. 1928 C. A. Nicholson *Hell & Duchess* II. vi. 286 'May I give Audrey your blessing?' 'Ra-ther and my love.' 1933 J. Buchan *Prince of Captivity* II. i. 153 'You've heard Kenneth speak of him.' 'Rather. I want to meet him.' 1975 *Listener* 10 Apr. 472/2 The producer had..prevented the willing son in the parable from saying 'Yes, ra-*ther*, Dad,' when asked to get on with the 'allotment'.

9. f. Ellipt. phr. *rather you than me* (or *I*): I would rather that you did or underwent something than I (used to convey admiration, commiseration, etc.).
1968 'C. Aird' *Henrietta Who?* xv. 192 A proper mix-up, isn't it?.. Rather you than me. 1969 V. Gielgud *Necessary End* xiii. 104 'You're a brave fellow. Rather you than I. Good luck.

rather (ræ·ðəɹ), *sb. U.S. dial.* [f. the adv.] A choice, preference. Cf. *druther, *ruther.
1903 *Dialect Notes* II. 326 'I would stay at home if I *had my rather*.' Also pronounced *ruther*. 1913 H. Kephart *Our Southern Highlanders* xiii. 283 'It matters not, so I've been told, Where the body goes when the heart grows cold; But,' she concluded, 'a person has a rather about where he'd be put.' 1930 W. Faulkner *As I lay Dying* 106 And if I had my rathers, you wouldn't be here at all. 1961 C. Himes *Black on Black* (1973) 46 If I had my rathers I'd make up my own band.

rathest, *adv.* Restrict † *Obs.* to senses 1 and 2 in Dict. and add: **3.** (Later example.)
1884 B. Potter *Jrnl.* 27 Oct. (1966) 107, I would rathest of all copy the raised plaques of Wedgwood.

rathite (rā·tait, ra·þait). *Min.* [ad. G. *rathit* (H. Baumhauer 1896, in Zeitschr. f. Kryst. und Min. XXVI. 594), f. the name of G. vom *Rath* (1830–88), German mineralogist: see -ite[1].] Any of a group of sulpharsenites of lead found in the Binnental in southern Switzerland.
1897 *Mineral. Mag.* XI. 225 Characteristic of rathite, but not invariably present, are parallel systems of very fine striæ..on the crystal faces and on the fractured faces. 1905 [see *hutchinsonite]. 1953 *Amer. Mineralogist* XXXVIII. 330 Weissenberg and precession camera studies on baumhauerite, dufrenoysite and rathite, lead to the following new structural data... Rathite-II is monoclinic... The unit cell dimensions do not agree with previous structural or morphological data on rathite, although they were obtained on a fragment of the original material. 1965 *Zeitschr. für Kristallogr.* CXXII. 434 Rathite-I, $(Pb,Tl)_3As_4(As,Ag)S_{10}$, is a mineral of a sulfosalt group, to which rathite-II, rathite-III, rathite-IV, dufrenoysite, baumhauerite and scleroclase belong. *Ibid.*, Rathite-I and -III form two modifications of a single species and should perhaps have a name different from rathite-II... In the Lengenbach quarry rathite-II is frequently found, whereas rathite-I occurs rarely, and then usually polysynthetically twinned. 1974 *Neues Jahrb. für Mineral.: Monatshefte* 530 Several kinds of minerals which belong to a group of lead sulfarsenites called 'rathite group' have been found in Lengenbach quarry (Ct. Wallis, Switzerland) and crystallographically investigated. The existence of two common periods of about 7·9 Å and 8·4 Å is a marked characteristic of them.

Rathke (rā·tkə). *Anat.* [The name of Martin Heinrich *Rathke* (1793–1860), German anatomist, who described the structure in 1838 (*Arch. für Anat., Physiol. und Wissenschaftl. Med.* 482).] *Rathke's pouch* or *pocket* (also *pouch of Rathke*): a diverticulum of the oral cavity, which in developing vertebrates forms the anterior lobe of the pituitary body.
1889 A. Macalister *Textbk. Human Anat.* 620 The stomodæum joined the oral end of the mesenteron at about the fourteenth day and from its upper part is projected the hypophysial pouch of Rathke. 1892 E. L. Mark tr. *O. Hertwig's Textbk. Embryol.* xiv. 285 Even after the rupture of the pharyngeal membrane there is retained, in front of its attachment, a small pit, which constitutes Rathke's pocket. 1906, 1915 [see *hypophysis 3]. 1945 W. J. Hamilton et al. *Human Embryol.* x. 153 An ectodermal diverticulum, Rathke's pouch, arises from the stomatodaeal roof. *Ibid.* xii. 253 Rathke's pouch loses its attachment to the pharyngeal roof..and develops into the various subdivisions of the anterior lobe of the definitive hypophysis. 1962 [see *pars intermedia]. 1973 F. Beck et al. *Human Embryol. & Genetics* xvii. 188 The thickening in the anterior wall of Rathke's pouch becomes the *pars anterior* of the adult pituitary gland.

ra·t-hole, *sb.* [Rat *sb.*[1] 7.] **1.** A hole used by a rat for passage or abode. Also *fig.*, a cramped or squalid building, room, or the like; a refuge or hiding-place. Also *attrib.*
1812, 1879 [see Rat *sb.*[1] 7 a]. 1879 [see *junk-shop* s.v. *junk *sb.*[3] 5]. 1912 *Dialect Notes* III. 581 He wouldn't know enough to pound sand in a rat-hole; so don't get him. 1922 D. H. Lawrence *Let.* 30 Apr. (1962) II. 701 Yet I don't believe in Buddha—hate him in fact—his rat-hole temples and his rat-hole religion. 1941 M. U. Schappes *Let.* 21 July in *Lett. from Tombs* (1941) 85 The rat-hole: a hotel where state's-witnesses are kept! *a* 1944 K. Douglas *Alamein to Zem Zem* (1946) 133 We stayed in our positions, like a terrier at a rat-hole. 1976 'B. Shelby' *Great Pebble Affair* 164 The warehouse is..a dummy warehouse designed to mask our rathole. 1976 N. Thornburg *Cutter & Bone* iii. 80 Mo was not just frigid,..she was dead, a cadaver with a welded womb and a cunt like a rathole, full of dust and bits of straw and feathers from old nests.
2. *Oil industry.* **a.** A shallow hole drilled near a well hole to accommodate the kelly or a pipe joint when it is not in use. **b.** A hole of smaller diameter drilled at the bottom of a larger hole.
1921 W. H. Jeffery *Deep Well Drilling* v. 209 Some rotary drillers drill a shallow well at a point midway and in front of the slush pumps, in which they rest the drill stem when not in use or when waiting to set in another joint of drill pipe. This is termed the 'rat hole'. 1939 D. Hager *Fund. Petroleum Industry* ix. 210 When the kelly is deep enough for a joint of drill stem, the kelly and bit are pulled out. The kelly is set to one side, fitting in a hole (the 'rat-hole') cut in the floor of the derrick. 1972 L. M. Harris *Introd. Deepwater Floating Drilling Operations* xvi. 164 During and immediately following the unloading of the water cushion and the rathole mud, the surface pressure is determined by the amount of fluid to be unloaded. 1975 G. Anderson *Coring* v. 89 Most diamond coring is now done in the full diameter of the borehole. Whether coring is done this way or in a rat hole depends on the full gauge-keeping ability of the rock bits used prior to initiating the coring operation.
3. *N. Amer.* A seemingly bottomless hole; used *fig.* or allusively of something that demands excessive expenditure.
1961 in Webster. 1975 D. Lambro *Federal Rathole* vi. 43 The contractor was refused..additional loan money because..it would be 'pouring money down a rathole'. 1976 *Globe & Mail* (Toronto) 21 Dec. 7/1 The committee will examine..Minaki Lodge, the rathole in northwestern Ontario down which increasing quantities of public money seem to be disappearing. 1977 *Time* 19 Dec. 13/2 Since the B-1 bomber will not be part of our military inventory, to build two more airplanes would simply amount to pouring half a billion dollars down a rathole.

ra·t-hole, *v.* [f. prec.] **1.** *intr.* and *trans. Oil Industry.* To drill a hole of smaller diameter at the bottom of (one of larger diameter).
1922 L. C. Sands in D. T. Day *Handbk. Petroleum Industry* 268 If the oil stratum should inadvertently be penetrated before the casing has been set and all water excluded from overlying formations, it will be expensive, if not impossible, to make a successful producer of the well. To forestall this danger it is common practice to 'rat-hole' ahead at intervals. 1939 D. Hager *Fund. Petroleum Industry* ix. 228 If the sand is expected, the hole can be reduced and 'ratholed' ahead, a smaller hole being drilled.
2. To hide or store; not to invest or spend (money).
1948 *Amer. Speech* XXII. 220 Stashing was overt and frequently a co-operative enterprise, whereas to *rat-hole* implied that an individual was storing something of possibly public property for his own use. 1971 *Wall St. Jrnl.* 17 June 1/5 Everyone ratholed their money last year. Then they saw the (economic) situation hasn't changed so they decided if they're going to buy, now's the time to do it. 1977 *New Yorker* 4 July 55/1 Speaker said he was 'ratholing' the Jif jar. He was not ready to pay the Internal Revenue Service any portion of the gold's value (his privilege until it is sold).

Rathskeller, rathskeller (rā·tskelə.ɹ). Also **ratskeller.** [ad. G. *ratskeller* (formerly *rathskeller*), f. *rat* council as in *Rathaus + -*s* gen. ending + *keller* cellar.
The form *rathskeller* was preferred to avoid the phonetic association with Rat *sb.*[1] + Cellar *sb.*]
a. A cellar in a German town hall in which beer or wine is sold. **b.** An underground beer-hall or restaurant. Also *transf.*
1900 Ade *More Fables* 159 Mr. Byrd..happened to be in a Rathskeller not far away. 1903 *Current Lit.* Apr. 495/2 The first rathskeller was established in New York in 1863 by Fred. Hollander. 1916 *Harper's Mag.* Dec. 18 He likes the religious cool of the rathskeller... He takes his ease in his inn. 1929 *Papers Mich. Acad. Sci., Arts & Lett.* X. 318/1 *Rathskeller*, a dugout. 1934 S. R. Nelson *All about Jazz* vi. 121 They used to work the *Rathskellers* or night-clubs of the lower grades. 1946 E. O'Neill *Iceman Cometh* (1947) I. 27 It's the No Chance Saloon. It's Bedrock Bar, The End of the Line Café, The Bottom of the Sea Rathskeller! 1969 *Courier-Mail* (Brisbane) 30 June 3/4 The ratskeller will be a 'cellar, with barrels of wine and beer, where people will dine at leisure'. 1975 *Country Life* 11 Dec. 1676/1 The *Rathaus* or town hall..stands in the cobbled market square with the usual *Ratskeller* for refreshments.

raticide (ræ·tisəid). Also **ratticide.** [f. Rat *sb.*[1] + -i- + -cide.] **a.** One who, or that which, kills rats, esp. a chemical substance used as a rat poison. **b.** *rare.* The killing of rats.
1847 W. J. Broderip *Zool. Recreations* 319 The celebrated ratticide Billy has long since gone to that bourne whence neither rats, dogs, nor travellers ever return. 1908 W. R. Boelter *Rat Problem* iv. 146 Raticide. This preparation.. is manufactured in the United States, where it is sold under the name of Azra. 1922 *Jrnl. R. Sanitary Inst.* XLII. 316 Practical experience..supports the view that the safety margin for Squills is greater than for any other practical raticide at present in use. 1936 *Times* 21 Aug. 11/5 The end of real research into effective raticides. 1941 *Nature* 1 Mar. 263/1 Guidance in production, research and application as relating to insecticides, fungicides, raticides and repellents for all purposes. 1976 *New Scientist* 26 Aug. 454/2 William Kotzwinkle's *Doctor Rat* is a comic triumph. About half is narrated by a mad laboratory rat, whose perceptions into the relationship between mass ratticide and grantsmanship do make delightful satire.

ratification. (Earlier and later examples of *ratification meeting*.)
1848 *Campaign* (Washington, D.C.) 21 June 64/2 A great democratic ratification meeting was held at New Orleans on the night of the 8th June. 1904 *N.Y. Even. Post* 23 June 2/6 The first Roosevelt ratification meeting..will be held this evening..when the Republican Club of the Thirty-first Assembly District will endorse the nominees of the Chicago convention.

ratifica·tionist. [f. Ratification + -ist.] One who favours ratification (of a treaty, etc.).
1921 *Glasgow Herald* 23 Dec. 7/3 On the ratificationists' side it had been suggested that there should be a time limit to speeches.

ratine (rǎti·n). Also ‖ **ratiné** (ratine). [Fr., pa. pple. of *ratiner* to frieze.] A clothing fabric of rough open texture.
1913–14 *T. Eaton & Co. Catal.* Fall & Winter 17 The materials for Suits are very beautiful..Whipcords, Corduroys, Boucle Tweeds, Ratines. 1922 *Daily Mail* 15 Nov. 1 (Advt.), All-silk ratine. For frocks, children's wear, furnishings. 1923 *Westm. Gaz.* 22 Mar. 9/2 Mrs. Almond's blue ratine looked as if it had just descended on her and had been made in heaven. 1934 Webster, *Ratiné*... Also *ratine.* 1966 *Vogue* Dec. 71 White ratine tunic dress. 1967 *Times* 21 Feb. 9/1 Ratiné..was launched, or rather relaunched, at the French prêt-à-porter fair last year. 1969 N. W. Parsons *Upon Sagebrush Harp* xi. 57 We noticed a long cobweb floating over our heads... Long silvery strands moving through the motionless air from west to east. Each was like a white ratiné string, fine as silk except for knots at irregular intervals.

rating, *vbl. sb.*[1] Add: **2. b.** (Later examples.)
1906 *Westm. Gaz.* 6 June 6/2 The cruiser..landed fourteen officers, 290 ratings, and eighty-five marines. 1925 *Nation* 26 Sept. 756/2 Numerous ground ratings are needed to handle airships in and out of the hangars in addition to a very considerable repair staff of specially skilled ratings. 1932 *Daily Express* 27 Jan. 1/2 Many of the ratings belong to Portsmouth, and some are married and have families. 1955 *Times* 12 May 7/5 Two naval ratings, both aged 19, were arrested on board H.M.S. Ark Royal. 1979 *Jrnl. R. Soc. Arts* CXXVII. 545/2 Six scholarships for deck ratings were awarded.
3. The value of a property which is claimed to be standard or limiting for a piece of equipment or a material, or to be necessary for its optimal or standard use; a rated value (*rated *ppl. a.*[1] 2).
Quot. 1893 repr. the action of *rate *v.*[1] 8 rather than the result.
1893 *Trans. Amer. Inst. Electr. Engineers* X. 260 This wire was quite uniform in diameter, and gave evidence of careful rating. The curve of rating should be of the same general equation as that of fusion, since both represent isothermal conditions. 1905 *Jrnl. Inst. Electr. Engineers* XXXV. 367 The 'normal carrying capacity' or 'rating' of a fuse wire may be defined as: The maximum current which the fuse is capable of carrying continuously without deterioration or undue heating. 1916 *Standardization Rules Amer. Inst. Electr. Engineers* 25 A transformer of given kv-a rating must be capable of delivering the rated output at rated secondary voltage. 1941 E. Molloy *Aero-Engine Pract.* 57 The rating of an aero-engine is prepared not with the mere hope that it may be suitable for some aircraft or other, but with the intention of fulfilling a definite specification, whether civil or military. 1967 G. Arnold *Re-Wiring House* 19 Cables used in house wiring usually operate at below their maximum current rating. 1967 M. Chandler *Ceramics in Mod. World* iv. 123 This enables the insulator to be operated at higher ratings than would otherwise be possible. 1968 Miller & Sawers *Techn. Devel. Mod. Aviation* iii. 94 The fuel used in the 1920's was no better than that sold for cars, with an octane rating of about 50. 1969 *Jane's Freight Containers 1968–69* 12/1 *Rating*, means the maximum gross weight and is the maximum permissible combined weight of the freight container and of its contents. 1974 *Homes & Gardens* Apr. 150/2 The fuse rating is given in the accompanying instruction book as well as being clearly marked on the appliance rating plate. 1978 *Amateur Photographer* 2 Aug. 131/1 It is important that the total current drawn from all sockets should not exceed the rating of the supply socket—13 or 15-amp.

4. a. An assessment or measure (of a person's achievement, behaviour, skill, status, etc.); a grade, category or standing.
1921 A. W. Procter *Princ. Public Personnel Admin.* viii. 162 A rating of individual efficiency is intended to be a measurement of the value of the services rendered throughout a given period of time by an individual employee. 1939 [see *dating *vbl. sb.* c]. 1948 B. G. M. Gundkler *Bantu Prophets S. Afr.* iv. 88 The farmer has certain definite rules for social rating. 1951 M. McLuhan *Mech. Bride* (1967) 58/2 Culture ratings à la Emily Post are not often made in accordance with the consumer mentality. *Ibid.* 59/2 Woe to the indigent intellectual who accidentally acquires a 'high' rating without the economic appendages. 1964 M. Argyle *Psychol. & Social Probl.* xi. 138 Suppose a firm selects personnel by interviewing applicants. We can find out how effective this is by comparing ratings (or the rank-order) of candidates made by the interviewer with their subsequent success at the job. 1968 *Listener* 6 June 748/2 There is to be a ballot during the interval to determine which of three works..should be repeated after the interval. One doesn't have to think very long to discover how invidious this could turn out to be. First performances are quite bad enough for the composers concerned without their having to worry about Instant Popularity Ratings as well. 1974 *Plain Dealer* (Cleveland, Ohio) 26 Oct. 4-D/4 Many sports publications may have underestimated the Cavaliers in the pre-season ratings.
b. *Broadcasting.* Usu. *pl.* orig. **Crossley rating** [from Archibald M. *Crossley* who in 1930 began the regular reports of the Co-operative Analysis of Broadcasting]: an estimate, based on statistical sampling, of the size of the audience of any particular radio or television programme; its popularity so assessed.
1939 *Business Week* 25 Feb. 36/2 You've heard some radio comedian crack, after getting off a poor gag, 'There goes my Crossley'. He is referring to his popularity rating with one of the services that measure..the size of radio listening audiences. 1940 *Time* 29 Jan. 50/2 In the radio business, 'Crossley ratings' are the official box-office count. Crossley's boss is the Cooperative Analysis of Broadcasting, instituted eleven years ago..and now subscribed to by 635 sponsors. 1941 G. Marx *Groucho Lett.* (1967) 47 The Tommy Riggs show, which I spurned, now has a Crosley rating of seventeen... This will give you an idea of how little I know about audiences and what they want. 1947 *Billboard* 1 Nov. 3 The majority of new shows on the air—so far at least—are failing to show evidence of any particular rating strength. This is especially true when ratings are related to talent budgets. 1952 H. V. Grohmann *Advertising Terminol.* 64 *Rating*, the popularity test for a specific show. 1959 *Daily Mail* 11 Aug. 8/8 According to the ratings, the only court of public opinion that counts in these circles, Hylton shows are successful. 1962 *Times* 19 May 4/3 Public-service broadcasting, free from 'the tyranny of the ratings'. 1971 *Daily Tel.* 2 Aug. 7/4 Though the two BBC channels between them still share the ratings with ITV on a 50–50 basis, the trend is sufficiently strong to alarm the major companies. 1977 *Time* 7 Feb. 40/3 The hokiness of Hollywood fame got to him too. He would say, 'Even my friendships are related to ratings.'
5. *attrib.* and *Comb.*, as (sense 1) *rating area, authority*; (sense 2) *rating badge*; (sense *3) *rating point, scale*; (sense *4 b) *ratings battle, issue, terms, war.*
1928 *Britain's Industr. Future* (Liberal Industr. Inquiry) v. 408 Rating areas should be drastically revised with a view to a more equal distribution of the burden of rates. 1962 L. Golding *Dict. Local Govt.* 328 Rating authorities, i.e., the local authorities responsible for levying and collection of rates, are the councils of county boroughs and county districts. 1910 *Our Navy* (U.S.) Apr. 19/1 A following the sea Is plenty good enough for me, Since I've got a ratin' badge tacked on my arm. 1921 *Sea Bag* (U.S.S.S. Oklahoma) 22 June 2/1 How many men suddenly wake up to the fact that they haven't watch-marks on their jumpers or rating-badges on their sleeves? 1976 *Sun* (Baltimore) 16 May A12/4 Since it [*sc.* Washington] is a bigger TV market than Baltimore, it costs more for a political advertisement. The price in Washington is $90 a rating point. In Baltimore it is $50 a rating point. In Baltimore, a rating point represents about 7,500 homes. 1974 *Times* 15 Oct. 16/6 The silly ratings battle..claims about the number of viewers watching each channel. 1927 *Scribner's Mag.* Apr. 418/2 One institution, in its confidential rating scale to be filled out by the school principal,..and others.., asks..as to 'moral earnestness, loyalty,..modesty'. 1967 M. Argyle *Psychol. Interpersonal Behaviour* v. 102 Another method [of measuring competence] is to use ratings: a number of rating scales are devised covering various aspects of a task, and are filled in by special observers. 1958 *Wall St. Jrnl.* 28 Nov. 1/6 Behind the ratings issue is the bigger question of whether the big TV networks, including the Columbia Broadcasting system, National Broadcasting Co. and American Broadcasting Co., should come under Federal regulation. 1977 *TV Times* (Brisbane) 13 Aug. 11/1 Australian ratings terms have been revised to incorporate the introduction of colour TV and consequent multi-set use in many homes. 1980 *Times* 25 Jan. 19/6 The ratings war— the battle to win as many viewers as possible for a programme.

ratio. 1. a. Delete † *Obs. rare* and add: spec. *Law.* Reason or rationale upon which a juridical decision is based; = *ratio decidendi* (see sense *1 b).
1964 G. Abrahams *Police Questioning & Judges' Rules* i. 20 Therein, it is submitted, is the *ratio* of that decision against the admissibility of statements. 1971 *Mod. Law Rev.* XXXIV. vi. 691 The judge considered..that *Torquay Hotel Co.* v. *Cousins* was authority, 'that this tort of interference with contracts applies not only where there is interference with contracts already made but where there is interference with contracts to be made in accordance with a regular course of dealing.' Is this the *ratio* of the *Torquay* case? 1976 J. M. Kelly *Stud. Civil Judicature of Roman Republic* v. 112, I have, by negative implication, foreshadowed certain characteristics of the *unus iudex* inasmuch as the factors which form the *ratio* of the *centumviri* or *recuperatores* will not be present in his case. 1977 *Times* 23 Dec. 18/5 That restricted definition was unnecessary for the ratio of that decision.
b. In Lat. phrases, esp. *Philos.* and *Law*: **ratio cognoscendi,** that in virtue of which knowledge of something is possible; that in virtue of which something is known to exist; **ratio decidendi** (pl. **rationes decidendi**), rationale of judgment; a principle underlying and determining a judicial decision; **ratio essendi, existendi,** that in virtue of which something exists.
1830 W. Hamilton in *Edin. Rev.* LII. 178 The existence of external things, which is given only *through* their intuition, it admits; the intuition itself, though the *ratio cognoscendi*, and *to us* therefore the *ratio essendi* of their reality, it rejects. 1862 in C. Clark *House of Lords Cases* VIII. 392 The observations made by Members of the House, whether law Members or lay Members beyond the *ratio decidendi* which is propounded and acted upon in giving judgment, although they may be entitled to respect, are only to be followed in as far as they may be considered agreeable to sound reason and to prior authorities. 1865 S. H. Hodgson *Times & Space* vii. 488 Now both the cause, or ratio existendi, and the reason, or ratio cognoscendi, in every particular case must be given by actual experience. 1877 *Law Rep. Exchequer Division* II. 233 The ratio decidendi in these cases does not appear in the reports. 1890 W. James *Princ. Psychol.* I. x. 337 But if the brand is the *ratio cognoscendi* of the belonging, the belonging, in the case of the herd, is in turn the *ratio existendi* of the brand. 1902 J. W. Salmond *Jurisprudence* viii. 176 A precedent, therefore, is a judicial decision which contains in itself a principle. The underlying principle which thus forms its authoritative element is often termed the *ratio decidendi*. 1903 G. E. Moore *Principia Ethica* iv. 127 Kant..admits that Freedom is the *ratio essendi* of the Moral Law, whereas the latter is only *ratio cognoscendi* of Freedom. 1923 C. D. Broad *Sci. Thought* viii. 267 Sensa are..in some way the *ratio cognoscendi* of the physical world, whilst the physical world is.. the *ratio essendi* of sensa. 1948 *Law Q. Rev.* LXIV. 463 The court is always at liberty to propound alternative rules of law, each of which it may elevate to the status of a ratio decidendi. 1970 *Internat. & Compar. Law Q.* XIX. I. 37 Which of these *rationes decidendi* possesses binding force? 1972 *Evangelical Q.* XLIV. 241 These activities compose the *ratio cognoscendi* of the pardoning grace of God; they make us aware of God's activity which is the *ratio essendi* that calls forth man's response. 1977 *Law Q. Rev.* XCIII. 378 The importance of the *ratio decidendi* is that it is the rule of law for which a case is authority.
5. Special Comb.: **ratio detector** *Electronics*, an F.M. detector whose two output voltages are such that their sum is constant and their ratio, rather than their difference, is proportional to the ratio of the two applied frequency-dependent voltages, so that its insensitivity to changes in amplitude is not confined to the carrier frequency. Also *ratiometer.
1947 *RCA Rev.* VIII. 201 A new circuit for f-m detection known as the ratio detector is coming into wide use. 1965 *Wireless World* July 8 (Advt.), Printed circuit for I.F. amplifiers and ratio detector. 1974 Harvey & Bohlman *Stereo F.M. Radio Handbk.* v. 99 Normally, in a ratio detector of the balanced type, the centre-point of the load is earthed and the output signal is taken from the tertiary winding.

ratio (rēi·ʃio, rēi·ʃio), *v.* [f. the sb.] *trans.* To enlarge, amplify, or reduce by a certain ratio. So **ra·tioed** *ppl. a.*, **ra·tioing** *vbl. sb.*
1943 H. T. U. Smith *Aerial Photographs* viii. 196 Each print which departs from the average scale or shows any apparent tilt is rectified and 'ratioed', or corrected for scale, by means of a projection printer. *Ibid.* 352 *Ratioed print*, a print prepared at a predetermined scale by photographic enlargement or reduction. 1970 J. A. Howard *Aerial Photo-Ecol.* ix. 105 Aerial photographs..sometimes.. are enlarged or reduced in size for a special purpose. These photographs are then known as ratioed prints or ratioed photographs. 1978 *Nature* 9 Mar. 142/2 This ratioing operation reduces the effective amplitude fluctuations of the laser by more than 100-fold.

ratiocinable (ræʃi-, rætiǫ·sinăb'l), *a. nonce-wd.* [f. Ratiocinate *v.* + -able.] Arrived at or deducible by reasoning.
1916 G. B. Shaw *Androcles & Lion* Pref. p. lxxxvii, An inveterate Roman Rationalist, always discarding the irrational real thing for the unreal but ratiocinable postulate.

ratiocinate, *v.* Add: Also with pronunc. (rætiǫ·sinēit). Also (*rare*) *trans.* and *refl.* So **ratio·cinated** *ppl. a.*
1900 G. B. Shaw *Let.* 30 Dec. (1972) II. 214 The conventional, factitious, ratiocinated motives & conclusions of his characters. 1926 Fowler *Mod. Eng. Usage* 483/2 *Ratiocinate* & its derivatives..may fairly be pronounced răti- rather than răshĭ-;..the OED, however, gives only răshĭ-. 1934 C. P. Snow *Search* II. iv. 180 Ratiocinating myself into honesty about my posturings.

ratiocinatively (ræʃi-, rætiǫ·sinĕtivli), *adv.* [f. Ratiocinative *a.* + -ly[2].] By the process of reasoning; by ratiocination.
1965 J. Lawlor in J. Gibb *Light on C. S. Lewis* 68 Some of the inconsistencies we were after could be approached ratiocinatively, and examined for logical contradiction; but the deeper kinds of awareness were to be reached intuitively rather than through rationalizations.

ratiocinator. (Later example.)
1971 Heath & Prendergast tr. J. Kristeva in *Signs of Times* 4 His [*sc.* Leibniz's] attempt to construct a 'calculus ratiocinator'.

ratiometer (rēiʃiǫ·mitəɹ). [f. Ratio + -meter.] A device for measuring the ratio of two electrical quantities.
1925 R. O. Kapp *G. Kapp's Transformers* (ed. 3) ix. 220 The ratio of a transformer is tested by an apparatus known as a ratiometer. This is really an autotransformer of which the windings are so subdivided that any ratio within very fine limits is obtainable. 1942 *Jrnl. Sci. Instruments* XIX. 23/2 A suitable current ratiometer has been constructed.. consisting of two coils on the same spindle maintained in a non-uniform field by a torsionless suspension. 1976 *Physics Bull.* Nov. 509/1 It has two amplifiers, phase sensitive detection or 'lock-in' and a ratiometer in one unit.

ratiomorphic (ræʃiomǭ·ɹfik), *a. Psychol.* [f. Ratio + Gr. μορφ-ή + -ic.] (See quots. 1954 and 1966.) Hence **ratiomo·rphous** *a.*
1954 E. Brunswik in *Acta Psychologica* XI. 109 Perception and thinking thus emerge as different forms of imperfect inferences regarding the environment, subsumable to a common behavior model patterned upon reasoning ('ratiomorphic' reduction, if this Latin-Greek hybrid be permitted). 1966 K. R. Hammond *Psychol. E. Brunswik* i. 38 By 1955, however, he [*sc.* Brunswik] had coined the term 'ratiomorphic' to represent the organism's process of coordinating uncertain data in order to make an inductive inference from them. Perception is a '..ratiomorphic subsystem of cognition..'. The process is, in other words, 'reasoning-like'. 1971 R. Martin tr. *Lorenz's Stud. Animal & Human Behaviour* II. 302 All kinds of constancy apparatus are—in principle—'ratiomorphic' in the most rigorous sense of the term, since all incorporate processes analogous to those of both induction and deduction. 1977 R. Taylor tr. *Lorenz's Behind Mirror* vii. 119 Egon Brunswik coined the term 'ratiomorphous' to describe..all these sensory and nervous processes [which] take place in areas of our nervous system which are completely inaccessible to our consciousness and our self-observation. *Ibid.* 162 These unconscious processes are what Egon Brunswik called 'ratiomorphous' computing mechanisms.

ration. Now with pronunc. almost always (ræ·ʃən). Add: **3. a.** (Further *pl.* examples.)
1862 R. Henning *Let.* 19 Oct. (1966) 110 Biddulph or Mr Hedgeland goes to the out-stations with provisions, or rations, as they call them. 1917 A. G. Empey *Over Top* 305 *Rations*, various kinds of tasteless food issued by the Government to Tommy, to kid him into the fact that he is living in luxury, while the Germans are starving. 1919 W. H. Downing *Digger Dial.* 41 'Wet rations':..Cooked foods etc...'Dry rations':..Uncooked food..'Iron rations': ..Emergency rations. 1922 C. E. Montague *Disenchantment* ii. 15 A little famished London cab-tout, a recruit, still rectilinear as a starved cat even after a month of army rations.
c. *esp.* an officially limited allowance for civilians in time of war or shortage. Hence phr. *off* (*the*) *ration*, in addition to the allowance; unrestricted.
1917 *Times* 28 Feb. 10/3 Captain Bathurst, replying to Mr. Faber, said the Food Controller, since the issue of voluntary rations, had been in communication with the War Office, and an Army Council instruction was issued last week limiting the sugar ration for civilian and combatant prisoners of war to 7 oz per week. 1919 E. H. Starling *Feeding of Nations* vi. 127 Each individual can buy of it according to his desire and satisfy his Calorie needs above those supplied in the rations. 1922 H. W. Clemesha *Food Control in North-West Division* ii. 34 There must have been many families who were unable to afford the additional rations of meat which the cards of children would have enabled them to obtain. 1928 W. H. Beveridge *Brit. Food Control* x. 230 The wealthy classes in particular suffered from the reduction of the meat ration. 1944 *Ourselves in Wartime* vii. 154/2 More often than not, they forfeited their personal sweet ration, which amounted to 3 ozs. of sweets or chocolate a week in 1943, for the sake of the children. 1948 *Ann. Reg. 1947* 246 Food shops were compelled to limit strictly the amount of 'off ration' foods which customers could buy at one time. 1950 N. Streatfeild *Mothering Sunday* 31 She was the last person to look for extras off the ration. 1959 in I. & P. Opie *Lore & Lang. Schoolch.* vii. 105 We are three spivs of Trafalgar Square Flogging nylons tuppence a pair, All fully fashioned, all off the ration, Sold in Trafalgar Square. 1960 J. Rae *Custard Boys* I. iv. 41 A little shop..where the old lady would sell us sweets and chocolate off the ration. 1975 S. Briggs *Keep Smiling Through* 150 Some articles off the ration could add an exotic touch to the menu. There was turbot in 1940 and whalemeat in 1942. *Ibid.* 161 'Off the ration' foods..like salt cod were publicized as 'grand for children as well as grown-ups and what a bargain!'
d. *transf.*
1850 Browning *Christmas Eve* ii. 7 Still, as I say, though you've found salvation, If I should choose to cry— as now—'Shares!'—See if the best of you bars me my ration!
e. Mil. slang phr.: *come up* (or *be given*) *with the rations*, to be awarded automatically (used deprecatingly, of military medals and decorations, to imply that they have not been earned).
1925 *N. & Q.* 25 July 71/2 Came up with the rations. 1928 H. Williamson *Pathway* xvii. 378 'Did *you* get the Military Cross?' asked Mrs. Ogilvie. 'Yes, it came up with

the rations.' 'Oh!' 'A soldier's joke, Mrs. Ogilvie.' 1937 J. A. Lee *Civilian into Soldier* 204 'Bit of decoration. Congratulations.' 'Came up with the rations.' He took the ribbon. But if he joked he was pleased in his soul. 1957 J. Braine *Room at Top* xviii. 162 Lampton has no decorations apart from those which all servicemen who served his length of time are given, as they say, with the rations. 1973 A. Price *October Men* xv. 210 The British Military Cross.. didn't come up with the rations.
4. a. *ration bag, -beef* (earlier example), *boot, party, rum, -sugar, -tea*; **ration book,** a book entitling its holder to a ration; **ration card, coupon,** a card or coupon entitling its holder to a ration; **ration sheep** *Austral.*, the sheep to be killed for food for the workers on a station; also *ration-sheep paddock, ration paddock*; **ration strength,** the number of men in an armed force, estimated by the rations supplied to them; **ration ticket** = *ration card.
1862 R. Henning *Let.* 5 Sept. (1966) 103 The 'ration bags' contained flour, sugar, tea, sardines, bacon, cheese, salt beef and salmon and jam. 1917 A. G. Empey *Over Top* 305 *Ration bag*, a small, very small bag for carrying rations. 1835 J. E. Alexander *Sketches in Portugal* iv. 101 Into an upper room marched two troopers, with a camp-kettle between them containing water, followed by two others with another kettle, containing a savoury mess of ration-beef, boiled with bread and onions. 1918 *Times* 1 Nov. 3/2 The Ministry of Food wish to remind the public that persons registered with retailers for tea must renew their registration as soon as possible by depositing with the retailer the 'spare counterfoil 2' on leaf 7 of their new ration books. 1939 *New Statesman* 18 Nov. 700/1 The ration books ..have now been distributed. 1973 *Country Life* 20 Dec. 2120/1 As these notes are written, ration books are being issued and motorists seem anxious to use them. 1902 'Coldstreamer' *Ballads of Boer War* vii. 70 If you find a time to suit, Just cop 'im with a ration boot. 1882 H. Vizetelly *Paris in Peril* II. vii. 35 At some establishments strangers were politely informed that dinners were only served to the regular clientèle, who had handed over their ration-cards to the proprietor. 1922 H. W. Clemesha *Food Control in North-West Division* ii. 39 When the ration cards and the vouchers had all been distributed they were lodged either with retailers or wholesalers. 1940 *Economist* 9 Mar. 415/1 A general census is to be taken preparatory to the issue of ration-cards [in France]. 1975 T. Allbeury *Special Collection* v. 32 The ration cards were no longer honoured. The food wasn't there any more. 1944 *Sun* (Baltimore) 12 Dec. 12/8 A 'red market' in meat—collection of ration coupons for point free cuts. 1935 G. L. Meredith *Adventuring in Maoriland* v. 41, I sometimes have to do the slaughtering. There is no 'ration paddock'. 1917 A. G. Empey *Over Top* 305 *Ration party*, men detailed to carry rations to the front line. 1928 Blunden *Undertones of War* xiii. 145 At the ration-party's rendezvous.., our hearty Quartermaster Swain..was guarding..our issue of rum. 1918 E. A. Mackintosh *War* 94 Punch concocted out of ration rum. 1911 C. E. W. Bean *'Dreadnought' of Darling* xxxiv. 293 In some of the Western towns they find a convenient substitute even for ration sheep. 1914 H. B. Smith *Sheep & Wool Industry Australasia* vi. 34 On a station the first sheep that are usually shorn are the ration sheep. These are the sheep that are to be killed for household and shearers' use. 1946 F. Davison *Dusty* viii. 81 From then on he took him [the pup] out whenever he had to bring in the ration sheep for a killing. The ration sheep grazed in a small paddock near the homestead. *Ibid.* x. 105 The fence bounding the ration-sheep paddock. 1931 W. S. Churchill *World Crisis* VI. xxi. 323 Out of 425,000 men comprising the entire manhood of the country, borne on the ration-strength of the Serbian army at the beginning of October, over 100,000 had been killed or wounded. 1965 B. Sweet-Escott *Baker St. Irregular* vii. 201 The ration strength of A.F.H.Q. at this time was..something like that of a fighting division. 1892 *Missing Friends* iii. 54 The most inferior goods in the market are called *ration-tea* and *ration-sugar*. 1938 N. MacOwan *Glorious Morning* II. ii. 59 There's food to get. Our ration tickets are only available today.

rational, *a.* (*adv.*) and *sb.*[1] Add: **A.** *adj.* **1. a.** (Later examples.)
1908 *Jrnl. Abnormal Psychol.* III. 166 Everyone feels that as a rational creature he must be able to give a connected.. account of himself. 1975 J. Plamenatz *K. Marx's Philos. Man* i. 17 He always speaks of man as a self-conscious, rational, active being who can make choices and can initiate change deliberately.
c. *rational psychology, psychological science*: psychology, or the science of mind, as studied by deduction from general principles, and distinguished from an empirical approach. Now *Obs.*
1817 Coleridge *Biog. Lit.* I. ix. 139 These delusions were such, as might be anticipated..from his ignorance of rational psychology. 1849 L. P. Hickok *Rational Psychol.* 21 Those a priori conditions which give the necessary and universal laws to experience, and by which intelligence itself is alone made intelligible, are the elements for a higher Psychological Science which we term *Rational*. 1861 J. S. Mill *Let.* in A. Bain *John Stuart Mill* (1882) iv. 118 It will enable me..to do the kind of service which I am capable of to rational psychology. 1892 W. James *Coll. Ess. & Rev.* (1920) xx. 321 We certainly need something more radical than the old division into 'rational' and 'empirical' psychology, both to be treated by the same writer between the covers of the same book.
3. c. *rational mechanics*: mechanics as deduced logically from first principles.
[1687 Newton *Philosophiæ Naturalis Principia Mathematica* p. v, *Mechanicam* vero duplicem Veteres constituerunt: *Rationalem* quæ per Demonstrationes accurate procedit, & *Practicam*. *Ibid.*, Quo sensu *Mechanica rationalis* erit Scientia Motuum qui ex viribus quibuscunq; resultant, & virium quæ ad motus quoscunq; requiruntur, accurate proposita ac demonstrata. 1729 A. Motte tr. *Ibid.* I. p. vii, The ancients considered Mechanics in a twofold respect; as rational, which proceeds accurately by demonstration, and practical.] *Ibid.* p. ix, In this sense Rational Mechanics will be the science of motions resulting from any forces whatsoever and of the forces required to produce any motions, accurately proposed and demonstrated. 1902 J. W. Gibbs *Elem. Princ. Statistical Mech.* p. ix, Nothing will more conduce to the clear apprehension of the relation of thermodynamics to rational mechanics..than the study of the fundamental notions and principles of that department of mechanics to which thermodynamics is especially related. 1928 R. De Villamil *Rational Mech.* p. vi, I have been asked: Why I call this a book on 'Rational Mechanics'? and Do I not consider all Mechanics as being 'Rational'? 1952 *Jrnl. Rational Mech. & Anal.* I. p. ii, The *Journal of Rational Mechanics & Analysis* nourishes mathematics with physical applications, aiming especially to close the rift between 'pure' and 'applied' mathematics and to foster the discipline of mechanics as a deductive, mathematical science in the classical tradition. 1958 *Science* 4 Apr. 729/1 In the United States..rational mechanics is not a recognized science. Indeed, there are some who disbelieve in its existence. 1977 C. Truesdell *First Course Rational Continuum Mech.* i. 4 Rational Mechanics is the part of mathematics that provides and develops logical models for the enforced changes of position and shape which we see everyday things suffer.
5. a. (Further examples.)
1901 H. B. Fine *College Algebra* II. i. 86 An expression is called rational if it does not involve an indicated root of an expression in which a variable letter occurs. 1917 T. M. MacRobert *Functions of Complex Variable* v. 89 The ratio of two polynomials is called a Rational Function. 1940 C. C. MacDuffee *Introd. Abstract Algebra* iii. 82 A quadratic equation with rational coefficients which does not have a rational root. 1946 A. A. Albert *College Algebra* iv. 111 A rational function of several symbols is any algebraic expression obtainable by formally applying a finite number of the rational operations of addition, subtraction, multiplication, and division to the symbols and numbers. 1966 R. E. Johnson *University Algebra* i. 20 There exist consecutive integers n and $n+1$, but there do not exist consecutive rational numbers.
c. (With admixture of sense 4.) *Physics.* The epithet given by O. Heaviside to electrical units and equations now described as *rationalized.
1882 O. Heaviside in *Nature* 24 Aug. 391/1 If..electricity stands foremost amongst the exact sciences, it follows that its unit measures should be determined with the utmost accuracy. Yet, twenty years ago very little advance had been made toward the adoption of a rational system. 1892 *Ibid.* 28 July 293/1 If we let the rational practical units be the same multiples of the 'absolute' rational units as the present practical units are of *their* absolute progenitors, then [etc.]. 1905 *Proc. R. Soc.* A. LXXVI. 551 The calculations would have been the same if e had been measured in 'rational' electric units instead of those in common use, but we should then have had an experimental value equal to 4π times that mentioned above. 1905 [see *Giorgi]. 1911 *Encycl. Brit.* XXVII. 744/2 If the filament is an endless or poleless iron filament magnetized uniformly by a resultant external magnetic force H, the flux density will be expressed in rational units by the equation $B = I + H$. 1925 W. H. Timbie *Elements of Electricity* (ed. 2) vi. 171 In order to have Ohm's Law hold with these units, the rational oersted must be equal to 1·26 C.G.S. units. 1942 *Phil. Mag.* XXXIII. 487 All the above formulæ, whether those of the rational system or of the ordinary system, are entirely independent of any choice of units.
7. Descriptive of methods of analysis and planning that make use of calculation to bring about a projected result, esp. in economic or social organization.
1915 M. Epstein tr. *Sombart's Quintessence of Capitalism* xii. 182 He is ever ready to adopt a newer method if it is more rational, whether in the sphere of organization, of production, or of calculation. 1926 E. Grossmann *Methods Econ. Rapprochement* 30 The most important economies will follow not from the simplification of the machinery of distribution but from a rational organization of production itself. 1930 T. Parsons tr. *Weber's Protestant Ethic* 21 But in modern times the Occident has developed..a very different form of capitalism which has appeared nowhere else: the rational capitalistic organization of (formally) free labour. *Ibid.*, Rational industrial organization, attuned to a regular market..is not, however, the only peculiarity of Western capitalism. 1943 J. A. Schumpeter *Capitalism, Socialism & Democracy* xi. 122 The rational attitude presumably forced itself on the human mind primarily from economic necessity. *Ibid.*, When the habit of rational analysis of.. the daily tasks of life has gone far enough, it turns back upon the mass of collective ideas and criticizes..them by way of such questions as why there should be kings..or tithes or property. *Ibid.* 123 Capitalist practice turns the unit of money into a tool of rational cost-profit calculations, of which the towering monument is double-entry bookkeeping. 1969 Simon & Stedry in Lindzey & Aronson *Handbk. Social Psychol.* (ed. 2) V. xl. 272 The classical economic theory of markets with rational agents and perfect competition is a deductive theory that requires almost no contact with empirical data..to establish its propositions. 1977 A. Giddens *Stud. in Social & Polit. Theory* v. 206 Weber's characterization of modern capitalism as involving above all the 'rational' organization of resources geared to the accumulation of profit is unsatisfactory.
C. *sb.*[1] **3.** Delete † *Obs. rare.* and add later examples.
1958 D. E. Littlewood *University Algebra* (ed. 2) ix. 150 The set of integers K can be embedded in a field of quotients. This field of quotients is defined as the rationals. 1971 D. G. H. B. Lloyd *Mod. Syllabus Algebra* vi. 110 Between any two rationals a third rational can be inserted. In this respect the set of rationals differs completely from the set of naturals or the set of integers.

rationalism. 2. Restrict *Theol.* to a, b in Dict. and add: **c.** The view that reason is the only guide leading to the improvement and progress of the human race and that adherence to religious or other 'non-rational' beliefs is out-dated.
1876 (*title*) Constructive rationalism. 1897 *Agnostic Ann.* I. 18 In my progress from Rome to Rationalism many other considerations have influenced me. 1923 J. S. Huxley *Ess. Biologist* 231 It is the task of Rationalism to see that religion, this fundamental and important activity of man, shall neither be allowed to continue in false or inadequate forms, nor be stifled or starved, but be made to help humanity in a vigorous growth that is based on truth and in constant contact with reality. 1968 A. J. Ayer *Humanist Outlook* 3 A broader movement of Rationalism or Free Thought, which was not merely anti-clerical but hostile to any form of religious belief. 1973 C. Campbell in *Rationalism in 1970s* 81 If the aim of rationalism is merely to attack and demolish the myths that we and others hold then there is more than enough work to keep us busy for a very long time. 1973 C. Macy (*title*) Rationalism and humanism in the new Europe.
3. (Earlier and later examples.)
1831 *Edin. Rev.* Sept. 247 The fundamental principles of Rationalism we take to be these:—That human reason, or the reasoning faculty, is the sole arbiter as to what is to be received as truth, and what is to be regarded as error, by the human mind; that facts recognized by sense or consciousness form the materials on which the reasoning faculty is to be exercised. 1967 *Encycl. Philos.* VII. 69/1 The philosophical outlook..which stresses the power of a priori reason to grasp substantial truths about the world and correspondingly tends to regard natural science as a basically a priori exercise... The spirit of rationalism in this sense is particularly associated with..Descartes, Spinoza, and Leibnitz.
5. The principle or practice of effecting assessment, planning, or organization in the economic or social sphere by rational (sense *7) methods.
1915 M. Epstein tr. *Sombart's Quintessence of Capitalism* xii. 182 Absolute rationalism is the first [principle]. Economic activities are ruled by cold reason, by thought. *Ibid.* xxv. 325 Economic rationalism owes much of its growth to technical rationalism. 1930 R. H. Tawney in T. Parsons tr. *Weber's Protestant Ethic* 1 (e) The word 'rationalism' is used by Weber..to describe an economic system based..on the deliberate and systematic adjustment of economic means to the attainment of the objective of pecuniary profit. 1935 *Encycl. Social Sci.* XIII. 114/1 In social and historical life the power of rationalism derives from the confidence which individuals and societies place in reason. 1958 G. Myrdal *Value in Social Theory* vii. 135 Basic to the eagerness in trying to drive valuations underground is the rationalism of our Western culture.

rationalist, *sb.* and *a.* Add: **1. b.** One who applies scientific methods of reasoning or calculation to social and economic life.
1958 G. Myrdal *Value in Social Theory* iii. 57 The American has..started to measure, not only human intelligence,..and personality traits, but moral leanings and the 'goodness' of communities. He is a rationalist. 1969 A. Etzioni in Lindzey & Aronson *Handbk. Social Psychol.* (ed. 2) V. 547 In effect, the rationalists advocate an approach that maximises conflict and makes nuclear war more likely. 1977 T. Parsons *Social Syst. & Evol. of Action Theory* i. 71 Phrasing it as 'the problem' I hope makes clear that I have not been a naïve rationalist.
2. a. Also, one who believes that reliance upon human reason does away with the need for religion. Cf. *rationalism 2 c.
1876 *Constructive Rationalism* 5 The destruction of orthodox Christianity being accomplished, there remains for the Rationalist much more to do. He has to frame a code which shall rule in the place of the code of Moses and of Jesus. 1897 *Agnostic Ann.* I. 42 Why do eminent Rationalists, Freethinkers, and Agnostics stand aloof. 1908 [see *ethicist]. 1942 E. T. Kerr in *Why I am a Rationalist* 33 Few Rationalists..owed to their schools..any kind of.. approach to independent thinking. 1954 *Rationalist Ann.* 73, I cannot possibly make my probabilities into certainties, and this is what some Rationalists appear..to do. 1973 C. Campbell in *Rationalism in 1970s* 81 It is an important part of a rationalist's duty to clear himself of myths..as well as to expose the myths of others.
b. (Further examples, esp. in sense of *rationalism 2 c and 5.)
1921 G. B. Shaw *Back to Methuselah* p. lxxiii, They banish the Bible from their houses, and sometimes put into the hands of their unfortunate children Ethical and Rationalist tracts of the deadliest dullness, compelling these wretched infants to sit out the discourses of Secularist lecturers. 1942 J. A. Schumpeter *Capitalism, Socialism & Democracy* (1943) xi. 122 There is however one more point about the concept of rationalist civilizations that I will mention here. *Ibid.*, The rationalist attitude may go to work with information and technique so inadequate that [etc.]. 1955 P. Edwards *Logic of Moral Discourse* ii. 52 The Rationalist view of causation. 1973 F. A. Hayek *Law, Legislation & Liberty* i. 33 The desire to remodel society after the image of individual man..since Hobbes has governed rationalist political theory. 1974 *New Humanist* May 16/1 Clerical pressure continued to be applied to booksellers, discouraging them from displaying Rationalist books. 1977 A. Giddens *Stud. Social & Pol. Theory* vii. 239 Durkheim replied by asserting the existence of a radical distinction between 'egoism' and 'rationalist individualism'.

rationalistically, *adv.* (Earlier and later examples.)
1854 Geo. Eliot tr. *Feuerbach's Essence Christianity* xxv. 245 Zwinglius only expressed..rationalistically..

what the others declared mystically. **1910** W. Temple *Faith & Mod. Thought* iii. 79 Why rationalistically minded folk should suppose that in a moment of almost intolerable joy people are going to have an exact memory for dates and places I cannot conceive. **1934** A. C. Ewing *Idealism* v. 258 The more rationalistically inclined philosophers have.. erred in putting forward extravagant claims to certainty.

rationality. Add: **2. b.** (Further examples.) Also *attrib.*

1908 *Jrnl. Abnormal Psychol.* III. 166 Any act..is immediately justified by distorting the mental processes concerned and providing a false explanation that has a plausible ring of rationality. **1933** J. L. Gillin *Social Path.* xxvi. 452 Capitalism is characterized by rationality. By that we mean a tendency to long range planning, careful consideration of the adaptation of means to ends, and cold and careful calculation of what measures will bring the greatest gain. **1961** H. M. Johnson *Sociol.* ix. 204 The expression 'economic rationality'..is perhaps confusing, since purely technical rationality in production is also 'economic', although not necessarily economical. **1969** Simon & Stedry in Lindzey & Aronson *Handbk. Social Psychol.* (ed. 2) V. xl. 272 The first [principle] is the assumption of objective rationality, which permits strong predictions to be made about human behavior without the painful necessity of observing people. **1975** T. McCarthy tr. *Habermas's Legitimation Crisis* (1976) II. iii. 46 Output crises have the form of a rationality crisis in which the administrative system does not succeed in reconciling and fulfilling the imperatives received from the economic system. *Ibid.* v. 62 A rationality deficit can arise because contradictory steering imperatives..are then operative within the administrative system. **1976** H. Leibenstein *Beyond Economic Man* v. 73 Suppose, but only for a moment, that rationality is interpreted as 'calculatedness'.

rationalizable, *a.* Add: Hence **ra:tionalizabi·lity.**

1936 Wirth & Shils tr. *Mannheim's Ideology & Utopia* iii. 124 The intuitional approach..conceives of knowledge and rationalizability as somewhat uncertain.

rationalization. Add: **1. b.** *Psychol.* The justification of behaviour to make it appear rational or socially acceptable by (subconsciously) ignoring, concealing, or glossing its real motive; an act of making such a justification.

1908 *Jrnl. Abnormal Psychol.* III. 166 Two different groups of false explanations can be distinguished..according as they are formed mainly for private or mainly for public consumption. The former of these I would term 'evasions', the latter 'rationalisations'. **1924** J. Riviere tr. *Freud's Coll. Papers* I. 341 His [*sc.* Adler's] theory does what all patients do and what our waking thought in general does—namely, makes use of a rationalization, as Jones has called it, in order to conceal unconscious motives. **1947** E. F. Frazier in *Amer. Sociol. Rev.* XII. 271 A dynamic sociological theory of race relations which will discard all the rationalizations of race prejudice. **1953** R. F. C. Hull tr. *Jung's Coll. Wks.* VII. 214 My patient clung to his intellectual world and defended himself with rationalizations against what he regarded as his illness. **1961** H. Bonner *Psychol. of Personality* ix. 281 The term 'rationalization' has become a household word. It is essentially a method of self-justification. It is motivated by the fear of criticism and disapproval by others. **1977** F. J. Bruno *Human Adjustment* ii. 67 The defense mechanism called rationalization is used when the real motive for one's behavior is unacceptable to the ego.

2. b. *Physics.* The reformulation of the equations and definitions of electromagnetism so that the factor 4π is removed from those relating to systems without spherical symmetry.

1891 O. Heaviside in *Electrician* 16 Oct. 656/2 When.. the real advantages of the rational system become widely recognized and thoroughly assimilated, then will come a demand for the rationalisation of the practical units. **1942** *Phil. Mag.* XXXIII. 486 The simplification of formulæ due to rationalization can be illustrated by considering a parallel plate condenser. **1951** *Electr. Engin.* LXX. 332/2 Note that, in effect, a 4π was inserted in the denominator of the classical expression for force between parallel wires to accomplish rationalization, and that the elimination of 4π from the magnetomotive force relation followed without any additional arbitrary insertion of 4π. **1969** L. Young *Systems of Units Electr. & Magn.* 197 The appearance or disappearance of 4π's in the equations of electromagnetism upon rationalization can be interpreted in two ways... Just as our point of view has been described as 'rationalization of units', so the other point of view has been called 'rationalization of quantities'.

c. *Econ.* and *Sociol.* The process of applying rational (sense *7) methods, esp. of standardization and simplification, to the planning and organization of economic enterprises or the administration of social groups in order to achieve a particular result such as maximum profit or efficiency; an example of this.

[**1905** M. Weber in *Archiv f. Sozialwissenschaft u. Sozialpolitik* XX. 29 Alsdann nun wiederholte sich, was immer und überall die Folge eines solchen „Rationalisierungs"-Prozesses ist.] **1921** E. & C. Paul tr. *Rathenau's In Days to Come* III. i. 125 The general wellbeing of the country is doubled or trebled by the setting of idle hands to work and by the rationalisation of production. **1926** D. Houston *Memorandum on Rationalisation in U.S.* 3 The term rationalisation as used in Europe today includes, I take it, the three elements of stabilisation, standardisation and simplification of industry or of individual enterprises. **1934** P. & I. Petroff *Secret of Hitler's Victory* iii. 38 The soullessness of modern labour, which had reached its climax in consequence of the rationalization, became the outstanding feature of the whole period. **1936** H. A. Phelps *Princ. & Laws of Sociol.* xx. 424 A compensating general trend is the increasing rationalization of social life. **1939** H. Hodge *Cab, Sir?* 259 The would-be Napoleon..puts his rationalisation schemes into the waste-paper basket. **1947** Henderson & Parsons tr. *Weber's Theory Social & Econ. Organization* i. 112 One of the most important aspects of the process of 'rationalization' of action is the substitution for the unthinking acceptance of ancient custom, of deliberate adaptation to situations in terms of self-interest. **1959** *Listener* 31 Dec. 1147/2 A rationalization proposal [in the U.S.S.R.] is a technical improvement using known means which lacks the degree of originality demanded of a patentable invention. **1971** J. J. Shapiro tr. *Habermas's Toward Rational Society* vi. 99 Until then [*sc.* the 19th century] modern science did not contribute to the acceleration of technical development nor, consequently, to the pressure toward rationalization from below. **1972** W. J. Mommsen *Age of Bureaucracy* iv. 80 Another secular force of social change is found, namely rationalization..by which tradition-bound or value-oriented forms of political and social organization are gradually replaced by purely instrumentally rational institutions. **1976** *Star* (Sheffield) 20 Nov., The company had announced 'rationalisation' plans meaning the closure of the Dronfield works.

rationalize, *v.* Add: **1. a.** (Earlier and later examples.)

1803 *Lett. Miss Riversdale* II. 79 This interesting sentiment [*sc.* friendship]..secures the permanence of happiness, by rationalizing (if I may use such a word) its origin. **1935** *Encycl. Social Sci.* XIII. 116/1 The problem was to rationalize human social life on the basis of self-evident and universal principles. **1965** W. J. Harvey in Geo. Eliot *Middlemarch* 11 Bulstrode..who rationalizes his worldly success as an example of divine providence.

c. *Psychol.* To give plausible reasons for (one's behaviour) that ignore, conceal, or gloss its real motive. Also *absol.* or *intr.*

1922 H. Somerville *Pract. Psycho-Anal.* i. 14 It is clear that patient is rationalising, and that as a matter of fact he is eaten up with jealousy. **1925** J. Riviere tr. *Freud's Coll. Papers* III. 330 The patient's consciousness naturally misunderstands them and puts forward a set of secondary motives to account for them—rationalizes them, in short. **1932** H. G. Wells *Work, Wealth & Happiness of Mankind* vii. 279 To rationalize has one meaning in psychology, another meaning in the sociological writings of Max Weber, and quite another in the loose discussions of modern politicians and business men. **1966** *Word Study* Dec. 3/2, I think we all rationalize with the thought that the free democratic society which produced us..had a right to be presented (*sic*) in such a fashion.

2. b. *Physics.* To subject (the units or equations of electromagnetism) to rationalization (sense *2 b).

1892 *Nature* 28 July 292/2 It is..very desirable that the practical units themselves should be rationalized. **1899** *Electrician* 29 Dec. 325/2 If we take the permeability of ether to be 4π units instead of unity, we rationalize at one stroke all our present units except the units of magnetic force and magnetic pole strength. **1973** J. Yarwood *Electricity & Magnetism* ii. 32 In rationalising electrical units the object..is to avoid the occurrence of 4π in systems without spherical symmetry and of 2π where cylindrical symmetry is absent.

c. To organize (economic production or the like) according to rational or scientific principles so as to achieve a desired or predictable result; esp. to reduce the number of (personnel, industrial plants, etc.) in such a way that the remainder are more efficiently deployed.

1926 E. Grossmann *Methods Econ. Rapprochement* 30 International cartels will be able to rationalise production in a way impossible in the present state of affairs. **1931** *Ann. Reg. 1930* II. 26 The Lancashire cotton industry. The steps taken to 'rationalise' the industry. **1953** J. B. Carroll *Study of Lang.* iv. 127 A recent attempt to rationalize an artificial language by making maximal use of elements common to the most widely used natural languages is Interlingua. **1962** *Listener* 22 Mar. 509/2 Their numbers go down: they are 'rationalized'. In 1920 there were nine evening newspapers in London; now there are two. **1977** *R.A.F. News* 11–24 May 7/2, I am..aware..of the need to rationalise reporting systems to reduce paperwork.

rationalized, *ppl. a.* Add: *spec.* in *Physics,* applied to units, equations, and definitions in electromagnetism that are formulated so that the factor 4π appears only when a system with spherical symmetry is involved. Cf. *rational *a.* 5 c.

1933 [see *Giorgi]. **1951** *Electr. Engin.* LXX. 332/1 The MKS Rationalized system seems now to be replacing the others rapidly and may well come to be the accepted system for all theoretical work in electricity. **1969** L. Young *Systems of Units Electr. & Magn.* vi. 78 The rationalized form of the CGS practical system is theoretically possible but is never used. **1973** J. Yarwood *Electricity & Magnetism* xvi. 608 The rationalised mksA units in electricity form a part of the wider SI system of units.

rationalizing, *vbl. sb.* (Earlier and later examples.)

1865 J. S. Mill *Auguste Comte* 54 The way to a complete rationalizing of those sciences..has been shown nowhere so successfully as there. **1927** A. Huxley *Let.* 25 Feb. (1969) 284, I can't see that there's anything to distinguish his rationalizings of religious emotions from those of anyone else. **1971** P. Gresswell *Environment* 105 Footpaths in some parishes need reorganising and rationalising for today's needs.

rationative, *a.* Delete † *Obs.* and add later example.

a **1966** 'M. na Gopaleen' *Best of Myles* (1968) 195 An issue too imponderable for rationative evaluation.

rationing, *vbl. sb.* (In Dict. s.v. Ration *v.*) (Later examples.) *rationing by the purse,* raising the price of a commodity so as to restrict the number of people who can afford to buy it. Similarly *rationing by price.*

1917 *Times* 1 May 7/6 The German Government now knows all about rationing, but while it has been learning the German people has eaten up its supplies. **1924** E. M. H. Lloyd *Experiments in State Control* xxiii. 290 As for the argument that rationing increases consumption, this was not true of the articles most severely rationed in Great Britain. **1930** *Economist* 22 Mar. 637/1 In the last resort, a rationing of credit was the only expedient left to central banks. **1940** *Times* (Weekly ed.) 10 Jan. 9/3 Rationing may have to be used to a much greater extent than in the last war. **1947** *People* 22 June 2, I find school lunches a great help with the rationing problems. *Ibid.*, It is the end of the rationing period. **1950** *Hansard Commons* 24 Apr. 617 Is this not a case of rationing by the purse? **1975** S. Briggs *Keep Smiling Through* 149 The great wartime invention, borrowed from the Germans, was points rationing. This widened choice as much as it could be widened within a rationing system. You could even..choose where to shop without being tied to the grocer where you were registered for basic rations. **1979** W. Safire in *N.Y. Times Mag.* 9 Sept. 16/1 *Rationing by price,* a system in which economic goods go to the people who are most willing to pay for them. This is the normal way of distributing goods in capitalist countries, and is increasingly used in Communist countries. Economists generally consider it an efficient system, but some politicians consider it immoral and prefer a system of rationing by political pull.

‖ **ratissage** (ratisɑ̃ʒ). *Econ.* [Fr., lit. 'scraping, raking in of stakes'.] A device whereby the Bank of France calls in a portion of its country's commercial banks' foreign currency balances, and so temporarily improves its reserves of foreign exchange.

1957 *Economist* 28 Dec. 1144/1 The technique of *ratissage* dates from March of this year even though it is only now that it has been broadened to include dollars as well as EPU currencies. **1959** *Ibid.* 10 Jan. 155/2 A substantial part of the EPU deficit is ascribed to further *ratissage* by the Bank of France. **1959** *Times* 7 Apr. 17/1 The French recovery has now gone far enough for the French authorities to hope that their so-called 'ratissage' operations may be liquidated this month.

ratite, *a.* Add: (Later examples.) Also as *sb.*, a bird belonging to this group.

1911 J. A. Leach *Austral. Bird Bk.* 52 'Discontinuous distribution' as applied to land animals, e.g...ratite birds in South America..implies a land connexion. **1939** J. Fisher *Birds as Animals* iii. 32 Sometimes..it [*sc.* the hallux] is altogether absent, as it is in auks, bustards and ratite birds. *Ibid.* vi. 73 Ostriches flock, so do the other ratites. **1978** *Sci. Amer.* July 102/2 Among the other ratites are the ostrich, the emu and the extinct giant moa of New Zealand. **1979** *Nature* 14 June 633/2 The supposed presence of a ratite bird, a member of a group otherwise restricted to the Southern Hemisphere, in the Upper Cretaceous of Mongolia is very doubtful.

ratlin(e, ratling. 1. *ratline stuff* (earlier and later examples).

1833 Poe *MS. found in Bottle* in *Gift 1836* (1835) 80 A pile of ratlin-stuff and old sails. **1883** *Man. Seamanship for Boys' Training Ships R. Navy* (Admiralty) (1886) 312 Hitch your rattling stuff round the third shroud from aft. **1954** Bradford & Quill *Gloss. Sea Terms* 154/1 *Ratline stuff,* twelve- or fifteen-thread but usually eighteen-thread right-handed, tarred rope, used for ratlines, heavy lashings and heavy lines.

Rato (rēɪ·to). *Aeronaut.* Also **rato.** [f. the initial letters of *r*ocket-*a*ssisted *t*ake-*o*ff.] A jato in which the engine is a rocket engine.

1953 P. C. Berg *Dict. New Words* 134/2 *R.A.T.O., Rato,* rocket-assisted take-off. **1962** *Aeroplane* CIII. 26/1 Designed to carry outsize cargo, the Stratocruiser is to have an increased payload of 60,000 lb and rato units will be used.

ratomorphic (ræ:tomǭ·ɹfik), *a. rare.* [f. Rat *sb.*[1] + *-omorphic,* after Anthropomorphic *a.*] (See quot. 1964.) Also **ra·tomorph,** one who holds a ratomorphic view of human behaviour.

1964 Koestler in *Listener* 14 May 786/2 It refuses to attribute to man any mental processes which cannot be shown to occur in lower animals. In other words, for the anthropomorphic view of the rat, Behaviourism has substituted a ratomorphic view of man. **1967** *Economist* 21 Oct. 295/3 The author [*sc.* Koestler] presents a trenchant.. criticism of behaviourism and experimental 'ratomorphic' views of men. **1977** H. G. Burger in B. Bernardi *Concept & Dynamics of Culture* 460 In trying to overlook symboling, those who oversimplify subhuman abilities are 'ratomorphs', and have overlooked two and a half million years of hominid evolution.

ratoon, *v.* Add: **b.** *trans.* To cut down (plants) to induce them to send up new shoots. So **ratoo·ned** *ppl. a.*

1925 *Glasgow Herald* 23 Apr. 14/2 Reports indicate that ratooned cotton has suffered. Ratooned plants produce a

much earlier crop than new plants and..Zululand had ratooned a considerable quantity this year.

ratooner (rătū·nəɹ). [f. prec. + -er.] A plant that ratoons.

1922 *Chambers's Jrnl.* Dec. 800/2 A second crop can be obtained from the dwarfed stumps of the trees after the first crop has been picked, but the ochro is a bad ratooner.

rat race. Also **rat-race.** [f. Rat *sb.*[1]] † **1.** A dance. *U.S. slang. Obs.*

1937 *Amer. Speech* XII. 74/2 C.C.C. speech... Terms for recreations: *rat-race,* dance of low-grade nature. **1948** Mencken *Amer. Lang.* Suppl. II. 707 The vocabulary of the jazz addict... A dance is a *rat-race* or *cement-mixer.*

2. A fiercely competitive struggle or contest; a struggle to maintain one's position in work or life. orig. *U.S.*

1939 C. Morley *Kitty Foyle* xxvi. 261 Their own private life gets to be a rat-race. **1940** *Time* 16 Dec. 26/3 Veteran fliers blanched when they saw the hourly, crowded 'rat race' at Randolph—the close-packed stream of trainers, gliding in to land and take on fresh cadets and instructors. **1946** *War Report* (B.B.C.) 350 Our armour is now 'swanning' as they say in the British Army, or in American parlance, 'the rat race is on'. **1947** J. Steinbeck *Wayward Bus* 214 He was afraid of his friends and his friends were afraid of him. A rat race, she thought. **1954** Wodehouse & Bolton *Bring on Girls* 219 'Is anything the matter with you?' 'Just the rat-race. I don't quite know why I've been doing it.' **1956** R. Fuller *Image of Society* iii. 70 A boy's got to have guts to make his way in this rat race of a modern world. **1958** *Spectator* 19 Sept. 381/2 Modern economic life is more like a rat-race than a rational way of life. **1959** *Observer* 8 Mar. 17/7, I don't like this rat-race for promotion. **1959** *Spectator* 2 Oct. 435/2 A realism that encourages in its popular press a rat-race morality in the guise of room at the top. **1960** *Daily Tel.* 18 May 17/7 A spirited criticism of 'the daily rat race' to get to work in London. **1967** G. F. Fiennes *I tried to run Railway* iv. 31 It became a rat race to see who could get teams in first so as to improve recruitment. **1973** C. Bonington *Next Horizon* iv. 68 Another artist, who had abandoned the rat-race and settled in Coniston. **1976** F. Zweig *New Acquisitive Society* I. v. 52 The shedding of middle-class values and style of life in the younger generation..is an outright negation of middle-class existence, defined in such derogatory terms as 'rat race'. **1978** D. A. J. Seargent *UFO's* v. 111 A motor car—the prime symbol of our motorised, machine-orientated, rat-race society.

Hence as *v. trans.* and *intr.*, to take part in a 'rat race'; **ra·t-racer, ra·t-racing** *vbl. sb.* and *ppl. a.*

1937 *Clarionette* (Univ. of Denver) (St. Patrick ed.) 18 Mar. 1/3 If you're off for a little body-swaying to music, you are..'rat-racing'. **1960** *Listener* 7 Jan. 41/2 A scarifying glimpse of rat racing in local government. **1962** *Guardian* 15 Oct. 7/2 A new modern figure—the new kind of ambitious rat-racer, the Snopes in the grey flannel suit. **1968** *Ibid.* 5 Oct. 6/5 Literary people in this country seem to have been..rat-racing each other to the nearest vacant editor's chair. **1968** *New Scientist* 21 Nov. 418/3 Looking into an aquarium..is just the medicine for a chap who has spent all day rat-racing against a computer. **1969** E. Lemarchand *Alibi for Corpse* ix. 116 I'm a damn sight saner than people who spend their lives rat-racing and jabbering their beads off. **1971** *New Scientist* 1 July 5/1 The belief among rat-racers that the physical exercise delays thrombosis. **1971** *Guardian* 3 July 8/6 Middle aged Frank who wants to be a drop-out from rat-racing society. **1977** D. Morris *Manwatching* 124 Eccentricity of dress and behaviour is commonplace for them and they enjoy social freedoms unknown to other rat-racing citizens.

ratsbane. Add: **1.** (Further example.)

1877 'Mark Twain' in *Atlantic Monthly* Dec. 723/1 What was that cat's name that eat a keg of ratsbane by mistake over at Hooper's?

Hence **ra·tsbany** *a.*

1937 Blunden *Elegy* 24 And sets pot to mouth And once again moistens his ratsbany drouth.

rat's-tail. Add: **2. d.** (Earlier example.)

1810 J. Lambert *Trav. Lower Canada* I. ix. 162 The dress of the Habitant is simple, and homely;..His hair is tied in a thick long queue behind, with an eelskin; and on each side of his face a few strait locks hang down like, what are vulgarly called, 'rats' tails'.

f. = *rat's-tail fescue* (see 3 below).

1950 *N.Z. Jrnl. Agric.* Sept. 219/3 In a way the rapid ingress of ratstail into pastures during the 1930's was a good thing.

3. rat's tail cactus = *rat-tail cactus* s.v. *rat-tail 5; **rat's-tail fescue,** an annual grass, *Vulpia myuros.*

1957 *Dict. Gardening* (R. Hort. Soc.) I. 146/2 Rat's Tail Cactus. Stems weak, pendent, slender. **1958** S. H. Scott *Observer's Bk. Cacti* 61 *Aporocactus flagelliformis.* Usually referred to as the 'rat's tail cactus' because of its pendent method of growth. **1858** G. Bentham *Handbk. Brit. Flora* 602 Rat's-tail Fescue... A tufted annual, usually about a foot high. **1917** S. F. Armstrong *Brit. Grasses* vii. 100 Rat's-tail Fescue... An annual, occurring chiefly in waste places. **1944** W. J. Stokoe *Caterpillars Brit. Butterflies* 224 Rat's-tail Fescue Grass... This small tufted grass grows mostly in dry pastures. **1954** C. E. Hubbard *Grasses* 137 (*heading*) Rat's-tail Fescue.

rat-tail. Add: **3. b.** A rat-tailed spoon (see Rat-tailed *a.* 2). **c.** = Rat's-tail 2 d.

1974 *Country Life* 14 Nov. 1447/1 Spoons ranging from 1661 to 1718..include..split-ended rat-tails or treffid spoons. **1977** J. Wainwright *Day of Peppercorn Kill* 121 The woman's hair was in sodden rat-tails... Her shoes squelched water.

4. Substitute for def.: A deep-water marine fish belonging to the family Coryphænoididæ, esp. one of the genus *Macrurus,* characterized by a long, tapering tail. Also *attrib.* (Later examples.)

1905 D. S. Jordan *Guide to Study of Fishes* I. xii. 209 In the deep-sea allies of the codfishes, the grenadiers or rat-tails (*Macrouridæ*), the numbers [of vertebrae] range from 65 to 80. **1928** Russell & Yonge *Seas* iv. 91 That curious fish of the cod family known as the Macrurus or rat-tail.. spends the greater part of its life in the cold dark depths over the abyssal plain. **1936** P. S. Barnhart *Marine Fishes California* 24 Rat-tails. The fishes of this family are deep-water fishes, and are seldom seen or taken. **1956** J. L. B. Smith *Old Fourlegs* ix. 92 In this they [*sc.* seals] are like humans, who will not eat those perfectly wholesome but.. unfortunately named 'Rat Tails'. **1975** *New Yorker* 19 May 32/1 There might be a few rattails (bottom feeders related to the shark family) near the ocean floor. **1975** *Sci. Amer.* Oct. 86/3 The bait was visited by only a few eelpouts, brotulids and rattail fish (grenadiers).

5. (Further examples in sense 2 of Rat-tailed *a.*) **rat-tail cactus,** a pendent or creeping cactus, *Aporocactus flagelliformis,* native to central America and having spiny stems bearing scarlet flowers; **rat-tail comb,** a comb with a long tapering handle at one end; **rat-tail file** (earlier and later examples); **rat-tail radish** (example); cf. *rat-tailed *a.* 1 d.

1744 in *Maryland Hist. Mag.* (1926) XXI. 251, 6 Ratt Tale Files. **1867** *Gardeners' Chron.* 3 Aug. 807/1 (*heading*) Raphanus caudatus, or Rat-tail Radish. **1895** *Army & Navy Co-op. Soc. Price List* 754 (*caption*) Queen Anne or Rat-tail [spoons]. **1900** L. H. Bailey *Cycl. Amer. Hort.* I. 283/2 Rat-tail Cactus. Creeping or pendent, slender and very branching. **1904** *Daily Chron.* 20 July 5/6 The bride's father presented her with a superb tiara of diamonds and pearls, and a canteen of rat-tail silver. **1925** S. T. Warner *Espalier* 77 The rat-tail spoons, The china dishes. **1945** *Sun* (Baltimore) 30 Jan. 6-0/3 It was made in 1898 from a rat-tail file which had been used in a bicycle factory. **1946** M. Free *All about House Plants* xvii. 203 Rat-tail Cactus..has long flexible stems about ½ inch in diameter covered with bristly hairs. **1972** F. Perry *Flowers of World* 60/1 *Aporocactus flagelliformis,* the Rat-tail Cactus, is well named for the long flexible stems..hang down all round the plant. They are covered with bristly hairs and carry masses of crimson flowers. **1978** *Chatelaine* (Canada) Dec. 67/1 Tuck ends in with a rat-tail comb.

rat-tailed, *a.* **1. b.** For 'tail' substitute 'flexible respiratory organ resembling a tail'. In full, *rat-tailed maggot.* (Later examples.)

1895 L. C. Miall *Nat. Hist. Aquatic Insects* ii. 198 The Rat-tailed Maggot, a common inhabitant of stagnant pools. **1935** *Discovery* July 212/1 The problem [of breathing under water] had been solved long before by the rat-tailed maggot, with its telescopic tube reaching to the surface. **1952** J. Clegg *Freshwater Life Brit. Isles* ii. 27 The larvae of one or two insects, such as the Rat-tailed Maggot,..are well adapted for living in the black mud at the bottom of these unwholesome waters. **1968** *Oxf. Bk. Insects* 130/2 They swim freely, breathing through their long tails, which can be extended to 6 inches to reach the water surface—hence their name Rat-tailed Maggots.

d. In the names of certain plants, esp. *rat-tailed radish,* an Asian radish, *Raphanus caudatus,* cultivated for its edible fruit. Cf. Rat-tail 5 in Dict. and Suppl.

1867 *Gardeners' Chron.* 3 Aug. 807/1, I shall continue to grow the Rat-tailed Radish. **1885** W. Robinson tr. *Vilmorin-Andrieux's Veg. Garden* 499 Rat-tailed Radish... The edible part of this Radish is not the root, but the silique or seed-vessel, which is gathered before it is fully grown. **1949** *Nat. Geogr. Mag.* Aug. 213/2 In India the rat-tailed radish..is grown for its fleshy, edible seed pods. **1969** *Oxf. Bk. Food Plants* 170/2 The Rat-tailed Radish..is grown in southern Asia. The part eaten is not the root but the fruit, which reaches a length of 8 to 10 inches.

2. b. Of a comb: having a long, tapering, handle at one end. Cf. *rat-tail comb* s.v. *rat-tail 5.

1973 *Daily Colonist* (Victoria, B.C.) 15 July 1/3 It turned out to be an ordinary, black rat-tailed comb.

rattan, ratan, *sb.*[1] Add: **4.** *rattan chair, furniture, mat, rocker, rope, screen, ware.*

1879 *Harper's Mag.* July 211 In the large parlor..with rattan chairs galore..presided Karl Whitaker. **1925** W. S. Maugham *The Letter* i. 9 The room is..quite simply furnished with rattan chairs, in which are cushions. **1972** D. Bloodworth *Any Number can Play* xv. 140 Comfortable rattan chairs. **1895** *Montgomery Ward & Co. Catal.* Spring & Summer 617/3 (*heading*) Rattan and reed furniture. **1966** D. Forbes *Heart of Malaya* ii. 31 The old kind, built on stilts like a Malay house, with wide verandahs and rattan furniture, is still the best. **1925** W. S. Maugham *The Letter* i. 9 Rattan mats on the floor. **1895** *Montgomery Ward & Co. Catal.* Spring & Summer 617/3 We show a larger assortment of rattan rockers in our special Furniture Catalogue. **1900** W. W. Skeat *Malay Magic* 172 Six or eight coils of rattan rope..are placed on a triangle formed with three rice-pounders. **1902** Conrad *Youth* 205 The straggling building of bricks,..resounded with the incessant flapping of rattan screens. **1971** K. Hopkins *Hong Kong* 247 Industries in which collective agreements have been signed include.. rattan ware.

Hence **ratta·nning,** chastisement with rattan sticks.

1847 H. Melville *Omoo* xxix. 110 The ratanning of the young culprits..may also be considered as in some measure characteristic of the [French] nation.

rat-tat, *sb.* Add: **c.** *Comb.*, as **rat-tat ginger** = *rat-a-tat ginger* s.v. *rat-a-tat.

1962 M. Duffy *That's how it Was* ii. 26 Rat-tat ginger, two door-knockers tied together and a piece of black cotton leading round the corner so I could knock on two doors at once.

rat-tat-tat, etc.: (further examples); also freq. used to represent the noise of reports from fire-arms.

1793 S. E. Phillips *Let.* 6 May in F. Burney *Jrnls. & Lett.* (1972) II. 109 The dear Postman is just arrived... A loud *rattat tattoo* was heard at the door. **1907** G. Manington *Soldier of Legion* iii. 127 The sombre background was punctuated again and again..by lightning like red flashes. Rat! tat! tat! tat!... These were Winchesters. **1957** L. MacNeice *Visitations* 42 Rat-tat-tat-tash of shields upon Ida. **1972** *Angling Times* 6 Apr. 14/5 It is only a matter of time before I get that tiny rat-tat-tat on the rod tip. **1974** M. Butterworth *Man in Sopwith Camel* i. 12 Rat-tat-tat-tat, his rear gunner was spraying a libation back on to Braithwaite's riddled body.

Hence **rat-ta·t(-tat, etc.)** *v. intr.* (and *trans.*), to (cause to) make a sharp rapping sound, to knock.

1910 *Daily Chron.* 14 Apr. 9/5 The lady rat-tat-tatted for half an hour. Then the housekeeper..sternly asked the visitor to be so good as to go away. **1916** H. S. Walpole *Dark Forest* i. vii. 188 A machine gun 'rat-tat-tat-tated' close to us. **1953** 'N. Blake' *Dreadful Hollow* 106 Nigel rat-tat-tatted an imaginary tommy-gun at them. **1966** A. Cavanaugh *Children are Gone* ii. vi. 49 'Well, g'night.' Her heels rat-tatted down the hall.

ratten-, ratting-crook. (Later examples.)

1866 E. L. Linton *Lizzie Lorton of Greyrigg* I. xii. 268 Two large iron 'ratten-crooks' and several smaller ones, for cauldrons and kettles, hung from the 'rannel balk'. **1974** *Country Life* 17 Oct. 1149/2 A *ratten-crook* is a pot crane hanging from the *rannel-balk* which is the wooden beam across and above an open hearth.

ratter. 1. For 'a dog' in Dict., read 'a dog, cat, or other animal', and add earlier and later examples.

1857 S. H. Hammond *Wild Northern Scenes* xi. 121 There was an assemblage of all the cats in that part of the town... Off at the right was an old spotted ratter. **1946** E. O'Neill *Iceman Cometh* iv. 233 With that..line of bull, you ought to be able to sell skunks for good ratters! **1972** R. Adams *Watership Down* xxv. 190 The farm cat was bewildered by the speed and fury of Bigwig's charge. It was no weakling and a good ratter.

3. *Austral.* One who steals opal from another's mine.

1931 I. L. Idriess *Prospecting for Gold* xxvi. 239 Ratters are men, a gang as a rule, who work your opal out for you while you sleep. **1964** W. C. Eyles *Bk. of Opals* vii. 83 When the miners..went down the shaft, they found the ratters had cleaned the place out entirely. **1976** *Nat. Geographic* Oct. 564/2 The mine ratters (thieves) are here, but that's been going on since King Solomon's mines.

ratticide, var. *raticide.

rattinet. For **1832** read **1838** and add earlier and later examples.

Appears to be *Obs.*

1811 *Weekly Reg.* 21 Sept. 46/1 Rattinets..can only be made of wool long enough to be combed. **1836** H. Manwaring *Tailor's New Guide* 15 Velvetteen Jacket, body and skirt, may be lined with rattinet.

ratting, *vbl. sb.* **1.** (Later examples.)

1946 *Sun* (Baltimore) 17 June 18/3 His turning long-distance state's evidence, his ratting, so to speak, to the cops..strikes at the very core of the latest design to maintain 'power' in the hands of a people's government. **1948** *Richmond* (Va.) *Times-Dispatch* 13 Feb. 30/1 McKeever.. performed the most odious job of 'ratting' college football has known.

3. *Comb.*, as **ratting canoe** *Canad.*, a small native boat designed for hunting muskrats in swamps and marshes.

1962 R. Slobodin *Band Organization of Peel River Kutchin* 14 A child armed with a .22 rifle, paddling or portaging the light, narrow ten-foot ratting canoe. **1968** R. M. Patterson *Finlay's River* 74 A—gave an initial display of his dexterity by upsetting a small ratting canoe with Butler and himself in it.

rattle, *sb.*[1] Add: **II. 4. a.** Also **rattle-rattle.** (Further examples.) *fig.* phr. *with a rattle*: with sudden or unexpected rapidity (orig. *Horse-racing*).

1888 *Daily Chron.* 10 Dec. 6/2 Bachelor came on with a rattle and won by a length and a half. **1909** in J. R. Ware *Passing Eng.* 206/2 The only approach to a sensation was caused by Warrington and Kettleholder, the former coming 'with a rattle' in the morning to the price taken about him in the excitement caused by his forward running in the Cesarewitch. **1926** E. Bowen *Ann Lee's* 251 Only the rattle-rattle of my bicycle. **1928** D. H. Lawrence *Lady Chatterley's Lover* ii. 11 She heard the rattle-rattle of the screens at the pit. **1977** *Evening Gaz.* (Middlesbrough) 11 Jan. 13/5 The Merryweather crew came with a rattle to level at the penultimate end 15–15 [in Bowls].

e. *N. Amer.* A succession of small, noisy waterfalls forming rapids; a fast-moving stream.

1776 G. Cartwright *Jrnl.* 14 Aug. (1779) 200 We fished in the stream below the rattle, and also in the lower pool, and killed seventy-eight fish. **1861** L. de Boilieu *Recoll.*

Labrador Life xiii. 166 In the different bays are brooks, and in these brooks are 'rattles', as they are termed, or, more properly speaking, 'falls', though none are of any great magnitude. **1907** J. G. Millais *Newfoundland* iii. 70 We had only to unload twice in passing 'rattles', as they called the strong rapids. **1925** *Dialect Notes* V. 339 *Rattle,*..a swift brook. **1975** *Canad. Antiques Collector* Mar.–Apr. 23/1 From the sealhunt we have:..rattle, river rapids, and so on.

f. The rustling quality of a sheet of finished paper when handled, indicative of its hardness and density.

1900 Cross & Bevan *Paper-Making* (ed. 2) v. 137 As a consequence, it adds the quality of 'wetness' to the pulp, which again confers the quality of hardness and 'rattle' upon the finished paper. **1962** F. T. Day *Introd. to Paper* ii. 24 Starch is added to paper furnishes and serves either as a sizing agent or to give the paper more substance and better handle for its use imparts stiffness and 'rattle' to the finished sheet.

g. *Hunting.* A particular note on the horn.

1908 L. C. F. Cameron *Otters & Otter Hunting* 203 *Rattle,* the note sounded on the horn at the 'worry'. **1954** J. I. Lloyd *Beagling* 143 *Rattle,* an exciting, vibrant sounding of the horn. **1976** *Shooting Times & Country Mag.* 16–22 Dec. 25/2 A rattle on the horn had hounds racing to the spot.

5. b. (Later example.)

1842 C. Ridley *Let.* in *Cecilia* (1958) ix. 111 Wells..is tiresome again... I wish I had courage to give her a good rattle, but if I did I think she would not bear it.

6. b. (Earlier example.)

1748 Richardson *Clarissa* III. 127 Sir, said I, I see what a man I am with. Your rattle warns me of the snake.

7. (Earlier and later examples.)

1716 D. Ryder *Diary* 17 May (1939) 235, I was vexed to see her so long entertained with such a rattle as he. **1969** *N.Y. Rev. Bks.* 2 Jan. 3/4 Editor of a biographical history of philosophy yet welcomed as a rattle and raconteur..Lewes stands in these pages like a wax effigy. **1971** E. Mavor *Ladies of Llangollen* i. 33 Great confidante, greater rattle, she was ever recording..what she was pleased to call 'boosey' whist parties beneath the Woodstock oaks.

9*. *Naut.* phr. *in the rattle*: on the commander's report of defaulters; in confinement; in trouble.

1914 'Bartimeus' *Naval Occasions* ii. 10 'In the bloomin' rattle, I am,' explained the disturber of traffic. **1919** W. Lang *Sea-Lawyer's Log* xii. 152 Ordinary Seaman Oldroyd spent the first dog-watch last night..washing his undergarments, but, having done so, he hung the same up to dry in the fore ammunition lobby, where they were subsequently discovered by the Gunner, who promptly placed Oldroyd 'in the rattle', hence his appearance as a defaulter. **1942** *Penguin New Writing* XV. 13 He was taken off, bawled out, put in the rattle. **1951** H. Hastings in *Plays of Year 1950* IV. 72 You ain't gonna put him in the rattle on account of a bit of leg-pull? **1964** J. Hale *Grudge Fight* vi. 91 The Andrew, that had taken him round the world a few times, given him his good conduct stripes and removed them when he'd been in the rattle. **1973** 'B. Mather' *Snowline* xviii. 212 The Old Man..let the others out, but..your bloke is back in the rattle.

10. rattle-box, (*c*) (earlier and later examples); (*d*) *transf.*, applied to a conveyance or machine; **rattle-free** *a.*, devoid of rattles; **rattleproof** *a.*, capable of preventing rattling; hence **rattleproofing** *vbl. sb.*; **rattle-weed,** (*a*) substitute for def.: the bugbane, *Cimicifuga racemosa*; (*c*) = Loco[1]; (*d*) = *rattle-box* (*c*) in Dict. and Suppl.; (earlier and later examples).

1817 A. Eaton *Man. Bot.* 80 *Crotalaria..sagittalis,* (rattle-box) leaves lance-oblong. **1835** J. E. Alexander *Sk. in Portugal* viii. 179 In May, the fleet of her Most Faithful Majesty consisted of the following ships:—..18, Audax,.. Fine, stout brig, but very ugly. 16. Providenza, .. Ditto, a perfect rattle-box. **1884** [see *loco *sb.*[1] a]. **1929** M. A. Gill *Underworld Slang, Rattle box,* machine gun. **1943** R. Holt *George Washington Carver* 199 He would caution stockmen against the rattlebox (*Crotalaria*). **1972** G. Beine *Land of Coyote* 90 These are rattleboxes, and there, some rabbitsfoot clover. **1973** 'H. Howard' *Highway to Murder* vii. 87 He was crowding ninety and so was his rattlebox. **1962** *Times* 3 May 19/4 It [*sc.* a car] is impressively quiet throughout..completely rattle-free and draughtproof. **1924** *Motor* 21 Oct. 626/1 Table utensils held in rattleproof devices. **1976** *Norwich Mercury* 19 Nov. 8/5 (Advt.), But Ziebart is rustproofing and soundproofing..and squeak-proofing and rattleproofing. **1791** *Trans. Amer. Philos. Soc.* III. 114 American Bane-berry, Black Snake-root, Rattle-weed. **1851** R. Glisan *Jrnl. Army Life* (1874) vi. 70 The rattle-weed..derives its name from the fact that its pod is full of loose seed, and makes a rattling noise when dry. **1864** *Rep. Maine Board Agric.* 45 Last year nothing grew on the field where it had been applied but rattle-weed. **1931** W. N. Clute *Common Names Plants* 110 *Crotalaria sagittalis*..is frequently known as rattle-box or rattle-weed.

rattle, *v.*[1] Add: **I. 4. a.** Also with *about, around,* esp. *transf.* and *fig.*, implying the occupation of an area or space larger than that which is comfortable, necessary, or desirable.

1869 L. M. Alcott *Little Women* II. iii. 43, I saw you two girls rattling about in the what-you-call-it [*sc.* charabanc], like two little kernels in a very big nutshell. **1926** M. J. Atkinson in J. F. Dobie *Rainbow in Morning* (1965) 81 He rattles around in his office like one pea in a pod. **1967** T. Stoppard *Rosencrantz & Guildenstern are Dead* III. 95 We can move,..change direction, rattle about, but our movement is contained within a larger one that carries us along. **1973** *Washington Post* 13 Jan. A23/6, I don't want that kind of power rattling around inside the bureaucracy.

II. 5. d. *Cricket.* To bowl *down* the opposing team's wickets speedily and cheaply; to skittle *out* batsmen in a similar manner.

a **1842** B. Aislabie in P. Norman *Scores & Ann. W. Kent Cricket Club* (1897) 370 M was a Morgan, who rattled them down. **1862** *Baily's Mag.* Apr. 259 Caffyn and Bennett rattled down their wickets..for 20 runs. **1873** *Ibid.*, July 409 In the second innings the two fast bowlers..rattled out the Marylebone men in grand style. **1898** G. Giffen *With Bat & Ball* vii. 94 On the sticky wicket..Hearne and Poughet 'rattled' us out. **1926** H. S. Altham *Hist. Cricket* xviii. 207 He..saw Kent rattled out by Painter and Roberts for 76.

6. d. To fire (bullets) rapidly; to carry off (a person) by firing.

1890 Kipling in *Scots Observer* 12 July 200/2 If a beggar can't march, why, we [*sc.* machine-guns] kills 'im an' rattles 'im into 'is grave. **1916** 'Boyd Cable' *Action Front* 198 He rattled off burst after burst of fire.

7. a. Delete † *Obs.* and add earlier and later examples.

1542 N. Udall *Erasmus's Apophthegmes* sig. K5, How Diogenes ratleed & shooke vp couetous persones. **1931** S. W. Ryder *Blue Water Ventures* xvi. 217 He should have rattled his officer-of-the-watch for slackness.

9. *to rattle off* (further examples); also spec. *Cricket*: to score or 'knock off' with ease (the runs necessary for victory); *to rattle up* (chiefly *Cricket*): to score rapidly, within a certain time, or before enforced retirement.

1860 *Baily's Mag.* Sept. 427 Captain Bathurst, in the fine old family style, rattled up 10 and 21. **1875** *Ibid.* June 108 Ultimately the South were left with about 40 to get to win, and Mr. W. G. Grace and Jupp rattled off these without difficulty. **1896** G. B. Shaw *Let.* 15 Feb. (1965) I. 597, I do not make a third of the income expected by men who rattle off their copy at anything from 20/- to 40/- a thousand. **1926** H. S. Altham *Hist. Cricket* xviii. 208 Jackson and Sellars rattled up 24 in a quarter of an hour. **1973** *Advocate-News* (Barbados) 20 Feb. 14/5 Such an 'uncertainty' would take the form of a dramatic batting collapse, giving the Australians enough time to rattle up a good second innings score. **1976** *0–10 Cricket Scene* (Austral.) 30/2 And to show he has lost none of his zest for runs, he rattled off scores of 171 not out, 12, 114 not out and 36 in the World Cup series in England.

10. (Further examples.)

1977 J. Laker *One-Day Cricket* 66 The Sri Lankans rattled the score along. **1977** *Sunday Times* 9 Jan. 28/6 They rattled their reply of 240 for four to the Bangladesh score of 266 for nine declared, at more than four runs an over.

11. For *U.S.* read 'orig. *U.S.*' and add earlier and later examples. Also, to irritate, to 'nettle'.

1869 J. R. Browne *Adventures Apache Country* xxviii. 282, I think he was slightly rattled by the formidable appearance of our escort. **1904** F. Lynde *Grafters* xxviii. 360 For once in a way the ex-district attorney was too nearly rattled to be fully alert to his surroundings. **1905** *Pall Mall Mag.* Nov. 546/1, I don't see you need be rattled. **1927** M. de la Roche *Jalna* xxii. 276 Don't be a duffer... The more Piers sees he can rattle you the more he'll do it. **1928** E. Wallace *Double* iv. 52 Why the devil are they bothering me? There's something about this business that is rattling me. **1936** P. Fleming *News from Tartary* 65 But I had the empty satisfaction of seeing that I had (slightly) rattled Pai. **1959** E. H. Clements *High Tension* v. 82 Trust a woman to put her oar in! That's got Alister nicely rattled! **1977** J. F. Fixx *Compl. Bk. Running* p. xviii, I was less easily rattled by unexpected frustrations.

rattled (ræ·t'ld), *ppl. a.* [f. Rattle *v.*[1] + -ed[1].] Agitated, confused, frightened. Also *Comb.*

1910 *N.Y. Even. Post* 10 Feb. 8/1 The plight of Ohio's rattled Republicans is enough to win grimy tears from the stony basilisk. *a* **1974** R. Crossman *Diaries* (1976) II. 413 'Oh God!' I thought, 'Harold's in such a rattled state, with Wigg on one side and poor Marcia and Gerald on the other.' **1977** W. M. Spackman *Armful of Warm Girl* 62 It had seemed to her that he, rather, had been rattled-sounding.

rattle-mouse. 1. (Later example.)

1960 M. Burton *Wild Animals Brit. Isles* 48 In the Isle of Wight it [*sc.* the serotine bat] is known as rattle-mouse.

rattler. Add: **1. c.** (Earlier and later examples.)

1709 W. King *Useful Trans. Philos.* II. sig. A 2[v], Nothing could be more useful than a full..Inspection of Human Tongues... It is hop'd that if any Persons know themselves to be..Tongue-Padds, Spokesmen, Rattlers, Bouncers, &c. they would..bequeath their Tongues to be dissected. **1959** *She* May 21/3 *Rattler,* great talker.

2. b. (*a*) (Further example); (*b*) *gen.*, any (rattling) form of transport, esp. a train.

1829 P. Egan *Boxiana* 2nd Ser. II. 674 Boscoe made his appearance in a rattler, with four prime prads. **1871** *Lakeside Monthly* Oct. 323/1, I am going on the rattlers tomorrow to nick a lot of flats and molls. **1903** A. M. Binstead *Pitcher in Paradise* viii. 193 On the followin' Saturday afternoon I took the rattler down to Aldershot. **1904** 'No. 1500' *Life in Sing Sing* 252/1 *Rattler,* a car. **1922** R. Parrish *Case & Girl* 333 We caught another rattler two hours later, and got off at Patacne. **1924** D. H. Lawrence *England, my England* 102 Miss Stokes had a punchture. 'Let me wheel the rattler,' said Albert. **1936** I. L. Idriess *Cattle King* xli. 348 'Well, sir, for last year you certainly paid your fare on the railways.' 'I've never jumped the 'rattler' in my life.' **1951** *Collier's* 17 Nov. 8/2 We're rolling across the country in a very luxurious rattler. **1966** 'L. Lane' *ABZ of Scouse* II. 88 *Rattler,* a tram or street-car. **1977** 'J. Fraser' *Hearts Ease* v. 38 'Where's the ambulance?' 'We sent it away... The first one was an old rattler. We've sent for the Daimler which has better springing.'

c. For *U.S.* read 'orig. *U.S.*' (Later examples.)

1909 *Chambers's Jrnl.* July 431/2 Many of the little snakes of the tropics are as poisonous as the dreaded rattler. **1918** W. Cather *My Antonia* i. ii. 17 She had killed a good many rattlers. **1949** G. B. Shaw *Buoyant Billions* iv. 48 You cannot charm the rattlers and gaters as I can. **1956** L. M. Klauber *Rattlesnakes* I. i. 11 Some believe any snake that vibrates its tail when angry or alarmed to be a rattler. **1963** D. P. Mannix *All Creatures Great & Small* xi. 190 A friend sent me a very fine Mexican green rattler nearly six feet long. **1978** P. Theroux *Picture Palace* ii. 8, I was moving round the room, hunched like a cowboy that hears a rattler.

3. a. (Earlier example in sense of 'storm'.)

1835 T. Power *Jrnl.* 25 Mar. in *Impressions Amer.* (1836) ii. 266 Our breeze freshened gradually all the evening, until by midnight it blew a rattler.

b. Also *gen.*, anyone or anything remarkably good or able, esp. with regard to speed.

1853 F. Gale *Public School Matches* 13 The first ball is well pitched and comes in a rattler to the middle stump. **1883** 'Mark Twain' *Life on Mississippi* xxiv. 271 That 'Cyclone' was a rattler to go, and the sweetest thing to steer that ever walked the waters. **1886** M. Thompson *Banker of Bankersville* ix. 134 Your partner is a whole team, hain't he? He's a rattler! **1894** 'Mark Twain' in *St. Nicholas* Mar. 395 It was a rattler, that caravan, and a mighty fine sight to look at. **1917** H. Garland *Son of Middle Border* xxiv. 290 You may consider yourself hired for as long as you please to stay. You're a rattler.

d. A long, resounding word.

1865 'Mark Twain' in *Californian* 18 Mar. 8/1 One of them rattlers with a clatter of syllables as long as a string of sluice boxes.

5. *attrib.* and *Comb.*, as (sense 2 c) *rattler hatband*; **rattler-jumper,** one who jumps (Jump *v.* 6 b) a train; so **rattler-jumping** *vbl. sb.*

1978 *Detroit Free Press* 5 Mar. 23/1 'One day when we were in a local shop, the owner remarked that he would pay $10 for a rattler hatband.' That casual remark launched the couple into a business. **1934** *Bulletin* (Sydney) 7 Mar. 33/2 It looks as though the Queensland Government will have to run special trains to cope with 'rattler'-jumpers, who nowadays travel in packs. **1933** *Ibid.* 3 May 20/1 Of all vocations rattler-jumping is the least easy.

rattlesnake, *sb.* Substitute for first part of def.: A venomous American pit viper belonging to the genus *Sistrurus* or *Crotalus* of the family Crotalidæ. (Later examples.)

1910 *Encycl. Relig. & Ethics* III. 143/1 The rattlesnake-doctor, who cured or prevented the bite of the rattlesnake, was usually distinct from other medicine-men. **1932** W. Faulkner *Light in August* ii. 29 He carried with him his own inescapable warning, like a flower its scent or a rattlesnake its rattle. **1956** L. M. Klauber *Rattlesnakes* I. i. 13 All rattlesnakes have rattles, and no other kind of snake has them. **1975** R. L. Beals *Peasant Marketing System of Oaxaca, Mexico* iii. 31 The rattlesnake was associated with the rain cult.

b. rattlesnake fern (earlier and later examples); **rattlesnake grass** (examples); **rattlesnake herb** (earlier examples); **rattlesnake leaf** (see quots.); **rattlesnake('s) master** (examples); **rattlesnake orchid,** an epiphytic orchid of the genus *Pholidota,* esp. *P. imbricata,* which is native to parts of south-east Asia and bears pendant racemes of light brown flowers; **rattlesnake plantain** (earlier and later examples); **rattlesnake root,** (*c*) one of several other plants believed to help cure the effect of rattlesnake bites; (later examples); **rattlesnake weed** (*c*) = **rattlesnake root* (*c*); (later examples).

1814 F. Pursh *Flora Amer.* II. 656 *Botrychium virginianum*..is known by the name of Rattle Snake Fern. **1931** W. N. Clute *Common Names Plants* 109 The spore-cases of one of our ferns are borne in spikes that so strongly suggest the rattles of the rattlesnake that it is commonly known as the rattlesnake fern. **1814** J. Bigelow *Florula Bostoniensis* 25 Rattlesnake grass... A large grass found in meadows and readily recognized by its swelling spikelets. **1878** J. B. Killebrew *Grasses of Tennessee* 232 Rattlesnake Grass..resembles quaking grass very much. **1736** B. Franklin *Poor Richard 1737* 3 (*caption*) Rattle-Snake Herb. **1763** tr. *L. du Pratz's Hist. Louisiana* II. 43 The Rattle-snake-herb has a bulbous root like that of a tuberose, but twice as large. **1822** A. Eaton *Man. Bot.* (ed. 3) 294 *Goodyera pubescens,* rattle-snake leaf, scrophula-weed. **1829** A. H. Lincoln *Familiar Lect. Bot.* 288 *Goodyera pubescens,* rattle-snake leaf. **1806** *Farmer's Calendar* sig. D4, Notwithstanding a free use of sweet oil, plantane, hoarhound, *prenanthet alba,* called here rattlesnake's master, &c. the swelling and pain progressed. **1836** M. Holley *Texas* v. 103 A root called rattlesnake's master grows abundantly in the pine woods and is said to be an efficient remedy. **1843** F. Marryat *Narr. Trav. & Adv. M. Violet* II. vi. 134 Close to my feet I beheld five or six stems of the rattlesnake master weed. **1899** H. B. Cushing *Hist. Choctaw Indians* 229 They [*sc.* the Choctaws] possessed an antidote for the bite and sting of snakes and insects, in the root of a plant called rattle snake's master. **1943** D. C. Peattie *Great Smokies* 189 Thus a little orchid, the rattlesnake plantain, with net-veined leaves, looks enough like a snakeskin to suggest that it may be a 'rattlesnake master' or cure for snake bites. **1887** G. Nicholson *Illustr. Dict. Gardening* III. 105/1 *Pholidota*..Rattlesnake Orchid. **1903** H. J. Chapman *Watson's Orchids* (ed. 2) 430 It [*sc. Pholidota*] is commonly known as the Rattlesnake Orchid. **1965** A. D. Hawkes *Encycl. Cultivated Orchids* 369/1 The persistent or deciduous, large, concave bracts which occur in most species—often

almost hiding the blossoms—give the common name of 'Rattlesnake Orchid' to at least some of the cultivated Pholidotas. **1778** J. Carver *Trav. Interior Parts N. Amer.* 482 The Rattle Snake Plantain, an approved antidote to the poison of this creature. **1943** [see **rattlesnake master*]. **1972** *Islander* (Victoria, B.C.) 16 Apr. 16/3 Rattlesnake plantain ..is a denizen of the woods. **1889** *Cent. Dict.* 786/3 Cancer-weed, the rattlesnake root, *Prenanthes alba,* of the United States, a milky-juiced composite having an intensely bitter root. **1941** R. S. Walker *Lookout* 48 Among the wild plants once employed as antidotes for the bites of poisonous reptiles are..Virginia snakeroot, button snakeroot, and rattlesnake-root. **1885** *Outing* Nov. 180/1 A pretty thing sends a creeping feeling down our backs, because it is rattlesnake weed. **1936** G. A. Reichard *Navajo Shepherd & Weaver* 45 A yellow-green commonly seen is made by brewing the leaves and stems of one of the goldenrods (Bigelovia) called by some Whites 'tall rattlesnake weed'.

II. *transf.* **2.** In full, *rattlesnake cocktail, whisk(e)y.* A potent alcoholic drink or cocktail.

1862 [see *cobbler 3]. **1867** T. C. Baker *Hist. U.S. Secret Service* xix. 246 It is hardly worth the while to present to the Government a bill for a few decanters and rattlesnake whisky. **1903** A. M. Binstead *Pitcher in Paradise* iv. 110 He went from bar to bar drinking 'rattlesnake cocktails'. **1930** *Savoy Cocktail Bk.* 132 Rattlesnake Cocktail... So called because it will either cure Rattlesnake bite, or kill Rattlesnakes, or make you see them. **1947** *Daily Progress* (Charlottesville, Va.) 3 Sept. 8/5 A bottle of 'rattlesnake whiskey' was seized in a recent Chinatown raid by the Federal Bureau of Narcotics, which offers this recipe for the concoction: Place a live rattlesnake in a large jar; add rice wine, dried toads, soy beans, sliced deer antlers and a handful of dried sea horses. Age six months. **1953** T. Shane *Bar Guide* ii. 31 Rattlesnake. 2 dashes Pernod. 1 tsp. Lemon Juice. ½ tsp. Powdered Sugar. ½ Egg White. 1 oz. Rye. Shake with cracked ice and strain.

rattlesnake (ræ·t'lˌsnēɪk), *v. rare.* [f. prec.]

1. *trans.* To deceive or trick, as with the cunning of a serpent.

1818 Keats *Let.* 3 Feb. (1958) I. 223 We must cut this, and not be rattlesnaked into any more of the like.

2. *trans.* and *intr.* To snake (see Snake *v.*[1] 4) with a rattling sound; to travel like a rattlesnake.

1961 M. Spark *Prime of Miss Jean Brodie* ii. 23 The evening paper rattle-snaked its way through the letter box. **1981** J. Barnett *Firing Squad* ii. 155 Messerschmitts came in at low level, rattlesnaking along the beaches.

rattling, *ppl. a.* Add: **4. c.** (Later examples.)

1885 *Punch* 4 July 4/1 You do see some rattling pretty, fresh faces. **1930** A. G. Hays in W. E. Weeks *All in Racket* 13 This is a rattling good story. **1978** *Jrnl. R. Soc. Arts* CXXVI. 636/1 Herkomer's *The Last Muster* is a rattling good picture.

6. *Comb.*, as *rattling-boned* adj.

1933 E. Sitwell *Eng. Eccentrics* ii. 42, I am afraid the ancient and rattling-boned gallant rather gloried in this fall from grace.

rat-trap. Add: **1. a.** Also *attrib.* in sense 'resembling a rat-trap'.

1904 *Westm. Gaz.* 3 Mar. 1/3 A gaunt man with a rat-trap face. **1907** *Ibid.* 20 Sept. 4/2 His [*sc.* a pike's] rat-trap jaws. **1978** R. Westall *Devil on Road* xxiii. 217 He had a rat-trap jaw and little deep-set eyes.

b. *transf.* A shabby or ramshackle building or dwelling. *colloq.*

1838 De Quincey in *Blackw. Mag.* XLIII. 21/2 Ay; but mind you,—put case that he or that she should die in this rat trap before sentence is passed. **1876** H. T. Williams *Pacific Tourist* 205 The following are among the..oddities which have, through miners' freaks and fancies, been used to denote settlements and camps and diggings, small or large:..Rat-Trap Slide. **1892** 'Mark Twain' *Amer. Claimant* iv. 56 It wouldn't have occurred to anybody else to name this poor old rat-trap Rossmore Towers. **1974** *Amer. Speech 1971* XLVI. 77 *Shabby hotel,*..flop joint, joint, rat trap.

2. (Further examples.)

1931 H. W. Bartleet *Bartleet's Bicycle Bk.* 33 Every practical cyclist who saw my pedals praised the scheme of rattrap and rubber in combination. **1974** *Sumter* (S. Carolina) *Daily Item* 24 Apr. (Western Auto Advts. Suppl.) 4 Chromed rat trap racing pedals! **1976** *Billings* (Montana) *Gaz.* 28 June 2-c (Advt.), Rat trap pedals... Big 'knobby' tread tires.

3. *Building.* Applied *attrib.* to a form of Flemish bond (Bond *sb.*[1] 13 a) in which the bricks are laid on edge and the 'headers' span the whole thickness of wall, dividing the wall cavity into square spaces.

1932 E. Gunn *Economy in House Design* ix. 31 Brickwork beneath the tile hanging should be built in rat trap bond. **1939** —— *Building Technique* 39/1 Rat-trap or box-bond... This form of walling, which may be briefly described as Flemish-bond with all bricks laid on edge, is well-recognised in Sussex and adjoining counties where tile-hung walls are common. **1974** *Bricks* (Brick Development Assoc.) i. 33 (*heading*) Rat-trap bond.

ratty, *a.* Add: **2. a.** (Earlier and later examples.)

1867 'Mark Twain' *Notebk.* (1935) viii. 99 Village of Bethany... It is fearfully ratty—some houses—mud. **1884** —— *Huck. Finn* ix. 78 We got..a ratty old bed-quilt off the bed. **1962** J. D. MacDonald *Girl* xii. 185 Pooty-Tat sat on a ratty couch. **1969** E. B. White *Let.* 2 Nov. (1976) 586 When he received me, in a ratty apartment in the West Seventies or thereabouts, he was wearing toga and sandals. **1970** 'D. Halliday' *Dolly & Cookie Bird* iv. 51 A ratty half-dozen people had spilled out of the bar-café... A lot of money changed hands rather quickly. **1974** R. M. Pirsig *Zen & Art of Motorcycle Maintenance* III. xxvi. 317 John always kept his BMW spic and span. It really did look nice, while mine's always a little ratty, it seems.

b. *colloq.* Ill-tempered, irritated, angry.

1909 M. B. Saunders *Litany Lane* xvi. 215 Shut up. She's ratty. **1913** H. S. Walpole *Fortitude* i. iv. 53 All right, you needn't be ratty about it! **1929** W. P. Ridge *Affect. Regards* 226 Have I ever got ratty with you, Elsie? **1976** T. Heald *Let Sleeping Dogs Die* vi. 122 I'd simply have asked her what the hell she was so ratty about.

3. *Austral.* and *N.Z. colloq.* Mad, eccentric, silly. Phr. *to be ratty over*: to be infatuated with.

1900 H. Lawson *On Track* 75 Trav'lers and strangers failed to see anything uncommonly ratty about him. **1906** E. Dyson *Fact'ry 'Ands* xiv. 184 Already the Beauties had decided that Connie was 'as ratty as rabbits'. **1922** A. Wright *Colt from Country* 86 There was a rough-up in a pub; he got a knock, had a fit, and went real ratty, and that was the end of him. **1941** Baker *Dict. Austral. Slang* 59 *Ratty,* stupid, silly. *Ratty over* (a person), infatuated with.

4. *Comb.*, as *ratty-looking* adj.

1884 'Mark Twain' *Huck. Finn* xix. 182 Both of them had big, fat, ratty-looking carpet-bags.

Hence **ra·ttily** *adv.*, ill-temperedly, irritably.

1977 T. Heald *Just Desserts* vii. 169 'We're supposed to be buddies,' said Bognor, rattily. **1980** *Times Lit. Suppl.* 4 July 754/2 Compliments fly and are quoted in abundance in this symposium. One reads on, rattily, in the hope of meeting some unusual personal habit,..some outstanding trait of character or behaviour.

‖ **ratu** (ratu·). Also **ratoo.** [Indonesian.] The title of a petty monarch or native regional ruler in parts of Indonesia and in Fiji.

1798 S. H. Wilcocke tr. *Stavorinus's Voy. E. Indies* I. ii. i. 214, I..appoint the said *pangorang,* to be *pangorang ratoo,* or hereditary prince, and heir to the crown and the whole empire of Bantam. **1820** J. Crawfurd *Hist. Indian Archipelago* III. viii. i. 22 The genuine native term for king in Javanese is *Ratu,* which is the same word that is written *Datu* in some other languages. **1821** J. Leyden tr. *Malay Annals* 157 The toddy-man's son became ratu of Majapahit. **1880** *Encycl. Brit.* XIII. 607/2 Ratu Loro Kidul is princess of the southern sea, and has her seat among the caves and fiords of the southern coast. **1907** *Daily Chron.* 30 Nov. 4/6 They are captained by Ratu Kadavu Levu, the grandson of the last native monarch of Fiji, Thakombau. Ratu is Fijian for Prince. **1937** M. Covarrubias *Island of Bali* (1972) viii. 250 Finally it is time for the prince, the *ratu,* to appear; the *patih* recites his praises and..begs him to enter. **1966** *Economist* 16 Apr. 232/1 General Suharto, Indonesia's strong man, has found a way of remaining loyal both to his guru..and to his ratu (king). **1977** *South China Morning Post* (Hong Kong) 14 Apr. (Business Suppl.) 11/3 Ratu Sir Kamisese Mara of Fiji, representing the developing countries, said in reply [etc.].

‖ **rature** (ratür). *rare.* [Fr.] A scribal erasure or deletion.

1931 T. S. Eliot *Charles Whibley* 5 He..composed rapidly..and made very few *ratures* or corrections.

Raudive (rǭdī·v). The name of Konstantin *Raudive* (1909–74), Latvian psychologist, used *attrib.* in connection with a phenomenon involving tape recordings of sounds said to represent voices of paranormal origin.

1971 *Psychic News* 20 Mar. 8/6 The Raudive story appears to have started when a Swede, Frederick Jurgenson wrote a book in 1969 describing how when he was in the woods recording birdsong, faint voices appeared on his tape. Some years later, Dr. Raudive contacted Jurgenson and discussed the phenomena. **1972** P. Bander *Carry on Talking* vii. 85 It would require a fevered imagination to convert it to Raudive-type voices. **1973** *Times* 4 Dec. 17/7 There is nothing new in PK (psychokinesis) and telepathy, or even the seemingly incredible Raudive voices. **1974** C. F. Panati *Supersenses* iv. 109 Professor..Dean..feels that the Raudive phenomenon is real enough and that the explanation is simple and straightforward. **1975** C. Wilson *Mysterious Powers* vi. 94 As a result of the publicity surrounding Dr. Raudive's work, voice phenomena have been dubbed 'Raudive voices'.

Raudixin (raudi·ksin). *Pharm.* [f. *rau(wolfia + -*dixin.*] A proprietary name for a hypotensive preparation containing the dried root of *Rauvolfia serpentina* (*rauwolfia).

1953 *Trade Marks Jrnl.* 9 Sept. 811/1 *Raudixin*... Anti-hypertensive agents... E. R. Squibb and Sons Limited. **1953** *Official Gaz.* (U.S. Patent Office) 10 Nov. 317/1 Mathieson Chemical Corporation, New York..*Raudixin*... Claims use since May 4, 1953. **1954** *Jrnl. Amer. Med. Assoc.* 16 Oct. 736/1 About two months prior to the incident ..he had received Raudixin (a whole rauwolfia root), 50 mg. twice a day. **1956** N. M. Ferguson *Textbk. Pharmacognosy* viii. 207 There are several proprietaries on the market which contain rauwolfia or its constituents. Chief among these are Raudixin, Rauwiloid and Serpasil. **1967** [see *rauwiloid].

rauli (rau·li). [Amer. Sp., f. Mapuche *ruili.*] A deciduous tree resembling a beech, *Nothofagus procera,* belonging to the family Fagaceæ and native to temperate regions of Chile and Argentina; also, the reddish hardwood timber of this tree.

1908 Elwes & Henry *Trees Gt. Brit. & Ireland* III. 555 *Nothofagus procera,* known as Rauli, is less common than *N. obliqua.* **1943** Record & Hess *Timbers of New World* 168/1 Rauli..has about the same range as Roble, but its timber is much more highly esteemed. **1955** *World Timbers* (Timber Devel. Assoc.) III. 97 Rauli works easily with both hand and machine tools. **1979** *Timbers of World* (Timber Res. & Devel. Assoc.) 214 Rauli occurs in pure stands on rich soil. *Ibid.*, There is also a tendency for rauli to show a pore ring on the tangential face.

raunch (rǭnʃ). *colloq.* (orig. *U.S.*). [Back-formation from next.] **a.** Shabbiness, grubbiness, dirtiness. **b.** Crudeness, vulgarity, licentiousness; boisterousness, earthiness.

1964 *Time* 21 Feb. 46/2 Presley made his pelvis central to his act, and the screams of his admirers were straight from the raunch. **1967** *Time* 18 Aug. 63 Calvin Coolidge High is an actual Manhattan school building, its rust and raunch unretouched for the camera. **1975** *Manch. Guardian Weekly* 2 Aug. 20 Bette Midler is..no Streisand, her material is blue and her songs are old. Yet she's been camped out at one of Broadway's biggest theatres for several months now, making raunch respectable in a sellout review called Clams on the Half Shell. **1976** *N.Y. Times* 9 July C19 There are bars that are all elegance, and bars that are all raunch, and bars that breathe both elegance and raunch and therefore are considered chic. **1978** *Maclean's Mag.* 4 Dec. 65/1 The result is a 200,000-word flop, in which raunch doesn't work and the highfalutin philosophy sinks without a trace. **1979** *Guardian* 14 Mar. 13/6 Her co-producer wanted to raise the raunch-quotient by having her perform in a garter belt.

raunchy (rǭ·nʃi), *a. colloq.* (orig. *U.S.*) [Origin unknown: cf. *ranchy *a.*] **1.** Inept, incompetent, sloppy; unpleasant, contemptible, mean, disreputable; dirty, grubby.

1939 *Forum & Century* July 45/1 Depending on how good or how 'raunchy' we [*sc.* Air Force cadets] were, we drilled from one to three hours in the torrid heat. **1949** *Cavalier Daily* (Univ. of Va.) 22 Oct. 4/1 This situation could become embarrassing—if the writer in question happened to be well-known as a somewhat raunchy character in reality. **1953** Berrey & Van den Bark *Amer. Thes. Slang* (1954) § 759/2 *Raunchy flying,* clumsy flying technique. **1965** D. E. Westlake *Fugitive Pigeon* 170, I suddenly felt raunchy. Still in the same slacks I'd been wearing when this thing started. **1968** *Amer. Speech 1967* XLII. 229 *Raunchy,*..a pejorative adjective used to modify anything which the speaker wishes to denigrate, with the general connotation of 'stinky, grubby, scabby, dirty, or cheap'. **1971** *Daily Colonist* (Victoria, B.C.) 30 Nov. 18/7 I'll bet..the girls would boycott guys with dirty, tangled hair, filthy jeans, raunchy sweat shirts and bare feet. **1979** *Now!* 14 Sept. 87/3 Millgarth police station, down the rough, raunchy end of Leeds's city centre.

2. Of persons, their actions, etc.: boisterous, earthy, sexually provocative, aggressively licentious, suggestive. Also in extended uses, esp. of language, humour, songs, etc.: bawdy, salacious, smutty; tending to excite sexual feeling.

1967 'E. Queen' *Face to Face* iv. 17, I fell in love with him. In a raunchy sort of way he's beautiful. **1969** *Sat. Rev.* (U.S.) 31 May 44/2 A blend of raunchy humor, unpleasant perversity, and..sickening brutality. **1970** *Melody Maker* 12 Sept. 7 Most of the songs were too twee, and the rest seemed to be too raunchy. **1971** *Sunday Australian* 7 Nov. 27/6 Russell now has two albums of his own on which to disport his raunchy rock compositions. **1973** *Daily Colonist* (Victoria, B.C.) 23 Nov. 23/1 A drunk at the next table was singing some raunchy songs. The songs kept getting dirtier and dirtier. **1974** T. P. Whitney tr. *Solzhenitsyn's Gulag Archipelago* I. i. i. 21 They had noticed two raunchy broads going to bathe. **1976** *Times* 4 Feb. 14/4 Jurors are asked to pronounce judgment on a particularly raunchy book, while lawyers make suggestive jokes. **1977** *Gay News* 7–20 Apr. 38/3 (Advt.), Interested in meeting guys in tight raunchy levis to show me around. **1977** D. Anthony *Stud Game* xxiii. 145 If you mean *Couplings,* I liked it... I happen to like raunchy films.

Hence **rau·nchily** *adv.*, in a raunchy manner; **rau·nchiness.**

1972 *Time* 17 Apr. 66/3 They are a raunchily genteel exercise. **1975** *New Yorker* 20 Jan. 62/3 A shaggy-dog tale of a raunchiness Tolstoyan in scale, if not in tone, is related with single-minded, uninterruptible passion by one of the male guests. **1977** D. O'Sullivan in D. Marcus *Best Irish Short Stories* II. 96 No..customs and excise officer ever streaked from his intimations of mortality as raunchily as Emily Brontë from the stool. **1980** *Observer* 13 Jan. 36/2 The language is nearly devoid of metaphor, but doesn't shirk a beguiling raunchiness.

raupo. Add: (Further examples.) Also *ellipt.*, a hut built of raupo.

1851 V. Lush *Jrnl.* 25 Apr. (1971) 75 Reached the Lusks' *raupo* about 9. **1863** 'Pakeha Maori' *Old N.Z.* vi. 79 My house was a good commodious *raupo* building. **1897** [see *korupe]. **1905** W. Baucke *Where White Man Treads* 145 Here and there a patch of stunted raupo standing listless in its sour and stagnant ooze. **1920** J. Mander *Story N.Z. River* xxvi. 317 There was suddenly a ghostly movement in the rapoo and the reeds. **1933** *Bulletin* (Sydney) 9 Aug. 21/2 The pollen and roots of the raupo..were..regularly eaten. **1944** *Coast to Coast 1943* 96 They [*sc.* fleas] came from the dust under the raupo mats. **1960** *Guardian* 9 Dec. 6/3 It was five months on voyage and a tent or a raupo (rush) hut at the end of it. **1975** *Turangi* (N.Z.) *Chron.* 2 Apr. 1/1 Prior to the tailrace being established the area had been raupo swamp.

raurekau (raure·kau). Also **raureka.** [Maori.] A small evergreen tree, *Coprosma australis,* belonging to the family Rubiaceæ, native to New Zealand, and bearing small white flowers and red berries. Also *attrib.*

1905 W. Baucke *Where White Man Treads* 254 Pork.. alternated with stacks of eels enclosed in wrappings of 'raurekau' leaves. **1928** Cockayne & Turner *Trees N.Z.* 36 Raurekau. A low, bushy tree..or tall shrub, with dark-coloured bark. **1949** *Landfall* III. 31 He stumbled through a jungle of rareka and gorse. **1963** Poole & Adams *Trees & Shrubs N.Z.* 173 Raurekau. Small tree reaching 7 m... Flowers in fascicles.

rauriki (rau·riki, ra·riki). Also **rariki.** [Maori.] = *puha.
1944 *Mod. Jun. Dict.* (Whitcombe & Tombs) 331 Rauriki, raddiky..The Maori word for sowthistle. **1949** E. de Mauny *Huntsman in his Career* II. 122 Weeds grew in profusion, rariki and nettle. **1958** A. Wall *Queen's English* ix. 48 Sowthistle..is actually edible and much used by the Maoris as *rauriki*, corruptly 'raddiky'. **1966** *N.Z. Encycl.* II. 785/2 The juice of rauriki.., a latex, was also used.

‖ **rauschpfeife** (rau·ʃpfəifə). *Mus.* Pl. **-n.** [Ger., = reed-pipe.] **1.** (See quot. 1964.)
1876 Stainer & Barrett *Dict. Mus. Terms* 374/2 *Rauschpfeif* [*sic*],..a stop in old organs of two ranks of pipes, consisting of a twelfth and fifteenth, or a fifteenth and octave twelfth. **1964** S. Marcuse *Mus. Instruments* 436/1 *Rauschpfeife*,..2..organ stop first mentioned by Arnold Schlick in 1511. It seems to have consisted originally of reed pipes with conical resonators; it became transformed in the course of the c. to a 2-rank stop of flue pipes. Since the mid-17th c. the stop was treated as a mixture, often 3-rank... In the 18th c. it was enlarged further... The original meaning of the word had long been forgotten: to Praetorius already it was a 'rustling' pipe (from [G.] *rauschen*, to rustle). **1976** D. Munrow *Instruments Middle Ages & Renaissance* vii. 60/1 The 'manual', the main part of the instrument [*sc.* a Renaissance organ], with eleven registers, was composed of reeds and flue stops, including a *Zink*, *Regall*, and *Rauspfeiffen.*
2. A reed-cap shawm of the Renaissance period.
1939 A. Carse *Mus. Wind Instruments* xi. 128 Two instruments of the shawm type figure in one of Burgkmair's famous series of woodcuts 'Kayser Maximilians I Triumph' (c. 1516) and are there named *rauschpfeiffen.* **1964** S. Marcuse *Mus. Instruments* 436/1 *Rauschpfeife* (? MHGer. *Rusch*, rush), 1. family of Ger. Renaissance reed-cap shawms, with wide conical bore.., terminating in a bell, the double reed concealed in a wooden cap. **1968** *Observer* 19 May 40 David Munrow..has a collection of more than 100 historic woodwind instruments with engaging names like..the rauschpfeife. **1976** D. Munrow *Instruments Middle Ages & Renaissance* vi. 50/4 *Rauschpfeifen* and *schreierpfeifen*.. are reed-cap shawms... Of the two, rauschpfeifen seem to have been more common. **1978** *Early Music* Apr. 253/1 No rauschpfeifen..are preserved, and yet no one denies they existed.

Rauwiloid (rau·wiloid). *Pharm.* [f. *rauw(olfia + *-iloid.] A proprietary name for a hypotensive preparation containing a number of alkaloids extracted from *Rauvolfia serpentina* (*rauwolfia).
1953 *Trade Marks Jrnl.* 15 July 616/2 *Rauwiloid*... Medicinal tablets for the treatment of hypertension. Riker Laboratories Inc... City of Los Angeles. **1953** *Official Gaz.* (U.S. Patent Office) 13 Oct. 295/1 *Rauwiloid*... Claims use since Oct. 28, 1952. **1954** *Jrnl. Amer. Med. Assoc.* 17 July 1027/1 The treatment of hypertension of varying degrees of severity with alseroxylon (Rauwiloid). **1956** [see *raudixin]. **1967** H. Beckman *Dilemmas in Drug Therapy* 175/1 Usual doses of the Rauwolfia preparations employed in treating hypertension are..alseroxylon (Rauwiloid), 2–4 mg. daily; ..rauwolfia (Raudixin, Rauserpa, Rauval), 200–400 mg. daily in divided doses.

rauwolfia (rauwǫ·lfiă, -vǫlfiă). Also **rauvolfia** and with capital initial. [mod.L. (P. C. Plumier *Nova Plantarum Americanarum Genera* (1703) 19), f. the name of Leonhard *Rauwolf* (d. 1596), German physician, botanist, and traveller + -ia¹.] **1.** A tropical shrub or small tree of the genus so called, belonging to the family Apocynaceæ and bearing clusters of small white flowers and red or black berries; *esp.* a shrub of one of the several species cultivated for the medicinal drugs obtained from their roots.
1752 P. Miller *Gardeners Dict.* (ed. 6) s.v. *Rauvolfia*, Four-leaved Rauvolfia, with narrow Leaves. **1823** *Curtis's Bot. Mag.* L. 2440 (*heading*) Three-leaved Rauwolfia. **1902** L. H. Bailey *Cycl. Amer. Hort.* IV. 1503/2 The Rauwolfia flourishes with great luxuriance in the shade of other shrubs. **1955** *Sci. Amer.* Oct. 81/1 Reserpine is an alkaloid extract from the snakeroot plant (named Rauwolfia for a 16th-century German physician). **1962** N. Maxwell *Witch-Doctor's Apprentice* i. 1 Many types of rauwolfia were employed by jungle shamans centuries before our medical men thought of tranquillizers. **1976** *Hortus Third* (L. H. Bailey Hortorium) 942/1 Rauvolfias are cultivated as ornamentals and for curiosity. **1976** W. A. R. Thompson *Herbs that Heal* ix. 148 This unleashed the flood-gates of the pharmaceutical industry, whose scouts started scouring the earth for rauwolfia.
2. *Pharm.* Also **Rauwolfia Serpentina.** The dried roots of *Rauvolfia serpentina* or related species, or an extract therefrom, containing a number of alkaloids (notably reserpine) and used medicinally, esp. to treat hypertension.
[**1949** *Brit. Heart Jrnl.* XI. 350/2 This overwhelming body of support in favour of regarding *R. Serpentina* as the remedy of choice. *Ibid.* 354/1 The hypotensive action of *R. Serpentina*.] **1952** *Ann. Internal Med.* XXXVII. 1149 In our clinic we have relied chiefly upon various combinations of hydrazinophthalazine, Rauwolfia and veratrum, principally because these drugs appear to be the safest..of any medicinal regimen we have tried. **1954** *Brit. Pharmaceutical Codex* 649 Rauwolfia has a depressant action on the central nervous system. **1957** H. W. Youngken in R. E. Woodson et al. *Rauwolfia* ii. 32 The drug Rauwolfia or Rauwolfia Serpentina consists of the dried root of *Rauwolfia serpentina.* .. The commercial sources of the drug..have been India, Pakistan, Ceylon, Burma, and Siam. **1966** *New Scientist* 27 Jan. 236/2 Physicians and pharmacists..are inclined to think that only a few vegetable drugs such as..digitalis, penicillin and rauwolfia are important in the present day *materia medica.* **1977** Lewis & Elvin-Lewis *Med. Bot.* vii. 187/1 *Rauvolfia* acts synergistically with other hypotensive drugs, and in the more severe cases of hypertension it is used in combination with *Veratrum viride* or protoveratrines A and B.
3. *attrib.*, as *rauwolfia alkaloid, berry.*
1942 *Biol. Abstr.* XVII. 117/2 The various effects suggest that the *Rauwolfia* alkaloids probably act on the vasomotor system and also directly on plain muscles of the blood vessels and intestines. **1977** S. Loebl et al. *Nurse's Drug Handbk.* 252 The rauwolfia alkaloids decrease blood pressure and have a sedative effect accompanied by bradycardia. **1932** *Discovery* July 231/1 Three kinds of starlings come with the great blue pigeons to the Rauwolfia berries.

rav (rɒv). Also **rov.** [Yiddish.] A rabbi; freq. prefixed to personal names.
1892 I. Zangwill *Childr. Ghetto* I. I. xiv. 314 'Ah, you will become a Rav!'.. 'What's that about a Rav?.. Does he want me to become a Rabbi?' **1893** —— *Ghetto Tragedies* 4 The great Rav Rotchinsky from Brody was to deliver a sermon. **1962** 'E. McBain' *Empty Hours* iii. 115 'I know who killed the *rov*.'.. 'She says she knows who killed the rabbi.' **1967** C. Potok *Chosen* xiv. 238 From one to three we would have the actual Talmud session itself, the shiur, with Rav Gershenson. **1973** *Jewish Chron.* 19 Jan. 34/2 The daughters and family of the late Mrs B—. C—...wish to thank the Rav, rabbonim..and friends for their visits..and numerous letters of sympathy.

ravanastron (rā:vānā·strǫn). Also **ravanastra.** [Origin unknown: freq. associated with the legendary King Ravana (see quots.); cf. Skr. *rāvaṇahasra* a kind of stringed instrument.] An ancient Hindu stringed instrument played with a bow.
1864 C. Engel *Music of Most Anc. Nations* ii. 81 The Hindoos maintain that the *ravanastron*, one of their old instruments played with the bow, was invented about five thousand years ago by Ravanen, a mighty king in Ceylon. **1876** Stainer & Barrett *Dict. Mus. Terms* 374/2 Ravanastron, a stringed instrument played with a bow in use among the Buddhists. **1896** H. Saint-George *Bow* ii. 7 Of existing bowed instruments the Ravanastron..most certainly seems to be the oldest. **1903** R. Hughes *Mus. Guide* I. 249/1 *Ravanastron*, a primitive violin with one or two strings... It is still used by the Buddhists. **1953** S. Beckett *Watt* 71 A ravanastron hung, on the wall, from a nail, like a plover. **1965** S. Krishnaswami *Mus. Instruments of India* 16 One of the earliest stringed instruments played with a bow was called the *ravanastra.* This instrument was associated with Ravana. **1969** N. Deane tr. *Bachmann's Origins of Bowing* i. 8 The quest for rudimentary types of bowed instruments led to the Welsh crwth and the Indian *ravanastron*, both regarded as the ancestor of our violin.

rave, *sb.*² **1. b.** (Examples.)
1847 *Rep. Comm. Patents 1846* (U.S.) 81 The raves are carried in front in such a form as to furnish a frame for the dash-board. **1851** J. S. Springer *Forest Life* v. 106 It was astonishing to see how he [*sc.* a teamster] had gnawed the rave of the sled. **1895** *Montgomery Ward Catal.* Spring & Summer 594/1 Bob Sleigh Gearing..Bob Knees..Bob Raves ..Bob Rollers.

rave, *sb.*³ Add: **2.** *slang.* **a.** A passionate (and usu. transitory) liking for or infatuation with a person or thing; a sudden display of extreme enthusiasm or popularity, a 'craze'. Also, one who or that which excites feelings of this kind.
1902 Farmer & Henley *Slang* V. 380/2 *Rave*,..a strong liking; a craze: as 'X has a *rave* on Miss Z.' **1924** G. B. Stern *Tents of Israel* xvi. 240 Even if Jeanne-Marie had a rave on me, I'm not responsible. **1927** L. Mayer *Just between us Girls* xv. 91 He said I was simply a rave in these pajamas. **1941** L. Eyles *For my Enemy Daughter* vii. 161 That, too, is a bit schoolgirlish, isn't it? Getting a 'rave' on a woman I admire. **1949** N. Marsh *Swing, Brother, Swing* iii. 38 Carlisle remembered the confidences that Félicité had poured out in her convent days, concerning what she called her 'raves'. **1958** *People* 4 May 14/7 Kitza Kazacos, Greek singer, who became a rave on B.B.C. but was allowed to languish without follow-up dates, now tells me from New York that she is on three TV programmes. **1959** C. MacInnes *Absolute Beginners* 70 The newest of the teenage singing raves. **1962** L. Davidson *Rose of Tibet* 8 T.L. had been having at the time one of his not uncommon raves; on this occasion for the mental-disciplinary benefits of a classical language.
b. A highly enthusiastic or laudatory review or notice of a book, play, film, etc. Also in extended use, a favourable opinion; a strong recommendation. Freq. *attrib.* (passing into *adj.*). orig. *U.S.*
No earlier definite example of this use has been traced in *Variety* (see quot. 1926).
1926 *Amer. Mercury* Dec. 464/2 One of the paper's [*sc. Variety*'s] coinages should be officially embraced by the dictionary and bred into the language. It refers to a flattering, enthusiastic review by a sycophantic critic as a *rave.* **1935** E. E. Cummings *Let.* 31 Jan. (1969) 135 Have been epistling with Pound, whom yessed in Paris for a full ½ hour under lurid misapprehension that his 'Douglas' rave intended 'South Wind'. **1936** *Amer. Speech* XI. 221 The producer waits for the early editions of the morning papers. He scans them avidly for the notices. They may be raves, in which case the critics have reviewed the show in glowing superlatives. **1942** *Melody Maker* 4 July 5/4 Raves coming thick and fast for George Auld's new powerhouse band now at the Arcadia Ballroom, N.Y. **1943** D. Powell *Time to be Born* i. 6 The critical raves and the big sale. **1951** Wodehouse *Old Reliable* ix. 114 Of course he can open the safe. He's an expert. You should have read what the papers said of him at the time of the trial. He got rave notices. **1958** *Listener* 27 Nov. 898/1, I yield to none in my admiration for this pianist, whose first London notice I had the honour to write long before the war (a 'rave' in case you think I am always wrong). **1961** *John o' London's* 12 Oct. 423/1, I don't suppose *The Young Doctors*..will collect rave tributes. **1969** Gish & Pinchot *Lillian Gish* xv. 222 Richard Barthelmess also received raves for his sensitive portrayal of Cheng. **1972** [see *pan *sb.*¹ 10]. **1974** *Publishers Weekly* 26 Aug. 299/2 The later work is distinctly finer than the earlier (though that was good enough to draw a rave foreword, here reprinted, from Ford Madox Ford). **1977** M. Kenyon *Rapist* v. 50 His music..had opened to raves in New York. **1979** *Tucson Mag.* Apr. 68/3 These three-day bus tours..have received rave notices from all who have gone along. **1980** *Times Lit. Suppl.* 10 Oct. 1131/1 Enzensberger's rave review of the novel is reprinted.
c. A lively party; a rowdy gathering.
1960 *News Chron.* 16 Feb. 6, I wandered around to a rave I knew was going on in Covent Garden. **1963** *Sunday Times* 8 Sept. 29/3 A rave, blast or orgy—all synonyms for party. **1964** C. Dale *Other People* iv. 96 A man who..thought that parties—'raves'—ended at ten-thirty. **1965** G. Melly *Owning-Up* vii. 75 We..organized all-night raves. **1968** *Listener* 7 Nov. 606/1 Have you heard, the Touch-Paceys are economising this year by combining their children's bonfire party with their annual fancy dress rave?
Hence **rave-in** (see *-in³), **rave-up** = sense 2 c above.
1967 *Melody Maker* 21 Jan 10 Pop enthusiasts have been treated to rave-ups featuring such world-class stars as the Four Tops. **1967** *New Statesman* 17 Mar. 356/3 Last week police arrested scores of teenagers at a rave-in, and left-wing Catholics staged a pray-in. **1972** 'Miss Read' *Tyler's Row* vii. 88 'Well, let's have this rave-up of a meal now,' suggested Peter. **1973** H. Miller *Open City* ix. 89 Phyllis McBain is invited to an old-style rave-up, knickers and husbands optional. **1974** R. Rendell *Face of Trespass* ii. 29 Some Victorian pretence that a simple Westbourne Grove rave-up was really a conference. **1977** *Rolling Stone* 7 Apr. 20/1 (Advt.), When Argent, that legendary British band of so many hit songs and rave-up performances, began to wane, these three musicians stepped out of its shadow and into the fierce light of their own creative genius.

rave, *v.*¹ Add: **3. a.** (Earlier and later examples with *about*.)
1816 Jane Austen *Emma* II. xv. 282, I quite rave about Jane Fairfax. **1921** *Collier's* 16 Apr. 20/2 He began to 'rave', about the place he had built on Long Island for him and Delores. **1978** J. Updike *Coup* iv. 150 So you're the young man my daughter has been raving about.
b. *slang.* To give oneself over to enjoyment; to 'live it up'; to depart rowdily or with the intention of having a good time. Cf. *rave *sb.*³ 2 c.
1961 *New Statesman* 26 May 830/2 When we got there, most of the art-student element had raved off to some shindig. **1965** G. Melly *Owning-Up* vii. 75 The word 'rave', meaning to live it up, was as far as I know a Mulligan-Godbolt invention. **1965** *Sunday Times* (Colour Suppl.) 19 Sept. 13/3 He started out by raving at weekends to Bridlington.

ravel, *sb.*¹ Add: **1.** Also, a cluster.
1913 D. H. Lawrence *Sons & Lovers* xiii. 380 There was a lovely yellow ravel of sunflowers in the garden.

ravel, *sb.*³ (Earlier example.)
1805 J. Austin in *Trans. Soc. Arts* XXIII. 242 An universal ravel or sniffle, useful at the beaming of all kinds of webs.

ravel, *v.*¹ **2.** (Later *fig.* examples.)
1956 *Essays in Crit.* III. 320 The discussion then ravels out into a note on such secondary Virgilian sources as the Saturnian prophecy. **1963** Oglesby & Hewes *Highway Engin.* (ed. 2) xvii. 544 The roads raveled rapidly and in the worst instances became during a single season merely a pile of loose stones.

ravel (ræ·v'l), *v.*⁴ Var. Rabble *v.*³ in Dict. and Suppl.
1923 *Discovery* Nov. 291/2 The other furnace..in which the flames actually play over the surface of the mundic which is 'ravelled' from the side by a pole some eight or nine feet long.

Ravelian (rave·liǎn), *a.* Also **Ravellian.** [f. the name of Maurice *Ravel* (1875–1937), French composer.] Of, pertaining to, or characteristic of the works of Ravel. Also as *sb.*, an exponent of Ravel's music.
1933 M. D. Calvocoressi in *Listener* 22 Feb. 305/1 There is no mistaking the thoroughly 'Ravel-ian'—and therefore French—quality of the mind and imagination which is at work. **1937** *Times* 29 Dec. 12/3 The charm of the early quartet and septet will keep them fresh when some of the more purely Ravellian works have been discarded as too dry. **1946** *Penguin Music Mag.* Dec. 32 The progress of Walton from his early and somewhat Ravelian *Facade* to the Violin Concerto. **1952** B. Ulanov *Hist. Jazz in Amer.* (1958) xxii. 310 Johnny Richards' excursions into Debussyan and Ravelian pasture. **1965** *Listener* 28 Oct. 680/2 The two syllables '*Ma-man*', pronounced as a falling fourth (a characteristic Ravelian fingerprint), strike an almost sentimental note. **1970** L. Davies *Ravel Orchestral Music* 46 It culminates in a more virtuoso episode which leans too obviously on skimming Ravelian scales. **1981** *Times* 26 Feb. 13/6 Manuel Rosenthal, a Ravelian to his fingertips.

ravelling, *vbl. sb.* **2.** (Later *fig.* example.)
1903 B. Potter *Tailor of Gloucester* 22, I am worn to a ravelling.

raven, *sb.*¹ Add: **4. a.** Similative, as *raven-shadowing* adj.; instrumental, as *raven-covered* adj.
1895 W. B. Yeats *Wanderings of Usheen* in *Poems* 7 We think on Oscar's pencilled urn, And on the heroes lying lain, On Gabhra's raven-covered plain. **1950** C. Day Lewis in *Penguin New Writing* XXXIX. 22 A driven heart, a raven-shadowing mind Loom above all my pastorals.
b. raven-tree, a tree in which ravens build their nests.
1904 *Westm. Gaz.* 23 July 13/1 The 'raven tree' is all that remains..to remind one of the former existence of these birds in those localities. **1908** *Chambers's Jrnl.* Apr. 284/1 Nearly every parish had its 'raven-tree'.
c. raven's duck (examples).
1761 *Newport* (Rhode Island) *Mercury* 28 Apr. 4/3 Just Imported..Russia and ravens duck. **1775** in *New Hampsh. Hist. Soc. Coll.* (1863) VII. 4, 120 Tents, to be made of Raven's duck. **1868** G. G. Channing *Recoll. Newport* 200 A miller called one day at the store to purchase a piece of ravensduck, with which to make or to repair sails for his windmill. **1931** *Sun* (Baltimore) 12 Jan. 6/6 Hemp sails, known as raven's duck, were used, the cotton duck being unknown at that time.

Raven (rē¹·v'n). The name of J. C. *Raven*, 20th-cent. psychologist, used *attrib.* and in the possessive with reference to non-verbal intelligence tests devised by him to measure Spearman's *g* factor in the ability to understand abstract relationships, solve problems, etc., and designed to be especially useful where language disadvantages exist; esp. *Raven('s) Progressive Matrices (Test).*
1948 *Psychometrika* XIII. 28 The results obtained by means of the Raven tests are indicative of the fact that these ..fulfill most of the requirements needed for testing normal and handicapped subjects. *Ibid.* 34 Apart from Raven B.. the remaining tests..are not loaded in factor *L.* **1954** A. Anastasi *Psychol. Testing* x. 270 The..Raven Progressive Matrices Test provides a promising tool for this purpose. **1964** M. Critchley *Developmental Dyslexia* xiv. 82 On Raven's progressive matrices he scored 32 out of 60: his I.Q. was estimated to be 93. **1972** *Jrnl. Social Psychol.* LXXXVII. 69 Intelligence scores were available only for the controls, who had been administered the Raven's Progressive Matrices Test. **1973** B. B. Wolman *Introd. Gen. Psychol.* xxxii. 666/1 One of the most widely used nonverbal tests is the Raven Progressive Matrices Test. *Ibid.*, A vocabulary test is also available to accompany the Raven Matrices. **1976** H. M. Proshansky et al. *Environmental Psychol.* (ed. 2) ix. 129/2 The..tests of spatial skills were..a short form (six items) of the Embedded Figures Test..and the Ravens Matrices.

Ravenna (rave·nă). The name of a town in northern Italy, used *absol.* or *attrib.* in *Ravenna grass* to designate *Erianthus ravennæ*, a large ornamental grass native to southern Europe and distinguished by greyish leaves and spikes of purplish-grey flowers.
1900 L. H. Bailey *Cycl. Amer. Hort.* II. 540/1 Ravenna Grass. A tall, hardy grass, 4–7 ft. high, very ornamental. **1929** J. W. Bews *World's Grasses* vi. 243 'Ravenna' is a Mediterranean species, extending as far north as upper Italy. **1976** G. S. Thomas *Perennial Garden Plants* x. 308 Ravenna Grass. In hot summers this grass is a striking addition to the garden.

ravenously, *adv.* (Later examples.)
1907 G. B. Shaw *Major Barbara* II. 217 Shirley (looking at it ravenously but not touching it..). *Ibid.*, (He turns to the table and attacks the meal ravenously). **1915** W. S. Maugham *Of Human Bondage* xliv. 211 She could not have eaten more ravenously if she were starving. **1951** C. S. Lewis *Prince Caspian* iii. 37 He would have made much more fuss about this if he had not by now been so ravenously hungry.

Ravenscroft (rē¹·vĕnzkrǫft). [The name of George *Ravenscroft* (1618–81), English glassmaker.] An article made of the flint-glass or lead-glass devised by George Ravenscroft. Also *attrib.* or as *adj.*
1929 W. A. Thorpe *Hist. Eng. & Irish Glass* I. iv. 128 Threads of trailed glass applied on the surface..are evident in several marked Ravenscroft glasses. **1948** E. B. Haynes *Glass through Ages* I. ix. 127 This..piece must be a Ravenscroft glass, perhaps the first example to be found with the plain seal. **1961** E. M. Elville *Collector's Dict. Glass* 87/1 (*caption*) Ravenscroft roemer bearing the seal of the raven's head..circa 1677. **1970** G. Savage *Dict. Antiques* 348/1 (*caption*) A rare Ravenscroft showing signs of criselling... Lead glass. 17th century. **1975** *Country Life* 2 Jan. 11/3 (*heading*) A Syllabub Jug in Ravenscroft Glass. *Ibid.*, This jug..is typical Ravenscroft of, I suppose, the late 1680s. **1976** *Times* 4 May 14/6 A documentary item of the beginning of English glass, a Ravenscroft silver-mounted and engraved decanter jug,..sold for £2,420.

raver. Add: **b.** A passionate enthusiast for a particular thing, idea, or cause; a fanatic. Also, one who likes to 'live it up' or have a wild time esp. in sexual relationships (cf. *rave *v.*¹ 3 b). *slang.*
1959 C. MacInnes *Absolute Beginners* 63, I did actually begin to be a raver for those weekly meetings. **1960** [see *mod *sb.*² and *a.*]. **1961** *Guardian* 6 Mar. 3/4 The bearded ravers usually associated with the more esoteric ranges of HiFi. **1968** Busby & Holtham *Main Line Kill* v. 44 There's a bloke I know makes his own LSD. Some of the ravers are giving it a try. **1971** *Cape Herald* 15 May 7/4 She looks like a raver But I could never please her. **1976** P. Cave *High Flying Birds* iii. 31 She was neither attractive nor plain; not a raver or a ragbag. **1976** *Listener* 22 Jan. 84/2 Things hot up when a raver and his girl swoop in by motor-bike. **1978** *Sunday Mail* (Brisbane) 24 Sept. 34/4, I have never analysed why, but many pop musicians are ravers—people who like to live it up—with a strong self-destructive streak.

ravers (rē¹·vəɪz), *pred. a. slang.* [f. Rav(ing *ppl. a.*¹; cf. *crackers *pred. a.*] Raving mad, delirious. Also, in weakened sense, furious, angry.
1938 N. Marsh *Death in White Tie* xxii. 242 Bart has driven me stark ravers, he's been so awful. **1939** —— *Overture to Death* xvii. 189 In Henry's..opinion Miss Prentice is practically ravers. **1951** E. Hyams *Sylvester* xxviii. 150 'You said you wanted to meet Sylvester Green. Well, here I am.'.. 'Stark ravers. I served for two years with Green. This man isn't even much like him.' **1967** N. Marsh *Death at Dolphin* iv. 100 Jeremy..will probably go stark ravers if they're sold out of the country.

‖ **ravigote** (ravigǫ·t). Also **ravigotte.** [Fr., f. *ravigoter* to invigorate.] (See quot. 1877.)
1830 R. Dolby *Cook's Dict.* 435/2 *Ravigote*, shred.. chervil, chives, pimpernel, and tarragon; this latter ought to predominate; (the mixture of these articles constitutes the *ravigote*); take some *velouté*,..add..butter and the *ravigote*; stir.., and serve. **1861** Mrs. Beeton *Bk. Househ. Managem.* 240 Ravigotte, a French Salad Sauce. **1877** E. S. Dallas *Kettner's Bk. of Table* 373 *Ravigote*, pick-me-up.. from the French verb *ravigoter*, to cheer or strengthen... The French give the name of Ravigote to an assemblage of four herbs—tarragon, chervil, chives, burnet—minced small or used as a faggot, and supposed..to have a rare faculty of resuscitation. *Ibid.*, *Ravigote sauce* is simply the English butter sauce to which a ravigote is added. **1943** D. Powell *Time to be Born* x. 255 A dish of crab meat ravigote. **1951** *Good Housek. Home Encycl.* 594/1 Serve with a ravigote sauce flavoured with chopped chives, chervil and tarragon. **1964** A. Launay *Caviare & After* 142 *Ravigotte*, a highly-seasoned white sauce which is served either hot or cold. **1975** *Amer. Speech* 1969 XLIV. 94 There is one instance of *ravigote*, but with two *t*'s, in *crabmeat ravigotte*..; it is 'a sauce or dressing coloured green with spinach purée and seasoned with vinegar and a mixture of herbs'.

ravin¹, **raven**². **2. a.** (Example in spelling *raven*.)
1935 W. Empson *Poems* 4 Nor heeds if the core be brown with maggots' raven.

ravine, *sb.* Add: **3.** (Earlier and later examples.) Also *fig.*
1781 G. Washington *Diary* 30 Sept. (1925) II. 263 We also began two inclosed works on the right of Pidgeon Hill—between that and the Ravine above Moves Hill. **1926** [see *peak *sb.*³ 5 c]. **1930** R. Campbell *Adamastor* 51 The phosphorescent whales..Bore through the gloom their long ravines of gold.
4. *ravine-gully*; *ravine-wrinkled* adj.
a **1930** D. H. Lawrence *Etruscan Places* (1932) 22 A modest, Italian sort of ravine-gully. **1950** C. Day Lewis in *Penguin New Writing* XXXIX. 20 Earth's face grew rapidly older, ravine-wrinkled.

ravinement (rävi·nmĕnt). *Geol.* [a. F. *ravinement* gullying.] An unconformity in river or shallow marine sediments caused by interruption of deposition by erosion.
1921 L. D. Stamp in *Geol. Mag.* LVIII. 109 A 'ravinement' may be defined as an irregular junction which marks a break in sedimentation... Although one of the commonest of geological phenomena, there is no English word which expresses quite so aptly the relationship. *Ibid.*, The period between two successive 'ravinements' constitutes a 'cycle of sedimentation'. **1923** —— *Introd. Stratigr.* i. 17 When such an unconformity is traced laterally into a continuous series of marine deposits it is frequently found that, although the bedding of the upper series is parallel to that of the lower, there are signs that the sea, in depositing the upper series, has slightly eroded the top of the lower series. Such an erosion line is termed a 'ravinement'. **1969** Bennison & Wright *Geol. Hist. Brit. Isles* i. 8 In the field an unconformity can..be proved because the erosion surface or ravinement is conspicuous.

ravioli (rævi,ōᵘ·li), *sb. pl.* [It., pl. of *raviolo* in the same sense: see Raviol.] Small square pasta cases filled with meat or vegetables.
1841 Thackeray in *Fraser's Mag.* June 721/2 For the same money, I might have had..a heap of macaroni, or ravioli. **1846** Dickens *Pictures from Italy* 47 Real Genoese dishes, such as Tagliarini; Ravioli..with fresh green figs. **1898** L. Merrick *Actor Manager* 66 Oliphant was duly introduced to ravioli. **1947** H. Innes *Lonely Skier* i. 23 'No need to try and catch their eyes,' Joe Wesson said through a mouthful of *ravioli*. **1956** A. Wilson *Anglo-Saxon Att.* II. iii. 363 *Ravioli* was in preparation for dinner. **1969** B. Malamud *Pictures Fidelman* 160 The deaf woman was up and down to supply the glass blower with ravioli, cheese, bread. **1972** H. C. Rae *Shooting Gallery* iv. 247 She cooked up a tin of Ravioli, spiced it with Parmesan and mixed herbs. **1975** J. Gores *Hammett* xvi. 110 Soup to start. Ravioli. Salad after.

ravish, *v.* **2. b.** (Later examples.) Also *absol.*
1939 G. B. Shaw *Geneva* III. 70 Am I to allow him to kill me and ravish my wife and daughters? **1981** *Sunday Times* (Colour Suppl.) 8 Mar. 104 He ravished and pillaged...left sons to hate him, women to fight over his wealth.

ravished, *ppl. a.* **2.** (Later examples.)
1901 G. B. Shaw *Caesar & Cleopatra* I. 102 A man comes from the south with stealing steps, ravished by the mystery of the night. **1953** R. Lehmann *Echoing Grove* 22 Ravished, startled, they watched the apparition wave up and down,.. with rapid wing beats, low above the terraces.

raw, *a.* (*sb.*) Add: **A.** *adj.* **1. a.** Also, *raw milk.*
[**1743** W. Ellis *Mod. Husbandman* July x. 48 If we make raw Milk Cheese.] **1871** *N. & Q.* 4th Ser. VIII. 415, I think that 'rammilk' is *rahm* milk—*i.e. cream* milk and not raw milk. **1950** *N.Z. Jrnl. Agric.* Mar. 221/2 Some doctors say raw milk is better for health than pasteurised milk, so who is to be believed? **1979** A. Parker *Country Recipe Notebk.* viii. 103 Raw milk ('farm milk') is at present officially described as untreated milk.
f. Applied to the taste of tea: harsh, not mellow.
1881 *Tea Cycl.* III. 220/1 To obtain a raw, rasping and pungent flavor I am compelled to underferment, the indication of which is that the colour of infused leaves are of a greenish brown tint. **1892** J. M. Walsh *Tea, its Hist. & Myst.* vii. 170 Ceylon and Javas are either 'raw', 'uncooked', .. or sour in flavour. **1933** C. R. Harler *Culture & Marketing of Tea* xiv. 278 The infused leaf of tea made from under-withered leaf is generally *greenish.* The infusions from such leaf are usually *raw* and *rasping.* **1958** T. Eden *Tea* xiv. 176 *Tea-Tasting Terms..Harsh, Raw, Rasping.* Bitter, due to the presence of unfermented polyphenols; a common defect of non-wither teas.
g. *raw humus*, vegetable matter not yet fully decomposed; incompletely formed humus.
1891 W. Schlich *Man. Forestry* II. i. 32 (*heading*) Accumulation of raw humus. **1926** Tansley & Chipp *Study of Vegetation* vii. 117 In cold, moist soils poor in mineral salts and acid in reaction..the leaf litter and other plant debris remain on the surface very little changed and often form a thick layer which is called raw humus. *Ibid.* 132 The soil is covered with a thick layer of raw humus. **1935** *Forestry* IX. 43 Raw humus is characterized by its excessive accumulation (slow decomposition), expandibility, and frequently by the presence of some structural remains of plants... [It is] characterized also by an extremely low base content. **1952** S. A. Waksman *Soil Microbiol.* v. 136 In evergreen forests, the largely organic surface layers are usually not mixed with the inorganic soil layers; the former are referred to as the 'raw humus' or 'duff'. *Ibid.* 144 The surface layer of the raw humus soil may undergo considerable leaching. **1975** *Soil Sci.* XX. 25/1 The raw humus..has been extracted successively with hexane, ether, and ethanol.
h. *to come the raw prawn*: see *prawn *sb.* 3 c.
2. a. *raw silk*, also, a fabric of spun silk; also *attrib.*
1866 A. D. Whitney in *Our Young Folks* Feb. 104 Two pairs of bright brown raw silk stockings..completed the mountain outfit. **1953** M. McCarthy in *Harper's Mag.* Mar 42/1, I was wearing a bright apple-green raw silk blouse. **1965** D. Mackenzie *Lonely Side of River* i. 18 Raw-silk summer curtains rustled in the drawing room. **1978** *Observer* 29 Jan. 25/4 There are lots of clothes around made in what is loosely termed 'raw silk'. This is a misnomer as raw silk is actually the silk before it has been woven into fabric and what we call raw silk is actually a slub silk.
c. *rawhide* (earlier and later examples). Hence *rawhiding*, a whipping; also *fig.*; *rawhide* vb. trans., to whip; also *fig.*
1829 *Massachusetts Spy* 16 Sept. 2/4 She..took down a raw hide..and..kept the whip moving. **1848** *Knickerbocker* XVIII. 519 The editor, it was predicted, would catch a raw-hiding before sun-set. **1858** *Spirit of Times* 6 Feb. 356/3 One of our citizens was rawhided in the street.. by a Mr. Huntington. **1935** W. Faulkner *As I lay Dying* 109 Like as not you got to take a rawhiding for thinking they meant it. **1935** H. L. Davis *Honey in Horn* viii. 100 He had been rawhided into a hunt that showed up his lack of endurance. **1944** H. Evatt *Snow Owl's Secret* 85 The huskies do like the sound of the singing rawhide. **1949** *Sat. Even. Post* 7 May 103/1 Joe went along as packer, rawhiding a string of bony horses up into the brownie country. **1979** *Tucson Mag.* Apr. 28/2 Sometimes the whole door was of rawhide.
attrib. (Earlier and later examples.) Also *fig.*
1841 G. Catlin *Lett. on N. Amer. Indians* I. x. 71 The raw-hide thong, with which it was tied to a stake. **1883** Sweet & Knox *On Mexican Mustang through Texas* i. 18 I'm just pining away for a fight. I'm a rawhide Texan, *I* am. **1897** *Slocan* (B.C.) *Pioneer* 8 May 1/2 A rawhide and pack trail has been constructed from the town of Brandon to the Two Friends mine. **1940** *Chambers's Techn. Dict.* 704/1 *Rawhide hammer*, a hammer the head of which consists of a close roll of hide projecting from a short steel tube; used by fitters to avoid injuring a finished surface. **1957** J. Kerouac *On Road* (1958) iii. 21 Here came this rawhide old-timer Nebraska farmer. **1973** J. Wainwright *Devil you Don't* 14 The expensive, rawhide shirt. **1976** A. Murray *Stomping Blues* iv. 51 Down from the cloudlike realms of abstraction and fantasy to the bluesteel and rawhide textures of..the everyday struggle for existence.
d. Undistilled (water): delete † and add later examples; also, not filtered or otherwise treated; unrefined or partly refined (sugar),

undeveloped (land), (*N. Amer.*), untreated (sewage).
1797 Raw sugar [see Sugar *sb.* 1 b]. **1868** *Chem. News* 20 Nov. 248/2 Several accidents happened; one..by which no less than 150,000 gallons of raw sewage were pumped into a tank holding 430,000 gallons of purified sewage. **1882** C. G. W. Lock et al. *Sugar Growing & Refining* p. vii, Sugar-cane..is extensively cultivated, and the manufactured product, under the name of 'raw sugar', forms the staple produce of many of our colonies. **1883** Sweet & Knox *On Mexican Mustang through Texas* xxi. 282 [He] came to Atascoso County, Texas, and bought a piece of raw land. **1925** G. Fairrie *Sugar* vii. 151 At the commencement of the nineteenth century the methods of converting the juice of the cane into raw sugar and the process of refining the raw sugar were very different from what they are to-day. **1930** *Engineering* 25 July 121/1 The net quantity of raw water distilled and passed into the feed line as make-up amounts to 1,138 tonnes per day. **1939** *Sun* (Baltimore) 11 Apr. 3/2 His agency is not interested in the price paid for the 'raw land' on which such developments were built. *Ibid.* 28 Sept. 12/1 The work might be done in part by carrying raw sewage lines to Colgate creek. **1956** *Jrnl. Amer. Water Works Assoc.* XLVIII. 1281 (*heading*) Relation of treatment methods to limits for coliform organisms in raw waters. **1958** Raw sewage [see *recirculate *v.*]. **1972** *Works Engineer* Nov./Dec. 32/1 Raw feed may contain domestic detergents, in which case, antifoaming agents must be added to the treated water. **1973** *Daily Colonist* (Victoria, B.C.) 7 July 3/1 Speculation in raw land is a major contributor to high housing costs..according to Mayor Art Phillips. **1976** *Chem. in Brit.* XII. 375/3 A recent exercise which nicely illustrates the applications of radiotracers in large systems was carried out, in a raw-water reservoir of capacity 3.10^6 m³. **1978** *Daily Tel.* 7 July 19/1 The slide in the daily price of raw sugar on the London futures market continued yesterday. **1978** *Oxford Times* 15 Dec. 4/6 Raw sewage has been bubbling up through manhole covers.
e. *raw material* (further examples).
1864 J. H. Newman *Apologia* vii. 392 The raw material of human nature. **1930** R. Campbell *Poems* 12 Taking as raw material for his lays The good old English beer he loves to praise. **1971** I. G. Gass et al. *Understanding Earth* i. 26/2 The processes of erosion, which provide the raw materials of the sedimentary rocks.
f. Of measurements, data, or the like: not yet subjected to a process giving them significance; unadjusted; naïvely calculated.
1904 *Amer. Jrnl. Psychol.* XV. 263 Weight..has a raw correlation of 0.34, after correction we eventually get 0.43. **1920** Yoakum & Yerkes *Mental Tests* iii. 78 The result of examination alpha is expressed in a total score which is the sum of the raw scores of the several tests. **1945** *Jrnl. Exper. Psychol.* XXXV. 46 Only the general problems involved in the evaluation of the raw data will be treated in this paper. **1950** *Sun* (Baltimore) 11 May 4/3 McCarthy's frequent statements that the proof of his charges of Communist infestation in the State Department lies..in the 'raw files, of the FBI. **1954** A. Anastasi *Psychol. Testing* ii. 24 The 'raw score' on the test..may be expressed as number of correct items, time required to complete a task, number of errors, or some other objective measure appropriate to the content of the test. Such a raw score is meaningless until evaluated in terms of a suitable set of norms. **1971** *World Archaeol.* III. 120 Naroll's formula..shows too much variation in raw numbers of population and square meters. **1974** *Nature* 1 Nov. 27/1 A raw spectrum was obtained by averaging the values for a given grating position weighted according to the reciprocal of their variances. **1975** *Ibid.* 31 Jan. 327/2 The raw magnetic field data are translated to a Jupiter centred spherical coordinate system. **1977** *Time* 4 Apr. 13/2 The console operators do not see a raw radar picture. The information is translated into digital bits and then filtered through complicated computer programming. **1978** *Daily Tel.* 16 Jan. 2/1 Sir Charles..said he had been given warnings as far back as April about the deteriorating situation but had not been prepared to release what he felt were 'raw' forecasts about losses.
g. Of manufactured material: unused.
1917 Bennett & Heron *Guide to Kinematogr.* i. 12 Raw stock is divided broadly into two classes, Ordinary and Non Flam. **1934** *Tit-Bits* 31 Mar. 12/2 Exposed film is 'stuff'; unexposed film is 'raw stock'. **1968** *Globe & Mail* (Toronto) 3 Feb. 27/3 She paints on the floor of her..studio, beginning with raw canvas. **1971** W. G. Salm *Stereo in your Home* xii. 168 Prerecorded open-reel tape production line in Ampex plant records all four tracks simultaneously, taking raw tape from large blank pancake. **1973** *Center City Office Weekly* (Philadelphia) 9 Oct. 5 We could not shoot today. No money to buy the raw stock film. **1979** *N. & Q.* Aug. 348/2 Print-outs from the raw text tapes.
h. Of a glaze: (see quot. 1934).
1934 Webster, *Raw glaze*, a glaze made from materials which need no preparation, but can be bought ready for use. **1964** H. Hodges *Artifacts* ii. 46 Any glaze in which the raw materials are simply ground up and applied in this way is called a raw glaze.
3. a. *raw sienna, umber*, sienna and umber which have not been calcined; also, the colours of these pigments; *raw deal*: see *deal *sb.*² 4 c; *raw edge*, the unfinished edge of a cut piece of fabric; also *fig.*; (cf. *raw-edged*, sense 9 in Dict. and Suppl.).
1869 *Bradshaw's Railway Man.* XXI. 460/1 (Advt.), Raw Turk. Umb. **1886** H. C. Standage *Artists' Man. Pigments* iv. 43 *Yellow ochres* (these include Jaune de Mars, Sienna, or Raw Sienna). *c* **1890** tr. *T. de Dillmont's Encycl. Needlework* 6 Rounded seam.—Back-stitch your two edges together.. then..roll the outer one in, with the left thumb, till the raw edge is quite hidden, hemming as you roll. **1895** *Montgomery Ward Catal.* Spring & Summer 252/3 Artists tube oil colors...Prussian blue. Raw sienna. Raw umber. Roman ochre. **1906** R. Fry *Let.* 17 Apr. (1972) I. 263, I did the wood-work in one coat, pure raw umber and white over a burnt sienna stain. **1908** M. Morgan *How to dress Doll* ii. 20 Overcasting is only used to keep raw edges on a seam.. from fraying. **1948** F. A. Staples *Watercolour Paintings* (1951) iv. 49 *Raw sienna.* Bright yellow with slight reddish tone. Transparent. **1951** R. Mayer *Artist's Handbk.* ii. 59 *Raw umber*... Its composition is similar to that of sienna but it contains no manganese. A dark brown, its tones vary from greenish or yellowish to violet-brown. **1978** *Detroit Free Press* 5 Mar. D9/1 Stitch a one-inch hem on each side, turning under raw edge. **1979** A. V. Badgley *Rembrandt Decisions* (1980) viii. 108 Hot cups of coffee..slowly salved the raw edges of Duncan Forbes' departure.
c. Psychol. *raw feel*, a term for the immediate impression evoked by a stimulus, prior to conscious evaluation.
1932 E. C. Tolman *Purposive Behavior in Animals & Men* 250 The dyed-in-the-wool mentalist will again protest. Such discrimination-box experiments..will not and cannot convey what may be called the 'raw feel' of these discriminanda. *Ibid.* 452 *Raw feel*, a name for the peculiar *quale* of experience. **1950** *Mind* LIX. 174 What a psychology of discriminations leaves out..he calls 'raw feels'. **1956** Meehl & Sellars in Feigl & Scriven *Minnesota Stud. Philos. Sci.* I. 249 To suppose that 'raw feels' as we shall call them, will be found to be emergent..is to suppose that raw feels..are the *a*'s and *b*'s in the *generalized* function. **1969** H. D. Lewis *Elusive Mind* ix. 181 There seems, in short, to be some 'immediate data of first person experience..(e.g. directly experienced sensations, thoughts, feelings..etc.)'. These are also described in many places as 'raw feels', a somewhat inelegant but suggestive term made popular, I believe, by Professor R. W. Sellars.
6. e. Of a person: naked (esp. when sleeping). *colloq.*
1931 D. Runyon in *Hearst's International* May 64/2 He puts her in the 'Vanities' and lets her walk around raw. **1952** M. R. Rinehart *Swimming Pool* xx. 185 Or maybe she sleeps raw. **1962** J. F. Straker *Coil of Rope* vii. 69 Did I shock you? I always sleep raw. **1974** H. Waugh *Parrish for Defence* (1975) lxvii. 309 She didn't own any night-gowns. She slept raw.
9. *raw-edged* (earlier and later examples), *-jawed, -ribbed* (later example), *-seamed, -skinned, -smelling* adjs.
1828 W. Carr *Dial. Craven* (ed. 2) II. 75 *Raw-edg'd*, not hemmed, without a selfedge. **1920** E. Sitwell *Wooden Pegasus* 105 Where raw-edged shadows sting forlorn As dank dark nettles. **1972** *Ulster* (Sunday Times Insight Team) ix. 151 How raw-edged the relationship was..was demonstrated by..the first Army 'victory' in Ulster. **1932** *Flynn's* 24 Dec. 136/1 They..resort to what they call a 'cold-turkey' heel or a 'raw-jawed clout'... They refer to the act of going into a store and carrying out several articles without using any finesse at all. **1967** R. Lowell *Near Ocean* 13 The chinook Salmon..Raw-jawed, weak-fleshed. **1922** Blunden *Shepherd* 81 The young black heifer and the raw-ribbed mare. **1957** T. Hughes *Hawk in Rain* 51 Suddenly he awoke and was running-raw In raw-seamed hot khaki. **1922** Joyce *Ulysses* 233 A rawskinned crown, scantily haired. **1906** *Macmillan's Mag.* Apr. 476 Next morning I woke in the raw-smelling dawn, feeling like a corpse.
B. *sb.* **2. a.** *on the raw* (further *fig.* examples). *in the raw* (see quot. 1934); also, naked.
1915 W. S. Maugham *Of Human Bondage* 71 He had a knack of saying bitter things, which caught people on the raw. **1926** A. Bennett *Lord Raingo* II. lxxiv. 341 What got 'im on the raw was Tommy Hogarth going against 'im in that business. **1934** Webster s.v. Raw n., *In the raw*, in one's natural or crude state; hence, in one's or its true nature or character; in naked truth; as, to present life in the raw. **1941** B. Schulberg *What makes Sammy Run?* viii. 188 To go swimming in the raw. **1942** *R.A.F. Jrnl.* 27 June 24 There is a long tale of other victims of nature in the raw. **1944** E. Waugh *Diary* 16 Apr. (1976) 561 Auberon surprised her in her bath and is thus one of the very few men who can claim to have seen his great-great-grandmother in the raw. **1959** *Times* 9 Nov. 6/7 That is an argument which gets me very much on the raw. **1961** *New Eng. Bible Acts* v. 33 This touched them on the raw, and they wanted to put them to death. **1970** V. Canning *Great Affair* iv. 68 As Xavier's pyjamas were much too small for me I slept in the raw. **1972** L. P. Davies *What did I do Tomorrow?* vii. 93 My, my. Village life in the raw.
c. *the raws*, the bare fists. *slang.*
1899 C. Rook *Hooligan Nights* ii. 27 The average Hooligan ..has usually done a bit of fighting with the gloves... But he is better with the raws.
3. Also *transf.*
1928 *Daily Mail* 16 Aug. 19/3, I am not at all sure that here is not a star in the raw.
4. a. Also *Comb.* in **raw bar** *U.S.*, a bar selling raw oysters.
1943 *Sun* (Baltimore) 5 Oct. 16/6 The boys at the raw bar in the end of Bill's place last night said the way oysters are this season a feller'll have to eat shells and all to get a mess. **1973** *Washington D.C. Yellow Pages* 1314 Chuck O'Brien's Riverboat. Featuring fine seafood and steaks. Informal raw bar. Cocktail lounge.

Rawang (rǎwæ·ŋ). [Native name.] A Tibeto-Burman language.
1934 J. T. O. Barnard *Handbk. of Rāwang Dial. of Nung Lang.* p. v, This is the first book on the Nung language, which has many dialects, of which, however, Rāwang may be taken as the one most commonly spoken. **1954** E. R. Leach *Polit. Syst. Highland Burma* iii. 45 *Nung*—several distinct dialects. Rawang and Daru dialects said to be mutually unintelligible. **1964** E. A. Nida *Toward Sci. Transl.* ix. 202 In Rawang, a language of Burma, a somewhat similar distinction between dead and alive is employed, but with special restrictions. **1976** *Sci. Amer.* Oct. 140/1 A translator has had to restate most of them, from the Japanese or the Xhosa or the Rawang ('just one of hundreds of Tibeto-Burman languages spoken').

raw-boned, *a.* Add: Also *transf.*
1802 C. Wilmot *Let.* 19 Oct. in T. V. Sadleir *Irish Peer* (1920) 102 A cold wild desolate country bare and rawboned. **1886** W. Morris *Let.* 23 June in Mackail *Life Morris* (1899) II. xvi. 161 Stirling..a very raw-boned town.

raw-head¹. (Earlier and later *attrib.* examples.)
1823 Scott *St. Ronan's Well* II. vi. 110 Tell a raw-head-and-bloody-bone story about a footpad. **1918** [see *funk-hole].

rawin (rē¹·win). *Meteorol.* [f. *ra(dar + win(d *sb.*¹] A determination of the atmospheric wind speed and direction made by tracking a balloon-borne target with radar; also *transf.*, the instrument itself.
1946 *Bull. Amer. Meteorol. Soc.* XXVII. 371/1 A system of obtaining winds-aloft reports by electronic means known as rawins is now gaining great favor. **1948** T. A. Blair *Weather Elements* (ed. 3) iii. 70 By the use of radar methods developed during World War II, a balloon carrying a radar target (reflector) can be followed through and above the clouds, making possible the determination of upper-air wind direction and force in all kinds of weather... Such soundings are known as rawins. **1951** *Jrnl. Meteorol.* VIII. 126/1 Monthly resultant rawins for 24 United States stations were obtained for the layer surface to 10,000 ft and for the 10,000-ft level. **1967** R. W. Fairbridge *Encycl. Atmospheric Sci.* 581/1 It [*sc.* a radiosonde] consists of a small radio transmitter sent aloft by a helium or hydrogen-filled balloon which transmits the values of the meteorological elements in code to ground stations. If the instrument is tracked by radar to determine wind speed and direction aloft, it is called a rawin. If the two are combined in one, it is called a rawinsonde. **1979** C. Kilian *Icequake* vi. 103 A few tractors and a collapsed rawin tower were all that was left of the station.

rawinsonde (rē¹·winsǫnd). *Meteorol.* [f. prec. + *sonde.] A balloon-borne device comprising a radiosonde and a radar target which both transmits meteorological data to ground stations and permits rawin observations to be made, freq. applied to the balloon and instrument package combined.
1946 *Bull. Amer. Meteorol. Soc.* XXVII. 371/1 The recent trend in practice is to combine radiosonde observations and winds-aloft observations in one operation, a Rawinsonde. **1955** *Sci. News Let.* 24 Sept. 197/1 Equipment the Weather Bureau plans to purchase includes:..sixty-five new rawinsondes, to measure winds aloft, including the 200-mile-per-hour river of air known as the jet stream. **1959** *Jrnl. Geophysical Res.* LXIV. 1835 Because of the great altitude of the core of the 'polar-night' jet stream, only isolated rawinsonde observations have penetrated the core. **1970** *Jrnl. Atmospheric Sci.* XXVII. 420/1 Sufficient information content exists within the operational U.S. rawinsonde network to resolve the three-dimensional structure of frontal zones. **1975** *Q. Jrnl. R. Meteorol. Soc.* CI. 336 The rainfall patterns are interpreted within a framework provided by routine upper air data supplemented by long sequences of nominally 1-hourly rawinsondes.

rawky, *a.*² (later examples.)
1935 E. R. Eddison *Mistress of Mistresses* x. 194 The air between the cliffs, ruffled in mists and rawky vapours. **1936** J. G. Horne *Flooer o' Ling* 22 Or rawky day creeps up the sky.

Rawlplug (rǭ·lplʌg), *sb.* and *v.* Also **Rawl-plug**, and with small initial. [f. the name of J. J. and W. R. *Rawl*ings, English electrical engineers, who introduced it + Plug *sb.*] **A.** *sb.* A proprietary name for a kind of thin cylindrical plug, made of fibre or plastic, which can be inserted into a hole in masonry, etc., in order to hold a screw or nail. Also applied *loosely* to any plug of this type.
1912 *Trade Marks Jrnl.* 30 Oct. 1648 *Rawlplug*... A wall plug for electric wiring made of fibre. Rawlings Bros., Limited, 82, Gloucester Rd., London, S.W.; Electrical engineers. **1923** *Radio Times* 28 Sept. 35/1 For any job connected with Wireless where you use a screw..always use Rawlplugs. **1941** M. Treadgold *We couldn't leave Dinah* vii. 124 The shelf for brushes that Nick Lindsay had fixed up with rawl-plugs. **1947** 'G. Orwell' in *Tribune* 7 Feb. 12/1 The amateur handyman, with his tack hammer and his pocketful of rawlplugs. **1960** Willmott & Young *Family & Class in London Suburb* ii. 24 Her role is..to stand at the bottom of a ladder handing up his power-tool..or a box of Rawlplugs. **1962** H. Thurston *Where is thy Sting?* iv. 55 A long panel of looking-glass which Philippa had fixed with rawlplugs on one side of the fireplace. **1972** D. Haston *In High Places* i. 9 The climber drills a hole in the rock, hammers in an expansion bolt (something like the domestic Rawlplug), attaches a carabiner and proceeds in normal fashion.
Also **Rawl**, a proprietary term used *attrib.* or as a prefix in names of tools, screws, bolts, and related accessories.
1937 *Trade Marks Jrnl.* 29 Sept. 1151 *Rawl*... Cutlery and edge tools. The Rawlplug Company Limited. **1958** *Engineering* 14 Feb. 54 (Advt.), The holes are drilled with a Rawltool to the exact size, the Rawlbolts dropped in and after the machine has been positioned the bolts are tightened. **1971** C. Bonington *Annapurna South Face* 246 American rawl stud bolts for hard rock. **1976** *Shooting Mag.* Dec. 9/1 (Advt.), The cabinet has been designed for fixing to the wall by means of three Rawlbolts through strengthening bars.

B. *v. trans.* To attach by means of a Rawlplug or the like; to drill a hole in (a wall, etc.) and insert a Rawlplug. Hence **ra·wlplugging** *vbl. sb.*
1960 'A. Burgess' *Right to Answer* i. 5 He'd rawlplugged his pictures..deep into the walls. **1964** E. & M. A. Radford *Hungry Killer* xiv. 133 The bookcase, Rawlplugged to the wall. **1971** *Ideal Home* Apr. 52/1 Brass hooks rawlplugged to the wall hold it in place. **1972** R. Quilty *Tenth Session* 92 You should have Rawl-plugged the wall. You're cracking the plaster. **1974** M. Butterworth *Man in Sopwith Camel* i. 27 Plastering, paperhanging, rawlplugging, joinering.

rawly, *adv.* Add: **4.** (Later examples.)
1955 E. Blishen *Roaring Boys* ii. 99 For..two years I had been rawly warring with my classes. **1979** *Chatelaine* Jan. 64/3 The secret realm in which their love flowered—so rawly, with such unanticipated greed!
5. So as to be bare or exposed.
1924 'L. Malet' *Dogs of Want* ix. 270 Every nerve of his body seeming rawly outside his skin instead of normally and decently covered by it.

raxed, *ppl. a.* (Later example.)
1898 N. Munro *John Splendid* xv. 147 A raxed shoulder he had met with at Dumbarton.

raxing, *vbl. sb.* and *ppl. a.* (Later examples.)
1876 C. C. Robinson *Gloss. Mid-Yorks.* 110/2 A person will tell of 'a nasty raxin' pain' he is subject to. **1893** R. L. Stevenson *Catriona* I. xiii. 143 My craig'll have to thole a raxing. **1935** A. J. Cronin *Stars look Down* I. ix. 67 What he did mind was the bother when a tub ran off; it nearly killed him, the raxing and straining to lift it back upon the line.

ray, *sb.*¹ Add: **I. 1. a.** Fig. phr. (*little*) *ray of sunshine*, a person (freq. a young woman) who enlivens or cheers another; a happy or vivacious person. Cf. Sunshine *sb.* 2 a.
1915 A. Bennett *These Twain* (1916) xx. 485 You're a little ray of sunshine, and all that, and I'm the first to say so. **1929** J. B. Priestley *Good Companions* II. iv. 364 Why are you now our little ray of sunshine? **1959** M. Scott *White Elephant* v. 56 Are you two in this to make money or just to be little rays of sunshine? **1972** C. Fremlin *Appointment with Yesterday* iv. 31 Milly rather fancied herself in the rôle of little ray of sunshine to brighten his declining years. **1978** 'M. M. Kaye' *Far Pavilions* xxxvii. 540 He hasn't exactly been a ray of sunshine up to now.
5. c. Chiefly *Science Fiction.* A supposed destructive beam of energy emitted by a ray-gun or similar device. Cf. *death-ray* s.v. *death *sb.* 19.
1898 H. G. Wells *War of Worlds* vi. 39 Only the fact that a hummock of heathery sand intercepted the lower part of the Heat-Ray saved them. **1919** G. B. Shaw *Heartbreak House* 39, I will discover a ray mightier than any X-ray: a mind ray that will explode the ammunition in the belt of my adversary before he can point his gun at me. **1926** G. Hunting *Vicarion* xiii. 215 I'm glad they never perfected that ray they used to talk about for disposing of an enemy at a distance without betraying the disposer. **1940** Graves & Hodge *Long Week-End* vi. 93 An inventor..claimed to have produced a ray that would set fire to anything inflammable. **1969** E. von Daniken *Chariots of Gods?* ii. 25 They will hammer and chisel in the rock pictures of what they had once seen: Shapeless giants,..staves from which rays are shot out as if from a sun.
8. c. (Later examples.)
1925 Eames & MacDaniels *Introd. Plant Anat.* vii. 176 The ray is more or less like a brick wall, the individual cells representing the bricks. **1953** K. Esau *Plant Anat.* xi. 252 The dicotyledons typically contain only parenchyma cells in the rays.
9*. *Astr.* Any of the long bright lines of pale material that can be seen to radiate from some lunar craters.
1838 J. P. Nichol *Phenomena & Order Solar Syst.* II. vi. 171 The most remarkable circumstance connected with this variety in the Moon's shining power is those rays issuing chiefly from craters and extending over a large space. **1873** R. A. Proctor *Moon* iv. 253 The telescope..has discovered numerous small craters of varying depth in the midst of many of the rays, and it reveals the fact, that these small craters..do not penetrate through the matter we are examining, inasmuch as there comes from their bases always the same kind of light that characterizes the ray. **1895** T. G. Elger *Moon* 27 The rays emanating from Tycho surpass in extent and interest any of the others. **1922** H. S. Jones *Gen. Astron.* iv. 102 From some of the craters, under favourable conditions of illumination, bright rays or streaks can be seen radiating radially in all directions. **1962** *Listener* 1 Feb. 223/2 The mysterious lunar rays issuing from Tycho, Copernicus, and other craters also fit better into an igneous theory. The rays cross mountains, walled formations, ridges, and seas without marked deviation.
10. a. *ray-shorn* adj.
1872 Geo. Eliot *Middlem.* II. xxxvii. 265 The other great dread—of himself becoming dimmed and for ever ray-shorn in her eyes.
b. In sense 8 c, as *ray cell, initial, tracheid.*
1907 D. P. Penhallow *Man. N. Amer. Gymnosperms* v. 83 Pits on the lateral walls of the ray cells are an invariable feature of all investigated species of..Coniferales. **1933** *Forestry* VII. 93 It is essential to study..the development of the ray cells in the wood. **1953** K. Esau *Plant Anat.* vi. 126 The ray initials give origin to the ray cells. **1975** *Sci. Amer.* July 102/2 Among the components of the cambium are what are called ray initials; the continuation of a ray initial down into the sapwood of a stem, a branch or a trunk is known as a wood ray. **1907** D. P. Penhallow *Man. N. Amer. Gymnosperms* vi. 88 In the higher Coniferæ the medullary ray is distinguished by the presence of an element which differs materially in its structure from the associated parenchyma cells. These elements have been designated as ray tracheids. **1940** Brown & Panshin *Comm. Timbers U.S.* vii. 128 Ray tracheids attain their best development in the genus *Pinus.* **1956** F. W. Jane *Structure of Wood* v. 91 Ray tracheids often form the marginal cells of the rays.
c. In sense 9 a, *ray-finned* adj.
1933 A. S. Romer *Vertebr. Paleontol.* iv. 85 That [*sc.* the history] of the later ray-finned fishes has no such interest. **1968** [see *lung-fish* s.v. *lung *sb.* 7]. **1970** R. M. Black *Elements Palaeont.* xvii. 249 The ray-finned fish have had an expansionist evolution.
d. ray blight, a fungus disease of chrysanthemums caused by *Ascochyta chrysanthemi*, which attacks the flowers, causing discoloration and shrivelling of the petals; **ray diagram**, a diagram showing the paths of light rays through an optical system; **ray-fin**, a fish belonging to the subclass Actinopterygii, to which most living bony fish belong and which includes those having thin fan-like fins with dermal rays; **ray fleck**, the marking caused by the exposure of a ray in sawn timber; **ray gun**, a hand-held device that can be made to emit rays, esp. (in *Science Fiction*) destructive or harmful ones; **ray therapy treatment**, the treatment of disease with radiation; radio-therapy; **ray-tracing**, the calculation of the path taken by a ray of light through an optical system.
1907 F. S. Stevens in *Bot. Gaz.* XLIV. 241 The Chrysanthemum Ray Blight... The common name chosen for the disease..is taken from the most conspicuous symptom of the malady, a blighting of the corolla. **1961** *Amat. Gardening* 21 Oct. 6/2 Ray blight is much less common than the other two bloom diseases. **1965** Nakajima & Young *Art of Chrysanthemum* vii. 81 If the ray blight is not checked, it may continue on to destroy all blooms in the immediate area. **1980** J. W. Hill *Intermediate Physics* xii. 123 Draw two ray diagrams to show how a real and virtual image may be obtained of an object placed the same distance away from two different mirrors. **1945** A. S. Romer *Vertebr. Paleontol.* (ed. 2) v. 89 Most of the more characteristic Paleozoic ray-fins were once assigned to *Palaeoniscus.* **1963** P. H. Greenwood *Norman's Hist. Fishes* (ed. 2) xvii. 306 The Bony Fishes can be divided into three main groups or subclasses: Actinopterygii (ray-fins), Crossopterygii (fringe-fins) and Dipneusti (lung-fishes). **1934** Brown & Panshin *Identification Comm. Timbers U.S.* 211 Ray fleck: a portion of a ray as it appears on the quarter surface. **1940** —— *Comm. Timbers U.S.* viii. 201 Some woods possess low, closely spaced, but relatively conspicuous ray flecks. **1968** *Canad. Antiques Collector* July 26/1 Quarter sawed figure is characterised by the annual growth rings appearing as parallel stripes and by the appearance of rays on the surface. In such woods as oak and chestnut these rays are called ray fleck or flake. **1931** *Amazing Stories* Dec. 804/1 The rayguns of the battlecraft, being of superior range, melted down the mortars of the fort at the magazine. **1951** A. C. Clarke *Sands of Mars* iv. 40 It was a modified air pistol... 'If you say it's like a ray-gun I'll certify you.' **1957** [see *bug-eyed *a.*]. **1958** *Spectator* 19 Sept. 379/1 But as a space-veteran who once triggered a ray-gun with Flash Gordon, let me advise you to read on. **1967** *Autocar* 28 Dec. 29/3 As the car nears each set of lamps a patrolman..points the ray gun at the cell situated between the two lamps. A beamed radio signal from the gun activates the fog warning lamp switch. **1977** W. McIlvanney *Laidlaw* xxvi. 116 It was a beautiful smile... It hit Harkness like a ray-gun and he felt his concentration atomise. **1928** *Daily Express* 20 Dec. 8/3 When the phrase 'ray-therapy' crept into one of the royal bulletins, I heard educated persons explaining that it meant treatment by wireless! **1943** *Gloss. Terms Electr. Engin.* (*B.S.I.*) 144 *Radio-therapy*, [deprecated synonym] *ray therapy*, the treatment of diseases by radiation. **1918** L. Silberstein *Simplified Method of tracing Rays* p. v, Our purpose is not to treat the whole subject of geometrical optics, but..that part of it which is called by the short name of 'ray tracing'... Given the ray incident upon any system of lenses..find the emergent ray. **1943** D. H. Jacobs *Fund. Optical Engin.* xxiv. 381 Ray-tracing equations are all derived from one *exact* law: Snell's law. **1974** W. T. Welford *Aberrations of Symmetrical Optical Syst.* iii. 41 This process of finding a ray path in terms of the numerical values of the incidence heights and convergence angles at each surface in turn is called raytracing. **1904** *Science Siftings* 12 Mar. 320/2 The Finsen light concentrates as much violet rays as can be found in a hundred square feet of sunlight. The same principle enters into all ray treatment. **1905** *Westm. Gaz.* 4 May 12/2 Six patients suffering from skin diseases..died after the ray-treatment.

Ray (rē¹). The name of the English naturalist, John *Ray* (1627–1705), used in the possessive to designate **Ray's bream**, a deep-bodied, dark brown and silver, European, marine fish, *Brama brama* (formerly *B. raii*), of the family Bramidæ, which was named in his honour by M. E. Bloch (*Ichtyologie* (1797) VII. 75).
1836 W. Yarrell *Hist. Brit. Fishes* I. 117 Ray's Bream.. appears to have been less perfectly known to the older writers than might have been expected. **1925** J. T. Jenkins *Fishes Brit. Isles* 74 Ray's Bream is a fish of characteristic appearance, with a body elevated and compressed, and a very long dorsal fin. **1959** A. C. Hardy *Open Sea* II. x. 208 This species should not be confused with Ray's bream.. which has also been called the black sea-bream. **1969** A. Wheeler *Fishes Brit. Isles & N.-W. Europe* 339/2 Ray's bream has little commercial value in northern waters though its flesh is very palatable.

ray, *v.*¹ Add: **4. a.** (Later examples.)
1899 P. H. Wicksteed tr. *Dante's Paradiso* 341 A point I saw which rayed forth light. **1922** E. R. Eddison *Worm Ouroboros* xxx. 372 Yellow flames of candles..on either side of the mirror rayed forth tresses of tinselling brightness.
6. *trans.* To treat with, or examine by means of, X-rays or other invisible radiation.
1921 *Science* 23 Sept. 278/1 The total number of offspring by the pairs in which the females were rayed..was 2460. **1933** *Discovery* Feb. 46/2 Tissues taken from an animal which had been rayed [with doses of Gamma rays]. **1955** *Proc. Nat. Acad. Sci.* XLI. 155 The *Tradescantia* microspore chromosomes react..as double threads when rayed at prophase.
raying *vbl. sb.*¹ (further examples, without *out*).
1921 *Science* 23 Sept. 278/1 Eggs which were laid during the first six days after raying and mating. **1933** *Discovery* Feb. 46/2 The dose of gamma rays needed to kill a culture at once..is enormous, and as the dose decreases the interval between the raying and the death of the culture becomes larger. **1955** *Proc. Nat. Acad. Sci.* XLI. 150 Accurate analysis of chromosomal aberrations could not be made until about 6 hours after raying.

rayed, *ppl. a.*¹ Add: **2.** = *irradiated *ppl. a.* 1 c.
1921 *Science* 23 Sept. 277/2 Wild type (red-eyed) females ..were X-rayed... Sisters of the rayed females were used as controls. **1938** Hevesy & Paneth *Man. Radioactivity* xxiv. 245 The rayed rock-salt assumes a blue-violet colour.

rayed, *ppl. a.*² (Later *arch.* or *poet.* examples.)
1905 W. H. Hunt *Pre-Raphaelitism* I. 163 From the depth of this rayed region we ascended to the further margin of the mist lake into the crystal air. **1918** W. Stevens in *Others* Dec. 11 We hang like warty squashes, streaked and rayed.

ray·ing, *ppl. a.* [f. Ray *v.*¹ + -ing².] **a.** Moving in rays. **b.** Emitting rays; radiating.
1891 G. Meredith *One of our Conq.* III. vii. 131 Popular artists..have figured in scenes of battle the raying fragments of a man from impact of a cannon-ball on his person. **1905** *Westm. Gaz.* 25 Apr. 2/3 The day That crowns us royal with the raying sun.

rayl (rē¹l). *Acoustics.* [f. *Rayl(eigh.] A unit of specific acoustic impedance equal to one dyne-second/cm.³ (in the C.G.S. system) or one newton-second/m.³ (in the S.I.).
1954 L. L. Beranek *Acoustics* i. 11 The specific acoustic impedance is the complex ratio of the effective sound pressure at a point of an acoustic medium or mechanical device to the effective particle velocity at that point. The unit is newton-sec/m³, or the mks rayl. (In the cgs system the unit is dyne-sec/cm³, or the rayl.) **1971** W. W. Seto *Schaum's Outl. Theory & Probl. Acoustics* ii. 40 At standard atmospheric pressure and 20 °C..the characteristic impedance of air is..415 rayls.

Rayleigh (rē¹·li). *Physics.* [The title of J. W. Strutt, 3rd Lord *Rayleigh* (1842–1919), English physicist.] **1.** Used, usu. *attrib.*, to designate various concepts, devices, and phenomena he invented or investigated, as **Rayleigh('s) criterion**, the criterion by which adjacent lines or rings of equal intensity in a diffraction pattern are regarded as resolved when the central maximum of one coincides with the first minimum of the other; **Rayleigh disc**, a lightweight disc suspended by a fine thread so that when it is placed at an angle to incident sound waves their intensity can be calculated from the measured torque on the disc; **Rayleigh instability** (see quot. 1977); also called *Rayleigh–Taylor instability* [Sir Geoffrey *Taylor* (1886–1975), English mathematician]; **Rayleigh limit**, the upper limit of a quarter of a wavelength placed on the difference between the optical paths of the longest and shortest rays of those going to form an image in order that the definition shall be close to the ideal (which corresponds to no path difference); **Rayleigh number**, a dimensionless parameter that is a measure of the instability of a layer of fluid due to differences of temperature and density at the top and bottom (see quot. 1950); **Rayleigh scattering**, the scattering of light by particles small compared with its wavelength, the intensity of the scattered light being inversely proportional to the fourth power of the wavelength (and therefore much greater for blue light than for red); so **Rayleigh-scattered** *a.*; **Rayleigh wave**, a type of wave that travels over the surface of a solid with a speed independent of its wavelength, the motion of the particles being in ellipses so that the surface undulates.
1937 Jenkins & White *Fund. Physical Optics* v. 123 Extending Rayleigh's criterion for the resolution of diffraction patterns..to the circular aperture, two patterns are

said to be resolved when the central maximum of one falls on the first dark ring of the other. *Ibid.* vii. 159 The Rayleigh criterion for resolving of images. **1970** D. W. Tenquist et al. *University Optics* II. v. 197 The chromatic resolving power of a prism, defined as $\lambda/\delta\lambda$ where $\delta\lambda$ is the smallest change of wavelength discernable [*sic*] in accordance with the Rayleigh criterion at a mean wavelength λ, is given by [etc.]. **1913** *Physical Rev.* I. 309 (*heading*) A method of producing known relative sound intensities and a test of the Rayleigh disk. **1972** J. M. Taylor tr. *Meyer & Neumann's Physical & Appl. Acoustics* vi. 209 The Rayleigh disk..is practically never used any more to determine particle velocity, which can be derived much more quickly and conveniently from electroacoustic sound pressure measuring devices. **1961** S. Chandrasekher *Hydrodynamic & Hydromagnetic Stability* x. 428 An important special case ..is that of two fluids of different densities superposed one over the other (or accelerated towards each other); the instability of the plane interface between the two fluids, when it occurs (particularly in the second context), is called Rayleigh-Taylor instability. **1971** I. G. Gass et al. *Understanding Earth* xix. 277/1 A Rayleigh instability..does not necessarily depend upon the existence of a density inversion. .. Where the depth of the fluid is very large, the fluid at the bottom is compressed by the overlying fluid and, in many cases, an instability develops before the temperature is high enough to produce a density inversion. **1977** *Sci. Amer.* Oct. 144/2 The instability at the interface between a denser fluid overlying a lighter fluid, when the interface is otherwise in hydrostatic equilibrium, is called a Rayleigh instability (or sometimes a Rayleigh–Taylor instability). **1923** Glazebrook *Dict. Appl. Physics* IV. 216/1 The adoption of the Rayleigh limit thus makes it possible considerably to increase the aperture of a lens system of any given type and to come close to the full theoretical resolving power with systems which, judged geometrically, would appear hopelessly over- or under-corrected. **1976** *Sci. Amer.* Aug. 77/2 Ideally a lens should be at the Rayleigh limit for light of all wavelengths. **1950** O. G. Sutton in *Proc. R. Soc.* A. CCIV. 298 The existence of a sustained convective regime depends upon the value of the non-dimensional quantity $Ra = -\beta g\alpha h^4/\kappa\nu$, which we shall call the Rayleigh number. **1980** *Sci. Amer.* July 82/2 Convection begins when the Rayleigh number exceeds a critical value. **1971** *Physics Bull.* July 387/1 More detailed studies of the linewidths of Rayleigh scattered lines can provide hitherto inaccessible information on the behaviour of fluids near their critical point or in the neighbourhood of phase transitions. **1937** Jenkins & White *Fund. Physical Optics* xii. 280 The first quantitative study of the laws of scattering by small particles was made in 1871 by Lord Rayleigh, and such scattering is frequently called Rayleigh scattering. **1973** C. Sagan *Cosmic Connection* (1975) xiii. 90 The beauty of the sunset, the sky, and distant landscapes are all due to Rayleigh scattering. **1920** A. E. H. Love *Treat. Math. Theory Elasticity* (ed. 3) xiii. 313 The waves travel over the surface with a velocity, which is independent of the wave-length $2\pi/f$, and slightly less than the velocity of equivoluminal waves propagated through the body. Waves of this kind are often called 'Rayleigh-waves'. **1956** J. C. Jaeger *Elasticity, Fracture & Flow* iii. 135 Rayleigh waves are not the only simple type of surface wave which can be predicted and identified. **1971** I. G. Gass et al. *Understanding Earth* xxiv. 336/1 (*caption*) A Rayleigh wave from an earthquake in Columbia recorded in Montana.

2. *Astr.* Also **rayleigh.** A unit of luminous intensity equal to one million photons per square centimetre per second.

1956 D. M. Hunter et al. in *Jrnl. Atmospheric & Terrestrial Physics* VIII. 345 We suggest that $4\pi B$ be given the unit of 'rayleigh' (symbol *R*), where *B* is in units of 10^6 quanta/cm^2 sec sterad. Hence $1R = 10^6$ quanta/cm^2 (column) sec. (The word 'column' is often inserted into these units to convey the concept of an emission-rate from a column of unspecified length). **1970** *Nature* 2 May 435/2 In the direction of maximum intensity the Lyman-α flux is 160 Rayleighs, which can be regarded as typical for the direction of the solar apex. **1974** *Science* 25 Jan. 317 A two-channel ultraviolet photometer aboard Pioneer 10 has made several observations of the ultraviolet glow in the wavelength range from 170 to 1400 angstroms in the vicinity of Jupiter. Preliminary results indicate a Jovian (1216 angstrom) glow with a brightness of about 1000 rayleighs and a helium (584 angstrom) glow with a brightness of about 10 to 20 rayleighs.

Rayleigh–Jeans (rēⁱ·li,dʒī·ns). *Physics.* The name *Rayleigh and that of Sir James *Jeans* (1877–1946), English physicist and astronomer, used *attrib.* with reference to an approximation to Planck's law (see *Planck) that is valid at long wavelengths, according to which the flux of radiant energy from a perfect radiator, at any particular wavelength, is proportional to its temperature divided by the fourth power of the wavelength.

1930 Ruark & Urey *Atoms, Molecules & Quanta* iii. 57 We obtain the radiation density, $\rho_\lambda d\lambda = 8\pi kT\lambda^{-4}d\lambda$. This is called the Rayleigh–Jeans distribution law. *Ibid.* 59 Planck's law approaches the Rayleigh–Jeans law if *h* approaches zero. **1948** Worthing & Halliday *Heat* xiii. 445 Rayleigh erroneously included a factor of 8 in his evaluation of c_1. The error was pointed out by the English physicist James Jeans (1877–). For that reason, the law is often referred to as the Rayleigh–Jeans law. Just the reverse of what occurred in connection with the Wien law, the Rayleigh law predicts correct values in the long-wavelength region but fails elsewhere. **1970** G. K. Woodgate *Elem. Atomic Struct.* iii. 36 The approximation corresponds, for thermal radiation, to..$\bar{n} \approx kT/h\nu \gg 1$, which in eq. (3.16) or (3.13) leads at once to the Rayleigh–Jeans approximation for the energy density per unit frequency range.

rayless, *a.* Add: †**2. c.** *Physics.* Not accompanied by or emitting alpha, beta, or gamma rays. *Obs.*

1904 *Proc. R. Soc.* LXXIII. 493 The first change is a 'rayless' one, *i.e.*, the transformation is not accompanied by the appearance of α, β, or γ rays. **1906** [see *Actinium 2]. **1907** N. R. Campbell *Mod. Electr. Theory* ix. 210 It appears that many changes which are usually classed as 'rayless' are accompanied by the emission of β rays without α rays, but the liberation of energy in such changes is so small compared to those in which α rays are emitted that it seems desirable to make the distinction implied by the term.

raymonder (rēⁱ·mǫndəɹ). *Cricket.* = *Ramrod 2 a.

1870 [see *ramrod 2 a]. **1878** H. C. Adams *Wykehamica* xxiii. 431 *Raymonder*, a ball bowled underhand, in a series of hops along the ground, (traditionally said to be derived from one Raymond, who bowled after this fashion). Sometimes it was pronounced 'ramroder'.

Raynaud (rēⁱ·nōᵘ). *Med.* [The name of Maurice *Raynaud* (1834–81), French physician, who described various cases displaying Raynaud's phenomenon in 1862 (*De l'Asphyxie Locale et de la Gangrène Symétrique des Extrémités*).] *Raynaud's disease* or *syndrome*: an ill-defined disease or syndrome characterized by *Raynaud's phenomenon*, in which spasm of the arteries of the digits (often due to low temperature or vibration) leads to pallor, pain, and numbness, and in severe cases to gangrene.

1883 *Trans. Clin. Soc.* XVI. 179, I have watched three cases which came within the category of Raynaud's disease. **1901** J. Hutchinson in *Med. Press & Circular* CXXIII. 403/1 The expression 'Raynaud's disease' would imply that there is some one malady complete in itself, and having all the symptoms the same in all cases which is suitably denominated by that name. That is not the case. *Ibid.*, I would rather speak of Raynaud's phenomena than of Raynaud's disease, for the former are things which we understand and are the same in all cases... What do we mean by Raynaud's phenomena?.. Local syncope, local asphyxia, symmetrical gangrene of the extremities are synonymous terms. **1925** Raynaud's disease [see *ganglionectomy]. **1932** *Amer. Jrnl. Med. Sci.* CLXXXIII. 188 The increasing amount of literature attests to the tendency to utilize the terms 'Raynaud's disease' or 'Raynaud's syndrome' as a general depository for a heterogeneous group of cases far removed from the condition originally described by Raynaud. **1936** *Q. Jrnl. Med.* XXIX. 399 For more than sixty years the term 'Raynaud's disease' was used as a convenient label for case after case of obscure aetiology in which pallor, cyanosis, pain, or gangrene of hands, feet, nose, or ears, happened to be symptoms, prominent or otherwise. *Ibid.* 401 'Raynaud's Phenomenon' may, therefore, be defined as 'Intermittent pallor or cyanosis of the extremities, precipitated by exposure to cold, without clinical evidence of blockage of the large peripheral vessels and with nutritional lesions, if present at all, limited to the skin'. **1937** Raynaud's syndrome [see *ganglionectomy]. **1946** E. V. Allen et al. *Peripheral Vascular Dis.* vii. 185 The predilection of Raynaud's disease for the female is one of the outstanding etiological factors. *Ibid.* viii. 206 Raynaud's phenomenon may occur primarily as in Raynaud's disease or it can occur secondarily in association with a number of conditions and diseases. **1973** *Times* 26 May 3/3 The name for the unusual affliction was Raynaud's Phenomenon, Mr Alan Lipfriend, Mr Lambert's counsel, told Mr Justice Mocatta. It was also known as vibratory white finger. The fingers went white, numb and stiff. **1975** *Daily Colonist* (Victoria, B.C.) 18 Nov. 2/1 My doctor says I have Raynaud's disease... It is just like I am allergic to cold... My hands and feet are affected and hurt.

rayograph (rēⁱ·ogrɑf). Also with capital initial. [f. the name of Man *Ray* (1890–1976), U.S. artist and photographer + -o + -graph.] A type of photograph made without a camera by arranging objects on light-sensitive paper which is then exposed and developed. Cf. *photogram 3. Also **ray·ogram.**

1932 *N.Y. Times* 17 Apr. viii. 11/5 Julien Levy Gallery—Photographs by Man Ray. A fine retrospective, including a group of 'rayograms'. **1937** *Photography 1839–1937* (Museum of Mod. Art, N.Y.) 68 Man Ray refers to his shadowgraphs as 'rayographs' or 'rayogrammes'. **1942** P. Guggenheim *Art of This Century* 106 Man Ray..took up photography in 1917. Invented Rayograph technique, 1921, and explored other possibilities of photography, especially in making Dada and Surrealist compositions. **1951** J. I. H. Bauer *Revolution & Tradition Mod. Amer. Art* iii. 28 Dada attracted only one distinguished American follower, the painter and photographer Man Ray, much of whose life was spent in Paris where he produced without the use of a camera his extraordinary 'rayographs'. **1956** *Focal Encycl. Photogr.* 836/2 Further developments came about 1921 when Man Ray and L. Moholy-Nagy..made their 'rayographs' and photograms, using not merely opaque flat objects but also three-dimensional and translucent ones. **1972** C. W. E. Bigsby *Dada & Surrealism* ii. 19 The photomontage consisted of a collage of photographs. Like the Rayogram (a photographic process devised by Man Ray) it was a joke at the expense of realism. **1974** *Encycl. Brit. Macropædia* XIV. 320/2 Man Ray was supporting himself by making fashion photographs in 1922 when he accidentally set a glass funnel, a graduate, and a thermometer on a piece of photographic paper, thus producing 'Rayographs'. **1975** *New Yorker* 19 May 13/1 (Advt.), The prices are more out of the ordinary: $7,500 is asked for Ray's rayograph 'Egg-beater and Abstracted Segment of Living Space'.

rayon¹. Add: **3. a.** Any of the class of fibres and filaments composed of or made from regenerated cellulose; also, fabric or cloth made from these. Formerly known as *artificial silk*.

1924 *Drapers' Record* 14 June 685/2 'Glos' having been killed by ridicule, the National Retail Dry Goods Association of America has made another effort to produce a suitable name for artificial silk. This time their choice has fallen on 'rayon'. **1925** *Glasgow Herald* 26 Mar. 15/1 The Viscose Company states that it will discontinue the use of wood pulp as a base for rayon when its wood pulp contracts expire. **1927** T. Woodhouse *Artificial Silk* 1 The sight of almost any article made from artificial silk (or Rayon, as it is also called) is sufficient to arouse admiration. **1951** *Good Housek. Home Encycl.* 230/2 Rayons are classified according to the highly technical processes by which they are manufactured. **1966** [see *man-made a.]. **1969** *Encycl. Polymer Sci. & Technol.* XI. 844 High-tenacity rayons are consumed by industry as reinforcing cords for manufacturing all types of rubber tires, drive belts, high-pressure hoses, and straps and tapes. **1973** H. McCloy *Change of Heart* ii. 18 Her stockings were real silk, not flimsy nylon..or coarse rayon.

b. *attrib.* and *Comb.*, as *rayon damask, gabardine, jersey, satin, stocking, taffeta, yarn*; *rayon-containing, -corded* adjs.

1930 *Daily News Rec.* 17 Feb. 19/2 The manufacture of rayon-containing fabrics normally is a highly competitive business. **1964** *Economist* 26 Sept. 1254/1 The rayon-corded SP tyres. **1952** M. Laski *Village* iv. 66 The rayon-damask couch of the three-piece suite. **1930** *Silk & Rayon Directory & Buyer's Guide of Gt. Brit.* 296/2 (*heading*) Gabardine, Rayon. **1947** *Sun* (Baltimore) 31 Oct. 3/7 (Advt.), Confident of its own good looks, this rayon gabardine wins your heart at once. **1965** *Which?* Mar. 94/2 *Rayon garberdine*, a fabric with a diagonal rib effect. *Ibid.* 94/3 *Rayon jersey*, a soft stretch, knitted fabric. Drapes well. **1973** *Guardian* 19 June 15/1 Matte viscose rayon jersey, long evening dress. **1977** B. Pym *Quartet in Autumn* v. 51 The dressing gown was a jazzy rayon satin. **1929** *Rayon Record* III. 587/1 The lower temperature up to about 110° is utilized for silk or rayon stockings. **1948** 'J. Tey' *Franchise Affair* xi. 119 Fawn-grey rayon stockings. **1974** M. Kelly *That Girl in Alley* iv. 70 She was wearing..beige rayon stockings. **1952** M. Laski *Village* viii. 137 A counterpane of rayon taffeta machine-embroidered with flowers. **1929** *Rayon Record* III. 411/2 A few samples of yarns and fabrics illustrating the decorative value of rayon yarns..have been received. **1947** *British Rayon Man.* x. 168 Much attention was given to the question of the best kind of package for rayon yarns.

‖ **rayon²** (rayǫ·n). Also **raion.** [a. Russ. *raión*.] In the U.S.S.R., a small territorial division for administrative purposes.

1936 [see *oblast]. **1948** J. Towster *Polit. Power in U.S.S.R.* iv. 66 All the units are divided into districts (*raions*). **1959** *Economist* 14 Mar. 946/1 In at least two of Moscow's fifteen raions, the chaps at the local Agitpunkts seem to have been lying down on the job. **1964** S. P. Dunn tr. *Levin & Potapov's Peoples of Siberia* 9 The creation in 1931–1932 of nomadic and rural soviets, rayons and national okrugs on a territorial basis finally undermined the importance in the social structure of the peoples of the North, of their former clan and tribal organizations and of the social elements which headed them. **1976** [see *okrug].

Rayonism, Rayonnism (rēⁱ·ǫniz'm). Also rayon(n)ism, ‖ **rayonnisme.** [ad. F. *Rayonnisme*, f. *rayon* Rayon¹ + *-isme* -ism; cf. Russ. *luchizm*, f. *luch* ray.] A style of abstract painting developed *c* 1911 in Russia by M. Larionov (1881–1964) and N. Goncharova (1881–1962), in which projecting rays of colour are used to give the impression that the painting floats outside time and space. Hence **Ray·on(n)ist** *a.*, of or pertaining to Rayonism; also as *sb.*; **Rayoni·stic** *a.*

1922 *Encycl. Brit.* XXXII. 9/1 Gontcharova's setting for the 1914 production of the 'Coq d'Or' and Larionov's 'Les Contes Russes' of 1915 mark the invasion of the theatre by cubist ideas. The colour scheme was still that of Russian peasant art; but the design was based on abstract forms, and aimed at a rhythm in harmony with the music and the dances. To this development the name of *rayonnisme* has been given. **1956** B. S. Myers *Encycl. Painting* 295 *Larionov, Michel* (1881–), Russian abstract painter who in 1909 developed a type of painting known as Rayonism, a dynamic form of space penetration consisting of rays of light and suggesting in some ways the work of the Futurists. **1956** Lake & Maillard *Dict. Mod. Painting* 241/1 *Rayonism*,..launched by Michael Larionov in 1911–12... A Rayonist canvas must give the impression of gliding out of time. **1968** D. Barran tr. *Veronesi's Into Twenties* iii. 76 Larionov founded the rayonnist movement, loosely based on the concepts of the futurist movement. **1969** Denis & de Vries *World's Art* II. xi. 224 Rayonism in Russia was of the same nature, (*Rayonistic Manifesto*, 1912, by Larionov). **1972** C. W. E. Bigsby *Dada & Surrealism* ii. 10 In some ways it [*sc.* Dada] was a part of that artistic re-examination which spawned such schools as impressionism, cubism, futurism and, more exotically, suprematism, rayonism, plasticism, vorticism and synchronism. **1975** *Physics Bull.* Feb. 60/3 The art world was no less fertile with the cubists, the futurists, vorticists, rayonnists and the Blauer Reiter group all active. *Ibid.* 61/1 Rayonnism was a style of painting invented by Mikhail Larionov and used by him and Natalia Goncharova around 1912–14. **1977** *New Yorker* 2 May 31/3 What makes it unique is the inclusion of some dazzling experimental pictures from the early twentieth century—Cubist, Futurist, Rayonist, and Suprematist.

‖ **rayonnant** (rẹyonaṅ), *a.* Also fem. **rayonnante** (-aṅt); pl. **rayonnants.** [Fr.] Beaming,

radiant. Also **rayonnant de joie**, radiant with joy.

The form with *-te* in quot. 1825 is erron.

1821 M. W. Shelley *Let.* 2 Apr. in P. B. Shelley *Lett.* (1964) II. 278 Yesterday he came *rayonnant de joie*—he had been ill for some days, but he forgot all his pains. **1825** H. Wilson *Mem.* II. 79 The next evening found us all quite rayonnante, waiting for our dinner. **1831** C. C. F. Greville *Mem.* (1874) II. xiii. 111 The Ministers were *rayonnants*. **1965** Lady Birkenhead *Illustrious Friends* ix. 92 Her ladyship was rayonnante... She sallied forth in a blue silk ball-dress and lively spirits.

‖ **rayonné,** *a.* Add: **3.** Of a person: radiant, effulgent.

1860 Queen Victoria *Let.* 20 June in R. Fulford *Dearest Child* (1964) 260 She is so embellié and rayonné as to look like a young girl.

‖ **rayonnement** (rẹyonmaṅ). [Fr.] Radiance, effulgence; influence (of culture, etc.).

1910 Wyndham Lewis *Lett.* (1963) 45 The benevolence and rayonnement that is the sign and beauty of a fine nature shines on faults without hiding them. **1966** *Economist* 23 Apr. 340/2 Nor are the producers allowed to show the film outside France, because it might damage the reputation of 'communities' which contribute to the 'cultural and humanitarian *rayonnement* of France'.

Rayonnism, Rayonnist: see *Rayonism, Rayonnism.

‖ **Raza** (rā·sa). [Mexican Sp., a. Sp. *raza* race.] Usu. in phr. *La Raza*, the race, designating the strong sense of racial and cultural identity held by Mexican-Americans.

1964 W. Madsen *Mexican-Americans of S. Texas* iii. 15 The Mexican-American thinks of himself as both a citizen of the United States and a member of *La Raza* (The Race). **1968** *Economist* 8 June 54/1 The preservation of *La Raza* within the dominant American culture means that Mexican-American families observe the Roman Catholic faith, speak Spanish and yield to the male the authority associated with Mediterranean custom. **1969** *Time* 4 July 14/2 *La Raza*, the race, meaning all Mexicans and Mexican Americans, and derived from the mystical theory of the 19th century philosopher, José Vasconcelos, that people of mixed race will inherit the earth. At best, it is a rallying cry betokening a mild form of cultural nationalism; at worst, it connotes outright racism. **1973** *Black Panther* 7 Apr. 4/3 The Oakland Mexican-American 'Raza' community continues its boycott of public schools.

‖ **Razakar** (razākā·ɹ). Also **Razakhar.** [Urdu *razākār*.] A Muslim who voluntarily pledges to fight in defence of his religion; hence, a member of a fanatical semi-military faction with this end. Also *attrib.*

1948 *Keesing's Contemp. Archives* 31 July–7 Aug. 9421 Hyderabad's Moslems formed the Razakar (volunteer) movement which, in recent months, has in effect become the private army of the Moslem party in Hyderabad. **1957** P. Griffiths *Mod. India* xii. 107 The Ittehad-ul-Muslimeen, with its semi-military organisation known as the Razakars, took a bitterly communal line. Each Razakar vowed to 'fight to the last to maintain the supremacy of the Muslim power in the Deccan'. **1968** H. Gray in M. Weiner *State Politics in India* viii. 402 Kasim Razvi organized a voluntary group of fighters named 'Razakars', who provided protection to landowners and the government administration during a Communist-led uprising. **1970** R. Wingate *Ismay* viii. 180 Attlee replied in a long personal letter [*c.* Sept.–Oct. 1948] pointing out that in fact the Nizam had not been a free agent, but in the hands of the 'Razakars' (the extreme Muslim party). **1971** *Guardian* 29 Oct. 13/1 They [*sc.* Pakistani army units] have left in the countryside a patchwork of police and Razakhar regimes. **1971** *Peace News* 29 Oct. 5/1 The mukti fouj attacked a radio station occupied by the Pakistani army and their civilian hirelings, the *razakars*. **1972** M. Shakir *Muslims in Free India* iv. 56 The emergence of the Razakars was a logical corollary of the Ittehad's political doctrine of collective Muslim sovereignty... It is the biggest exclusive Muslim Party in Hyderabad. **1974** *Encycl. Brit. Macropædia* IX. 76/1 Immediately after Indian independence a fanatical Muslim faction, the Razākārs, fomented tensions in the state and the city [*sc.* Hyderabad].

razamataz(z, varr. *Razzmatazz, razzamatazz.

razee, *sb.* Add: Also † 8 **raze.** (Earlier example.)

1794 R. F. Greville *Diary* 14 Sept. (1930) 335 Two large Ships razes which are line of Battle Ships cut down & mounted with very heavy guns 24 Pdrs.

razee, *v.* Add: **2.** (Earlier example.)

1820 *Deb. Congress U.S.* 28 Jan. (1855) 1008 It would not follow that they should have power to *razee* a State..by depriving the admitted State of equal rights.

razeed *ppl. a.* (earlier U.S. examples).

1847 *Knickerbocker* June 496 The 'Chicken Mauma' was persecuting the Cherokee advocate with her *razeed* (i.e., reduced,) offers in reference to the sale of the 'funny chickens'. **1867** *Harper's Mag.* Oct. 679/2 This 'mittimus' of the Squire's was a razeed, square-topped old chaise.

razer, var. Raser¹ in Dict. and Suppl.

‖ **razet** (raze). *Bullfighting.* [Provençal *raset*.] In southern France: a contest in which teams of combatants compete to snatch a rosette from between the bull's horns.

1932 R. Campbell *Taurine Provence* ii. 43 The finest thing in the French arena..is the razet, or the course of the cocarde-bearing bulls. **1967** McCormick & Mascareñas *Compl. Aficionado* vi. 210 Confusion with Spanish toreo arises in the sport which the French call the *course de cocardes* (or *razet*), in which the athlete, unarmed with cape or muleta, attempts to snatch from between the bull's horns the rosette (*cocarde*), or divisa, of the owner.

‖ **razeteur** (razetȫɹ). [Provençal, f. prec.] A member of a bullfighting team which engages in a *razet*.

1932 R. Campbell *Taurine Provence* ii. 43 They [*sc.* the bulls] know every habitual razeteur by sight. **1961** *Times* 8 July 10/6 There are two classes, a *tourneur* or decoy, whose function is to turn the bull in order to favour the chances of his partner, the *razateur* (sometimes *crocheteur*). **1963** E. & A. Heimann tr. *Droit's Camargue* iii. 28 The razéteurs... Called thus because they pass so close to the bull that they literally graze, or shave, by him. **1976** N. Roberts *Face of France* ix. 106 The *raseteurs*, the young men who get their name from the *razet*, or running half circle, which they describe in their efforts to snatch the [bull's] *cocarde*.

razmataz(z, varr. *Razzmatazz, razzamatazz.

razoo¹ (răzū·). *N. Amer. slang.* Also **razzoo, razzooh.** [Prob. alteration of *Raspberry 3* (cf. *Razzberry, *Razz *sb.*) with arbitrary suffix *-oo*, perh. after Kazoo.] Ridicule; the arousing of indignation or the like, provocation; a sound of contempt, a 'raspberry'. Also in phr. *to give the razoo*: to ridicule. So as *v. trans.*, to arouse or provoke; to manhandle.

1890 *Grip* (Toronto) 18 Jan. 40/1 Shall I razoo old Mowat on the Separate School business? *Ibid.* 19 Apr. 265/1 What is all this racket about the Independence of Parliament?.. It is dependent on the presence of Members..on the whips razoo round. **1908** H. Green *Maison de Shine* 208 Can't a man take a flat o' beer wit' out gittin' the razoo? **1926** *Flynn's* 16 Jan. 639/2 The ginny with th' poke gave th' fly th' razoo an' we split a bunch of nifty kale. **1939** R. Chandler *Big Sleep* xxvi. 235 My information is Apartment 301, but all I get there is the big razzoo. **1942** Berrey & Van den Bark *Amer. Thes. Slang* § 297/1 Ridicule; banter, ..*razoo*. *Ibid.* § 297/4 (*Ripe*) *raspberry* or *razzberry, razoo*, a sound of contempt by vibrating the tongue between the lips, loosely any expression of derision or ridicule, hence ridicule. *Ibid.* § 297/5 Ridicule; banter,..*give the razz* or *the razoo*. **1944** H. Wentworth *Amer. Dial. Dict.* 496/1 *Razoo*, to manhandle, use roughly. **1959** *Washington Post* 22 Dec. c-18/5 Yesterday's hero, Fidel Castro, now gets the lustiest Bronx razzoohs since Adolf Hitler was flipping his wig for the cameras.

razoo² (rā·zū). *Austral.* and *N.Z. slang.* Also **rahzoo, razhoo.** [Origin uncertain.] A (non-existent) coin of trivial value, a 'farthing'. Also in phr. *brass razoo*. Used in neg. contexts only.

1930 *Bulletin* (Sydney) 5 Nov. 21/1 The useless graft on patch and flat! They never think a bloke has earned a darned razoo for that. **1931** W. Hatfield *Sheepmates* xxx. 268 Richards never has a rahzoo. **1940** F. D. Davison *Woman at Mill* ii. 151, I found myself on the streets again, without a brass razoo. **1943** *Coast to Coast 1942* 118 Up till the present he hadn't a brass razoo towards the seven and sixpence. **1947** J. Morrison *Sailors belong Ships* 187, I wouldn't give you a razhoo for anything between there and Charmian Road. **1964** J. Cleary *Flight of Chariots* viii. 361, I wouldn't give a brass razoo for his chances out there. **1968** R. Clapperton *No News on Monday* vii. 80 He isn't rolling in the stuff—he hasn't got two brass razoos to rub together. **1976** *Sunday Mail* (Brisbane) 25 Apr. 16/7 Last week he signed a contract for the new $356,000 building and then cheerfully announced: 'I haven't a razoo.'

razor, *sb.* Add: **1. b.** (Further examples.) Also *Ockham's razor*.

1901 T. C. Allbutt *Science & Medieval Thought* 57 Now this scientific economy, perhaps first formulated, or effectively used, by William Ockham, in the phrase 'entia non sunt multiplicanda'—known as 'Ockham's rasor'—is what is called now-a-days 'materialism'. **1907** Ld. Curzon *Frontiers* 7 Frontiers are indeed the razor's edge on which hang suspended the modern issues of war or peace, of life or death to nations. **1936** J. Buchan *Island of Sheep* xii. 235 In the Norlands life had always been on a razor's edge. **1944** W. S. Maugham (*title*) The razor's edge. **1960** A. Huxley *Let.* 17 July (1969) 894 Perhaps Ockham's razor isn't a valid scientific principle. Perhaps entities sometimes ought to be multiplied beyond the point of the simplest possible explanation. For the world is doubtless far odder and more complex than we ordinarily think. **1976** A. White *Long Silence* vi. 49 He was living on a razor's edge. Sooner or later, the Germans were going to begin to suspect. **1977** M. Goulder in J. Hick *Myth of God Incarnate* iii. 60 Natural explanations, where they are at all plausible, are surely to be preferred on the basis of Occam's razor.

3. a. *razor blade* (further examples), *strop* (earlier and later examples).

1936 *Discovery* Aug. 255/1 Glass razor blades can be ground to powder under foot when used. **1945** 'G. Orwell' *Animal Farm* viii. 64 Making cocks fight with splinters of razor-blade tied to their spurs. **1977** *Jersey Even. Post* 26 July 18/3 Also reported stolen is a silver razor-blade-shaped pendant. **1759** *Newport* (Rhode Island) *Mercury* 26 June 4/3 Hones, Razor-strops &c. **1866** *Harper's Mag.* Nov. 788/2 Packwood, some fifty years ago, led the way in England of..systematic advertising, by impressing his razor-strop indelibly on the mind of every bearded member of the kingdom. **1946** G. Millar *Horned Pigeon* i. 1, I only heard the noise of a man's razor strop.

b. *razor-edged* (later example), *-shaped* (earlier example); *razor-keen, -sharp, -thin; razor-like*.

1972 K. Bonfiglioli *Don't point that Thing at Me* xiii. 98 Even the shadows, razor-edged, purple and green, were painful to look at. **1955** *Times* 11 May 14/6 Political interest is razor keen. **1978** *Detroit Free Press* 5 Mar. (Parade Suppl.) 14D/1 (Advt.), With special Holder hands never come near razor-keen stainless steel blades. **1842** Poe in *Gift* 148 The razor-like crescent. **1977** *Rolling Stone* 19 May 93/2 The sultry title cut, with its razorlike clarion guitar lead. **1829** P. Egan *Boxiana* 2nd Ser. II. 299 He had now not the slightest chance with Curtis, who..drew streams of blood from his razor-shaped nose, and knocked him down. **1921** R. Hichens *Spirit of Time* v. 80 Something of it he must have seen—but what?.. The suggestion of a razor-sharp silhouette? **1975** J. Grady *Shadow of Condor* viii. 132 She carried..a flat, thinly sheathed razor-sharp knife taped to her stomach. **1979** *N.Y. Rev. Bks.* 25 Oct. 48/1 (Advt.), A witty, razor-sharp satire on monogamy. **1971** C. Bonington *Annapurna South Face* viii. 95, I..peered over the top, to see that the ridge was now razor-thin and looked even more difficult beyond the point I had reached. **1973** P. Evans *Bodyguard Man* viii. 64 He cut razor-thin slices through the most congested areas of traffic.

c. razor clam (later examples); **razor-cut** *v. trans.*, to cut (hair, etc.) with a razor; also *fig.*; hence as *sb.*, a haircut effected with a razor instead of scissors; also as *ppl. a.*; **razor-cutting** *vbl. sb.*; **razor-edge,** also *attrib.*; (further *fig.* examples); **razor gang,** (*a*) a gang of thugs armed with razors; (*b*) *Railway slang* (see quots. 1966 and 1970); **razor-grass,** substitute for def.: a West Indian sedge belonging to the genus *Scleria*, esp. *S. pterota*, which has sharp-edged leaves; (earlier and later examples); **razor-man,** a thug armed with a razor; **razor plug, point,** a power-point for plugging in an electric razor; **razor-slasher,** one who slashes another (usu. across the face) with a razor; a member of a razor gang; hence **razor-slash** *v. trans.* [back-formation], to slash with a razor; also as *sb.*, the action of lacerating thus; a wound so made; **razor-slashing** *vbl. sb.*; **razor strop fungus,** the birch polypore, *Piptoporus betulinus*; **razor toe,** a pointed toe on a shoe; an (outmoded style of) shoe with a razor toe.

1935 J. C. Lincoln *Cape Cod Yesterdays* 48 The dictionary ..even mentions the 'razor clam' among them. **1960** M. Sharcott *Place of Many Winds* ix. 162 Commercial crab fishermen often use clams, particularly razor clams, as bait. **1964** F. Warner *Early Poems* 77 Cruelty razor-cut my arteries. **1965** *Family Circle* Oct. 60 Hair as dark as this ideally goes into a sleek and sophisticated styling of the very short tapered razor cut. **1969** J. N. Smith *Is he dead, Miss ffinch?* vi. 27 I'd had time for a hair-do..a razor-cut, tapered down to the neck. **1971** R. Falkirk *Chill Factor* vi. 57 His hair was razor-cut. **1974** R. B. Parker *God save Child* (1975) iv. 33 He was dark-skinned with longish black hair carefully layered with a razor cut. **1976** Scott & Koski *Walk-In* (1977) xii. 66 Their hair was razor cut to just above the collar line. **1968** J. Ironside *Fashion Alphabet* 197 Razor cutting became popular in the 1950s, first in men's barber shops and later in women's hairdressers. The use of a razor means that the hair can be layered and thinned when wet and shaped more effectively. **1977** *Oxford Consumer* June 6/1 Razor cutting is now a rarity,..and only 7 people had had a perm recently. **1927** D. H. Lawrence *Mornings in Mexico* 61 The instant moment is forever keen with a razor-edge of oblivion. **1941** B. Schulberg *What makes Sammy Run?* iii. 49 With one razor-edge phrase he had cut me down to his level. **1962** H. O. Beecheno *Introd. Business Stud.* i. 7 Ours is a razor-edge economy and maintaining our balance of payments.. becomes a matter of overriding importance. **1976** 'A. York' *Dark Passage* xiii. 152 His finances..were in a razor edge state... He lived like a millionaire,..but there was no cash around. **1957** *Essays in Criticism* VII. 311, I suppose that Mr. Conquest would not consider deliverance from the caprice of motorists, or even of wide boys and razor gangs, altogether undesirable for the free mind. **1966** H. Sheppard *Dict. Railway Slang* (ed. 2) 10 *Razor gang*, economy men from Headquarters. **1970** F. McKenna *Gloss. Railwaymen's Talk* 38 *Razor gang*, an investigating committee, searching rosters and rotas for 'unproductive time'. **1977** *Times* 4 May 10/6 There were razor gangs on our race-courses. **1864** A. H. R. Grisebach *Flora Brit. W. Indian Islands* 787 Razor-grass: *Scleria scindens*. **1922** *Blackw. Mag.* July 11/1 The great sweep of razor-grass rustled golden. **1954** *Farmer's Guide* (Jamaica Agric. Soc.) 587 Razor-grass... At least nine different kinds of Scleria occur in Jamaica, and all of them can be unpleasant weeds due to the cutting edges of the leaves. **1969** S. M. Sadeek *Windswept & Other Stories* 29 The cart rolled on..into the savannah of..beezie-beezie reeds and razor-grass. **1958** *New Statesman* 5 Apr. 436/3 The razor-men arrive at his door. **1977** E. W. Hildick *Loop* vii. 36 Noah..was..a 'painter' or razorman with some northern racetrack gang. **1961** *Times* 26 May 9/6 Putting razor plugs in the bathrooms. **1969** C. Hodder-Williams *98.4* i. 7 There was a razor point so I went out to the car and fetched my shaver. **1978** *Cornish Guardian* 27 Apr. 15/1 (Advt.), 21 letting bedrooms (basins, razor points). **1958** M. Procter *Man in Ambush* xiv. 162 This girl had reason to be afraid. She had been razor-slashed once. **1959** N. Mailer *Advts. for Myself* (1961) 292 In the worst of perversion, promiscuity, pimpery,..rape, razor-slash, bottle-break.., the Negro discovered a morality of the bottom. **1963** T. Tullett *Inside Interpol.* xi. 160 A razor-slash across

the face. **1976** R. Hill *Another Death in Venice* i. iii. 55 Dunkerley the pimp, razor-slashing young prostitutes who wouldn't pay. **1980** P. Ableman *Shoestring's Finest Hour* ii. 30 The pimp..is called..Ted the Slash because he's got a razor slash on his cheek. **1951** S. Spender *World within World* iv. 213 Some of the recruits turned out to be a gang of Glasgow razor-slashers. **1961** *John o'London's* 6 July 24/2 Greene's slum *Faust* was articulate in a way unlikely in the most intelligent razor-slasher. **1938** F. D. Sharpe *Sharpe of Flying Squad* xix. 204 Warfare between the gangs was confined to individual beatings-up and razor-slashings. **1979** W. J. Fishman *Streets of E. London* 106/2 His face.. resembling the cross lines of a railway complex as a result of razor slashing. **1923** J. Ramsbottom *Handbk. Larger Brit. Fungi* 129 The name 'razor-strop fungus' is often given to *P[olyporus] betulinus*, as up till the early part of last century it was used for making strops. **1966** F. H. Brightman *Oxf. Bk. Flowerless Plants* 116/2 *Piptoporus betulinus*..has also been recommended for stropping razors, and is sometimes referred to in books as the 'Razor Strop Fungus'. **1895** *Montgomery Ward Catal.* Spring & Summer 509/1 The Razor Toe... This style shoe is becoming very popular on account of the long narrow toe, and patent tip. **1897** C. T. Davis *Manuf. Leather* (ed. 2) xxii. 303 The pedestrian or runner avoids 'razor toes'.

razor, *v.* Add: **a.** Also, to cut *out* (with a razor blade); to shave *away, off*.

1974 *Globe & Mail* (Toronto) 24 July 5/1 Articles taken out of magazines in the libraries..'I'll just say they were razored out. Definitely.' **1975** M. Kenyon *Mr Big* xviii. 175 He..had razored off the moustache. **1977** D. Seaman *Committee* 42 A roughness on the chin each morning that had to be razored away.

b. To slash or assault with a razor.

1937 E. Ambler *Uncommon Danger* viii. 110 By the time I'd finished with the beggar he would have..razored his own father and mother if I'd told him to. **1954** 'N. Blake' *Whisper in Gloom* vii. 91 They might..terrorise a suspected informer—beat him up or razor him.

Hence **ra·zoring** *vbl. sb.*

1950 W. Sansom in *Penguin New Writing* XL. 44 It was not the kind of shop one would have expected of Sally—and perhaps this proved a key to the outcome of that night's razoring. **1963** *Times* 5 June 16/1 Mr. C. Lindsay, for the defence, said Osborne had been afraid of possible razoring by the barons over a debt of £1 and four or five ounces of tobacco.

razor-back, *sb.* and *a.* Add: **A.** *sb.* **1.** (Earlier example.)

1823 W. Scoresby *Jrnl. Voy. Northern Whale-Fishery* 143 Several razor-backs (Balaena physalis) had been seen, but no whales.

2. (Earlier and later examples.)

1849 J. Barrow *Facts Texas* iii. 57 Hogs are a very numerous family, but they are of very indifferent breed, and receive the appellation of 'razor backs', which is significant enough of their appearance. **1878** C. Hallock *Amer. Club List & Sportsman's Gloss.* p. ix, *Razorback*, a domestic hog which runs wild in the woods of the Southern States. **1941** *Arkansas: Guide to State* 99 Outside the imagination, a true razorback probably does not exist. **1976** N. Thornburg *Cutter & Bone* xiii. 302 It has come to him now, what it was about the razorback.

3. A narrow ridge-like back in cattle and horses.

1844 H. Stephens *Bk. of Farm* II. 164 A high narrow shoulder is frequently attended with a rigid back bone, and low-set narrow hooks, a form which gets the appropriate name of razor-back. **1908** *Animal Managem.* 25 The 'razor' back may..be due only to want of muscle which judicious rest, food, and work will produce. **1941** I. L. Idriess *Great Boomerang* vii. 51 Fine upstanding beasts... No 'razor-backs' going away with nothing behind. These were 'table-tops'; you could throw your blanket on any beast and camp on his back.

4. Chiefly *Austral.* and *N.Z.* A steep-sided, narrow ridge of land.

1874 W. M. Baines *Narr. E. Crewe* xi. 247 From a high 'razor-back', I had a magnificent view. **1889** *Trans. N.Z. Inst.* 110 Supposing the traveller to be standing on a narrow spur, or razorback, leading to the mountain-top. **1902** [see *cow-track* s.v. *cow *sb.*¹ 7]. **1911** *Chambers's Jrnl.* Dec. 30/1 Twice the way led along a real 'razor-back'. On both sides the mountain sloped precipitously. **1957** P. White *Voss* vi. 153 Presently the path, which had reached a razorback..wound suddenly..and plunged down.

5. *U.S. Circus slang.* A circus hand; *spec.* one who loads and unloads the wagons.

1904 *Everybody's Mag.* X. 658/1 That night it took the Old Man 'n Early Jim both to keep a razorback from carvin' up Ibree. **1909** *Youth's Compan.* LXXXIII. 289/4 There was too much worth seeing outside. The loaders—'razorbacks', in circus language—were putting the great clanking parade wagons on the flat cars. **1926** R. E. Sherwood *Here we are Again* 162 Canvasmen or 'razorbacks', as they are known in the slang of the circus, are rarely in funds. **1975** *New Yorker* 13 Oct. 38 Some people..were watching the roller coaster... I went up to the razorback who ran the controls.

B. *adj.* (Further examples.)

1851 G. S. Cooper *Jrnl. Expedition Overland* 110 Gullies.. ran down from each side of the razor-back ridge. **1896** [see *hump-backed a.]. **1899** B. Tarkington *Gentleman from Indiana* iv. 44 A squad of thin, 'razor-back' hogs. **1924** J. Masefield *Sard Harker* iii. 126 It was one of the half-wild razor-back hogs which the negroes allowed to stray in the woods there. **1976** N. Thornburg *Cutter & Bone* xiii. 302 As it fell open Bone was able to see his T-shirt underneath, and the emblem on it: a red Arkansas razorback hog, name and symbol of the state university's sports teams.

razor-backed *a.* (further example).

1904 *Daily Chron.* 12 May 5/3 A fierce struggle ensued for the possession of two razor-backed ridges above which runs the main Peking road.

razor-grinder. 1. (Earlier examples.)

1789 *Boston Directory* 184 Fillis William, razor-grinder. **1798** D. Wordsworth *Jrnl.* 22 Feb. (1941) I. 9 Met a razor-grinder with..a boy to drag his wheel.

razor-shell. Add: (Further examples.) Also *attrib.*

1792 J. Belknap *Hist. New Hampshire* III. 183 The Razor-shell clam '*Solen ensis*'. **1901** E. Step *Shell Life* ix. 155 We now reach the Razor-shell family, characterised by having the valves of the shell of equal length. **1928** Russell & Yonge *Seas* ii. 38 The long razor-shells (*Solen*) may occasionally be dug at low tide. **1971** *Oxf. Bk. Invertebrates* 80 The razor shells, looking like the old-fashioned cut-throat razors, are among the most specialized of the burrowing bivalves.

razz (ræz), *sb. slang* (orig. *U.S.*). [Short for *Razzberry.] = *Raspberry 3*.

a **1919** C. Briggs *Oh Man!*, She'll prob'ly give me the razz for being out late last night! **1920** S. Lewis *Main Street* xxiii. 282 The Red Swede got the grand razz handed to him. **1921** *Collier's* 15 Jan. 20/1 The mob gave him the razz. **1926** N. V. Lindsay *Going-to-the-Stars* 52 Let us think of the Irish flute in the morn,..And forget our jazzes and our razzes and our hates. **1935** *Punch* 27 Feb. 248/2, I wasn't asked parties; I got no rise..; the girls gave me the razz—and all for the reason I'd no badge to show. **1960** E. W. Hildick *Jim Starling & Colonel* viii. 62 That band chap blew him the razz! **1961** *Punch* 18 Jan. 129 What say, honey?—let's give this communal living the razz and just go off somewhere, the two of us. **1961** *Spectator* 9 June 835/1 He selects one of them for punishment.., delivers a sonorous 'razz' and pretends to cane him. **1967** J. D. R. McConnell *Eton* 61 Offenders may be summoned to the Library for a 'razz'. **1977** *Times Lit. Suppl.* 29 Apr. 534/5 Even the peppiest, most two-fisted and up-and-coming borough librarian would get the razz for buying it.

razz (ræz), *v. slang* (orig. *U.S.*). [f. the sb.] *trans.* To hiss or deride; to make fun of (a person). Hence **ra·zzing** *vbl. sb.*

1921 *Collier's* 19 Feb. 5/3 It [*sc.* a crowd]..will razz its local favorite with as much enthusiasm as it will the visitin' boxer at the first sign of foul fightin'. **1921** *Sat. Even. Post* 18 June 65/2 I'd of rather took fifty socks on the jaw than the razzing the crowd give Bat. **1924** P. Marks *Plastic Age* 52 The fellows razzed the life out of me. *Ibid.* 60, I don't mind the razzing myself,..but I don't like the things they said to poor little Wilkins. **1932** J. Lawson *Man's Life* xvi. 161 The person who never could appreciate institutional life..is always with us... His chief hobby when at home has been razzing the wife, or his mother, because the bacon is too fat or too lean..and anyhow he doesn't like bacon at all. **1939** L. Jacobs *Rise Amer. Film* 378 He turned out a series of domestic comedies that caused him to be hailed for his 'razzing' of American foibles. **1941** J. McCormack in L. A. G. Strong *John McCormack* x. 168, I have seen a great deal of baseball in America, but I have never been able to reconcile myself to the continuous razzing of the pitcher. **1956** B. Holiday *Lady sings Blues* (1973) iii. 33 When I came to work the other girls used to razz me, call me 'Duchess' and say, 'Look at her, she thinks she's a lady.' **1968** *Punch* 24 July 129/2 Ilya Ilf, pooh-poohing purges, Razzed Red Russians with wry stories. **1975** A. Bergman *Hollywood & Le Vine* ii. 29, I continued on down the street... I anticipated the razzing of the Dead End Kids. **1977** *TV Times* (Austral.) 20 Aug. 29/1 My kids will get razzed about it at school the next day. No one knows more about my mistakes than I do.

razzamatazz, var. *Razzmatazz.

razzberry (ræ·zbĕri). *N. Amer. slang.* Also **razbery.** [Var. of Raspberry.] = *Raspberry 3*.

1922 *Collier's* 15 July 4/3 No matter if all the rest of the crowd gives me the razzberry, why they'll be at least two guys pulling for me. **1927** [see *bird *sb.* 5 b]. **1928** C. Sandburg in *Woman's Home Compan.* Aug. 112/3 Hand 'em the razzberries. **1948** *Daily Ardmoreite* (Ardmore, Okla.) 27 May 8/3 Here in the home of the Bronx jeer it usually is rewarded with a noisy razberry. **1975** E. Iglauer *Denison's Ice Road* ii. 35, I sure got the razzberries from the boys.

razzia. Add: **b.** (Earlier and later examples.)

1855 *Poultry Chron.* 4 Apr. 98/1 The owners of manors.. carried out a 'razzia' on the enemy's territory of Leadenhall market. **1965** C. D. Eby *Siege of Alcázar* (1966) v. 100 Small bands had been stealing out of the fortress at night to scavenge in the houses near by. The purpose of these *razzias* was to bring back food for the infirmary.

So **ra·zzia** *v. intr.*, to maraud.

1846 R. Ford *Gatherings from Spain* iv. 34 The object of these border *guerrilla*-warfares was..to 'harry', to 'razzia'.

razzle (ræ·z'l), *sb. slang.* [Short for Razzle-dazzle.] **a.** A 'good time', a spree; usu. in phr. *on the razzle*.

1908 A. Bennett *Old Wives' Tale* iv. i. 435 'What puzzles me most is what the devil you were doing in a place like that. According to your description, it must be a ——.' 'I went there because I was broke,' said Matthew. 'Razzle?' Matthew nodded. **1915** W. S. Maugham *Of Human Bondage* 249 We won't 'alf go on the razzle. **1927** *Daily Express* 2 June 6/4 Its heroine..is a Frenchman's idea of a great English lady out on the razzle. **1930** J. B. Priestley *Angel Pavement* v. 213 And now we're going on the razzle. **1943** —— *Daylight on Saturday* xxvi. 201 I've got three [absentees]... One's off on a razzle. **1968** 'J. Le Carré' *Small Town in Germany* 210 Your wife was in England, and you went on the razzle with Leo. **1978** 'L. Black' *Foursome* vii. 56 He loved making new friends, joining up with them for a razzle in the nightspots.

b. = Razzle-dazzle *sb.* b in Dict. and Suppl.

1969 *Wall St. Jrnl.* 30 Sept. 1/1 His specialty is 'Razzle', a game that in one form or another has entranced fair-goers since ancient times.

ra·zzle, *v. slang.* [f. the sb.] *intr.* To live a life of pleasure, to enjoy oneself; to go 'on the razzle'.

1908 G. B. Shaw *Lett. to Granville Barker* (1956) 120 He will probably put it to you whether, as a gentleman, you can ask for a salary when you have been doing nothing but razzling in America. **1951** E. Bagnold *Loved & Envied* iii. 39 We ought to be fairly flush... It's not an expensive island. We ought to be able to razzle a bit, if there's anywhere to razzle.

razzle-dazzle, *sb.* Add: **a.** (Earlier and further examples.) Also, deception, fraud; extravagant publicity.

1889 *Gallup* (New Mexico) *Gleaner* 18 Mar. 4/2 A Kansas paper..recently told of a 'regular old razooper, who, having got a skate on, indulged in a glorious razzle-dazzle'. **1898** G. B. Shaw *Mrs. Warren's Profession* i. 175, I don't bet much and I never go regularly on the razzle-dazzle as you did when you were my age. **1928** *New Yorker* 15 Dec. 24/1 Suspecting some sort of razzle-dazzle, the wiser of the two men said he would buy the seats at their box-office value. **1938** *Sun* (Baltimore) 20 Jan. 3/2 With such razzle-dazzle financing the general practice, it is not surprising to find oil derricks sprouting up on the grounds of the State capitol. **1962** 'K. Orvis' *Damned & Destroyed* vii. 51 The razzle-dazzle I had handed the two drug-ring musclemen. **1969** *New Yorker* 29 Nov. 47/1, I want models, I want a private plane, I want this, I want that, I want some razzle-dazzle. **1977** *Time* 19 Sept. 25/1 Lance ran for the Democratic gubernatorial nomination, using the financial razzle-dazzle that later was to become such a liability. **1978** *New York* 3 Apr. 17/3 It [*sc.* a musical] has pizzazz and razzle-dazzle, bursts of energy and invention, music and laughter.

b. (Later examples.) (See also quot. 1935).

1896 [see Switchback *a.*]. **1935** *Amer. Mercury* June 230/2 *Razzle-dazzle*, kelsy [i.e., a prostitute]; also used by the public in reference to carnival rides, although not so used by carnies themselves. **1968** D. Braithwaite *Fairground Archit.* iii. 34 The steam swing and 'Razzle Dazzle' drew inspiration from mechanisms in the spinning frame, [etc.]. *Ibid.* 60 Four years before his death in 1897, Savage patented the 'Razzle Dazzle', otherwise known as 'Whirligig' or 'Aerial Novelty'.

c. *attrib.* or as *adj.* Of, pertaining to, or characterized by, razzle-dazzle; dazzling, spectacular.

1889 *Road* (Denver, Colorado) 28 Dec. 5/1 Clint Butterfield incloses us a razzle-dazzle card of some kind that has a very neat little design of a nightmare etched in blood red and India ink. **1946** *N.Y. Times Bk. Rev.* 4 Aug. 5/1 A great many people are reading Mr. Wakeman's razzle-dazzle novel these days. **1951** M. McLuhan *Mech. Bride* (1967) 10/2 The newsreel is provided with a razzle-dazzle accompaniment. **1965** *Economist* 22 May p. xxiv/1 A front page [of a newspaper] full of short items and a Gallic profusion of typefaces, alike in text and headlines—..what the Americans call 'circus' or 'razzledazzle' make-up. **1971** *Daily Tel.* 19 Oct. 19 Mr Thorley's chairmanship is, or was, intended to be an interregnum between the razzle-dazzle rule of Sir Derek Pritchard and the accession of the Showering dynasty. **1974** *Plain Dealer* (Cleveland) 13 Oct. c. 12/6 Freshman halfback Pat Healy scored three touchdowns, the last on a razzle-dazzle 32-yard pass from quarterback Bruce Basile. **1977** *Time* 13 June 39/1 Erdman comes to his subject with the sure hand of one who knows, from the inside, what lurks in the hearts of financial razzle-dazzle artists.

razzle-dazzle, *v.* (Later example.)

1976 *Houston* (Texas) *Chron.* 22 Sept. 7–4/4 Lady Bird eats it in an orange print pants suit and that Texas smile that razzle-dazzles 'em.

razzmatazz, razzamatazz (ræ:zmătæ·z, ræ:zămătæ·z). *colloq.* (orig. *U.S.*). Also **razamataz(z, razmataz(z, razz-ma-tazz,** etc. [Origin unknown; perh. alteration of Razzle-dazzle.] **a.** A type of rag-time or early jazz music; old-fashioned 'straight' jazz; sentimental, 'corny' jazz; hence anything old-fashioned; stuff, rubbish. **b.** Noisy, showy publicity; meretricious or extravagant display; an event surrounded by such publicity or display; fuss, commotion, garishness. Also *attrib.* or as *adj.*

In quot. 1899 the sense is uncertain but may be 'up-to-date, stylish' or 'cultured, superior'.

1899 G. Ade *Fables in Slang* 37 It would be a Big Help to the Poor and Uncultured to see what a Real Razmataz Lady was like. **1901** T. D. Collins (*title of piano music*) Raz-a-ma-taz. **1901** W. H. Smith (*title of piano music*) Raz-ma-taz. **1936** *Amer. Mercury* XXXVIII. p. x/2 *Rooty-toot*,—unadulterated corn; razz-ma-tazz. **1937** *Amer. Speech* XII. 48/1 *Razmataz band*, a band which plays in an outmoded style. **1938** *Brit. Empire Mod. Eng. Illustr. Dict.* 1257/1 Razz-ma-tazz (*Am.*), old-fashioned jazz. **1942** Berrey & Van den Bark *Amer. Thes. Slang* § 579/2 'Straight Jazz.' (Old-fashioned jazz, which reproduces the score faithfully, as distinguished from 'swing')..*razzmatazz*. **1947** *Sat. Rev. Lit.* (U.S.) 25 Oct. 65/2 Expert horsing of the old razzmatazz style by an expert horsewoman. **1950** C. Coben *Old Piano Roll Blues* (song), And while we kiss, kiss, kiss away all our cares, The player piano's playin' razzamatazz, I wanna hear it again. **1953** Berry & Van den Bark *Amer. Thes. Slang* (1954) (ed. 2) § 233/5 Something old-fashioned, ..*razzmatazz*. **1958** *People* 4 May 8/2 She will, from next

Friday when she flies to Cannes, be getting the full razzama-tazz, big-star build-up. **1958** *Spectator* 1 Aug. 174/3 Don't you remember *anything* about the Twenties but crime, booze, flappers, religious razzmatazz? **1959** J. Wain *Travelling Woman* x. 148 The enormous selling bonanza that was going on about him, in its astonishing flood of genuine goodwill, even a grain here and there of genuine piety, with unscrupulous salesman's razzmatazz, heightened his sense of living in a dream. **1961** *Sunday Times* 26 Nov. 48/3 Barbara Murray, a girl who is entirely wasted on rats, retorts and all that razzmatazz. **1963** *The Beatles* 9 Though some of our material is a bit out of the way for a razzamatazz chap like him. **1965** G. McInnes *Road to Gundagai* iv. 59 The great wide streets have an air of grandeur which even the razz-ma-tazz of neon cannot wholly mar. **1969** *Listener* 17 Apr. 544/1 *Oh! What a lovely war* (Paramount) is a razz-ma-tazz spectacular. **1970** *Times* 9 Mar. 13/1 He turned, as might have been expected, a fairly serious event into a razamataz. **1971** *Morning Star* 8 Mar. 4/7 Some of the hotels and centres can be a bit razzamataz and noisy, especially at night. **1972** *Daily Colonist* (Victoria, B.C.) 16 Feb. 32/1 There was no need to go through 'all this razzmatazz'. The replacing of white centre lines with yellow ones wasn't all that difficult to comprehend. **1973** *Daily Tel.* 8 Nov. 5 (Advt.), We thought the car good enough not to need any launch gimmicks or razmataz. **1973** *Daily Colonist* (Victoria, B.C.) 23 Nov. 22/3 Opening day Thursday had all the razzmatazz of a revival meeting. **1974** *Time Out* 27 Sept. 23/1 There is the host, resplendent in white satin, razmatazz shirt. **1977** *Listener* 20 Oct. 508/3 This programme included a razzmatazz of presentational devices which seemed better suited to a giveaway quiz show. **1979** *Guardian* 1 May 30/1 In keeping with the showbiz, razzmatazz side of the election, the most glamorous transport belongs to the media.

razzo (ræ·zo). *slang.* [Prob. alteration of Raspberry.] The nose.

1899 [see *lower *v.* 1 e]. **1936** [see *acid *sb.* 3].

razzoo(h, varr. *Razoo¹.

R-boat. [Partial tr. G. *R-boot*, abbrev. of *räumboot* minesweeper.] In the war of 1939–45, a German minesweeper.

1942 *Times* 10 June 4/3 An R-boat is stated to be an armed motor minesweeper. **1945** P. Scott *Battle of Narrow Seas* ii. 5 Besides E boats there were flotillas of R boats (corresponding roughly to our M.Ls.), used for minesweeping and defensive patrolling along the occupied coasts. **1961** Granville & Kelly *Inshore Heroes* vii. 71 The Raumboot (R-boat) was a patrol vessel of between 85 and 115 feet, with a speed of about 20 knots. **1978** F. Maclean *Take Nine Spies* v. 183 A whole group of German R-boats was sent from Sicily to the Aegean.

RDX (ā̆ɹ,dī,e·ks). [f. *R*esearch *D*epartment (Woolwich, England) E*x*plosive.] = *Cyclonite.

1941 *Newsweek* 8 Dec. 43/2 One [explosive], developed in cooperation with the British and identified with the stuff used in Britain's 'superbombs', is known as RDX and credited with 40 per cent more bursting power than TNT. **1947** *Times* 9 July 5/7 RDX, the main high explosive development of the war, was yet another chemical contribution to victory—opening up, incidentally, a new field of organic chemistry. **1974** *Encycl. Brit. Macropædia* VII. 89/1 The torpedo warhead torpex..is a cast mixture of RDX, TNT, and aluminum.

re, *sb.*² Add: (Later examples.) Now freq. apprehended as a preposition, and used in weakened senses to mean 'about, concerning'.

The use as a preposition has freq. been condemned: see Fowler *Mod. Eng. Usage* (1926) s.v. *illiteracies* and A. P. Herbert (examples below).

1926 in H. W. Fowler *Mod. Eng. Usage* 484/1 Dear Sir,—I am glad to see that you have taken a strong line re the Irish railway situation. *Ibid.* 484/2 Reference had been made in a former issue to some alleged statements of mine re the use of the military during the recent railway dispute. **1935** A. P. Herbert *What a Word!* iii. 80 We herewith enclose receipt for your cheque £4 on a/c *re* return of commission *re* Mr. Brown's cancelled agreement *re* No 50 Box Street top flat. **1939** [see *inclusivity]. **1976** *Time* 27 Dec. 2/2 Re your article on legitimized gambling..and specifically state lotteries: the inefficiency of revenue collection is horrendous and the odds for winning are unconscionable. **1977** *Time* 7 Feb. 1/3 Re my archaeological explorations in Syria: it is not true that '..the Italian archaeologists have been slow to publicise their discoveries'. **1979** *Verbatim* Summer 5/2 G. Bocca's observations re public signs.

re, abbrev. of Rupee.

1913 W. T. Rogers *Dict. Abbrev.* 164/1 *Re.* (money), rupee. **1962** *Housewife* (Ceylon) Apr. 10 These courses are practical, economical (Re. 1/- for 3 lessons). **1971** *Hindustan Times Weekly* (New Delhi) 4 Apr. 11/2 Gramdal was better by Re 1 on good offtake.

re-, *prefix.* Add: **4. b.** (Later examples.)

1885 G. B. Shaw *Let.* 14 Dec. (1965) I. 146, I re-return the cheque, and if you re-re-return it I will re-re-re-return it again. **1922** Joyce *Ulysses* 526, I rererepugnosed in rererepugnant. **1954** *New Biol.* XVI. 43 Under the title 'Vital Blarney'..I reviewed, or to be pedantic I re-reviewed, Bernal's book *The Physical Basis of Life*.

5. a. *re-abolish* vb., *-alliance, -apportion* vb. (further examples), *-apportionment* (further examples), *-bandage* vb., *-beam* vb., *-biff* vb., *-break* vb. (earlier and later examples), *-cable* vb., *-calibrate* vb., *-calibration, -canalization, -canalize* vb., *-canvass* vb., *-carve* vb., *-centrifuge* vb., *-certification, -certify* vb., *-chromatograph* vb., *chromatography, -clean* vb., *-clone* vb. (hence *-cloning* vbl. sb.), *-codify* vb., *-conceptualization, -conceptualize* vb., *-configure* vb., *-conscript* vb., *-contamination, -contrast* vb., *-cool* vb., (hence *-cooling* vbl. sb.), *-debit* vb., *-decontaminate* vb., *-decontamination, -demarcation, -differentiate* vb. (further examples), *-differentiation, -enrich* vb., *-enrichment, -equilibrate* vb., *-equilibration* (further examples), *-estimate* vb. and sb., *-evocation, -exploration, -explore* vb., *-expose* vb., *-fabricate* vb., *-feature* vb., *-fecundate* vb., *-flush* vb., *-foliate* vb., *-foliation, -format* vb., *-forward* vb. (hence *-forwarding* vbl. sb.), *-incubate* vb., *-infarction, -initialize* vb., *-input* vb., *-intensify* vb., *-isolate* vb., *-license* vb., *-list* vb., *-lubricate* vb., *-mapping* vbl. sb., *-nucleation, -orchestrate* vb. (further examples), *-orchestration, -origination, -originator, -pattern* vb. (hence *-patterning* vbl. sb.), *-peg* vb. (hence *-pegging* vbl. sb.), *-phosphorylate* vb., *-pile* vb., *-postpone* vb., *-proportion* vb. (further examples), *-punch* vb., *-punctuation, -rat* vb. (later example), *-recovery, -remember* vb., *-remembrance, -riddle* vb., *-scrutinize* vb., *-scrutiny, -sex* vb. (further example), *-shade* vb., *-show* vb. (hence *-showing* vbl. sb.), *-stack* vb., *-stage* vb., *-structuration, -suspend* vb., *-suspension, -suture* vb. and sb., *-synthesis, -synthesize* vb., *-tailor* vb., *-target* vb., *-tightening* vbl. sb., *-time* vb., *-triangulate* vb., *-triangulation, -uptake, -walk* vb., *-winded* ppl. a., *-zip* vb.

1963 Auden *Dyer's Hand* 461 The distinction between the things of God and the things of Caesar is reabolished. **1847** Webster, Realliance. **1973** *Jrnl. Genetic Psychol.* Mar. 137 Their [*sc.* neo-Freudians'] realliance will contribute something to the explanation of the latter theory. **1967** M. E. Jewell *Legislative Representation in Contemp. South* v. 124 The Kentucky legislature was one of the first to reapportion both houses substantially on a population basis. **1971** C. A. Auerbach in N. W. Polsby *Reapportionment in 1970s* ii. 90 All state legislatures will be reapportioned according to the principle of one vote, one value. **1931** *Times Lit. Suppl.* 18 June 476/4 There should be a reapportionment of seats. **1974** *Anderson* (S. Carolina) *Independent* 19 Apr. 1B/1 Members of the House-Senate conference committee asked ..for free conference power that would allow them to rewrite district lines in the House reapportionment bill. **1920** C. H. Stagg *High Speed* x. 180 Dan helped him rebandage his hands. **1979** *Sunday Express* 28 Jan. 3/8 Within an hour of starting that wound had been stitched and rebandaged. **1919** E. Pound *Quia pauper Amavi* 16 The infant beams at the parent, The parent re-beams at its offspring. **1934** Blunden *Choice or Chance* 53 Ye men of England, hear the clarion. If Inferior nations biff you, them rebiff. **1805** R. W. Dickson *Pract. Agric.* II. 943 And when the weather is bad these cocks are never re-broken out, being only lightened up to let the air pass through them more freely. **1905** *Daily Chron.* 31 July 4/7 The leg was badly set, and had to be re-broken. **1943** V. Sackville-West *Eagle & Dove* i. xv. 89 Her left arm..had had to be re-broken and re-set most painfully several times. **1908** *Daily Chron.* 7 Apr. 1/7 Chicago, Monday... This afternoon..an alleged interview with Hackenschmidt is re-cabled from a London newspaper. **1909** *Cent. Dict.* Suppl., Recalibrate. **1971** *Nature* 6 Aug. 391/2 We are currently recalibrating our Dobson ozone spectrophotometers. **1978** *Sci. Amer.* Feb. 34/3 In the 1980's shuttle-recoverable instruments that can be recalibrated and still better instruments in high orbit may answer the questions. **1911** Webster, Recalibration. **1977** *Nature* 6 Jan. 18/1 Node markers..can thus easily be reset by 93 yr periodic recalibration observations of maximum northerly midwinter full moonrise azimuth. **1961** R. D. Baker *Essent. Path.* v. 82 New blood vessels form in the lumen (recanalization). **1943** *Amer. Speech* XVIII. 222 General semantics..is offered as a means of recanalizing those responses..that cause morbid over-excitation of the nervous system. **1962** *Punch* 12 Sept. 366/2 Water conservation..to the extent of recanalising the water. **1925** T. Dreiser *Amer. Trag.* II. iii. xxvi. 329 The twelve men.. re-canvassing for their own mental satisfaction the fine points made by Mason. **1924** J. Masefield *Sard Harker* 4 Men remembered this rhyme, and plied that it should be recarven. **1956** *Nature* 7 Jan. 45/2 The homogeneous supernatant was recentrifuged once or twice. **1976** *Ibid.* 15 Jan. 114/2 The supernatant was recentrifuged at 20,000*g* for 10 min. **1976** P. R. White *Planning for Public Transport* i. 20 Maintenance facilities may be very limited, problems of major overhaul and recertification being handled by sale of a vehicle to a dealer who provides a reconditioned vehicle in part exchange. **1978** *Jrnl. R. Soc. Med.* LXXI. 13 Such developments..could be more effective and acceptable than some form of periodic recertification in maintaining standards in practice. **1934** Webster, Recertify. **1976** *National Observer* (U.S.) 17 July 11/3 The Israelis were recertifying their credentials as a people of almost unbelievable resourcefulness and courage. **1977** *Proc. R. Soc. Medicine* LXX. 58/2 Only the American Board of Family Practice is putting the idea into practice by an MCQ recertifying exam from October 1976. **1945** Rechromatograph [see **rechromatography*]. **1948** *Amer. Scientist* XXXVI. 511 If either of these two zones is cut out, eluted, and rechromatographed on a fresh column, it will form a single zone. **1971** *Nature* 16 Apr. 456/2 After 24 h dialysis against 0·01 M phosphate buffer..the residual fibre was rechromatographed, the peak fractions were pooled and reduced to a final volume of 1 ml. **1945** *Jrnl. Biol. Chem.* CLVII. 327 Rechromatography was usual, especially when two zones were bordering upon one another. In such instances they were cut out as one and rechromatographed. **1950** L. Zechmeister *Progress in Chromatogr.* xii. 162 The peptides were characterized by the ratio, total nitrogen/amino nitrogen. Such ratios observed were not altered by rechromatography. **1978** *Nature* 14 Dec. 735/2 Rechromatography on Sephadex G-50 gave only a single peak. **1921** *Daily Colonist* (Victoria, B.C.) 29 Oct. 8/1 (Advt.), Fine recleaned currants. **1960** *Farmer & Stockbreeder* 29 Mar. 5/1 The following are wholesale prices for recleaned seed per cwt ex-store unless otherwise stated. **1971** *Nature* 30 July 313/2 Six clones were mixed (*GdH/GdD*), two of these were recloned, and 106 out of 107 of these sub-clones showed either *GdH* or *GdD* while one was again mixed. **1962** *Cold Spring Harbor Symp. Quantitative Biol.* XXVII. 410 Recloning of these clones gives rise to all converted clones. **1973** *Listener* 20 Dec. 846/3 We shall not avoid increasing dislocation..unless we can recodify large areas of international behaviour. **1961** Webster, Reconceptualization. **1973** *Sci. Amer.* Sept. 120/2 With a reconceptualization of the hospital as a therapeutic community.., many of the chronic inpatients were able to be returned to the community. **1977** Fontana & Van de Water in Douglas & Johnson *Existential Sociol.* iii. 126 Understanding the world in this manner demands a thorough-going reconceptualization of our usual notions of truth and progress in knowledge. **1961** Webster, Reconceptualize. **1977** A. Giddens *Stud. in Social & Polit. Theory* ii. 118 Let us at this juncture reconceptualize 'structure' as referring to *generative rules and resources* that are both applied in and constituted out of action. **1964** M. McLuhan *Understanding Media* ii. xxxi. 313 The viewer of the TV mosaic, with technical control of the image, unconsciously reconfigures the dots into an abstract work of art. **1946** L. B. Lyon *Rough Walk Home* 17 Only his singular, re-conscripted breath Could fan to a purpose all that pyre his death. **1961** Webster, Recontamination. **1962** *Economist* 19 May 706/2 The gas must be protected against re-contamination through leaks. **1966** D. G. Brandon *Mod. Techniques Metallogr.* 187 The gas used must be extremely pure if immediate recontamination of the surface is to be avoided. **1957** R. N. C. Hunt *Guide to Communist Jargon* xlviii. 160 In connection with the Brest-Litovsk treaty [Lenin] recontrasted those who were 'revolutionaries out of sentiment' with 'real revolutionaries'. **1934** Webster, Recool. **1969** *Gloss. Terms Water Cooling Towers* (*B.S.I.*) 5 *Recooled water temperature*, average temperature of the circulating water entering the basin. **1968** C. G. Kuper *Introd. Theory Superconductivity* v. 93 These nucleation centres are remarkably stable—they often survive the heating of the specimen to room temperature and subsequent recooling. **1934** S. W. Rowland *Hughes-Onslow's Lawyer's Man. Book-keeping* (ed. 3) i. 9 The bank, for its part and from its point of view, credited when the cheque was paid in. Consequently when the cheque is found to be worthless, it redebits. **1968** *Lebende Sprachen* XIII. 87/2 The bank may redebit the account. **1935** A. P. Herbert *What a Word!* vi. 187, I do not think that she [*sc.* the Ship of State] was ever..'re-decontaminated'. *Ibid.* i. 21 The answer from high places was: 'A process of redecontamination would be advisable.' **1969** *P.E.N.* IX. 48 He recalled that at the beginning of the 1939 War the use of the word 'contaminate' for a gas attack had seemed comic, particularly when it involved 'decontamination' and 'redecontamination' stations. **1938** *Times* 17 Jan. 11/5 The chief violations [of the Soviet constitution] have been the redemarcation of internal frontiers and the formation of new territorial and administrative units. **1960** *Observer* 20 Mar. 1/4 The Ghana Government claimed that 14 people arrested last week..had conspired to conduct a campaign of violence and civil disturbance there to provoke 'foreign intervention' and the redemarcation of frontiers. **1911** *Cornh. Mag.* Apr. 497 It is as if John Brown on his death-bed were to have his tissues pass into a state of flux, and then get simpler and simpler, until you would have to say, This is no longer a man, but merely a mass of man's protoplasm, and as if finally this mass were to redifferentiate up again. **1960** *New Biol.* XXXI. 90 A second possibility is that tissue cells undergo an apparent de-differentiation to form the young regenerate or the bud but, like cells in tissue culture, retain their tissue specificity and later re-differentiate into tissues of the same kind as those from which they came. **1970** *Jrnl. Gen. Psychol.* LXXXII. 182 He must again re-differentiate these boundaries. **1889** *Cent. Dict.*, Redifferentiation. **1921** *Discovery* Feb. 28/2 Such a process, which we may style dedifferentiation followed by redifferentiation, is clear evidence of the possibility of reversing development. **1960** *New Biol.* XXXI. 89 Some cells..normally change their shapes and functions in the fulfilment of their proper roles in the organism's economy. Such reversible changes have been called 'modulations' by Weiss, and the distinction between them and more profound re-differentiations may seem rather arbitrary. **1951** *Sci. Amer.* Nov. 18/2 The gas is.. cycled back into the reservoir several times to be re-enriched. **1976** *Ibid.* Dec. 33/3 If the uranium is to be returned to the gaseous-diffusion plants for reenrichment, it is converted into uranium hexafluoride. **1971** I. G. Gass et al. *Understanding Earth* i. 27/1 Once an igneous rock has completely solidified, however, the absence of a fluid phase and the reduction of temperature make it very difficult for the minerals to re-equilibrate to new assemblages which would be stable at lower temperatures. **1970** G. Germani in I. L. Horowitz *Masses in Lat. Amer.* xvi. 591 It satisfied their need for re-equilibration through the emphasis on 'order, discipline, hierarchy', and through the demobilization of the lower classes. **1971** I. G. Gass et al. *Understanding Earth* i. 33/1 The process of re-equilibration..is materially aided by the introduction of water in the environment of weathering. **1934** Webster, Re-estimate, *v.t.* **1952** S. Spender *Shelley* 44 Not so much a re-estimate, as a restoring of some sort of balance. **1964** K. G. Lockyer *Introd. Critical Path Analysis* ix. 89 An alternative is to insert the actual (or re-estimated) times. **1924** S. Joyce in J. Joyce *Lett.* (1966) III. 104 This re-evocation and exaggeration of detail by detail and the spiritual dejection which accompanies them are purely in the spirit of the confessional. **1952** C. P. Blacker *Eugenics* 138 The re-evocation of the repressed memory, though painful like an incision, cured the sufferer. **1977** *Proc. R. Soc. Med.* LXX. 385/2 One re-exploration was done over the same period as 36 cholecystectomies overall. **1933** *Proc. R. Soc.* A. CXLII. 350 For this reason we have not re-explored this region, since we could not hope to detect the presence of groups of such weak intensity. 1977 Proc. R. Soc. Med. LXX. 385/2 Over the same period another 22 patients were reexplored after operation elsewhere. 1946 Nature 28 Dec. 946/1 The slides are located in their former position and re-exposed. 1950 F. E. Zeuner Dating Past (ed. 2) 264 A Final wet phase, of a very minor character re-exposes by stream erosion the levels containing Middle Stone Age. a 1942 B. Malinowski Sci. Theory of Culture (1944) 164 In a small farcical form, such a charter has been refabricated in the Blut und Boden doctrine of modern Naziism. 1976 Dumfries & Galloway Standard 25 Dec. 12/2 The policy favoured at present is to re-process all nuclear fuel in a few politically-stable countries, return the re-fabricated fuel to the country of origin, and retain the wastes for 'safe' storage. 1922 Joyce Ulysses 554 The face of Martin Cunningham, bearded, refeatures Shakespeare's beardless face. 1957 L. Durrell Justine III. 199 The resonance of this one phrase refecundated his powers of feeling. 1880 'Mark Twain' Tramp Abroad xlvii. 495 The tints remained during several minutes..paling almost away for a moment, then re-flushing,—a shifting, restless, unstable succession of soft opaline gleams. 1937 Discovery Aug. 246/2 Wintering, refoliating, flowering, and seeding. 1963 Times Lit. Suppl. 22 Feb. 143/1 The Book of Durrow..was taken to pieces,..rearranged, refoliated, repaired, reconditioned and..superbly rebound. 1956 Nature 31 Mar. 619/2 Oidium heveae is most prominent in Malaya at the time of refoliation after 'wintering' of the trees. 1977 J. L. Harper Population Biol. Plants xii. 398 Defoliation is often complete and is followed by refoliation. 1967 E. R. Lannon in Cox & Grose Organiz. Bibliogr. Rec. by Computer iv. 88 The user may initially employ his own Preprocessor to edit and reformat his data. 1973 Computers & Humanities VII. 214 The cards were built onto a disk file by a program that reformatted the material into fixed-length records. 1911 Mrs. H. Ward Case of Richard Meynell xxiii. 484 Hester's telegram, sent originally to Upcote and reforwarded, had reached Meynell in Paris. 1957 M. Lowry Let. 29 Apr. (1967) 407 Your letter of March 12..Cape sent it back to Canada again, so that it had to get reforwarded again from B.C. before I received it. 1947 J. Hilton So well Remembered iv. 265 George's last two letters had never reached Charles... (They did arrive, eventually, after a series of fantastic re-forwardings.) 1962 H. L. Kern et al. in A. Pirie Lens Metabolism Rel. Cataract 386 The lenses were subsequently removed, dipped in saline containing antibiotics, and reincubated at 37 °C. 1961 Lancet 22 July 213/2 Absence of reinfarction. 1972 Computers & Humanities VI. 282 As before, the user must establish output procedures and appropriate tests for upper and lower limits and reinitialize counters. 1973 C. W. Gear Introd. Computer Sci. vi. 246 It is very easy to make the mistake of transferring back to the start of the loop, forgetting to re-initialize variables when it is required. 1964 C. Dent Quantity Surveying by Computer vi. 88 The items are queried and re-input, except for zero items, which will not be required to appear in the bill in any case. 1967 J. D. Dews in Cox & Grose Organiz. Bibliogr. Rec. by Computer II. 24 The tape..can then be corrected and re-input to correct the file. 1963 Daily Tel. 1 Nov. 14/2 To these torments must be added that of reintensified bombing. 1967 E. Chambers Photolitho-Offset ix. 133 Reintensify if added contrast is required. 1946 Nature 14 Sept. 379/1 Leaf infection of onion seedlings was obtained by ascospore inoculation, and the fungus was re-isolated from the lesions. 1977 J. L. Harper Population Biol. of Plants xi. 348 The bacterial agent must be re-isolated from the experimentally infected plant. Ibid., The re-isolated micro-organism and that originally inoculated must be tested for identity. 1934 Webster, Relicense. a 1974 R. Crossman Diaries (1977) III. 870, I then said that we must break this relicensing operation into two stages. 1977 Times 11 Oct. 4/3 Dr Lemon..did not favour relicensing all pilots who had suffered heart attacks. 1963 Times 29 May 7/2 The practice of 'stop-listing', 'delisting', and then 'relisting' areas can be a powerful deterrent to industrialists. 1976 Times 25 Oct. 14/7 Save in cases of nullity, the jurisdiction to relist depended on the likelihood of an injustice having been done. 1967 Karch & Buber Offset Processes x. 474 Old grease should be removed and the gears re-lubricated every one million impressions. 1965 G. J. Williams Econ. Geol. N.Z. viii. 99/1 A detailed remapping of the area will reveal the importance of deep pre-Second Period weathering. 1933 H. G. Wells Shape of Things to Come III. 261 The need for a planned 'renucleation' in the social magma that arose out of this dissolution. 1934 —— Exper. Autobiogr. II. vii. 481 Socialism, if it is anything more than a petty tinkering with economic relationships is a renucleation of society. 1901 Westm. Gaz. 13 May 4/3 The 'Marseillaise' has just been reorchestrated by order of the Minister of War. 1975 New Yorker 19 May 85/1 It seems to me that the actual personalities and events of the Nez Perce war were possibly even more interesting..than were the respectfully created counterparts..reorchestrated for us today. 1940 L. MacNeice Poems 251 Smuggling over the frontier Of fact a sense of value, Metabolism of death, Re-orchestration of world. 1975 New Yorker 16 June 97/1 In the late sixties, the opera was quite often given..but always in an edition by Claudio Abbado marred by many cuts, by some reorchestration, and, most gravely, by the recasting of Romeo as a tenor. 1854 Thoreau Walden 163, I occasionally observed that he was thinking for himself and expressing his own opinion, a phenomenon so rare that I would any day walk ten miles to observe it, and it amounted to the re-origination of many of the institutions of society. 1832 J. S. Mill Let. 22 Oct. in Coll. Wks. (1963) XII. 128 They were the reoriginators of any belief among us. 1935 L. MacNeice Poems 32 The basic facts repatterned without pause. 1952 C. P. Blacker Eugenics x. 246 The gene-complex has a holistic or integrative action of its own, a capacity to undergo changes, to adjust itself to a repatterning of its constituent elements. 1972 Guardian 28 Oct. 24 It is open to Mr Barber..to repeg the pound at an exchange rate far above the level to which it has now fallen. 1978 N.Y. Times 30 Mar. D1/6 The yen..has actually been revalued upward by 38.43 percent since it was repegged at the Smithsonian rate of 308 to the dollar in December 1971. 1938 Sun (Baltimore) 12 July 14/5 Accompanied by rumors..of a possible 'repegging' at its old ratio of between $4.86 and $4.87, the British pound sterling declined today..to its lowest point in more than a year. 1964 G. H. Haggis et al. Introd. Molecular Biol. vi. 182 (caption) An enzyme at the inner boundary rephosphorylates the diglyceride to phosphatidic acid, in interaction with ATP. 1965 Phillips & Williams Inorg. Chem. I. xvii. 638 The ADP is then rephosphorylated via various sugar-phosphates and the oxidation of glycogen. 1877 Nature 27 Sept. 468/2 In repiling and reheating this iron several times this defective appearance is gradually removed. 1947 Penguin New Writing XXX. 104 The lame boy stayed behind and helped me re-pile the tins. 1956 D. Gascoyne Night Thoughts 37 To swell the roar that rises with each climax repostponed. 1967 Karch & Buber Offset Processes iv. 125 Modification is possible to condense, expand,..reproportion height and width. 1969 P. L. Berger Rumor of Angels v. 121 The openness and the reproportioning this attitude entails have a moral significance, even a political significance, of no mean degree. 1963 Rep. Comm. Inquiry Decimal Currency xiv. 138 Ancillary machine costs:..re-punching card and tape records. 1965 Math. in Biol. & Med. (Med. Res. Council) II. 48 A start has now been made on re-punching the British Columbia marriage records for 1946–55 in a form suitable for testing such a system. 1887 G. B. Shaw Let. 7 Feb. (1965) I. 162 The American printer..has taken upon himself the repunctuation of 'Cashel Byron'. 1966 Mod. Lang. Q. Sept. 256, I was, I believe, responsible for most of the detailed examination of poems in A Survey of Modernist Poetry—for example showing the complex implications of Sonnet 129 before its eighteenth-century repunctuations. 1975 D. W. S. Hunt On Spot iv. 54 As I heard him say over the lunch table once, 'to rat is difficult; to re-rat..' and he broke off as though to show that to find a description of a second change of party was beyond even his eloquence. 1938 Times 22 Jan. 5/1 The prospects of a re-recovery in the United States. 1922 Joyce Ulysses 671 With greater difficulty remembered, forgot with ease, with misgiving reremembered. 1923 D. H. Lawrence Birds, Beasts & Flowers 24 For we are on the brink of re-remembrance. 1922 Joyce Ulysses 133 But my riddle! he said. What opera is like a railway?—Opera? Mr. O'Madden Burke's sphinx face reriddled. 1973 Nature 6 Apr. 377/1 The role of postgraduate students may well be rescrutinized. 1963 Punch 6 Feb. 182/3 The whole business ..deserved re-scrutiny. 1955 Auden Shield of Achilles ii. 45 Re-sex the pronouns, add a few details. 1951 L. MacNeice tr. Goethe's Faust I. 14 The little god of the world, one can't reshape, reshade him. 1961 Webster, Reshow. 1976 Times Lit. Suppl. 21 May 620/5 Pabst's maligned film, which is still frequently reshown. 1977 Listener 24 Mar. 385/2 Most programmes are not re-shown. 1976 K. Benton Single Monstrous Act iii. 17 Let's go and see that film at the local. It's a re-showing of The Godfather. 1841 Civil Engin. & Arch. Jrnl. IV. 341/2 Scintling (removing and restacking the bricks in the hacks). 1892 Daily News 26 Feb. 5/7 A patch of brickwork, six feet by three feet, had been taken out of the chimney to be re-stacked. 1923 Daily Mail 30 May 7 The combat will be restaged at the forthcoming pageant. 1929 Daily Express 16 Jan. 6 A famous comedian ..suddenly decided to alter his programme and not restage his old, worn-out material. 1962 M. Cook tr. Dia's Afr. Nations & World Solidarity ix. 97 It will be vain to hope for a profound change toward a progressive economy..without a bold restructuration incompatible with an exaggerated desire to spare the former capitalist structures. 1970 B. Brewster tr. Althusser & Balibar's Reading Capital (1975) III. iii. 259 Reproduction appears to be the general form of permanence of the general conditions of production, which in the last analysis englobe the whole social structure, and therefore it is indeed essential that it should be the form of their change and restructuration, too. 1972 Times 26 June 12/4 The restructuration of a basic sector of European economy. 1900 Daily News 27 Sept. 7 The Count..is hard at work directing the repairs to the airship, which has already been resuspended. 1946 Nature 5 Oct. 486/2 The mixture is..re-suspended in saline to give a concentration of 2–5 per cent. 1976 Nucl. Engin. Internat. Nov. 37/2 Eventually the Pu settles to the ground and may be resuspended by being blown up by winds. 1884 Pall Mall Gaz. 16 Oct. 5/2 The Exchequer, however, is still in want of funds, and the resuspension is expected before long. 1978 Nature 27 Apr. 754/2 The lack of response to fire in the pollen input could be due to..the resuspension and resedimentation which occurs in the lakes. 1884 Practitioner XXXIII. 289 (heading) Resuturing of granulating wounds. 1901 Brit. Med. Jrnl. 2 Feb. 263/1 Should any difficulty be met with in replacing the bowel, the opening can be enlarged and then resutured without causing any weakening in the lower part of the abdominal wall. 1961 Lancet 19 Aug. 402/2 After resuture of the wound the patient's general condition gradually improved. 1977 Ibid. 1 Jan. 28/1 One of these is abdominal wound disruption—a complication fatal at worst and a serious nuisance at best, for, should all go well directly after resuture, there is both a prolonged hospital stay and a much increased risk of incisional hernia. 1927 Haldane & Huxley Animal Biol. v. 120 The actual process of contraction may have an efficiency of 90 to 100 per cent., but an amount of energy greater than the work done in contraction is wasted as heat during the re-synthesis of the lactic and phosphoric acids, so that the whole process has an efficiency of only about 40 per cent. 1956 Nature 7 Jan. 22/2 There is little to support the notion that phospholipids act as intermediates in the resynthesis of triglycerides during fat absorption. 1978 Ibid. 12 Oct. p. ix/1 (Advt.), Components of cells in all living organisms undergo continual breakdown and resynthesis. 1928 A. B. Callow Food & Health 21 The fats are split up into their component parts, fatty acids and glycerol. These substances are absorbed by certain body-cells and again resynthesized into fats. 1964 G. H. Haggis et al. Introd. Molecular Biol. v. 137 All cells require new proteins during growth and division, and are subsequently constantly resynthesizing their enzymes and other proteins. 1928 Daily Express 9 Aug. 14/1 (Advt.), O'coats turned and retailored. 1949 M. Mead Male & Female xii. 260 The reactionary and the cynic make common cause in suggesting that..laws and ideals should be re-tailored to recognize the deviations and discrepancies between ideal and practice. 1977 Times 27 Apr. 23/8 On the bonus side value-added tax machinery is retailored to harmonize with EEC directives. 1966 Aviation Week & Space Technol. 18 Apr. 31 This was the retargeted date after the March 30 attempt. 1893 Pall Mall Gaz. 25 Jan. 7/1 The wedding of the Princess Sophie to the Crown Prince of Greece implied a further re-tightening of the bonds of friendship between Copenhagen and Berlin. 1960 Times 1 June 18/7 But the production as a whole needs re-timing. 1967 Karch & Buber Offset Processes x. 468 The sheets should be ¼ inch away from the guides at this point. If not, the feeder must be retimed to the press. 1977 Dædalus Summer 109 He persuaded the Survey of India to retriangulate part of the region. 1927 Blackw. Mag. Feb. 256/1 It would mean the retriangulation of the whole area, and endless delays and doubts as to the fixing of the position of soundings already taken off the coast. 1975 J. B. Harley U.S. Maps i. 7 Supplementary work on the retriangulation continues. 1974 M. C. Gerald Pharmacol. v. 101 Norepinephrine is taken back into the presynaptic neurons (re-uptake process), thus removing it from the receptor area. 1977 Lancet 21 May 1081/1 The amount of catecholamine entering the blood is dependent upon a complex process of neuronal release and reuptake into nerve terminals. c 1864 E. Dickinson Bolts of Melody (1945) 246 One rewalks a precipice. 1969 J. T. Burtchaell Catholic Theories of Biblical Inspiration since 1810 vii. 303 The individual Christian can and certainly should rewalk the route from paganism to Christ. 1877 G. M. Hopkins Poems (1967) 68, I hear the lark ascend, His rash-fresh re-winded new-skeinèd score In crisps of curl. 1970 'D. Halliday' Dolly & Cookie Bird xi. 168 He..pulled back a zipper..re-zipped it. 1974 'M. Underwood' Pinch of Snuff xi. 97 Brian watched him re-zip the bag.

b. *re-brighten* vb., *-catholicization*, *-centralization*, *-civilianization*, *-civilianize* vb., *-institutionalization*, *-institutionalize* vb., *-modernize* vb., *-phonemicization*, *-phonologization*, *-phonologize* vb., *popularize* vb., *-solemnize* vb., *-stabilization*, *-Stalinization*, *-Stalinize* vb., *-standardization*, *-sterilization*, *-sweeten* vb., *-vascularization*, *-vascularize* vb., *-volatilization*.

1925 E. Sitwell Poetry & Crit. 23 Miss Stein..breaks down the predestined groups of words..; then she rebrightens them,..and builds them into new and vital shapes. 1949 Scottish Jrnl. Theol. II. 241 The pressure for visible unity, for a recatholicisation of the Churches, has ..[come] from the 'evangelical' movements burdened with fulfilling the mission of the Gospel to the whole world. 1925 Glasgow Herald 5 Oct. 11 Only Scotland and the Northern (Newcastle) [administration] will continue to function, but the official policy is definitely recentralisation because of the enormous fall in the volume of the work. 1957 C. P. McVicker Titoism p. x, Decentralization as Re-centralization. 1967 Economist 14 Oct. 149/2 There is not necessarily a contradiction between the drive for reform and the temporary re-centralisation of investment policy, plus a restrictive incomes policy. 1947 Manch. Guardian 18 July, Certain foreigners are eligible for 'recivilianisation'. 1962 S. E. Finer Man on Horseback xi. 190 How far it is possible for a military régime, starting with quasi-civilian institutions,..ultimately to re-civilianize itself. Ibid. 197 Could the régime not pass..from a quasi-civilianized military régime..to a re-civilianized one? What would be the criteria of such re-civilianization? Ibid. 198 In both these cases it has taken a long time for the 're-civilianization' to take root. 1978 Sci. Amer. Feb. 50/2 Most patients are placed in nursing homes.., a process the Department of Health, Education, and Welfare has labeled 'reinstitutionalization', since most homes have more than 100 beds (and yet offer only custodial care). 1976 Guardian Weekly 14 Nov. 7/2 The first task of the Carter administration must be to reinstitutionalise American foreign policy. 1973 Courier & Advertiser (Dundee) 7 Aug. 1/3 (Advt.), Remodernised 4-apartment house. 1976 S. Wales Echo 27 Nov. 9/4 (Advt.), Remodernised fur coats. 1951 Z. S. Harris Methods in Structural Linguistics ix. 93 The following not infrequent situation is also a special case of resegmentation of a segment for purposes of rephonemicization. 1972 R. Jakobson in A. R. Keiler Reader in Hist. & Compar. Linguistics 135 Within the framework of an isolated functional dialect, one cannot speak either of an increase or a reduction of a phonological system, but only of a restructuring, that is, of its rephonologization. 1975 Amer. Speech 1971 XLVI. 266 Change cannot be incorporated within Saussurian idealized homogeneity of structure; that does not change, it is merely rephonologized. 1906 Westm. Gaz. 11 Aug. 12/2, I do not intend to repopularize the stereoscope. 1959 I. & P. Opie Lore & Lang. Schoolch. ii. 38 The following favourite accumulates on the principle of 'The House that Jack built'. It was repopularized by the American folk-singer Burl Ives in 1953, but had been current in Britain for anyway forty years before his visit. 1949 'J. Nelson' Backwoods Teacher xvi. 173 There before an assemblage they wrote their names in the Bible and took each other for better or worse—the same to be re-solemnized later when the preacher should come through. 1953 P. L. Fermor Violins of Saint-Jacques 28 A phoenix's funeral that was resolemnised each evening. 1921 Nineteenth Cent. Feb. 231 A disastrous blow to the restabilization of Europe. 1958 A. J. Toynbee East to West xxv. 75 The population started to increase at an accelerating rate (it is still increasing today, though re-stabilization is now in sight). 1956 Washington Post 19 Nov. A21/1 Even today, after the unspeakable horror of the blood bath in Hungary, the betting is still somewhat against a 're-Stalinization'. 1974 R. J. Osborn Evolution Soviet Politics v. 195 Just as with Khruschev's anti-Stalin campaign, Brezhnev's cautious but firm 're-Stalinization' has been a weapon for dealing with current problems. 1968 Russian Rev. XXVII. 309 All this evidence might appear to be leading to the conclusion that the party leadership is trying to re-Stalinize, but is forced to proceed very gradually in the face of opposition. 1911 Webster, Restandardization. 1938 Mind XLVII. 103 It is to be urgently hoped that a similar restandardization will be attempted for the present scale. 1952 C. P. Blacker Eugenics 204 The Pintner-Patterson scale (1917) and the Arthur Performance scale, the latter a restandardization of ten of the tests used in the former, are now in common use. 1900 A. H. Buck Ref. Handbk. Med. Sci. (ed. 2) I. 567/2 The best silk sponges are expensive so that resterilization would be necessary. 1966 Auden About House 29 Shrines.. Rose again Resweetened the hirsute West. 1924 Jrnl. Path. & Bacteriol. XXVII. 205 Revascularisation of the lobule takes place. 1954 Martin & Hynes Clin. Endocrinol. (ed. 2) i. 4 If the grafts were placed at a distance away from the

portal trunks revascularization from other vessels was not followed by active function. 1963 Lancet 12 Jan. 89/2 Given sufficient time..dead bone can become revascularized, and a case that starts off as a non-union with no callus can ultimately unite. 1971 Nature 23 July 279/2 The question of how skin grafts are revascularized is still not resolved. 1866 Q. Jrnl. Geol. Soc. XXII. 441 The volatile matters removed by distillation..have found a reception where no revolatilization could occur. 1938 R. W. Lawson tr. Hevesy & Paneth's Man. Radioactivity (ed. 2) xviii. 170 The condensation and revolatilization of bismuth hydride.

c. *re-cane*, *-fenestrate*, *-lampshade*, *-litter*, *-mine*, *-neck*, *-pew* (earlier example), *-staff*, *-washer* vbs.

1971 Islander (Victoria, B.C.) 16 May 5/4 There was also a small chair that went with the set but he was having the seat re-caned. 1970 H. Braun Parish Churches xvii. 214 Few early churches escaped being at least in part refenestrated during the High Gothic era. 1971 Country Life 3 June 1366/3 He did not make great changes..other than to refenestrate the east and south sides. 1918 A. Bennett Roll-Call I. ii. 24 The lampshade craze increasing in virulence, they had between them re-lampshaded the entire house. 1775 W. H. Marshall Minutes Agric. (1778) sig. D3, Shovelling the gangways, and re-littering them with long dung. 1960 Farmer & Stockbreeder 16 Feb. 145/3 The whole place is then thoroughly cleaned,..and rested a week before being re-littered. 1917 J. Masefield Old Front Line iii. 38 It was all mined, countermined, and re-mined. 1967 Guardian 2 Mar. 6/6 A genuine theorbo, built as a lute in 1584 and renecked in 1730. 1978 Early Music Oct. 531/2 It seems to have been cut down all round and re-necked in the 18th century to make a seven-string viol. 1839 F. Witts Diary 20 May (1978) 160 The church has been recently repewed with deal. 1898 Westm. Gaz. 4 Apr. 7/2 The hospital committee accepted the resignation, and a special meeting has been called to consider the subject of restaffing the institution. 1898 Daily News 16 Apr. 9/5 The Management Committee..decided to close the Institution..and restaff the place with as little delay as possible. 1960 Times 16 Sept. 13/5 Recently a tap over a wash hand basin in my office..required re-washering. Ibid., The local water company re-washer water taps free of charge in order to save water. 1974 J. Wainwright Evidence I shall Give xx. 92 The handyman who re-washers a tap.

're (-ə1, -1), contraction of Are, pl. pres. ind. of Be, as *you're, they're*. Freq. used in the representation of speech (and for metrical reasons in verse). Cf. Yare, Y'are.

1591 Shaks. Two Gent. v. iv. 44 O, 'tis the curse in love, and still approv'd, When women cannot love where they're belov'd. 1676 G. Etherege Man of Mode iv. i. 56 The women indeed are little beholding to the young men of this age; they're generally only dull admirers of themselves. 1738 Swift Polite Conv. 2, I hope, you're never the worse. But where's your manners? 1836 Dickens Pickw. (1837) ii. 12 'They're beginning up stairs,' said the stranger—'hear the..fiddles tuning.' 1978 J. Updike Coup (1979) i. 39 Whoever the hell you are, you're the best thing I've seen today.

reable, *v.* Restrict †*Sc. Obs.* to sense in Dict. and add: **2.** (rī,ēi·b'l) *trans. Med.* To rehabilitate (a patient).

1945 Lancet 24 Mar. 387/2 For some year or two I have suggested to friends who write or teach that we should say 'reable' instead of 'rehabilitate'. It therefore gives me pleasure..to read of 'reabling' and 'reablement' in your last three issues. 1947 E. Spriggs in I. Brown Say Word 101, I am glad to see that the words Reable and Reablement are now being frequently used in the Lancet and other journals.

Hence **rea·blement**, rehabilitation.

1945 Lancet 3 Mar. 275/2 The committee believe that a revised programme of reablement should be administered by a senior tuberculosis officer. 1945, 1947 [see above]. 1955 Brit. Jrnl. Physical Med. XVIII. 119/1 Reablement is seen as an intricate problem which has to be considered from very various aspects. 1968 Lancet 2 Nov. 985/2 To be comprehensive and successful any reablement programme must be built on well-organised services for health, education, social work, and vocational training.

reabsorb, *v.* Add: **b.** *intr.* for *pass.* To be reabsorbed.

1916 Galsworthy in Scribner's Mag. Jan. 17/1 In one's heart rose an ecstasy of love for this..earth which breeds us all, and into which we reabsorb.

reabso·rptive, *a.* [Re- 5 a.] Having the quality of reabsorbing.

1946 Nature 23 Nov. 730/1 The concept of competition for secretory and reabsorptive mechanisms in the renal tubules has proved very fruitful. 1975 Ibid. 27 Feb. 747/1 Mammalian bladder possesses, in addition to an exceptionally high electrical resistance, an aldosterone-stimulated Na^+ reabsorptive mechanism.

reaccommodate, *v.* (Later example.)

1920 Nineteenth Cent. Oct. 629 It will take time before the Jews can again reaccommodate themselves to the local conditions.

reach, *sb.*[1] Add: **I. 4. b.** (Later examples.) Also, a course that is approximately at right angles to the wind.

1949 Sun (Baltimore) 20 June 16/1 From the start to Sandy Point, the skippers had to face a headwind, the next leg..was a reach, while the trip to Poole's Island Light resulted in another beat. 1976 Oxf. Compan. Ships & Sea 695/1 Reach, the point of sailing of a vessel which can point her course with the wind reasonably free and her sails full throughout. A broad reach is the same but with the wind abeam or from slightly abaft the beam. 1977 Modern Boating (Austral.) Jan. 14/1 Destiny II will go like a rocket on a reach or a run.

II. 5. *spec.* in *Cricket*, the extent to which a batsman can play forward without moving his back foot; in *Boxing* (see quot. 1958).

1851 F. Lillywhite's Guide to Cricketers 18 A good length ball depends entirely upon the size and reach of a batsman. 1897 K. S. Ranjitsinhji Jubilee Bk. Cricket iii. 67 Batsmen vary greatly as to their 'reach'—that is, the distance they can safely play forward or advance the bat in making a drive. 1951 Sport 30 Mar.–5 Apr. 11/3 He quickly found that O'Hara, his opponent, was longer in the reach. 1958 F. C. Avis Boxing Ref. Dict. s.v. Reach: the distance from the extremity of one hand to that of the other when the arms are fully held out sideways.

III. 12. c. (Later examples.)

1955 Times 9 May 3/3 The action therefore slows almost to a standstill in the middle reaches of the play. 1971 J. B. Carroll et al. Word Frequency Bk. p. xxxvi, Hapax legomena..come from almost anywhere in the lower reaches of the theoretical type distribution.

IV. 16. *attrib.* and *Comb.*, as (sense 13 a) *reach land*; **reach rod,** a connecting rod for transmitting manual motion to a remote part of a mechanism.

1795 T. Chapman Jrnl. 4 Nov. in Hist. Mag. Amer. (1869) V. 359 They appear contented and Happy, having Plenty of fine reach Land. 1909 Cent. Dict. Suppl., Reach-rod, in a locomotive, a rod which connects the reverse-lever in the cab to the bell-crank on the reverse-shaft of the valve-gear. 1972 L. M. Harris Introd. Deepwater Floating Drilling Operations 238 Some compartments will have reach rods to bilge drainage valves.

17. Used *attrib.* to designate a fork-lift truck whose fork can be moved forward and backward as well as up and down.

1962 Engineering 2 Nov. 584/1 Until the reach truck appeared, the provision of two hydraulic functions in a fork truck was considered difficult. 1963 Times Rev. Industry May 91/3 Reach types—The fork carriage is movable in a horizontal plane, with forks reaching beyond the front wheels to withdraw the load to a position within the wheelbase.

reach, *v.*[1] Add: **I. 6. c.** *U.S. slang.* To bribe.

1906 A. H. Lewis Confessions of Detective 72 I'd been squared; it was known that I could be reached. 1912 A. Train Courts, Criminals & Camorra III. ix. 234 In America, if the criminal can 'reach' the complaining witness or 'call him off' he has nothing to worry about. 1929 C. F. Coe Hooch! v. 105 You could reach the..Attorney without tippin' your hand to him at all. 1967 L. Katcher Earl Warren xvi. 124 It is impossible..to open a big, notorious gambling operation without buying off public officials... This does not necessarily mean a sheriff or a District Attorney or a chief of police is being reached. 1968 W. Safire New Lang. Politics 373/1 Reached, bought off; corrupted. A public official may be approached with no implication of commitment on his part; when he has been spoken to, the implication is that he is neutralized or partially persuaded pending a final decision; when he is reached, however, he has been bought and sold.

8. b. (Later examples.) Now *esp.*, to communicate with (a person).

1919 E. O'Neill Moon of Caribbees 113 Ut is only from your chance meetin' wid Harry..that I happen to know where to reach you. 1967 D. Francis Blood Sport v. 55 The two drivers, reached by telephone, met us by appointment. 1973 R. Ludlum Matlock Paper xiii. 122 I'm off the phone now. Would you like to try reaching Miss Ballantyne?

II. 12. c. (Later examples.) *to reach for the roof, sky*, etc. (orig. *U.S.*), of a person held at gunpoint: to raise the hands above the head; *to reach for one's gun*, etc., used *fig.* for: to react with extreme hostility.

The statement in quot. 1953 is commonly attributed to the German Nazi leader H. Goering (1893–1946), but it has been traced (in a slightly different form) to the nationalistic play Schlageter (1933) by the German dramatist H. Johst (1890–): see Oxf. Dict. Quotations (ed. 3, 1980).

1904 L. Tracy King of Diamonds iii. 39 He reached over for the stone. 1910 W. M. Raine B. O'Connor ii. 25 Now reach for the roof. 1927 Ladies' Home Jrnl. Dec. 6/3, I reached for the stars pronto, without even turning my head. 1931 W. James Sun Up 42 'Stick em up,' I says... One of 'em flinches some but finally reaches for the sky. 1953 in Oxf. Dict. Quotations (ed. 2) 223/2 When I hear anyone talk of Culture, I reach for my revolver. 1959 Spectator 9 Oct. 480/2 Cherwell was not a cultured man; indeed he seems almost to have reached for his gun when 'culture' was in the air. 1967 Guardian 12 Oct. 9/6 This is the point at which Laing's critics reach for their guns. 1968 R. F. Adams Western Words (ed. 2) 245/2 Reach for the sky!, a gunman's command to raise the hands in the air. 1980 Listener 14 Aug. 200/1 Members of PRO Dogs..will..be reaching for their choke-chains because..criticism of..dogs is a treasonable activity.

e. (Later example.)

1935 W. G. Hardy Father Abraham IV. i. 349 But Abraham strode along and snuffed the air and was tender to Sarai as he helped her on and felt his whole heart reaching out to her.

f. To make an unwarranted inference; to jump to a conclusion; to guess. orig. and chiefly *U.S.*

1960 'E. McBain' Give Boys Great Big Hand x. 106 This may be reaching, but here it is anyway, for what it's worth. 1963 'G. Bagby' Murder's Little Helper xii. 149 Anderson moaned. 'Inspector,' he said, 'Isn't that reaching?' 1964 L. Treat in H. Waugh Merchants of Menace (1969) 11 The way he saw things, the Lieutenant was sure reaching for it—far out. 1973 G. Moffat Lady with Cool Eye xi. 130 Dawson..might very well be visiting Mrs Wolkoff..but no: as the Americans say, we're reaching. 1978 R. Ludlum Holcroft Covenant xviii. 212 'Aren't you reaching, Miles?'.. 'I said it was a theory, but not without some support.'

14. b. (Later example.)

1935 B. Malinowski in M. Black Importance of Lang. (1962) 90 A word rich in associations and reaching out in many directions.

c. (Later example.)

1951 M. McLuhan Mech. Bride (1967) 132/2 Some interesting assumptions that reach back more than two centuries.

16. a. Also *absol. spec.* in *Baseball*: to reach first base.

1891 Young Man Apr. 128/1, I sent letter upon letter after him, but they don't seem to reach. 1974 Anderson (S. Carolina) Independent 20 Apr. 6A/6 The Cavs tied it in their half of the inning as Duval White reached on an error, stole second, went to third on Dan Berstein's first of four hits and scored on Jonathan Williams' fly ball. 1976 Billings (Montana) Gaz. 6 July 1-C/3 Dan Fuchs reached on an error, Rich Popp singled to score Knudtson, and Mike Klunder brought Fuchs home on a sacrifice fly.

reachable, *a.* Add: **2. a.** (Later examples.) Also, as *sb.*, one who may be reached (in quot. 1974, denoting a person who can be bribed).

1914 R. & E. Shackleton Four on Tour in Eng. xviii. 186 Few visitors go to Runnimede [sic], because it has not been a readily reachable place by rail. 1974 Publishers Weekly 27 May 31 (Advt.), The secret deals behind the rip-off, put-down and sell-out of the American people..to Big Money interests..with the help of the reachables in office! 1978 Detroit Free Press 5 Mar. A4/1 Most buds and twigs and other reachable food are gone. If snow is deep enough, deer cannot escape.

b. *Math.* That may be reached (from a specified point) by passing along the lines of a graph.

1959 Sociometry XXII. 143 The matrix M contains all these paths and also a path of length l from each point to itself. Thus each point is reachable from itself. 1965 F. Harary et al. Structural Models ii. 43 Point v is reachable from u in a digraph D if and only if there is a sequence from u to v. 1978 D. D. Šiljak Large-Scale Dynamic Syst. iii. 149 A reachable set $V_l(v_j)$ of a point v_j is a set of points v_l reachable from v_j.

Hence **reachabi·lity** *Math.*, the possibility of reaching one point of a graph from another; freq. *attrib.*

1959 Sociometry XXII. 141 We present a logical sequence of theorems that characterize strengthening and weakening group members... We..find it convenient to introduce the concept of the 'reachability matrix' of a group. 1965 F. Harary et al. Structural Models ii. 43 Reachability is reflexive since every point of V is reachable from itself by a path of length 0. Ibid. 45 The difference between reachability and joining is that in the latter instance we ignore the direction of the lines. Ibid. v. 117 The reachability matrix $R(D)$ whose entries are denoted r_{ij} and defined as follows: $r_{ij} = 1$ if v_j is reachable from v_i; otherwise $r_{ij} = 0$. 1978 D. D. Šiljak Large-Scale Dynamic Syst. iii. 149 The reachability concept of digraphs.

reacher. Add: **3.** *Yachting.* (See quot. 1903.)

1903 Outing XLII. 646/2 Intermediate between the baby and the balloon are various jib-topsails called 'reachers' and numbered 1, 2, 3, according to the size. 1977 P. Johnson Offshore Manual Internat. 152 Under the rule 'jibs' includes all fore and after headsails such as genoas, staysails, big-boy/bloopers, reachers and drifters.

reaching, *vbl. sb.*[1] Add: **3.** (In sense 15 c of the vb.) *reaching foresail, jib, sail, staysail.*

1948 R. de Kerchove Internat. Maritime Dict. (ed. 2) 581/2 Reaching foresail, a triangular sail which sets on the forestay... Also called Genoa foresail, Genoa jib. 1924 G. H. P. Muhlhauser Cruise of Amaryllis vii. 273 There is now rather more wind and we are carrying reaching jib and staysail. 1962 D. F. Southern in A. Garrett Roving Commissions 11 It was sunny, warm and most pleasant running under our large reaching sail. 1924 G. H. P. Muhlhauser Cruise of Amaryllis vii. 273 At sunset I changed to the working jib but left the reaching staysail.

reaching, *vbl. sb.*[2] (Later example.)

1791 J. Woodforde Diary 7 June (1927) III. 275 She was..very much swelled in the face by reaching and very weak.

reach-me-down, *a.* and *sb.* Add: **A.** *adj.* **b.** *transf.* and *fig.* Ready-made, stock; derivative, inferior.

1907 G. B. Shaw Let. 11 July (1972) II. 700 You are not.. one of these reach-me-down people. 1914 —— in New Statesman 14 Nov. 19/2 Europe has a stock of ready-made constitutions... It is therefore quite possible that a reach-me-down constitution proposed..by an international congress..might prove acceptable. 1932 V. Woolf Common Reader 2nd Ser. 230 There are a dozen occasions on which a reach-me-down character will satisfy him well enough. 1951 C. Day Lewis Lyrical Poetry of Thomas Hardy in Proc. Brit. Acad. XXXVII. 158 Often..Hardy seems to lose all touch with his medium, and will dress up his subjects in the shoddiest, reach-me-down verse. 1962 Times 14 Jan. 12/5 The clergyman full of reach-me-down psychology. 1975 Listener 4 Dec. 760/1 The role he has chosen is that of a reach-me-down James Bond. 1976 Glasgow Herald 26 Nov. 5/5 Daniel Wayenberg's programme of works by Brahms, Bartok, Ravel and Mussorgsky proved an admirable one with not a reach-me-down sonata in sight.

B. *sb.* **1. a.** (Earlier and later examples.)

1862 F. Sinnett Acct. Colony S. Australia 53 Waxen dummies, in their model reach-me-downs. 1922 Mrs. A. Sidgwick Victorian i. 11 She has evidently bought a reach-

me-down at one of the cheap shops, and as for her hat it's the limit. 1926 A. Bennett Lord Raingo I. xxxv. 165 He had said stiffly that he would enter the House of Lords in no hired reach-me-down. 1928 D. Byrne Destiny Bay i. 117 A small rat-faced man in a suit of reach-me-downs. 1931 R. Church High Summer iv. vi. 322 'Buy a..ready-made suit'.. 'Not even the prospect of saving Filosilk from irretrievable ruin could induce me to wear a reach-me-down.' 1981 Times 12 Mar. 10/8 The costumes..look..like tattered reach-me-downs from the Royal Shakespeare Company.

b. *transf.* and *fig.*

1916 G. B. Shaw Pygmalion 191 The rest of the story.. would hardly need telling if our imaginations were not so enfeebled by their lazy dependence on the ready-mades and reach-me-downs of the ragshop in which Romance keeps its stock of 'happy endings' to misfit all stories. 1958 Spectator 31 Jan. 136/3 This second-hand reach-me-down of a musical comedy. 1963 Times 28 Jan. 6/3 Ministers realize..that there would be acute political difficulties in dressing the shop window with the cast-off reach-me-downs of policy from years ago. 1977 Gramophone Dec. 1015/1 My particular ration of reviews usually spares me the necessity of sitting in judgement on the familiar reach-me-downs of the classical repertoire, and the past year has been no exception.

2. *pl.* Trousers.

1877 [in Dict.]. 1905 Westm. Gaz. 20 Apr. 2/1 There is a gentleman in pegtop reach-me-downs (I believe this is the correct method in America of describing that portion of gentleman's attire which a lady is never supposed to notice). 1979 G. Mitchell Mudflats of Dead ix. 96 Tatty old reach-me-downs..and a gosh-awful dirty sweater.

reachy, *a.* Delete *rare* and add later examples.

1902 Rep. Evolution Committee R. Soc. I. 96 Though not so 'reachy' as a fine Indian Game, they never have the general appearance of Leghorn. 1952 A. W. Hunter Leighton's Compl. Bk. Dog xvi. 227 Neck [of Kerry Blue Terrier]: Strong and reachy.

reacquaint, *v.* (Later examples.)

1973 M. Amis Rachel Papers 20, I wondered if there were any important lies I had told her which it would be worth reacquainting myself with, but could think of none. 1977 National Observer (U.S.) 1 Jan. 6/5 One purpose of the tests is to give owners fun with their dogs and reacquaint them with their pets.

reacquisi·tion. [f. Re- 5 a + Acquisition.] The action of reacquiring; a thing which or a person who is reacquired.

1904 Lancet in Westm. Gaz. 30 Sept. 10/1 The man on his holiday..rises to the occasion by the reacquisition of powers which belong not to the town. 1959 Times 10 Jan. 8/5 He paid tribute to the Socratic wisdom which had rendered Mr. Forster so valuable a reacquisition to the college. 1977 Word 1972 XXVIII. 194 It looks as if we have a reacquisition here.

react, *v.*[1] Add: **1. c.** *trans.* To cause to react chemically or immunologically *with* or *together*.

1944 [see *adduct sb.]. 1949 [see *oxo(-) 2 c]. 1962 J. C. Wright Metallurgy in Nuclear Power Technol. v. 91 In the Degussa process fine beryl is reacted with excess lime at 1,500°C. 1963 Listener 14 Mar. 459/2 A team of workers.. showed that xenon and fluorine could readily be reacted together to produce xenon-tetrafluoride. 1971 Nature 21 May 195/1 These, and several antisera previously made against Australasian marsupial sera, were reacted with a range of sera from Australasian and American marsupials in immunoelectrophoretic tests. 1972 Sci. Amer. Aug. 34/1 It is fairly easy to react unsaturated molecules with a variety of chemical reagents.

2. (Later examples.) Also, of a person: to respond; to behave in response *to* an event, a statement, etc.

1913 H. James Small Boy & Others xv. 212 What tenuity of spirit it argues that I should neither have enjoyed nor been aware of missing..a space wider than the schoolroom floor to react and knock about in. 1921 G. B. Shaw Back to Methuselah p. xlviii, The assumption that he may safely cross Oxford Street in a state of unconsciousness, trusting to his dodging reflexes to react automatically and promptly enough to the visual impression produced by a motor bus. 1928 E. O'Neill Strange Interlude I. 25 My heart pounding!..seeing Nina again!..how sentimental..how she'd laugh if she knew! and quite rightly..absurd for me to react as if I loved..that way. 1931 —— Haunted II, in Mourning becomes Electra (1932) 251 Lavinia... I have a right to love! Orin (reacting as his father had—his face grown livid—with a hoarse cry of fury grabs her by the throat). You—you whore! 1953 Times 13 May 3/5 Mr. Selwyn Lloyd said the Siamese Government..was reacting to this threat by taking precautionary measures on the north-eastern frontier. 1966 Guardian 11 Aug. 8/2 Pictures not only of the member addressing the House..but of other members reacting to his speech. 1979 Economist 28 Apr. 28/2 A firework lobbed by a demonstrator caused the police to react.

4. *spec.* of share prices: to fall after rising.

1896 [see Dict.]. 1908 Daily Chron. 10 Jan. 1/7 In one or two directions Stock Markets were again heavy, and prices further reacted. 1927 Daily Tel. 22 Nov. 2 Cairn line reacted 9d and P. and O. fell 2. 1972 Ibid. 29 Apr. 17/6 The shares reacted to 222p before rallying to 228p, a net loss of 2.

reactance (ri,æ·ktāns). *Electr.* [ad. F. *réactance* (É. Hospitalier 1893, in *L'Industrie Électrique* 10 May 210/1), f. *réaction* Reaction: see -ance and cf. Eng. *resistance, impedance.*]

1. a. The non-resistive component of impedance, arising from the effect of inductance or capacitance or both and causing the current to be out of phase with the e.m.f. causing it. Cf. Impedance in Dict. and Suppl.

1893 Trans. Amer. Inst. Electr. Engineers X. 413 The term 'impedance' is suggested for the denominator..and for the quantity enclosed between the brackets, the term 'reactance' is proposed. 1896 F. Bedell Princ. Transformer iv. 68 The reactance is, accordingly, equal to the component of the impressed electromotive force at right angles to the current, divided by the current. Reactance is measured in ohms. 1902 W. G. Rhodes Alternating Currents 62 We propose to determine the equivalent resistance R, and the equivalent reactance S, of the combination. 1940 Amateur Radio Handbk. (ed. 2) i. 10/2 The reactance of a condenser increases as the frequency of alternating current decreases... The reactance of an inductance decreases as the frequency decreases, which is the opposite effect to that of a condenser. 1970 D. F. Shaw Introd. Electronics (ed. 2) ii. 33 The general definition of complex impedance is $Z = R + jX$, where R is the resistance and X is the reactance of the circuit. 1975 D. G. Fink Electronics Engineers' Handbk. VII. 8 The reactance of a capacitor is given by $X_c = 1/2\pi fC = 1/\omega C$ (ohms) where $f = \omega/2\pi$ is the frequency in hertz.

b. = *Reactor 2 a.

1923 Mod. Wireless I. 263 Try reversing the connections of the Armstrong inductance or reactance. 1947 R. Lee Electronic Transformers & Circuits vi. 150 To avoid introducing losses and attenuation in the transmission bands, reactances as nearly pure as practicable are used in the elements of a wave filter. 1961 [see *reactor 2 a].

2. *Mech.* and *Acoustics.* The imaginary component of a mechanical or acoustic impedance, producing a phase difference between a driving force and the resulting motion but no dissipation of energy.

1925 Physical Rev. XXV. 91 In the acoustic filters the impedances were combinations of the reactances, $iM\omega$ and $1/C\omega$, ω being 2π times the frequency, M the inertance and C the capacitance. 1932 W. West Acoustical Engin. i. 3 Any impedance may be expressed as the vectorial sum of two components, a resistance and a reactance, whose phases differ by 90°. Ibid. iv. 54 The impedance comprises a resistance component, which depends on the rate of absorption of sound by the walls of the cavity..and an elastic reactance, in parallel, which depends on the volume of the cavity. 1950 Kinsler & Frey Fund. Acoustics viii. 202 The acoustic reactance x of a sound medium..is the component that results from the effective mass and stiffness of the medium. 1958 Condon & Odishaw Handbk. Physics II. iii. 22/2 The reactance is the sum of the two contributions: $-ik/\omega$, the elastic reactance, and $i\omega m$, the inertial reactance. 1968 H. J. Pain Physics of Vibrations & Waves ii. 38 Mass, like inductance, produces a positive reactance, and the stiffness behaves in exactly the same way as the capacitance.

3. *Psychol.* A term sometimes used for the type of response aroused in a person who feels his freedom of choice is threatened or impeded.

1966 J. W. Brehm Theory of Psychol. Reactance i. 2 Since this hypothetical motivational state is in response to the reduction (or threatened reduction) of one's potential for acting, and conceptually may be considered a counterforce, it will be called 'psychological reactance'. 1970 Jrnl. Personality & Social Psychol. XIV. 18 If the communication arouses reactance, the individual will tend to reject its position in order to demonstrate his freedom to decide for himself. 1972 Harvard Law Review LXXXVI. 399 'Reactance' is conceived to be an aversive state created in the individual by a threatened loss of his freedom to act or choose.

4. *attrib.* and *Comb.*

1921 Wireless World IX. 187/1 Reactance coils are used in conjunction with valve sets. 1964 R. F. Ficchi Electr. Interference x. 203 A reactance-grounded system is one in which a reactance is inserted in the connection to ground. 1967 Electronics 6 Mar. 186/1 For slow scan applications, however, the reactance components would become excessively large and difficult to stabilize with temperature. 1972 L. S. Wrightsman Social Psychol. in Seventies x. 306 Reactance theory concerns itself with situations where one's freedom of choice is threatened. 1976 Tedeschi & Lindskold Social Psychol. viii. 360/1 Numerous experiments have shown reactance-like effects in the laboratory. Ibid. 361/1 Reactance studies in which subjects are told they are being evaluated for leade ship.

reactant (ri,æ·ktänt), *sb.* (*a.*) *Chem.* [f. React *v.*[1] + -ant[1].] A reacting substance or species. Also *attrib.* or as *adj.*

1928 Chem. Rev. V. 67 In hydrogenation there is an optimum ratio to the partial pressures of the reactants. 1936 T. J. Webb Elem. Princ. Physical Chem. viii. 253 In general the rate of a given reaction depends upon the instantaneous concentration of the reactants. 1948 Ann. Rep. Progr. Chem. XLV. 10 The immediate consequence of absorption of some of the energy of the incident radiation is the conversion of the absorbing reactant molecule into the product. 1950 Sci. News XV. 65 Once a layer of oxide has formed, subsequent reaction must take place by the reactants.. diffusing through it. 1971 Jrnl. Oil & Colour Chemists' Assoc. LIV. 849 The necessary reactants, ammonia and sulphur dioxide, appeared to be absorbed from the atmosphere by droplets of water condensed on the paint surface. 1973 J. G. Tweeddale Materials Technol. II. ii. 23 With the majority of materials suitable for melting and freeze casting, gases such as oxygen, hydrogen and nitrogen..are much more soluble in, or reactant to, the liquid state of a material than they are to the solid state.

reaction. Add: **2. b.** (i) (Further examples.) Also, any chemical change. Also extended to transformations of atomic nuclei and other particles.

1926 H. G. Rule tr. J. Schmidt's Text-bk. Org. Chem. 131 The Hofmann method, in which bromine and potassium hydroxide are brought into reaction with acid amides. 1936 Nature 29 Feb. 344/1 The typical features of nuclear reactions are therefore perhaps most clearly shown by neutron impacts. 1950 N. V. Sidgwick Chem. Elements I. 397 The reaction of boron trichloride with an alcohol is quite violent. 1956 A. H. Compton Atomic Quest i. 53 The amount of U-235 needed to bring about a chain reaction was many times smaller than earlier measurements had indicated. 1972 R. A. Jackson Mechanism i. 5 Unimolecular reactions occur as a result of reorganization of the bonds within a molecule, with or without rupture into fragments. Ibid. 6 Bimolecular reactions take place during a collision of two molecules. 1977 J. Narlikar Struct. Universe ii. 36 The star's nuclear reactions generate energy which goes towards providing the necessary pressure to support the star against its tendency to contract with gravitation.

(ii) *spec.* (usu. with qualifying adj.) a reaction characteristic of acid or alkali; hence, acidity, neutrality or alkalinity.

1856 W. A. Miller Elem. Chem. II. x. 710 A salt which affects neither the blue of litmus nor the yellow of turmeric is said to have a neutral reaction. 1899 Jrnl. Physiol. XXIV. 289 On the addition of acid a dispersion of the flakes of coagulum occurs with production of an opalescent fluid having an acid reaction. 1900 Proc. R. Soc. LXVI. 116 The effect of the acid or basic reaction of the salt on the hydrosol. 1938 Jrnl. Inst. Brewing XLIV. 466/1 The highest grade of asbestos is a completely insoluble material, neutral in reaction and tasteless. 1956 K. Imhoff et al. Disposal of Sewage iv. 23 The 'Reaction' of aqueous wastes may be described as acid, neutral, or alkaline, depending on the concentration of hydrogen ions. Pure water is neutral in reaction. 1968 Passmore & Robson Compan. Med. Stud. I. vi. 2/2 It follows that the reaction of an aqueous solution may be expressed either in terms of $[H^+]$ or $[OH^-]$.

3. c. *Psychol.* A response to a stimulus which can be observed, estimated, or measured.

1887 Jrnl. Mental Sci. XXXII. 604 The reaction is found to occupy less time in the insane than in the sane. 1943 C. L. Hull Princ. Behav. xviii. 326 Many reactions.. approach the all-or-none type, which differs appreciably from the galvanic skin reaction,..etc. Ibid., Reaction is a joint multiplicative function of habit strength and drive. 1968 R. F. C. Hull tr. Jung's Analytical Psychol.: Tavistock Lectures ii. 54 Other characteristic disturbances are:.. reaction expressed by facial expression, laughing, movements of the hands or feet or body..insufficient reactions like 'yes' or 'no'.

d. In general use: a response (to an event, a statement, etc.); an action or feeling that expresses or constitutes a response.

1914 H. James Notes of Son & Brother v. 117 His [sc. William James's] letters..mark the beginning of those vivacities and varieties of intellectual and moral reaction which were for the rest of his life to be the more immeasurably candid and vivid. 1922 Joyce Ulysses 650 Did Bloom discover common factors of similarity between their respective like and unlike reactions to experience? 1932 Kipling Limits & Renewals 134 Only the speed of my reactions saved me from bumping into Bunny when he pulled up without warning beside a lorry. 1946 E. O'Neill Iceman Cometh I. 89 They all hoot him down in a chorus of amused jeering. Hugo is not offended. This is evidently their customary reaction. 1980 D. Brierley Blood Group O 43 Reaction to the death of Desnos had set in.

e. *Radio.* A former name for positive feed-back. Also *ellipt.*, a reaction coil.

1917 R. D. Bangay Elem. Princ. Wireless Telegr. II. 205 We now come to the final and most efficient method of utilising the properties of the magnifying valve for the purposes of reception of spark signals. It is known as the reaction method. 1920 Wireless World 24 July 327/2 The reaction may cause some howling if it is not in phase with the frame. 1922 A. C. Lescarboura Radio for Everybody i. 32 This small coil is frequently called the 'tickler' or reaction. 1923 Radio Times 28 Sept. 22 Two-valve long range receiving set with anode tuning and reaction. 1933 Boys' Mag. XLVII. 117/2 To modernize an old set.. swinging reaction can easily be altered to Reinartz reaction. 1943 C. L. Boltz Basic Radio xi. 184 Other names for positive feed-back are reaction and regeneration. Ibid. xiv. 222 Reaction in the hands of the inveterate knob twiddler is a curse to everybody.

4. a. (Earlier and later examples.)

1792 A. Young Trav. France I. 557 A most curious political combination, which seems to shew, that..where evils are of the most alarming tendency, there is a re-action, an undercurrent, that works against the apparent tide, and brings relief. 1965 New Statesman 7 May 724/2 In spite of his achievements as a social reformer, he became a symbol of black reaction to organised labour. 1969 A. G. Frank Latin Amer. (1970) xxii. 341 After eliminating popular leadership in labor, student, and other organizations..reaction settled back into comfort. 1974 tr. Sniečkus's Soviet Lithuania 29 Operating under conditions of extreme political reaction at home, the Communist Party of Lithuania skilfully combined legal and illegal forms of political activity.

b. *Econ.* A downward movement (of share prices, etc.) following an upward one.

1841 [see *inflation 6]. 1925 Scribner's Mag. Sept. 338/1 The mercurial American..sees either a coming 'business boom' or a season of 'reaction', and he usually bases his belief on visible tendencies in trade and industry. 1930 Daily Express 8 Sept. 10/2 It is safer to buy on reactions than on the top of a rise.

5. *reaction drive, experiment, force, mechanism, period* (earlier example), *potential, rate, speed, threshold, time* (later examples), *velocity, vessel, word*; **reaction chamber,** (*a*) a vessel in which a chemical reaction occurs, esp. in an industrial process; (*b*) the combustion chamber of a rocket; **reaction circuit,** that part of the anode circuit of a thermionic valve

which produces positive feedback in the grid circuit; **reaction coil,** an inductance coil included in the anode circuit of a thermionic valve so as to cause positive feedback in the grid circuit; **reaction engine,** substitute for def.: any engine in which motive power is derived from the reaction exerted by a jet of escaping fluid (usu. gas); (further example); **reaction formation,** a term originally used by Freud for the tendency of a repressed wish or feeling to be expressed in a contrasting or opposite form; **reaction index** *Broadcasting* (see quots.); **reaction jet,** a jet engine used to provide intermittent thrust for changing or correcting the velocity of a craft or the like; **reaction machine** (further example); **reaction motor** = *reaction engine*; **reaction pattern,** an assumed pattern of behaviour or response established in the nervous system; **reaction propulsion,** any form of propulsion which utilizes the reaction exerted by escaping fluid as the source of motive power; *spec.* jet or rocket propulsion; **reaction rim** *Petrol.*, a zone of one mineral enclosing another and formed by reaction of the latter with its surroundings; **reaction shot** *Film* and *Television*, the photographing of a person responding to an event or to a statement made by another; **reaction time,** the time taken by a person (or any living organism) to respond to a stimulus; **reaction turbine,** a turbine which is driven by the pressure drop experienced by the working fluid in passing across or through the rotor; **reaction type,** a physical or personality type whose members are expected to react in specific ways; **reaction wood,** modified wood that forms in branches and leaning trunks and tends to restore upward direction of growth.

1924 Chem. Abstr. XVIII. 2426 Liquid hydrocarbons are heated under pressures of 100–300 lbs. per sq. in. at temps of 350–550° and after partially cooling are passed through a series of vaporizing or reaction chambers of successively diminishing pressure. 1952 E. Burgess Rocket Propulsion iii. 61 The combustion chamber in which the burning takes place is sometimes known as the 'blast' or 'reaction' chamber. 1966 H. O. Ruppe Introd. Astronautics I. i. 10 Usually, an oxidizer and a fuel are stored in separate tanks, and transported by a feed system to the reaction chamber. 1969 H. T. Evans tr. G. Hägg's Gen. & Inorg. Chem. xxi. 529 The nitrose is pumped up to a denitrating and concentrating tower..where it is combined with the chamber acid pumped up from the reaction chamber. 1919 Sci. Abstr. B. XXI. 291 The case of reaction circuits is also dealt with, and conclusions are drawn as to the audion characteristics of importance in a particular arrangement. 1931 Boys' Mag. XLV. 125/1 A simple capacity coupled inductive reaction circuit, better known as 'swinging' reaction. 1917 R. D. Bangay Elem. Princ. Wireless Telegr. II. 207 The amount of this extra E.M.F. induced in the grid coil by the reaction coil can be controlled by adjusting the coupling between the two windings. 1943 C. L. Boltz Basic Radio xiv. 221 With telephones on head hold the reaction coil in the hand and move towards the aerial coil. 1957 'T. Sturgeon' Thunder & Roses 122 They fired up the reaction drive and began to move toward the sun. 1967 N. E. Borden Jet-Engine Fundamentals 9 All gas-turbine engines, together with pulsejets, ramjets, and rocket motors, belong to a class of power plants called reaction engines. 1893 Amer. Jrnl. Psychol. VI. 242 Is the sensorial-muscular difference entirely conditioned by the technique of the ordinary reaction-experiment? 1933 J. C. Flugel Hundred Years Psychol. viii. 186 The reaction experiment was..a legacy both from the personal equation problems of the astronomers and from Helmholtz's measurement of the speed of the nervous impulse in the sensory nerves. 1965 C. N. Van Deventer Introd. Gen. Aeronaut. ix. 201/2 The reaction force, which is equal to the acceleration force, propels the balloon in the westerly direction shown. 1910 A. A. Brill tr. Freud's Three Contributions to the Theory of Sex iii. 83 A sub-species of sublimation is the suppression through reaction-formation, which..begins even in the latency period of infancy. 1924 [see *character-trait]. 1934 Mind XLIII. 113 How rare is that love of virtue which does not bear the trace of neurotic reaction-formation. 1960 Encounter XV. v. 25 A 'reaction-formation' in the psychoanalytic sense. 1974 W. B. Arndt Theories of Personality xvii. 349 Another common example of reaction formation is where repressed hostility for a person, say a mother's hostility toward her child, is converted by the ego into excessive concern for his well being. 1967 Listener 17 Aug. 195/3 The reaction index? This is supposed to measure what the audience thought about the show. 1972 Ibid. 7 Sept. 295/3 If radio had existed in the mid-1820s, a series of late Beethoven quartets would have had..a 'Reaction Index' (that's the measure of appreciation) too low to be mentioned. 1972 P. Moore Can You speak Venusian? xv. 153 Dr. Barber..proposes to fit large reaction jets to the tops of mountains on opposite sides of the world... When the tilt begins, one simply switches on the jets. 1978 Nature 5 Oct. 378/2 Reaction jets are fired once or twice daily to reduce wheel momentum that accumulates due to disturbance torques on the spacecraft. 1972 D. G. Shepherd Aerospace Propulsion i. 4 It should also be recognized that the propeller operates by the same basic principle... Thus all propulsion systems are reaction machines, although this is often popularly attached only to jet engines. 1972 R. A. Jackson Mechanism v. 80 Stereochemical studies allow us further insight into the details of a reaction mechanism by providing information about the direction of approach of a reagent or of the movement of groups during a reaction. 1935 C. G. Philp Stratosphere & Rocket Flight xi. 54 In its simplest terms the reaction motor most favoured at present takes the form of a rocket. 1962 F. I. Ordway et al. Basic Astronautics ii. 13 Archytas devised a wooden pigeon propelled by a steam reaction motor. 1923 K. Dunlap in Psychol. Rev. XXX. 94 A feeling is always a real stimulus pattern which is the beginning of a reaction pattern. 1926 W. McDougall Introd. Social Psychol. (ed. 20) 417 Those who..imply that they can explain alleged instinctive behaviour by postulating in vague general terms a 'reaction pattern' in the nervous system corresponding to every movement and attitude displayed..have never succeeded in demonstrating the validity of such interpretation in any single case of instinctive behaviour. 1946 F. P. Chisholm in W. S. Knickerbocker 20th Cent. English 177 Understanding the nature of language-situations is itself a very powerful developer of more mature reaction-patterns, and hence of more adequate 'command of the language'. 1959 Times Lit. Suppl. 27 Mar. 181/3 A psychic organism that..is bound up with instincts and emotions and is the source of innate reaction-patterns which go much farther to determine the behaviour of the individual than he cares to recognize. 1887 Jrnl. Mental Sci. XXXII. 604 Guicciardi and Tanzi have made a series of observations on the 'reaction period' in fourteen cases where there were hallucinations of hearing. 1952 C. L. Hull Behavior System i. 12 Absolute zero of reaction potential. 1972 G. R. Lefrancois Psychol. Theories Human Learning vi. 129 Reaction potential, sometimes called excitatory potential, is a measure of the potential that a stimulus has for eliciting a specific response. 1935 C. G. Philp Stratosphere & Rocket Flight ii. 5 Reaction propulsion may be said to date from 1919, in which year Professor Goddard..announced to the world the results of his researches made with rockets. 1962 F. I. Ordway et al. Basic Astronautics ii. 18 Konstantin Tsiolkovsky..seems to be the first scientist who realized that it is only possible to travel in space by means of reaction propulsion. 1971 P. J. McMahon Aircraft Propulsion iii. 80 Combining the thermodynamic and propulsion efficiencies gives the overall efficiency of the reaction propulsion engine. 1946 Nature 6 Apr. 439/1 This is strong evidence for the conclusion that the decrease in the bimolecular reaction-rate as the reactive carbon atom changes from primary to secondary is due to an increase in steric hindrance. 1972 Weston & Schwarz Chem. Kinetics i. 3 Experimental studies of reactions generally involve the determination of reaction rates as a function of several variables: chemical composition, temperature, pressure, or volume. [1886 Bull. U.S. Geol. Survey No. 28. 52 Wherever it comes in contact with the olivine, the peculiar reactionary rims of amphibole described by Törnebohm..are finely developed.] 1892 Amer. Jrnl. Sci. CXLIII. 516 The attention of the writer was repeatedly attracted by a fibrous growth around olivine, that resembles..the reaction rims that have been described as existing between garnet and serpentine. 1921 Mineral. Mag. XIX. 145 If the corona can be shown to be due to alteration or modification of the nucleus the term reaction rim is preferred. 1960 Turner & Verhoogen Igneous & Metamorphic Petrol. (ed. 2) vi. 145 The very existence of solid-solution series in so many groups of igneous minerals, and the frequent development of reaction rims (coronas) of one mineral around central cores of another, are evidence of reaction. 1953 Reaction shot [see *off screen, off-screen adv. (phr.) and a.]. 1966 Guardian 11 Aug. 8/2 The Select Committee on televising the House of Commons..was understandably cautious about 'reaction shots', by which the committee meant pictures not only of the member addressing the House..but of other members reacting to his speech. 1930 G. R. de Beer Embryol. & Evolution xv. 105 Atavism is..due to the reproduction of a set of conditions (a definite system of reaction-speeds) which obtained in the ancestor. 1964 A. Edel in I. L. Horowitz New Sociology xiv. 224 The comparable discovery of color blindness or different reaction-speed or influence of drugs or alcohol affects the concept of a reliable observer in other fields. 1970 W. Sahakian Psychol. Learning xi. 230 The concept of the reaction threshold is well established. 1879 Reaction-time [see *psychometry 2]. 1883 Mind Apr. 177 (heading) Reaction time and attention in the hypnotic state. 1908 J. M. Cattell in Essays in Honor of W. James 574 The incoming currents and the pre-existing structure of the centres cause the discharge, and the perception, in my opinion, usually follows the discharge in time. This time order I pointed out more than twenty years ago in the case of the reaction-time. 1923 R. A. Knox Memories of Future v. 73, I went round to his house, where he looked at my tongue, X-rayed me, felt my pulse, took my reaction-times, and shook his head importantly. 1971 J. W. Kling et al. Woodworth & Schlosberg's Exper. Psychol. (1972) ix. 309/2 The simple reaction time of human subjects..is one form of latency measure. 1972 K. Bonfiglioli Don't point that Thing at Me xiii. 96 Another shot rang out, followed one fifth of a second later by the bang of the car door... There is still nothing wrong with the Mortdecai reaction-time. 1881 Encycl. Brit. XII. 524/1 Turbines in which part only of the available energy is converted into kinetic energy, before the water enters the turbine wheel, may be termed Reaction Turbines. 1929 T. M. Naylor Steam Turbines i. 3 The only pure reaction turbines in use are of very small power, and are used for fluids other than steam, e.g. the garden sprinkler.., distributors on sewage filter beds, [etc.]. 1969 W. Thomson Thrust for Flight 111 The reaction turbine is one in which the whole of the pressure drop and the velocity increase occur as the gas flows through the rotor blade passages. 1904 Psychol. Bull. I. 230 The chief directions in which Kraepelin has stimulated his associates to work psychologically, are a number of more biological than purely psychological reaction-types, such as retardation and inhibition, flight of ideas, etc. 1917 C. E. Long tr. Jung's Analytical Psychol. iii. 158 Relatives, and especially related women, have therefore on the average, resemblance in reaction-type. 1917 A. Meyer Coll. Papers (1951) II. 44 The organic reaction types. 1922 [see *ecophene]. 1934 E. B. Strauss tr. Kretschmer's Text-bk. Med. Psychol. xiii. 184 In our description of the various reaction-types we shall make use of psychopathological instead of normal material. 1953 H. Read True Voice of Feeling II. iii. 262 The artist.. reveals an inward unity and concentration of personality in marked contrast with the extraneous dissipations and diversities of the average reaction-type. 1904 Brit. Med. Jrnl. 10 Sept. 564/2 The reaction velocity of the chemical or physical processes that result in the agglutination of bacteria, is a very variable factor. 1926 Jrnl. Inst. Petroleum Technologists XII. 202A When the apparatus is shut down for cleaning, the reaction vessel is opened and steam or water is pumped through the tubes. 1973 Materials & Technol. VI. viii. 504 The mixture is stirred in a reaction vessel and the monomer becomes suspended as an emulsion of tiny droplets. 1949 H. P. Brown et al. Textbk. Wood Technol. I. xii. 288 The existence of internal forces in living trees, independently of the presence of reaction wood, has also been suggested. 1980 Family Handyman Sept. 70/3 The fruitwoods (cherry, apple) are more difficult [to dry out], depending upon the size of the log and the presence or absence of reaction wood. 1920 T. P. Nunn Educ. v. 44 The subject's memory does not throw up a number of suggestions from which a suitable reaction-word is consciously selected.

rea·ctionarily, *adv.* [f. Reactionary *a.* + -ly[2].] In a reactionary manner.

1966 Economist 23 July 329/1 Mr Wilson's Government is now following a right-wing policy of deflation more resolutely, more ruefully, more reactionarily..than any of its predecessors since the war. 1970 R. A. H. Robinson Origins of Franco's Spain 301 Prieto, leader of the Bilbao Socialists,..called the Statute reactionarily anti-liberal: 'Spain cannot tolerate that territory turning itself into a Vaticanist Gibraltar.'

rea·ctionariness. *rare.* [f. Reactionary *a.* + -ness.] Reactionary character.

1923 U. L. Silberrad Lett. Jean Armiter vi. 149, I believe he felt almost kindly..towards me and my antiquated reactionariness.

reactionary, *a.* and *sb.* Add: **A.** *adj.* **2.** (Earlier and later examples.) Also, in Marxist use, unfavourably contrasted with *revolutionary*.

1840 J. S. Mill in London & Westm. Rev. Mar. 276 The philosophers of the reactionary school—of the school to which Coleridge belongs. 1953 R. Lehmann Echoing Grove iii. 156 The British Government's iniquitous non-intervention policy playing into the hands of Franco's abominable reactionary conspiracy. 1957 Economist 28 Dec. 1134/1 Hungarians are to have both the 'true freedom' and the bananas. Mr Kadar's masters were shrewd enough to realise that, in this reactionary country, the latter is appreciated more than the former. 1974 tr. Sniečkus's Soviet Lithuania 23 The bourgeois nationalists set up various kinds of reactionary armed organisations to combat the revolutionary movement. 1975 J. Plamenatz K. Marx's Philos. of Man xi. 299 Ideas and practices adopted by the bourgeois when they put themselves forward as the champions of society as a whole against irresponsible power and a reactionary nobility can become a nuisance to them once they are installed as the dominant class and their position comes to be challenged.

B. *sb.* (Later examples.) Also, in Marxist use, an opponent of communism.

1908 Nation 1 Oct. 302/2 Part of Wall Street will vote for Taft because it believes him at heart a 'reactionary', or, at least, the less of two evils. 1921 G. B. Shaw Back to Methuselah II. 52, I can find you hundreds of the most sordid rascals, or the most densely stupid reactionaries, with all these qualifications. 1951 G. Mikes Down with Everybody! I. i. 15 'A reactionary and a foreign spy is a man who is not pleased when his wages are cut or when he is asked to work more for the same wages and who is not pleased to starve.' 'Very good, Comrade Barna, very good indeed.' 1967 tr. Quotations from Chairman Mao Tsetung (ed. 2) 75 The reactionaries in all countries are fools of this kind... Their persecution of the revolutionary people only serves to accelerate the people's revolutions. 1981 L. Deighton XPD ii. 5 Literally pogoni means epaulette, but for a citizen of the USSR..it is a symbol of the hated reactionary.

rea·ctionaryism. Also **reactionarism.** [f. Reactionary *a.* and *sb.* + -ism.] = *Reactionism. Hence **rea·ctionarist.**

1911 Galsworthy Patrician I. i. 11 His common sense continually impelled him, against the sort of reactionaryism of which his son Miltoun had so much, to that easier reactionaryism, which..makes what material capital it can out of its enemy, Progress. 1922 Contemp. Rev. Aug. 240 His party had previously been coloured by monarchism and reactionarism. 1924 Glasgow Herald 22 May 7 The advent of the Labour Government was heralded by the reactionarists as the end of all things. 1965 Listener 4 Nov. 723/1 Reactionaryism, whether in politics or art, is never pleasant to contemplate.

rea·ctionism. [f. Reaction + -ism.] Reactionary principles or practice.

1891 J. M. Robertson Mod. Humanists 91 For the Bentham group Burke finally represented sheer reactionism. 1930 A. I. Nazaroff Tolstoi vi. 90 This lack of 'social sensibility' shocked them. They even mistook it for the mortal sin of 'reactionism'.

reactionist, *sb.* Add: (Earlier and later examples.) Also, a person who reacts *against* something.

1857 J. F. Maguire Rome: its Ruler & its Institutions p. ix, I trust I have done sufficient to enable the reader..to estimate, at their right value, the accusations which have been made against him [sc. Pope Pius IX], as a reformer of the one day, and a reactionist of the next. 1861 Q. Rev. CIX. 294 The utter weariness of spirit which this unresting scepticism has bred in most minds of the highest order of thought; the deep study into which it has driven the noble reactionists who have arisen there..have entirely altered the whole tone of religious feeling amongst our Teutonic brethren. **1900** *Dublin Rev.* Jan. 35 Reactionists..signatories to the Royal supremacy in 1559, who in 1579 appeared in prison as recusants. **1902** G. B. Shaw *Mrs. Warren's Profession* p. xxx, Clergymen's sons are often conspicuous reactionists against the restraints imposed on them in childhood. **1909** [see *baby *sb.* 1 e].

reactivate (ri,æ·ktĭvĕ¹t), *v.* [Re- 5 a.] *trans.* To make active or operative again.

1903 *Med. Rec.* (N.Y.) 14 Feb. 251/2 The serum can be reactivated by a little fresh serum not only from a normal rabbit, but from the goat and the rat. **1949** M. Mead *Male & Female* viii. 176 Menstruation..is itself believed to be reactivated by marriage. **1953** *Times* 13 May 3/5 The importance of maintaining the base there in such condition that it could be reactivated immediately in case of war. **1957** *Observer* 27 Oct. 12/2 An American recession may also come in the wake of reactivating the dollar problem. **1958** *Engineering* 14 Mar. 352/3 In this building humidity control is accomplished automatically by passing the air required for ventilation, plus a percentage of recirculated air, through desiccant driers automatically reactivated by gas. **1976** A. White *Long Silence* xx. 179 Please reactivate the plan to surround the station. **1977** *Living with Tanker Surplus* (Shell Internat. Petroleum Co.) 6 A ship in lay-up cannot easily or quickly be reactivated.

Hence **reactiva·tion**, the action or result of reactivating.

1903 *Med. Rec.* (N.Y.) 14 Feb. 251/1 It has been found that this 'reactivation' of the serum, as it has been called, can often be brought about by the serum of various animals. **1919** *Lancet* 10 May 799/1 Professor Chauffard and M. L. Girard brought forward a new and striking proof of the tuberculous nature of erythema nodosum—the reactivation of [illegible] nearly extinct by injection of tuberculin. **1926** *Glasgow Herald* 1 May 4 Very characteristic of spring is what may be called reawakening, if we may use the word to include reactivation from a rest that is deeper than sleep. **1967** *Times* 18 Nov. p. i/1 Sophisticated arguments about the 're-activation' of bank deposits or the velocity of circulation. **1970** H. Trevelyan *Middle East in Revolution* 12 The Agreement provided for the reactivation of the Base for defence against the invasion of any Arab State or Turkey by a State other than Israel.

reactive, *a.* Add: **3. a.** (Further examples.) Also *reactive formation* = *reaction formation* s.v. *reaction 5; *reactive inhibition*, the inhibiting effect of the supervention of fatigue or boredom on the response to a stimulus.

1925 J. Strachey tr. *Freud's Case of Hysteria* in *Coll. Papers* III. 67 Repression is often achieved by means of an excessive reinforcement of the thought contrary to the one which is to be repressed. This process I call *reactive* reinforcement... *Ibid.* 68 The thought which asserts itself exaggeratedly in consciousness and (in the same way as a prejudice) cannot be removed I call a *reactive thought*. **1943** C. L. Hull *Princ. Behavior* xviii. 327 Uniform time intervals between reinforcements great enough to prevent the accumulation of appreciable amounts of reactive inhibition. **1968** *Globe & Mail* (Toronto) 3 Feb. 11/6 These young people can join 'negative identity' groups and become..therapeutically channelled into some reactive kinds of activity. **1974** Chaplin & Krawiec *Systems & Theories of Psychol.* vi. 263 In and of itself fatigue will generate a form of inhibition which Hull calls reactive inhibition, and the animal will cease to respond. **1975** *Way* Suppl. No. 25. 67 Dynamic psychology might interpret this as evidence for reactive formation, which is a technical term suggesting the control of an unacceptable impulse by exaggerating the opposite tendency. **1977** G. H. Sage *Introd. Motor Behav.* (ed. 2) xvi. 359 The first of these is an inhibition (reactive inhibition) caused by a reluctance to repeat a stimulus.

b. Also, characterized by reaction to a stimulus.

1927 C. Spearman *Abilities of Man* iv. 47 Stern's division of persons into the 'spontaneous' and the 'merely reactive'. **1952** M. K. Wilson tr. *Lorenz's King Solomon's Ring* xi. 171 A few birds—usually old, strongly reactive ones—take off, emitting 'Kiaw' cries and thereby provoking the whole flock to leave the ground with them.

c. *Psychol.* Of mental illness: thought to be caused by reaction to environmental stress; exogenous; so *reactive schizophrenic*, a person with reactive schizophrenia (cf. *process schizophrenia* s.v. *process *sb.* 13 a).

1924 A. A. Brill *Bleuler's Textbk. Psychiatry* xii. 537 Reactive depressions, which become aggravated to a mental disease, are quite rare in the light of present views. **1955** McCarthy & Corrin *Med. Treatment of Mental Dis.* xix. 326 Reactive depressions are the mild, brief depressions, the result of situations depressing in themselves, and not out of proportion to the circumstances. **1958** H. Weiner in L. Bellak *Schizophrenia* iv. 157 Whether such reactive psychoses actually imply a different series of etiologic contributing factors, i.e., 'non-organic', is highly speculative. **1962** *Lancet* 10 Jan. 105/1 Two were middle-aged men with anxiety states... The third was a middle-aged lady with a reactive depression. **1962** *Psychol. Bull.* LIX. 329/1 Reactive schizophrenia indicates a good prognosis..with notable stress precipitating the psychosis. *Ibid.* 332/1 If the Rorschach diagnosis is offered by them, it then appears that reactive schizophrenics are not psychotic. **1965** J. Pollitt *Depression & its Treatment* 95 The terms 'endogenous' and 'reactive'..on the basis of presence or absence of an external cause. **1966** I. B. Weiner *Psychodiagnosis in Schizophrenia* vi. 94 A group of process schizophrenics made no more frequent errors than the reactive schizophrenics. **1971** *Brit. Med. Bull.* XXVII. 77/2 One school of thought..allocates such cases to a special category of 'reactive psychoses'. **1974** M. C. Gerald *Pharmacol.* xvi. 310 When the cause of depression can be identified,..we term this reactive or exogenous depression. **1976** L. J. West et al. *Treatment of Schizophrenia* vi. 278 The good premorbid/bad premorbid, and process/reactive dimensions do seem to identify significant characteristics.

5. a. *Chem.* Readily susceptible to chemical reaction.

1888 *Nature* 22 Mar. 503/1 The unsaturated hydrocarbons..are..more reactive than the paraffins. **1942** G. Wendt *Chem.* vii. 179 As in the other families of elements we shall meet, the lightest of the group—in this case fluorine—is the most reactive. **1957** G. E. Hutchinson *Treat. Limnol.* I. xi. 708 It is not clear, however, in what form the reactive iron is actually present in the lake water. **1972** R. A. Jackson *Mechanism* iv. 78 A trapping or inhibiting agent may be insufficiently reactive towards a particular radical to divert it from its normal course of reaction, or the agent may react in such a way as to regenerate a reactive intermediate.

b. Of a process: involving chemical reaction.

1950 *Chem. Abstr.* XLIV. 8190 (*heading*) Reactive diffusion in metals. **1960** *Ibid.* LIV. 23646 The reactive diffusion in a system Nb-B is accomplished by diffusion of B atoms through the reaction products toward the metal. **1971** *New Scientist* 8 Apr. 96/1 Reactive evaporation..is often used for depositing metal oxides. **1973** R. L. Burwell in Basolo & Burwell *Catalysis* II. 72 Ethylene is adsorbed by reactive adsorption. Perhaps it first adsorbs as π-ethylene and then reacts to monoadsorbed ethane. **1978** Felder & Rousseau *Elem. Princ. Chem. Processes* x. 384 In this chapter we show..how calculated enthalpies of reaction are incorporated in energy balances on reactive processes.

c. Of a dye or other colouring material: designed to react chemically with the substrate, usu. in order to become fixed.

1957 *Jrnl. Soc. Dyers & Colourists* LXXIII. 238/2 The search for more practical methods of using reactive dyes has continued, and this work has now culminated in the production of a new range of dyes—the Procion (ICI) dyes—which are water-soluble but contain a reactive group capable of combining with cellulose under alkaline conditions. **1973** *Materials & Technol.* VI. iv. 320 The direct, basic, sulphur, vat, azoic and reactive dyes are used also for colouring the fibre [*sc.* rayon]. *a* **1977** *Harrison Mayer Ltd. Catal.* 39/1 *Reactive colours*, a new range of leadless colours for underglaze, on-glaze or in-glaze decoration... On firing, these colours react with the glaze to give interesting and variegated effects.

d. Of coke or coal: having a high reactivity (sense *b).

1963 *Economist* 21 Dec. 1284/1 Reconciling the clean air policy with the British love of open fires means trying to produce enough smokeless fuel of the sort that can be burnt in the fireplaces of smokeless zones. But the Minister of Power now estimates that there will be a gap..between the demand for such 'reactive' smokeless fuels and the supply. **1974** *Encycl. Brit. Macropædia* IV. 783/2 Coals of medium rank are used for low-temperature carbonization because they yield reactive cokes.

6. a. *Electr.* Possessing or pertaining to electrical reactance; *spec.* applied to the vector component of an alternating current (or voltage) which is 90° out of phase with respect to the associated voltage (or current), i.e. wattless; *reactive power*, the product of the voltage and the reactive current, or of the current and the reactive voltage; *reactive volt-ampere*, a unit of reactive power.

1892 J. A. Fleming *Alternate Current Transformer* II. ii. 269 Reactive Coil or Dimmer.—This is a device used in alternating current work to serve the same ends as rheostats in direct current plants. **1892** S. P. Thompson *Dynamo-Electr. Machinery* (ed. 4) xxii. 628 We have here two electromotive-forces, the impressed O A and the reactive A E.., with their resultant the effective electromotive-force O E, with an angle E O A..between them. **1914** J. H. Morecroft *Continuous & Alternating Current Machinery* vi. 205 The component of voltage 90° out of phase with the current we shall call the reactive component of the voltage or the reactive voltage. **1916** *Standardization Rules Amer. Inst. Electr. Engineers* 15 *Reactive volt-amperes*, the product of the reactive component of the voltage by the total current, or of the reactive component of the current by the total voltage. **1920** *Whittaker's Electr. Engineer's Pocket-Bk.* (ed. 4) 222 The reactive current changes from leading to lagging as the field excitation is reduced. **1951** W. V. Lyon *Alternating Currents* ix. 225 In an induction motor, neglecting heating losses, the active component provides the useful mechanical power delivered by the motor, while the reactive component is associated with the production of the alternating magnetic field necessary for the operation of the motor. **1956** C. S. Siskind *Electr. Circuits* xi. 274 The reactive component—the quadrature component—is..kept small, the latter, acting adversely to lessen the power factor, is expressed as the reactive volt-amperes, abbreviated kvar. **1961** *Listener* 9 Nov. 768/2 There is also the possibility, on direct current lines, of providing what is called 'reactive power'—power where the current is out of step with the voltage—for the operation of converter equipment. **1973** J. Earl *Tuners & Amplifiers* iv. 83 The source itself can be mostly resistive or mostly reactive. **1978** *Gramophone* May 1958/1 It is claimed to have an output of 100 watts RMS working into an 8-ohm inductive or reactive load.

b. *Mech.* and *Acoustics.* Possessing or pertaining to mechanical or acoustic impedance.

1934 Olson & Massa *Applied Acoustics* ii. 31 (*caption*) Resistive and reactive air load per unit area on one side of a vibrating piston of radius R centimeters set in an infinite baffle. **1963** [see *impedance 2]. **1976** A. H. Benade *Fund. of Musical Acoustics* xxi. 457/2 The other register hole on my special clarinet can be called a reactive register hole, i.e., an aperture in which the flow depends mainly on the inertia of the air moving through it.

7. *Gram.* (See quot.) *rare.*

1957 R. W. Zandvoort *Handbk. Eng. Gram.* v. ii. 224 To express the speaker's reaction to a previous statement by the person addressed which is repeated by the speaker (*reactive questions*)... You can't catch me.—I can't, can't I? said Philip.

reactivity. Add: **a.** (Further examples.)

1907 *Chem. Abstr.* I. 1395 Transformation of open chains to cyclic compounds enhances their reactivity. **1934** *Jrnl. Theol. Stud.* XXXV. 402 The author would establish that religion consists in simultaneous reactivity of the affective, conative and cognitive elements of mentality. **1949** *Brit. Jrnl. Psychol.* Dec. 86 When the individual's P.G.R. [psychogalvanic reflex] scores are obtained for a given attitude they must be expressed for each person relative to his own general P.G.R. reactivity. **1950** *Sci. News* XV. 70 Consideration of the chemical reactivity of crystals will obviously be focussed primarily on the surfaces. **1956** *Nature* 10 Mar. 477/1 It is assumed that the reactivity of the anhydride is not increased by adsorption. **1972** DePuy & Chapman *Molec. Reactions & Photochem.* vii. 137 Electronic effects are more important than the steric effects in determining the relative reactivity of dienophiles.

b. The rate of reaction of a coke or coal with (orig.) carbon dioxide, or with oxygen, under specified conditions.

1926 *Rep. Fuel Res. Board 1925* (Dep. Sci. & Industr. Res.) 23 A method has been designed which gives an empirical figure for reactivity to carbon dioxide. **1930** *Engineering* 22 Aug. 251/4 A convenient distinction has been drawn between the combustibility of a coke, denoting the rate at which it burns in oxygen or air, and its reactivity, meaning the rate at which it reacts with carbon dioxide and produces the reducing atmosphere in which the metallic iron is extracted. **1947** G. W. Himus *Elem. Fuel Technol.* xv. 296 The 'Reactivity Value' of coke is defined by the Fuel Research Board as 'the number of millilitres of carbon monoxide formed from 100 millilitres of carbon dioxide' under certain definitely specified conditions. **1971** *Materials & Technol.* II. x. 601 Gas coke..has a lower reactivity than most coals and is not suitable for use on open fires. *Ibid.* 604 The reactivity of coke with respect to oxygen is often measured by the Critical Air Blast (C.A.B.) method.

c. *Nuclear Physics.* A measure of the extent to which a reactor (or part of it) deviates from a steady critical condition (see quot. 1962).

Orig. in a broader sense.

1945 H. D. Smyth *Atomic Energy* viii. 89 The reactivity of the pile was so far above expectations that it would have been beyond the capacity of the control rods to handle if the remainder of the heavy water had been added. **1947** *Physical Rev.* LXXII. 17/1 The reactivity of a chain-reacting pile depends critically on the balance of neutron production, absorption, and leakage. **1954** R. Stephenson *Introd. Nuclear Engin.* vii. 270 In order to operate at any appreciable power, some excess reactivity must be available to overcome the various factors which act to reduce the ability of the chain reactor to multiply neutrons. **1960** *Gloss. Atomic Terms* (U.K.A.E.A.) 44 At any steady state of operation the reactivity is zero. **1962** *Gloss. Terms Nuclear Sci.* (B.S.I.) 97 *Reactivity*, the value of the expression $(k_{eff}-1)/k_{eff}$, which is a measure of the departure of a reactor from the critical condition... k_{eff} is the multiplication constant (effective). **1963** B. Fozard *Instrumentation Nuclear Reactors* xii. 149 A sudden change in reactivity in either direction is liable to produce thermal shock in the fuel elements. **1974** *Encycl. Brit. Macropædia* XIII. 314/2 The reactivity of the core increases when the rods..are withdrawn.

reactor (ri,æ·ktəɪ). Also (*rare*) **reacter.** [f. React *v.*¹ + -or.] **1.** A person, animal, or organism that reacts to a stimulus, esp. under test or experimental conditions; *spec.* one showing an immune response to a specific antigen.

1890 W. James *Princ. Psychol.* I. xi. 433 One must bear in mind that in these experiments the reacter always knew in a general way the *kind* of question which he was to receive. *Ibid.* xiii. 525 The reacter awaits the signal and reacts if it is of one sort, but omits to act if it is of another sort. **1895** E. B. Titchener in *Amer. Jrnl. Psychol.* VII. 83 Reagent *or* reactor. **1907** *Psychol. Rev.* (Monogr. Suppl.) VIII. iii. 314 In Series III. the reaction consisted in stopping a vertical movement... The reactor moved the pencil up and down against this vertical guide. **1928** *Daily Tel.* 6 Nov. 7/7 Out of 835 animals..122 reacted to the double intradermal test, and 94 of these reactors proved tuberculous at autopsy. **1932** *Amer. Rev. Tuberc.* XXV. 367 In this case the reactor was the sanatorium chef, who has developed tuberculosis. **1932** [see *Schick-negative adj.]. **1961** *Listener* 30 Nov. 933/1 There may be some value in studying people's reactions to events and also the appearance of these reactors at any given moment. **1969** Myers & Steele *Bovine Tuberculosis* vi. 351 Since 1961, the [N.Z.] Government has provided the funds to pay for the tuberculin testing costs, and compensation for reactors sent to slaughter. **1973** *Black Panther* 10 Nov. 3/4 They are not actors, but reactors, and not leaders, but followers. **1976** *Nature* 13 May 144/1 Among the other categories of patients tested, the percentage of negative reactors varied from 3 to 14%.

2. a. *Electr.* A coil or other piece of equipment which provides reactance in a circuit.

1913 H. B. Dwight *Constant-Voltage Transmission* iii. 14 The name 'synchronous condenser'..is not quite so appropriate when the machine is used with a constant-voltage transmission line, because for a large share of the time the current in the machine is not leading, but lagging, and the machine at that time does not behave as a condenser, but would more accurately be called a 'synchronous reactor'. **1920** *Chem. Abstr.* XIV. 2756 The cast-in-concrete air-core reactor is recommended for furnace work. **1951** *Engineering* 9 Nov. 584/2 Each pair of ignitrons is capable of supplying a rectified current of 650 amperes at 600 volts, the 30 per cent ripple in which is reduced by the smoothing reactor. **1958** J. Shepherd et al. *Higher Electr. Engin.* xvi. 389 Current-limiting reactors may be connected in series with

each generator, in series with each feeder, or between each bus-bar section. **1961** G. F. Tagg *Pract. Electr. Engin* I. 33 One method of reducing short-circuit current is by the employment of reactances or inductance coils which, for arge a.c. power systems, are known as reactors.

b. A vessel or apparatus in which substances are made to react chemically, esp. one in an industrial plant.

1935 *Industr. & Engin. Chem.* Sept. 1072/2 In the pyrolysis of ethane and propane best results were obtained when using a reactor consisting of a helical coil of KA2S tubing... This reactor was placed in a radiant-type electrically heated furnace. **1939** *World Petroleum* May 42/2 The plant consists essentially of an assembly of three polymerization reactors, debutanizer, rerunning column for separation of polymer into dimer..and trimer.., with two nickel catalyst hydrogenation reactors for conversion of the dimer to finished iso-octane. **1974** *Daily Tel.* 17 July 19/4 The kettle he should have been watching was a three-ton reactor heating synthetic resin. The reactor..was just two degrees under 100 degrees centigrade when it was seen and could have cracked..or exploded. **1975** *Sci. Amer.* Nov. 107/1 Using air to remove the graphite from the diamond surface enables us to carry out both operations in a single reactor. **1978** *Nature* 22 June 582/1 The accident happened in a reactor used for the production of trichlorophenol.

c. = *nuclear reactor* s.v. *nuclear *a.* 4. Freq. *attrib.*

1945 [see *nuclear reactor]. **1947** *Newsweek* 8 Sept. 76/3 The tight-lipped Atomic Energy Commission did not tell all it knows about the new 'reactor'. The active substance is plutonium. **1950** *Chemical Engin. Progress* XLVI. 110/1 It behoves this nation to keep in the forefront of reactor technology. *Ibid.*, The enormous cost of the reactor program. **1957** *Economist* 19 Oct. 256/1 A maximum limit of 20 per cent enrichment with uranium 235 ..is usually imposed on uranium exports from the US, and the limit is exceeded only in such special cases as the fuel core for Harwell's Dido research reactor. **1963** B. Fozard *Instrumentation Nuclear Reactors* xii. 144 Most of the neutrons in a reactor core..are produced at the instant of fission as the products of a reaction of great violence. **1971** *Materials & Technol.* II. xii. 734 Reprocessing..is essential if plutonium is to be utilized in a mixed national power system of fast and thermal reactors. **1975** *Nature* 16 Oct. 525/3 The air in the reactor room is maintained at a lower pressure than the air outside, so that if radioactivity were released from the reactor, it would not immediately be dispersed into the atmosphere.

read, *sb.*² Add: (Earlier and later examples.) Also *transf.*, something for reading, esp. with ref. to its value as entertainment or information (freq. with qualifying adj.).

1825 Jamieson Suppl. s.v., Will ye gie me a read of that book? **1902** J. Milne *Epistles Atkins* i. 8 The soldiers..have 'another good read'. **1958** *Observer* 9 Feb. 15/3 A.G.'s usual solid, lively read. **1961** *John o' London's* 21 Sept. 327/3 *My Friend Sandy* can be hugely recommended..as a pleasantly light, bright sophisticated read. **1963** T. Parker *Unknown Citizen* v. 136 He'd come back to the prison, had his tea, and gone to bed to lie down and have a read. **1975** L. Trilling in *Times Lit. Suppl.* (1976) 5 Mar. 250/4 Was she [*sc.* Jane Austen] perhaps to be thought of as nothing more than a good read?.. Now that we have before us that British locution, which Americans have lately taken to using, the question might be asked why the phrase should have come to express so much force of irony and condescension. **1977** J. I. M. Stewart *Madonna of Astrolabe* xviii. 256 *Tamburlaine* is a tolerable read... As a stage play it is pretty hopeless. **1981** *Times* 2 Mar. 12/6 The labels are informative to the point of saturation. If you do not like the wine, you might at least enjoy the read.

read, *v.* Add: **5. a.** Also, to understand (musical notation); *spec.* = *sight-read* s.v. *sight *sb.*¹ 17.

1792 H. Newdigate *Let.* Mar. in A. E. Newdigate-Newdegate *Cheverels* (1898) ix. 133 Her Voice was not strong but..they are quite astonish'd with her knowledge of Music & facility in reading it. **1894** G. B. Shaw in *Fortn. Rev.* Feb. 258 To do half-a-dozen things much more difficult than reading music. **1918** —— in *Nation* 22 June 308/1 To wile away the time by reading at sight a bundle of band parts and vocal scores of a rather difficult opera. **1938** D. Baker *Young Man with Horn* i. v. 56 Jeff's band didn't play from music, though they could all read music. **1974** *Listener* 24 Jan. 106/3, I could read the music and be able to make it work right away with five minutes' rehearsal.

b. *spec.* to peruse books, newspapers, etc., for quotations suitable for inclusion as illustrative examples in a dictionary.

1876 J. A. H. Murray *Let.* 29 Nov. in K. M. E. Murray *Caught in Web of Words* (1977) vii. 146, I dont for words of that kind believe in the quotation test at all..because you know that not one millionth of current literature is read, & that it is the veriest chance or succession of chances which has caught *carriageless*..& missed a thousand others as good. **1961** R. W. Burchfield in *Essays & Studies* XIV. 39 A large number of literary sources..are being systematically read against an Oxford dictionary. **1977** K. M. E. Murray *Caught in Web of Words* xii. 235 Lowell's book of literary essays, *My Study Windows*, was one of those read for the Dictionary.

c. (Later examples.)

1921 P. L. Haworth *Trailmakers of Northwest* 206 As Brennan had lost one eye and could not see any too well out of the other, he was glad to have one of us ride in his canoe and read water for him. **1932** L. Golding *Magnolia St.* ii. ii. 300 In the little town in..Lancashire where she was born quite as many people read tea-leaves as read their ABC. **1951** E. Rickman *Come racing with Me* iii. 19 We are talking about 'reading' a race, which is the practice on the spectator's part of a comprehensive and discriminating view of a field of horses from start to finish, so that the performance of all or most of the runners, and their relative positions at various stages, are intelligently observed and memorised. **1965** Priestley & Wisdom *Good Driving* xi. 81 You get into the habit of registering mentally all the signs..which enable you to 'read' the road in front of you. **1967** *Boston Sunday Herald* 26 Mar. iv. 3/8 An optical scanner..may eliminate the sorting machines by 'reading' the zip code on the letter and dispatching it accordingly. **1967** Karch & Buber *Offset Processes* ii. 20 An optical system 'reads' the photograph, and a heated stylus is directed to penetrate the plate to be printed, producing halftone dots. **1969** R. Welsh *Beginner's Guide Curling* xvii. 120 The ability to read strange ice..and knowing exactly when to sweep are other qualities of a good skip. **1970** G. F. Newman *Sir, You Bastard* vi. 159 Ambition drowning the man was how she would read his promotion. **1971** *Sunday Express* (Johannesburg) 28 Mar. 17/1 You read a putt, stroke it properly along the line you have chosen, and then the ball breaks off in the opposite direction. **1972** *Daily Tel.* 5 May 3/3 A meter reader rang the bell and told my wife he wanted to read the meter in the garage. **1974** *Times* 19 Feb. 15/3 Most people are not used to 'reading' plans..and have only slightly less difficulty with architectural photographs. **1977** *Time* 14 Nov. 48/1 They broke down and then analyzed the RNA in the archaebacteria's ribosomes, the structures that 'read' the message of the master molecule DNA and produce the protein necessary for life. **1978** *Monitor* (McAllen, Texas) 12 Feb. 1-B/1 This generation of Gypsies..will wake up to modern life and give up many of the old customs... My job was supposed to be reading palms. **1979** *SLR Camera* Jan. 36/3 Like the now discontinued EF the AE-1 uses a silicon photocell to read the light.

e. To interpret (a design) in terms of the setting up needed to reproduce it on a loom. Also with *in.*

1839 Ure *Dict. Arts* 267 In both modes of manufacture, the piece is mounted by reading-in the warp for the different leaves of the heddles. **1895** T. F. Bell *Jacquard Weaving & Designing* i. 9 The straight-edge EE..will slide up and down in the frame, to mark the line on the design paper that is next to be read by the lasher. **1897** [see *reading-machine* s.v. Reading *vbl. sb.* 10 b]. **1924** T. Woodhouse *Jacquards & Harness* iv. 107 Before describing the remaining parts of the machine, it will..be best to indicate how the design is read. **1958** A. Hindson *Designer's Drawloom* xi. 105 The weaver can tie up the pattern single-handed, but it can be done more easily and quickly if there is a helper to read the pattern draft.

f. To study (a subject, a 'school') at a university; to read for (a degree). Cf. sense 15 c.

1884 [see *Greats* s.v. Great C. 10]. **1955** *Times* 23 May 6/1 Agriculture is no longer a subject to be ashamed of; it produces no inferiority complex in those who read it. **1966** *Rep. Comm. Inquiry Univ. Oxf.* II. 49 Graduates reading first degrees. *Ibid.* 85 Women undergraduates reading arts. **1970** [see *English *sb.* 3 d]. **1977** *Professional Careers Bull.* Autumn 1/1 Partially it has been due to an ever increasing demand from sixth formers to read law.

g. Phr. *to read one's shirt* (see quot. 1925). *slang.*

1918 *Nat. Geogr. Mag.* June 499 They..speak of 'reading their shirts'. **1925** Fraser & Gibbons *Soldier & Sailor Words* 237 *To read a shirt*, to search it for lice. **1931** 'D. Stiff' *Milk & Honey Route* xiii. 144 It is said, for instance, that the hobo spends a great deal of his time reading his shirt, seeking certain animals known as 'seam squirrels'.

h. *Computers.* To copy or extract data on or in (any storage medium or device); to copy, extract, or transfer (data). Also const. *into*, *out of.*

1940 W. J. Eckert *Punched Card Methods Scientific Computation* 4 The number are..read into the machines by..electrical contacts made through the holes. **1945** *Jrnl. Franklin Inst.* CCXL. 277 When the punched tapes are ready, the problem is placed on the machine by automatic controls which 'read' the first tape and make the specified assembly. **1948** *Math. Tables & Other Aids to Computation* III. 123 The speeds at which words can be read (or written) by the machine will be much less than the speeds at which the machine can transfer words internally. *Ibid.* 124 When additional instructions are received they can be read into the machine from an instruction tape. **1950** *High-Speed Computing Devices* ix. 151 The tape reader automatically reads punched tape..and transcribes the data represented by the holes in the tape to a deck of cards. Other equipment can perform the reverse operation, reading the holes punched in the cards and producing a tape. **1959** E. M. McCormick *Digital Computer Primer* ix. 135 The tape is then connected into the computer system and the information read from it to the computer. **1964** F. L. Westwater *Electronic Computers* i. 2 The card or tape is then 'read'. This may be done by allowing the holes to pass under tiny wire brushes. *Ibid.* iv. 59 To read a word out of the store we have to open a gate at the end, and this permits pulses to escape. **1964** *Ann. N.Y. Acad. Sci.* CXV. 654 The length of time required to read information from or store information into one of the 1,024..12-bit memory locations. **1970** O. Dopping *Computers & Data Processing* xiv. 226 The computer time for file maintenance..is often mainly determined by the time for reading and writing magnetic tape. **1972** *Computer Jrnl.* XV. 201/1 The commonest way of reading a file into the system. **1972** *Guardian* 14 Aug. 10/3 Computers can already 'read' a high speed disc-store at around 500,000 characters a second. **1978** J. K. Atkin *Basic Computer Sci.* vii. 92 To read a bit from the memory it is necessary to interrogate a particular core by sending current pulses..along the appropriate x- and y-wires.

i. To receive and understand the words of (a person) by radio or telephone, to hear; to detect (an object) by sonar; *transf.*, to understand the words or intentions of (a person).

1956 *Amer. Speech* XXXI. 228 [U.S.A.F. slang] *Do you read me?* As in conversation by radio, this means 'Do you understand me?' The answer might be, 'Yes, *five by five*', meaning loud and clear. **1956** 'E. McBain' *Cop Hater* (1958) ix. 85 'Are you stoned now, or can you read me?' 'I hear you,' Ordiz said. **1960** *Master Detective* July 83/1 Static-laced code crackle sounded from the speaker. 'Poelzell. I read you. Keep the Dodge in sight.' **1963** *Times* 25 May 10/7 'Does *anyone* read poor Philip?' A comforting voice from a glider, still airborne: 'Humphrey to Philip. Loud and clear.' **1967** R. J. Serling *President's Plane is Missing* (1968) ix. 164 'Don't be so oversolicitous, Rod. It's as bad for a marriage as being too inconsiderate. Do you read me?' 'I read you, Nancy.' **1968** R. Severn *Game for Hawks* x. 120 'How d'you read her, Cass?' he asked, sourly. 'Could she be taking you for a ride?' **1970** B. Knox *Children of Mist* v. 103 'If you can hear..this is an emergency call.'.. Thane pressed the microphone button. 'Fenn, we read you.' **1972** J. Porter *Meddler & her Murder* x. 131 The girl friend listening?.. Oh, I read you. Well, I'll make it short and sweet. **1974** L. Deighton *Spy Story* xviii. 193 A couple of conventional subs steaming a parallel course... We read them on the sonar and ranged them. **1977** D. Bennett *Jigsaw Man* xi. 203 As from the end of this call, this number will be discontinued. I am reading you back for the fast time.

6. c. (*fig.* example.)

1956 E. H. Hutten *Lang. Mod. Physics* vi. 262 The empiricist prejudice..is, indeed, very strong, but it is obviously not true that we simply read off our hypotheses from data.

e. *to read up*: to study (a subject, a topic, etc.) intensively and systematically; to familiarize oneself with (a subject) by reading.

1842 J. S. Mill *Let.* 22 Aug. in *Wks.* (1963) XIII. 542, I began to *read up* the subject. **1856** C. M. Yonge *Daisy Chain* ii. xxvii. 657, I dread reading up all I must read presently. **1869** 'Mark Twain' *Innocents Abroad* xv. 147, I shall throttle down my emotions hereafter, about this sort of people, until I have read them up. **1894** 'R. Andom' *We Three & Troddles* xvii. 149 Those miserable, hollow shams who read up the cricket news..in the evening papers. **1915** R. Brooke *Coll. Poems* p. cxxxvii, I've been peacefully reading up the countryside all the morning. **1921** K. Macaulay *Dangerous Ages* v. 103 You should read it up beforehand, and try if you can understand it. **1962** D. Lessing *Golden Notebk.* ii. 280 Those Russians, they're pretty well up in my field, I read them up. **1977** F. Branston *Up & Coming Man* xiv. 152 He would have covered his interests by reading up the minutes of all committees.

f. Computers. *to read out*, to extract (data); to transfer from internal storage; so *to read in.*

1946 *Ann. Computation Lab. Harvard Univ.* I. 62 The storage counter cams..control the number impulses for reading out either from a switch or from a storage counter. Figure 26 shows the circuits for a read-out. *Ibid.* 159 The number of columns shifted is recorded in a counter and a predetermined number of significant digits and a power of ten are read out. **1957** [see *off-line *a.* and *adv.* A. 2]. **1959** E. M. McCormick *Digital Computer Primer* ii. 7 When the problem is completely solved..the calculation is stopped, and the output, or answer, is read out. **1961** [see *read-out 1 a]. **1968** *Times* 10 Dec. 6/8 On each orbit the storage system reads out the information to a ground station. **1970** O. Dopping *Computers & Data Processing* xiv. 222 When all the records have been read-in, all that is needed then is to print the contents of the 50 cells. **1971** *Physics Bull.* Mar. 158/3 It covers those devices in which information can be stored for a limited or controlled time and then read out leaving the device capable of repeated use.

7. b. (Later examples.) Also, to interpret or comprehend.

1962 *Listener* 22 Nov. 886/3 When East removed the double into One Spade, West read his partner for a psychic opening and bid Three No Trumps. **1967** *Ibid.* 28 Dec. 846/3 He..wants..the celebration of the Eucharist (so I read him) to take the form of a prayer meeting. **1970** *Sunday Tel.* 20 Dec. 21/7 Gleeson mesmerises batsmen unable to read him, not into error but into strokelessness.

8. b. Also with *in.*

1903 *Westm. Gaz.* 13 Nov. 7/2 The learned counsel argued that his lordship must read in a negative... In a contract for personal service you must have in it a negative, express or implied. **1919** 'C. Dane' *Legends* 96 She said to me once that the critics had 'read in' things that she had never dreamed of—that it made her doubt her own motives. **1966** G. N. Leech *Eng. in Advertising* xv. 141 In 'lovely, oveny biscuits', 'oveny' can only be made denotatively meaningful by reading in something extra. **1979** E. H. Gombrich *Sense of Order* iv. 99 Finding it difficult, if not impossible, to tell at any point where we see elements and where texture..; where we are reading and where we are 'reading in'.

9. a. Also, to convey (a statement) when read; to say. Cf. sense 18 b below.

1894 [see *it *pron.* 3 f]. **1904** G. Parker *Ladder of Swords* xvi. 229 A footman..came to Angèle, bearing a note which read: 'Your friend is very ill, and asks for you.' **1916** G. B. Shaw *Androcles & Lion* p. lxvi, Your examination paper will read 'The time of Jesus was worth nothing... Dr. Crippen's time was worth, say, three hundred and fifty pounds a year. Criticize this arrangement.' **1946** Bible (Rev. Standard Version) *Mark* xv. 26 And the inscription of the charge against him read, 'The King of the Jews.' **1961** New English Bible *Rom.* xii. 19 There is a text which reads, 'Justice is mine, says the Lord, I will repay.'

11. e. Phr. *to take* (something) *as read*: to treat (a statement, a subject, etc.) as if it has been agreed, without having a discussion about it; to take for granted. Occas. with other introductory vbs.

1886 G. M. Hopkins *Lett. to R. Bridges* (1955) 244 Objections on your part, if any, are now too late and will be 'taken as read'. **1928** D. L. Sayers *Unpleasantness at Bellona Club* iii. 22 Don't let's harrow our feelings. Take it as read. **1930** E. M. Brent-Dyer *Chalet Girls in Camp* x. 151 'It's really I who ought to say 'sorry', you know.'.. 'We'll take it all as read,' put in Miss Wilson hastily. **1938** M. Allingham *Fashion in Shrouds* xxi. 397, I think we can almost take that as read, don't you? **1959** 'M. M. Kaye' *House of Shade* v. 59 I'll take it as read. **1973** H. Miller *Open City*

xvii. 187 You can regard your complaint of boorishness..as read.

12. a. Also const. *with.*

1885 A. Edwardes *Girton Girl* I. iii. 68 Geoffrey Arbuthnot, B.A. Cantab., is willing to read classics and mathematics with Miss Bartrand. Terms, five shillings an hour.

13. a. Also *transf.*

1977 F. Branston *Up & Coming Man* xiv. 152 What could be more natural than for a keen new councillor to read himself in on past decisions?

b. *to read out of* (earlier and later examples); also *refl.*; *to read in*: to admit or induct formally; to make (a person) a member of an armed service, to conscript.

1836 W. Dunlap *Thirty Years Ago* I. xxi. 201 By the death of his parents, he was left in possession of some property, which he dissipated even before he was 'read out of meeting'. **1841** *Congress. Globe* 30 June 133/2 Mr. Alford concluded by warning the 'tariff boys' of the South, that instead of their reading him out of church, if they did not mind he would read them out of church. **1890** *Harper's Mag.* Feb. 349/2 They said I wasn't no Christian; and so they got together and read me out o' the church. **1915** F. Hopwood *Let.* 20 May in M. Gilbert *Winston S. Churchill* (1972) III. Compan. ii. 920 It is plain that a First Sea Lord remains in office until the new Patent appointing his successor is passed the Seal & he is 'read in' at the Board. **1938** C. S. Forester *Ship of Line* viii. 110 Excellent, Mr. Bush. Read 'em in. **1942** E. Paul *Narrow St.* xxii. 186 Men..who may be read into the army and shot as traitors if they try to strike, are difficult material. **1976** *Time* 27 Dec. 14/2 Arafat also warned that any Palestinian group that rejected the idea..must read itself out of the P.L.O.

II. 15. a. Also, to read (sense *5 a) music.

1889 G. B. Shaw in *Star* 24 May 4/2 The few who are really able to read at sight. **1976** *Star* (Sheffield) 3 Dec. 16/7 (Advt.), Organist required for a 9-day period at Christmas. Must be able to read.

c. (Earlier and further examples.)

1823 M. Wilmot *Let.* 1 Oct. (1935) 202, I trust that Edwᵈ may obtain the gold medal which he is reading for. **1863** Mrs. Gaskell *Dark Night's Work* xv. 271 You knew him at Hamley, I suppose? I remember his reading there with Mr. Ness. **1874** J. Codman *Mormon Country* i. 2 They buy a big book..and, having read up thoroughly, fill the cavities of their minds with details from these to supply what they did not learn from their extended visit of half a day. **1890** C. M. Yonge *More Bywords* 129 The two sisters are reading up for the Oxford exam. **1911** J. London *Let.* 17 Aug. (1966) 350, I should advise you..to read up on socialism. **1938** E. C. Lodge *Terms & Vacations* iv. 51 Those of us who read for University schools went to the different Colleges to lectures. **1962** [see *literæ humaniores]. **1976** 'M. Albrand' *Taste of Terror* xviii. 103 Why don't you read up on it in the *Britannica*?

d. To act as a publisher's reader.

1850 Thackeray *Pendennis* II. iii. 27 Warrington artfully inspired the two gentlemen who 'read' for Messrs. Bacon and Bungay with the greatest curiosity regarding 'Walter Lorraine'. **1891** G. Gissing *New Grub Street* I. vi. 122 She..liked to know who 'read' for the publishing-houses. **1956** P. Scott *Male Child* ii. i. 112, I..asked her whether she had ever read for a publisher. **1978** S. Hodge *Gollancz* iii. 64 They went on reading for him.

e. To receive and understand a message by radio, telephone, etc.

1930 *Amer. Speech* V. 289 The receiving operator 'receives', 'copies' or 'reads'. **1962** J. Glenn in *Into Orbit* 213 *Schirra*: 'John, leave your retro-package on through your pass over Texas. Do you read?' *Glenn*: 'Roger'. **1963** *Times* 25 May 10/7 'Philip to Kitty—do you read?' No reply. **1966** D. Francis *Flying Finish* ii. 26, I said, 'Port Ellen tower this is Golf Alpha Romeo Kilo November, do you read?'

16. c. *you wouldn't read about it*: exclamation used to express a mixture of incredulity and disgust. *Austral.* and *N.Z. colloq.*

1950 J. Cleary *Just let me Be* xiv. 135 Everything I backed ran like a no-hoper. Four certs I had, and the bludgers were so far back the ambulance nearly had to bring 'em home. You wouldn't read about it. **1962** D. Cusack *Picnic Races* xxi. 249 He drew a deep breath. 'You wouldn't read about it.' **1973** H. Williams *My Love had Black Speed Stripe* x. 69 You wouldn't read about it. A bloke his missus reckons was a doctor of philosophy, whatever that was, and just about the biggest dill you could meet.

18. a. (Later examples.)

1887 A. Birrell *Obiter Dicta* 2nd Ser. viii. 260 When the dish is served, we only ask, Is it good?..when the book comes out, Does it read? **1931** *N. & Q.* 17 Oct. 287/2 The translation is oddly unequal. It is often conspicuously clear and vigorous; sometimes it is halting and dull; occasionally, it does not 'read'.

b. Also, to convey a statement when read.

1891 F. H. Williams *Atman* v. 270 The letter reads as follows.

d. Of a measuring instrument: to have a graduated scale of a specified kind.

1862 *Catal. Internat. Exhib., Industr. Dept., Brit. Div.* II. No. 2941, Standard barometer on Fortin's principle, reading from an ivory zero point in the cistern.

19. e. Of an actor: to audition *for*, to rehearse *for* a role. Also, *to read in for*, to take the part of (another actor) at a rehearsal.

1943 S. Lewis *Gideon Planish* 71 'And will you read for it?'.. 'You mean try and see if I can act one of the parts?' 'Professionally, we call it "read for a part".' **1966** A. E. Lindop *I start Counting* ii. 44 She'd had a letter..asking her to report to the Jubilee Hall, where she was to read for the Amateur Dramatic Society. **1968** J. Bingham *I love, I Kill* iii. 32 Shirley, you read in for Sarah... You'll be understudying her anyway. **1970** E. Berckman *She asked for It* ii. 18 'How's it done?' 'You ring your agent, and say you'll read for the part.' *Ibid.* 19, I stood over her, literally, till she'd rung and said she'd read for the producer. **1971** *Guardian* 10 Dec. 10/5 The [play] reading is on again... Somebody can read in for Graham.

III. 22. *Computers.* The infin. used *attrib.* and in *Comb.* with the sense 'reading'.

1953 A. D. & K. H. V. Booth *Automatic Digital Calculators* xii. 115 Mounted close to the drum are a series of read/record heads. **1960** *Proc. Inst. Electr. Engineers* VII. B. 56 (*heading*) A digital computer store with very short read time. **1964** F. L. Westwater *Electronic Computers* iv. 77 The resulting change of flux as the core switches will cause an electromotive force in the read wire which we can recognise. **1965** *Wireless World* July 340/1 The read/write cycle time is 600 nsec. **1971** J. H. Smith *Digital Logic* vi. 126 In practical core memories the *X* and *Y* co-ordinates are only used for identifying the core required and each core in the plane is threaded with additional 'read' and 'write' wires. **1979** J. E. Rowley *Mechanical In-House Information Syst.* i. 64 Read-write heads..give direct access to the opposite track, and then this track may be searched sequentially.

23. read-around ratio *Computers*, the number of times that a particular bit in an electrostatic store can be read without degrading bits stored nearby.

1953 *Math. Tables & Other Aids to Computation* VII. 112 The inherent coupling between adjacent storage locations on the face of any cathode ray tube places a limit on the number of times any point on the raster may be consulted before its neighbors are regenerated. For the Institute machine this number, called the 'read around' ratio, is in the neighborhood of 30. **1969** P. B. Jordain *Condensed Computer Encycl.* 415 A high read-around ratio means greater reliability.

read, *ppl. a.* Add: **2.** Also *read up.*

1873 'Mark Twain' *Gilded Age* xi. 112, I *am* better read up in most sciences, maybe, than the general run of professional men. **1883** H. E. Manning *Eternal Priesthood* xx. 277 He is a welcome visitor..a ready and amusing guest, read up in the newspapers, and full of the events of the day.

readability. Add: (Earlier and later examples.) Also in extended sense, the quality of, or capacity for, being read with pleasure or interest, considered as measured by certain assessable factors, as ease of comprehension, attractiveness of subject and style.

1843 S. Coleridge *Let.* July in *Mem. & Lett.* (1873) I. xi. 281 If bad arrangement in S.T.C. is injurious to readibility, in S.C. it will be destructive. **1899** J. Southward *Mod. Printing* II. i. 4 A book..is not a good book because it is a pretty one; readability is of far more importance than picturesqueness. **1935** Gray & Leary *What makes Book Readable* ii. 38 The weight of opinion of all judges is that an informal.., non-technical.., adult..vocabulary is an important contribution to readability, whereas a vocabulary limited to 1,000–1,500 words.., is not essential. **1948** R. Flesch in *Jrnl. Appl. Psychol.* XXXII. 228 To measure the readability ('reading ease' and 'human interest') of a piece of writing, go through the following steps. **1953** *Journalism Q.* XXX. 417/1 One can think of cloze procedure as throwing all potential readability influences in a pot, letting them interact, then sampling the result. **1958** H. B. & A. C. English *Dict. Psychol. & Psychoanal. Terms* 441 The criteria brought forward for determining readability suggest that it is not a single variable but a combination of at least three or four. **1963** R. Morris *Success & Failure in Learning to Read* iv. 97 The attempts that have been made to implement [E.L.] Thorndike's 'readability-control' programme have added considerably to the resources available to the teachers of reading today. **1975** *Language for Life* (Dept. Educ. & Sci.) xiv. 218 The writers would be able to refashion their stories to adjust readability levels. **1977** *N.Y. Rev. Bks.* 23 June 4 (Advt.), One might define a great essay as a short excursion which has infinite readability.

readable, *a.* (and *sb.*) Add: **2. a.** (Earlier examples.)

1771 *Monthly Rev.* Dec. 493 The real Author of these Letters..chose to turn it to what literary advantage he might make of a couple of very *readable* volumes. *a* **1817** Jane Austen *Northanger Abbey* (1818) I. vi. 74 Sir Charles Grandison! That is an amazing horrid book... I thought it had not been readable.

b. (Later example.)

1977 *New Yorker* 1 Aug. 13/3 Sales clerks..stacked giant displays of discounted readables.

readapt, *v.* Add: **b.** *intr.* To become adapted anew.

1961 Webster, *Readapt*.., to become adapted again. **1971** N. Freeling *Over High Side* iii. 226 The little flat..had become 'working woman alone' instead of 'married couple'. .. 'I'm glad to see you're re-adapting with no trouble.'

reader. Add: **2. c.** (Examples.) Also, one similarly employed by a theatre to read plays offered for production.

1829 H. Foote *Compan. Theatres* 146 Drury-Lane.—Season 1828–9. Lessee and Manager—B. Price, Esq... Reader of Plays—Mr. Frederick Reynolds. **1859** E. Fitzball *Thirty-Five Years Dram. Author's Life* I. vii. 262, I have been dramatic reader myself, in the Theatre Royal, Covent Garden,..some years. **1891** G. Gissing *New Grub Street* II. xiii. 4 One of Mr Jedwood's 'readers'..was expressing a doubt whether Fadge himself was the author of the review. **1895** G. B. Shaw in *Sat. Rev.* 2 Mar. 280/2 The Lord Chamberlain's reader is not selected by examination either in literature or morals. **1924** J. Galsworthy *White Monkey* ii. vi. 168 Here were manuscripts, of which the readers to Danby and Winter had already said: 'No money in this.' **1956** P. Scott *Male Child* ii. iv. 153 He told me..he'd get another reader..in your place. He thought your illness had impaired your judgment. **1976** M. Green *Children of Sun* v. 171 J. B. Priestley..became..a publisher's reader for John Lane.

d. One who reads designs in weaving (see *read *v.* 5 e).

1839 Ure *Dict. Arts* 267 The weaving of imitation shawls is executed, as usual, by as many shuttles as there are colours in the designs, and which are thrown across the warp in the order established by the reader. **1932** L. Hooper *New Draw-Loom* ii. ix. 82 The reader, looking carefully at the line, No. 1 of the design at the side next to the numerals, must count the number of dark squares with which it begins ..and call 'Take 1'. **1970** *Classification of Occupations* (Office of Population Censuses & Surveys) 70/1 (Index), Reader: design; textile.

e. One who reads music; a sight-reader.

1947 G. B. Shaw in *Mus. Times* Jan. 10/1 It takes years of practice to train a group of good readers to sing in tune not only passably but exactly. **1977** *Grimsby Even. Tel.* 5 May 3/3 (Advt.), Pianist wanted also Trombonist by local rehearsal dance band. Must be readers.

3. a. (Examples in Jewish context.)

1872 *Minutes S. Manch. Hebrew Congreg.* 29 Sept. in I. W. Goldberg *South Manch. Hebrew Congreg.: 80 Years of Progress* (1952) 8 That the Reverend H. D. Marks be elected Reader, Stipendiary Secretary and Minister to the Congregation. **1973** *Jewish Chron.* 2 Feb. 43/1 A memorial service for Mr. Victor Schiller, honorary reader of the Lecton Synagogue..was held at the synagogue.

b. reader-aloud, one who reads (a literary text, etc.) aloud, esp. to an audience. Also *reader-alouder*. Cf. Read *v.* 11 a.

1938 *Times* 16 Sept. 13/4 Fountains are less trouble in bedrooms than readers-aloud or raconteurs. **1952** G. Raverat *Period Piece* viii. 145 Aunt Etty was the best reader-aloud I have ever known. **1952** *Sat. Rev.* (U.S.) 13 Sept. 6/3 Hemingway is a reader-alouder, it appears. **1977** *Listener* 10 Nov. 624/4 Lots of subordinate clauses can make life very difficult for the reader-aloud.

6. a. (Earlier examples.)

1718 C. Hitching *Regulator* 20 A reader, *alias* pocket-book. **1789** G. Parker *Life's Painter Varieg. Char.* xv. 151 *Reader.* Is a pocket-book; a person cannot be too careful of this article, particularly if he should have..any rum screens in it, that is, bank notes.

b. (Further example.)

1977 'L. Egan' *Blind Search* iv. 57 McAllister was a gambler... This is a deck of readers—marked cards.

c. *U.S. Criminals' slang.* (See quot. 1926.)

1926 *Clues* Nov. 162/1 *Reader*, a circular notifying police officers to arrest the party described thereon. **1955** *Publ. Amer. Dial. Soc.* xxiv. 150 Sometimes there is a 'detainer' issued for a thief... This is called a *reader* or a *dipsy*.

7. A device for obtaining data stored on tape, cards, or other media (usu. converting the data into coded electrical signals).

1946 *N.Y. Times* 15 Feb. 16/3 When the problem is punched on the cards they are dropped into a slot in a 'reader'. **1964** F. L. Westwater *Electronic Computers* iv. 80 Even with the faster types of card reader it is difficult to exceed 800 digits per second. **1968** *Brit. Med. Bull.* XXIV. 205/1 It may be possible to eliminate the stage of transfer onto punched cards by using an optical reader, for there is now a rapid development in this type of device. **1972** M. Woodhouse *Mama Doll* xi. 145 Some people at Admiralty ran the tape through a five-hole reader for us, and gave us back seven hundred and eighty-four groups of digits.

8. A machine for producing on a screen a magnified, readable image of any desired part of a microfilm or other microform.

1950 *Amer. Documentation* I. 141/2 A new reading machine just announced..holds much promise. This reader giving a clear, sharp image..is relatively inexpensive. **1962** A. Günther *Microphotogr. in Lib.* (Unesco) 7 Micro-opaque cards..may be readily filed. However, they need much more light for projection and, therefore, a more complicated and more expensive reader, which must be equipped with a blower for cooling. **1975** P. G. New *Reprography for Librarians* iv. 48 The librarian committed to exploiting micro materials must not only consider investing in a multitude of portable readers for loan, but must also ensure that his library is fully equipped with..viewers for use on the premises.

9. *attrib.* and *Comb.*, as *reader group, participation, response*; *reader-contributor, -writer*; (sense 2 c) *reader's report*; **reader-printer,** a reading machine (sense *(*b*)) that can also produce enlarged, readable copies.

1946 *R.A.F. Jrnl.* May 146 The success of the new magazine will depend on the continuance of the excellent reader-contributor relationship which was fostered. **1951** M. McLuhan *Mech. Bride* (1967) 112/2 These magazines, carefully geared to both the purse and heart strings of their respective reader groups, feature houses and rooms in which almost nobody ever lives. *Ibid.* 5/2 This kind of newspaper invites reader participation in its triumphs. **1959** H. W. Ballou *Guide to Microreproduction Equipment* 167 Thermo-Fax Brand Microfilm Reader-Printer... Special Features: Reader and Printer combined in one machine for automatic push-button copying or reading. **1971** *Ann. Rep. Curators Bodl. Libr. 1969–70* 47, 4,011 prints were made on the microfilm reader-printer. **1940** *Kenyon Rev.* II. 274 The reader-response has been altered through a lessening of the pleasure with which the utterance is received. **1979** *Maledicta* III. 83 Among those critics who use psychoanalytical theory there is little agreement over what one can say legitimately about 'reader response'. **1897** 'S. Grand' *Beth Bk.* (1898) xlvii. 460 Mr. Kilroy took the manuscript himself to a publisher..who..accepted it... Beth..heard the reader's report. **1978** E. Tidyman *Table Stakes* ii. v. 241 Each morning a mailboy would arrive with a stack of scripts..Attached were the readers' reports. **1951** S. Spender *World within World* 310 Reader-writer walk together in a real-seeming dream-alliance leading into gardens inhabited by Stephen Daedalus and Marcel.

readership. Add: **3.** The total number of (regular) readers of a periodical publication, as a newspaper or magazine; all, or a section, of such readers considered collectively. Also *attrib.* orig. *U.S.*

1923 O. G. Villard *Some Newspapers & Newspapermen* 189 The appeal of the *News* to the masses has been so successful that it now has a readership of some forty thousand. **1947** C. L. Allen (*title*) A readership study of 3 typical Wisconsin hometown dailies. **1951** *Sunday Times* 2 Dec. 1/3 Mr. Stephen's..experienced counsel and reflections [will] become available to the whole Sunday Times readership. **1958** *New Statesman* 30 Aug. 241/1 It holds its vast circulation..by grace of Mr. Gilbert Harding, whose weekly column (according to readership surveys) is the *People's* biggest pulling feature of all. **1963** *Guardian* 10 Apr. 7/2 Another variation, reflecting different readership, is the background of the characters. **1971** *Nature* 2 Apr. 310/3 In view of the intended readership the selection of topics seems reasonable enough. **1979** *London Rev. Bks.* 25 Oct. 2/3 The obvious difference will relate to the subjects generated by the nationality of the *London Review*'s readership, and by that of its contributors.

read-in (rī·d,in). *Computers.* [f. vbl. phr. *to read in* (*read *v.* 6 f).] The input of data to a computer or storage device.

1946 *Ann. Computation Lab. Harvard Univ.* I. 63 If a quantity is standing in the counter at the time of read-in so that addition must be performed, the carry circuits are utilized. **1952** [see *matrix 6 d]. **1959** E. M. McCormick *Digital Computer Primer* ii. 8 When the storage..comprises relays and vacuum tubes,..special read-in equipment is required so that the information will be in a form the computer can use. **1970** *Computers & Humanities* IV. 166 Only the read-in statements of the program set would need to be changed to accommodate this programming.

readiness. Add: **4. b.** *Psychol.* The stage of physiological or developmental maturity at which an organism is able to take in new learning with ease. Also *attrib.*

1948 E. R. Hilgard *Theories of Learning* ii. 22 There is another kind of readiness familiar to educators. This is illustrated by the use of such a term as 'reading readiness' to refer to the child's reaching a maturity level appropriate to the beginning of reading. **1956** H. C. Lindgren *Educ. Psychol.* ix. 236 Many schools postpone the teaching of reading for first-graders who do not demonstrate 'reading readiness' on standardized tests. **1967** J. C. Nunnally *Psychometric Theory* iii. 77 In schools, predictive validity is at issue in measures of 'readiness'. **1976** *Woman's Day* (U.S.) Nov. 58/2 Clare Pederson..takes a particularly dim view of readiness tests... 'Why don't school systems take the money they waste on readiness tests and spend it on books instead?' **1977** P. R. Ammon in Hom & Robinson *Psychol. Processes in Early Educ.* vii. 184 The question of readiness pertains not only to *how* children can learn, but also to *what* they can learn.

Reading (re·diŋ). [The name of the county town of Berkshire.] **1.** *Reading Beds*, a set of beds of sand, clay, and gravel of fluviatile origin underneath the London clay in the London and Hampshire basins.

1817 *Trans. Geol. Soc.* IV. 283 In many parts of this great valley or trough of chalk [*sc.* the Thames valley] we recognized our Reading beds in their proper place, as the inferior strata of the plastic clay formation. **1854** *Q. Jrnl. Geol. Soc.* X. 164 The leaves preserved in the Reading beds. **1882** A. Geikie *Text-bk. Geol.* 844 The Woolwich and Reading Beds, or 'Plastic Clay' of the older geologists, consist of lenticular sheets of plastic clay, loam, sand, and pebble-beds. **1923** [see *Landenian *a.*]. **1969** Bennison & Wright *Geol. Hist. Brit. Isles* xv. 336 The Reading Beds facies is the most extensive of the three in the London Basin.

2. a. *Reading onion*, a variety of onion developed by the firm of Sutton & Son (formerly of Reading). Also *absol.*

1845 E. Acton *Mod. Cookery* (ed. 4) xxii. 508 The Reading onion is the proper kind for pickling. **1885** W. Robinson tr. *Vilmorin-Andrieux's Veg. Garden* 366 White Spanish, or Reading, Onion... Bulb quite flat, 3 to 4 inches in diameter. **1963** *Sutton's Seed Catal.* 66/1 Sutton's Improved Reading. A maincrop onion.

b. *Reading sauce*, a sharp sauce flavoured with onions, spices, and herbs.

1861 Mrs. Beeton *Bk. Househ. Managem.* 240 *Reading sauce*,..walnut pickle..shalots..Indian soy..bruised ginger [etc.]... *Seasonable.*—This sauce may be made at any time. **1862** 'L. Carroll' in *College Rhymes* III. 114 There are epithets That suit with any word—As well as Harvey's Reading Sauce With fish, or flesh, or bird. **1878** M. Jewry *Warne's Model Cookery* 46/1 A large spoonful of sauce, either Worcester or Reading. **1907** *Yesterday's Shopping* (1969) 33/2 *Sauces* (Other Makers')... Reading.

3. Designating the gypsy caravan of traditional design, supposedly first built in Reading (see quots.). Chiefly in *Reading wagon.*

1940 F. G. Huth in *Jrnl. Gypsy Lore Soc.* XIX. 117 The Reading Waggon, or old type of Gypsy *vardō*, with large wheels running outside the body of the van. **1951** *Archit. Rev.* CIX. 317 There are five distinct types of gypsy wagon in use on the roads today... Reading wagon. Originally built by the Dunton family at Reading, this is probably the oldest type of van and is generally accepted as the 'gypsy shape'. **1972** Ward-Jackson & Harvey *Eng. Gypsy Caravan* ii. 43 The Reading type of van had certainly been evolved by the 1860's. **1975** *Country Life* 2 Oct. 840/1 We thought we might buy an old gypsy wagon... I found the ruin of a 'Reading' or showman's wagon. **1976** *Horse & Hound* 3 Dec. 68/4 (Advt.), Gypsy Caravan of Reading style for sale.

reading, *vbl. sb.* Add: **1. e.** *Computers.* The copying, extraction, or transfer of data. Also with *in*, *out*. Also *transf.* Freq. *attrib.* Cf. *read *v.* 5 h, 6 f.

1949 E. C. Berkeley *Giant Brains* iv. 44 The reading of a hole in a column of a punch card is done by a brush of several strands of copper wire pressed against a metal roller. **1950** *High-Speed Computing Devices* ix. 155 The input to the Tabulator is from a single feed with reading stations examining the two most advanced cards simultaneously. **1964** F. L. Westwater *Electronic Computers* iv. 65 When the magnetised spot of wire passes under the reading head there will be a change of magnetic flux through the coils on the head. **1964** J. Z. Young *Model of Brain* xiii. 217 We can thus say that the vertical lobe system is necessary for 'reading-out' of the memory as well as for 'reading-in'. **1970** O. Dopping *Computers & Data Processing* viii. 122 After a few milliseconds, the beginning of the tape block reaches the reading head and the buffer register in the channel starts to be filled with information.

2. a. Also with *aloud, out.*

1936 F. R. Leavis *Revaluation* ii. 44 Here, if this were a lecture, would come illustrative reading-out—say of the famous opening to Book iii. **1960** C. Day Lewis *Buried Day* iv. 64, I don't remember much reading-aloud, before I could read to myself.

b. Also with *aloud.*

1960 C. Day Lewis *Buried Day* vii. 149 These were H.R.K.'s incomparable readings-aloud from Jane Austen, Thackeray, Dickens or the poets. **1974** *Listener* 14 Mar. 347/2 The least we can ask from a reading-aloud of poetry we know is that it adds to our gain from private reading.

e. (Earlier and later examples.)

1787 J. Cobb *Eng. Readings* 5 But tell me, Kitty, how did this *rage* for English Readings reach a town so far from London? **1813** M. Edgeworth *Let.* 16 May (1971) 55 We have been to one of Mrs. Siddons readings—*Measure for Measure*... In settling with Sheridan she came short 10 or 12 thousand pounds and her Readings are to make up this defalcation. **1916** M. B. Lowndes *Diary* 12 Apr. (1971) 71, I went to the most remarkable Poets' Reading I have ever attended... I was moved by Mr de la Mare reading five poems of great beauty. **1953** *Ann. Reg. 1952* CXCIV. 377 The Hell scene in Shaw's *Man and Superman* had been staged with elaborate simplicity as a 'reading'.

7. c. An extract from a previously printed source; in *pl.* freq. denoting a particular selection of such extracts intended to be read at one time or as a unit.

1835 C. Fry (*title*) Daily readings. Passages of Scripture, selected for social reading. **1865** Ctess. of Cawdor (*title*) Short Sunday evening readings selected and abridged from various authors. **1908** Robinson & Beard (*title*) Readings in modern European history. A collection of extracts from the sources. **1931** W. L. Valentine *Readings in Exper. Psychol.* p. xiv, The original purpose was to include as a single reading only a single experimental paper. **1947** *Mind* LVI. 278 This joint work is intended to be read in conjunction with a companion volume of 'readings' culled from the classics of ancient and modern philosophy. **1972** *Sci. Amer.* Feb. 117/1 *The Science of Matter* offers more than 160 samples, averaging a couple of pages each; *A History of Medicine*..gives us a couple of dozen readings some 10 or 12 pages in length.

9. (Further examples.)

1814 *Morning Herald* 14 Mar. in J. Agate *These were Actors* (1943) 31 Mr. Kean thought fit to leave out the whole of the first line in this declaration... This, in the saucy jargon of the day, may be called 'a new reading'. **1865** Dickens *Mut. Friend* II. iii. x. 94 By-the-by, that very word, Reading, in its critical use, always charms me. An actress's Reading of a chambermaid, a dancer's Reading of a hornpipe, a singer's Reading of a song, a marine-painter's Reading of the sea, the kettle-drum's Reading of an instrumental passage, are phrases ever youthful and delightful. **1929** A. Carse *Orchestral Conducting* iii. i. 96 The personality of a conductor, the individuality of his readings ..count for more than technical correctness. **1945** H. Wood *About Conducting* 105 Every aspirant to a conductor's career should..make himself acquainted with the traditional readings of the classical repertoire. **1969** *Listener* 13 Mar. 360/2 The structure of his film implies one reading of Isadora's life, while its content implies a quite contrary interpretation. **1977** M. Allen *Spence in Petal Park* iv. 16 Someone turned him over..after death, I would say. The pathologist will tell us for sure, but that's my reading.

10. a. *reading-circle, clinic, habit, -lamp* (earlier and later examples), *light, list, material, matter* (earlier and later examples), *party* (earlier and later examples), *rate, -readiness, scheme, society* (earlier and later examples), *table* (earlier examples).

1871 Mrs. Stowe *Pink & White Tyranny* xi. 124 They would get up their reading-circles, and he would set her to improving her mind. **1926** R. Macaulay *Crewe Train* ii. v. 118 A reading circle. You all study some book together, and meet and talk about it. **1963** R. I. McDavid *Mencken's Amer. Lang.* 320 Minton points out the spread of technical medical terminology to education, as *clinic* (yielding *reading clinic* and *speech clinic*). **1975** *Language for Life* (Dept. Educ. & Sci.) xxvi. 514 There should be a reading clinic or remedial centre in every L.E.A. **1940** R. S. Lambert *Ariel & all his Quality* v. 131 You could hear complaints..that broadcasting was undermining the reading habit. **1963** D. Pryce-Jones in Sissons & French *Age of Austerity* 212 The war may have enlarged the reading habits of a great many people. **1975** *Language for Life* (Dept. Educ. & Sci.) xxi. 304 The reading habit should be established early. **1782** *Catal. Stock in Trade Benjamin Martin* 14 A reading lamp, with magnifying glass and shade. **1908** Mrs. H. Ward *Diana Mallory* ii. x. 212 She was bending over the fire..a reading-lamp beside her. **1960** T. Cooper *Winter's Day* ii. v. 134 Do..switch off the big light; this reading-lamp on the desk is ample. **1936** M. Allingham *Flowers for Judge* xix. 273 The green reading light..shining down upon his papers. **1945** Wilson & Wright *Tomorrow's House* xi. 116/2 Getting a decent reading light is by no means a matter of setting a table lamp on the night table. **1981** L. Deighton *XPD* xli. 329 He had a small reading light by which to read the documentation. **1925** *Scribner's Mag.* July 61/1 Books on fishing..should, in my opinion, have a place on every reading list. **1981** *Times Lit. Suppl.* 6 Feb. 136/4 The names of these daunting authors..make an occasional modest appearance on reading-lists. **1961** *Educ. in Scotland 1960* (HMSO) 44 A welcome increase in the provision of supplementary reading material. **1975** *Language for Life* (Dept. Educ. & Sci.) xvii. 253 Her first task is to assess the attainment level of every child and provide each with reading material of the right level of readability. **1848** Thoreau in *Union Mag.* Aug. 79/2 An odd leaf of the Bible,..Emerson's Address on West India Emancipation..an odd number of the Westminster Review... This was the readable, or reading matter, in a lumberer's camp in the Maine woods. **1923** R. Macaulay *Told by Idiot* ii. i. 68 Wise men and women would derive such pleasure as they could from the writings of others, without putting themselves to the trouble of providing reading matter in their turn. **1972** 'E. Ferrars' *Breath of Suspicion* xi. 185 I'm leaving the choice of some reading-matter for you to Bernard. **1785** B. Sheridan *Jrnl.* (1960) i. 43 Yesterday evening we spent at Mr Vesey's—a sort of conversatione—and reading party. **1930** J. S. Huxley *Bird-Watching & Bird Behaviour* iii. 61, I was spending some of the spring vacation with a reading-party on the coast of North Wales. **1980** D. Newsome *On Edge of Paradise* ii. 59 A meeting there with a reading-party was usually the prelude to some summer expedition abroad. **1960** *Bookseller* 17 Dec. 2330/3 A 'reading-rate controller'.., an inverted T-square with the handle part moving down the page of a book,..is attached to a sloping desk... A ready reckoner shows the number of..[lines] read in a minute, and the pupil can set the speed of this and then try to read faster. **1975** *Broadcast* 28 July 11/3 Our videodisc player's reading rate is 30 million bits per second. **1948, 1956** Reading-readiness [see *readiness 4 b]. **1964** M. Critchley *Developmental Dyslexia* iv. 15 Bound up with the problem of when a child should first receive formal instruction in reading is the notion of a state of 'reading-readiness'. **1976** *Woman's Day* (U.S.) Nov. 58/2 How can reading-readiness scores have meaning when reading experts are still debating what skills are needed for beginning reading? **1974** *Education & Community Relations* Jan. 3 Several multiracial primary schools foresaw a major change 'in the selection of reading schemes and supplementary readers'. **1975** *Language for Life* (Dept. Educ. & Sci.) vii. 104 The reading scheme is at the centre of this material in most young children's early experience of reading. **1775** T. Campbell *Diary* 21 Mar. (1947) 58 Strolled into the Chapter Coffee-house..remarkable for a large collection of books, & a reading Society &c—I subscribed a shilling for the right of a years reading. **1828** M. O'Brien *Jrnl.* 28 Oct. (1968) iii. 21, I hope we shall manage the reading society, though we can only muster three members at Present. **1890** G. B. Shaw in *Star* 28 Feb. 2/4, I repaired to the London Institution to see 'The Shakespere Reading Society' recite 'Much Ado'. **1794** T. Sheraton *Cabinet-Maker & Upholsterer's Drawing-Bk.* II. iii. Pl. 44 (*caption*) A Reading & Writing Table. **1855** Trollope *Warden* ix. 134 A huge arm-chair fitted up with candlesticks, a reading table, a drawer, and other paraphernalia.

b. reading age, reading ability expressed in terms of the age (during the period of development) for which a comparable ability is calculated as average; **reading chair,** a chair designed to facilitate reading; *spec.* one equipped with a book-rest upon one arm; **reading copy,** a copy of a book that is usable although in less than perfect condition; **reading-glass,** (*b*) in *pl.*, a pair of spectacles for use when reading; **reading-machine,** (*b*) a device for producing an enlarged, readable image from microform; (*c*) a device for automatically producing electrical signals corresponding to the characters of a text; **reading notice** *U.S.* (see quot. 1909).

1921 C. Burt *Mental & Scholastic Tests* iii. iii. 271 Consequently, a score of sixty words indicates a mental age for reading at ten;..according to the formula:—Reading Age $= (4+\frac{\text{Words}}{10})$ years. *Ibid.*, The reading ages of four and five pretend to little more than a conventional significance. **1945** F. J. Schonell *Psychol. & Teaching of Reading* i. 21 There is always a great increase in eye movements as the reading material increases in difficulty for particular reading ages. **1952** Anderson & Dearborn *Psychol. of Teaching Reading* i. 10 If the reading age is appreciably below the mental age, the child is regarded as a reading problem. **1961** *Guardian* 28 Apr. 13/3 He looks a dissipated 20... His reading age is 8·2. **1975** *Language for Life* (Dept. Educ. & Sci.) ii. 11 There are at least a million adults with a reading age of below 9·0 who cannot read simple recipes. **1803** T. Sheraton *Cabinet Dict.* 17 *Arm-chair* for a library, or a reading chair... These are intended to make the exercise easy, and for the convenience of taking down a note or quotation... The reader places himself with his back to the front of the chair, and rests his arms on the top yoke. **1853** A. J. Downing *Architect. Country Houses* xii. 426 Fig. 218 is a reading-chair of a simple and good form,..having a desk for a book on one arm, and a stand for a candle on the other—both being..easily lifted out.., when not in use. **1951** E. Paul *Springtime in Paris* iii. 54 There was a long table, and ranged on both sides, good reading chairs. **1977**

READING

J. Hodgins *Invention of World* iii. 44 The tall green reading chair that had recipes..shoved under its cushion. **1952** J. Carter *ABC for Book-Collectors* 164 Reading copy, a usually apologetic, but occasionally slightly defiant, term meaning that the book is not in collector's condition. **1977** J. Wilson *Making Hate* iv. 52 The *Just So Stories*. I had an early edition, a torn reading copy, but quite clean. **1853** Dickens *Bleak House* xli. 405 The green lamp is lighted, his reading-glasses lie upon the desk. **1972** G. Bill *Villains Galore* i. 1 Clara..needed reading glasses for all but the largest print. **1937** M. L. Raney *Microphotogr. for Libraries* 76 There is today plenty of work for reading machines to do, since the entire contents of great libraries that have filming cameras lie open to order in so far as copyright allows. **1940** A. Huxley *Let.* 14 Oct. (1969) 461, I would like to have..micro-photographs suitable for reading by means of a reading machine. **1959** *Library Resources & Technical Services* III. 90 The average library user does not meet the microcopy until he has to use it on the reading machine. **1964** *Litho-Printer* Aug. 34/2 Even optical reading machines, which are now entering the field of practicability cannot quite dispense with human work: they need clean copy, at least re-typed from edited manuscripts. **1965** R. R. Karch *Graphic Arts Procedures* (ed. 3) xiii. 338 Specially-designed figures printed at the bottoms of bank checks are printed with ink capable of being magnetized and read by electronic reading machines for routing the checks to proper places. **1980** J. Drummond *Such a Nice Family* viii. 38 Would you like us to fix up a reading-machine for you?.. It'll throw up an enlargement of the text. **1909** Webster, *Reading-notice*, in a newspaper or periodical, a paid advertisement so set up as to have the appearance of regular news or editorial or contributed matter. **1970** R. K. Kent *Lang. Journalism* 109 *Reading notice*, an advertisement in a newspaper or magazine that is set in body type and in columns so as to appear the same as editorial matter.

reading, *ppl. a.* **2.** (Further examples of (*the*) *reading public*.)

1877 M. W. Chapman in *Harriet Martineau's Autobiogr.* III. 99 The reading public..were longing to express their grateful acknowledgements. **1916** E. Pound *Let.* 17 Nov. (1971) 99 That many-eared monster with no sense, the reading public. *a* **1936** Kipling *Something of Myself* (1937) iii. 47 Our reading public..were..as well educated as fifty per cent of our 'staff'. **1962** M. McLuhan *Gutenberg Galaxy* 132 There was no reading public in our sense... Under manuscript conditions an author would..have no public. An advanced scientist today has no public. **1975** S. Schoenbaum *W. Shakespeare* xi. 120 A dramatist had least to say about..publication... He strove, after all, to please audiences in the theatre, not a reading public.

readjustment. Add: **2.** *Comb.*, as **readjustment rule** *Linguistics* (see quot. 1972).

1968 Chomsky & Halle *Sound Pattern Eng.* i. i. 10 The 'readjustment rules' relating syntax to phonology make various other modifications in surface structures. **1972** R. A. Palmatier *Gloss. Eng. Transformational Grammar* 141 *Readjustment rule*,..one of a set of special rules which prepare the syntactic surface structure of a sentence for inputting to the phonological component; one of the types of phonological rules..which determine admissible, or possible, and inadmissible, or impossible, classificatory matrices. **1977** *Canad. Jrnl. Linguistics 1976* XXI. ii. 215 He also includes the suggestion, discussing Trager's similar analysis of short *a* in New Jersey English in 1940.., that such factors may be handled as a type of readjustment rule in the historical development of the phonology of a language.

readmire, *v.* (Later example.)

1930 O. W. Holmes *Let.* 12 May (1953) II. 1246, I finished it [*sc.* a book] a few days ago. I readmired the Rousseau and Machiavelli and believed without adequate knowledge what you say about foundations.

read-mostly (rīdˌmōᵘ·stli), *a.* *Computers.* [f. Read *v.* + Mostly *adv.*] Applied to a memory whose contents can be changed, though not by program instructions, but which is designed on the basis that such changes will be very infrequent compared with the number of occasions when the contents are read.

1971 *New Scientist* 1 Apr. 29/3 Initially, read-mostly memories will be used for semi-permanently stored computer instructions, such as in process control computers. **1977** *Sci. Amer.* Sept. 139/1 Another variation on the read-only memory is the read-mostly memory... Read-mostly memories have two forms. The commonest is the optically erasable read-only memory. *Ibid.*, An alternative form of read-mostly memory is the electrically alterable read-only memory.., which can be altered without the necessity of erasing the entire array.

read-only (rīdˌōᵘ·nli), *a.* *Computers.* [f. Read *v.* + Only *adv.*] Applied to a memory whose contents cannot be changed by program instructions but which can usually be read at high speed; also *ellipt.* Abbrev. ROM s.v. *R II. 2 a.

1961 H. R. Foglia et al. in *IBM Jrnl. Res. & Development* V. 67/1 Another form of memory can be used that may be read at these high speeds, with capability of being changed in a few minutes. Memories of this type, in which the fast-read cycle is of prime importance, may be called read only memories. **1969** P. B. Jordain *Condensed Computer Encycl.* 416 Read-only storage is used for rapid access to information of a permanent nature. *Ibid.*, Read-only stores are not to be confused with nondestructive read-out memories, memories that may be altered but can be repeatedly read without loss of data. **1970** O. Dopping *Computers & Data Processing* x. 135 Read-only memories are sometimes used in the internal organization of a computer, but in some machines they are also used for the permanent storage of standard programs. **1977** *Engineering Materials & Design* Aug. 9/3 The function program for the device is held in a 16 kilobit read only memory. **1979** R. Mutch *Gemstone* viii. 96 Probably only need one read-only.

read-out (rī·dˌaut). Also **readout.** [f. vbl. phr. *to read out* (*Read *v.* 6 f).] **1. a.** *Computers.* The extraction or transfer of data from a storage medium or device. Also *transf.*

1946 [see *Read *v.* 6 f]. **1952** [see *Matrix 6 d]. **1961** *IBM Jrnl. Res. & Development* V. 67/1 During readout the voltage pulse will be applied to a selected conductor corresponding to the word to be read out. **1970** O. Dopping *Computers & Data Processing* x. 151 Read-out is effected when a desired word passes through the pulse-shaping circuits. **1971** J. Z. Young *Introd. Study Man* xxii. 302 This reprogramming of the read-out of the DNA may serve to bring into play a new complex of enzyme systems.

b. The display of data by an automatic device in an understandable form. Also *transf.*

1961 *Aeroplane & Astronautics* CI. 573/2 The range indicator displays have a six-figure readout correct to the nearest 25 m. **1965** *Wireless World* July 35 (Advt.), For rapid transistor measurements, it can be set up for direct meter readout. **1971** *Engineering* Apr. 44/2 (*caption*) The new Sangamo extensometer which offers direct readout plus great sensitivity. **1971** *Nature* 13 Aug. 443/3 Displays have been selected during evolution to provide an accurate 'read out' of an animal's internal state to another animal.

c. (See quot. 1966.)

1966 *Britannica Bk. of Year 1965* 807/2 *Readout*, the radio transmission of data or pictures from a space vehicle either immediately upon acquisition or later by means of playback of a tape recording. **1970** N. Armstrong et al. *First on Moon* iii. 66 There was still work to do, starting with spacecraft-to-ground readouts.

2. A device for extracting or displaying data.

1954 *IRE Trans. Electronic Computers* Dec. 12 (*heading*) A radio-frequency nondestructive readout for magnetic core memories. *Ibid.* 15/1 It was possible to sense any core in this array plane using the rf readout. **1967** *Electronics* 6 Mar. 114/1 (Advt.), Our all-new nixie tube—the industry's lowest-cost electronic readout, and one sure to usher in a whole new generation of low-cost digital instrumentation. **1969** A. C. Marin *Rise with Wind* i. 4 Any efforts..to create an identity for the Director always fell flat, killed by the ever-present awareness of the computerlike brain... You could almost sense the logic circuits working, the tapes running on the reels, the punch cards appearing on the read-out. **1972** *Physics Bull.* July 418/3 An easy to read 2½ digit led readout reduces error when interpreting the measurements, particularly in low light conditions. **1976** M. Machlin *Pipeline* liii. 538 She was pleased to see that Schultheiss was alone in the control shack, which was a large building filled with a complicated set of flashing lights, meters, and digital readouts.

3. A record of its output produced by a computer or scientific instrument.

1967 *Economist* 11 Feb. 542/1 The computer read-out is fed into the draughting machine which should then draw the correct shape of the ship for the designer to check. **1970** H. Harrison *Captive Universe* 142 Just look at these figures... These are from a machine, a readout. **1971** *Daily Colonist* (Victoria, B.C.) 11 Dec. 17/5 The echogram—the read-out of an electronic device that is useful in diagnosing such abnormalities as tumours. **1977** *Daily Tel.* 25 Jan. 15/3, I left behind at the institute about 100-yards of paper read-outs, my metabolic processes recorded in squiggly lines.

4. *attrib.*

1965 *Wireless World* July 352/3 The average readout time is two seconds but for maximum readout it is five seconds. **1966** *Times Rev. Industry* Oct. 52/1 The units in the range—sample injection systems, oven and temperature controllers, columns, detectors and read-out units—are of modular design. **1968** H. Harris *Nucleus & Cytoplasm* iv. 68 Our text-books are full of diagrams showing ribosomes rolling along a messenger tape and thus fulfilling their role as a 'read-out' mechanism. **1976** *Offshore Platforms & Pipelining* 57/1 Readout gauges on the control panel will display tension in each line. **1976** B. Bova *Multiple Man* (1977) i. 16 The computer was humming to itself, lights flickering on its read-out console.

rea·d-through. Also **readthrough, read through.** [f. Read *v.*] **1.** An act of reading through; spec. *Theatr.*, an initial rehearsal at which actors read their parts from scripts.

1961 A. Wilson *Old Men at Zoo* ii. 95 She has to go to a read-through whatever that may be. **1963** E. Humphreys *Gift* ii. i. 209 He was late for the read through... Started talking loudly to the actress next to him and greeting everybody when we had already begun to read. **1966** 'O. Norton' *School of Liars* vii. 113, I..came back for the first read-through, and the acting notes. **1971** *Guardian* 3 July 9/3 'This script is a mess..' said Adrienne Corri the other day at a read-through for my TV series. **1976** J. Grenfell *Joyce Grenfell requests Pleasure* xvi. 230 Our spirits went up and down..after the usual enthusiastic read-through when the material seemed so original and imaginative.

2. *Biochem.* The continued transcription of genetic material by RNA polymerase that has overrun a termination sequence.

[**1970** Goff & Minkley in L. Silvestri *RNA-Polymerase & Transcription* 134 One interesting interpretation of these results is that the largest RNA class is produced when rho fails to function at the first termination signal used *in vivo* and the polymerase reads through to a second 'stop signal'.] **1970** A. A. Travers in *Nature* 14 Mar. 1011/2 (*caption*) Fig. 3. Readthrough by *E. coli* σ initiated RNA polymerase into sequences transcribed by σ[T4] initiated polymerase. **1971** *Ibid.* 15 Dec. 207/2 These sequence data strongly suggest that the IIb protein is a read-through product resulting from polypeptide chain elongation. **1978** *Ibid.* 30 Mar. 417/2 One of these..is the only oligonucleotide consistent with transcriptional 'read-through' of the termination site.

readva·nce, *sb.* Chiefly *Geol.* [f. the vb.] A renewed advance.

1879 *Geol. Mag.* Decade II. VI. 250 The recurring glacial conditions were too insignificant to cause a re-advance of the ice-sheet. **1927** Peake & Fleure *Apes & Men* 29 With the re-advance of the temperate forest in Europe, as the climate improved, most of them [*sc.* the mammals] finally disappeared. **1975** J. G. Evans *Environment Early Man Brit. Isles* ii. 45 In the north of the British Isles, three readvances during the general retreat of the ice are marked by 'moraines'.

readvertise, *v.* Add: Now *spec.* to give further notice of (a job vacancy). Hence **rea·dvertised** *ppl. a.*; **rea·dvertising** *vbl. sb.*; **readve·rtisement.**

1934 Webster, Readvertisement. **1963** *Times* 19 Apr. 7/1 The decision to readvertise the post of Chief Constable of Glamorgan was 'reluctantly' agreed to by the standing joint committee of the county council today. **1964** A. Battersby *Network Analysis* x. 153 The table could obviously be extended to include reprinting, re-advertising, and later royalty payments. **1971** *Engineering* Apr. 124/2 Previous candidates for the second post (now re-advertised) will be reconsidered if they so indicate. **1977** *Times Educ. Suppl.* 21 Oct. 46/1 (Advt.), Headteachers; (1) Balderston Comprehensive Community School (readvertisement). **1978** *Nature* 9 Nov. 109/1 It would like to readvertise the head of technology post inviting both engineers and astronomers to apply.

ready, *a.*, *adv.*, and *sb.* Add: **A.** *adj.* **I. 1. e.** *U.S. slang.* Excellent, first-rate; mature, fully competent. Chiefly of music or musicians.

c **1938** N. E. Williams *His Hi De Highness of Ho De Ho* 35/2 When an individual or a piece of music is high class or greatly admired, we indicate it by saying, 'He's ready!' or 'That's ready!' **1944** C. Calloway *Hepsters Dict.*, *Ready*,.. 100 per cent in every way. Ex., 'That fried chicken was ready.' **1945** *Tomorrow* June 27/3 This time he was *ready*, so to speak, for it was on this second sojourn..that he began to impress his musical contemporaries. **1968** in R. Russell *Jazz Style in Kansas City* (1971) 183 When he came back, several months later, he was a new musician. *He was ready.*

f. *ready room*, a room in an aircraft-carrier where pilots are briefed and await orders to fly. *U.S. Mil.*

1945 in Webster Add. **1953** P. C. Berg *Dict. New Words* 134/2 *Ready room*,..the room of an aircraft-carrier where pilots ready for flight assemble to receive their briefing. **1971** W. H. Cracknell in *Profile Warship* III. 55/2 Below the flight deck..was the gallery deck. Here were a rudimentary combat intelligence centre, squadron ready rooms, and other air department offices. **1977** *Time* 2 May 1/1 Stamped indelibly inside my head is what used to be on Navy ready-room walls—'Flying itself is not inherently dangerous, but like the sea, it is unmercifully unforgiving of human error.'

6. d. *U.S. Blacks.* (See quots.)

1967 J. Horton in *Trans-Action* Apr. 5/2 One either knows 'what's happening' on the street, or he is a 'lame', 'out of it', 'not ready' (lacks his diploma in street knowledge), a 'square'. **1970** C. Major *Dict. Afro-Amer. Slang* 96 *Ready*, hip; receptive. **1973** T. Kochman *Rappin' & Stylin' Out* 163 Another term such as 'ready' is descriptive of the person who 'has his diploma in street knowledge', which means knowing what's happening, taking advantage of opportunities, avoiding pitfalls, and being prepared to move where the action is.

IV. 16. a. (Further examples.)

1767 J. Woodforde *Diary* 24 July (1924) I. 64 My father sent me down a couple of fowls ready roasted. **1775** P. Freneau *Poems* (1902) I. 169 How could the wretches help but marching on, When at their backs your swords were ready drawn? **1796** J. Woodforde *Diary* 15 Oct. (1929) IV. 314 They have let their House ready furnished to a Revd. Mr. Beevor. **1868** *E. Acton's Mod. Cookery* (rev. ed.) xxx. 590 Never purchase it [*sc.* coffee] ready ground unless compelled to do so. **1930** H. Nicolson *Let.* 2 Jan. (1966) 40, I suppose that I shall get into the way of finding these paragraphs leaping ready-armed to the mind. **1952** J. B. Oldham *Eng. Blind-Stamped Bindings* 3 Sale of books ready-bound. **1960** *Farmer & Stockbreeder* 23 Feb. 123/3 Sold, ready-cooked in foil containers, the pies are in two sizes. **1976** *Glasgow Herald* 26 Nov. 16/3 Food can be bought ready prepared if not for the table at least for the oven or the saucepan.

b. *ready-carved, -cooked, -folded, -ground, -prepared* (further examples), *-roasted, -shelled, -sliced, -traced, -trained, -written* (later example). (See also *Ready-mixed *a.*)

1803 M. Wilmot *Let.* 31 May in Londonderry & Hyde *Russian Jrnls.* (1934) i. 13 Then ready carved bouillé, ready carved fricasées etc. **1974** L. Deighton *Spy Story* xi. 112 She put some ready-cooked pizzas into the oven. **1964** *McCall's Sewing* xiii. 234/2 *Ready-folded braid*, these braids are of a woven bias construction. **1960** *Farmer & Stockbreeder* 15 Mar. (Suppl.) 11/2 Put the almonds through a nut mill (neither minced nor ready-ground almonds will do.) **1875** T. Seaton *Fret Cutting* vii. 71 The high price he has had to pay..for ready-prepared wood. **1959** *Times* 9 Mar. (Suppl.) p. vii/7 Ready-prepared vegetables and other fresh produce are now being offered in the supermarkets. **1754** Richardson *Grandison* IV. xviii. 144 He makes her..become herself the cat's paw to help him to the ready-roasted chestnuts. **1909** *Daily Chron.* 14 Dec. 6/3 Pound boxes of ready-shelled walnuts at 1s. 3d. **1958** *Times Lit. Suppl.* 23 May 281/2 Bud Floyd and his wife Debbie tortured in their antiseptic, ready-shelled, air-conditioned inferno. **1960** *Times* 11 Feb. 3/5 Every day on television one can see the new styles ingeniously used to advertise instant coffee and ready-sliced bread. **1977** J. Wainwright *Day of Peppercorn Kill* 17 Ready-sliced bread—because if she ever handled a bread-knife she'd cut her damn-fool fingers off. **1967** E. Short *Embroidery & Fabric Collage* iv. 94 As the firm also sold ready traced materials and supplied the threads for working them, Morris's influence was widespread. **1946** *Nature* 5 Oct. 491/1 Sir Reginald pointed out first that in research on the problems of an old traditional industry there are usually no ready-trained scientific workers. **1977** *Belfast Tel.* 22 Feb. 4/1 (Advt.), With the full library of ready-written, ready-to-use application packages available in the U.K., the System Ten 220 Series can be quickly harnessed to your work.

d. Const. with infin. or with *for* and following noun, and used as attrib. phrases expressing preparedness for the action indicated, as *ready-to-eat, -use,* etc. (See also *Ready-to-wear *a.* (and *sb.*).)

1887 G. M. Hopkins *Let.* 12 May (1956) 379 Publishers 'tapped a stratum'..of almost untouched reading or ready-to-read public. **1897** *Sears, Roebuck Catal.* 294/3 (*heading*) Ready-to-use table cloths. **1907** *N.Y. Times* 14 Sept. 4 Through this store's efforts a new attitude toward ready-for-service clothing has been adopted by hundreds of men. **1909** H. N. Casson *C. H. McCormick* 237 Certain ready-to-eat foods are now being made from wheat. *c* **1938** *Fortnum & Mason Price List* 72/1 'Ready to serve' dishes. **1959** *Times* 9 Mar. (Suppl.) p. vi/3 Preparing cartons of frozen ready-to-cook chickens. **1972** *Listener* 7 Sept. 292/3 A range of ready-to-serve fish dishes with shrimp and lobster sauces. **1976** N. Roberts *Face of France* v. 66 A civilisation of leisure will be raring for ready-to-eat pork products. *a* **1977** *Harrison Mayer Ltd. Catal.* 98/2, 1 of 0·28 litre pack of the following liquid, ready-to-use glazes.

e. With sbs. used attrib. in sense 7, as *ready-reference, -use*.

1928 G. Campbell *My Mystery Ships* xiv. 250 A lucky shot from her might 'touch off' any of the ready-use ammunition which was at the guns. **1955** 'N. Shute' *Requiem for Wren* iii. 69, I give her a coconut out of the ready-use locker. **1963** *Amer. N. & Q.* Jan. 77/1 The book..can be used as a ready reference tool to answer questions about nearly all basic biological research in modern times. **1971** *Engineering* Apr. 122/2 Here's the tough, expanding, ready-reference file.

17. a. *ready-hearted*; also *ready-smiling*.

1937 A. L. Rowse *Sir Richard Grenville* ii. 31 So ready-hearted, so busy and generous about life's affairs. **1940** Blunden *Poems 1930–40* 252 Bright-tressed, ready-smiling, April-eyed.

C. *sb.* **1.** Also in *pl.*, bank notes.

1937 Partridge *Dict. Slang* 690/2 *Readies*, money in bank..notes. **1968** [see *Green *sb.* 7 d]. **1974** D. Francis *Knock Down* xiv. 157 He sort of winks at me and gives me a thousand quid in readies. **1977** *Private Eye* 4 Mar. 9/1 Send £50 to the address below, preferably in readies.

2. a. Now usu. in phr. *at* (*the*) *ready*. Also *transf.* (Further examples.)

1931 A. Curtayne *St. Anthony of Padua* viii. 78 Galloping full tilt with visor down and lance couched at the ready. **1955** *Times* 22 July 10/5 Others were more cautious, with fur capes or dark coats over their frocks and umbrellas at the ready. **1978** J. Carroll *Mortal Friends* i. iii. 30 The troops in the lorries were standing, rifles at ready.

b. *U.S. colloq.* The condition of being prepared to start (something). Freq. in phr. *to get a good ready* and varr., to assume a favourable stance or position for this. Also *transf.* Now *rare*.

[**1855** 'Q. K. Philander Doesticks' *Doesticks; what he Says* xviii. 153 The music got 'good ready' for a fair start, and at the word 'go' they went.] **1872** 'Mark Twain' in *Buyers' Man. & Business Guide* 76, I could have ketched them cats if I had had on a good ready. **1878** B. F. Taylor *Between Gates* 71 A time hardly long enough for a century plant to get a good ready for blossoming. **1897** A. H. Lewis *Wolfville* i. 2 So we begins to draw in our belts an' get a big ready.

3. *Rope-making.* A strand in a rope or cable.

1857 R. Chapman *Treat. Ropemaking* 84 The only method to be obtained is to give one turn or twist to the strand or readie, while the machine draws it a certain length. **1883** *Man. Seamanship for Boys' Training Ships* (Admiralty) (1886) 125 You now commence to form the long-splice, by unlaying one strand, and filling up the space it leaves with the opposite strand next to it..these strands being composed of three small strands, which are called readies. **1957** *Encycl. Brit.* XIX. 546/1 As the strand is twisted it is wound on a large reel and appears as a smooth, round strand composed of a number of individual yarns. This is known as the 'ready'.

4. *ready-up*, a conspiracy or swindle; a case of fraudulent manipulation; a fake. Cf. Ready *v.* 4 b. *Austral. slang.*

1924 *Truth* (Sydney) 27 Apr. 6 *Ready up*, a fake. **1926** J. Doone *Timely Tips for New Australians* 7 *Ready-up*, a conspiracy. **1945** Baker *Austral. Lang.* xv. 267 *Ready-up*, a case in which illegal methods are used to influence the outcome of a decision or an action. **1961** H. R. F. Keating *Rush on Ultimate* v. 85, I don't accept all the pretences and ready-ups you people put out.

ready, *v.* Add: **1. b.** *intr.* or *absol.* To make oneself ready or prepare in any way. *U.S.*

1967 *Wall St. Jrnl.* 12 Dec. 1 Machinists Union President Roy Siemiller, readying for aerospace bargaining, and Steelworkers chief I. W. Abel feel they must match the big rubber and auto settlements. **1972** *Time* 17 Apr. 22 (*caption*) In a cloud of catapult steam, a U.S. jet readies to attack Viet Nam.

3. a. For 'Now only *dial.*' in Dict. read 'Now chiefly *N. Amer.*' and add further examples. Also const. *up*.

Perh. influenced in early use by Redd *v.*¹ 6, to put in order, make neat.

a **1849** J. Keegan *Legends & Poems* (1907) 111 Hould your gob, and go ready the room for the dacint boys to sit down. **1864** *Harper's Mag.* Apr. 616/1 The pot ought to be a-bilin' for dinner, and the Kitchen to be readied up. **1902** H. Lawson *On Track* 73 Anyway, she'll have one readied up somehow. **1926** *Amer. Mercury* Dec. 464/2 The report that a certain man 'authored the show, which will be *readied* next Fall, when it will be *hoked* up (from hokum)'. **1934** *Bulletin* (Sydney) 25 July 38/1 This has invited the All-Blacks to call in on their way back from their tour of Britain next year and get what is being readied up for them. **1937** V. McNabb *God's Way of Mercy* xii. 185 You can make your home a world to ready your soul for the doings of that great new world. **1958** *Observer* 2 Mar. 6/6 It can be readied in 15—in less if we have some warning—not in two hours. **1959** H. E. Watters *Check List Canad. Lit.* p. xii, To the Editor, Miss Francess G. Halpenny, who readied the manuscript for the printers, I am heavily indebted. **1968** 'E. Lathen' *Stitch in Time* xvi. 141 The tax people were readying an attack on Martin. **1969** *Sun* 22 July 38 American scientists are readying the first space station for lift-off into earth orbit in 1971. **1971** J. Z. Young *Introd. Study Man* iii. 28 These..go back, readying the system for further signalling. **1978** J. A. Michener *Chesapeake* 509 She asked her slaves to ready the small sloop. **1979** *Arizona Daily Star* 5 Aug. A 10/2 An army of scientists and engineers readied a 2-mile floating defense line yesterday.

c. Of a person. Also const. *up*. Now chiefly *N. Amer.*

1846 *Swell's Night Guide* 78 He's to be readied at any downey move, and knows how to work it. **1895** J. Barlow *Strangers at Lisconnel* 205 Nothin' else 'ud suit them except gettin' all readied up for us to be skinkin' out in the evenin' late. **1904** H. Lawson *Over Sliprails* 161 The girl's relations..had a parson readied up, and they were married the same day. **1924** J. Galsworthy *White Monkey* II. ix. 198 I'll put you wise about our authors, and ready you up to go before the Board. **1940** *Sun* (Baltimore) 27 Feb. 12/4 Johnny Paychek already has been matched with him. Lee Savold is being readied for a shot at him. **1968** *Globe & Mail* (Toronto) 17 Feb. 4 (Advt.), An executype..to ready you for a party, or extra work.

4. a. (Further example.)

1925 P. Wallace *Major* iv. 58 He sat..in his office..deploring inwardly the tendency of owners to 'ready' their horses for Epsom.

b. (Further example.)

1933 *Bulletin* (Sydney) 15 Nov. 33/1 All readied-up, I thought, though not bad fun.

ready-made, *ppl. phr.*, *a.*, and *sb.* Add: **3. b.** (Earlier example.)

1777 P. Thicknesse *Year's Journey* II. iii. 153 The principal manufacture of the city [*sc.* Paris] is *ready-made love*.

5. Also *transf.* and *fig.* (Earlier and later examples.)

1831 M. Edgeworth *Let.* 29 Mar. (1971) 501 Then she is so fond of..her own family. She seems as if she was a ready-made family for Fanny. **1905** *Daily Chron.* 22 Nov. 10/2 Wholesale manufacturers and confectioners rejoice greatly as they see their trade in Christmas 'ready-mades' annually swelling. **1933** C. Spring *Fatality in Fleet St.* viii. 86 He looked like a film Cossack..ready-made. **1967** *Economist* 10 June 1101/2 The typical Italian climbing shop cannot afford to carry a big enough range to demonstrate the advantages of readymades.

b. The term introduced by Marcel Duchamp (1887–1968), French artist, to denote representatives of a dadaistic art-form created by him, in which simple manufactured objects are exhibited as works of art; the art-form itself.

[**1915** M. Duchamp in A. Schwarz *Marcel Duchamp* (1969) 54 Précises les 'Readymades'. En projetant pour un moment à venir..'d'inscrire un readymade'.] **1933** D. Gascoyne *Short Survey of Surrealism* ii. 28 Such was Marcel Duchamp's disgust for 'art' that he invented a new form of expression, which he called *Ready-Made*. A Ready-Made was any manufactured object that the artist liked to choose. **1958** *Times* 30 May 3/7 The 'ready-mades' in which Marcel Duchamp parodied the exhibition work of art, signing his name on the specially chosen article—a bicycle wheel, or a snow shovel. **1968** *N.Y. City* (Michelin Guide) 17 Marcel Duchamp..created his provocative 'ready-mades' which consist of simple objects, which the artist declared to be works of art (having intervened only to give them a name) thus, to be displayed. A Fountain, which was..a can bought in a department store. **1972** C. W. E. Bigsby *Dada & Surrealism* ii. 11 The ready-mades were those objects which he [*sc.* Marcel Duchamp] himself had selected as being commonplace.

rea·dy-mix, *a.* (and *sb.*) orig. *U.S.* [f. Ready *a.* + Mix *sb.*¹] = *Ready-mixed *a.* (and *sb.*) Also *fig.*

1950 M. McCarthy in *Reporter* 18 July 39/1 The ready-mix cake turns out 'terrific'. **1958** [see *Instant *a.* 4 c]. **1963** *Punch* 28 Aug. 322/3 A studio audience complete with ready-mix audio-visual sound programme. **1966** *Guardian* 28 Apr. 4/2 Factory food, in the form of ready-mixes, was certainly better than that produced by 'the multitude of indifferent cooks..now wandering around the country'. **1976** E. Ward *Hanged Man* iii. 13 A ready-mix concrete truck marked Metro Concrete jolted along the access road. **1980** *Times Lit. Suppl.* 30 May 608/4 Germaine is the principal new ingredient in the Barthian narrative ready-mix.

ready-mixed, *a.* (and *sb.*) orig. *U.S.* [f. Ready *a.* 16.] Of paints, concrete, and other artificial compound substances, having some or all of the constituents already mixed together. Also *ellipt.* as *sb.*

1895 *Montgomery Ward Catal.* Spring & Summer 254/2 Seroco Ready Mixed Paints and Enamels. **1908** *Sears, Roebuck Catal.* 71/1, 85 cents per gallon buys our guaranteed Seroco Ready Mixed House Paint. **1930** *Engineering* 16 May 631/3 The practice of selling ready-mixed concrete was introduced in the United States about ten years ago. **1960** *Farmer & Stockbreeder* 22 Mar. (Suppl.) 10/2 You can make up your own flooring compounds, but probably the simplest way is to buy a tin of a ready-mixed material. **1973** M. Truman [illegible] The Ready-Mixed Concrete Company which in the past had been used almost exclusively by contractors paving Jackson County roads. In Judge Truman's 125-mile road-building program, only three-fourths of a mile were paved with Ready-Mixed. **1977** J. I. M. Stewart *Madonna of Astrolabe* iii. 51 To what extent should we be justified in pouring unexpected wealth, as if it were so much ready-mixed concrete, into shoring up a mere sense-of-beauty-occasioning object?

ready-to-wear, *a.* (and *sb.*) orig. *U.S.* Also **ready-for-wear.** [f. *Ready *a.* 16 d.] **1.** Of clothing: = Ready-made *a.* 2, 4.

1895 *Montgomery Ward Catal.* 276/1 (*heading*) Ready to wear clothing. **1905** *Daily Chron.* 27 Feb. 8/3 A more exclusive type of ready-to-wear hat is the..sailor turban, toque, or narrow boat shape. **1930** *Times* 17 Mar. 9/4 The ready-to-wear sections bring the new styles within the reach of modest purses. A new spring catalogue giving illustrations of their ready-for-wear clothes has been prepared. **1963** *Amer. Speech* XXXVIII. 45 ..announced the premature birth of Neiman-Marcus as 'the South's finest and only exclusive women's ready-to-wear shop'. **1977** M. Sokolosky in *R. Merle's Virility Factor* 81 ..He has a Ready-Mixed ..ready-to-wear business. **1981** *Country Life* 12 Feb. 414/2 The ready-to-wear collections for spring and summer.

2. as *sb.* (An article of) ready-made clothing. Chiefly *pl.*

1933 *Blackw. Mag.* Dec. 502/2 There was a young person, looking quite chic in a suit of ready-to-wears. **1973** J. Ashford *Double Run* iii. 21 His clothes had the hanging look of cheap ready-to-wears which had gone half. **1974** *Country Life* 31 Jan. 54/3 He has private clients, and is..selling ready-to-wear in boutiques. **1977** *South China Morning Post* (Hong Kong) 13 Apr. 14/8 A vice-president of Bloomingdales..commented that the ready-to-wear is wonderful for the film industry.

reafference (riæ·fĕrĕns). [ad. G. *reafferenz* (von Holst and Mittelstaedt 1950, in *Naturwissenschaften* XXXVII. 464).] Cf. Afferent *a.* Sensory stimulation in which the stimulus changes as a result of the individual's movements in response to it.

1965 *Sci. Amer.* Nov. 85/1 A key to its operation is the availability of 'reafference'. This word was coined by the German physiologists Erich von Holst and Horst Mittelstaedt to describe neural excitation following sensory stimulation that is systematically dependent on movements initiated by the sensing animal. **1973** J. S. Bruner *Beyond Information Given* (1974) xiv. 249 The crucial issue in the regulation of intentional action is the corrective comparison of information that is fed back as a result of action that has been initiated, using the difference between the two as a basis of correction. It is immediately apparent that this is the concept of reafference as a source of regulation in behavior. **1973** R. Martin tr. von Holst's *Behav. Physiol. of Anim. & Men* I. iii. 141 Rather than asking about the relationship between a given afference and the evoked efference (i.e. about the reflex), we set out *in the opposite direction*..asking: What happens in the CNS with the afference (referred to as the 'reafference') which is evoked through the effectors and receptors by the efference? **1973** C. D. Kernig *Marxism, Communism & Western Soc.* VII. 103/1 As early as 1955 Anokhin..had discovered a phenomenon corresponding to the reafference principle. **1979** *Sci. Amer.* Sept. 138/3 Holst..pointed out..it has been suggested that reafference is necessary for the development of accurate visually guided spatial behavior.

re-a·fferent, *a.* [f. Re- + Afferent *a.*] Of or pertaining to reafference.

1965 *Sci. Amer.* Nov. 87/1 This kind of movement, with its continued reafferent visual stimulation, is the critical factor in compensating for displaced visual images. **1965** M. Cronkey & [illegible] *Theory of Types* i. 5 Under certain circumstances reafferent stimulation (that is, stimulation resulting from muscular activity) is a prerequisite to the development of a concept of visual space. **1971** R. Martin tr. *Lorenz's Stud. in Anim. & Hum. Behav.* II. 300 Such sensory signals directly produced by inherent movements of the organism are referred to as *reafferent signals*, and the signal relating to the corresponding motor command is transmitted to the perceptual mechanism is the efferent signal copy. **1973** R. Martin tr. von Holst's *Behav. Physiol. of Anim. & Men* I. iii. 149 There is absence of the reafferent 'stop'..which normally arrests the correcting movement. **1975** L. Ganz in A. H. Riesen *Developmental Neuropsychol. of Sensory Deprivation* vi. 198 They believe that the reafferent perception of a moving part of the body is essential for the development of coordination.

Reaganism (rēⁱ·gănizᵉm). [f. the name of Ronald W. *Reagan* (b. 1911), American Republican politician, Governor of California 1967–75, and President of the United States 1981– + -ism.] The policies and principles advocated by Ronald Reagan; adherence to or support of these. Also **Rea·ganite,** a supporter of Reagan; also *attrib.* or as *adj.*

1966 *Punch* 7 Dec. 832/3 Reaganism is really rampant now. **1975** *Financial Times* 20 Nov. 5/8 Wits have suggested that 'Reaganism is extremism in defence of Fordism'. **1976** *Times* 21 Feb. 6/7 The Reaganites..have moved on... If organization was the key to success, Mr Ford would never stand a chance. **1977** *Time* 3 Jan. 12 (*caption*) Reaganite at G.O.P. convention. **1981** *Economist* 31 Jan. 13/2 The right way lies through greater scope for bodies like the International Monetary Fund..which lends its money tied to almost Reaganite advice. **1981** *Times* 31 Jan. 13/3 Mrs Thatcher approximates on occasions to the tones of Reaganism.

reagent. Add: **1.** Now applied to any substance employed in chemical reactions (cf. sense 2). (Further examples.)

1904 L. W. Riggs *Elem. Man. Chem. Laboratory* III. 82 Any ion..when separated from other cations..will exhibit with reagents certain phenomena characteristic of that ion. **1929** C. R. Hayward *Outl. Metall. Pract.* xxii. 487 Recarburizing..may..be accomplished by adding the reagents to the converter. **1948** Currier & Rose *Gen. & Appl. Chem.* xix. 151 Nitric acid..is a very important substance: in the laboratory as an oxidizing reagent, and in industry for the manufacture of explosives. **1964** N. G. Clark *Mod. Org. Chem.* v. 65 Traditional reagents and reaction conditions rarely affect the paraffins, but they exhibit useful reactivity with the fluorocarbons. **1973** *Materials & Technol.* VI. viii. 325 To make a bonded joint, the PTFE is first etched with a reagent.

3. *Comb.* **reagent grade,** a grade of commercial chemicals characterized by a high standard of purity; freq. *attrib.*; **reagent paper,** paper treated with a reagent for use in chemical tests.

1935 *Jrnl. Amer. Chem. Soc.* LVII. 408/2 Baker and Adamson $CuSO_4.5H_2O$, Reagent Grade, was recrystallized twice from distilled water. **1945** *Ibid.* LXVII. 1096/1 Reagent grade ammonium sulfate was heated at 75° for several days. **1966** A. S. Prasad *Zinc Metabolism* xv. 275 Six..(the patients) received reagent grade zinc sulfate..and three..received reagent grade ferrous sulfate. **1973** *Jrnl. Amer. Chem. Soc.* XCV. 1913/1 All reagents and solvents used were reagent grade and, unless noted otherwise, were used without further purification. **1908** *Practitioner* Mar. 410 A pea-sized piece being rubbed up with 2 c.c. of water, and the reagent-paper dipped into this.

reaggregate, *v.* Add: **b.** *intr.* To come together again.

1962 H. Blumenthal et al. in Pirie *Lens Metabolism Rel. Cataract* 300 The subunits..reaggregate after removal of urea.

reagin (riˌēⁱ·dʒin). *Immunol.* [a. G. *reagin*, f. *reag-ieren* to react + -in -in¹.] **a.** The complement-fixing substance in the blood of persons with syphilis which is responsible for the positive response to the Wassermann reaction.

1911 R. W. Matson tr. *A. Wolff-Eisner's Clin. Immunity & Sero-Diagnosis* iii. 33 To avoid errors, it is..best to use the term 'reactive substances' (reagins) rather than 'antibodies', since the latter implies a neutralization in the sense of an antitoxin. **1915** J. E. R. McDonagh *Biol. & Treatm. Venereal Dis.* x. 71 Owing..to the fact that a positive Wassermann reaction may be obtained in conditions other than syphilitic ones, the reaction ceases to be a specific reaction. Therefore the third factor ought not to be called an antibody, since it is in no wise specific, hence it is best called reacting substance, or Reagin. **1937** H. Eagle *Lab. Diagnosis Syphilis* i. 24 There is reason to believe that Wassermann's first theory was correct, and that the active component of syphilitic serum, so-called reagin, may well be an antibody to *Spirochaeta pallida* despite its reactivity with normal tissue lipoids. **1942** *Jrnl. Lab. & Clin. Med.* XXVII. 729 It seems controversial as to whether reagin is an antibody to lipid haptens of the host..or an antibody to the spirochete. **1976** A. E. Wilkinson in Catterall & Nicol *Sexually Transmitted Dis.* 215 Although the function of reagin is still uncertain, its level seems to be thoroughly related to the amount of tissue reaction by the host, rising rapidly with increasing numbers of treponemes in early syphilis and later falling as the number of organisms declines with developing immunity.

b. The antibody which is involved in allergic reactions, causing the release of histamine and similar agents when it combines with antigen in tissue and capable of producing sensitivity to the antigen when introduced into the skin of a normal individual.

1925 [see *Atopy]. **1963** *Advances in Immunol.* III. 181 Reagin still represents a nebulous concept to many immunologists, some doubting the legitimacy of its classification as an antibody. **1969** R. S. Weiser et al. *Fund. Immunol.* xv. 163 The antibodies responsible for P-K type reactions, the so-called 'reagins' or P-K antibodies, have long been a mystery. **1977** *Nature* 16 June 618/1 Allergic diseases such as hay fever, extrinsic asthma, drug hypersensitivities and some forms of urticaria are mediated by allergen-specific antibodies of the IgE class, known also as reagins.

Hence **reagi·nic** *a.*, of, pertaining to, or being (a) reagin.

1931 A. F. Coca *Asthma & Hay Fever in Theory & Pract.* I. xii. 332 Bona fide reaginic reactions are indicative of..past, present or potential sensitivities in the atopic individual. **1945** *Vet. Rec.* LVII. 339/2 (*heading*) Reaginic allergy in cattle. **1975** *Nature* 6 Feb. 475/1 In the rat, reaginic antibodies were reported and their possible significance in *Schistosoma mansoni* infection considered. **1977** A. M. Denman in Holborow & Reeves *Immunol. in Med.* xi. 295 IgE constitutes the major class of reaginic antibody but it seems likely that some IgG..antibodies also contribute.

real, *a.*², *adv.*, and *sb.*³ Add: **A.** *adj.* **I. 1. c.** *real money*: (*b*) *colloq.* A large sum of money.

1918 R. W. Lardner *Treat 'em Rough* 120, I could go out and pitch baseball and make real money. **1939** A. Huxley *After Many Summer* i. iv. 46, I did some business this morning... Might make a lot of money. *Real* money. **1964** L. Deighton *Funeral in Berlin* iii. 21 'Whom do you feel like?' I liked that 'whom'—you've got to pay real money these days to get a secretary that could say that.

(*c*) *colloq.* The coinage or currency in which one habitually reckons, freq. as opp. to foreign currency.

1973 L. Meynell *Thirteen Trumpeters* iv. 50 So I'm paying one thousand seven hundred and ten lire for my Pimms?.. What's it mean in real money? **1977** *Vole* No. 2. 17/2 Just before the demise of real money and the introduction of decimal coinage, the officials of Gloucester Shoveha'penny League invested £10 in old-style halfpennies. **1977** *Zigzag* Mar. 7/1 They charged me three hundred francs. Well, that's..quite a lot in real money.

d. (Further examples.)

1875 *Encycl. Brit.* I. 544/2 Every quadratic equation has always two roots, real or imaginary. **1910** *Ibid.* I. 613/1 The development of the theory of equations leads to the amplification of real numbers, rational and irrational, positive and negative, by imaginary and complex numbers. **1952** S. C. Kleene *Introd. Metamath.* i. 6 That there are infinite sets considered in mathematics which cannot be enumerated was shown by Cantor's famous 'diagonal method'. The set of the real numbers is non-enumerable. **1965** Patterson & Rutherford *Elem. Abstract Algebra* iii. 85 The real number *a* is called the real part of the complex number (*a*, *b*) and the real number *b* is called the imaginary part. **1972** S. W. P. Steen *Math. Logic* iii. 178 Having defined the integers we can then define rational numbers as triplets of integers, then real numbers as Dedekind Sections of rational numbers and lastly complex numbers as ordered pairs of real numbers.

f. *real time*, the actual time during which a process or event occurs, esp. one analysed by a computer, in contrast to time subsequent to it when computer processing may be done, a recording replayed, or the like.

1953 *Math. Tables & Other Aids to Computation* VII. 73 With the advent of large-scale high-speed digital computers, there arises the question of their possible use in the solution of problems in 'real time', i.e., in conjunction with instruments receiving and responding to stimuli from the external environment. The criteria for satisfactory operation in such real-time service are different from those generally encountered. **1964** *Listener* 19 Nov. 784/1 A higher speed in computers means that their complexity can increase very rapidly, too, and that they can more easily engage in activities in what we call 'real time'. That is to say, they can calculate at the actual speed of the events taking place. **1968** *Times* 10 Dec. 6/8 The data gathered by the telescopes are stored on board the satellite by magnetic tapes and discs... The Smithsonian experiment can also be used in real time, transmitting information as it gathers it. **1970** *Nature* 20 June 1110/2 The data are telemetered to ground-based stations which record the information on magnetic tape and provide a digital print-out in real time. **1973** *Sci. Amer.* May 115/2 It is wrong to detail a suspense plot, even though we all recall from real time how *Apollo 13* limped back safely. **1973** *Nature* 12 Oct. 294/1 As we are working in scientific 'real time', we have to ask at what stage the work will be when filming is in progress. **1979** R. Hawkey *Side-Effect* xi. 83 The Real Time was three hundred milliseconds, but it was shot in slow motion.

attrib.

1953 [see above]. **1960** *N.Y. Times* 17 July 13/4 As an experiment, Air Force and Weather Bureau meteorologists attempted to use the pictures to make 'real time' forecasts of the weather—forecasts fresh enough to be useful. **1968** *Times* 1 Nov. 23/6 Computers have been slow to conquer the real-time control of industrial processes and traffic flows. **1970** O. Dopping *Computers & Data Processing* vi. 96 An example of a real-time process is a cheque account system in a bank where all transactions, e.g. withdrawals, are reported to the computer before they are finished. **1972** *Guardian* 9 Feb. 3/8 We do think we know how to develop a satellite with a near real-time (instantaneous) capability. **1975** *Offshore Engineer* Sept. 52/2 Sea & Storm is also..showing a wave data processing unit..which will give virtually realtime treatment of data from wave buoys. **1977** *Navy News* June 44 (Advt.), To undertake training of our customers' engineers/programmers/technicians on all aspects of software applicable to real-time radar systems.

2. a. Phr. *real life, real world* (passing into senses 3 and 4). Also *attrib.*

1771 T. Jefferson *Let.* 3 Aug. in Koch & Peden *Life & Selected Writings* (1944) viii. 358 Considering history as a moral exercise, her lessons would be too infrequent if confined to real life. **1801** M. Edgeworth *Belinda* I. iii. 70 Nothing is more unlike a novel than real life. **1836** Dickens *Pickw.* (1837) v. 44 A curious manuscript..curious as a leaf from the romance of real life. **1838** J. S. Mill in *Westm. Rev.* XXXI. 28 The writers and readers..in France have..a thirst for something which shall address itself to their real-life feelings. **1852** [in Dict.]. **1876** C. M. Yonge *Womankind* v. 34 Insolence to a governess is an old stock complaint. In real life, I never heard of it from anyone by birth and breeding a lady. **1909** *Daily Chron.* 16 Apr. 3/5 Jocelyn Johnstone..showed..humour in her sketches of..'real life' scenes. **1923** C. D. Broad *Sci. Thought* xiii. 536 Now, in real life, there are no examples of pure creation. **1937** 'G. Orwell' *Road to Wigan Pier* ix. 182 One could..give everything away, change one's name and start out with no money... But in real life nobody ever does that kind of thing. **1957** P. Suppes *Introd. Logic* xii. 286 Textbook problems (as opposed to real-life problems). **1963** *Amer. Speech* XXXVIII. 296 The instances in which its selection depends on real world context. **1966** *Listener* 19 May 727/1 The Vice Chancellor of Lancaster University strongly believes 'that the university must keep contact with the real world outside'. May I take this opportunity to ask..: (*a*) what is real about the real world? (*b*) why it is always outside? **1977** *National Observer* (U.S.) 15 Jan. 13/1 The roles each of us plays on the revolving stage of the real world have been well described. **1978** P. Marsh et al. *Rules of Disorder* ii. 33 In the perception of our non-academic pupils, school is..a waste of time..not..a part of their 'real' lives.

4. a. spec. *Econ.* Reckoned by purchasing power rather than monetary or nominal value.

1775 Johnson *Journey to Western Islands* 368 Lesley..related so punctiliously, that a hundred hen eggs, new laid, were sold in the Islands for a peny... Posterity has since grown wiser; and having learned, that nominal and real value may differ, they now tell no such stories. **1776** Adam Smith *Wealth of Nations* I. i. v. 39 Labour, like commodities, may be said to have a real and a nominal price. Its real price may be said to consist in the quantity of the necessaries and conveniences of life which are given for it; its nominal price, in the quantity of money. **1882** R. Bithell *Counting-House Dict.* 208 The nominal value of a coin is that value which is assigned to it by law, and often differs very materially from its *real* or *metallic* value. **1885** J. L. Joynes tr. *Marx's Wage-Labour & Capital* 10 The real wage expresses the price of labour in relation to the price of other commodities... Real wages may remain the same, or they may even rise, and yet the relative wages may none the less have fallen. **1929** *Soc. Sci. Abstracts* 23 The close similarity of the general price level..substantiates its use as a measure of 'real income'. **1936** K. A. H. Egerton *Dict. Econ. Terms* (ed. 2) 131 Real wages at such times change at a very different rate, and sometimes in the opposite direction, from nominal or money wages; being based on the purchases a wage at any given time will make. **1964** Gould & Kolb *Dict. Soc. Sci.* 454/1 If a series of national product estimates for several years is divided by a price index, each year's national product being divided by the price index for that year, the resulting series is known as *deflated* or *real* national product, or national product in *real terms*. **1976** *Glasgow Herald* 26 Nov. 1/6 Real earnings have fallen in the past few years and there is no way we can agree to any further reduction in the purchasing power of our members. **1981** *Sunday Times* 26 Apr. 13/4 Despite an urgent maintenance and restoration programme, it [*sc.* the National Trust] is spending less in real terms on looking after its property than it was two years ago.

b. *real ale*, a name for draught (or bottled) beer brewed from traditional ingredients, matured by secondary fermentation in the container from which it is dispensed, and served without the use of extraneous carbon dioxide; *real coffee*, coffee made directly from ground coffee beans, as opposed to 'instant' coffee.

1964 L. Deighton *Funeral in Berlin* iv. 281 Could you find us a little cup of real coffee? **1972** *What's Brewing* Oct., Mr A— B—..is ripping out the keg taps and replacing them with real ale from wooden barrels. **1973** C. Hutt *Death of Eng. Pub* i. 25 The beer-drinker who feels strongly about the declining quality of his pint has two organisations he can turn to..—the long-established Society for the Preservation of Beers from the Wood, and the more recently formed, more militant CAMRA (Campaign for Real Ale). **1974** N. Freeling *Dressing of Diamond* 201 'Where's the patrol, Gilbert?' 'Be back any minute.' 'Then you might make us some real coffee.' **1974** *Good Beer Guide* (CAMRA) 2 The real ale we are talking about has to stand up to three tests; in the way it is brewed, the way it is stored and the way it is served. **1976** *Evening Standard* 29 Dec., The most popular of about a dozen real ales brought in from distant parts to a growing number of pubs in the capital. **1980** *Times* 23 Sept. 8/3 In the 1970s..the 'real ale' fashion took hold.

c. (*a*) (Further examples.)

1846 *Punch* 20 June 272/2 You, who will not subscribe to the real thing; come, pull out your purse to the *name*:..although you know that you ask for the 'Ragged Schools',..beg subscriptions for the 'Youths of Limited Circumstances'. **1884** *Art Amateur* Dec. 23/1 Those persons who indulge in..having..Japanese rooms in their houses, but have only a ludicrous imitation, will be interested in seeing here the real thing. **1902** T. W. H. Crosland *Outlook Odes* 31 My tobacco merchant, who sells me two ounces of the real thing every week. **1939** *War Illustr.* 2 Dec. 365 The 'stand-by' atmosphere of the first few weeks of war may be lost at any moment in the urgency of the 'real thing'. **1977** *Time* 22 Aug. 40/2 But the copied Coke may not work. India's soft-drink fanciers have learned to distinguish between ersatz Coke, which is peddled everywhere on the Indian market, and the Real Thing.

(*b*) *spec.* True love as distinct from infatuation, flirtation, etc.

1857 C. M. Yonge *Dynevor Terrace* I. xi. 173, I could not part with you where we were not sure the 'real thing' was felt for you. **1906** J. Galsworthy *Man of Property* III. iii. 302 This was none of those affairs of a season that distract men and women about town... This was the real thing! **1919** Wodehouse *Damsel in Distress* v. 61 It had come at last. The Real Thing. George had never been in love before. Not really in love. **1931** J. Cannan *High Table* x. 152 He was afraid that she would think he was just flirting—that it wasn't the Real Thing. **1941** M. McCarthy in *Partisan Rev.* VIII. 327 All that conjugal tenderness had been a brightly packaged substitute for the Real Thing. **1955** E. Waugh *Officers & Gentlemen* II. ii. 189, I thought of you at the last. Ever since we met I've known I had found the real thing. **1960** *Woman's Own* 19 Mar. 17/2 Once these phases are over, you should be ready for the Real Thing..the man who will be exactly right for *you*. **1973** G. Scott *Water Horse* (1974) xvi. 109 A girl..whom she knew to be looking for the Real Thing in Spain.

d. *the real McCoy*: see *McCoy.

e. *real tennis* = Tennis *sb.* 1. Also *attrib.*

The usage distinguishes the original game from the modified form which became the more popular after 1874; see

Lawn-tennis and Tennis *sb.* 2. Derivation from Real *a.*¹ is a folk etymology.

1880 [see *jeu de paume* s.v. *Jeu c]. **1902** [see *royal tennis* s.v. *Royal *a.* 15 a]. **1954** A. S. C. Ross in *Neuphilol. Mitteilungen* LV. 22 The games of real tennis and piquet..are still perhaps marks of the upper class. **1966** *Oxford Mag.* Michaelmas, No. 8, p. 149 A splendid exhibition of real tennis was given at the Merton Street court on Sunday. **1972** *Daily Tel.* (Colour Suppl.) 14 Jan. 25/4 There are 17 Real Tennis courts in use in the country. **1975** *Country Life* 30 Jan. 258/1 Today tennis means to most people lawn tennis, while its..ancestor..has survived under the title 'real tennis'... Only one public school boasts a real tennis court... Canford in Dorset.

5. c. (Later examples.) Also *loosely*, aware of, or in touch with, real life.

1961 *Noble Savage* Fall 12 He [*sc.* Seymour Krim] alludes to something called 'direct writing', and he finds that criticism gets in the way of his 'truer, realer, imaginative bounce'. **1964** *Sunday Express* 1 Mar. 22/5 Most [actors]..are so insincere... Albie..is an exception... He's a real person. **1966** *New Statesman* 17 June 873/2 This was a realler America than I had known in the past, hitching on this or that bandwagon or presidential campaign. **1967** P. Welles *Babyhip* (1969) xxviii. 179 Sometimes I wish I were back in Paris. The people seemed realer. **1969** *Newsweek* 9 June 95 Why suffer all the bad hotels and rotten food and accountants and taxes if you waste the opportunity on stage to be real? **1973** *Scotsman* 7 Aug. 8/5, I notice..the editor-designate of the much discussed 'Scottish International' review telling us that Glasgow is a 'realler' city than Edinburgh. **1976** *New Yorker* 1 Mar. 35/1 'Ellen. Be real for once. I said we'd get together.' 'In your letter you said we'd have dinner.' **1977** *Time* 25 July 45/2 Billy is very sweet and very gentle and very real.

6. c. *real estate*: also *attrib.*

1840 *Spirit of Times* 25 Jan. 562/1 A negro, the holder of a ticket in the grand real estate lottery.., came pushing into a lottery office in great excitement. **1843** *Niles' Nat. Reg.* 4 Mar. 5 Real estate bank... A committee of the legislature of Arkansas have reported the facts connected with the management of this institution. **1849** *Knickerbocker* XXXIII. 174 His father had recently made some heavy real-estate purchases. **1854** H. D. Thoreau *Walden* 88 This experience entitled me to be regarded as a sort of real-estate broker by my friends. **1880** *Harper's Mag.* Sept. 562 This region was ..seized upon by real-estate speculators. **1892** Kipling *Lett. of Travel* (1920) 85 The packed real-estate offices; the real-estate agents themselves. **1903** *Westm. Gaz.* 11 Sept. 2/3 The law might almost be forgiven for making no provision for dealing with real-estate-owning paupers. **1965** H. T. Ansoff *Corporate Strategy* (1968) vi. 104 A company which primarily buys and sells..may be an investment trust, a pension fund, or a real estate syndicate. **1969** *Sydney Morning Herald* 24 May 30/1 (Advt.), The Real Estate Institute of New South Wales..will commence the next evening course of lectures in Real Estate and Valuation Practice. **1972** *Accountant* 17 Aug. 193/2 The cannibalization of assets, particularly of real estate subsidiaries. **1978** S. Brill *Teamsters* vi. 208 He sincerely believed that real-estate investments were the gold mines of the future.

III. 11. *real-seeming* adj.

1948 E. Bowen *Why do I Write?* 24 You and I, by writing a story, impose shape—on fictitious life, it's true, but on life that is real-seeming enough to be familiar and recognisable. **1951** S. Spender *World within World* 310 Reader-writer walk together in a real-seeming dream-alliance leading into gardens inhabited by Stephen Daedalus and Marcel.

B. *adv.* **1.** (with adjs.) For 'chiefly *Sc.* and *U.S.*' read 'orig. *Sc.* and *U.S.*' and add further examples.

Not common in standard use in southern England.

1939 *War Illustr.* 28 Oct. 219/1 If I had not been on fire I could easily have shot down two more. It was real bad luck, but my pals accounted for three besides the one I hit. **1943** K. Tennant *Ride on Stranger* viii. 77 He's real clever. **1959** J. Ludwig in *Tamarack Rev.* Summer 7 Some day she'd get real tough with her son Sidney. **1968** *Globe & Mail* (Toronto) 17 Feb. 50/3 (Advt.), Austin Healey Sprite black, radio, a real nice car. **1968** K. Weatherly *Roo Shooter* 111 It was real heavy going, and I must have dried the flamin' plugs and points twenty times. **1976** *Daily Mirror* 18 Mar. 24/4 I'm havin' a rest—I feel real listless.

2. (with advbs.) *colloq.* (chiefly *N. Amer.* and *Austral.*).

1893 H. A. Shands *Some Peculiarities of Speech in Mississippi* 52 Real down... Used by cultivated whites to mean *exceedingly* or *extremely*. A thing that is extremely nice is said to be *real down* nice. **1924** J. C. French *Writing* x. 290 Avoid: They live *good* in that camp (say live *well*), I *sure* will write *real* soon (say *surely* will, *really* soon). **1933** R. L. Pooley in *Amer. Speech* VIII. 61/2 One such [grammarian], commenting on the sentence, 'I will write *real* soon,' corrects *real* to read *really*. This is utter nonsense. No one ever says *I will write really soon*... It simply isn't English. **1942** Z. N. Hurston in A. Dundes *Mother Wit* (1973) 225/1 De man looked at me real hard for dat. **1947** K. Tennant *Lost Haven* xix. 317 Everyone said she was lucky... Everything fell out 'real nice' for her. **1959** *Weekly Times* (Melbourne) 30 Sept. (Advt.), How about picking up your phone and asking your B.F.E. dealer to arrange a free demonstration of a '35' on your property real soon? **1967** G. Jackson *Let.* 13 July in *Soledad Brother* (1971) 121, I felt real bad about that. **1975** D. Lodge *Changing Places* ii. 57 You and I must have lunch together real soon.

C. *sb.*³ **4.** *Math.* A real number.

1866 W. R. Hamilton *Elem. Quaternions* I. i. 10 Such scalars are..simply the reals (or real quantities) of Algebra. **1940** W. V. Quine *Math. Logic* 273 The fact that every bounded class of reals has a least bound is the basic formal difference between the reals and the ratios. **1967** Condon & Odishaw *Handbk. Physics* (ed. 2) I. ii. 22/1 Apart from the reals and the complex numbers they are the only associative hypercomplex systems with real coefficients and no divisors of zero. **1979** *Proc. London Math. Soc.* XXXVIII. 367 Let x, h, Q denote large reals with $x \geqslant h > Q$.

D. In colloq. phr. *for real* (orig. *U.S.*). **a.** as *adj. phr.* Genuine, (in) earnest, true, sincere.

1956 [see *Pick *v.*¹ 20 p]. **1972** M. J. Bosse *Incident at Naha* 55 Was that kid for real? I mean, he didn't look as if he could fight his way out of a paper bag. **1973** *Black Panther* 17 Mar. 8/1 This is no 'scare tactic', it is for real. **1973** W. Fairchild *Swiss Arrangement* vii. 78, I told him I was staying with Mom and that's for real. **1977** B. Randall *Fan* 34 Dear Mr. Breen. Are you for real?.. Give us a rest.

b. *adv.* Really, truly, actually; in reality.

1962 J. Glenn in *Into Orbit* 183 Everyone seemed to sense that we were going for real this time. **1964** M. McLuhan *Understanding Media* (1967) II. xxx. 320 It was Hitler who gave radio the Orson Welles treatment for real. **1970** E. Bullins *Theme is Blackness* (1973) 179 Yeah! For real! Shoot me! That's what she says. **1972** J. E. Franklin in W. King *Black Short Story Anthol.* 355 'We'll let her in on our team, see?' 'For real?' 'Not for real, just play-like.' **1977** *R. Air Force News* 22 June–5 July 11/2 Two pilots have in fact done the job 'for real'—both Sqn Ldr Marshall and Flt Lt Dave Fischer have put Harriers down on the deck of HMS Bulwark.

real estate: see Real *a.*² 6 c in Dict. and Suppl.

‖ **realia** (reˌā·liă, riˌēⁱ·liă). [neut. pl. of late L. *reālis* actual, real.] **1.** *Educ.* Objects which may be used as teaching aids but were not made for the purpose. *N. Amer.*

1950 J. S. Kinder *Audio-Visual Materials & Teaching Techniques* xiii. 333 Realia include such items as objects, specimens, samples, relics, artifacts, souvenirs, and even models and dioramas. **1962** *Library Jrnl.* LXXXVII. 819/1 They ignore the miniature models, facsimiles, mock-ups, and realia which one used in construction of exhibitions and bulletin boards. **1970** J. Riddle et al. *Non-Book Materials* (prelim. ed.) 44 Realia are entered under title. The title may be taken from the box or accompanying data. Where no title can be found, the catalogues will supply one, e.g. [Robin's egg] (Realia). **1975** *Ontario Libr. Rev.* LIX. 243 The study encompassed seventeen types of audio and visual materials;..microforms, overhead projectors, sculpture replicas and realia, media kits and simulation games.

2. Real things, actual facts, esp. as distinct from theories about them.

1952 G. Sarton *Hist. Sci.* I. xiv. 362 The *realia*, we should always remember, cannot be learned in books but only in the practice of a scientific profession. *Ibid.* xix. 480 There is a real need of new translations..fully explained by a scholar..acquainted not only with philologic details but with all the realia expressed or implied. **1962** Y. Malkiel in Householder & Saporta *Probl. Lexicogr.* 5 Attention is focused on..strictly lexical data (at the expense of *realia*, proper names, and the like). **1975** *Times Lit. Suppl.* 25 Apr. 464/1 Takes Christian scholars to task for being too theological about Judaism, for ignoring its *realia*.

realign (riˌălai·n), *v.* [f. Re- 5 a + Align *v.*] **a.** *intr.* for *refl.* To fall into line again; to return to previously aligned positions.

1923 *Glasgow Herald* 14 Aug. 7 If France does not meet us half way, enabling the Allied front to realign, there will be a separate reply to Germany.

b. *trans.* To align again or anew. Also *fig.*

1957 V. W. Turner *Schism & Continuity in African Soc.* v. 131 Private intrigues provide means whereby individuals seek to realign the social structure in pursuit of their own advantage. **1967** Cox & Grose *Organization & Handling Bibl. Rec. by Computer* iv. 93 There is a Restructure program which realigns and copies the data base files. **1973** *Computers & Humanities* VII. 139 The texts are realigned at the place where the match was found.

realignment. For 'Chiefly *U.S.*' read 'orig. *U.S.*' and add further examples.

1929 P. Hughes *Catholic Question 1688–1829* III. ii. 197 A re-alignment in English political life suddenly brought uncertainty to an end. With that re-alignment the Catholic Question emerges. **1935** A. P. Herbert *What a Word!* iv. 92 Comment in the Lobbies centred round the re-alignment of, etc., etc. **1957** *Times Lit. Suppl.* 20 Dec. 766/2 Clear and virtually immeasurable material benefits, widespread re-alignments of jobs, changes of thinking and attitudes, and alteration of ways of working. **1959** *Economist* 3 Jan. 11/2 The chance thus arose of the concerted realignment of European currencies. **1964** *Ann. Reg. 1963* 78 Externally, political realignments with the Pacific and South East Asia continued, a number of official visits being exchanged. **1979** *Time* 2 Apr. 10/2 In the scramble for realignment in the face of the impending treaty, Hussein has even reconciled with Yasser Arafat..with whom he has been at bloody odds since 1970.

realism. Add: **1. a.** (Earlier and later examples.)

1826 R. Whately *Logic* 299 Nothing, perhaps, has contributed more to the error of Realism than inattention to this ambiguity. *Ibid.* 300 All these absurdities are in fact but the extreme and ultimate point of Realism. **1828** J. S. Mill in *Westm. Rev.* IX. 155 Their [*sc.* Aristotelian logicians'] classification of names according to the *mode of their signification* (of which the doctrine of the Predicables forms a part) when purified from the taint of Realism which adheres to the expression but without infecting the substance, constitutes a prodigious step in the theory of naming. **1948** C. Hartshorne *Divine Relativity* iii. 122 Even if one takes the 'conceptualist' solution of the problem of nominalism and realism, one need not therefore deny that God may have something corresponding to concepts. **1970** N. Wolterstorff *On Universals* vii. 170 Our position is that of realism and 'nominalism'. Is rapprochement between two such ancient armies possible?

b. (*a*) (Earlier examples.)

1830 W. Hamilton in *Edin. Rev.* LII. 169 If the veracity of consciousness be unconditionally admitted..the doctrine is established which we would call the scheme of Natural Realism or Natural Dualism. *Ibid.* 180 The scheme of Natural Realism, which it is Reid's immortal honour to have been the first..to embrace.

(*b*) In the 20th century, applied to philosophical theories reacting against 19th-century idealism which, while they agree in affirming that external objects exist independently of the mind, differ in their accounts of appearance, perception, and illusion. More recently (opp. to *verificationism), the theory that the world has a reality that transcends the mind's analytical capacity, and hence that propositions are to be assessed in terms of their truth to reality, rather than in terms of their verifiability. Cf. also *naïve realism* s.v. *Naïve *a.* 1 b, *New Realism.

The quotations are chosen to give some idea of the range and diversity of views in modern theories of realism.

1906 *Mind* XV. 308 Some of the leading supporters of the new Realism (especially Mr. Moore and Mr. Russell) connect it with an extremely nominalistic type of Logic... This Logic, however, seems to be quite capable of recognising types such as those of Plato. **1920** D. Drake et al. *Ess. Critical Realism* p. vi, Our realism is not a physically monistic realism, or a merely logical realism... To find an adjective that should connote the essential features of our brand of realism seemed chimerical, and we have contented ourselves with the vague, but accurate, phrase *critical realism*. **1920** J. Laird *Stud. in Realism* i. 13 Any realism of this kind, even if it defends common sense, defends a common sense which is very sophisticated indeed. **1938** G. D. Hicks *Critical Realism* p. xiii, Realism, as Professor Perry has defined it, stands for the principle that 'things may be, and are, directly experienced without owing either their being or their nature to that circumstance'. **1954** M. R. Cohen *Amer. Thought* ix. 271 The practical consequence of Peirce's realism is his sharp distinction between what is useful and what is true. **1963** J. Macquarrie *Twentieth-Cent. Relig. Thought* xvii. 259 In German realism, which stands rather apart from the Anglo-American brand, the influence of phenomenology is noticeable. **1967** *Encycl. Philos.* VII. 80/2 Representative realism also accounts for illusions, dreams, images, hallucinations, and the relativity of perception... [It] is the easiest inference from the scientific account of the causal processes up to the brain in all perceiving and fits other scientific evidence. **1969** L. W. Forguson in K. T. Fann *Symposium on J. L. Austin* III. 328 Austin's purpose with regard to both the sense-datum theory and philosophic realism was entirely negative... In his view, the problems..either aren't problems at all or..aren't philosophical..but scientific problems. **1972** K. R. Popper *Objective Knowl.* p. vii, While I am prepared to uphold to the last the essential truth of commonsense realism, I regard the commonsense theory of knowledge as a subjectivist blunder. **1973** M. Dummett *Frege: Philos. Lang.* xiii. 466 The fundamental tenet of realism is that any sentence on which a fully specific sense has been conferred has a determinate truth-value independently of our actual capacity to decide what truth-value is. **1979** *Sci. Amer.* Nov. 139/1 Realism can be stated formally as the belief that a mere description of data is not all that should be required of a theory.

2. c. orig. *U.S.* (*a*) In legal theory, the doctrine that the law is to be discovered by studying actual legal decisions and procedures, rather than by recourse to enactments or statutes; (*b*) more loosely in political theory, the view that actual political power is the subject-matter of politics, as opp. to doctrine, law, rights, or justice.

1930 J. Frank *Law & Modern Mind* v. 42 (*heading*) Legal realism. **1930** K. N. Llewellyn in *Columbia Law Rev.* XXX. 449 A sophisticated reversion to a sophisticated realism. Gone is the ancient assumption that law is because law is. *Ibid.* 461 (*heading*) Realism as to 'society'. **1951** J. H. Herz *Polit. Realism & Polit. Idealism* ii. 24 Political Realism has at all times insisted that the nature of politics is fundamentally determined by the struggle for power. **1959** B. Crick *Amer. Sci. of Polit.* v. 85 The popular character of pragmatic realism as reformism can be seen most vividly in the literature of the era. **1960** G. Schubert *Public Interest* iv. 148 A second thread of Legislative Realism emphasizes the effect of each chamber..in conditioning the behavior of the individual members. **1961** O. Kirchheimer *Polit. Justice* v. 216 Through the psychological variant of legal realism we have become conscious of all the numerous factors in the judge's personality structure which might become determinative factors of judicial action. **1962** B. C. Borning *Polit. & Soc. Thought C. A. Beard* ii. 19 The very spirit of science that stimulated Beard and others to aim at realism was closely related to the premise that man by using reason could not only understand but also control his environment. **1968** W. Twining *Karl Llewellyn Papers* I. ii. 5 At Oxford, American Realism was just another Aunt Sally. **1977** M. Clanchy in E. Attwooll *Perspectives in Jurisprudence* x. 176 Realism has been given a special meaning in jurisprudential thinking to distinguish between real rules of law and paper rules.

3. (Further examples.)

1894 C. L. Morgan *Psychol. for Teachers* ix. 203 Realism ..involves the introduction of such details as shall assimilate the representation to actual fact, and the incorporation of the results of generalisation in individual persons or concrete things. **1912** Laran & Gaston-Dreyfus *Courbet* 51 Gautier was astonished at seeing Realism in a shed. **1924** [see *Expressionism]. **1937** H. Read *Art & Society* iv. 180 My underlying contention, that there is an inherent contradiction between art and vulgarism (or, to confine ourselves to æsthetic terms, between art and realism). *Ibid.* vii. 247 There are, in fact, only these three basic modes—realism, idealism and expressionism. **1957** B. S. Myers *Art & Civilization* xxvi. 645 The late-nineteenth- and early-twentieth-century realism reappeared in the United States during the depression years after 1929 as a school of Social Realism. **1957** *Observer* 3 Nov. 14/2 Realism was Courbet's answer to the quarrelling schools of French Classicism (Ingres) and Romanticism (Delacroix). **1970** F. Martí-Ibáñez *Adventure of Art* xii. 571/1 Some interpreters regard personal intimacy as the catalyst for Picasso's return to realism. **1977** T. Neville *Challenge of Mod. Thought* xix. 117 It is precisely by means of this extreme realism that Kafka points the inadequacy of the known facts as a guide to ultimate truth.

realist, *sb.* (and *a.*). Add: **3. c.** One who adheres to or is influenced by principles of realism (sense *2 c).
1930 K. N. Llewellyn in *Columbia Law Rev.* XXX. 463 The problem calls for exploration, from the realist's angle, by cautious study of detail. **1954** M. R. Cohen *Amer. Thought* ii. 64 The realists insist that any theory of value that is not arbitrary must be based on actual experience. **1960** G. Schubert *Public Interest* iv. 142 Political scientists generally..appear to agree with those Realists who insist that there must be a general consensus to accept the decisions of public officials, if a democratic polity is to exist. **1971** Chambliss & Seidman *Law, Order & Power* i. 4 The American legal realists, who insisted that we must study the *law in action* as well as the *law in the books*. **1977** M. Clancy in E. Attwooll *Perspectives in Jurisprudence* x. 176 The historians of law will tend to be a realist.
4. (Later examples.)
1931 R. Pound in *Harvard Law Rev.* XLIV. 697, I approach the subject of the call for a realist jurisprudence..with some humility. **1959** Hart & Honoré *Causation in Law* iv. 92 The general scepticism as to the possibility of framing rules which developed into the 'Realist' movement of the 1930's. **1977** *Daedalus* Summer 58 We may discover that the realist paradigm, which stresses the primacy of foreign policy, has to be seriously amended, not only for the present but for the past.

realistic, *a.* Add: **1. a.** (Earlier and later examples).
1829 H. C. Robinson *Diary* 13 Aug. (1967) 102 [Goethe] repeated the remark which is one of his fixed ideas that it is by..facts that even a poetical view of nature is to be..authenticated... It is this which had made Goethe a *realistic* poet, as opposed to the idealism of such poetry as Wordsworth's. **1943** *Ulster: Brit. Bridgehead* (H.M.S.O.) 6 (*caption*) This realistic picture shows British troops in training in Northern Ireland. **1971** *Hi-Fi Sound* Feb. 67/3 High fidelity stereo at its most successful is wide-ranging and realistic, analytical and rich in detail.
2. a. (Later examples.)
1936 *Sun* (Baltimore) 26 Feb. 1/5 Mr. Eden, although doubting his wisdom, wished Chamberlain success in his 'realistic' search for lasting peace. **1962** *Listener* 12 Apr. 626/1 Neither a minimum property qualification nor even a simple educational test..would be realistic or just in the present situation of South Africa. **1963** *Observer* 3 Nov. 33/1 'Realistic' can vary from its theatre meaning of 'with damp washing' to its place at the top of the conjugation 'I am realistic, you have compromised, he has sold out'. **1973** *Howard Jrnl.* XIII. 321 Realistic payment of prisoners is a novel idea to students of the British penal system.
3. (Later examples.)
1843 J. S. Mill *Logic* I. i. viii. 197 The philosophers who overthrew Realism..retained long afterwards, in their own philosophy, numerous propositions which could only have a rational meaning as part of a Realistic system.

reality. Add: **7.** *attrib.* and *Comb.*, as *reality content, control, -revealer, value*; *reality-based, -centred* adjs.; **reality principle,** the principle propounded by Freud that the actual conditions of living modify the pleasure-seeking activity of the libido; **reality-testing,** the testing of an emotion or thought in a real-life context: also *attrib.*; hence **reality-test** *sb.* and *v.*, **-tested** *ppl. a.*
1960 L. Pincus *Marriage* i. 25 A challenge to move forward to fuller and more reality-based relations. **1964** *Listener* 19 Apr. 683/1 These concepts themselves keep the child's thinking 'reality-centred'. **1951** J. M. Fraser *Psychol.* x. 112 When we find someone in whose life..phantasy achievements occupy a very large place, we are probably justified in thinking that the reality-content of his motivation is a little low. **1949** 'G. Orwell' *Nineteen Eighty-Four* i. 37 Whatever was true now was true from everlasting to everlasting... All that was needed was an unending series of victories over your own memory. 'Reality control', they called it: in Newspeak, 'doublethink'. *Ibid.* 54 It's merely a question of self-discipline, reality-control. **1932** R. Macaulay *Dangerous Ages* xii. 236 Your eye is at present in..an impermanent stage in its struggle towards the adult level of the reality-principle. **1922** C. J. M. Hubback tr. *Freud's Beyond Pleasure Princ.* i. 5 Under the influence of the instinct of the ego for self-preservation it [*sc.* the pleasure-principle] is replaced by the 'reality-principle'. **1954** D. Riesman *Individualism Reconsidered* xxii. 345 By the reality principle alone, mankind could not be governed. **1957** N. Frye *Anat. of Criticism* (1971) ii. 75 In literature, what entertains is prior to what instructs,..the reality-principle is subordinate to the pleasure-principle. **1968** *Listener* 22 Feb. 244/1 The real world—where there are limits on what is possible..where..the reality principle operates. **1976** S. Hynes *Auden Generation* vi. 185 It is the existence of Europe, and not any political doctrine that is the reality principle here. **1962** A. Huxley *Island* ix. 136 Murugan calls it dope... We, on the contrary, give the stuff good names—the *moksha*-medicine, the reality-revealer, the truth-and-beauty pill. *Ibid.* 141 'Which is the easy way?' Will asked. 'Education and reality-revealers.' **1925** J. Riviere tr. *Freud's Papers on Metapsychol.* in *Coll. Papers* IV. 20 Their entire disregard of the reality-test. **1968** N. N. Holland in Levine & Madden *Art Victorian Prose* 333 The more we reality-test a work of literature, the more we become aware of the reality of ourselves as separate beings. **1960** L. Pincus *Marriage* i. 19 Some remnants of ego (reality-tested experience) may..be repressed. **1925** J. Riviere tr. *Freud's Papers on Metapsychol.* in *Coll. Papers* IV. 16 One mode of thought-activity was split off; it was kept free from reality-testing and remained subordinated to the pleasure-principle alone. **1953** J. Strachey tr. *Freud's Interpretation of Dreams* II. in *Compl. Wks.* V. 566 In other words it becomes evident that there must be a means of reality-testing (i.e. testing things to see whether they are real or not). **1955** H. Lasswell *Struct. of Metaphysics* ii. 72 These then contribute their psychic charges, which both intensifies our feeling of disquietude and weakens our reality-testing abilities. **1961** L. Pincus *Marriage* i. 20 The more healthy process of reality-testing and reality acceptance. **1974** S. A. Rennison *Psychol. Needs & Polit. Behav.* iv. 44 The principle of adequate reality testing allows us to come to grips with this problem. **1923** Reality-value [see *Image sb. 5 a]. **1961** *Listener* 23 Nov. 856/2 From the point of view of emancipation, as opposed that is to the point of view of truth or (as he called it) reality value, Freud was deeply affected by two considerations.

realizability. (Example.)
1975 *Nature* 20 Mar. p. xiv (Advt.), The emphasis throughout is on the theoretical foundations of optimum source synthesis, including conditions for physical realizability and mathematical methods for satisfying them.

realization. Add: **1. b.** (Later examples.)
1966 *Guardian* 1 Sept. 7/3 Ralph Ortiz, American and destroyer of pianos..doesn't call them happenings any more, he dreams them 'realisations'. **1976** *Southern Even. Echo* (Southampton) 6 Nov. 7/8 Another well-made film, John Huston's realisation of the Rudyard Kipling short story 'The Man Who Would Be King', is showing at the end of next week at the Palace, Bordon.
c. *Math.* An instance or embodiment of an abstract group as the set of symmetry operations or the like of some object or set.
1954 Beaumont & Ball *Introd. Mod. Algebra & Matrix Theory* iv. 133 Find a realization in geometry of the group consisting of the elements e, a, a^2, a^3, a^4 where $a^5 = e$... **1965** Patterson & Rutherford *Elem. Abstract Algebra* ii. 52 A particular member of an equivalence class is called a realization of the corresponding abstract group. **1979** *Proc. London Math. Soc.* XXXVIII. 260 But this coequalizer, if computed in [illegible] would yield precisely the geometric realization of [illegible].
d. *Statistics.* A particular series which might be generated by a specified random process.
1957 Kendall & Buckland *Dict. Stat. Terms* 242 A realization of a stochastic process $\{x_t\}$ is one of the series of values $(\ldots x_{n-2}, x_{n-1}, x_n, \ldots)$ to which it may give rise. The realization may be regarded as a 'member' of the process in the same way that an individual observation is regarded as a member of a population. **1965** Cox & Miller *Theory Stochastic Processes* i. 9 The realization illustrates the very irregular behaviour of the system. **1975** D. R. Brillinger *Time Series* ii. 18 ..will be described as a realization, trajectory, or sample path of the time series.
e. *Linguistics.* The phonetic, phonological, graphic, or syntactic manifestation of a linguistic unit, structure, or set of features. Also *attrib.*
1954 Pei & Gaynor *Dict. Linguistics* 6 *Actualization*, the perceptible result of the articulation of the phonemic variants or of the archiphoneme... Also called *realization*. **1966** *Language* XLII. 187 A transformation places an element in position, and a realization rule provides a spelling in terms of the alphabet of the neighboring component. **1971** R. Fowler in *Archivum Linguisticum* II. 131 ..misleadingly influenced by accidents of surface-structure realization. *Ibid.* 133 TG theory..offers a pattern for what I call 'realization rules'—rules which effect the transition between combinations of feature sets and strings of morphemes. **1971** B. Mafeni in J. Spencer *Eng. Lang. W. Afr.* 102 ..are spelt alike, but they are actually two different lexical and grammatical items with different phonetic realizations. **1974** *Language* L. 344 All speakers clearly simplify to a certain extent the phonetic realization of words derived from their own languages, and yet pronounce them with their sound structures more intact than do non-native speakers. **1977** *Canad. Jrnl. Linguistics* XXII. 198 Important questions remain about Halliday's model, particularly as regards the precise relationship between meaning potential and its realization at the level of form and the nature and details of the realization rules which connect the two.
4. *Mus.* The action of completing or enriching the texture of a piece of music left sparsely notated by a composer; the texture of music so completed or enriched.
1911 E. Newman tr. *Schweitzer's J. S. Bach* II. xxxv. 447 The only original instrument to be considered in connection with the realisation of the thorough-bass is the organ. **1946** *Penguin Mus. Mag.* Dec. 92 There is an atmosphere about these 'realisations' by Nadia Boulanger which is usually absent from our choral recordings. **1954** *Grove's Dict. Mus.* (ed. 5) VII. 69/1 *Realization*, a useful modern term for the setting forth of a thorough-bass in full harmony, with more or less elaborate textures, from a continuo part, either at sight in performance or in editing old music. **1958** *Listener* 4 Dec. 964/3 His [*sc.* Roman's] realization of Leo's *Dixit*, which he conducted in 1747, shows how far his predilection for the Italians went. **1959** *Collins Mus. Encycl.* 534/1 *Realization*, the act of completing the harmony of a 17th- or 18th-cent. work by providing a keyboard accompaniment based on the indications afforded by the figured bass. **1966** *Listener* 10 Mar. 364 Deryck Cooke's fine realization of Mahler's tenth symphony hardly received its due on Wednesday. **1980** *Early Music* Jan. 109/2 The continuo realization by Claire Caillard follows the principles of Saint Lambert.
Hence **realiza·tional** *a.*
1965 L. Northbourne tr. *Schuon's Light on Ancient Worlds* 66 The Biblical, mystical and 'realizational' character of Christianity. **1968** *Language* XLIV. 574 Working with [illegible], found the rewrite format surprisingly easy to set up. The realizational format was rather more difficult. **1972** *Ibid.* XLVIII. 373 The realizational rule that, in his analysis, provides the input for the vowel-reduction rule..would have to re-apply in order to convert *u* to *ü* again. *Ibid.* 408 The present article sets the frame for a wider study of suffixal *i* in Germanic by linking the realizational rules of underlying *ī* in two Germanic dialects—Old English and Gothic. **1979** *Trans. Philol. Soc.* 203 Vowel-invertors are much less varied than those of the consonants, but there is one realizational difference which suggests interesting patterns of historical change.

realize, *v.*[2] Add: **1. b.** (Earlier example.)
1769 A. Ferguson *Inst. of Moral Philos.* vi. v. 257 Providence..tends to realize imaginary wealth.
d. *Mus.* To complete a piece of music left sparsely notated by a composer; to enrich the texture of a work, esp. by orchestrating music written for a single voice or instrument.
1911 E. Newman tr. *Schweitzer's J. S. Bach* II. xxxv. 451 Our ears are different from those of Bach's day. Orchestra and chorus are much larger..if we realise the thorough-bass in the same scale it sounds too loud. **1947** A. Einstein *Mus. in Romantic Era* iv. 58 To interpret the role of the piano in orchestral form, to 'realize' it. **1962** A. Jacobs *New Dict. Mus.* 304 *Realize*, to work out in full and artistically such music as was originally left by its composer in a sparsely-notated condition... Though lacking the advantage of being self-explanatory, 'realize' is superior to 'arrange' in this context since it avoids the suggestion of altering the composer's intentions. **1980** *Early Music* Jan. 111/2 Other reconstruction work has involved realizing short score into full score (as in parts of the Overture in D).
3. b. Also, to become aware of the presence of (a person).
1926 H. S. Walpole *Dark Forest* ii. iv. 269 The moment I realized him I felt afraid.
realized *ppl. a.*, *spec.* in phr. *realized eschatology* Theol. (see quot. 1946).
1936 C. H. Dodd *Apostolic Preaching & its Devel.* iii. 156 This promise of a second coming is realized in the presence of the Paraclete..in the life of the Church... The evangelist, therefore, is deliberately subordinating the 'futurist' element in the eschatology of the early Church to the 'realized eschatology' which..was from the first the distinctive and controlling factor in the *kerygma*. **1946** E. L. Mascall *Christ, Christian, & Church* vi. 102 In recent years much stress has been laid upon the notion of 'realized eschatology'—the view that..the last Day and the Final Judgement are actually present in Christians now. **1977** G. W. H. Lampe *God as Spirit* i. 17 It has been argued convincingly that T. W. Manson, C. H. Dodd, and other exponents of 'realized eschatology' offered a one-sided interpretation of the N.T. evidence.

reallocate (riæ·lǝkeɪt), *v.* [f. Re- 5 a + Allocate *v.* 1.] *trans.* To apportion or assign again.
1957 *Economist* 7 Sept. 863/1 When the French Government reallocates next month the oil prospecting rights over a 25,000 square mile tract of the Sahara Desert, several major British and American oil companies are likely to be committed to carry out intensive exploration in the area. **1964** A. Battersby *Network Analysis* i. 9 It [*sc.* network analysis] reduces the examination of a large project to three stages:..(c) Re-allocating money or other resources to improve the schedule.

realloca·tion. [f. Re- 5 a + Allocation.] **a.** The action of apportioning or assigning again. **b.** A case or example of this.
1958 E. Gowers *Plain Words* 48 We might find that, even though the Board of Trade could still not resist announcing that certain surplus government factories are now 'available for re-allocation', they would at least put it that, 'In short,' they would add in a burst of confidence, 'they are to be re-let'. **1959** *Farmer & Stockbreeder* 16 Feb. 121/3 Should there be any re-allocation of profitable pennies from corn, milk and sugar beet, the cattle producer is first in the queue. **1962** A. Battersby *Guide to Stock Control* 108 No figure in the third column is less than any in the fourth column; this means any further re-allocation of orders must of necessity increase the total stock value. **1964** *Ann. Reg. 1963* 124 In 1963 this amounted to 81,563 tons, with sizeable additional reallocations. *a* **1974** R. Crossman *Diaries* (1975) I. 636 We soon found that some of the 199 sites available were not using up their allocation and we had some very nice re-allocations to do in order to make sure that our total didn't fall down.

really, *adv.*[2] Add: In adj. phr. **really truly** chiefly *N. Amer. Children's speech*, authentic, genuine.
1908 L. M. Montgomery *Anne of Green Gables* xi. 114 They all had puffed sleeves..it was awfully hard there among the others who had really truly puffs. **1909** M. Diver *Candles in Wind* xxxiii. 348 Such a really truly knight! **1911** G. Stratton-Porter *Harvester* xvi. 356 There are fairies! Really truly ones! They have found the remainder of the willow dishes. **1911** T. Dreiser *Jennie Gerhardt* 249 She thinks you are her really truly uncle. **1942** *Post* (Morgantown, W. Va.) 2 May 5/7 The [family] have one of the prize sites with a really, truly beach.

realm. 2. b. (Further example.)
1924 W. B. Selbie *Psychol. Relig.* 80 Though the term unconscious is used very loosely by Freudians it generally means the realm of [illegible] various emotions which have from time to time been repressed.

‖ **Realpolitik** (re,ā·lpoliti·k). Also **realpolitik.** [Ger.] Practical politics; policy determined by practical, rather than moral or ideological, considerations. Also *transf.* Cf. *practical politics* s.v. *Practical *a.* (*sb.*) 6.
1914 G. B. Shaw in *New Statesman* 14 Nov. (Suppl.) 5/2 He [*sc.* Friedrich von Bernhardi] proposes that we, his Imperial masters in *Realpolitik*, will do precisely what our Junkers have just made us do. **1915** E. B. Holt *Freudian Wish & its Place in Ethics* iv. 151 This science is '*Realpolitik*', the Politics of Reality. **1920** *Times* 10 Jan. 13/2 An ever-growing share..might not altogether suit the *Realpolitik* of this country. **1946** A. Huxley *Testing Pilate* iv. 275 Freudism became the *realpolitik* of psychology and philosophy. **1958** C. H. Dodd *Authority of Bible* xii. 266 In the last days of the monarchies Israel became involved in..the large '*Realpolitik*' of the time. **1930** *Times Lit. Suppl.* 4 June 433/2 The conflict between these two ideals—*Realpolitik* and a policy founded upon principles of justice and morality. **1948** R. Robinson *Jaeger's Aristotle* vi. 113 The letter that we possess is the solemn record of this practical pact between *Realpolitik* and theoretical schemes of reform. **1952** J. D. Mackie *Earlier Tudors* x. 351 [Thomas Cromwell] had little belief in the omnipotence of the papacy and pinned his faith to *Realpolitik*. **1968** *New Statesman* 19 Apr. 494/2 But the bare-faced hypocrisy with which they have attempted to conceal their military realpolitik, and which has now been devastatingly exposed, is a serious tactical error. **1962** *Listener* 17 Apr. 731/2 Writing in the eighteen-fifties—the decade which saw the birth of the name and the concept of *Realpolitik*—Mommsen was imbued with the sense of need for a strong man. **1970** G. Greer *Female Eunuch* 109 Even the best educated of them [*sc.* women] know that arguments with their men-folk are [illegible] realpolitik. **1979** *N.Y. Rev. Bks.* 25 Oct. 49/2 Soviet policy toward Iran comes from revolutionary ideology not from traditional *Realpolitik*.
Hence **realpoli·tiker,** one who believes in, advocates, or practises *Realpolitik*.
1930 C. Sforza in *Time & Tide* 4 Apr. 435/2 The United States of Europe? asserted the *real politikers*, whom, by a strange legerdemain, the defeat of Hohenzollern Germany has conjured up again in France. **1931** *Times Lit. Suppl.* 22 Jan. 55/1 Both [Cavour and Bismarck] were *Realpolitiker*, endowed with an extraordinary capacity for gauging the forces with which they had to deal. **1962** *Times* 14 June 8/5 In all this he [*sc.* Pierre Flandin] took the line of a French *Realpolitiker*. **1969** *Observer* 2 Dec. 21/4 ..not in the sentimentally brutal way supposed by the realpolitikers. **1976** *Survey* Winter 16 Czechoslovakia may look even more remote than Angola..but its fate counts in the over-all balance, whatever the *Realpolitiker* may think.

‖ **Realschule** (re,ā·lʃūːlǝ). Also **realschule.** Pl. **Realschulen.** [Ger.] In Germany and Austria, a secondary school with a modern rather than classical academic curriculum. Cf. *real school* s.v. Real *a.*[2] 10 c.
1833 [see Real *a.*[2] 10]. **1879** *Encycl. Brit.* X. 471/1 Of more recent growth is the system of *realschulen*, where Latin is the only ancient language taught, the other branches being modern languages, especially French and English, mathematics and natural philosophy, geography and modern history. **1949** R. K. Merton *Social Theory & Social Structure* xiv. 343 Herbert, who first actually organized a *Realschule*. **1969** *Listener* 10 Apr. 480/2 Experimental step towards comprehensive education: class in a Realschule. **1980** *Jewish Chron.* 23 May 14/2, I later taught in the Realschule, founded by his grandfather, Rabbi Samson Raphael Hirsch.

Realtor (rī·ǝltǝ(r)). *U.S.* Also **realtor.** [f. Realty[1] + -or.] A proprietary term in the U.S. for a real-estate agent or broker who belongs to the National Association of Realtors (formerly the National Association of Real Estate Boards). Also *gen.*, an estate agent.
1916 C. N. Chadbourn in *Nat. Real Estate Jrnl.* 15 Mar. 111/2, I propose that the National Association adopt a professional title to be conferred upon its members which they shall use to distinguish them from outsiders. That title is the copyrighted word *realtor*, defined by the National Association against misuse... I therefore, propose that the National Association adopt and confer upon its members, dealers in realty, the title of realtor (accented on the first syllable). **1922** S. Lewis *Babbitt* xiii. 157 We ought to insist that folks call us 'realtors' and not 'real-estate men'. Sounds more like a reg'lar profession. **1925** O. W. Holmes *Let.* 17 Dec. in *Holmes–Laski Lett.* (1953) I. 807 These realtors, as they call themselves, I presume are influential. **1929** *Sun* (Baltimore) 8 Jan. 26/3 (*heading*) Realtors doubt plan for Fox Theater here. **1931** *Evening Standard* 25 Apr. 15/2 (*heading*) 'Realtor' recommends Surrey. **1934** E. Pound *Eleven New Cantos* xxxv. 23 His Wife now acts as his model and the Egeria Has, let us say, married a realtor. **1942** *Amer. Speech* XVII. 209/2 The ambitious realtor's favorites, the overworked [street names] Grand, Broadway, and Inspiration. **1948** *Official Gaz.* (U.S. Patent Office) 14 Sept. 340/2 National Association of Real Estate Boards, Chicago, Ill... Service Mark. *Realtors.* For services in connection with the brokerage of real estate... Claims use since Mar. 31, 1916. **1962** R. Buckminster Fuller *Epic Poem on Industrialization* 139 The organized religions The world's premier realtors. **1969** *Parade* (N.Y.) 14 Dec. 18/2 The realtor who sold most of the property to the hippies has had her office windows smashed. **1970** *Globe & Mail* (Toronto) 25 Sept. 40/2 (Advt.), Metro wide established realtor with country wide referral contacts. **1973** R. C. Dennis *Sweat of Fear* ix. 59 The realtor said..'Let me point out some of the features of this lovely, lovely home.' **1979** *Tucson Mag.* Apr. 33/3 Included are..bankers and lawyers; social and political activists; professors and artists, renovators and historians, journalists and realtors.

realty[1]. Add: **4.** Also *attrib.*
1888 A. Randall-Diehl *Two Thousand Words* 175 *Realty-man*, a dealer in real estate. **1948** E. Wharton *Hermit & Wild Woman* 135, I chanced on a record of the transaction in the realty columns of the morning paper. **1934** E. Pound *Eleven New Cantos* xi. 48 Beecher's church organized by realty agents. [illegible] They walked like Men viii. 47 People were throwing realty promises in a mad attempt to find a place to live. **1968** *Globe & Mail* (Toronto) 17 Feb. 51 (Advt.), Lawyer to manage litigation, realty transactions by property development corporation. **1973** J. Gores *Dead Skip* xv. 100 The tract home had been rented from the realty office by phone. **1975** *New Yorker* 5 Sept. 20/3 The realty company that is the agent for U.S. Steel gave us thirty days to get out.

ream, *sb.*[2] Add: **4.** (See quots.)
1962 *Gloss. Terms Glass Industry* (B.S.I.) 39 *Ream*, a non-homogeneous layer in flat glass. [illegible] II. vi. 408 In the drawing of sheet glass the outer layers of glass may have come from the glass originally on the surface..and may be somewhat deficient in alkali compared with the main glass... This results in a type of inhomogeneity known as 'ream', in which the inhomogeneity is in a plane at right angles to the plane of the glass.

ream, *sb.*[3] Add: **d.** *transf.* in *pl.* A large quantity.
1931 D. H. Lawrence *Love Poems & Others* 54 Eh, what a shame it seems As some should ha'e hardly a smite o' trouble An' others has reams. **1927** J. S. Huxley *Relig. without Revelation* iv. 113 This simple personal fact illustrates, better than could whole reams of argument, the extreme complexity of religion. **1976** *San Francisco Examiner* 30 May (This World Suppl.) 19/1 Spacecraft sent there in recent years have [illegible] reams of sound, ordered data, yet the charisma of Mars remains.

ream, *v.*[3] Add: **I. 3. a.** Also, to clear (something) of an obstruction; to excavate. Also without *adv.* *U.S.*
1967 'T. Wells' *Dead by Light of Moon* x. 99 The toilet..flowed over and Mr. Hawthorne had to come up and ream out the pipes. **1978** J. A. Michener *Chesapeake* 641 At the bottom of the Chesapeake..this primeval riverbed existed, sixty feet deeper than the shallow waters surrounding it [illegible].
II. f. 4. *U.S. slang.* To cheat, to swindle. Cf. *rim *v.*[2]
1914 'High Jinks, Jr.' *Choice Slang* 17 *To Ream one* (*to*), to swindle one. **1938** A. J. Liebling *Back where I came From* 84 He had invented a new technique for reaming the customers. **1942** Berrey & Van den Bark *Amer. Thes. Slang* 131/8 Cheat; defraud..ream. **1952** S. Kauffmann *Philanderer* (1953) xvii. 106 [illegible] they're liable to ream you good.
5. Usu. with *out*. To reprimand. *U.S. colloq.*
1950 E. Hemingway *Across River* xii. 117 [illegible] **1972** R. Busby *Reasonable Man* xvi. 145 [illegible] **1973** J. Ryder *Trevayne* (1974) xi. 89 I'll get my ass chewed... I'll get reamed anyway for letting you make this tour. **1976** 'A. Hailey' *Overload* iv. xvii. 380 A half-wit in my department has been sitting on the thing all morning. I'll ream her out.

reamed (rīmd), *ppl. a.* [f. Ream *v.*[3] + -ed[1].] Of a hole: enlarged by reaming.
1900 *Westm. Gaz.* 29 Nov. 5/1 Two bolts..engage in two carefully reamed holes in the pivot and lever, and are secured by castellated nuts and split pins.

reamer (rī·mǝɹ), *v.* [f. the sb.] *trans.* To ream; to clear out with a reamer.
1934 Webster, *Reamer*,..*v.t.*, to cut with a reamer, as in enlarging a diamond dies. **1935** tr. *H. Mignet's Flying Flea* (Air League Brit. Empire) xiii. 237 It may happen..that the platinum points..have seized up... Reamer this out, a very little, with a metal rod covered with a strip of emery cloth. **1938** J. Healey *Metal Aircraft for Mechanic* ii. 14 Tubular rivets are a reamer fit, so having reamered the hole to size, radius the edge of the metal slightly.

reaming, *vbl. sb.* Add: **2.** *fig.* A reprimand. *colloq.*
1973 M. Woodhouse *Blue Bone* xi. 111 One major stink... Massive reamings are being handed out. **1976** 'J. Charlton' *Remington Set* xxiii. 119 You're bloody cheerful..for a bloke that's headed for a number one reaming from the CO.

re-a·nalyse, *v.* [Re- 5 a.] *trans.* To analyse again.
1934 in Webster. **1946** *Nature* 17 Aug. 243/2 As commercial penicillins recently have contained much more penicillin *K* than was formerly the case, the statistics of the Johns Hopkins Hospital were re-analysed as regards the time factor. **1964** K. G. Lockyer *Introd. Critical Path Anal.* ix. 88 By taking the original network and inserting into it the *actual* times..it is simple to re-analyse the network. **1973** J. J. McKelvey *Man against Tsetse* iii. 200 Ehrlich rushed to Liverpool, confirmed the results of Thomas and Breinl, returned to Germany, and with his colleague Alfred Bertheim reanalyzed atoxyl. **1977** *Word 1972* XXVIII. 93 Yet, because of changes elsewhere in the adjective system, they must be reanalyzed synchronically as NPs.

reana·lysis. [Cf. prec. and Analysis.] A second, or further, analysis.
1934 in Webster. **1962** H. A. Gleason in Householder & Saporta *Probl. Lexicogr.* ii. 92 In some cases no kind of analysis or reanalysis can force the material into the sort of model that descriptive linguistics find appropriate to grammatical statements. **1968** Chomsky & Halle *Sound Pattern Eng.* i. i. 10 If the syntactic component were to be connected to an orthographic rather than a phonetic output system, the reanalysis into phonological phrases would be unnecessary. **1977** *Times* 19 Nov. 14/4 The Bancroft Library..decided to submit the plate to detailed reanalysis.

reanneal (ri,ănī·l), *v.* [Re- 5 a.] *trans.* To anneal again; in *Biochem.*, to change (single-stranded nucleic acid) back into a double-stranded form; also *intr.*, to change from a single-stranded to a double-stranded form. Hence **reannea·ling** *vbl. sb.*
1963 J. Osborne *Dental Mech.* (ed. 5) x. 212 If a considerable amount of bending and re-bending has to be done the wire must be re-annealed from time to time. **1971** *Nature* 25 June 503/1 Mouse satellite DNA..consists of highly repetitive nucleotide sequences which therefore reanneal relatively rapidly after denaturation with alkali. **1980** *Sci. Amer.* Apr. 77/3 The first step of this process is called denaturation, the second step reannealing. *Ibid.*, When the DNA is denatured and reannealed to radioactive RNA, only the remains of those colonies that contained a plasmid whose sequence matches the messenger become radioactive.

reanswer, *v.*[2] For *Obs.*[−1] read *rare* and add later examples.
1933 J. Clayton *Sir Thomas More* v. 87 From the time of S. Anselm..the most profound and subtle philosophical questions had been raised and answered, and again reconsidered and reanswered. **1977** *Word 1972* XXVIII. 78, I am grateful for her patience in answering and reanswering countless questions and either producing or approving almost all the sentences included in this article.

reap, *sb.*[2] Add: **3.** *Judo.* (See quot. 1968.) Cf. next.
1968 K. Smith *Judo Dict.* 167 *Reap*, an action of the leg or foot to sweep away the legs or feet of an opponent in execution of a throw. **1975** R. Butler *Where All Girls are Sweeter* ii. 8, I..locked his arm and gave him what the judo boys call a 'reap' and his arm cracked loudly as he went down on his back.

reap, *v.*[1] Add: **B. 2. e.** *Judo.* To sweep (one leg or both legs) from under one's opponent.
1950 E. J. Harrison *Judo* iii. 56 When reaping your opponent's leg..you should turn your head..and gaze upwards at the ceiling. **1954** E. Dominy *Teach Yourself Judo* vii. 73 Now bring your right hip past his right and reap his leg away as already described. **1956** K. Tomiki *Judo* iii. 74 Making a sickle of your leg, apply the back of your right knee to that of *uke*'s left knee, crosswise. Sharply reap his left leg toward your right oblique back corner.

reaper. Add: **1. b.** fig. *the* (*Great, Grim, Old*) *Reaper*: Death personified.
The expression arises from the iconographic portrayal of Death wielding a scythe: cf. Scythe *sb.* 2.
1839 Longfellow in *Baltimore Lit. Monument* May 17/1 O, not in cruelty, not in wrath, The reaper came that day. **1931** *N. & Q.* 5 Sept. 180/2 One is startled by the inroads which the great reaper has made in the ranks of the Knights since the 15th Edition. **1940** A. Upfield *Bushranger of Skies* xiv. 161 That he happened to be the seventh son of a seventh son..was said..to account for his escapes from The Reaper. **1976** R. Lewis *Witness my Death* v. 182 The old house had been silent, waiting with him for the Old Reaper to come again. **1977** *New Statesman* 2 Sept. 304/1 The Grim Reaper has been rather too vigorously at work among us... At least ten of the best have gone.
2. a. (Earlier and further examples.)
1844 *Let.* 8 Nov. in *Ohio Cultivator* (1845) 15 Mar. 47, I intended..to have written you immediately after harvest, respecting the performance of your Reaper. **1845** *Ohio Cultivator* 1 Jan. 8/3 (*heading*) McCormick's Virginia Reaper. This is another valuable invention for wheat growers. **1847** *Monthly Jrnl. Agric.* June 583 In what consists the difference between their [*sc.* Hussey's and McCormick's] Reapers we are not exactly advised. **1868** *Iowa State Agric. Soc. Rep. 1867* 237 By the folding arrangement the driver can fold the reaper-bar..rendering it perfectly portable. **1923** R. Herrick *Homely Lilla* xvi. 238 But there are tractors and reapers and sprayers. **1981** *Country Life* 7 May 1266/1 Meadows were mown by horse-drawn reaper.
b. reaper-(and-)binder = Self-binder.
1895 C. D. Warner *Golden House* ix. 116 He's got no more feeling in business than a reaper-and-binder. **1901** G. B. Shaw *Three Plays for Puritans* p. xxxii, Fools who go laboriously through all the motions of the reaper and binder in an empty field. **1915** C. Mackenzie *Guy & Pauline* 238 Close at hand was the hum of a reaper-and-binder. **1953** E. Hyams *Gentian Violet* 54 A reaper-binder was cutting oats.

reaping, *vbl. sb.* Add: **1. b.** *Judo.* The action of reaping (*Reap *v.*[1] 2 e) the leg or legs of one's opponent.
1954 E. Dominy *Teach Yourself Judo* vii. 70 The Major Outer Reaping. This is one of the most effective and popular throws in judo. **1956** K. Tomiki *Judo* iii. 68 *O-soto-gari* (Major Outer Reaping Leg Throw). **1976** *Oxf. Compan. Sports & Games* 547/2 The most successful throws have proved to be..*o-soto-gari* (major outer reaping throw), [etc.].

reapprai·sal. [Re- 5 a: cf. next.] A second appraisal; a reassessment (of something) esp. in the light of new facts.
1911 in Webster. **1953** [see *agonizing reappraisal* s.v. *Agonizing *ppl. a.* 1 b]. **1959** *Listener* 17 Dec. 1063/1 The Government of Signor Segni has found it necessary to make a reappraisal of the Vanoni plan. **1971** C. M. Kernan *Lang. Behavior in Black Urban Community* i. 2 It promotes an informed reappraisal of the linguistic abilities of Black students. **1976** *Morecambe Guardian* 7 Dec. 15/4 Local government was reorganised by them without any re-appraisal of the whole basis of local government finance.

reapprai·se, *v.* [Re- 5 a.] *trans.* To make a fresh valuation of, to revalue; to reassess, freq. in the light of new facts. Hence **reapprai·sed** *ppl. a.*, **reapprai·sement, reapprai·ser.**
1895 *U.S. Customs Guide* 124 As I consider the appraisement made by the United States appraisers too high..I have to request that the same may be reappraised..with as little delay as your convenience will permit. *Ibid.* 125 Re-appraisement should take place immediately. **1903** *Daily Chron.* 3 Nov. 5/3 Mr. Low..arranged to have the rental reappraised every twenty-five years. **1906** *Westm. Gaz.* 1 Sept. 2/1 The August circular issued by the United States Government, and dealing with 'Reappraisements of Merchandise by U.S. General Appraisers'. *Ibid.*, Autograph bats specially selected, entered at 15*s.* 6*d.*, reappraised at 21*s.* per each... Entered value is net. Reappraised value less 20 per cent. and 5 per cent. *Ibid.*, The appraisers put a higher value upon them; the reappraisers decide that the true value is 21*s.*, less 20 per cent. and 5 per cent. **1961** *Lancet* 22 July 213/1 There is singularly little evidence that sufficient thought has been devoted to the problem of re-appraising the treatment. **1976** *New Yorker* 8 Mar. 127/1 His subject—finding appropriate uses for technology, and reappraising the engineer's role in society—is important.

reap-silver. For †*Obs. rare* read *Obs.* exc. *Hist.* and add later examples.
1843 Carlyle *Past & Present* ii. x. 123 The Lakenheath eels cease to breed squabbles between human beings; the penny of reap-silver to explode into the streets of the Female Chartism of St. Edmundsbury. **1929** F. M. Powicke in *Cambr. Med. Hist.* VI. vii. 229 The definition of the competence in jurisdiction of the monastic cellarer and the borough reeve, the wrangles about reapsilver and other dues.

rear, *sb.*[3] (and *a.*[1]). Add: **I. 2. b.** The buttocks or backside. *colloq.*
1796 *True Briton* 26 Oct. 3/3 Lord Camelford can boast of a power which rivals that of the First Lord of the Admiralty. He has made Captain Couver a *yellow rear*. **1851** H. Melville *Moby Dick* I. xxi. 159 He put his hand upon the sleeper's rear. **1876** 'Mark Twain' *Adv. Tom Sawyer* ii. 28 In another moment he was flying down the street with his pail and a tingling rear.., and Aunt Polly was retiring from the field with a slipper in her hand. **1949** N. R. Nash *Young & Fair* i. ii. 16 Just once is enough, Baby. (*She slaps her on the rear*) Come on—get to work. **1965** H. Gold *Man who was not with It* vi. 49 You used to have some fat, some curves there. Quite a rear you used to have—quite a rear.
3. a. *spec.* The back part of a motor vehicle.
1966 *Publ. Amer. Dial. Soc.* 1964 xlii. 8 *Rear*,..the aft suspension of a car; the differential of a car; the entire aft of an automobile. **1976** *Evening Post* (Nottingham) 16 Dec. 8/2 The 38-ton Bedford TM costs £16,887 for the tractor business end and trailers or huge van-type rears are £1,100 to £1,700 extra.
b. A (public or communal) water-closet, lavatory, or latrine. Also *pl.* (const. as *sing.*). orig. *School* and *University slang*.
1902 Farmer & Henley *Slang* VI. 4/2 *Rear*..(University), a jakes. **1907** H. Nicolson *Let.* 31 Apr. in J. Lees-Milne *Harold Nicolson* (1980) I. ii. 29 The usual bad rears with its hook and eye lock. **1940** [see *Lat[2]]. **1946** B. Marshall *George Brown's Schooldays* xliii. 170 And now let's raid the rears and rout out any of the other new swine that are hiding there. **1969** Visct. Buckmaster *Roundabout* ii. 30 We also had to know a Latin description of the rear, which we called Foricas.
II. 7. a. *rear-line* (later example), *-link*.
1971 C. Bonington *Annapurna South Face* iii. 40 Lieutenant Bishnuparsad, our rear-link wireless operator, was already installed there. *Ibid.*, He was to stay here at the pension paying-post throughout the expedition, acting as rear link and also handling all our mail. **1974** T. P. Whitney tr. *Solzhenitsyn's Gulag Archipelago* i. iv. 167 They were vehement in their rear-line wrath (the most intense patriotism always flourishes in the rear).
b. (Later examples.)
1920 Rear pocket [see *custard pie* s.v. *Custard 2 b]. **1920** *T. Eaton & Co. Catal.* Spring & Summer 395/3 Rear Tire Carrier suitable for all models of Ford touring cars. **1925** F. Scott Fitzgerald *Great Gatsby* i. 10 Then there was a boom as Tom Buchanan shut the rear windows and the caught wind died out about the room. *Ibid.* 12 All the cars have the left rear wheel painted black as a mourning wreath. **1931** E. S. Gardner in *Detective Fiction Weekly* 7 Mar. 325/1 One of the officers..ensconced himself in the rear seat. **1951** *Catal. Exhibits, Festival of Britain* p. xxix, This new Foden rear-engine chassis has revolutionised normal design practice. **1952** V. Canning *House of Seven Flies* 5 A second sailor opened the rear door of the car for him. **1964** V. J. Chapman *Coastal Vegetation* vi. 170 Whether one is investigating fore-, mid- or rear-dunes, it will be found that the water is fresh. **1966** 'A. Hall' *9th Directive* xx. 184 The car..was gathering speed..when I..got the rear door open and lurched inside. **1968** *Listener* 26 Dec. 868/3 Following the coffee-table book comes the rear-window book: the huge unread, unreadable volume that lies on the shelf behind the back seat. **1969** B. Knox *Tallyman* vi. 120 Rear-wheel skids should be steered into, said the rule-book. **1973** *Country Life* 1 Mar. 540/2 Rear-seat passengers are not too badly off for leg room. **1975** *Ibid.* 2 Jan. 32/2 A real omission here is a heated rear window... Rear wipers are likewise unknown. **1976** P. R. White *Planning for Public Transport* iii. 56 The rear-engine layout was also adopted for single-deckers. **1978** *Dumfries Courier* 20 Oct. 22/1 (Advt.), All are quality cars with spacious reclining seats, fitted carpets,..heated rearscreen, radial tyres, etc.
8. a. *rear-facing*.
1978 *Cornish Guardian* 27 Apr. 24/5 (Advt.), 1973 Volvo 145 D/L Estate... Rear-facing child seats.
b. *rear-driven* (so *-drive*), *-lit*; *rear-illuminate, -project* vbs.
1904 Rear-driven [Listed in Dict.]. **1961** *Twentieth Century* Feb. 124 Rear-lit cloths become more common [in the theatre]. **1970** *Nature* 19 Dec. 1217/1 A number of test-areas in the form of circular holes in a metal plate are uniformly rear-illuminated to a supra-threshold luminance. **1972** *Country Life* 26 Oct. 1060/3 The rack and pinion steering is responsive yet without quite the feel of a rear-drive car. **1973** *Jrnl. Genetic Psychol.* June 255 The stimuli..were rear-projected onto a 27·9 cm[2] opaque glass screen. **1977** *Lancashire Life* Jan. 79/1 No rear drive Citroen has been made since the 1930s.
c. At the rear, as *rear-engined, -mounted*.
1933 *Motor* 10 Oct. 524/1 The rear-engined Trojan. **1957** *Sci. News Let.* 23 Mar. 190/1 Rear-engined cars are here to stay. **1960** *Farmer & Stockbreeder* 19 Jan. 90/1 In this country the accepted method of handling silage has been by means of a rear-mounted buckrake. **1975** *Drive* New Year 102/2 The protesting chatter from the air-cooled rear-mounted engine is more a symptom of asthma than mechanical stress. **1976** P. R. White *Planning for Public Transport* iii. 56 The higher maintenance costs and poorer availability of the rear-engined models.
9. rear echelon *U.S. Mil.*, that section of an army concerned with administrative and supply duties; also *transf.*; **rear end,** (*a*) the back part or section (of anything, esp. a vehicle); (*b*) *slang*, the backside or buttocks (of a person); hence as *v. trans.* (N. Amer.), to collide, or cause (one's vehicle) to collide, with the rear end of another vehicle; **rear-ender,** a rear-end collision; **rear gunner,** a member of the crew of a military aircraft who operates a gun from a compartment or turret at the rear of the aircraft; **rear-lamp, -light,** a (usu. red) lamp at the rear of a vehicle which can be switched on to serve as a warning light in the dark; **rear mirror,** a rear-view mirror (see *rear-view* attrib.); **rear pillar** (see quot. 1930); **rear projection** = *back projection* s.v. *Back-B; **rearsight,** a part of a camera viewfinder, situated at the back, to which the eye is applied; **rear-view** *attrib.*, giving a view to the rear; *spec.* of a mirror inside a motor vehicle in front of the driver.
1934 Webster, Rear echelon. **1947** *Amer. Speech* XXII. 55 *Rear echelon commando*, a soldier assigned to the rear. **1967** *Boston Sunday Herald* 7 May iii. 14/2 The number [of servicewomen] in Vietnam will remain small, chiefly because there is no large 'rear echelon' setup of the kind maintained in Europe in World War II. **1977** P. Johnson *Enemies of Society* xii. 165 This shaky argument, of the type which convinced the rear echelons of the Gadarene swine, carried the day. **1868** Rear end [see sense 7 b in Dict.]. **1926** *Daily Colonist* (Victoria, B.C.) 19 Jan. 5/3 Two passengers were killed and fifty injured today in a rear-end collision of..two subway trains. **1937** J. Weidman *I can get it for you Wholesale* xxviii. 268 She's a pain in the rear end. **1961** *Amer. Speech* XXXVI. 273 *Rear end*,..the differentials of a tractor. **1967** G. Kelly in *Coast to Coast 1965–6* 95 Blokes my age are sitting on their rear-ends ordering the rest..around. **1976** *Islander* (Victoria, B.C.) 1 Aug. 3/4 A driver came to an abrupt stop in front of me. I slithered all over the wet road but did I rearend her? Of course not. **1978** *Detroit Free Press* 2 Apr. 3A/1 The men, who were on a chartered city bus traveling to a football game in 1972 when it was rear-ended by another bus, rejected a settlement of $500 apiece and took their case to the jury. **1932** *Erie Railroad Mag.* Apr. 46/1 With all his fast running I never knew of him piling them up, of any but a few derailments and never a rear-ender. **1920** *Flight* XII. 11/1 A central passage leads through to what in the military machine was the rear gunner's cockpit, which is now occupied by the 'postman'. **1944** 'N. Shute' *Pastoral* i. 3 He had developed into a very good rear-gunner in the Wimpey. **1977** *R.A.F. News* 27 Apr.–10 May 8/2 The aircraft was hit again and again and the rear gunner was wounded. **1907** *Westm. Gaz.* 17 Sept. 4/2 When the compulsory carrying of rear-lamps has been suggested the proposal has always been violently resisted. **1937** *East London Rubber Co. Ltd. Motor Catal.* 154/2 Top covers for rear lamps. **1918** A. Quiller-Couch *Foe-Farrell* iii. 54 The car purred and glided away... We watched the rearlight turn the corner. **1967** N. Freeling *Strike out where not Applicable* 159 There is nothing that looks so like the rear lights of a car as the rear lights of another car. **1968** Rear mirror [see G.T. s.v. *G III. f]. **1930** *Motor Body Building* LI. 105/1 *Rear pillar*, a vertical frame member at the back corner of the body. **1977** Rear pillar [see *Pillar *sb.* 2 c]. **1960** *Practical Wireless* XXXVI. 316/2 A team of demonstrators who operated the sequence of exhibit animations, rear projection films and synchronised sound and provided a live commentary. **1976** Botham & Donnelly *Valentino* xii. 93 Working behind the screen, with rear-projection to help them follow the story. **1971** *Amateur Photogr.* 13 Jan. 50/3 The rearsight is quite large, has a permanently attached rubber eye-cup and is adjustable between +1 to −4 dioptres to suit individual eyesight. **1978** *Ibid.* 2 Aug. 79/2 Accessories included: carrying strap, body cap, rearsight rubber eyecup, etc. **1926** Rear-view [see *Driving *vbl. sb.* 3 a]. **1959** H. Nielsen *Fifth Caller* xiii. 195 His face had been in the rearview mirror. **1969** G. Macbeth *War Quartet* 17 For the moment they were framed In rear-view mirrors. **1974** V. Nabokov *Look at Harlequins* (1975) iv. iv. 173, I see today..the rearview reflection of that sweet wild past.

rear, *v.*[1] Add: **II. 9. c.** Also const. *into*.
1871 Browning *Prince Hohenstiel-Schwangau* 52 To play at horticulture, rear some rose Or poppy into perfect leaf and bloom.
III. 10. b. *fig.* phr. *to rear its* (*ugly*) *head*, and varr. = *to raise its* (*ugly*) *head* s.v. *Raise *v.*[1] 17 e.
1857 Trollope *Barchester T.* II. viii. 124 Rebellion had already reared her hideous head within the [bishop's] palace. **1946** K. Tennant *Lost Haven* (1947) vii. 96 Another problem reared its ugly head. **1966** B. Kimenye *Kalasanda Revisited* 21 Scandal of even the mildest type failed to rear its head. **1971** *Daily Tel.* 5 July 1 The problem of broken rails is rearing its ugly head again in the current spate of railway accidents. **1976** *0–10 Cricket Scene* (Austral.) 21/1 They crumbled as their inexperience reared its ugly head.

rear-guard[1]. Add: **2. b.** *rear-guard action*, a defensive stand by the rear-guard of a retreating army. Also *fig.*
1898 [in Dict., sense 2]. **1946** *Ess. & Stud.* XXXI. 46 Yet this scene is a magnificent rear-guard action by Cleopatra. **1954** F. C. Avis *Boxing Ref. Dict.* 93 *Rearguard action*, defensive boxing. **1977** *World of Cricket Monthly* June 29/1 It was left to Murray and Roberts to provide a defiant rearguard action which postponed the end.

reargue, *v.* Add: Also *absol.* or *intr.*
1972 *N.Y. Law Jrnl.* 31 Oct. 15/5 The informal motion to reargue is granted. **1973** *Ibid.* 31 Aug. 2/5 A letter from counsel for South Wall Associates, which shall be deemed a motion to reargue.

rear-horse. (Earlier and later examples.)
1869 *Rep. U.S. Comm. Agric. 1868* 308 The *Mantes* or 'rear-horses' prey upon other insects. **1900** *Everybody's Mag.* July 21 Most people are acquainted with the praying mantis—otherwise known as the 'rearhorse'. **1935** H. T. Fernald *Appl. Entomology* (ed. 3) xvi. 87 They [*sc.* mantids] are often called rearhorses, devil-horses,..or mule killers.

rearing, *vbl. sb.* Add: **5. b.** (Further example.)
1967 K. M. Smith *Insect Virol.* x. 184 The hot medium is poured into rearing jars.
6. rearing-house, a building in which young chickens are kept.
1824 J. H. Barlow *Hatching Poultry by Steam* 15 The Rearing House is of the same size and dimensions as the Hatching House, in which Fowls are kept until they are three weeks or a month old. **1948** L. Robinson *Mod. Poultry Husbandry* xi. 227 When growing stock are first placed in arks or other types of range rearing-house, a temporary run of wire netting should be erected. **1962** D. de Saulles *Pictorial Poultry Keeping* (ed. 2) This pullet is just about ready to be taken from the rearing house and put into laying quarters. **1975** T. Allbeury *Special Collection* xxiii. 168 The lights were on in the rearing house... He could hear the shrill noise of eight thousand week-old chicks.

rearing, *ppl. a.* Add: **3.** = *raring* adj. s.v. *Rare *v.* 1 b.
1926 G. Frankau *My Unsentimental Journey* viii. 104 A good many of the eight hundred other diners were 'raring' (Anglicé—rearing) 'to go'. **1947** H. Walsh *Fourth Point of Star* 34, I am rearing to talk this business over with her. **1963** *Listener* 28 Feb. 393/3 Inside most liberal Lincolnographers there seems to be a Lincolnolater rearing to genuflect.

rearing crew. *N. Amer.* [app. f. Rear *sb.*[3]] = *rear-crew* s.v. Rear *sb.*[3] 9.
1963 R. D. Symons *Many Trails* xviii. 184 The last men..will form the nucleus of the 'rearing crew'—that is, the crew which will bring up the rear of the drive, taking care to roll into the water any logs left stranded on the banks. **1971** *Daily Progress* (Charlottesville, Va.) 6 May 7-D/5 Freeing stranded logs is the job of a 32-man 'rearing crew', which works seven days a week with power boats, winches and occasionally a stick of dynamite.

rearmament. (Later examples.)
1905 *Daily Chron.* 14 Mar. 1/7 The artillery rearmament scheme accounts for £1,213,000. **1935** *Hansard Commons* 22 May 363 Re-armament that could not fail to cause all lovers of peace considerable anxiety. **1938** *Encycl. Brit. Bk. of Year* 60/2 In Great Britain the economic consequences of re-armament are..rather intangible and remote. **1938**, etc. [see *Moral Rearmament* s.v. *Moral *a.* 7 f]. **1961** *Daily Worker* 25 May 2 Without Britain's tolerance German rearmament in the air would be jeopardised.

rearou·sal. [f. Re-arouse *v.*] A second or further arousal.
1934 in Webster. **1949** M. Mead *Male & Female* iv. xiv. 293 The demon to be avoided..is lack of potency, defined in a number of quantitative ways—frequency, time, interval before rearousal.

re-arrange, *v.* (Earlier example.)
1824 De Quincey in *London Mag.* Jan. 5/2, I have therefore abstracted, re-arranged, and in some respects..have improved, the German work on this subject.

re-arti·culate, *v.* [Re- 5 a.] To articulate for a second or further time. So **re-arti·culated** *ppl. a.*
1963 *Economist* 21 Sept. 987/2 A wholly rearticulated Office of External Relations. **1964** *Language* XL. 78 Verbs with rearticulated root vowel often appear in apocopated form in rapid speech. **1964** E. Palmer tr. *Martinet's Elem. General Linguistics* ii. 45 The necessity..of re-articulating a foreign mode of experience to conform to the model which is familiar to us. **1973** Matias & Willemen tr. M. Cegarra in *Screen* Spring/Summer 177 It is clearly the whole notion of film rhetoric which needs to be re-articulated.

rearward, *adv.* Add: **1. c.** *Comb.*, as *rearward-facing, -hinged* adjs.
1955 *Times* 23 June 4/4 It had been decided in Australia to have rearward-facing seats. **1959** *Economist* 13 June 1042/1 The engine drives a ducted fan, which supplies compressed air to two concentric rings of jet openings facing downward..while rearward-facing jets propel it. **1967** *Jane's Surface Skimmer Systems 1967–68* 25/1 There is a rearward-hinged car-type door on each side of the cabin.

rearwardly, *adv.* (Later examples.)
1888 *Engineer* 4 May 374/3 Having a handle..extending rearwardly beyond the suction tube. **1934** *Canad. Patent Office Rec.* 30 Oct. 2464/1 Beams extending..rearwardly of the rear axle. **1946** A. W. Judge *Mod. Petrol Engines* vi. 237 The cooling air..leaves the engine cowling through a rearwardly inclined exit. **1973** *Times* 30 Oct. 17/8 A reasonable proportion of the water is thrown rearwardly.

rearwardness (rɪə·ɹwǝɹdnės). *rare.* [f. Rearward *a.* + -ness.] The state of being in the rear or in arrears.
1903 *T.P.'s Weekly* 16 Oct. 621/2 It is advantageous to keep oneself quite a year behind contemporary literature; this rearwardness saves both time and money.

re-asce·nded, *ppl. a.* [f. Reascend *v.* + -ed[1].] That has ascended again.
a **1838** R. Grant *Sacred Poems* (1839) 8 Oh! from earth to heav'n restor'd, Mighty re-ascended Lord. **1906** *Westm. Gaz.* 14 Apr. 6/2 Rise, O saints..Round your re-ascended Sun circling soar!

reason, *sb.*[1] Add: **II. 5. c.** Phr. *for reasons best known to oneself*, for seemingly perverse reasons.
1638 W. Chillingworth *Relig. Protestants* 84 Yet it hath pleased God (for Reasons best known to himselfe) not to allow us this convenience. **1743** Fielding *Jonathan Wild* iv. xiii. 383 Indeed those, who have unluckily missed it, seem all their Days to have laboured in vain to attain an End, which Fortune, for Reasons only known to herself, hath thought proper to deny them. **1847** A. Brontë *Agnes Grey* xiii. 191 If they chose to 'take' me, I went; if, for reasons best known to themselves, they chose to go alone, I took my seat in the carriage. **1894** Somerville & 'Ross' *Real Charlotte* III. xli. 133 Removing his pipe and the hat which, for reasons best known to himself, he wore while at work. **1938** G. Graham *Swiss Sonata* vi. 250 She tried very hard to adopt me, but my father, for reasons best known to himself, wouldn't give me up.
III. 10. e. *the age of reason*, (*a*) (freq. with capital initials) the late seventeenth and eighteenth centuries in Western Europe, during which cultural life was characterized by faith in human reason; the Enlightenment (cf. Enlightenment 2); (*b*) *R. C. Ch.*, the age at which a child is capable of discerning right from wrong and can be held responsible for his or her actions; also *loosely*.
(*a*) **1794** T. Paine (*title*) The age of reason; being an investigation of true and fabulous theology. **1902** Chesterton *Twelve Types* 129 Carlyle..denied every one of the postulates upon which the age of reason based itself. **1926** R. H. Tawney *Relig. & Rise of Capitalism* i. 61 The sanguine optimists of the Age of Reason. **1971** R. J. White *Second-Hand Tomb* xviii. 200 You medieval scholars suffer from a double dose of spiritual pride where the Age of Reason is concerned.
(*b*) **1884** Addis & Arnold *Cath. Dict.* 17/1 The age of reason is generally supposed to begin about the seventh year... At that time a child becomes capable of mortal sin. **1947** E. Sutton tr. *Sartre's Age of Reason* viii. 126 You have..reached the age of reason, my poor Mathieu..but you try to dodge that fact too, you try to pretend you're younger than you are. *Ibid.* xviii. 360 He yawned again as he repeated to himself: 'It's true, it's absolutely true: I have attained the age of reason.' **1955** tr. *Maupassant's Compl. Short Stories* 1237, I am seven years old today. As it is the age of reason, I want to thank you for having brought me into this world. **1974** *Oxf. Dict. Chr. Ch.* (ed. 2) 24/1 The attainment of the age of reason is commonly marked by a child's First Communion.
V. 23. *reason-monger*; *reason-contained, -giving, -wrought* adjs.
1973 *Art Internat.* Mar. 75/2 The frost of the passage, its chill, reason-contained fire is even more remarkable. **1855** Bagehot in *Nat. Rev.* July 36 The strong analytic, comprehensive, reason-giving powers..were utterly foreign to his mind. **1933** Dell & Baynes tr. *Jung's Mod. Man in Search of Soul* viii. 186 In works of art of this nature..we cannot doubt that the vision is a genuine, primordial experience, regardless of what reason-mongers may say. **1906** Hardy *Dynasts* ii. i. ii. 152 Here, then, ends My hope for Europe's reason-wrought repose!

reason, *v.* Add: **6. b.** (Later example.)
1900 *Outrageous Fortune* x. 117 There is little need now to recapitulate those arguments with which I reasoned down the dictates of my better nature.
c. To drive *away* or *off* by reasoning.
1839 E. A. Poe in *Burton's Gentleman's Mag.* (Philadelphia) Sept. 150, I struggled to reason off the nervousness which had dominion over me. **1854** M. L. Charlesworth *Ministering Children* ix. 139 Let the sinner then beware how he reasons away and rejects the awful Word of God. **1866** M. C. Harris *Christine* xiv. 74 It was very natural, the doctor said to himself, trying to reason away the pain he felt.

reasonability. (Further example.)
1973 *Atlantic Monthly* June 58/2 In those early days of knowing him, I still believed in reasonability.

reasonable, *a.* Add: **8.** *reasonable-sized* adj.
1965 E. Jutikkala in Glass & Eversley *Population in Hist.* xxiii. 554 The only reasonable-sized city in Finland, Turku,..must be discussed separately from the surrounding province.

reasonably, *adv.* Add: **5.** *Comb.*, as *reasonably-priced, -sized* adjs.
1902 *Daily Chron.* 22 June 3/3 See that reasonably-priced meals are provided. **1960** *Farmer & Stockbreeder* 8 Mar. 77/3 A reasonably-priced flowmeter was necessary. **1968** Fox & Mayers *Computing Methods for Scientists & Engineers* v. 76 That is if the elements of the inverse matrix A^{-1} are large for reasonably-sized A.

reasoned, *ppl. a.* Add: **a.** (Later example.)
1904 E. F. Benson *Challoners* vii. 146 He would sooner have mated her with a thief or an adulterer..than with a reasoned atheist.
b. *reasoned amendment*, an amendment to a bill in Parliament that seeks to prevent a further reading by proposing reasons for the alteration or rejection of the bill.
1909 *Times* 9 Nov. 10/3 Such an amendment would..have obvious advantages over the longer form which is known as a reasoned amendment. **1975** *Daily Tel.* 23 July 1 The Conservative 'reasoned amendment'..supported the Government's belated commitment to reduce the disastrous rate of inflation.
c. *reasoned bibliography* [cf. *Catalogue raisonné], a descriptive survey of relevant books and articles appended to an essay or the like.
1958 *Listener* 11 Sept. 392/3 The fourth essay..which..is accompanied by a reasoned bibliography of formidable fullness. **1969** *Archiv* CCV. 490 Her reasoned bibliography ranges as far afield as Calcutta and Hiroshima.

re-aspire, *v.* (Later example.)
1936 R. Campbell *Mithraic Emblems* 17 From the sacred flames they feast In hymns of incense re-aspire To praise His throne of silver fire.

reassemble, *v.* Add: **2.** Also *fig.*
1945 P. Larkin *North Ship* 9 The meadows are bright With the coldest dew; The dawn reassembles.
reasse·mbled, reasse·mbling *ppl. adjs.*
1904 *Westm. Gaz.* 21 Oct. 2/3 The first sitting of the reassembled Chambers. **1906** *Macmillan's Mag.* Apr. 438 To my reassembling senses..came the realisation of a greater tragedy. **1977** W. M. Spackman *Armful of Warm Girl* 88 Nicolas collected his wits enough to follow, with a re-assembled smile.

reassertion. (Earlier example.)
1843 Mill *Logic* I. ii. i. 221 That cannot be called reasoning or inference which is a mere reassertion in different words of what had been asserted before.

reassignment. (Examples.)
1884 [see Re- 5 a]. **1960** *Amer. Speech* XXXV. 216 Hoenigswald recognizes the following types of sound change:..split from reassignment of noncontrasting phones. **1976** *National Observer* (U.S.) 18 Dec. 2/4 The Appeals Court..ordered reassignment of 20,000 to 30,000 students—up to half the student total—to achieve racial balance. **1978** *N.Y. Times* 29 Mar. B4/4 Sent..Sam Perlozzo, second baseman, to minor leagues for reassignment.

reassimilate, *v.* (Example.)
1952 G. Sarton *Hist. Sci.* I. xvi. 395 Another result of the long wars was the existence of a relatively large body of veterans who..could not be easily reassimilated.

reassociate, *v.* (Later example.)
1964 G. H. Haggis et al. *Introd. Molecular Biol.* xi. 293 When the pH is brought back to neutrality the α and β pairs reassociate into the four-unit molecule.
Hence **re:associa·tion.**
1923 J. S. Huxley *Essays of Biologist* iv. 152 Dissociation in most cases is not complete;..now and again re-association of the parts [of the mind] occurs. **1953** Hinsie & Campbell *Psychiatric Dict.* (ed. 2) 738/2 *Reassociation*, a process of renewed or refreshed association occurring in hypnoanalysis of the war neurosis, during which the patient relives the traumatic event with emotional vividness. Such forgotten experiences will then become a part of his normal personality and consciousness. **1977** *Proc. R. Soc. Med.* LXX. 560/1 Nucleic acid reassociation experiments indicate that at least a proportion of tumour cells carry virus genetic material.

reassortment. (Later examples.)
1959 B. Wootton *Social Sci. & Social Path.* xi. 330 He may break down under the stress of the eleven-plus re-assortment. **1971** D. J. Cove *Genetics* i. 8 The formation of a zygote by the gametes of the two different strains, followed by the production of ascospores must provide an opportunity for the shuffling or reassortment of genetic information affecting different characters.

reassurer. (Earlier example.)
1787 J. A. Park *Marine Insurances* xv. 315 Re-assurance..may be said to be a contract, which the first insurer enters into, in order to relieve himself from those risks which he has incautiously undertaken, by throwing them upon other underwriters, who are called re-assurers.

reassu·ring, *vbl. sb.* [f. Reassure *v.* + -ing[1].] The action of the verb Reassure.
1865 in Webster. **1902** 'Mark Twain' in *Harper's Weekly* 6 Dec. 5/2 Alfred..did what he could..to respond with some show of heart to the Major's kindly pettings and reassurings.

reasty, *a.* For '†*Obs.* exc. *dial.*' read 'chiefly *dial.*' and add: **1.** (Later examples.)
1914 C. Mackenzie *Sinister Street* II. iv. ii. 859 A luxurious mournfulness was in the view, and he leaned out over the sill scenting the reasty London air. **1959** *Listener* 10 Dec. 1035/2, I paid for all mistakes with drops of sweat Strained from the reasty gammon of my pain. **1964** J. Hale *Grudge Fight* v. 76 What you'll do is shut your trap and get back to your reasty pit. **1974** P. Wright *Lang. Brit. Industry* xvii. 166 In some homes..bacon with a strong nasty taste is not rancid but *reasty*. **1978** *Jrnl. Lancs. Dial. Soc.* xxvii. 13 Please, Looard, send summat good to eyt; Not reeasty bacon or fatty meyt.

reattach, *v.* Add: **2.** (Later examples.)
1921 O. E. Inglis *Burchard's Textbk. Dental Path. & Therapeutics* (ed. 6) vi. xxviii. 777 The first indication [of when a tooth has been forced out by accident] has ever been acted upon by parents who have quickly pressed teeth into their places where they became reattached. **1952** W. H. Archer *Manual of Oral Surg.* i. 41/2 The tooth became re-attached, and circulation into the pulp tissue was re-established. **1957** G. Ryle in C. A. Mace *Brit. Philos. in Mid-Century* 254 The notion of meaning had been..partly detached from the notion of naming and re-attached to the notion of saying.
refl. **1953** I. Glickman *Clin. Periodontol.* xxxvii. 609 The problem hinges upon the question of whether the periodontal tissues reattach themselves to tooth surface previously denuded by disease..or whether they are attached at the level of the pre-existent pocket. **1980** D. Williams *Murder for Treasure* v. 45 The guard..had re-attached himself to Treasure's entourage.
Hence **reatta·ched** *ppl. a.*
1928 H. K. Box *Treatm. Periodontal Pocket* iii. 117 The arrangement of the reattached tissues on the curetted cemental surface.

reattachment. Add: **b.** *Dentistry.* The re-establishment of the connections between a tooth and the jaw.
1908 O. E. Inglis *Burchard's Textbk. Dental Path. & Therapeutics* (ed. 3) iv. xv. 410 Kirk records a case of immediate replantation in early life, followed in old age by root resorption. The tooth when extracted contained secondary dentine, which could only have formed as the result of a reattachment of the pulp. **1928** H. K. Box *Treatm. Periodontal Pocket* iii. 103 Gentle digital pressure..appeared to be the only precaution taken to insure reattachment. **1953** I. Glickman *Clin. Periodontol.* xxxvii. 609 The term 'reattachment' is to be used here to connote the restoration of gingival contour effected by the re-embedding of the periodontal membrane into newly formed cementum, and attachment of gingival epithelium to tooth surface denuded by periodontal disease. **1962** Blake & Trott *Periodontol.* x. 96 By combined cementum and soft tissue curettage, it is possible to obtain a degree of attachment of soft tissue to cementum with a consequent real reduction in pocket depth... This is..called 'reattachment'.

reattempt, *v.* (Later examples.)
1861 M. Arnold *On Translating Homer* iii. 102, I think that the task of translating Homer into English verse both will be re-attempted, and may be re-attempted successfully. **1895** Hardy *Jude the Obscure* iii. vi. 198 He had begun to sit in his parlour during the dark winter nights and re-attempt some of his old studies.

Réaumur. Delete ‖ and add after 'physicist': René-Antoine Ferchault de *Réaumur*.
a. Used *ellipt.* (in Dict.: further examples).
1799 Malthus *Jrnl.* 30 May (1966) 39 In the summer of 1793 the therm was 25 Reaumur. **1832** S. Austin tr. *H. Pückler-Muskau's Tour in Germany, Holland & England* III. 196 The room..is regularly heated to a temperature of fourteen degrees Reaumur. **1950** 'G. Orwell' *Shooting Elephant* 18, I had been unequal to translating Réaumur into Fahrenheit, but I know that my temperature was round about 103.
b. Used in the possessive and *attrib.* to designate various processes, materials, and concepts developed by him, as **Réaumur malleable iron,** a name for white-heart malleable cast iron; **Réaumur's porcelain,** a devitrified form of glass produced by prolonged exposure to heat near but below the fusion temperature, formerly used for chemical

vessels; **Réaumur process**, a process for annealing iron leading to the production of white-heart malleable iron, first published by him in 1722; **Réaumur('s) scale**, his temperature scale (see Dict.).

1912 W. H. HATFIELD *Cast Iron* xiii. 195 Sample J..was a..sample of English Réaumur malleable cast iron. It consisted mainly of well-laminated pearlite, in which was immersed the remaining annealing carbon, with a skin of well-developed ferrite. **1832** G. R. PORTER *Porcelain & Glass* xvi. 317 M. Reaumur,..in..1739, communicated the result of these to the Royal Academy of Sciences in Paris. The subject becoming by this means more generally known, glass, when thus converted, obtained, and has since kept the name, of Reaumur's porcelain, a designation which it owes to its appearance rather than to its real properties. *Ibid.* 325 Glass has been converted into Reaumur's porcelain during volcanic eruptions, by being enveloped in burning lava. **1868** C. L. EASTLAKE *Hints Househ. Taste* x. 219 In the production of what is called Réaumur's porcelain, the formation of crystals may be determined by the application of heat lower than that necessary to effect the perfect fusion of the glass. **1918** P. MARSON *Glass* iii. 17 'Réaumur's Porcelain' is a glass in a devitrified state, and is used for pestles and mortars, devitrified glass being less brittle than ordinary glass and similar to vitrified porcelain. **1911** B. STOUGHTON *Metall. of Iron & Steel* (ed. 2) xiii. 356 In the original Réaumur process the castings are packed in iron oxide during the anneal, and this operation is carried on at a temperature of about 800 to 875°C..for a period of four or five days. **1949** J. E. GARSIDE *Process & Physical Metall.* xxi. 374 Malleable cast-irons are made by two methods—..(2) The Whiteheart process, also known as the European or Réaumur process. **1795** W. NICHOLSON *Dict. Chem.* II. 935 In Reaumur's scale, the number of degrees between these two points is 80. **1863** E. ATKINSON tr. *Ganot's Elementary Treatise on Physics* vi. i. 194 Besides the centigrade scale two others are frequently used—Fahrenheit's scale and Réaumur's scale. **1958** S. PETTERSSEN *Introd. Meteorol.* (ed. 2) ii. 20 The Reaumur scale was much used in Central Europe until the beginning of this century but has since gone out of use.

reave (rīv, rē[1]v), *sb. Archæol.* [Origin unknown: perh. f. OE. *rǣw* REW *sb.*[1]] A long low bank or wall found on Dartmoor.

1848 J. H. MASON in S. Rowe *Perambulation of Dartmoor* 130 In tracing the northernmost *reave* from Hamildon..we lost it in a tin-work. **1976** *Current Archaeol.* V. 250/2 Four major walls on the moor... The walls were known as reaves in local dialect (pronounced 'raves'). *Ibid.* 252/1 The lynchets seem to show that the reaves..were built by a people who already had fields laid out in parallel strips. **1978** *Antiquity* Mar. 26/1 Dartmoor reaves..(the word derives from the Old English *raew*, meaning a row) are long, low banks, constructed mainly of stone, and often covered in vegetation. These may run for any distance up to 15 km, and they may reach 0·5 m or more in height.

reb[1]. For *U.S.* read 'Chiefly *U.S.*' and add: (Further examples.) Also *spec.* = *REBEL *sb.*[1] 1 e; cf. *Johnny Reb* s.v. *JOHNNY, JOHNNIE 3. Also *attrib.*

1897 KIPLING *Captains Courageous* iii. 62 Earnin' my bread on the deep waters, an' dodgin' Reb privateers. **1904** —— *Traffics & Discoveries* 31 Then we got into the Colony [*sc.* Orange Free State], and the rebs—ministers mostly and schoolmasters—came round the cars with fruit and sympathy and texts. **1909** 'O. HENRY' *Options* (1916) 81 The alleged aristocratic superiority of a 'reb' ought to be visible to him at once. **1916** 'TAFFRAIL' *Pincher Martin* v. 73 They wanted me to order the rebs. to shove off. **1928** S. V. BENÉT *John Brown's Body* 181 'Hello, Charley,' he said, 'Where you been?'.. 'Out hearing the Rebs,' he said. **1938** J. DANIELS *Southerner Discovers South* xxxv. 346 The boy.. insisted in the hospital that he was the best alligator catcher on the coast of Georgia. Perhaps he was. Maybe still one Reb can beat ten Yankees. **1947** *Sierra Club Bull.* May 80 The stars and stripes, the flag of our now united country raised in honor of our visit, and I, an old battle-scarred and weather-worn Reb. **1963** *Amer. Speech* XXXVIII. 45 *Reb*, a Southern driver or a Southern trucking firm. **1979** S. SHEPPARD *Four Hundred* i. 4 A certain scheme for separating some of the Rebs from their money.

reb[2] (reb). Also **rebb** and with capital initial. [Yiddish, *abbrev.* of *rebbe* *REBBE.] A traditional Jewish courtesy title prefixed to a man's first name or surname.

1882 [see *KEHILLA]. **1892** I. ZANGWILL *Childr. Ghetto* I. 147 'Well, how goes it, Reb Moshé?' said Reb Shemuel.., noticing Moses loitering. He called him 'Reb' out of courtesy and in acknowledgement of his piety. The real 'Reb' was a fine figure of a man. **1932** L. GOLDING *Magnolia Street* I. v. 84 The beard of Reb Berel, the beadle, was in attendance. **1967** C. POTOK *Chosen* i. 15 Reb Saunders ordered them never to lose because it would shame their yeshiva. **1975** *Publishers Weekly* 3 Feb. 75/3 A ragged stranger stands by the door of the synagogue... The boy whispers to the woman that the stranger, reb Naftoli, is a *badchen*, a merrymaker, who speaks in rhymes. **1978** I. B. SINGER *Shosha* ix. 170 We had a preacher in our family—Reb Zekele Preacher, they called him.

rebab (rībæ·b). Also **rabab(a, rabap, rebaba, rubabah**, etc. [a. colloq. Arab. *rebāb*, classical Arab. *rabāb* in the same sense: cf. RIBIBE *sb.*] A plucked or bowed stringed instrument of Arabian origin, now in use in North Africa and the Middle East, and among the Islamic populations of the Indian sub-continent, Malaysia, and Indonesia.

1738 [see *OUD]. **1802** *Jrnl. F. Horneman's Trav.* iii. 72 The song of these Fezzan girls is Soudanic. Their musical instrument is called *rhababe*. **1836** E. W. LANE *Acct. Manners & Customs Mod. Egyptians* II. v. 73 A curious kind of viol, called rabab, is much used by poor singers, as an accompaniment to the voice. **1856** R. F. BURTON *Personal Narr. Pilgrimage to El-Medinah* III. ii. 76 They have the one-stringed Rubabah, or guitar. **1925** *Blackw. Mag.* Feb. 217/2 Zakka Khel the *Subedar* played cunningly, on the rebab, an instrument like a mandoline. **1929** F. STARK *Let.* 22 Dec. (1974) I. 226 Mahmud..got his *rebaba*, played with the fingers and a bow on one string, and they [*sc.* Bedouin tribesmen in Iraq] showed us the four different modes of their music. **1937** M. COVARRUBIAS *Island of Bali* (1972) viii. 213 There are other instruments, such as..the two-string violin (*rebab*), which are used mainly as a lead for the melody. **1958** O. CAROE *Pathans* xix. 309 There were performers who chanted to the *rabap*. **1960** G. E. EVANS *Horse in Furrow* xiii. 177 In Arabia the poet singer to the 'rebab' (lute with one string and a bow) is a recognised authority who dare not for his life deviate by a word from the known facts. **1967** I. DIGS *Bedouin Boyhood* xi. 104 Long minutes later I heard some broken tunes sent up by the *rababa*. **1972** M. SHEPPARD *Taman Indera* iii. 57 (*caption*) A [Malay] maker of musical instruments, working on a *rebab*. **1976** *Listener* 17 June 784/2 A musician in eastern Afghanistan who played a rebab, a bowed string instrument he had made out of a biscuit tin and a broom handle.

reback (rībæ·k), *v.* [f. RE- 5 a + BACK *v.* 2.] *trans.* To replace the damaged spine of (a binding or book). So **reba·cked** *ppl. a.*, **reba·cking** *vbl. sb.*

1901 D. COCKERELL *Bookbinding* xxii. 305 Re-backing. Bindings that have broken joints may be re-backed. **1952** J. CARTER *ABC for Book-Collectors* 150 Re-backed. This means that the binding of the book has been given a new backstrip or spine. **1970** *Bodl. Libr. Rec.* VIII. 192 Although it [*sc.* an eighteenth-century New Testament] has been rebacked and repaired..it is now very worn. **1981** *Sotheby's Catal. Printed Bks.* 16 Mar. 34, 179 Lambarde (W.), Dictionarium Angliae Topographicum & Historicum, portrait, calf, rebacked, 1730.

re-bar (rī·bāɹ). *U.S.* [f. RE(INFORCING *ppl. a.* + BAR *sb.*[1]] A steel reinforcing rod in concrete.

1961 *Publ. Amer. Dial. Soc.* xxxvi. 29 *Re-bar*,..abbreviated term for reinforcing steel, which is used to strengthen nearly all concrete structures. **1974** *Spartanburg* (S. Carolina) *Herald* 24 Apr. A1/6 Reinforcing steel bars, known in the industry as re-bars, are used to reinforce concrete in the construction of highways, bridges, buildings and other structures. **1979** *Civil Engineering* Nov. 39/2 A rebar bolt has exponential load transfer characteristics.

rebarbarization. (Further example.)

1949 WELLEK & WARREN *Theory of Lit.* 340 For the 'rebarbarization' of literature, cf. the brilliant article, 'Literature' by Max Lerner and Edwin Mims.

rebarbative, *a.* Delete *rare* and substitute for def.: Repellent, forbidding; unattractive, dull; unpleasant, objectionable. (Further examples.)

1927 *Observer* 1 May 20 The small minority that is not put off by the dry and rebarbative quality of the tuition. **1946** BEERBOHM *Mainly on Air* 67 Even A. B. Walkley..found Ibsen rather rebarbative. **1958** I. MURDOCH *Bell* iv. 48 Still, everyone appeared to be extremely nice, except that that Dr. Greenfield man was a trifle rebarbative. (This was a word which Toby had recently learnt at school and could not now conceive of doing without.) **1958** J. PRESS *Chequer'd Shade* i. 12 Unless he commands sufficient poetic authority to compel the general acceptance of his minted coinage his work will inevitably be judged rebarbative and obscure. **1963** *Listener* 7 Mar. 436/1 Some rebarbative club member does seem to have opened one or two windows in the smoking-room. **1971** P. D. JAMES *Shroud for Nightingale* iii. 63 Her face matched her personality, rebarbative and defensive. **1976** *Gramophone* Aug. 266/1 An accessible musical language can be equally deceptive and pose as many (though of course different) problems as an entirely novel and rebarbative idiom.

Hence **reba·rbatively** *adv.*; **reba·rbativeness**; **rebarbati·vity.**

1947 I. BROWN *Say Word* 102 Is Max our only dealer in rebarbativity? **1966** *Punch* 16 Nov. 754/3 A veritable drummer boy, all knobbly North Country rebarbativeness and almost incomprehensible obsolete dialect. **1968** I. MURDOCH *Nice & Good* vii. 57 Beyond Uncle Theo were some alien holiday-makers,..of whom this part of the beach happily attracted few, because of its rebarbatively stony nature. **1975** M. AMIS *Dead Babies* lvii. 214 During the Americans' twenty-minute absence from the sitting-room Celia joined in her husband's wholly successful attempt..to moderate Roxeanne's rebarbativeness to the odd aside. **1976** *Listener* 22 Apr. 495/1 A rejection of formal logic and with it of a rebarbatively technical foundation and medium for philosophical thinking.

rebat(e, varr. *RABAT *v.*

rebate, *sb.*[1] Add: Now usu. with pronunc. (rī·bē[1]t). (Further, incl. *attrib.*, examples.) Also *Comb.*

1869 *Bradshaw's Railway Man.* XXI. 448 Rebate account..£1,348. **1907** *Daily Chron.* 9 Dec. 4/3 They cannot dislodge the Welsh makers [of tin-plates] from their hold of the rebate trade. **1908** *Times* 1 Feb. 5/2 The rebate-taker, the franchise-trafficker, the manipulator of securities..and the man-killer all alike work at the same web of corruption. **1908** *Daily Chron.* 1 Feb. 5/3 The President [*sc.* Roosevelt] attacks by name..the Atchison and Santa Fé Railway.. for its rebate practices and intervention in the money market. **1955** *Times* 10 May 15/3 An interesting feature of German price lists for the outside world is the appearance of rebates to shipbuilders—rebates for indirect exports, which recall the cartel. **1957** CLARK & GOTTFRIED *University Dict. Business & Finance* 292 In current usage, a rebate is distinguished from a discount in that the former is not taken out or deducted in advance, but is handed back after payment of the full amount. **1965** H. K. COMPTON *Gloss. Purchasing & Supplies Managem. Terms* 115 *Rebate*, an allowance (or discount) on price, usually given after the completion of the contract, and most frequently based on some relationship with the business turnover. **1965** *McGraw-Hill Dict. Mod. Econ.* 426 Rebates to large and more favored shippers were used extensively in the U.S. railroad industry in the nineteenth century as a form of price discrimination. **1976** *Milton Keynes Express* 2 July 21/2 The decision to ban rebate tenants at Fishermead was revealed in this month's issue of the magazine 'City Limits'.

rebate *v.*[1] Add: **2. d.** To pay back (a sum of money) as a rebate; to give a rebate on.

1957 CLARK & GOTTFRIED *University Dict. Business & Finance* 292/2 Under customs regulations..import duties paid on goods which are later re-exported may be rebated in part or in full. **1977** *Time* 21 Feb. 49/1 Much of the energy tax would have to be rebated in some form to the poor to help them meet higher living expenses.

Hence **rebatable** (rībē[1]·tăb'l, rī·bē[1]:tăb'l), *a.*

1972 *Daily Tel.* 15 Feb. 17 The CBI's two main recommendations are a 2½ p.c. cut in corporation tax..and a halving of the rebatable heavy oil duty.

rebb, var. *REB[2].

rebbe (re·bə). Also with capital initial. [Yiddish, f. Heb. *rabbī* RABBI *sb.*] A rabbi; *spec.* a Chasidic religious leader.

1881 *Encycl. Brit.* XIII. 681/2 The Zaddikim (or 'righteous') and Rebbes, as their leaders are called, live in magnificence upon the contributions of the most ignorant of the people. **1882** [see *CHEDAR]. **1965** J. A. MICHENER *Source* (1966) 45 Two others prayed in loud voices on a line of their own, while the old rebbe, incredibly ancient, Cullinane thought, mumbled prayers that no one else could have heard. **1967** *N.Y. Times* 22 Mar. 29 (Advt.), The Lubavitcher Rebbe, Rabbi..Schneerson, has issued his annual call to world Jewry to observe the Purim festival. **1978** I. B. SINGER *Shosha* xiv. 243 To me..you are my rebbe. Your every word is filled with wisdom and love of God as well.

rebbitzin (re·bitsin). Also **rebbetzin, rebbitzen** and with capital initial. [Yiddish, fem. of *REBBE.] The wife of a rabbi.

1892 I. ZANGWILL *Childr. Ghetto* I. iv. 110 'His third wife was Kitty Green..,' persisted the Rebbitzin. **1926** S. ASH *Kiddush Ha-Shem* II. ii. 109 She was standing in the next room together with her mother, the *rebbetzin*. **1965** J. A. MICHENER *Source* (1966) 670 For the next ten years Leah, the young rebbetzin of Eliezer bar Zadok, knew only the Judenstrasse. **1966** H. KEMELMAN *Saturday Rabbi went Hungry* (1967) xxxvii. 236, I just called the hospital and they told me the rebbitzen had a boy. **1972** —— *Monday Rabbi took Off* v. 43 Believe me, the *rebbitzin* is real class, a Wellesley graduate. **1978** I. B. SINGER *Shosha* xii. 212 Mother's old-fashioned clothes brought a condescending expression from the rebbitzen.

rebec. Now the usual form of REBECK *sb.*[1] a (in Dict. and Suppl.).

1755 [see REBECK *sb.*[1] a]. **1915** A. DOLMETSCH *Interpretat. Mus. 17th & 18th Cent.* vii. 463 A little band consisting of a bass-viol, a tenor viol, a violin or rebec..and a flute. **1932** R. DONINGTON *Work & Ideas A. Dolmetsch* 15 It is probably the rebecs that are the nearest relatives of the violins... A number of fine rebecs have recently been finished... Now that the difficulties of the preliminary research work have been overcome..the rebecs are not difficult or costly to make. **1955** *Times* 20 May 3/5 A concert for children will consist of recorder and harpsichord pieces..and 'French brawls' for rebec and tambourin. **1980** *Early Music* Jan. 87/3 Progress Instruments, by applying modern plastics to make the backs of rebecs and plucked stringed instruments, are seeking to bring them within everyone's purse without sacrificing their authenticity of tone.

Rebecca. Add: **a.** (Further examples.)

1843 *Times* 10 Jan. 6/3 There has been..a mob of lawless depredators..who assembled nightly for the purpose of destroying the turnpike gates... These ruffians are headed by a very tall man, dressed for disguise as a female who goes by the name of Rebecca, and, as many of his associates are likewise dressed as females, the whole gang have been christened 'Rebecca and her daughters'. *Ibid.* 22 June 5/5 (*heading*) Rebecca riots.

c. Var. *REBEKAH.

Rebeccaite (earlier examples); also *transf.*

1843 A. SEDGWICK *Let.* 23 June in *Life* (1890) II. i. 56 Dress me like one of the Welsh Rebeccites [*sic*], and call me a water-nymph. **1878** F. KILVERT *Diary* 25 Nov. (1940) III. 434 A large party of Rebaccaites being out spearing salmon below Rhayader Bridge.

rebeck, *sb.*[2] For 'Now only *Hist.* or *poet.*' read 'Chiefly *Hist.*' See *REBEC.

re-become, *v.* (Further examples.)

1920 A. HUXLEY in E. Sitwell *Wheels* (Fifth Cycle) 35 And the stretched gargoyles re-become Women and men! **1938** L. MACNEICE *I crossed Minch* II. ix. 131 It is useless to ask you..to re-become your ancestor the Irish peasant.

re-beget, *v.* (Later example.)

1935 W. EMPSON *Poems* 19 Searching the cave gallery of your face My torch meets fresco after fresco ravishes Rebegets me.

re-begin, *v.* (Later example.)

1930 AUDEN *Poems* 44 And love's worn circuit re-begun.

Rebekah (rībe·kă). *U.S.* Also **Rebecca.** [The form, in the Authorized Version of the Bible, of the name *Rebecca*, used in allusion to Gen. xxiv. 60.] A member of a society or 'order' of women, founded in Indiana in 1851 as a complementary organization to that of the Odd Fellows. Also *attrib.*

1860 C. E. DE LONG *Jrnl.* 27 July in *Calif. Hist. Soc. Q.* (1931) X. 252 Met in the Lodge Room. Rebeccas & Brothers & had a free & easy. **1913** *Chicago Record-Herald* 16 Mar. v. 6/5 The staff of Maple Leaf Rebekah Lodge, No. 369, will confer the Rebekah degree on a large class of candidates. **1930** *Randolph Enterprise* (Elkins, W. Va.) 16 Jan. 5/4 They sure have a fine bunch of Odd Fellows and Rebekahs down there. **1949** *Milwaukie* (Oregon) *Rev.* 28 July 6/3 Mrs. Wanda Million and Mrs. Esther Lineagar were initiated into the Milwaukie Rebekah lodge at last week's meeting in the city hall. **1972** *Fairbanks* (Alaska) *Daily News-Miner* 3 Nov. 7/2 (*caption*) Rebekahs—Members of the Rebekahs are interested in forming an Independent Order of the Odd Fellows in Fairbanks.

rebel, *a.* and *sb.*[1] Add: **A.** *adj.* **1. a, b.** (Further attrib. examples.)

1861 O. W. NORTON *Army Lett., 1861–65* (1903) 24 The rebel camps are within two miles of us. **1937** *Granta* 3 Feb. 219/1 Seeing the contrast between the military efficiency of the rebel army and the unpreparedness of the people for war [at the time of the Spanish Civil War 1936–39]. **1944** H. FAST *Freedom Road* 167 It was not essentially a problem of reconstruction, not even a problem of readmission of the rebel states into the Union. **1963** *Times* 11 May 11/5 Dubbed the Rebel City for espousing the cause of Perkin Warbeck in 1492, Cork has always maintained a fighting reputation. **1980** *Times* 3 Jan. 1/7 The sources said..that the Russians were using sophisticated M124 helicopter gunships against rebel strongholds.

2. a. (Later examples.)

1931 *Ann. Reg. 1930* II. 49 The Prime Minister's speech, as was to be expected, was considered unsatisfactory by the 'rebel' group in the Labour Party. **1976** *Southern Even. Echo* (Southampton) 18 Nov. 17/5 Action would now be concentrated on Albert Johnson quay where most of the rebel dockers worked.

B. *sb.*[1] **1. a.** (Further examples.)

1902 KIPLING *Traffics & Discoveries* (1904) 32, I shot my Bible full of bullets after Bloemfontein went... Take it and pray over it before we Federals help the British to knock hell out of you rebels. **1966** *BBC Handbk. 1966* 97 Stanleyville radio was in the hands of the rebels in the Congo (Leopoldville) Republic. **1976** *Daily Tel.* 20 July 4/2 Armed men, believed to be Moslem rebels, ambushed a bus in the southern Philippines on Sunday.

d. A supporter of the American cause during the War of Independence (1775–83). **e.** A supporter of the Southern, or Confederate, cause during the American Civil War (1861–5); hence, by extension, used colloq. for SOUTHERNER 2. Chiefly *U.S.*

1775 *Massachusetts Spy* 3 May 3/1 The commanding officer accosted the militia in words to this effect, 'Disperse you damn'd rebels!' **1788** *Ann. Reg. 1776* I. 181 Gen. Clinton, with two brigades of British..were sent to make an attempt upon Rhode Island... The rebels having abandoned the island at their approach, they took possession of it without the loss of a man. **1847** *Knickerbocker* XXIX. 54 Mrs. Mowatt is..a great granddaughter of one of those old 'rebels' who signed the Declaration of Independence. **1861** E. COWELL *Diary* 15 Apr. in *Cowells in Amer.* (1934) 290 [The] proceeding caused the necessary diversion of 'the rebels'' course. **1864** [in Dict., sense 1 a]. **1895** W. H. CHAMBLISS *Diary* 305 The malignant epithets, 'Yankee' and 'Rebels',..were invented by fanatics and foreigners to aggravate our interstate quarrel. **1905** A. C. RICE *Sandy* 123 'Was he a rebel?'.. 'He was a Confederate, sir! I never knew a rebel.' **1929** *Amer. Speech* IV. 344 *Rebel*, a Southerner. **1938** *Oklahoma Supreme Ct. Rep.* CLXXXIII. A. 509/2 The Northern man is often referred to as 'Yankee' and the Southern man as 'Rebel'. **1959** W. PETERS *Southern Temper* xiii. 211 [The] publisher of the Augusta *Courier*.. exudes pleasantness and good will, even when his caller is a Northern newspaperman or writer. 'Well,' he said to one such not long ago, 'I guess you've come down here to give us rebels hell.'

2. a. (Later example.)

1955 *N.Y. Times* 27 Oct. 28/2 Mr. Dean..is a mixed-up rebel because his father lacks decisiveness and strength.

b. *rebel without a cause*: the title of a cinematographic film released in the U.S. in 1955, applied to a (young) person whose aggressive behaviour is attributed to feelings of frustration or insecurity rather than to loyalty to a particular cause.

[**1955** *N.Y. Times* 27 Oct. 28/2 It is a violent, brutal and disturbing picture of modern teenagers that Warner Brothers presents in its new melodrama..'Rebel Without a Cause'. Young people neglected by their parents or given no understanding and moral support by fathers and mothers who are themselves unable to achieve balance and security in their homes are the bristling heroes and heroines.] **1963** *Times* 30 May 15/6 It is a story of a rebel-without-a-cause.. and of the gradual disillusioning of a younger man..with the romantic image that at first glance he presents.

3. *rebel-hearted, -held* adjs.; **rebel yell**, a characteristic shout or battle-cry uttered by Confederate soldiers during the American Civil War; also *transf.*

1926 C. DAY-LEWIS in *Oxford Poetry* 19 What sense Have they the pioneer-minded, the rebel-hearted, If man's fulfilment rest on no 'perhaps' Outside him? **1966** *BBC Handbk. 1966* 98 The Service's East African Unit..became..the only source of news from rebel-held territory [in the Congo Republic]. **1863** A. J. L. FREMANTLE *Jrnl.* 2 July in *Three Months in Southern States* 265 The Southern troops, when charging, or to express their delight, always yell in a manner peculiar to themselves... The Confederate officers declare that the rebel yell has a particular merit, and always produces a salutary and useful effect upon their adversaries. A corps is sometimes spoken of as a 'good yelling regiment'. **1868** *Harper's Mag.* Sept. 488/1 A tall woman..uttered a long, piercing cry, which Humphreys afterward described as 'a rebel yell', and Alec as 'a keen whoop'. **1936** M. MITCHELL *Gone with Wind* vi. 121 Stuart Tarleton's voice rose, in an exultant shout, 'Yee-aay-ee!' as if he were on the hunting field. And she heard for the first time..the Rebel yell. **1945** S. LEWIS *Cass Timberlane* (1946) xxix. 193 No dance at the Heather Country Club was canonical without the presence of Jay Laverick, emitting the rebel yell. **1974** P. RUELL *Death takes Low Road* x. 126 For God's sake, lassie, can you no' keep your voice down to less than a rebel yell?

rebel, *v.* **2.** For † *Obs. rare*[—1] read *rare* and add later example.

1908 A. S. M. HUTCHINSON *Once aboard Lugger* IV. ii. 219 To-day the empress sway of conventionality is rarely rebelled.

rebeldom. 1. (Further examples.)

1887 J. D. BILLINGS *Hardtack & Coffee* 280 Afterwards [soldiers] went by thousands into other sections of Rebeldom. **1893** [see *CHINCAPIN].

† **rebelism** (re·bĕliz'm). *U.S. Obs.* [f. REBEL *sb.*[1] + -ISM.] Adherence to the principles or practice of the Confederates during the American Civil War (1861–5).

1862 *Constitution* (Middletown, Conn.) 26 Mar. 4/1 There is a good deal of rebelism in the *Old Bailie*. **1864** *Harper's Mag.* July 271/1 He..is silent or pretends to rebelism when his mistress..is by. **1867** *Congress. Globe* 10 Dec. 103/3 The action of Congress can have no other effect than to embarrass the work of reconstruction,..to feed the spirit of rebelism and incite insubordination.

rebelly, *a.* Delete *rare* and add later examples.

1936 'F. O'CONNOR' *Bones of Contention* 179 The child and grandchild of rebelly men. **1959** *Manch. Guardian* 2 July 5/4 Finsbury..seems always to have been a fairly rebelly quarter.

rebid (rībi·d, rī·bid), *v. Bridge.* [RE- 5 a.] *trans.* and *intr.* To bid (*BID *v.* 3 c) again. Hence **re·bid** *sb.*; **rebi·ddable** *a.*

1923 *Daily Mail* 6 Oct. 6/4 Z has called '3 clubs'. A bids '2 spades'. Y or Z may point out the insufficiency or may in turn re-bid 3 clubs. **1927** M. C. WORK *Contract Bridge* iii. 27 It is not the one-bids, or the ones jumped to twos, which produce the heavy sets in Contract; it is the game-going jumps and the rebids to reach game-going declarations that do the damage. **1929** —— *Compl. Contract Bridge* iv. 25 One of them being strong enough to bid two of a major, and East not being able to rebid. **1945** PHILLIPS & REESE *How to play Bridge* vi. 45 If your hand is weak, but your suit is rebiddable, you should bid Three of it. **1958** *Listener* 9 Oct. 572/3 The Diamonds are not rebiddable. **1971** *Daily Tel.* 21 Aug. 8/3 East contested the part score with 3 diamonds, judging that his two honours would be adequate support opposite the rebid suit. **1975** *Times* 11 Jan. 7/5 By rebidding his suit he imagined that he was asking his partner to choose between Hearts and Spades. **1976** *Cumberland News* 26 Nov., In many respects rebids are more important than opening bids or first responses for very rarely do these bids give precise information about the hands.

rebind, *v.* (Earlier and later examples.)

1820 *Rep. Comm. Public Rec. 1800–1819* 525 The total sum of £1,000 per Annum..to the Purposes of maintaining ..and..occasionally re-binding..the Public Records. *Ibid.* 527 Many different Series of Records..have been carefully repaired and suitably rebound. **1901** D. COCKERELL *Bookbinding* i. 18 Nearly all librarians complain that they have to be continually rebinding books. *Ibid.* xxii. 306 When the sewing cords or threads of a book have perished it should be re-bound. **1946** H. J. PLENDERLEITH *Preservation of Leather Bookbindings* 10 Some [volumes of a catalogue] have actually had to be rebound. **1963** B. C. MIDDLETON *Hist. Eng. Craft Bookbinding Technique* ii. 8 Thousands of old books..were rebound in modern morocco.

rebinding, *vbl. sb.* (Earlier and later examples.)

1857 D. LAING *Penni Worth of Witte* p. iii, It must have suffered in the rebinding, by being rather unsparingly cut in the edges. **1931** A. ESDAILE *Student's Man. Bibliogr.* vi. 178 Most bibliographers are aware of the great mass of important knowledge which has been destroyed by careless re-binding.

‖ **reblochon** (rəblɔʃoṅ). [Fr.] The name of a soft French cheese made originally and chiefly in Savoy.

1908 DOANE & LAWSON *Varieties of Cheese* 43 Reblochon. This is a soft French cheese weighing 1 to 2 pounds. It is made from fresh whole milk which is curdled with rennet at a temperature of 80 °F. or above, the time allowed being about thirty minutes. **1931** F. COLCHESTER-WEMYSS *Pleasures of Table* vii. 70 Reblochon is a delightful cheese, and not at all awkward as regards keeping. **1935** O. BURDETT *Little Book of Cheese* iv. 59 Though not often to be had in England, and, perhaps, not very often beyond the borders of Savoy, *Reblochons* must be mentioned. It makes a convenient postscript to Brie, for it is a smallish round, flat cheese, very creamy and rich in flavour. **1967** *House & Garden* Apr. 96/3 Fertilizers mask the taste of grass... Reblochon is losing its hint of gentian. **1971** *Sunday Times* 28 Mar. (Colour Suppl.) 34/4 *Reblochon*, the most distinguished of the cheeses of the Haute Savoie. Made with the yield of the second milking of the fawn Tarentais cow (*reblochе* in the local patois means 'second milking'). **1973** J. M. WHITE *Garden Game* 44 He cut off a sliver of *reblochon* and sampled it... He waved the cheese-knife.

reboa·ntic, *a.* Chiefly *poet.* [f. REBOANT *a.* + -IC.] = REBOANT *a.*

1903 KIPLING *Five Nations* 124 When the Conchimarian horns Of the reboantic Norns Usher gentlemen and ladies With new lights on Heaven and Hades. **1907** *Academy* 23 Mar. 298/1 Even then the Norns couldn't be reboantic, supposing that word to be correctly derived from *reboo* = to bellow back.

reboard, *v.* Delete *rare* and add later examples. Also *absol.*

1906 *Westm. Gaz.* 24 Mar. 10/2 Her bulwarks kept her afloat, and the crew reboarded her and brought her safely to Milford Haven. **1977** *R.A.F. News* 5–18 Jan. 3/2 He was winched to safety and later put down on the tanker to..reboard the catamaran. **1977** *Western Living* (Vancouver) Apr. 33/2 The conductor ordered all to reboard.

rebody, *v.* Restrict †*Obs.*[—1] to sense in Dict. and add: **2.** *trans.* To furnish (a motor vehicle, etc.) with a new body. Hence **rebo·died** *ppl. a.*

1963 *Times* 19 Jan. 3/6 His car had to be rebodied. **1977** *Custom Car* Nov. 30/3 Their Rio was slated by the press, who saw it as nothing more than a rebodied Dolomite.

reboiler (rīboi·ləɹ), *v.* [RE- 5 c.] *trans.* To fit with a new boiler.

1889 [in Dict. s.v. RE- 5 c]. **1908** *Westm. Gaz.* 25 Apr. 2/3 The Banter Line decided to reboiler the *Caesar* and go for the Atlantic record. **1961** *Guardian* 5 June 10/6 One of the two original engines..was..reboilered just a year before the line closed.

reboiler (rīboi·ləɹ), *sb.* [f. REBOIL *v.*[2] + -ER[1].] A heater for vapourizing the liquid at the bottom of a fractionating column.

1956 MCCABE & SMITH *Unit Operations of Chem. Engin.* xii. 685 The liquid feed flows down the column to the still, which in this type of plant is called the 'reboiler', and is subjected to rectification by the vapor rising from the reboiler. **1966** *McGraw-Hill Encycl. Sci. & Technol.* XIII. 185/2 Many [stripping] processes employ a combination of all three; that is, after absorption at elevated pressure, the solvent is flashed to atmospheric pressure, heated, and admitted into a stripping column which is provided with a bottom heater (reboiler). **1977** F. G. SHINSKEY *Distillation Control* ix. 273 Multiple reboilers are generally used in severe fouling service, so that they may be individually cleaned without interrupting operation of the column.

re-book, *v.* (Further examples.)

1898 G. GISSING *Let.* 27 Jan. in R. A. Gettman *G. Gissing & H. G. Wells* (1961) 85 The trunk..will have to be..rebooked to Rome. **1906** *Westm. Gaz.* 6 Sept. 8/3 Such a bylaw..compels a passenger to get out and rebook, even though he miss his train. But he would still have to rebook if going by a later train. **1978** 'S. WOODS' *Exit Murderer* 55 I've re-booked your room for you.

rebop: see *BEBOP.

rebore (rī·bō[ə]ɹ), *sb.* [f. the vb.] A re-boring of one or more cylinders of an internal-combustion engine; also, an engine which has had its cylinders re-bored.

1954 'TORRENS' *Motor Cyclist's Workshop* (ed. 6) xx. 126 When your piston is 'worn out', your cylinder is sorely in need of a rebore or..regrinding. **1967** J. MILLS *Low-Cost Car Repairs* iii. 75 The check is to determine whether a recon-set can be used or whether a rebore is needed. **1970** K. BALL *Fiat 600, 600D Autobook* i. 13 A rebore will require new pistons and rings a size larger than the original and these will usually be supplied by the garage..undertaking the rebore. **1977** *S. Wales Echo* 18 Jan. (Advt.), Exchange crankshafts, camshafts, short engines, rebores, pistons.

reborrowing (rībǫ·ro[u]iŋ), *vbl. sb.* [f. REBORROW *v.* + -ING[1].] The action of the vb. REBORROW (esp. in *Philol.*); also, that which is reborrowed.

1869 *Bradshaw's Railway Man.* XXI. 221 Pay off and create in lieu of reborrowing. **1933** L. BLOOMFIELD *Language* xxv. 452 The same re-borrowing of this last word appears in Old English. **1953** K. JACKSON *Lang. & Hist. Early Brit.* II. 332 Förster notes that the AS. form of Nene is spelt five times in Peterborough documents as *Nyn*, and regards this as a re-borrowing. **1979** *Amer. Speech* LIV. 29 Of the 6 pure reborrowings, 4 are from *cent*: Indonesian *sen*, Samoan *sene*, Tongan *seniti*, and Swahili *senti*.

rebound, *sb.* Add: Now usu. with pronunc. (rī·baund). **3. a.** *spec.* in *Basketball, Football*, etc., the return of a ball from the backboard, goal-post, etc., after an unsuccessful shot; a ball that rebounds in this manner. Also *attrib.*

1922 W. E. MEANWELL *Basket Ball for Men* vi. 61 The opposing attack is following in hard for the rebound. *Ibid.*, The hook pass..permits the guard to reverse and present his back to the oncoming rebound man. **1948** *Sun* (Baltimore) 7 Jan. 11/1 Jimmy Kirby..tapped in four field goals when taking rebounds. **1954** *Basketball* ('Know the Game'

Series) 35 When a shot is missed obtaining the rebound is vital... Although defensive players usually start with the best position for the rebound offensive, players must make every effort to beat the defender to the best rebound positions. **1955** *Sun* (Baltimore) 4 Feb. 15/1 He is an outstanding rebound man and the team's third leading scorer. **1969** *Eugene* (Oregon) *Register-Guard* 3 Dec. 1D/3 'It was the board play that killed us,' continued Belko, pointing to Wichita's 62–43 edge in rebounds. **1971** *N.Y. Times Guide Spectator Sports* iii. 78 A man much taller than the rest..had the best chance to retrieve the 'rebound' over the heads of smaller players. **1976** *Western Mail* (Cardiff) 22 Nov. 16/3 Preece was on hand to fire home the rebound from close range. **1978** *Dumfries Courier* 13 Oct. 4/6 Milligan levelled the scores for Girvan connecting with a rebound after a shot had been blocked in the goalmouth.

c. (Examples in *Med.*) Freq. *attrib.*

1951 *Jrnl. Clin. Endocrinol.* XI. 235 (*heading*) Spermatogenic rebound phenomenon after administration of testosterone propionate. **1954** *Jrnl. Nerv. & Mental Dis.* CXX. 46 (*heading*) Abstention [from addictive drugs], rebound, and readjustment. **1974** M. C. GERALD *Pharmacol.* vi. 115 The relief produced by the shrinkage of swollen nasal membranes is merely temporary and is followed by rebound congestion. *Ibid.* xv. 288 More often death has been attributed to secondary causes such as..suicides arising from rebound depression during periods of drug withdrawal or abstinence. **1977** *Lancet* 9 Apr. 774/1 Thrombocytopenia caused by alcohol is reversible after alcohol withdrawal, and is followed by rebound thrombocytosis. **1978** M. G. HARMATZ *Abnormal Psychol.* vii. 162/1 Patients felt even more deeply depressed when drug action wore off than before they had treatment (a rebound effect). Modern antidepressants..avoid these side effects.

d. Restrict † *Obs.* to phrases in Dict. and add: Also † *in* or *on the rebound*: during a period of reaction following an emotionally disturbing experience, esp. a broken engagement or a refusal of marriage. Also used without *prep.*

1853 MRS. GASKELL *Ruth* III. iv. 129 His first rebound to Jemima. **1859** *Harper's Mag.* Aug. 341/1 Ellen Bond caught his heart 'in the rebound', as somebody says. **1861** C. M. YONGE *Young Step-Mother* xxi. 295 We may steer her safely through, above all, if one of the six cousins will but catch him in the rebound. **1864** G. J. WHYTE-MELVILLE *Queen's Maries* II. xxxviii. 234 It is an old saying that 'many a heart is caught on the rebound'. *a* **1899** V. S. LEAN *Collectanea* (1904) IV. 41 Many a heart is caught in the rebound, *i.e.* after a repulse by another. **1921** GALSWORTHY *To Let* III. x. 290 Ah! it was strange—this marriage. The young man, Mont, had caught her on the rebound, of course, in the reckless mood of one whose ship has just gone down. **1931** H. S. WALPOLE *Judith Paris* III. i. 428 She has but accepted him on the rebound from her trouble with Beaminster. **1969** A. GLYN *Dragon Variation* ix. 283 Maybe the girl had had a fight with Carl, and had turned to Jeff on the rebound. **1977** A. MORICE *Murder in Mimicry* I. ix. 91 Her passion for Gilbert was already on the wane... The rebound was in full swing. **1978** *Lancashire Life* Sept. 47/2 Perhaps it was on the rebound from this disappointment that he contracted his disastrous first marriage.

rebound, *v.* Add: **1. b.** (Further examples.)

1955 *Times* 9 Aug. 12/2 Equities rebounded sharply yesterday after their reaction last week although they did not always fully hold their gains. **1979** *Arizona Daily Star* 1 Apr. D2/3 Coal prices just don't seem to want to rebound.

e. *Basketball.* To catch a rebound (*REBOUND *sb.* 3 a).

1954 *Sun* (Baltimore) 20 Dec. B18/2 Pollard is the best balanced basketball player I have ever seen;..he can rebound with the best big men in the business. **1974** *State* (Columbia, S. Carolina) 3 Mar. 1-D/1 Kelley went to the line for a one-and-one at 1:04 and missed. Stewart rebounded and McCurdy failed to connect twice.

reboundant (rībau·ndănt), *a. Her. Obs.* (exc. *Hist.*). [f. REBOUND *v.* + -ANT[1].] = REVERBERANT *a.* 1.

1688 [see REVERBERANT *a.* 1]. **1828** W. BERRY *Encycl. Heraldica* I., *Reboundant*, or *Rebounding*, a term sometimes used, in ancient heraldry, for the tail of a lion, when turned up like the letter S, the end of the tail being outwards, which was called *reverberant, beaten back*, or *reboundant*, as if the animal had beaten it to its back and it had rebounded; the usual way of turning the end of the tail, in ancient times, being inwards; but it is now seldom borne otherwise than outwards, and, therefore, not noticed in the blazon. **1878** B. BURKE *General Armory* p. xliv, *Reboundant*, an ancient term for the tail of a lion when turned up and bent in the form of a letter S with the point outwards, the ancient way of depicting the tail was usually with the point turned towards the back, unless blazoned reboundant. **1910** W. A. COPINGER *Heraldry Simplified* 368/1 *Reboundant*, or Rebounding, formerly applied to the tail of a lion when turned up like the letter S.

rebounder. Add: **b.** *Basketball.* A player who is skilled in catching rebounds (*REBOUND *sb.* 3 a).

1949 *Sun* (Baltimore) 28 Dec. 14/6 Dolhou goes on the injured list with..a growing reputation as an adept floorman, rebounder and set shot. **1954** *Ibid.* 20 Dec. B18/3 What more do you want..shooter, playmaker, rebounder, durability. **1974** *Spartanburg* (S. Carolina) *Herald-Jrnl.* 21 Apr. B1/1 He started on the basketball team..playing the point position as playmaker and still being one of the leading rebounders for Coach Bill Hinson's Viking team. **1979** *N.Y. Times* 25 Nov. 1. 10/3 In winning the Southeastern Conference regular-season title and posting a 23–8 mark, Dale Brown had size, quickness, shooters, ballhandlers, rebounders and experience.

rebounding, *vbl. sb.* Add: **b.** *spec.* in *Basketball*, the action of catching a rebound (*REBOUND *sb.* 3 a). Also *attrib.*

1954 *Basketball* ('Know the Game' Series) 35 (*heading*) Rebounding. **1957** *Encycl. Brit.* III. 181 *Rebounding*,..the fundamental ability to get proper position, time the jump and properly retrieve rebounds is a vital part of the game. **1969** *Eugene* (Oregon) *Register-Guard* 3 Dec. 1D/4 The rebounding was that close only because the hosts had trouble getting going on the backboards. **1972** *N.Y. Times* 4 June 7/4 Bill Moore, who holds most of Susquehanna University's rebounding records, has been appointed assistant basketball coach at the university. **1976** *Springfield* (Mass.) *Daily News* 22 Apr. 39/1 We lost this game chiefly for two reasons: offensive rebounding (16–10 Celts with Cowens pulling nine) and turnovers (the Braves coughing up 30 to 17).

rebours, *a.* and *sb.* Restrict † *Obs.* to senses in Dict. and add: **B.** *sb.* Also in phr. ‖ *à rebours* (a rəbūr), in the wrong way, perversely; through perversity.

1906 W. JAMES *Let.* 3 Apr. in R. B. Perry *Thought & Character W. James* (1935) II. 393 How lamely I, for one, must have expressed myself to be taken so *à rebours*. **1939** T. S. ELIOT *Idea Christ. Society* i. 19 It [*sc.* a dislike of everything maintained by Germany or Russia] may..lead us to be mere imitators *à rebours*, in making us adopt uncritically almost any attitude which a foreign nation rejects. **1949** I. DEUTSCHER *Stalin* iv. 96 In making a fetish of the underground, in shying away from the broader opportunities for action, they tended to reduce the resolution to impotence. They were liquidators *à rebours*. **1951** M. LOWRY *Let.* 25 Aug. (1967) 262 While he took no human action at all.. some principle of tyrannic yet thwarted force in his feeling has worked against him, *à rebours*: now he does take action ..mysteriously the thing begins to work for him.

rebozo, reboso. (Earlier and later examples.)

1807 J. PINKERTON *Mod. Geogr.* (rev. ed.) III. 185 The Mexican ladies..when they are at home, or go out in a carriage,..wear what is called the *rebozo*, or muffler, like the shawls now used at Madrid. **1844** J. GREGG *Commerce of Prairies* I. 216 A Mexican female is scarcely ever seen without her rebozo or shawl. **1927** W. CATHER *Death comes for Archbishop* I. ii. 25 She was dressed in black, with a white apron, and a black reboso over her head, like a Mexican woman of the poor. **1947** J. STEINBECK *Wayward Bus* 235 The little modest dark girls in blue *rebozos*. **1956** R. BRADBURY *October Country* 19 Behind him came women in black rebozos. **1977** *Guider* July 310/3 The dancers wear white dresses with coloured belts and rebozos tied from hand to hand.

rebranch (rībra·nʃ), *v.* [f. RE- 5 a.] *intr.* To ramify; to branch again. Hence **rebra·nching** *vbl. sb.*

1888 J. LE CONTE *Evolution* I. i. 14 A growing tree branches and again branches in all directions... Even so the tree of life, by the law of differentiation, branches and rebranches continually. **1895** W. H. HUDSON *Introd. Philos. H. Spencer* i. 19 That doctrine of the gradual branching and rebranching of species..went somewhat vaguely by the name of the development hypothesis. **1935** HUXLEY & HADDON *We Europeans* v. 161 Evolutionary branches may ..unite again after they have diverged and then either rebranch or remain united.

rebreathe *v.* Add: Hence also **rebrea·thing** *vbl. sb.* (freq. *attrib.*).

1897 W. MARCET *Contrib. Hist. Respiration in Man* 109 Curves were taken showing the effects produced by the rebreathing of six litres of air during two minutes. **1935** *Sun* (Baltimore) 17 May 5/2 Their faces are covered with four layers of gauze, muffling nose, mouth, eyes and ears. One purpose of the gauze is to increase slightly the rebreathing of exhaled carbon dioxide. **1960** *Brit. Jrnl. Anaesthesia* XXXII. 256 (*heading*) A rebreathing technique for the determination of arterial P_{CO_2} in the apnoeic patient. **1976** H. MACINNES *Death Reel* xxiii. 184 He was pulled in..by a diver using re-breathing apparatus.

rebroadcast (ri-), *v.* [RE- 5 a.] *trans.* To broadcast again; *spec.* to broadcast (a programme received from another station).

1923 *Daily Mail* 14 Aug. 5/3 A special orchestral concert ..will be relayed to all the broadcasting stations in Britain, and thence be re-broadcast by them on their own particular wave-lengths. **1939** *Sun* (Baltimore) 9 Feb. 1/2 Several hours after the vessel's first distress call, she rebroadcast her SOS appeal. **1948** *John o' London's Weekly* 10 Dec. 598/3 The difficulty in receiving the Scottish station makes it unlikely that more than a handful of Scott's English admirers will be able to judge how *Waverley* sounds on the air, but perhaps the serial will be re-broadcast later on another wavelength. **1951** A. C. CLARKE *Sands of Mars* viii. 94 It was a live programme, beamed to Mars..picked up and rebroadcast. **1965** *Economist* 9 Oct. 152/3 Plays supplied by the networks to their provincial affiliates, and already censored in Madrid, must be recensored before being rebroadcast. **1974** P. GZOWSKI *This Country* 12/2 What I said live to the Maritimes was recorded in Toronto and rebroadcast an hour later, then recorded in Winnipeg and so on.

Hence **rebroa·dcast** *ppl. a.*, **rebroa·dcasting** *vbl. sb.*; also **rebroa·dcaster**, a station that rebroadcasts material received from elsewhere.

1956 *Nature* 10 Mar. 451/2 There are fuller accounts of the External Services and of the re-broadcasting of the B.B.C. programmes throughout the world. **1957** *BBC Handbk.* 38 Programmes of this kind may be conveyed to the rebroadcaster either by short-wave transmission or as recordings by sea or airmail. **1962** *Ibid.* 102 The local station may record the BBC transmission for rebroadcasting later. **1969** *Listener* 17 July 91/3 The rebroadcast *Morals and Medicine* debate on abortion was a sterilised, factual affair by comparison. **1973** *Ibid.* 13 Sept. 348/3 No one could tell which programmes might one day be required..for rebroadcasting. **1974** *BBC Handbk.* 273/1 Every programme chosen for distribution to re-broadcasters is of the highest quality.

rebroa·dcast (rī-), *sb.* [f. the vb.] A repeat broadcast, esp. one of a programme received from another station; also, the action of broadcasting again.

1927 *Observer* 24 July 4/5 Rebroadcasts can be made on both sides of the Atlantic and..though they are interesting the quality is usually painful and always bad. **1939** *Sun* (Baltimore) 9 May 2/4 Thousands of British subjects heard the message as broadcast by the French stations or picked up through the short-wave rebroadcast from the United States. **1940** *Ibid.* 23 Feb. 1/4 Explaining the experiment, involving receipt of National Broadcasting Company programs from New York city by a receiver 130 miles distant, and simultaneous rebroadcast by Station W2XB, Dr. Baker said [etc.]. **1943** R. CHANDLER *High Window* x. 79 This ball game is a studio re-broadcast. **1963** *Guardian* 19 Apr. 12/2 The National Aeronautics and Space Administration had promised to allow some broadcasts direct from the capsule as well as the rebroadcast of television tapes. **1965** *Economist* 9 Oct. 152/2 Even [Spanish] commercials taped for rebroadcast day after day must be vetted daily. **1974** *Radio Times* 14 Mar. 30/3 Re-broadcasts of programmes from a fortnightly series on the arts.

rebuff (rī₁bʊ·f), *v.*[2] *rare.* [f. RE- 5 c + BUFF *a.* or *BUFF *v.*[2] 2.] *trans.* To restore to a buff colour.

1924 GALSWORTHY *White Monkey* III. x. 281 On both sides flat houses, recently re-buffed.

rebuffer (rībʊ·fəɹ). *rare.* [f. REBUFF *v.* + -ER[1].] One who rebuffs others.

1950 O. NASH *Family Reunion* (1951) 109 The medica tyrant, the social rebuffer.

rebuild, *v.* Add: (Further examples.)

1927 *Scribner's Mag.* Apr. 356/1 Another boy..wore a black silk handkerchief across his face because he had no nose then and his face was to be rebuilt... They rebuilt his face, but..they could never get the nose exactly straight. **1961** M. BEADLE *These Ruins are Inhabited* (1963) ii. 19 The fire..had gone out... While George rebuilt it, I set about frying bacon and eggs.

rebuild *sb.* (further examples); **rebuilt** *ppl. a.* (further examples); hence also **rebui·lded** *ppl. a.* (*poet.*).

1924 R. CAMPBELL *Flaming Terrapin* iii. 48 Sodom, rebuilded, scorns the wilting power. **1934** WEBSTER, *Rebuilt*.., of factory products, as typewriters, disassembled, reconstructed with new parts replacing those that are worn, and refinished. **1959** J. THURBER *Years with Ross* vii. 124 A rebuilt typewriter I had been using. **1960** K. M. KENYON *Archæol. in Holy Land* iv. 106 Sometimes the collapsed wall survived in the shape of an inverted U, and the rebuild consisted of a capping which might have foundations on the same level as the original wall. **1972** *World of Wild Wheels* (Custom Car) 10/1 As the moulding drops off they tend to leave it off. It's less work in the re-build. **1973** *Country Life* 1 Mar. (Suppl.) 11 Modern house being a rebuild of the original manor house. **1976** *Evening Post* (Bristol) 23 Apr. (Advt.), Bodywork, resprays, engine re-build, reasonable prices. **1978** *Gramophone* July 173/3 The Canterbury organ was ailing and unfit (in fact a rebuild is about to begin).

rebu·kative, *a. rare.* [f. REBUK(E *v.* + -ATIVE.] Disapproving, rebuking. So **rebu·katively** *adv.*

1924 'O. DOUGLAS' *Pink Sugar* iii. 30 Miss Dickson.. asked if we had called yet..and added, rather rebukatively, that Mr. M'Clandish had called at once. **1950** —— *Farewell to Priorsford* 219 Do you suppose she meant gangsters or only what Jane Austen called 'less worthy females'? No, Janet, I'm not backbiting, so don't look so rebukative.

rebunk (rībʊ·ŋk), *v.* [f. RE- 5 a + *BUNK *sb.*[4], after *DEBUNK *v.*] *trans.* To restore the reputation of or regard for (a person who has been debunked). So **rebu·nking** *vbl. sb.*

1960 *Times Lit. Suppl.* 6 May 291/4 We live in an age of rebunking, when it has become fashionable once more to take the Victorians almost as solemnly as they took themselves. **1962** *Ibid.* 11 May 341/1 Mr. Swanberg sets out neither to debunk nor to rebunk 'the Chief', but to describe him..as a social and political phenomenon.

rebunker (rībʊ·ŋkəɹ), *v.* [f. RE- 5 a + *BUNKER *v.* 1 b.] *intr.* To take in a further supply of coal or oil for consumption on a voyage.

1899 C. J. C. HYNE *Further Adv. Capt. Kettle* iii. 60 But wood, as compared with coal, is bulky stuff to carry, and as the stowage capacity of these stern wheelers is small, they have to make frequent calls to rebunker. **1929** *Daily Express* 14 Jan. 11/1 The Wimpole rebunkered and provisioned here and sailed for Grimsby on the midday tide.

rebu·rgeoning, *vbl. sb.* [f. REBURGEON *v.* + -ING[1].] A renewed budding or sprouting (in quot. *fig.*).

1929 'M. B. ELDERSHAW' *House is Built* ix. 210 The reburgeoning of life throughout the country did not leave this household quite untouched.

rebu·rial. [RE- 5 a.] A second interment (of a corpse).

1922 Z. GREY *To Last Man* viii. 185 It struck Jean as singular that neither Esther Isbel nor Mrs. Jacobs suggested a reburial of their husbands. **1980** J. BARNETT *Palmprint* v. 50 I've arranged for a re-burial service by the Reverend Solomon.

rebus, *sb.* (Later *attrib.* examples.)

1928 O. JESPERSEN *Internat. Lang.* II. 170 The number of roots admitted in primitive Esperanto was extremely small, and a good deal of ingenuity was used to express as much as possible by means of compounds and derivatives... The great number of these rebus-words..has deterred many intelligent people from Esp. **1969** A. PARPOLA et al. *Decipherment of Proto-Dravidian Inscriptions of Indus Civilization* ii. 8 It now appears that it [*sc.* the Indus script] is a purely logographic script, based on the so-called rebus principle. This means that each sign represents a whole word, which may comprise one or more syllables, and that a given word is expressed by a clearly recognizable picture of a quite different thing, which has, however, the same phonetic value.

re-bush (rību·ʃ), *v.* [f. RE- 5 c + BUSH *v.*[3]] *trans.* To provide with a replacement bush. Hence **re-bu·shing** *vbl. sb.*

1885 C. G. W. LOCK *Workshop Receipts* 4th Ser. 331/2 It is important..that enough metal is left round these holes to admit of their being re-bushed if necessary. **1970** K. BALL *Fiat 600, 600D Autobook* ix. 110/2 If this clearance is exceeded, replace either the roller shaft or rebush the shaft bore. *Ibid.* 112/1 Rebushing must be followed by reaming to a uniform internal diameter of ·867 inch ± ·0004 inch.

‖ **rebus sic stantibus** (rē[1]·bɒs sik stæ·ntibɒs), *phr.* [mod.L.] Things standing thus; provided that conditions have not changed; *spec.* in *International Law* used of the principle that a treaty lapses when conditions are substantially different from those which obtained when it was concluded; *clausula rebus sic stantibus* [*CLAUSULA], a clause to this effect.

[**1598** A. GENTILI *De Iure Belli Libri Tres* III. xiv. 599 Et tandem facto, ac silentio hic quoque, vt in priuatis, dices ratificari foedus... Item excipitur de rerum statu immutatio. Si mutatio nequit praeuideri. Etiam & adiecto iureiurando subintelligi clausula, *Rebus sic stantibus*. **1760** tr. *de Vattel's Law of Nations* I. II. xvii. 230 The only state of things, on account of which the promise is made, is essential to it, and the change of that state alone can lawfully hinder or suspend the effect of that promise. This is the sense which ought to be given to that maxim of the civilians, *Conventio omnis intelligitur rebus sic stantibus*.] **1849** R. WILDMAN *Institutes Internat. Law* I. iv. 175 The doctrine, that every treaty implies a condition of defeasance on any material change of circumstances, which is usually called the rule de rebus sic stantibus, is rejected by Grotius and Bynkershoek as calculated to destroy the obligation of all compacts. **1883** *Wharton's Law-Lex.* 693/2 *Rebus sic stantibus*, at this point of affairs. **1927** *Amer. Jrnl. Internat. Law* XXI. 509 The principle of *rebus sic stantibus* was invoked by Austria-Hungary in 1908 as a justification..for the annexation of Bosnia and Herzegovina. **1939** E. H. CARR *Twenty Years' Crisis* xi. 233 International lawyers evolved the doctrine that a so-called *clausula rebus sic stantibus* was implicit in every treaty, i.e. that the obligations of a treaty were binding in international law so long as the conditions prevailing at the time of the conclusion of the treaty continued, and no longer. **1973** I. M. SINCLAIR *Vienna Convention on Law of Treaties* iv. 105 All international lawyers are aware of the pitfalls surrounding the application of the *clausula rebus sic stantibus* and the controversies which have raged as to its admissibility as a ground for the unilateral denunciation or termination of a treaty. **1975** D. FISICHELLA in S. E. Finer *Adversary Politics & Electoral Reform* III. 264 It seems probable that the second ballot would create '*rebus sic stantibus*' further incentives to manipulation from outside and therefore increased factionalization.

rebuttable, *a.* (Later examples.)

1908 *Westm. Gaz.* 28 Jan. 8/3 Will you state when you think it is rebuttable and when you think it is irrebuttable? **1971** *Daily Tel.* 25 May 3/3 If the gift is substantial..there should be a rebuttable presumption that the witness exercised undue influence on the testator.

rebuttal. Add: Also *attrib.*

1925 *North Western Reporter* CCII. 898/2 Rebuttal evidence properly is that which explains away, contradicts, or otherwise refutes the defendant's evidence 'by any process which consists merely in diminishing or negativing the force of' it. **1971** *N.Y. Law Jrnl.* 23 Nov. 18/3 Petitioner's rebuttal affidavit..was received by the court after the court's decision had been published. **1976** *Billings* (Montana) *Gaz.* 17 June 3-H/8 Smith's attorney..presented John L. Agro..as a rebuttal witness. **1976** *Southern Even. Echo* (Southampton) 11 Nov. 4/4 So bitter was the resentment in Southampton of the criticism contained in the late Sir John Hodsoll's report, that the Public Records Office took the unique step of including rebuttal evidence in the State Archives.

rec, rec. (rek), colloq. abbrev. of RECREATION[1]. Also *attrib.* and *ellipt.* (= *recreation ground*).

1929 'R. CROMPTON' *William* iv. 86 Somehow or other the Outlaws got through the lessons before 'rec'. **1931** E. RAYMOND *Mary Leith* v. 170 You saw the mounted specials clear the Rec. **1948** PARTRIDGE *Dict. Forces' Slang* 153 *Rec space*, the ratings' recreation spaces on board ship. **1950** P. TEMPEST *Lag's Lex.* 176 *Rec*, used in some prisons instead of the word 'association'. The prisoners' recreation time. **1960** J. R. ACKERLEY *We think the World of You* 46 The only open space, besides the Rec., in the neighbourhood. **1962** W. GRANVILLE *Dict. Sailors' Slang* 94/2 *Rec room*, *recreation room* at Osborne, now closed. **1967** *Boston Sunday Globe* 23 Apr. B 41/1 (*heading*) A rec hall for patients. **1972** *Sat. Rev.* (U.S.) 24 June 18/3 The dance was held in the games room of the rec center. **1975** 'E. LATHEN' *By Hook or by Crook* xiii. 129 The wedding presents were supposed to go on the Ping-Pong table in the rec room. **1977** J. SAVAGE *Nemesis Club* vi. 75 He often goes to the rec. after school.

recado. Add: **2.** Also, a saddle-cloth.

1961 G. DURRELL *Whispering Land* vii. 180 Using our saddles and the woolly sheepskin saddle-cloth, called a *recado*, as back-rests.

recalcitrant, *a.* and *sb.* Add: **B.** *sb.* Also *transf.*

1918 *Oxf. Mag.* 21 June 343/2 The American Universities have generally adopted the 'Elective System'... There is, however, a distinguished recalcitrant in the University of Princeton.

Hence **reca·lcitrantly** *adv.*

1976 *Gramophone* Apr. 1603/1 A dialogue between God and a recalcitrantly sinful soul.

reca·lculation. [f. RECALCULATE *v.*: see -ATION.] The action of recalculating.

1848 M. SOMERVILLE *Phys. Geogr.* (1849) App. II. 425 The height here assigned to the Peak of Aconcagua differs 700 feet from that given by Captain Fitzroy. A re-calculation..of his elements has led us to adopt a much greater elevation for the giant of the Chilian Andes. **1872** H. WRAY *Some Applic. of Theory to Practice of Construction* 150 This gives a weight rather less than was provided for, but the difference is not enough to render recalculation necessary.

recalescence. Add: The temporary generation of heat associated with a change in crystal structure when a solid is cooled through a transformation temperature (sometimes sufficient to produce a brief rise in temperature). (Earlier and further examples.) So **recale·scent** *a.*

1873 W. F. BARRETT in *Phil. Mag.* XLVI. 477 Wherever the momentary expansion of the wire is feeble or absent, there likewise this recalescence, as it might be termed, is also feeble or absent. **1912, 1916** [see *DECALESCENCE]. **1973** J. G. TWEEDDALE *Materials Technol.* II. ii. 15 Once nucleation is really started solidification may proceed quite quickly and is then likely to be associated with release of thermal energy as recalescence.

recall, *sb.*[1] Add: Also with orig. U.S. pronunc. (rī·kǭl). **1.** (Further examples.)

1869 *Sphinx* 27 Nov. 274/2 Even the gods forgot to applaud—about the highest compliment which could be paid to the actor, and worth fifty recalls. **1964** *Financial Times* 3 Mar. 13/1 Several hundred were called back yesterday, and the recall will continue this week and next. The final assembly tracks for bicycles..are being restarted. **1969** T. PARKER *Twisting Lane* 197 What's more I did a Borstal recall as well; that's like an extension of it. **1972** *Sat. Rev.* (U.S.) 17 June 6/3 The previous December, General Motors had announced the biggest recall ever—6·7 million Chevys were called in for breaking engine mounts. **1976** *Globe & Mail* (Toronto) 15 Dec. 5/1 More glass has been found in Coca-Cola bottles and..he expects the soft drink company will have to issue another recall.

c. (Earlier and later examples.)

1835 J. E. ALEXANDER *Sketches in Portugal* vii. 160 The bugle sounds the recal for the skirmishers. **1880** J. H. SHORTHOUSE *John Inglesant* xxii. 283 The horn below sounding the recall.

2. Delete *rare*. In *Psychol.*, the act of calling to mind something previously learned or experienced, esp. in memory tests, usu. distinguished from recognition; also *attrib.*; *total recall*, the ability to call to mind every detail that caught the attention.

1894 M. W. CALKINS in *Psychol. Rev.* I. 480 The likelihood of recall increases, therefore, by the recency of the position. **1901** *Ibid.* VIII. 363 A somewhat similar feeling had that day been generated by wholly different content.. and..there had been direct emotional recall through emotion. **1932** F. C. BARTLETT *Remembering* xv. 256 The matter of recall is mainly a question of interest, while the manner of recall is chiefly one of temperament and character. **1934** R. MACAULAY *Going Abroad* xvii. 134 Total Recall is always subject to the gaps in one's memory. **1953** *Columbia-Viking Desk Encycl.* 805/2 His complicated style seeks by total recall to recapture the minutest psychological and sensory detail. **1953** MILLER & SELFRIDGE in Saporta & Bastian *Psycholinguistics* (1961) 203/1 The same data are replotted to show the relation of the recall-score to the length of the list. **1963** J. D. SALINGER *Seymour: an Introduction* 238, I was not..the least bit intoxicated by my own powers..of almost total recall. **1970** *Jrnl. Gen. Psychol.* Oct. 236 Part-whole learning interferes with motor recall just as it does with verbal recall. **1971** *Ibid.* Oct. 215 Recall-memory tests call for reproduction of previous input-output events. **1976** GHATALA & LEVIN in Levin & Allen *Cognitive Learning in Children* iii. 62 The possibility has been raised that the processes and/or types of stored information utilized in recognition decisions differ from those utilized in recall. **1977** M. W. EYSENCK *Human Memory* ii. 35 On a subsequent, unexpected recall test, semantically processed material was much better recalled than phonemically or structurally processed material. **1977** *Lancashire Life* Feb. 70/3 Clara's memory may be less than the total recall she would have wished.

3. c. *U.S.* Removal of an elected government official from office by a system of petition and vote; this method of terminating a period of office. So *recall election*.

1902 *Arena* XXVIII. 470 If the Recall was in force in any locality, whenever a petition signed by any number over one-half of the registered voters of that locality for the recall of any officer of that locality was filed, that office would become vacant. **1911** *Ann. Amer. Acad. Pol. & Soc. Sci.* XXXVIII. 163 (*heading*) Popular control under the recall. **1970** *Internat. & Compar. Law Q.* 4th Ser. XIX. II. 188 Electoral laws..or the right of recall, or the right to be elected to public office. **1976** *Billings* (Montana) *Gaz.* 16 June 3-B/1 *Philadelphia* (AP)—Critics of Mayor Frank L. Rizzo..filed petitions Tuesday asking for his recall. **1978** *Detroit Free Press* 16 Apr. 8A/4 Five suburban politicians have lost their jobs because they supported a MSHDA-backed housing program, and three more are facing a recall election in Birmingham next month. **1979** *Tucson* (Arizona) *Citizen* 3 Oct. 5A/1 A near-loss in a bitter recall election.

4. *attrib.* and *Comb.*, as *recall clause, coverage, signal, telegram*.

1976 *Sunday Times* 30 May 8/3 If by June 15 they can collect 145,448 signatures..then Rizzo has 10 days to resign or face a referendum... Hence, the procedure is known as the 'recall clause'. **1971** *Wall St. Jrnl.* 11 Aug. 28/1 Practically every week now brings news that a potentially faulty or dangerous product is being recalled... Many manufacturers are increasingly interested in.. product recall coverage. **1904** *Daily Chron.* 11 June 5/4 The eight blocking ships saw the recall signal right enough, but.. disregarded it. **1916** H. G. WELLS *Mr. Britling sees it Through* I. i. 27 He wished he knew of somebody who could send a recall telegram from London.

recall, *v.*[1] **3. c.** (Earlier example.)

1671 LOCKE *Essay Draft B* (1931) 264 The mind can repeat..those ideas..by the power it has of recalling and bringing in view any of its own ideas.

Récamier (rē[1]kæ·myē[1], ‖ rekamye). [Name of Jeanne Françoise Julie Adélaïde *Récamier* (1777–1849), French hostess.] **a.** (Also *Madame Récamier* and with small initial.) Used *attrib.* and *absol.* to designate a *chaise-longue* of the type on which Madame Récamier is portrayed reclining in a painting by David. **b.** Used *attrib.* and *Comb.* to designate a reclining position similar to that in which Madame Récamier is portrayed.

1924 M. J. MURRY *Voyage* ix. 168 Mrs. Tancred got up from her Récamier and stretched out both hands to him. **1936** H. NICOLSON *Let.* 19 Feb. (1966) 244 Draped curtains and Madame Récamier sofas and wall-paintings. **1938** E. BOWEN *Death of Heart* I. iii. 66 Anna, on the sofa in a Récamier attitude. **1944** 'P. QUENTIN' *Puzzle for Puppets* i. 9 A Madame Récamier couch and a huge mirror. **1964** H. GRISEWOOD *Last Cab on Rank* i. 7 A charming récamier sofa. **1966** J. AIKEN *Trouble with Product X* v. 88 She herself was disposed, Recamier-fashion, on one of the twin beds, against a pile of pillows. **1971** M. SMITH *Gypsy in Amber* vii. 53, I don't understand..how a man of your obvious good taste could put a récamier in this room. **1972** J. AIKEN *Butterfly Picnic* x. 180, I had adopted a sort of Récamier posture, reclining on my elbows. **1974** H. MCCLOY *Sleepwalker* iv. 49 She..sat on a Récamier couch. **1979** *N.Y. Times Mag.* 16 Dec. 142/3 The smaller studio at the north end of the space is more formal, primarily because of the major furnishings—a matching pair of *récamiers* (Empire-style lounges).

recap (rī·kæp), *sb.*[1] orig. *U.S.* [f. RECAP *v.*] A pneumatic tyre that has been recapped.

1939 in WEBSTER Add. **1940** in *Amer. Speech* (1944) XIX. 63/2 In recapping it is necessary to make a distinction between a full recap and a top-cap. **1943** *Sun* (Baltimore) 28 Apr. 7/1 The 'recaps for all' announcement had given the public the impression we are over the hump in the rubber situation. **1968** *Wanganui* (N.Z.) *Chron.* 15 Nov. 10/6 (Advt.), If it's tyres or Recaps you need, then try our 10-minute exchange service. **1969** *Truth* (Melbourne) 12 July 31/7 (Advt.), Guaranteed recaps from only $6.50.

recap (rī·kæp), *sb.*[2] Colloq. abbrev. RECAPITULATION[1] I.

1950 in WEBSTER Add. **1955** POHL & KORNBLUTH *Space Merchants* ii. 22 A brief recap of the extensive plans his lecture agent had made. **1959** *Times Lit. Suppl.* 2 Oct. 561/1 Comments and 'recaps', hints of tragedies to come or to be overcome, helped to give to oral narrative both atmosphere and shape. **1969** *Morning Star* 1 Nov. 3/8 Three films by John Lennon and Yoko Ono were shown to the Press this week—a recap of a recent New Cinema programme and a foretaste of two new John–Yoko programmes. **1973** *Times* 13 July (Motor Racing Suppl.) p. vi/1 A quick recap on my instructor's words. **1977** *N.Y. Rev. Bks.* 24 Nov. 42/4 Instant observation, like sports' reporting, is seldom more than re-cap, resumé, or the usual proclamation of personal preference.

recap, *v.*[1] (Earlier and further examples.)

1856 'STONEHENGE' *Brit. Sports* I. ii. 21 The cases may easily be recapped, and used many times. **1967** KARCH & BUBER *Offset Processes* ix. 375 Fill the fountain bottle with the fountain solution and re-cap it.

2. *trans.* To renew (a worn pneumatic tyre) by cementing, moulding, and vulcanizing a strip of camel-back on the tread. orig. *U.S.*

1939 in WEBSTER Add. **1943** *Sun* (Baltimore) 28 Apr. 7/1 Any civilian can get his tires recapped. **1946** W. H. GROUSE *Automotive Mechanics* xxvii. 579 Worn tires that have good casings without separated plies or broken or damaged cord can be recapped so that they will have considerably longer life.

Hence **reca·ppable** *a.*, that can be recapped; **reca·pper,** (*b*) one who recaps tyres; **reca·pping** *vbl. sb.*

1939 WEBSTER Add., Recappable. 1940 Recapping [see *RECAP sb.[1]]. 1942 *Sun* (Baltimore) 9 Apr. 26/5 Camelback would be released to retreaders and recappers for passenger tires. 1959 L. E. ONEACRE in M. Morton *Rubber Technol.* iv. 104 The main advantage of an electric mold is that the recapper does not have to install a steam boiler to supply heat that would be needed for steam molds. 1967 H. J. STERN *Rubber* (ed. 2) ix. 401 Many of the tyres produced are retreaded when worn. Sometimes the process is described as recapping, or remoulding, according to the external area which is renewed. 1973 *Black Panther* 24 Nov. 11/3 Rebozo was the largest tire re-capper in Florida. 1976 *Billings* (Montana) *Gaz.* 30 June 7-E/4 (Advt.), Plus ·32 to ·58 federal excise tax each and recappable tire.

recap (rī·kæp), *v.*[2] Colloq. abbrev. RECAPITULATE *v.* 1. Also *absol.*

1950 in WEBSTER Add. 1958 'N. BLAKE' *Penknife in my Heart* ii. 36 Let me recap. the situation. 1959 H. HOBSON *Mission House Murder* xviii. 183 Let us recap, Mr. Ford, and assess your position. 1971 *New Scientist* 1 July 44/1 The computer résumé they use when she recaps the results to the other scientists. 1976 'A. HALL' *Kobra Manifesto* ix. 129 'Want to recap anything?' 'No, I've got it.'

recapacitate, *v.* Restrict *rare* to sense in Dict. and add: **2.** *trans. Physiol.* To restore potency to (a spermatozoon). Cf. *CAPACITATE *v.*, *CAPACITATION.

1959 M. C. CHANG in C. W. Lloyd *Rec. Progress Endocrinol. & Reproduction* 153 Decapacitated rabbit sperms can be recapacitated again. *Ibid.* 161 The capacitated sperms can be decapacitated by treatment with seminal plasma.. but can be recapacitated if they remain longer in the female tract. 1969 *New Scientist* 31 July 234/1 Such sperm were said to have suffered decapacitation, and could regain fertilizing ability (be 'recapacitated') by spending a second period in the female tract.

Hence **re:capacita·tion.**

1970 [see *DECAPACITATED *ppl. a.*]. 1974 *Jrnl. Reproduction & Fertility* XLI. 243 It was possible to demonstrate sperm recapacitation.

reca·pitalize, *v.* [RE- 5 b.] *trans.* and *intr.* To capitalize (shares, etc.) again. Also *fig.* So **re:capitaliza·tion; reca·pitalized** *ppl. a.*

1927 *Sunday Express* 11 Sept. 1/1 The 'recapitalisation' bubble..is still crowding northern Bankruptcy courts with failures affecting thousands of people. 1928 *Observer* 15 July 20/3 Does anybody suppose that the recapitalised mills are paying their shareholders any such return? 1945 *Richmond* (Va.) *News-Leader* 1 Aug. 13/2 Thomas E. Wilson, chairman of the Board of Directors of Wilson & Co., Inc., today announced plans to recapitalize the company's 274,085 outstanding shares of $6 preferred stock. 1947 J. HAYWARD *Prose Lit. since 1939* 25 The break with the pre-war period was irreparable, and..a reorientation of intelligence and sensibility, a recapitalisation of experience, was inevitable. 1979 *Beautiful British Columbia* Spring 16/1 Bishop declared bankruptcy, re-capitalized, and Secrets of Chinatown was released the following year.

recapitulant (rīkăpi·tiŭlănt), *a. rare.* [f. RECAPITULATE *v.*: see -ANT[1].] = RECAPITULATORY *a.*

1929 W. FAULKNER *Sound & Fury* 378 He repeated his story, harshly recapitulant. 1932 —— *Light in August* xii. 269 The boys' diction was slow now, recapitulant. 1939 —— *Wild Palms* 244 You expressed gratitude almost tediously recapitulant.

recapitulate, *v.* Add: **1. c.** *Mus.* To restate, usu. in similar but distinct form, a musical theme which has been developed in an intervening section. Cf. *RECAPITULATION[1] 1 c.

1959 D. COOKE *Lang. Music* v. 250 The second subject of the finale picks up the six-note chromatic scale..; when it is recapitulated, the same notes are used in the bass.

recapitulation[1]. Add: **1. b.** (Further examples.)

1904 G. S. HALL *Adolescence* I. p. viii, Realizing the limitations and qualifications of the recapitulation theory in the biologic field, I am now convinced that its psychogenetic applications have a method of their own. 1919 J. B. WATSON *Psychol.* vii. 266 The recapitulation theory.. holds..that ontogeny repeats phylogeny—that the developing child must pass through all the stages that the race has passed through. 1924 R. M. OGDEN tr. *Koffka's Growth of Mind* ii. 48 He can dismiss both the recapitulation- and the utility-theories. 1957 P. HALMOS *Towards Measure of Man* i. 19 Today, the theory of recapitulation is, on the whole, discredited.

c. *Mus.* The section of a composition or movement (esp. one in sonata form) in which some or all of the themes presented in the exposition are repeated, usu. in a modified form. Also *attrib.*

1879 GROVE *Dict. Music* I. 551/1 In the recapitulation of his [*sc.* Beethoven's] subjects,..there is a growing tendency to avoid the apparent platitude of repeating them exactly as at first. 1898 G. B. SHAW *Perfect Wagnerite* 3 In classical music there are, as the analytic programs tell us, first subjects and second subjects, free fantasias, recapitulations, and codas. 1934 C. LAMBERT *Music Ho!* II. 127 His [*sc.* Satie's] unusual employment of what might be called interrupted and overlapping recapitulations. 1947 A. EINSTEIN *Music in Romantic Era* vii. 70 Beethoven had been criticized for having held too fast to the sonata form in his third *Leonore* Overture, or—more precisely—that he had not foregone the recapitulation. 1947 *Penguin Music Mag.* Dec. 29 After the climax Mendelssohn duly modulates to his 'recapitulation-section'. 1959 *Collins Music Encycl.* 613/2 The procedures denoted by the names of its [*sc.* sonata form's] three sections—Exposition (*i.e.* presentation), Development (*i.e.* discursive treatment), and Recapitulation (*i.e.* return)—were present in earlier music. *Ibid.* 614/1 The Recapitulation may differ from the Exposition in details, *e.g.* in orchestration and in the use of new accompaniments to themes. 1959 *Listener* 23 July 152/1 The first theme of the sonata-form opening movement, expounded in three-four time, is transformed in the abbreviated recapitulation to four-four. 1979 C. DEXTER *Service of all the Dead* iii. 22 He had sedulously drilled the set works into them—their themes, their developments and recapitulations.

2. For *rare*[−1] read *Theol.* and add later examples.

1913 E. GRUBB *Doctrine of Person of Christ* iv. 28 This idea of 'recapitulation'..is one of the deepest and most pregnant thoughts contributed by Irenæus. 1957 *Oxf. Dict. Chr. Ch.* 1142/1 The conception of recapitulation was elaborated esp. by St. Irenaeus, who interpreted it as both the restoration of fallen humanity to communion with God through the obedience of Christ and as the summing-up of the previous revelations of God in past ages in the Incarnation. 1969 J. ATKINSON in *Dict. Chr. Theol.* 285/2 The recapitulation in Christ connotes the total work of God for man's redemption.

reca·rbon, *v.* [f. RE- 5 c + CARBON *sb.*] *trans.* To fit (an arc-lamp) with new carbon electrodes. So **reca·rboning** *vbl. sb.*

1902 W. J. DIBDIN *Public Lighting* 445 There are two pairs of carbons in each lamp, such being necessary to allow of the lamp burning throughout the whole of a winter night without being re-carboned. 1902 *Encycl. Brit.* XXVIII. 86/2 The cost of carbons and the labour of recarboning. 1917 C. N. BENNETT *Guide to Kinematogr.* xi. 185 If someone will..kindly re-carbon the arc for us and turn on the electric current we shall be ready to start showing.

recarbonate (rīkā·ɹbŏnē[i]t), *v.* [f. RE- 5 c + CARBONATE *v.*[1]] To charge (water) with carbon dioxide after softening. Hence **reca·rbonated** *ppl. a.*; **re:carbona·tion.**

1907 CROSS & BEVAN *Paper-Making* (ed. 3) xv. 356 In order to retain this carbonate in a more permanently soluble form the Archbutt-Deely plant is provided with means for re-carbonating the softened water by the injection of carbon dioxide. 1926 *5th Ann. Rep. Ohio Conf. Water Purification* 60 (*heading*) Methods of recarbonation of lime-soda softened water. 1936 E. S. HOPKINS *Water Purification Control* ix. 163 The lime-softened water may be satisfactorily stabilized at low cost by recarbonating it with carbon dioxide gas. 1969 *Civil Engin.* (N.Y.) June 40/1 The recarbonated water then flows through two ballast ponds in series. 1978 R. WALKER *Water Supply, Treatment & Distribution* xv. 183 The recarbonation chamber must be sized and baffled to provide between 15 and 30 minutes retention time.

recarburization (rīkā:ɹbiŭrəizē[i]·ʃən) *Steel-making.* [RE- 5 c.] Addition of carbon (in the form of coke, anthracite or a carbon-rich alloy such as pig iron) to steel to produce the desired composition after refining.

1888 *Jrnl. Iron & Steel Inst.* 331 The recarburisation of the bath of iron gives rise, in the case of basic or neutral open hearth furnaces, to a very considerable loss of metal. 1924 E. L. RHEAD *Metallurgy* (ed. 3) xi. 222 Recarburisation is now often effected by the use of anthracite. This is put at the bottom of the ladle or thrown into the stream of metal as it runs from the furnace. 1940 SIMONS & GREGORY *Steel Manuf. simply Explained* xv. 105 Anthracite coal is used if the recarburization is carried out in the ladle.

Also **reca·rburize** *v. trans.*; **reca·rburizing** *vbl. sb.*

1888 *Jrnl. Iron & Steel Inst.* 332 This recarburising material consisted in the case of the first charge of 0·145 ton of ferro-manganese and 0·030 ton of ferro-silicon. 1929 C. R. HAYWARD *Outl. Metall. Pract.* xxii. 475 Deoxidize and recarburize the bath. 1950 *Engineering* 30 June 738/3 When making high-carbon steels..it was formerly common practice to run the carbon down in the furnace and then to recarburise to the composition required by the addition of coke or anthracite to the ladle. 1974 *Encycl. Brit. Macropædia* XVII. 645/1 For a long time, high-carbon steels such as rails were made by recarburizing the steel with hot metal. 1974 D. M. CONSIDINE *Chem. & Process Technol. Encycl.* 654/1 Deoxidizing, recarburizing, or alloying additions are made to the ladle during steel tapping from the furnace to adjust the final composition of the steel to the desired specification.

re-ca·se, *v.* [RE- 5 c.] *trans.* To furnish with a new case; to case again; *spec.* to rebind (a book) using its original case or a new one. Hence **reca·sed** *ppl. a.*; **reca·sing** *vbl. sb.*

1853–8 [see RE- 5 c in Dict.]. 1920 T. J. WISE *Bibliogr. Writings J. Conrad* I. 4 It is clear that ex-library copies of *Almayer's Folly*..have recently been made attractive and marketable by the apparently simple process of re-casing them in cloth. *Ibid.*, when the binding looks very green and new..the book is probably a re-cased one. 1952 J. CARTER *ABC for Book-Collectors* 150 A book which, being shaken or loose, has been taken out of its covers and re-settled in them more firmly is said to be re-cased. *Ibid.*, Most re-casing is not hard to detect. 1977 *Shotton's* (Durham) *Catal.* Oct., Recased into orig. cloth.

recast, *v.* **1.** Delete *rare*[−1] and add later examples.

1894 E. FAWCETT *New Nero* 17 He recast a sudden look upon Fanshawe and his face drearily brightened. 1899 P. H. WICKSTEED tr. *Dante's Paradiso* ii. 21 Now thou wilt urge that the ray here is darkened rather than in other parts, because here it is recast from further back.

2. c. *Theatr.* To assign (an actor) to another part; to cast (a role, etc.) again. Also *fig.* Cf. CAST *v.* 48.

1951 N. MARSH *Opening Night* vii. 163 A lively and almost cosy discussion about recasting had developed. 1962 L. PAYNE *Too Small for his Shoes* viii. 162 I'll have to recast the part and re-shoot every sequence he was in. 1979 A. SCHOLEFIELD *Point of Honour* 80 My father was recast as a hero..saviour of lives in the face of the enemy. 1981 N. J. CRISP *Festival* vi. 143 Should the play..justify..a possible transfer to the West End, it will..have to be recast.

recategorize (rīkæ·tĕgŏrəiz), *v.* [RE- 5 a.] *trans.* To assign to another category; to reclassify. So **recategoriza·tion; reca·tegorizing** *vbl. sb.*

1949 M. MEAD in M. Fortes *Social Structure* 32 We may find ..a recategorizing of cultural experience into such characteristic American forms as fraternities. 1957 PARTRIDGE *English gone Wrong* i. 12 Those..horrifics, *recategorize* and *recategorization*, have..fascinated the combatant services and the Civil Service. 1970 *Jrnl. Gen. Psychol.* Apr. 151 The meaning properties tend to be exhausted rather rapidly by the recategorization instructions. 1977 *Guardian Weekly* 19 Sept. 6/3 He claims there are no political detainees in Chile. (Those left in gaol have been recategorised as common criminals.)

recce (re·ki), *sb.* orig. *Mil. slang.* Also **reccy.** Shortened form of RECONNAISSANCE. Freq. *attrib.* Cf. *RECCO.

1941 *Illustr. London News* CXCVIII. 802 (*caption*) A motor-cycle 'reccy' unit (special reconnaissance troops of armoured force). 1942 E. WAUGH *Put out more Flags* 151 The C.O. has just gone forward with his recce group to make his recce. 1944 *R.A.F. Jrnl.* Aug. 260 We aren't likely to do any serious fighting unless..we are faced by recce patrols by the enemy. 1951 R. CAMPBELL *Light on Dark Horse* iv. 71 My lads became truly proficient in practical recce-work. 1958 G. USHER *Death in Bag* vi. 56, I did a bit of a recce. 1961 W. VAUGHAN-THOMAS *Anzio* iii. 45 No recce plane tried a quick dash to capture a tell-tale photograph. 1968 *Listener* 19 Dec. 812/1 In a side-street we stopped. 'I'll just do a bit of a recce,' the Fiddler said. 1974 *Observer* (Colour Suppl.) 28 Apr. 25/4 We were interrupted by a knock at the door of the mews flat in Belgravia, where Deighton had set up temporary headquarters during his London recce. 1977 'E. CRISPIN' *Glimpses of Moon* xii. 235 He had had it off all right, thanks..to making careful reccys. 1979 'A. HAILEY' *Overload* I. xvii. 102, I sometimes think about two guys in Korea, close buddies of mine. We were on a recce patrol near the Yalu river.

recce, *v.* orig. *Mil. slang.* Also **reccy.** [Shortened f. RECONNOITRE *v.*] *trans.* and *intr.* To reconnoitre (a place, etc.).

1943 J. H. FULLARTON *Troop Target* xxiv. 175 We're even reccying alternative positions twenty miles back to withdraw to. 1944 M. STANDING in *War Report* (B.B.C.) (1946) II. iv. 106 Do not unload until waders have recce'd the water in front of your door. 1945 *Comment from Italy* (Three Arts Club) 48 So I find four bottles and start out across the fields to recce for a farmhouse. 1945 E. WAUGH *Brideshead Revisited* 299 I'm going out myself with the adjutant to recce training areas. 1958 P. SCOTT *Mark of Warrior* I. 59 I'll know better to-day when I've recced it. 1959 'J. CHRISTOPHER' *Scent of White Poppies* x. 156 These two recce'd up here, and found them gone. 1968 *Listener* 11 July 40/3 He was..recce-ing the new tourist 'firsts' like the Canadian North-West Passage. 1972 *Shooting Times & Country Mag.* 4 Mar. 21/2 'Lofters'..come over to recce the food situation in the more populated areas. 1976 *Listener* 8 Apr. 438/1 Neither of us had seen any of his architecture in Virginia, so I set off to recce Richmond and Charlottesville.

recco (re·ko). *Mil. slang.* Abbrev. of RECONNAISSANCE. Also *attrib.* Cf. *RECCE *sb.*

1917 A. G. LEE *Let.* 24 May in *No Parachute* (1968) 19 They were still flying two-seater Nieuports, doing..reccos and photography. 1934 *Flight* 18 Jan. 51/1 A shorter 'recco' is also done to the north, if trouble is brewing. 1942 in Forbes & Allen *Ten Fighter Boys* 60 Green section were detailed to..proceed to Chatham, where a recco-machine was hanging around. *Ibid.* 155 My next trip was a scramble to intercept some Hun machines on recco off Dungeness. 1942 W. SIMPSON *One of our Pilots is Safe* ii. 19 That was the last 'recco.' flight we made, and for months we had to content ourselves with mock air battles. 1943 *Life* 9 Aug. 100/2 (*caption*) Armed recco over Rabaul. 1966 P. DERRIG *Pride of Green Berets* 61 Civil Guard recco units.

reccy, var. *RECCE *sb.* or *v.*

recd. Also **rec'd.** Abbrev. of *received* pa. t. and pa. pple. of RECEIVE *v.*

1599 J. DAY in P. Henslowe *Henslowe's Diary* (1904) I. 57 Rec[d] of m[r] Hinchloe more in ernest of The Tragedy of Thomas Merrye 20[s] Joh. Day. W Haughton. 1775 D. GARRICK *Let.* 1 Nov. (1963) III. 1044 In the middle of the play last night I rec[d] your very extraordinary Note. 1851 J. CHAPMAN *Diary* 23 Apr. in G. S. Haight *Geo. Eliot & J. Chapman* (1940) 158 Rec[d] a letter from Susanna giving a better account of Beatrice. 1876 W. WHITMAN *Daybks. & Notebks.* 12 Mar. (1978) I. 6 Sent & recd. 1934 E. POUND *Eleven New Cantos* xxxiv. 17 And on his return was recd. by Gouverneur Morris and Mr Astor with a pubk. dinner at Tammany Hall. 1965 H. GOLD *Man who was not with It* xi. 86 It had cost me to go back to Pittsburgh, although I had rec'd value.

recede, *v.*[1] Add: **1. e.** Of a colour: to appear to be more distant from the eye than another in the same plane; = RETIRE *v.* 3 b. Cf. *ADVANCE *v.* 2 b.

1935 A. H. RUTT *Home Furnishing* iv. 35 Advancing and receding qualities in colors are a reality, as psychologists have proved. The warm hues seem to advance and the cool ones to recede. 1951 *Good Housek. Home Encycl.* 152/1 The cool tints..tend to 'recede' and will give a feeling of space.

5. a. (Earlier example.)

1788 E. SHERIDAN *Let.* 27 July in *Betsy Sheridan's Jrnl.* (1960) iv. 107 And now to *recede*—I had just sent off my letter yesterday when Mrs Angelo call'd, as usual all life and spirits and full of news.

Hence **rece·ded** *ppl. a.*

1909 M. B. SAUNDERS *Litany Lane* I. vi. 69 Her attendant lady.., Augusta of the receded fringe.

receding, *ppl. a.* (Further examples.)

1895 A. W. PINERO *Second Mrs. Tanqueray* III. 103 A man..with a low forehead, a receding chin, a vacuous expression. 1956 *Jrnl. Theol. Stud.* VII. 18 The incidence of the gradually receding preposition παρά supports the impression. 1958 [see *ADVANCE *v.* 2 b]. 1977 *Transatlantic Rev.* LX. 183 If I have learned anything it is that the past is not a receding dream but an ever burgeoning presence at our backs that sustains us.

receipt, *sb.* **VI. 17.** *receipt tax* (earlier example).

1787 DUKE OF DORSET in O. Browning *Despatches from Paris* (1909) I. 217 It is fear'd that the Duty is intended to include Stamp-receipts after the plan of the Receipt-Tax in England.

receipt, *v.*[1] Add: **2.** Also *fig.*

1938 E. BOWEN *Death of Heart* III. i. 327 She receipted Portia's remark with an upward jerk of the chin.

3. (Earlier and later examples.)

1832 *Sen. Doc. 23rd U.S. Congress I Sess.* No. 512 (1835) II. 829 [Stock] will be delivered..to an issuing officer,..who will receipt therefor. 1862 O. L. JACKSON *Colonel's Diary* (1922) v. 96 Major Lyford..receipted to me for the safe delivery of the cargo. 1913 J. LONDON *Valley of Moon* 503 These two assistants had..been receipted for by the local deputy sheriff.

receipted *ppl. a.* (further example).

1979 *Daily Tel.* 21 Nov. 3/3 Drivers would need to present full log sheets of receipted bills to support their case for relief.

receiptor. (Earlier and later examples.)

1814 *Mass. Supreme Court Rep.* XI. 319 The receiptors are precluded, by their own act, from calling in question the validity of the attachment. 1914 F. RAWLE *Bouvier's Law Dict.* III. 2824/2 The officer taking the goods often.. delivers them to some third person, termed the 'receiptor', who gives his receipt for them.

receivable, *a.* Add: **A.** *adj.* **1. d.** Of a broadcast signal or a broadcasting station.

1962 *Rep. Comm. Broadcasting 1960* 196 in *Parl. Papers 1961–2* (Cmnd. 1753) IX. 259 Both the BBC and the ITA.. accept it as their duty to see that their present services are as nearly as possible available to everybody in the country. Not only should they be receivable; they must also be technically acceptable.

B. *sb. pl. Comm.* Debts owed to a business, esp. regarded as assets. Cf. *PAYABLES *sb. pl.*

1863 'E. KIRKE' *My Southern Friends* xxii. 231 When I went home..we had only nineteen thousand in bank. I had exhausted all our receivables. 1947 *Sun* (Baltimore) 30 Sept. 19/1 A move to..take care of a steadily expanding volume of receivables acquired. 1955 *Times* 10 May 17/2 Long-term receivables have been separated from current receivables, and a new item, short-term borrowings, appears under current liabilities. 1978 *Daily Tel.* 21 Mar. 3 (Advt.), The HP 3000 will run your payroll, process invoices, print out receivables and take care of general ledger work.

receive, *v.* Add: **I. 4. a.** (Further example.)

1861 D. G. ROSSETTI tr. *Dante's Vita Nuova* 299 The sonnet has two parts... It might well receive other divisions also.

d. Also of other, non-recording, apparatus.

1957 *Encycl. Brit.* XXI. 912I/2 The television signal is received on a short wave antenna and carried by a transmission line to the receiver.

e. Of a radio or television set: to reproduce the sound or picture transmitted by (a station or a person). Also *transf.*, of the user, and *absol.*

1908 *Rep. Brit. Assoc. Adv. Sci. 1907* 621 In this way it is possible to receive at Hythe from Elmers End. 1930 *Morn. Post* 18 Aug. 3/4 In order to receive foreign stations consistently it is necessary to employ a powerful receiver. 1969 M. PUGIT *Last Place Left* xxv. 187 Get what I mean, now? You receiving me, strength four? 1970 J. EARL *Tuners & Amplifiers* iii. 75 If you are proposing to try receiving distant stations..you will need a tuner of the best possible selectivity.

II. 13. a. Also in phr. *to receive in marriage* = *to take in marriage* s.v. TAKE *v.* 14 b.

1835 O. PRATT in *Utah Gen. & Hist. Mag.* (1938) XXIX. 34, I baptized Sarah Marinda Bates, near Sacketts Harbor, whom I received in marriage upwards of one year after.

14. b. (Later example.)

a 1817 JANE AUSTEN *Persuasion* (1818) III. iv. 58 He thought it a very degrading alliance; and Lady Russell.. received it as a most unfortunate one.

IV. 24. Also, to receive visitors.

1902 H. JAMES *Wings of Dove* IX. xxxii. 499 'I'm commissioned to ask you from her to go and see her.'.. He was.. bewildered. 'Then she can receive—?'

25. The infin. used, usu. *attrib.*, to designate the receiving mode, controls, etc., of a radio or telecommunication system.

1920 *Wireless World* 7 Aug. 356/1 A send–receive switch. 1966 'A. HALL' *9th Directive* xiii. 121 Loman was buzzing for me so I switched the radio to 'receive'. 'Do you hear me, Quiller?' 1970 *New Scientist* 24 Dec. 554/3 The Applications Technology Satellite 3..provided two transmit and two receive voice channels.

received, *ppl. a.* Add: **1. a.** (Further examples.) Also *received idea* = *idée reçue* s.v. *IDÉE. Cf. also *received text* s.v. TEXT *sb.*[1] 1 d.

1608 E. TOPSELL *Serpents* 219 Suidas followeth the common received opinion, that the Salamander quencheth the fire. 1959 *Times Lit. Suppl.* 20 Feb. 94/4 The error he has discovered in the received reference books..deserves special note from all those concerned with the period. *Ibid.* 20 Mar. 159/1 Her neat and tidy balance sheets are summaries of received ideas into which she ventures to inject no new thought. 1960 J. BAYLEY *Characters of Love* ii. 52 The appearance of Courtly Love..has left an immense legacy of received ideas about sex and society. 1973 *Howard Jrnl.* XIII. 330 In former days, the received view was accepted more readily and unquestioningly. 1973 *Times* 13 Dec. 13/4 It is a received idea that television is the most powerful medium ever devised. The proposition is hardly ever questioned. 1976 E. MACLAREN *Nature of Belief* iv. 39 His formal education may help him to be critical or sceptical about quite a lot of 'received opinion'.

b. Of language or pronunciation: *received pronunciation*, the pronunciation of that variety of British English widely considered to be least regional, being originally that used by educated speakers in southern England; also, the 'accepted', standard pronunciation of any specified area, Received Standard; *Received Standard* (*English*), the spoken language of a linguistic area (usu. Britain) in its traditionally most correct and acceptable form. Hence in other derived uses.

[1818 *Trans. Amer. Philos. Soc.* I. 259 According to its most generally received pronunciation, it is more properly a diphthong.] 1869 A. J. ELLIS *On Early Eng. Pronunc.* I. 13 The alphabet required for writing the theoretically received pronunciation of literary English. 1874 —— *Ibid.* IV. 1095/1 The tip of the tongue for received English is not so advanced towards the teeth or gums, as for the continental sound. 1882 —— in *Trans. Philol. Soc.* 21 We say they are dialectal forms of the received *down.* 1889 [see *RP* s.v. *R II. 2 a]. 1890 *Dialect Notes* I. 26 For the study of pronunciation the received spelling is very ill adapted. 1913 H. C. WYLD in *Mod. Lang. Teaching* IX. 261/2 When he speaks of *Standard English*, he is, I believe, referring to what I now call *Received Standard.* 1914 M. MONTGOMERY in *Ibid.* X. 11/2 Yet in that country [*sc.* Germany], as time goes on, a process of assimilation towards a single 'Received Standard' is said to be growing more, rather than less, marked. 1932 D. JONES *Outl. Eng.* xviii. 148 In Received English there are six affricates which may be represented phonetically by.. diagraphs. 1932 *S. P. E. Tract* XXXVII. 542 These authors.. define the 'Received Pronunciation' as that of 'the great public schools, the Universities, and the learned professions'. 1936 *Trans. Philol. Soc.* 80 My own recollection of this opposition to Received Speech is that the dialect speaker acquires a consciousness of 'correctness' in speech accompanied by a powerful objection to being caught..'talking fine'. 1937 D. JONES in *Le Maître Phonétique* Apr.–June (Suppl.), I take the view that foreigners learning English should be free to choose whatever pronunciation they prefer. Many naturally choose what has been termed 'received' pronunciation (R.P.), as being a widely understood type of English. 1940 J. H. JAGGER *Eng. in Future* i. 15 The influence of the various forms of Modified Standard—to accept Professor Wyld's terms—upon each other and upon Received Standard. 1962 A. C. GIMSON in R. Quirk *Use of Eng.* 281 *Received Pronunciation*, or RP, suggesting..the result of a collective social judgment rather than of a conscious, prescriptive agreement. 1964 C. BARBER *Ling. Change Present-Day Eng.* ii. 20 The influence of the mass-media and of mass-education..does not necessarily produce speakers of Received Standard English. 1969 S. POTTER *Changing Eng.* i. 14 Other cities, notably Edinburgh and Dublin, have their received pronunciations, and so have other regions of the English-speaking world. 1973 G. W. TURNER *Stylistics* v. 147 Even now 'received pronunciation' will help its user to obtain credit when ordering goods by telephone. 1974 J. I. M. STEWART *Gaudy* x. 179 There is no such thing as an Oxford accent, since what phoneticians call Received Standard English came into existence without the university's playing any very identifiable part in the process.

receiver[1]. Add: **1. c.** (*a*) *Amer. Football*, an offensive player eligible to catch a pass; also, a defender designated to receive a kick-off or punt, a safety man; (*b*) *U.S. Baseball*, a catcher. Cf. *wide receiver* s.v. *WIDE *a.*

1897 *Encycl. Sport* I. 421/2 The direction of a pass must depend upon where the prospective receiver is. 1908 *Baseball Mag.* Aug. 16/1 Flint was a wonderful catcher and the amount of work he could stand would make some of the receivers of today take notice. 1921 *Outing* Jan. 156/3 (*caption*) Crangle of Illinois was..a sure receiver of the forward pass. 1935 L. LITTLE *How to watch Football* vii. 140 The passer..would be helpless without the receiver who is able to get into position. 1940 D. HILL *Football through Years* 64/2 The accompanying illustrations.. show what a split second can mean to the receiver of the ball in fast brilliant play. 1957 *Encycl. Brit.* IX. 474/2 Eligible receivers were the players at the end of the line of scrimmage. 1967 *Ebony* June 128/3 Battey has been Howard's chief rival as the American League's premier receiver. 1972 J. MOSEDALE *Football* i. 7 A great receiver like Paul Warfield..loafs through his pass patterns until the defender relaxes. 1981 *Sports Illustr.* 12 Feb. 28 The Irish set up one touchdown by Walker when their kickoff receivers got their signals crossed.

2. c. (Further example.)

1977 *N.Z. Herald* 8 Jan. 1-3/10 Meanwhile, the Nelson receiver of wreck, Mr R. K. Watson, said no legal action would be taken against the three men who boarded the boat and claimed salvage rights.

d. An official of the Metropolitan Police Force (see quot. 1966).

1829 *Act* 10 *Geo. IV.* c. 44 § 10 It shall be lawful for His Majesty to appoint a proper Person to receive all Sums of Money applicable to the Purposes of this Act, who shall be called 'The Receiver for the Metropolitan Police District'. 1902 *Encycl. Brit.* XXXI. 818/1 The county council of any county within the Metropolitan Police District has to transfer to the receiver of police a sum bearing..proportion to the police rate. 1928 *Daily Mail* 7 Aug. 17/2 The Receiver is concerned with equipment and so forth..of..the Metropolitan Police. 1966 J. D. DEVLIN *Police Procedure, Administration & Organisation* iii. 20 The police authority of the Metropolitan Police Force is the Home Secretary, and the official responsible for police property, buildings and finance is the Receiver for the Metropolitan Police.., who is appointed by the Crown. *Ibid.*, In provincial forces, the duties and functions of the Receiver fall on the police authority.

4. d. (See quot. 1970.)

1938 M. LANGLEY *Refuelling in Flight* 18, I insisted that the tanker (giver) should formate on the liner or bomber (receiver). 1970 *Gloss. Aeronaut. & Astronaut. Terms* (*B.S.I.*) x. 17 *Receiver aircraft*, an aircraft which is being refuelled in the air.

7. b. (Further examples.) Also occas. applied loosely to the complete telephone receiving-unit.

1897 *Sears, Roebuck Catal.* 472/2 (Advt.), The improved long distance battery telephone of the regular Bell telephone style with..compound pole receiver. 1918 S. LEACOCK *Frenzied Fiction* iv. 52 'Great-grandfather,' I said, as I hung up the receiver in disgust, 'you are a Mutt!' 1936 DYLAN THOMAS in *New Verse* Dec. 17 The parting of hat from hair, Pursed lips from the receiver. 1971 *Daily Tel.* 11 June 32/6 The amount of telephoning which has to be done does not justify two receivers at the charity's office. 1980 A. N. WILSON *Healing Art* xvi. 195 He had a telephone receiver to his ear.

c. An apparatus for receiving radio or other signals transmitted as electromagnetic waves; now *spec.* a combined tuner and amplifier (without a loud-speaker).

1891 *Rep. Brit. Assoc. Adv. Sci. 1890* 757 To calculate the force between two neighbouring Hertzian receivers. 1912 *Chambers's Jrnl.* Jan. 60/2, I had got our receiver into 'tune' with the transmitter on board a steamer some miles away. 1927 *Radio Assoc. Official Handbk.* 50 The ultimate Television receiver will be a simple piece of apparatus. 1930 *Morning Post* 18 Aug. 3/4 With any good receiver..several foreign stations may be regularly well received. 1933 [see *LOOKER *sb.* 1 d]. 1955 *Radio Times* 22 Apr. 3/2 The F.M. receiver is designed to take advantage of the full frequency range transmitted. 1966 *McGraw-Hill Encycl. Sci. & Technol.* XI. 200/1 In most pulse-radar systems a single antenna serves for both transmission and reception... The duplexer protects the sensitive receiver by disconnecting it from the antenna during the presence of the powerful transmitter pulse. 1973 *Daily Tel.* (Colour Suppl.) 12 Oct. 31/4 Other products not listed include radio tuners and tuner/amplifiers (known now as 'receivers'). 1976 A. WHITE *Long Silence* 31 By the time Dick was thirteen, he'd built his own radio receiver and transmitter. 1978 *Gramophone* May 1958/1 Trio (B. H. Morris) plan to demonstrate several new amplifiers, tuners, receivers and turntables.

d. A detector of sound or other compressional waves.

1920 *Physical Rev.* XV. 178 A pair of receivers mounted on a horizontal rod which may be rotated..is an efficient device for getting the direction of a source of sound. 1931 STEWART & LINDSAY *Acoustics* x. 261 The earliest type of acoustic receiver was the so-called Broca tube, consisting.. of a sphere or nipple *C* of rubber or sheet metal attached to the end of a listening tube *T*. 1957 NOLTINGK & TERRY in E. G. Richardson *Technical Aspects of Sound* II. ii. 111 Magnetostrictive and piezoelectric receivers may be designed. 1973 *Nature* 30 Nov. 297/1 This communication describes the results of a reversed seismic refraction line carried out with explosives and seabed receivers in the median valley of the Mid-Atlantic Ridge.

receiver-general. Add: †**2.** *transf.* **a.** *slang.* A prostitute. *Obs.* **b.** *Pugilists' slang.* (See quot. 1903.) *Obs.*

1811 *Lexicon Balatronicum*, *Receiver General*, a prostitute. 1821 P. EGAN *Boxiana* III. 356 It was evident M'-Dermot was doomed to be a Receiver-General; although he had nobbed Purcell over the right eye..Purcell had the best of it. 1829 *Ibid.* 2nd Ser. II. 180 Dick was now a receiver-general, and his mug was severely bruised. 1903 FARMER & HENLEY *Slang* VI. 5/1 *Receiver-general*..(pugilists'), a boxer giving nothing for what he gets.

receivership. Add: **1.** Also in extended use.

1934 H. G. WELLS *Exper. Autobiogr.* II. ix. 732 Lenin's reconstructed Communist Party was a much more effective step towards an organized receivership.

2. (Further examples.)

1929 *Times* 30 Oct. 14/2 After the close of the market it became known that the receivership in the Cuba Cane Sugar Company had been made permanent. 1967 R. STEIN *Great Cars* 222/2 Other troubles piled up. In 1921, Lincoln was forced into receivership, and Ford bought the company. 1976 F. ZWEIG *New Acquisitive Society* II. v. 113 The sinking enterprise finally ends up in receivership.

receiving, *vbl. sb.* Add: **1.** (Later examples.)

1931 *Writer's Digest* Oct. 28 (*To*) *fall for receiving*, to be convicted to having stolen property in one's possession... 'He caught me with the rocks, so I fell for receiving.' 1956 E. GRIERSON *Second Man* ii. 37 He was also in trouble again with the police: a little receiving. 1979 *Tucson* (Arizona) *Citizen* 20 Sept. 11D/7 Paul Jones of Cal, who caught 10 passes against Arizona last Saturday, leads in receiving with 13.

2. a. (Further examples of *receiving order.*)

1930 *Daily Express* 30 July 2/7 Receiving Orders are announced in the 'London Gazette'. 1977 *Private Eye* 1 Apr. 4/1 Harry Alan's financial affairs last attracted attention in 1973 when Air Express Travel obtained a receiving order against him for £700.

b. *receiving country, depot, home, -house* (earlier and later examples), *office* (examples), *pen, -room* (further examples), *-ship* (further examples), *-station* (further examples), *-yard* (further example).

1938 *Washington Post* 21 Dec. 6/1 Refugees would be allowed to leave Germany for a 'receiving country'. 1958 J. J. SPENGLER in B. Thomas *Economics of Internat. Migration* ii. 17 (*title*) Effects produced in receiving countries by pre-1939 immigration. 1970 *Soviet Weekly* 13 June 11 Our laundry has 32 receiving depots and washes for 150,000 people. 1967 *U.S. Supreme Court Reports* CCCLXXXVII. 27 The fact of the matter is that, however euphemistic the title, a 'receiving home'..for juveniles is an institution for confinement. 1973 *Washington Post* 13 Jan. A5/2 Under the plan children in 'predisposition status' and under Receiving Home authority will be transferred into one of five categories. 1824 E. WEETON *Jrnl.* May (1969) II. 280, I wished to see the General Post Office, so..I took that letter all the way there, instead of putting it into one of the receiving Houses nearer at hand. 1854 E. E. HALE *Kanzas & Nebraska* ix. 224 A boarding-house or receiving-house, in which three hundred persons may receive temporary accommodation on their arrival. 1900 S. A. NELSON *ABC of Wall St.* 157 *Receiving houses*, houses which make a business of receiving and selling cash grain. 1908 *Daily Chron.* 21 Apr. 1/6 He should..see that the receiving house clause was given the very fullest effect to. *c* 1865 in R. Whitehouse *London Album* (1980) Pl. 70, Midland Railway, Receiving Office. 1885 *List of Subscribers, Classified* (United Telephone Co.) 2 The Midland Railway Company will also receive Goods at the following Receiving Offices. 1972 *Classification of Occupations* (Dept. Employment) II. 366/2 *Receiving office assistant*, receives from customers articles requiring service and returns finished articles at a receiving office. 1931 Receiving pen [see *CRUSH *sb.* 4 c]. 1846 G. DODD *Brit. Manuf.* VI. 184 [The] sail-cloth is taken to a 'receiving-room', where it is examined, freed from lumps and irregularities, measured, and weighed. 1978 M. PUZO *Fools Die* v. xxix. 333, I noticed that most of the men threw their car keys on the table in the first receiving room. 1901 *Daily Colonist* (Victoria, B.C.) 3 Nov. 2/3 A third class ensign on the receiving ship Columbia lying at the New York Navy Yard. 1978 K. BONFIGLIONI *All Tea in China* IV. xiii. 180 The only other vessels in the anchorage were the receiving ships. 1939 *War Illustr.* 28 Oct. 219/1 From his bed in an R.A.F. medical receiving station hidden away in the woods 'somewhere in France'. 1977 G. W. H. LAMPE *God as Spirit* ii. 54 Divine communications at the subconscious level, for which the actual recipients had acted simply as passive receiving stations. 1923 Receiving yard [see *FORCING *vbl. sb.* 3 a].

c. Of things: designed for the reception of radio signals or the like, as *receiving aerial, apparatus, circuit, set* (also *fig.*), *station.*

1923 E. W. MARCHANT *Radio Telegr. & Teleph.* iii. 24 The exact arrangement of the receiving aerial may be varied within fairly wide limits. 1962 A. NISBETT *Technique Sound Studio* 267 This highly inefficient transformer action is improved if the receiving aerial is so constructed as to resonate at the desired frequencies of reception. 1908 Receiving apparatus [see *PICK *v.*[1] 20 f]. 1925 *Times* 28 May 20/4 Such a standard must obviously be based upon a consideration of two factors—namely, the limitations of the present transmitting and receiving apparatus; and, secondly, the æsthetic element. 1923 E. W. MARCHANT *Radio Telegr. & Teleph* iv. 38 When the coherer was first used in connection with receiving circuits, it was connected between the aerial and the ground. 1955 *Radio Times* 22 Apr. 3/2 A special type of receiving circuit is necessary for F.M. 1916 *Lit. Digest* (N.Y.) 1 Jan. 13/2 His outfit comprised only a cheap home made receiving set! But it did the work, just the same. 1937 *Discovery* Nov. 334/2 His [*sc.* the shrew's] bewhiskered snout is a receiving-set fitted to pick up any broadcast interesting to a shrew. 1953 A. HUXLEY *Let.* 9 Aug. (1969) 682 Receiving sets grafted into the tissues of animals, so as to make them robots responsive to the radioed will of their masters. 1975 *Listener* 4 Dec. 738/2 A fee required of each owner of a receiving set. 1923 E. W. MARCHANT *Radio Telegr. & Teleph.* iii. 23 A number of vertical wires are attached to a suspended horizontal wire and brought down in a fan-shape to the receiving station.

3. Special combinations. **receiving barn** *U.S.*, a stable in which horses are placed before a race to prevent tampering; **receiving blanket** *N. Amer.*, a soft blanket in which to wrap a baby (cf. RECEIVER[1] 6 a); **receiving line** orig. *U.S.*, a row of persons by whom guests are greeted in turn on arrival.

1946 *Sun* (Baltimore) 26 Oct. 8/4 Refusal of George D. Widener to allow his equine star, Lucky Draw, to compete in the Pimlico Special simply because of the receiving barn is to be regretted. 1949 *Ibid.* 29 Apr. 18/4 There has been no doping case since the rules governing the receiving barn were revised last fall. 1974 TILLEY & PLOWDEN *This is Horse Racing* 181 *Receiving barn*, facility where horses are isolated for certain period before post time, to minimize chances of tampering. 1926 *Infants' Dept.* Oct. 5028/1 *Layette*... 1 Receiving blanket. 1944 K. HARDY *Sewing for Baby* viii. 214 You will need two cotton receiving blankets. 1970 *Globe & Mail* (Toronto) 26 Sept. 12/8 (Advt.), Since mid-summer we have paid out Four Hundred and Twenty-one Dollars for the basic essentials in layettes; shirts, diapers, gowns, and receiving blankets. 1933 H. L. ICKES *Diary* 17 Nov. (1955) I. 125 He told Anna in the receiving line last night that he was going to order me down there. 1971 M. LEE *Dying for Fun* xl. 189 Millie Panhard Geltzer went to the Woman of the Year Luncheon... She shook hands with the receiving line.

receiving, *ppl. a.* Add: **2.** *receiving end*, the position that receives a transmitted signal or discharged object. Usu. *fig.*, used loosely in colloq. phr. *to be on* (or *at*) *the receiving end*, to be the (unfortunate) recipient of some action, event, etc.; to bear the brunt, to suffer.

1933 [see *INVERT *v.* 2 g]. 1937 H. L. ICKES *Diary* 2 Oct. (1955) II. 219, I shall refuse to be at the receiving end of any more brickbats. 1942 BERREY & VAN DEN BARK *Amer. Thes. Slang* § 674 *Catcher's position*, receiving end. 1946 B. PEMBERTON in S. H. Adams *Alexander Woollcott* xiv. 142 But it [*sc.* the nature of Woollcott's criticism] was hell for those on the receiving end. 1949 J. SZIGETI *With Strings Attached* xxxiii. 302 It cannot be the love of music pure and simple, for we share this with multitudes who either are on the receiving end alone or, not content with this, also aspire to the satisfaction of making music themselves. 1955 W. C. GAULT *Ring around Rosa* vi. 78 Jan had just brought a right hand from right field and the wrestler had been on the receiving end. 1958 P. KEMP *No Colours or Crest* viii. 151 His experience of guerrilla warfare had been, as it were, on the receiving end; for he had served before the war on the North-West Frontier and in Palestine. 1962 A. NISBETT *Technique Sound Studio* x. 176 Dialling is rather more of a performance than being on the receiving end of a phone call. 1968 G. JONES *Hist. Vikings* III. ii. 202 As in Scandinavia, so at the receiving end in Europe the times were favourable to the art and practice of *viking.* 1976 J. SNOW *Cricket Rebel* 134 The wrist had been on the receiving end of a Dennis Lillee bouncer.

re-cement, *v.* Add: Hence **re-cemented** *ppl. a.*

1946 *Nature* 13 July 58/1 The lower part of the profile consists, according to the district, either of re-cemented chalk or compact sand, both of which are very water-retentive. 1965 G. J. WILLIAMS *Econ. Geol. N.Z.* iii. 26/2 Re-cemented brecciated quartz is occasionally seen.

recency. Add: **a.** (Further examples.)

1882 C. E. DUTTON *Tertiary Hist. Grand Cañon District* v. 83 Even here where historic antiquity merges into geologic recency the one gives us no measure of the other. 1948 H. NICOLSON *Diary* 23 Sept. (1968) 149 Frank Pakenham tells me that I was on the list for peerages last December... I might have come up again had it not been for the recency of Croydon. 1964 *Language* XL. 204 The recency and taxonomic character of paralinguistics and kinesics are not sufficient cause for overlooking them. 1976 *Nature* 5 Feb. 395/2 Genetic similarity may serve as direct measure of the recency of the cladistic event which separated the two compared lines of descent.

b. *Psychol.* The fact of being recent as it affects the facility with which learned material or an experience is recalled; freq. *attrib.* Cf. *PRIMACY 1 b.

1894 M. W. CALKINS in *Psychol. Rev.* I. 482 The influence of recency, too, can be studied. 1916 J. B. WATSON in *Psychol. Bull.* XIII. 77 Other investigators hold that the stamping in of a successful act depends upon the principles of *recency* and *frequency* and is not dependent upon the pleasantness or unpleasantness resulting from the activity. 1929 K. S. LASHLEY in C. Murchison *Found. Exper. Psychol.* xiv. 555 It seems probable that primacy and recency are effective only when they increase the intensity or stimulating value of the situations. 1938 R. S. WOODWORTH *Exper. Psychol.* ii. 38 The wrong name recalled acquires recency value and blocks the correct name. 1948 E. R. HILGARD *Theories of Learning* vii. 183 As it applies to memory, the law of proximity becomes a law of recency. 1964 W. K. ESTES in A. W. Melton *Categories of Hum. Learning* 98 The suggestion has been put forward..that the negative recency function results largely from response tendencies the *Ss* bring with them to the experiment. 1971 *Sci. Amer.* Aug. 85/1 There is considerable evidence that the recency effect is due to retrieval from short-term storage.

recensionist. (Example.)

1962 *Listener* 29 Nov. 920/1 Mr Edel is a recensionist, reconstructing the life of his hero from a myriad documents.

recent, *a.* Add: **1.** (Later examples.)

1966 *Listener* 29 Sept. 479/2 Gerhard's Op. 1..its beauty [is] marred only by a lack of variety and a tendency to go on too long. Nearly fifty years more recent, the *Duo Concertante* for violin and piano shows neither quality. 1976 *Encounter* June 72/2 Professor Daiches' method, then, consists partly of summarising the results of recent research on the ancient Near East, and partly of commentary on the Pentateuch. 1976 *Daily Tel.* 20 July 1/1 The four countries had agreed on the ban during their recent economic summit in Puerto Rico.

2. b. (Further example.)

1877 *Encycl. Brit.* VI. 134/2 If not set when either moist or recent, they [*sc.* beetles' legs] may be softened by being placed for a night in any small vessel containing a layer of wet sand.

4. b. Also = *HOLOCENE *a.* (Later examples.)

1882 A. GEIKIE *Text-bk. Geol.* 883 Above them [*sc.* glacial deposits] lie younger accumulations such as river-alluvia, peat-mosses, lake-bottoms,..raised lacustrine and marine terraces, which, merging insensibly into those of the present day, are termed Recent or Prehistoric. 1927 [see *HOLOCENE *a.*]. 1959 J. D. CLARK *Prehist. Southern Afr.* ii. 49 A.. method of correlating the succession of events in Africa with those in Europe during the Pleistocene and early Recent times. 1975 *Nature* 20 Mar. 209/2 Sterols..have been identified in Recent and ancient sediments.

recently, *adv.* **b.** (Further examples.)

1922 JOYCE *Ulysses* 123 What is it?..—A recently discovered fragment of Cicero's. 1937 *Granta* 3 Feb. 219/1 He was off with the five survivors of our English group to join the recently formed English Battalion. 1960 *Farmer & Stockbreeder* 19 Jan. 102/3 Females changed hands at prices ranging to 130 gs, paid for a recently-calved C.M. cow. 1979 *Tucson* (Arizona) *Citizen* 20 Sept. 3C/6 It already has been replaced by the recently-completed 254-bed West Wing.

recep. (rĭse·p). **1.** Abbrev. of *reception room(s)* s.v. RECEPTION 10. (Freq. used in advertisements.)

1920 *Evening News* 2 Mar. 7/2 (Advt.), Exchange tenancy houses,..4 bed. 2 recep. bath. 1926 [see *BED *sb.* 1 e]. 1950 *Evening Standard* 21 Mar. 12/3 (Advt.), At Harrow... Charm. mod. hse... 2 recep., kit. 1976 'Z. STONE' *Modigliani Scandal* I. iv. 41 The house..was small—three bedrooms, two recep. and a study.

2. *Theatr.* See *RECEPTION 5 c.

reception. Add: **2. d.** (Further examples.) Now usu. a party at which guests are formally greeted, esp. after a wedding.

1865 LD. BROUGHTON *Recoll. Long Life* (1911) VI. ix. 54 On March 5 [1842] I dined at Lord Palmerston's... Lady Palmerston had a reception afterwards. 1906 *Mrs. Beeton's Bk. Househ. Managem.* 1680 The orthodox wedding breakfast seems likely to become a thing of the past, so much has it been superseded by the tea and reception which usually follow afternoon weddings. 1907 *St. Nicholas* Oct. 1119/1 The 'happy couple'..will hurry back to Red Feather's house to the reception. 1928 J. SYKES *M. A. Disraeli* 3 Lord Broughton ('Recollections of a Long Life') says it was in Lady Palmerston's day that an evening party at a Minister's house began to be called a reception. 1944 J. LEES-MILNE *Prophesying Peace* (1977) 72 We went together to the Dorchester reception given by the bride's mother. 1965 *Proc. Classical Assoc.* LXII. 35 During the Reception a gavel was presented for the use of the Association by Miss D. J. Wood, Joint Hon. Secretary of the Manchester and District Branch. 1977 *Times* 14 July 16/6 *Receptions*... Lord Grey of Naunton, accompanied by Lady Grey, entertained members of the Royal Over-Seas League at a reception in the House of Lords last night.

e. A place where guests register on arriving at a hotel, etc., or where an organization's clients, etc., are received. Usu. without article and with capital initial.

1917 E. FENWICK *Diary* 13 Nov. in *Elsie Fenwick in Flanders* (1981) 183 The reception was a regular Hades of women and children and men..all terribly wounded. 1930 E. WAUGH *Vile Bodies* x. 169 'Bless you,' said the woman at the counter marked 'Reception', 'all our rooms have been booked for the last six months.' 1958 'CASTLE' & 'HAILEY' *Flight into Danger* viii. 107 Those cars.. were promptly waved..to parking spaces well clear of the entrance to Reception. 1969 D. CLARK *Nobody's Perfect* iii. 110 Not one gets past me or Bert. We hold them in reception, phone up the one they want to see, and make them wait till they're fetched. 1970 G. F. NEWMAN *Sir, You Bastard* v. 148 A sister met him in the reception of the decaying building. 1975 M. DRABBLE *Realms of Gold* IV. 246 Frances had a call from Reception telling her that it would take at least two hours to put a call through to London.

f. Shortened f. *reception room* s.v. RECEPTION 10. Cf. *RECEP. 1.

1929 *Daily News* 13 Sept. 11/5 Semi-detached Houses... Attractive Elevations... Large hall, 2 reception, 3 bedrooms. 1977 *Evening Post* (Nottingham) 27 Jan. 14/3 (Advt.), An extremely good two double bedroomed semi-detached house with garage, two receptions, kitchen, bathroom, gardens. 1980 *Daily Tel.* 29 July 1/8 The 18th century house has four receptions, nine bedrooms and eight bathrooms.

5. c. *Theatr.* An ovation granted a popular actor on taking the stage. Also abbrev. *recep.*

1847 F. A. KEMBLE *Let.* 15 Feb. in *Records Later Life* (1882) III. 160, I wish I could avoid my 'reception', as it is called, because any loud sound shakes me from head to foot. 1936 W. H. LANE CRAUFURD *Murder to Music* vii. 104 'Did you hear my little recep.!' he cried ecstatically.

d. The receiving of broadcast signals; the efficiency with which this is done, as regards audibility, picture quality, freedom from interference, etc.

1907 *Rep. Sel. Comm. Radiotelegr. Convention* 129/2 in *Parl. Papers* VIII. 1 Receivers tuned for the reception of waves of such lengths..can be rendered quite immune from influence by..longer and more powerful waves. 1923 E. W. MARCHANT *Radio Telegr. & Teleph.* ix. 104 For the long wave-lengths the change in the spacing produces less effect on the efficiency of reception than it does for shorter wave-lengths. 1943 C. L. BOLTZ *Basic Radio* xiii. 203 We are here concerned with the reception of C.W., I.C.W., and amplitude-modulated carriers. 1949 *Radio Times* 15 July 6/3, I expected the television to be perfection, but..the same excuses everywhere—'Of course, reception isn't so good here because of the electrical appliances in the district.' 1962 [see *RECEIVING *vbl. sb.* 2 c]. 1968 A. MARIN *Clash of Distant Thunder* (1969) ix. 63 We can get the Television Français... The reception is not very good. 1972 *Times* 21 Sept. (Ireland Suppl.) p. ii, At that time television was only just getting off the ground..and many areas still did not have reception.

6. (Further example in *Amer. Football.*)

1976 *Billings* (Montana) *Gaz.* 16 June 5-C/3 Bell is expected to have a chance at several receptions, with two top passing quarterbacks on hand.

10. (sense 2) *reception area, camp, -clerk, committee, counter, day* (earlier examples), *desk, evening, hall, night, room* (further examples), *town*; **reception centre,** a centre for the reception of newcomers or visitors; *spec.* a hostel providing temporary accommodation for the destitute; **reception class,** the lowest class in an infant school, into which children going to school for the first time are admitted; **reception statute** *U.S. Law*, a statute passed by an American state after Independence providing that the common law of England be received as binding in that state, subject to repeal and local interpretation.

1939 *Times* 2 Nov. 8/7 While anxiety was expressed about the effects of breaking up home life, tributes were paid to householders and others in the reception areas for their friendly helpfulness. 1971 F. FINLAY *Boy in Prison* vi. 79, I was handed over to one of the Grendon officers, who greeted me pleasantly and took me down an open pathway to the reception area. 1979 *Tucson Mag.* June 40/1 There was..a wedding performed in an airplane followed by a sky dive to the reception area. 1918 W. OWEN *Let.* 13 Sept. (1967) 576 Write to the 2nd Man., not this Recep. Camp as I'll be joining about Sunday. 1954 *Encounter* Feb. 39/2 It is when the fugitive succeeds in moving to the reception camps of Western Germany, and..back into civil life, that his enthusiasm turns to disgust. 1942 *Nation* 27 Apr. 41 Men passing the 'screening test' will be ordered promptly to reception centers for final examination and induction. 1948 *Act* 11 & 12 *Geo. VI.* c. 29 § 17 It shall be the duty of the [National Assistance] Board to make provision whereby persons without a settled way of living may be influenced to lead a more settled life, and the Board shall provide and maintain centres, to be known as reception centres, for the provision of temporary board and lodging for such persons. 1957 H. ROOSENBURG *Walls came tumbling Down* ix. 199 We were waiting for a truck to take us to a DP reception centre. 1978 C. A. BERRY *Gentleman of Road* ix. 69 Normal folk who live on the other side of the soup enjoy reception centres close to their doors as little as they favour Salvation Army hostels or prisons. 1972 *Where* Jan. 10/1 If a few children in each reception class could already read well, teachers would be forced to make more flexible arrangements. 1980 *Guardian* 24 June 11/2 Children with problems are very often referred as priority cases to..reception classes in infants schools. 1934 T. F. TWEED *Blind Mouths* ii. 11 The reception-clerk, who had been placidly reading his newspaper, dropped it and became suddenly alert. 1981 L. DEIGHTON *XPD* xxvii. 220 [He] walked quickly across to.. the lift... The reception clerk looked up. 1851 W. K. NORTHALL *Before & Behind Curtain* 89 We believe Mr. Marks consulted some members of the reception-committee. 1920 [see *EXTREMIST]. 1960 C. MACINNES *Mr. Love & Justice* 96 If you've got any ideas of seeing a lawyer, or having any sort of reception committee for me, that's up to you. 1978 *Navy News* Oct. 2/1 Among the 'reception committee' were the mayor and treasurer of the town of Ajax, Ontario. 1975 N. LUARD *Robespierre Serial* xv. 133 The hotel manager was behind the reception counter. 1853 E. TWISLETON *Let.* 16 Jan. (1928) iv. 67 It was Mrs. Crawford's reception day, and I was struck with her excellent, easy manners. 1884 G. MEREDITH *Let.* 30 May (1970) II. 738, I.. hope to present my respects on one of her reception days. 1936 'J. TEY' *Shilling for Candles* xxiv. 263 He shot into the darkness of the great lounge and across it to the green light of the reception desk. 1940 E. GILL *Autobiogr.* vi. 180, I went to the hotel and managed to make myself understood at the Reception Desk. 1978 S. SHELDON *Bloodline* xliii. 363 Alec followed the waiter out of the large dining room into the small office behind the reception desk. 1846 J. K. POLK *Diary of President* (1929) 70 These informal reception evenings are very pleasant. 1938 D. DU MAURIER *Rebecca* iii. 17 A certain sofa..midway between the reception hall and the passage to the restaurant. 1976 *Washington Post* 19 Apr. C17/6 (Advt.), First floor has a very large reception hall, spacious living room [etc.]. 1848 S. THORNE *Jrnl. Boy's Trip* (1936) 30 To night being reception night I went to see the President and his lady again. 1846 S. S. MAGOFFIN *Diary* 7 Aug. in S. M. Drumm *Down Santa Fé Trail* (1926) 70 My dirt-floored chamber, dining-room, parlour, reception room &c. &c. is quite dessolate. 1906 C. H. B. QUENNELL *Mod. Suburban Houses* p. vii, The accommodation generally required is the three usual reception rooms—sometimes Drawing and Dining Rooms will suffice with what is known to the House Agent as the Sitting Hall in addition. 1961 *Times* 18 Oct. 20/2 More often than not, reception-room heating is inadequate. 1931 *Columbia Law Rev.* XXXI. 416 The so-called reception statutes vary in form, in the date chosen as a deadline, and as to the implications of the extent of reception. 1956 E. H. POLLACK *Fundamentals of Legal Research* i. 10 The 'reception' statutes, enacted by most American states in the eighteenth and nineteenth centuries, gave special recognition to the English common law and statutes..as of 1607, the date of the earliest English settlement at Jamestown. 1976 J. K. LIEBERMAN *Milestones!* i. 19 Reception statutes did not freeze into place the particulars of the common law of England... What was received..was the *process* of the common law..not the particular results that English judges had reached. 1958 *Times Rev. Industry* Feb. 24/1 Some 70,000 people are to be 'overspilled' from Glasgow City over the next 10 to 15 years into the new towns..and other reception towns.

receptionist. Add: **b.** A person employed by a hotel, medical surgery, or other organization, to receive clients, etc. Cf. *RECEPTION 2 e.

1901 *Girl's Own Paper* 12 Jan. 234/1 She answered an advertisement of a Kensington photographer who needed a 'receptionist' and saleswoman—shop-woman in fact. 1927 *Daily Express* 19 Feb. 2/4 An operation which she underwent..for the purpose of advertising the [plastic surgery] business, and on condition that she would continue to be employed as secretary and receptionist. 1932 D. L. SAYERS *Have his Carcase* iii. 41 'Ai'm afraid,' said the receptionist..'that all our rooms are engaged.' 1956 *Times* 21 Jan. 7/5 Such a qualification would ensure that the secretary ..could, in her employer's absence, conduct responsible interviews on his behalf, as opposed to receptionist duties. 1960 M. SPARK *Bachelors* vii. 99 Dr Lyte sat in his consulting room after the last of the evening surgery had departed and his receptionist had locked up and gone home. 1976 B. BOVA *Multiple Man* (1977) vi. 66 The reception lobby was equally quiet... A curved desk with all the paraphernalia of a busy receptionist... But no people.

receptive, *a.* Add: **1. a.** Also of sensory processes in an organism.

1906 C. S. SHERRINGTON *Integrative Action of Nerv. Syst.* iii. 90 The flexion-reflex has a 'receptive skin-field' which though extensive is characteristic for it. 1934 A. FORBES in C. Murchison *Handbk. Exper. Psychol.* iii. 175 When the receptive substance is excitatory the combination causes contraction of the muscle cell. 1975 M. & N. SAMUELS *Seeing with Mind's Eye* xi. 152 Receptive visualization provides us with the means for getting in touch with images from our inner center.

c. *Med.* and *Psychol.* Affecting or relating to the comprehension of speech or writing, esp. as impaired by a brain disorder.

1926 H. HEAD *Aphasia* I. II. iii. 204 Closer observation showed that the 'receptive' aspect of her use of language had not in reality escaped. *Ibid.* 207 These defects..are entirely inexplicable on any theory..of separate loss of the 'emissive' and 'receptive', 'motor' and 'sensory' functions, which are supposed to accompany speech. 1955 R. JAKOBSON in Saporta & Bastian *Psycholinguistics* (1961) 423/2 The classical distinction between the so-called emissive (or expressive) aphasia on the one hand, and the receptive (or sensory) aphasia on the other. 1963 OSGOOD & MIRON *Approaches to Study of Aphasia* iii. 60 Patients displaying breakdown..in the comprehension of speech and word meanings, but without impairment of hearing per se (i.e., the classical 'sensory' or 'receptive' aphasia), are most likely to have lesions in the left temporal zone. 1961 R. BRAIN *Speech Disorders* vii. 93 Central aphasia is characterized by both receptive and expressive disturbance. 1973 J. W. BROWN *Pick's Aphasia* p. xi, With regard to 'receptive' disturbances, Pick assumed a complementary mental structure underlying the transition of heard sound to thought. 1977 W. H. PERKINS *Speech Path.* (rev. ed.) v. 132/2 By the same token, receptive aphasia, sensory aphasia, posterior aphasia, and fluent aphasia identify essentially the same language disorder as Wernicke's aphasia, or pragmatic aphasia. *Ibid.* xv. 341 The North-western Syntax Screening Test..of receptive and expressive grammatical ability has become available.

receptivity. Add: (Further examples.) Also, in the philosophy of Kant, a passive quality contrasted with spontaneity or activity.

1796 F. A. NITSCH *View of Kant's Princ.* 74 That a given variety can occur in our perceptions, knowledge, &c. supposes a..Receptivity which is totally Passive. *Ibid.* 75 The Receptivity, as far as it receives varieties of the first description, may be called external sense, and as far as it receives varieties of the second description, internal sense. 1950 M. PEAKE *Gormenghast* xxiv. 159 This poem..shall be addressed to as many as are here present in all the variance of their receptivity, status and acumen. 1956 C. MORRIS *Var. of Human Nature* i. 17 Receptivity should be the keynote of life. *Ibid.* ix. 191 In Factor D receptivity is directed toward nature, toward the man-cosmos relation, or to the needs of other persons. 1958 B. BERNSTEIN in J. A. Fishman *Readings in Sociol. of Lang.* (1968) 234 It is now necessary to show how this mode of perceiving and the attendant structuring of receptivity conflicts with and induces a resistance to formal education. 1973 CHURCHILL & AMERIKS tr. *Husserl's Experience & Judgment* I. i. 79 This phenomenologically necessary concept of receptivity is in no way exclusively opposed to that of the activity of the ego. 1974 A. E. BLUMBERG tr. *Schlick's Gen. Theory of Knowl.* iii. § 40. 371 The introduction of the antithesis between spontaneity and receptivity—in modern terms, between activity and passivity—is entirely inappropriate at this point.

receptor. 3. a. For def. read: The region of an antibody molecule which shows specific recognition of an antigen. Also *attrib.* and *Comb.*, as *receptor group*; *receptor-destroying* adj. (Further examples.) [The sense is due to P. Ehrlich, who coined the G. *receptor* (*Berlin klin. Wochenschr.* (1900) 21 May 453/2.)]

Ehrlich's theory of receptors originally took a broader form than that in which it is now accepted; hence in the first part of this century there was some overlap between this sense and sense c below.

1903 J. COATS *Man. Path.* (ed. 5) 151 By careful and increasing dosage the protoplasm of the cell may be gradually stimulated to form more and more receptor groups. 1935 F. P. GAY *Agents of Dis. & Host Resistance* xix. 377 The entire receptor hypothesis of Ehrlich rests on a purely imaginary misconception... There is no particular objection in referring to this absorption as due to certain chemical units that one may call the 'receptors' of the susceptible cell. .. It is another matter to assume..that the hypothetical receptors.., when injured or destroyed, are reproduced in excess and poured into the blood stream as specific antibodies to those foreign incitants that have engendered their production. 1941 KOLMER & TUFT *Clin. Immunol. Biotherapy & Chemotherapy* iii. 63 This [*sc.* Ehrlich's side-chain theory] is now regarded as untenable because of the physiologic improbability that there can exist a sufficient number of specific receptors for an innumerable number of antigens. 1951 WHITBY & HYNES *Med. Bacteriol.* (ed. 5) xii. 221 The enzyme is best known..as the receptor-destroying enzyme (RDE) which has helped to elucidate the phenomenon of hæmagglutination by viruses of the mumps–influenza group. 1967 *Cold Spring Harbor Symp. Quantitative Biol.* XXXII. 431/1 We assume first that nothing except antibody recognizes antigen, and we must therefore assume that the receptor for antigen is antibody already present at a site, in or on the cell, prior to exposure to antigen. 1970 FENNER & WHITE *Med. Virol.* iii. 45 The importance of glycoprotein receptors for the attachment of influenza virus has been demonstrated in experiments in which they have been destroyed by bacterial neuraminidase ('receptor-destroying enzyme'). 1973 *Sci. Amer.* June 82/3 The essential initial step in the immune response is the contact of an antigen—a foreign substance—with receptors on the surfaces of lymphocytes derived from the bone marrow, the so-called *B*-lymphocytes.

b. *Biol.* Any organ or structure which on receiving stimuli of a certain kind from its environment generates nerve impulses that convey information about that aspect of the environment. Freq. *attrib.*

1906, etc. [see *EFFECTOR 2]. 1906 C. S. SHERRINGTON *Integrative Action Nerv. Syst.* i. 13 Electrical stimuli applied to receptor organs are..efficient excitors of reflexes. *Ibid.* ix. 309 The branching at the receptive end places it in communication not with one but with several receptor cells. 1919 W. M. BAYLISS *Introd. Gen. Physiol.* iv. 112 The eye may be said to be the most accurately adjusted of all our receptor organs. 1920 T. P. NUNN *Education* xiii. 170 To pick out and distinguish the different elements and qualities of which the world is composed..is more possible for higher animals by the enormous development of the receptor-system. 1934 *Nature* 22 Sept. 445/1 Normal vision may be due to a receptor which gives rise to a red sensation, one which gives rise to a blue sensation and one which gives rise to a not blue, not red sensation which, of course, corresponds to a green sensation. 1962 *Listener* 8 Nov. 779/2 The use of the eye instead of the ear as the principal receptor for information probably has produced some consistent concomitant psychological changes. 1971 *Jrnl. Gen. Psychol.* LXXXV. 87 Reason..has proposed that susceptible persons are more highly receptive to incoming stimuli, and that this receptivity is global rather than being confined to one receptor system alone. 1974 D. & M. WEBSTER *Compar. Vertebr. Morphol.* x. 197 A blow on the head mechanically stimulates the eye's receptor cells. 1975 *Sci. Amer.* July 108/2 There are other receptors on the antennae of mosquitoes that respond to chemical stimuli.

c. *Physiol.* A region of a neurone or other tissue which specifically recognizes and responds to a neurotransmitter, hormone, or other substance. Freq. *attrib.*

1912 E. H. STARLING *Princ. Human Physiol.* vi. 312 The existence of..a 'receptor' substance, as he calls it, has been furnished by Langley. *Ibid.* 314 Receptor substances may act as intermediaries in every case of propagation of an impulse across a synapse of whatever description. 1939 M. A. GOLDZIEHER *Endocrine Glands* ii. 7 The organotropic hormones of the pituitary..are supposed to act upon the cells of the individual endocrine glands which constitute their receptor organs. 1955 *Biochim. & Biophys. Acta* XVI. 268 The picture which has emerged, assumes that acetycholine is stored in an inactive bound form. Stimulation releases the ester which combines with a receptor protein. 1961 R. D. BAKER *Essent. Path.* iii. 38 Materials from the infarct stimulate the bone marrow and cause leukocytosis, and their influence on central nervous system receptors may cause fever. 1973 *Nature* 20 Apr. 497/2 Receptors can be defined as components of a tissue which specifically react with a drug or hormone. *Ibid.*, The meaning of 'receptor' is becoming very diffuse and a clearer definition of the word is necessary. 1975 *N.Y. Times* 27 Mar. 9/1 The test shows whether the tumor cells contain receptors, or 'landing sites', for estrogen molecules, which are then taken into the cells and stimulate their growth.

recess, *sb.* Add: Now also with pronunc. (rī·ses). **3. a.** Chiefly *N. Amer.* exc. in Parliamentary use. (Earlier and further examples.)

1620 *Jrnls. House of Lords* 22 Mar. 61/1 They [*sc.* the Commons] humbly desire to know the Time of the Recess of this Parliament, and of the Access again, as they may accordingly depart and meet again at the same Time their Lordships shall. 1851 K. QUENTIN *Reisebilder & Studien* II. 58 Um 12 Uhr verliess ich mit den Kindern die Schule. Sie haben eine Pause (Recess) von einer Stunde. *c* 1860 E. DICKINSON *Poems* (1955) I. 117 Whose Beryl Egg, what School Boys hunt In 'Recess'—Overhead! 1913 A. HUXLEY *Let.* 3 Feb. (1969) 47, I had a very good vacation, or do you call it Ree-cess?, as one (female) American asked me. 1942 *Amer. Mercury* July 91 Must be a recess in Heaven—pretty angel like that out on the ground. 1951 E. PAUL *Springtime in Paris* ii. 27, I could not possibly hope to reach the cage before the bank closed at 12 o'clock noon, for the two-hour lunch recess. 1975 *Weekend Mag.* (Montreal) 1 Nov. 22/1, I watched him carefully as he won game after game at recess one day.

9. e. *Criminals' slang.* The lavatory in a prison. Usu. *pl.*

1950 P. TEMPEST *Lag's Lexicon* 177 *Recess*, the lavatory and urinal, which are generally situated in a recess (two cells knocked into one). 1958 F. NORMAN *Bang to Rights* III. 103 The recesses are give [*sic*] a good clean out. 1974 *Observer* (Colour Suppl.) 10 Feb. 17/1 Locked in their cells [*sc.* in Winson Green Prison, Birmingham] at 5.30., with one opening later to go to the recesses (lavatories) and to have a hot drink.

f. = *recess printing* below.

1971 D. POTTER *Brit. Eliz. Stamps* iii. 36 This different-

RECESS

size stamp, printed by recess, interrupted the unity of the set.

11. Also *Comb.* *recess time* (earlier and later examples); **recess printing**, a method of printing used in the production of postage stamps (see quot. 1951); hence **recess-print** *v.* (usu. as pa. pple.).

1930 *Times Educ. Suppl.* 26 July p. iv/1 All [stamps] are recess-printed in designs appropriate to the occasion. 1976 *Times* 10 Aug. 6/7 The first issue, containing a finely drawn head of Queen Victoria...and recess printed by Bradbury Wilkinson. 1914 A. B. Creeke *Stamp-Collecting* iii. 66 *Recess-printing*: The design is cut into the plate, and the ink stands up slightly on the stamp. 1951 R. J. Sutton *Stamp Collector's Encycl.* 190 Recess Printing: Strictly speaking, any process where the inked image is below the plate surface of the plate, cliché, block or cylinder; but in modern philatelic parlance refers to the present-day machine-printed, photo-mechanically engraved plate method of reproduction, which in its essentials is similar to the line-engraving by which most of the first and early stamps were printed. A 'recess' printed stamp has a distinct raised image. 1869 Mrs. Stowe *Oldtown Folks* xxxiii. 431 At recess-time she strolled out with me into the pine woods back of the school-house. 1946 G. Wilson *Fidelity Folks* iii. 84 A half dozen biscuits soaked in it ought to keep starvation away until recess time.

recess, *v.*² Add: Now also with pronunc. (rĭse·s). **3. a.** For *U.S.* in Dict. read 'Chiefly *U.S.*' and add further examples.

1933 W. J. Abbot *Watching World go By* xvii. 316 The convention was thrown into confusion. It recessed almost in a riot. 1943 *Sun* (Baltimore) 12 Feb. 17/1 Tomorrow's holiday on which the country's major securities and commodities exchanges will recess. 1970 *Daily Tel.* 6 July 9 The French Parliament recessed for its three-month summer holiday last week. 1977 *Ibid.* 4 May 19/1 A Turin court trying 53 Leftist 'Red Brigade' guerrillas recessed indefinitely yesterday for lack of citizens willing to serve on the jury. 1977 *New Society* 7 July 23/2 The inquiry recessed earlier this month so that the contending parties would have a chance to wade through transcripts. 1977 *Time* 26 Dec. 22/1 Congress recessed last week for a month-long holiday without enacting his energy bill.

b. *trans.* Chiefly *U.S.* To put (a meeting, etc.) into recess; to adjourn, suspend.

1954 W. Faulkner *Fable* 80 It takes more ammunition to recess a war for ten minutes than to stop a mere offensive. 1967 *Guardian* 12 June 8/4 Hans Tabor recessed the meeting and told them to 'stand by' for any urgent call. 1970 W. Wager *Sledgehammer* xix. 173 When Gillis recessed the proceedings...not a single juror had been picked. 1978 *Detroit Free Press* 2 Apr. 10A/5 Exhausted negotiators agreed to recess formal talks for the weekend and meet in private.

recession, *sb.*¹ Add **1. d.** *Philol.* The transference of accentuation towards or on to the first syllable of a word.

1886 *Amer. Jrnl. Philol.* VII. ii. 246 A tendency existed to recession from the end of the word. 1929 *S.P.E. Tract* xxxii. 388 This condition, which lightens the syllable, allows and even invites loss and recession of accent.

4. *Econ.* A temporary decline or setback in economic activity or prosperity.

1929 *Economist* 2 Nov. 806/1 The material prosperity of the United States is too firmly based, in our opinion, for a revival in industrial activity—even if we have to face an immediate recession of some magnitude—to be long delayed. 1930 *Engineering* 3 Jan. 21/2 The paramount problem is now whether this recession is yet at an end. 1938 R. Ahrens *Cause for Alarm* i. 16 'Trade recession' they called it... As far as I could see there wasn't a great deal of difference between a trade recession and a good old-fashioned slump. 1958 *Spectator* 30 May 676/2 This is partly due to the inability of the United States to pull itself out of recession. 1976 F. Zweig *New Acquisitive Society* ii. 99 The private sector, particularly in the throes of recession, is limited in its ability to pay by the discipline of the market system. 1981 *Times* 11 Mar. 19/4 The economy is now in deep recession.

recessional, *a.* and *sb.* Add: **A.** *adj.* **1. a.** (Further examples.)

1932 Chesterton *Coll. Poems* (1933) 45 We fancied heaven preferring much Your rowdiest song...to such Very recessional repentance. 1973 *Times* 15 Nov. 6/6 Timothy Farrell, sub organist of Westminster Abbey, thundered recessional music.

3. *recessional moraine* (Geol.), a form of moraine which is deposited during a temporary halt or minor readvance of a receding glacier or ice sheet, similar in appearance to a terminal moraine.

[1897 *Jrnl. Geol.* V. 427 To account for the moraines of recession by any scheme of ups and downs of the solid earth.] 1907 R. D. Salisbury *Physiogr.* v. 277 As the edge of the ice was melted back, it sometimes halted for a time far back from the position of its maximum advance. Beneath the edge in such positions, terminal moraines were made. Such terminal moraines are sometimes called recessional moraines. 1925 W. J. Miller *Introd. Physical Geol.* viii. 255 Recessional moraines, forming a great succession of curving ridges, are wonderfully displayed to the south of Lakes Michigan and Erie. 1957 G. E. Hutchinson *Treat. Limnol.* I. i. 82 The terminal or recessional moraines of valley glaciers may persist, in certain circumstances, in a sufficiently well-preserved state that they can dam the stream that replaces the glacier. 1976 C. L. Matsch *N. Amer. & Great Ice Age* vi. 84 Recessional moraines are valuable aids to the glacial geologist because their spacing might allow the calculation of rates of melting if appropriate materials are available for radiocarbon dating.

recessionary (rĭse·ʃənări), *a.* [f. Recession *sb.*¹ + -ary¹.] Of, pertaining to, or characterized by (economic) recession.

1958 *Times* 24 Nov. 15/2 Moreover, there are several factors that have so far provided a cushion against recessionary forces. 1970 *Times* 22 May 8 He admits himself baffled by the combination of high inflationary symptoms in prices and interest rates with recessionary symptoms in output. 1974 *Daily Tel.* 19 July 17 The outlook is clearly recessionary, but it is still doubtful if the Chancellor could afford to put back more than £500 million into the economy. 1979 *Nat. Westm. Bank Q. Rev.* Aug. 4 The desire to keep the economy at a high level of activity in general world wide recessionary conditions.

recessive, *a.* Add: **A.** *adj.* **1. a.** Also of persons, retiring, reserved. *rare.*

1925 T. Dreiser *Amer. Tragedy* I. i. xiii. 95 She, for her part, felt recessive and thence evasive.

b. *Philol.* *recessive accent*, stress transferred towards or on to the first syllable of a word.

1879 W. W. Goodwin *Elem. Gr. Gram.* (ed. 2) 13 When a word throws its accent as far back as possible...it is said to have recessive accent. 1926 Fowler *Mod. Eng. Usage* 168/1 He [sc. the latinist] has still to reckon with the recessive-accent tendency, which has so great a vogue in English. 1933 S. Potter *Mod. Ling.* Aug. 82/3 Add to these specimens the 'recessive accent'—stressing of the first syllable of a word which has previously been accented on the second or third syllable. 1973 A. H. Sommerstein *Sound Pattern Anc. Gr.* v. 172 Action nouns...have recessive accent if masculine but final accent if feminine.

c. *Philol.* That tends to recede from use or fall into desuetude.

1936 *Ore. Mich. Publ. Lang. & Lit.* XIII. 14 Inasmuch as the -*th* plural was a recessive characteristic, we can infer that the -*eth* form as well as the -*th* form occurring south of this line. 1962 *Amer. Speech* XXXVII. 272 Possibly coal bed is a recessive term in the Banks. 1972 M. L. Samuels *Linguistic Evol.* vi. 92 The loopholes for a given feature may or may not shift... If it does shift, there may be complete levelling resulting in the elimination of the recessive form.

2. *Biol.* [f. G. *recessiv* (Mendel, *Versuche über Pflanzenhybriden*, in *Verh. d. Naturforsch. Ver. Brünn* (1865) IV. 10).] Applied to a hereditary trait which is not perceptibly expressed in heterozygotes, being masked by a dominant allele of the gene that determines it; hence applied also to an allele which can affect the phenotype only in the absence of some other, dominant, allele. Const. *to* the dominant allele.

1900, etc. [see *dominant *a.* 7]. 1930 *Glasgow Herald* 21 Aug. 4 The characters of tallness and dwarfness were thus separated out. One of them, dwarfness, had temporarily disappeared in the first filial generation, and for that reason was called by Mendel the recessive character. 1930 R. A. Fisher *Genet. Theory Nat. Selection* iii. 52 The pronounced tendency of the mutant genes to be recessive, to the gene of wild type from which it arises, calls for explanation. 1950 *Science News* XV. 124 The research worker may find blood group inheritance a convenient field for study... Recessive characters, even in the presence of dominant ones, are quite easily recognised, for the heterozygotes can be distinguished from the homozygotes by using the appropriate antibodies. 1964 M. Argyle *Psychol. & Social Probl.* vi. 78 A single recessive gene produces the phenomenon of skipping generations and of carriers who do not show the condition themselves; certain kinds of mental deficiency are like this. Schizophrenia is thought by some to be inherited by two or more recessive genes. 1971 D. J. Cove *Genetics* ii. 14 If a heterozygous strain...resembles the strain homozygous for the *A* allele...thus the *A* allele is said to be dominant to the *a* allele, or conversely the *a* allele is said to be recessive to the *A* allele.

B. *sb.* *Biol.* **a.** An individual in which a particular recessive allele is expressed. **b.** A recessive allele or character.

1900 W. Bateson in *Jrnl. R. Hort. Soc.* XXV. 58 Mendel discovered that in this generation the numerical proportion of dominants to recessives is approximately constant, being in fact as three to one. 1902 *Rep. Evolution Committee R. Soc.* I. 8 The recessives are thenceforth not only apparently but actually pure, and if allowed to fertilise themselves give rise to recessives only, for any number of generations. 1905 R. C. Punnett *Mendelism* 15 This condition behaves as a single recessive to the normal state. 1916 W. Bateson *Probl. Genetics* ix. 32 We find that...a diversity of recessives may appear within a moderately short period. 1931 E. B. Ford *Mendelism & Evolution* iii. 38 The first gene to be discovered in *Drosophila simulans* was that for yellow body-colour. It is a sex-linked recessive. 1949 W. C. Allee et al. *Princ. Animal Ecol.* xxxiv. 655/1 Deleterious autosomal recessives are strikingly abundant in certain wild populations of *Drosophila*. 1973 J. Murray *Genetic Diversity & Natural Selection* ii. 5 The lowered fitness is the result of the unmasking of deleterious recessives.

recessiveness (rĭse·sivnĕs). *Biol.* [f. prec. + -ness.] The state or property of being recessive. Opp. *dominance 2.

1909 W. Bateson *Mendel's Princ. Heredity* 71 Basing his procedure on a knowledge of the dominance or recessiveness of each character the breeder may thus guide his operations with certainty. 1938 J. Rice *Genetics & Clinician* ii. 17 In the case of the red and white four o'clocks, the dominance of 'red' is but 50%, as is also the recessiveness of white, i.e., neither character is definitely dominant nor recessive. *Ibid.*, The dominance of some characters in the echinoderm may be changed to recessiveness by altering the chemical nature of the sea water. 1975 J. B. Jenkins *Genetics* iii. 83 Bateson...proposed his presence-and-absence hypothesis... which stated that dominance is due to the presence of a particular gene and that recessiveness results from the loss of that gene.

recharge, *sb.* **1.** Delete *rare* and add: *spec.* in *Hydrology*, the replenishment of the water content of an aquifer as a result of the absorption of water into the zone of saturation (freq. induced artificially by sinking wells into the aquifer); the water so added. Also (*rare*), the action of recharging a battery. (Further examples.)

1928 M. Arndt *Storage Batteries* vii. 146 This is a common feature of all high rate recharges, as the more dense electrolyte formed by the electrolytic action within plate pores and at plate surfaces does not have time to diffuse throughout the electrolyte during the earlier part of the recharge period. 1931 *Trans. Amer. Geophysical Union* XII. 208 At eight of the plats some recharge was indicated, but some of these had been irrigated in the previous summer. 1940 O. E. Meinzer *Hydrol.* x. 404 The recharge is increased if the intake area receives not only the local precipitation but also the surface flow of a tributary catchment area. 1966 R. G. Kazmann *Mod. Hydrol.* v. 139 When the supply of soil moisture in a given place is fully replenished, any additional water received from the surface is carried downward under the influence of gravity, either directly to the water table or to the intermediate belt of the zone of aeration: this phenomenon is termed the 'recharge' of an aquifer. 1976 J. D. Bredehoeft et al. in J. C. Rodda *Facets of Hydrol.* ix. 241/2 Variations in recharge change the inflow as well as the saturated thickness of the aquifers to the west of Barstow.

4. Special Comb. in *Hydrology*: **recharge area**, an area of ground surface through which is absorbed the water that will percolate into a zone of saturation in one or more aquifers; **recharge basin**, an artificially constructed basin, freq. in sandy material, used to collect water for artificial recharge of an aquifer; **recharge well**, a well used to inject water into an aquifer by artificial recharge.

1951 H. E. Thomas *Conservation of Ground Water* ii. 29 The ground-water phase of the hydrologic cycle is one of movement from the places where water enters the aquifer—the 'recharge' areas—to the place where the water is discharged from the ground. 1958 Betson & Arkis in M. J. Kirkby *Hillslope Hydrol.* viii. 308 The impact upon streamflow of paving over a thin-soil primary source area...would be far less than a similar amount of paving over a primary recharge area. 1951 H. E. Thomas *Conservation of Ground Water* iii. 143 The city has recharge basins totaling 65 acres which permit infiltration of up to 4 million gallons a day into the ground. 1970 *Daily Tel.* 18 Sept. 5/1 Long Island is combating the threat [of drought] by constructing 'recharge basins' capable of retaining about 10 per cent. of the water now being lost to the sea. 1951 H. E. Thomas *Conservation of Ground Water* iii. 125 Permits were issued with the proviso that water used for cooling and air-conditioning be returned to the same aquifer through recharge wells. 1976 Raudkivi & Callander *Analysis of Groundwater Flow* iii. 58 Water is pumped through the aquifer from the recharge well to the well.

recharge, *v.* **1. b.** Add to def.: *spec.* to replenish the water content of (an aquifer). (Further examples.)

1942 A. C. Swinnerton in O. E. Meinzer *Hydrol.* xiv. 698 It is possible to recharge basalt by running water down drilled wells. 1966 *Daily Tel.* 29 Sept. 3/4 Where infiltration is most vigorous, underlying aquifers may be recharged. 1976 Raudkivi & Callander *Analysis of Groundwater Flow* v. 108 If the well is at the centre of an island,...the aquifer is recharged from the lake surrounding the island.

5. a. *trans.* To restore an electric charge to (a battery). **b.** *intr.* Of a battery: to acquire an electric charge again, to become recharged.

1876 *Int. Dict.*, sense 1 b. 1893 J. T. Niblett *Portable Electricity* iii. 188 As a rule the user of a secondary battery knows...what rate his battery is intended to be recharged. 1928 M. Arndt *Storage Batteries* vii. 143 A partial charge... may be given to recharge a battery sufficiently to meet some special demand. 1960 *Farmer & Stockbreeder* 22 Mar. (Suppl.) 8/2 The battery runs for 18 hours on a single charge and is recharged for a few pence with a trickle charger. 1974 *Sci. Amer.* Nov. 134/2 The batteries had to be recharged by the generator,... in about 40 hours. 1976 *Country Life* 22 Jan. 200/3 Batteries for a lawnmower...have been recharged.

c. In *fig.* phr. *to recharge one's (the) batteries*, to restore fitness and mental composure by means of changed circumstances, esp. rest; also *absol.*

1921 W. S. Churchill *Let.* 9 Feb. in M. Soames *Clementine Churchill* (1979) xiii. 194 Subordinate everything in yr life to regathering yr nervous energy, and recharging yr batteries. 1971 A. Price *Alamut Ambush* x. 127 His London existence had been frenetic, and Fifie was where he recharged his batteries. 1976 J. Snow *Cricket Rebel* 83, I felt that I was at the halfway stage of my career and needed the rest to recharge the batteries. 1976 G. Moffat *Short Time to Live* xii. 136 London's...overwhelming... You come home to recharge, and go back.

Hence **recha·rgeable** *a.*, that may be recharged; **recharging** *vbl. sb.* (further examples).

1893 J. T. Niblett *Portable Electricity* iii. 187 Nearly all cases of failure of secondary batteries are due to lack of sufficient knowledge on the part of the user for properly conducting the operation of recharging. *Ibid.* 188 The rate at which a secondary battery receives the recharging current may be allowed to vary considerably without fear of damage to any of its parts. 1926 C. Connolly *Let.* 8 June in

Romantic Friendship (1975) 139, I need recharging. I expect we both do. 1942 Partridge *Usage & Abusage* 170/2 The invention of either new words or new senses (i.e., the re-charging of old words). 1949 *Econ. Geol.* XLIV. 523 In considering sites for water storage, the hydrologic engineer and the economic geologist should not overlook rechargeable aquifers. 1964 T. L. Kinsey *Audio-Typing & Electric Typewriters* iii. 16 The batteries, like accumulators, are rechargeable. 1972 *New Scientist* 11 May 321/1 These newcomers have provided...new electrical devices—such as...a rechargeable battery with an energy-to-weight ratio some 10 times higher than that of the familiar lead and sulphuric acid system. 1977 *Sci. Amer.* May 27/2 More expensive techniques are to build infiltration pits and recharging wells, by means of which water is pumped into the ground rather than drawn from it. 1978 *Nature* 15 June p. xv/1 The recorders are...supplied with rechargeable fibre tipped pens.

‖ **réchaud** (reʃo). [Fr., f. stem of *réchauffer* *Réchauffer *v.*] A receptacle in which food is warmed or kept warm.

1925 B. Rackham tr. *E. Hannover's Pottery & Porcelain* III. viii. 185 The factory hardly created much in the way of new forms...beyond a series of '*Réchauds*' (foodwarmers). 1955 H. Newman in *Apollo* Feb. 55 The typical French 'réchaud' (food-warmer) is usually cylindrical or slightly conical, and is of greater diameter than that of a tea-warmer. 1968 G. Godden *Greengage Summer* ix. 206 The rose-coloured wine, the réchaud flame, the lights were reflected in the windows over and over again.

réchauffe, *v.* Delete † *Obs.* and add: Now in form **rechauffe**. Also *fig.* Cf. *Réchauffer *v.*

1836 Disraeli in Monypenny & Buckle *Life Disraeli* (1910) I. App. B. 387 Canning irritated by Copley's re-chauffing in a speech Philpotts' pamphlet. 1931 R. Campbell *Georgiad* i. 18 His melancholy recipes (*sic*) For 'happiness'... How to 'rechauffe' the stock-pot of desire.

réchauffé. Add: Also **rechauffe, réchauffée.** **A.** *sb.* (Further *lit.* and *fig.* examples.)

1851 E. Ward *Jrnl.* 5 Feb. (1951) 123 Took tea with the Godleys, met Mr Roswilo, and had a *réchauffé* both of the ball supper and the ball gossip. 1936 *Times Lit. Suppl.* 16 May 337/2 The main objection to the book is that it is a 'rechauffe'... Stevenson's contribution now bears a 'family' resemblance to his piece then'. 1977 *Times* 5 Sept. 10/5 Cris de Meynan...is a useful bottle for casual meals of cold game or rechauffes.

B. *adj.* Of food: reheated. Also *fig.*, rehashed.

1909 Webster, *Réchauffé*...p.a. masc., *réchauffée*...p.a. fem....Warmed over;—of a dish food. Also *fig.* 1921 *Sat. Westm. Gaz.* 23 Sept. 14/1 Professor Wendell...frequently inserts what the dust-cover as jacket of the English edition denominates his 'humanity' between a hackneyed quotation and a platitude tastefully rechauffé. 1977 *Gramophone* Feb. 1307/1 These, then, are humdrum, *rechauffé* performances full of gestures by rote. 1979 *Broadcast* 29 Dec. 16/4 Canned laughter: Artificial sauce used to season rechauffé wells.

Hence **rechauffeed** *ppl. a.* = sense B above.

1883 E. W. Hamilton *Diary* 3 Feb. (1972) II. 395 The programme, it may be taken in addition to the unexciting non-contentious, rechauffeed bills, in Metropolitan Government and Local Government for Ireland.

‖ **réchauffer** (reʃofe), *v.* [Fr.: see Réchauffé.] Usu. as *inf.*, to warm up again; *fig.*, to rehash. Hence ‖ **réchauffage** (-ɑ̃ʒ) [Fr. suff. -*age*: cf. -age], a rehash.

1965 *Punch* 31 Mar. 473/3 This plan to store up the dead at minus two hundred degrees C and then *rechauffer* them to life when medical science has learned how to put right whatever it was they died of. 1967 *Punch* 28 June 961/3 In telling the story of her career Lady Summerskill has fallen into the Politician's Pitfall of believing that it is possible to *réchauffer* old Parliamentary speeches. 1972 E. Lucie-Smith in Cox & Dyson *20th-Cent. Mind* III. xvi. 466 Most of what they produced was a mere *réchauffage* of pre-war ideas.

re·check, *sb.* [f. the vb.] A renewed or second examination or investigation.

1946 *Amer. Jrnl. Med. Sci.* CLXXI. 851 A recheck showed that no fermentation test had been done on the urine, which vitiated the value of the test. 1972 T. Aedies *This Suitcase is going to Explode* xvii. 190 Those rechecks on the refugee scientists were supposed to be thorough. 1977 J. M. Johnson in Douglas & Johnson *Existential Sociol.* viii. 243 Upon receiving the reports and memo, the supervisor of Unit Two did not return the forms to the workers for a recheck.

recheck, *v.* [Re- 5 a.] *trans.* and *intr.* To check again. Hence **reche·cking** *vbl. sb.*

1902 'Mark Twain' in *Harper's Weekly* 6 Dec. 4/2 Your luggage...rechecked, fare-ticket and sleeper changed. 1957 J. S. Huxley *Relig. without Revelation* iii. 51 Constant checking and rechecking against fact. 1962 L. Deighton *Ipcress File* xi. 71, I decided to recheck his security clearance. 1967 Karch & Buber *Offset Processes* x. 469 Retighten the sprocket and recheck by rotating the cylinder. 1977 'E. Trevor' *Theta Syndrome* ii. 27 A young intern rechecked the patient. 1977 P. G. Winslow *Witch Hill Murder* ii. 215 All that's left is to check and re-check on every aspect of the case.

recherché, *a.* (Later examples.)

1939 Joyce *Finnegans Wake* (1964) 149 The fiery goodmother Miss Fortune (who the lost time we had the pleasure we have had our little *recherché* brush with, what, Schott?). 1946 A. Christie *Come, tell me how you Live* viii. 148 We have a very delicious and recherché lunch with the French Commandant. 1955 *Times* 16 May 11/4 The more recherché of whose oratorios ought to be remembered for future eighteenth-century festivals. 1970 I. Murdoch *Fairly Honourable Defeat* i. v. 55 The smell...was fresh and bitter and at the same time nauseating. Hilda wondered if it were not caused by some extremely recherché form of dry rot. 1978 *Dædalus* Fall 4 Our perception of a lifetime has become increasingly recherché.

‖ **recherche du temps perdu** (rəʃɛrʃ dü tɑ̃ pɛrdü). [ad. F. *à la recherche du temps perdu* (also used), lit. 'in search of the lost time', used by Marcel Proust (1871–1922) as the title of a reminiscent novel (1913–27).] The remembrance of things past; the narration or evocation of one's early life.

1946 L. P. Hartley *Sixth Heaven* xi. 233 A feeble poor-spirited attempt to revive the joys of childhood, a journey à la recherche du temps perdu. 1952 *Spectator* 19 Dec. 854/2 Like so many sensitive novels by contemporary women novelists, it is a *recherche du temps perdu*. 1966 M. Steen *Looking Glass* x. 117, I...listened to the two old friends sharing their *recherche du temps perdu*. 1975 *Broadcast* 3 Nov. 14/3 Old Times by Harold Pinter...is, of course, a play about *recherche du temps perdu*.

‖ **Rechtsstaat** (re·χ᾽tsʃtɑt). Also (*erron.*) **Rechtstaat**. [Ger., f. *rechts* gen. of *recht* Right *sb.*¹ + *staat* State *sb.*] A country in which the rule of law prevails.

1934 J. D. Lewis *Genossenschaft-Theory of Otto von Gierke* vi. 69 Being a *Rechtsstaat*, the modern state, the German state for example, stands within law and recognizes legal limitations upon its own sovereign will. 1944 F. A. Hayek *Road to Serfdom* vi. 54 The early nineteenth-century discussions in Germany about the nature of the *Rechtsstaat*. 1948 [see *kulturstaat s.v. *Kultur]. 1963 *Economist* 5 Oct. 51/1 A *Rechtsstaat* (state based on law) only the judgements of a competent court are valid and binding. 1977 *Times* 21 Apr. 5/6 Boyo Try as the lawyers might...to make Franco Spain a *Rechtsstaat*.

‖ **recibiendo** (reθivie·ndo). *Bullfighting.* [Sp., lit. 'receiving', f. *recibir* (see next).] A method of killing the bull by which the bullfighter receives the charging bull on the point of his sword. Also as quasi-*adv.*

1900 *Encycl. Brit.* XXVI. 460/1 If the *matador* remains without moving, or rather moving only his body to avoid the horns, the thrust is known as *recibiendo*. 1932 E. Hemingway *Death in Afternoon* ix. 88 He did kill several times recibiendo, receiving the bull on the sword in the old manner. 1976 E. P. Benson *Bulls of Ronda* iv. 30 'He's going to receive the bull. What they call a *recibiendo*.' ...The bull...propelled itself against the waiting man... The sword plunged deep into the entrails of the bull.

‖ **recibir** (reθivi·r). *Bullfighting.* [Sp., lit. 'to receive'.] The action on the part of a bullfighter of receiving a charging bull while remaining stationary.

1838 *Q. Rev.* LXII. 411 The picador, holding his lance under his right arm, pushes to the right, and pulls his horse to the left; the bull is thus turned from his plunge, and passes on to the next horseman. This is called 'recibir', 'hoc habet'. 1932 R. Campbell *Taurine Provence* iii. 76 There are several forms of *estocada*... Some...are attacks on stationary bulls, others, as in the Recibir, are performed by the stationary man on the charging bull.

recidivism. (Later examples.)

1971 *Sci. Amer.* May 51/1 There is no strong evidence... that recidivism increases with leniency of sentence. 1977 *Listener* 15 Dec. 803/2 But though the problem of recidivism and rehabilitation still remains, it can hardly be the chief cause of today's overcrowded prisons.

recidivist. Add: **a.** (Later examples.)

1931 *Sun* (Baltimore) 20 Jan. 10/2 This creates a body of recidivists who are being constantly released and as constantly returned. 1964 *Listener* 26 Mar. 507/1 The Reader in criminology at Oxford has given me the following definition of a recidivist: A recidivist is the offender who neither becomes his ways spontaneously nor learns to avoid detection, and who is neither deterred by the experience of conviction nor reformed by any of the methods in the courts' repertoire. 1981 W. Ebersohn *Divide Night* ii. 26 Old recidivists who felt at home only in jail and would be back again and again.

b. *attrib.* or as *adj.*

1920 *Contemp. Rev.* Nov. 684 It is the source of depraved, unchastened, even of recidivist, tendencies, as well as of those which point towards a wider and more perfect life. 1931 J. S. Huxley *What dare I Think?* iii. 88 Sterilization has been suggested, but this seems disproportionate save in recidivist cases of philoprogenitiveness which seem otherwise incurable. 1950 *Chambers's Jrnl.* 230/1 Often...he had the opportunity of seeing the overcrowded masses of recidivist (frequent offender) prisoners, many of them unprepossessing and hardened-looking roughs, with whom he was glad he had not to mix. 1962 *Lancet* 15 Dec. 1278/1 My work as a probation officer among recidivist alcoholics in Pentonville Prison.

recidivity. (Example.)

1937 G. Frankau *More of Us* xii. 123 Man asks you...not to blame him for his recidivities.

recipe, *v.* *imper.* and *sb.* Add: **B.** *sb.* **2.** (Earlier and later examples.) Also *fig.*

1716 T. Cave *Let.* 5 Oct. in M. M. Verney *Verney Lett. 18th C.* (1930) II. xxii. 45 Sister Lovett and I greatly admire the Ink you wrote last with,...but dare not wish for the Recipe it being no doubt a Secrett. 1947 [see *promoting *vbl. sb.* 2 b]. 1969 *Times* 12 May 16/3 The D.N.A. molecules in the cells of normal people contain a chemical recipe for manufacturing ordinary haemoglobin. 1981 *Times* 9 Apr. 9/7 Another recipe made especially for Easter is fritters of soft ricotta cheese.

4. *Comb.*, as **recipe-book**, a book containing recipes, a cookery-book; also *fig.*

1872 *Young Englishwoman* Dec. 662 Recipes of practical utility will appear...to be neatly copied into their recipe-books. 1952 M. Laski *Village* iii. 62 To have read the recipe-books and produced appetising meals for her family would have meant, to her, willing acceptance of...servitude. 1964 P. Strevens in D. Abercrombie et al. *Daniel Jones* 113 Teaching grammars in the future...will be superior...to those of the past which have simply been grammar-teaching recipe-books. 1976 L. Deighton *Twinkle, twinkle, Little Spy* xiv. 140 Bessie Mann was asking me how many kids we were going to have, and I found myself looking at recipe books and baby carriages.

reci·pher, *v.* Also **recypher.** [Re- 5 a.] *trans.* To encipher again (a message that is already in cipher). Hence **reci·pher** *sb.*; **reci·phering** *vbl. sb.*

1961 in Webster. 1963 *Times* 21 May 7/4 He said the purpose of such a table would be to convert plain language letters or punctuation into figures which could then be subsequently reciphered on a reciphering pad. 1975 C. Mott-Radclyffe *Foreign Body in Eye* ii. 28, I remember one telegram, all in re-cypher, beginning, 'I saw the Minister of Foreign Affairs on return from my leave yesterday. His Excellency remarked how sunburnt I was looking after my recent holiday in Kenya.'

recipiangle. (Later example.)

1939 *Nature* 22 Apr. 673/2 The seventy objects which comprise Mr. Court's most recent benefaction include an Italian 'recipiangle'...instrument...is an instrument used for measuring angles in surveying, and it is also marked with lines and scales which can be used for making calculations.

recipience. Add: **b.** Recipient state or condition.

1902 in Webster. 1906 S. S. Laurie *Synthetica* I. 7 The subject-reality...appears to be a mere potency of recipience and reaction. 1923 *Times Lit. Suppl.* 29 Nov. 801/2 The [*sc.* the neo-realists'] insistence on the independent reality of the physical world and the passive recipience of the mind in knowing it.

recipient, *a.* and *sb.* Add: **B.** *sb.* **4.** *Linguistics.* The indirect object of a verb or the complement of an adjective.

1937 O. Jespersen *Analytic Syntax* xxxviii. 156 In conventional grammar the term *object* is further used for what is governed by an adjective. Here, too, I have avoided the term object...and have adopted instead the term *Recipient* abbreviated R... R is seen to be chiefly used where Latin, German, etc. have a dative which is not governed by a transitive verb. 1972 R. Quirk et al. *Gram. Contemp. Eng.* vii. 350 The most typical function of the indirect object is that of *recipient* (or 'dative' participant); *ie* of animate being passively implicated by the happening or state: I've found *you* a place. *Ibid.* 353 The subject may also have a recipient (or 'dative') role with verbs such as *have, own,* [etc.] ..., as is indicated by the following relation: Mr Smith has bought/given/sold his son a radio→So now his son has/owns/possesses the radio.

reciprocal, *a.* and *sb.* Add: **A.** *adj.* **3. c.** *reciprocal innervation* or *inhibition* (Physiol.), an arrangement of nerve stimulation as a result of which contraction of one muscle or group of muscles to produce movement is accompanied by simultaneous inhibition of an antagonistic muscle or group of muscles whose contraction would tend to produce the opposite movement.

1897 C. S. Sherrington in *Jrnl. Physiol.* XXII. 327 It has revealed to me an almost unexpectedly significant number of examples of depressor effect generally, perhaps always, in combination with pressor effects, that is to say, in the form of reciprocal innervations. 1906 —— *Integrative Action Nervous Syst.* iii. 83 (*heading*) Reciprocal inhibition. 1942 Johnston & Whillis *Gray's Anat.* (ed. 28) 942 Experimental evidence shows that all the fundamental reflexes of posture..., the reciprocal innervation of antagonists, and the elementary combination of synergists, can still occur without the cerebellum. 1960 W. B. Crow *Synopsis of Biol.* lxv. 445 Neurons connect up two or more efferent paths so that two or more muscles come into action together, or one is inhibited when another comes into operation (reciprocal innervation). 1971 *Sci. Amer.* Aug. 75/1 One principle Sherrington discovered was 'reciprocal innervation'. 1977 G. H. Sage *Introd. Motor Behav.* (ed. 2) viii. 159 The flexion reflex consists of a contraction of the flexor muscles while reciprocal connections with the antagonistic extensor muscles cause reciprocal inhibition.

d. *Genetics.* Of each of a pair of crosses: complementary to another in that the male parent in each is of the same kind as the female parent in the other.

1902 W. Bateson *Rep. Evolution Comm. R. Soc.* I. 21 Reciprocal crosses...showed clearly that of the two pairs of antagonistic characters, the violet colour is dominant. 1909 —— *Mendel's Princ. Heredity* xi. 203 When the cross is in the form pure single ♀ × double-throwing ♂, all the F_1 plants give a mixture of doubles and singles in F_2; but when the reciprocal cross is made, namely double-thrower used as ♀ × pure single used as ♂, it is found that the F_1 plants are of two kinds. 1948 H. P. Riley *Genetics & Cytogenetics* vi. 88 The cross curved × wild type is known technically as the reciprocal of the cross wild type × curved. *Ibid.* vii. 96

When genes are in autosomes reciprocal crosses normally give identical results, but when genes located in the X chromosome are dealt with, the results of reciprocal crosses are different. 1971 D. J. Cove *Genetics* viii. 111 If some genetical information is also carried in the cytoplasm, the unequal contribution of cytoplasm to the zygote by the two gametes should provide a way of detecting it. The technique used for this is the reciprocal cross.

e. *Physics.* In the names of quantities defined as reciprocals of standard physical units, as *reciprocal centimetre, ohm, second* (see quots.).

1934 H. E. White *Introd. Atomic Spectra* i. 6 Wave numbers...are units with the dimensions of reciprocal centimeters, abbreviated cm^{-1}. 1960 Brand & Speakman *Molecular Struct.* i. 8 The passage of energy between matter and radiation occurs in quanta of magnitude ΔE, where $\Delta E = h\nu$. When ΔE is expressed in terms of the erg and ν as reciprocal seconds, Planck's constant, *h*, has the value $6{\cdot}6256 \times 10^{-27}$ erg sec. 1960 A. D. Cross *Introd. Pract. Infra-Red Spectroscopy* i. 2 Band positions are quoted in units of wave number (ν) which are expressed in reciprocal centimetres (cm^{-1})... However, the true unit of frequency ($\bar{\nu}$) is given in reciprocal seconds (sec^{-1}). 1978 *Sci. Amer.* Dec. 66/1 The maximum conductivity is somewhat greater than 2,000 reciprocal ohms per centimeter (equivalent to a resistivity of less than ·0005 ohm per centimeter).

f. *reciprocal course*: a path followed by a person or craft which is opposite in direction to one with which it stands in relation, e.g., to the one desired or the one which was followed immediately before.

1946 *Happy Landings* (Air Ministry) July 5/3 The immediate action is to turn on the reciprocal course. 1958 'N. Shute' *Rainbow & Rose* ii. 57, I...flew out to sea...on the reciprocal course, and then turned in again and flew towards the coast on 110°. 1961 B. Fergusson *Watery Maze* ii. 49 All these craft had lately been 'degaussed'...with unexpected effects on their compasses, and some of the shoreward-bound parties were steering on a reciprocal course.

7. Phonetics. *reciprocal assimilation* (see quot. 1972).

1915 G. Noël-Armfield *General Phonetics* ix. 32 If the adjacent sounds act upon each other more or less equally the influence may be called reciprocal. *Ibid.* 33 Reciprocal assimilation is common in diphthongs. 1939 L. H. Gray *Foundations Lang.* iii. 68 It [*sc.* assimilation] may be reciprocal, when the modification is mutual. 1972 Hartmann & Stork *Dict. Lang. & Linguistics* 21/2 If two sounds influence each other mutually the term *reciprocal assimilation* (alternative term: *coalescent assimilation*) is used, e.g. English *seven* pronounced as [sebm̩] where the labio-dental [v] has become bilabial [b] which in turn influences the alveolar nasal [n] changing it to the bilabial nasal [m].

B. *sb.* **3.** Restrict † *Obs.* to sense in Dict. and add: **b.** A noun, pronoun, or verb that expresses mutual action or relationship.

1961 R. B. Long *Sentence & its Parts* xvi. 357 Reciprocals can refer only to plural nouns, pronouns, or nounal units; reflexives can refer to singulars as well. 1965 *Canad. Jrnl. Linguistics* Spring 175 Especially reflexives, reciprocals, iteratives, passives, and the Eyak progressive almost always take the vocalic forms.

reciprocate, *v.* Add: **2. b.** (Later examples.)

1903 G. B. Shaw *Man & Superman* iii. 97 He...is peevish and sensitive when his advances are not reciprocated. 1922 Joyce *Ulysses* 717 Hospitality extended and received in kind, reciprocated and reappropriated in person. 1924 C. Mackenzie *Old Men of Sea* viii. 131, I was rather disappointed to hear that she was unlikely to reciprocate Dick Duffy's affection.

3. b. Also *gen.*, to return love or liking.

1883 'Mark Twain' *Life on Mississippi* iv. 542 He had loved one 'too fair for earth', and she had reciprocated. 1916 G. B. Shaw *Androcles & Lion* p. xxx, Unless you love your neighbor as yourself and he reciprocates you will both be the worse for it. 1922 Joyce *Ulysses* 679 In what manners did she reciprocate? She remembered... She provided... She admired. 1936 L. C. Douglas *White Banners* xi. 235 Hannah liked her, and felt that the girl heartily reciprocated.

reciprocating, *ppl. a.* **2. b.** Add to def.: Applied *spec.* to engines in which the working fluid drives an oscillating piston. (Earlier and later examples.)

1822 J. Robison *Syst. Mech. Philos.* II. 577 When water is to be driven along a main by the strokes of a reciprocating engine, it should be forced into an air-box. 1901 *Trans. Inst. Engineers & Shipbuilders in Scotland* XLV. i. 25 An ordinary reciprocating engine performs one complete cycle of operations...in two strokes of the piston. 1911 *Encycl. Brit.* XXV. 824/1 The rapid development of the marine steam turbine makes it probable that it will displace the reciprocating engine in all large and fast ships. 1948 *Daily Tel.* 9 Apr. 4/4 The aircraft developed in recent years in America were...of conventional design. Their engines were reciprocating engines, and they were designed to fly at normal speeds. 1973 *Times* 13 Dec. 35/2 The Wankel still relies on petrol and unfortunately uses rather more of it than a comparable reciprocating engine.

reciprocator. Add: **b.** A reciprocating engine forming part of a composite power plant.

1907 *Westm. Gaz.* 16 Sept. 4/1 Not an ounce of steam will be wasted, the principle being to carry the exhaust steam from the high-pressure reciprocators to the low-pressure turbines. 1931 *Times* 24 Jan. (Trade & Engin. Suppl.) p. iv/3 Another alternative, in which the exhaust steam from the reciprocator is led to a turbo generator which delivers its current to a motor on the propeller shafting, has been applied to the existing 'City' ships of the Ellerman Line. 1952 Fox & McBirnie *Marine Steam Engines & Turbines* xi. 197 Only the reciprocator is used while entering or leaving port.

reciprocitarian. Add: (Later example.) Also *attrib.* or as *adj.*

1906 *Athenæum* 3 Feb. 134 To these he appeals on reciprocitarian lines by offering a reduction of our wine duties. 1932 *Times Lit. Suppl.* 1 Dec. 909/4 Dilke asserts that he [*sc.* Joseph Chamberlain] was a strong 'Reciprocitarian' and had already the notion of a 'British Zollverein'.

reciprocity. Add: **1. a.** Also in *Social Science* (see quots. 1960, 1972).

1904 W. T. Mills *Struggle for Existence* xx. 262 Reciprocity is a new word in politics, but it expresses an old fact in real life. 1952 T. M. Newcomb *Social Psychol.* ix. 308 The last stage in acquiring roles involves recognizing reciprocity between oneself and others. 1960 G. W. Allport *Personality & Soc. Encounter* xi. 175 The child understands that members of other countries are as attached to their own lands as he is to his; this is the principle of reciprocity. 1967 E. A. Hoebel in P. Bohannan *Law & Warfare* ii. 187 In Comanche society reciprocity is not developed to an exaggerated degree. 1971 E. Ardener *Soc. Anthrop. & Lang.* p. lv, Maussian systems of exchange and reciprocity are analogous to systems of communication, of which language is also one. 1972 *Jrnl. Social Psychol.* LXXXVII. 89 Gouldner has postulated the existence of a universal norm of reciprocity which stipulates that...people should help those who help them.

2. b. (Earlier example.)

1838 F. Haywood tr. *Kant's Critick Pure Reason* i. ii. ix. 76 Both the Judgments, the relationship of which forms the hypothetical judgment..., in whose reciprocity likewise the disjunctive consists.

3. *attrib.* and *Comb.*, as *reciprocity-monger, technique, treaty*; **reciprocity failure** *Photogr.*, departure from adherence to the reciprocity law, found with all emulsions, in which greater exposure than that predicted by the law is required at both very low and very high light intensities; **reciprocity law** *Photogr.*, the statement that the degree of blackening of an ideal emulsion is constant for a given incident energy, i.e., for a given product of light intensity and exposure time; **reciprocity theorem** *Physics*, a theorem which states that the response of a given physical system is unchanged under interchange of the locations of a constant excitation and of the measured response; also called the Onsager principle; *spec.* (*a*) *Electr.* (see quot. 1957); (*b*) *Nucl. Physics*, the statement that time-reversal leaves the transition rate for a nuclear reaction unchanged.

1923 *Jrnl. Optical Soc. Amer.* VII. 1110 In order to show in a different way the magnitude of the reciprocity failure, the values of sensitivity have been computed. 1966 D. G. Brandon *Mod. Techniques Metallogr.* 13 Emulsions produced specifically for metallography are designed to retain their speed for long periods, that is, to have a low reciprocity failure, while emulsions for astronomical applications, involving exposure times of several hours, show almost no reciprocity failure at all. 1973 *Sci. Amer.* Dec. 122/3 Lowering the temperature of the emulsion tends to suppress the recombination of the ions and hence to suppress reciprocity failure. 1979 *SLR Camera* May 11/1 Unless the meter takes into account reciprocity failure then the film cannot be correctly exposed. [1900 *Astrophysical Jrnl.* XI. 89 Scheiner, in 1891, proved that the increase in the number of fainter stars on prolonging the exposure fell far below what would be expected according to the law of reciprocity.] 1907 Sheppard & Mees *Investigations Theory of Photogr. Process* ii. vi. 214 A reversing action of the released bromine may...be deduced from the failures of the Bunsen and Roscoe 'reciprocity law'. 1942 C. E. K. Mees *Theory of Photogr. Process* vi. 236 Bunsen and Roscoe laid down a general law for photochemical reactions which states that the product of a photochemical reaction is dependent simply on the total energy employed... From the reciprocal relation between time and intensity in the Bunsen–Roscoe expression, it was called the reciprocity law. 1974 Dainty & Shaw *Image Sci.* ii. 35 For very long or very short exposure times the process of latent image formation is less efficient, and this is known as reciprocity law failure. 1885 A. Crump *Formation Polit. Opin.* 198 The declamations of the Fair Traders and the reciprocity-mongers...fail to disturb the convictions of the sound thinkers in the country. 1973 *Times* 18 Oct. (Brazil Suppl.) p. ii/4 There is a dependence on imported technology and design, as well as insufficient use of reciprocity techniques. [1876 *Proc. R. Soc.* XXV. 118 Although the principle of reciprocity appears to be firmly grounded on the theoretical side, instances are not uncommon in which a sound generated in the open air at a point A is heard at a distant point B, when an equal or even more powerful sound at B fails to make itself heard at A.] 1938 G. P. Harnwell *Princ. Electricity & Electromagnetism* xiii. 451 The reciprocity theorem is a consequence of the symmetry of the determinant and is generally stated in a limited form involving the current in one branch and the emf. in another. 1952 Blatt & Weisskopf *Theoret. Nucl. Physics* x. 529 We obtain the reciprocity theorem... This theorem states that the probability for a transition proceeding one way in time is equal to the probability for the same transition but with the sense of time reversed. 1957 B. I. & B. Bleaney *Electr. & Magnetism* iii. 69 We have the Reciprocity Theorem, which states that a given e.m.f. in the *p*th branch will produce the same current in the *q*th branch of a circuit as the same e.m.f. in the *q*th branch would produce in the *p*th branch. 1968 [see *Onsager]. 1970 I. E. McCarthy *Nuclear Reactions* ii. 31 The reciprocity theorem or time-reversal invariance is an essential property of nuclear systems. 1847 H. Clay in *Whig Almanac 1848* 22/1 Out of these acts have sprung a class...of treaties,... commonly called Reciprocity Treaties.

recirculate, *v.* [Re- 5 a.] To circulate again. **a.** *trans.* To make available for reuse. **b.** *intr.* Of material: to take part in recirculation.

1916 Harding & Willard *Mech. Equipment Buildings* I. xv. 349 The air is simply re-circulated, no fresh air being taken into the heating system from the outside. 1934 W. Trinks *Industrial Furnaces* (ed. 3) I. vi. 392 In several designs of fuel-fired furnaces, burned gases are drawn from the heating chamber by a fan and recirculated, mixing with the fresh combustion gases. 1958 *Jrnl. Amer. Water Wks. Assoc.* L. 1025/2 Treated sewage can be recirculated from the bottom of the final tank to the wet well where it is mixed with the incoming raw sewage. 1959 *Motor Manual* (ed. 36) v. 107 The balls recirculate (thereby reducing the rate of wear) and their use reduces friction in the mechanism to a minimum. 1964 *Proc. R. Soc.* B. CLIX. 262 The few large lymphocytes which re-circulate from the blood to the lymph do so for only a brief period. 1967 E. Chambers *Photolitho-Offset* ix. 127 The machine has a unique roller-jet agitation...allowing for the developer to be recirculated. 1977 *Jrnl. Neuropath. & Exper. Neurol.* XXXVI. 471 Microglial cells are continually replaced by circulating mononuclear cells and may, themselves, also recirculate.

Hence **reci·rculated** *ppl. a.*, **reci·rculating** *vbl. sb.*

1916 Harding & Willard *Mech. Equipment Buildings* I. xvi. 442 This apparatus is designed for use where maximum cooling of the air by evaporation in addition to air cleansing is desired with recirculated spray water. 1947 T. N. Adlam *Radiant Heating* 5 It can be arranged to discontinue recirculating and admit all fresh air after the correct room temperature has been attained. 1967 *Punch* 5 July 22/2 Once he has overcome in training the psychological objection to drinking recirculated fluids, the astronaut will not be bothered by such trivialities.

reci·rculating, *ppl. a.* [f. prec. + -ing².] **1.** Circulating again or continuously, in senses of the vb.

1916 Harding & Willard *Mech. Equipment Buildings* I. xiv. 337 The use of recirculating ducts, through which part or all of the air delivered to the rooms is returned to the furnace to be reheated, will greatly reduce the cost of operation. 1954 G. M. Fair et al. *Water Supply & Waste-Water Disposal* xxx. 908 The recirculating system generally includes pumps, hair-catchers, chemical feeds..., filters, disinfecting equipment..., and heaters. 1964 *Proc. R. Soc.* B. CLIX. 262 Some of the large lymphocytes in thoracic duct lymph divide...but...the number of new small lymphocytes which they contribute to the total re-circulating pool is extremely small. 1975 *Offshore* Aug. 3/1 (Advt.), Even the initial lead slurry, when you use our patented re-circulating mixer, will be to the desired weight.

2. *recirculating ball*, a ball-bearing running in a closed ball race; usu. *attrib.* with reference to a form of automotive steering mechanism in which a half-nut containing an eccentrically mounted ball race can be made to move along a helical cam by rotation of the cam.

1946 W. H. Crouse *Automotive Mech.* xxiii. 494 The recirculating ball-and-nut type of steering-gear mechanism that uses a grooved ball nut, ball bearings, and a gear sector. 1956 E. Molloy *Automobile Engineer's Ref. Bk.* xv. 37 Screw-and-nut gears incorporating balls between the screw and nut, and known as recirculating ball steering gears, are widely used on all classes of vehicles above the 12-h.p. range. 1960 *Farmer & Stockbreeder* 23 Feb. 45/2 Light, responsive, re-circulating ball type steering. 1970 *A.A. Bk. of Car* 151/4 (*caption*) Recirculating balls are used to reduce friction between a rotating worm and a nut moving along it. 1980 *Times* 9 May 26/2 Instead of rack and pinion Toyota has decided to stick with recirculating ball steering.

recirculation. Add to def.: *esp.* the process of making available for reuse waste products or other material. (Further examples.)

1916 Harding & Willard *Mech. Equipment Buildings* I. xiv. 330 All of the air is brought in from the outside to warm the building, and no recirculation is allowed. 1934 W. Trinks *Industr. Furnaces* (ed. 3) vi. 394 (*caption*) Furnace designed for recirculation of products of combustion. 1954 G. M. Fair et al. *Water Supply & Waste-Water Disposal* xxx. 905 Modern swimming pools are designed for the recirculation of pool water and its continuous repurification, reheating, and redisinfection. 1964 *Proc. R. Soc.* B. CLIX. 280 It seems likely that a re-circulation through the lymph nodes is a feature of the life-history of small lymphocytes among mammals generally. 1975 *Nature* 20 Mar. 175/2 Recirculation of metal and other wastes could save an increasing proportion of the total energy used in Sweden.

recirculatory (rīsɜ·ɹkiŭlētəri, rīːsɜɹkiŭlēi·tɔri), *a.* [Re- 5 a.] Involving recirculation; recirculating.

1951 *Engineering* 23 Nov. 664/1 The steering gear is of the recirculatory-ball pattern. 1959 *Motor* 23 Sept. 179/2 Available as an extra is a recirculatory interior heater.

‖ **récit** (resi). [Fr.] **1.** *Mus.* (See quots.)

1884 F. Niecks *Conc. Dict. Mus. Terms* 201 *Récit* (Fr.), (1) what is performed by one singer or one instrumentalist, a vocal or instrumental *solo*. (2) The principal part in a piece of concerted music. (3) One of the manuals and corresponding stops of the organ, the Swell Organ. 1924 L. J. De Bekker *Black's Dict. Mus. & Musicians* 538/1 *Recit*, Fr. solo part; principal of several parts. 1954 *Grove's Dict. Mus.* (ed. 5) VII. 72/1 *Récit*..., a 17th-century term for a declamatory melody supported by simple accompaniment,...but also, by analogy, an instrumental piece, such as an organ piece with the tune played on a solo stop. 1968 A. Niland *Introd. to Organ* 109 Cavaillé-Coll never really took to the swell box, being content to enclose only the récit expressif, even in a five manual organ. 1975 *New Yorker* 28 Apr. 133/2 The *Récit* [*sc.* a rank of organ pipes] is in mysterious darkness, behind the louvres of the box that encloses it.

2. *Lit.* Narrative, account (freq. opposed to dialogue); a relating of events. Also, a book or passage consisting largely of narrative.

1944 'G. Orwell' in *Horizon* X. 237 The whole book, *récit* as well as dialogue, is written in the American language. 1959 *Times Lit. Suppl.* 27 Mar. 173/1 The *récit*...studies always to keep an uncluttered foreground where only one or two characters are permitted to disport themselves; for the most part it sticks to narrative and reported speech. 1960 *Ibid.* 18 Nov. 737/3 The four *récits* which compose his latest work are set as far apart as Ethiopia, Spain, Italy and the Midi. 1964 *New Statesman* 17 Apr. 610/2 Its form is French: a flowery *récit* by an ageing recluse describing his attempts to make life and thought converge. 1978 *Observer* 26 May 30/1 The rhythms of his *récit*—his dialogue is a different matter—are terribly unvarious.

recital. Add: **3. b.** Now also in wider sense, a performance of instrumental music or of music and songs, freq. from the works of several composers.

1929 *Radio Times* 8 Nov. 417/2 A Recital of Gramophone Records. 1962 *Amer. Speech* XXXVII. 19 New York City boasts of many fine museums, art galleries, and recital halls. 1981 *Early Music News* Mar. 2 Gillian Weir...will give a joint harpsichord and organ recital, conceived as a salute to Couperin and Bach.

recitation. Add: **3.** (Earlier examples.)

1770 P. Fithian *Let.* 30 Nov. in *Jrnl. & Lett.* (1900) 8 At nine the Bell rings for Recitation, after which we study till one, when the Bell rings for Dinner. 1780 E. Parkman *Diary* 9 Feb. (1899) 208 He has been absent from ye Recitations so long, that he ought to be one of ye first that returns at this Term.

4. (sense 3) *recitation bench* (further example), *-method, -room* (earlier and later examples); (sense 2) *recitation music, -note* (earlier example).

1949 'J. Nelson' *Backwoods Teacher* 48 Likely-looking ones I placed on the table...beside the four, long, recitation benches. 1899 W. James *Talks to Teachers* i. 4 Traditions of instruction...evolved from the older American recitation-method. 1927 *Grove's Dict. Mus.* (ed. 3) III. 371/2 A long series of 'Recitation Music', *i.e.* compositions for piano intended to accompany the declamation of various well-known poems, was written by Stanley Hawley. 1844 W. B. Heathcote *Canticles* p. iii, The first half of the verse is said on the recitation-note G. 1827 *Harvard Reg.* Sept. 202 We hurry to the Chapel, and then crowd to the recitation room. 1899 W. James *Talks to Teachers* xiv. 161 The flowing life of the mind is sorted into parcels suitable for presentation in the recitation-room.

recitative, *a.*¹ and *sb.* Add: **B.** *sb.* **3. b.** Also *fig.*

1957 M. Spark *Comforters* iii. 42 It said: On the whole she did not think there would be any difficulty with Helena... There seemed, then, to have been more than one voice: it was a recitative, a chanting in unison.

4. *Comb.*, as *recitative-like* adj.

1947 A. Einstein *Mus. in Romantic Era* xi. 143 Characteristically Lisztian recitative-like interjections. 1963 *Times* 28 Feb. 16/3 It was especially in his perfect shaping of the recitative-like passages that Mr. Frank demonstrated his cultivated musicianship.

recitative, *v.* **2.** (Later example.)

1932 [see *preludingly *adv.*].

recitativo. Add: **2.** Used in certain Italian phrases designating varieties of recitative, as **recitativo accompagnato** (akǫmpanʸā·to): in which the vocalist is accompanied by an orchestra; **recitativo secco** (se·ko) [lit. 'dry recitative']: in which the vocalist has little or no musical accompaniment; **recitativo stromentato** (strǫmentā·to) [lit. 'accompanied recitative'] = *recitativo accompagnato* above.

[1771 C. Burney *Present State of Mus. France & Italy* 285 Signor Rinaldo di Capua has at Rome the reputation of being the inventor of accompanied recitatives. 1801 T. Busby *Dict. Mus.*, *Recitative accompanied*, a *recitative* is said to be *accompanied* when besides the bass there are parts for other instruments.] 1866 [see *recitativo stromentato* below]. 1947 A. Einstein *Mus. in Romantic Era* ix. 96 The genuine melodrama, in which the spoken word, usually an emotional monologue, was given an orchestral background in the manner of a *recitativo accompagnato*. 1963 *Listener* 14 Feb. 313/1 *Recitativo accompagnato* began to encroach more and more on the traditional *secco*. 1828 T. Busby *Mus. Man.* 144 *Recitativo secco*, (Ital.) unaccompanied recitative. 1876, 1883 [see Secco *a.*]. 1891 G. B. Shaw in *World* 13 May 27/2 Start on the *recitativo secco* by entirely expunging the first two lengths after the duel. 1955 E. Dent in H. Van Thal *Fanfare for E. Newman* 105 It is practically a reversion to *recitativo secco* in a more pretentious form. 1963 Auden *Dyer's Hand* 524 Ariel is song; when he is truly himself, he sings. The effect when he speaks is similar to that of *recitativo secco* in opera. 1977 *Listener* 26 May 692/1 The vocal line is written in the customary *recitativo secco* style. [1828 T. Busby *Mus. Man.* 144 *Recitativo istromento*, (Ital.) accompanied recitative.] 1837 J. A. Hamilton *Dict. Two Thousand Italian, French, German, Eng. & Other Mus. Terms* 58 *Recitativo stromentato* (Italian) recitative accompanied by the orchestra. 1866 *Chambers's Encycl.* VIII. 140/2 It is termed *recitativo accompagnato, strumentato*, or *obbligato*. 1905 E. J. Dent *Alessandro Scarlatti* ii. 45 Here we have an unmistakable *recitativo stromentato*, the earliest (1686) that I have been able to find.

recite, *v.* **7.** (Earlier Amer. examples.)

1742 E. A. Holyoke *Diary* 22 Aug. (1911) 32 This day began to recite to Mr Flynt & Mr Appleton in Tully and virgill & Greek testament. 1759 S. Gardner *Jrnl.* 21 Mar. in *Essex Inst. Hist. Coll.* (1913) XLIX. 6 Finished reciting, the Dr. gave us good advice. 1815 *Niles' Weekly Reg.* IX. 18/1 Those not immediately engaged in reciting to some one or other of the professors, remain in their own chambers.

Reckitt (re·kit). Also *erron.* **Reckett.** [See quot. 1877.] Used in the possessive as the proprietary name of a blue (*blue *sb.* 2 c) for laundry use; also as the name of the colour of this substance, a clear cobalt blue, esp. *transf.* Hence **Reckitt's bluebag**, the bag in which this product is marketed.

1877 *Trade Marks Jrnl.* 19 Feb. 471 Reckitt's Blue in squares will be found far more beautiful & much more economical than any other... Francis Reckitt, of Hull, Yorkshire, on behalf of self and partner, James Reckitt, trading as I. Reckitt and Sons, at...London, and as Isaac Reckitt and Sons, at Hull, Yorkshire; starch manufacturers. 1893 S. R. Crockett *Stickit Minister* 46 The book...was bound in a peculiarly deadly blue, of a rectified Reckitt tint, which gave you dazzles in the eye. 1898 G. B. Shaw *Let.* 25 Apr. (1972) II. 37 Mud & fog are not so fatal as sciroccos [*sic*]...& Reckitt's blue. 1920 E. Sitwell *Wooden Pegasus* 83 Then (Recketts' blue) a puff of wind. 1925 'H. H. Richardson' *Way Home* ii. i. 113 As for the sky, Mahony declared it made him think of a Reckitt's bluebag. 1930 E. V. Lucas *Down Sky* 138 A bunch of yellow-haired girls lying on marble slabs looking at a sea of Reckitts. 1969 R. Harris *Nat. Hist. Collecting* 49 Some sponges are brightly coloured. One is called the Reckitt's blue, not without reason. 1971 *Guardian* 6 Feb. 3/3 The flies zoom between the Reckitt's Blue walls. 1976 *Trade Marks Jrnl.* 1 Dec. 2497/2 Reckitt's Bag Blue. Ready for use... Laundry blue for sale in the United Kingdom and for export to the Irish Republic. Reckitt & Colman Products Limited, trading as Reckitt & Colman,...Hull, Yorkshire; manufacturers and merchants.

Recklinghausen's disease, var. *von Recklinghausen's disease.

reckon, *v.* Add: **I. 5. d.** *colloq.* To rate highly, to esteem. Usu. in negative phrases.

1957 *Evening News* 12 Nov. 6/4 If...an East Ender wants to say that he does not consider the character of another to be worth while he says 'I don't reckon him'. 1977 *Sunday Times* 52/3, I don't reckon the chances of *Young Scientists of the Year*...against *Just William*.

6. a. Now usu. *colloq.*, esp. in the U.S. (formerly chiefly in southern States).

1810 M. Dwight *Jrnl.* 9 Nov. in *Journey to Ohio in 1810* (1912) 37 The people here talk curiously, they all reckon instead of expect. 1863 *Congress. Globe* 7 Feb. 783/3 If you can take this property by compact, I reckon you cannot take it against the consent of the owners. 1893 H. A. Shands *Some Peculiarities of Speech in Mississippi* 53 *Reckon*,...this word is almost always used in the ordinary conversation of our best educated people for *think* or *suppose*, and corresponds to a like use of *guess* in the Northern States. 1900 *Cosmopolitan* Feb. 389/1 She met Sam on the way out, and says she: 'Sam, what do you reckon? My quilt took the premium.' 1963 *Social Problems* Spring 367/1, I reckon it'll always be lucky. 1977 I. Shaw *Beggarman, Thief* iii. ii. 193, I don't reckon I've had a fare there for more than ten years.

II 11. b. (Further examples.)

1902 Kipling *Five Nations* (1903) 63 We reckon not with those Whom the mere Fates ordain. 1945 T. S. Eliot *What is a Classic?* 8 We may say confidently that it [*sc.* the definition] must be one which will expressly reckon with him.

Hence as *sb.*, an act of consideration; a 'think'. *colloq.*

1949 F. Sargeson *I saw in my Dream* vii. 57 He reckons he's going to... Then he'd better go and have another reckon, the girl said.

reckonable, *a.* Delete *rare* and add: Also, admissible for the purposes of reckoning. (Further examples.) Hence **reckonabi·lity**, the quality of being reckonable.

1905 *Daily Chron.* 21 July 4/4 Its only drawback is that it is called 'Hardy's country' by those fiction-enthusiasts who now form so reckonable a section of Weymouth guests. 1967 *Encycl. Philos.* VII. 94/1 (*heading*) Reckonability. 1971 *Daily Tel.* 15 Sept. 6/8 This means that for every £1,000 of reckonable earnings the pension offered must amount to at least £10 for each year of service. 1973 *Times* 29 Nov. 16/3 There were bound to be practical difficulties and problems of definition in reopening at this late stage the question of reckonable service of teachers who joined the profession immediately after the war.

reckoning, *vbl. sb.* Add: **9.** *reckoning tablet.*

1930 T. S. Eliot tr. *St.-J. Perse's Anabasis* 67 He who has spread on the ground his reckoning tablets.

reclaim, *sb.*¹ Restrict 'Now *rare*' to senses in Dict. and add: **II. 5.** = *reclaimed rubber.

1935 Dawson & Porritt *Rubber* 208/2 Although reclaim as compared with new rubber definitely leads to inferior mechanical properties in vulcanisates,...it has important applications in rubbers where mechanical strength and abrasion resistance are of secondary significance. 1954 H. J. Stern *Rubber* vi. 200 Small differences in temperature from one part of the rubber to the other greatly affect the uniformity of the reclaims. 1971 R. Singleton in C. M. Blow *Rubber Technol. & Manufacture* vi. 207 First-quality reclaim made from whole tyres contains about 45% rubber hydrocarbon by weight. The remaining 55% consists of valuable carbon black, a little mineral filler, and softeners.

reclaim, *v.* Add: **I. 3. d.** (Further examples.)

1903 G. B. Shaw *Man & Superman* p. xi, Far beyond mere lovemaking into politics, high art, schemes for reclaiming new continents from the ocean. 1922 Joyce *Ulysses* 68 Reclaim the whole place. Grow peas in that corner there. 1943 J. S. Huxley *TVA* vi. 30 Gullied and badly eroding land reclaimed in the Valley. 1966 *Listener* 26 May 751/2 You could take advantage of this fact by reclaiming a large area of the Wash itself.

e. To recover (rubber) for reuse by freeing it from impurities and rendering it plastic again; more widely, to make (re)usable (what has been used or rendered unusable). Also *absol.*

1895 *Sci. Amer.* 26 Oct. 267/1 Methods of reclaiming rubber. 1898 *India-Rubber & Gutta-Percha & Electr. Trades' Jrnl.* XVI. 184/1 Prior to that time [*sc.* 1870] the use of rubber reclaimed from fibrous wastes had been confined practically to one large factory in Boston and one near New York. 1937 H. Barron *Mod. Rubber Chem.* xxi. 254 Miller carried out four cycles of reclaiming starting from a reclaim. That is, he vulcanised, reclaimed, added sulphur, revulcanized, reclaimed, etc. 1937 *Iron Age* 5 Aug. 38/3 Many new parts of Monel rendered unserviceable as a result of being turned undersize or bored oversize in the process of manufacture are readily reclaimed by spraying. 1962 A. Nisbett *Technique Sound Studio* vi. 113 The tape is finally wiped and reclaimed. 1970 [see *reclamation 2 d]. 1972 P. W. Allen *Natural Rubber & Synthetics* v. 121 About 90 per cent of the world's supply of new rubbers is not reclaimed but ends up as unwanted waste products. 1976 *Conservation News* Nov./Dec. 18/2 It cost twice as much to reclaim bottles as it did to buy new ones. 1977 *Lancashire Life* Dec. 92/4 It is particularly annoying to see stones from old property bulldozed aside instead of being reclaimed for future use.

reclaimed, *ppl. a.* Add **1.** (Further examples.)

1961 D. M. Disney *Mrs. Meeker's Money* vii. 77 A reclaimed island of land-scaped brick and glass on the fringe of the business district. 1974 W. J. Burley *Death in Stanley St.* ix. 164 The new waste-disposal unit, pride of the city, had been built on a promontory of reclaimed land.

2. Rendered reusable; *reclaimed rubber*, rubber obtained from used vulcanized rubber by treating it to remove free sulphur and foreign substances and to render it plastic, chiefly used mixed with crude rubber in low-grade rubber goods.

1897 *Sci. Amer.* 25 Sept. 196/2 The reclaimed rubber of commerce is obtained by steaming or devulcanizing old rubber waste, generally shoes, freed more or less perfectly from fiber. 1913 B. D. Porritt *Chem. Rubber* v. 66 The consumption of reclaimed rubber...is probably equal to the world's production of raw rubber. 1942 *Time* 2 Feb. 63/3 Reclaimed rubber, ordinarily bypassed by manufacturers because it wears out faster than natural rubber. 1958 *Jrnl. Amer. Water Works Assoc.* L. 1021 (*heading*) Emergency use of reclaimed water for potable supply at Chanute, Kan. 1971 R. Singleton in C. M. Blow *Rubber Technol. & Manufacture* vi. 203 Reclaimed rubber has become widely accepted as a raw material which possesses processing and economic characteristics that are of great value in the compounding of natural and synthetic rubber stocks. 1976 *Lancs. Evening Post* 7 Dec. 13/5 (Advt.), Demolition materials. Reclaimed timber (cut to your requirements)... Reclaimed stone.

reclaiming, *vbl. sb.* Add: **a.** (Further examples.)

1892 *Sci. Amer.* 7 May 293/3 These [*sc.* old boots and shoes] are sorted roughly, put up in bales, and shipped to the companies who make a business of reclaiming. 1913 B. D. Porritt *Chem. Rubber* v. 60 'Reclaiming' at the present time is an industry by itself, several large works being exclusively engaged on this work. 1971 R. Singleton in C. M. Blow *Rubber Technol. & Manufacture* vi. 203 The raw material for reclaiming is scrap rubber in a wide variety of forms, but tyres...form the major quantity.

b. (Further examples.)

1895 *Sci. Amer.* 26 Oct. 267/1 Rubber reclaiming factories. 1954 H. J. Stern *Rubber* vi. 200 Peptising agents accelerate the reclaiming process and are useful with both natural and synthetic rubbers. 1957 *Times* 20 Dec. 17/6 In our reclaiming plant overtime working was in force from October.

reclamation. Add: **2. c.** (Earlier and later examples.) Cf. also *land reclamation* s.v. *land *sb.*¹ 10 a.

1848 J. S. Mill *Let.* 3 Feb. in W. Ward *Aubrey de Vere* (1904) iv. 132, I look much more than you do to reclamation of waste lands. 1955 *Times* 24 May 8/3 The United Nations Relief and Works Agency is soon to present...a report on the feasibility of an ambitious reclamation scheme known as the Sinai strip. 1970 *New Yorker* 15 Aug. 32/1 The Dutch... knew...how difficult and dirty any sort of reclamation work could be.

d. The action or process of reclaiming used or unusable objects or materials.

1937 *Iron Age* 5 Aug. 36/2 This...has been largely responsible for the extended use of the spray process in applying this metal both for the protection of new parts and for the reclamation of worn or unserviceable items. 1954 *Publ. Calif. State Water Pollution Control Board* No. 9. i. 13 Implicit in the study of waste water reclamation and utilization is the idea that eventually it will be of economic value as a water source. 1962 A. Nisbett *Technique Sound Studio*

vii. 133 In BBC practice reclamation also includes not only checking through the tape to remove spacers, trailers, etc... but also removing all temporary joints and replacing them by cemented joints. **1970** *New Society* 5 Mar. 387/3 The reclamation industry reckons it saves Britain £1,000 million by reclaiming otherwise imported material.

6. Special *Comb.*: **reclamation disease** [tr. G. *urbarmachungskrankheit* (B. Sjollema 1933, in *Biochem. Zeitschr.* CCLXVII. 151)], a disease affecting crops, esp. cereals, grown on reclaimed land, caused by a deficiency of copper and distinguished by discoloured leaves and the failure of affected plants to produce seed.

1937 F. T. Heald *Introd. Plant Pathol.* xviii. 365 The curative value of boron has also been demonstrated for the 'reclamation or bog disease', a trouble characteristic of swampy heath soils in European countries. **1949** Butler & Jones *Plant Pathol.* ix. 312 Among the group of crop disorders..owing to the active measures taken during modern times to reclaim peat moor, swamp, and polder soils, that which eventually became widely termed 'reclamation disease' is one of the most important. **1961** W. Stiles *Trace Elements in Plants* (ed. 3) iii. 99 Reclamation disease.. affects oats and other cereals.

réclame. Add: (Earlier example.) Also, popular acclaim, notoriety, glory, fame.

In quot. 1870 the sense seems to be 'an advertisement'.

1870 O. Logan *Before Footlights & behind Scenes* xxii. 255 Perhaps you think I mean this as a *reclame* for the Sherman House. **1906** R. Fry *Let.* 16 Mar. (1972) I. 257 Pictures.. which he offered at ridiculous prices for the *réclame* of getting into the Museum. **1945** R. Hargreaves *Enemy at Gate* 105 The effulgent *réclame* of the conqueror of Wurmser and the Archduke Charles. **1977** *Times Lit. Suppl.* 25 Feb. 200/2 The author of a novel..which the serious papers denounced as prurient, so adding both to the number of copies sold and to his réclame among his colleagues.

reclassify (rīklæ·sifəi), *v.* [Re- 5 a.] *trans.* To classify again; to alter the classification of.

1920 in Webster. **1928** *Daily Tel.* 27 Nov. 8/2 They will have an opportunity of reclassifying their institutions. **1946** *Nature* 28 Sept. 439/1 Existing roads should be reclassified and the design of new roads should not be attempted before their purpose was clearly determined. **1953** *Times* 8 June 7/4 Whether the Government will consider the advisability of reclassifying crash helmets as motor-cycle accessories. **1972** W. McGivern *Caprifoil* iii. 47 'Those files aren't available.' 'They've been reclassified?' **1977** *Evening Post* (Nottingham) 27 Jan. 2/8 Re-classify their status—they are part retired, but certainly not wholly unemployed.

reclinable, *a.* (Further example.)

1957 *Archit. Rev.* CXXII. 351 (*caption*) In both, seats are reclinable and the windows are double-glazed.

recline, *v.* Add: **1. a.** (Later example.)

1972 *Daily Tel.* (Colour Suppl.) 25 Aug. 19/2 A back-row, next-to-bulkhead seat is often fixed, i.e. the backrest cannot be reclined.

3. d. Of a seat: to admit of mechanical inclination of the back to a reclining or recumbent position.

1972 *N.Y. Law Jrnl.* 24 Oct. 3/2 The company is engaged in the manufacture and sale of upholstered furniture, principally medium priced chairs that recline. **1974** *Trafford's Catal.* 963/3 Multiposition metal reclining chair... Can recline to many positions.

recliner. Add: **2.** A chair in which one may comfortably recline; a reclining chair or seat. Also *Comb.*, as *recliner chair, seat.* orig. *U.S.*

1928 E. O'Neill *Strange Interlude* ix. 275 There is a stone bench at center, a recliner at right. **1948** Partridge *Dict. Forces' Slang* 153 *Recliners*, Navy 'issue' armchairs. (Ward-room.) **1970** *Globe & Mail* (Toronto) 25 Sept. 31/2 (Advt.), The comfort of hushed travel in recliner seats. **1977** E. Leonard *Unknown Man No. 89* xx. 192 Jay Walk, in his desk-chair recliner, had his shoes off. **1978** *Lancashire Life* Oct. 115/2 Buoyant Upholstery have just introduced the Wellbeck Chameleon range of fourteen interchangeable items, made up of Chesterfields, unit pieces, settees, recliner chair, wing chair, [etc.].

reclining, *ppl. a.* (Further examples.)

1883 L. M. Mitchell *Hist. Anc. Sculpture* xix. 354 By a recent correction in the placing of the reclining figure..lines of unexpected beauty in the composition of Pheidias have been revealed to us. **1966** D. Hall *Henry Moore* iv. 72 He exploited his stone-carving breakthrough in a series of female figures: upright busts, reclining figures.

reclining, *vbl. sb.* Add: **reclining chair** (examples); **reclining seat,** a seat which may be adjusted to a reclining position, esp. in a motor vehicle or aeroplane.

These combinations can equally be seen as examples of the *ppl. a.*

1863 Geo. Eliot *Let.* 26 Dec. (1956) IV. 124 Another munificent friend has given me the most splendid reclining chair conceivable. **1907** *Yesterday's Shopping* (1969) 276/2 Improved portable suspensory Reclining Chair, with leg rest in canvas. **1976** B. Bova *Multiple Man* (1977) xiv. 148 We sat side by side in the most luxurious reclining chairs I'd ever flown in. **1943** S. C. Menefee *Assignment: U.S.A.* i. v. 117 She settled her ample proportions into the reclining seat next to me. **1974** 'D. Craig' *Dead Liberty* xix. 108 Boxanford arranged..to change his car... He wanted reclining seats.

reclude, *v.* Add: **2. b.** Also *refl.*

1911 M. Beerbohm *Zuleika Dobson* ii. 22 No woman who knows that of herself can be rightly censured for not recluding herself from the world.

reclusage. Restrict † *Obs.* to sense in Dict. and add: **2.** Retirement, reclusion.

1960 *Times Lit. Suppl.* 18 Nov. 742/3 For more than half a lifetime..he had enjoyed a voluntary reclusage on the Riviera di Levante.

recluse, *a.* and *sb.* Add: **c.** *Comb.*, as *recluse-like* adj.

1946 E. Blunden *Shelley* xvii. 213 The fashionable round ..did not prevent her from falling under the spell of the recluse-like Shelley.

reclusion. 1. a. (Further examples.)

1908 E. Wharton *Hermit & Wild Woman* 33 In a life of penance and reclusion her eyes might be opened to her iniquity. **1971** T. Merton *Contemplation in World of Action* ii. v. 300 It must not be imagined that these problems of order rose exclusively from a lack of legislation and from a too-free development of 'charisms' of pilgrimage, hermit-solitude or reclusion.

reclusive, *a.* Add: Now freq. of persons. (Further examples.)

1965 *Listener* 16 Sept. 426/2 A reclusive New Englander who wrote but did not flourish in the literary climate of Transcendentalism. **1971** *Wall St. Jrnl.* 13 Aug. 1/6 Esquire ..ran a cover showing the reclusive Howard Hughes. **1979** *Daily Tel.* 1 Sept. 21/3 Equal partnership deals are not common in the business career of the reclusive Mr Ludwig.

Hence **reclu·sively** *adv.*, in the manner of a recluse; **reclusiveness** (further examples).

1925 C. Connolly *Let.* 25 Jan. in *Romantic Friendship* (1975) 56 The last week of Minehead..left me with an intense reclusiveness. **1963** *Punch* 13 Feb. 249/1 His life was spent uneventfully, reclusively almost. **1976** *National Observer* (U.S.) 14 Feb. 15/2 My symptom was reclusiveness. **1979** *N. & Q.* June 240/2 W. S. W. was both a Scholar and subsequent Fellow of Trinity College, Cambridge, where he lived reclusively.

recode (rīkōᵘ·d), *v.* [Re- 5 a.] *trans.* To put into another or different code; *spec.* in *Psychol.*, to rearrange mentally (information presented by a problem, situation, or test).

1951 [implied in *recoding *vbl. sb.*]. **1957** J. S. Bruner *Contemp. Approaches in Psychol.* 59 All the measurements can be recoded into a simple rule. **1964** M. McLuhan *Understanding Media* (1967) viii. 90 The human ear can be compared to a radio receiver that is able to decode electromagnetic waves and recode them as sound. **1971** *Jrnl. Gen. Psychol.* LXXXV. 213 There are often possibilities for S to encode the information in one of the languages not intended or to recode the information in a second language. **1977** J. M. Scandura *Problem Solving* vii. 301 In order to recover the original elements..they must be recoded. *Ibid.* 309 Extraneous processes may reduce memory load..where two digits are recoded..as one unit.

So **reco·ding** *vbl. sb.*

1951 G. A. Miller *Lang. & Communication* xi. 233 The task also illustrates something we can call recoding, and in many problems it can be shown that the restructuring process is, in whole or in part, a matter of coding the information in a new form. **1957** J. S. Bruner *Contemp. Approaches to Cognition* 60 Once a system of recoding has been worked out whereby information is condensed into more generic codes, the problem of mastery becomes one of mastering the recoding system. **1964** J. Z. Young *Model of Brain* ii. 21 The conversion from one sort of physical system to another is called re-coding (e.g. speech into writing). Information is thus that feature of the system that remains invariant under re-coding. **1970** B. Milner in Pribram & Broadbent *Biol. of Memory* 42 This maze has 28 choice-points, so that, even with recoding, the sequence of turns to be remembered cannot be encompassed within the span of immediate memory. **1977** J. M. Scandura *Problem Solving* vii. 304 Even on simple tests of memory span, recoding and rehearsal processes tend to be highly dependent on individual preference.

recogitate, *v.* **1. b.** (Later examples.)

1920 in Webster. **1932** H. Crane *Let.* ?Feb. (1965) 401, I had to spend the rest of the day and evening cogitating and recogitating.

recognition. Add: **7. c.** *Psychol.* In the study of thinking and memory, the mental process whereby things are identified as having been previously apprehended or as belonging to a particular known category, usu. distinguished from the process of recall.

1894 Creighton & Titchener tr. *Wundt's Hum. & Anim. Psychol.* xx. 297 The simplest case of assimilation is the cognition of an object; the simplest case of successive association, its recognition. **1894** *Psychol. Rev.* I. 608 There were some incidental illustrations of false recognition. **1923** C. Spearman *Nature of Intell.* xix. 313 Recognition..is often traceable to nothing more than an awareness of similarity. **1951** G. A. Miller *Lang. & Communication* vi. 121 It is a general rule of verbal learning that recognition is easier than recall. **1965** E. E. Harris *Foundation of Metaphys. in Sci.* xix. 380 There are two kinds of problems..in attacking which the cybernetic approach has been used... The second are problems of transmission and of recognition. **1965** K. M. Sayre (*title*) Recognition: a study in the philosophy of artificial intelligence. *Ibid.* i. 33 The task of achieving mechanical recognition of letter-patterns brings up problems of both sorts. **1973** A. J. Pomerans tr. *Piaget & Inhelder's Memory & Intelligence* 1 It is difficult to decide whether his [*sc.* the subject's] recognition is based on the remembrance or conservation of perceptive schemata..or whether it reflects the organization of the sense data by these schemata.

d. *out of* (or *beyond*) *recognition*, to such a degree as to be unrecognizable.

1901 G. B. Shaw *Three Plays for Puritans* 202 The world, instead of having been improved in 67 generations out of all recognition, presents, on the whole, a rather less dignified appearance. **1916** —— *Androcles & Lion* p. xli, Jesus is refined and softened almost out of recognition. **1964** M. Drabble *Garrick Year* ii. 33 After she was born,..things improved out of all recognition. **1977** *Rolling Stone* 5 May 30/5 Futuristic explorers..returning to their own world to find it changed beyond recognition.

8. (sense 7 a) *recognition-call, scene*; (sense *7 c) *recognition habit, learning, memory, schema, test, vocabulary, word*; **recognition colour**: also *transf.* in *Mil.* use; **recognition grammar** *Linguistics*, a grammar based on the analysis of given sentences in a corpus (opp. generative grammar); **recognition mark** (later examples); **recognition marking**: also *transf.* in *Mil.* use; **recognition picketing** *U.S.*, the picketing of an employer to obtain union recognition; **recognition signal** *Mil.* (see quot. 1963).

1911 J. A. Thomson *Biol. Seasons* ii. 155 Love-calls and song probably had their roots in the simple recognition-call or characteristic signal of the species. **1944** *Return to Attack* (Army Board, N.Z.) 32/2 Three tanks, displaying British recognition colours, climbed the hill. **1966** A. F. R. Brown in *Automatic Transl. of Lang.* (NATO Summer School, Venice 1962) 49 A recognition grammar will turn out to be a thousand times more complicated than a conventional descriptive grammar. **1968** J. Lyons *Introd. Theoret. Ling.* vi. 230 We have put the categorial system in the form of a 'recognition' grammar and the 'rewrite' system in the form of a 'production' grammar. **1920** T. P. Nunn *Educ.* xiii. 169 Learning to read involves, in fact, building up recognition-habits. **1970** M. R. Amato *Experim. Psychol.* xii. 550 The simplest case of recognition learning is verbal discrimination in which an arbitrarily selected 'correct' item is to be identified from an accompanying, but incorrect, item. **1906** M. C. Dickerson *Frog Bk.* 26 These brilliant colours..may act as recognition marks for others of the same species. **1939** A. S. Pearse *Anim. Ecol.* (ed. 2) iii. 31 He [*sc.* E. S. Poulton] cites the conspicuous white tails of the rabbit and antelope as examples of recognition marks. **1960** M. Burton *Wild Animals Brit. Isles* 121 A patch of white around the short tail [of the red deer] furnishes a 'recognition mark', common to most of the deer family. **1977** T. I. Storer et al. *Elements Zool.* (ed. 4) xiii. 220/2 Recognition marks and other signals are often important in intraspecific communication. **1940** in *Brit. Aviation Colours of World War Two* (R.A.F. Museum Series) (1976) III. 9 (*heading*) Aircraft colouring and recognition markings. **1975** Gander & Chamberlain *German Tanks of World War 2* vi. 53/2 Perhaps the most universally applied markings used on German tanks was the tactical national recognition marking. This was usually a black cross outlined in white. **1955** H. E. Garrett *Gen. Psychol.* x. 396 Students do not always distinguish between those facts which should be learned for recall and those for which recognition memory is sufficient. **1973** J. G. Greeno in B. B. Wolman *Handbk. Gen. Psychol.* viii. 150/2 The agreement between the data and the theory demonstrates that it is appropriate to analyze recognition memory in terms of a concept of trace strength. **1960** *U.S. Statutes at Large 1959* LXXIII. 542 (*heading*) Boycotts and recognition picketing. **1962** N. S. Falcone *Labor Law* xi. 345 Recognition picketing is generally defined as picketing an employer's establishment to force the employer to recognize and bargain with the union. **1932** T. S. Eliot *Selected Ess.* 194 The Recognition Scene, so important in Shakespeare's later plays. **1971** *Jrnl. Gen. Psychol.* Jan. 166 Recognition schema operating on coded features are entirely possible. **1958** P. Scott *Mark of Warrior* ii. 149, I want you to set up your recognition signals on the D[ropping] Z[one]. itself. **1963** *Dict. U.S. Mil. Terms* (U.S. Dept. Defense) 180 *Recognition signal*, any prearranged signal by which individuals or units may identify each other. **1978** R. V. Jones *Most Secret War* v. 48 You have to shoot your opponent out of the ocean..if he does not make the right recognition signal. **1923** P. B. Ballard *New Examiner* vii. 81 The third and last test was a Recognition Test... The candidate..had to underline the one word or phrase which would make each statement true. **1966** J. M. Brown et al. *Applied Psychol.* xii. 418 Recognition tests..were used to evaluate the memorability of advertising messages. **1966** J. Derrick *Teaching Eng. to Immigrants* ii. 99 Most stories will contain far more material than the pupils are expected to reproduce themselves (i.e. relying on and helping to build up their 'passive' or recognition vocabulary). **1977** P. Strevens *New Orientations Teaching Eng.* v. 62 Recognition vocabulary..can lie outside the confines of controlled vocabulary, grammar, etc., as long as the learner understands it when he meets it. **1957** Partridge *Eng. gone Wrong* ii. 44 *Monolithic*, especially perhaps in *monolithic unity*, is a recognition-word, a keyword, a badge.

recognitive, *a.* (Further examples.)

1930 *New Statesman* 16 Aug. 593/1 Its function appears to be mainly critical and recognitive. **1977** *Maledicta* Summer 33 The relatively passive attitude of acceptance of good or evil we call *recognitive.*

recognitory, *a.* Add: (Further example.) Now *rare.*

1964 R. Perry *World of Tiger* iii. 42 It is difficult to think of any recognitory purpose this marking could serve.

recognizability. (Further examples.)

1938 [see *extruder 2]. **1979** *Sci. Amer.* June 42/2 The transformation of an embryo into a fetus..is a transformation from external recognizability only as human to increasing recognizability as an emergent person.

recognize, *v.*[1] Add: **4. a.** Also *absol.*

1974 M. Pendyre *Breach of Security* xxiii. 132 Our customary criteria for recognition are that the government should control the whole country... But..those raving Afro-Asian countries have recognized already. **1977** D. Beaty *Excellency* v. 66 The State Department isn't going to recognize until they *do* have the details.

recognized, *ppl. a.* (Further examples.)

1945 *Guide to Educ. System Eng. & Wales* (Min. of Educ.) 60 *Recognised Efficient School*, independent school inspected by H.M.I.s and regarded as efficient by the Ministry. **1966** *Rep. Comm. Inquiry Univ. Oxf.* I. i. 47 We recommend that the existing university category of 'Recognized Student' should be revised. **1974** G. Hubbard *Quaker by Convincement* iv. i. 180 Until recently we were a Recognized Meeting, part of the Preparative Meeting of Kingston upon Thames, some three miles away.

recognizer. Add: **2.** A device which can interpret speech by identifying the sounds and assigning them the correct meaning.

1952 *Jrnl. Acoustical Soc. Amer.* XXIV. 637 The recognizer discussed will automatically recognize telephone-quality digits spoken at normal speech rates by a single individual, with an accuracy varying between 97 and 99 percent. **1958** *Listener* 11 Dec. 984/2 One aspect of this work..is that of Drs. Ahmed and Fatehchand..on the direct recognition of the spoken word [by a computer]... It is safe to predict that a recognizer of 95 per cent. accuracy could be built within five years. **1973** *Physics Bull.* May 281/1 Ideally the recognizer should perform this action irrespective of the speaker and the acoustic environment in which he is speaking. This means that the machine not only has to recognize the speech sounds it receives, but it also has to ignore those facets of the signal that convey information irrelevant to the task of recognizing the speech. **1976** W. A. Ainsworth *Mechanisms of Speech Recognition* x. 104 An automatic speech recognizer may be defined as any mechanism, other than the human auditory system, which decodes the acoustic signal produced by the human voice into a sequence of linguistic units which contain the message that the speaker wishes to convey. *Ibid.* 111 If the world of discourse of the speech recognizer is sufficiently restricted it is sometimes possible to employ semantic information to choose between words or phrases which seem equally likely on phonetic, syntactic or other grounds.

recoil, *sb.* Add: **3*.** *Nucl. Physics.* The result of a collision between two sub-atomic particles, or of spontaneous decay of a single particle, in which the two resulting particles move in opposite directions with speeds determined by conservation of momentum.

1909 *Nature* 24 June 490/1 Rutherford..suggests the possibility of the phenomenon being due to a recoil effect rather than to a volatility possessed by the product radium B. **1912** *Phil. Mag.* XXIV. 622 It is well known..that the emission of α particles from radioactive substances is accompanied by a vigorous recoil of the residual atoms. **1933** *Discovery* Apr. 107/2 The energy of recoil is greatest when it [*sc.* a neutron] strikes a hydrogen nucleus, and the recoiling atom may travel 30 c.m. or more in air before it is brought to rest. **1964** J. B. Hasted *Physics of Atomic Collisions* iii. 107 Since the angle of recoil is related to the velocity, a suitable positioning of slits should serve to select atoms of a certain velocity.

4. a. Further names of devices intended to diminish or absorb recoil, as *recoil box, pad, reducer, spring* (in Dict.: examples); also, names of devices used to measure the force or energy of the recoil of a firearm, as *recoil gauge, machine, recorder*; also *recoil action.*

1908 *Westm. Gaz.* 13 Feb. 4/2 The recoil action of the spring..closes these valves. **1892** W. W. Greener *Gun & its Development* (ed. 5) 480 Recoil breeching of rope is..the simplest gear for taking the recoil. Others in use are..the indiarubber breeching, or the recoil box of Mr. E. T. Booth. **1890** *Field* LXXVI. 461/1 Ascertaining, by chronograph and recoil-gauge, what the forward velocity of the shot and the backward movement of the gun respectively amount to. **1896** W. W. Greener *Gun & its Development* (ed. 6) xii. 316 (*caption*) New mechanical gun-rest and recoil-gauge. **1900** G. T. Teasdale-Buckell *Experts on Guns & Shooting* xii. 138 We cannot follow Mr. Toms..in his remarks about the recoil gauge. **1896** *Field* 28 Mar. 457/1 There would thus be an increase of the back-thrust registered on the recoil machine. **1931** G. Burrard *Mod. Shotgun* I. vi. 145 A recoil pad can be a great boon to those who are at all sensitive to recoil or to almost anyone who uses a gun in a hot climate where one frequently has to shoot in nothing but a thin shirt. **1974** R. Dunlap *Gun Owner's Bk. of Care, Repair & Improvement* xiii. 207 We have the obvious changes in original wood fittings for firearms: fitting recoil pads to shotguns and rifles, which more or less combines with the desire for shortening or lengthening buttstocks. **1976** *Shooting Times & Country Mag.* 18–24 Nov. (Advt.), Beavertail fore-end in walnut, pistol grip stock with recoil pad. **1906** H. Sharp *Mod. Sporting Gunnery* v. 123 The sportsman does not appear to have attached very much importance to these recoil recorders, possibly for the reason that machines devised on the above lines tell only the weight of recoil. **1942** *R.A.F. Jrnl.* 3 Oct. 29 The compensator or recoil reducer was very effective. **1975** G. T. Garwood *Shotguns & Cartridges for Game & Clays* (ed. 3) xvi. 155 In the USA various proprietary 'recoil reducers' are advertised, the best known being the Edwards. This is a sealed device, and the working principle is not disclosed. **1859** H. C. Folkard *Wild-Fowler* xxii. 126 The best plan of all is Colonel Hawker's invention of a steel spiral recoil-spring. **1882** 'Marksman' *Dead Shot* (ed. 5) 331, I have recently seen several attached to heavy breech-loaders, in which the recoil-spring and fittings were so short and cramped, as to be only twelve or fourteen inches in length. **1971** G. T. Garwood *Gough Thomas's Second Gun Bk.* xi. 211 If it were not for the friction device and the recoil spring, the parts of the gun with which the shooter makes contact..would not commence to recoil at all.

a*. In *Nucl. Physics.* (cf. sense 3* above), as *recoil atom, electron, energy, momentum, nucleus, proton, ray, track.*

1912 *Phil. Mag.* XXIV. 622 Recoil atoms produce a strong ionization in the gas they traverse. **1942** Pollard & Davidson *Appl. Nucl. Physics* iii. 40 A cloud chamber for observation of alpha particles, protons, and heavy recoil atoms is not hard to construct. **1923** *Physical Rev.* XXI. 483 The velocity of secondary β-rays excited in light elements by γ-rays agrees with the suggestion that they are recoil electrons. **1966** S. E. Liverhant *Outl. Atomic Physics* iv. 111 In an experimental arrangement designed to measure the coincidences between the scattered photon and the recoil electron in Compton scattering, the detectors are to be placed symmetrically about the direction of the incident X-ray beam. **1949** Friedlander & Kennedy *Introd. Radiochem.* xi. 253 Neutron capture is always followed by γ-ray emission, and the nucleus receives some recoil energy in this process. **1963** *Radiochem. Man.* (Radiochemical Centre) ii. ii. 5 When an atom in a chemical compound captures a neutron, by an (n, γ) reaction, the atom recoils with an energy usually greater than that of the chemical binding forces: recoil energies are usually in the range of a few MeV whilst chemical bond energies are usually only a few eV. **1950** D. Halliday *Introd. Nucl. Physics* iii. 106 The recoil momentum of a disintegrating nucleus will be influenced in magnitude and direction by the presence of a neutrino. **1962** Semat & Albright *Introd. to Atomic & Nuclear Physics* (ed. 4) xiv. 470 The mass to which the recoil momentum is transferred can be considered infinite in comparison with that of an atom, so that the velocity of recoil is zero. This phenomenon is sometimes called recoilless emission of radiation. **1934** *Proc. Cambr. Philos. Soc.* XXX. 99 The ionisation due to recoil nuclei of the energies here concerned is almost entirely primary. **1949** O. Oldenberg *Introd. Atomic & Nuclear Physics* xxi. 237 The two tracks so produced, that of the original alpha particle deflected by the collision and that of the 'recoil nucleus', are both well defined. **1942** Pollard & Davidson *Appl. Nucl. Physics* iii. 41 The recoil protons caused by neutrons in a gas containing hydrogen can..be detected with such equipment. **1949** O. Oldenberg *Introd. Atomic Physics* xix. 195 When the cloud chamber is filled with CH_4 and subjected to neutron bombardment, the short, straight tracks of recoil protons show up. **1913** E. Rutherford *Radioactive Substances* iv. 178 By observing the deflections of a pencil of recoil rays, both in a magnetic and electric field, the velocity and value of e/m of the recoil atoms can be deduced. **1926** R. W. Lawson tr. *Hevesy's Man. Radioactivity* vi. 59 In consequence of their smaller velocity, the phenomenon of scattering occurs in a much more marked degree with recoil rays than with α-particles. **1927** *Proc. R. Soc.* A. CXVI. 664 This paper describes the measurement of the mobility of a single atom of actinium A immediately after it reaches the end of its recoil track. **1930** E. Rutherford et al. *Radiations from Radioactive Substances* vi. 155 At ordinary pressure, the recoil track is shown by a knob at the end of the track. As the pressure is reduced, the recoil track becomes longer and often shows evidence of a marked scattering.

b. recoil escapement (earlier example); **recoil gear** *Mil.* (see quot. 1940); **recoil starter,** a device for starting a small internal-combustion engine in which a cord, wound round a pulley, is rewound by a spring after being pulled for the starting cycle.

1838 *Penny Cycl.* XII. 299/1 [This] motion is called the recoil, and this escapement is thence called the recoil escapement. **1904–5** *Jrnl. R. Artillery* XXXI. 303 Recoil gear. **1911** H. A. Bethell *Mod. Artillery in Field* i. 26 (*heading*) Hydropneumatic recoil gear. **1940** *Chambers's Techn. Dict.* 706/2 *Recoil gear* (Artillery), The whole recoil mechanism, embracing both buffer and recuperator. **1960** *Farmer & Stockbreeder* 16 Feb. 106/2 (Advt.), Petrol-engine model with..two-stroke engine;..automatic recoil-starter eliminating the use of loose starting rope. **1972** P. Dempsey *How to repair Small Petrol Engines* vii. 235 Rewind starters. Sometimes called recoil starters, these devices are found on outboards, lawnmowers, go-karts.

recoiling, *ppl. a.* (Examples in *Nucl. Physics.*)

1911 *Ann. Rep. Progress of Chem.* VII. 272 A coating of silver $20\mu\mu$ in thickness stopped the recoil completely, while $10\mu\mu$ allowed some 60 per cent of the recoiling atoms to pass through. **1950** D. Halliday *Introd. Nucl. Physics* iii. 107 Allen tried to detect recoiling Li^7 ions formed during the disintegration by *K*-capture of Be^7. **1963** Bowen & Gibbons *Radioactivation Analysis* vii. 109 The recoiling atom travels for a short distance before it gives up all its excess energy, and may undergo various chemical reactions in the process.

recoilless (rīkoi·l,lės), *a.* Also **recoil-less.** [f. Recoil *sb.* + -less.] Having no recoil.

a. *Mil.* Applied to a firearm in which recoil is reduced or eliminated by deflection of much of the combustion gas to the rear.

1948 *Jrnl. British Interplanetary Soc.* July 163 There were two types [of rocket-firing guns] scheduled for development, one static.., the other a portable 'recoilless' model on a wheeled chassis. **1953** *Times* 28 May 5/4 Both the United States and the French armies have produced admirable recoilless guns since 1946. **1957** *Economist* 7 Sept. 855/1 Recoilless anti-tank rifles could knock out any tank built in the Soviet Union. **1972** M. Kenyon *Shooting of Dan McGrew* i. 9 You'll be wanting a recoil-less rifle to defend your honour. **1975** *Times* 10 Apr. 8/7 Mortar and recoilless rifle fire has been used against the international airport.

b. *Nucl. Physics.* Applied to transitions occurring in an atomic nucleus bound in a crystal lattice in which a photon is emitted from the nucleus without recoil (the Mössbauer effect: see *Mössbauer).

1960 *Physical Rev.* CXX. 1093/1 Mössbauer's observation rests on the fact that in the case of a nucleus bound in a crystal, a γ ray can be emitted or absorbed without any energy transfer to and from the lattice. The probability of such a recoilless transition is, in most cases, small. **1962** [see *recoil momentum* s.v. *recoil *sb.* 4 a*]. **1971** Greenwood & Gibb *Mössbauer Spectrosc.* i. 11 Mössbauer experiments usually utilise the recoilless emission of γ-rays by a radioactive source followed by their subsequent resonant recoilless reabsorption by a non-active absorber. **1974** *Nature* 19 Apr. 638/2 The 13·3-keV transition to the ground state has long been one of the most attractive candidates for high resolution experiments because..the low energy of the γ photon assures that a high probability of recoil-less transitions will result even at room temperature.

recollating, *vbl. sb.* (Further example.)

1881 P. Fitzgerald *World behind Scenes* iii. 177 This [painting] represents Farren & Farley, but the recollating with Zoffary's work makes this picture comparatively feeble.

recolonization. Add: **2.** *Ecol.* The return of an animal, plant, or other organism to an area once inhabited by the species or group concerned.

1923 *Jrnl. Ecol.* XI. 242 In places re-colonisation by *Saxifraga oppositifolia* of such 'blow-outs' was observed. **1956** *Nature* 11 Feb. 282/1 By recolonization, the yield of a sprayed plot may be diminished. **1958** *New Biol.* XXVII. 47 The 'homing' instinct of the injected cells is so remarkable that the recolonization hypothesis was not at first accepted. **1973** *Nature* 3 Aug. 254/2 Factors inhibiting the recolonization by plants of colliery spoil..were reviewed by American and British workers.

recolonize, *v.* Add: **1.** (Later example.)

1976 *Listener* 18 Mar. 329/2 The Europeans are back, recolonising Africa.

2. *Ecol.* Of a plant, animal, or other organism: to return to (a former habitat of the species or group concerned).

1943 J. S. Huxley in *Discovery* Jan. 9/1 Mountain regions..have become re-colonized since the retreat of the ice. **1954** *New Biol.* XVII. 18 The best that can be done is to allow the elephant-grass, *Pennisetum*, to recolonize a fallow plot. **1958** *Ibid.* XXVII. 47 The appropriate cells.. then find their way into the damaged tissue to recolonize it. **1961** *Times* 19 Apr. 14/7 The hares have recolonized the fields. **1963** *Lancet* 19 Jan. 133/1 This pattern suggested that the patients' skin had been recolonised by normal commensal organisms. **1967** *Oceanogr. & Marine Biol.* V. 320 It [*sc.* a sea-urchin]..became extinct during the glacial periods, though surviving in Australia, and recolonizing New Zealand in interglacial periods.

recolt. Substitute for entry:

‖ **récolte** (rekǫlt). Also 8 **recolt.** [Fr.] A harvest or crop. (Chiefly in France.)

1788 [in Dict. s.v. Recolt]. **1865** M. Eyre *Lady's Walks* xxix. 311 Chesnuts are also a *récolte*, they are..commonly sold ready roasted and stripped of the husk in the markets. **1971** *Country Life* 2 Dec. 1557/2 Much of this surplus..has now been re-classified to help the meagre *récolte* in 1971.

recombi·nable, *a.* [f. as next + -able.] Capable of recombining or being recombined.

1964 D. Michie in G. H. Haggis *Introd. Molecular Biol.* x. 272 The most important conclusion from this work concerns the attempt to relate the size of the recombinable units, as measured genetically, to the dimensions of the DNA molecule. **1970** J. S. Bruner in K. Connolly *Mechanisms Motor Skill Development* 79 It is in the altered nature of failure that one sees most vividly the differentiation of a gross act into a set of recombinable constituents.

recombinant (rīkǫ·mbinănt), *a.* and *sb.* *Genetics.* [f. Recombin(e *v.* + -ant[1].] **A.** *adj.* Formed by recombination.

1942 *Jrnl. Genetics* XLIII. 320 Double and higher recombinant types were neglected. **1960** *New Biol.* XXXI. 71 A daughter chromosome might be formed by copying first the ab fragment and then the C portion of the original chromosome so that we now have a recombinant chromosome, abC. **1971** D. J. Cove *Genetics* iii. 21 If a yellow-conidiospored strain is crossed to a strain requiring the vitamin biotin, it is found that a considerable excess of parental types over recombinant types is obtained. **1975** *Nature* 18 Dec. 562/3 The hazards associated with cloning recombinant DNA molecules can only be speculated about, since there is no experimental evidence to prove or deny that they exist. **1977** *Time* 7 Mar. 51/1 Should Harvard and M.I.T. be permitted to go ahead with experiments in so-called recombinant DNA-experiments involving the implantation, in cells of a common bacterium, of alien DNA-borne genes? **1978** *Daedalus* Spring 39 Much of the discussion about recombinant DNA research has centred on whether the work is likely to create hazardous organisms.

B. *sb.* A recombinant organism or cell.

1951 *Jrnl. Gen. Microbiol.* V. 59 Produce a double infection and..obtain in the population of virus units resulting a proportion of recombinants. *Ibid.*, The existence of recombinants can..only be demonstrated if conditions can be so arranged as to favour their selective proliferation. **1969** A. M. Campbell *Episomes* ix. 124 Crosses between lambda and 434 fail to produce any recombinants that generate one immunity and respond to another. **1976** *Nature* 1 July 2/3 Foreign genes are inserted into the bacterium by splicing them into a plasmid..and reintroducing the recombinant into the bacterium.

recombinase (rīkǫ·mbinēⁱz). *Biochem.* [f. Recombin(ation + *-ase.] An enzyme or enzyme system which promotes genetic recombination.

1964 A. W. & P. B. Kozinski in *Proc. Nat. Acad. Sci.* LII. 211 We postulated..that recombination between T_4DNA molecules requires a specific enzyme, 'recombinase'. **1969** A. M. Campbell *Episomes* ii. 18 These two proteins seem to function in a recombinase system. **1970** *Austral. Jrnl. Biol. Sci.* XXIII. 1237 The product of $rec\text{-}w^+$ could be a regulator specifically controlling the recombinase which initiates recombination at the *cog* locus.

recombination. Add: **1.** (Earlier examples.)

1828 in Webster. **1847** A. de Morgan *Formal Logic* xi. 218 It is good against those who confound analysis and recombination of existing materials with introduction of them.

2. *Physics.* The recombining of ions and electrons to form neutral atoms. Freq. *attrib.*

1897 *Phil. Mag.* XLIV. 424 When a gas is acted on by the Röntgen rays a steady state is reached when the rate of production of the ions by the rays is equal to their rate of recombination. **1942** J. D. Stranathan *'Particles' Mod. Physics* i. 8 Let us suppose that there are *n* pairs of ions present per cc. at any time, *n* positive ions and *n* negative ions... The number of recombinations *R* per cc. per second is then given by $R = \alpha n^2$ where α is a constant called the coefficient of recombination. **1962** *Guardian* 10 July 9/5 Atoms in the atmosphere would be broken up in extremely large numbers, so that this recombination light would be visible even to the naked human eye. **1969** J. J. Sparkes *Transistor Switching* i. 13 It will first have to supply the recombination current for any charge already present. **1974** *Encycl. Brit. Macropædia* XIV. 506/2 Other forms of radiation met with in plasma physics include line and recombination radiation.

3. *Genetics.* **a.** The formation by a sexual process of genotypes that differ from both the parental genotypes.

1903 *Proc. Cambr. Philos. Soc.* XII. 53 Since the resolution of a compound character may be spoken of as an analysis leading to a distribution of the components among the gametes, the term synthesis should surely be reserved for a recombination that has taken place in such a way that the gametes become bearers of the compound character again, as they were in the pure compound form. **1909** W. Bateson *Mendel's Princ. Heredity* iii. 71 These cases of novelties resulting through a re-combination of the factors brought in by the original pure types are striking because it is not at first sight evident how the novelty has been produced. **1941** J. S. Huxley *Uniqueness of Man* iv. 107 Recombination—*i.e.*..reshuffling of old genes in new constellations owing to independent assortment after a cross. This accounts for most of the differences observed between brothers and sisters in the same family. **1976** *Times Lit. Suppl.* 6 Aug. 985/2 Recombination was an idea that Darwin had lacked in his attempt to explain how natural selection and breeding were connected.

b. The formation by crossing-over of chromosomes that differ from both the chromosomes from which they derive. Also *attrib.*

1923 Bridges & Morgan *Third-Chromosome Group of Mutant Characters of Drosophila Melanogaster* i. 9 If Dichæte is crossed to pink, and the F_1 female is back-crossed to a pink male, most of the flies are of the two original types, Dichæte or pink; but a small number of the offspring are both Dichæte and pink or neither (i.e., wild-type). These two latter classes are called 'recombination classes' and the 'percentage of recombination' may be found... The use of the term 'recombination' in this technical sense is a shortening of the full term 'recombination of linked characters'. **1939** C. D. Darlington *Evolution of Genetic Systems* xiv. 77 This recombination we now see is more profound than Weismann imagined. It extends beyond the chromosomes to the genes. The number of units capable of recombination is not five or even fifty, but five thousand or fifty thousand. *Ibid.*, Taking the sum of the haploid number of chromosomes and of the average chiasma frequency of all the chromosomes in a meiotic cell as a recombination index. **1940** *Jrnl. Genetics* XL. 429 Let *x* be the recombination frequency. **1943** *Biol. Rev.* XVIII. 50 In one set of individuals AB and ab may be more favoured than Ab and aB, the reverse may be true elsewhere and, as recombination is the only means short of mutation of changing the arrangement, this inconstancy of advantage must favour some degree of recombination. **1955** *Jrnl. Gen. Microbiol.* XIII. 346 Genetic recombination has now been demonstrated amongst several viruses. **1965** *Proc. Nat. Acad. Sci.* LIII. 457 The term 'recombination' when used in the context of bacterial genetics connotes to many either the process of DNA transmission known as conjugation or the formation by conjugation of any progeny which inherit phenotypic traits derived from both parents. It can, however, be used more strictly to denote the series of physical and chemical events which serve to link genes derived from one parental DNA with those derived from another parental DNA. **1976** *National Observer* (U.S.) 5 July 6/1 The object of these guidelines is to ensure that experimental DNA recombination will have no ill effects on those engaged in the work, on the general public, or on the environment. **1977** A. W. F. Edwards *Foundations Math. Genetics* viii. 94 Linkage is not complete, and its magnitude is measured by the recombination fraction, *r*, between the two loci.

recombina·tional, *a.* *Genetics.* [f. prec. + -al.] Of or pertaining to recombination.

1959 *Nature* 14 Nov. 1593/1 (*heading*) Recombinational lethals in a polymorphic population. *Ibid.* 1594/1 A recessive lethal can be produced by recombinational instead of mutational change. **1969** *Genetics* LXI. 298 The problem of recipient culture variability during recombinational analyses. **1977** *Jrnl. Protozool.* XXIV. 27/2 The duplex forms could have arisen as a consequence of a prior mutational or recombinational event.

Hence **recombina·tionally** *adv.*, by or as a result of recombination.

1969 W. D. Stansfield *Theory & Probl. Genetics* ix. 184 Recombinationally separable forms of a gene within a cistron are referred to as heteroalleles.

recombina·tionless, *a.* *Genetics.* [f. as prec. + -less.] That does not show recombination.

1969 L. Levine *Biol. of Gene* v. 101/2 They were able to isolate two recombinationless (Rec) mutants in an F strain of *Escherichia coli* K12. **1974** *Nature* 4 Jan. 44/1 Recombinationless bacterial mutants are highly sensitive to near-ultraviolet, suggesting that DNA may be an indirect target of the action of near-ultraviolet.

recombine, *v.* **2.** (Earlier and later examples.)

1859 Mill *Liberty* ii. 85 With what a salutary shock did the paradoxes of Rousseau explode like bombshells in the midst..of onesided opinion..forcing its elements to recombine in a better form. **1910** W. M. Wheeler *Ants* viii. 131 These characters..are relatively stable in particular races or varieties and have a tendency to combine and recombine in endless permutation. **1942** J. D. Stranathan *'Particles' Mod. Physics* i. 27 Ions formed in a gas have a tendency to recombine. **1974** *Encycl. Brit. Macropædia* IX. 811/2 Nitrogen ions may recombine similarly.

recombinogenic (rīkǫ:mbinodʒe·nik), *a.* *Biol.* [f. Recombin(ation + -o + *-genic.] Tending to cause genetic recombination.

1965 *Genetics* LII. 107 (*heading*) The recombinogenic effect of thymidylate starvation in *Escherichia coli* merodiploids. **1971** *Nature* 12 Nov. 71/3 It is not unreasonable to assume that the protein plays an important part in some facet of meiosis, and in view of the T_4 evidence, neither is it unreasonable to suppose that it has some recombinogenic activity.

recommend, *sb.* For '*dial.* and *U.S.*' read '*colloq.* (orig. *U.S.*)' and add earlier and later examples.

1806 L. Dow *Travels* I. iv. 110 This morning, I went on shore, having no proper recommends with me. **1832** J. J. Strang *Diary* 19 Feb. in M. M. Quaife *Kingdom of St. James* (1930) 202 There is no complaint against me and they offer me a good recommend. **1892** B. Potter *Jrnl.* (1966) 227 Miss Emmet..wedged in a recommend of farm-house lodgings of her cousins. **1908** *Practitioner* Nov. 731 The Committee pays for out-patient 'recommends' at the rate of one guinea for six. **1924** J. Galsworthy *White Monkey* i. viii. 65 They'll give you a good recommend, won't they? **1967** [see *book-society* s.v. *book *sb.* 18]. **1977** *Listener* 30 June 867/4 William McIlvanney's *Laidlaw* comes with a recommend from Ross Macdonald.

recommend, *v.*[1] **7. c.** (Earlier example.)

1813 Jane Austen *Pride & Prej.* I. xviii. 217 Let me recommend you, however, as a friend, not to give implicit confidence to all his assertions.

recommendation. Add: **4. a.** Also, that which is recommended; a proposal or suggestion.

1911 G. B. Shaw *Doctor's Dilemma* 299 How this was effected may be gathered from the recommendations finally agreed on. **1929** *Star* 21 Aug. 19/1 It is interesting to record that some of our recommendations have duly improved in capital value. **1976** *Daily Tel.* 20 July 2/3 A report following a public enquiry into the disaster made a number of observations and recommendations.

recommended, *ppl. a.* Add: Also, advised, prescribed.

1968 *Globe & Mail* (Toronto) 3 Feb. 10/7 Sulphur dioxide levels in the smelter were 40 times the recommended safe level. **1977** *Sniffin' Glue* July 17 Recommended reading: Anything by Colin Wilson.

recommission, *v.* Add: **b.** *intr.* for *pass.* Of a ship. Hence **recommi·ssioning** *vbl. sb.*

[**1909** *Army & Navy Gaz.* 1 May 431/2 Fleet Surg. H. B. Marriott to Doris on recommiss.] **1922** *Daily Mail* 3 Nov. 12/5 Naval Appointments..to Emperor of India on recommissioning. **1928** *Observer* 15 July 12/4 It was intended that she should return home at the end of the present cruise to re-commission. **1977** *Navy News* June 15/1 The Arethusa recommissioned at Portsmouth in April.

recommittal. (Earlier example.)

1837 *Second Rep. Inspectors of Prisons* i. 90 in *Parl. Papers* XXXII. 1 Judging from the immense number of recommittals, it would almost seem that the effect produced by imprisonment..is not such as materially to deter from the commission of crime.

recompensive, *a.* For † *Obs. rare*⁻¹ read *rare* and add later example.

1924 *Brit. Weekly* 21 Aug. 443/2, I am glad to tell that I am having recompensive explorations here.

recomposition. Add: **b.** *Linguistics.* (See quots.)

[**1933** J. Marouzeau *Lexique Terminol. Linguistique* 158 *Recomposition*..Procédé par lequel on restitue à l'un des éléments d'un composé la forme qu'il avait à l'état autonome; ainsi quand on donne à lat. *recludo* la forme *reclaudo* d'après le simple *claudo*.] **1935** T. Hudson-Williams *Short Introd. Study Compar. Gram.* 8 *Recomposition* is a species of analogy; the form of a compound verb is affected by that of the simple verb; the simple form is restored or, occasionally, retained consciously in the compound; e.g. *sē+paro* should give..*sēpero*; but the literary dialect reformed it to *sēparo*... *Decomposition* is the opposite process, the simple verb being affected by the compound. **1964** A. Martinet *Elements of General Linguistics* iv. 126 An element like *tele-*..which today combines freely with monemes and syntagms that exist outside the combinations in question.. behaves in fact like an affix... Perhaps in the case where a new syntagm is formed we might speak of 'recomposition' from elements which are extracted by analysis. **1972** Hartmann & Stork *Dict. Lang. & Linguistics* 192/2 *Recomposition*, the process or result of using a borrowed element as an affix to form new words, e.g. *tele* in *telecast, teleview, teleprinter.*

recompress (rīkǫ̆mpre·s), *v.* [Re- 5 a.] *trans.* To compress again; to increase again the pressure of air or other gas in (a vessel) or acting on (a person); *esp.* to subject (an aircraft pilot, diver, etc.) to increased pressure again.

1945 *Jrnl. Gen. Physiol.* XXVIII. 220 The frogs..were recompressed within 30 seconds. **1950** *Ibid.* III. 255 The chamber was recompressed at free-fall rate. **1951** A. R. Behnke in J. F. Fulton et al. *Decompression Sickness* iii. 87 Should symptoms recur.., recompress the diver to a depth giving relief. **1967** P. D. Griffiths in R. I. McCallum *Decompression of Compressed Air Workers* 230 The patient must be observed constantly and recompressed at once should symptoms return, preferably to the minimum effective pressure. **1969** Haxton & Whyte in Bennett & Elliott *Physiol. & Med. of Diving* i. 12 It is sometimes necessary to recompress a patient to a pressure slightly higher than that at which he has been working.

recompression (rī:kǫ̆mpre·ʃən). [Re- 5 a.] **1. a.** The state of being compressed again. **b.** The action of compressing again, *esp.* by exposure to increasing air pressure (e.g. during descent from a high-altitude flight without pressurization or following decompression after a dive).

1939 *Jrnl. R. Aeronaut. Soc.* XLIII. 822 After the shock the speed again becomes subsonic, so that the consequent divergence of the streamlines on the back part of the upper surface generates a recompression. **1943** M. A. & F. A. Hitchcock tr. *Bert's Barometric Pressure* I. ii. iv. 501 M. Bucquoy mentions..the cure of muscular swellings by recompression. **1951** Ferris & Engel in J. F. Fulton et al. *Decompression Sickness* ii. 23 The pain of bends is immediately relieved by recompression of 3000- to 6000-foot equivalents and recurs upon reascent to the original altitude. **1969** A. R. Behnke in Bennett & Elliott *Physiol. & Med. of Diving* xi. 227 Hoppe-Seyler (1857) described blockage of pulmonary vessels by nascent bubbles and the inability of the heart to propel blood under these conditions. He proposed recompression to absorb the liberated gas and re-establish circulation. **1973** *Nature* 21–28 Dec. 523/1 These did not occur at maximum altitude, but during the re-compression. The other runs gave clear heart signals throughout. **1977** *Hongkong Standard* 14 Apr. 16/3 Mr Ng was later sent to Queen Mary Hospital, after the specialists were satisfied there was no need for further recompression.

2. recompression chamber or **lock,** a chamber in which a person can be subjected to an air pressure above that of the atmosphere.

1951 A. R. Behnke in J. F. Fulton et al. *Decompression Sickness* iii. 87 Individuals believed to be 'cured'..have been rushed to the recompression chamber in a state of collapse. **1967** P. D. Griffiths in R. I. McCallum *Decompression of Compressed Air Workers* 229 An important factor is that medical recompression locks often have a safe working pressure..of only 45–50 p.s.i.g. **1976** *Daily Tel.* 27 Jan. 15/1 A 14-year-old boy, critically ill with carbon monoxide poisoning, recovered consciousness yesterday in the Royal Navy's recompression chamber at Rosyth dockyard. **1977** *Proc. R. Soc. Med.* LXX. 503/2 This involves being able to carry out a detailed examination of the patient, and to put up infusions, give injections or insert catheters in the confined, noisy and pressurized space of a recompression chamber.

recon (rīkǫ·n), *v.*[2] *U.S. Mil. slang.* *trans.* and *intr.* Abbrev. of Reconnoitre *v.*

1966 *National Observer* (U.S.) 26 Dec. 1/4 We launched a small operation and while reconning the area, saw a bunch of color near a tree line. **1969** I. Kemp *Brit. G.I. in Vietnam* v. 96 Our orders are to recon only, and avoid all contact with the enemy whatsoever.

recon (rīkǫ·n), *sb.*[1] *U.S. Mil. slang.* Abbrev. of Reconnaissance; a reconnaissance unit. Freq. *attrib.*, as *recon company, unit* (etc.). Cf. *recce *sb.*

1918 E. M. Roberts *Flying Fighter* 337 *Long Recon*, a trip of from 20 to 80 miles behind the Hun lines to gather information. **1942** *Yank* 25 Nov. 21 He was temporarily with the recon. **1943** J. Goodell *They sent me to Iceland* ii. 31 Convoyed by jeeps and recon cars we sped through the town. **1946** *Sun* (Baltimore) 27 July 12/1 (*heading*) Ex-Patton recon unit to parade in Cumberland. **1948** N. Mailer *Naked & Dead* i. ii. 20 The men in recon looked small and lost in comparison to the other platoons. **1950** 'D. Divine' *King of Fassarai* xiv. 108 Should have the recon reports by now. Get through to Air Command again. **1968** *Globe & Mail* (Toronto) 13 Feb. 4/1 It is recon's bad luck to live in an area bordered by an ammunition dump, a flightline loading area and the 26th Marine Regiment's command post. *Ibid.* 4/3 The survivors of the recon company are frightened but uncowed. **1975** A. Price *Our Man in Camelot* v. 93 'He was a pilot in recon.'..'Photographic reconnaissance,' he explained. **1977** 'E. McBain' *Long Time no See* xiii. 208 Our recon patrol found an enemy base camp.

recon (rī·kǫn), *sb.*[2] *Biol.* [f. Rec(ombination + *-on[1].] A piece of genetic material which can be exchanged but not divided by genetic recombination; thus the shortest piece which can be so exchanged.

1957 S. Benzer in McElroy & Glass *Symposium on Chem. Basis of Heredity* 71 The unit of recombination will be defined as the smallest element in the one-dimensional array that is interchangeable..by genetic recombination. One such element will be referred to as a 'recon'. **1969** A. M. Campbell *Episomes* iii. 38 The 'unit factor' of the classical geneticist is replaceable by the muton, the recon, the cistron, or even a collection of linked cistrons, each in the appropriate operational context. **1978** N. Jardine in Hookway & Pettit *Action & Interpretation* 122 The operons, cistrons, recons and mutons of the molecular geneticist.

reconcentration. (Further examples.)

1956 *Nature* 21 Jan. 126/2 Kunkel and Tiselius were not able to demonstrate the heterogeneity of serum albumin by their method, and the ability of the method here described to do this may also be due to this reduction of diffusion by reconcentration. **1972** *Times* 26 June 12/3 Biological reconcentration of filter feeding organisms..invalidates the dilution hypothesis.

reconcile, *v.* Add: **I. 5. b.** (Further example.)

1840 *Act 3 & 4 Vict.* c. 52 § 6 If..Prince Albert shall..be reconciled to or shall hold Communion with the See or Church of Rome.

II. 10. b. *Accountancy.* To establish the consistency of (one account) *with* another, esp. by allowing for transactions made or begun but not yet fully recorded (as when a cheque has been issued but not yet presented for payment). Cf. *reconciliation 4 b.

1900 W. W. Snailum *Fifteen Studies in Book-Keeping* xi. 122 At the end of each financial period it will be necessary to 'reconcile' the bank account... This is effected by means of a 'reconciliation statement'. **1930** A. Palmer *Munro's Book-Keeping & Accountancy* (ed. 10) 26 The Bank Pass Book..would show a balance at the credit of £174, which would be reconciled as follows. **1947** [see *reconciliation 4 b]. **1970** R. W. Wallis *Accounting* v. 66 Reconciliations may also establish the accuracy of the different parts of the accounting system within an organization, for example by reconciling the debtors' control account in the ledger with the total of the individual accounts in the sales (debtors) ledger.

reconciliate, *v.* (Later example.)

1922 *Glasgow Herald* 26 Apr. 11/5 The question of reconciliating the two wings of the army.

reconciliation. Add: **4. b.** *Accountancy.* The action or practice of rendering one account consistent with another by balancing apparent discrepancies; *reconciliation statement*, a statement of account whereby such discrepancies are adjusted.

1895 J. Thornton *Man. Bookkeeping* xi. 187 See that all Banker's charges..are duly entered in your own books, or you will have difficulty with your Reconciliation. **1929** L. C. Cropper *Book-Keeping & Accounts* iv. 46 In order to explain this divergence it is necessary to construct a statement, known as a 'Reconciliation Statement'... A specimen example is appended showing how this 'reconciliation' is arrived at. **1947** F. H. Jones *Jordan's Mod. Book-Keeping* I. iii. 33 In order to reconcile the Cash Book balance with the statement of Account balance a Bank Reconciliation Statement is compiled. **1957** W. W. Bigg *Cost Accounts* xiii. 231 Assuming that it is desired to keep the Cost and Financial Accounts entirely distinct it is still imperative that they be rendered capable of reconciliation one with the other. **1973** A. & E. E. Fieldhouse *Elem. Book-Keeping* 80 If the balances of the two books should disagree a Reconciliation Statement should be made out.

recondi·tion, *v.* [Re- 5 a.] *trans.* **1.** To restore to a proper, habitable, or usable condition; to repair or rehabilitate.

1920 *Glasgow Herald* 29 Apr. 7/1 The Agamemnon..is being reconditioned at the Brooklyn Navy Yard. **1922** *Flight* XIV. 366/1 In the name of economy, the R.A.F. has had to be content with machines built during the War and 'reconditioned', or, at best, with designs got out during the War. **1930** S. Runciman *Hist. First Bulgarian Empire* ii. i. 53 A strong line of Imperial fortresses..had probably been reconditioned by Constantine Copronymus. **1935** *Punch* 27 Mar. 346/2, I see with shame that H.M. Gov. propose in a White Paper to 'recondition' the Navy. But I see with delight and surprise that for once the *Shorter O.E.D.* does not acknowledge the existence of the filthy verb. **1966** J. S. Cox *Illustr. Dict. Hairdressing* 125/2 *Recondition* (Hair),.. to restore hair by means of suitable substances and/or treatments to its normal condition.

2. *Forestry.* To reduce warping and collapse in (timber) by heating in a steam-filled atmosphere for several hours.

1932 *Rep. Forest Products Res. Board 1930* iv. 42 Experiments have been made during the year to ascertain the possibility of reconditioning or restoring collapsed Tasmanian oak. **1938** H. E. Desch *Timber* x. 127 Timber that is badly warped or cupped, without showing any visible signs of collapse, may also be successfully re-conditioned. **1979** J. G. Wilkinson *Industr. Timber Preservation* vii. 198 Conventional kilns can also be used to: 1. Recondition timber which has collapsed. 2. Destroy fungal growth and insect infestations.

3. *Psychol.* To alter the responses of (a person) by means of conditioning techniques; to replace (existing responses) in this manner.

1935 [implied in *reconditioning *vbl. sb.* 3]. **1942** *Sun* (Baltimore) 4 Nov. 9/5 The army has been consistently uninterested in taking the rejects into conditioning battalions and reconditioning them. **1957** W. Sargant *Battle for Mind* x. 220 It is very difficult indeed to condition or recondition such persons [*sc.* psychopaths], some of whom are criminals, until later in life when their brainwave patterns become much more normal. **1967** J. A. Hadfield *Introd. Psychotherapy* xiv. 88 The objection then to treatment by re-conditioning the symptoms alone is..that we are not dealing with the real illness.

Hence **recondi·tioned** *ppl. a.*

1932 Kipling *Limits & Renewals* 374 Our pernicious system of employing reconditioned souls on such delicate duties. **1933** *Sun* (Baltimore) 21 Apr. 19/3 (Advt.), Guaranteed reconditioned cars. **1944** Auden *For Time Being* (1945) 52, I moved the vices out of the city into a chain of re-conditioned lighthouses. **1957** L. F. R. Williams *State of Israel* ix. 155 Israel's first elected Parliament, which met in a reconditioned cinema. **1977** *Western Mail* (Cardiff) 5 Mar. 14/3 (Advt.), Massey Ferguson 35, 4-cylinder Tractor. Reconditioned engine.

recondi·tioning, *vbl. sb.* [f. prec. + -ing[1].] The action or process of the vb. **1.** Restoring to proper or adequate condition; rehabilitation, repair. Also, conversion or modernization (of houses, etc.).

1920 *Sphere* 27 Mar. 339 (*heading*) Reconditioning. A present striking phase of the great British shipping industry. *Ibid.* (*caption*) The word at the top of this page—'Reconditioning'—may be unfamiliar to the general public, but it is to-day well known to all the shipping world. **1926** *Manch. Guardian Weekly* Feb. 104/2 Measures are being considered for the improvement and reconditioning of existing rural cottages. **1936** *Discovery* Apr. 117/1 Systematic re-conditioning of working-class houses throughout the country. **1944** M. Laski *Love on Supertax* x. 95 You really ought to let me give your hair a thorough re-conditioning.

2. *Forestry.* The steaming of timber to reduce warping and collapse (see *recondition *v.* 2).

1932 *Rep. Forest Products Res. Board 1931* ii. 11 The treatment, which has been called re-conditioning, consisted essentially of warming the timber, which was first dried to a moisture content of 15 per cent.,..to 210 °F in saturated air. **1948** *New Biol.* IV. 89 The kiln load is given a stress-relieving or reconditioning treatment. **1979** J. G. Wilkinson *Industr. Timber Preservation* vii. 198 Reconditioning typically involves heating defective boards for between four and eight hours at 100 °C in a steam-filled atmosphere.

3. *Psychol.* The replacement through conditioning of one conditioned response by another; the re-establishing of a conditioned response after its extinction.

1935 J. E. Wallin *Personality Maladjustments* xi. 461 Such bonds must be loosened or dissipated by substituting other emotional bonds that are more potent by a process of emotional reconditioning. **1940** Hilgard & Marquis *Conditioning & Learning* 349/1 *Reconditioning*, the re-establishment of a conditioned response after it has been diminished by extinction or forgetfulness. **1957** W. Sargant *Battle for Mind* x. 221 The need to vary methods of conditioning and reconditioning according to the different temperaments is clearly shown by a study of the way prison sentences affect various types. **1967** J. A. Hadfield *Introd. Psychotherapy* xiv. 88 One could not deal with this guilt by re-conditioning because neither she nor we knew of its existence. **1972** J. W. Kling et al. *Woodworth & Schlosberg's Exper. Psychol.* xiv. 570/2 Reconditioning: if extinction trials are followed by a single presentation of the CS–US combination much or all of the effects of extinction will be overcome.

reco·njure, *v.* [Re- 5 a; cf. F. *reconjurer* (Cotgr.).] *trans.* To conjure again; to reconstruct in imagination; to recall.

1611 Cotgrave *Dict.*, *Reconjurer*, to reconiure, to coniure againe. **1904** *Edin. Rev.* Jan. 53 Nor can the antiquarian reconjure their image in the past from their ruins of today. **1915** C. Mackenzie *Guy & Pauline* i. 57 There was neither passion nor sentiment in the music..yet in solitude when Guy reconjured the sound afterward, it returned to his memory like fire.

reconnaissance. Add: **1.** (Further examples.)

1944 *Return to Attack* (Army Board, N.Z.) ii. 7 When he [*sc.* Rommel] attacked, Maryland bombers of the RAF caught his tanks... He withdrew, calling his attack a 'reconnaissance in force', a phrase which he was to use again.

attrib. **1950** 'D. Divine' *King of Fassarai* vii. 46 One of the big four-engined aeroplanes that came past them occasionally on reconnaissance flights. **1966** *Daily Tel.* 19 Apr. 32/5 A multi-million-pound order for a British EMI 'reconnaissance pod'. **1976** H. Tracy *Death in Reserve* xxi. 167 They should have a light aircraft going to take reconnaissance photographs.

reconnoitre, *v.* Add: Also *U.S.* **reconnoiter.**

2. a. (Later example.)

1948 N. Mailer *Naked & Dead* iii. v. 523 He could not reconnoiter the pass.

reconnoitrer (earlier example).

1860 C. M. Yonge *Hopes & Fears* III. xiii. 254 Tearfully she thanked the trusty reconnoitrer.

reconnoitring, *ppl. a.* Add: (Earlier example.)

1759 Ld. G. Sackville *Let.* 16 June in *Rep. MSS. Mrs. Stopford-Sackville* 307 in *Parl. Papers 1904* (Cd. 1892) XLVII. 1 Prince Ferdinand's and Marshal Contades' reconnoitering partys met this morning.

Hence **reconnoi·tringly** *adv.*, in a reconnoitring manner. *rare.*

1924 A. D. Sedgwick *Little French Girl* iv. iii. 324 Giles could almost see him nibbling reconnoitringly at the edge of the stained oak mantelpiece.

reconstitute, *v.* Add: **1.** *trans.* **a.** (Further examples.)

1944 J. Lees-Milne *Prophesying Peace* (1977) 68 Thurstan wants to pull down some Georgian dwellings of little intrinsic worth, and to reconstitute some medieval overhangs and fronts. **1977** *Church Times* 5 Aug. 16/3 Bishop Runcie said he thought that, if the Orthodox were dissatisfied with the reactions of the Conference, the dialogue might have to be reconstituted.

b. *spec.* to restore the previous constitution of (dehydrated food) by the addition of liquid.

1917 *Nat. Food Jrnl.* 24 Oct. 59/1 The total amount of dried milk (full cream) that may be available equals approximately 2,200 tons, and would represent nearly 4,000,000 gallons of whole milk when reconstituted. **1945** *ABC of Cookery* (Min. of Food) vii. 22 It is not always necessary to reconstitute dried eggs before use. **1951** L. Nicholls *Trop. Nutrition & Dietetics* (ed. 3) x. 202 Vegetables, fish or meat dried by this method are easily reconstituted with good retention of flavour and colour. **1960** *Times Rev. Industry* July 18/1 Fruits..stored..in the dried state and..reconstituted..by soaking in water and cooking. **1972** *Homes & Gardens* Aug. 98 Ice cream powders..can be reconstituted with water, or with milk and cream.

2. *intr.* To undergo or take part in reconstitution.

1974 *Sci. Amer.* June 77/3 Both sodium and potassium reconstitute into feldspars during metamorphism or igneous melting.

reco·nstituted, *ppl. a.* [f. Reconstitute *v.* + -ed[1].] That has been constituted or formed anew; applied *spec.* to food which has been dehydrated and subsequently made ready for use by adding liquid. *reconstituted stone* = *reconstructed stone* s.v. *reconstructed *ppl. a.* a.

1850 H. Martineau *Hist. Peace* II. iv. xii. 157 The first act of the reconstituted government was to carry a new Coercion Bill. **1925** *Chem. Abstr.* XIX. 545 Acidity of the reconstituted milk was a little lower than that of raw milk. **1928** *Daily Mail* 30 July 13/3 Reconstituted and synthetic cream. **1946** *Daily Tel.* 30 Jan. 2/1 Dark fruit cake... 6 oz fruit, 2 reconstituted eggs, [etc.]. **1951** *Good Housek. Home Encycl.* 348/2 Reconstituted dried egg may be used in making batter. **1966** D. Francis *Flying Finish* vi. 77 The floors were some sort of reconstituted stone heated from underneath. **1969** *Guardian* 20 Sept. 9/7 Mother and father eating fish fingers and reconstituted potatoes in gloomy silence. **1976** A. Davis *Television* ii. 32 The BBC converter ..works by storing the signals relating to each line of the picture and releasing them at the right moment to take their place in a reconstituted picture at the new standard. **1977** A. Hallam *Planet Earth* 159/3 Reconstituted stone, where the natural material is crushed and recast in blocks simulating the original in color and texture, with the advantage that the blocks are of uniform size.

reconstitution. Add: **1. a.** (Earlier example.)

1848 E. A. Poe *Eureka* 141 The regathering of..Matter and Spirit will be but the re-constitution of..God.

2. The restoration of dehydrated food to its original constitution by the addition of liquid.

1920 *Nat. Food Jrnl.* II. 595/1 The Clauses..prohibiting the addition to milk of colouring matter or water, the reconstitution of milk [etc.]..will remain in force. **1945** *ABC of Cookery* (Min. of Food) xvii. 59 If dried eggs are used they may be added dry with the other dry ingredients and the water needed for reconstitution added with the mixing liquid.

reconstruct, *v.* Add: **2.** (Further examples.) Also *spec.* in *Philol.*, to reform hypothetically (part of) a protolanguage by deduction from (later) recorded languages.

1917 *Science Progr.* XI. 682 Not only is the past retrieved in fragments; in some museums and exhibitions, and to a certain extent in historical plays, it is actually reconstructed. **1930** *Language* VI. 164 It is a well-known fact that it is impossible to reconstruct a complete paradigm of Indo-European personal pronouns. **1965** *Ibid.* XLI. 19 The younger protolanguages which we can reconstruct within the Indo-European family..cannot be placed in an identical frame of reference with Proto-Indo-European. **1976** E. Maclaren *Nature of Belief* ii. 11 We can reconstruct how the process must have gone on.

Hence **reconstru·ctable, -ible** *adjs.*, capable of being reconstructed.

1961 Webster, Reconstructible. **1965** *Language* XLI. 19 [Proto-Indo-European] is reconstructible only on the basis of internal evidence. **1978** *Ibid.* LIV. 285 Morphology is harder to recapture, and syntax is even more slippery—if indeed it is reconstructable at all beyond a rather shallow level. **1978** *Nature* 13 Apr. 605/2 By using chronologies recording different elements of climate a greater range of climatic variables will be reconstructable.

reconstructed, *ppl. a.* Add: **a.** (Further examples.) *reconstructed stone* (see quot. 1950).

1909 W. G. Renwick *Marble & Marble Working* xv. 175 Reconstructed Sicilian marble was selected for lining the walls of the operating-rooms at the Manchester Royal Infirmary, opened last year. *Ibid.* 176 (*caption*) The first building to be erected in Reconstructed Portland Stone.

1933 Bloomfield *Language* 302 Students of the Romance languages reconstruct a Primitive Romance ('Vulgar Latin') form before they turn to the written records of Latin, and they interpret these records in the light of the reconstructed form. 1935 *Specification* XXXVII. 245/1 Reconstructed stone is natural stone—reconstructed, and is to be distinguished from artificial stone, which may be described as high-grade concrete. 1950 *Ibid.* LII. 343/1 Reconstructed stone is natural stone crushed and moulded into the required shape after it has been formed into a plastic mass by the addition of cement and water. 1951 E. E. Evans-Pritchard *Social Anthrop.* iii. 43 The reaction against the attempt to explain social institutions by their reconstructed past..came at the end of last century. 1957 *Gloss. Terms Stone in Building* (*B.S.I.*) 30 *Reconstructed stone*, a building material manufactured from cement and natural aggregate for use in a manner similar to and for the same purpose as natural building stone. 1959 *Language* XXXV. 425 We..agree on the essential artificiality of Reconstructed Proto-Indo-European.

b. *U.S.* Converted from (a form of) Communism.

1966 *New Statesman* 14 Oct. 549/1 (Vietnam) As reconstructed peasants sleep Upon their AID-assisted beds. [1973 R. Hayes *Hungarian Game* viii. 63 All 10 seem to be unreconstructed Stalinists, somewhat to the fanatical left of both Rákosi and Gerő.]

reconstruction. Add: **1. b.** Usu. with capital initial. Also, the period during which this process occurred. (Further examples.)

A fuller treatment of this sense (also in derivative forms) may be found in M. M. Mathews *Dict. Americanisms* (1951).

1890 C. L. Norton *Political Americanisms* 93 *Reconstruction*. After the Civil War the question of restoring the lately seceded States..became the leading civil problem of the time. The measures introduced into Congress were popularly known as Reconstruction Bills. 1949 D. S. Freeman in B. A. Botkin *Treas. S. Folklore* p. x, The existing general pattern of Southern folklore probably was set in late 'slave days' and during the Reconstruction. 1967 *Freedomways* VII. 133 In history the horrors of slavery are watered down and sketchily covered so as not to enrage the complacent black student, while the period following Reconstruction is covered as if the Negro had strangely disappeared from the face of the earth. 1978 *Names* Mar. 106 In much of the South it [sc. the township] is an artifact of Reconstruction governments after The War.

c. The rebuilding of an area devastated by war. Also, the restoration of economic stability to such an area.

1925 A. J. Toynbee *Survey of Internat. Affairs 1920–23* I. 39 Austrian reconstruction (scheme drafted; protocols signed). *Ibid.* 40 Hungarian reconstruction (collaboration decided on). 1933 *Radio Times* 14 Apr. 75/1 Several grandiloquently named new departments of State, such as the Ministry of Supply and the extremely short-lived Ministry of Reconstruction. 1940 *Economist* 31 Aug. 280/1 Those who did not find work before August 1st were assembled in reconstruction camps. 1946 *R.A.F. Jrnl.* May 172 All the women were free to return to their interrupted training, to resume their pre-war occupation or to help with the gigantic task of reconstruction. 1953 P. C. Berg *Dict. New Words* 135/1 *Reconstruction area*, an area which has to be redeveloped as a whole, on account of war damage. 1974 tr. *Sniečkus's Soviet Lithuania* 51 As soon as the nazis had been expelled, the working people plunged into the task of reconstruction.

2. (Later examples.)

1930 *Language* VI. 185 The scarcity of material for dual forms to substantiate the I[ndo-]E[uropean] reconstruction is not surprising. 1937 'M. Innes' *Hamlet, Revenge!* II. vi. 166 You must do that reconstruction all over again. 1959 *Language* XXXV. 423 No reputable linguist pretends that Proto-Indo-European reconstructions represent a reality. 1977 M. Goulder in J. Hick *Myth of God Incarnate* iv. 65 As with the account I have given of Jesus, we are forming a reconstruction of history, and such reconstructions can never be more than probable.

Hence **reconstru·ctional** *a.*, of or pertaining to reconstruction; **reconstructionary**; **reconstructionist**, (*b*) one who reconstructs the past mentally (see Reconstruct *v.* 2); also *attrib.* or as *adj.*

1864 M. H. Throop *Future* 119 The act of secession extinguished the Union party as soon as it was adopted, except as a *reconstructionist* party. 1920 G. S. Gordon *Let.* 21 Jan. (1943) 133 Now—in this reconstructional mood—Professors are no longer Scholars or Professors, but 'Heads of Depts.'. 1925 E. F. Norton *Fight for Everest: 1924* vi. 131 These were undoubted signs of reconstructional work. 1949 Wellek & Warren *Theory of Lit.* iv. 32 We must, these literary reconstructionists argue, enter into the mind and attitudes of past periods. 1958 *Archivum Linguisticum* X. 1. 8 The..prism of literature..may twist the actual sequence of events, as the reconstructionist is tempted to envision it. 1966 M. Scriven in W. H. Dray *Philos. Analysis & Hist.* 255 The difference between this analysis and the reconstructionist approach. 1978 *Language* LIV. 470 The reconstructional changes he suggested have withstood the test of time. 1979 *Jrnl. R. Soc. Arts* Apr. 262/2 Rationality, objectivity, universalism and abstract analysis are features of the reconstructionist approach.

reconstructive, *sb.* Add: (Further example.) *rare.*

1927 *Daily Express* 24 Nov. 5 (Advt.), The prescribed reconstructive which creates the good red blood in which no germ can obtain a footing.

reconte·st, *v.* [Re- 5 a; cf. F. *recontester* (Cotgr.).] *trans.* To contest again.

1611 Cotgrave *Dict.*, *Recontester*, to recontest; make new protestation of, or complaint vnto. 1897 W. J. Locke *Derelicts* xviii. 239 The old man vaunting the ancients and Joyce defending the moderns, until a veritable Battle of the Books was recontested. 1922 *Glasgow Herald* 11 Nov. 9 Both are recontesting the seats they have occupied in Parliament.

reconvene, *v.* Add: (Later examples.) Hence **reconve·ning** *vbl. sb.*, a renewed convening.

1903 *Westm. Gaz.* 20 June 7/1 A prominent delegate..urges the reconvening of the Joint High Commission. 1906 *Ibid.* 11 Apr. 5/1 The anthracite operators..made a counter-proposal—namely, to re-convene the Coal Commission. 1972 *Maclean's Mag.* Sept. 10/2 The Select Committee..which he reconvened. 1977 *Time* 17 Jan. 12/2 To break the deadlock when and if the talks reconvene..Richard last week made firm Britain's readiness to accept a major role in the transition period from white to black rule.

reconversion. Add: **1. b.** (Further examples.) *spec.* alteration (of industry, etc.) to peacetime requirements after war. In recent use, conversion by adaptation of function, modernization; also, an object so converted.

1944 *Sun* (Baltimore) 21 Jan. 7/2 'Reconversion' in the foreign field.., like the reconversion of domestic industry from war to peace production, is one of our major problems. 1946 *News Chron.* 2 Mar. 2/2 The difficulties of demobilisation and of industrial re-conversion. 1956 *Planning* XXII. 239 Reconversion in industry is an essential consequence of the introduction of the common market... Reconversion as defined in the [Spaak] report means modernisation or rationalisation of production methods to cope with increased competition. 1976 *National Observer* (U.S.) 28 Aug. 14-B/4 (Advt.), He covers reconversions as well as new buildings.

reconvey, *v.* **2.** (Earlier *absol.* example.)

1838 W. Bell *Dict. Law Scotl.* 67 When he [sc. the assignee] reconveys to the cedent, it [sc. the deed] is called a *retrocession*.

reconviction. (Further examples.)

1909 *Rep. Commissioners of Prisons* I. 15 in *Parl. Papers* (Cd. 4847) XLV. 133 The highest proportion of re-convictions is in this class no less than 40 per cent. 1968 *Economist* 3 Feb. 16/2 Reconvictions bring those under suspended sentences into prison after all. 1976 *Howard Jrnl.* XV. 1. 7 Striking variations in reconviction rates.

record, *sb.* Add: **I. 1. a.** (Further example.)

1870 'Mark Twain' in *Galaxy* Oct. 575/1 That verdict is of record, and holds good to this day.

2. *on record* (further examples). Also, *to go on record*: to give oneself a place on a formal record, to be recorded (*as* favouring a given course of action, etc.); to express one's opinion. Also, *to be on record, to put* (oneself, etc.) *on record*. orig. *U.S.*

1900 *Congress. Rec.* 11 Jan. 785/1, I would be perfectly contented if Senators would put their vote on record. 1920 H. G. Wells *Outl. Hist.* 169/1 Greece had suddenly begun to produce literature, and put itself upon record as no other nation had ever done hitherto. 1930 E. M. Brent-Dyer *Chalet School & Jo* v. 64 It is on record that seventeen people had to go and remake theirs [sc. beds]. 1930 *Daily Express* 8 Sept. 10/4 President Parsons of Woolworths has also gone on record with the statement that [etc.]. 1940 *Publishers' Weekly* 1 May 3007/1 The Association has previously been on record against legislation of this kind. 1967 *N.Y. Times* (Internat. Ed.) 11–12 Feb. 1/1 West German Vice Chancellor Willy Brandt put the Bonn Government on record today as being willing to bar any nuclear explosions. 1975 J. Grady *Shadow of Condor* vii. 116 I've hardly heard anything of what's being done..and I would like to go on record right now to that effect. 1978 *Lancashire Life* July 43/2 It is on record that soldiers from the 6th Manchester Regiment in Egypt provided enough eager players to hold a Lancs. v Cheshire match shortly before going on to Gallipoli.

II. 5. d. (Further examples.)

1887 M. Shearman *Athletics & Football* v. 143 T. G. Little and J. H. T. Roupell..tied at 5ft. 9in., a height which remained the 'record' for the next five years. 1924 C. Mackenzie *Heavenly Ladder* xxiii. 286 The various 'records' we've broken..were not of our own seeking. 1955 N. & R. McWhirter (*title*) Guinness book of records. 1975 *Oxf. Compan. Sports & Games* 734/1 Thin air, a handicap in the endurance events, contributed to the shattering of world records in 11 track and field events with a premium on explosive effort.

e. (i) A disc or, formerly, a cylinder from which recorded sound or television pictures can be reproduced. Occas. also, a recording made on magnetic tape.

1878 *Design & Work* 19 Jan. 72/2 The vibrations resulting when a voice..utters certain words or other sounds, instead of being caused to transmit corresponding vibrations to a distance, are caused to produce a material record... The record can also be multiplied precisely as a photographic portrait can be multiplied. 1878 *Cassell's Family Mag.* June 443/1 Mr. Edison is now engaged in devising a finished instrument capable of storing up speeches and music of all kinds, and of allowing the records to be sent by post. 1892 W. Gillet *Phonograph* ii. 11 Previous to taking a record the cylinder has a sheet of tinfoil carefully wrapped round it. *Ibid.*, The record being finished, to reproduce it we have but to bring the cylinder back to the point of starting, and again rotate it. 1897 *Sears, Roebuck Catal.* 485/2 One graphophone talking machine... 12 Musical and Talking Records, your own selection. 1919 H. L. Wilson *Ma Pettengill* i. 20 With a..hired help to bring him his breakfast in bed and put on another record and minister to his lightest whim. 1919 Wodehouse *Damsel in Distress* xxiii. 278 I've been dancing to your music for years! I've got about fifty of your records on the Victrola at home. 1949 Frayne & Wolfe *Elem. Sound Recording* xxix. 601 The making of very high-quality magnetic records has become possible commercially. 1966 *Listener* 3 Nov. 646/1 Musicians' Union..objects to any new broadcast popular music programmes relying mainly on records. 1975 G. J. King *Audio Handbk.* vii. 154 Although there are still a few mono releases, the basic record is cut for two-channel stereo. 1976 *National Observer* 13 Nov. 8/2 Video-disc players, which attach to your television and broadcast shows from records, are being developed by several major manufacturers. 1978 *Gramophone* June 136/3 The programme is recorded digitally using the well proved PCM (pulse code modulation) encoding system... Philips are forecasting that players and disc records will be available in the early 1980s.

(ii) *fig.* (See quots.)

1926 Maines & Grant *Wise-Crack Dict.* 12/2 *Put on a new record*, change the subject. 1976 W. Goldman *Magic* III. xii. 207 'There's something crazy here..and I care.' 'Oh Jesus, I'm sick of that record.' 1977 Partridge *Dict. Catch Phr.* 177/2 *Put another record on!* and *change the record!*.. Addressed to..anyone..'going on about something'.

f. An account of a person's conduct in a particular sphere, preserved for reference; *spec.* a record (or history) of criminal convictions or prison sentences. orig. *U.S.*

1901 *Land of Sunshine* Apr. 234 In that crowded hall were many men with 'records'. a 1911 [see *pedigree sb. 2 d]. 1918 A. Woods *Crime Prevention* vii. 87 The convict has a hard enough row to hoe when he gets out to overcome his record, even if he is mechanically capable. 1938 [see *arm sb.[1] 2 b]. 1952 M. Allingham *Tiger in Smoke* x. 168 Not one of you has got a real record..and you don't want to spoil it, eh? 1954 *Manch. Guardian Weekly* 18 Mar. 3 McCarthy had described her as a 'code clerk' whose 'Communist record' was known to the country. 1969 [see *moderate sb. a]. 1973 'E. Ferrars' *Foot in Grave* viii. 144 'You didn't know..that he'd got a record.' 'A police record?' 'Yes.'

g. In various phrases: *off the record* (orig. *U.S.*): unofficially, confidentially; also as adj. phr.; also, *for, on the record*, for the sake of having the facts recorded or known; also, *to put* (*set*, etc.) *the record straight*: to achieve a proper record of the facts; to correct a misapprehension.

1933 H. L. Ickes *Secret Diary* 24 Mar. (1953) I. 9 He met and answered every question, although in some instances his answers were off the record. 1935 *Time* 2 Sept. 16/2 Only a very few Canadian tycoons took a calmer off-the-record view. 1939 *Time* 16 Oct. 101/1 By such slightly off-the-record stunts as burglarizing the plane factory..the Major sleuths out a sabotage gang. 1943 M. Asquith *Off Record* 10 If it is an Americanism, all the better. They are our allies, and if no one else understands the meaning of 'Off the Record', they will. 1949 *Manch. Guardian Weekly* 31 Mar. 2/3 Professor Schuman, who is a tidy-minded man, wanted to keep the record straight. 1949 R. Chandler *Little Sister* xxxi. 226 Off the record—we were always sure. We just didn't have a thing on him. 1951 *N.Y. Times Book Rev.* 22 Apr. 35/4 (*heading*) For the record. 1953 A. C. Clarke *Prelude to Space* viii. 43, I thought you might like to come along. For the record, you can be one of our legal advisers. 1965 A. Nicol *Truly Married Woman* 34 Look here, Olu, do you chaps hate us?.. Of course, this is all off the record. We are speaking as man to man. 1967 N. Freeling *Strike out where not Applicable* 30, I said we'd get the gendarmerie to look at things, just to get the record straight, what? 1971 *Daily Tel.* 24 Apr. 9/6 Our Saxon forebears..[regarded] all smithy conversations as off-the-record and therefore not slanderous at law. 1972 *N.Y. Times* 3 Nov. 35/1 Sir Rudolf is a spirited and independent man who feels he is obligated to put the record straight. 1973 D. Westheimer *Going Public* i. 15, I wouldn't..advise you to lower your standards. I'll say for the record I'm not advising that. 1973 *Times* 17 Dec. 14/4 An unattributable criticism of the oil companies by one minister was followed next day by an on-the-record reversal. 1976 J. Snow *Cricket Rebel* 76, I was impressed by the fact that he had bothered to get in touch with me to put the record straight. 1977 *Oxford Consumer* Mar. 5/1 Mr. Shergold made further investigations and sent us the following letter which should help to set the record straight. 1978 *Church Times* 1 Sept. 10 I'd be surprised if they didn't get down to some pretty forthright talking about women priests—not necessarily on the record, but among themselves. 1978 R. V. Jones *Most Secret War* iv. 41 Our discussion, which he had assured me was 'off the record', was reported back to the Air Ministry.

h. *pl.* Used *absol.* and *attrib.* (with capital initial) to designate a place where official records are kept; *spec.* a criminal records office or department (cf. sense *5 f).

1934 *Discovery* Nov. 319/2 The buildings under observation were the Great Temple, the Records Office and the Police Barracks. 1937 M. Allingham *Dancers in Mourning* xxvi. 314 Yeo had become a new man since the message from the [Police] Records Department. 1958 'J. Byrom' *Or be he Dead* iii. 52 'I'll send you a copy..so that you can bring your files up to date.'.. 'I'm sure Records will be much indebted to you.' 1973 'C. Aird' *His Burial Too* iv. 38, I did a person check with Records before I left the Station... Criminal Records Office have no knowledge of him.

i. *Computers.* A number of related items of information which are handled as a unit.

1957 *Proc. Western Joint Computer Conf.* 215/1 A record might be defined as all of the individual items of information (or words) about a given file unit. 1963 *Communications* (Assoc. Computing Machinery) VI. 267/1 Suppose each record of an input tape contains up to 50 fields.., some of which may be missing in any given input record. 1964 T. W. McRae *Impact of Computers on Accounting* vi. 189 A tape reel holds a certain number of records just as a ledger holds

a certain number of accounts. 1966 C. J. Sippl *Computer Dict. & Handbk.* 265/1 The most basic subdivision of a record is called a field. 1970 O. Dopping *Computers & Data Processing* i. 14 We know that the first word in this record is always the last name of the subscriber, the second one is his first name, etc. 1973 C. W. Gear *Introd. Computer Sci.* i. 13 To avoid confusion we will refer to the set of information on each card (name and phone number) as a record.

IV. 13. a. (sense 5 a) *record-keeping*; (sense 5 d) *record-breaker, -breaking* (further examples), *-holder, -setting, -smasher, -smashing*; (sense *5 e) *record-buying, -collecting, collector*.

1894 *Rep. Vermont Board Agric.* XIV. 93 Raising trotters for sporting men..has been done with the one idea of producing a record breaker, regardless of every other qualification. 1929 A. Huxley *Do what you Will* 145 Modern record-breakers have been ready to undergo..hardships for the sake of money. 1976 *Liverpool Echo* 7 Dec. 7/3 A Birkenhead company is making a name for itself as record breakers in the shipping world. 1929 A. Huxley *Do what you Will* 147 The record-breaking was to have a numerous audience. 1937 C. Isherwood *Sally Bowles* 66 Huge contracts for Sally, record-breaking sales for the novels I should one day write. 1950 *Sport* 22–28 Sept. 20/4 Not satisfied with his record-breaking average of 23 m.p.h., Joy rode again last weekend in vain hope of lowering his hundred mile time. 1949 L. Feather *Inside Be-Bop* i. 5 Lester's introduction to the record-buying jazz public. 1970 *Guardian* 7 Aug. 8/2 The two orchestras..are..little known..by the record-buying public. 1956 M. Stearns *Story of Jazz* (1957) xvii. 216 To the record-collecting converts known as 'moldy figs', however, this was not 'authentic' jazz. 1932 *New Yorker* 14 May 57/1 Mme. Ljungberg, hitherto accessible to record-collectors principally in opera albums, introduces herself as a singer of Strauss. 1946 *Penguin Music Mag.* Dec. 91 The record collector, who has built up his library.., really gets to know the music to which he listens. 1934 *Discovery* Dec. 352/1 The record holders of the fastest time between England and Australia. 1963 *Times* 30 Jan. 4/3 That graceful and intelligent runner, Johnson, is still the United Kingdom recordholder over 800 metres. 1977 *Whitaker's Almanack 1978* 584/2 Terry Paine,..the Football League's appearance record-holder, played his..final League match against Southampton. 1965 D. E. C. Eversley in Eversley & Glass *Population in Hist.* I. 34 Where central legislation..enforces record-keeping, the change tends to be abrupt. 1977 J. M. Johnson in Douglas & Johnson *Existential Sociol.* viii. 246 Those typifications of the daily work of intake workers that did not result in an official 'case' for record-keeping purposes. 1969 *Jane's Freight Containers 1968–69* 112/2 Time required to unload a trailer has been reduced from 10 minutes to a record-setting 90 seconds. 1972 J. Mosedale *Football* ii. 21 The Eagles won 14–0, a record-setting two playoff shutouts in a row. 1928 *Daily Sketch* 7 Aug. 22/3 Arne Borg, the record-smasher at all distances in the swimming world, was at work again yesterday. 1889 *Puck* (N.Y.) 7 Aug. 399/2 We will soon have as many record-smashing ocean-steamers as we now have champion pugilists.

b. (sense 5 a) *record book, card, office* (further example); (sense *5 e) *record cabinet, case, company, deck* (*deck *sb.*[1] 3 f), *groove, label* (*label *sb.*[1] 7 c), *library, needle, rack, shop, storage, store.*

1961 *Evening Standard* 6 Feb. 22/3 A further step toward deathless record-book fame in its dreariest form. 1976 *0–10 Cricket Scene* (Austral.) 5/1 That was the start of the climb..which was to re-write the record books and lift Australia to the peak of performance. 1967 H. Pinter *Basement* 70 *Law* goes to the record cabinet. He examines record after record. 1934 *Burlington Mag.* Sept. 142/1 Prolonged study of the data which record-cards may provide. 1960 M. Spark *Bachelors* i. 6 The specialist himself would possibly remember only the gist, and then only with the aid of his record cards. 1977 *News of World* 17 Apr. 7/2 Details..noted down by teachers on their pupils' record cards. 1908 *Sears, Roebuck Catal.* 201/1 Disc Record Cases..made from wood covered with black seal grain imitation leather... No. 1 holds 50 7-inch disc records. 1949 D. Smith *I capture Castle* xiii. 245 A wireless and a gramophone combined..[and] a record case to match. 1938 D. Baker *Young Man with Horn* iv. vii. 278 The record company went broke before any of their records were issued. 1978 *Lancashire Life* Apr. 29/3 Life after death is a tall order, but this month a record company comes near to meeting it by supplying the next best thing. 1976 A. Hope *Hi-Fi Handbk.* 116 If you intend giving your record deck a fairly hard working life [etc.]. 1946 E. Hodgins *Mr. Blandings* i. 9 He lowered the rusty tone arm, complete with needle, on to the record groove. 1976 M. Maguire *Scratchproof* xi. 172 'The record labels too, I suppose.' 'Steaming off and switching?' 1977 *Listener* 25 Aug. 247/2 Elvis had been doing it..already, on a little Tennessee record label. 1961 *Times-Picayune* (New Orleans) 19 June III. 11/3 A stereo series called 'Adventures in Music', which is an instructional record library for elementary schools. 1974 E. Ambler *Dr. Frigo* I. 50 On shelving built along the inner wall was..hi-fi equipment and a record library. 1918–19 *T. Eaton & Co. Catal.* Fall & Winter 369/2 Record Needles, 100 for 15c. 1904 G. B. Shaw *Let.* 23 Nov. (1972) II. 467, I was startled to hear that Edith Livia was getting..twenty-five shillings a week for work at the Record Office. 1973 A. Roy *Sable Night* ii. 22 The bookcase and record-rack yielded nothing. 1975 R. Butler *Where All Girls are Sweeter* ii. 17, I looked through his record racks... There was a nice mixture from Bach to Vivaldi. 1960 *Twentieth Cent.* Apr. 341 Record-shops..have mushroomed all over Britain in the last five years. 1975 *Guardian* 22 Jan. 14/5 The top 40 singles are based on local record shop returns. 1939–40 *Army & Navy Stores Catal.* 838/2 Radiogramophone..with five record albums and..extra record storage. 1974 *Times* 8 Apr. 13/3 A record storage cabinet in a Queen Anne style. 1949 *Billboard* 2 Apr. 34 Albums listed are those classical and semi-classical albums selling best in the nation's retail record stores. 1977 D. Westlake *Nobody's Perfect* (1978) 107 A branch of a major department store..the record stores, the shoe stores, the ladies' clothing stores.

c. (Further examples.)

1912 Chesterton *Manalive* II. i. 194 Smith was one of the University's record men for shooting. 1922 *Brit. Med. Jrnl.* 2 Sept. 412/1 During 1919,..3,420 new students were registered, being 1,105 greater than the 'record' entry of 1891. 1937 *Discovery* Sept. 264 His 'dash'..was accomplished in record time. 1948 'J. Tey' *Franchise Affair* viii. 86 If you could have seen your face when I introduced you to her... It cured me of her in record time. 1960 *Farmer & Stockbreeder* 16 Feb. 83/1, 27,000 gn record-priced bull. 1978 *Lancashire Life* Apr. 69/2 Visitors numbered a record 114,000 in 1977, and it is hoped that this trend will continue.

14. Special combinations. **record album**, † (*a*) a holder for gramophone records, *obs.*; (*b*) = *album[1] 6; **record changer**, a device for automatically placing another record on the turntable of a gramophone when the preceding record has ended; also **record-changing** *a.*; **record club**, a society which enables members to purchase selected gramophone records at reduced prices; **record hop** *slang*, a dance at which the music is provided by gramophone records; a place where such entertainment is held; **record jockey** *U.S. slang* (immediately supplanted by *disc jockey*: see *disc 8 f); **record linkage**, the process of combining items of information or sets of data relating to the same subject; **record player**, orig. a turntable and pick-up unit designed to be plugged into and played through a radio; now usu. a gramophone; **record sleeve**, a stiff envelope in which a gramophone record may be stored (cf. *sleeve *sb.* 7); **record token**, a voucher exchangeable in a shop for a gramophone record or records; **record type**, a type-face including special sorts reproducing the contractions or particular letter forms found in mediæval manuscripts.

1925–6 *T. Eaton & Co. Catal.* Fall & Winter 391/1 These Record Albums are made with strong cardboard covers... Each album will hold 12 records. 1945 *Billboard* 24 Mar. 18/2 (*heading*) Best-selling record albums by classical artists. 1955 Keepnews & Grauer *Pictorial Hist. Jazz* viii. 87/2 When people got around to..dissecting it [*sc.* a variety of jazz] in books and record-album notes,..it became known as 'Chicago style'. 1931 *N.Y. Times* 12 July 4/8 The Capehart 400 series..has several novel features... The record changer is equipped with a special constant-speed electric motor which operates both the turn-table and the record-shifting mechanism. 1947 *Gramophone* Nov. 88/1 The amplifier and record changer are contained in a small chair-side consol. 1977 *Times* 18 Apr. (Gramophone Suppl.) p. i/2 A radio-gramophone, with automatic record changer, was installed in our living room. 1931 *Wireless World* 23 Sept. 349/3 H.M.V.'s ace set is the model 531, being a nine-valve superheterodyne with automatic record-changing radio-gramophone. 1943 *Gramophone* Dec. 107/1 The spindle and turntable move left and down, which leaves the area free for the first record to drop upon the 'floor' of the record-changing unit. 1958 *Manch. Guardian* 21 Jan. 6/6 If anything the record club is likely to spread an interest in records of serious music rather than reduce the profits of the big companies. 1961 G. Smith *Business of Loving* xi. 229 We started as a record club... The first discs..were fifteen shillings. 1960 *Punch* 9 Mar. 345/1 Akin to the juke-box joints are the record-hops. 1966 *Wall St. Jrnl.* 25 July 1 'Record hops'..are dances often organized by a disc jockey and plugged by him over the air, as a means of supplementing his income. 1940 *Variety* 3 Apr. 39/3 [Quoting J. Kapp] The name bands are come on for the record jockeys who ride herd over not only Decca records but all the others. 1946 H. L. Dunn in *Amer. Jrnl. Public Health* XXXVI. 1412/2 In the process of record linkage the uniting of the fact-of-death with the fact-of-birth has been given a special name, 'death clearance'. 1959 *Science* 16 Oct. 954/1 The term record linkage has been used to indicate the bringing together of two or more separately recorded pieces of information concerning a particular individual or family. 1968 *Brit. Med. Bull.* XXIV. 208/2 If birth and marriage records are included in the system, it is possible to use record linkage to assemble sibships, parent-offspring groups and eventually pedigrees. 1934 *Wireless World* 5 Jan. 10/3 (*caption*) The Collaro record player incorporated in a radio-gramophone cabinet. 1939 *New Regal-Zono Records* Feb. 4 (Advt.), The thousands already sold of the Columbia electric record-player prove conclusively how many fully appreciate the facility of playing their records through their radio sets. 1958 *Sunday Times* 3 Aug. 3/6 Now that the station has bought a new record-player and gets all its music in prerecorded tapes, the personal touch is all but gone. 1960 *Practical Wireless* XXXVI. 377 (Advt.), Turns any gramophone into a first-class tape-recorder and back into a record-player in a moment. 1973 M. Amis *Rachel Papers* 68 You mean to tell me that it's only made in stereo?..What about the people who don't own stereo record-players? 1977 *Gramophone* July 241/1 The humble record player of earlier years could still be found in a few places. 1954 *Gramophone Record Rev.* July 415 On the record sleeve, the user is extolled to play the record at full room volume. 1963 L. Deighton *Horse under Water* xviii. 74 The brightly coloured record sleeves that are the folk art of the new world. 1978 P. Porter *Cost of Seriousness* 47 An old woman, So the record sleeve denotes, Is singing of death In a young world. 1958 M. Kelly *Christmas Egg* II. 76 The envelope that had held his brother's record token. 1977 *Radio Times* 26 Nov.–2 Dec. 56/1 EMI Record Tokens..can be exchanged at over 5,000 leading record shops. 1886 F. W. Maitland *Let.* 24 Apr. (1965) 19 As regards mode of printing:—The use of 'record type' seems undesirable. 1934 V. H. Galbraith *Introd. Use of Public Records* v. 77 A century ago the Record Commission in its publications tried by means of 'record type' virtually to reproduce the document, with all its abbreviations. 1972 C. R. Cheney in Campbell *Charters of Rochester* p. vi, Originals and early copies are reproduced as faithfully as is possible, without going to the extreme of using 'record type'.

record, *v.* Add: **III. 9. c.** To convert (sound or visual scenes, esp. television pictures) into a permanent form from which they can afterwards be reproduced by machine. Also *absol.*

1892 W. Gilbert *Phonograph* II. 31 The instrument is now ready to register any sound... While the handle is steadily turned.., speak slowly and distinctly the words you wish to be recorded. 1902 *Encycl. Brit.* XXXI. 680/2 When the phonograph records the sound of an orchestra, it does not record the tones of each instrument. 1935 H. C. Bryson *Gramophone Record* iv. 70 A good modern record is designed to record frequencies between 250 and 5,000 without discrimination. 1960 *How TV Works* 37/1 A video-tape machine recording the opening of Parliament. 1967 S. Beckett *Eh Joe* 15 Joe's opening movements followed by camera at constant remove... No need to record room as a whole. 1972 *Daily Tel.* 29 Jan. 3 The full proceedings of an inquest were recorded for broadcasting for the first time yesterday. 1978 *Radio Times* 9–15 Dec. 81/4 Most people who record and play back BBC television programmes using videocassette equipment do not realise they infringe copyright. 1980 S. Brett *Dead Side of Mike* xiii. 145 I'd better go. We're about to record.

d. *intr.* Of a performer or instrument: to be suited to sound recording.

1923 O. Mitchell *Talking Machine Industry* viii. 88 It has been complained by some that, in technical phrase, she does not record well. 1925 P. A. Scholes *Second Bk. Gramoph. Rec.* p. xviii, The Piano, as an instrument, records less well than other instruments.

e. *trans.* and *intr.* Of a performer: to give a performance, or a performance of (a work), that is recorded.

1927 *Daily Tel.* 12 Feb. 7 She has recorded the Ave Maria from 'Otello'..as if she had recorded all her life! 1930 *Melody Maker* Feb. 155/3 The band..has recently jumped into prominence, having already recorded for Edison Bell. 1966 *Listener* 10 Mar. 345/1 Mostly they recorded at the end of their careers, and violinists seem to deteriorate much earlier than pianists.

V. 12. The infin. used in the sense 'recording'. **a.** *attrib.* (often as a name of a part), as *record button, head*; also *record–reproduce* adj.

1950 Godfrey & Amos *Sound Recording & Reproduction* (1952) vi. 162 When the *Record* button is depressed, current is fed to the wiping head. 1973 *Times* 28 Nov. 6/7 Miss Woods..pressed the 'record' button (it was next to the 'stop' button). 1975 P. G. Winslow *Death of Angel* ix. 186 He pushed the record button. But the heads of his machine were too old. They didn't erase the music. 1946 *Trans. Amer. Inst. Electr. Engineers* LXV. 216/2 The playback head is located a few feet along the wire from the record head so that in listening tests it is possible to get a quick comparison between a few bars of music as recorded and as played back. 1950 G. A. Briggs *Sound Reproduction* (ed. 2) xix. 135 For reproduction, the erase and record heads are switched off and the play-back head is brought into use. 1976 I. R. Sinclair *Master Stereo Cassette Recording* i. 9 Though some high quality reel-to-reel machines use separate record and replay heads,..it is usual on cassette recorders to have only one head used for both recording and for replay. 1951 *Bell Syst. Technical Jrnl.* XXX. 1146 The ring-type record–reproduce head. *Ibid.*, The process was repeated..for several record–reproduce speeds.

b. Not *attrib.*

1950 G. A. Briggs *Sound Reproduction* (ed. 2) xix. 136 By suitable switching of output circuits, one amplifier may be arranged to function for both record and replay. 1958 C. N. G. Matthews *Tape Recording* iv. 36 During recording, to reproduce it [*sc.* the tape] is kept in close contact with the heads by pressure pads or by its own tension. *Ibid.* 38 Equalization is switched automatically as the machine is switched from record to playback. 1971 *Hi-Fi Sound* Sept. 49/2 (Advt.), 4 track stereo record and replay using popular Compact Cassettes. 1976 K. Bonfiglioli *Something Nasty in Woodshed* x. 119 Even on virgin tape I still got the ghostly muttering if it was played through on 'record'..at a nil recording level.

recordable, *a.* Add: (Further examples.) Also, worthy of record, memorable.

1917 *Wireless World* May 87 It might prove useful when signals of recordable strength are obtainable. 1971 *Jrnl. Gen. Psychol.* Jan. 168 Psychologists sometimes deal with body process in either recordable terms or in hypothetical body-process terms. 1974 'M. Innes' *Appleby's Other Story* viii. 62 Everything recordable about this room had been recorded.

Recordak (rĭkǭ·ɹdæk). Also with small initial. The proprietary name of an apparatus manufactured by Kodak Ltd. for producing a photographic record of a series of documents, such as bank cheques, etc.

1929 *Encycl. Brit.* XVII. 805/1 G. L. McCarthy..has designed a camera intended to photograph upon a strip of motion picture film all the checks passing through a bank. This camera, which is known as the 'Recordak', provides a permanent record which greatly diminishes the risk of fraud. 1937 *Trade Marks Jrnl.* 31 Mar. 360/2 Recordak... Photographic apparatus and parts thereof,..and films prepared for exhibition. Kodak Limited,..London,..; manufacturers. 1942 *R.A.F. Jrnl.* 2 May 11 The Recordak machine has replaced the clumsy methods of the 'Tours' post office. The airgraph service is now established. 1948 E. L. Irey *Tax Dodgers* xiii. 251 It was the only bank in Atlantic City which used a recordak. 1967 Cox & Grose

Organization & Handling Bibl. Rec. by Computer 185 In 1956 we installed a 'magnetic tape to microfilm' device..and decentralized the whole file, using about forty Recordak readers, each containing all of the 160 million names.

recordant, *a.* For *rare*[—1] read *rare* and add further example.

1877 Ruskin *St. Mark's Rest* II. iv. 42 They are merely shapes of amphora..usefully recordant of different ages of the wine.

recordation. Add: **2.** (Later example.)

1881 Meredith *Tragic Comedians* I. iii. 40 She was prepared to express her recordation of the circumstance in her diary.

4. (Further examples in *U.S. Law.*)

1924 G. W. Thompson *Commentaries on Mod. Law of Real Property* VIII. liv. 274 The majority rule holds to the tenet that the recordation of an instrument void on its face is not of itself constructive notice. 1938 *U.S. Statutes at Large* LII. 1006 Recordation of Aircraft Ownership. *Ibid.*, Every such conveyance so recorded..shall be valid as to all persons without further recordation. 1948 *Columbia Law Rev.* Dec. 1248 Recordation of title to chattels is normally governed by local laws. 1962 *Iowa Law Rev.* Winter 227 Was his recordation at that time at such place in the records that it should be concluded [etc.]? 1976 *Washington Post* 19 Apr. c10/4 All costs incident to settlement and conveyancing, including..recordation costs and taxes..will be at the cost of the purchaser.

recorded, *ppl. a.* Add: **1. b.** Also, *recorded delivery*, a Post Office service whereby the safe delivery of an item of mail is recorded in a register signed by the recipient; also *ellipt.*, a letter or package sent by recorded delivery.

1961 *Use Recorded Delivery* (G.P.O. Leaflet P.L. 140) 2 Recorded delivery is designed for the customer who wants to be able to prove, if necessary, that a letter or packet has been delivered. 1968 R. Petrie *MacLurg goes West* I. iv. 31 There was this package, see. Registered. No. I'm wrong there; it was that other thing. Recorded delivery, with a little orange sticker on the top. 1969 W. J. Burley *Death in Willow Pattern* i. 8 The post—two recorded deliveries among the rest. 1977 'E. Crispin' *Glimpses of Moon* xii. 256 A letter marked in one place 'Special Delivery', and in another, 'Recorded Delivery'.

2. a. Of sounds or images: converted into a durable form (e.g. on disc, tape, or film) from which the original can be reproduced by suitable apparatus.

1932 *Radio Times* 29 July 242/1 The B.B.C...[has] decided to supplement with recorded programmes the proposed broadcasting service from the new Empire short-wave transmitter at Daventry. 1949 *Ibid.* 15 July 13/1 Stand Easy... (Wednesday's recorded broadcast). 1958 M. Kelly *Christmas Egg* III. 106 He was addicted to the collection of recorded music. 1962 A. Nisbett *Technique Sound Studio* 167 Recorded effects consist principally of those which cannot conveniently be created in the studio: cars, aircraft, birdsong,..and so on. 1975 *Sci. Amer.* May 45/1 A 'videodisc' system that presents recorded pictures and sound on a standard television set will be put on the home-appliance market next year.

b. Of a recording medium: bearing a recording.

1962 A. Nisbett *Technique Sound Studio* 243 A little of the bias signal can still be heard if recorded tape is pulled slowly over the reproducing head.

recorder[1]. Add: **1.** (Later examples.) Now in England and Wales, a part-time judge presiding over certain Crown Courts (see quot. 1971).

1955 *Times* 3 May 6/3 At the first trial before the Recorder (Sir Gerald Dodson) earlier in the session the jury had failed to agree. 1965 *Modern Law Rev.* XXVIII. v. 563 The Recorder of Manchester..interrupted a case..to address us on the iniquitous way in which his court was being used. 1971 *Act* 19 & 20 *Eliz. II.* c. 23 §21 Her Majesty may from time to time appoint qualified persons, to be known as Recorders, to act as part-time judges of the Crown Court and to carry out such other judicial functions as may be conferred on them under this or any other enactment... Every appointment of a person to be a Recorder shall be of a person recommended to Her Majesty by the Lord Chancellor, and no person shall be qualified to be appointed a Recorder unless he is a barrister or solicitor of at least ten years' standing. 1972 *Daily Mail* 2 Aug. 6/4 Judges' itineraries can be changed, solicitors can sit as recorders. 1979 T. Skyrme *Changing Image Magistracy* 219 New-style Recorders differed from their predecessors in that they were not affiliated to any borough, or to any other local government area, but might be required to sit in any part of the Circuit. They were also required before appointment to undertake to sit on not less than twenty days a year.

4. (Further examples.) Also, † the recording part of an early gramophone or phonograph; now *spec.* a tape-recorder.

1899 *T. Eaton & Co. Catal.* Spring & Summer 191/2 A Columbia Graphophone, with clockwork motor, recorder, reproducer, hearing tube, speaking tube and horn. *Ibid.* 191/3 Recorder, with sapphire point, $5.00. Reproducer, with sapphire point, $5.00. 1902 *Encycl. Brit.* XXXI. 680/1 The recorder describes a series of spirals diminishing from the circumference to the centre of the disc. 1908 *Sears, Roebuck Catal.* 201/3 Recorder, for..our special home graphophone. 1914 *Cassier's Engineering* XLV. 414/2 Typical thermoelectric recorders were..described. 1935 *Discovery* Nov. 324/1 The necessary constancy of motion of cinematographic film in recorders, reproducers, printers, and associated apparatus. 1948, etc. [see *flight recorder* s.v. *flight *sb.*[1] 15]. 1957 *Times* 13 Dec. 18/2 The Minifon pocket recorder is..regarded as an essential item for every-day use in the office. 1959 W. S. Sharps *Dict. Cinemat.* 109/1 A magnetic recorder is equipment incorporating an electromagnetic transducer..for recording electric signals as magnetic variations in the medium. 1971 *Engineering* Apr. 105/1 (Advt.), One of Bell & Howell's portable, multi-channel recorders is the answer. 1973 A. Broinowski *Take One Ambassador* iii. 32 She cocked an eyebrow over her mini-cassette recorder.

5. *attrib.* and *Comb.*, as *recorder–reproducer*.

1937 *Jrnl. Soc. Motion Picture Engineers* XXIX. 217 (*caption*) Magnetic recorder–reproducer. 1970 *Proc. IEEE* LVIII. 886 (*heading*) Signal recorder–reproducer using a coherent light source and a photographic film record.

recorder[2]. Restrict '*Obs.* exc. *Hist.*' to senses b, c in Dict. and add: **a.** (Later examples.)

The popularity of the instrument spread in the twentieth century after its revival by Arnold Dolmetsch in 1919.

1920 *Glasgow Herald* 10 Aug. 6 One was able to understand why the Greeks went into battle to the soft strains of solemn music rendered on flutes and recorders. 1932 R. Donington *Work & Ideas A. Dolmetsch* 16 The first group of early instruments to regain something of its original popularity has been the family of recorders, or English flutes. Many hundreds of Dolmetsch recorders are already in use. 1958 M. Dolmetsch *Personal Recoll. A. Dolmetsch* viii. 88 Our broken consorts now, for the first time, included the recorder, the instrument employed being, of course, the ancient boxwood and ivory recorder which Arnold had brought over from England. It was played by his first recorder pupil, namely the Harvard Professor Peabody. 1962 E. Hunt *Recorder & its Music* 7 The present-day meteoric return to popularity of the recorder—whose seductive tone charmed the ears of Henry VIII, Shakespeare and Pepys—is a development unparalleled in the history of any other musical instrument. 1976 D. Munrow *Instr. Middle Ages & Renaissance* i. 14/1 The essential features of the recorder are its beak-shaped mouthpiece and the number of its finger-holes: seven finger-holes plus a thumb-hole.

recordership. (Later example.)

1975 *Daily Tel.* 5 Nov. 12/5 (*heading*) Recordership for woman solicitor.

recording, *vbl. sb.* Add: **3. b.** The action or process of recording sound or television pictures.

1904 S. R. Bottone *Talking Machines & Records* 67 Male voices generally come out more true to the singer's timbre than ladies', the delicate overtones of these latter being more altered in recording. 1923 O. Mitchell *Talking Machine Industry* ii. 15 Mr. Fenby,..in 1863, took out a patent for the electrical recording and reproducing of sound. 1935 *Discovery* Oct. 309/1 The air plant has to be shut off for a few minutes during an actual recording, on account of the noise of the air being forced through the ventilators. 1976 I. R. Sinclair *Master Stereo Cassette Recording* i. 5 Magnetic recording..is almost as old in concept as disc recording, but has had to wait for modern technology to be developed to the stage at which it could be used successfully.

4. A representation of sounds or pictures in a form from which the original can be reproduced by suitable apparatus.

1932 *New Yorker* 11 June 46/3 Accepting the recordings as accurate, I venture that Mme. Leider is a musicianly performer with a fine sense of text. 1949 *Radio Times* 15 July 13/1 Sandy Macpherson takes you to 'The Chapel in the Valley' (BBC recording). 1958 J. Moir *High Quality Sound Reproduction* vii. 130 The standards adopted for 78-r.p.m. recordings have proved unnecessarily robust for electrical reproducers. 1968 *Listener* 18 July 91/1 The Seekers, lugubriously watching a recording of themselves singing 'The Carnival is Over'. 1977 *Times* 24 Aug. 14/4 The basis of the library will be its gradual acquisition of recordings by poets of their verse at the time of its publication.

5. *attrib.* and *Comb.*, as (sense 3 a) *recording fee*; (sense *3 b) *recording deck* (*deck *sb.*[1] 3 f), *right, room, session, studio, tape, van*; **recording amplifier**, one provided to amplify the signals supplied to the cutter (in disc recording) or to the recording head (in tape recording); **recording channel**, a circuit or set of equipment used for sound or video recording; **recording engineer**, an engineer responsible for the technical aspects of recording when a sound or video recording is made; **recording head**, a head (*head *sb.* 11 g) for recording on to magnetic tape or wire; **recording level**, a measure of the average strength of a recorded signal.

In some cases not clearly distinguishable from the ppl. adj.

1934 *Wireless World* 5 Jan. 9/3 The recording amplifier is specially designed to amplify the currents delivered by the microphone or 'A' amplifier to a degree suitable for the electrical conditions as determined by the recording head. 1964 A. A. McWilliams *Tape Recording & Reproduction* ix. 185 The purpose of the recording amplifier is to deliver sufficient current to the record head to magnetize the tape correctly over the working frequency range. 1975 G. J. King *Audio Handbk.* vii. 160 The difference signals are finally mixed with the sum signals and pass with the latter through the RIAA equaliser and hence to the cutter head, via the recording amplifier. 1938 *Motion Pict. Sound Engin.* (Acad. Motion Pict. Arts & Sci.) v. 69 A stage recording channel uses one or more transmitters on the stage to initiate the electrical energy necessary for recording either on film or on disc. 1949 Frayne & Wolfe *Elem. Sound Recording* xi. 184 A complete recording system is known as a recording channel. 1975 G. J. King *Audio Handbk.* x. 237 The controlled amplifier in the recording channel is adjusted in gain by rectified signal from a control amplifier fed from the recording amplifier. 1977 *Times* 24 Aug. 14/2 Three superior recording decks would cost £8,000. 1962 A. Nisbett *Technique Sound Studio* 268 *Recording engineer* (BBC), professional engineer whose job includes the technical recording of sound and the editing of tape, but not microphone balance and control. 1977 *Gramophone* Mar. 1453/1 It is indifferently balanced, with the recording engineer conspicuously experimenting with his levels. 1898 Recording fee [in Dict., sense 3]. 1934 *Wireless World* 5 Jan. 9/2 The 'recording' and 'reproducing' heads are each provided with a micrometer adjustment for controlling the separation of the pole pieces. 1971 *Physics Bull.* June 359/1 Magnetic tapes of the kind used in computers or as video-tapes in television vary in abrasiveness so that some tapes cause excessive wear of expensive recording heads. 1934 *Wireless World* 5 Jan. 10/2 Provided that the maximum recording level of 10 dbs. below 1 milliwatt is not seriously exceeded when recording, the distortion due to non-linear magnetic effects is hardly noticeable aurally. 1975 G. J. King *Audio Handbk.* vii. 157 Recording level may be given in terms of amplitude or velocity. 1962 Recording right [see *gear *v.* 4 a]. 1907 *Westm. Gaz.* 12 Dec. 9/4 Employees..assisted many of the frightened girls to cross from the blazing building to the Gramophone recording-rooms. 1975 *Language for Life* (Dept. Educ. & Sci.) xxv. 425 Almost a quarter of the schools had a projection room and 14 per cent a recording room. 1927 *Melody Maker* Sept. 923/1 All their days..appear to be occupied with recording sessions. 1962 *Times* 5 July 15/4 The raw product of a recording session is a magnetized length of tape. 1928 *Gramophone* Apr. 451 (*heading*) Round the recording studios. 1958 [see *cut *v.* 23 d]. 1977 *Times* 18 Apr. (Gramophone Suppl.) p. iv/2 Contemporary popular music is to a large extent a child of the recording studio. 1960 *Guardian* 9 Nov. 11/2 The biggest manufacturer of recording tape in Europe. 1977 *Times* 24 Aug. 14/1 Imagine 10 shelves..holding 380 reels of 10½ in polystyrene recording tape. 1940 P. Fleming *Flying Visit* i. 17 Wires..poured into the streamlined flanks of a..lightly armoured recording van.

recording, *ppl. a.* Add: **a.** (Further examples.)

In some cases not clearly distinguishable from the vbl. sb. used *attrib.* (see prec., sense *5).

1892 W. Gilbert *Phonograph* III. 82 Adjust the screw on the guide rod until the recording stylus makes a slight groove on the revolving wax cylinder. 1937 'M. Innes' *Hamlet, Revenge!* III. iii. 249 Phonetic nicety apart..the machine..combined recording and reproducing units in an unusually compact way. 1949 *Radio Times* 15 July 14 Richard Dimbleby, with the BBC Mobile Recording Unit. 1962 A. Nisbett *Technique Sound Studio* 260 The Peak programme meter (PPM) is used by the BBC and most recording companies.

b. Of measuring instruments: able to produce a record of readings obtained.

1873 [in Dict.]. 1904 R. M. Walmsley *Electr. in Service of Man* I. ix. 357 A recording wattmeter, which would record the number of watts at every instant, so that..the total energy could be measured up or calculated. 1930 C. J. Stewart *Aircraft Instruments* viii. 146 It is sometimes necessary to record the temperature of the air during flight, and for this purpose recording thermometers are used. 1961 R. Rawlinson in G. F. Tagg *Pract. Electr. Engin.* III. 363 Situations demanding the use of a recording ammeter are fortunately not common.

recordist (rĭkǭ·ɹdist). [f. Record *v.* + -ist.] One who records or makes recordings.

1931 L. Cowan *Recording Sound for Motion Pictures* 384 *Recordist*, person engaged in recording sound. 1938 *Nature* 5 Feb. 226/2 That highly skilled, imaginative, ingenious, agile and patient collaborator in film-making who bears the regrettable though comprehensible title of 'recordist'. 1956 *B.B.C. Handbk.* 1957 65 The Corporation has made steady progress in securing world-wide sources of news in pictures... It maintains a staff of newsreel film cameramen and recordists in this country. 1958 *Times* 2 Dec. 3/2 (Advt.), Electro-Encephalography Recordist (single-handed) required. 1960 *Guardian* 9 Nov. 11/1 The federation issued a policy statement for the guidance of amateur recordists. 1966 V. C. Lewis *Bird Recognition* III. 3/1 The recordist of the 'songs' and 'calls' illustrated herewith is of the opinion that the use of sound concentrators..tends to detract from 'naturalness' of the recorded sounds. 1978 *Daily Tel.* 23 Sept. 1/3 All BBC television programme production is threatened from today by an unofficial overtime ban by cameramen and sound recordists.

Recordite (re·kǭɹdait), *sb.* (and *a.*) *Obs.* exc. *Hist.* [f. the name of the *Record* newspaper + -ite[1].] One who subscribes to views represented by the evangelical Church of England newspaper, the *Record* (1828–1949). Also *attrib.* or as *adj.*

1853 W. J. Conybeare in *Edin. Rev.* XCVIII. 284 This exaggeration of Evangelicanism, sometimes called the Puritan, sometimes, from its chief organ, the Recordite party, we shall now endeavour to describe. *Ibid.*, Thus from *justification by faith* the Recordite infers the worthlessness of morality. 1875 F. Arnold *Our Bishops & Deans* I. i. 11 We suspect we may supply the ellipse by the words 'Protestants' or 'Recordites'. 1965 W. R. Ward *Victorian Oxford* x. 224 Gladstone arranged with Vice-Chancellor Jeune to delay the by-election till the last possible moment, but his legal adviser Phillimore was still on tenterhooks that 'some Recordite would be put up to poll two or three votes and claim to be returned before a committee'.

recordless, *a.* (Earlier example.)

1854 'G. Greenwood' *Haps & Mishaps* 116 That old, old city of a forgotten and recordless past.

re-cork, *v.* Add: (Later examples.) Also *fig.*

1906 *Daily Chron.* 3 Apr. 7/4 While he was recorking the bottle Dunstan picked up the glass. 1922 Joyce *Ulysses* 523 (He uncorks himself behind: then, contorting his features, farts loudly.) Take that! (He recorks himself.) 1969 E. Penning-Rowsell *Wines of Bordeaux* xviii. 271 The leading châteaux re-cork their reserves of old wines roughly every twenty-five years.

‖ **recorte** (reco·rte). *Bull-fighting.* [Sp., lit. 'cutting, trimming'.] A pass by which the torero cuts short the bull's charge.

1925 E. Hemingway in *This Quarter* I. II. 217 A series of acceptable veronicas ending in a very Belmontistic recorte. 1932 —— *Death in Afternoon* vii. 67 A recorte is any pass with the cape that..stops him brusquely. 1932 R. Campbell *Taurine Provence* iii. 71 The cape, the use of which varies from the vulgar vuelta and recorte to the most perfect designs. 1967 McCormick & Mascareñas *Compl. Aficionado* ii. 54 The banderillero..performs a rough recorte to fix the toro gently before he disappears behind a burladero.

recount, *sb.*[1] For † *Obs.* read *rare* and add later example.

1905 *Daily Chron.* 20 July 3/1 We..are not bored by the intolerable recount of flukey rounds.

recounter, *sb.*[1] (Later example.)

1953 Partridge *Shaggy Dog Story* ii. 17 But not even the most cavalier of casual recounters may omit two extremely significant and pertinent literary examples.

recoup, *sb.* Add: Also *attrib.*

1904 H. G. Turner *Hist. Colony Victoria* II. ix. 276 The too general use of the recoup system under which public works were authorised to be paid for out of future loans. 1966 *Public Administration* XLIV. 411 If the recoup mechanism is to do its job.

recoup, *v.* Add: **3. b.** Also *absol.* or *intr.*

1906 L. J. Vance *Terence O'Rourke* I. iii. 19 Each..had seemed to be broken in fortune, and..ready to seize upon any chance to recoup. 1976 C. Bermant *Coming Home* II. vii. 218, I had..acquired so many debts that if I didn't return to England to recoup, we might have to run for it.

c. *intr.* To regain lost health, vitality, etc.; to recuperate.

Prob. arising from confusion with *recuperate*.

1939 M. Spring Rice *Working-Class Wives* iv. 79 We have sent her away for two years running to help her to recoup. 1955 W. Gaddis *Recognitions* I. ii. 66 The Society recouped: found its own Marguerite.

recoupe, *sb.* **b.** For 1835 read 1810 as date of quot.

recouperation (rĭkūpĕrē[i]·ʃən). *nonce-word.* [App. blend of Recoup *v.* and Recuperation.] = Recoupment.

1904 G. B. Shaw *Let.* 23 Nov. (1972) II. 466 It seems to me that unless you were living quite madly beyond your income in the old days, the economies of the last few years ought by this time to have produced some degree of recouperation.

recouple, *v.* Add: Hence **recou·pling** *vbl. sb.*, a repeated joining in pairs or linking.

1890 W. James *Princ. Psychol.* I. ii. 76 The multiplicity of emotional and instinctive reactions in man, together with his extensive associative power, permit of extensive recouplings of the original sensory and motor partners.

recoupment. (Earlier and later examples.)

1839 *N.Y. State Supreme Court Rep.* (Wendell XXII, 1840) 156 The offer came under the third category, *recoupment.* 1905 *Daily Chron.* 18 July 5/5 The full effect of 'recoupment' and 'betterment' would reduce it [*sc.* net cost]. 1953 *Words & Phrases* (St. Paul, Minn.) XXXIX. 28/2 A 'recoupment' is a counterclaim arising out of the contract sued upon, whereas a 'set-off' is a counterclaim arising from an independent transaction. 1977 *Times Educ. Suppl.* 21 Oct. 3/1 A nationwide recoupment scheme to equalize the burden of financing advanced further education.

recourse, *sb.*[1] Add: **4. c.** *Law.* Phr. *without recourse*, used to indicate that the endorser of a bill, etc., shall not be held liable for its non-payment.

1800 *U.S. Supreme Ct. Rep.* (1816) VII. 160 Pay the within to James Welch, on order, without any recourse whatever. 1805 *Ibid.* (1807) III. 203 The words *without recourse* do not imply *without value*. 1839 J. Bouvier *Law Dict.* I. 499/1 The words commonly used are *sans recours*, or without recourse. 1878 M. D. Chalmers *Digest of Law of Bills of Exchange* iv. 97 It is held in America that an indorser 'without recourse' is responsible..*e.g.*, where the bill is a forgery. 1948 D. Richardson *Simple Guide Bills of Exchange Acts* II. 56 When a party adds to his signature the words *sans recours* (*i.e.*, without recourse to me)..he means that, should the bill be dishonoured, the holder cannot look to him for repayment.

recover, *sb.* **4. a.** Delete † *Obs.* and add later example.

1915 Kipling *New Army in Training* ii. 10 The squads at bayonet-practice had their balance, drive, and recover already.

b. (Earlier example.)

1818 'T. Brown' *Brighton* III. ii. 123 'I hold it that a prime coachman's a better fellow than a paltry—' peer he was going to say, but he knew how to pull up to a hair; so, making a recover, he added—'man of fortune'.

recover, *v.*[1] Add: **I. 1. f.** To remove (certain substances) from industrial waste in order that they may be reused.

1906 R. W. Sindall *Paper Technol.* iv. 37 About 75 to 85 per cent. of the soda used in the treatment of esparto can be recovered. 1929 Clapperton & Henderson *Mod. Paper-Making* xxii. 315 In almost all mills a large quantity of fibre and clay may be recovered from the back waters of the machines. 1929 *Industr. & Engin. Chem.* May 446 (*caption*) Air filters recover dust from gyratory crushers... This dust is sold as agricultural limestone. 1941 *Coke & Smokeless-Fuel Age* III. 285/1 All coke producers are required to recover benzole from their gas. 1969 D. Stewart *Paper* 33 The caustic soda is recovered in the following way.

8. b. For † *Obs.* read *rare* and add earlier and later examples.

1602 Jonson *Poetaster* III. i. D3, Did you neuer heare any of my verses? *Horace.* No, Sir; but I am in some feare, I must, now. *Crisp.* Ile tell thee some (if I can but recouer 'hem). 1957 L. Durrell *Justine* IV. 233 It is strange when everything about Alexandria is so vivid that I can recover so little of that lost period.

9. b. (Later examples.)

1816 Jane Austen *Emma* III. iii. 39 A young lady who faints, must be recovered. 1940 W. Faulkner *Hamlet* III. i. 179 This legal dollar which would be little enough compensation, not for the time he had spent recovering the cow. 1967 *Listener* 7 Sept. 302/3 Well, I think Mr Wilson would put one well on the way to convalescence, and the Queen would recover one completely.

III. 18. a. Also *fig.*

1921 H. Crane *Let.* 19 Sept. (1965) 64 *The L[ittle] R[eview]*, I was informed..is recovering, will shortly reappear as a quarterly under Pound, Picabia, etc.

recoverability. (Further examples.)

1964 N. Chomsky *Current Issues in Linguistic Theory* ii. 46 But if there are several alternative designated elements, the comments on 'recoverability' must be slightly revised. 1965 G. J. Williams *Econ. Geol. N.Z.* xviii. 286/1 The recoverability of coal from a thick seam by underground mining is, of course, low. 1968 *Language* XLIV. 234 It will consist of..deletion of the subject of the embedded sentence, subject to the recoverability condition (i.e. only in those cases where it is identical with that of the matrix sentence). 1979 *Sci. Amer.* Jan. 29/3 Estimates of ultimate recoverability vary from one coalfield to another depending on the accessibility of the coal.

recoverable, *a.* Add: **1. a.** (Further examples.)

1952 F. H. Norris *Paper & Paper Making* xiv. 222 Water derived from boiler washings and the rag washers is not recoverable. 1964 N. Chomsky *Current Issues in Linguistic Theory* ii. 41 A deleted element is, therefore, always recoverable. 1974 *Canad. Jrnl. Linguistics* XIX. 147 It is clear that not every instance of the operation of vowel deletion in Canadian French results in a recoverable vowel.

c. Designating mineral reserves which by reason of their location and purity may be extracted economically.

1950 E. Ayres in L. M. Fanning *Our Oil Resources* vii. 234 The oil-shale regions of Colorado, Utah, and Wyoming..contain the bulk of recoverable shale oils as well as the bulk of total shale oils. 1959 D. L. Katz et al. *Handbk. Natural Gas Engin.* xi. 462/2 The initial gas content [of a natural reservoir] minus the content at a selected abandonment pressure gives the recoverable gas. 1973 E. N. Tiratsoo *Oilfields of World* xii. 336 Those volumes of hydrocarbons technically recoverable, to a high degree of certainty, but the exploitation of which is deemed uneconomic. 1976 *Conservation of Resources* (Chem. Soc.) 20 So far we have used up some 16 % of total possible recoverable oil reserves, and only about 4 % in the case of coal. 1979 *N.Y. Rev. Bks.* 17 May 14/4 The world's recoverable coal reserves are several times those of oil.

2. b. (Later example.)

1962 *Times Lit. Suppl.* 28 Sept. 759/1 It was realized that many severe mental disturbances were more recoverable than had previously been supposed.

recovered, *ppl. a.* Add: (Further examples, in sense *1 f of the vb.) † *recovered rubber* = *reclaimed rubber* s.v. *reclaimed *ppl. a.* 2.

1892 *Sci. Amer.* 7 May 293/2 Of all the materials used in rubber compounding, none was found to be as effective as recovered rubber, and this for the simple reason that when carefully prepared it is rubber. 1897 *India-Rubber & Gutta-Percha & Electr. Trades' Jrnl.* 12 Apr. p. vi (Advt.), G. W. Laughton & Co., manufacturers of recovered rubbers. *Ibid.* p. vii (Advt.), The Recovered Rubber Works, Ltd., Clayton, Manchester. 1902 *Chambers's Jrnl.* Oct. 683/2 This rejuvenated substance is known in New York as 'recovered' rubber. 1906 R. W. Sindall *Paper Technol.* iv. 37 The organic matter is burnt off, and the mass left behind consists mainly of impure carbonate of soda. This residue is known as 'recovered ash'. 1929 Clapperton & Henderson *Mod. Paper-Making* xxii. 317 Most of the recovered stuff will have to be used in a paper of a lower grade. 1952 F. H. Norris *Paper & Paper Making* xiv. 225 The most important point is the final recovery cost as compared with the value of the recovered fibre.

recoveree. Restrict 'now *rare* or *Obs.*' to sense in Dict. and add: **2.** One who is recovering from a disease or an illness.

1957 *Times* 15 Oct. 3/4 There was the Asian influenza casualty..who was replaced gallantly by an influenza recoveree, Mr Robert Harben. 1973 *Washington Post* 5 Jan. B5/2 The survey also showed that employers will hire wheelchair people, ex-convicts, T.B. recoverees, the deaf and the blind ahead of epileptics.

recoverer[1]. **1.** (Later examples.)

1884 Tennyson *Becket* III. iii. 135 Our recoverer and upholder of customs hath in this crowning of young Henry by York and London..violated the immemorial usage of the Church. 1978 A. Pratt *Directory of Waste Disposal & Recovery* 107 This list does not cover in great depth the more traditional areas of recovery—metals, paper, textiles etc. In these fields there are many recoverers and merchants.

recovering (rĭkʌ·vəriŋ), *vbl. sb.*[2] [f. Re-cover *v.*[2] + -ing[1].] The action of Recover *v.*[2]

1904 *Daily Chron.* 28 July 8/5 His umbrella..may want re-covering. 1930 A. D. Stubbs *Pastimes that Pay* 45 When your dining room chairs require re-covering there is no need to send them to the upholsterer. 1967 M. Holford *Photogr. Handbk.* 147 Send them to the local umbrella recovering shop.

recovery. Add: **I. 2. d.** In general use, the act of regaining an original position, esp. after rhythmic movement.

1876 [see *recurb]. 1949 Shurr & Yocom *Mod. Dance* 173 Practice slowly at first. Fall: *one, two, three, four*: Recovery. *Ibid.* 190 *Recovery*, a series of movements used in order to return to position after a full sequence.

e. Reversion of a material, object, or property to a former condition following removal of an applied stress or other influence.

1885 P. G. Tait *Properties of Matter* xi. 218 All elastic recovery in solids is gradual. 1895 *Proc. R. Soc.* LVIII. 132 Recovery of elasticity which the overstrained material undergoes with the mere lapse of time. 1939 Williams & Homerberg *Princ. Metallography* (ed. 4) iii. 73 Complete recovery cannot be attained in polycrystalline metals. 1966 C. R. Tottle *Sci. Engin. Materials* vii. 164 Recovery can also take place by annihilation of positive and negative dislocations, with or without the added movements in climb or cross-slip. 1975 E. R. Trotman *Dyeing & Chem. Technol. of Textile Fibres* (ed. 5) vi. 129 Recovery from strain [of polynosic fibres] gives good dimensional stability to fabrics which contain them.

3. b. Also, cf. *on the mend* s.v. Mend *sb.* 5. (Further example.)

1834 H. Evans *Diary* 27 July in *Chrons. Oklahoma* (1925) III. 206 We found this Camp in a desolate situation... The sick some little on the recovery.

II. 4. a. (Later *attrib.* example.)

1951 Koestler *Age of Longing* I. i. 9 Your hand..is on lend-lease to a vicious old man; it is my recovery grant, or whatever you call it.

5. c. (Later examples.)

1922 Joyce *Ulysses* 710 The dun for the recovery of bad and doubtful debts. 1964 W. D. Park *Collection of Debts* (ed. 2) iii. 37 The court has power, in any action for the recovery of a debt or damages, to order interest at such rate as it thinks fit on the whole or part of the debt for the whole or part of the period it has been due.

e. The extraction of reusable substances from the waste produced by a process; also, the original extraction of a useful substance from a mixture, raw material, etc.

1885 *Jrnl. Iron & Steel Inst.* I. 216 (*heading*) Recovery of residuals from furnace gases. 1906 R. W. Sindall *Paper Technol.* iv. 37 (*heading*) Soda recovery. 1923 S. J. Truscott *Textbk. Ore Dressing* 3 With copper,..though ore containing as little as 2·5 per cent may exceptionally be successfully smelted, a better recovery is obtained when the content is higher. 1923 W. H. Walker et al. *Princ. Chem. Engin.* vii. 228 (*heading*) Recovery of waste heat from furnaces. 1951 K. K. Landes *Petroleum Geol.* xi. 621 Modern production methods..result in a much higher percentage of ultimate recovery and, conversely, a lower percentage of residual oil left underground. 1962 F. T. Day *Introd. Paper* iv. 40 The excess water which falls through the machine wire bed flows away for recovery. 1978 J. Updike *Coup* (1979) vi. 247 Engineer's my title; recovery's my racket... Better recovery in the established fields is the name of the game... It's a miracle, what you can squeeze out of a rock if you know where to pinch it.

f. The return or capture of a ringed or tagged animal after its release; = *return *sb.* 15 b.

1909 *Brit. Birds* III. 180 Turning now to..the recovery of marked birds, it is as yet too early perhaps to expect many results of interest. The most interesting recovery that has as yet been reported is ring No. 4308. 1940 H. F. Witherby et al. *Handbk. Brit. Birds* III. 23 Twenty-five recoveries of British ringed nestlings indicate movements of comparatively short distances only. 1959 *Listener* 19 Feb. 321/1 We have been marking young seals with identity tags since 1951..and we have had some good recoveries. 1965 P. Wayre *Wind in Reeds* iii. 36 Of the 284 ducks I ringed at Mileham, forty-five were recovered... Thirteen recoveries were from abroad. 1976 L. Brown *Brit. Birds of Prey* ix. 114 Most of the recoveries are of birds which died unnatural deaths.

g. An amount recovered, usu. in contrast with that expended or initially available.

1931 *Economist* 17 Jan. 112/2 Total recoveries for December were, for the first time for over three years, slightly (30,000,000 francs) below the expected level. 1958 *N.Z. Timber Jrnl.* May 56/1 *Recovery*, the ratio of final product to log volume in timber conversion. 1973 J. L. Gregoire in V. S. White *Mod. Sawmill Techniques* v. 118 Figure 5.2 shows typical recovery on a 9-inch-diameter log 16 feet long with 1/16-inch taper per foot... From this log we are able to obtain a 1×4 14 feet long and six pieces of 2×4 16 feet long.

h. The retrieval of a satellite or spacecraft after a flight. Freq. *absol.* and *attrib.*

1949 *Jrnl. Brit. Interplanetary Soc.* VIII. 197 Two possibilities for increasing the chances for physical recovery after impact seemed worthy of investigation. 1960 D. E. Bailey in K. W. Gatland *Spaceflight Technol.* 220 The main problems of satellite recovery are associated with deceleration, heating and tracking. 1961 *Ann. Reg.* 1960 385 Subsequently the U.S. Air Force made two further recoveries of the smaller Discoverer satellites by snatching them in mid-air as they floated down towards the sea by parachute. **1962** D. Slayton in *Into Orbit* 23 The recovery techniques which we would put into play to find and rescue the Astronaut and his capsule after they had landed. **1967** *Technol. Week* 20 Feb. 10/3 Orbital telemetry indicated that the capsule battery should have sufficient charge to operate the radio beacon and flashing light that serve as recovery aids.

7. a. (Later examples.) Freq. in economic contexts.

1932 A. Salter *Recovery* iv. i. 282 Here what is needed is a moratorium of several years, say four or five, to cover the depression and a period for Germany's recovery after it. **1940** G. Crowther *Outl. Money* iii. 113 Prices rose with staggering rapidity, until..they were over one million million times the pre-war level. This is 'inflation without recovery'. **1958** *Times* 28 June 11/7 The pound staged a remarkable recovery against the United States dollar. **1974** B. Pearce tr. *Amin's Accumulation on World Scale* II. iv. 497 In order to explain world recovery, all that remains is to analyze the effects of new techniques.

c. The restoration to working condition of a disused mine.

1932 *Trans. Inst. Mining Engineers* LXXXII. 452 The recovery of two separate underground districts which had been sealed off for several years due to the occurrence of fire.

III. 10. *attrib.* and *Comb.*, as (sense 3) *recovery area, room, school, unit, ward*; (sense 5) *recovery airfield, area, crew, fleet, line, ship, team, vehicle*; (sense 5) *recovery furnace, plant*; (sense 7) *recovery area, party*; **recovery time,** (*a*) the time required for an object or material, esp. an item of electronic equipment, to return to some specified condition following an action, e.g. the passage of a current; (*b*) *Railways*, time allowed in a schedule in excess of that which would be required in normal running.

1963 *Dict. U.S. Mil. Terms* 11 It is not expected that combat missions would be conducted from a recovery airfield. **1965** *Guardian* 23 Aug. 1/2 It was taken for granted that the astronauts would be brought down in the Bermuda 'recovery area' at 12 22 a.m. **1971** *Ibid.* 2 July 1/5 The Soyuz made a..soft landing in the expected recovery area. **1976** *Scotsman* 27 Dec. 2/7 An attempt to have South Ayrshire designated as a recovery area. **1977** *Times* 1 Apr. 3/4 After the transfer to the recovery area he [sc. a baby] was left in the care of a pupil nurse. **1971** *Guardian* 1 July 1/5 A helicopter-borne recovery crew..found the cosmonauts.. without any signs of life. **1976** *Daily Times* (Lagos) 26 Aug. 9/3 Landings in darkness are generally avoided by Soviet space controllers as they hamper the work of recovery crews. **1968** *Guardian* 23 Sept. 1/2 The splash-down appears to have been some way from the recovery fleet. **1942** G. S. Witham *Modern Pulp & Paper Making* (ed. 2) viii. 225 The Wagner Recovery Furnace..embodies an attempt to greatly increase the efficiency of heat utilization in the recovery furnace over what is possible with the traditional rotary furnace. **1963** R. R. A. Higham *Handbk. of Papermaking* v. 107 There are various designs of rotary recovery furnaces although these are generally of the horizontal type. **1976** J. D. Lee *Nuclear Power*..1. 8 One of the deck crewmen started paying out the recovery line, and the rubber boat bobbed away from the submarine. **1933** *Sun* (Baltimore) 7 Oct. 1/8 McKee named his ticket the 'Recovery Party'. **1929** Clapperton & Henderson *Mod. Paper-Making* xxii. 318 The water goes to waste, or back into use, to the recovery plant. **1970** *Adv. in Chem.* XCVII. 223 The performance of butadiene recovery plants improves as solvent selectivity increases. **1926** S. S. Goldwater in *Trans. Amer. Hospital Assoc.* XXVIII. 276 A large ward designed for an acute surgical service should have recovery rooms, where postoperative cases may be cared for. **1951** *Anesthesiology* July 476 The use of the recovery room has without question saved lives. **1964** G. L. Cohen *What's Wrong with Hospitals?* iii. 51 She wanted a post-operative recovery room. **1979** *Arizona Daily Star* 1 Apr. K10/1 Hospital auxiliary needs volunteers to make infant sweaters,..blankets and stuffed toys for children coming from the recovery room. **1909** *Westm. Gaz.* 11 Sept. 9/4 The establishment of open-air recovery schools. **1962** D. Slayton in *Into Orbit* 23 The Navy stuck most of its recovery ships out in that part of the world. **1976** J. Jackson *Flameout* (1977) vi. 108 Red Cross volunteers were still carrying coffee and Coke to the recovery teams. **1944** *Proc. Radar* (Mass. Inst. Technol. Radar School) xi. 18 The time required for elimination of these free ions after disappearance of the main pulse signal is referred to as the recovery time of the device. **1959** G. R. Partridge *Princ. Electronic Instruments & Instrumentation* xix. 373 This interval from the end of the dead time to the moment when another full-size pulse can be produced is known as the recovery time. **1961** *Trains Illustrated* Nov. 684/1 Time regaining soon began and by Doncaster, with the joint help of 4 min recovery time and a top speed of 85 m.p.h. on the level at Moss, 6 min had been picked up. **1964** *Proc. IEEE* LII. 1301/1 The time required for full recovery of a gap between silver contacts in vacuum..ranged from about 1 μ sec to 30 μ sec... This short recovery time is to be contrasted with the much longer recovery time in gases which is of the order of milliseconds. **1967** R. K. Richards *Electronic Digital Components & Circuits* ii. 31 The time recovery of the minority carriers is called the 'recovery time' of the diode. **1977** *Modern Railways* Dec. 483/2 Nothing more than 3 min 18 sec is sufficient for even time to Stowmarket and the ensuing sharp 54 min to Diss was just kept from where recovery time should have balanced the arrears. **1965** *Nursing Times* 5 Feb. p. iv/2 (Advt.), Recovery Unit. *a* **1944** K. Douglas *Alamein to Zem Zem* (1946) 10 Field workshops with huge recovery vehicles and winches. **1974** A. Douglas *Noah's Ark Murders* i. 6 The recovery vehicle was positioned directly opposite the car. **1965** *Nursing Times* 5 Feb. p. lxii (Advt.), Enrolled Nurses..For Recovery Ward to do full-time day duty. **1970** H. McLeave *Question of Negligence* i. 14 Cameron strolled around the recovery ward to inquire, as always, about the patients he had done that day.

recreance[2]. (Later example.)

1897 F. Thompson *New Poems* 103 To give the pledge, and yet be joined That a pledge should have force to bind, This, O Soul, too often still Is the recreance of thy will!

recreate, *v.*[1] Add: **5.** (Later examples.) Now chiefly *U.S.*

1978 *Verbatim* Winter 6/1 The President plans to recreate on Labor Day. **1979** *Sunset* Apr. 16/3 (Advt.), Recreate. It's fun in Colorado. For the best in summer fun take yourself and your family away.

recreation[1]. Add: **5.** *attrib.* and *Comb.*, as *recreation area, centre, ground, hall, home, leader, leadership, league, officer, ramble, room, tent, therapy, time, vehicle.*

1962 *Recreation* Dec. 531/1 Recreation areas are never too large for the future. **1978** *N.Y. Times* 29 Mar. 82/3 Ocean Beach Park, a recreation area on Long Island Sound. **1943** J. S. Huxley *TVA* 17 Guntersville, formerly a sleepy market town..is now becoming..an important recreation centre. **1974** M. G. D. Dixey *Local Recreation Centres* i. 10 Most local authorities recognise the need for indoor recreation centres. **1859** Recreation ground [see sense 3 a in Dict.]. **1898** E. Howard *Tomorrow* xiv. 147 These wretched slums will be pulled down, and their sites occupied by parks, recreation grounds, and allotment gardens. **1969** I. & P. Opie *Children's Games* xii. 341 The merry-go-rounds a type of swing placed in the recreation ground by the local council. **1936** J. B. Hilton *Playground of Death* iv. 42, I learned on an Eltrent Recreation Ground. **1943** J. S. Huxley *TVA* 75 This method of sectional prefabrication has now been successfully applied to larger buildings, such as..recreation halls. **1976** *National Observer* (U.S.) 10 July 6/2 Add to this the boom in recreation and leisure homes and you have Ozark hills and hollows chock full of people. **1923** *Playground* Apr. 35/2 A recreation leader should be active in the social work program of his city. **1936** H. R. Clarke *Playground Man.* ii. 32 Discuss the future possibilities for playground and recreation leaders. **1963** H. D. Corbin *Recreation Leadership* ii. 20 The recreation leader is responsible for the organization, direction, and supervision of recreational activities. **1924** *Playground* Apr. 118 The marked increase in employed recreation leadership. **1976** *National Observer* (U.S.) 22 May 14/4 (Advt.), Liberal Arts Career Studies... Recreation Leadership. *Ibid.* May 17/5 We were told to scrounge up our own games through local recreation leagues to get as much experience as possible. **1976** *Evening Times* (Glasgow) 1 Dec. 7/2 The Carsluith pattern who died attempting to defend recreation officer Neil McLellan. **1853** Recreation ramble [see sense 3 a in Dict.]. **1854** Recreation room [see *coco* 2]. **1890** *Harper's New Monthly Mag.* Feb. 342/1 An excellent canteen and a recreation-room are, however, now provided in almost every barrack. **1978** J. Wainwright *Thief of Time* 16 In the Recreation Room. Playing Ludo. **1930** E. M. Brent-Dyer *Chalet Girls in Camp* v. 73 The two big tents for the commissariat and recreation tents had ridge-poles. **1977** M. Eiseman *Prel. Lang.* iv. 60 If they play volleyball, that is recreation therapy. **1909** 'Mark Twain' *Is Shakespeare Dead?* iv. 45 It seriously shortened his..recreation-time. **1974** *State* (Columbia, S. Carolina) 25 Feb. 17/1 (Advt.), Home sites. Mobile Homes. Recreation vehicles. **1977** *Globe & Mail* (Toronto) 27 Apr. 25/2 He predicted sales of RVs will continue to rise, 'because the recreation vehicle looks solid when compared with other vacation forms'.

recreational, *a.* Add: **a.** (Later examples.) Also used, for or as a form of, recreation; concerned with recreation.

1890 *Century Mag.* June 176 The recreational section has been a most unequivocal Success, and has already proved a boon to East London. **1946** *Q. Jrnl. Forestry* XL. 2 The Americans have a word for it: the 'recreational use' of land. This is primarily what National Parks are for. **1946** *R.A.F. Jrnl.* May 167/1 There is an extensive technical and recreational library. **1956** A. H. Compton *Atomic Quest* 110 The 'Recreational' Plans for the great protected nature reserve..are being discussed by Deal town council with..recreational societies. **1976** *Billings* (Montana) *Gaz.* 20 June 1-F/1 We used to hear about recreational shopping—it was never such a fad as it is now—but I think now it is a pretty grim, serious business. **1977** *Age* (Melbourne) 18 Jan. 24/1 (Advt.), Caravan Owners! Recreational vehicle reports required to include a 24 ft. Glendale, prononounced ideal for a young couple's first home.

b. *recreational mathematics*, mathematics studied or indulged in for pleasure or amusement.

1940 Kasner & Newman *Math. & Imagination* 156 Researches in recreational mathematics sprang from the same desire to know, were guided by the same principles, and required the exercise of the same faculties as the researches leading to the most profound discoveries in mathematics and mathematical physics. Accordingly, no branch of intellectual activity is a more appropriate subject for discussion than puzzles and paradoxes. **1973** *Sci. Amer.* Sept. 176/1 It would..be hard to imagine two problems in combinatorial point-set geometry more remote from foreseeable practical applications unless one thinks of recreational mathematics (with its two virtues: amusement and pedagogy) as a branch of applied mathematics.

Hence **recrea·tionalist,** one who advocates or promotes the provision of facilities for recreation; **recrea·tionally** *adv.*

1970 *Daily Tel.* 31 Mar. 14 Landowners, farmers, residents, recreationalists, preservationists, conservationists, naturalists, sportsmen and many others have to be considered. **1975** *Nature* 18 Sept. 185/1 Extracts of the poppy plant have been used since the days of the Homeric epics medically and recreationally. **1977** *Jrnl. R. Soc. Arts* CXXV. 257/1 The NCC [sc. Nature Conservancy Council] is trying to achieve a revolution in attitudes reconciling the needs of the recreationalists and the conservationists. **1979** *Sci. Amer.* Feb. 32/2 The diver's work is far more demanding than might be supposed by someone who has dived recreationally in clear, warm water.

recreationist (rekri,ēi·ʃənist). [f. Recreation[1] + -ist.] **a.** One who pursues a recreation. **b.** = *Recreationalist. (See also quot. 1952.)

1904 *Daily Chron.* 13 Sept. 6/6 The campers had relaxed their vigilance,..and a sudden startled shout of 'Police,' the Recreationists discovered helmeted and uniformed stalwarts advancing on them. **1952** D. Riesman in *Antioch Rev.* Dec. 420 Students of leisure—'recreationists' perhaps we'd better call them. **1963** *Times* 6 Nov. 13/3 A useful exercise..to bring poachers..round a table with gamekeepers..together with naturalists, architects, town planners, industrialists, and recreationists (if the word will serve). **1977** *Daily Colonist* (Victoria, B.C.) 8 July 13/4 Cougar Air..will be relying heavily on lumber and mining companies for business as well as recreationists.

recredence (rīkrī·dĕns). *rare*[−1]. [Prob. back-formation on Recredential *a.* and *sb.*, infl. by Credence *sb.* 4 b.] In phr. *letters of recredence* = Recredential *sb. pl.*

1855 E. C. Grenville Murray *Embassies & Foreign Courts* xii. 345 In this audience the ambassador presents his letters of recall to the sovereign, and usually makes a farewell speech... He receives at the same time his letters of recredence.

recredential, *a.* and *sb.* For † *Obs.* read *Obs.* exc. *Hist.* and add: **B.** *sb.* (Later example.)

1909 *Eng. Hist. Rev.* XXIV. Apr. 256 He obtained his recredentials on 18 October 1666.

recrudency. For † *Obs. rare*[−1] read *rare* and add later example.

1930 *Jrnl. Hellenic Stud.* XLIV. p. lxii, This success is all the more notable as it synchronizes with the recrudency of efforts to circumscribe Greek studies.

recrudescence. Add: **2.** *transf.* A revival or rediscovery (of something regarded as good or valuable).

1906 H. W. & F. G. Fowler *King's Eng.* i. 15 A literary tour de force, a *recrudescence*, two or three generations later, of the very respectable William Lamb. **1973** *Times Lit. Suppl.* 3 Aug. 900/2 The first fruit of this act of recrudescence was the catalogue of drawings.

recruit, *sb.* Add: **I. 3. a.** (Later example.)

1960 *Harper's Mag.* Sept. 677/1, I had gone in by way of recruit of clocks—for you must know I'm a clock peddler.

c. *Ecol.* An animal which has recently reached the size that qualifies it to be counted as a member of the population to which it belongs.

1938 *Jrnl. du Conseil* X. 269 It has been assumed that the same number of recruits would be found. **1948** M. Graham *Rational Fishing of Cod of N. Sea* iii. 86 Codling from the deeper areas of the North Sea grow more slowly and will not enter as recruits until they are..2–3 years old. **1977** J. L. Harper *Population Biol. of Plants* v. 122 Order and organization appear in populations because of feedback from existing populations to new recruits and because these recruits themselves interact.

II. 8. *recruit drill, training.*

1904 A. Huxley *Let.* 14 Dec. (1969) 32, I have,i n company with Gielgud, accomplished forty recruit drills this half and am now very nearly a full fledge[d] territorial. **1914** (*title*) Recruit training (infantry). **1914**: as to set up I felt just like you went to his recruiting-schooner for you. **1940** *Whig Almanac* & *U.S. Reg. for* 1850 26/2, 23 cents per day, hardships in war, and no hope at all, requires the aid of a recruiting sergeant. **1948** *Contact Books* xi. 62/1 When first, as a boy, I got to know boxers, unemployment and hardship were the greatest recruiting sergeants for the ring. **1846** J. C. Frémont *Rep. Exploring Exped. Rocky Mts.* 160 The bottoms of this river (the Bear), and some of the creeks.. form a natural resting and recruiting station for travellers. **1907** G. B. Shaw *John Bull's Other Island* p. xxxiv, Every school is a recruiting station; every church is a barrack.

recruit, *v.* Add: **I. 3. c.** To become a member of (a natural population). Also *intr.* (const. to). Cf. sense 6 d below.

1960 *Oceanogr. & Marine Biol.* III. 357 The stock to which an individual recruits is solely determined by the time at which it attains the critical size. **1967** *Ibid.* V. 415 Before 1950 only a part of each year-class recruited the fishery as three years of age, the remainder recruiting it at age four, but in later years most, if not all, of the members of each year-class recruited at three years of age.

6. a. (Later *transf.* examples.)

1936 [see *montaña]. **1961** *New English Bible Acts* xvii. 5 The Jews recruited some low fellows from the dregs of the populace. **1974** *Economist* 11 May 36/2 The neo-fascists among whom many of the stewards at his meetings were recruited.

b. *U.S.* To (attempt to) induce (an athlete) to sign on as a student at a college or university.

1913, etc. [implied in *recruited *ppl. a.*]. **1974** *Time* 21 Jan. 62/1 With impressive speed he recruited new talent and turned out a winner his first season. **1979** *Arizona Daily Star* 1 Apr. (Parade Suppl.) 23/1 Like her fellow junior from West Virginia, Earl Jones.., Ostrowski already is being heavily recruited by the country's top college coaches.

c. *Physiol.* To bring (additional muscle fibres or muscular activity) into play by the recruitment of their neurones.

1938 *Amer. Jrnl. Physiol.* CXXII. 49 The manner of recruiting mechanical energy during hyperpnea was extremely variable. **1979** *Sci. Amer.* Sept. 148/3 Slow-twitch units, resistant to fatigue and generating relatively little tension, are the first to be recruited.

d. *Ecol.* Of a natural population: to acquire by recruitment (sense *2 c). Cf. sense 3 c above.

1977 J. L. Harper *Population Biol. of Plants* v. 126 Even in a controlled, 'homogeneous' environment the numbers of seeds that are recruited into a germinating population are determined by the individual properties of each seed.

II. 9. (Later examples.) Also with *up*.

1860 H. J. Hawley *Diary* 10 May in *Wisconsin Mag. Hist.* (1936) Mar. 336 We..stoped [*sic*] giving the teams a rare advance to recruit up a little which they need. **1896** E. Dowson *Let.* 19 Mar. (1967) 346, I believe I have recruited a little since I came here.

d. *Anat.* The incorporation into a tissue or region of cells from elsewhere in the body.

1973 *Laboratory Investigation* XXVIII. 56/1 Small bronchioles, bronchi, and tracheas showed cells distributed within all three layers which could be recruited by movement of cells from capillaries through the epithelium to airway lumina. **1978** *Nature* 2 Feb. 403/1 These compartments have the property that once the boundary has formed, the cells from the neighbouring compartments can never cross it, so each develops by further subdivision but not by recruitment of cells from outside.

recruital. 1. (Later example.)

1889 W. R. Smith *Relig. Semites* i. 12 The urban population is maintained only by constant recruital from the country.

recruited, *ppl. a.* (Later examples in sense *6 b of the vb.)

1913 *Collier's* 1 Mar. 20/2 None of them belongs to the class of 'recruited' athletes, so common in our colleges twenty years ago. **1974** *Anderson* (S. Carolina) *Independent* 19 Apr. 36/3 One of the most heavily recruited South Dakota schoolboys in history. **1979** *Tucson* (Arizona) *Citizen* 28 Apr. 1B/2 He was one of the state's most highly recruited athletes this season, with every major college swimming power after him.

recruiter. 1. For † *Obs.* read *Obs.* exc. *Hist.* and add later example.

1954 Brunton & Pennington *Members of Long Parl.* i. 2 The constituency concerned was short of a member until the election of the 'Recruiters'—that is, the members elected from 1645 onwards.

2. (Later examples.) Also, one who recruits employees.

1899 A. Conan Doyle *Duet* 301 Frederick William, the half-mad recruiter of the big Potsdam grenadiers. **1944** *Living Off Land* viii. 159 The recruiter has been to District Headquarters. **1970** *Wall St. Jrnl.* 20 Mar. 1/1 One reason for the recent increase in recruiting is that recruiters seem to be their value in haggling over perks. **1977** *Graduate* 9 Dec. 15/3 It is also, incidentally, the recruiter's intention to deal with all applicants in an efficient and courteous manner.

recruiting, *vbl. sb.* Add: **b.** *recruiting bill, campaign, drive, -market, office, poster, -schooner, sergeant, station.*

1708 Addison *Let.* 24 Jan. (1941) 89 The next day,..they cramped the former Recruiting Bill by a new clause. **1976** *Sunday Mail* (Glasgow) 28 Nov., There have been recruiting campaigns for both police and teachers since regionalisation. **1955** *Railway Mag.* May 345/1 A big recruiting drive for staff..has been launched by British Railways. *a* **1971** R. Whytt in J. Burnett *Useful Toil* (1974) i. 126 A recruiting drive was on, and her brothers might have to go. **1901** *Macmillan's Mag.* Apr. 476/1 The recruiting-market is in direct competition with all other avenues of employment. **1848** J. R. Lowell *Biglow Papers* 1st Ser. viii. 125 He looked through the dirty pane of the recruiting office window. **1919** G. B. Shaw *Augustus does his Bit* 244 This is the Town Hall Recruiting Office. Give me Colonel Bogey, sharp. **1909** *Regulations for Recruiting for Regular Army & Special Reserve* (War Office) i. 8 Illustrated recruiting posters, large and small, have been ordered to be supplied for display in public buildings. **1966** G. Orwell *Inside Whale* 184 After the bombs and the food-queues and the recruiting posters, a human voice! **1971** P. D. James *Shroud for Nightingale* i. 23 She was attractive enough for a recruiting poster. **1934** 'R. Dacy' *Enchanted Isl.* i. 16 But soon heard his recruiting-schooner for you. **1840** *Whig Almanac & U.S. Reg.* 1850 26/2, 23 cents per day, hardships in war, and no hope at all, require the aid of a recruiting sergeant. **1948** *Contact Books* xi. 62/1 When first, as a boy, I got to know boxers, unemployment and hardship were the greatest recruiting sergeants for the ring. **1846** J. C. Frémont *Rep. Exploring Exped. Rocky Mts.* 160 The bottoms of this river (the Bear), and some of the creeks.. form a natural resting and recruiting station for travellers. **1907** G. B. Shaw *John Bull's Other Island* p. xxxiv, Every school is a recruiting station; every church is a barrack.

recruitment. Add: **2. a.** (Later examples.)

1945 in *Amer. Speech* (1946) XXI. 78/2 The material included the..recruitment programs..to the advertisements from the company. **1958** *Times* 26 Mar. (Careers in Industry Suppl.) p. xviii, They may study the recruitment literature of many employers. **1971** M. E. Ray *Recruitment Advertising* i. 9 Recruitment advertising is but a small part of the wide and varied affairs of a Personnel Officer.

c. *Ecol.* Increase in a natural population as progeny grow and become recruits (sense *3 c); the extent of such increase.

1938 *Jrnl. du Conseil* X. 266 A stock will be in equilibrium with fishing when: C = A+G−M where C is capture, A is recruitment, G is growth, M is natural mortality. **1938** H. G. Champion in Champion & Trevor *Man. Indian Silviculture* i. v. 146 The extension over the regeneration area as a whole of the light conditions appearing most favourable to regeneration is..the correct first step, always bearing in mind the possibility of different requirements for recruitment, i.e. new seedlings, and establishment, i.e. the further development of seedlings already present. **1954** W. E. Hiley *Woodland Managem.* xx. 357 A number of trees will be included, which have grown..but were omitted from the earlier enumeration because they had not then reached this size. These are called 'recruited trees'... It has been claimed that the volume of these trees—the recruitment—should be omitted from the calculation. **1965** *Oceanogr. & Marine Biol.* III. 357 A problem which has recently received considerable attention is the mechanism underlying recruitment of adolescent fish to the various North Sea Summer–Autumn spawning stocks. **1970** S. H. Gardiner tr. *Assmann's Princ. Forest Yield Study* 459 The lower the threshold of measurable diameter the smaller is the amount of recruitment. **1977** J. L. Harper *Population Biol. of Plants* v. 144 Changes in the composition of light after it has passed through a leaf canopy may be one of the critical factors hindering the recruitment of seedlings under vegetation.

3. *Physiol.* **a.** The involvement of successively more motor neurones in response to an unchanging stimulus.

1923 Liddell & Sherrington in *Proc. R. Soc.* B. XCV. 335 The several forms assumed by the course of the contraction indicate the various time relations exhibited by the progressive involvement of additional motoneurones during the development of the reflex. That process may, for convenience of statement, be designated 'recruitment'. **1937** Best & Taylor *Physiol. Basis Med. Pract.* iv. 1278 Many reflex arcs gradually increase to a maximum when a stimulus of *unaltered* intensity is merely prolonged. This is due to the activation of a progressively greater number of motoneurones. The phenomenon is called recruitment and is figuratively spoken of as a 'inertia' by Sherrington. **1975** A. Vander et al. *Human Physiol.* (ed. 2) viii. 210 The tension of the muscle can be controlled by the recruitment of additional motor units.

b. The phenomenon shown by an ear which, while having a relatively high threshold for the perception of quiet sounds, perceives louder sounds with undiminished intensity, i.e. increases in objective intensity of sound result in abnormally great increases in perceived loudness.

1937 E. P. Fowler in *Arch. Otolaryngol.* XXVI. 517 When there is no recruitment of loudness in the poorer ear, i.e. no change in the difference in hearing at the thresholds required to balance the loud sounds binaurally, it means that there is an impedance (conduction) lesion in the poorer ear. **1948** *Proc. R. Soc. Med.* XLI. 517 The deafness of the affected ear present at threshold disappears at higher intensities, and this in its simplest terms constitutes the phenomenon of Loudness Recruitment. **1960** *Jrnl. Speech & Hearing Res.* III. 15/1 The classical technique for the direct measurement of loudness recruitment in subjects with unilateral hearing loss is performed by having the subject equate the loudness of a pure tone on one ear with the loudness of a pure tone of identical frequency on the other ear. **1971** D. E. Rose *Audiol. Assessment* x. 332 Recruitment came to be viewed..as the distinguishing feature of an ear with a cochlear lesion.

recruity. (Earlier example.)

1890 [see *once *adv.* A. 8].

recrystallization. Add: **1.** (Earlier and further examples.) *spec.* in *Metallurgy*, a rearrangement of the crystalline structure of a metal at high temperatures which tends to reduce distortion of the lattice.

1793 *Trans. Soc. Improv. Med. & Chirurg. Knowledge* 30 (*heading*) A process for preparing pure emetic tartar by recrystallization. **1925** *Jrnl. Iron & Steel Inst.* CXII. 474 A formula is derived for establishing the relation between the progress of recrystallization and the degree of deformation in the hot state. **1937** *Discovery* Feb. 45/2 The firn consists of granular ice, formed by partial recrystallization of the snow. **1947** J. C. Rich *Materials & Methods Sculpture* vi. 60 The setting of plaster of Paris after it has been mixed with water is also referred to as a rehydration and occasionally also as a recrystallization of the calcined gypsum back to its original hydrated rock form. **1952** J. Hazard tr. *Becker's Organicum* iii. 38 The most important method for purifying solids is recrystallization. **1960** M. C. Nutt *Metallurgy & Plastics for Engineers* v. 78 Recrystallization occurring at elevated temperatures is the process of the formation and growth of unstrained grains, supplanting entirely the cold-worked structure.

2. Special Comb.: **recrystallization temperature** *Metallurgy*, the temperature at or above which recrystallization of the crystal lattice of a metal takes place (see quots.).

1927 *Carnegie Scholarship Mem.* XVI. 166 No systematic investigation has been made..into the relation between the degree of cold-work and the recrystallization temperature. **1948** J. E. Garside in H. W. Baker *Mod. Workshop Technol.* I. vi. 133 More precisely, cold working is defined as that process of deformation which is conducted below the recrystallisation temperature of the metal or alloy. **1961** G. E. Dieter *Mech. Metallurgy* v. 155 For practical considerations a recrystallization temperature can be defined as the temperature at which a given alloy in a highly cold-worked state completely recrystallizes in 1 hr. *Ibid.*, The recrystallization temperature decreases with increasing purity of the metal. **1976** M. C. Nutt *Metallurgy & Plastics for Engineers* v. 78 When a metal is heated, the internal force changes only slightly, but the movement of the atoms increases sharply. In time a temperature is reached at which movement of the atoms can start, and they can arrange themselves into lattices that are not distorted. This degree of heat is known as the recrystallization temperature.

recrystallize, *v.* Add: Also *intr.* (Further examples.)

1935 G. E. Doan *Princ. Physical Metallurgy* iii. 103 Lead ..recrystallizes spontaneously at room temperature, even if only slightly deformed. **1946** *Nature* 5 Oct. 482/2 The hard rolled silver specimens when heated to about 900 °C. recrystallized very rapidly. **1956** *Ibid.* 3 Mar. 429/2 Tyrosine was recrystallized from hot water.. **1963** D. W. & E. E. Humphries tr. *Termier's Erosion & Sedimentation* xvii. 338 Certain rocks are dissolved and recrystallized so readily under the effect of differential pressures, that they can flow as plastic substances toward zones where the pressure is lowest. **1971** I. G. Gass et al. *Understanding Earth* iii. 60/2 The mineral can recrystallize to different and denser crystalline structures.

Hence **recry·stallizable** *a.*, that may be crystallized again; **recrystallized** *ppl. a.* (further examples); **recry·stallizing** *ppl. a.*

1859 *Proc. R. Soc.* IX. 653 The gold-salt is a bright yellow crystalline precipitate, difficultly soluble in boiling water, and not recrystallizable without some alteration. **1908** [see *idioblastic *a.*]. **1962** Simpson & Richards *Junction Transistors* iii. 48 This recrystallized germanium forms a single crystal of the same orientation as the germanium disk and, except for the *p*-type impurity, is indistinguishable from it. **1976** Cotterill & Mould *Recrystallization & Grain Growth in Metals* iii. 40 The recrystallized grain size depends chiefly on the degree of deformation and to a lesser extent on the annealing temperature. *Ibid.* 54 'Recovery' will continue until the entire specimen has been consumed by the recrystallizing grains.

rect, *a.* Add: **d.** *fig.* Upright.

1890 E. Johnson *Rise Christendom* 102 A rect and good and great soul, what is this but God sojourning in the body of man?

Hence **re·ctly** *adv.*, directly. *rare*[−1].

1922 Joyce *Ulysses* 208 Swiftly rectly creaking rectly rectly he was rectly gone.

rectal, *a.* Add: *rectal gland*, a gland that excretes into the rectum; *esp.* in cartilaginous fishes, a gland that excretes salt so as to maintain the osmotic balance.

1887 Marshall & Hurst *Pract. Zool.* xi. 217 The rectal gland is a thick-walled tube, about three quarters of an inch long, lying in the body-cavity dorsal to the rectum. **1925** R. E. Snodgrass *Anat. & Physiol. Honeybee* vi. 165 Nothing definite is known of the function of the rectal glands. **1974** D. & M. Webster *Compar. Vertebr. Morphol.* xvii. 442 Excess salt is actively excreted by the rectal, or digitiform gland [of chondrichthyeans].

Hence **re·ctally** *adv.*, by way of the rectum.

1906 *Practitioner* Nov. 645 This serum was injected rectally. **1977** *Lancet* 16 Apr. 857/2 We have measured the effect of rectally administered cimetidine on acid secretion in the canine stomach.

rectangle, *sb.* Add: **1. a.** Also, something that has the shape of a rectangle.

1898 G. B. Shaw *Arms & Man* i. 8 For an instant the rectangle of snowy starlight flashes out. **1901** 'Linesman' *Words by Eyewitness* iii. 60 The indistinct rectangles of the companies shook themselves out into a single rank. **1925** F. Scott Fitzgerald *Great Gatsby* vii. 174 A small rectangle of light which I guessed was the pantry window. **1940** W. Faulkner *Hamlet* iii. i. 168 As if a rectangle of opaque glass had been set into nothing's self. **1965** A. Lurie *Nowhere City* (1977) ix. 103 The shop windows had been painted over in irregular rectangles of red, blue, green and white.

rectangle, *a.* (Later example.)

a **1796** Burns *Caledonia* in *Works* (1800) IV. 356 Rectangle-triangle, the figure we'll chuse.

rectangular, *a.* **1. a.** (Later *Comb.* examples.)

1960 *Farmer & Stockbreeder* 8 Mar. (Suppl.) 10/3 Rectangular-shaped pods which should be gathered when 1 to 1½ in long. **1976** *Private Eye* 24 Dec. 8/1 They must be true, rectangular-type icebergs, without cracks or crevasses.

rectangularism (rektæ·ŋgiŭlăriz'm). *rare.* [f. Rectangular *a.* + -ism.] A tendency towards or preference for rectangular forms.

1954 *Archit. Rev.* CXVI. 88/2 But in Holland there was no convulsion change; *de Stijl* won common acceptance by compromise..with existing styles, as in the case of Dudok's romantic rectangularism.

‖ **recte** (re·kte), *adv.* [L., lit. 'in a straight line, rightly'.] **1.** Correctly: used to indicate that the word or phrase following it within a parenthesis is the correct version of that which immediately precedes the insertion.

1886 *Trans. Philol. Soc.* 621 Leg. *contini* (recte *cointinni*) gen. of *cointinn* s.f. 'strife, controversy'. **1934** *Times Lit. Suppl.* 3 May 325/2 'Tithreks Saga' (*recte* 'Thithriks Saga af Bern'). **1939** Joyce *Finnegans Wake* (1964) iii. 543 The villa of the Ostmanorum to Thorstan's, *recte* Thomars Sraid. **1979** *Trans. Philol. Soc.* 184 Kent translates the portion after my square brackets 'This indeed..(is) my activity (*recte* physical-dexterity).'

2. *Mus.* In phr. (*per*) *recte et retro* [med.L., in the right way and backwards], applied to the movement of a canon cancrizans (see *cancrizans *a.*).

[**1801** T. Busby *Compl. Dict. Mus.*, *Recte*,..a word signifying forwards and particularly pertaining to the *Canon*.]

1836 *Penny Cycl.* VI. 243/1 The canon *Recte et Retro* has but one peculiarity, and pretends to only one merit, namely, that it may be sung either forwards or backwards. **1876** [see *cancrizans *a.*]. **1909** R. Dunstan *Composer's Handbk.* x. 155 A Canon 'per Recte et Retro' is one that may be sung *forwards* and *backwards* at the same time, producing two parts in one. **1922** R. O. Morris *Contrapuntal Technique* 50 In each pair of voices the part of the lower is that of the upper begun at the end and sung backwards, i.e. a canon *per recte et retro.* **1954** *Grove's Dict. Mus.* (ed. 5) V. 79 Stainer wrote a hymn-tune 'per recte et retro' in 1898.

rectenna (rekte·nă). [f. *rect*(*ifying ant*)*enna.*] A unit combining a receiving aerial and a device for rectifying the current it produces.

1975 *Pop. Sci.* Sept. 66/2 Transmitting electric power from outer space to earth... The rectifying units could be coupled to the receiving antenna or built into it—a combination now called a rectenna. **1977** *Indian Express* 18 Aug. 6 A micro-wave beam is directed to rectenna on the ground. **1978** *Times* 8 Sept. 19/6 The ground antenna which will receive the microwave beam and rectify this energy to direct-current electricity—the name 'rectenna' has been coined—will consist of a grid of wires carried on insulated posts.

rectification. Add: **1. a.** Also *spec.* in Chinese communism, the correction of errors in ideology and practice within the communist party. Also *attrib.*, as *rectification campaign, drive, movement.*

1956 tr. *Sel. Wks. Mao Tse-Tung* IV. 111 In the rectification campaign of 1943 all bureaux..of the Central Committee of the Party, all regional and district Party Committees should..endeavour to gain experience. **1959** *New Statesman* 17 Jan. 64/3 We make sure that..any professor or student who needs rectification is encouraged to volunteer for three or four weeks of unskilled labour in the steel works. **1962** E. Snow *Red China Today* (1963) xlvii. 376 It was the opening shot of an official 'rectification' or *cheng-tung tso-feng* movement. **1967** tr. *Quotations from Chairman Mao Tsetung* (ed. 2) 4 As we used to say, the rectification movement is 'a widespread movement of Marxist education'. Rectification means the whole Party studying Marxism through criticism and self-criticism. **1971** J. J. Taylor in D. J. Dwyer *China Now* (1974) xxii. 425 The Cultural Revolution is to the international united front what the rectification drives within the CCP in 1941 and 1942 were to the anti-Japanese front in China: an effort to maintain purity and fervour in the revolutionary nucleus. **1977** *Time* 21 Mar. 22/3 Mao's.. willingness periodically to shake up the bureaucracy in 'rectification campaigns'.

c. *Photogrammetry.* The process of preparing a plan view from an aerial photograph taken at an oblique angle.

1920 M. N. MacLeod *Mapping from Air Photographs* ii. iv. 54 The rectification, or plotting of photos by comparison with fixed points, is at best a roundabout method. **1921** *Geogr. Jrnl.* LVII. 141 The theory of the rectification of photographs that should have been taken vertically, but are really tilted several degrees in an unknown direction, involves propositions in the theory of perspective which are not readily accessible in convenient form. **1928** *Ibid.* LXXI. 589 Whether the tilt is to be eliminated by photographic 'rectification', or by transferring the map grid to the photograph in perspective,..it is first necessary to find the tilt and the height at the moment of exposure. **1944** P. G. McCurdy et al. *Man. Photogrammetry* ii. 54 Optical rectification is the process of projecting the image of a tilted photograph into a horizontal plane. **1968** *Times* 1 Nov. 6/8 A process known as rectification is already standard practice. This is an automatic method of giving correct positions in plan to elevated features such as hills.

4. The process or act of permitting an electric current to flow preferentially in one direction; *esp.* in *Electr.*, the conversion of an alternating current into a direct current; also in *Physiol.*, the action of nerve membranes in allowing electrical impulses to be conducted preferentially in one direction.

anomalous rectification (Physiol.), the phenomenon of permitting current to flow preferentially from low to high ionic concentrations.

1895 *Electrician* 9 Aug. 488/1 A commutator..to which current is led and from which it is taken, after rectification. **1905** *Proc. R. Soc.* LXXIV. 485 The rectification is less complete in proportion as the temperature of the carbon filament increases. **1922** *Encycl. Brit.* XXXII. 1027/2 The remainder of the plate voltage is created by the rectification by the valve of the speech currents induced in the secondary circuit. **1941** *Jrnl. Gen. Physiol.* XXIV. 562 The most convenient specification of the rectification characteristics of the membrane. **1949** A. L. Hodgkin et al. in *Arch. Sci. Physiol.* III. 139 There is a marked rectification but it evidently takes some time to develop... We therefore propose the use of the term 'delayed rectification' to describe this effect. **1962** *Jrnl. Physiol.* CLXIII. 61 A fall in the membrane conductance when the membrane current flows from the inside to the outside of the fibre, and rise in conductance when the current is in the opposite direction has been called anomalous rectification. **1965** *Math. in Biol. & Med.* (Med. Res. Council) vi. 258 In 1941, Cole and Curtis showed that, in the steady state, the membrane of squid nerve has a much lower resistance to outward (depolarizing) currents than to inward (hyperpolarizing) currents. By analogy with the behaviour of electrical rectifiers this phenomenon was called rectification and the term 'delayed rectification' was introduced—emphasizing the fact that the changes in resistance take an appreciable time (a few milliseconds) to occur. *Ibid.*, Rectification can occur in the absence of [ionic] concentration gradients.., and in skeletal muscle fibres..and in cardiac muscle fibres..rectification in the opposite direction to that predicted from simple changes in membrane ion concentration occurs. **1966** R. G. Kloeffler *Electron Tubes* x. 209 For the purpose of analysis of rectification by diodes, a clear concept may be obtained by assuming that both the device and the load are linear. **1970** J. Shepherd et al. *Higher Electr. Engin.* (ed. 2) xxv. 798 The semiconductor diode rectifier is now replacing the mercury-arc rectifier for polyphase rectification in all applications except those involving the highest voltages.

rectified, *ppl. a.* Add: **1. b.** *Photogrammetry.* Designating a plan or photograph which has been corrected for errors of perspective (cf. *rectify *v.* 1 c).

1920 M. N. MacLeod *Mapping from Air Photographs* i. iv. 27 A 'rectified' print of the negative is obtained which is true to scale all over and can be traced on to the map. **1969** G. C. Dickinson *Maps & Air Photographs* xv. 245 All five or nine photographs are printed fused together into one image, the obliques being 'transferred' or rectified into vertical views before printing.

3. b. (Examples.)

1892 S. P. Thompson *Dynamo-Electric Machinery* (ed. 4) iii. 38 The currents are now 'rectified', or 'redressed', as our continental neighbours say, but are not continuous. *Ibid.* (*caption*) Curve of rectified or commuted alternating current. **1910** G. W. Pierce *Princ. Wireless Telegr.* xviii. 197 The rectified current obtained by applying the alternating voltage *V* could be read on the galvanometer. **1947** R. Lee *Electronic Transformers & Circuits* iii. 48 With large values of capacitance, the rectified voltage..increases to within a few per cent of the peak voltage. **1958** *New Scientist* 10 July 342/2 A wire 'tickled' a crystal of silicon, and when the alternating radio signal was applied to the combination the current would flow only in one direction: this 'rectified' current could then be used to work a Morse receiver. **1965** *Wireless World* July 335/1 The rectified current does not then cause potential changes in the source.

4. Of tulip flowers: having variegated colouring caused by a virus affecting the plant.

1659 T. Hanmer *Garden Bk.* (1933) 21 When they [*sc.* the colours of tulips] streame away, as they doe ever wholly in flowers perfected, or rectified (as the tearme is) the leaves retaine not the least blew or yellow in them. **1850** *Beck's Florist* 23 Some say, that as they broke or became rectified, another number was given them. **1881** *Encycl. Brit.* XII. 259/2 The breeder bulbs and their offsets may grow on for years producing only self-coloured flowers, but after a time.. some of the progeny 'break', that is, produce flowers with the variegation which is so much prized. The flower is then said to be 'rectified'. **1911** [see *Rembrandt]. **1929** A. D. Hall *Bk. Tulip* v. 99 This change [in colouring] is called 'breaking', the flower is termed 'broken' or 'rectified', while the original form is known as a 'breeder'. **1948** [see *Rembrandt].

rectifier. Add: **1. b.** *Photogrammetry.* A device for preparing, by optical or other means, a plan view from an oblique aerial photograph.

1921 *Geogr. Jrnl.* LVII. 141 The construction of the photographic 'rectifier'..embodies some curious properties of the lens with a flat field. **1932** McCaw & Cazalet tr. *O. von Gruber's Photogrammetry* xi. 277 If a photograph is projected on a plane (map plane) oriented in a definite manner such that the projection also indicates a definite scale, it is customary to describe the process as rectification and the apparatus therefore as a rectifying camera or simply a Rectifier. **1962** *Photogrammetric Record* IV. 84 The great increase in setting accuracy and definition available in the SEG V (and other modern rectifiers). **1964** *Exhib. Guide 10th Internat. Congr. Photogrammetry* 76 Wild E4 Rectifier-enlarger... Rectification elements: Inclination of table in *x* and *y* directions or across the diagonals.

2. b. Substitute for def.: A device or substance which permits an electric current to flow preferentially in one direction; *esp.* in *Electr.*, a device for converting an alternating current into a direct current; also applied in *Physiol.* to a nerve membrane which conducts electrical impulses preferentially in one direction. (Earlier and further examples.)

1895 *Electrician* 9 Aug. 488/1 An efficiency of 96 per cent. is claimed for the rectifier. **1911** *Encycl. Brit.* XXVII. 835/1 A common type of rectifier is another tube containing gas at a low pressure. **1926** R. W. Hutchinson *Wireless* 119 Fig. 96 shows a method of using a crystal as a detector or rectifier. **1941** *Proc. Soc. Exper. Biol. & Med.* XLVIII. 293 In a nerve fiber membrane, current may pass more easily in one direction than in the other,..the nerve fiber membrane behaves as a rectifier rather than a simple resistance. **1955** *Amer. Jrnl. Physiol.* CLXXXIII. 671/1 (*heading*) Rectifier properties of Purkinje fibers. **1958** *Times* 11 Feb. 15/2 Thirty years ago we started with the copper oxide rectifier, and 10 years later we introduced the Westalite selenium rectifier. Last year we added the germanium-type rectifier to our range. **1961** [see *inverter 2 a]. **1970** J. Shepherd et al. *Higher Electr. Engin.* (ed. 2) xxv. 795 In its single-phase form the mercury-arc rectifier consists of a graphite or carbon-coated iron anode and a mercury-pool cathode enclosed in an envelope from which all air has been removed.

4. rectifier (**photo-**)**cell, rectifier photo-electric cell,** a photovoltaic cell.

1933 *Sci. Proc. R. Dublin Soc.* XX. 538 The selenium rectifier cell of Bergmann. **1935** *Discovery* July 214/2 The so-called rectifier cell has been developed, with its great convenience of being able to dispense with batteries. **1936** [see *light-sensitive* s.v. *light *sb.* 15]. **1952** *Jrnl. Sci. Instrum.* XXIX. 137 Correcting the deviation from the theoretical value of the response of selenium rectifier photo-cells to obliquely incident light.

rectify, *v.* Add: **1. c.** *Photogrammetry.* To correct errors of perspective in (an oblique aerial photograph, or a position derived from one) in order to obtain a plan view.

1919 *Geogr. Jrnl.* LIII. 390 This method..is..not so satisfactory as our method of rectifying the print in a camera. **1928** *Ibid.* LXXI. 591 The photographed positions of two control points are rectified. **1944** P. G. McCurdy et al. *Man. Photogrammetry* x. 440 (*heading*) Control and computation to rectify the individual photographs. **1979** *Sci. Amer.* Apr. 28/3 Essentially an orthophoto is an aerial photograph which has been rectified and on which contour lines, spot heights, and other information are superimposed.

7. c. Substitute for def.: To permit (an electric current) to flow preferentially in one direction; *esp.* in *Electr.*, to convert (an alternating current) into a direct current. (Earlier and further examples.)

1892 S. P. Thompson *Dynamo-Electric Machinery* (ed. 4) xxiii. 652 A commutator, which rectified the alternations. **1895** *Electrician* 9 Aug. 488/1 The town supply of current at a pressure of 3,000 volts is transformed down to 65 volts and rectified. **1922** *Encycl. Brit.* XXXII. 1024/2 It [*sc.* the valve] can..be used to separate out the two constituents of a high frequency alternating current and 'rectify' them into a direct current. **1962** *Jrnl. Physiol.* CLXIII. 111 The 2–3 membrane was assumed to be permeable only to potassium and to rectify anomalously: the extent of this rectification is illustrated. **1962** A. Lytel *Industr. Electronics* ii. 38 These rectifiers are two-element tubes and are used in power supplies to rectify or convert alternating current to direct current. **1964** B. V. Rollin *Introd. Electronics* v. 65 To rectify a signal by mixing it with a reference voltage of the same frequency and observing the resulting d.c. output.

rectifying, *vbl. sb.* (Later examples.)

1845 G. Dodd *British Manufactures* V. iii. 63 This distillation is the rectifying of the spirit, by which a certain portion of essential oil is removed from it. **1928** *Geogr. Jrnl.* LXXII. 383 When we come to what should be the most interesting chapter on map-plotting, contouring, and rectifying, we find it composed of chunks taken mostly from German propagandist pamphlets. **1960** *Nature* 5 Nov. 495/1 (*heading*) Rectifying properties of heart muscle. **1962** D. F. Shaw *Introd. to Electronics* xii. 246 The rectifying property of a semi-conductor diode is a consequence of the asymmetrical conduction across the contact between a metal and a semi-conductor.

rectifying, *ppl. a.* Add: **1.** (Further examples.)

1892 S. P. Thompson *Dynamo-Electric Machinery* (ed. 4) xxiii. 659 A rectifying commutator. **1906** *Proc. Physical Soc.* XX. 182 Wehnelt..proposed to employ vacuum-tubes with one electrode covered with..oxides and heated, as rectifying valves for alternating currents. **1932** *Geogr. Jrnl.* LXXX. 463 The Panorama-camera..consists of nine lenses working through prisms, and, in conjunction with a special rectifying printing apparatus, produces the equivalent of a photograph taken with a very wide-angled lens. **1944** P. G. McCurdy et al. *Man. Photogrammetry* ii. 54 Rectifying apparatus is often designed so that a change in scale and tilt removal can be made simultaneously. **1947** R. Lee *Electronic Transformers & Circuits* iv. 86 This circuit requires more rectifying tubes.

2. *rectifying column*, a distillation column in which the distillate is subjected to successive stages of purification by continually condensing and redistilling the vapour.

1891 S. P. Sadtler *Handbk. Industr. Org. Chem.* vi. 216 An improved Savalle rectifying column as used generally in French and Belgian distilleries. **1923** W. H. Walker et al. *Princ. Chem. Engin.* xvii. 557 An apparatus in which this direct interchange of heat and consequent condensation and evaporation can take place is called a rectifying column. **1946** *Nature* 20 July 105/1 The work..included the design and construction of..a liquid methane rectifying-column. **1955** Coulson & Richardson *Chem. Engin.* II. xviii. 611 A column apparatus may be used for batch or continuous operation. In the former case the column is mounted on a large boiler or still which holds the charge, and the column is said to be a rectifying column.

rection. Delete *rare* and add later examples.

1953 [see *intraverbal* s.v. *intra-[1]]. **1968** J. Lyons *Introd. Theoret. Linguistics* vi. 241 Concord..is usually distinguished from *government* (or 'rection', in the usage of some authors).

Hence **re·ctional** *a.*

1938 *Trans. Philol. Soc.* 115 Where..a given declension implies a given gender, the inflexion may be said to express a rectional category in addition to the categories of case and number. **1949** *Archivum Linguisticum* I. ii. 167 Notionally noun compounds are either attributive or rectional (governing).

rectitudinous, *a.* Add: (Later examples.) Also as *sb.*

1906 F. S. Oliver *Alexander Hamilton* v. ii. 381 The rectitudinous inquisition that is enjoyed under the freedom of the press. **1966** 'W. Cooper' *Mem. New Man* i. ii. 27 We were both wearing soberly rectitudinous dark clothes. **1978** *N.Y. Rev. Bks.* 18 May 23/1 It is not so much an appetite for hypocrisy as for the sententious and rectitudinous—for 'expressin' right'—that is a vital impulse in the American folk character.

‖ **rectius** (re·ktiŏs), *adv.* [L., compar. of *recte *adv.*] More correctly: used similarly to *recte *adv.* 1.

1932 *N. & Q.* 6 Feb. 103/1 'Gallinatia': (*rectius* 'Galimatia(s)')... The more correct spelling is Galimatia(s). **1965** J. S. Roskell *Commons & their Speakers* i. 12 A London chronicle says that he was..replaced by William (*rectius* John) Doreward. **1980** *Daily Tel.* 13 Aug. 14, I am by no means sure that Mr Anthony Powell should be allowed to rebuke the editors..for 'oddly' referring to Garter King at Arms (*rectius*, Garter King of Arms).

recto, *sb.* and *adv.* Add: **A.** *sb.* Also in *Palæography*, the front of a leaf of manuscript.

1849 D. Rock *Ch. of Fathers* I. i. iii. 280 The verses, in a very old hand, at the recto of fol. 258. **1964** F. Bowers *Bibliogr. & Textual Crit.* iii. vi. 84 A textual critic can find the most desirable combination of recto and verso settings. **1978** *Bodl. Libr. Record* IX. 324 The writing exercises..are confined to the rectos of the pages.

recto-. Add: Also more widely in *Surg.* with the sense 'of or pertaining to the rectum', as **re·ctopexy** [*-pexy], the fixation of a prolapsed rectum; **re·ctoscope** [-scope], an instrument for use in rectoscopy; **recto·scopy,** visual examination of the rectum.

a **1898** *Syd. Soc. Lex.*, Rectopexy. **1902** J. P. Tuttle *Treat. Dis. Anus* xvii. 691 (*caption*) Rectopexy for procidentia recti—the incision. **1977** *Lancet* 22 Jan. 170/2 On 10 patients a transabdominal rectopexy was performed, using a modified Ripstein procedure. **1890** Billings *Med. Dict.* 442/2 *Rectoscope*, speculum for the rectum. **1906** P. L. Mummery *Sigmoidoscope* i. 7 There are several patterns of pneumatic sigmoidoscope or rectoscope now in use. **1977** *Time* 17 Jan. 51/1 The surgeon cuts directly through the urethra with a marvellous combination of scalpel and fibre-optics looking glass called a rectoscope. *a* **1898** *Syd. Soc. Lex.*, Rectoscopy. **1909** *Index-Catal. Library Surgeon-General's Office, U.S. Army* XIV. 343/1 Rectoscopy. *See* Rectum (Exploration of). **1967** N. S. Kapany *Fiber Optics* vii. 171 The application of flexible fiberscopes not only to gastroscopy, but also to..rectoscopy.

rector. Add: **3. a.** Now also in the Church of England, the leader of a team ministry. In the Roman Catholic Church, a parish priest.

1923 S. Kaye-Smith *End of House of Alard* ii. 117 They came to the cottage where the Rector lived, instead of in the twenty-five roomed Rectory. **1927** *Catholic Times* 11 Feb. 21/2 In 1901 he became rector of St. Joseph's, Birkenhead. **1972** *Daily Tel.* 7 Aug. 10/5 Only the leader of the team, usually called 'Rector', is the beneficed freehold incumbent. **1977** Macmorran & Elphinstone *Handbk. for Churchwardens* (new ed.) vii. 66 In the context of a team ministry.. the incumbent of the benefice or benefices to which a team ministry extends is always styled 'rector'... The other members of the ministry are styled 'vicars'.

4. a. (Later examples.)

Now used more widely in Eng. to designate the principal of a higher educational institution, as the Royal College of Art, the Imperial College of Science and Technology, Liverpool Polytechnic, etc.

1916 Joyce *Portrait of Artist* (1969) i. 55 All the fellows would make fun and talk about young Dedalus going up to the rector to tell on the prefect of studies. **1950** *Chambers's Encycl.* XIII. 658/2 [Sir H. T. Tizard] was..rector of the Imperial College of Science and Technology 1929–42. **1973** *Stornoway Gaz.* 3 Mar. 1/4 The Nicolson [School] has had its share of brilliant rectors, teachers and pupils.

rectoral, *a.* Delete '(Said only of God.)' and add further examples.

1754 *Session Papers, Petition T. Tullideph* (Court of Session, Scotland) 5 Mar. 4 The University Meeting requested the Rector to hold a Rectoral Court against the next Day [at St. Andrews]. **1919** A. Gordon *Cheshire Classis* 121 Some of the above provisions cannot fail to remind us of Richard Baxter's 'rectoral' theory of the ministerial office.

rectorial, *a.* Add: **B.** *sb. Sc.* A rectorial election.

1899 *Student* (Edin. Univ.) 2 Nov. 41 One student writes protesting against the enormities of the Rectorial. **1920** *Glasgow Herald* 27 Nov. 6 The Scottish Universities, to whose noisy 'Rectorials' Viscount Bryce made reference. **1923** *Ibid.* 26 July 6 St David's Day..is March 1, the day of the Rectorial. *Ibid.*, The torchlight procession on the night of the Rectorial. **1968** *Guardian* 30 Oct. 16/4 Edinburgh has never known a rectorial like this.

rectorite (re·ktŏrait). *Min.* [See quot. 1891 and -ite[1].] An aluminosilicate of sodium that is a clay mineral of the montmorillonite group and occurs as large, soft white leaves or plates.

1891 Brackett & Williams in *Amer. Jrnl. Sci.* CXLII. 16 The second hydrous silicate of alumina..is found in the Blue Mountain mining district in Marble Township, Garland county [Arkansas]... We propose the name Rectorite for this,..in honor of Hon. E. W. Rector, of Hot Springs, Ark., who originated and has so unceasingly supported..the bills providing for the Geological Survey of Arkansas. **1950** *Amer. Mineralogist* XXXV. 590 The structural scheme of rectorite consists of contiguous pairs of pyrophyllite-like units separated by pairs of layers of water molecules. An equally apt description would be the alternation of one pyrophyllite unit with one vermiculite unit. **1970** *Clays & Clay Minerals* XVIII. 239 Pyrophyllite is widespread in pelitic rocks of the Manning Canyon Shale in north central Utah, and the association of this mineral with other clay minerals, especially rectorite, is related to the origin. The regular mixed-layer clay mineral rectorite seems to form as a result of the alteration of muscovite–paragonite during late stages of diagenesis... Pyrophyllite subsequently formed from the alteration of rectorite during advancing metamorphism.

rectory. 1. For † *Obs. rare*[−1] read '*Obs.* exc. *Hist.*' and add later example.

1919 J. E. H. Thomson *Mem. T. Dunlop* ii. 19 All that remained of the Rectory or Pedagogy, that in pre-Reformation days represented the later University.

rectosigmoid (rektosi·gmoid), *sb.* and *a.* *Med.* [f. Recto- + Sigmoid.] **A.** *sb.* The region of the junction of the rectum and the sigmoid. **B.** *adj.* Of, pertaining to, or being this region.

1912 *Trans. Amer. Surg. Assoc.* XIII. 159 Carcinoma of the rectum and rectosigmoid remains a local condition until a late stage. **1913** *Jrnl. Amer. Med. Assoc.* 18 Oct. 1489/1 (*heading*) Villous polypus of recto-sigmoid juncture removed by ligation and clamp. **1961** *Lancet* 16 Sept. 624/2 Barium-enema showed a narrow rectosigmoid, with the calibre increasing near the level of the iliac crest. **1962** *Ibid.* 5 May 951/2 Neoplasm of the rectosigmoid region and sigmoid colon. **1977** *Proc. R. Soc. Med.* LXX. 273/1 Anterior resection of the rectum..for..adenocarcinoma of the rectosigmoid.

Also **re:cto-sigmoi·dal** *a.*

1902 J. P. Tuttle *Treat. Dis. Anus* xvii. 678 Any neoplasm of the sigmoid or upper portion of the rectum may induce a gradual descent until the growth reaches a resting-place in the ampulla of the rectum... Thus, unusual contracture at the recto-sigmoidal juncture..will cause the arrest of the faecal masses. **1914** *Brit. Jrnl. Surg.* I. 683 (*caption*) Small but well-marked recto-sigmoidal anastomosis.

rectress. Add: **3.** = Rectoress 2. *rare.*

1906 *Month* July 66 The rector and rectress, and their two delicate-looking, perfectly-dressed daughters.

‖ **rectus** (re·ktŏs). [L., = straight, right.] **1.** *Mus.* In a fugal composition, the version of a theme performed in the basic or original, as opposed to the reversed or inverted, order. *rectus et inversus* [L., 'inverted'] = *recte et retro* s.v. *recte *adv.* 2.

1931 D. F. Tovey *Compan. to 'Art of Fugue'* 31 The next two fugues are *tours de force*, being compositions that can be inverted note for note from beginning to end. For purposes of comparison Bach writes his *Inversus* under his *Rectus.* **1938** *Oxf. Compan. Mus.* 134/2 The Canon Cancrizans.. other names for it are Canon Recte et Retro or Rectus et Inversus. **1959** *Listener* 30 July 189/3 The English Rite puts the *Gloria* at the end. Rubbra starts it as an elaborate double fugue *rectus et inversus.* **1960** *Times* 30 Nov. 17/3 Fugue, *rectus et inversus*, canon, ostinato..shall these dry bones live? Britten..uses them all. **1962** *Listener* 27 Dec. 1109/2 The rectus versions of the 'mirror' fugues XII and XIII.

2. *Law.* Phr. *rectus in curia* [lit. 'right in court'], innocent, acquitted, set right in point of law.

[*a* **1135** in B. Thorpe *Anc. Laws & Inst. Eng.* (1840) I. 240 Omni domino licet submonire hominem suum, ut ei sit ad rectum in curia sua.] *a* **1577** T. Smith *De Republica Anglorum* (1583) ii. iii. 46 Yet with a clause, *modo stet rectus in curia*, that is to say, that no man obiect against the offendor. **1645** J. Howell *Lett.* iii. vii. 59 He is now come to be again *rectus in curia*, absolutely acquitted and restor'd to all things. *a* **1706** Evelyn *Diary* an. 1680 (1955) IV. 229 Saturday came other Witnesses of the Commons.. who tooke off all the former days objections, & set the Kings Witnesses *recti in Curia.* **1816** *Edin. Rev.* XXVII. 122 He must come *rectus in curia*, and swear to the falsehood of the libel. **1866** J. G. Murphy *Crit. Comm. Exodus* 310 We meet with the propitiation or atoning sacrifice, by which they become *recti in curia*, right in point of law. **1934** E. Pound *Eleven New Cantos* xxxii. 7, I pray you place me rectus in curia in this business with the Emperor.

recueillement. Delete *rare* and add later examples.

1897 G. du Maurier *Martian* i. 35 The deep stillness and studious *recueillement* that brood over the scene. **1903** E. Wharton *Sanctuary* II. v. 150 The silence, the *recueillement*, about her. **1931** R. Fry in W. Rose *Outl. Mod. Knowledge* 936 It succeeds in arousing a mood of *recueillement* not unlike that which emanates from some of Giorgione's compositions. **1977** *Times Lit. Suppl.* 20 May 610/4 Every writer required his *recueillement* (a word that appears several times in his letters), his time of in-gathering and collecting.

‖ **reculer pour mieux sauter** (rəküle pur myö sote), *phr.* [Fr., lit. 'to draw back in order to leap better'.] Making use of a withdrawal or setback in such a way as to advance or succeed all the more.

[*c* **1500** J. D'Arras *Mesuline* (1895) xx. 113 Alwayes wyse men goo abacke for to lepe the ferther.] **1820** Lady Granville *Let.* 30 Aug. (1894) I. 170 'Yes,' he said, 'with people in general it is *reculer pour mieux sauter*, with her *sauter pour mieux reculer*.' **1907** G. B. Shaw *Let.* 21 Mar. (1972) II. 675 The sales that are influenced by my name go up steadily from year to year; and all the apparent slumps are cases of 'reculer pour mieux sauter'. **1920** D. H. Lawrence *Women in Love* i. 6, I think my coming back home was just *reculer pour mieux sauter.* **1951** R. F. Harrod *Life J. M. Keynes* x. 431 This was, in his mind, a case of *reculer pour mieux sauter.* **1972** *Times* 30 May (Hongkong Suppl.) p. v/4 The hesitation in currency circulation in 1967–1968 following the fall in deposits was a matter of *reculer pour mieux sauter*: growth..took off again.

recultivation. (Earlier example.)

1850 *Proc. Philol. Soc.* IV. 208 It is now just about a hundred years ago that Klopstock paved the way to the recultivation of German.

recumb, *v.* For † *Obs.* read *rare* and add later examples.

1906 M. Dods *Later Lett.* (1911) 213 Will you excuse pencil, as I am in a run down condition and my doctor bids me 'recumb' as much as I can. **1925** O. W. Holmes in *Holmes–Laski Lett.* (1953) I. 693 Now I shall recumb with a hot bottle at my back.

recumbent, *a.* (and *sb.*) Add: **A.** *adj.* **1. b.** (Later examples.) *recumbent stone circle*, in *Archæol.*, a stone circle characterized by the presence of one large stone lying flat flanked by two tall uprights.

1933 V. G. Childe in *Proc. Soc. Antiquaries Scotland* LXVII. 51 Recumbent Stone Circles may have been erected in late Hallstatt times. **1943** J. & C. Hawkes *Prehist. Britain* iii. 59 Their adaptation of a passage-grave tradition in the remarkable 'Recumbent Stone' circles of north-east Scotland. **1962** Gordon & Lavoipierre *Entomol. for Students of Med.* xx. 134 Like the head and thorax the abdomen is covered with long hairs; on the dorsal aspect of the abdomen the hairs may either lie flat (a condition described as recumbent) or they may be raised. **1963** *Field Archaeol.* (Ordnance Survey) (ed. 4) 40 There is an eccentric type found in North-east Scotland... This is the 'Recumbent' stone circle which has more than seventy examples in this area.

3. Geol. *recumbent fold*, a fold whose axial plane is nearly horizontal; so *recumbent anticline, syncline.*

1909 *Summ. Progr. Geol. Surv.* 1908 52 A discrepancy between the two limbs of the recumbent fold. **1910** *Q. Jrnl. Geol. Soc.* LXVI. 617 The sliding is not confined to the lower limbs of recumbent anticlines. *Ibid.*, The schists of the Highlands of Scotland are disposed in a succession of recumbent folds of enormous amplitude. **1922** [see *nappe 2]. **1937** [see *fan *sb.*[1] 10 f]. **1962** Read & Watson *Introd. Geol.* I. viii. 451 A recumbent anticline..faces laterally and a recumbent anticline in which the hinge-region sags or droops faces downwards. **1964** W. C. Putnam *Geol.* vi. 131/1 In an extreme case, the whole fold may be forced over on its side so that its axial plane..is horizontal, or very nearly so... Such a structure is called a recumbent fold. **1969** H. Robinson *Morphology & Landscape* iii. 34 Sometimes the pressure exerted upon a recumbent fold is sufficiently great to cause it to be torn from its roots and to be thrust forward.

recuperate, *v.* Add: Also with pronunc. (rĭkū·pĕrēɪt). **1. a.** Delete † *Obs.* and add later examples.

1896 J. A. H. Murray *Let.* 22 Apr. in K. M. E. Murray *Caught in Web of Words* (1977) xiv. 279 We reached the summit however, & recuperated our energies. **1977** *Guardian Weekly* 7 Aug. 11/5 No plant in the world has shown that it can recuperate plutonium on an industrial scale from oxide-bearing fuel.

b. (Later example.)

1924 *Proc. Classical Assoc.* 13 Both these Associations have fully recuperated any loss which they had made during the war.

Hence **recu·perating** *vbl. sb.* and *ppl. a.*

1894 'R. Andom' *We Three & Troddles* xix. 174 We each mentioned our favoured recuperating localities. **1979** B. Parvin *Deadly Dyke* xxiv. 128 Recuperating firemen stood watching the thick palls of smoke.

recuperation. Add: **3.** *Gunnery.* The action of a recuperator (sense *4).

1922 *Encycl. Brit.* XXXI. 1185/2 As the buffer flow-space is greatest at the termination of recuperation, some check is required to prevent a metal-to-metal blow. *Ibid.* 1186/1 Recuperation may be by means of steel springs or compressed air.

recuperative, *a.* (and *sb.*) Add: **A.** *adj.* **5.** Of, pertaining to, or being a recuperator (sense 2), or an air heater using the same principle.

1906 A. L. J. Queneau tr. *Damour's Industr. Furnaces* x. 142 The volume..of the recuperative chambers should be calculated to suit the exchange of the calories to be effected, according to the specific heats of the recuperating refractory bricks. **1923** *Iron Age* CXI. 1782 The recuperative installation is adopted where blast furnace or coke oven gas is available. **1930** *Engineering* 31 Jan. 155/2 Two methods of transferring heat from a hot gas to a cold one were in use, and might be distinguished as belonging, respectively, to the recuperative and to the regenerative type. In the recuperative type, the cool and the hot gases were separated by a conducting wall through which the transfer of heat took place. **1938** *Jrnl. Iron & Steel Inst.* CXXXVIII. 327P Of the 848 pit holes surveyed,..6·2 % are of the one-way-fired recuperative type, 2·1 % are of the bottom-fired recuperative type. **1962** [see *recuperator 2]. **1971** B. Scharf *Engin. & its Language* xiv. 204 Air heaters are classified as recuperative or regenerative, and the recuperative heaters are further subdivided into tubular or plate-type heaters.

recuperator. Add: **2.** (Further examples.) Now restricted to a form of heat exchanger in which hot waste gases, being conducted continuously along a system of flues, impart heat to incoming air or gaseous fuel flowing in the opposite direction in parallel flues by conduction through the dividing walls. Cf. Regenerator 2.

1906 A. L. J. Queneau tr. *Damour's Industr. Furnaces* ii. 47 Two systems are still in practice—the Siemens recuperator with inversion, and recuperation without inversion, by parallel counter currents. **1911** F. W. Harbord *Metallurgy of Steel* II. xxi. 541 Gorman's furnace..formed a further step in advance, as the waste gases on their way to the chimney passed around horizontal fireclay pipes or 'recuperators'. **1938** H. Etherington *Mod. Furnace Technol.*

vii. 314 In a recuperator, the hot waste gases and the cold air are led through separate channels in close contact.. Regenerators operate on a different principle. **1953** D. J. O. Brandt *Manuf. Iron & Steel* xxviii. 203 Soaking pits are of two kinds, regenerative pits, which are fired in two directions, being reversed at intervals, and recuperative which are fired in one direction only, the heat in the outgoing gases being as far as possible transferred to the incoming air and fuel in a recuperator. **1962** G. R. Bashforth *Manuf. Iron & Steel* IV. ii. 39 In the recuperative type of soaking pit, the flow of fuel and air is maintained in one direction... The waste products of combustion pass through a recuperative chamber... These recuperators may either be of the refractory or metallic type.

3. That which restores one's health or spirits.

1905 *Smart Set* 17 Sept. 24 A/2 (Advt.), A day trip on these steamers is calculated to brace the entire system, and the jaded business man will find them a splendid recuperator.

4. *Gunnery.* (See quot. 1922.)

1918 E. S. Farrow *Dict. Mil. Terms* 498 *Recuperator gauge*, in artillery, a gauge for verifying the charge of the recuperator, in liquid and in compressed gas. **1922** *Encycl. Brit.* XXXI. 1184/1 The recuperator returns the gun to the firing position after it has come to rest under the action of the recoil resistance. **1925** *Jrnl. R. Artillery* LII. 38 The recuperator question was taken in hand early on in the war,a nd by the end of 1918 all springs had been replaced by air recuperators. **1962** *Ordnance Techn. Terminol.* (U.S. Army Ordnance School) (AD 660 112) 86/1 *Counterrecoil mechanism*, a hydraulic, pneumatic, or mechanical system that returns a gun into battery, or firing position, after recoil; a recuperator.

Hence **recu:perato·rial** *a.*

1976 J. M. Kelly *Stud. in Civil Judicature of Roman Republic* ii. 47 If then we discard the dominant theory, how are we to explain the special recuperatorial jurisdiction otherwise?

recurb (rĭkɒ̄·ɹb). *rare*⁻¹. [f. Re- + Curb *sb.*: cf. F. *recourber* vb., L. *recurvāre* Recurve *v.*] The curved shape produced at the repeated climax of systematic oscillation.

1876 G. M. Hopkins *Wreck of Deutschland* xxxii, in *Poems* (1967) 62 The recurb and the recovery of the gulf's side, The girth of it and the wharf of it and the wall.

recurrable (rĭkɒ̄·răb'l), *a.* [f. Recur *v.* + -able.] That can recur.

1935 E. Pound *Let.* 17 Apr. (1971) 273, I don't know that I have been clear enough re *recurrable* epithets—either to be simple and natural so that repeat don't worry one, or else strange and part of definite intended stylization.

recurrence. Add: **6.** *attrib.* and *Comb.*, as *recurrence frequency, interval*; **recurrence formula, relation** *Math.*, an expression which defines the general member of a series in terms of the preceding members; **recurrence surface** [tr. Sw. *rekurrensyta* (E. Granlund 1932, in *Sveriges Geologiska Undersökn.* Ser. C. No. 373. viii. 73)], a horizon in a peat bog between highly decomposed and slightly decomposed peat, indicating the commencement of a period of active peat growth; **recurrence time** *Math.*, the time between two successive occasions when a Markov process enters any given state.

1902 E. T. Whittaker *Course Mod. Analysis* x. 210 The recurrence-formulae. We proceed to establish a group of formulae which connect Legendre functions of different orders. **1925** *Biometrika* XVII. 165 (*heading*) Recurrence formulae for the moments of the point binomial. **1965** *Wireless World* Sept. 431/1 It remains now to provide a suitable pulse generator of variable recurrence frequency to fire the thyristor. **1965** R. G. Kazmann *Mod. Hydrol.* iv. 76 Statistical studies made to determine the recurrence interval of this design-flood resulted in figures ranging from 1000 to 90,000 years. **1933** *Biometrika* XXV. 420 (*heading*) On a recurrence relation connected with..double Bessel functions. **1961** M. M. Nicolson *Fund. & Tech. Math. for Scientists* xvi. 369 A set of formulae relating Legendre polynomials of different orders *n*; such relations are called recurrence relations. **1979** Page & Wilson *Introd. Computational Combinatorics* ii. 5 If such a recurrence relation can be produced, it can usually be made the basis of an algorithm for computing values of the desired function. [**1934** *Irish Naturalists' Jrnl.* V. 134 To look for Granlund's 'rekurrenz-surfaces'.] **1938** *New Phytologist* XXXVII. 452 Granlund suggests that such layers are due to slowing up of bog growth by unfavourable conditions, and to the level marking the sudden renewal of growth he gives the name 'Rekurrenzflache [*sic*]', which has been translated by Jessen as Recurrence-surface. **1956** H. Godwin *Hist. Brit. Flora* iii. 34/2 In his work on the raised bogs of Scania, Nilsson..has been able to identify no fewer than nine recurrence surfaces between *c.* 3500 B.C. and the present day. **1975** J. G. Evans *Environment Early Man Brit. Isles* iv. 77 Resumption of peat growth, leading to the formation of 'recurrence surfaces', takes place when conditions of high rainfall return. [**1943** *Rev. Mod. Physics* XV. 54/2 (*heading*) The average time of recurrence of a state of fluctuation in which the molecular concentration in a sphere of air of radius *a* will differ from the average value by 1 percent.] **1949** *Trans. Amer. Math. Soc.* LXVII. 99 A new method of finding the second moment of the recurrence times of finite or infinite Markov chains. **1971** R. A. Howard *Dynamic Probabilistic Syst.* I. v. 287 θ_{ii} is the number of transitions between a departure from state *i* and the first return to *i*; it is called the first passage time from state *i* to state *i*, or the recurrence time of state *i*.

recurrency. **2.** For † *Obs.* read *rare* and add later example.

1928 H. Poutsma *Gram. Late Mod. Eng.* (ed. 2) I. I. i. 83 The principal verbs used to express recurrency, i.e. the iterative aspect of an action..are *can, to use* and *will*.

recurrent, *a.* and *sb.* Add: **A.** *adj.* **3. d.** Math. *recurrent relation* = **recurrence relation*.

1896 *Phil. Trans. R. Soc.* A. CLXXXVII. 522 These recurrent relations between the functions for different values of *n* hold for general complex values of *m* and *n*. **1931** E. W. Hobson *Theory Spherical & Ellipsoidal Harmonics* ii. 67 Recurrent relations..between the functions $Q_n(\mu)$ for different values of *n*.

recurring, *ppl. a.* and *vbl. sb.* The stressed syllable now usually has the same sound as the infinitive, not that of *recurrent*.

recurringly (rĭkɒ̄·riŋli), *adv.* [f. Recurring *ppl. a.* + -ly².] In a recurring manner; repeatedly.

1915 'A. Hope' *Young Man's Year* xxix. 272 It pointedly and recurringly reminded him that there were more women than one in the world. **1923** *Daily Mail* 10 Aug. 4/2 A bogus manager..is 'a manager who engages artists and recurringly fails to pay their salaries'.

recursion. Restrict 'Now *rare* or *Obs.*' to sense in Dict. and add: **2. a.** The application or use of a recursive procedure or definition; *primitive recursion* [tr. G. *primitive rekursion* (R. Péter 1934, in *Math. Ann.* CX. 613)], definition of a function of natural numbers by induction on a single argument or (equivalently) by simple recursion formulæ; *recursion formula*, an equation relating the value of a function for a given value of its argument (or arguments) to its values for other values of the argument(s).

[**1871** *Math. Annalen* IV. 113 Man hat also für die Funktion $R^{m,\nu}$ folgende Recursionsformel: $2(\nu-1)/z \cdot R^{m,\nu} = R^{m+1,\nu-1} + R^{m-1,\nu+1}$.] **1930** *Proc. London Math. Soc.* XXX. 267 For other values of *r* we define *f*(*r*, *n*, *k*) by recursion formulae. **1933** *Ann. Math.* XXXIV. 863 The recursion formulas, $m+1 = S(m)$, and $m+(k+1) = S(m+k)$. **1934**, **1974** [see *recursive *a.* 2 a]. **1943** *Trans. Amer. Math. Soc.* LIII. 42 Schema (I) introduces the successor function,..and Schema (V) the schema of primitive recursion. **1961** *Commun. Assoc. Computing Machinery* IV. 65/1 The growing extent and direction of application of recursion in programming research. **1964** E. Bach *Introd. Transformational Gram.* iii. 46 Care must be taken to ensure that unwanted recursion (looping) does not occur. **1967** Klerer & Korn *Digital Computer User's Handbk.* i. 167 Even if recursive procedures are explicitly outlawed,..recrsuion can take place unwittingly. **1972** R. A. Palmatier *Gloss. Eng. Transformational Gram.* 142 Recursion is restricted to the transformational component of the grammar. **1973** C. W. Gear *Introd. Computer Sci.* v. 232 Fortran does not allow recursion. **1975** F. R. Palmer in W. F. Bolton *Eng. Lang.* i. 34 The structure of language involves 'recursion' of the kind illustrated by 'This is the house that Jack built', 'This is the mouse that lived in the house that Jack built' and so on—if necessary *ad infinitum*.

b. A recursive definition.

1936 *Math. Ann.* CXII. 727 There are other definitions of this sort, e.g. certain recursions with respect to two or more variables simultaneously, which cannot be reduced to a succession of substitutions and ordinary recursions. **1963** W. V. Quine *Set Theory* § 11. 79 There are the familiar so-called recursive definitions or recursions. **1966** N. Chomsky *Topics Theory Generative Gram.* ii. 33 An utterly fantastic proposal, namely, that a grammar should contain no recursions in its system of rules. **1971** *Computers & Humanities* V. 155 Algol is more powerful in that it allows recursions, has block structure, and permits expressions in many places.

recursive (rĭkɒ̄·ɹsiv), *a.* [f. L. *recurs-* (see Recursant *a.*) + -ive.] **1.** Periodically or continually recurring. Now *rare* or *obs.*

1790 *Loiterer* 13 Mar. 7 Till your ear be so attuned to one particular measure, that your ideas may be spontaneously absorbed into the same revolving eddy of recursive harmony.

2. a. *Math.* and *Logic.* [after similar uses of G. *rekurrent* (D. Hilbert 1904, in *Verhandl. des dritten Internat. Math. Kongr.*), *rekursiv* (K. Gödel 1931, in *Monatshefte f. Math. u. Physik* XXXVIII. 179).] Involving or being a repeated procedure such that the required result at each step except the last is given in terms of the result(s) of the next step, until after a finite number of steps a terminus is reached with an outright evaluation of the result; *recursive definition*, a definition (of a function) which is either primitive recursive or (now usu.) general recursive; *recursive function*, a function which has or which may be given a recursive definition; *recursive relation*, a property of, or relation between, natural numbers whose truth value for all arguments is a recursive function; *recursive set*, a set of natural numbers whose defining property is recursive; *general recursive* adj. phr., applied to a function or relation which is recursive and is defined for all natural number values of its arguments; *partial recursive* adj. phr., applied to a function defined by a recursive process which for some or all values of the arguments does not terminate, leaving the value of the function undefined; *primitive recursive* adj. phr., applied to a function or relation which can be generated by primitive recursion and substitution from the zero, successor, and identity functions.

1934 Kleene & Rosser *Gödel's Undecidable Propositions Formal Math. Syst.* (typescript) 3 We define the class of recursive functions to be the totality of functions which can be generated by substitution..and recursion..from the successor function *x*+1, constant functions.., and identity functions. *Ibid.*, A relation R shall be recursive if the representing function is recursive. *Ibid.*, Recursive functions have the important property that, for each given set of values of the arguments, the value of the function can be computed by a finite procedure. Similarly, recursive relations (classes) are decidable in the sense that, for each given set of natural numbers, it can be determined by a finite procedure whether the relation holds or does not hold... The functions *x*+*y*, *xy*, x^y and *x*! are clearly recursive. **1936** *Math. Ann.* CXII. 727 In this paper we offer several observations on general recursive functions, using essentially Gödel's form of the definition. *Ibid.*, Ordinary or 'primitive' recursive functions. *Ibid.* 729 A recursive function (relation) in the sense of Gödel..will now be called a primitive recursive function (relation). **1938** *Jrnl. Symbolic Logic* III. 151 If we omit the requirement that the computation process always terminate, we obtain a more general class of functions, each function of which is defined over a subset (possibly null or total) of the *n*-tuples of natural numbers... These functions we call partial recursive. **1940** *Mind* XLIX. 240 Preliminary considerations, such as..the exact specification of the rules for the use of recursive definitions. **1943** *Mind* LII. 268 Quite elementary theorems, requiring for their proofs recursive arguments to take care of the indefinite number of variables involved. **1943** *Trans. Amer. Math. Soc.* LIII. 44 A system E of equations defines recursively a general recursive function of *n* variables if, for each set $x_1, \ldots, x_n$ of natural numbers, an equation of the form $f(x_1,\ldots, x_n) = x$, where f is the principal function symbol of E, and where $x_1, \ldots, x_n$ are the numerals representing the natural numbers $x_1, \ldots, x_n$, is derivable from E..for *exactly* one numeral *x*. **1944** *Bull. Amer. Math. Soc.* L. 285 In the present paper, 'recursive function' means 'general recursive function'. *Ibid.* 288 Closely related to the technical concept [of a] recursively enumerable set of positive integers is that of a recursive set of positive integers. This is a set for which there is a recursive function *f*(*x*) such that *f*(*x*) is say 2 when *x* is a positive integer in the set, 1 when *x* is a positive integer not in the set. We may also make this the definition of the set being recursively soluble. For 2 and 1 may be regarded as the two possible truth-values, true, false, of the proposition 'positive integer *x* is in the set'. **1962** R. B. Braithwaite in B. Meltzer tr. *Gödel's Formally Undecidable Propositions* 12 An arithmetical function is recursive if it is the last term in a finite sequence of functions in which each function is recursively defined by a rule involving two functions preceding it in the sequence (or is the successor function or a constant or obtained by substitution from a preceding function). **1964** E. Mendelson *Introd. Math. Logic* 125 Relations obtained from primitive recursive (or recursive) relations by means of the propositional connectives and the bounded quantifiers are also primitive recursive (or recursive). **1965** Herman & Plassman tr. *H. Hermes' Enumerability, Decidability, Computability* i. 29 Today it is generally believed that *every* system of algorithms can be defined by recursive functions. This gives a deeper meaning to Gödel's result. *Ibid.* iii. 82 The essence of Ackermann's proof of the existence of a computable function which is not primitive recursive consists in defining a computable function which increases in a certain sense faster than any primitive recursive function. **1967** *Encycl. Philos.* VII. 92/1 This Herbrand–Gödel–Kleene notion of general recursive function can be put in the context of instructions and computations discussed above. **1970** *Nature* 19 Dec. 1234/1 Turing formulated his concept of an abstract computing machine; the functions computable by these machines are exactly the recursive functions. **1974** A. Kenny tr. *Wittgenstein's Philos. Gram.* 34 Is there a further step from writing the recursive proof to the generalization? Doesn't the recursion schema already say all that is to be said?

b. *Linguistics.* Applied to a grammatical feature or element which may be involved in a procedure whereby that feature or element is repeatedly reintroduced; applied to a grammatical rule in which part of the output serves as input to the same rule.

1955 N. Chomsky *Logical Struct. Linguistic Theory* (microfilm, Mass. Inst. Technol.) vi. 248 We will find many other reasons to question the validity of the extension of the notion of *production* to *recursive production*. **1957** —— in *Janua Linguarum* IV. 57 Bar-Hillel has suggested..that Pike's proposals can be formalized without the circularity that many sense in them by the use of recursive definitions. **1968** J. Lyons *Introd. Theoret. Linguistics* vii. 326 The adverb is a recursive category..in the sense that one adverb may modify another. **1970** —— *Chomsky* viii. 90 It will be observed that rules (2), (3) and (4) are recursive, but in different ways. Rule (2) is left recursive; rule (3) is right recursive; and rule (4) is self-embedding. **1972** R. A. Palmatier *Gloss. Eng. Transformational Gram.* 142 A recursive rule is a rule which reapplies indefinitely to its own output... The recursive power of a grammar, which resides entirely in the syntactic component, is its ability to generate an infinity of sentences... The recursive mechanism is the system of rules which account for the infinite properties of language... A recursive element is one from which strings can be derived that contain the same element... The recursive property of a grammar..is its provision for embedding sentences within other sentences. **1977** *Word 1972* XXVIII. 336 Logicians of the first half of the century had developed and used recursive grammars with such clarity that Chomsky's 'application' can hardly be regarded as a *tour de force*.

c. *Computing.* Applied to a statement, definition, subroutine, or the like, some part of which makes use of the whole of itself, so that its explicit interpretation requires in general many successive executions; applied also to languages, compilers, etc., which allow of such techniques.

Quot. 1958 uses the word in a context where 'iterative' would now be usual.

1958 *Commun. Assoc. Computing Machinery* Aug. 10 The idea of recursive curve fitting has been in use for some time as a graphical technique for fitting curves 'by eye' to observational data. **1959** *Numerische Math.* I. 45 The definition of expressions, and their constituents, is necessarily recursive. **1960** *Ibid.* II. 312 It is then impossible to call in a subroutine while one or more previous activations of the same subroutine have not yet come to an end... We intend to describe..a means of removing the..restriction..; hence the name 'recursive programming'. **1973** C. W. Gear *Introd. Computer Sci.* v. 233 We can understand recursive procedures by imagining that many different copies of the procedure are available. **1979** Page & Wilson *Introd. Computational Combinatorics* vi. 136 Since backtrack programming is closely related to tree searching we can consider using recursive techniques in our implementations.

3. *Phonetics.* A term sometimes used to refer to consonants accompanied by glottal closure or implosion. Also as *sb.*

1924 R. L. Turner in *Bull. School Oriental Stud.* III. 304 According to one of my informants, an *m* accompanied by glottal closure and distinguished from ordinary *m*, exists in Magarkurā, one of the Mongolian languages of Nepal. Prince Troubetzkoy refers to consonants in the Caucasian languages accompanied by complete closure of the glottis. These he calls 'recoursives', a convenient term I have anglicized as 'recursives'; he indicates them by a dot above or below the letter. **1934** Webster, *Recursive*, adj.,.. formed with an inward movement of air caused by lowering the larynx with closed glottis;—said of certain consonants in Sindhi (*g*, *j*, *d*, *b*). **1974** *Encycl. Brit. Macropædia* IX. 448/1 One major feature distinguishing Sindhi from the rest of the northwest group is the development of a series of imploded stops (also called suction stops and recursive stops), for *b*, *d*, *j*, and *g*.

recursively (rĭkɒ̄·ɹsivli), *adv.* [f. prec. + -ly².] In a recursive manner: esp. *recursively defined*, having a recursive definition; *recursively enumerable*, (of a set of natural numbers) generated by a general recursive function having one parameter which ranges through all possible values.

1934 Kleene & Rosser *Gödel's Undecidable Propositions Formal Math. Syst.* (typescript) 21 These are arithmetic propositions which involve only recursively defined functions. **1943**, **1944** [see *recursive *a.* 2 a]. **1944** *Bull. Amer. Math. Soc.* L. 285 A set of positive integers is said to be recursively enumerable if there is a recursive function *f*(*x*) of one positive integral variable whose values for positive integral values of *x*, constitute the given set. **1961** *Commun. Assoc. Computing Machinery* IV. 10/2 Many of the constituents of the Algol language are defined recursively. **1962** B. Meltzer tr. *Gödel's Formally Undecidable Propositions* 46 A number-theoretic function $\phi(x_1, x_2, \ldots x_n)$ is said to be recursively defined by the number-theoretic functions $\psi(x_1, x_2, \ldots x_{n-1})$ and $\mu(x_1, x_2, \ldots x_{n+1})$ if for all $x_2, \ldots x_n$, *k* the following hold:

$$\phi(0, x_2, \ldots x_n) = \psi(x_2, \ldots x_n),$$
$$\phi(k+1, x_2, \ldots x_n) = \mu(k, \phi(k, x_2, \ldots x_n), x_2, \ldots x_n).$$

1964 *Mem. Amer. Math. Soc.* LI. 8 An index of ϕ_2 is obtainable primitive recursively from an index of ϕ_1, and vice versa. **1968** [see *nest *v.* 4 b]. **1968** *Language* XLIV. 571 The set of all sets of integers is non-denumerable; the set of all Turing machines is denumerable; each recursively enumerable set corresponds to a Turing machine; therefore there are sets of integers which are not recursively enumerable. **1970** *Ibid.* XLVI. 787 Sentence margins are slots where nuclear sentence patterns may recursively be embedded within other sentences. **1970** Eilenberg & Elgot *Recursiveness* 84 The class of all recursively enumerable sets is countable. **1974** Kernighan & Plauger *Elem. Programming Style* iii. 54 Learning to think recursively takes some effort, but that is repaid with smaller and simpler programs... Use recursive procedures for recursively-defined data structures.

recursiveness (rĭkɒ̄·ɹsivnės). [f. as prec. + -ness.] The property of being recursive.

1936 *Amer. Jrnl. Math.* LVIII. 346 Since the results of the present paper were obtained, it has been shown by Kleene.. that analogous results can be obtained entirely in terms of recursiveness, without making use of λ-definability. **1962** R. B. Braithwaite in B. Meltzer tr. *Gödel's Formally Undecidable Propositions* 11 The notion of recursiveness has played a central part in metamathematics since Gödel's work on it. *Ibid.* 12 The importance of recursiveness for metamathematics in general lies in the fact that recursive definition enables every number in a recursively defined infinite sequence to be *constructed* according to a rule, so that a remark about the infinite sequence can be construed as a remark about the rule of construction and not as a remark about a given infinite totality. **1964** E. Bach *Introd. Transformational Gram.* ii. 16 In order to provide for an infinite number of terminal strings, a grammar must have a basic property called recursiveness. **1970** *Nature* 19 Dec. 1234/1 The concept of recursiveness was formulated in Gödel's sensational work in 1931 on the limitations of mechanical languages. **1975** G. Sampson *Form of Lang.* v. 89 As soon as we allow 'S' to appear on the right-hand side of one of the rules, the grammar acquires the property of 'recursiveness'.

re·curve, *sb.* *Archery.* [f. the vb.] A backward-curving end of the limb of a bow; a bow designed with this feature. Also *attrib.*

1961 E. Burke *Archery* i. 10 (*caption*) Named parts of the bow..recurve..upper limb..bowsight..lower limb.. nock. **1962** G. H. Gillelan *Young Sportsman's Guide to Archery* ii. 19 The other important bow design is the recurve, so named because its tips have a reverse curl. **1979** *Country Life* 26 July 287/1 The 'Bowhunter' style involves any type of bow, usually of a recurve construction, that is a combination of glass-fibre and wood..to buy a new bow works out as follows: long bow, £20–£50; recurve, £150–£180. **1980** *Hunting Ann. 1981* 81/3 For targets of opportunity, a recurve or long bow can be handled faster.

recurved, *ppl. a.* Delete 'in 19th c.' and add later examples.

1925 E. H. Wilson *Lilies E. Asia* 23 Flowers white, fragrant, funnel-shape, dilated, more or less recurved at the apex. **1936** D. Wilkie *Gentians* vi. 33 The calyx is tubular, and the lobes..are ovate and recurved. **1961** J. E. Collin *Brit. Flies* VI. 1. 71 Proboscis not very stout nor much recurved. **1980** *Plantsman* I. 251 Both [lilies]..have delicate pendant flowers with slightly recurved petals.

recusal (rĭkiū·zăl). [f. Recuse *v.* + -al.] An objection to a judge as prejudiced.

1958 *Manch. Guardian* 5 Aug. 5/7 The submission to recusal was based on reports that incorrectly stated the facts. **1980** *Times* 30 Oct. 6/8 They should not feel that by making their applications for his recusal they had prejudiced their case.

recyclable (rīsəi·klăb'l), *a.* [f. *recycle *v.* + -able.] Capable of being recycled.

1971 *New Yorker* 4 Dec. 177 Publishers might do well to encase their pages and bindings in some sort of ersatz (but recyclable) horn. **1972** *Guardian* 30 Oct. 12/1 The shape of cars to come—a disposable commuter car which would be 'recycleable' after perhaps two years. **1976** *Nature* 6 May 66/2 Evidence suggests that net water flow from the rectal lumen to haemolymph results from increases in intercellular space osmotic pressure due to the transport across lateral cell membranes of recyclable intracellular solutes. **1979** *Observer* 30 Dec. 3/8 Theoretically they [*sc.* big plastic bottles] are recyclable, but actually they are constructed so as to make reclamation highly improbable.

Hence **recy:clabi·lity.**

1973 *Sci. Amer.* July 1 In terms of its attributes, its utility and recyclability, glass is a natural. **1978** *Jrnl. R. Soc. Arts* CXXVI. 609/2 Lifetime and recyclability of the product.

recycle (rīsəi·k'l), *v.* Also **re-cycle.** [Re- 5 a.] **1.** *trans.* **a.** To reuse (a material) in an industrial process; to return to a previous stage of a cyclic process.

1926 [implied in *recycling *vbl. sb.*]. **1928** *Jrnl. Inst. Petroleum Technologists* XIV. 766 It is economically more advantageous to stop cracking in the first cycle when coke formation begins and produce more gasoline by re-cycling those fractions which do not form great quantities of coke during cracking. **1929** *Proc. R. Soc.* A. CXXIV. 43 It ought to be possible to obtain nearly the theoretically possible yield by returning to the reaction chamber or 'recycling' all the products formed except the gasoline. **1945** H. D. Smyth *Gen. Acct. Devel. Atomic Energy Mil. Purposes* ix. 100 Any given sample of material is recycled many times. **1958** *Times* 17 Oct. 5/1 It is envisaged that plutonium produced in the working of the reactor will later be recycled through it. **1964** N. G. Clark *Mod. Org. Chem.* iv. 62 Using only a small volume of solvent, which is continually recycled, it is possible to carry out the equivalent of many hundreds of separate extractions. **1972** *Sci. Amer.* Oct. 69/1 Their new process is the first closed-loop, spray-etching system that electrolytically reverses the chemical reaction of etching. It continuously recycles cupric chloride and has reduced the cost of etching wiring boards by over 90%. **1980** *Times* 7 Mar. 25/3 The uranium is recycled back to an enrichment plant to make new thermal-reactor fuel, and the plutonium is stored.

b. *spec.* To reuse (a waste material), to convert (waste) into or *into* a usable form; also, to reclaim (a material) from waste.

1960 *Aeroplane* XCIX. 521/2 It has systems which reduce all organic waste to a small amount of ash and recycle urine and waste water into drinkable water. **1967** *Technology Week* 23 Jan. 34/3 It would allow us to economically desalt sea and brackish water, recycle water from sewage. **1971** *Sci. Amer.* May 95/1 (Advt.), You bring us the cans and we'll recycle them. **1971** *New Yorker* 16 Oct. 33 What you ecology-minded ladies don't realize is that before a bottle can be recycled it has to be emptied. **1973** *Guardian* 22 Mar. 15/1 The Liberals of Kew..have been recycling paper, and have managed to scrape a regular £25 a month. **1974** *Listener* 28 Feb. 278/1 Such a plant would recycle steel, aluminium, zinc, lead and copper from scrap. **1979** *China Now* Mar./Apr. 31/3 The report covers all methods of recycling organic materials.

c. *transf.* in connection with natural processes. Usu. in *passive*.

1965 G. J. Williams *Econ. Geol. N.Z.* i. 2/2 These [beds of sediment] are of considerable interest to economic geologists for through them much detrital gold was recycled within and beyond the primary gold-bearing areas. **1970** *Nature* 17 Oct. 273/2 The annual discharge of dissolved sodium in rivers is about 20 × 10⁷ tons, of which 9 × 10⁷ tons have been recycled from the sea through the atmosphere. **1971** I. G. Gass et al. *Understanding Earth* iii. 68/2 Much of the ocean will be recycled in the ocean-floor spreading process. **1973** *Sci. Amer.* Apr. 61/1 Stars continually recycle their material through the interstellar medium.

d. *fig.*

1969 *Guardian* 12 May 1/5 (*heading*) Bankers find way to recycle hot money. **1970** *Nature* 25 July 321/2 It is not possible to recycle the output of the secondary schools without there being some intermediate opportunity for broadening the intellectual experience of the young men and women concerned. **1973** *Ibid.* 2 Mar. 4/2 A further five [cases] may be the result of the virus being recycled in swill. **1973** *Black Panther* 4 Aug. 7/3 Those workers finding themselves without jobs..are re-cycled back to their former jobs at the reduced wages. **1974** *Weekend Mag.* (Montreal) 16 Mar. 2/2 The kids are appropriating the Fifties, proving once more that fads (like garbage) can be recycled. **1974** *Newsweek* 7 Oct. 52/1 A new international banking system to recycle OPEC funds into loans to the poorer nations. **1978** *Washington Post* 8 Aug. C4/5 Many juveniles, he adds, are repeat offenders, 'recycled' through the system.

2. a. *trans.* To repeat (a process) on a computer or counting device; also *absol.* **b.** *intr.* Of a computer: to repeat a procedure.

1962 A. Shepard in *Into Orbit* 103 Walt decided to recycle the count—or set it back—to allow for this delay. **1973** *Sci. Amer.* May 110/3 The three input terminals of a NAND gate connected to Q_e, Q_d and Q_a of a series of five flip-flops would cause the apparatus to recycle on the count of 25 (11001). **1970** A. Cameron et al. *Computer & O.E. Concordances* 47, I made a preliminary run and found a large number of keypunch errors that I had missed originally... I decided therefore to recycle.

3. *intr.* To undergo recycling.

1970 *Nature* 28 Nov. 856/2 The inability of most newly formed lymphocytes to recycle from blood to lymph could explain their truncated life span. **1975** *Ibid.* 24 July 247/1 The PhD degree is a relatively easy target to attack because its recipients do not seem to confer on society the same sort of benefits as, say, medical doctors. More recycle into the educational system than go elsewhere. **1978** *Amat. Photographer* 29 Nov. 128/3, I had noticed the unit appeared to be taking longer than usual to recycle after each shot, but assumed the battery was getting low.

Hence **recy·cled** *ppl. a.*; **recy·cling** *vbl. sb.*; *recycling time* (Photogr.), the time required to recharge the capacitor of a flash unit.

1926 *Petroleum Devel. & Technol. 1925* 338 With the use of higher pressures and temperatures permitting the ultimate cracking of this cut, more recycling of this fraction may be practiced. **1958** *Amat. Photographer* 31 Dec. 2/2 (Advt.), Angle 60°, recycling time 6 sec. **1964** M. Gowing *Britain & Atomic Energy 1939–1945* ix. 258 There were heavy losses of the product and of time in recycling and in washing and cleaning the machines. **1969** *Focal Encycl. Photogr.* (rev. ed.) 636/2 The recycling time and the number of flashes obtainable from a set of batteries..are also the subject of specific measurement methods. **1970** C. S. Russell et al. *Drought & Water Supply* viii. 68 Only 46 percent ..indicated a willingness to drink recycled domestic water. **1970** *Daily Tel.* 12 Sept. 5/7 They lived off their recycled body wastes. **1971** *Ibid.* 29 Apr. 14/6 The more than 9·1 million aluminium cans represent about 460,000 lb of litter and solid waste removed for complete recycling. **1972** *Guardian* 6 June 15/4 Recycling enthusiasts are..collecting ..old bottles and tin cans. **1975** *Nature* 17 Jan. 149/2 There is little commercial future for recycled glass in high grade uses at present. **1975** *N.Y. Times* 25 Oct. 27/3 The recycling boom is waning because of unfavorable economic conditions. **1977** *Financial Times* 4 June 5/3 Retreading is a form of recycling, which should be encouraged. **1978** D. Bloodworth *Crosstalk* xxiii. 180 He had driven the Deputy Director..half mad with his hesitation, his recycled arguments for accepting and not accepting.

recycle (rīsəi·k'l), *sb.* Also **re-cycle.** [f. the vb.] The operation or process of recycling a material, etc.; also, the material itself. Orig. and freq. *attrib.*, usu. denoting material subjected to or set aside for recycling.

1926 *Petroleum Devel. & Technol. 1925* 339 The gas oil was returned to the cracking system as recycle charging stock. **1936** W. L. Nelson *Petroleum Refinery Engin.* iii. 19 Recycle stock has about the same boiling-range and.. physical characteristics as gas oil. **1939** *World Petroleum* Mar. 104/3 The asphalt bottoms produced are released through exchangers against reduced crude. They are blended with the cracked recycle gas oil and thence through the tar coolers to tankage. **1946** *Nature* 30 Nov. 800/1 The gradual deterioration in the quality of re-cycle benzene is due to the preferential accumulation of paraffins. **1961** *Engineering* 2 June 781/1 Reduce the cost of fuel burned by using plutonium recycle or spikes of fully enriched uranium. **1966** *McGraw-Hill Encycl. Sci. & Technol.* XIII. 185/2 The rich solution from the absorption step must be stripped in order to permit recovery of the absorbed solute and recycle of the solvent. **1975** *Nature* 13 Feb. 496/3 Recycle, or reuse of materials, is an important aspect of the proper management of these resources. *Ibid.* 2 Oct. 369/1 There is a great variety of possible schemes for incorporating a converter, or heat engine, into the heat recycle.

recycler (rīsəi·kləɹ). [f. *recycle *v.* + -er¹.] One who or that which recycles (waste products, etc.).

1973 *Nature* 13 Apr. 483/3 Exhaust gas recyclers which cause an increase in the fuel consumption. **1974** *Ibid.* 19 Apr. 641/3 Developing..new manufacturing designs congenial to recyclers. **1976** *National Observer* (U.S.) 13 Mar. 8/2 With the emergence of the week-end home mechanic, the industry has tried to change its image from that of 'junk-yards keepers' to 'recyclers'.

recyclist (rīsəi·klist). *rare.* [f. as prec. + -ist.] An advocate of the recycling of waste products; a recycler.

1973 *Times* 1 Aug. 12/1 Perhaps pop artists were the first Recyclists.

red, *a.* and *sb.*¹ Add: **A.** *adj.* **I. 1. b.** *fig.* (Later examples.)

1795–1804 W. Blake *Four Zoas* in *Compl. Writings* (1957) 336 Red rage redounds. **1892** W. B. Yeats *Countess Kathleen* ii. 34 God's red anger seize them. **1938** —— *Herne's Egg* iii. 28 The Great Herne himself And he in a red rage. **1952** Dylan Thomas *Coll. Poems* p. x, Of fear, rage red, manalive.

g. *red, white, and blue*: the colours of the Union Jack, hence, the flag itself; also *attrib.* or as *adj.*, patriotic, devoted to the service of Britain.

1855 D. T. Shaw *Britannia, Pride of Ocean* 1 May the Service United ne'er sever, And both to their Colours prove true, The Army and Navy for ever! Three cheers for the Red, White and Blue! **1912** R. Brooke *Lett.* (1968) 387 Aren't you, perhaps, going to lecture..about the British Empire, on 'Heart-Cries under the Red White and Blue', or some such title? **1971** *Scope* (S. Afr.) 19 Mar. 30/1 They were all that he was not; British in tradition; red-white-and-blue in sentiment. **1972** P. Lovesey *Abracadaver* xv. 191 Our careers are *dedicated* to the red, white and blue. There is no need to remind us where our duty lies. **1974** *Times* 24 Aug. 2/2 Anyone joining his organization had his background checked 'to avoid communist infiltration... If the man has a red, white and blue background, then he is okay'. **1977** *Sniffin' Glue* July 21 He just averts his gaze to the red white and blue and exchanges nothings with the silly mayor.

h. *to paint the town red*: see Paint *v.*¹ 9.

3. (Further examples.)

1865 Swinburne *Chastelard* v. i. 141 The men of Pharaoh's, beautiful with red And with red gold. **1892** W. B. Yeats *Countess Kathleen* i. 18, I am half mindful to go pray to him To cover all this table with red gold. **1931** M. Allingham *Look to Lady* xiii. 144 The real Chalice..is made of English red gold.

b. (Further example.) Also *red 'un*, a sovereign.

1816 Scott *Antiq.* I. xv. 325 It's a red half-guinea to him every time he mounts his mare. **1890** in Barrère & Leland *Dict. Slang* II. 175/1 The youth, her wish obeying, placed a coin down—gently saying—'There's a red 'un—or in other words "a quid!"' **1899** C. Rook *Hooligan Nights* ii. 25 Honest work..will bring in but a few shillings a week; and what is that compared to the glorious possibility of nicking a red 'un? **1901** G. B. Shaw *Capt. Brassbound's Conversion* II. 265 E'll give huz fawv unnerd red uns. **1905** *Hackney & Kingsland Gazette* 15 Sept. 3/7 He said 'Here comes a German with a red lot (gold chain, etc). If you have heart, pull it.' **1981** A. Hewins *Dillen* iii. 20, I don't think much o' that stone you got. I'll give you a nice red un for it.

c. For *U.S.* read 'orig. *U.S.*' (Earlier and later examples.) Also (*U.S.*) in phr. *nary* (*a*) *red* (*cent*): see *nary *a.*

c **1839** J. S. Jones *People's Lawyer* (1856) I. i. 8 It would not have cost you a red cent. **1900** W. Archer *Let.* 1 Feb. in C. Archer *William Archer* (1931) xii. 263 We have never agreed about plays, and we never will... I have never given a red cent for the ideas in plays. **1904** Kipling *Traffics & Discoveries* 23 I'd turned in every red cent on the Zigler. **1943** K. Tennant *Ride on Stranger* xvii. 188 'To think of it,' groaned George Benson. 'We don't get a red cent, not a flaming red cent.' **1958** J. Carew *Black Midas* ix. 193 He will pay you seven dollar..and not a red cent extra. **1976** T. Sharpe *Wilt* xiii. 135 'I'll alimony you for all the money you've got.' 'Fat chance. You won't get a red cent.' **1979** *Tucson Mag.* Apr. 34/3 In ten years, the city has not spent one red cent from any federal funds for Barrio Historico.

5. a. (Later example.)

1922 Joyce *Ulysses* 23 You know that red Carlisle girl, Lily?

c. *Red Indian*: see Indian *sb.* 2 b in Dict. and Suppl.

7. esp. in phr. *red face*, a sign of embarrassment or shame.

1937 Partridge *Dict. Slang* 692/1 *Red face* (or *neck*), *have a*, to be ashamed. **1973** *Listener* 14 June 786/1 Mediterranean weather caused red faces among long-range weathermen who had to confess they'd got June wrong so far. **1977** *Listener* 30 June 865/1 The celebrated Samuel Palmer fakes ..that have left so many red faces in the world of fine art. **1980** B. Parvin *Death in Past* v. 30 She..grabbed me and said: 'It's true—I'm going to have a baby!' Was my face red! **1981** L. Deighton *XPD* iii. 13 There was secret material..[that] would have caused a few red faces here in Whitehall.

9. b. (Earlier and further examples.) Also, Bolshevik, communist; freq. *spec.* of or pertaining to the U.S.S.R.; *red revolution*, a socialist or communist revolution.

1848, etc. [see *Red Republic]. **1917** [see *Red Guard 1 a]. **1919** *Times* 7 Oct. 4/3 That I was prepared to create a Red Revolution in England..is something which I have never said. **1920** *Blackw. Mag.* Sept. 404/2 The Red Government, still bent upon the destruction of Europe, was..recognised. **1924** R. Macaulay *Orphan Island* xix. 252 It is mainly a catalogue of grievances, together with rousing addresses... 'What we call Red journalism.' **1926** *Brit. Gaz.* 12 May 3/6 After an attempt to hold a 'Red' Meeting in Edgware-road, a crowd of about 2,000 people was said to have collected and arrests were made by the police. **1927** W. E. Collinson *Contemp. Eng.* 85 The spread of the Bolshevistic propaganda has led to the fear, lest Labour should go red. **1929** J. Buchan *Courts of Morning* I. 129 The Scotsman had become their special intimate... Judson, who seemed to have known him before, called him Red Geordie. **1934** *Discovery* Feb. 55/2 All along that frontier, every three hundred yards, there are Red soldiers with rifle and machine-gun. **1940** W. Empson *Gathering Storm* 49 Revolt and mercy fired no sparks In the Red argument at all. **1948** E. B. White *Let.* 24 Jan. (1976) 290 My desk got so deep in Red literature that I had to fumigate myself every night before going home. **1951** *Sun* (Baltimore) 19 June 7/1 Count Wolf von Westarp, cofounder of the band of neo-Nazis,..has indignantly denied any Red ties. **1958** *Spectator* 6 June 723/2 There are still hundreds of writers in gaol all over the Red Empire, not to mention Franco's or Salazar's prisons. **1965** B. Pearce tr. *Preobrazhensky's New Econ.* 189 The red managers, proletarian engineers, and business executives have no monopoly of the means of production. **1970** M. O'Brine *Crambo* lii. 193 He is still a Red Navy man. He has the right to be buried at sea. **1972** D. Bloodworth *Any Number can Play* ii. 9 He infuriated the communists because he ran too just and egalitarian a kingdom to suit the sacred cause of red revolution. **1976** G. Mansell *Why External Broadcasting?* 18 Other totalitarianisms, whether of the red or the black variety. **1981** *Times* 3 Mar. 13/2 Anything is better than the horrors of nuclear war..better red than dead. **1981** *Time Out* 24–30 Apr. 7/4 Rosenthal and her fellow-candidates will be..hoping that Sir Horace Cutler's deepest fears of a 'red' London are realized.

II. 12. *red-blood* (fig.), *-fire, -roof, -rose* (later examples), *-wine* (later examples).

1915 J. London *Let.* 5 Nov. (1966) 463, I go ahead content to be admired for my red-blood brutality. **1925** V. Woolf *Common Reader* 262 The high-brow public and the red-blood public. **1943** Wyndham Lewis *Let.* 8 Aug. (1963) 360 The vulgarly red-blood American attitude (the lady and gentlemen complex). **1976** A. J. Russell *Pour Hemlock* (1979) ii. 19 I'm not their kind of people... This is a Redblood administration, I'm a Mollycoddle. **1918** D. H. Lawrence *New Poems* 15 As it guards the wild north cloud-coasts, red-fire seas running through The rocks. **1913** —— *Love Poems & Others* 33 The subtle, steady rush..of advancing God..Is heard..In the tapping haste of a fallen leaf, In the flapping of red-roof smoke. **1895** W. B. Yeats *Poems* 234 The red-rose-bordered hem. **1942** W. Stevens *Notes toward Supreme Fiction* 36 The channel slots of rain, the red-rose-red. **1877** E. S. Dallas *Kettner's Bk. of Table* 376 *Matelote Relish*, small onions and mushrooms in a red-wine sauce. *Ibid.* 483 It is difficult to procure the mild red-wine vinegar in London. **1943** E. M. Almedingen *Frossia* iv. 192 A nice plump partridge, red wine sauce, and cranberry jelly. **1971** *Vogue* 15 Sept. 125/2, I had chicken in red wine sauce with mushrooms and bacon.

13. *red-black, -gold* (earlier and later examples), *-golden, -pink, -purple* (further examples), *-rose, -white* (later example), *-yellow* (later example).

1910 *Westm. Gaz.* 25 Jan. 5/2 The material employed is the finest red-black rubber. **1975** R. H. Rimmer *Premar Experiments* II. 174 Even before I touched her, her nipples were engorged, red-black and demanding. **1607** E. Topsell *Four-footed Beasts* 661 This beast is of red-gold-colour. **1871** Swinburne *Songs before Sunrise* 237 Till the red-gold harvest-rows, Full-grown, are full of the light. **1923** D. H. Lawrence *Birds, Beasts & Flowers* 98, I have..seen..His red-gold, water-precious, mirror-flat bright eye. **1973** J. Cleary *Ransom* i. 21 She was..beautiful, with that red-gold hair that was a sensation on colour television. **1962** I. Murdoch *Unofficial Rose* 12 Her red-golden hair. **1880** E. Glaister *Needlework* ix. 101 If the flowers be another colour than yellow, say red-pink, or blue, the darning may be the same colour. **1951** E. Paul *Springtime in Paris* xv. 268 The pharmacy had large old-fashioned globes of coloured liquid, red-pink like Corsican wine and transparent blue-green. **1851** *Southern Planter* (Richmond, Va.) July 197/2 Improved Red Purple Straw on corn land. **1929** A. Clarke *Pilgrimage & Other Poems* 15 Vats of red-purple dye. **1917** G. Frankau *Inn of Thousand Dreams* in *City of Fear* 26 Once more I press..Your finger-tips against these lips Your own red-rose lips knew. **1920** J. Masefield *Enslaved* 9 Little red-white blossoms flecked me. **1937** V. Woolf *Years* 333 There was a red-yellow glow... The sun was sinking through the London dust.

14. a. *red-belted* (see also sense 14 b in Dict.), *-bordered, -carpeted, -checked, -clayed, -cloaked* (further examples), *-coloured* (later example), *-eared* (see also sense 14 b below), *-eaved, -edged, -ensigned, -flowered* (see also sense 14 c in Dict.), *-furred, -hatted* (earlier and later examples), *-labelled, -lipped* (further example), *-mouthed* (see also sense 14 b in Dict.), *-rimmed, -screened, -striped, -tabbed, -tied, -tiled* (later example), *-toothed, -veined* (see also sense 14 c in Dict.). See also *red-blooded etc. as main entries in Suppl.

1925 F. Scott Fitzgerald *Great Gatsby* iv. 81 A glimpse of red-belted ocean-going ships. **1922** Joyce *Ulysses* 657 Dry them..in a long redbordered holland cloth passed over a wooden revolving roller. **1905** *Westm. Gaz.* 6 Sept. 6/2 There were waiting on the red-carpeted platform.. officials representing the railway company. **1922** Joyce *Ulysses* 484 They appear on a redcarpeted staircase. **1948** M. Laski *Tory Heaven* viii. 106 He climbs the red-carpeted steps under the gay awning. **1976** N. Roberts *Face of France* xvi. 165 A red-carpeted dais. **1973** M. Amis *Rachel Papers* 170 My sister, a swirl of red-checked nightie, flew through the doorway. **1978** R. Ludlum *Holcroft Covenant* xxiv. 277 Running across the fronts of the booths were brass rods holding red-checked curtains. **1913** J. Masefield *Daffodil Fields* 2 Some short-grassed fields begin, Red-clayed and pleasant. **1910** W. B. Yeats *Green Helmet* 21 A tall red-headed red-cloaked man stands upon the threshold. **1980** *Jrnl. R. Soc. Arts* Mar. 241/2 Max Ernst's red-cloaked, bird-masked lady in *The Robing of the Bride*. **1942** S. Spender *Life & Poet* 12 Since we believe socialism to be just, novels should preach socialism and see everything through red-coloured spectacles. **1900** W. B. Yeats *Shadowy Waters* 46 A red-eared hound follows a hornless deer. **1881** O. Wilde *Poems* 80 The dusky red-eaved sheds. **1918** G. Frankau *Judgement of Valhalla* 6 (*title*) The song of the red-edged steel. **1922** Joyce *Ulysses* 221 Father Conmee..took his rededged breviary out. **1942** E. Sitwell *Street Songs* 7 Man's threatening shadow Red-edged by the sun like Cain, has a changing shape. **1965** G. McInnes *Road to Gundagai* xiii. 239 The navy blue red-edged flag. **1892** D. Sladen *Japs at Home* (ed. 2) xxvi. 283 While in the distance looms the harbour of Yokohama, full of the mighty red-ensigned steamers of the England he pined for night and day. *a* **1915** Joyce *Giacomo Joyce* (1968) 16 Poised on its edge a woman's hat, red-flowered. **1932** Blunden *Face of England* 66 Where the sheep's parsley tops a red-furred stem. **1863** Geo. Eliot *Romola* I. xiii. 222 The boy-cardinal Giovanni de Medici, youngest of the red-hatted fathers. **1918** A. Bennett *Pretty Lady* x. 52 The young red-hatted officer. **1930** Blunden *Poems* 185 Where the red-hatted cranks Have fixed a portcullis With notice-board—thanks! **1922** Joyce *Ulysses* 95 The redlabelled bottle on the table. **1881** O. Wilde *Poems* 70 An amorous red-lipped boy. **1913** Kipling *Songs from Books* 239 And red-mouthed shadows racing By, that thrust me from my food. **1934** Dylan Thomas *Let.* *c* 26 May (1966) 133 I've wasted some of my tremendous love for you on a lank, redmouthed girl with a reputation like a hell. **1916** Joyce *Portrait of Artist* (1969) 252, I fear his redrimmed horny eyes. **1962** I. Murdoch *Unofficial Rose* xii. 115 The red-rimmed eyes. **1977** P. Hill *Liars* vii. 91 Her eyes were red-rimmed, as if she had been crying. **1930** Blunden *Poems* 134 The red-screened windows of schoolhouse and inn. **1955** E. Pound *Classic Anthol.* II. 94 Fang Shu's black-dappled team of four Drew his red-screen'd car to the war. **1865–6** W. Whitman *Sequel to Drum-Taps* 19 The red-striped artilleryman. **1940** Blunden *Poems 1930–40* 209 He.. damned, at each pause, red-tabbed Brigade, Whose orders for grimness more than the frost-spell made us shiver. **1948** W. Fortescue *Beauty for Ashes* xxvii. 207, I accosted a red-tabbed English officer who directed me to it at once. **1977** J. Cleary *High Road to China* iv. 107 Johnny Silversmith, red-faced and red-tabbed, came to our table. **1911** G. K. Chesterton *Innocence of Father Brown* iv. 105 The red-tied youth. **1960** D. Potter *Glittering Coffin* iii. 38 Red-tied adolescent poets. **1977** H. Osborne *White Poppy* viii. 69 An iron bedstead on a red-tiled floor. **1881** O. Wilde *Poems* 178 The red-toothed lightning. **1925** Blunden *Eng. Poems* 123 Through red-toothed nettles. **1877** C. Patmore *Unknown Eros* xi. 51 He had put, within his reach, A box of counters and a red-vein'd stone. **1907** *Daily Chron.* 18 Mar. 6/2 His face is clear, with the red-veined cheeks of a sailor. **1956** R. Finlayson in C. K. Stead *N.Z. Short Stories* (1966) 22 He was a thickset florid man with a red-veined nose.

b. *red-eared* (see also sense 14 a above), *-lipped*.

1937 *Jrnl. Tennessee Acad. Sci.* XII. 45 (*caption*) The Red-Eared Sunfish. Only the males have the red tip on the opercular flap. **1952** A. Carr *Handbk. Turtles U.S. & Canada* II. 251 The red-eared turtle can usually be recognized..by the long, oval expansion of the broad supra temporal stripe, which is usually bright red. **1910,** etc. Red-lipped snake [see *herald *sb.* 5].

c. *red-leafed* (later examples; also *fig.*).

1911 E. Pound *Canzoni* 1 Ah! red-leafed time hath driven out the rose. **1923** Blunden *To Nature* 47 There shone the Ancre, red-leafed woods above it.

15. a. *red-decked, -flushed, -gilded, -lined* (earlier and later examples; see also *red-line *v.*), *-lit, -polished, -struck, -washed*.

1923 D. H. Lawrence *Birds, Beasts & Flowers* 66 Red-decked socialists, Hibiscus-breasted. **1871** W. Whitman *Passage to India* 18 The red-flush'd cheeks, and perfumes. **1943** V. Woolf *Haunted House* 124 The red-flushed clouds. **1949** E. Pound *Pisan Cantos* lxxix. 77 The mountain forest is full of light The tree-comb red-gilded. **1820** Keats *Lamia & Other Poems* 57 Why were they proud? Because red-lin'd accounts Were richer than the songs of Grecian years? **1921** W. de la Mare *Crossings* 31 Unlocks the trunk and pushes back its red-lined lid. **1966** *Sunday Times* (Colour Suppl.) 4 Dec. 73/3 [GI Jargon] *Red-lined*, cancelled or classified unserviceable. **1979** *N.Y. Times* 24 Jan. B 18/1 Redlining has an undeniable racial component whereby redlined neighborhoods often coincide with nonwhite neighborhoods. **1930** R. Campbell *Gum Trees*, Along the red-lit rim of space In lofty cadences they rhyme. **1934** V. G. Childe *New Light Most Anc. East* iv. 89 Household vessels, always the most sensitive indicator of ethnic change, are radically altered. Though Black-topped and Red-polished ware continue to be manufactured they are no longer the vehicle for new shapes. **1977** *Jrnl. R. Soc. Arts* CXXV. 476/2 The most important items found in the tombs of *Kotchati* and acquired by the Cyprus Museum are two clay (Red Polished ware) models of sanctuaries. **1923** D. H. Lawrence *Birds, Beasts & Flowers* 82 Since the Lamb bewitched him with that red-struck flag His fortress is dismantled. **1932** Blunden *Face of England* 50 Red-washed cottages.

b. *red-glowing, -panting*.

1936 R. Campbell *Mithraic Emblems* 24 A sombre grape, whose heart, Red-glowing to the hilted dart, Seems a lit furnace that he fans. **1922** Joyce *Ulysses* 47 Unheeded he kept by them.., a rag of wolf's tongue redpanting from his jaws.

III. 16. e. Applied to hearts and diamonds in a pack of cards.

1764 [in Dict., sense 16 a]. **1908** R. F. Foster *Auction Bridge* 50 Here is an example of a hand which is not a good red declaration. **1910** W. Dalton *'Saturday' Bridge* vii. 100 Doubling an original red-suit declaration cannot be recommended on anything very short of a certainty. **1973** *Times* 29 Sept. 11/7 He needed to find both red aces on the left... His game bid was against the odds.

f. Applied to the representation of British territories on maps: see sense *1 e of the sb.

1916 J. Buchan *Greenmantle* iv. 52 You see that map... South Africa is coloured green. Not red for the English, or yellow for the Germans. **1934** A. Huxley *Beyond Mexique Bay* 36 The non-existent young lady in fancy dress would be mortally offended by the suggestion that the place [*sc.* British Honduras] should be painted anything but red on the map. **1964** *Critical Q.* Winter 320 You shook your finger at the map and said..'Africa, I want it red.' **1975** *Listener* 4 Sept. 297/1 At the beginning of the 20th century, practically every country exporting spices was marked red on the map. **1977** A. Wilson *Strange Ride R. Kipling* iv. 212 Rhodes arranged for Rudyard to make a visit up to the territory of Rhodesia, which he saw as the first step in his dream of an all-red British route from Cape Town to Cairo.

17. a. *Animals*, as **red buck**, substitute for def.: = *impala, *rooibok; (earlier and later examples); **red bug**, (*b*) substitute for def.: = Jigger *sb.*² in Dict. and Suppl.; (examples); **red cat** *S. Afr.* = *rooikat; **red crab** (earlier and later examples); **red crab** (earlier example); **red dog**, the dhole, *Cyon* (or *Cuon*) *alpinus*; **red fox** (*b*) (earlier and later examples); **red hare**, (*b*) a southern African hare belonging to the genus *Pronolagus*, distinguished by speckled buff and black fur, with reddish fur beneath the body and a red-brown tail; **red hartebeest**, a variety of hartebeest, *Alcelaphus bucelaphus caama*; **red howler**, a howler monkey, *Alouatta seniculus*, found in forested areas of South America and distinguished by long red-brown fur; **red mite**, (*a*) a blood-sucking mite, *Dermanyssus gallinæ*, which attacks poultry; (*b*) = Red spider; **red panda** = Panda; **red river hog**, a West African race of the bush pig, *Potamochœrus porcus*; **red setter**, an Irish setter belonging to the breed sometimes so called, distinguished by a long, silky, dark red coat, drooping ears, and a long feathered tail; **red squirrel**, (*a*) substitute for def.: a small North American squirrel, *Sciurus hudsonicus*, also called the chickaree; (earlier and later examples); (*b*) the common European squirrel, *Sciurus vulgaris*, now relatively rare in Britain; **red wolf**, (*b*) substitute for def.: a North American wolf, *Canis rufus*, native to parts of the south-western states, where it is rare; (later examples); (*c*) a variety of the common wolf, *Canis lupus*.

1813 J. Campbell *Jrnl.* 30 Oct. in *Trav. S. Afr.* (1815) xl. 484 The following are the number of creatures killed by our people during the journey..Redbucks..6..Rhebucks..3. **1965** Red-buck [see *impala]. **1804** D. McKinnen *Tour Brit. W. Indies* x. 171 The red bug..has stained the cotton so much in some places this year as to render it of little or no value. **1827** J. L. Williams *View W. Florida* 29 Red bugs are numerous, especially in mossy woods. **1856** *Rep. Comm. Patents 1855: Agric.* (U.S.) 104 The 'red-bugs', or..'cotton-stainers', generally make their appearance about August. **1909** Red bug [see *bête rouge]. **1939** *Sun* (Baltimore) 18 Aug. 11/2 The chiggers—'red bugs' to some—were terrific. **1955** *Sci. News Let.* 16 July 42/1 Chiggers, called red bugs down South, cause the most exquisite itching. **1731** G. Medley tr. *Kolb's Present State Cape Good-Hope* II. 127 A few that are call'd Wild Red Cats..have a streak of bright Red running along the ridge of the back. **1947** *Cape Times* 3 May 14 Buck which used to be plentiful have been almost exterminated by wild red cats. **1966** E. Palmer *Plains of Camdeboo* x. 180 There could have been lynx or red cats in the mountain behind us... Even in a zoo these animals are wonderful, the size of a small leopard, not spotted but brick red with jet-black pointed ears and emerald eyes. **1825** C. Waterton *Wanderings S. Amer.* 285 Amongst the bare roots of the trees..a red crab sometimes makes its appearance. **1862** *Chambers's Encycl.* III. 528/2 The name Dhole is extended to some other very similar species or varieties, natives of Ceylon, Nepaul, and other parts of the East, to which the common name Red Dogs has been sometimes applied. **1894** Kipling *Second Jungle Bk.* 178 'What moves?' said Phao..'The dhole, the dhole of the Dekkan—Red Dog, the Killer!' **1957** P. J. Darlington *Zoogeogr.* vi. 394 *Cuon* (the Dhole or Red Dog), widely distributed in southern and eastern Asia. [**1637** T. Morton *New Eng. Canaan* II. v. 79 The Foxes are of two colours; the one redd, the other gray.] **1778** in *Essex Inst. Hist. Coll.* (1913) XLIX. 109 Sold..38 red fox skins. **1917** H. E. Anthony *Mammals Amer.* 72/2 The Red Fox mates in February or early in March. **1974** *Harper's & Queen* Sept. 37/3 (*caption*) Red Fox Jacket £450. **1844** J. Backhouse *Narr. Visit Mauritius & S. Afr.* xxviii. 485 The Red Hare or Roode Haas..is smaller than the Common Hare. **1912** J. Stevenson-Hamilton *Animal Life Afr.* xvi. 252 The red hare..has only recently acquired the dignity of a separate genus. **1939** [see *klipbok]. **1971** C. M. van der Westhuizen in D. J. Potgieter et al. *Animal Life S. Afr.* 396/1 The red hares (*Pronolagus* spp.) are peculiar to Southern Africa, where they inhabit elevated and hilly country. **1947** J. Stevenson-Hamilton *Wild Life S. Afr.* xii. 80 The Cape or red hartebeest..is now found only in the remoter parts of the north-west of the Cape Province and portions of the deserts of Bechuanaland and South-West Africa. **1966** E. Palmer *Plains of Camdeboo* viii. 134 The red hartebeest was almost certainly here for it was once one of the commonest antelope in the Cape. **1979** Delany & Happold *Ecol. Afr. Mammals* vii. 152 Blue wildebeest, red hartebeest and eland..are elements of a fauna found further to the south. **1865** Red howler [see Howler 1 b]. **1894** H. O. Forbes *Hand-bk. Primates* I. 194 The Red Howlers always travel in large companies. **1958** J. Carew *Wild Coast* viii. 109 The forest was full of noises—the roar of red howlers. **1894** *Rep. Vermont Board Agric.* XIV. 176 A little kerosene on the roosts will destroy the red mites that are so troublesome. **1912** J. H. Robinson *Princ. & Pract. Poultry Culture* xx. 342 Red mites..secrete themselves about the roosts. **1950** *N.Z. Jrnl. Agric.* Feb. 146/3 Some vegetable crops, including beans, are liable to become severely infested with red mite. *Ibid.* 182/3 Infestations of body lice, red mite, and intestinal worms all weaken the birds' constitution. **1976** Walters & Parker *Keeping Chickens* vi. 61 Red mites. These are in fact greyish in colour, about the size of a pinhead and live in cracks and joints in the woodwork of the house. **1955** F. Bourlière *Mammals of World* vi. 184 The Red Panda is a solitary animal. **1971** L. H. Matthews *Life of Mammals* II. ix. 266 The red or lesser panda..inhabits parts of western China and the slopes of the Himalayas. **1868** Red river hog [see River *sb.*¹ 5]. **1953** G. M. Durrell *Overloaded Ark* i. 38 A fully grown pair of Red River Hogs fled... They were the most vivid orange colour with long white tufts on their ears, and a flowing mane of white hair along their backs. [**1872** 'Idstone' *Dog* xii. 108 The Irish Setter should be of a pure rich mahogany red.] **1885** [see Setter *sb.*¹ 11 a]. **1893** R. B. Lee *Hist. & Descr. Mod. Dogs (Sporting Div.)* xvii. 343 One cannot say that the Irish red setter, the Irish terrier, and the water spaniel of Ireland, came at any recent date from one stock. **1912** A. Huxley *Let.* 23 June (1969) I. 46 For the past week the..beautiful red setter has been ill, refusing nourishment. **1954** M. K. Wilson tr. *Lorenz's Man meets Dog* viii. 78 A Red Setter or a dog of a similar long-haired, long-eared breed. **1977** *Irish Press* 29 Sept. 16/2 (Advt.), Red Setter pups 6 weeks old, male and female, parents F.T.C. [**1637** T. Morton *New Eng. Canaan* II. v. 81 There are Squirils of three sorts, very different in shape and condition; one..is red, and hee haunts our howses, and will rob us of our Corne.] **1682** T. Ash *Carolina* 22 There are.. the Red, the Grey, the Fox and Black Squirrels. **1795** *Stat. Acct. Scotland* XV. xxi. 439 The red squirrel..has become extremely common of late years. **1902** W. D. Hulbert *Forest Neighbors* 102 Other sounds there were..the scolding of the red squirrel, disturbed and angry. **1935** [see *grey squirrel]. **1971** *Country Life* 17 June 1538/1 It is not within my memory when red squirrels were about in fair numbers. **1972** *Ecology* LIII. 1142/1 Pasture juniper fruits provide winter food for red squirrels. **1942** G. M. Allen *Extinct & Vanishing Mammals* 229 The typical form of red wolf was slightly the smallest of the three races. **1964** [see *grey wolf* s.v. *grey, gray *a.* 8 b]. **1969** J. Fisher et al. *Red Bk.* 75/2 It [*sc.* the giant panda] appears to be without natural enemies, with the possible exception of the leopard and the red wolf. **1978** B. H. Lopez *Of Wolves & Men* 279 My wife and I raised two hybrid red wolves.

b. *Birds*, as **red bishop (bird)**, an African weaver belonging to the genus *Euplectes*, esp. *E. orix*; **red grouse** (examples); **red mavis** *U.S.*, the common ground thrush, *Toxostoma rufum*; = *brown-thrasher* s.v. Brown *a.* 6.

1884 E. L. Layard *Birds S. Afr.* (ed. 2) 462 Red Bishop Bird..though not an uncommon bird, is certainly a very local one. **1939** *Nature* 1 Apr. 566/1 The red bishop has never been found to have more than three wives in his large territory. **1955** Mackworth-Praed & Grant *Birds E. & N.E. Afr.* II. 951 Zanzibar Red Bishop..differs from the South African and Tanganyika Territory races of the Red Bishop. **1966** E. Palmer *Plains of Camdeboo* xii. 197 Here would be red bishop birds in summer plumage, the grenadier of Barrow... Sita saw a bird like a jewel sail past her... It was a red bishop bird, black and glowing scarlet, on its way to the reeds with its drab-coloured harem. **1776** T. Pennant *Brit. Zool.* (ed. 4) I. 269 Grous..Red... The plumage on the head and neck is of a light tawny red. **1910** Malcolm & Maxwell *Grouse* i. 1 We shall be almost exclusively concerned with the red grouse. **1927** S. Gordon *Days with Golden Eagle* xiv. 93 The grey or hooded crow is a far more deadly enemy to red grouse than the golden eagle. **1971** *Country Life* 12 Aug. 390/1 It [*sc.* heather] provides the main food of that highly famed bird, the Scottish red grouse. **1854** Thoreau *Walden* 171 Upon the topmost spray of a birch sings the brown-thrasher—or red mavis, as some love to call him. **1858** *Atlantic Monthly* Dec. 869/2 The Red Mavis..has many habits similar to those of the Cat-Bird. **1917** T. G. Pearson *Birds Amer.* III. 179 Brown Thrasher... Other names..Mavis; Red Mavis; Song Thrush.

c. *Fishes*, etc., as **red bream** (later examples); **red drum** (earlier and later examples); **red groper**, a red-skinned form of the blue groper, *Achœrodus gouldii*, a marine fish found off the coast of southern Australia; **red grouper**, delete 'Mexican' and add: found off the coast of south-eastern North America; (earlier and later examples); **red snapper**, substitute for def.: one of several important marine food fishes belonging to the family Lutjanidæ, esp. *Lutjanus campechinus* of eastern North and Central America; (earlier and later examples).

1924 *Truth* (Sydney) 27 Apr. 6 *Red bream*, name given to young schnapper. **1969** *Man* (Austral.) Mar. 87/2 Another prize for the table is the snapper or red bream. **1709** J. Lawson *New Voy. Carolina* 156 Black Drums are a thicker-made fish than the Red Drum. **1969** *Daily Progress* (Charlottesville, Va.) 25 May 5 B/1 A big Channel Bass, or Red Drum as it is called in some circles, lay on the charmingly beautiful wild-beach. **1893** J. D. Ogilby *Edible Fishes & Crustaceans N.S.W.* 134 (*heading*) Red Groper. **1962** L. Wedlick *Fishing in Austral.* iv. 159 The red groper is now considered to be the female of the species. **1829** Red grouper [see Grouper 1]. **1976** *National Observer* (U.S.) 23 Oct. 19/4 This is the place for exotic fish eating, with surprises like..Florida red grouper. **1775,** etc. [see Snapper *sb.*¹ 6 b]. **1973** *Nature* 6 July 49/1 The red snapper, *Etelis marshi* (an Indian Ocean fish). **1978** *Detroit Free Press* 16 Apr. (Detroit Suppl.) 28/2 There is a respectable number of fish dishes, ranging from the irreproachable Florida red snapper to frog legs, scallops and deviled crab. **1978** *Times* 4 Nov. 13/3 Red Snapper baked in a sauce of fresh tomatoes, onions, celery.

d. *Plants*, as **red alder** = *red els*, *rooiels; **red ash**, (*a*) for *pubescens* substitute *pennsylvanica*; (earlier and later examples); (*b*) (later example); **red bay**, for *carolinensis* substitute *borbonia*; (earlier and later examples); **red bean** (see sense 19 below); **red beech**, (*a*) substitute for def.: the common North American beech, *Fagus grandifolia*; (earlier and later examples); (*c*) a southern beech of New Zealand, *Nothofagus fusca*; **red birch** (earlier and later examples); **red elm**, substitute for def.: one of several elms, esp. the American slippery elm, *Ulmus fulva*; (earlier and later examples); **red fir**, (*b*) substitute for def.: a fir of western North America belonging to the genus *Abies*, esp. *A. magnifica*; (examples); (*c*) add = *Douglas fir* s.v. *Douglas¹; (examples); **red grass** *S. Afr.*, one of several reddish pasture grasses, esp. *Themeda triandra*; **red iron bark** (earlier and later examples); **red mahogany** (earlier and later examples); **red maple** (earlier and later examples); **red oak**, substitute for def.: a North American oak, *Quercus borealis* (or *Q. rubra*), or a closely related species; (earlier and later examples); **red osier**, substitute for def.: (*a*) *N. Amer.* in full, *red osier dogwood*; one of several species of dogwood, esp. *Cornus stolonifera*; (earlier and later examples); † (*b*) the basket willow, *Salix* × *rubra*; **red pine**, (*a*) (earlier and later examples); (*b*) (later examples); (*c*) (earlier and later examples); (*d*) the Japanese pine, *Pinus densiflora*; cf. *matsu; **red sandal wood** (earlier example); **red spruce**, for *rubra* substitute *rubens*; delete rest of def.; (earlier and later examples); **red willow**, one of several North American willows with reddish bark, esp. *Salix lævigata*; also = *red osier*.

1907 T. R. Sim *Forests & Forest Flora Cape of Good Hope* xiv. 217 Red Alder or Red Els. **1784** *Mem. Amer. Acad.* I. 492 *Fraxinus*..The White Ash. The Red Ash. The Black Ash. The Prickley Ash. **1965** *Austral. Encycl.* I. 165/2 The red almond (also called red ash, white-leaf, leather-jacket and cooper's wood), is a smooth-barked tree. **1969** T. H. Everett *Living Trees of World* 286/2 The red ash forms an irregular head with stout, erect branches. **1731** M. Catesby *Nat. Hist. Carolina* I. 63 The Red Bay... The wood is fine grain'd, and of excellent use for Cabinets. **1734** [see *hand-board* s.v. *hand *sb.* 63]. **1938** M. K. Rawlings *Yearling* iv. 35 The red bay thicket seemed impenetrable. [**1637** T. Morton *New Eng. Canaan* II. ii. 63 Beech there is of two sorts, redd and white.] **1810** F. A. Michaux *Hist. Arbres Forestiers de l'Amérique Septentrionale* I. 27 Red beech.. dans les Etats du nord. **1894** *Amer. Folk-Lore* VII. 99 *Fagus sylvatica*, white beech, red beech. **1928** Cockayne & Turner *Trees N.Z.* 83 *Nothofagus fusca*..Red-beech. A tall, massive tree. **1970** R. M. Lockley *Man against Nature* x. 204 The native red beech *Nothofagus* takes a hundred.. years to mature. **1785** H. Marshall *Arbustrum Americanum* 19 Red birch..grows to a pretty large size. **1918** N. Duncan *Battles Royal* I. ii. 41 The crew was gathered close about a roaring red birch fire. **1949** Collingwood & Brush *Knowing your Trees* 172 River birch is also called red birch. **1805** M. Lewis *Jrnl.* 12 Apr. in *Orig. Jrnls. Lewis & Clark Exped.* (1904) I. vii. 299 Some timber..consists of Cottonwood red Elm, with a small proportion of small Ash. **1810** [see *moose elm]. **1956** *Handbk. of Hardwoods* (Forest Prod. Res. Lab.) 88 Elm, English..Other Names. Red elm, nave elm. *Ibid.* 92 Commercial white elm may include some slippery elm (*Ulmus fulva*) also known as red elm. **1844** Lee & Frost *Ten Years in Oregon* 81 The red fir constitutes the greater part of the timber of the country. **1884** [see *Douglas¹]. **1884** [see *mountain hemlock* s.v. *mountain 9 d]. **1949** Collingwood & Brush *Knowing your Trees* 106 In close stands the trunks of red fir are clear of branches from sixty to eighty feet. **1957** M. Hadfield *Brit. Trees* 32 Douglas fir,..Oregon pine, red fir. **1974** *Country Life* 12 Dec. 1855/1 A fine Red fir (*Abies amabilis*) a rather rare tree from Washington State and British Columbia. **1929** J. W. Bews *World's Grasses* vi. 253 The most important grass in enormous areas of African savanna..is the 'Rooi gras' or 'Red grass' of S. Africa. **1955** J. Cope *Fair House* ii. 37 He [*sc.* a horse] likes the red-grass down here. **1884** A. Nilson *Timber Trees N.S.W.* 65 *E*[*ucalyptus*] *leucoxylon*.—Red-flowering Ironbark; Black Ironbark; Red Ironbark. **1944** *Living off Land* vii. 137 Slow growers with a dense grain.. include..Snow Gum, Red Ironbark. **1884** Red mahogany [see *forest mahogany* s.v. *forest *sb.* 5]. **1965** *Austral. Encycl.* III. 404/2 Phillip's 'gum-tree' was the red mahogany, *E*[*ucalyptus*] *resinifera*. **1770** J. R. Forster *Trav. N. Amer.* I. 167 The red Maple, or *Acer rubrum*, is plentiful in these places. **1955** [see *pace *v.* 2 a]. **1976** *Sci. Amer.* Nov. 111/3 There were other valley species, among them..the sugar maple..and its relatives the silver maple.., the red maple (*A. rubrum*) and the box elder. **1634** W. Wood *New Englands Prospect* 16 Of Oakes there be three kindes, the red Oake, white, and black. **1663** *Rec. Town of Plymouth* (Mass.) (1889) I. 64 Lott lyeth on the easterly side of the fourth lott and att the south end bounded with a Rid oake stake. **1901** H. Robertson *Inlander* 310, I des gwine down to de branch to git me some red-oak bark. **1958** G. A. Petrides *Field Guide Trees & Shrubs* 294 The barks of many red oaks are dark in color. **1976** *Sci. Amer.* Nov. 112/1 The red oak and the white oak, which put down a deep tap root, seldom survive the process of transplantation. **1807** F. Pursh *Jrnl. Bot. Excursion* (1923) 48 Cornus several sorts, among which is the Osier rouge or Red Osier. **1857** Thoreau *Maine Woods* (1864) 174 There grew..*Cornus stolonifera*, or red osier. **1864** *Ibid.* 314 *Cornus stolonifera* (red-osier dogwood), prevailing shrub on shore of West Branch. **1946** T. M. Stanwell-Fletcher *Driftwood Valley* 112 The moose browsed on young twigs of willow and red-osier dogwood. **1971** *Islander* (Victoria, B.C.) 30 May 3/1 A red osier dogwood..shaded our tents. **1972** Fredericksen & East

Silence of North xx. 182 His [*sc.* a bear's] broad burly rump was vanishing in a thick tangle of red osier halfway down the slope. 1809 E. A. Kendall *Trav. U.S.* III. 145, I have referred the sapling of the lumberers to the yellow, red or Norway pine. 1829, etc. [see *Norway¹]. 1884, etc. [see *matsu]. 1884 A. Nilson *Timber Trees N.S.W.* 81 *F*[*renela*] *Endlicheri.*—Red Pine; Black Pine. 1900 A. H. Kent *Veitch's Man. Coniferæ* (new ed.) 145 Two of the species are of great importance in their native countries on account of their valuable timber, viz., the Rimu or Red Pine of New Zealand, *Dacrydium cupressinum*, and the Huon Pine. 1911 [see *Moreton Bay pine* s.v. *Moreton Bay a]. 1916 E. Pound *Lustra* 89 The red-pine-tree god looks on him and wonders. 1970 R. M. Lockley *Man against Nature* x. 204 The rimu or red pine *Dacrydium* [takes] two hundred years to mature to the milling stage. 1840 J. Pereira *Elem. Materia Medica* II. 1142 Red Sandal or red Sander's wood (*lignum santali rubri*; *lignum santalinum rubrum*) is imported in roundish or somewhat angular billets, which are blackish externally, but of blood-red internally. 1777 *Quebec Gaz.* 17 Apr. 2/1 The logs covering the sleepers, shall be of ash or red spruce. 1943 R. Peattie *Great Smokies* 157 The red spruce..crowns only our highest peaks. 1977 J. L. Harper *Population Biol. Plants* xx. 622 At least 50% of the trees in the upper canopy were red spruce. 1784 *Mem. Amer. Acad.* I. 491 Salix... The White Willow. The Red Willow. The Rose Willow. The Dogwood. 1855 Longfellow *Hiawatha* i. 12 The bark of the red willow. 1895 *Outing* XXVII. 211/1 The lake..was covered with a growth of red willows and rushes. 1969 T. H. Everett *Living Trees of World* x. 96/1 Among lower-growing American willows the following are of tree size: the red or polished willow [etc.].

e. *Minerals* etc., as **red clay,** (*a*) in general use; (*b*) a fine-grained, red or reddish-brown, abyssal clay of diverse origins, containing windblown particles, meteoric and volcanic dust and debris, and insoluble organic remains; **red earth,** † (*a*) (in Dict., sense 19); (*b*) a red soil of the tropics and sub-tropics, usu. clayey and highly leached, and coloured by iron; cf. *red loam* (next entry); **red loam,** a red soil of the tropics, usually friable and highly leached; cf. *red earth* (*b*) above; **red mud,** (*a*) a marine mud of terrigenous origin, found on continental shelves and in other shallow waters, and coloured by iron oxides; (*b*) a residue from the extraction of alumina from bauxite, coloured red by ferric impurities; **red rock,** (*a*) *U.S.* (see quot. 1904); (*b*) *Geol.*, a predominantly or wholly granophyric rock of red colour associated with some large gabbroic masses; **red soil,** a general name for leached soil of the tropics and sub-tropics, coloured red by ferric compounds; cf. sense A. 1 f in Dict.

1387 Red clay [in Dict.]. 1827 J. L. Williams *View of W. Florida* 89 In the gulf [of Mexico], the..red clay lands approach within eighteen or twenty miles of the coast. 1874 *Proc. R. Soc.* XXII. 427 The bottom consists of 'Globigerina-ooze' or of the red clay produced by the decomposition of the shells of Foraminifera. 1916 H. F. Cleland *Geol.* vi. 242 Radiolarian ooze and red clay shade into each other in certain places, the deposit being called radiolarian ooze when these organic remains constitute 25 per cent. of the mass. 1926 *Jrnl. Geol.* XXXIV. 140 The *terra rosea* of the Istrian Peninsula, Dalmatia, and Greece consists of red clay, residual from a limestone basement. 1964 W. C. Putnam *Geol.* xiv. 371/2 Some workers once thought the red clay consisted mostly of meteoritic dust... Now the evidence appears more convincing that the red clay has a land-derived origin, and that it consists for the most part of the very finest clay and related particles. 1976 *National Observer* (U.S.) 23 Oct. 9/3 But Cooper and his boys are a long way from the red-clay farm he describes so vividly in *Families*. a 1977 *Harrison Mayer Ltd. Catal.* 14/2 *Red clay*, a term applied to all the ferruginous bodies, high in iron and manganese. Typified by the clay known as Etruria Marl. 1877 C. W. Thomson *Atlantic* I. iv. 315 Wherever, throughout the islands, a section of the limestone is exposed of any depth, it is intersected by one or two horizontal beds of an ochre-like substance, called locally 'red earth'... This 'red earth', mixed with varying proportions of decayed vegetable matter and coral-sand, forms the surface layer of vegetable soil. 1889 *Bull. U.S. Geol. Survey* No. 52. 25 The red earth of the southern portion of the Great Appalachian Valley is apparently identical, both in composition and in the method of accumulation, with the 'terra rossa' of southern Europe, the 'laterite' of India, and the 'red earth' of Bermuda. 1932 [see *red loam* below]. 1958 C. Achebe *Things fall Apart* xxv. 184 There was a small bush behind Okonkwo's compound. The only opening into this bush from the compound was a little round hole in the red-earth wall through which fowls went in and out. 1966 *Official Yearbk. Australia* LII. 878 Red earths associated with old land surfaces are widely distributed throughout the semi-arid areas. 1932 H. Greene tr. *Vageler's Introd. Tropical Soils* v. 163 The younger the soil, the greater the predominance of unchanged siallitic material, and the soil has then the character of a more or less plastic red loam. 1932 G. W. Robinson *Soils* xiii. 271 P. Vageler draws a distinction between red loams and red earths... In comparatively young soils..the clay is of a siliceous type and the soil is described as a red loam. In the red earths, removal of silicic acid or accession of sesquioxides has proceeded sufficiently to give a weathering complex of a predominantly sesquioxidic character. 1885 *Rep. Sci. Results Voy. H.M.S. Challenger: Narrative* I. xxi. 918 Red muds. 1926 *Jrnl. Geol.* XXXIV. 140 The red muds which Murray found in such quantities in the Atlantic Ocean off the mouth of the Amazon River. 1936 *Metals Handbk.* (Amer. Soc. Metals) 902 The Bayer process is almost universally employed for the purification of bauxite. In this process the bauxite is digested with caustic soda solution under pressure and the alumina dissolved out as a solution of sodium aluminate. The residue, known as red mud, contains the oxides of iron, silicon, and titanium [etc.]. 1972 *Daily Tel.* (Colour Suppl.) 27 Oct. 25/1 There are tentative hopes for a new industry based on 'red mud'. This is the unpleasant-looking residue of the aluminium process. .. Red mud may yet be pay dirt. 1880 J. F. Carll *Geol. Oil Regions* vi. 72 In this record we have two important facts to work upon—the top of the conglomerate and the presence of red rock beneath it and not far below its base. 1893 *Bull. U.S. Geol. Survey* No. 109. 23 The red rock..is found in three distinct though indefinitely outlined areas... It occupies a position between the gabbro and the fragmental rocks. 1904 *Dialect Notes* II. 387 *Red rock*, the drillers' name for the red shale underlying the Panama Conglomerate. 1908 *Jrnl. Geol.* XVI. 774 The 'red rock dike' of analysis No. 12 is called gabbro-aplite by the author. 1918 *Ibid.* XXVI. 632 The 'red rock' has purposely been left out of the discussion of variations from the gabbro,..because its geologic relations are very different... The gray gabbro rapidly gives place to a bright red rock... The 'red rock' has become widely known under this name because of its brilliant color and the difficulty of giving it a more accurate classification... The rock here discussed is intrusive and granitoid. 1969 Bennison & Wright *Geol. Hist. Brit. Isles* xi. 268 Red rocks of Permian and Triassic age outcrop in south-west England. 1889 *Bull. U.S. Geol. Survey* No. 52. 29 It closely resembles the similar red soils of Virginia and of many other regions. 1906 E. W. Hilgard *Soils* iii. 34 The 'red' soils formed from the so-called granites and slates of the western slope of the Sierra Nevada of California. 1932 G. W. Robinson *Soils* xvi. 323 The red soils which occupy most of southern India outside the black cotton area..are probably similar to the red loams and red earths of East Africa. 1940 G. Taylor *Australia* v. 77 All rocks in the west which are not very poor in iron (and not situated in a basin) yield red soils, since the paucity of organic matter allows for a rapid oxidation of the iron in the clay. *Ibid.* viii. 181 The red soils indicate that the material has been peroxidized under a hot sun in an arid climate not subjected to periodic flooding. 1970 *E. Afr. Standard* (Nairobi) 23 Jan. 16/8 (Advt.), Freehold red-soil plot in the most desired residential area.

f. Combined with other colours, as **Red and White Friesian,** a cow or bull belonging to the breed so called, distinguished by its red and white coat from the black and white animals of the older Friesian breed; also *ellipt.*

1962 *Guardian* 23 Oct. 2/5 A breed of cow making its first appearance at the [Royal Dairy] show this year is the Red and White Friesian. *Ibid.*, One of the Red and Whites has given 8·62 gallons in three milkings at the show. 1975 *N.Z. Jrnl. Agric.* Sept. 65/2 Breeders began experimenting by crossing their purebred cattle with Red and White Friesian and Danish Red bulls.

18. Prefixed to the name of a part (or some distinctive feature) used to denote the whole: **a.** of persons, as **red-beard** (later example); **red-hat,** (*b*) a staff officer (*Mil. slang*).

1868 W. Morris *Earthly Paradise* I. 194 Thou laughest—hast thou never heard Of this same valorous Red Beard, And how he died? 1916 W. Owen *Let.* 3 July (1967) 398 Red-Hats gallup up to us at startling speed.., but they never stay long, or criticise. 1918 A. Bennett *Pretty Lady* xxii. 146 It was the red hat put me off. 1919 W. Deeping *Second Youth* xxxiv. 288 When the real job's finished we just throw up our caps and shout. I wish the red-hat element would try to understand that. 1978 A. Waugh *Best Wine Last* xv. 179 A number of very high-ranking officers were invited... The visiting red hats were not impressed.

b. *spec.*, forming the names of certain birds, fishes, plants, etc., as **red-ear,** (*a*) in full, *red-ear sunfish*; a North American freshwater fish, *Lepomis microlophus*, which has a red patch on its operculum; (*b*) in full, *red-ear turtle*; a small North American turtle, *Pseudemys scripta elegans*, distinguished by a reddish stripe behind the eye; also called a slider; **red fin,** (*c*) the Australian name for the English perch, *Perca fluviatilis*, which was introduced into Australia in 1868; **red-root,** (*a*) (earlier and later examples). See also *red-heart etc. as main entries in Suppl.

1948 *List Common Names Fishes U.S.* (Amer. Fisheries Soc.) I. 16 Redear Sunfish *Lepomis microlophus*. 1957 M. B. Trautman *Fishes of Ohio* iv. 518/1 Three years later the offspring of these 14 Red ears appeared to be..numerous. 1958 R. Conant *Field Guide Reptiles & Amphibians* 58 Baby Red-ears, commonest of all pet turtles, are sold in enormous numbers. 1977 *N.Z. Herald* 8 Jan. 4–9/5 (Advt.), Miniature red ear turtle, tank, element and thermostat. 1979 *Arizona Daily Star* 1 Apr. c 11/4 A state-record redear sunfish has been reported from Parker Canyon Lake... The big redear was identified by Ft. Huachuca fisheries biologist Bruce Halsted. 1951 T. C. Roughley *Fish & Fisheries Austral.* 152 English perch (Redfin—*Perca fluviatilis*). 1969 *Southerly* XXIX. 127 Twice, in the dusk, he caught a redfin. 1709 J. Lawson *New Voy. Carolina* 78 The Red-Root whose Leaf is like Spear-mint, is good for..sore Mouths. 1941 R. S. Walker *Lookout* 59 The commonest shrub..is New Jersey tea or red-root.

19. a. With miscellaneous sbs., as **red alert,** a warning of danger; an instruction to be prepared for an emergency, or, in hospitals, to admit only emergency cases; also, a state of readiness for an emergency; also *fig.* and (with hyphen) *attrib.*; **red anchor,** used *attrib.* to designate that period in the history of the Chelsea porcelain factory during which it produced high-quality porcelain with a distinguishing red anchor mark; also applied to porcelain of this period; **Red Arrow,** a familiar name for the nightly express train from Moscow to Leningrad; **red-arse** *Mil. slang*, a recruit; **Red Astrachan,** a red-skinned variety of eating apple; **red atrophy** *Med.*, a later stage of massive necrosis of the liver, in which the organ is red rather than yellow; **red ball** *U.S. slang*, a fast freight train or truck; high priority freight; also *attrib.*; **red-band** *Prison slang* (see quot. 1950); **Red Bank,** the name of an oyster bed in Co. Clare, Ireland, used attrib. in *Red Bank oyster*; **red banner** = *red flag 1, 3 a; also used in the title of various distinctions and orders in the U.S.S.R.; **red bean,** (*a*) the red-skinned seed of one of several legumes; (*b*) an Australian timber tree, *Dysoxylum muelleri*, of the family Meliaceæ; **red beds** *Geol.*, add: more widely, sedimentary strata deposited at any period in a continental environment, composed largely of sandstone, siltstone, or shale, and coloured red by iron compounds which usu. coat individual grains; also *attrib.* in *sing.*; (earlier and further examples); **red-berry,** (*a*) substitute for def.: one of several North American plants, esp. the red baneberry (*Actæa rubra*); (earlier and later examples); (*b*) (later example); **red biddy** *colloq.*, a drink consisting of methylated spirits and cheap red wine; also, inferior red wine; **red blanket** *Austral. slang* (see quot.); **red blood cell** or **corpuscle** = *red cell; **red board** *U.S. slang*, (*a*) a stop signal on a railway; (*b*) (see quot. 1935); **red body,** also, the gland that this aggregation supplies, or both structures together; = *red gland* below; (earlier and later examples); **Red Branch** [tr. Gael. *Craebh Ruaid*], in Irish epic tradition, the name of the most famous of the royal houses of Ulster; **Red Brigade(s),** a left-wing extremist terrorist group operating in Italy from the early 1970s; hence **Red Brigader,** a member of this group; **red carpet,** the carpet of this colour traditionally laid down on formal occasions to greet important visitors, used *fig.* to indicate a ceremonial welcome or lavish reception; freq. *attrib.*, as *red carpet treatment*, etc.; **red caviar,** the red roe of fish other than the sturgeon; **Red Centre,** the remote interior of Australia; cf. *centre *sb.* 11 h; **Red Chamber,** the Senate chamber of the Canadian Parliament Building, Ottawa; hence, the Senate itself; **red channel,** at a port, airport, etc., the channel through which passengers should pass who have goods to declare; † **red children,** North American Indians considered as under the guardianship of a white person or agency (*obs.*); **red-cooking** *vbl. sb.*, a form of Chinese cookery in which meat is fried quickly and then stewed in soya sauce; hence (as a back-formation) **red-cook** *v. trans.*; **red core,** a disease of strawberries caused by the fungus *Phytophthora fragariæ*, which attacks the roots, staining the central part of them and making affected plants wilt; **red corpuscle** = *red cell; **red country,** large tracts of red sand, *spec.* in Australia; **Red Crescent,** also, the equivalent of the Red Cross in other Muslim countries; (earlier and later examples); **red Devon,** a large red-brown bull or cow of the breed so called, usually kept for the production of beef; = *Devon (*a*); **red duster** *slang* = *red ensign*; **red dwarf** *Astr.*, an old, relatively cool star lying on the main sequence; also *attrib.*; **red emperor,** an Australian marine fish, *Lutjanus sebæ*, found off parts of the northern coast; **red ensign** (examples); **red figure** *Archæol.*, used *attrib.* and *absol.* to designate a technique of vase painting devised in Athens in the late sixth century B.C. in which figures and patterns are outlined and detailed with lines of black paint on a red clay ground and the background filled in with black; so **red-figured** *a.*; **red flannel,** flannel dyed red and formerly used esp. for making underwear, nightwear, etc.; also used *colloq.* in *pl.* to designate clothing made from red flannel; also *fig.*; **red flannel hash** *U.S.*, a hash (Hash *sb.* 1) made with beetroot; **red giant** *Astr.*, a relatively cool giant star; **red gland** *Zool.*, a gland in the wall of a swim bladder which secretes gas into the bladder so as to increase the buoyancy of the fish; also, esp. formerly, the rete mirabile that supplies the gland, or both structures together; **red hardness** *Metallurgy*, the property, exhibited by some steels used for machine tools, of retaining a high degree of hardness up to a low red heat; **red hat,** the symbol of a cardinal's office; (see also sense 18 a); **red Indian,** (*a*) (see Indian *sb.* 2 b); (*b*) an Australian marine fish, *Patæcus fronto*, which resembles a blenny; **red judge,** *spec.* a high court judge (later examples); **Red Kaffir** *S. Afr.* (see quot. 1904); **red lamp,** (*b*) = *red light 2; **Redland** *sb.* and *a.*, the Soviet Union, Russian (*slang*); **red lane** (earlier example); **red lantern** = *red light 2; **red lead** *Naval slang*, (*a*) tomato ketchup; (*b*) tinned tomatoes; **red leg** *Zool.*, a bacterial disease of frogs causing hæmolytic septicæmia and a red flush on the ventral surfaces of the hind legs; **red meat,** dark-coloured meat, as beef or lamb (as opposed to chicken, veal, etc.); also *fig.*; **red menace,** the political or military threat regarded as emanating from the Soviet Union; **red morocco** (later examples); **Red Ned** *Austral.* and *N.Z. slang*, inferior red wine or other similar drink; **red noise** (see quots. 1961); **red nucleus** *Anat.*, each of a pair of nuclei in the tegmentum of the midbrain, dorsal to the substantia nigra, which form part of the extra-pyramidal motor system; **Red Paint,** applied to an ancient North American Indian people known from burials in which large quantities of red ochre were used; **red palm oil,** palm oil having a red colour, obtained by boiling the fruit in water instead of by fermentation; **red-pencil** *v. trans.*, to mark (in red) as erroneous or unacceptable; to correct (a piece of written work); **red peril** = *red menace* above; **red planet** (with def. article), the planet Mars; **red poley** = Redpoll² in Dict. and Suppl.; **Red Prince,** nickname of Prince Frederick Charles of Prussia (1828–85); **red rain,** rain that is red or reddish in colour because of suspended dust or, rarely, red algæ; **red reflex** *Ophthalm.*, a red glow, seen in ophthalmoscopy when the interior of the eye is illuminated, caused by the light reflected from the fundus having passed through the choroid; **red ribbon,** (*a*) also **riband** (later example); (*c*) *U.S.* as a symbol of temperance; **Red Riding Hood** [f. the fairy tale *Little Red Riding Hood*], used *attrib.* to designate a type of cape with a hood; **red robin,** (*c*) *U.S.*, a perennial herb, *Castilleja coccinea*, which has bright red bracts surrounding small greenish-yellow flowers; **red rock fault** *Geol.*, a fault in Permian red beds, forming part of the boundary of the Cheshire Basin; **Red Rover,** a children's chasing game; also, the child who is 'it' in this game; **red sable,** the fur of the Japanese mink (kolinsky), used esp. for artists' brushes; a brush made from this fur; **red squill** (see Squill); **red star,** a symbol of the Soviet Union; **red steer** *Austral. slang* (see quot. 1941); **red-stone** (later examples); **red tabby,** a cat, esp. a long-haired one, with a reddish-orange coat patterned in a deeper red; **red tag,** name of an artificial fly used in angling; **Red Terror,** the persecution of opponents by the Bolsheviks after the Russian revolution of 1917; also *transf.*; **red tide** = *red-water 4; **Red Tory** *Canada*, one of a political group who, while maintaining some conservative principles, yet support many liberal and socialist policies; so **Red Toryism**; **red vision** *Ophthalm.* = *erythropsia* s.v. *erythro-; **red warning** = *red alert* above.

See also as main entries in Suppl.: *Red Army, *red belt, *red book, etc.

1961 Webster, Red alert. 1962 'K. Orvis' *Damned & Destroyed* xxvii. 203 His every move will be under red-alert watch. 1967 *Guardian* 29 Dec. 1/1 The emergency bed service..has put out a 'red alert' to more than 200 hospitals in the London area. 1970 *Daily Tel.* 26 Sept. 1/6 Fearing that the Arabs will attempt to seize another British plane..the Government sent out a 'red alert' to airlines yesterday, warning them to exercise stringent anti-skyjacking precautions. 1972 'G. Black' *Bitter Tea* (1973) xii. 188 It looked as though a phone call to me had resulted in a red alert. 1973 M. Amis *Rachel Papers* 98 Fortunately, my room was in a state of red alert nowadays and Rachel's telephone call hadn't caught me with my pants down. 1975 *Times* 26 Nov. 1/4 The health authorities at Croydon yesterday put out a 'red' alert of hospital beds... Hospitals in its area will take only emergencies. 1981 W. Safire in

N.Y. Times Mag. 22 Feb. 9/1 The red alert flashed here a few weeks ago—warning of incoming semantic missiles from the new Secretary of State, Al Haig. 1957 Mankowitz & Haggar *Conc. Encycl. Eng. Pott. & Porc.* 49/2 Work of the 'raised' (i.e. embossed) and 'red anchor' periods enjoys the highest esteem. *Ibid.*, The resulting quality..was perhaps inferior to 'red anchor' Chelsea. 1966 *Daily Tel.* 26 Oct. 16/4 A Chelsea figure of a Chinaman, Red Anchor period, was sold for £4,200. 1975 *Oxf. Compan. Decorative Arts* 140/2 This factory's [*sc.* Chelsea's] figures of the 'red anchor' period (1748–55 when the factory mark was an anchor painted in red) are among the loveliest in the whole range of European porcelain. 1973 J. Shub *Moscow by Nightmare* viii. 87 The Krasnaya Strela—the Red Arrow—leaves Moscow at 11.53 every night. 1974 A. Williams *Gentleman Traitor* v. 94 You'll be given your ticket on the Red Arrow Express for Leningrad. 1946 R. Grinstead *They dug Hole* I. i. 13 And so it goes on. The everlasting bickering between old sweat and red-arse! 1947 D. M. Davin *Gorse blooms Pale* 193 You were only a bloody redarse in those days. 1847 J. M. Ives *New England Bk. Fruit* 36 Red Astracan.—This beautiful apple is of medium size, of a round and rather flat form. 1860 R. Hogg *Fruit Man.* 21 Red Astrachan... Flesh white, and richly flavoured. 1876 J. Burroughs *Winter Sunshine* 128 The red astrachan [is] an August apple. 1948 *Newsweek* 30 Aug. 32/1 The best and most popular American apples are descended from Russian apple trees—Borominka, Titovka, Red Astrakhan, Alma Ata—imported into the United States a hundred years ago. 1977 *N.Z. Herald* 8 Jan. 1–5/5 One store was selling a New Zealand apple, red astrachan, at 33c a pound. 1849 E. Sieveking tr. *Rokitansky's Man. Path. Anat.* II. I. ii. 122 Atrophy of the liver, independent of the *marasmus senilis* of the organ, appears in various forms. We first draw attention to two distinct forms which have not been remarked hitherto... Owing to their distinctive colouring, they may be appropriately termed yellow and red atrophy. 1961 R. D. Baker *Essent. Path.* xvi. 409 Massive necrosis... If the patient lives several weeks after the onset of jaundice subacute red atrophy is found at autopsy. 1927 *Amer. Speech* II. 388/2 Fast freights are known as *red balls*. 1934 *Sun* (Baltimore) 3 May 12/6 Several who have worked on these 'red ball' runs told me that after a man has been on a truck twenty-four hours he's tired and unstrung. 1944 *N.Y. Times* 8 Oct. iv. 5/4 The famous Red Ball highway—a belt of one-way roads for truck convoys that actually kept pace with General Patton's advance. *Ibid.* 5/5 It was assumed that bullets and butter, and gasoline as important as either, would be delivered. It was—once the Red Ball got rolling. 1968 *T.V. Times* (Austral.) 29 May 18/3 In railway language 'Red Ball' means top priority freight. 1950 P. Tempest *Lag's Lexicon* 178 *Redband*, a privileged prisoner. He is allowed to travel freely about the prison in pursuance of his duties. 1952 J. Henry *Who lie in Gaol* x. 143 That prisoners could walk through the house unaccompanied by a red-band or an officer seemed to her little short of madness. 1976 A. Miller *Inside Outside* 3, I would then have a cup of coffee brought to me by a 'red band'. 1876 *Encycl. Brit.* V. 803/1 Near Pooldoody is the great Burren oyster bed called the Red Bank,..from which a constant supply of the excellent Red Bank oysters is furnished to the Dublin and other large markets. 1922 Joyce *Ulysses* 325 He spat a Red bank oyster out of him. 1935 S. & B. Webb *Soviet Communism* II. ix. 759 The Red Banner of Toil is awarded 'by special decision' of the Central Executive Committee (TSIK) of the All-Union Congress of Soviets. 1957 *Encycl. Brit.* XXII. Pl. iv, following p. 704, *Order of the Red Banner*: For conspicuous bravery or self-sacrifice in time of war, special capacity for leadership or some action contributing decisively to the success of soviet arms (military or civil). 1966 tr. Lin Piao in *Quotations from Chairman Mao Tse-tung* (Foreword), The most fundamental task in our Party's political and ideological work is at all times to hold high the great red banner of Mao Tse-tung's thought. 1974 tr. *Sniečkus's Soviet Lithuania* 28 New fighting patriots took the places of those that were killed or jailed, keeping the red banner of revolution flying. 1977 *N.Y. Rev. Bks.* 26 May 24/3 The pilgrims move in ranks, by sections, red banners flying. 1892 T. F. Garrett *Encycl. Pract. Cookery* 93/2 Red Haricot Beans à la Bourguignonne... Take 1 qt. of Red Beans, pick out any stones. 1895 [see *black bean* s.v. *black a. 19]. 1908 J. H. Maiden *Forest Flora N.S.W.* III. 115 'Red bean' is, however, the commonest name..because it is supposed to resemble the timber of the Black Bean.. except in colour. 1931 E. Sherson *Bk. Vegetable Cookery* viii. 161 Boil the little red beans in the usual way. 1932 [see *pencil cedar* s.v. *pencil *sb.*² 7 b]. 1965 *Austral. Encycl.* I. 468/2 Other trees to which the name [bean] is applied are the red bean.., walnut bean.., and yellow bean. 1977 *Sunday Times* (Colour Suppl.) 4 Dec. 19/1 *Afters*: 'toffee apple', red-bean pancakes. 1980 *Times* 21 June 11/6 Red-bean salad in a sweet, spicy dressing. 1849 *Q. Jrnl. Geol. Soc.* V. 25 The appearance at certain points of the series of stratified deposits of red sandstones and other rocks coloured by the peroxide of iron, in regions where the older formations contain comparatively few red beds, is a fact observed in many countries. 1922 *Bull. Geol. Soc. Amer.* XXXIII. 107 The thickest beds of such [iron] ore in the Appalachian district are in areas where the underlying red beds of Upper Ordovician age are either very thin or are entirely eroded away. 1946 L. D. Stamp *Britain's Structure & Scenery* xii. 124 In early Permian times conditions repeated those found in early Old Red Sandstone times. The older geologists who used the term New Red Sandstone for the red beds of the Permian and Trias introduced this very useful comparison. 1974 *Encycl. Brit. Macropædia* XVIII. 694/1 Continental sediments, especially of red-bed facies often associated with evaporite deposits, are especially widespread and characteristic of the Triassic. Throughout Eurasia north of Tethys, such rocks are a conspicuous part of the rock record. 1785 G. Washington *Diary* 28 Jan. (1925) II. 338, I discovered.. the red berry of the Swamp. 1805 M. Lewis *Jrnl.* 12 Apr. in *Orig. Jrnls. Lewis & Clark Exped.* (1904) I. vii. 299 The under brush is willow, red wood,..the red burry [*sic*], and Choke cherry. 1951 *Dict. Gardening* (R. Hort. Soc.) IV. 1754/1 Rhagodia..Australian Red Berry. 1973 E. Goudie *Woman of Labrador* p. xix, Well fortified with her never-ending supply of tea and redberry pie. 1928 *Daily Express* 5 Dec. 13/2 Glasgow has not relaxed its war-time drink restrictions..but nothing is being done to make the sale of this horrible 'Red Biddy' punishable by prison. 1939 Joyce *Finnegans Wake* (1964) I. 39 Blotto after divers tots of hell fire, red biddy, bull dog, blue ruin and creeping jenny. 1950 E. Hyams *From Waste Land* 204 Pamphlets issued by the French wine trade..to persuade the customer that the Red Biddy he is drinking is something very special and fine indeed. a 1953 Dylan Thomas *Adventures Skin Trade* (1955) 85 'I suppose he thinks red biddy's like bread and milk,' Mr. Allingham said. 1961 C. Willock *Death in Covert* iii. 67 Any idea where we could get any of the hard stuff? This flipping red biddy's burning a hole in my stomach. 1977 M. Kenyon *Rapist* v. 58 Next time it'd be Majorca..and what if he was not a red biddy man? At five bob a bottle he bloody soon would be. 1926 A. Giles *Exploring in Seventies* 127 Tinned meat in 6 lb tins ('red blanket' we called it). The tins were painted red without labels or description of contents. 1846 *Phil. Trans. R. Soc.* CXXXVI. 66 Professor Rudolph Wagner was the first to point out the circular form of the red blood-corpuscle of the Lamprey, but he does not appear to have noticed the existence of a nucleus. 1910 H. W. Armit tr. *Ehrlich & Lazarus' Anæmia* i. 3 Tarchanoff proposed that by determining the loss of water during profuse sweating, and by comparative red blood cell counts both before and after the sweating, an estimate of the quantity of blood could be arrived at. 1950 *Sci. News* XV. 87 Thanks to the hæmoglobin in them, the red blood corpuscles of our blood carry oxygen from lungs to muscles and brain, which all the time use it up. 1971 W. M. Dougherty *Introd. Hematol.* v. 135/2 The hematocrit (packed cell volume) is the determination that equates what volume of a given unit of blood is composed of red blood cells (erythrocytes and reticulocytes). 1929 *Bookman* (U.S.) July 527/2 *Red board*, when..a train has to stop for orders. 1935 A. J. Pollock *Underworld Speaks* 96/1 *Red board*, board facing the grandstand on which a horse race is declared official by the judges. 1946 *Sun* (Baltimore) 5 Apr. 18/1 After each race there was much 'red board' speculation and betting as to which horse had won. 1968 *Wall St. Jrnl.* 31 Jan., 'One thing about Sam,' he says. 'He never bet the red board.' (In track jargon, to bet the red board is to claim you picked the winner—after the race is over.) 1973 *Amer. Speech* 1969 XLIV. 259 *Red board*, stop signal on an overhead signal bridge. 1785 A. Monro *Struct. & Physiol. Fishes* iii. 28 A red-coloured organ..is found on the inner side of the air-bag of the cod, haddock, etc.: but in those fishes where the air-bag communicates with the alimentary canal, this red body is either very small and simple..or entirely wanting. 1911 *Proc. Zool. Soc.* I. 184 In the vast majority of cases 'red body' includes both rete mirabile and gas gland. 1963 L. Birkett tr. *Nikolsky's Ecol. Fishes* i. 7 In the two red bodies in the swim-bladder of the eel, there are 88,000 venous and 116,000 arterial capillaries. 1723 D. O'Connor tr. *Keating's Gen. Hist. Ireland* 91 The Lodge of *Teagh na Craoibhe Ruadhe*, which signifies in English the House of the *Red Branch*, where the most renowned Champions lodged their arms... The Champions.. were distinguished by the Title of Champions of the Red Branch. 1772 S. O'Halloran *Introd. Study Hist. Ireland* I. v. 40 Long before the birth of Christ we find an *hereditary* order of chivalry in Ulster, called *Curaidhe na Craoibhe ruadh*, or the Knights of the Red Branch, from their chief seat in Emania, adjoining to the palace of the Ulster kings, called Teagh na Craoibhe ruadh, or Academy of the Red Branch. 1879 *Encycl. Brit.* IX. 75/1 Ulster, whose warriors of the *Craebh Ruaid* or Red Branch are the most prominent figures in the Heroic period, had no Fenians. 1889 W. B. Yeats *Wanderings of Oisin* III. 41 Came by me the Kings of the Red Branch with roaring of laughter and songs. 1892 —— *Countess Kathleen* iv. 73 And on tales Of Finian labours and the Red-Branch Kings. 1970 N. Chadwick *Celts* ix. 268 The ruling king in the [Tain] cycle is Conchobar mac Nessa of the House of the Red Branch at Emain Macha in Ulster. 1973 *Times* 11 Dec. 1/2 Messages found this morning in a telephone booth said that Dr Amerio would be kept in a 'people's jail' as a reprisal against dismissals at Fiat. The messages claimed that the kidnapping was the work of the 'Red Brigades'. 1977 *Time* 18 July 29/1 Some unemployed university graduates..have joined in the terrorism of groups like the 'Red Brigades'. 1978 *Ann. Reg.* 1977 136 Terrorist action with a political flavour was also carried on by other extremist groups, in particular by the left-wing Red Brigade. *Ibid.*, In their efforts to spread terror the Red Brigaders resorted to shooting in the legs or kneecap a number of fairly prominent persons. 1979 *Rolling Stone* 11 Jan. 79 Outside..the Red Brigades are blowing people's knees off. 1934 S. Lewis *Work of Art* 72 He's got to be a certified public accountant,..or one-night-stand lecturer that blows in and expects to have the red carpet already hauled out for him. 1938 N. Marsh *Artists in Crime* ii. 28 Be sure to have the red carpet out. 1952 *N.Y. Times* 21 Aug. 21/6 (*heading*) Englewood rolls out red carpet for little baseball league teams. 1960 *Daily Mail* 13 Dec. 7/8 A champagne party was laid on for Mr. Steven Mueller, 2,000,000th passenger to leave London in a TWA jet airliner... There was a hitch when Mr. Mueller arrived. He is just two years and four months old. The red carpet treatment went to his mother. 1966 J. A. Morris *Bird Watcher* (1968) ii. 30 We're rolling out the red carpet for him. Nothing will be too good for good old Congressman Herper. 1969 'G. Black' *Cold Jungle* viii. 122 A complete breakdown in red carpet reception arrangements. 1975 *Evening Standard* 23 Sept. 12/1 (Advt.), May we give you the red carpet treatment? Why get footsore and weary comparing notes at all the carpet stores? 1977 *New Yorker* 10 Oct. 67/2 Eaton's interest in the Soviet Union—to which he has made eight red-carpet journeys since 1958—was indirectly whetted by John D. Rockefeller, Sr. 1894 T. F. Garrett *Encycl. Pract. Cookery* 324/1 *Red Caviare*, this is a very inferior quality, made from the roe of any fish, such as the grey mullet, or carp. 1927 A. Martineau *Caviare to Candy* iii. 31 *Red Caviare*. This is made from the roe of a hen pike. 1946 G. Stimpson *Bk. about Thousand Things* 80 Virtually all the red caviar..is displayed as of Russian origin. 1957 V. Nabokov *Pnin* i. 10 Those stupendous Russian ladies..infuse a magic knowledge of their difficult and beautiful tongue..in an atmosphere of Mother Volga songs, red caviar, and tea. 1964 A. Launay *Caviare & After* i. 21 Red caviare (*Keta*) is made from the salmon roe..bright orange in colour. 1974 M. G. Eberhart *Danger Money* (1975) v. 57 They bought..some red caviar. 'It would be good with rye bread,' Susan suggested. 1935 H. H. Finlayson (*title*) The red centre: man

and beast in the heart of Australia. *Ibid.* ii. 22 The Luritja Country—the south-west portion of Central Australia and contiguous tracts in the adjoining States—..might well be known as the Red Centre. Sand, soil, and most of the rocks are a fiery cinnabar. 1979 *Jrnl. R. Soc. Arts* Apr. 293/1 Nearly a decade of good rains have turned the famous Red Centre into something approaching a Green Centre. 1905 *Eye Opener* (Calgary, Alberta) 25 Feb. 1/6 The innocent hawbuck who imagines that the red chamber is full of dignity and high thoughts has never listened to the debates from the galleries. 1948–9 *Parl. Affairs* II. 50 The Senate Chamber at the east end, known from its bright leather upholstery as the 'Red Chamber', is similar in design to the Commons Chamber but is much smaller and has no side galleries. 1955 *Chatelaine* Apr. 13 Canada's first woman senator is Mrs. Norman F. Wilson, who shattered a fifty-year-old tradition that had preserved the Red Chamber as an exclusively men's club when she stepped over the threshold in 1930. 1965 *Globe & Mail* (Toronto) 6 Dec. 6/5 Mr. Pearson..could make no better beginning than to appoint to the Red Chamber 12 such Canadians. 1968 Red channel [see *green a. 1 i]. 1979 *Guardian* 2 Aug. 1/8 Customs staff will examine in detail the baggage of all passengers going through the red channel and carry out more spot checks..on those using the nothing-to-declare green channel. 1801 B. Hawkins *Lett.* (1916) 379 Your father is desirous that his red children would consent to establish houses of entertainment and ferries on these roads. 1855 J. H. Chambers in *Montana Hist. Soc. Contrib.* (1940) X. 136 Col Vaughan gave his red children a talk. 1871 *Weekly Manitoban* 5 Aug. 2/4 She wishes her Red children, as well as her White people, to be happy and contented. 1972 K. Lo *Chinese Food* I. 23 We red-cook it [*sc.* fish], quick-fry it, clear-simmer it, deep-fry it, steam it, and hot-plunge it. 1956 B. Y. Chao *How to cook & eat in Chinese* I. vi. 65 Red-cooking is stewing with soy sauce, some materials needing pre-frying, some not. It is so-called because the soy-sauce juice gives a reddish colour. 1972 K. Lo *Chinese Food* I. 12 In red-cooking the meat or poultry is first quick-fried and then simmered in broth or water along with soya sauce and other constituents of the soya herbal sauce. 1936 *Ann. Rep. E. Malling Res. Station 1935* 144 The roots of the affected [strawberry] plants showed the 'red core'. 1952 E. Ramsden tr. *Gram & Weber's Plant Dis.* III. 364/1 No method of soil treatment has proved effective against red-core disease. 1970 *Countryman* Spring 111 Red core was first known as a disastrous disease in 1921 in Scotland. 1846 *Phil. Trans. R. Soc.* CXXXVI. 64 The well-known red corpuscle of the blood of the Frog. 1871 *Q. Jrnl. Microsc. Sci.* XXII. 361 The chemical and formal structure of the red corpuscle. 1911 C. E. W. Bean *'Dreadnought' of Darling* x. 89 The river came down in flood. People had been forced to clear out of some of the 'frontages' and camp back in the red country. *Ibid.* xv. 142 The mallee..covers thousands of square miles of red country in Victoria, New South Wales, and South Australia. 1936 A. Russell *Gone Nomad* viii. 60 Picture great wind-scoured plains of red sand..; a sun glaring wanly day after day from a sky reddened with the dust of incessant sand-storm..—and you will see the great 'Red Country' of the West Darling in the big seven years' drought of 1897–1903. 1877 H. Ponsonby *Let.* 6 Sept. in A. Ponsonby *Henry Ponsonby* (1942) 376 The Red Cross or Crescent is a valuable association in its attempts to protect the sick and wounded in time of war. 1959 *Chambers's Encycl.* XI. 552/1 The League of Red Cross Societies (founded 1919) is a federation of national Red Cross and Red Crescent societies. 1970 *Times* 3 Apr. (Arab League Suppl.) p. viii/3 Their own Palestinian Red Crescent facilities. 1971 *Shankar's Weekly* (Delhi) 4 Apr. 10/4, I think it will be a better idea to get some business men and others contribute money, buy medicines and other equipment with that and donate it to the East Pakistan Red Crescent or some other organisation. 1976 G. Seymour *Glory Boys* ii. 23 The big tanks had rumbled into..Nablus. He recalled..the wail of the Red Crescent ambulances. 1912 R. Lydekker *Ox & its Kindred* v. 101 Attention may be directed to the well-known red Devons. 1979 V. Canning *Satan Sampler* ix. 197 A colour gravure of a red Devon heifer. 1925 Fraser & Gibbons *Soldier & Sailor Words* 237 *Red duster, the*, the Red Ensign of the Mercantile Marine. 1928 *Daily Express* 10 Aug. 15/1 His papers have not yet come through allowing him to fly the White Ensign, so, meanwhile, the Vita sails under the 'red duster'. 1944 *Times* 7 June 6/7 A glance at the ships, with their different flags, the Red Duster of the Merchant Navy, the Stars and Stripes, [etc.]. 1977 *Jrnl. R. Soc. Arts* CXXV. 216/1 Today even the Red Duster is to some extent a flag of convenience. [1916 *Proc. Nat. Acad. Sci.* II. 17 Differing from the dwarf red stars most conspicuously in density, dimensions, and total brightness.] 1929 J. H. Jeans *Universe around Us* v. 293 The great gulf which lies between the red giants and the red dwarfs. 1959 *Listener* 26 Feb. 370/1 Instead of cooling steadily down towards the Red Dwarf stage..the Sun is becoming more energetic as it ages. 1978 *Nature* 22 June 645/1 The energies of even the weak radio flares emitted by a red dwarf star are several orders of magnitude greater than those from large and infrequent solar outbursts. 1951 T. C. Roughley *Fish Austral.* 68 The red emperor or king snapper..grows to a weight of at least 40 pounds. 1956 M. West *Gallows on Sand* xiii. 145 Fillets of red emperor, caught while we were at the bottom of the sea. 1965 *Austral. Encycl.* IX. 41/1 The juvenile form of the red emperor.. bears a red broad-arrow mark, from which it derives its name of government bream. 1730 *Royal Navy Orders & Instructions* 12 Merchant-Ships are to wear a Red Ensign; with the Union Jack in a Canton at the upper End next the Staff; and a white Jack, with a red Cross, commonly called *St. George's Cross*, passing quite through the same. 1910 *Encycl. Brit.* X. 459/1 The red ensign is the distinguishing flag of the British merchant service. 1961 B. Fergusson *Watery Maze* vi. 123 Red Ensign ships—merchant ships with Merchant Navy crews, requisitioned for Admiralty service—were administered by the Director of Sea Transport. 1976 *Oxf. Compan. Ships & Sea* 695/2 The Red Ensign ..is today flown by all British merchant vessels and also by many yachts. 1893 P. Gardner *Catal. Greek Vases in Ashmolean Mus.* iv. 22/1 The natural advantages..of the red-figure method caused it to speedily supersede the older style. 1899 R. Glazier *Man. Hist. Ornament* 77 The Transitional period (B.C. 500–470), when the black silhouette figures on a red ground gave way to the red figure period on a black ground. 1918 J. D. Beazley *Attic Red-Figured Vases in Amer. Museums* i. 5 The earliest red-figure vases are mostly amphorae or cups. 1936 *Burlington Mag.* May 253/1 Amongst the fifth-century red-figure vases there are many of high artistic merit. 1960 R. G. Haggar *Conc. Encycl. Continental Pott. & Porc.* 211/1 Greek pottery is generally classified by style: Geometric, Oriental influence, Black-figure, Red-figure and mixed styles. 1975 J. Boardman *Athenian Red Figure Vases: Archaic Period* ii. 11 Red figure is the reverse of black figure. *Ibid.* 12 Relief line is so important in early red figure that..a word about it is called for. 1978 K. J. Dover *Greek Homosexuality* I. 10 An Attic red-figure vase of the early fifth century depicts a man at a dinner party. 1890 Red-figured [in Dict., sense 14 a]. 1918 J. D. Beazley *Attic Red-Figured Vases in Amer. Museums* p. v, The Andokides painter, one of the first artists to use the red-figured style. 1919 J. C. Hoppin (*title*) Handbook of Attic red-figured vases. 1960 E. H. Gombrich *Art & Illusion* i. 40 The Greek vase painters made use of this principle of reversal when they switched over from the earlier black-figured technique ..to the red-figured style. 1977 *Jrnl. R. Soc. Arts* CXXV. 96/2 It [*sc.* the *anthemion*] was a favourite border design, too, with the painters of Greek pottery of the period, particularly of the red-figured Attic ware. 1848 *Santa Fé* (New Mexico) *Republican* 28 June 2/4 A proportionate lot of..Cinto laces, Red flannel shirts, [etc.]. 1860 *Times* 15 Sept. 10/1 Most of us wear no linen.., the red flannel shirt answering all purposes of outward and inward raiment. 1906 E. Nesbit *Railway Children* vi. 137 'How lucky we *did* put on our red flannel petticoats!' said Phyllis. 1940 L. I. Wilder *Long Winter* (1962) ix. 83 Her red flannel underwear was so hot. *Ibid.* 84 It's too hot for my red flannels, Ma! 1943 *Sun* (Baltimore) 24 Nov. 17/5 It is time for Kent countians to 'get into their red flannels'. 1978 Walkley & Foster *Crinolines & Crimping Irons* ii. 46 Both natural wool and red flannel became favourite materials for underwear and nightwear. 1907 *Dialect Notes* III. 248 *Red flannel hash*,.. beet hash. 1951 E. Graham *My Window looks down East* vii. 59 She had a real hankerin' for red-flannel hash. 1977 J. Cheever *Falconer* 60 His mother..served the red flannel hash with poached eggs. 1916 *Proc. Nat. Acad. Sci.* II. 17 The brightest stars in the cluster are the red and yellow giants. 1929 J. H. Jeans *Universe around Us* v. 272 The stars of large diameter shewn in the table..are red and have very high luminosities; they are red giants. 1966 *Random House Dict.* 142/3 *Betelgeuse*, a first magnitude red giant star in the constellation Orion. 1977 J. Narlikar *Struct. Universe* ii. 39 The red-giant reactions do not take as long to complete as the reactions during the main-sequence stage. 1896 *Jrnl. Anat. & Physiol.* XXX. 550 The red glands occupy about the anterior half of the internal surface of the ventral wall of the swim-bladder. 1926 H. M. Kyle *Biol. Fishes* xi. 276 The fish is able to exercise some selection of gases, and the 'red glands' or retia mirabilia of the Physoclists are evidently used for this purpose. 1931 J. R. Norman *Hist. Fishes* ix. 174 The walls of the bladder are richly supplied with fine blood-vessels, and at certain areas these are accumulated to form the so-called red bodies or red glands, masses of interlacing and tightly packed arteries and veins. 1974 D. & M. Webster *Compar. Vertebr. Morphol.* xv. 368 The rete mirabile here [in teleosts] produces gas, which the red gland secretes into the gas bladder. 1910 H. P. Tiemann *Iron & Steel* 312 Red hardness is the name they give to the property of a tool when it maintains its cutting edge after its nose is red hot. 1925 *Trans. Amer. Soc. Steel Treating* VIII. 693 Molybdenum when present in sufficient amounts imparts to steel the properties of 'red hardness'. 1937 *Discovery* May 153/1 The forerunner of self-hard steel,..Musket self-hard steel, employed [carbon and tungsten]..to confer the then new so-called red hardness... Tungsten contributes to high speed steel the property of red hardness. 1819 *Orthodox Jrnl.* May 175/2 [He] laboured afterwards most earnestly to counteract its contents, and was honoured with the red hat by his holiness. 1863 Geo. Eliot *Romola* I. i. 36 Men who love to see avarice and lechery under the red hat and the mitre. 1969 R. H. Bainton *Erasmus of Christendom* (1970) i. 44 To receive the red hat he must go to Rome. 1975 R. Player *Let's talk of Graves* vi. 217 Cardinal Cavalle..had got his red hat from Pius the Ninth. 1934 *Bulletin* (Sydney) 16 May 20/3 Someone identified the thing as *Patæcus fronto*, better known as the 'Red Indian' in N.S.W., where it is sometimes caught on the reefs. 1965 *Austral. Encycl.* VII. 395/2 Red Indian fish (*Pataecus fronto*), a fish of the southern Australian rocky shore-lines. Its high dorsal fin forms a crest like the feathers of a Red Indian's head dress. 1963 'J. Prescot' *Case for Hearing* viii. 125 There's precious little point in letting him out on bail when a red judge is going to send him back again for at least a couple of years. 1972 *Times* 20 May 3/7, I would very much like to see more work being committed for trial by the red judge. 1977 *Daily Tel.* 19 Oct. 6/5 Since May there have been no visits by 'Red judges' because of a shortage of accommodation. Arrangements have been made for the more important criminal cases—murder and manslaughter—to be tried either at the Old Bailey or Maidstone. 1812 W. J. Burchell *Jrnl.* 27 May in *Trav. Interior S. Afr.* (1822) II. viii. 160 They had intended going to the *Roode Kaffers* (Red Caffres). 1879 R. M. Ballantyne *Six Months at Cape* iii. 44 This red-Kafir is in truth a savage. 1904 D. Kidd *Essential Kafir* i. 31 Red ochre and oil are rubbed into the skin, and frequently into the blanket. When this latter is done by a tribe the people are called Red Kafirs. 1846 *Blackw. Mag.* Nov. 595/1 Almost any serial will give hints enough to an acute boy, [and]..guide him to the door with the red lamp. 1927 W. E. Collinson *Contemp. Eng.* 96 The red lamp as the sign of a '*maison tolérée*'. 1929 J. L. Hodson *Grey Dawn* II. ii. 159 You should have seen the queue at the Red Lamp. 1962 H. Myers tr. *Pingaud's Holland* 51 The dirty little canals near the harbour in the 'red lamp' district. 1942 Berrey & Van den Bark *Amer. Thes. Slang* §49/1 Redland, Russia. 1966 J. Gardner *Amber Nine* iii. 44 If Redland have got a finger in the pie then that part of the world could be warmish. 1969 W. Garner *Us or Them War* i. 15 Morton picked up the camera... He said, 'Exacta. Made in Dresden. East Germany. A favourite with Redland agents.' 1977 C. Wood *James Bond* iv. 40 'You can imagine who the first suspects are?' Bond could. 'Redland.' 1821 M. Wilmot *Let.* 17 Jan. in *More Lett.* (1935) 92 Melodious clang of knives forks and plates to flourish *down the red Lane* the most magnificent supper that Gourmands ever guttled. 1958 L. Durrell *Mountolive* xv. 286 The quarter lying beyond the red lantern belt. 1973 *Whig-Standard* (Kingston, Ontario) 11 Aug. 7/3 He was on the loose and cutting a swathe in the red lantern district. 1918 R. W. Kauffman *Our Navy at Work* 6 'Red lead' is catsup, which it hugely resembles. 1919 G. M. Battey *70,000 Miles in Submarine Destroyer* 261 Beans and 'red lead' for breakfast. 1945 'Tackline' *Holiday Sailor* v. 55 Everything went into a pot-mess—meat, spuds, peas, beans, rice, oxo, 'red-lead' (tinned tomatoes)—and the result was invariably good. 1959 W. L. Corzine *Sailors in Nightgowns* vii. 100 We were having those delicious beans for breakfast, which we always camouflaged with red lead (catsup). 1905 Emerson & Norris in *Jrnl. Exper. Med.* VII. 34 The name often given to the disease in these letters is 'red-leg', and this is also the name used by the frog-catchers. 1964 G. Durrell *Menagerie Manor* iii. 75, I sent them [*sc.* toads] away for post-mortem, and the report came back that they were suffering from an obscure disease called red-leg. 1974 *Amer. Jrnl. Vet. Res.* XXXV. 1243 Since 1890, septicemic frog disease (red-leg) has had a devastating effect on frogs kept under laboratory conditions. 1898 Red meat [see sense 8 c]. 1933 E. O'Neill *Ah, Wilderness!* I. 24 Poetry's his red meat nowadays, I think—love poetry—and socialism, too, I suspect. 1972 *Times* 3 July 12/2 Colin Carr chose..a lollipop, a Popper Polonaise, rather than good red meat, but it served to show off an enviable fluency on the instrument. 1977 *Jrnl. R. Soc. Arts* CXXV. 369/1 Dairy products, cereal products and red meat account for 60 per cent and 75 per cent respectively of the energy and protein intake of the population. 1925 B. Coan *Red Web* 6 It is time, right now, to get down to cases about this thing we hear called the 'red menace'. 1932 J. F. Carter *What we are about to Receive* xviii. 204 But once the election is over..we shall quietly lay aside our witch hunting, put the Red Menace in cold storage. 1934 R. V. C. Bodley *Japanese Omelette* xvii. 174 The substance of his words did not confirm the rather pessimistic views of my soldier friends in Changchun on the subject of the 'Red Menace'. 1977 *Time* 14 Feb. 17/1 Young..was quick to warn Smith that his efforts to gain U.S. support by invoking the 'Red menace' would not succeed. 1819 J. Taylor *Naturales Curiosæ* 129 Adonis—Red Morocco... Its flowers are of a bright scarlet, with a black spot or eye at the bottom, and are frequently sold in London under the name of red-morocco. 1931 M. Grieve *Mod. Herbal* I. 389/1 'Red Morocco' was a somewhat strange old English name for this plant [*sc.* pheasant's eye]. 1941 Baker *Dict. Austral. Slang* 59 *Red Ned*, cheap red wine. 1941 —— *N.Z. Slang* vii. 62 Such terms for strong drink as.. *red Ned*. 1972 I. Hamilton *Thrill Machine* xxvi. 120 Jo clutched the glass of Red Ned that I thrust at him. 1961 Ward & Shapiro in *Jrnl. Meteorol.* XVIII. 642/1 The only other characteristic of these spectra is their resemblance to 'red noise', that is, generally higher variance at the lower frequencies (longer periods). [*Note*] The term 'red noise' was suggested by Prof. E. N. Lorenz to describe this phenomenon. *Ibid.* 646/1 The spectrum of a time series of random numbers, which has been modified so that there is a moderate correlation between successive values, exhibits a characteristic damping of the higher frequencies. This type of spectrum, having more power (or variance) at lower frequencies can, by analogy to the spectrum of light, be called a 'red noise'. 1963 *Jrnl. Atmospheric Sci.* XX. 182/1 A general suppression of relative variance at higher frequencies and consequent inflation at lower frequencies, as compared to the even distribution of relative variance across all frequencies shown by the 'white noise' spectrum. Following a suggestion by E. N. Lorenz, Shapiro and Ward have called this phenomenon 'red noise'. 1979 *Nature* 23 Aug. 672/1 The spectra were calculated using a fast Fourier algorithm, the spectral estimates were smoothed, and a white or red noise null continuum was assumed in a significance testing. 1890 W. H. Vittum tr. *Edinger's Twelve Lect. Struct. Cent. Nerv. Syst.* vi. 93 Below the thalamus is a rounded ganglion,—the nucleus ruber, the red nucleus of the tegmentum. 1942 F. A. Mettler *Neuroanat.* xiii. 302 The red nuclei are important extrapyramidal relay stations. 1972 M. L. Barr *Human Nerv. Syst.* vii. 110/1 The red nucleus is..involved in pathways through which the cerebellum is able to influence motor function. 1974 D. & M. Webster *Compar. Vertebr. Morphol.* xii. 294 These fibers decussate in the midbrain and then form a large capsule around the red nucleus. 1917 W. K. Moorehead *Stone Ornaments used by Indians in U.S. & Canada* 53 Oval forms occasionally found in the Red Paint People's graves in Maine are much weathered and appear very old. 1947 R. P. T. Coffin *Yankee Coast* 225 Before the dawn and the Dawn People, there were the Red Paint Men... They had been gone so long now that not even the teeth of them are left, only the red paint, color of life, they smeared their bodies with. 1970 S. Trueman *Intimate Hist. New Brunswick* iii. 43 'Red Paint' Indian burial ground dating back 3,500 to 4,000 years; in the bottom of each grave is a covering of red ochre, bright red oxide. 1933 *Discovery* May 158/1 Red palm oil and some samples of maize oil are good sources [of vitamin A]. 1975 *Sci. Amer.* June 126/1 Cereal and vegetable oils are generally poor in carotene—except for red-palm oil, which is very rich. 1959 *Encounter* Dec. 29/1, I have been red-pencilling student papers for a good many years. 1966 *Eng. Stud.* XLVII. 116 There are many teachers of English who look upon the adjective *corny* as a word to be red-pencilled whenever it turns up in a student paper. 1979 *Verbatim* Summer 2/2 To red-pencil is 'to censor or correct'. 1927 *Observer* 4 Dec. 13/1 We have to guard against the Red Peril on our borders. 1973 *Sat. Rev. Society* (U.S.) Mar. 48/3 At the end of the Fifties the red peril slunk off. 1873 *Punch* 4 Jan. 1/2 Mars. The Red Planet salutes you. But you are a slow lot. 1894 J. E. Gore tr. *Flammarion's Pop. Astron.* iv. iv. 374 The red planet varies in brightness according to its position in the sky. 1972 *Nature* 4 Feb. 251/2 They show clearly that the red planet is far from being dead, and that weathering and volcanic activity are taking place on Mars to a significant degree. 1977 *Time* 17 Oct. 45/1 The two Viking landers and their orbiters have spent much of the 15 months since the arrival on Mars snapping pictures of the Red Planet. 1941 *Coast to Coast* 22 The mounted trooper found a couple of red poley steer skins in Jo Wiggins's slaughter-yard. 1878 Disraeli *Let.* 13 June in Monypenny & Buckle *Life Disraeli* (1920) VI.

ix. 318 Lord B. mistook His Royal Highness for the father of the bride, who soon appeared as 'The Red Prince'. 1888 *Random Recoll. Courts & Society* vi. 140 The wedding of the Princess of Dessau with Prince Frederick Charles—the Red Prince—in September, 1857, was solemnized in the evening at the Palace with all the customary formalities. 1958 *Everyman's Encycl.* V. 494/1 Frederick Charles of Prussia.., known as the 'Red Prince' because of the uniform he usually wore. 1885 Etheridge & Seeley *Phillips' Man. Geol.* (ed. 2) I. ii. 18 In the Arctic regions minute spherical particles of iron are sometimes brought down from the air in snow, as though the earth occasionally entered clouds of meteoric dust. A like cause must account for the red rain which fell at Blankenburg in 1819, and owed its colour to cobalt chloride. 1904 G. S. West *Treat. Brit. Freshwater Algæ* 189 *Sph*[*ærella*] *lacustris*..is abundant all over the country in ditches, rain-pools, and bog-pools. The cells..frequently become brick-red in colour owing to the presence of hæmatochromin... The curious phenomenon known as 'Red Rain' owes its colour in a few instances to the presence of this Alga. 1933 E. Hawks *Bk. of Air & Water Wonders* iv. 102 Red rain, accompanied by sand, fell on March 10, 1901, at Vienna and in Italy. 1884 H. E. Juler *Handbk. Ophthalmic Sci. & Pract.* xiv. 364 If the mirror be held at a considerable distance from an emmetropic eye no image of any details of the fundus is seen, but only a red reflex. 1954 S. Duke-Elder *Parsons' Disease of Eye* (ed. 12) vii. 94 If the fundus reflex is seen as a uniform red glow (the red reflex), the eye is emmetropic or approximately so. 1971 *Brit. Med. Bull.* XXVII. 69/1 The 'red reflex' coincides with a very early stage of development of a retinal neoplasm. 1853 Disraeli in *Hansard Commons* 30 June 1045 The very next day his Sovereign elevated him [*sc.* Lord Ellenborough] in the Peerage, and decorated him with the Red Riband. 1879 'Mark Twain' *Lett.* (1917) I. 355 He couched his lance and ran a bold tilt against total abstinence and the Red Ribbon fanatics. 1908 *Costumes Classical & Fashionable* (Liberty & Co.) 16 Red Riding Hood Cloak in cloth. 1936 N. Streatfeild *Ballet Shoes* vii. 110 Pauline wore shorts and a shirt, and Petrova an apron and a red-riding-hood cloak over her frock. 1964 M. Laski in S. Nowell-Smith *Edwardian England* iv. 204 White dresses with sashes..topped, for transit, by a red velvet Red Riding Hood cloak. 1966 M. Steen *Looking Glass* i. 14, I was taken to parties..in my Red Ridinghood cape. 1826 W. Darlington *Florula Cestrica* 72 *E*[*uchroma*] *coccinea*... Painted Cup. Red Robin. 1855 J. Phillips *Man. Geol.* viii. 190 One [fault] stated to cause a dislocation to the extent of 1,000 yards, is called the 'red rock fault', north of Pendleton near Manchester. 1942 E. M. Anderson *Dynamics of Faulting* v. 76 The 'North Staffordshire Boundary Fault'.. is not a single fault, as two separate north-north-easterly fractures bound the coalfield... The northern of these is known as the Red Rock Fault. 1969 Bennison & Wright *Geol. Hist. Brit. Isles* xi. 268 The thickness of the beds in the Cheshire Basin indicates a great amount of downwarping which continued into Triassic times. The basin may have been bounded by faults, such as the Red Rock Fault, which continued to move during Permian times. 1891 *Amer. Folk-Lore* IV. 224 Red Rover. The boy who is 'it' is called the 'Red Rover', and stands in the middle of the street, while the others form a line on the pavement on one side. 1898 A. B. Gomme *Trad. Games* II. 107 The players, except one, take their stand at one side, and one stands at the other side in front of them. When all are ready, the one in front calls out 'Cock', or 'Caron', when all rush across to the other side, and he tries to catch one of them in crossing. This game is called 'Red Rover' in Liverpool... 'Red Rover' is shouted out by the catcher when players are ready to rush across. 1974 J. Keats *Of Time & Island* v. 82 The little children played Red Rover. 1892 A. G. Thornton *Illustr. Catal. Drawing & Surveying Instruments & Materials* 107 Finest Red Sable Brushes, Round Black Handles, plated ferrules. 1899 M. Marks *Cycl. Home Arts* 62/2 The red sables are somewhat too strong for water-colour. 1910 *Encycl. Brit.* XI. 350/2 The fur [of the kolinsky] has often been designated as red or Tatar sable. 1948 F. A. Staples *Watercolour Painting* i. 3 The best red sable brushes should be used. 1970 *Oxf. Compan. Art* 169/2 The best soft brush for water-colour is the red 'sable', made from the fur of the Siberian mink. 1927 M. Dekobra tr. *Wainwright's Madonna of Sleeping Cars* xiv. 186 The gorilla with the pallid brow, marked with a Red star. 1969 G. Macbeth *War Quartet* 73 In his coat He wore the red star. 1979 J. Barnett *Backfire is Hostile!* xiv. 158 This strange aeroplane ..with a red star on its tail. 1941 Baker *Dict. Austral. Slang* 59 *Red steer, the*, fire, esp. a bush-fire. 1963 J. Cantwell *No Stranger to Flame* 12 The cane-cutters, made negro by sun and by soot from fires (Red Steers, they called them). 1971 F. Hardy *Outcasts of Foolgarah* 118 Like the bushfires: hadn't he patented the special extinguisher to end the blight of the red steer for all time? 1947 J. C. Rich *Materials & Methods of Sculpture* viii. 222 Many kinds [of sandstones] receive their names from their colors, i.e. bluestone, brownstone, and redstone. 1959 N. Sluman *Blackfoot Crossing* 38 On it Sikimi placed a fine revolver, some otter skins, a redstone pipe, and several small sacks of rare pigments. 1976 *Burnham-on-Sea Gaz.* 20 Apr. (Advt.), An attractive redstone detached cottage conveniently close to the town centre. 1876 G. Stables *Domestic Cat* vi. 51 The first cat of the Tabby kind which claims our attention is the Red or Sandy Tabby. 1903 F. Simpson *Bk. of Cat* xxv. 288/2 Red tabbies..are one of the difficult varieties to obtain. 1948 P. M. Soderberg *Cat Breeding* 248 Red Tabbies cannot compete with several other breeds for popularity. 1977 D. S. Richards *Handbk. Pedigree Cat Breeding* vii. 107 Ginger cats, or more correctly red tabbies, will invariably be male. 1898 Red tag [see Zulu *sb.* and *a.* 3]. 1923 *Daily Mail* 11 Aug. 7 Dace..have been caught with fly..and upper parts of the Lea should yield some good specimens of these fish to the black gnat, Zulu, red tag, and coachman. 1918 in J. Degras *Soviet Documents on Foreign Policy* (1951) I. 130, I wish to emphasise that the so-called 'Red Terror'—which is grossly exaggerated and misrepresented abroad—was not the cause but the direct result and outcome of Allied intervention. 1922 'Sapper' *Black Gang* xvi. 267 Experts of the Red Terror..butchers of women and children. 1930 *Morning Post* 13 Aug. 13 Never before, even in 1918–19, when the Red Terror was at its height, have persecutions reached such a terrible level. 1957 *Encycl. Brit.* XIX. 713/1 The beginning of the Red terror coincided with the period of greatest food shortage, before the harvest. 1977 *Socialist Press* 2 Mar. 7/1 His Committee of Public Safety directs accelerating repression (the 'red terror') against 'enemies of the people'. 1947 *Sun* (Baltimore) 3 Sept. 3/4 A tiny sea creature was blamed today for the 'red tides' which destroyed fish. 1970 T. D. Brock *Biol. Microorganisms* xix. 655 The red tide, an occasional occurrence in inshore areas, results from extensive blooms of red-pigmented dinoflagellate species. Red tides, which probably develop when the seawater becomes unusually enriched with nutrients, are of practical significance because some dinoflagellates produce fish toxins that may cause extensive and unsightly fish kills. 1980 *N.Y. Times* 13 Sept. 8/2 The red tide..usually occurs in much milder form each year in late summer or fall. 1975 *Globe & Mail* (Toronto) 22 Jan. 29/8 But the choice is not merely between the two levels of government; it is between two kinds of conservatism. Red Tory, blue Tory. 1976 *Weekend Mag.* (Montreal) 10 Jan. 2/1 All good Canadians are on the side of the Red Tories, the Dalton Camp forces who favor family allowances and the welfare state, Canadian nationalism, immigration, bilingualism and multiculturalism, the obviously good and liberal things that keep Canada great. 1974 *Globe & Mail* (Toronto) 29 Oct. 5/3 David Smith could outpace Alderman Anne Johnston in Ward 11, depending on how that ward buys her Red Toryism as opposed to his knee-deep Liberalism. [1879 X. Galezowski in *Recueil d'Ophtalmologie* I. 534 (*heading*) Sur la vision rouge des opérés de cataracte.] 1883 *Ophthalmic Rev.* II. 281 Purtscher suggests that red-vision would probably be heard of more frequently..after cataract operations, were it not for the careful protection of the eyes. *Ibid.* 278 He went into the open air and the red vision disappeared. 1959 S. Duke-Elder *Parsons' Dis. Eye* (ed. 13) xxiv. 373 Erythropsia (red vision) occurs particularly after cataract extraction if the eyes are exposed to bright light. 1940 Red warning [see *alert *sb.* 1 b]. 1963 *Times* 22 Jan. 10/3 Commander J. R. E. Langworthy, secretary of the Emergency Bed Service, said last night that if the cold spell continues a 'red warning' might have to be considered. Under this hospitals are asked to stop admitting any but emergency cases. 1969 *Daily Tel.* 14 Jan. 1/6 The next step, if the position worsened, would be a red warning. This stops all but very urgent admissions, reserves of local nurses are called in, and extra beds are put up in wards.

b. With adjs. (and derived sbs.), as **red-green** *a.*, pertaining to or affecting the ability to distinguish between red and green; **red-raw** *a.*, rubbed or irritated until the flesh is exposed and inflamed; also *fig.*; **red-sensitive** *a.*, sensitive or responding to the colour red; **redward** *a.* and *adv.* (further examples); also **redwards.**

1888 Red-green-blindness [in Dict.]. 1935 *Discovery* Aug. 231/1 According to this theory red-green blindness is due to the red-green nerve being atrophied, leaving the yellow-blue nerve still in action. 1956 C. Auerbach *Genetics in Atomic Age* 56 A good example for a sex-linked recessive gene in man is red-green colour-blindness. This abnormality is much more frequent in men than in women. 1958 *Listener* 6 Nov. 730/1 He, too, was a red-green colour-defective. 1964 S. Duke-Elder *Parsons' Dis. Eye* (ed. 14) xxiv. 364 The red-green cases fall into two main groups, protanopes and deuteranopes. 1971 J. Z. Young *Introd. Study Man* xxxviii. 553 Those who are red-green blind cannot identify ripe or rotten fruit or even see red berries among the leaves! 1924 A. J. Small *Frozen Gold* 184 A red-raw panic. 1957 T. Hughes *Hawk in Rain* 13 All day he stares at his furnace With eyes red-raw. 1936 *Discovery* May 151/2 This..was an advantage in the days of blue-sensitive materials when actinometers were first put forward, but has lost this advantage now green-sensitive and red-sensitive materials are so universally employed. 1967 Karch & Buber *Offset Processes* v. 149 Although it [*sc.* orthochromatic film] is insensitive to red, dyes can be added to make it red-sensitive. 1927 *Publ. Allegheny Observatory* VI. 136 The redward shifts of the solar lines. 1946 *Nature* 10 Aug. 205/2 The departure from the normal redward shift must be due to changes in the sun, and may readily be attributed to movements of the sodium vapour. 1973 *Sci. Amer.* May 118/1 A scarlet pigment, mercuric sulfide, is in fact a low-frequency reflector, reflecting about equally everything redward of a half-reflecting point in the orange. 1979 *Nature* 19 Apr. 719/1 The IR line moved redwards by ~150 Å and the red line bluewards by ~70 Å.

B. *sb.*¹ **1. c.** For 'in billiards' read 'in billiards and related games'. (Earlier and later examples.)

1857 J. E. Ritchie *Night Side of London* 128 'Good stroke'—'Bad flewke'—'On the red'. 1895 [see *cush]. 1928 C. Bergener *Contrib. Study of Conversion of Adjectives into Nouns* 135 The reds must be potted before you take the colours [in snooker]. 1974 *Rules of Game* 79 Three points if the cue ball hits the red into a pocket. 1977 *Cleethorpes News* 6 May 29/4 Hood potted the last red and this left Barnes in trouble, needing all the colours and a snooker to boot to pull off a win.

e. The red colour conventionally used in map-making to represent British territories.

1899 *Manch. Guardian* 2 May 7/1 The destruction of the Transvaal's independence..would blot out from the mass of red on the map of South Africa a spot of brown. 1966 *Observer* (Colour Suppl.) 27 Feb. 5/1 Red on the map tends to be spots, not splashes nowadays—but there are still more than 1,500 British islands. 1975 P. Mason *Kipling* vi. 150 A busy talkative man..preaching war..wanting to paint the map red.

f. *to see red*: to get very angry; to lose self-control.

[1900 J. K. Jerome *Three Men on Bummel* xiii. 292, I began, as the American expression is, to see things red.] 1901 'L. Malet' *Hist. Sir R. Calmady* i. v. 39 Happily violence is shortlived, only for a very little while do even the gentlest persons 'see red'. 1923 *Daily Mail* 19 June 15 It maddened me, I think, and I saw red—and before I knew what I was doing I stabbed him. 1937 A. Christie *Death on Nile* xi. 119 Why—? Because she thinks I'm not her social equal! Pah—doesn't that make you see red? 1953 J. Wain *Hurry on Down* x. 221 Instead of answering he leaned across and snatched at the packet Charles held in his hand. Charles saw red. His livelihood was in danger. 1974 *Times* 31 Jan. 2/6 'The village was incensed when a woman was left to die in her bath because an ambulance man on a go-slow refused to come out,' he said. 'We saw red and said we would form an action group to drive ambulances and cars.' 1977 *Daily Mirror* 15 Mar. 2 (*heading*) MPs see red over soaring prices.

g. The colour conventionally (now less commonly) used to indicate debit items and balances in accounts, used esp. in phrases *in the red*: in debt, overdrawn, losing money (also *fig.*); *out of the red*: in credit, making a profit. Hence, debt, an overdraft. Cf. *black *sb.* 2 d.

1926 Maines & Grant *Wise-Crack Dict.* 10/1 *In the red*, losing money in show parlance. 1927 *Scribner's Mag.* Apr. 380/2 'We've got to put forth our best efforts from now till the end of the month, or we'll be in red on the books,' he announced. 1928 *Publisher's Weekly* 10 Nov. 1957/2 About 966 copies more and the title will be out of the red. 1931 F. L. Allen *Only Yesterday* viii. 212 The Philadelphia Sesquicentennial was sinking deeper and deeper into the red. 1949 *Harper's Mag.* Mar. 62/2 The corporation was nearly a million dollars in the red. 1955 *Times* 28 June 3/3 With Tordoff and Saeed opening Somerset's second innings with commendable vigour, Leicestershire went further into the red. 1960 *Times* 15 Feb. 11/6 The British Transport Commission is already in the red to the tune of at least £30m. 1966 O. Norton *School of Liars* i. 5, I don't think the manager at Barclays has ever heard of the Married Women's Property Act,..my red is Andrew's red. 1977 D. Williams *Treasure by Degrees* xviii. 169 A quarter of a million pounds..would be more than sufficient to keep the College out of the red for the foreseeable future. 1978 S. Brill *Teamsters* vii. 268 CCC has never run in the red.

h. A red light, lamp, etc., meant as a signal to stop. Also *fig.*

1970 'W. Haggard' *Hardliners* xiv. 155 At the top of Whitehall he jumped his first red, slipping left to the Mall against the signal. 1972 D. Bloodworth *Any Number can Play* xvi. 153 'We go by the position of the lights, not the colours,' soothed Ivansong, as they roared through a red. 1976 'P. B. Yuill' *Hazell & Menacing Jester* vi. 67, I was doing over fifty and jumping reds.

3. b. (Later examples.)

1927 E. Hemingway *Men without Women* 207 You tried the red?..we'll have a round of the red. 1961 J. B. Priestley *Saturn over Water* viii. 113 We drank a bottle of Chilean red, and she made coffee. 1969 *Listener* 27 Mar. 417/3 A bottle of red with this, which lasted right through the half-pound of Brie cheese that followed. 1974 K. Millett *Flying* (1975) iii. 286 Nell goes off to seek another bottle of red.

5. b. (Further examples.) Also, a variety of potato bearing red-skinned tubers or the tubers themselves.

1829 G. Griffin *Collegians* II. xxx. 333 The English reds are a nate piatie. 1902 A. Bennett *Grand Babylon Hotel* xxii. 251 The 'Spanish reds' from Catalonia, including the dark 'Tent' so often used sacramentally. 1926 R. N. Salaman *Potato Varieties* v. 28 A red,..when selfed, gives rise to three plants bearing red tubers to one bearing white. 1929 W. Faulkner *Sartoris* ii. v. 125 Got an old red we been saving for you... John would have enjoyed that fox. 1961 *Guardian* 24 Oct. 8/6 Some Russian dry whites and reds, selling at as little as 7s. 6d. a bottle. 1968 *Ibid.* 23 Feb. 9/6 No Swiss reds are of real distinction. 1968 K. Weatherly *Roo Shooter* 8 A number of roos were resting. The big buck was typical of the reds, standing on his tips about seven feet. 1972 E. Hargreaves *Fair Green Weed* vii. 93 I'm to buy in more cattle, good reds, in the herd book. 1976 *Southern Evening Echo* (Southampton) 18 Nov. 16/6 Least wastage was 1½oz. from a pound of 'Reds'. 1977 *Age* (Melbourne) 18 Jan. 13/2 Let me forecast in no uncertain terms that this policy can lead only to a severe shortage of high quality reds in five years' time. 1977 *Grimsby Even. Tel.* 24 May 12/4 Principal sorts were: Cod.., coley.., rockfish.., reds.

6. b. (Earlier example.) Also, a Russian Bolshevik; a communist, or extreme socialist. Phr. *reds under the bed*, used to denote an exaggerated or obsessive fear of the presence and harmful influence of communist sympathizers in a particular society, institution, etc.

1851 *Punch* XX. 245/2, I dreamt that I stood in the Crystal Halls, With Chartists and Reds at my side. 1922 S. Lewis *Babbitt* v. 56 Say, juh notice in the paper the way the New York Assembly stood up to the Reds? 1928 D. L. Sayers *Lord Peter views Body* iii. 44 I'm a Tory, if anything. I'm certainly not a Red. Why should I help to snatch the good gold from the Primrose Leaguers and hand it over to the Third International? 1931 P. Hodges *Britmis* i. 20 The Orenburg Army,..had been operating against the Reds south of the Trans-Siberian Railway. 1940 [see *liberal *sb.* 1 c]. 1947 *Partisan Rev.* XIV. 354 'All those guys,' he said, 'are just sore because they are not rich. Give any one of them a million dollars and they would forget all about being reds.' 1957 *Economist* 7 Dec. 882/2 Dr Villeda..has been at pains to show the Americans that he is no red. 1972 *Times* 24 May 16/3 This sort of 'reds under the bed' scare.. could only be counter-productive. 1974 *Socialist Worker* 23 Nov. 16/1 The question now is whether the Broad Left leadership has the guts to campaign openly against the Social Contract and for a £30 a week claim in the face of mounting hostility from the government and an increasingly bitter Reds under the Bed campaign from the forces of reaction inside and outside the NUM leadership. 1976 C. Bermant *Coming Home* i. i. 16 There came the depredations of the Russian civil war, first from the Reds then the Whites. 1976 *West Lancs. Evening Gaz.* 15 Dec., Apparently it is usual practice for Tory writers, if they think an election is imminent, to re-hash previous pre-election writings, same methods, reds under the bed, Communists, or other disguises becoming members of the Labour Party. 1977 'J. Le

Carré' *Hon. Schoolboy* xiii. 305 There's a story that you people had some local Russian embassy link... Any Reds under your bed..if I may ask?

7. chiefly *U.S.* A red cent (see sense 3 c of the adj. in Dict. and Suppl.).

1849 *Alta California* (San Francisco) 12 July 2/5 Silver is not Plenty on the Pharaoh and his host's Tables, and any body can eat it, and bet a red on any card he chuses. **1856** 'Ouesticks' & 'Doesticks' *Hist. & Rec. Elephant Club* 244 Judge—'Have you got ten dollars?' Mr. W—'Tis true, I hain't a red.' **1869** 'Mark Twain' in *Harte & Twain' Sk. Sixties* (1926) 199 Greely would sure up money on his own as long as he had a red. **1903** J. London *Let.* 1 June (1966) 175, I don't care a red how much the Lazar-sheets roast me. **1922** Joyce *Ulysses* 151 Didn't cost him a red. **1936** J. A. McKenna *Black Range Tales* 267 Many who came into Frisco had not a dad-blasted red left to their name.

8. ellipt. for *red alert* (sense *19 a of the adj.).

1943 D. Nixon *Raiders Overhead* iii. 28 Every night, and all night, there were raids. On the evening of the 16th the 'red' came up at 8.5 p.m. **1943** G. Greene *Ministry of Fear* iv. i. 223 Yellow's up.. About time for the Red I should think.

9. *Naut.* The port side of a ship. Also quasi-*adv.*

1948 Partridge *Dict. Forces' Slang* 153 *Red, the,* the port side of a ship. It shows a red light. **1956** 'Taffrail' *Arctic Convoy* xi. 103 Someone shouted: 'There they are, sir! Bearing red nine-oh!'—otherwise ninety degrees on the port beam. **1958** W. King *Stick & Stars* 66 Object bearing red five oh.

10. = *red-bird 2, *red devil 3. *slang.*

1967 W. Murray *Sweet Ride* vii. 107 It's pills, mostly. Reds, goofballs, all kinds. And grass, of course. **1969** *Oz* May 21/1 Mixing 'reds' & alcohol can lead to a one way trip because the two drugs potentiate each other, i.e. 1+1 = more than 2. **1972** J. Wambaugh *Blue Knight* (1973) xvi. 233 What've you got, boy? Bennies or reds? Or maybe you're an acid freak?

11. *Comb.*, as (sense *6 b) *red-hunting* vbl. sb. and ppl. a. See also *red-baiting vbl. sb.

1927 U. Sinclair *Oil!* 313 Sure thing! He's nuts on this red-hunting business, and the pinks are worse than the reds, he says. **1935** H. L. Ickes *Secret Diary* (1953) I. 401 He feels about Red hunting just as I do and thinks it is absurd to deny communists an opportunity to express themselves or to have a ticket on the ballot. **1962** M. McCarthy *On Contrary* 37 Such Red-hunting publications as *Counterattack.*

redacter, var. Redactor.

1816 Scott *Tales of my Landlord* 1st Ser. I. 8, I am not the writer, redacter, or compiler, of the Tales of my Landlord.

rédacteur. (Further examples.)

1848 J. G. Lockhart *Let.* 4 Jan. in *N. & Q.* (1946) 9 Mar. 90, I wrote only yesterday to thank him for the Life of the Chancellor..tho' not to congratulate him on his redacteur. The book is awfully ill done. **1883** *Daily News* 2 Oct. 5/6 Other rédacteurs of the once famous *Journal des Débats.* **1962** *Economist* 27 Jan. 334/3 In the French tradition, the *Député*..hands the big news of the day to a star rédacteur who comments on the story as he tells it.

redaction. Add: **2. a.** (Earlier example.)

1785 T. Jefferson *Writings* (1894) IV. 68 The English of which is, that the redaction of the paper had been taken from the improved culprit, and given to another.

b. Also, an adaptation; a shortened form, an abridged version.

1948 *Observer* 30 May 3/3 Finally, we have..what is described as a 'redaction' or compression—this dangerous device grown in popularity—of Lytton's *The Last Days of Pompeii*, by S. Fowler Wright. **1977** *New Yorker* 8 Aug. 2/3 Vinnette Carroll's singing-and-dancing redaction of the Book of Matthew. **1978** *Amer. N. & Q.* Mar. 103/2 In 1661 Samuel Smithson produced a prose redaction of the old metrical romance of Guy of Warwick.

4. *attrib.* and *Comb.*, as **redaction criticism** (see quot. 1976); hence *redaction critic, redaction-critical adj.*

1970 N. Perrin *What is Redaction Criticism?* i. 22 Although he does not use the term, Lightfoot was actually the first redaction critic. **1976** *Christian Believing* 27 The redaction critic returns by way of the work of form criticism to the Synoptic Gospels as wholes. **1976** *Times Lit. Suppl.* 8 Oct. 1285/3 As far as I know, all redaction criticism begins by considering how Luke or 'Matthew' has altered the Marcan narrative. **1968** D. M. Barton tr. *Rohde's Rediscovering Teaching of Evangelists* i. 15 Various basic theological ideas in the individual gospels were presented through redaction-critical work on the synoptic gospels. **1970** N. Perrin *What is Redaction Criticism?* i. 27 Bornkamm's article..is the first thoroughgoing redaction-critical investigation of the theological peculiarities and theme of Matthew's Gospel. **1966** Keck & Martyn *Stud. in Luke-Acts* (1968) i. 65 At the present time the method called redaction-criticism is luring us into a one-sided concentration on the work of editors. **1968** D. M. Barton tr. *Rohde's Rediscovering Teaching of Evangelists* ii. 37 Redaction criticism..endeavours to understand the gospels in their entirety against the background of a definite theological situation in the church. **1970** N. Perrin *What is Redaction Criticism?* i. 1 Redaction criticism is an attempt to represent in English the German word *Redaktionsgeschichte*, which Willi Marxsen proposed as the designation for a discipline within the field of New Testament studies. **1976** *Times Lit. Suppl.* 8 Oct. 1285/3 What is called redaction criticism, the attempt to discover the theological and cultural presuppositions of the gospel writers by examining how they have edited ('redacted') their material.

redactional *a.* (later examples); also, of or belonging to a particular redaction.

1968 *Language* XLIV. 15 In this theory also, RV [sc. Rigvedic] *dayām* would probably be redactional for original *dayām. **1971** *Ibid.* XLVII. 65 Emeneau's hunch was that some forms beginning with *dy-* which occur in environments interdicted by the Sievers-Edgerton theory might actually be puristic redactional substitutions for Middle Indic forms in *j-*. **1971** *New Testament Abstracts* XV. 285 A minute analysis of vocabulary in this parable shows that before redactional activity it dealt with a king (God) who acted like a shepherd who separated the sheep from the goats with sure judgment.

redactoral (rĭdæ·ktŏrăl), *a.* [f. Redactor + -al.] = Redactorial *a.*

1970 R. S. Foster *Redaction of Israel* v. 110 The first redactoral addition in [2 Kings xvii.] v. 24–28 explains that the Samaritans were a mixed race.

red alert: see *red a. 19 a.

Red Army. [see Red *a.* 9 b in Dict. and Suppl.] **a.** The name given to the Russian Bolshevik army and later to that of the Soviet Union.

1918 *Manch. Guardian* 13 Dec. 7/4 The Bolshevik Government..is engaged in creating a Red Army of over a million men. **1935** N. Mitchison *We have Been Warned* ii. 172 In Moscow..on May Day there are great processions..and marching by the Red Army. **1942** J. B. Priestley *Daylight on Saturday* viii. 51 The Red Army is still showing what a workers' republic can do. **1976** 'M. Barak' *Secret List H. Roehm* viii. 81 After the Revolution he remained an officer in the Red Army. **1978** *Detroit Free Press* 16 Apr. (Record) 15/1 At last, in January 1944 the Red Army broke through the German ring. **1980** M. Bar-Zohar *Deadly Document* x. 164 The band chimed in with..the Red Army song.

b. Adopted as the name of the army in other, esp. communist, countries.

1936 *Encycl. Brit.* II. 393/1 The [Hungarian] Red Army was organised, primarily to ensure the maintenance of the dictatorship. **1954** tr. *Mao Tse-Tung's Red China* 17 The Chinese Soviets and their Red Army have grown out of the development of the agrarian revolution, which liberates the masses of the peasants from oppression and exploitation. **1965** M. Michael tr. *J. Myrdal's Report from Chinese Village* i. 4 The peasants of northern Shensi..set up their own soviet republic and formed their own Red Army. **1974** tr. *Snezhkus' Soviet Lithuania* 20 Many joined the ranks of the Lithuanian Red Army. **1979** *Times Lit. Suppl.* 14 Feb. 165/1 Horthy came to power in the winter of 1919–20..after Romania's military victory over the Hungarian Red Army. **1980** *Times* 31 Mar. 16/4 The Communists' Red Army had just completed its epic Long March from the Southeast to the new headquarters at Yenan in remote northern Shensi province.

c. The name of a left-wing extremist terrorist organization in Japan.

1972 *Sat. Rev.* (U.S.) 24 June 30/2 The Red Army, an extremist terrorist group in Japan, sponsored and trained by the radical Popular Front for the Liberation of Palestine. **1974** *Times* 7 Feb. 5/1 The Popular Front for the Liberation of Palestine and the Red Army of Japan in a statement issued yesterday sought to justify the attack on the Singapore oil storage tanks—where it is claimed 15 tanks were blown up. **1976** K. Benton *Single Monstrous Act* iii. 19 The Libyans..seem to have doled out money to every revolutionary group there is, from the Japanese Red Army to the Provos.

d. Red Army Faction, the name of a terrorist organization of West Germany.

1977 *Time* 19 Sept. 8/3 The initials represent the now familiar Red Army Faction, widely described as the Baader-Meinhof gang, the group that murdered both Buback and Ponto. **1979** R. Perry *Bishop's Pawn* ix. 174 The terrorists..were definitely operating under the Red Army Faction umbrella.

reda·te, *v.* [Re- 5 a.] *trans.* To change the date of; to assign a new date to. Hence **reda·ting** *vbl. sb.*

1611 Cotgrave *Dict.*, *Redater*, to redate, or adde a new date unto. **1864** *Spectator* 31 Dec. 1508 Instead of rewriting or redating the previous part of my letter I prefer to send it as it was written. **1935** Huxley & Haddon *We Europeans* ii. 54 A recently propounded re-dating of a fragment of a skull. **1962** *Early Music* Jan. 103/2 The new madrigal was not really published in appreciable quantity until the late 1590s, and the re-dating of a central group of manuscript sources shows that dissemination in manuscript was the main way in which the repertory circulated during the years 1520–40.

red-back. **1.** (In Dict. s.v. Red *a.* 18 b.)

2. *Austral.* In full, *red-back spider*; = *jockey spider* s.v. *jockey sb. 9.

1933, etc. [see *jockey spider* s.v. *jockey sb. 9]. **1936** K. C. McKeown *Spider Wonders Austral.* xi. 152 The Red-back Spider..has adapted itself to a life in close association with man. **1953** A. Upfield *Murder must Wait* xiv. 130 Five red-back spiders..lying in wait to inject their poison. **1956** *Coast to Coast 1955* 59 Look, there's a red-back in it. **1978** *Telegraph* (Brisbane) 11 Aug. 4/4 The six were victims of a wolf spider—of a type previously unknown and potentially as dangerous as a redback. **1979** *Ibid.* 22 Feb. 3/1 It's a battle to the death. Redback spider versus poisonous dugite snake.

red-backed, *a.* For the Latin name of the red-backed sandpiper substitute *Erolia alpina*; also, the red-backed mouse or vole (*Clethrionomys gapperi*), and the Australian red-backed spider (= *red-back 2). (Earlier and further examples.)

1709 J. Lawson *New Voy. Carolina* 126 The Red-back'd Snake [is found in Carolina]. **1897** *Proc. Biol. Soc. Washington* XI. 113 The following brief synopsis of the Red-backed Voles is based on a study of specimens in the collection of the U.S. Biological Survey. **1934** *Bulletin* (Sydney) 6 June 21/2 In the Chillagoe..district the red-backed spider is known as 'the jumping red-back'. **1936** D. McCowan *Anim. Canad. Rockies* viii. 71 The Red-backed mouse is a forest dweller..of medium size and has a coat that is marked by a fairly broad belt of chestnut brown hair over the spine. **1940** Gabrielson & Jewett *Birds Oregon* 285 The handsome Red-backed Sandpiper is a common migrant on the coast. **1942** [see *red-backed shrike* s.v. *Shrike]. **1961** R. Murie *Two in Far North* ii. iii. 117 Olaus was interested in learning all he could about the distribution of the red-backed mouse. **1963** *Times* 5 June 14/4 There was still to come the red-backed shrike. **1966** 'J. Hackston' *Father Clears Out* 134 Safe from..red-backed spiders, and spiders without red backs. **1977** J. L. Harper *Population Biol. Plants* xv. 465 The seed was collected and buried mainly by..red-backed voles.

red-bait (re·dbeit), *sb. S. Afr.* [tr. Afrikaans *rooiaas.*] A large sea-squirt, *Pyura stolonifera*, an ascidian which is used as bait by anglers.

1895 *Agricultural Jrnl.* (Dept. of Agric., Cape Colony) 912 The bait most in use is crayfish, and 'rooi aas' (red bait) a species of Zoophyte. **1905** J. Gilchrist in Flint & Gilchrist *Science in S. Afr.* iii. iv. 192 One of the features of the rocky parts of the coast line from Cape Point eastwards is the clusters of 'rooias' or 'red-bait'. **1930** C. L. Biden *Sea-Angling Fishes of Cape* xii. 217 The name 'red-bait' is an English interpretation of the Dutch 'rooi aas'. **1945** *Cape Times* 10 Oct. 7 Red bait could be cut by the sackful. **1957** S. Schoeman *Strike!* iii. 29 Not a single fish would take redbait or mussel. **1970** *Std. Encycl. S. Afr.* III. 205/2 'Sea-squirts', including the common 'red-bait' (*Pyura*) and the transparent *Ciona*, live permanently attached to rock surfaces.

Re·d-baiting, *vbl. sb.* orig. *U.S.* Also with small initial. [*red sb.[1] 6 b.] Harassment of those of known or suspected communist sympathies; also *attrib.* So (as a back-formation) **re·d-bait** *v. trans.* and *intr.*, to oppose, thwart, or persecute because of communist associations; **Re·d-baiter, red-baiter,** one who seeks out and harasses supposed communists.

1929 *Nation* 2 Oct. 343/1 Red-Baiters... Mr. Woll and his underlings have become..obsessed with an anti-Communist hysteria. **1934** M. H. Weseen *Dict. Amer. Slang* 386 *Red-baiting*, looking for communists. **1937** *Nation* 14 Aug. 167/1 A new red-baiting campaign. **1939** *Sun* (Baltimore) 22 Mar. 12/1 It was well-nigh impossible to say a word for capitalism and against the vagaries of Communism without being accused of Red-baiting. **1940** G. Seldes *Witch Hunt* p. xi, Nevertheless, that program was red-baited into compromise and failure. **1943** F. Scully *Rogue's Gallery Hollywood* 83 Fugitives from red-baiting America, they still rallied 'round leaders like Pirandello. **1946** *Sun* (Baltimore) 19 July 20/1 The War Department hasn't gone off half-cocked to 'witch hunt, red bait or to bust' unions. **1950** *Manch. Guardian Weekly* 15 June 15 The 'Amerasia' case is something that the Red-baiters have wanted to open up again ever since the conviction of Alger Hiss. *Ibid.* 16 Nov. 9 The defeat of the Liberals..suggests that vociferous 'Red-baiting' is, however silly, good politics. **1962** D. Lessing *Golden Notebk.* iv. 483 The fanciest bit of red-baiting I've heard in a long time. **1969** *Listener* 13 Feb. 214/2 One might sneer away the Red-baiting of those days and shrug off the ceaseless cries of 'Fascists!' from the Left, and yet be uneasy about the future. **1974** *Socialist Worker* 2 Nov. 12/5 The methods could be used to red-bait or victimise workers involved in the present disputes. **1976** *Times Lit. Suppl.* 13 Jan. 89/1 Professor Nove may be in danger of justifying the charge of red-baiters of being an apologist for Stalin. **1977** *Time* 7 Feb. 46/3 Boyle, 40, stars as the Red-baiting chairman of the Senate Permanent Subcommittee on Investigations.

red-bellied, *a.* (Earlier and later examples.)

1709 J. Lawson *New Voy. Carolina* 126 Red-bellied Land-Snakes. **1731** M. Catesby *Nat. Hist. Carolina* I. 19 The Red-bellied Wood-pecker... The belly near the vent, is stained with red. **1917** T. G. Pearson *Birds Amer.* II. 162/2 The Red-bellied Woodpecker..evinces a decided taste for fruit. **1934** Dylan Thomas *Let.* 15 Apr. (1966) 108 It there must be a sorry sort of thing to be: a fat, red-bellied one you could tell no one about. **1965** Mrs. L. B. Johnson *White House Diary* 3 June (1970) 281 Most startling of all was a fish whose every scale stood out..the red-bellied angelfish. **1973** M. Crowell *Greener Pastures* 109 It is the red-bellied woodpecker that visited the feeder last week. **1977** C. McCullough *Thorn Birds* iv. 75 Of snakes the variety was almost endless..red-bellied black snakes.

red belt. **1.** [*red a. 9 b.] **a.** Territory under the political control or influence of the U.S.S.R. **b.** Elsewhere, an area of communist strength or influence.

1947 *Sun* (Baltimore) 22 Dec. 2/1 The 'Red Belt' states of Europe are moving into a fairly solid bloc. **1966** M. R. D. Foot *SOE in France* ix. 257 The French communist organizations in the 'red belt' round Paris. **1969** P. Allum in Henig & Pinder *European Political Parties* 233 The DC is strongest in the 'white provinces' where it regularly polls half the vote and it is weakest in the 'red belt' where its vote is only around 30 per cent. **1977** *Time* 4 Apr. 22/3 But much as Moscow might like a Red belt across Africa, even Angola and Mozambique are not anxious to be totally under Soviet control. **1978** *Sunday Star* (Toronto) 26 Feb. A4/3 When voters of the Paris red-belt suburb of Kremlin-Bicetre attend election meetings, the gloves are off.

2. [Red *a.* 16.] A belt worn by one who has attained a certain degree of proficiency in judo or karate; also, a person qualified to wear such a belt.

1952 E. J. Harrison *Judo* i. 14 A black belt is worn in the first five Dan grades..and a red belt in the tenth and higher Dan grades. **1968** *Radio Times* 7 Feb. 9/4, I am only a Red Belt myself..a solid red belt is worn by the highest grade masters. **1971** *Rand Daily Mail* 27 Mar. 6/8 Executive members of the South African Division of the Japanese Karate Association put their heads together and came up with a new belt—a red one symbolising a junior brown belt. **1976** D. Jacks *Judo* 8 The adult beginner wears a white or red belt and the grades then progress.

red-belted: see *red a. 14 a, red a. 14 b.

red-berried, *a.* (Earlier and later examples.)

1731 P. Miller *Gardeners Dict.* s.v. *Sambucus*, The Mountain red-berry'd Elder. **1930** J. Masefield *Wanderer of Liverpool* 19 Red-berried blackthorn. **1972** *Hilliers' Man. Trees & Shrubs* 360 *Sambucus racemosa* L. 'Red-berried Elder'. A medium-sized to large shrub.

red-bird. **1.** For *Piranga æstiva* substitute *Piranga rubra*; for *P. rubra* substitute *P. olivacea*. Add Latin names of the Baltimore oriole (*Icterus galbula*) and the cardinal (*Richmondena cardinalis*). (Earlier and later examples.)

1709 [see *Baltimore]. **1929** W. Faulkner *Sartoris* iv. iv. 337 They saw redbirds darting like arrows of scarlet flames. **1938** M. K. Rawlings *Yearling* ix. 81 A red-bird sang in an arc across the sink-hole. **1952** E. B. White *Let.* Feb. (1976) 459 It's a nice place to be, what with the..red-bird saying 'Portugee, Portugee'.

2. The drug secobarbital (Seconal); also, a tablet of this drug (coloured red). *slang.*

1969 Lingeman *Drugs from A to Z* 219 *Seconal*... Slang names: red birds, red devils, reds. **1976** *Billings* (Montana) *Gaz.* 11 July 4-E/1 In order on DAWN's list of drugs most frequently recorded in crisis situations—in which a drug user sought help or died—were heroin, marijuana, aspirin, LSD, secobarbital (marketed as Seconal and known as red devils, Mexican reds and red birds) [etc.].

re·d-blooded, *a.* [Red *a.* 14 a.] **1.** Having red blood.

1804, 1840 [see Red *a.* 14 a].

2. Restored to health and strength after weakness or exhaustion.

1877 Tennyson *Harold* iv. iii. 131 Sit down, sit down, and eat, And, when again red-blooded, speak again.

3. *transf.* Virile, vigorous, full of life, spirited.

1881 A. A. Hayes *New Colorado* xi. 155 [Nothing] can be conceived more exasperating to a strong red-blooded man than to..have a villain take his watch and money. **1888** W. Whitman in *Cent. Mag.* (1911) Dec. 254/2 John's letter appeals to me..because of its uncompromising red-blooded espousal of the book. **1914** E. R. Burroughs *Tarzan of Apes* xix. 257 Tarzan..did what no red-blooded man needs lessons in doing. **1923** *Daily Mail* 18 Feb. 10 (Advt.), It's a rip-roaring, red-blooded yarn that no woman will be able to read unmoved. **1947** B. Schulberg *What makes Sammy Run?* viii. 151 The play was about two red-blooded guys who are always scrapping. **1966** Wodehouse *Plum Pie* i. 11 A redblooded loony doctor under the influence of the divine passion ought surely to have put the thing through months ago. **1978** M. Puzo *Fools Die* xxiii. 387 To me he sounded like any red-blooded American or Venetian businessman.

re·dbone. *U.S.* A hound belonging to a variety distinguished by a red or red and tan coat, which was once used to hunt racoons. Also *attrib.*

1916 W. H. Miller in *Field & Stream* XXI. 177/1 Good Redbone, Pennsylvania or Portsmouth hound stock is what we want up North. **1929** — *Amer. Hunting Dog* iii. 57 Two strains are well known, the Redbone, an ancient breed of Southern coon-hound, and the J. E. Williams dogs. **1948** W. Faulkner *Intruder in Dust* i. 5 A true rabbit dog, some hound, a good deal of hound, maybe mostly hound, redbone and black-and-tan. **1975** E. Wigginton *Foxfire 3* 18, I had a good stock a' dogs—blue tick and redbone mix.

red book, red-book. Add: **4.** *Little Red Book*: a popular name used in Western countries for 'Quotations from Chairman Mao Tse-tung' (published in English in 1966). Also in extended and allusive use.

1967 [see *Maoism]. **1970** G. Jackson *Let.* 17 Apr. in *Soledad Brother* (1971) 225 Burn it; all the fascist literature, burn that too. Then equip yourself with the Little Red Book. **1971** D. Bagley *Freedom Trap* viii. 193 Communist Albania has ceased to hew to the Moscow line. Enver Hoxha, the Albanian party boss, has read the Little Red Book and thinks the thoughts of Mao. **1974** *Times* 21 Oct. 12/8 A hotelier in..Lübeck has arranged to provide in all bedrooms..a German translation of Chairman Mao's little red book. **1976** J. Crosby *Nightfall* x. 47 If I write my own little Red Book, that will be in it. Hit the pricks—or they get ideas. **1977** *N.Y. Rev. Bks.* 12 May 22/3 Photos showed rampaging teenagers waving their little red books of Mao quotations.

redbreast. Add: **3.** *U.S.* The long-eared sunfish, *Lepomis megalotis.*

1888 G. B. Goode *Amer. Fishes* 66 In Pennsylvania it [sc. the long-eared sun-fish] is called 'Sun Perch' and 'Red Headed Bream', elsewhere it is the 'Red Breast'. **1948** *New Hanover Fishing Club Prize List* 30 Annual prizes for fresh-water fish..Red Breast.

red-brick. Also **red brick, redbrick.** **1.** A red building-brick. Freq. *attrib.*

1712 J. Mortimer *Whole Art Husbandry* II. 150 The black Mould..will in time degenerate into a red-brick Earth. **1835** J. Romilly *Diary* 11 Mar. (1967) 70 They are nasty red-brick churches, in the worst stile of 1760. **1839** [in Dict. s.v. Red *a.* 16 a]. *a* **1847, 1888** [in Dict. s.v. Red *a.* 12 a]. **1916** E. F. Benson *David Blaize* v. 101 His horizon and aspirations stretched no farther than this red-brick arena. **1943** 'B. Truscot' *Redbrick University* 17 The material used in them [*sc.* universities] was..a hideously cheerful red brick suggestive of something between a super council-school and a holiday home for children. **1960** J. Betjeman *Summoned by Bells* v. 46 But for me, Less academic, red-brick Chalfont Road Meant great-aunt Wilkins, tea and buttered toast. **1977** *Western Mail* (Cardiff) 5 Mar. 7/3 For the Opposition leader, it was a nostalgic return to the red brick establishment of Kesteven and Grantham girls' school which she left 34 years ago.

2. (Also with capital initial.) Used *attrib.* or quasi-*adj.* to denote a British university founded in the late nineteenth or early twentieth century in a large industrial city, with buildings of red-brick, as distinct from the older universities (esp. Oxford, Cambridge, the ancient universities of Scotland, and some of the London colleges) built predominantly in stone, and also as distinct from the new universities founded after the 1939–45 war; of or pertaining to such a university; also *ellipt.*, a red-brick university; *collect.*, such universities in general. Also *transf.*

1943 'B. Truscot' *Redbrick University* 18 The range of interests represented in a Redbrick staff common-room. *Ibid.* 19 It may be natural enough for him to go on to Red-brick, but to..enter Oxbridge is something infinitely more exciting. **1944** 'H. Ashton' *Yeoman's Hospital* ix. 197 Marriner took his professorship at that frightful red-brick university. **1950** *Times Educ. Suppl.* 10 Mar. 183 (*heading*) Redbrick criticized. **1958** *Times Lit. Suppl.* 17 Jan. 30/4 Talk of..'the red-brick intellectuals', though no Movement founder-member had done more than *teach* at one of the provincial universities. **1958** *Observer* 16 Feb. 12/3 One of the new 'redbrick actors', neither actorish in aspect nor conventionally po-voiced. **1958** *New Statesman* 22 Feb. 233/1 Under education, the correct entry is: 'Educated Thomas Cooks, American Express, Wayfarers, etc., etc.' Europe has been my Redbrick. **1960**, etc. [see *Oxbridge]. **1966** G. Sinstadt *Whisper in Lonely Place* iii. 33 He's a research engineer, degree from one of the red-bricks, middle twenties. **1975** D. Lodge *Changing Places* i. 9 Rummidge..had lately suffered the mortifying fate of most English universities of its type (civic redbrick). **1976** M. Hinxman *End of Good Woman* i. 13 They kept introducing her to eligible mates. Revolutionaries at the London School of Economics, posh chums from Oxford..budding scientists and engineers at the 'red-brick' universities. **1977** *Jrnl. R. Soc. Arts* CXXV. 670/1 Some of the best safety managers I know left school at 14 or 15 years of age. Conversely, we have seen people who come from red bricks and grey stones but are quite unable to do the job at all.

red-bud. Add: **a.** (Earlier and later examples.) Also *attrib.*

1705 R. Beverley *Hist. & Present State Virginia* iv. 56 They dish up [roots, herbs, etc.] various ways, and find them very delicious Sauce to their Meats;..such are the Red-buds, Sassafras-Flowers, Cymnels, Melons, and Potatoes. **1709** J. Lawson *New Voy. Carolina* 100 The Red-Bud-Tree bears a purple Lark-Heel. **1931** W. Faulkner *Sanctuary* xviii. 164 Lilac and wistaria and redbud..had never been finer. **1946** D. C. Peattie *Road of Naturalist* v. 58 The red-bud trees begin to bloom. **1977** *New Yorker* 2 May 51/2 The dogwoods and redbuds blossomed.

redcap, red-cap, red cap. Add: **4.** *Mil. slang.* A military policeman.

1919 *Athenæum* 1 Aug. 695/1 In your July 18 issue a correspondent mentions 'red-hat' as an army policeman. I have always found 'red-cap' to be the more familiar term. **1949** G. Cotterell *Randle in Springtime* 7 'Mind yourself, there are some redcaps in that jeep.' 'I seen 'em,' the driver grinned, slowing down to below forty miles an hour, as another jeep, containing three Military Policemen, approached and passed them. **1964** J. Hale *Grudge Fight* viii. 127 The redcaps and the R.A.F. police. **1976** J. O'Connor *Eleventh Commandment* iv. 53 She used to take me to night-clubs tucked away which no officers or redcaps knew about.

5. Chiefly *U.S.* A porter at a railway station.

1919 S. Lewis *Free Air* xxiv. 245 A factory illuminated by arc-lamps,—the baggage—the porter..red caps. **1929** M. de la Roche *Whiteoaks* x. 145 A 'redcap' darting into the throng, the bag clutched in his hand. **1931** W. Faulkner *Sanctuary* xxi. 227 A man shouted 'taxi' at them; a redcap tried to take Fonzo's bag. **1942** *Sun* (Baltimore) 10 Oct. 10/2 There are only thirty-five red caps to help the anxious passengers with their luggage. **1960** B. Keaton *Wonderful World of Slapstick* iii. 59 Kelly didn't explain that there were no redcaps at Victoria Station. **1969** J. A. McPherson in A. Chapman *New Black Voices* (1972) 156 He had redcaps in the Chicago stations telling the soldiers who to ask for on the train. **1977** *Times* 19 Apr. 15/7 In France and Switzerland the larger stations have 'red cap' porters who shift baggage on a prescribed tariff.

red carpet: see *red a. 19 a.

red cedar. **a.** Substitute for def.: A North American species of juniper, esp. *Juniperus virginiana*, or the western conifer, *Thuja plicata*. Also *attrib.* (Earlier and further examples.)

1682 S. Wilson *Acct. Province Carolina in Amer.* 12 This Country hath..divers sorts of lasting Timber, that England hath not, as Cedar white and red, Cypress, Locust, Bay and Laurel Trees. **1797** *Deb. Congress U.S.* 10 Feb. (1849) 2113 It would be expedient..to secure some of the lands in South Carolina and Georgia, well clothed with live oak and red cedar timber, for the purpose of building ships of war. **1851** *Knickerbocker* XXXVII. 377 The country-bred traveller..inhales the odor of the red-cedar buckets. **1884** C. S. Sargent *Rep. Forests N. Amer.* 7 The hemlock and the red cedar (*Thuya*) are still important elements of the forests. **1904** E. Step *Wayside & Woodland Trees* 81 The Virginian Juniper..or 'Red Cedar', as it is called on the American continent, is..frequently planted in our parks and gardens. **1958** G. A. Petrides *Field Guide Trees & Shrubs* 22 The junipers (including the Red Cedar) may bear either scaly or hollowed 3-sided needles. **1969** T. H. Everett *Living Trees of World* iv. 63/1 Western red-cedar..is native along the Pacific Coast of North America. **1972** *Ecology* LIII. 1141/2 'Fruits' of pasture juniper and red cedar are important foods for red squirrels.

red cell. A blood cell containing hæmoglobin; an erythrocyte; = *red blood cell, red corpuscle*, both s.v. *red a. 19 a.

1885 Delafield & Prudden *Handbk. Path. Anat. & Histol.* (ed. 2) ii. 50 In the extravasation of blood by diapedesis, the white blood-cells may pass through the walls of the vessels..; the red cells, on the other hand, are carried passively through the walls by minute amounts of fluid. **1896** *Boston Med. & Surg. Jrnl.* CXXXV. 131/2 Nucleated red cells have usually been classified as microblasts, normoblasts, megoblasts..and those with dividing nuclei. **1936** *Lancet* 11 July 88/2 A..study of the permeability of red cells. **1968** Passmore & Robson *Compan. Med. Stud.* I. xxvi. 2/1 The red cells are by far the most numerous of the blood cells; for every white cell there are about 500 red cells and about 30 platelets.

2. *attrib.*

1917 C. Price-Jones *Blood Pictures* i. 11 Assuming the average red-cell count of a woman to be 4,450,000 per c.mm. **1941** *Science* 24 Jan. 87/1 (*heading*) Red cell volume circulating and total as determined by radio iron. **1947** *Radiology* XLIX. 303/2 After acutely toxic doses of such agents as Sr[89] and Pu[239], the minimum red cell count occurs in survivors at a period later than the time of death of non-survivors. **1968** Passmore & Robson *Compan. Med. Stud.* I. xxvi. 5/1 Red cell precursors normally account for one third to one tenth of the bone marrow cells. **1976** *Med. Clin. N. Amer.* LX. 945 Pure red cell aplasia is a disorder in which patients cease making new red cells. **1976** I. Chanarin et al. *Blood & its Dis.* ii. 11 The normal..red cell count at sea level is..$10^6/\mu l$ $5{\cdot}0 \pm 0{\cdot}6$ [for men].

red cent: see *red a. 3 c in Dict. and Suppl.

red-cheeked, *a.* (Earlier and later examples.)

1602 Marston *Antonio's Revenge* v. iv. sig. K, Red cheekt Bacchus. **1872** Geo. Eliot *Middlemarch* II. xxxvii. 254 Pratt, a red-cheeked man. **1923** E. O'Neill *Anna Christie* i. 7 He is a boyish, red-cheeked, rather good-looking young fellow of twenty or so. **1960** *Farmer & Stockbreeder* 16 Feb. (Suppl.) 3/1 From the car he waved goodbye to her. Red-cheeked, in her felt shoes, one hand clutching her dress at the neck, she waved back.

Red China. [*red a. 9 b.] Communist China; the People's Republic of China. Hence **Red Chinese** *sb.* and *adj. phr.* Cf. *mainland China* s.v. *mainland 2 b.

1934 tr. Mao Tse-Tung (*title*) Red China. **1937** E. Snow *Red Star over China* i. 17 There has been perhaps no greater mystery among nations, no more confused an epic, than the story of Red China. **1966** 'G. Black' *You want to die, Johnny?* viii. 151 You couldn't stop junks landing..I'm thinking of Red Chinese. **1967** R. J. Serling *President's Plane is Missing* (1968) vii. 138 Is there any reason to suspect the Red Chinese or Russians? *Ibid.* 139 Suppose, sir, Red Chinese agents had somehow kidnapped Jeff Haines. **1971** J. Henderson *Copperhead* (1972) xv. 295 Targets inside Russia and Red China. **1972** J. Ball *Five Pieces of Jade* ii. 27 A systematic new campaign by the Red Chinese to pump narcotics..into the United States. **1973** 'G. Black' *Bitter Tea* (1973) iii. 44 Plenty of people in Malaysia still hate Red China. **1978** *Listener* 8 June 730/3 The entry permit into what we used to call 'Red China' is stamped on your passport.

red-circle, *v.* To separate out by circling in red ink; usu. *fig.*; *spec.* (see quots. 1974 and 1977). Hence **red-circling** *vbl. sb.*

1965 E. Brown *Big Man* xxiii. 202 The night had been red-circled for pleasure and for entertainment. **1973** *Maclean's Mag.* July 63/1 This was not a problem peculiar to the Trudeau regime, but it grew worse when the Treasury Board approved a plan to 'red circle' my 28 senior officers, the heart of my department. *Ibid.*, He suggested that if a committee were set up, the impending red circling might be avoided. **1974** *Globe & Mail* (Toronto) 14 Feb. 1/2 [The secretaries] fear their position will be further downgraded by a reclassification scheme undertaken by the Treasury Board whereby secretaries could be red-circled—held in their current pay bracket until the new secretarial classifications catch up to them. The possibility of red-circling has stalled salary negotiations between the Treasury Board and the Public Service Alliance of Canada. **1976** *Ibid.* 15 Sept. 1/2 The alliance said red-circling is not permitted and that an employee can only be demoted if he is proved to be incompetent or incapable of performing the duties of his position. **1977** *Spare Rib* May 22/1 Certain men at Trico-Folberth's were 'red circled', placed in a special category to preserve their higher rate of pay. (It is called 'red circling'

when workers moved on to a job with a lower rate of pay take their previous higher rate with them.)

redcoat, red-coat, red coat. Add: **1. c.** A steward at a Butlin's holiday camp.

1950 L. Blair *Butlin Holiday Bk. 1949–50* 66 The snow eventually disappeared and the 'Redcoats' prepared to return to the Holiday Villages. **1962** R. North *Butlin Story* v. 61 Charlie was a Redcoat at Filey for four seasons. **1966** P. J. Kavanagh *Perfect Stranger* iii. 23 To counteract my snobbism, he sent me to Butlin's Holiday Camp to do a month as a Holiday Uncle, or Redcoat. **1979** *Daily Tel.* 2 Aug. 3/3 A holiday camp 'redcoat'..claimed a world record at Brighton yesterday, by eating 100 peanuts in 46 seconds.

d. The title of a particular attendant at the door of the House of Lords.

1972 *Times* 22 July 12/5 Redcoat is the only attendant dressed in red in the House, a reminder that his was a royal appointment of Charles II's originally. The King, visiting the Lords and finding no one to greet him, made his own appointment on the spot. **1974** *Daily Tel.* (Colour Suppl.) 6 Dec. 29/2 C. D. Maxted, known as Red Coat, is a familiar figure to all peers as they arrive at the Lords.

3. (Later example.)

1906 *Westm. Gaz.* 6 Sept. 2/1 The British markets want large, bright apples, preferably of the red-coat type.

red cross, red-cross. Add: **3. a.** *red cross flag* (further example).

a **1850** Wordsworth *Prelude* (1959) x. 385 The proud fleet that bears the red-cross flag.

b. *Red Cross work.*

1914 W. Owen *Let.* 23 Sept. (1967) 284 There are already too many ladies offering to help with red cross work.

re·d-cross, *v. rare.* [f. the sb.] *trans.* To mark with a red cross.

1869 Browning *Ring & Bk.* IV. xi. 128 You would have ..forced me..find my way submissive to the fold. Be red-crossed on the fleece, one sheep the more.

red-crossed, *a.* (Later examples.)

1900 W. S. Churchill in *Morning Post* 17 Feb. 8/1 White-hooded, red-crossed ambulance waggons. **1916** 'Boyd Cable' *Action Front* 165 Another [ambulance wagon] was overturned,..and in the Red-Crossed canvas tilts of others gaped huge tears and rents. **1935** C. S. Forester *Afr. Queen* ii. 30 The Mediterranean squadron..with the red-crossed Admiral's flag in the van. **1962** J. B. Priestley *Margin Released* ii. iv. 113 The starched and red-crossed debs.

red-currant. Add: **c.** *red currant jam, tart.*

1788 J. Woodforde *Diary* 8 July (1927) III. 36 We had for Dinner to Day some Peas and Beans, a Piggs Face..and black and red currant Tarts. **1861** Mrs. Beeton *Bk. Househ. Managem.* 771 Red-currant jam. **1866** J. Blackwood *Let.* 10 Sept. in *Geo. Eliot Lett.* (1956) IV. 307 My little boy..declined red currant tart. **1958** R. Page *Let.* 11 Oct. in R. McDouall *Clubland Cooking* (1974) 166 Another member had a weakness for Red Currant and Raspberry Tart.

redd, *sb.*[1] Add: **1.** Also with *up.* (See also quot. 1893–4.)

1893–4 R. O. Heslop *Northumb. Words* II. 569 By inversion, 'A fine *red* up' is sometimes used to indicate a scene of disorder. **1917** 'H. H. Richardson' *Fortunes R. Mahony* II. ii. 105 She herself, in proper wifely fashion, proposed to give her little house a good red-up, in its master's absence.

redd, *sb.*[2] **2.** (Earlier and later examples.)

1808 Jamieson s.v. Red, Redd. With their snouts they form a hollow in the bed of the river, generally so deep, that, when lying in it, their backs are rather below the level of the bed. This is called the redd. **1913** F. M. Halford *Dry-Fly Man's Handbk.* III. i. 307 An observant man will detect the heaps of clean gravel or redds where the ova have been deposited by the trout... If there are salmon in the river, their redds too will be visible. **1916** *Trans. Inverness Sci. Soc.* VIII. 324 Salmon and all kinds of trout are very much alike in their spawning habits. The spawning bed, often called a 'redd', is composed of gravel or rough sand. **1960** *New Scientist* 2 June 1392/1 A study of the nature of redds—the gravel banks chosen by the female trout to receive her eggs—has shown that an essential feature is the presence of water currents. **1971** W. Hillen *Blackwater River* xi. 105 The alevins emerge from the spawning nest, or redd, in late winter. **1977** *New Yorker* 2 May 47/2 Everywhere, in fleets, are the oval shapes of salmon. They have moved the gravel and made redds.

redd, *v.*[2] Add: **6. a.** Also in *U.S.* and general use.

1842 *Spirit of Times* (Philad.) 12 Aug. (Th.), I never used to red up their chamber without thinking of it. **1896** E. Higginson *Flower that grew in Sand* 120 'You got your front room red up, Emarine?' 'No; I ain't had time to red up anything.' **1909** A. Quiller-Couch *True Tilda* xix. 258 They tumbled out and redded up the place in a hurry. **1912** Mulford & Clay *Buck Peters* i. 19, I guess you two men can take care of each other while I red up. **1951** L. Craig *Singing Hills* xix. 181 You take this baby while I redd up the room. **1977** J. Aiken *Five-Minute Marriage* ix. 141 The rooms..are all clean and redd up, sir.

redden, *v.* Add: **2. d.** Of a pullet: to acquire a deeper shade of red in the comb and wattles as the bird approaches maturity and prepares to begin laying.

1909 T. W. Sturges *Poultry Man.* vii. 106 When a pullet is about to redden up and develop her comb previous to laying, the change from one pen to another will check this development. **1950** *N.Z. Jrnl. Agric.* Oct. 332/3 If any [pullets] appear unlikely to start 'reddening up' for a month or more, they should..be sold immediately. **1967** T. R. Morris in T. C. Carter *Environmental Control in Poultry Production* ii. 27 Once the flock has begun to 'redden up' it is usually too late to alter sexual maturity.

reddening, *vbl. sb.* [f. Redden *v.* + -ing[1].] The action of making or becoming red. Also *attrib.*

1847 T. T. Stoddart *Angler's Compan. Scotl.* 116 Worms on their transference to the moss-jar still undergo the process of scouring..that of toughening, and..the further one of reddening. *Ibid.*, The reddening matter..is a species of high-coloured earth, reduced to a powder. **1927** Haldane & Huxley *Animal Biol.* vii. 150 Impulses..run directly to the local vessels, which open up, causing reddening of the skin. **1978** Pasachoff & Kutner *University Astron.* xxiii. 579 Traditionally, studies of reddening and extinction have been used to find the distances to stars.

red devil. **1.** A type of Italian hand grenade. Also *attrib.*

a **1944** K. Douglas *Alamein to Zem Zem* (1946) vi. 44 The little tin 'red devil' grenades, bombastic little crackers that will blow a man's hand off and make a noise like the crack of doom. **1967** *Sunday Times* (Colour Suppl.) 10 Sept. 45/4 *Red devils*, Italian hand grenades painted red. They made a lot of noise but caused little damage.

2. *the Red Devils*: popular name for the Parachute Regiment of the British Army.

1943 in G. G. Norton *Red Devils* (1971) ii. 24 General Alexander directs that I Para Brigade be info[rmed] that [they] have been given name by Germans of 'Red Devils'. **1948** M. Packe *First Airborne* vii. 78 They..inspired the German paratroops..to christen them the Red Devils. **1974** *Times* 19 Apr. 15/4 The Red Devils free-fall team is giving another display. **1977** *Times* 8 June 4/1 The Red Devils, the Army's daring freefall parachute team.

3. = *red-bird 2.

1967 *Boston Sunday Herald* 26 Mar. i. 12/2 Friday's 'goof ball' raid in a South End apartment (where 3,600 so-called 'red-devils' were confiscated) was the result of three months investigatory work. **1971** 'D. Shannon' *Murder with Love* (1972) iv. 67 Quite a collection of the pills..the Blue Angels and Red Devils and Yellow Submarines. **1974** M. C. Gerald *Pharmacol.* xi. 205 Short-acting barbiturates such as..secobarbital ('red devils').

redding, *vbl. sb.*[1] Add: **2.** *redding-out*: the process of undergoing or experiencing a red-out.

1933 *Jrnl. R. Aeronaut. Soc.* XXXVII. 407 The phenomenon of 'redding out' is essentially and solely ocular in origin due to postural congestion of the vascular retina. **1951** Nayler & Ower *Flight To-day* (ed. 3) i. 24 Accelerations..in the other direction, i.e., upwards towards the pilot's head, of more than 2g [are common] in bunts, which lead to risks of 'redding out' in contrast to 'blacking out'. **1961** R. L. Christy in H. G. Armstrong *Aerospace Med.* xvi. 250/2 Occasionally there may be a temporary loss of vision and there have been a few scattered reports that objects appear red and produce the phenomenon commonly referred to as 'redding out'.

reddingite (re·diŋəit). *Min.* [f. the name of *Redding* township, Fairfield County, Connecticut, where the mineral was first found: see -ite[1].] A hydrated phosphate of ferrous iron and bivalent manganese (the latter predominating), $(Mn,Fe)_3(PO_4)_2.3H_2O$, which forms an isomorphous series with phosphoferrite and occurs as pinkish, yellowish, or colourless, translucent or transparent, orthorhombic crystals.

1878 Brush & Dana in *Amer. Jrnl. Sci.* CXVI. 35 In addition to the above minerals, as original constituents of the same deposit, are amblygonite (hebronite), and a phosphate of manganese isomorphous with scorodite which we shall describe under the name reddingite. **1955** [see *phosphoferrite* s.v. *phospho-]. **1964** *Amer. Mineralogist* XLIX. 1122 Landesite, a rare hydrated manganous-ferric phosphate, is an alteration product of reddingite which, in turn, is derived from the hydration and alkali-leaching of lithiophilite-triphylite at the Berry Quarry, Poland, Maine. **1971** *Mineral. Abstr.* XXII. 18/1 Iron-manganese phosphate minerals, new to the Congo are..reddingite-phosphoferrite, bermanite, phosphosiderite, [etc.].

reddish, *a.* Add: **1. b.** (Further examples of insect names.)

1889 *Cent. Dict.* 5018/3 *Reddish light-arches*, a British noctuid moth. **1907** R. South *Moths Brit. Isles* 1st Ser. 279 The Reddish Light Arches..occurs in beech woods. *Ibid.* 321 The Reddish Buff... The female is much smaller than the male. **1968** *Oxf. Bk. Insects* 82/1 It is easy to confuse this moth with the less common Reddish Light Arches.

2. a. *reddish-blue, -brown* (earlier and later examples), *-purple, -violet.*

1934 *Webster*, Reddish-blue. **1962** I. Murdoch *Unofficial Rose* xviii. 179 A dark reddish-blue sky. **1964** S. Duke-Elder *Parsons' Dis. Eye* (ed. 14) xxi. 296 The anterior ciliary veins are dilated, and a reddish-blue zone surrounds the cornea. **1629** J. Parkinson *Paradisi in Sole* l. 518 The scales..are eyther of a reddish browne, whitish, or greenish colour. **1855** J. Phillips *Man. Geol.* 411 Over all is a continuous widely spread reddish-brown clay. **1980** *Catal. Fine Chinese Ceramics* (Sotheby, Hong Kong) 16 The unglazed stoneware burnt reddish-brown in the firing. **1629** J. Parkinson *Paradisi in Sole* viii. 55/2 A white [flowring Tulipa], speckled with a reddish purple, more or lesse, of diuers sorts, with white, yellow, or blew bottomes. **1963** A. Lubbock *Austral. Roundabout* 108 The granite ridges round Cloncurry, reddish-purple and bare. **1971** L. A. Boger *Dict. World Pott. & Porc.* 278/2 Inscriptions were often painted on the black background in red (or reddish purple). **1980** *Catal. Fine Chinese Ceramics* (Sotheby, Hong Kong) 92 Covered with a bright reddish-purple glaze attractively streaked and mottled in milky blue. **1856** *Rep. Comm. Patents 1855: Agric.* 273 The *Red-striped* [sugar] cane,..and the *Violet* or *Reddish-violet*, which is only a variation from the former,..will generally prosper..[in] the Southern States. **1964** S. Duke-Elder *Parsons' Dis. Eye* (ed. 14) x. 102 The ciliary form for the most part a diffuse reddish-violet blush.

Hence **re·ddishly** *adv.*

1946 R. Capell *Simiomata* ii. 47 He is donnish, tall and reddishly fair.

reddition. Add: **1. d.** Recompense or restitution.

1929 R. Bridges *Testament of Beauty* iv. 143 And for her soilure make Reddition to Nature.

3. (Later example.)

1950 J. N. D. Kelly *Early Christian Creeds* ii. 36 This solemn rehearsal, or reddition, of the creed before baptism was universally observed in the West.

red dog. **1.** (In Dict. s.v. Red *a.* 19 a.)

2. A low grade of flour.

1889 in *Cent. Dict.* **1931** *Hearings U.S. Congress House Comm. Ways & Means* 131 That would probably include 'red dog' flour, which is a very low grade flour which is considered feed. **1946** *Sun* (Baltimore) 14 Feb. 14/1 'Red dog' is fine bran particles and small quantities of wheat flour.

3. Either of two card games (see quot. 1934).

1930 *Sun* (Baltimore) 18 Aug. 6/1 Playing red dog for money. **1934** *Webster*, *Red dog*... 3. a. A game in which players hold each five cards and bet, for a pool, that they hold a higher card in the same suit than the top card of the stock. b. A variety of stud poker played with seven cards, the first two and the last one being dealt face down, in which all buried red cards are wild. **1935** *Encycl. Sports* 400/1 *Red dog.* This is a card game for any number of players from three to eight. **1938** J. D. Carr *Crooked Hinge* xxi. 290 It was, in the terms of an American pastime called Red Dog, 'high, low, jack, and the goddam game'. **1945** A. A. Ostrow *Compl. Card Player* 47 (*heading*) Red Dog (Also known as high-card pool). *Ibid.* 48 Six-Spot Red Dog... In this variant of red dog each player receives 3 cards. **1974** *Hoyle's Mod. Encycl. Card Games* 289 Despite its simplicity, red dog can build up to high stakes.

4. A manœuvre in American football, in which an opponent rushes the player who is passing the ball. Also as *v. trans.*, to rush (a player) in this way.

1959 *Washington Post* 17 Nov. C1/3 A variety of defenses which stress red-dogging the passer. **1959** *Time* 30 Nov. 56/2 Huff is at his rugged best when he knifes through the line and 'red-dogs' a quarterback as he fades to pass. The crash of Huff's tackle can stir the Giant bench to bellowing glee. **1966** Rote & Winter *Lang. Pro Football* iii. 132 *Red dog*,..surprise defensive maneuver where one or more linebackers..charge across line of scrimmage after ball carrier.

reddy, *a.* Add: **b.** (*b*) *reddy-brown* (further examples), *mauve.*

1946 G. Millar *Horned Pigeon* i. 1 A bedside light shone into her reddy-brown curls. **1968** D. Ireland *Chantic Bird* i. 5 Ma's photos..were in a flat, wooden box, covered with flowered paper. Reddy brown. **1970** A. Draper *Swansong for Rare Bird* vi. 45 My best shirt was reddy mauve. **1977** 'M. Underwood' *Murder with Malice* ii. 24 The reddy-brown stain on the mushroom-pink carpet showed where her head had lain.

reddy, var. Ready *sb.* 1 in Dict. and Suppl.

1962 R. Cook *Crust on its Uppers* i. 24 Not enough reddy in it in my case. *Ibid.* iii. 39 'Loot!'..'In reddy?' *Ibid.* viii. 65 Reddies which should be sailing into her African kick.

rede, *sb.*[1] Add: **5.** *Comb.* *rede-craft* (see quot.).

1880 W. Barnes (*title*) An outline of rede-craft (logic) with English wording.

redeal, *v.* [Re- 5 a.] To deal again. Also as *sb.*, a fresh deal.

1935 *Sun* (Baltimore) 11 Jan. 1/2 The New Deal is to be 'redealt'. **1959** T. Reese *Bridge Player's Dict.* 182 If the pack is deficient in any respect there must always be a redeal. **1964** *Official Encycl. Bridge* 287/2 When there is a redeal, the current deal is canceled; the same dealer deals again, unless he was dealing out of turn. *Ibid.* 450/1 Hands are never redealt at duplicate except in special cases.

red-eared: see also *red a. 14 a, 14 b.

redeemed, *ppl. a.* Add: **b.** Of land: reclaimed. *rare.*

1838 H. Colman *1st Rep. Agric. Mass.* (Mass. Agric. Survey) 37 From one acre of redeemed meadow 4½ tons of English Hay were weighed and sold in 1836–7.

redefe·ction. [Re- 5 a + Defection.] Return to a country from which one has previously defected. Hence (as a back-formation) **redefe·ct** *v. intr.*; **redefe·ctor.**

1957 *Britannica Bk. of Year* 814/1 *Redefection*, n, the returning of exiles to their native country. **1959** *Washington Post* 7 Apr. A 16/2 Many months before the Hungarian Revolution, the Communist propaganda mills began an elaborate campaign to encourage the 'redefection' of homesick or otherwise dissatisfied or disillusioned refugees from

satellite countries. *Ibid.*, The destiny that awaited these 'redefectors' was of course easily predictable. **1963** *Punch* 13 Feb. 249/3 Defected Western scientist who wishes to redefect. **1963** *Daily Progress* (Charlottesville, Va.) 7 Aug. 36/2 West German social workers concede the justice of complaints by many of the re-defectors that they are treated as 'outcasts' in West Germany. **1974** 'J. Le Carré' *Tinker, Tailor* xxi. 179 What's Tarr supposed to be doing now: redefecting to us?

redefine, *v.* (Further examples.)

1946 *Nature* 14 Dec. 885/1 Since inequality in the number of filaments in *Spirographis* was the only positive character distinguishing it from *Sabella*, that distinction is no longer valid, and both should be united under *Sabella*, as re-defined. **1951** R. Firth *Elem. Social Organiz.* v. 163 Let us..re-define the notion of primitive. **1964** *Ann. Reg. 1963* 231 A Trades Union Council re-defined the duties of unions on 3 July. **1979** *Listener* 16 Aug. 212/3 The many attempts ..to get round the decline of religion and worship by re-defining them in various Pickwickian senses.

redefinition. Add: (Further examples.) Hence **redefini·tional** *a.*

1944 J. S. Huxley *On Living in Revolution* 13 A re-definition of the status of colonies. **1949** *Mind* LVIII. 146 It will be convenient to call a statement in which a word is used in a high or in a low sense a *redefinitional statement.* **1956** *Nature* 25 Feb. 370/2 Many, and perhaps all, scientific advances involve the re-definition of terms. **1970** S. L. Barraclough in I. L. Horowitz *Masses in Lat. Amer.* iv. 125 Another effect of changing technology is to force a redefinition of the traditional relations between peasants and management. **1977** *New Society* 5 May 225/1 There is a very close concern with the meanings of words and their definition or re-definition.

redemption. Add: **8. redemption yield,** the yield of a stock calculated as a percentage of the redemption price and allowing for any capital gain or loss which that price represents relative to the current price.

1948 G. Crowther *Outl. Money* (ed. 2) ii. 74 If the price of this bond in the market is 101, that means that for every £101 invested now, the purchaser will get £6 in interest, but he will also lose £1 in capital value—that is to say, the 'redemption yield' is 2½ per cent per annum. **1972** *Daily Tel.* 2 Dec. 24 The redemption yield is defined as that rate of interest at which the present price is made to equal the present discounted value of the future stream of interest payments plus the present discounted value of the eventual capital repayment. The rate of interest used for the purpose of discounting these future interest and capital values is the redemption yield rate of interest itself. **1973** *Ibid.* 7 Mar. 21 At the issue price of £99½ the 1980 stock will give a running yield of 9·05 p.c. and a gross redemption yield of 9·10 p.c. to 1980.

redemptionless, *a.* (Earlier example.)

1846 J. Brown *Let.* 12 Aug. (1912) 93 By the bye, is not he a redemptionless devil that Sir Robert?

redemptivism (rĭde·mᵖtiviz'm). *rare.* [f. Redemptive *a.* + -ism.] The desire to redeem. So **rede·mptivist** *a.*

1924 C. Mackenzie *Heavenly Ladder* xxiii. 289 You are obsessed by redemptivist phantoms... I perceive you the victim of an absurd idea that you have to save other people. Even your own God Jesus Christ made no attempt to do that. *Ibid.* 290 You misunderstand what you call my redemptivism... I regard myself as an automatic purveyor of Almighty God's bounteous Grace by administrating His Sacraments.

redemptor. Add: **3.** = Trinitarian *sb.* 1.

1880 Mrs. Oliphant *Cervantes* ii. 25 The friar, Jorge Olivar, one of the Brothers of Mercy, and official Redemptor for the province of Aragon.

redemptorial (rĭdĕmᵖtōᵊ·riăl), *a. rare.* [f. as Redemptory *a.* + -al.] = Redemptory *a.*

1900 R. W. Barbour *Thoughts* 6 He pleads His own crucified person. His very redemptorial existence is His plea.

Redemptorist. **1.** (Earlier and later examples.)

1833 J. B. Robertson in *Schlegel's Philos. Hist.* I. p. xlii, In conjunction with..some of the Redemptorists—a most able, amiable, and exemplary body of ecclesiastics at Vienna—he established in 1820, a religious and political journal. **1840** J. R. Hope-Scott *Diary* in R. Ornsby *Mem. J. R. Hope-Scott* (1884) I. xii. 227 Viewed as instruments of the Church, he thinks that most of the orders are now of little use. Those of the Jesuits and Redemptorists, with one or two more of the same kind as the latter, he conceives to be the only effectual orders. **1915** *Encycl. Relig. & Ethics* VIII. 68/1 The Redemptorists have remained sturdily faithful to their primary work of giving missions and retreats. **1975** *Church Times* 27 June 14/5 It [sc. the Roman clerical collar] is still worn by Redemptorists, Passionists and Oratorians.

Redemptoristine. Add: (Further examples.) Also *attrib.* or as *adj.*

1886 T. Livius tr. *Saintrain's Our Lady of Perpetual Succour* (ed. 3) I. xxiv. 211 In the public chapel attached to the convent of the Redemptoristine Nuns, St. Alphonsus, Dublin, the Holy Picture has a beautiful shrine. **1931** *Tablet* 8 Aug. 183/2 The Redemptoristine nuns from Clapham went into their new home at Chudleigh in 1925. *Ibid.*, The Redemptoristines are established also in a number of convents in Canada, Italy, Austria, France, Belgium, Spain, Holland and Brazil. **1969** *Observer* (Colour Suppl.) 13 Apr. 54/1 *Redemptoristines.* Founded by St Alphonsus Liguori at Amalfi, Italy, in 1731.

redemptress (rĭde·mᵖtrĕs). [After Redemptor.] A female redeemer; = Redemptrice.

1865 [see Restoress].

redeploy·, *v.* [Re- 5 a.] **a.** *intr.* To carry out redeployment; to change employment to increase overall efficiency. **b.** *trans.* To move (troops, labour, materials, etc.) from one area of activity to another.

1945 *Sunday Times* 27 May 7 Re-deploying to crush Japan. **1945** *Sat. Rev. Lit.* (U.S.) 16 June 12/1 Others stated that the plan to redeploy troops from Europe was a great mistake. **1948** *Picture Post* 3 Apr. 11/2 Labour forces had to be redeployed. **1949** *Manch. Guardian Weekly* 28 Apr. 3/3 A very extensive proportion of this industry could be redeployed within six months. **1958** *Times Lit. Suppl.* 22 Aug. 473/2 Organized labour cannot be induced to redeploy over the day-shift, thus releasing labour for a second day (and sometimes a third night) shift. **1966** *Hansard Commons* 20 June 628 Action is needed equally to deal with the problem of internal demand, public and private, and to redeploy resources, both manpower and capacity, according to national priorities, and check inflation. **1970** *New Society* 5 Mar. 389/1 The unification of hitherto separate services.. should make it easier to recruit and redeploy social service staff. **1971** *Brit. Jrnl. Industrial Relations* July 160 Some believed that there was a vast pool of under-used labour and that a short sharp burst of deflation would force companies to release surplus manpower which could then be redeployed into the essential export industries. **1980** *Times* 29 Feb. 19 Hawker's philosophy since aerospace nationalization has been to redeploy its resources into the electrical and mechanical engineering business it knows well.

Hence **redeploy·able** *a.*, available for redeployment, able to be redeployed.

1946 I. Shaw in *New Yorker* 2 Feb. 24/2 'I'm redeployable,' Olson sang.

redeploy·ment. [Re- 5 a.] Movement or reallocation (of troops, labour, resources, etc.); reorganization for greater efficiency; transfer to alternative employment.

1945 *Time* 12 Feb. 17/2 The new blueprint for U.S. redeployment calls for an army of 6,500,000 men to defeat Japan. **1945** *Newsweek* 28 May 28/3 The redeployment of three and four star generals from the European theater. **1949** *Manch. Guardian Weekly* 28 Apr. 3/3 Only by redeployment of labour can higher wages, greater productivity, and lower costs be achieved. **1955** *Times* 7 July 13/1 The growth of deposits had been checked (partly, as with Martins, by the redeployment of funds by large customers). **1959** *Economist* 3 July 42/1 Redeployment of liquid resources in this way, incidentally, can also keep directors and managers in the powerful style to which they are accustomed. **1966** *Daily Tel.* 2 Nov. 1/2 Temporary bridging finance for house purchase also has a special importance at a time when the redeployment of labour needs to be encouraged. **1970** *New Scientist* 24 Sept. 613/2 If they all have similar reserves of manpower, there should soon be no shortage for the Soviet Union's developing economy. The question may then be one of retraining and redeployment. **1974** *Financial Times* 20 Mar. 18/8 If they cancel the entire venture now, some 21,000 workers on the programme in Britain and a similar number in France will face either redundancy or redeployment.

redeposit, *v.* Add: (Further examples.) Also *absol.* Hence **redepo·sited** *ppl. a.*, **redepo·siting** *vbl. sb.*

1905 *Westm. Gaz.* 8 Feb. 9/3 The Bill which the Board proposes to redeposit in Parliament. *Ibid.*, The Board had decided to take an early opportunity of redepositing their bill. **1946** F. E. Zeuner *Dating Past* vi. 169 Re-deposited Younger Loess (probably result of ploughing), with Neolithic. **1963** A. J. Hall *Textile Sci.* vi. 282 At the same time that soil is passing out of the fabric into the main bulk of detergent liquor some of this soil is passing from this liquor into the fabric and there being re-deposited on the fibres. **1965** G. V. Williams *Econ. Geol. N.Z.* xix. 334/2 These rocks consist of conglomerates, black shales, re-deposited sandstones, mudstones and pseudo-tillites. **1975** *New Yorker* 17 Nov. 37/2 'I never heard of a bank saying "Do not redeposit",' she said. **1977** *Antiquaries Jrnl.* LVII. 235 There are tips of redeposited sands and loams..associated with features in the timber building sequence.

rederi·ve, *v.* [Re- 5 a.] To derive again.

1968 P. A. P. Moran *Introd. Probability Theory* vii. 293 It has probably been independently rederived by more writers than almost any other result. **1968** C. G. Kuper *Introd. Theory Superconductivity* xiv. 237 Thus all the results of Chapter 12 may be rederived.

redesign, *v.* Add: (Further examples.) Hence **redesi·gned** *ppl. a.*, **redesi·gning** *vbl. sb.*

1914 *Jrnl. Inst. Electr. Engineers* LIII. 94/2 It would be safe to re-design on a scientific basis all the cables mentioned in the British Standard Specification. **1930** G. B. Shaw *Apple Cart* p. xxix, Wren was not content to re-design and rebuild St Paul's: he wanted to redesign London as well. **1943** *Mind* LII. 63 Though minor improvements and modifications of traditional language and its categories are permissible, any radical 'redesigning' of them must necessarily distort reality and be ultimately unintelligible. **1946** *Nature* 21 Dec. 897/2 Many of the re-designed features were not amenable to accepted strength computation methods. **1969** *Jane's Freight Containers 1968–69* 185/1 This berth was re-designed and re-constructed to handle the London-Gothenburg passenger, car and cargo service of the England Sweden Line. **1973** *Sci. Amer.* Apr. 17/1 A further need is to redesign medical education so that physicians and other health personnel are trained and challenged to provide primary care and to deal with the chronic diseases. **1978** *Dumfries Courier* 20 Oct. 10/3 (*caption*) The 900 series has a 5 cm. longer wheelbase and a completely redesigned chassis.

redesi·gn, *sb.* [f. the vb.] A fresh design; designing again.

1930 *Daily Express* 6 Oct. 1/4 The re-design of the airship's hull. **1946** *Nature* 28 Sept. 439/1 Redesign of existing roads and the planning of new roads must serve two purposes, circulation and safety. **1963** *Guardian* 15 Mar. 8/6 This is a redesign..of the regular Tuscan pattern. **1973** *Sci. Amer.* Sept. 69/1 Most of the work being done on the problem of air pollution resulting from the internal-combustion engine is focused on redesign of the engine or on adding emission-control devices to it. **1977** *Times* 15 Aug. 5/2 Our last redesign came in 1971.

redevelop, *v.* Add: **a.** (*intr.* examples.)

1967 *Oceanogr. & Marine Biol.* xiv. 363 After a short time polarity in growth redevelops. **1978** *SLR Camera* Nov. 9/2 The image will then re-develop to a warm brown tone.

b. *spec.* in *Town Planning.*

1936 E. E. Finch & C. G. Eve *Rep. Town Planning Scheme* (Corporation of London Publ. Health Dept.) 13 It therefore appears that nearly a quarter of the building site area has been redeveloped in the last 30 years. **1941** H. J. Manzoni in F. E. Towndrow *Replanning Britain* 97 (*heading*) Relating the redeveloped units. **1947** *Act* 10 & 11 *Geo. VI.* c. 51 § 5(3) For the purposes of this section a development plan may define as an area of comprehensive development any area which in the opinion of the local planning authority should be developed or re-developed as a whole. **1956** N. Lichfield *Econ. Planned Devel.* iv. 76 Where an area is to be re-developed, demolition and clearance may be required. **1971** P. Gresswell *Environment* 91 *Comprehensive* development covers a large area, usually in towns—perhaps including a number of streets which are to be redeveloped to one comprehensive plan. **1976** *Star* (Sheffield) 29 Oct. 10/2 Our re-developed city has been advertised as a tourist and conference centre—and yet the horrible menace of litter continues.

redevelopment. Add: *spec.* in *Town Planning.* Also *attrib.*

1935 *Act* 25 & 26 *Geo. V* c. 40 §13 It shall be the duty of the local authority to cause the area to be defined on a map and to pass a resolution declaring the area so defined to be a proposed re-development area. **1938** *New Statesman* 25 June 1060/1 As to the great city itself, no satisfactory redevelopment of it is possible so long as population and business pour into it and its overall density is increased. **1940** *Economist* 28 Dec. 794/2 The committee recommends that in cases where private ownership presents an obstacle to planned development, redevelopment or conservation on an economical basis, the areas should be bought outright by planning authorities. **1944** in D. Tyerman *Ways & Means of Rebuilding* iii. 59 Mr Cadbury had been considering plans of redevelopment schemes at thirty to the acre. **1951** L. Keeble *Princ. & Pract. Town & Country Planning* xvi. 345 In towns where a great deal of redevelopment overspill will have to be accommodated..it will often be possible to plan some completely new neighbourhoods for development at later stages. **1964** J. B. Cullingworth *Town & Country Planning* xii. 241 Housing, comprehensive redevelopment and traffic. **1966** N. Lichfield *Cost Benefit Anal. Town Planning Cambridge* iv. 26 The projects comprise a group of interrelated redevelopment areas, roads and car parks. **1972** R. Quilty *Tenth Session* 136, 61 Sainsbury Road is right in the middle of a redevelopment area. One of those decaying streets back of Kentish Town. **1977** *Grimsby Even. Tel.* 13 May 1/3 The 'shocking conditions' they have to live in while their area goes through a redevelopment phase.

red-eye. Add: **3.** (Examples.)

1857 Thoreau *Maine Woods* (1864) 172 The birds sang quite as in our woods,—the red-eye, red-start, veery, wood-pewee, etc. **1917** T. G. Pearson *Birds Amer.* III. 103 The Job..photographed several times a female Red-eye solicitously feeding two voracious young Cowbirds. **1953** D. A. Bannerman *Birds Brit. Isles* II. 252 The red-eye has been called 'preacher-bird' because of its unceasing vocal efforts.

4. (Earlier and later examples.)

1819 J. A. Quitman *Diary* in J. F. H. Claiborne *Life & Corr. J. A. Quitman* (1860) I. 42 Whiting and I had to treat to 'red-eye' or 'rot-gut', as whiskey is here called. **1838** *Yale Lit. Mag.* III. 12 An Indian tribe that..seldom ever passed the prairie except to sell their skins, and purchase 'red-eye'. **1903** *Sun* (N.Y.) 15 Nov., Ben made for the nearest red-eye plant, and inside an hour he was roaring and shooting up the town. **1910** in J. Lomax *Cowboy Songs* 305 Drink that rot gut, drink that rot gut, Drink that red eye, boys. **1911** C. E. Mulford *Bar-20 Days* viii. 95 Anybody'd think you was full of red-eye, th'way you act. **1949** A. Hynd *We are Public Enemies* ii. 44 Barrow put down a slug of red eye and walked up to her. **1957** J. Steinbeck *Short Reign of Pippin IV* 69 Serving red-eye in shot-glasses. **1976** *Observer* 5 Dec. (Colour Suppl.) 18 (Advt.), Most of the liquor to be had [in New Orleans] at that time [sc. 1865] was known as 'redeye'. Because that's what it did.

5. *U.S. slang.* Tomato ketchup.

1923 G. H. McKnight *Eng. Words* 56 *Red eye* for 'catsup'. **1947** *Sun* (Baltimore) 7 Aug. 10/3 Red-Eye..that great disguise, ketchup. **1960** Wentworth & Flexner *Dict. Amer. Slang* 423/2 *Redeye*.., ketchup. *W.W.I and W.W.II Armed Forces use.*

6. *Austral.* A cicada, *Psaltoda mœrens.* Also *attrib.*

1925 [see *floury a. d]. **1945** Baker *Austral. Lang.* xii. 214 Most noted among our appellations for cicadas are: baker, floury baker,..red eye and double drummer. **1958** *Austral. Encycl.* II. 379/2 (*caption*) Red-eye cicadas. *Ibid.* 380/1 Other well-known cicadas are the red-eye..and the aptly named cherry-nose.

7. *red-eye gravy*: gravy made by adding liquid to the grease from cooked ham or other lean meat. *U.S.*

1947 *Reader's Digest* Apr. 130/1 Pinky brown slices of cured ham that almost floated in red-eye gravy. 1949 *Newsweek* 11 July 6/2 Truman had..'good Missouri hams, red-eye gravy, and hominy grits'. 1959 *Washington Post* 29 Oct. D5/1 To the folks in the hominy grits and red-eye gravy belt there is only one game this week—Louisiana State vs. Mississippi. 1977 *Time* 24 Oct. 27/2 Dennis serves up his baked ham and red-eye gravy, grits, green beans, carrots, buttermilk biscuits and coffee.

8. *U.S. colloq.* Used *attrib.* to designate an aeroplane flight on which the traveller is unable to get adequate sleep because of the hour of arrival or because of differences in time zones.

1968 Mrs. L. B. Johnson *White House Diary* 31 Mar. (1970) 642 Lynda was coming in on 'the red-eye special' from California, about 7 A.M., having kissed Chuck good-by at Camp Pendleton last night as he departed for Vietnam. 1972 'J. Lange' *Binary* 19 They all looked tired... Phelps had brought them out to California on a red-eye flight, let them sleep a few hours, then dragged them up for a meeting. 1973 *Time* 25 June 8/3 He took two 'red-eye' flights from Seattle to the capital. 1976 *National Observer* (U.S.) 4 Sept. 5/1 Schweiker..and Newhall took the red-eye special back to Washington that same night. Newhall just wanted to sleep, but Schweiker was, in Newhall's words, 'euphoric'.

9. *Canada.* A drink made from beer and tomato juice.

1973 *Daily Colonist* (Victoria, B.C.) 29 Aug. 2/2, I did manage to acquire a fair liking for 'red eye'..a mixture of beer and tomato juice. 1975 'S. Marlowe' *Cawthorn Jrnls.* (1976) xiii. 107 'I'll have a redeye,' Lester told the barman, who mixed tomato juice and beer for him, half and half. 1976 *Maclean's Mag.* 22 Mar. 51/2 The red-eye, a murky combination of beer and tomato juice.

red-eyed, *a.* Add: **1.** (Later examples.)

1911 J. Masefield *Everlasting Mercy* 47 Old parson, red-eyed as a ferret. 1934 C. Carmer in B. A. Botkin *Treas. S. Folklore* (1949) III. ii. 490 Two-Toe is a red-eyed 'gator and about fourteen feet long.

2. b. In the names of certain insects.

1934 *Nat. Geogr. Mag.* May 612/2 Spiders and red-eyed flies are numerous. 1937 C. Longfield *Dragonflies Brit. Isles* 170 The Red-eyed Damsel-fly should be easy enough to tell by its eyes alone.

Red Fed (re·d fed). *N.Z. colloq.* Also **redfed.** [f. Red *a.* 9 b (in Dict. and Suppl.) + Fed(eration.] A member of the Federation of Labour (founded 1909); now *gen.*, one who rebels against the established order, a left-winger. Hence **Re·d-Fed-ism.**

1914 *Evening Post* (Wellington, N.Z.) 14 Jan. 10/5 The charge you prefer against the 'Red Feds'. 1916 *Chrons. N.Z. Expeditionary Force* 15 Nov. 134/1 I'll poke the fellow in the jaw that calls me a 'Red Fed'. 1931 N. E. Coad *Such is Life* 24 They are on the look-out for Bolshevists, Red-Feds and I.W.W.s. 1948 D. W. Ballantyne *Cunninghams* I. xxxi. 152 That crazy young redfed..who landed himself in trouble over at the works with his gabbing. 1959 K. Sinclair *Hist. N.Z.* iii. 199 In contrast to the United Labour Party, the 'Red Feds' eschewed political action. *Ibid.*, The 'Red Fed's' rejection of the principle of industrial arbitration. 1959 *N.Z. Listener* 26 June 4/1 Those dark days when many a university lecturer and many a student were being hauled across the coals for radicalism, Communism, red-Fed-ism, and all sorts of other nonsense. 1969 F. Sargeson *Joy of Worm* iv. 134 And when besides he's Red Fed, Maoriland worker, not to mention *Das Kapital* all rolled into one—well, it's a toss-up whether he can continue to survive. 1970 D. M. Davin *Not Here, not Now* VI. i. 280 He sounded a bit of a red-fed, though. No time for Baldwin or any of that lot over in England.

Redfern (re·dfɜːn). [f. the name *Redfern*, Maddox.] (See quot. 1909.) Also *attrib.*

1909 J. R. Ware *Passing Eng.* 207/2 *Redfern* (*Soc.*, 1879), perfectly-fitting lady's coat or jacket. From the vogue obtained, 1879 on, by Redfern, Maddox, W. Regent Street, whose lady's tailoring became celebrated over the whole world. 1932 N. Coward *Cavalcade* I. iii. 24, I am now going home to have a bath and put on my new Redfern model.

Red Fife (red fəif). Also **Red Fyfe.** [f. Red *a.* + the name of David *Fife* (1804?–77), Canadian botanist.] A rust-resistant variety of spring wheat, developed in Canada during the 1870s by David Fife.

[1851 *Watchman* (St. Thomas, Ontario) 1 Feb. 2/4 In another column will be found an article on the 'Fife Wheat' which is held in such high estimation..by all who have tried it.] 1883 *Prince Albert* (Saskatchewan) *Times* 3 Oct. 6/2 Try it and be convinced, and by all means give the 'red fyfe' the preference. 1889 *Experimental Farms: Rep. 1888* (Canada Dept. Agric.) 29 One sample of the Red Fyfe was grown in Ontario. 1900 *Westm. Gaz.* 28 Dec. 2/1 The quantity and quality of Red Fife..is as fine as ever. 1932 *Discovery* Mar. 73/2 The suitable wheat [for prairie farming], Red Fife, was discovered by accident. 1936, 1965 [see *marquis 5]. 1973 H. Robertson *Grass Roots* iii. 56 Homesteaders were able to purchase high-quality Red Fife wheat.

red-fish. Add: **2. b.** After 'American' insert 'or North Atlantic'. (Earlier and later examples.)

1763 tr. *Le Page du Pratz's Hist. Louisiana* II. 26 This Gulf abounds with delicious fish; as..red fish, cod, sturgeon, ..and many other sorts. 1843 *Southern Lit. Messenger* I. 121/2 The waters too, furnished their finny..treasures,—the red fish, buffalo [etc.]. 1897 [see **channel bass*]. 1955 S. A. Grau *Black Prince* 190 He had not lifted his head from the redfish he was cleaning. 1962 K. F. Lagler et al. *Ichthyol.* viii. 247 The giant redfish (*Arapaima gigas*) of the Amazon is one of the largest freshwater bony fishes. 1969 A. Wheeler *Fishes Brit. Isles & N.-W. Europe* 480/1 The redfish is widely distributed in the North Atlantic.

4. *Austral.* = *nannygai.

1951 T. C. Roughley *Fish & Fisheries Austral.* 26 The nannygai, or redfish..occurs round the southern half of Australia. 1966 [see *nannygai]. 1979 *Verbatim* Summer 7/2 Australian fishermen found that redfish sold better than nannygai though the difference is entirely linguistic.

red flag. 1. (See Red *a.* 4 b.)

2. As a sign of danger, a warning, or a signal to stop. Also *attrib.* and *fig.*

1777 P. Thicknesse *Year's Journey* I. iii. 23 There is a red flag hoisted gradually higher and higher, as the water flows into the harbour [at Calais]. 1856 *N.Y. Herald* 12 Jan. 1/4 James Flood is road master of his section; any obstruction being on the track it is the duty of the flagman to exhibit his red flag. 1885 C. M. Yonge *Nuttie's Father* I. xiii. 153 They went into a hole and stuck fast, while the red-flag traction engineman prodded her with an umbrella. 1908 Kipling in *Flag* (Union Jack Club) 8 Thou didst flee up Cheepe, calling..for a red flag. 1968 J. Updike *Couples* ii. 146 You're sore as hell about some silly thing, maybe Harold's snubbing you, maybe you have the red flag out, but you're right there. 1973 'C. Aird' *His Burial Too* iii. 32 The roadmen went and got into a muddle with their flags... One of them..apparently gets a power complex every time anyone puts a red flag into his hand. 1976 *Billings* (Montana) *Gaz.* 17 June 2-A/4 Time and time again, Eizenstat addressed the members and urged them to avoid 'red flag' words or issues—such as homosexual rights, amnesty, opposition to capital punishment, full federalization of welfare costs. 1977 *Time* 17 Oct. 39/3 He..warned present and future White House aides to be on the alert for 'red flags' of moral dilemmas that may arise while serving a President.

3. a. As a symbol of revolution, socialism, or communism, freq. *spec.* of Soviet Russian communism. Also *attrib.* and *fig.*

1848 A. H. Clough *Let.* 26 Feb. in T. Arnold *N.Z. Lett.* (1966) 78 The Red Flag flying at Paris. 1857 C. M. Yonge *Dynevor Terrace* I. xx. 322 Muskets and pikes were here and there seen, and once he recognized the sinister red flag. 1878 *Indianapolis Sentinel* 23 May 4/6 We denounce the red flag communism imported from Europe. 1888 E. Bellamy *Looking Backward* xxiv. 353 They were paid by the great monopolies to wave the red flag and talk about..blowing people up. 1889 J. Connell in *Justice* 21 Dec. 3/2 Then raise the scarlet standard high! Within its shade we'll live and die! Though cowards flinch, and traitors jeer, We'll keep the Red Flag flying here! 1891 [see Red *a.* 4 b]. 1909 *Westm. Gaz.* 19 Oct. 2/1 The Socialists who in such circumstances would be returned in many constituencies..would be the real thing, of the red-flag order. 1914 Chesterton *Flying Inn* xxiii. 274 A very coarse strip of red rag..had been tied round the wooden sign-post by way of a red flag of revolution. 1922 W. J. Locke *Tale of Triona* xvii. 199 The only positive ideal in England at the present moment is Bolshevism. The only flag waved..is the red flag. 1957 *Encycl. Brit.* XIX. 711/1 In the urban centres the victory was won under the red flag of class warfare. 1967 *Guardian* 4 Dec. 1/4 The purpose of the visit is political—to show the Red Flag over Cairo. 1973 *Times* 21 Nov. 6/8 Red flag on Cunarder. The former Cunard liner Franconia..left Southampton yesterday flying the hammer and sickle. She was bought..by..the Far Eastern Steamship Company, of Vladivostock. 1974 P. Gore-Booth *With Great Truth & Respect* 210 The country was infested with various hostiles, Red Flag Communists, White Flag Communists, Karens or simply dacoits. 1977 *Time* 4 Apr. 24/3 There was little of the tedious Red-flag waving 'revolutionary culture' with which visitors to Peking are entertained.

b. *the Red Flag*: a socialist song by James Connell (see quot. 1889 above).

1909 R. Brooke *Let.* 2 Jan. (1968) 154 Yes! yes! Herbert Samuel..is a Socialist... He used to sing *The Red Flag* after dinner every night instead of grace. 1935 H. Nicolson *Let.* 1 Nov. (1966) 222 He..left the hall while they sang the *Red Flag.* 1968 A. Diment *Bang Bang Birds* ix. 160 A covey of West German millionaires who were singing the Red Flag in a hypnotised dirge. *a* 1974 R. Crossman *Diaries* (1976) II. 507 As an outgoing member I went down to the last debate and took part in singing the 'Red Flag' and 'Auld Lang Syne'. 1976 *Whitaker's Almanack 1977* 361/1 Labour M.P.s in their jubilation sang 'The Red Flag', shouted, and waved order papers in the air.

4. = *red ensign* s.v. Red *a.* 19 a. See Ensign 5.

1901 W. C. Russell *Ship's Adventure* v. 78 He..had begun the sea life in the Royal Navy as midshipman, but..had quitted the white for the red flag.

Hence **red fla·gger,** one who carries a red flag, a communist; **red fla·ggery,** communist doctrines.

1920 R. Macaulay *Potterism* II. i. 54, I hate red-flaggery, and all other flaggery. 1921 *Times* 4 Feb. 11/6 The common enemy..what Belfast of today calls the 'red-flaggers'. 1923 *Glasgow Herald* 23 June 8/3 That modest pattern which ..Mr. Kirkwood is finding it so difficult to assimilate with Red Flaggery. 1934 G. B. Shaw *On Rocks* II. 231 That's the way to dish these Labor chaps and Red flaggers.

red-flowered: see *Red *a.* 14 a, Red *a.* 14 c.

Red Fyfe, var. *Red Fife.

Red Guard. [Red *a.* 9 b in Dict. and Suppl.] **1. a.** A member of an organized detachment of workers during the Russian Bolshevik revolution of 1917; also, such units collectively.

1917 *Times* 12 Nov. 8/5 In Moscow, the 'Red Guard' was defeated. *Ibid.* 13 Dec. 8/5 The fighting at Tamarovka seems to have been between detachments of shock battalions..and local troops, with sailors, 'Red Guards', infantry, and armoured cars. 1943 E. M. Almedingen *Frossia* x. 407 At the corner of Sredny a red guard picket awaited them. 1957 *Encycl. Brit.* XIX. 710/2 Definitely expecting a crushing defeat of Trotsky's Red guards, the committee of public defense gave orders to the cadets of the military schools to arrest the military revolutionary committee. 1961 *Everyman's Conc. Encycl. Russia* 388 Kerenskiy's Provisional Government..was overthrown..with the aid of the Red Guards. 1977 *N.Y. Rev. Bks.* 9 June 46/3 The role of the Petrograd Red Guard in the October seizure of power has been greatly exaggerated.

b. A member of a paramilitary group of Russian soldiers and Finnish communist sympathizers who seized power in Finland in 1918; the group itself.

1922 *Encycl. Brit.* XXXI. 74/1 Whole trainloads of revolutionary *soldateska* arrived from Petrograd... They entered the so-called Finnish 'Red Guards', and ransacked the country. The reactionaries..organized the 'White Guards'. 1956 A. G. Mazour *Finland between East & West* iii. 43 In this manner there came into being the so-called Red Guard.

2. A name given to (*a*) the armed units of village people in the Second Revolutionary Civil War in China, 1927–37, (*b*) a youth movement during the Cultural Revolution in the People's Republic of China, 1966–76; also, a member of one of these movements.

1966 *Guardian* 24 Aug. 9/7 The teenage demonstrators, who are known as the 'Red Guard'..appeared to have taken over permanently the churches serving Peking's Christians. 1966 *Economist* 27 Aug. 813/1 The rioters were clearly identified as a new youth group of secondary school and university student activists called the Red Guard. The Red Guard made its first appearance at the monster rally on August 18th, when the students were publicly congratulated for their revolutionary zeal by Mao Tse-tung. 1967 *Listener* 19 Jan. 80 The Red Guards damaged a number of temples and old buildings, destroyed old books. 1971 T. W. Robinson *Cultural Revolution in China* p. vii, The Red Guard phase of the Great Proletarian Cultural Revolution began in the late summer of 1966. 1973 M. Lindsay in Yuan-li Wu *China* vi. 142 Mao insisted that the Red Guards (local village militia) had an important function in harassing small enemy forces and assisting the Red Army. 1977 *N.Y. Rev. Bks.* 12 May 22/3 To purge the party bureaucrats Mao mobilized these adolescents as Red Guards, but millions of them later had to be dispersed to the countryside. 1978 *China Now* July/Aug. 25/1 During the Cultural Revolution in the late 1960's, the influx of Red Guards from all over the country swelled the numbers to nine million.

3. *transf.* Applied to various other radical groups and their members.

1966 *Observer* 25 Sept. 1 In his winding-up speech to the Liberal Assembly..Mr Grimond yesterday hit back at the party's youthful 'Red Guard'. 1966 *Time* 4 Nov. 35 Japan's Red Guards are members of the Socialist opposition—aided by Communists and the *Komeito* (Clean Government Party). 1967 *Guardian* 17 July 14/3 The growing anger of the Liberal Party's leadership at the outspoken criticism by the party's young 'Red Guards'. 1968 *N.Y. Times* 10 Feb. 2 As in all such movements there is an extremist group, called 'Red Guards' in Turin, that seeks to eliminate virtually all traditional authority, to elect professors and to confer marks based on the findings of student committees. 1968 *Listener* 10 Oct. 482/2 A young Red Guard of British music recently expressed to me his concern about form in pop music. 1969 *Guardian* 20 Nov. 3/5 West Bengal's Red Guards, as the Marxist volunteers have been nicknamed. 1970 *Ibid.* 8 Apr. 12/3 India's most turbulent politician, Jyoti Basu, was addressing maybe 100,000 Marxists on the Maidan in Calcutta. He was surrounded by his own Red Guards; tough, supple, and enormously well-disciplined young men.

Hence **Red Guardism,** the Chinese Red Guard movement.

1967 *N.Y. Times* 5 Mar. 2 By November last year Red Guardism was in full cry, and zealous youths were swarming across China..holding aloft little red books containing quotations from Mr. Mao.

red gum[2]. Add: **2.** esp. *E. camaldulensis* or one of a group of closely related species. (Later examples.)

1911 E. M. Clowes *On Wallaby* ix. 249 Farther eastward iron bark and stringy bark prevail, and red gum follows the course of the Murray and its tributaries. 1920, etc. [see *marri]. 1955 M. R. Jacobs *Growth Habits of Eucalypts* vi. 219 The red gums form one of the most widespread and stable groups of the eucalypts.

2*. *U.S.* The sweet gum, *Liquidambar styraciflua*, an important timber tree.

1839 *Southern Lit. Messenger* V. 113/2 Dislodge the rackoon from its lofty hole in the red-gum tree. 1916 E. T. Seton *Woodcraft Man.* 288 Sweet Gum, Star-Leaved, or Red Gum, Bilsted, Alligator Tree, or Liquidambar. 1942 W. M. Harlow *Trees Eastern U.S.* 193 Now that veneered furniture has so largely replaced solid pieces, redgum wood has come into prominence.

red-headed, *a.* **2. b.** (Earlier and later examples.)

1744 W. Ellis *Let.* 11 Aug. in *Mod. Husbandman* (1750) VI. xxi. 161 The red-headed Thistle is growing on the Hills. 1901 E. Mounts *Islands in Ocean of Memory* 58 You may throw the bait out within three feet of the boat and a red-headed hag or gony will make a dive at it. 1947 J. Stevenson-Hamilton *Wildlife S. Afr.* xxx. 249 The red-headed squirrel (*Paraxerus palliatus*), is distinguished by the rufous colour of the head and underparts.

re·d-heart. 1. A cherry belonging to a variety bearing heart-shaped fruit with red flesh; also, the fruit of a tree of this kind. Also *attrib.*

[1664 J. Evelyn *Kalendarium Hortense* 68 June.. Cherries,..Heart, Black, Red, White.] 1707 [see *black heart* s.v. *black *a.* 19]. 1764 [see Red *a.* 12 b]. 1833 H. Barnard *Let.* 25 May in *Maryland Hist. Mag.* (1918) XIII. 377 Here were..numerous trees of ripe cherries, black hearts and red hearts. 1887 *Harper's Mag.* Dec. 31/1 Under the largest of two red-heart cherry-trees sat a girl shelling pease. 1904 E. A. G. Glasgow *Deliverance* 238, I used to cut round old Fletcher's pasture..to keep from passin' by his red-heart cherry-tree.

2. One of several trees with reddish bark or wood, esp. the western North American *Ceanothus spinosus*, an evergreen shrub or small tree belonging to the family Rhamnaceæ and bearing clusters of blue or white flowers.

1926 J. Masefield *Odtaa* xiv. 231 He saw a footmark in some soft earth close to a red-heart. 1937 *Range Plant Handbk.* (Forest Service, U.S. Dept. Agric.) B39 The branches of a number of species [of *Ceanothus*], such as.. redheart, or spiny myrtle (*C. spinosus*), end in spines. 1951 W. L. Jepson *Flowering Plants Calif.* 620 Red-heart. Straggling shrub 5 to 10 ft. high, or forming a small tree up to 24 ft. 1965 *Austral. Encycl.* V. 225/2 *Dissiliaria baloghioides* (hauer or redheart). 1973 G. M. Chippendale *Eucalypts W. Austral. Goldfields* 93/1 Redheart [sc. *Eucalyptus decipiens*] is a spreading, twisted, gnarled tree. 1973 *Stand. Encycl. S. Afr.* IX. 266/1 Red-heart..(*Acacia nilotica* = *A. benthami*) Thorn-tree with a spreading rounded canopy.

red herring. 2. b. (Earlier and later examples.)

1884 *Liverpool Daily Post* 11 July 5/4 The talk of revolutionary dangers is a mere red-herring. 1928 *Manch. Guardian Weekly* 10 Aug. 105/1 Both the Opposition parties are trying to drag in the Protectionist red herring in the vain hope of causing dissension. 1956 [see *chivvy *v.*[1]]. 1967 G. F. Fiennes *I tried to run Railway* iv. 48 The Coroner's opinion that the detonators were 'something of a red herring'. 1975 M. Russell *Murder by Mile* xi. 116 This could be a side-issue or red herring designed to..turn me from the genuine scent. 1976 *Southern Even. Echo* (Southampton) 13 Nov. 9/3 He accused Mr. Deacon of introducing a red herring into the issue.

red horse, red-horse. Add: **3.** *slang* (orig. *Mil.*). Corned beef.

1864 I. Jackson *Let.* 28 June in *Some of Boys* (1960) 184 Supper..is coffee & Red Hoss. 1905 J. Bowe *With 13th Minnesota in Philippines* 24 Of bean-soup, hard-tack, and red horse..we have had our fill. 1920 W. B. Ellington *Company 'A', 23rd Engineers* 27 We have red horse and rice pudding for dinner. 1941 J. Smiley *Hash House Lingo* 46 *Red horse*, corned beef.

red-hot, *a.* (and *sb.*). Add: **2. a.** Also, outstanding, uninhibited, lively, sexy, passionate; esp. in phr. *red-hot momma*, (*a*) a woman who sings in a particular earthy style; (*b*) a girlfriend, lover. Also *transf.* in jazz; cf. *hot *a.* 8 g.

1888 J. Runciman *Chequers* 116 You take the fellows in town that make their living after dark... There's some red-hot ones up—you know where—in Piccadilly. 1926 Whiteman & McBride *Jazz* viii. 169 A red hot mama song. 1934 S. R. Nelson *All about Jazz* ii. 58 Dorsey is a red-hot stylist and technician [on the saxophone]. 1934 C. Lambert *Music Ho!* III. 210 The negro associations of jazz, the weary traveller, the comforting old mammy, the red-hot baby, have become a formula of expression only, as empty and convenient as the harlequin and columbine of the nineteenth century. 1935 *Time* 21 Jan. 58 Sophie Tucker, famed as 'the last of the red hot mamas'. 1936 Wodehouse *Laughing Gas* ii. 24 The bride-to-be is probably some frightful red-hot mamma. 1940 O. Nash *Face is Familiar* 87 Affection..leads to breach of promise If you go round lavishing it on red-hot momise. 1942 Z. N. Hurston in A. Dundes *Mother Wit* (1973) 223/2 A red hot pimp like you *say* you is, ain't got no business in the barrel. 1950 A. Lomax *Mister Jelly Roll* 69 A red-hot bass player, seventy-nine years old, a proud Creole. 1957 R. Hoggart *Uses of Literacy* v. 132, I first heard 'Paper Doll' sung in the 'red-hot' fashion by an American star crooner. 1976 in D. Villiers *Next Year in Jerusalem* 204 (*caption*) 'The Last of the Red Hot Mommas', Sophie Tucker. 1977 J. Wainwright *Do Nothin'* xi. 183 It was jive and blues; either red-hot or smoochy.

b. Also, sensational, lively, exciting, intense. (Further examples.)

1887 *Lantern* 19 Feb. 6/1 A red-hot newsy journal. 1891 'Mark Twain' in *Harper's Mag.* Dec. 97 Suddenly a red-hot new idea came whistling down into my camp. 1904 J. C. Lincoln *Cap'n Eri* xi. 205 '"Fightin' Fred Starlight, the Boy Rover of the Pacific",' he read aloud. 'Humph! Is it good?' 'Bet your life! It's a red-hot story.' 1915 Wodehouse *Psmith Journalist* v. 33 My idea is that *Cosy Moments* should become red-hot stuff. I could wish its tone to be such that the public will wonder why we do not print it on asbestos. 1955 *Times* 16 May 3/3 Local propaganda on the virtues of nationalization has resembled rather the cooing of lethargic doves than the strident militancy of red hot Socialism. 1969 *John Edwards Mem. Foundation Q.* V. II. 60 An urban audience more accustomed to 'red hot' fox trots than to barn dances. 1977 *Belfast Tel.* 24 Jan. 18/6 The clinching of these red-hot finals must be hailed as another major breakthrough for the sport.

d. *Austral. slang.* Unfair, unreasonable.

1896 H. Lawson *While Billy Boils* 281 When..she paused for breath, he drew a long one, gave a short whistle, and, said: 'Well, it's red-hot!' 1907 A. Wright *Keane of Kalgoorlie* 107 'It's red 'ot,' put in Dave, 'th'way these 'ere owners makes er pore man give 'em a lump in th' sweep. 1941 Baker *Dict. Austral. Slang* 59 A red hot price.

3. *red-hot poker*, substitute for def.: a tall perennial herb of the genus *Kniphofia* (formerly *Tritoma*), esp. *K. uvaria*, belonging to the family Liliaceæ, native to southern or tropical Africa, and bearing spikes of red, yellow, or white flowers. (Earlier and later examples.)

1887 'F. Anstey' *Talking Horse* (1892) 216 The dahlias and 'red-hot pokers' and gladioli..burnt with a sinister glow. 1916 M. Hampden *Flower Culture* xvii. 206 Red-hot pokers are not over [in November]. 1934 G. A. R. Phillips *Aristocrats of Flower Border* xiii. 191 Far more descriptive of its vivid beauty are the common names of torch lily and red hot poker. 1971 H. Evans *How to cheat at Gardening* xi. 168 The Red Hot Poker..looks marvellous mixed with Sea Holly.

4. as *sb.* A frankfurter; a hot dog. *U.S. slang.*

1892 *Chicago Figaro* VI. 157/2 The appetizing savors of 'red hots'. 1934 J. T. Farrell *Calico Shoes* 46 Don leaned against the thrown-together red hot stand, munching at a hot-dog sandwich. 1971 B. Malamud *Tenants* 143, I got this redhot with mustard on it I'm gonna eat my meat.

redi·al, *v.* [f. Re- 5 a + *dial *v.* 4.] *intr.* and *trans.* To dial again.

1961 'E. Lathen' *Banking on Death* (1962) viii. 70 His daughter broke the connection. He started to redial then put the phone down. 1966 'A. Hall' *9th Directive* xiv. 132 All three lines were busy and I began redialling the numbers. 1973 'E. McBain' *Let's hear It* xiii. 193 In as long as it took for the caller to re-dial, the phone began to ring again. 1976 'D. Craig' *Faith Hope & Death* xvi. 111, I re-dialled and still engaged, so it was not a wrong number.

redictate, *v.* Add: (Later example.) Also *absol.*

1964 T. L. Kinsey *Audio-Typing & Electric Typewriters* iii. 12 After a recording has been transcribed the dictator need only re-dictate over the existing work and it is automatically erased, the new dictation being recorded in its place. 1974 R. Crossman *Diaries* (1975) I. 15, I decided therefore to re-dictate this whole first transcript in plain intelligible English.

rediffu·se, *v.* *Broadcasting.* [Re- 5 a.] *trans.* To disseminate, broadcast, or rebroadcast by rediffusion. So **rediffu·sed** *ppl. a.* Also **rediffu·ser,** a person who or company which rediffuses a programme.

1931 *Times Educ. Suppl.* 1 Aug. p. iv/1 In order to receive such rediffused signals the listener requires an ultra-short wave adapter. 1932 *B.B.C. Year-bk. 1933* 70 Apart from the purely copyright aspect..the BBC..may very reasonably question the equity of the public rediffuser being able to use its programmes for the same..fee as is paid by the private listener. 1948 *Architect Rev.* CIV. 131/2 A public address system enables broadcast programmes to be received and rediffused throughout the school. 1950 *Sport* 7–11 Apr. 22/4 It may not be long before television programmes are rediffused through cinemas.

rediffusion (rīdifiū·ʒən). *Broadcasting.* [Re- 5 a.] **a.** The dissemination, broadcasting, or rebroadcasting of a programme by (*a*) reproduction on loudspeakers and screens in public places, (*b*) transmission by a broadcasting company which was not responsible for making it, or (*c*) publication by other media of items from a radio or television programme.

1927 *Observer* 13 Nov. 19/5 The best programmes of the British service will be available for re-diffusion throughout the Continent as well as throughout the Empire and the rest of the world. 1933 *B.B.C. Year-bk. 1934* 28 It was found possible to apply the [Copyright] Act..to public loud-speaker 'rediffusion'. 1948 *Daily Tel.* 9 June 6/5 The Association for the Protection of Copyright in Sport set out to secure certain safeguards, particularly against general 'rediffusion' of television on big cinema screens. 1950 *Sport* 7–11 Apr. 22/4 Sporting events will, I think, be among the most popular types of rediffusion. 1967 W. Soyinka *Kongi's Harvest* 2 Who but a lunatic Will bandy words with boxes With government rediffusion sets Which talk and talk and never Take a lone word in reply.

b. *spec.* The distribution of radio or television transmissions within a community by cable from a single receiver.

1935 *Nature* 2 Feb. 196/1 'Rediffusion' is a method of distributing a broadcast programme over an independent line network to a number of subscribers. 1968 Bethell & Burg tr. *Solzhenitsyn's Cancer Ward* I. xix. 295 Vadim..was happily surprised to discover there was no radio... (The reason for this omission was that for years they had been planning to move the clinic into better-equipped quarters, and the new place..was going to be wired with rediffusion points throughout.) 1975 C. Stuart in J. Reith *Diaries* ii. 179 Wireless or relay exchange (also called rediffusion) was the practice of wiring broadcast programmes to individual subscribers by commercial companies operating under licence from the Post Office.

redi·g, *v.* [Re- 5 a.] *trans.* To dig again. So **redi·gging** *vbl. sb.*

1907 *Church Q. Rev.* July 470 This division seems to entail some redigging of ground already trenched. 1922 G. Bell *Let.* 16 Feb. (1927) II. xxii. 633, I must tell you the Yusufiyah is one of the oldest canals in the world... Julian sailed down it to Ctesiphon and the Abbasids re-dug it. 1963 *Times* 4 June 12/5 A year later the tree can be lifted out after the trench is redug. 1965 J. A. Michener *Source* 892 Couldn't they have redug the tunnel?

Red Indian: see Indian *sb.* 2 b in Dict. and Suppl.

redingote. (Earlier and later examples.)

1793 F. Burney *Let.* *c* 9 Feb. (1972) II. 13 He was quite wet through his *redingotte*. 1802 C. Wilmot *Let.* 3 Jan. in *Irish Peer* (1920) 22 Benches..where servants are generally stretch'd in 'Redingotes' (Great coats) and cock'd hats. 1890 *Athenæum* 28 June 838/2 The Emperor [Napoleon], buttoned up in his legendary grey *redingote* and seated on a white horse stands motionless on a small rise of ground. 1930 *Times* 17 Mar. 9/4 Some of the new redingotes are made to give the effect of a coat and skirt with pleated lingerie vest. 1939 *Country Life* 11 Feb. p. xxxviii/2 This Matita two-piece redingote and dress is in a tone-on-tone effect in light and dark grey. 1965 [see *cloqué]. 1973 *Times* 14 Nov. 16/8 The Queen will wear a classic redingote of blue silk.

redingtonite (re·diŋtənait). *Min.* [f. the name of the *Redington* mine, Knoxville, Napa Co., Calif. + -ite[1].] A hydrous sulphate of iron, magnesium, chromium, and aluminium occurring as fibrous masses of a white, yellowish, or purple colour.

1888 G. F. Becker in *Monogr. U.S. Geol. Survey* No. 13. 279 A hydrous chromium sulphate occurs in fissures in silicified serpentine... It seems appropriate to give the name redingtonite to this hitherto unknown mineral. 1965 G. J. Williams *Econ. Geol. N.Z.* x. 149/1 He found deep lilac or purple earthy redingtonite associated with the folia of the fuchsite.

red ink. 1. *slang.* Cheap red wine; also applied to some other inferior alcoholic drinks. Chiefly *U.S.*

1919 *Red Cross Mag.* Nov. 22/3 He at once took ten of his fellow students to a sixty-cent 'red-ink' and spaghetti dinner down on Tenth Street. 1926 J. Black *You can't Win* xii. 153 Barrels of the deadly 'foot juice' or 'red ink', as the winos called it. 1930 *N.Y. Times Mag.* 16 Feb. 19 Today's word 'rum' used in a broad sense to designate all kinds of forbidden liquors, may refer to..the 'red ink' of Greenwich Village. 1942 H. W. Van Loon *Lives* 631 The wine problem was easily settled. Any kind of 'red ink'—any kind of that cheap Chianti..would be satisfactory. 1952 E. O'Neill *Moon for Misbegotten* III. 140 You'd lie awake..with..the wine of passion poets blab about, a sour aftertaste in your mouth of Dago red ink! 1976 W. H. Canaway *Willow-Pattern War* iii. 28 Lunch..was a real workaday snack this time: *raclettes* and *rösti* with a half-bottle of red ink.

2. *U.S. colloq.* The debit side of an account: cf. *Red sb.[1] 1 g. Also in extended use.

1929 *Century* Mar. 605/2 Red ink returns were as prolific as asparagus, which meant you..dug deep for the freight money. 1939 S. Bent *Newspaper Crusaders* ii. 35 The long-drawn crusade whereby St. Louis was taken out of political red ink and put on the credit side of the electoral ledger merits examination. 1948 *Sun* (Baltimore) 31 Jan. 1/3 We cannot play with red ink when we're financing a great government. 1967 *Boston Sunday Herald* 7 May IV. 9/2 Give us enough red ink and make money plentiful enough and the economic skies will soon clear. 1977 *Time* 28 Feb. 26/1 Carter's projected $57·4 billion deficit is an improvement over the $68 billion in red ink anticipated for fiscal 1977, which ends on Sept. 30. 1979 *Financial Rev.* 19 Oct. 27/1 The company would report a loss of more than $US 3 million for the third quarter, bringing the red-ink figure for nine months close to $US 6 million.

redintegration. 2. c. (Later examples.)

1912 B. Dumville *Fund. Psychol.* x. 208 All suggestion of things not present is due to a process of redintegration; things found or put together in past experience tend to call one another up. 1920 H. L. Hollingworth *Psychol. Functional Neuroses* ii. 19 Redintegration is to be conceived as that type of process in which a part of a complex stimulus provokes the complete reaction that was previously made to the complex stimulus as a whole. This is not precisely Hamilton's use of the term 'redintegration', but the process is so similar that the term may be used here without injustice. 1938 G. W. Allport *Personality* xix. 525 One variation of the associational theory is the doctrine of redintegration. 1947 G. Murphy *Personality* viii. 172 If the term canalization marked off no specific kind of event but were purely an alternative for such terms as conditioning, positive adaptation, or redintegration, there would be no justification in using it.

redintegrative, *a.* Delete $rare^{-1}$ and add further examples. Hence **redi·ntegra:tively** *adv.*

1870 S. H. Hodgson *Theory of Practice* I. 370 States of consciousness which the redintegrative activity has the tendency to produce. 1890 W. James *Princ. Psychol.* I. xiv. 581 The forms of its transitions, whether redintegrative, associative, or similar, are due to unknown regulative or determinative conditions. 1920 H. L. Hollingworth *Psychol. Functional Neuroses* ii. 18 The redintegrative mechanism, whereby a part reinstates a previous whole, is one of the most enlightening concepts ever offered to psychology. 1933 G. Murphy *Gen. Psychol.* xviii. 352 All about us are objects to which we do not respond redintegratively because this would involve dozens of *conflicting* action patterns. 1941 *Brit. Jrnl. Psychol.* Oct. 167 Three of the six CO stimuli aroused reactions of a redintegrative character. Some of them went back to situations of childhood and early life. 1946 C. Morris *Signs, Lang. & Behavior* 292 The requirement for redintegrative (or 'part-whole') efficacy. 1960 *Encounter* Jan. 81/1 This book has been for me..a remarkable redintegrative experience.

rediscount, *sb.* Add: Also *attrib.*, as *rediscount rate.*

1927 *New Republic* 21 Sept. 108/2 The action of the Reserve Board in ordering the Chicago bank to reduce its rediscount rate from 4 to 3½ percent has resulted in a direct challenge of the Board's authority to compel such a change. 1929 *Times Lit. Suppl.* 21 Mar. 218/3 This..should be corrected by the raising of the Reichsbank rediscount rate and the forcing down of German prices. 1951 *N.Y. Herald Tribune* 15 Nov. 20/2 The rise in the Bank of England rediscount rate from 2 to 2½ per cent. 1970 G. Jackson *Let.* 17 Apr. in *Soledad Brother* (1971) 223 The missionaries, with the benefits of christendom, school us on the value of symbolism, dead presidents, and the rediscount rate.

rediscou·ntable, *a.* [Re- 5 a.] That may be discounted again.

1964 *Economist* 8 Aug. 573/1 Restrictions on the eligibility of rediscountable paper.

redi·scounting, *vbl. sb.* [f. Rediscount *v.* + -ing[1].] The action of discounting again. Also *attrib.*

1931 *Economist* 2 May 946/2 This, in its turn, has occasioned a very decided increase in rediscounting. 1961 *Ann. Reg. 1960* 491 Gerrard and Reid, discount brokers, were granted borrowing and rediscounting facilities at the Bank of England. 1974 B. Pearce tr. *Amin's Accumulation on World Scale* II. iii. 429 The commercial banks in the under-developed countries do without rediscounting by the bank of issue.

redisperse, *v.* (Later examples.)

1946 *Nature* 28 Dec. 946/2 Elutriation methods of size separation redisperse coagulæ into their ultimate particles, thereby producing erroneous results. 1956 *Ibid.* 17 Mar. 521/2 The precipitates were then..redispersed by high-speed stirring.

redissolve, *v.* **a.** (Later examples.)

1941 J. S. Huxley *Uniqueness of Man* 98 The mineral framework of the bones is redissolved to be used up by the living cells. 1946 *Nature* 7 Sept. 350/2 The precipitates are.. redissolved in a 3 per cent solution of sodium dihydrogen phosphate. 1965 Phillips & Williams *Inorg. Chem.* I. xv. 567 It dissolves in an excess of the base, and can then be reprecipitated by addition of thionyl chloride, but excess of this acid does not, however, redissolve the aluminium sulphite.

redistil, *v.* Add: Hence **redisti·lled** *ppl. a.*

1930 *Engineering* 21 Mar. 394/1 A parcel of redistilled magnesium of exceptional purity. 1956 *Nature* 21 Jan. 130/2 Using the normal procedure of shaking infective fluid with 20 per cent of freshly redistilled ether and leaving overnight.

redistillation. (Earlier example.)

1666 Boyle *Orig. Formes & Qual.* 281 Neither do liquors, that have already been distill'd, obtain that colour upon redistillation.

redistributive, *a.* Add: Also of wealth, etc.

1931 G. B. Shaw *Fabian Ess.* p. viii, Redistributive taxation within Capitalist limits means dole for idleness. 1971 *Morning Star* 2 Sept. 3 Taxation structure should be much more progressive and redistributive and should include effective taxation of wealth, capital gains and gifts and a heavier tax on unearned income. 1974 *Guardian* 27 Mar. 15/4 The Budget is redistributive in favour of the working man.

redistrict, *v.* Add: (Further examples.) Hence **redi·stricting** *vbl. sb.* Also *attrib.*

1870 *Trans. Illinois Agric. Soc.* VII. 510 Mr. Flag moved that the subject of redistricting the State be referred to a special committee of nine. 1890 *N.Y. Weekly Tribune* 22 Oct. 12/3 Democratic rascalities in redistricting and in voting and counting will not prevent but will hasten and insure the passage of a bill to secure fair Congressional elections in future. 1949 *Illinois State Register* (Springfield) 1 Feb. 6/4 States throughout the nation are eyeing the Illinois plan for redistricting schools which has proved to be a tax saver as well as a more efficient means of educating youths. 1973 *Time* 25 June 18/3 He supports busing and the redistricting of the Richmond school system to achieve racial balance. 1977 *Time* 21 Mar. 57/2 The mock rebellion is a protest against a redistricting plan under which Martha's Vineyard will lose the seat that it has had in the Massachusetts legislature for 285 years. 1980 *Christian Sci. Monitor* (Midwestern ed.) 4 Dec. 5/1 The shift will be reflected in a gain of congressional seats by the region, when the US House of Representatives is redistricted on the basis of the 1980 census.

‖ **redivivus** (redivī·vŏs), *a.* Also fem. **rediviva,** fem. pl. **-æ.** [L., see Redivive *a.*] Come back to life; = Redivive *a.*

Always placed after the sb.

1675 R. Head (*title*) Proteus redivivus: or the art of wheedling. 1681 H. Nevile (*title*) Plato redivivus. 1843 H. James *Let.* 11 May in R. B. Perry *Tht. & Char. W. James* (1935) I. 47, I believe Jonathan Edwards *redivivus* in true blue would, after an honest study of the philosophy that has grown up since his day, make the best possible reconciler and critic of philosophy—far better than Schelling *redivivus*. 1856 C. M. Yonge *Daisy Chain* II. xxii. 589 He walked round and round his friend, called him Nicholas Randall *redivivus*, quoted Dogberry. 1937 *Mind* XLV. 492 Paton.. has written the sort of defence of Kant against his critics which a Kant *redivivus* might have written in self-defence. 1939 Joyce *Finnegans Wake* (1964) III. 490 She's write to him she's levt by me, Jenny Rediviva! 1974 *Listener* 28 Feb. 284/3 The belligerent women of all parties, tricoteuses *redivivæ*. 1975 *Times* 5 May 15/2 Some still believe in Stormont Redivivus. 1979 *Amer. N. & Q.* June 167 There isn't much more to say about this first volume of GW [sc. *Gesamtkatalog der Wiegendrucke*] *redivivus*.

Redjang, var. *Rejang. **red lead**: see also *Red *a.* 19 a.

red-legs, red-leg. Add: **3.** Also = Redshanks 3 a. (Later examples.)

1886 Britten & Holland *Dict. Eng. Plant-Names* 399 Red Legs. From the general redness of the stems. (1) *Polygonum bistorta*... (2) *Polygonum Persicaria*. 1960 *Oxf. Bk. Wild Flowers* 126/1 Spotted Persicaria or Red-legs..can usually be easily recognized by the red stems and the dark blotch in the middle of the leaves. 1971 *Countryman* Winter 126 The redleg..was flowering in late summer.

4. *U.S. Mil. slang.* An artilleryman.

1900 P. Revere *Cleveland in War with Spain* 164 The battery marched down the street... Cleveland's 'red-legs' were off. *Ibid.*, For once the 'dough boys' were envied by the 'red legs'. 1927 *Amer. Legion Monthly* July 16 Reilly's redlegs..admired the advance guard of Bengal lancers. 1969 S. N. Spetz *Rat Pack Six* 71 Anyway, you'll get a chance to cool it down there, just guarding a bunch of Red Legs.

red letter. Add: **2. b. red-letter day** (earlier and later examples).

1704 S. Knight *Diary* in *Amer. Speech* (1940) XV. 231/2 Red letter day. 1740 *Mock Campaign* 18 Their empty Eccho only now displays Great —'s Power on Red Letter Days. 1905 Proctor & Frere *New Hist. Bk. Common Prayer* (ed. 3) ix. 338 It is difficult to see clearly the motive which determined the selection of the black letter Saints' Days. In the case of the red letter days it was clearly the desire to bring the festivals to the test of the Bible. 1919 *Granta* 1 May 4/1 January 18th, 1889 should be a red-letter day in the history of Cambridge University, for it was on that date that *The Granta*, like a new planet, swam into our ken. 1965 C. E. Pocknee *Parson's Handbk.* (ed. 13) xvi. 145 The 1928 Book classifies days as holy, special, and ordinary. To *Holy* (or *Red-letter*) *Days* 1928 has added St. Mary Magdalene, 22 July, and the Transfiguration of our Lord, 6 August.

red-letter *v.* (later example).

1940 A. Upfield *Bushranger of Skies* xiv. 160 The history of his life was red-lettered with luck.

red light. 1. A warning light, esp. one instructing traffic to stop. Hence *fig.*, a sign of danger; a warning; a signal to pause or desist in some course of action or thought; esp. in phr. *to see the* (or *a*) *red light.*

1849 C. Brontë *Shirley* III. iii. 44 He is one of Mrs. Yorke's warning-examples—one of the blood-red lights she hangs out to scare young ladies from matrimony. 1862 Anon. *Railway Traveller's Handy Bk.* 99 *Danger, to stop*, is shown by a red light fixed upon a pole being turned full upon the line... *Caution, to go slowly*, is shown..by a green light. *All right, to go on*,..by a white light. *c* 1864 Brough & Halliday in M. R. Booth *Eng. Plays of 19th Cent.* (1973) IV. 239 There, that's our signal—that's the red light on our railway, and means 'danger'. 1907 A. Quiller-Couch *Major Vigoureux* xxiii. 234, I fancy the man has begun to see the red light. 1927 *Daily Tel.* 15 Nov. 9/3 Men see a red light when they find that things they have called their own—like intelligence—may be given to women too. 1931 E. Wallace *On Spot* x. 120 If Con had been better acquainted with Tony he would have seen the red light. 1938 *Mag. Digest* Jan. 66 (*heading*) A red light for the pugnacious. 1946 *Sun* (Baltimore) 29 Jan. 11/1 What in fact the Associated Press has done is to put up the red light. 1948 G. V. Galwey *Lift & Drop* vii. 190 Roberts' carelessness in getting pinched showed them the red light and they ceased operations. 1958 *Spectator* 28 Feb. 251/1 The French Government may see the red light. 1974 E. Ambler *Dr. Frigo* II. 97 He's hit by a drunk driver running a red light and killed. 1977 *New Yorker* 24 Oct. 52/3 He drove through a red light in Pennsylvania and..was picked up by police.

2. The sign of a brothel. Freq. *attrib.*, as *red light district*, etc.

1900 *N.Y. Jrnl.* Nov. 19 (*caption*) Children of the 'red light district'. 1900 *Boston Transcript* 4 Dec. 14/3 The disorderly houses in the 'red-light' district were all closed last night. 1925 H. L. Foster *Trop. Tramp with Tourists* 276, I was amazed to find that about two thirds of every city consisted of red-light district. 1928 *Daily Express* 26 Sept. 11/1 It is further asserted that he has elevated the 'red light' houses to the level of a business in New York. 1947 [see *geisha]. 1951 E. Paul *Springtime in Paris* xi. 205 Strangers in cities of Spain or Italy, who want to locate the red-light district, are advised to seek a point equidistant between the principal seminary and the main barracks of the Guardia Civil. 1962 *Coast to Coast 1961–62* 131 Harry was rough, tough and hairy, and he knew where to find a swy game, a sly-grog joint or a red light in every capital city. 1967 O. Wynd *Walk Softly* vi. 91 He is reported to have sizeable interests in the Naboshima Red Light district, and many bars. 1973 *Islander* (Victoria, B.C.) 17 June 16/3 At one time there were about 2,000 men in the camp, and except for the red-light girls at the other end of town, only two women. 1978 *Detroit Free Press* 16 Apr. 11A/2 Besides, she said, a legal red-light district would probably be bad for business. 1981 *Observer* 2 May 3/4 The trial will seem far removed from the red-light districts and suburbs where Sutcliffe struck.

3. A children's game in which one participant turns his back on the others, who try to sneak up on him without being seen moving.

1953 P. G. Brewster *Amer. Nonsinging Games* 35 Red Light (Mississippi). The player who is 'It' counts rapidly to ten and then cries, 'Red Light!'.. At the cry of 'Red Light!' each must stop and hold the position in which he was when he heard it. 1969 I. & P. Opie *Children's Games* vi. 195 Other names [for 'Peep behind the Curtain'] include.. 'Red Light' (Liverpool, Blackburn, Spennymoor, Peterborough, Helensburgh, and Edmonton, Alberta), [etc.]. 1975 *New Yorker* 10 Mar. 38/1 Evenings, screen doors bang behind children rushing out to meet each other for hide-and-seek, giant steps, dead dog, red light.

Hence **red-li·ghter,** a prostitute.

1913 A. Stringer *Shadow* 37 He could hobnob with bartenders and red-lighters.

red line, *sb. phr.* Used to describe the British army, esp. in phr. *thin red line*; also *transf.*

1855 F. Duberly *Let.* 8 June in E. E. P. Tisdall *Mrs. Duberly's Campaigns* (1963) v. 147 They advance, supported by the impenetrable red line, our infantry. 1877 W. H. Russell *Brit. Exped. Crimea* III. ii. 156 The Russians ..dashed on towards that *thin red line tipped with steel.* 1890 Kipling *Departmental Ditties* (U.S. ed.) 61 It's 'Thin red line of 'eroes' when the drums begin to roll. 1935 'G. Orwell' *Clergyman's Daughter* iv. 226 Napoleon Buonaparte..soon found that in the 'thin red line' he had more than met his match. 1971 *Guardian* 24 Feb. 10/1 Trade unionists..are beginning to give physical support to the Union of Post Office Workers as their 'thin red line'. 1974 *Times* 19 Apr. 14 Home of the thin red (or khaki) lines... 'Home of the British Army', say the road signs leading into Aldershot.

re·d-line, *v.* Also **redline, red line.** To circle or mark in red ink; freq. *fig.* (see quots.). Hence **re·d-lining** *vbl. sb.* See also *red-lined* s.v. Red *a.* 15 a in Dict. and Suppl.

1942 *Yank* 23 Sept. 14/1 Who is it the yardbird sees when he gets red-lined on the payroll for signing his name wrong? 1945 *Amer. Speech* XX. 261 To *redline* a soldier is to cross off his name on the payroll for a particular month because of an improper signature or some other irregularity on anyone's part. 1961 *Richmond* (Va.) *Times-Dispatch* 18 Jan. 4/1 The American Automobile Association may 'red-line' Prince George county because of its policy toward traffic violators. 'Red-lining' is a method used by the AAA to warn motorists of possible speed traps. 1973 *Times* 7 Feb. 20/4 They found that Laurelton had been 'red-lined' by the bank, which meant that it was not possible to get a normal mortgage. 1973 *Black Panther* X. XXVIII. 4/4 'Red-lining' is a corrupt scheme whereby real estate developers..arrange unofficial agreements with banks to have them refuse to grant improvement loans for particular buildings. Eventually the homes fall into a state of such disrepair that the owner is willing to sell it. 1976 *In Common* VI. II. 4/1 Common Cause Governing Board..voted CC support of legislation aimed at discouraging 'redlining'—the practice by lending institutions of discriminating arbitrarily against certain city neighborhoods in making home mortgage loans. 1977 *Listener* 9 June 763/1 The policy of so-called 'red-lining', that is, drawing a line round certain undesirable areas and refusing to lend money on property within the red line. 1979 *Verbatim* Summer 2/2 To redline an aircraft, thus, is 'to ground' it.

redly, *adv.*[1] Add: **b.** *Comb.*, as *redly-lipped*, *-squirting* adjs.

1910 J. Masefield *Ballads & Poems* 91 Maids that were redly-lipped and comely-skinned. 1930 R. Campbell *Adamastor* 90 Its steel-shot bulk with redly-squirting Nose and lolling tongue.

red man. Add: **2. b.** One of the extinct Beothuk people of Newfoundland.

1955 L. E. F. English *Historic Newfoundland* 9/2 They [sc. Beothuks] were described as of ordinary height—yet there are recorded instances of giant Red Men of seven feet in stature. 1969 H. Horwood *Newfoundland* xi. 78 The original Red Men, who, because of their attachment to red ochre, gave their nickname to all the other native tribes of North America.

4. = *Roman *sb.*[3] 2.

1966 E. Palmer *Plains of Camdeboo* xiv. 235 The Red Men come into the house at night... Of medium size—perhaps three inches with the legs fully spread..orange-red, hairy, fast, with great snapping beak-like jaws which, most horrifyingly, they can move at the same time up and down and from side to side.

Redmondite (re·dmǫndəit), *a.* and *sb.* [f. the name *Redmond* (see below) + -ite[1].] **A.** *adj.* Of or pertaining to the Irish politician John Edward Redmond (1856–1918) or his nationalist ideas. **B.** *sb.* A supporter of Redmond or his policies.

1895 *Notes from Ireland* 2 Feb. 47 The Redmondite manifesto. *Ibid.*, The past week has seen the issue of two familiar appeals..one from the McCarthyites, and one from the Redmondites. 1905 T. D. Sullivan *Recoll. Troubled Times in Irish Politics* xxxvii. 339 At the first blush one might suppose that the Tory party would scoff at the Redmondite motion. 1910 *Westm. Gaz.* 9 Feb. 2/2 He cannot even pretend he is entitled to a Redmondite summons. 1913 *Times* 30 Dec. 3/2 Mr John MacNeill denies that his friends are actuated by hostility towards the Redmondites. 1928 T. M. Healy *Lett. & Leaders of my Day* II. xxxi. 391 In the middle of the negotiations with the Liberal Cabinet as to the Home Rule Bill—which the Redmondites were certain to denounce as inadequate—Dillon tried to get control of the *Freeman*. 1962 Greene & Laurence in G. B. Shaw *Matter with Ireland* 142 Sinn Féin, which..had already won two by-elections against Redmondite candidates. 1973 S. Levenson *James Connolly* xvii. 208 He accused the Redmondites of accepting partition without a struggle. 1978 D. Murphy *Place Apart* ii. 17 My mother, whose ancestry was Redmondite at best.

Hence **Re·dmondism,** the ideas or policies associated with Redmond.

1914 R. Casement *Let.* 29 July in R. MacColl *Roger Casement* (1956) vii. 133 Redmondism has no real support here at all.

red-mouthed: see *Red *a.* 14 a, Red *a.* 14 b.

re·dneck. Also **red-neck, red neck. 1.** *U.S.* **a.** A member of the white rural labouring class of the southern States; one whose attitudes are considered characteristic of this class; freq., a reactionary.

Originally, and still often, derogatory, but now also used with more sympathy for the aspirations of the rural American.

1830 A. Royall *Southern Tour* I. 148 This may be ascribed to the *Red Necks*, a name bestowed upon the Presbyterians in Fayetteville. 1893 H. A. Shands *Some Peculiarities of Speech in Mississippi* 53 *Red-neck*,..a name applied by the better class of people to the poorer inhabitants of the rural districts. 1904 *Dialect Notes* II. 420 *Redneck, n.*, An uncouth countryman. 'The hill-billies came from the hills, and the *rednecks* from the swamps.' 1913 J. Davis *Life & Speeches* iii. 42 If you red-necks or hill billies ever come to Little Rock be sure and come to see me—come to my house. 1936 W. Faulkner *Absalom, Absalom!* 122 Rich and poor, aristocrat and redneck. 1959 *Times Lit. Suppl.* 28 Aug. 491/4 The ugly faces and, under prompting or provocation, the uglier actions of a handful of red-necks, crackers, tar-heels and other poor white trash here and there in the South. 1960 *Spectator* 15 Jan. 83/2 The old patrician families who are opposed to the graft, blackmail and demagogy by which the Boss, the tribune of the rednecks, keeps himself in power. 1969 *Observer* 7 Dec. 25/3 They [sc. communes] all shared two experiences: the search for new values, and attention from local rednecks and the police. 1971 J. Bishop *Days of Martin Luther King, Jr.* IV. 329 The fearful Southern red-neck, committed to the credo that the black man is a bridge between the animal kingdom and the human, derided the speech as typical 'coon shouting'. 1973 *Black World* Mar. 56 Carload of rednecks came with the darkness to Slim's house. Blew the horn until Slim's daddy opened the door. 1975 *Daily Tel.* 15 Oct. 17/7 Was it because they might think his [sc. Governor George Wallace's] reputation as a Right-wing 'red neck' a political embarrassment? 1976 *Time* 27 Sept. 47/1 That was the point Carter was attempting to make when he said in 1970 that Maddox 'has compassion for the little man', and when he said that a Humphrey-Wallace ticket in 1972 'would do well in the South', and when he called himself 'basically a redneck'. 1977 D. James *Spy at Evening* x. 71 Middle-class rednecks like you..get all worked up about it. 1978 J. Updike *Coup* v. 192 Her momma's a washrag and her daddy's a redneck.

b. *attrib.* or as *adj.*

1961 D. Alexander *Bloodstain* xi. 134 You should never have come out here alone. This is redneck country. Every man in every one of these houses is a Night Rider. 1965 *Listener* 20 May 730/2 His general manner and accent suggest a person who might hold the racist views of a red-neck Southern bigot. 1971 B. Malamud *Tenants* 60 'I' grows up in redneck Mississippi in pure black poverty. 1972 R. Bloch *Night-World* (1974) vii. 43 See how far you can march through Georgia today before some redneck sheriff busts you for vagrancy. 1973 *Freedomways* XIII. 52 Even Faulkner's ability was distorted by the pervasive racism of his redneck traditionalism. 1974 *New Yorker* 25 Feb. 102/3 He seems Southern redneck—a common man who works outdoors in the sun—to the soul. 1976 *National Observer* (U.S.) 17 July 4/1 Quite possibly Mississippi's only self-avowed redneck Republican. 1979 *Arizona Daily Star* 22 July 1. 5/4 Carter..ran on a virtually redneck platform for the 1970 nomination... After running a redneck campaign, [he] pledged an end to discrimination in his inaugural address.

2. (See Red *a.* 18 a.)

3. *S. Afr.* = *rooinek.

1900 A. H. Keane *Boer States* p. xviii, *Rooinek*, 'Red-neck', in reference originally to some merinos introduced by an English farmer into the Free State, and marked with a red brand on the neck. These were spoken of as *red-necks*—an expression afterwards extended to the English themselves, and then as a term of contempt to the British troops in red uniform. 1921, etc. [see *rooinek]. 1936 R. Campbell *Mithraic Emblems* 111 To find a red-neck cheap upon this day You do not need to wander far away. 1972 J. McClure *Caterpillar Cop* ii. 18 What's with this Red-neck?.. Another bloody English immigrant?

red-necked, *a.* Add: **a.** (Further examples.)

1782 J. Latham *Gen. Synopsis Birds* I. II. 558 (*heading*) Red-necked W[oodpecker]. 1847 P. H. Gosse *Birds of Jamaica* 334 The red-necked gaulin—*Egretta ruficollis*. 1926 Red-necked wallaby [see *brush wallaby* s.v. *brush *sb.*[1] 4]. 1971 *Country Life* 27 May 1292/3 The red-necked phalarope ..though still Ireland's rarest breeding bird, has shown an increase.

b. Holding redneck views; characteristic of a redneck (sense *1); conservative. orig. *U.S.*

1960 *Washington Post* 30 Apr. A6 'Uncle Earl' cavorted in typically 'red-necked' style at a fashionable luncheon. 1973 *Publishers' Weekly* 2 Apr. 58/2 A villain (red-necked political appointee with shady family connections to an equally red-necked but powerful Senator). 1977 *Arab Times* 13 Dec. 6/4 The white, red necked, blue collared worker is more interested in better wages, lower taxes, improved Social Security than justice in the Middle East.

redness. Add: (Further examples.) Also, the state or quality of being politically 'red'.

Under Chairman Mao the concept of 'redness' as opposed to 'expertness' was of great significance in the People's Republic of China.

1940 R. S. Lambert *Ariel & all his Quality* iii. 84 Press campaigns against alleged BBC 'redness'. 1973 R. Taylor *Educ. & Univ. Enrolment Policies in China, 1949–1971* 39 The subordination of expertise to ideology (or 'redness'). 1973 T. R. Tregear *Chinese* iii. 57 In the industrial field 'expertness' assumed greater importance than 'redness'. 1975 I. C. Y. Hsü *Rise of Mod. China* (ed. 2) xxvi. 796 The government in 1957 initiated a 'socialist education movement' among the industrial and agrarian population, followed by the dispatch of military and civil leaders to physical labor as an example to the people. The importance of 'redness', i.e., ideology over expertise, was very much emphasized. 1976 *Lancs. Evening Post* 7 Dec. 3/9 He grabbed witness by the collar, causing scratches to the neck and redness. 1981 G. Hammond *Revenge Game* xii. 137 In Keith's mind, a set of vague ideas solidified into a herring of most extraordinary redness.

‖ **redningskoite** (re·dniŋzkoi:tə, re·dniŋzʃoi:tə). Also **redningschoite, redningsshöite,** and with capital initial. [Norw., f. *redning* rescue + *skøyte* a type of fishing vessel.] A kind of Norwegian lifeboat. Also *attrib.*

1906 H. W. Smyth *Mast & Sail* 434 *Redningskoite*, a Norwegian sea-keeping lifeboat for assisting the Northland fishing fleets in bad weather. 1925 *Yachting Monthly* XXXVIII. 135/1 The original Redningschoites were designed, by the late Colin Archer, to be the best heavy weather boats in the world. 1935 C. E. T. Lewis *Lifeboats & their Conversion* ii. 39 The cruising lifeboats of Scandinavia and Russia as typified by the Norwegian ketch-rigged redningsshöites. The main duty of these craft is to keep in touch with the herring fleets throughout the winter, and render any help required. 1937 *Yachting Monthly* LXII. 508/1, I wanted to try out the conventional *redningskoite* sail plan. 1962 D. Phillips-Birt *Fore & Aft Sailing Craft* vii. 177 The redningskoite is illustrated in Figs 89, 90, and 91. The Norwegian term is rendered literally as 'rescue sailing ship'.

re·do, *sb.* [f. the vb.] A doing over again (in various senses); a repetition.

a 1953 Dylan Thomas *Quite Early One Morning* (1954) 79 The decorators were in at the mortuary, giving the old home a bit of a re-do like. 1961 D. B. Shields in *Webster* s.v., No pleasanter prospect than a redo of our South American trek. 1977 *Surgery* LXXXI. 41 (*heading*) 'Redo' surgery after operations for aneurysm and occlusion of the abdominal aorta.

redondite (rēdǫ·ndəit). *Min.* [f. the name of *Redonda* Island, W. Indies + -ite[1].] A hydrated phosphate of aluminium and iron, $(Al,Fe)PO_4.2H_2O$, occurring as whitish amorphous masses.

1870 C. U. Shepard in *Amer. Jrnl. Sci.* C. 96 Barrandite has gr. = 2·576. Redondite gives gr. = 2·019. 1964 *Amer. Mineralogist* XLIX. 445 X-ray powder data are given on redondite from Ponikla and Listenec, Czechoslovakia, from Redonda Island, and for tangaite from Tanganyika... Redondite (and tangaite) give patterns that are identical and distinct from those of variscite and clinovariscite... The authors feel that the name redondite should be used for all material of this type with Al > Fe and the name tangaite is superfluous. 1975 *Mineral Abstr.* XXVI. 225/2 Variscite, crandallite, and redondite, relatively rare phosphate minerals, are found in phosphatic rocks associated with carbonatites and syenites in the Grande islet (Brava island) Cape Verde.

redouble, *sb.*[2] *Bridge.* [f. next.] A call that redoubles a bid.

1906 'Cut-Cavendish' *Compl. Bridge Player* 98 The redouble is the *rara avis* of the Bridge world. 1910 J. B. Elwell *Auction Bridge* 103 The laws of Auction, as embodied for club play, limit the doubling feature to one double and one re-double. 1925 [see *business 21 d]. 1964 *Official Encycl. Bridge* 450/1 Ill-judged doubles of game or slam contracts may lead to redoubles. 1975 *Times* 20 Dec. 10/7 The Double of a No trump is primarily for a penalty: what does the opener mean to convey by a Redouble?

redouble, *v.*[1] Add: **b.** *spec.* in Bridge, to double again (a bid which an opponent has already doubled). Also *absol.* or *intr.*

1894 'Boaz' *Pocket Guide to Bridge* 6 The leader has then to ask the adversaries whether he or they wishes to redouble. 1898 [see *double *v.* 1 g]. 1910 J. B. Elwell *Auction Bridge* 102 The partner..may be well satisfied with the double, and, perhaps, in a position to redouble... The general Auction laws do not limit the number of times that a declaration may be doubled and re-doubled. 1921 *Sat. Westm. Gaz.* 1 Oct. 17/1 The player who doubles would not lose much if the rule was that *his* double could not be redoubled. 1980 *Times* 12 Jan. 10/6 South doubles for a take-out and West redoubles.

redoubled, *ppl. a.* Add: **1. c.** *spec.* in Bridge.

1964 *Official Encycl. Bridge* 450/1 When the standard of play is high, redoubled contracts are rare.

redoubler. 1. For $rare^{-0}$ read *rare* and add example.

1959 T. Reese *Bridge Player's Dict.* 182 At the slam level the odds in favor of the redoubler are better yet.

redoubling, *vbl. sb.* Add: **1. c.** *spec.* in Bridge.

1899 A. Dunn *Bridge* 63 Doubling and Redoubling... After the trump suit is announced, the adversaries may 'double' the value of the suit selected... The dealer and his partner may redouble. 1908 *Laws of Auction Bridge* §55 Doubling and redoubling affect the score only, and not the value in declaring. 1963 G. F. Hervey *Handbk. Card Games* 134 Doubling and redoubling do not..increase the size of a contract.

redoubt, *sb.* Add: **1. b.** (Later examples.)

1955 *Times* 16 May 3/4 Labour's most south-westerly redoubt is in acute danger. 1963 *Times* 30 May 13/2 It would be too embarrassing for all concerned if Ebbw Vale and the other Labour redoubts were to be presented on nomination day with official Labour candidates on top of the sitting 'independents'. 1970 R. Lowell *Notebk.* 72 Let's face it, English is a racist redoubt. 1977 *Time* 21 Mar. 10/3 Like his predecessors, he has tried—and failed—to overcome the age-old linguistic dispute that makes Belgium the staunchest redoubt of tribalism in Western Europe.

3. (Earlier example.)

1787 [see *assemblée].

red-out (re·daut). [f. Red *a.* and *sb.*; cf. *black-out.] A reddening of the vision resulting from an accumulation of blood in the head when the body is accelerated downwards.

1942 *Richmond* (Va.) *Times-Dispatch* 21 Dec. 13/1 The doctors learn airplane ambulance work... They learn about grayouts, blackouts and redouts. 1943 *Effects of Flight* (U.S. Navy Training Div., Bur. Aeronaut.) iv. 70 High numbers of *g*'s acting upward through the vertical axis of the body can be produced..and when these are excessive there may be the sensation of everything turning red, followed by unconsciousness. This has been called 'red-out'. 1946 R. A. McFarland *Human Factors in Air Transport Design* ix. 363 Values of −2 or −3*g* produce fullness and a throbbing pain in the head. The vessels of the eyes become congested, and reddening of vision, or 'red-out', may occasionally occur. 1962 F. I. Ordway et al. *Basic Astronautics* xii. 463 If the acceleration is applied in the opposite direction, then a condition known as red-out occurs... The blood..increases in weight to the point where the heart can no longer pump it. 1970 *Daily Tel.* (Colour Suppl.) 10 July 25/2 With a severe bout of negative G you get a red-out, when so many vessels burst that your vision becomes crimson and you see through a haze of blood. 1980 *Verbatim* Winter 19/1 Too many positive G's may cause loss of consciousness—a *red-out*.

Hence **red ou·t** *v. intr.*, to undergo or experience a red-out. See also *redding *vbl. sb.*[1] 2.

1955 M. Relfer *Dict. New Words* 175 *Red out*, to experience a red field of vision, congestion of the face and, in particular, the eyes, pressure in the head, etc., as a result of blood rushing to the head in certain aerial maneuvers, such as a rapid climb, inverted spins, outside loops, etc. 1980 *Verbatim* Winter 19/1 A pilot is in danger of redding out when an aerial maneuver puts a strong positive-G stress on his body.

redowa. In etym. read 'ad. Czech *rejdovák*, f. *rejdovati* to steer, manipulate (as with a carriage pole), to wheel about'. Substitute for def.: A Bohemian folk dance, in Western Europe developed into a dance in relatively quick triple time; the music for such a dance. (Earlier and later examples.)

1845 J. R. Planché *Graciosa & Percinet* II. 9 Charming Polka—Redowa Polka! Pink of Polkas thou'rt to me! 1855 E. R. Smith *Araucanians* vi. 65 The schottisches and redowas of the modern ball-room. 1885 A. Dodworth *Dancing* viii. 62 *Redowa*. When first introduced this dance had the time of a polka mazourka... Our beautiful waltz of to-day is a subdued redowa. 1903 W. Lamb *How & what to Dance* vii. 70 The Redowa, when first introduced, began with a promenade movement, but it is now generally commenced with the circular figure. 1960 P. J. S. Richardson *Social Dances of 19th Cent. Eng.* ix. 99 As late as 1894 or 1895 I can clearly remember being present at a popular assembly in London when the Redowa was announced, but only about three veteran couples..were able to perform it.

redox (rī·dǫks, re·dǫks). *Chem.* [f. Red(uction + Ox(idation.] A reversible reaction in which one species is oxidized and another reduced; usu. *attrib.*, indicating (some connection with) such a reaction, or a simultaneous oxidation and reduction, as *redox couple, electrode, indicator, potential, reaction*, etc.

1928 Michaelis & Flexner in *Jrnl. Biol. Chem.* LXXIX. 689 The success met on application of the theoretical fundamentals of oxidation-reduction potentials (hereinafter to be referred to by us as 'redox' potentials)..to organic redox systems..has given great impetus to more elaborate study of physiological materials. 1938 *Nature* 15 Oct. 723/1 Increasing use of redox indicators is being made with marked success in analytical chemistry. 1940 Glasstone *Text-bk. Physical Chem.* xii. 932 A reversible oxidation-reduction electrode..sometimes abbreviated to.. 'redox' electrode. 1959 *Engineering* 2 Jan. 25/2 A mercaptan modifier and redox catalyst are employed to obtain the desired molecular structure in the shortest possible reaction time. 1965 Phillips & Williams *Inorg. Chem.* ix. 312 Since galvanic cells consist essentially of two redox couples opposed one against the other, it is convenient to choose one couple arbitrarily and regard it as having zero potential. 1970 Ambrose & Easty *Cell Biol.* ii. 66 Note..that a favourable redox potential does not necessarily mean that a particular reaction will take place, but simply that it is thermodynamically feasible. 1971 *Nature* 4 June 311/2 An internal redox apparently occurs, and the electron moves from cation to anion. *Ibid.* 12 Nov. 89/1 The red colour of Mars suggests the presence of ferric iron which is not readily explained by the redox state implied by the second hypothesis. 1978 P. W. Atkins *Physical Chem.* xii. 362 In a redox titration a reduced form of an ion (e.g., Fe^{4+}) is oxidized by the addition of some oxidizer (e.g., Ce^{4+}).

redpoll[2], -polled. (Earlier and later examples.)

1891 J. Macdonald *Stephens's Bk. Farm* (ed. 4) III. VI. 416/1 The Norfolk and Suffolk Red Polled breed stands highest for dairying purposes. The Red Polls are handsome symmetrical animals of medium size. 1949 *Caribbean Q.* I. II. 36 A small herd of grade red poll cattle was maintained. 1970 G. E. Evans *Where Beards wag All* viii. 95, I have kept

Red Polls, because they are harmless, and the Suffolk breed for a Suffolk man.

red rag, red-rag. Add: **3.** (Earlier and later examples.)
1873 C. M. YONGE *Pillars of House* II. xviii. 151 Jack will do for himself if he tells Wilmet her eyes are violet; it is like a red rag to a bull. **1875** 'MARK TWAIN' in *Atlantic Monthly* Feb. 219 'What do you suppose I told you the names of those points for?' 'Well, to—to be entertaining, I thought.' This was a red rag to a bull. **1965** *Listener* 25 Nov. 874/3 Professor Allen has adopted more an 'inquiry programme' style to examine some of the sacred cows and red rags of American life.
4. *Naut. slang.* = *red ensign*, *red duster.
1910 D. W. BONE *Brassbounder* 129 Pluggin' a Dutchman's naethin'; it's th' 'Rid Rag' that Kelly's doon oan. **1929** D. J. MUNRO *Roaring Forties* 23 Up went Old Glory... We followed suit with the 'Red Rag' waving in defiance.

redraw, *v.*[2] Add: **4.** *trans.* To draw up again, to compose again.
1815 A. CONSTABLE *Let.* 26 Apr. in J. Constable *Corr.* (1962) I. 126 My father has had his *will redrawn*, leaving *out* the objectionable clause.

Red Republic. Also with small initials. [RED *a.* 9 b.] A republic based on socialist principles, *spec.* the French Second Republic, proclaimed in 1848. So **Red Republican,** one who holds radically republican views and advocates the use of force to realize them, esp. a supporter of the European revolutions of 1848; **Red Republicanism,** the principles and views of Red Republicans.
1848 *Illustr. London News* 1 July 415/1 The 'Red Republicans' have justified their name. They have filled the streets of Paris with blood... The working classes, or 'Red Republicans', were imbued with the doctrines of Communism. **1848** *Tablet* 2 Sept. 566/4 The red republic is now in a situation somewhat analogous to that of the Jacobins under Bonaparte. **1850** (*title of newspaper*) The Red Republican. **1850** E. P. WHIPPLE *Washington & Principles of Revolution* 29 We are proposing all those intricate problems which red republicanism so swiftly solves. **1857** C. M. YONGE *Dynevor Terrace* I. xx. 327 Raise our barricade for the rights of the Red Republic, and cry *La liberté, l'égalité, et la fraternité.* *Ibid.* II. ii. 23 All I am clear about is, that even a Red Republican is less red than he is painted. **1858** *N.Y. Tribune* 11 Jan. 2/5 When Mr. Bigler was in Kansas last summer, he was the known, open and enthusiastic advocate of what some hereabout call Red Republicanism. **1874** TROLLOPE *Phineas Redux* I. xxxiii. 276 One..advocates the personal government of an individual ruler, and the other that form of State which has come to be called a Red Republic. **1875** P. K. O'CLERY *Hist. Ital. Revolution* v. 185 The other [was] a soldier and nothing more, knowing nothing of politics beyond a rabid red-republicanism. **1960** R. K. WEBB *Harriet Martineau* xi. 322 Only Cavour was capable..of steering his way between the reactionary agents and the red republicans. **1972** T. COLEMAN *Passage to Amer.* xiv. 221 Kossuth came to America in 1851... He came to be seen as a continental red republican.

redressive, *a.* For *rare*[—1] read *rare* and add later example.
1965 H. KAHN *On Escalation* ii. 45 The first use of nuclear weapons..is likely to be less for the purpose of destroying the other side's military forces..than for redressive, warning,..or deterrence purposes.

redri·ll, *v.* [f. RE- 5 a + DRILL *v.*[4]] *trans.* To drill (crops or ground) again in the same season. So **redri·lling** *vbl. sb.*
1959 *Times* 5 Aug. 4/1 (Advt.), Aldrin wettable powder, used as a seed dressing, has proved outstandingly successful against the Black Cutworm.., eliminating time and money-wasting re-drilling and reducing handling and application costs. **1961** *Times* 24 July 6/6 Wheat bulb fly is in some years a very serious pest. In 1953 the estimated total loss in this country, taking into account crops which had to be redrilled or patched, was estimated at £1,250,000.

Red River. *Canad.* The name of a river flowing from North Dakota, U.S.A., to Lake Winnipeg, Manitoba, Canada, used *attrib.* in **Red River cart** *Hist.* (see quot. 1875); **Red River fever,** typhoid fever; **Red River frame** *Hist.*, used in a particular style of architecture (see quot. 1921); **Red River jig,** a fast, intricate jig.
1857 J. PALLISER *Jrnl.* 14 July (1863), The Red River cart is one admirably suited to the exigencies of the country. **1875** J. CARNEGIE *Saskatchewan & Rocky Mts.* ii. 13 Red River cart, a stout two-wheeled vehicle of the toughest quality though entirely made of wood, wheels, body, shafts, and all, being fitted together without a single particle of iron. **1968** [see *ox-train* s.v. *OX 5]. **1975** *Whig-Standard* (Kingston, Ontario) 21 Nov. 25/1 I've been here [*sc.* at Calgary] for three days and I haven't seen one buffalo, or a Red River cart or a redcoated Mountie or a saloon shoot-out. **1880** D. CURRIE *Lett. of Rusticus* 62/1 Red River fever was very prevalent at the Portage. **1889** J. G. DONKIN *Trooper & Redskin in Far North-West* iii. 41 In fact it is the typho-malarial scourge known as Red River fever in Manitoba, jungle fever in India,..and Rocky Mountain fever in British Columbia. **1945** K. M. HAIG *Brave Harvest* 40 Aunt Alice's heart sank. The Red River fever! **1966** M. KAVANAUGH *Assiniboine Basin* xxxv. 223 Fisher spent six weeks in the hospital suffering from typhoid—sometimes called Red River fever. **1882** *Edmonton Bull.* 29 July 4/1 They are of the style known as Red River frame and are for use as storehouses. **1921** *Beaver* Feb. 15/2 The 'big house' was a two-and-a-half storey building, with a large kitchen behind, built from the same plan as the officers' dwellings in Fort Garry, and known as a Red River frame building. **1963** MACLEOD & MORTON *Cuthbert Grant of Grantown* 93 There on the western limit of his seigniory, Cuthbert Grant built the great log house, in the Red River frame style. **1872** *Canad. Monthly* Oct. 305/1 The principal dance, in fact the only one, is called a Red River jig, which somewhat resembles a horn-pipe, male and female participating in it. **1930** L. MUNDAY *Mounty's Wife* iii. 44 The jigging by the men, known as the Red River jig, is really done very cleverly and is so swift. **1965** G. SHEPHERD *West of Yesterday* x. 74 Two compatriots came out on the floor and danced a Red River jig.

red rot. [f. RED *a.* + ROT *sb.*[1]] **a.** Fungal decay of standing trees or of timber characterized by red-brown rotted tissue; *esp.* that caused by *Trametes pini* in conifers.
1894 W. SOMERVILLE tr. *Hartig's Text-bk. Dis. Trees* 1 The damping-off of seedling beeches,..the red-rot of the spruce, &c., were known to foresters more than a hundred years ago. **1925** HAWLEY & HAWES *Man. Forestry* (ed. 2) vii. 127 The practice..of leaving for seed purposes trees diseased with red rot is unsafe because it tends to perpetuate the disease, not through the seed of the old trees, but by the fungus upon them. **1934** [see *CONK *sb.*[2]]. **1950** R. MOORE *Candlemas Bay* 119 The house was starting in to go right now, the way all the fine old houses went, as surely as if red rot had got into its solid beams and were eating it hourly down.
b. A fungal disease of sugar cane characterized by red patches within the canes.
1907 *Bull. Hawaiian Sugar Planters' Assoc. Exper. Station* vii. 25 The fungus is *Colletotrichum falcatum*, and the disease it causes is called the 'red rot' of the cane. **1928** F. S. EARLE *Sugar Cane* v. 154 The red rot fungus gains entrance to the stalks through borer holes and other injuries. **1944** *Phytopathology* XXXIV. 210 The lesions produced on the inoculated leaves were in all ways similar to lesions produced by virulent red-rot cultures. **1975** *Mycologia* LXVII 56 Perithecia produced by *C[olletotrichum] graminicola* most closely resemble those of *Glomerella tucumanensis*, the causal agent of the red rot of sugarcane... Carvajal and Edgerton..connected the conidial state, *C. falcatum*, with *P[hysalospora] tucumanensis*. The perfect state of this fungus was transferred by von Arx and Müller..to the genus *Glomerella*, and the name generally accepted today is *G. tucumanensis*.

red-shank(s, redshank. Add: **2. a.** *to run* (etc.) *like a redshank*: also in other dialects and *N.Z.* (further examples).
1873 E. WAUGH *Snowed-Up* v. 88 They're off like red-shanks! An' they'n come noan back to-neet, noather. **1891** G. CHAMIER *Philosopher Dick* 466 Without another thought or look behind us we were off like red shanks. **1901** W. F. BARRY *Wizard's Knot* vi. 53 Don't be running from us that way like a redshank. **1903** *N.Z. Illustr. Mag.* VIII. 93 They would be off into the bush like redshanks.

red shift, *sb.* Chiefly *Astr.* Also **red-shift, redshift.** [f. RED *a.* and *sb.* + SHIFT *sb.*] Displacement of spectral lines towards the red end of the spectrum; increase in the wavelength of electromagnetic radiation. Also *fig.*
1923 A. S. EDDINGTON *Math. Theory Relativity* v. 164 The red shift in the spiral nebulae. **1936** *Five Halley Lectures* 17 This telescope should penetrate to distances where the effects of red-shifts are so great that the requisite data will lie well above the threshold. **1958** *Listener* 11 Dec. 971/2 When we look at these distant regions we find that the light is reddened, indicating that the galaxies are receding from us. As far as we can see, the red shift of the most distant nebulae is still increasing linearly with distance. **1967** *Daily Tel.* 30 Jan. 10/4 The gigantic red-shift of quasars is therefore considered proof by some astronomers that they may be as much as six to eight thousand million light years away from our earth. **1971** *Sci. Amer.* May 56/3 Red shifts are commonly expressed as a fraction or percentage obtained by dividing the measured displacement of a line by the wavelength of the undisplaced line. **1973** A. GARNER *Red Shift* 155 Next time it'll be all right, every time, and it isn't. Next time will make up for him—and me. Never... Galactic. Red shift. The further they go, the faster they leave. The sky's emptying. God, this wind's cold. **1976** *Pract. Electronics* Oct. 793/3 During 1975 two teams of astronomers in the United States recorded a red-shift of $z = 0{\cdot}5240$ in the optical spectrum of a quasar. **1979** *Nature* 11 Oct. 498/1 Hyperpolarisation of the membrane causes a blueshift of the spectrum when probe is applied to the inside and a redshift for probe bound to the outer surface.
Hence **red-shifted** *a.*, exhibiting a red shift; also *transf.*; (as a back-formation) **red-shift** *v. trans.*
1963 *Nature* 16 Mar. 1041/1 Some broad lines..may be red-shifted hydrogen lines. **1964** *Listener* 20 Aug. 266/2 Matthews and Greenstein were able to identify the lines of 3C 48, which they found to be red shifted by 37 per cent. **1973** *Nature* 3 Aug. 264/1 As the 3 cm radiation propagates through an expanding Universe it will be redshifted. *Ibid.* 21/28 Dec. 517/1 The report..seems to provide a rare parallel to the visual pigments of the freshwater Osteichthyes, and to raise the question whether its pigment is a porphyropsin or a redshifted rhodopsin. **1976** *Pract. Electronics* Oct. 793/3 They set up their apparatus to scan for the 21 cm line at what could be its redshifted wavelength. **1979** *Nature* 5 July 20/1 The apparent colours of the faint galaxy identifications (presumably at redshifts of about 0·5) are considerably bluer than expected from redshifting the spectrum of nearby giant ellipticals such as M87.

red shirt, redshirt. 1. a. A supporter of Garibaldi, esp. one of the thousand who sailed with him in 1860 to conquer Sicily.
1864 YOUNG & STEVENS *Garibaldi: Life & Times* lxxv. 200 Naples had gone mad with joy: men, women, ragamuffins, priests, Redshirts, ex-Bourbon *sbirri*, *lazzaroni*,..—all lent their voices to..the general cry of '*Viva Garibaldi!*' *Ibid.* lxxvi. 202 Some of them mended their lives when Garibaldi came, and fought well in the ranks of the Red-shirts before Capua. [**1868**: in Dict. s.v. RED *a.* 4.] **1948** F. FRENAYE tr. *Levi's Christ stopped at Eboli* xvii. 168 When King Franceschiello had to leave Naples.., Garibaldi and his Red Shirts set out to attack him. **1979** *Guardian* 12 June 9/4 It took a march on Rome..—echoes of Garibaldi's thousand Red Shirts—to get water and elementary sewerage installed.
b. In more general use, a revolutionary, an anarchist, a communist.
1889 [in Dict. s.v. RED *a.* 18 a]. **1905** *Daily Chron.* 12 Sept. 3/2 Because I made a stand in my native town for municipal ownership of public utilities, I was branded a 'red-shirt', a 'dynamiter', and an 'Anarchist'. **1911** H. S. HARRISON *Queed* xviii. 234 Queed wrote a stinging little article..holding up to public scorn journalistic red-shirts who curry-combed the masses. **1934** T. S. ELIOT *Rock* i. 42 Enter *redshirts* in military formation. **1940** G. GREENE *Power & Glory* I. i. 13 You remember this place—before the Red Shirts came?
c. *spec.* A member of a Pathan nationalist organization formed in North-West Province in 1921 and lasting until the creation of Pakistan in 1947; also *attrib.* or as *adj.*
1930 *Civil & Milit. Gaz.* (Lahore) 1 June 1/5 In Mardan and Charsadda..the activities of the 'Red Shirts' have again increased. **1932** *Ann. Reg. 1931* 156 An organisation closely allied with Congress, the 'Red Shirt' Army under Abdul Ghaffar Khan, was a source of much anxiety in the North-west Frontier Province. **1948** G. CUNNINGHAM *Diary* in N. Mitchell *Sir George Cunningham* (1968) vii. 152, I could tell he felt he was on rather weak ground in talking about the Red Shirt activities by the twiddling of his bare toes. **1968** N. MITCHELL *Sir George Cunningham* v. 87 He records a recruiting meeting at Swabi..on 11th February, and the fact that four notorious ex-Red Shirts had publicly given him purses towards any war fund purpose.
2. *U.S.* A college athlete whose course is extended by a year during which he does not take part in university events, in order to develop his skills and extend his period of eligibility at this level of competition. Also *attrib.*, as *redshirt year*.
1955 *Life* 5 Dec. 144/2 Although he is what the pros call 'redshirt', a player with one more year of college eligibility, five pro clubs are eyeing him. **1970** *Time* 7 Dec. 78 He worked even harder in his sophomore year as a 'redshirt', practicing with the varsity but not playing in any games—so that he would have an additional year of eligibility. **1976** *Honolulu Star-Bull.* 21 Dec. H-2/1 Crowe was coming off a redshirt year last season while Bonup was going through one.
So as *v. trans.*, to keep out of university competition for a year for the above reasons; so **re·d-shirting** *vbl. sb.*
1950 *Birmingham* (Ala.) *News* 27 Sept. 35/1 He coached all the juniors and senior linemen and the boys red shirted. *Ibid.* 19 Nov. C1/2 There are not enough players to have a 'B' squad or red shirt promising sophomores. **1958** *Tuscaloosa* (Ala.) *News* 8 Jan. 6/8 An anticipated argument over 'red shirting'—the practice of holding athletes out of competition to prolong their eligibility—failed to develop. **1963** *San Francisco Chron.* 3 Dec. 44 He could have been red-shirted but he was an uncertain commodity last year. **1966** *Time* 14 Oct. 49 They [*sc.* the Big Ten] also are forbidden to 'red-shirt' prospects—putting them on a five-year program, keeping them out of action as sophomores in order to beef them up. **1968** *Daily Progress* (Charlottesville, Va.) 10 Apr. A10 Edwards supported continuation of red-shirting, a common practice that permits an athlete to use four years to complete three years of varsity eligibility.

redskin. Add: **2.** A variety of potato.
1908 *Chambers's Jrnl.* Oct. 702/1 The chief products grown are..potatoes... Tasmanian 'redskins' are..exported. **1973** *Courier & Advertiser* (Dundee) 21 Feb. 2/8 Scandinavia demands Redskin..and Maris Piper.

red snow. 1. (Further examples.)
1820 *Edin. Phil. Jrnl.* III. 307 (*heading*) Observations on the red snow of Mount St. Bernard. *Ibid.*, Some imperfect observations on the red snow of the Alps were made by M. Saussure in 1778, and the result of them appeared in the 3d volume of his Travels. **1894** J. W. MOORE *Meteorol.* xx. 236 Red snow and green snow have been observed in the Arctic Regions and elsewhere. **1912** *Q. Jrnl. R. Meteorol. Soc.* XXXVIII. 220 The red snow first attracted attention because the hoof-prints of the pack animals..were observed to be 'splotched with red as if the snow-crust had cut the mules feet and dyed the snow with drops of blood'. **1933** E. HAWKS *Bk. of Air & Water Wonders* vii. 145 Red snow was known to Pliny, and was attributed by him to a dust with which the snow became covered after it had lain for some time on the ground... The phenomenon of red snow is due to the presence of some genus of algæ, scientifically known as *Protococcus nivalis*. **1973** *Islander* (Victoria, B.C.) 17 June 5/2 The three explorers saw white ptarmigan, and the unique 'red snow' of the area.

red spot. 1. *Astr.* = *great red spot* s.v. *GREAT *a.* 20.
1879 *Monthly Notices R. Astron. Soc.* XL. 86 The very remarkable red spot which has attracted the attention of every observer during the present opposition. **1962** *Listener* 26 July 136/1 After 1882 the Red Spot began to fade, and since then it has undergone various changes. **1977** *Times Educ. Suppl.* 21 Oct. 21/2 Neither is it true to say that with regard to the Red Spot on Jupiter, 'scientists do not know what causes it'.

2. A defect of cheese in which there are fine red spots throughout.
[**1900** *Bull. N.Y. Agric. Exper. Station* No. 183. 189 The evidence seems to be conclusive that the red spots are produced by the growth of a minute plant which finds its way into the curd before it is put to press.] **1932** *Discovery* Feb. 59/2 The trouble in cheese known as 'open-ness' is being investigated and red spot in cheese and oiliness in butter have been studied with some success. **1955** J. G. DAVIS *Dict. Dairying* (ed. 2) 662 The peculiar fault known as 'red spot' in Cheddar cheese is due to an organism biochemically resembling the mastitis streptococcus which apparently lives a saprophytic existence in the udders of certain cows.

redstart. 1. a. Substitute for first part of def.: A small European and North African bird belonging to the genus *Phœnicurus* of the family Turdidæ, esp. *P. phœnicurus.* (Later examples.)
1925 C. E. RAVEN *In Praise of Birds* vi. 76 Not far off was a Redstart's nest in a piece of old iron piping. **1950** J. BUXTON *Redstart* xi. 132 Am I to describe as redstarts only those species which are placed in the genus *Phoenicurus*? **1973** T. SOPER *New Bird Table Bk.* iii. 31 Redstarts and woodpeckers are hole-nesters.
2. (Earlier and later examples.)
1731 M. CATESBY *Nat. Hist. Carolina* I. 67 The Red-Start... These Birds frequent the shady Woods of Virginia. **1947** R. T. PETERSON *Field Guide to Birds* (ed. 2) 208 The Redstart is one of the most butterfly-like of birds.

red-tape, red tape. Add: **b.** (Earlier and later examples.)
1736 LD. HERVEY *Poet. Epistle to Queen* in *Mem.* (1848) II. 156 Let Wilmington, with grave, contracted brow, Red tape and wisdom at the Council show. **1837** C. G. F. GORE *Stokeshill Place* I. vii. 142 My dear, you mistake John Barnsley... Dearly as he loves a bit of red tape, you never saw him try to inspire any other man with the love of business. **1938** *Daily Progress* (Charlottesville, Va.) 18 Aug. 4/1 The trial time has been about cut in half by the preliminary clearing away of red tape. **1956** A. WILSON *Anglo-Saxon Attitudes* I. i. 4 He resigned from the Labour Party and the House of Commons to fight your battle without the restraints of red tape. **1977** *Time* 14 Feb. 24/3 Straw-bossing the operation was Energy Chief James Schlesinger, who helped draft the natural gas bill and cut red tape to get gas supplies moving.
red-taped *a.*, also, restricted by red-tape (further examples); **red-ta·p(e)y** *a.* [-Y[1]] = RED-TAPISH *a.*
1889 *Columbus* (Ohio) *Dispatch* 21 Jan., Whether the newspaper reports are extravagant, or the official reports are too red-tapy or timid,..it is hard to say. **1905** D. SLADEN *Playing the Game* iv. 37 He'll be more red-tapey than ever, so as not to let the Japanese suspect anything. **1928** E. E. CUMMINGS *Enormous Room* iv. 61 The whole rotten red-taped Croix Rouge. **1954** *Sun* (Baltimore) 19 July 26/6 Red-taped—Jimmy Kennedy..wants to work but can't. **1975** *Daily Tel.* 21 July 5/8 There is..'a strictly Christian name approach' so that they are as different as possible to the 'red-tapey concept' of the social services.

red tapeworm. *joc.* [Blend of RED-TAPE and TAPEWORM.] Red-tapism or a red-tapist regarded as parasitic or as a disease of society.
1917 *Times* 26 Oct. 8/3 Send the papers to the Christchurch Museum in New Zealand as an example of the Red Tape-worm of England. **1918** *Studies* Dec. 664 The State ought to provide it; but the mechanism is clogged by the red-tapeworm. **1939** *Amer. Speech* XIV. 6 Others refer disparagingly to 'poluticians',..and 'red tapeworms'.

red-top. Add: **2.** (Earlier and later examples.) Substitute for def.: One of several pasture grasses, esp. *Agrostis stolonifera* or one of its varieties. Also *attrib.*
1790 S. DEANE *New-England Farmer* 115/1 The red top grass is so natural to every soil in this country, that all our old fields..are full of it. **1856, 1889** [see *FINETOP]. **1891** M. E. WILKINS *Humble Romance* 92 The whole yard..[was] covered with a tall waving crop of red-top. **1937** *Range Plant Handbk.* (Forest Service, U.S. Dept. Agric.) G8 It seems preferable to use redtop, as a generic name for most of the native range species of *Agrostis*. **1958** J. G. MACGREGOR *North-West of 16* v. 65 In bays, mostly on the west side and shaded from the afternoon sun, grew Red Top, the glory of the wild hay meadows. In places it was six feet high. **1972** *Daily Colonist* (Victoria, B.C.) 9 Mar. 5/3 The birds [*sc.* prairie chickens] would survive when redtop grass was raised, since it was combined in midsummer.

reduce, *v.* Add: **II. 17. c.** In mod. use, the opposite of OXIDIZE *v.* in Dict. and Suppl.; to cause to undergo reduction. (Further examples.)
1872 *Mining Mag. & Rev.* I. 250 When oxides are heated with carbon, the oxygen they contain combines with the carbon to form two invisible gases..which pass away into the air, the ore being thus 'reduced'. **1890** W. JAGO *Inorg. Chem.* iii. 48 There are other examples of reduction in which the bodies are simply reduced to a lower stage of oxidation. **1935** J. W. MELLOR *Comprehensive Treat. Inorg. & Theoret. Chem.* XIV. lxvii. 608 Cobaltic fluoride in hydrogen at 200°..is reduced to cobaltous fluoride. **1955** *Sci. News Let.* 7 May 297/2 Using heats as high as 3,100 degrees Fahrenheit, quartzite rock is reduced with coke and charcoal. **1971** *Nature* 1 Jan. 13/1 The resulting electrons are used to reduce carbon dioxide to carbohydrate. **1974** *Sci. Amer.* Dec. 68/3 The molecule that has lost electrons is said to have been oxidized; the one that has received them is said to have been reduced. Thus in photosynthesis water is oxidized and carbon dioxide is reduced.
III. 21. f. *Photogr.* To decrease the density of (a negative or print).
1889 E. J. WALL *Dict. Photogr.* 158 Bromide prints may be reduced in exactly the same way as negatives. **1903** A. WATKINS *Photogr.* 118 Do not throw away paper prints which are too dark from over printing. They can be reduced. **1956** *Focal Encycl. Photogr.* 950/1 Before starting to reduce a negative or print it should be well soaked in water. **1963** P. MOYES *Murder à la Mode* iv. 65 'What does "reducing" mean, exactly?' 'Makin' the print lighter... If the neg's too contrasty, like you can't get the light part to print without the dark's too dark. So you reduce it. By nibbin' the dark part with cyanide.'
22. (Later examples without const.)
1941 *Sun* (Baltimore) 17 Feb. 18/4 He was a top sergeant there. His cooking career began in France in July, 1918, when he was 'reduced' and made a mess sergeant. **1948** PARTRIDGE *Dict. Forces' Slang* 154 *Reduced*, reduced in rank. (Services' colloquialism verging on jargon.) **1953** K. TENNANT *Joyful Condemned* xxiii. 220 The deputy-governor..had been reduced..because of some trouble in a gaol of which he had been governor.
IV. 26. a. (Later examples.)
1903 E. A. ROSS in *Amer. Jrnl. Sociol.* IX. 197 There never has been a good reason for supposing we shall be able to reduce everything social to a single element. *Ibid.* 198 It is certain, nevertheless, we cannot reduce the whole man to a 'cell' in a 'social organism'. **1920** *Psychol. Rev.* XXVII. 71 The psychological simplification of human behavior, which reduces instinctive conduct to the functioning of psychical dispositions or impulses.
c. Also, to condense, come down *to*. (Further examples.)
1924 *Times Lit. Suppl.* 6 Nov. 704/2 His success or failure hangs..on the degree of intensity with which he fuses his material—and perhaps the old distinction between fancy and imagination reduces in the end to that. **1953** J. B. CARROLL *Study of Lang.* iii. 78 The problem of describing verbal behavior..reduces to the problem of describing the strengths..of verbal responses under various stimulus conditions. **1956** E. H. HUTTEN *Lang. Mod. Physics* iii. 109 The equations of motion for a material particle as given by the general theory reduce to the equations of motion of Newton, when we consider the simplest case of a Euclidean, limited, region of space. **1971** *Ideal Home* Apr. 69/1 The size of houses in sq. ft. has tended to reduce quite rapidly over the last few years. **1973** *Daily Tel.* 15 May 19 (Advt.), After only 8 years the amount you need to pay in cash will reduce and if present conditions continue you pay nothing after 10 years. **1978** *Amer. Polit. Sci. Rev.* Sept. 964/1 Over time those claims reduce to nothing more than rationalizations to maintain power. **1979** *Daily Tel.* 1 Dec. 27 (Advt.), The Company invests 98% to 113% of each payment (depending on your starting age), except in the first two years when these figures reduce to 73% to 89%.
d. To articulate (a speech sound) in a way requiring less muscular effort; to form (a vowel) in a more neutral, centralized articulatory position; to weaken, obscure.
1874 A. J. ELLIS *Early Eng. Pronunc.* IV. 1099/1 Reducing (r_0) from a consonant to a pure glide. *Ibid.* 1315 So that (oo') often falls into the juncture (AA), or else (ee', oo') are reduced to two syllables. **1892** W. W. SKEAT *Primer Eng. Etymol.* ii. 25 The *day* in *Monday* has been reduced to *-dy* (di) in familiar speech. **1909** D. JONES *Pronunc. Eng.* I. 46 Cases occur in which almost all other vowels may be reduced to ə when unstressed. **1934** C. DAVIES *Eng. Pronunc.* 12 Back-rounded vowels were unrounded and reduced to an indistinct sound similar to [ʌ] or [ə]. **1957** E. J. DOBSON *Eng. Pronunc. 1500–1700* II. x. 871 The unstressed back vowels seem not to have been reduced to [ə] as early as ME *ĕ*. **1962** A. C. GIMSON *Introd. Pronunc. Eng.* vii. 143 In present RP the secondary accent has been lost and the former [e] or [ɛ:] reduced to [ə] or elided.
e. *intr.* To lessen one's weight, to slim.
1926 MAINES & GRANT *Wise-Crack Dict.* 12/2 *Perpendicular your outline*, reduce a little. **1929** E. LINKLATER *Poet's Pub* xii. 145 'And how did they reduce?' asked Jean... 'They perspired without shame,' said the professor. **1958** *Times Lit. Suppl.* 13 June 334/5 A commendably simple and, at the same time, reliable guide for those who wish to 'reduce' without too much trouble. **1963** R. WOLFF *I, Keturah* (1964) II. xvii. 230 Miss Hawthorne said abruptly 'I think you ought to reduce, Keturah.' So she bought a book on dieting. **1971** *Homes & Gardens* Sept. 65/3, I try continually to reduce, but you cannot take a couple out for a gay evening and be on a diet.

reduced, *ppl. a.* Add: **4. e.** Mathematically modified to a more convenient form.
1862 E. ATKINSON tr. *Ganot's Elem. Treat. Physics* X. ix. 724 The resistance offered by the element and galvanometer is equal to the resistance of 4·08 yards of such copper wire, and this is said to be the reduced length of the element and galvanometer in terms of the copper wire. **1916** W. C. McC. LEWIS *Syst. Physical Chem.* II. iii. 90 It is more convenient to make use of the reduced form [of van der Waals' equation], i.e. pressures, volumes, and temperatures will be expressed as fractions..of their critical values. **1930** L. BRAND *Vectorial Mech.* xiv. 445 This is the same as the equation..for a simple pendulum of length $l = k^2/b$. For this reason l is called the reduced length of the pendulum. **1934** H. E. WHITE *Introd. Atomic Spectra* ii. 34 The preceding equations will apply to two masses m and M rotating about their center of mass, if m be replaced by $mM/(m+M) = m/(1+m/M) = \mu$. μ is called the reduced mass and approaches m as $M \to \infty$. **1950** CORBEN & STEHLE *Classical Mech.* iv. 54, μ is called the reduced mass of the system and is given by $\mu = m_1m_2/(m_1+m_2)$. **1973** *Sci. Amer.* July 25/2 The 'reduced', or simplified, form of the horn equation shows similarly that at any point in the horn the acoustic wavelength depends on the square root of the difference between the squared frequency and a 'horn function' U that depends in a rather simple way on the nature of the horn flare.
6. b. *Phonetics.* Of a vowel sound: articulated less distinctly than a stressed vowel; weakened and centralized; become more obscure than the vowel of which it is a reflex.
Much used in descriptions of the reduced grade of ablaut theory in Indo-European philology.
1894 V. HENRY *Short Compar. Gram. Eng. & German* I. iii. 76 To a normal grade *ĕy*, *ĕw*, there corresponds a reduced grade *ĭ*, *ŭ*. **1909** O. JESPERSEN *Mod. Eng. Gram.* I. xv. 423 Besides this 'full' [ə] we have a reduced [ə]. **1938** *Language* XIV. 41 Statements that the reduced grade preceded the accent and the zero grade followed it conflict with some of the best established and most pervading ablaut schemes. **1957** E. J. DOBSON *Eng. Pronunc. 1500–1700* II. x. 868 Gil uses *ė* as the symbol for the reduced vowel from ME *ŭ* in *oner* 'honour'. **1962** A. C. GIMSON *Introd. Pronunc. Eng.* vii. 120 As the great variety of spellings indicates, /ə/ may represent the reduced (obscured, 'schwa') form of any vowel or diphthong in an unaccented position. **1964** *Language* XL. 156 Indo-European reconstructions would require a consideration of laryngeals or reduced-grade vowels or both. **1968** CHOMSKY & HALLE *Sound Pattern Eng.* 28 The segment represented as ə will be referred to as the 'reduced vowel'. **1972** A. A. PRINS *Hist. Eng. Phonemes* i. 38 Indo-European had the following vowel system:..Reduced Vowels: $ə_1$, $ə_2$.
c. *spec.* Lowered in price.
1939 JOYCE *Finnegans Wake* (1964) I. 166 A real fur, reduced to 3/9. **1941** E. BOWEN *Look at all these Roses* 8 She came up to London..and bought reduced coats and shoes for the little girls. **1975** M. KENYON *Mr Big* xviii. 174 It's *reduced.* And it really fits.

reduceless (rĭdiū·slės), *a.* [f. REDUCE *v.* + -LESS.] Incapable of reduction, that cannot be lessened.
c **1864** E. DICKINSON *Poems* (1955) II. 641 As an Estate perpetual Or a reduceless Mine. **1954** W. FAULKNER *Fable* 16 They wore..the same regimental numerals, to the rest of the regiment which had not only preceded them by that reduceless gap but which had even seemed to be fleeing from them.

reducend. For † *Obs.* read *rare* and add later examples.
1847 G. BOOLE *Math. Anal. of Logic* 45 Reducend Mood, *Baroko.* **1880** W. H. S. MONCK *Introd. Logic* xvi. 181 Both extremes of the Reducend Syllogism occur in its conclusion.

reducer. Add: **1.** (Further example.)
1923 W. S. MAUGHAM *Our Betters* I. 23 He's the great reducer... What does he reduce?.. The Duchess of Arlington told me he'd taken nine pounds off her.
2. c. For *Photogr.* read *Chem.* (cf. sense 2 e below). (Further examples.)
1935 J. N. FRIEND *Text-bk. Physical Chem.* II. vii. 302 When E_0 is negative the ion in the lower state of oxidation..functions as a reducer or reductant. **1973** *Sci. Amer.* Oct. 128/2 The oxidant, hydrogen peroxide, is stored in a plastic bag within the pressure vessel and mixed in the exit opening with reducer in the soap.
e. *Photogr.* A chemical used to reduce the density of a print or negative.
1897 E. J. WALL *Dict. Photogr.* (ed. 7) 500 Belitski's Reducer. This is the most convenient as it is one solution. *Ibid.* 501 If the negative has been dried, soak in water till wet, then cover with the reducer. **1905** *Westm. Gaz.* 25 Feb. 14/2 This [*sc.* Farmer's] reducer is compounded of potassium ferricyanide solution, and a solution of hypo. **1911** [see *BLEACHER 4]. **1960** G. A. GLAISTER *Gloss. Bk.* 342/2 Proportional reducers exert a uniform action on the negative, thus preserving the tone. **1977** J. HEDGECOE *Photographer's Handbk.* 242 Persulfate reducer..makes a better reducer for negatives which are over-developed and therefore need less density and contrast.
f. A means of reducing one's weight.
1903 E. SANDYS in *Athletics for Women* 99 For those able to dispense with a few pounds of surplus adipose tissue, it [*sc.* swimming] is one of the best of reducers.
g. *Printing.* (See quot. 1968).
1963 KENNEISON & SPILMAN *Dict. Printing* 161 *Reducers*, the addition of varnish, boiled oils, etc., to printing inks to enable them to work more easily. **1968** *Gloss. Terms Offset Lithogr. Printing* (*B.S.I.*) 26 *Reducer*. 1. A liquid, miscible with ink, used to reduce its consistency. 2. A substance for addition to a printing ink to reduce its colour strength without necessarily affecting its consistency.

reducibility. Add: spec. in *Logic*, as *axiom of reducibility* (see quot. 1952).
1910 WHITEHEAD & RUSSELL *Principia Math.* I. 168 (*heading*) The hierarchy of types and the axiom of reducibility. *Ibid.* 174 We assume, then, that every function of one variable is equivalent, for all its values, to some predicative function of the same argument. This assumption seems to be the essence of the usual assumption of classes; at any rate, it retains as much of classes as we have any use for... We will call this assumption the *axiom of classes*, or the *axiom of reducibility*. **1930** L. S. STEBBING *Mod. Introd. Logic* xxiii. 463 Ramsey has suggested a reconstruction of the system of *Principia Mathematica* in which the axiom of reducibility is no longer needed. **1942** D. D. RUNES *Dict. Philos.* 266/2 As an indication or rough description of the axiom of reducibility, it may be said that it cancels a large part of the restrictive consequences of the prohibition against impredicative definition..and, in approximate effect, reduces the ramified theory of types to the simple theory of types. **1952** S. C. KLEENE *Introd. Metamath.* iii. 44 To escape this outcome, Russell postulated his *axiom of reducibility*, which asserts that to any property belonging to an order above the lowest, there is a coextensive property (i.e. one possessed by exactly the same objects) of order 0. **1963** W. V. QUINE *Set Theory & its Logic* xi. 251 The axiom of reducibility regales us after all with attributes unspecifiable except by quantifying over attributes whose order is as high as their own.

reducible, *a.* **4. a.** (Later examples.)
1932 LEWIS & LANGFORD *Symbolic Logic* ix. 282 Functions like this one, which can be expressed in equivalent form by means of functions of a lower degree of generality, will be said to be *reducible*. **1967** R. A. GEORGE tr. *Carnap's Logical Struct. of World* I. 6 An object (or concept) is said to be reducible to one or more other objects if all statements about it can be transformed into statements about these other objects.
b. (Later examples in *Chem.*)
Absence of a const. is usual in this sense.
1903 H. C. JONES *Elem. Inorg. Chem.* xxii. 281 Cobaltous oxide..is a greenish powder, easily reducible to the metal. **1957** G. E. HUTCHINSON *Treat. Limnol.* I. xi. 707 Some of both oxidizable and reducible iron was no doubt in organic combination. **1976** *Nature* 15 Jan. 147/1 These bulk membranes, being less dense than water, separated aqueous reducing agents and potentially reducible substrates.

reducing, *vbl. sb.* Add: **2.** (in sense *26 e of the vb.) *reducing belt, pill, treatment*; **reducing gear** = *reduction gear* s.v. *REDUCTION 13; **reducing machine,** an apparatus for producing scale models.
1928 A. HUXLEY *Point Counter Point* v. 69 No idea how comfortable those rubber reducing belts are till you've tried them. **1917** E. BUTLER *Transmission Gears* iii. 66 The first use of a planetary reducing gear as a transmission between a petrol motor and the driving wheel of an automobile, was made in 1888–9 by Edward Butler on his petrol cycle. **1947** J. C. RICH *Materials & Methods of Sculpture* vi. 191 In using platinum or palladium for medals, the original model is first cast in a hard bronze, and by means of a reducing machine a small steel die is cut of the precise size of the medal. With this die the copies are struck. **1968** *Canad. Antiques Collector* July 10/1 Benjamin Cheverton's 'reducing machine' had been employed for the scale model. This ingenious machine kept proportions exact, so that a six-inch statuette would lose nothing of the artistry of the original. **1955** W. GADDIS *Recognitions* II. vii. 608 I'm taking *scads* of these marvelous reducing pills that simply take your appetite away. **1907** F. H. BURNETT *Shuttle* xix. 195 Mina is growing fat, and spends her days in taking reducing treatments.

reductant (rĭdʌ·ktănt). *Chem.* [f. REDUCT(ION + -ANT[1], after OXIDANT.] A reducing agent.
1925 *Chem. Rev.* II. 128 The heroic efforts to measure the potential of a pure solution of a reductant are evidence of the tenacity of a preconception. **1935** [see *REDUCER 2 c]. **1968** J. MARCH *Adv. Org. Chem.* xix. 853 In some cases both the oxidant and reductant are organic. **1976** *Sci. Amer.* July 71/2 The process is termed gaseous direct reduction or solid direct reduction depending on the state of the reductant.

reductase (rĭdʌ·ktēⁱz). *Biochem.* [ad. F. *réductase* (M.-E. Pozzi-Escot 1902, in *Bull. de la Soc. Chim. de Paris* XXVII. 559), f. *réduct-ion* REDUCTION: see *-ASE.] **a.** Any enzyme which promotes chemical reduction.
1902 *Jrnl. Chem. Soc.* LXXXII. I. 655 The author [*sc.* Pozzi-Escot] has discovered a new class of diastases which he calls 'reductases'. **1914** [see *DEHYDRASE a]. **1938** *Ann. Rev. Biochem.* VII. 112 The protein of acetaldehyde reductase has been obtained in a crystalline condition. **1974** *Nature* 13 Dec. 579/1 The enzymes responsible for the first step in nitrate assimilation and for nitrate respiration are the nitrate reductases, both these processes involving the conversion of nitrate to nitrite. **1977** *Proc. R. Soc. Med.* LXX. 617/1 Thurnham..measured the erythrocyte glutathione reductase activity.
b. reductase test, a method of estimating the bacterial content of a sample of fluid, usu. milk, by measuring its reducing power.
1910 *Analyst* XXXV. 207 (*heading*) Reductase test for milk. **1932** *Discovery* Feb. 59/2 In the north of Europe..general use is made of the 'reductase' test, which takes the time required to decolourize a definite solution of methylene blue as a measure of the purity of milk. **1964** *Biol. Abstr.* XLV. 3876/1 Application of the reductase test to 261 randomly selected staphylococcus strains showing a positive plasma coagulase test yielded reductase times of 15 minutes or less in 96·6% of the cases.

‖ **reductio** (rĭdʌ·ktiō). Pl. **reductiones** (rĭdʌktiōᵘ·nīz). [L., = REDUCTION.] Used in various Latin phrases: **1. reductio ad impossibile,** reduction to the impossible: a method of proving a proposition by drawing an absurd or impossible conclusion from its contradictory.
1552 T. WILSON *Rule of Reason* (ed. 2) f. 56 The other croked waye (called of the Logicians, *Reductio ad impossibile*) is a reduccion to that, whiche is impossible. **1843** J. S. MILL *Logic* I. II. iii. 265 We shall thus discover the error in our generalisation, by what the schoolmen termed a *reductio ad impossibile*. **1869** W. S. JEVONS *Substitution of Similars* 44 This indirect or negative method is closely analogous to the *indirect proof*, or *reductio ad absurdum*, so frequently used by Euclid and other mathematicians, and a similar method is employed by the old logicians in the treatment of the syllogisms called *Baroko* and *Bokardo*, by the *reductio ad impossibile*. **1884** J. N. KEYNES *Formal Logic* III. iv. 181 This method of reduction is called *Reductio ad impossibile*, or *Reductio per impossibile*, or *Deductio ad impossibile*, or *Deductio ad absurdum*. **1962** W. & M. KNEALE *Devel. Logic* i. 8 What Aristotle attributed to Zeno was presumably the discovery of the use of the *reductio ad impossibile* in metaphysics.
2. reductio ad absurdum: reduction to the absurd (see REDUCTION 9 b). Also with superl. **reductio ad absurdissimum** (the most absurd, the greatest absurdity).
1741 I. WATTS *Improvement of Mind* I. xiii. 181 *The Respondent* may be attack'd either upon a Point of his own Concession, which is call'd *Argumentum ex concessis*, or by reducing him to an Absurdity, which is call'd *Reductio ad absurdum*. **1824** J. S. MILL in *Westm. Rev.* II. 34 This we admit: and we regard it as a decisive *reductio ad absurdum* of his own argument. **1865** —— *Exam. Hamilton's Philos.* iv. 44 Hegel..has fairly earned the honour..of having logically extinguished transcendental metaphysics by a series of *reductiones ad absurdissimum*. **1896** G. B. SHAW *Our Theatres in Nineties* (1932) II. 170 Madame Sarah Grand's position is a *reductio ad absurdum* of our whole moral system. **1931** *Times Educ. Suppl.* 10 Jan. 9/4 A *reductio ad absurdissimum* is seen in the fact that an aspirant for employment in a business house, who has actually qualified for exemption from matriculation by virtue of the subjects endorsed on his school certificate, is not infrequently informed that his application cannot be entertained unless he produces the actual matriculation certificate. **1939** *Canadian Forum* July 126/2 It..may prove to be no more than the *reductio ad absurdum* of his own introspectiveness. **1955** *Times* 23 June 11/3 When the House of Commons regards even nationalized boards as little more than civil servants, and piles a committee on top of their committee we approach the *reductio ad absurdum*. **1963** [see *Christocentricity* s.v. *CHRISTO-]. **1969** L. RUBY *Art of Making Sense* (rev. ed.) viii. 84 Here is a more complex sample of the reductio ad absurdum. **1976** *Times Lit. Suppl.* 3 Dec. 1522/4 Whether the result represents anything more than a reductio ad absurdum of the traditional musicological pastime of theme-spotting. **1977** *Time* 7 Feb. 51/1 *Reductio* is always *ad absurdum*.

reduction. Add: **II. 6. e.** *transf.* The process of explaining behaviour, social or mental activity, etc., by reducing it to its component factors or to a simpler form; also by ascribing a complex result to the operation of a few or one of its factors.
1916 A. A. BRILL tr. *Freud's Wit & its Relation to Unconsc.* ii. 28 The briefest reduction of the meaning by which one could replace this joke would be..[etc.]. **1927** W. M. WHEELER in *Proc. 6th Internat. Congr. Philos., 1926* 34 The reduction of these new properties to those of the parts in the sense of identification, and the finding of a causal determination also in this sense is impossible. **1928** H. G. & C. F. BAYNES tr. *Jung's Contrib. to Analytical Psychol.* 56 When the unsuitable structures have been reduced, and the natural course of things restored, the possibility of a normal life being thus attained, reduction is not to be pushed further. **1950** H. HARTMANN *Ess. Ego Psychol.* (1964) vi. 112 We may refer to such simplifications as 'theories by reduction'. They see one specific phase..as the sole causative factor for a character type. **1960** J. STRACHEY in *Freud's Compl. Wks.* VIII. 23 Here and elsewhere in this work Freud uses the word 'reduction' in the sense of taking something back to its original form. **1960** R. F. C. HULL tr. *Jung's Gen. Aspects of Dream Psychol.* in *Coll. Wks.* VIII. 240 Obviously this reduction is quite unsatisfying from the scientific point of view... The discovery of a single antecedent is by no means sufficient.
f. *Computers.* The transformation of data into a simpler or more amenable form.
1958 *Jrnl. Assoc. Computing Machinery* V. 89 (*heading*) Special purpose analog machine for data reduction. **1969** P. B. JORDAIN *Condensed Computer Encycl.* 423 Reduction can take several forms: changing the encoding to eliminate redundancy, or extracting significant details from the data and eliminating the rest or choosing every second or third out of the totality of available points.
7. b. *reduction sentence*: a sentence giving conditions for the use of a concept less strict than a definition.
1936 R. CARNAP in *Philos. of Sci.* III. 441 We shall call R_1 and R_2 reduction sentences for 'Q_3' and '$\sim Q_3$' respectively. **1949** [see *non-dispositional* s.v. *NON- 3]. **1963** A. PAP *Introd. Philos. of Sci.* ii. 32 The virtue of the reduction sentence, then, is that it permits us to ascribe a disposition *D* to an object only if the relevant experiment has been performed and found to have a positive outcome. **1965** P. CAWS *Philos. of Sci.* viii. 54 Carnap..closes the loophole in the definition by rewriting it as a *reduction sentence*.
10. a. Also in *Chem.*, the opposite of OXIDATION (senses a and b in Dict. and Suppl.): the removal of oxygen from, or addition of hydrogen to, a compound; partial or complete donation of an electron to an atom or molecule; a decrease in the proportion of electronegative constituents in a molecule or compound. (Further examples.)
1900 W. A. SHENSTONE *Inorg. Chem.* 177 The terms oxidation and reduction are no longer confined to changes in which oxygen plays a part... The term reduction may be applied to any change which involves a decrease in the relative amount of the negative radicle present in a compound. **1913** J. B. COHEN *Org. Chem. Adv. Students* II. ii. 100 The difference between the two catalysts is also brought out in the case of heptine C_7H_{12}, copper giving heptene C_7H_{14} and polymerisation products.., and nickel effecting complete reduction to heptane. **1930** L. B. FLEXNER tr. *L. Michaelis's Oxidation-Reduction Potentials* 10 We shall simply collect together as equivalent processes (1) the addition of oxygen, (2) the loss of hydrogen, and (3) the loss of electrons and call them all oxidations, and their converses, reductions. **1950** N. V. SIDGWICK *Chem. Elements* II. 1327 Ferrous compounds are formed either from the metal or by the reduction of the ferric. **1964** N. G. CLARK *Mod. Org. Chem.* x. 177 Aldehydes readily undergo reduction to alcohols. **1970** AMBROSE & EASTY *Cell Biol.* ii. 62 Reduction is regarded conversely as involving a gain of electrons, and substances which have the characteristic of giving up electrons to other substances are called reducing agents. **1979** *Archaeology* July–Aug. 21/2 When there is a lack of oxygen in the 'reduction' the iron oxide constituent of the earthenware clay remains in its ferrous or black state.
d. *Philos.* In phenomenology, the process of reducing an object of consciousness or an idea to its pure essence through elimination of all reference to extraneous things, in particular by eliminating (or 'bracketing') all reference to the real world of material objects; esp. as *eidetic, phenomenological, transcendental reduction*. (See quot. 1943.)
1914 *Mind* XXIII. 590 He [*sc.* Husserl] deals in a most valuable section with the relation of consciousness to natural reality, with the province of pure consciousness, and with the phenomenological reductions. **1924** *Monist* XXXIV. 520 To carry out the 'phenomenological reduction', i.e., to isolate an object from its existential or systematic connections, is equivalent to considering it as it is originally given, without the distorting influences of 'theory'. **1931** W. R. B. GIBSON tr. *Husserl's Ideas* 44 The corresponding Reduction which leads from the psychological phenomenon to the pure 'essence', or,..from factual ('empirical') to 'essential' universality, is the eidetic Reduction. *Ibid.* 114 We propose to speak..of phenomenological reductions... From the epistemological viewpoint we would also speak of transcendental reductions. **1943** M. FARBER *Found. Phenomenol.* i. 20 The 'reduction' opens up a universal field for philosophical investigation which is free from all pre-judgements and assumptions, hence its crucial methodological importance. Husserl is careful to distinguish eidetic reduction (proceeding from fact to essence) from transcendental reduction, according to which the phenomena are characterized as 'irreal'... The method of phenomenological reduction is applied in order to achieve the presuppositionless field of philosophy. **1966** A. GURWITSCH *Stud. Phenomenol. & Psychol.* v. 111 Performance of the transcendental without the eidetic reduction discloses the flow of 'my' transcendentally purified mental states in phenomenal time. **1970** A. GIORGI *Psychol. as Human Sci.* iii. 148 By means of the phenomenological reduction, i.e., by a change in attitude, the world can be considered as phenomenon. **1972** H. SPIEGELBERG *Phenomenol. in Psychol. & Psychiatry* ii. 76 He did not believe in a strict separation between the world of essences and the world of facts, thus rejecting Husserl's 'eidetic reduction'.
11. b. *spec.* of the size of a copy or photographic image in photography, microphotography, etc.
1889 E. J. WALL *Dict. Photogr.* 158 Whereas in enlargements the greater distance is between lens and sensitive surface, in the case of reduction the greater distance must be between the lens and negative. **1959** F. LUTHER *Microfilm* 1 Reductions greater than those now in common use were employed to produce microfilms that could meet the rule-of-thumb test of quality, that is, be enlarged back to original size without substantial loss of definition or legibility. **1962** A. GÜNTHER *Microphotogr. in Libr.* 5 The advantages of the use of photographic reduction in recording documents are so obvious that it is not surprising that the first microphotograph was made shortly after the invention of photography. **1965** *Focal Encycl. Photogr.* II. 1255/1 Reduction in printing, in copying and graphic arts work applies to reproduction at a scale of less than 1:1 or same size. **1973** D. A. SPENCER *Focal Dict. Photogr. Technol.* 518 A 1:3 reduction is a copy that is one-third of the linear size of the original.
c. *Photogr.* Diminution of the density of a print or negative.
1889 E. J. WALL *Dict. Photogr.* 158 Reduction will proceed in proportion to the amount of ferridcyanide [*sic*] present. **1902** A. WATKINS *Photogr.* 93 Where a rapid plate (of poor quality) does not seem to give sufficient contrast before it fogs over, a knowledge of reduction and intensifying will often give a good negative. **1956** *Focal Encycl. Photogr.* 951/2 The object of reduction is to make a very dense negative easier to print, or to lighten undesirably black areas of a print. **1977** J. HEDGECOE *Photographer's Handbk.* 242 Assess the progress of reduction by removing the negative at frequent intervals.
d. *Mus.* Transcription of a full orchestral score for a smaller number of instruments, esp. for piano; the reduced score thus produced.
1884 F. NIECKS *Dict. Mus. Terms* 273 *Riduzione*, reduction. **1966** [see *piano reduction* s.v. *PIANO *sb.*[2] 2 d]. **1973** L. LOCKWOOD in A. Tyson *Beethoven Stud.* 118 (*caption*) Reduction of *Sehnsucht*, bars 3–6. **1979** [see *open score* s.v. *OPEN *a.* (*adv.*) 22 c]. **1980** *Times* 28 June 9/7 Mahler songs, not in their familiar orchestral settings but in piano versions. Some are originals, some reductions, and some fall in between.
e. *Cytology.* The halving of the number of chromosomes per cell that occurs at one of the two anaphases of meiosis (cf. *POST-REDUCTION, *PREREDUCTION); chiefly *attrib.* in **reduction division,** the meiotic cell division during which reduction occurs. [The senses are due to A. Weismann, who used G. *reduktion, reduktionstheilung* (*Über die Zahl der Richtungskörper und über ihre Bedeutung für die Vererbung* (1887) i. 14, ii. 35).]

1891 *Jrnl. R. Microsc. Soc.* 461 There is a 'reduction-division', for twelve chromosomes are found in each new cell. **1896** E. B. WILSON *Cell* v. 182 The process of reduction is very obviously a provision to hold constant the number of chromosomes characteristic of a species. **1906** *Rep. Brit. Assoc. Adv. Sci. 1905* 570 Weissman predicted that a transverse division of the chromosomes would be found to take place by which the reduction would be brought about. *Ibid.*, A true reduction division is found to occur in the heterotype stage. **1927** HALDANE & HUXLEY *Animal Biol.* ii. 59 Instead of the chromosomes dividing, the members of a pair come to lie side by side; and at division one whole chromosome of a pair is separated from the other. This process is called the reduction of the chromosomes, for owing to it, each of the two cells produced at this division possess only half of the ordinary number of chromosomes for the species. **1931** E. B. FORD *Mendelism & Evolution* i. 12 Each chromosome must contain many factors. These will be inherited together for, at the reduction division, they will pass into the same germ cell without the opportunity of random assortment. **1948** H. P. RILEY *Genetics & Cytogenetics* iv. 65 The reduction division reduces the *number* of chromosomes and centromeres. **1971** D. J. COVE *Genetics* i. 7 This type of cell division whereby a diploid cell can give rise to haploid cells is called reduction division. **1979** *Sci. Amer.* Feb. 104/1 Meiosis, the 'reduction division' whereby one male germ cell divides to form four sperm cells, each of which has half the normal complement of chromosomes.
f. *Phonetics.* Weakening; obscuring (of a vowel); substitution of a sound which requires less muscular effort to articulate.
1909 O. JESPERSEN *Mod. Eng. Gram.* I. ix. 260 Weak /iu/ has in some words kept both sounds, though with an early reduction of /i/ to /j/. **1953** K. JACKSON *Lang. & Hist. Early Brit.* 293 The reduction of pretonic *ǫ* to *ö* in Welsh. **1959** C. L. WRENN *Word & Symbol* (1967) 39 *Gerrans*..shows the regular Cornish reduction of the final *t* to *s*. **1962** A. C. GIMSON *Introd. Pronunc. Eng.* vii. 120 This reduction of unaccented vowels, typical of a stress-accent language such as English, has been a feature of the English sound system for over a thousand years. **1970** B. M. H. STRANG *Hist. Eng.* vi. 342 The reduction of vowels is a sign that these formal distinctions were no longer functionally important.
g. Reducing or limiting the use of addictive drugs. Usu. *attrib.*
1914 JACKSON & HELLYER *Vocab. Criminal Slang* 69 *Reduction*.., the reduction cure for a 'habit'. Example: 'The only sensible way of getting off is on the reduction'. **1953** W. BURROUGHS *Junkie* iii. 39 The 'thirty-day cure'. This is not a reduction cure. They don't give any junk... All they offer the addict is thirty days' detention. *Ibid.* viii. 73, I have never known one of these self-administered reduction cures to work. **1962** 'K. ORVIS' *Damned & Destroyed* ii. 18 You've tried everything with your daughter, haven't you?.. Forced her to take cold-shock and reduction treatments.
13. reduction division: see sense 11 e above; **reduction gear** *Engin.*, a system of gear wheels in which the driven shaft rotates more slowly than the driving shaft; so **reduction gearing**; **reduction negative, print** *Photogr.*, a negative or print made from a larger original; so **reduction printing.**
1896 E. T. CARTER *Motive Power & Gearing* xxviii. 532 The double reduction gear used on the Frankfort-Offenbach tramcars. **1942** *R.A.F. Jrnl.* 3 Oct. 18 The port propeller amd reduction gear casing were wrenched off. **1971** B. SCHARF *Engin. & its Lang.* xii. 161 Any system of gears in which the speed of the driven shaft is lower than the speed of the driving shaft may be described as a reduction gear. **1896** E. T. CARTER *Motive Power & Gearing* 617/1 (Index), Reduction gearing. **1934** *Jane's Fighting Ships* 94 Turbines with single reduction gearing. **1942** J. LISTON *Aircraft Engine Design* viii. 146 The inherent tendency for the propeller efficiency to drop at high speeds can be offset by suitable reduction gearing. **1945** *Jrnl. Soc. Motion Picture Engineers* Apr. 290 This process [*sc.* reduction printing] is commonly used in making 16-mm negatives or prints from 35-mm originals. Film thus made is referred to as a reduction negative or reduction print, as the case may be. **1943** *Ibid.* Dec. 507 An effort was made to learn if the wows introduced into the reduction prints by the printer itself could be reduced by an increase in the speed of the printer. **1973** D. A. SPENCER *Focal Dict. Photogr. Technol.* 420 The assembly is also used to make reduction prints—e.g. by projecting 35 mm film on to 16 mm raw stock in the camera. **1945** *Jrnl. Soc. Motion Picture Engineers* Apr. 290 Reduction printing is the process of producing and recording photographically a smaller image, usually on a smaller film, from a larger image.
reductional *a.* (examples in *Cytology*: cf. sense 11 e above); **redu·ctionally** *adv.*
1903 *Biol. Bull.* IV. 266 Van Winiwarter..considers it probable that one of these divisions is reductional. **1905** *Proc. Acad. Nat. Sci. Philadelphia* LVII. 195 There is no evidence that chromosomes divide in different ways in the first maturation mitosis, some equationally and some reductionally. **1914** *Jrnl. Morphol.* XXV. 622 The univalent chromosomes..conjugate first in the equator of the first maturation spindle and there separate reductionally. **1920** etc. [see *EQUATIONAL *a.* 3]. **1939** SANSOME & PHILP *Rec. Adv. Plant Genetics* (ed. 2) vi. 183 Muller's Theory implies that the whole chromosome divides reductionally at the first division of meiosis. **1975** *Nature* 8 May 111/1 Normally fertilisation stimulates both the completion of the reductional first meiotic division and the subsequent equational second division.

reductionism (rĭdʌ·kʃəniz'm). [f. REDUCTION + -ISM.] In philosophy, the practice of trying to show that certain entities may be eliminated by reducing all reference to them to reference to some other entities. In more general use, the practice of describing a phenomenon (particularly one involving human thought and action) in terms of an apparently more 'basic' or 'primitive' phenomenon, to which the first is then said to be equivalent; for example, the practices of describing mental states in terms of the behaviour that expresses them, of describing organic processes in terms of the physico-chemical reactions which underly them, of describing social and political transformations in terms of the economic changes which engender them. In each case it is supposed that 'reduction' both explains, and also simplifies; 'reductionism' is therefore often used as a term of abuse for those theories which simplify too much, by reducing one phenomenon to another that is too basic to explain it.
1948 A. L. KROEBER *Anthropol.* (rev. ed.) xv. 576 The whole problem of the double aspect of our phenomena can also be seen as hingeing on how far we wish..to carry or not to carry what might be called intellectual reductionism. **1952** R. M. HARE *Lang. Morals* 180 Nor am I committing the sin of 'reductionism' which, because of its excessive prevalence, has become a fashionable target for philosophical heresy-hunters. **1953** W. V. QUINE *From Logical Point of View* 20 The other dogma is *reductionism*: the belief that each meaningful statement is equivalent to some logical construct upon terms which refer to immediate experience. **1960** B. G. ANDREAS *Exper. Psychol.* viii. 212 In theory construction reductionism refers to the use of constants and laws from one scientific discipline to explain the relationship found in another realm of investigation. **1965** *Jrnl. Politics* XXVII. 783 Reductionism. In its belief that every major irritation in society has a simple cause, a simple explanation and a simple solution, the radical right reduces such irritations to Communism. **1969** *Daily Tel.* 1 Nov. 7/7 It [*sc.* the minute dissection of music's anatomical structure] can easily succumb to the kind of 'reductionism' which..claims that any organism can be fully explained by a careful analysis of its smallest components. **1969** *Times Lit. Suppl.* 20 Nov. 1341/1 Reductionism..assumes that all properties of organisms..are ultimately reducible to physics and chemistry. **1971** R. F. MURPHY *Dialectics of Social Life* (1972) ii. 68 Durkheim..directed his critique against theories that sought to reduce social explanation to areas of psychology and biology that were not known at the time... Such 'reductionism' to the unknown is indeed a shallow gambit. **1976** *Nature* 3 June 439/1 Reductionism rests on the belief that the whole can be fully explained in terms of the parts. **1979** *Bull. Amer. Acad. Arts & Sci.* Apr. 20 What would be needed instead is what I once heard Etienne Gilson describe as a dogmatic basis for dogmatic tolerance. For that we must go beyond the reductionism that has shaped all of us to a historical awareness of the deepest issues that have divided the two communities. **1981** *Times Lit. Suppl.* 6 Feb. 137/4 Structuralism..has offered certain threats to reductionism.

reductionist. Add: **b.** An advocate of reductionism; one who attempts to analyse or account for a complex theory or phenomenon by reduction. Also *attrib.* or as *adj.*
1934 in WEBSTER. **1943** *Mind* LII. 129 The behaviour field which, as a whole, is molar from the reductionist's point of view. **1953** K. BRITTON *John Stuart Mill* iii. 86 Here it was Comte who maintained a reductionist view, holding that all the laws of psychology must find their ultimate explanation in terms of bodily changes. **1956** A. J. AYER in H. D. Lewis *Contemp. Brit. Philos.* 60 The reductionist's hero is the average man who is patently a logical construction. **1960** J. COHEN *Chance, Skill & Luck* i. 13 It does not follow that psychology lacks the status of an independent science and must be 'reduced' to neurophysiology, though this is what latter-day reductionists in effect demand. **1964** I. L. HOROWITZ *New Sociol.* 9 It succeeds in re-tooling the 'culture-lag' doctrine by making it conform to its own reductionist image of society. **1974** M. WILES *Remaking of Christian Doctrine* i. 5 Reductionists who in the face of the evolution of the human species want to say that man is really nothing but his animal ancestry in another form. **1977** *Times Lit. Suppl.* 1 Apr. 409/2 While the interpretation may be reductionist, it is a plausible one.

reductioni·stic, *a.* [f. REDUCTION + -ISTIC.] = REDUCTIONIST *a.* Hence **reductioni·stically** *adv.*
1960 B. G. ANDREAS *Exper. Psychol.* viii. 213 In psychology a theory may be considered reductionistic if, in addition to describing behavioral events, it employs constructs from physiology. **1965** *Jrnl. Politics* XXVII. 783 The list of problems and events treated reductionistically is enormous. **1971** *Jrnl. Gen. Psychol.* LXXXIV. 152 It is quite easy to see how each generation of reductionistically oriented theorizers was so easily seduced into using the most current, exciting, interesting, and potentially useful technology as the basis for its ideas. **1974** B. F. SKINNER *About Behaviorism* xiv. 240 A science of behavior has been said to dehumanize man because it is reductionistic. **1976** SMYTHIES & CORBETT *Psychiatry* i. 7 Current sociological theory is just as fragmented, disorganized and reductionistic as Freudian psychology.

reductive, *a.* and *sb.* **A.** *adj.* Delete 'Now *rare*' and add: **1. b.** *Psychol.* That leads back to an earlier state.
1928 H. G. & C. F. BAYNES tr. *Jung's Contrib. to Analytical Psychol.* 58 If now, by means of a reductive procedure, we uncover the infantile pre-stages of an adult psyche, we find as the ultimate foundation the infantile seeds. **1950** J. A. HADFIELD *Psychol. & Mental Health* xvi. 410 We style our method direct reductive analysis: it is reductive in that we analyse back to the deep-seated and predisposing causes as well as the more recent and precipitating causes. **1962** *Listener* 29 Mar. 568/2 Psychoanalysis is primarily historical and reductive, deducing the analysand's present situation from his past history. **1967** J. A. HADFIELD *Introd. Psychotherapy* xxii. 182 By reductive analysis I mean all those systems of psychotherapy whose methods are to go into the past to discover the causes of the neurosis in childhood experiences.
2. a. (Further examples.)
1924 C. E. MONTAGUE *Right Place* iii. 29 Calvin cast his reductive shadow over the naturally high spirits of Scotland. **1957** G. E. HUTCHINSON *Treat. Limnol.* I. ix. 626 Reductive organic sediments. **1969** *Listener* 14 Aug. 203/3 A British cinema..need not be one that seeks to indoctrinate foreigners in some reductive British image. **1974** *Times* 19 Nov. 9/4 Mr. Wilson..is not a reductive writer: he wants to recreate Crowley, not to explain him away. **1979** *Quarto* Oct. 3/4 There may be something reductive in Wolfe's constant harping on the rivalry and the petty resentments of the astronauts and their wives.
b. That tends to reduce, or is connected with reduction, esp. *REDUCTION 6 e and 7 b. (See also *REDUCTIONISM.)
1937 T. PARSONS *Struct. Social Action* v. 181 He [*sc.* Pareto] is thus free at the outset at least from the 'reductive' tendencies so prominent in the older positivism. **1957** P. LAFITTE *Person in Psychol.* v. 62 Reductive tendencies in social psychology can now be considered. The first of these is the tendency to see the person as an object which is pushed around in a field of force. **1966** O. WOJTASIEWICZ tr. *Kotarbinski's Gnosiology* IV. i. 221 All reasonings are usually divided into deductive and reductive, the former being from reason to consequence, and the latter, from consequence to reason. **1977** R. WILLIAMS *Marxism & Lit.* I. iv. 62 What is in fact idealized, in the ordinary reductive view, is 'thinking' or 'imagining'. **1979** *Dædalus* Summer 96 The search for a common denominator can prove fruitful—if it does not degenerate into a reductive maneuver.
c. *Art.* = *MINIMAL *a.* c.
1967 *New Yorker* 25 Feb. 99 To judge by art magazines and museum programs, nothing new has been done in the past few years but Happenings, optical displays, and so-called primary structures and reductive paintings. **1970** *Britannica Bk. of Year 1969* 798/3 *Rejective art*, a simplified and often depersonalized art (as painting or sculpture) based on the principle of the artist rejecting the various options open to him; called also *reductive art*, *reductivism*, *rejectivism*.
B. *sb.* For † read *rare* and add later examples.
1871 T. D. HAYE tr. *Taine's On Intelligence* I. II. i. 53 It is the *special reductive*, that is to say, the contradictory sensation, which fails in the conflict, and, instead of depriving the image of its externality, becomes itself effaced. **1890** W. JAMES *Princ. Psychol.* II. 125 The usual explanation of hypnagogic hallucinations is that they are ideas deprived of their ordinary reductives.

reductivism (rĭdʌ·ktiviz'm). [f. REDUCTIVE *a.* and *sb.* + -ISM.] **1.** *Art.* = *MINIMALISM.
1967 *Listener* 17 Aug. 220 Bernard Cohen's *White Plant* dates from the period when his earlier 'linguistic' style had degenerated into a hothouse aestheticism; the rather self-conscious reductivism of his recent exhibition was far less cloying. **1970** [see *REDUCTIVE *a.* 2 c].
2. = *REDUCTIONISM.
1972 *Village Voice* (N.Y.) 1 June 74/3 Dylan had been moving toward Duchamp's brand of Cartesian reductivism and public withdrawal. **1975** *Times Lit. Suppl.* 23 May 566/1 The reductivism implied in this enterprise—the reducing of ideas to another level of meaning or set of causes.
So **redu·ctivist** *a.* and *sb.* = *minimalist* s.v. *MINIMAL *a.* c.
1967 *Listener* 13 July 45/3 Harold and Bernard Cohen were the two foremost British painters during the early 'sixties who were trying to evolve a visual language to correspond to what 'the artist thinks'. Now both seem to have given this up as a bad job and fallen in line with current reductivist tendencies. **1967** *Sat. Rev.* (U.S.) 23 Sept. 23 New York..is dominated by large numbers of artists who swim in one or two schools producing closely related works—lately, the reductivists and the remainders of the Pop people.

red 'un: see *RED *a.* 3 b.

redund (rĭdʌ·nd), *v. rare.* [Shortened f. REDUND(ANT *a.* and *sb.*] **a.** *intr.* To be redundant; to contain a redundancy. **b.** *trans.* To make redundant.
1905 *Daily Chron.* 16 Jan. 4/7 The phrase 'inadvertently forgotten' does not redund so much as appears. **1959** *Guardian* 23 Dec. 4/1 'I've worked here before,' said another, 'but I was redunded in May.'

redundancy. Add: **2.** *spec.* **a.** *Engin.* The presence in a framework of more members than are needed to confer rigidity.
1904 J. B. JOHNSON et al. *Theory & Pract. Mod. Framed Structures* (ed. 8) I. xxv. 260 Another common example of redundancy is where two diagonals are used in the same quadrilateral. **1923** W. L. MARSH *Internat. Air Congr., London, 1923* 828 Owing to the extreme redundancy of the structure [of the airship] the calculations cannot be tackled by the graphical methods employed on most types of girder work construction. **1950** J. C. GRASSIE *Elem. Theory Structures* ix. 129 (*heading*) Conditions for internal and external redundancy in structures. **1966** J. L. MERIAM *Statics* iii. 81 For a truss that is statically determinate externally, there is a definite relation between the number of its members and the number of its joints necessary for internal stability without redundancy.
b. The condition of having more staff in an organization than is necessary. Hence, the

state or fact of losing a job because there is no further work to be done; a case of unemployment due to reorganization, mechanization, loss of orders, etc.

1931 *Economist* 11 Apr. 780/1 Such economics create redundancy of staff and unemployment rather than increased employment. 1934 *Planning* II. xxvi. 3 The shipbuilding and wool textiles industries have succeeded in establishing common instruments with which to combat redundancy. 1952 *Economist* 12 July 77 The strike against redundancy is a comparatively new phenomenon in industrial relations. 1955 *Times* 31 Aug. 4/6 The men stopped work after a dispute..over the way to handle redundancy at the works because of a reorganization scheme. 1956 *Economist* 7 July 12/2 The unions now appear ready to lay rather more emphasis on bargaining for higher severance pay, and rather less on demonstrations against the fact of redundancies. 1957 *Observer* 8 Sept. 9/4 This [*sc.* Gloucester] is a rather troubled city, with 15,000 people—nearly a third of the insured population—in the aircraft industry, and the prospect of substantial redundancy as defence contracts run out. 1972 *Accountant* 5 Oct. 420/1 Should a staff surveyor become redundant, redundancy pay would be considered, the maximum benefit being limited to one month's pay (based on salary at the date of redundancy) for every completed year's service. 1976 J. R. L. Anderson *Redundancy Pay* i. 11 The terms of the final merger called for heavy redundancies. 1977 M. Drabble *Ice Age* ii. 240 There isn't any work. There's large-scale unemployment. Redundancies everywhere. 1977 I. Shaw *Beggarman, Thief* iii. vi. 267 We live in the age of what the British call redundancy.

c. *Linguistics.* The element or degree of predictability in a language arising from knowledge of its structure; the fact of superfluity of information in a piece of language.

1948 *Bell Syst. Techn. Jrnl.* XXVII. 398 The redundancy of ordinary English, not considering statistical structure over greater distances than about eight letters, is roughly 50%. 1954 G. A. Miller et al. in J. S. Bruner *Beyond Information Given* (1974) iii. 59 When missing or ambiguous portions of a stimulus pattern can be supplied correctly..on the basis of the context alone, the missing portions carry little or no information. This fact is referred to as the redundancy of the language. 1972 J. L. Dillard *Black English* vii. 283 What the linguist calls redundancy—a technical term..which explains how we are able to understand sentences which we have not heard plainly or to read paragraphs in which a great deal of the print has been scrambled. 1977 A. Sheridan tr. *J. Lacan's Écrits* iii. 86 This notion of redundancy in language originated in research that was all the more precise because a vested interest was involved. 1979 E. H. Gombrich *Sense of Order* iv. 104 It is easy to understand how our grasp of ordinary language profits from high redundancies. We can afford to miss or mishear individual sounds or even words without losing the meaning.

d. *Engin.* The incorporation of extra parts in the design of a mechanical or electronic system in such a way that its function is not impaired in the event of a failure.

1962 J. Glenn in *Into Orbit* 38 The engineers had a word for this insistence on inserting backups into the system. They called it the principle of 'redundancy'. 1972 L. M. Harris *Introd. Deepwater Floating Drilling Operations* viii. 84 Redundancy implies that alternate methods of well control and operating subsea equipment will be available in the event of failure of any one component or group of components. 1972 *Sci. Amer.* Jan. 46/2 Perhaps the most unusual feature of the grand-tour spacecraft will be a computer called STAR ('self-test and repair'), provided with enough redundancy to operate for at least 10 years.

3. *attrib.* and *Comb.*, as (sense *2 b) *redundancy agreement, pay, payment, scheme*; (sense *2 c) *redundancy rule*; **redundancy check** *Computers*, a check on the correctness of processed data that involves a comparison with accompanying data derived from them prior to processing.

1951 *Public Administration* XXIX. 374/1 Examples, in the years between the two wars, of competitive industries being turned into monopolies, sometimes with the help of the State—the agricultural schemes, or the redundancy agreements, as in tinplate and shipbuilding. 1969 *Guardian* 3 July 12/1 The dockers and the port employers should start thinking about the sort of redundancy agreement that would meet their needs. 1962 *Gloss. Terms Automatic Data Processing (B.S.I.)* 33 *Redundancy check, redundant check*, a check that uses extra digits, which do not themselves fully represent the data concerned. 1970 O. Dopping *Computers & Data Processing* ii. 49 The most common form of redundancy check is the parity check. 1969 *Times* 7 Nov. 21/7 (*heading*) Redundancy pay. 1976 J. R. L. Anderson *Redundancy Pay* i. 11 The terms of the final merger called for heavy redundancies... There would be a bit of redundancy pay. 1980 *Times Lit. Suppl.* 31 Oct. 1240/4 Gus Baedecker, the London adman of *Events Beyond the Heartlands*, who uses his redundancy pay to take Kate and the children away from it all to a cottage on the Welsh coast. 1965 *Act* 13 & 14 *Eliz. II* c. 62 § 1 Where on or after the appointed day an employee who has been continuously employed for the requisite period—..is dismissed by his employer by reason of redundancy,..then,..the employer shall be liable to pay to him a sum (in this Act referred to as a 'redundancy payment'). 1966 *Listener* 17 Mar. 391/2 We carried out too, our pledge to introduce redundancy payments for those who were temporarily out of a job through the speeding of the process of industrial change. 1972 M. Jones *Life on Dole* ii. i. 98 The Redundancy Payments Act of 1965..compels the employer to pay out a lump sum..to each employee who is dismissed. 1965 N. Chomsky *Aspects of Theory of Syntax* 214 More generally the *phonological redundancy rules*, which determine such features as voicing of vowels.., can be supplemented by analogous syntactic and semantic redundancy rules. 1972 *Archivum Linguisticum* III. 14 A morphological feature.. must be recognized as being syntactically relevant, by means of a lexical redundancy rule. 1976 *Language* LII. 296 Lexical redundancy rules define the set of possible underlying morphemes in a language, in addition to minimizing the feature specifications required in the lexicon. 1969 *Guardian* 3 July 12/2 A generous redundancy scheme for dockers could save money.

redundant, *a.* and *sb.* Add: **A.** *adj.* **1. d.** *Engin.* Of a component of a framework, or a force or moment on it: capable of being removed without causing loss of rigidity. Hence of a framework: containing more than the minimum number of components necessary for rigidity.

1890 *Jrnl. Assoc. Engin. Societies* IX. 242 (*heading*) Deflection of framed structures and the distribution of stresses over redundant members. 1908 E. S. Andrews *Theory & Design of Structures* xi. 290 Redundant frames have the following disadvantages:—(1) Any stress in one member caused by bad fitting or change of temperature causes stress in all the other members. 1929 Niles & Newell *Airplane Structures* xiv. 316 In any redundant structure, the distribution of stresses will be such that not only are the conditions of equilibrium satisfied, but also that the deformations of all parts of the structure will be consistent with respect to each other. 1953 C.-K. Wang *Statically Indeterminate Structures* i. 4 In analyzing indeterminate structures it is necessary to have as many extra conditions, in addition to those of statics, as there are redundant reactions. 1976 A. C. Palmer *Structural Mech.* vii. 147 Determine the number of redundant forces (or moments) for each of the plane frames in Figure 7.22.

e. Of a person: no longer needed at work; unemployed because of reorganization, mechanization, change in demand, etc.

1928 *Britain's Industr. Future* (Liberal Industr. Inquiry) xxv. 358 We reach, finally, the pressing, but difficult, problem of the redundant workers. When everything possible has been done..there is little doubt that we shall still have to deal with a large surplus of labour in the coal-mining industry. 1934 J. B. Priestley *Eng. Journey* x. 346 You may do a good stroke of work by declaring the Stockton shipyards 'redundant', but you cannot pretend that all the men who used to work in those yards are merely 'redundant' too. 1956 *Times* 21 July 7/5 Redundant workers..workers dismissed on the score of redundancy. 1958 *Spectator* 30 May 713/3 Over five thousand other men were rendered redundant. 1969 H. E. Bates *Vanished World* xii. 156 Nowadays,..it would no doubt be said that I became redundant. I prefer the old way: I was unexpectedly sacked. 1974 C. Hill *Scorpion* 49, I rang his office... He doesn't work there any more. He was made redundant about two months ago. 1976 *Milton Keynes Express* 30 July 9/3 He had been made redundant and needed money.

f. Of a language: containing material which is predictable from context or a knowledge of its structure; also of a language feature, predictable in this way.

1954 G. A. Miller et al. in J. S. Bruner *Beyond Information Given* (1974) iii. 59 If a language is highly redundant, the relative information per symbol is much lower than it would be if successive symbols in a message could be chosen independently. 1965 W. S. Allen *Vox Latina* 78 Towards the end of a word sounds tend to become more 'redundant', i.e. predictable in terms of what has already been uttered. 1979 E. H. Gombrich *Sense of Order* iv. 104 If the message reads that the meeting was suspended for lack of a q.u.o.r.u.m. every successive letter can be said to be increasingly redundant.

B. *sb.* Restrict † *Obs.* to sense in Dict. and add: **1. b.** *Engin.* A redundant component of a framework (see sense A. 1 d above).

1953 C.-K. Wang *Statically Indeterminate Structures* i. 5 When the equations are solved and the redundants found, they can be put back on the given indeterminate structure and the remaining reactions solved by the equations of statics. 1976 A. C. Palmer *Structural Mech.* vii. 138 Although the frame was three times redundant, this extra piece of information enabled us to reduce the number of unknown redundants to two.

2. A person who leaves his job because of redundancy (sense *2 b); = *REDUNDANTEE.

1975 *Times* 4 Aug. 12/1 A call for volunteer redundants has not fallen on deaf ears.

Hence **redu·ndantize** *v. trans.*, to make (a person) redundant.

Fortunately *rare*.—Ed.

1949 *Picture Post* 19 Nov. 22/1 The blameless little men, so many of whom have been and will be 'redundantised'.

redundantee (rīdʌ·ndăntī·). *rare.* [f. REDUNDANT *a.* + -EE[1].] A person who has been made redundant.

1963 *Times* 8 May 13/7 Could not redundancies due to the Beeching plan be handled in a similar way? I am sure that industrialists would cooperate to try to find work for these men, possibly on the basis of one redundantee (?) to so many regular employees. 1971 *Selling Today* Sept. 8/2 Much is offered by way of genius. Like the recent 'redundantee' who offered a world of experience, expertise, exuberance and enthusiasm.

reduplicate, *v.* **2.** (Later example.)

1973 *Sci. Amer.* Feb. 59/1 The manner in which two-syllable adjectives reduplicate is different... A verb reduplicates by the entire word, but the adjective reduplicates in terms of its constituent syllables.

reduviid, *a.* and *sb.* Insert in etym. after *Reduvius* (J. C. Fabricius *Systema Entomologiæ* (1775) 729). (Earlier and later examples.)

1888 J. S. Kingsley *Riverside Nat. Hist.* II. 267 These are strongly suggestive of certain tropical forms of Reduviids. 1909 *Lancet* 20 Nov. 1495/2, I have recently suspected a reduviid bug..as having some relationship with the causation of kala-azar in Madras. 1962 Gordon & Lavoipierre *Entomol.* xxxix. 237 Triatomines may be distinguished from other reduviid bugs. 1965 B. E. Freeman tr. *Vandel's Biospeleol.* xii. 182 Certain reduviid bugs are frequently found in the entrances of tropical caves.

redux, *a.* Add: **2.** Brought back, restored.

[1662 Dryden (*title*) Astraea Redux. A poem on the happy restoration and return of His Sacred Majesty.] 1873 Trollope (*title*) Phineas redux. 1971 J. Updike (*title*) Rabbit redux.

red-veined: see *RED *a.* 14 a, RED *a.* 14 c.

redward, *a.* and *adv.*: see RED *a.* 19 b in Dict. and Suppl.

red ware[2]. Add to def.: Also, a type of fine, glazed pottery. (Earlier and later examples.)

1699 M. Lister *Journey to Paris* 139 As for the Red Ware of China, that has been and is done in England, to a far greater perfection than in China, we having as good Materials. 1934 *Discovery* June 166/2 Plain burnished red ware. 1959 *Chambers's Encycl.* XI. 137/2 Unglazed redware was made till about 1770... Thomas Astbury (1688–1743) is supposed to have been the first to glaze this finely potted red ware and to decorate it with pads of white clay stamped in patterns. 1965 E. Tunis *Colonial Craftsmen* v. 121 Redware as it came from firing was no harder than a modern flowerpot and it was just as porous. To make it useful, the potters glazed it. 1975 *Country Life* 26 June (Suppl.) 56/2 Christie's... Fine English Porcelain and Pottery..Astbury glazed redware..bell-shaped mug. 1977 *New Hampshire Times* 27 July 17/2 The potter had only one type of clay available, the local glacial clay underlying most of the coastal area. This clay fires in a kiln to a red color and therefore is called redware. Redware collapses past a certain firing temperature.

red-water. Add: **3.** (Earlier example.)

a 1759 N. Owen *Jrnl. Slave-Dealer* (1930) 30 If they are found out they are obliged to drink a large quantity of poyson, comonly caled red watter, which soon puts an end to thier days.

4. A mass of water made red by pigmented plankton, esp. dinoflagellates.

1856 *Edin. New Philos. Jrnl.* IV. 264 Alphonse Albuquerque..saw, from the stern of his vessel, issuing from the strait, and expanding outside, a stream of red water, which flowed towards Aden. 1902 *Amer. Naturalist* XXXVI. 189 The 'red water' occurred for two hundred miles..along the coast, from the region of Santa Barbara to San Diego. 1933 *Science* 7 July 13/1 *Provocentrum* is more frequently prominent in production of 'red water' than reports have indicated. 1948 *Jrnl. Marine Res.* VII. 57 The presence of this [human respiratory] irritation was reported as associated with 'red water', dying fish and onshore winds. *Ibid.* 60 'Red water' containing 56×10^6 dinoflagellates per liter.

red, white, and blue: see *RED *a.* 1 g.

redwing. 1. a. Substitute for def.: A European thrush belonging to one of the subspecies of *Turdus musicus* (or *T. iliacus*), distinguished by red patches on the flanks and under sides of the wings. (Later examples.)

1894 A. Newton *Dict. Birds* 778 The notes of the Redwing are indeed pleasing in places where no better songster exists. 1954 D. A. Bannerman *Birds Brit. Isles* III. 192 To these islands the Continental redwing is a winter visitor. 1977 J. L. Harper *Population Biol. Plants* ii. 47 Redwings..eat the fleshy receptacles [of rose hips].

b. For *Agelæus* substitute *Agelaius*. (Earlier and later examples.)

1778 J. Carver *Trav. N. Amer.* 474 The second sort [of blackbird] is the red wing, which is rather smaller than the first species. 1947 *Chicago Tribune* 2 Sept. 7/3 The grackles and redwings also are having a high old time these days roaming around the country. 1974 A. Dillard *Pilgrim at Tinker Creek* xi. 201 By the creek, where..redwings scatter.

c. (Earlier and later examples.)

1878 T. J. Lucas *Camp Life & Sport S. Afr.* vi. 85 We had a sprinkling of 'red wing', but the game grey partridge was more predominant. 1962 Mackworth-Praed & Grant *Birds Southern Third Afr.* 208 Red-wing... Top of head brown with darker centres to feathers.

redwood. Add: **1. a.** (Earlier and later examples.)

1619 W. Phillip tr. *Schouten's Relation Wonderfull Voiage* 37 In each Canoe..there lay two whole broad planckes of fayre redde wood. 1957 *Handbk. Softwoods* (Forest Prod. Res. Lab.) 42 Timber of this species [sc. *Pinus sylvestris*] imported from the Continent is commonly called redwood, red deal, or simply 'red'. 1963 [see *KAPUR].

b. Delete *Sc.*; = *compression wood* s.v. *COMPRESSION 6. (Later example.)

1925 [see **compression wood*].

Redwood[2] (re·dwud). [The name of Sir Boverton *Redwood* (1846–1919), British chemist.] **a.** *Redwood viscometer*: either of two types of viscometer (differing in the ranges of viscosity for which they are suitable), which were designed by Redwood and are used esp. to measure the viscosity of petroleum and its products.

[1886 *Jrnl. Soc. Chem. Industry* 29 Mar. 131/1 He had been in the habit of using one of Mr. Redwood's instruments

for determining viscosities.] 1896 B. Redwood *Petroleum* II. ix. 605 The Redwood viscometer..is a modification, designed by the author in 1885, of the instrument formerly used at the Battersea Works of Price's Patent Candle Company. 1931 G. Barr *Monogr. Viscometry* iv. 96 Liquids for which the Redwood No. II viscometer is specified..are difficult to free from suspended impurities. 1949 A. C. Merrington *Viscometry* v. 58 The calibration of a Redwood viscometer with a number of oils using an apparatus of the Thorpe and Rodger type. 1972 Harker & Allen *Fuel Sci.* vii. 98 The instrument most commonly used in the United Kingdom to measure viscosity is the Redwood Viscometer... The two standard instruments are the No. 1 and the No. 2 Redwood viscometers. The former is used for thin oils having viscosities of less than 2,000 Redwood seconds, and the latter for more viscous oils. 1973 A. L. Mills in Hobson & Pohl *Mod. Petroleum Technol.* (ed. 4) xx. 730 These instruments are now rarely used except perhaps for black oils... In fact, the method of viscosity determination using the Redwood viscometer has been deleted from the Institute of Petroleum 'Standard Tests'.

b. *Redwood second* (also *second Redwood*): a unit of viscosity used in conjunction with Redwood viscometers and equal to one second of the time required for a given quantity of fluid to pass through a capillary in the instrument. So *Redwood time, unit, viscosity*, etc., and with ellipsis of second word.

1913 *Petroleum World* June 272/1 The remaining three columns respectively give their colour, density at 20° C. and viscosity at 20° in Redwood units. 1930 *Engineering* 5 Sept. 308/1 The fuel used throughout this test was Mexican boiler oil with..a viscosity of 200 secs. Redwood. 1949 A. C. Merrington *Viscometry* v. 57 The results are normally expressed as 'Redwood seconds' at the temperature of the test. *Ibid.* A Redwood time of T_R seconds. 1967 A. S. Brunjes in Bland & Davidson *Petroleum Processing Handbk.* xii. 32 The Kinematic, Saybolt, Universal, redwood [*sic*] No. 1, and Engler scales. 1973 P. J. King et al. in Hobson & Pohl *Mod. Petroleum Technol.* (ed. 4) vi. 215 Reference to..viscosities expressed in terms of Redwood seconds..is still found in the literature. 1973 J. G. C. Pope in *Ibid.* xviii. 654 The viscosity of relatively mobile oils is recorded as so many seconds Redwood I @ 100° F. 1973 W. H. Thomas in *Ibid.* xxv. 860 When the Redwood viscosity requirements are quoted in specifications the usual procedure is to determine viscosity in kinematic or dynamic units and to convert these into Redwood by means of a conversion chart.

red-worm. Add: **3.** A parasitic nematode worm belonging to the family Strongylidæ, esp. to the genus *Strongylus*, which infests the intestine and other organs of many vertebrate animals, causing severe anæmia and general debility.

1891 R. Wallace *Rural Econ. Austral. & N.Z.* xxviii. 374 Sheep also suffer from the red worm..in their fourth stomach. 1951 *Chambers's Jrnl.* Oct. 587/1 Red-worms belong to the Strongyle family. 1970 Miller & West *Black's Vet. Dict.* (ed. 9) 768/1 Thiabendazole is a useful drug for the removal of red worms in horses.

reed, *sb.*[1] Add: **I. 1. b.** (Further examples.)

c 1593 T. Deloney *Garland of Goodwill* (1631) iii. sig. B1, But senselesse man, what de I meane, Upon a broken reede to leane. 1617 J. Chamberlain *Let.* 20 Dec. (1939) II. 123 Yf you trusted to him you trusted to a rotten reede who wold have failed you in the end. 1621 *House of Lords Jrnls.* 30 Apr. 101/1 Their lordships..reported, That they.. demanded of his lordship [*sc.* F. Bacon] whether it were his Hand..who answered 'My lords, it is my Act, my Hand..I beseech your Lordships, be merciful unto a broken Reed.' 1757 Smollett *Reprisal* i. i. 7 You lean upon a broken reed if you trust to their compassion. 1926 R. H. Tawney *Relig. & Rise of Capitalism* ii. 108 Human efforts, social institutions, the world of culture, are at best irrelevant to salvation, and at worst mischievous. They distract man from the true aim of his existence and encourage reliance upon broken reeds. 1961 I. Murdoch *Severed Head* xiii. 118 A nervous shrinking which was not exactly dislike made me hesitate to probe the motives of such a being. Therewith some vague yet powerful train of thought led me to say, 'I'm a broken reed after all.' 1973 *Times Lit. Suppl.* 23 Mar. 311/3 The history of the opposition shows what bruised reeds the generals were.

II. 8. a. *double reed* (earlier and later examples); also (with hyphen) *attrib.*

1876 Stainer & Barrett *Dict. Mus. Terms* 137/2 *Double reed*,..the vibrating reed of instruments of the oboe class. 1879 Grove *Dict. Mus.* I. 151/2 *Bassoon*.., a wooden double-reed instrument of eight-foot tone. 1931 G. Jacob *Orchestral Technique* iii. 26 The bassoon also agrees well with its double-reed cousin the oboe. 1961 A. Baines *Musical Instruments* ix. 233 The European shawm reed is of harder material prepared like all Western double reeds by folding over a strip of seasoned cane, shaping and binding the ends together, and paring down and finally separating the tip. 1974 *Encycl. Brit. Macropædia* XIX. 848/1 The human voice..may be classified as a double-reed aerophone in which the vocal chords act as a double reed.

d. (Earlier and later examples.) *double reed*: see sense 8 a.

1838 C. Fox *Jrnl.* 5 June (1972) 50 Professor Wheatstone ..then played the Chinese reed, one of the earliest instruments constructed. 1871 H. Calderwood *Let.* 23 June in Calderwood & Woodside *Life H. Calderwood* (1900) 216 The orchestra mostly reeds and strings. 1877 G. B. Shaw *How to become Musical Critic* (1960) 26 The strings and reeds were a little better than usual. 1926 Whiteman & McBride *Jazz* ix. 199 In the double reeds, I am planning to add a bassoon. 1939 Joyce *Finnegans Wake* (1964) 408 Brass and reeds, brace and ready! 1959 'F. Newton' *Jazz Scene* vi. 107 Three trumpets, three trombones, four reeds, piano. 1961 J. A. MacGillivray in A. Baines *Musical Instruments* x. 244 The clarinet..marked (like the oboe among the double reeds) the arrival of the fully lip-controlled instrument. 1974 *Encycl. Brit. Macropædia* XIX. 855/1 Shawms were a particularly important family of loud double reeds. 1975 *New Yorker* 19 May 6/3 Joe Muranyi on reeds, and Bobby Pratt on trombone.

12. (Later example.) *reed-and-tie*, used of a style resembling reeds bound together.

1875 T. Seaton *Man. Fret Cutting & Wood Carving* vi. 68 Make a little reed round the uncarved or T part of the bracket and the support... This will form a neat reed, and give a pretty finish to your work. 1960 H. Hayward *Antique Coll.* 235/1 *Reed-and-tie moulding*, an ornament composed of contiguous parallel convex mouldings bound together by straps simulating ribbons. 1971 *Country Life* 1 Apr. 766/1 The grandiloquence of Louis XVI's France, with heavy reed-and-tie borders..also had a place at fashionable West-End silversmiths.

13. a. *reed boat, -swamp, -whistle*; (sense 8) *reed cap, section*; (sense 10) *reed hook, space*.

1902 *Encycl. Brit.* XXV. 377/1 The catamaran and the reed boat were known to the Peruvians. 1977 *Time* 28 Nov. 60/1 Now Heyerdahl is about to take a reed boat down the Tigris River. 1964 S. Marcuse *Musical Instruments* 441/1 *Reed cap*, a small wooden cap with a blowhole on top; it enclosed the reed of some 16th-c. double-reed instrs. 1976 Reed-cap [see *RAUSCHPFEIFE]. 1910 L. Hooper *Hand-Loom Weaving* (1920) 328 *Reed hook*, hook for entering reed. 1914 H. Nisbet *Preliminary Operations of Weaving* I. ix. 359 The reacher, with the right hand, then proceeds to select the warp threads from a bunch held in the left hand, and delivers them in consecutive rotation to a reed-hook which is inserted through successive eyes of the harness by the drawer-in. 1957 Simpson & Weir *Weaver's Craft* (ed. 8) viii. 97 (*caption*) Reed hooks. 1939 D. Baker *Young Man with Horn* iii. i. 117 Rick..started setting chairs together the way they should go, in threes: reed section, brass section, rhythm section, and the extras one on top of another. 1975 *New Yorker* 21 Apr. 8/3 Billy Harper, a young and very exciting tenor saxophonist..steams up the reed section of the Thad Jones-Mel Lewis band. 1919 *Brit. Manufacturer* Nov. 35/2 Wide hand looms of high reedspace scarcely require more effort than those for narrower weaving. 1971 *Nature* 11 June 364/2 Here the invasion of reedswamp from the north and west was incomplete. 1975 J. G. Evans *Environment Early Man Brit. Isles* iii. 58 The vegetation.. passes through a variety of stages—reed swamp, carr..and raised bog. 1864 J. A. Grant *Walk across Afr.* xi. 245 On his arm he carried a reed-whistle three inches long, but it seemed to be more for ornament than use. 1962 R. P. Jhabvala *Get Ready for Battle* ii. 101 A toyman with toys stuck on the end of a long pole.., blowing on a reed-whistle.

b. *reed-cutter* (later example); *reed-cutting, -drawing* (earlier and later examples).

1974 *Country Life* 3 Oct. 922/1 In winter..the reed cutters took the harvest that served for thatch all over Britain. 1973 R. Adams *Watership Down* xxxiii. 260 The 'boat' was a miniature punt, used for reed-cutting. 1874 Hardy *Far from Madding Crowd* I. vi. 74 Oak seized the cut ends of the sheaves, as if he were going to engage in the operation of 'reed-drawing'. 1946 N. Wymer *Eng. Country Crafts* v. 50 The preparation of the straw—variously known as yelming, reed-drawing, or gabbling—consists of removing all unsuitable pieces and arranging the strands level.

c. (Also parasynthetic) *reed-bottomed, capped, -choked, -encumbered, -fringed, -stemmed, -throated.*

1835 C. Mathews *Let.* 7 Feb. in A. Mathews *Mem. Charles Mathews* (1839) IV. 343 Then behold six reed-bottomed, ragged, ricketty chairs. 1977 *Early Music* July 342/2 A rauschpfeife, a relatively easy (i.e. non-embouchure) instrument, presumably derived from a reed-capped bagpipe chanter. 1952 V. Canning *House of Seven Flies* viii. 125 The narrow, reed-choked mouth of an old cut. 1892 W. B. Yeats *Countess Kathleen* iii. 55 Leave marshes and the reed-encumbered pools. 1906 A. B. Cooper *Flood-Tides* 4 By wold and wilderness, by reed-fring'd lake. 1952 V. Canning *House of Seven Flies* viii. 124 Flat, reed-fringed islands. 1942 W. Faulkner *Go down, Moses* 266 She held a reed-stemmed clay pipe but she was not smoking it. 1914 W. B. Yeats *Responsibilities* 76 From that reed-throated whisperer Who comes at need.

14. reed bat = *reed legget* below; **reed-horn,** (*a*) a fog-horn in which the sound is produced by a current of air blowing on a reed (sense 8 c); (*b*) *slang*, a saxophone; **reed-knife,** a knife-like instrument used in tuning a reed-organ; **reed legget** = *LEGGET; **reed-man,** (*a*) a player of a reed instrument; (*b*) one who works with reeds; **reed-mark** (see quots.); **reed-marked** *a.*, of cloth, having the warp threads lying unevenly; **reed relay** *Electr.*, a small, high-speed, switching device consisting of a pair of contacts, enclosed in a glass tube, which can be brought together by an external magnetic field.

1969 Reed bat [see *LEGGET]. 1902 *Encycl. Brit.* XXX. 266/2 At the Trinity House experiments with fog signals at St. Catherine's (1901) several types of reed-horn were experimented with. 1936 *Metronome* Feb. 61/2 *Reed horn*, sax. 1876 Stainer & Barrett *Dict. Mus. Terms* 339/2 An organ is tuned by means of hollow cones and reed-knives. 1961 *Thatcher's Craft* (Rural Industries Bureau) vii. 205/1 (*caption*) Norfolk reed leggett used for dressing reed into position. 1872 Reed-man [see *brass-man* s.v. *BRASS *sb.* 6]. 1938 D. Baker *Young Man with Horn* i. v. 47 There was the band playing 'Home, Sweet Home' as a one-step with the reed man getting into clear and going absolutely wild on a clarinet. 1951 Wallace & Bagnall-Oakley *Norfolk* vii. 84 The old villages..began as trading places for the eel-fishers, the reed-men and the smugglers from the sea. 1977 *New Yorker* 6 June 128/2 It consists of eleven Laurence studio performances (about forty minutes in all), backed by two reedmen (Paul Quinichette or the late Bobby Jaspar) and two rhythm sections. 1931 E. Midgley *Techn. Terms Textile Trade* I. 261 *Reed marks*, a type of defect in woven fabrics due to the warp threads running in 'twos' or 'threes'. 1961 Blackshaw & Brightman *Dict. Dyeing* 145 *Reed marks*, marks or streaks running the warp of a cloth and caused by defects in the functioning of the reed during weaving. 1894 T. W. Fox *Mechanism of Weaving* iii. 37 Sometimes warp threads are allowed to run in pairs throughout the piece without being looked upon as a serious defect; such material is said to be reed-marked, or without cover. 1947 *Electr. Engin.* LXVI. 1104 (*heading*) Glass enclosed reed relay. 1966 *Times* 16 Dec. 11/6 The key component in the Ambergate exchange is a miniature reed relay. 1975 Fink & McKenzie *Electronics Engineers' Handbk.* xxiii. 41 Figure 23–46 shows the reed relay, combining small size and high reliability. A magnetic field induced by an external coil follows the path of the encapsulated contact arm, causing a force to pull the two arms together.

reed, *v.* Add: **4.** *Weaving.* To pass (warp threads) through the splits of a reed.

1894 T. W. Fox *Mechanism of Weaving* ii. 17 It will be noticed that the threads from shaft 4 are reeded two in a dent, and those from the remaining shafts three in a dent. 1957 *Textile Terms & Definitions* (Textile Inst.) (ed. 3) 79 *Reed, v.*, to draw ends through a reed (local, to sley, to bob the reed or to enter the reed).

reeded, *ppl. a.* Add: **2.** (Earlier example.)

1778 J. Bampfylde *Sixteen Sonnets* 16 Counting the frequent drop from reeded eaves.

3. (Earlier and later examples.)

1829 H. Foote *Compan. to Theatre* 36 The upper circle and tiers, including both the slips and lower gallery, are each supported in part by 14 slender shafts, reeded, of iron. 1935 *Archit. Rev.* LXXVIII. 33 The window is glazed with reeded glass. 1952 [see *AGBA]. 1978 R. Rendell *Sleeping Life* ii. 14 No one came when they rang the bell on the neighbouring front door, a far more trendy and ambitious affair of wrought iron and reeded glass.

5. Of wood: having a specified kind of reed.

1839 Ure *Dict. Arts* 972 A quantity of well-seasoned and clean reeded deal is required for forming the joints.

6. *fig.*

1926 T. E. Lawrence *Seven Pillars* (1935) lxxxix. 495, I knew that Sherif Abd el Main should be still at Shobek, so rode boldly up the silent street in the reeded starlight, which played with the white icicles.

reediness. (Earlier and later examples.)

1844 H. Stephens *Bk. Farm* II. 365 The straw of the former kind is strong and inclined to reediness. 1931 E. Midgley *Techn. Terms Textile Trade* I. 20 The warp threads..must lie in the cloth an equal distance apart, or a defect known as 'reediness' is created.

re-edit *v.*, **re-editing** *vbl. sb.* (Examples in *Cinemat.*) Cf. *EDIT *v.* 2 d, *EDITING *vbl. sb.* b.

1953 K. Reisz *Technique Film Editing* ii. 168 The picture and words were slightly re-edited to fit in with the music. 1975 *Listener* 21 Aug. 242/2 Not as severely compromised by studio re-editing and reshooting as some of his earlier films had been.

reedmergnerite (rīdmɜ̄·ɹgnərəit). *Min.* [See quot. 1954 and -ITE[1].] A colourless triclinic silicate of sodium and boron, $NaBSi_3O_8$.

1954 C. Milton et al. in *Bull. Geol. Soc. Amer.* LXV. 1286 Cores and cuttings from the Green River formation in Utah contain two new minerals, reedmergnerite and eitelite,..in dolomitic shale. Reedmergnerite..occurs in many wells in Duchesne and Uintah counties as crystals rarely more than 1 mm long, colorless, triclinic..; habit stubby prismatic... The name honors Frank S. Reed and John L. Mergner, technicians of the Geological Survey. 1974 *Amer. Mineralogist* LIX. 79/1 This study was undertaken to determine more precisely the bond lengths and angles of danburite.. for comparison with topologically similar paracelsian ($BaAl_2Si_2O_8$) and hurlbutite ($CaBe_2P_2O_8$) and with structurally similar feldspars: anorthite ($CaAl_2Si_2O_8$)..; albite ($NaAlSi_3O_8$); and reedmergnerite ($NaBSi_3O_8$).

Reed–Sternberg (rīd stɜ̄·ɹnbəɹg). *Path.* [The names of Dorothy M. *Reed* (1874–1964), U.S. pathologist, and C. *Sternberg* (1872–1934), Austrian pathologist, who described the cell in 1902 and 1898 respectively.] *Reed–Sternberg cell*: a binucleate or multinucleate giant cell characteristic of Hodgkin's disease.

[1937 *Surg., Gynecol. & Obstetr.* LXIV. 466/1 Hodgkin's disease,..with its Sternberg Reed cells, fibrosis, necrosis, and eosinophilic infiltration, needs no introduction.] 1947 Jackson & Parker *Hodgkin's Dis. & Allied Disorders* i. 7 The fact that the Reed-Sternberg cells are frequently scattered, isolated, and often separated widely by cells of other cell types favors an inflammatory process rather than a neoplasm. 1980 *Brit. Med. Jrnl.* 29 Mar. 903/1 The dermis was infiltrated by lymphocytes,..mononuclear Hodgkin's cells, and classical binucleate Reed-Sternberg cells.

re-educate, *v.* Add: (Further examples.) Now often *spec.* with the object of changing political beliefs or social behaviour.

1947 *Hansard Commons* 15 Dec. 1434 The work of the *Kulturbund zur demokratischen Erneuerung Deutschlands* in re-educating the German people in the spirit of democracy and international understanding. 1955 *Treatment of Brit. P.O.W.'s in Korea* (H.M.S.O.) 4 'Re-educating' the prisoners. The Chinese technique of 're-education' embraced every phase of daily life in the prison camps. 1967 *Listener* 18 May 653/1 The aim..is to re-educate the prisoner rather than to punish him. 1975 *Chinese Econ. Stud.* VIII. iv. 3 Chairman Mao teaches us that 'it is necessary for educated

youths to go to the countryside to be reeducated by the poor and lower-middle peasants'. 1976 *New Yorker* 26 Jan. 110/2 The Chinese considered him sufficiently important to be spared and 'reëducated', or brainwashed. 1976 W. H. Canaway *Willow-Pattern War* vi. 64 Agricultural communes which specialized in re-educating professors and other intellectuals as labourers.

re-education. Add: (Further examples: cf. prec.) Also *attrib.*

1906 *Trans. Assoc. Amer. Physicians* XXI. 724 Re-education is undoubtedly one of the most important factors in producing lasting cures. 1944 J. S. Huxley *On Living in Revolution* 151 This gang has succeeded in imposing its ideas on a considerable minority of the German people, and..this constitutes a grave problem of re-education. 1945 'G. Orwell' *Animal Farm* iii. 26 He formed..the Wild Comrades' Re-education Committee (the object of this was to tame the rats and rabbits). 1945 *Times* 8 May 7/5 It is becoming clear that the 're-education' of Germany by the allies will not be a pious aspiration, but an unavoidable duty. 1951 Koestler *Age of Longing* ix. 337 You don't like revolutionary vigilance,..and discipline, and re-education camps. 1974 N. Freeling *Dressing of Diamond* 133 A year of physiotherapy..in a re-education centre. 1976 Scott & Koski *Walk-In* (1977) xiii. 79, I am once again politically unreliable. I once again face re-education. 1977 *Time* 9 May 22/1 Also targetted for resettlement are most of the 30,000 political prisoners the regime admits are still interned in 're-education' camps.

reedy, *a.* Add: **3. a.** (Later example.)

1890 G. Meredith *Let.* 26 Mar. (1970) II. 993 A reedy state of health forbids my going to Dinners.

b. (Earlier example.)

1743 W. Ellis *Mod. Husbandman* Dec. viii. 410 As they make good Part of their Rent by the Sale of their Wheat Straw, they are very careful to preserve it as reedy or long as they can.

e. Of cloth: having the warp threads unevenly distributed.

1931 E. Midgley *Techn. Terms Textile Trade* I. 261 *Reedy*, a term applied to a cloth which shows reed marks.

5. *Comb.* (Later example.)

1905 *Westm. Gaz.* 1 Nov. 1/3 The valley of the reedy-voiced little Ervola.

reef, *sb.*[1] Add: **1.** (Earlier and later *transf.* examples.)

1846 *Swell's Night Guide* 48 Ruttum turned out a quid as big as a moke's egg, took a reef in his patter trap. 1884 'Mark Twain' *Huck. Finn* xxxi. 316, I lit out, and shook the reefs out of my hind legs. 1903 Somerville & 'Ross' *All on Irish Shore* i. 2 'I dunno, Master Freddy; it might be 'twas a hare,' returned Patsey, taking in a hurried reef in the strap that was responsible for the support of his trousers. 1924 E. Pound *Let.* 3 Dec. (1971) 190 Am also letting out another reef in my long job. Installment of which should soon be inspectable.

3. *reef-earing* (further examples); **reef-knot,** (*b*) (later example); **reef-knot** *v.* (earlier example); **reef net** *N. Amer.*, a type of net used for catching salmon; also *attrib.*; hence **reef netter,** a fisherman who uses a reef net.

1883 *Man. Seamanship for Boys' Training Ships R. Navy* (Admiralty) (1886) 51 Reef-earrings are pieces of rope, in size according to the size of the leech-rope, as when a topsail is reefed the reef-earring, when passed is supposed to bear the same amount of strain as the leech-rope. 1974 P. Wright *Lang. Brit. Industry* xv. 148 Over the mainsail came, broke all the reef-ear-rings, an' then we'd full sail on. 1974 *Maclean's Mag.* Nov. 10/1 Show the other lads the difference between a reef knot and a granny. 1883 *Man. Seamanship for Boys' Training Ships R. Navy* (Admiralty) (1886) 87 For a topgallant sail or royal,..[a roband-hitch] is ..not clove-hitched, the two nearest robands being reef-knotted together. 1917 *Pacific Fisherman Yearbk.* 60 (*heading*) The Siwash reef net. *Ibid.*, Reef net fishing was confined to the flood tide. 1970 *National Fisherman* Feb. 21-B/1 In 1969 there were 63 pairs of reefnet vessels registered in Puget Sound. 1974 B. & R. Hill *Spirit in Stone* iii. 35 A man whom we will call the ritualist and several assistants are fishing for salmon with a reef net. 1939 *Pacific Fisherman* June 45/2 With the increase in gear, considerable friction arose between purse seiners and reef netters.

reef, *sb.*[2] Add: **2. a.** (Earlier and later examples.)

1857 in *Occasional Papers Univ. Sydney Austral. Lang. Res. Centre* (1966) No. 9. 21 On this gold-field the word reef shall be taken to mean any seam of quartz, the average thickness of which..shall exceed three (3) feet. 1939 C. W. Towne *Her Majesty Montana* 114 Even before the end of placer mining, Butte prospectors had located quartz on a black-stained reef. 1955 *Times* 9 May 18/3 The total development footage driven was 48,295 ft., and of the 21,085 ft. on basal reef and sampled 12,990 ft., or 62 per cent., proved payable. 1966 'J. Hackston' *Father clears Out* 16 An offshoot from our old reef provided the quartz.

c. *S. Afr.* (With capital initial.) = *RAND *sb.*[2] 1 b. Also (usu. with small initial), rock in a mine which is not gold- or diamond-bearing.

1893 T. Reunert *Diamonds & Gold S. Afr.* i. 21 The surface shales and basalt surrounding the pipes are called 'Reef'. *Ibid.* 22 In the upper levels of the mines intrusive masses of shale and igneous rock are met with, called 'Floating Reef'. They are destitute of diamonds. *Ibid.* 28 The encasing rock of the mine (or the 'Reef', as the diggers called it) being exposed by the removal of the diamondiferous ground, began to disintegrate, and fall into the mine... The reef troubles..more than once threatened to involve the whole mine in ruin. 1905 L. Phillips *Transvaal Probl.* ii. 49 Meetings took place along the Reef from Boksburg to Krugersdorp. 1926 S. G. Millin *S. Africans* 77 The richest road in the world, whose sixty miles run over the gold-mines of the Reef. 1970 W. Smith *Gold Mine* xvi. 44 Free gold.. rapidly worked its way down..its journey accelerated by the vibration of the conveyor and bin as mine reef was dropped. 1975 'D. Jordan' *Black Account* xvii. 89 One of the houses with the Reef's pre-war style of flat roofs and enormous bay windows.

4. (sense 1) *reef-forming, -making, -strewn* adjs.; (senses 2 a and *2 c) *reef development, town, value*; **reef-break** *Surfing* (see quot. 1970); **reef-builder,** for 'coral insect' read 'cœlenterate or other marine organism' (earlier and later examples); **reef flat,** the horizontal upper surface of a reef; **reef-knoll,** a hillock, usu. of limestone, formed from ancient coral; **reef-limestone,** limestone which was formed in reefs.

1966, 1968 Reef break [see *point break* s.v. *POINT *sb.*[1] B. 14]. 1970 *Studies in English* (Univ. of Cape Town) I. 26 A reef break, surf breaking over a reef, will provide a good, fast ride. 1869 *Amer. Naturalist* III. 352 We could find no evidence that the reef-builders at the present time.. are working upon so high a northern line. 1972 *Sci. Amer.* June 54/1 The chief animal reef-builders today are the corals, but many other marine invertebrates are important members of the reef community. 1971 *Daily Tel.* 11 Oct. 17 A limited amount of reef development in the lower western portion of the mine yielded reasonable values. 1886 *Trans. R. Soc. Edin.* XXXII. 557, I..came upon the coral rock exposed in flat surfaces resembling those of the ordinary reef-flat. 1931 J. S. Gardiner *Coral Reefs & Atolls* ii. 35 Such rock masses as are visible on the reef flat do not stand up above the high tide level. 1976 R. C. Selley *Introd. Sedimentol.* viii. 297 There are three main morphological elements to a reef: the fore-reef, the reef flat and the back-reef. *Ibid.* 299 Tidal channels..traverse the reef-flat. 1967 *Oceanogr. & Marine Biol.* V. 330 Tethyan corals include such reef-forming genera as *Stylina, Isastraea*, and *Thamnastrea*. 1890 R. H. Tiddeman in *Rep. Brit. Assoc. Adv. Sci. 1889* 602 At the foot of these mounds, or reef-knolls as I would call them, we have in many places a breccia formed of fragments of the limestone. 1969 Bennison & Wright *Geol. Hist. Brit. Isles* ix. 211 Extensive sheet reefs, not necessarily primarily organic in origin, as well as reef-knolls are widespread, the reef-knolls occurring on the flanks of massifs. 1893 P. Lake tr. *E. Kayser's Textbk. Compar. Geol.* iii. 225 In these reef limestones..the greater part of the rock is formed not by corals, but by the rock-building algae..*Gyroporella* and *Diplopora*. 1938 M. Black *Hatch & Rastall's Petrol. Sedimentary Rocks* (ed. 3) viii. 163 The term 'reef limestone' has been used in geological literature with varying significance. In this discussion, shelly or structureless, unbedded limestones which show no clear connection with sessile benthonic organisms will be left out of consideration. 1956 W. Edwards in D. L. Linton *Sheffield* 6 Shirley and Horsfield..have described the reef-limestones of the northern fringe of the main outcrop near Castleton. 1855 J. Phillips *Man. Geol.* xvi. 491 The reef-making madrepores are seldom found below 100 feet. 1961 *Times* 14 Dec. 17/5 Ice and reef-strewn channel. 1938 N. Devitt *Spell of S. Afr.* 185 At a military court held in a certain Reef town, a civilian was charged with murder. 1955 *Times* 3 May 17/2 Reef values to the south-west of the fault are expected to be similar to those encountered in the President Brand and Western Holdings mines.

reef, *v.*[1] Add: **1. b.** (Further examples.) *Criminals' slang*, to pull up (the lining of a pocket) so as to steal the contents; to pick (a pocket); hence, to steal or obtain dishonestly in any fashion; also more *gen.* to remove, to take or strip *off*, to pull *down*.

1899 [see *LEATHER *sb.* 2 e (*a*)]. 1901 M. Franklin *My Brilliant Career* xvii. 142 She was the only one who bothered with a bathing-dress. The rest of us reefed off our clothing. 1903 Farmer & Henley *Slang* VI. 10/1 *Reef*..(thieves'), to draw up a dress-pocket until the purse is within reach of the fingers. 1926 *Variety* 29 Dec. 7/4 The cleverest wire who ever reefed an insider would be astonished to hear that a 'milk man' was a hambo, who stole more bows than the applause warranted at the finish of his act. 1938 *Surg., Gynecol. & Obstetr.* LXVI. 200/2 An attempt was made to shorten the quadriceps by reefing the tendon and fastening the aponeurosis of the internal vastus to the patella under tension. 1944 L. Glassop *We were Rats* xviii. 102 'Where'd you get all the smash?' asked Pat. 'The Harday organization,' said Gordon, 'works fast. I reefed it off a few Parsees like steam.' 1949 —— *Lucky Palmer* xiv. 124 Mugs deserve to have their dough reefed off them. 1953 K. Tennant *Joyful Condemned* xxiii. 223 They vowed it [*sc.* a magpie] ran squawking to inform on anyone who was reefing down a bit of lightning conductor to make an aerial. 1955 *Publ. Amer. Dial. Soc.* xxiv. 95 Some careful tools reef every score. 1959 *Economist* 7 Feb. 505/1 Where public servants..feather their nests when they are not reefing money off honest citizens. 1967 K. Giles *Death in Diamonds* vi. 104 If I go near the car pool they'll reef it off me. 1976 *Courier-Mail* (Brisbane) 17 July 24/9 Collins 'reefed' his $140 watch from his left hand. 1977 *Times* 13 July 5/4 As the talent suckers chummy, the wire reefs his leather... A slick pickpocket team has a private language for its dirty work.

c. To feel the genitals of (a person). *coarse slang.*

1962 Parker & Allerton *Courage of his Convictions* i. 33, I enjoyed reefing girls much more than lessons. The girls enjoyed it too. 1972 B. Rodgers *Queens' Vernacular* 101 Cop a feel..reef (Brit gay sl); take somebody's pulse.

B. *Comb.* **reef-topsail,** used *attrib.* to designate a breeze of a strength in which topsails are reefed; also *fig.* ? *obs.*

1840 R. H. Dana *Two Yrs. before Mast* xxxi. 235 We had a steady 'reef-topsail breeze' from the westward. 1849 H. Melville *Redburn* 1648 By night it was a reef-topsail-breeze. 1909 B. Lubbock *Deep Sea Warriors* 16 The sail-maker's reef-topsail voice drowned my question.

reef, *v.*[2] Add: So **ree·fing** *vbl. sb.*[2]; also *attrib.* (Further examples.)

1865 *Mining Surveyors' & Registrars' Rep.* (Dept. Mines, Victoria) Sept. 46 The southern or Gipps Land slope of the Great Dividing Range..will become one vast reefing district. 1874 A. Bathgate *Colonial Experiences* viii. 95 Quartz crushing for gold..gives abundant promise for the future, notwithstanding that the interest of the speculating public has been somewhat shaken in 'reefing'. 1874 C. Holloway *Jrnl. Visit N.Z. 1873–75* I. 121 (typescript), These Block's [*sic*] are distant about 20 miles from the rising reefing district of Lyell. 1906 J. M. Bell in P. Galvin *N.Z. Mining Handbk.* 5 Reefing is being carried out at a number of places.

reefable (rī·făb'l), *a.* [f. REEF *v.*[1] + -ABLE.] Capable of being reefed.

1909 *Westm. Gaz.* 26 Oct. 5/1 Instead of having two rigid planes or wings, set one on either side of the body, it has a single transversal span of canvas which is reefable, like that of the sail of a ship.

reefed, *ppl. a.* Add: **2.** Also with *up*, and *transf.*

1874 K. H. Digby *Temple of Memory* iv. 78 The windmills with the reef'd-up sails. 1962 *Into Orbit* 245 *Reefed*, the condition of a parachute which is not fully deployed, in order to reduce the initial stress.

reefer[1]. Add: **1.** (Earlier and later examples.)

1818 'A. Burton' *Adventures J. Newcome* i. 40 'Hoy! Reefers! Reefers!—with your sport you seem to make a *Dover Court*.' 1939 Joyce *Finnegans Wake* (1964) 323 Reefer was a wenchman.

2. (Earlier and later examples.) Also (*N Amer.*), an overcoat.

1878 C. Hallock *Amer. Club List & Sportsman's Gloss.* p. ix/2 *Reefer*, a short jacket worn by sailors. 1921 [see *covert cloth*]. 1935 A. J. Pollock *Underworld Speaks* 96/1 *Reefer*, an overcoat. 1947 *Words: New Dict.* 480/1 *Reefer, n. Slang.* 2. A short, double-breasted overcoat. 1968 [see *LOUNGE *sb.* 2 c]. 1970 J. H. Gray *Boy from Winnipeg* 47 Our winter overcoats—'reefers' we called them—lasted so well that one might serve as many as three boys before being discarded.

attrib. (Further examples.)

1901 G. B. Shaw *Capt. Brassbound's Conversion* ii. 241 Sprawl supine on the floor, with their reefer coats under their heads. 1928 R. Macaulay *Keeping up Appearances* xv. 170 Cary spoke with a hint of nervous defiance, thrusting her hands into the pockets of her reefer coat. 1936 N. Streatfeild *Ballet Shoes* x. 161 On the day of the interview, Nana cleaned Pauline's reefer coat, and blue beret. 1955 *Times* 20 Aug. 3/3 He was wearing a reefer jacket and uniform trousers under a blue naval raincoat. 1969 N. W. Parsons *Upon Sagebrush Harp* i. 4 Rena and I shivered in our white serge reefer coats. 1978 *Jrnl. R. Soc. Arts* CXXVI. 702/2 Pugin shocked his contemporaries by attending important meetings with aristocratic clients dressed in a seaman's jersey and reefer jacket.

3. [Or perh. ad. Mexican Sp. *grifo* marijuana, one who smokes marijuana.] A cigarette containing marijuana; marijuana; one who smokes marijuana. Also *attrib. slang* (orig. *U.S.*).

1931 [see *CAMP *a.* (and *sb.*[5])]. 1932 *Melody Maker* Sept. 749/2 'Song of the Weed', 'Got the South in my Soul', 'I Heard', and 'Reefer Man' are all worth your half-crowns. 1933 *Chicago Defender* 2 Dec. 5 The humble 'reefer', 'the weed', the marijuana, or what have you by way of a name for a doped cigarette has moved to Park Ave. from Harlem. 1940 R. Chandler *Farewell, my Lovely* xxxiii. 256 He sold reefers... With the right protection behind him. 1946 S. Jackson *Indiscreet Guide to Soho* 120 'Reefers' (cigarettes made from marihuana) used to sell in thousands in the West End before the war. 1952 *Amer. Speech* XV. 335/2 One who smokes is a nicotine-hound or a reefer (especially a smoker of marihuana). 1952 M. Tripp *Faith is Windsock* ix. 137 Got any reefers?..Gimme some skin man. 1956 'N. Shute' *Beyond Black Stump* 279 And then we got to smoking those reefers... Those cigarettes! 1959 *News Chron.* 26 Aug. 1/2 'Reefer' cigarettes, made up from hashish, are sold in the West End and in Notting Hill at 5s. apiece. 1967 M. M. Glatt et al. *Drug Scene* iv. 49 Quite a few were on heroin and cocaine, most smoked reefers, but not too many were on Purple Hearts. 1972 W. Labov *Lang. in Inner City* p. xxii, Reginald then makes another disruptive move, suggesting that they get a bag of reefer (marijuana). 1976 *Milton Keynes Express* 2 July 9/6 [He]..was fined £100 after three ounces of cannabis, enough to make 240 'reefers', was found at his home. 1979 *High Times* Mar. 25 Louisiana state cop displays a pot of pot discovered among 30 tons of Columbian reefer.

4. *Criminals' slang.* (See quots.)

1935 N. Ersine *Underworld & Prison Slang* 62 *Reefer*,..a pickpocket. 1941 Baker *Dict. Austral. Slang* 59 *Reefer*... (2) A pickpocket's accomplice.

reefer[2]. Add: **1.** (Earlier example.) Also *N.Z.*

1859 *Adelong Mining Jrnl.* 15 Apr. 4/2 On Monday last he made known to a few of our old reefers his discovery. 1940 Baker *N.Z. Slang* iv. 28 Gold-fields brought [to N.Z.] the reefer, the deep lead, the gutter, the monkey shaft.

3. *Austral.* One associated with the Great Barrier Reef.

1951 J. Devanny *Travels N. Queensland* xv. 78 We plunged into the water up to our knees..holding the sticks which experienced 'reefers' invariably carry.

ree·fer[3]. Alteration of REFRIGERATOR. Usu. = *refrigerator car* or *ship*. Also *attrib.*, and as *adj.* = REFRIGERATED *ppl. a.*

1914 *Wells Fargo Messenger* III. 39 Ten thousand halibut must be packed in Wells Fargo 'reefers' between sun-up and

sunset. 1924 'Digit' *Confessions 20th Cent. Hobo* 12 *Reefer*, a refrigerator box car for perishable goods with an ice-box at each end. 1926 *Amer. Speech* I. 652/2 *Reefer*, refrigerator. 1951 *Manch. Guardian* 27 Sept. 14/2 Denmark also provides examples of English, or rather American naming, in the African Reefer and the Indian Reefer—'Reefer' being an Americanism for a vessel carrying refrigerated cargo. 1953 *Sun* (Baltimore) 7 Nov. (B ed.) 6/3 Then..to San Francisco for 'reefer' cargo—refrigerated fruit. 1958 J. Kerouac *On Road* i. 19 We didn't know..what boxcars and flats and de-iced reefers to pick. 1961 *Amer. Speech* XXXVI. 273 *Reefer box*, a refrigerated trailer. 1963 *North* (Ottawa) May–June 14/1 Price had been showing them how to can the local fruit and prepare it for freezing in the Indian Affairs reefer. 1963 T. Pynchon *V.* i. 22 To the mezuzah nailed up over the vegetable reefer and the Zionist banner hanging in back of the salad table Da Conho added this prize. 1965 R. B. Oram *Cargo Handling* (1969) vi. 99 Refrigerated space is now commonly referred to as 'reefer' space. 1968 P. Durst *Badge of Infamy* iv. 32 Steaks are in the reefer, the french fries are all out and in the wire basket. 1971 *Maclean's Mag.* Sept. 34/1 Reefers are insulated vans with Thermo King refrigerated units on them. They can carry anything from ice cream to corpses. 1976 *Times* 6 Oct. 21 The underlying strength of the refrigerated ship (reefer) business. 1978 *Jrnl. R. Soc. Arts* CXXVI. 186/1 At present in the Dry Cargo Fleets of the World the United Kingdom is first in Reefers, second in Containers and, it would seem, is leaving Bulkers and General Cargo to others. Reefer and Container ships certainly need high grade officers who are good navigators and capable ship handlers in traffic.

reefing, *vbl. sb.*[1] Add: **b.** *reefing breeze* (later example), *gear, hook, spindle, wheel.*

1956 A. F. Loomis *'Hotspur' Story* 118, I was sailing in a reefing breeze. 1911 J. Barten *Compl. Naut. Pocket Dict.* 156/2 *Reefing gear*, Mechanismus zum Segelreffen. 1961 F. H. Burgess *Dict. Sailing* 169 *Reefing gear*, patent roller fittings used in some small sailing boats, to dispense with the use of reef points. 1860 'Vanderdecken' *Yarns* 36 The topmast shrouds..should have..reefing hooks and thimbles. 1962 *Roving Commissions 1961* 43 Mr 'Christie' Mahoney arranged for our reefing spindle to be mended. 1840 *Civil Engin. & Archit. Jrnl.* III. 104/1 This vessel..has the reefing wheels after Mr. Hall's patent.

Reek (rīk), *sb.*[3] *Ireland.* [Var. of RICK *sb.*[1]: cf. REEK *v.*[3]] A mountain, used *spec.* in *pl.* in *Macgillicuddy's Reeks* (also, *the Reeks*), county Kerry. Also in *sing.*, as pop. name for Croaghpatrick, county Mayo. *Reek Sunday*, the last Sunday in July, on which pilgrimages are made to Croaghpatrick.

1780 A. Young *Tour in Ireland* i. 381 Nothing stops the eye till Mangerton and Macgilly Cuddy's Reeks point out the spot where Killarney's lake calls for a farther excursion. *Ibid.* ii. 3 Mangerton, and the Reeks, in Kerry; the Galbies in Corke..these are the principal in Ireland. 1808 J. Milner *Let.* 22 Sept. in *Inquiry concerning Ireland* (ed. 2, *c* 1810) 326 The forked, cloud-capped Reeks, overlooking the Atlantic Ocean. 1870 P. W. Joyce *Irish Local Names* 36 Croagh; *Cruach*, a rick or stacked up hill... Croaghpatrick; St. Patrick's rick or hill. 1871 T. C. Pope *Council of Vatican* 236 He required a period of nearly six weeks to complete the remaining portion of the journey to Rome. He commenced the ascent of the Alpine reeks on a Friday. 1922 Joyce *Ulysses* 290 From the streamy vales of Thomond, from M'Gillicuddy's reeks the inaccessible and lordly Shannon the unfathomable. 1930 *Irish Rosary* May 321 From our drawing-room windows one had a perfect view of the Reek. 1959 D. D. C. P. Mould *Peter's Boat* iii. 31 Carrauntual, the 3414 foot height at the western end of the great ridgeway of the MacGillycuddy's Reeks. *Ibid.* 32 The ridge of the Reeks springs up suddenly, a wall of rock, from the plains of Kerry. 1960 [see *GARLAND *sb.* 9]. 1964 B. Whelpton *Unknown Ireland* vii. 98 The Gap of Dunloe between the Reeks and the Purple Mountains.

reek, *v.*[1] Add: **3. b.** (Later examples.)

1961 *Newark Evening News* 21 Nov. 12 The day before Election Day, to which we are entitled as a legal holiday, we were informed to report to our respective polls to work as 'workers of the party'.... Such tactics reek of totalitarianism! 1969 *Listener* 3 Apr. 467/3 The plot fairly reeks of the confessional.

reel, *sb.*[1] Add: **2. c.** *off the reel* (further examples); also, immediately, quickly; also *right* (or † *sharp*) *off the reel*.

1825 J. Neal *Bro. Jonathan* I. vi. 156 So then, says he to me, says he; sharp off the reel;—as 'cute a feller, that, as ever you seed. 1833 J. K. Pauling *Banks of Ohio* II. v. 78 I'd as good a mind as I ever had to shoot a wild deer, to have a fight with him off the reel, and settle the right of soil at once. 1835 *Gentleman's Vade-Mecum* (Philadelphia) 14 Feb. 3/1 Where's my old man—tell me that..where's Tom Bloomberg—tell us right off the reel. 1900 Ade *Fables in Slang* 27 He could tell you quick—right off the reel. 1927 *Daily Tel.* 3 May 17/2 The Eton XI..were undefeated, having scored four victories off the reel. 1941 J. Smiley *Hash House Lingo* 40 *Off the reel*, added to an order to signify a 'rush order'. 1946 *Sunday Dispatch* 8 Sept. 6/5 Won six races off the reel for Wembley Cubs. 1951 F. Yerby *Woman called Fancy* ii. 44 'You want to bring me back?' Court said. 'Why, little Fancy?' 'Can't answer that —not right off the reel. Don't rightly know myself.'

3. b. *spec.* The flanged cylinder or core on which magnetic tape or punched paper tape is wound; also *transf.*, a length of tape wound on such a cylinder. (Further examples.)

1939 *Wireless World* 19 June 611/2 (*caption*) The necessary controls, together with the reels of steel tape, are mounted on top of the [recording] cabinet. 1953 E. T. Canby *Home Music Syst.* xiii. 227 The reel size is seven inches. 1956 *RCA Rev.* XVII. 366 The tape is unwound from a reel..and after passing over the video heads, is pulled by the capstan. 1958 H. G. M. Spratt *Magnetic Tape Recording* iii. 55 The tape..when wound on the normal type of spool employed, will result in a reel ranging from 5–12 in. in diameter. 1964 *Communications Assoc. Computing Machinery* VII. 630/1 This standard covers the physical dimensions of take-up (or storage) reels, with either fixed or separable flanges, so that reels of perforated tape may be interchanged among machines of various manufacturers. 1977 T. Allbeury *Man with President's Mind* ix. 101 The video tapes..were..sent in a special top priority diplomatic bag to Moscow. The priority was so high that the reels were not even rewound before despatch.

c. *Cinemat.* and *Photogr.* A flanged cylinder on which film is wound; usu. *transf.*, a length of film wound on such a cylinder; *loosely*, a (long) portion of a motion picture; also, † the spool on which photographs were mounted in a mutoscope (see quot. 1901).

In early usage in Cinemat., *reel* was restricted in signification to a fixed length of film, normally one thousand feet at 35 mm. gauge, complete films being called *two-reelers*, etc. (see *REELER[1] 3). The word is now used of other fixed lengths, of standard lengths of film at other gauges, and also without regard to length.

1896 Queen Victoria *Jrnl.* 3 Oct. (1980) 222 The new cinematograph process,—which makes moving pictures by winding off a reel of film. 1901 *Everybody's Mag.* Aug. 230/2 Ordinary photographs are printed from the negative film and mounted on a central spool, from nine hundred to twenty-seven hundred pictures to a 'reel', as it is called. 1912 *Maclean's Mag.* Apr. 634/1 It is comprised in three 'reels', which means that there are 3,000 feet of film, requiring a full hour to run. 1915 *Chicago Herald* 1 Nov. 8/5 'The Sentimental Lady' is five reels of whipped cream lightness and frothy texture. 1916 F. H. Richardson *Motion Picture Handbk.* (ed. 3) 198 There has been some inclination to increase the size of reels to two and even three thousand feet... One thousand feet of film has been and should continue to be the standard reel of film. 1921 B. Schulberg *What makes Sammy Run?* iv. 51 People..become characters in Hollywood movies... In the last reel the good brother has to be killed off so that the bad brother can be regenerated. *Ibid.* xi. 193 Two-reel horse-operas. 1968 *Tamarack Rev.* Spring 9 He was waiting for the others to arrive so they could run through the second reel together. 1972 *Daily Tel.* 19 June 24/4, I don't know what I took. I know I shot three reels—36 pictures on each—but don't ask me what is on them.

d. *reel-to-reel* attrib. phr.: applied to a form of tape-recorder in which tape passes between two reels which are mounted separately on the recorder (cf. *open-reel* s.v. *OPEN *a.* 22 c; contrast *CASSETTE d and *CARTRIDGE 1 d (iii)); also to the tape used in such a machine; also *absol.*

1961 *N.Y. Times* 10 Sept. ii. 15/3 'Reel-to-reel' machines move the tape..from an open supply reel to a take-up reel. 1967 *Tape Recording Mag.* Jan. 12/2 Transferring recording material on to an orthodox quarter-inch tape reel-to-reel model. 1975 *Gramophone* Jan. 1389 (Advt.), It will make going in for tape much more worthwhile—cassette or reel-to-reel or both. 1976 *Broadcast* 29 Nov. 15/1 Capitol Records..wants..listeners..to send cassettes or reel-to reel tapes of their vocal efforts. 1977 *Design Engin.* July 97/2 A range of two and four channel cassette tape recorders offers recording facilities usually associated with reel-to-reel instruments.

6. *reel line* (earlier example), *stand*; **reel-backing** *U.S. local* (see quot. 1976); **reel barge,** a barge, carrying extended lengths of pipe coiled on a reel, which is used to lay submarine pipelines; also *attrib.*; **reel boy,** a boy attending to the reeling of yarn, etc.; **reel-fed** *a. Printing*, using reeled paper; **reel-land** *nonce-wd.*, the world of the cinematograph; **reel man,** a sailor who holds the reel from which a log-line depends; **reel-room,** the room in a cinema where reels of cinematographic films are kept; **reel ship,** a self-powered ship performing the function of a reel barge; **reel timing,** a method of playing certain kinds of slot machine (see quot.).

1959 W. Faulkner *Mansion* xii. 346 His uncle had the gun..a black strong small-gauge length of reel-backing running from the trigger through a series of screw eyes to the sash of the window screen. 1976 C. S. Brown *Gloss. Faulkner's South* 161 *Reel-backing*.., a heavy (often old) length of fishing-line wound as the first layer on a fishing reel, and hence the last piece if the entire line is reeled out. 1972 *Study of Potential Benefits Offshore Oil & Gas Devel.* (Internat. Managem. & Engin. Group of Brit. Ltd.) x. 80 There is also the highly promising reel-barge technique, which, however, has so far been limited to comparatively small diameter pipe. 1975 *Offshore* Aug. 121/1 Santa Fe International Corp's reel barge Chickasaw has successfully laid a 10-in pipeline in more than 1,000 feet of water. 1975 [see *PIPE-LAY *sb.*]. 1975 *Offshore Progress* (Shell Internat. Petroleum Co.) 17 Another technique which has been used for some time in the Gulf of Mexico for small-diameter pipelines is the reel barge: the pipe is welded together on-shore (thereby minimizing expensive offshore time) and coiled on a reel. The barge is then pulled along the right of way, unreeling pipe behind it. 1918 *Nation* (N.Y.) 7 Feb. 130/1 Defeating the man in whose flax mill he had worked as a reel boy. 1946 V. S. Ganderton in H. Whetton *Pract. Printing & Binding* xii. 153/1 Multi-colour rotaries and methods have been brought to a high degree of perfection... They are either sheet- or reel-fed, but the delivery is flat. 1971 D. Potter *Brit. Eliz. Stamps* xiv. 150 Reel-fed printing takes the paper into the press on a continuous reel, and the printed sheets are later divided. 1926 *Chambers's Jrnl.* Aug. 605/1 Some..would fain have treated her much in the style of those who at the present day mob the stars of reel-land. 1837 J. Kirkbride *Northern Angler* 3 With regard to the reel and reel-line I need say nothing. 1851 H. Melville *Moby Dick* III. xxxix. 226 The towering resistance of the log caused the old reelman to stagger strangely. 1928 *Daily Express* 8 Oct. 2/1 Hundreds of children..filed from the..Cinema..while the staff tackled a fire in the reel-room behind the gallery. 1976 *Offshore Platforms & Pipelining* 143/2 Santa Fe expects to launch its new reel ship in 1977. 1889 *Cent. Dict.*, Reel-stand. 1961 *World's Press News* 6 Jan. 7 (Advt.), A..Rotary Newspaper Press. The machine will be arranged with..six 3-arm magazine reelstands. 1969 E. H. Pinto *Treen* 320 The silk on the spool of the cocoon winder was then transferred to the lead-weighted reel stand. 1964 A. Wykes *Gambling* iii. 71 But another method, called 'reel timing', was not easy to track down... First, the player had to determine the exact number of seconds that each reel spun before coming to rest after the handle was pulled. Secondly, he had to memorize the sequence of all 60 symbols on the three reels... He had to be able to count..a certain number of seconds between the insertion of the coin and the pulling of the lever.

reel, *sb.*[3] Add: **1.** (Further examples.) *Virginia reel*: see also VIRGINIA 1 d.

1745 in R. Forbes *Lyon in Mourning* (Scott. Hist. Soc. Ser.) (1895) I. 208 He..took his share in several dances, such as minuets, Highland reels (the first reel the Prince called for was, 'This is not mine ain house'). 1788 J. O'Keefe (*title*) The Highland reel. 1814 [see FOURSOME *a.* 2]. 1818 B. Dun *Nine Quadrilles* Pref., There are two kinds of music to which the Scotch reel is danced, viz. the reel properly so called, and the strathspey. 1840 Lytton *Money* iii. v. 85 Do you remember her dancing the Scotch reel with Captain Macnaughten? 1843 [see EIGHTSOME *a.* or *adv.*]. 1913 *Times* 3 June 11/3 The old Scotch reel is rarely danced today, as the young folk prefer eight-somes as more 'romping'. 1950 *Oxf. Jun. Encycl.* IX. 278/2 The chief dances performed at the [Highland] Games are the Sword Dance, the Foursome Reel, the Reel of Tulloch, the Highland Fling, and the Seann Triubhas. 1955 *Highland Dancing* ii. 61 During the Highland Reel the distance between the points may be reduced. 1964 W. G. Raffe *Dict. Dance* 414/2 *Reel o' Tulloch*... This and other Scottish Reels in 4/4 time, come from ancient periods. *Ibid.*, This Reel is a variant of the Foursome Reel. 1971 *Country Life* 23 Dec. 1790/1 Get up and take a whirl in an eightsome reel. 1974 *Encycl. Brit. Micropædia* VIII. 468/3 Popular reels include the Irish Sixteenhand Reel and the Scottish reels Maury's Wedding and the Duke of Perth.

2. (Later examples.) Also *U.S. dial.*, a song for such a dance.

1818 [see sense 1 above]. 1883 Grove *Dict. Mus.* III. 92/1 The Irish reel is played much faster than the Scotch. 1964 *Amer. Folk Music Occasional* i. 61 From such a man you will hear ballads, breakdowns, reels. 1968 J. Arnold *Shell Bk. Country Crafts* 320 In the form of the reel, strathspey, or pibroch it has a primitive echo which is evocative for all its convention and sophistication.

reel, *v.*[2] Add: **2. b.** Also const. *out* (*rare*). Also (const. *off*), to cover (a distance, etc.) rapidly; to accomplish or perform without pause or effort.

1870 'Mark Twain' in *Galaxy* Dec. 883/2 The hands [of my watch] would straightway begin to spin round and round... She would reel off the next twenty-four hours in six or seven minutes. 1872 —— *Roughing It* iv. 46 We reeled off ten or twelve miles. 1928 *Granta* 30 Nov. 172/2 Milton just reeled out bits about Christmas when he was up here. 1961 *Trains Illustr.* Nov. 685/2 The 11 miles between posts 137 and 148 were reeled off at an average of 93.8 m.p.h. 1972 J. Mosedale *Football* iv. 50 The Rams became the most exciting team in football, reeling off six straight victories. 1976 *0–10 Cricket Scene* (Austral.) 33/2 In a fine performance in which he reeled off 35.5 overs, he captured 5–148 in the Third Test in Australia at Adelaide in 1974.

3. a. (Later examples.) Also, to draw *in*, as with a reel.

1942 *Tee Emm* (Air Ministry) II. 57 A large passenger aircraft was struck while its trailing aerial was still reeled out. 1975 *Offshore* Aug. 121/1 The test pipeline was reeled from the Chickasaw into water depths of approximately 1,000 to 1,040 feet. 1978 *Sci. Amer.* Feb. 158/1 To launch a kite hold it about 20 degrees forward of the vertical and in approximately the correct flying attitude, release it as a gust of wind passes and slowly reel out line. *Ibid.* 158/2 To bring the kite down you probably will reel in the kite line. 1979 *Amat. Photographer* 10 Jan. 73/2 It's also dangerous, and illegal, to have permanent power points in a bathroom, so you must reel in an extension cable each time so you can use your enlarger, safelights, etc.

b. *intr.* Const. *out*: to become uncoiled from a reel. So **ree·l-out** *sb.*

1975 *Daily Tel.* 11 Sept. 3/7 Three cases of seat belts reeling out with no load being taken by the belt until all the webbing was off the reel, and four cases of excessive reel-out.

re-elect, *v.* Add: Hence **re-ele·cted** *ppl. a.*

1838 J. L. Adolphus in J. G. Lockhart *Life Scott* VII. vi. 221 He spoke very beautifully and warmly of the re-elected candidate who sat by him. 1974 *Times* 4 Mar. 2/7 Re-elected Conservatives are anxious to make their views known.

reeler[1]. 2. Delete *Obs. rare*[−1] and add: In mod. use, a machine which winds paper, yarn, etc., on to reels. (Later examples.)

1906 [see *RE-REEL *v.*]. 1907 [see *cop-winder* s.v. *COP *sb.*[1] 8]. 1929 Clapperton & Henderson *Mod. Paper-Making* xvi. 246 Many defects in the paper..will..cause a break at the reeler, owing to the high tension of the paper as it passes from the roll to the reeler bar. 1952 F. H. Norris *Paper & Paper Making* xvii. 246 The reeler is equipped with a yardage counter and may have four winding drums. 1955 S. C. Gilmour *Paper* vi. 59 On most modern machines..a drum reeler is more common.

3. *Cinemat.* Used with a qualifying number, as *two-reeler* or *two reeler*, to designate a film consisting of the given number of reels. orig. *U.S.* Cf. *REEL *sb.*[1] 3 c.
1916 *Chicago Herald* 17 Feb. 3/4 Essanay will make an international release of the eight-reeler. 1922 H. L. Wilson *Merton of Movies* v. 90, I got another two reeler to pull off after this one. 1938 F. H. Ricketson *Managem. Motion Picture Theatres* 121 Coming-attraction trailers can be spotted after the newsreel. A two-reeler, if the length of the feature permits, or a one-reel comedy can follow. 1976 L. Kennedy *Presumption of Innocence* ii. 88 A kind of mad, surrealistic quality, like an early Chaplin two-reeler.
Hence ree·lerman, one who operates a reeler (sense 2).
1929 Clapperton & Henderson *Mod. Paper-Making* xvi. 247 The yardage is kept on two different tickets by the reelerman.

reeler[2] (rī·lǝɹ). [f. Reel *v.*[1] + -er[1].] A stagger; esp. in slang phr. *to cop a reeler*, to get drunk. **b.** One who sways or staggers; a drunken person.
1937 J. Curtis *You're in Racket, Too* v. 60 Make him swear blind he'll be quiet as he comes up the stairs, see? Of course, if he's copped a reeler you'll have to skip it. 1960 A. Clarke *Later Poems* (1961) 76 Though every farewell has been banned, Student or reeler from a band Flung it.

reeling, *vbl. sb.*[1] Add: **1. a.** (Further examples.) Also *concr.*, reeled yarn or the like.
1894 *Cassell's New Technical Educator* IV. 369/1 The reelings are then weighed and made up into bundles. 1906 W. Macfarlane *Princ. & Pract. Iron & Steel Manuf.* iv. 47 Bars for certain purposes are straightened by reeling. 1952 F. H. Norris *Paper & Paper Making* xvii. 246 There are also the faults, which in turn will add their quota of troubles in supercalendering and reeling. 1973 J. G. Tweeddale *Materials Technol.* II. iv. 95 A simplified form of a two-high mill of this kind can be used for straightening rolls and tubes by causing spiral flexture [sic] in the cold condition, a process called reeling. 1974 *Encycl. Brit. Macropædia* XVIII. 173/2 Reeling is the process of unwinding raw silk filament from the cocoon directly onto a holder.
b. *reeling drive, machine* (later examples).
1962 G. A. T. Burdett *Automatic Control Handbk.* vii. 9 This is the basis of a large number of electronic control schemes embracing..coiling and reeling drives [etc.]. 1964 Harbord & Hall *Metallurgy of Steel* xxxii. 506 Rods are passed through the reeling machine. This consists of a pair of conical rolls, revolving both in the same direction, and lying side by side, their axes being placed, not horizontally, but inclined to the horizon a few degrees in opposite directions, so as to cross each other at a slight angle in the middle of their length. 1926 J. R. Walker *Story of Steel* xii. 217 The next step is to pass the tube through what is known as the reeling machine... In this operation any mill-scale is removed; the tubes are given a smooth, burnished surface. 1971 W. K. V. Gale *Iron & Steel Industry: Dict. Terms* 168 *Reeling machine*, a machine which straightens round steel bars by passing them between specially shaped rollers which induce reverse bending.

reely, reelly (rī·li). Representing a vulgar pronunciation of Really *adv.*[2]
In quot. 1792 representing the pronunciation of a German speaker.
1792 F. Burney *Jrnl.* May (1972) I. 153 Mrs. Schwellenberg exclaimed 'But, Miss Berner, I hear it bin reelly true you will Marry?' 1910 H. G. Wells *Hist. Mr. Polly* vi. 158 Thought my bicycle was on fire..all right reely. 1933 E. A. Robertson *Ordinary Families* x. 222 'Is it a very good boat?'..'Oh, reely, Ted.' 1939 Joyce *Finnegans Wake* (1964) ii. 527 Of course it was downright very wicked of him, reely meeting me disguised. 1967 N. Marsh *Death at Dolphin* v. 'Well—I don't know, reely, if we've anybody free at the moment,' said the clerk.

re-embarkment. Add: Also 8 re-im-. (Earlier and later examples.)
1728 G. Carleton *Mem. Eng. Officer* 95 The heavy artillery landed for the siege was return'd aboard the ships, and everything in appearance prepar'd for a re-imbarkment. 1915 J. Churchill *Jrnl.* 21 Mar. in M. Gilbert *Winston S. Churchill* (1972) III. Compan. I. 722 The re-embarkment would take longer.

re-embo·died, *ppl. a.* [f. Re-embody *v.* + -ed[1].] Reincarnated. So **re-embo·diment**.
1901 'A. Hope' *Tristram of Blent* xiii. 175 That re-embodiment or resurrection of her in the girl who moved and talked and sat like her, who had her ways though not her face. 1924 W. B. Selbie *Psychol. Relig.* 271 In Indo-European folk-lore, dogs, wolves, and hares represent such re-embodied spirits.

re-embroi·der, *v.* [Re- 5 a.] *trans.* To ornament with additional embroidery. So **re-embroi·dered** *ppl. a.*; **re-embroi·dering** *vbl. sb.* (also *attrib.*).
1927 *Daily Express* 8 Apr. 5 The gown is of ivory silk lace, re-embroidered with small china beads. *Ibid.*, Interesting example of the re-embroidering vogue. In this case an embroidery of coloured taffeta and bugle beads was applied to black Spanish lace. 1963 *Times* 24 Jan. 12/4 Re-embroidered white lace is used. 1968 J. Ironside *Fashion Alphabet* 235 *Re-embroidered*: This is a lace with an all-over motif outlined and emphasised by re-embroidering either with silk thread, ribbon, braid or metallic yarns. 1974 *Times-Picayune* (New Orleans) 15 Aug. v. 6/1 The bride..wore a peau de soie gown styled with a sculptured yoke of re-embroidered lace and a cameo neckline.

re-emission. (Further examples.)
1955 Friedman & Weisskopf in W. Pauli *Niels Bohr* 136 These states..have a finite lifetime, since they can decay by the re-emission of the incident particle. 1968 G. M. B. Dobson *Exploring Atmosphere* (ed. 2) iii. 62 As a result of this complicated process of constant absorption and re-emission of radiation,..the ground..absorbs about half of the incoming solar radiation.

re-emit, *v.* (Further examples.)
1924 *Proc. Physical Soc.* XXXVI. 422 It is possible that the α-particle is in some way attached to the residual nucleus. Certainly it cannot be re-emitted with any considerable energy, or we should be able to observe it. 1955 *Bull. Atomic Sci.* Mar. 92/2 Several neutrons are re-emitted when a uranium atom is exploded by one neutron. 1969 *Times* 19 Feb. 13/6 Atoms and molecules of the atmosphere absorb sunlight and then reemit the energy at wavelengths which are characteristic of the particular type of atom or molecule.

re-e·mphasize, *v.* [Re- 5 a.] *trans.* To emphasize again, to place renewed emphasis on. So **re-e·mphasis**.
1857 E. B. Browning *Aurora Leigh* I. 26 From many a volume, Love re-emphasised. 1894 J. R. Illingworth *Personality, Human & Divine* i. 18 This intimacy and immediacy of possible union between the soul and God..had long vanished from the popular religion. Luther re-emphasized it. 1934 Webster, Re-emphasis. 1948 J. Towster *Polit. Power in U.S.S.R.* viii. 174 The changes or re-emphases that took place. 1971 *Nature* 23 Apr. 490/3 He also reemphasizes his original contention that an anomalous situation does exist.

Reemy, var. *R.E.M.E., Reme.

re-enclo·se, *v.* [Re- 5 a.] To enclose again.
1598 Florio *Worlde of Wordes* 307/2 *Racchiudere*, ..to re-enclose or shut vp againe. 1849 E. A. Poe *Let.* 13 Jan. (1949) II. 418 Please re-enclose me the printed papers. 1870 'Mark Twain' *Let. to Publishers* (1967) 30, I re-enclose the Express letters. 1907 R. Brooke *Let.* 29 Oct. (1968) 114 You probably know all this already from Mrs Leon. I re-enclose her letter.

re-encounter, *sb.* **b.** (Later examples.)
1904 *Daily Chron.* 28 July 8/5 If she is wise she will..avoid the exhausting re-encounters in the flesh. 1948 *Times Lit. Suppl.* 18 Sept. 526/3 Mr. Sassoon..sets down his personal experience of the poet. 1974 Frith & McLaughlin in R. K. Harris *Nuclear Magnetic Resonance* (Chem. Soc. Specialist Periodical Rep.) III. xii. 367 This..broadens our definition of cage recombination to include reaction of the original partners in the radical pair on re-encounter after their initial diffusive separation.

re-enforcer. [In Dict. s.v. Re-enforce *v.*] (Later example.)
1914 W. McDougall *Social Psychol.* 404 The energy of the new impulse..may function as a re-enforcer of purely intellectual activities.

re-engine, *v.* Add: (Later examples.) Also of an aeroplane. Hence **re-e·ngined** *ppl. a.*; **re-e·ngining** *vbl. sb.*
1941 *Sun* (Baltimore) 20 Jan. 2/7 Reengined and rearmed Hurricanes, new Tornadoes..and improved Defiants. 1955 *Times* 23 Aug. 5/1 The Grants Scheme should be extended to include the re-engining and re-conditioning of suitable trawlers. 1967 *Jane's Surface Skimmer Syst.* 1967–68 36 [caption] A re-engined version of the SR.N5. 1973 *Stornoway Gaz.* 2 June 2/5 The cruising speed will be 14 knots—she has not been re-engined.

re-engineer, *v.* [Re- 5 a.] *trans.* To design and construct anew; also *transf.* and *fig.*, to arrange or contrive anew.
1944 *Sun* (Baltimore) 14 Dec. 4-0/3 Management and labor should decide now..whether jobs re-engineered for women will be restored to provide work for servicemen. 1946 *Jrnl. Inst. Electr. Engineers* XCIII. IIIA. 59/2 It was decided therefore to re-engineer the system using aerial arrays. 1958 J. K. Galbraith *Affluent Society* xii. 194 The first task of the Madison Avenue man is to 're-engineer' his image to include something besides the production of goods. 1974 W. Rees-Mogg *Reigning Error* v. 104 The most fundamental relationships of human life have been re-engineered in order to prevent sufferings which turned out to represent the unavoidable limitations of human existence. 1977 *Economist* 5 Feb. 43 The new GT55P arm has been completely re-engineered.
So **re-enginee·red** *ppl. a.*, **re-enginee·ring** *vbl. sb.*
1962 *Rep. Comm. Broadcasting* 1960 337 in *Parl. Papers* 1961–2 (Cmnd. 1753) IX. 259 The importance..of planning the use of Bands IV and V (and when the time comes the re-engineered Bands I and III)..as an integrated whole. 1973 C. Sagan *Cosmic Connection* (1974) xxii. 150 Our motivations for planetary re-engineering must be clear. 1977 *Time* 18 Apr. 46/2 What would happen, they ask, if by accident or design, one variety of re-engineered *E. coli* proved dangerous?

re-enli·ster. [f. Re-enlist *v.* + -er[1].] A person who enlists again.
1908 *Daily Chron.* 8 May 8/2 The majority of the re-enlisters 'are the discontented, the thriftless, the criminal class'.

re-enlistment. Add: Also *attrib.*
1953 R. Wells (*song-title*) Re-enlistment blues. 1970 *Times* 28 May 7/7 A source of much resentment is the iniquitous system whereby the Army maintains a high re-enlistment rate.

re-enslavement. (Earlier example.)
1859 Ld. Lyons *Let.* 6 Dec. in Ld. Newton *Lord Lyons* (1913) I. ii. 19 The reenslavement of all the emancipated negroes.

re-enter, *v.* Add: **1. c.** (Later example.) Also in wider contexts.
1803 G. Colman *John Bull* iv. i. 56 Re-enter Dan booted. 1937 'M. Innes' *Hamlet, Revenge!* II. iii. 137 And so—just conceivably—re-enter the spies.
2. c. *spec.* of spacecraft. Also *absol.*
1961 *Times* 6 May 8/2 Seven minutes after take-off the report came through that the capsule was beginning to reenter the earth's atmosphere. 1962 S. Carpenter in *Into Orbit* 36 The capsule would not re-enter too quickly or the deceleration will be too great. 1968 *Times* 16 Dec. 7/3 A second crucial phase begins when Apollo 8 re-enters the Earth's atmosphere. 1977 G. Scott *Ult Pursuit* x. 90 The Americans..spotted it... It looked as though it was about to reenter.
d. In the drilling of offshore oil wells and similar holes: to enter (a borehole) again with a drilling bit.
1961 W. Bascom *Hole in Bottom of Sea* xii. 244 Two parallel guide lines, stretched between the ship and fittings on the bottom, had a sliding crossbar to guide the bit so that the hole could be re-entered at will. 1967 *Ocean Industry* June 15/1 Sonar system for hole-re-entry which makes it possible to re-enter a small diameter borehole in three-mile-deep water. 1968 *Proc. Offshore Exploration Conf.* 241 The drilling personnel have found it necessary to remove the drill string and reenter the hole. 1975 *Proc. Offshore Technol. Conf.* I. 27/1 One beacon..is attached to the object being re-entered (well head) and the other beacon..is attached to the re-entry tool.
4. (Further examples.)
1838 W. Howitt *Rural Life Eng.* I. i. v. 81 On applying to the steward he found that he was actually re-entered as tenant to the farm.
Hence **re-e·nterable** *a.*, capable of being re-entered.
1969 P. B. Jordain *Condensed Computer Encycl.* 423 A reenterable program can service several tasks concurrently by switching from one task to another task at high speed.

re-entering, *ppl. a.* **2.** (Further examples.)
1958 *Punch* 17 Sept. 365/1 Re-entering nose-cones. 1959 *Daily Tel.* 14 Apr. 17 The proposed American attempt to catch a re-entering satellite in a very long shot indeed. 1974 *Encycl. Brit. Macropædia* XV. 938/2 In the early 1960s a new technique..was developed, using materials similar to those employed as heat shields for re-entering space vehicles.

re-e·ntrancy. [f. Re-entrance + -y[3].] **a.** *Electr.* The state of being re-entrant; also, a measure of the number of complete turns required to trace out an armature winding (see *re-entrant *a.*[2] a).
1901 Sheldon & Mason *Dynamo Electr. Machinery* iii. 47 Any closed-coil winding, single or multiple, may be singly or multiply re-entrant, the re-entrancy being reckoned as great as that of any single winding on the armature. 1902 *Jrnl. Inst. Electr. Engineers* XXXI. 933 A few writers..take the re-entrancy as being the number of times we must go around the armature in tracing out the whole winding. *Ibid.* 935 Condition of Re-entrancy.—The first condition to be fulfilled by a proposed winding is that it should re-enter upon itself. 1907 Hobart & Ellis *Armature Construction* viii. 157 Multiplicity and re-entrancy of multiple windings.
b. The capability of being entered again.
1976 P. C. Sanderson *Minicomputers* iv. 72 For some applications, hardware should allow re-entrancy. This allows the same procedures to be performed on different blocks of data. 1976 *Nature* 11 Mar. 176/1 There is an absence of detailed comment on..modern programming techniques which are important for memory-starved minicomputers, such as subroutine re-entrancy.

re-entrant, *a.* (and *sb.*) For b read B and add: **A.** *adj.* **1. a.** (Further examples.)
1967 M. Chandler *Ceramics in Mod. World* iv. 122 It is difficult to form re-entrant shapes from two dies. 1973 J. G. Tweeddale *Materials Technol.* II. 34 Notches, flaws, or re-entrant angles..are likely to give rise to stress concentrations. 1975 D. G. Fink *Electronics Engineers' Handbk.* xii. 49 Magnetron oscillators are single-port devices. Both the slow-wave circuit and the electron stream are reentrant; i.e., the circular geometry is always used.
b. *Mus.* Designating a form of tuning of the open strings of the citole, cittern, and ukulele, in which the fourth course is tuned to a higher pitch than the third, as e', d', g, b or e', d', g, a for the cittern.
1948 *Galpin Soc. Jrnl.* I. 48 The cittern's curious re-entrant tuning gives simply-fingered versions of all the chords commonly used in contemporary music. 1961 A. Baines *Mus. Instruments* vii. 166 Its [*sc.* the cittern's] tunings..were re-entrant, with the fourth course higher in pitch than the third, as on the modern ukelele. 1976 D. Munrow *Instruments Middle Ages & Renaissance* iv. 27/1 The instrument which Tinctoris describes [*c.* 1487] is unquestionably the ancestor of the renaissance cittern, with..a re-entrant tuning for its four metal strings. *Ibid.*, The earliest account of the tuning of any stringed instrument, that of Jerome of Moravia (*c.* 1250), described three fiddle tunings, one of which is re-entrant.
2. a. Of or pertaining to something which returns upon itself, as in *Electr.*, applied to a

form of armature winding (see quot. 1901); in *Acoustics*, applied to a form of horn loudspeaker in which the bore is divided and folded upon itself before expanding to the flare, in order to reduce space.
1901 Sheldon & Mason *Dynamo Electr. Machinery* iii. 46 A singly-re-entrant winding is one in which, by successive regular advances, all the coils have been laid when an advance of 360° has been made. To be doubly-re-entrant would mean that after advancing between successive coils, in the order of their winding, is doubled; and the whole winding is not complete until the armature has been gone around, angularly, twice, i.e., through an advance of 720°. 1902 *Jrnl. Inst. Electr. Engineers* XXXI. 933 A winding is re-entrant if it comes back to the starting point and is then complete. 1928 *Gramophone* Jan. 345/1 There are now listed three models (called 're-entrant') in which a relatively broad acoustic system is, by means of embodying a double reflexion of tone, enabled correctly to expand to quite a wide-mouthed horn in no greater depth from front to back than is allowable in the relatively shallow American pattern cabinet. 1940 *Chambers's Techn. Dict.* 710/1 The majority of windings are singly re-entrant. 1960 *Practical Wireless* XXXVI. 395/2 Speakers of the re-entrant type will be found most suitable. 1961 Briggs & Cooke *A to Z in Audio* 79 One of the earliest applications of this principle [*sc.* that of the exponential horn] to sound reproduction was probably the re-entrant gramophone produced by HMV in 1927.
c. *Computers.* Of, pertaining to, or designating a program or subprogram which may be called or entered many times concurrently from one or several programs without alteration of the results obtained from any one execution.
1964 *Proc. Fall Joint Computer Conf.* I. 45 (*heading*) Method of control for re-entrant programs. *Ibid.* 45/1 A routine which permits unlimited multiple entrances and executions before prior executions are complete is called a re-entrant routine. 1970 O. Dopping *Computers & Data Processing* xiv. 221 A form of programming which allows re-entry into a partially used subroutine is called re-entrant programming. 1976 H. D. Baecker in *Virtual Storage* (Infotech International Ltd) 195 Allocation of and access to local variables in recursive or re-entrant environments. *Ibid.*, Re-entrant programs are not only a good thing because of their alleged economy of space.
B. *sb.* **b.** *Geogr.* A prominent, angular intrusion into a landform, such as an inlet between two coastal promontories or a valley extending into a hill or mountain side.
1894 [see *pocket-hole s.v. *pocket *sb.* 13]. 1899 R. T. Hill *Geol. Jamaica* i. 18 The interior mountains are marked by deeply etched knife-edged salients and angular re-entrants. 1936 *Bull. Amer. Assoc. Petroleum Geologists* XX. 1224 The profound reentrant between the escarpment of the Sierra das Furnas and that of the Serra do Tuba. 1942 J. Onslow *Bowler-Hatted Cowboy* v. 53 Dense spruce spread from the valley upwards, following the big re-entrants. 1973 J. Bonington *Next Horizon* xviii. 248 The road was like a switchback gone mad, as it backed from valley floor, over spurs, round re-entrants and down again.

re-entry. Add: **2. b.** *card of re-entry*; also, in Bridge, a card which by winning a trick gives the lead (in his own hand or in dummy) to a player who has previously had it. Also *re-entry card* and *ellipt.* as *re-entry*.
1899 A. Dunn *Bridge* 22 If the long suit hand has no card of re-entry, he will be prevented from bringing in his suit. 1901 [see *suck *v.* 6]. 1908 R. F. Foster *Auction Bridge* 84 In planning the play of a no-trumper, the declarer must be careful to provide for re-entry cards. 1958 *Listener* 9 Oct. 572/1 If he had a suit headed by K Q J and a re-entry he would pass, but overcall. 1967 Cohen & Barrow *Bridge Players' Encycl.* 405/1 *Re-entry*, a card by which a player who had the lead (including the opening lead) can regain it.
4. *Astronautics.* The return of a spacecraft into the earth's atmosphere.
1948 *Jrnl. Brit. Interplanetary Soc.* VII. 34 The technique of atmospheric re-entry will be of great importance..during excursions into space. 1961 *Guardian* 6 May 1/3 The vapour trail caused by the re-entry of the capsule. 1968 *Times* 16 Dec. 7/2 The reentry of the first rocket stage into the atmosphere. 1970 *Guardian* 18 Apr. 1/5 The spacecraft..appeared to be badly scarred by the heat of re-entry. 1977 G. Scott *Ult Pursuit* x. 89 Clearly it had not been programmed to come down where it did; you've seen how it was damaged in re-entry and landing.
5. The act of re-inserting a drilling bit into a borehole during the drilling of an offshore oil well or similar hole.
1961 W. Bascom *Hole in Bottom of Sea* xiii. 271 There would be no riser pipe or other means of hole re-entry. If the bit were once withdrawn, the hole would be lost. 1969 *Design of Deep Ocean Drilling Ship* (U.S. Nat. Research Council) 64 A riser pipe may be the best means of hole re-entry... When re-entry is attempted—after a bit change, for example—as the pipe tip approaches bottom a sensing sonde will be lowered into position at the bit which will be able to sense the tip of the projecting casing. 1968 *Proc. Offshore Exploration Conf.* 242 Reentry was attempted by two engineers with only one reentry for a number of attempts. 1974 *Geotimes* Dec. 16/2 The casing broke just below the reëntry core, and was displaced downward..after 4 successful reëntries, eventually preventing further reëntry.
e. *Surfing.* (See quot. 1968.)
1968 W. Warwick *Surfriding in N.Z.* 14/1 A re-entry is a roller coaster, taken a stage further. To perform this tactic, your board must be moving roughly parallel to the top of the wave and as the wave begins to break, kick your board to the top so that it is sitting on top of the curl with virtually no water covering its deck. Your board will appear to have stalled at the top of the wave for a second then it will be sucked back down with the curl. 1970 *Surf* I. x. 11/2 Finishes it off with a re-entry.
f. Return to one's usual place or mode of living.
1972 *Listener* 31 Aug. 270/1 The standard pattern of war veterans..a difficulty in what is called 're-entry', in getting back into the civilian society. 1974 *N.Y. Times* 8 July 1/3 'Country' is where you go..to escape from your city responsibilities, and when country-time ends, re-entry can be traumatic.
4. b. *Philately.* (See quots.)
1912 P. J. Melville *Postage Stamps in Making* I. ix. 96 The varieties of the first British stamps which are best described as having had the roller applied twice, or 're-entries' known to have existed on Plate 145 of the 1d., Small Crown, imperforate, and other plates of Great Britain. 1951 J. Sutton *Stamp Collector's Encycl.* 191 *Re-entry*, duplication of part of a stamp design due to a first impression having been inadequately erased, and thus enabling traces of its 'entry' to appear in conjunction with the new impression, causing a doubling of a part of the image... Known in the U.S.A. as the 'shift'. 1971 D. Potter *Brit. Eliz. Stamps* xiv. 159 On recess-printed stamps the entry die may momentarily make contact in part, twice, and slight doubling of the design results. This is known as a re-entry. 1972 D. & M. Patrick *Hodder Stamp Dict.* (1973) 204/1 *Re-entry*, a second attempt to rock in the stamp design on a steel plate for engraving postage stamps... A perfect re-entry leaves no trace on the stamps. 1975 B. Gunston *Philatelist's Compan.* 227 *Re-entry*, characteristic doubling or thinning of portions of the design of a line-engraved stamp caused by the impression having been entered more than once on the plate.
5. *attrib.* (*a*) in sense *2 c, as *re-entry angle, black-out, capsule, heating, parachute, problem, vehicle*; (*b*) in sense 'giving permission for re-entry into a country', as *re-entry permit, visa*; *re-entry card*: see sense 2 b above; *re-entry point Computers* (see quots.).
1970 *Daily Tel.* 17 Apr. 6/1 Without this burn, the spacecraft would miss by 100 miles the critical re-entry angle, between 5·6 and 7·2 degrees. 1966 *Electronics* 14 Nov. 54 Reentry blackout has plagued every flight in the Mercury and Gemini series. 1967 *Technology Week* 20 Feb. 10/2 Biosatellite I's re-entry capsule containing biological specimens..is believed to have re-entered in or near Australia. 1962 F. I. Ordway et al. *Basic Astronautics* i. 4 Overview of the problems of reentry heating. 1970 P. Cave *High Flying Birds* i. 11 Originally designed as a possible re-entry parachute device for space capsules, the flying sail designed by Francis M. Rogallo in the early sixties has come a long way. 1958 F. Frenaye tr. *Levi's Christ stopped at Eboli* 123 Before they are aware of it..their re-entry permit has expired, and they have lost their chance. 1972 R. Perry *Fall Guy* iv. 73 Wondering whether I'd be granted a re-entry permit. 1922 *Gloss. Terms Automatic Data Processing* (B.S.I.) 11. 39 The point at which a routine is re-entered from a subroutine is a re-entry point. 1971 *Gloss. Terms Data Processing* (B.S.I.) 91 *Re-entry point*, the address or the label of the instruction at which the computer program that called a subroutine is reentered from the subroutine. 1957 *Times* 9 Nov. 6/5 The President's positive claim that the United States has solved the so-called re-entry problem of bringing space missiles back to earth. 1965 *New Scientist* 2 Dec. 638/1 Further improvement is being made with re-entry vehicles which embody penetration aids against anti-missile defence. 1973 *Times* 26 Feb. (Arms for Peace Suppl.) p. ii/4 The United States, by possessing the ability to put multiple independently targeted reentry vehicles on her launchers, has many more nuclear warheads than the Soviet Union. 1973 J. Snow *Mission by Nightmare* xi. 125 I'm not coming back... They've cancelled my re-entry visa. 1977 *Times* 11 May 7/8 The Soviet authorities had given him a permanent reentry visa so that he can return to Moscow.

re-equip, *v.* (Later examples.)
1919 J. L. Garvin *Econ. Foundations Peace* viii. 158 The industrial centres, wherever existing, have to be re-equipped. 1944 J. S. Huxley *On Living in Revolution* 148 Neither the profit motive nor political considerations should be allowed to interfere with the job of re-equipping Europe. 1960 *Farmer & Stockbreeder* 19 Jan. (Suppl.) 6/1 Currently engaged in re-equipping many leading breeding and research farms.

re-e·scalate, *v.* [f. Re- 5 a + *escalate *v.* 2.] *trans.* To escalate (a war or conflict) again. Also *absol.*
1965 H. Kahn *On Escalation* xii. 237 Further bargaining is rarely successful unless the conflict is re-escalated. 1965 *Ibid.* 244 Up and down this ladder nations may escalate, de-escalate and re-escalate. 1972 *Sat. Rev.* (U.S.) 6 May 34/1 President Nixon has always kept for himself the option of re-escalating the war.
So **re-escala·tion**, the act or process of re-escalating. Also *attrib.*
1965 H. Kahn *On Escalation* xii. 231 De-escalation dominance might also involve being in a good position to resume fighting if the other side forced further action. The latter property could also be called 're-escalation dominance'. *Ibid.* 234 Of course, de-escalation cannot guarantee that re-escalation will not occur.

re-estate, *v.* Delete † *Obs.* and add later example.
a 1945 E. R. Eddison *Mezentian Gate* (1972) ix. 82 Kallias's meaning was by this alliance to re-estate his power in the Meszrian Marches.

reet (rīt), *a.* U.S. dial. var. Right *a.* (usu. in sense 8).
Also a common dial. var. in the U.K.: see *Eng. Dial. Dict.*
1934 in Webster. 1942 *Amer. Mercury* July 85 Jelly got into his zoot suit with the reet pleats. 1943 *Crisis* July 201/2 Negro youths will crack at anyone of any race who is nice looking. They will say, 'A fine queen..a reet chick'. 1946 B. Treadwell *Big Bk. Swing* 125/1 *Reet*, fine, O.K. 1977 *Hot Car* Oct. 42/3 People I've spoken to who went last year said it was reet good and very alcoholic.

re-eva·luate, *v.* [Re- 5 a.] *trans.* To evaluate again.
1945 *Physiol. Rev.* XXV. 126 Those same changes which were originally considered slight and of no consequence have more recently been re-evaluated in the light of new experiments and are now spoken of as 'worthy of emphasis'. 1964 M. A. K. Halliday et al. *Linguistic Sci.* p. xii, As theories develop so must applications thereof be re-propounded, and books continually re-evaluated. 1970 *Jrnl. Social Psychol.* LXXXVII. 145 The modified semantic differential was translated into Afghan colloquial speech and reevaluated according to the criteria of excellence. 1976 *Nature* 29 Apr. 740/2 It asked the Secretary of the Department of Health, Education and Welfare (of which NIDA is a part) to reevaluate the project and determine whether Rubin is suitably qualified to conduct the research. 1979 *Time* 2 Apr. 2/3 Their music has the power to force people to reevaluate their ideas and institutions.
Hence **re-evalua·tion**, a second or further evaluation.
1946 *Nature* 16 Nov. 689/2 In view of the great progress made in recent years on the pancreatic enzymes, there is no doubt that much of the older work on the peptidases, etc., of the intestinal canal will need re-evaluation. 1959 *Encounter* Sept. 58/1 These re-evaluations called into question the accepted idea of the populist basis of American radicalism. 1970 *Publishers' Weekly* 8 June 152/1 This capability is worth noting; it may cause a reevaluation of short-run publishing. 1972 *Maclean's Mag.* Sept. 12/2 It is too much to hope that it will cause a re-evaluation in the selection of officers from the top down. 1976 *Nature* 29 Apr. 740/2 The reevaluation was carried out on March 11 at a closed meeting.

reeve, *sb.*[1] **2. d.** (Earlier and later examples.)
1853 S. Strickland *27 Yrs. Canada West* II. 271 The Counties..choose their reeves, and deputy reeves where the population admits of it, and these form the county council. 1884 *Brandon* (Manitoba) *Blade* 17 Jan. 8/3 The Reeve, in a few well chosen remarks, dwelt on the duties and responsibilities devolving on them as servants of the people. 1945 G. W. Brown *Canad. Democracy in Action* vii. 89 The town council consists of a mayor, a reeve, and two or three councillors elected for each of the wards. 1965 *Victoria* (B.C.) *Daily Times* 20 July 11/8 The reeve said the general principle of regional planning is good but the mechanics need improving. 1968 *Globe & Mail* (Toronto) 5 Feb. 1/3 Deputy Reeve Wilfred Shaw said 61 families had been removed from their houses by firemen and other town employees.

reeve, *v.*[1] Add: **3. d.** *fig.* To gather together.
1876 G. M. Hopkins *Wreck of Deutschland* xii, in *Poems* (1967) 55 Yet did the dark side of the bay of thy blessing Not vault them, the millions of rounds of thy mercy not reeve even them in?

Reeves (rīvz). The name of John Reeves (1774–1856), English naturalist, used *absol.*, *attrib.*, or in the possessive in **Reeves('s) pheasant** to designate a long-tailed Chinese pheasant, *Syrmaticus reevesii*, introduced to Europe by him and named in his honour by J. E. Gray in 1829.
1829 J. E. Gray in E. Griffith et al. tr. *Cuvier's Animal Kingdom* VIII. 25 (*heading*) Reeves' Pheasant, *Phasianus Reevesii*. 1834 G. Bennett *Wanderings in N.S.W.* II. iv. 55 In the aviary [at Macao], the beautiful..Reeves's Pheasant, was seen. 1922 W. Beebe *Monogr. Pheasants* III. 146 Then came my first view of a live Reeves in its wild home. 1926 J. S. Huxley *Ess. Pop. Sci.* 47 If a female Reeves pheasant is crossed with a male of another race, the males among the hybrid offspring show many characters of the male Reeves pheasant. 1951 J. Delacour *Pheasants of World* ix. 227 Reeves's Pheasants are found on all the higher wooded hills of Central China north of the Yangtze River. 1975 *Islander* (Victoria, B.C.) 19 Oct. 4/3 We now have..the rare Reeves pheasant. 1976 *Shooting Times & Country Mag.* 18–24 Nov. 28/2 He is, however, incorrect in his remarks about the Reeves pheasant.

reevesite (rī·vzəit). *Min.* [f. the name of Frank *Reeves*, Australian geologist + -ite[1].] A hydrated basic carbonate of nickel and iron, $Ni_6Fe_2CO_3(OH)_{16}.4H_2O$, occurring as yellowish plates.
1967 J. S. White et al. in *Amer. Mineralogist* LII. 1182 Reevesite: This mineral occurs as bright yellow fine-grained aggregates lining cavities and cracks in the weathered meteorites. *Ibid.* 1193 The mineral is named reevesite in honor of Dr. Frank Reeves, who was responsible for the discovery of the Wolf Creek meteorite crater in 1947. 1971 *Ibid.* LVI. 1077 Reevesite..has been found in the nickel ore from the Bon Accord area in the Barberton Mountain Land, South Africa.

reeving, *vbl. sb.*[1] (Further *attrib.* examples.)
1969 E. H. Pinto *Treen* 266 The pear-shaped, elm item is a reeving block. 1971 *Gloss. Terms Materials Handling* (B.S.I.) v. 18 *Reeving thimble*, a thimble..of sufficient internal length and breadth for one thimble to pass through another. 1975 *Offshore* Sept. (Dutch Suppl.) 49–12/1 The actual pulling wires, also 57-mm diameter, ran from the Pontra Maris to a sheave pontoon anchored 2 km away by eight 7·2-ton capacity delta anchors. The reeving system allowed a total pulling force of 600 tons.

re-excite, *v.* Add: (Earlier example.) Hence **re-exciting** *vbl. sb.* and *ppl. a.*
1697 J. Sergeant *Solid Philos.* 195 Some short time must be allow'd for the coming of Impressions from without ..and the Re-exciting them in the Fancy. *Ibid.* 438 Such Sounds, thro' the use of the Words are apt to re-excite the Memory. 1964 J. Z. Young *Model of Brain* xiii. 211 Here we may notice that because it contains re-exciting circuits it could provide an increase in the 'command to attack' by what amounts to a positive feed-back.

re-existence. (Later example.)
1973 *Times Lit. Suppl.* 14 Dec. 1536/2 He builds up imaginary examples of transmigration, re-existence, etc.

re-export, *sb.* **2.** (Earlier examples.)
1761 J. Glen *Descr. S. Carolina* 48 The Exports of South Carolina Produce are inserted in one Account, and the Re-exports of imported Commodities and Manufactures in another. 1775 *Jrnls. Continental Congress U.S.* (1905) III. 502 The reëxport employs ships, sailors [etc.].

ref (ref), *sb.*[1] Colloq. abbrev. of Referee *sb.* 3 b.
1899 R. H. Barbour *Halfback* xxii. 233 De Farge (the referee) is awfully down on holding and off-side plays. Last year he penalized us eight times during the game. But he's all right... He's the finest little ref that ever tossed a coin. 1939 *War Illustr.* 9 Dec. 394/3 Prisoners taken from U-boats and merchant ships are spending a happy hour kicking the ball about... An armed guard stands by, but not to protect the 'ref'. 1941 *London Opinion* Sept. 45/2 The referee goes to examine the eye while the crowd roars. Can he go in? The ref. moves away. He can. He does. 1957 I. Cross *God Boy* (1958) 108 You're a pretty good ref, Sister. 1962 *Observer* 9 Sept. 16/2 A mob stormed outside the club offices shouting: 'We want the ref.' 1966 F. Shaw et al. *Lern Yerself Scouse* 51 *Buy a bewk, ref!* The referee appears to have forgotten the rules of the game. 1972 A. Draper *Death Penalty* iii. 18 He detested referees who were continually blowing their whistles... A ref could make or mar a game. 1976 *Listener* 29 Jan. 117/1 Adam is able to make good jokes about Cambridge..and there is no ref to blow the whistle on him.

ref (ref), *sb.*[2] Colloq. abbrev. of Reference *sb.* (in Dict. and Suppl.) **a.** In sense 3.
1926 F. M. Ford *Let.* 9 Mar. (1965) 168 Yours of the 23d ult. Ref German Translation of *No More Parades*. 1967 Wodehouse *Company for Henry* ix. 149 'I want his advice.' 'With ref. to what?' 'Oh, something that's coming up'. 1971 —— *Much Obliged, Jeeves* xi. 108 It's with ref to that book you pinched from the Junior Ganymede.
b. In sense *6 b.
1901 [see *bovrilize *v.*]. 1907 Westbrook & Wodehouse *Not George Washington* xvii. 183 Your refs. must be A1, or you don't stand an earthly. 1934 H. G. Wells *Exper. Autobiogr.* I. iv. 161 Such questions seemed to me already of far more importance than satisfying J.K. or securing a satisfactory 'ref.' when my apprenticeship was up. 1974 P. Wright *Lang. Brit. Industry* xii. 102 Refs (references).

ref (ref), *v. trans.* and *intr.* Colloq. abbrev. of Referee *v.*
1929 R. C. Sherriff *Journey's End* ii. i. 50 *Raleigh.* Did you play Rugger? *Osborne.* Yes. But mostly reffing at school in the last few years. 1964 J. Hale *Grudge Fight* viii. 125 A scrum developed while Windy who was supposed to be reffing blew his whistle and went red in the face and didn't dare come too close. 1968 *Punch* 2 Oct. 457/3 Who says the game was badly reffed? The sending-off of Nobby Stiles, For nothing, was supremely deft. 1975 *Times* 4 Jan. 12/3 Muhammad Ali..was fighting Mildenberger and Teddy Waltham was reffing. 1977 *Gay News* 24 Mar. 2/2 Norman has recently been booked to ref games in California.

reface, *v.* Add: **3.** To face (a person, a concept) again.
1906 *Daily Chron.* 18 Apr. 3/4 Rather than re-face Mag McGhie..David prefers to 'face an angry Maker'. 1979 R. Rendell *Make Death love Me* xii. 106 It would teach her to assume responsibility and re-face reality.

refained (rifē[i]·nd), *a.* Also **refaned, refayned, refeened** (-fī·nd). Repr. an affected pronunc. of Refined *ppl. a.* (with reference to sense 2). Freq. joc. or derogatory.
1930 A. Huxley *Brief Candles* 25 Altogether too much the lady—refained; you know the type. A Governess;..the genteel, Jane Eyre, daughter-of-clergyman kind. 1932 *New Statesman* 9 Jan. 36/2 Few audiences could take seriously a performance of *Hamlet* in which..Horatio [spoke] in what is known as the 'refaned' accent. 1939 R. Campbell *Flowering Rifle* ii. 46 The most 'refayned' of all that breed. 1940 W. de la Mare *Pleasures & Speculations* 137 His [*sc.* the advertiser's] tone is usually genteel and refaned. 1940 H. G. Wells *New World Order* 53 A friendly adviser..protests against 'the wombs of associated labour'... My adviser produces.. 'the lap of social labour', which is more refained but pure nonsense. 1940 *John o' London's* 3 May 149/3 Edinburgh being very 'refained', the word [*sc.* keelie] is seldom used in the more polite society. 1941 V. Woolf *Between Acts* 122 The old lady..looked too refined. 'Refeened'—Mrs. Manresa qualified the word to her own advantage... She could span the old lady's 'refeenment'. 1961 *Radio Times* 27 Apr. 63/4 *The Kilt Is My Delight* is indeed a delightful programme, but..may I suggest that it sometimes takes on a less 'refaned' air? Always these immaculately dressed ladies and gentlemen prance around in some 'stately home'. 1962 N. Marsh *Hand in Glove* i. 28 Nicola..wondered if Mr. Period would find the phrase 'refeened,' a word he often used with humorous intent. 1962 *Punch* 14 Feb. 297/3 An appalling gold-digger with a refained accent. 1969 M. O'Brine *Mills* xi. 43 The Sangsters were much too 'refained to argue with someone they were meeting for the first time. 1972 A. MacVicar *Golden Venus Affair* v. 53 Her accent was 'refained', but it betrayed a Glasgow East End origin. 1976 *Listener* 25 Mar. 386/4 The blowzy, mini-skirted divorcee with a loutish son and an impossibly 'refeened' mother.
Hence **refai·nment, refee·nment** (with reference to Refinement 2).
1933 H. Matheson *Broadcasting* iii. 63 Universal education may mean a universal hybrid speech..often overlaid with a veneer of what can only be called 'refainment'. 1941 [see above]. 1960 J. Mitford *Hons & Rebels* i. 17 Nancy [Mitford], even in those early days preoccupied with U and Non-U usage, made up a poem illustrative of the main 'refainments' of Miss Broadmoor's speech.

refan, *v.* Add: **2.** [Re- 5 c.] *Aeronaut.* To fit (a turbo-fan) with a new fan. So **refa·nned** *ppl. a.*, **refa·nning** *vbl. sb.*
1973 *Internat. Aerospace Abstr.* XIII. 2222/1 (*heading*) Refanned commercial gas turbine engines. 1974 *Ibid.* XIV. 2236/1 (*heading*) Reduction of JT8D powered aircraft noise by engine refanning. 1975 *Times Lit. Suppl.* 21 Mar. 318/3 Improved engine design has produced a new generation of quieter aircraft, earlier and noisier types can be refanned and further progress in the direction of peace and quiet may be expected. 1978 *Flight Internat.* 18 Feb. 429/1 United Airlines' flirtation with the refanned 727-300B in 1975 marked the end of this transport's major development.

refan (rī·fæn), *a.* and *sb.* *Aeronaut.* [f. prec. vb.] **A.** *adj.* **a.** = *refanned *ppl. a.* **b.** = *refanning *vbl. sb.* (used *attrib.*). **B.** *sb.* A refanned engine.
1974 *Internat. Aerospace Abstr.* XIV. 2236/1 This would be accomplished by retrofitting the existing fleet with quieter refan engines and new acoustically treated nacelles. 1975 *Ibid.* XV. 2764/2 The objective of the refan program is to demonstrate the technical feasibility of substantially reducing the noise levels of existing JT8D powered aircraft. 1978 *Flight Internat.* 18 Feb. 428/2 The principal advantage of refans (whether cropped or sized up) over new engines can be summarized as lower development, acquisition and ownership costs; [etc.].

refect, *v.* Add: **2.** *trans.* To eat (fæcal pellets).
1960 M. Burton *Wild Animals Brit. Isles* 60 The droppings refected are different from those discarded. 1964 R. M. Lockley *Private Life of Rabbit* x. 102 Termites may refect food as much as six times.

refection, *sb.* Add: **2. d.** The eating of fæcal pellets, practised by rabbits and some other animals.
1939 *Nature* 10 June 982/1 The pellets frequently constitute more than one third of the stomach contents [of the rabbit] and refection to such a degree seemed too improbable. 1952 L. H. Matthews *Brit. Mammals* vi. 136 In 1939 the habit of 'refection' was rediscovered in the rabbit. 1964 R. M. Lockley *Private life of Rabbit* x. 102 Many animals, including insects as well as hares and rabbits, have this habit of refection. 1973 *Bk. Brit. Countryside* (Automobile Assoc.) 367/2 Feeding is by refection, a similar method to chewing the cud. Food is eaten then excreted in semi-digested form as soft moist pellets. These are eaten again and passed through the intestines to be fully digested.

refectory, *sb.* Add: (The pronunc. re·fĕktəri is still used by some Roman Catholics.) **b.** *refectory table* (also with ellipsis of *table*): see quots. 1948, 1960.
1913 L. V. Lockwood *Furniture Collector's Gloss.* 51/2 *Refectory*, an early long, narrow table upon which was served a meal. 1923 H. Stanley-Barrett *Old World Galleries A.B.C. Hist. Eng. Antique Furnit.* (ed. 2) 118 The Elizabethan trestle refectory table usually had heavily carved bulbous legs. 1928 *Daily Express* 18 Apr. 4. Refectory tables ..were the principal pieces of furniture in medieval and Tudor times. 1948 *Antique Collector* Aug. 127/1 In the late 16th and 17th centuries the common dining-table was an oblong one with either four or six turned legs connected by square sectioned stretchers. In contemporary inventories it was usually called a 'long table', but in order to conjure up a picture of jovial monks dining, the long table has been renamed a 'refectory table', which inaccurate term is often used today. 1960 H. Hayward *Antique Coll.* 235/1 *Refectory table*, popular modern term for a long table of the type in use in the second half of the 16th cent. until the Restoration. 1971 D. Francis *Bonecrack* iv. 51 We sat..with our feet up on a sixteenth century Spanish walnut refectory table. 1976 *Cumberland News* 3 Dec. 29/5 (Advt.), We are most interested in old oak furniture—dressers, court cupboards—kitchen presses, bedding chests, kitchen and refectory tables as well.

refeed (rifī·d), *v.* Also **re-feed.** [Re- 5 a.] *trans.* To feed again, esp. after a period of starvation.
1884 [in Dict. s.v. Re- 5 a]. 1943 *Nutrition Abstr. & Rev.* XII. 637/2 [The birds] were then refed on maize until the original weight was regained. 1971 *Jrnl. Nutrition* CI. 1564/2 Rats were starved 48 hours and were refed the high carbohydrate diet for 0, 1, 2, 3, 4, and 7 days.
So **refe·d** *ppl. a.*, **refee·ding** *vbl. sb.*
1932 *Biol. Abstr.* VI. 410/2 (*heading*) Re-feeding after starvation. 1950 *Arch. Biochem.* XXVII. 177 The percentage of nitrogen in the extracts obtained from the livers of the rats refed with a high-carbohydrate diet was significantly lower than the percentage of nitrogen in the extracts obtained from the pellet-refed and the fat-refed rats. 1964 *Proc. Soc. Exper. Biol. & Med.* CXV. 441/1 The data for refed rats reflect a marked increase in liver lipid upon refeeding. 1967 M. Kenyon *Whole Hog* viii. 94, I put them [*sc.* laboratory animals] on controlled feeding after a period of starvation and their reactions to the re-feeding.. would be similar to human reactions.

refer, *v.* Add: **7. f.** To postpone the passing of (a candidate) in an examination or the acceptance of (an application for a degree), provision being made for re-examination at a later date.
1907 *Practitioner* June 795 A student, who should venture to put upon an examination paper what is the only logical outcome of the teaching of text-books, ought, without hesitation, to be referred. 1908 A. S. M. Hutchinson *Once Aboard Lugger* i. i. 32 'I had forgotten. Your examination?' ..'I failed. I was referred for three months.' 1927 *Univ. Oxford Examination Statutes* vi. 227 The examiners shall have power..to recommend the Board to refer the Student's application for leave to supplicate back to him in order that he may present himself for re-examination. 1976 *Daily Times* (Lagos) 3 Nov. 31/2 Twenty-four students drawn from the states and the Armed Forces passed the prescribed test while four students were referred. 1979 *Jrnl. R. Soc. Arts* Dec. 9/1 The candidates for this session's examination were exclusively restricted to those who had been referred in a previous examination or who were resitting in order to obtain higher grades.
8. a. (Further examples.) Also const. *back to*.
1927 W. E. Collinson *Contemp. Eng.* 123 The elements of committee English which I had picked up in the Debating Society were soon reinforced by the constantly heard expressions: standing orders, terms of reference (defining the scope of a committee's labours), to refer back (to a committee for further consideration). 1934 G. B. Shaw *On Rocks* ii. 267, I must really refer you back to him for further consideration and report. 1961 New Eng. Bible *Luke* xxiii. 15, I have myself examined him in your presence and found nothing in him to support your charges. No more did Herod, for he has referred him back to us. 1976 [see sense 8 d below].
d. To send or direct (a person) to a medical consultant or institution for specialist treatment.
1961 *Lancet* 2 Sept. 517/2 We are indebted to Dr. J. F. O'Connell for referring this patient. 1970 H. McLeave *Question of Negligence* xxii. 186 She did consult me, but she was referred by someone else. 1973 *Guardian* 9 Mar. 13/2 The slum Doctor round the corner who never examines his patients before referring them. 1973 *Listener* 19 Apr. 507/1 They're all referred by the GP... They're all psycho-geriatrics. 1976 *Women's Report* Sept./Oct. 2/1 Yet the clinic makes use of the NHS because all clients needing surgery or expensive treatment have to be referred back to NHS hospitals or clinics.
referring *vbl. sb.* and *ppl. a.* (later examples).
1950 P. F. Strawson in *Mind* LIX. 320 We very commonly use expressions of certain kinds to mention or refer to some individual person or single object or particular event or place or process... I shall call this way of using expressions the 'uniquely referring use'. *Ibid.* 326 'Mentioning', or 'referring', is not something an expression does; it is something that some one can use an expression to do.

referability (re:fĕrăbi·liti). *rare.* [f. Referable *a.* + -ility.] The fact or quality of being referable.
1964 R. H. Robins *Gen. Linguistics* vi. 233 Non-favourite sentences of class I have no such referability to a longer sentence in which they may be incorporated.

referee, *sb.* Add: **3. b.** (Earlier and further examples.)
1840 *Spirit of Times* 25 Jan. 559/3 He was a general referee and umpire, whether it was a horse swap, a race, a rifle match, or a cock fight. 1856 *Porter's Spirit of Times* 6 Dec. 229/1 In [baseball] matches, an umpire is chosen on each side, and a referee to decide, when the umpires cannot agree. 1887 G. A. Hutchison *Football* ii. 11 In case of infringement, the referee shall..order a scrummage to be formed. 1906 *Daily Chron.* 7 Sept. 9/4 No great blame attaches to the referee, who probably did not notice the forward till he was apparently off-side. 1936 H. B. T. Wakelam *Game goes On* 17 Other innovations during the Daring 'Eighties were the appointment of neutral referees and the provision of whistles for those functionaries. 1951 F. N. S. Creek *Soccer for Boys* iii. 18 The duration of the game is mainly the responsibility of the referee. 1977 *Times* 16 Mar. 12/1 The main topic of conversation..was the alleged attempt to bribe the Danish referee before the first leg.
c. A person appointed to examine a scientific or other learned work and comment on its suitability for publication.
1884 *Proc. London Math. Soc.* XV. 160 The original paper has been divided into two..at the suggestion of the referee. 1926 A. E. Housman *Let.* 14 Oct. (1971) 242 A report of mine ..decided the Syndics not to accept a treatise of Richmond's. ..A. W. Ward..told him that I was the referee. 1970 *Physics Bull.* Jan. 2/2 The majority of authors expressed their appreciation of the value of the constructive criticism of the referees in improving the quality of their papers. 1971 *Nature* 22 Oct. 571/3 Each paper was carefully scrutinized by one senior referee and by one of the two distinguished editors-in-chief.
4. A person who may be referred to for information or guidance on the character or other qualities of someone, *spec.* of an applicant for employment, for an academic or other award, or the like.
1862 H. Mayhew *London Labour* IV. 12 Classification of the workers and non-workers of Great Britain... Referees, or those who give characters to professional beggars when a reference is required. 1882 *Sydney Slang Dict.* 7/2 *Referees*, those who give characters to enable dishonest persons and thieves' accomplices to obtain situations. 1944 *Oxf. Univ. Gaz.* LXXIV. 316/1 Candidates are requested to send in their names with eight copies of any statement that they may

wish to make..giving the names of not more than three referees. 1971 *Reader's Digest Family Guide to Law* 689/2 One way an employer can assess the abilities of an applicant is to ask for references from former employers. It may also be useful to telephone the referee, who may be prepared to give more information informally than he can provide in writing. 1972 *Library Assoc. Record* Nov. 224/1 On three occasions lately I have sent for an application form and job description only to find that these did not arrive until two or three days before the closing date, this making it very difficult..to arrange referees, etc. 1976 *Oxf. Univ. Gaz.* CVII. 209/1 Applications, including a *curriculum vitae*,..and the names of two academic referees, should be sent to the Secretary of the Marjory Wardrop Fund.

referee, *v.* Add: **2.** *trans.* and *intr.* To examine and evaluate (a scientific paper, thesis, or book); to act as referee (in sense *3 c of the sb.). **refereeing** *vbl. sb.* (later examples). Hence **referee·d** *ppl. a.*
1966 *Rep. Comm. Inquiry Univ. Oxf.* II. 452 Editing or refereeing for journals. 1970 *Physics Bull.* Jan. 2/2 (*heading*) Refereeing of research papers. *Ibid.* 3/1 If a referee is unable to referee a paper himself, he is invited to pass it to an appropriate colleague. 1970 *Computers & Humanities* IV. 312 All submitted papers will be refereed. 1971 *Nature* 24 Sept. p. xvi/2 (Advt.), All papers are scrupulously refereed and the journal is guided by an editorial board of distinguished scientists whose activities cover all aspects of polymer research. 1975 *Ibid.* 6 Nov. 1/1 The Scientific Information Committee of the Royal Society has recently put forward a set of guidelines for the refereeing of papers for publication. 1978 *Maledicta* II. 10 Unfortunately for them, our journal is not a 'refereed' academic publication.

reference, *sb.* Add: **3. d.** *Logic* and *Linguistics.* The act or state of referring through which one term or concept is related or connected to another or to objects in the world; also as *objective reference*, and *attrib.* as *reference class, property*.
1883 F. H. Bradley *Princ. Logic* I. i. ii. 55 Judgment is not the synthesis of ideas, but the reference of ideal content to reality. *a* 1914 C. S. Peirce *Coll. Papers* (1933) III. xix. 366 Dyadic relations..which can only subsist between two subjects of different categories of being..may advantageously be termed a reference. 1927 Ogden & Richards *Meaning of Meaning* (ed. 2) i. 9 It is Thought (or, as we shall usually say, *reference*) which is directed and organized, and it is also Thought which is recorded and communicated. 1946 C. I. Lewis *Knowledge & Valuation* x. 270 This property ϕ may be called the *reference property*. *Ibid.* 271 The class of things having it [*sc.* a property] may be called the *reference class*. 1951 G. Humphrey *Thinking* viii. 228 In its original form as the 'problem of meaning' the question has to be obscured by the invention of new descriptive terms such as *transcendent reference, objective reference* or *context* in order to satisfy the scientific conscience. 1956 G. Ryle in A. J. Ayer *Revol. in Philos.* 7 Both [Frege and Bradley] saw that it is..intrinsic to a thought to be true or false, or to have 'objective reference'. 1959 K. R. Popper *Logic Sci. Discovery* viii. 155 This class α, which is assumed to be *non-empty*, serves, as it were, as a frame of reference, and will be called a (finite) *reference-class*. 1972 *Language* XLVIII. 446 In order to understand problems of semantic theory, it is crucial to understand why most philosophers think of the notion of reference as the key element in such theories. Reference is the relation between singular term and bearer, as well as between general predicate and the entities of which the predicate is true. With the notion of reference go the notions of naming, describing, and—therefore—truth. 1974 P. F. Strawson *Subject & Predicate* ii. 47 Hearer and speaker should each understand the name..as having a certain unique reference; and..the reference should be the same for each of them.
e. *Sociol.* and *Psychol.* The process by which or the extent to which an individual establishes a relation with elements in society as a standard for comparing status and values (see also *frame of reference* s.v. *frame *sb.* 4 d (ii)). Freq. *attrib.* (see also *reference group* in sense 8 a below).
1937 G. Murphy et al. *Exper. Social Psychol.* (rev. ed.) iv. 220 A laboratory situation is set in which social factors determine a reference frame which must be used by the subject in perceiving. 1947 Sherif & Cantril *Psychol. of Ego-Involvements* vi. 137 Judgments of the physical characteristics of others..become..ego-involving judgments in which an individual uses himself..as a central point of reference. 1948 M. & C. Sherif *Outl. Social Psychol.* xviii. 621 This ordering of responses held whether a respondent made as many as nineteen consensual references or as few as one. 1956 Gardner & Thompson *Social Relations & Morale* v. 23 Displacements of the distributions of scores along the reference continuum. *Ibid.* vii. 43 The most general reference population..would be: 'All the persons (living or dead) you have..known in any way'.
6. a. *spec.* = *referee *sb.* 4.
1837 Dickens *Let.* 24 Feb. (1965) I. 238, I have..taken the liberty of mentioning your name, among those of other references, to testify to my being 'sober and honest'. 1865 [in Dict.]. 1934 D. L. Sayers *Nine Tailors* ii. ii. 95 'Did he give you any references?'..'Yes..he did. He give me the name of a garridge in London..and..said if I was to write to the boss, he'd put in a word for him.'
b. A (usu. written) report produced by a referee (*referee *sb.* 4); a testimonial.
1895 in *Funk's Stand. Dict.* 1924 Galsworthy *White Monkey* ii. ix. 197 That was my first job since the war, so I can whistle for a reference. 1936 *Punch* 21 Oct. 467/1 'I have references, excellent references... Here is one from a lady in Eaton Square.' (Hands it to the employer.) 1940 G. D. H. & M. Cole *Counter-Point Murder* viii. 85 Corcoran said he wrote to us taking up the reference, and got back a letter speaking very highly of the firm. 1976 'P. B. Yuill' *Hazell & Menacing Jester* ii. 27 'Did Thornton get another job?' 'Nobody's asked me for a reference.'
8. a. *reference librarian, library* (further examples), *point*; **reference book**, (*a*) a book used for reference purposes; cf. *book of reference* in sense 7; (*b*) *S. Afr.*, an identity document or group of documents officially introduced in 1952 or the name for a pass (see *pass *sb.*[2] 8 f), regulating movement in particular areas, which all non-white residents must carry (replaced in 1977 by 'travel documents': see quot. 1977 s.v. *pass *sb.*[2] 8 f); **reference electrode** *Electr.*, an electrode the potential of which can be accurately maintained and reproduced, and in relation to which other potentials can be measured; **reference frame** = *frame of reference* s.v. *frame *sb.* 4 d; occas. *transf.*; **reference group** *Sociol.* and *Psychol.*, a group to which a person may or may not belong but which he, perhaps subconsciously, refers to as a standard in forming his attitudes and behaviour; **reference tube** *Electr.*, a cold-cathode gas-filled tube which can maintain an accurately fixed voltage across itself for long periods.
1889 *Cent. Dict.*, Reference book. 1952 [see *pass *sb.*[2] 8 f]. 1954 L. G. Green *Under Sky like Flame* xii. 174 On the Gold Coast cooks carry a reference book. 1960 *Observer* 27 Mar. 16/4 Failure to produce a reference book on demand is a criminal offence. 1967 [see *author *v.*]. 1969 *Golden City Post* July 3 The thugs..robbed him of his money, reference book and personal documents. 1970 C. L. Cline *Lett. George Meredith* I. p. xxxii, The inaccessibility of reference books in provincial towns may also have contributed to the slothfulness of the editing. 1971 *Drum* Mar. 4 They walked purposefully towards the women who searched themselves for their reference books. But it was not the reference books the policemen wanted. 1971 *Rand Daily Mail* 27 Mar. 2/2 The vast majority of short-term prison sentences are for minor statutory offences connected with Bantu influx control, taxation and reference book regulations. 1972 *Physics Bull.* Apr. 225/2 For a reference book the index is inadequate. 1926 *Jrnl. Amer. Chem. Soc.* XLVIII. 34 The effect of temperature upon the potential of reference electrodes. 1948 Glasstone *Textbk. Physical Chem.* (ed. 2) xii. 940 There is no reliable method known for determining the absolute potential of a single electrode. The only sound procedure is to combine the electrode with a reference electrode of known potential difference and to measure the E.M.F. of the resulting cell. 1975 M. R. Jenkins in Williams & Wilson *Biologist's Guide to Princ. & Techniques Pract. Biochem.* vii. 202 Calomel electrodes..are the most common reference electrodes and they consist of a solution of mercurous chloride (calomel) and potassium chloride in contact with solid mercurous chloride and mercury. 1921 J. M. Bird *Relativity & Gravitation* ii. 36 (*heading*) The reference frame for space. 1940 C. S. Sherrington *Man on his Nature* ii. 62 There was as yet no reference-frame of natural law, of chemistry or physics, by help of which to orientate the natural fact. 1967 [see *frame *sb.* 4 d (i)]. 1970 *Nature* 17 Oct. 272/2 In deriving the equations of the electromagnetic field in a rotating reference frame, Schiff used the explicit transformation to rotating co-ordinates. 1942 H. H. Hyman *Psychol. of Status* ii. 37 Satisfaction with status is consequently also a function of the reference group, since the reference group is a variable of the judgment. 1957 *Jrnl. Abnormal & Social Psychol.* LV. 360/1 Attention has also been given to the influence of his reference groups: the groups in which he aspires to attain or maintain membership. 1969 *Times* 14 Apr. 7/3 The two groups of farmers have different reference groups, a different set of individuals with which they will compete for status and recognition. 1970 C. T. Restrepo in I. L. Horowitz *Masses in Lat. Amer.* xiv. 542 The absence of contacts—and hence the lack of visibility of reference groups—has kept the peasants unaware even of their own needs. 1973 'J. Patrick' *Glasgow Gang Observed* xiii. 115 Tim's reference group for clothes was the teenage record scene. 1977 R. Holland *Self & Social Context* v. 119 The area of social theory most closely related to that of role, namely 'reference group' research. 1951 L. I. Edwards in J. D. Stewart *Reference Librarian* iii. 61 Theoretically the reference librarian should know something about everything because of the varied nature of the queries received. 1978 W. White in *W. Whitman's Daybks. & Notebks.* I. 87 For this information and other help I am grateful to George A. Masterton, reference librarian at Wayne State University. 1860 Reference library [see *art gallery* s.v. *art *sb.* V]. 1976 *Nature* 1 Apr. 466/3 This book is a 'must' and it certainly deserves space on a shelf of every reference library. 1977 *Evening Post* (Nottingham) 24 Jan. 3/2 Now there is a reference library, a craft work base, and two other areas into which children can move for private reading and study. 1884 Reference-point [see *bench-mark b]. 1936 M. Sherif *Psychol. of Social Norms* vi. 96 This subjectively established standard or norm serves as a reference point with which each successive experienced movement is compared and judged to be short. 1977 *Rolling Stone* 5 May 33/1, I really do think Rod Stewart, Bruce Springsteen and the Who are the most appropriate reference points for rock and roll. 1960 *Electronic Engin.* XXXII. 218 Studies were confined to neon-filled tubes..having molybdenum anodes and cathodes and employing high-stability reference-tube manufacturing techniques. 1962 G. A. T. Burdett *Automatic Control Handbk.* vii. 23 The primary duty of the reference tube..is to provide voltage which can be accurately maintained within close limits for long periods.
b. In extended scientific and technical use denoting an object, property, value, or the like, used as a basis for comparative measurement or standardization. Also *absol.*
1878 J. W. Draper *Sci. Mem.* ii. 57, I had previously passed through the slit a beam of sunlight reflected from a mirror, so as to have a reference spectrum with fixed lines. 1901 *Shop & Foundry Pract.* III. xxvii. 23 Reference gauges are gauges that represent either an accurate subdivision of the imperial yard, or some arbitrary size or shape adopted for some purpose and required to be preserved. 1941 C. O. Fairchild et al. *Temperature* 305 It is not always possible to maintain the reference junctions..at a desired temperature during the calibration of a thermocouple. 1950 K. Henney *Radio Engin. Handbk.* (ed. 3) xxi. 775 A convenient and consistently accurate method of measuring the amplitude of the signals is required, as well as a reference level common to the entire system. 1952 Markus & Zeluff *Electronics for Communication Engin.* ix. 342/1 This cavity has a high Q and is used as the frequency reference. 1953 Amos & Birkinshaw *Television Engin.* I. i. 20 D.c. restoration is only possible provided the picture signal contains a reference signal related to black level and the television waveform therefore includes such a signal. 1966 Williams & Fleming *Spectrosc. Methods Org. Chem.* iv. 81 The positions of proton resonances in an NMR spectrum are measured relative to the resonance position of the twelve equivalent protons of an arbitrary reference substance, tetramethylsilane. 1975 D. G. Fink *Electronics Engineers' Handbk.* xix. 3 The reference pressure for sounds in air, corresponding to 0 dB, has been defined as a sound pressure of 0·0002 microbar. 1978 P. W. Atkins *Physical Chem.* xxix. 968 The potentiometer reading is used in the normal way to find its new potential relative to the reference.

reference, *v.* Add: **1. b.** To relate (a measurement) to a defined base or zero level.
1971 *Nature* 3 Sept. 51/2 A complete separation of the explosion population from the earthquake population is obtained when referenced to the arbitrary decision line. 1972 *Science* 23 June 1349/3 The intensity of the sound was as high as 80 db (referenced to 0·0002 dyne/cm²).
2. (Further examples.)
1971 *Nature* 3 Sept. 71/3 Each chapter is very fully referenced. 1972 *Physics Bull.* May 295/3 The book is well produced and well referenced. 1975 *Nature* 3 Apr. p. iv (Advt.), Published as a two-book set for easier handling, each part is fully referenced and illustrated. 1977 *Jrnl. R. Soc. Arts* CXXV. 451/2 One BCS paper is referenced on page 35 of the White Paper Cmd. 6354. 1978 *Sci. Amer.* Jan. 28/3 The version we see (not explicitly referenced) is probably from an edition of about 100 years ago or even a 20th-century reprint. 1980 *Encounter* May 16/2 It is difficult to remember that Ferrar and Debenham's formulation of the sequence of Beacon rocks is still referenced today.
Hence **re·ferenced** *ppl. a.*, **re·ferencing** *vbl. sb.*
1884 Referencing [see sense 3 in Dict.]. 1971 *Nature* 30 Apr. 602/1 It is a very good book indeed—but it could have been excellent if just a little extra thought had gone into the original plan and an extra month into the final editing and referencing. 1972 *Science* 5 May 503/3 The authors..give an additional line of the frequently quoted but never referenced turbulence poem by Richardson (1922) beginning 'Big whorls have little whorls'. 1978 *Nature* 14 Dec. 739/2 The referencing, which includes entries up to the first half of 1976, is impressive.

referend (re·fĕrĕnd, refĕre·nd). [ad. L. *referend-um*, gerund or neut. gerundive of *referre* to Refer.] That by which or, more commonly, to which reference is made, *spec.* that which is signified by a particular sense of a word.
1925 *Monist* XXXV. 427 By the content of a judgment is meant the referend plus that which is predicated of the referend, and by the referend is meant that to which the judgment refers. 1930 L. S. Stebbing *Mod. Introd. Logic* ii. 13 We shall find it convenient to use the word 'referend' to stand for that which is signified. *Ibid.*, It is perhaps unfortunate to have to introduce new terminology, but the word 'object' is not suitable for the purpose for which we need the technical term 'referend'. The referend is that which is being referred to. 1939 *Trans. Philol. Soc.* 74 How can we disentangle ourselves from the close meshwork of our native language and find a *tertium quid* or set of neutral referends to serve as a measure for both languages compared? 1940 *Kenyon Rev.* 269 The monosign is referential in the sense that what it means.., its referend, is something distinct from itself. 1941 *Mind* L. 151 He thinks he is holding fast to some identical referend throughout, whereas in truth he has only an identical symbol which is changing its referend. 1956 J. Whatmough *Language* 262 Referend. That which is symbolized or referred to by a verbal symbol, e.g. the referend of *rain* is 'the moisture of the atmosphere condensed and falling in visible drops'. 1957 S. Potter *Mod. Linguistics* vii. 141 The bird, the living creature that flies with our eyes, we may call the *referend*, and the picture of it that we have in our minds as we speak..may be called the *image*. 1977 *Word* 1972 XXVIII. 162 The auxiliary is invariably present when the referend is not the subject of the clause.

referendum. Delete ‖ and add earlier and further examples. Pl. **referendums, -enda.**
In terms of its Latin origin, *referendums* is logically preferable as a modern plural form meaning ballots on one issue (as a Latin gerund *referendum* has no plural); the Latin plural gerundive *referenda*, meaning 'things to be referred', necessarily connotes a plurality of issues. Those who prefer the form *referenda* are presumably using words like *agenda* and *memoranda* as models. Usage varies at the present time (1981), but *The Oxford Dictionary for Writers and Editors* (1981) recommends *referendums*, and this form seems likely to prevail.
1847 G. Grote *Let.* 25 Sept. in *Seven Lett. concerning Politics Switzerland* (1847) iv. 81 The clergy made efficient use of their influence over the popular *referendum*. 1870 *Sat. Rev.* 7 May 602/1 The *veto* or *referendum* has this much in common with the Bonapartist *plebiscitum*, that it is submitted to a body which is not an assembly and which cannot discuss. 1870 *Rep. Mass. Bureau of Statistics of Labor* I. 358

We want the referendum. **1889** F. O. ADAMS *Swiss Confederation* vi. 77 In Federal matters there are now two Referendums. **1911** W. S. CHURCHILL *Let.* 29 Mar. in R. S. Churchill *Winston S. Churchill* (1969) II. *Companion* II. xiv. 1061 The collapse of the Referendum policy in the House of Lords was the subject of comment in the Lobbies yesterday. **1945** —— in *Times* 22 May 4/1 If you should decide to stand on with us, all united together until the Japanese surrender is compelled, let us discuss means of taking the nation's opinion, for example, a referendum, on the issue whether in these conditions the life of this Parliament should be further prolonged. **1945** C. ATTLEE in *Ibid.* 4/2, I could not consent to the introduction into our national life of a device so alien to all our traditions as the referendum, which has only too often been the instrument of Nazism and Fascism. Hitler's practices in the field of referenda and plebiscites can hardly have endeared these expedients to the British heart. **1965** H. V. WISEMAN *Britain & Commonwealth* II. iii. 86 Referenda have been held in New Zealand on the question of prohibiting the sale of intoxicating liquors. **1975** *Referendum on U.K. Membership of European Community* (Cmnd. 5925) i. 3 The present White Paper is concerned only with the organisation of the referendum. *Ibid.* iii. 5 The Government propose to ensure that the postal and proxy voting facilities which are available for general elections are also available for the referendum poll. **1975** *Times* 10 Apr. 5/1 The Liberals think that Referendum Day should be a public holiday. **1975** *Act Eliz. II* c. 33 § 1 A referendum shall be held on the question whether the United Kingdom is to remain a member of the European Economic Community. **1976** H. WILSON *Governance of Brit.* iii. 75 There was great interest in constitutional circles in my announcement on 23 January 1975 that collective responsibility would be relaxed for the period of the referendum campaign on membership of EEC—the famous 'agreement to differ'. **1976** *Ann. Reg. 1975* 53 Two referenda had already been held in Wales in 1975. **1976** *Times* 22 Dec. 11/7 Since we are now likely to hear much about referendums, could we ask the BBC not to continue calling them referenda? **1977** *Daily Tel.* 16 Feb. 1/3 Proposed referenda on the plan for Scottish and Welsh Assemblies should be consultative only and not binding on Parliament. **1977** *Times* 17 Mar. 19/3 They did not tell us at the time of the referendum that Brussels was to reform the English language. **1977** *Time* 21 Nov. 28/2 The many referendums on the ballots reflected a growing public demand for more efficient and less meddlesome government.

referent, *sb.* and *a.* Delete *rare* and add: **1.** (Further example.)

1921 *Contemp. Rev.* Mar. 315 The whole administration is conducted by the provincial government in Bratislava (Pressburg), under the Minister for Slovakia and his thirteen 'Referents' or State Secretaries.

2. c. *sb.* That to which something has reference; *spec.* that which is referred to by a word or expression. Also in *Comb.* (*appositively*), as *referent-object*.

1923 OGDEN & RICHARDS *Meaning of Meaning* i. 13 The word 'thing' is unsuitable for the analysis here undertaken, because in popular usage it is restricted to material substances—a fact which has led philosophers to favour the terms 'entity', 'ens' or 'object' as the general name for whatever is. It has seemed desirable, therefore, to introduce a technical term to stand for whatever we may be thinking of or referring to. 'Object', though this is its original use, has had an unfortunate history. The word 'referent', therefore, has been adopted. **1931** F. C. S. NORTHROP *Sci. & First Principles* ii. 49 This theory [*sc.* the physical theory of nature] is untenable unless there is a referent for atomicity and motion in something other than the microscopic particles. **1937** *Harper's Monthly Mag.* Dec. 49/2 Knowledge about technological unemployment..is not advanced by the syllogisms of classical economists. The classicists treat the term as a thing-in-itself without finding the referents which give it meaning. **1938** *Mod. Lang. Rev.* Oct. 547 The reinterpretation of the term [*baroque*] has sprung, not.. from the feeling that it could be better applied to some other type of art, but from the revaluation of its original referent. **1950** *Papers Mich. Acad. Sci., Arts & Lett.* XXXVI. 323 The same question may..be phrased by asking how to bridge the gap between..sign and referent. **1958** S. STUBELIUS *Airship, Aeroplane, Aircraft* 7 As a rule, there has been no difficulty in ascertaining the referents in my linguistic material, with the aid of context. **1964** *Eng. Stud.* XLV. 385 Gender distinctions had always been functional in English as indicating either the sex of the referent or the non-significance of sex-distinction of the referent. **1964** *Language* XL. 229 The much-studied relations of words to their referent-objects. **1968** [see *DENOTATUM]. **1970** *New Society* 5 Mar. 394/2 Clearly, if the ongon is to have any meaning at all, there must be some way of knowing at what level the symbols are to be interpreted, bearing in mind that each drawing or material object will have several referents. **1973** *Times Lit. Suppl.* 26 Oct. 1306/3 Canada often provides the landscape for his fable or the referents of his argument. **1976** *Archivum Linguisticum* VII. 17 Most Castilian loanwords from other languages have been borrowed to refer to referents, or fulfil functions, when Latin had originally an apparently good word for the same purpose. **1979** *Dædalus* Summer 97 'Official' discourse is reproached for being without a referent in psychic or social reality.

3. *Logic.* Any member of the class of all terms bearing a given relation to any term. (The correlative of *relatum*.)

1903 [see *DOMAIN *sb.* 4 f]. **1933** L. S. STEBBING *Mod. Introd. Logic* (ed. 2) vii. 111 Every relational proposition has a converse, which consists in interchanging the terms with, or without, a change in the relation asserted to hold between them. The term *from* which the relation proceeds is called the referent; the term *to* which it proceeds is called the relatum. **1947** H. REICHENBACH *Elem. Symbolic Logic* iii. 115 There will be more than one relatum with respect to Peter if Peter has other children... The class of referents is also called the domain of function, and the class of relata is called the converse domain. **1967** R. A. GEORGE tr. *Carnap's Logical Struct. of World* III. 60 A relation extension is called one-many if, for each relatum, there exists only one referent.

referential, *a.* Add: **b.** Of or pertaining to a referent (sense *2 c); *spec.* of language or symbolism: that indicates a referent or has a referent as object.

1884 [in Dict.]. **1923** OGDEN & RICHARDS *Meaning of Meaning* i. 13 Besides this referential use which for all reflective, intellectual use of language should be paramount, words have other functions which may be grouped together as emotive. *Ibid.* ix. 318 Unless the referential and the affective-volitional aspects of mental process are clearly distinguished, no discussion of their relation is possible. **1922** T. C. POLLACK *Nature of Lit.* IX. 195 In *phatic communion*, one person uses words to come into relation with another. In *referential symbolism*, one person uses words to direct the attention of another to certain referents... In *evocative symbolism*, one person uses words to evoke a controlled experience in another. **1946** C. MORRIS *Signs, Lang. & Behavior* 60 The current distinction between 'referential' and 'emotive' terms, a basic distinction in the work of C. K. Ogden and I. A. Richards. **1946** H. JACOB *On Choice of Common Lang.* 116 The never entirely separable functions of language roughly classified as 'emotive' and 'referential'. **1964** [see *EXTRASOMATIC *a.*]. **1964** S. JACOBSON *Adverbial Positions in Eng.* i. 38 Several adverbs and adverbial phrases, especially such as express restriction or particularization, refer to, i.e. direct attention to, some particular constituent within the sentence. They may be called referential adverbials, and together with the constituent referred to, which may be a word, phrase, or clause, they form a structure of reference... Thus..'Only a Person can forgive'.., where *only* refers to *a Person*. **1964** R. H. ROBINS *Gen. Linguistics* 29 Word translation, or the finding of lexical equivalents, is easiest.. with words in other languages which are such as to have a referential meaning more or less uniform in all cultures. **1971** *Archivum Linguisticum* II. 40 Nor is there any compulsion to believe in the primacy of any subdivision of meaning, for instance in the referential 'table'-ness of 'table'. **1980** *Mind* LXXXIX. 601 The truth of the speaker's belief q is neither necessary nor sufficient for the referential use of his definite description.

referentially *adv.* (examples).

1922 W. E. JOHNSON *Logic* II. 120 Those so-called constants which are dependent upon context are only referentially constant. **1963** J. LYONS *Structural Semantics* iii. 39 Each of the [colour-]terms is referentially, or denotationally, vague in the sense that it denotes an area of the spectrum whose boundaries are not fixed precisely. **1966** *Philos. Rev.* LXXV. 11 Some of Socrates' substitutions are within intensional or referentially opaque contexts. **1975** *Nature* 10 Apr. 510/1 [He] is not bound by referentially constrained situations in his use of VIC [*sc.* Visual Communication, a therapeutic programme]. **1980** *Mind* LXXXIX. 599 Has the definite description been used referentially? No doubt the students took it in the attributive way... They do not even realize that the teacher intends his description referentially.

referral (rĭfɜ̄·răl). [f. REFER *v.*: see -AL II. 5.] **a.** The act of referring; *spec.* the referring to a third party of personal information concerning another.

1934 in WEBSTER. **1943** *Sun* (Baltimore) 27 Nov. 6/3 The publication by Senator Butler of his report and his referral of it to the Truman Committee. **1968** *Globe & Mail* (Toronto) 15 Jan. 24/7 (Advt.), Television and newspaper advertising will supply starting work followed by constant referral jobs. **1969** *Ithaca* (N.Y.) *Jrnl.* 27 Nov. 32/2 The name of the woman who was two months pregnant was apparently given secretly to the..company by one of her friends, who was paid for the 'referral' with a vaporizer or other small appliance. **1971** *Nature* 30 Apr. 545/1 The Unisist committee asks that associated countries should survey information services of national, regional or international scope and make plans to integrate these into a 'world referral network'. **1971** *N.Y. Law Jrnl.* 23 Nov. 20/4 After referral to me by order of Mr. Justice McInerney, this motion for a withdrawal order is granted. **1981** *Times* 10 Mar. 18/6 The senior management of these businesses comes out well from these referrals.

b. The referring of an individual to an expert or specialist for advice; *spec.* the directing (usu. by a general practitioner) of a patient to a medical consultant for specialist treatment (see *REFER *v.* 8 d). Also *attrib.*

1955 D. M. DEED in C. Morris *Social Case-Work in Gt. Brit.* (ed. 2) iii. 70 At the time of referral [to a family caseworker] the child was in a temporary hostel. **1958** *Times* 8 Nov. 7/5 The public..is in turn increasingly unwilling to trust themselves to the care of their N.H.S. practitioner, thus reinforcing the habit of unnecessary referral to hospitals. **1960** in L. Pincus *Marriage* II. 58 At the time of their referral to the Bureau, they had been married for nine and a half years. **1966** *Lancet* 24 Dec. 1403/1 The referral system, whereby consultants see patients only at the request of the family doctor, came into being at the end of the 19th century. **1968** *World Medicine* 26 Nov. 35/1 The referral diagnosis was Addison's disease. **1972** *Where* May/June 136/3 He can go to his family doctor who will either examine him and take the necessary tests to establish the diagnosis before giving treatment, or will send him to the nearest hospital clinic with a letter of referral. **1976** *Proc. R. Soc. Med.* LXIX. 949/1 Our surgical unit is no less busy than most in the country: it has its share of medical ward referrals, urgent admissions and emergencies. **1978** *Times* 2 Nov. 14/2 They [*sc.* the Samaritans] can, and do, act as a medical or psychiatric referral agency.

reffo (re·fo). *Austral. slang.* [Abbrev. REFUGEE *sb.*] A European refugee; *spec.* a refugee who left Germany or German-occupied Europe before the war of 1939–45. Now *Obs.* exc. *Hist.*

1941 BAKER *Dict. Austral. Slang* 59 *Reffo*, a refugee from Europe. **1951** CUSACK & JAMES *Come in Spinner* 278 'The woman's a Viennese.' 'Oh, a reffo?' **1955** J. CLEARY *Justin Bayard* x. 137 She talked even now in her letters of the Dagoes and Balts and reffos, the New Australians, who were taking over the country. **1960** *Times Lit. Suppl.* 3 June 349/3 Stefan is the refugee—from Bulgaria... Lionel.. considers Stefan 'a bloody reffo' and a nuisance. **1961** P. WHITE *Riders in Chariot* 221 He was..a blasted foreigner, and bloody reffo, and should have been glad he was allowed to exist at all. **1965** *Listener* 2 Sept. 339/2 Australians..had their full quota of terms such as wogs, dagos, and so on, and refugees were commonly referred to as 'bloody reffos'. **1976** *Australasian Post* 8 Apr. 54/2 Several decades ago, the normal expression for migrants was 'reffo', which was then officially changed to 'migrant'.

refill, *sb.* Add: **b.** The renewed contents of a glass; a second or further drink.

1929 'E. QUEEN' *Roman Hat Mystery* iii. 53, I asked her if it was time to go in for my orangeade refills. **1960** C. MACINNES *Mr Love & Justice* 79 The star ponce beckoned for refills... The girl..brought the glasses. **1966** M. SHARMAN *Seeds of Violence* ii. 11 'I think we'd better have refills.' He busied himself with the drinks. **1968** 'R. RAINE' *Night of Hawk* xxxiii. 156 Drink up and I'll give you a refill. **1977** *Rolling Stone* 30 June 25/1 She lets go with a loud, decidedly unsentimental laugh that startles a room-service waiter trying to set down refills on a coffee table hopelessly cluttered with empty glasses and Heineken bottles.

c. *adj.* That requires or serves as a refill.

1907 *Yesterday's Shopping* (1969) 345/1 Nickel Pencil.. 4 in. long, with refill leads. **1918** W. OWEN *Let.* 9 Sept. (1967) 574 I'd like a parcel with re-fill battery, cigarettes, & chocolate. **1921** *Edin. Rev.* Jan. 158 Perhaps in time each party will provide not only its own refill ministers and kitchenmaids but even its own king. **1961** *Lebende Sprachen* VI. 103/2 *Refill cartridge*, die Nachfüllmine..für Kugelschreiber.

refillable (rĭfi·lăb'l), *a.* [f. REFILL *v.* + -ABLE.] Capable of being refilled.

1920 in WEBSTER. **1961** R. A. FORESMAN in H. R. Shepherd *Aerosols* iii. 48 Although disposable aerosols are the only real market factor at this writing, the refillable units were of some consumer importance as late as 1958. **1962** *Harper's Bazaar* Aug. 65 An elegant new black and gold refillable atomiser. **1971** *Nature* 23 July 274/1 For electrophoresis, refillable glass cannulae containing 25 % KCl and a coiled silver wire were implanted chronically. **1977** *Private Eye* 13 May 23/3 (Advt.), Refillable ballpens and cartridge pens that look good and feel good.

refi·nance (*also* -finæ·ns), *v.* [RE- 5 a.] *trans.* To finance again; to provide with further capital. So **refi·nanced** *ppl. a.*; **refi·nancing** *vbl. sb.* and *ppl. a.*

1908 *Standard* 6 Apr. 6/4 Plans have been drafted for the refinancing of the Erie Railway Company. **1921** *Glasgow Herald* 6 July 8/2 The first important refinancing operation, since the Funding and Victory issues, was the Conversion Loan announced with the Budget statement. **1922** *Daily Mail* 26 Oct. 3 The conditions of the Government guarantee and the details of the scheme for re-financing the company were first officially announced. **1930** *Times Lit. Suppl.* 30 Oct. 893/1 Oscar, generously refinanced through 'Gamy', is sailing..safely into harbour. **1934** *Planning* I. xix. 3 Refinancing in the post-war boom had left an evil legacy: the practices of local finance, of the issue of part-paid shares and of calling up unpaid capital had intensified the problem and to a large extent control had passed into the hands of the banks. **1971** *Flying* Apr. 25/2 The R. J. Enstrom Helicopter Company..is being reactivated and refinanced as a wholly owned subsidiary of Franklin Capital Corp. of Michigan. **1976** *Economist* 16 Oct. 102/2 Instead of forcing foreigners to convert their sterling holdings into other reserve assets once and for all, such a scheme would allow Britain to refinance any rundown in its sterling balances, both private and official, as and when it occurred. **1979** *Jrnl. R. Soc. Arts* CXXVII. 428/1 There would be danger, if interest rates fell substantially, that mortgages would be redeemed and refinanced.

Hence **refi·nance** *sb.*, renewed or additional finance; chiefly *attrib.* in *refinance credit* (see quot. 1970); **refina·nceable** *a.*, capable of being refinanced.

1959 *Economist* 28 Feb. 817/1 The authorities have not lifted the ban on 'refinance' credits... These are usance credits, usually of three months' duration, opened in order to meet sight drafts payable in London. **1965** *Ibid.* 19 June p. ix/1 What of 'refinanceable credits'—the portions of medium-term advances for export credits which now allow the banks to grant such advances with no impact on their liquidity ratio? **1970** *Penguin Dict. Commerce* 276 *Refinance credits*, credits obtainable by an overseas buyer, where the exporter cannot provide credit and the buyer does not wish to pay cash. **1976** *Bank of England Q. Bull.* XVI. 194 Cash flow may well be negative in the earlier years, thus requiring refinance at uncertain interest rates.

refinedness. Delete ? *Obs.* and add further examples.

1717 *Censor* II. sig. A4ᵛ, The sensible Part of the World in their Pleasures, as well as graver Conduct, are proud of being influenc'd by Examples that give them the Credit of Discernment, and a Refinedness of Taste. **a 1945** E. R. EDDISON *Mezentian Gate* (1972) xxxviii. 227 'Your son, I said. There are other names for bastards.' 'I have always admired the refinedness of your language.'

refiner. Add: **1. b.** *spec.* A machine used in paper-making in which knots and lumps in the pulp are broken down by scissoring between blades or discharged by centrifugal action.

1902 *Encycl. Brit.* XXXI. 458/2 One form of beater [for paper making] has already been referred to..but engines of quite a different construction are now used largely in

American mills, and also to some extent in Great Britain. These are known as 'refiners'. *Ibid.* 459/1 By the use of the refiner the time occupied in the beater can be reduced by nearly one half. **1929** CLAPPERTON & HENDERSON *Mod. Paper Making* vi. 68 The final clearing of knots, and the reduction of the longer fibres to a uniform length, are often performed by a refiner or perfecting engine. **1963** R. R. A. HIGHAM *Handbk. Papermaking* ii. 55 The Bauer refiner consists of two discs which rotate in opposite directions, the bars of which form a variety of cutting surfaces according to the character of the stock and the grade of paper to be manufactured.

refinery. Add: **1.** (Later examples.) Also *attrib.*

1939 *Thorpe's Dict. Appl. Chem.* (ed. 4) III. 348/2 The largest refinery plant in this country is at Prescot, where copper is refined..from blister copper. **1951** DYLAN THOMAS *Let.* Jan. (1966) 351 Today I was taken to see a great new black-towered hissing and coiling monster, just erected in the middle of the refinery. **1954** *Thorpe's Dict. Appl. Chem.* (ed. 4) XI. 187/1 Refined sugar presupposes treatment with bone charcoal in refineries as distinct from factories dealing with sugar beet or sugar cane. **1959** *Listener* 2 Apr. 582/1 Refinery gases, oil, and now refrigerated methane. **1970** W. G. ROBERTS *Quest for Oil* ix. 93 The newest refineries use desulphurisation processes of this kind to treat a wide cut from the light end of the crude oil right up to kerosine or gas oil. **1981** J. SIMPSON *Moscow Requiem* II. v. 159 The coast of Saudi Arabia and the peninsula where stood the second largest oil refinery in the world.

refit, *sb.* (Further examples.)

1945 *Jane's Fighting Ships 1944–45* 432 A contract was signed..for the repair and refit of this battle-cruiser. **1955** *Times* 7 May 8/1 An explosion occurred last night in the diesel room of H.M.S. Daring under refit in Devonport dockyard. **1975** *Drive* New Year 28/2 Is it more economical to write off a damaged car and pay the client out or make him wait for a re-fit? **1976** *Gramophone* Sept. 453/3 The Canterbury instrument is now in need of a refit.

reflate (riflē^i·t), *v.* [f. RE- 5 a after *DEFLATE *v.* 3, *INFLATE *v.* 4.] **a.** *absol.* To raise the pressure of demand (in an economy) after a period of falling pressure. Also *trans.*, to expand (the money supply or the flow of expenditure) or raise (prices) after a period of contraction or reduction. **b.** *intr.* for *pass.* Of an economy: to be affected by or subject to reflation. Hence **refla·ting** *vbl. sb.*

1932 *Sun* (Baltimore) 12 Apr. 10/2 There are plenty of ways to inflate or 'reflate' without putting the Federal Government $2,000,000,000 more into debt to accomplish the purpose. **1939** *Ibid.* 18 Dec. 13/6 The Administration was pleased to see farm prices on the rise for the New Deal dedicated a large part of its recovery promotion energies to schemes for 'reflating' commodity quotations. **1958** *Engineering* 21 Mar. 354/2 The rest of the world is afraid that the American determination not to reflate is going to start a deflation abroad. **1960** *Economist* 15 Oct. 220/2 Britain's present poor export performance makes it imperative that we should not reflate too early. **1966** *Daily Tel.* 2 Nov. 1/1 It is not a sign that the Government has begun to reflate the economy to halt unemployment and start industrial production moving forward. **1971** *Ibid.* 10 Sept. 19 The banks have 'saved' £600 million and..the authorities will probably want a substantial slice of it locked up again in case the economy reflates too fast again. **a 1974** R. CROSSMAN *Diaries* (1976) II. 134 If we wait until the officials tell us it's wise to reflate, the results won't show for years ahead and we'd have no prospect of winning the election. **1975** *Washington Post* 31 Aug. c7/4 The French and Germans are now having to reflate their economies by massive government expenditures. **1979** *Dædalus* Spring 43 Governments are..taking pains not to reflate the economy enough to wipe out unemployment for fear of rekindling inflation and a balance of payments crisis.

reflation (riflē^i·ʃən). [f. RE- 5 a after *DEFLATION 3, *INFLATION 6.] The process of reflating or taking measures designed to allow an expansion in economic activity to be resumed.

1932 *Economist* 20 Feb. 394/2 Its purpose has been aptly described as 'reflation', to prevent further deflation..and to undo some of the present extreme deflation. **1932** *Sun* (Baltimore) 12 Apr. 10/2 It does seem in order to ask why, if inflation or 'reflation' or whatever they may choose to call it is really the principal concern, they do not..come forward with a straight inflation or 'reflation' project. **1932** *Hansard Commons* 20 Apr. 1582, I propose a different thing altogether [from inflation]. I would describe it as reflation, which I would define as controlled expansion of the note issue to keep pace with increased production. **1933** *Daily Tel.* 24 Apr. 11/3 The fight in Congress to beat depression by inflation—reflation, as some call it—grows more heated. **1940** G. CROWTHER *Outl. Money* iii. 116 The custom has grown up of referring to what happens during the upward phase of the normal trade cycle not as 'inflation' but as 'reflation'. For the present, we can think of 'reflation' as being restricted to a rise of prices that merely restores the *status quo ante*—the position before the start of the preceding deflation—and inflation to any further rise in prices after this point. **1959** *Economist* 17 Jan. 194/2 Every pound's worth of income tax relief this year would bring less risk to sterling than every ten shillings' worth of reflation made in any other way. **1971** *New Scientist* 10 June 606/1 A reflation stimulus is urgently needed now to fill the yawning gap in consumer demand. **1975** *Evening Standard* 24 July 36/2 (*heading*) It's reflation time says EEC. **1977** *Times* 29 Aug. 9/3 An electorally opportunistic government..might be strongly tempted to chance some reflation in the run up to a general election next year.

reflationary (riflē^i·ʃənări), *a.* [f. *REFLATION + -ARY¹.] Characterized by, suggestive of, or tending to reflation.

1932 *Times Lit. Suppl.* 8 Dec. 931/3 The 'reflationary' policy of the American Government will in the end, he thinks, set prices rising again. **1940** *Economist* 7 Dec. 704/2 As long as the bulk of the gold inflow in the United States was the counterpart of a movement of capital there was no inherent reflationary or inflationary virtue to be found in it. **1957** *Ibid.* 26 Oct. 286/1 No small reflationary measures taken by Britain now, however, would make the situation then any better. **1969** *Indian Econ. Rev.* Feb. 32 As it proved, what expansion of the money supply there was in the 1930s was not sufficient to be reflationary... that is, it served to increase national income without inducing a significant rise in prices. **1969** *Daily Tel.* 28 Oct. 16 An election just after a reflationary Budget would be cynically received. **1974** *Guardian* 27 Mar. 15/3 The TUC wanted a mildly reflationary Budget and Mr Healey has given them.. a mildly deflationary one. **1976** LD. SELWYN-LLOYD *Mr. Speaker, Sir* vii. 150 Many heads had already been shaken at the reflationary measures of the first two years or so of the Parliament and the idea of bursting through to growth and success.

reflationist (riflē^i·ʃənist). [f. *REFLATION + -IST.] One who supports or advocates a policy of reflation.

1959 *Economist* 31 Jan. 393/1 The next questions that eager reflationists put to him should presumably be about the date of that budget and about investment allowances. **1972** *Guardian* 24 Jan. 6/8 Mr Macleod was a natural reflationist. **1980** *Spectator* 1 Nov. 4/2 What is unique to Britain is the great strength of the reflationists in the party of the Right.

reflect, *v.* Add: **I. 4. b.** Also used with reference to other forms of wave or radiation. (Further examples.)

1902 *Chem. News* 24 Jan. 47/2 (*heading*) Rays capable of being reflected in radiation emitted by a mixture of chlorides of radium and barium. **1909** *Proc. R. Soc.* A. LXXXII. 495 The fraction of the incident α-particles which are reflected. **1937** *Discovery* Jan. 3/1 The ionosphere—that region in the upper atmosphere where free electrons reflect wireless waves. *Ibid.* 4/1 Pulses are radiated in all directions: some reach the receiver by travelling direct along the ground, others by travelling high up into the ionosphere, where they are reflected downwards by one or more of the conducting layers. **1950** D. HALLIDAY *Introd. Nuclear Physics* vi. 147 A substance with a positive scattering length should..reflect neutrons totally at small external glancing angles. **1960** N. N. TOOS *Theory Mech. Vibration* iv. 308 If the bar has an end, the disturbance wave will be reflected as it reaches that end. In a certain subsequent time period the reflected wave and the incident wave coexist in the bar. Afterward, only the reflected wave remains. The manner in which the reflection takes place depends on the end condition. **1974** S. W. FLAX et al. in R. S. Reneman *Cardiovascular Applications Ultrasound* ii. 19 Another factor is how blood cells reflect ultrasound.

c. *Physiol.* To give out (an impulse) along a motor nerve, in response to one received along a sensory nerve. Usu. *pass.*

1833, 1855 [in Dict., sense 4]. **1869** J. C. DALTON *Treat. Human Physiol.* (ed. 5) II. 314 The function of the gray matter is..to receive the impulse conveyed to it, and to reflect or send back another. **1906** H. W. SYERS tr. *J. P. Morat's Physiol. of Nerv. Syst.* ii. 218 The impulse is reflected from the posterior roots to the tracts of the spinal cord, in conscious impressions. **1931** H. G. WELLS et al. *Sci. of Life* I. iii. 86/1 The impulse may be wholly reflected in a reflex or pass on in part and more or less modified to the hemispheres. **1950** P. D. F. MURRAY *Biol.* viii. 68 The term 'reflex'.. refers to the manner in which the impulse, having passed in to the central nervous system, is 'reflected' outwards from it.

7*. With direct statement, question, or exclamation as obj. (For indirect uses, see sense 12 b in Dict.)

1862 MRS. H. WOOD *Channings* II. vi. 102 'No, no; it would not be right of him to make me his wife now,' she reflected. **1926** 'J. M. H. RIDDELL *Senior Partner* II. xi. 223 'She has the Poussett kind of talk,' he reflected, 'and the same uppish way with her.' **1956** E. PHILLPOTTS *Portreeve* I. vi. 30 'Let what will come, there's amusement in it,' she reflected. 'Tis hunting of a sort. Fox-hunting—man-hunting—what more has life for me?' **1919** V. WOOLF *Night & Day* xxxiii. 506 Even if she started this very moment she would be too late for..at least, he reflected, she would not be home till Tuesday night.

II. 12. **a.** Also occas. with *over*.

1906 W. S. MAUGHAM *Bishop's Apron* ix. 137 Winnie reflected over this for a moment.

14. b. (Further example.)

1979 *Nature* 11 Jan. 84/1 If the flight control centre was indeed 'astounded', this surely reflects on the rate at which the Soviet team get access to the data from US missions.

reflectance (rĭfle·ktăns). *Physics.* [f. REFLECT *v.* + -ANCE.] The proportion of the light incident upon a surface, which is reflected or scattered by it; *spec.* a complex number whose modulus is the proportion of the radiant flux (at some specified wavelength or range of wavelengths) which is reflected, and whose argument indicates the change of phase undergone by the reflected light. Cf. **reflection coefficient, factor*. Also *attrib.*

1926 *Jrnl. Optical Soc. Amer.* X. 178 Reflectance..is the ratio of reflected to incident radiant energy. **1932** [see *INFRA-RED *a.* and *sb.* A. 1]. **1956** *Nature* 14 Jan. 74/2 The assessment of particular reflectance characteristics of cotton may be made on the automatic Nickerson-Hunter cotton colorimeter. **1957** V. J.-R. KEHOE *Technique Film & Television Make-Up* viii. 96 Black velvet has approximately 2 per cent reflectance, while some white paper is as high as 90 per cent. **1960** *Illuminating Engin.* LV. 228/1 The concepts of transmittance and reflectance can be extended to cover the separation of specular and diffuse components which are functions of angles. **1975** D. H. BURRIN in Williams & Wilson *Biologist's Guide Princ. & Techniques Pract. Biochem.* v. 138 Reflectance spectrophotometers, which measure the radiation absorbed when a light beam is reflected by the sample, allow the determination of absorption spectra of pastes and suspensions of microorganisms which are too opaque to transmit radiation. **1977** J. HEDGECOE *Photographer's Handbk.* 132 (*caption*) Meter reads direct off mid-gray (ideally 16 percent reflectance) card which simulates subject mid-tone.

reflected, *ppl. a.* Add: **2. b.** Also used of other waves and radiations (cf. *REFLECT *v.* 4 b). (Further examples.)

1830 *Encycl. Metrop.* II. 753 A tendency in the reflected Sound to deviate from the direction which a ray of Light regularly reflected at the echoing surface would follow. **1862** R. MALLET *First Princ. Observational Seismol.* II. xii. xvi. 356 Ottajano and Somma..sustained the subordinate shock, of reflected waves from the N.E. flank of Vesuvius. **1909** *Proc. R. Soc.* A. LXXXII. 497 For β-particles the number of reflected particles..decreases with the atomic weight of the reflector. **1924** *Year-bk. Wireless Teleg. & Telephony* 13 A system of telegraphy employing reflected beams was not only possible, but possessed very many advantages. **1960** [see *REFLECT *v.* 4 b]. **1975** LEOPOLD & ASHER *Fund. Abdominal & Pelvic Ultrasonography* ii. vi. One can obtain information from all areas of the body by suitably observing the reflected ultrasound pulses.

reflecting, *ppl. a.* Add: **1. a.** Also, that reflects waves or radiation of other kinds (cf. *REFLECT *v.* 4 b). (Further examples.)

1850 W. & R. CHAMBERS *Nat. Philos. Acoustics* 16 The reflected [sound] waves have the same form as if they diverged from a point on the other side of the reflecting surface, directly opposite to the origin of the waves and equally distant from the surface. **1869** J. TYNDALL *Sound* (ed. 2) i. 13 Like sound also, light and radiant heat, when sent through a tube with a reflecting interior surface, may be conveyed to great distances with comparatively little loss. **1937** *Discovery* Jan. 3 The reflecting ionosphere layer. **1953** REED & RUSSELL *Ultra High Frequency Propagation* 313 The amplitude of the reflected energy is dependent.. upon the dynamic mechanics of the reflecting media.

reflection, reflexion. Add: **2. a.** Also, the similar action of surfaces on other waves and radiations (cf. *REFLECT *v.* 4 b). (Further examples.)

1902 *Chem. News* 24 Jan. 47/2 Certain rays existed in the radiation emitted by certain radio-active bodies which were capable of reflection. **1909** *Proc. R. Soc.* A. LXXXII. 497 The diffuse reflection of the α-particles is a consequence of their scattering. **1929** *Jrnl. Sci. Instruments* VI. 34 Sound has all the properties of a wave motion..and exhibits the phenomena of reflection, interference, diffraction and resonance. **1941** A. B. WOOD *Textbk. Sound* (ed. 2) iii. 31 The direct reflection of a sound of short duration from a surface at some distance such as the wall of a building or a cliff is generally described as an echo. **1960** [see *REFLECT *v.* 4 b]. **1969** *Times* 28 Aug. 3/3 Seismic reflections indicate the thickness of the rocks.

9*. *Cryst.*, *Math.*, *Physics.* The conceptual operation of inverting a system or event with respect to a plane, each element being transferred perpendicularly through the plane to a point the same distance the other side of it. Freq. *attrib.*

1899 W. J. LEWIS *Treat. Crystallogr.* iii. 18 We shall often express the relation of two planes, or two lines, to a plane of symmetry bisecting the angle between them by the statement that they are reciprocal reflexions in the plane. **1910** *Nature* 26 May 368/1 Its 880 known solutions (8 × 880, if we admit reversals and reflections of the square with respect to the diagonals and with respect to the two medial lines) are 'different'. **1955** W. PAULI *Niels Bohr* 30 The mathematical group was further amplified by including the reflections of space and time. *Ibid.* 33, I am restricting myself..to the discussion of the reflection of all coordinates simultaneously while I do not consider the reflection of space or time separately. **1965** A. F. BROWN tr. *Zhdanov's Crystal Physics* v. 144 Symmetry groups containing only the operations of reflection, rotation and inversion, and not containing any translations, are called point groups. **1965** *Sci. Amer.* Dec. 28/1 Until December, 1956, they [*sc.* physicists] had assumed that if an event is possible, its mirror image is also possible, and that if one looks at some real event in a mirror, what one sees could also actually happen. This was known as reflection symmetry. **1971** I. G. GASS et al. *Understanding Earth* i. 19/1 The stereogram of zircon..shows a four-fold rotation axis in the centre and also shows a number of reflection planes. **1972** F. J. BUDDEN *Fascination of Groups* xxvi. 507 The two-dimensional pattern of fig. 26.051 contains translations and glide reflections, but no rotations.

10. *reflection oscillator*; *reflection-reducing* adj.; **reflection coefficient, factor** *Physics* = *REFLECTANCE; **reflection nebula** *Astr.*, a nebula which is visible only by virtue of the light which it reflects; **reflection profiling** *Geol.*, profiling (sense *3) by means of reflection shooting; **reflection shooting** *Geol.*, seismic prospecting in which shock waves generated at the earth's surface are detected up to a mile away after having been reflected at the interface between strata, the depth of which is deduced from their time of arrival.

1942 A. HUND *Frequency Modulation* i. 139 The reflection coefficient for horizontal polarization. **1959** BORN & WOLF *Princ. Optics* xiii. 627 The complex reflection and transmission coefficients of the film may immediately be evaluated. **1975** E. HEIGHT *Optics* iii. 43 Determine the values of the amplitude reflection coefficients for light incident at 30° on an air-glass interface. **1920** Reflection factor [see *REFLECTOMETER]. **1971** E. SKUDRZYK *Acoustics* xv. 302 The amplitude reflection factor represents the ratio of the reflected to the incident pressure wave with respect to magnitude and phase. **1936** *Astrophysical Jrnl.* LXXXIV. 219 (*heading*) Reflection nebulae. **1974** *Sci. Amer.* Oct. 34/3 The reflection nebulae are useful for studying the properties of the interstellar dust grains, but they are distinguished from the true emission nebulas, which shine as a result of the atomic processes going on within them. **1958** B. MCCOLLUM & J. E. Duntan et al. *Sci. of Petroleum* I. viii. 396/2 (*heading*) Equipment used in reflection profiling... **1960** Moore in van Andel & Shor *Marine Geol. Gulf of Calif.* 193 The sedimentary structure of the continental terrace of the Costa de Nayarit..has been investigated geophysically by means of continuous acoustic-reflection profiling. **1971** I. G. Gass et al. *Understanding Earth* xvi. 243/2 Reflecting horizons..can be mapped over vast areas by continuous reflection profiling. **1962** CORSON & LORRAIN *Introd. Electromagn. Fields* xi. 406 On reflection-reducing coatings on lenses improve the transmission significantly. **1929** *Trans. Amer. Inst. Mining & Metall. Engineers* LXXXI. 606 The distances are short in reflection shooting compared with those in refraction shooting. **1951** K. K. LANDES *Petroleum Geol.* ii. 48 Reflection shooting has recently become important again as a method of detailing rock structure where refraction shooting is not practicable. **1973** R. E. CHAPMAN *Petroleum Geol.* ii. 45 More detailed structural information is obtained from reflection shooting, in which the elastic waves are partly reflected by surfaces of contrasting density. **1971** *Physics Bull.* June 331/1 Reflection spectroscopy..concerns the measurement of the frequency dependence of the specular reflectivity of the material to determine either the positions and strengths of features in the absorption spectrum or its optical constants.

reflectionless *adv.*, also as *adj.*, not giving rise to any reflection.

1961 *Rev. Sci. Instruments* XXXII. 828/1 A reflectionless wave-guide termination. **1956** *Nature* 25 Feb. 392/1 A thin transverse film having a surface resistivity equal to the wave impedance of the waveguide forms a reflexion-less termination when [etc.].

reflectious (rĭfle·kʃəs), *a. nonce-wd.* [f. REFLECT *v.* + -IOUS. Cf. REFLEXIOUS *a.*] = REFLECTIVE *a.*

1874 HARDY *Far from Madding Crowd* II. xv. 311 'Justice is come to weigh him in the balance,' I said in my reflectious way.

reflective, *a.* and *sb.* Add: **A.** *adj.* **4. c.** (Further examples.)

1942 M. BODKIN *Archetypal Patterns in Poetry* 314 An hypothesis..that archetypal patterns, or images, are present within the experience communicated through poetry, and may be discovered there by reflective analysis. **1961** *Manas* 5 Apr. 1/2 Mr. Lyford gives voice to a temper that represents..an achieved plateau of reflective thinking.

reflectivity. Add: Also (*a*) the degree to which anything incident on a surface is reflected; (*b*) the degree to which a surface reflects what is incident upon it. (Further examples.)

1910 *Physical Rev.* VIII. 152 The reflectivity of metal atoms striking surfaces of the same metals at room temperature [is lower] is zero. **1936** *Discovery* Aug. 237/1 Steel mirrors of high reflectivity. **1946** [see *ALUMINIZE *v.* 4]. **1960** *McGraw-Hill Encycl. Sci. & Technol.* XI. 395/1 Typical curves of the reflectivity of the polarized components versus angle of incidence are given in Fig. 2. **1974** *Mineral. Abstr.* XXV. 267/1 Reflectivities of pyrite..and sphalerite were measured with POH microscope and SPS-1 microscope photometer.

reflectometer. Substitute for def.: Any of various instruments for measuring quantities associated with reflection; *spec.* (*a*) one for measuring the critical angle of a transparent solid so that its refractive index may be calculated; (*b*) one for measuring the intensity of light reflected or scattered by a surface so that its reflectance may be calculated. (Earlier and later examples.)

1891 *Jrnl. Chem. Soc.* LX. 513 (*heading*) Measurement of refractive indices at high temperatures by means of the total reflectometer. **1920** *Sci. Papers U.S. Bureau of Standards* XVI. 415 The use of an incomplete sphere as a reflectometer furnishes two new absolute methods for the determination of diffuse reflection factors. **1936** *Sci. Abstr.* A. XXXVIII. 1223 (*heading*) Ultrasonic total reflectometer to measure speed of sound and elastic constants of solids. **1962** R. H. KAY in J. Thewlis *Encycl. Dict. Physics* VI. 240/1 Instruments primarily designed to measure reflecting power of materials are known as reflectometers and usually compare the test specimen against such a standard as magnesium oxide or aluminium. **1977** *Sci. Amer.* Feb. 88/3 (Advt.), A time domain reflectometer (TDR) works in a manner similar to radar. It generates repetitive pulses of energy that are sent down a cable and displayed on a cathode-ray tube screen. Any cable faults (impedance changes) cause pulse reflections. **1977** BOXALL & VON FRAUNHOFER *Conc. Paint Technol.* ix. 193 The contrast ratio is the reflectometer reading obtained over the black tile after the reflectometer has been set to 100 % over the white tile.

Hence **reflecto·metry**.

1967 *Jrnl. Optical Soc. Amer.* LVII. 445 (*heading*) Terms, definitions, and symbols in reflectometry. **1977** *Sci. Amer.* Feb. 88/2 (Advt.), I knew that the TV industry used time domain reflectometry a lot.

reflector. **1.** Delete † *Obs.* in Dict. and add further examples.

1790 T. WILKINSON *Mem.* IV. 132 By which means [*sc.* comparing cast lists of 1747 and 1789] the unprejudiced reflector may draw a fair conclusion. **1921** W. C. BOOTH *Rhetoric of Fiction* I. vi. 157 It was not until authors had discovered the full uses of the third-person reflector that they could effectively show a narrator changing as he narrates.

4. *a.* *spec.* in *Geol.*, a stratum or interface that reflects seismic waves.

1933 *Bull. Amer. Assoc. Petroleum Geologists* XVII. 258 The geologic section contains many strata which act as good reflectors of wave energy. **1954** C. H. DIX *Seismic Prospecting for Oil* xi. 217 In areas where the reflectors are almost planes with small dip and small variations of dip, the reflections carry across the records with little change in character or time. **1965** *Bull. Amer. Assoc. Petroleum Geologists* XLIX. 350/1 Lateral continuity of the reflectors appears to be large compared to those within the..aprons and on the basins. **1978** *Nature* 29 June 744/1 The tilt of the terraces parallels a northwards dip of subsurface reflectors toward a sedimentary basin off Hudson Strait.

b. Also, something designed to reflect other forms of radiation (as radio waves or neutrons); e.g. part of a nuclear reactor designed to reflect escaping neutrons back into the core. (Further examples.)

1897 *Strand Mag.* Mar. 277/1 How far have you sent a telegraphic despatch on the air? Did you use a reflector? **1909** *Proc. R. Soc.* A. LXXXII. 496 The zinc sulphide screen S..was fixed behind the lead plate P, in such a position that no α-particles could strike it directly. When a reflector was placed in the position RR at about 1 cm. from the end of the tube, scintillations were at once observed. **1921** E. W. MARCHANT *Radio Teleg. & Telephony* ii. 11 The strength of the signal received, when reflectors were used, was estimated to be about ten times as great as when there were no reflectors. **1943** *Gloss. Terms Telecomm.* (*B.S.I.*) 66 A passive aerial placed behind an active aerial is usually called a reflector. **1946** *Chemical Age* 27 Oct. 490/1 And the escape of neutrons from the system can be reduced (relatively) by increasing the size of the system, or by a reflector (e.g., a layer of graphite). **1958** *Times* 1 July 7/3 It consists of an arrangement of small rods of uranium fuel in ordinary water and will be used to investigate..the use of different materials as 'reflectors' to reflect neutrons, which might otherwise escape, back into the core. **1977** N. FRENCH *Dictionary* (ed. 2) Cat. a near-crit. mass... Put a high-class reflector round it. **1978** *Nature* 16 Feb. 497/3 Chain reaction is unlikely, since the mass of uranium used in such reactors is normally subcritical, and special methods, e.g. a reflector, must be used to keep sufficient neutrons within it to maintain a chain reaction.

c. A piece of reflective material, now commonly a red disc, mounted at the rear of a vehicle or by the roadside so as to show its presence by reflecting the light from headlights.

1900 *Cyclist Touring Club Gaz.* Dec. 544/1 If a rearward indication is considered desirable, the Lea Reflex reflector and lens does everything needful... It throws back the light of the overtaking vehicle in a red glow through a wide angle. **1931** *Highway Code* 13 If you do not use a red rear lamp remember to keep your red reflector clean and properly adjusted. **1940** *Which? Car Suppl.* Oct. 133/1 We checked the condition of all lamps and reflectors. **1973** A. PRICE *Colonel Butler's Wolf* vii. 68 'He came directly down the road... His headlight had picked up the first of the reflectors. Even my bicycle light picks 'em up'.. 'Then supposing a car came up—it must have come in as he was approaching it—could it have cut off the reflectors and then blinded him?' 'Mmm—it could have, I suppose—but it would have lit 'em all up first and warned him there was a corner here.' **1978** *Highway Code* 35 Make sure your cycle is safe to ride. At night you must have front and rear lamps and a red rear reflector.

Hence **reflecto·red** *a.*, of a lamp: fitted with a reflecting surface or surfaces.

1916 A. BENNETT *These Twain* xix. 429 The glittering light of the latest triple-jetted and reflectored gas-lamps which the corporation..had placed in Crown Square. **1978** *Amer. N. & Q.* Dec. 65/2 Pierre Patte's early advocacy of reflectored lights in the auditorium.

reflectorize (rĭfle·ktərəiz), *v.* orig. *U.S.* [f. REFLECTOR + -IZE.] *trans.* To treat or coat with a substance that reflects light. Chiefly as **refle·ctorized** *ppl. a.*

1944 *New Hampshire Highways Signs & Road Marking* 56 Route markers 24″ × 24″ white on black. Plain or reflectorized. **1947** *Sun* (Baltimore) 22 May 2/5 A bus accident in Texas has resulted in one of several attempts by that state to reflectorize curb surfaces to enhance driving safety. *Ibid.*, A traffic official there..the reflectorized surface now warns fast drivers that the road does not go straight on. **1951** *Richmond* (Va.) *Times-Dispatch* 15 Feb. 7/6 More and more 'reflectorized' paint—the kind that gleams brilliantly under the headlights at night—is being used to mark traffic lanes on Virginia highways. **1959** *B.S.I. News* Mar. 4/2 (*caption*) Left to right: sign with button reflectors, sign reflectorized all-over, painted sign. **1961** *Evening Star* (Dunedin, N.Z.) 28 Mar. 1/7 They signalled with torches and skipped out of the path of the oncoming traffic which even knocked down large reflectorised stop signs. **1967** *New Scientist* 8 Jan. 225/2 A great improvement on the type of number plate now in use in Britain would be one with black letters on a white, 'reflectorized' background. **1973** *Houston Chron.* 21 Oct. 3 (Advt.), Bicycle has black saddle and grips. 10″ × 1 3/8″ tires. Fully reflectorized for safety. **1978** *Highway Code* 15 Watch out for blind people who may be carrying white sticks (white with two red reflectorised bands for deaf/blind people).

reflectoscope (rĭfle·ktŏskōᵘp). [f. REFLECT *v.* + -O + -SCOPE.] An instrument for investigating opaque bodies by transmitting ultrasound into them and measuring its reflection.

1944 *Sun* (Baltimore) 26 May 9/1 The invention, credited to Dr. Floyd A. Firestone, physicist and acoustics expert, is known as the 'supersonic reflectoscope'... The machine can gauge thickness of metal when one side is inaccessible. **1964** I. EDLER in D. Gordon *Ultrasound as Diagnostic & Surg. Tool* xiv. 124 The apparatus used for ultrasound-cardiography is an ultrasound reflectoscope.

reflex, *sb.* Add: The pronunc. (rī·fleks) is now standard. **2. c.** *Linguistics.* A form (word, sound unit, etc.) corresponding to, or derived from, another comparable form.

1890 S. PRIMER in *PMLA* V. II. 196 It is doubtful whether it [*sc.* the *a*-sound] is a reflex of the older pronunciation. **1945** Y. MALKIEL in *Univ. Calif. Publ. Linguistics* I. IV. iii. 51, -entia seems to have taken root in Italy more than anywhere else, as follows from the..reflexes of *absentia, haerentia, licentia, negligentia, scientia*. **1965** A. ZETTERSTEN *Stud. Dial. & Vocab. Ancrene Riwle* 67 It is necessary to emphasize that this Mercian sound coalesced with the reflexes of /æ/ and /a/ in ME. **1970** *Publ. Amer. Dial. Soc.* LIV. 4 This chapter presents the regionally divergent reflexes of Middle English parent phonemes. **1971** J. ANDERSON in A. J. Aitken et al. *Edin. Stud. Eng. & Scots* 110 The permutation of *may* and the subject [in *May your cabbages wither away*] is the superficial reflex of this underlying structure [*I wish that your cabbages may wither away*]. **1975** *Language* LI. 983 The assumption is..that the different reflexes of the same unit are due to phonetic change, either in the lending or the borrowing language, and that therefore the borrowings took place at different times.

6. a. Also in literary use.

1921 G. B. SHAW *Back to Methuselah* v. 238 Martellus: Control your reflexes, child. The Newly Born: My what! Martellus: Your reflexes. The things you do without thinking.

b. *attrib.*, as *reflex apparatus, mechanism, movement, stimulus, therapy, time*; **reflex arc**, the connected set of nerves concerned in the production of a reflex action. Cf. REFLEX *a.* 5.

1924 R. M. OGDEN tr. *Koffka's Growth of Mind* iii. 77 This statement..agrees..with what we have already learned about the reflex-apparatus. **1882** B. BRAMWELL *Dis. Spinal Cord* i. 24 Every half segment of the spinal cord with its sensory and motor nerve roots is, in theory and probably also in fact, a perfect reflex arc. *Ibid.* 25 The multipolar nerve cells of the anterior cornua probably constitute the centre of this reflex arc. **1924** R. M. OGDEN tr. *Koffka's Growth of Mind* iii. 69 Always beginning with a sensory neurone and ending with a motor neurone, this apparatus is called a reflex-arc. **1976** *Radiol. Clin. N. Amer.* XIV. 432/2 The reflex arcs controlling normal esophogeal motility are interrupted by disease of the medullary nuclei. **1885** *Encycl. Brit.* XIX. 29/1 Stimulation of a sensory surface may simultaneously produce, by a reflex mechanism, movement, secretion and consciousness. **1924** R. M. OGDEN tr. *Koffka's Growth of Mind* iii. 69 A reflex-mechanism is.. conceived as a pre-determined, inherited connection between afferent (receptor) and efferent (effector) pathways. *Ibid.* 91 A stimulus excites a reflex-movement. **1937** *Discovery* Nov. 341/1 When the reasoning powers are great, as in man, the mind is less responsive to outside influences or reflex stimuli. **1956** A. HUXLEY *Adonis & Alphabet* 31 Unorthodox medicine tries to influence the autonomic system by direct mechanical action in the form of osteopathy, chiropractic, reflex therapy and acupuncture. **1913** *Amer. Jrnl. Physiol.* XXXI. 309 How soon, after administering a minimal and also stronger dose of alcohol, does a change in reflex time appear? **1964** L. MARTIN *Clin. Endocrinol.* (ed. 4) iii. 135 Delayed tendon reflexes are characteristic, and particularly the ankle jerks. Sherman *et al.* (1963) measured the reflex-time in 50 hypothyroid cases and considered that the increase had a diagnostic significance.

7. *Photogr.* A reflex camera. Cf. *REFLEX *a.* 7.

1926–7 *Army & Navy Stores Catal.* 969/2 Reflex cameras. The 'Mentor' Folding Reflex is constructed..of the finest materials. **1940** *Wall's Dict. Photogr.* (ed. 15) 548 A twin-lens reflex consists of a rigid box which forms the camera, on top of which is mounted a structure such as that shown in the sketch, but with the mirror fixed in position. **1948** H. S. NEWCOMBE *Twin-Lens Camera Compan.* 15 Other people also find it difficult to hold a normal reflex steady. **1958** *Oxford Mail* 19 May 7/3 In the single-lens reflex you actually focus by looking through the camera lens with a mirror. **1977** J. HEDGECOE *Photographer's Handbk.* 14/1 The single lens reflex is the most highly developed and deservedly popular camera for advanced work.

reflex, *a.* Add: **5. a.** *reflex action*: also in extended or *fig.* use.

a 1846 B. R. HAYDON *Autobiogr.* (1927) III. xv. 278 He [*sc.* Wordsworth]..had a portion of the spirit of the mighty ones..but..did not possess the power of using that spirit otherwise than with reference to himself and so as to excite a reflex action only. **1917** KIPLING *Diversity of Creatures* 159 Only the Lord can understand..How much is reflex action and How much is really sin.

7. *Photogr.* Applied to a camera in which the image from the main lens (or from a duplicate of it) is reflected by means of a mirror on to a glass screen and can be seen and adjusted up to the moment of exposure.

1895 W. DE W. ABNEY *Instantaneous Photogr.* p. v (Advt.), The 'Reflex' Manufacturing Company..Patentees and Sole Manufacturers of the 'Reflex' Camera... The actual working lens also serves for the finder. **1911** *Encycl. Brit.* XXI. 505/2 Although reflex cameras are rather heavy and bulky as hand cameras, they have many advantages over the ordinary hand camera. **1946** R. J. C. ATKINSON

Field Archaeol. v. 157 The chief advantage of the reflex type of camera is the full-sized focusing-screen, which is a valuable aid to composing the photograph and to exact focusing. **1976** *Daily Mail* (Hull) 30 Sept. 2/6 (Advt.), Edixa SL, 35 mm. Reflex Camera, flash gun, quartz, cine camera, lenses, etc; £32. **1976** J. TATE tr. *A. Bodelsen's Operation Cobra* v. 27 As neither his mother nor his father earned so much money now, he knew he could not reckon on getting a reflex camera.

8. *Electronics.* **a.** Applied to a circuit, amplifier, etc. in which the same valves or transistors are used for amplification of both high- and low-frequency signals (usu. the radio and audio frequencies respectively); also applied to the action of such a device.

1923 *Radio News* Feb. 1455/1 Very much discussion has been carried on as to the practicability of the so-called reflex circuit—a circuit designed, developed and patented in April and November of 1917 by Mr. Marius Latour, a French radio engineer. **1924** W. JAMES *Construction Two-Valve & Crystal Reflex Receiver* 3 The first valve is connected to operate as a dual or reflex amplifier; that is, it gives high-frequency and low-frequency-amplification. **1928** LAUER & BROWN *Radio Engin. Princ.* (ed. 2) vii. 163 (*heading*) Reflex amplification. **1934** *Pract. Wireless* 1 Sept. 723/1 Reflex circuits are rarely used at the present time, although they were extremely popular between 1922 and 1924. *Ibid.*, It is possible to obtain the same output from two valves wired in a reflex arrangement as from three valves connected in a more conventional circuit. **1957** R. F. SHEA *Transistor Circuit Engin.* xi. 375 When size and cost are of extreme importance, reflex circuits may be used. **1961** J. M. CARROLL *Design Man. for Transistor Circuits* viii. 154 Reflex circuits in which i-f and a-f gain are achieved in the same transistor stage have recently been incorporated into economy broadcast receivers.

b. Applied to an oscillator, esp. a klystron, in which the same resonant cavity serves to modulate the electron beam and to produce an amplified microwave signal.

1942 J. G. BRAINERD et al. *Ultra-High-Frequency Techniques* x. 339 A reflex klystron oscillator, in which a single resonator acts both as the buncher and the catcher. **1945** *Proc. IRE* XXXIII. 112/2 Reflex oscillators can be considered as oscillators in which an electron stream passes through a longitudinal radio-frequency field across a 'gap' between two electrodes, then into a drift space in which there is a retarding electric field produced by a negative repeller electrode, and finally returns through the radio-frequency field across the gap. **1969** *IEEE Trans. Industr. Electronics* XVI. 103/1 The oscillation frequency of a reflex klystron is determined by..the size of the cavity resonator. **1975** D. G. FINK *Electronics Engineers' Handbk.* IX. 31 Reflex Klystrons are used as test signal sources, receiver local oscillators, pump sources for parametric amplifiers, and low-power transmitters for FM line-of-sight relays.

9. Applied to a photocopying process in which the original document is illuminated by light passing through a piece of sensitized paper placed in contact with it, a negative image being formed on the paper according to the amount of light reflected by the original; also applied to equipment or materials connected with such a process.

1943 *Jrnl. Sci. Instruments* XX. 18/1 The Kodak Reflex plate has been introduced to solve the problem of the exact preservation of scale in copying maps, machine drawings and other line diagrams. *Ibid.*, The latter [distortion] is eliminated by the nature of reflex copying. *Ibid.*, Reflex printing has been carried out with photographic paper for a number of years. **1947** *Jrnl. R. Aeronaut. Soc.* LI. 318/1 The exposure is made through the paper with a yellow reflex screen. *Ibid.*, A plasticising solution is applied to the cellulose film and the exposed reflex paper is then laminated on to the material. **1956** 'C. BLACKSTOCK' *Dewey Death* vii. 149 We then take a piece of..reflex contact document paper—the shiny side is the business end. **1958** T. LANDAU *Encycl. Librarianship* 267/2 The principle of reflex copying was established by Albrecht Breyer of Berlin,..who in 1839 produced reflex prints by placing silver chloride papers in contact with printed pages. **1972** A. TYRRELL *Basics of Reprography* xiii. 210 Reflex printing by the diffusion-transfer technique has been very popular... Attempts have been made to use diazo-sensitizers in reflex copying.

re·flexing, *vbl. sb. Electronics.* [f. REFLEX *a.* + -ING¹.] The use or action of a reflex circuit.

1925 A. H. MORSE *Radio* ii. 60 The modern practice of 'reflexing', or amplifying both audio and radio-frequency in a single triode. **1939** *A. W. A. Techn. Rev.* IV. 37 Reflexing does not appreciably reduce the higher audio note response. **1961** J. M. CARROLL *Design Man. for Transistor Circuits* viii. 154/1 The reflexing circuit, normally used as a second i–f amplifier and first audio amplifier, can provide gain from a single transistor only a few db less than the gain obtained from two transistors in conventional circuits.

reflexive, *a.* and *sb.* Add: **A.** *adj.* **2.** Restrict † *Obs.* to senses in Dict. and add: **c.** *Social Sciences.* Applied to that which turns back upon, or takes account of, itself or a person's self, esp. methods that take into consideration the effect of the personality or presence of the researcher on the investigation.

1934 G. H. MEAD *Mind, Self & Society* xxi. 173 Cooley and James..endeavor to find the basis of the self in reflexive affective experiences, i.e., experiences involving 'self-feeling'. **1957** P. LAFITTE *Person in Psychol.* 17 All learning depends on the reflexive interpretation of one's experience together with the experience of others. *Ibid.* 21 The psychologist's reflexive judgements will be limited by his knowledge of himself in some of the ways that his subjects' reports are limited. **1970** A. GOULDNER *Coming Crisis of Western Sociol.* xiii. 489 The historical mission of a Reflexive Sociology is to transcend sociology as it now exists. *Ibid.* 490 A Reflexive Sociology means that we sociologists must.. acquire the ingrained *habit* of viewing our own beliefs as we now view those held by others. **1972** M. LANDAU *Polit. Theory & Polit. Sci.* i. 32 A reflexive prediction is one in which the prediction is itself a factor which may materially alter the projected or anticipated outcome. **1977** DOUGLAS & JOHNSON *Existential Sociol.* p. xiii, Our emphasis on the problematic and situated nature of meaningful experience contrasts..with the structuralism of Alvin Gouldner's 'reflexive sociology'. **1977** R. HOLLAND *Self & Social Context* v. 82 In both cases the person producing the theory is included within the subject matter he attempts to understand. The usual term for this kind of approach is 'reflexive', a word which has begun to appear in the human sciences.. but which has long been implicit in social theory.

5. (Further example.) Also of phrases, esp. in *Linguistics*.

1933 L. BLOOMFIELD *Language* xii. 193 In English we say *he washed him* when actor and goal are not identical, but *he washed himself* (a *reflexive* form) when they are the same person. **1979** *Trans. Philol. Soc.* 11 We might of course propose to handle such facts by a purely local reflexive deletion rule.

6. (Later examples in *Physiol.*) Also in extended use.

1927 *Jrnl. Nerv. & Mental Dis.* LXV. 463 We..succeeded in producing in apes..a reflexive contraction of the adductor muscles of the thigh. **1971** L. KOPPETT *N.Y. Times Guide Spectator Sports* i. 13 To have any chance at all, the batter, whose action must be entirely reflexive, needs protection from additional trickery. **1971** *Sci. Amer.* Aug. 74/1 A number of biologists..adopted the radical hypothesis that animal behavior was almost wholly reflexive. On this view the continually changing array of stimuli that an animal encounters as it moves through its environment was thought to produce a large part of the animal's repertory of behavior by reflexes and their mutual interactions. **1975** *New Yorker* 13 Jan. 30/2 At the end, the audience rose in an ovation—but at concerts like this one standing ovations have become reflexive. **1976** *Nature* 1 Apr. 392/1 To suggest ..that the evolutionary considerations which determine the mating systems of mammals and birds have any light.. to shed on the tensions and asymmetries commonly observed in human sexual relationships is to invite reflexive dismissal as a 'sexist'.

7. *Math.* and *Logic.* Applied to any relation which always holds between a term and itself. [The sense is due to G. Vailati, who used It. *riflessività* reflexivity (*Rivista di Matematica* (1891) I. 134).]

1903 B. RUSSELL *Princ. Math.* xix. 159 All kinds of equality have in common the three properties of being reflexive, symmetrical, and transitive. **1937** R. CARNAP *Logical Syntax of Lang.* iv. 261 Conditions which require for symmetrical, reflexive, and transitive relations the property of non-emptiness. **1953** A. A. FRAENKEL *Abstract Set Theory* i. 34 Any set is equivalent to itself... Equivalence is a reflexive relation. **1972** F. J. BUDDEN *Fascination of Groups* xx. 374 Conjugacy is a relation between the elements of a group. It is evidently reflexive (since $y = 1y1^{-1}$).

reflexively, *adv.* Add: **b.** = REFLEXLY *adv.*; in the manner of a reflex action, automatically. *rare.*

1952 *Mind* LXI. 252 As is usual with such reflexively repeated dicta, this one is true but never practical to follow. **1966** D. F. GALOUYE *Lost Perception* xvi. 172 Instantly the three men twisted round. The closest was Wellford. Reflexively, he loosed a laser discharge.

reflexiveness. (Further examples.)

1903 B. RUSSELL *Princ. Math.* xxvi. 219 The property of a relation which insures that it holds between a term and itself is called by Peano *reflexiveness*. **1933** *Mind* XLII. 36 In the *Principia Mathematica* $p \supset p$ assigns to the relation ⊃ the property of reflexiveness. **1968** *New Scientist* 16 May 339/1 Three properties of conditions are frequently mentioned: reflexiveness, symmetry and transitivity... A reflexive condition is a relation between an idea and itself... An example is equality.

reflexivity (further examples).

1940 W. V. QUINE *Math. Logic* iii. 138 A natural extension of the notions of reflexivity and commutativity which were applied to a statement composition. **1950** L. M. HAMMOND et al. tr. *Hilbert & Ackermann's Princ. Math. Logic* iv. 135 The properties of reflexivity, symmetry, and transitivity of dyadic predicates. **1965** *Canad. Jrnl. Linguistics* X. 175 They [*sc.* classifiers] refer most clearly to the voice of the verb:..passivity, reflexivity. **1977** DOUGLAS & JOHNSON *Existential Sociol.* v. 172 Reflexivity refers to the mutual interdependence of observer or knower to what is seen or known.

reflexivization (rĭfle:ksivəizē^i·ʃən). *Linguistics.* [f. next + -ATION.] The action of making (a verb, noun phrase, etc.) reflexive; the process or fact of being made reflexive. Cf. REFLEXIVE *a.* 5 in Dict. and Suppl.

1965 N. CHOMSKY *Aspects of Theory of Syntax* iii. 145 In a sentence such as..'the boy hurt the boy', the two phonetically identical Noun Phrases are necessarily interpreted as differing in reference; sameness of reference requires reflexivization of the second Noun Phrase. *Ibid.*, The reflexivization rule can be formulated as an erasure operation that uses the Noun Phrase to delete another. *Ibid.* 146 The reflexivization rule..will apply only when the integers assigned to the two items are the same. **1966** P. M. POSTAL in F. P. Dinneen *Rep. 17th Round Table Meeting Linguistics & Lang. Stud.* 182, I view the process of reflexivization as a complex of a number of partially independent operations. **1972** *Language* XLVIII. 390 The interactions of other pronominals with indirect object, shown in reflexivization properties for example, suggest that postpositions are two-place predicates rather than one-place (intransitive) predicates. **1978** *Studies in Eng. Lit.: Eng. Number* (Tokyo) 65 Reflexivization, which converts the simple personal pronoun into the compound refl. pronoun in *-self*, is apparently optional in ME.

reflexivize (rĭfle·ksivəiz), *v. Linguistics.* [f. REFLEXIVE *a.* + -IZE.] **a.** *trans.* To make (a verb, noun phrase, etc.) reflexive. **b.** *intr.* To become reflexive. Hence **refle·xivizing** *ppl. a.*

1965 S.-Y. KURODA *Generative Gram. Stud. in Japanese Lang.* (Mass. Inst. Technol. thesis) v. 144 The object of a subordinate clause which is coreferential with the matrix subject will be Provincialized rather than Reflexivized. **1967** J. R. ROSS *Constraints on Variables in Syntax* (Mass. Inst. Technol. thesis) vi. 254 It would be expected that the leftmost occurrence of *Bill* would be able to reflexivize the rightmost. **1971** *Language* XLVII. 160 It appears that in Japanese the subject can optionally reflexivize (with *zibun*) an identical NP [= noun phrase] in an embedded sentence, if that embedded sentence is a part of the VP. **1976** J. S. GRUBER *Lexical Struct. Syntax & Semantics* I. iii. 73 The expression of Accompaniment must, like the expression of Location, not be generated in construction with the verb. This follows from phrase order and also the fact that they do not reflexivize. **1978** *Language* LIV. 142 Level III DA usages in our data always include *on-* along with the reflexivizing apparatus and reverential elements.

reflexly, *adv.* Add: Now usu. stressed **re·flexly**. (Further examples.)

1932 S. ZUCKERMAN *Social Life Monkeys & Apes* ix. 146 It seems reasonable to surmise that when the sexual skin is swollen with oedema, its sensory nerve endings are stimulated by pressure and that this reflexly rouses the animal to sexual behaviour. **1961** *Lancet* 12 Aug. 367/2, I thought of..Superdiagnostex, lying in the cupboard where I hurled it reflexly when a salesman left it on trial last month. **1974** *Sci. Amer.* Oct. 100/1 A coordinated motor performance could be described as being differentiated into many parts along the time dimension, so that each part is reflexly triggered by the sensory components of its predecessor.

reflexogenous, *a.* Add: (Further examples.) **reflexogenic** *a.* (example).

1933 *Ann. Rep. London Co. Council* IV. III. 133 Sometimes the epileptic fit is intimately bound up with the appropriate stimulus... Much depends on the nature of the reflexogenous stimulus. **1969** *Jrnl. Compar. Neurol.* CXXI. 124/2 Stimulation of the skin of the head and of the rostral half of the belly elicited reflex movements of the ipsilateral forelimb, and this area is referred to as the forelimb reflexogenous zones. **1973** *Nature* 3 Aug. 310/1 These calculations ..predict very high pressures in the aorta, and imply reflexogenic control over cerebral perfusion pressures.

reflexology (rī:fleksǫ·lŏdʒi). *Psychol.* [f. REFLEX *sb.* + -OLOGY; cf. G. *reflexologie*.] **a.** The theory that the behaviour of organisms is made up of established patterns of simple or complex reflex responses; the scientific study of reflex action as it affects behaviour.

[**1912** W. VON BECHTEREW in *Deutsche Med. Wochenschr.* 8 Aug. 1481 (*heading*) Was ist Psychoreflexologie?] **1927** *Psychol. Abstr.* I. 590 Thanks to its objective method, reflexology has rendered service to applied sciences, psychiatry and pedagogy. **1933** E. & W. MURPHY tr. A. Gerver in *Bechterev's Princ. Human Reflexology* 8 Bechterev..founded this Institute in 1918, and its chief aim was the study of the anatomy and physiology of the central nervous system and also the study of the principles of reflexology, a new branch of science created by him... The Institute is now called the Reflexological Institute for the Study of Brain. **1957** T. KILMARTIN tr. *Aron's Opium of Intellectuals* vi. 198 'Refloxology' does not solve the riddle of existence any more than materialistic sociology. **1973** C. D. KERNIG *Marxism, Communism & Western Society* VII. 144/1 (*heading*) Results of recent research in reflexology. **1975** C. BURT *ESP & Psychol.* iv. 54 The physiological psychologists (materialistic monists) who adopted Huxley's reflexology.

b. A technique for relaxing nervous tension through a method of foot massage.

1976 M. SEGAL *Reflexology* 1 Known also as compression foot massage, reflexology is a natural art of healing. *Ibid.*, Like acupuncture, reflexology has been used by the Chinese for 5,000 years. **1976** *Seed* V. v. 31 (Advt.), Reflexology (compression massage of the feet). **1977** D. BERKSON *Foot Bk.* ii. 13 Various kinds of pressure-point therapy and reflexology as well as shiatsu.

Hence **re:flexolo·gical** *a.*; **re:flexo·logist**.

1927 *Psychol. Abstr.* I. 650 Therapeutics ought to investigate..the methods of reflexological orthopedics applicable to the cure of mental diseases. **1933** Reflexological [see sense a above]. **1933** E. & W. MURPHY tr. A. Gerver in *Bechterev's Gen. Princ. Human Reflexology* 13 Not only have psychiatrists, reflexologists, and neuropathologists been in close touch with him [*sc.* Bechterev], but also teachers, both rural and urban, have always given him a ready ear. **1938** *Brit. Jrnl. Psychol.* Jan. 337 Behaviourists and reflexologists who do not believe that a rat has any mental life involving anticipations or foresights which govern his behaviour. **1957** *Listener* 31 Oct. 689/1 Another analogy was suggested by the experiments of the Russian reflexologists, Bechterev and Pavlov. **1976** M. SEGAL *Reflexology* i. 3 We, as reflexologists, do the same by trying to relax the patient and relieve nerve tension.

refloat, *v.*¹ Add: **a.** Also *transf.* in *Econ.* (Further examples.)

1973 *Oxf. Mag.* 1 June 1/2 He gallantly accepted the responsibility of re-floating it [*sc.* the *Magazine*]. **1974** B. PEARCE tr. *Amin's Accumulation on World Scale* II. iii. 434

If a slump in cotton sales should occur, the central bank would refloat the producers through an additional issue. **1977** *Economist* 3 Sept. 67/1 Portugal refloats the escudo downwards.

b. *intr.* To float again.

1906 *Daily Chron.* 23 June 6/3 On the rising tide the Talisman refloated.

reflow, *sb.* **b.** *fig.*

1969 *Daily Tel.* 4 Sept. 1/4 The re-flow of funds into London after the wave of speculation which followed devaluation was not as large as the initial outflow. **1975** *Washington Post* 19 Feb. A15/2 May I ask at that point if you have had an opportunity to examine the reflow in the purchase of goods and services.

reflower, *v.* Add: **2.** Also *fig.* (Examples.)

1878 Swinburne *Poems & Ballads* 2nd Ser. 178 Out of the herbs on the walls reflowering. **1977** *Arab Times* 13 Dec. 7/3 This technology has reflowered in the sports hall, where highly specialised surfaces are needed.

Hence **reflo·wered** *ppl. a.*, covered with flowers again.

1907 E. Nesbit in *Daily Chron.* 19 Feb. 6/7 Hark to the sigh of the reflowered tomb: 'Ah, live, live, live, for Spring goes by, goes by!'

reflux. Add: **1. c.** Also, of the contents of the stomach or intestine.

1937 R. Schindler *Gastroscopy* xiv. 275 The continuous unregulated reflux of intestinal juice containing bile and pancreatic juice may play the greatest role. **1978** T. C. Jewett in E. Lebenthal *Digestive Dis. Children* 414 The redundant mucosa which forms a rosette at the cardio-esophageal area may also act as a barrier to reflux of gastric contents up the esophagus.

d. *Chem.* The condition, process, or action of refluxing; also *concr.*, the condensed vapour involved in this. Freq. in adverbial phr., as *at* or *under reflux*, in a vessel fitted with a reflux condenser.

1897 *Jrnl. Chem. Soc.* LXXI. 1036 The operation is carried out under reflux. **1923** H. M. Bunbury *Destructive Distillation of Wood* xii. 213 It is sometimes necessary to make a total reflux, i.e. to return all the condensed vapours to the column. **1936** W. L. Nelson *Petroleum Refinery Engin.* xv. 278 The reflux flows down the column but it changes composition from plate to plate so that all of the material that is originally put into the top of the column as reflux is vaporized and returns to the product storage tank. **1958** F. H. Garner et al. in H. W. Cremer *Chem. Engin. Pract.* V. vii. 422 Liquid-liquid extraction with reflux is usually applied to the separation of homologous substances where both the separation factor and distribution coefficient are low. **1959** *Petroleum Handbk.* (Shell Internat. Petroleum Co.) (ed. 4) 172 The light product returned in this manner to the top of the column is called the reflux. **1962** J. T. Marsh *Self-Smoothing Fabrics* xxii. 372 Melamine..may be determined gravimetrically after hydrolysing a known weight of the treated fabric in 0·2N HCl for 30 min. under reflux. **1968** *Inorg. Syntheses* XI. 20 The mixture is stirred at reflux for 6 hours. *Ibid.* 38 The reflux and stirring are continued for an additional 5–6 hours. **1975** *Jrnl. Chem. Soc.: Perkin Trans.* I. 297/1 The mixture..was heated at reflux under nitrogen. *Ibid.* 344/2 Ethyl cyclo-oct-4-enylideneacetate (1·0 g) was added to sodium hydroxide solution (20 ml; 2N) and brought to reflux.

2. reflux condenser *Chem.*, a condenser so mounted or designed that condensed vapour runs back into the stock of boiling liquid; **reflux œsophagitis** *Med.*, œsophagitis caused by the flow of fluid into the œsophagus from the stomach or intestine; **reflux valve** = *check-valve* s.v. *check-.

1891 F. W. Streatfield *Pract. Work in Org. Chem.* 64 The operation is most conveniently conducted in a tubulated retort to the neck of which is attached a reflux condenser. **1939** *Inorg. Syntheses* I. 82 The mixture is placed in a small flask fitted with a reflux condenser and kept at a temperature of about 40 °C. for approximately 48 hours. **1962** J. T. Marsh *Self-Smoothing Fabrics* xxii. 368 The flask is then cooled and the reflux condenser washed down with about 50 cc of distilled water. **1952** R. Belsey in F. A. Jones *Mod. Trends Gastro-Enterol.* I. vii. 144 Reflux oesophagitis leads to thickening of the oesophageal wall, not penetration or perforation. **1980** *Brit. Med. Jrnl.* 29 Mar. (Advt. facing p. 946) 'Tagamet' represents a mark of reliability in reducing gastric acid, and has revolutionised the treatment of disorders such as peptic ulcer and reflux oesophagitis, where acid plays a part. **1888** *Lockwood's Dict. Mech. Engin.* 282 *Reflux valve*,..a flap valve used for the purpose of taking off the pressure of a head of water acting in a backward direction against a set of pumps. **1893** Turner & Brightmore *Princ. Waterworks Engin.* vi. 375 'Re-flux' valves are introduced at such points as the inlets to reservoirs or towards the 'delivery' sides of deep depressions in the pipe-line. **1962** L. B. Escritt *Pumping Station Equipment & Design* vii. 72 If a reflux valve sticks open.. water will rush backwards from the rising main.

reflux (rī·flɒks), *v. Chem.* [f. the sb.] **a.** *intr.* Of a liquid: to boil in circumstances such that the vapour returns to the stock of liquid after condensing. **b.** *trans.* To boil (a liquid) in this way, esp. in a flask fitted with a reflux condenser; also *absol.*

1923 H. M. Bunbury *Destructive Distillation of Wood* xii. 229 The steam is then almost completely shut off and the contents allowed to reflux gently for about two hours. **1926** Lowy & Baldwin *Lab. Bk. Elem. Org. Chem.* 45 After all the alcohol or acetone has been added, reflux the mixture on a water bath for half an hour. **1938** A. A. Morton *Lab. Technique in Org. Chem.* iv. 83 The stopcock is closed, and the liquid allowed to reflux until the thermometer records the lowest temperature possible. **1958** A. I. Vogel *Elem. Pract. Org. Chem.* xvii. 420 Place 0·5 g. of the dry acid..into the flask, add 2·0–2·5 ml. of redistilled thionyl chloride and reflux gently for 30 minutes. **1965** Adams & Raynor *Adv. Pract. Inorg. Chem.* v. 49 Reflux equimolar quantities of vanadyl acetylacetonate and pyridine in ether. **1973** *Nature* 9 Mar. 113/2 The reincorporation of 1 molecule of H_2O into the partially dehydrated talc was achieved by refluxing the material with 50 volumes of benzene/water azeotropic mixture.

Hence **re·fluxing** *vbl. sb.* and *ppl. a.*

1923 H. M. Bunbury *Destructive Distillation of Wood* xii. 213 The bulk or the whole of the condensate from the rectifier, which contains most of the less volatile constituent, is returned to the column; this operation is termed 'refluxing'. **1959** R. J. Hengstebeck *Petroleum Processing* iii. 49 If bubble-cap trays are used, actual tray requirements are usually estimated by means of empirical correlations of 'tray efficiency' against the properties of the refluxing liquid. **1962** J. T. Marsh *Self-Smoothing Fabrics* xxii. 368 Refluxing for 30 min. is generally enough to hydrolyse the amino-aldehyde. **1975** L. F. & M. Fieser *Reagents for Org. Synthesis* V. 381 Raney nickel which had been deactivated by treatment with refluxing ethanol.

refocus, *v.* Add: *trans.* and *intr.* (Further examples.)

1943 A. W. Judge *Automobile Electr. Maintenance* (ed. 2) viii. 222 Provided the correct bulb is used as a replacement it should be unnecessary to re-focus after bulb renewal. **1971** S. Hill *Strange Meeting* i. 55 Then, he seemed to come to abruptly, and his eyes re-focused. **1981** J. Gardner *Licence Renewed* vi. 70 Bond just had time to refocus his glasses. The horses were off.

reforest, *v.* Add: = Reafforest *v.* 2. Chiefly *N. Amer.* (Further examples.)

1918 *Jrnl. Forestry* XVI. 335 It has been assumed that 75 per cent of the State-owned land must be reforested artificially. **1939** *Geogr. Jrnl.* XCIV. 178 The French have spent 317 million francs on reforesting over a million acres. **1969** S. M. Jepsen *Trees & Forests* i. 27 During 1965, more acres..of national forest land were reforested by planting and seeding than in any previous year.

reforestation (further examples).

1918 *Jrnl. Forestry* XVI. 889 It is absolutely essential to ameliorate the rivers, especially the Hun Ho, in the mountainous collecting basin, by reforestation. **1976** T. Walker *Spatsizi* xi. 121 Natural re-forestation was well advanced in the burn with pines ten feet high.

reform, *sb.* Add: **1. c.** *ellipt.* (with capital initial), the Reform Club (see 6 a below).

1853 *London Clubs* 51 *The Reform Club.* Next [to Boodle's] in order amongst political clubs stands the *Reform*, although we are not sure that it is not surpassed in seniority by its great rival..the *Carlton*. Both had their origin in the exciting era of 1830, and the Reform Bill. **1860** A. J. Munby *Diary* 18 Mar. in D. Hudson *Munby* (1972) 56, I feel no interest in..Reform and Carlton conflicts. **1886** B. Potter *Jrnl.* 14 Jan. (1966) 163 It was rumoured yesterday, 13th, that Morley was going to leave the Reform. Harrison has left the Athenaeum. **1940** H. Nicolson *Let.* 14 July (1967) 102, I dined..at the Reform and we listened afterwards to Winston. **1978** G. Greene *Human Factor* III. i. 97 They made a habit of lunching alternately at the Reform and the Travellers once a month on a Saturday.

2. b. *reform through labour* [tr. Chinese *láodòng gǎizào*], in China, an element of ideological reformation whereby criminals and dissidents are made to work as a part of their political reeducation.

1957 P. S. H. Tang *Communist China Today* v. 247 The theoretical basis for the policy of 'reform through labor' was stated by Mao Tse-tung in his 1949 report *On People's Democratic Dictatorship*..promising that 'reactionaries' who desisted from counter-revolutionary activities would not be put to death but would be given work in order to 'reform themselves through labour so as to become new men'. **1962** E. Snow *Red China Today* (1963) xxi. 156 He inspects reform-through-labour farms. *Ibid.* xlvii. 361 No one is entitled to assume that 'reform through labor' in China is administered by humanitarians. **1977** *China Now* July/Aug. 13/3 Bao spent seven years undergoing Reform Through Labour (Lao Dong Gai Zao) having been arrested in 1957.

4. a. Also (with capital initial) preceded by designating adj.

1873 Lady G. Fullerton *Life L. de Carvajal* II. viii. 243 Her beloved friends..were both nuns of the Augustinian Reform. **1893** *Mod. Lang. Notes* VIII. 344 (*heading*) History and texts of the Benedictine reform of the tenth century. **1911** A. Brennan *St. Lawrence of Brindisi* xvi. 147 The Fathers of St. Giles, who belonged to the Alcantarine Reform, were delighted to receive him [*sc.* Father Lawrence] as their guest. **1953** K. Sisam *Stud. Hist. Old Eng. Lit.* vi. ii. 106 The second half of the tenth century was the period of the Benedictine Reform.

6. a. (Also with initial capital.) *reform Convention, Democrat, mayor, movement, party* (earlier and later examples), *politician*; **Reform Club,** a club instituted to promote (usu. political) reform; *spec.* the name of a London club in Pall Mall founded in 1836; **Reform(ed) Neutral** *Philol.*, an international language developed by Rosenberger and de Wahl from *Idiom Neutral* (see *idiom 5); **reform school** orig. *U.S.*, a reformatory for young persons.

1835 *Times* 16 June 5/6 People have heard of the Middlesex Reform Club. It seems that this formidable 'corporation' is on the eve of dissolution. **1837** *Times* 27 Nov. 5/1 A numerous meeting of Whig-Radical members of the House of Commons was held to-day (Saturday), at 12 o'clock, at the Reform Club-house, Pall-mall. **1877** *Public Acts Michigan* 42 Reform club temperance societies may be incorporated in pursuance of the provisions of this act. **1884** B. Potter *Jrnl.* 22 Feb. (1966) 68 Lord Rollo..is a member of the Reform Club, but seldom goes. *a* **1974** R. Crossman *Diaries* (1975) I. 42 One of the Reform Club dinners at which Charles Snow took the chair and captivated Harold Wilson. **1851** *Documentary Hist. Amer. Industr. Society* (1910) VIII. 317 A National Reform Convention is however ever to be held. **1887** *Courier-Jrnl.* (Louisville, Kentucky) 8 Feb. 1/2 (*heading*) The Reform Democrats manage to get Mr. Randall into very deep water. **1968** *Listener* 5 Sept. 290/1 His machine had just taken a terrible beating at the hands of a reform mayor. **1839** J. S. Mill in *Westm. Rev.* XXXII. 476 The question is not now about particular reforms, but how to carry on the Reform movement. **1922** A. L. Guérard *Short. Hist. Internat. Lang. Movement* II. vi. 139 Reformed-Neutral of 1907 looked more natural than the primitive form. The restitution of international *c* wherever it had been replaced by *s* or *k* greatly improved the appearance of the language. But one of the most obvious blemishes of the Idiom was not corrected: the accumulation of final consonants as in *nostr*. *Ibid.* viii. 172 He [*sc.* Prof. Peano] is, like Dr. Molenaar, and like Messrs. Rosenberger and de Wahl in their Reform-Neutral, a radical, a posteriorist. **1946** H. Jacob *On Choice Common Lang.* iii. 27 Idiom Neutral, the early product of the Akademi, under the influence of Rosenberger and de Wahl, soon became Reform-Neutral, and when Rosenberger died de Wahl carried on his studies until, in 1922, he published his own system, Occidental. **1839** J. S. Mill in *Westm. Rev.* XXXII. 477 No reformer can hope to realize any reforms of importance, but by means of a strong and united Reform party. **1970** D. Goldrich et al. in I. L. Horowitz *Masses in Lat. Amer.* v. 189 It is much more common for 'reform' parties to symbolize the *peasant* as the forgotten. **1904** A. French *Barrier* iii. 22 The reform politicians, those bees who buzzed continually and occasionally stung, had been after the young man. **1859** Reform school [in Dict.]. **1860** C. E. DeLong *Jrnl.* 7 Dec. in *Calif. Hist. Soc. Q.* (1931) X. 258 In the afternoon a crowd of us went out to view the site of the [state] Reform School [for boys]. **1913** J. London *Valley of Moon* i. 3 An' her with seven, an' two of 'em in reform school. **1958** *New Statesman* 6 Sept. 294/3 A reform-school boy whose criminal side has been straightened up by the analysts. **1973** J. Cleary *Ransom* i. 12 Even the Police Commissioner's wife didn't go to finishing school—some of us reckon he found her in a reform school.

b. Reform Judaism, a liberalizing movement initiated in Germany by the philosopher Moses Mendelssohn (1729–86), to accommodate the Jewish faith to the European intellectual enlightenment. Also in various related *attrib.* collocations, as *Reform Jew, party, Synagogue*, etc. Occas. *ellipt.* as predic. adj.

In Britain, Reform Judaism occupies a middle position between Orthodox and Liberal Judaism.

1843 *Voice of Jacob* 27 Oct. 21/2 The Frankfort Reform Association..meets with great sympathy among a large portion of the Jews here... The Anti-reform party..seriously proposes to counteract the..'reformers' in an honourable manner. **1844** *Southern Q. Rev.* Apr. 325 The Reform party maintain that the old written law,—the law of Moses and the Prophets, is the only divine..law, but that the Talmud and the decisions of the Rabbins..are not divine. *Ibid.* 333 An important effort was made in London..establishing a Reform Synagogue. **1845** *Voice of David* 15 Aug. 219/1 The 'Berlin Reform Association'..is said to have declared..that should the Synod not agree with its (the Reform Association's) views, the latter would independently pursue its own course. **1860** *N.Y. Times* 6 Aug. 8/3 The innovations of the 'reform party'..which he [*sc.* Rabbi Isaacs] attributed to religious pride... The congregation..remained uncontaminated by these pretended reforms. **1870** R. D-C. Lewin *What is Judaism?* 8 The signal triumphs which have attended the efforts of the Reform School of Judaism. *Ibid.*, Charitable institutions which are so largely supported by Reform Jews. **1892** I. Zangwill *Childr. Ghetto* III. 36 The paper was founded to inculcate..the principles of true Judaism... But this is rank Reform; it's worse than the papers we came to supersede. *Ibid.* 143 The Reform Synagogue, though a centre of culture and prosperity, was cold, crude, and devoid of magnetism. **1916** H. Sacher *Zionism & Jewish Future* 48 Why should not all Jews recover their spiritual unity through Reform Judaism? **1959** *Tamarack Rev.* Summer 12 Try to convince Gershon that she'd joined the Orthodoxes in protest against her sons' becoming Anglican-like Reform Jews. **1966** 'A. Blaisdell' *Date with Death* xvi. 208, I never was very religious, we were Reform but not much given to..keeping up with temple. **1977** H. Fast *Immigrants* IV. 238 What about the Reform Jews? The capacity of the rich is always larger than the capacity of the poor. **1980** *Times* 18 July 4/3 It is intended to broadcast a service from a Reform synagogue in London..the first occasion that an established non-Christian religious service has been included in the regular output.

reform, *v.*[1] Add: **12.** Also re-form. To subject (petrol, hydrocarbons, etc.) to *reforming *vbl. sb.* 2.

1924 *Proc. 31st Ann. Convention Pacific Coast Gas Assoc.* 724 The artificial gas portion of the commercial mixture may be manufactured by 'reforming' natural gas rather than by producing this gas from oil. **1931** *U.S. Bureau of Mines Techn. Paper* No. 483. 1 There is a definite demand for a means and process for re-forming hydrocarbon gases, including refinery gas. **1941** W. L. Nelson *Petroleum Refinery Engin.* (ed. 2) xxiv. 527 These authors conclude that it is not economical to top the light gasoline and reform the naphtha separately. **1966** *McGraw-Hill Encycl. Sci. & Technol.* XI. 403/2 In a typical operation, a Pennsylvania straight-run gasoline of 44 octane number may be reformed to give a product with an octane number of 80..with a yield of 66%. **1974** *Sci. Amer.* Oct. 67/2 The carbohydrates decay exoergically to form fossil fuels such as methane..which can then be re-formed endoergically to yield hydrogen gas.

reformabi·lity. *rare.* [f. Reformable *a.*: see -ity.] Capacity for being reformed.

1904 *Daily Chron.* 29 Nov. 6/1 The Council has made provision for all women sent from London courts, regardless of the question of reformability.

reformate (rīfǭ·ɹmēɪt). [f. Reform(ing *vbl. sb.* + -*ate*, after *distillate, filtrate*, etc.] The end-product of the process of reforming petroleum products.

1949 *Industr. & Engin. Chem.* Oct. 2185/1 The crude reformate was condensed at 32° F. **1951** *World Petroleum* Nov. 73/2 In recent weeks two new processes have been announced for separating and purifying the aromatics from reformate. **1958** W. L. Nelson *Petroleum Refinery Engin.* (ed. 4) xxi. 813 Reformate normally has an end point 15 to 30°F higher than that of the feedstock. **1973** Hadley & Turner in Hobson & Pohl *Mod. Petroleum Technol.* (ed. 4) xii. 442 Catalytic reforming of straight-run material yields a C_6–C_8 reformate rich in the aromatic hydrocarbons, benzene, toluene, ethylbenzene, and the xylenes.

reforma·tionist. [f. Reformation + -ist.] One who supports or advocates reformation.

1906 *Macmillan's Mag.* June 589 One who is proved incorrigible..may indeed be secluded, but..if the reformationists are right, his seclusion should have no penal character. **1928** *Sunday Express* 17 June 12/5 The chaos in the Church may harden into bitter anarchy and disintegrating conflict between Reformationists, Adorationists, and Modernists. **1969** *Listener* 17 July 87/3 'To make them better citizens,' says the Reformationist.

reformatory, *a.* and *sb.* Add: **A.** *adj.* (Later examples.)

1932, etc. [see *approved *ppl. a.* 5]. **1933** J. Masefield *Conway* II. 61 H.M.S. reformatory-ship *Clarence* was fired and burned by the boys on board her.

B. *sb.* (Earlier and further examples.) Also *N. Amer.*, a reforming institution for women or for first offenders.

1834 J. S. Mill in *Monthly Repos.* VIII. 735 He proposes that those who are convicted of offences..should be no otherwise ill-treated than by being compelled to live as a community apart... If all who, in any manner violated the laws, were removed into such a place of reformation, the inhabitants of the reformatory would speedily outnumber the remainder of the community. **1870** *N.Y.* (State) *Laws* I. 320 The action of the commissioners..in locating said prison or industrial reformatory..at Elmira..is hereby approved. **1878** *Harper's Mag.* Dec. 109/1 Our reformatories..do not check the first steps in wrong-doing. **1912** M. Nicholson *Hoosier Chron.* iii. 53 They were going to cut down the Reformatory's appropriation last winter. **1950** *Times* 20 Mar. 5/3 As the result of the laudable impulse for reformatory treatment, the expectation may have been fostered among criminals that they will find that the salutary discomforts of prison have been reformed away. **1970** *Globe & Mail* (Toronto) 26 Sept. 2/2 [He] was sentenced to two years less a day definite and 18 months indeterminate in reformatory after admitting that he stole about $50,000 from Sunnybrook Hospital. **1975** *Washington Post* 27 Feb. B1/7 Michael Craddock..thinks the reformatory is a danger, a 'public nuisance'. **1977** *Ibid.* 25 Feb. A2/1 About 200 inmates at the Ohio State Reformatory hurled trays and food and fought guards during a disturbance today in the institution dining room.

reformed, *ppl. a.* and *sb.* Add: **A.** *ppl. a.* **1. a.** (Further examples.) Also *U.S.*

With capital initial still used widely to denote Calvinist (as opp. Lutheran) Protestantism.

1794 T. Coxe *View U.S.* ix. 373 There are and have been in the legislative, executive and judicial branches of the general government, persons of the following denominations—Episcopalian,..Reformed, Roman, and probably others. **1837** J. M. Peck *Gazetteer Illinois* (ed. 2) 73 A Seminary is about being established in a settlement of Reformed Presbyterians. **1844** I. D. Rupp *He Pasa Ekklesia* 466 The Reformed Methodists took their origin from a feeble secession from the Methodist Episcopal Church, in..1814. **1847** R. Davidson *Hist. Presbyterian Church Kentucky* viii. 216 Campbellites..affected the title of Reformers, or Reformed Baptists, and spoke of 'The Reformation' as if there had never been any Reformation before. **1928** W. D. Brown *Hist. Reformed Church in Amer.* i. 7 The Reformed Church in America is the direct outgrowth of the emigration from the Netherlands. **1954** *Collier's* 20 Aug. 21/3 Three allocations were decided upon—..the third to an Evangelical Reformed Church. **1967** R. McA. Brown *Ecumenical Revolution* viii. 142 In 1934 the Evangelical Synod of North America merged with the Reformed Church in the United States to form the Evangelical and Reformed Church, while a few years later both of these new groups merged to form what is now the United Church of Christ. **1969** T. F. Torrance *Theol. Sci.* ii. 87 So far we have been thinking of this mainly in terms of strictly Reformed theology but in some respects it had an even greater development in Lutheran theology. **1977** *Washington Post* 18 Mar. D18/3 The signers included Billy Graham and leaders of the..Reformed Church in America. **1978** *Church Times* 29 Dec. 1/3 The consultation is proposing to the sponsoring bodies that a dialogue programme at world level be implemented between the Anglican and Reformed traditions.

d. *Judaism.* (With capital initial.) Subscribing to, or characteristic of, Reform Judaism (see *reform *sb.* 6 b).

1844 *Voice of Jacob* 19 July 188/2 There is a 'reformed Synagogue at Liverpool'..that..is to take no part in the election of a Chief Rabbi. **1859** *N.Y. Times* 30 Sept. 1/5 At the tabernacle of the Reformed Society..Dr. Adler preached in German. **1876** Geo. Eliot *Dan. Der.* II. iv. xxxii. 298 He was affectionately directed by a precocious Jewish youth, who entered cordially into his wanting not the fine new building of the Reformed but the old Rabbinical school of the orthodox. **1898** W. J. Locke *Idols* vi. 70 Think of Simeon Goldberg, a good friend, a man..of the Reformed faith. **1918** H. Barnett *Canon Barnett* II. xxxiv. 65 Minister of Reformed Synagogue in New York. **1971** *Guardian* 16 Nov. 8/6 Grandfather had been a warden of a reformed synagogue. **1977** *Church Times* 12 Aug. 5/5 On our last day we went with a party to a service in the Reformed synagogue in Haifa.

3. b. Also **re-formed.** Of petroleum products: subjected to or obtained by *reforming *vbl. sb.* 2.

1924 *Proc. 31st Ann. Convention Pacific Coast Gas Assoc.* 725 Reformed natural gas. **1931** *U.S. Bureau of Mines Techn. Paper* No. 483. 2 The natural gas employed was composed almost entirely of methane (CH_4) and ethane (C_2H_6), but the re-formed product comprised methane..and hydrogen..as the chief combustible constituents. **1952** Kirk & Othmer *Encycl. Chem. Technol.* VIII. 793 Generally a greater part of the natural gas is re-formed, and a smaller amount is mixed with air and added to the re-formed gas in order to control the specific gravity and burning characteristics of the mixture. **1966** *Petroleum Handbk.* (Shell Internat. Petroleum Co.) (ed. 5) 73/2 Reformed natural gas..can usually be blended to give a gas interchangeable with manufactured gas.

reformer[1]. Add: **3. b.** (Earlier example.)

1780 G. Selwyn in *15th Rep. R. Comm. Hist. Manuscripts* App. vi. 443 in *Parl. Papers* 1897 (C. 8551) LI. 1 My best and ablest friends here are dead; their survivors supine and superannuated; their connections new Whiggs and Reformers, and Associators.

c. *U.S.* A member of one of the reformed sects of various Protestant denominations in the 19th century, *esp.* the Campbellite Baptists. *Obs.* exc. *Hist.*

1831 J. M. Peck *Guide for Emigrants* 258 The Reformers, or Methodist Protestant church, have several societies and preachers in the State [of Illinois]. **1834** [see *Campbellite 1]. **1871** E. Eggleston *Hoosier Schoolmaster* xii. 101 Squire Hawkins..had become a member of the 'Reformers'..who now call themselves 'Disciples', but whom the profane will persist in calling 'Campbellites'. **1931** W. W. Sweet *Relig. on Amer. Front* I. ii. 26 Between 1829 and 1832, something like 10,000 Kentucky Baptists withdrew to form the Disciples Church. Besides the Campbell followers, who were known as *Reformers*, there were several thousand anti-mission Baptists in Kentucky.

d. An advocate or adherent of Reform Judaism (cf. *reform *sb.* 6 b).

1855 *Jewish Chron.* 20 July 245/3 The two parties, orthodox and reformers. **1870** *N.Y. Times* 3 Apr. 3/3 The tenets of Mr. Lewin represent the most advanced opinions of the Reform School, and are therefore shared in full by a minority only of the Reformers themselves. **1892** I. Zangwill *Childr. Ghetto* III. II. iv. 39 'By worshipping bare-headed, and by seating the sexes together, they have defiled Judaism.' 'Stop..who told you the Reformers do this?' **1934** *Times Lit. Suppl.* 10 May 334/3 The 'Reformers', too, are unlikely to have much material earlier than their foundation, though, no doubt, their papers throw light on the schism and their *ritualia* are of great beauty. **1976** B. Williams *Making of Manchester Jewry* iv. 105 To Reformers the future of Judaism..appeared to depend upon..a degree of accommodation to the values of the surrounding milieu.

5. An installation or apparatus for the reforming of petroleum products (*reforming *vbl. sb.* 2).

1934 *Jrnl. Inst. Petroleum Technologists* XX. 347 One of the new refineries in France..includes a cracking still, a viscosity breaker and a naphtha reformer. **1958** *Times Rev. Industry* May 24/3 The pilot scale installation incorporates..a catalytic reformer in which gaseous hydrocarbons interact with steam to produce the hydrogen. **1971** *Daily Tel.* 6 Dec. 6/7 The board found it necessary to retain specialist workers employed on a hydro-carbon reformer plant. **1974** *Times* 22 Mar. 21/4 A 'reformer' unit which processes the fuel (natural gas or propane) to produce a hydrogen-rich feed for the fuel cell. **1978** *Trends in Oil & Gas Refining* (Shell Internat. Petroleum Co.) 3 In a simple refinery there are few secondary units—perhaps..a catalytic reformer (for the improvement of octane quality of motor gasoline).

reforming, *vbl. sb.* Add: **2.** Also re-forming. The treatment of hydrocarbons so as to produce changes in composition; *spec.* (i) increasing the octane number of petrol by heating it under pressure over a catalyst (the major effects being an increase in the proportions of aromatic and other unsaturated cyclic compounds, and loss of hydrogen); (ii) partially or completely converting gaseous hydrocarbons to carbon monoxide and hydrogen by heating with steam over a catalyst.

Not all the early reforming processes used a catalyst.

1924 *Proc. 31st Ann. Convention Pacific Coast Gas Assoc.* 724 The 'reforming' or 'cracking' of natural gas is a process of much interest to several of the Southern California gas utilities. **1932** *Jrnl. Inst. Petroleum Technologists* XVIII. 262 In the cracking or 're-forming' of gasoline, one company uses a De Florez furnace, heating the gasoline to 925° F. at about 1000 lb. pressure, obtaining a recovery of 86 to 88 per cent. of gasoline of high octane number from a straight-run paraffin type gasoline. **1941** *Oil & Gas Jrnl.* 27 Mar. 87/1 A catalytic process for the 'reforming' or converting of low-octane to high-octane gasolines at high temperatures in the presence of hydrogen gas. **1966** *Kirk-Othmer Encycl. Chem. Technol.* (ed. 2) X. 417 A basic problem in catalytic steam reforming and cracking is that the pyrolysis reactions..compete with the steam-hydrocarbon reactions. **1971** *Sci. Amer.* Dec. 57/2 During the 1950's new platinum-alumina-halogen catalysts were introduced to carry out the catalytic reforming of low-octane oil fractions. **1974** *Ibid.* Oct. 68/2 In the U.S. methane is the customary commercial source of hydrogen. In the process called re-forming it is treated with high-temperature steam, producing carbon monoxide (CO) and molecular hydrogen.

reformism (rīfǭ·ɹmiz'm). [f. Reform *sb.* + -ism.] A policy of social, political, or religious reform, *spec.* in *Politics*, the theory that socialism can be established in an evolutionary way by reforms within a country's existing legislative system rather than by revolution. Cf. *revisionism.

1904 R. C. K. Ensor *Mod. Socialism* p. xxvi, The germs of all Von Vollmar's reformism may be found in his own speeches before Bernsteinism appeared. *Ibid.* p. xxvii, The concrete spirit of reformism, which is careful of national peculiarities in its domestic politics, cannot overlook them wholly in foreign affairs. **1920** *Contemp. Rev.* Dec. 872 Your Right leaders are favouring reformism on those questions which I have called fundamental and decisive. **1926** *Socialist Rev.* Jan. 317 Industrial Conferences where working men..are learnedly lectured by middle-class apostles of reformism. **1937** 'C. Caudwell' *Illusion & Reality* xii. 309 Since this Kingdom of Heaven was to be achieved by non-resistance, by heavenly forces and a general change of heart, it was bound to become mere reformism and end as a machine for tying the oppressed of the Empire to the throne of Constantine. **1957** R. N. C. Hunt *Guide to Communist Jargon* xli. 139 Reformism was the view..that the evils of the capitalist system could be exorcised by reforms. **1964** P. G. Casanova in I. L. Horowitz *New Sociol.* 72 The transformation of social structures in the twentieth century has brought both sides to make partial concessions to reformism. **1970** F. C. Weffort in I. L. Horowitz *Masses in Lat. Amer.* xi. 403 Reformism was conceived within an ideological framework of consecration of the State as the only solution to social and economic problems. **1973** C. D. Kernig *Marxism, Communism & Western Society* VII. 154/2 On the international level contemporary reformism is represented by the Socialist International founded in 1951. **1976** *Times* 9 Aug. 11/3 Their convictions, constancy and devotional practices form in fact a needed counterweight within the church to the prevalence of reformism.

reformist. Add: **1. b.** *spec.* an advocate or supporter of *reformism. Also as *adj.*

1904 R. C. K. Ensor *Mod. Socialism* 164 We are revolutionaries, because..we are not at all sure, Citizen Millerand, of attaining our desired solution by the reformist method. **1906** M. Minturn tr. *Jaurès's Stud. in Socialism* p. vi, The situation reached its climax in 1899 with the entrance of the Reformist Millerand into the Waldeck–Rousseau coalition cabinet. **1913** V. G. Simkhovitch *Marxism versus Socialism* 292 Whether they call themselves revisionists, reformists, laborites or plain socialists..the overwhelming majority of the socialists of today are tending to be reformers. **1920** *19th Cent.* Aug. 206 England..allowed the right of private judgment to her middle classes in the seventeenth century and allowed every kind of Reformist literature to enter the country. **1927** H. J. Laski *Communism* i. 39 The growth of capitalism..seemed to suggest that the day of its end was far distant. Everywhere there grew up reformist socialism. **1941** Koestler *Scum of Earth* xi. 113 The sectarian hatred between Stalinists, Trotskyists, and Reformists still existed. **1950** E. H. Carr *Bolshevik Revolution* I. i. 12 The 'Economists'..reached the same practical conclusion as the legal Marxists that it was necessary to postpone to an indefinite future the revolutionary socialist struggle of the proletariat and to concentrate meanwhile on a reformist democratic programme in alliance with the bourgeoisie. **1969** *Daily Tel.* 1 Feb. 19/3 An article published in Soviet newspapers yesterday blamed 'reformist' journalists for the 'moral tenor' in Czechoslovak life. **1974** tr. *Sniečkus's Soviet Lithuania* 12 A reformist opportunist trend..had developed on the basis of the petty-bourgeois nationalistic ideology. **1977** *Time* 14 Nov. 16/3 The situation in Italy is too critical for a reformist policy like the one the Communists propose.

reformulate, *v.* Add: (Further examples.)

1962 E. E. Evans-Pritchard *Ess. Social Anthropol.* 9 Our knowledge has increased and some of our theoretical propositions have had to be reformulated accordingly. **1968** C. G. Kuper *Introd. Theory Superconductivity* iv. 55 These questions must be carefully reformulated before they can be answered. **1970** *Times* 20 Apr. 4/4 It might be necessary to reformulate existing trace mineral mixtures where this material was used. **1970** G. Germani in I. L. Horowitz *Masses in Lat. Amer.* xvi. 586 Class and mass theories must be reformulated within a more general framework.

Hence **refo·rmulated** *ppl. a.*

1965 *Mod. Law Rev.* XXVIII. 536 Alternative policies..embodied in a reformulated principle.

reformulation. (Later examples.)

1922 A. G. Hogg *Redemption from this World* iv. 135 We are keeping well within the limits of legitimate reformulation when we say [etc.]. **1951** E. E. Evans-Pritchard *Social Anthropol.* v. 86 The theories have been shaped and reshaped by this steady growth in knowledge and they have.., in each reformulation, directed observation into deeper layers. **1957** J. S. Huxley *Relig. without Revelation* (rev. ed.) ix. 220 The crude distinction in terms of ethical absolutes like 'good' and 'evil' requires reformulation in the light of psychology and history. **1968** Fox & Mayers *Computing Methods for Scientists & Engineers* x. 195 For this purpose we would prefer a reformulation of the problem. **1970** G. Germani in I. L. Horowitz *Masses in Lat. Amer.* xvi. 586 Such reformulation has been suggested by the mass and 'national-popular' movements in Latin America. **1979** *Dædalus* Winter 38 The deeper cause of this popular success surely lies in the new philosophy's reformulation of the antipolitics of May.

refract, *v.* Add: **1.** (Examples relating to waves other than light.)

1874 *Proc. R. Soc.* XXII. 532, I have dealt with the effect of the atmosphere to refract sound upwards. **1944** A. Holmes *Princ. Physical Geol.* xiv. 284 The waves advance more rapidly through the deeper water opposite a bay than through the shallower water opposite a headland... The waves thus become curved or refracted..towards parallelism with the shore line. **1966** *McGraw-Hill Encycl. Sci. & Technol.* XI. 408/1 Waves propagated through the solid earth are refracted by changes of material or changes of density. **1974** Harvey & Bohlman *Stereo F.M. Radio Handbk.* vii. 144 The sky-wave..may be refracted back towards the earth by the layers of ionized gas.

c. *intr.* To undergo refraction.

1964 *Oceanogr. & Marine Biol.* II. 84 If waves reach the beach unaffected by offshore underwater topography one has an easier task than if waves are breaking and refracting around an offshore bar.

4. *Ophthalm.* To measure the focusing characteristics of (an eye) or of the eyes of (a person). Also *absol.*

1897 J. Thorington *Retinoscopy* v. 45 To give a patient thus refracted with the retinoscope his emmetropic correction..an allowance must always be made, in all meridians, of one diopter. **1904** —— *Refraction & how to Refract* (ed. 3) ix. 235 If a young subject must be refracted without drops, then the fogging method must be followed. *Ibid.*, The [manifest] method by which the eyes of patients past forty-five years of age are refracted. **1953** N. Bier *Contact Lens Routine & Pract.* iv. 56 If the measurements..in the corneal fit were 8·75/13, do not refract with a trial lens of 8 mm radius and 13 mm diameter. **1968** *Sci. Jrnl.* Dec. 21/1 Normally when 'refracting' human patients an ophthalmologist relies on verbal reports and applies corrective lenses until the subject says that he sees most clearly.

refraction. Add: **2.** More widely, change in direction of propagation of any wave as a result of its travelling at different speeds at different points along the wave front. (Further examples.)

1874 *Proc. R. Soc.* XXII. 532 This refraction explains the well-known difference which exists in the distinctness of sounds by day and by night. **1914** [see *absorption 4 (*d*)]. **1944** A. Holmes *Princ. Physical Geol.* xiv. 297 (*caption*) Diagram to illustrate the development of a hooked spit by the refraction of oblique waves. **1971** *Nature* 12 Feb. 452/2 T. D. Krishna Kartha describes his work on the variation of velocity (refraction) of microseisms approaching Cochin in southern India. **1974** Harvey & Bohlman *Stereo F.M. Radio Handbk.* vii. 144 Refraction of the radio wave occurs due to the effects of the varying density of the gas layers in which the wave is travelling.

6*. *Ophthalm.* Measurement of the focusing characteristics of eyes. Also *attrib.* Cf. *refract *v.* 4.

1900 J. Thorington *Refraction & how to Refract* ix. 229 The great danger in any refraction..is an overcorrection. **1928** W. S. Duke-Elder *Pract. Refraction* xxi. 302 Test..the depth of cycloplegia by testing the accommodation before the refraction is done. **1953** N. Bier *Contact Lens Routine & Pract.* iv. 56 The practitioner's contact lens refraction set. *Ibid.*, An alteration of 0·5 mm in the corneal radius is approximately equivalent to 3·00 D.S. in refraction. **1961** *Lancet* 30 Sept. 760/2 An ophthalmologist holds refraction clinics weekly in the surgery. **1975** M. Ruben *Contact Lens Pract.* iv. 67/2 The cylinder found by refraction is −3·00 D.

7. refraction profiling *Geol.*, profiling (sense *3) by means of refraction shooting; **refraction shooting** *Geol.*, seismic prospecting in which shock waves generated at the earth's surface are detected at several points along a line some miles long, the relation between the time of arrival at each point and its distance giving information about the nature and depth of the underlying strata.

1929, 1963 Refraction profiling [see *profiling *vbl. sb.* 3]. **1929, 1951** Refraction shooting [see *reflection shooting* s.v. *reflection 10]. **1960** C. Gatlin *Petroleum Engin.* iii. 37/1 Refraction shooting is used primarily as a reconnaissance tool to select areas and obtain interpretative data for the more detailed reflection method. **1978** *Nature* 27 Apr. 789/1 Seismic refraction shooting near the Isle of Lewis in the Outer Hebrides has shown the presence of major sedimentary units.

refractive, *a.* Add: **3.** (Later examples corresponding in sense to *refraction 6*.)

1953 N. Bier *Contact Lens Routine & Pract.* iv. 58 [Temporary spasm of accommodation] may persist throughout the refractive examination. **1969** J. R. Gregg *How to communicate in Optometric Pract.* iii. 31 At key spots along the refractive procedure, build in ways of showing confidence and understanding.

refractometry (rīfræktǫ·metri). [f. Refract *v.* + -o + -metry.] The measurement of refractive indices of media.

1902 *Encycl. Brit.* XXX. 239/2 Refractometry by total reflection. **1950** *Engineering* 10 Feb. 143/2 Increased attention has been devoted to methods of high-precision refractometry. **1958** *Oxf. Univ. Gaz.* 23 Apr. 892 Studies in refractometry of living cells with particular reference to the investigation of osmoregulatory function. **1974** *Encycl. Brit. Macropædia* IX. 633/2 Of the many laboratory principles employed in the process analytical instruments, the three in widest use today are refractometry, absorption spectroscopy, and gas chromatography.

Hence **refractome·tric** *a.*, of or pertaining to a refractometer; made by means of refractometry; **refractome·trically** *adv.*

1904 *Nature* 4 Feb. 334/2 A simple thermostat for use in connection with the refractometric examination of oils and fats. **1920** *Amer. Jrnl. Physiol.* LI. 278 Reading refractometrically the serum non-protein increase after the intravenous injection of a known amount of acacia or gelatin solution. **1929** *Canad. Jrnl. Res.* I. 13 The total solid content was determined refractometrically. **1937** *Discovery* June 180/2 The application of refractometric work to observe the progress of essential processes in the production of plastics may not be so commonly known. **1970** *Nature* 26 Dec. 1269/1 The relative values [of purity] given by the four methods were: Lowry, 1·0; gravimetric, 0·93; refractometric, 0·87; ultraviolet absorption, 0·32. *Ibid.*, The protein concentration..was also determined refractometrically and gravimetrically.

refractor. Add: **4.** *Geol.* A stratum, or an interface between strata, detected in refraction shooting.

1946 *Geophysics* XI. 40 The production in refraction shooting varies widely, depending on the surface conditions and also on the depth of the refractors. **1976** W. M. Telford et al. *Appl. Geophysics* iv. 365 Where a single refractor is being followed, a series of short refraction profiles are often shot rather than a long profile.

refractoriness. Add: **1. c.** *Physiol.* Temporary inability to respond fully to nervous or sexual stimuli.

1932 W. Burridge *Excitability* xxi. 172 Refractoriness is here defined as a condition of inexcitability of an excitable tissue which follows the receipt of an adequate stimulus. **1937** *Wilson Bull.* XLIX. 251 Birds invariably passed the climax of activity after a time and underwent regression. This was due to 'throwing out of gear' or development of refractoriness at some part of the sexual mechanism. **1949** *Ibid.* LXI. 221 This refractoriness must 'wear off' before external stimuli can induce a new gonadal activation. **1963** S. Ochs in E. E. Selkurt *Physiol.* ii. 28 In the alpha group of A fibers of the frog sciatic nerve, this period of absolute refractoriness lasts only a little longer than 1 msec.

refractory, *a.* and *sb.* Add: **A.** *adj.* **5.** *Physiol.* Temporarily unresponsive or not fully responsive to nervous or sexual stimuli; *esp.* in *refractory period*, a period of reduced responsiveness following a response to such a stimulus.

1879–80 *Jrnl. Physiol.* II. 400 The same absence of response was observed whenever the second excitation occurred 'during the commencement of the systole of the ventricle' (termed by Marey the 'refractory period'). **1900** J. Burdon-Sanderson in E. A. Schäfer *Text-bk. Physiol.* II. 449 This rhythmicity is attributed to the liability of the heart to be 'refractory' for a certain period after each excitation. **1937** L. V. Heilbrunn *Outl. Gen. Physiol.* xxxvi. 415 In skeletal muscle, the refractory period is much shorter than in cardiac muscle. **1950** *Nature* 16 Dec. 1034/2 During this time, experimental birds are 'refractory'—they cannot be forced into spermatogenesis by means of photostimulation. **1952** *Jrnl. Physiol.* CXVII. 534 After the earliest stimulus the membrane potential falls again with hardly a sign of activity, and the membrane can be said to be in the 'absolute refractory period'. The later stimuli produce action potentials of increasing amplitude, but still smaller than the control; these are in the 'relative refractory period'. **1967** J. L. McCary *Human Sexuality* xi. 176 After orgasm, the man enters the refractory period..; the sexual stimulation that was previously effective and pleasurable now becomes unavailing and distasteful. Women, on the other hand, usually do not go into a refractory period. **1971** M. B. V. Roberts *Biol.* xviii. 268 The importance of the refractory period is that, together with transmission speed, it determines the frequency at which an axon can transmit impulses. **1972** *Nature* 18 Feb. 366/1 A substance from the male accessory glands..rendered the female permanently refractory to further insemination. **1974** P. Svendsen *Introd. Animal Physiol.* xi. 119 The very long action potential and refractory period in cardiac muscle are of great importance. They ensure that..two successive contractions cannot add together without a period of relaxation between them.

B. *sb.* **1.** (Later example.)

1860 Dickens *Uncomm. Trav.* in *All Year Round* II. 394/1 The Refractories were picking oakum... The oldest Refractory was, say twenty; youngest Refractory, say sixteen.

2. More widely, any refractory material. (Further examples.)

1907 *Jrnl. Iron & Steel Inst.* LXXIII. 384 The plant, which has only recently been laid down by the American Refractories Company at Joliet, represents the first successful attempt at manufacturing refractory materials in the west of America. **1931** *Daily Express* 15 Oct. 14/1 For many months manufacturers of refractories have had to struggle against the dumping of large quantities of Belgian sand. **1957** *New Scientist* 9 May 44/1 It was inevitable that factories making fire-bricks—the original refractory—should spring up alongside iron- and steel-works. **1962** *Science Survey* III. 344 The tundish that evens out the flow between the ladle and mould is usually a simple refractory-lined box. **1967** M. Chandler *Ceramics in Mod. World* v. 140 A high proportion of refractories must serve also as structural materials.

refra·cture, *sb.*[2] [Re- 5 a.] Renewed fracture (of a bone).

1908 *Practitioner* Oct. 535 Instances of refracture in long bones, at an old site of previous similar injury.

refresh, *sb.* Delete 'Now *colloq.*' and add: **3.** The process of renewing the data stored in a memory device or displayed on a cathode-ray tube. Usu. *attrib.*

1967 *Technology Week* 20 Feb. 22/3 The complete refresh memory is made up of 16 parallel magnetostrictive delay-line loops that store all of the picture elements for one frame. **1972** D. Lewin *Theory & Design of Digital Computers* vii. 264 To produce a steady picture on the CRT the contents of the display file must be periodically cycled through (a word at a time) and passed to the display unit; this procedure is known as the refresh cycle. **1977** J. C. Boyce *Digital Computer Fundamentals* viii. 211 Typically the refresh operation must be performed about every 2 milliseconds. **1977** *Sci. Amer.* June 57/2 (Advt.), Every [memory] board is fast. With 'hidden refresh' and *no* 'wait state'.

refresh, *v.* Add: **1. a.** (Further examples.) Also, to plunge (cooked vegetables, etc.) into cold water as part of the cooking process.

1861 Mrs. Beeton *Bk. Househ. Managem.* xxv. 591 Let the herbs be as fresh as possible for a salad, and, if at all stale or dead-looking, let them lie in water for an hour or two, which will very much refresh them. **1877** E. S. Dallas *Kettner's Bk. of Table* 45 Some..prefer to eat them [*sc.* asparagus] with oil and vinegar. In this case they are, as the French say, to be *refreshed* with cold water. **1972** *Guardian* 18 Aug. 11/3 Lasagne... Boil the pasta for ten minutes then drain and run under the cold tap, and drain again... Cannelloni: these are boiled like lasagne, and refreshed in the same way.

5. (Further examples.)

1977 A. P. Malvino *Digital Computer Electronics* vii. 180 Because capacitor charge leaks off, the stored data must be refreshed every few milliseconds. **1977** *Sci. Amer.* Sept. 139/3 A memory used to refresh the information presented in a conventional video display, which is scanned point by point in a repeating linear pattern, does not require a memory with random access.

refresher. Add: **3. a.** (Earlier examples.)

1826 F. Reynolds *Life & Times* I. iv. 148 For the cause, after refresher on refresher, came on within the space of a few months. **1831** —— *Playwright's Adventures* vi. 108 He also knew that barristers..can only be kept alive by refreshers.

4. *attrib.*, applied to training or instruction provided as a review of material previously studied or to instruct a person in new developments, techniques, etc., esp. in *refresher course*; *refresher leave*, leave granted for the purpose of attending a refresher course.

1907 *Interim Rep. War Office Comm. Provision of Officers* 10 Given a short term of liability, and short periods of recall to the Colours for 'refresher' training, many officers..would remain therein. **1914** Hamel & Turner *Flying* x. 209 Others..go direct into the Royal Flying Corps Reserve..where they are available for periodical 'refresher' courses and for employment in the event of war. **1930** *Times Educ. Suppl.* 20 Sept. 403/4 Refresher courses for teachers. **1945** *Jrnl. Amer. Med. Assoc.* 12 May 141/1 A clinical refresher training program for..medical officers. **1959** *Listener* 17 Dec. 1085/1 This is not an introduction to Kipling, but a splendid refresher course. **1972** *Accountant* 17 Aug. (Suppl.) 14/3 (Advt.), Terms of appointment include generous refresher leave periods. **1976** H. Wilson *Governance of Britain* vi. 121 A major conference would involve one or two, or more, full briefing meetings at No. 10..with refresher briefings to review the progress of the conference and decide tactics to meet a changing situation. **1977** *New Yorker* 12 Sept. 103/1 A refresher training course in the basic sciences that I was teaching at the Armed Forces Institute of Pathology in the fall of 1969. **1980** R. Perry *Grand Slam* v. 42 [He] was making a complete pig's ear of shadowing me... He was in dire need of a refresher course.

refreshing, *vbl. sb.* Add: **1.** Also, the plunging of cooked vegetables, etc., into cold water.

1961 S. Beck et al. *Mastering Art of French Cooking* viii. 422 A second important French technique is that of refreshing. As soon as green vegetables have been blanched..they are plunged for several minutes into a large quantity of cold water.

refreshing, *ppl. a.* **3.** (Earlier example.)

1716 *Parish Rec. Kenilworth* in *Mod. Lang. Rev.* (1951) XLVI. 327 Paid to Mr. Palmer a Refreshing fee at Christmas Sessions.

refreshment. Add: **7.** *refreshment bar, counter, room* (earlier and later examples), *saloon, stand, station, stop, table, tent*; **Refreshment Sunday** (later examples).

1860 Dickens *Uncomm. Trav.* in *All Year Round* II. 418/2 Crowds of us had sandwiches and ginger-beer at the refreshment-bars..in the Theatre. **1889** E. Dowson *Let.* 17 Mar. (1967) 50, I..searched through the Law Court refreshment bars. **1973** 'B. Mather' *Snowline* vi. 68 The first class refreshment bar at Sealdah Station. **1908** *Busy Man's Mag.* Jan. 89/2 The dance over, he took them to the refreshment counter for a cup of coffee and a sandwich. **1979** *National Trust* Spring 14/3 Nowhere perhaps is the transformation more apparent than in the Tea Room, now freed from the unsightly refreshment counter. **1835** Dickens in *Evening Chron.* 7 Mar. 3/4 The Militia-man..repaired to Bellamy's kitchen—a refreshment room where persons who are not members [of the House of Commons] are admitted on sufferance, as it were. **1966** G. W. Turner *Eng. Lang. in Austral. & N.Z.* vii. 157 Because the narrow-gauge lines necessary for a mountainous country preclude corridors, and therefore dining cars, in our trains, the New Zealand *refreshment rooms* on railway stations are notable. **1976** *Flintshire Leader* 10 Dec. 3/5 Mrs Parry worked in the refreshment room at Chester railway station. *a* **1828** J. Bernard *Retrospections of Stage* (1830) II. x. 318, I..purchased a cottage..surrounded by..meadowland; the former being small enough for a refreshment saloon. **1936** A. Russell *Gone Nomad* ii. 8, I..sauntered down the corridors to the refreshment saloon at the first suggestion of thirst. **1976** *Honolulu Star-Bull.* 21 Dec. A-3/1, I have two gripes about the refreshment stand operation at Blaisdell Center Arena. **1860** Dickens *Uncomm. Trav.* in *All Year Round* II. 513/2, I travel by railroad... I am hungry when I arrive at the 'Refreshment' station where I am expected. **1977** C. McCullough *Thorn Birds* iii. 67 Our next refreshment stop is a place called Blayney. **1912** *Encycl. Relig. & Ethics* V. 770/2 The fourth Sunday in Lent, when the Gospel for the day narrates the Feeding of the Five Thousand, has long been called *Dominica Refectionis*, or 'Refreshment Sunday'. **1974** [see *Lætare]. **1977** *Church Times* 1 Apr. 5/4 'The God of love my Shepherd is' set in *The English Hymnal* for Refreshment Sunday. **1860** Dickens *Uncomm. Trav.* in *All Year Round* II. 514/2 You are going off by railway, from any Terminus... You present to your mind, a picture of the refreshment-table at that terminus. **1885** A. Edwardes *Girton Girl* I. xiv. 276 Let us bend our steps to the refreshment tent. **1928** E. Waugh *Decline & Fall* I. ix. 94 The refreshment tent looked very nice.

refrigerant, *sb.* Add: **3.** Also, a substance used as the working fluid in a refrigerator. (Further examples.)

1901 [see **champagne gas*]. **1926** *Encycl. Brit.* III. 319/1 In the refrigerating cycle, the refrigerant is made to pass into the evaporating coils so as to enable heat to be absorbed from the commodity to be cooled. **1964** *Listener* 7 May 776 (Advt.), ICI salesmen are today successfully selling..refrigerants to Icelanders (where you'd think it would be cold enough already). **1970** *Times* 16 June 2/7 The water is mixed with a liquid hydrocarbon freezing agent such as butane. The refrigerant takes heat from the water which in turn produces ice crystals and concentrates unwanted solids and salts into a brine slurry.

refrigerate, *v.* Add: **1. c.** (Later examples.)

1957 *Times* 1 Nov. 11/7 Operations which cannot be undertaken at normal temperature may be performed if the body is refrigerated. **1979** *Arizona Daily Star* 5 Aug. J 5/2 Refrigerate overnight before using.

2. (Earlier example.)

1559 P. Morwyng tr. *Gesner's Treasure of Evonymus* 151 A man must put les wyne to new routes then to dry: and perauentur, les also to them whiche ought to refrigerat and coule.

refrigerated, *ppl. a.* Add: Also applied, by extension, to the container in which food is kept, displayed, transported, etc., in a refrigerated condition.

1943 J. S. Huxley *TVA* xv. 128 TVA, in collaboration with the University of Tennessee, designed and built refrigerated barges..to encourage the fruit and vegetable freezing industry in the Valley. **1958** *Brit. Standard Specification* No. 3053 (*title*) Open and closed refrigerated display cabinets (for the retail sale of frozen packaged foods in temperate climates). **1962** [see *gondola 4 d]. **1967** *Economist* 7 Jan. 29/2 Bulgaria has bought a fleet of..long-distance refrigerated trucks (built on the standard American pattern)... There has been heavy investment in refrigerated rail cars bought from Hungary and Poland. **1967** *Commercial Fisheries Rev.* Dec. 53/1 The catch of the refrigerated tuna boats was 7,985 tons in 1966. **1976** *Southern Even. Echo* (Southampton) 3 Nov. (Advt.), Cunard have bought 10 fast refrigerated ships.

refrigerating, *vbl. sb.* Add: Also *attrib.*

1909 *Chambers's Jrnl.* Jan. 23/1 The refrigerating-engineer..claims to play the chief part in the successful maintenance of the overseas trade in chilled meat.

refrigeration. Add: **3.** *attrib.* and *Comb.*, as *refrigeration company, machinery, unit.*

1976 *National Observer* (U.S.) 28 Aug. 7/2 Les Whitely, who owns a refrigeration company in the San Francisco Bay area. **1943** J. S. Huxley *TVA* 30 The production of new refrigeration machinery. **1969** *Coast to Coast 1967–68* 103 The Natwicks sat on the front veranda to watch the traffic..the semi-trailers and refrigeration units, the decent old-style sedans, [etc.]. **1979** *Tucson* (Arizona) *Citizen* 20 Sept. 1B/4 We have two refrigeration units in our house.

refrigerator. Add: **2.** (Further examples.)

1841 C. Cist *Cincinnati in 1841* (Advt.), Refrigerators or Ice Chests. **1958** *Times* 13 Jan. 11/2 Only 10 per cent. of the population of this island have refrigerators, against 90 per cent. in the United States. **1975** *N.Z. Jrnl. Agric.* Sept. 39/1 My first impression was that the New Zealand farm equipment manufacturers were attempting to 'sell refrigerators to the Eskimos', because, among the massive displays by the British manufacturers, a comparative handful of New Zealand firms were trying to enter an overcrowded market.

b. *refrigerator engineer, -freezer, -maker, ship, truck.*

1909 *Westm. Gaz.* 6 Sept. 5/3 The second refrigerator-engineer..informed us that the boats had put off. **1960** M. Spark *Ballad of Peckham Rye* ii. 27 Humphrey Place, refrigerator engineer of Freeze-eezy's. **1963** *Which?* 6 Feb. 42/1 The combined refrigerator-freezer..will hold more frozen food than a refrigerator's freezing compartment. **1976** *Woman's Day* (U.S.) Nov. 125 Self-defrosting refrigerator-freezer has built-in energy-saving condenser. **1950** *Manch. Guardian Weekly* 7 Dec. 15/3 The Sullivan tunes in the interest of the butcher and baker and refrigerator-maker. **1877** in *Sci. Amer.* (1977) Jan. 14/3 A despatch from M. Tellier to the French Academy of Sciences announces the arrival of the refrigerator ship *Frigorific* at Pernambuco, Brazil. **1921** *Daily Colonist* (Victoria, B.C.) 21 Oct. 12/5 The Moliere, another refrigerator ship..is at Seattle loading. **1976** *Islander* (Victoria, B. C.) 12 Sept. 10/1 A refrigerator ship carrying, eggs, bacon and steel to England. **1971** P. O'Donnell *Impossible Virgin* x. 211 Get Brunel boxed up and put in the refrigerator truck. **1974** R. B. Parker *God save Child* vi. 48 A big refrigerator truck lumbered by on the highway.

refry (rīfrəɪ·), *v.* [Re- 5 a.] *trans.* To fry again. So **refrie·d** *ppl. a.*, esp. in *refried beans* (U.S.) [Sp. *frijoles refritos*], a dish consisting of pinto beans boiled and fried in advance and refried when required; also *fig.*

1957 *House Beautiful* Sept. 126/2 Main course is a barbecue, Yucatan fashion, accompanied by refried beans. **1960** *N.Y. Times Mag.* 1 May 72/1 The sales of such canned items as tortillas, refried beans and green chilies..have shown marked increases in recent months. **1967** V. Bennett *Compl. Bean Cookbk.* iv. 117 To refry the beans: heat them in the balance of the bacon drippings. **1976** *Sat. Even. Post All-Amer. Cookbk.* 130/2 California refried beans with cheese. This is one of the best and heartiest dishes for which we have to thank the Spanish and the Mexicans. *Ibid.*, The beans are ready to refry at your convenience. **1977** McKnight & Tobler *Bob Marley* ix. 113 Of the ten songs, three are re-fried oldies.

refuel, *v.* Add: **1.** Also, to fill up the fuel-tank of (a car, an aircraft, etc.); now the usual sense.

1973 D. Robinson *Rotten with Honour* 186 He refuelled the middle-aged Saab. **1974** *Guardian* 25 Jan. 13/1 The Italian attitude has been one of reciprocity: you refuel our aircraft and we'll refuel yours. **1976** C. Egleton *State Visit* iii. 31 A tanker would roll out of the BP hangar to refuel a Trident which was on turn round for Amsterdam.

2. *absol.* or *intr.* To take on more fuel.

1940 N. Monks *Squadrons Up!* i. 18 They [*sc.* the pilots] dropped down to reload, refuel, and grab a sandwich. **1958** *Sat. Even. Post* 20 Sept. 129/1 The first waves of F-100 and F-101 fighters have already passed through.., the second waves are air-refueling. **1973** D. Kyle *Suvarov Affair* xiv. 173 He..decided to drop down to the seaplane station at Port Hardy, refuel and report. **1977** *R.A.F. News* 11–24 May 4/5 Flt Lt Hermer then flew to the Olna to refuel... He was unable to shut down his aircraft while refuelling.

refu·elling, *vbl. sb.* [f. prec. + -ing[1].] The action of the verb. Freq. *attrib.*

1917 W. S. Churchill in M. Gilbert *Winston S. Churchill* (1977) IV. Compan. I. 93 A sheltered anchorage with refuelling facilities, the whole properly netted and surrounded by an extensive system of mines, will have been created. **1930** *Engineering* 9 May 612/1 If we permit refuelling in flight. **1935** *Jrnl. R. Aeronaut. Soc.* XXXIX. 267 The refuelling system in the centre portion of the wing is an interesting time saving device. **1954** W. Tucker *Wild Talent* (1955) xv. 203 They're setting up refuelling bases and they hope to send that ship completely around the world. **1955** *Times* 12 July 7/5 An R.A.A.F. Canberra jet bomber will make a round-Australia flight to-morrow, making only two refuelling stops. **1968** J. Sangster *Touchfeather* xviii. 205 The other crewman was standing with the refuelling mechanic. **1978** R. Ludlum *Holcroft Covenant* 9 It was a refueling station never detected by Allied Intelligence.

refuge, *sb.* Add: **3. a.** Also *spec.*, an establishment that offers shelter to a woman who has been physically ill-treated by her husband (or another man with whom she has cohabited).

1976 *Lancaster & Morecambe Guardian* 7 Dec. 27/3 (Advt.), Battered Women's Refuge—Women's Information Centre. **1977** *New Society* 25 Aug. 389/1 The Hastings Refuge Group began campaigning in January 1976 in the hope of getting a house from the council to use as a refuge. Meanwhile they set up an emergency service for battered wives.

c. (Earlier and later examples.)

1869 [see *island *sb.* 2 c]. **1930** V. Sackville-West *Edwardians* iii. 122 Standing upon the refuge waiting to cross Park Lane, he had seen her drive out of Stanhope Gate. **1976** *Cumberland & Westmorland Herald* 4 Dec. 1/1 It was ludicrous that Belisha beacons could not be erected. People would be running the gauntlet from the pavement to the central refuge.

d. A mountain hut in which climbers and walkers can shelter.

1817 H. C. B. Campbell *Jrnl.* 4 Sept. in G. de Beer *Journey to Florence* (1951) 62 We were eight hours in reaching the top of this wondrous mountain... We stopped at the Refuge No 2. **1873** *Young Englishwoman* Nov. 524/2 Napoleon appointed that ten or twelve 'refuges' should be built for storm-harassed travellers here. **1899** G. Bell *Let.* 28 Aug. (1927) I. 51 We..walked up to the Refuge de l'Alpe in two hours. Two German men turned up at the Refuge. **1933** G. D. Abraham *Mod. Mountaineering* i. 8 The hut is situated less than an hour below the Shoulder... To use this refuge for ordinary mountaineering is an abuse; it is only intended and equipped for special parties *in extremis*. **1956** G. Rébuffat *Mont Blanc to Everest* 42 'There's the refuge. Look, they're signalling to us with lanterns!' How I wished they were. But I knew that the Solvay hut was at least 600 feet lower. **1967** 'G. Carr' *Lewker in Tirol* vi. 84 He had not been inside an Alpine hut for years, but this one was very different from the penitential refuges he remembered in Haute Savoie. **1973** *Guardian* 25 Apr. 1/8 The boys had only to be a short distance off route to walk right past their assessors, who were waiting in the safety of the mountain refuge on Foel Grach's summit.

e. *Biol.* A region in which a natural population can survive through a generally unfavourable period.

1929 *Bull. Geol. Soc. Amer.* XL. 663 On the west coast of Norway the ice seems to have reached out into the sea, but it was probably so thin that many islands and high peninsulas rose above it and formed places of refuge for arctic plants and animals. **1946** *New Phytol.* XLV. 235 The tundra refugees could survive only in the tundra refuges but not in those offered by the coastal mountains. **1954** A. J. Cain *Animal Species* v. 60 Its [*sc.* the arid period's] effect was to confine forest-living animals of all sorts to several refuges where rainfall was sufficient to maintain the sort of habitat necessary for them. **1979** *Guardian* 28 Aug. 3/2 The Dartford warbler..suffered population crashes seven times between 1860 and 1945, but recovered each time because a breeding population of several dozen pairs remained in a habitat refuge.

f. *U.S.* A bird sanctuary.

1933 *Sun* (Baltimore) 9 May 14/5 The refuge..is under the supervision of Mrs. R. L. Duke, who since 1931 has played host to the waterfowl which make it a winter 'resort'. **1956** Peterson & Fisher *Wild Amer.* xxix. 320 At the Refuge headquarters they told us that their census of nesting grebes on Tule Lake..showed 3500 pied-bills. **1976** *National Observer* (U.S.) 24 Jan. 9/4 The conservation group..succeeded in setting up a 150-acre refuge in central Wisconsin... It is a roosting site where eagles have been stopping..as they migrate southward.

5. refuge room, a gas-proof room.

1938 *Times* 10 Mar. 11/1 An internal passage will form a very good refuge-room if it can be closed at both ends. **1940** [see *goof *v.* 1 a].

refuge, *v.* Add: **2.** (Further *fig.* example.)

1929 R. Bridges *Testament of Beauty* III. 85 What grave lore had refuged with the Ishmaelite was stealing back from exile to its western home.

refugee, *sb.* Add: **1. d.** Someone driven from his home by war or the fear of attack or persecution; a displaced person. Also *fig.*

1914 E. A. Powell *Fighting in Flanders* vii. 190 The road from Antwerp to Ghent..was a solid mass of refugees. **1926** T. E. Lawrence *Seven Pillars* (1935) VI. lxxix. 436 Then there were the guests and refugees whom we might expect so soon as the news of our establishment was rumoured in Damascus. **1944**, etc. [see *displaced *ppl. a.* b]. **1957** L. Durrell *Justine* I. 39 You are a mental refugee of course, being Irish. **1976** *National Observer* (U.S.) 6 Mar. 18/4 Robert MacNeil, a 44-year-old refugee from NBC and the British Broadcasting Corp., opens with a succinct summary of the program's topic.

3. a. *refugee family, scholar*; **refugee capital** = *hot money* s.v. *hot *a.* 12.

1936 Refugee capital [see *hot money* s.v. *hot *a.* 12]. **1950** *N.Y. Times* 12 Sept. 11/2 Since the Korean invasion began, as much as $300,000,000 to $400,000,000 in 'refugee capital' has flowed here from Europe. **1940** *Manch. Guardian Weekly* 23 Feb. 147 Contracts have just been signed admitting 500 refugee families from Germany and Poland to San Domingo. **1936** *Discovery* Apr. 98/1 The most distinguished of the refugee scholars. **1978** P. Sutcliffe *Oxf. Univ. Press* VII. iv. 260 Some of the refugee scholars eventually made their way to America.

b. Also, of or pertaining to a refugee or refugees, as *refugee camp, centre, colony, project, style, train.*

1902 J. Buchan in J. Adam Smith *J. Buchan & his World* (1979) 38/2 [Milner] has turned over to me..the Boer refugee camps. **1906** *Westm. Gaz.* 23 Apr. 7/1 In the refugee camps yesterday rude altars were erected. **1953** *News Chron.* 2 June 1/4 The Mall looked like a gigantic refugee camp. Over 30,000 people were bedding down along the pavements. **1975** O. Sela *Bengali Inheritance* ii. 18 Kids, spawned in refugee camps, brought up in squalor. **1941** Koestler *Scum of Earth* 177 Refugee-centre besieged by crowd but said there is some British ambulance in Périgueux. **1940** *Manch. Guardian Weekly* 23 Feb. 147 This is the first step in an ambitious plan for a large refugee colony..that will be one of the most important refugee projects in the New World. **1864** K. Cumming *Jrnl. Hospital Life* (1866) ix. 146/2 His two daughters were with him, and were keeping house in *two* rooms, refugee style. **1888** M. Grigsby *Smoked Yank* xxvi. 224 General Hazen asked me to take charge of the refugee train that was assigned to his division.

refugee, *v.* Substitute for def.: **a.** *trans.* To cause (someone) to become a refugee. **b.** *intr.* To be or become a refugee; to depart or live as a refugee. Chiefly *U.S.* Hence **refugee·ing** *vbl. sb.*

1750 [in Dict.]. **1806** in B. Hawkins *Lett.* (1916) 429 It will be some time before the Greek young will get rid of the remains of that alloy which debased the agents and refugeed their associates. **1862** K. Stone *Jrnl.* 25 Aug. in *Brokenburn* (1955) 139 The planters generally are moving back to the hills as fast as possible. There are two families refugeeing in our neighbourhood. **1864** K. Cumming *Jrnl. Hospital Life* (1866) x. 157/2 Many of the citizens of Mobile..had *refugeed* from fear of an attack. **1866** W. Reid *After War* 250 Many of his pupils were..negroes that had been 'refugeed' from the Red River country. **1874** L. Collins *Hist. Sk. Kentucky* I. 162/1 [There have been] about 1,200 deaths..among the negroes refugeed at Camp Nelson. **1904** R. E. Lee *Recoll. & Lett. Gen. R. E. Lee* xv. 270 In the early years of the struggle, my mother and sisters, when 'refugeeing' had boarded..at his home. **1936** M. Mitchell *Gone with Wind* xviii. 324 The exodus of women, children and old people from the city began... Many..who took the train that night had already refugeed five and six times before. **1942** J. Lees-Milne *Ancestral Voices* (1975) 43 The two were very entertaining about their refugeeing with the niece of an old friend near Chedworth. **1965** 'Han Suyin' *Crippled Tree* xxix. 412 The house in which Aunt Number One refugeed herself when her husband, Uncle Liu, wanted to give her baby number twelve.

refugee·dom. [f. Refugee *sb.* + -dom.] The condition of (being) a refugee.

1967 P. E. H. Durston *Mortissimo* xv. 120 He had come to dislike the paraphernalia of refugeedom—votive masses for the unlikely deliverance of the homeland. **1968** *Guardian* 20 Sept. 9/3 Today, the possibilities of social refugeedom are just as great..but few if any parents are worth running away from.

refugium (rĭfiū·dʒiɒm). *Biol.* Pl. -ia. [a. L. *refugium* place of refuge (see Refuge *sb.*).] A refuge (sense *3 e), *spec.* one in which a species survived a period of glaciation.

1955 *Canad. Jrnl. Bot.* XXXIII. 442 (*heading*) Nature of the evidence in support of glacial refugia. 1967 M. E. Hale *Biol. Lichens* ix. 142 It occurs as a disjunct in the celebrated unglaciated refugium of south-western Wisconsin. 1976 *Islander* (Victoria, B.C.) 16 May 7/1 The presence of endemic mammals on the Island suggests that suitable survival areas (refugia) existed there during the peaks of the last two glaciations.

refund, *v.*[1] **1. c.** For † *Obs.* read *rare* and add later example.

1920 A. S. Pringle-Pattison *Idea of God* i. 9 If any one prefers to use the term universe for the sum of created or dependent beings, he may, of course, refund the universe into God as its creative source.

refuse, *sb.*[2] Add: **2.** *refuse bin, can, cart, collection, collector, disposal, heap, sack, tip, tipping.*

1959 J. Kirkup tr. *de Beauvoir's Mem. Dutiful Daughter* III. 212 In the evenings, there would be the refuse bin to empty. 1976 'W. Trevor' *Children of Dynmouth* i. 14 The wind..rattled the refuse-bins on the ornamental lamp-posts. 1955 W. Gaddis *Recognitions* II. v. 539, I knew it, said Mr. Sinisterra, standing behind a refuse can. 1974 J. Wainwright *Hard Hit* 33 Along the street, the refuse cart is collecting the empties. 1945 *Listener* 12 July 35/1 For three months now there has been no refuse collection of any kind [in Berlin]. 1974 *Listener* 19 Sept. 368/3 There's a restaurant in Kensington Park Road. It needs six refuse collections a week. 1958 *Daily Mail* 25 Oct. 5/2 It happened to the rat-catcher (*he's now a rodent operator*), the dustman (*refuse collector*), and the sweeper (*street orderly*). 1976 Botham & Donnelly *Valentino* iii. 24 A string of unskilled jobs. Messenger, refuse collector, dishwasher and laundry assistant. 1906 *Westm. Gaz.* 10 Jan. 2/3 Owing to the narrow limits of Manhattan Island the problem of refuse-disposal is far more difficult in New York than in any other great city in the world. 1972 *Country Life* 28 Dec. 1790/2 Whereas refuse disposal will be a county function, refuse collection will be that of the district. 1816 W. Phillips in *Trans. Geol. Soc.* III. 112 In 1805, I noticed some crystals of the oxyd of uranium on the refuse heaps of Tin Croft mine. 1921 R. A. S. Macalister *Text-bk. European Archaeol.* I. x. 556 Most Danish archaeologists..call these remains *affaldsdynger* (refuse-heaps) or *skaldynger* (shell-heaps). 1972 *Police Rev.* 1 Dec. 1557/2 Does he, at the moment of picking up your dustbin or refuse-sack, become the owner..of its contents? 1977 *Cornish Times* 19 Aug. 7/2 (Publ. Notice) Refuse sacks will be provided for premises which would normally receive a collection on these days. 1969 M. Pugh *Last Place Left* xviii. 132 We reached a quarry where we buried Nell's craft lightly in the refuse tip. 1981 *Observer* 17 May 3/3 Refuse tips are probably the richest wildlife refuges in cities. 1974 *Country Life* 3 Oct. 940/3 This valley..one of the loveliest in the south west..is threatened by refuse tipping.

refusenik (rɪfiū·znik). Also **refusnik.** [Partial tr. Russ. *otkáznik*, f. stem of *otkazát'* to refuse: see *-NIK.] A Jew in the Soviet Union who has been refused permission to emigrate to Israel.

1975 *Nature* 31 Jan. 297/2 If, as is often the case with scientists, the initial application is rejected, one may spend months or years as a 'refusnik', with neither the opportunity nor the necessary time to keep up one's reading or think about one's own research. 1976 *Listener* 26 Aug. 237/1 Hundreds of people all over Britain make regular telephone calls to refuseniks every week. 1978 *Daily Tel.* 19 Dec. 11/4 The couple..have recently been putting on a satirical show ..mainly to keep up the morale of their 'refusenik' friends. 1980 *Jewish Chron.* 18 July 18/1 The dissidents languishing in exile, in prison camps and insane asylums, and the refuseniks cut off from family and friends and..from their sources of livelihood. 1980 *Radio Times* 29 Oct. 63/4 Tonight Avital talks about her life since she left Russia, a life of waiting and campaigning to free her husband and other Jewish refusniks from jail in the USSR.

refusing, *vbl. sb.* (Later example.)

1901 *Chambers's Jrnl.* Sept. 663/2 He will, if not instantly checked, learn a lot of bad tricks, such as..slipping his head-collar at night, and 'refusing' in the hunting-field.

refusnik, var. *REFUSENIK.

refutability. For *rare*—[1] read *rare* and add later example.

1957 C. A. Mace *Brit. Philos. in Mid-Century* 160 One can sum up all this by saying that falsifiability, or refutability, is a criterion of the scientific status of a theory.

refute, *v.* Add: ¶ **5.** *trans.* Sometimes used erroneously to mean 'deny, repudiate'.

1964 C. Barber *Ling. Change Present-Day Eng.* v. 118 For people who still use the word in its older sense it is rather shocking to hear on the B.B.C., which has a reputation for political impartiality, a news-report that Politician A has *refuted* the arguments of Politician B. 1978 *Observer* 7 May 4/9 Mr O'Brien, who was first elected general secretary three years ago, refutes the allegations. 1979 *Daily Mail* 17 Feb. 15/3 He refuted allegations that she took her own life because of police harassment. 1980 *Bookseller* 19 July 257/1, I refute Mr Bodey's allegation that it is our policy not to observe publication dates, and to display new titles in newsagents immediately on receipt from the publisher.

Reg.[1], abbrev. of *REGINA.

1792 W. Boscawen *Treatise on Convictions on Penal Statutes* 48 *Reg.* v. *Matthews.* 1848 E. W. Cox *Reports Cases in Criminal Law* II. 422 *Reg.* v. *Hawkes*..settles that question, about which some doubt had been previously entertained. 1976 *Law Rep. Queen's Bench Div.* 417 (*heading*) Reg. v. Michael (Crown Ct.).

reg[2] (reg). *Physical Geogr.* [N. African Arab.] A flat area of desert covered with gravel or boulders; stony desert.

1904 A. Knox *Gloss. Geogr. & Topogr. Terms* 324 *Reg*, firm level ground, generally without vegetation, a barren, naked plain. 1926 *Chambers's Jrnl.* June 341/1 Beyond the harbour,..away to the east, lies open stony 'Reg', and thence the vast, empty desert. 1963 D. W. & E. E. Humphries tr. *Termier's Erosion & Sedimentation* ii. 38 *Regs* and *serirs* are planed areas with a covering of boulders, which tumble from the surface of the hamadas or from the plains below them. The term reg is generally reserved for the low plains used by caravans. Moreover, this term is applied commonly to all bouldery ground which has been subjected to deflation. 1966 M. Woodhouse *Tree Frog* xxvi. 196, I was somewhere near the edge of a *reg*, one of those huge flat plains of gravel which are, more than anything, the true desert. 1976 L. Deighton *Twinkle, twinkle, Little Spy* xxiii. 226 The going changed to the gravelly surface of the 'reg' and then to rough 'washboard'.

reg[3] (reg), colloq. abbrev. of Regulation.

1952 M. Shaara in *Mag. of Fantasy & Sci. Fiction* Oct. 4 Wisher had decided..to follow the regs without question. For without the regs, the Mapping Command was a death trap. 1971 J. Sangster *Your Friendly Neighbourhood Death Pedlar* iv. 86 I'm sorry I can't do what you ask. Company regs. 1977 *Hot Car* Oct. 53/2 In Germany it will possibly do well because of their strict regs about modding a car.

regai·n, *sb.* [f. the vb.] **1.** An act of regaining; recovery. Also, an amount regained or recovered.

1927 *Observer* 2 Oct. 19/5 Take into consideration..wages cost, depreciation and interest on working capital, general expenses, discount, regain, and waste. 1927 *Morning Post* 4 Oct. 4/4 Progress of time will see..a regain of position of the horse in the ranks of industry.

2. The weight of moisture in a textile fibre or fabric expressed as a proportion of the weight of the material when thoroughly dry.

1904 J. M. Matthews *Textile Fibres* iii. 46 The amount of normal wool is obtained by adding to the dry weight of the wool the amount of moisture supposed to be present in the air-dried material under normal conditions of humidity and temperature. The added amount is termed 'regain', and is officially fixed by the conditioning house. The permissible percentage of regain varies with the form of the manufactured wool. 1941 *Nature* 4 Oct. 408/2 Dry wool shows strong polarization under an applied potential, but as its regain is increased its conductivity increases exponentially. 1961 A. F. W. Coulson *Man. Cotton Spinning* II. i. ii. 285 The changes in moisture regain depend on temperature.

regal, *a.* and *sb.*[1] Add: **A.** *adj.* **4.** *regal lily* = *REGALE *sb.*[1] 4; *regal pelargonium*, a house plant belonging to a group of varieties of *Pelargonium × domesticum*, flowering in spring and early summer.

1925 E. H. Wilson *Lilies E. Asia* 38 The bulbs of the Regal Lily are often part yellow-brown or orange-coloured. 1939 D. T. Macfie *Lilies* vii. 102 It is difficult to be moderate in the choice of words when talking of the regal lily. 1980 *Observer* 4 May 44/7 Charles Lyte..discovered in his research the extraordinary exploits of the men who introduced delights like the..Regal Lily into Britain. 1903 T. W. Sanders *Amateur's Greenhouse* 311 Pelargonium.. Decorative and Regal Kinds. These are grown in great quantities for Covent Garden Market. 1951 J. E. Cross *Bk. Geranium* xii. 104 The plant which still bears the name Pelargonium in commerce is *Pelargonium domesticum*, known in England generally as Show or Regal Pelargonium. 1955 *Times* 26 May 12/3 At Chelsea this year are several large groups of pelargoniums, both the zonal forms for outdoor bedding and the regal types for the greenhouse. 1962 R. Page *Educ. Gardener* ix. 260 Pots of palest pink regal pelargoniums..line the steps. 1966 Rochford & Gorer *Rochford Bk. Flowering Pot Plants* viii. 116 The Regal Pelargoniums..are generally known simply as Pelargoniums.

5. *regal* (*walnut*) *moth*, a large brown and yellow moth, *Citheronia regalis*, found in the eastern United States.

1854 E. Emmons *Agric. N.Y.* V. 238 Regal Walnut-moth..feeds on the walnut. 1887 S. W. Denton *Pages from Naturalist's Diary* (1949) 121, I have..caught eight sphinx moths, and many others, one like the regal walnut moth. 1912 *Country Life in Amer.* 1 Aug. 38 The blue horned hickory devil..turns into the Regal moth. 1972 Swan & Papp *Common Insects N. Amer.* xix. 270 Regal moth..Also known as hickory horned moth and royal walnut moth.

B. *sb.* **3. a.** (Later Hist. example.)

1905 R. H. Benson *King's Achievement* III. xi. 482 He noticed for a moment a wonderful red stone on the thumb, and recognized it. It was the Regal of France that he had seen years before at his visit to St. Thomas's shrine at Canterbury.

regal, *sb.*[2] **2.** (Later examples.)

1944 W. Apel *Harvard Dict. Mus.* 633/2 The reed stops of the later organs are frequently called 'regal'. 1976 *Gramophone* Nov. 837/2 This is instanced by his almost spooky use of the 16-foot regal from the top manual coupled to the pedals.

regale, *sb.*[1] Add: **4.** The specific epithet of *Lilium regale*, used to designate a fragrant, white-flowered lily of the species so called, which was discovered in China by E. H. Wilson in 1903 and named by him in 1912 (*Horticulture* XVI. 110). Also *attrib.*

1935 Woodcock & Coutts *Lilies* i. 3 Then came the epoch-making introduction of that choice representative of the genus, justly called *regale*. 1949 H. Nicolson *Let.* 15 June (1968) 171 Out of this jungle growth I wish regale to rise. I know it means keeping regale seeds each year. 1962 R. Page *Educ. Gardener* viii. 236 The regale lilies open their cream-pink trumpets. 1963 W. Blunt *Of Flowers & Village* 174 The Madonna lily is to the Regale lily as is the Parthenon to the Mansion House. *Ibid.*, Regale smells like a Bond Street hairdresser's.

regale, *sb.*[2] Add: **1. a.** (Later examples.)

1897 E. Coues *New Light Early Hist. Greater Northwest* I. 8 All were merry over their favorite regale, which is always given on their departure, and generally enjoyed at this spot, where we have a delightful meadow to pitch our tents, and plenty of elbow-room for the men's antics. 1922 E. R. Eddison *Worm Ouroboros* xxxiii. 429 That night was supper set in Lord Juss's private chamber: a light regale, yet most sumptuous.

b. (Earlier example.)

1791 F. Burney *Jrnl.* Aug. (1972) I. 46 There was a grand regale of sweetmeats, fruits, & cakes.

regalia[1]. Add: **2.** Also erron. as sing.

1953 *Times* 29 May 15/4 The regalia which will be used at the Coronation..is that which is normally used at the Coronations of our kings.

regalia[2]. (Earlier example.)

1819 H. Busk *Dessert* 379 Amber ginseng, and purified eringoes, Regalia's, and imperial's, and *maringoe's*.

regality[1]. **1. b.** For † *Obs. rare*—[1] read *rare* and add later examples.

1966 *New Statesman* 22 July 140/2 Her firmness is deeply satisfying and, in the final act, that excessive regality was lost in the gentle girlhood..which she can present when she chooses. 1979 *Daily Tel.* 4 Dec. 15/1 She is a narrator whose regality, though it is all natural style and never affectation, proves oddly inhibiting to those she interviews.

regard, *sb.* Add: **II. 10. c.** (Later examples.)

1847 Dickens *Dombey* (1848) xxvi. 265 'Your regards, Edith, my dear?' said Mrs Skewton, pausing, pen in hand, at the postscript. 1978 W. J. Burley *Wycliffe & Scapegoat* ix. 160 Give my regards to your father and tell him not to worry.

IV. 16. *attrib.*, as **regard ring** (see quots.).

1889 in *Cent. Dict.* 1890 W. Jones *Finger-Ring Lore* viii. 414 'Regard rings', of French origin, were common even to a late period, and were thus named from the initials with which they were set forming the acrostic of these words: Ruby Emerald Garnet Amethyst Ruby Diamond Lapis lazuli Opal Verd antique Emerald. 1912 O. M. Dalton *Franks Bequest Catal. Finger Rings* 302 A 'regard ring', so called from the fact that the initial letters of the gems composing the bezel form that word. 1951 M. Flower *Victorian Jewellery* 253 *Regard ring*, a ring set with a row of small stones of different kinds, the initial letters of which spell a word. 1973 *Country Life* 29 Nov. (Suppl.) 56/1 A tiny antique 'Regard' ring, the word spelt by the first letters of the stones: ruby, emerald, garnet, amethyst, ruby and diamond. 1978 *Illustr. London News* Nov. 129/2 (*caption*) Early Victorian 'regard' ring, £140.

regard, *v.* Add: **I. 7. c.** Also, *as regarding.*

1884 Browning *Ferishtah's Fancies* 111, I am in motion, and all things beside That circle round my passage through their midst,—Motionless, these are, as regarding me.

regardless, *a.* Add: **1. c.** *ellipt.* (passing into *adv.*) for 'regardless of expense' or 'regardless of consequences', used postpositively. orig. *U.S.* Phr. *to press on regardless*: see *PRESS *v.*[1] 15 a.

1872 'Mark Twain' *Roughing It* xlvii. 334 We are going to get the thing [*sc.* a funeral] up regardless, you know. 1896 *Advance* (Chicago) 30 July 150 Miss Bond got herself up regardless, and came in resplendent in ruby velvet and white swansdown. 1898 J. D. Brayshaw *Slum Silhouettes* 40 Who do yer think is down 'ere, got up regardless? D'Arcy's mash, Daisy Chapman. 1911 H. Quick *Yellowstone Nights* xi. 289 We got a bulletin from his doctors and messages from him to rush S.F. 41144 to its passage, regardless, or he'd accept a bid he'd got for the Bottle Imp. 1920 W. J. Locke *House of Baltazar* xvii. 205 I've a jolly good mind to set him up regardless, like a pre-war nut. 1928 T. E. Lawrence *Let.* 16 Apr. (1938) 587, I thought some plutocrat publisher was backing you, regardless: the price he offered was so fantastic. 1928 E. O'Neill *Strange Interlude* II. 58 *Evans.* (*Blundering on regardless now.*) I know it's hardly the proper time—. 1940 M. Dickens *Mariana* iii. 63 'It's a shilling. Is that too much?' He laughed at her. 'I told you—we're dining out regardless to-night.' 1960 *Observer* 24 July 17/6 What a marvellous feeling when you find the boat is sailing on regardless. 1962 *Listener* 19 Apr. 687/2 The microphone picks everything up, and transmits it regardless. 1980 J. Scott *Gospel Lamb* vii. 103 San fairy anne, as the Frogs used to say. He was clobbered, regardless.

regasify (rigæ·sifəi), *v.* Also **re-gasify.** [Re- 5 b.] *trans.* To convert back into gas.

1926 [implied in regasified *ppl. a.* below]. 1940 *Gas Age* 24 Oct. 47/1 The liquefied natural gas would be stored in special insulated tanks, to be withdrawn..and regasified when need arose. 1946 *Nature* 20 July 105/2 The methane would be extracted and stored as liquid at periods of low demand, and re-gasified to enrich water gas at periods of high demand. 1967 *Sci. Amer.* Oct. 32/1 Natural gas is liquefied, stored and regasified for use in a city on days of peak demand.

So **rega·sified** *ppl. a.*; also **regasifica·tion,** the action or process of regasifying.

1926 R. W. Lawson tr. *Hevesy & Paneth's Man. Radioactivity* xxiii. 167 The uncondensed impurities (nitrogen, hydrogen, rare gases) are then pumped off, and after removal of the liquid air, the regasified emanation is allowed to stream into the vessel in which it is to be used. 1940 *Oil & Gas Jrnl.* Oct. 51/1 We began an intensive study of natural-gas liquefaction, its storage and subsequent regasification. 1967 *Sci. Amer.* Oct. 32/1 Regasification involves heating liquid natural gas (*LNG*), which is at a temperature of −259 degrees Fahrenheit, so that it vaporizes and is raised to a safe temperature of 40 degrees F. 1978 *Liquefied Natural Gas* (Shell Internat. Petroleum Co.) 8 The heat required for re-gasification can be taken from a variety of sources—sea water, power station effluent or gas-fired heaters.

regathering, *vbl. sb.* (Later example.)

1955 *Times* 9 Aug. 10/2 They are kept together by a rule of life, by monthly meetings on the mainland, and a 10-day regathering on the island each June.

regatta. Add: **3.** A cotton fabric, usu. made in twill; a striped garment made in this fabric. Also *attrib.*, esp. as *regatta shirt*, an informal light-weight striped shirt.

1861 T. Baines *Jrnl.* 21 July in *Explorations S.-W. Afr.* (1864) 59, I gave Hendrick a couple of regatta shirts and elastic braces. 1910 *Westm. Gaz.* 7 Feb. 5/5 (Advt.), Woven Ginghams, Zephyrs, Regattas, and heavier grades, good for tub frocks. 1910 *Encycl. Brit.* VII. 278/1 *Regatta* is a stout, coloured shirt cloth similar in make to a jeanette. It was originally made in blue and white stripes and was used largely and is still used for men's shirts. 1962 *B.S.I. News* Jan. 8/2 The cloths are coloured-woven regatta fabrics, one of which is all cotton and the other a blend of cotton with one-third rayon staple. *Ibid.* 9/1 The blend has been evolved more recently; requirements for it have been based on regattas submitted to comprehensive trials by hospitals and local authorities. 1963 *New Yorker* 1 June 115 *These* shirts..in regatta, sky blue..navy or red. 1972 *Canad. Antiques Collector* Nov.–Dec. 10/1 The Langley records list substantial quantities of Chesterfields, frock coats, Scotch tweeds,..regatta shirts.

Régence[2] (reʒaŋs). [Fr.: see Regence[1].] Used, chiefly *attrib.*, to designate the style of costume, furniture, and interior decoration, characteristic of the first third of the eighteenth century in France, during which occurred the French Regency (see Regency 5).

1919 H. Nicolson *Let.* 1 Feb. in J. Lees-Milne *H. Nicolson* (1980) vii. 114 A huge Régence writing table. 1930 R. Cuthill tr. *J. Schober's Silk & Silk Industry* iv. 272 *Regeance* [sic], a Jacquard fabric with small designs and also stripes in the warp direction; usually in striking colours. 1930 *Morning Post* 18 July 14/5 A Regence settee and four fauteuils. 1963 *House & Garden* Mar. 79/2 The term Régence ..is used for the period 1710 to 1730. 1968 *Ibid.* May 47 (*caption*) The sitting-room, showing one of the bergères, Régence overmantel and Austrian clock. 1973 *Country Life* 10 May 1307/1 Slender *Régence* ribbonwork displaced the more plastic forms of the Baroque. 1977 *Times* 16 June 18/6 An ormolu-mounted Régence commode at 53,000 francs..or £6,235.

regency. Add: **3. b.** (Later examples.) Also in *transf.* or extended senses: a group that manages or administers in the absence of a manager or political leader; an inside group or clique controlling affairs.

1940 *Sun* (Baltimore) 13 Aug. 15/5 Secretary Knox named a three-man 'regency' to publish and edit the *News* while he holds his Cabinet post. 1956 *New Republic* 9 Jan. 2/3 There's some evidence the Administration 'Regency' is using Ike's absence to even old scores. 1963 *Daily Tel.* 12 Jan. 15/3 Leading members of the Parliamentary Labour party have been angered by a suggestion..that a 'Regency' has been set up to rule in the absence of Mr. Gaitskell.

7. a. Regency point, a kind of lace (see quots.); also *attrib.*

1865 F. B. Palliser *Hist. Lace* xxx. 364 A 'point' lace, with the 'cloth' or 'toilé' on the edge, for many years was in fashion, and in compliment to the Prince, was named..'Regency Point'. It was a durable and handsome lace. 1900 E. Jackson *Hist. Hand-Made Lace* 193 Regency Point Lace. This lace, made in Bedfordshire, was in great demand during the Regency early in the nineteenth century. The edge is thick, the ground, a complicated réseau, or hand-made mesh. 1930 T. Wright *Romance of Lace Pillow* xv. 219 During the Regency (1810–1820) there was made in Northamptonshire a striking lace, with fillings of a bold character, which was called Regency Point.

b. Passing into *adj.* Applied to styles of architecture, clothing, furniture, etc., characteristic of the English Regency (sense 5 in Dict.), and, more generally, the late eighteenth and early nineteenth century.

The isolated early use in quot. *c* 1793 may possibly reflect the public controversy surrounding the Regency Bill of 1788.

c 1793 Jane Austen *Volume Third* in *Minor Works* (1954) 211 'She sends me a long account of the new Regency walking dress Lady Susan has given her... She says nothing indeed except about the Regency.' 'She *must* write well thought Kitty, to make a long letter upon a Bonnet & Pelisse.' 1880 E. Glaisher *Needlework* viii. 86 Shield-backed chairs of the Regency fashion. 1887 Regency cap [in Dict.]. 1909 *Cent. Dict.* Suppl., Regency style. 1918 A. Bennett *Pretty Lady* vi. 25 He had furnished his flat in the Regency style of the first decade of the nineteenth century. *Ibid.* vii. 33 Regency furniture and china. 1936 *Discovery* Oct. 321/2 The short 'Brutus' curls of regency mock-classical beauties. 1938 C. Day Lewis *Overtures to Death* 18 We gaze At a Regency terrace, curved Like the ritual smile. 1954 J. Betjeman *Few Late Chrysanthemums* 43, I pulled aside the thick magenta curtains—So Regency, so Regency, my dear—. 1958 —— *Coll. Poems* 250 It's for Regency now I'm enthusing So we've Regency stripes on the wall. 1963 N. Freeling *Because of Cats* iii. 47 There were Regency-striped silk cushions. 1973 P. Moyes *Curious Affair of Third Dog* v. 57 The Regency-stripe wall-paper. 1976 *Denbighshire Free Press* 8 Dec. 16/7 (Advt.), Curtains... Regency stripe/gold, 2 width each side, length 53 in. £20. 1977 C. McCullough *Thorn Birds* iii. 64 The Louis Quinze sofa and chairs, the Regency escritoire. 1977 *S. Wales Echo* 18 Jan. (Advt.), Only four remaining on a small development of just 10 Regency style Detached Houses.

regenerable (rɪdʒe·nĕrăb'l), *a.* [f. Regener(ate *v.* + -able.] Capable of being regenerated.

1920 in Webster. 1927 A. Kocourek *Jural Relations* x. 140 Regenerable relations.. are mesonomic relations which are convertible into zygnomic relations by means of a jural act or event. 1976 *Nature* 24 June 660/1 Microparticles were needed in the thermally regenerable ion-exchange process to ensure rapid rates of ion-exchange.

regenerant, *a.* Restrict *rare*—[1] to sense in Dict. and add: **B.** *sb.* A regenerating agent.

1961 in Webster. 1963 *Engineering* 13 Sept. 338/3 The regenerant for the anion column can be either sodium carbonate or sodium hydroxide.

regenerate, *v.* Add: **2.** Also in *Biol.* (Further examples.)

1895 *Arch. für Entwicklungsmech. der Organismen* II. 122 The power of an adult animal to regenerate lost parts. 1909 R. H. Lock *Rec. Progress Study of Variation* iii. 61 The power of regenerating a lost part must clearly often be of service to the creatures which possess it. 1959 [see *regeneration 3]. 1961 Lenhoff & Loomis *Biol. Hydra* 409 Hydra treated with $10^{-4}M$ lipoic acid for short periods immediately after removal of their hypostomes and tentacles completely lose the capacity to regenerate those structures. 1970 Ambrose & Easty *Cell Biol.* i. 21 The capacity to regenerate certain tissues, possessed by most embryonic animals, is still present in some mature animals and plants.

3. (Later example.)

1962 *Which?* Oct. 294/1 After a time, the resin [in a water softener] has no sodium left, and has to be 'regenerated' by adding sodium chloride.

c. *Chem.* and *Textiles.* To re-precipitate (a natural polymeric substance, as cellulose, proteins) following chemical processing, esp. in the form of fibres; to make (fibres) in this way. Cf. *regenerated *ppl. a.* 2.

1925 *U.S. Patent 1,528,219*, I..have invented certain new and useful Improvements in a Process of Regenerating Cellulose from Viscose. 1948 J. T. Marsh *Textile Sci.* i. 8 It has not been possible to regenerate fibres from wool, but successful attempts have been made with silk. 1950 R. W. Moncrieff *Artificial Fibres* 90 Some of the sodium cellulose xanthate decomposes, regenerating cellulose which is maintained in emulsion form by that part of the sodium cellulose xanthate which is still undecomposed. 1955 Cockett & Hilton *Basic Chem. of Textile Preparation* iv. 82 Attempts have been made to regenerate both silk and wool in which the protein raw material is in a linear or near linear form. 1972 M. A. Taylor *Technol. of Textile Properties* 30 Azlon is the generic term given..to fibres regenerated from natural protein, such as casein from milk. 1973 *Materials & Technol.* VI. iv. 277 The extruded filaments were injected into a bath of dilute sulphuric acid to re-precipitate, or 'regenerate', the original cellulose and form textile threads.

5. a. Also in *Biol.* (Further examples.)

1901 T. H. Morgan *Regeneration* i. 20 A piece of hydra regenerates without the formation of new material. 1928 J. S. Huxley *Ess. Pop. Sci.* 251 When small pieces of a planarian regenerate, they exhibit what we may call polarity. 1971 [see *regulate *v.* 5].

regenerate, *ppl. a.* and *sb.* Add: **A.** *ppl. a.* **4.** *Biol.* Formed or modified by regeneration.

1952 *Q. Rev. Biol.* XXVII. 169/2 Intimacy of morphological relation between the regenerate and the adult tissue has demanded that study of the process of regeneration be made against the background of the anatomy and physiology of adult tissues.

B. *sb.* Restrict † *Obs.* to sense in Dict. and add: **2.** *Biol.* A limb or other part formed by regeneration.

1952 *Q. Rev. Biol.* XXVII. 169/2 The histology of the regenerate emphasizes the continuity and interrelation between adult and regenerating tissue. *Ibid.*, The regenerate..is nourished by the adult blood stream. 1960 [see *dedifferentiation]. 1964 [see *melanogenesis* s.v. *melano-]. 1977 *Sci. Amer.* July 69/3 A graft between a proximal level of a host cockroach leg and a distal level of a donor leg gives rise to a normally oriented intercalary regenerate and forms a normal leg segment.

regenerated, *ppl. a.* Add: **2.** *Chem.* and *Textiles.* Of natural polymeric materials (as cellulose, proteins): re-precipitated (esp. in the form of fibres) following chemical treatment. Of fibres: prepared from a substance in this way.

1904 *Jrnl. Soc. Chem. Industry* 29 Feb. 177/1 The next operation..is one for the purpose of denitrating the cellulose, in order that the fibre may ultimately consist of what might be termed 'regenerated' cellulose. 1933 *Trans. Faraday Soc.* XXIX. 230 Regenerated celluloses are more highly activated than cotton which has merely been swollen. 1941 *Thorpe's Dict. Appl. Chem.* (ed. 4) V. 121/2 Viscose and regenerated cellulose rayons in general show a much higher moisture adsorption at all relative humidities than do native cellulose fibres. 1948 J. T. Marsh *Textile Sci.* i. 8 Regenerated fibres may be classified according to the nature of the parent material, cellulose or protein. 1963 A J. Hall *Textile Sci.* ii. 66 Several types of protein are satisfactory for the production of regenerated protein fibres and among those most used are casein..and the natural proteins..extracted from groundnuts and soya beans. 1964 N. G. Clark *Mod. Org. Chem.* xvii. 363 Regenerated fibres are derived from naturally occurring fibrous material by first converting it into a soluble derivative, forcing a solution of this through a minute jet to give a 'thread' of solution, and finally recovering the original or its derivative from solution as a solid thread or fibre. 1973 *Materials & Technol.* VI. iv. 277 Wool and silk are both protein fibres, and it is not surprising that attempts have been made to produce regenerated protein fibres.

regeneration. Add: **1. c.** *Forestry.* The natural regrowth of a forest which has been felled or thinned. Freq. *attrib.*

1888 E. E. Fernandez *Man. Indian Sylviculture* i. 6 The name regrowth will be specially given to the new crop obtained by coppice regeneration. 1909 P. T. Maw *Pract. Forestry* ix. 183 A Seed Felling or Regeneration Felling is made when a good seed year has come. As its name implies, it is the felling made for the actual regeneration of the area. It consists in the removal of all the trees except a few, which are left as mother trees, to seed the whole area; and also, to form a light canopy or shelter wood for the young crop. 1928 R. S. Troup *Silvicultural Syst.* iv. 38 Where regeneration is sufficiently well advanced it should be freed from overhead cover. *Ibid.* 45 The latter will be retained as regeneration areas under the revised working plan. 1979 *Biol. Abstr.* LXVII. 2139/2 The conditions of regeneration and growth of seedlings varied with stands and plots.

3. (Further examples.) Also in *Biol.*

1901 T. H. Morgan *Regeneration* i. 23 The word 'regeneration' has come to mean, in general usage, not only the replacement of a lost part, but also the development of a new, whole organism, or even a part of an organism, from a piece of an adult, or of an embryo, or of an egg. 1959 W. Andrew *Textbk. Compar. Histol.* xii. 478 While asexual reproduction is not common among the echinoderms, a high power of regeneration is present and a single arm can regenerate a whole starfish. 1978 *Nature* 27 July 374/1 Can re-innervation take place not only by regeneration of the original axons but also by collateral sprouting of..undamaged fibres?

4. *Electronics.* Positive feedback (see *feedback, feed-back *sb.* a).

1922 *Proc. IRE* X. 244 The effect of regeneration (that is, the supplying of energy to a circuit to reinforce the oscillations existing therein) is equivalent to introducing a negative resistance reaction in the circuit. 1943 [see *reaction 3 e]. 1957 *Practical Wireless* XXXIII. 694/1 The hole is used..in the next stage for a regeneration control. 1969 J. J. Sparkes *Transistor Switching* iii. 74 Regeneration will only commence provided the loop gain A_v..is greater than 1.

5. *Chem.* and *Textiles.* The action or process of regenerating fibres, etc. Cf. *regenerate *v.* 3 c.

1925 *U.S. Patent 1,528,219* The regeneration of cellulose from viscose solutions. 1950 R. W. Moncrieff *Artificial Fibres* ii. 20 The original cotton had consisted of short, hairy, nearly opaque fibres, and after regeneration it consisted of very long, smooth, transparent filaments, but it was still the same essential material. 1953 *Chem. Abstr.* XLVII. 643 (*heading*) Regeneration of egg albumins under pressure. 1972 M. A. Taylor *Technol. Textile Properties* 28 The regeneration and polymerisation of the cellulose occurs after the filament has been coagulated and stretched.

regenerative, *a.* (and *sb.*) Add: **2. b.** Applied to a principle or technique of refrigeration by which the uncooled portion of the working fluid loses some heat prior to the major cooling step by exchange with the cooled portion.

1896 *Proc. Chem. Soc.* XI. 222 In all continuously working circuits of liquid gases used in refrigerating apparatus the regenerative principle applied to cold first introduced by Siemens in 1857..has been adopted. *Ibid.* 231 If..hydrogen, previously cooled by a bath of boiling air, is allowed to expand at 200 atmos. over a regenerative coil..a liquid jet can be seen. 1922 Glazebrook *Dict. Appl. Physics* I. 565/1 The usual process [for the commercial liquefaction of gases] is a regenerative one, first successfully developed by Linde, in which the Joule-Thomson effect of irreversible expansion in passing a constrictive orifice..serves as the step-down in temperature, and a cumulative cooling is produced by causing the gas which has suffered this step-down to take up heat in a thermal interchanger from another portion of gas that is on its way to the orifice. 1961 F. E. Hoare et al. *Exper. Cryophysics* i. 4 Dewar was employing regenerative cooling to produce a jet of cold hydrogen gas which could be used for cooling other systems.

c. *Astronautics.* Applied to a method of cooling the walls of a rocket engine by circulating the fuel through them.

1947 *Amer. Jrnl. Physics* XV. 131/2 In the motor, between 2 and 3 percent of the heat due to combustion passes through the chamber and nozzle walls into the coolant, which returns again to the combustion chamber when regenerative cooling is utilized. 1949 G. P. Sutton *Rocket Propulsion Elements* vi. 142 In regenerative cooling the motor parts are cooled by means of a built-in jacket or cooling coil in which the oxidizer or the fuel are used as the coolant fluid. 1962 F. I. Ordway et al. *Basic Astronautics* x. 413 This regenerative cooling method serves two purposes. It cools the walls of the thrust chamber and adds thermal energy to the propellant. 1974 *Encycl. Brit. Macropædia* XV. 938/2 The conventional method of cooling [in rocket engines] is known as regenerative cooling.

3. Applied to any method of braking in which energy is extracted from the parts braked, to be stored and re-used.

1904 *Electrical Mag.* I. 600/1 The regenerative braking action comes into play automatically. 1930 *Engineering* 6 June 722/2 Regenerative braking had been adopted on the majority of the electric locomotives recently placed in service. 1958 *Ibid.* 14 Mar. 340/1 A bus using the regenerative transmission... In this system, when the vehicle is braked..energy is absorbed in accelerating a..flywheel... Then when the vehicle is restarted the energy of the flywheel is used to accelerate it, resulting in a saving of fuel. 1973 *Sci. Amer.* Dec. 23/2 A regenerative braking system would employ the vehicle's electric motors as generators during braking or downhill driving, thus putting the kinetic energy of the vehicle back into the storage system.

4. *Electronics.* Pertaining to or employing positive feedback (see *feedback, feed-back *sb.* a); *regenerative feedback*, positive feedback.

1915 *Proc. IRE* III. 231 It is always better practice to use the cascade circuits for the radio frequencies, even if the regenerative circuits are not employed with each individual audion system. 1919 *Wireless World* Aug. 250/2 By using regenerative feed back much higher amplification can be realized, but the operation becomes less stable. 1922 *Sci. Amer.* Sept. 160/1 Armstrong's regenerative receiver, now so widely employed, is ever so much more sensitive than the ordinary vacuum tube receiver. 1947 R. Lee *Electronic Transformers & Circuits* ix. 254 The next pulse occurs when the negative grid voltage decreases sufficiently so that regenerative action starts again. 1969 J. J. Sparkes *Transistor Switching* iii. 59 The cross-coupling resistor..can be shunted by a capacitor..to speed up the regenerative switching of the circuit. 1971 *Physics Bull.* July 385/2 The high spectral intensity results from the fact that, since the laser is a regenerative oscillator, the oscillation linewidth decreases with increasing laser power—in contrast to the behaviour of any thermal source. 1975 G. J. King *Audio Handbk.* ii. 35 Positive feedback means that the phase of the signal fed back is coincident with the phase of the source or input signal. This is regenerative feedback which results in sustained oscillation.

regeneratively *adv.* (further examples).

1947 *Amer. Jrnl. Physics* XV. 131/1 The coolant liquid absorbs heat as it circulates in ducts around the motor and is then injected into the combustion chamber (regeneratively cooled type). 1949 G. P. Sutton *Rocket Propulsion Elements* vi. 142 The German Me 163 motor has a steel cooling jacket in which fuel cools the motor regeneratively. 1969 J. J. Sparkes *Transistor Switching* iii. 74 When T_1 is conducting, raising V_1 until it is about equal to V_{B_2} turns T_1 on so that T_1 is switched off regeneratively.

regenesis. (Later U.S. example.)

1973 *Black World* June 90/2 Sister [Sherley Anne] Williams breaks her [book [*sc. Give Birth to Brightness*] down into three major parts. Part 1 is called 'Regenesis'.

regent, *sb.* Add: **3. d.** (*a*) (Earlier and later examples.)

1813 *Niles' Reg.* V. 79/2 The regents of the university, expressly endeavored to effect this important object. 1969 *Morning Star* 13 Oct. 5/3 The Director of Afro-American Studies declares her sacking raises grave doubts about the Regents' desire to encourage black participation. 1976 *New Yorker* 26 Apr. 32/2 One of the Smithsonian's regents..is chairman of the House Appropriations Committee. 1977 *Detroit Free Press* 11 Dec. 11-D/1 Regents for Oklahoma State University Friday honoured Terry Miller by retiring his No. 43 football jersey and approving a commendation to be awarded at the next regents meeting.

4*. A variety of potato.

1846 [see *early *sb.* a]. 1868 M. Jewry *Warne's Model Cookery & Housekeeping Bk.* 14 Potatoes.—*We* think the best are..the regents for winter use. 1892 I. Zangwill *Childr. Ghetto* II. 6 'Kidneys or regents, my child?'..said Guedalyah the greengrocer. 1927 T. P. McIntosh *Potato* ii. 20 Not much appears to be known about Regent, which was a later introduction [*sc.* after 1836].

4.** A chairman of a branch of the Daughters of the American Revolution.

1890 *Constitution & Bye-Laws, Daughters Amer. Revolution* 4 When twelve or more members of the Society shall be living in one locality they may organize a Chapter. They may elect a presiding officer whose title will be Regent. 1928 *Harper's Mag.* Oct. 529/2 The Daughters upheld Mrs Brousseau and the contention of the Massachusetts State Regent. 1946 *Nat. Historical Mag.* Mar. 144/2 Please read over the foregoing statement again, Madam Regent. 1974 *Marlboro Herald-Advocate* (Bennettsville, S. Carolina) 18 Apr. 4/2 Mrs. Walter Hughes, local regent, also attended the Congress.

5. a. regent honeyeater, a bird, *Zanthomiza phrygia*, of the family Meliphagidæ, having black plumage with yellow bars and spots and found in the eucalyptus forests of south-east Australia.

1913 G. M. Mathews *List Birds Austral.* 270 *Zanthomiza phrygia phrygia.* Regent Honey-eater. 1967 A. Rutgers *Birds Austral.* 262 Regent Honey-eaters make a lot of noise and have a loud laughing call.

reggae (re·gēⁱ). Also **Reggae, Reggay.** [Origin unknown; perh. connected with Jamaican English *rege-rege* quarrel, row (in Cassidy and Le Page, *Dict. Jamaican Eng.* (1967) 380/1).] A kind of popular music, of Jamaican origin, characterized by a strongly accentuated off-beat and often a prominent bass; a dance or song set to this music. Also *attrib.*

1968 (*song-title*) Do the Reggay. 1969 *Daily Mirror* 10 Oct. 19/1 *Reggae*, West Indian music. 1969 *Observer* 23 Nov. 25/8 The visiting American executives ..dancing the Reggae, Jamaica's successor to the Ska. 1969 *Listener* 25 Dec. 905/2 A very dapper and jaunty Reggae group called the Pioneers. 1970 *Melody Maker* 3 Oct. 25/6 If I ever did reggae again, it would have to be darned good reggae, and there's not much of that around. It's such a blank type of music. 1971 *Advocate-News* (Barbados) 17 Sept. (Guyana Suppl.) p. vi (Advt.), A rum punch..served to an atmosphere of reggae, calypso and steelband music. 1973 G. Sims *Hunters Point* xiii. 120 I'm a reggae fan. West Indian music. And early Beatles like 'From Me to You'. 1973 *Black World* Jan. 77/2, I heard the Rastas credited with starting everything from the island's most popular dance, 'Reggae', to the embryonic Black Consciousness movement. 1975 *Globe & Mail* (Toronto) 16 July 7/3 The reggaes..should be viewed as songs of social protest in which the dispossessed describe their personal experience and comment upon the social injustice of the system. 1976 *Telegraph* (Brisbane) 28 Apr. 58/4 As reggae grows in popularity, ever more artists are performing material with a reggae flavour. 1977 McKnight & Tobler *Bob Marley* iii. 42 So we come to reggae, which the British initially found difficulty in pronouncing, let alone understanding. 1978 *Sunday Times* 29 Jan. 43/2 Althea and Donna met at a reggae festival in Ochos Rios. 1979 *Spectator* 1 Dec. 13/3 The bulk of the reggae-blacks were born here and yet feel themselves to be foreigners.

Regge (rē̍ⁱ·dʒe). *Nuclear Physics.* The name of T. E. *Regge* (b. 1931), Italian physicist, used *attrib.* to designate certain concepts in the theory of the scattering of sub-atomic particles, as **Regge pole,** a pole of a complex function relating the scattered amplitude of partial waves to angular momentum; **Regge trajectory,** a path traced in the complex angular momentum plane by a Regge pole as the energy varies; esp. a plot of spin against the square of the rest mass for a group of particles.

1961 *Physical Rev. Lett.* VII. 394/2 We may satisfy Feynman's principle therefore by postulating that all poles of the *S* matrix are of this type (Regge poles). 1962 *Ibid.* VIII. 41/2 Each point is supposed to lie on a Regge trajectory. 1962 *Physical Rev.* CXXVI. 2204/2 This perturbation theory behavior is very different from that of the Regge case. *Ibid.*, Strongly interacting particles may exhibit the Regge behavior. 1973 [see *pole *sb.*[2] 9*]. 1973 B. H. Bransden et al. *Fundamental Particles* viii. 163 (*caption*) The Regge trajectories of some meson states. Mesons differing in spin by one unit appear to lie on the same Regge trajectory. 1973 L. J. Tassie *Physics Elem. Particles* xii. 170 Most work on Regge theory is concerned with describing collision processes, and in this respect the Regge pole model is not a theory with a high predictive power. 1975 *Sci. Amer.* Feb. 62/3 The Regge trajectories turn out on observation to be nearly linear, meaning that the angular momentum of the particles on a particular trajectory is given to a good approximation by a linear function of the mass of the particle squared. 1977 P. D. B. Collins *Introd. Regge Theory* ii. 69 When such a Regge pole occurs for a physical integer value of *l* it will correspond to a physical particle or resonance. 1977 *Nature* 21 July 207/2 Hadrons on the same Regge trajectory have a remarkably simple relation between mass and angular momentum: $J = \alpha' M^2 + \alpha_0$, where *J* is the total angular momentum of the hadron, *M* is its mass and α' and α_0 are called the Regge 'slope' and 'intercept' respectively.

Reggeization (re:dʒeᵢəizēⁱ·ʃən). *Nuclear Physics.* [f. prec.: see -ization.] Treatment or modification in accordance with Regge theory.

1964 *Rev. Mod. Physics* XXXVI. 641/1 We have throughout considered the theory of spin ½ fermions, which as stated above shows the factoring property which is necessary for the success of the Reggeization procedure. 1975 *Physics Bull.* Jan. 25/2 Schnitzer's study of the Reggeization of non-abelian gauge theories is now seen to have been prophetic.

So **Re·ggeized** (also r-) *a.*

1971 N. Dombey in Cumming & Osborn *Hadronic Interactions of Electrons & Photons* ii. 37 Now assume instead that the pion is Reggeized; i.e. S_J has a moving pole at $J = \alpha_\pi(t)$. 1973 *Physics Bull.* Feb. 99/1 Reggeized baryon exchange models give poor quantitative agreement with the results. 1973 *Jrnl. Physics* A. VI. 506 A reggeized absorption model with no free parameters..is applied to spin-2⁺ production reactions.

Reggeon (re·dʒeɒn). *Nuclear Physics.* [f. as prec. + *-on[1].] (A particle represented by a) Regge pole or trajectory, or a virtual particle regarded as exchanged in the type of scattering they represent. Hence **reggeo·nic** *a.*

1964 *Physics Lett.* IX. 269/1 Mandelstam has given some arguments that moving branching points may appear in a relativistic theory as a result of singularities to the right in the *j*-plane for particles with spin. These new singularities correspond to the production threshold of resonance states (reggeons) with negative orbital momenta. *Ibid.* XII. 153/2 If this fact is correct it would modify the reggeonic branch points and the elastic scattering asymptotic amplitude. 1974 *Physics Bull.* May 206/4 High energy backward scattering is studied by (i) covariant reggeization techniques, (ii) the use of a super multiplet reggeon propagator model to generate polynomial residues. 1977 P. D. B. Collins *Introd. Regge Theory* ii. 71 The power behaviour expected from the exchange of a Regge trajectory (sometimes called 'Reggeon')..may be contrasted with that from a fixed-spin (elementary) particle. 1978 *Nature* 19 Jan. 214/2 Reggeons with arbitrarily high spin can be exchanged with equanimity.

regiculture (re·dʒikɒ:ltiŭɹ). *rare.* [f. L. *rēgi-, rex* king + Culture *sb.*] Honour or homage to kings.

1880 Swinburne in T. H. Ward *Eng. Poets* III. 281 For all her evil report among men on the score of passive obedience and regiculture.

‖ **regidor** (reχido·r). Pl. regidores, regidors. [Sp. *regidor* alderman, f. *regir* to rule.] In Spain and the former Spanish dominions in America, a member of a cabildo or municipal council; a councillor; a village official.

1622 J. Mabbe tr. *Aleman's Rogue* I. i. iii. 33 Thus it fared with a Regidor, who being espied by an old man,.. call'd him unto him. 1755 Smollett *Cervantes' Don Quixote* I. p. xiv, Dedicated to the alcaides, regidors, and gentlemen of the noble town of Argamasilla. 1834 A. Pike *Prose Sketches & Poems* 170 The Regidor, or Assistant Alcalde, Miguel Sena, has only perjured himself three times. 1848 E. Bryant *What I saw in California* xxii. 283 The first of these pueblos is governed by its corresponding body of magistrates, composed of an alcalde or judge, four regidores or municipal officers, a syndic and a secretary. 1895 G. E. King *New Orleans* vii. 115 Instead of a superior council, there was a cabildo, with regidores, alcaldes, [etc.]. 1934 *Hist. Soc. Southern California Publ.* XVI. 142 He was *regidor* of Los Angeles in 1838–39. 1950 G. Brenan *Face of Spain* vi. 143 He introduced himself as the *regidor* of the village municipality. 1969 *Femina* (Bombay) 26 Dec. 41/1 One of the labourers summoned the *regidor*, a village official, to the scene. 1974 *Encycl. Brit. Micropædia* II. 422/1 In local affairs, each municipality in Hispanic America was governed by its *cabildo*, or city council... Its members, *regidores* (councillors) and *alcaldes ordinarios* (magistrates), along with the local *corregidor* (royally appointed judge), enjoyed considerable prestige and power.

‖ **régie** (reʒi). Also with capital initial. [Fr., f. *régir* to rule.] In France and certain other countries: a government department that administers a state-controlled industry or service; formerly *esp.*, one responsible for taxation, customs and excise, etc.; a government monopoly used as a means of taxation, *esp.* the tobacco monopoly in the former Turkish Empire. Also *attrib.*

1791 Ld. Gower in *Despatches Earl Gower* (1885) 61 The 4th [article of a decree] allows tobacco in leaves to be stored, for a year, in the ware-houses of the *Régie*. 1879 *Encycl. Brit.* IX. 738/1 Unfortunately, he [*sc.* Frederick the Great] adopted the French ideas of excise, and the French methods of imposing and collecting taxes,—a system known as the Regie. 1883 *Pall Mall Gaz.* 9 May 5/1 The Turkish tobacco régie..is designed to include a company having the exclusive right of preparing tobacco for home consumption and of selling it to the public. 1884 *Ibid.* 5 Sept. 6/2 Ladies.. smoke the strong régie cigar with evident enjoyment. 1890 *Athenæum* 11 Oct. 474/3 All the frequenters of a country inn [in France]..consume the tobacco of the *régie*. 1923 *Glasgow Herald* 26 Feb. 10 The exploitation of the railways of the Ruhr and the Rhineland by a Franco-Belgian 'regie' is believed to have been decided. 1929 W. Ray tr. *Hegemann's Frederick the Great* 122 The King indeed was thoroughly well satisfied with his Régie escapades... The patient Prussians had barely two more years to wait before death came to rid them of their great king, the extortions of the French *Régie*, [etc.]. 1958 G. T. Matthews *Royal Gen. Farms 18th Cent. France* I. ii. 43 Prior to 1548 the various salt taxes and commercial concessions constituting the *grandes gabelles* were partly farmed to individual tax-farmers and partly managed by government *régies*. 1964 Ridley & Blondel *Public Administration in France* II. vii. 181 Finally we come to the revenue or, as the French call them, fiscal divisions. Until recently there were four more or less autonomous services (or *régies*). These had remained virtually unchanged since the Revolution and corresponded roughly to the main sources of state revenue: direct taxes, indirect taxes, customs duties, and registration fees, stamp duties and the national domain... After the war it was decided that the four *régies* should be transformed into two divisions of the ministry. *Ibid.* x. 239 Traditionally there were two ways of organizing a public service, the *régie* and the concession; the former operated by a government department or a local authority, the latter on contractual terms by private enterprise. 1977 S. J. & E. K. Shaw *Hist. Ottoman Empire & Mod. Turkey* II. iii. 233 In 1883 the Public Debt Commission turned the tobacco monopoly over to a private German-French company called the *Régie cointéressée de tabacs de l'Empire Ottoman*, which paid a fixed annual fee..in return and then divided the profits with the Ottoman treasury. The Régie had the sole right to buy and process all tobacco sold in the empire and regulate its cultivation... The tobacco..was stored in the Régie warehouses.

régime, regime. Add: **1.** (Later examples.)

1908 A. Bennett *Old Wives' Tale* I. iii. 45 She was a shrivelled little woman, capable of sitting twelve hours a day in a bedroom and thriving on the *régime*. 1943 *Ann. Allergy* I. 33 Others in whom the psychic element is important are nevertheless improved by a hygienic régime or by symptomatic medication. 1973 *Daily Tel.* 13 Feb. 16 This is not a diet to enter upon without medical prescription... To embark on this régime without due regard to the consequences may delay diagnosis of other disorders.

2. a. (Later examples.) Now freq. applied disparagingly to a particular government or administration.

1955 *Times* 2 May 8/3 But none of us is prepared, either, to bolster up the aging régime of Chiang Kai-shek. *Ibid.* 11/5 Only King Saud and the régime in the Yemen (which recently survived in undiminished medieval splendour an abortive *coup d'État*) remain patently faithful to Egypt. 1973 *Guardian* 16 Apr. 1/6 The Smith regime in Rhodesia.

b. *old regime* (earlier and later *transf.* examples).

1816 W. Scott in *Q. Rev.* XIV. 192 A crime against sentiment which no author, of moderate prudence, would have hazarded under the old *régime*. 1842 Geo. Eliot *Let.* 30 Aug. (1954) I. 144 There ought to be..a few spectral clingers to the memory of the old régime in the era of political regeneration. 1912 F. A. Talbot *Moving Pictures* xii. 136 Under the old *régime* darkness prevailed from one end of the programme to the other, save, perhaps, during a short interval. 1971 R. Bendix in A. Bullock *20th Cent.* xv. 352/2 Their overthrow of an 'old regime' fulfils the first task of their [*sc.* revolutionary movements'] ideological mission. 1976 J. B. Hilton *Gamekeeper's Gallows* xv. 159 'Take her back home again tomorrow.'.. The old regime was over.

3. *Physical Geogr.* **a.** The condition of a watercourse with regard to changes that may be occurring in its form or bed and the possibility of an equilibrium in which there is neither erosion nor deposition; = *regimen 5.

[1779 P. L. G. Du Buat *Principes d'Hydraulique* I. iv. 73 Ainsi, par le terme régime, nous entendons proprement la vitesse du courant, comparée à la résistance du terrain qui forme le lit.] 1856 *Min. Proc. Inst. Civil Engineers* XV. 241 The case of the River Clyde, at Glasgow, should be carefully examined, when considering any measure for the *régime* of the Thames. *Ibid.* 242 To regulate the low-water *régime*, by removing the shoals below London Bridge. 1895 *Ibid.* CXIX. 282 Observations were made at thirty sites... Each was known by long local experience to have been in a state of permanent regime, the canal having been flowing for years on its self-silted bed. 1925 F. Reeves *Notes & Data Rly. Engin.* 30 One frequently sees the results of this absence of accurate knowledge of the *régime* of the stream in washaways, bridges of unnecessary size, etc. 1927 *Min. Proc. Inst. Civil Engin.* CCXXIII. 268 The conditions of great rivers in unstable regime, presenting every kind of irregularity of flow. 1957 *New Scientist* 26 Dec. 30/3 The regime theory of canals was originally developed in India.. and stemmed from field observations of the self-adjusting character of these artificial alluvial canals. *Ibid.*, From the regime viewpoint the behaviour of a river is visualised as fluctuation about equilibrium or 'regime' dimensions. 1965 A. Holmes *Princ. Physical Geol.* (ed. 2) xviii. 543 The whole régime of sandbanks and inner channels eventually reaches an all-over width that meets the requirements of all but the very greatest floods.

b. The condition of a body of water with regard to the rates at which water enters and leaves it.

1874 *Chem. News* 27 Feb. 101/2 (*heading*) Pluvial régime of the torrid zone in the basin of the Atlantic Ocean. 1933 *Geogr. Jrnl.* LXXXII. 174 While some writers have thought the régime of the lake (the balance between gains and losses) to depend almost entirely on the precipitation on and evaporation from the lake-surface, Mr. Gillman finds that the mean inflow from tributary basins is by no means negligible. *Ibid.*, Theeuws held that the old régime of the lake was changed once and for all by the *débâcle* of about 1875.

4. The set of conditions under which a system occurs or is maintained.

1890 *Rep. Brit. Assoc. Adv. Sci. 1889* 502 We should expect that, after the change of loads has been frequently repeated so that a cyclic *régime* is established, the wire will, for any value of load between the two extremes, be longer during unloading than during loading. 1920 A. Fage *Airscrews in Theory & Exper.* xii. 176 The study of the working régime of a helicopter. 1942 *Electronic Engin.* XIV. 665/3 It has been found that the duration of this low voltage régime may be increased to..20–30 microseconds by connecting an additional condenser directly between anode and cathode. 1957 J. K. Charlesworth *Quaternary Era* II. xlviii. 1410 Pluvial conditions over vast areas of the world.. were replaced by a régime of desiccation. 1971 *Sci. Amer.* Sept. 118/1 Without altering the horticultural regime of keeping 90 percent of the land fallow the Tsembaga's 1,000 best acres might have supported a population of 200 or more per square mile. 1978 *Nature* 29 June 752/1 Anemones were..maintained in circulating seawater at 10 °C for 6 months before experimentation in a 12-h light and 12-h darkness regime.

regimen. Add: **5.** *Physical Geogr.* = *régime, regime 3 a.

1810 *Encycl. Brit.* XVIII. 65/1 We shall..learn the mutual action of the current and its bed, and the circumstances which ensure the stability of both. These we may call the regimen or the conservation of the stream, and may say that it is in regimen or in conservation. 1851 *Min. Proc. Inst. Civil Engineers* X. 231 Experiments and observations were made on the velocity and regimen of the stream. 1966 *McGraw-Hill Encycl. Sci. & Technol.* XI. 584/2 Most natural streams are in regimen. 1971 R. F. Flint *Glacial & Quaternary Geol.* iii. 47 It will be useful to follow the practice of engineers in reference to streams of water, and refer to the system or activity of the glacier as a whole, based on its meteorology, economy, rate and possible type of flow, and fluctuation, as the regimen of the glacier. The term, applied to glaciers as well as streams, is not quantitatively precise; it is broadly descriptive.

regimental, *a.* Add: **1.** (Earlier example.)

1659 J. Jones *Let.* 1 Dec. in J. Mayer *Inedited Lett. Cromwell & Other Regicides* (1861) 112 But crosse windes stayed the messenger at the water side till saturday last, soe that the tyme of meeteing at whitehall is come upon us before wee canne have any regimental meetings of o[u]r offic[e]rs.

2. *Mil. slang.* Maintaining or observing strict discipline.

1919 *Athenæum* 1 Aug. 695/1 *Regimental*, an Old Army adjective for a strict disciplinarian. 1948 Partridge *Dict. Forces' Slang* 154 *Regimental*... As an adjective, applied to an officer or N.C.O. who was a stickler for details. 'So-and-so's too regimental for words.'

regimentation. Add: (Later examples.) Also with ref. to a whole society.

1936 *Sun* (Baltimore) 6 July 8/4 Let us..take the word 'regimentation'... One dictionary has its meaning as 'enforced socialism'. 1937 *Liberty* 10 July 4/2 The same sort of regimentation that we find in Russia. 1943 J. S. Huxley *TVA* i. 7 The possibility of obtaining the efficiency of a coordinated plan without totalitarian regimentation. 1958 A. Huxley *Let.* 15 Feb. (1969) 845 Uniformity and tidiness..are so admirable in a work of art or a scientific theory but..in human life spell regimentation.

Regina (rɪdʒəi·nă). [L. *rēgina.*] A queen; used to designate the prosecution in criminal proceedings during the reign of a queen, as in law reports.

1717 W. Salkeld *Rep. Cases adjudg'd in Court of King's Bench* I. 460 Domina Regina *versus* Wigg. 1792 W. Boscawen *Treat. on Convictions on Penal Statutes* 36 In the earliest case, indeed, upon the statute of 5 An. viz. *Regina* v. *Matthews*, the Court seem to have thought otherwise. 1976 *Law Rep. Queen's Bench Division* 372 Regina v. Kellett.

region. Add: **5. b.** A relatively large subdivision of a country for economic, administrative, or cultural purposes that freq. implies an alternative system to centralized organization; *spec.* one of the nine local government areas into which the mainland of Scotland has been divided since 1975, when the former system of counties was abolished. *Standard* (*administrative*) *region*: one of the eight (formerly nine) areas into which England is divided for industrial planning, demographic surveying, etc.

1921 G. D. H. Cole *Future of Local Govt.* ii. 15 What is really needed is..a systematic scheme of development including both towns and rural areas over the whole of a wide Region. 1933 H. Finer *Eng. Local Govt.* vii. 160 The largest area of government..which would comprehend the main large-scale services, to be managed or regulated by a Council popularly elected for the whole of the Region. 1950 Ormrod & Walker *Butterworth's Annotated Legislation Service: Statutes Suppl. No. 63* I. 10 England and Wales are divided into fourteen hospital regions, each administered by a Regional Hospital Board. Each region is based on a town in which there is a University Medical School. 1958 *Britain: Official Handbk. 1959* (H.M.S.O.) i. 14 Table 2 shows the distribution of the population by urban and rural districts and the populations of the standard administrative regions, of the seven major conurbations and of 16 large cities. 1959 W. Isard et al. *Industrial Complex Anal.* i. 6 Until the 'ultimate' is achieved in social science theory, analysts must be content with sets of regions—or hierarchies of sets of regions—which tend to differ from problem to problem. 1966 *Census 1961: Occupation Tables* p. xix, (*heading*) Areas for which statistics are given. Standard Regions. The constitution of the Standard Regions of England and Wales used in this volume is as follows. 1973 *Times* 23 May 16/6 The Government were confident of their ability to continue giving the regions the assistance they needed. 1973 *Act* 21 & 22 *Eliz. II*. c. 65 § 1 Scotland (other than Orkney, Shetland and the Western Isles) shall be divided into local government areas to be known as regions. 1976 *Scottish Daily Express* 23 Dec. 6/6 Even in booming Aberdeen and the thriving Grampian Region there are troubles between the two councils.

c. An area of the world made up of neighbouring countries that, from an international point of view, are considered socially, economically, or politically interdependent.

1925 W. S. Culbertson *Internat. Econ. Policies* App. ix. 549 Regional or world conferences..to establish in regions, or with respect to subjects in the agenda, 'a workable basis of coöperation among the nations of the earth'. 1948 G. A. Johnson in K. M. Panikkar et al. *Regionalism & Security* 45 There are..yet few who would..divide the world into regions, each a federation. 1959 W. Isard et al. *Industrial Complex Anal.* i. 5 For a long time economists, geographers, sociologists, political scientists, city and regional planners, and other social scientists have been concerned with the concept of 'region'. 1970 Cantori & Spiegel *Internat. Politics of Regions* i. 1 We will consider regions to be areas of the world which contain geographically proximate states forming, in foreign affairs, mutually interrelated units. 1977 M. Hudson *Global Fracture* xiii. 167 Regional consolidation is occurring within five broad geopolitical blocs... Each of these regions is characterized by a broad range of complementary products.

d. *Broadcasting.* A part of the country covered by a particular programme service or broadcasting company; *transf.*, the company itself.

1929 *Radio Times* 8 Nov. 442/1 The Northern Region—Manchester 2ZY 797 kc/s. (376·4 m.) 1949 *Ibid.* 15 July 12/4 In Other Regions—Midland (296·2 m)..North (449·1 m), [etc.]. 1956 *B.B.C. Handbk. 1957* 124 The expansion of television resources in the Regions. 1968 *Listener* 25 July 123/2 This was a good effort by a small ITV region—Anglia —and the old companies of ITV have been putting on quite a brave show. 1974 *B.B.C. Handbk. 1975* 56/2 The Manchester site will eventually accommodate network production centre, television region and local radio station.

6. c. Phr. *in the region of*: round about, approximately.

1966 'A. Hall' *9th Directive* x. 97 The breech-pressure is in the region of 20 tons p.s.i. 1972 *Country Life* 5 Oct. (Suppl.) 7/2 Delightful house... Offers in the region of £40,000. 1979 *Solihull News* 26 May (Classified Section) 24/1 (Advt.), A superb and spacious centrally heated residence... Price: offers in the region of: £29,500.

8. (In sense *5 b) *region-planning.*

1921 G. D. H. Cole *Future of Local Govt.* ii. 14 What is really needed is plainly not mere town-planning, but region-planning. 1931 W. A. Robson *Devel. of Local Govt.* I. ii. 143 Similar problems are presented in connection with other fundamental region-planning requirements.

regional, *a.* Add: **1. b.** Geol. *regional metamorphism* [tr. F. *métamorphisme régional* (G. A. Daubrée *Études et Expériences Synthetiques sur le Métamorphisme* (1860) II. ii. 59)]: metamorphism affecting rocks over an extensive area as a result of the large-scale action of heat and pressure.

[1859 T. S. Hunt in *Q. Jrnl. Geol. Soc.* XV. 489 We must commence by distinguishing between the local metamorphism which sometimes appears in the vicinity of traps and granites and that normal metamorphism which extends over wide areas and is apparently unconnected with the presence of intrusive rocks.] 1871 —— in *Amer. Naturalist* V. 494 The problem to be solved in regional metamorphism is the conversion of sedimentary strata..into aggregations of crystalline silicates. 1937 Wooldridge & Morgan *Physical Basis Geogr.* x. 133 Much more important are the great masses of metamorphic rock which have resulted from what is often called regional metamorphism, *i.e.* deep burial of rock masses..due to earth-movement. 1971 I. G. Gass et al. *Understanding Earth* i. 35/1 Regional metamorphism is accompanied by more or less intense deformation.

c. Of, pertaining to, or connected with a region (esp. in senses *5 b and c). So *regional board, planning*, etc.

1921 G. D. H. Cole *Future of Local Govt.* ii. 15 There is an overwhelming case, from the standpoint of public convenience and efficiency of service, for the regional planning of publicly owned road transport services. 1927 *Rep. Comm. Publ. Libraries Eng. & Wales* 151 in *Parl. Papers* (*Cmd. 2868*) XII. 231 Our conclusion is that a national system should be built up on..the grouping of public libraries round regional centres, which will generally be the great urban libraries. 1933 H. Finer *Eng. Local Govt.* vii. 164 If the regional solution were adopted it would..furnish an area large enough to include most of those services for which a large area has been found necessary. 1941 *Economist* 18 Jan. 68/2 Lord Nuffield..envisaged regional boards throughout the country which would co-ordinate hospital finance and policy. 1943 in J. S. Huxley *TVA* 5 The Tennessee Valley Authority initiated regional planning on a scale never before attempted in history. 1956 G. N. Flemming *Organisation of Technical Coll.* 1 There are at present three main types of technical college,..described in this circular as local, area and regional colleges. 1957 G. L. Goodwin *Brit. & United Nations* 459 The prospects of world peace will turn not on the United Nations but on the effectiveness of global and regional balances of power. 1959 W. Isard et al. *Industrial Complex Anal.* i. 6 The concept of regional structure has come to be relativistic. 1962 *Lancet* 27 Jan. 219/2 The unit would be organised on regional-board level. 1965 Haas & Schmitter *Polit. of Econ. in Lat. Amer. Regionalism* i. 1 One of the first prolonged attempts at regional economic integration between independent developing countries. 1969 *Times* 3 Feb. 10/8 Farming is not a subject which often enters the realm of regional planning. 1972 W. Isard et al. *Ecologic-Economic Anal. for Regional Development* p. xv, We constantly assert that no longer can regional development and regional planning be treated in their traditionally narrow contexts. 1976 C. A. Smith *Regional Anal.* I. p. xi, Careful field studies of markets and regional economics.. would have identified regional patterns not predicted by central-place theory..thereby stimulating the development of new regional system theories. 1977 M. Hudson *Global Fracture* xiv. 178 The currency-debt of a nation or regional grouping of nations not politically associated with the creditor central bank. 1978 *Dumfries Courier* 13 Oct. 2/6 The modernisation project at Templand had been praised by the Regional Planning Committee.

d. Of or pertaining to a broadcasting region (sense *5 d). Also, designating a B.B.C. radio service which operated during the 1930s.

1929 *Radio Times* 8 Nov. 403/3, 2BE *Belfast*..9.0 Regional News. 1930 *B.B.C. Year-bk.* 30 The Regional Stations of Daventry, London, and Manchester will cover about 75 per cent. of the population. 1962 *Rep. Comm. Broadc. 1960* 157 The four major companies having been appointed [by the ITA], this consideration prompted the appointment of the relatively large number of relatively small 'regional' companies. *Ibid.* 221 The present sound programmes of the BBC are..basically national programmes. Within them, regional programmes are accommodated. 1965 [see *national *a.* 1 d]. 1968 *Writing for BBC* (ed. 2) 12 A particularly worth-while opportunity for would-be contributors exists in the Regional editions of programmes.

B. *sb.* **1.** A B.B.C. radio service which operated during the 1930s.

1936 J. Reith *Diary* 20 Jan. (1975) iii. 185 We got it [*sc.* a statement] out at 9.38 as we had to collect all the Regionals and Empire. 1938, 1971 [see *national *a.* 1 d]. 1978 *Broadcast* 27 Feb. 4/2 A balance between the majors and regionals has emerged.

2. The part of a gravity anomaly or magnetic anomaly that is due to deep features and varies only gradually from place to place.

1940 L. L. Nettleton *Geophys. Prospecting for Oil* xii. 222 If this regional is properly estimated and removed,..the local features will show up in their proper form and relief. 1954 *Geophysics* XIX. 1 The problem of regionals and residuals arises in all geophysical methods which are based on measurement of a 'potential' field. 1967 *Bull. Amer. Assoc. Petroleum Geologists* LI. 2388/1 The 13th-order regional fits closely the observed data.

3. In general use, *ellipt.* for *regional* (*stock*) *exchange, newspaper, stamp*, etc.

1958 *Gibbons Stamp Monthly* 1 Sept. 2/3 Pictorials can be captioned—to most people these Regionals will be confusingly anonymous. 1965 *Time* 3 Sept. 58 Because of this growth..the cost of seats on the regionals has been rising steadily. 1969 *Times* 5 May (Wall Street Suppl.) p. v/7 Many of the regionals even encouraged institutional, or mutual fund, business... The major brokerage houses handling institutional transactions could thus buy a relatively inexpensive seat on a regional exchange. 1971 D. Potter *Brit. Eliz. Stamps* iv. 59 Jersey and Guernsey stamps were withdrawn on 1 October 1969, and the two islands issued their own stamps. But the regionals remained valid for postage elsewhere in the United Kingdom. 1974 *State* (Columbia, S. Carolina) 27 Feb. 3-B/4 The South Atlantic regional at Norfolk, Va. 1975 B. Gunston *Philatelist's Companion* 247 *Regional*, a stamp issued for use in only part of the territory under the authority of a postal administration (eg Scotland, in the case of the UK). 1975 *Times* 13 Aug. 12/1 While national newspaper managements were still thinking in the fifties of hot metal.. modernization ran across Britain and the regionals profited.

regionalism. Add: **1.** (Further examples.) Also, on a national or international scale: the theory or practice of regional rather than central systems of administration, or of economic, cultural, or political affiliation; the study of such phenomena as they relate to geographic factors.

1919 Geddes & Brandford in C. B. Fawcett *Provinces of England* p. ii, 'Regionalism' was, indeed, first a French word: and this not merely in geography, but also in politics, and long before the war. From Brittany to Provence its studies have been long preparing. 1923 G. M. Trevelyan *Manin & Venetian Revolution* xiv. 244 He abandoned his Republican faith and his Venetian 'regionalism' in view of the new circumstances of Italy. 1931 E. C. Mower *Internat. Govt.* v. 89 Regionalism, it is claimed, is justified both from the defensive value of natural frontiers and on sound economic principles, having regard to the needs of modern industrial life. 1934 *Encycl. Soc. Sci.* XIII. 208/2 Regionalism has been called a manifestation of 'world federalism' and an intermediate stage between administrative decentralization and federalism. 1936 *Columbia Univ. Quart.* Mar. 268 Regionalism can..be defined as the study of the relation of man to geographic areas, and the potentialities which this relation presents in terms of human welfare and progress. 1948 G. A. Johnson in K. M. Panikkar *Regionalism & Security* 45 Of these claims on behalf of regionalism the United Nations Charter is concerned directly with only one. 1959 A. H. Robertson *Europ. Institutions* i. 4 As the idea of universalism waned,..that of regionalism developed. 1962 L. Golding *Dict. Local Govt.* 332 When War broke out in 1939 regionalism was applied in practice by dividing the country for purposes of civil defence and the administration of other emergency services into twelve large areas, each of which was placed in the charge of a Commissioner. 1965 Haas & Schmitter (*title*) The politics of economics in Latin American regionalism. 1970 Cantori & Spiegel *Internat. Polit. of Regions* i. 1 Sometimes 'regionalism' has been studied exclusively in terms of regional organization. 1977 M. Hudson *Global Frontiers* xv. 195 (*heading*) The new regionalism.

2. A regional word, phrase, or peculiarity of pronunciation which is not part of the standard language of a country; regional distinctiveness in literature.

1953 S. A. Brown in A. Dundes *Mother Wit* (1973) 40/1 We go then to what is called the New Negro Movement, then to Regionalism. 1954 F. G. Cassidy *S. Robertson's Development Mod. English* v. 126 The third [sc. *you-all*] is a regionalism. 1955 *Times* 7 May 9/4 The regionalism of American writing falls into place beside that of Scotland or Ireland. 1964 *Language* XL. 93 Intellectual leaders of the Seicento..did not hesitate to use, as nonce-forms, regionalisms like *parapaglia* 'butterfly' (based on Bolognese *parpaja*..) and *sfragaro* 'wastrel' (Calabrese-Sicilian). 1974 R. A. Hall *External Hist. Romance Lang.* 216 Some lexical regionalisms have been inevitable in films made in, say, Mexico or Argentina. 1978 *Amer. Speech* LIII. 13 The layman applies the term imprecisely to a large body of lexemes including true slang, jargon, regionalisms, and colloquialisms, which are vaguely perceived as slang by such groups as college students.

regionalist. Add: Further examples with reference to geographical, administrative, or international regionalism. Also *attrib.*

1919 Geddes & Branford in C. B. Fawcett *Provinces of Eng.* p. iii, The most discerning..regionalists of to-day are also among the most appreciative of truly comprehensive politics. 1925 L. Mumford in *Survey* 15 Apr. 151/1 The regionalist attempts to plan such an area so that all its sights and resources..may be soundly developed. 1934 *Encycl. Soc. Sci.* XIII. 216/1 There was a revival of regionalist feeling in Italy after the World War. 1937 F. Borkenau *Spanish Cockpit* i. 59 The Basque provinces, naturally, voted for the Basque regionalists. 1941 J. Masefield *In Mill* 112 The Regionalist novel-writers. 1944 A. Brecht in *Regionalism & World Organization* 11 The universalists want a world-wide organization; the major-regionalists recommend federations of continental scope; the minor-regionalists propose federal groupings of smaller countries; and the ideological unionists advocate a confederation of democracies or a league of the United Nations. 1977 *Economist* 23 Apr. 56/1 The three major regionalist parties [in Belgium]—the Walloon Rally, the Volksunie and the Francophone Front. 1977 G. P. Atkins *Lat. Amer. in Internat. Polit. System* i. 9 Regionalists were primarily concerned with security arrangements for and keeping the peace in delineated geographical areas.

regionality (ridʒənæ·liti). [f. Regional *a.* + -ity.] Nature or character connected with or pertaining to a region.

1961 in Webster. 1966 *New Statesman* 8 Apr. 510/1 The zone of time-space that middle-class western man mostly inhabits is the continuous present of the Western World: the particularities of the particular place he happens to live in are often little more than a backdrop to home and work and leisure—unless he's feeling jolly, when he is sometimes prepared to wear his regionality as a mask or humour. 1976 *Amer. Speech* 1973 XLVIII. 282 For obvious reasons, the concept of 'regionality' in regard to language is a preoccupation of the editors of the *Dictionary of American Regional English.* 1979 *Dictionaries* I. 27 Many of the statements about regionality are qualified in some way.

regionaliza·tion. [f. Regional *a.* + -ization.] The action of adapting economic, political, social, or cultural organization to a geographical or administrative region.

1920 A. R. Orage in C. H. Douglas *Credit Power & Democracy* 152 The suggested regionalisation of the administration of the industry may be regarded as acceptable to the Miners' Federation. 1930 *Aberdeen Press & Jrnl.* 3 Nov. 5/6 We have just completed..what might be called the first try-out of programme regionalisation. 1938 Odum & Moore *Amer. Regionalism* I. ix. 216 The present regionalization of the country. 1952 *Property Owners' Jrnl.* Jan. 10/2 The other major factor is..regionalisation. Bricks are expensive things to carry... Consequently, brick houses must be built as near the brick-fields as possible. 1963 *Times* 23 Apr. 13/3 The fashion for regionalization is strengthening in Britain today. 1970 *Nature* 26 Dec. 1250/2 In regions..which have urban populations large enough for regionalization of intake to be a real possibility, less than half the students accepted in 1969 came from the region in which the universities are situated. 1975 *Church Times* 18 July 12/3 It may be believed that SCM is committing itself to increased regionalisation and student work from within. 1978 *Nature* 2 Feb. 403/2 The development of the adult fly involves first the regionalisation of the embryo into a number of territories.

regionalize (rí·dʒənălaiz), *v.* [f. Regional *a.* + -ize.] *trans.* To bring under the control of a region for administrative purposes; to divide into regions; to organize on a regional basis. So **re·gionalized** *ppl. a.*, **regionali·zing** *vbl. sb.*

1921 G. D. H. Cole *Future of Local Govt.* xii. 112 May it not be possible to escape the disadvantages of central ownership and control by regionalizing instead of nationalizing many industries and services. 1938 Odum & Moore *Amer. Regionalism* I. viii. 188 We have 'regionalized' our nation and subregionalized and districted our states. 1962 *Times* 23 Jan. 4/2 The draw for the second round of the F.A. Amateur Cup—to be played on February 3 and no longer regionalized. 1962 *Daily Tel.* 8 May 1/7 Miners' M.P.s repeatedly voiced suspicion that Lord Robens's real intention ..is to 'regionalise' the coal industry by creating autonomous boards in Scotland, Lancashire and elsewhere. 1972 *Times of India* 28 Nov. 1/2 Various service cadres will be regionalised. 1978 *Jrnl. R. Soc. Arts* CXXVI. 218/1 It costs just as much to broadcast to half a million people as to broadcast to fifty million, so that the more you regionalize your output the more expensive it becomes. 1978 *Radio Times* 11–17 Mar. 15/4 In future years Ceefax will offer regionalized news.

regionally, *adv.* (In Dict. s.v. Regional *a.*) (Later examples.)

1886 *Science* 10 Sept. 233/2 The preservation of rock-oils in every formation, of every geological age, all over the world; subject, however, locally or regionally, to subsequent change or destruction. 1962 *Rep. Comm. Broadc. 1960* 151 We have examined the BBC on its allocation of money between its sound and television services, both nationally and regionally. 1974 *Nature* 1 Nov. 28/1 Viewed regionally, the area of greatest regressive tendency within the depositional regime can be identified as the central and northern North Sea.

‖**regisseur** (rɛʒisör). *Theatr.* and *Ballet.* [Fr.] A stage manager or artistic director. Also *transf.* in *Cinemat.*

1828 J. Ebers *Seven Yrs. of King's Theatre* ii. 58 He had been a kind of manager of the Opera at Bologna, and subsequently Regisseur of the Théâtre Italien at Paris. 1925 *Daily Herald* 20 May 9/3 The three main streams of the revolutionary theatre..derive from the three *regisseurs.* 1935 A. Revusky *Jews in Palestine* x. 173 Piscator, the leading theatrical regisseur of Germany. 1949 *Ballet Ann.* III. 27 *Prince Igor*, revised by that experienced and excellent *régisseur*, Nicolas Beriosoff. 1954 'E. Box' *Death in Fifth Position* i. 10, I was introduced to..the *regisseur* or director of the [ballet] company. 1965 *New Statesman* 16 Apr. 621/2 Shaw, Chekhov and Brecht wrote for a producers' theatre, intervening when necessary as their own *régisseurs* to bully their actors into subordination to their texts. 1968 *Listener* 14 Mar. 357/1 Zeffirelli is a *régisseur* in the grand 19th-century naturalistic manner. 1977 *New Yorker* 16 May 79/3 Serge Grigoriev, who was Diaghilev's regisseur, restaged it for the Royal Ballet.

register, *sb.*^{1} Add: **I. 1. a.** Also, a record of attendance at a school.

1887 C. D. Warner *Their Pilgrimage* (1888) vi. 165 Mr. King discovered by the register that the Bensons had been there. 1888 C. M. Yonge *Our New Mistress* ii. 14 She called over the names... The registers had got into a muddle, and there was no knowing who had left school and who was only absent. *a* 1930 D. H. Lawrence *Phoenix II* (1968) 22 One day my bread-stealer arrived at half past two, when the register was closed. 1955 E. Blishen *Roaring Boys* iv. 183, I called the register... The ginger-haired boy answered to the name of Grange. 1961 M. Spark *Prime of Miss Jean Brodie* iii. 59, I must mark the register for today before we forget. There are two new girls. 1978 R. Mills *Comprehensive Educ.* 44 His lessons..began with the calling of the register.

c. A person's face, regarded as an indication of feeling or emotion. *slang.*

1899 'J. Flynt' *Tramping with Tramps* II. iv. 271, I hain't seen your register for many a day.

5. b. A quantity recorded or registered.

1904 T. Holdich *India* xii. 351 At this point the rainfall is extraordinary, 50 or 60 feet being a not unusual register at Cherra Punji on the edge of the plateau.

II. 8. c. *Art.* One of a number of bands or sections into which a design is divided.

1937 *Discovery* Sept. 287/1 As a rule these plant designs [on Jhukar pottery] were painted in black, or a deep purple, the red being used for the broad bands separating the registers. 1966 Roniger & Dunn tr. *Lazarev's Old Russ. Murals & Mosaics* i. 39 In the middle register of the apse is the great monumental composition of the *Eucharist.* 1977 *Times Lit. Suppl.* 4 Feb. 137/3 The outside [of a conical drinking-horn] is decorated in paint or in enamel. There are two upper registers with scenes of animals and hunters. 1980 *Catal. Fine Chinese Ceramics* (Sotheby, Hong Kong) 222 Two groups of nine bosses arranged in three registers within rectangular enclosures, the registers alternating with bands of repeated curlicues.

d. *Linguistics.* A variety of a language or a level of usage, *spec.* one regarded in terms of degree of formality and choice of vocabulary, pronunciation, and (when written) punctuation, and related to or determined by the social role of the user and appropriate to a particular need or context.

1956 T. B. W. Reid in *Archivum Linguisticum* VIII. 32 He will on different occasions speak (or write) differently according to what may roughly be described as different social situations: he will use a number of distinct 'registers'. 1962 *Canadian Jrnl. Linguistics* VII. 69 Interference may also vary according to the social role of the speaker in any given case. This is what the Edinburgh School has called *register.* 1966 G. N. Leech *Eng. in Advertising* vii. 68 Varieties of English distinguished by use in relation to social context are called *registers.* 1971 P. Young in J. Spencer *Eng. Lang. W. Afr.* 173 A novel, then, can be seen as an amalgam of registers within a wider register of literary endeavour. 1972 *N. & Q.* Dec. 446/2 Chaucer must therefore have used what was, for the London of his time, a more formal, possibly more archaic, register. 1977 P. Strevens *New Orientations Teaching of Eng.* x. 119 They are aware..of the idea of 'varieties' of English, and they probably know the term 'register'—a variety related to a particular use of the language, a particular subject or occupation.

e. *Phonetics.* A type of phonation, essentially controlled by the larynx, but distinct from tone, employed contrastively in some languages (e.g. Cambodian).

1964 J. C. Catford in D. Abercrombie et al. *Daniel Jones* 34 'Register' differences..are associated with tone-differences in several S.E. Asian languages. 1967 D. Abercrombie *Elem. Gen. Phonetics* 101 In Cambodian, for example, every syllable is spoken with one of two registers, which are mainly distinguished from each other by the position of the larynx in the throat. The same is true of Gujerati spoken in Surat, the difference here being between 'tight' and 'breathy' phonation.

9. a. (Further examples.)

Now used outside the U.S. in the sense 'perforated or open-work plate by which warm air is admitted'.

1920 E. Frost *Let.* Mar. in *Lett. R. & E. Frost* (1972) 86, I am writing to you with a pencil generally these days, because I can sit and warm my feet over the register at the same time if I use a pencil. 1950 R. Moore *Candlemas Bay* ii. 86 Two rooms were warmed by hot-air registers through the kitchen ceiling. 1957 V. Nabokov *Pnin* vi. 145 A cranky-looking oil furnace in the basement did its best to send up its weak warm breath through registers in the floors. 1967 *Gloss. Terms Gas Industry* (*B.S.I.*) 91 *Register*, a fitment equipped with a damper or movable louvres which permit adjustment or closure. 1970 *Daily Tel.* (Colour Suppl.) 25 Sept. 14 (Advt.), Built-in ducts waft warm air to each room through small, skirting-level registers.

c. *Rope-making.* A disc containing concentric circles of holes through which the component yarns of a strand pass, the rotation of the disc serving to twist them together. Also *register plate.*

1793 J. Huddart *Brit. Patent* 1952 5 The register is calculated to form the strand into shells of yarns, and therefore they must be made of different sizes. 1846 G. Dodd *Brit. Manuf.* VI. 199 The system for attaining any required intensity of twist is called the 'register'. 1855 W. Cotton *Brief Memoir Capt. Joseph Huddart* 20 His great principle (concentric circles) was accomplished by what he called a register plate. 1950 A. E. Haarer *Ropes & Rope-Making* vii. 43 This..is what we call a register plate. Behind it you can see the yarns passing together into a short tube. 1957 D. Himmelfarb *Technol. of Cordage Fibres & Rope* v. 127 Register plates are generally heavy castings of curved cross section, with holes approximately ½-inch to 1-inch in diameter.

10. b. = *cash register* s.v. *cash *sb.*^{1} 3 a. chiefly *U.S.*

1895 in *Funk's Stand. Dict.* 1911 *Daily Colonist* (Victoria, B.C.) 29 Apr. 13/4 Two robbers..took $160, all the money in the register, and made good their escape. 1976 'E. McBain' *Guns* vii. 148 Colley wishes he could see into the open drawer of the register. 1977 *Transatlantic Rev.* LX. 40 He..then counted the cash in the register.

c. In mechanical calculators, a device in which numbers representing data or the results of arithmetical operations are stored or displayed; in an electronic computer or calculator, a location in store having a small capacity but negligible access time and used for a specific purpose (hence with qualifying sbs., as *address, control, storage register*).

1928 *Monthly Not. R. Astron. Soc.* LXXXVIII. 451 Their principal deficiencies were the absence of tens transmission in the multiplier register,..and the excessive labour of zeroising or clearing the registers. 1946 *Math. Tables & Other Aids to Computation* II. 151 The Brunsviga Dupla.. had two product registers..and had red and white figures as in the multiplier register. 1947 *Proc. I.R.E.* Aug. 759/1 The counter advances one stage on receiving a pulse, and hence is an adder as well as a register. 1956 [see *address *sb.* 7 c]. 1959 *Commun. Assoc. Computing Machinery* Oct. 3/2 Besides

the internal memory, the arithmetic unit has four fast access cells. In these cells, the words are stored in dynamic form, in registers. 1964 F. L. Westwater *Electronic Computers* i. 13 This number, when placed in a special register called the control register, will cause the machine to obey the instruction. 1973 *Sci. Amer.* Aug. 102/2 (Advt.), In addition to its computer-like operational stack, the HP-35 has a constant storage register which lets you store any number and recall it as often as you want for repeat operations, without ever having to re-enter it. 1977 *Ibid.* Sept. 86/3 In microprocessors registers are employed for the temporary storage of data, of partial results, of instructions and of the addresses where other data or instructions are to be found.

11. b. (Further examples.) More widely, exact coincidence of position of superimposed images, esp. in colour printing; so *in, out of register.* Also *transf.*

1683–4 J. Moxon *Mech. Exerc. Printing* (1962) 348 Out of register, bad register. 1771 P. Luckombe *Hist. & Art of Printing* 500 *Out of register*, when pages are not worked even on each other. 1907 [see *duotone]. 1915 [see *print *v.* 14 a]. 1947 *Electronics* Jan. 75/2 Color fringing..is also not present in the simultaneous system [of colour television], but a similar effect due to lack of register among the three simultaneous images may be present. 1950 *Proc. R. Soc.* A. CCI. 189 It [sc. a diffraction grating] is then cut in half, perpendicular to the rulings, and the two halves are put together face to face in register. 1966 H. Williamson *Methods Bk. Design* (ed. 2) xxii. 366 Colour printing usually costs more per colour than does black printing because of the laborious work of getting and maintaining register. 1967 V. Strauss *Printing Industry* xi. 735/2 Printers distinguish several kinds of register, depending on the intricacy of a job. No-register means that the several color areas are completely independent of each other, loose register that minor variations in their relations are inconsequential; tight, close, or hairline register indicates that these relations must be quite exact. 1975 J. B. Harley *O.S. Maps* ix. 138 Road casings were omitted..to assist with colour register when printing. 1975 *Nature* 25 Sept. 331/2 It is a rod-shaped, coiled-coil molecule, about 410 Å long, composed of two parallel α-helical chains which are in register. 1978 *Amat. Photogr.* 11 Jan. 69/2 A bas relief is made by printing through a negative and a positive, sandwiched together slightly out of register.

12. (sense 1) *register clerk*; **register board,** a flat surface with pegs or guides such that sheets of paper or film placed on it may be brought into the same relative position; **register mark** *Printing* (see quots.).

1967 Karch & Buber *Offset Processes* viii. 331 The register board..performs a very important part in obtaining accurate registration by providing a mechanism to jog each sheet into exactly the same position for entry into the head of the press. 1977 J. Hedgecoe *Photographer's Handbk.* 245 (*caption*) Take a negative 2¼ ins sq..and attach it to a strip of film which has been punched to fit the register board. 1887 C. D. Warner *Their Pilgrimage* (1888) iii. 63 The register clerk stood fingering the leaves of the register with a gracious air. 1927 H. Hubbard *Colour Block Print Making* 208 *Register marks*, in colour printing, marks for controlling the position of the paper in printing to ensure register. 1937 *Discovery* Oct. 300/2 Register marks are drawn on the stones just off the edge of the design. These the printer uses for placing the paper in exactly the right place on the stones used in subsequent printings. 1971 D. Potter *Brit. Eliz. Stamps* xv. 174 Autotron marks, long bars, provide the electronic check for colour registration. Register marks..serve a similar purpose.

register, *sb.*^{2} (Later examples in official titles.)

1948 *Daily Progress* (Charlottesville, Va.) 7 Apr. 13/7 He is the Register of Wills. That's what the state constitution calls him. 1972 *Publishers' Weekly* 21 Aug. 58/1 Now Acting Register of Copyrights on leave to UNESCO, Miss Ringer has been opposing appointment of George D. Cary as Register.

register, *v.* Add: **1. a.** (Later *fig.* examples.)

1891 'L. Malet' *Wages of Sin* I. iii. 53 He was always thinking, doing, feeling, experiencing something... Always registering impressions, making observations. 1946 D. C. Peattie *Road of Naturalist* iv. 49 All I could do was.. register on the one hand my sympathy with Abie and, on the other, the isolation of my own interests. 1955 W. Heisenberg in W. Pauli *Niels Bohr* 22 The observer has.. only the function of registering decisions. 1972 *Daily Tel.* 31 Jan. 7 Those responsible for the television serialisation of Heinrich Mann's *Man of Straw*..worked hard to register the idea that the novel..prophesied the rise of the Nazis.

2. a. (Later *absol.* example.)

1930 N. R. Stephenson *Nelson W. Aldrich* iii. 48 The Senate passed the bill, Aldrich and Platt registering against it.

c. *trans.* and (now chiefly) *intr.* (for *refl.*) To enter the name of (a guest or visitor) in the register of a hotel or guest-house; to enter one's name in such a register. orig. *U.S.*

1848 *Lit. Amer.* 14 Oct. 237/1 Sixty miles down the Monongahela brought us to Pittsburgh, and about half past 7 P.M., I was registered at the 'Monongahela House'. 1850 M. Reid *Rifle Rangers* I. v. 52 Take your supper, engage a snug room, and wait for me. Don't register till I come—I'll attend to that. 1891 'Mark Twain' in *Harper's New Monthly Mag.* Dec. 96/2, I arrived in Washington, registered at the Arlington Hotel, and went to my room. 1905 A. Bennett *Tales of Five Towns* II. 264 'You haven't registered,' Nina called to him... He advanced to sign. 1922 H. Titus *Timber* xv. 136 She..stopped her car at the Commercial Hotel where she registered and was given a room. 1936 G. B. Shaw *Millionairess* IV. 187 You have allowed my husband to bring a woman to my hotel and register her in my name. 1967 Beavis & Medlik *Man. Hotel Reception* ii. 12 The receptionist should hand the pen to the guest when asking him to register. 1977 *Rolling Stone* 30 June 81/1 We then registered at the Airport Inn.

d. *intr.* (for *refl.*) To enter oneself or have one's name recorded in a list of people (freq. as a legal requirement), as being of a specified category or having a particular eligibility or entitlement.

1940 *Economist* 26 Oct. 521/2 Special delivery certificates have been issued for householders who must register with a single coal merchant. 1941 M. B. Lowndes *Diary* 7 Nov. (1971) 225, I registered for my rations, sugar, bacon, butter, etc. at Fortnum's, where quality is *excellent.* 1952 B. Pym *Excellent Women* xxvi. 243, I told them of a laundry, a grocer and a butcher where they might register. 1965 *Listener* 10 June 875/3 To intimidate Negroes who might be tempted to register as voters. 1973 'B. Mather' *Snowline* vii. 85, I tried to get him to register..as an addict. You get a scrip to buy the damned stuff on prescription. 1975 S. Briggs *Keep smiling Through* 149/2 You could even..choose where to shop without being tied to the grocer where you were registered for basic rations. 1976 *Southern Even. Echo* (Southampton) 13 Nov. 8/4 The next day he registered as unemployed. 1977 *Time* 17 Jan. 24/2 This presumably would include all those civilians who fled the country to avoid the draft, simply failed to register or refused to submit to induction.

3. b. Of a person: to indicate or express (a particular feeling or emotion), esp. by facial expression.

1901 'L. Malet' *Hist. Sir R. Calmady* VI. viii. 568 The brightness died out of Honoria's face. She registered sharp annoyance against herself. 1915 Wodehouse *Something Fresh* iii. 56 A stage-director of a moving-picture firm would have recognized the look; Lord Emsworth was 'registering' interest. 1925 A. P. Herbert *Laughing Ann* 32 For I don't have no adventure in the street, Men don't register emotion when we meet. 1977 *Private Eye* 29 Apr. 3/3 On being told, her face registered shock and horror.

c. *intr.* Of a person (orig. and esp. a film actor): to portray a particular role with conviction. Also of the ideas or feelings concerned: to communicate themselves successfully, to be convincing. Hence *gen.* of feelings, thoughts, utterances, etc.: to produce the desired effect, to make an appropriate impression on the person intended. Freq. const. (*up*)*on, with.*

1913 Esenwein & Leeds *Writing Photoplay* 24 It is sometimes said that an effect, a bit of business, or an emotion which an actor is endeavoring to portray, 'will not register', meaning that it will not 'get across' or be understood by the audience in the way intended by the producer. 1915 *N.Y. Times* 1 Nov. 11 This new movie star 'registers', as the film folk have it. 1928 *Sunday Dispatch* 16 Dec. 14/4 It looks..as though the producers had not been willing to risk spending money on it in case Miss Eagels did not register well. 1934 H. G. Wells *Exper. Autobiogr.* II. ix. 704 He never did as he intended or the hint was too feeble to register upon our minds. 1939 *Punch* 6 Sept. 255/1, I give a cough. A significant cough... The cough registers. Deep silence ensues. 1951 N. Balchin *Way through Wood* iv. 60 Even that didn't register. You see I didn't know *where* Joe had been knocked down. 1951 M. McLuhan *Mech. Bride* (1967) 141/2 The slick-chick and the corporation executive, as they now register on the popular imagination, are already inside the totem machine. 1964 'A. Gilbert' *Knock, knock, who's There?* i. 14, I couldn't help seeing the name... I looked sharply at Ted, wondering if it was going to register with him. 1966 *Listener* 17 Feb. 253/1 Sixteen-year-old Alexandra can only hope to register with her mother, so she finds out sadly. 1977 *Daily Mirror* 16 Mar. 5/3 With the five-year-old it did not register.

d. *intr.* To appear or produce a response on a recording or measuring instrument.

1947 *Math. Tables & Other Aids to Computation* II. 356 When two pure imaginaries are multiplied together a minus sign will register. 1974 *Nature* 6 Sept. 19/1 The ion energies were too small to register.

4. b. (Further example.)

1976 *Physics Bull.* May 200/1 The images of the projected mask and the structure on the silicon wafer are superimposed and alignment is accomplished by registering the two images.

c. *Mil.* To adjust a gun in relation to (its target); to align (artillery) with its target.

1958 *Observer* 9 Feb. 11/4 The American Polaris..will free still further the Western nuclear deterrent from dependence on large static bases..which can be registered in advance. 1958 *Listener* 11 Sept. 386/3 The position had been liberally registered by Russian gunners from the city; hence the cannon balls. 1959 H. MacLennan *Watch that ends Night* IV. vii. 166, I had to spend ten hours in that hole with the body, for the machine guns were registered so close to the ground a rat couldn't have escaped.

6. *trans. Rope-making.* To form (a strand) by the use of a register. Cf. *register *sb.*^{1} 9 c. Also *absol.*

1793 J. Huddart *Brit. Patent* 1952 3 The spindle is turning in, registering the strand. 1800 *Remarks on Patent Registered Cordage* (Huddart & Co.) 3 He has invented a method of manufacturing cordage, whereby every yarn holds a situation in the strand, in which it bears its proportion of the strain of the rope. This is termed registering the strands. 1855 W. Cotton *Brief Memoir Capt. Joseph Huddart* 26 In order to render them impervious to water, it was necessary to register them at a higher angle. 1968 W. Tyson *Rope* I. iii. 10 In 1799 Huddart patented..a means of registering the strands at a short length from the tube and winding up the rope as made, thus preserving a uniformity of twist.

registered, *ppl. a.* Add: **1.** *spec.* of postal items recorded at the point of dispatch and indemnified against loss or damage. Also *ellipt.* as *sb.*

1837 *Act* 1 *Vict.* c. 34 § 25 All registered Letters shall be delivered to the Post Office, and also be delivered by the Post Office, under all such Regulations in every respect as the Postmaster General shall from Time to Time appoint. 1855 Mrs. Gaskell *Let.* 21 July (1966) 359, I am going to send it by registered parcel post. 1864 D. G. Rossetti *Let.* 5 Feb. (1965) II. 498 It shall..fly in a registered letter to you as soon as may be. 1874 [in Dict.]. 1921 *Daily Colonist* (Victoria, B.C.) 9 Apr. 17/7 Three bandits..last night held up a United States mail truck here and robbed it of three pouches of registered mail. 1930 J. B. Priestley *Angel Pavement* xi. 553 Don't forget you've got three registered there; bring me the receipts in the morning. 1946 W. Stevens *Let.* 12 Nov. (1967) 537 The proofs..are being returned by registered mail today. 1962 *John o' London's* 11 Jan. 27/3 *Postman*..'Ere, I got a registered for yer. 1975 E. Rudinger *Wills & Probate* ii. 85 He enclosed a short covering letter listing the enclosures, and sent it by registered post. 1976 *New Yorker* 8 Mar. 90/3 She found waiting for her on the kitchen table the mail in response to the registered letters she had sent to the United States senators. 1981 *Guardian* 15 May 4/3 The registered envelopes, posted in Dublin and Waterford, have been handed to police but most people have held on to the money.

2. *Rope-making.* Formed by means of a register. Cf. *register *sb.*^{1} 9 c.

1800 *Remarks on Patent Registered Cordage* (Huddart & Co.) 5 But this loss of proportional strength, increases with the number of internal yarns contained in the first strands, while the registered strands bear a proportional strength to their number of yarns. 1846 G. Dodd *Brit. Manuf.* VI. 199 A registered strand, or the strand produced by twisting the yarns together by this machine, is a smooth, uniform piece of cordage, all the yarns twisting round in one direction.

3. *registered nurse*, a nurse who has been entered on an official register. See also *state-registered nurse* s.v. *state *sb.* 41 a.

1896 *Brit. Med. Jrnl.* 18 Jan. 158/1 The precedent of the General Medical Council should be followed, and an analogous body created, composed of representatives of the nursing interests, the medical profession, the Privy Council, and the registered nurses. 1903 *Outlook* 14 Mar. 604/1 Such a person shall be given a Regents' certificate of proficiency and be privileged to bear the title of 'Registered Nurse' (R.N.). 1905 *Rep. Sel. Comm. Registration of Nurses* iv. in *Parl. Papers* VII. 733 No person should be entitled to assume the designation of 'Registered Nurse' whose name is not upon the Register. 1949 *Reader's Digest* June 91/1 They passed with flying colors the examinations for registered nurse. 1976 L. Hockey *Women in Nursing* v. 50 After the age of twenty-five years, registered nurses are likely to move into a variety of other designations, mostly probably becoming ward sisters.

register office. Add: *spec.* the office of a registrar of births, marriages, and deaths.

The officially correct form in the U.K. at the present time (1981). Cf. *registry office 2.

1893 [in Dict.]. 1954 T. S. Eliot *Confidential Clerk* III. 121 We'd meant to be married very quietly In a register office. 1976 *Southern Even. Echo* (Southampton) 2 Nov. 6/5 Colours are gradually creeping into the dresses in shades of pale pinks and blue satin ribbon edging, and cream is a favourite for register office weddings. 1980 *Times* 13 Aug. 1/8 Couples [in Russia] intending to get married apply to the local register office.

registrable, *a.* Add: Also **registerable.** (Later examples.)

1918 *Brit. Med. Jrnl.* 31 Aug. 212/2 In speaking of the minimum period, it is to be remembered that that is..only sufficient to gain a registrable qualification. 1960 *Times* 28 Apr. 3/3 (Advt.), Qualifications: Medical qualification registerable in New South Wales. 1962 *Economist* 20 Jan. 246/1 These agreements, which may become registerable. 1971 *Daily Tel.* 1 July 2/6 Places where Jews and Quakers celebrated marriage should become registerable in the interests of inter-religious harmony. 1976 *Sunday Times* (Lagos) 26 Sept. 11/1 At least, in some alert official's estimation, 4,000 registrable voters might just call it a centre. 1977 *Nature* 6 Jan. p. xiv/3 (Advt.), Applicants must have a veterinary qualification which is registerable in the United Kingdom.

registral, *a.* Delete *rare* and add later examples.

1967 E. Salzman *20th-Cent. Music* xiv. 173 Tiny, cell-like structures which retain their..identity through every kind of registral, rhythmic, dynamic, and color shift. 1970 *Language Sciences* Oct. 11/1 Rules that change this basic competence into dialect forms, into registral forms. 1980 *Dædalus* Spring 189 The expansion of registral sonority and the appoggiatura, G-sharp, strongly stress the arrival of the A.

registrant (re·dʒistrănt). orig. *U.S.* [f. Register *sb.*^{1} or *v.* + -ant^1.] One who registers (in various senses), esp. one who thereby gains a particular entitlement.

1890 in Webster. 1928 *Index Trade-Marks U.S. Patent Office 1927* 5 (*heading*) Alphabetical list of registrants of trade-marks for the year 1927. 1942 *Nation* (N.Y.) 27 Apr. 481 Up to Jan. 1, 1942, the Army rejected 8 per cent of the registrants passed as physically fit by Selective Service. 1955 W. W. Denlinger *Compl. Boston* 65 The many registrants carrying this name in the studbooks since 1934. 1964 *Economist* 24 Oct. 360/2 The registrants are more likely to be Democrats. 1976 *Century of Trade Marks* (Patent Office) 8/2 The definition specifically includes marks 'proposed to be used'; though the Courts have held that this means that the registrant must have a 'present intention' to use the mark. 1977 E. Ambler *Send no more Roses* ii. 18 The subject of that particular seminar was of fairly general interest and the number of registrants was high. 1979 *Rescue News* Mar. 2/6 The registration fee for the symposium is £4... Registrants will receive the definitive Symposium programme.

registrar. Add: (Also with pronunc. redʒistrā·ɹ.) **1.** *spec.* the title of (*a*) a senior officer with administrative responsibility in certain universities; (*b*) a local official responsible for maintaining an index of births, marriages, and deaths in the area under his authority.

(*a*) 1756 *Reply to Dr. Huddesford's Observations relating to Delegates of Press* 4 A Convocation being appointed to be held in the Theater on the second of July, the Vice-Chancellor gave directions to the Registrar to prepare the forms of nomination. 1797 [in Dict.]. 1870 D. P. Chase *Registrarship of University* 5 The Registrar has been relieved of a great amount of labour connected with the University accounts. 1900 *Statuta et Decreta Univ. Oxon.* 283 The Registrar of the University shall be elected in Convocation. He..is required to attend..all meetings of the Houses of Congregation and Convocation and of the Congregation of the University,..and generally to perform all duties necessary for carrying on the business of the Houses. 1943 'B. Truscot' *Redbrick Univ.* iii. 59 After a pause for breath we come to the whole of the University [of Bristol] Council, the Deans of Faculties, the Professors and Professores Emeriti, the Librarian, the Registrar, twenty-nine representatives of Convocation, [etc.]. 1953 K. Amis *Lucky Jim* i. 16 He'd been passing behind the Registrar's chair.., had stumbled and had knocked the chair aside just as the other man was sitting down. 1975 J. Mann *Captive Audience* i. 10 The crowd of students..far from being calmed by the duplicated communication which the registrar had delivered had become wild and agitated. 1980 *Times* 1 Aug. 15/7 (Advt.), In view of the forthcoming retirement of the present Registrar, applications are invited for the post of Registrar of the University of Wales.

(*b*) 1876 C. M. Yonge *Three Brides* II. xiii. 242 They put up their banns at the Union at Brighton, and were married by the Registrar. 1880 A. Trollope *Duke's Children* II. xxvii. 325 None of your private chaplains... Just the registrar, if there is nothing better. 1892 I. Zangwill *Childr. Ghetto* II. i. xxv. 218 Let us be married honestly by a registrar. 1967 *Guardian* 1 Aug. 4/3 The shot-gun marriages tend to take place in the registrar's office under the mistaken impression that the church does not marry pregnant brides.

3. A doctor of a certain grade in a hospital: orig. a junior doctor whose duties included the maintenance of a register of patients; now usu. a senior officer undergoing training as a specialist or consultant. Cf. *resident *sb.* 3.

1862 *Med. Times & Gaz.* 18 Oct. 411/2 Besides these there are a Resident Medical Officer, or Physician's Assistant..; a Medical and Surgical Registrar at a salary of £25 a year; two House Surgeons. 1894 *Brit. Med. Jrnl.* 10 Nov. 1089/1 Rayner, Herbert E., F.R.C.S.Eng., appointed Surgical Registrar and Anæsthetist to the Hospital for Sick Children, Great Ormond Street. 1937 *Ibid.* 4 Sept. 470/2 Qualified students of the school can obtain appointments as house-physicians and house-surgeons, obstetric assistants, surgical, gynaecological, and medical registrars. 1961 *Lancet* 29 July 264/2 There would seem to be intra-professional divisions in which interests do not quite coincide—e.g., the unplaced registrars and the established consultants. 1965 P. Ferris *Doctors* iii. 59 What senior registrars want is to be appointed consultants. 1977 *Western Morning News* 30 Aug. 3/3 Some new patients have to wait as long as two to three years before they are seen, because the consultant surgeons spend so much of their time with follow-up cases; these could be handled easily and effectively by registrars. 1980 *Times Lit. Suppl.* 1 Aug. 879/2 In interviews with residents (in Britain, registrars) she found that they expressed strong preference for the middle-class patient.

registrarship. Add: (Also stressed *registra·rship.*) (Further examples, with reference to *registrar 3.)

1889 *Brit. Med. Jrnl.* 9 Nov. 1077/1 London Hospital,.. —Surgical Registrarship. Salary £100 per annum. 1937 *Ibid.* 4 Sept. 467/2 In addition, the following appointments are open to all qualified students of the hospital:..two medical registrarships at £100 per annum. 1963 *Lancet* 12 Jan. 117/2 After he was demobilised in 1946 he held registrarships in Bristol at Southmead Hospital and the Children's Hospital.

registration. Add: **1. b.** Also (occas. without article) *ellipt.*, = **registration number* below.

1973 J. Pattinson *Search Warrant* vii. 111 A blue Chrysler... It had a New York registration. 1973 'I. Drummond' *Jaws of Watchdog* xii. 152 His car..was a Ferrari but with British registration. 1976 L. Deighton *Twinkle, twinkle, Little Spy* x. 111 'The same registration!' said Mann excitedly. 'That makes four times the same number.'

c. *registration fee*; also (with reference to the registration of motor vehicles) *registration book, number, plate.*

1922 *Michelin Guide Gt. Brit.* III. 711 When disposing of his car the motorist must fill in the name and address of the new owner in the registration book. 1968 'D. Rutherford' *Skin for Skin* ii. 17 The Morris 1000 was parked at the kerb-side... I had the registration book and the cover note ready. 1869 *Bradshaw's Railway Man.* XXI. 14 Certificates are required for transfers. Registration fee 2*s.* 6*d.* each deed and seller. 1967 *Post Office Guide* 85 The registration fee must be paid by postage stamps affixed to the cover. 1977 J. Bingham *Marriage Bureau Murders* ii. 24 There would be a down payment, a registration fee..and further annual payments. 1903 *Act* 3 *Edw. VII* c. 36 § 6 A person driving a motor car shall,..if an accident occur to any person,..or to any horse or vehicle..owing to the presence of the motor car on the road, stop and, if required, give his name and address, and also..the registration mark or number of the car. 1911 *Motor Man.* (ed. 13) viii. 260 A registration number once issued cannot be transferred to another car. 1959 M. Gilbert *Blood & Judgement* xii. 128 Are you the owner of a blue Riley saloon car, registration number GKR 692? 1977 B. Pym *Quartet in Autumn* v. 39 The car was an important status symbol and large sums of money could be paid for particularly desirable registration numbers. 1956 Registration plate [see *all over *advb. phr.* 1 b]. 1974 F. Nolan *Oshawa Project* xviii. 170 A black Volkswagen with a Darmstadt registration plate. 1977 'D. Cory' *Bennett* v. 133 Relatively few cars, and fewer still with foreign registration plates.

2. (Further examples.) Also used with reference to other keyboard instruments, esp. the harpsichord.

1921 G. A. Audsley *Organ-Stops* 1 Haphazard methods of registration must be shunned. 1961 R. Russell in A. Baines *Mus. Instr. through Ages* iv. 80 Undue preoccupation with such things as registration..tends to obscure the fundamental musical requirements of the instrument [sc. the harpsichord]. 1966 *Listener* 19 May 737/3 He is the most characterful harpsichord player since Landowska. Purists may question his frequent changes of registration. 1974 *Daily Tel.* 18 Feb. 11/1 The excesses to which Bach's 'Goldberg' Variations can easily lend themselves were strictly avoided by George Malcolm..on the harpsichord... Registration was also kept within reasonable limits. 1976 *Gramophone* May 1761/2 Chorzempa's [organ] registration balances happily with the orchestra.

3. Substitute for def.: The state of being in register (Register *sb.*^{1} 11 b in Dict. and Suppl.), or the action of obtaining this.

1901 *Chambers's Jrnl.* June 364/1 The skilled attendant replaces them in the clip, one upon another, taking a little care to ensure perfect 'registration'..and, lo! there is a finely painted lantern slide! 1949 *Electronics* Dec. 69/2 The three color images in camera and picture tube must be very precisely aligned, both electrically and optically, to secure accurate registration. 1959 Halas & Manvell *Technique Film Animation* III. xix. 218 Background artists should know what registration manipulation is possible under the camera. 1962 W. H. Stevens in G. A. T. Burdett *Automatic Control Handbk.* viii. 61 Separate panels are now cut from the multiple boards, and the fixing and component holes are drilled or punched... Registration of the holes with the circuit is achieved by means of the pilot holes. 1967 *Listener* 30 Mar. 424/2 The three pictures are equally focused and accurately in registration one with the other. 1971 D. Potter *Brit. Eliz. Stamps* iii. 37 The line was printed in two operations as part of the tricolour production, and the very accurate registration required was not forthcoming. 1975 J. B. Harley *O.S. Maps* i. 11 A second sheet of plastic material carrying an opaque coating is placed in exact registration with the first.

registrative, *a.* For *rare*$^{-1}$ read *rare* and add further example.

1878 W. James in *Jrnl. Specul. Philos.* XII. 11 At one time, 'scientific' thought, mere passive mirroring of outward nature, purely registrative cognition..would seem to be his [sc. Spencer's] ideal.

registree (redʒistrī·). [f. Register *v.* + -ee^1.] One who is registered (in various senses).

1923 G. B. Shaw in *Daily News* 18 Dec. 6/1 My refusal to credit the trade union known as the General Medical Council with the power to confer Omniscience and Infallibility on its registrees. 1966 *Punch* 28 Dec. 945/3 Miss Shuter at the desk had hysterics when an irate would-be registree broke into her glassy sanctuary and shook her by the shoulders.

re·gistry o:ffice. 1. = Register office; *spec.* † a place where a register of positions in domestic employment is kept (*obs.* exc. *hist.*).

1728 Swift *Let. c* 10 May in *Works* (1766) XVII. 169, I will take up the bones, and make of it a skeleton, and put it in my registry office. 1834 J. S. Mill in *Monthly Repos.* VIII. 439 Prying into the records of the..registry office. 1836 [in Dict.]. 1839 J. Romilly *Cambr. Diary* 16 Sept. (1967) 178 Lucy went..to the Registry Office..to get a place for Frances Wilderspin. 1892 C. M. Yonge *Cross Roads* xv. 157 She was in communication with the registry office there; but she would not take what the matron of the lodge called 'rackety situations'. 1910 E. M. Forster *Howards End* vii. 59 Would you come round with me to the registry office? There's a housemaid who won't say yes but doesn't say no. 1964 M. Laski in S. Nowell-Smith *Edwardian England* iv. 144 Registry offices abounded, but for really high-class servants the best method was..use of the advertisement columns of the *Morning Post.*

2. = *register office.

1911 G. B. Shaw *Getting Married* 236 Marriages gave place to contracts at a registry office. 1931 J. S. Huxley *What dare I Think?* vi. 205 The marriage ceremonial among most primitive peoples..contains a religious motive, just as much as does a Christian wedding ceremony (and just as little as does a wedding in a registry office). 1974 'R. Tate' *Birds of Bloodied Feather* iii. 67 Can't we just make a date? It only means slipping into a registry office. 1976 *Daily Times* (Lagos) 27 Aug. 16/4 Workers at the registry office explained to Ajar that according to the law his wife could retain her maiden name. 1980 J. Cartwright *Horse of Darius* ii. 27 They..were married in the registry office.

reg'lar (re·glăɹ), *a.* and *adv.* Repr. a colloq. pronunc. of Regular *a.* and *adv.*

1842 Dickens *Let.* 17 Feb. (1974) III. 69 The Newhaven serenade was not so good; though there were a great many voices, and a 'reg'lar' band. 1843 —— *Martin Chuzzlewit* (1844) xxv. 306, I says 'my half a pint of porter fully satisfies; perwisin', Mrs. Harris, that it is brought reg'lar, and draw'd mild.' 1899 F. W. Maitland *Let.* 10 Oct. (1965) 201 Wherever I go I shall expect my E.H.R. 'reglar'. 1950 *Collier's* 6 May 56/2 (*caption*) I'm a 'real pal'—a 'reg'lar fellow'—a 'good scout'!

reglementary, *a.* (Later example.)

1937 M. Covarrubias *Island of Bali* iii. 58 The independent village is called a *desa*, a term we shall employ to designate the legal, 'complete' village that has the three reglementary temples.

regnancy. Restrict *rare*$^{-1}$ to sense in Dict. and add: **2.** *Psychol.* In the sense of *regnant *ppl. a.* 2 c.

1938 [see *regnant *ppl. a.* 2 c]. 1963 S. R. Maddi in Wepman & Heine *Concepts of Personality* vii. 185 Murray..with his concept of regnancy, has gone a bit further than Allport in attempting to conceptualize relevant brain processes. 1964 Gould & Kolb *Dict. Soc. Sci.* 352/2 The authors point to Allport's biophysical traits and Murray's regnancies as illustrations of hypothetical constructs.

regnant, *ppl. a.* Add: **2. c.** *regnant process* (Psychol.): in the theory of personality, a hypothesis that dominant brain processes exist which determine behaviour (see quot. 1938).

1938 H. A. Murray *Explor. Personality* ii. 45 It may prove convenient to refer to the mutually dependent processes that constitute dominant configurations in the brain as *regnant* processes; and, further, to designate the totality of such processes occurring during a single moment..as a *regnancy.* 1974 W. B. Arndt *Theories of Personality* xii. 237 We must infer the characteristics of regnant processes from the behavior of organisms.

regolith (re·goliþ). *Geol.* [erron. f. Gr. ῥῆγο-ς rug, blanket + -lith.] The unconsolidated solid material covering the bedrock of a planet.

1897 G. P. Merrill *Treat. Rocks* v. 299 This entire mantle of unconsolidated material, whatever its nature or origin, it is proposed to call the regolith, from the Greek words ῥηγος, meaning *a blanket*, and λιθος, *a stone.* 1935 *Jrnl. Geol.* XLIII. 745 'Regolith' was introduced by Merrill to include all unconsolidated surficial material and therefore embraces far more than residual weathered rock. 1949 F. J. Pettijohn *Sedimentary Rocks* ix. 282 Residual soils (regolith of Merrill, saprolith of Becker, and sathrolith of Sederholm) are the products of weathering formed *in situ.* 1970 *Nature* 24 Jan. 321/2 The solid rocks at Tranquillity Base are covered by a 4–6 m thick regolith or dust layer composed of local rock fragments..and spheres or fragments of glass. 1976 J. Kleczek *Universe* iv. 155 The solid lunar globe is covered by a layer of loose broken rock material called regolith. 1977 A. Hallam *Planet Earth* 16/1 Meteorite debris [on the moon] amounts to only about 2% of the sampled regolith.

Hence **regoli·thic** *a.*

1955 *Trans. R. Soc. N.Z.* LXXXII. 1015 Under the soil of the upland surface of the Belmont plateau there are arrested streams or sheets..of formerly regolithic debris now forming deposits of head on slopes. 1977 *Nature* 6 Jan. 38/2 Since the returned sample is of regolithic materials it could also contain basalt fragments.

regosol (re·gosǫl). *Soil Sci.* [erron. f. as prec. + *-sol.] A poorly developed soil without definite horizons, overlying and formed from deep, unconsolidated deposits such as sand or loess.

1949 Thorp & Smith in *Soil Sci.* LXVII. 120 Soon after the definition of Lithosols was published in the 1938 Yearbook, it was realized that many weakly developed soils occur in deep soft-rock deposits, like loess and sand, that are not *stony* in the ordinary sense of the word. These nonstony soils were called lithosols for a time, but a practical need was felt for distinguishing deep soft soil materials from very stony ones. Hence the proposal of the name *Regosol.* This new term is not yet fully established in the literature, and it may prove desirable eventually to give the concept some other status than that of a great soil group. 1968 R. W. Fairbridge *Encycl. Geomorphol.* 523/1 Soil scientists sometimes use 'regosol' for soils developed without distinct horizons over deep unconsolidated 'rock', e.g., a mature alluvial formation. 1976 *Sci. Amer.* Sept. 174/3, ·7 billion hectares of sandy, undifferentiated soils called regosols are nonarable.

Hence **regoso·lic** *a.*

1956 *Proc. Soil Sci. Soc. Amer.* XX. 268/2 The Iowan loess is separated from the Farmdale loess by a regosolic buried soil.

regrade, *v.*^{3} (Earlier and later examples.)

1826 A. Macomb *Let.* 18 Nov. in *Reg. Deb. Congr. U.S.* (1829) III. 1572 The road..is to be regraded. 1869 *Rep. Comm. Agric. 1868* (U.S. Dept. Agric.) 362 They may be readily and rapidly leveled in the construction of a new road, or regraded when displaced by wear. 1960 *Farmer & Stockbreeder* 26 Jan. 70/2 The pigs are..regraded by the Association's own system of selective grading. 1977 *Times* 18 Mar. 4/5 Unless the corporation offers to regrade the cameramen there is little chance that their union..will allow the programme to go ahead.

Hence **regra·ded** *ppl. a.*, **regra·ding** *vbl. sb.*

1920 *Glasgow Herald* 1 Sept. 6 They have agreed to a conference to consider regrading. 1923 *Ibid.* 28 Mar. 10 Heavy extraordinary expenditure has to be budgeted for, including three-quarters of a million for arrears of regraded salaries. 1956 D. L. Linton *Sheffield* 41 The regrading consequent on these changes would favour the Froggatt stream. 1962 A. Battersby *Guide to Stock Control* 105 Transferring from one stock to another, e.g. re-grading.

regra·ss (ri-), *v.* [Re- 5 c.] To put (land) under grass again. So **regra·ssing** *vbl. sb.*

1901 *Yearbk. U.S. Dept. Agric.* 30 Experiments in regrassing were undertaken at Tucson, Ariz., in cooperation with the agricultural experiment station. 1940 *Advisory Bull. War Food Production* No. 1. 1 Ploughable grassland

too poor to be initially cereal-worthy should..be ploughed and immediately re-grassed. *Ibid.* 28 Re-grassing properly and methodically undertaken enormously extends the effective grazing season.

regress, *v.* Add: **2. b.** *Psychol.* To return in one's mind to an earlier period or stage of life as a result of mental illness or through hypnosis or psychoanalysis. Also *trans.*, to induce regression in (a person). See also *regression 4 d.

1926 J. I. Suttie tr. *Ferenczi's Further Contrib.* xi. 137 Now the stage to which these two neurotics regressed seems to be the infantile stage of the first year of life. 1950 *Psychoanalytic Q.* XIX. 501 The immutability of a constant, passive environment forces him to adapt, i.e., to regress to infantile levels. 1956 Ambrose & Newbold *Handbk. Med. Hypnosis* vii. 146 If a child can be hypnotised and regressed with suitable suggestions, causing him to re-live the actual traumatic episode, much tension can be overcome. 1957 P. Lafitte *Person in Psychol.* vi. 75 The person changes.. for the worse, perhaps regressing directly to infantile behaviour. 1960 *Times Lit. Suppl.* 3 June 356/3 Harry was later hypnotized by a friend of Dr. Puharich and 'regressed' through his life memories to see if he had any knowledge of Egyptian history, language, or religion. 1970 T. X. Barber *LSD, Marihuana, Yoga & Hypnosis* vi. 259 When regressed hypnotically to the time of the original conditioning, all subjects again manifested the eye-blink response. 1976 F. H. Frankel *Hypnosis* v. 67 He was then regressed in time, and referred to business difficulties that he had experienced earlier that day and in recent weeks. 1978 Gris & Dick *New Soviet Psychic Discoveries* ix. 104 A girl student..insisted that she be regressed by a co-student.

3. *intr. Genetics.* To tend or evolve towards the mean value for the population; to display regression to the mean (*regression 4 b).

1885 *Nature* 24 Sept. 509/2 The type is an ideal form towards which the children of those who deviate from it tend to regress. *Ibid.* 510/1 The stability of a type would, I presume, be measured by the strength of its tendency to regress. 1892 F. Galton *Finger Prints* i. 21 There is a constant tendency in the offspring to 'regress' towards the parental type. 1909 *Westm. Gaz.* 21 Apr. 5/1 There is a tendency for children of exceptional parents to regress towards the average stock. 1953 Srb & Owen *Gen. Genetics* xxiii. 497 Instead of showing the average value of their selected parents, the progeny regress from this value toward the original population mean, and in fact average only a little better than the population from which their parents came. 1975 A. Smith *Human Pedigree* iii. 67 The reason why we have not ended up as uniform as tailor's dummies is that there is only a tendency to regress to the mean... The various genes involved in such a character as weight will combine, from time to time and to confound the general rule, in a manner that is unexceptional [*sic*]. A child will then be heavier than both its parents. It will not have regressed.

4. *trans. Statistics.* To calculate the coefficient(s) of regression of (a variable) *against* or *on* another variable. *colloq.*

1971 *Nature* 8 Oct. 407/1 These parameters were regressed on measurements made of site factors on five sampling sites from the fives dated flows. 1977 D. M. Smith *Human Geogr.* vii. 170 Katzman..estimated several education production functions by regressing measures of output against measures of school input and local socio-economic status of residents.

regressed (rĭgre·st), *ppl. a.* [f. prec. + -ed^1.] That has regressed or been regressed.

1948 *Jrnl. Nerv. & Mental Dis.* CVII. 443 Lewis's observation goes back to 1926 when he saw a regressed schizophrenic patient reacquire a dorsiflexor response to plantar stimulation. 1965 B. E. Freeman tr. *Vandel's Biospeleol.* xxvi. 419 The variations in the regressed eyes of the mole may be noted.

regression. Add: **4. a.** (Later examples.)

1917 *Jrnl. Genetics* XXVII. 117 What is frequently called reduction in evolution, or, to use a less ambiguous term, regression. 1950 *Sci. News* XV. 136 They pointed out that the regression of tumours caused by Compound E did not generally last indefinitely. The tumours usually recurred. 1965 B. E. Freeman tr. *Vandel's Biospeleology* xxvi. 417 Regression of the eyes is more marked when it is phyletically ancient.

b. *Genetics.* The tendency for the mean value of a partially inherited quantitative character, among any class of relatives of an individual or a group chosen for their values of that character, to lie between the (mean) value for that individual or group and the mean value in the general population. Esp. as *regression to the mean.*

1885 F. Galton in *Nature* 24 Sept. 507/1 The experiments showed further that the mean filial regression towards mediocrity was directly proportional to the parental deviation from it. 1889 [in Dict., sense 4]. 1912 J. A. Thomson *Heredity* (ed. 2) ix. 321 The amount of the regression affords a useful measure of the intensity of the inheritance. If the regression is slight, it means that the intensity of the inheritance is high. 1952 Srb & Owen *Gen. Genetics* xxiii. 497 This phenomenon, in which the progeny of selected parents slip back toward the average of the population from which the parents were chosen, has long been known. In this connection the phenomenon was called regression. Today, the term regression has a broader statistical connotation, but it is still applicable in its original sense to the problems we are discussing. 1975 A. Smith *Human Pedigree* iii. 66 At this stage it is necessary to refer to the phenomenon known as the regression to the mean. Where random mating exists, and where there is no evolutionary pressure favouring any characteristic that is controlled by many genes, the offspring will have a tendency to be nearer average for that characteristic than their parents.

c. *Statistics.* The relationship between the mean value of a random variable and the corresponding values of one or more other variables; **coefficient of regression** = *regression coefficient* in sense 8 below.

1897 K. Pearson in *Phil. Trans. R. Soc.* A. CLXXXVII. 259 The coefficient of regression may be defined as the ratio of the mean deviation of the fraternity from the mean offspring to the deviation of the parentage from the mean parent. *Ibid.*, From this special definition of regression in relation to parents and offspring, we may pass to a general conception of regression. Let A and B be two correlated organs (variables or measurable characteristics) in the same or different individuals, and let the sub-group of organs B, corresponding to a sub-group of A with a definite value *a*, be extracted. Let the first of these sub-groups be termed an array, and the second a type. Then we define the coefficient of regression of the array on the type to be the ratio of the mean-deviation of the array from the mean B-organ to the deviation of the type *a* from the mean A-organ. 1917 *Phil. Mag.* XXXIV. 205 When the regression of the first variable on the remaining $n-1$ variables is linear, the multiple correlation coefficient measures the dependence of the first variable on the others. 1925 R. A. Fisher *Statistical Methods for Res. Workers* v. 114 The following qualitative examples are intended to familiarise the student with the concept of regression. 1943 M. G. Kendall *Adv. Theory Statistics* I. xiv. 328 In this chapter we shall mainly be concerned with the case in which regressions are linear or very nearly so. 1952 C. G. Lambe *Elem. Statistics* vii. 56 This straight line which gives an estimate of the average value of y associated with any value of x is called the line of regression of y on x and p/σ_x^2 is called the coefficient of regression of y on x. 1972 T. H. & R. J. Wonnacott *Introd. Statistics for Business & Economics* xiii. 287 Multiple regression is the extension of simple regression, to take account of the effect of more than one independent X variable on the dependent variable Y.

d. *Psychol.* The process of regressing, or a tendency to regress, in the sense of *regress *v.* 2 b; *spec.* the tendency of the libido, under the stress of frustration, to return to a simpler and more satisfying stage of development; also, the state of returning mentally to an earlier period, esp. in hypnosis and psychoanalysis.

1910 tr. S. Freud in *Amer. Jrnl. Psychol.* XXI. 214 The flight from the unsatisfying reality into what we call.. disease, but which is never without an individual gain in pleasure for the patient, takes place over the path of regression, the return to earlier phases of the sexual life, when satisfaction was not lacking. 1913 C. G. Jung *On Psychoanalysis* in *XVIIth Internat. Congr. Med.* § xii. 68 [Freud] called this phenomenon of reactivation or secondary exaggeration of infantile reminiscences 'Regression'. 1920 *Challenge* 21 May 44/3 The *libido*..in its regression to the collective unconscious, gives rise to the similation of archaic psychical adaptations. 1948 *Jrnl. Nerv. & Mental Dis.* CVII. 443 In regression to infantile levels the subject assumed the sleeping posture of an infant. 1961 *Economist* 11 Mar. 962/1 The poor layman who has laboriously got on to nodding terms with infantile sexuality, regression, Oedipal conflicts, displacement and the rest. 1970 T. X. Barber *LSD, Marihuana, Yoga & Hypnosis* vi. 255 Under regression to infancy, the hypnotized person does not topple from his chair. 1971 *Jrnl. Gen. Psychol.* Apr. 208 In psychology the term 'regression' refers to a primitivization of behavior. 1978 Gris & Dick *New Soviet Psychic Discoveries* ix. 105 Let me amplify on regressions. Whatever people think, their previous lives are not individual experiences.

7. *Geogr.* A retreat or withdrawal of the sea from the land.

1908 W. J. Sollas et al. tr. *Suess's Face of Earth* III. 364 Every new transgression (regression), in so far as the encroaching line of breakers itself has not denuded the land, will encounter an altered relief. 1937 *Bull. Amer. Assoc. Petroleum Geologists* XXI. 1436 Rhythmic transgressions and regressions of the sea continued throughout the period of Jackson sedimentation, evidenced by the interwedging of marine and non-marine sediments, as the strand line moved..back and forth. 1975 *Sci. Amer.* Feb. 90/3 The stratification of sedimentary deposits suggested successive marine transgressions onto the continents and regressions from them. The regressions could be attributed to the subsidence of the ocean basins and the transgressions to the partial filling of the basins with sediment eroded from the continents.

8. *attrib.* and *Comb.*, as (sense *4 c) *regression analysis, formula, function, theory*; **regression coefficient,** a coefficient in the regression equation; *esp.* the first-order coefficient, which is estimated by the covariance of the two variables divided by the variance of the independent variable; **regression curve,** a graph of the expected value of the dependent variable plotted against the value of the independent variable(s); **regression equation,** an equation which gives the expected value of the dependent variable as a function of the value(s) of the independent variable(s); **regression line** = *regression curve* above.

1948 *New Biol.* IV. 36 We can then use the technique of regression analysis to determine to what extent we can account for the variation in yield in terms of variation in rainfall. 1976 *National Observer* (U.S.) 19 June 18/3 It may seem absurd to label the probability of murder as 'P(M)' and subject it to the technique of regression analysis. 1903 *Phil. Trans. R. Soc.* A. CC. 20 Not only the slope (regression coefficient) of the line, but its position is identical. *Ibid.* 21 If an organ has been modified only by indirect selection, then its partial regression coefficients on any complex of other organs, however large or small, provided it includes all the directly selected organs, will remain unchanged by the selection. 1925 R. A. Fisher *Statistical Methods Res. Workers* v. 114 The regression coefficients are of interest and scientific importance in many classes of data where the correlation coefficient, if used at all, is an artificial concept of no real utility. 1964 R. von Mises *Math. Theory Probability & Statistics* xi. 576 The correlation coefficient is the geometrical mean of the two regression coefficients. 1905 *Res. Mem. Drapers' Co.* XIV. 21 Yule's method of approaching the problem from the form of the regression curves is available and capable of very great extension. 1925 R. A. Fisher *Statistical Methods Res. Workers* v. 114 The function which represents the mean height at any age is termed the regression function of height on age; it is represented graphically by a regression curve, or regression line. 1943 M. G. Kendall *Adv. Theory Statistics* I. xiv. 327 The means of arrays will in general lie more or less closely round smooth curves... Such curves are called regression curves and their equations..are called regression equations. 1972 G. A. Beaumont *Elem. Math. Statistics* xii. 152 We begin with a certain minimal property of the regression curves. 1897 *Proc. R. Soc.* LX. 480 The characteristic or regression equations which we have to find. 1943 Regression equation [see *regression curve* above]. 1978 *Nature* 18 May 184/2 The proportion of non-scientific staff [*A*]..is determined largely by the total staff employed (*S*) and the number of addresses amongst which it is dispersed (*D*) according to the regression equation $A = 22{\cdot}23 \log S - 1{\cdot}86D - 8{\cdot}77$. 1971 *World Archaeol.* III. 115 The regression formula predicts that a compound with five adults will have approximately [illegible] huts. 1925 Regression function [see *regression curve* above]. 1904 *Biometrika* IV. 139 The actual degree of resemblance of our brothers being equally variable, is measured by the steepness of this regression line. 1925 Regression line [see *regression curve* above]. 1971 *World Archaeol.* III. [illegible] Variation away from the regression line is a function of wealth that is not used in this way. 1967 *Times Rev. Industry* Feb. 111/3 The lay reader should not be put off by the complicated language of regression theories, as Professor Kaldor's ideas are fascinating enough, and well enough argued in this lecture to be quite comprehensible to the non-economist.

regressive, *a.* Add: **1. c.** Also *spec.* of a tax, that bears proportionately harder on persons with lower incomes; *regressive assimilation* (examples).

1889 *Cent. Dict.*, Regressive assimilation. 1924 F. M. Stenton in Mawer & Stenton *Introd. Survey Eng. Place-Names* ix. 174 Some of these names, which enter into local nomenclature in considerable numbers, may be due to 'regressive assimilation'. 1939 [see *progressive *a.* 2 c]. 1964 C. Barber *Ling. Change Present-Day Eng.* iii. 62 Regressive assimilation..in which the sound exerting the influence comes later in the word than the one influenced. 1976 *Hansard Commons* 9 June 1597 The [licence] fee is a poll tax and it is regressive. It bears very hard on the worse-off.

d. *Psychol.* Of, pertaining, or relating to psychological regression.

1926 J. I. Suttie tr. *Ferenczi's Further Contrib.* xi. 137 Besides this regressive trait that fetters the patients to their bed..there may also be at work..the 'secondary' function of the neurosis. 1957 P. Lafitte *Person in Psychol.* xi. 161 Concentration camp life has plenty of examples of exceptional, as well as of regressive, behaviour. 1969 *Listener* 22 May 736/2 It [sc. *The Boston Strangler*] seeks to entertain by feeding us on clinical information about behaviour of the most regressive kind. 1970 R. F. Bales *Personality & Interpersonal Behav.* iii. 49 To..teach in such a way presumably helps one's normal defenses by providing in one's overt behavior a good example for the more regressive inner self.

4. *Geogr.* Of, pertaining to, or being a regression of the sea.

1937 *Bull. Amer. Assoc. Petroleum Geologists* XXI. 1436 Near the close of this regressive movement the Loma Novia, Government Wells, and Chernosky Sand members were deposited. 1950 *Ibid.* XXXIV. 284 The regressive type of bioherm or reef may contain within it back-reef types of sediments such as red shale and anhydrite, but the transgressive type does not. 1968 D. L. Eicher *Geologic Time* v. 49 Transgressive and regressive sequences generally do not contain a complete sedimentary record of all environments that prevailed laterally at the time. 1978 *Nature* 29 June 749/2 There seems little reason to invoke oscillations in sea level..to account for any other transgressive or regressive sequences observed.

regressiveness (earlier example).

1853 W. Bagehot in *Prospective Rev.* IX. 421 There was a want of prospectiveness and a superfluous amount of regressiveness.

regressi·vity. [f. Regressive *a.* + -ity.] The state of being regressive; regressiveness.

1904 G. S. Hall *Adolescence* I. ii. 55 Retarded development of an organ..is an indication of regressivity. 1972 *Times* 23 Oct. 18/6 VAT will have small regressivity.

regressor (rĭgre·səɹ). *Statistics.* [f. Regress *v.* + -or.] Any of the independent variables in a regression equation. Also *regressor variable.*

1956 *Jrnl. R. Statistical Soc.* B. XVIII. 230 Two multiple correlations based on the same numbers of regressors. 1961 Kendall & Stuart *Advanced Theory Statistics* II. xxviii. 355 We understand by 'linear regression' that the conditional mean value of y is a linear function of the regressors $x_1, \ldots, x_p$. 1978 *Jrnl. Econometrics* VIII. 307 (*heading*) Posterior distribution for the multiple correlation with fixed regressors.

regret, *sb.* **2. b.** (Earlier U.S. example.)
1851 T. A. Burke *Polly Peablossom's Wedding* 177 The invitations went out, and strange to say, not a single 'regret' was sent in; but all came.

regret, *v.* Add: **3.** *absol.* or *intr.* To feel regret.
1853 Mrs. Gaskell *Ruth* II. x. 281 Those who had umbrellas were putting them up; those who had not were regretting and wondering how long it would last. **1883** H. Conway *Called Back* vi. 77 'Do you regret, Mr. Vaughan?' 'No—not if there is a chance.'
regretting *vbl. sb.* (further example).
1907 G. B. Shaw *John Bull's Other Island* iv. 105 No more neglect, no more loneliness, no more idle regrettings and vain-hopings.

regretfully, *adv.* Add: **2.** It is to be regretted (that); = Regrettably *adv.* (A regrettable use, prob. after *hopefully *adv.* 2.)
1976 *New Statesman* 20 Aug. 237/1 Regretfully, that is no ground for leniency towards him. **1977** *N.Z. Woman's Weekly* 10 Jan. 36/4 Regretfully I'm one of those who suffer from an odd compulsion. **1977** *Times Lit. Suppl.* 15 Apr. 468/4 The investigators, who must regretfully remain anonymous, have produced..a richness of archaeological potential which it will take years to absorb and assess. **1977** *Jrnl. R. Soc. Arts* CXXV. 336/2 Regretfully, however, the editorial staff may justifiably be thought to stand criticized for what has been omitted.

regre·ttableness. [f. Regrettable *a.* + -ness.] Regrettable character.
1913 *Eng. Hist. Rev.* July 555 The regrettableness of the lapses from what might have been.

regrind (rī·graind), *sb.* [f. the vb.] An act of regrinding.
1952 L. J. St. Clair *Design & Use of Cutting Tools* x. 187 Tool θ_p would require a metal removal of ·037" from the end (regrind). **1958** *Engineering* Apr. 8/1 (Advt.), You know a Dormer drill right down to the last regrind. **1975** *Drilling Technol. & Collet Chuck* (Bristol Erickson Ltd.) 9 'Stubbing' means that for short holes 3 diameters of drill plus one diameter for chip clearance gives four times more drill life with feeds increased by at least 33⅓%, equals more holes per re-grind.

regrou·nd, *v.* [Re- 5 c.] *trans.* To furnish with a new ground or basis for etching, painting, etc. So **regrou·nding** *vbl. sb.* Cf. Ground *sb.* 6 and *v.* 8.
1831 J. Constable *Let.* 4 Dec. (1966) IV. 360, I know very well it can be blotched up, with dry point burr, regrounding, &c, but that is hateful. *Ibid.* 28 Feb. (1966) IV. 368 It is necessary..to reground a plate. **1937** *Discovery* Mar. 76/2 To add to the plate he cleans it thoroughly, regrounds it, draws the new work and continues as before.

regroup, *v.* Add: Also *intr.*
a **1944** K. Douglas *Alamein to Zem Zem* (1946) 69 We had all seen the enemy so disorganized that it did not seem possible he could regroup enough to give us much trouble. **1976** *National Observer* (U.S.) 22 May 10/3 It was very important for me to find day care so I could regroup psychologically. **1981** *Times* 25 Apr. 12/3 The host and the other men said their evening prayers. They regrouped around the fire.

regrou·pment. [f. Regroup *v.* + -ment.] Rearrangement in groups; a rearranged group. Also *attrib.*
1920 *Glasgow Herald* 1 July 6/4 A quiet continuance of the existing regime will allow..a regroupment without any definite break in development. **1961** *Encounter* Jan. 9/2 Villagers from the hills..have been 're-grouped' in new areas where they could be better protected and controlled—but only a third of these re-groupments have been completed. **1961** *Guardian* 12 May 7/4 Algerians..who have been herded by the French into 'regroupment centres'. **1963** *Economist* 17 Aug. 565/1 India is thinking of rounding up the ..Nagas into 'regroupment villages'.

regrou·t, *v.* [Re- 5 c.] *trans.* To furnish with grouting again.
1967 *Do it Yourself* Nov. 1330/2 The joints between ceramic tiles can be re-grouted with a proprietary cement. **1974** *Sunday* (Charleston, S. Carolina) 28 Apr. 1-C/2 The tiles will be 'thoroughly cleaned and regrouted and restored to their original newness', he said.

regrow, *v.* Add: Also *trans.* (Further example.) **regrowth,** add def.: the phenomenon of growing or increasing again, *esp.* the renewed growth of vegetation after partial destruction by harvesting, fire, etc.; also *concr.*, the new vegetation that results; (earlier and later examples).
1741 W. Ellis *Mod. Husb.* May v. 85 Folding Sheep..will likewise prevent the Regrowth of the Trefoil. **1920** A. S. Pringle-Pattison *Idea of God* 72 The Tubularia, a kind of sea-anemone, re-grows its flower-like head. **1944** *Forestry Terminol.* (Soc. Amer. Foresters) 60/1 *Regrowth*, herbage that grows after grazing or after the plants have gone through a period of dormancy. **1956** Peterson & Fisher *Wild Amer.* i. 17 Other great stretches, greener and fresher than their surroundings, showed where self-sown regrowth had reclaimed old cleared areas. **1977** J. L. Harper *Population Biol. Plants* xiv. 438 The sward was then rested from grazing until the regrowth had approached 12·5 cm. *Ibid.* xx. 630 This form of regrowth may be more significant than seedlings as a means of recovery after fire. **1977** *Listener* 2 June 710/3 The National Union of Tailors and Garment Makers..see evidence of the regrowth of the sweat-shop.

regula. 2. For *rare*⁻¹ read *rare* and add later example.
1870 S. H. Hodgson *Theory of Practice* II. iv. 255 Logic is the regula of the sequences in meanings..; Grammar the regula of sounds and language.

regular, *a.*, *adv.*, and *sb.* Add: **A.** *adj.* **2. a.** (Further examples in *Astr.*)
1811 *Phil. Trans. R. Soc.* CI. 298 The arguments which I have given in the foregoing article, where only nebulæ of an irregular round figure were considered, need not be repeated when a regular circular form is presented to our view. **1922** *Astrophysical Jrnl.* LVI. 164 This state of affairs led Sir John Herschel to avoid the discussion of physical distinctions among nebulae and to elaborate his father's formal classification in an ingenious manner. All nebulous objects were divided into regular and irregular, and the latter alone into nebulae and clusters. **1926** *Ibid.* LXIV. 324 The characteristic feature of extra-galactic nebulae is rotational symmetry about dominating non-stellar nuclei. About 97 per cent of these nebulae are regular in the sense that they show this feature conspicuously. **1973** L. Oster *Mod. Astron.* xx. 303 We can do little more than speculate at present as to why some spiral galaxies are barred and others are 'regular'.

d. *Math.* In various senses (see quots.).
1893 A. R. Forsyth *Theory Functions Complex Variable* i. 12 A function which is monogenic, uniform and, except at poles, continuous, is called a meromorphic function. [*Note*] Sometimes regular, but this term will be reserved for the description of another property of functions. *Ibid.* iii. 52 A point *a* in the plane may be such that a function of the variable has a determinate finite value there, always independent of the path by which the variable reaches *a*; the point *a* is called an ordinary point [*Note.* Sometimes a regular point.] of the function. *Ibid.* viii. 163 The singularities..in the vicinity of which each branch of the function is uniform. [*Note*] These singularities will, for the sake of brevity, be called regular. **1908** H. Hilton *Introd. Theory Groups Finite Order* ii. 8 A permutation with the same number of symbols in each cycle—such as (1 4 3) (2 5 7) (9 6 8)—is called regular. **1968** P. A. P. Moran *Introd. Probability Theory* iii. 117 It is convenient to use the words 'regular' and 'positively regular' for the cases where the vector $P^n p(o)$ converges to a vector independent of p(o), and where in addition this vector has all its elements positive. **1972** A. G. Howson *Handbk. Terms Algebra & Anal.* xxxi. 154 Some authors weaken the definition of an analytic function on a domain *S* by asking only that the function should be analytic (in the above sense) at all but a finite number of points of *S*. A function which is analytic in the stronger sense would be described by them as being regular on *S*. *Ibid.* xxxvi. 181 Those points..possess a tangent plane..and are called regular points. Points which are not regular are said to be singular. **1978** *Sci. Amer.* Oct. 96/2 A prime is regular if and only if it does not evenly divide the numerator of any of the first *p* − 3 numbers in the series of fractions called the Bernoulli numbers... Of the primes smaller than 100 all but 37, 59 and 67 are regular.

3. d. Also *spec.* of a long-standing client or customer.
1841 Dickens *Barn. Rudge* xi. 296 The regular Maypole customers..each..in..his allotted seat in the chimney corner. **1911** G. Stratton-Porter *Harvester* vii. 108, I have orders to fill for regular customers. **1966** H. Mills *In Pursuit of Evil* vi. 59 'Is that a regular customer of yours?' I said to the..woman who owned the shop. **1973** 'H. Howard' *Highway to Murder* x. 125 Most of the ladies who patronize my salon are regular clients.

e. orig. *U.S.* Designating size or quality: average, medium; standard. (In quot. 1952 as *sb.*)
1952 *Amer. Speech* XXVII. 266 With regard to sizes of suits there are three basic divisions: regulars—for men of average height and weight; shorts..and longs. **1977** *Guardian* 11 June 14/7 If a Big Mac is too big for you then you might settle for a regular hamburger. **1978** *N.Y. Times Mag.* 23 July 22/2 In gasoline 'regular' has changed from meaning 'without tetraethyl lead' to its opposite—'with lead'.

6. c. (Further examples.)
1956 [see *Belsen]. **1960** J. Rae *Custard Boys* II. xiii. 155 You're becoming a regular creeping Jesus. **1977** *Globe & Mail* (Toronto) 2 Mar. 6/6 Here we are today, being urged to go on a regular orgy of appeasement.

d. *regular fellow* (or *guy*), an agreeable, ordinary, or sociable person. Freq. as a term of mild approbation; a 'decent chap'. *colloq.* (chiefly *U.S.*).
A fortuitous juxtaposition in quot. 1840: cf Guy *sb.*² 2.
[**1840** Barham *Ingoldsby Legends* 1st Ser. 44 Did you see her..With her knees to her nose, and her nose to her chin, Leering up..You'd lift up your hands in amazement, and cry, '—Well! I never *did* see such a regular Guy!'] **1920** F. Scott Fitzgerald *This Side of Paradise* ii. 52, I know I'm not a regular fellow, yet I loathe anybody else that isn't. **1924** *Amer. Mercury* Jan. 51/2 He was just one of so many mute and inglorious Babbitts preparing to qualify as regular fellows. **1930** E. H. Lavine *Third Degree* xiii. 161 So he [*sc.* a policeman] usually decides to become a 'regular guy'. **1936** C. S. Lewis *Allegory of Love* iv. 173 Chaucer.. was not a 'regular fellow', *un vrai businessman*, or a rotarian. He was a scholar, a courtier, and a poet. **1936** H. L. Mencken *Amer. Lang.* (ed. 4) 254 When G. K. Chesterton made his first visit to the United States he was much upset when an admiring reporter described him as a regular guy. **1969** P. Wiles in Ionescu & Gellner *Populism* 167 W. J. Bryan was to a degree exceptional even in the USA, a 'regular guy'. **1977** *Zigzag* June 17/1 He seems to be a regular, normal guy.

8. *Astr.* Of a satellite: (see quot. 1951).
1948 D. ter Haar in *Kgl. Dansk. Vid. Selsk. Mat.-Fys. Medd.* XXV. No. 3. 63 This group has orbits which are all approximately in the equatorial plane of the primary and whose eccentricities are small. We shall call these satellites the 'regular' satellites. **1951** G. P. Kuiper in J. A. Hynek *Astrophysics* viii. 357 The satellite systems vary from the beautifully regular case of Uranus to a completely irregular system like Neptune... 'Regularity' is measured by low relative inclinations, low inclination with respect to the planetary equator, small orbital eccentricities, a common sense of motion—the same as that of planetary rotation—and some degree of regularity in the distances to the planet. **1960** *Jrnl. Brit. Astron. Assoc.* LXX. 35 Regular satellites, eighteen in number, travelling in almost circular orbits in the plane of the equator of the parent planet.

C. *regular-built* (earlier example).
1817 J. K. Paulding *Lett. from South* I. 105, I can do this without forfeiting my character as a 'regular built' traveller, whose duty it is to tell all he sees, and more besides.

D. *sb.* **4. a.** (Example in *sing.*)
1917 W. Owen *Let.* 23 Nov. (1967) 509 The C.O. is a terrible old 'Regular'.
b. (Later U.S. example.)
1894 'Mark Twain' *Let.* 28 Jan. (1917) II. 606 When the mind-curist is done with you, you *have* to call in a 'regular'.
c. (Earlier and later examples.)
a **1852** [see *casual *sb.* 3 a]. **1872** 'Mark Twain' *Roughing It* xliii. 299, I struck up friendships with the reporters of the other journals, and we swapped 'regulars' with each other and thus encouraged work. 'Regulars' are permanent sources of news, like courts, bullion returns, 'clean-ups' at the quartz mills, and inquests. **1937** S. L. Bernstein in C. Davy *Footnotes to Film* iv. 229 He can take the risk of alienating his 'regulars' in order to satisfy the requirements of the few. **1949** S. Gibbons *Matchmaker* xx. 240 Mr. Waite was not a Regular at The Peal of Bells. **1959** 'A. Gilbert' *Death takes Wife* ix. 112 A woman entered his shop... She wasn't one of his regulars. **1970** *Daily Tel.* 28 Sept. 2/2 The Ministry of Technology does not plan at present to introduce coal rationing..but it believes that local coal merchants will limit supplies to 'regulars only'. **1978** *Dumfries Courier* 20 Oct. 16/3 Their popularity with the Edenbank regulars was demonstrated last week when the couple were presented with a framed colour aerial photograph of the hotel.

5. (Further examples.) Also in phr. *to go regulars*, to share profits. Now *Obs.*
1840 H. Cockton *Valentine Vox, Ventriloquist* ii. 5 I'll hire the large concert room upon the Market Hill, and you shall go reg'lars in the profits. **1882** *Sydney Slang Dict.* 7/2 *Regulars*, a thief's share of plunder. *Ibid.* 10/1 A cross-cove, who had his regulars for stalling, cried 'Cop bung', as a pig was marking.

6. (Later example.)
1973 *Bodl. Libr. Rec.* IX. 11 Lines 1–2 are a list of solar regulars, sometimes called ferial regulars.

regularity. Add: **2.** *attrib.*
1925 C. D. Broad *Mind & its Place* x. 457 The two conditions..are *not* jointly sufficient..to cause a memory even on the most extreme form of the regularity-theory of causation. **1935** *Aristot. Soc. Suppl. Vol.* XIV. 47, I accept the rationalist view, because it seems to me the only alternative to the so-called regularity view—that which reduces causal inference to a mere psychological habit or instinct. **1951** A. C. Ewing *Fund. Questions Philos.* (1968) viii. 162 All this should make one hesitate very much before accepting the regularity theory merely because it is the simplest and keeps closest to what is empirically observed. **1954** A. J. Ayer *Philos. Ess.* vi. 145 The same is true even on a 'regularity' view of causation. **1965** *Language* XLI. 186 The second breakthrough [in linguistics] was achieved in the 1870's, in the emergence of what I shall call the regularity hypothesis. **1977** M. Mandelbaum *Anat. Hist. Knowl.* iii. 50 Hart and Honoré drew a sharp contrast between the plain man's notion of causation..and the regularity view, which they accepted as being..applicable in the sciences.

regularization. (Earlier example.)
1853 G. H. Lewes *Comte's Philos. Sci.* 289 The regularization of habitual or continuous intercourse.

re·gularizer. *rare.* [f. Regularize *v.* + -er¹.] A person or thing that produces regularity.
1921 Ld. Shaw *Lett. to Isabel* xvii. 98 'There is a world elsewhere.' That is the secret... That it is which is the great regularizer.

regulate, *v.* Add: **5.** *refl.*, and *intr.* for *refl.* *Biol.* To exhibit regulation (sense *1 b).
1902 *Archiv für Entwicklungsmech. der Organismen* XV. 228 Pieces which are more active may be expected to regulate more widely. **1926** J. S. Huxley *Ess. Pop. Sci.* 235 The portion of substance which in its normal position would have developed into a half, has the power, if isolated, of regulating itself and its internal structure so as to give rise to a whole. **1971** *Nature* 24 Sept. 233/1 If half the optic tectum is removed and the optic nerve regenerates, the system regulates to preserve a retinotopic projection. **1977** *Sci. Amer.* July 67/1 Parts of the early embryo of various animals can be removed and the remaining parts will embryonically regulate to form a normal whole.

regulated, *ppl. a.* Add: **a.** Also *regulated tenancy*, a tenancy the rent of which is regulated by the terms of the Rent Acts (see quot. 1965).
1965 *Act Eliz. II* c. 75 § 1 In this Act 'regulated tenancy' means—a) a tenancy to which the Rent Acts apply by virtue of this section; or b) a statutory tenancy arising on the termination of such a tenancy as is mentioned in paragraph (a) of this subsection. **1970** *Internat. & Compar. Law Q.* 4th Ser. XIX. ii. 206 Unfurnished tenancies under

the Rent Acts being either (a) 'controlled' or (3) 'regulated tenancies'.
c. Proverb.
1819 'P. Atall' *Hermit in America* i. 29 Accidents will happen in the best regulated families. **1850** Dickens *David Copperfield* xxviii. 291 'My dear friend Copperfield,' said Mr. Micawber, 'accidents will occur in the best regulated families; and in families not regulated by that pervading influence..of Woman.' **1864** C. M. Yonge *Trial* II. iii. 60 Accidents will happen in the best regulated families. **1939** W. S. Maugham *Christmas Holiday* x. 285 Accidents will happen in the best regulated families... If you find you've got anything the matter with you, go and see a doctor right away. **1961** M. Kelly *Spoilt Kill* iii. 159 Foul play... Even in the best regulated families.

regulating, *ppl. a.* Add: **3.** *Biol.* Of developing organisms or tissues (cf. *regulate *v.* 5).
1902 *Archiv für Entwicklungsmech. der Organismen* XV. 231 When a regulating piece is supplied with abundant food, growth and regulation are..very intimately connected. **1926** J. S. Huxley *Ess. Pop. Sci.* 259 This passage from a regulating to a non-regulating condition takes place during gastrulation, before the first structural differentiation, in the shape of the medullary plate, is visible at all.

regulation. Add: **1. b.** *Biol.* The property whereby a living organism can adapt the form of its body to accommodate for changes made or damage done to it, and whereby, in the normal course of development, the nature and growth of the various parts are so interrelated as to produce an integrated whole. Also *attrib.* [a. G. *regulation* (H. Driesch 1898, in *Ergebnisse d. Anat. und Entwicklungsgeschichte* VIII. 718).]
1902 *Archiv für Entwicklungsmech. der Organismen* XV. 187 The term regulation is employed here in the sense given by Driesch in including not only the actual regeneration of organs but any other changes, *e.g.*, changes in the general form or outline and proportions of the body which may accompany or follow the replacement of lost parts. **1928** J. S. Huxley *Essays Pop. Sci.* 261 For Haldane, regulation places organisms in a different category from any non-living systems: for Driesch, it demands the intervention of vitalistic 'forces' such as his hypothetical entelechy. *Ibid.* 280 Once the tissues of the organism have become differentiated and it is capable of function, extraordinary powers of regeneration and regulation are developed. **1948** *New Biol.* V. 121 When the two cells resulting from the first cleavage division of a newt's egg are separated, at least one and often both form a complete though half-size individual. And conversely, two two-celled embryos can be put together to form a single large individual. This phenomenon of regulation, the adjustment of the entire developmental mechanism to disturbances of this sort, is a characteristic of the early stages of development of many kinds of animals. **1970** Ambrose & Easty *Cell Biol.* xiii. 422 Eggs of this type, which have the capacity to redevelop normally after a disturbance, are known as regulation eggs.

2*. *Electr.* The degree to which the output (or some other property) of an apparatus remains the same when the load varies, expressed as the percentage change in the former for a given change in the latter.
1899 Franklin & Williamson *Elem. Alternating Currents* [illegible] 129 A transformer of which the secondary e.m.f. [illegible] out little with increase of current is said to have good regulation. **1900** M. A. Oudin *Standard Polyphase Apparatus & Systems* iii. 42 A certain three-phase unit [illegible] machine of large output gave a regulation of 6¾ per cent. from full load to no load [illegible]. **1947** *Proc. IRE* XXXV. 444/2 The operations can be performed to any desired degree of precision, providing power supplies of excellent regulation and circuit components of high precision are used. **1977** *Physics Bull.* June 247/2 (advt.), The 227 output is dependable, too. Excellent regulation of 0·005%, stability to 0·05% and low noise combine to assure high precision output. **1977** *Design Engin.* July 81/1 Output regulation is specified as less than 0·1% for max to min load and for ±10% line voltage.

regulative, *a.* Add: **2.** *Biol.* Pertaining to or being regulation (sense *1 b); whose development is guided by regulation; opp. *mosaic *a.*¹ 5.
1902 *Archiv für Entwicklungsmech. der Organismen* XV. 229 Recognition of the fact that certain regulative processes are wholly or in part mechanical in nature is of importance for the future study of regulation. **1933** [see *mosaic *a.*¹ 5]. **1933** J. H. Woodger tr. *L. von Bertalanffy's Mod. Theories Devel.* x. 143 In the regulative eggs the cleavage occurs before segregation, so that every blastomere still contains the requisites for the formation of the whole organism. **1936** *Jrnl. Exper. Zool.* LXXIV. 91 The ability of the separated half blastomeres following the first cleavage to produce a one-half of normal size—so-called mosaic eggs—or a whole of their own size—so-called regulative eggs. **1947** L. B. Arey *Developmental Anat.* (ed. 5) ix. 161 All gradations exist between determinative, mosaic eggs and indeterminate, regulative eggs. *Ibid.*, Even the mosaic egg of a tunicate is regulative before fertilization occurs. **1973** R. Searls in S. J. Coward *Developmental Regulation* ix. 241 A number of experiments indicate that the limb mesenchyme is completely regulative.

regulatively *adv.* (earlier and later examples).
1854 H. L. Mansel *Man's Concept. Eternity* 10 A conception which is *speculatively* untrue may be *regulatively* true. **1952** *Mind* LXI. 554 Principles which function regulatively in ethical inquiry.

regulator. Add: **4. b.** *Econ.* A change in the rate of taxation which the Chancellor of the Exchequer may use to manipulate the economy between budgets; the power to operate such an alteration.
1961 *Daily Tel.* 18 Apr. 24/6 (*heading*) Economy 'regulators'. Changes without budget. *Ibid.*, This is the power he is taking to operate two new 'regulators' of the economy at any moment the Government thinks fit. *Ibid.* 7 July 1/8 Labour spokesmen maintained their objection to the Bill mainly on the grounds that the two 'economic regulators' were left in the hands of the Government. **1968** *Times* 29 Nov. 2/4 The activation of the 10 per cent regulator has effectively doubled the tax..since the spring. **1976** *Daily Tel.* 1 Nov. 16 Full use of the regulator by itself would raise an extra £1,100 million of revenue in a full year and add 1½ p.c. to prices. The regulator currently permits variations of up to 10 p.c. in VAT rates and in excise duties such as those on drink, tobacco and petrol between Budgets. **1977** *Whitaker's Almanack 1978* 354/1, I have decided to use my regulatory powers to increase by 10 per cent the rate of the duties on tobacco [illegible].

5. regulator gene *Genetics* [tr. F. *régulateur* (Jacob & Monod 1959, in *Compt. Rend.* CCXLIX. 1282)], a gene which codes for a polypeptide which can act as an operator to modify the frequency of initiation of transcription, so as to inhibit or stimulate the synthesis of mRNA (and hence of enzyme) on the structural genes of the operon.
1961 Jacob & Monod in *Jrnl. Molecular Biol.* III. 334 A new type of gene, which we shall call a 'regulator gene'... A regulator gene does not contribute structural information to the proteins which it controls. The specific product of a regulator gene is a cytoplasmic substance, which inhibits information transfer from a structural gene (or genes) to protein. In contrast to the classical structural gene, a regulator gene may control the synthesis of several different proteins: the one-gene one-protein rule does not apply to it. **1966** R. M. Campbell *Episomes* ix. 117 The regulator gene product has since been isolated. **1973** J. B. Jenkins *Genetics* xii. 327 There is some evidence for regulator genes and repressor substances in higher organisms, although their mode of operation appears to be quite different from those discussed under the operon concept.

regulatory, *a.* Add: **2.** *Biol.* Pertaining to, being, or involving regulation (sense *1 b).
1902 *Archiv für Entwicklungsmech. der Organismen* XV. 227 Morgan has proposed the name 'morphallaxis' for certain regulatory form-changes occurring in..lower animals, in consequence of which a piece gradually assumes more or less exactly the proportions of the whole organism. **1935** J. S. Huxley *Ess. Pop. Sci.* 102 This regulatory function of higher mechanisms is seen also in the domain of pure physiology. *Ibid.* 280 This period of non-regulation not only succeeds one regulatory phase, but precedes another. **1948** *New Biol.* V. 122 Many animals (especially those which undergo spiral cleavage) have little trace of profound regulatory power from the fertilised egg onwards. **1964** *New Perspectives in Biol.* IV. 146 A considerable amount is known about regulatory mechanisms in bacteria. *Ibid.* 170 The regulatory mechanisms that control cell differentiation in multi-cellular organisms may, in the not too distant future, cease to be one of the main unknown areas in biology.

Regulo (re·giŭlo). Also **regulo.** [a. L. *rēgulō* first pers. sing. pres. indic. of *rēgulāre* to regulate.] **a.** The proprietary name of a thermostatic control for a domestic gas oven.
1922 *Trade Marks Jrnl.* 30 Aug. 1998 Regulo... Taps and valves of ordinary metal... Radiation, Limited, Birmingham; manufacturers. **1926** *Economist* 28 Mar. 731/1 The 'New World' cooker, with the 'Regulo', revolutionised gas cooking. **1935** E. Waite *Good English Food* ii. 113 This oven..must be the right heat... Gas Companies provide the Regulo at the side. **1968** S. E. Ellacott *Everyday Things in England 1914–68* ii. 33 About 1930 an oven-control dial was introduced for gas cookers. This was named 'Regulo'.
b. *attrib.*, esp. followed by a numeral indicating one of a scale of temperature settings marked on a Regulo. Also *fig.*
1940 *Army & Navy Stores Catal.* 1011 'Regulo-controlled' gas cookers..fitted with the Regulo oven heat controller. **1956** [illegible] Cook in a fairly hot oven (Regulo Mark 6). **1958** *Spectator* 6 June 733/1 There was the Guildhall, where..a temperature approximating to Regulo 7, caused such breathlessness among woodwind. **1958** *Times* 6 Oct. 13/5 Place in well greased roasting tin and cook for 45–60 minutes, according to size of bird. Regulo Mark 7. **1968** S. E. Ellacott *Everyday Things in England 1914–68* ii. 33 Cookery books..added the appropriate Regulo readings for their recipes. **1971** 'D. Halliday' *Dolly & Doctor Bird* vii. 160 The permanently at Regulo 5. **1977** *New Society* 6 Oct. 22/1 The inevitable Tory tactic of keeping things low key by a regulo one campaign, which would let apathy take its toll. We went for regulo nine.

regulon (re·giŭlǫn). *Biol.* [f. Regul(ate *v.* + *-on¹.] A unit comprising all the genetic material whose transcription is regulated by a single inducer or repressor.
1964 Maas & Clark in *Jrnl. Molecular Biol.* VIII. 365 The term *regulon* is proposed to describe such a system in which the production of all enzymes can be controlled by a single repressor substance. Two types of regulons can be distinguished, those in which the structural genes for the enzymes are adjacent to each other (histidine, tryptophan) and which thus consist of single operons, and those in which they are not (arginine) and which thus consist of several operons. **1976** *Ann. Rev. Microbiol.* XXX. 549 Because a single repressor regulates the specific expression of these operons with G3P as the inducer, the system has been referred to as the *glp* (for glycerophosphate) regulon. **1978** *Jrnl. Molecular Biol.* CXXIV. 359 (*heading*) Dominant constitutive mutations in *malT*, the positive regulator gene of the maltose regulon in *Escherichia coli*.

regur (re·gəɹ, rē·gəɹ). [ad. Hind. *regar* black soil, ad. Telugu *rē-gaḍa*, *rē-gaḍi* clay.] Rich, dark, calcareous soil rich in clay, formed mainly by the weathering of basaltic rock and occurring extensively on the Deccan Plateau of India. Cf. *black cotton soil* s.v. *black *a.* 19. Also *attrib.*
1828 *Edin. New Philos. Jrnl.* VI. 119 Immense deposits of a black alluvial clay are met with in various parts of India. It is denominated cotton ground, from the circumstance of that plant being always cultivated upon it. It is the regur soil of the ryuts. **1858** [see *black cotton ground* s.v. *black *a.* 19]. **1879** Medlicott & Blanford *Man. Geol. of India* I. xviii. 429 Regur, in its most characteristic form, preserves the constant characters of being highly argillaceous and somewhat calcareous, of becoming highly adhesive when wetted, [illegible]. **1906** E. W. Hilgard *Soils* [illegible]. **1963** A. Gourevitch [illegible] regur [illegible].

regurgitate. 1. a. (Earlier *fig.* example.)
1753 Smollett *Count Fathom* II. lxii. 233 Renaldo's grief seemed to regurgitate with redoubled violence.

rehab (rī·hæb), *sb. slang.* (Not used in the U.K.) [abbrev. Rehabilitation.] **1.** = *rehabilitation 2 d. Also *attrib. Austral.*, *N.Z.*, and *Canad.*
1948 K. Stronach in A. E. Woodhouse *N.Z. Farm & Station Verse* (1950) 182 (*title*) Rehab. **1949** L. Peterson *Chipmunk* 33 The car was completely theirs though; he'd paid cash for it out of his rehab money. **1953** M. Scott *Breakfast at Six* i. 12 This block's a Soldiers' Settlement. You know the sort of thing. Under Rehab. *Ibid.* 27 It was the Rehab officer in charge of these settlements or others. **1959** C. C. Slatter *Gone in my Hand* (1960) xi. 141 Rehab was the caper you joker. **1964** *Canad. Weekly* 28 Nov. 12/3 By then I had a wife and child and my army rehab allowance was $90 a month. **1966** *N.Z. News* 27 Apr. 9/2 Rehabilitation assistance to returned war servicemen and women, a service known and respected throughout New Zealand as 'Rehab', officially ended on March 31. **1966** G. W. Turner *Eng. Lang. in Austral. & N.Z.* viii. 172 A house may be bought with a 'rehab loan' or a State advances loan.
2. = *rehabilitation 2 c. Also *attrib. U.S.*
1961 Partridge *Dict. Slang Suppl.* 1246/2 *Rehab*, a rehabilitation ward or department in a hospital: since ca. 1945. **1970** *New York* 16 Nov. 45/1 The storefront drug-rehab center. **1976** *Amer. Speech* 1973 XLVIII. 208 Afterward, alcoholics go to rehab 'the rehabilitation center' to readjust to a life free from the use of alcohol. **1977** *Chicago Tribune* 2 Oct. xii. 68/8 (Advt.), Rehab Nurse. Immediate position available in skilled nursing facility for Rehab Nursing Supervisor. **1978** *Tucson Mag.* Dec. 26/3 Having young people off to 're-hab' centers against their will may also be illegal.
3. = *rehabilitation 2 a. *U.S.*
1975 R. H. Nimmer [illegible] (1976) 125 If I can persuade him to sell for ten thousand dollars each, I'll toss this house in at ten thousand dollars. That will give us fifteen thousand dollars more for rehab. **1978** *New York* 3 Apr. 36/3 [illegible] rehab [illegible].
Hence as *v. trans.*; also **reha·bber.**
1977 *Archit. Rec.* Mar. 13/3 We can and should shorten time for developers and rehabbers. **1977** *Sat. Rev.* 23 July 9/2 The rehabilitation, into 16 apartments and several common rooms, of an abandoned tenement at 1186 Washington Avenue... 'rehab-ing' additional tenements in the vicinity of 1186. **1978** *Harper's Mag.* June 4/3 [illegible] rehabbed [illegible]. **1979** *Arizona Daily Star* 1 Apr. 31/3 Rehabbing an older city home involves some special problems.

rehabilitate, *v.* Add: **3. b.** To restore (a disabled person, a criminal, etc.) to some degree of normal life by appropriate training.
1944 *New Statesman* 27 May 353/1, I think Dr. Rogerson's Patient would have had a very different outlook, had he been properly Rehabilitated. **1951** [illegible] As soon as the wounded were rehabilitated they joined the new men, because of their valuable experience. **1968** *Listener* 19 Dec. 816/1 They want to rehabilitate us, but we think it is the rehabilitation officers who need rehabilitating, not us. **1978** *Lancashire Life* 51/1 The emphasis today is to rehabilitate old folk to enable them to remain active for as long as possible.
c. *absol.* for *refl.* To return from military to civilian status or purpose.
1945 *Daily Tel.* 29 June 4/3 Such assistance will be a precious aid to the first victims of the Axis—a people of 15,000,000 struggling to rehabilitate and develop, despite the grievous loss produced by the systematic murder by the Italians of the trained personnel and educated youth.

Also **rehabi·litating,** *vbl. sb.*
1924 *Glasgow Herald* 18 Mar. 6/3 The rehabilitating of Austria was carried a stage further.

rehabilitation. Add: **2. a.** (Further example.)
1973 *Detroit Legal News* 30 Aug. 13/8 Owner or interested party who appeared verbally granted a period of two weeks in which to..secure a building permit, and to immediately begin rehabilitation.

c. Restoration (of a disabled person, a criminal, etc.) to some degree of normal life by appropriate training. Cf. *rehabilitate *v.* 3 b.
1940 M. J. MacDonald in *Hansard Commons* 17 Oct. 867 There is one..aspect of the healing of the wounded—..which I should like to mention,..it is the secret of the maximum cure possible for the patient. It is the process known as rehabilitation. It is not sufficient that the wound should be healed; the wounded part of the patient must be enabled to function again so that he may once more play his part in society as a worker... I have appointed an adviser on rehabilitation. **1941** *Ann. Reg. 1940* 150 The principal questions for expert study being the..rehabilitation of men disabled in the war. **1952** *Rehabilitation* July (verso front cover). The British Council for Rehabilitation was founded in 1944... Rehabilitation was defined as 'the whole range of services from the time of the onset of the individuals' disability to the point at which he is restored to normal activity or the nearest possible approach to it'. **1974** *Science* 2 Aug. 423/2 People who are down on jails believe that the institutional setting is too dehumanizing for any meaningful rehabilitation to take place. **1979** *Internat. Rehabilit. Med.* I. 73/1 Mair..defined rehabilitation as implying the restoration of patients to their fullest physical, mental and social capability.

d. The retraining of a person, or the restoration of industry, the economy, etc., after a war or a long period of military service.
1941 *New Statesman* 15 Feb. 161/1 A military victory would be followed by the economic and democratic rehabilitation of France and Germany. **1941** *Times* (weekly ed.) 30 July 16/2 The possession of these..assets will ease the task of the Allied Governments when the time comes for the rehabilitation of European finance after the war. **1946** *R.A.F. Jrnl.* May 170 The career-finding agency..was inaugurated as one of the primary aids for rehabilitation of members. **1950** *N.Z. Jrnl. Agric.* May 458/2 (*heading*) Rehabilitation of Ex-servicemen. *Ibid.*, Land for the rehabilitation of returned servicemen has been..plentiful in Canada.

3. *attrib.*, as *rehabilitation area, camp, centre, counselling, grant, officer, programme, studies.* Also **rehabilitation medicine** (see quot. 1971).
1977 *Detroit Free Press* 11 Dec. 18-B/3 Improving city neighborhoods are excellent buys—especially rehabilitation areas where urban pioneers have turned the neighborhood around but buildings are still available at reasonable prices. **1967** W. Soyinka *Kongi's Harvest* 31 All the prostitutes were sent off to a rehabilitation camp. **1978** R. Ludlum *Holcroft Covenant* xx. 228 They were shocked beyond anything we can imagine when they learned about the 'rehabilitation camps'. Auschwitz, Belsen—it blew their minds. **1944** *Ourselves in Wartime* iii. 41 The Ministry of Labour..laid its plans for the rehabilitation of injured citizens. Men and women were trained at one of the Ministry of Labour's rehabilitation centres. **1967** R. Rendell *Wolf to Slaughter* vi. 64 What d'you do..? Start screaming like an addict in a rehabilitation centre? **1977** *Wandsworth Borough News* 16 Sept. 1/5 Government plans to erect a rehabilitation centre for 125 homeless men on the site of the old Battersea General Hospital. **1976** *Laurel* (Montana) *Outlook* 9 June 13/4 Graduate students received 94 master of science degrees, with 46 being in education, 31 in special education, and 15 in rehabilitation counselling. **1956** T. H. Raddall *Wings* 30 After I got my discharge from the army I took a forestry course at U.N.B., on my rehabilitation grant. **1969** *Jrnl. Amer. Med. Assoc.* 6 Jan. 137 (*heading*) Rehabilitation medicine's challenge for the 1970's. **1971** *Lancet* 27 Nov. 1207/2 A few years ago the University of New York announced that..they would no longer use the term 'physical medicine' in the context of rehabilitation but would replace it with the term 'rehabilitation medicine'. *Ibid.*, The following..relates to a World Congress to be held in Sydney, Australia, in August, 1972: 'Rehabilitation Medicine is a special area of medical practice traditionally concerned with the problems of the severely disabled and with the task of restoring them to a place of independence and dignity in society.' **1977** *Rehabilitation* Jan.–Mar. 8/1 With the recent international tendency to use the term 'Rehabilitation Medicine' instead of 'Physical Medicine' and, therefore, bringing in socio-logical and psychological factors, the pressing need for co-ordinated post-graduate study became the Council's challenge. **1979** *Internat. Rehabilit. Med.* I. 44/1 The ultimate goal is to establish rehabilitation medicine alongside diagnosis and treatments, as one of the three activities of all practising doctors. **1968** Rehabilitation officer [see *rehabilitate *v.* 3 b]. **1950** *N.Y. Times* 20 Apr. 1/2 President Truman approved..a bill authorizing an economic rehabilitation programme for the Navajo and Hopi Indian tribes. **1973** *Brit. Med. Jrnl.* 17 Mar. 687/3 (*heading*) Chair of rehabilitation studies. **1974** *Ibid.* 15 June 622/1 Dr. Cairns Aitken, senior lecturer in the department of psychiatry, Edinburgh University, has been appointed to the newly founded chair of rehabilitation studies at that university.

rehabilitative (rīhăbi·lĭtātiv), *a.* [f. Rehabilitat(e *v.* + -ive.] Of or pertaining to rehabilitation; designed to rehabilitate.
1958 *Times* 8 Oct. 11/3 The rehabilitative process should begin with a man's arrival and should not stop short at the moment of release. **1963** T. & P. Morris *Pentonville* vii. 172 Many inmates are full of ideas about how prison could be improved, both as a deterrent and a rehabilitative technique. **1973** *Black Panther* 22 Sept. 5/2 There are no rehabilitative services offered in the Texas prisons. **1979** *Arizona Daily Star* 8 Apr. (Parade Suppl.) 4/1 This is short-term alimony, where the husband pays his former wife a monthly stipend for a limited period, generally two years, or even a lump sum. The term used by lawyers is 'rehabilitative alimony'. The idea is for the ex-wife to take advantage of the alimony period to return to school, to take a brushup course, or to launch a career. **1979** *Internat. Jrnl. Sociol. of Law* VII. 329 She feels that with the turn away from the rehabilitative model, indeterminate sentences and parole, we may inadvertently end up with a less discriminating system.

rehabilitee (rĭ:hăbĭlĭtī·). [f. Rehabilit(ate *v.* + -ee¹.] One who is (being) rehabilitated.
1972 *Rehabilitation* July–Sept. 7 Only in recent years would government seem to have begun to appreciate the importance of social aspects and the quality of life available for the rehabilitee, as opposed to concentrating on industrial aspects. **1978** *Jrnl. R. Soc. Med.* LXXI. 449 Two beds in the Lonsdale Unit were reserved for rehabilitees.

reha·logenize, *v.* *Photogr.* [f. Re- 5 b + Halogen + -ize.] *trans.* To convert the metallic silver in a developed image back to a silver halide (with the silver or the image as obj.). So **reha·logenizing** *vbl. sb.*
1940 F. J. Mortimer *Wall's Dict. Photogr.* (ed. 15) 549 *Rehalogenising*, the re-conversion of the silver image in a finished print or negative into silver chloride, bromide, or iodide. **1969** M. J. Langford *Advanced Photogr.* xi. 235 This bleach 're-halogenises' the silver image—the ferricyanide oxidises the silver which reacts with the bromide and reverts to a silver halide. **1977** J. Hedgecoe *Photographer's Handbk.* 324 *Bleach*, chemical bath capable of rehalogenizing black metallic silver.
Hence **reha:logeniza·tion,** the process of rehalogenizing.
1956 *Focal Encycl. Photogr.* 967/1 *Rehalogenization*, conversion of a silver image back into one of silver halide. **1958** *Newnes Compl. Amat. Photogr.* 323 As this process involves rehalogenisation of the silver image it is essential that the print and utensils used are completely free from hypo. **1967** E. Chambers *Photolitho-Offset* iv. 48 An improvement in the dot etching properties of the positives can be effected by tanning the gelatin in proportion to the silver image, resulting in a greater degree of lateral etching without loss of density. This process is known as rehalogenisation, or metallising.

rehash, *v.* Add: **2.** Chiefly *U.S.* To consider, mull over, discuss (an idea, performance, etc.) afterwards.
1965 Mrs. L. B. Johnson *White House Diary* 14 Dec. (1970) 340 Our houseguests..were all gathered around Lyndon rehashing the events of the evening. **1974** *Los Angeles Times* 13 Oct. iii. 9/1 The Dodgers were anything but downtrodden as they rehashed the game. **1977** *Time* 31 Jan. 48/3 If he meets someone he knows after a session he may stop him on the street to rehash it.

rehear, *v.* **2.** (Earlier and later examples.)
1799 F. Burney *Jrnls. & Lett.* (1973) IV. 297 He has taken the amazing trouble & toil of copying the whole, from the pleasure the interview gave him! though he may always re-hear it *de vive voix!* **1947** *Observer* 28 Dec. 2/5, I long to rehear such great contrapuntal machines as the *Quoniam* and *Amen* choruses.

rehearsal. Add: **1.** *spec.* in *Psychol.*, the intentional repetition (mentally or verbally) of information in order to keep it temporarily in the memory.
1935 *Amer. Jrnl. Psychol.* XLVII. 66 Reminiscence..has not yet been demonstrated to occur independently of review, either intentional rehearsal or casual revival, during the interval between the two tests of retention. **1951** S. S. Stevens *Handbk. Exper. Psychol.* xvii. 654/1 Color naming has commonly been used during the rest interval in an attempt to reduce rehearsal. **1960** O. H. Mowrer *Learning Theory & Behavior* x. 365 A telephone number will be 'remembered', without rehearsal, for a short time after it is seen in a directory (or heard spoken), but will then usually be lost quite completely. **1976** [see *rehearse *v.* 1 d]. **1979** W. A. Wickelgren *Cognitive Psychol.* viii. 235 The classification of memory into learning, storage, and retrieval leaves the cognitive process of rehearsal somewhat in limbo.

2. a. Also, the act of practising any ceremony, e.g. a wedding or a state occasion. (Further examples.)
1953 Rodgers & Pearce *Altar Bound* 45 The rehearsal dinner takes place immediately before or after the rehearsal, and..is given by either the parents of the bride or groom. **1963** A. Vanderbilt *New Compl. Bk. Etiquette* ix. 93 It is becoming more and more popular for the wedding rehearsal to be held in the late afternoon the day before the wedding, followed by a rehearsal dinner, which may be scheduled for six-thirty or seven o'clock. In some sections of the country, mainly the South and Midwest, it is customary for the groom's parents to give the rehearsal dinner. **1977** *Times* 24 Mar. 2 (*heading*) Ulster trip by Princess becomes rehearsal for Queen's visit.

3. *attrib.*, as *rehearsal break, complex, hall, pianist, room*; **rehearsal band,** a band that meets to practise jazz, dance music, etc.; **rehearsal dinner** *U.S.*, a dinner held after a wedding rehearsal (usually in the evening before the wedding); **rehearsal script** (see quot.).
1969 *Down Beat* 17 Apr. 19/3 There have been any number of rehearsal bands around New York in the past 10 years. **1970** *New Yorker* 23 May 78/2 The so-called 'rehearsal' bands of Thad Jones–Mel Lewis, Clark Terry, [etc.]. **1957** *Sentinel* (Milwaukee) 14 Nov. ii. 6/3 'That six minutes is beginning to look like an awful long time,' shivered Junie, chewing nonchalantly on a coffee cup during a rehearsal break. **1977** *West Briton* 25 Aug. 10/2 The dawning of a new and promising era for Redruth Amateur Operatic Society... The society took over its new rehearsal complex at Plain-an-Gwarry. **1953, 1963** Rehearsal dinner [see sense *2 a above]. **1974** *State* (Columbia, S. Carolina) 3 & 4 Mar. 6-B/6 The groom gifts his attendants at the rehearsal dinner or bachelor dinner. **1960** 'E. McBain' *Give Boys Great Big Hand* x. 107 Unfurnished rehearsal halls and the cubbyhole offices of music publishers. **1976** *Listener* 29 July 121/2 The last 17 hopefuls..gather together in the rehearsal hall. **1977** S. Brett *Star Trap* i. 11 There was a guy..who was rehearsal pianist for the show. **1959** M. Summerton *Small Wilderness* i. 18 Reading a part in a rehearsal room. **1979** *Jrnl. R. Soc. Arts* CXXVII. 500/1 The school housed an excellent theatre, a number of rehearsal rooms, a swimming pool, [etc.]. **1960** Rehearsal script [see *master-scene* s.v. *master *sb.*¹ 29].

rehearse, *v.* Add: **1. d.** *Psychol.* To repeat, either mentally or orally (what it is desired to keep temporarily in the memory); to engage in rehearsal (sense *1).
1917 *Arch. Psychol.* xl. 103 Frequent reviews..throw into relief the portions [of memorised material] that are hazy, inexact and confused..because they fix more clearly in mind the material that is rehearsed. **1951** S. S. Stevens *Handbk. Exper. Psychol.* xvii. 654/1, 84 per cent of her younger subjects and 70 per cent of the older ones reported that they had rehearsed in the interval between learning and recall. **1964** *Jrnl. Exper. Psychol.* LXVIII. 414/2 The present experiment is designed to investigate the consequences for short-term memory of instructing Ss to rehearse a sequence of digits in groups of one, two, three, four, and five during presentation. **1976** G. R. & E. F. Loftus *Human Memory* iv. 56 Rehearsal can do two things: it can keep the information in short-term store for as long as we continue rehearsing, and it can also act as a mechanism by which information is transferred from short-term to long-term store. **1979** W. A. Wickelgren *Cognitive Psychol.* iii. 84 Time yourself reciting the alphabet verbally. Then time yourself while you go through the alphabet again, this time imaging each letter as if it were projected on a screen. It will take you about three times as long..as it did to rehearse each letter verbally.
rehearsed *ppl. a.*, restrict † to sense in Dict. and add: (*b*) that has been practised beforehand.
1871 L. Lockhart *Fair to See* I. vi. 167 He turned to confront them, on their entrance, with a carefully rehearsed mien.

reheat (rī·hīt), *sb.* [f. the vb.] **1. a.** The action or an instance of reheating; *spec.* artificial or spontaneous heating of the working fluid in a turbine taking place between stages. Also *attrib.*
1913 [see *reheat factor* below]. **1918** *Engineering* 6 Sept. 245/1 The case of a turbine consisting of two equal stages which, in order to avoid complications due to 'reheat', will be assumed to be operated by water. **1938** *Van Nostrand's Sci. Encycl.* 950/1 Many industrial processes in which heat plays a part, employ reheating, sometimes to the extent of several 'reheats'. **1953** Jennings & Rogers *Gas Turbine Analysis & Practice* iii. 104 One of the factors contributing to this lower efficiency is the..deleterious effect of reheat in the compressor. **1959** *Motor* 27 May 562/3 The resultant gases pass through the high-speed turbine, then a second, 're-heat' combustion chamber followed by the power turbine. **1965** *Economist* 17 Apr. 326/1 The main items needed..are another reheat furnace for the slabbing mill and a cold reduction line. **1966** *McGraw-Hill Encycl. Sci. & Technol.* XI. 425/2 Under suitable conditions of initially high steam pressure and superheat, one or two stages of reheat can be advantageously employed to improve thermodynamic efficiency of the cycle. **1975** J. B. Woodward *Marine Gas Turbines* ii. 44 The reheat step consists of a constant-pressure heating interposed between two expansion processes.
b. *Aeronaut.* = *after-burning *vbl. sb.* 2. Also, an afterburner.
1949 [see *after-burning *vbl. sb.* 2]. **1950** *Engineering* 6 Oct. 295/1 The performance of a jet engine may be changed by 'reheat' i.e., the burning of additional fuel with residual oxygen remaining in the combustion gases after they have passed through the turbine. **1957** *Ann. Reg. 1956* 1. 14 The hero of the occasion was Mr. L. P. Twiss flying a Fairey Delta 2 research aircraft, powered by a Rolls Royce Avon turbo-jet fitted with reheat. **1959** *Spectator* 4 Sept. 295/1 A fighter making a 700-mile-an-hour run a few feet from the spectators, with re-heat ablaze, used to be a grand spectacle. **1972** D. Hart-Davis *Spider in Morning* ii. 21 The reheats were in and burning fuel at a terrifying rate. **1976** *Farnborough Internat. Exhibition* (Official Programme) 11/2 Such is the engines' thrust when reheat is on that fuel is burnt up at an incredible 20 tons an hour.
2. Special Comb.: **reheat factor,** a measure of the performance of a multistage steam turbine, usu. expressed as the ratio of the measured efficiency of the turbine and the (lower) efficiency expected on the assumption of adiabatic expansion of the steam.
1913 H. M. Martin *Design & Constr. of Steam Turbines* v. 44 It will be noted that after an expansion of one hundred-fold or so, the ratio ϵ/η becomes practically constant. This ratio..is known as the 'reheat factor'. **1950** J. K. Salisbury *Steam Turbines & their Cycles* i. 34 The sum of the stage available energies is greater than the turbine available energy by a small amount. The ratio of these two quantities is..the reheat factor. **1961** Fox & McBirnie *Marine Steam Engines & Turbines* (ed. 2) xviii. 438 The significance of the

reheat factor is that in a multi-stage steam turbine any particular stage benefits somewhat from the inefficiency of the preceding stages.

reheated, *ppl. a.* Add: *spec.* in *Aeronaut.*, equipped with or augmented by afterburning. Cf. *reheat *sb.* 1 b.
1961 F. K. Mason *Hawker Aircraft since 1920* 74 Work started on a development using a fifty degree swept wing and a re-heated Avon of greatly increased power. **1976** *Farnborough International Exhibition* (Official Programme) 43/1 With a reheated take-off thrust of 38,000 lbs., this engine entered airline service in Concorde.

Reheboth, var. *Rehoboth.

Rehoboam (rīhŏbō^u·ăm, rī,ŏ-). [f. the name of *Rehoboam*, son of Solomon, King of Judah (I Kings xii–xiv).] †**1.** A shovel hat. *Obs. rare*⁻¹.
1849 C. Brontë *Shirley* I. i. 11 A personage of short stature..bearing on broad shoulders a hawk's head,..the whole surmounted by a Rheoboam [**1850** Rehoboam] or shovel-hat, which he did not seem to think it necessary to lift.
2. A large bottle for wine or spirits, bigger than a Jeroboam and smaller than a *Methuselah 2.
1895 *Brewer's Dict. Phr. & Fable* (new ed.) 1050/1 A rehoboam of claret or rum is a double jeroboam. **1959** *Gloss. Packaging Terms* (*B.S.I.*) 28 *Rehoboam*, a wine bottle—capacity 6 reputed quarts. **1960** *Times* 11 Jan. 17/2 A vigorous passing movement across the ground by three Frenchmen with a dummy rehoboam of champagne was ended only by a flagrant knock-on. **1972** [see *Methuselah 2].

Rehoboth (rī·,ŏbǫþ). *S. Afr.* Also erron. **Reheboth.** [a. Heb. *rĕhōbhōth* wide places.] A Biblical place-name (Gen. xxvi. 22), applied to a river, town, and district in Namibia (South West Africa), and used as the name of a people of mixed African and European descent. Also *attrib.* Hence **Re·hobother.** So **Rehobo·thian** *sb.*
1875 C. J. Andersson *Notes of Trav. in S. Afr.* vi. 88 The cattle, etc., might have been retaken had our friends, the Rehobothians, at once accompanied me in pursuit of the marauders. **1926** S. G. Millin *South Africans* vii. 198 To-day self-governing half-caste tribes like the Bondelswarts and the Rehoboths still exist in South-West Africa to trouble the souls of their white neighbours. **1930** C. G. Seligman *Races of Africa* ii. 34 The old Hottentot population of the Cape has become largely absorbed by racial admixture with incoming Europeans and East Indian slaves, and has thus constituted the basis of the present.. 'Rehoboth' half-breeds. **1935** L. G. Green *Great African Mysteries* x. 125 The half-caste Rehoboths..set up nothing less than a civilised republic in a savage land. **1937** *Decisions High Court S.-W. Africa Jan. to Dec. 1936* 59 *Held*, that in this Territory the term 'Baster' when it is ascribed to a person's race is well known to refer to the members of the Reheboth Bastard Community. **1960** *State of Union 1959–60* lxv. 430 The *Rehobothers* whose number is estimated at 8,900, are of mixed origin, having immigrated from the Cape. **1970** *Standard Encycl. Southern Afr.* II. 191/2 The majority are..the descendants of early Basters and they insist on being called Rehoboth Basters, a term they reserve strictly to themselves. **1973** *Observer* (Colour Suppl.) 2 Sept. 56/1 Hereros, Ovambos, Damaras, Kakaovelders, Bushmen and Rehobothers (until recently known as the Bastards).

reho:spitaliza·tion. [f. Re- 5 a + *hospitalization.] The act or the state of being admitted again to hospital.
1974 M. C. Gerald *Pharmacol.* xvi. 305 The remaining patients retain varying degrees of psychopathology, with many requiring periodic rehospitalization.

rehouse, *v.* (Later examples.)
1904 G. B. Shaw *Common Sense of Municipal Trading* viii. 75 The municipality bargains with the Local Government Board as to how many people it must rehouse. **1935** *Scrutiny* IV. 134 The problem of re-housing the inhabitants after slum-clearance is at present dealt with unsatisfactorily. **1966** *Listener* 6 Oct. 511/1 The first few years after the war.. saw the re-housing of Turner in the glory that he deserved. **1978** *Lancashire Life* Apr. 105/1 This trolley can be moved freely on its castors, then re-housed in its own base unit. **1980** *Jrnl. R. Soc. Arts* Mar. 186/2, 5¾ per cent of London's total built up area was bought and developed for railway use without any statutory responsibility to re-house the displaced residents.
Hence **rehou·sed** *ppl. a.* Also *absol.* or as *sb.*
1927 *Scots Observer* 26 Feb. 2/5, I have seen the re-housed in their new environment. **1940** Harrisson & Madge *War begins at Home* xii. 331 Many of the re-housed families would go back to their old homes if they had not been pulled down.

rehousing, *vbl. sb.* (Later examples.)
1904 G. B. Shaw *Common Sense of Municipal Trading* viii. 76 The displaced have solved the rehousing problem by crossing the river into Battersea. **1936** 'G. Orwell' *Diary* 27 Feb. in *Coll. Essays* (1968) I. 189 The re-housing is almost entirely the work of the Corporation. **1977** G. Scott *Hot Pursuit* iii. 31 Singapore's progressive rehousing policies.

rehydra·te, *v.* [Re- 5 a.] **a.** *intr.* To absorb water again, esp. after dehydration.
1923 J. W. Mellor *Comprehensive Treat. Inorg. & Theoret. Chem.* III. 763 Crystals of the hemihydrate which practically re-hydrated on cooling. **1968** *Physics Bull.* Dec. 430 Technical advances..have resulted in good products which rehydrate quickly.
b. *trans.* To add water to again after dehydration; *esp.* to restore (dehydrated food) to a palatable state by the addition of water.
1962 F. I. Ordway et al. *Basic Astronautics* xiii. 521 The water necessary to rehydrate these processed foods can be recovered. **1965** *Listener* 10 June 875/1 We saw the astronauts learning to rehydrate their plastic bagfuls of strawberry cereal. **1969** *New Scientist* 16 Jan. 128/2 The food was shown to be fully dehydrated; and when later rehydrated it tasted excellent. **1971** *Sci. Amer.* Aug. 18/3 The patient is first rehydrated by intravenous injection, if necessary, and is then fed the oral solution at the rate at which fluid is lost by diarrhea. **1978** *Jrnl. R. Soc. Med.* LXXI. 223 He was given broad-spectrum antibiotics and rehydrated.
Hence **rehydra·table** *a.*, that may be rehydrated; **rehydra·ted** *ppl. a.*
1956 *Nature* 4 Feb. 239/2 When a flake of magnesium vermiculite, which has been partially dehydrated to the phase containing single sheets of interlayer water molecules .., is allowed to become rehydrated, a dark line is observed to enter the flake at the edges and gradually converge on the centre. **1969** *New Scientist* 16 Jan. 128/3 The rehydrated product should as far as possible be identical with the fresh product. **1970** N. Armstrong et al. *First on Moon* vi. 127 U.S. choice beef was better for rehydratable meat cubes than prime, which had too much fat. **1975** *Radio Times* 12–18 July 40 Menu [for U.S. astronauts], rehydratable romaine soup,..rehydratable coffee. **1975** *Daily Colonist* (Victoria, B.C.) 12 July 1/2 The cosmonauts will dine on steak in plastic pouches, rye bread, cheese spread, rehydrated strawberries and tea with lemon and sugar [etc.].

rehydra·tion. [Re- 5 a.] The process of adding or taking up water again, usu. after dehydration.
1866 [in Dict. s.v. Re- 5 a]. **1923** J. W. Mellor *Comprehensive Treat. Inorg. & Theoret. Chem.* III. 763 Special precautions are needed to remove water rapidly before it cools much below 130°, or rehydration sets in. **1936** *Discovery* Oct. 329/1 In summer influenza the protein particles of the blood plasm are subject to hydration, in early winter dehydration takes place while rehydration in late winter is followed closely by an epidemic in February and March. **1947** J. C. Rich *Materials & Methods of Sculpture* iv. 67 The setting of plaster of Paris after it has been mixed with water is also referred to as a *rehydration* and occasionally as a *recrystallization* of the calcined gypsum back to its original hydrated rock form. **1965** *New Scientist* 26 Aug. 498/3 Natural mayonnaise would not be used, since the emulsion separates on freezing giving an oil which, it was feared, would hinder re-hydration. **1979** *Nature* 29 Mar. 389/1 The possibility of oral rehydration has provided Third World health authorities with a very valuable breakthrough in the treatment of diarrhoeal diseases.

‖ **Reich** (raiχ^y, raik). Pl. **Reiche.** [Ger., = kingdom, realm, state: see Riche, rike.] Chiefly during the period 1871–1945, the German state or commonwealth; also, one of a sequence of empires or régimes in Germany, esp. the *Third Reich.
Apart from *Third Reich*, collocations with an ordinal are rare and do not constitute recognized English historical terminology.
1921 *Times* 19 Jan. 11/2 All the States proclaim..their loyalty to the Reich. **1924** *Hansard Commons* 16 Jan. 152 We have always pointed out that,..if any part of the German Reich wished to set up an autonomous area for itself, they must utilise their own constitutional machinery. **1933** *Times* 15 Mar. 15/2 During the past week the Nazi steam-roller has passed over every one of the seventeen Federal States of the Reich. **1946** *Britannica Bk. of Year* (U.S.) 341/2 Germany was deprived not only of all Hitler's annexations of territory, but also of all former German reich territory east of the Oder and western Neisse rivers. **1972** F. Forsyth *Odessa File* ii. 46, I have commandeered this ship in the name of the Reich. **1973** *Nature* 14 Sept. 107/2 A penetratingly sketched background of the fortunes and misfortunes of the Second and Third *Reiche*. *Ibid.*, The Jews had been emancipated in Prussia in 1812 and..had come to occupy high places in the Second Reich. **1974** J. White tr. *Poulantzas's Fascism & Dictatorship* vii. v. 338 The dismembering of the State apparatus was also expressed in contradictions between the central authority of the Reich and the provincial authorities. **1976** *New Yorker* 8 Mar. 130/2 The villains here are a secret 'Comrades Organization' of ex-Nazis, who live in Brazil and hope to establish a Fourth Reich.
Also (f. the gen. sing.) **Rei·chsbank** [G. *bank* bank], the name of the central bank of the German Reich, 1875–1945; **Rei·chsbanner** [G. *banner* banner], a republican para-military organization in Germany from 1924 to 1933; **Rei·chsmark** [G. *mark* Mark *sb.*² in Dict. and Suppl.], the monetary unit of the German Reich, replaced in 1948 by the *Deutsche mark; **Rei·chsma:rschall, -ma:rshall** [Ger., in full *Reichsmarschall des Grossdeutschen Reiches* Marshal of the Greater German Reich], a title bestowed on Hermann Göring (1893–1946) in 1940 by Adolf Hitler; **Rei·chsrat** (formerly **-rath**) (-rāt) [G. *rat(h* council: cf. *Bundesrat], (*a*) the parliament of the Austrian part of the Habsburg Empire; (*b*) the council of the federated states of Germany between the end of the 1914–18 war and 1933; **Rei·chswehr** (-vēr) [G. *wehr* defence], the name of the German army between 1919 and 1935. See also *Reichstag.
All usu. with capital initial in Eng. as in German.
1879 *Encycl. Brit.* X. 466/1 The Imperial Bank (Reichsbank) ranks far above the others in importance. **1977** *New Yorker* 3 Oct. 85/1 Nazi archives at the Reichsbank.. yielded substantial data. **1924** *Times* 6 Dec. 11/1 Some scattered organizations..were amalgamated to be the nucleus of a united Republican guard, known as the Reichsbanner Black–Red–Gold.., the 'Great Germany' idea being invariably pushed into the foreground at Reichsbanner demonstrations. **1935** C. Isherwood *Mr. Norris changes Trains* viii. 130 The newspapers were full of death-bed photographs of rival martyrs, Nazi, Reichsbanner and Communist. **1964** *Listener* 25 June 1038/2 The Reich authorities began..to condemn the Reichsbanner, an organization which existed solely to defend the Republic. **1874** *Anglo-Brazilian Times* (Rio de Janeiro) 23 Dec. 5/3 The business on Hamburg has been done at 448 reis per reichsmark at 90 days and 454 reis for sight drafts. **1924** *Times* 7 Nov. 11/6 The new banknotes of 10, 20, 50 and 100 Reichsmarks, which are part of the reformed and stabilized German currency..were exhibited at the Reichsbank this morning. **1978** *Time* 3 July 8/1 The old reichsmarks of the Nazi regime had become worthless. **1940** *R.A.F. in Action* 16 On the 20th May, 1940, the Reichsmarshal..was forced to promise that more guns would be provided immediately. **1976** J. Wheeler-Bennett *Friends, Enemies & Sovereigns* ii. 36, I have always regretted that I was not present to see his defeat of Goering on that celebrated occasion when the Reichsmarschall had 'made a monkey' out of the chief American prosecutor, Robert Jackson, and David had deflated him. **1858** *New Amer. Cycl.* II. 392/2 The council of state, or *Reichsrath*, composed of 12 members, is a body coördinate to the ministry, and communicating immediately with the emperor. **1905** *Spectator* 11 Feb. 201 Meanwhile the Pan-German delegates in the Austrian Reichsrath are agitating for economic severance from Hungary and alliance with Germany. **1919** *German Constitution* (H.M.S.O.) i. 11 A Reichsrat is formed for the representation of the German States in Federal legislation. **1943** S. H. Thomson *Czechoslovakia in European Hist.* ix. 180 The Vienna Parliament (Reichsrat) which functioned as the legislative body for the Austrian half of the Empire. **1969** E. Wall *Europe Unification & Law* iii. 72 Though the Reichsrat, representing the federated Länder, could object to a bill, the objection could be overridden by a two-thirds majority of the Reichstag or..by referendum. **1920** *Times* 30 Oct. 9/3 He depicted the Reichswehr as mainly a Royalist institution. **1934** *Ann. Reg. 1933* 182 A Heimwehr patrol [in Austria] shot dead a German Reichswehr soldier. **1976** J. Wheeler-Bennett *Friends, Enemies & Sovereigns* iv. 100, I had had first-hand experience of the Seekt and Schleicher periods of the *Reichswehr* and of its brief honeymoon with Hitler which began to wane with the Night of the Long Knives in June 1934.

Reichert (rəi·χəɹt). The name of Emil *Reichert* (1838–94), German food scientist, used in the possessive and *attrib.* (freq. in *Comb.* with another name, as *Meissl, Polenske, Wollny*) with reference to standard procedures for determining the proportion of volatile water-soluble fatty acids present in butter, fats, and oils (one of which he published in *Zeitschr. f. anal. Chem.* (1879) XVIII. 68).
1885 *Analyst* X. 103 Reichert's process possesses every advantage over Hehner's, which is becoming replaced by the former. **1887** *Ibid.* XII. 203 (*heading*) On Reichert–Meissl's method of butter analysis. **1892** *Ibid.* VII. 171 Reichert..proposed to saponify 2·5 grammes of butter with caustic soda and alcohol, evaporate off the alcohol, add 50 c.c. of water and 2 c.c. dilute sulphuric acid, and to distil 50 c.c. in a weak current of air. This method, although Reichert himself calls it Hehner's method, is now known as the Reichert process. *Ibid.* 175 The Reichert–Wollny method is largely adopted in every country except France, and may be considered a standard method. **1894** *Ibid.* XIX. 189 Filtered butter-fat will give a constant Reichert–Meissl number for many weeks. **1901** *Ibid.* XXVI. 71 (*heading*) Note on the Reichert value of butter and other fats. **1906** [see *Polenske]. **1928** [see *Kirschner]. **1936** *Analyst* LXI. 404 As the original Reichert process, using 2·5 g. of fat, and as the Reichert–Meissl process, using 5 g., have been obsolete since Wollny modified the Meissl process nearly 50 years ago, and as the name Reichert is common to the different forms, it may now be used alone in place of the indiscriminate use of the hyphenated forms, Reichert–Meissl, Reichert–Meissl–Wollny, Reichert–Wollny and Reichert–Polenske, when applied to the soluble volatile acids. **1957** *Encycl. Brit.* IV. 469/2 The Reichert–Meissel [*sic*] (Reichert–Wollny) value..is a valuable characteristic in butterfat analysis. **1973** [see *Polenske].

Reichian (rəi·χiăn), *sb.* and *a.* [f. the name of Wilhelm *Reich* (1897–1957), Austrian psychologist + -ian.] **A.** *sb.* A supporter of the theories or practices of Wilhelm Reich, esp. those relating to sexual energy as vital energy (cf. *orgone), to its effect in determining character and mental health, or to his hypothesis that authoritarian regimes emerge in cultures that are sexually repressive. **B.** *adj.* Of, pertaining to, or following Reich or his theories.
1959 *Partisan Rev.* XXVI. 51 The Reichians want to believe in Socialism again. **1959** N. Mailer *Advts. for Myself* (1961) 295 The Yoga's *prana*, the Reichian's orgone, Lawrence's 'blood'. **1969** P. A. Robinson *Freudian Left* i. 10 The true Reichian is convinced that Reich's greatest contributions lay..in biophysics and astronomy. **1970** R. Lowell *Notebook* 247 Such cures the bygone Reichian

prophets swore to. **1976** *Listener* 8 Jan. 4/2 One of the unique features of present-day Portugal is the concurrence of political and sexual revolution. It provides a laboratory for Reichian radicals who see a causal relationship between sexual repression and totalitarianism, on one side, and sexual liberation and revolution, on the other. **1978** *N.Y. Rev. Bks.* 23 Feb. 29/4 The Victorian idea of TB as a disease of low energy..has its exact complement in the Reichian idea of cancer as a disease of unexpressed energy.

Reichstag (rəi·χʸsˌtăk, rəi·ksˌtāg). [Ger., f. gen. sing. of *REICH + *tag* diet: cf. *BUNDESTAG.] The diet or parliament of the German Empire (1871–1918) (formerly also, that of the North German Confederation) and of post-Imperial Germany until 1945; the building in Berlin in which this parliament met. Also *transf.*

1867 *Times* 3 Jan. 10/1 It is proposed to exclude Government functionaries, not from the future Diet of the Confederation (Reichstag), but only from the Parliament which is about to assemble. **1870** GEO. ELIOT *Let.* 3 Apr. (1956) V. 87 We went to the Reichstag one morning, and were so fortunate as to hear Bismarck speak. **1889** M. H. VAN DE VELDE *Cosmopolitan Recollections* I. v. 167 Will he attend the Reichstag? **1909** M. A. VON ARNIM *Caravaners* xiii. 220 She began to talk to me..about..our Reichstag. **1935** C. ISHERWOOD *Mr. Norris changes Trains* xvi. 263 We ought to be grateful to van der Lubbe, because the burning of the Reichstag had melted the snow. **1944** J. S. HUXLEY *On Living in Revol.* 138 In 1928..the Nazis had secured less than 2 per cent. of the seats in the Reichstag. **1975** *Times* 4 Mar. 1/2 From the Tory benches, there was a claim that the Commons was being turned into a Reichstag.

2. *attrib.*, as **Reichstag Fire**, a fire which destroyed the Reichstag building on 27 Feb. 1933, believed to have been engineered by the Nazi party in order to facilitate their seizure of power; **Reichstag Trial**, the subsequent trial of the alleged incendiary, Marinus van der Lubbe, and others; also *transf.*, as the type of a staged trial.

1933 *Times* 1 Mar. 13/2 A communication about the Reichstag fire was issued by the Prussian authorities. **1976** S. HYNES *Auden Generation* v. 143 The Reichstag fire appears as an emblem of public terror. **1968** D. HOPKINSON *Incense Tree* xii. 156 The trial was a travesty in the tradition of the Reichstag Trial at Leipzig. **1970** *Peace News* 17 Apr. 8/1 Bobby Seale has been sentenced to *four years* for contempt of court at Attorney General Mitchell's 'Reichstag Trial' in Chicago.

reide·ntify, *v.* [RE- 5 a.] **a.** *trans.* To identify again or in a new way; also *absol.* **b.** *intr.* To identify oneself with something again.

1934 WEBSTER, Reidentify, v.t. **1959** P. F. STRAWSON *Individuals* i. 32 We cannot attach one occasion to another unless, from occasion to occasion, we can reidentify elements common to different occasions. *Ibid.* 55 A condition..of the possession of a single, continuously usable framework..was the ability to *re*identify..some elements of the framework in spite of discontinuities of observation. **1964** C. M. WISE in D. Abercrombie et al. *Daniel Jones* 208 The second element..has been reidentified as a voiced postvocalic allophone. **1966** 'HAN SUYIN' *Mortal Flower* xii. 258 Others are longing to go back, to reidentify, but they are afraid of changing.

reification. Add: Also (rēˈˌifikēˈ·ʃən). (Further examples.) Also, depersonalization, esp. such as Marx thought was due to capitalist industrialization in which the worker is considered as the quantifiable labour factor in production or as a commodity.

1937 T. PARSONS *Struct. Soc. Action* xiii. 476 Positivistic empiricism has been predominantly a matter of the 'reification' of theoretical systems. **1941** H. MARCUSE *Reason & Revol.* II. i. 279 Marx's early writings are the first explicit statement of the process of reification (*Verdinglichung*) through which capitalist society makes all personal relations between men take the form of objective relations between things. **1954** H. J. EYSENCK *Psychol. Politics* viii. 262 Freud's reification of mental mechanisms is a literary rather than a scientific device. **1962** MACQUARRIE & ROBINSON tr. *Heidegger's Being & Time* I. i. 72 The Thinghood itself which such reification implies must have its ontological origin demonstrated. **1971** J. J. SHAPIRO tr. *Habermas's Toward Rational Soc.* iii. 39 The active assault upon culture is based on the same reification as the fetishism of those students who believe that by occupying university classrooms they are taking possession of science as a productive force. **1976** G. THERBORN *Sci., Class & Soc.* i. 26 The ugly consequences, in Friedrich's view, result from a 'reification' of the current epistemological stance of science. **1979** E. H. GOMBRICH *Sense of Order* v. 143 To see the [wavy] line as water, mountains or, perhaps, a fluttering ribbon might be described as 'reification', to see it as a living serpent as 'animation'.

reificatory (rĭ-, rēˈˌifikēˈ·tŏri), *a.* [f. REIFICAT(ION + -ORY².] Of, pertaining to, or characterized by, reification.

1951 Z. S. HARRIS *Methods in Structural Ling.* ii. 18 The reificatory question of what parts of human behavior constitute language. **1969** D. TRIESMAN in Cockburn & Blackburn *Student Power* 148 The International Student Conference..coalesced around the fundamental and reificatory tenet that it would only discuss problems of the 'student as such'.

reify, *v.* Add: Also with pronunc. (rēˈ·ifəi). (Later examples.)

1931 M. R. COHEN *Reason & Nature* III. iii. 390 There is..a fundamental philosophic issue: the extent to which the principle of unity should be hypostatized or reified (I wish the use of the word *thingified* were more common). **1953** C. E. OSGOOD *Method & Theory in Experim. Psychol.* xvi. 680 The second hindrance to objectivity is the ubiquitous tendency to reify the word, to assume the word itself somehow carries its own meaning. **1971** *Times Lit. Suppl.* 31 Dec. 1619/3 To look upon them [*sc.* economic laws] as objective necessities, as bourgeois economists do, is to reify them. **1979** E. H. GOMBRICH *Sense of Order* x. 282 The temptation to 'reify' the shield into the open mouth of a gaping mask..proved as irresistible as did the opportunity of turning spiralling volutes into suggestions of eyes.

Hence **re·ified** *ppl. a.*, **re·ifying** *vbl. sb.* and *ppl. a.*

1941 H. MARCUSE *Reason & Revol.* iv. 115 Lordship and bondage result of necessity from certain relationships of labor, which are, in turn, relationships in a 'reified' world. **1962** MACQUARRIE & ROBINSON tr. *Heidegger's Being & Time* II. vi. 487 Why does this reifying always keep coming back to exercise its dominion? **1965** B. PEARCE tr. *Preobrashensky's New Economics* 47 One can..understand its laws in the spirit of vulgar economics, that is, by offering in the guise of science mere superficial description, complete with the reified relations of commodity production. **1969** R. BLACKBURN in Cockburn & Blackburn *Student Power* 207 An alienated society naturally encourages a re-ifying vocabulary. **1979** E. H. GOMBRICH *Sense of Order* ix. 242 It is surely not far-fetched to interpret its coiling frame as a reified flourish on a reified support.

reign, *sb.* Add: **4.** *attrib.*, as **reign mark,** a mark on a piece of oriental ceramic ware indicating in whose reign it was made; **reign name, title,** the symbolic name adopted by a Japanese or (formerly) Chinese ruler, by which his reign is known and dated.

1936 *Burlington Mag.* Jan. 10/2 Distinguished by a reign mark in blue or pink enamel. **1980** *Catal. Fine Chinese Ceramics* (Sotheby, Hong Kong) 6 Where a reign mark is given after the measurement no attribution to the period of this reign is intended unless the words '*and period*' are added. [**1834** C. GUTZLAFF *Sketch of Chinese Hist.* I. iv. 89 The Han Dynasty... The characters given after the emperors' names are the kwŏ-haou, 'national designations' of the emperors during their reigns. **1848** S. W. WILLIAMS *Middle Kingdom* II. xvii. 229 Kwoh Hiau, or Reigning Title.] **1935** C. P. FITZGERALD *China* xxiii. 457 Every Ming Emperor retained the same reign title for the full duration of his reign. **1974** *Encycl. Brit. Macropædia* X. 78/1 Mutsuhito, who took the reign name Meiji ('enlightened rule', 1863–1912). **1976** *Times* 10 Nov. 17/1 Reign titles disappeared from China with the fall of the empire in 1911: in Japan they still survive.

reigner. For † *Obs.* read *rare* and add later example.

1908 *Daily Chron.* 13 Feb. 4/4 Louis XIV., the record reigner, was never in his life clean all over from the natural exhalations of a monarch's skin.

Reil (rəil). *Anat.* [The name of Johann Christian *Reil* (1759–1813), German anatomist.] *island of Reil*: an area of the cerebral cortex which overlies the corpus striatum but is concealed within the lateral sulcus (the fissure of Sylvius).

Described by Reil in 1809 (*Arch. f. Physiol.* IX. 144).

1840 G. V. ELLIS *Demonstrations of Anat.* 31 The fissure [of Sylvius] divides above into two portions, one of which passes before, and one behind some small convolutions which constitute the island of Reil. **1888** W. R. GOWERS *Man. Dis. Nervous Syst.* II. iv. 6 Within the fissure of Sylvius lie the small convolutions of the island of Reil, or insula, four or five in number, which spread out like a fan. **1939** O. LARSELL *Textbk. Neuro-Anat.* XVIII. 219 The lateral fissure (fissure of Sylvius) opens to the hidden cortical surface of the island of Reil. **1961** A. R. BUCHANAN *Functional Neuro-Anat.* xxi. 173 The insular part of the cerebral cortex or island of Reil can be readily exposed by removal of the opercular portions of the frontal, parietal and temporal lobes of the cerebrum.

Reilly, var. *RILEY.

reim. Add: See also RIEM in Dict. and Suppl. (Earlier and later examples.)

1852 M. B. HUDSON *Feature in S. Afr. Frontier Life* I. v. 127 Their minister had with a reim round his neck Been dragged from his station by Bowker's command. **1862** *Colburn's United Service Mag.* III. 212 They have always what the Dutch call a 'forelouper'..who goes in front of the two leading oxen to pull them on with a strap or *reim*, as the Dutch call it. **1947** P. J. PRETORIUS *Jungle Man* vii. 86 When I secured a captive I took a reim, tied it round the neck of the animal, and the boy was then sent off to camp leading it.

reimbibe, *v.* Add: Also *fig.*

a **1960** E. M. FORSTER *Maurice* (1971) iii. 15 Then he would reimbibe the face and the four words.

reimbursabi·lity. [f. REIMBURSABLE *a.*] The quality of being reimbursable.

1978 *Nature* 16 Nov. 201/3 What this means is that Congress is limiting the reimbursability of salaries of the best faculty, the stars, the Nobel Prize winners, those people who make our institutions great.

reimmerse, *v.* Add: Also *fig.*

1905 *Speaker* 1 Apr. 18/1 It is to the desert that yearly pilgrimages bring hosts of the faithful..to re-immerse themselves in the original enthusiasm.

reimpje, reimpie, var. *RIEMPIE. (See also RHEIMPY.)

1891 B. MITFORD *Romance of Cape Frontier* xxiii. 419 Just slip off these bits of *reimpje*, Tambusa; and give me an assegai and a stick or something. **1923** O. SCHREINER *Thoughts on S. Afr.* 17 The great clean, bare 'voorhuis' (front room) with its mud floors and its chairs and sofa with reimpje seats. **1926** *Brit. Weekly* 27 May 158/4 Men and women sat on reimpje-seated chairs. **1927** *Glasgow Herald* 14 May 11/4 The sitting-room with its beautiful chairs and settee of native wood with 'reimpje' bottoms (criss-crossed thongs of home-cured leather), all in a simple old Dutch style, looked strangely unfamiliar.

reimplant, *v.*, **reimplantation.** (Later examples of both.)

1919 *Jrnl. Amer. Med. Assoc.* 26 July 301/1 Bonnefon's experiences confirm that if a small pathologic process can be cut out completely and then reimplanted in the old site, the environment being normal, it will be invaded by normal cells and lose its pathologic characteristics. **1955** *New Biol.* XVIII. 32 The reimplantation of stored infant tissues into the same animal, and..the observation of the reciprocal influences of host and implant, are technically practicable and established procedures. **1973** *Sci. Amer.* Feb. 28/3 We can transfer the eye into a tissue culture and change its orientation when we reimplant it in a host embryo, and correlate the result with the results of direct-transplantation experiments.

reimportation. (Earlier example.)

1857 MILL *Pol. Econ.* (ed. 4) II. III. xxiv. 229 The Bank reserves can replenish themselves without any re-importation of the gold.

reimpregnate, *v.* (Later example.)

1953 R. LEHMANN *Echoing Grove* 29 But still the stones seemed rocked, the unsterile mounds, reimpregnated, exhaled dust's fever.

reimpression. 2. (Later example.)

1924 W. B. SELBIE *Psychol. Relig.* 89 Every one is aware of unaccountable recollections of this kind. Such a re-impression of familiar things may take place even though there is no conscious recollection.

reimschoon, var. *REMSKOEN.

rein, *sb.*¹ Add: **4.** *rein-ring*; **rein-hand** (earlier example).

1843 *Ainsworth's Mag.* IV. 436 Our well-fed 'Phaeton' pulled his team together.., dropped his rein-hand, [etc.]. **1955** E. POUND *Classic Anthol.* III. 188 A leathered front-board with tiger-fell And metal rein-rings as well. **1968** J. ARNOLD *Shell Bk. Country Crafts* xxiii. 267 Attached to the hames are tug-hooks and rein-rings.

reincarna·tionism. [f. REINCARNATION + -ISM.] A belief in, or doctrine of, reincarnation.

1907 W. DE MORGAN *Alice-For-Short* viii. 75 This is an entirely unwarranted speculation, based upon no data; a neotheosophical reincarnationism without so much as a single Himalayan Brother to back you up! **1938** *Jrnl. Theol. Stud.* XXXIX. 192 We shall find ourselves emphasizing the subordinationism and the reincarnationism to such an extent that the Origenian theology may well appear to be merely one more form of Gnosticism.

reindeer. Add: **2.** *reindeer meat, skin* (later examples), *steak*; **reindeer lichen** (earlier example); **reindeer tongue,** the tongue of a reindeer, usu. smoked, considered as a delicacy.

1770 G. CARTWRIGHT *Jrnl.* 11 July (1792) I. 8 They [*sc.* caribou] find there many extensive tracts of land destitute of wood, and covered with plenty of Reindeer Lichen. **1926** *Daily Colonist* (Victoria, B.C.) 5 Jan. 2/4 Reindeer meat was the principal attraction on the menu. **1933** Reindeer-skin [see *KAMIK]. **1977** *Country Life* 13 Jan. 80/1 His wife makes..the soft reindeer-skin shoes for winter. **1958** W. BICKEL tr. *Hering's Dict. Class. & Mod. Cookery* 502 *Reindeer steak*,.. steak cut from a tender loin, sautéd in butter. **1973** D. FRANCIS *Slay-Ride* v. 67 She gave us reindeer steaks in a rich dark sauce. **1788** *Times* 1 Jan. 4/3 (Advt.), Smoked Salmon and Dutch Herring, Fine New French Olives, and New Rein Deer Tongues. **1857** J. H. WALSH *Economical Housekeeper* II. iii. 32 Reindeer Tongues, Pork Pies, and a whole host of similar commodities, are sold at the 'sausage shops' in London. **1935** M. MORPHY *Recipes of All Nations* 499 Reindeer are eaten in Norway, and smoked reindeer tongues are considered a great delicacy. **1973** J. FLEMING *You won't let me Finish* xvi. 129 I'm told I am to have some smoked reindeer tongue for my lunch.

reinduction. Delete † and add later example in sense of *INDUCTION 4 d.

1944 N. MAILER in *Cross-Section* 336 A half-year later, he was made (after reinduction school) a captain.

‖ **reine** (rɛn). *Cookery.* [Fr., lit. 'queen'.] Chiefly in phr. *à la reine* 'in the fashion of a queen', used to designate dishes prepared in some special way. Also used alone following the name of the dish.

1845 E. ACTON *Mod. Cookery* i. 37 (*heading*) Rabbit soup à la reine. Wash and soak thoroughly three young rabbits. **1884** MADAME VALERIE *Cookery for Amateurs* ii. 25 *Soups*...

À la Reine. Although this potage has a fine name it is easily made if you have the remains of cold fowl or turkey. **1930** H. BELLOC *New Cautionary Tales* 13 Turbot à la Reine, and Ices. **1958** W. BICKEL tr. *Hering's Dict. Class. & Mod. Cookery* 254 *Sole..queen style*, à la reine: poached, covered with creamed fish velouté, garnished with truffle slices and small fish dumplings. **1962** *Listener* 26 July 155/3, I had this dish in a tiny restaurant in the Dordogne Valley. It is called *plaice reine*.

Reinecke (rəi·nekə). *Chem.* The name of A. *Reinecke*, 19th-c. German chemist, used in the possessive or *attrib.* to designate (*a*) a red crystalline complex salt, ammonium diamminetetrakis(isothiocyanato)chromate(III), $NH_4[Cr(NCS)_4(NH_3)_2].H_2O$, which is used esp. in *Biochem.* to precipitate large cations, and (*b*) the parent acid of this salt, $H[Cr(NCS)_4(NH_3)_2]$, which can also be isolated as red crystals.

Reinecke described the preparation of the salt in 1863 (*Ann. d. Chem. u. Pharm.* CXXVI. 113).

1892 *Jrnl. Chem. Soc.* LXII. II. 798 Reinecke's salt crystallises in rectangular tables; it is insoluble in absolute ether, but dissolves in water to a ruby-red solution and in alcohol. **1928** *Chem. Abstr.* XXII. 764 The urine is slightly acidified and evapd. to ½ its vol., purified with charcoal, and the creatinine pptd. with Reinecke Salt. **1928** *Brit. Chem. Abstr.* A. 542/2 The mixed potassium ammonium salt of Reinecke's acid, obtained by melting together potassium dichromate and ammonium thiocyanate. **1933** *Biochem. Jrnl.* XXVII. 157 Reinecke's salt, $[(NH_3)_2Cr(CNS)_4]NH_4$, is proving to be a valuable precipitant for a variety of basic substances, and its use is likely to extend. **1935** *Org. Syntheses* XV. 75 The undissolved residue from the second extraction consists chiefly of Morland salt (the guanidine salt of the Reinecke acid). **1965** tr. Hein & Herzog in G. Brauer *Handbk. Preparative Inorg. Chem.* II. xxiv. 1377 The total yield of air-dry Reinecke salt amounts to 250–275 g. *Ibid.* 1378 $H[Cr(SCN)_4(NH_3)_2]$... Synonym: Reinecke acid. **1966** PHILLIPS & WILLIAMS *Inorg. Chem.* II. xxvi. 322 The anion of Reinecke's salt..and that of K_2PtCl_4 are used to precipitate large organic cations.

Hence **reineckate** (rəi·nĕkēˈt) [ad. G. *reineckat*: see -ATE¹], (a salt of) the anion present in Reinecke's salt, $[Cr(NCS)_4(NH_3)_2]^-$.

1928 *Brit. Chem. Abstr.* A. 526/2 The cuprous 'Reineckate' is removed by filtration. **1939** *Thorpe's Dict. Appl. Chem.* (ed. 4) III. 115/1 Reinecke's salt... Its solution gives precipitates with the heavy metals and organic bases, alkaloids yielding characteristic crystalline compounds—reineckates. **1955** J. A. LOVERN *Chem. Lipids of Biol. Significance* ii. 61 Choline is estimated in a variety of ways. The simplest and most frequently used methods depend on the formation of a sparingly-soluble complex with the reineckate radical. **1957** *Jrnl. Antibiotics* X. 188 The antibiotic was..crystallized as its reineckate. **1964** *Oceanogr. & Marine Biol.* II. 152 The reineckate ion is removed with silver nitrate.

‖ **Reine Claude** (rɛn klod). [Fr., perh. a. the name of *Claude* (1499–1524), daughter of Louis XII and wife of François I.] = GREENGAGE. Also *attrib.*

1731 P. MILLER *Gardeners Dict.* s.v. Prunus 16. La Reine Claude, i.e. Queen Claudia. This is a small round Fruit, of a yellowish Colour,..and its Juice is richly sugar'd. **1860** R. HOGG *Fruit Man.* 252 Reine Claude. See Green Gage. **1929** E. A. BUNYARD *Anat. Dessert* 118 In France it [*sc.* the greengage] is always known as Reine Claude, and the legend runs that it is thus named after the wife of François I. **1941** MRS. BELLOC LOWNDES *I, too, have lived in Arcadia* xvi. 299 A large old *Reine Claude*, of which the sweet luscious fruit was famed. **1962** *Harper's Bazaar* Aug. 37 Delicious though the *reine claude* can be. **1973** *Guardian* 20 Jan. 3/4 It is the land of..the honeyed Reine-Claude greengages.

reined, *ppl. a.* Add: **1.** Also *reined-back*.

1905 W. H. HUNT *Pre-Raphaelitism* II. vii. 174 At first acquaintance with the poet, I thought that later in my knowledge of him I should see some phases of the reined-back pose of Woolner's bust, but this I was unable to do. **1961** *New Statesman* 21 July 92/1 The reined-back rhythms of this verse are especially pleasing.

reinfect, *v.* (Later examples.)

1889 G. A. SMITH *Bk. Isaiah* (ed. 2) I. 422 Not only to find it [*sc.* sin] 'hindering, disturbing, complicating all', but reinfecting with the lust and odour of sin the will which gave it birth. **1928** L. E. H. WHITBY *Med. Bacteriol.* xxi. 210 Finally, the envelope of the corpuscle bursts and the merozoites are set free. These in turn reinfect red corpuscles and develop into trophozoites, thus carrying on the cycle. **1960** *Farmer & Stockbreeder* 12 Jan. 12 Loose smut re-infects growing corn and so perpetuates the disease.

reinfesta·tion. [RE- 5 a.] A second or further infestation.

1911 in WEBSTER. **1946** *Nature* 2 Nov. 636/2 One application of 0·5 per cent 'Gammexane' dust has been generally found to..prevent re-infestation for some time. **1968** *Times* 30 Oct. 12/2 A central portion of the sewer system was cleared of rats..and no attempt was made to prevent re-infestation.

reinforce, *v.* Add: **2. d.** *Psychol.* To strengthen (a response), usu. by repetition of a stimulus, esp. one that is painful or rewarding.

1906 C. S. SHERRINGTON *Integrative Action Nervous Syst.* v. 175 These widely separate reflex-arcs therefore re-inforce one another. **1927** G. V. ANREP tr. *Pavlov's Conditioned Reflexes* vii. 117 The second method consisted in contrasting the single definite conditioned stimulus..with different neighbouring stimuli which were never reinforced. **1951** S. F. NAGEL *Found. Social Anthropol.* iv. 58 To remain effective, however, the conditioning must be 'reinforced'. **1970** *Jrnl. Gen. Psychol.* July 3 Bugelski..notes that in the orthodox stimulus-response view of learning based on the reinforcement of an instrumental response, it is essential that the behavior appear first and then be..'reinforced by the psychologist (or someone) immediately'. **1973** *Howard Jrnl.* XIII. 281 A points system was used to pay the men for desirable behaviours that would be likely also to be reinforced in life outside the institution.

reinforced, *ppl. a.* Add: **2.** Special collocations: **reinforced concrete,** concrete with steel bars or network embedded in it to increase its tensile strength; **reinforced plastic,** plastic strengthened by the inclusion of a layer of fibre (esp. glass).

1902 *Min. Proc. Inst. Civil Engineers* CXLIX. 297 Reinforced concrete is extremely economical..where an imposing building is not required. **1906** *Daily Chron.* 27 Apr. 6/2 There is undoubtedly a great future for reinforced concrete. **1910** *Encycl. Brit.* VI. 837/2 The introduction of steel concrete (also known as ferroconcrete, armoured concrete, or reinforced concrete) is generally attributed to Joseph Monier, a French gardener. **1958** *Engineering* 14 Mar. 350/1 A sufficient number of reinforced-concrete buildings have now been in use for more than 50 years to show that, properly designed, reinforced concrete is as durable material as is likely to be required for most purposes. **1966** *McGraw-Hill Encycl. Sci. & Technol.* II. 337/1 Modern bridge abutments are usually made of reinforced concrete. **1974** *Encycl. Brit. Macropædia* III. 184/1 The towers are of reinforced concrete and the cables built up of strands of twisted wire. **1940** *Brit. Plastics* Aug. (Advt. section) 5 Two-piece housing in reinforced plastic material for electric hair drying unit. **1947** W. J. BROWN *Fabric Reinforced Plastics* iv. 76 The increasing use of reinforced plastics for engineering applications. **1959** *Engineering* 16 Jan. 86/1 The principal advantage of the reinforced plastics hull is the much reduced maintenance work and its easy repairability. **1971** *Nature* 30 July 305/1 If the exceptional properties of carbon fibres are to be utilized for engineering purposes they must be fabricated as a 'reinforced plastic'.

reinforcement. Add: **3. b.** *spec.* Increase in the intensity or amplitude of sound.

1879 [in Dict., sense 3]. **1937** A. T. JONES *Sound* viii. 198 When the stem of a vibrating tuning fork is placed on top of a wooden table or other extended wooden surface there is also a reinforcement of the sound. **1959** E. PULGRAM *Introd. Spectrography of Speech* vii. 58 This gain in amplitude is obtained in exchange for loss of duration, for a tuning fork thus placed for reinforcement will cease to operate more quickly than one not so placed. **1969** L. F. YERGES *Sound, Noise, & Vibration Control* 70 Today..deliberate 'electrical' amplification is the more significant means of sound reinforcement. *Ibid.* 145 Reinforcement is usually necessary in the following spaces: 1. Legitimate theaters with more than 1000 seats. 2. Lecture halls with more than 300 seats. 3. Almost all gymnasiums, arenas, and large assembly halls.

c. *Psychol.* (An act of) strengthening or establishing of a response, esp. in learning theory through the repetition of a rewarding or painful stimulus, or the satisfaction of a need; also *attrib.*

1876 W. JAMES *Coll. Ess. & Rev.* (1920) 31 The whole question of its predetermination relates to the intensity of the degree of reinforcement with which the triumphant representation occurs. **1906** C. S. SHERRINGTON *Integrative Action Nervous Syst.* v. 175 This reinforcement is significant of the solidarity of the whole spinal mechanism. **1927** G. V. ANREP tr. *Pavlov's Conditioned Reflexes* vii. 117 The first method consisted in repeating the definite conditioned stimulus a great number of times always accompanied by reinforcement. **1948** E. R. HILGARD *Theories of Learning* iv. 84 Primary reinforcement..is identified with diminution of need. Secondary reinforcement is mediated by a stimulus which has been closely and consistently associated with the need reduction. *Ibid.* xii. 347 Doubts about this basic pattern for reinforcement theory have been expressed frequently throughout the preceding chapters. **1953** C. E. OSGOOD *Method & Theory in Experim. Psychol.* ix. 376 The two crucial conditions for learning [in Hull's system] are thus contiguity and reinforcement. **1960** J. B. CARROLL in Saporta & Bastian *Psycholinguistics* (1961) 333/1 A purely Pavlovian or Watsonian view of language learning has been supplanted generally by some variety of reinforcement theory. **1963** *Listener* 7 Feb. 238/1 The pleasures of gambling, right up to the time the ruined rake rises from the table to shoot himself, are..due to 'the principle of intermittent reinforcement'. **1973** *Howard Jrnl.* XIII. 269 If reinforcement techniques can circumvent this double-bind situation, they may be justified.

5. The strengthening structure or material employed in reinforced concrete or plastic.

1905 G. J. FIEBEGER *Civil Engin.* xxi. 405 A beam may have its reinforcement on the tension side only, or on both the tension and compression sides of the neutral axis. **1958** *Engineering* 14 Mar. 350/1 Adequate cover to the reinforcement, including binding wires and stirrups, is essential, as is the elimination of all unnecessary steelwork. **1973** *Materials & Technol.* VI. viii. 521 The resin and the reinforcement, cut to size, are introduced separately into the mould. **1974** *Encycl. Brit. Macropædia* XIV. 519/1 Reinforcements [for plastics] include cotton and asbestos flocks; glass fibres, chopped or in the form of rovings, mats, or monofilaments; carbon fibres; and mineral whiskers.

reinforcer. Add: **b.** *Psychol.* That which serves to reinforce or strengthen a response.

1958 L. KRASNER in Saporta & Bastian *Psycholinguistics* (1961) 84/2 'Good' and a head shake were effective reinforcers. **1967** *Listener* 12 Jan. 55/2 By making reinforcers contingent on behaviour we can alter behaviour in a very effective way. **1974** B. F. SKINNER *About Behaviorism* 39 The behavior is said to be *strengthened* by its consequences, and for that reason the consequences themselves are called 'reinforcers'.

reinforcing, *vbl. sb.* Add: **2.** *concr.* = *REINFORCEMENT 5.

1966 *McGraw-Hill Encycl. Sci. & Technol.* XI. 426/1 As reinforcing for concrete, steel in several of the following forms may be used.

reinforcing *ppl. a.* (further examples).

1906 C. S. SHERRINGTON *Integrative Action of Nervous Syst.* v. 175 Gentle stimuli to the skin of a limb exerted a reinforcing influence on closely following stimuli applied to the limb region of the cortex of the brain. **1938** B. F. SKINNER *Behav. of Organisms* vi. 244 It is possible to show that an emotional or reinforcing stimulus..is effective without regard to various minor properties. **1948** E. R. HILGARD *Theories of Learning* iv. 108 Any stimulus which has been associated with need-reduction may itself serve as a reinforcing agent. **1949** POSTMAN & EGAN *Experim. Psychol.* xiv. 298 The instrumental response and the classical response have both been established through the same reinforcing stimulus—food. **1970** [see *homopolymer* s.v. *HOMO-]. **1973** *N.Y. Law Jrnl.* 31 Aug. 1/5 Three major steel companies..were indicted yesterday..on charges of violating anti-trust law in the sale of reinforcing steel bars in Texas. **1973** P. DICKINSON *Gift* v. 75 The men and machines had dug a vast, rectangular wound in the clay;..half was still hummocked mud,..littered with grids of reinforcing rods.

reinfuse, *v.* Add: Hence **reinfu·sion.**

1963 *Lancet* 5 Jan. 61/1 We reviewed the problem of blood-transfusion for Jehovah's Witnesses and discovered that they will accept immediate reinfusion of their own blood.

‖ **Reinga** (reˌi·ŋa). Also **reinga, Re-i-nga;** † **Treaingha.** [Maori, = 'place of leaping'.] In Maori tradition, the place where departed spirits make their way into the next world; hence, the land of departed spirits.

1822 *Proc. Church Missionary Soc.* 364 They say, that, at the death of a Chief, his soul goes to the Treaingha, at the North Cape. **1830** *New Zealanders* x. 236 Réinga signifies, properly, the place of flight; and is said, in some of the accounts, to be a rock or a mountain at the North Cape, from which, according to others, the spirits descend into the next world through the sea. **1884** M. A. MARTIN *Our Maoris* vi. 79 The natives in the north of the island still point out the cliff from which the spirits [of the dead] made their descent into the sea on their way back to the Island of Hawaii, from whence their forefathers came. This cliff was called the Re-i-nga, *i.e.*, the leaping-place. **1938** R. D. FINLAYSON *Brown Man's Burden* 48 Depart, O father, to the Reinga, to the far Hawaiki, to the Lord of the Dead. **1949** P. BUCK *Coming of Maori* IV. iv. 516 Thus death closed the account of the body, and the soul (*wairua*) entered the spirit land (*reinga*) with a clean sheet and without apprehension.

reinge·stion. [RE- 5 a.] = *REFECTION *sb.* 2 d.

1956 THOMPSON & WORDEN *Rabbit* iii. 27 It is difficult to believe that reingestion has not in fact been observed many times. **1964** H. N. SOUTHERN *Handbk. Brit. Mammals* 254 Utilization of food [by rabbits] assisted by reingestion, or refection, in which soft faecal pellets (mainly from caecum, where microbial digestion occurs) are swallowed. **1964** R. M. LOCKLEY *Private Life of Rabbit* x. 103 Reingestion was observed both out of doors by day and below ground.

reinikaboo: see *RANNYGAZOO.

reink, *v.* (Later examples.)

1937 *Discovery* Oct. 300/2 The stone being kept damp, and re-inked for each print. **1955** J. RYDER *Printing for Pleasure* v. 61 After taking the first print, and without re-inking, print onto the top sheet of the platen packing.

reinsman. Add: Also *Austral.* and *N.Z.*, spec. in *Trotting*. (Further examples.)

1904 *N.Y. Times* 13 Dec. 7 A number of well-known amateur reinsmen started from the Harlem River Speedway. **1930** W. BANNING in W. & G. H. Banning *Six Horses* 361 A reinsman was a master driver who..was able to drive each span of his complement wholly independent of the other. **1969** *Sydney Morning Herald* 24 May 27/1 Western Districts reinsman Gordon McWilliam lost a battle of tactics behind the hot favourite, Cocky Raider. **1977** *N.Z. Herald* 8 Jan. 1-9/4 The Cambridge reinsman R. F. Mitchell..had a farewell present at Cambridge last night when he drove Pompano Prince to victory.

reinstatement. Add: **1. c.** *Mil.* Re-establishment of a serviceman in a previously held civilian job after demobilization. Chiefly *attrib.*

1945 *Daily Herald* 17 May 4/3 The Reinstatement Act was working very smoothly—'and I believe I shall have very few cases to go to the reinstatement committees at all.' **1946** *R.A.F. Jrnl.* May 160 With the best will in the world, reinstatement regulations cannot help in such cases.

reinsure, *v.* Add: Also *intr.* or *absol.* (Earlier examples.)

1755 N. MAGENS *Ess. Insurances* I. 94 Such Insurer, his Executors, Administrators, or Assignees, may re-insure to the Amount of the Sum before insured. **1802** S. MARSHALL *Treat. Insurance* I. 115 At Marseilles..the insured, in such case, sues the insolvent insurer till he obtains a sentence authorizing him to re-insure at the expense of the insolvent.

reintegration. Add: **3.** (Later examples.)

It has supplanted *redintegration* as the usual form in this sense.

1973 *Times* 13 Feb. 7/2 Soviet consular officials have told them that their reintegration 'will take time'. **1977** *Sunday Times Mag.* (Perth, Austral.) 4/4 It was not the chattering of birds on the roof that set reintegration in process. **1979** *Internat. Rehabilit. Med.* I. 45/2 Intensive rehabilitation is indicated when..at least social reintegration into the family can be achieved.

reintegrative (rĭˌi·ntĭgrătiv), *a.* [f. REINTEGRATE *v.* + -IVE: see REDINTEGRATIVE *a.*] Tending to reintegration.

1957 V. W. TURNER *Schism & Continuity in Afr. Society* x. 303 A society continually threatened with disintegration is continually performing reintegrative ritual. **1974** *Gen. Systems* XIX. 67/2 But so brilliantly had Mayan..societies systematized their communications, that the patient's very gestures ('symptoms') had actuated the reintegrative liturgies. **1981** J. CAREY *John Donne* vii. 221 Donne's interest lies less in resurrection itself than in reintegration... His sole intent is to emphasize the power of the reintegrative act.

reinterpret, *v.* (Further examples.)

1920 A. S. EDDINGTON *Space, Time & Gravit.* ix. 141 It has been reinterpreted and has finally become merged in the conservation of energy. **1952** B. ULANOV *Hist. Jazz in Amer.* i. 7 The rhythmic base of music has been reinterpreted. **1979** *N.Y. Rev. Bks.* 25 Oct. 11/4 A seasoned cult member is assigned to the nascent convert..to reinterpret his old life according to the new beliefs.

reinterpretation (further examples).

1956 *Nature* 10 Mar. 443/2 For these reasons there should be a re-interpretation of humanism. **1974** R. A. HALL *External Hist. Romance Langs.* 6 The reinterpretation of an unfamiliar form or part of a form under the influence of another form with which it has been identified.

reinvention. (Further examples.)

1964 W. L. GOODMAN *Hist. Woodworking Tools* 38 It appears to be a re-invention of the Minoan and Roman tool. **1973** *Sci. Amer.* Apr. 85/1 James thought up the idea of the differential gear (actually a reinvention).

reinvert, *v.* (Later example.)

1907 J. H. PARSONS *Dis. Eye* iv. 43 Just as with a convex lens, the image is inverted. It is re-inverted psychologically in the brain.

reinvestment. (Further examples.)

1857 MILL *Pol. Econ.* (ed. 4) I. I. iv. 70 All property..is a part of capital, so soon as it..is set apart for productive reinvestment. **1931** *Economist* 3 Jan. 25/2 A fair amount of reinvestment business following the heavy dividend disbursements of this week sufficed..to give prices a firm tendency. **1980** W. ASH *Incorporated* vi. 61 Various proposals for the re-investment of the profits.

Reissner (rəi·snəɪ). *Anat.* The name of Ernst *Reissner* (1824–78), German anatomist, used in the possessive or with *of* to designate a thin vestibular membrane of the internal ear, separating the scala vestibuli from the central duct of the cochlea.

1872 [see *SCALA]. **1945** MCNALLY & STUART in C. & C. L. Jackson *Dis. Nose, Throat, & Ear* III. 366 (*caption*) Gross dilatation of the scala media has occurred with displacement of Reissner's membrane on to the wall of the scala vestibuli. **1974** D. & M. WEBSTER *Compar. Vertebr. Morphol.* x. 229 The scala media is separated from the scala vestibuli by a thin, epithelial structure, Reissner's membrane.

reissue, *sb.* Add: (Further examples.) Also, a reissued gramophone record.

1937 *Discovery* Oct. 320/1 Apart from this curious re-issue, Godwin's book was reprinted in the Harleian Miscellany in 1746. **1948** *Hansard Commons* 21 Jan. 217 American films..were good enough to go on attracting cinema-goers even on re-issue. **1966** *Listener* 5 May 664/3 No reservations about a wonderful bargain HMV reissue. **1977** *Rolling Stone* 24 Mar., More people acquaint themselves with Fleetwood Mac and dig back to old reissues.

reistafel, var. *RIJSTTAFEL.

Reiter (rəi·təɪ). *Path.* The name of Hans *Reiter* (1881–1969), German bacteriologist, used in the possessive to denote a disease or syndrome first described by him in 1916 (*Deutsche Med. Wochenschr.* 14 Dec. 1535) which is characterized by arthritis, conjunctivitis, and urethritis, typically affects young men, and is usually caused by bacteria of the genus *Chlamydia*.

1923 STEDMAN *Med. Dict.* (ed. 7) 859/2 *Reiter's disease*, a fever of a more or less remittent type, lasting about seventeen days, accompanied with pains in the joints, conjunctivitis, iritis, cystitis, and enlargement of the spleen; it has been observed in the tropics, but its causation is unknown. **1946** *Jrnl. Infect. Dis.* LXXIX. 134 (*heading*) The possible relationship of the pleuropneumonia-like organisms to Reiter's disease, rheumatoid arthritis and ulcerative colitis. **1962** *Lancet* 26 May 1111/1 The symptom-triad of arthritis, conjunctivitis, and urethritis, generally known as Reiter's syndrome, has a striking predilection for young men in military service. **1972** [see *keratoderma* s.v. *KERATO-]. **1978** *Jrnl. R. Soc. Med.* LXXI. 335 In Reiter's disease, acute uveitis is usually asynchronous.

Reithian (rī·þiăn), *a.* Also **Reithean.** [f. the name of J. C. W. *Reith* (1889–1971), 1st Lord Reith of Stonehaven, Director-General of the British Broadcasting Corporation (1927–38) + -IAN.] Of, pertaining to, or characteristic of Reith or his principles, esp. relating to the responsibility of broadcasting to enlighten and educate public taste.

1961 *Guardian* 26 Oct. 10/5 The Reithian conception of broadcasting was barely in evidence at the time [*sc.* in 1923]. **1963** *Times* 12 Jan. 4/5 An elaborate compromise between past and present balancing Reithian principles against television's need to survive as show business. **1969** *Observer* (Colour Suppl.) 21 Dec. 38/1 From the great Reithean days the BBC had come to be taken for granted as a pillar of enlightened orthodoxy. **1973** *Times* 9 June 10/4 Nowhere has the Reithian prediction—offer the public what it wants and it will want what it gets—been vindicated more awesomely. **1977** *Punch* 31 Aug.–6 Sept. 327/3 The BBC..still retains at least something of the old Reithian sobriety or an image of established quality.

‖ **reja** (rē·ha). [Sp.] In Spain, a wrought-iron screen or grille used to protect windows, chapel tombs, etc.

1845 R. FORD *Hand-bk. Travellers in Spain* I. 252/2 The *coro*..is railed off by a fine *reja*, the work of Sancho Muñoz, 1519. **1870** R. H. BUSK *Patrañas* 147 Then night came: the *maja* stood at her *reja*, looking out for her serenader. **1914** BYNE & STAPLEY *Rejeria of Span. Renaissance* p. vii, Renaissance Architecture in Spain could not be fully appreciated without examining the towering wrought-iron grilles, or Rejas, of the period. **1924** J. MASEFIELD *Sard Harker* iv. 289 The windows..were covered with old iron rejas. **1969** S. SITWELL *Gothic Europe* xii. 139 One remembers Spanish cathedrals very notably on account of their *rejas* or wrought iron screens which are a feature peculiar to Spain.

Rejang (rēˈdʒa·ŋ). Also **Redjang.** [Native name.] An Indonesian people of southern Sumatra; a member of this people. Also, their language.

1783 [see *BATTA² *a.* and *sb.*]. **1839** T. J. NEWBOLD *Straits of Malacca* II. xiv. 227 The Rejangs, a people inhabiting the interior of Sumatra. **1932** W. L. GRAFF *Lang. & Languages* xi. 423 The population of *Indonesian* tongue amounts to about 50 millions... About eight geographical groups can be distinguished... In the *Sumatra* group, *Achinese, Battak, Rejang, Lampong, Malay, Mentaway*. **1955** P. VOORHOEVE *Crit. Survey Stud. Lang. Sumatra* v. 20 We find in the language of Rejang another language which is undoubtedly independent although closely related to Malay.

reject, *sb.* Now with pronunc. (rī·dʒekt). **1.** Restrict † *Obs.* to sense in Dict. and add: **b.** [f. the vb.] One who is rejected or discarded by others, esp. as unsuitable for some activity (orig. for military service).

1925 *Glasgow Herald* 13 Mar. 8/7 Probably the large proportion of rejects is not so much a symptom of national decadence as a result of the fact that the right sort of men are not coming forward in sufficient numbers. **1942** *Sun* (Baltimore) 4 Nov. 9/4 He said that..the 'army has been consistently uninterested in taking the rejects into conditioning battalions and reconditioning them'. **1971** *Sunday Express* (Johannesburg) 28 Mar. 17/2 Time and again I have heard members at the turnstiles say that they would prefer to watch South African-born players rather than overseas rejects. **1974** E. TIDYMAN *Dummy* vi. 87 The misborn and the unwanted..society's rejects. **1979** R. RENDELL *Make Death love Me* ii. 21 He knew someone who..was also a reject of the University of Kent.

2. (Further examples.)

1935 H. C. BRYSON *Gramophone Record* ix. 234 If rejects are kept below 15 per cent. with rigid examination, then efficiency is high. **1949** J. DEKETH *Fund. Radio-Valve Technique* vii. 61 If this fusing method were applied to values of the dimensions of the A-technique..there would be a higher percentage of rejects in manufacture. **1952**, etc. [see *export reject*]. **1969** [see *CULLER 1 b].

3. *attrib.* **a.** Appositive.

1955 H. KURNITZ *Invasion of Privacy* (1956) ii. 17 This pioneer used a war surplus Eimo camera and 'reject' film which he developed in his bath tub. **1963** R. R. A. HIGHAM *Handbk. Papermaking* ii. 71 The lighter reject materials..are ejected through a special automatically controlled V-notch slide valve. **1965** *Wireless World* July 22 (Advt.), *Antex* reduces operational fatigue, with resulting drop in reject output. **1977** 'M. YORKE' *Cost of Silence* ix. 69 His friends..had seen him with Madge and made a few cracks about reject models.

b. General attrib. uses.

1958 *Times* 26 Feb. 8/4 The reject figures of Army recruits has [*sic*] given some cause for concern here. **1963** R. R. A. HIGHAM *Handbk. Papermaking* ii. 71 Heavy material which will not pass through the screen is continuously forced downwards into a reject trough and is removed from a heavy reject box connected to the bottom of the volute trough.

reject, *v.* Add: **I. 6. d.** *Psychol.* Of a parent or guardian: to spurn (a child) by denying it the normal emotional relationship between parent and offspring.

1931 *Smith Coll. Stud. in Soc. Work* I. 407 Case histories are presented showing the attitude toward their parents.., husbands, and children of twelve mothers who rejected their children. **1932** *Ibid.* II. 237 This type of relationship cannot exist when a mother rejects her child. **1961** H. C. SMITH *Personality Adjustment* xviii. 513 Children raised in negligent and understaffed orphanages are not actively rejected but suffer severe deprivation of warmth and affection. **1973** A. JANOV *Primal Scream* vii. 74 But to feel really rejected means to..feel utterly alone and unwanted as that child.

10. *Med.* To show an immune response to (a transplanted organ or tissue) so that it fails to survive in the body of the recipient. Also *absol.*

1953 *Nature* 3 Oct. 603/1 Embryonic cells transplanted into embryos of different genetic constitutions may survive into adult life, although their hosts would almost certainly have rejected them if transplantation had been delayed until after birth. **1968** *Observer* 7 Jan. 1/1 Although he is now entering the crucial period where his body could begin to reject the implanted heart, today's hospital bulletin said there were no signs of rejection or infection. **1969** *Daily Progress* (Charlottesville, Va.) 12 Jan. A2/3 'The heart rejects like crazy,' Dr Shumway comments. **1974** R. M. KIRK et al. *Surgery* ii. 35/1 A graft that will be ultimately rejected at first appears to be accepted by the host tissues.

II. 11. The infin. used *attrib.*, designating a part of a record player by means of which the turn table is made to stop (and the pick-up arm usu. returned to its rest) before a side has ended. Also stressed (rī·dʒekt).

1947 *Gramophone* Dec. p. xi/2 Automatically plays eight 10-in. or 12-in. mixed records. Repeat and reject switch provided. **1975** J. GRADY *Shadow of Condor* ii. 34 Strains from *Carmen* came through the speakers. Malcolm..hit the reject lever. **1976** R. L. SIMON *Wild Turkey* xviii. 129 He shuffled over to the turntable and pushed the reject button.

rejecting *vbl. sb.* (later example); also as *ppl. a.*

1931 *Smith Coll. Stud. in Soc. Work* I. 407 The purpose of the study was to test part of the hypothesis..by a comparison of the case histories of a group of rejecting and non-rejecting mothers. **1939** P. M. SYMONDS *Parent–Child Relationships* i. 24 Attempts to define rejecting behavior are rare. **1970** H. EDELSTON *Found. & Growth of Character* III. ii. 117 We hear a great deal of the rejecting mother: not quite so much of the over-demanding child.

rejected, *ppl. a.* Add: **a.** Also spec. *Psychol.*, refused or denied the normal relationship between parent and child. Cf. *REJECT *v.* 6 d.

1931 *Smith Coll. Stud. in Soc. Work* I. 407 The problems for which the rejected child was referred were more frequently of the aggressive, rebellious type. **1961** H. C. SMITH *Personality Adjustment* xviii. 513 Such severely rejected children tend to develop a general apathy..to all human relationships.

c. Gram. *rejected condition* (see quots.).

1947 PARTRIDGE *Usage & Abusage* 80/2 Those sentences in which the principal clause speaks of what would be or would have been, and in which the *if*-clause states, or implies, a negative. Grammarians call this: Rejected Condition, as in 'If wishes were horses, beggars would ride'. **1957** R. W. ZANDVOORT *Handbk. Eng. Gram.* v. ii. 218 Clauses expressing a condition that is not, or is not likely to be, realized [I should not mind so much, if I was not so busy] are called clauses of *rejected condition*.

rejectee (rīdʒekti·). *U.S.* [f. REJECT *v.* + -EE¹, after *draftee*.] One who is rejected as unfit for military service. Also *transf.* Cf. *REJECT *sb.* 1 b.

1941 *Sun* (Baltimore) 18 June 3/1 Twenty-six youths rejected by selective service officials..began taking physical examinations..for admittance to the first camp in the United States to be established for 'rejectees'. **1942** *Nation* 27 Apr. 41 (*heading*) When the draftee becomes a rejectee. **1977** *Time* 20 June 48/2 Determined to become a doctor, Braun did what an increasing number of rejectees do each year: he looked abroad. **1978** J. A. MICHENER *Chesapeake* 637 'Grandpop, take a look at the kind of boys we want,' and when Cudjo continued pestering him, he pointed to rejectees half the old man's age.

rejection. Add: **1. c.** *Psychol.* The refusal or inability to accept emotionally the fact of being a parent to one's child; the state of rejecting a child or of being rejected by a parent. Cf. *REJECT *v.* 6 d.

1931 *Smith Coll. Stud. in Soc. Work* I. 407 Thirty-five cases of rejection were chosen in which staff members unanimously agreed to that diagnosis. **1939** P. M. SYMONDS *Parent–Child Relationships* i. 10 Such concepts and terms as rejection and overprotection seem to have emerged into common use out of the child guidance movement. *Ibid.*, Newell..reports on 33 children diagnosed as cases of maternal rejection. **1957** L. C. STECKLE *Probl. Human Adjustment* (rev. ed.) vi. 132 Rejection is most effective in building uncertainty. *Ibid.* 133 Parental rejection usually results in aggressively hostile behavior in the child. **1973** A. JANOV *Primal Scream* vii. 74 Once that is felt, there are no more feelings of 'rejection'.

2. *Electronics.* The process of attenuating an unwanted electrical signal. Freq. *attrib.*

1940 *Chambers's Techn. Dict.* 712/2 Rejection filter. **1950** LAWSON & UHLENBECK *Threshold Signals* xii. 346 Besides automatic biasing, rejection filters in the i-f amplifier can be used to reduce the effects of c-w interference. **1953** [see *DETECTION 3]. **1957** R. W. LANDEE et al. *Electronic Designers' Handbk.* xvi. 29 If the feedback network by itself does not have a complete null, the depth of the null for the rejection amplifier will be less than the depth of the null for the null network taken by itself. **1967** W. A. STOVER *Circuit Design for Audio, AM/FM, & TV* xiii. 227 The IF rejection is enhanced by placing a resonant circuit or filter near the tuner input.

3. *Med.* Failure of transplanted tissue to survive or function in the body of the recipient as a result of the immune response it evokes in the latter.

1954 *Proc. R. Soc.* B. CXLIII. 43 Incompatibilities (falling short of rejection) became apparent when homografts were exchanged between members of separate sublines which..stood only eight to twelve generations apart. **1974** *Times* 5 Apr. 18/3 Professor Shumway believes that heart-transplant patients, like those given kidney transplants, gradually develop a tolerance to the grafted organ and so

become less likely to have to cope with severe rejection episodes. **1974** M. C. GERALD *Pharmacol.* i. 7 We are optimistic that..more effective drugs to prevent the rejection of organ transplants will be discovered in the present decade.

4. *Comb.*, as **rejection form** *rare* = *rejection slip* below; **Rejection Front,** an alliance of Arab groups, who refuse to consider a negotiated peace with Israel (see *REJECTIONIST); **rejection slip,** a formal notice sent by an editor or publisher to an author with a rejected MS.

1907 WODEHOUSE *Not George Washington* II. ii. 42, I papered the walls with editorial rejection-forms, of which I was beginning to have a representative collection. **1917** 'W. N. P. BARBELLION' *Jrnl. Disappointed Man* (1919) 296, I used to file..rejection forms and meditated writing a facetious essay on them. **1975** *Financial Times* 23 Dec. 4/8 The 'Rejection Front', which is led by the PFLP, stands strongly opposed to efforts at a Middle East settlement and to conservative regimes in the Middle East, especially in Saudi Arabia and Iran. **1978** *Radio Times* 28 Jan.–3 Feb. 15/4 The Arab 'Rejection Front' nations of Iraq, Algeria, Libya and the People's Democratic Republic of Yemen were prepared to give finance, training facilities and arms to his group. **1906** J. LONDON *Let.* 19 Nov. (1966) 223, I have just received from you, along with a rejection-slip, two poems..which have evidently been submitted to you over my name. **1933** DYLAN THOMAS *Let.* 9 May (1966) 15 Forget the 'annihilative reverse' of the rejection slip. **1953** H. MILLER *Plexus* II. xiii. 217 If they were thin envelopes it meant rejection slips, with a request to forward postage for the return of the scripts. **1978** P. SUTCLIFFE *Oxf. Univ. Press* IV. i. 108 He, Gerrans, and Doble soon settled upon a formula, a terse but dignified communication that might be a little more comforting to the author than a bare rejection slip, the use of which the Press has always eschewed. **1979** F. ADCOCK *Inner Harbour* 2 'Please send future work'—Editor's note on a rejection slip.

rejectionist (rĭdʒe·kʃənist). [f. REJECTION + -IST.] An Arab who refuses to accept a negotiated peace with Israel. Also *transf.* Also *attrib.* or as *adj.* Hence **reje·ctionism,** the policy of a rejectionist.

1976 *Guardian* 2 June 2/2 If Mr Kosygin wants to engineer the format of an Arab 'rejectionist' coalition to confront American penetration of the area, he is unlikely to get much encouragement from President Assad. **1976** *Guardian Weekly* 19 Dec. 8/4 Condemned to death by the mainstream Arafat-led guerrilla leadership—'Fascist' he calls it—Abu Nidal is Palestinian rejectionism at its most uncompromising. **1977** *Time* 10 Jan. 22/1 Last week a prominent member of the rejectionist Popular Front for the Liberation of Palestine..and his wife..were found dead in their West Beirut apartment. **1977** *Listener* 18 Aug. 195/1 If..the peace moves collapse..he [*sc.* King Hussein] will be..helpless in a stormy sea of Arab rejectionism. **1977** *Times* 29 Nov. 15/4 It offends no previously declared principles—except those of the 'rejectionists' on both sides. **1979** *Economist* 1 Dec. 14/1 Khomeini would have remained a voice in the wilderness if his austere rejectionist doctrines had not caught the mood of a people whose religion is still young and vigorous.

rejective, *a.* (Examples.)

1957 *Publ. Amer. Dial. Soc.* XXVIII. 87 The other clause must not be 'rejective' with regard to it [*sc.* the fronted clause]. Examples: ..He likes it you think? ..Is he coming do you doubt?.. (*doubt, forbid, disagree*, etc. are rejective). **1967** *New Yorker* 25 Feb. 108 Besides being called minimal art, it is known as..'reductive art', 'rejective art', [etc.]. **1970** *Britannica Bk. of Year 1969* 798/3 *Rejective art*, a simplified and often depersonalized art (as painting or sculpture) based on the principle of the artist rejecting various options open to him; called also *reductive art, reductivism, rejectivism*.

rejector. Add: **1. b.** *Electronics.* = *rejector circuit* below.

1923 *Wireless World* 7 July 441/1 Signals with the frequency of the aerial circuit will pass through the ordinary tuning device, and little will pass through the rejector. **1946** *Electronic Engin.* XVIII. 45/1 Methods of bass compensation in common use involving arrangements of chokes, condensers, tuned acceptors or rejectors..all fail where high fidelity is required. **1977** L. J. GIACOLETTO *Electronics Designer's Handbk.* (ed. 2) xxiv. 115 (*caption*) Complex frequency characteristic of..the admittance of a single-tuned circuit to which a rejector is coupled.

2. Special Comb.: **rejector circuit,** a circuit consisting of a capacitor and an inductor connected in parallel and having values chosen such that the combination offers a very high impedance to signals of a particular frequency.

1923 *Wireless World* 7 July 441/1 The smaller the damping of the rejector circuit, the better it will perform. **1929** *Daily Express* 7 Nov. 14/3 A rejector circuit, which acts as a by-path for unwanted stations. **1952** *Electronic Engin.* XXIV. 314/1 As the selectivity of the 50 kc/s amplifier is not sufficient to reject the 48 kc/s frequency completely, a separate rejector circuit is used. **1969** NELKON & HUMPHREYS *Electronics & Radio* viii. 170 The parallel *LC* circuit is known as a 'rejector' circuit because the main current is zero at f_0.

re·jig, *sb.* [RE- 5 c.] Reorganization, rearrangement.

1965 *New Statesman* 23 Apr. 630/1 The *Sunday Citizen*, for all its admitted demerits (which may yet be rectified if still another rejig..is accomplished effectively). **1974** *Guardian Weekly* 10 Aug. 5/1 The idea of the late-night front page rejig is unknown.

reji·g, *v.* [RE- 5 c.] *trans.* To refit or re-equip; to mend. Also *fig.*, to rearrange, refashion, alter.

1948 *Daily Express* 22 Apr. 1/6 Britain will send experts to help rejig French factories. **1958** *Spectator* 2 May 558/1 To alter the period of the action [of *Twelfth Night*], to rejig the entrances and exits of the characters..is really a kind of forgery. **1962** *Economist* 22 Sept. 1084/1 Schemes for rejigging the conditions of press competition. **1972** *Times Lit. Suppl.* 14 Apr. 419/4 Current attempts to rejig Spanish sixteenth and seventeenth-century literature in terms of racial influences, art history, and so on. **1976** A. WHITE *Long Silence* iv. 34 Three [weeks]..he spent with Jean Duclerc, helping re-jig the wireless. **1979** *Economist* 13 Oct. 81/3 Last year, faced with slower-than-expected increases in electricity demand, Hydro-Québec rejigged part of the project to reduce overall capacity.

Hence **reji·gged** *ppl. a.*; **reji·gging** *vbl. sb.*

1960 *Times Lit. Suppl.* 30 Sept. 635/3 The subject is fascinating and the setting never dull, and re-jigged (as the technicians say) it might make an enthralling story. **1969** *Daily Tel.* 6 Oct. 1/1 Whitehall's structure will be considerably changed as a result of the Prime Minister's 'rejigging' of the machinery of government. **1972** *Guardian* 15 June 15/2 Who will take over the re-jigged RIBA? **1977** *New Statesman* 17 June 809/1 His rejigged Radio 4 *Today* programme is now packing the listeners in. **1980** *Jrnl. R. Soc. Arts* Feb. 152/1 This leads me to suggest that the industrial planning process needs re-jigging.

rejoicing, *vbl. sb.* Add: **1. c.** *Rejoicing of* (or *over*, etc.) *the Law* [tr. Heb. *Simchat Torah* (in Suppl.)], the Jewish feast at the conclusion of the Feast of Tabernacles, celebrating the gift of the covenant of the Law.

1861 J. T. BANNISTER *Temples of Hebrews* 390 Tisri.. [Day] 23. The rejoicing for the law, a solemnity in memory of the covenant that the Lord made with the Hebrews, in giving them the law by the mediation of Moses. **1892** I. ZANGWILL *Childr. Ghetto* I. 134 It was the Rejoicing of the Law, and the Sons of the Covenant had treated him to rum and currant cake. **1903** W. ROSENAU *Jewish Ceremonial Institutions & Customs* v. 101 On the Day of Rejoicing Over the Law special inducements are held out to the younger members of the congregation to participate actively in the public service. **1925** *Jewish Encycl.* XI. 364/2 *Simhat Torah* ('The Rejoicing over the Law'): Name given to the second day of *Shemini 'Azeret*; it falls on the 23d of Tishri and closes the Feast of Sukkot. **1953** T. H. GASTER *Festivals Jewish Year* vi. 100 Not impossibly, the custom of celebrating the Rejoicing in the Law as a *wedding* was inspired by the idea of sublimating to a more spiritual plane the..staging of a mock wedding at harvest festivals. **1962** B. ABRAHAMS tr. *Life Glückel of Hameln* iii. 52 Her companions were not to return before *Simchat Torah*.. Festival of the Rejoicing of the Law, celebrated immediately after that of Tabernacles. **1978** J. SACKS in P. Moore *Man, Woman, & Priesthood* iii. 39 On *Simchat Torah*, the festival of 'Rejoicing in the Law', adults and children forget decorum and dance and sing around the synagogue in celebration of the ending of the yearly cycle of Torah-reading and the beginning of the new.

‖ **rejon** (rehō·n). *Bull-fighting.* Pl. **rejones.** [Sp. *rejón* lance, spear, f. *rejo* pointed iron bar, *reja* ploughshare (L. *rēgula* straight piece of wood, f. *regere* to keep straight).] A wooden-handled spear, usu. placed from horseback.

1838 *Q. Rev.* LXII. 397 These noble 'Bestiarii' fought with the '*rejon*', a short projectile spear, about four feet long. **1893** CHAPMAN & BUCK *Wild Spain* v. 57 The knight, mounted on fiery Arab steed, was armed only with the *rejon*, or short sharp lance of those days, five feet in length, and held at its extreme end. **1932** *Times Lit. Suppl.* 7 Jan. 11/1 Nobles and gentlemen, on horseback with long heavy spears (*rejones*)..awaited the bull's attack. **1934** R. CAMPBELL *Broken Record* viii. 184 His *rejon* is nearly always mortal. **1957** A. MACNAB *Bulls of Iberia* x. 99 The *rejón* act..is more popular in Portugal, but is also quite common in Spain. **1967** MCCORMICK & MASCAREÑAS *Compl. Aficionado* i. 19 The rider plays the bull with the horse itself, placing long banderillas, darts known as *rejones*, and killing with a long lance, also called a *rejón*. **1973** *Times* 5 Dec. 15/1 They attempt to finish him [*sc.* the bull] off with a *rejon*, a long-bladed, wooden-handled spear.

‖ **rejoneador** (rehoneadō·r). *Bull-fighting.* Pl. **-es**; fem. **-a.** [Sp.] A mounted bull-fighter who places *rejones* (see prec.).

1926 *Blackw. Mag.* Sept. 290/2 No small skill in horsemanship is required to be a successful rejoneador. **1932** E. HEMINGWAY *Death in Afternoon* iii. 31 A man who kills them [*sc.* bulls] on horseback with a javelin, using trained thoroughbred horses, is called a rejoneador or a caballero en plaza. **1957** R. CAMPBELL *Portugal* vi. 114 The great Peruvian horsewoman, the beautiful Conchita Cintron, earned immortal fame as a *rejoneadora*. **1957** A. MACNAB *Bulls of Iberia* x. 100 The *rejoneadores*..are technically deemed amateurs. **1967** MCCORMICK & MASCAREÑAS *Compl. Aficionado* i. 19 Rejoneo has little to do with the true art of toreo on foot; it is too showy, and the danger is risked by the horse, not the man (*rejoneador*). **1973** *Times* 5 Dec. 15/1 Women do fight in Spanish bull rings..but exclusively as *rejoneadoras*, from the backs of..horses.

‖ **rejoneo** (rehonē·o). *Bull-fighting.* [Sp.] The art of bull-fighting on horseback with *rejones* (see prec.).

1961 *John o' London's* 609/2 *Rejoneo*, the art of bull-fighting on horseback. **1965** *Pix* (Austral.) 13 Feb. 36/2 Conchita Cintron, first woman to master both rejoneo and the classic Spanish style on foot, was a matadora in her own right. **1967** MCCORMICK & MASCAREÑAS *Compl. Aficionado* i. 19 Alvarez cites further instances of corridas organized to celebrate marriages, all antedating the late mediaeval and renaissance practice of *rejoneo*, or knightly jousting against toros bravos. **1978** *Radio Times* 17–23 June 25/3 In the bull-ring he practises 'rejoneo', the aristocratic form of fighting bulls from horseback.

rejuvenate, *v.* Add: **2.** *Geol.* To restore to a condition characteristic of a younger landscape.

1903 H. LE R. FAIRCHILD *Le Conte's Elements Geol.* (ed. 5) ii. 23 If the land area be lifted up the graded streams are rejuvenated—that is, their grade and power are increased. **1944** A. HOLMES *Princ. Physical Geol.* xi. 195 When a river that has already established a flood-plain is rejuvenated, it cuts through its own deposits into the underlying rocks. **1954** W. D. THORNBURY *Princ. Geomorphol.* vi. 143 During the low sea levels of the glacial ages streams presumably were rejuvenated in their lower courses.

rejuvenated *ppl. a.*, **rejuvenation** (examples of each in *Geol.*).

1936 *Geogr. Jrnl.* LXXXVII. 20 Before rejuvenation the river had developed an open mature valley in marked contrast to its present rugged canyon. **1954** W. D. THORNBURY *Princ. Geomorphol.* vi. 142 Mature or old-age topography is likely to have superposed upon it youthful features as a result of rejuvenation. **1969** G. C. DICKINSON *Maps & Air Photographs* xiii. 209 If, for reasons such as rise of the land, fall in sea level, or glacial overdeepening of the main valley, a river begins to cut down into its valley floor, a rejuvenation head will form at the upper limits of this downcutting, working its way progressively upstream. **1970** R. J. SMALL *Study of Landforms* iii. 105 In geologically active areas (such as New Zealand) it is possible that both fault- and fault-line scarps exist together. Indeed, it has been suggested that what are termed 'composite' and 'rejuvenated' scarps may be developed in these circumstances. **1971** *Nature* 19 Feb. 539/1 Further local rejuvenations of the main graben and step faults continued until mid-Pleistocene times, deepening the Lake Naivasha and Lake Natron sectors.

rejuvenatory (rĭdʒū·vĭnătŏri), *a.* [f. REJUVENATE *v.* + -ORY².] Tending to cause rejuvenation.

1971 *Nature* 25 June 530/2 A reverse transcriptase would have 'rejuvenatory' potential if it could restore lost or masked primary information from secondary copies. **1972** *Ibid.* 15 Dec. 414/1 This may represent a special rejuvenatory function of sexual processes.

rejuvenescent, *a.* **2.** (Later example.)

1899 R. FRY *Let.* 9 June (1972) I. 173, I have very rarely had a more rejuvenescent visit to Cambridge than this last.

‖ **relâche** (rəlāʃ). [Fr.] A period of rest, an interval; a break *from* something.

1863 A. J. MUNBY *Diary* 10 Apr. in D. Hudson *Munby* (1972) 155 Caldwell's is one of the few public dancing rooms in London, which is frequented by respectable women... When we arrived, the relâche was begun. **1905** G. BELL *Let.* 24 Oct. (1927) I. xi. 226 Tonight he has asked Yves Guyot to dinner because I said I wanted to see him, so we shall have a little 'relâche' from archaeology. **1910** W. S. BLUNT *Let.* 19 Mar. in R. S. Churchill *Winston S. Churchill* (1969) II. Compan. II. xv. 1159 Wd you rather have a *relâche* from politics?

relapsing, *ppl. a.* **6.** *relapsing fever*, substitute for def.: either of two similar kinds of fever characterized by relapses, caused by spirochætes of the genus *Borrelia* and transmitted respectively by lice and by ticks. (Earlier and later examples.)

1849 *Dublin Q. Jrnl. Med. Sci.* VIII. 50 This fever has been well called a relapsing fever; that is, it was made up of two parts, crisis being very generally present at the termination of each. **1936** *Lancet* 22 Feb. 448/1 Recent investigation of a small outbreak of relapsing fever in Kfar Vitkin, south of Hedera in the coastal plain, showed that all infections could be traced to a cave infested with *Ornithodorus papillipes*. **1966** DUNLOP & ALSTEAD *Textbk. Med. Treatment* (ed. 10) 218 Tetracycline..is the drug of choice, although subsequent relapses of tick-borne relapsing fever due to *Bor. duttoni* may be experienced. **1974** PASSMORE & ROBSON *Compan. Med. Stud.* III. xii. 75/1 The spirochaetes responsible for louse-borne relapsing fever is *Borr*[*elia*] *recurrentis* and for the tick-borne form, *Borr. duttoni*. *Ibid.*, Louse-borne relapsing fever is a disease of cold weather which occurs in epidemic form usually in the wake of disasters such as wars or earthquakes.

relata: see *RELATUM.

relatable, *a.* Add: **b.** Now usu. with *to*. Also, that may be shown to possess mutual relation. Hence **relatabi·lity.**

1937 *Burlington Mag.* July 38/2 The 'relatability' of pictorial forms..follows on the artist's realization of their basis, undifferentiated essentials. **1956** *Jrnl. Theol. Stud.* VII. 88 It is found that 38·5 per cent. of the elements of psalm language is certainly not relatable to psalm contents; the relatability of a large part of the remainder thus becomes questionable. **1964** *Language* XL. 244 A proposal for the relatability of two languages..has been traditionally based on the discovery of systematic sound correspondences between certain of their forms. **1975** T. F. MITCHELL in W. F. Bolton *Eng. Lang.* iv. 165 *Blackboard* is less obviously relatable to *black board* than *blackbird* is to *black bird*.

relate, *v.* Add: **II. 9. c.** To feel affectively involved or connected with someone or something; to have an attitude of personal and sympathetic relationship *to*.

1950 *Childhood Educ.* Nov. 115/1 Group formation such as takes place in the classroom tends to be adult-centered and dependent upon the varying ways children relate to the teacher. **1965** *Listener* 10 June 864/2 Attitudes to other people at the unconscious level appropriate to an early stage of infancy, of the time before we learnt the capacity to relate. to whole persons. **1966** *New Statesman* 14 Oct. 549/1 The Civic Action now begins As friends and former foe relate. **1968** *Globe & Mail* (Toronto) 13 Feb. 33/8 (Advt.), Candidates should..be able to relate to senior officers of the University. **1969** C. DAVIDSON in Cockburn & Blackburn *Student Power* 361 If we only relate to on-campus issues, we run the risk of laying the counter-revolutionary groundwork. **1971** *Guardian* 7 Jan. 8/3 Married people can still relate. **1971** M. SPARK *Not to Disturb* iii. 89 'What do you mean, I don't relate?' she says. 'When you relate you don't ask what you mean. There's such a thing as a trend.' 'Who do you think you are, you—Chairman Mao?' **1977** J. L. HOULDEN *Patterns of Faith* ii. 20 It is possible to relate to him [*sc.* God] and..a Christian is one who finds that the relating is best done in ways that bear on the figure of Jesus.

relation, *sb.* Add: **3. d.** *Logic.* A constituent of a proposition or propositional function that connects two terms (a dyadic relation) or more (triadic, *n*-adic, etc.).

1870 C. S. PEIRCE *Coll. Papers* (1933) III. iii. § 47. 28 *Inclusion in* or *being as small as* is a *transitive* relation. **1885** W. JAMES in *Mind* X. 32 No relation-expressing proposition is possible except on the basis of a preliminary acquaintance with such 'facts'..as this. **1910** WHITEHEAD & RUSSELL *Princ. Math.* I. § 30. 245 Functions of this kind always mean 'the term having such and such a relation to *x*'. **1940** W. V. QUINE *Math. Logic* v. 201 Relations in the sense here considered are known, more particularly, as dyadic relations. **1956** H. REICHENBACH *Direction of Time* ii. 26 When the points are in a linear order, or serial order, they are governed by an asymmetrical and transitive relation. **1965** HUGHES & LONDEY *Elem. Formal Logic* xxxix. 272 Such relations are said to be non-symmetrical relations. Examples are: 'implies', 'brother of'... Such relations are said to be non-transitive relations. Examples are: 'one mile distant from', 'first cousin of'. *Ibid.* 274 Every dyadic relation must be either reflexive or irreflexive or non-reflexive. **1978** C. KIRWAN *Logic & Argument* i. 23 A binary relation such as hating, which holds from some but not all things to themselves, is neither reflexive nor irreflexive. Likewise many binary relations are neither transitive nor intransitive and many are neither symmetrical nor asymmetrical.

e. *Philos.*, esp. as *external relation*, a connection existing between one thing and another which is not intrinsic to the identity of the first; *internal relation*, a connection between one thing and another which is intrinsic to the identity of the first.

1893 F. H. BRADLEY *Appearance & Reality* iii. 31 Every quality in relation has..a diversity within its own nature, and this diversity cannot immediately be asserted of the quality. Hence the quality must exchange its unity for an internal relation. *Ibid.* iv. 40 This solid unit, existing only by virtue of external relations, is forced to expand. **1922** G. E. MOORE *Philos. Stud.* 288 Yet this last, according to me, is one of the things which the dogma of internal relations denies. **1935** A. J. AYER in *Aristotelian Soc. Suppl. Vol.* XIV. 179 The connexion between the proposition which Mr. Ryle mistakes for the dogma of internal relations and the dogma of internal relations as we understand it, is that they both follow from the proposition that all a thing's characters are intrinsic to it. **1956** R. A. WOLLHEIM in A. J. Ayer *Revolution in Philos.* ii. 22 In logic this view is known as the theory of 'internal relations'. All the relations in which an object stands are rooted in its nature as firmly as triangularity is rooted in the nature of the triangle. **1975** HARGREAVES & WHITE tr. *Wittgenstein's Philos. Remarks* iii. 63 The essential difference between the picture conception and the conception of Russell, Ogden and Richards, is that it regards recognition as seeing an internal relation, whereas in their view this is an external relation.

5. d. In phr. *no relation*, denying relationship by blood or marriage despite having the same surname.

1930 E. M. BRENT-DYER *Chalet Girls in Camp* xii. 175 Except..Ruth Wynyard, Lilli van Huysen, and Greta Macdonald—no relation!—all of them had been her [*sc.* Mrs. Macdonald's] pupils. **1977** *Private Eye* 13 May 14/1 We shall see much more of it now that Mr Moss Evans (no relation) has been elected to one of the two highest offices in the land, as General Secretary of the TGWU.

7. *Comb.*, as **relation-axis** *adj. phr. Gram.*, involving or consisting of a word expressing a relation and another with respect to which the relation holds; **relation-word** *Gram.*, a word expressing relation between other words or groups of words, e.g. a preposition or conjunction.

1933 L. BLOOMFIELD *Language* 267 They [*sc.* English substantive expressions] occur in the position of axis in the relation-axis construction (*beside John*), with a positional meaning of, say, 'center from which a relation holds good'. **1964** E. A. NIDA *Toward Sci. Transl.* iv. 57 In the phrases *through the house*, *behind the store*, and *in the shed*, the relationship between the prepositions *through*, *behind*, and *in* and the following immediate constituents (consisting of the noun with preposed determiner *the*) may be described as 'relation-axis'. **1925** GRATTAN & GURREY *Our Living Lang.* xii. 79 [The work of a preposition] is to show the relation in which a noun stands to some other part of the sentence... For this reason it is also known as a *Relation-word*. **1962** J. SÖDERLIND in F. Behre *Contrib. Eng. Syntax* 117 *Of*-groups where *of* is a pure relation-word.

relational, *a.* Add: **2.** (Further examples in *Gram.*)

1930 J. T. HATFIELD et al. *Curme Vol. Ling. Stud.* 37 The immaterial adnominal cases are the following;..relational—a man *in stature*. **1946** *Language* XXII. 219 A *relational phrase* has two immediate constituents. **1967** *Child Devel.* XXXVIII. 841 This study is concerned with the ability of preschool children to use the relational terms 'more', 'same', and 'less' when comparing the number, length, and weight of objects. **1979** *Trans. Philol. Soc.* 215 The element *-ai/ei-* is also found in association with one of the so-called 'relational particles'.

Hence (*Gram.*) as *sb.*, a conjunction or preposition considered as a relation-word; a relational particle.

1964 E. A. NIDA *Toward Sci. Transl.* iv. 62 There are four principal functional classes of lexical symbols: object words, event words, abstracts, and relationals. **1969** *Language* XLV. 485 Relationals are any units which function primarily as markers of relationships between other terms, e.g. *at*, *by*, *because*, *and*, *or*. **1978** *Ibid.* LIV. 80 Some English prepositions correspond to Japanese genitive constructions with *no* plus relationals.

relationism. a. (Further examples.)

1958 W. STARK *Sociol. Knowl.* viii. 338 By the concept of relationism he [*sc.* K. Mannheim] means that if we formulate a truth, we should not do so in abstract and absolute terms, but must always include in the formula the concrete conditions to which it is related. **1975** *Nature* 1 May p. iv (Advt.), The book is concerned with space and time as abstract relations which hold between objects and events (relationism), and as aspects of nature with causal properties of their own (absolutism).

relationist. Add: Also, one who holds that events are connected in a relative way. Also *attrib.* or as adj.

1885 W. JAMES in *Mind* X. 31 And may not the 'relationists' be right after all? **1923** C. D. BROAD *Sci. Thought* iii. 89 The former alternative is taken by the Relationist... Time just consists of the relations of before and after among events. **1974** L. SKLAR *Space, Time, & Spacetime* iii. 167 According to the relationist, the postulation of space, time, or spacetime as entities..is simply a confusion. *Ibid.*, The prerelativistic and relativistic versions of the relationist position. **1975** *Nature* 6 Feb. 485/3 Wishing to retain a pure relationist stance, he proposes that the statement 'is absolutely accelerated' is not a relational term, giving it instead the status of a complete assertion.

relationless, *a.* Add: **b.** Bereft of relation in general.

1889 'SCOTUS NOVANTICUS' *Metaph. Nova et Vetusta* (ed. 2) II. 86 The datum is not atomic or relationless. **1911** *Proc. Mus. Assoc.* May 121 Oneness and homogeneity could be evolved from such apparently relationless short-cut phrases.

relationship. Add: Also, an affair; a sexual relationship.

1944 M. LASKI *Love on Supertax* viii. 81 'Were you going to marry Lou?' asked Clarissa... 'We hadn't got further than a relationship,' Sid said. **1974** J. GARDNER *Corner Men* v. 41 Bob and I weren't hallo young lovers. We had a relationship, but I wasn't in love with him. **1975** R. RENDELL *Shake Hands for Ever* viii. 76 'Did Mr Hathall have a special friendship with any girl here?'..'Do you mean a relationship? D'you mean, was he *sleeping* with anyone?' **1977** *Rolling Stone* 30 June 62/2 People don't fall in love anymore, they have relationships. **1981** C. R. LAJEUNESSE *Dead Man Running* iii. 11 Rowena and I had a relationship at first, which had been a no-strings-attached affair. Then.. she became serious and I had shied away.

relative, *a.* and *sb.* Add: **A.** *adj.* **9.** Special collocations: *relative address* (Computers), an address (sense *7 c) which is defined only in relation to some other address; hence *relative addressing*, the practice of using relative addresses; *relative density* = *specific gravity* s.v. GRAVITY 4 c; normally defined using water or (for a gas) hydrogen as standard; (cf. quot. 1704 in sense A. 4 b); *relative deprivation* (Sociol.), deprivation as experienced by a person in respect of opportunities, standard of living, etc., which is relative to the circumstances of the group or society of which he is a member; *relative humidity* (Meteorol.): see HUMIDITY 1; *relative pitch*: (Mus.), the pitch of a note in relation to another; the ability to recognize or reproduce this; also in extended use in *Phonetics*; *relative sexuality* (Biol.) [tr. G. *relative sexualität* (M. Hartmann 1909, in *Arch. f. Protistenkunde* XIV. 325)], the phenomenon shown by those species of which an individual or gamete may act as either female or male, according as it is less or more male than its mate.

1956 BERKELEY & WAINWRIGHT *Computers* viii. 352/2 Relative addresses are translated into absolute addresses by adding some specific 'reference' address. **1970** O. DOPPING *Computers & Data Processing* xix. 312 The relative addresses should be tagged to show that they will later have to be modified. **1966** C. J. SIPPL *Computer Dict. & Handbk.* 268/1 Relative addressing is a feature of great significance in multiprogramming, time-sharing, and real time operations. **1967** KLERER & KORN *Digital Computer User's Handbk.* i. 20 Relative addressing is done with addresses that are generated relative to some preset location whose relative address is 0. **1879** J. D. EVERETT *Units & Physical Constants* iv. 30 The relative density of water at various temperatures.., the density at 4° C. being taken as unity. **1892** G. F. BARKER *Physics* III. i. 315 Relative density is the ratio of the absolute density of a gas or vapor to that of air or of hydrogen. **1957** A. EFRON *Basic Physics* I. ix. 105 The relative density of lead is 11·3. **1963** A. F. ABBOTT *Ord. Level Physics* ix. 115 The relative density of a substance is the ratio of the mass of any given volume of it to the mass of an equal volume of water. Numerically, specific gravity and relative density are identical. **1974** FOLIVI & GOLMAN *New Certif. Physics* ii. 73/1 The ratio of the density of a substance divided by the density of water is the relative density of the substance. **1949** S. A. STOUFFER et al. *Amer. Soldier* I. iv. 125 To help explain such variations in attitude by education, age, and marital condition, a general concept would be useful. Such a concept may be that of relative deprivation. *Ibid.* 126 The concept of relative deprivation may seem..not to be applicable to the educational differentials in attitude. **1966** *New Statesman* 8 July 55/2 [illegible] expression 'relative deprivation', as Runciman [illegible] deprivation is largely imaginary and the emphasis [illegible] relativity. **1972** DOWSE & HUGHES *Polit. Sociol.* [illegible] When they cannot achieve these values..dissatisfaction, anger and often aggression occur. This type of situation is quite usual one in any complex society and is termed 'relative deprivation', which may be defined as 'a perception that develops from a discrepancy between the "ought" and the "is" of collective value satisfaction'. **1926** D. C. MILLER *Sci. of Mus. Sounds* vii. 216 Many writers on the subject have held that the quality of a vowel, as well as that of a musical instrument, is characterized by a particular set of overtones accompanying a given fundamental, the pitches of these overtones varying with that of the fundamental so that the ratios remain constant; this is the *relative-pitch* theory. **1929** *Melody Maker* Mar. 300/3 A person enjoying the ability to determine the interval between two or more musical sounds is said to possess Relative Pitch. **1933** L. BLOOMFIELD *Language* vii. 117 The Japanese language is said to distinguish two relative pitches, normal and high. **1949** R-M. S. HEFFNER *Gen. Phonetics* vii. 213 It is the relative pitch of speech sounds which is a linguistic factor in differentiation between meanings. **1969** H. L. SMITH in A. A. Hill *Linguistics Today* ix. 94 The four relative levels of stress in English.., the four relative pitch heights, and the four *junctures* or *transitions*..form three independent but interdependent systems of *prosodic* or *suprasegmental phonemes*. **1977** *Proc. R. Soc. Med.* LXX. 134/1 There is a continuum of skills which ranges from 'tone deafness' through 'relative pitch', to 'absolute pitch'... People who have relative pitch, in so far as they are able to tell, when given a certain reference tone, that a second sound is higher or lower in pitch. **1948** F. E. FRITSCH *Struct. & Reprod. Algae* 327 Relative sexuality, in which one gamete [of *Spirogyra*] behaves respectively as male and female towards others, is also on record. **1967** E. STEINER tr. *Esser & Kuenen's Genetics of Fungi* ii. 96 In the light of recent work certain older data are no longer to be interpreted as relative sexuality.

relativism. Add: **a.** (Earlier example.) Also, a name given to theories or doctrines that truth, morality, etc., are relative to situations and are not absolute.

1865 J. GROTE *Exploratio Philosophica* I. xi. 229 The notion of the mask over the face of nature is exactly that which I am sure Dr Whewell does not wish to fall into—it is what I have called 'relativism'. If 'the face of nature' is reality, then the mask over it, which is what theory gives us, is so much deception, and that is what relativism really comes to. [**1934** C. MORRIS in G. H. Mead *Mind, Self & Society* p. xix, Philosophically the position is that of an objective relativism: qualities of the object may belong to it only relative to a conditioning organism.] **1941** H. MARCUSE *Reason & Revolution* II. ii. 353 According to Comte, relativism is inseparable from the conception that sociology is an exact science dealing with the invariant laws of social statics and dynamics. **1959** A. BRECHT *Polit. Theory* vii. [illegible] I do not intend to minimize the extent to which Comte's positivism actually contributed to preparing the ground for modern Scientific Method and Value Relativism. **1976** W. J. STANKIEWICZ *Aspects Polit. Theory* vii. [illegible] What is logically excluded is relativism as a methodology: a methodology demands fixity of purpose; a fixed purpose excludes relativism.

b. Special collocations: (*a*) *historical relativism*, the view that there can be no objective standard of historical truth, as the interpretation of data will be affected by subjective factors characteristic either of the historian or of the period in which he lives; (*b*) *ethical relativism*, the view that there are no universal or objective ethical standards; that each culture develops the ethical standards that it finds acceptable and that these cannot be judged by the ethical standards of another culture; (*c*) *cultural relativism*, the view that there are no objective standards by which to evaluate a culture; that a culture cannot be understood except from the point of view of its own values or customs; the practice of studying a culture from such a standpoint.

(*a*) **1937** T. PARSONS *Struct. Soc. Action* xiii. 480 In place of a theory of dialectic evolution on the Hegelian model there emerges a complete historical relativism. **1945** K. R. POPPER *Open Society* II. xxii. 191 But this so-called 'historical relativism' by no means exhausts the historical character of the Marxist theory of morals. **1956** W. KLUBACK *Dilthey's Philos. Hist.* iii. 58 The value of any age was true for that age but could not with validity be applied to other ages. For Dilthey historical relativism did not imply pessimism. On the contrary, it made man aware of his place in history. **1977** M. MANDELBAUM *Anat. Hist. Knowl.* vi. 150 Some of the conventional arguments for historical

relativism, and against the objectivity of historical knowledge, lose much of their force.

(*b*) **1937** T. PARSONS *Struct. Soc. Action* xi. 447 He [*sc.* Durkheim] was forced to define normality with reference to the social type alone, thus ending in a complete ethical relativism. **1944** *Brit. Jrnl. Med. Psychol.* XX. 113/1 This [empirical] point of view is distinct both from ethical absolutism and ethical relativism. **1964** M. RADER *Ethics & Human Community* ix. 236 There is a kind of incongruity in combining the two kinds of relativism. The methodological type requires tolerance..the ethical type condones the most intolerant of societies. **1968** *Internat. Encycl. Soc. Sci.* V. 158 The 'reductionist' form of ethical relativism, which presents the ethical beliefs of a people as functionally dependent on their other beliefs and practices.

(*c*) **1958** F. M. KEESING *Cultural Anthropol.* ii. 47 The scientific habit of looking at each people's standards and values objectively, seeing them as 'relative' to the particular view of life fostered within the culture concerned, has led some thinkers to a philosophic position often called 'cultural relativism'. **1968** *Internat. Encycl. Soc. Sci.* III. 543/2 The methodology of cultural relativism rests on the assumption that the ethnologist is able to transcend, or to eliminate for the moment, his own cultural conditioning and values and to assume the subjective..mentality of an adherent of..the culture. **1976** T. EAGLETON *Crit. & Ideology* iv. 134 Imperialism..bred an awareness of cultural relativism at precisely the point where the absolute cultural hegemony of the imperialist nations needed to be affirmed.

relativist, *sb.* Add: **1.** (Further examples.)

1935 K. KOFFKA *Princ. Gestalt Psychol.* 347 The relativist's argument rejects in the first place the distinction between the picture as a geographical and a behavioral object. **1953** M. GINSBERG *Ess. Sociol. & Soc. Philos.* I. vii. 124 Relativists generally stress the great diversity of morals. Yet the similarity is much greater. **1967** *Encycl. Philos.* III. 75/1 Writers who call themselves relativists always accept the first and second and sometimes accept the third of the theses. **1976** W. J. STANKIEWICZ *Aspects Polit. Theory* v. 97 If relativists are right in assuming that the analysis of values is complete when the latter are declared to be tastes, wishes or attitudes, it would seem impossible for moral statements to form any kind of pattern for either the individual or society.

2. A student or proponent of the theory of relativity.

1914 [implied in sense 3 below]. **1919** *Nature* 11 Dec. 374/2 The out-and-out relativist will not admit an absolute measure of acceleration any more than of velocity. **1922** A. S. EDDINGTON *Theory of Relativity* 16 The reason why the relativist resurrects this ancient truism is because it is only in this undissected combination of four dimensions that the experiences of all observers meet. **1968** *Amer. Jrnl. Physics* XXXVI. 1109/1 His [*sc.* Einstein's] great passion for the physical explanation of the laws of nature resulted in his abandoning ether and absolute time, thus radically modifying long-established Newtonian space-time. Thus, he was the first true relativist. **1977** *Listener* 24 Mar. 365/2 Newton's definitions of time and space..which were found to be implausible with the coming of the relativists at the end of the 19th century.

3. *attrib.* or as *adj.*

1914 C. D. BROAD *Perception* v. 286 It is no special objection to the relativist theory. **1921** *Nature* 8 Dec. 467/1 The differential equations which the relativist mathematicians use. **1939** V. A. DEMANT *Relig. Prospect* i. 12 Man has no criteria by which to guide himself in the movements of time. He resorts to beating back the forces that oppose what he conceives to be the most advanced phase. This attitude is completely relativist. **1949** *Scrutiny* XVI. I. 26 It is commonly held that the essential point about totalitarian morality is the denial of a universal moral law binding on all mankind and its replacement by a relativist view of morals. **1962** *Listener* 10 May 821/1 We are usually too timid, too relativist, to be so vehement in our defence of righteousness today.

relativistic, *a.* **1.** Add: Of, pertaining to, or characterized by relativism or relativity. (Further examples.)

1917 A. S. PRINGLE-PATTISON *Idea of God* 212, I applied this specially to the case of the secondary qualities which are usually regarded as the stronghold of the relativistic theory. **1937** T. PARSONS *Struct. Soc. Action* xiii. 480 On the empirical plane one of the first radical representatives of this relativistic view is Dilthey. **1959** A. BRECHT *Pol. Theory* vi. 249 Roscoe Pound..called it futile to wait for a statement of absolutes, and recommended practical work on the basis of our present civilization—again an activity entirely legitimate from the relativistic viewpoint. **1964** M. JACOBS *Pattern in Cultural Anthropol.* xii. 297 No anthropologist lacks admiration for northwestern art work, no matter how ridiculously relativistic his dogma. **1979** *Sci. Amer.* Mar. 94/2 It seems that a much more relativistic process is going on in the visual system. The boundary between each target square and its immediate background gives only the relation between the light reflected by each of these areas. **1981** *Times Lit. Suppl.* 13 Feb. 176/4 In *Gulliver's Travels* Swift chose the most 'relativistic' form, the travel book, in order to attack the root of relativism.

2. *Physics.* **a.** Pertaining to or based on the theory of relativity; modified or formulated according to the assumptions or consequences of the theory of relativity.

special relativistic, general relativistic adjs.: based on or taking account of the special theory (only) or the general theory, respectively, of relativity.

1914 L. SILBERSTEIN *Theory of Relativity* iv. 94 It requires, according to the relativistic view itself, some essential, though numerically slight, modifications. **1926** *Physical Rev.* XXVIII. 1070 The relativistic theory of the hydrogen atom is apparently incomplete. **1938** *Ann. Reg. 1937* 355 Heitler and Bhabha..used relativistic quantum mechanics to determine the number of secondary positrons and electrons produced when a fast primary electron passes through matter. **1958** CONDON & ODISHAW *Handbk. Physics* ii. 19 The constancy of v_s implies a constant ratio between classical time *t* and relativistic proper time τ. **1970** *Nature* 17 Oct. 273/1 The rectilinear Galilean transformation was discarded by Einstein in favour of the special-relativistic (Lorentz) transformation to a uniformly moving frame. **1972** *Ibid.* 18 Feb. 361/2 The theory of black holes..may perhaps be considered as one of the aspects of general-relativistic physics which is better understood. **1974** G. REECE tr. *Hund's Hist. Quantum Theory* vii. 99 Sommerfeld applied his theory, which took account of the relativistic correction, to the X-ray term.

b. Characterized by or designating circumstances, esp. those involving speeds approaching that of light or large gravitational potentials, in which discrepancies between the predictions of the theory of relativity and of Newtonian mechanics or classical electromagnetism become significant.

1934 *Discovery* Oct. 285/2 He examined the principles of thermodynamics as they apply in a relativistic universe. **1964** *Astrophysical Jrnl.* CXXXIX. 925 It has been argued in the past that the energy of the relativistic particles associated with the Crab cannot be greater than the kinetic energy of the filaments; otherwise the nebula would expand faster than it is observed to do. This argument is only correct, however, if the relativistic particles are trapped within the filaments. **1967** *Ibid.* CL. 1005 To extend these results on neutron and supermassive star models to arbitrarily relativistic stars rotating with arbitrary angular velocity is a problem which..is numerically complicated. **1973** C. W. MISNER et al. *Gravitation* xxiv. 633 The relativistic instability occurs far outside the Schwarzschild radius when the star is very massive... Rotation can stabilize it against relativistic collapse for a while. **1978** *Astrophysical Jrnl.* CCXXIII. 14 Fragmentation of a relativistic shock wave in either the free expansion or the self-similar blast wave solutions probably cannot be avoided.

Hence **relativi·stically** *adv.*

1947 *Physical Rev.* LXXII. 340/1 Relativistically, ν should be replaced by $c\alpha$. **1955** L. D. LANDAU in W. Pauli *Niels Bohr* 52 Perturbation theory has been reconstructed in a relativistically invariant way. **1970** *Nature* 17 Oct. 271/1 To discuss the recoil relativistically we must speak of 4-momentum.

relativitist, var. *RELATIVIST *sb.* 2.

1931 *Sci. Progress* XXV. 632 As the relativitist would point out, only the resultant symmetry of our chemical molecules, inspected in the aggregate, is observable. **1939** *Mind* XLVIII. 62 The relativitist, they tell us, has *discovered* that what we think of as a ruler of fixed length, has in fact different lengths according to its position.

relativity. Add: **2.** The quantitative dependence of observations on the relative motion of the observer and the observed object; that branch of physics which is concerned with the description of space and time allowing for this dependence.

The modern theory of relativity, developed largely by Albert Einstein (1879–1955), is an extension and generalization of the corresponding principles in classical, or Newtonian, mechanics.

The *principle of relativity*, in its restricted form, is the postulate that the laws of nature have the same form in all inertial reference frames; in its more general form, it states that the laws of nature, when expressed in a suitable ('covariant') form, have the same form in all reference frames, whether inertial or not.

The *special theory of relativity* (1905), based on the restricted principle of relativity and the hypothesis of the constancy of the speed of light *in vacuo* as seen by observers in any inertial frames, resulted in a theoretical framework for the unification of space and time in a four-dimensional continuum and for the equivalence of mass and energy, and showed how the uniform relative motion of observers affects their measures of length and time.

The *general theory of relativity* (1915), essentially a theory of gravitation, is based on the general principle of relativity, the postulated equivalence of inertial and gravitational mass, and the assumption that the results of the special theory must be valid in the limiting case of zero gravitational potential; it leads to a new set of equations of motion and the result that space-time is curved by the presence of gravitational fields.

1876 J. C. MAXWELL *Matter & Motion* vi. 84 Our whole progress up to this point may be described as a gradual development of the doctrine of relativity of all physical phenomena. Position we must evidently acknowledge to be relative. *Ibid.* 85 (*heading*) Relativity of force. **1882** J. B. STALLO *Concepts & Theories of Mod. Physics* xii. 204 The same considerations which evince the relativity of motion also attest the relativity of its conceptual elements, space and time. **1905** W. J. G. tr. *Poincaré's Sci. & Hypoth.* v. 76 The state of the bodies and their mutual distances at any moment will solely depend on the state of the same bodies and on their mutual distances at the initial moment, but will in no way depend on the absolute initial position of the system and of [*read* on] its absolute initial orientation. This is what we shall call, for the sake of abbreviation, the law of relativity. [**1905** *Sci. Abstr.* A. VIII. 2277 A. Einstein... The relativity of lengths and times.] **1906** J. W. YOUNG tr. H. Poincaré in *Bull. Amer. Math. Soc.* XII. 243 The principle of relativity, according to which the laws of physical phenomena must be the same for a stationary observer as for one carried along in a uniform motion of translation. *Ibid.* 247 Let us consider the principle of relativity; this principle is not only confirmed by our daily experience,..but it appeals to our common sense with irresistible force. And yet it also is being fiercely attacked. **1912** *Phil. Mag.* XXIII. 375 An acceptance of the Einstein theory of relativity necessitates a revision of the Newtonian system of mechanics. **1916** *Monthly Notices R. Astron. Soc.* LXXVI. 701 These considerations have led Einstein to his postulate of general relativity, which requires the laws of nature to be invariant for *all* transformations of co-ordinates. **1920** R. W. LAWSON tr. *Einstein's Relativity* vii. 20 As a result of an analysis of the physical conceptions of time and space, it became evident that in reality there is not the least incompatibility between the principle of relativity and the law of propagation of light, and that by systematically holding fast to both these laws a logically rigid theory could be arrived at. This theory has been called the special theory of relativity. **1921** *Nature* 1 Dec. 434/2 The result is valid for both the special and the general theory of relativity. **1922** E. P. ADAMS tr. *Einstein's Meaning of Relativity* iii. 68 We shall be true to the principle of relativity in its broadest sense if we give such a form to the laws [of physics] that they are valid in every such four-dimensional system of co-ordinates, that is, if the equations expressing the laws are co-variant with respect to arbitrary transformations. **1928** *Times* 3 Dec. 8/2 The whole point of the theory of relativity is the discovery of invariants, or absolute quantities, the same to all observers, and identical throughout the universe. **1932** W. T. STACE *Theory of Knowl.* xiv. 389 The space-time of modern relativity mechanics. **1959** *Listener* 9 Apr. 631/2 In everyday experience, special relativity gives virtually the same results as Newtonian theory. **1968** *Amer. Jrnl. Physics* XXXVI. 1109/1 Although the principle of relativity is subject to a possible experimental disproof in the future, the importance of the postulational approach is that it freed relativity from electrodynamics as a basis and made special relativity more universal. **1973** L. J. TASSIE *Physics of Elementary Particles* 203 An important result of the theory of special relativity is time dilatation, or the slowing down of moving clocks. **1974** *Encycl. Brit. Macropædia* XV. 584/2 The general theory of relativity derives its origin from the need to extend the new space and time concepts of the special theory of relativity from the domain of electric and magnetic phenomena to all of physics and, particularly, to the theory of gravitation. **1976** M. G. BOWLER *Gravitation & Relativity* p. vii, Einstein's theory of gravitation, general relativity, has been verified at the one per cent level. **1978** *Sci. Amer.* Feb. 131/1 The present understanding of the fundamental laws of nature arose from three principles: special relativity, general relativity and quantum mechanics.

3. The relative grading of posts or salaries, usu. considered within one business (*internal*) or in comparison with others (*external*). Freq. *pl.*

1962 *Rep. Comm. Broadc. 1960* 192 The BBC's policy on the payment of its staff takes careful account of internal and external relativities. In assessing internal relativities, the broad aim is to define the difficulty and responsibility of posts at each level over a very wide range of professions... To maintain external relativities, the Corporation draws comparisons with a suitable range of different employment. **1966** *New Statesman* 21 Jan. 71/1 Union A makes a claim in January, on the grounds that they have fallen behind B and C. When A's claim is agreed, B makes a claim in February, because relativity has been destroyed. **1971** *Nature* 20 Aug. 513/1 The institution pressed for the use of internal relativities to determine salaries. **1974** *Times* 9 Feb. 1/2 The new principles and procedures for adjusting wage relativities, ..since the Government agreed to set up machinery inside the Pay Board to examine major relativity claims.

relativization (relătivəizē¹·ʃən). [f. next + -ATION.] **1.** *Physics.* A relativistic treatment of a problem or theorem. *rare.*

1921 H. L. BROSE tr. *Moszkowski's Einstein the Searcher* vii. 162 Now, the conception of time has been entirely revolutionized by Einstein himself... We thus approach a relativization of causality... Something physiological that ultimately..resolves itself into a relativization of time.

2. The action of making relative; the fact or process of being made relative. Freq. in *Philos.* and *Linguistics.*

1942 *Mind* LI. 237 This is not exclusive subdivision, but relativization or canalization. **1945** *Polish Sci. & Learning* VI. 19/1 The relativization of the definability of a sign '*a*' to a set of propositions *X* becomes obvious. **1948** L. SPITZER *Linguistics & Lit. Hist.* 73 His [*sc.* Cervantes'] humor, which admits of many strata..of relativization and dialectics—bears testimony to his high position above the world. **1959** K. R. POPPER *Logic Sci. Discovery* 346, I only learned from Rényi's paper how fertile this relativization could be. **1968** *Language* XLIV. 55 Relativization, Relative reduction and Modifier inversion were set up to handle other constructions. **1977** M. DUMMETT *Elem. Intuitionism* v. 206 The relativization property guarantees that the logical laws which hold good whenever the individual variables are taken as ranging over any admissible domain also hold good when they are confined to some inhabited subdomain which can be characterized by a predicate of the language.

relativize (re·lătivəiz), *v.* [f. RELATIVE *a.* + -IZE.] **1.** *Physics.* To render or treat according to the principles and results of the theory of relativity.

1935 J. DOUGALL tr. *Born's Atomic Physics* iv. 84 Before Einstein, no one ever hesitated to speak of the simultaneous occurrence of two events... Einstein proved that this concept must be 'relativized', since two events may be simultaneous in one frame of reference, but take place at different times in another. **1956** E. H. HUTTEN *Lang. Mod. Physics* iii. 108 If we can make a uniform motion into an accelerated one, with a stroke of the pen so to speak, it means that the concept of force becomes relativised.

2. To render relative; to make something relative to, or dependent on, something else. Freq. in *Philos.* and *Linguistics.*

1937 T. PARSONS *Struct. Soc. Action* xi. 447 His [*sc.* Durkheim's] theory of religion, by associating it with the social type, relativized another great body of phenomena. **1948** L. SPITZER *Linguistics & Lit. Hist.* 81 The pun is a bifocal manner of expression which relaxes and relativizes the firmness with which language usually appears to speaking man. **1966** J. J. KATZ *Philos. Lang.* ii. 14 The philosopher of language..need not restrict his philosophical solutions and claims by relativizing them to the conceptual systems. **1976**

Language LII. 285 Kuhn proposes to relativize the notion of science. **1978** F. BURTON *Politics of Legitimacy* iii. 87 The raid that evening served to relativize the type of criticism that Jimmy was making.

Hence **re·lativized** *ppl. a.* (examples in *Linguistics*).

1972 *Language* XLVIII. 306 This [*sc.* receiving primary stress] should happen, for example, if the relativized NP were the subject of an embedded intransitive sentence. **1977** *Ibid.* LIII. 94 Ross formulates the relativization rule in such a way that it always involves movement of the relativized NP out of the sentence dominating the rest of the relative clause.

relator. Add: **4.** *Linguistics.* A sentence-element (usu. a preposition) serving to relate one phrase to another.

1933 [see *DESCRIPTOR]. **1953** W. J. ENTWISTLE *Aspects of Lang.* v. 157 The *relatum* of a language corresponds to the logical category of substance and finds its pure expression in proper nouns; the *descriptum* corresponds to quantity and has its pure expression in numerals; the *descriptor* with quality and is pure in adverbs; the *relator* with relation and is pure in prepositions. **1965** *Language* XLI. 73 Thus the whole string is a relator-axis phrase of which *on* manifests relator and the rest manifests axis. **1978** *Language* LIV. 353 *Relator* is assigned to constituents that serve to introduce embedded clauses (e.g. *that*), and is thus partially co-extensive with the complementizer of recent TG.

‖ **relatum** (rĭlā·tŏm, -ē¹·tŏm). Pl. **-ata.** [a. L. *relātum*, neut. pa. pple of *referre* REFER.] **a.** *Logic.* = RELATE. *sb.* 2. **b.** *Linguistics.* (Brøndal's term for) the substantival member of a prepositional phrase (see *RELATOR 4).

1872 G. GROTE *Aristotle* iii. 101 Habit, disposition, perception, cognition, position, &c., are all *Relata*. *Ibid.* 102 The *Relatum* and its Correlate seem to be *simul naturâ*. If you suppress either one of the pair, the other vanishes along with it. **1893** W. MINTO *Logic* iii. 118 In mediæval logic the term *Relata* was confined to these perfect cases, but the Category had a wider scope with Aristotle. **1903** B. RUSSELL *Princ. Math.* ii. 24 The class of terms to which some term has the relation *R*..I call the class of *relata*. Thus if *R* be paternity..the relata will be children. **1933** [see *DESCRIPTOR]. **1946** *Language* XXII. 219 The *relatum* is most commonly a noun or other type of substantive expression. **1953** [see *RELATOR 4]. **1974** L. SKLAR *Space, Time, & Spacetime* iii. 167 For temporal and spatiotemporal relata the idealization is that of the instantaneous event.

relaunch, *v.* Add: (Further example.) Also *fig.*

1964 *Yearbk. Astron. 1965* 135 The heavier landing vehicle was the penalty to be paid for taking the entire propulsion machinery down on to the lunar surface and re-launching it. **1971** *Daily Tel.* 27 May 1/5 The company was relaunched nearly two years ago... It was originally part of Henry Bowen-Davies' £8m Davies' Investment group which collapsed in 1967. **1980** *Listener* 3 Jan. 6/3 Rupert Murdoch relaunched the *Sun* as a down-market tabloid.

re·launch, *sb.* [f. the vb.] A renewed launch. Freq. *fig.* of a business or commodity.

1970 *Daily Tel.* 2 Mar. 21/6 The re-launch had the desired effect and Vim's share of the market increased from 33 p.c. to 38 p.c. **1979** *Guardian* 14 Nov. 20/1 (Advt.), The relaunch of Times Newspapers Limited signals the return of the most challenging selling opportunities around.

relax, *sb.* Restrict † *Obs.* to sense 2 and add: **1.** (Further examples.)

1853 C. BRONTË *Villette* II. xxiii. 141 That bustle and business to which, till five p.m., there was no relax. **1925** A. S. M. HUTCHINSON *One Increasing Purpose* I. xxii. 137 That..sigh she gave,..and that relax into his arms. **1961** *Times* 2 Nov. 16/2 Miss Brodie herself, who is always arousing her headmistress's suspicions because of her damning of braces and blessing of relaxes.

relax, *v.* Add: **I. 1. b.** *spec.* in *Ent.*

1902 W. J. HOLLAND *Butterfly Bk.* 41 When butterflies or moths have been put up in papers or mounted on pins without having their wings expanded and set it becomes necessary, before setting them, to relax them. **1939** DUNCAN & PICKWELL *World of Insects* xix. 389 Before such stored specimens can be mounted they must be relaxed. **1976** P. W. CRIBB *Lepidopterist's Handbk.* vii. 85, I have just relaxed and set some tortrices without too much trouble.

II. 7. c. Of persons: to become less tense or anxious. Freq. imp., 'stop worrying!', 'calm down!'.

1935 A. J. POLLOCK *Underworld Speaks* 135/2 *You can relax*, the person playing the hand in contract informing his partner that the contract will be made. **1941** *Men Only* July 70 (*caption*) All right, relax. I'm just watching it [*sc.* the baby] for someone! **1954** T. S. ELIOT *Confidential Clerk* I. 18 As you're here, Eggers, I can just relax. **1959** *Woman* 4 Apr. 48/2, I patted his knee. 'Relax, darling. Our problem is soon to be solved.' **1976** C. WOLFF *Older Love* i. 12, I relaxed over a *crème caramel* and was happy.

10. Chiefly *Physics.* To return towards a state of equilibrium.

1959 G. TROUP *Masers* iii. 45 Interactions occur between the lattice vibrations and the molecules, which enable the molecular system to 'relax' to thermal equilibrium. **1972** *Physics Bull.* Aug. 451/3 The electronic spins, initially unpolarized, relax slowly towards their equilibrium polarization which, in the high field applied and at a very low temperature of 0·4 K, is nearly 100%. **1973** *Nature* 24 Aug. 496/1 We have thus assumed that the observed strains result from a single system which relaxed exponentially after the start of the eruption with a time constant of 7·5 d. **1978** *Sci. Amer.* Sept. 124/2 Regular patterns of differential extinction..have occurred as the supersaturated faunas of 13 species of small flightless mammals have relaxed toward the smaller number of species that are appropriate to particular mountaintops.

relaxant, *a.* and *sb.* **A.** *adj.* (Later example.)

1977 *Lancet* 24–31 Dec. 1332/1 It may be that these neurons are involved in mediating relaxant effects of the intestine in response to a food bolus.

B. *sb.* (Later example.)

1977 *Sci. Amer.* May 99/2 One of its active alkaloids is the basis of drugs that are important in modern surgery as muscle relaxants.

relaxation. Add: **6.** *Engin.* and *Math.* A method of solving a set of simultaneous equations (originally *spec.* ones describing the equilibrium of a rigid load-bearing structure) by guessing a solution and successively modifying it to accord with whichever equation or constraint is currently least closely satisfied. Freq. *attrib.*

1935 *Proc. R. Soc.* A. CLI. 60 The method of systematic relaxation... Imagine that one constraint is relaxed, so that one joint is permitted to travel slowly through a specified distance in some specified direction. **1940** R. V. SOUTHWELL *Relaxation Methods in Engin. Sci.* i. 11 The relaxation procedure is a means whereby simultaneous equations may be solved, not exactly, but with steadily increasing approximation. **1957** L. FOX *Two-Point Boundary Problems* iii. 39 In most problems of the type suitable for relaxation the equations can be arranged so that the biggest coefficient in any row lies in the diagonal. **1972** *Physics Bull.* May 273/1 During the war, Southwell and his team had been extending relaxation methods from redundant pin-jointed frameworks to the stress analysis of the continuum.

7. Chiefly *Physics.* The gradual return of a system towards equilibrium; *esp.* the reduction of stress caused by gradual plastic deformation in material held at constant strain. Freq. *attrib.*, as **relaxation time**, the time taken for a system to return to a state of equilibrium; *spec.* (in cases in which the process of return is exponential), the time taken for the deviation from equilibrium to be reduced by a factor *e*.

1867 J. C. MAXWELL in *Phil. Trans. R. Soc.* CLVII. 53 A time T, which may be called the 'time of relaxation' of the elastic force. **1908** J. JEANS *Math. Theory Electricity & Magnetism* x. 349 The time..in which all the charges in the dielectric are reduced to $1/e$ times their original value, is called the 'time of relaxation', being analogous to the corresponding quantity in the Dynamical Theory of Gases. The relaxation-time admits of experimental determination. **1937** *Trans. Amer. Soc. Mech. Engineers* LIX. 451/2 There are many reasons why relaxation tests at constant extension are useful and significant. **1949** *Aircraft Engin.* Jan. 2/1 The field of research offered by the plastic, creep and relaxation properties of metals under general stress systems at high temperatures is yet only partially explored. **1958** *Jrnl. Iron & Steel Inst.* CXC. G3/1 The experiments on relaxation here described based on 10,000 h duration have given some quantitative characteristics for relaxation for 4 types of steel at 410° and 470° C. **1959** G. TROUP *Masers* iii. 35 We shall see that collisions are in fact a form of 'relaxation process' (process tending to restore the system to equilibrium). **1962** CORSON & LORRAIN *Introd. Electromagn. Fields* v. 191 The free charge density ρ therefore decreases exponentially with time at a rate such that after a time..called the relaxation time, it is reduced to $1/e$ or 36·8% of its original value. **1969** C. O. SMITH *Sci. of Engin. Materials* xiii. 367 The relaxation test is usually performed by maintaining total strain (elastic plus plastic) at a constant level and measuring the decrease in load (or stress) as a function of time. **1971** *Nature* 8 Jan. 93/1 Many phenomena, for example, may be assigned their typical relaxation times—the average time for an effect to fade away... Thus, a fit of temper may have a relaxation time of a few minutes, the satiation of hunger by a meal lasts for a few hours. **1972** *Ibid.* 22 Dec. 447/1 Physicists are also interested in such phenomena as the changes in the qualities of the varnish on a Stradivarius violin, these being relaxation phenomena demonstrating both temporal and irreversible properties.

8. Special Comb.: **relaxation oscillator** *Electr.*, a form of oscillator in which the period and resulting waveform are determined by the slow charge and rapid discharge of a resistor–capacitor or inductor–capacitor circuit. See also sense 7 above.

1942 E. WILLIAMS *Thermionic Valve Circuits* v. 128 The simplest relaxation oscillator is perhaps the series connexion of a d.c. supply voltage, a resistance and a condenser, a neon lamp being shunted across the condenser. **1943** *Electronic Engin.* XV. 412 In general the time-base will be derived from a relaxation oscillator producing a 'saw-tooth' waveform. **1966** *McGraw-Hill Encycl. Sci. & Technol.* XI. 438/2 One of the most widely used forms of relaxation oscillator is the astable multivibrator..which generates a rectangular or square wave.

relaxed, *ppl. a.* Add: **3.** esp. (in sense 7 of vb.) informal, leisurely, at ease; unanxious, free from constraint or tension.

1958 *Listener* 19 June 1032/2 All the speakers sounded relaxed and informal. *Ibid.* 14 Aug. 249/2 What struck me most was the leisurely, richly human world—'relaxed' we should call it now, when nobody is relaxed—that it evoked. **1961** *Times* 16 Mar. 15/4 'Relaxed' has taken the place of the outmoded 'bronzed and fit'. **1972** M. WOODHOUSE *Mama Doll* ix. 121 He'll be fine, Bottle. Very relaxed character. **1978** *Times* 23 Apr. 12/4 You want a meal at the right price in a relaxed atmosphere.

relaxedly *adv.* (further examples); **relaxedness** (further examples).

1952 *Scrutiny* XVIII. IV. 275 What we have here, of course, is not relaxedness or distraction. **1957** Relaxedness [see *COMPÈRE *v.*]. **1974** M. Z. LEWIN *Enemies Within* xxxiii. 147, I drove a lot faster and a lot less relaxedly. **1977** *Times* 22 Sept. 8/1 Edward Heath..didn't have anything so relaxedly Edwardian as a confidant.

relaxer. Restrict *rare* to sense in Dict. and add: **2.** One who applies the method of relaxation (sense *6).

1957 L. FOX *Two-Point Boundary Problems* iii. 39 There are many tricks a skilled relaxer can use to accelerate the convergence of the process and generally lighten his work. **1959** A. M. OSTROWSKI in R. E. Langer *Numerical Approximation* 4 One of the fields where this difficulty is felt to a particularly high degree is that of relaxation. The practical relaxer has it 'in his fingertips' how to steer the successive relaxations.

relaxin (rĭlæ·ksin). *Physiol.* [f. RELAX *v.* + -IN¹.] An ovarian hormone first found in rodents, in which it relaxes the pelvic ligaments and softens the cervix of the uterus.

1930 H. L. FEVOLD et al. in *Jrnl. Amer. Chem. Soc.* LII. 3341 The only physiological property thus far discovered for this hormone is its action on the pelvic ligaments and for this reason we propose the name 'Relaxin'. **1968** PASSMORE & ROBSON *Compan. Med. Stud.* I. xxxvii. 45/2 The polypeptide hormone, relaxin, is probably also produced by the placenta. **1978** *Amer. Jrnl. Obstetrics & Gynecology* CXXX. 473/1 The role of human relaxin in pregnancy has not been determined, but relaxin extracted from human luteal tissue is active in the guinea pig pubic symphysis assay.

relaxing, *vbl. sb.* Add: **2.** *attrib.* in *Ent.*, applied to containers in which dead insects are relaxed, containing a pad soaked with fluid.

1907 *Yesterday's Shopping* (1969) 699/3 Entomologists' requisites..Killing and relaxing boxes. Each..1/6. **1912** H. ROWLAND-BROWN *Butterflies & Moths* I. v 53 The nervures and wing-attachments must be softened to permit of their being rearranged. I have not found that the 'relaxing-boxes' are much good for this purpose with other than recently killed specimens. **1952** E. F. DAGLISH *Name this Insect* p. xiii, After some hours in a relaxing chamber the legs and wings will be as easy to arrange as are those of fresh-killed specimens. **1963** R. L. E. FORD *Pract. Entomol.* 32 Specimens may be left in a relaxing tin for about twenty-four hours when they will be ready to set. **1976** P. W. CRIBB *Lepidopterist's Handbk.* vii. 84 A relaxing box is a clean plastic box with a layer of clean absorbent material in the bottom.

rela·xity. *rare.* [irreg. f. RELAX *v.* + -ITY; cf. LAXITY.] Relaxedness, freedom from restraint or tension; the state of being relaxed.

a **1784** S. JOHNSON *Sermons* (1788) I. xiii. 271 Men have ever been persuaded, that by doing something, to which they think themselves not obliged, they may purchase an exemption from such duties as they find themselves inclined to violate: that they may commute with heaven for a temporary fine, and make rigour atone for relaxity. **1908** *Daily Mail* 30 June 9/5 The great secret of voice production is relaxity.

relay, *sb.* Add: **2 d.** A series of motor vehicles intended to cover a prescribed route (usu. in sequence); an operation involving this.

1942 *R.A.F. Jrnl.* 27 June 6, I chartered a relay of cars which got us to Beirut..up the coast road. **1971** M. TAK *Truck Talk* 129 *Relay*, a procedure commonly used in companies to keep as many trucks as possible moving over the road. **1973** *Amer. Speech 1969* XLIV. 207 *Relay*, driving operation in which the driver takes his unit from one terminal to another, where a new driver takes over to deliver it to the next terminal, and so on. **1975** *Drive* Nov.–Dec. 110/1 The travellers soon continued their journeys—care of AA Relay. *Ibid.* 110/2 Relay's south-east team had to..cover a Rolls-Royce..from east London and take it back to ..Chelsea.

e. *Bridge.* In full, *relay bid*. (See quot. 1964.)

1959 T. REESE *Bridge Player's Dict.* 183 The relay method is used in some systems played by European teams. In certain sequences the responder does not try to give a picture of his own hand but makes a series of relay-bids at the lowest level so that he can learn more about his partner's hand. **1961** *Times* 30 Aug. 11/5 If the partner responds the minimum in the suit immediately above the opening bid...this response is either negative (discouraging) or natural (with more than 10 points). In either event it is known as a 'relay' bid and is forcing; it does not indicate a real suit but invites the opener to disclose what values he holds. **1962** *Listener* 27 Sept. 494/2 In this auction North's 1 NT and 2 NT were 'relay bids', just asking partner to describe his hand. **1964** *Official Encycl. Bridge* 452/1 *Relay*, a minimum bid unrelated to the bidder's hand, aimed simply at keeping the bidding open so that the bidder's partner can describe his hand. **1980** *Times* 12 July 7/4 After One Club—One Diamond–One Heart—the usual rebid by responder is One Spade. This is a 'relay' bid, asking opener to clarify his hand.

3. b. In full, *relay race*. A race of runners in sequence; *spec.* one run by teams of four athletes, a baton being passed in each team from one runner to the next. Also (in quot. 1920), each of the four sections of a relay race. Also in other sports, e.g. Swimming,

where members of a team perform in sequence. Also *attrib.*

1898 M. SHEARMAN *Athletics* x. 301 So popular has this form of racing become that within the last year a number of athletic meetings have been held at which there were a series of these relay races. **1908** T. A. COOK *Olympic Games* 187 Relay Race 1600 Metres... Teams of four with four reserves. **1908** *Daily Chron.* 18 Apr. 5/6 It looked as if its representatives would carry off the prize for the one mile relay race, in which four runners run one lap, carrying a flag each. At the end of the lap the flag is handed to a relay walker, and in his turn the walker hands the flag to a cyclist who completes the race. **1920** *Isis* 13 Oct. 2/2 Ten yards is allotted each side of the starting line in which to pass the baton to the next competitor, for every relay subsequent to the one which begins the race. **1922** F. W. H. NICHOLAS *Handbk. Athletics for Beginner* (ed. 2) x. 44 Relay 100 yards and relay hurdles may be run up and down. **1927** W. DEEPING *Kitty* xxiv. 310 To him life was like a relay-race: you snatched the baton from the failing hand of the past, and sped ahead without looking back till some other racer took the baton from you. **1929** G. M. BUTLER *Mod. Athletics* ii. 8 Names are taken..and made up into senior and junior relay teams of four each. *Ibid.* 9 Juniors are under no circumstances allowed to compete in the senior relays. **1939** *Encycl. Brit. Bk. of Year* 650/2 Peter Fick lowered the world's record for the 400-metre relay..while Ralph Flanagan swam 400 metres free style in 4 min. 46·2 sec. **1950** *Oxf. Jun. Encycl.* IX. 454/1 There are also two kinds of team relay race: the medley, in which there are exponents of all three styles of swimming; and the free-style, in which all members swim the same stroke. **1952** ARMBRUSTER & MOREHOUSE *Swimming & Diving* (ed. 2) x. 201 The types of relays in swimming are usually of two kinds. *Ibid.* 204 The swimming take-off in relay racing differs from that in back relay racing. **1955** R. BANNISTER *First Four Minutes* ii. 18 He had helped Hungary on two occasions to capture the 4 × 1,500 metres relay World record. **1958** *Times* 13 Aug. 2/6 The main hopes in this country lie in the men's four by 100 and four by 400 metres relays. **1958** [see *BATON *sb.* 2 b]. **1974** *Country Life* 14 Feb. 292/3 The England women's only gold medal in athletics..came in the 4 × 400 metres relay. **1976** *Liverpool Echo* 6 Dec. 18/1 Visibility on the course, however, was too poor to permit the senior relay and a three miles race was substituted. **1978** G. WRIGHT *Illustr. Handbk. Sporting Terms* 150/3 *Relays*, events in which teams of swimmers swim in sequence... A relay team usually consists of four swimmers, but occasionally larger numbers are used.

4. a. In mod. use, any electrical device, usu. incorporating an electromagnet, whereby a current or signal in one circuit can open or close another circuit. Also *transf.* (Further examples.)

1907 *Cornhill Mag.* Mar. 363 The difficulty is overcome by using the partly exhausted current to move a special kind of 'switch', or key, called a 'relay'. **1923** E. W. MARCHANT *Radio Telegr. & Teleph.* v. 71 The telephone may be replaced by an ordinary Post Office relay, such as is used for working on the ordinary telegraph line. **1935** MONSETH & ROBINSON *Relay Systems* i. 1 The function of protective relays in modern power systems is to initiate the operation of devices to isolate transmission circuits and apparatus when trouble develops. **1956** G. A. MONTGOMERIE *Digital Calculating Machines* x. 211 For adding numbers, three sets of relays are used, designated as A, B, and C; they are wired together so that, if two numbers are sent respectively to A and B, the sum of the two numbers appears on C. **1968** *Brit. Med. Bull.* XXIV. 200/2 The sensory relay region in the thalamus. **1969** *Times* 16 Jan. 4/7 Relays are instruments used to switch electrical circuits on and off automatically. They usually consist of an electromagnet which, when activated by an electrical signal, opens or closes a switch in another circuit.

b. An installation or satellite which receives, amplifies, and retransmits a radio transmission so that it may be received over a wider area. Freq. *attrib.*

1921 *Wireless World* 10 Dec. 575/1 These men have banded themselves into a relay organisation. They have laid out in definite form certain traffic routes.., and messages..are broadcasted across the country any time of the night... These relay routes enable the transmission of personal messages from coast to coast, and from the Canadian border to the Mexican border. **1923** *Radio Times* 28 Sept. 26/3 The engineers of the British Broadcasting Company will employ a wireless relay across the Thames. **1945** *Wireless World* Oct. 305 (*heading*) Extra-terrestrial relays. Can rocket stations give world-wide radio coverage? **1962** *Rep. Comm. Broadc. 1960* 257 in *Parl. Papers 1961–2* (Cmnd. 1753) IX. 259 The relay companies are prohibited from originating any programmes of their own. **1966** *McGraw-Hill Encycl. Sci. & Technol.* XIII. 469/2 By far the largest number of television circuit miles is provided by microwave radio relay. **1966** *Electronics* 14 Nov. 47 The company has developed an antenna that allows a plane, say flying over North America, to communicate with a relay satellite orbiting about the equator. **1975** D. G. FINK *Electronics Engineers' Handbk.* XXII. 61 The communications satellite is a radio relay, consisting of a receiver and transmitter, plus a command receiver and transmitter, to control the satellite.

c. A radio transmission or programme which has been relayed.

1929 *Radio Times* 8 Nov. 395/3 We were testing all the arrangements for the Schneider Trophy relay, making sure that the loud-speaker system at various points round the coast could pick up our broadcast. **1929** *B.B.C. Yearbk. 1930* 383 Listeners can..expect to find a number of relays of Central European stations included in the British programmes. **1965** *Listener* 25 Nov. 873/2 *Don Carlos* (Third Programme, November 20) was a direct relay. In spite of all the disadvantages—in timing, indifferent quality of reception, and applause—this kind of broadcast has the incalculable quality of excitement and immediacy that no recording or tape can hope to equal.

5. Special Combs.: **relay rack**, a rack or frame on which relays are mounted, usu. used in a telephone exchange; **relay station**, a radio station that serves as a relay; also *fig.*; **relay valve** *Engin.*, a fluid valve in which the main flow is controlled by a diaphragm actuated by a weep derived from the main flow.

1908 *Daily Chron.* 8 Apr. 3/6 Each girl sits in front of a relay rack, fitted with a bewildering number of small holes, each of which represents a subscriber. **1930** [see *RACK *sb.*[2] 5 g]. **1970** *Jrnl. Gen. Psychol.* LXXXII. 58 A desk-type relay rack with two standard rack panels and an enclosed back stood on a table in the experimental room. **1923** *Radio Times* 28 Sept. 2/2 The proposed relay stations.. will have a power of 100 to 150 watts. **1969** *Times* 26 Feb. 8/7 Light..is converted to a train of nervous impulses which are transmitted down the optic nerve to a relay station known as the lateral geniculate body (L.G.B.) and from there to the striate cortex. **1974** *B.B.C. Handbk. 1975* 21/1 Savings on capital expenditure in 1974–5 were achieved through postponement of work on the proposed Caribbean relay station. **1939** R. N. LE FEVRE *Man. Pract. Gas Fitting* xix. 390 The relay valve..is made in a variety of sizes to suit particular gas rates and pipe connections. **1970** MILES & PINKESS *Gas Appliance Control* ii. 41 It is..possible to have a leak in a joint in the weep pipe or in the cover of the relay valve itself which would pass sufficient gas to hold the relay valve open, even if the control were shut. **1977** R. PRITCHARD et al. *Industr. Gas Utilization* ix. 414 Relay valves are used to control gas or air flows using a small actuating valve, which may be a solenoid valve or a thermostat in the weep line.

relay, *v.* Add: **4. a.** *trans.* To pass on or retransmit (telephonic or broadcast signals received from elsewhere); *loosely*, to transmit.

1878 *Telegr. Jrnl.* VI. 274/2 They have finally solved the important problem of *relaying* telephone sounds. **1904** *Marconigram* July 16/2 With a telegraphone in Chicago, one may telephone from New York, have the telegraphone record his message and repeat it over another wire to St. Louis, where another machine relays it to Denver. **1923** *Glasgow Herald* 22 Mar. 9/2 Little progress has so far been made, as the experiments have only recently been commenced, but last night a Birmingham concert was relayed for London with some success. **1923** *Daily Mail* 14 Aug. 5/3 A special orchestral concert which will be relayed to all the broadcasting stations in Britain. **1958** *Radio Times* 14 Feb. 3/3 The sensitive receiving equipment is also used for relaying programmes from the Commonwealth and the U.S.A. **1969** *Times* 16 July 4/1 The television pictures to be relayed back to earth will be taken by a camera fixed on a special attachment. **1974** *B.B.C. Handbk. 1975* 20/2 Prokofiev's opera *War and Peace* was relayed from the New Sydney Opera House in Australia. **1977** *Rep. Comm. Future of Broadcasting* (Cmnd. 6753) iii. 21 We saw a cable company in Toronto relaying programmes on 24 channels: but several of them were relaying the same programme.

b. *transf.* To pass on (a message or information).

1956 A. H. COMPTON *Atomic Quest* ii. 117 These men were thoroughly acquainted with our wartime methods of bomb construction and relayed the techniques to Russia. **1974** *State* (Columbia, S. Carolina) 13 Feb. 5-A/1 President Nixon relayed word through a spokesman Tuesday that he has no plans to visit the Middle East.

Hence **re·layed** *ppl. a.*; **re·laying** *vbl. sb.*

1904 *Marconigram* July 16/1 The steel belt machine will transmit a record..by relaying, to great distances. **1949** *Radio Times* 15 July 6/1 [We] presented an electrophone to our aged father on October 5th, 1908... I have a vivid recollection of..listening to a relayed programme.

relea·rning, *vbl. sb.* [RE- 5 a.] Learning again.

1961 'E. FENWICK' *Friend of Mary Rose* ii. 20 In the new house..he would have an immense amount of relearning to do. **1978** *Dædalus* Fall 33 Nothing is ever learned beyond the need for relearning.

releasable, *a.* Delete *rare* and add later examples.

1936 *Nat. Geogr. Mag.* LXIX. 93/2 Heavy items of apparatus to be carried in releasable form on the outside of the gondola. **1950** *Manch. Guardian* 15 Sept. 7/3 At Church Fenton.., to quote from an Air Ministry News Service message, will be 'a miniature bombing range and miniature aircraft complete with releasable bombs to amuse the children'. **1980** *Nature* 31 Jan. 488/1 The content of LH-RH in the synaptosomal pellet obtained after centrifugation showed a decline with time, indicating depletion of a releasable pool of the neuropeptide.

release, *sb.*[1] Add: **1. c.** *Psychol.* Liberation from emotional or physical tension. Also *attrib.* and in gen. use.

1915 E. B. HOLT *Freudian Wish* i. 20 Just what shall happen depends on the relative strengths of the suppressed wish and of the censor, and on the amount of release which the joke affords as well as on the degree of violence which it does to the censor. **1933** E. & W. MURPHY tr. *Bechterev's Gen. Princ. Hum. Reflexol.* xxv. 272 All processes of release from inhibition are based on the retention—in the centres—of the traces of reflexes. **1934** E. B. STRAUSS tr. *Kretschmer's Text-bk. Med. Psychol.* i. 12 Are they [*sc.* fits] not really a symptom of cortical paralysis (produced by the lesion) in the sense that they result from a release of sub-cortical activities from cortical control? **1936** P. T. YOUNG *Motivation of Behav.* v. 247 Tension and release occur constantly in the trivial events of daily life. **1948** L. KANNER *Child Psychiatry* (ed. 2) xvii. 245 The term 'release therapy' indicates that the ventilation of specifically oriented feelings of hostility, guilt, and anxiety constitutes a main therapeutic facet. **1954** A. H. MASLOW *Motivation & Personality* xi. 187 It is very likely that catharsis, as originally defined by Breuer and Freud, is in essence a more complex variant of release behavior. **1959** *Times* 4 Apr. 7/6 As the American male is said to approach his car as a form of self-expression, so the German sees it as an instrument of release. **1966** *Listener* 4 Aug. 174/1 In some quarters this loss of binding traditions has resulted in the retreat beyond all controls—the anti-culture of 'release'. **1978** M. T. ERICKSON *Child Psychopathology* vi. 117 The therapeutic effect of release therapy is based on the child's acting out or talking about a traumatic event that is the source of the disturbance.

2. d. A written authorization or permission for publication, esp. from an owner of copyright or a person depicted in a photograph.

1965 *Tamarack Rev.* Winter 13 Don't touch anything, and be sure to get a release. **1966** K. GILES *Provenance of Death* i. 6 Publishing your photo in an ad. without a release could be libel. **1970** C. WHITMAN *Death out of Focus* ix. 130 A photographer would be a damned fool to come in here with a print for which he had no release. **1979** R. COX *Auction* vii. 186 The late Herr Schneider bequeathed it [*sc.* a picture] to me and I was given a release by Herbstein as a result.

6. c. *Teleph.* The action of freeing for further use apparatus or circuitry which has been engaged. Freq. *attrib.*

1892 Release key [see sense 6 b in Dict.]. **1919** J. POOLE *Pract. Telephone Handbk.* (ed. 6) xxi. 368 Its armature short-circuits the no-voltage release magnet and the switch arm falls back to its open position. **1921** W. AITKEN *Automatic Telephone Systems* I. 185 The calling receiver may be replaced just before the register is connected and cause a premature release. **1969** S. F. SMITH *Telephony & Telegr. A* ii. 39 If a relay were required to have a high value of release current, it would need as many springs as possible of maximum thickness. **1970** N. N. BISWAS *Princ. Telephony* iii. 80 This alarm circuit becomes a necessity in all exchanges where the release of the entire switching stages is controlled by the calling subscriber.

d. *Phonetics.* The action or manner of relaxing or terminating the obstruction involved in articulating a stop consonant.

1920 in WEBSTER. **1951** Z. S. HARRIS *Methods in Structural Linguistics* 44 In some English dialects perhaps the sequence [tr] (post-dental [t] plus voiceless spirant release), are each composed of smaller segments. **1964** J. C. CATFORD in D. Abercrombie et al. *Daniel Jones* 34 Variations in vocal fold thickness..produce qualitative variations in..the release-sound of glottal stop. **1969** *English Studies* L. 328 This implies that the difference of total duration between [ptk] and [bdg] is very nearly equivalent to the difference of duration of the release stage. **1978** *Amer. Speech 1975* L. 295 In the style of pronunciation favored by barbershoppers, final voiced stops like those in the key words have a release that gives the impression of an indistinct vowel.

e. *Jazz.* A passage of music that serves as a bridge between repetitions of a main melody. Chiefly *U.S.*

1936 L. DOWLING tr. *Panassié's Hot Jazz: Guide to Swing Music* 18 The group of eight measures designated by the letter *b* is called the 'middle part' because it makes the first appearance in the middle of the tune. [*Translator's note*] Also called, quite poetically, 'the release'. **1937** *New Republic* 24 Nov. 69/1 But then the band comes down to the release and Benny holds up one finger and Jess nods. **1946** MEZZROW & WOLFE *Really Blues* (1957) 344 We played a more staccato style on the release. **1949** L. FEATHER *Inside Be-bop* II. 67 In the release there is another beautiful sweeping phrase. **1959** AVAKIAN & PRINCE in M. T. Williams *Art of Jazz* (1960) xvii. 184 Charlie..develops a series of riffs through the first sixteen bars (tension); then, in the eight-bar release, he contrasts this by playing melodic lines characteristically made up mainly of even eighth notes (relaxation), then returns to eight bars of riffs (tension). **1972** A. WILDER *Amer. Popular Song* ii. 56 The conventional *A-A-B-A* structure (main strain: its virtual repetition: a release, almost always new material: and finally, a literal, varied, or extended restatement of the main strain) was used in *Ol' Man River*.

7. The action of releasing information or other material for publication or public showing; the information or material released.

a. (The releasing of) a news item or official statement, usu. to the press. orig. *U.S.*

1907 *N.Y. Even. Post* (semi-weekly ed.) 15 July 4 The report was given to the press associations..labelled 'confidential', with a fixed date for 'release', before which no part of it was to be used. **1927** G. ADE *Let.* 31 May (1973) 120, I will be..up to my eyes in the weekly release grind. **1931** F. L. ALLEN *Only Yesterday* ix. 276 Press agents distributed their canned releases. **1932** *Atlantic Monthly* Mar. 269/1 The press agents..did not pour forth their releases to a..coöperating press. **1957** [see *BACKGROUNDER]. *a* **1974** R. CROSSMAN *Diaries* (1975) I. 343 The release wasn't ready until a few minutes before I had to deliver the speech.

b. The action of making a film available to cinemas or a gramophone record to purchasers; also, the film or record itself.

1912 *Motion Picture Ann.* 25 An Essanay release called 'Sunshine'. **1927** *Daily News* 8 June 4/4 Some of the recent 'releases' show that Hollywood and Germany are being challenged seriously in the matter of production. **1929** '*His Master's Voice' New Records* Mid-June 13 Theme songs from two great American films that are scheduled for release in the autumn. **1932** *New Yorker* 14 May 57/2, I have not seen it [*sc.* a gramophone record] on any official list and it seems to be a special release. **1949** *St. Paul* (Minnesota) *Pioneer-Press* 19 June 11/3 Busiest year on record for total releases is 1921 when American moviegoers had their choice of 854 different features. **1966** *Illustr. London News* 30 July 31/2 Perhaps this could be quietly excised before the film goes out on general release. **1966** *Guardian* 22 Dec. 4/7 In the pop/folk field the best new release is by The Incredible String Band. **1974** *Times* 19 Oct. 9/1 New Releases... Piano Music by Erik Satie. **1977** *Time* 4 July 4 (Advt.), There's a film to watch—a recent release—8 tracks of stereo to listen to, free naturally, and plenty of room to stretch out or stroll about.

8. Special Combs.: **release agent**, a substance which is applied to a surface in order to prevent adhesion to it, esp. in food packaging and concrete construction; **release date**, a date fixed for the release of information or other material (see sense 7 above); **release group**, a group of servicemen due for release from conscripted service; **release note**, a note authorizing the release of (part of) an aircraft as fit for service; now also in extended use.

1960 A. E. Bender *Dict. Nutrition* 107/2 *Release agents*, substances applied to tinned or enamelled surfaces or plastic films to prevent the food adhering; e.g. fatty acid amides, microcrystalline waxes, petrolatum, starch, methyl-cellulose. 1965 W. H. Taylor *Concrete Technol. & Pract.* vii. 160 An ideal release agent..should produce a clean stripping action with a minimum of surface defects on the hardened concrete. 1974 Briston & Katan *Plastics in Contact with Food* iii. 61 Silicone resins are also used as release agents. The baking industry, for instance, uses silicone resins to coat bread baking pans and hundreds of releases from a single coating of resin have been reported. 1910 *Moving Picture World* 26 Mar. 488/1 (*heading*) Independent release dates. 1934 L. C. Douglas *Forgive us our Trespasses* (1937) xv. 306 He decided not to take another look at the gripping letter until he had done at least one essay. He always tried to keep about three weeks ahead of the release date. 1965 *Amer. N. & Q.* Mar. 103/2 Its fine appendix of 'Serials from 1912 to 1930', showing title, director, cast, release date, releasing company. 1945 *News Chron.* 18 Apr. 2/4 We think it would have been much fairer to lower the release group age, such as all men over 45 in Group One and so on, and let some of the youngsters who have been in so-called deferred jobs have a turn. 1946 *R.A.F. Jrnl.* May 129 W.A.A.F. personnel whose release groups have appeared in an advance promulgation are invited to apply for vacancies. 1930 *Air Ann. Brit. Empire* 234 The firm must issue with every consignment they deliver a release note certifying that all inspection has been carried out. 1963 *Times Rev. Industry* Mar. 51/1 When a motor dealer asked a customer from whom he bought a second-hand Wolseley car to sign a 'release note', which turned out to be a guarantee of a third party's commitments under a hire-purchase agreement, the customer was not liable on the guarantee.

release, *v.*[1] I. 4. d. Of a public or military authority: to make available (requisitioned or otherwise withheld items) to the public; to return (land or property) to civilian use.

1917 *Globe* 21 Feb. 4/4 Only this morning a daily paper of some standing remarked that the Government had not 'released' any Colonial mutton..last week. 1945 *Daily Tel.* 27 July 3/3 (*heading*) R.A.F. & Navy to release houses. *Ibid.*, The Admiralty and Air Ministry are to do all they can to alleviate the housing situation by releasing property.

II. 6. a. (Later example with *of*.)

1979 *Petroleum Rev.* XXXIII. 675/3 To release the diver of this chore, remote-controlled systems are being developed.

c. *U.S.* To make (an employee) redundant. *euphem.*

1976 *National Observer* (U.S.) 24 Jan. 1/4 The two most difficult things I ever had to do were, one, tell 23 teachers we were going to release them [etc.]. 1977 *Time* 12 Dec. 54/2 He closed 1,700 stores, released 100,000 employees, borrowed heavily to revamp and enlarge the remaining 1,932 supermarkets.

7. To make available for publication or public showing; to publish (printed matter, recorded material or the like). orig. *U.S.*

1904 *N.Y. Times* 25 July 5 Chairman Cannon's speech and President Roosevelt's response are completed. The latter is in the hands of the press associations, and will be released Wednesday afternoon. 1912 *Motion Picture Ann.* 42 List of Licensed Pictures. Regularly released during the year 1912. 1916 'B. M. Bower' *Phantom Herd* v. 71 We've just got to release films the market calls for. 1937 A. Thirkell *Summer Half* xi. 298 If a film gets to Eastchester it means it's been released for simply months. 1957 *Essays & Stud.* X. 5 Among words that incur reproach are..release (the expression 'to release a film' is denounced by a bishop as 'an abominable Americanism'). 1962 *Sunday Times* (Colour Suppl.) 10 June 7 This is also true of American records, a great many of which are only released because companies have to take them to get some really lucrative artist. 1972 *Daily Tel.* 18 Jan. 9/5 Rehearsals have already started and the record is expected to be released some time in the Autumn. 1980 *Time Out* 21–27 Nov. 49/3 Films considered by their multinational distributors as too 'difficult' to release conventionally.

releaser. Add: **b.** *Dairying.* A device which removes milk from the vessel in which the output of a milking machine accumulates. Freq. *attrib.*

1950 *N.Z. Jrnl. Agric.* Apr. 378/1 Probably the most important part of any milking shed is the releaser room, as it is here that milk or cream can most easily become affected by unsatisfactory conditions. 1950 *Ibid.* Oct. 369 Up-to-date assembly of releaser, cream separator, skimmed-milk pump, and cream cooler. 1967 Harvey & Hill *Milk* (ed. 4) xiii. 224 Where milk pipe-lines are provided to transmit the milk directly to the dairy, as in parlours, bails or with milk lines in cowsheds, a releaser is required to remove the milk from the system. Sufficient milk accumulates in the releaser jar which operates valves which seal off the vacuum system and allow the milk to be discharged. 1977 D. N. Akam in Thiel & Dodd *Machine Milking* iii. 82 A design of diaphragm releaser milk pump that is available in the UK is vacuum driven using a pulsator operating at 50 pulsations/min.

c. *Biol.* [tr. G. *auslöser* (K. Lorenz 1935, in *Jrnl. für Ornithol.* LXXXIII. 143).] A sign stimulus (see *sign *sb.* 12); restricted by some writers to one that acts between animals of the same species. Freq. *attrib.*

1937 K. Lorenz in *Auk* LIV. 249 All such devices for the issuing of releasing stimuli, I have termed releasers (*Auslöser*), regardless of whether the releasing factor be optical or acoustical, whether an act, a structure or a color. 1953 J. S. Huxley *Evol. in Action* iv. 96 The only definite releaser known in man is the pattern made by a mother's smile to her infant. 1953 N. Tinbergen *Herring Gull's World* xiv. 216 Ritualisation is the result of a secondary evolutionary process which is closely linked to the releaser-function. 1964 *Nature* 8 Aug. 528 The red patch on the bill seems to be a genuine social releaser. 1965 *Listener* 9 Aug. 207/2 Because animal signal codes are uniform within each species and fixed for long periods, special signal structures may evolve, and these are called releasers. 1971 *Nature* 16 Apr. 431/2 Releaser pheromone effects exist in man, at least in larval forms, and some involve pheromones of other mammals (musk, civetone). 1975 J. Alcock *Animal Behavior* vi. 153 The first concept we shall examine is the sign stimulus or releaser, that portion of the total stimulus configuration which acts as the effective cue in releasing a specific behavior pattern. 1980 A. P. Brookfield *Animal Behaviour* vii. 58 Sign stimuli which elicit behaviour in members of the same species are called releasers.

releasing, *vbl. sb.* Add: **2.** Special Comb.: **releasing factor** *Physiol.*, any of several oligopeptides, released from the hypothalamus into the pituitary portal system, which promote the release from the adenohypophysis into the bloodstream of some specific peptide hormone.

1955 *Endocrinology* LVII. 443 Posterior pituitary extracts contain a corticotropin-releasing factor (CRF) that stimulates the release of ACTH from rat anterior pituitary tissue *in vitro*. 1961 *Ibid.* LXXVII. 609/1 In 1959, Shibusawa et al...claimed to have prepared a thyrotropin releasing factor (TRF) from dog hypothalamic extracts and from urine. 1966 *Brit. Med. Bull.* XXII. 266/2 On this view, various hormonal agents (now called releasing factors) are liberated from nerve-endings (of hypothalamic nerve tracts) into the capillaries (primary plexus) of the portal vessels in the median eminence. 1974 M. C. Gerald *Pharmacol.* xxiii. 415 The supreme command post of the endocrine system directs the activity of the anterior pituitary by neurosecretory mediator substances called releasing factors. 1977 *Time* 14 Oct. 42/1 Andrew Schally..isolated identified and synthesized three separate hormones—'releasing factors'—by which the hypothalamus directs the release of key hormones from the pituitary.

relegate, *v.* Add: **2. d.** *Sport.* To reallocate (a team) to a lower division of a league. Cf. *relegation 1 c.

1923 *Times* 28 Apr. 12/5 Norwich County..will..be relegated to the Second Division next season. 1934 *Times* 7 May 4/5 Everton, when they were relegated for the first time in their history, climbed back immediately. 1981 *Times* 16 May 19/1 After a trying beginning, that saw the club relegated to the second division.

relegation. Add: **1. c.** *Sport.* The demotion of a team to a lower division of a league; *spec.* in *Assoc. Football*, the reallocation to a lower division of the Football League of an agreed number of teams scoring the fewest points in a division in the course of a season's play. Also *attrib.*

1924 *Times* 5 May 6/6 Fractions in goal averages decided promotion and relegation. 1928 *Daily Express* 10 Aug. 13/7 Their supporters have recovered from the bitter disappointment felt when relegation became inevitable. 1949 *Times* 9 May 6/5 There was the question about relegation from the Championship. 1951 *Sport* 6–12 Apr. 6/2 Key man in the successful battle now being waged by West Bromwich Albion to steer clear of the First Division relegation zone is Jack Vernon. 1969 [see *inject *v.* 2]. 1969 *Listener* 1 May 625/3 On Saturday, more than 250 million people are estimated to have been watching the end of the season. 1977 *Daily Mirror* 12 Apr. 26/4 We are out of the relegation zone now.

relentment. a. (Later examples.)

1922 Joyce *Ulysses* 404 The prolongation of labour pains in advanced gravidancy by reason of pressure on the vein, the premature relentment of the amniotic fluid (as exemplified in the actual case) with consequent peril of sepsis to the matrix. 1925 C. E. Montague *Disenchantment* iv. 65 Great are the forces of decent human retirement after a hearty let-out with the temper.

re-let, *v.* (Earlier example.)

1780 A. Young *Tour in Ireland* I. 53, I found rents in general at 20s. an acre, with much relet at 30s.

re-let (rī·let), *sb.* [f. the vb.] A property that is let again.

1969 *Guardian* 29 Oct. 1/8 A vast increase in the number of 're-lets' among existing corporation houses. 1971 *Daily Tel.* 5 Aug. 10/7 Relets could be made to young people and earn £40 to £50 a week for the landlord instead of perhaps £10. 1976 *Times* 7 Jan. 13 Even allowing for the substantial numbers of relets from the existing stock, the magnitude of the loss of this source of housing in the new communities is evident.

relevance. Add: (Further examples.) Also *spec.* in recent use, pertinency to important current issues (as education to one's later career, etc.); social or vocational relevancy.

1949 *Poetry* (Chicago) Feb. 299 Tate holds that the poem is autonomous, and that the only relevance the subject-ideas have is to each other within the formal meaning of the work itself. 1955 *Bull. Atomic Sci.* Apr. 126/1 Relevance is another one of these non-assessable quantities which circumstances require to be assessed. 1970 *Time* 30 Nov. 40 The impetus came largely from student demands for 'relevance', especially for the overdue admission of more minority-group students. Activism has also done much to curb the old absurdities of trivial research and needless Ph.D.s. 1975 *Language for Life* (Dept. Educ. & Sci.) ix. 129 We have heard the case for 'relevance' carried to the point of excluding fantasy or any stories with settings or characters unfamiliar to the pupils from their first-hand experience. 1976 *Times* 12 Feb. 11/7 Hal [sc. a novel]—while laudable in its social intentions—is little more than a piecing together of stock responses to the current demand for 'relevance'. 1977 *Chem. in Brit.* Mar. 105/1 It may seem anomalous in these days of 'relevance' philosophy in tertiary education that the average student of chemistry gets little inkling from his teachers of the vast practical importance of disperse systems in industry. 1978 *New Scientist* 21 Sept. 850/2 'Relevance' in research implies both social efficacy and psychic commitment by the research worker.

relevancy. Add: **1. b.** (Later examples.)

Now less common than *relevance* in general use.

1961 *Jrnl. Physical Chem.* LXV 317/1 We are reporting these investigations..because of their relevancy to problems of the study of apparently simple exchange reactions of chlorine. 1980 *Times Lit. Suppl.* 30 May 600/1 A tendency to confuse relevancy with recency.

2. A relevant remark. (*Nonce use* influenced by Irrelevancy.)

1895 'Mark Twain' in *N. Amer. Rev.* July 20 Conversations consisted mainly of irrelevancies, with here and there a relevancy, a relevancy with an embarrassed look, as not being able to explain how it got there.

relevant, *a.* Add: **1.** (Later examples.) Cf. *relevance.

1948 D. Cecil *Two Quiet Lives* II. 140 To learn everything that could possibly be thought relevant to the subject. 1969 *Harper's Mag.* Nov. 86 Either we can commit ourselves to changing the institutions of our society that need to be changed, to make them—to use a term which I hate—'relevant'..or we can sit back and try to defend them. 1970 *N.Y. Times* 1 July 44 Museums should have a more involved or relevant public role. 1976 *Listener* 20 May 627/3 The ultimate sin of the broadcaster is to keep off the air, because of his political or social prejudices, subjects which are relevant and significant. 1978 S. Braden *Artists & People* p. xvii, What actually makes a work of art relevant to people? It has been said that relevance is achieved when artists meet the real observations of their public.

relevé (rəlәve). [Fr., lit. 'raised up'.] **1.** = Remove *sb.* 2 c.

1825 Lady Blessington *Jrnl.* Dec. in E. Clay *Lady B. at Naples* (1979) 141 The fragments of *entrées* and *relevés*. 1846 *Jewish Manual, or Pract. Information Jewish & Mod. Cookery* p. xv, *Releves*, or *Removes*, are top and bottom dishes, which replace the soup and fish. 1889 [see *main course* s.v. *main *a.* 11]. 1906 *Mrs. Beeton's Bk. Househ. Managem.* lxii. 1668 *Relevé*.., the remove. A course of a dinner, consisting of large joints of meat, four-footed game, and sometimes joints of fish. 1961 Froud & Turgeon tr. *Larousse Gastronomique* 805/2 Remove. Relevé—Dish which in French service *relieves* (in the sense that one sentry relieves another) the soup or the fish.

2. *Ballet.* (See quot. 1957.)

1930 Craske & Beaumont *Theory & Pract. Allegro in Classical Ballet* 66 Execute two *Petits Battements* with a *relevé* and *dégagé*. 1953 *Ballet Ann.* VII. 83 That infinitesimal moment of holding the breath in a *relevé* on point in a held pose. 1957 G. B. L. Wilson *Penguin Dict. Ballet* 227 *Relevé* or *temps relevé*, lit. a lifted step. The raising of the body on half- or full-point or points. 1976 *New Yorker* 29 Mar. 92/3 He has an immobile thick torso, a heaviness in plié and relevé.

rele·vel, *v.* [Re- 5 a.] *trans.* To level again.

1928 *Daily Tel.* 14 Aug. 14/3 Gutters have been filled in, dangerous curb-stones removed, level crossings relevelled. 1975 J. B. Harley *O.S. Maps* i. 7 Each area is relevelled in a cyclic system,.. the interval depending on the character of the country. Mountain and moorland areas are relevelled every forty years.

relexification (rīːleksifikēɪ·ʃən). *Linguistics.* [f. Re- 5 a + Gr. λέξι-ς word + -fication.] The process of replacing a word or group of words in one language with a corresponding word or group of words from another language, without grammatical adjustment of the items introduced.

1962 W. A. Stewart in F. A. Rice *Stud. Role Second Languages in Asia, Africa, & Lat. Amer.* 46 The vocabulary derived from one source language has been largely replaced ..by a more recent vocabulary derived from another language, while the original grammatical structure is preserved... This process of relexification seems to be the converse of restructuralization. 1965 *Amer. Speech* XL. 172 William Stewart has hypothesized that this Portuguese-based pidgin was re-lexified to yield the English, Dutch, and French pidgins which are the progenitors of the Creoles. This idea of a re-lexification which left the original syntax almost intact is a very attractive one. 1968 *Word* XXIV. 263 The effect of this process is a kind of continuous and massive 'relexification'. 1971 [see *initiator]. 1972 J. L. Dillard *Black English* iii. 121 Within the Negro community, the use of Africanisms has been demonstrably larger in the past; allowing for relexification, we can still see a great deal of indirect influence. *Ibid.* 303 Relexification is the replacement of a vocabulary item in a language with a word from another, without a change in the grammar. If I

change the sentence I am very tired to I am très tired I have in a sense relexified the English sentence. A 'Latin' sentence like ego amo tu is of course simply a relexification of I love you with Latin words. 1974 R. A. Hall *External Hist. Romance Lang.* 33 According to certain..theories, these two varieties..would have been the predecessors of West African Pidgin Portuguese, from which all other modern pidgins and creoles would have sprung by a process of 'relexification'. 1975 *Language* LI. 685 If all of a group of PC's [*sc.* pidgin or creole languages], such as those usually embraced in relexification hypotheses.., have a common ancestor, then the extension of the use of 'mouth' in that ancestor would account for 'mouth' having the added sense in all those PC's.

relexify (rīle·ksifəi), *v.* *Linguistics.* [f. as prec. + -fy.] *trans.* To introduce into (a language) vocabulary taken from another language without grammatical adjustment of the items introduced. Hence **rele·xified** *ppl. a.*

1962 W. A. Stewart in F. A. Rice *Stud. Role Second Languages in Asia, Africa, & Lat. Amer.* 46 If a language A can be shown to derive its vocabulary from language B and its grammatical structure from language C, then language A can be both 'restructured B' and 'relexified C' at the same time. *Ibid.*, It is ..possible to consider them all..as relexified forms of some prior language. 1965 *Orbis* XIV. 521 We know that the Philippine creoles were also relexified very rapidly. 1965, 1972 [see prec.]. 1972 J. L. Dillard *Black English* iii. 122 Hawaiian Pidgin English *pau* (relexified with a Hawaiian word), Melanesian Pidgin English *finish* constitute other relexifications of *cabá*. 1979 *Amer. Speech* LIV. 296 A uniform 'plantation creole'..later 'relexified' into something nearer to the speech of whites.

reliability. Add: **2.** *Statistics.* The extent to which a measurement made repeatedly in identical circumstances will yield concordant results.

1904 *Amer. Jrnl. Psychol.* XV. 238 The reliability with which any system of measurement represents any particular form of intelligence. 1925 F. C. Mills *Statistical Meth.* xvi. 561 By the study of successive samples, and by the testing of the subordinate elements in a given sample when broken up into significant sub-groups, much more may be learned as to the reliability of a given measure..than by unquestioning acceptance and uncritical employment of the usual mathematical formulas for probable errors. 1938 A. E. Waugh *Elem. Statistical Meth.* vii. 138 We can increase the reliability of the mean by studying more cases, and..the reliability is greater also when the variation among the original figures is small. 1950 J. P. Guilford *Fundamental Statistics in Psychol. & Educ.* (ed. 2) xvii. 473 Tests of differences and correlation coefficients may often prove to be insignificant merely because the measures used were lacking in reliability. 1978 R. J. Jesson *Statistical Survey Techniques* i. 15 In considering reliability we shall be referring to a measure of the closeness of each observation to its own average over repeated trials.

3. *attrib.*, as *reliability engineer, race, test, trial*; **reliability coefficient**, any of various measures of statistical reliability; freq. the coefficient of correlation between two sets of measurements made of the same set of quantities.

1910 C. Spearman in *Brit. Jrnl. Psychol.* III. 281 A very convenient conception is that of the 'reliability coefficient' of any system of measurements for any character. By this is meant the coefficient between one half and the other half of several measurements of the same thing. 1930 *Psychol. Rev.* XXXVII. 140 The reliability coefficient of a variable, *X*, is a special type of correlation coefficient which indicates the degree to which individuals systematically differ from each other in the trait as measured. 1954 *Psychol. Bulletin* LI. 229/1 The several types of reliability coefficient do not answer the same questions and should be carefully distinguished. 1972 *Jrnl. Social Psychol.* LXXXVII. 48 The split-half method was employed and resulted in a reliability coefficient for the instrument of ·83. 1969 *Word Study* Apr. 3/2 The reliability engineers, on the other hand, did not want to avoid taboo words; they were chiefly interested in alarming the program to potential failures. 1977 *Chicago Tribune* 2 Oct. XII. 57/4 (Advt.), Reliability Engineer, to direct and perform component reliability studies, coordinate with system requirements, and function as reliability consultant. 1907 *Strand Mag.* Nov. 491/2 A result extraordinarily interesting should be worked out from this thousand-mile [car] reliability race. 1904 *Technics* Aug. 114 As a 'reliability test', the car was driven from London to Newport (Mon.), a distance of about 160 miles. 1929 *Even. News* 18 Nov. 16/4 [He] crashed on his motor-cycle while taking part in a reliability test on Portsdown-hill. 1902 *Car* 3 Sept. 43/1 The cars entered for the Automobile Club's Reliability Trials which are being held this week began to arrive at the Crystal Palace at a very early hour. 1904 *To-Day* 18 May 58/2 The Automobile Club has arranged to hold a reliability trial for motor boats. 1963 P. Drackett *Motor Rallying* i. 9 But the true progenitor of the rally was the reliability trial. 1970 *Which?* July 199/1 We have not done any extended reliability trials on the single samples of television sets we tested.

reliable, *a.* Add: **b.** (Further examples.) Also, a reliable person or animal.

1908 [see *peacherino]. 1910 W. M. Raine *Bucky O'Connor* (1920) ii. 20, I hate to have you take that gun, though. I meant to run you down with that same old Colt's reliable. 1911 R. D. Saunders *Col. Todhunter* xii. 171 'You never can tell about these old reliables,' said Tom. 'Solomon might take it into his head to get frisky any minute.' 1950 *Western Folklore* Apr. 138 The cowboy's six-shooter speaks a language universally understood. Familiar epithets for the revolver were *equalizer, shootin' iron,..Old Reliable*. 1972 *Village Voice* (N.Y.) 1 June 19/3 One of the Governor's old reliables, Assemblyman Robert Kelly, sponsored the bill in the lower house and told everyone how great it was. 1970 E. Snow *Red China Today* (1976) 33 The 'three-way alliance' of mass organizations, Party 'reliables', and army political work teams which had completed the Party purging.

2. *Statistics.* Yielding concordant results when repeated.

[1892 *Analyst* XVII. 228 When the Babcock test is made according to the instruction given with the machine, strictly reliable results are obtained.] 1932 *Jrnl. Gen. Physiol.* XVI. 23 Under such conditions it might be expected that volumetric measurements be somewhat less reliable than in the simple case first examined. 1942 J. P. Guilford *Fundamental Statistics in Psychol. & Educ.* xiv. 273 By a perfectly reliable test, we mean one that is free from errors of measurement. 1970 D. W. Matheson et al. *Introd. Exper. Psychol.* ii. 26 A sampling technique is reliable if several samples from the same population yield similar data. *Ibid.* vi. 66 If a test is reliable, a subject will receive approximately the same score each time he takes the test.

relic. Add: **2. c.** An old person. *colloq.*

1869 'Mark Twain' in *Buffalo Express* 21 Aug. 1/3, I came upon a noble Son of the Forest sitting under a tree, diligently at work on a bead reticule... I addressed the noble relic as follows. 1902 —— in *Harper's Mag.* Dec. 15/1 'How much of it can you two undertake?' 'All of it!' burst from both ladies at once... 'You do ring true, you brave old relics!' 1981 B. Healey *Last Ferry from Lido* ix. 161 So far as he's concerned the Ca' Silvestro and the old lady are just a pair of ancient relics.

4. c. phr. *relic of barbarism*, a survival or reminder of bad conditions or practices.

1852 *Harper's Mag.* Dec. 126/2 Railing against the church, against society, against institutions, against 'relics of barbarisms'. 1870 J. H. Newman *Grammar of Assent* iv. 75 When Mr. Wilberforce, after succeeding in the slave question, urged the Duke of Wellington to use his great influence in discountenancing duelling, he could only get from him in answer, 'A relic of barbarism, Mr. Wilberforce'. 1919 W. T. Grenfell *Labrador Doctor* iv. 68 After giving a talk on psychical influence he had the jacket removed as 'a relic of barbarism'. 1921 T. Wolfe *Let.* 2 Sept. (1956) 16 This 'point system' of selecting teachers is a relic of barbarism.

d. *Linguistics.* The survival of an archaic form; an instance of this. (See also *relict *sb.* 6.)

1943 *Language* XIX. 257 Nowhere..was there an indication of the genuine vitality of this set of suffixes, which, divested of any specific function, had become mere meaningless relics. 1951 *Amer. Speech* XXVI. 252 The occurrences of *clabbered milk* in the northern counties of the state are probably explained as a sporadic relic. 1962 *Ibid.* XXXVII. 170 In the word *one*, the mid-central vowel is occasionally replaced by /u/—a relic usage evidently related to the pronunciation of *home* as /hum/.

3. c. *Biol.* A relict species.

1947 [see *epibiotic *a.* and *sb.* 1]. 1965 B. E. Freeman tr. *Vandel's Biospeleology* vii. 70 *Troglochaetus* would seem to be a marine relic. 1974 *New Phytologist* LXXIII. 974 Thistles, mulleins and foxgloves..appear as the stemless relics of the pachycaul inflorescences.

7. (in sense *4 d) *relic form*; **relic area**, a region noted for the survival of old or otherwise archaic language forms.

1953 *Language Learning* IV. 104 Relic areas, on the other hand, are those whose geographical or cultural isolation, and relative lack of prestige, has caused the retention of older forms or prevented the spread of forms characteristic of these areas. 1962 *Amer. Speech* XXXVII. 171 The regional words used by the Ocracokers are the regional words of the North Carolina coast, especially the relic area which lies around Albemarle Sound. 1972 H. Kurath *Studies Area Linguistics* i. 2 He [*sc.* the area linguist] will reserve judgment and recommend further investigation when the linguistic variants exhibit a complicated and apparently erratic dissemination as in certain transition zones or relic areas. 1933 Relic form [see *hyperform* s.v. *hyper- IV]. 1951 *Amer. Speech* XXVI. 13 The preservation of relic forms is made possible by geographical or cultural isolation. 1972 M. L. Samuels *Linguistic Evolution* vi. 92 The receiving system itself becomes *less* divergent from its neighbour than before, retaining only relic forms from its antecedent.

B. *adj. Geogr., Geol.*, and *Biol.* = *relict *a.* 4.

1894 J. Geikie *Great Ice Age* (ed. 3) xxxi. 488 In many of the Swedish lakes there occur certain forms of life which appear to be a relic-fauna of the Yoldia Sea. 1926 W. H. Twenhofel *Treat. Sedimentation* v. 369 The Salton Sink of California is probably an example of a relic sea which appears to have been severed from the Gulf of California in the building of the Colorado River delta. 1940 *Jrnl. Genetics* XL. 72 At this stage it is usual for some or all of the chromosomes to show 'relic' coils or spirals. These coils are to be regarded as the remains of the spirals of the previous division. 1966 Mrs. L. B. Johnson *White House Diary* 2 Apr. (1970) 379 He described the 'relic forest' of maple, quaking aspen, Douglas fir, and ponderosa pine with huge trunks. By some strange mystery of nature, they have survived from a much earlier time, when the climate was different here. 1976 H. M. French *Periglacial Environment* v. 95 The presence of obviously relic pingos..in present-day periglacial environments complicates attempts to identify the conditions for present-day pingo growth. 1978 *Nature* 7 Sept. 19/1 Appropriate physical conditions for the origin of life could exist on the relic regolith grains.

relict, *sb.* Add: **1.** (Further example.)

1884 'Mark Twain' *Huck. Finn* xxxvii. 384 Things that was valuable..on account of them being relicts.

3. b. Also *transf.* of a person. (Further example.)

1928 *Daily Express* 3 July 10/2 Our British boards of railway directors are like an ante-room to a museum. They are crowded with relicts of the easy pre-war age for whom the world has never changed.

d. *Biol., Geogr.*, and *Geol.* A relict species, structure, etc.

1905 F. E. Clements *Res. Methods in Ecol.* 321 *Relict*, a species belonging properly to an earlier type of succession than the one in which it is found. 1950 *Jrnl. Ecol.* XXXVIII. 294 A few relicts of the former open fen are to be seen in the form of scattered plants of *Cirsium palustre*..and *Angelica sylvestris*. 1971 *Nature* 5 Feb. 377/2 Rare plant species are often relicts surviving in restricted ecological niches. The opportunities for these species to spread to other, suitable habitats are minimal. 1977 A. Hallam *Planet Earth* 17/2 The breccias are revealing to us the relicts of an original lunar crust that formed and was reconstituted several times.

6. *Linguistics.* Used *attrib.* or in *Comb.* to denote language or vocabulary which is a survival of otherwise archaic or old forms. Cf. also *relic 4 d, 7.

1934 Priebsch & Collinson *German Lang.* vii. 364 Often we find a wedge of linguistic innovation along the rivers and highways with relict-areas preserving ancient forms on the high moors and along the wooded hills. 1947 *Ibid.* (ed. 2) i. 14 The Logudoresian dialect retains certain relict-words with affinities in Basque. 1948 *Trans. Phil. Soc.* 1947 14 It now appears that the original Germanic language of the Low German area was not in any essential matter distinguished from Frisian, but that it was afterwards High Germanized, leaving Frisian as a relict language of its original state. 1963 H. C. Darby in Brown & Foote *Early English & Norse Stud.* ii. 9 Another example of relict names is found on Dunsmore Heath.

relict, *a.* Restrict † *Obs.* to senses in Dict. and add: **4.** *Geogr., Geol.*, and *Biol.* Surviving from a previous age or in changed circumstances after the extinction or disappearance of related forms or structures.

In origin prob. an *attrib.* use of the sb. rather than a revival of the adj. (cf. quot. 1901).

1898 J. Geikie *Earth Sculpture* xvi. 274 The direction, and to a large extent the shape or form of relict mountains, are thus mainly determined by the geological structure. 1901 *Ann. & Mag. Nat. Hist.* VII. 315 Those [animals] remaining in the old place formed a zonally-disposed relict-fauna. 1932 E. W. Sinnott *Plant Sociology* iv. 63 Besides these characteristic relict species there are others whose fidelity is due to a narrowly specialized adaptation to definite physico-chemical relations of the habitat. 1939 W. H. Twenhofel *Princ. Sedimentation* xii. 459 Relict seas are bodies of water that have become separated from the parent body by diastrophic, depositional, or volcanic causes. *Ibid.*, Well-known relict seas are..the Caspian Sea, Lake Nicaragua, and Lake Baikal. 1945 M. J. D. White *Animal Cytol. & Evolution* xiii. 293 [*Saga serrata*] seems to be a 'relict' species, since the localities in which it occurs are very discontinuous and the individual populations of very small size. 1954 W. D. Thornbury *Princ. Geomorphol.* xvi. 413 Relict features attributable to former existence of periglacial conditions have been described at many places in Europe and North America. 1974 J. D. Millman *Recent Sedimentary Carbonates* I. vii. 221 Coralline algae in continental shelf sediments may be relict but others are modern. 1977 *Birds* Winter 19/1 Fowlmere is a small, relict fen, lying in a hollow of the Gog Magog hills. 1978 *Sci. Amer.* Sept. 111/1 The lobe fins were far less successful as fishes (they survive only as lungfishes and a few relict forms). 1978 T. Rowley *Villages in Landscape* ii. 49 Village plans may include relict features of early defensive structures, such as the alignment of Roman town walls.

5. *Astr.* Remaining from the 'big bang'.

1971 *Nature* 3 Sept. 36/2 The discovery in 1965, by Penzias and Wilson, of background radiation which may well be relict radiation from this fireball. 1978 *Sci. Amer.* July 54/3 Encounters between cosmic rays and photons of the relict radiation would severely drain the energy of the cosmic rays above some energy threshold.

relief[2]. Add: **3. a.** (Further examples.) Also, financial assistance afforded to those in need by the state under other legislative provisions.

1921 *Daily Colonist* (Victoria, B.C.) 22 Mar. 1/3 More than $400,000 has now been expended by the city in providing relief, it will be pointed out. 1957 R. Hoggart *Uses of Literacy* ii. 40 At my grandmother's we were not living 'on relief' but, like many around us, we were 'a bit short'. 1965 Mrs. L. B. Johnson *White House Diary* 12 Aug. (1970) 309 Their world is so narrow, and in their homes—frequently broken homes with one parent or an aged grandparent on relief—the vocabulary is often limited to grunts or profanity. 1966 G. Jackson *Let.* 3 Mar. in *Soledad Brother* (1971) 95 I've heard men brag about..taking money from black women who are on relief. 1978 S. Brill *Teamsters* vii. 285 He didn't make a lot... But we never went on relief.

5. b. (Earlier example.)

1709 Swift in *Tatler* 10 Sept., Little Parson Dapper, who is the Common Relief to all the lazy Pulpits in Town.

c. (Earlier example.)

1788 W. Dyott *Diary* 1 Sept. (1907) I. 53 We had ninety dishes, fifty-five the first course and relieves, and thirty-five the second course.

7. Delete *rare* and add: *spec.* remission of income-tax due on a proportion of earned income.

1889 A. Chapman *Income-Tax Grievances & their Remedy* i. 9 Appellants who..prove..that their profits have not been equal to the sum at which they were assessed and have paid duty, are entitled to relief and repayment of tax. 1916 [see *income-tax]. 1931 *Economist* 28 Feb. 456/2 A resident British holder of 'Kaffirs' is subject to income tax on dividends at the full rate of British tax (4s. 6d. in the £), less Dominion tax relief (at present 2s. 3d., namely, half the

British rate). He is thus liable at the 'reliefed' rate of 2s. 3d. 1969 *Times* 2 May 25 Tax relief can reduce the cost of your investment by up to £16.10.0 per £100 of premium. 1972 *Accountant* 17 Aug. 191/1 The strict ban against relief for part-time directors and employees. 1973 P. O'Donnell *Silver Mistress* iv. 72 If it's a phony charity account..they probably get tax relief. 1977 *Money Which?* Mar. 123/3 If you become entitled to tax relief on a new outgoing or allowance,..tell the taxman straightaway.

9. b. *relief agency, committee, fund* (earlier and later examples), *organization, party, team, work, worker*; **relief road**, a road designed to divert traffic from congested areas; **relief roll** *U.S.*, a list of people receiving state relief; **relief ticket**, a small sum of money given to alleviate hardship; **relief well**, a hole drilled to intersect an oil or gas well in which there is a fire or a blow-out, so as to provide a route for water or mud to stop it.

1951 T. Sterling *House without Door* i. 7 A Jewish relief agency..which trained refugee Jews in manual skills. 1971 Piven & Cloward in M. Edelman *Polit. Lang.* (1977) iii. 53 Relief agencies are..compelled to invent rituals of degradation and to subject their clientele to them. 1842 *Picayune* (New Orleans) 23 Jan. 2/5 The Relief Committee of the Firemen's Charitable Association, will meet..at the Firemen's Insurance Office. 1862 *Times* 14 Apr. 11/4 Proud men..who go before 'relief committees' and submit to be questioned about their wants. 1892 J. C. Harris *On Plantation* 139 Where they lived remote from the relief committees, the families of the soldiers were not so well provided for as they had a right to expect. 1842 S. Bamford *Passages in Life of Radical* II. xxi. 104 He had some money in hand belonging to the relief fund. 1863 *Observer* 26 Apr. 5/4, I cannot..recommend too strongly..to your lady readers' kind consideration, the 'Cracow Ladies' Committee', who are connected with the 'Ladies' Relief Fund Committee' in London. 1914 Relief fund [see *flag-day* s.v. *flag *sb.*[4] 7]. 1952 M. McCarthy *Groves of Academe* (1953) vi. 120 We're not yet relief organisations, you must admit. 1974 *Whig-Standard* (Kingston, Ontario) 11 Jan. 7/1 A..graduate..with many years of experience with relief organizations. 1978 *Internat. Relations Dict.* (U.S. Dept. State Library) 15/2 U.S. agricultural surpluses are donated to 'friendly governments' through non-profit relief organizations. 1933 J. Buchan *Prince of Captivity* I. iii. 85 Now he has gone and lost himself and..they're talking of a relief party. 1940 *Gloss. Highway Engin. Terms (B.S.I.)* 9 *Relief road*,..a road to enable through traffic to avoid congested areas or other obstructions to movement. 1959 *Oxford Mag.* 26 Feb. 276/1 A relief road is invented, and it must then be guessed how much of the flow along each existing route will be diverted into it. 1960 *Oxford Mail* 10 Oct. 4/6 The idea of relief roads to link the suburbs with the centre is one that deserves to be considered. 1976 *S. Wales Echo* 27 Nov. 5/7 At present it is planned to join the relief road to Hirwaun Road near Tudor Terrace. 1937 C. Himes *Black on Black* (1973) 127 Remembering suddenly the time the Belle Vernon Milk Company dumped hundreds of gallons of milk into the gutters of Cedar Street when the relief rolls in Cleveland were the highest they'd ever been. 1938 *Sun* (Baltimore) 16 Apr. 8 The President himself has said that road building will 'take very few people off the relief rolls'..Spending should be limited to relief. 1976 *National Observer* (U.S.) 17 Jan. 5/2 Americans who are elderly, blind, disabled, or who have impoverished dependent children—generally, Americans who are on the relief rolls. 1977 M. Edelman *Polit. Lang.* v. 80 Social work counseling..apparently has little or no effect on client satisfaction, behavior, or the size of relief rolls. 1970 *Guardian* 28 Jan. 3/8 Relief teams are racing against the rainy season, due in the Nigerian delta. 1972 *Ibid.* 22 May 10/6 There is a drought at present in the villages... The relief teams have moved in. 1848 *United Irishman* 20 May 224/2 A beautiful and fertile island..became gradually poorer and poorer... Millions of men, who toiled their lives through from morning to night, found at length they had no *rights* but a right to public alms, and had realized, with all their toiling, nothing but the chance of a relief-ticket. 1976 J. O'Connor *Eleventh Commandment* i. 21 He got a bit of beer money and came away laughing; relief tickets, they call them, and worth a couple of quid... The shops that took relief tickets would always overcharge the working class. 1925 A. B. Thompson *Oil-Field Explor. & Devel.* I. vii. 299 On reaching a rich gas sand the Gleason well ran wild at a rate of about 15,000,000 cub. ft. per day... A relief well failed to effect its object, although sunk only 135 ft. away. 1939 D. Hager *Fund. Petroleum Industry* ix. 222 The relief well was deflected so that it reached the burning hole at a depth of 7046 ft. 1975 *Petroleum Rev.* XXIX. 238/3 Just one month after the fire had started, the first relief well commenced injecting sea water. 1979 *Tucson* (Arizona) *Citizen* 3 Oct. 3 A/3 Two relief wells are being drilled at an angle to Ixtoc 1 in an effort to rechannel the gusher to a controlled well. 1879 *Good Words* 495/2 There is another young engineer superintending his dam, a relief work about two miles long. *Ibid.* 566/2 A collector, after having tried to induce some Mhars to go to a famine relief work close by, who refused, looking over a wall saw two of them devouring a dead dog. 1895 Kipling *Day's Work* (1898) 170 They've gone as far as to admit extreme local scarcity, and they've started relief-works in one or two districts. *Ibid.* 202 Then in the evening he pitches in a twenty-page demi-official to me, saying that the people where he is might be 'advantageously employed on relief-work'. 1915 R. Fry *Let.* 28 July (1972) II. 388 I've been a long time in France—went to see the Quaker relief work. 1921 *Daily Colonist* (Victoria, B.C.) 22 Mar. 1/3 At the end of its resources as far as relief work is concerned, the city of Vancouver will urge the Dominion and Provincial Governments to carry a burden which it believes should rightfully be borne by them. 1938 R. D. Finlayson *Brown Man's Burden* 16 They were relief workers and flax-cutters, working hard and making good money. 1973 *Guardian* 9 June 13/4 Latham belongs to that new breed of white man—the relief worker.

relief[3]. Add: **4.** *relief-carving, construction, decoration, panel, -polish* (sb. and vb.), *-polishing, process* (examples); **relief map**, a map that indicates the relief of the land, either by the analogous form of its surface or by a system of colouring, shading contour lines, or the like; **Relief nib** [*Relief*, proprietary name], a manufacturer's name for a special kind of nib.

1892 E. Rowe *Hints on Chip-Carving* 60 In relief-carving the teaching must be individual, and consequently fewer students can be taught by one teacher. 1970 *Oxf. Compan. Art* 960/2 Such sculpture is not properly relief carving (e.g. the figures of gods and giants on the Great Altar of Zeus at Pergamum). 1962 *Times* 28 Feb. 5/1 At the same gallery Mr. Michael Rothenstein shows a small collection of his new relief-constructions. 1960 *Connoisseur's Handbk. Antique Coll.* 236/1 There are various ways of producing relief decoration: by freehand modelling, free-incising or piercing or, more frequently, by pressing soft clay in plaster moulds. 1960 R. G. Haggar *Conc. Encycl. Cont. Pott. & Porc.* 380/1 *Relief decoration*, figures, flowers, and ornamental decoration formed in the mould, or moulded or modelled separately and luted to the ware with slip. 1876 *Nature* 11 May 23/1 Relief-maps and Models illustrating Geological Phenomena all over the world. 1880 'Mark Twain' *Tramp Abroad* xxxiii. 358 He showed us the whole thing on a relief map. 1934 J. Bygott *Introd. Map Work & Pract. Geogr.* vi. 43 After inspecting even a small-scale relief map of Northern England..we realise that the longest rivers flow from the eastern slopes towards the North Sea. 1971 R. W. Purton *Let's look at Maps & Mapmaking* 16 It is possible to buy relief maps moulded in plastic, on which the physical features are raised as..on a model. [1908 *Trade Marks Jrnl.* 4 Mar. 347 Relief..Pens..Esterbrook Steel Pen Manufacturing Company..New York..21st November 1907.] 1920 A. Huxley *Limbo* 144 He selected a pen—with a Relief nib he would be able to go on for hours without getting tired—and a large square sheet of writing-paper. 1938 E. Bowen *Death of Heart* II. iii. 214 Today..she made the following purchases... Half a dozen Relief nibs. 1960 *Twentieth Century* Oct. 343 The pen tray filled with compact sheaves of new relief nibs. 1937 *Burlington Mag.* Feb. 59/2 The relief-panels of the Pisa pulpit. 1961 *Times* 21 Dec. 3/3 He makes relief-panels and sculpture from charred wood. 1933 Greaves & Wrighton *Pract. Microsc. Metallogr.* (ed. 2) xiv. 237 Relief polish is obtained..so that micro-constituents are visible under the microscope. 1968 E. Stach in Murchison & Westoll *Coal & Coal-Bearing Strata* i. 4 Nowadays coals are normally relief-polished and then examined under vertically incident light using immersion objectives. 1902 C. Salter tr. *H. F. V. Jüptner's Siderology* II. iii. 121 Relief polishing shows up the hardest constituents, especially cementite, in relief. 1924 Greaves & Wrighton *Pract. Microsc. Metallogr.* vi. 48 Both 'relief polishing' and 'polish attack' may be used to display the microstructure of steels containing cementite. 1889 *Cent. Dict.*, *Relief processes*, those processes in mechanical or 'process' engraving by which are produced plates or blocks with raised lines. 1940 *Chambers's Techn. Dict.* 713/2 *Relief process* (*Photog.*), any colour process using matrices. 1965 Zigrosser & Gaehde *Guide Coll. Orig. Prints* iv. 53 *Relief processes*, a general term that includes woodcuts, wood engravings, linoleum cuts, [etc.].

reliefer[2] (rīlī·fəɹ). [f. Relief[2] + -er[1].] One who receives state relief (sense *3 a).

1936 *Harper's Mag.* Jan. 203/2 Reliefers don't of course live by themselves and form a compact group with relief as their only topic. 1938 *Sun* (Baltimore) 6 Sept. 8/4 Baltimore streets and alleys could do with a little attention from 'grateful reliefers'. 1947 *Ten Eventful Years* IV. 629/1 *Reliefer*, an unemployed person on federal government relief rolls. 1973 B. Broadfoot *Ten Lost Years* xx. 225 No, we never thought of the poor people. The reliefers.

relieve, *v.* Add: **I. 3. b.** Also *refl.*, to defecate or urinate, and *fig.*

1931 S. Tremayne *Trial A. A. Rouse* 184, I wanted to relieve myself. 1952 Bible (Rev. Standard Version) *I Sam.* xxiv. 3 And he came to the sheepfolds by the way, where there was a cave; and Saul went in to relieve himself. 1956 H. Gold *Man who was not with It* (1965) xii. 99 There's a stomach ache of music..; it churns and stretches, trying to relieve itself. 1960 V. Nabokov *Invitation to Beheading* xiv. 141 The bliss of relieving oneself, which some hold to be on a par with the pleasure of love. 1961 *Encounter* Feb. 25/1 It [*sc.* a kitten] learned to go down into the alley to relieve itself in the dirt there. 1977 *Sunday Times* 30 Jan. 30/3, I urgently wished to be alone to relieve myself (a serious problem in winter orienteering).

5. c. (Later examples.) Also, *euphem.*, to dismiss *from* a position, to deprive *of* membership.

1875 'Mark Twain' in *Atlantic Monthly* June 733/1 He was 'relieved' from duty when the boat got to New Orleans. Somebody expressed surprise at the discharge. 1952 E. O'Neill *Moon for Misbegotten* III. 111 He relieves her of the pitcher and tumblers as she comes down the steps. 1972 *Newsweek* 10 Jan. 11/3 Its present chief..has expelled some 3,000 members (Lascorz was relieved of his membership during a previous clean-up in 1969).

relie·vedly, *adv.* [f. Relieved *ppl. a.* + -ly[2].] In a relieved manner, with relief from anxiety.

1911 R. Brooke *Let.* 22 Dec. (1968) 327, I rather grasp relievedly at them, after I've beaten vain hands in the rosy mists of poets' experiences. 1925 *Glasgow Herald* 1 Aug. 6/3 The country relievedly witnesses the passing of the crisis. 1951 M. Leinster in D. Knight *100 Yrs. Sci. Fiction* (1969) 200 Sometimes he was able to thrust aside..the fact that Jane was dead. Now he grappled relievedly with the question of his sanity or lunacy.

reliever. Add: **1. d.** *N. Amer.* A pitcher who relieves the opening pitcher in a baseball match.

1967 *Boston Herald* 8 May 16/2 Fregosi homered in the fifth..off reliever Bob Humphreys. 1976 *Washington Post* 19 Apr. D1/4 Los Angeles chased reliever Roger Moret in the seventh with a five-run explosion. 1979 *Arizona Daily Star* 5 Aug. C2/2 Craig Swan combined with reliever Neil Allen on an eight-hitter as New York stopped Montreal's five-game winning streak.

relieving, *ppl. a.* Add: **2. a.** *relieving officer* (earlier and later examples).

1836 *Falmouth Pkt.* 23 Sept. 5/2 Application for relief is made to the relieving officer. 1841 *Punch* 23 Oct. 170/2 The family..told me they were literally dying of hunger, and that they had applied to the vestry, who had referred them to the..relieving officer. 1850 C. Kingsley *Alton Locke* II. xiv. 210 In the midst of all the rout, the relieving officer stood impassive, jotting down scraps of information. 1980 G. M. Fraser *Mr American* ix. 167 She has an order for medical attendance from the relieving officer.

relievo[2] (rili·vo). Also **relievio.** [prob. f. Relieve *v.* + *-o[2].] A children's seeking game in which a captured player may be released by another member of his or her side; the call effecting the release.

1888 S. O. Addy *Gloss. Words Sheffield* 296 *Bedlams* or *relievo*, a game played by boys. *Ibid.* 297 If..one of the boys out at field runs through the *den* shouting '*Relievo*', without being caught by the *tenter*, the prisoner is allowed to escape. 1912 J. Stephens *Crock of Gold* v. 39 'It's a nice game,' said the Leprechaun, 'and so is Cap-on-the-back, and ..Relievo, and Leap-frog.' 1913 —— *Here are Ladies* 261 Tip-and-Tig, Horneys and Robbers, Relievo we played. 1969 I. & P. Opie *Children's Games* ii. 110 Names which reflect the rescue element are: 'Release', 'Releaster', 'Reliev-i-o', 'Tig and Relievo', [etc.]. *Ibid.* iv. 172 In Scotland, Wales, and the northern half of England, 'Relievo' is the principal seeking game with two sides. *Ibid.* 173 Commonly it is enough for the 'releaser' to shout 'Relievo' or 'Rallio' or perhaps 'Bish-Bash' as he rushes through the den. 1970 *Daily Colonist* (Victoria, B.C.) 19 Aug. 2/1 Summer streets full of youngsters playing Hoist The Sails, Relievo or Giant Step. 1974 *Amer. Speech* 1971 XLVI. 83 Line and running games: crack-the-whip, fly-the-whip, follow-the-leader, leap frog, redman, red rover, rolla-rolla, relievio, [etc.].

relifting, *vbl. sb.* (Further example.)

1904 H. Belloc *Avril* III 105 The repose and the re-lifting of musical notes.

re·light, *sb.* *Aeronaut.* [f. the vb.] A re-ignition of a jet engine in flight. Usu. *attrib.*

1945 J. Grierson *Jet Flight* iv. 101 The fuel should be shut off for a minute or so in order to blow surplus fuel out of the engine before the relight. 1955 Mangham & Peace *Jet Engine Man.* vi. 100 The relight button..enables the ignition circuit to be energised without the starter motor. 1976 B. Jackson *Flameout* (1977) II. iv. 69 A thorough examination of the relight systems, every fuel valve, and every fuel feed pipe. 1977 [see *relighting *vbl. sb.*]. 1977 *R.A.F. News* 11–24 May 11/1 Flt Lt Harry Apiafi..suffered a flame out in the Canberra's starboard engine... While attempting a relight, Harry lost the other engine as well.

reli·ghting, *vbl. sb.* [f. Relight *v.* + -ing[1].] The action of the vb.; *spec.* in *Aeronaut.*, the re-ignition of a jet engine in flight.

1955 Mangham & Peace *Jet Engine Man.* v. 98 Re-lighting is more positive at low engine-windmilling speeds and at lower altitudes. 1960 H. Zeffer *Princ. & Pract. Aircraft Electr. Engin.* xvii. 517 A typical situation in which re-lighting would be necessary is one where a jet engine has been stopped by an accretion of ice entering the air intake. 1965 *Gloss. Mining Terms* (B.S.I.) VII. 10 *Re-lighting station*, a place in a mine at which safety lamps can be re-lighted under controlled conditions. 1977 *R.A.F. Yearbk.* 34/1 The relighting drills cover three contingencies: the 'hot' relight speaks for itself, the unassisted relight involves diving the aircraft to get the engine rpm sufficiently high for a cold relight attempt, and the assisted relight uses the GTS in exactly the same way as starting the engine.

religieuse. Add: Also 8 *erron.* **religieux** (*pl.*).

1. (Earlier and later examples.)

1694 Ld. Perth *Let.* 17 Sept. (1845) 43 Lady Lucy is a most excellent religieuse. 1777 P. Thicknesse *Year's Journey* II. l. 142 This virtuous, and..amiable society of *religieux* [ed. 3, 1789, *religieuses*]. 1959 *Times* 25 July 9/1 They [*sc.* béguinages] are communities of *religieuses* who live in separate houses.

2. (See quot. 1968.)

[1929 E. J. Kollist *French Pastry, Confectionery & Sweets* iv. 57 Gâteau Religieuse... Set on clean baking-sheet eclairs from pâte à choux paste... In between each eclair decorate with whipped cream.] 1954 C. Turgeon *Tante Marie's French Cakes & Pastries* 50 Little Nuns. *Religieuses*. .. This pastry is perhaps the most popular in all France. 1961 A. Wesker *Kitchen* II Pastry called 'Religieuse'. 1968 *Guardian* 17 Feb. 9/8 A Religieuses [*sic*] is a gorgeously opulent edifice of éclairs bound with a mortar of chocolate and cream.

religio-. Add: *religio-ethnic, -historical, -musical, -mystical, -philosophic, -philosophical* (example), *-political, -psychiatric* (also *-psychiatry*), *-sexual*.

1966 J. E. Hofman in J. A. Fishman *Readings Sociol. of Lang.* (1968) 626 Our hypothesis that whenever religio-ethnic concentrations and certain other factors coincide, a situation is created which enhances the ideological climate suitable to retentiveness of the ethnic mother tongue. 1953

W. R. Trask tr. *Auerbach's Mimesis* i. 17 The reader is at every moment aware of the universal religio-historical perspective which gives the individual stories their general meaning and purpose. 1959 'F. Newton' *Jazz Scene* iii. 45 'Shouting' sects..have made the most powerful single religio-musical contribution to jazz. 1976 *Listener* 22 July 92/2 George Sand..turned religio-mystical..with a spate of earnest, spiritual books. 1926 Fowler *Mod. Eng. Usage* 393/1 We must take account of religio-philosophic speculations. 1931 *Times Lit. Suppl.* 21 May 408/2 The world-wide religio-philosophical movement known as Theosophy. 1928 *Weekly Dispatch* 3 June 10/5 One of the most remarkable contributions to the religio-political discussion on record. 1979 M. A. Screech *Rabelais* vi. 215 The direct religio-political propaganda. 1964 S. Z. Klausner *Psychiatry & Relig.* i. 1 (*heading*) Religio-psychiatry: a social institution. *Ibid.*, The religio-psychiatric movement is born through several thousand similar encounters. 1968 *Internat. Encycl. Soc. Sci.* XII. 632/2 Religio-psychiatry is a twentieth-century movement whose participants are concerned with the relation between religious and scientific approaches to mental, emotional, or spiritual healing. 1946 D. C. Peattie *Road of Naturalist* iv. 49 The Poles snarled back a stream of commingled religio-sexual obscenity, to show how dirty they could talk if they pleased.

religion. Add: **4. c.** *religion of nature*: the worship of Nature in place of a more formal system of religious belief.

1902 W. James *Var. Relig. Exper.* iv. 91 In that 'theory of evolution' which..has within the past twenty-five years swept so rapidly over Europe and America, we see the ground laid for a new sort of religion of Nature, which has entirely displaced Christianity from the thought of a large part of our generation. 1961 D. G. James *Matthew Arnold* i. 22 The essay itself is given up chiefly to a warm exposition of her religion of nature.

8. *religion-arousing* adj.; *religion-complex*; *religion-game*.

1957 J. S. Huxley *Relig. without Revelation* (rev. ed.) vii. 174 Potential religion-arousing objects. 1922 *Brit. Jrnl. Psychology* Oct. 117 Such complexes clearly exist in the normal mind with perfectly free access to consciousness, *e.g.* the 'religion complex'. 1961 J. Wilson *Reason & Morals* ii. 120 Thus J. R. Lucas..even puts in a good word for the religion-game.

religionist. Add: (Later examples.) Also, one professionally occupied with religion; a minister or preacher.

1870 O. Logan *Before Footlights & behind Scenes* xl. 603 While clergymen and religionists, as now, stand afar off and denounce the theatre. 1895 *Wales* Aug. 361/2 The antagonism that some classes of religionists have shown to the pastoral care of the church is to be attributed, most frequently, to one of two things. 1939 Wyndham Lewis *Jews* v. 55 The Jews are..the great religionists of the West. 1958 *New Statesman* 6 Sept. 304/3 Lately in the United States *religionist* has taken on a new and definite meaning. When *religionists* are referred to in the current press, it is clear that this term includes anyone who is professionally occupied with religion, of whatever church, movement or status—that is, anyone from Billy Graham to Reinhold Niebuhr. 1966 *Ibid.* 13 May 683/1 Secularists are not demanding what religionists demand and obtain, namely public subsidy for their own propaganda institutions. They deplore all propagandising of children. 1977 *Private Eye* 1 Apr. 8/1 Mr Arthur Blessit..a Los Angeles religionist, was assaulted by a number of similarly moved but differently associated people.

religionless, *a.* Add: **A.** *religionless Christianity* [tr. G. *religionsloses Christentum*], Christianity dissociated from many of the doctrines and practices of conventional religion.

1953 R. H. Fuller tr. *Bonhoeffer's Lett. & Papers from Prison* v. 123 If religion is no more than the garment of Christianity—and even that garment has had very different aspects at different periods—then what is a religionless Christianity? 1963 *Times* 7 May 13/4 Archbishop Heenan is reported in today's Sunday press to have said that Anglican discussions about religionless Christianity are an embarrassment to Roman Catholics who work for Christian unity. 1969 A. Richardson *Dict. Chr. Theol.* 288/2 It is in the light of Barth's denunciation of religion that Bonhoeffer's plea for 'religionless Christianity', by which he meant unpietistic, unchurchy Christianity, should be understood. 1974 *Oxf. Dict. Chr. Ch.* (ed. 2) 187/1 Though writers of the Death of God school have taken up his [*sc.* Bonhoeffer's] idea of religionless Christianity, his teaching represents a search for the beyond in the midst, and a demand for a radical reform of the Church, which in its existing form he thought to have no message for the present day. 1977 *Church Times* 29 Apr. 6/4 Nor is it one of the newer-style jobs in which the intransigent element in Christian faith is dissolved away in..religionless Christianity.

B. as *sb. collect.*, people without religious belief.

1964 *New Statesman* 14 Feb. 254/3 Literature which Arnold was quite right to suggest would come to occupy the importance of religion for the religionless.

religiose, *a.* (Later examples.)

1932 [see *beglamour *v.*]. 1966 I. Jefferies *House-Surgeon* ix. 168 It was only later, when they were improving, that they would complain about their neighbours, or become religiose, or mutter. 1971 *Times Lit. Suppl.* 22 Oct. 1319/2 Donald's parents—the dying, religiose father and steely, moralizing mother—inhabit an area of real, alarming unsavouriness. 1975 J. Nicoll *Dante Gabriel Rossetti* ii. 48 An unhappy love affair with Rossetti's religiose sister Christina.

‖ **religioso** (relidʒiō·so), *a.* and *sb.* *Mus.* [It., = religious.] **A.** *adv.* As a direction to the performer: in a devotional manner. **B.** *sb.* A devotional effect; a passage to be played devotionally. **C.** *adj.* Having a devotional quality.

1837 J. A. Hamilton *Dict. Mus. Terms* (ed. 4) 58 *Religioso, religiosamente* (Italian), with religious feeling, in a devotional manner. 1876 Stainer & Barrett *Dict. Mus. Terms* 377/1 *Religiosamente, religioso*, in a religious or devotional manner. 1941 W. C. Handy *Father of Blues* v. 63, I was featuring *The Holy City* as a cornet solo and these saxophones contributed wonderfully to the religioso. 1961 *Times* 4 Dec. 14/5 The tremulous, *religioso* registrations with which the organist coloured the recitatives. 1975 *Sunday Times* 14 Dec. 31/1 In the Prologue to 'The Golden Legend'.. Sullivan *religioso* sounded no more inspired than the average Mus.D. exercise of a century ago. 1977 *Gramophone* Apr. 1541/2 In the slow movement there is little sense of *religioso* in the bare chordal writing for strings at the start. 1980 *Country Life* 17 Jan. 177 Even more *religioso* than Gounod at times was the Abbé Liszt.

religious, *a.* and *sb.* Add: **A.** *adj.* **3. a.** (Further examples.)

1769 *Account of Society for promoting Relig. Knowl.* 5 The design of this Society being to promote Religious Knowledge among the Poor. 1800 H. More *Let.* 11 Sept. (1925) 177, I knew that every Anti-Abolitionist in the world was.. an enemy to religious instruction at home. 1809 M. Waring (*title*) A diary of the religious experience of Mary Waring. 1836 *Introd. Discourse & Lect. Amer. Institute of Instruction 1835* 105 The parent who neglects the religious education of his child might as well suffer him to wander filthy and ragged in the streets. 1850 C. Kingsley *Alton Locke* I. xi. 178 'Schooling hasn't made wages rise, nor preaching neither.' 'But surely..all this religious knowledge ought to give you comfort.' 1858 Geo. Eliot *Scenes Clerical Life* II. 193 It may be that some of Mr. Tryon's hearers had gained a religious vocabulary rather than religious experience. 1872 *Q. Rev.* CXXXII. 534 The people will have to decide at a general election upon this great question of Religious or Secular Education. 1914 G. B. Shaw *Parents & Children* p. c, The last ray of art is being cut off from our schools by the discontinuance of religious education. 1960–61 *Where?* Winter 16/2 *Religious instruction* (*RI*), the only subject which state schools are obliged to teach by law. 1961 *Regulations G.C.E. Examinations* (Univ. London) 22 *Religious Knowledge* Ordinary Level. There will be one paper of 2½ hours. 1968 *Guardian* 28 Nov. 6/3 It is thought that a new attitude in the schools would encourage student teachers to take religious education as a subsidiary subject at colleges of education. 1973 *Listener* 23 Aug. 251/1 To disbelieve in God's existence is..a matter of distrusting the testimony of others or lacking a religious experience oneself.

c. Special collocations. *religious philosophy*: the philosophical study of religion; philosophy that accepts the concept of an omnipotent God; hence *religious philosopher*; *religious psychology*: psychology which accepts that a religious context is basic to man's personality and behaviour.

1840 J. S. Mill in *Westm. Rev.* XXXIII. 297 Of Coleridge as a moral and religious philosopher..there is neither room, nor would it be expedient for us to speak more than generally. *Ibid.* 298 We must be looking for a religious philosophy, and our main hope ought to be that it will be such a one as fulfils the conditions of a philosophy—the very foremost of which is, unrestricted freedom of thought. 1902 W. James *Var. Relig. Exper.* iv. 105 An interpretation of Christ's message which in these very Gifford lectures has been defended by some of your very ablest Scottish religious philosophers. *Ibid.* xviii. 431, I doubt if dispassionate intellectual contemplation of the universe, apart from inner unhappiness and need of deliverance on the one hand and mystical emotion on the other, would ever have resulted in religious philosophies such as we now possess. 1912 R. B. Perry *Pres. Philos. Tendencies* vii. 148 The English school of idealists..has from the outset offered a religious philosophy based on the supremacy of consciousness. 1927 J. S. Huxley *Relig. without Revelation* iv. 120 Those who, through study or profession, are brought into contact with religious psychology. *Ibid.* viii. 290 Thouless, who writes on religious psychology from the standpoint of a psychologist who is also a professing Christian. 1951 E. A. Burtt *Types of Relig. Philos.* (rev. ed.) p. vii, An exposition of the main points of view in religious philosophy. *Ibid.* i. 7 What significant comparisons may we make between the religious psychology of individuals who participate in quite different cultures? 1960 D. A. Lowrie *Rebellious Prophet* xiv. 196 Berdyaev..is inclined to consider him [*sc.* Bulgakov] a religious philosopher rather than a theologian. 1974 B. A. Brody *Philos. of Relig.* p. vii, For centuries, a principal issue in traditional religious philosophy had been whether one could prove the truth or falsity of a variety of fundamental doctrines.

5. Of a horse: **a.** (See quot. 1788.) **b.** *U.S.* 'Having no vicious traits' (D.A.E.).

1788 Grose *Classical Dict. Vulgar Tongue* (ed. 2) sig. Z4*, *Religious horse*, one much given to prayer, or apt to be down upon his knees. 1869 *Overland Monthly* III. 127 It is amusing to hear one ask of another, when about to purchase a horse: 'Is he religious?'

6. *Comb.*, as *religious-mad, -minded, -sane* adjs.

a 1930 D. H. Lawrence *Apocalypse* (1931) vi. 98 Men were religious-mad: not religious-sane. 1888 C. M. Yonge *Beechcroft at Rockstone* II. xx. 153 Thoroughly religious-minded,..his aspirations had been blighted by his father's death. 1954 A. Seton *Katherine* xxvi. 447 Religious-minded Katherine had never been... This strict penitential garb and talk of pilgrimage were surely some passing derangement.

B. *sb.* **2. b.** Delete † *Obs. rare* and add later examples.

In recent use not a gallicism.

1651 T. Matthew *Life Lady L. Knatchbull* (1931) II. i. 87 Dame Mary Roper..was a much younger Religious. 1922 Joyce *Ulysses* 706 Anal violation by male religious (fully clothed, eyes abject) of female religious (partly clothed, eyes direct). 1939 R. Godden *Black Narcissus* xxx. 273 You've forgotten who you are. You're a religious. A nun. 1948 W. S. Maugham *Catalina* xxix. 189 It behoved Catalina to become a religious. 1980 I. Murdoch *Nuns & Soldiers* i. 56 When she was being converted she was already purposing to be a religious.

religiously, *adv.* Add: **5.** *Comb.*, as *religiously-minded*.

1935 B. Russell *Relig. & Sci.* vi. 144 The sacred history related in the Bible and the elaborate theology of the ancient and mediaeval Church have become less important than formerly to most religiously minded men and women.

reline, *v.*[1], **relining**, *vbl. sb.* Add: **a.** (Further examples.)

1921 *Automobile Engineer* XI. 168/1 It is necessary to remove the rear-hubs from their tapers..in order to gain access to the internal brake for relining. 1933 *Radio Times* 14 Apr. 121/2 Have your brakes tested..if they need relining—specify Ferodo... The Ferodo guarantee is tied to the steering wheel of your car when you have the brakes relined with Ferodo. 1976 J. Drummond *Funeral Urn* v. 21 She..asked when she might have the car. He grunted dourly, 'Gotta re-line the brakes.'

b. *spec.* in *Art.* To attach a new backing canvas to (a painting).

1911 M. J. Gunn *Print Restoration & Picture Cleaning* viii. 146 Nothing but re-lining will often save a valuable picture from perishing. 1948 G. L. Stout *Care of Pictures* v. 96 The painting..had been relined at least once. 1957 *Encycl. Brit.* XVII. 68E/2 *Relining*, the procedure of attaching a new or secondary canvas at the back of a canvas support when that fabric has become too weak to serve its purpose or when the ground or paint has become loosened from it. *Ibid.*, If it [*sc.* the painting] has been relined before, the old relining canvas is removed. 1978 *Daily Tel.* 9 Dec. 14/4 He has cleaned, restored or relined over 85 pictures.

reli·ner. [f. Reline *v.*[1] + -er[1].] **1.** A person who provides oil-paintings with fresh linings.

1905 W. H. Hunt *Pre-Raphaelitism* I. 183 The reliner decided that the varnish was neither mastic nor copal. 1911 M. J. Gunn *Print Restoration & Picture Cleaning* viii. 147 In case the services of a picture re-liner and restorer should be needed.

2. Material providing a fresh lining, as for the brakes of a motor vehicle.

1920 T. Eaton & Co. Catal. Spring & Summer 395/1 *Tire Reliners* for Ford cars. Made of several layers of heavy tire fabrics. 1945 *Sun* (Baltimore) 8 Nov. 14/3 Motorists could recognize a sound used tire if it had a tread design at least every four inches, no emergency patches on the inside such as boots and reliners, [etc.].

reliquiæ, *pl.* Add: **1.** (Earlier and later examples.) Also, literary remains; unpublished or uncollected writings.

1654 E. Gayton *Pleasant Notes Don Quix.* III. vii. 114 A sort of these Theeves are now redivivous, (the Reliquiæ I believe of Knight-Errantry) who goe by the name of Spirits. 1933 *Times Lit. Suppl.* 2 Nov. 746/2 These reliquiae of Lytton Strachey, collected from periodicals and other sources.. belong to all times of his life as a writer. 1948 *Mind* LVII. 517 Scarcely less important..are the Jena manuscripts [of Hegel] published partly by Lasson in 1923 and partly by Hoffmeister in 1931–2. Armed with these *reliquiae* a scholar could approach the making of a commentary with fair confidence.

relish, *sb.*[1] Add: **3. b.** (Earlier and further U.S. examples.) Also *attrib.*

1797 W. Priest *Travels in U.S.A.* (1802) 32 About eight or nine in the morning they breakfast on tea and coffee, attended always with what they call *relishes*, such as salt fish, beef-steaks, sausages, broiled fowls, ham, bacons &c. 1826 J. F. Cooper *Last of Mohicans* I. vi. 72 Glad to eat their venison raw, and without a relish too. Here..we have plenty of salt. 1963 R. I. McDavid *Mencken's Amer. Lang.* iii. 120 *Rolliches*, pickled rolls of meat, are still occasionally made in the Hudson Valley, and sometimes anglicized to *relishes*. 1978 *Chicago* June 221/1 The $4.50 to $7.95 dinners include a relish plate (crisp vegetables, cheddar cheese spread, and scoop of homemade liver pâté).

re·lished, *ppl. a.* [f. Relish *v.*[1] + -ed[1].] Liked (as food); enjoyed, appreciated.

1901 *Yearbk. U.S. Dept. Agric.* 1900 433 There was an outbreak of seventeen-year cicadas, which afforded an abundant and greatly relished food supply.

reload, *v.* Add: **2. b.** Also, to load (a camera, cassette, etc.) again.

1888 *Judge* Christmas Number 43/1 One Hundred Exposures may be made without 're-loading' the camera. 1897 *Sears, Roebuck Catal.* 473/3 The camera..can be reloaded.. [in] any place from which the light is excluded. 1940 *Chambers's Techn. Dict.* 713/2 *Reload*, to remove exposed film and insert unexposed film in a camera or magazine in a dark-room or under light-tight conditions (e.g. in a changing bag). 1977 J. Hedgecoe *Photographer's Handbk.* 45 Instead of buying 35 mm film in cassettes you can purchase a bulk length..and then keep reloading cassettes yourself.

reload (rī·lōᵘd), *sb.* [f. the vb.] That which serves to reload anything, as a film placed in a camera, etc.

1928 *Daily Express* 14 Dec. 14/6 (Advt.), He can use the rod... Shaving Stick as a reload. 1958 *Newnes Compl. Amat. Photogr.* 75, 35 mm. film is also supplied as daylight-loading or darkroom-loading reloads for cassettes. 1961 *Guardian* 16 Mar. 6 He found..a small Minox pocket camera..and, in a brown paper bag, two Minox reloads. 1976 *Shooting Mag.* Dec. 47/2 Despite the budget price it does not make second-quality reloads. 1977 J. Wainwright *Nest of Rats* I. vii. 46 A shooter; a thirty-eight Colt 'Agent' revolver... There was a box of re-loads included in the parcel.

reloa·der. [f. Reload *v.* + -er[1].] That which or one who reloads.
1909 *Cent. Dict.* Suppl., *Reloader*,..a self-loading conveyer used to collect and transport coal from a coal-storage yard or pocket and to deliver it to railroad-cars or vessels or to place it in other near-by storage-places. 1973 'A. Hall' *Tango Briefing* vii. 94 This man's forte was fast use of the automatic reloader... He was using something like a ·44 Magnum. 1976 *Shooting Times & Country Mag.* 18–24 Nov., I would very much appreciate advice from reloaders of home made cartridges. 1976 *Shooting Mag.* Dec. 46/2 No reloaders..should have anything to do with powder of unknown vintage that has been subjected to unknown storage conditions.

reloca·table, *a.* [f. Relocat(e *v.* + -able.] That can be relocated.
1872 'Mark Twain' *Roughing It* xli. 290 At midnight.. the ledge would be 'relocatable'. 1976 *Milton Keynes Express* 11 June 34/1 (Advt.), Heating Engineer... To initiate sales leads, progress enquiries including those for Intercities Relocatable Boiler House, design systems and negotiate quotations through to acceptance. 1977 *Gloss. Terms Data Processing* (*B.S.I.*) vii. 9/1 *Relocatable program*, a computer program that is in such a form that it may be relocated. 1979 *Personal Computer World* Nov. 69 (Advt.), Relocatable linkable output.

relocate, *v.* For '*U.S.*' read 'orig. *U.S.*' Add: **1. c.** To move to another place; to resettle; to change the location of.
1834 A. Lincoln et al. in I. D. Tarbell *Early Life A. Lincoln* (1896) xvii. 198 To view and relocate a part of the road..we have made the said relocation on good ground. 1866 *Rep. Indian Affairs* (U.S.) 76 If the Indians could be removed to some remote place equally fertile, and there relocated, it would no doubt be to their advantage. 1908 *Pacific Monthly* Feb. 204/2 The section east of here..has been practically all relocated, in places the new track being miles away from the original location. 1936 *Sun* (Baltimore) 21 July 1/3 Families which could not be supported by a shifting of land to moisture-holding grass production for cattle raising..would be 'relocated'. 1956 H. Foster in D. L. Linton *Sheffield* 245 The city does not face what would have been the almost insuperable problem of re-locating its vast heavy industries. 1964 T. W. McRae *Impact of Computers on Accounting* i. 23 This was a somewhat laborious business..causing a whole battery of 'words' to be re-located to different addresses. 1970 *Globe & Mail* (Toronto) 25 Sept. b2/8 Each of the three Canadian companies..has been looking at the possibility of relocating their data centres. 1978 *N.Y. Times* 30 Mar. b3/5 Paving the way for the Nestlé Company to relocate its White Plains headquarters on the property.
2. (Earlier and further examples.) Freq. without const.
1841 in Webster. 1851 C. Cist *Cincinnati* 143 [This] determined the company to re-locate on higher ground. 1864 *Congress Globe* 9 Mar. 1018/2 In a larger number of cases these persons having taken homesteads, and again desiring to sell and relocate,..have paid for the lands. 1957 [see *runaway *sb.* (and *a.*) 7]. 1964 *New Statesman* 3 Apr. 533/2 It offers an enjoyable evening out in Jaguarthreepointfoursville where the hero has relocated from his Scottish tenement by selling cash registers. 1968 'E. Lathen' *Stitch in Time* xvi. 134 He was relocating, he explained..because of a sudden desire to specialize in dermatology. 1971 *Daily Mail* 16 Mar. 22/1 (Advt.), Applicants should be prepared to work in our London Office initially, and relocate to new premises in Basingstoke..by 1973. 1978 *New York* 3 Apr. 37/2 The company had to relocate for technical reasons. 1979 *N.Y. Rev. Bks.* 25 Oct. 55/1 (Advt.), Lady author/lecturer who can easily re-locate seeks male counterpart.

relocation. Add: **2.** (Earlier U.S. and later examples.)
1837 A. Lincoln Let. 5 Aug. in *McClure's* (1896) Mar. 316/1, I also tacked a provision on to a fellow's bill, to authorize the relocation of a road. 1873 *Trans. Illinois Dept. Agric.* X. 371 The court shall appoint three viewers to examine the necessary re-location. 1901 S. E. White *Claim Jumpers* 232 Under the terms of a relocation, we can use the old stakes and 'discovery'. 1948 *Sierra Club Bull.* Dec. 5/1 A general relocation of the road was thereupon planned, including a higher crossing of Yosemite Creek. 1963 C. R. Cowell et al. *Inlays, Crowns, & Bridges* viii. 89 Re-location of the copper ring may be inaccurate in such an elastic material. 1967 *Boston Sunday Globe* 23 Apr. 9/5 A private company has sponsored refugee relocation. 1973 M. Mann (*title*) Workers on the move: the sociology of relocation. 1979 *Navajo Times* (Window Rock, Arizona) 24 May 1/2 He recognized that relocation would be difficult for the Navajos. 1981 J. Sutherland *Bestsellers* i. 29 A move towards bestsellerism..would seem to presage a general relocation of the bestseller.
3. *attrib.*, as *relocation allowance, assistance, cost, director, expense, grant*; **relocation centre** *U.S.*, an internment camp to which persons of Japanese birth or origin were committed during the war of 1939–45.
1958 *Observer* 12 Jan. 18/3 (Advt.), Liberal relocation allowances will be given successful candidates. 1968 *Globe & Mail* (Toronto) 3 Feb. b6/3 (Advt.), Generous relocation allowance. *Ibid.* 17 Feb. b6 (Advt.), Excellent advancement potential and benefits program. Relocation assistance will be offered. 1976 *Star* (Sheffield) 20 Nov. (Advt.), Relocation assistance will be provided. 1943 S. Menefee *Assignment: U.S.A.* 68 Hearst reporters got anti-Japanese statements from Mayor Fletcher Bowron and other prominent figures in Los Angeles and played up the Dies Committee's 'exposures' of the relocation centers. 1967 *Economist* 22 Apr. 354/2 In the days just after the bombing of Pearl Harbour, many Americans vented their panic..on the Nisei... The brunt fell most heavily on the large communities on the West Coast... These were evacuated en masse to inland 'relocation centres'. 1970 *Internat. & Compar. Law Q.* 4th Ser. XIX. ii. 237 A sparsely populated Coast where relocation costs are nominal. 1963 *Freedomways* Summer 425 A relocation director,..whose job is to help relocate families displaced from areas where new housing is being constructed. 1961 *Times* 27 Jan. 3/3 (Advt.), Relocation expenses guaranteed for selected Engineers. 1977 *Navy News* July 35 (Advt.), Relocation expenses will be considered where appropriate. 1977 *Times Educ. Suppl.* 21 Oct. 52/3 (Advt.), Relocation grants available in approved cases.

relucence, -ency. Restrict †*Obs.* to *relucency* and add later example of *relucence*.
1926 *Spectator* 15 May 849/2 The mystical life..found again and again inspiration and relucence from the poets.

relucent, *a.* (Later *fig.* example.)
1897 F. Thompson *New Poems* 33 The relucent song take for thy sacred meeds!

reluctance. 1. b. (Later examples.) Also *attrib.*
1967 Kurrelmeyer & Mais *Electricity & Magnetism* xii. 291 An air gap of 0·1 mm has the same reluctance as 1 m of iron of permeability 10,000. 1968 *New Scientist* 11 Jan. 63/2 The reluctance motor is a synchronous machine; its speed is determined entirely by the frequency of the ac supply. 1977 *Gramophone* Aug. 366/1 The moving-iron (variable reluctance) principle is used with a larger than usual fixed magnet.

relu·ctantism. *rare.* [f. Reluctant *a.* + -ism.] A reluctant state or condition; reluctance.
1906 *Century Mag.* Feb. 552/2 The incisive coldness of Miss Lamb's demeanor..was sufficient to chill..her youthful admirers into a state of objectified reluctantism.

reluctivity. Substitute for def.: The reciprocal of the magnetic permeability. (Later example.)
1917 C. M. Smith *Electr. & Magn. Measurements* x. 277 Reluctance and its reciprocal permeance are characteristics of the circuit. Reluctivity and permeability are characteristic of the given material.

rem, *a.* Slang abbrev. of Remanded *ppl. a.*
1887 [see *bust *sb.*[3] d].

rem (rem), *sb.*[1] Pl. **rem, rems.** [f. initial letters of *roentgen equivalent man.*] Orig., a quantity of ionizing radiation having the same effect on human tissue as one roentgen of X-rays. The dosage in rems is now calculated by multiplying the dosage in rads by the relative biological effectiveness. Cf. *rep[7].
1947 *Nucleonics* Oct. 38/2 The rep and rem units were introduced by Dr. H. M. Parker. *Ibid.* 39/2 Roentgen-equivalent-man (or mammal), rem... One rem is the estimated amount of energy absorbed in tissue which is biologically equivalent in man to 1 r of gamma- or X-rays. By definition: 1 rem = 83/RBE erg/gm tissue. 1957 *Encycl. Brit.* XVI. 591/1 The biological effects of radiation are not solely dependent on the amount of energy released; it is also a question of how highly localized the energy is. Such considerations have led to another unit, the rem, supposed to be that dose of radiation which has the same biological effect as 1 rad of X-radiation. 1958 W. D. Claus *Radiation Biol. & Med.* xviii. 431 The dose in rem is obtained by multiplying the physical dose in rads by the RBE appropriate to the situation. 1958 *Observer* 5 Oct. 15/4 The most active watches..could deliver a dose of five rem (units of radioactive dose measurement) within five years. The International Commission on Radiological Protection has recommended that no one should receive an accumulated dose of more than five rem by the age of 30. 1975 *Nature* 27 Mar. 278/2 Fifteen rem is equivalent to 1·5 rad of alpha radiation (since alpha radiation is considered 10 times as carcinogenic as gamma radiation). 1976 *Sci. Amer.* Nov. 31/2 If the fresh fission products from one megaton of fission were spread uniformly over a perfectly flat area of 1,000 square miles, the gamma-ray dose rate one meter above the ground would be about 250 rems per hour after 10 hours.

REM (rem, ɑːɪiːe·m), *sb.*[2] Also **rem.** Abbrev. of *rapid eye movement* (see *rapid *a.* 2 a). Freq. *attrib.*, designating a distinctive type of sleep that occurs at intervals throughout the night and is characterized by such eye movements, more dreaming and bodily movement, an increased pulse rate, and faster breathing.
1957 Dement & Kleitman in *Jrnl. Exper. Psychol.* LIII. 340/1 In most of the remaining text the following abbreviations will be used: REM's (rapid eye movements) and NREM's (no rapid eye movements). 1969 *Sunday Times* (Colour Suppl.) 16 Feb. 21/3 Subjects regularly deprived of the opportunity to dream by being woken at the onset of REM periods began to show psychological disturbance after a few nights. 1972 F. R. Freemon *Sleep Research* i. 4 This second type of sleep, called the rem state, has low voltage EEG activity mixed with bursts of theta waves and frequent conjugate eye movements. 1976 Smythies & Corbett *Psychiatry* xiv. 265 During REM sleep the brain shows intense metabolic activity. 1977 S. Dunkell *Sleep Positions* ii. 38 As we continue our journey through the night, the duration of each successive REM phase increases.

remagnetize, *v.* (Earlier example.)
1839 Advt. in G. S. Haight *George Eliot & John Chapman* (1969) 259 Compasses made to order and remagnatized [*sic*].

remain, *v.* Add: **2. c.** *it* (or *that*) *remains to be seen*: it is not yet known or certain.
1796 Ld. Glenbervie *Diary* 16 Oct. (1928) I. 88 It remains, however, to be seen what will be the ultimate result in the present instance of a struggle as yet perhaps but in its infancy. 1828 *Athenæum* 12 Feb. 103/1 Whether or not the 'Life of Columbus' will restore it, remains yet to be seen. 1859 *Times* 4 Feb. 9/4 That remains to be seen. 1866 Mayne Reid *Headless Horseman* xvi. 88 It remains to be seen how we shall get over it. 1938 H. L. Mencken *Let.* 23 Apr. (1961) 427 Whether I'll write anything for publication remains to be seen. 1967 *Listener* 6 July 20/2 How far or how quickly the new government can get anywhere.. remains to be seen. 1976 *Southern Even. Echo* (Southampton) 13 Nov. 3/6 It would remain to be seen to what extent it would be practical or desirable to build houses there.
4. a. Also, with *on*.
1912 J. Joyce *Let.* 23 Aug. (1966) II. 311 Tomorrow I must pawn my watch and chain in order to remain on a little longer. 1939 H. Nicholson *Diary* 11 Apr. (1966) 397 Harold Macmillan is enraged that Chamberlain should remain on.
6. b. Delete †*Obs.*—[1] and add later examples.
1872 Hardy *Under Greenwood Tree* I. i. vi. 78 The tunes they that morning essayed remained with him for years. 1899 'Mark Twain' *Man that corrupted Hadleyburg* in *Harper's Mag.* Dec. 30/2 A remark which he made to me has remained with me to this day, and has at last conquered me. *a* 1927 I. Duncan *My Life* (1928) viii. 78 Another, even greater impression, that has remained with me all my life was the 'Rodin Pavillon'. 1977 B. Pym *Quartet in Autumn* vi. 54 She had once noticed an old woman with a lost expression peering through one of the surrounding hedges and that impression had remained with her.

remainder[1]. Add: **3. b.** (Further U.S. examples.)
1872 'Mark Twain' *Roughing It* l. 357 There'll be a double-barreled inquest here..and your remainders will go home in a couple of baskets. 1885 —— *Huck. Finn* viii. 62, I was having a good enough time seeing them hunt for my remainders.
5. (Earlier and later examples.) Also *transf.*, an unused portion of goods, unused material; = Remnant *sb.* 4 b.
1757 *Monthly Rev.* Sept., C. Henderson, Bookseller, under the Royal Exchange, having purchased the remainder of the impression of the following very entertaining book.. proposes to sell them for 4s. only. 1854 *Gowans'* (115 Nassau St., N.Y.) *Catal.* No. 13. 6 Remainders of editions by other publishers. 1865 *N. & Q.* VII. 510/2 (Advt.), 'Remainders' of valuable books, all in new condition, at greatly reduced prices. 1914 J. Leatham *Daavit* 69 My dear good old mother bocht a remander fae Johnnie Hitcheon, and took it an' me ti Saunders ti be mizhur't. 1926 C. N. Bennett *Photogravure* 121 Paper makers, like drapers, have their remnants, though the name for them in the paper making industry is 'remainders'. 1930 J. H. Appel *Business Biogr. John Wanamaker* viii. 104 'Bargain Room' opened—'a place where remainders of lots are sold at smaller prices'.
6. *remainder biscuit* (further example); (sense 5) *remainder binding, list, -shop.*
1899 *Sketch* 1 Nov. 62/1 The poor evening paper cannot afford this. It must..be content with the 'remainder biscuit' of the morning's telegrams. 1912 *Chambers's Jrnl.* Dec. 773/2 It is pitiful to see the rows of discarded books in circulating libraries and remainder-shops. 1931 *Times Lit. Suppl.* 10 Sept. 688/2 The unsold sheets of a published book are re-issued with a cancel title or a new preface, or in a remainder binding. 1977 *Gay News* 24 Mar. 21/3, I use anything that's cheap on the remainder list.
7. Special Comb.: **remainder theorem** *Math.*, the theorem that if a polynomial $f(x)$ is divided by $(x-a)$ the remainder will be $f(a)$.
1886 G. Chrystal *Algebra* I. vii. 134 (heading) Results of the application of remainder theorem. 1933 R. W. Brink *College Algebra* xix. 295 Without performing the divisions, by means of the Remainder Theorem find the remainder after each of the following divisions. 1971 Willerding & Hoffman *College Algebra* ix. 267 By the Remainder Theorem, the remainder when x^3+7x^2+3x+3 is divided by $x+1$ is 6. As a corollary of the remainder theorem we have the Factor Theorem.

remai·nder, *v.* [f. Remainder[1].] *trans.* To dispose of (an unsold part of an edition of a book) at a reduced price; to treat as a remainder (sense 5). Also *transf.* So **remai·ndered** *ppl. a.*, **remai·ndering** *vbl. sb.*
1904 *Heffer & Sons' Catal.* 2 As the History of 'Remaindered' Books would almost prove, it might be said that no Book was really great until it had been 'Remaindered'. 1906 *Times* 17 Nov. 9/3 How many books do we see every year produced by publishers who..'remainder' them at a few pence a copy? 1907 *Times* 25 Mar. 12/1 There is no doubt now that the boycott is not meant to stop remaindering at low prices. 1910 *Library* I. 46 The plays in question were printed in the years with which they are dated and unsold copies..remaindered in 1619. *Ibid.* 49 A nineteen-year-old edition was then being remaindered. 1932 *John o' London's Weekly* 25 June 428 He told me he had bought them when they were remaindered by publishers, at 9d. a copy. 1959 *Daily Tel.* 29 Dec. 6/2 Swift turnover for cash, sometimes of goods specially ordered for the sales, and the 'remaindering'

of clothes, carpets, furniture and whatnot which might not otherwise be sold so quickly—or even, where fashion is important, at all. 1968 C. M. Vines *Little Nut-Brown Man* x. 155 He liked his books to be in short supply, thus perhaps appearing better sellers than they were; or he disliked the thought of being remaindered. 1981 *Country Life* 1 Jan. 34/1 Picture-books seem to end up by being sold off cheap as remaindered volumes.

remainer[2]. (Further example.)
1922 Joyce *Ulysses* 688 How did the centripetal remainer afford egress to the centrifugal departer?

re·make, *sb.* [f. the vb.] **1.** A second formation of a gold-bearing reef. *Austral.*
1865 *Mining Surveyors' Rep.* (Mining Dept., Victoria) Mar. 74 The lode was however very thin, and ran completely out at 70 feet deep, leaving no track. However a party are now prospecting this ground, to discover if a remake of this reef exists.
2. (Also **re-make.**) A remaking of a film or of a script, usually with the rôles played by different actors; an adaptation of the theme of a film.
1936 *Variety* 24 June 4/4 James Melton assigned the lead in Warners' remake of 'Desert Song'. 1940 *Time* 22 Jan. 76/3 The result is not just another remake, for Director Hawks's weird idea was also to remake the sex of his leading character. 1948 *Sunday Pictorial* 18 July 11/4 'If Winter Comes' (Empire) is a re-make of the famous weepie novel. 1952 *Time* 2 June 92/3 *Lovely to Look at* (M-G-M), a re-make of the old Broadway musical *Roberta* (filmed in 1935 with Fred Astaire and Ginger Rogers). 1957 *Observer* 1 Sept. 11/7 The romance is a remake by director Leo McCarey of his 'Love Affair', which seemed a good film when Irene Dunne and Charles Boyer did it twenty years ago. 1960 *Times* 23 Feb. 4/1 A Hollywood company has undertaken a Western remake of Mr. Akira Kurosawa's famous Japanese costume drama. 1977 *New Statesman* 2 Sept. 312/2 The technicolour remake of the talkie remake of some..silent Hollywood goodie.

remand, *sb.* Add: **2.** (Later example.)
1970 G. F. Newman *Sir, You Bastard* i. 35 The door used by remands, down near the witness box.
3. *attrib.*, as *remand prisoner, warrant*; **remand centre,** an institution to which young persons between the ages of 14 and 21 years are remanded to await trial or sentence; since 1967, such an institution for a person of any age; *Canad.*, such an institution for adults; **remand home,** an institution to which young persons between the ages of 8 and 14 years are remanded or are committed for detention.
1948 *Criminal Justice Act* 11 & 12 *Geo. VI* c. 58 s. 48(1)(a) Remand centres, that is to say places for the detention of persons not less than fourteen but under twenty-one years of age who are remanded or committed in custody for trial or sentence. 1967 *Criminal Justice Act Eliz. II.* c. 80 s. 66(1) Notwithstanding that a remand centre is provided under section 43 of the Prison Act 1952 for the detention of persons of or over the age of fourteen but under the age of twenty-one who are remanded or committed in custody for trial or sentence, any person required to be detained in an institution to which the Act applies may be detained in a remand centre for any temporary purpose or for the purpose of providing maintenance and domestic services for that centre. 1970 G. Greer *Female Eunuch* 180 His impudence in courtrooms and remand centres. 1974 *Globe & Mail* (Toronto) 4 Sept. 1/1 The Calgary remand centre—the first in Alberta—is open for business. The centre will house men held in custody between court appearances pending their trials. 1976 *Southern Even. Echo* (Southampton) 10 Nov. 9/2 A Southampton schoolboy was remanded in custody for seven days to Winchester remand centre after the Magistrates decided he was 'unruly'. 1902 *Times* 13 Jan. 9/4 The Children's Committee reported that the three remand homes at Pentonville-road, Harrow-road, and Camberwell-green were opened for the reception of children on January 1. 1933 *Act* 23 & 24 *Geo. V* c. 12 s. 108(2) References in any Act to places of detention provided under Section one hundred and eight of the Children Act, 1908, shall be construed as references to remand homes provided under this Act. 1934 'J. Spenser' *Limey breaks In* ii. 21 The policeman who took me to the remand home led me into a restaurant and gave me a good feed before he handed me over. 1963 M. Duggan in C. K. Stead *N.Z. Short Stories* (1966) 101, I came in for a couple of remand home stares, bread and water and solitary and take that writ on his eyeballs. 1972 G. Sereny *Case of Mary Bell* i. iii. 46 With her father's agreement, she was taken to stay at Fernwood Remand Home, a Newcastle County Council Children's Home for girls. 1897 Remand prisoner [in Dict., sense 1]. 1977 *Belfast Telegraph* 19 Jan. 4/8 Two visitors to republican prisoners in Crumlin Road jail had been..attacked by loyalist gangs.., a remand prisoner claimed today. 1963 J. N. Harris *Weird World Wes Beattie* (1964) i. 12 Wes sits in the Psychiatric Hospital on an attorney general's remand warrant.

remanence. 2. Delete *rare*—[1] and add later examples in *Theol.*
1964 J. Stacey *John Wyclif & Reform* v. 104 The next assertion was a doctrine of Remanence. If annihilation was denied then, in his view, the bread and wine remained bread and wine. 1964 R. H. Bainton *Hist. Christianity* viii. 238/1 This was not to say..that Christ is not in the sacrament. He is there, in addition to and along with bread and wine, whose substance remains. This doctrine is called remanence. 1974 *Encycl. Brit. Macropædia* XIX. 1051/1 He [*sc.* Wycliffe] sought to replace it [*sc.* the doctrine of transubstantiation] with a doctrine of remanence (remaining).

3. *Physics.* Residual magnetism, *spec.* *retentivity 1 (but see quot. 1962).
1917 G. D. Shepardson *Telephone Apparatus* iv. 279 When the current has been reduced to zero, there still exists a more or less permanent magnetization such as indicated by *OE*, the power of holding this residual magnetization being sometimes called the 'remanence' of the iron, sometimes expressed as a percentage of the maximum magnetization. 1924 C. R. Underhill *Magnets* xxv. 435 The remanence is the structural flux density of a permanent magnet, sometimes called the residual induction. 1947 *Electronic Engin.* XIX. 379/1 The wire originally used..was a medium carbon steel having a remanence of 6,000/7,000 gauss. 1948 [see *retentivity]. 1962 M. McCaig in D. Hadfield *Permanent Magnets & Magnetism* ii. 26 The hysteresis loop of largest area is known as *the* hysteresis loop. The values of remanent magnetism..and coercive force..for this hysteresis loop are known as the remanence and coercivity respectively. This definition of remanence conforms with the usage recommended by the British Standards Institution... In the U.S.A.... the same quantity is called residual magnetism, while the word remanence is used to describe the state of an actual magnet after magnetization. Owing to its own self-demagnetizing field such a magnet operates at a point in the top left-hand quadrant of the hysteresis loop. As the British Standard refers to the flux density in such a magnet as 'residual magnetism' there is a complete interchange of meanings of the terms 'remanence' and 'residual magnetism' on the two sides of the Atlantic. 1973 J. G. Tweeddale *Materials Technol.* I. iv. 93 When an electromagnetically induced field is changing rapidly, as it might do in a piece of electrical or electronic equipment, it is obvious that a very low remanence is desirable if energy loss and generation of heat is to be avoided. 1976 *Nature* 5 Feb. 381/1 Many of the intrusions have been sampled during our new study and the palaeomagnetism of those possessing stable remanences after a[lternating] f[ield] cleaning is reported here.

remanent, *a.* Restrict 'Now *rare*' to senses in Dict. and add: **4.** *Physics.* Of magnetism: remaining in a substance or specimen after removal of the magnetizing field.
1866 E. Atkinson tr. *Ganot's Elem. Treat. Physics* (ed. 2) X. v. 678 The iron used for the electromagnet..must be pure, and be made as soft as possible... If this is not the case the bar retains, even after the passage of the current a quantity of magnetism which is called the remanent magnetism. 1912, 1931 [see *retentivity]. 1939 L. F. Bates *Mod. Magnetism* viii. 258 In general, the remanent magnetism possessed by a ferromagnetic which has been placed in a strong field is sufficient to give satisfactory deflections of a sensitive astatic magnetometer system. 1944 *Proc. IRE* XXXII. 667/2 The remanent flux will go through a series of values corresponding to the sum and difference frequencies between the recording signal and the supersonic signal. 1962 [see *remanence 3]. 1971 I. G. Gass et al. *Understanding Earth* xvi. 237/1 The intensity of this remanent or permanent component of magnetisation in basalts is invariably greater than that induced by the present Earth's field.

‖ **remanié** (rəmaniˌe), *a. Geol.* and *Geogr.* [a. F. *remanié*, pa. pple. of *remanier* to re-handle, reshape.] Derived from an older stratum or structure.
[1866 *Q. Jrnl. Geol. Soc.* XXII. 237 If the Diestien beds be divided into upper sandy ooze and lower muddy ooze, the sections show that where the first has been removed and 'remanié', the resulting Scaldésien beds take the form of 'Crag gris'.] 1870 *Ibid.* XXVI. 72 In this 'remanié' deposit there appears to exist an assemblage of species peculiar to two distinct epochs. 1894 J. Geikie *Great Ice Age* (ed. 3) xii. 160 The marine organisms..may indicate an inter-glacial submergence to the extent of 300 or 400 feet, but on the other hand the deposits in which they occur may be *remaniés*. They may well have been dragged forward by the ice from a lower level. 1913 *Rep. Brit. Assoc. Adv. Sci.* 1912 622 Excavation (some 8 feet deep) in 'rubble-drift' material, mostly *remanié* stuff from the Boulder Clay which caps the hill above. 1957 J. K. Charlesworth *Quaternary Era* I. iv. 87 If the lateral glacier is steep it may fall on a trunk glacier as a remanié glacier. 1964 V. J. Chapman *Coastal Vegetation* vi. 137 If a dune undergoing erosion is stabilized, it represents a moderate remanié form. 1969 Bennison & Wright *Geol. Hist. Brit. Isles* i. 10 A bed may contain fossils of a previous geological age known as derived or remanié fossils. 1978 *Nature* 16 Nov. 258/2 At Hamilton in western Victoria, a bed of calcareous clay containing phosphatic nodules rests unconformably on limestone and clay of early and middle Miocene age and contains abundant *remanié* foraminifera from both formations.

‖ **remaniement** (rəmanimɑ̃). [Fr.: see prec.] A rearrangement, a reconstruction.
1920 *Glasgow Herald* 3 Aug. 5 Much more..needs study in the latest 'remaniement' of the Turkish Ministry... The Grand Vizier..is determined that the new Cabinet shall be composed of moderate men. 1933 E. K. Chambers *Eng. Folk-Play* 87 The play is said to have been given as far back as 1807, but to me it suggests a literary *remaniement*.

remap, re-map (ri-). *U.S.* [Re- 5 a.] = *redistricting *vbl. sb.* Also *attrib.*
1962 *Nashville Tennessean* 2 Sept. 2-b/1 (*heading*) State remap is still issue. 1962 *Economist* 1 Dec. 908/1 Tennessee has approved a constitutional convention on what is called 're-map'. 1974 *State* (Columbia, S. Carolina) 15 Feb. 4-a/3 The department rejected a bitterly disputed remap plan that would have divided the state into 28 districts drawn to maintain the integrity of county lines.

remargin, *v.* (Later example.)
1952 J. Carter *ABC for Book-Collectors* 153 When one or more of the three outer margins of a leaf has been restored, it is said to be re-margined.

remark, *v.* Add: **3. a.** Quot. *a* 1704 to read: 1694 Locke *Ess. Hum. Und.* (ed. 2) iii. ix. 275 It is easie to observe, what has been before remarked, [etc.].
b. Also without *prep.*, to make remarks.
1845 J. Ruskin *Let.* 17 June in H. I. Shapiro *Ruskin in Italy* (1972) 118 Perhaps..it is an English cheesemonger & his wife, who come in, and remark, as happened to me the other day while I was looking at the gates of Ghiberti.

remarkable, *a.* and *sb.* Add: **A.** *adj.* **1.** Also as quasi-*adv.*
1779 J. Woodforde *Diary* 6 Feb. (1924) I. 245 Mr. Ferman and myself went to see a remarkable large Pigg. 1818 W. Sewall *Diary* 13 Mar. (1930) 39/1 The scholars appeared remarkable well. 1871 E. Eggleston *Hoosier Schoolmaster* iv. 39 He uses sech remarkable smart words. 1890 Kipling *Barrack-Room Ballads* (1892) 8 We aren't no thin red 'eroes.. But single men in barricks, most remarkable like you.
B. *sb.* (Later U.S. example.)
1946 *Richmond* (Va.) *Times-Dispatch* 15 Jan. 10/2 (*heading*) Religious remarkables.

remarque. (Later examples.)
1890 F. G. Kitton *Charles Dickens by Pen & Pencil* I. iv. facing p. 49 Charles Dickens, his wife and her sister—1843... Remarque: Miss Mary Hogarth. 1925 C. Morley *Thunder on Left* vii. 80 If they're girls, how mothers hurry to drill and denature those bright dreaming wits. They love them chiefly because they make so pretty a vignette in the margin of their own self-portrait—like a *remarque* in an engraving.

rema·ster (ri-), *v.* Also **re-master.** [Re- 5 a.] *trans.* To make a new master of (a record); to issue (a recording) from a new master: see *master *sb.*[1] 9* a. Hence **rema·stering** *vbl. sb.*
1967 *Punch* 25 Jan. 132/3 Both have been remastered, and only nostalgia entitles you to prefer the original 78s. 1970 *Soviet Weekly* 20 June 4/2 They include speeches which he recorded in the earliest years of Soviet power skilfully remastered and transferred to tape. 1975 *Gramophone* Aug. 322/1 They..will be most grateful to A. C. Griffith for the skill with which the transfer and re-mastering has been carried out. 1977 *Rolling Stone* 19 May 23/1 Jeffreys' best-known song (first released in 1973 as a single on Atlantic and now remastered and included on the new LP).

re·match, *sb.* [Re- 5 a.] A return match.
1941 *Sun* (Baltimore) 10 Apr. 17/1 Abraham Simon, originally the May opponent in a rematch of a 13-round go with the Bomber in Detroit, in June in New York. 1972 J. Mosedale *Football* v. 62 They lost the playoff in a rematch with Los Angeles. 1973 *Times* 15 May 14/8 'I didn't hit out enough,' Mrs Court said, depressed over her showing but willing to play a rematch. 1978 *N.Y. Times* 30 Mar. d22/1 He chose to give Muhammad Ali a rematch before taking on Norton.

re:mate·rialize, *v.* [Re- 5 a.] *intr.* To materialize again. Hence **re:materializa·tion**; **re:mate·rialized** *ppl. a.*
1907 W. De Morgan *Alice-For-Short* xxvii. 280 It was as nothing to Moses to cease to exist when hunted for, and to re-materialize when convenient. 1921 *Glasgow Herald* 29 Jan. 13/1 The fate of the Manchester Repertory and the failure of the Glasgow one to rematerialise warn us [etc.]. 1928 *Sunday Express* 8 Apr. 5/1 He comes back to each of the characters in the way in which they remember him, and ..because of their glimpses of his rematerialised self, [etc.]. 1956 R. M. Lester *Towards Hereafter* xiv. 169 Levitation, of course, is often a form of dematerialization and rematerialization. 1959 *Times* 13 Oct. 16/4 Their lot is to be transfixed..sawn in half, dematerialized and rematerialized. 1962 *Punch* 26 Dec. 926/3 It's just a simple matter of Rematerialisation. 1978 *Sci. Amer.* Mar. 54/1 The virtual photon rematerializes into any one of a very large number of possible combinations of new particles.

‖ **remboîtage** (rɑ̃bwatɑ̃ʒ). [Fr., f. *remboîter* to re-case (a book).] (See quot. 1952.)
1952 J. Carter *ABC for Book-Collectors* 153 *Remboîtage* means the transferring of a book from its own binding to another more elegant, more nearly contemporary, more appropriate—anyway, more desirable; or, alternatively, the transferring into a superior binding of a text more interesting or more valuable than the one for which it was made. 1968 C. P. Bracken *Roman Ring* iv. 27, I defy anyone to detect our remboîtages... Many early books were rebound anyway.

Rembrandt (re·mbrænt). The name of the Dutch painter *Rembrandt* (1606–69) used *absol.* or *attrib.* to designate a Darwin tulip with streaked or variegated flowers.
1902 *Jrnl. R. Hort. Soc.* XXVI. 975 Tulips, Rembrandt... A striped-flowered section raised from the 'Darwin', by MM. Krelage & Son, Haarlem. 1908 *Ibid.* XXXIII. 233 Rembrandts..are broken Darwins, and..often beautifully marked and feathered. 1911 J. Weathers *Bulb Bk.* 440/1 What are now known as 'Rembrandt' Tulips are broken or rectified Darwin Tulips. 1929 A. D. Hall *Bk. Tulip* v. 103 The broken forms [of Darwin tulip] are now given a class to themselves and called 'Rembrandts'. 1948 J. C. Wister *Bulbs for Home Gardens* xi. 122 Rembrandt tulips..are striped varieties, mostly 'rectified' or 'broken' Darwins. 1974 A. Huxley *Plant & Planet* xviii. 197 Those tulips called Rembrandts, Bybloems and Bizarres by the fanciers, in which the flowers are streaked or feathered in another colour.

Rembrandtesque, *a.* (Earlier and later examples.)
1863 'Ouida' *Held in Bondage* I. xi. 242 Look at that little Venus Anadyomene, Arthur, with the fire-light shining on

her; quite Rembrandtesque, isn't it? 1934 *Burlington Mag.* May 213/2 The artistic Rembrandtesque pen-and-ink sketches then in vogue. 1952 M. Allingham *Tiger in Smoke* iii. 55 One of the old naphtha flares..making Rembrandtesque clouds above them. 1976 *Listener* 12 Aug. 174/1 There is not a single illustration from the Book of Job, although one might have expected it to be full of Rembrandtesque motives.

Rembra·ndtian, Rembra·ndtic, *adjs.* = Rembrandtesque *a.*
1863 *Miss Jemima's Swiss Jrnl.* (1963) i. 8 The white mob-caps of the old women having quite a Rembrandtic effect. 1967 *Listener* 16 Feb. 232/1 He uses a swirl of off-white paint to give the face a Rembrandtian nose.

R.E.M.E., REME (rī·mi). Also **Reemy.** [Acronym f. initials of *R*oyal (Corps of) *E*lectrical and *M*echanical *E*ngineers.] A Corps of the British Army, formed on 1 June 1942, which handles the repair and maintenance of military machinery. Also *attrib.*
1942 *Daily Tel.* 1 Sept. 4/4 R.E.M.E.—you may pronounce it 'Reemie'—marks an important step forward in Army organisation and a break with a tradition which goes back to Creçy and Poitiers. 1943 Hunt & Pringle *Service Slang* 55 *Reemy*, the Royal Electrical and Mechanical Engineers, formed in 1942 for the repair of Army tanks. 1944 A. Jacob *Traveller's War* xiii. 219 It could not have functioned without R.E.M.E. and the enormous industrial undertaking in the rear which this corps operated. 1965 A. Nicol *Truly Married Woman* 16 He had had an outing.. with Higgins, a REME Lieutenant, to see some ruined, Roman fortifications. 1975 C. Mott-Radclyffe *Foreign Body in Eye* xiii. 209 My son is a corporal in the REME and is working as a fitter in one of the ordnance sheds at Tel-el-Kebir. 1978 R. V. Jones *Most Secret War* xlix. 484 We had two other R.E.M.E. officers, Majors K. G. Dobson and R. A. Fell.

reme:diabi·lity. *rare.* [f. Remediable *a.*: see -ity.] = Remediableness.
1964 A. O. J. Cockshut *Unbelievers* 25 Speculations about the remediability of life..are scattered through his writings.

remedial, *a.* Add: **2.** *Educ.* **a.** Designating or pertaining to special classes, teaching methods, etc., in basic educational skills to help schoolchildren who have not achieved the proficiency necessary for them to be able to learn other subjects with their contemporaries.
1924 E. M. Paulu (*title*) Diagnostic testing and remedial teaching. 1927 *Psychol. Abstr.* I. 217 A group of 29 teachers..took a course in remedial reading... They applied to themselves the remedial techniques about which they were studying. 1942 F. Schonell *Backwardness in Basic Subj.* x. 204 If remedial work with backward readers is to be effective, a teacher should have..detailed information from a thorough diagnosis... Remedial methods must have therapeutic as well as pedagogical value. 1975 *Language for Life* (Dept. Educ. & Sci.) xviii. 270 Two groups of children were given remedial education, one in the Remedial Centre and one elsewhere. 1977 *Cork Examiner* 8 June 4/9 The most urgent point here was the need to eliminate overcrowded classes by reducing the pupil–teacher ratio to not more than 35 pupils per teacher and the appointment of more remedial teachers. 1978 *Times Lit. Suppl.* 1 Dec. 1394/2 Caught up in the whole nasty mess is the one other child character in the book, Manjit Mirza, a Sikh girl who shares Ronnie's remedial reading lessons.
b. Of a child: receiving or requiring remedial teaching.
1966 *New Statesman* 22 Apr. 575/2 The youngest, according to her passport, is 11 years old. She looks and acts more like a remedial eight-year-old. 1969 *Word Study* Feb. 1/1, I first became aware that something was amiss, neologically, when I heard a teacher refer to one of her students as 'a remedial reader'. 1975 *Language for Life* (Dept. Educ. & Sci.) xxv. 408 The average weekly time lay between 3 hrs. 12 mins. and 3 hrs. 22 mins. for all except the 12 year old 'Remedial' pupils. 1976 *Cumberland News* 3 Dec. 10/2 And then there was the problem of finding somewhere for the remedial children, who need special teaching.
3. Concerned with or aimed at the overcoming of muscular disabilities or postural defects by means of special exercises.
1925 H. E. Stewart *Physiotherapy* xxii. 321 Faulty postures and weight-bearing..will readily deform one [*sc.* a child] whose tissues are subnormal. It is of vital importance to discover these conditions early and institute proper remedial measures. 1943 O. F. G. Smith *Rehabilitation, Re-Education & Remedial Exercises* i. 9 When the patient has learnt to use the appliance,..whether a piece of needlework or a remedial apparatus, some personal responsibility should be put on him to continue the work in his own time. 1951 F. Charlesworth *Chiropodial Orthopædics* viii. 132 Progressive remedial exercises, Faradism and other form[s] of physiotherapy are a necessary adjunct to the treatment. 1974 *Times* 11 Jan. 15/1 This will be discussed..at a meeting with the Councils of the Association of Occupational Therapists, the Chartered Society of Physiotherapy and the Society of Remedial Gymnasts. 1975 Arnheim & Sinclair *Clumsy Child* vi. 50/2 The most effective remedial program for the clumsy individual is one founded on the principles of psychomotor development.

remediate, *a.* Delete †*Obs.*—[1] and add further *arch.* example.
1906 G. G. Coulton tr. *Pearl* 33 And washed me in blood remediate.

reme·diate, *v.*[2] [Back-formation from Remediation.] *trans.* To remedy or redress.
1969 *Word Study* Feb. 2/1, I encountered the phrase 'teachers *remediating* speech difficulties' (my italics again). Another patent neologism, I thought, derived obviously from a transitive weak verb *to remediate*, itself a back-formation on the analogy of the often-heard *to orientate* 'back-formed' from the noun *orientation*. 1973 *Black World* Mar. 31 Without supportive services to..remediate their academic deficiencies,..the students experienced an extremely high failure rate the first year. 1976 *Canad. Jrnl. Linguistics* Spring 92 The effect of expansion was to increase the time allowed for segment analysis, thus remediating such falling behind.

remediation. Delete *rare* and add further examples. *Esp.* the giving of remedial teaching or remedial therapy (see *remedial *a.* 2 a, 3). Freq. *attrib.*
1954 L. J. Cronbach *Educ. Psychol.* vii. 215 (*heading*) Remediation of emotional difficulties. 1969 *Language* XLV. 599 The report was originally designed to provide descriptive information on which to base linguistic retraining or 'remediation' programs. 1970 H. Osser in S. Rogers *Children & Lang.* (1975) v. 304 Most research has been concerned with diagnosis of problems rather than with their remediation. 1975 *Language for Life* (Dept. Educ. & Sci.) xvii. 263 The school psychological service has provided a series of in-service training, workshop courses in the assessment and 'remediation' of specific reading difficulties. 1975 Arnheim & Sinclair *Clumsy Child* vi. 50/1 Because of the inconspicuousness of symptoms and the difficulties inherent in diagnosing clumsiness, delay in remediation has..been the rule rather than the exception. 1978 M. T. Erickson *Child Psychopathol.* x. 220 Educational remediation is usually conducted by a teacher who has been trained in a special educational program. *Ibid.*, Most remediation teachers are employed by school systems.

remedy, *sb.* Add: **4.** Also *attrib.*
1920 *Act* 10 *Geo. V* c. 3 § 1(1), As though for the figure '4' in the column relating to the remedy allowance in respect of millesimal fineness there were substituted the figure '5'.

remeet, *v.* Add: **1.** (Further example.)
1953 J. S. Huxley *Evolution in Action* iii. 71 A number of forms..which then remet when the ice retreated.
2. *trans.* To meet (a person or thing) again.
1928 *Observer* 24 June 8/6 (Advt.), So tersely..does he tell his tale that within 300 pages we re-meet the classical heroes of three generations. 1970 I. Petite *Meander to Alaska* i. viii. 77 As usual, in travelling north to Alaska, I had the feeling of remeeting spring.

remember, *v.* Add: **I. 1.** Also *transf.* Cf. *memory *sb.* 1 c, d, 2 d.
1933 *Boys' Mag.* XLVII. 170/1 If two cars pass over the detectors simultaneously then right of way is given to one and the arrival of the other is 'remembered', the right of way being accorded to it as soon as the first is safely through. 1958 *Engineering* 21 Mar. 358/1 Upon playback the machine 'remembers' the original picture without loss of detail. 1980 'D. Rutherford' *Turbo* ix. 130 They bend on impact instead of breaking but the material remembers its original shape and goes back to it.
4. a. (Later examples, const. *about.*)
1847 G. P. R. James *Whim* III. ix. 164 Remember about the burning of the will. 1891 W. Morris *News fr. Nowhere* ix. 67, I *do* remember about that strange piece of baseless folly. 1919 G. B. Shaw *Heartbreak House* i. 9 *Nurse.* Youve actually remembered about the tea! (*To Ellie*) O, miss, he didnt forget you after all!
b. (Further U.S. examples.)
1862 M. D. Colt *Went to Kansas* x. 150, I then remembered of reading of such a practice among Southern ladies. 1903 *Profitable Advertising* Nov. 500 'Do you remember of ever making a purchase as the result of an advertisement?' asked the writer. 1923 B. Hecht *Florentine Dagger* xiii. 224 She remembers dimly, she says, of striking him with a dagger. 1948 *Amer. Speech* XXIII. 237 Of the two or three thousand local [Pennsylvania Dutch] people whose speech the present writer has heard during the past seventeen years, no native has ever said, 'I remember it', but always, 'I remember of it'.
5. b. (Later example.)
1890 C. M. Yonge *Slaves of Sabinus* ii. 22 He remembered him of snow-capped Hermon.
II. 7. a. (Later examples.)
1922 Joyce *Ulysses* 398 Would to God that foresight had remembered me to take my cloak along! 1935 E. R. Eddison *Mistress* (1967) xiii. 229 And while he felt about for firm ground then Lessingham again, most courtly and submissive, remembering Derxis of that former passage with Alquemen.

remembering, *vbl. sb.* (Further examples.)
1918 W. Stevens in *Poetry* (Chicago) May 63 Wait now; have no rememberings of hope, Poor penury. 1969 K. H. Pribram in *Sci. Amer.* Jan. 73 (*title*) The neurophysiology of remembering.

remembrance, *sb.* Add: **7. d.** *Garden of Remembrance* (also with small initials), a garden commemorating the dead, esp. those killed in the world wars of 1914–18 and 1939–45.
1954 J. Betjeman *Few Late Chrysanthemums* 46 They'll catch me coming..Across the Garden of Remembrance? No, That would be blasphemy. 1959 *Listener* 22 Jan. 166/1 Here is a statue of Byron... Here are cenotaphs commemorating other philhellenes of several nations... The visit to what is now a garden of remembrance was a profoundly moving experience. 1973 J. Rossiter *Manipulators* iv. 48 With any luck..I'll find the bastard dead and scattered over a garden of remembrance.

9. a. (Later example.)
1845 C. Dickens *Let.* 2 Oct. (1977) IV. 396, I send you the claret jug. But for a mistake, you would have received the little remembrance almost immediately after my return from abroad.
10. *attrib.* and *Comb.*, as *remembrance-banquet, wreath*; **Remembrance Day,** the Sunday nearest to 11 Nov., kept in remembrance of those killed in the world wars of 1914–18 and 1939–45, and since 1945 combined with Armistice Day; **Remembrance Service,** a service held on Remembrance Day; **Remembrance Sunday** = *Remembrance Day*; **Remembrancetide,** the period immediately preceding Remembrance Day, considered as if part of the liturgical year.
1930 R. Graves *Ten Poems More* 9 A fresh remembrance-banquet to forestall The Knight turned hermit. 1921 *Times* 11 Nov. 12/2 We have received a number of appeals which may be specially associated with 'Remembrance Day'. 1929 *Radio Times* 8 Nov. 440/2 Remembrance Day, Nov. 11. Wear a Flanders Poppy. 1946 [see *Armistice Day*]. 1974 P. McCutchan *Call for Simon Shard* i. 5 The old soldiers had brought a Remembrance Day wreath. 1964 L. Deighton *Funeral in Berlin* i. 312 You have an invitation. It's the Remembrance Service. 1977 *Belfast Tel.* 14 Feb. 4/6 He would like to tell these people who were sniping at the Legion that its Remembrance services would continue. 1942 C. Milburn *Diary* 8 Nov. (1979) 157 Remembrance Sunday, and great news today! American troops have landed in North Africa. 1946 *Glasgow Herald* 31 Oct. 4/4 Remembrance Sunday, when Christians gather to remember the young and brave of two generations who died for freedom and the hope of a better world. 1954 R. Macaulay *Let.* 7 Nov. in *Last Lett. to Friend* (1962) 175 Remembrance Sunday. A poppy on my coat, another on my car. 1977 B. Pym *Quartet in Autumn* v. 42 The only services that drew congregations of any size were Harvest Festival, Remembrance Sunday and the Carol Service at Christmas. 1970 *Sussex Life* Nov. 73/1 Remembrancetide this year will occupy the period November 2–8. 1977 *Daily Tel.* 5 Nov. 14 The 20th century has seen the creation of new commemorative rituals—and those of Remembrancetide are not the only ones. 1977 *Lancs. Life* Nov. 74/3 They have invited..a local boy who served in the RAF..to place the remembrance wreath on their memorial.

remembrancer. Add: **2.** (Further examples.) Also, a memoirist, a chronicler.
1951 [see *continuity 6]. 1957 *Times* 8 Aug. 8/2 Both [wrote] personal portraits and reminiscences. In both cases Posterity will need to bear in mind that they are not always reliable remembrancers. 1968 G. Jones *Hist. Vikings* iv. ii. 356 Byrhtnoth's brave but..foolhardy stand..found no remembrancer among the victors.

remembrancing, *vbl. sb.* (Further example.)
1952 *Essays in Criticism* II. ii. 150 Many religious poems in English had a Latin refrain which gave the recurrent gesture of authority and devotional remembrancing.

remen (re·men). [Ancient Egyptian.] An ancient Egyptian measure of length (see quots.).
1934 F. Petrie *Measures & Weights* 5 Remen (⅘ of 29·2 in.). The *Remen* doubled was the diagonal of the square cubit, 29·161 [in.]. 1959 *Chambers's Encycl.* IX. 183/1 The Egyptian royal cubit (20·63±0·2 in. or 524±5 mm)... From this cubit was formed the double remen, the length of the diagonal of a square with sides equal to the royal cubit. Thus, the double remen was equal to $\sqrt{2}\times 20{\cdot}63$ in. (29·16 in., 740·66 mm). This was the basis of the ancient Egyptian land measure. 1969 *Listener* 18 Dec. 859/3 The Royal cubit is $\sqrt{2}$ Egyptian remens, i.e. the diagonal of a 1×1 remen square. The 'Megalithic yard' is thus $\sqrt{5}$ remens, or the diagonal of a 2×1 remen rectangle.

remicle (re·mik'l). *Ornith.* [f. L. *rēmi(g-, rēmex* Remex: see *-cle* s.v. -cule.] A smaller outermost primary wing feather in some birds.
1887 R. S. Wray in *Proc. Zool. Soc.* 344 [In the wild-duck's wing] the distal predigital (11) is always small and is designated the remicle. 1924 *Bull. Amer. Museum Nat. Hist.* L. 316 In the three specimens of *Gavia stellata* seen in the flesh the normal number of quills, ten large ones and the remicle, were present. 1964 A. L. Thomson *New Dict. Birds* 665/2 In most non-passerine species there are—not counting a remicle, if present..—10 primaries in normal individuals.

re·migatory, *a. rare.* [f. L. *rēmigāt-*, ppl. stem of *rēmigāre* to row + -ory[2].] Pertaining to or connected with rowing.
1911 J. Munro *F. J. Furnivall: a Record* p. xvii, A special providence seems to have guarded over Furnivall on his remigatory excursions.

remi·litarize (ri-), *v.* [Re- 5 a.] *trans.* To re-arm (a country or territory that has earlier been disarmed or demilitarized). So **remilitariza·tion.**
1937 *Nation* 28 Aug. 215/1 Since the remilitarization of the Reich. 1939 Webster Add., *Remilitarize*.., to prepare or equip again with military forces, defenses, etc. 1944 D. Thomson *French Foreign Policy* (Oxf. Pamphlets World Affairs lxvii) 5 In 1936..Hitler paved the way for all further aggressions by occupying and remilitarizing the left bank of the Rhine. 1969 *Daily Tel.* 5 Feb. 24/6 The explicit intention to re-militarise Sinai and encourage terrorist groups. 1976 *Survey* Summer–Autumn 13 The status quo.. trend in Soviet policy between 1964 and 1973 was obscured by..remilitarization of the stagnant economy.

remind, *v.* Add: **2. a.** Also *absol.*, and with direct speech as obj.
1891 Kipling *Light that Failed* xiv. 291 It will recall and remind and suggest and tantalise, and in the end drive you mad. 1966 D. F. Galouye *Lost Perception* ii. 24 'Manuel sent the last two messages,' Gregson reminded. *Ibid.* iv. 44 Forsythe withdrew from his sightless isolation long enough to remind, 'Next week's Thanksgiving.' 1976 B. Freemantle *November Man* vii. 95 'The details.. indicated criticism of the Soviet Union,' reminded Kodes.

reminding, *ppl. a.* (Earlier example.)
1872 Geo. Eliot *Middlemarch* II. iv. xxxvii. 271 Even the pale stag seemed to have reminding glances.

remi·nding, *vbl. sb.* [f. Remind *v.* + -ing[1].] The act of reminding; a reminder.
1836 J. S. Mill *Let.* Feb. in *Wks.* (1963) XII. 294 The things..were not done even after numerous remindings. 1865 —— *Comte* 129 Everything that he can do without the aid of incessant remindings from other thinkers, is merely provisional, and will require a thorough revision.

Remington (re·miŋtən). [The name of Eliphalet *Remington* (1793–1861) and his son Philo (1816–89), gunsmiths of Ilion, New York, the original manufacturers.] A proprietary term for a make of firearms and typewriters.
1865 S. Bowles *Across Continent* iii. 23 Perhaps he had intuitive knowledge of our brave hearts and our innumerable Colts', Smith and Wessons', Remingtons', Ballards', and double-barreled shot-guns. 1871 W. W. Greene *Mod. Breech-Loaders* 192 The Remington Rifle...has been extensively used in America, France, Denmark, and Austria. 1888 *Official Gaz.* (U.S. Patent Office) 23 Oct. 350/2 Type-writing machines.—Standard Typewriter Manufacturing Company, New York, N.Y... Used since 1880. 'The word "Remington".' 1895 G. B. Shaw *Let.* 27 Aug. (1965) I. 551, I typewrite on a Bar Lock; but they now all imitate the Remington type so closely that there is no telling. 1895 W. S. Churchill in *Daily Graphic* 27 Dec. 4/3 The rebels, who use Remingtons, fired independently. 1897 *Sears, Roebuck Catal.* 570/2 The Remington Semi-Hammerless Single Barrel Breech Loading Shot Gun... You take no risk in buying the old and reliable Remington. 1906 *Official Gaz.* (U.S. Patent Office) 4 Dec. 1683/1 Remington Arms Company, New York, N.Y... Used ten years. *Remington*... Shotguns, Pistols, Revolvers, and Rifles. 1926 *Trade Marks Jrnl.* 19 May 1177 *Remington*... Typewriters, accounting typewriters and portable typewriters and accessories... Remington Typewriter Company..New York, U.S.A.; manufacturers. 1935 N. Marsh *Enter Murderer* xvi. 193 'Is there a typewriter?' 'There is. A Remington.' 1949 *Lubbock* (Tex.) *Morning Avalanche* 23 Feb. ii. 1/1 Among the rare pistols are an 1850 double-barreled dueling pistol.. and a five-shot Remington pistol. 1959 A. K. Lang in *Alfred Hitchcock's Mystery Mag.* Feb. 77/2 The blonde found herself staring down the muzzle of his Remington. 1973 'A. Hall' *Tango Briefing* ix. 114 A Remington ·410 across his knees..just the one shot. 1975 *Country Life* 9 Oct. 920/4 A link between typewriters and women's emancipation. The first Remingtons were produced in 1873.

reminisce, *v.* Add: **2.** Also with direct speech as obj.
1961 *Dallas Morning News* 10 Sept. vi. 6 'I bought my first dress from him when I was still a struggling young actress,' she reminisces. 1969 A. Glyn *Dragon Variation* viii. 233 'I remember when the whole thing was eighteenth century,' he reminisced. 'Chandeliers, brocades.' 1978 *Daily Tel.* 28 Aug. 3/1 'I remember the teacher asking what we wanted to be,' Signor Santo Del Bon reminisced.

reminiscence. Add: **4.** *Psychol.* An improvement in the memory or performance of something partially learned, occurring after the learning has ceased.
1913 P. B. Ballard in *Brit. Jrnl. Psychol.* Monograph Suppl. I. ii. 17 As obliviscence is a gradual process of deterioration in the capacity to revive past experiences, so is reminiscence a gradual process of improvement in that capacity. *Ibid.* 31 In the case of very young children the interval which secures maximal reminiscence seems to be three days. 1935 *Amer. Jrnl. Psychol.* XLVII. 89 The results of the two experiments..indicate that reminiscence..occurs independently of casual revival or intentional review. 1951 S. S. Stevens *Handbk. Exper. Psychol.* xvii. 653/1 Although most of the work on reminiscence had used verbal material, the appearance of reminiscence is not restricted to verbal learning. 1978 E. Gulian et al. in M. M. Gruneberg et al. *Practical Aspects Memory* ii. 596 Training sessions following closely together produce no marked improvement in performance and..clear-cut progress shows-up only after a training gap. This finding is akin to the phenomenon of reminiscence, which is described in a wide range of learning studies.

remini·scer. = Reminiscencer.
1966 *Punch* 30 Nov. 825/1 The revenants and reminiscers: Hearne, Tuckwell, Mozley, Gunning, and the rest—chirpy, inconsequent, lords of anecdote, abbots of unreason.

remini·scing, *vbl. sb.* [f. Reminisc(e *v.* + -ing[1].] The action of the verb Reminisce.
a 1910 'Mark Twain' *Autobiogr.* (1924) II. 204 A deal of pretty jolly reminiscing was done. 1929 E. W. Springs *Above Bright Blue Sky* 239 She wanted to do a lot of reminiscing, but I cut her short.

remise, *sb.*[2] Add: **4.** A specially planted shelter for partridges. Also *attrib.*
1905 *Kynoch Jrnl.* Jan.–Mar. 45 An instance where this 'remise' system has been carried out most successfully and on a large scale. *Ibid.* 46 When the natural food is exhausted, in hard weather a few handfuls of small corn..are scattered about inside the 'remise', which gives employment to many coveys who are hard pressed for food, and they also serve as a sort of headquarters to which all partridges in the neighbourhood can retire if disturbed. 1939 *Country Life* 11 Feb. p. xxii/2 It is this danger which is also one of the dangers of laying out a partridge *remise* as part of a plantation scheme.

remish (rī·miʃ), slang abbrev. of Remission (sense 4 b).
1958 F. Norman *Bang to Rights* i. 15 That is if I dont get nicked for nothing and get a few days chokey and a few days remish.

remit, *sb.* Restrict 'Chiefly *Sc.*' to senses in Dict. and add: **3. b.** (rī·mit) *N.Z.* An item submitted for consideration at a conference, etc.
1916 *Maoriland Worker* 12 July 4/7 Messrs. Hutchison and Harper moved that the Order Paper Committee put on a remit dealing with Conscription.—Carried. 1918 *Conf. United Federation of Labor* (N.Z.) 4/1 Mr. B. Martin moved the Auckland District Council remit: 'That capitation to the National Executive from the District Councils under clause 12 be reduced to 3d per member.' 1958 *N.Z. Listener* 5 Sept. 8/2 We might see that New Zealand would put a remit up to the International Board and it would be turned down. 1963 *Manawatu Standard* 9 Apr. 12/9 A Canterbury remit that the Government should appoint a Minister of Road Transport..could not meet with the general approval of the Associated Chambers of Commerce delegates in Rotorua. 1966 G. W. Turner *Eng. Lang. in Austral. & N.Z.* viii. 174 Policies of influential bodies are a good deal determined by remits from below.
c. A set of instructions, a brief.
1963 *Guardian* 30 May 8/6 The remit given to Sir Gilbert Flemming, who is considering the possibility of the dispersal of Government departments. 1971 *New Scientist* 25 Feb. 407/2 The remit is essentially to produce a scenario of nutritional developments. 1973 M. Mackintosh *King & Two Queens* xv. 209 Your self-imposed remit in America was to find out..what he was covering up. 1977 *Undercurrents* June–July 12/3 The answer lies simply in the fact that the DHSS is a bureaucratic department which does not have the remit to support publications other than official government documents.

remitless (rī·mi-tlės), *a. rare.* [f. Remit *v.* + -less.] Without remission; unpardoned; ceaseless.
a 1907 F. Thompson *Works* (1913) I. 198 Meek guides and daughters to the blinded heaven In Œdipean, remitless wandering driven.

remittance. Add: **2.** *remittance man* (further examples); also *fig.*
1903 [see *palouser]. 1959 T. S. Eliot *Elder Statesman* iii. 93 Everyone would sneer at the fellow from London, The limey remittance man for whom a job was made. 1969 *Listener* 9 Jun. 42/2 I'd arrived at the end of the line already; the proclaimed remittance man of an obsolete social system. 1975 C. Aird *Slight Mourning* vii. 69 'Last heard of in the backwoods of Queensland.'..'A remittance man, I'll be bound.'

remi·ttence. *rare.* [f. as Remittent *a.* + -ence; cf. F. *rémittence*.] = Remittency (example in *Path.*).
1901 *Practitioner* Mar. 311 When the paroxysms return in the evening and are prolonged so that the intermission or remittence takes place in the morning, this fever may be easily mistaken for typhoid.

‖ **remittitur** (rīmi·titʊ̈ɪ). *Law.* [a. L. *remittitur*, third pers. sing. pass. of *remittere* to Remit.] **1.** The remission of excessive damages awarded to a plaintiff, or a formal statement of this.
1770 G. Wilson *Rep. Cases King's Courts, Westm.* I. 30 The court said plaintiff might take judgment *de melioribus damnis* where several damages are given, or enter a *remittitur*. 1792 B. J. Sellon *Pract. Courts King's Bench* I. xi. 500 Where a verdict was given for a greater sum than the amount of the damages laid in the declaration, court will suffer amendment to be made by plaintiff, entering a *remittitur* of the *extra* sum. 1848 J. J. S. Wharton *Law Lexicon* 579/2 *Remittitur damna*, where a jury gives greater damages than a plaintiff has declared for, it may be rectified by entering a *remittitur* for the excess; or, if a plaintiff have signed judgment for the greater sum, the court will give him leave to amend it, by entering a *remittitur* for the excess.
2. The act of sending the transcript of a case back from an appellate to a trial court for record or further work; a formal notice of this.
1794 W. Tidd *Pract. Court King's Bench* II. xli. 718 Where a writ of error determines in the Exchequer Chamber, by abatement or discontinuance, the judgment is not again in this court, till there be a *remittitur* entered. 1796 B. J. Sellon *Pract. Courts King's Bench* II. xx. 526 On the hearing, the lords either affirm or reverse the judgment on which the clerk of the Parliaments draws a *remittitur*, by which the transcript of the record is remanded into the King's Bench, with the affirmance or reversal to be entered of record. 1820 *Tomlins's Law-Dict.* (ed. 3) II. s.v. Remittitur, In cases of appeal, the Record itself, or a transcript thereof, is sent from the Court of *B.R.* to the Exchequer-Chamber, or House of Lords; when judgment is given in the superior Court, or the Writ of Error abates, or is discontinued, the record or transcript is returned (*Remittitur*, sent back), to the Court of *K.B.*, and the entry of this circumstance is termed a *Remittitur*. 1848 J. J. S. Wharton *Law Lexicon* 579/2 *Remittitur of record*, formerly when a writ of error, in the Exchequer Chamber, abated, or was discontinued, the transcript must have been remitted, and a *remittitur* entered, before a defendant could sue out execution. 1972 *N.Y. Law Jrnl.* 10 Oct. 2/1 Motion to amend the remittitur granted, the return of the remittitur is requested and, when returned, it will be amended.

remix, *v.* Add: Hence **remi·xing** *vbl. sb.*
1956 *Nature* 10 Mar. 490/1 The undiminished fertility of the inbred plants when pollinated by bees promotes rapid remixing of genetic material under favourable conditions. 1975 *McGraw-Hill Yearbk. Sci. & Technol.* 256/1 After extraction, the interaction between the feed stream and the enriched product must be such that isotopic remixing, or scrambling, does not occur.

remnant, *sb.* Add: **2. b.** Also *spec.*, in allusion to Isa. x. 22, a small number of Jews that survives persecution, in whom future hope is vested.
1859 'L.N.R.' *Missing Link* xiii. 171 The Exiled Remnant ..refuse to snuff a candle or poke the fire, but impatiently call, 'Shuboth-guy—Shuboth-guy,' as the stoker passes. 1892 I. Zangwill *Childr. Ghetto* II. xv. 16 The rest of the 'remnant' that was met to save Israel looked more commonplace. 1914 J. Hastings *Encycl. Relig. & Ethics* VII. 607/2 The function of Judaism is to fulfil the Isaianic ideal of a missionary 'Remnant'. *Ibid.*, Judaism is to be the religion of a Remnant. 1932 C. Roth *Hist. Marranos* i. 16 It had been only a weak remnant which had accepted baptism as the alternative to death. 1969 *Guardian* 18 Sept. 8/6 (*heading*) Germany's remnant. *Ibid.*, There are only around 30,000 Jews in Germany today, where once there were 600,000. 1972 C. Raphael *Feast of Hist.* i. 32 In the immediate aftermath of the Holocaust, Jewish history.. seemed to offer a message of bleakness... In the Holy Land, the remnant faced enmity and restriction.
d. *Geomorphol.* = *residual *sb.* 5.
1893 [see *monadnock]. 1896 *Ann. Rep. State Geologist N.J. 1895* 10 This eastern belt of remnants, which are really outliers of the continuous portion of the Pensauken.., runs through Camden..and Salem counties. 1907 *Amer. Jrnl. Sci.* CLXXIV. 470 At the present time there are remaining only few traces of these old bolson surfaces. Most of these remnants have been preserved only on account of being covered by extensive lava sheets. 1942 [see *berm 1 b].
6. *attrib.* and *Comb.*
1864 E. G. White *Testimonies* (1871) I. 467 All who have a desire to draw away from God's remnant people..should have the privilege. 1885 —— *Testimony for Church* xxxii. 228 The remnant church will be brought into great trial and distress. 1905 *Westm. Gaz.* 21 Oct. 18/2 'It is not, then, a curious fact,' I said, 'that there should be so many comparatively new books on your remnant stall.' 1905 'O. Henry' *Trimmed Lamp* (1907) 115 Did you ever notice me leaning on the remnant counter or peering in the window of the five-and-ten? *a* 1936 Kipling *Something of Myself* (1937) iii. 75 The *Pioneer* had made as much out of its share in this remnant-traffic as it had paid me in wages. 1972 N. Znamierowski *Rugmaking* 17/1 Remnant counters..are.. excellent sources.

remnantal (re·mnăntăl), *a. Geol.* [f. Remnant *sb.* + -al.] Of or pertaining to a remnant.
1907 *Amer. Jrnl. Sci.* CLXXIV. 470 Farther south at Paraje..and at El Paso, the same remnantal levels are noted. 1942 [see *berm 1 b].

remobiliza·tion. [Re- 5 a.] The action of mobilizing again; a further mobilization.
1919 J. L. Garvin *Econ. Found. Peace* viii. 152 German workers might then have the sympathy of their class in other lands to a degree making quite impossible the effective re-mobilisation of the Grand Alliance as a debt-collecting agency. 1977 J. L. Harper *Population Biol. Plants* 392 The demand is met by remobilization of other materials.

remo·bilize, *v.* [Re- 5 a.] *trans.*
1. *Geol.* To make fluid or plastic again.
1954 R. L. Parker tr. *Niggli's Rocks & Mineral Deposits* xiv. 519 Old deposits may be remobilized in depth and may give rise to solutions that ascend to higher levels. 1965 G. J. Williams *Econ. Geol. N.Z.* xii. 178/2 Mr Wood's descriptions seem to leave little doubt as to a genetic association of the Moke Creek greenschist and the enclosed copper ore, even though the latter must have been remobilized during metamorphism. 1971 I. G. Gass et al. *Understanding Earth* xxi. 313/1 Continental mountain ranges are..principally composed of original continental crust (in great part remobilised).
2. *Mil.* To recall to active service.
1963 *Times* 26 Feb. 7/7 The prosecutor, M. Gerthoffer, who is a magistrate remobilized with the rank of general, wound up his two-hour final speech.

re·model, *sb. Arch.* [f. the vb.] The act of modelling or constructing a building again; a remodelled building.
1956 *Archit. Rev.* CXX. 119/3 Rethinking is needed on the difference between surface grime on a noble building and a cut-rate remodel which might well show a fundamental lack of architectural conviction or emotive power. 1974 R. C. Dennis *Conversations with Corpse* ii. 15 A house, French Regency, I think... I imagine it's a remodel. 1978 *Tucson Mag.* Dec. 90/3 In a remodel (existing house) we have to be neater, and work slower.

remonetization. (Earlier U.S. example.)
1877 *Rep. U.S. Monetary Comm. 1876* (44th Congr., 2nd Sess., Senate Rep. No. 703) i. 90 It is not a particular silver coin, the remonetization of which is demanded, but it is the metal silver, in whatever denominations of coin the law may authorize.

REMONETIZE

remonetize, *v.* (Earlier U.S. example.)
1877 *N.Y. Tribune* 16 Nov. 8/1 They regard the ultimate passage of a bill of some kind, remonetizing silver, as a certainty.

remonstrantly, *adv.* (Earlier example.)
1872 Geo. Eliot *Middlem.* IV. lxxvi. 240 'But when she saw the good that might come of staying ——' said Dorothea, remonstrantly.

remonstrate, *v.* Now usu. with pronunc. (re·mǫnstrēᶦt).

remontant, *a.* and *sb.* Add: Also used of strawberry plants bearing fruit for a longer period than usual. (Later examples.)
1923 J. H. McFarland *Rose in Amer.* ii. 21 The Hybrid Perpetual roses are also called Remontant. Both designations are misnomers so far as bloom is concerned. **1965** E. B. Le Grice *Rose Growing Complete* xii. 170 Single, coarse, once-flowering climbers had, at least a thousand years ago, become many-petalled, or dwarf, or remontant (repeat-flowering). **1968** R. Hay *Gardener's Round* 78 Plant the 'remontant' or perpetual strawberries to have a crop in the autumn. **1969** *Oxf. Bk. Food Plants* 74/2 The perpetuals or remontants are an interesting group [of strawberries], which flower successively during the summer and produce fruit from July till October. **1979** *Guardian* 13 Oct. 15/5 Cover remontant strawberries with cloches.

remo·ralize, *v.* [Re- 5 b.] *trans.* To make moral again; to re-instil with morals. So **remo:raliza·tion.**
1967 *Guardian* 16 Oct. 6/5 Violence and pain still provide an evil satisfaction which the remoralisation of sex has not yet exorcised. **1974** *Daily Tel.* 21 Oct. 6/8 We are able to remoralise whole groups and classes of people, undoing the harm done..by permissiveness in television, in films, on bookstalls. *Ibid.* 21 Oct. 6/8 We shall need intellectual as well as moral courage to grapple with the dilemmas inherent in the remoralisation of public life.

remorse, *sb.* Add: **7.** *remorse-stricken* adj.
1973 M. Amis *Rachel Papers* 56, I couldn't resist taking a certain fascinated pleasure in his remorse-stricken face.

remo·rtgage, *v.* [Re- 5 a.] *trans.* To mortgage anew; to change the terms of a mortgage on (a property). So **remo·rtgage** *sb.*; **re-mo·rtgaging** *vbl. sb.*
1960 *Farmer & Stockbreeder* 15 Mar. 125/1, I have tried to raise capital by various means, including re-mortgaging, but without success. **1961** Benjamin & Atholl *How to borrow Money* vii. 77 A practical alternative to offering a second mortgage as the security may be to re-mortgage the house. *Ibid.*, The probability is that you have had the house for a number of years so that if you could effect the re-mortgage, the amount you would have to repay on your present mortgage would be substantially less. *Ibid.* 78 The comparative costs of re-mortgaging and raising a loan on a second mortgage..should be compared. **1976** *Milton Keynes Express* 25 June 33/5 (Advt.), Deposit loans, personal loans, remortgages, second mortgages, business finance. **1977** *S. Wales Echo* 18 Jan. 11/1 (Advt.), Building society re-mortgages and second mortgages arranged. **1978** *Cornish Guardian* 27 Apr. 34/2 (Advt.), Also available—First mortgages, re-mortgages and personal loans for tenants.

remote, *a.* (and *sb.*) and *adv.* Add: **A.** *adj.* **3. f.** Situated, occurring, or performed at a distance (not necessarily great); *remote control*, control of apparatus, etc., at a distance; also (with hyphen) *attrib.*; so *remote-controlled* ppl. adj., *remote-control* vb. trans. and intr. (also *fig.*).
1904 L. Andrews *Electricity Control* i. 8 It is probable..that for installations of a few thousand horse-power only, some simple method of mechanical remote control will be generally preferred. **1920** *Wireless World* 7 Aug. 356/1 Pilot's and mechanic's cockpits are not very roomy compartments and therefore it has become standard practice to employ 'remote control', that is to say the main portion of the wireless apparatus..are [*sic*] fitted in one or two boxes which can be suspended in any convenient part of the main fuselage of the machine; these circuits being controlled by a small unit..which may be fitted on the dashboard of the machine. **1933** *Times* 16 May 9/2 A remote control device for the selection of several alternative wireless programmes will soon be made available to the public. **1943** *Gloss. Terms Electr. Engin.* (*B.S.I.*) 84 *Remote-controlled substation*, a substation the operation of which is controlled at a distance. **1956** *Nature* 28 Jan. 160/2 The remote-handling device for removal of the collectors containing the enriched product without exposure to air. **1957** *Economist* 9 Nov. 525/2 Because of their radioactivity, none of the materials can be handled normally. All operations are carried out painstakingly by remote control. **1961** G. Millerson *Telev. Production* iii. 28 (*caption*) Lens turret,..rotated by rear handle..or remote switching. **1966** P. O'Donnell *Sabre-Tooth* xv. 203 Two transmitters..were remote-controlled from the H.Q. section. **1967** Cox & Grose *Organization & Handling Bibl. Rec. by Computer* iv. 95 The use of these direct access devices also paves the way for remote-terminal inquiry. **1970** O. Dopping *Computers & Data Processing* vi. 96 Remote processing of data..normally requires multiprogramming. In remote processing, input and output goes via communication lines. **1970** 'B. Mather' *Break in Line* xv. 187, I wondered if he were still in Calcutta or was remote-controlling from London. **1970** *New Scientist* 6 Aug. 286/1 The study defines remote-access computing as the use of computers where the main computer installation is at a distance from the user, who employs a terminal device to communicate with the computer over telephone or other links. **1972** *Times* 11 Sept. (Botswana, etc., Suppl.) p. vi/2 (*caption*) Remote sensing, a development of aerial photography, can point to possible indications of mineral deposits. **1973** C. W. Gear *Introd. Computer Sci.* iv. 162 Many computer systems have low speed input/output devices, called remote terminals, attached to the central computer. **1974** *Harrods Christmas Catal.* 69/1 Remote-control Gantry Crane, battery operated... 28" high. £8·50. **1977** *Nature* 6 Jan. 34/2 Until this year, the most accurate means of studying the atmospheric pressure at the surface of Mars were provided by remote-sensing from fly-by or orbiting spacecraft. **1978** R. V. Jones *Most Secret War* viii. 68 The German Navy was said to have developed remote-controlled rocket-driven gliders of about three metres span. **1981** *Oxford Jrnl.* 15 May (Advt.), 20" Colour TV. Remote control. *Ibid.*, 14" Colour Portable TV with infra-red remote control hand unit.
6. *not the remotest*: also *ellipt.*
1928 D. L. Sayers *Unpleasantness at Bellona Club* xvii. 205 'Was the quantity marked on the bottle?' 'I haven't the remotest. You'd better ask her.' **1969** E. Stewart *Heads* (1970) 94 'Why do you think he was trailing Father Fields?' 'I haven't the remotest,' Greg said.
B. *sb.* **c.** *U.S. Broadcasting.* An outside broadcast (see quots.). Cf. *nemo.
1937 *Amer. Speech* XII. 100 A *remote pickup* or simply a *remote* means a program brought from some point other than station studios. **1937**, etc. [see *nemo]. **1947** *Billboard* 1 Nov. 17 First Remote on War Dead's Arrival... What is believed to be the first video broadcast by a remote unit from a moving object will be essayed tomorrow. **1962** *Sat. Rev.* 1 Sept. 17/2 Accent is not only a low-budget show; it is a 'remote', A 'remote' shoots on location with video-cruiser facilities, as distinct from a studio show. **1967** *Boston Globe* 30 Mar. 3/1 CBS said in future days Cronkite may be seen in some news remotes while Zenker remains at the desk. **1976** *Listener* 15 July 53/1 'Remotes' are what American television technicians call outside broadcasts.
C. *adv.* In comb. with a pple. forming an adj., = *remotely *adv.* 2 b.
1943, etc. Remote-controlled [see sense A. 3 f above]. **1959** H. Barnes *Oceanogr. & Marine Biol.* iv. 200 Various systems could be devised, some complicated and expensive using remote-indicating compasses, but we have merely mounted an ordinary liquid compass in the field of view of the camera. **1976–7** *Sea Spray* (N.Z.) Dec./Jan. 94 The unit is available either directly mounted to a Borg–Warner marine reverse transmission unit or remote mounted and coupled to engine or reverse gear by a universal joint shaft.

remotely, *adv.* Add: **2. b.** At or from a distance (not necessarily great). Freq. in comb. with a pple. forming an adj.
1957 *Railway Mag.* Nov. 758/2 The remotely-controlled signalbox, normally unstaffed, is retained. **1967** *Jane's Surface Skimmer Systems 1967–68* 122 Take-off (4 hp continuous) for remotely-driven accessory box. **1971** *Physics Bull.* July 395/3 The appropriate parts of the projector are light proof and the shutter is operated remotely so that photographic records may be taken in a lit room. **1973** *BBC Handbk. 1974* 246/2 The network control rooms handle..remotely controlled studios, such as the news studio at Westminster.

remoteness. Add: **b.** A remote region; = Remote *sb.* b. *nonce-use.*
1880 'Mark Twain' *Tramp Abroad* xxxii. 345 Switzerland, and many other regions which were unvisited and unknown remotenesses a hundred years ago, are in our days a buzzing hive of restless strangers every summer.

remo·tivate, *v.* [Re- 5 a.] *trans.* To motivate anew. Hence **remo·tivating** *vbl. sb.* and *ppl. a.*
1974 *Listener* 28 Feb. 271/1 They try, in a favourite word of probation officers, to 'remotivate' men who have been through the penal system. **1976** *Archivum Linguisticum* VII. 28 Writers tend to be consistent with themselves if not with each other, although rarer forms are probably remotivated on each occasion. **1977** *Spare Rib* Jan. 8/1 Towards the end of my career, interviewers were invited to a Re-motivating Lecture with supervisors and regional controllers. **1977** D. Morris *Manwatching* 184 (*caption*) Re-motivating Actions succeed by replacing a companion's unwanted mood with a new, more attractive mood.

remou: see *remous.

‖ **remoulade** (remulād). *Cookery.* Also **ré-**. [a. F. *rémoulade.*] A French salad dressing (see quots.).
1845 E. Acton *Mod. Cookery* iv. 135 (*heading*) Remoulade. This differs little from an ordinary salad dressing. **1861** Mrs. Beeton *Bk. Househ. Managem.* 241 (*heading*) Remoulade, or French salad-dressing. *Ibid.*, 4 eggs, ½ tablespoonful of made mustard, salt and cayenne to taste, 3 tablespoonfuls of olive-oil, 1 tablespoonful of tarragon or plain vinegar... Green remoulade is made by using tarragon vinegar instead of plain. **1877** E. S. Dallas *Kettner's Bk. of Table* 376 *Remoulade*..may be..described as a Mayonnaise made with hard-boiled yolks of eggs. **1939** A. Simon *Conc. Encycl. Gastron.* I. 46/1 *Rémoulade*, a salad dressing consisting of the yolks of hard-boiled eggs, oil and vinegar, salt and pepper. Mustard is sometimes added. **1961** *Listener* 20 Apr. 719/2 There is always a little of it [*sc.* tarragon] in smooth French sauces, such as Béarnaise, tartare, and rémoulade. **1966** N. Freeling *Dresden Green* I. 21 He..stopped at the dairy..for a piece of cheese, celery remoulade salad. **1978** G. Vidal *Kalki* iii. 79 The preparation of a shrimp remoulade.

re·mould, *sb.* [f. the vb.] A worn tyre on to which a new tread has been moulded. Also *attrib.* Cf. *retread *sb.*
1956 C. Willcock *Death at Flight* iii. 35, I asked the firm's transport department to change both front tyres not three weeks ago. And I told them no remoulds. **1960** *Farmer & Stockbreeder* 29 Mar. 21/1 (Advt.), New and remould tyres on terms! Also the new extra grip remould for town and country use. **1972** *Practical Motorist* Oct. 157/1 A remould uses the carcass of a tyre that has already done a lifetime of service. **1973** *Times* 28 Apr. 4/2 Mr Assender claimed yesterday that £6·50 was a fair price for a 'remould quality' 145-13 tyre. **1976** *Drive* Sept.–Oct. 77/1 *Remould* or *remould quality tyres* should by now have disappeared from the forecourt vocabulary, being replaced by the two official designations.

remount, *sb.* **1. a.** (Earlier example.)
1781 R. F. Greville *Diary* 5 Aug. (1930) 11 This was a favorable opportunity to take a ride, & try a new mare I had lately purchased, & one of a remount, made within a short time of my Appointment.

† **remous** (rəmū·). *Aeronaut. Obs.* Pl. **remous** (with *erron. sing.* **remou**). [Fr., = 'eddy, ship's wash'.] (See quot. 1916.)
1911 *Aeroplane* 8 June 8/1 Brooklands has three constant *remous* or eddies, two downward and one upward. *Ibid.*, The only way to get ——'s 'bus into the air is to 'taxi' to the sewage farm *remou* and get pulled off the ground by it! **1914** G. Hamel *Flying* viii. 167 An attempt has been made, by a well known military pilot, to classify remous as 'rollers', 'half-rollers', and 'wulliwas'. **1915** G. Bacon *All about Flying* vi. 106 The little eddies known as 'remous' are more entertaining than annoying. **1916** H. Barber *Aeroplane Speaks* 140 *Remou*, a local movement or condition of the air which may cause displacement of an aeroplane.

removability. (Earlier examples.)
1789 *Deb. Congress U.S.* 16 June (1834) 464, I am not satisfied that removability shall be acquired only by impeachment. *Ibid.* 6 Aug. (1834) 679 The Senate..insisted on the amendment to the Treasury bill, respecting the removability of the Secretary by the President.

removal. 4. (Further *attrib.* examples.)
1939 M. B. Lowndes *Let.* 23 Oct. (1971) 183 The removal man..told me some interesting things about the art of moving and storing furniture. **1962** J. G. Bennett *Witness* xviii. 218 One of the removal men asked him if a sofa was to go 'up the apples'. **1973** *Times* 28 Dec. 16/1 (Advt.), Assistance with removal expenses if necessary. **1974** M. Gilbert *Flash Point* xiii. 115 They wore corduroy trousers and jackets belted at the waist... They looked like removal men. **1979** *Homes & Gardens* June 77/2 They used his pension to buy an old removal van.

removalist (rīmū·vălist). *Austral.* [f. Removal + -ist.] A person or firm engaged in household or business removals.
1959 S. J. Baker *Drum* (1960) 139 *Removalist*, a person or firm engaging in the shifting of household or business effects. **1966** —— *Austral. Lang.* (ed. 2) i. 4 There is a good deal of evidence to suggest that..*removalist* (a person or firm engaged in moving furniture, etc.)..is an Australian original. **1971** *Classified Telephone Directory* (Brisbane) Pink Pages 251/1 (Advt.), Approved Government contractor for removals and storage. A. F. Palmer Removalists. **1972** D. Williamson *Removalists* (1973) 61 A self assured removalist in a dust coat enters. The dustcoat is emblazoned with the emblem 'Aussie Removalists'.

remove, *sb.* Add: **2. c.** (Earlier example.)
1771 B. Franklin in M. Farrand *Benjamin Franklin's Mem.* (1949) 124/1 Every Man at the first Remove, found under his Plate an Order on a Banker.
6. d. *Printing.* The number of sizes by which the type of a footnote or side-note is smaller than that of the text; hence, the note itself.
1890 C. T. Jacobi *Printing* v. 70 Footnotes are nearly always set in type two sizes (or removes, as they are called) smaller than text... Side-notes are frequently put into three or four removes smaller. **1898** J. Southward *Mod. Printing* I. xxxvii. 224 The usual type for notes is two or three removes from the text. **1934** V. Steer *Printing Design & Layout* xvi. 293 Footnotes are explanatory notes at the foot of the page usually set in type two removes from the size used for the text. **1960** G. A. Glaister *Gloss. Bk.* 345/2 *Removes*, quotations or notes set at the foot of a page and in smaller type than that of the text.

remove, *v.* Add: **I. 1. f.** *Cricket.* Of a bowler or ball: to dismiss (a batsman).
1969 *Wisden's Cricketers' Almanack* 300 Underwood..accounted for Redpath and Walters, each getting an inside edge to the ball that removed him. **1976** *Eastern Even. News* (Norwich) 22 Dec. 14/2 With the fourth ball of his second over Lever removed Venkataraghavan, the ball brushing the batsman's glove before passing through to wicketkeeper Alan Knott. **1977** *Evening Post* (Nottingham) 24 Jan. 16/3 Selvey removed Sivaramakrishnan with his fourth ball.

‖ **remplaçant** (rańplasań). Also fem. **-e.** [Fr.] One who replaces another; a substitute.
1850 Lady Eddisbury *Let.* 23 Mar. in N. Mitford *Ladies of Alderley* (1938) 280 At 5 Macaulay sent word he was too ill to come, Ed. could get no remplaçant & so at ¼ to 8, our dinner having dwindled, we sent for Mama. **1880** E. W. Hamilton *Diary* 23 May (1972) I. 14 The defence of the Government is..that as he [*sc.* Sir B. Frere] is more conversant than any '*remplaçant*' could be with the important question of confederation, he had better be left to carry that through. **1915** W. J. Locke *Jaffery* xiii. 177 'We've settled nothing about a remplaçante for Mrs. Considine.'..'No one can replace Mrs. Considine.'

‖ **remskoen** (re·mskun). *S. Afr.* Also **reimschoon, remschoen, riemschoen, rimschoen**; pl. also **-e.** [Afrikaans:—Du.

remschoen, f. *rem* brake + *schoen* Shoe.] = Skid *sb.* 3 a. Also *fig.*
See J. Smuts et al. *Voorloper* (1976) for additional variant forms.
1816 G. Barker *Diary* 13 Feb. (MS.) in J. Smuts et al. *Voorloper* (1976) 671 My box was set upon the rim-schoen to keep it dry. **1822** W. J. Burchell *Trav. Southern Afr.* I. 151 The remschoen (lock-shoe or skid), is a log of wood, generally about eight inches square, and nearly two feet long, having a groove in it to receive the felly of the wheel. **1835** A. Steedman *Wanderings & Adventures in S. Afr.* I. I. ii. 121 On regaining the track, we found the *reimschoon*, or iron slipper, which had fallen from the waggon, lying in the road. **1898** *Cape Argus* 2 Feb. 36, I am pleased to find that my frequent allusion to the backward element in the legislative Council as a *riemschoen party* has gone home. **1912** *East London Dispatch* 2 May 5 Riemschoen Party.—The name applied a few years back to that party in Cape politics which appeared to be averse from progress; the word Riemschoen is applied in other directions with the same meaning, e.g. 'Riemschoen Districts'. **1949** L. G. Green *In Land of Afternoon* x. 126 Voortrekker wagons were equipped with wooden axles and the *remskoen* instead of brakes. **1957** *Cape Times* 8 Aug. 9/1 The wagons..with *remskoene* on, slithered down the steep slopes.

‖ **remuage** (rəmuaʒ). *Wine-making.* [Fr., lit. 'moving about'.] The periodic turning or shaking of bottled wine (esp. champagne) to move sediment towards the cork before disgorgement.
1926 P. M. Shand *Bk. Wine* v. 154 The bottles are now stacked in wooden racks..for the delicate operation of *remuage*. **1935** Schoonmaker & Marvel *Compl. Wine Bk.* i. 34 Here takes place the curious process known as the 'shaking', or *remuage*. **1958** D. Morris *French Vineyards* ii. 34 When the *remuage* is finished, the bottles are left in darkness. **1967** A. Lichine *Encycl. Wines & Spirits* 430/2 *Remuage* (moving around) is a term used in the process of making Champagne. Bottles placed in specially built racks are turned or shaken a little every day for about four months before they are shipped, so that the sediment may move down towards the cork. **1977** T. Heald *Just Desserts* viii. 187 Along the walls were countless bottles top downwards in racks 'Ready for the *remuage*... Gets the sediment down to the cork.'

‖ **remuda** (rəmū·dă). [Amer. Sp., a. Sp., exchange, replacement.] A herd or collection of saddle-horses kept for remounts.
1892 *Dialect Notes* I. 251 *Remuda*, a 'bunch' of horses, about a score. Usually applied to geldings only. **1903** A. Adams *Log of Cowboy* 9 The *remuda*, under Bill Honeyman as horsewrangler, numbered a hundred and forty-two, ten horses to the man. **1907** S. E. White *Arizona Nights* v. 92 In a moment the first of the remuda came into view, trotting forward with the free grace of the unburdened horse. **1924** W. M. Raine *Troubled Waters* xi. 113 Presently he got up and strolled toward the remuda. **1927** *Blackw. Mag.* Nov. 650/1 In the feeble flare remudas keep on passing. **1955** R. Hobson *Nothing too Good* vii. 61, I knew this was the horse remuda, the advance guard of the drive. **1972** T. A. Bulman *Kamloops Cattlemen* ii. 15 They usually brought with them a pretty fair *remuda* of horses.

‖ **remueur** (rəmuö̃ɪ). *Wine-making.* [Fr., lit. 'mover'.] One who engages in *remuage* (see quots.).
1926 P. M. Shand *Bk. Wine* v. 154 The *remueur's* task is gradually to work down all the sediment. **1965** O. A. Mendelsohn *Dict. Drink & Drinking* 279 *Remueur*, the craftsman who daily twists and slightly shakes the bottles containing champagne in the making. **1976** N. Roberts *Face of France* xxv. 229 The man who does the tilting and turning [of champagne bottles] is called a *remueur*... He can handle 30,000 bottles a day.

remu·ster, *v.* orig. *Services'.* [Re- 5 a.] *intr.* **a.** with pass. sense. To be assigned to other duties. **b.** for *refl.* To assemble again. Hence **remu·stering** *vbl. sb.*
1942 *R.A.F. Jrnl.* 3 Oct. 13 Because Bill Snooks is unfit for air crew duties, he should be allowed to re-muster to..a sedentary trade. *Ibid.* 14 A.C. 2 So-and-So has certain qualifications which make him suitable for re-mustering or training. **1963** *Times* 5 June 14/1 No. 500 (County of Kent) Squadron, Royal Auxiliary Air Force, disbanded in 1957, will remuster for one day to receive its squadron standard from Lord Avon. **1966** *Punch* 6 July 16 Modern football is a managers' game... Attack is based on the counter which passes the opposing defence before it can remuster after its own attack. **1977** 'J. Herriot' *Vet in Spin* (1978) xviii. 166 Normally when an aircrew is grounded he remusters on the ground staff, but yours is a reserved occupation.

remu·tiny, *v.* [Re- 5 a.] *intr.* To mutiny again.
1895 Hardy *Jude* I. iii. 20 He anxiously descended..trying not to think of..the captain with the bleeding hole in his forehead, and the corpses round him that remutinied every night on board the bewitched ship.

Remy Martin (remi martæń). Also **Rémy Martin.** [Name of the shippers.] The proprietary name of a cognac; a drink of this.
[**1951** T. E. Carling *Compl. Bk. Drink* v. 42 Principal Cognac Producers..Remy Martin.] **1961** C. Willock *Death in Covert* i. 9 A large shot of Rémy Martin in a balloon glass. **1963** *Official Gaz.* (U.S. Patent Office) 26 Feb. TM 137/1 *Remy Martin* for Cognac. First use 1884. **1965** L. Meynell *Double Fault* I. iii. 29 Evelyn Barker had a Rémy Martin in front of her. **1965** P. D. Wall *Trio* 1966) iv. 55 He promptly loaded his briefcase with Pernod and Remy Martin. **1975** D. Bloodworth *Clients of Omega* xxv. 242 Sipping his Rémy Martin with its lacing of java. **1976** *Trade Marks Jrnl.* 14 Apr. 783/2 Remy Martin..Brandy. E. Remy Martin and Co... 29th August, 1973.

remytho·logize, *v.* [Re- 5 b, after *demythologize *v.*] *trans.* To provide with a new mythological system; to reinterpret the elements of (an older mythology) in terms of a newer one. Hence **remythologiza·tion.**
1964 K. G. Grubb *Layman looks at Church* v. 156 The Bible..has to be 'demythologised' and then remythologised. **1973** R. Slotkin *Regeneration through Violence* ii. 36 Both [myth and art] serve as means of ordering and explaining a chaotic and threatening environment. The remythologization of the West began with attempts by French and Spanish Jesuits and English Puritans to order the chaos of the New World. **1974** *Canadian-Amer. Slavic Stud.* VIII. 492 The updating, transformation and 'remythologization' of these legends constituted a form of justification of the validity of their world-view. **1976** H. Montefiore in *Christian Believing* 148, I may expect to 'translate' or 're-mythologize' its thought forms and imagery.

Renaissance. Add: Also with small initial.
1. c. (Further examples.)
1842 Queen Victoria *Jrnl.* 14 Sept. (1980) 37 We..saw the fine greenhouse the Duke has built, all in stone, in the Renaissance style. **1930** R. Fry *Let.* 12 Sept. (1972) II. 650 [Montrésor] has..a very ambitious and rather good Renaissance Gothic church. It's odd what a really good and convenient style that makes—in fact it does Gothic much better with less fuss than Gothic itself. **1963** A. Lubbock *Austral. Roundabout* 190 Airy, Renaissance-style stucco arches. **1976** *Early Music* IV. 512/2 (Advt.), Renaissance viols from 16th-century models. **1980** I. Murdoch *Nuns & Soldiers* i. 40 A programme of Renaissance music.
d. Special Combs. **Renaissance humanism** = Humanism 4; **Renaissance man**, one who exhibits the virtues of an idealized man of the Renaissance; also *fig.*
1906 W. H. Woodward *Stud. Educ. Renaissance* vii. 128 That the Frenchmen in their King's train should be profoundly impressed with the Renaissance man as they found him declared in Rodrigo Borgia, and his enigmatic son, in Ludovico Sforza or Ercole d'Este, is no cause for wonder. **1948** W. K. Ferguson *Renaissance in Hist. Thought* iii. 71 Bayle..interpreted Renaissance humanism as an enlightened revolt against barbarism. *Ibid.* v. 128 The discontented rebels against the restrictions of contemporary bourgeois society..took the lead in the idealization of the Renaissance man, combining the cult of genius with that of free, egoistic personality. **1955** P. O. Kristeller *Classics & Renaissance Thought* i. 10 Renaissance humanism was not as such a philosophical tendency or system, but rather a cultural and educational program which emphasized and developed an important but limited area of studies. **1970** E. Pace *Saberlegs* (1971) xiv. 132, I knew your father... A fine man. So many-sided. What I believe you would call a Renaissance man. **1975** *Language* LI. 443 Renaissance humanism was responsible for the most successful system of syntactic analysis to be conceived prior to the advent of explicit syntactic theorizing in the 20th century. **1977** *Time* 8 Aug. 32/3 At 50, Hood is the Renaissance man of sailing; he designed, cut the sails and outfitted *Independence*, the first man in history to control every aspect of a 12-tonner from drawing board to helm.
2. a. (Further examples.)
1925, etc. [see *Negro Renaissance* s.v. *negro 7]. **1969** A. Cockburn in Cockburn & Blackburn *Student Power* 18 The astonishing works of Mao Tse Tung..bear witness to the flowering of the May 4 Movement which..has justly been called the Chinese Renaissance. **1969** *Physics Bull.* June 221/1 The 'renaissance' in optics, one of the oldest disciplines in physics, has been brought about mainly by the advent of the laser. **1973** *Black World* Sept. 95 Arna Bontemps was not of the 'Harlem Renaissance'... His first novel and his poems..appeared just when..the Renaissance flopped. **1975** *Nature* 3 Apr. 391/1 A renaissance occurred in 1969 when Adler proved that bacteria have specific chemoreceptors.
Hence **Renai·ssancer**, one who participates in a renaissance; = next; **Renai·ssancist**, an advocate or student of a renaissance; also *attrib.* or as *adj.*
1895 J. M. Falkner *Lost Stradivarius* 261 Neo-Platonism..has enthralled..many minds from Proclus and Julian to Augustine and the Renaissancists. **1899** G. B. Shaw *Let.* 17 Oct. (1972) II. 113 The mosque [of Sulieman]..is a successful attempt to take St Sophia and give it refined grandeur in the spirit of Brunelleschi and the early dignified Renaissancers. **1949** Renaissancer [see *Brahmsian *a.* and *sb.*]. **1973** *Compar. Stud. Soc. & Hist.* XV. 473 That a near-century of scholarship..should fail indeed to validate even the concept of a Renaissance, would appear to have little if any bearing on..the prosperity of the guild of Renaissancists in our time. *Ibid.* 478 In characteristic Renaissancist fashion.

renaissant, *a.* Add: **2.** = Renascent *a.*
1972 *Times* 3 Jan. 15 Rapidly rising output and renaissant business confidence and investment are normally a time at which profits rise. **1972** E. Longford *Wellington* II. xxi. 331 This was all very difficult for a renaissant Tory party which meant to win and win soon.

renal, *a.* Add: **2.** *renal dialysis*, dialysis performed artificially as a substitute for normal kidney function; freq. *attrib.* to denote a device to do this; *renal dwarfism, infantilism, osteodystrophy, rickets*, osteodystrophy due to the failure of the kidneys to convert dietary vitamin D to a more active form.
1912 O. May in *Univ. Coll. Hosp. Mag.* II. 99 (*heading*) A case of 'renal infantilism'. **1920** H. Barber in *Lancet* 3 Jan. 18/1 Until some standard text-book gives a full account of this condition it is not easy to select a suitable name; but as the kidney disease not infrequently has a very insidious onset, and many of the cases seek advice for the first time for want of development or bone deformity, some name such as renal dwarfism may be used. **1926** G. V. Ashcroft in *Jrnl. Bone & Joint Surg.* VIII. 279 Renal rickets is a disease not mentioned in medical text-books. *Ibid.* 288 It is to the association of the typical clinical picture, the typical X-ray picture, and deficient renal function that the term Renal Rickets has been applied. **1929** Thursfield & Paterson *Garrod's Dis. Children* (ed. 2) iii. 123 The primary cause of renal rickets is the inability of the diseased kidneys properly to excrete phosphorus. **1930** *Lancet* 10 May 1002/2 Two conditions associated with bone deformities have been differentiated from late rickets—namely, cœliac rickets and renal rickets. *Ibid.*, Cases of cœliac rickets and renal infantilism only occasionally show rickets. *Ibid.*, Rickets occasionally complicates renal dwarfism. **1943** Liu & Chu in *Med.* XXII. 103 The term 'renal osteodystrophy' seems to be a suitable generic name to include cases of osseous disorder associated with renal insufficiency, while the exact nature of the pathological process in the skeleton is still undetermined. **1957** *Brit. Med. Bull.* XIII. 57/2 The terms 'renal dwarfism', 'renal infantilism' and 'renal rickets', although obviously not invariably applicable, were apt in their original use. **1960** *Jrnl. Amer. Med. Assoc.* 24 Dec. 2124/1 Renal dialysis proved successful in treating previously intractable heart failure. **1962** *Lancet* 2 June 1169/1 More common..is the form associated with renal failure which was at one time called 'renal rickets'..but which is now more elegantly referred to as 'renal osteodystrophy'. **1963** *Ibid.* 5 Jan. 16/1 Renal-dialysis units mostly use complex and powerful artificial kidneys which require experienced surgical, medical, and biochemical supervision. **1974** *Times* 17 Apr. 2/8 If kidneys for transplant were available many people maintained by expensive renal dialysis machines could be given a fuller life and many of the 5,000 who die each year from kidney failure could be saved.

renardite (rĕnā·ɹdəit). *Min.* [a. F. *renardite* (A. Schoep 1928, in *Bull. de la Soc. Française de Min.* LI. 247), f. the name of A. F. *Renard* (1842–1903), of the University of Ghent: see -ite[1].] A hydrated basic phosphate of lead and uranium, $Pb(UO_2)_4(PO_4)_2(OH)_4 \cdot 7H_2O$, found as minute, yellow orthorhombic crystals.
1929 *Amer. Mineralogist* XIV. 244 Renardite... Found as minute crystals with quartz, torbernite and clay from the Kasolo Mine, Katanga, Belgian Congo. **1956** [see *dewindtite]. **1971** *Mineral Abstr.* XXII. 267/1 The geological, mineralogical, and metallogenic character of the uraniferous schists of the Salamanca province of western Spain are described... Secondary minerals are represented by gummite,..renardite [etc.].

renascence. 1. (Later examples.)
1912 E. St. V. Millay in F. Earle *Lyric Yr.* 185 Renascence... O God, I cried, give me new birth, And put me back upon the earth! **1973** *Nature* 20 July 184/3 *The Serengeti Lion*..has greater significance in that it reflects the renascence of animal study in Africa.

rena·tionalization. [Re- 5 b.] The action of removing (a formerly nationalized industry, etc.) from private ownership and bringing it under national control again.
1923 W. P. Livingstone *Galilee Doctor* iv. ii. 250 It had given them a charter of renationalization. **1957** *Economist* 2 Nov. 437/1 For months the threat of renationalisation kept steel shares subdued and they have been consistent laggards in the market. **1958** *Engineering* 28 Feb. 271/2 Steel leaders are hardening in their opposition to re-nationalisation. **1971** *Guardian* 5 Mar. 13/5 The campaign for 'Renationalisation without compensation'.

rena·tionalize, *v.* [Re- 5 b.] **1.** *absol.* To reinvest with national character.
1927 *Scots Observer* 26 Feb. 15/3 Professor M'Fadyen,..has given us..a book,..based on his rich experience as a missionary professor in India... The argument that missions denationalise the native is noted, and reasons are given for the contention that missions really renationalise.
2. *trans.* To transfer (a formerly nationalized industry) from private to national ownership again.
1954 *Ann. Reg. 1953* 22 Steel and transport were to be renationalized. **1959** *Daily Tel.* 13 Mar. 19/7 Its proposals not only to renationalise steel and road transport but also to nationalise by one means or another half of British industry. **1980** *Times* 30 Sept. 4/6 The next Labour Government would renationalize..transport activity.

renatura·tion. [Re- 5 a.] The process of restoring the nature or properties of what has been denatured.
1940 *Nature* 31 Aug. 301/1 The reversion of heat denaturation in proteins..has been demonstrated... The present method of renaturation is of interest owing to the well-defined conditions under which the reversion is achieved. **1965** Peacocke & Drysdale *Molecular Basis Heredity* iv. 37 A quite different approach to this problem has now become available with the discovery that, under carefully chosen conditions, the two strands of DNA of micro-organisms and bacteriophage may be separated and then subsequently reunited to restore the original helical structure and biological activity. This process of 'renaturation' may be followed by centrifugation of DNA. **1978** *Nature* 28 Sept. 352/1 Kunitz demonstrated by enzymatic and physico-chemical methods that renaturation of the inhibitor yielded a product indistinguishable from the original protein.

rena·ture (rī-), *v.* [Re- 5 a.] **a.** *trans.* To restore the nature or properties of (what has been denatured).
1946 *Nature* 30 Nov. 768/2 Some fairly close system of supervision by the international authority will be essential to ensure that the denatured material..is not being 're-natured' so as to make it suitable for use in a bomb. **1977** *Sci. Amer.* Apr. 44/2 Stewart found that after cooling and the removal of the detergent the interferon recovered its original biological activity: it was renatured.
b. *intr.* To undergo renaturation.
1965 Peacocke & Drysdale *Molecular Basis Heredity* iv. 38 Such a close correlation clearly has great potentialities in the assessment of the relationship between micro-organisms but cannot be used for higher plants and animals since their DNA are more heterogeneous and do not 're-nature'. **1973** *Nature* 11 May 55/2 They demonstrated that the centromeric regions..renature the most rapidly after denaturation.
Hence **rena·tured** *ppl. a.*, **rena·turing** *vbl. sb.*; also **rena·turable** *a.*
1955 *Bull. Atomic Sci.* Jan. 12/1 If reconversion of the denatured nuclear explosives is not a protracted process, the threat of seizure, renaturing, and atomic rearmament would remain indefinitely. **1964** G. H. Haggis et al. *Introd. Molecular Biol.* xii. 315 Such a renatured DNA regains its biological activity and can be used to transform bacteria. **1970** *Nature* 26 Sept. 1310/1 Renaturable DNA is defined as DNA that after denaturation..rapidly reverts to a duplex when the denaturing conditions are removed.

rench (rentʃ), var. Rinse *v.* (Now chiefly *U.S. dial.*)
1591, etc. [see Rinse *v.* 2 aβ, 4 aβ in Dict.]. *a* **1841** J. Guild *Jrnl.* in *Proc. Vermont Hist. Soc.* (1937) V. 263 She would..go down to a brook about forty rods and stand in the brook and rench her close. **1859** J. C. Hotten *Dict. Slang* 81 *Rench*, vulgar pronunciation of *rinse*. '*Wrench* your mouth out,' said a fashionable dentist one day.—*North.* **1919** H. L. Mencken *Amer. Lang.* iii. 91 The Yankees..still clung, in their common speech, to such forms as..rench for rinse,..and the employment of precisely the same forms by thousands of Irish immigrants..gave them a certain support. **1941** E. P. O'Donnell *Great Big Doorstep* i. 2 How many time I'm gunna tell you..to come and rench out the diaper? **1960** V. Williams *Walk Egypt* iv. ii. 257 Then you best rench your hands.

rencontre, *sb.* Add: **1. c.** An organized but informal meeting of scientists.
1975 *Chem. in Brit.* XI. 145/1 One approach is to organize small informal meetings—*rencontres*—at which chemists can meet, be educated by (and educate) representatives from the other sciences. **1975** *Physics Bull.* Dec. 515/3 A few months ago the SRC organized a *rencontre* in Aberdeen on 'Combinatorics' which was a get-together for mathematicians, physicists and chemists. **1977** *Chem. in Brit.* XIII. 105 Theoretical research horizons in colloid science formed the theme of a recent *rencontre* sponsored by the Science Research Council.

rendered, *ppl. a.* Add: **2.** Of a brick or stone surface: covered with a render (cf. Render *sb.*[2] 5).
1971 *Country Life* 25 Feb. 447 (Advt.), It can be applied to brick, concrete, rendered or roughcast surfaces with equal success. **1973** *Nation Rev.* (Melbourne) 31 Aug. 1450/1 The white rendered, modernised..premises..have been an inconspicuous residential for the past three years. **1978** M. & N. Ward *Home* 90 Stucco-rendered walls.

rendering, *vbl. sb.* **3. c.** (Earlier and later examples.) Also *concr.*
1792 G. Cartwright *Jrnl. Labrador* I. p. xiii, *Rendering oil*, a sealer's term for melting fat into oil. **1945** *ABC of Cookery* (Ministry of Food) xii. 46 Rendering means melting to extract the fat from surrounding tissues. **1979** N. & I. Lyons *Champagne Blues* 174 We cook the steak in renderings of pork belly.

rendezvous, *sb.* Add: **5. c.** The pre-arranged meeting (and usu. docking) of two or more spacecraft in space; an instance of this.
1959 *ARS Jrnl.* Aug. 592/1 Many proposed space missions will require achieving rendezvous of two bodies in an orbit about a planet. **1962** F. I. Ordway et al. *Basic Astronautics* ix. 385 Orbital operations involving rendezvous with satellites or space stations. **1962** *Listener* 29 Nov. 901/2 The vehicle has to be put into a transfer-orbit which will take it from the Earth inward to the orbit of Venus, meeting the planet at a pre-selected rendezvous. **1965** *Times* 16 Dec. 10/1 The Americans achieved the first rendezvous of man in space today. **1969** *Guardian* 22 July 18/3 About ninety minutes after lift-off Eagle began the complex series of manoeuvres leading to rendezvous. **1973** C. Sagan *Cosmic Connection* xix. 139 Rendezvous and docking maneuvers are reasonably well developed in manned missions even now.

rendezvous, *v.* Add: **1. d.** Of a spacecraft or its crew: to effect a meeting in space, *spec.* to dock with another spacecraft.
1960 *IRE Trans. Aeronaut. & Navig. Electronics* VII. 112/2 The system will eventually rendezvous at R_p. **1966** *Punch* 12 Jan. 68/2 Under bright Uranus We'll rendezvous in space. **1966** *Electronics* 3 Oct. 134, 4 [*sc.* a computer] helped the crew rendezvous and link up with the Agena target on the first orbit. **1969** *Observer* 20 July 7/2 Collins had a difficult time 'space-walking' to an Agena rocket with which they had rendezvoused.

rendezvousing, *vbl. sb.* Add: (Later example.) Also as *ppl. a.*
1965 K. W. Gatland *Spacecraft & Boosters* II. 90/1 Saint or Satellite Inspector..was intended to provide the capability of rendezvousing with an unidentified satellite orbiting the Earth. **1973** *Daily Colonist* (Victoria, B.C.) 28 Jan. 22/3 Most [clocks] have served rendezvousing couples and time watchers well for many years.

rendingly (re·ndiŋli), *adv.* [f. Rending *ppl. a.* + -ly[2].] In a rending or heart-breaking manner; painfully.
1926 H. Crane *Let.* 19 Aug. (1965) 273, I have made up a kind of friendship with that idiot boy... He is rendingly beautiful at times.

rendition. 3. a. For *U.S.* read 'orig. *U.S.*' and add later examples.
1922 Joyce *Ulysses* 237 Ben Dollard does sing that ballad touchingly. Masterly rendition. **1939** N. Monsarrat *This is Schoolroom* III. xvii. 385 No account of twentieth-century culture would be complete without reference to the impact of the dance-band world..as well as strange words and phrases like 'rendition'. **1975** *Radio Times* 3 Apr. 17/1 It's comedian Roy Hudd, strumming the strings..with a rendition of 'Auntie Maggie's Remedy'.
b. Visual representation *of* anything.
1959 E. Pulgram *Introd. Spectrography of Speech* xiii. 89, I chose a sustained sound... The sketch of Fig. 11 is a complete and exact spectrographic rendition of it. **1972** *Sci. Amer.* Nov. 45/2 (Advt.), If you are interested in additional dimensions of photo-optical performance—rendition of corners,..stray-light shielding—this chart..may provide the detailed answer. **1978** *Amateur Photographer* 2 Aug. 109/2 A polarising filter may be used to darken skies without affecting the rendition of foreground detail.

rendu (rańdü), *a.* [Fr., = rendered, delivered.] Of imported goods: designating the price on delivery, including tariffs and delivery costs. Cf. *franco *a.*
1957 Clark & Gottfried *Dict. Business* 275/2 *Rendu price*, one on imported goods meaning that the price includes the cost of the goods themselves, freight insurance, landing fees, tariffs, and the costs of delivering the goods direct to the buyer's place of business. **1959** E. E. Nemmers *Dict. Econ.* 254 *Rendu price*, an import *delivered price*. The price of imported goods including all charges for tariff and freight. **1962** [see *franco *a.*].

rendzina (rendzī·nă). *Soil Science.* [a. Russ. *rendzina*, ad. Polish *rędzina*.] A fertile lime-rich soil which occurs typically under grass or open woodland on relatively soft calcareous bedrock (e.g. chalk and some limestones) and has a dark, friable, humus-rich surface layer above a softer pale calcareous layer formed by the breakdown of the underlying rock. Also *attrib.*
1927 C. F. Marbut tr. *K. D. Glinka's Great Soil Groups of World* 34 The humus carbonate soils such as the Rendzinas..constitute a good example of the influence of the parent rock on the soil forming process. **1928** C. L. Whittles tr. *E. Ramann's Evolution & Classification of Soils* v. 91 Recently under the influence of Russian soil workers the term 'rendzina' has been applied to all soils which have developed from the weathering of calcareous rocks. **1932** G. W. Robinson *Soils* xiv. 285 The writer is occupying a debatable position in assigning the chalk soils of England to the rendzina group. **1946** Lutz & Chandler *Forest Soils* xi. 386 Highly calcareous materials frequently give rise to immature soils called rendzinas. **1955** *Proc. Prehistoric Soc.* XXI. 53 Vegetation covering the surface of the loess produced soils due to chemical weathering which can be classified as podsols, chernozems, rendzinas, terra rossas and others. **1971** *Nature* 13 Aug. 453/1 The flora has survived through the post-glacial forest and blanket bog intervals in Teesdale partly on rendzina soils associated with rotted crystalline marble on Widdybank and Cronkley fells. **1973** *Country Life* 29 Nov. 1787/3 *The Rural Landscape of Kent*..is strictly for the diligent reader who can disentangle rendzinas from stagnogley soils. **1976** *Interim* IV. III. 14 Chalk rendzinas go straight from the A horizon to the C.

renegader (re·nīgēᶦdəɹ). [f. Renegade *sb.* or *v.* + -er[1].] = Renegade *sb.* 2.
1846 J. R. Lowell in *Boston Courier* 17 June 2 Haint they cut a thunderin' swarth, (Helped by Yankee renegaders).

renegadism. (In Dict. s.v. Renegade *sb.*) Also **renegadeism.** (Earlier and later examples.)
1826 B. R. Haydon *Jrnl.* 25 Feb. in *Autobiogr.* (1853) II. 115 The Academy is certainly modified, but still John Bull never pardons an appearance of renegadeism. **1939** A. J. Toynbee *Study of Hist.* VI. 104 The..profanity of Jason..gave Hellenism such a vogue and Renegadism such an impetus.

renego·tiate, *v.* [Re- 5 a.] *trans.* To negotiate a second or further time. Hence **renego·tiated** *ppl. a.*
1934 in Webster. **1962** *Economist* 16 June 1096/2 The annexes to these agreements, which determine the air services to be operated, are normally re-negotiated every six months. **1969** *Listener* 26 June 894/3 When you propose to renegotiate housing subsidies, this really means higher rents for a lot of people. **1975** *Times* 19 Feb. 2/8 Mr Crosland..is expected to support the argument..when Mr Callaghan..brings back the final renegotiated terms. **1976** *Film & Television Technician* Dec. 1/2 Among the key mechanisms in achieving this, have been the re-negotiated cost of living clauses which the employers have unilaterally abandoned.

renegotia·tion. [Re- 5 a.] A second or further negotiation.
1934 in Webster. **1945** *Britannica Bk. of Year* 771 The renegotiation by the government of original contracts to bring them more in line with actual costs as revealed by experience. **1963** *Ann. Reg. 1962* 274 The renegotiation of the U.S.–Spanish Bases agreement, due for renewal in 1963. **1976** H. Wilson *Governance of Britain* 10 The final Cabinet tally, after the re-negotiations of the terms of entry, was publicly announced to be seventeen in favour, seven against. **1980** *Boston Globe* 30 Mar. 76 Today the 25-year-old..insists on a renegotiation of his contract.

renegue, *v.* Add: Now more commonly with pronunc. (rīne·g)or (rīnēᶦ·g). Also (*U.S.*) **renig.**
1. (Later examples.)
1914 Joyce *Dubliners* 163 'There's one of them, anyhow,' said Mr Henchy, 'that didn't renege him.' **1922** —— *Ulysses* 324 We fought for the royal Stuarts that reneged us against the Williamites and they betrayed us.
4. b. (Earlier example, in form *ranague.*)
a **1849** J. Keegan *Legends & Poems* (1907) 64 Amn't I to undherstand that..Peggy is goin' to ranague you for Micky Gorman?
c. To change one's mind, to recant; to break one's word; to go back *on* a promise or undertaking or contract; to disappoint expectations. orig. and chiefly *U.S.*
Now the dominant sense, and freq. in spelling *renege.*
1784 A. Ellicott *Diary* 24 Nov. in C. V. Mathews *Andrew Ellicott* (1908) i. 27 The Hussey immediately Reniged and reclaimed the Bed. **1866** C. H. Smith *Bill Arp* 153 When the Secretary read out my name all mixed up with the Republic, I felt I was obleged to renig. **1906** 'O. Henry' *Four Million* 123 It might brace her up and keep her from reneging on the proposition to skip. **1917** H. Franck *Vagabonding down Andes* 32 Hays renigged at the last moment, but I accepted the invitation issued to the 'general public'. **1935** A. Squire *Sing Sing Doctor* ix. 141, I was afraid our man might renege on his contract. **1936** 'N. Blake' *Thou Shell of Death* xiii. 236 She turned very calm and quiet... 'I'll never renege. I'll write to Jack... He must come back.' **1946** *Time* 21 Oct. 99/1 The picture begins to renege on its early promise. **1951** A. R. Lewis *Naval Power & Trade in Mediterranean* v. 150 The naval assault was a success, but Hugh reneged on his side of the bargain. **1962** J. McCabe *Mr. Laurel & Mr. Hardy* iii. 82 Anderson made a few more films with Stan before he reneged on a contract detail regarding payment. **1968** *Daily Tel.* 13 Dec. 1/2 The Minister is equally annoyed that the National Federation of Building Trade Operatives appears to have reneged on its earlier promise to accept a 1d-an-hour wage cut. **1973** R. Ludlum *Matlock Paper* v. 43 'You're offering me a chance to renege?' 'Of course. You're under no obligation to us.' **1977** *Time* 7 Mar. 23/1 He was given certain undertakings from other people that they subsequently reneged on. **1981** *Times* 27 Jan. 12/8 Labour's record on immigration has been almost identical to the Conservatives'—it introduced stringent controls and reneged on Britain's commitment to the East African Asians who had United Kingdom passports.

reneguing, *vbl. sb.* (Later example.)
1921 G. B. Shaw *Back to Methuselah* p. lxv, There was no Prime Minister to whom such renagueing or trafficking would ever have occurred.

renewability. (Example.)
1976 *Nature* 27 May 350/2 The renewability of forests is a key point in the book.

renewable, *a.* Add: **2.** Of a source of energy: not depleted by its utilization.
1971 *Sci. Amer.* Sept. 43/2 (*caption*) Continuous, or renewable, energy supply can be divided into two categories: solar and nonsolar. **1972** Rocks & Runyon *Energy Crisis* 8 We have already expanded in numbers and living standard far beyond the capacity of our most accessible and renewable energy source [*sc.* flowing water] to sustain us even in the present. **1975** Ruedisili & Firebaugh *Perspectives on Energy* iv. 295 Geothermal, tidal, and wind energy are three..renewable sources of energy that have been harnessed to a greater or lesser extent. **1978** *Nature* 20 Apr. 661/1 Renewable sources of energy—wind, wave, sun, geothermal heat and the like—are..taken seriously at the United Kingdom Atomic Energy Authority.
B. as *sb.* A renewable source of energy.
1974 *Oceanus* XVII. 20 (heading) Using two renewables. **1980** *Times* 22 Aug. 10/2 The CEGB decision to take the first commercial steps for wind-powered electricity makes it easier..to take renewables seriously.

renewal. Add: **a*.** A planned urban re-development. Also in phr. *urban renewal* s.v. *urban *a.*
1965 *Economist* 6 Feb. 544/1 For its size Boston is the most renewal-minded city in the country. Mr. Slayton, the federal Urban Renewal Commissioner, recently called its eleven integrated projects..'a laboratory demonstration of renewal techniques'. **1967** *Boston Sunday Herald* 26 Mar. I. 9/6 The BRA leadership is pushing to exempt 'planned unit developments' in renewal areas. **1978** *Jrnl. R. Soc. Arts* CXXVI. 595/2 In urban renewal schemes where roadways and footpaths are altered the lower voltage system usually requires replacement.
b. renewal theory, the branch of probability theory which considers populations of objects which fail and need renewal after randomly distributed intervals.
1940 G. A. D. Preinreich *Present Status of Renewal Theory* 25 Lotka has a method of mathematically describing renewal theory, using the Hertz method of approximation to the solution. **1958** *Jrnl. R. Statistical Soc.* B. XX. 243 The study of renewal theory has its origins in the discussion of self-renewing aggregates and the non-stochastic treatment of questions of population growth. **1966** S. Beer *Decision & Control* x. 216 Renewal theory concerns itself with strategies for replacing worn out parts, machines, aircraft or anything else that has to be renewed.

reng (reŋ). [ad. Pers. *rang*, Skr. *raṅga* colour, hue.] A colouring, esp. a hair-dye.
1929 Redgrove & Foan *Blonde or Brunette?* xi. 73 The leaves, dried and powdered, of *Indigofera argentea*, L. ('reng'), cultivated in Persia, constitute the best form in which to employ indigo. **1934** *S.P.E. Tract* XLI. 16 *Reng* is perhaps now in sufficiently common use by hair-dressers to be accorded a place in the national vocabulary. **1966** J. S. Cox *Illustr. Dict. Hairdressing* 79/1 *Henna-Reng*, a hair dye consisting of a combination of Henna and Indigo.

renga (re·ngă). Also **renge, renka.** [Jap., = linked (verse).] A form of Japanese verse established by the 15th century and consisting of a series of half-tanka, contributed by different poets in turn.
See also note s.v. *haiku.
1877 W. G. Aston *Gram. Jap. Written Lang.* (ed. 2) x. 198 *Renka*..is where one person composes part (commonly the second part) of a *tanka*, the remainder being added by some one else. **1890** B. H. Chamberlain *Things Japanese* 272 A favourite game at these tournaments called *renge*, wherein one person composes the second hemistich of a verse and another person has to provide it with a first hemistich, seems to date from the eleventh century. **1911** —— *Jap. Poetry* iv. 159 This was termed *Renga*, lit. 'linked verses'. **1968** E. Miner *Introd. Jap. Court Poetry* 163 *Renga*... A form dating from about the thirteenth century; several authors would compose a sequence, usually of a hundred sections or stanzas, alternating 5, 7, 5 syllable lines with 7, 7 syllable units, any two of which formed a complete poem.

rengas (re·ŋgas). [Malay.] Any of several East Indian trees of the family Anacardiaceæ, esp. *Gluta renghas*, containing a sap that often produces allergic reactions in those touching it; the wood or sap of this tree. Also *attrib.*
1836 J. Low *Diss. Soil & Agric. Penang* iv. 200 *Runggas*—a lofty tree, the juice of which is deleterious to the human frame, creating swellings over the whole body. **1935** I. H. Burkill *Dict. Econ. Products Malay Penin.* I. 1079 Furniture of 'rĕngas' may certainly be used without inconvenience. **1939** A. Keith *Land below Wind* xiii. 221 One disadvantage of wearing shorts was the long expanse of leg which was left bare, with the result of..patches of *rengas* poisoning from rubbing against the leaves of the *rengas* tree. **1940** E. J. H. Corner *Wayside Trees of Malaya* I. 116 Most *Rengas*-trees are large and not a few are among the tallest and finest in the forest. **1956** *Nature* 25 Feb. 366/1 The Colonial hardwood, rengas, has been shown to contain two pigments. **1965** R. McKie *Company of Animals* xi. 164 Jim..confirms the story of the Rengas tree rash. Some people are allergic to this tree.

renguerra (rengwē·ə·rā). *Vet. Sci.* Also **renguera.** [S. Amer. Sp. *renguera* limping, lameness, f. *renguear* (Sp. *renquear*) to limp.] = *swayback 2.
1917 S. H. Gaiger in *Jrnl. Compar. Path. & Therapeutics* XXX. 209 Renguera is a new and hitherto undescribed disease of lambs, occurring in the Peruvian Andes. **1938** *Nature* 5 Mar. 400/1 The resulting lesions of demyelination are anatomically related to those observed in the enzootic paraplegia (renguerra) reported to occur among lambs bred in certain areas in the Peruvian Andes. **1966** A. Robertson *Internat. Encycl. Vet. Med.* II. 698 Synonyms [for copper deficiency]. Enzootic ataxia; swayback;..renguerra.

renicky-boo: see *rannygazoo.

renierite (rənʸē·ə·rəit). *Min.* [ad. F. *reniérite* (J. F. Vaes 1948, in *Ann. de la Soc. géol. de Belgique* LXXII. B22), f. the name of A. *Renier*, 20th-c. Belgian geologist: see -ite[1].] A sulphide of copper, germanium, and other metals $(Cu,Fe)_3(Fe,Ge)S_4$, occurring as yellowish tetragonal crystals and granular masses.
1949 *Mineral Mag.* XXVIII. 737 Renierite... Sulphide of Cu, Fe, Ge (7·75 %), Zn, As... Named after Prof. Armand Renier, Director of the Geological Survey of Belgium. Near germanite. **1966** *Mineral Abstr.* XVII. 537/2 Germanium-bearing (250 p.p.m. Ge) sphalerite occurring with galena in mineralized Cambrian limestone and dolomite in..southern Sardinia, is found to contain numerous inclusions of renierite up to 10μ in size.

renig, U.S. var. Renegue *v.* in Dict. and Suppl.

renin (rī·nin). *Physiol.* [f. L. *rēnes* kidneys + -in[1].] † **1.** A substance extracted from animals' kidneys and used in medicine. *Obs.*
1894 G. M. Gould *Illustr. Dict. Med., Biol. & Allied Sci.* 940/2 Extracts have been prepared from nearly every organ in the animal body;..cerebrin, from the brain,..ossin, from bones, renin, from the kidneys. **1900** Dorland *Med. Dict.* 565/1 *Renin*, a therapeutic extract prepared from the kidneys of animals.
2. A proteolytic enzyme secreted by and stored in the kidneys, which acts in the blood to convert angiotensinogen (hypertensinogen) to angiotensin (hypertensin). [Coined in this sense as G. *renin* (R. Tigerstedt, in *Compt. Rend. 12me Congrès Internat. de Médecine 1897* (1899) II. II. 29).]
1906 *Lancet* 19 May 1375/2 The pressor substance, to which these workers give the name 'renin', is not dialysable. **1938** *Proc. Soc. Exper. Biol. & Med.* XXXIX. 214 Undialyzed renin (0·2 cc) caused moderate vasoconstriction. *Ibid.* 215 Renin is an enzyme-like substance which is activated by a kinase-like material contained in the protein fraction of plasma and whole blood. **1959** [see *hypertensin]. **1965** [see *hypertensinogen]. **1968** Passmore & Robson *Compan. Med. Stud.* I. xxxiii. 20/2 Renin is an enzyme which on reaching the blood activates an α_2-globulin called angiotensin formed in the liver, making angiotensin I, a polypeptide containing ten amino acids. Another enzyme converts angiotensin I to angiotensin II by removing two amino acids. This last substance is the most potent pressor agent known. **1977** Freeman & Davis in J. Genest et al. *Hypertension* vi. 211/1 Renin is synthesized and stored..in the granules of JG [*sc.* juxtaglomerular] cells which are located primarily in the renal afferent arteriole, although these granular cells have also been identified in the efferent arteriolar wall..and in the mesangial cells.

renitence. For † *Obs.* read *rare* and add later example.
1917 C. R. Payne tr. *Pfister's Psychoanal. Method* viii. 168 The result of this renitence consists mostly in the continuance of those symptoms of disease which depend on the repression.

renka, var. *renga.

renminbi (renminbi). Also **jenminpi, renminpi, Renminbi.** [Chinese *rénmínbì*, f. *rénmín* people + *bì* currency.] **a.** The name of the currency introduced in China in 1948. **b.** Occas. used for yuan, the basic unit of this currency.
1957 *Encycl. Brit.* V. 546 B/1 In 1953 the official currency was the jenminpi or People's bank note on the mainland. **1971** [see *jiao]. **1973** *Times* 21 Mar. (China Trade Suppl.) p. iii/6 The basic unit of renminbi—which is abbreviated to RMB—is the yuan. **1974** *China Reconstructs* July 14/3 The Chinese currency, the Renminbi, is stable. **1975** *Ann. Reg. 1974* 320 More than 60 countries were already using the Chinese renminpi as the trading currency with China. **1979** *Fortune* 21 May 110/2 Its young tellers..eagerly explain to a visitor the tax advantages of converting his money into Chinese renminbi and keeping it in Peking.

renneting (re·nĕtiŋ), *vbl. sb.* *Cheese-making.* [f. Rennet *sb.*[1] + -ing[1].] The action or process of adding rennet in order to curdle milk. Also *attrib.*
1894 J. Oliver *Milk, Cheese & Butter* xi. 172 If no other heating than to obtain renneting temperature, this commenced as soon as last milk has arrived. **1917** Walker-Tisdale & Woodnutt *Pract. Cheesemaking* xiii. 108 Renneting. Having ripened the milk and regulated it to the renneting temperature, the rennet is added. **1932** R. H. Leitch *Cheddar Cheese-Making* vii. 41 The standard temperature of renneting Cheddar cheese is 86 degrees Fahrenheit; under average conditions there is no advantage in a higher renneting temperature. **1937** Harvey & Hill *Milk Products* iv. 200 Renneting is carried out at a temperature of 80°–91° F. **1950** J. G. Davis *Dict. Dairying* 100 Renneting. When the correct acidity has been reached rennet is added at the rate of about 1 ml. per gal. of milk... The renneting acidity is one of the crucial points in cheese-making. **1976** —— *Cheese* III. xxi. 498 There are four crucial stages in cheesemaking. These are the acidity at renneting, the stage of pitching, the point at which the whey is removed from the curd, and the time of milling, salting and putting to press.

reno- (rīno), comb. form of L. *rēnes* kidneys (now more usual than Reni-), as in Renopericardial *a.*, *renography, etc.

renogram (rī·nogræm). *Med.* [f. as next + -gram.] A graphical record of the varying radioactivity of a kidney into which a radioactive substance has been injected; also, a radiograph or autoradiograph of a kidney.
1954 P. G. Smith in M. Campbell *Urol.* I. II. ii. 189 The exposure is made as the last few cc. of the contrast medium are being injected. The needle is immediately withdrawn and a second film is made as rapidly as the cassettes can be changed. This last film is known as a renogram or nephrogram. **1964** C. C. Webster in J. F. Glenn *Diagnostic Urol.* x. 190 (*caption*) The radioisotope renogram and blood clearance tracing are produced with three scintillation probes connected through three rate meters and three recorders. **1974** Passmore & Robson *Compar. Med. Stud.* III. xxii. 16/1 When hippuran (sodium ortho-iodohippurate) labelled with ^{131}I is given intravenously the substance is secreted rapidly into the proximal tubular fluid. Scintillation counters placed over the kidneys measure the radiation emitted by the isotope and the activity/time curve which is recorded is called the renogram. *Ibid.* 16/2 Patients with unilateral disturbances in renal function show differences in the shape and amplitude of the two renograms.

renography (rīnǫ·grăfi). *Med.* [f. *reno- + -graphy.] Renal angiography or autoradiography.
1911 *Brit. Med. Jrnl.* 1 Apr. 748 (*heading*) A lecture on renography. **1964** C. C. Webster in J. F. Glenn *Diagnostic Urol.* x. 189 Since its inception in 1955, radioisotope renography has been increasingly utilized for the external measurement of individual kidney function. **1971** *Nature* 17 Sept. p. xi (Advt.), There is an active research programme in renography carried out in conjunction with the renal unit.

Renoiresque (renwāre·sk), *a.* [f. the name of Pierre Auguste *Renoir* (1841–1919), French painter + -esque.] Of, pertaining to, or characteristic of Renoir or his work.
1958 *Listener* 5 June 934/2 He succeeded in creating a Renoiresque atmosphere of the turn of the century. **1961** *Times* 18 Jan. 15/2 In her Renoiresque paintings Morisot lost a little of her freshness. **1971** R. A. Carter *Manhattan Primitive* xvi. 150 An oddly put-together woman in a..black scarf top from which her Renoiresque bosom kept emerging.

renominate, *v.* For 'a second term of office' read 'a further term of office' and add later examples.
1927 *New Republic* 21 Sept. 122/2 It declares that Mr. Coolidge must, shall and will be renominated. **1975** J. P. Morgan *House of Lords & Labour Govt.* vii. 180 They illustrated the possible influence of systems of renominating voting Peers at successive General Elections. **1981** *Times* 28 Apr. 15/2 Devices to make it difficult if not impossible for moderate candidates to be renominated.

reno·rmalizable, *a.* *Physics.* [f. *renormaliz(e *v.* + -able.] That permits of renormalization.
1955 L. D. Landau in W. Pauli *Niels Bohr* 66 This phenomenon, not being renormalizable, cannot be considered within the limits of the theory. **1968** D. Lurié *Particles & Fields* vi. 266 Such theories, in which all divergences can be absorbed in the coupling-constant and mass renormalizations and thereby ignored, are known as renormalizable field theories. **1978** *Sci. Amer.* Feb. 132/2 For many years it seemed there was no convincing renormalizable theory of the weak interactions.
So **reno:rmalizabi·lity.**
1955 L. D. Landau in W. Pauli *Niels Bohr* 54 This renormalizability of the theory is in reality only approximate. **1975** *Nature* 11 Sept. 95/2 In 1971..a Dutch graduate student, 't Hooft, opened the flood-gates by giving convincing arguments for the renormalisability of a rather special type of Yang–Mills gauge theory.

re:normaliza·tion. *Physics.* [Re- 5 a.] A method used in quantum mechanics of removing unwanted infinities from the solutions of equations by redefining parameters such as the mass and charge of subatomic particles. Freq. *attrib.* Cf. *normalize *v.* 3 a.
1948 *Physical Rev.* LXXIV. 1430/1 The divergent terms in the line shift problem can be thought to be contained in a renormalization of the mass of a free electron. **1954** W. Heitler *Quantum Theory of Radiation* (ed. 3) vi. 277 What is observable is the total mass and the total charge of the electron and these include the self-mass and self-charge. Although it is still a major unsolved difficulty of the theory that these quantities turn out to be infinite, they should, whatever their value, be combined with the 'original' mass and charge (i.e. the theoretical mass and charge when no interaction with the radiation field existed at all). For the original plus the self-mass and charge the observed finite values of mass and charge should then be substituted. This procedure will be called the re-normalization of mass and charge. *Ibid.* xxx. 310 This is the relativistic form of the 'renormalization terms'. **1954** *Physical Rev.* XCV. 1329/1 Ever since the overwhelming success of the applications of renormalization technique in quantum electrodynamics, the problem of understanding this renormalization procedure without the use of perturbation methods has been of great interest. **1962** N. R. Hanson in A. B. Pippard et al. *Quanta & Reality* v. 88 Theoreticians..invented a technique to diminish the number of possible solutions into something which practising physicists could manage. The result is a rather arbitrary procedure called 'renormalization'. It rejects as physically unpromising most solutions of any wave equation. **1977** L. Streit in Price & Chissick *Uncertainty Principle & Foundations of Quantum Mechanics* xviii. 353 Virtually every second calculation of quantum electrodynamics included the process of throwing away an infinite term and interpreting the remainder as the 'correct result'. These procedures were formalized in the renormalization theory of Feynman, Dyson and Schwinger. **1979** *Sci. Amer.* Mar. 67/2 In the 1950's it became apparent that the aims of the renormalization procedure can be achieved by a large family of mathematical transformations; these make up the renormalization group.

reno·rmalize, *v.* *Physics.* [Re- 5 a.] *trans.* To apply renormalization to.
1955 L. D. Landau in W. Pauli *Niels Bohr* 54 The results hereby obtained can be renormalized, that is, if the physical charge of the electron is defined by its interaction with quanta of zero frequency and its mass, as the physical mass of the electron, the undetermined constant Λ disappears from the formulae for the physical effects. **1972** *Sci. Amer.* Nov. 50/1 The second obstacle to a satisfactory theory is that many calculations involving the weak force quickly lead to infinite results. In electromagnetic theory similar divergences are handled by the process called re-normalization. Until recently no one could see how to re-normalize the divergences presented by the weak force.
So **reno·rmalized** *ppl. a.*, calculated with the use of renormalization.
1954 *Physical Rev.* XCV. 1329/1 A rather unexpected and quite surprising feature is obtained by comparing the re-normalized coupling constant with the unrenormalized coupling constant. **1977** M. E. Fisher in U. Landman *Statistical Mechanics & Statistical Methods in Theory & Application* 17 Renormalized field-theoretic perturbation theory.

renosterbos, -bush, varr. *RHENOSTERBOS.

renounceable, *a.* Delete *rare*$^{-1}$ and add further examples.

1955 *Times* 8 July 14/2 Renounceable allotment letters and application forms for additional shares have been posted. 1979 *Daily Tel.* 19 Jan. 13/2 A three-for-one capitalisation issue will then produce for Midland 86 million renounceable shares which it can sell free of stamp duty.

renovascular (rīnovæ·sciŭlăɹ), *a. Med.* [f. *RENO- + VASCULAR *a.*] Pertaining to the blood vessels of the kidneys.

1961 *Medicine* XL. 347 (*heading*) Functional characteristics of renovascular hypertension. 1974 PASSMORE & ROBSON *Compan. Med. Stud.* III. xx. 8/1 Neutralization of the activity of angiotensin II by specific angiotensin antibodies does not always prevent or reverse hypertension induced by renovascular means.

renovated, *ppl. a.* Add: *renovated butter* = *process butter* s.v. *PROCESS *sb.* 13.

1899, etc. [see *process butter* s.v. *PROCESS *sb.* 13]. 1906 *Daily Chron.* 13 Sept. 5/2 Mr. Hehner went on to explain the nature of American renovated butter. He said that sometimes butter 'went off', and it was then melted down and the sour milk run off and replaced by pure milk and cream. 1937 HARVEY & HILL *Milk Products* iii. 178 Renovated butter. Butter which has become unfit for human consumption is treated in many countries by a process which is said to render it suitable for such consumption. Such methods, however, are not practised in Great Britain.

renovize (re·novəiz), *v. U.S. rare.* [A blend of RENOV(ATE *v.* + MODERN)IZE *v.*] *trans.* To restore and modernize.

1933 *Daily Progress* (Charlottesville, Va.) 25 Jan. 3/6 A 'renovize Philadelphia' campaign has met marked success in its early stages. 1935 A. P. HERBERT *What a Word!* ii. 53 A North American warrior tells me that she has seen 'to renovise'! 1965 *Amer. Speech* XL. 303 This suffix has been attached to..verbs (*renovize, flavorize*).

‖ **renseignement** (rañsęnʸmañ). [Fr.] (A piece of) information; also, a letter of introduction.

1841 E. EVERETT *Let.* 30 Dec. in Dickens *Lett.* (1974) III. 4/1 At Washington you will be able to get abundance of letters & *renseignement* for every part of the interior. *c* 1863 MRS. GASKELL *Lett.* (1966) 931, I am..sending my courier to you in hopes that you will most kindly give him 'renseignements' on one or two points. 1873 W. JAMES *Let.* 25 May in R. B. Perry *Thought & Char. W. James* (1935) I. xx. 346, I take up the pen today mainly on a matter of business, that is, to get at as early a date as possible certain *renseignements* which may affect my choice of how to spend next winter. 1875 LADY C. SCHREIBER *Jrnl.* 20 Nov. (1911) I. 390 Called on the Consul, saw some fine Oriental dishes at his house, got from him various renseignements. 1921 *Glasgow Herald* 17 May 3/8 Given, however, accurate renseignements, properties of high potential value are to be acquired.

Renshaw[1] (re·nʃǭ). *Tennis.* The name of William Charles *Renshaw* (1861–1904) and his twin brother Ernest (1861–99), used *attrib.* in *Renshaw smash* to denote a kind of fast overhead volley with which they were associated.

[1882: see SMASH *sb.*[1] 1 b.] 1883 *Field* 7 July 11/1 Deuce was called in five of the games, and the fifth game was won by four of the 'Renshaw smashes'. 1889 W. M. BROWNLEE *Lawn-Tennis* v. 18 This return very soon was called the Renshaw smash'. 1975 *Oxf. Compan. Sports & Games* 835/1 He took the ball early—the 'Renshaw smash' was celebrated.

Renshaw[2] (re·nʃǭ). *Physiol.* [Name of Birdsey *Renshaw* (1911–48), U.S. neurologist, who investigated such cells.] *Renshaw cell*: a nerve cell in the spinal cord that is innervated by collaterals from a motor neurone and forms synapses with that and adjacent motor neurones so as to provide an inhibitory feedback path.

1954 J. C. ECCLES et al. in *Jrnl. Physiol.* CXXVI. 533 A detailed study of the interneuronal discharges has established that these interneurones form a specialized group mediating the inhibitory path from motor axons. They may appropriately be given the distinguishing title of 'Renshaw cells'. 1974 M. C. GERALD *Pharmacol.* xv. 277 The Renshaw cells exert an inhibitory influence on the flow of nerve impulses along motor neurons in the spinal cord. 1976 W. R. INGRAM *Rev. Anat. Neurol.* i. 23 Some investigators have questioned the validity of the Renshaw cell concept.

rent, *sb.*[1] Add: **2. b.** *fair rent*, the amount of rent which a tenant may reasonably be expected to pay for the use of specified land or property; *spec.* that officially registered by a Rent Office for a particular tenancy.

1886 *Act* 49 & 50 *Vict.* c. 29 § 6 The landlord or the crofter may apply to the Crofters Commission to fix the fair rent to be paid by such crofter to the landlord for the holding. 1926 *Act* 16 & 17 *Geo. V.* c. 52 § 2 The expression 'full fair rent' in relation to a small holding means the rent which a tenant might reasonably be expected to pay for the holding if let as such and the landlord undertook to bear the cost of structural repairs. 1965 *Act* 13 & 14 *Eliz. II.* c. 75 § 27 In determining..what rent is or would be a fair rent under a regulated tenancy of a dwelling-house regard shall be had,..to the age, character and locality of the dwelling-house and to its state of repair. *a* 1974 R. CROSSMAN *Diaries* (1975) I. 263, I had to deal with all the questions about rateable value and with the fair-rent clauses. 1976 *Southern Even. Echo* (Southampton) 13 Nov. 6/5 The rent officer in determining fair rents can consider under his brief the property and then fix his figure with reference to similar properties.

c. (Later examples.)

Still *Obs.* except in *U.S. dial.* use.

c 1847 J. S. COYNE in M. R. Booth *Eng. Plays of 19th Cent.* (1973) IV. 186 You used not to wear such waistcoats as that when you lived in Fuller's Rents. *a* 1902 S. BUTLER *Way of All Flesh* (1903) lv. 254 A rag and bottle merchant in Birdsey's Rents. 1926 *Dialect Notes* IV. 388 *Rent*,..apartment or rentable house. 1943 [see *rent-hunter* in sense 4 b below].

d. *Pol. Econ.* (See quot. 1817.)

[1777 J. ANDERSON *Enquiry Nature Corn-Laws* 45 It is not, however, the rent of the land that determines the price of its produce, but it is the price of that produce which determines the rent of the land.] 1815 T. R. MALTHUS *Inquiry Nature & Progress Rent* 1 The rent of land may be defined to be that portion of the value of the whole produce which remains to the owner of the land, after all the outgoings belonging to its cultivation..have been paid, including the profits of the capital employed, estimated according to the usual and ordinary rate of the profits of agricultural stock at the time being. 1817 D. RICARDO *Princ. Pol. Econ.* ii. 49 Rent is that portion of the produce of the earth, which is paid to the landlord for the use of the original and indestructible powers of the soil. 1848 MILL *Pol. Econ.* I. II. xvi. 500 The rent, therefore, which any land will yield, is the excess of its produce, beyond what would be returned to the same capital if employed on the worst land in cultivation. 1884 J. RAE *Contemp. Socialism* ix. 455 No part of Ricardo's theory is more elementary or more unchallenged than this, that the rent of land constitutes no part of the price of bread, and the high rent is not the cause of dear bread, but dear bread the cause of high rent.

e. Money or cash, esp. that acquired by criminal activity or in exchange for homosexual favours; hence *ellipt.* (as quasi-*adj.*), = *rent boy* in sense 4 c below.

1828 W. T. MONCRIEFF *Tom & Jerry* I. 20 Blunt, my dear boy, is..to be able to flash the screens—sport the rhino—shew the needful—post the pony—nap the rent... Money, money, is your universal good. 1925 FRASER & GIBBONS *Soldier & Sailor Words* 242 *Rent*, money: cash. 1936 J. CURTIS *Gilt Kid* xii. 127, I haven't done anything since I've come out of the nick and the old rent's running a bit low. 1967 A. WILSON *No Laughing Matter* III. 306 I've been rent myself once... I just gave what they paid me for. 1977 *Gay News* 24 Mar. 15/3 A word of warning about the Strand Bar in Hope Street... It's rough and some of the people there are rent.

f. *Econ.* The financial advantage or gain regarded as emanating from a particular skill or ability, *spec.* in phr. *rent of ability* (see quot. 1929).

1879 A. MARSHALL *Econ. Industry* II. xii. 144 Rent of rare natural abilities is a specially important element in the incomes of business men. 1905 G. B. SHAW *Irrational Knot* p. xv, There is an important economic factor, first analyzed by an American economist (General Walker), and called rent of ability. 1929 S. E. THOMAS *Elem. Econ.* (ed. 4) xvii. 261 We may say that there is a rent element in both profits and wages, and that this element depends on the natural or acquired gifts of the employer or worker concerned. Where the differential payment is due to differences of ability, it may be suitably and correctly described as a rent of ability. 1930 *Times* 6 May 12/2 The Fabian Society formerly pleaded for the rent of ability, but the plea fell on deaf ears.

4. b. *rent-hunter* (in sense 2 c in Dict. and Suppl.), *-owner, -receiver, -yielding.*

1943 *Boston Herald* 28 July 12/1 A recent..cartoon showed two weary rent-hunters walking past the White House. 1844 MILL *Ess. Pol. Econ.* iii. 89 All which is produced beyond this, whether it be in the hands..of any of the numerous varieties of rent-owners, may be taken for immediate enjoyment. 1943 in E. Blunden *Return to Husbandry* 12 The squire has become an absentee landlord, a mere rent-receiver. 1848 MILL *Pol. Econ.* I. III. v. 565 Selling at a scarcity value..never is, nor has been, nor can be, a permanent condition of any of the great rent-yielding commodities.

c. *rent allowance, assessment, book, contract, control,* (hence *rent-controlled* adj.), *man, office, officer, rebate, restriction, tribunal*; **rent boy** *slang*, a young male homosexual prostitute; **rent car** *U.S.*, a hire-car; **rent party** *U.S.* = *house-rent party* s.v. *HOUSE *sb.*[1] 23; **rent strike,** a refusal to pay rent, usu. by a number of people as a form of protest; **rent table** (see quot. 1952).

1947 *Rep. Assistance Board 1946* iv. 21 The Statutory Regulations provide for the addition of a 'rent allowance' according to the circumstances of each individual case. 1974 COOTE & GILL *Women's Rights* vii. 259 The council gives rent rebates to council tenants and rent allowances to other tenants. 1965 *Act* 13 & 14 *Eliz. II.* c. 75 § 25 There shall be constituted rent assessment committees in accordance with the provisions of Schedule 2 to this Act. 1970 *Internat. & Compar. Law Quarterly* 4th Ser. XIX. II. 208 Generally, the working of the rent assessment committees was commended. 1830 M. EDGEWORTH *Let.* 4 Nov. (1971) 427 The rent to us is to be from his commencement the raised rent. See Rent book. 1973 *Courier & Advertiser* (Dundee) 1 Mar. 6/4, I haven't seen my rent book for three years. 1978 *Lancashire Life* Apr. 67/1 He inspected Aunt Clara's rent-book and asked her for the names of her grocer and butcher. 1969 Rent-boy [see *KEPT *ppl. a.* 1 a]. 1975 *Daily Tel.* 24 July 3/6 Many of the boys became male prostitutes... They became known as 'rent boys'. 1976 DEAKIN & WILLIS *Johnny go Home* iii. 56 Between the ages of fifteen and twenty he had been a rent boy, a boy prostitute living and working in the West End. 1932 W. FAULKNER *Light in August* xv. 338 They went straight to the garage where Salmon keeps his rent car. 1970 *Islander* (Victoria, B.C.) 6 Dec. 7/1 Ben quickly turned his machine into a 'rent' car. 'They call them taxis now,' he explained. 1906 *Chambers's Jrnl.* Jan. 118/1 Evidences of the long-continued disturbance of rent-contracts in Ireland. 1940 W. FAULKNER *Hamlet* I. i. 14 But then I hear tell he always makes his rent contracts later than most. 1931 *Rep. Inter-Departmental Committee Rent Restrictions Acts* (Min. Health) xiv. 46 Some of us, if we had had the task of devising the original system of rent control..would perhaps have proposed the setting up of rent courts. 1965 *Listener* 20 May 727/1 Local authority subsidies, rent control of private rented houses. *a* 1974 R. CROSSMAN *Diaries* (1975) I. 48, I am engaged on rent controls already. 1946 M. B. LOWNDES *Let.* 9 Jan. (1971) 270 Luckily Susan's flat is rent controlled. 1971 B. MALAMUD *Tenants* 6 The building was rent-controlled, and from the District Rent Office..Harry had learned he was a statutory tenant. 1978 I. B. SINGER *Shosha* viii. 152 He had a rent-controlled apartment for which he paid no more than thirty zlotys a month. 1943 L. HUGHES in *Poetry* Sept. 312 The rent man knocked. 1969 *Punch* 1 Jan. 15/1, I do worry about the effect which its long illness is having on its general standing with the butcher, the baker, and the rentman. 1977 COOTE & GILL *Women's Rights* vii. 257 Check the rent register at your local Rent Office. 1965 *New Statesman* 9 Apr. 561/3 Widely publicised rent scales would..make it easier for rent officers. 1973 E. BERCKMAN *Victorian Album* 176 Before we ever took this lease, I went to the Rent Officer. 1976 *Southern Even. Echo* (Southampton) 13 Nov. 6/5 The rent officer in determining fair rents can consider under his brief the property and then fix his figure with reference to similar properties. 1926 C. VAN VECHTEN *Nigger Heaven* ix. 150 There were..the modest rent-parties. 1956 [see *house-party* s.v. *HOUSE *sb.*[1] 23]. 1968 P. OLIVER *Screening Blues* vi. 203 One of the most frequently heard songs in the rent-party repertoire was *The Boy in the Boat.* 1936 G. WILSON *Rent Rebates* 10 There is a not inconsiderable body of opinion which has already expressed itself in favour of the adoption of rent rebate schemes. 1971 *Reader's Digest Family Guide Law* 132/2 To apply for a rent rebate, a tenant must give details of his own income. 1977 *R.A.F. News* 11–24 May 4/6 Some 4,036 soldiers and 59 leading aircraftsmen were on rent rebate. 1921 A. W. BOON *Rent Restriction Act, 1920* 5 The Rent Restriction Act is very intricate and much involved. 1940 *Economist* 5 Oct. 422/2 Profiteers have been threatened with the Rent Restriction Acts. 1952 A. CHRISTIE *Mrs. McGinty's Dead* ii. 17 Under the Rent Restriction Act the landlord couldn't get the old woman out. 1976 *Evening Post* (Nottingham) 17 Dec. 32/8 Rent increases amounting to £2,025..were not included because of the failure to act on the partial lifting of rent restrictions in November, 1973. 1970 *N.Y. Times* 5 Feb. 38/6 The student organization also is lending moral and organizational support to..a widespread local rent strike. 1973 *Freedom* 26 May 1/2 Let the Trade Union movement now show its regard for the value of education by giving..support to student rent-strikes. 1977 *Transatlantic Rev.* LX. 120 In English class he'd go off on a crusade for Food Co-ops and Rent Strikes. 1927 MACQUOID & EDWARDS *Dict. Eng. Furnit.* III. 241 A..type, known as a 'Rent-Table', was introduced about this time [*sc.* 1750]. 1952 J. GLOAG *Short Dict. Furnit.* 387 *Rent Table*, a type of office table made during the second half of the 18th century, with a round or octagonal top, with drawers immediately below. 1961 'J. WELCOME' *Beware of Midnight* i. 9 The furniture, too, was solid and respectable. There was an octagonal rent table. 1973 V. CANNING *Finger of Saturn* i. 4 He..set his briefcase on the round rent table. 1945 *Daily Herald* 20 Apr. 4/3 Unanimous proposals of the Committee are: The establishment of 198 rent tribunals for England and Wales. 1973 E. BERCKMAN *Victorian Album* 28 It means spending half your life before the Rent Tribunal. 1974 COOTE & GILL *Women's Rights* vii. 256 You may be able to recover the extra amount by going to the Rent Tribunal.

rent, *v.*[1] Add: **2. a.** Also *absol.*

1671 W. BERKELEY in E. D. Neill *Virginia Carolorum* (1886) viii. 335 In Virginia about forty thousand persons..have come to settle and rent. 1911 M. W. OVINGTON *Half Man* 44 Not only were they unable to rent in neighbourhoods suitable for respectable men and women. 1979 *N.Y. Rev. Bks.* 17 May 6/2 New Yorkers rent. They don't buy.

b. To obtain money from (someone) by criminal means or in exchange for homosexual favours. *slang.*

1898 O. WILDE *Let.* 11 May (1962) 738 Bosie..is devoted to a dreadful little ruffian aged fourteen... Every time he goes home with Bosie he tries to rent him. 1956 C. MACKENZIE *Thin Ice* xii. 172 'I reckon you thought I was trying to rent Mr. Fortescue, eh?' 'To do what?' I asked in astonishment at such an expression. 'Get money out of him.'

5. (Earlier and later examples.) Also const. *for.* Now chiefly *N. Amer.*

1784 G. WASHINGTON *Diary* 15 Sept. (1925) II. 292 The Plantation on which Mr. Simpson lives rented well—viz. for 500 Bushels of Wheat. 1805 *New-England Palladium* (Boston) 26 July 3/3 Two convenient Tenaments, for small families, that will rent at 12 pr. cent of what they will be sold for. 1947 *Chicago Daily News* 25 Feb. 1/4 (*caption*) 4-room apartment to rent for $120. 1974 *Whig-Standard* (Kingston, Ontario) 11 Jan. 7/2 The smallest 'bedsitter' apartment in central London rents for about $25 per week.

7. a. Used (with reference to sense 2) in the form *rent-a-* prefixing a noun (usu. with hyphens) to designate the rental of the thing specified, orig. and chiefly a motor vehicle.

Rentacar forms part of a proprietary term in the U.S.

1921 *Chicago Central Business & Office Building Directory* 531/1 *Rent-a-Ford* (*Inc.*) 1450 S. Michigan av. 1924 *Official Gaz.* (U.S. Patent Office) 19 Feb. 503/1 The Rentacar Company, Toledo, Ohio..Rentacar U-Drive..Automobiles.

Claims use since Aug. 6, 1921. 1935 *Arch. Dermatol. & Syphilol.* XXXII. 78 A man..who owned a 'rent-a-car' business. 1963 *Fortune* Sept. 78 (Advt.), Avis is only No. 2 in rent a cars. So we have to try harder. 1966 J. GARDNER *Amber Nine* xi. 161 Martin..was at the wheel of the Rent-a-Car Merc. 1969 'G. BLACK' *Cold Jungle* ix. 137 The rentacar Zephyr was still waiting on grass. 1971 *E. Afr. Standard* (Nairobi) 13 Apr. 13/6 Rent-a-train and unit-train operations across North America would siphon more than 4 million tons of cargo from the Seaway. 1972 'G. BLACK' *Bitter Tea* (1973) ii. 26 The key to my rentacar was in one of my damp pockets. 1976 *National Observer* (U.S.) 21 Aug. 7/1 Rent-a-horse service is available from a riding school next door. 1977 *Rolling Stone* 19 May 11/2 His is a typical L.A. rent-a-home.

b. In various extended and fanciful uses, as *rent-a-crowd, rent-a-mob*, etc., to denote the spontaneous acquisition or instant availability (usu. for some transitory purpose) of the thing specified. Also *transf.* in *concr.* senses.

1961 *Daily Tel.* 21 Dec. 8/6 Dictators!!! When you liberate a territory or mop up a colonialist enclave, are you disappointed and upset to receive only a tepid welcome from the people? Let *rentacrowd* help you! We can supply cheering crowds for all occasions. 1964 C. DRIVER *Disarmers* x. 233 The phenomenon which Peter Simple of the *Daily Telegraph* cruelly christened 'Rentacrowd': London's instantly available progressive claque ready..to demonstrate on a whole range of causes. 1968 *Guardian* 22 Aug. 8/6 Ali..is the only man in Britain at the moment who can summon up a sizable Rentacrowd fast. 1970 *Peace News* 5 Sept. 7/1 One of the chief rentacops was reportedly bitten in the genitals by one of his own dogs. 1970 *Guardian* 27 Oct. 11/5 The strategy was based upon a tactic which Oxford students called Rentamob..a hard core of rioters who could turn a demonstration into a confrontation. 1972 M. JONES *Life on Dole* i. 11 In Merthyr Tydfil it was the day for the demonstration against unemployment...There are no professional marchers—no Rentacrowd—here. 1976 *Times* 27 Jan. 4/1 Squatters in London are reported to be using children in a 'rent-a-kid' system, as a means of being rehoused. 1977 *New Society* 7 July 15/1 Trouble was caused not by ordinary workers, still less by management, but by Rent-a-picket... 'There's always the Rent-a-crowd element that hangs on to strikes.' 1979 *Daily Tel.* 17 Nov. 2/1 Sir John denied that the trade union movement was contaminated by the 'rent-a-mob' philosophy.

rentability. Delete *rare*$^{-1}$ and add later examples. *spec.* (see quot. 1964).

1922 W. SCHLICH *Man. Forestry* (ed. 4) I. II. 107 These changes should not, in the long run, seriously affect the rentability of the forest industry. 1964 *Financial Times* 3 Mar. 5/2 Rentability is the ratio of an undertaking's profits to its capital. 1976 F. ZWEIG *New Acquisitive Society* II. iii. 107 Public enterprise..is also supposed to achieve productivity and rentability.

rental, *sb.* Add: **2. b.** A house, flat, car, etc., let out for rent. Chiefly *N. Amer.*

1952 *Sat. Even. Post* 22 Nov. 25 Sometimes a mob of hoodlums in a rental forced off the road by pursuing police would get tangled up in their own arsenal. 1968 *Globe & Mail* (Toronto) 3 Feb. 3/3 Habitat, Expo's futuristic housing complex, has started a new life as an ordinary rental development. 1970 *New York* 16 Nov. 42/3 They began breaking up the mills and foundries into rentals for small-time manufacturers. 1972 J. GORES *Dead Skip* xi. 76 The house was a rental, and rental properties meant landladies. 1981 *Nordic Skiing* Jan. 47/1 The ski shop carries complete rentals in all sizes and reservations on rental equipment are accepted.

3*. The fact or process of renting (in the sense of RENT *v.*[1] 3).

1915 *Nat. Real Estate Jrnl.* Nov. 332/2 The duty of the agent to owners whose property he has in charge for rental, requires him to make untiring effort to promptly secure desirable tenants. 1928 *Publishers' Weekly* 12 May 1951 The rental of children's books has not so far been well tested out, as only four of the stores replying have collections of children's books. 1977 *Grimsby Even. Tel.* 31 May 11/2 Advt.), The property is on rental.

4. *rental agency, agent, car, house*; **rental library** chiefly *U.S.*, a library at which a charge is made for the loan of books.

1947 S. L. MCMICHAEL *How to operate Real Estate Business* XXVII. 221 The rental agency must use considerable tact in fitting tenants into the right locations. 1972 J. PHILIPS *Vanishing Senator* III. iv. 136 Couldn't you try rental agencies? 1978 R. LUDLUM *Holcroft Covenant* xiii. 153 The rental agency was not amused, but Holcroft gave them no choice. 1915 *Nat. Real Estate Jrnl.* Nov. 332/2 (*heading*) Duties of rental agents. 1967 *Boston Sunday Globe* 23 Apr. B43/7 The rental agent is Martin Cerel, Natick realtor. 1979 *N.Y. Rev. Bks.* 17 May 37 (Advt.), Britain's most experienced rental agent is here to give you every assistance with your plans for accommodation. 1962 *Time* 7 Dec. 84/3 Cut-rate rental cars are generally as clean and well-serviced as the big three. 1978 S. SHELDON *Bloodline* xxxiii. 315 It took almost three hours for Max to drive to Lesgets in a Volkswagen, the cheapest rental car he could find. 1953 A. UPFIELD *Murder must Wait* ix. 85 Rental houses are few..and the demand for them is heavy. 1928 *Publishers' Weekly* 14 July 169 His basement book-store..is now the home of an unusually successful rental library. 1934 G. CONKLIN *How to run Rental Library* i. 11 The rental library..is strictly limited to that type of book-renting business which is organized for the purpose of profit. 1946 R. CHANDLER *Let.* 9 Jan. (1966) 136 The publishers have co-operated in the rental library swindle over a period of years.

rentalsman (re·ntălzmæn). *Canad.* [f. RENTAL *sb.* by analogy with *OMBUDSMAN.] An official with responsibility for the equitable letting and administration of rented property.

1970 *Deb. & Proc. Legislative Assembly Manitoba* 30 June 3525 In my view, the singularly most unique feature of the proposed amendments is the establishment of the office of Rentalsman. 1971 *Canad. Labour* Mar. 15/3 This rating was based on: a single easy-to-read law, containing a broader statement of tenant rights than the other provinces; appointment of a 'rentalsman' (rental ombudsman); and a public education program on the new legislation. 1974 *Globe & Mail* (Toronto) 13 Apr. 8/2 Ceilings for rent increases..may be breached if the landlord can convince a new Government official called a rentalsman that higher increases are necessary because of costs.

rent-charge. Add: **2.** *attrib.*, as *rent-charge, bank, stock.*

1909 *Daily Chron.* 14 Sept. 5/6 Rent-charge banks were formed to aid the peasants in redeeming these charges. 1869 *Bradshaw's Railway Man.* XXI. 290 The debenture holders having refused to accept the 4½ per cent. rent-charge stock at par. 1909 *Gt. Central Railway Co. Rep.* 6 Aug. 13 The South Yorkshire Rent-Charge Stocks.

‖ **rente** (rant). [Fr.] Stock, esp. government stock; the interest or income accruing from such stock. Cf. RENT *sb.*[1] 3 b.

1873 R. BROWNING *Red Cotton Night-Cap Country* III. 185 Lying stretched on straw, The produce of your miserable *rente*! 1920 J. GALSWORTHY *In Chancery* III. ix. 283 She had, he knew, but one ambition—to live on her '*rentes*' in Paris. 1926 D. L. SAYERS *Clouds of Witness* v. 116 There were substantial dividends from capital invested in French *rentes.* 1927 *Financial Times* 13 July 5/5 French Banks and Rentes hardened, recovering the ground lost recently. 1931 S. JAMESON *Richer Dust* xvii. 495, I must find a job... I can't just live on my *rentes.*

rented, *ppl. a.* Add: **3.** *lower-rented.*

1801 JANE AUSTEN *Let.* 3 Jan. (1952) 100 It used to be lower rented than any other house in the row. 1939 M. S. RICE *Working-Class Wives* viii. 195 The shortage will not have been overtaken by 1941 even if the process of 'filtering-up', (i.e. people who can afford to do so vacating the lower-rented houses and moving into unsubsidized higher-rented houses) were steady and complete.

Rentenmark (re·ntənmăɹk). [Ger., f. *renten* securities: see MARK *sb.*[2] in Dict. and Suppl.] A unit of currency introduced in Germany in November 1923 and tied to the nation's industrial and agricultural resources; in 1924 it was replaced by the *reichsmark* (see *REICH).

1923 *Times* 8 Nov. 12/4 The issue of Rentenmarks is to be delayed till it can take place simultaneously throughout the country. 1940 G. CROWTHER *Outl. Money* i. 19 The Rentenmark note was not itself land, nor was there any method by which the holder of a Rentenmark note could possess himself of the land that was supposed to be behind his note. 1967 T. STOLPER tr. *G. Stolper's German Econ.* iv. 91 In the case of the Rentenmark collateral consisted of the 'real-estate debts' of agriculture and of..industrial companies. 1974 *Spectator* 21 Dec. 803/1 The Rentenmark was a new unit of currency supposedly backed by the land of the Reich.

renter, *sb.*[1] **4. b.** Delete † *Obs.* and add later examples.

1882 SWEET & KNOX *Sk. Texas Siftings* 51 The joyful glee of farming with negro renters 'on the shares'. 1938 *Mississippi* (Fed. Writers' Project) 104 *Renters*, who hire land for a fixed amount to be paid either in crop values or in cash. 1970 J. BLACKBURN *Land of Promise* viii. 119 He was on his way to see a renter on one of his farms south of town.

6. A male prostitute. *slang.*

1893 O. WILDE *Let.* Mar. (1964) 336, I would sooner be blackmailed by every renter in London, than have you bitter, unjust, hating. 1895 M. BEERBOHM *Let.* 3 May (1964) 103 It was horrible leaving the court day after day and having to pass through a knot of renters. 1969 *Jeremy* I. III. 22/2 *Renter*, male prostitute. 1972 D. SUTTON *Lett. R. Fry* I. 5 In many cases 'affairs' were more idealistic than that practised by the 'renters' of Piccadilly.

7. One who organizes the distribution of films to exhibitors.

1908 *Variety* 16 May 11 There are other and larger questions pressing the attention of the exhibitor, renter and manufacturer. 1911 D. S. HULFISH *Cycl. Motion-Pict. Work* II. 112 The film industry is definitely separated into three branches: *manufacturer, renter,* and *exhibitor.* The renter owns the picture films. 1920 I. P. GORE in L. Carson '*Stage' Year Bk. 1920* 52 The success which attended the efforts of the exhibitors, renters, and manufacturers, to combat the peril with which they were confronted. 1927 *Melody Maker* Aug. 820/2 The renters could render far more assistance..if they would take more interest in the showing of their films at every cinema where they are booked. 1940 *Economist* 13 July 43/1 To ensure that a minimum number of films are made, renters must produce or acquire one British film.

rentering, *vbl. sb.* Delete † and add later examples. Also *attrib.*

1901 P. N. HASLUCK *Tailoring* 21 There are three kinds of absolutely invisible stitches which are used to repair tears... They are stoating, fine-drawing, and rentering. 1921 [see *INVISIBLE *a.* 1 f]. 1955 J. E. LIBERTY *Pract. Tailoring* (ed. 2) iii. 24 Seaming and rentering... this is used in place of stoting when the material does not lend itself to being stoted.

‖ **rentier** (rañti,e). Also fem. **rentière** (-iʸę̄r). [Fr.] One who makes an income from property or investment. Cf. *RENTE.

1847 in WEBSTER. 1885 *Instructions Clerks Classifying Occupations* (Census Eng. & Wales, 1881) 100 Persons without specified occupations..returned by property, rank, &c... Rentier. 1885 A. EDWARDES *Girton Girl* III. xv. 265 We are private citizens—rentières, living on our means. 1921 R. H. TAWNEY *Acquisitive Society* v. 68 If it [*sc.* a society] is to..avoid the creation of a class of *rentier*, it must not use for current consumption the whole of the wealth annually produced. 1948 A. HUXLEY *Let.* 25 Jan. (1969) 579 Maria's uncle..is a rentier, living on an unelastic income. 1954 M. BERESFORD *Lost Villages* II. vi. 205 Any ex-demesne land which *rentier* lords were prepared..to lease. 1964 T. B. BOTTOMORE *Elites & Society* iii. 45 Pareto's two types of elite..which he also refers to as the 'speculators' and the 'rentiers'. 1969 *Listener* 6 Feb. 173/2 This grumbling, sinisterly superior *rentière* lived opposite Virginia Lodge, Parkwood Hill, the Dales' suburban residence. 1973 'D. JORDAN' *Nile Green* xxxi. 145 It's oil sheiks and Latin American generals and Lebanese rentiers who are going to buy your bonds. 1976 M. GREEN *Children of Sun* viii. 316 Brian..fulminated against *rentiers* and Money Men..but..he remained a hedonist and a snob.

renting, *vbl. sb.*[1] Delete *rare* and add: **1.** (Later examples.)

1946 E. O'NEILL *Iceman Cometh* p. vii, The renting of rooms on the upper floors, under the Raines-Law loopholes. 1974 *Howard Jrnl.* XIV. 41 Renting to the poor became increasingly unprofitable.

2. *slang.* In the sense of *RENT *v.*[1] 2 b.

1956 C. MACKENZIE *Thin Ice* xiv. 183 Mr. Jack Shore has done quite a lot of blackmailing, or in his own elegant phraseology a lot of renting, during his inglorious career. 1963 —— *Life & Times* II. 255 At this date [*sc.* 1899] the cant word among homosexuals for their proclivities was 'so'. That seems to have vanished completely from current cant, though 'renting' for male prostitutes and 'camping' to express the way they attract the soliciting male still survive.

3. *attrib.*, as *renting firm, house.*

1920 I. P. GORE in L. Carson '*Stage' Year Bk. 1920* 52 In the same way the big renting firm came to the rescue of its small rival. 1927 *Melody Maker* Aug. 819/2 Every renting house of any standing requires a capable and experienced cinema musician permanently on its staff.

‖ **rentrée** (rañtre). [Fr.] A return, esp. a return home after an annual holiday.

1892 E. DOWSON *Let.* 22 Aug. (1967) 240 Many thanks for your note, which I found on my rentrée last week. 1896 BEERBOHM in *Yellow Bk.* XI. 20 His *rentrée* into the still silent town strengthened his..resolves. 1913 —— *Fifty Caricatures*, (*caption*) Rentrée of Mr. George Moore into Chelsea. 1961 *Guardian* 22 Sept. 16/1 The Rentrée is the return to the great cities, the beginning of a new school year. 1977 *Times* 30 Aug. 4/1 Life is about to begin again in France... The word of the moment is *la rentrée.*

renunciant, *sb.* (Earlier and later examples.)

1848 T. ARNOLD *Let.* 2 Jan. (1966) 216 Even now there is a little band of Renunciants scattered over the world. 1931 E. WILSON *Axel's Castle* viii. 257 All were pessimists, renunciants, resignationists, 'tired of the sad hospital' which earth seemed to them.

renversement. Delete † *Obs.* and add later examples. Now *spec.* in *Aeronaut.* (orig. *U.S.*), an aeroplane manœuvre consisting of half-loop effected simultaneously with a half-turn (see also quot. 1956).

1954 W. FAULKNER *Fable* 343 The car making the last *renversement* because now it could go no farther. 1956 *U.S. Air Force Dict.* 435/2 *Renversement*,..any airplane maneuver or performance in which the airplane is made to reverse direction, as in a chandelle, Immelmann turn, or wingover. 1966 E. V. RICKENBACKER *Rickenbacker* (1968) vi. 106, I immediately put my Spad into a *renversement*—pulled the stick straight back to start a loop and simultaneously rolled it over in a half turn. 1973 *Times* 2 Jan. (Forward into Europe Suppl.) p. xii/6 Perhaps the Henry Moore retrospective..and the Francis Bacon exhibition..herald a *renversement.*

‖ **renvoi** (rañvwa). *Law.* [Fr., f. *renvoyer* to send back: cf. RENVOY *sb.*] (See quots.)

1897 J. T. B. SEWELL tr. Labbé in *Outl. French Law as affecting Brit. Subjects* iv. 69 In order to justify the system which repudiates the doctrine of 'renvoi'..it suffices to define exactly what is the intention of the law which authorizes the foreign law to decide. 1898 *Law Q. Rev.* XIV. 231 An English specialist versed in the problems raised by the conflict of laws may well never have heard the name of the *Renvoi.* 1904 J. P. BATE *Doctrine of Renvoi* iv. 53 The basis of the Renvoi-theory is the doctrine that when a conflict-rule refers a matter to a foreign law, the foreign law is referred to *in its totality.* 1905 J. WESTLAKE *Priv. Internat. Law* (ed. 4) ii. 25 In its substance the subject [*sc.* private international law] has been no less deeply affected by the substitution..of political nationality for domicile as the criterion of personal law, and by the controversy which has consequently sprung up about the doctrine called the *renvoi.* 1935 G. C. CHESHIRE *Priv. Internat. Law* vi. 133 The second solution is to give a decision on the assumption that the doctrine of *renvoi* is recognized by English law. This famous doctrine, which presupposes that a reference to the law of the domicil means a reference to the whole of that law, may be explained as follows in connexion with our case of X. The English Court, upon referring to the Private International Law of France, finds that it is referred back to English law as being the law f X's nationality. There is a *renvoi* or a remission to English law. 1959 E. JOWITT *Dict. Eng. Law* II. 1525/1 *Renvoi*, a term employed in private international law to denote the sending, or determination, of a matter to or according to the law of a tribunal outside the jurisdiction where the question arose. 1970 *Internat. & Compar. Law Q.* 4th Ser. XIX. I. 36 The generally accepted English view has been that tort cases are outside the scope of the doctrine of *renvoi.*

reoccu·rrence. [RE- 5 a.] A further occurrence; a recurrence.

1817 D. O'CONNELL *Let.* 24 June (1972) II. 152 There will not be a reoccurrence even of those wretched squabbles.

1964 B. TRNKA in D. Abercrombie et al. *Daniel Jones* 188 The re-occurrence of voiceless consonants in the same morpheme is found more frequently than that of the corresponding voiced consonants.

reorder, *v.* Add: **3. b.** (Later example.)

1967 COX & GROSE *Organization & Handling Bibl. Rec. by Computer* vi. 162 Not all delayed books are going to be re-ordered.

reordered *ppl. a.*, **reordering** *vbl. sb.* (examples in sense 3 b of vb.).

1938 *New Statesman* 21 May 860/2 Ordered, counter-ordered and reordered machines. 1962 A. BATTERSBY *Guide to Stock Control* v. 42 The boundary between them remains as a mark which indicates the Re-ordering Level (ROL). 1969 *Jane's Freight Containers 1968–69* 416/1 Consistent production rates possible with more frequent re-ordering cycles.

reo·rder, *sb.* [RE- 5 a.] A renewed or repeated order for goods.

1901 *Scotsman* 8 Apr. 9/7 Current trade is confined to small reorders. 1928 *Publishers' Weekly* 9 June 2376 It is not easy on placing advance orders..to get the full value of the suggestions of the clerks, but on the reorders this is simpler. 1977 *Time Out* 21 Jan. 3/3 The chap behind the counter is often the one who chooses the wine he sells and he is there to take your re-orders when you come back.

reo:rganiza·tional, *a.* [f. REORGANIZATION + -AL.] Of or pertaining to reorganization.

1972 *Guardian* 26 Jan. 4/3 Yugoslavia's League of Communists united today behind party reorganisational changes.

reo:rganiza·tionist. [f. REORGANIZATION + -IST.] One who favours (political) reorganization; *spec.* a member of a radical Chinese faction (fl. 1930: see quot. 1975). Also *attrib.* or as *adj.*

1930 *Times* 25 Mar. 23/5 Nanking was faced by a revolt from the so-called 'Reorganisationists', a group mainly consisting of the political leaders of the left who had been excluded from office. 1967 J. ISRAEL in A. Feuerwerker et al. *Approaches to Mod. Chinese Hist.* 292 Many Wuhan figures who cast their lot with Wang Ching-wei (exiled leader of the Reorganizationist faction) argued that the revolution remained 'unfinished'. 1975 I. C. Y. HSÜ *Rise Mod. China* (ed. 2) xxiii. 653 Wang and his left-wing followers were out of office. The latter group retaliated by accusing Chiang of betraying the principles and ideas of Sun, and demanded a reorganization of the KMT in the spirit of the 1924 manifesto—hence their nickname 'The Reorganizationists'.

reorganize, *v.* Also *intr.* for *refl.*

1857 J. HYDE *Mormonism* vii. 183 They completely re-organized in May, 1857. 1972 'E. LATHEN' *Murder without Icing* i. 14 The time has come for him to reorganize. He will be selling some earlier ventures.

Hence **reo·rganized** *ppl. a.*

1892 'MARK TWAIN' *Amer. Claim.* xiv. 129 The rude impact of the thought of these people upon his reorganized condition of mind. 1929 P. HUGHES *Catholic Question, 1688–1829* III. iii. 283 In the re-organised Catholic Committee, O'Connell had from the first been a force.

reo·rient, *v.* [RE- 5 a.] **1.** *trans.* To rearrange, give a new orientation or direction to (ideas, etc.); to help (a person) to find his bearings again; to redirect (a thing).

1933 *Times Educ. Suppl.* 25 Feb. 57/4 Russia's children are suffering in the grim struggle; but they are not having to be reborn, to reorient their whole lives. 1939 *Sun* (Baltimore) 25 July 8/7 'Britain has *re-Oriented* her China policy.' I don't know whether that pun was Mr. Byas' own. 1951 G. HUMPHREY *Thinking* ix. 306 Psychic 'scaffolding', 'hypotheses', by which the original *Aufgabe* is reoriented. 1956 A. H. COMPTON *Atomic Quest* 32 His [*sc.* Einstein's] quantum and relativity theories were reorienting physics thought. 1966 D. F. GALOUYE *Lost Perception* vii. 74 'I feel so out of touch,..as though I've stood still for two whole years while everything passed me by.' 'We'll reorient you,' Forsythe assured. 1975 *Nature* 10 July 109/1 On day 123 the satellite was reoriented to allow the four Ariel-V experiments which view along the spin axis to observe Cyg X-1, placing it outside the field of view of this experiment. 1977 D. BENNETT *Jigsaw Man* iv. 71 I'll give you a sort of running commentary,' she said. 'Reorient you.'

2. a. *refl.* To adjust (oneself) *to* something; to come to terms with something; to adopt a new direction or relation.

1937 *Jrnl. Compar. Psychol.* XXIV. 296 The animals did recognize a new situation in the tests... Many took much longer about the first jump in the test trials, and appeared to be re-orienting themselves to the whole environment. 1955 M. LOWRY *Let.* July (1967) 381 What with this eye business I have to revise entirely my method of writing and..reorient myself to it. 1962 *Lancet* 8 Dec. 1229/2 Some will reorient themselves in their new surroundings in about three weeks, recognising their new ward as their home. 1963 *Economist* 19 Jan. 232/2 If the railways can really re-orient themselves.

b. *intr.* for *refl.*

1960 C. D. SIMAK in *Mag. Fantasy & Science Fiction* June 104 He had to reorient, he knew. He had to come to..terms..with this situation. 1974 *Nature* 6 Sept. 16/1 The other spectra are so similar to V that Watkins also assigns them to zinc vacancies. But the V[1] centres do not reorient under uniaxial stress. 1977 *Jrnl. Protozool.* XXIV. 29/2 Some conjugating pairs subjected to antiserum treatment separate belatedly, after a partial cortical fusion has occurred and when the conjugants have reoriented into a heteropolar configuration.

reo·rientate, *v.* [RE- 5 a.] **1.** *trans.* = *REORIENT *v.* 1.

1933 *Planning* 6 June 2 The most urgent necessity..for a common and intensive effort by all parties to reorientate their various points of view. 1958 *Times Lit. Suppl.* 21 Nov. 675/1 How does one re-orientate one's philosophy and economic policy? 1959 B. WOOTTON *Soc. Science & Soc. Pathol.* viii. 267 Glueck's object in treating criminality as a disease is to concentrate attention..on the practical problem of how his [*sc.* the offender's] behaviour may be re-orientated. 1981 P. MCCUTCHAN *Shard calls Tune* xv. 163 [He] came to a small town... He knew where he was... He was re-orientated nicely.

2. *refl.* = *REORIENT *v.* 2 a.

1940 *Scrutiny* IX. 196 We may perhaps hope that Scotland will be able to reorientate itself with the end of the epoch that has seen its cultural disintegration. 1959 M. SUMMERTON *Small Wilderness* xiv. 174 Struggling to reorientate myself..I tried to place myself in his shoes. 1979 J. WAINWRIGHT *Duty Elsewhere* xvi. 47 Cooke had reorientated himself a little. He was still dizzy.

reo:rienta·tion. [RE- 5 a.] The action or process of reorienting; a fresh orientation.

1920 *Contemp. Rev.* July 6 There will be needed a great collaboration of wisely directed effort at home, combined with an entire reorientation of attitude and policy on the part of the Allies. 1938 *Burlington Mag.* May 248/2 He seems to minimize the imposition of sartorial forms by commercial dictation, surely more often the cunning contrivance of change for change's profitable sake than the concrete expression of social reorientations. 1942 L. B. NAMIER *Conflicts* 16 Far-seeing statesmen discerned the need of such a reorientation. 1951 R. FIRTH *Elem. Soc. Organization* iii. 110 This is represented..by a re-orientation of resources in goods or labour power. 1967 G. STEINER *Lang. & Silence* 83 Many reorientations, many ways of ordering and choosing are available to scholarship and the imagination. 1977 P. BAELZ *Ethics & Belief* vi. 74 Different religions give different answers to this question, all of them involving both a reorientation of insight and understanding and a new pattern of action.

reovirus (rī·ovəiᵊrɒs). *Biol.* [mod.L., f. initial letters of *r*espiratory, *e*nteric, *o*rphan (see *orphan virus* s.v. *ORPHAN *a.* 2) + VIRUS.] Any of a group of related double-stranded RNA viruses that are sometimes associated with disease in animals, including respiratory and enteric infection in man.

1959 A. B. SABIN in *Science* 20 Nov. 1388/1 Studies..have provided the defininitive information required for establishing these strains as members of a new group of viruses, for which I am proposing the name 'reoviruses'. 1970 *Nature* 28 Mar. 1209/1 *Vaccinia* virus and reovirus assemble their mRNA synthesizing enzymes within the mature virion. 1973 R. G. KRUEGER et al. *Introd. Microbiol.* xix. 532/1 Three serological types of reovirus have been identified. The structure of the particles..consists of an RNA core some 40 nm in diameter and an icosahedral capsid 30–50 Å in width. No envelope is present. The 92 capsomeres are closely packed and appear as pentagonal or hexagonal columnar prisms (possibly truncated pyramids) arranged in 5:3:2 symmetry. 1979 *Sci. Amer.* Nov. 65/1 In mice..reovirus attacks the acinar cells of the pancreas.

reo·xidize, *v.* [RE- 5 b.] **a.** *trans.* To oxidize again.

1940 GLASSTONE *Textbk. Physical Chem.* xii. 1020 The ion, e.g., Ti^{++++}, is first reduced at the cathode, e.g., to Ti^{+++}, which is the active reducing agent; in the process it is re-oxidized to Ti^{++++}, and is again reduced at the cathode and so on. 1957 *Science* 12 Apr. 691/3 Partially reduced ribonuclease..was reoxidized. 1973 *Chem. Soc. Rev.* II. 43 The unwanted epimer can be reoxidized to (78) for recycling.

b. *intr.* To take up oxygen again.

1966 *Economist* 16 Apr. 289/3 The zinc vapour is stripped out of the gases by a spray of molten lead—a vital part of the new process, since it removes the zinc before it can re-oxidise. 1976 *Sci. Amer.* July 78/2 The product of the Purofer process is a briquette that does not reoxidize readily because of its small surface area.

reoxygenate, *v.* Add: Hence **reo:xygena·tion.**

1957 G. E. HUTCHINSON *Treat. Limnol.* I. ix. 593 The full reoxygenation of a very deep lake..presents real difficulties.

rep[1]**.** Substitute for def: Colloq. abbrev. of REPUTATION. Common in early 18th c.; now chiefly *U.S.* (Later examples.)

1873 J. H. BEADLE *Undevel. West* xix. 367 Of the town proper, a majority of citizens were negroes, with them a few whites of doubtful 'rep', and perhaps a dozen Indians. 1910 E. A. WALCOTT *Open Door* xii. 155 An' me a white man, too, even if me rep. is off color. 1935 'E. QUEEN' *Spanish Cape Mystery* iii. 68 Got a rep as a bad customer. 1956 'E. MCBAIN' *Cop Hater* (1958) x. 90 We got a big rep as it is. Ain't nobody in this city who ain't heard of The Grovers. 1978 M. PUZO *Fools Die* iv. 51 He was a legitimate bad guy with an obvious rep in Vegas.

rep[4]**.** (Further examples.)

1906 [see *CON d]. 1930 E. M. BRENT-DYER *Chalet School & Jo* iv. 57 'There's only rep left, and I've arithmetic and French to write!' protested Cornelia.

rep[5]**.** Abbrev. of REPRESENTATIVE *sb.*; esp. a sales representative.

1896 T. EYTON *Rugby Football* 10 Joe..has annually played and led the Tauranga reps. against the Auckland teams. 1933 L. G. D. ACLAND in *Press* (Christchurch, N.Z.) 18 Nov. 15/7 *Rep*... Under the arbitration award it is laid down that 'the shearers shall elect a representative', who deals with the employer..whenever any dispute or question arises. 1938 E. AMBLER *Cause for Alarm* ii. 28 No travellers seen except on Tuesdays and Thursdays... Reps.., Tuesdays and Thursdays. 1959 T. GIRTIN *Unnatural Break* xxviii. 101 The local rep of the League of Empire Loyalists got up in the middle and made a protest. 1959 *New Statesman* 19 Dec. 874/1 One young sales rep, whom I met quite early in the month, was already discovering his soul and finding the process painful. 1969 T. LLOYD in R. Blythe *Akenfield* xiv. 218, I am the only member of this club who isn't a farmer's son..or a rep. 1977 *Jrnl. R. Soc. Arts* CXXV. 679/1 We can have three thousand safety reps for a labour force of thirty five thousand. 1979 *Tucson* (Arizona) *Citizen* 20 Sept. 2A/2 The House turned down an amendment by Rep. Eldon Rudd.

rep[6]**.** Colloq. abbrev. of REPERTORY 3 (occas. of REPERTOIRE); a repertory company or theatre. Freq. *attrib.*

1925 *Amer. Speech* I. 36/2 A rep show is made up of players with a repertoire of plays. 1929 J. B. PRIESTLEY *Good Companions* II. ii. 283 Each member of the troupe prided himself..on having a large répertoire, known always as a 'rep'. 1933 S. O'CASEY *Let.* 24 Sept. (1975) I. 465 A Rep or a Stage Society production. 1948 H. L. MENCKEN *Amer. Lang.* Suppl. II. 690 *Rep company*, a company presenting a répertoire of plays on the road. 1959 *Times* 5 Jan. 12/2 While the productions are out of Oxford we let the theatre to local societies, visiting reps or small-cost tours. 1959 *Manch. Guardian* 30 Jan. 7/1 She has returned to 'weekly rep.', producing for a sound but as yet undistinguished company which must perform potboilers for most of the year. 1971 *Guardian* 27 Sept. 10/1 Faithful and ageing Rep-goers. 1977 R. BARNARD *Death in High C's* ii. 18 You're back with..the same old rep production. 1977 *Radio Times* 12–18 Nov. 15/1 After the war Major Bates joined Worthing rep as stage manager.

† **rep**[7]**.** *Obs.* [f. initial letters of *roentgen equivalent physical.*] A quantity of ionizing radiation that will release the same amount of energy in human tissue as one rad (formerly roentgen) of X-rays. Cf. *REM *sb.*[1]

1947 *Nucleonics* Oct. 38/2 If the energy lost by ionization in the tissues is the same as the energy loss for one roentgen of gamma radiation absorbed in air, the dose is spoken of as one roentgen-equivalent-physical (abbreviated 'rep'). [*Note*] The rep and rem units were introduced by Dr. H. M. Parker. 1955 *Bull. Atomic Sci.* June 211/1 On a conservative estimate, a dose of 200 reps, such as many Hiroshima survivors must have received, would probably have caused each of their offspring to inherit, on the average, at least one mutation produced by the exposure. 1962 F. I. ORDWAY et al. *Basic Astronautics* xii. 494 The rep or roentgen equivalent physical is the unit that equals the ionization produced by other forms of radiation to that of x-rays and gamma-rays. The rep is defined as the amount of particulate radiation that produces an energy absorption of 93 ergs/g in human tissue.

repacification. For *rare*$^{-1}$ in Dict. read *rare* and add later example.

1937 *Times* 17 Nov. 17/4 Each costly 'pacification' of Wazirs, Mahsuds, or Mohmands is to be followed by their equally costly repacification at regular and almost predictable intervals.

repacker. (Example.)

1976 *Shell in Industr. Chemicals* (Shell Internat. Petroleum Co.) 5 Group companies sell Teepol to blenders and re-packers.

repaint, *sb.* Add: **1.** (Further examples.)

1901 R. FRY *Let.* 10 Dec. (1972) I. 182, I am at present cutting off the sky with a knife and have found a marvellous landscape underneath, entirely concealed by repaints. 1935 *Burlington Mag.* Nov. 202/1 The picture was..then almost entirely covered with re-paint.

2. (Later examples.) Also *fig.* and *attrib.*

1964 G. MARX *Let.* 27 Oct. (1967) 258, I will be in London in April, doing a repaint job on my old quiz show. 1968 S. JAY *Sleepers can Kill* vii. 74 The car was a black Opel in need of a repaint. 1970 J. WAINWRIGHT *Prynter's Devil* iii. 50 Vac the room first, kiddo. Then start the repaint job.

3. Something repainted, esp. a golf-ball.

1922 WODEHOUSE *Clicking of Cuthbert* i. 16 Why, you are a pearl among women, the queen of your sex... You make the rest look like battered repaints. 1931 M. ALLINGHAM *Look to Lady* xiv. 148 'How's the car business?'.. 'I sold a lovely repaint in Norward last week... We faked it up lovely—registration book and everything.' 1936 *Sun* (Baltimore) 1 Aug. 817, I should like to know whether they [*sc.* golf balls] are new ones or repaints. 1955 *Times* 23 Aug. 4/4 Repaints cost as much as new balls used to.

repair, *sb.*[2] Add: **3.** *repair bill, kit, -man* (chiefly *U.S.*), *outfit, -ship, -shop* (earlier and later examples), *station, time, work*(*s*).

1908 *Westm. Gaz.* 7 Jan. 4/2 The effects of wear and tear would be reduced to a minimum, and the repair bill..would be kept very low. 1970 *Observer* 1 Feb. 31/6 Ten new pence..now buys a 'Panti-hose and Tights Repair Kit' from Woolworths. 1975 *Times* 28 Aug. 11/8, I do not regard a song and dance as the infallible, all-purpose dramatic repair kit. 1871 W. S. HUNTINGTON *Road-Master's Assistant* ii. 9 It is a common practice for repairmen, when replacing mended iron, to squeeze it in perfectly tight. 1928 *Sat. Even. Post* 4 Feb. 140/1 If your car suffers from any of these common motor ills, take it to your car dealer or repairman. 1958 *Times* 13 Sept. 7/7 Such dilemmas as that of the Los Angeles couple whose T.V. stops one night and the husband must go out and find the repairman. 1976 *Washington Post* 19 Apr. A22/4 The tenants refuse to let repairmen into their apartments. 1908 *Sears, Roebuck Catal.* 517/1 Traveler's big complete repair outfit. 1976 J. R. L. ANDERSON *Redundancy Pay* i. 17 He..bought a pump, repair outfit, and a torch-battery cycle lamp. 1905 *Westm. Gaz.* 16 Nov. 6/2 The King's repair-ship 'Assistance'

was floated off at Tetuan yesterday. 1866 *Harper's Mag.* Sept. 543/1 In the repair-shops of the Columbus and Indianapolis Railroad. 1979 *Jrnl. R. Soc. Arts* July 466/1 Somehow the hospital must contrive to be both repair-shop and home. 1906 *Westm. Gaz.* 27 June 7/3 The mechanic..hung on to the radiator from the starting-line to the repair-station. 1934 *Discovery* Nov. 326/1 Its main depot and repair station is at Lunghwa near Shanghai. 1962 *Autom. Data Proc.* (*B.S.I.*) 52 *Repair time*, time spent outside the periods allocated to routine maintenance and supplementary maintenance in diagnosing and clearing faults, equipment testing and maintenance. 1962 D. R. COX *Renewal Theory* vii. 80 Suppose that a machine is subject to stoppages and call the time necessary to restart a stopped machine a repair-time. 1906 *Westm. Gaz.* 22 Aug. 10/1 Important repair work is..being undertaken..at the cathedrals of Winchester, Gloucester, York, and Canterbury. 1907 *Ibid.* 21 Mar. 9/1 With no repair works and with insufficient and sometimes incompetent staff, they ran their omnibuses as many hours as they could anyhow be kept on the road. 1969 *Gloss. Landscape Work* (*B.S.I.*) v. 30 *Repair work*, the treatment of incisions, bruises and other wounds or injuries [in trees etc.].

repairability (rīpēᵊ:răbi·lĭti). [f. REPAIRABLE *a.* + -ITY: cf. REPARABILITY.] The state or quality of being repairable.

1969 *Jane's Freight Containers 1968–69* 464/1 The ship lines are worried about repairability, particularly in foreign ports.

repairable, *a.* (Further examples.)

1890 [see *ADOPTION 2 d]. 1936 [see *AMENITY 1 a, 3 b]. 1950 [see *DILLY *sb.*[5]]. 1972 *Police Rev.* 10 Nov. 1447/1 There is a reasonable chance that the tyre will be repairable. 1976 *Sci. Amer.* Sept. 73/1 The Shuttle will also provide new capabilities in hauling multiple satellites in a single mission..performing in-orbit servicing and returning repairable systems.

repai·rableness. [f. REPAIRABLE *a.* + -NESS.] Capacity for being repaired.

1909 *Daily Chron.* 21 Aug. 6/6 The [hosepipe] tyre succumbed to the superior repairableness of the detachable.

repairing, *vbl. sb.*[2] **2.** (Later examples of *repairing lease.*)

1935 E. FARJEON *Nursery in Nineties* 522 The house..was on a repairing lease, and some hundreds of pounds were demanded if we left at once. 1972 C. DRUMMOND *Death at Bar* ii. 45 The house agent..rubbed his hands at the prospect of getting some free work done on the place for it was not a repairing lease.

repaper, *v.* Add: (Earlier and later examples.) Also *absol.* Hence **repa·pering** *vbl. sb.*

1854 MRS. GASKELL *North & South* (1855) I. vii. 96 The landlord..had relented from his expressed determination not to repaper. 1857 —— *Life Charlotte Brontë* II. xiii. 313 Some re-papering and re-painting in the Parsonage. 1964 L. DEIGHTON *Funeral in Berlin* xxxvi. 207, I scarcely recognized the office. It had been re-papered. 1974 —— *Spy Story* xiv. 134 You said repapering the sitting-room would be for my birthday.

reparation. Add: **5. d.** *pl.* Compensation for war damage owed by the aggressor.

In the *Treaty of Peace* (1919) the English heading of Part VIII is 'Reparation', but the French is 'Réparations'.

1921 *Glasgow Herald* 28 Oct. 11 The mere purchase of foreign securities to meet reparations..simply means the transference of worthless papers from one body of financiers to another. 1931 F. L. ALLEN *Only Yesterday* ii. 24 Lodge..wanted Germany to be disarmed, saddled with a terrific bill for reparations, and if possible dismembered. 1947 *Sun* (Baltimore) 2 Apr. 10/2 Reparations to Russia must be paid out of current German production. 1976 C. BERMANT *Coming Home* II. v. 184 Israel..partly with the help of German reparations..was experiencing something of an economic miracle.

7. (Further examples in sense 5.)

The sing. and the pl. forms are both found.

1919 J. M. KEYNES *Econ. Conseq. Peace* v. 139 The endless controversy and intrigue between the Allies themselves..culminated in the presentation to Germany of the Reparation Chapter in its final form. 1919 *Treaty of Peace* VIII. Art. 234 The Reparation Commission shall after May 1, 1921, from time to time, consider the resources and capacity of Germany. 1930 *Economist* 4 Jan. 11/2 The British delegation has left for the Hague to attend the resumed Reparations Conference. 1931 *Times Lit. Suppl.* 24 Sept. 717/3 The impossibility of real Reparations payments. 1968 *Tamarack Rev.* Spring 12 'We have restored Jewish pride.' By taking German reparation money?' 1977 *Time* 10 Jan. 46/1 More than $20 billion of foreign capital has poured in: mostly gifts from Jews abroad, reparations payments from West Germany and U.S. aid.

repartee, *v.* Add: **1.** (Later example.)

1910 G. B. SHAW *Let.* 9 Dec. (1972) II. 957, I spent an hour and a half shouting, bullying..& reparteeing until I was as one in a Turkish bath.

re-parti·tioned, *ppl a.* [RE- 5 a.] That has been partitioned afresh.

1921 N. ANGELL *Fruits of Victory* iii. 100 The new states of repartitioned Europe seem..either unable or unwilling to help their neighbours.

repat (rīpæ·t), colloq. abbrev. of *REPATRIATE *sb.* (also REPATRIATION). Also *attrib.*

1946 BRICKHILL & NORTON *Escape to Danger* xxii. 197 Typical of the repat. boys was 'Chuck' Lark. 1948 *R.A.F. Rev.* June 5/1 (*heading*) Family camp (for repats). 1968 *Punch* 12 June 854/2 The emotions attendant on leaving had the same admixture as before: the expat and the repat were, after all, the same person. 1974 P. FLOWER *Odd Job* iii. 27 He'd got along on a Repat. pension.

repatriate, *v.* Add: Now also with pronunc. (rīpæ·t-). **1. b.** *transf.* of money.

1909 *Westm. Gaz.* 6 Aug. 10/4 A definite step is about to be taken to repatriate the United States silver coin which circulates in the Dominion of Canada. 1940 *Economist* 13 Jan. 64/2 A considerable amount of French capital remained in London last September and has for the most part been repatriated over the past four months. 1966 *Wall St. Jrnl.* 14 Nov. 24/4 In addition much of their overseas profits are 'repatriated' thus cutting the balance of payments deficit. 1978 *Whig-Standard* (Kingston, Ont.) 19 Jan. 7/7 Unless the province proposes to seize it [*sc.* the Sun Life Building] as a part of the $400 million to be repatriated to Quebec, it has no legal power to do anything.

c. *Canad.* To devolve or return (legislation) to the constitutional authority of an autonomous country. Cf. *PATRIATE *v.*

1961 *Ann. Reg. 1960* 73 The Federal Government suggested that the British North America Act should be completely repatriated, making it entirely amendable in Canada rather than on application to the British Parliament. 1968 *Globe & Mail* (Toronto) 5 Feb. 8/5 Federal and provincial governments had been unable..to agree on a formula for repatriating and amending the constitution. 1978 *Independencer* (Ottawa) Jan./Feb. 5/1 Any attempt to repatriate our constitution would be divisive and not help National Unity.

repatriated *ppl. a.* (further example).

1966 *Wall St. Jrnl.* 14 Nov. 24/4 Pfizer International's repatriated earnings and exports to affiliates have totaled more than $420 million since 1951.

repatriate (rīpæ·t-, rīpēᶦ·triĕᶦt), *sb.* [f. the vb.] A repatriated person.

1921 *Glasgow Herald* 15 Nov. 5/3 The majority of these repatriates have the choice of living on totally inadequate means or entering the workhouse. 1945 *Daily Mirror* 27 Sept. 1/4 Seven ships have been named as bringing repatriates home to Britain. 1973 *Times* 17 Nov. 4/8 Mrs Meir, the Prime Minister, and..the Defence Minister, were among those at Lod airport today to give the repatriates a heroes' welcome.

repatriation. Add: Now also with pronunc. (rīpæ·t-). **1.** (Further *attrib.* examples.)

1945 *Daily Mirror* 27 Sept. 1/4 The last batch of liberated prisoners and internees in Singapore boarded a repatriation ship yesterday. 1951 R. CAMPBELL *Light on Dark Horse* 1 It was through an erroneous repatriation-order that I obtained this last panorama of my early home. 1973 *Bulletin* (Sydney) 25 Aug. 17/3 The Repatriation Commission has amassed a fortune... The Repatriation Act empowers the commissioners to act as guardians of these men's affairs... These pensioners are visited regularly by repatriation officers.

2. *Canad.* Devolution or return of legislation to the constitutional authority of an autonomous country. Also *fig.* Cf. *PATRIATION.

1961 *Ann. Reg. 1960* 73 Justice Minister Fulton proposed a two-stage process beginning with repatriation [of the British North America Act] and followed by the working-out of a method of amendment. 1968 *Globe & Mail* (Toronto) 5 Feb. 8/5 Mr. Pearson was asked..why the federal Government had not discussed repatriation of the constitution from Britain. 1976 *Maclean's Mag.* 17 May 45/2 The arguments for and against the repatriation of the Canadian culture take on a national scope.

repayment. Add: **3.** *Comb.*, as **repayment mortgage** (see quots.).

1965 *Legal Side of buying House* (Consumers' Assoc.) ii. 31 There are two main differences between a fixed mortgage combined with an endowment policy.., and a repayment mortgage. First, the insurance premiums..are eligible for tax relief... Secondly, if the borrower dies before the mortgage is paid off..the endowment assurance policy..pays all that is due. 1968 B. D. COLEMAN *Money—How to save it, spend it, & make It* iv. 25 This annuity mortgage (also known as a 'repayment' mortgage), ..is simply a mortgage loan on a property to be bought, the principal (or capital borrowed) and interest to be repaid together by means of fixed instalments over a fixed period. 1974 *Listener* 24 Jan. 98/3 A building society prefers its borrowers..to stick to the simple repayment mortgage: paying back the advance over a period by monthly instalments of capital and interest combined.

repeal, *sb.* Add: **4.** *Comb.*, as **Repeal Warden** *Irish Hist.*, a local official of the Loyal National Repeal Association. Cf. sense 2 b.

1841 D. O'CONNELL in P. S. O'Hegarty *Hist. Ireland under Union* (1952) xiv. 103 The Office of Repeal Warden..must be purely ministerial. They must not..be considered, as Representatives or Delegates. 1903 M. MACDONAGH *Life Daniel O'Connell* xx. 392 The article further suggested that the Repeal wardens should be instructed in the military uses and abuses of railways. 1966 L. J. MCCAFFREY *Daniel O'Connell & Repeal Year* ii. 70 The most important cogs in the Repeal machinery on the local level were the priests. They organized and participated in Repeal meetings, and they selected the parish Repeal Wardens.

repea·lless, *a. rare.* [f. REPEAL *sb.* + -LESS.] Without any cancellation or repeal; from which nothing is erased.

c 1862 E. DICKINSON *Poems* (1955) I. 318 God can summon every face On his Repealless—List.

repeat, *sb.* Add: **1. c.** In *U.S.* phr. *and repeat*, used to denote the return of a horse or the like back over the distance it has just come. Cf. *RETURN *sb.* 1 g.

1819 *Va. Herald* (Fredericksburg) 19 May 4/5 Second day two miles and repeat, free for all ages. 1856 *Trans. Mich. Agric. Soc.* VII. 276 Trotting horses shall be tested in harness, by going at least one mile and repeat. 1903 A. ADAMS *Log of Cowboy* ix. 131 A race horse can't beat an ox on a hundred miles and repeat to a freight wagon.

d. *Broadcasting.* A repetition of a programme which has already been broadcast.

1937 *Printers' Ink Monthly* May 40/3 *Repeat*, a term denoting the second broadcast of a regular studio program broadcast for those stations not served by the original broadcast due to time differences. 1941 *B.B.C. Gloss. Broadcasting Terms* 28 *Repeat*, repetition (as distinct from reproduction) of a programme which has been broadcast, either live or recorded, on one or more previous occasions. 1959 HALAS & MANVELL *Technique Film Animation* xvi. 144 Animation seemed to be too elaborate a process to undertake for the limited number of repeats possible in the television medium. 1965 *Spectator* 5 Mar. 289/2 In the current fortnight no fewer than twenty-seven of the BBC's programmes are 'repeats'. 1973 *Listener* 6 Dec. 798/2 A true interstice piece was the repeat of E. M. Forster's talk on Crabbe, in the interval of *Death in Venice.* 1976 *Weekend Echo* (Liverpool) 4/5 Dec. 2/3 If B.B.C. and ITV have to put repeats on, why don't they show those that were on 15 to 20 years ago.

5. a. In general *attrib.* (or *adj.*) use, designating a further example or instance of the specified sb.; repeated, occurring again; esp. as *repeat order, performance.*

1888, 1891 [in Dict., sense 4 c]. 1908 A. W. MYERS *Compl. Lawn Tennis Player* xv. 237 'E.R.' comes back in a few minutes for a 'repeat order'. 1935 E. F. BENSON *Lucia's Progress* vii. 206 'Went like hot cakes, ma'am,' said the proprietor, '..and I've just telephoned a repeat order.' 1949 *Radio Times* 15 July 13/4, 9.50 p.m. A repeat performance of *Thais* by Massenet. 1961 D. A. BANNERMAN *Birds Brit. Isles* IX. 15 By 10th May all normal clutches have been laid, but repeat-clutches are begun till the last days of May. 1974 J. WAINWRIGHT *Hard Hit* 68 It is a repeat performance of last night; a staring up at the ceiling..a haunting—a remembering. 1978 *Lancashire Life* Nov. 89/1 Postal orders rolled in, followed by repeat orders, together with letters from delighted winners testifying to the efficiency of the pills. 1980 *Times* 6 Sept. 13/5 Last year over half our visitors to London were on repeat visits.

b. Special Combs.: **repeat buying,** the persistent buying of brands with which a shopper is familiar; **repeat fee,** a fee paid to a radio or television artist each time his performance is re-broadcast; **repeat pattern** = sense 4 b; **repeat-sign** *Mus.* = sense 2 d.

1972 A. S. C. EHRENBERG (*title*) Repeat-buying: theory and applications. 1973 *Nature* 3 Aug. 316/1 A longstanding assumption in the theory of repeat buying has now been explained by results on consumers' brand switching behaviour. 1969 *Daily Tel.* 6 Mar. 18 The principle of the repeat-fee for the repeated employment of an artistic work has been established in the entertainment world for more than 50 years. 1975 *Broadcast* 21 July 12/2 Repeat fees had to be avoided. 1959 *Listener* 9 Apr. 629/1 There are even numbers of schools where the bastard activity of hand-painted repeat-pattern-making is still practised. 1967 E. SHORT *Embroidery & Fabric Collage* i. 6 Symmetrical motifs, and repeat patterns, come into their own. 1946 A. L. BACHARACH *Brit. Music of Our Time* iii. 62 Ostinato bass-figures..could be indicated by repeat-signs.

repeat, *v.* Add: **I. 1. b.** Used in radio communication, dictation, etc., to emphasize or clarify an important part of the message. Often combined with a negative. Also *transf.*

1938 W. BULLITT Cable 19 Mar. in R. W. Clark *Freud: Man & Cause* (1980) xxiii. 507, I can make available immediately $10,000: but can not (repeat not) be responsible for more. 1943 F. J. BELL *Condition Red* xvi. 259 We are not—— repeat——not——a Jap. 1952 *New Statesman* 24 May 612/2 We must not, repeat not, call *Dragon's Mouth*..a play. 1957 'J. WYNDHAM' *Midwich Cuckoos* iii. 23 A notification from the R.A.F. was received in Trayne of some unidentified flying object, not, repeat not, a service machine, detected by radar in the Midwich area. 1961 B. PYM *No Fond Return of Love* xi. 104 A notice..which said 'Nobody, repeat *nobody*, is to tamper with the electric heating apparatus in here'. 1973 R. HAYES *Hungarian Game* xxxv. 209 The request was for numbers of people on staff, *repeat*, numbers on staff. 1978 *Guardian Weekly* 8 Oct. 10/2 If— repeat, if—the security forces have been tapping the home telephone of the editor of the Economist.

3. c. With direct speech as obj.: to say or utter again (something that has just been said by oneself or another).

1766 O. GOLDSMITH *Vicar of Wakefield* I. xii. 119 'A groce of green spectacles!' repeated my wife in a faint voice. 1866 C. M. YONGE *Dove in Eagle's Nest* I. ii. 58 'Ah! if the steeple of the Dome Kirk were but finished, I could not mistake it,' said Christina... 'Dome Kirk?' repeated Ermentrude; 'what is that?' 1956 'C. BLACKSTOCK' *Dewey Death* xii. 278 'He wasn't really bad,' said Barbara... 'Bad!' repeated Mr. Dodds. 'What does that mean?' 1976 H. MACINNES *Agent in Place* xx. 218 'Yes,' Tom repeated, 'he knew he had been tricked.'

II. 6. b. Also, to broadcast (a radio or television programme) again.

1923 *Radio Times* 28 Sept. 12/1 Why is it apparently not thought advisable to repeat the 'Request Nights', which..are so popular? 1955 *Ibid.* 22 Apr. 28/3 *Music and Movement*..To be repeated on Friday at 9.55 a.m. 1974 *Listener* 29 Aug. 277/3 There could be no better celebration of the art of standing up and holding forth than the late Dr

Bronowski's *The Ascent of Man*, the last episode of which was repeated over the weekend as a tribute.

c. (Later examples.)

1965 *Listener* 10 June 867/2 A certain rugged, irregular shape tends to repeat throughout the picture. **1967** E. Short *Embroidery & Fabric Collage* i. 33 An allover pattern in embroidery differs from one that is printed in that it does not necessarily have to repeat exactly.

d. *Educ.* (orig. *U.S.*). To undertake (a course or period of instruction) again. Cf. *Repeater 5 d.

1945 C. V. Good *Dict. Educ.* 342/2 *Repeater*, a pupil who has repeated or is currently repeating the work of a grade or part of a subject at some designated level of difficulty. **1973** *Sun-Herald* (Sydney) 26 Aug. 83/1 A suggestion has come that he should repeat third year as he is so young. **1976** *National Observer* (U.S.) 28 Aug. 6/3 Make them repeat the course, repeat a year, drop a grade in rank, anything short of expulsion. **1977** *Rolling Stone* 5 May 45/3 Mark had to repeat first and second grades.

7. d. (Further examples.)

1954 E. B. White *Let.* 28 July (1976) 398 At my age, Miss T., a writer repeats like an onion. **1981** P. Hansford Johnson *Bonfire* I. vii. 71, I hope these aren't cucumber sandwiches... Cucumber always repeats.

8. b. Also *transf.*

1965 *Listener* 9 Sept. 393/2 It contradicts most cogently the persistent accusation that Strauss repeated himself.

repeatability (rĭpītăbi·lĭti). [f. Repeatable *a.* + -ity.] Capacity for being repeated; *spec.* the extent to which consistent results are obtained on repeated measurement (cf. *Reproducibility).

1920 *Music & Lett.* Oct. 289 Repeatability is thus in music an element of the beautiful. **1951** G. Humphrey *Thinking* iv. 108 The criterion of repeatability [of experiments] is not fulfilled. **1961** A. Fleur *Hume's Philos. of Belief* 209 The ultimate warrant for accepting these new scientific ideas lies in their implicit open general challenge to falsification and in their implicit open general promise of repeatability. **1965** *Wireless World* July 338/2 The problems of obtaining good stability and repeatability of resistance value. **1972** *Physics Bull.* May 286/1 By using advanced measurement techniques and controlling the loading procedure a short term repeatability of ±1 part in 20 000 (±0·005%) can be achieved. **1976** G. C. Spivak in J. Derrida *Of Grammatology* p. lxxxvi, Denying the uniqueness of words, their substantiality, their transferability, their repeatability, *Of Grammatology* denies the possibility of translation.

repeatable, *a.* Add: *spec.* of a scientific experiment or result.

1935 [see *operation 5 a]. **1949** *Monthly Notices R. Astron. Soc.* CXIII. 396 The reason why so many experiments are approximately repeatable is that we take infinite pains to select them from the others. **1955** R. O. Kapp *Facts & Faith* 45 The precision with which experiments are repeatable does not prove that it is in the nature of matter to behave in an orderly manner but only that it is in the nature of scientists to do so. **1969** *Listener* 6 Mar. 301/1 An American botanist..and his wife..threw up their careers to devote themselves to evolving a repeatable experiment which could incontrovertibly demonstrate ESP. **1977** *Theology* LXXX. 196 We are here neither in the world of sheer unaccountable miracle nor in that of repeatable experiment.

repeater. Add: **3. a.** (Earlier examples.)

1725 C. Mordaunt *Let.* in E. Hamilton *Mordaunts* (1965) vii. 141 It [*sc.* a watch] is a silent Repeator. **1766** H. Brooke *Fool of Quality* I. vii. 290 She did further rob the said right hon. &c. of a large purse of money, his gold repeater, snuff-box, diamond-ring.

b. (Earlier example.)

1782 S. Hood *Let.* 30 Apr. (1895) 135 Sir George..took the Eurydice, Admiral Drake's repeater, to carry his *duplicate* despatches.

d. (Earlier and later examples.) Also in *Teleph.*

1859 T. P. Shaffner *Telegr. Man.* xxxv. 486 If the line be 600 miles long, and the battery arrangements fail to charge it sufficient for telegraphing, it is the practice to operate it by..the application of an apparatus called a repeater. **1923** *Sci. Amer.* Feb. 106/2 The development of the vacuum tube repeaters..put an entirely different aspect on the problems which have confronted the telephone engineer in the past. *Ibid.* 106/3 These repeaters are placed at regular intervals along the line and as the currents become weakened they pick them up, and..deliver back into the line a current many times stronger. **1958** *Times* 1 July 8/3 The idea behind the work now in hand is to make possible the inclusion of submerged repeaters at more frequent intervals along the cable, which would proportionately increase the capacity of the communications system. **1972** *Sci. Amer.* Sept. 102/2 Each repeater used in coaxial cables and each relay station used in microwave links adds some noise, mostly from its input circuits.

e. = *Relay *sb.* 4 b. Freq. *attrib.*

1936 *R.C.A. Rev.* I. 26 The modulations are passed on to the distant terminal via the repeater stations. **1940** *Ibid.* V. 36 In order to choose the proper amplifying system it becomes necessary to know the amount of gain to be incorporated in each repeater amplifier. **1946** *Jrnl. Brit. Interplanetary Soc.* VI. 72 Yet three repeater stations circling the Earth could provide a steady, reliable service from Pole to Pole with little more power output than the present London transmitter. **1947** *Proc. IRE* XXXV. 1226/1 In communications systems involving a number of similar repeaters, the distortion permissible in a single repeater is very small. **1959** *Aeroplane* XCVII. 542/1 (*caption*) The 500-lb. repeater satellite proposed by the Space Electronics Corporation. **1965** *New Statesman* 30 Apr. 674/1 Early Bird is an active repeater satellite. That is, it receives signals from powerful ground stations, amplifies them, and re-broadcasts them to the ground. **1972** *Sci. Amer.* Feb. 15/1 Microwaves do not bend with the curvature of the earth, so that for long links it is necessary to use repeaters. **1979** *Ibid.* Jan. 62/3 One example of a 'next generation' circuit that could be built with existing technology is a repeater station in a fiber-optics communication link.

5. For *U.S.* in Dict. read Chiefly *U.S.* **a.** (Earlier and later examples.)

1868 [see *colonist 3]. **1871** *Scribner's Monthly* I. 366 Repeaters changed their coats and hats after every vote. **1904** [see *colonizer 2].

b. (Further examples.) Also, one who repeats an offence; a recidivist.

1899 J. Flynt *Tramping* iv. 386 'Revolver' or 'repeater', is both a tramp and a criminal term for the professional offender, who is continually being brought up for trial. **1938** *Encycl. Brit. Bk. of Yr.* 185/1 These young felons are what prison language describes as 'repeaters', young 'old offenders', who have previously, almost continuously, served prison sentences. **1954** *Daily Mail* 10 Mar. 5/6 As regards the 'repeaters', if a child sees his name in the papers it may well be an incentive..to future wrongdoing. **1965** Mrs. L. B. Johnson *White House Diary* 18 July (1970) 303, I asked Nick about repeaters among young criminals. He used some horrifying figure—I believe it was 70 percent. **1977** *Time* 11 July 35/1 After stronger juvenile laws were enacted and violent repeaters were finally jailed in New Orleans, teen-age homicides declined from 29 in 1973 to five in 1975.

c. Also *gen.*, one who repeats an achievement or success.

1944 *Sun* (Baltimore) 13 Jan. 11/2 Mr. Fetterman and Mr. Huffer..got..certificates for their suggestions. Mr. Fetterman is a repeater. He..isn't sure just how many citations have come from the War Production Board for his ideas.

d. *Educ.* A student who undergoes a course or period of instruction again.

1912 *Jrnl. Educ. Psychol.* June 328, 4640 of the children were 'repeaters'... There is nothing to show whether the per cent. thus promoted consists of repeaters regaining their lost grade or of bright children who were skipping a grade. **1945** [see *repeat *v.* 6 d]. **1976** *National Observer* (U.S.) 6 Nov. 17/3 Repeaters are assigned to schools and remedial classes according to age as well as grade.

e. One who returns repeatedly, esp. to a hotel.

1970 *Globe & Mail* (Toronto) 26 Sept. 31/6 (Advt.), The Bremen probably has the largest number of repeaters on her cruises. **1971** *New Yorker* 4 Dec. 183/2 (Advt.), We're a small hotel... Almost all our guests are repeaters. **1977** *Time* 30 May 21/1 By last week the number of visitors had passed 60,000 (including repeaters), even though news accounts of the 'miracle' cloth have been spotty.

repeatered (rĭpī·təɹd), *a.* *Telegr.* and *Teleph.* [f. Repeater + -ed[2].] Equipped with repeaters.

1932 *Telegraph & Telephone Jrnl.* XVIII. 120/2 The post-war development, as the standard form of trunk line-plant, of repeatered cables which will not carry direct current, drastically limited the progress of further trunk mechanisation. **1964** *Discovery* Oct. 46/1 (*caption*) Repeaters are inserted at intervals along submarine cables to amplify the signal about a million times. A typical trans-oceanic repeatered cable may now carry up to 138 circuits.

repeating, *ppl. a.* Add: **1. f.** Telegr. and Teleph. *repeating coil*, a type of transformer used to transmit a signal from one circuit to another without alteration.

1889 *Telephone* I. 494/2 In connection with one or more of the local circuits on the board are placed repeating coils which terminate in single lines in the local exchanges. **1958** J. R. G. Smith *Elem. Telecomm. Pract.* vi. 105 A repeating coil is a special type of transformer in which the ratio of the windings is equal,..and is used to 'repeat' speech currents from one part of a circuit to another.

4. Of a pattern: repeated or recurring uniformly over a surface.

1959 *Listener* 16 Apr. 679/1 Stuffy repeating patterns, 'folksy' craftwork. **1967** E. Short *Embroidery & Fabric Collage* iii. 74 Initials could be designed as a separate motif or incorporated into a repeating design.

‖ **repêchage** (re·peʃɑ̄ʒ). *Sport* (orig. *Rowing*). Also **repechage**. [a. F. *repêchage*, f. *repêcher* to fish out, rescue; to give an examination candidate a second chance to pass.] An extra contest in which the runners-up in the eliminating contests compete for a place in the final. Also *attrib.*

1928 *Daily Express* 7 Aug. 12 M. Bernasconi, their representative in the single sculls, met Joe Wright..in the repechage—second chance—contests for Saturday's second-round losers. **1948** *Call-Bulletin* (San Francisco) 3 July 5/7 Harvard, upset by Cornell in the first trial heat, got back into the running by the 'repechage' or second-trial system. **1955** *Times* 25 Aug. 2/7 On Friday there will be repechages for teams beaten in the opening heats. **1959** *Times* 20 Apr. 3/1 Those teams knocked out in the first round took part in a *repêchage*. **1976** *Yachts & Yachting* 20 Aug. 375/3 The following day there is a repechage in the same waters. **1978** *Times* 30 June 21/6 The Poles won by virtue of the 'repêchage' principle, which provides for one of the 61 defeated teams to reenter the competition by beating the other defeated teams.

repellant, *a.* and *sb.* Add: **A.** *adj.* **1. c.** = Repellent *a.* 2 b.

1897 *Sears, Roebuck Catal.* 274 [Wrap] made of imported black repellant cloth.

d. = *Repellent *a.* 2 d.

1944 *Living off Land* v. 111 Repellant cream should be smeared thoroughly on all parts of the skin which are unprotected by clothing.

B. *sb.* (Further examples in sense of *Repellent *sb.* 4.)

1908 *Jrnl. Econ. Entomol.* I. 83 He had tried repellants against the cotton boll weevil, including lemon, cinnamon, tar and clove oil. **1945** *Tee Emm* (Air Ministry) V. 51 Use the shark repellant sparingly. **1958** Moth-repellant [see *moth *sb.*[1] 3]. **1958** *Sunday Times* 20 July 16/5 Simple dressings..and an insect repellant are obvious necessities.

repellent, *a.* and *sb.* Add: **A.** *adj.* **2. d.** Causing certain insects or other animals not to settle or approach.

1971 'G. Black' *Time for Pirates* i. 15 The air reeked from ..mosquito-repellent smudge. **1979** D. Kyle *Green River High* x. 131 We were smothered in repellent cream, but that didn't stop them [*sc.* insects].

B. *sb.* **4.** A substance that causes certain insects or other animals not to settle or approach. Freq. in *Comb.* preceded by the name of the animal, as *insect repellent* (see *Insect *sb.* 4 a), etc.

1908 *Jrnl. Econ. Entomol.* I. 81 (*heading*) Experiments with repellents against the corn root-aphis. **1923** *Ibid.* XVI. 222 A very effective repellent for practical use is a mixture of one part furfural to four parts pine tar oil. **1942**, etc. [see *moth *sb.*[1] 3]. **1949** *Consumer Reports* July 311/1, 38 brands of insect repellents. **1950** 'N. Shute' *Town like Alice* 43 If they were to spend another night upon the veranda she must get hold of some mosquito repellent. **1955** *Sci. Amer.* Aug. 76/3 It is neither an attractant nor a repellent to unconditioned salmon, and would have meaning only to those conditioned to it. **1963** 'F. Richards' *First come, First Kill* v. 60, I probably smell to high heaven of insect repellent. **1968** C. Helmericks *Down Wild River North* I. xv. 234 Covering myself..with canvas against the angry insects blown back from the horses' backs, and bathing my hooded face with repellent. **1979** R. Perry *Bishop's Pawn* viii. 144 This left the insects free to concentrate on me and the repellent I was using hadn't matured with age.

repercuss, *v.* Restrict ?*Obs.* to senses in Dict. and add: **3.** [Back-formation from Repercussion.] *intr.* To cause or admit of repercussions (sense *6 a, *fig.*); to have an unwanted or unintended effect; to reflect or rebound *on* something.

1923 [see *extra-organismal* s.v. *extra- 1]. **1969** F. Halliday in Cockburn & Blackburn *Student Power* 323 There are also examples where an initially political campaign by students repercusses back into the campus and detonates an internal revolt within higher education. **1972** *Guardian* 18 Feb. 13/1 The public crucifixion of a mandarin looks likely to repercuss for years. **1975** J. De Bres tr. *Mandel's Late Capitalism* vii. 243 The tendency towards thorough planning and organization within the companies or enterprises of late capitalism necessarily repercusses on the structure of the bourgeois class. **1976** *Daily Tel.* 1 Dec. 3/3 It is a script which the plaintiffs feel cannot do anything but repercuss poorly on their reputation if it is thought that 'King Kong' is associated with that.

repercussion. Add: **6. a.** Also *fig.*, a resulting effect or implication; an unwanted or unintended reverberation. Freq. *pl.*

1906 *Pall Mall Gaz.* 22 Jan. 1 The disasters of Tsardom in the Japanese war have had a repercussion all over Europe. **1935** *Times* 5 July 15/3 The direct effects and indirect repercussions of any projected action. **1948** *Hansard Commons* 26 Jan. 673 All practical measures will be adopted..to minimise repercussions upon other unconvertible European currencies. **1969** T. F. Torrance *Theol. Sci.* ii. 85 The inclusion of that fact in the Reformation doctrine of the Grace of God had immense repercussions. **1978** *Lancashire Life* Oct. 96/1 If the strike could be expected to 'bite' anywhere, with anarchic repercussions, Merseyside was the place.

repercussive, *a.* Add: **4. b.** *fig.* Of an action, decision, etc.: having repercussions (sense *6 a).

1974 *Daily Tel.* 11 May 17/5 He said that because of the decision to go ahead with the tour he was worried about the repercussive effect on British and international sport. **1975** *Financial Times* 27 Oct. 17/4 Britain will in an important sense continue to be 'reliant' on other sources herself, since she cannot escape repercussive consequences in her own industry and economy whenever Western Europe suffers. **1979** *Jrnl. R. Soc. Arts* CXXVII. 554/2 The repercussive effects of pay policy.

reperforator (rīpə·ɹfŏrē[i]təɹ). *Telegr.* [f. Re(ceiving *ppl. a.* + Perforator.] A machine which perforates paper tape in accordance with telegraphically received signals.

1916 *Papers Inst. P.O. Electr. Engin.* No. 59. 22 Parment ..proposes re-perforators at the receiver end for *x* messages, receiving a printed slip simultaneously. **1948** *Annals Computation Lab. Harvard Univ.* XVI. 61 The reperforator and the printer operate on a time division basis. **1973** Goacher & Denny *Teleprinter Handbk.* iii. 10/1 Reperforators are used to store teleprinter signals on punched paper tape so that they may be retransmitted later by means of a suitable tape reader.

repertoire. Add: **b.** *transf.*

1872 E. Braddon *Life in India* vi. 201 A Lascar crossing-sweeper whose native dialect is Bengali or Tamil, and from

whose linguistic *répertoire* Oordoo and Hindoo have been wholly omitted. **1959** R. Postgate *Good Food Guide* 211 Latest additions to his marvellous repertoire are Honey Duck..and a poussin stuffed with mushroom butter and herbs, encased in a very thin pastry and baked. **1961** Webster, A small but dependable repertoire of jokes designed to amuse the young—Frank Sullivan. **1965** *Listener* 20 May 753/1 Easily reached from Dublin, New Grange itself, with its rich repertoire of geometric art, is the showpiece of Irish prehistory. **1971** *Nature* 13 Aug. 443/2 The most striking aspects of an animal's behavioural repertoire are often the 'displays' it gives in sexual or aggressive encounters. **1973** *Archivum Linguisticum* IV. 55 The analysis of *repertoires*, namely, what that community knows of and does with the languages concerned.

repertorial (repəɹtō[ə]·riăl), *a.* [f. Repertory + -al.] Of or pertaining to (a) repertory.

1898 J. London *Let.* 6 Dec. (1966) 8 Worth far more than five dollars, at the ordinary repertorial rate of so much per column. **1912** G. B. Shaw *Let.* 1 May in *Lett. to Granville Barker* (1956) 181 To follow a year of Shaw with yet another Shaw is not very repertorial. **1928** *Observer* 1 Apr. 15/3 The producer's laudable desire to deliver Ibsen's humour from the old repertorial gloom was most happily realized in some of the minor parts.

repertorily (re·pəɹtŏrĭli), *adv.* *rare.* [f. Repertory + -ly[2].] In the manner of repertory.

1928 *Observer* 22 Jan. 13/4 Miss Margot Drake's Ann catches fire in the later phases of the play, but some of the other parts are somewhat repertorily done.

repertory. Add: **3. b.** A type of theatrical presentation in which the plays performed by a company are changed at regular short intervals; repertory theatres collectively.

1896 [see *repertory theatre* below]. **1910** G. B. Shaw *Let.* 30 Apr. in *Lett. to Granville Barker* (1956) 164 Producing a lot of plays merely to ascertain which draws the most money, and running that and dropping the rest is not Propagandist Repertory. **1926** [see *co-star *sb.*]. **1951** *Oxf. Compan. Theatre* 664/2 The pioneer work of all these theatres stimulated an ever-growing interest in Repertory. **1974** *Encycl. Brit. Macropædia* XVIII. 229/1 The change from repertory to the single play and the rise of realistic production also shifted artistic control from the actor to the manager.

c. *ellipt.* for *repertory company*.

1933 P. Godfrey *Back-Stage* ix. 134 The number of small stock companies, calling themselves resident repertories.. continued to consolidate their positions with provincial audiences.

4. *attrib.*, as (sense *3 b) *repertory acting, actor, actress, company, movement, play, player, system, theatre.*

1917 J. Agate *Buzz, Buzz!* ii. 146 It is in this way that Repertory acting gets its revenge. **1951** *Oxf. Compan. Theatre* 664/1 Glasgow audiences became acquainted with Repertory acting and production of a high standard. **1917** J. Agate *Buzz, Buzz!* ii. 145 Let us recall the Repertory actor who, desponding of intellectual success, decided to 'go back to the profession'. **1951** *Oxf. Compan. Theatre* 665/2 John Drinkwater, a Repertory actor and dramatist. **1979** K. O'Hara *Searchers of Dead* vi. 64 Noel was..a sound hard-working repertory actor. **1917** J. Agate *Buzz, Buzz!* ii. 146 The Repertory actress sometimes succeeds in sending you away from the theatre concerned for the character she has been representing. **1977** 'J. le Carré' *Hon. Schoolboy* xiii. 290 She frowned..like a repertory actress doing Forgetfulness. **1909** G. B. Shaw *Let.* 29 Dec. in *Lett. to Granville Barker* (1956) 160, I may shortly doubt whether he will throw himself into the repertory company to be cast for anything you please. **1926** *Scribner's Mag.* Aug. 224/1 Mr. Ames showed what could be done with a first-class repertory company. **1967** *Oxf. Compan. Theatre* (ed. 3) 796/2 There have been many plays, first produced by a repertory company, which have then been transferred to London. **1977** J. Aiken *Last Movement* i. 13 He was..highly experienced; he had been in different repertory companies since the age of sixteen. **1951** *Oxf. Compan. Theatre* 663/2 It is impossible to name any one person as the sponsor of the Repertory Movement in England, but no one can deny that it owes much to the vision and courage of J. T. Grein. **1955** *Radio Times* 22 Apr. 7 The famous Liverpool Playhouse..was founded in 1911 by Miss Horniman, who had started the repertory movement in England. **1974** *Encycl. Brit. Macropædia* XVIII. 231/1 Miss A. E. F. Horniman (pioneer of the British repertory movement). **1903** G. B. Shaw *Let.* 12 Jan. (1972) II. 302 Much Ado.. would come in on tour as a Shakespearean repertory play. **1933** —— in E. J. West *Shaw on Theatre* (1958) 235 All the players in the country, whether they are British Drama League players or Repertory players or regular professional players. *c* **1913** D. McCarthy *Drama* (1940) 60 The repertory system is certainly a means to getting good acting. **1974** *Encycl. Brit. Micropædia* VIII. 514/1 Major English companies using the repertory system include the Royal Shakespeare Theatre in Stratford-on-Avon and London and the National Theatre Company. **1896** W. Archer *Theatrical 'World' of 1895* 390 A repertory theatre, where unbroken runs shall be forbidden by the articles of association. **1897** G. B. Shaw *Our Theatres in Nineties* (1932) III. 273 What we want in order to get the best work is a repertory theatre with alternative casts. **1909** *Times* 9 June 8/6 It may..take a little time for the London public to grow used to the frequent changes of bill which a repertory theatre implies. **1976** E. Dewhurst *After Ball* i. 5 The newly-formed Frensham repertory theatre club had come to the end of its first meeting.

repetend (re·pĭtend), *a.* *rare.* [ad. L. *repetendus*, gerundive of *repetĕre* to repeat.] That is to be repeated.

1929 R. Bridges *Test. Beauty* iv. 181 Taketh repetend life and exuberant difformity of disorder'd growth.

‖ **répétiteur** (repetitö̆r). Also **repetiteur**. [Fr., = tutor, coach.] **a.** One who teaches musicians and singers, esp. opera singers, their parts.

1938 *Oxf. Compan. Mus.* 792/2 *Répétiteur*.., choirmaster of an opera-house. **1941** L. A. G. Strong *John McCormack* iv. 70 Covent Garden in those days was blessed by the possession of a master répétiteur at the piano. **1948** *Penguin Mus. Mag.* June 79 He went to his first post as *répétiteur* at Cologne when he was barely 17. **1961** *Times* 13 July 5/3 A stolid arc of singers gathered round the répétiteur. **1970** *Guardian* 24 Apr. 8/2 The opera-house..needed someone to do three jobs; a *repetiteur* was needed, an oboist and an assistant conductor. **1974** *Courier-Mail* (Brisbane) 3 Aug. 15/10 (Advt.), The Australian Opera has vacancies for experienced repetiteurs. **1977** R. Barnard *Death on High C's* i. 11 Little Mr Pettifer, the repetiteur, was seated at the piano... The cast was bustled into position around him.

b. One who supervises ballet rehearsals, etc.

1952 *Ballet Ann.* 1953 143/1 The Sadler's Wells Ballet... Professor of Dancing and Repetiteur: Harijs Plucis. **1964** W. G. Raffe *Dict. Dance* 416/2 The *répétiteur* is often a private tutor; but in Theatre he is in charge of..the full preparation for the show; he may also be the ballet-master. **1977** *Times* 16 May 8/6 Then to the Royal Ballet..eventually becoming..principal *répétiteur*.

repetition[1]. Add: **6.** *repetition choice, device, phenomenon, rate*; **repetition compound** (see quot.); **repetition compulsion** *Psychoanal.*, a term first used by Freud to describe behaviour that is caused by a more powerful instinct than that of pleasure, whereby a response is repeated regardless of the result; also *transf.*

1934 *Brit. Jrnl. Psychol.* Jan. 254 This difference in repetition choice correlated with a difference in teachers' ratings on the trait of pride. **1957** R. W. Zandvoort *Handbk. Eng. Gram.* ix. i. 286 *Repetition Compounds*.., a type of compound which consists in the repetition of the word constituting its first element: *goody-goody, pretty-pretty.* **1925** A. Strachey tr. *Freud's The Uncanny* in *Coll. Papers* IV. 391 We are able to postulate the principle of a repetition-compulsion in the unconscious mind, based upon instinctual activity and probably inherent in the very nature of the instincts—a principle powerful enough to overrule the pleasure-principle. **1941** L. Trilling in D. Lodge *20th Cent. Lit. Crit.* (1972) 288 [Freud] first makes the assumption that there is indeed in the psychic life a repetition-compulsion which goes beyond the pleasure principle. **1953** A. Koestler in *Encounter* I. ii. 28/2 British foreign policy..and French internal politics..seem to be dictated by this kind of repetition-compulsion. **1961** J. A. C. Brown *Freud* i. 4 This phenomenon, described by Freudians as the repetition compulsion, is met with most frequently clinically ..in the choice of a mate where the same personality type is selected each time. **1974** S. Arieti *Amer. Handbk. Psychiatry* (ed. 2) III. 164/2 In the hyponoic and sometimes.. anoetic qualities of the hypnoid state, the dominance of repetition compulsion becomes apparent. **1941** L. MacNeice *Poetry of W. B. Yeats* 164 The twentieth century suspected most poetic repetition-devices. **1954** A. H. Maslow *Motivation & Personality* xi. 188 (*heading*) Repetition phenomena; persistent, unsuccessful coping; detoxification. **1940** *Chambers's Techn. Dict.* 714/1 *Repetition rate*, the number of times repetition is demanded in a telephone conversation, this being related to the line or transmitter noise, [etc.]. **1948**, etc. [see *pulse repetition rate* s.v. *pulse *sb.*[1] 5]. **1969** *Times* 4 Feb. 13/3 It [*sc.* pulsar NP 0532] has the fastest repetition rate of all known pulsating stars.

repetitional, *a.* (Later example.)

1965 *Eng. Studies* XLVI. 160 It is..harder to ascertain cases of *amphibolia* than repetitional figures or double constructions.

repetitive, *a.* Add: **B.** as *sb.* = *repetition compound* s.v. *Repetition[1] 6.

1961 R. B. Long *Sentence & its Parts* xvii. 383 The category of repetitives..includes..a few words with components repeated without change. *Poohpooh, tomtom*,.. and *hushhush* are repetitives of this kind.

repetitor (rĭpe·titǫɹ). [Ger.] A private tutor, esp. in Law, at a German university or college.

1770 *Diary* (MS., Eng. Coll., Rome), Wed 7..in time of Repetition..to y[e] R.C. [Roman College] a little Theological act perform'd by Padre Angeleni one of y[e] Repetitors in y[e] German Colledge. **1886** in Webster. **1895** H. Rashdall *Universities Europe Middle Ages* I. iv. 250 A *Repetitio* in Medicine and Arts [at Bologna]..was, as a rule, not given by the Master himself but by a 'Repetitor', who attended the lecture and then repeated it to the students afterwards and catechized them upon it. **1968** *Listener* 30 May 699/3 The *repetitor*, a lawyer—not a recognised university teacher and often looked at askance by academic lawyers—sets up house in a university town and acts as a kind of 'crammer', compensating for the frequent lack of suitable classes in the legal faculties and other teaching deficiencies.

repha·se, *v.* [Re- 5 a + *Phase *v.* 2.] *trans.* To phase again; to readjust the proposed timing of. So **repha·sing** *vbl. sb.*

1957 *Economist* 2 Nov. 423/2 Rather coyly, the Chancellor added his 1958 and 1959 figures together.., remarking only that the revision 'will mean a re-phasing of the nuclear power programme'. **1957** *New Scientist* 7 Nov. 38/2 It is thought necessary to 'rephase' the plan for modernising the railways. **1970** *Daily Tel.* 22 Jan. 21/3 The CEGB may well rephase the programme to avoid 'bunching' and the 'feast and famine' cycle. **1971** *Nature* 19 Feb. 517/2 BP Chemicals International has announced..the rephasing of the construction of a chemical plant at Baglan Bay, South Wales. *a* **1974** R. Crossman *Diaries* (1976) II. 584 Harold [Wilson] and Peter Shore now feel that the whole closure programme over the next eighteen months should be rephased and slowed down.

rephra·se, *v.* [Re- 5 a.] *trans.* To put into different words; to express in an alternative way. Also *absol.* and *fig.* Hence **rephra·sing** *vbl. sb.*

1895 [in Dict. s.v. Re- 5 a]. **1949** M. Mead *Male & Female* viii. 176 The extreme ingenuity with which man has rephrased his own physiology. **1952** M. R. Rinehart *Swimming Pool* xi. 103 Perhaps I'd better rephrase the question. **1953** *Essays in Criticism* III. 109 A good opportunity to rephrase and to say more properly what I had to say. **1961** R. B. Long *Sentence & its Parts* xv. 342 Careful and formal styles..resort to rephrasings that avoid the problem. **1966** Ogilvy & Anderson *Excurs. Number Theory* i. 5 When somebody else comes along and rephrases the question or perhaps asks a new one, a breakthrough results. **1967** Cox & Grose *Organization & Handling Bibl. Rec. by Computer* iv. 98 The second major objective is to develop a man–machine dialoguing capability which will permit real time rephrasing of input queries. **1981** 'J. Ross' *Dark Blue & Dangerous* xiv. 80 I'll rephrase what I said. I *know* that you knew Sergeant Proctor.

replace (rĭplē[i]·s), *a.* *rare.* [f. the vb.] Designed to replace something that is worn out or is being discarded.

1927 *Daily Tel.* 10 May 4/5 The life of the first tracks was about 2,000 miles... The replace tracks..embody such obvious improvements that they will undoubtedly give a much longer life.

replaceabi·lity. [f. Replaceable *a.*: see *-bility.] The state, property, or condition of being replaceable.

1890 in Webster. **1907** A. W. Pollard *Bks. in House* 37 As to what should be sold and what kept, the one sovereign test is that of replaceability. **1959** P. F. Strawson *Individuals* v. 161 Replaceability by quantifier and variable.

replaceable, *a.* Add: **b.** *absol.* in *Chem.*, denoting those hydrogen atoms in an acid which may be replaced by base.

1895 W. A. Tilden *Introd. Study of Chem. Philos.* (ed. 8) xv. 140 Tartaric acid is a case of similar kind. Its molecular formula cannot be less than $C_4H_6O_6$, on account of the existence of the double tartrates, which prove that the acid contains two replaceable basic hydrogen atoms. **1930** W. R. Anderson *School Cert. Chem.* iv. 47 With sulphuric acid we can get a salt by turning out half the hydrogen present, but this salt still contains replaceable hydrogen. **1962** Parkes & Harrison *Basic Physical & Inorg. Chem.* xv. 202 A normal salt is one in which all the replaceable hydrogen of the acid and the hydroxyl (or oxygen) of the base have reacted to form water.

replacement. Add: **1. b.** *absol.* in *Min.* The dissolution of one mineral and the simultaneous deposition of another in its place. Freq. *attrib.*

1906 *Econ. Geol.* I. 839 As a general term synonymous with 'metasomatism', 'replacement' is preferable to 'substitution'. **1911** *Ibid.* VI. 534 Replacement ore-bodies are generally associated with fissures..capable of conducting solutions from considerable distances. **1928** W. Lindgren *Min. Deposits* (ed. 3) ii. 27 Metallic ores are often formed by replacement. *Ibid.* xxviii. 739 The quartz monzonite contains a great number of replacement veins carrying much tourmaline. **1965** G. J. Williams *Econ. Geol. N.Z.* iii. 23/2 The ore-shoots are typical replacement-bodies of quartz, mullock and pug along narrow shears. **1970** K. C. Jackson *Textbk. Lithol.* iv. 196 Crystal growth, particularly where replacement is important, often results in inclusions of unincorporated minerals in the growing crystal. **1972** M. H. Battey *Mineral. for Students* vi. 160/1 Deposition is also influenced by the nature of the country-rocks. The fluids react with these..to produce replacement deposits in the neighbourhood of the vein fissures.

2. Something which or someone who replaces another.

1894 *Q. Jrnl. Geol. Soc.* L. 383 The hypothesis that the rock is a siliceous replacement of a limestone. **1934** H. G. Wells *Exper. Autobiogr.* I. ii. 62 He..sold little, I think, but jam-pots and preserving jars to the gentlemen's houses round about, and occasional..table glass and replacements. **1944** *Yank* 26 May 3 At the Rapido some replacements couldn't tell the difference between our fire and Jerry's. **1954** W. Faulkner *Fable* 4 The original regiment had been raised in this district... And most of its subsequent replacements had been drawn from this same district. **1973** *Times* 15 Feb. 12/6 There are, however, two significant changes in the list of substitutes (whom rugby officials insist on calling 'replacements' to differentiate between them and the fellows used in other sports for tactical reasons without anyone having gone to hospital).

3. *attrib.* and *Comb.*, as *replacement cost, price, thrust*; **replacement theory** (see quot. 1979); **replacement therapy**, therapy aimed at making up a deficit of a substance normally present in the body.

1928 *Britain's Industrial Future* (Liberal Industr. Inquiry) ii. vi. 67 The capital plant of most..undertakings is relatively small, but the roads are an exception, of which the replacement cost must exceed £1,300,000,000. **1936** J. M. Keynes *Gen. Theory Employment* xi. 135 The price which would just induce a manufacturer newly to produce an additional unit of such assets, *i.e.* what is sometimes called its *replacement cost*. **1963** *Rep. Comm. Inquiry Decimal Curr.* xiv. 141 The true 'replacement cost' is the present value to the user of the future flow of services he could expect from the old machine were it not prematurely replaced. **1977** *Time* 24 Jan.

44/3 Proponents of replacement-cost accounting argue that the machine should be carried on the books at the price of a new machine. **1974** *Terminol. Managem. & Financial Accountancy* (Inst. Cost & Managem. Accountants) 15 *Replacement price*, the price at which material could be purchased, identical to that which is being replaced or revalued. **1957** C. W. Churchman et al. *Introd. Operations Research* xvii. 482 (*heading*) Relevant costs in replacement theory considerations. **1969** J. Argenti *Managem. Techniques* 226 (*heading*) Replacement Theory... Problem... When to replace plant, machinery, etc. **1979** *Gloss. Terms Work Study* (*B.S.I.*) 12 *Replacement theory*, a body of mathematical theory connected with the problems of determining the most economical time to replace or repair a piece of equipment. **1962** H. Burn *Drugs, Med. & Man* xiii. 138 In replacement therapy extracts of glands such as the thyroid gland, the pancreas, the parathyroid gland and the adrenal cortex were prepared from the glands of animals and were given to patients whose own glands were deficient. **1977** *Lancet* 14 May 1048/1 A young woman with von Willebrand disease who asked for a termination of pregnancy and tubal ligation had had..only one bleeding episode in her life requiring replacement therapy. **1971** I. Butykai tr. *Lukovich's Electric Foil Fencing* i. 20 Fencers show a definite preference for angular attacks, replacement thrusts and ripostes.

replacer (rĭplē[i]·səɹ). [f. Replace *v.* + -er[1].] A person or thing that replaces another; a substitute. Also *attrib.*

1895 in *Funk's Stand. Dict.* **1913** G. B. Dibblee *Newspaper* 110 One may perhaps grumble at the rather obvious significance of the new 'replacers'. **1960** *Farmer & Stockbreeder* 9 Feb. 73/1 Early weaning is done at three weeks, the piglets being moved on to a home mixed replacer meal, compounded for £67 10s a ton. **1965** *Language* XLI. 280 The pronominal replacers are marked for low stress.

repla·cive, *a.* and *sb.* [f. Replace *v.* + -ive.] **A.** *adj.* That replaces something else; substitutive; *spec.* in *Linguistics*, of a morph or morpheme.

1948 *Language* XXIV. 440 Morphemes may be classified.. as (1) additive, (2) replacive, (3) additive and replacive, and (4) subtractive. **1949** E. A. Nida *Morphology* (ed. 2) 72 In English replacive morphemes are abundantly illustrated in the verbs which undergo a change of syllabic in the past-tense. **1965** *Canad. Jrnl. Linguistics* X. 139 It [*sc.* the Giamina language] became extinct not through replacive bilingualism with Spanish or English but with one of the Yokuts languages. **1974** P. H. Matthews *Morphology* vii. 122 *Men*, for example, would be said to consist of the regular allomorph *man* of the morpheme man plus a 'replacive morph' ('replace *a* with *e*' or '*a* →*e*') which was assigned as yet another allomorph of plural. **1977** *Word* 1972 XXVIII. 193 It seems possible to classify Welsh metanalysis into three main types: *additive, subtractive*, and *replacive*.

B. *sb.* Something which replaces or substitutes for something else; *spec.* in *Linguistics*, a replacive morph or morpheme.

1948 *Language* XXIV. 441 The shift of stress in related nouns and verbs in English..is also a type of replacive. **1949** E. A. Nida *Morphology* (ed. 2) 55 In the example *feet* as a plural of *foot* we may describe the replacement as /iy←u/. Such morphemes are called 'replacives'. **1954** *Word* X. 224 A 'replacive'..is not by any stretch of the imagination composed of phonemic material. **1962** H. A. Gleason in Householder & Saporta *Probl. Lexicography* 87 The 'replacive'..quite artificially makes an affix out of a process. **1977** *Word* 1972 XXVIII. 203 Complicated series of alterations: some replacives (e.g. *Glama*), but also syncope, assimilation, and so on.

repla·n *v.* [Re- 5 a.] *trans.* To plan again. Hence **repla·nning** *vbl. sb.*

1888 [see Re- 5 a]. **1943** J. S. Huxley *TVA* xv. 129 Replanned so as to provide docks and terminals..Guntersville has become transformed. **1946** *Nature* 28 Sept. 438/2 No schemes for reconstructing and replanning London will be satisfactory without drastic adjustments to existing facilities for transport. **1960** *Farmer & Stockbreeder* 22 Mar. 120/3 Farm manager, Frank Stevens, played an important part in replanning the farm. **1976** S. R. Simpson *Land Law & Registration* ix. 170 Similar considerations apply to physical planning or replanning and to 'land reform'. **1978** P. Boardman *Worlds of Patrick Geddes* vii. 244 The Viceroy opened a competition with a prize of £500 for the best replanning scheme for the city [*sc.* Dublin].

replantation. Add: **2.** *Med.* Permanent reattachment to the body of a part which has been removed or severed.

1870 [in Dict.]. **1976** *Daily Tel.* 26 Nov. 17/6 One of the first reports of a successful replantation of a hand is published in the current issue of the *Journal of Bone and Joint Surgery*. **1980** *Times* 12 Aug. 11/5 Microsurgical replantation of limbs is carried out throughout the North American continent, Australia and many European countries.

replate, *v.* Add: **2.** *trans.* and *intr.* (See quot. 1961.)

1961 H. B. Jacobson *Mass Communications Dict.* 283 *Replate*, to recast a page of type to insert an important but late story. **1967** *Punch* 18 Jan. 91/1 This..was replated between editions to alter a reference to the *Guardian's* sales. **1967** M. Shulman *Kill 3* i. viii. 54 Since it's a London story, let's hold it till as late as possible. The opposition [newspapers] will have to re-plate if they want to pick it up. **1980** 'L. Black' *Eve of Wedding* v. 58 'How late will you stay open?' 'Until about three for fudging. We can replate to about two.'

replay, *v.* Add: **2.** To play (a gramophone record or a tape) again, or to play back; to reproduce (what has been recorded).

1922 *Daily Mail* 18 Nov. 8 Each instrument is fitted with our special 'Repeater' which automatically replays records when desired without the operator's attention. **1962** A. Nisbett *Technique Sound Studio* 241 Tape which is replayed on the same head as was used for recording does not exhibit faults which would be at once apparent if the tape were replayed on most other machines. **1973** *Sci. Amer.* Jan. 117/1 We could replay the recorded sounds at leisure as many times as necessary to make an accurate comparison with the frequencies of our standard disk. **1973** L. Cooper *Tea on Sunday* xi. 93 He recalled the people... So often by running the first interviews through again as if they were a section of a film being replayed he picked up some clue. **1976** Dexter & Makins *Testkill* 140 One of Byron's cover drives, replayed later on TV in slow motion as a textbook stroke. **1977** *Rolling Stone* 19 May 96/2 The Betamax enables you to record (on tape) your favourite TV programs for replaying later.

replay, *sb.* (In Dict. s.v. Replay *v.*) Add: Now usu. with pronunc. (rī·plē[i]). **1.** (Later examples.)

1932 *St. George's Hosp. Gaz.* XXVIII. 25 The re-play took place at Chiswick House, St. Thomas's winning by 7 wickets. **1947** *Sporting Mirror* 7 Nov. p. iii/1 They reached the Junior Cup Final, but after a drawn game at Maidenhead, lost the replay to Reading Albion. **1951** *Sport* 27 Jan.–2 Feb. 4/3 Sunderland were the visitors to St. James' Park in a 6th round replay. **1966** *Listener* 20 Jan. 88/2 The less exhausted players and survivors of replays in the interminable F.A. Cup. **1978** *Morecambe Guardian* 14 Mar. 10/2 In the old days, Chorley were Morecambe's traditional cup rivals and the Shrimps will do well even if they force a replay.

2. The action or an instance of replaying a sound recording, piece of film, etc. Freq. *attrib.*, denoting equipment used for this.

1953 E. S. Gardner *Case of Green-Eyed Sister* viii. 117 You had insisted on a replay of the tape. **1958** S. Ellin *Eighth Circle* ii. xvii. 132 He put Berrigan's 'I Can't Get Started' on the phonograph and set it for replay. **1962** A. Nisbett *Technique Sound Studio* vii. 130 The facilities for mixing—requiring two replay decks in addition to a mixer and recorder—are generally beyond the scope of the amateur. **1972** *Guardian* 24 Aug. 10/1 It would have been helpful if an echo machine could have produced for the President a replay of his acceptance speech in the same hall four years ago. **1974** *Cleveland* (Ohio) *Plain Dealer* 13 Oct. c. 2/1 The scoreboards will be placed on each side and the instant replay screens at each end. **1975** O. Sela *Bengali Inheritance* xv. 124 Now get the mike connected... He.. pressed the replay tab. **1976** *Daily Tel.* 16 July 3/3 A video-tape machine is to be used by London Transport to run 'replays' of violence at underground stations. **1978** S. Brett *Amateur Corpse* xv. 137 Gerald spooled through till nearly the end of the tape... The replay button was pressed.

3. *transf.* and *fig.*

1975 P. Fussell *Great War* ix. 317 And the economic ruin uncompleted by the Great War was finished by the Second, which necessitated a replay, but much magnified, of immense indebtedness to the United States. **1976** W. H. Canaway *Willow-Pattern War* xv. 153, I lay awake..doing an involuntary replay of that horrible dream. **1977** *Time* 30 May 20/2 As Poland approaches the first anniversary of the 1976 riots, an occasion that could invite a replay of last year's protest, the Party Chief is under pressure from Moscow to keep the lid on dissent. **1977** D. Anthony *Stud Game* xxiii. 142 Dusty Gordon's party would be a replay of the Hollywood parties Paul Sherwood had dragged me to.

replenishment. Add: **4.** *attrib.*, as *replenishment tanker*.

1963 *Times* 5 Feb. 10/1 The Admiralty had placed an order for three replenishment tankers, to be built on Tyneside. **1976** *Southern Even. Echo* (Southampton) 3 Nov. 2/3 Captain Averill, Master of the RFA replenishment tanker Olwen..was supporting Royal Navy frigates on duty off Iceland.

replete, *a.* **2. a.** (Later U.S. example.)

1973 *N.Y. Law Jrnl.* 5 June 4/4 Statutes are replete with misplaced commas.

replete (rĭplī·t), *sb.* [f. the adj.] Something that is replete; *spec.* an ant which is distended with food.

1908 W. M. Wheeler in *Bull. Amer. Mus. Nat. Hist.* XXIV. 379 In most cases, as McCook has shown, it is the major workers that most readily tend to become repletes. **1923** *Jrnl. Proc. Roy. Soc. W. Austral.* IX. 47 The impulse to develop repletes is probably due to the brief and temporary abundance of liquid food..in arid regions. **1929** *Encycl. Brit.* XX. 885/2 Since ants..have not the art of making receptacles, they [*sc.* honey ants] have adopted the curious method of using the crops of certain workers or soldiers for the purpose of food storage... Individuals thus functioning are termed *repletes*... When hungry the ants stroke the repletes and receive from them droplets of regurgitated honey-dew collected during times of plenty. **1979** *National Geographic* Nov. 630 The swollen worker ants, called repletes, have been fed by other workers until their abdomens are nearly grape size.

reple·vining, *vbl. sb.* [f. Replevin *v.* + -ing[1].] The action of being replevined. (In quot. *fig.*)

a **1953** H. Belloc *Farewell to Juliet* in *Sonnets & Verse* (1954) 99 One that was pledged, and goes to his Replevining.

replica. Add: **1. b.** (Later examples.) *spec.* in *Linguistics* (see quots. 1956 and 1966). Also *attrib.*

1956 E. Haugen in *Publ. Amer. Dial. Soc.* xxvi. 39 The speakers of language B have *borrowed* it from A... The item as pronounced by speakers of A we shall call the *model* and the diffused item as pronounced by speakers of B we shall call the *replica*. **1963** M. Frayn in Sissons & French *Age of Austerity* xv. 336 The orange-girls, dressed up as replica Nell Gwyns. **1966** R. A. Hall *Pidgin & Creole Languages* i. 5 The European would conclude that it was useless to use 'good language' to the native, and would reply to him in a replica of the latter's incomplete speech.

c. *Mus.* A repeat.

1740 J. Grassineau *Mus. Dict.* 198 *Replica, Reditta*, or *Riditta*, a *repetition*, that is, when one part after a silence repeats or runs over the same notes and intervals, and in fact the same song, which some part had gone over before it, during that silence. **1952** P. A. Scholes *Conc. Oxf. Dict. Mus.* 493/1 *Replica* (It.), repeat.

2. *Comb.* **replica method** = *replica technique* below; **replica plate** *Microbiol.*, a plate of culture medium which has been simultaneously inoculated with numerous microbial clones by holding it against a piece of velvet or similar material which has previously had a plate of grown colonies of micro-organisms held against it; so **replica plating**, the technique of making replica plates, usu. with culture media that contain various antibiotics or lack various nutrients, so that unusual clones of micro-organisms can be recognized; **replica technique**, a method of producing a model of an etched metallic surface for subsequent examination in an electron microscope, used when it is impracticable to take a thin slice of the metal.

1941 *Jrnl. Appl. Physics* XII. 695/2 One basic advantage of the replica methods as compared to the direct methods of surface observation with the electron microscope. **1951** V. E. Cosslett *Pract. Electron Microscopy* ix. 214 Electron microscopy has to be content with the indirect alternative of replica methods, in which an impression is taken from a surface on to a thin film which may then be examined by transmission in the usual way. **1952** J. & E. M. Lederberg in *Jrnl. Bacteriol.* LXIII. 399 (*heading*) Replica plating and indirect selection of bacterial mutants. *Ibid.* 400 Replica plating is used to facilitate routine tests involving repetitive inoculations of many isolates on different media. *Ibid.* 401 A single initial plate may be used to imprint more than one fabric if carry-over from one replica plate to another vitiates serial transfer. **1958** *Times* 31 Oct. 10/7 'Replica plating'..enables strains of bacteria resistant to particular antibiotics, for example, streptomycin, to be quickly isolated. This is a simple and beautiful technique, which has proved of extreme value in research. **1970** D. A. Hopwood in Norris & Ribbons *Methods in Microbiol.* IIIa. vi. 404 A particularly interesting application of replica plating is in isolating bacterial variants of changed potentiality for sexual reproduction. **1977** *Physiologia Plantarum* XXXIX. 140/2 The cells..then grow with the same arrangement on the replica plate as on the master plate. **1943** *Jrnl. Appl. Physics* XIV. 24/1 The direct replica technique consists in casting a thin film of plastic on the prepared surface. **1955** *Jrnl. Iron & Steel Inst.* CLXXIX. 392/1 A simple two-stage replica technique, using an intermediate dry-stripped Formvar film on to which carbon is evaporated is described. **1966** [see *replicate *v.* 2 b]. **1966** D. G. Brandon *Mod. Techniques Metallogr.* 50 Surface replica techniques..increase the resolution available to the metallographer by a factor of 100 over that obtainable by optical microscopy but have a much reduced sensitivity to changes in surface tilt.

replicabi·lity. [f. next + -ity.] The state, property, or condition of being replicable.

1957 *Psychiatry* XX. 80/2 Replicability—six investigators can listen to the same tape, and, within the limits of human error, apply the same analytic categories..and come out with the same transcription. **1971** *Jrnl. Gen. Psychol.* LXXXIV. 304 If that were the case, replicability would depend upon the degree that the important personality variables were adequately sampled. **1972** *Nature* 17 Mar. 99/2 Ascites tumours with high replicability are, felt Dr Steel, toys for the experimentalist. **1978** J. Dunn in Hookway & Pettit *Action & Interpretation* 150 Seeking to render human performance in an idiom in which replicability and inter-observer reliability are at a premium.

replicable, *a.* Restrict † *Obs. rare*[−1] to sense in Dict. and add: **2.** That may be repeated experimentally.

1953 J. B. Carroll *Study Lang.* ii. 34 Reasonably consistent and replicable descriptions of the phonemes of a language. **1973** H. J. Eysenck *Inequality of Man* iii. 108 Even if we regard the observed slight difference as replicable. **1974** *Encycl. Brit. Macropædia* XIII. 1004/2 Replicable phenomena that can be demonstrated with certainty.

So **re·plicably** *adv.*

1964 Crystal & Quirk *Prosodic & Paralinguist. Features in Eng.* iv. 50 Shorter pauses than those of unit length appear to be replicably distinctive.

replicase (re·plikē[i]z, -s). *Biochem.* [f. Replic(ate *v.* + *-ase.] An enzyme which synthesizes a complementary RNA molecule on an RNA template.

1963 Spiegelman & Hayashi in *Cold Spring Harbor Symp. Quantitative Biol.* XXVIII. 162/1 The RNA-dependent-RNA-polymerase which replicates RNA will be called a

replicase. **1971** *New Scientist* 20 May 436/3 In making the complementary RNA the replicase observes a strict one-way system; it moves from the so-called 3′ end of the RNA to the 5′ end. **1973** C. Weissmann et al. in F. T. Kenney et al. *Gene Expression & its Regulation* ii. 20 Viral replicases show a very high template specificity for the homologous, intact viral RNA.., as well as for the complementary minus strand.., to the exclusion of all other, unrelated viral RNAs and most other RNAs examined.

replicate, *sb.* Add: **2.** *Science.* A repetition of an experiment or trial; each of a number of similar parts or procedures which constitute an experiment or trial. Cf. *Replication 4 c.

1929 *Jrnl. Agric. Sci.* XIX. 213 In the earlier quantitative experiments..the precision of the results left much to be desired, since only four replicates could be used. **1953** *New Biol.* XIV. 85 It was decided that ten replicates, each of 50 ears [of wheat], would probably be sufficient to reveal important changes in population from year to year. **1970** *Sci. Jrnl.* May 65/1 Earliness and qualities of the varieties could always be reproduced, even if many replicates were made. **1976** *Jrnl. Heredity* LXVII. 204/2 The germinability from one microscope field..in each drop constituted one replicate. *Ibid.*, Values given represent data from a single pollen sample germinated in replicates as stated.

replicate, *a.* Add: **3.** *Science.* Being a replicate (*Replicate *sb.* 2).

1961 *Lancet* 29 July 231/2 Replicate assays (i.e. repeat assays on different days). **1972** *Science* 26 May 914/3 Three to five replicate chemical analyses were made for each major tissue. **1978** *Nature* 3 Aug. 459/1 The Institute of Petroleum method recommends that 24 replicate runs are carried out to obtain reasonable statistics.

replicate, *v.* Add: **2. b.** (Further examples.)

1964 R. D. Heidenreich *Fund. Transmission Electron Microsc.* iii. 71 The intensity distribution bears a close relation to the topography of the surface being replicated. **1966** D. G. Brandon *Mod. Techniques Metallogr.* 47 The original specimen surface may be preserved and the potential resolution of the carbon and silicon monoxide replica techniques still obtained by replicating the surface with a plastic material and then replicating this plastic negative of the surface with carbon or silicon monoxide. **1973** *Sci. Amer.* Jan. 102/1 The simplest method of studying snow crystals is to replicate them by letting them fall into a thin layer of dilute solution of plastic and solvent. The solvent evaporates rapidly, leaving a thin plastic cast of the snow crystal.

c. *Science.* To repeat (an experiment or trial) and obtain a consistent result.

1923 *Biometrika* XV. 283 We may obtain an estimate of what the variability would be if the conditions of any one trial could be replicated in a number of experiments with the same variety. **1969** *Sci. Jrnl.* Dec. 49/2 Beveridge.. replicated Thouless' experimental finding of differential cultural susceptibility to phenomenal regression. **1970** T. Lupton *Managem. & Social Sci.* (ed. 2) iii. 74 The studies have since been replicated with similar groups and different groups. **1973** J. L. Fleiss *Statistical Methods for Rates & Proportions* iii. 26 One often undertakes a study in order to replicate (or refute) another's research findings.

d. *Biol. intr.*, *trans.*, and *refl.* Of genetic material or a living organism: to reproduce or give rise to a copy of (itself).

1957 J. Lederberg in McElroy & Glass *Chem. Basis Heredity* vii. 743 The very interesting statement that the bacterial nucleus apparently does not replicate by a non-dispersive mechanism. **1958** *New Scientist* 10 July 341/1 A characteristic of living matter is its ability to replicate itself. **1960** *New Biol.* XXXI. 20 They [*sc.* the genes] may continue for millions of years exactly replicating their complex structure. **1965** *Listener* 2 Sept. 332/1 It would be fallacious..to suppose that the major code-bearing molecules, the nucleic acids, can..replicate themselves in isolation. **1968** H. Harris *Nucleus & Cytoplasm* i. 7 The other possibility which could account for the persistence of this information in the enucleate cell is that the relevant RNA might be replicated in the cytoplasm. **1969** *Listener* 10 July 34/1 Arthropods survive, replicate, live off their environment. **1972** *Proc. R. Soc.* B. CLXXXI. 29 DNA is replicated at fork-like growing points. **1977** *Sci. Amer.* Nov. 54/3 Attributes by which we identify living things—their capacity to replicate themselves, to repair themselves, to evolve and to adapt. **1979** *Jrnl. R. Soc. Arts* CXXVII. 645/1 The bacterium rapidly divides and replicates.

e. To imitate; to make or be a model or replica of.

1958 *Word* XIV. 365 It is worth considering whether a formalized investigation replicating the game [of 'Twenty Questions'] would not produce a valid and economical description of a vocabulary. **1966** J. J. Katz *Philos. Lang.* iv. 99 Verbal exchanges between the computers replicate the publicly observable phenomena that occur when human speakers communicate in a natural language. **1967** M. Argyle *Psychol. Interpersonal Behaviour* x. 194 Other activities may be used including..setting up groups in cooperation or competition to replicate organizational problems. **1971** *Sci. Amer.* Oct. 53/2 Synthetic melts of lunar composition can replicate the texture seen in the granular moon rocks. **1977** *Times* 19 Nov. 14/4 Striations that could be replicated with modern but not ancient tools. **1979** *Times* 24 Dec. 10/8 The gallery is to have three rooms that replicate rooms at Hutton Castle.

f. To copy exactly.

1970 *Computers & Humanities* IV. 233 The index entries that have been generated in this way, using the computer's ability to replicate strings of characters, are sorted into alphabetic sequence. **1971** *Sci. Amer.* June 88/3 Since binary data are generally consumed within calculation centers, the ability to replicate the data for future manipulations is essential. **1976** *Ibid.* June 63/1 The signal transmitted from the earth is received by the spacecraft's radio system, which faithfully replicates the received phase, increases its frequency by 10 percent and transmits it back to the earth.

replicated *ppl. a.* (further scientific examples); hence also **re·plicating** *vbl. sb.* (usu. *attrib.*) and *ppl. a.*

1926 *Jrnl. Min. Agric.* XXXIII. 506 A replicated experiment provides a valid estimate of error. **1957** F. H. C. Crick in McElroy & Glass *Chem. Basis Heredity* vii. 747 In a replicating structure one expects the bases to pair rather accurately, so that mistakes are rare. **1960** *New Biol.* XXXI. 23 Some persistent change has therefore occurred in the replicating system for proteins. **1961** M. E. Haine *Electron Microscope* x. 245 The replica..has one surface flat and the other following the topography of the replicated surface. *Ibid.*, Replicas were formed in plastic and were limited by the large molecular size of the replicating material. **1971** *Nature* 13 Aug. 502/1 Cellulose acetate replicating tape is used to prepare negative replicas of the etched surfaces. **1976** *National Observer* (U.S.) 17 July 15/3 It appears easy to come up with organic compounds and water. But the big gap is, How often do you get replicating organisms? That, if you will, is the missing link in the origin of life. **1977** *Jrnl. Agric. Sci.* LXXXVIII. 127 (*heading*) Response to family selection based on replicated trials. **1978** *Nature* 20 July 212/2 The combined nicking-closing and negative supercoiling activity of gyrase are presumed to act at some site at or beyond the replicating fork to relieve the positive supercoiling strain which builds up during replication and to aid unwinding.

replication. 4. Restrict † *Obs.* to senses in Dict. and add: **c.** *Science.* Repetition of an experiment or trial so as to test the trustworthiness of its conclusion.

1926 *Jrnl. Min. Agric.* XXXIII. 506 The method adopted is that of replication. **1953** *New Biol.* XIV. 85 More than one sample must always be taken in order to discover the natural variation between individual samples. This is termed 'replication'. **1971** *Jrnl. Gen. Psychol.* LXXXV. 200 The stability of the means can be seen to be high in the replications.

6. (Further examples.)

1951 V. E. Cosslett *Pract. Electron Microscopy* ix. 223 Most materials for replication are initially either too rough or too smooth for the purpose. **1966** D. G. Brandon *Mod. Techniques Metallogr.* 45 A major advance in replication technique was made when Bradley discovered that an evaporated carbon film was virtually structureless and would faithfully follow surface contours down to a limit of resolution of about 20 Å. **1969** *Computer & Humanities* III. 193 Gutenberg's press in mid-fifteenth century made multiple replication of the visual expression of ideas practicable. **1972** *Language* XLVIII. 346 The vocalic sub-hierarchy assigns markedness values exclusively by means of replication, whereas the consonantal sub-hierarchy assigns markedness values by means of complementation *and* replication, depending on the feature specification. **1975** *New Yorker* 3 Mar. 96/3 The real triumph of 'Female Friends' is the gritty replication of the gross texture of everyday life, placed in perspective and made universal.

b. *Biol.* The process by which genetic material or a living organism gives rise to a copy of itself.

1948 *Nature* 5 June 872/2 Replication seems commonly to mark the end of active cell division and the onset of differentiation. **1955** *Sci. Amer.* Oct. 70/1 Sometimes, because of a mistake in some step of the replication process, a daughter cell gets a gene carrying a garbled message; that is, it does not bear precisely the same information as its original counterpart. **1971** D. J. Cove *Genetics* v. 73 (*caption*) M. Meselson and F. W. Stahl's experiment to demonstrate the semi-conservative replication of DNA. **1972** *Proc. R. Soc.* B. CLXXXI. 29 Replication of chromosomal DNA occurs in sections arranged in tandem. **1977** *Sci. Amer.* Dec. 94/2 In viral replication (and microbial replication in general) mutants appear at a frequency of about one in a million particles, depending on the viral strain and the conditions of culture.

replicative, *a.* Add: **2.** *Biol.* Of, pertaining to, or involved in replication.

1960 *New Biol.* XXXI. 130 A hint that some divisions of the plant kingdom have replicative mechanisms that either derange more easily or are more tolerant of novelty than others. **1971** *Sci. Amer.* Aug. 52/1 As has been abundantly demonstrated over the past two decades, DNA is the replicative molecule of the cell. **1975** *Ibid.* May 28/3 We have isolated a structure reflecting that stage: several plus strands of graduated lengths partially bonded to the minus strand on which they are being synthesized; we call such a structure the replicative intermediate. **1976** *Nature* 28 Oct. 731/1 Plasmids can be viewed as primitive bacteriophages that have not yet acquired those specialised functions necessary for a complex replicative cycle.

replicatively, *adv.* (Later example.)

1957 R. K. Merton *Student-Physician* App. C. 304 The second criterion is that results must be *replicatively* consistent.

replicator (re·plikē[i]tǫɹ). [ad. F. *réplicateur* (Jacob & Brenner 1963, in *Compt. Rend.* CCLVI. 298), f. as next + *-eur* -or.] **1.** *Biol.* A postulated section of nucleic acid at which replication is initiated and away from which it proceeds in one or both directions.

1963 *Cold Spring Harbor Symp. Quantitative Biol.* XXVIII. 331/1 A unit capable of independent replication or replicon would carry two specific determinants... A structural gene controlling the synthesis of a specific initiator... An operator of replication, or replicator, i.e., a specific element of recognition upon which the corresponding initiator would act, allowing the replication of the DNA attached to the replicator. **1969** A. M. Campbell *Episomes* viii. 107 Effectively, the F factor would have taken over the normal replicator function for the major part of the bacterial chromosome. **1973** *Virology* LIV. 270/1 According to the replicon theory, some replication-deficient mutants should be found in the postulated replicator gene. **1978** *Bull. Amer. Acad. Arts & Sci.* Feb. 17 In cases where more than two X chromosomes were present there were multiple late replicators.

2. That which replicates (in any sense).

1964 *Listener* 15 Oct. 575/1 Looking as far into the technological future as I dare, I would like to describe the invention to end all inventions. I call it the replicator; it is simply a duplicating machine. It could make, almost instantly, an exact copy of anything. **1972** *Science* 27 Oct. 359/1 Adult liver..has been described as a 'discontinuous replicator'..that divides at a low 'wear and tear' replacement rate.

replicon (re·plikǫn). *Biol.* [ad. F. *réplicon* (Jacob & Brenner 1963, in *Compt. Rend.* CCLVI. 298), f. *réplic-ation* Replication: see *-on[1].] **a.** A piece of genetic material which replicates as a unit, beginning at a single site within it.

1963 *Cold Spring Harbor Symp. Quantitative Biol.* XXVIII. 330/2 A genetic element such as an episome or a chromosome (of a bacterium or of a phage) constitutes a unit of replication or replicon. Such a unit can only replicate as a whole... The capacity to behave as a replicon must depend upon the presence and activity of certain specific determinants. In other words, the properties of such units require that they set up specific systems of signals allowing, or preventing, their own replication. **1971** *Jrnl. Molecular Biol.* LVIII. 873 It has been hypothesized that in mammalian cells DNA replication proceeds *via* two replication forks per replicon, which proceed in opposite directions. **1972** *Bacteriol. Rev.* XXXVI. 365/1 The hypothesis was proposed that those DNA molecules that are capable of replication (termed 'replicons') are circular in structure and carry at least two gene loci controlling their replication; at one locus on the replicon is located a regulator gene which produces a diffusible substance (initiator) acting upon the second locus, an operator of replication (replicator), to permit DNA replication to be initiated from that point. **1974** *Nature* 9 Aug. 467/2 Each chromosome contains many tandemly arranged units of replication (replicons), and each replicon is comparable to the whole chromosome of a bacterium or a virus in that its replication is usually bi-directional. **1976** *Ibid.* 29 Jan. 281/1 Each of these components is capable of autonomous replication in certain host strains of bacteria; that is, each is a 'replicon'.

replo·t, *v.* [Re- 5 a.] *trans.* To plot or represent again. So **replo·tter**; **replo·tting** *vbl. sb.*

1896 *Rep. Board of Ordnance & Fortification* (U.S.) 18 A replotting arm for attachment to the Lewis position-finder. **1897** *Ann. Rep. Chief of Ordnance to Sec. of War 1896* (U.S.) 581 A photograph and description of the Lewis replotter. *Ibid.* 582 With this device it is possible to correctly replot the position of a moving target..at intervals of ten seconds of time without hurry or confusion. **1902** *Encycl. Brit.* XXXI. 367/1 A complete automatic replotter is carried on the table of each instrument, by the use of which the observer can instantly convert the range and direction of the target as read from the instrument into the corresponding range and direction from the gun itself. **1965** Phillips & Williams *Inorg. Chem.* I. xvii. 633 The change in ease of oxidation down the series is brought out by replotting some of the values. **1968** *Economist* 6 July 54/1 If an essential invention occurs out of order, the time-scale has to be revised and the forecast path of advance replotted.

replumb, *v.* Restrict † *Obs.*[−0] to sense in Dict. and add: **2.** [A separate formation on Re- 5 a.] *trans.* To redo or replace the plumbing in (a building). Chiefly as **replu·mbed** *pa. pple.* Also **replu·mbing** *vbl. sb.*

1909 H. G. Wells *Tono-Bungay* iii. ii. 291 My uncle distinguished himself by the thoroughness with which he did the repainting and replumbing. **1973** *Irish Times* 2 Mar. 26/1 (Advt.), Re-roofed, re-wired and re-plumbed in recent years. **1976** *Newmarket Jrnl.* 16 Dec. 39 (Advt.), These properties have been fully modernised including re-roofing, re-wiring, re-plumbing, etc. **1977** [see *see *v.* 22]. **1978** *Morecambe Guardian* 14 Mar. 23/3 (Advt.), The house has been re-wired, replumbed, had a new bathroom installed, etc.

reply, *sb.* Add: **1. d.** *reply-paid* adj. (later examples).

1928 E. Wallace *Double* xviii. 272 It was evidently, from the indicator, a reply-paid message. **1973** *Times* 14 Mar. 4/6 In the present poll this outcome may well have been achieved by the retiring conservators distributing reply-paid proxy forms.

2. b. A pleading by the plaintiff after the delivery of the defence; the final speech of Counsel in a trial.

1837 in Carrington & Payne *Rep. Cases Nisi Prius* VII. 676 The counsel for the prosecution may re-examine the witness, and after the prisoner's counsel has addressed the jury, will be entitled to the reply. **1837** *Rex* v. *Stannard* in *Ibid.* 675 C. Phillips waived his right of reply. **1875** *Act* 38 & 39 *Vict.* c. 77 Sched. i. § xxiv. 55 A plaintiff shall deliver his reply, if any, within three weeks after the defence or the last of the defences shall have been delivered, unless the time shall be extended by the Court or a Judge. **1898** *Criminal Evidence Act* 61 & 62 *Vict.* c. 36 s. 3 In cases where the right of reply depends upon the question whether evidence has been called for the defence, the fact that the person charged has been called as a witness shall not of itself confer on the prosecution the right of reply. **1961** L. F. Sturge *Basic Rules Supreme Court* xxiii. 62 The position is further confused by the fact that the Rules give the name 'reply' to what the legal profession is accustomed..to call a 'defence to counterclaim'... The modern practice is to head the pleading 'Reply and Defence to Counterclaim' and to head each part respectively 'Reply' (meaning the equivalent of the common law replication) and 'Defence to Counterclaim'. **1964** Ld. Evershed et al. *Atkin's Encycl. Court Forms* (ed. 2) XXXII. 35 Pleadings subsequent to the reply still bear their ancient names: rejoinder, surrejoinder, rebutter and surrebutter, although modern rules do not refer to them by name. **1975** I. H. Jacob *Bullen & Leake's Precedents of Pleadings* (ed. 12) ix. 109 No pleading subsequent to a reply or a defence to counterclaim may be served except with the leave of the court.

4. A signal sent by a transponder in response to interrogation. Also *attrib.*, as *reply pulse.*

1945 [see *interrogate *v.* 4 a]. **1947** L. N. Ridenour *Radar System Engin.* viii. 263 The replies may be made more complicated in a variety of ways for the purpose either of identifying the beacon or of using it as part of an auxiliary communication system. **1963** R. S. H. Boulding *Princ. Radar* (ed. 7) xxii. 471 Measurement of the time from the commencement of the interrogating pulse to the receipt of the beginning of the reply pulse enables the distance between the aircraft and the beacon to be determined. **1965** R. S. Berkowitz *Mod. Radar* I. ii. 12 When the radar interrogates a beacon and receives a reply whose power is fixed by the characteristics of the beacon transmitter, separate calculations are necessary for the out and back paths.

reply, *v.* Add: **I. 1.** (Further *fig.* examples.)

1714 Pope *Rape of Lock* III. 24 The Nymph exulting fills.. the sky; The Walls, the Woods, and long Canals reply. **1785** Cowper *Task* VI. 231 There is in souls a sympathy with sounds... Some chord in unison with what we hear is touched within us, and the heart replies. **1930** R. Campbell *Poems* 9 Clear as a glass the day replies To every feature save her eyes.

3. (Later examples.)

1849 J. L. Campbell *Chief Justices Eng.* II. xxxiii. 401 Lord Mansfield hesitated long about making the right to reply depend upon the giving of evidence by the defendant. **1923** W. G. Russell *Treatise on Crimes* (ed. 8) II. 1835 If the defendant is undefended there is no right to sum up or reply if he calls no witnesses, whether he himself does or does not give evidence: but there is a right to reply if he calls a witness.

repolariza·tion. [Re- 5 b.] A renewed polarization (in the senses of Dict. and Suppl.); the action of repolarizing.

1922 *Public Opinion* 11 Aug. 136/2 What is aimed at is the repolarisation of the individual items of society. **1958** Cranefield & Hoffman in *Jrnl. Gen. Physiol.* XLI. 633 (*title*) Propagated repolarization in heart muscle. **1968** *Economist* 20 July 32/1 It also caused him to predict a repolarisation of South African politics. **1973** *Nature* 9 Feb. 400/1 The immediate consequence of contact of ovum and spermatozoon is an action potential, a depolarization and repolarization including a transient reversal of polarity.

repo·larize, *v.* [Re- 5 b.] *intr.* To polarize again.

1958 J. H. Burn *Lect. Notes Pharmacol.* (ed. 5) 37 Normally a second response can be produced when the intracellular potential has repolarized to only 2/3 of its full resting negativity. **1965** *Math. in Biol. & Med.* (Med. Res. Council) VI. 262 When curve C is reached the only remaining point at which the ionic current is zero is the new value E_r, so that the membrane repolarizes. **1973** *Nature* 9 Feb. 400/2 The eggs then repolarized, somewhat slowly, to the −60 mV typical of the end of phase III. **1980** *Sci. Amer.* May 74/3 In an unmyelinated fiber the entire axonal membrane must depolarize and then repolarize.

repo·lish (rī-), *sb.* [f. the vb.] A renewed polishing.

1905 *Daily Chron.* 9 Sept. 2/6 The floor is waxed and polished, so only needs dusting and a very occasional repolish.

report, *sb.* Add: **2. e.** A teacher's official statement in writing about the work and behaviour of a pupil at a school.

1873 C. M. Yonge *Life J. C. Patteson* I. i. 16 The half-yearly reports often lament his want of zeal and exertion. **1906** R. Brooke *Let.* 1 Apr. (1968) 47 My term's report.. has come in, & is very bad. Result: the family are shocked. **1973** 'M. Innes' *Appleby's Answer* ii. 19 It was a mark I'm simply bound in conscience to put into a pupil's report.

3. e. A statement in which an accusation is made against (a sailor, etc.); the charge itself; esp. in phr. *on report*, on a charge.

1850 H. Melville *White Jacket* II. xxxviii. 246 The names of such offenders shall be put down on the report. **1915** *Recruiter's Bull.* (U.S. Marine Corps., N.Y.) June 17/2, I was in that Corps for fifteen years and never saw a man on report. **1948** Partridge *Dict. Forces' Slang* 154 *In the report*, a colloquial synonym of 'in the *rattle*'. (Navy.) **1963** T. & P. Morris *Pentonville* vi. 126 When an officer observes a prisoner committing an offence he must decide whether or not to place the prisoner on Governor's Report... Reports are heard on every day except Sundays and public holidays. **1969** *Punch* 5 Mar. 350/2 We're just coming up to the conference point with my sergeant, fifteen hundred hours outside Queensway tube. I'll be on report if he catches me exceeding three miles per hour.

f. *weather report*: see Weather *sb.* 6 a in Dict. and Suppl.

8. *attrib.* and *Comb.*, as *report sheet, stage*; **report card,** (*a*) *U.S.*, a document comprising a school report; also *attrib.* and *transf.*; (*b*) *Austral.* (see quot. 1969).

1929 W. Faulkner *Sound & Fury* 213 But to have the school authorities think that I have no control over her, that..I didn't even know she had a report card. **1952** B. Ulanov *Hist. Jazz in Amer.* (1958) xvi. 202 He set up a rating system for bands, based on the report-card letters A to D. **1953** *Manch. Guardian Weekly* 12 Nov. 2/1 The Republicans would be less doleful this morning if they had not seconded so volubly the Democrats' contention that these four elections would constitute 'a report card' on the Eisenhower Administration. **1969** Eagleson & McKie *Terminology of Austral. Natl. Football* iii. 7 *Report card*, a card on which at the termination of a match, umpires record particulars of any charge(s) they may have against players. **1977** *Time* 22 Aug. 11/3 Carter's early forcefulness..drove six Latin countries..to reject U.S. military assistance rather than agree to prepare 'report cards' for Washington on human rights. **1980** J. Irvine *Cais Obsessions* xi. 195 Erin's baby bootees, her silver food pusher, her first report card. **1957** C. Smith *Case of Torches* xiv. 189 He wrote it all down... He doodled on his report sheet. **1966** P. O'Donnell *Sabre Tooth* i. 7 His report-sheet shows that he was a good man. **1966** Freeman & Abbott *A.B.C. of Parliamentary Procedure* 74 Reports from Committee of the Whole House ('Report Stage')...When a Bill is committed *pro formâ* to enable the member in charge to introduce numerous amendments..the Bill so amended is reported and recommitted for a future day. **1976** *Liverpool Echo* 6 Dec. 5/8 We expect concessions to be made during the committee and report stages of the Bill.

report, *v.* Add: **I. 2. d.** To say factually. Also with direct speech as object.

1929 M. A. Gill *Underworld Slang* 15/1 Words that can be used in place of 'said'. Reported. **1977** B. Freemantle *Charlie Muffin* iv. 45 'Completely misread the interview,' he reported.

3. a. Also with *back*. Phr. *to report out* (a bill), of a committee of Congress: to return (a bill) to the legislative body for debate.

1883 *Rep. U.S. Bureau Indian Affairs* p. xxix, The bill as read and referred was reported back by the Senate Committee. **1948** *Sun* (Baltimore) 31 May 8/1 The bill recently was reported out favorably by the House Armed Services Committee. **1965** Mrs. L. B. Johnson *White House Diary* 10 Apr. (1970) 257 The Committee reported out the Civil Rights Bill, quicker and stronger than ever expected. **1976** *N.Y. Rev. Bks.* 15 Apr. 20/3 The Senate Antitrust Subcommittee may well report out a 'vertical divestiture' bill, which will then go before the full Judiciary Committee. **1976** *Daily Tel.* 25 Sept. 10/4 The society..has arranged for its ticket selling committee to meet next Tuesday to report back their ideas to the society. **1979** *Sci. Amer.* July 68/1 The bill was reported out of committee, but when it was pointed out that enforcing such a statute would cost money and that the bill should be referred to the appropriations committee, support waned and the bill did not come to a vote.

II. 5. b. (Later examples.) Also used in less formal contexts of a journalist or broadcaster. Phr. *to report back*: to return with a formal report (to one's principal).

1961 *Providence* (Rhode Island) *Jrnl.* 20 July 5 He would study the correct method and report back to the council. **1966** *Rep. Comm. Inquiry Univ. Oxf.* I. 371 An inferior body receives its policy from the superior body and can then be told to do its work and decide within the limits of the policy laid down, only reporting back in cases of real doubt or difficulty. **1968** *Listener* 21 Nov. 669/1 That resolution gave us only 70 hours to report back to the Assembly on the organisation of the force. **1973** *Guardian* 19 Mar. 22/3 The good and bad of compensation: Judy Hillman and Malcolm Stuart report on page six. **1974** *Radio Times* 28 Feb. 35/2 What does a holiday cruise offer you? *Gladys Nicol reports*. *Ibid.* 35/5 The World Tonight: News—Douglas Stuart reporting.

c. Also with *in*.

1969 I. Kemp *Brit. G.I. in Vietnam* viii. 159 'I'm reporting in,' I told the sergeant. **1977** D. Beaty *Excellency* viii. 99 He murmured something about having to report in at the African Airways counter.

reportabi·lity. [f. Reportable *a.*: see -ity.] The quality or state of being reportable.

1960 Bruner & Klein in J. S. Bruner *Beyond Information Given* (1974) vii. 119 The work of Blackwell in America, and of Dixon in England, suggests strongly that the reportability of a stimulus or awareness of a subject depends upon the nature of response requirements.

reportable, *a.* Add: **2.** That should or must be reported to some authority.

1942 L. D. Kitchin *Road Transport Law* 29/1 No notice need be given..to a test and inspection to be conducted within 48 hours of a 'reportable' accident in which the vehicle has been involved. **1976** *Sci. Amer.* Oct. 28/3 Smallpox was made a reportable disease. **1976** G. Thurston *Coronership* iii. 54 Death in custody in a police station is not, strictly, reportable unless it is unnatural or of unknown cause.

reportage. Add: Also with pronunc. (|| rəportā3). **3.** The describing of events (usu. by an observer); *spec.* the reporting of events for the press or for broadcasting, esp. with reference to its style; an instance of this, a piece of journalistic or factual writing. Also *transf.* and *attrib.*

1891 E. Dowson *Let.* 7 Feb. (1967) 184 Howells..writes in dialect, in Yankee and that's not realism (even Zola doesn't go so far) it's reportage. **1931** *Times Lit. Suppl.* 19 Feb. 129/2 Ludwig's works written before the war, plays, novels, *reportage*. **1938** C. Connolly *Enemies of Promise* xvi. 173 The article has a future, especially in the form of the critical essay..and the skilled 'reportage'. **1942** E. Paul *Narrow St.* xix. 155 What L'Hibou suggested..that Madame Absalom should do with her ears, had best be omitted from this reportage. **1954** Koestler *Invisible Writing* xv. 169 Németh's book was planned as a long psychological novel, Kunzi's book as a straightforward reportage. **1959** L.-H. Liang tr. *Ting Yi's Short Hist. Mod. Chin. Lit.* xv. 217 During the early period of the War of Resistance, reportage ..became the most popular of all literary forms among both the writers and the broad reading public. **1961** *Listener* 28 Dec. 1115/3 Where is your black-and-white photograph without sun and sparkle?... A big enough illustration, one would think, to render suspect almost any black and white reportage. **1966** *Listener* 8 Apr. 525/1 In the fourth issue..there was too much reportage. **1969** *N.Y. Rev. Books* 2 Jan. 28/3 Reportage like this account of a journey through the Sudetenland as the Germans entered. **1978** *Poland* May 31/2 Young people need a confirmation of their knowledge about their own country..and that..is precisely what they find in a realistic reportage. **1979** *London Rev. Bks.* 25 Oct. 8/4 His study of the Hyde Park orators might have been taken as a masterly piece of reportage.

reportative (rīpō·ɹtātiv), *a.* [f. Report *v.* + -ative.] That presents or introduces reported speech.

1973 *New Society* 6 Sept. 580/2 Containing in fact only one four letter word used in a reportative context. **1975** *Language* LI. 804 Normally, a sentence containing the quotative clitic translates with 'one says', 'they say', or 'it is said', i.e. a reportative verb having an unspecified subject.

reportedly, *adv.* (In Dict. s.v. Report *v.*) (Further examples.)

1901 *Daily News* 28 Nov. 13/6 Some chance remarks reportedly made by Prince Akihito..sent the match-makers searching farther afield. **1969** *Listener* 6 Aug. 199/1 He was..strongly criticized for reportedly kissing an African woman in public. **1972** *New Yorker* 4 Mar. 85/1 What their constituents have said to them reportedly is not appreciative. **1979** *Time* 2 Apr. 23/1 This probe reportedly is ready to produce indictments against the man that Carter made director of the Office of Management and Budget.

reporter. Add: **1. a.** (Later example.)

1952 P. Edwards in *Shakespeare Survey* V. 35 If the 1609 Quarto of *Pericles* is reconstructed, or 'reported', we should infer, from the suggestion that two hands are at work on the manuscript, that the text was compiled by two 'reporters'.

c. Also (*Sc. Law*), one who receives reports (on juvenile offenders).

1968 *Social Work (Scotland) Act* c. 49 s. 36(1) For the purpose of arranging children's hearings and for the performance of such other functions in relation to the children's panel or to children's hearings as may be assigned to him by this Part of this Act, a local authority shall, in accordance with the provisions of this section, appoint an officer, whole-time or part-time, to be known as the reporter. **1976** *Howard Jrnl.* XV. i. 31 The key figure in the new system is the reporter. It is his function to decide, on the basis of reports, whether the child referred to him by the police, social worker or education department is in 'need of compulsory measures of care'.

2. b. (Earlier and later examples.) Also one who does similar work for other kinds of journal, or for radio or television. Also *reporter-at-large*.

1798 *Deb. Congress U.S.* 21 Mar. (1851) 1289 The House ought to render the reporters as independent..as they could be. **1946** McCarthy *Let.* in *Publics* Nov. Mr. Hersey.. is *The New Yorker*'s reporter-at-large. **1968** *Listener* 12 Sept. 321/1 Some of the strikers, including some sports reporters, had proposed a return to work... The radio reporters of *France-Inter* returned ten days later..and TV reporters..agreed to resume work in mid-July.

c. In the titles of newspapers and periodical publications.

1797 (*title*) The reporter, or the general observer. **1853** (*title*) St. Helens newspaper and midweek reporter. **1870** (*title*) Cambridge University reporter. **1956** (*title*) Surrey county reporter. **1961** (*title*) Rating and valuation reporter.

4. *Chem.* In full *reporter group*. A group whose spectroscopic properties are sensitive to its chemical environment and well characterized and which is used as a means of obtaining structural information about a system or molecule in which it occurs.

1970 *Nature* 21 Mar. 1103/1 The method is based on the anisotropy of signals from the nitroxide radical which thus acts as a 'reporter' of molecular motions (for example, whether the spin label is in a region of free or unrestricted movement). **1974** *Sci. Amer.* Mar. 76/1 The method involves attaching a 'reporter' group, usually a nitroxide group that has an unpaired electron, to the carbon chain of a test molecule's fatty-acid tail.

5. *attrib.*, in appositive use, and *Comb.*, as *reporter-director, -material, -photographer, politician, -researcher*; *reporter-like* adj.

1909 C. S. Peirce *Let.* 14 Mar. in R. B. Perry *Tht. & Char. of W. James* (1935) II. 440 So it is reported by my rather reporter-like memory. **1889** 'Mark Twain' *Connecticut Yankee* 108 It was my purpose..to start a newspaper... So I wanted to..be finding out what sort of reporter-material I might be able to rake together. **1978** W. F. Buckley *Stained Glass* ii. 15 One reporter-photographer from *L'Humanité* pressed for admittance. **1894** G. B. Shaw *Let.* 2 Dec. (1965) I. 464 Surely so fine a spirit could have been

rescued from the reproach of being..an ignorantly contemptuous reporter-politician? **1976** *Time* 27 Sept. 3/1 More than 70 *Time* correspondents, writers, reporter-researchers and editors set out to assess the South as it is today.

reporterage (rīpō·ɹtərēd3). [f. Reporter + -age.] = *Reportage 3.

a **1936** Kipling *Something of Myself* (1937) v. 131 Yet the book was not all reporterage. **1939** Joyce *Finnegans Wake* (1964) I. 70 Making his reporterage on Der Fall Adams for the Frankofurto Siding.

reporting, *vbl. sb.* Add: **b.** (Further examples.) *reporting company Canad.*: (see quot. 1973).

1886 *Encycl. Brit.* XXI. 842/1 Numerous mechanical reporting machines have been invented. **1973** *Daily Colonist* (Victoria, B.C.) 5 Oct. 9/2 It is a reporting company (new term for what was formerly called a public company) with 5,000 shareholders, mostly in B.C. **1974** *BP Shield Internat.* Oct. 7/3 Prince Philip's masterly handling of the final reporting-back session of the conference.

repo·rtless, *a.* [-less.] Unknown, not reported; without repute.

c **1865** E. Dickinson *Poems* (1955) II. 740 Reportless Subjects, to the Quick Continual addressed—But foreign as the Dialect Of Danes, unto the Rest. *c* **1884** *Ibid.* III. 1120 Still thou art What surgeons call alive—Though slipping—slipping I perceive To thy reportless Grave.

reportorial, *a.* (Earlier U.S. and later examples.)

1858 *82nd Anniv. Amer. Independence* (Boston, Mass.) 6 As far as reportorial observation could extend, the best possible temper prevailed. **1926** [see *dated *ppl. a.* 3]. **1949** E. B. White *Let.* 20 Nov. (1976) 315 Boyer wrote..some reportorial articles that experienced editors..regarded as thoroughly non-objective. **1955** *Bull. Atomic Sci.* Sept. 249/1 The stifling of routine reportorial inquiry. **1977** M. Lipper in Bond & McLeod *Newslett. to Newspapers* I. 73 This is especially striking when one keeps in mind..the role of coffeehouses or clubs as forerunners of modern reportorial staffs.

reporto·rially, *adv.* [f. Reportorial *a.* + -ly[2].] In a reportorial manner; as a newspaper reporter.

1862 *N.-Y. Tribune* 22 Apr. 1/4 At headquarters this morning—I mean those of General Heintzelman, to which I am reportorially attached—I found things quiet enough. **1901** *Pop. Sci. Monthly* Feb. 382 Unfortunately, the weather will not let the newspaper alone, and so..the newspaper must keep pegging away at it, editorially and 'reportorially', until the present anomalous state of things is developed. **1972** J. G. Vermandel *Last seen in Samarra* xvii. 114 Safer than admitting to firsthand knowledge, but reportorially a flop, as he kept complaining bitterly to Alex. **1981** *Times Lit. Suppl.* 8 May 512/3 Because Wilkinson remains so reportorially self-conscious,..he's unable to comprehend the tramps among whom he lives.

repo·rtship. *rare.* [f. Report *sb.* + -ship.] An instance of reporting for a newspaper.

1912 G. Frankau *One of Us* xiii. 129 It was indeed a triumph of reportship: They gave the artiste's *rôles*, her lap-dog's photo.

repose, *sb.* **5. b.** (Later example.)

1909 *Chambers's Jrnl.* Oct. 664/1 The workmen lost the large conception of their ancestors, the patterns [of carpets] lacked repose.

reposing, *vbl. sb.* (Later *attrib.* example.)

1936 [see *funeral home* s.v. *funeral *sb.* 6].

reposit, *v.* **2.** (Earlier example.)

1800 C. B. Brown *Arthur Mervyn*, II. iv. 69 The grave was covered, the spade reposited under the shed, and my seat in the kitchen resumed.

reposition, *v.* Delete rare[−1] and add: **a.** (Later examples.)

1959 *Times* 2 Oct. 11/3 Arm rests have been redesigned and repositioned. **1967** Cox & Grose *Organization & Handling Bibl. Rec. by Computer* II. 46 To do so requires that the space left be rapidly and accurately repositioned at the 'exposing' position. **1981** S. Brett *Situation Tragedy* i. 9 They rearranged their cameras, repositioned their sound-booms.

b. *intr.* To adjust or alter one's position.

1947 A. C. Douglas *Gliding & Soaring* 48 If the tow is to consist of more than two sailplanes the manner in which they shall release or reposition should be decided beforehand. **1977** *Chicago Tribune* 2 Oct. IV. 3/2 The Prisendam will make a second visit in April while repositioning from Singapore to Vancouver.

Hence **reposi·tioned** *ppl. a.*, **reposi·tioning** *vbl. sb.*

1968 P. A. P. Moran *Introd. Probability Theory* ix. 408 Thus by repositioning there is no restriction in supposing that the distribution of *X* is confined to the points 0, 1, 2,.., zero being its true left extremity. **1969** *Gloss. Terms Dentistry* (*B.S.I.*) 69 Repositioned flap procedures. (1) Apically repositioned flap operation... (2) Apically displaced flap operation... (3) Laterally repositioned flap operation. **1975** T. Allbeury *Special Collection* ix. 60 Air experts had noticed..the repositioned undercarriage members. **1977** *Proc. R. Soc. Med.* LXX. 432/1 No changes in the surgically repositioned bone. **1980** A. Coppel *Hastings Conspiracy* xxxviii. 230 A major repositioning of large sections of the Soviet Army.

repository, *sb.* **1. c.** (Earlier examples.)

1759 A. Murphy *Let.* 22 July in D. Garrick *Private Corr.* (1831) I. 101 But yours is Beaver's Repository, and there you must judge whether they [*sc.* the horses] are marketable, or likely to tire before they come to the winning-post. **1767** J. Wedgwood *Let.* 23 May (1965) 53, I spent a great part of the day in search of a Room for my repository.

repossess, *v.* Add: (Later examples.) Also *spec.*, to regain possession of or seize (goods being bought by hire-purchase) when a purchaser defaults on his payments.

1964 *Reading Teacher* Dec. 210/2 One might think..of a reading program that would enable the children to investigate more widely on their own the worlds of Robin Hood, the cowboys, the spacemen..when the television is being repaired or repossessed. **1969** *Rolling Stone* 28 June 28/2 The starting point was having their car repossessed in Nashville ten years ago. **1972** *New Society* 12 Oct. 98/3 The record company repossessed the amplifiers and so Bolan, of necessity, switched back to acoustic guitar. **1977** *Field* 13 Jan. 40/1 These committees..are to consider cases where a farmer needs to repossess a cottage for an incoming employee.

reposse·ssed, *ppl. a.* Chiefly *U.S.* [f. prec. + -ed[1].] That has been regained or seized back, esp. by a vendor; second-hand.

1933 *Sun* (Baltimore) 21 Apr. 19/2 (Advt.), Repossessed Car Corp., 31 W. North Ave. **1936** H. L. Mencken *Amer. Lang.* (ed. 4) vi. 293 For the former [*sc. second-hand*] the automobile dealers..have substituted *reconditioned, rebuilt, repossessed* and *used*. **1957** O. Nash *You can't get there from Here* 148 When he was knocked down by a repossessed scooter and the Boy Scouts administered Second Aid. **1974** *Sumter* (S. Carolina) *Daily Item* 24 Apr. 14B/5 (Advt.), Save by getting a repossessed home now. **1977** *Times* 12 Dec. 26/9 (Advt.), Sugar daddy offers—repossessed pastel mink coat..lynx collar..£2,000.

repossession. Add: **1.** (Later examples.) Also *spec.*, the recovery of goods being bought by hire-purchase when a purchaser defaults on his payments; legal proceedings to effect this. Also *attrib.*

1938 *Sun* (Baltimore) 26 Feb. 18/1 Some used-car dealers ..employ thugs to beat up customers, if necessary, in repossession activities. **1972** *Mod. Law Rev.* XXXV. I. 24 This balance of power could be achieved if retailers' claims for debt are abolished, leaving repossession or an adverse credit report as the sanctions against non-payment. **1977** *Field* 13 Jan. 40/2 If an occupier refuses an offer of suitable housing, it may provide ground for repossession by the farmer.

repo·st (rī-), *v.*[2] [f. Re- 5 a + Post *v.*[1]] *trans.* To post (a letter, etc.) again.

1963 *Times* 9 Jan. 4/7 Having stolen those postal packets, counsel said, the defendant extracted the passbooks and then reposted them to the building societies. **1977** *Lancs. Life* Nov. 83/3 There were four letters in the box, none of which was addressed to us. Three I re-delivered.., the other I re-posted.

So **repo·stage.**

1855 D. G. Rossetti *Let.* 23 Jan. (1965) I. 241, I am asked by William to request from you the re-postage of *Athenaeums* when quite done with.

repo·sting (rī-), *vbl. sb.* [f. Re- 5 a + Posting *vbl. sb.*[3]] Appointment to a new post.

1970 R. Wingate *Ismay* vii. 159 In India itself regiments with a century or more's traditions of war, service and comradeship had to be split, inevitably reducing their efficiency during the reposting period to zero. **1972** D. Bloodworth *Any Number can Play* xix. 200 Pawkinson-Convoy has been recalled for immediate consultation to London and probable reposting. **1981** J. Bingham *Brook* 16 He had applied for a re-posting to Melford.

repoussé. b. (Examples.)

1858 Queen Victoria *Let.* 1 Apr. in R. Fulford *Dearest Child* (1964) 83 A vase by Veité—silver repoussé, really very fine. **1911** [see *crusta f]. **1977** *Jrnl. R. Soc. Arts* CXXV. 485/1 The part [of the helmet] which covered the forehead is plain, but above it, in *repoussé*, there is an arched ridge round which curly hair is shown.

|| **repoussoir** (rəpūswār). [Fr., f. *repousser*: see Repoussé *a.*] An object in the foreground of a painting serving to emphasize the principal figure or scene. Also *transf.* and *fig.*, and *attrib.*

1873 H. James in *Galaxy* Mar. 427 Mr. Casaubon is an excellent invention: as a dusky *repoussoir* to the luminous figure of his wife he could not have been better imagined. **1890** W. James *Princ. Psychol.* II. xxvi. 513 The relative motion felt by the retina is assigned to that one of its components which we look at more in itself and less as a mere repoussoir. **1906** *Westm. Gaz.* 24 Mar. 2/2 A cool, tranquilly pleasing background is degraded to mere dulness in consequence of the gaudy gowns in front of it. Has the word *repoussoir* any meaning to her? **1925** A. Huxley *Along Road* III. 169 His exquisitely subtle use of *repoussoirs* and that extraordinary mastery of colour. **1936** *Burlington Mag.* May 208/1 The strong repoussoir character of the trees on the left. **1948** L. Spitzer *Linguistics & Lit. Hist.* v. 235 Since there had to be the Muse..she had to become the *repoussoir*; the personification of the Greek culture which had to be rejected. **1970** T. Hilton *Pre-Raphaelites* v. 150 Brown's..landscapes of the 1850s..shunning the usual devices of *repoussoir* trees and the conventions of aerial perspective. **1974** *Times Lit. Suppl.* 15 Mar. 261/3 The sitter's shoulders sometimes compressed into the narrowing oval of the frame, a *repoussoir* for the all-important face. **1977** 'M. Innes' *Honeybath's Haven* v. 46 The traveller who approaches Hanwell Court by the main drive has the advantage of first viewing the mansion disposed beyond a gigantic *repoussoir* known to art historians as the *Poseidon urging the Sea-Monster to attack Laomedon*.

repper (re·pəɹ). *slang. rare*[−1] [f. Rep(utation) + *-er[6].] = Reputation 4 a.

1910 R. Brooke *Let.* 7 Nov. (1968) He will find — in his Bed one night: and then she will force him to marry her to save her Repper.

repple depple (re·p'l de:p'l). *U.S. Mil. slang.* Also **reppo depot.** [f. Repl(acement + Dep(ot modified by reduplication and rhyme.] A replacement depot (see quot. 1945).

1945 *Amer. N. & Q.* Dec. 136/2 *Repple Depple*, overseas replacement depot where soldiers are assembled before sailing back to the United States. **1945** *N.Y. Times Mag.* 9 Dec. 20/3 One of the last phases of the European war's end—the overseas replacement depot or 'repple depple', where Yanks slated for the long voyage home are gathered together... Repple depples, in short, are dreary places. *Ibid.* He talked out of turn to the repple depple Powers-That-Be. **1947** *Amer. Speech* XXII. 216 *Repple Depple* had a less popular form, *Reppo Depot*. A person who never approached the fighting front was a *Reppo Depot Ranger*. **1973** S. Alsop *Stay of Execution* (1974) III. 283, I suppose in the American army we would have been sent to a Replacement Depot, or Repple Depple. **1978** J. A. Michener *Chesapeake* 771 He was sent to Korea, not with a formed unit but to a replacement depot, a repple depple he explained to his parents.

represent, *v.* Add: **9. c.** *Math.* To act as a representation of (a group).

1897 [see *primitive *a.* 5 d]. **1971** D. Gorenstein in Powell & Higman *Finite Simple Groups* ii. 77 We conclude that $H/O_p(H)$ is faithfully represented as a linear group on the Frattini factor group of $O_p(H)$.

representability. (Examples.)

1879 W. James *Coll. Ess. & Rev.* (1920) 95 The craving for clear representability..leads often to an unwillingness to treat any abstractions whatever as if they were intelligible. **1977** M. Cohen *Sensible Words* i. 14 What these men share is a confidence in the visual representability of meaning.

representation. Add: **2. e.** *Math.* The image of a homomorphism from a given (abstract) group to a group or other structure having some further meaning or significance; such a homomorphism.

1897 W. Burnside *Theory of Groups of Finite Order* ii. 22 As long as we are dealing with the properties of a group *per se*, and not with properties which depend on the form of representation, the group may, if convenient, be replaced by any group which is simply isomorphic with it. **1908** H. Hilton *Introd. Theory of Groups Finite Order* xv. 180 One and the same group of linear substitutions may give rise to two or more representations of *G*. **1940** D. E. Littlewood *Theory of Group Characters* iv. 48 To several elements of the group may correspond identical matrices, so that the representation is not simply, but multiply isomorphic with the group. **1949** S. Kravetz tr. *Zassenhaus' Theory of Groups* ii. 35 A representation is said to be faithful if the homomorphy induced by the representation is an isomorphy. **1971** D. Gorenstein in Powell & Higman *Finite Simple Groups* ii. 76 A faithful, irreducible representation of an abelian group on any vector space is necessarily cyclic. **1974** *Encycl. Brit. Macropædia* I. 752/2 Technically, a representation of a group is a homomorphism of it into another group, most commonly into the group of invertible linear transformations (or matrices) on some linear space. **1980** *Sci. Amer.* May 68/2 The way mathematicians construct a group depends to a large extent on whether the group has a natural representation as the transformations of some geometrical object.

representational, *a.* Add: **b.** *spec.* in *Art.* (See quots. 1961, 1962.)

1923 [see *non-representational* s.v. *non- 3]. **1934** C. Lambert *Music Ho!* II. 115 The repetitions of..[an] underlying curve in an abstract or representational picture have no dramatic content. **1956** R. Macaulay *Towers of Trebizond* xxii. 256, I could see he [*sc.* an ape] was going to be a painter of the abstract type...I thought Suliman ought to try and be a little representational too. **1959** Halas & Manvell *Technique Film Animation* 12 Though there is still..some useful place in normal draughtsmanship and painting for the exactly representational illustration as against a good photograph, there would seem to us to be no argument in favour of an exactly representational animated picture. **1961** M. Levy *Studio Dict. Art Terms* 96 *Representational art*, that kind of painting or sculpture which tries..to reproduce the physical appearance of objects, persons, or other subjects. As distinct from non-representational art where the interest in surface appearances is of little or no account. **1962** R. G. Haggar *Dict. Art Terms* 286/2 *Representational*, describes art in which figures and objects are depicted as they appear to the eye, or as they are known to be, in contradistinction to abstract or non-representational art. **1965** *New Statesman* 9 Apr. 566/2 The academy is now positively soliciting exhibits from Pop artists and abstractionists. However,..there's no cause for alarm. The R[oyal] A[cademy] will continue to show the best of British representational art, but it aims, also, to provide an annual cross-section of everything which is being done by British artists.

representationalism (further examples). So **representa·tionalist** *a.* and *sb.*

1846 J. D. Morell *Hist. View Philos.* I. ii. 232 The great aim of Reid's philosophy..was..to controvert the

representationalist hypothesis. **1921** A. Huxley *Crome Yellow* xii. 115 One could admire representationalism in the Old Masters... But in a modern? **1934** C. Lambert *Music Ho!* II. 113 It is all very well to hammer out a theory, however mistaken, that applies to an art functioning in space: it is quite another matter to apply this to an art that functions in time. Most of the modern fallacies about abstraction, literary sentiment, representationalism, romantic contamination, etc. in music are due to ignoring this elementary distinction. **1937** R. I. Aaron *John Locke* II. iii. 121 These accounts would have been the same if Locke had never adopted the representationalist position... Though nominally Locke remains representationalist in his explanation of the knowledge we have of our minds, actually he proceeds as if we know ourselves and our operations directly. **1976** *Jrnl. R. Soc. Arts* CXXIV. 567/1 The Brocks were among the last great representationalists, preservers of a world of recognizable human types in clearly defined historical settings. **1978** *N. & Q.* Feb. 91/2 He felt attracted to this best ally of representationalism.

representative, *a.* and *sb.* Add: **A.** *adj.* **1. d.** (Later example.)

1934 A. C. Ewing *Idealism* vi. 283 By the term 'representative theory of perception' I mean to cover *any* theory of perception which admits the existence of physical objects in the realist sense but is not a direct theory.

e. *representative fraction*: the ratio of a distance on a map to the distance it represents on the ground. Cf. *R. F.* s.v. *R II. 2 a.

1886 H. D. Hutchinson *Mil. Sketching made Easy* i. 2 If the Representative Fraction is marked on a sketch, the scale can be understood, and the sketch can be used, by anyone, even though it be a foreign one. **1969** G. C. Dickinson *Maps & Air Photographs* vii. 99 Whether written as a fraction or a ratio, this means of expressing the scale is called the representative fraction (R.F. for short) of the map.

2. b. (Later examples.)

1879 Gladstone *Gleanings* I. viii. 214 We have, *proh pudor!* found no better method of providing for peace and order in Jamaica..than by the hard and vulgar, even where needful, expedient of abolishing entirely its representative institutions. **1921** H. Samuel *Let.* 8 May in M. Gilbert *Winston S. Churchill* (1977) IV. Compan. III. 1461 The very early establishment of representative institutions. **1975** J. Plamenatz *K. Marx's Philos. Man* xv. 464 How.. would they [*sc.* the workers] ensure that this system was not as much of a sham as bourgeois representative government was, according to Marx?

B. *sb.* **2. a.** (Earlier and further examples.)

1635 *Essex Inst. Hist. Coll.* (1862) IV. 93/1 By the towne rep[re]sentative, 22 of the 12th moneth. **1787** *Constitution U.S.* i. §2 No person shall be a representative who shall not ..be an inhabitant of that state in which he shall be chosen. **1977** *Time* 18 July 10/3 Young proved himself a sensible and reasonable Representative—and also an independent one. **1979** *Daily Tel.* 8 Jan. 8 Members of Parliament are representatives and not delegates.

b. Also, similar legislative bodies in Australia and New Zealand.

1852 *Act* 15 & 16 *Vict.* c. 72 §32 There shall be within the Colony of New Zealand a General Assembly, to consist of the Governor, a Legislative Council, and House of Representatives. *Ibid.* §40 For the Purpose of constituting the House of Representatives of New Zealand it shall be lawful for the Governor,..to summon and call together a House of Representatives in and for New Zealand. **1891** *National Australasian Convention Debates* 23/1 This Convention [of Sydney, March 1891] approves of the framing of a federal constitution, which shall establish,—1. A parliament, to consist of a senate and a house of representatives. **1930** L. G. D. Acland *Early Canterbury Runs* 1st Ser. ii. 27 He was a member of the House of Representatives and of the Provincial Council. **1965** *Austral. Encycl.* VII. 9/1 The Speaker's chair in the House of Representatives is also of interest.

4. a. (Further examples.)

1957 Clark & Gottfried *Dict. Business & Finance* 299/1 *Representative*, in selling, a salesman, either one employed by the seller or operating as an agent. **1961** P. F. Payne *Brit. Commercial Institutions* v. 70 The manufacturer's agent carries out functions similar to those of the wholesaler's representative. **1976** *Gramophone* June 37/3 Raymond Cooke and Robert Cox of KEF demonstrated the excellence of their loudspeakers and there were recitals by record industry representatives.

representativity (re:prīzentāti·vīti). [f. Representative *a.* + -ity.] Representative character; representativeness.

1901 *N. Amer. Rev.* Apr. 632 By far the most signal instance of Professor Wendell's open-mindedness is his recognition of Mark Twain's..representativity as a Westerner.

representee. Restrict †*Obs.* to senses in Dict. and add: **3.** *Law.* One to whom a representation is made.

1911 G. S. Bower *Law Actionable Misrepresentation* I A representation is a statement made by, or on behalf of, one person (hereinafter called 'the representor') to, or with the intention that it shall come to the notice of, another person (hereinafter called 'the representee'), which relates, by way of affirmation, denial, description, or otherwise, to a matter of fact. **1971** R. A. Percy *Charlesworth on Negligence* (ed. 5) ii. 34 In *Jones v. Still* an honest misrepresentation, even if negligent, was held to give no cause of action, since no duty to take care in making the statement arose because there was no contract, fiduciary relationship or reliance by the representee upon the special skill, knowledge or training possessed by the representor.

representor. Delete †*Obs.* and add: *spec.* in *Law*: one by or on behalf of whom a representation is made.

1911, 1971 [see *representee 3].

repress, *v.*[1] Add: **3. c.** *Psychol.* [tr. G. *verdrängen* (used in this sense by Breuer & Freud 1893, in *Neurol. Centralbl.* XII. 10).] In *Psychoanalysis*, of a patient or person who is the object of study: to keep out of the conscious mind, or suppress into the unconscious (unacceptable memories or desires). Also *absol.*

1909 A. A. Brill tr. *Freud's Sel. Papers on Hysteria* i. 7 The patient has not reacted to psychic traumas because the nature of the trauma..concerned things which the patient wished to forget and which he therefore intentionally inhibited and repressed from the conscious memory. **1919** M. K. Bradby *Psycho-Analysis* III. vii. 82 He believes the unconscious to be exclusively composed of contents repressed from the conscious. **1920** *Discovery* Mar. 69/2 The motive for repression is one's personal comfort. One represses to preserve one's peace of mind. **1943** J. S. Huxley *Evolutionary Ethics* ii. 15 The impulses whose thwarting generated the guilty hate may themselves become coloured with guilt, or be repressed. **1977** R. A. Baron et al. *Psychol.* x. 337 This unconscious mechanism can begin to create new problems... The individual loses some control over the situation when he represses his awareness of it.

repressed, *ppl. a.* Add: (Later examples in sense of *repress *v.*[1] 3 c.) Also *absol.*

1904 *Psychol. Bull.* I. 357 The theory of Freud, that dreams are disguised realizations of repressed desires. **1919** M. K. Bradby *Psycho-Analysis* I. iii. 34 It [*sc.* the unconscious] also contains the repressed. **1923** J. S. Huxley *Ess. Biologist* v. 187 The publication of Darwin's *Origin of Species* was to them what psycho-analysis is (or may be) to a patient with a repressed complex. **1954** D. Riesman *Individualism Reconsidered* (1955) xxii. 336 It was not easy to find convincing evidence for the existence of repressed Oedipal desires in every adult whom he analyzed. **1960** M. Spark *Bachelors* vi. 81 Repressed homosexuality is a meaningless term because no one can prove it. **1973** J. G. Starke *Validity of Psycho-Anal.* ii. 14 For psycho-analysts, perhaps the most important component is *repressed* material.

represser. Add: **1.** (Later example in the sense of *repress *v.*[1] 3 c.)

1951 *Jrnl. of Personality* XIX. 472 Repressers..are patients who avoid contact with such emotional material. They are apt to block in the presence of sexual or aggressive stimuli.

2. = *repressor 2.

1957 [see *repressor 2]. **1971** *New Scientist* 22 July 182/1 Two classes of DNA-switching molecules are known so far in bacteria, one class acts negatively—repressers—and one acts positively, but indirectly—sigma factors.

repressible, *a.* Restrict *rare*[−0] to sense in Dict. and add: **2.** *Biochem.* That may be inhibited by the action of a repressor; susceptible to the action of a repressor.

1957, 1964 [see *repressor 2]. **1974** *Nature* 10 May 110/1 The interactions which control repressible and inducible operons in bacteria are now well understood in principle.

repressing (rīpre·siŋ), *vbl. sb.*[2] Also **re-** (with hyphen). [f. Re- 5 a + Pressing *vbl. sb.*[1]: cf. *Repess *v.*[2]] **1.** A new impression made from an old matrix of a sound recording.

1960 'I. T. Ross' *Murder out of School* iv. 38 Old records.. collectors' items now, that had somehow never come out as well in the long-playing repressings the companies had issued. **1975** *Daily Tel.* 17 Mar. 10/3 If you can only afford one of the three recent issues, it must be 'Chopin 3' of RCA's collection of re-pressings.

2. The action of pressing again.

1967 M. Chandler *Ceramics in Mod. World* v. 142 Shaping methods for fireclay bricks include extrusion and wire-cutting (often followed by re-pressing).

repre·ssing, *ppl. a.* [f. Repress *v.*[1] + -ing[2].] That represses. So **repre·ssingly** *adv.*

1872 Geo. Eliot *Middlem.* II. III. xxxii. 152 Solomon put his hand before her repressingly. **1909** A. A. Brill tr. *Freud's Sel. Papers on Hysteria* vii. 161 What becomes conscious as an obsession and obsessive affect and substitutes the pathogenic memory in the conscious life, are compromise formations between the repressed and the repressing ideas. **1951** *Jrnl. of Personality* XIX. 472 We hypothesized that intellectualizing patients..would show higher accuracy for threatening material than the repressing type of patient. **1975** I. M. Blanco *Unconscious as Infinite Sets* vi. 82 The unrepressed unconscious of the ego is, in the Freudian conception, the *repressing* aspect of the ego.

repression. Add: **2. c.** *Psychol.* The action, process, or result of suppressing into the unconscious or keeping out of the conscious mind unacceptable memories or desires. Also *attrib.*

1909 A. A. Brill tr. *Freud's Sel. Papers on Hysteria* iv. 88 If I could now make it probable that the idea became pathogenic in consequence of the exclusion and repression, the chain would seem complete. **1910** S. Freud in *Amer. Jrnl. Psychol.* XXI. 193, I called this hypothetical process 'repression' (*Verdrängung*), and considered it was proved by the undeniable existence of resistance. *Ibid.*, One of my cases, in which the conditions and the utility of the repression process stand out clearly enough. **1930** R. Lehmann *Note in Music* VII. 274 Gerald was not free, not calm and balanced; quite the reverse—a tangle of passionate conflicts and repressions. **1939** T. S. Eliot *Family Reunion* i. 40, I always said his [Lordship Suffered from what they call a kind of repression. **1954** R. F. C. Hull tr. *Jung's Devel. of Personality* in *Coll. Wks.* XVII. iv. 115 No breaking down of repressions can ever destroy true creativeness. **1973** J. G. Starke *Validity of Psycho-Anal.* ii. 14 In essence, repression is the mental process of rejecting and excluding material from consciousness.

3. *Biochem.* The inhibition of enzyme synthesis by the action of a repressor on an operon.

1957 [see *repressor 2]. **1959** *Jrnl. Molecular Biol.* I. 176 It now appears to be a general rule, for bacteria, that the formation of sequential enzyme sequences involved in the synthesis of essential metabolites is *inhibited* by their end product. The convenient term 'repression' was coined by Vogel to distinguish this effect from another, equally general, phenomenon: the control of enzyme *activity* by end products of metabolism. **1973** R. G. Krueger et al. *Introd. Microbiol.* xv. 437/2 Repression is distinct from feedback inhibition in that the former results in the cessation of enzyme synthesis whereas the latter leads to inactivation of an existing enzyme.

repressive, *a.* (Later examples in sense of *repress *v.*[1] 3 c.)

1921 R. Macaulay *Dangerous Ages* xii. 234 His phrases drifted over Mrs Hilary's head. '..a deterrent force residing in the ego and preventing us from stepping outside the bounds of propriety..conflict with the progress of human society..inhibitory and repressive power of the censor.' **1944** J. S. Huxley *On Living in Revolution* 57 It is much harder to feel strongly about social problems such as malnutrition or unemployment, because the connection with the repressive mechanism is not so automatic. **1968** A. Heron *Towards Quaker View of Sex* i. 7 This still repressive and inhibited outlook towards sex..has brought difficulties to the serious student of human behaviour. **1970** *Nature* 26 Sept. 1371/2 Whose tortured minds are nourished by absurd slogans such as 'creative vandalism' and 'repressive tolerance'.

repressor. Restrict *rare* to sense in Dict. and add: **2.** *Biochem.* A substance which by its action on an operon can inhibit the synthesis of a specific enzyme or set of enzymes.

1957 H. J. Vogel in McElroy & Glass *Chem. Basis Heredity* ii. 286 In order to facilitate further discussion, the following terminology will be used hereafter: a relative decrease, resulting from the exposure of cells to a given substance, in the rate of synthesis of a particular apo-enzyme is termed 'enzyme repression'; the substance thus decelerating enzyme synthesis is a 'represser' ('repressor'); an enzyme-forming system that can be antagonized by a represser is 'repressible'; and, under conditions of repression, the formation of the enzyme is 'repressed'. **1964** G. H. Haggis et al. *Introd. Molecular Biol.* x. 273 Either inducible cells contain built-in repressors, which are inactivated by added inducers, or repressible enzyme synthesis depends on intracellular inducers inactivated by external repressors. There is evidence that the first of these alternatives is the correct one. **1971** J. Z. Young *Introd. Study Man* iii. 54 The repressor is produced by a regulator gene. **1973** B. J. Williams *Evolution & Human Origins* vi. 91/1 Diffusable repressor substances that form complexes with specific operator regions prevent the mRNA transcription.

repressory (rīpre·sōri), *a.* [f. Repress *v.*[1] + -ory[2].] Having the qualities of a repressor; designed to repress.

1905 W. J. Locke *Morals M. Ordeyne* (1906) xi. 117 But what do I know of the repressory methods employed in seminaries for young ladies? **1954** D. Riesman *Individualism Reconsidered* xxii. 347 The repressory forces must draw their energies from the great energy reservoir of the id.

re-pre·ssuring, *vbl. sb.* [f. Re- 5 a + Pressure + -ing[1].] The pumping of fluid into an oil well so as to increase or maintain the pressure in the oil-bearing strata, allowing more oil to be extracted.

1929 *Jrnl. Inst. Petroleum Technologists* XV. 430 If a flush field is produced inefficiently and re-pressuring is left until the field is almost commercially exhausted the ultimate yield of the pool must be very considerably less than when properly controlled back-pressures have been used from the beginning and re-pressuring operations commenced in the early life of the field. **1940** *Sun* (Baltimore) 24 Sept. 6/8 An oil field at Walters, Okla., has been revived by a method known as repressuring. **1961** *Economist* 2 Dec. 955/2 This pipeline..will eventually carry..salt water in the opposite direction, to be pumped into the underground structure in order to keep up underground pressure and keep the oil flowing up and out to the coast. This 're-pressuring' project is one of the largest ever undertaken outside the United States.

repre·ssurize, *v.* [Re- 5 a.] *trans.* To pressurize again; to renew pressure in.

1953 *Jrnl. Inst. Electr. Engineers* C. II. 646/2 After completion of the modifications, the system was repressurized, but when the routine 66-kV d.c. 15-min test was applied, failure occurred on two phases. **1962** J. Glenn in *Into Orbit* 41 You have one large handle for repressurizing the cabin with oxygen in case of a bad leak. **1971** *Daily Tel.* 26 July 22/5 He and Irwin will then repressurise their cabin, take off their spacesuits..and then get seven hours rest.

repricing (rīprəi·siŋ), *vbl. sb.* [Re- 5 a.] The act of changing, usually by increasing, a price.

1959 *Times* 1 June 11/2 If anomalies are to be avoided.. this 'repricing' should be undertaken systematically. **1974** *Daily Colonist* (Victoria, B.C.) 8 Sept. 7/7 There would be re-pricing, probably several times, during the model year.

reprime, *v.*[2] Add: **b.** *absol.*

1976 *Shooting Mag.* Dec. 65/1 (Advt.), Lee Load-All 12 G... 2nd Station reprimes with a push from inside the shell to prevent concave heads.

reprint, *sb.* Add: **1. b.** *attrib.*

1928 *Publishers' Weekly* 30 June 2603 Bookstores have in a new form a problem that confronted them twenty-five years ago when the reprint fiction began to appear in cloth binding. **1951** A. C. Clarke *Sands of Mars* xi. 135 She immediately sold the second reprint rights of Gibson's latest series. **1951** M. McLuhan *Mech. Bride* (1967) 23/2 The current effort to make almost every reprint cover look lustier than the next has brought them all to a dead level of fleshliness. **1952** E. Mannin *Let.* in *Manch. Guardian Weekly* 16 Oct. 13/1 May I, as an English author who is a victim of the deplorable American publishing habit of farming-out their cheap editions to what are called 'reprint houses', be allowed to point out some important points. **1961** T. Landau *Encycl. Librarianship* (ed. 2) 323/1 *Reprint series*, a number of publications, being reprints, not necessarily related in subject or treatment, issued by a publisher in uniform style and assigned a collective series title. **1964** F. Bowers *Bibliogr. & Textual Crit.* vi. i. 163 The normal reprint transmission of variants is disrupted by the annotator's correction of error. **1981** J. Sutherland *Bestsellers* i. 14 The [American] bookclubs..have in the past been much less reprint affairs than in the UK.

reprint, *v.* Add: **1. a.** Also *absol.*

1934 H. G. Wells *Exper. Autobiogr.* II. ix. 646 Macmillan's, my English publishers, were caught unawares by the demand and had sold out the first edition before they reprinted.

3. *intr.* for *pass.*

1821 R. Southey *Let.* 11 Jan. in *N. & Q.* (1975) Sept. 402/1 Do not bind your set, till I send you some corrections and additions for the first volume, which is now reprinting. **1942** *World Rev.* Apr. 17 In this war he is a Home Guard officer and the author of four training manuals, which, despite the paper shortage, reprint every few months. **1967** *Listener* 12 Jan. 68/3 It will be for these [colour photographs] that *Slowly Down the Ganges* has to reprint. **1980** *Daily Tel.* 21 Aug. 14 The book has sold 10,000 copies since May. It is now reprinting.

reprisal, *sb.* Add: **I. 4. a.** Also *attrib.*, as *reprisal attack, raid.*

c **1945** *Hutchinson's Pict. Hist. War* XII. 157/1 The vicious battering of the industrial and military bases of Germany provoked none of the great so-called 'reprisal' raids which characterised April and May. **1947** R. W. Cooper *Nuremberg Trial* 81 Everything was to be prepared to carry out reprisal attacks on London. **1956** H. Nicolson *Diary* 29 Oct. (1968) 311 Israel has launched against Egypt an attack that seems more serious than a reprisal raid.

reprise, *sb.* Add: Now usu. with pronunc. (rəprī·z).

The renewed influence of Fr. *reprise* is apparent in the modern pronunc. and development of sense 7.

II. 7. d. *transf.* in *Linguistics.* The repetition of a word or word-group occurring in a preceding phrase; a restated element. Also *attrib.* as *reprise construction.*

1950 *Archivum Linguisticum* II. 144 The aim of this article is to establish the frequency with which reprise constructions occur in *Chanson de Roland*. **1955** [see *grammaticalization]. **1959** M. Schlauch *Eng. Lang. in Mod. Times* iv. 99 Reprise constructions (called resumptions by Partridge) in formal discourse employed the appositional pronoun to recall a noun separated from its verb by a long series of interrupting modifiers. **1963** F. T. Visser *Hist. Syntax Eng. Lang.* I. i. 53 Type 'He, Alexander, cwæð.'.. The following 'reprise construction' is remarkable for the subject's being expressed four times. **1971** *Catholic Biblical Q.* XXXIII. 218 This structure makes sense out of *didaskalos* (vs. 10) as a reprise, for, far from ending the preceding unit (as it would as an inclusion), it introduces the theme of vss. 9–15, *Jesus as teacher*.

e. The repetition of a theatrical performance; a restaging or rewriting of a play (esp. for television), a repeated showing of a (piece of) film; a rerun, a replay. Also in extended use, a further performance of any kind; a reconstruction, a repeat.

1953 *Sun* (Baltimore) 9 Oct. 10/2 Mr. Moore came on the screen bubbling over with a joke which he was eager to share with the viewers. It required a reprise from the previous week's performance. **1955** *Times* 20 May 3/4 It is the season for *reprise* at this club theatre. Formerly, in shows of this kind, the menu has included so many delightful dishes of the past served cold by inexpert chefs as to be somewhat gruesomely indigestible. **1961** *Guardian* 1 Feb. 9/4 A polished reprise of his campaign sermons. **1968** *Globe & Mail* (Toronto) 17 Feb. 39 Nancy last night watched a television reprise of her dramatic victory in amazingly fast time in the Olympic giant slalom. **1972** *Publishers' Weekly* 11 Sept. 51/1 The author spins a good tale on the level of a reprise of what life was like for a young girl in a small town 30 years ago. **1977** *Time* 30 May 25/1 Carter headed next to the studios of KNXT-TV for a locally televised reprise of his successful national call-in program.

reprise, *v.* Restrict †*Obs.* (exc. *Arch.*) to senses in Dict. and add: **1. f.** (With pronunc. rəprī·z.) To repeat (a theatrical performance, song, etc.); to restage or rewrite. Cf. *reprise *sb.* 7 e.

1965 *Observer* 5 Dec. 24/3 The theatre rocks as Dolly reprises more times than is artistically justifiable the chorus of this..infectious title-song. **1970** *Ibid.* 29 Nov. 29/1 The song is tearfully reprised off-screen. **1975** *Listener* 5 June 735/1, I decided that..it would be necessary to reprise this scene in a heightened form.

repristinate, *v.* Delete *rare* and add further examples. Hence **repri·stinated** *ppl. a.*

1969 R. H. Bainton *Erasmus of Christendom* (1970) i. 14 The glory of Greece, the grandeur of Rome, the grace of Galilee should repristinate society and revivify the Church. **1970** *Jrnl. Ecumenical Stud.* VII. 804/1 The disavowal of the doctrine..and its displacement by the repristinated authorities of Scripture and Christian experience. **1977** *Theology* LXXX. 200 The power exercised by the remembered and interpreted past has been deliberately conserved and repristinated by methods which have not excluded secular rituals.

repri·vatize, *v.* [Re- 5 b; cf. G. *reprivatisieren.*] *trans.* To make private again; to denationalize. Also *absol.* Hence **reprivatiza·tion.**

1950 Webster Add., Reprivatize v.t. **1959** *Economist* 4 Apr. 53/1 A whole series of political and legal hurdles will have to be taken before the way is clear to denationalise, or reprivatise, in earnest. **1963** *Ibid.* 12 Jan. 136/3 He has promised to continue the work of 'reprivatisation' (denationalisation). **1980** *Evening News Mag.* 18 Jan. 22/1 The ugliest word yet coined by Whitehall is its latest—'reprivatisation', meaning the selling back to private enterprise of parts of nationalised industries. **1980** *Economist* 6 Dec. 18/2 Reprivatisation of the profitable naval yards.

repro (rī·pro). Colloq. abbrev. of Reproduction. **1.** *Printing* and *Photogr.* = Reproduction 1 e. Chiefly *attrib.* in, or *ellipt.* for, *repro proof* = *reproduction proof* s.v. *Reproduction 3.

1946 Melcher & Larrick *Printing & Promotion Handbk.* 247/1 The offset or other printer..should..be encouraged to reject any proofs..of poor quality and ask for better 'repros'. **1948** R. R. Karch *Graphic Arts Procedures* viii. 231 Care must be taken to see that repro proofs are well printed. **1952** R. W. & E. W. Polk *Practice of Printing* (ed. 4) viii. 62 There are two kinds of reproduction proofs (commonly [known as repros) used for the making of plates for letterpress, offset, and gravure. **1967** E. Chambers *Photolitho-Offset* iii. 31 Reproduction proofs on paper can be transferred into same-size negatives... The process..consists in making a repro-proof and covering with a sheet of special film. **1972** *Screw* 12 June 28/4 What's the chance of getting a repro of that shot? **1973** D. A. Spencer *Focal Dict. Photogr. Technol.* 528 Repro pulls can be made on baryta coated paper, or ink-accepting transparent or translucent film. **1977** J. Hedgecoe *Photographer's Handbk.* 17 (*heading*) A repro copy stand.

2. A reproduction or copy, usu. of a piece of furniture (cf. Reproduction 2 a in Dict. and Suppl.). Chiefly *attrib.* or as *adj.* Also *Comb.*

1958 *Spectator* 8 Aug. 193/1 Her Tudorbethan villa, with its Repro-Jaco interior. **1967** *House & Garden* Mar. 79/1 Good traditional design (*not* 'repro') marries brilliantly with good contemporary. **1967** *Sunday Times* 14 May 12/7 The total environment is compromised..by the 'repro' furniture introduced into the flats. **1970** 'D. Halliday' *Dolly & Cookie Bird* ii. 13 We have a workshop of our own for repairs and a bit of repro work. **1973** *Daily Tel.* (Colour Suppl.) 16 Mar. 41/3 You may ask what is the difference between a fake and a 'repro'. **1976** *New Society* 13 May 363/1, I can't imagine..a Rembrandt on the wall opposite, even if it were an unframed repro. **1978** *Morecambe Guardian* 14 Mar. 27/8 (Advt.), Good antique and repro furniture.

reproach, *v.* **2. a.** (Later *absol.* example.)

1961 B. Fergusson *Watery Maze* xv. 364 Like the good soldier he was, he never reproached or repined.

reproachable, *a.* **1.** (Later example.)

1972 *Sunday Tel.* 30 Apr. 14/2 It is here that 'The Green Flag' is reproachable. The history of a rebellion is incomplete if it gives hardly a clue as to the nature and attitudes of the power at which the rebellion was directed.

reprobance. For †*Obs. rare*[−1] in Dict. read '*rare* (only in allusive use of quot. 1604)' and add later example.

1878 Swinburne *Poems & Ballads* 2nd Ser. 213 Like Absalom with locks luxurious, Or like Judas fallen to reprobance.

repro·cess, *v.* [Re- 5 a.] *trans.* To subject (something) to a special process again. So **repro·cessed** *ppl. a.*; **repro·cessing** *vbl. sb.*

1939 *Sun* (Baltimore) 21 Feb. 14/4 'Reprocessed'..should be applied to wool made into a fibrous state for reuse. **1944** *Richmond* (Va.) *Times-Dispatch* 16 Jan. 9/1 A 'reprocessing' routine has been made an important part of the discharge procedure. **1948** *News Chron.* 15 Jan. 2/2 Production can be increased by reprocessing what we used to throw away. **1956** *Nature* 3 Mar. 400/1 The United Kingdom is likely..to re-process the burnt elements from reactors. **1962** E. Godfrey *Retail Selling & Organization* xix. 190 When an item is sold, the punched ticket is returned to the warehouse, where it is re-processed for sales records to be made. **1977** *New Yorker* 9 May 142/2 The Japanese are.. troubled..by our objections to the use of reprocessed nuclear fuel. **1979** *Bull. Amer. Acad. Arts & Sci.* May 44 Participants discussed the time necessary for a nation to convert a reprocessing plant to weapons production and how that time might be lengthened without giving up nuclear energy.

reproduce, *v.* Add: **3. d.** *trans.* To cause to be heard (sound originating elsewhere or on another occasion); also *absol.* Freq. with advbs.

1899 *T. Eaton & Co. Catal.* Spring & Summer 191/1 A graphophone..a perfect machine, reproduces perfectly. **1924** *Radio Times* 19 Dec. 620/1 (Advt.), He can 'pick' up the entertainment being sent out from any British or Continental Station and reproduce it at Loud Speaker strength. **1961** G. A. Briggs *A to Z in Audio* 21 If you can reproduce the bottom note of a double bass properly, you are not doing badly. **1978** *Gramophone* July 272/1 Totally enclosed headphones..can usually reproduce extreme bass frequencies more easily.

reproduced, reproducing *ppl. adjs.* (examples in sense *3 d of vb.).

1941 *B.B.C. Gloss. Broadcasting Terms* 28 *Reproducing desk*, table carrying one or more turn-tables and other equipment for playing gramophone records, disc recordings, or pressings. **1946** *Penguin Music Mag.* Dec. 93 Reproducing instruments are improving as much as recording has improved. **1964** *Listener* 17 Sept. 442/2 Recordings made for the reproducing piano by Josef Lhévinne. **1970** J. Earl *Tuners & Amplifiers* iii. 69 Power so liberated..detracts from the reproduced signals.

reproducer. Add: **2.** (Further examples.) Also, any device for reproducing recorded sound.

1899 *T. Eaton & Co. Catal.* Spring & Summer 191/3 The price of the Universal Graphophone, with a long run clockwork motor recorder, reproducer, hearing, speaking tubes and horn, is $50. **1937** *Jrnl. Soc. Motion Picture Engineers* XXIX. 218 With a high-quality microphone, a high-quality reproducer, and a suitably corrected amplifier, the response curve can be made uniform. **1961** G. A. Briggs *A to Z in Audio* 169 The console includes a tape reproducer with 7½", 15" and 30" speeds. **1978** *Gramophone* Aug. 399/3 The Stanton 681EEE, which continues in production, is already a very fine reproducer.

3. *Computers.* A machine for making copies of punched cards or tape.

1940 W. J. Eckert *Punched Card Methods* ii. 20 The High Speed Reproducer. This machine is used to transfer information from one card to another. **1949** [see *interpreter 5 a]. **1964** F. L. Westwater *Electronic Computers* vi. 100 A machine called a reproducer..automatically reproduces the information in a pack of old cards into new ones. **1970** A. Chandor et al. *Dict. Computers* 291 *Paper tape reproducer*... Also known as a reperforator.

reproducibility (rī:prŏdiū·sĭbi·lĭti). [f. next + -ity.] The capacity to be produced again; the quality of being reproducible; the extent to which consistent results are obtained when produced repeatedly. Cf. *repeatability.

1936 H. J. Paton *Kant's Metaphysic of Experience* I. xix. 371 Kant assumes that experience presupposes the reproducibility of appearances. **1939** *Jrnl. Amer. Chem. Soc.* LXI 3336/2 The agreement between two such sets of points indicates the reproducibility. **1946** *Nature* 7 Sept. 347/2 In many circuits increased reproducibility of electrical conditions does not indicate an increased reproducibility of light-emitting characteristics. **1953** *Industr. & Engin. Chem.* Feb. 465/1 Within the reproducibility of the data, no effect of change in catalyst quantity or flow rate was noted. **1962** *Listener* 6 Dec. 952/1 This accuracy, and the reproducibility that goes with it, lies at the heart of mass production. **1969** *Physical Rev. Lett.* XXIII. 1402/2 The accuracy obtained for the energy is better than ±0·1% and the reproducibility is about ±0·05%. **1977** J. D. Douglas in Douglas & Johnson *Existential Sociol.* i. 54 This oversimplifies the problem of objectivity by assuming that reproducibility constitutes objectivity and that providing a record of research makes it possible to reproduce results. **1981** *Times Lit. Suppl.* 13 Feb. 169/4 The illustrations to Milton..might suggest that one thing Blake liked about printmaking was the sheer fact of reproducibility, for a surprising proportion of the Milton pictures exist in two or more copies.

reproducible, *a.* Add: (Further examples.)

1949 E. P. Abraham et al. in H. W. Florey et al. *Antibiotics* II. VIII. xv. 637 The initial experiments were concerned with finding reproducible conditions for obtaining penicillin. **1972** *Physics Bull.* Jan. 15/1 Atomic clocks now provide a unit of time reproducible to 1 part in 10^{12}.

Hence **reprodu·cibly** *adv.*

1961 *Jrnl. Physical Chem.* LXV. 317/2 The round cells were reproducibly positioned in the light beam which entered the thermostated mineral oil-bath through a window. **1974** *Nature* 13 Dec. 589/1 A survey of various brain regions of the rat revealed that cell-free fractions from the corpus striatum contained adenyl cyclase which was reproducibly activated by low concentrations of D-LSD.

reproduction. Add: **1. c.** (Earlier example.) Also *attrib.* as *reproduction rate.*

1782 J. Wesley in *Arminian Mag.* Oct. 545 He [*sc.* Buffon] substitutes for the plain word *Generation*, a quaint word of his own, *Reproduction*, in order to level man not only with the beasts that perish, but with nettles or onions **1928**, etc. Reproduction rate [see *gross *a.* 6 a]. **1936** *Discovery* Sept. 298/1 It is the net reproduction rate which measures whether the population is maintaining itself or not.

e. (Earlier and later examples.)

1856 Kay & Johnson *Rep. Cases in Chancery* II. 285 Having regard to the international treaties, the Plaintiff reserves his right of reproduction, which is a sufficiently apt word in this case. **1870** W. A. Copinger *Law of Copyright in Works of Lit.* vi. 101 Copyright may be infringed..by reproduction under an abridged form. **1923** H. Crane *Let.* 21 June (1965) 137 He is offering..$25.00 for the original, and the reproduction rights without any payment. **1967** Karch & Buber *Offset Processes* iv. 71 The cost of reproduction (photographable) proofs..brought about the obvious question.

f. *Econ.* In Marxist theory, the process by which given capital is maintained for further production by the conversion of part of its product into capital; *simple reproduction*, reproduction in which the amount of capital remains constant, any surplus value being consumed; *enlarged, expanded* (etc.) *reproduction*, reproduction in which the amount of capital is increased by conversion of part of the surplus value into additional means of production. Also *attrib.*

1887 Moore & Aveling tr. *Marx's Capital* II. xxiii. 578 The conditions of production are also those of reproduction. *Ibid.* 579 If this revenue serve the capitalist only as a fund to provide for his consumption..then..simple reproduction will take place. *Ibid.* 582 The value of the capital advanced divided by the surplus-value annually consumed, gives the number of years, or reproduction periods [etc.]. **1939** *Rev. Econ. Stud.* VII. 32 According to Marx's analysis, in the case of enlarged reproduction, if gross investment..is not larger than c_1+c_2..there would be no net accumulation of capital at all. **1955** M. Dobb *Econ. Theory & Socialism* xvi. 266 These examples were designed to show the relations which would need to hold for expanded reproduction (i.e. a process of annual net investment) to take place and continue of its own momentum. **1965** B. Pearce tr. *Preobrazhensky's New Economics* 62 The consciously adopted economic policy of the State is quite often not a reaction to the difficulties encountered in practice in developing socialist reproduction. **1970** B. Brewster tr. *Althusser & Balibar's Reading Capital* III. iii. 266 The analysis of reproduction destroys the appearance..of a 'free' contract between the worker and the capitalist. **1975** *Chinese Econ. Studies* VIII. iv. 79 In the course of reproduction, the scale of operation would not have expanded... This reproduction based on the original scale is called simple reproduction.

g. The process of reproducing sound; the degree of fidelity with which this is done.

1908 *Sears, Roebuck Catal.* 195/2 The Type FH Harvard Disc Talking Machine... Perfectly uniform speed, essential to perfect reproduction, is obtained. **1924** *Radio Times* 19 Dec. 619/2 (Advt.), Sound reproduction that is very near perfection. **1946** *Penguin Music Mag.* Dec. 93 The standard of reproduction with which people appear to be content, even very musical people, is far from satisfactory. **1962** *Times* 5 July 15/4 It would be foolish to assume that sound reproduction has reached a stage of ultimate perfection;..research is constantly in progress both on ways of improving reproduction from discs and on alternative means of reproduction. **1978** *Gramophone* July 275/1 If headphones are what you want, you will not be disappointed. The reproduction of classical music or pop, piano, voices, orchestra—all are splendid.

2. a. Also, in more recent use, an article of furniture, etc., in a style reproduced from an earlier period. Also *attrib.*

1925 C. Camp in *Scribner's Mag.* Sept. 318 People who call your best pieces reproductions when you know that they are not. **1925** *Scribner's Mag.* Oct. 15 (Advt.), A reproduction whale-oil lamp which measures 18½ inches to the top of the bulb. **1964** *Times Rev. Industry* Mar. 37/3 Sales [of furniture] are growing (unfortunately, almost entirely of 'reproduction' styles, which do not help our attempts to project an image of a new Britain). **1975** M. Kenyon *Mr Big* xx. 190 Some reproduction Chippendale dining-chairs. **1977** *Whitaker's Almanack 1978* 887/1 The footwear industry is based on Florence, reproduction furniture at Cascini and Poggibonsi. **1981** J. B. Hilton *Playground of Death* vi. 71 There were one or two expensive reproduction pieces—a corner cupboard (for drinks) and a Jacobean footstool.

3. Special Combs.: **reproduction constant** or **factor** *Nuclear Physics* = *multiplication *constant* or *factor*; **reproduction proof** *Printing*, a printed proof for use as an original for further, photographic, reproduction.

1962 *Newnes Conc. Encycl. Nucl. Energy* 729/1 *Reproduction constant.* This is an alternative, and less-frequently used, name for the multiplication factor..of a reactor. **1945** H. D. Smyth *Gen. Acct. Devel. Atomic Energy Mil. Purposes* v. 35 The whole success or failure of the uranium project depended on the multiplication factor k, sometimes called the reproduction factor. **1947** *Science* 10 Jan. 28/1 Usually, *k* is called the 'reproduction factor' of the system. A self-sustaining chain reaction evidently is possible only when $k > 1$. **1952** Glasstone & Edlund *Elem. Nucl. Reactor Theory* iv. 79 A multiplication factor or reproduction factor, defined as the ratio of the number of neutrons of any one generation to the number of corresponding neutrons of the immediately preceding generation. **1948** R. R. Karch *Graphic Arts Procedures* 370/1 (Index), Reproduction proofs. **1949** Melcher & Larrick *Printing & Promotion Handbk.* 246/2 Reproduction proofs must be perfect—if anything, more perfect than would be acceptable in finished copy, since every slightest flaw will be duplicated in the whole run. **1967** Reproduction proof [see *repro 1].

reproductive (rīprŏdʌ·ktiv), *sb. Zool.* [f. the adj.] A reproductive insect.

1934 C. A. Kofoid *Termites* i. 8 The alates become functional reproductives only after they leave the parent colony and, in isolation, start a new one. **1971** E. O. Wilson *Insect Societies* (1972) xix. 370/2 Some of the workers are inseminated and serve as the usual reproductives. **1977** *Nature* 2 June 395/3 At a critical time..approximately one generation time before the season ends, the entire effort is thrown into producing reproductives (queens and males).

reproductory, *a.* For *rare*⁻⁰ in Dict. read *rare* and add example.

1962 *Sci. Survey* XIX. 289 Reproductory behaviour and its approximate synchronisation depend to a large extent on the action on the central nervous mechanisms of hormones.

repro·file, *v.* [Re- 5 c.] *trans.* To give a new profile to; to reface. So **repro·filed** *ppl. a.*; **repro·filing** *vbl. sb.*

1963 *Times* 24 May (London Underground suppl.) p. xii/5 Wheels where the steel 'tyres' are worn away may be changed for 're-profiling'. *Ibid.*, A worn pair of wheels is readily replaced by a newly reprofiled (or refaced) pair. **1964** J. Summerson *Classical Lang. Archit.* iv. 29 The Mannerist idiom of Vignola whose famous cornice—reprofiled—you will at once recognize.

repro·gram, repro·gramme, *v.* [Re- 5 a.] **1.** *trans.* To program differently; to supply with a new program. Also *transf.* and *fig.*

1963 [implied in reprogramming *vbl. sb.* below]. **1964** *Ann. Reg. 1963* 390 If any alteration had to be made to the map this could be done simply by re-programming the tape instead of having to redraw the map. **1972** *Listener* 6 July 13/2 A small clockwork mouse..[has] been very fully wound up and pointed in a certain direction, but there doesn't seem to be any means of re-programming it on the way. **1973** *Publishers Weekly* 18 June 9 (Advt.), Shows the hitherto unsuccessful dieter how to 'reprogram' himself for lifetime slimness. **1974** *News & Courier* (Charleston, S. Carolina) 21 Apr. C-10/2 The birds were taken from their nests as babies and were raised as pets. Ben David now is trying to reprogram the birds to fly free and kill to eat and live. **1977** D. Ramsay *You can't call it Murder* i. 63 He resumed..questions. Bam, bam, bam. As if he were reprogramming a computer.

2. To allocate to different spending programmes.

1971 *Hearings Comm. Armed Services, U.S. Senate 92nd Congress 1st Sess.* (S. 939) III. 2093 Fiscal year 1967 Navy funds were reprogramed to match maritime administration funds for program initiation.

So **re·programming** *vbl. sb.*; also **repro·grammable** *a.*

1963 *Rep. Comm. Inquiry Decimal Currency* xiv. 138 in *Parl. Papers 1962–3* (Cmnd. 2145) XI. 195 Ancillary machine costs:..re-programming of computers. **1971** J. Z. Young *Introd. Study Man* xxii. 302 This reprogramming of the read-out of the DNA may serve to bring into play a new complex of enzyme systems. **1974** *Jrnl. Politics* XXXVI. 77 One instrument for executive spending flexibility is 'reprogramming' of funds within an appropriation. **1978** *Sci. Amer.* Feb. 72/2 Since the machines are reprogrammable, it is feasible to apply them to the assembly of products manufactured in families of models.

reprographer (rīprǫ·grăfəɹ). [f. as next + -er¹, as *photographer*, etc.] One who makes facsimile copies of documents.

1967 *Britannica Bk. of Year* 804/1 *Reprography*,..facsimile reproduction (as by photocopying) of graphic matter (as documents); *reprographer*,..*reprographic, adj.* **1969** S. Cuadra *Ann. Rev. Information Sci.* IV. vi. 178 Old-style reprographers not only mixed their own 'soup', but they also made their own work schedules. **1972** *Libr. Resources* Spring 150 The education of library reprographers is a permanent concern.

reprographic (rīprŏgræ·fik), *a.* and *sb.* [f. as next + -ic: see -graphic.] **A.** *adj.* Of or pertaining to reprography.

1961 *Engineering* 17 Nov. 633 Development to improve the efficiency of reprographic services. **1963** *Ibid.* 3 May 606 'Reprographic' has become recognized as the omnibus adjective for the many processes and methods now used to produce copies of documents of all kinds. **1966** *Times* 11 May 17/3 Admel International, of Weybridge, Surrey, a leading maker of what is called reprographic equipment,..is to be sold to an American group. **1973** *Nature* 31 Aug. 535/3 As reprographic technology progresses, interlibrary affiliates grow, and..publishers are likely to find the economic realities of their business approaching prohibitiveness. **1977** *Times* 24 Oct. 3/3 Its [sc. HMSO's] empire includes printing presses, binderies, warehouses, bookshops, reprographic units and laboratories.

B. *sb. pl.* (const. as *sing.*) = next.

1967 *Financial Times* 15 Mar. 21/2 The expansion in reprographics has had to call on many branches of technology to cope with the demand for better machines. **1969** *Daily Tel.* (Colour Suppl.) 31 Jan. 13/1 (Advt.), Reprographics: a word coined little over five years ago to describe the means used to reproduce the enormous volumes of drawn, typed and written material which has to be circulated in industry, commerce and government today. **1975** B. J. Perry in Barr & Line *Ess. Information & Libraries* 119 Information science research is a complex area that requires the application of skills from many different disciplines, including..reprographics. **1980** *Financial Rev.* 5 Mar. 21/4 A 30 per cent drop in world demand for reprographics products.

reprography (rīprǫ·grăfi). [ad. G. *reprographie*, f. *repro-duktion* Reproduction + *photo-graphie* Photography.] The branch of technology concerned with the copying and reproduction of documentary and graphic material.

1961 *Bibliogr., Documentation, Terminol.* I. 105 The term 'reprography' is today used to cover all processes for the photographic reproduction and multicopying of documents. **1962** *Archives* V. 235 *Reprography.* This new and convenient, though not very likeable, word is a collective term for the processes of facsimile reproduction of documents, whether by means of photographic, electronic or other methods. **1963** *Special Libraries* Dec. 646/1 The term 'reprography', coined about ten years ago, is intended to describe all methods of facsimile reproduction of documentary materials with the exception of conventional printing. Although little used in the United States, the term is achieving considerable popularity in Europe. **1967** *Financial Times* 15 Mar. 21/1 The almost unpronounceable word 'reprography' is in. It covers a field with ill-defined boundaries, edging at its simplest into pencil-making and at its most sophisticated into desk-top photosetting. **1977** *Author* Summer 67 Cassettes and other forms of electronic storage are examples of what is becoming known as reprography.

reptation. (Later example.)

1947 R. L. G. Irving tr. *Casteret's My Caves* iii. 54 Reptation involves an attitude so seldom adopted by the human race that it appears to be the fate allotted to a few inferior creatures.

reptilarium: see *reptiliary.

reptile, *sb.*¹ Add: The pronunc. (re·ptail) is now standard in the U.K. **2.** (Later example.)

1974 Wodehouse *Aunts aren't Gentlemen* v. 45 She spoke as follows, her manner and diction similar to those of a sergeant-major addressing recruits. 'What's the matter with you, you poor reptile?'

3. *reptile man* (later example).

1975 J. McClure *Snake* v. 64 The reptile man at the museum..said..that method of preservation had been abandoned.

reptile (re·ptail), *sb.*² *Math.* Also **rep-tile.** [f. *rep(licating) tile* with a pun on Reptile *sb.*] A two-dimensional figure of which two or more can be grouped together to form a larger figure having the same shape.

1966 S. W. Golomb in *Jrnl. Combinatorial Theory* I. 281 Certain polyominoes are 'rep-tiles', i.e., they can be used to tile enlarged scale models of themselves. **1972** C. S. Ogilvy *Tomorrow's Math* (ed. 2) iv. 73 The following are unproven conjectures... Every rep-tile also tiles a parallelogram... A rep-tile with five or more sides cannot be convex. **1977** *Sci. Amer.* Jan. 110/3 Another kind of nonperiodic tiling is obtained by tiles that group together to form larger replicas of themselves. Solomon W. Golomb calls them 'reptiles'.

reptiliary (repti·liări). [f. Reptile *sb.* + -ary¹, after *aviary*.] A building or enclosure in which reptiles are kept, as for display. Also **reptila·rium** [-arium]; **repti·llery** [-ery].

1928 *Daily Tel.* 11 Sept. 14 This is the new open-air reptiliary. **1938** *Times* 30 Apr. 9/4 Before long a reptilarium would be installed at Whipsnade. **1976** *Southern Even. Echo* (Southampton) 10 Nov. 1/4 A reptillery has been in existence at Holiday Hill, near Lyndhurst, for about 15 years. **1978** *Autoworld* No. 68. 23/1 You can visit a wild goat park.., a butterfly reserve,..a reptiliary in the New Forest.

reptilious, *a.* For *rare*⁻¹ in Dict. read *rare* and add further example.

1936 L. C. Douglas *White Banners* xv. 317 Considering with what reptilious patience she had waited.

reptillery: see *reptiliary.

republic, *sb.* (and *a.*) Add: **2. a.** Now also applied loosely to any state which claims this designation.

1947 E. Waugh *Scott-King's Mod. Europe* 4 Out of it [*sc.* history] emerged the present republic of Neutralia, a typical modern state, governed by a single party, acclaiming a dominant Marshal, supporting a vast ill-paid bureaucracy whose work is tempered and humanised by corruption. **1976** *Whitaker's Almanack 1977* 829/2 Republic of Burundi... Burundi became independent as a Constitutional monarchy but this was overthrown on November 28, 1966. The Constitution and Parliament were also abolished. The President rules through a Cabinet of Ministers and the UPRONA party apparatus. Burundi is a one-Party State.

republican, *a.* and *sb.* Add: **A.** *adj.* **3. a.** (Earlier and later examples in modern sense.)

The Republican party formed in 1854 is now a predominantly conservative party, favouring agricultural, commercial, and financial interests and a limited central government (see M. M. Mathews *Dict. Americanisms*, 1951, and Sperber & Trittschuh *Amer. Polit. Terms*, 1962, for further details of the development of this sense).

1854 A. E. Bovay *Let.* 26 Feb. in F. Curtis *Republican Party* (1904) I. vi. 177 Urge them..to band together under the name I suggested to you at Lovejoy's Hotel in 1852. I mean the name 'Republican'. **1856** *Porter's Spirit of Times* 4 Oct. 71/1 New Jersey..[was] discovered by the late Republican Convention, in their explorations for a candidate for Vice President. **1905** *Baltimore Amer.* 7 Mar. 4/1 Republican Senators..cannot find desks on the Republican side, as there are more Republicans in the Senate than there are desks for Republican Senators. **1976** *Columbus* (Montana) *News* 1 July 6/3 Charles Eckels attended the Republican Convention in Helena last week as chairman from this District.

b. Used as a distinguishing epithet by political parties outside U.S.

1958 *Listener* 27 Nov. 865/1 One [new French party] is the Republican Centre. **1975** *Financial Times* 31 Oct. 6/8 Mr Ecevit, chairman of the main Opposition, the Republican People's Party (RPP). **1980** S. J. Burki *Pakistan under Bhutto* ii. 22 The formation of the landlord-dominated Republican Party.

B. *sb.* **2. b.** *transf.* Cf. Republic *sb.* 3 b.

1816 I. D'Israeli *Inquiry Lit. & Polit. Char. James I* 3 His other brothers, the republicans of literature, want a heart to admire the man.

republicanization. (Later example.)

1970 R. A. H. Robinson *Orig. Franco's Spain* i. 53 'Republicanisation' of local councils took place on a large

scale; possibly this was as much the result of voluntary changes of allegiance..as of arbitrary measures.

republicanly, *adv.* For *rare*⁻¹ in Dict. read *rare* and add later example.

1837 J. S. Mill *Let.* June in *Works* (1963) XII. 339 Twelve pages on the *immediately present* state of French politics.. written republicanly by Thibaudeau.

Republicrat (ripʌ·blikræt). *U.S. politics.* Also **Republocrat.** [Blend of Republican *sb.* 3 + Democrat.] A member of a political faction that includes both Republicans and Democrats. Also, a conservative Democrat with Republican sympathies.

1940 *Better Eng.* Oct. 55/1 A republocrat, as 'Time' uses it, is a republican or a democrat who will have anyone but Mr. Hoover. **1944** *N.Y. Post* 14 Apr. 19/4 The 'Republicrats', meaning reactionary Democrats in league with the Republicans, want to discredit F.D.R. **1946** *Time* 22 Apr. 21/1 Would you be good enough to give us the names of the 'republocrats' in the House and Senate who have organized for the purpose of defeating President Truman's legislative program? **1949** *Southern Farmer* July 3 All we ask is that the Hindu philosophers tell us how to make the Southern Republicans climb up a rope and disappear.

Hence **Republicra·tic** *a.*

1944 *New Republic* 19 June 801/1 The Republicratic gang engineering the supposed Southern 'revolt' against the New Deal appears well satisfied with the results of the Mississippi convention.

repudiant (ripiū·diănt), *a. rare.* [See Repudiate *v.* and -ant¹.] Characterized by repudiation; = Repudiative *a.*

1954 W. Faulkner *Fable* 144 The owner..still..invincibly repudiant.

repudiationist. Add: Also in *gen.* use. Also *attrib.* or as *adj.*

1930 *Times* 11 Nov. 15/4 That is the most effective way in which he can support the Government and repudiate the repudiationists. **1931** *Star* 8 May 16/1 The small Lang 'Repudiationist' rump came to Mr. Scullin's assistance. **1932** *New Statesman* 16 Jan. 53/2 Germany is repudiationist to a man.

repulsion. Add: **3. c.** *Genetics.* The condition of two genes, in an individual heterozygous at each of two linked loci, when the dominant allele of each occurs on the same chromosome as the recessive allele of the other. Opp. *coupling *vbl. sb.* 6 e.

1908 *Science* 15 May 786/1 When in F_1 the two dominants, femaleness and the *grossulariata* factor coexist, there is a repulsion between them, such that each gamete takes one or other of these two factors, not both. **1911** *Proc. R. Soc.* B. LXXXIV. 3 We were therefore led to recognize—A. A system of partial coupling where the two factors are generally associated. B. A system of complete repulsion (or as we have sometimes called it 'spurious allelomorphism') under which two factors are never associated in the same gamete. **1926, 1970** [see *coupling *vbl. sb.* 6 e]. **1977** Mather & Jinks *Introd. Biometrical Genetics* viii. 203 The associated distribution will lead to coupling linkage in F_1 and so may be denoted by K, while the dispersed distribution will give repulsion linkage and so may be denoted by R.

5. *attrib.* **repulsion motor** *Electr.*, an a.c. commutator motor for single-phase operation in which current is supplied to the stator only, the armature being short-circuited through the brushes and its current induced from the stator winding.

1904 E. R. Raymond *Alternating Current Engin.* iv. 187 The repulsion motor..can be used as an electric brake. **1940** *Whitaker's Electr. Engineer's Pocket-Bk.* (ed. 4) 229 The repulsion motor is the simplest and commonest of all a.c. commutator motors. **1972** C. C. Barnes et al. *Electrics 72/73* 1 A repulsion motor has a single-field winding and a wound rotor with a centrifugal device to short-circuit the commutator.

reputational (repiūtē·iʃənăl), *a.* [f. Reputation + -al.] Of or pertaining to reputation.

1936 in *Shaw & Mrs. Campbell* (1952) 117 This alarms me to see people blundering away..the reputational chances of their lifetime. **1967** *Listener* 19 Jan. 103/3 [T. F.] Powys is now at the nadir of the twenty-year reputational trough that follows death.

reputed, *ppl. a.* **1.** Delete † *Obs. rare* in Dict. and add: Now used after an adverb, as *internationally reputed*, etc.

Not often found in good sources.

1928 *Daily Express* 8 Aug. 15/3 Dr. Hanslick, the universally reputed professor of musical history. **1969** *Daily Tel.* 4 Aug. 17/6 An internationally reputed geologist.

2. b. *reputed pint, quart*, etc.: (see quot. 1904). Also, the amount of liquid contained by such a measure.

1904 Makins & Lambert *Licensed Victuallers' Handbk.* (rev. ed.) xiii. 216 Reputed quart means a bottle containing the sixth of a gallon, and a reputed pint a bottle containing the twelfth of a gallon. An Imperial Pint is an eighth part of a gallon. **1935** [see *Methuselah 2]. **1959** *Gloss. Terms Packaging* (B.S.I.) 29 The normal bottle of wine should contain ⅙ of a gallon..and is known as a 'reputed quart'... Variants of the above are known as 'mock quarts'..and may be of less capacity.

request, *sb.*¹ Add: **I. 2. b.** *spec.* A letter, etc., asking for a particular record, song, etc., to be played on a radio programme, often accompanied by a personal message; a record played or a song, etc., sung, either over the radio or to a live audience in response to a request.

1928 *Radio Times* 12 Oct. 79/1 The B.B.C. can never promise to comply with requests, for suitable opportunities may not arise for weeks or even months. **1949** *Ibid.* 15 July 24/1 Listeners' requests played by Sandy Macpherson at the BBC theatre organ. **1966** *Listener* 4 Aug. 181/3 Judging by the requests, the classical and pre-classical composers are out. **1977** *Zigzag* Mar. 8/2 My sister..saw Roy Eldridge..playing requests in a bar.

3. b. (Further example.)

1976 *Daily Tel.* 20 July 2/7 A list should be provided on request.

II. 11. *request item, night, number, programme, session, week*; **requestman** *Naut.*, a seaman who makes a written request to an officer; also *pl.*, applied to the occasion appointed for the presentation of such requests; **request stop**, a stop at which a bus will halt only on request from a passenger or intending passenger.

1943 *Radio Times* 18 Sept. 17/1 Special Request Items. **1974** 'J. Black' *Bigoted Apprentice* i. iii. 19 The pluggers kept the initiative by inventing the request item. **1916** 'Taffrail' *Pincher Martin* vii. 116 'Request-men an' defaulters—'shun!' bawled the master-at-arms. **1951** H. Hastings *Sea Gulls over Sorrento* i. in J. C. Trewin *Plays of Year* IV. 73 You'd better write out a request... You know the routine... Requestmen in at 0900 tomorrow morning and not before! **1961** *Times* 17 Feb. 14/6 Shortly after breakfast C.P.O. Coleshill knocked on my cabin door and entered with his list of requestmen and defaulters. **1923** *Radio Times* 28 Sept. 12/1 Why is it apparently not thought advisable to repeat the 'Request Nights', which are so popular? **1971** D. Halliday *Dolly & Doctor Bird* iii. 37 The small coloured orchestra..suddenly broke..into a request number for young tango. **1889** *Cent. Dict.*, Request-program.. A program made up of the numbers the performance of which has been requested by the audience. **1955** *Radio Times* 22 Apr. 10/2 A request programme for 'troopies' in the operational areas aim to keep spirits high. **1959** M. Shadbolt *New Zealanders* 76 The crackling voice of the request session announcer. **1943** G. Greene *Ministry of Fear* iv. i. 223 Buses still quickly past the Request stops. **1960** *Daily Express* 25 Nov. 1/2 (caption) Five years' continuous service without once stopping at a request stop. **1973** *Times* 15 Oct. 17/5 Hiding behind another bus at a request stop so as to avoid seeing the uplifted hand. **1928** *Radio Times* 7 Dec. 649/2 (heading) Children's Hour request week... The Fourth Request Week will begin on January 7, 1929.

requester. (Later example.)

1973 *Nature* 13 Apr. 465/2 How many requesters, before writing, saw my paper?

‖**requeté** (rekete·). *Hist.* Also **Requetè.** [Sp.] A member of a Carlist militia that took the Nationalist side during the Spanish Civil War of 1936–39.

1936 *Times* 26 Aug. 12/4 The troops engaged on the insurgent side have consisted of *requetes*, Navarrese volunteers devout to a man. **1938** *Times Lit. Suppl.* 27 Aug. 558/3 The first, longer and better of the two stories.. describes how a Carlist *requeté* in North Spain deserted his side to secure the reprieve of his father. **1944** Wyndham Lewis *Let.* 1 Jan. (1963) 374 I was glad to learn from Augustus that he had exchanged his requeté uniform for that of the Home Guard. **1957** P. Kemp *Mine were of Trouble* ii. 21 The Requeté movement drew its main strength from the Basque provinces, especially from Navarre. **1979** D. Robinson *Eldorado Network* xvii. 129 The Requetés..were a sort of Basque militia... Fought for Franco.

requirant. (Later example.)

1821 in G. E. Cory *Rise of S. Afr.* (1910) I. vii. 213 The R.C. Pastor Scholtz finds himself obliged to declare that he has not any ground for further action against the requirant.

required, *ppl. a.* Add: **b.** *required reading*, literature which one is required to read for an educational course or which must be read in order to gain an understanding of some subject.

1921 H. J. Laski in *Holmes-Laski Lett.* (1953) I. 370 They are quite intolerable—pushing little professors full of pedantic details, nosing into the dull routine of unimportant matters, pushing their little quack remedies, interested in getting the wrong books on to lists of required reading. **1930** *Publishers' Weekly* 15 Mar. 1547/1 Some of the important courses in our colleges and universities cannot be taught successfully..because of the lack of a sufficient number of books for required reading. **1954** *N.Y. Times Bk. Rev.* 31 Jan. 1/1 Here is a book that should be required reading for Democrats in 1954. **1962** *Listener* 17 May 873/2 They genuinely stimulate thought and thus become required reading for social critics, amateur and professional. **1976** J. Wainwright *Walther P.38* 62 The usual sort of stuff which..was 'required reading' for any moderate education.

Hence **requi·redness**, the fact or quality of being required.

1938 W. Köhler *Place of Value in World of Facts* ii. 35 At the bottom of all human activities are 'values', the conviction that some things 'ought to be' and others not. Science, however, with its immense interest in mere facts seems to lack all understanding of such 'requiredness'. **1946** C. Morris *Signs, Lang. & Behavior* iii. 63 A command such as 'Come here!' may signify with high consistency the requiredness of the response which it prescribes. **1977** A. Ecclestone *Staircase for Silence* iv. 78 Péguy spoke often of it as invincible anxiety; while Abraham Heschel called it a sense of requiredness, in the language of the Bible: what is required of me?

requisite, *a.* and *sb.* Add: **A.** *adj.* **b.** *requisite variety*, the variety necessary in a system for it to be able to control another system in which there is variety.

1956 W. R. Ashby *Introd. Cybernetics* xi. 207 This is the law of Requisite Variety. To put it more picturesquely: only variety in R can force down the variety due to D; only variety can destroy variety. **1966** S. Beer *Decision & Control* xii. 281 But however many interacting sub-systems of preys and predators, big fleas and little fleas, are involved, it is none the less evident that the balance of animal populations would be grossly upset very rapidly unless the law of requisite variety held in general throughout nature. **1975** R. M. Glorioso *Engin. Cybernetics* v. 72 As a minimum requirement, the number of distinguishable states of the system must equal the number of possible states of the environment, although the proper states of the system must be available as well. Thus, to reduce the variety in the environment, the system must be capable of achieving the 'requisite variety'.

requisitely, *adv.* Delete † *Obs.* and add later example.

1976 *Daily Tel.* (Colour Suppl.) 13 Feb. 29/2 Neither in England nor elsewhere could Jack Scala expect ever to rise to the top earning bracket until he is over 30 and projects the requisitely aggressive and reliable image.

requisition, *sb.* Add: **5.** *attrib.*, as *requisition form, note, paper, slip.*

1929 T. H. Burnham *Engineering Econ.* xv. 206 Tools may only be issued against a requisition form signed by a duly authorized official. **1926** G. Frankau *Peter Jackson* (1925) 226 A faked requisition note. **1974** P. Gore-Booth *With Great Truth & Respect* 150 We..were alarmed by the appearance of requisition notices on the doors of adjoining houses. **1931** *Daily Colonist* (Victoria, B.C.) 23 Apr. 1/7 Burns said he would start tonight for Los Angeles. Governor Marshall having honored requisition papers. **1936** L. Bemelmans *Life Class* III. ii. 221 A stack of requisition slips, for the carpentry department, for the painters,..for the plumber.

requisitioner (rekwizi·ʃənəɹ). [f. Requisition *v.* + -er¹.] = Requisitionist.

1968 M. Guybon tr. *Solzhenitsyn's First Circle* lxii. 392 After Christmas they were going to send you out all for it the cities, and the requisitioners would starve the peasants.

requisitionize (rekwizi·ʃənaiz), *v. rare.* [f. Requisition *sb.* + -ize.] *trans.* To request or require (one to do something) by written requisition.

1845 J. Keble *Let.* 6 Feb. in G. Battiscombe *John Keble* (1963) xiii. 254 If it is thought well to requisitionize the Proctors to veto the affair.. you may put my name down to it.

requotation. [Re- 5 a.] A new or revised quotation (of the price of a share).

1964 *Economist* 29 Aug. 838/1 A requotation for their shares. **1973** *Daily Tel.* 14 Feb. 1/6 Pergamon Press..is back in profit and looking for an early requotation on the Stock Exchange.

requote, *v.* Add: Also, to quote a new price for (a share).

1967 *Economist* 27 May 946/4 The shares were requoted after the financial troubles.

re-ra·diate, *v.* Also **reradiate.** [Re- 5 a.] *trans.* To radiate again (what has been absorbed or received). Also *absol.*

1923 F. W. Raynes *Heating Systems* xv. 184 The arrangement and grouping of the heating surfaces affect the radiant heat transmitted, in that a more or less percentage is rendered ineffective through being simply re-radiated from surface to surface. **1928** *Daily Tel.* 10 Jan. 16 A short-wave receiver of a sensitive type which will not re-radiate, and so interfere with neighbouring receivers. **1942** J. D. Stranathan *Particles Mod. Physics* vi. 255 Electrons of the scattering material are supposed to be set into vibration by the varying electric field. These, in turn, re-radiate energy in all directions. This re-radiated energy represents the scattered X-rays. **1952** *Archit. Rev.* CXI. 18/2 Any type of screen will itself be heated by the sunshine falling upon it, and will re-radiate this heat into the building. **1976** *Sci. Amer.* Sept. 75/3 The earth intercepts a vast amount of solar energy... About 60 percent is reflected without interacting further and most of the remainder is absorbed by the atmosphere or by oceans and landmasses and is promptly re-radiated.

Hence **re-ra·diated** *ppl. a.*

1942 [see above]. **1974** Harvey & Bohlman *Stereo F.M. Radio Handbk.* vii. 155 This reradiated signal is picked up by the dipole. **1978** R. V. Jones *Most Secret War* xxi. 175 We could in principle re-radiate this already re-radiated signal back to the aircraft.

re:-radia·tion. [Re- 5 a.] The action of re-radiating; also *concr.*

1881 [in Dict. s.v. Re- 5 a]. **1915** W. H. Eccles *Wireless Telegr.* 132 The Joulean wastage in the surrounding ground

is usually so great that it is well to limit the re-radiation, and to seek rather to transfer the collected energy somewhat rapidly to the detector. **1924** *Times* 31 Jan. 7/4 The first intelligible programmes began to come through in time for successful re-radiation here before the close of the old year. *Ibid.*, A large number of letters have been received..from listeners..commenting on the surprisingly good quality of the re-radiation. **1976** *Nature* 2 Sept. 15/2 Solar radiation is absorbed by changing energy levels of atoms or molecules, producing excitation. Fluorescence or reradiation then follows.

rerail, *v.* Add: (Examples of the simple verb.) Also *fig.*

1895 *Funk's Stand. Dict.*, *Re-rail*.., to put on the track again; to cause to take the rails again, as derailed rolling-stock. **1914** W. de Morgan *Ghost meets Ghost* xi. iii. 435 'And where else did you go?' said the Earl, to rerail the conversation. **1967** G. F. Fiennes *I tried to run a Railway* ii. 6, I learned how to re-rail wagons and engines without sending for the crane.

reraise, *v.* Add: (Later example.) Hence **rerai·sing** *vbl. sb.*

1937 *Jrnl. Eng. & Germanic Philol.* XXXVI. 2 The raising (resp. re-raising) of a (< a) > æ. **1937** *Times* 31 Dec. 12/1 The object of making it impossible for such points of friction to be reraised in future.

re-read, *sb.* [Re- 5 a.] An instance of reading again.

1973 M. Amis *Rachel Papers* 185 Finally, I belaboured one of Gerard Manley Hopkins's sleazier lyrics, implying (a last-minute reread made clear) that it was high time we burned all extant editions of the little fag's poetry. **1977** T. Allbeury *Man with President's Mind* ii. 13 The report got a routine re-read.

re-rea·dable, *a.* [Re- 5 a.] Capable of being re-read; capable of being read about for a second time with pleasure.

1948 F. R. Leavis *Great Tradition* i. 9 His books..have a permanent life as light reading—indefinitely re-readable. **1967** *Punch* 22 Feb. 283/3 He makes the Paris of the 'twenties, Joyce and Fitzgerald re-readable, something of a feat.

re-reader. [Re- 5 a.] One who re-reads.

1957 'G. Edwards' *Talking of Books* 190 It can be argued that all these are specialists or exceptional classes of re-readers. **1965** K. Tillotson *Mid-Victorian Stud.* xxiv. 309 Dickens..is perhaps chiefly a re-reader—going back also to that shelf of boyhood favourites lovingly enumerated in *David Copperfield*.

re-reco·rd, *v.* Also **rerecord.** [Re- 5 a.] *trans.* To record again. Also *absol.*

1930 *Proc. IRE* XVIII. 1335 Before sound film can be recorded a matched print must be made. **1932** *Times Educ. Suppl.* 17 Dec. p. iii/2 The records made by the great tenor..have been re-recorded by the Company. **1955** L. Feather *Encycl. Jazz* 25 He..restored him to the music world by rerecording *Honky Tonk Train Blues*. **1964** *Oceanogr. & Marine Biol.* II. 429 The background characteristics of low speed recordings may be somewhat improved by re-recording at higher speeds. **1977** *New Yorker* 29 Aug. 67/1 The music of long-dead musicians long ago rerecorded for in-flight listening. **1978** W. F. Buckley *Stained Glass* xi. 110 It was recorded and rerecorded that in her youth she had actually studied under Clara Schumann.

Hence **re-recorded** *ppl. a.*, **re-recording** *vbl. sb.*

1930 *Proc. IRE* XVIII. 1333 Once it was discovered that re-recording was practical a great many other reasons for its use were discovered. **1962** A. Nisbett *Technique Sound Studio* v. 104 It is now possible to compare the original and re-recorded signals directly. **1975** P. G. Winslow *Death of Angel* ix. 180 It seems to be a re-recording of another tape. **1977** *Gramophone* Aug. 357/2 The incredibly high copying speeds needed to make the exercise economically viable (32 or 64 times normal speed) put severe limits on the re-recorded sound quality.

re-ree·l, *v.* [Re- 5 a.] *trans.* To wind again on to a reel, or from one reel to another. Hence **re-ree·ling** *vbl. sb.*; **re-ree·ler** (see quot. 1964).

1906 R. W. Sindall *Paper Technol.* vi. 75 (caption) Double-Drum Reeler. The reels from the paper-machine are re-reeled, slit, and finished off on this machine. **1929** Clapperton & Henderson *Mod. Paper-Making* vii. 246 (caption) Four-drum winder for re-reeling newsprint. **1937** E. J. Labarre *Dict. Paper* 211/1 *Re-reeling machine*, the apparatus on which the web of paper is re-reeled after passing through the operations of tub-sizing..and drying. **1964** *Gloss. Letterpress Rotary Printing Terms* (B.S.I.) 17 *Re-reeler*, an auxiliary unit to rewind the web or webs for subsequent operations. **1967** E. Chambers *Photolitho-Offset* xv. 237 The paper can be either cut into sheets or re-reeled for finishing elsewhere.

re-refine, *v.* Add: (Later examples.) Hence **re-refi·ned** *ppl. a.*; **re-refi·ner**; **re-refi·ning** *vbl. sb.*

1971 *Daily Colonist* (Victoria, B.C.) 10 Feb. 21/5 The re-refined oil sells for 26 cents a quart. **1973** *Sci. Amer.* Feb. 48/3 Today only 100 to 150 million gallons of the 1,100 million gallons of oil annually drained from the nation's crankcases are re-refined. **1975** *Conservation of Energy* (Shell Internat. Petroleum Co.) 17 Shell companies are examining the most effective means of collecting and re-refining used lubricating oils, including re-refining. *Ibid.*, It is possible that 20 per cent of the world demand for lubricants will be met by re-refined oils in the early 1980s. **1977** *Laconia* (New Hampshire) *Evening Citizen* 21 July 5/5 Your car's dirty motor oil might soon be 'refined' through an environmentally safe process which will help conserve..crude oil. *Ibid.*, A lot of refiners' raw material is being burned as fuel. **1977** *Lubricants Business* (Shell Internat. Petroleum Co.) 8 Re-refining is likely to be primarily attractive in a number of developing countries where savings in foreign exchange could be an important factor.

re-relea·se, *v.* [Re- 5 a.] *trans.* To release (a film, record, etc.) again. So **re-relea·se** *sb.*; **re-relea·sable** *a.*; **re-relea·sed** *ppl. a.*

1948 *Daily Mail* 7 Feb. 2/5 'Mrs. Miniver'..turned up..on 're-release', and lasted only seven days. **1968** *Listener* 26 Sept. 416/2 The Disney product had to be aimed at all ages, it had to be 'endlessly re-releasable'. **1975** *N.Y. Times* 4 Mar. 38/1 The new D. W. Griffith Theater, which has re-released Cocteau's 'Les Enfants Terribles'. **1977** *It* May 28/1 There are the re-released 'Jimmy Weatherspoon and Ben Webster' 'Stitt Palysbird'. **1977** *Guardian Weekly* 28 Aug. 20/1 There are currently 38 Presley albums available in this country..and, with re-releases, he has had 117 singles on sale here.

re-representation. (Later example.)

1932 *Brain* LV. 459 These impulses..may be..passed on to the main part of the lateral nucleus by short internuclear connections where they acquire a re-representation.

re-reveal, *v.* (Earlier example.)

a **1631** Donne *Poems* (1635) 367 Davids Successors, in holy zeale, In formes of joy and art doe re-reveale To us so sweetly and sincerely too, That I must not rejoyce as I would doe.

re-ri·de, *v.* [Re- 5 a.] *trans.* To ride (a route, contest, etc.) again. Hence **re·ride** *sb.*

1884 *Cyclists' Touring Club Monthly Gaz.* Nov. 335/2 Favourable stretches of high-way are ridden and re-ridden with wearying iteration upon safety bicycles, which are 'safe' only in name. **1903** *Daily Chron.* 28 Sept. 8/7 The tie will have to be re-ridden. **1976** *Billings* (Montana) *Gaz.* 17 June 1-H/3 Judges gave one cowboy a fourth reride.

rero·be, *v.* [Re- 5 c.] *trans.* To dress in a fresh robe; to clothe in a robe again. So **rero·bing** *vbl. sb.*

1849 C. Brontë *Shirley* II. v. 126 She immediately took the book from her and, with her own hands, commenced the business of disrobing and rerobing her. **1934** Dylan Thomas *18 Poems* 32 Second Rise of the skeleton and Re-robing of the naked ghost.

re-roll, *v.* (Later examples.) Hence **re-ro·ller**, one who or that which rolls iron or steel again; **re-ro·lled** *ppl. a.*

1931 *Times* 16 Mar. 19/7 The re-rollers are feeling the pinch. **1955** *Times* 14 July 15/3 Difficulties were experienced due to the national shortage of re-rolled steel products. **1963** *Punch* 13 Feb. 242/1 It is one of the largest 're-rollers' in the industry. **1967** Karch & Buber *Offset Processes* viii. 357 Reroll the screens and reassemble them on the pump. **1976** *Woman's Day* (N.Y.) Nov. p. H/1 On lightly floured surface roll out each half 3/16 inch thick. Cut in 3-inch squares (reroll scraps).

re-rou·te, *v.* [Re- 5 a.] *trans.* To set upon a new route; to re-direct. Also *fig.* So **re-rou·t(e)ing** *vbl. sb.*

1929 *Daily Express* 5 Jan. 2/1 Post Office officials..said that it was the first move in a general 're-routing' of mails. **1936** J. Steinbeck *In Dubious Battle* xiii. 203 The police stood, re-routing the traffic. **1961** 'E. Lathen' *Banking on Death* xi. 94 Arthur was on a Chicago-Boston flight which was rerouted to Washington because of the blizzard. **1971** *Daily Tel.* 21 Oct. 14/5 A council is to spend £250 on re-routeing a drainpipe to save an asparagus bed. **1978** S. Herzel in P. Moore *Man, Woman, & Priesthood* viii. 108 Any real understanding of sexuality and spirituality must be re-routed through a new awareness of the metaphoric and personal character of language itself.

re-ro·w, *v.* [Re- 5 a.] *trans.* To row (a race) again. Also *absol.* Hence **re-·row** *sb.*

1892 *Pall Mall Gaz.* 13 June 3/2 The C.U.B.C. had the race re-rowed this morning. **1901** *Daily Chron.* 19 July 8/3 Kingston and London were ordered to re-row. *Ibid.* 12 Aug. 8/2 After a foul and a re-row the Senior Pairs fell to H. U. Gould and C. M. Steele. **1979** *Oxford Mail* 31 May 16/8 Division II (6–10): Subject to rerow 11.45 today.

re-ru·bber, *v.* [Re- 5 c.] *trans.* To provide (a tyre) with a fresh covering of rubber. So **re-ru·bbering** *vbl. sb.*

1908 *Westm. Gaz.* 2 June 4/2 At the end of three months.. the first set are sent to be re-rubbered... Re-rubbering [costs] £30. **1923** *Daily Mail* 16 Feb. 5 (Advt.), Tyre re-rubbering and general Tyre repairs. **1973** *Times* 28 Apr. 4/2 A wornout tyre that has been rerubbered with a new tread.

re-run, *sb.* [f. the vb.] **1.** A repeat showing of a motion picture; also, the film itself; also *transf.*, of broadcast or printed material. Also *fig.*

1934 in Webster. **1955** *Times* 28 July 9/7 Film costs rather more than 'live' television, but it has a great advantage for re-run purposes. **1965** *Times Lit. Suppl.* 25 Nov. 1070/4 Re-runs of old silent films. **1968** *Guardian* 22 Mar. 10/6 *Around the World in 80 Days* starts a rerun in 70 millimetre at the Coliseum Cinerama Theatre. **1968** P. Marlowe *Hire me a Hearse* ix. 137 It didn't take very long for Hazard to run the spools through again. In the rerun some of it sounded repetitious. **1971** *Wall St. Jrnl.* 22 July 1/5 Liberty magazine, which was around from 1924 until 1950, makes a comeback with reruns of its old articles. **1978** G. Vidal *Kalki* vi. 151 On the wall opposite the window a row of TV monitors showed us..a rerun of *I love Lucy*. **1979** *Dædalus* Summer 15 The current quarrel..is just the latest rerun of that earlier script.

2. The repeated performance of a computation or computer program. Usu. *attrib.*

1946 J. B. Conant *Man. Operation Automatic Sequence Controlled Calculator* ii. 50 Starting, stopping and rerun instructions. **1958** [see *generator 4 c]. **1980** R. Longbottom *Computer Syst. Reliability* vii. 105 A further category of rerun time is incurred when it is found that previously created files have been lost... The time to restore files is again application dependent.

3. A repeated occurrence or attempt.

1976 H. Wilson *Governance of Britain* iii. 65 A tendency developed in the 1960s for a defeated minister almost automatically to seek for a re-run at Cabinet. **1977** *Daily Tel.* 17 Mar. 1/1 The Left-wing appeared to be in no mood to have a re-run of last year's rebellion, when the Government was defeated by 28 votes.

re-run, *v.* Add: (Further examples.) *spec.* to show (a motion picture) again (also *fig.*); to subject again to an experimental or computational procedure.

1962 'D. Shannon' *Extra Kill* vii. 103 There was a picture Fox made... Last week I saw it's being rerun at a neighborhood house. **1967** E. Chambers *Photolitho-Offset* XV. 239 The product cannot be re-run through the machine. **1971** J. B. Carroll et al. *Word Frequency Bk.* p. xxxvii, The discrepancy was caused by technical problems in making final corrections..in the computer tape... Discrepancies in individual subject categories were so small..that it was not considered worthwhile to rerun the distributions. **1973** *Sci. Amer.* Dec. 45/2 Our entire simulation of the Antennae (with $n = 350$) could be rerun in less than five minutes on any fast modern computer. **1976** M. Maguire *Scratchproof* viii. 115 Re-run it... Slow down the projection speed. **1976** E. Maclaren *Nature of Belief* ii. 16 You can't re-run a piece of history with different variables the way you can repeat a chemistry experiment without the chlorine. **1977** *Virology* LXXVIII. 207/2 The DNA isolated by this procedure was then rerun in a CsCl density gradient.

Hence **re-run** *ppl. a.*

1929 *Amer. Speech* IV. 386 The term *smoke* is comparatively new..and means an inferior grade of whitish, cloudy alcohol, usually *re-run alky* at that. **1968** A. Diment *Gt. Spy Race* vii. 94 The excuses running through their minds like re-run television shows.

res¹ (rēⁱz). Pl. **res.** [L., = thing.] **1.** orig. in *Law* (see quots. 1851, 1854); hence *gen.*, the condition of something; the matter in hand, the point at issue, the crux.

[**1684** G. Mackenzie *Institutions Law of Scotland* i. ii. 10, I do resolve, first, to lay down what concerns the Persons of whom the *Law* treats: *Secundo*, what concerns the things themselves treated of, such as rights, obligations &c. *Tertio*, the actions whereby these rights are pursued, which answers to the Civilians, *objecta juris, viz. Personae, res, & Actiones*.] **1851** P. Colquhoun *Summary Roman Civil Law* II. 1 *Res*..is used in contradistinction to *persona*... In another and more restricted sense it signifies those objects of rights which are neither *personae, actiones*, nor *facta*. **1854** P. Cumin *Man. Civil Law* 59 *Q.* Define *Res. A.* All physical and metaphysical existences, in which persons may claim a right. **1923** *Law Rep. King's Bench Div.* II. 439 If the res—the thing actually and directly in dispute—has been already adjudicated upon, of course by a competent Court, it cannot be litigated again. **1947** Wodehouse *Full Moon* vi. 126 Do you mind if we get back to the *res*. Time presses. **1949** W. Stevens in *Trans. Connecticut Acad. Arts & Sci.* Dec. 166 The poem is the cry of its occasion, Part of the res itself and not about it. **1966** Wodehouse *Plum Pie* i. 23, I saw that I had better come to the *res* without delay.

2. Used in a number of Latin (*esp.* legal and philos.) phrases, as: **res co·gitans** *Philos.*, the concept of man as that of a thinking being.

[**1641** Descartes *Meditationes* ii. 23 Sed quid igitur sum? res cogitans: quid est hoc? nempe dubitans, intelligens, affirmans, negans, volens, nolens, imaginans quoque, & sentiens.] **1904** J. Iverach *Descartes, Spinoza & New Philos.* iii. 63 It is not possible to take the mind as a thing among other things; a mere *res cogitans* can apprehend nothing but thoughts or ideas. **1962** M. McLuhan *Gutenberg Galaxy* 247 The mental *res cogitans* and the material *res extensa*. **1972** Z. Vendler (*title*) Res cogitans: an essay in rational psychology.

res commu·nis *Law*, common property; something incapable of appropriation.

[*a* **1259** Bracton *De Legibus* (Rolls Ser., 1878) I. 54 Extra patrimonio autem dicuntur res sacræ, et religiosæ, et communes.] **1704** T. Wood *New Institute Imperial or Civil Law* ii. i. 65 *Res communes* are those in which no Person has a Property, neither can any one be Master of them, or deprive others of the use of them. **1854** P. Cumin *Man. Civil Law* 65 *Q.* When an island rises in the Sea, who is proprietor?.. *A.* The first occupier: until occupation, no one, it is a *res communis*. **1923** W. S. Holdsworth *Hist. Eng. Law* (rev. ed.) II. iii. iii. 273 He [*sc.* Bracton] then discourses upon..res publicæ and res communes. In his treatment of the latter topic there are traces that he had read Azo hastily. **1970** *Internat. & Compar. Law Q.* XIX. 237 It has been suggested that Antarctica has, in fact, become *res communis*.

res exte·nsa *Philos.*, a material thing considered as extended substance.

1940 *Philos. & Phenomenol. Res.* I. 181 This cosmos of meaning and unmeaning into which man is born, this rich structure of all kinds of relationships, is levelled down to the status of *res extensa*. **1962** [see *res cogitans].

res ge·stæ, (an account of) things done, achievements; an account of a person's

career; events in the past; in *Law*, the facts of a case, used esp. with reference to evidence that includes spoken words; also in *sing.* **res gesta.**

1616 G. Carew *Let.* 24 Jan. (1860) 27 In this gazette you may not expect any more than *res gestae*. **1696** W. Nicolson *Eng. Historical Libr.* I. 213 The like Scruples I have upon me as to some other *Res Gestæ* of this King, which are said to have been written by Robert Bale, sometime Recorder of London. **1794** *Trial of J. H. Tooke* in T. B. Howell's *State Trials* (1818) XXV. 440 That letter your lordships have received, and, I believe, without any objection from this side of the table, probably upon the ground, that as it is an answer to an act which is charged against the prisoner, it is fit to be received as part of the *res gesta* upon the subject. **1815** S. M. Phillipps *Treat. Law of Evidence* i. vii. 202 Hearsay is often admitted in evidence, as part of the *res gestae*; the meaning of which seems to be, that where it is necessary, in the course of a cause, to inquire into the nature of a particular act and the intention of the person who did the act, proof of what the person said at the time of doing it is admissible evidence, for the purpose of shewing its true character. **1930** Burrows & Cahn *S. L. Phipson's Law of Evidence* vi. 54 Acts, declarations, and incidents which constitute, or accompany and explain, the fact or transaction in issue, are admissible, for or against either party, as forming parts of the *res gesta*. **1936** *Mind* XLV. 518 Signor Gentile..roundly denies any distinction between *res gestae* and *historia rerum gestarum*, maintaining that all history is contemporary history. **1951** S. F. Nadel *Foundations Social Anthropol.* i. 14 Events in the past, which, being as it were closed chapters, *res gestae*, and having had their particular consequences, exemplify more typically the just-so happenings which limit our search for regularities. **1959** A. G. Woodhead *Study of Greek Inscriptions* 56 Cross-references concerning offices held or *res gestae* in general. **1969** *N. Dakota Law Rev.* Winter 208 Children talk best in their native habitat. What they say when pressures of emotion and strangeness are absent is more apt to be true, somewhat analogous to *res gestae*.

res i·ntegra *Law* (see quot. 1959); also *transf.*

1754 in F. Vesey *Reports* (1814) I. 11, I confess, if this had been *res integra*, I should doubt, whether the testator's declaration is a proper execution within the fifth clause.. but I find myself bound by such a number of former precedents, that I must give way to their superior weight. **1760** G. Gilbert *Cases in Law & Equity* 250 And if the Matter had been *res integra* and undetermined, he should have held it ill if it had been brought by the other Name. **1834** J. Ram *Science of Legal Judgment* xiv. 126 If the matter were entire.., *res integra*, a new case or point.., it might admit of difficulty. *a* **1873** Mill *Three Ess. Relig.* (1874) 203 The question whether it is so or not is *res integra*, untouched by any of the results of human knowledge and experience. **1959** Jowitt *Dict. Eng. Law* II. 1533/2 *Res integra*, a point governed neither by any decision nor by any rule of law, which must therefore be decided upon principle. **1961** *Times* 15 Feb. 15/3 If the matter were *res integra*, his Lordship saw great force that in 1843 the words 'voluntary contribution' would be understood as intended to cover an annual subscription.

res i·psa lo·quitur *Law*, a principle that the proven occurrence of an accident implies the negligence of the defendant unless he provides another cause; also *transf.* and *attrib.*

1659 H. Grimston tr. *Croke's Reports King James* 508 It is apparent, that the money was lent for Interest, and is more than the Statute permits; Wherefore being usury apparent, the Court shall adjudg it accordingly: And..if the corrupt agreement be not expressed in the verdict, and the matter is apparent to the Court to be usury, there the Jury needs not to shew that it was corruptly, for *res ipsa loquitur*. **1864** in Hurlstone & Coltman *Exchequer Reports* II. 725 There are certain cases of which it may be said res ipsa loquitur, and this seems one of them. **1872** *Wharton's Law-Lexicon* (ed. 5) 846/1 *Res ipsa loquitur* (the thing speaks for itself), a phrase used in actions for injury by negligence where no proof of negligence is required beyond the accident itself, which is such as necessarily to involve negligence. **1908** *Times Law Rep.* XXIV. 551/2 *Res ipsa loquitur* does not mean, as I understand it, that merely because at the end of a journey a horse is found hurt, or somebody is hurt in the streets, the mere fact that he is hurt implies negligence. **1927** Ld. Hewart in A. P. Herbert *Misleading Cases* p. v, '*Res ipsa loquitur*', as the man in the street said when a sack of flour, in the best manner of the attic declension, fell upon him from an upper room. **1954** *Cambr. Law Jrnl.* Apr. 132 In *Ybarra* v. *Spangard*..the Californian District Court of Appeal held that the plaintiff might sue all the defendants and recover a joint judgment in a *res ipsa loquitur* situation. **1965** *Mod. Law Rev.* XXVIII. v. 623 When a textbook attains an eighteenth edition surely the maxim *res ipsa loquitur* could hardly boast a better exemplification. **1973** *N.Y. Law Jrnl.* 19 July 12/3 At the new trial plaintiff shall be permitted to produce an expert to establish a case under the doctrine of res ipsa loquitur.

res judica·ta *Law*, a matter that has been adjudicated by a competent court.

[*a* **1259** Bracton *De Legibus* (Rolls Ser., 1881) IV. 266 Item cadit assisa, si petens petat per assisam quod per judicium amisit, quia cadit assisa propter exceptionem rei judicatæ, & agat si voluerit de falso judicio.] **1693** Ld. Stair *Inst. Law Scotl.* (ed. 2) iv. xl. 675 *Res Judicata* is Relevant, not only being a Decreet between the Pursuer and the Defender; But it is sufficient, if it was between their Predecessors or Authors. **1867** *Wharton's Law-Lexicon* (ed. 4) 830/1 *Res judicata*, a point decided by authority. **1927** C. K. Allen *Law in Making* iii. 109 It was a commonplace with the Orators, and especially with Cicero, that the *res iudicata*, or *iudicatum*, was an integral part of the civil law. **1955** *Times* 14 July 11/3 All too often he spoke of *res judicatae* as if they were mere *obiter dicta*. **1972** *Mod. Law Rev.* XXXV. i. 96 The husband's solicitor maintained that the issue of cruelty was *res judicata*.

res non ve·rba, 'things not words'; material fact or concrete action as opposed to mere talk.

1949 E. Pound *Pisan Cantos* lxxxii. 116 And for all that old Ford's conversation was better, Consisting in *res non verba*. **1961** *Times* 26 Apr. 20/2 The strapping muleteer who saves every situation by his guiding principle of *res non verba*.

res nu·llius *Law*, no one's property; strictly, a thing or things that can belong to no one.

[*a* **1259** Bracton *De Legibus* (Rolls Ser., 1878) I. 60 Res quidem nullius esse dicuntur pluribus modis, natura sive jure naturali, ut feræ bestiæ, volucres, et pisces... Item tempore dicunter res in nullius bonis esse, ut thesaurus.] **1704** T. Wood *New Inst. Imperial or Civil Law* ii. i. 67 *Res nullius* (or things which are not the Goods of any Person or number of Men) are those that are of a *divine* Right. **1833** J. S. Mill in *Jurist* IV. 24 It matters not that the property has now become *res nullius*, and is therefore, properly speaking, our own. **1951** F. Schulz *Classical Roman Law* iv. i. 361 Wild animals (game, fishes, birds) were *res nullius* as long as they enjoyed their natural freedom. **1977** *Times Lit. Suppl.* 28 Jan. 110/2 To the South African government, South West Africa had fallen into a state of *res nullius*, and sovereignty was acquired by virtue of her occupation.

res² (rez). Abbrev. of Residence *sb.*¹

1882 W. Whitman *Daybks. & Notebks.* (1978) II. 288 Sam Long 614 Sansom St—res: 3210 Race. **1972** *Guardian* 17 Feb. 11/1 Her own little bijou res in Chelsea. **1972** J. Gores *Dead Skip* x. 68 I've got a res add on Hemovich 5-0-7 Nevada Street.

resaddle, *v.* (Earlier example.)

1856 V. Lush *Jrnl.* 17 Jan. (1971) 174 Giving them strict injunctions to take care of Rollo I left them to resaddle.

re·sail, *sb.* [f. the vb.] A race sailed again.

1947 *Sun* (Baltimore) 8 Sept. 14/6 In the resail, sailed Saturday, the breeze barely let the racers finish within the time limit. **1970** *Cape Times* 28 Oct. 25/8 Experience has shown that resails of cancelled races are not always successful.

resale. Add: Also with pronunc. (rī·sēⁱl).

2. *attrib.* **a.** **resale price**, the price at which a commodity is sold again; **resale price maintenance**, the determination by a manufacturer of a minimum price at which his goods may be sold to the consumer or ultimate user.

1919 C. T. Murchison *Resale Price Maintenance* x. 159 The rights of a patentee, by conditions of sale, to control the use of his article even after it has passed from him and is in the hands of the vendee, is..unqualifiedly upheld and the *dicta* of the Court are certainly not hostile to the idea of resale price maintenance. **1929** *Congress. Rec.* LXX. 2431 A communication from the chairman of the Federal Trade Commission, transmitting a report of that commission on Resale Price Maintenance. **1936** *Publishers' Weekly* 12 Dec. 2298/2 Compare report of the Federal Trade Commission on resale price maintenance, 70th Cong. 2d Sess. H. Doc. No. 546. **1940** *Economist* 14 Sept. 332/1 Most newspapers were generally in favour of resale-price agreements with boycotting clauses. **1945** *Economica* XII. 228 The trading stamp is primarily the product of resale price maintenance. **1967** *Listener* 27 July 104/3 Price-cutting of chocolate and sweets begins following abolition of resale price maintenance. **1973** *Guardian* 1 June 12/1 Mr Heath..made his reputation at the Board of Trade partly on the abolition of resale price maintenance.

b. In sense 'second-hand'.

1960 *Farmer & Stockbreeder* 29 Mar. (Suppl.) 13/2 Every one of your surplus jute sacks has a good re-sale value. **1967** *Boston Sunday Herald* 26 Mar. i. 38/8 The offices opposite Fisherman's Beach, Swampscott, will specialize in both new and resale homes. **1970** *Globe & Mail* (Toronto) 26 Sept. B2/3 The supply of money is good for conventional mortgages on resale houses. **1975** M. Bradbury *History Man* iii. 40 It's a very sound residential area.., you'd keep your resale value. **1977** *New Yorker* 24 Oct. 39/2 I'm conducting here a gents' resale business. *Ibid.* 40/3 You were in a gentlemen's resale shop.

resam (resæ·m). [Malay.] A Malaysian tropical fern, *Gleichenia linearis*, which has creeping rhizomes and leathery pinnate leaves. Also *attrib.*

1909 H. N. Ridley *Malay Plant Names* 219 Resam. *Gleichenia linearis*.. A common fern. **1939** J. D. Gimlette *Dict. Malayan Med.* 200/1 *Résam* leaves are used by Malays in making internal and external medicines for fever. **1964** M. E. D. Poore in Wang Gungwu *Malaysia* i. ii. 51 If the frequency of cultivation increases, the land deteriorates seriously..and forest is replaced by unproductive lallang.. or resam (*Gleichenia* spp.).

resanctifica·tion. [Re- 5 a.] A second or further sanctification.

1897 'Mark Twain' *Following Equator* xlix. 473 The Fort was built three centuries ago by a Mohammedan Emperor—a resanctification of the place in the interest of *that* religion.

resaw (rī·sǭ), *sb.* [f. Re- 5 a + Saw *sb.*¹ See also Resawing *vbl. sb.*] **a.** A machine used for the further cutting of sawn wood. **b.** Wood cut by such a machine. Also *attrib.*

1915 *Saw in History* (Henry Disston & Sons) iii. 36 Boards and thin Cants, or Planks, made by the log saw are run through the Resaws and are manufactured into two or more thin boards. **1958** R. H. Hordern *Woodworking Machinist* I. i. 7 The band mill to achieve the preliminary breaking-down and squaring, and the band or circular resaw to carry out the final resawing. **1971** *Timber Trades Jrnl.* 21 Aug. 38/1 The first band resaw machine to be installed in the district is a Stenner type VHM 36in. *Ibid.* 11 Sept. 43/1 The new production line was so successful that the additional machines were installed. These are a VHS slab resaw followed by an EDS double edger. **1972** J. Minifie *Homesteader* viii. 63 Re-saw is a half-inch board with little to recommend it but ductility and ease of handling.

resche·dule, *v.* [Re- 5 a.] *trans.* To replan in accordance with a different timetable; to change the time of (a planned event or activity); *spec.* to arrange a new scheme of repayments of (an international debt).

1966 A. Battersby *Math. in Managem.* viii. 198 It can be ..used to prepare an amended input for re-scheduling the product launch. **1968** *Economist* 17 Feb. 40/1 Requests to reschedule the country's debts for just one more year. **1970** *Daily Tel.* 14 Apr. 5/6 Fords will try to avoid lay-offs by re-scheduling production. **1973** *Times* 14 Dec. 4/5 Programmes will have to be rescheduled, and both ITV and BBC were working on that yesterday. **1977** P. Way *Super-Celeste* 18 We may be rescheduling the flight. **1978** *N.Y. Times* 30 Mar. A3/3 The United States was prepared to.. reschedule outstanding loans to developing countries. **1980** *Daily Tel.* 3 Sept. 19 The task of rescheduling the Brazilian debt will present international banks with their biggest challenge yet.

Hence **resche·duled** *ppl. a.*, **resche·duling** *vbl. sb.*

1968 S. E. Varner *Rescheduled School Year* 6 One plan for a rescheduled school year—the all-year school or rotating four-quarter plan—has been discussed. **1973** 'D. Jordan' *Nile Green* iv. 23 There's this debt rescheduling problem still unsolved... But..we could talk with the [World] Bank and do a joint deal. **1976** P. R. White *Planning for Public Transport* vii. 145 By slight rescheduling of existing runs, a fairly fast direct journey can be provided from the larger village(s)..into the market town.

rescission. 2. (Later examples.)

1931 *Daily Express* 16 Oct. 14/6 Rumours regarding bond interest, caused a sharp rise..in the Four per Cent. Rescission issues. **1964** *Mod. Law Rev.* XXVIII. iii. 269 Keeping the contract alive for as long as the employers would tolerate the breach without exercising their right of rescission. **1976** *Evening News* (Newburgh, N.Y.) 12 Sept. 12A/1 The rescission of funds in no way affects the proposed federal prison.

resco·re, *v. Mus.* [Re- 5 a.] *trans.* To score (a piece of music) again. Hence **resco·red** *ppl. a.*; **resco·ring** *vbl. sb.*, the action or an instance of scoring again; a rescored version.

1890 *Daily News* 16 Aug. 5/4 Bizet..once undertook the task of re-scoring Bellini's 'Norma'. *Ibid.* He had thrown the re-scored MS. into the fire. **1926** Whiteman & McBride *Jazz* xi. 227 Most of the rehearsing and discussing and rescoring was done in consultations outside. **1955** G. Abraham in H. Van Thal *Fanfare for E. Newman* 22 Small details and large patches of drastic rescoring alike show Wagner's finer mastery of the orchestra in the 1860s. **1962** *Times* 8 Oct. 16/5 Sir Malcolm Sargent has rescored the music. **1976** *New Yorker* 9 May 129/1 Whether they were played in Arne's own orchestration or in the rescoring by Charles Edward Horn..was not stated. **1981** P. Dickinson *Seventh Raven* iv. 47 He met this guitarist and rescored the whole bloody thing.

re-scree·n, *v.* [Re- 5 a.] *trans.* To screen again. So **re-scree·ning** *vbl. sb.*

1950 *Hansard Commons* 7 Nov. 770 In view of the replies made yesterday by colleagues of the right hon. Gentleman that the main screening took place in 1943, when circumstances were very different, would he look into the question of whether re-screening, now that circumstances have changed, should be instituted? **1957** Manvell & Huntley *Film Music* iii. 97 Creating the score by screening and re-screening each sequence where music is required. **1967** E. Chambers *Photolitho-Offset* xi. 160 These Canon optical tone filters..eliminate moiré by diffusing the screen pattern through interference effects during rescreening. **1967** Karch & Buber *Offset Processes* ii. 12 Letterpress printed halftones (pictures) in books can often be 'picked up' as line shots, or rescreened for printing by the offset-lithographic process. **1979** *Listener* 29 Nov. 739/1 The Mudd documentary, which CBS will not re-screen, will gradually recede in memory.

rescuable, *a.* Delete *rare* and add later examples. Hence **rescuabi·lity.**

1975 *Sci. Amer.* Dec. 34/2 The transition from Stage I (rescuable) to Stage II (killed) is a first-order, or 'one hit', reaction, whose rate is directly proportional to the amount of colicin. **1976** *Listener* 3 June 693/1 Men had taken the most obvious solace in alcohol, about the only rescuable commodity from many of the ruins. **1979** *Nature* 22 Nov. 382/2 The lack of rescuability of focus-forming activity from transformants caused by MSV DNA fragments remains to be resolved.

rescue, *sb.* Add: **1. b.** *Bridge.* = *rescue bid* in sense 3 c below.

1917 [see *overbid *v.* 2 c]. **1932** H. Phillips *One Hundred Contract Bridge Hands* 114 West's double is for 'business'... North does not attempt a 'rescue'.

3. a. *rescue bell, boat, capsule, car, company, co-ordinator, cradle, dinghy, man, mission, operation, party* (further example), *service, ship, squad, station, team, tube, work.*

1939 *Sun* (Baltimore) 26 May 8/6 The rescue bell was plunged down to be attached to one of the aft hatches. **1960** W. O. Shelford *Subsunk* vii. 80 *Squalus* sank in an area where there were no appreciable tides to delay the divers or

RESERVOIR

research, *v.*¹ Add: Also with pronunc. (ri-sɜ̄ɪtʃ). **1. a.** Delete '*rare* or *Obs.*' and add: Also, to engage in research upon (a subject, a person, etc.).

1942 R. Chandler *High Window* (1943) xxxiii. 219 Some I was told, some I researched, some I guessed. **1959** *Encounter* Dec. 32/2 I'll author an article about it after I've researched the matter further. **1965** *New Statesman* 23 Apr. 642/1 Bryan Magee must have worn cosy blinkers when he researched lesbianism for his TV programme. **1971** D. Potter *Brit. Eliz. Stamps* xv. 180 Collect, study and research your stamps. **1978** S. Sheldon *Bloodline* xiv. 175 She researched the guests, found out their likes and dislikes, what they ate and drank, and what type of entertainment they enjoyed. **1980** *Times* 1 Oct. 1/4 Union leaders..said they wanted time to 'research' the company's case before the next meeting.

b. (Earlier and later examples.) Also const. *in(to)*, *on*.

1781 H. Walpole *Let.* 27 Jan. (1858) VII. 505, I know, as Gray would have said, how little I have *researched*, and what slender pretensions are mine. **1935** D. L. Sayers *Gaudy Night* i. 23, I believe she's researching on the Bacon family. **1958** *Times Lit. Suppl.* 14 Nov. 653/2 The *Savage Affair* is about one Michael Savage..researching into Commonwealth Drama. **1975** *Nature* 6 Nov. 27/3 He is a biochemist who has researched in many areas. **1977** *Navy News* July 12/3 Former lieutenant-commander, John Winton has researched into sea ballads, ships' logs, and sailors' personal writings and reminiscences to produce the latest of his works. **1977** K. O'Hara *Ghost of T. Penry* xv. 142 If you'd ever learn to research properly..you'd know you're always civil.

c. *trans.* To engage upon research for (a book or the like).

1965 *Listener* 4 Mar. 343/3 The book has been thoroughly and conscientiously 'researched', as the Americans say. **1973** *Which?* Dec. 367/1 This continued to research monthly features for sale to the mass-circulation *Daily Mirror*. **1975** *Publishers Weekly* 2 June 49/3 Barbara Villet, ex-*Life* reporter, researched her book in five big New York hospitals.

researchable (risɜ̄·ɪtʃăb'l), *a.* [f. Research *v.*¹ + -able.] Worthy of being researched; suitable for methodical investigation.

1967 L. von Bertalanffy *Robots, Men & Minds* II. 86 There is a wealth of researchable and fascinating problems which..will open new perspectives and bring evolution into the framework of organismic and systems thinking. **1977** A. Giddens *Stud. in Social & Polit. Theory* iv. 169 Ordinary language is not therefore just a topic that can be made available for analysis, but is a resource that every sociological or anthropological observer must use to gain access to his 'researchable subject-matter'.

researched, *ppl. a.* Restrict † (*obs.*) to sense in Dict. and add: **2.** That has been subjected to research; that is the result of research.

1956 C. W. Mills *Power Elite* ii. 33 The old southern aristocracy, in fictional image and in researched fact, is indeed often in a sorry state of decline. **1966** *New Society* 31 Mar. 25/3 This is a thorough, well researched, and authorised biography. **1978** *Jrnl. R. Soc. Arts* CXXVI. 737/2 The project teachers were involved in researched teaching about race relations. That is to say, the majority were either tape-recording their work or teaching with observation by colleagues.

researching, *ppl. a.* (Later examples.)

1930 J. B. Priestley *Angel Pavement* viii. 393 He wants to come along early next week and bring his researching friend Jiggs or Hoggs or something. **1981** J. Sutherland *Bestsellers* xiii. 145 The most respected researching novelist in Britain is probably Len Deighton.

researchist (risɜ̄·ɪtʃist). [f. Research *v.*¹ + -ist.] = Researcher.

1923 *Chambers's Jrnl.* Feb. 95/1 In an age when the superstitious seek eagerly for a sign..House of Clays would have disappointed a psychical researchist. **1961** D. Angus *Descent of Venus* v. 106 He claims to be a researchist in human relationships.

‖ **réseau** (rezo). Also † 6 *Sc.* **rasour**; **reseau.** Pl. **-x.** [Fr., = net, web, etc.] **1. a.** A plain net ground used in lace-making.

1578 [in Dict. s.v. Rasour]. **1578** in F. B. Palliser *Hist. Lace* (1865) xxxiii. 395 Fyve litell vaills of wovin rasour of threde. **1865** F. B. Palliser *Ibid.* vii. 105 There were two kinds of ground used in Brussels lace, the bride and the réseau. **1911** *Encycl. Brit.* XVI. 41/2 To the period from 1620 to 1670 belongs the development of long continuous scroll patterns with *réseaux* and *brides*. **1953** M. Powys *Lace & Lace-Making* v. 40 The laces of Louis XVI have a lighter design, finally becoming little more than a border to the Réseau or net grounds on which are sprinkled small sprays. **1959** *Chambers's Encycl.* VIII. 294/1 Later, in the 18th century, Milan lace adopted the *réseaux* made popular in northern Europe. **1975** *Oxf. Compan. Decorative Arts* 524/1 Sometimes the *réseau* was bobbin-made and the threads attached to the open edges of the *toilé*, following the pattern.

b. With qualifying adjective or phrase, as *réseau à l'aiguille* (alęgwiy) [Fr. *aiguille*, needle], hand-made net ground; *réseau ordinaire* (ǫrdinēᵊɹ) [Fr., ordinary], standard machine-made net ground; *réseau rosacé* (rozase) [Fr., rosaceous] a mesh ground with a flower pattern.

1865 F. B. Palliser *Hist. Lace* vii. 106 Since machine-made net has come into use the 'réseau à l'aiguille' is rarely made, save for royal trousseaux. *Ibid.*, Machinery has now added a third [way of making the réseau], the tulle or Brussels net, 'réseau ordinaire', made of Scotch thread. **1900** E. Jackson *Hist. Hand-Made Lace* 218 *Réseau Rosacé*, the name given to the réseau ground in Argentan lace. **1911** *Encycl. Brit.* XVI. 41/2 Grounds composed entirely of varieties of *modes* as in the case of the *réseau rosacé*..were sometimes made then [*sc.* about 1700 to 1760]. **1953** M. Powys *Lace & Lace-Making* iv. 14 The ground is the Réseau Ordinaire and the central filling the Mignon.

2. A network or grid, esp. one superimposed as a reference marking on photographs in astronomy, surveying, etc.

1902 *Nature* 5 June 140/1 The réseau is hinged in front of the plate, its correct register being determined by geometrical contacts. **1906** *Athenæum* 27 Jan. 111/1 Prof. Turner showed specimens of photographic reproductions of *réseaux* for stellar photography made by M. H. Bourget. **1940** C. A. Hart *Air Photogr. applied to Surveying* ix. 245 Extreme accuracy of recording on a stereoscopic instrument becomes of value only when there is provided a stable basis of measurement, such as a reseau, a device commonly used in astronomical measurements from photographs. **1963** W. K. Kilford *Elem. Air Survey* i. 19 A squared réseau is also sometimes engraved on the plate. This will make apparent any distortions of the negative due to non-flatness in the focal plane. **1976** J. B. Garner et al. *Surveying* xiii. 228 The Principal Point is engraved on this glass, as are a number of small crosses, conventionally at 10mm centres in either direction, all of which are imaged at every exposure. These crosses form a reseau grid and enable the user to determine any subsequent distortion of the film to considerable accuracy.

3. A spy or intelligence network, esp. in the French resistance movement.

1960 G. Martelli *Agent Extraordinary* v. 82 This arrangement..enabled him to devote more time..to the running of the *réseau*. **1966** M. R. D. Foot *SOE in France* ix. 258 Various intelligence réseaux in Paris. **1973** L. Snelling *Heresy* II. i. 62 These agents made no connection between Graham and the escape *réseau*. **1974** T. Allbeury *Snowball* vii. 37 Paul Loussier had been an active member of one of the SOE *réseaux* in Paris when the Resistance was only measured in hundreds.

resect, *v.* Add: **2.** (Further examples.)

1924 R. Howard *Surg. Emergencies* iv. 71 The portion of the gut to be resected is brought well out of the wound. **1976** *Nature* 25 Mar. 351/1 Two weeks after the initial operation the correct nerve to the anconeus muscle was crushed, or cut and resected.

3. *Surveying.* To map by resection. **resected** *ppl. a.* (further example).

1888 W. H. Richards *Textbk. Mil. Topogr.* (ed. 2) iv. 38 Having determined stations by intersection, as described, the surveyor may, by reference to them, resect or find the point on his survey which corresponds to his position on the ground. **1913** A. R. Hinks *Maps & Survey* v. 108 Round about the intersected and resected points, the detail is sketched in by eye estimation. **1931** M. Hotine *Surveying from Air Photographs* vi. 92 The problem is..to resect the position of the perspective centre in space.

resection. Add: **2.** (Further examples, not referring to bone.)

1881 *Amer. Jrnl. Med. Sci.* LXXXII. 263 (*heading*) Resection of about six and a half feet of the small intestine, with recovery. **1926** *Jrnl. Amer. Med. Assoc.* 20 Nov. 1728/2 Prostatic resection is applicable to a large number of cases of prostatic hypertrophy commonly subjected to prostatectomy. **1944** W. W. Babcock *Princ. & Pract. Surg.* xxxiii. 650 Avulsion of the entire ganglion has been supplanted by resection or division of the posterior sensory root. **1967** H. S. K. Singha *Pocket Surg.* viii. 131 Treatment consists in the resection of all visible tumour together with a wide margin of normal tissue. **1980** *Brit. Med. Jrnl.* 29 Mar. 882/1 Some surgeons can produce primary union of the bowel anastomosis in all but 5–10% of patients..even for anterior resection of the rectum.

3. *Surveying.* The process of determining the position and orientation of a plane table from bearings of points already mapped, prior to mapping surrounding detail in relation to it.

1888 W. H. Richards *Textbk. Mil. Topogr.* (ed. 2) iv. 38 Finding the place by resection with the magnetic needle. **1923** D. Clark *Plane & Geodetic Surveying* I. 212 Resection is usually confined to small scale work. **1976** M. Gilbert *Night of Twelfth* vii. 86 He demonstrated on the map, 'that brings you into an area roughly here. You can't do an exact resection, of course.'

resectionist (rīse·kʃənist). *Surg.* [f. prec. + -ist.] One who carries out resection.

1934 *Jrnl. Amer. Med. Assoc.* 24 Feb. 648/1 Transurethral resection..can be done by competent resectionists in more than 95 per cent of obstructions [of the prostate]. **1972** J. P. Mitchell *Princ. Transurethral Resection & Haemostasis* xxvii. 225 This cumbersome attachment to the normal resectoscope impedes the resectionist.

resectoscope (rīse·ktŏskōᵘp). *Med.* [f. Resect *v.* + -o + -scope.] A surgical instrument for transurethral resection.

1926 M. Stern in *Jrnl. Amer. Med. Assoc.* 20 Nov. 1726/2 It has been possible to reduce the problem to a mere cystoscopic procedure by the evolution of a cutting current capable of operating in a water medium, and a cystoscopic instrument for its application. This instrument is provided with a small movable ring or loop of tungsten wire which, when actuated by a suitable current, is capable of removing longitudinal spaghetti-like sections of tissue... Of these instruments I have named..the latter the resectoscope. **1951** M. Campbell *Clin. Pediatric Urology* i. 74 Miniature resectoscopes are now available for transurethral removal of vesical neck obstruction..uretocele or small intravesical growths. **1962** *Lancet* 28 Apr. 914/1, I must admit to a certain amount of fear of producing incontinence in using the resectoscope on the female bladder neck, and prefer to use fulguration first, even if resection is necessary later. **1974** J. D. Maynard in R. M. Kirk et al. *Surgery* viii. 180 Provided IVP shows some renal function, treatment [of bladder neck obstruction] consists of posterior bladder neck resection per urethra with a resectoscope.

reseda. Add: **2.** (Earlier and later examples.)

1873 *Young Englishwoman* Mar. 130/1 A costume of réséda-coloured poplinette is composed of a skirt rasterre, and a redingote tunic. *Ibid.* Apr. 182/2 A dress of réséda grosgrain silk. **1927** F. B. Young *Portrait of Clare* 25 She forced herself to dwell upon the vision of the reseda shantung dress designed by Liberty. **1968** E. Brill *Old Cotswold* v. 85 Its tall spires can sometimes be seen on the banks of newly-widened roads, the pale reseda of its leaves conspicuous among the darker greens of coarser herbage. **1977** *Western Morning News* 30 Aug. 9/2 (Advt.), 1977 BMW 528. Manual. Reseda Green. Sunroof. Tint.

re:sedimenta·tion. *Geol.* [ad. It. *risedimentazione* (C. I. Migliorini 1950, in *Atti Soc. Tosc. di Sci. nat., Mem.* A LVII. 83): see Re- 5 a.] Movement of previously deposited sediment from one location to another by marine currents. So **rese·diment** *v. trans.*; **rese·dimented** *ppl. a.*

1957 *Jrnl. Geol.* LXV. 231/1 Migliorini coined the term 'resedimented rocks' which is not entirely unambiguous but is nevertheless useful as a short way of referring to the deposits of turbidity currents. *Ibid.*, Various types of markings on the sole of the graywackes..explaining them on the basis of resedimentation by turbidity currents. **1973** *Nature* 26 Jan. 267/2 Because most coarse material resedimented from modern stable continental shelves passes down submarine canyons we interpret major occurrences of coarse resedimented material down slope of the Cambro-Ordovician shelf break as evidence of submarine canyon activity. *Ibid.* 268/1 Episodes of resedimentation occurred from soon after the inception of the miogeocline in Lower Cambrian times to the changes in sedimentary polarity..in Middle and Upper Ordovician times. **1978** [see *resuspension* s.v. *re- 5 a].

reseed, *v.* Add: Also *absol.* (Further examples.) Hence **resee·ded** *ppl. a.*, **resee·ding** *vbl. sb.* (further examples). Also as *sb.*, an area which has been reseeded.

1940 R. G. Stapledon *War Food Production Advisory Bull. No. 1* 26 The field was twenty-five acres, of which fifteen were re-seeded. *Ibid.*, The re-seeded portion of the field had the best grass. *Ibid.* 27 By 1938 we considered the area..to be in a sufficiently good condition to plough and re-seed. **1945** 'G. Orwell' *Animal Farm* viii. 72 The pasture was exhausted and needed re-seeding. **1946** *Nature* 26 Oct. 587/1 Reseeded grassland. **1960** *Farmer & Stockbreeder* 5 Jan. 94/2 During the past 20 years vast acreages of old grassland..have been ploughed and re-seeded on official advice. **1966** *Economist* 24 Sept. 1208/3 Your attack on crofting ignores the remarkable Lewis reseeding schemes. **1970** *Watsonia* VIII. 193 A large area where the *Cerastium* used to grow had been re-seeded. **1973** *Stornoway Gaz.* 2 June 4/3 Maybe you're sleeping soundly o' nights, dreaming of all your frisky lambs and ewes on the surging reseeds. **1981** *Times* 11 Feb. 8/4 The reseeding of rubber plantations in southern Thailand.

rese·gregate, *v.* [Re- 5 a.] *trans.* To segregate again. So **resegrega·tion.**

1923 *Daily Mail* 24 May 6 As travelling facilities improved these races began to resegregate themselves. *Ibid.*, With the upheaval of the war, when..the population was in a state of flux, resegregation probably became more rapid and accentuated. **1954** *Newsweek* 27 Sept. 59/2 (*heading*) Now resegregation. *Ibid.*, In this fashion were Greenbrier County's desegregated schools resegregated. **1970** *Washington Post* 22 Jan. A20/1 'Resegregation' is what may occur when a formerly *de jure* segregated school system complies with the order to desegregate and..people move around in such a way as to 'resegregate' the schools—only on a *de facto* basis this time. **1976** *Time* 27 Sept. 57/3 Another pattern of resegregation occasionally takes place within desegregated schools when students are simply assigned to segregated classes.

resele·ct, *v.* [Re- 5 a.] *trans.* To select again. So **resele·ction.**

1940 F. D. Davison *Woman at Mill* i. 83 Lot 32..had been forfeited and was available for re-selection. **1953** *New Biol.* XIV. 47 In Peru a special type [of cotton] of *barbadense* has been acclimatized and reselected. *Ibid.* 58 In Egypt we attempted also a further refinement of purity, and even improvement, by continued reselections of single-plant origin. **1979** *Guardian* 5 July 24/3 The NEC..agreed to put reselection of MPs back on the agenda. **1981** *Observer* 31 May 12/3 One left wing MP..said after last week's National Executive had voted against MP's being reselected unopposed [etc.].

resemblance, *sb.*¹ Add: **1. e.** *Biol.* An evolved similarity in appearance between organisms of different species.

1862 H. W. Bates in *Trans. Linn. Soc.* XXIII. 502 Mimetic analogies..are resemblances in external appearance, shape, and colours between members of widely distinct families. **1902** *Encycl. Brit.* XXVII. 149/2 Mimetic resemblance is far commoner in the female than in the male, a fact readily explicable by selection, as suggested by Wallace. **1912** *Proc. Zool. Soc.* 361 Some of the cases of resemblance [of Blattidæ to Coleoptera] are so detailed and close that it is impossible to regard them as anything but examples of true mimicry. **1931, 1951** [see *mimicry 2]. **1968** R. D. Martin tr. *Wickler's Mimicry* ii. 22 One of the

most astonishing cases of one species resembling another (interspecific resemblance) occurs in a butterfly of the swallowtail family (Papilionidae) in the species *Papilio dardanus*. **1974** *Encycl. Brit. Macropædia* XII. 214/1 In mimicry the animate agent of selection..interacts directly with at least two of the similar forms and is deceived by their similarity... A convergent resemblance, on the other hand, results from the action of similar forces of natural selection..on unrelated organisms, which may be geographically or temporally isolated from each other.

rese·nsitize, *v.* [Re- 5 a.] *trans.* To sensitize again.

1951 M. Hynes *Med. Bacteriol.* (ed. 5) vii. 91 After an interval..antibodies are formed again in sufficient amounts to resensitize the animal. **1967** E. Chambers *Photolitho-Offset* i. 3 If additions are required to the image, it is necessary to 'resensitise' the stone by eliminating from it the insoluble products originating from the gum, which stop penetration by any greasy substance.

resentment. Add: **2. d.** *Social Psychol.* A term introduced by Nietzsche (as G. *ressentiment*) to describe an attitude which arises, often unconsciously, from aggressive feelings frustrated by the sensed inferiority of one's situation or personality and freq. results in some form of self-abasement. Cf. *ressentiment.

1899 Haussmann & Gray tr. *Nietzsche's Geneal. Morals* i. 33 The slave-revolt in morality begins by resentment itself becoming creative and giving birth to values—the resentment of such beings, as real reaction, the reaction of deeds, is impossible to, and as nothing but [*sic*] an imaginary vengeance will serve to indemnify. **1911** A. M. Ludovici tr. *Nietzsche's Ecce Homo* in *Compl. Wks.* XVII. 20 Freedom from resentment and the understanding of the nature of resentment—who knows how very much..I am indebted to my long illness for these two things. *Ibid.* 21 Nothing on earth consumes a man more quickly than the passion of resentment. **1943** G. A. Morgan *What Nietzsche Means* vi. 150 At bottom, Nietzsche thinks, resentment is caused by a desire to stun pain. It differs from healthy revenge particularly in that, being impotent to express itself by immediate action, it poisons and consumes within. **1957** H. E. Barnes tr. *Sartre's Being & Nothingness* ii. 47 Others so as to make the Not a part of their very subjectivity, establish their human personality as a perpetual negation. This is the meaning and function of what Scheler calls 'the man of resentment'. **1974** B. F. Skinner *About Behaviorism* x. 154 The controlling measures used by an authority make it more likely that a person will escape or counter-attack, and relevant conditions may be felt as resentment; at the same time the measures may generate compliant behavior.

resequent (ri·sikwĕnt, risi·kwĕnt), *a.* (*sb.*) *Geomorphol.* [f. Re- + *-sequent* in Consequent, Subsequent *adjs.*] **a.** Designating, or characterized by the presence of, a stream or streams having a course which follows the dip of strata in the manner of a consequent stream but is stratigraphically at a lower level than the original surface of the underlying geological formation. Hence as *sb.*, a resequent stream.

1906 W. M. Davis in *Bull. Amer. Geogr. Soc.* XXXVIII. 608, I suggested a few years ago that streams which, after spontaneous development aside from an original consequent course, come again to follow it, should be called resequent streams. *Ibid.*, In the third stage..a new series of similar streams—lateral resequents—is beginning to be developed on the crown of the then newly-uncovered hard-rock anticlines. **1937** Wooldridge & Morgan *Physical Basis Geogr.* xv. 212 A good example is afforded by the Central Weald of Kent and Sussex, where there are many good examples of synclinal valleys for which resequent origin must be deemed probable. **1954** W. D. Thornbury *Princ. Geomorphol.* v. 114 Resequent valleys drain in the same direction as the original consequent drainage but are at lower topographic levels and have developed with respect to new base levels of erosion. **1970** R. J. Small *Study of Landforms* iii. 91 Where inversion of relief has once been achieved, continued erosion can lead to a restoration of the original structure-relief relationship; the synclinal valleys and anticlinal ridges are then, strictly speaking, 'resequent'. *Ibid.*, vii. 235 Synclinal streams may even be 'resequents', which are traditionally regarded as the eventual outcome of drainage development in areas of folded rocks.

b. Of a fault-line scarp or a related feature: having a relief similar to that originally produced by the faulting; freq. *spec.* where such relief results from erosion of an obsequent scarp.

1913 W. M. Davis in *Bull. Geol. Soc. Amer.* XXIV. 198 If this change of form is a slope, more or less perfectly graded, descending to the relatively depressed side, it may be called a resequent scarp. **1941** C. A. Cotton *Landscape* xxiii. 265 Fault-line scarps are of two kinds, resequent and obsequent, according as they face in the direction the initial fault-scarp faced on the same line of fault or in the opposite direction. *Ibid.*, Resequent fault-line scarps..are commoner than obsequent. **1970** R. J. Small *Study of Landforms* iii. 105 It will be seen that consequent and resequent scarps are very similar, and indeed many would argue that to make a distinction between them is purely academic, on the grounds that it is invariably impossible to decide in actual cases whether or not an obsequent scarp has previously existed along a fault-line (and thus whether the scarp in question is resequent or not).

reserpine (risɜ̄·ɹpin). *Pharm.* [ad. G. *reserpin* (J. M. Müller et al. 1952, in *Experientia* VIII. 338), f. initial letter of *Rauwolfia + *-e-* + L. *serp-entina* (see below), fem. of *serpentinus* Serpentine *a.*: see -ine⁵.] A colourless crystalline alkaloid $C_{33}H_{40}N_2O_9$, which is obtained from the roots of several plants of the genus *Rauwolfia, notably *R. serpentina*, and is used to treat hypertension and as a sedative.

1952 [see *chlorpromazine]. **1954** *Jrnl. Amer. Med. Assoc.* 20 Mar. 1040/1 It seems that reserpine reduces the blood pressure by exerting an effect on the central nervous system. **1958** *Observer* 30 Nov. 12/4 New drugs such as chlorpromazine and reserpine make the minds of the mentally ill more accessible. **1966** *New Scientist* 24 Nov. 430/3 Reserpine interferes with the retention, mainly in the hypothalamus.., of a number of amines. **1973** *Sci. Amer.* Sept. 121/1 Once widely used, reserpine has been replaced by the phenothiazines because of the greater frequency of reserpine's serious side effects. **1974** *Times* 20 Sept. 3/7 There seems little doubt that this link between regular use of reserpine and an increased cancer risk is genuine.

Hence **rese·rpinized** *a.*, treated with reserpine; **rese:rpiniza·tion,** the process of, or condition resulting from, treatment with reserpine.

1960 *Federation Proc.* XIX. 266/1 (*heading*) LSD-25 and brain serotonin in reserpinized rat. **1962** *Lancet* 22 Dec. 1330/2 The accumulation and retention of ¹⁴C from the blood-glucose in vivo are greatly diminished in the heart of 'reserpinised' rats. **1963** *Federation Proc.* XXII. 364/2 The animal was 5 to 10 times more sensitive to these substances after reserpinization than before. **1973** *Nature* 13 Apr. In the absence of endogenous catecholamines (that is reserpinization) the transient contractile tension response to the ionophore was attenuated, but elevation of resting tension persisted. **1975** *Pediatric Res.* IX. 463 (*heading*) The chronically reserpinized rat as a possible model for cystic fibrosis.

reservable, *a.* (Later example.)

1971 *Daily Tel.* (Colour Suppl.) 29 Oct. 7/1 There were three seat prices—50p, 80p and £1.10. But only the most expensive were reservable in advance.

reservation. Add: **I. 3. b.** (Earlier and later examples.) Also, a tract of land similarly set apart for (*a*) Canadian Indians, (*b*) African Blacks, (*c*) Australian Aborigines.

1789 *Deb. Congress U.S.* 25 May (1834) 41 The reservation,..of six miles square round the fort at Oswego, is within the territory of the State of New York. **1792** *Mass. Hist. Soc. Coll.* (1806) 1st Ser. I. 287 The whole Six Nations live on grounds, called the State Reservations, and are intermediate spaces settled on both sides by white people. **1859** *Native Voice* (Vancouver) (1959) Feb. 5/4 Has the Government of this Island the power to remove the Indians (by purchase) from that piece of outside Victoria Harbour known as the Indian Reservation? **1953** D. Cushman *Stay away, Joe* 22 Won't have anything to do with the Injuns off the reservation. **1957** M. Banton *W. Afr. City* i. 15 A tangle relating to the ownership of land in the Kru Reservation. **1965** *Austral. Encycl.* I. 88/2 Western Australia maintains 167 native reservations. **1970** *Times* 24 Mar. 6/6 Port Elizabeth is close enough to the Transkei Bantustan (African reservation) to be affected by the complex restrictions designed to stop industry from attracting Africans out of the reservations. **1971** D. Heffron *Nice Fire & Some Moonpennies* ii. 18 He told me to remember that when you're off the Reservation, people take you not as an individual but as an Indian. **1973** *Black Panther* 8 Sept. 8/2 The Apaches, originally from Arizona, New Mexico and parts of Mexico, were sent to Florida by the U.S. Army after they refused to remain on a reservation in Arizona.

attrib. (earlier and later examples).

1866 *Rep. Indian Affairs* (U.S.) 100 The reservation Indian is under the protection of the general government. **1946** G. Foreman *Last Trek of Indians* 260 Eighty of them came to their agency and enrolled with the reservation Indians. **1977** *Listener* 13 Oct. 462/1 In the 1950s..a whole generation of reds..was squashed... A few of us survivors are kept on like reservation Indians to remind the winners of how tolerant they've become.

c. orig. *U.S.* The action or fact of engaging seats, rooms, places, etc., or of hiring a vehicle, in advance; something reserved in advance. Also *attrib.*

1906 F. Lynde *Quickening* xiii. 118 That sleeping-car reservation for Thomas Gordon—have you secured it? **1907** *Springfield* (Mass.) *Weekly Republican* 19 Dec. 16 A considerable number of New York and Boston people have made reservations at the Curtis hotel in Lenox for the holiday season. **1925** *Scribner's Mag.* July 32/1 (Advt.), Reservations for 1925–26 should be made as soon as possible to insure entrance. **1935** R. Macaulay *Personal Pleasures* 18 Do tickets, passports, money, travellers' cheques, packing, reservations, boat trains, inns, crouch and snarl before you like those surly dragons that guard enchanted lands? **1949** *Skyline Trail* (Montreal) Mar. 14/2 It is most important that hikers procure their hotel reservations well in advance. **1968** *Globe & Mail* (Toronto) 17 Feb. 35 (Advt.), For reservations, call or write Atlantic Car & Truck Rental. **1971** *Financial Mail* (Johannesburg) 26 Feb. 656/2 When his travel agent phoned the Cape Town reservation centre about that day's 19.15 flight to Johannesburg, a reservation clerk said: 'We have no idea how many seats are left.' **1977** E. Leonard *Unknown Man No. 89* xx. 207 Ryan kept going,..past the reservation desk to the foyer.

d. Exemption from military service because of an important civilian occupation. Also *attrib.*, as *reservation age*, etc.

1916 *List of Certified Occupations* (Local Government Board) 4 The *only* ground for making these reservations is that the men protected are engaged on work of national importance. **1940** *Economist* 20 July 73/2 He [*sc.* Ernest Bevin]..is transferring persons to war work by pressure and persuasion rather than by compulsion. An early example of this was the raising of the reservation age in certain industries while postponing the date, so that men below the new age limit could..transfer into war jobs where they would still be reserved. **1941** *Manch. Guardian Weekly* 26 Sept. 194 The test of reservation should be the work a man is actually doing, not his declared occupation. *Ibid.*, The reservation system cannot be said to operate fairly in our present critical state unless its categories are tightened. **1942** *Ann. Reg. 1941* 27 Hitherto all men above the reservation age had been exempted, whether they were actually occupied or not. **1944** *Manpower* (Ministry of Information) 11 Each occupation had its own age of reservation. The more important the job, the lower the age of reservation. The skilled tradesmen in the Armed Forces were drawn from men below the age of reservation in corresponding civilian occupations.

e. = *reserve *sb.* 5 e. In full, *central reservation*.

1937 *Memorandum on Lay-Out & Construction of Roads* (Ministry of Transport) 15 Gaps for vehicles in the central reservation should in no case be formed opposite the exits from the minor roads. *Ibid.* 17 The separation of dual carriageways should be effected by a reservation of the greatest width practicable. **1959** *Highway Code* 16 Do not reverse or turn in the carriageway or cross the central reservation. **1972** *Times* 10 June 2/2 Police are to investigate why a new M1 crash barrier near the Toddington service area failed to stop a lorry from crossing the central reservation. **1977** *Oxford Mail* 20 Apr. 1 Oxfordshire's ambulance chief has called for safety barriers down the central reservation of the new Witney Bypass before there is a serious crash.

reservationist (rezəɹvē¹·ʃənist), *sb.* (and *a.*) *U.S.* [f. Reservation + -ist.] One who makes reservations (senses *3 c and 4 a). Also *attrib.* or as *adj.*

1920 *Glasgow Herald* 18 Mar. 9 The reservationist Senators are..too much attached to their limitations and qualifications to abandon them for anything. **1933** W. S. Holt *Treaties defeated by Senate* x. 296 The reservationists did not constitute a majority. **1978** *N.Y. Times* 30 Mar. B20/1 (Advt.), Reservationist... 2 yrs ticketing exp.

reserve, *sb.* Add: **I. 1. b.** Also, that part of the profit of a joint stock company which is not distributed to shareholders. *hidden reserve*: see *hidden *ppl. a.* 1 d.

1930, etc. [see *hidden *ppl. a.* 1 d]. **1940** *Economist* 25 May 936/1 This has..involved a redefinition of reserves. In former years this item in our figures was a repository for a mixture: reserves provided against specific assets, and other items properly chargeable against profits were combined with 'free reserves' which were, in effect, merely undivided profits. Now the line is more tightly drawn to include the latter and exclude the former. **1974** *Terminol. Managem. & Financial Accountancy* (Inst. Cost & Managem. Accountants) 63 *Reserves*, undistributed or surplus profits. The creation and distribution of certain reserves are affected by company policy and legal considerations, e.g. the provisions of the Memorandum and Articles of Association of the company.

c. The amount of a mineral, or of oil or natural gas, which is known to exist in the ground in a particular region and to be capable of exploitation. Usu. *pl.*

[Cf. quot. 1860 in sense 5.]

1912 M. H. Burnham *Mod. Mine Valuation* i. 66 Ore in a developed mine lying below that which the engineer is willing to class as a reserve, and pay for, is 'possible' only. [*Note*] The above table classes as reserves only the ore lying above the deepest level. **1922** *Bull. Amer. Assoc. Petroleum Geologists* VI. 444 Within the last few years the necessity has developed for estimating recoverable reserves on oil and gas properties. **1945** J. A. Brown in L. M. Fanning *Our Oil Resources* i. 6 In the Maracaibo basin of Venezuela alone there are proved reserves of at least 5 billion barrels of crude oil. **1969** *Australian* 24 Oct. 18/2 Ore reserves fall into four general categories—proven, profitable, possible and indicated. Proven is ore that has been blocked out on four sides; probable is ore that has been opened on two or three sides, while possible ore has been opened on one side only; indicated ore is ore that has been outlined by diamond drilling but which has not been opened by underground work. **1979** *Nature* 26 July 261/1 What the board is actually talking about if it uses the word 'proven' is 'reserves': materials that have been mapped out sufficiently well that they can be the subject of mining by known methods.

d. *spec.* Extra energy; a supply of energy or resilience. Usu. *pl.*

1929 H. Crane *Let.* c 23 Oct. (1965) 347, I feel quite rested already, but I know that I need a little 'reserve'. **1934** H. G. Wells *Exper. Autobiogr.* I. vi. 369 Dr. Collins heard of my plight and wrote also. I detected a helpful motive and wrote among other things to assure him that I had 'reserves' for a year or so. **1941** *Lilliput* Mar. 206/1 We had lovely weather all the time. I was so glad. Most children don't feel it, but Jonathan is such a fragile little thing, no reserves at all.

2. c. (Further example.) Also *pl.*, the reserve or second team.

1961 *Daily Express* 14 Jan. 10 Aston Villa assistant manager Dick Taylor..saw 19-year-old Cheung bamboozle Villa reserves. **1976** E. Dunphy *Diary* in *Only a Game?* v. 147 All you have got to look forward to is Aldershot reserves away next Wednesday. **1976** *Evening Post* (Nottingham) 14 Dec. 18/5 The England reserve, who injured his back against East Midlands last week, is replaced by Nottingham Casuals' Ian Henry. **1977** *Belfast Tel.* 22 Feb. 29/3 He played for the reserves last night and if there is no reaction, he will be in the Thistle first team tonight.

RESET

disease there must be some reservoir of infection. *Ibid.*, The most important reservoirs of infection are human or animal cases or carriers. Plants may be the reservoir of infection in some of the mycoses. **1947** *Ann. Rev. Microbiol.* I. 353. A precise demon-stration of the mechanisms through which the reservior of the disease functions would constitute a great advance in constructing the biological pattern of influenza. **1965** B. E. Freeman tr. *Vandel's Biospeleol.* xv. 246 It would be possible that bats serve as a reservoir of histoplasmosis. **1977** *Sci. Amer.* Mar. 61/2 When no human being harbours the [smallpox] virus, there should remain only one reservoir: the stocks in research and diagnostic laboratories.

3*. ¶ *au reservoir*: see *Au revoir.

4. a. (Further examples.)

1884 *Queen* 16 Feb. (Advt.), The 'Victor' reservoir pen can be used with any good fluid ink, and any ordinary nib... Price 3s.6d. **1913** *Trans. Soc. Trop. Med. & Hygiene* VI. 269 The monkey is most probably the normal 'reservoir host' [for *Physaloptera mordens*].

c. Special Combs.: **reservoir engineering,** the study and exploitation of natural oil and gas reservoirs; so **reservoir engineer**; **reservoir rock,** rock (capable of) forming a reservoir for oil or natural gas.

1949 M. Muskat *Physical Princ. Oil Production* i. 28 While..the..properties of the reservoir will determine its inherent potentialities as an oil-producing system, there is still much left to the choice of the reservoir engineer with regard to the actual exploitation program to be undertaken. **1973** *Mod. Petroleum Technol.* (Inst. Petroleum) (ed. 4) v. 172 The primary aim of a reservoir engineer is to obtain maximum recovery at minimum cost. **1946** *Petroleum Engineer* Jan. 51/1 The war..has led to a greater realization than ever before of the need for applying the most advanced reservoir engineering principles of production and oil conservation to obtain maximum economic recoveries. **1954** *Mod. Petroleum Technol.* (Inst. Petroleum) (ed. 2) iv. 122 A balanced view of production operations calls for some knowledge of the fundamentals of reservoir engineering as well as of the more purely mechanical methods employed. **1977** *Times* 2 Nov. 3 (Advt.), Reservoir engineering is largely an art... Our job is to get information about..oil-bearing rock..below the sea bed. **1912** E. H. C. Craig *Oil-Finding* iii. 46 The relative porosity of strata is one of the determining factors in the movements of oil, and the selection of a reservoir rock. **1951** K. K. Landes *Petroleum Geol.* vii. 191 Throughout the world, sandstone is by far the most important reservoir rock. **1975** G. Anderson *Coring* i. 18 Permeability normally varies from one location to another vertically as well as horizontally in a reservoir rock.

reset, *v.*[2] Add: **I. 7. a.** To cause (a device) to return to a former state, esp. a condition of readiness.

1919 R. Mordin *Strowger Automatic Telephone Exchange* iii. 64 When the release magnet armature knocks the double dog out of engagement with the shaft it also, by means of the lever shown, re-sets the side switch. **1931** *Proc. R. Soc.* A. CXXXII. 301 The mechanical relay M_R is used only for extinguishing or resetting thyratron Q. **1977** D. Anthony *Stud Game* xxv. 160, I..reset the burglar alarm.

b. *Computers.* To set (a binary cell) to zero; to return (a counting device) *to* a specified value, esp. zero.

1947 *Proc. IRE* Aug. 759/1 The triggered flip-flop is the only one to respond..; as it is reset it gives out a positive pulse. **1956** G. A. Montgomerie *Digital Calculating Machines* iii. 47 If the wheels are left in engagement on the return stroke they are reset to their previous values. **1971** J. H. Smith *Digital Logic* i. 12 A binary divider is a modified toggle which has only one input. If electrical pulses are applied to this input the unit will 'set'. The second pulse will 'reset' the circuit.

8. To set (hair) again.

1932 E. Bowen *To North* xii. 120 She..had had, since lunch, her hair shampooed and re-set.

II. 9. a. *intr.* for *refl.* **b.** Of a device: to return to an initial state.

1895 W. J. Locke *Gate of Samaria* xvi. 194 She..noticed a look upon Thornton's face,—the after-light, as it were, of a sneer, before the features had time to reset. **1971** *Gloss. Electrotechnical, Power Terms (B.S.I.)* i. iii. 13 A relay resets when it returns to its initial position. **1980** *Sci. Amer.* May 49/1 When both currents are removed, the switched junction automatically resets, closing the loop.

resettable, *a.* Add: (Further examples). Hence **resettabi·lity.**

1971 *Physics Bull.* July 390/1 The smooth tunability and resettability (1:3×10⁸ which is an instrumental limitation) is not understood. **1973** G. J. King *Newnes Colour T.V. Servicing Man.* I. i. 35/1 Not all models employ a resettable cut-out. **1975** *Gramophone* June 118/2 Resetability [*sic*] of the slider controls could be improved if a cursor line were engraved on the rectangular knob. **1979** *Sci. Amer.* Aug. 131/1 A system of this type can be viewed as a mathematical strongbox with a resettable combination lock.

resettle, *v.* Add: **1. a.** Also without const. Also, to settle again in another place; *spec.* to establish (a homeless or evicted person, etc.) again in a house or community, or (*S. Afr.*) in another area. Cf. *Resettlement 1 c.

1937 [implied s.v. *resettled *ppl. a.*]. **1951** R. Firth *Elem. Social Organiz.* iv. 142 In the olden days in Ireland a working team..was drawn from the community..to resettle an evicted family. **1965** *Listener* 30 Sept. 482/2 About 1,500 [old people]..require something between forty and fifty days to be resettled back in the community. **1972** *Stand. Encycl. S. Afr.* V. 379/1 A total of 115 000 Bantu from the western areas of Johannesburg and elsewhere were resettled in new homes in new Bantu towns. **1978** C. A. Berry *Gentleman of Road* ix. 74 You're in a bad state and I'd like to resettle you. **1978** A. Brink *Rumours of Rain* 105 'How many hundreds and thousands of 'exceptions' do you think there are?' he asked angrily. 'Whole societies uprooted and resettled.' **1981** *Eastern Province Herald* (S. Afr.) 2 June 5/6 This will cause friction between those who are resettled and those who were there before.

rese·ttled *ppl. a.*

1937 *Sun* (Baltimore) 25 May 12/2 Nobody in Washington is able to indicate the means by which resettled farmers and sharecroppers can pay rent on these expensive houses. **1977** *Time* 4 Apr. 21/1 When Sanjay and his elder brother Rajiv visited a community of resettled slumdwellers, they were given a tumultuous welcome. **1981** *Eastern Province Herald* (S. Afr.) 2 June 5/4 The synod..urged the Diocesan Council to create a committee to co-operate with other concerned groups, and resettled people themselves, in planning a strategy to cope with the problems brought about by forced removal.

resettlement. Add: **1. b.** *spec.* The act of resettling demobilized servicemen into civilian life.

1918 *Labour Gaz.* May 175/1 (*heading*) Resettlement of officers and professional men... The report of the Committee appointed to consider the resettlement of officers has been approved by the Minister of Reconstruction. **1920** *Brit. Med. Jrnl.* 4 Sept. 340/1 Many doctors who served long with the forces have found the period following demobilization a time of uncertainty and financial stress. For them 'resettlement' has proved an uphill business. **1922** *Encycl. Brit.* XXX. 821/2 (*heading*) The first two years of resettlement. The success of the preparations which had been made..is indicated by the figures of re-absorption of men demobilized. **1945** *Yorkshire Post* 19 Apr. 3/2 The Minister of Labour has provided several admirably chosen lecturers..to tell Servicemen of the plans made for their release and resettlement when their term of service comes to an end.

c. *S. Afr.* The act of resettling Blacks in their supposed homelands.

In South Africa resettlement involves the enforced removal of Blacks from one area, often followed by a period in temporary resettlement camps, before their removal to the 'homelands'.

1954 *Bantu* Nov. 11 (*caption*) Compare these neat little houses with the unhygienic tin and brick shacks of Sophiatown..and you would understand why the resettlement of the Bantu had become imperative. **1976** A. Lemon *Apartheid* viii. 153 The further resettlement of up to half the existing number of Africans in white areas is being seriously contemplated. **1981** *Observer* 31 May 12/4 About four million have been moved since 1948 and at least another million are marked down for 'resettlement'.

3. *attrib.* and *Comb.*, as *resettlement area, board, camp, centre, grant, office, officer, plan, programme, project, scheme, village.*

1971 *Progress* (Cape Town) May 7/4 The resettlement areas..to which so many African urban people have been endorsed. **1954** *Bantu* Nov. 4 The views expressed in this article are personal and in no way bind the Department or the Natives Resettlement Board. **1973** Resettlement board [see *resettlement programme* below]. **1971** *Progress* (Cape Town) May 7/5 The valiant attempts of the women of these missions and resettlement camps to make some sort of living for themselves and their families. **1972** *Times* 19 Dec. 3/1 There are events planned in all the 14 resettlement camps. **1973** Resettlement centre [see *resettlement programme* below]. **1977** *R.A.F. News* 11–24 May 7/1 Even the resettlement grant does not sweeten the bitter pill. **1945** *Daily Tel.* 17 May 3/3 They would be welcomed at the employment exchanges and the resettlement offices if they had not got a job. **1922** *Encycl. Brit.* XXX. 821/1 A number of officers known as Resettlement Officers were appointed.., whose business was to travel the country and investigate the causes which impeded the turnover from war to peace. **1946** *Lancet* 2 Feb. 177/1 Our resettlement officer almost got the job, but like me he didn't have the Finnish, or the Norwegian, Swedish, or Danish, without which, out there, the execution of big business is apparently impossible. **1974** 'J. Le Carré' *Tinker, Tailor* xxxii. 286 The first person..to visit him was the resettlement officer, talking about a friendly teaching agency. **1945** *Daily Tel.* 17 May 3/3 (*heading*) Mr. Bevin explains resettlement plans. **1973** *Guardian* 16 Apr. 20/1 No more Uganda Asians will be allowed into the resettlement centres. An announcement.. by the Home Office..marks the first stage in the resettlement programme and has the full backing of the Uganda Resettlement Board. **1973** J. Hoagland *South Afr.* xi. 299 The Portuguese moved quickly to establish the resettlement program in Mozambique. **1936** *Sun* (Baltimore) 28 Jan. 8/7 The Government's resettlement project at..Reedsville. **1921** W. Sherren *Rights of Ex-Service Man & Woman* iii. 23 The Civil Liabilities Resettlement Scheme of 1918 was instituted to provide assistance to ex-officers and men who had suffered serious hardship by reason of their military service **1954** *Bantu* Sept. 44 Throughout the ages various Resettlement Schemes have been undertaken similar to the Scheme of the Western Areas of Johannesburg. **1952** *Times* 22 Sept. 5/1 Volunteer teams..after receiving a short course of instruction in spoken Malay..live in the new resettlement villages. **1973** J. Hoagland *South Afr.* xi. 299 There were 433,000 Africans living in resettlement villages.

reshape, *v.* (Later examples.)

1916 Joyce *Portrait of Artist* (1969) 67 The change of fortune which was reshaping the world about him into a vision of squalor and insincerity. **1951** E. E. Evans-Pritchard *Social Anthropol.* v. 86 The theories have been shaped and reshaped by this steady growth of knowledge. **1960** *Farmer & Stockbreeder* 29 Mar. 17/1 British economic policy has had to be reshaped to deal with an adverse trend in the balance of payments.

resha·per (ri-). [f. Reshape *v.* + -er[1].] One who or that which reshapes.

1961 Webster, *Reshaper*, one that reshapes something; *esp.* a worker who does the final blocking of hats. **1964** F. L. Westwater *Electronic Computers* iii. 49 The effect of passing through a gate causes a deterioration in the pulse shape. This can be corrected by some non-logical element such as an amplifier or reshaper.

reshaping, *vbl. sb.* (Further examples.)

1951 R. Firth *Elem. Social Organiz.* iii. 89 This has been particularly marked with the coming of industrialism, foreshadowing the destruction or radical re-shaping of their social structure. **1964** F. L. Westwater *Electronic Computers* iv. 89 The pulse train is fed back into the tube after amplification and re-shaping. **1969** *Gloss. Landscape Work (B.S.I.)* v. 30 *Reshaping*, a pruning operation directed to improving the shape of a tree.

Resh Galuta (rēʃ gālūtā·). Also **Resch Glutha** and with small initials. [Aramaic, lit. 'chief of the exile'.] = *Exilarch. Also *transf.*

1829 H. H. Milman *Hist. Jews* III. xix. 160 Babylon, where, during his days of splendour, the Resch-Glutha fixed his residence. *Ibid.* 162 The Court of the Resch-Glutha is described as equally splendid; in imitation of his Persian master, he had his officers, counsellors, and cupbearers. **1867** C. M. Yonge *Pupils of St. John* x. 159 A stately personage..known as the Resch Glutha, or Chief of the Captivity. **1903** *Jewish Encycl.* V. 293/1 This last story indicates that the resh galuta had by that time become the subject of Mohammedan legend. **1931** C. Roth *Jewish Bk. of Days* 6 Huna Mari, Prince of the Captivity, executed, 470. This high office (otherwise known as that of Exilarch—*Resh Galutha*) was enjoyed by the secular head of the 'exile' in Babylonia in virtue of his descent from the house of David. **1971** *Encycl. Judaica* VI. 1031/2 During subsequent periods the *nesi'im* of Yemen were referred to as *resh galuta*, although they had no connections with the Babylonian exilarch or the House of David.

reshoot, *v.* (Later examples.)

1900 G. B. Shaw *Let.* 4 Nov. (1972) II. 194 He has.. become too curious to see how their marriage will turn out to reshoot himself more efficiently. **1900** *Law Rep. Appeal Cases* 429 No doubt the action of the current upon the net is calculated more or less to shift the position of it, and so from time to time to make it necessary to gather it up and reshoot it, to restore it to its perpendicular position. **1955** H. Kurnitz *Invasion of Privacy* (1956) xiv. 88 It's a hundred-to-one he'll..reshoot the picture so that not a speck of what you want in it remains. **1956** *Nature* 21 Jan. 120/1 The information can be conveniently stored against the day when better interpretation methods allow one to make the fullest use of the recordings, without having to reshoot in areas which are often exceedingly difficult of access. **1975** R. L. Simon *Wild Turkey* (1976) xi. 72 Badass has had a change in shooting schedule... We're going to have to reshoot two days. **1976** A. Davis *Television* 28 Plays were still recorded in sequence... The maxim otherwise was to stop the tape and reshoot only for technical mishaps.

reshoo·ting, *vbl. sb.* [f. prec. + -ing[1].] The action of the verb Reshoot.

1960 L. Cooper *Certain Compass* 121 There was some re-shooting we could do. **1975** [see *re-edit *v.*, re-editing *vbl. sb.*]. **1978** *Times* 26 Aug. 7/2 We had another preview. The picture was obviously better... He did not think any reshooting was necessary.

Resht (reʃt). The name of a province and town in north-west Iran, used *attrib.* to designate patchwork made there.

1888 A. S. Cole *Descr. Catal. Coll. Tapestry & Embroidery S. Kensington Mus.* 314 Patchwork of cloth of various colours... Resht work. *Persian.* **1923** *Daily Mail* 26 Feb. 15 There is also a kind of Persian patch-work called 'Resht'. It is really an amazingly ingenious mosaic of cloth, only *one* thickness being used, and was generally arranged in four colours—red, green, grey, and white. **1960** H. Hayward *Antique Coll.* 237/1 *Resht patchwork*, a type of mosaic patchwork produced at Resht, Persia, during the 18th and 19th cent. for covers and prayer-rugs. The designs are inlaid in coloured felts with outlines and details worked in coloured silks in chain-stitch and couched-work. Inferior examples are often not true patchwork, as the small pieces are applied to the ground and not inlaid.

reshta, var. *Rishta.

reshuffle, *v.* (Later examples with reference to a redistribution of posts within a cabinet, etc.)

1963 *Ann. Reg. 1962* 117 Chief Akintola proposed to reshuffle the Western Nigeria Development Corporation's management. **1973** A. Mann *Tiara* iii. 18 The London correspondent of *Messaggero* thought the British Cabinet would be reshuffled.

reshuffle, *sb.* Add: (Later examples.) Used *esp.* to denote a redistribution of posts within a government or cabinet, etc.

1922 L. Woolf *Downhill all Way* (1967) i. 37, I feel that this is no time for a mere reshuffle of the ancient Conservative and Liberal Pack and for entrusting power to one or other of the two parties whose political principles and practice are directly responsible for the disastrous situation in which the country finds itself today. **1931** A. L. Rowse *Politics & Younger Generation* x. 272 It [*sc.* the Tyrol] should be returned to Austria..at the first reshuffle in Europe. **1940** H. Nicolson *Diary* 3 Apr. (1967) 65 We get over the telephone the final text of the Government reshuffle. **1947** *New Secondary Education* (Ministry of Educ.) 10 Such reorganisation as could be effected in existing buildings by a 'reshuffle' of the children. **1953** E. Simon *Past Masters* iv. i. 211 Nothing was ever done about the dry

rot... Two rooms are involved... So another grand reshuffle is at hand. **1955** *Daily Express* 14 Oct. 1/5 (*caption*), I *do* wish the P.M. would make up his mind about the Cabinet reshuffle. **1965** *Listener* 17 June 884/2 A reshuffle in the Fund's liabilities among the various depositors. **1976** *Liverpool Echo* 22 Nov. 17/2 Moore went off injured..and it took them a good quarter of an hour to adjust to the enforced reshuffle. **1978** J. R. L. Anderson *Death in Greenhouse* ii. 32 [He] was given his present job at the Foreign Office in the Government re-shuffle..last year.

reshu·ffling, *vbl. sb.* [f. Reshuffle *v.* + -ing[1].] The action of the verb Reshuffle.

1883 E. W. Hamilton *Diary* 28 May (1972) II. 440 His having to quit the Government would be a loss in itself, but what is worse it would involve a reshuffling of the cards. **1926** J. S. Huxley *Ess. Pop. Sci.* 28 Sexual reproduction means first a reshuffling of the factors, and then a recombination of them in new arrangements. **1934** *Mind* XLIII. 525 The mere repetition and reshuffling of changeless entities to which, according to M. Bergson, we should be condemned if we did not have intuition to tutor us. **1937** *Sun* (Baltimore) 7 Dec. 9/2 The proposed 'reshuffling' of operations and routes of American lines.

reside, *v.*[1] Add: **2.** (Later *transf.* example.)

1973 C. Sagan *Cosmic Connection* xiii. 88 Such a solution is liquid at the temperatures and pressures at which the Venus clouds reside.

3. b. (Later example.)

1972 *Incorporated Linguist* XI. 30 The significance of linguistic theorizing resides..in the fact that it provides intellectual training while..introducing the learner to problems of the functioning of linguistic systems.

residence, *sb.*[1] Add: **1. d.** *Anthrop.* The place in which it is customary for a couple to settle after marriage, according to the prevailing kinship system. Also *attrib.* Cf. *Matrilocal *a.*, *Neolocal *a.*, etc.

1865 J. F. McLennan *Primitive Marriage* viii. 154 Teadhloch and cuedichc.., Gaelic names for family, mean the first having a common residence; the second those who eat together. **1889** E. B. Tylor in *Jrnl. Anthrop. Inst.* XVIII. 247 Now, on looking out from the schedules the adhesions of this avoidance-custom, a relation appears between it and the customs of the world as to residence after marriage. **1924** W. H. Rivers *Social Organisation* v. 90 The last aspect of father-right and mother-right to be mentioned ..is one with which perhaps I ought to have dealt in the second chapter, namely, the place of residence in case of marriage. **1938** G. A. Riechard in F. Boas *Gen. Anthrop.* ix. 421 Among primitives residence is of even greater importance. **1968** Jacobson & Schoepf tr. *Levi-Strauss's Structural Anthrop.* xv. 309 The impact of residence on descent. **1972** E. A. Hoebel *Anthrop.* (ed. 4) xxi. 427/2 There are, then, five basic varieties of residence. *Ibid.*, It is not a simple..matter as to whether a given household represents one kind of residence pattern or another.

2. b. Used *spec.* with reference to a residential post held by an artist, poet, sculptor, writer, etc., within a community or institution for the purpose of teaching his craft or influencing communal life. Also *transf.* (freq. *joc.*). Cf. also *poet-in-residence* (*Poet 1 e).

1954 R. Jarrell *Pictures from Institution* iii. 94 How glad Dr. Rosenbaum was that he was only a Composer in Residence. **1965** *Economist* 30 Oct. 499/2 Its money will also be used..to help symphony orchestras and support artists-in-residence at universities. **1968** Mrs. L. B. Johnson *White House Diary* 30 Apr. (1970) 667 He [*sc.* Eric Hoffer] said, 'I call myself a conversationalist in residence.' **1970** *Times* 16 Apr. 8/5 No girls throwing smoke in our eyes at last night's Gold Leaf concert..instead a tame composer, a 'composer in residence'... Peter Patterson..is the lucky man commissioned to write pieces. **1972** W. King *Black Short Story Anthol.* 133 She [*sc.* Alice Walker] is presently writer-in-residence at Tougaloo College,..Mississippi. **1975** *New Yorker* 5 May 45/3 An artist-in-residence at WNET's Television Laboratory and a consultant on television to the Rockefeller Foundation. **1976** C. Bermant *Coming Home* i. v. 70 A few weeks later, a Jewish boy was billeted on us... I'm not sure if I welcomed it, for I had begun to enjoy my role as Jew in residence. **1979** *Guardian* 27 June 8/3 (Advt.), Photographer in Residence required... There are no specific teaching commitments but the photographer will be expected to participate in the life of the centre and be available to students and others on an informal basis. **1980** *Early Music Gaz.* Apr. 8 (Advt.), Lessons in solo and ensemble performance are conducted by an artist-in-residence each semester.

5. a. (*of*) *no fixed residence* = (*of*) *no fixed abode* s.v. *Abode *sb.*[1] 5. ? *Obs.*

1859 *Times* 1 Feb. 9/5 Robert Murry, 35, and Johanna Murry, 28, described as husband and wife, and travelling ventriloquists of no fixed residence, were charged with attacking and robbing Robert Hobbs on the highway of a purse containing a 5*l.* Bank of England note. **1863** W. Hardman *Let.* (1925) ? 24 Aug. 72 His books were curious, inasmuch as the customers had no fixed residences.

e. [tr. Russ. *residentura.] A group or organization of intelligence agents in a foreign country.

1969 H. H. Cooper *Cave with Two Exits* i. 69 In Rome he was met by a young man from the Residence... The Resident himself..was extremely secure... His cover was strictly diplomatic.

8. *residence address, permit, time*; **residence city** [tr. G. *residenzstadt*], a seat of a royal or princely court; cf. *Residenz; **residence counsellor** *U.S.*, a psychiatric adviser attached to a residential block in a university; **residence general** [tr. F. *résidence générale*], the official residency of the senior French representative in a French protectorate.

1890 'Mark Twain' *Lett. to Publishers* (1967) 264, I do not know your residence-address or Whitford's. **1961** L. Mumford *City in Hist.* xiii. 386 The chief new cities built from the sixteenth to the nineteenth century were 'residence cities' for kings and princes, like Versailles, Karlsruhe, and Potsdam. **1973** *Pennsylvania Voice* 10 Oct. 5/1 A residence counselor who has been dealing with a student may feel that he is not properly equipped to help the student resolve his problem. **1955** *Times* 9 May 8/4 The crowd..had tried to break into the residence-general. **1962** *Guardian* 14 Aug. 1/7 Bishop Roseveare was given the letter informing of the cancellation of his residence permit. **1977** *Times* 16 Apr. 12/7 Herr Klee got a job..which clinched his residence permit. **1962** A. Battersby *Guide to Stock Control* viii. 76 With an average stock of 200 tons and annual sales of 1,200 tons, the residence time is 2 months. The more commonly used 'turnover' is 6 times a year, which is the reciprocal of the residence time. **1969** *Gloss. Terms Vacuum Technol.* (*B.S.I.*) 13 *Residence time*, the average time for which a molecule is bound to a surface in a state of sorption. **1977** *Sci. Amer.* May 23/2 The period of time a particular amount of water spends in its cycle is termed its residence time... Deep in the oceans the residence time of a body of water may be more than 1,000 years. **1978** *Nature* 20 July 246/1 The concept of oceanic residence time has been used widely in marine chemistry; element residence times range from 10⁸ to < 100 yr, and provide a useful measure of the reactivity of an element in the ocean.

residency. 1. a. Delete † *Obs.* and add later N. Amer. examples. Also *attrib.*

1966 *Publ. Amer. Dial. Soc. 1964* xlii. 35 *Shanty-..Irish*, i.e., those who remain in the lower-class communities near the center of the city (or, irrespective of residency, preserve the social traits of the shanty Irish). **1970** *Daily Colonist* (Victoria, B.C.) 2 Apr. 30/1 The controversial [abortion] measure contains no residency requirement and would allow a pregnancy to be terminated at any time. **1981** *Times* 16 Feb. 15/7 Degrees for people who want to be more effective and secure in their Jobs or Professions... No residency required. (Advt. by an Amer. university.)

c. Of a musician or a band: the state of being permanently or regularly engaged at a club, etc.

1966 *Melody Maker* 7 May 5/3 Colin Smith..has left to take up a residency at London's Georgian Room. **1968** *Ibid.* 30 Nov. 6/1 The Nice..used to have a residency there. **1971** *Ibid.* 4 Sept. 21 She moved back to Chicago and played long residencies there..as a pianist vocalist. **1975** *Evening Herald* (Dublin) 8 May 12/4 (Advt.), *Popular vocalist*..seeks position with musicians..; preference for residency in lounge.

2. a. (Later examples.) Also *transf.* (in sense of Resident *sb.* 2 d) and *attrib.*

1958 L. van der Post *Lost World of Kalahari* v. 90, I spent the evening and night with the Resident and his wife..in the ample Residency. **1971** R. Russell tr. *Ahmad's Shore & Wave* xv. 159, I once met Lord Mountbatten myself in Kashmir at a Residency lunch. **1980** J. Hone *Flowers of Forest* i. 8 The annual reception for the Queen's Birthday at the old British Residency on the Nile.

b. = *Residence *sb.*[1] 5 e.

1970 K. Benton *Sole Agent* xviii. 199 [He is]..supplied by the Soviet Illegal Residency here, in Lisbon, to help Rogov to get rid of the body. **1977** 'J. Le Carré' *Hon. Schoolboy* iii. 53 What did he do up there..his networks blown to smithereens? His foreign residencies, his reptile fund frozen solid by the Treasury—they meant his operational accounts..and not a friend in Whitehall or Washington to call his own? **1981** J. Simpson *Moscow Requiem* ii. vi. 177 The KGB resident in Riyadh..made his last radio contact with the KGB Residency in Aden.

4. *N. Amer.* The position or station of a resident (*Resident *sb.*[1] 3); the period during which one holds this position.

1924 *Mod. Hospital* XXIII. 422/1 Residencies should also be offered in general hospitals. **1933** *Southern Med. Jrnl.* XXVI. 773/2, I am not in sympathy with the prolonged residency extending over more than two years. **1949** G. D. Wolf *Physician's Business* (ed. 3) i. 10 A promising young doctor, following a year's internship, could receive an assistant residency for two years. *Ibid.*, The ideal residency is one in which all phases of the specialty are stressed. **1958** F. G. Slaughter *Daybreak* i. 9 They were also competing for a residency on the Neurosurgical Service—a vitally important stepping-stone for an ambitious doctor on the last lap of his training. **1975** R. H. Rimmer *Premar Experiments* i. 66 I've got a friend who's an MD... He's doing his residency at Boston City Hospital. **1977** *Washington Post* 24 May 1/2 Georgetown University Medical School and Providence Hospital have announced their joint development of a residency program in..general, or family, practice. **1979** *Sci. Amer.* Oct. 16/3 After an internship and residency at the Hospital of the University of Pennsylvania he was Special Projects Associate in epidemiology at the Mayo Graduate School of Medicine in Rochester, Minn.

resident, *a.* and *sb.*[1] Add: **A.** *adj.* **2. a.** (Further examples.)

1812 *Laws & Ordinances City of New-York* 202 It shall be the duty of the said resident physician and health commissioner, to meet daily at the health office, from the thirty-first day of May to the first day of October. **1844** [see *grave-trap* s.v. *Grave *sb.*[1] 6]. **1899** Somerville & 'Ross' *Some Experiences Irish R.M.* i. 10 You wouldn't meet a Christian out of doors, unless it was..a resident magistrate. **1902** *U.S. Laws & Statutes* (1903) XXXII. i. 694 With the first meeting of the Philippine legislature..there shall be chosen, by the said legislature..two resident commissioners to the United States. **1919** *Bull. Johns Hopkins Hospital* XXX. 191/2 The assistant residents, even those who did not become chief resident physicians, often continued in office for several years. **1931** V. Palmer *Separate Lives* 40 I'm a Resident Magistrate now, you know. One of the biggest stations in the Territory. **1952** Granville *Dict. Theatrical Terms* 151 *Resident manager*, the manager of the theatre as distinct from the acting manager of a visiting company. He handles all the local business. **1962** Plano & Greenberg *Amer. Polit. Dict.* iii. 38 *Resident commissioner*, a delegate elected by the people of a territory to represent them in the House of Representatives. He may speak in the House and serve on committees, but he may not vote... Only Puerto Rico has a resident commissioner. **1963** J. Joesten *They call it Intelligence* ii. viii. 73 Abel..held the post of 'resident officer'... His job was to recruit and organize local spies. **1974** P. Gore-Booth *With Great Truth & Respect* 364 On 16 May 1967, at about 6.00 in the morning, my bedside telephone rang. The Resident Clerk at the Foreign Office said that President Nasser had ordered the United Nations Expeditionary Force to leave the Egyptian frontier with Israel. **1976** *National Observer* (U.S.) 22 May 22/1 The resident correspondent can suggest people to interview. **1977** *Belfast Tel.* 19 Jan. 3/9 The resident band..will be George Chambers and the Apex.

4. Delete † *Obs.* and add later example.

1971 *Nature* 12 Mar. 91/1 The non-polar amino-acids, for example in myoglobin and haemoglobin, are resident in the interior of the protein molecule, out of contact with the aqueous environment.

B. *sb.* **1. a.** Also, a guest staying one or more nights at a hotel or boarding-house.

1956 M. Stewart *Wildfire at Midnight* xii. 103 A little sitting-room beside the residents' lounge. **1969** 'L. Bruce' *Death with Blue Ribbon* vii. 71 Carolus ordered three doubles. He had stayed in the Residents' Lounge too long.

c. A bird belonging to a species that does not migrate but is found all the year round in a particular area.

1896 Kirkaldy & Pollard tr. *Boas's Text Bk. Zool.* 465 Woodpeckers..are 'residents', or wander about in a limited locality. **1920** H. F. Witherby *Pract. Handbk. Brit. Birds* I. 18 After nesting our residents [*sc.* rooks] are subject to partial and irregular movements—some probably emigrating to Continent. **1978** *Ibis* CXX. 496 In West Africa.. residents and migrants overlap throughout the dry season.

2. b, c, d. These senses now only *Hist.* (Further examples.)

1839 T. J. Newbold *Pol. & Statistical Acct. Straits of Malacca* I. iv. 126 The public property there was estimated ..by the British resident. **1958** [see *Residency 2 a]. **1965** A. Nicol *Truly Married Woman* 88 Ride to the white man who is Resident. **1971** R. Russell tr. *Ahmad's Shore & Wave* i. 13 The British Resident was confiding to the Finance Minister his distrust of the Prime Minister.

3. *N. Amer.* A medical graduate who has completed an internship and is engaged in specialized practice under supervision in a hospital, usu. as training for independent specialization. Also *attrib.* Cf. *Registrar 3.

1892 *Rep. Johns Hopkins Hospital* III. 21 Dr. William Osler [gave]..Books for Resident's Reading Room. **1914** *Mod. Hospital* II. 30/2 The sixth floor is primarily for residents' quarters. The chief resident will have his study, bed room, and private bath. **1931** E. G. Reid *Great Physician* vi. 116 Long term residents took the place of the usual short term internship. His first Resident Physician had under him three assistant residents. **1938** [see *Intern *sb.*]. **1970** *Globe & Mail* (Toronto) 26 Sept. B8/4 (Advt.), Hospital approved for interne and resident training, nursing school, [etc.]. **1977** 'E. Trevor' *Theta Syndrome* vi. 85 Never..disturb the senior resident when he's on his rounds, unless the place is on fire.

4. [tr. Russ. *rezidént*.] An intelligence agent (in a foreign country). Cf. also *Rezident.

1963 'J. Le Carré' *Spy who came in from Cold* viii. 74 The Resident in a particular country would make a requisition... The Resident could draw it [*sc.* money] himself and hand it to the agent. **1968** A. Diment *Bang Bang Birds* v. 63 Of course this Agency has a resident in Stockholm. **1975** *Times* 16 Dec. 7/5 Herr Guillaume soon became a 'resident'—the head of a group of spies.

residenter. Add: **2.** (Further and *transf.* examples.) *old residenter*, a pioneer in the U.S.

1827 *Western Monthly Rev.* I. 70 Hence arose a feud and a collision of authorities between the old and the new 'residenters'. **1880** W. H. Patterson *Gloss. Words Antrim & Down* 82 *Residenter*, *sb.* an old inhabitant. **1898** E. N. Westcott *David Harum* 253, I ain't what ye might call an old residenter. **1921** H. Guthrie-Smith *Tutira* xxii. 212 Such residenters are attracted..by the influx of sparrows, rats, and mice. **1967** *Buchan Observer* 5 Sept. 7 Residenters ..are becoming greatly concerned. **1971** M. Tak *Truck Talk* 130 *Residenter*, any old tractor still used over the road. **1975** *New Yorker* 29 Dec. 36/3 Well, here's an old guy in the audience—an old residenter, that's what they called them—with a long grey beard.

residential, *a.*[1] **1. a.** Delete 'Now rare or *Obs.*' and add later examples.

1886 *Fortn. Rev.* XXXIX. 24 It [*sc.* a medical college for women] has no residential hall, nor is it desirable perhaps that it should have any. **1910** *Bradshaw's Railway Guide* Apr. 1009 Imperial Hotel. First-class family and residential. **1923** W. J. Locke *Moordius & Co.* xxiii. 308 An untidy boarding-house in Torrington Square, Bloomsbury, which called itself a Residential Hotel. **1943** [see *neighbourhood unit* s.v. *Neighbourhood 7]. **1960** *Times* 21 Mar. 8/5 Oxford, Cambridge and the other residential universities enjoy special prestige compared with the civic universities. **1963** *Specification & Recommendations for Permanent Residential Caravans* (B.S. 3632) 4 The term 'Permanent residential caravan' means a structure with a wheeled chassis..designed for use as a permanent dwelling..and so constructed as to be movable by towing or other methods of

transport. Hereinafter the term 'residential caravan' will be used for brevity. **1969** *Sunday Times* (Colour Suppl.) 9 Nov. 85/4 The residential towers in flat units will next year house only 400 students. **1977** *Times Educ. Suppl.* 21 Oct. 56/2 (Advt.), Required..a head teacher for this newly opening..residential school for maladjusted boys of secondary age. **1981** J. Carey *John Donne* i. 23 The Inns of Court ..operated like residential clubs or hotels.

b. (Later examples.)

1900 H. A. Jones *Mrs. Dane's Defence* III. 49 My dear Sir Daniel, we live in a residential neighbourhood..and what possible occupation is there..except to discuss scandal. **1922** Joyce *Ulysses* 647 A phenomenally beautiful tenor voice like that..could easily..procure for its fortunate possessor..an *entrée* into fashionable houses in the best residential quarters. **1981** J. B. Hilton *Playground of Death* iv. 46 Below was the black tumour of a residential area.

2. (Further examples.)

1869 *Bradshaw's Railway Man.* XXI. 298 The moderate advance of rates..did not affect the 'residential' or periodical and season tickets. **1934** Webster, Residential zone. **1952** *Density of Residential Areas* (Ministry of Housing & Local Govt.) 1/1 The term 'residential density'..signifies the degree of closeness with which dwellings, and hence the people occupying them, are arranged in the residential areas of towns and villages. **1961** E. A. Powdrill *Vocab. Land Planning* v. 97 The residential zone will become an industrial and commercial zone and so creating [*sic*] new problems. **1974** *Drive* Autumn 110/2 Disc parking and residential parking schemes, plus a maze of one-way streets, make life harder for motorists. **1977** *Age* (Melbourne) 18 Jan. 8/3 Parents doing residential adoptions (i.e. going to the country of the child's origin) are approved by the host country.

B. as *sb.* A residential hotel.

1953 K. Tennant *Joyful Condemned* xxxvi. 351 Margot took down from Stella the address of the residential where she and Dorsie were staying. **1954** J. Symons *Narrowing Circle* xxix. 129 She had a connection with the Gongora Residential. **1973** *Nation Rev.* (Melbourne) 31 Aug. 1450/1 The..premises..have been an inconspicuous residential for the past three years.

reside·ntially, *adv.* [f. Residential *a.*[1] + -ly[2].] As a residence; according to residence; with the provision of residential accommodation.

1913 *Chambers's Jrnl.* Jan. 51/2 London contains no single palace residentially associated with our long line of sovereigns. **1922** *Daily Mail* 1 Dec. 8 It is sufficiently developed to be comfortable residentially. **1974** *Sci. Amer.* Sept. 125/1 The American high school has all the requisites of an institution designed to promote mating: single males and females are assembled in substantial and equal numbers, are residentially selected for social homogeneity, [etc.]. **1978** *Daily Tel.* 24 July 13/2 The conference was being held residentially at Lambeth.

‖ **residentura** (rezidentū²·rā). Also **rezidentura.** [Russ. *rezidentúra*: cf. *Resident *sb.* 4.] A group or organization of intelligence agents in a foreign country.

1963 A. Dulles *Craft of Intelligence* viii. 102 Within the embassy there would still be intelligence officers, but they would restrict themselves, except in emergencies, to 'clean' operations... This unit the Soviets call the 'legal *residentura*'. Outside the embassy..perhaps in a bookstore or a photography shop...was headquarters of the 'illegal' *residentura*. **1969** A. Marin *Rise with Wind* i. 8 He wouldn't like to be a Russian *Kaygaybehnik* confronted with Clay... The man was capable of bursting into any accessible *residentura* and slaughtering the comrades. **1977** 'R. Rostand' *Killing in Rome* i. 5 Chief of the entire *rezidentura* [in Rome] was the first secretary himself, a KGB colonel. **1979** *Daily Tel.* 15 Oct. 5/2 The deputy chief of the KGB Rezidentura in Tokyo..maintains numerous contacts among the staff and research fellows.

‖ **Residenz** (rezide·nts). Also **residenz.** [Ger., lit. 'residence'.] The building in which a German princely court resided before 1918; a town which was the seat of a princely court. Hence **Residenzstadt** (-ʃtat) [G. *stadt* town], the seat of a court.

1840 Thackeray in *Fraser's Mag.* Aug. 151/2 Twice as many people in the streets as you see at midday in a German *residenz* or an English provincial town. **1848** —— *Van. Fair* lxiii. 570 Troops of people..flock to the Residenz and share in the pleasures of the fair and the festivities there. **1915** F. H. Burnett *Lost Prince* xxii. 215 But I want to know who lives at the Residenz? **1961** L. Mumford *City in Hist.* Note to plate 4. These royal tomb cities were thus the earliest form of permanent Residenzstadt, like Versailles or Karlsruhe. **1965** *Observer* 9 May 40/7 The State apartments in the old Kings' Residenz..are being repainted. **1973** *Country Life* 15 Nov. 1542/1 The two vases in the Residenz in Munich. **1980** *Times* 9 Aug. 8/7 The Residenz, the baroque palace of the Von Schönborns.

residual, *sb.* Add: **1. c.** The difference between an observed or measured value of a quantity and its true, theoretical, or notional value.

1868 J. C. Watson *Theoret. Astron.* vii. 370 In the case of a limited number of observed values of *x*, the residuals given by comparing the arithmetical mean with the several observations will not..give the true errors. **1872** *Mem. R. Astron. Soc.* XXXIX. 100 The usual treatment of equations, by rendering a minimum the sum of the squares of the residuals, not only assumes equal weights for the observations, but also that positive and negative errors are equally probable. **1906** Wright & Hayford *Adjustment of Observations* (ed. 2) ii. 12 The sum of the positive residuals is equal to the sum of the negative residuals. **1923** Glazebrook *Dict. Appl. Physics* III. 647/2 The distribution of 174 residuals in a certain case is shown in Fig. 1. The residuals are the differences between observed and calculated monthly sea levels at three tidal observatories. **1932** *Human Biol.* IV. 478 Departures from the average arm girth and calf girth of children of the given four skeletal dimensions, we have termed residuals and have regarded them as indices of muscular development. **1967** *Oceanogr. & Marine Biol.* V. 12 When the residuals at a port are evaluated at say hourly intervals and then plotted against time, a graphical representation of the variations in the surge component at that port is obtained. **1972** *Nature* 17 Mar. 96/1 A major contribution to understanding of the seismic velocity variation necessarily comes from travel time residuals (these are differences between observed arrival times of P-waves at seismometers and the theoretical arrival time from a standard set of tables).

2. a. (Later examples.)

1967 *Economist* 28 Oct. p. xxiii/2 According to the recorded items on the balance of payments Sweden should by now be dead broke. The one thing that has kept it afloat is an unexplained positive 'residual' in the balance of payments. **1973** 'E. McBain' *Let's hear It* iii. 42 His first impression was one of total harmony... Her face and figure came as residuals to his brief course in art appreciation.

b. The part of a gravity anomaly or magnetic anomaly that remains after subtraction of the regional. Also *attrib.*

1949 *Geophysics* XIV. 45 Since the density of salt is less than that of the surrounding sediments, its residual is negative. *Ibid.* 516 Residual maps have been used extensively by geophysicists to bring into focus local features which tend to be obscured by the broad features of the field. **1965** Krumbein & Graybill *Introd. Statistical Models in Geol.* xiii. 325 Many residual maps contain geological information of value in exploration for natural resources. **1978** *Nature* 13 July 146/1 After removing the mean values we projected the residuals on to profiles transverse and parallel to the plate motion and averaged the values over 0·5° intervals.

3. Also *transf.*

1899 *Daily News* 24 June 4/1 The casual docker is often a residual—the driftwood of society.

4. A royalty paid to an actor, musician, etc., for a repeat of a play, television commercial, etc.

1966 *Guardian* 14 May 7/1 The 'residuals', or BBC fringe benefits, lie chiefly in adoption by the Transcription Service. **1971** *Daily Tel.* 25 Mar. 21/4 He [*sc.* Frank Sinatra] will continue to receive money from..record royalties, residuals of his television programmes and a share in the profits of his films. **1972** D. Ramsay *Little Murder Music* 78 'What makes a jingle date such an important affair?.. The money?' 'And how. I could wind up making more in residuals alone.' **1977** *Rolling Stone* 13 Jan. 19/1, I love doing jingles because I get residuals.

5. *Geomorphol.* A portion of rocky or high ground remaining after erosion.

1968 R. W. Fairbridge *Encycl. Geomorphol.* 587/1 Normal fluvial erosion begins to be established, and eventually, only residuals of limestone..are left standing. **1970** R. J. Small *Study of Landforms* v. 162 Above the peneplain, a few isolated hills..would remain. Such residuals were referred to by Davis as 'monadnocks'.

residual, *a.* Add: **2. a.** (Later examples.) *residual legatee* = *residuary legatee* s.v. Residuary *a.* 1 b; also *fig.*; *residual powers* = *residuary powers* s.v. *Residuary *a.* 2 b.

1919 G. B. Shaw in *Irish Statesman* 25 Oct. 428/1 When the enfranchisement of the Dominions began with Canada, the question on which freedom depends: namely, which party is to have residual powers, was hardly raised. Those were early days for democracy; and the residual powers were left technically to England. **1963** *Listener* 21 Feb. 334/1 The Liberals cash in..by claiming to be the residual legatees of the radical tradition. **1967** *Boston Sunday Herald* 30 Apr. III. 10/1 A specially interesting use of this guide's table of 'Life Expectancy' and 'residual values' of textiles is in evaluating garments and other textile items donated to charity. **1969** F. Halliday in Cockburn & Blackburn *Student Power* 296 The dominant 'mainstream' faction.. was labelled as 'Trotskyite', a residual term meaning that the group was Marxist but opposed to the official Japan Communist Party line. **1976** *National Observer* (U.S.) 3 July 8/6 'There shouldn't be any wholesale dumping of what we think are good risks into the residual market,' says the Insurance Information Institute spokesman. **1979** *Jrnl. R. Soc. Arts* Jan. 97/2 Alice was the residual legatee and executrix.

d. Applied to magnetism which is retained after the removal of the magnetizing force or in the absence of a magnetizing current.

1837 *Phil. Mag.* X. 195 If the interrupted keeper be applied to a compound magnet, it will be attracted only with a force equal to the quantity of residual magnetism, which being very small the attraction will be comparatively feeble. **1874** *Tyer's Block Telegraph & Electric Locking Signals* (ed. 5) 14 There is also..an electro-magnet fitted with 'horns' or 'keepers'.., so arranged as to..retain for an indefinite period the residual magnetism produced on the passage of each signal. **1886** [see *Retentiveness 6]. **1902** *Encycl. Brit.* XXX. 430/2 If a bar of hard steel is placed in a strong magnetic field, a certain intensity of magnetization is induced in the bar, but when the strength of the field is afterward reduced to zero, the magnetization does not entirely disappear. That portion which is permanently retained, and which may amount to considerably more than one-half, is called the residual magnetization. **1912,** etc. [see *Retentivity]. **1917,** etc. [see *Remanence 3].

e. Applied to the small charge which some capacitors remain capable of delivering after 'discharge'.

1838 Faraday in *Phil. Trans. R. Soc.* CXXVIII. 24 It is the assumption for a time of this charged state of the glass between the coatings in the Leyden jar, which gives origin to a well-known phenomenon, usually referred to the diffusion of electricity over the uncoated portion of the glass, namely, the residual charge. **1878** *Encycl. Brit.* VIII. 39/2 Kohlrausch called attention to the close analogy between the residual discharge and the 'elastic recovery'..of strained bodies... The instantaneous strain which follows the application of a stress is analogous to the initial charge of the jar, and the gradually increasing strain which follows to the gradual formation of the latent or residual charge. **1921** T. F. Wall *Electr. Engin.* iv. 62 The successive charges after the first discharge are known as residual charges. **1938** H. G. Mitchell *Textbk. Electr.* iv. 64 The residual discharge of the condenser can easily be explained if we assume that the charge penetrates the glass and cannot all reach the conducting plates when these are connected.

f. *Physical Geogr.* Applied to a deposit or feature formed *in situ* by the weathering of rock, and to a soil largely composed of such material.

1895 *Geogr. Jrnl.* V. 140 Most of the peneplains that I have examined..still possess residual elevations,..evidently to be regarded as unconsumed remnants of the denudation of the former cycle. **1906** E. W. Hilgard *Soils* i. 11 When soils have been formed without removal from the site of the original rock, by simple weathering, they are designated as sedentary, or residual soils. **1933** [see *Peneplain *sb.*]. **1937** Wooldridge & Morgan *Physical Basis Geogr.* xi. 150 While its origin has given rise to considerable controversy, all are agreed that it [*sc.* laterite] is essentially a residual deposit. **1944** A. Holmes *Princ. Physical Geol.* viii. 119 Ultimately the angular block is transformed into an onion-like structure of concentric shells of rusty and thoroughly rotted residual material. **1954** W. D. Thornbury *Princ. Geomorphol.* iv. 74 Soils were commonly divided into two major groups, residual and transported. Residual soils were classified according to the type of rock from which they were derived. **1972** J. G. Cruickshank *Soil Geogr.* ii. 57 Parent materials may be classified by their mode of formation as follows: 1 Weathered rock in place which produces residual soils.

g. *residual stress*, stress present in an object in the absence of any external load or force.

1931 A. Nádai *Plasticity* xxxviii. 259 Stresses of this kind, remaining after partial plastic flow, may be called residual stresses. **1958** *Engineering* 18 Apr. 498/1 The operation of welding gives rise to residual stresses, i.e. stresses which exist independently of the external loading. **1976** Lindberg & Braton *Welding & Other Joining Processes* xii. 454 A weldment heated to a temperature at which the yield strength is low..will relieve the residual stresses and increase fatigue life.

residuary, *a.* and *sb.* Add: **A.** *adj.* **1. b.** (Later and *transf.* examples.)

1962 S. E. Finer *Man on Horseback* viii. 123 The Justice party..in a sense was the residuary legatee of the old Democrats. **1967** E. Rudinger *Wills & Probate* 11 The wife is probably the residuary legatee.

2. b. spec. *residuary powers*, powers remaining with one political group after other powers have been allocated to another group, e.g. as between a federal government and a province.

1919 G. B. Shaw *Let.* in *Irish Statesman* 22 Nov. 536/2 The meaning of residuary powers and the importance of the Australian precedent in their bearing on the Irish national question. **1950** Theimer & Campbell *Encycl. World Politics* 163/2 In some federations..the powers of the federal government are named in the constitution and all other powers (the residuary powers) belong to the regions.

3. Applied to the Established Church of Scotland after the Disruption in 1843. *Obs.* or *arch.*

1843 *Witness* 19 May 2/7 (*heading*) Residuary Assembly... After the Evangelical Party had left the House [etc.]. **1845** J. Bright in *Hansard Commons* 16 Apr. 882/2 Even in Scotland..there were the Secession Church, the Relief Church, and the Free Church; that which the State upheld being called by the complimentary name of the Residuary Church. **1883** R. Cleland *Inchbracken* iii. 22 The 'Residuary' Presbytery, as you are pleased to denominate the church of your fathers.

B. *sb.* Also *fig.*

1920 E. Gosse in *Edin. Rev.* Jan. 47 He was the residuary of his own temperament.

residue. Add: **1. e.** *Sociol.* A term used by Vilfredo Pareto (1848–1923) for fundamental impulses which motivate human conduct, and which are not the product of rational deliberation.

1933 *Harper's Mag.* CLXVII. 573 In each of these two groups of phenomena we have found a constant element, the sentiment out of which both the actions and the explanations rise. The *expression* of such a sentiment is what Pareto calls a 'Residue'. **1935** Bongiorno & Livingston tr. *Pareto's Mind & Society* II. vi. 508 It might perhaps be advisable to give word-names to the things we have been calling *a*, *b*, and *c*... Suppose we call the things *a*, *residues*, the things *b*, *derivations*, and the things *c*, *derivatives*. *Ibid.* 509 Residues correspond to certain instincts in human beings, and for that reason they are usually wanting in definiteness, in exact delimitation. **1935** L. J. Henderson *Pareto's Gen. Sociol.* v. 23 Theology and metaphysics and parts of law consist, in great measure, of systematic and extensive derivations from certain very important residues like those involving the words justice, duty, sanctity, and absolute. **1939** F. Creedy *Human Nature writ Large* xix. 400 Public documents which have won high praise and wide acceptance..owe their acceptability to the fact that their Residues or fundamental assumptions coincide with those of the crowd. **1958** W. Stark *Sociol. of Knowl.* viii. 320 These residues or drives determine all the comings and goings on the stage of life. The 'derivations', or thoughts and beliefs

which, objectively speaking, derive from the residues, but which subjectively appear to the people who harbour them as 'their ideas', 'their convictions', 'their philosophies'. **1966** S. E. Finer *V. Pareto: Sociol. Writings* 38 Thus, if you strip from the theory its variable part you are left with a residuum. This is the constant element and Pareto therefore calls it a *residue*.

2. (Later example.)

1967 E. Rudinger *Wills & Probate* 9 After disposing of specified items and sums of money, you give the remainder —the residue, lawyers call it—to some named person... You could, of course, provide that your residue should be divided..among a number of people.

3. Restrict † *Obs.* to sense in Dict. and add: **b.** *Number Theory.* A remainder left when a given number is divided into some integer; also, a number congruent to a given number modulo a third number; *quadratic*, *cubic*, etc., *residue*, a remainder left when a given number is divided into the square, cube, etc., of some integer; *residue class*, the class of integers congruent to one another modulo a given number (e.g. 2, 5, 8, 11, etc., are members of a residue class modulo 3).

1860 *Rep. Brit. Assoc. Adv. Sci. 1859* I. 231 The set of numbers 0, 1, 2... P − 1 (or any set of P numbers respectively congruous for the modulus P to those numbers) is termed a complete system of residues for the modulus P. **1899** *Q. Jrnl. Math.* XXX. 363 The same representation of the coefficients also determines the residue when the coefficient is not a multiple of the modulus. **1939** Uspensky & Heaslet *Elem. Number Theory* vi. 134 Of two numbers congruent for the modulus *m*, each is called a 'residue' of the other modulus *m*. *Ibid.* x. 270 A number *a* is said to be a quadratic residue of another number *m* if the congruence $x^2 \equiv a \pmod{m}$ can be satisfied by some integer *x*. **1948** O. Ore *Number Theory & its Hist.* ix. 214 Since these are the numbers that correspond to the same remainder *r* when divided by *m*, we say that they form a residue class (mod *m*). There are *m* residue classes (mod *m*). **1963** W. W. R. Ball *Math. Recreations & Ess.* (ed. 12) ii. 60 There is an arithmetic of residues, closely analogous to the arithmetic of ordinary numbers. **1966** Ogilvy & Anderson *Excursions in Number Theory* iv. 43 If we identify every integer, positive, negative, or zero, with its remainder modulo *m*, we thus have all the integers partitioned into congruence classes, or residue classes, modulo *m*. **1977** *Sci. Amer.* July 127/2 Gauss used complex numbers of the form $a + bi$..to formulate and prove a version of the law of quadratic reciprocity for biquadratic residues. The number *k* is said to be a biquadratic residue of another number *m* if *k* is congruent modulo *m* to the fourth power of an integer. Thus the biquadratic residues of 10 are 0, 1, 5 and 6.

c. *Theory of Functions.* (See quots.)

1893 A. R. Forsyth *Theory of Functions* x. 223 The sum of the residues of a doubly-periodic function relative to a fundamental parallelogram of periods is zero. **1957** T. M. Apostol *Math. Analysis* xvi. 524 In many cases it is relatively easy to evaluate the residue at a point without the use of integration. **1959** G. & R. C. James *Math. Dict.* 334/1 If $f(z)$ is an analytic function of the complex variable *z* in the 'deleted' neighbourhood consisting of all *z* satisfying $0 < |z - z_0| < \epsilon$, then the residue of $f(z)$ at z_0 is

$$\frac{1}{2\pi i}\int_C f(z)dz,$$

where *C* is a simple closed rectifiable curve about z_0 in the 'deleted' neighbourhood. The value of the residue is..the coefficient of $(z - z_0)^{-1}$ in the Laurent expansion of $f(z)$ about z_0. **1973** Rauch & Lebowitz *Elliptic Functions, Theta Functions, & Riemann Surfaces* i. 19 A differential of the *first kind* is one whose set of Laurent expansions has no poles. Differentials of the *second kind* have poles but with zero residue at each pole while differentials of the *third kind* allow poles with non-vanishing residues.

5. (Earlier and later examples.) In mod. use applied to any molecule when incorporated without major alteration in a larger one; *esp.* in *Biochem.*, an amino-acid, sugar, or other molecule incorporated in a polymer such as a protein, carbohydrate, etc.

1852 H. Watts tr. *Gmelin's Hand-bk. Chem.* VII. 76 The compound thus formed by substitution contains therefore the residues of the two compounds united, that is to say, the first compound + the second, *minus* an equal number of H- and O-atoms. *Ibid.* 77 Oxamide..may be regarded as a compound of the two residues, $C^4H^2O^4$ and 2NH. **1886** E. F. Smith tr. *V. von Richter's Org. Chem.* 30 Ordinarily, radicals are groups containing carbon, while all others, like OH, SH, NH_2, NO_2, are residues or groups. **1903** *Jrnl. Chem. Soc.* LXXXIV. I. 214 The (albumin) molecule may be regarded as made up of 125 groups or residues. **1934** W. R. Fearon *Introd. Biochem.* xv. 231 The test is believed to depend on the hydrolytic unmasking of an aldehyde group in the hexose residue. **1955** H. R. Downes *Chem. Living Cells* v. 186 In one sample of lactoglobulin, for example, there were found 20 serine residues, 36 from aspartic acid, and 50 from leucine. **1975** *Nature* 27 Feb. 694/2 Even the smallest protein (say 50 residues) is extremely complicated.

resign, *sb.* For † *Obs.* read *rare* and add earlier and later examples.

1639 Chapman & Shirley *Chabot* v. sig. H3ᵛ, My free Resigne of title, office,..would buy my poore lives safety. **1971** J. V. Allen *Cowboy Lore* iv. 159 It was a pistol shot that laid Pete out, It was his last resign.

resign, *v.*[1] Add: **I. 2. c.** Quot. *a* 1704 to read: *c* **1698** Locke *Some Thoughts on Conduct of Understanding* (1881) 61 [etc.].

d. Also const. with *inf.*

1898 G. B. Shaw *You never can Tell* II. 265 She smiles in spite of herself, and resigns herself to indulge him a little. *a* **1953** E. O'Neill *More Stately Mansions* (1964) I. ii. 39 Resign myself to be a grandmother!

II. 5. b. For *U.S.* read 'orig. *U.S.*' and add later examples.

1926 G. B. Shaw *Translations & Tomfooleries* 235 My brothers said I ought to resign from my clubs. **1959** *Chambers's Encycl.* II. 68/1 In 1922 he resigned from office and was raised to the peerage as earl of Balfour. **1973** Howat & Taylor *Dict. World Hist.* 1619/2 In April 1951 he resigned from the government..in protest against an increase in social service charges. **1980** *Times* 20 Aug. 1/2 Mr A. J. P. Taylor resigned from the British Academy over what he described as a witch hunt by some members to remove Professor Anthony Blunt.

7. (Later Austral. example.)

a **1964** in *Penguin Bk. Austral. Ballads* (1964) 75 'Die or resign, Jack Donahoe!' they [*sc.* police] shouted in their joy.

re-sign, *v.* Add: **b.** *intr.* Of a sportsman, performer, etc.: to sign a contract for a further period. Also *trans.*, to cause (a person) to do this.

1938 C. E. Sutcliffe et al. *Story of Football League* iii. 19 The player claimed from the club a sum of £250 which..had been promised him..on re-signing in June. **1951** *People* 3 June 8/3 We are urging players not to re-sign for next season until we have consulted the Ministry. **1976** *Jrnl.* (Newcastle) 26 Nov., Berwick Rangers have re-signed centre-forward Billy Laing..who has been on a month's trial at Shielfield. **1977** *Rolling Stone* 21 Apr. 5/1 (*caption*) The Rolling Stones re-sign with Atlantic.

re-si·gnal (ri-), *v.* [Re- 5 c.] *trans.* To refurnish with railway signals. Hence **re-signalling** *vbl. sb.*

1928 *Observer* 10 June 5/4 (*heading*) Re-signalling London Bridge. **1970** *Railway Mag.* Oct. 577/1 A further stage of the resignalling of the Bristol region..was completed in mid-August.., following which the London–Reading–Bath–Bristol route is now entirely controlled by colour-light signals. **1977** *Modern Railways* Dec. 469/1 However, until the Victoria and Brighton area resignalling projects are completed..the practical scope for a general speed-up of train services is minimal.

resignation. Add: **1. c.** In *Chess.*

1969 A. Glyn *Dragon Variation* ix. 271 One or two of the inexpert protested at the resignation. ('He hasn't been mated, has he? Why does he give up? Why doesn't he play on?')

5. *attrib.*, as *resignation letter*, *rally*, *speech*.

a **1974** R. Crossman *Diaries* (1976) II. 713 When the word didn't come he was compelled to write his resignation letter. **1961** *Economist* 2 Dec. 906/1 Democrats are beginning to repudiate their party at public 'resignation rallies'. **1938** H. Nicolson *Diary* 3 Oct. (1966) 374 It begins by Duff Cooper making his resignation speech.

resigna·tionist. *rare.* [f. Resignation + -ist.] One who follows a philosophy of resignation, a believer in resignationism.

1931 [see *Renunciant *sb.*].

resignee. Restrict ? *Obs.* to sense in Dict. and add: **2.** = Resigner.

1761 *Second Let. to Earl of Bxxx* 1 A right honourable resignee. **1973** *Times* 15 May 17/3 The only resignee that Mr Nixon chose to criticize.

resilience. Add: **2.** *spec.* The energy per unit volume absorbed by a material when it is subjected to strain, or the maximum value of this when the elastic limit is not exceeded. (Further examples.)

1908 E. S. Andrews *Theory & Design of Structures* i. 27 The work done per unit volume of a material in producing strain is called resilience. **1965** J. A. Cormack *Definitions Strength of Materials* iii. 67 Show that resilience per cubic inch in direct tension or compression may be expressed in the form $f^2/2E$, where *f* is the intensity of stress induced and *E* is the modulus of elasticity. **1978** B. I. Sandor *Strength of Materials* iv. 79 The maximum value of the elastic strain energy in a unit volume that has not been permanently deformed is called the modulus of resilience.

resi·liently, *adv.* [f. Resilient *a.* + -ly[2].] In a resilient manner, esp. such that the original position is resumed after bending or other shock.

1946 *Jrnl. Inst. Electr. Engineers* XCIII. IIIA. 330/2 The components were insulated from mechanical shock by mounting them on a tray resiliently supported within the frame. **1977** *BSI News* Oct. 8/3 A straight extension of ISO 2373 for the larger machines is not necessarily possible because of the difficulty of resiliently mounting them.

resilin (re·zilin). *Biol.* [f. L. *resil-ire* to jump back, recoil, Resile + -in[1].] An elastic material formed of cross-linked protein chains that is found in the cuticles of many species of insect, notably forming the hinges and ligaments of wings.

1960 *New Scientist* 14 July 104/1 Resilin is a protein of a new kind in which the protein chains are cross-linked by means of bonds of a new sort which are resistant to tryptic digestion. **1960** T. Weis-Fogh in *Jrnl. Exper. Biol.* XXXVII. 889 The main conclusion is that the characteristic elasticity is caused by a peculiar protein, called resilin, which differs from other structural proteins also in respect of amino acid composition. **1961** Bailey & Weis-Fogh in *Biochimica & Biophysica Acta* XLVIII. 453 We therefore propose to name it resilin, derived from the Latin resilire, *i.e.* to spring back (pronounced rez'ilin). (We are indebted to Professor D. S. Robertson, Cambridge, who suggested the name.) **1969** R. F. Chapman *Insects* xxii. 435 Resilin contains various amino acids including two previously unknown ones which provide the links between the protein chains. **1970** *Nature* 26 Dec. 1338/1 Resilin is a significant structural component of the sucking pump in many different insects. **1973** *Sci. Amer.* Nov. 92/3 The flea's leap is powered not by muscle alone but is assisted by the elastic protein resilin.

resilium (rizi·liŏm). *Conch.* [mod.L., f. as prec. + *-ium*, neut. ending.] The resilient central part of the 'hinge' of a bivalve, which tends to force apart the two valves.

1906 G. Bourne tr. *Pelseneer's Mollusca* in E. R. Lankester *Treat. Zool.* V. v. 213 The ligament finally becomes external..or internal; in the latter case it is a 'resilium'. **1926** *Proc. Malacol. Soc.* XVII. 44 Nacreous shells.. possess glutinous hinge-joints consisting of conchyolin disposed in a dual fashion as 'resilium'—more or less in the centre of the hinge base—and 'ligament' to the right and left of it. **1962** *Science* 29 June 1121/3 (*heading*) Aragonite in the resilium of oysters. **1973** R. T. Abbott *Kingdom of Seashell* 16/1 These valves are brought together by one or two strong adductor muscles and forced ajar by a rubberlike wedge, or resilium, which acts much the same way as a rubber wedge placed in the hinged side of a closing door.

re-silver, *v.* (Earlier and further examples.)

1839 in G. S. Haight *Geo. Eliot & J. Chapman* (1969) 259 Sextants, Quadrants, &c., re-silvered and repaired. **1873** *Young Englishwoman* Jan. 52/1 Whether looking-glasses can be re-silvered... A very good furniture dealer will re-silver this for you. **1973** W. H. Hallahan *Ross Forgery* iii. 40 The mirror must have been resilvered not too many years ago.

resin, *sb.* Add: **2*.** Any synthetic material resembling a natural resin; now usu. any of a large and varied class of synthetic organic polymeric materials (solid or liquid) that are thermosetting or thermoplastic (see also quot. 1934) and are used esp. as plastics or their chief ingredients. Freq. with qualifying adj. or sb., esp. *synthetic.*

1883 *Amer. Chem. Jrnl.* V. 338 Concentrated sulphuric acid on a mixture of benzoic aldehyde and resorcin gave a reddish resin. **1909** *Chem. Abstr.* III. 1818 Process of manufacturing synthetic resins as substitutes for shellac, consisting in treating *o*-cresol with formaldehyde in the presence of an acid. **1934** *Chem. Rev.* XV. 123 The resinous plastic field may well be divided into two main divisions dependent upon properties which find reflection in use. (1) The resins, which are melted for flow and cooled for hardening into the finished shape... (2) The resinoids, which in molding are heated for flow and also heated for hardening effect. **1937** *Discovery* Jan. 27/2 A new series of synthetic resins, claimed to be as clear as optical glass and to be..non-shattering,..is being marketed in the United States. **1943,** etc. [see *Ion exchange]. **1951** *Engineering* 20 Apr. 469/3 The setting of synthetic-resin glues by high frequency heating. **1957** *Which?* Autumn 20/2 The actual processing of these fabrics consists of various ways of putting the resin in under heat and pressure. **1970** Gait & Hancock *Plastics & Synthetic Rubbers* iii. 60 Production of phenolic resins is still increasing. **1973** *Materials & Technol.* VI. i. 86 Much imported and all British-made plywood is manufactured with synthetic resin adhesives of one type or other. **1976** *McGraw-Hill Yearbk. Sci. & Technol.* 182/1 Probably the cheapest resin to use for plastic bottles is polystyrene.

3. *resin-based*, *-bonded*, *-finished*, *-scented*, *-tipped*, *-treated* adjs.; **resin-weed** (earlier example).

1959 *Times* 3 Mar. 7/7 Resin-based paint. **1940** 'Plastes' *Plastics in Industry* x. 147 Not only have a number of private houses been built of resin-bonded plywood, but also several garages and petrol stations. **1959** *Engineering* 16 Jan. 86/1 This year there are in evidence still more hulls either moulded or sheathed in resin-bonded glass fibre. **1978** *Lancashire Life* Oct. 125/2 Birchwood ply used for work tops, resin-bonded and waterproofed particle board, steel runners with nylon bearings, [etc.]. **1963** A. J. Hall *Textile Sci.* v. 231 It also has the effect of making the resin-finished fabric tear more easily. **1937** J. Betjeman *Continual Dew* 11 Drained dark the pines in resin-scented rain. **1922** Joyce *Ulysses* 653 A pyre of crosslaid resintipped sticks. **1962** J. T. Marsh *Self-Smoothing Fabrics* i. 5 It was soon observed that the resin-treated fabrics possess certain properties in addition to 'crease-resistance'. **1852** L. B. MacKinnon *Atlantic & Transatlantic Sk.* I. 268, I found that he had spoken the truth, and that the resin grass, or weed, had peculiar leaves which always grew in the same direction.

b. Substitute for def.: *attrib.*, in terms denoting vessels in plants that contain resin secreted by cells lining them, as *resin canal*, *duct*, *passage*, *reservoir*.

1854, etc. [in Dict.]. **1896** W. R. Fisher tr. *Schlich's Man. Forestry* V. 14 Turpentine is chiefly found in the resin-ducts. **1924** W. S. Jones *Timbers* iv. 24 The presence of resin canals in dicotyledonous woods is, as in the case of Conifers, of considerable diagnostic value. *Ibid.* v. 30 Resin ducts are usually absent from the wood of many genera of Conifers. **1938** H. E. Desch *Timber* ii. 20 Resin canals run vertically in the stem and horizontally in the rays, and are just large enough to be seen with the naked eye. **1967** N. T. Mirov *Genus Pinus* vii. 486 When the pine is wounded, the resin canals are severed. The oleoresin, squeezed from the epithelial cells into the resin canals, may be gathered in receptacles attached below the wound.

resinate, *v.* Add: (Examples.) Also, to impregnate with synthetic resin.

1945 C. S. Forester *Commodore* xviii. 198 The Governor had taken advantage of the campaigns in which he had served to study the foods of the different countries. Vienna and Prague had fed him during the Austerlitz campaign; he had drunk resinated wine in the Seven Islands. **1966** *New Scientist* 22 Sept. 667/2 One of the drawbacks in resinating cloth is the tendency to reduce the durability of garments by making fibres more brittle.

resined, *ppl. a.* (In Dict. s.v. Resin *v.*) Add: **b.** Of wood: from which resin has been extracted or collected.

1926 *Contemp. Rev.* May 640 Resined wood lasts better than wood not resined, or wood from the same tree above the limit of the cuts.

resinification. Add: **2.** A reaction in which a synthetic resin is formed; conversion into a synthetic resin.

1913 *Chem. Abstr.* VII. 1484, H_2O inhibits polymerization and resinification, but favors the production of larger quantities of AcOH and HCO_2H. **1928** *Industr. & Engin. Chem.* Aug. 796/1 If a mixture of phenol and formaldehyde is heated, with or without a catalyst, resinification occurs. *Ibid.*, When a certain degree of resinification is reached, the process is usually interrupted. **1938** *Jrnl. Physical Chem.* XLII. 351 A more active catalyst in the resinification might be expected to increase the number of cross linkages. **1973** *Materials & Technol.* VI. viii. 587 If phenol is added in excess monomethylol phenol..is mainly formed and this, in a secondary condensation or resinification reaction, is converted into polynuclear methylene phenols.

resinify, *v.* In both senses used in connection with synthetic resins. (Further examples.) Also **re·sinified** *ppl. a.*, **re·sinifying** *vbl. sb.*

1920 *Chem. Abstr.* XIV. 3672 The Na salt of the 163° acid completely resinified on illumination in H_2O. **1928** *Industr. & Engin. Chem.* Aug. 798/1 Refractive indices of very highly resinified mixtures. **1933** *Ibid.* June 646/1 Like all resinifying reactions, that which forms vinyl resins is imperfectly understood. **1933** *Chem. Abstr.* XXVII. 4110 The disaggregated mass may be resinified more deeply with CH_2O. **1936** *Trans. Faraday Soc.* XXXII. 388 For the resinifying reactions we refer again to the scheme given earlier. **1963** F. M. Dean *Naturally Occurring Oxygen Ring Compounds* i. 2 Naturally occurring furans generally resinify in these conditions.

resinized, *ppl. a.* Add: **b.** Containing resin.

1908 W. R. Fisher tr. *Schlich's Man. Forestry* (ed. 2) V. 706 Resinised wood, owing to its easy combustibility, is excellent for kindling purposes, and in mountain districts abroad is still employed for torches.

resinography (rezinǫ·grăfi). [f. Resin *sb.* + -ography.] The study of the morphology, internal structure, and related properties of synthetic resins. Hence **resino·grapher,** one who practises resinography; **resinogra·phic** *a.*, of or pertaining to resinography; **resinogra·phically** *adv.*, by means of resinography.

1946 Rochow & Gilbert in J. J. Mattiello *Protective & Decorative Coatings* V. v. 476 Resinography, as the name implies, is the graphic study of resins. The term is proposed by the authors to serve as an analogue of metallography and mineralography. *Ibid.* 484 The resinographer must explain that the effect of magnetite on electrical and other properties should be considered. *Ibid.* 485 Resinographic methods, employing the optical microscope and reflected light, are designed for those large classes of resinous materials which are too opaque for examination by transmitted light. *Ibid.* 503 So much information about the fillers can be gathered quickly from a molding..that Stafford and Williams..usually briquet a portion of the sample and examine it resinographically. **1949** *Analytical Chem.* XXI. 461/1 A resinographic examination of this type of plastic reveals the number of phases, their mode of association; particle sizes and shapes, and relative reflectivity. **1961** G. L. Clark *Encycl. Microscopy* 527/1 Demonstration of macromolecular boundaries is the least developed and most difficult part of resinography. **1970** *Encycl. Polymer Sci. & Technol.* XII. 81 Usually there is enough reflectivity in the surface of a resinographic specimen to reveal the topography. **1976** T. G. & E. G. Rochow *Resinography* i. 14 Besides information from these four levels of study,..the resinographer must make his own examination and arrive at his own conclusions about the identity and utility of a plastic or fiber.

resinoid, *a.* and *sb.* Add: **B.** *sb.* **2.** A synthetic resin; *spec.* one that is thermosetting, or is not permanently soluble and fusible.

1925 Baekeland & Bender in *Industr. & Engin. Chem.* Mar. 225 The resinous products..are..decidedly infusible. Furthermore, they are insoluble in ordinary solvents and are incomparably stronger and more resistant to chemical and physical agents than the natural resins or the artificial resins of the Novolak type; in order to differentiate them from all these, we feel warranted in designating them more accurately under the name of 'phenol resinoids'. **1934** [see *resin *sb.* 2*]. **1935** C. Ellis *Chem. Synthetic Resins* I. i. 14 The term 'resin' is restricted by some to thermoplastic resins, and the word 'resinoid' is used to designate the heat-hardening resins. **1936** H. W. Rowell *Technol. Plastics* i. 12 Phenolic and other synthetic resinoids now have an increasing use in the varnish industry. **1949** B. L. Davies *Technol. Plastics* xi. 188 The three stages in the polymerization [of phenol with formaldehyde] were first recognized by Baekeland,..all being called resinoids to distinguish them from the permanently soluble, permanently fusible, resins. **1959** *Times* 18 Nov. 12/5 The diamonds produced in the laboratory consist, in effect, of abrasive grit suitable for use in resinoid-bonded grinding wheels.

† **resinophore,** *a.* (*sb.*) *Chem. Obs.* [ad. G. *resinophor* (W. Herzog 1921, in *Österr. Chem.-Zeitung* XXIV. 77): see Resin *sb.* + -phore.] *resinophore group*, a group whose presence in a molecule was considered to be responsible for resinous properties or resin formation. Also *ellipt.* as *sb.*

1922 *Chem. Abstr.* XVI. 1671 The 'resinophore group', formed from ureas capable of being converted into resinous substances, is the complex —N=C=N—. **1935** C. Ellis *Chem. Synthetic Resins* II. xxv. 562 An investigation of the effect of heat on several mono- and di-arylidene ketones showed that the grouping —CH=CHCO— was responsible for resinification; that is, this group is a 'resinophore'. **1937** R. S. Morrell et al. *Synthetic Resins & Allied Plastics* xv. 375 Comparison between Herzog and Kreidl's resinophore groups and Standinger's chain theories shows that preference must be given to the latter in any explanation of the causes of resinification.

resinosis (rezinōᵘ·sis). *Forestry.* [f. Resin *sb.* + -osis.] The excessive production of resin (in conifers).

1922 F. Dorrance tr. *Sorauer's Man. Plant Dis.* I. xvi. 716, I have had the opportunity to observe resinosis as a constitutional disease, i.e. as the manifestation, even in old trees, of a tendency throughout the whole plant body, to form resin excessively. **1968** F. G. Browne *Pests & Dis. of Forest Plantation Trees* ii. 950 In New Zealand the species [sc. *Pythium undulatum*] is closely associated with a virulent resinosis of the roots, root collars, and stems of *Larix decidua* and *L. leptolepis*.

resinous, *a.* Add: **5.** (Earlier examples.) Now *Obs.*

1742 J. T. Desaguliers *Diss. Electr.* 41 The Air being electrical of a vitreous Electricity, and sulphur of a resinous Electricity. **1756** *Phil. Trans. R. Soc.* XLIX. 152 Some writers on electricity have said, that there were two kinds of electrical fire, the one resinous, and the other vitreous.

resist, *sb.* Add: **3. a.** *spec.* Such a composition used to provide protection against the etchant or solvent in photo-engraving, photogravure, or photolithography. (Further examples.)

1886 W. T. Wilkinson *Photo-Engraving on Zinc & Copper* v. 34 When it is judged that the etching has proceeded far enough, the resist of ink and resin is removed by the copious use of turpentine. **1933** T. S. Barber *Photo-Engraving, Electrotyping & Stereotyping* i. 17 The portions of the bichromated surface which were exposed to light, and already made insoluble, acted as an acid resist. **1960** [see *micro-machining* s.v. *micro- 2 a]. **1967** V. Strauss *Printing Industry* v. 213 Polyvinyl alcohol..can be dissolved in water and sensitized with bichromates. Such a solution..can serve as a photomechanical coating resulting in an acid resist. Coating, exposure, and development of polyvinyl alcohol resists do not differ from other resists. But polyvinyl alcohol resists have the advantage that they need not be heated as high as glue-enamel resists in order to make them sufficiently acid-resistant. **1975** Fink & McKenzie *Electronics Engineers' Handbk.* viii. 6 The [semiconductor] wafer surface to be masked is coated with a photosensitive coating known as photoresist, or resist. The masking plate is then..exposed with ultraviolet light... The photomasking step is followed by an etching step. **1978** *Sci. Amer.* Nov. 63/1 The wafer is first coated with an X-ray-sensitive organic polymer called an X-ray resist.

b. *Pottery.* A material (usu. wax-based) which is applied to pottery in order to prevent glaze or lustre from adhering to certain parts and is removed before or during firing. Orig. and freq. *attrib.*

1904 A. Hayden *Chats on Eng. China* xii. 231 The..method, with the design left in white, was produced in handsome and highly artistic styles, and there is a pattern, known as the 'Resist' pattern, which is much sought after. **1910** J. F. Blacker *ABC of Collecting Old Eng. Pott.* xxiii. 214 The commonest application of silver resist is used on a white or ivory ground. **1933** W. B. Honey *Eng. Pott. & Porc.* iv. 59 Designs of formal flowers and inscriptions..were also done in darker brown or in 'resist' so as to appear of a lighter colour. **1957** Mankowitz & Haggar *Encycl. Eng. Pott. & Porc.* 189/2 The decoration is painted upon the glazed surface of the ware in a 'resist', covered with metallic solution and fired, the infusible resist portion firing away during the process leaving a white decoration reserved against a bright metallic background. **1967** M. Chandler *Ceramics in Mod. World* iii. 108 (*caption*) The water-soluble resist is washed off before the plates are fired. *a* **1977** *Harrison Mayer Ltd. Catal.* 19/1 Wax resist decoration is very frequently used both under and over a glaze and in combination with stain decoration.

resistance. Add: **1. b.** *passive resistance.* See also as main entry in Suppl.

c. Organized covert opposition to an occupying or ruling power; *spec.* (usu. with def. article and capital initial) in the war of 1939–45, the underground movement formed in France in June 1940 with the object of resisting the authority of the German occupying forces and the Vichy government; any organization of this type with similar ends.

1939 *War Illustr.* 28 Oct. 217/1 Underground resistance to Hitler has been organized amongst the workers in all the big industrial centres of Germany. **1940** *Times* 19 June 6/3 General de Gaulle..broadcast from London a message to the French nation last night. The text of his speech..is as follows:..Whatever happens the flame of French resistance must not and shall not be extinguished. **1946** A. Huxley *Let.* 28 Mar. (1969) 541, I was sent a number of French books recently... Novels about the Resistance—half heroism, half unutterable moral squalor. **1959** *Listener* 17 Sept. 454/3 To judge by the atmosphere of this tale the resistance in Denmark must have been a less desperate affair than it was in many parts of Europe. **1967** *Freedomways* VII. 143 The men and women who have been notable in African history—not only the rulers and statesmen, but also the educators and scholars, writers and artists, religious leaders, heroes of resistance. **1976** H. Tracy *Death in Reserve* xxii. 170 In this area, *Sanglier* had been an important man in the Resistance. **1981** *Guardian* 2 June 6/7 In the many rural areas [of Afghanistan] under the control of the resistance and unchallenged by the Russians, the insurgents impose their own curfew.

2. a. (Further examples.)

1953 *Sun* (Baltimore) (B ed.) 9 Sept. 3/4 How germs develop resistance to the drugs. **1968** [see *resistant *a.* 2].

b. *Psycho-anal.* Opposition, freq. unconscious, to allowing memories or desires which have been repressed as unacceptable or disruptive to emerge into the conscious mind.

1905 *Psychol. Bull.* II. 256 Resistance (in the form of indifference, etc.), would greatly delay the inquiry. **1909** A. A. Brill tr. *Freud's Sel. Papers on Hysteria* vii. 167, I started with the presupposition that..this paranoia must contain unconscious thoughts and repressed reminiscences which have to be brought to consciousness..by overcoming a certain resistance. **1913** E. Jones *Papers on Psycho-Anal.* i. 6 Whenever an individual considers a given process as being too obvious to permit any investigation into its origin, and shows resistance to such an investigation, we are right in suspecting that the actual origin is concealed from him. **1924** J. Riviere tr. *Freud's On Psychotherapy* in *Coll. Papers* I. 254 It [sc. a technique of suggestion] does not permit us..to recognize the *resistance* with which the patient clings to his disease and thus even fights against his own recovery. **1936** M. M. Green tr. *Reik's Surprise & Psycho-Analyst* iv. 52 We have discovered resistance in analysis as the emotional expression of this opposition. **1951** P. M. Symonds *Ego & Self* xii. 175 In order to defend the ego against anxiety, resistances are built up against the recognition of these unconscious impulses. **1964** E. Krapf tr. *Caruso's Existential Psychol.* iii. i. 150 Placing the burden of resistance on the patient alone provides a good excuse for one's own incapacity and totalitarian one-sidedness. **1977** C. F. Monte *Beneath Mask* ii. 53 Resistance to recall is one evidence of motivated forgetting or repression.

c. In *Comb.* with a preceding sb., as *crease resistance*, resistance to creasing, etc.

1932 [see *drug-resistance* s.v. *drug *sb.*[1] 3]. **1947** [see *crease *sb.*[2] 5]. **1959** *Times* 12 Jan. 11/4 Gives shrink, stretch, and rot-resistance. **1959** [see *flame-resistance* s.v. *flame *sb.* 10]. **1966** *Economist* 8 Jan. 113/1 There are now two processes..for giving man-made fibre garments a permanent press. This builds up crease resistance in normal wear. **1973** R. G. Krueger et al. *Introd. Microbiol.* xv. 422/1 Transmissible drug resistance was discovered..around 1960.

3. d. *line*, etc., *of least resistance*: (see quot. 1871); also *fig.*, the easiest method or course of action.

1865 Mill *Auguste Comte* 101 In the play of antagonistic forces, the path it [sc. our intelligence] points out is (in scientific phraseology) the direction of least resistance. **1871** G. E. Voyle *Dict. Artillery Terms* (ed. 2) 185/2 *Line of least resistance*, in blasting or mining, is a line drawn from the centre of the charge perpendicular to the surface of the ground. *Ibid.* 213/1 By taking 1/10 of the cube of the line of least resistance in feet, the proper charge of powder..is given in pounds. **1903** G. B. Shaw *Man & Superman* iii. 134 This Life Force says to him 'I have done a thousand wonderful things unconsciously by merely willing to live and following the line of least resistance: now I want to know myself and my destination, and choose my path.' **1908** W. McDougall *Introd. Social Psychol.* vii. 179 He often seems to act, not in the line of least resistance, but in the line of greatest resistance. **1923** W. G. Boulton *Blasting with High Explosives* xv. 45 L = Line of least resistance in feet... The line of resistance is the shortest distance from the chamber to the surface. **1931** *Economist* 7 Feb. 300/2 This was no doubt the line of least resistance, when money had to be found. **1954** M. Sharp *Gipsy in Parlour* xxiv. 233 He took by nature the line of least resistance. **1967** C. V. Bark *See Living Crocodiles* xi. 184 It is taking the line of least resistance. It saves trouble.

4. a. *spec.* Resistance to an unvarying electric current. (Earlier and later examples.)

1760 J. Wesley *Desideratum* 17 While the electric fire, which is in all bodies, is left to itself, undisturb'd by any external violence, it is more or less dense... And there is some resistence to every endeavour of altering its density... This resistence is least in metals, minerals, water, quicksilver, animals and vegetables... In these bodies the resistence is greater, when their surface are polish'd. **1767** J. Priestley *Hist. Electricity* 116 The difference between electric and non electric bodies was owing to the different resistance..to the passage of the electric fluid. **1963** A. F. Abbott *Ordinary Level Physics* xxxvii. 479 The resistance of a wire depends on its dimensions and the material from which it is made. **1975** [see *resistor].

6. (sense *1 c) *resistance cell, club, fighter, figure, forces, group, hero, man, movement, network, plan, work, worker*; (sense 4 a) *resistance-capacitance, -capacity* (both used *attrib.*), *-coupling*; *resistance-coupled* adj.; **resistance furnace,** an electric furnace which is heated by passing a current through elements of high resistance; **resistance pyrometer,** a form of resistance thermometer suitable for use at high temperatures; **resistance thermometer,** a temperature-measuring device in

which the change in electrical resistance with temperature of a platinum or other metallic wire is measured; **resistance transfer factor** *Biol.*, a plasmid that promotes the transfer from one bacterium to another of a genetic determinant for drug resistance; **resistance welding,** a method of welding in which the heat to cause fusion of the metals is produced by the passage of an electric current through the contact resistance between the two surfaces, these being held together by mechanical pressure.

1942 E. Williams *Thermionic Valve Circuits* iii. 33 The condenser, C_2, is therefore omitted and the circuit becomes the orthodox Resistance-capacitance Coupling circuit. *Ibid.* 69 Consider the resistance-capacitance coupled circuit. **1962** Simpson & Richards *Physical Princ. Junction Transistors* ix. 208 A resistance-capacitance coupled amplifier. **1924** P. J. Risdon *Wireless* xxvii. 221 In the case of resistance or resistance-capacity coupling, instead of an inductance coil, a high resistance is employed for establishing the necessary differences of potential. **1962** Simpson & Richards *Physical Princ. Junction Transistors* v. 88 Many audio and low radio-frequency amplifiers, having resistance-capacity coupling between stages. **1972** D. Ramsay *Little Murder Music* 139, I succeeded in making my way to a resistance cell. **1945** *Sun* (Baltimore) 6 Oct. 2/5 German underground units, composed chiefly of Hitler youth and former prisoners of war, were springing up in all parts of Germany, with some secret 'resistance clubs' reviving the worst of Nazism. **1931** Moyer & Wostrel *Radio Handbk.* xiii. 684 Resistance-coupled amplifiers. A typical circuit in which the amplifier is connected by the method of resistance coupling. **1962** Simpson & Richards *Physical Princ. Junction Transistors* ix. 219 A relatively heavily loaded resistance-coupled stage operating over a reasonably long range of ambient temperature. **1921** J. Scott-Taggart *Thermionic Tubes* vi. 207 A form of coupling between successive valves in an amplifier which has been of considerable use is that known as 'resistance coupling'. **1963** *Listener* 14 Feb. 304/2 He joins up with various bands of multi-racial resistance fighters. **1976** H. Tracy *Death in Reserve* xxii. 172 The local *gendarmerie*, guided by the old Resistance fighters,..covered the flanks. **1958** *Times Lit. Suppl.* 21 Mar. 149/1 The sexual fantasies of various Resistance and non-resistant figures. **1944** Resistance forces [see *E.A.M.* s.v. *E. III]. **1959** *Listener* 16 Apr. 674/3 Fighting between Tibetan resistance forces and the Chinese. **1904** M. Maclean *Mod. Electr. Pract.* V. 239 To prevent more or less arcing in what is ostensibly a resistance furnace. **1928** *Jrnl. Iron & Steel Inst.* CXVII. 853 Materials of construction for electric resistance furnaces, and the advantages of the latter over fuel-fired furnaces, are described. **1957** Resistance group [see *courier *sb.* 1 b]. **1974** J. Thomson *Long Revenge* ii. 22 For reasons of security, Mercer wasn't put in direct touch with the local Resistance group. **1975** N. Freeling *What are Bugles blowing For?* vii. 45 Was he a Resistance hero then? **1945** *Daily Herald* 8 May 4/8 The Germans..started shooting at Dutch Resistance men. *Ibid.* 28 May 4/2 Armed forces of the Danish resistance movement have passed a resolution..demanding the arrests [*sic*] of all collaborators. **1953** *Encounter* Nov. 6/2 The rising sales of even the most expensive Japanese garments..were taken to be a kind of resistance-movement against Western-style attire. **1978** R. V. Jones *Most Secret War* xxiii. 189 It would be some time before the Resistance movements began to operate coherently. **1969** F. Halliday in Cockburn & Blackburn *Student Power* 317 Students also provided the cadres for the various resistance networks..which channelled arms, funds and information to the FLN. **1944** *Daily Tel.* 15 May 4 Mr. Wareing reveals..the existence of a Resistance plan to seize power. **1899** *Physical Rev.* VIII. 197 The failure of Siemens' resistance pyrometer was simply due to faulty construction. **1920** *Whittaker's Electr. Engineer's Pocket-bk.* (ed. 4) 290 Resistance pyrometers. For exact readings when temperature is steady... Range: minus 200° up to 1000° C. **1959** *Jrnl. Iron & Steel Inst.* CXCIII. 318/3 He then describes the four main types of pyrometer, namely, the optical pyrometer, the resistance pyrometer, the thermocouple, and the radiation pyrometer. **1887** *Phil. Trans. R. Soc.* A. CLXXVIII. 164 In comparing the platinum resistance thermometer with other instruments, it will be seen that it is essentially practical. **1920** *Whittaker's Electr. Engineer's Pocket-bk.* (ed. 4) 286 The Callendar compensated resistance thermometer consists of a pure platinum wire wound on a mica or fireclay frame and connected by stout copper or platinum leads to terminals in the thermometer head. **1959** Resistance thermometer [see *conductivity 2]. **1960** Watanabe & Fukasawa in *Biochem. & Biophys. Res. Communications* III. 664 Each resistance factor is not a transmissible factor in itself but..the resistance factors are carried by some transmissible factor... We propose to refer to this factor as 'resistance transfer factor (RTF)' tentatively. **1965** *Times* 26 Feb. 15/3 The resistance transfer factor, like the sex factor, may stimulate conjugation—so that resistance can sweep rapidly through a colony. **1970** Passmore & Robson *Compan. Med. Stud.* II. xviii. 57/2 Resistance transfer factors represent a world-wide problem which is becoming increasingly serious. **1914** *Proc. Inst. Mech. Engin.* 170 The best, the most controllable and the least likely method of injuring the material, which gave the most uniform satisfactory results, was electric resistance welding. **1927** [see *metallic *a.* 1 h]. **1975** Bram & Downs *Manuf. Technol.* ii. 61 Resistance-welding techniques are characterised by the absence of flux or filler. **1959** *Listener* 17 Sept. 454/3 They are engaged in sabotage and resistance work. **1962** L. Deighton *Ipcress File* 221 Grenade was a resistance worker in 1940 when..every one else in France was rooting for Marshal Pétain. **1974** J. Thomson *Long Revenge* ii. 23 A suspected French Resistance worker.

resistanceless (rĭzi·stānslės), *a.* [f. Resistance + -less.] Marked by a total lack of (electrical) resistance.

1968 *Physics Bull.* Dec. 410/2 The Meissner effect provided an important clue to the understanding of superfluidity, by showing that it was more than resistanceless flow. **1969** [see *excitonic *a.*]. **1978** *Sci. Amer.* Apr. 88/2 Certain properties of the glassy superconductors, such as the transition temperature where resistanceless flow begins, are significantly different from those in crystalline materials.

resistant, *a.* and *sb.* Add: **A.** *adj.* **2.** That is not overcome by some disease or drug. Const. *to*.

1897 Muir & Ritchie *Man. Bacteriol.* 291 An animal is shedding into the air..myriads of bacilli which may rapidly spore, and thus arrive at a very resistant stage. **1898** *Yearbk. U.S. Dept. Agric.* 1897 399 The..cross yielded the Golden Clairette, a valuable new sort [of vine], apparently highly resistant to Phylloxera. **1925** *Scribner's Mag.* July 1/1 Genetics has made possible better strains of livestock... Disease is less to be feared because of resistant stocks. **1968** *New Statesman* 5 Jan. 10/3 A germ which normally inhabits the intestines of a cow, for instance, but which normally doesn't harm man, could be rendered resistant, perhaps, to a whole group of antibiotics, and could then escape from the animal, and pass that resistance on to another type of germ which *was* dangerous to man.

3. In *Comb.* with preceding sb.

1902 [see *fire-resistant* adj. s.v. *fire *sb.* B. 2]. **1932** [see *drug-resistant* adj. s.v. *drug *sb.*[1] 3]. **1936** [see *crease *sb.*[2] 5]. **1947** [see *flame-resistant* adj. s.v. *flame *sb.* 10]. **1959** *Times* 12 Jan. 11/4 Porous, crush-proof, stress and tear-resistant and washable. **1963** *B.S.I. News* June 8/2 Bad news for burglars was the announcement in May of a BSI specification for thief-resistant locks. **1968** *Times* 12 Oct. 18/8 One of the organisms sometimes responsible for travellers' diarrhoea is now sulphonamide-resistant. **1978** *Cornish Guardian* 27 Apr. 34/6 (Advt.), White Ceiling Tiles all with F.R.A. (Fire Resistant Additive).

B. *sb.* **a.** Delete 'Now *rare*' and add: *spec.* a member of a resistance movement. Cf. *résistant.

1944 *Ann. Reg. 1943* 172 To the injury of the United Nations and the French resistants. **1948** W. Fortescue *Beauty for Ashes* xxxv. 273 *Mademoiselle*..had been one of the true Resistants, refusing even in the blackest hours of the war to admit even the possibility of defeat, or to allow pessimism among the peasants. **1959** *Sunday Times* 22 Mar. 23/7 A disguised and wounded Hungarian resistant.

‖ **résistant** (resistaṅ). Also *fem.* **résistante.** [Fr.] A member of the French Resistance (see *resistance 1 c).

1966 M. R. D. Foot *SOE in France* vii. 175 Mathilde Carré..proclaimed herself already an ardent résistante. **1967** *Listener* 16 Mar. 370/3 Pierre d'Harcourt was twenty-six when, after Dunkirk, he shed his cavalry uniform to become a pioneer *résistant*. Maybe the cavalry discipline helped to restrain his impetuosity. **1978** *Dædalus* Fall 29 Everyone in France was a 'résistant'.

resister. Add: **1. b.** *spec.* A member of a resistance movement; a resistant. Also with capital initial.

1952 *Chambers's Jrnl.* Feb. 66/2 But all three of them, when they were drunk, believed themselves to have been the most stout-hearted Resisters, members of the Underground. **1959** *Encounter* Aug. 45/1 The treatment accorded to the records of the *Résistance* is especially entertaining... Thousands of Resisters wrote their personal recollections. **1966** M. R. D. Foot *SOE in France* vi. 129 In Greece and in Yugoslavia SOE sought to back any anti-German bodies of resisters.

resistibility. Add: **1.** (Later example.)

1974 *Times* 4 Feb. 8/2 Irresistibility is simply not borne out by statistics. A survey conducted in 1965 showed not only that 81 per cent of Italian males objected to the closing of licensed brothels but that nearly three quarters made use of prostitutes, which suggests a high degree of resistibility.

resistible, *a.* (Further examples.)

1903 G. B. Shaw *Man & Superman* iii. 134 As to your Life Force, which you think irresistible, it is the most resistible thing in the world for a person of any character. **1966** *Listener* 9 Nov. 597/3 The most resistible of the new MPs..his voice, as usual, half-way between a splutter and a jeer. **1979** *Financial Times* 14 Apr. 21/3 The water babies themselves are frightful infants..given to singing resistible underwater ditties.

resistive, *a.* Add: **2.** *Electr.* Pertaining to, possessing, or resulting from electrical resistance; *spec.* in connection with resistance as a component of impedance (in this sense opp. *reactive *a.* 6 a and b).

1929 *Wireless World* 30 Jan. 116/2 The reactive component in each case is drawn at 90° to the resistive component. **1937** A. V. Eastman *Fund. Vacuum Tubes* v. 153 Since power considerations in a voltage amplifier are minor, the output load may be partly reactive instead of resistive. **1947** R. Lee *Electronic Transformers & Circuits* i. 6 With many sine wave electronic transformers, the transformer load is resistive. **1958** *New Scientist* 21 Aug. 658/3 The Stellarator uses resistive heating only in the first stage, to bring the plasma into the million degree region. **1964** R. F. Ficchi *Electr. Interference* 217 Resistive coupling is the association of two or more circuits with one another by means of resistance mutual to the circuits. **1969** L. E. C. Hughes et al. *Dict. Electronics & Nucleonics* 231/1 Reactive loads are made entirely resistive by adding inductors or capacitors in series or shunt. **1980** *Times* 23 June 18/7 He used the equations for capacitative and resistive loss in electric cables.

resistivity. Add: **1.** Now usu. defined as the resistance of a conductor of unit length and unit cross-sectional area. (Earlier and later examples.)

1885, 1895 [see *-ivity]. **1943** C. L. Boltz *Basic Radio* i. 26 Anyone capable of working out a formula can therefore find the resistance of any conductor by looking up the tables for specific resistance or resistivity. **1962** *Newnes Conc. Encycl. Electr. Engin.* 658/1 Copper of the highest purity has a resistivity lower than that of any other known material except silver. A standard value for the resistivity of annealed copper was established in 1913 by the International Electrotechnical Commission as 0·017 241 Ω per m length and mm² section. **1975** *Country Life* 6 Feb. 331/3 The scientific techniques used by the archaeologists, such as the resistivity meters used to detect areas where the soil is broken up by the tops of mine shafts. **1975** G. Anderson *Coring* vii. 121 The porosity of the sand is above 20% and the resistivity ranges from 0.3 ohm-m for saltwater sands to several ohm-meters for oil-saturated sands.

2. Special Combs.: **resistivity surveying,** measurement of the current passing between electrodes embedded in the ground at a series of positions over a given area, in order to identify regions of differing resistivity and hence locate buried structural features; so **resistivity survey,** a set or programme of such measurements.

1927 *Terrestrial Magnetism* XXXII. 49 The Department of Terrestrial Magnetism of the Carnegie Institution of Washington has undertaken a series of resistivity surveys in regions where records of earth-current potential gradient are being obtained. **1952** M. B. Dobrin *Introd. Geophysical Prospecting* xvi. 298 The interpretation of the actual data obtained in field resistivity surveys is usually a highly empirical and generally unreliable process. **1966** *McGraw-Hill Encycl. Sci. & Technol.* XI. 22/2 Electromagnetic and electrical resistivity surveys are used to locate deposits of metallic sulfides, which, except for sphalerite, are good electrical conductors. **1931** *Engin. & Min. Jrnl.* CXXXI. 325 (*heading*) Earth resistivity surveying. **1953** R. J. C. Atkinson *Field Archaeol.* (ed. 2) i. 38 Resistivity surveying requires at least two operators, one at the instrument and one to move the electrodes. **1963** *Times* 3 June 6/2 Resistivity surveying has been conducted by Mr. Whybrew, of the Ministry's test branch, and buried features have been detected.

resistor (rĭzi·stəɹ). *Electr.* [f. Resist *v.* + -or.] A passive device which impedes the flow of an electric current, used to develop a voltage drop across itself or to limit current flow.

[**1759**: see Resister 2]. **1905** *Sci. Amer. Suppl.* 27 May 24586/1 The resistance medium or 'resistor', when solid, usually consists of a core of carbon, coke, or graphite. **1930** *Engineering* 31 Jan. 128/3 A higher-temperature zone heated by non-metallic resistor rods. **1947** R. Lee *Electronic Transformers & Circuits* vi. 181 Tubes may require resistors in the plate and grid leads to damp out parasitic oscillations. **1965** *Wireless World* July 329/2 To monitor this current a 50 Ω resistor is inserted in the earthed lead of the recording head. A 1 kc/s signal source is then injected into the radio input of the recording amplifier and the voltage across the resistor measured with a valve voltmeter or oscilloscope. **1975** Fink & McKenzie *Electronics Engineers' Handbk.* vii. 4 Wire-wound resistors usually exhibit an increase in resistance with higher frequencies because of skin effects.

resi·t (rī-), *v.* [Re- 5 a.] *trans.* and *intr.* To sit (an examination) again, usu. after failing at a previous attempt; to retake (an examination), e.g. in order to improve one's grade.

1959 *Chambers's 20th Cent. Dict.* 1380/2 Resit,..*v.i.* to sit an examination again after failing. —— *n.* an opportunity or act of resitting. **1968** *Sunday Times* 30 June 15 So many students resit the engineering examination each year that an eventual pass rate of 80 percent..may occur. **1974** A. Fowles *Pastime* i. 6 He'd had to re-sit English to get the minimum the library required. **1979** *Jrnl. R. Soc. Arts* Dec. 9/1 The candidates for this session's examination were exclusively restricted to those who had been referred in a previous examination or who were resitting in order to obtain higher grades.

resit (rī·sit), *sb.* [f. the vb.] The act of resitting an examination; also, an examination held specifically to enable candidates to resit. Also *attrib.*

1959 [see *resit *v.*]. **1972** D. Haston *In High Places* ii. 30 For once he'd passed his exams in June and didn't have any resits in September. **1973** *Guardian* 14 June 7/2 Resit school breaks its 11-plus record. Out of 50 pupils..who had to resit their 11-plus examination..31 have passed. **1977** *Belfast Tel.* 19 Jan. 3/4 One wonders how your recent correspondents, who feel that an Eleven-Plus resit is unnecessary, can accept the position arising as a result of the supplementary test which has not been held.

resite (re·zəit), *sb. Chem.* [ad. G. *resit* (H. Lebach 1909, in *Zeitschr. f. angew. Chem.* XXII. 1601), f. L. *rēs-ina* Resin *sb.* + *-it* -ite[1].] A name given to the insoluble, infusible resins that are the final products of phenol-aldehyde copolymerizations.

1913 H. Lebach in *Jrnl. Soc. Chem. Industry* XXXII. 559/2 For the final product, whether it be called Bakelite 'C', Resinite, Condensite, or by some other name, I suggest the name 'Resite'. **1935** C. Ellis *Chem. Synthetic Resins* I. xiv. 307 Resites (C stage) are branched-chain macromolecules. **1950** *Thorpe's Dict. Appl. Chem.* (ed. 4) X. 24/2 The resin hardens on further heating and becomes insoluble and infusible. This is the 'C' or resite stage and consists of

macromolecules with numerous branched and cross-linked chains. **1973** [see *resitol].

re-si·te (rī-), *v.* [Re- 5 a.] *trans.* To place on another site; to relocate. Hence **re-si·ting** *vbl. sb.*

1955 *Times* 19 July 9/7 Many argue that the bridge will still be needed even when the tunnel has been built; then the question of re-siting it may arise. **1967** E. Chambers *Photolitho-Offset* viii. 106 These cameras..have facilities for the exact siting of films, plates, transparencies and masks with three-point lay and location pins for accurate re-siting and register of copy. **1969** *Jane's Freight Containers 1968–69* 555/1 They are completely mobile and can be pushed or towed for re-siting. **1976** P. R. White *Planning for Public Transport* x. 213 In so far as population re-sited from existing industrial cities is located in new or expanding towns of about 100 000 population, similar improvements could occur.

resitol (re·zitǫl). *Chem.* [a. G. *resitol* (H. Lebach 1913, in *Chem.-Zeitung* 19 June 734/1), f. *resit* *resite *sb.* + *res-ol* *resol.] A name given to the rubbery, insoluble resins produced in phenol-aldehyde copolymerizations at a stage intermediate between resol and resite.

1913 H. Lebach in *Jrnl. Soc. Chem. Industry* XXXII. 559/2 Besides these products, Baekeland distinguishes an intermediate polymerisation product, which he terms 'Bakelite "B"'... This intermediate condensation product I would suggest to call 'Resitol'. **1950** *Thorpe's Dict. Appl. Chem.* (ed. 4) X. 24/2 When heated at temperatures above 100° c. it melts and froths, becoming more viscous and eventually reaching the rubbery stage, the 'B' or resitol stage. **1973** *Materials & Technol.* VI. viii. 588 Phenol resins evolve in three stages, which are connected with the degree of condensation: 1. Resol or A-stage (beginning of condensation): the resin is fluid, soluble, and still contains much water. 2. Resitol or B-stage (continued condensation, slight cross-linking): insoluble, rubbery. 3. Resite or C-stage (final condition of the cured product): infusible and insoluble.

resi·tuate (rī-), *v. N. Amer.* [Re- 5 a.] = Relocate *v.* 1 a. Also *refl.*

c **1865** E. Dickinson *Poems* (1955) II. 734 Bore Death from Passion All His East He—sovreign as the Sun Resituated in the West And the Debate was done. **1974** D. Richards *Coming of Winter* i. 6 He could feel sharp blades of undergrowth so he resituated himself once or twice.

reslash, *v.* Add: Hence **resla·shing** *vbl. sb.*

1970 J. Ford et al. in H. W. Mulligan *African Trypanosomiases* xxvii. 548 He [sc. Morris]..checked regeneration by stumping and burning, and not by annual reslashing. **1973** J. J. McKelvey *Man against Tsetse* iii. 162 Unpaid community labor, which was recruited under the provisions of the 1937 Sleeping Sickness Ordinance, accomplished almost all of the subsequent slashing and reslashing of regrowth over the years.

reslush (rīslɒ·ʃ), *v. Paper-making.* [Re- 5 b.] *trans.* To convert (dry or semi-dry paper stock) into slush by the addition of water.

1963 *Economist* 11 May 555/2 Drying the pulp for shipping and then 're-slushing' it. **1967** E. Chambers *Photolitho-Offset* xvi. 247 A few mills in this country make pulp from waste paper by 'reslushing'.

resmethrin (re·zmīþrin). [-*ethrin* f. *pyr)ethrin.] A synthetic pyrethroid employed as an insecticide in the form of a spray.

1971 *Pesticide Sci.* II. 16/1 (*table*) 5-Benzyl-3-furylmethyl (±)-*cis-trans*-chrysanthemate (Resmethrin). **1971** *Bull. World Health Org.* XLIV. 329/1 Resmethrin was superior to the other synthetic pyrethroids and to the natural pyrethrins against all insects tested. **1975** *Daily Tel.* 25 Oct. 8/2 The answer is to keep a bottle of resmethrin..handy and give any new plants a going over with it.

Resochin (rī·sokin). *Pharm.* Also **resochin.** [a. G. *resochin*, f. *reso-rcinol* Resorcinol + *chin-olin* Quinoline.] A proprietary name for *chloroquine.

1946 *Trans. R. Soc. Trop. Med. & Hygiene* XL. 170 Atebrin, sontochin, resochin and quinine were observed to cause arrest of development and degeneration of ring and amoeboid forms. **1947** [see *chloroquine]. **1951** *Trade Marks Jrnl.* 12 Dec. 1135/1 *Resochin*... Pharmaceutical preparations. Farbenfabriken Bayer..; Manufacturers and Merchants. **1955** *Official Gaz.* (U.S. Patent Office) 19 July TM 157/1 Farbenfabriken Bayer Aktiengesellschaft, Leverkusen-Bayerwerk, Germany. Filed Dec. 15, 1954. *Resochin.* Applicant claims ownership of German Reg. No. 602,220, dated Nov. 2, 1950. **1958** J. H. Burn *Lect. Notes Pharmacol.* (ed. 5) 105 Chloroquine was also discovered by the Germans as resochin, but was tested and developed in America during the last war. **1970** W. Peters *Chemotherapy & Drug Resistance in Malaria* i. 5 From the time of its synthesis by Andersag..in 1934, chloroquine, or Resochin as it was first known to the German workers, faced a precarious existence.

resocializa·tion (rī-). *Social Psychol.* [Re- 5 b.] The action or process of (re)inducing conformity to accepted standards of social behaviour. Also *attrib.*

1964 M. Argyle *Psychol. & Social Probl.* i. 20 When we come to consider the re-socialization, or indoctrination, of adults, we shall introduce two further principles of learning: cognitive learning and learning by means of emotional arousal. **1970** P. Mayer *Socialization* p. xiii, Major social changes, such as ongoing urbanization or industrialization, involve the resocialization of people of all ages. **1974** M. C. Gerald *Pharmacol.* xii. 232 Some view their program [sc. of Alcoholics Anonymous] as a resocialization process attempting to develop individual maturity. **1976** *Guardian* 23 July 15/5 The regular presence of women in the unit, unthinkable in the old isolation system, is seen as an essential part of the resocialisation process by staff and psychiatrists alike. **1977** F. Musgrove *Margins of Mind* ix. 196 A period of 'desocialization' may be an important preliminary to the effective resocialization of adults.

reso·ften (rī-), *v.* [Re- 5 b.] To soften again. Hence **reso·ftening** *vbl. sb.*

1611 Florio *Worlde of Wordes* 439/3 *Rimollire*,..also to remollifie, to resoften or asswage. **1889** P. N. Hasluck *Model Engin. Handybk.* 136 By the appearance of the colours we are informed of the temperature of the steel, or in other words, how far, or to what extent the resoftening has progressed. **1894** Duke of Argyll *Burdens of Belief* 90 Come burnished autumn with thy wealth of flame,..Come leaves with tints too blended for a name, And lakes resoftening lights that come from you. **1963** J. Osborne *Dental Mech.* (ed. 5) i. 16 The impression is taken, chilled, the surface resoftened by flaming and the impression reinserted.

resol (re·zǫl). *Chem.* Also **-ole** (-ōᵘl). [a. G. *resol* (H. Lebach 1909, in *Zeitschr. f. angew. Chem.* XXII. 1601), f. L. *rēs-ina* Resin *sb.* + *-ol* -ol.] A name given to the alcohol-soluble, usu. fluid resins formed as the first stage in phenol-aldehyde copolymerizations and often prepared as precondensates in the manufacture of plastic.

1913 H. Lebach in *Jrnl. Soc. Chem. Industry* XXXII. 559/2 These initial stages of Bakelite have received various names... Four years ago I proposed to call the whole of this class 'Resoles'. **1933** *Ibid.* LII. 420T/2 Similar *o*-hydroxy-substituted substances are supposed to represent molecules of Bakelite A (resole). **1935** C. Ellis *Chem. Synthetic Resins* I. xiv. 306 The resols, or initial products of alkaline condensation are mono- or poly-alcohols. **1956** J. N. Anderson *Appl. Dental Materials* xxi. 260 The resin is in the 'resole' state and is modified by adding fillers and colouring matter. **1973** *Materials & Technol.* VI. viii. 590 Resols used as starting materials have a very limited storage life owing to the presence of reactive groups.

resole, *v.* Add: **b.** *intr.* To admit of being resoled. *rare.*

1922 *Daily Mail* 24 Nov. 11 (Advt.), You can get 'cheap' shoes which look worse every day you wear them, and won't re-sole once.

resolution. Add: **I. 2. c.** More widely, the act, process, or capability of rendering distinguishable the component parts of an object or closely adjacent optical or photographic images, or of separating measurements of similar magnitude of any quantity in space or time; also, the smallest quantity which is measurable by such a process. (Further examples.)

1902 *Encycl. Brit.* XXXII. 776 A resolving power of 100,000 would suffice for the resolution of the closest lines in the spectrum. **1931** *Proc. R. Soc.* A. CXXXII. 307 The 'resolution' (*i.e.*, the smallest interval of time by which two impulses could be separated and still be separately recorded) of the relay ring..was between 1/100th and 1/200th second. **1935** *Nature* 12 Oct. 592/2 The remarks on numerical aperture may give rise to confusion, as the term 'definition' is used, instead of the correct one, 'resolution'. **1958** *Engineering* 28 Mar. 389/2 The accuracy and resolution of the equipment are both equal to one digit in the fourth significant place. **1962** *Which?* Mar. 70/1 We measured the resolution. This is the ability of the projector to reproduce fine detail. **1968** *Brit. Med. Bull.* XXIV. 255/1 This provides 8-bit resolution for the input samples. **1971** J. Z. Young *Introd. Study Man* ii. 20 As they examine the world more and more minutely with instruments of ever higher resolution they come upon phenomena not previously described. **1972** *Sci. Amer.* July 19/3 A 1,000 kilohertz..sonar would provide a resolution of 30 centimeters on a target 200 meters away. **1973** *Ibid.* June 47/1 The time resolution of the shutter is about a nanosecond. **1978** *Ibid.* Mar. 144/2 (Advt.), A usable resolution of 0·001° C makes the HP 2804 an excellent choice for measuring minute temperature differences. **1978** *Nature* 18 May p. xviii/1 By means of a selector key the weighing range of 200 g, resolution to 0·01 g, or of 2 kg, resolution to 0·1 g, can be selected.

II. 9. (Earlier example.)

1785 T. Parkinson *Syst. Mech.* iv. 78 (*heading*) Composition and resolution of forces.

re-solution. Add: Also **resolution.** (Later examples.)

1957 G. E. Hutchinson *Treat. Limnol.* I. xi. 708 Evaporation to dryness and resolution liberates some of the previously unreactive iron..in a reactive form. **1966** Phillips & Williams *Inorg. Chem.* II. xxxii. 524 A second factor controlling the re-solution of the hydroxide in alkali would appear to be the ease of change of coordination number. **1968** R. W. Fairbridge *Encycl. Geomorphol.* 1051/2 In southern England, many large speleotherms are today being destroyed by drips from the original formative source ('re-solution').

resolvancy (rĭzǫ·lvānsi). *rare.* [f. Resolve *v.* + -ancy.] An outcome or solution.

1930 G. Greene *Two Witnesses* 84 How utterly grown-up we then become, and the crowded confused days have to be made to clear, have to reach their own resolvancies.

resolve, *v.* Add: **I. 2. d.** (Earlier example.)

1785 T. Parkinson *Syst. Mech.* iv. 80 One force..may be resolved into any number, either in the same, or different planes, producing the same effect with it.

e. More widely, to distinguish (things of similar magnitude or close together in time). Cf. *resolution 2 c.

1932 *Proc. R. Soc.* A. CXXXVI. 313 Two particles separated by as little as 1/500th second could be 'resolved' and correctly recorded.

11. a. (Later examples.)

1975 *New Yorker* 17 Nov. 94/2 Bress proposed the questions to be resolved by the jury—whether the subpoenas were properly served on the McSurelys, whether they refused to comply with a lawful command of Congress, and whether their refusal was willful. **1976** *Amer. Speech 1973* XLVIII. 248 Nor can *Atlas* materials resolve Kypriotaki's basic question of lexical or environmental control of initial syllable deletion.

resolvent, *a.* and *sb.* Add: **B.** *sb.* **4.** *Galois resolvent*: see *Galois.

resolver. Add: **4.** *Electr.* An electromechanical device which transforms the representation of an electric vector from polar to Cartesian coordinates (see quot. 1956); also, more widely, an electronic device which resolves an input signal into components.

1952 G. A. & T. M. Korn *Electronic Analog Computers* vi. 283 A special rotatable transformer called a resolver. **1956** Berkeley & Wainwright *Computers* iii. 101 A special form of transformer known as a resolver..built something like an electric motor... Turning of the rotor varies the coupling between the windings so as to produce the sine and cosine functions. *Ibid.* viii. 353/1 Resolver..a device for resolving a vector into two mutually perpendicular components. **1975** *Daily Tel.* (Colour Suppl.) 24 Oct. 29/1 Many find that they can get a very pleasant 'surround' effect by using a pair of modest rear speakers coupled to a simple unit called a resolver. The Neal resolver, for example, sorts out the signals from a stereo amplifier and passes to the rear loudspeakers the out-of-phase information—the peripheral elements of the sound. **1976** *Sci. Amer.* Feb. 78/1 The addition of servo control calls for feedback from sensors such as potentiometers, encoders and resolvers, which measure the position of each joint.

resolving, *vbl. sb.* Add: **2.** *Comb.* **resolving power,** the capability of an optical or photographic system to separate or distinguish closely adjacent images; also, the similar capability of a radio telescope; **resolving time,** the interval from the start of a counted pulse in a pulse counter to the time when another pulse can be detected and counted separately.

1879 Resolving power [in Dict.]. **1887** *Encycl. Brit.* XXII. 374/1 While the resolving power of a spectroscope with grating..is independent of the wave-length for each order, the resolving power of a spectroscope with prism will vary inversely as the third power of the wave-length. **1955** *Sci. Amer.* Mar. 38/1 The resolving power is measured by the sharpness of the peak; it is usually expressed as the 'half-power beam width', which means the width of the arc along which the antenna receives half or more of the maximum signal power. **1977** J. Narlikar *Struct. Universe* iv. 107 Because visual wavelengths are considerably shorter than radio wavelengths.., the optical telescope can achieve a better resolving power with an aperture of a few centimetres than the radio telescope can with an aperture of several metres. **1942** W. B. Lewis *Electr. Counting* viii. 83 The anode potential rises more rapidly after extinction and the resolving time is reduced. **1948** *Jrnl. Sci. Instruments* XXV. 37/1 The resolving time of the input circuit of the scaler was measured directly by feeding into the scaler pairs of pulses of variable separation. **1963** B. Brown *Experimental Nucleonics* 175 Due to the short resolving time much higher count rates are possible.

resonance. Add: **1. c.** Substitute for def.: The phenomenon of an oscillating signal (as an electric current or electromagnetic radiation) producing an effect upon an oscillating current of the same frequency; the condition in which a circuit or device produces the largest possible response to an applied oscillating signal, *esp.* when its inductive reactance balances its capacitative reactance. (Earlier and later examples.)

1886 *Electrician* 20 Aug. 296/2 By this to-and-fro reflection, or electrical reverberation or resonance, the amplitude of the received current may be made far greater than the strength of the steady-flow current from the same impressed force. **1897** A. Hay *Princ. Alternate-Current Working* xii. 159 This phenomenon of the neutralisation of an inductance by means of a capacity is generally referred to as electrical resonance. **1920** E. W. Stone *Elem. Radiotelegr.* i. 22 When these two reactances are equal, a state of resonance is said to obtain. **1920** *Whittaker's Electr. Engineer's Pocket-bk.* (ed. 4) 446 Resonance is rarely established with the fundamental frequency of the supply, but is generally due to harmonics. **1943** C. L. Boltz *Basic Radio* viii. 129 The phenomenon of a circuit responding most to one frequency is called resonance. **1959** K. Henney *Radio Engin. Handbk.* (ed. 5) vi. 46 It was realized that a hollow closed conducting box of arbitrary shape possessed electrical resonance properties similar to the conventional coil and capacitor circuit. **1964** R. F. Ficchi *Electr. Interference* v. 48 Past the point of resonance, the inductive reactance continues to increase and the capacitor is no longer effective as a bypass

filter. **1975** D. G. Fink *Electronics Engineers' Handbk.* xv. 30 One type of forced commutation uses resonance to generate an alternation which brings the current in a conducting thyristor to zero.

d. *Physics* and *Chem.* (i) Generally, a condition in which a particle is subjected to an oscillating influence (as an electromagnetic field or another particle) of such a frequency that a transfer of energy occurs or reaches a maximum; an instance of this; an exchange of energy occurring under such conditions.

1895 *Abstr. Physical Papers* (Physical Soc.) I. 355 It is thus impossible that resonance should obtain between the electric waves of Hertz and the molecules of a body, and consequently impossible for an ordinary prism to disperse electric waves. **1902** *Phil. Mag.* III. 396 A new type of light absorption, which it may be possible to refer to the electrical resonance of small metallic particles for waves of light. *Ibid.* IV. 428 The variable nature of the colour..makes it appear improbable that the action is similar to that of aniline dyes; namely, a resonance within the molecule. **1931** *Proc. R. Soc.* A. CXXX. 477 The possibility of the occurrence of line spectra due to a resonance between the α-particle and the nucleus. **1935** *Physical Rev.* XLVII. 751/2 The large cross sections [for neutron scattering] may thus be called a resonance effect, but the 'resonance' is very unsharp. **1959** *Physical Rev. Lett.* II. 427/1 Quite a narrow resonance (half-width ≤ 20 Mev) appears in these cross sections just below this threshold. **1966** Williams & Fleming *Spectrosc. Methods Org. Chem.* iv. 78 Since different protons in an organic molecule have varying electronic environments, the precise value of the magnetic field required to bring any one into resonance at constant frequency will vary slightly from proton to proton. **1970** I. E. McCarthy *Nuclear Reactions* I. i. 6 The reaction occurred preferentially at four different energies between about 4 MeV and 5·3 MeV. Resonance was said to occur in the α, Al^{27} system at these energies. **1974** *Accounts Chem. Res.* VII. 341/2 The Raman spectra are dominated by the porphyrin vibrational modes which are enhanced by resonance with the allowed electronic transitions in the visible and near-ultraviolet region. **1976** *Sci. Amer.* Jan. 50/2 The method of detection employed is an application of Werner Heisenberg's uncertainty principle and it consists in searching for an enhancement at a particular energy in the probability of interaction between known particles. Such an enhancement is called a resonance.

(ii) *Chem.* = *mesomerism b.

1927 *Sci. Abstr.* A. XXX. 84 (*heading*) The problem of several bodies and resonance in quantum mechanics. [Abstr. of paper by W. Heisenberg in *Ztschr. f. Physik* (1926) XXXIX. 499.] **1931** *Physical Rev.* XXXVII. 489 In some cases there may be several ways of drawing valence bonds in a given compound. In such cases, the real situation is again a combination of the various possibilities, and on account of resonance the energy is lower than it would otherwise be. **1939** L. Pauling *Nature of Chem. Bond* ii. 35 In case that the extreme covalent structure A:B and the extreme ionic structure A^+B^- correspond separately to the same bond energy value, then the two structures will contribute equally to the actual state of the molecule, and the actual bond energy will be greater than the bond energy for either structure alone by an amount equal to the interaction energy of the two structures; that is, resonance between the two structures will stabilize the molecule. *Ibid.* xii. 408 A substance showing resonance between two or more valence-bond structures does not contain molecules with the configuration and properties usually associated with these structures. **1950** N. V. Sidgwick *Chem. Elements* I. 525 From the crystal structure of sodium formate Zachariasen has shown that in the ion there is complete resonance between the two C—O groups. **1968** *Nature* 24 Aug. 801/2 This departure from planarity, by interfering with resonance, was expected to make the molecule so unstable that one hardly expected to find it. **1978** P. W. Atkins *Physical Chem.* xv. 494 These structures are less stable than the Kekulé forms, because the A—D bonds, etc., are long and weak; therefore, although they must be allowed to take part in the resonance they contribute more weakly.

(iii) *spec.* (also *magnetic resonance*) the transition of a particle possessing a magnetic moment between different quantum states in the presence of a magnetic field and electromagnetic radiation of the appropriate frequency; a spectroscopic technique (as *electron spin resonance, nuclear magnetic resonance*, etc.) in which such phenomena are observed.

1938 *Physical Rev.* LIII. 318/2 The experimental procedure is to vary the homogeneous field for some given value of the frequency of the oscillating field until the resonance is observed by a drop in intensity at the detector and a subsequent recovery when the resonance value is passed. **1942** Nuclear magnetic resonance [see *nuclear *a.* 4]. **1952** *Physical Rev.* LXXXVIII. 951/1 We have observed conduction electron spin resonance absorption in fine particles of metallic sodium. **1957** *Endeavour* Oct. 185/1 When atomic nuclei are placed in a constant magnetic field of high intensity and subjected at the same time to a radio-frequency alternating field, a transfer of energy takes place between the high-frequency field and the nucleus. This phenomenon is known as nuclear magnetic resonance. **1965** R. N. Dixon *Spectroscopy & Structure* viii. 182 Electron spin resonance spectrometers usually use fields of the order of 3000 G, and the resonance is then at microwave frequencies of about 9000 Mc/sec. **1967** Atkins & Symons *Struct. Inorg. Radicals* i. 7 Signals can be detected by nuclear magnetic resonance under conditions where lines are too broad to detect by electron spin resonance. **1977** *Nature* 17 Nov. 272/2 The spectrum of the enriched pigment contains one additional resonance at approximately 130·8 p.p.m. downfield from TSP-d_4.

(iv) *spec.* in *Nuclear Physics*, a short-lived particle, or an excited state of a particle, manifested as an increase, at certain well-defined energies, in the probability of interaction of other particles.

1964 *Physical Rev. Lett.* XIII. 64/2 We have searched for a possible spin alignment of the resonance by analyzing the strong decay into Ξ^* (1530) + π. **1965** *Science* 10 Sept. 1183/2 When the nucleon is exposed to any kind of high-energy beams, it is transformed into short-lived states of higher energy, which are known under various names, such as 'hyperons' or 'resonances'. **1969** *Times* 5 Feb. 13/7 The Xi resonances belong to the heavier class of nuclear particles known as baryons. **1972** G. L. Wick *Elem. Particles* iv. 60 Analyses of particle tracks also helped to reveal numerous short-lived particles called resonances. **1975** *Physics Bull.* Dec. 537/3 Excited states of the nucleons, so called resonances, can be produced by inelastic electron scattering.

e. *Mech.* (i) A condition in which an object or system is subjected to an oscillating force having a frequency close to that of a natural vibration of the object or system; the resulting amplification of the natural vibration.

1899 Franklin & Williamson *Elem. Alternating Currents* v. 59 Mechanical resonance. If a periodic force of given maximum value and given frequency acts upon the body..the body will be set vibrating at the same frequency as that of the force, and the violence of the motion will be greatest, for the given value of the periodic force, when the frequency of the force is equal to the proper frequency of the body. **1913** *Phil. Mag.* XXVI. 125 If the lower bob is of solid metal..its damping coefficient will be small and the resonance in consequence probably inconveniently sharp. **1935** J. E. Younger *Struct. Design Metal Airplanes* xv. 272 In the prevention of dangerous structural vibration, the first principle is to avoid resonance. **1952** D. E. Christie *Intermediate College Mech.* xii. 306 In practical engineering it is frequently desirable to keep forced oscillations well away from resonance. **1959** *Listener* 5 Feb. 252/1 Where the lengths of waves and ship are almost equal, we experience severe pitch and heave motions... This 'resonance' can be destroyed in two ways. **1974** *Encycl. Brit. Micropædia* VIII. 525/1 Mechanical resonance..is known to have built up to such large proportions as to be destructive, as in the case of the Tacoma Narrows Bridge.

(ii) Amplification of wave or tidal motion in a body of water when this motion has the same frequency as a natural vibration of the body of water.

1955 P. H. Kuenen *Realms of Water* ii. 37 Lake Baikal..has..no resonance whatever and the tide is less than one inch. *Ibid.* 42 In the Malay Archipelago..resonances in the separate basins with natural periods of oscillation. **1975** *New Yorker* 12 May 94/2 Certain harbors sometimes have problems with a phenomenon known as resonance, wherein waves that might be, say, two feet high on the outside build up energy within the harbor until waves in there stand ten feet high or higher.

f. *Astr.* The circumstance or phenomenon of the periods of revolution of two bodies about a single primary being, exactly or approximately, in the ratio of two small integers.

1923 *Astron. Jrnl.* XXXV. 70/2 From the mechanical point of view, the chief feature of the motion of an asteroid of the Trojan group is due to resonance. **1928** *Bull. Amer. Math. Soc.* XXXIV. 283 Here the millions of 'stones' or 'rocks' which must constitute those rings revolve round Saturn and resonances are caused by the action of its larger satellites. **1968** R. A. Lyttleton *Mysteries Solar Syst.* vii. 229 The adopted value implied a period of the unknown planet of some 218 years, which is not very different from three times the period of Uranus, namely 84 years. It would be sufficiently near in fact to give a mild resonance, because of the approximate 3:1 ratio. **1979** *Science* 5 Oct. 39/1 Mimas has been implicated because it orbits Saturn in exactly twice the time that any particles happening to orbit within the Cassini division would. This whole-number relationship of orbital periods is called resonance.

g. *fig.* and other *transf.* uses.

1607, etc. [see sense 1 a in Dict.]. **1892** J. Sully *Human Mind* II. xiv. 58 That the corporeal resonance does form an essential ingredient in emotion is abundantly proved by a variety of facts. **1925** J. Laird *Our Minds & Their Bodies* i. 21 It is the commonest thing in the world to say that we are 'not ourselves' when our bodies, so to say, ring differently to us from their normal resonance. **1939** *Scrutiny* VII. 441 In particular, his temperament was painfully out of resonance with his father's. **1956** Day & de la Warr *New Worlds beyond Atom* v. 34 A part of the wave-form emitted by the drug forms a discord with the radiations of the disease while another part of it is in resonance with the organ which is being treated. **1962** E. F. Haden et al. *Resonance-Theory for Linguistics* 48 Resonance in Linguistics is a bond, imagined as a hybrid or a wave, linking two language entities. **1965** J. M. Stephens *Psychol. of Classroom Learning* vii. 176/2 In real life many problems have been solved by this seemingly mysterious, unconscious 'resonance'. **1967** E. H. Lenneberg *Biol. Foundations Lang.* ix. 378 Perhaps a better metaphor still is the concept of resonance... Exposure to adult language behavior has an excitatory effect upon the actualization process much the same way a certain frequency may have..upon a specific resonator; the object begins to vibrate in the presence of the sound. In the case of language onset, the energy required for the resonance is, in a sense, supplied by the individual himself. **1976** *Southern Even. Echo* (Southampton) 16 Nov. 3/5 For much of the piece there is an extra resonance and significance about Gray's otherwise-familiar anti-hero. **1977** A. Sheridan tr. *J. Lacan's Écrits* iii. 86 What is redundant as far as information is concerned is precisely that which does duty as resonance in speech.

2. b. The enhancement of one colour by its proximity to another or others. Cf. *resonant *a.* 2 b.

1933 *Burlington Mag.* Jan. 3/1, I knew that Titian was a master of rich and sumptuous colour. I knew how splendidly he could evoke from his blues and crimsons their fullest and deepest resonance.

3. *resonance frequency, particle, vibration*; **resonance absorption** *Nuclear Physics*, absorption of energy or of a particle under conditions of resonance; *spec.* = next; **resonance capture** *Nuclear Physics*, absorption of a particle by an atomic nucleus which occurs only for certain well-defined values of particle energy; **resonance chamber** = Resonator 2; **resonance energy,** (*a*) an energy value at which resonance occurs; (*b*) *Chem.*, the extent of stabilization of a molecular structure attributed to mesomerism; **resonance fluorescence,** fluorescence in which the light emitted has the same wavelength as that which excites the emission; **resonance hybrid** *Chem.*, a molecular structure which is a mesomeric combination of a number of forms; **resonance radiation,** (the radiation emitted in) resonance fluorescence; **resonance Raman spectrum** [*Raman], a Raman spectrum excited by light having a frequency equal to that of a band in the absorption spectrum of the scattering substance (see quot. 1975[1]); so *resonance Raman effect, spectroscopy*, etc.; **resonance scattering** *Nuclear Physics*, elastic scattering of a particle by an atomic nucleus at an energy of the incident particle for which the scattering cross-section is large compared with that for adjacent values of the energy (cf. *potential scattering* s.v. *potential *sb.* 4 c; **resonance stabilization** *Chem.* = *resonance energy* (*b*).

1945 H. D. Smyth *Gen. Acct. Devel. Atomic Energy Mil. Purposes* iv. 24 The term 'resonance absorption' is used to describe the very strong absorption of neutrons by U-238 when the neutron energies are in certain definite portions of the energy region. **1952** [see sense 1 d (iii) above]. **1961** G. R. Choppin *Exper. Nuclear Chem.* viii. 114 In indium a resonance absorption occurs for neutrons with a kinetic energy of 1·44 ev. **1937** G. Gamow *Struct. Atomic Nuclei* xi. 224 We may have here a case of resonance-capture. **1964** M. Gowing *Britain & Atomic Energy 1939–45* i. 40 Various other ways of reducing the capture of neutrons by resonance capture..were considered. **1919** D. Jones *Outl. Eng. Phonetics* vi. 15 The effect of a resonance chamber in modifying quality of tone may be illustrated experimentally. **1962** A. Nisbett *Technique Sound Studio* iii. 59 The mouth, nose and throat cavities act as resonance chambers for sound coming from the vocal cords. **1931** *Jrnl. Amer. Chem. Soc.* LIII. 1368 The energy of the bond is largely the resonance or interchange energy of two electrons. **1941** in M. Gowing *Britain & Atomic Energy 1939–45* (1964) 430 Neutrons of certain critical, or resonance, energies are strongly absorbed by uranium without causing fission. **1965** Phillips & Williams *Inorg. Chem.* I. iii. 75 The difference between the computed binding energy of a hypothetical structure and that computed using all possible structures, i.e. the actual binding energy of the real molecule, is called the resonance energy. **1925** *Sci. Abstr.* A. XXVIII. 121 In the case of resonance fluorescence the re-emitted line possesses a greater Doppler width than the incident line. **1977** I. M. Campbell *Energy & Atmosphere* viii. 260 (*caption*) Schematic diagram of apparatus for the resonance fluorescence detection of hydroxyl radicals in air. **1921** W. H. Eccles *Continuous Wave Wireless Telegr.* iii. 172 When an external sine force acts upon a vibrator, whether at the resonance frequency or not, it gradually builds up a vibrating motion to a definite final amplitude. **1974** *Encycl. Brit. Macropædia* XIX. 101/2 Wind supplied the power causing the bridge to vibrate at one of its torsional resonance frequencies without sufficient damping. **1939** L. Pauling *Nature Chem. Bond* xii. 407 Each of these tautomers in its normal state is represented..by a resonance hybrid of this structure and others. **1964** N. G. Clark *Mod. Org. Chem.* xxv. 525 The true structure of benzene is..a resonance hybrid, to which the two Kekulé formulae..contribute equally. **1975** *Sci. Amer.* May 43/2 Other resonance particles decay by the 'strong' interaction in about 10^{-23} second. **1905** R. W. Wood in *Phil. Mag.* X. 514 Repeated efforts have been made..to detect a lateral emission of yellow light by sodium vapour when in the act of absorbing sodium light... This seems to be the first case found of the phenomenon, which it may perhaps be well to style resonance radiation, to distinguish it from fluorescence. **1928** [see *quench *v.* 3 d]. **1963** R. W. Ditchburn *Light* (ed. 2) xvii. 661 Sodium absorbs and re-emits as resonance radiation the two well-known yellow lines at wavelengths 5890 Å. and 5896 Å. **1960** tr. Shorygin & Krushinskii in *Soviet Physics Doklady* V. 793 The possibility of observing resonance Raman spectra is limited to a considerable degree by the loss of light due to absorption. **1962** *Pure & Appl. Chem.* IV. 87 The study of the resonance Raman effect can contribute not only to the extension of the technique of Raman spectroscopy, but also to an investigation of the nature of the interaction of light with matter. **1975** *Jrnl. Chem. Soc.: Dalton Trans.* 381/1 When a molecule..is excited with a laser beam whose wavenumber corresponds or closely corresponds with the band maximum of a strongly allowed electronic transition of the molecule, then a rigorous resonance Raman (r.r.) spectrum may be obtained. Such spectra are normally characterised by an enormous increase in the intensity of a totally symmetric fundamental of the molecule together with the appearance of long overtone progressions in this same fundamental. **1975** *Nature* 25 Dec. 770/1 The results of Lewis and Spoonhower using resonance Raman spectroscopy imply the existence of a strong complex between retinal and a tryptophan residue in rhodopsin. **1937** Bethe & Placzek in *Physical Rev.* LI. 462/1 Near resonance, the resonance scattering σ_r must be added to the potential scattering. **1955** A. E. S. Green *Nuclear Physics* xiii. 456

These deviations are attributed to the interference between the potential and resonance scattering associated with the *l*th partial wave. **1978** *Nature* 26 Oct. 730/1 Resonance scattering of light from a beam of free atoms is an ideal technique for making precise absolute measurements of the shift in wavelength of the light relative to the reference wavelength of the beam atoms. **1939** L. PAULING *Nature Chem. Bond* i. 10 Because the resonating system does not have a structure intermediate between those involved in the resonance, but..a structure which is further changed by the resonance stabilization, I prefer not to use the word 'mesomerism'..for the resonance phenomenon. **1952** [see *OXYANION]. **1978** K. YATES *Hückel Molecular Orbital Theory* iii. 118 In order to evaluate the resonance stabilization of benzene, as represented by the formulation (XI), a comparison should be made with cyclohexatrienes possessing the benzene geometry. **1909** *Westm. Gaz.* 4 Sept. 10/1 If the period of vibration of the two parts is the same 'resonance vibrations' are set up.

resonant, *a.* and *sb.* Add: **A.** *adj.* **1. b.** *Phonet.* Of consonants: liquid or nasal. Cf. the sb. in Dict. and Suppl.

1943 K. L. PIKE *Phonetics* vii. 144 The *sonorants* are nonvocoid resonants and comprise the lateral resonant orals and resonant nasals (e.g. [m], [n], and [l]). **1948** W. F. TWADDELL in *Language* XXIV. 141 Before the resonant consonants, only /e/, /a/, and /u/ occurred regularly.

2. b. Of colours: emphasizing each other by contrast.

1887 *Portfolio* XVIII. 233/2 His painting has ever become slighter, higher in tone and less full and resonant in colour.

4. a. Involving, exhibiting, or bringing about electrical resonance (RESONANCE 1 c in Dict. and Suppl.).

1888 *Electrician* 28 Sept. 663/2 Drawing the micrometer terminals so far apart that sparks can only be made to pass by means of resonant action. **1925** Resonant frequency [sense 5 below]. **1938** *Jrnl. Appl. Physics* IX. 654/2 Such a resonator may often be equivalent to a conventional resonant circuit. **1947** R. LEE *Electronic Transformers & Circuits* vii. 192 The heights of resonant peaks and frequency distance between peaks depend upon circuit *Q* and coefficient of coupling *k*. **1959** K. HENNEY *Radio Engin. Handbk.* (ed. 5) v. 27 Multiple resonance imposes practical limits on the way in which resonant circuits may be combined. **1964** R. F. FICCHI *Electr. Interference* v. 56 The isolation offered by the turns eliminates resonant effects. **1966** *New Scientist* 15 Dec. 627/1 The device, known as a resonant-gate transistor, was developed by a team of scientists in..Pittsburgh.

b. *Physics* and *Chem.* Pertaining to, involving, exhibiting, or taking part in any other kind of resonance.

1934 Resonant frequency [sense 5 below]. **1944** L. F. & M. FIESER *Org. Chem.* xx. 523 Results of the ozonization experiments cited above accord with the concept of two resonant Kekulé forms contributing in an equal extent to the structure. **1960** DICKE & WITTKE *Introd. Quantum Mechanics* xvi. 291 When the energy of the incoming particles corresponds to such a state, a resonant condition is said to occur, in which the scattering cross-section is markedly greater than for nonresonant energies. **1965** *Physics Lett.* XIV. 159/1 The enhancement observed [in pion-nucleon interaction] need not necessarily be identified with a resonant state. **1973** *Sci. Amer.* Nov. 125/3 The reason is that one second is about the period at which a 10-inch pendulum vibrates naturally, that is, one second is the frequency of vibration to which the pendulum is resonant. **1976** *Chem. Physics Lett.* XLI. 292/1 The resonant electronic transition is the axial $d_{z^2} \leftarrow d_{z^2}$ transition at $17\,450\ cm^{-1}$.

5. Special collocations: **resonant cavity,** a cavity resonator (see *RESONATOR 3 b); **resonant frequency,** a frequency at which resonance (of any kind) takes place; **resonant scattering** *Nuclear Physics* = *resonance scattering* s.v. *RESONANCE 3.

1945 *Nature* 15 Sept. 323/1 Dr. J. T. Randall applied the resonant-cavity technique to the relatively ineffective magnetron of pre-war days, and made of it a radically new and immensely powerful device which remains the heart of every modern radar equipment. **1955** *Science* 9 Dec. 1132/3 The deuteron beam..then enters the first of two 48-megacycle resonant-cavity accelerator sections. **1925** W. GREENWOOD *Text-bk. Wireless Telegr. & Teleph.* i. 19 The resonant frequency is..equal to the natural frequency when the damping is negligible. **1934** J. P. DEN HARTOG *Mech. Vibrations* ii. 52 The forced frequency coincides exactly with the natural frequency... This important phenomenon is known as 'resonance', and the natural frequency is sometimes called also the 'resonant frequency'. **1964** R. F. FICCHI *Electr. Interference* v. 49 The smaller the total inductance, the higher the resonant frequency. **1948** *Physical Rev.* LXXIV. 926/1 The present work was begun with the object of detecting nuclear resonant scattering from Mg^{24} nuclei which are excited by resonant radiation from the radioactive Na^{24} nuclei. **1955** [see *potential scattering* s.v. *POTENTIAL *sb.* 4 c]. **1963** W. E. BURCHAM *Nuclear Physics* xiii. 484 From the observed cross-section for the resonant scattering effect, with appropriate account taken of the thermal broadening,..a total width of $(2{\cdot}1 \pm 0{\cdot}4) \times 10^{-5}$ eV for the 412 keV level of ^{198}Hg was found.

B. *sb.* (Further examples.) Also, a liquid consonant.

1948 W. F. TWADDELL in *Language* XXIV. 141, /i/ before resonants was at best very rare. **1956** J. WHATMOUGH *Language* 37 Thus we have the class known as resonants (e.g. *l*, *m*). **1970** B. M. H. STRANG *Hist. Eng.* ix. 406 The third sub-system consists of resonants, i.e. sounds which may or may not have syllabic function (thus corresponding to approximants, liquids and nasals in P[resent-day] E[nglish]). **1976** *Archivum Linguisticum* VII. 167 The gemination becomes an additional argument for the reconstruction of laryngeals next to a resonant or *s*.

resonantly *adv.* (further examples).

1971 *Nature* 15 Oct. 469/2 From the observational viewpoint the soft X-ray enhancement..can be interpreted in terms of solar X-rays resonantly scattered in the Earth's atmosphere. **1977** *Gramophone* May 1737/2 Here if anything he is even more resonantly impressive than he was for Klemperer. **1978** *Sci. Amer.* June 124/1 The audio oscillator was tuned to resonantly oscillate the water..so as to cause vibrational standing waves to form on the surface.

resonate, *v.* Add: **1. a.** (Further examples.) Also *fig.*

1946 *Physical Rev.* LXIX. 37 It [*sc.* a resonant cavity] was adjusted to resonate at about 30 mc/sec. **1955** R. S. H. BOULDING *Radar Pocket Bk.* vi. 72 An artificial line can be charged from an a.c. supply without a rectifier by making a choke in the charging circuit resonate, at the supply frequency, with the capacitance of the line. **1956** *Jrnl. Chem. Physics* XXIV. 468/1 In NMR spectra nuclei which give signals at the same applied field or resonate with the same Larmor frequency may be considered magnetically equivalent. **1976** *Publishers Weekly* 27 Sept. 80/1 Prose..resonating with the illustrations.

b. *spec.* in *Chem.* To exhibit mesomerism (cf. *RESONANCE 1 d (ii)). Const. *among* or *between* different structures, as if a real physical alternation were occurring.

1933 PAULING & WHELAND in *Jrnl. Chem. Physics* I. 369 In the first step the C—C bond breaks and there are formed two phenyl-methyl radicals, which however can resonate between only the structures A and B of Fig. 3. **1957** G. E. HUTCHINSON *Treat. Limnol.* I. iii. 196 The molecule resonates between the following structures. **1965** G. W. WHELAND *Resonance in Org. Chem.* i. 7 Only by exercising the utmost care in the choice of words can one avoid the appearance of implying that the molecules..are oscillating back and forth among the several structures, and hence that these structures must possess real physical significance... The common statements that the hybrid *resonates* among the structures and that the structures *resonate* with one another almost unavoidably give this quite erroneous impression. **1976** *Sci. Amer.* Nov. 59/1 Perhaps the baryon resonates between these configurations, much as the benzene ring resonates between its various possible structures.

2. *trans.* To act as a resonator for; to amplify by resonance.

1904 *Physical Rev.* XVIII. 231 Having the radiating aërial resonating the primary circuit, it is now necessary to have a second primary circuit in tune with the first. **1975** *Sci. Amer.* July 48/2 Signals become convolved when sounds are reverberated or resonated or, in the case of photographs, when images are blurred.

resonating, *ppl. a.* Add: (Further examples). *resonating chamber* = RESONATOR 2.

1912 *Chem. Abstr.* VI. 3220 An extensive research on the properties and behavior of resonating gas mols. **1933** *Jrnl. Chem. Physics* I. 611 A negative carbon atom, with one unshared and three shared electron pairs, occurs rather often in resonating structures, as in carbon monoxide, [etc.]. **1938** Resonating chamber [see *RESONATOR 2 b]. **1959** E. PULGRAM *Introd. Spectrogr. Speech* vii. 55 The resonating tuning fork continues to operate for a while after the struck fork has been stopped. **1968** B. M. H. STRANG *Mod. Eng. Structure* (ed. 2) 35 The character of the sound produced can be varied over a considerable range by changing the shape of the resonating chambers. **1974** D. M. ADAMS *Inorg. Solids* vii. 230 In PbI_2, with its inert pair of *s*-electrons, bonding is via two resonating *p*-electrons. **1974** R. C. HIBBELER *Engin. Mech.* xxii. 958 Resonating vibrations can cause tremendous stress and rapid failure of parts.

resonator. Add: **2. b.** (Further examples.)

1938 *Oxf. Compan. Mus.* 994/2 The frontal sinuses..have apparently considerable value as resonators: it is said that the Australian natives have a great want of resonance in their speech due to the small size of these sinuses. All the resonating chambers or passages mentioned above..are in direct communication with the air which has passed through the vocal cords. **1970** A. C. GIMSON *Introd. Pronunc. Eng.* (ed. 2) 11 These cavities [*sc.* pharynx, mouth, nasal cavity] function as the principal *resonators* of the note produced in the larynx. *Ibid.* 19 The way in which the speaker's vibrator and resonators function together.

3. a. Also, any device which displays electrical resonance. (Earlier and later examples.)

1889 *Phil. Mag.* XXVIII. 125 For this proof we must give somewhat different dimensions to our apparatus in order to be able to introduce electric resonators into its interior. **1943** *Proc. IRE* XXXI. 448/1 The simple circuit of Fig. 14(a) is well known as the Hertzian oscillator and is readily recognized as a lumped capacitance associated with a single-turn inductor. At frequencies up to several hundreds of megacycles such a resonator is quite practical.

b. *spec.* (in full *cavity resonator*) a hollow enclosure with conducting walls which is capable of containing electromagnetic fields having particular frequencies of oscillation, and of exchanging electrical energy with them; such devices are used esp. for the amplification or detection of microwaves.

1936 *Proc. IRE* XXIV. 1320 [All of these terminals may act as more or less sharply resonant systems... They may..be thought of as electromagnetic analogues of the Helmholtz resonator.] *Ibid.* 1324 A cylindrical resonator attached to the hollow tube system. **1943** *Ibid.* XXXI. 447/2 The flexibility of the Klystron is seriously limited by the fact that cavity resonators are permanently attached to the grids and thus form an integral part of the tube itself. **1952** REINTJES & COATE *Princ. Radar* (Mass. Inst. Technol. Radar School Staff) (ed. 3) ix. 613 Cavity resonators are hollow metal-walled chambers fitted with devices for admitting and extracting electromagnetic energy. **1956** *Nature* 10 Mar. 470/2 During the past four or five years,..the Radiation Laboratory has concentrated much effort..on the construction of a linear accelerator of the cavity-resonator type. **1966** *McGraw-Hill Encycl. Sci. & Technol.* II. 577/2 Coaxial cavity resonators are often used as wavemeters, particularly for lower than microwave frequencies. **1975** D. G. FINK *Electronics Engineers' Handbk.* ix. 29 In the reflex Klystron a single resonator is used to modulate the beam and extract of energy from it, making the tube simple and easy to tune.

4. An object or system which resonates, in any other sense.

1897 *Abstr. Physical Papers* (Physical Soc.) III. 356 Next a resonator of a perfectly-conducting material is imagined to be introduced. This resonator, while absorbing energy from the incident radiation and radiating it again without loss, will, in general, change the character of the radiation, either in its frequency, the law of its damping, or both. **1914** J. H. JEANS *Rep. Radiation & Quantum Theory* ii. 9 Planck ..supposed that the emission and absorption were accomplished by 'resonators' of perfectly definite periods. **1949** KOESTLER *Insight & Outlook* p. ix, The theory of memory traces as selective resonator systems suggested in various forms by Jacques Loeb..and others. **1959** E. PULGRAM *Introd. Spectrogr. Speech* vii. 55 This effect of sympathetic vibration is called resonance, and the object activated by it is a resonator. **1974** G. REECE tr. *Hund's Hist. Quantum Theory* ii. 24 He [*sc.* Planck] constructed a model for the emission and absorption of radiation at the walls of a black body by assuming the presence of resonators. **1979** *Sci. Amer.* Mar. 81/1 It was soon learned that a particular wavelength could be selected by designing an optical resonator that will allow only the chosen wavelength to pass repeatedly through the amplifying dye cell... The resonator consisted of a partially reflecting glass plate at one end of the laser and a diffraction grating at the other.

resonatory, *a.* For *rare*$^{-1}$ read *rare.* (Further example.)

1961 L. F. BROSNAHAN *Sounds of Language* i. 2 The process of vocal sound production in language appears on analysis to consist basically of the production of an air-stream, the conversion of some of the kinetic energy of this airstream into acoustic energy in the form of a complex of sound waves, and the subsequent resonatory modulation of that complex.

resorb, *v.* Add: *esp.* in *Physiol.*, to absorb into the circulation (material already in the body, esp. material that has been digested or broken down). (Further examples.)

1901 J. S. MARSHALL *Princ. & Pract. Operative Dentistry* iv. 76 The absorbent organ secretes..a digestive fluid or soluble ferment which dissolves or digests the dental tissues and alveolar walls, and prepares them to be resorbed. *Ibid.* xxxiv. 509 Bone is resorbed through the action of the osteoclast cells. **1928** MOORE & KEY tr. *Leriche & Policard's Normal & Path. Physiol. of Bone* iv. 71 If..one brings about increase of circulation, bone in the epiphyses is resorbed to the extent of detaching itself from the investing cartilage and becoming very friable. **1967** M. E. HALE *Biol. Lichens* viii. 118 The acids are not resorbed by the hyphae. **1968** R. D. MARTIN tr. *Wickler's Mimicry in Plants & Animals* xi. 107 The flea will resorb its eggs if the female host miscarries.

Hence **reso·rbed** *ppl. a.*; **reso·rbing** *ppl. a.*, undergoing resorption; also **reso·rbable** *a.*, that may be resorbed.

1910 J. P. BUCKLEY *Mod. Dental Materia Medica* ii. 315 (*heading*) Resorbed root. **1928** MOORE & KEY tr. *Leriche & Policard's Normal & Path. Physiol. of Bone* iv. 85 In resorbing callus the cells were hypertrophied. **1967** *Archivum Immunol. et Therapiae Exper.* XV. 809 (*caption*) Section through a resorbable 256 Walker tumor. **1973** *Nature* 9 Mar. 135/2 Seven [voles] were pregnant but four had resorbing embryos. **1976** *Biomed.* XXV. 131/1 Porous resorbable ceramics have been found to be most suitable [as bone substitutes] since they encourage replacement of resorbed ceramic with new bone.

resorbent, *a.* (Later examples.)

1889 *Science* 29 Mar. 232/2 Not until the silicates are developed, and the granitic quartz begins to form, does the resorbent action discontinue. **1962** K. F. LAGLER et al. *Ichthyology* viii. 257 When the fish descends..the resorbent capillary network collapses so as to enlarge the area of contact of the gas gland for rapid secretion of gases into the bladder.

resorcinol. Substitute for def.: = RESORCIN; a dihydric phenol whose uses include the manufacture of phenol-formaldehyde resins. (Further examples.)

This, rather than *resorcin*, is now the usual name for the substance.

1949 *Thorpe's Dict. Appl. Chem.* (ed. 4) IX. 476/1 Of the dihydric phenols, resorcinol is industrially second to hydroquinone in importance, and is used in considerable amount. **1959** *Times* 27 Apr. (Rubber Industry Suppl.) p. vii/3 In this country all tire cord is impregnated by an aqueous process..using the mixed latices and resorcinol-formaldehyde resin. **1964** N. G. CLARK *Mod. Org. Chem.* xxi. 439 The three dihydroxybenzenes are called catechol, resorcinol, and hydroquinone (quinol). **1972** *Materials & Technol.* V. iv. 95 Resorcinol adhesives, with the added advantage of being neutral or nearly so, are superior to all others as adhesives for wood, giving exceptional durability.

resorption. Add: Cf. RESORB *v.* in Dict. and Suppl. (Further examples.)

1889 *Science* 29 Mar. 232/2 While the silicates are crystallizing in a molten mass, if porphyritic quartz is present it undergoes resorption. **1910** LAKE & RASTALL *Textbk. Geol.* xiii. 232 These phenocrysts have often undergone a certain amount of corrosion, or resorption. **1928** MOORE & KEY tr.

Leriche & Policard's Normal & Path. Physiol. of Bone iv. 71 When osteoclasts are present there is resorption. **1962** BLAKE & TROTT *Periodontology* viii. 82 In other cases there is resorption of the apices of vital teeth. **1962** K. F. LAGLER et al. *Ichthyology* viii. 247 In many rayfishes..gas resorption is also performed by..the oval organ situated in the posterior portion of the gas bladder. **1974** *Nature* 22 Mar. 343/1 This is accomplished by inhibition of bone resorption, of the tubular reabsorption of calcium, and of the intestinal absorption of calcium.

resorptive, *a.* Delete *rare* and add earlier and later examples.

1886 *Philadelphia Med. Times* XVI. 490/2 The removal of such substances is dependent upon the action of specialized cells, called resorptive or giant cells, and the process is physiological. **1951** *Jrnl. Bone & Joint Surg.* A. XXXIII. 936 All resorptive processes occur principally in relation to calcified matrices and not to osteoid or uncalcified cartilage. **1978** *Sci. Amer.* Feb. 112/3 My own observations of such 'flypaper' trappers as butterworts and sundews suggest that these plants have secretory and resorptive mechanisms that are quite different from the ones likely to operate in pitcher plants.

resort, *sb.* Add: **I. 7.** (Further examples.)

1883 *Harper's Mag.* Sept. 521/1 The bustle of arrival and departure..[animates] the village in the way peculiar to American towns near a 'resort'. **1893** K. D. WIGGIN *Polly Oliver's Problem* (1894) ii. 30 She would become the head of a summer resort, with a billiard-room and a bowling-alley. **1909** 'O. HENRY' *Roads of Destiny* xviii. 294 He was manager at different times of..a dozen hotels and summer resorts, an insurance company, and a district leader's campaign. **1936** [see *holiday resort* s.v. *HOLIDAY 4). **1976** *Morecambe Guardian* 7 Dec. 14/4 The man behind the proposal described Carnforth as 'a poor little town that could never become a health resort'.

III. 11. *attrib.* and *Comb.*, as (sense 7) *resort city, cottage, estate, hotel, -motel, -motor hotel, railroad station, town*; **resort clothes, -wear,** clothes suitable for wearing at a holiday resort.

1974 *Sumter* (S. Carolina) *Daily Item* 20 Apr. 1A/5 Police mounted a room-by-room sweep of hotels in this resort city. **1978** R. LUDLUM *Holcroft Covenant* viii. 93 In every resort city there was always one major shop that catered to the reading requirements of a specific nationality. **1974** *Country Life* 2 May 1097/2 Their Côte d'Azur collection of women's resort clothes. **1977** N. FREELING *Gadget* i. 24 Bring more than resort clothes... Evening things, feminine. **1971** *Jamaican Weekly Gleaner* 3 Nov. 3/1 A long debate on the merits or otherwise of licensing resort cottages. **1975** *New Yorker* 19 May 115 (Advt.), The Mountain View House is a resort estate with a charm only a century of family ownership can achieve. **1919** E. HOUGH *Sagebrusher* 49 A few passengers from the resort hotel back in the town began to appear. **1977** *Whitaker's Almanack 1978* 778/2 Tourism is the most important industry, with a good choice of resort hotels. **1963** *New Yorker* 1 June 123/1 (Advt.), America's most *fun*derful resort-motel. *Ibid.* 8 June 58 (Advt.), America's 3 great new resort-motor hotels. **1928** *Publishers' Weekly* 3 Nov. 1868/1 The..delay involved when a package must change trains in a resort railroad station. **1970** *Southerly* XXX. 124 Occasionally a car hummed along the bitumen..making..for one of the string of resort towns further along the coast. **1972** D. E. WESTLAKE *Cops & Robbers* (1973) xvi. 243 You could always tell a resort town, it ran much heavier to neon. **1965** *Punch* 10 Mar. p. xvi/2, Harvey Nichols's spring collection, which includes Italian and French resort-wear, can be seen on Mar 15. **1975** *Harper's & Queen* June 69 The latest collection of resort wear.

resorter. Add: (Later example.) Also (*U.S.*), one who runs a business in a resort.

1927 *Scribner's Mag.* Apr. 383/2 Take me along on a Southern trip to see what the rich resorters are wearin'? **1978** *Minneapolis Tribune* 4 Apr. 6A/1 The severe reduction of motorized opportunities..would devastate the resorters and outfitters of northeastern Minnesota.

resound, *sb.* (Later example.)

1835 H. EVANS in *Mississippi Valley Hist. Rev.* (1927) XIV. 213 Distinctly..the resound of guns was heard in quick succession.

resound, *v.* Add: **1. a.** Also with *to*.

1861 F. O. MORRIS *Rec. Animal Sagacity & Character* 121 He..lies buried..in those very shrubberies which had so often resounded to his joyous cries.

resounding, *ppl. a.* Add: Also *fig.*

1977 *Guardian Weekly* 4 Sept. 16/1 The Federal Energy Administration..says that in the month ending in mid-August, American oil consumption was up a resounding 8 per cent over the same time last year.

resource. Add: **1. a.** (Later *sing.* examples.)

1965 H. I. ANSOFF *Corporate Strategy* i. 6 A large majority of decisions must be made within the framework of a limited total resource. **1969** *Nature* 20 Dec. 1233/1 Satisfactory land use—that is, one that will provide a sustained yield of a resource—must take full account of the ecology. **1978** *Sci. Amer.* Nov. 82/3 The very best of them, such as W. F. Leopold's classic *Speech Development of a Bilingual Child*, continue to be a rich resource for contemporary investigators.

b. See also *natural resources* s.v. *NATURAL *a.* 6 e.

6. *attrib.* and *Comb.*, as (sense 1 a) *resource allocation, base, limit, zone*; *resource-based, -bound, -intensive, -limited, -poor, -supplying, -wasteful* adjs.; **resource aggregation** (see quot. 1968); **resource centre,** a library or other centre which houses a collection of *learning resources* (*LEARNING *vbl. sb.* 4); such a collection itself; also *attrib.*; **resource industry,** an industry of which the raw materials occur as natural resources; **resource profile** (see quot. 1967); **resource time,** the length of time a resource is required for a specific project.

1967 A. BATTERSBY *Network Analysis* (ed. 2) ix. 144 When Esso Petroleum transferred their accounts to a computer, they recognized 20 categories of clerical staff for resource aggregation. **1968** *Gloss. Terms Project Network Anal.* (*B.S.I.*) 9 *Resource aggregation*, the summation of the requirements of each resource, for each time period, calculated according to a common decision rule. **1964** K. G. LOCKYER *Introd. Critical Path Anal.* viii. 70 The name given by Production Controllers to this aspect of their work is *Loading*; regrettably, new names have been devised by some of the earlier CPA workers, amongst them *Manpower Smoothing*, and *Resource Allocation*. **1965** H. I. ANSOFF *Corporate Strategy* i. 6 A resource-allocation pattern which will offer the best potential for meeting the firm's objectives. **1977** P. JOHNSON *Enemies of Society* vi. 80 Classical economics had dealt with scarcity, value, choice, resource-allocation and efficiency. **1974** *Times* 8 Jan. (Europe Suppl.) p. xvi/5 How do you regard that oil? Is it British or is it European or is it just Shell's? I regard it as part of the resource base. **1967** *Times* 28 Feb. (Canada Suppl.) 32 Resource-based industries..have been characteristic of the economy. **1977** *Bull. Amer. Acad. Arts & Sci.* Oct. 18 Doubts about the continuity of past, present, and future..are expressed in current resource-bound models of growth. **1968** *Globe & Mail* (Toronto) 3 Feb. 47/1 (Advt.), The School has a gymnasium and central library with resource centre. **1971** in T. D. F. Barnard *New Directions in Librarianship* 44 There are new trends towards treating the library as a nexus for resource centre development. **1976** *Ann. Rep. Manpower Services Comm. 1975–76* ii. 20/1 The report therefore recommended that a resource centre should be established so that any organisation wishing to mount industrial relations training could be directed to appropriate sources of advice and teaching material. **1968** *Globe & Mail* (Toronto) 13 Jan. B5/1 Long on resource industry and lean in secondary industry, British Columbia continues to be especially vulnerable to the fluttering of foreign economies and markets. **1970** *Toronto Daily Star* 24 Sept. 1/8 This list..is likely to be lengthened to include some resource industries, pipelines, and possibly steel. **1976** *Conservation News* Sept.–Oct. 20/2 A planned transition to a less resource-intensive economy. **1970** S. L. BARRACLOUGH in I. L. Horowitz *Masses in Lat. Amer.* iv. 132 In many regions there is a sharp resource limit that would permit at most a doubling in the amount of productive land available. **1967** A. BATTERSBY *Network Analysis* (ed. 2) xii. 208 Before running the programme a 'duration limited' run or a 'resource limited' run must be specified. **1977** *Jrnl. R. Soc. Arts* CXXV. 389/1 The processing of many materials could..become energy-limited rather than resource-limited. **1969** *New Scientist* 2 Oct. 18/1 The flow of food and raw materials to the developed nations (almost all of which are resource-poor) will slowly dry up. **1973** *Listener* 20 Dec. 846/2 The resource-poor countries. **1967** A. BATTERSBY *Network Analysis* (ed. 2) ix. 141 Fig. 9.1 is called a 'resource profile' of a project, and it is obtained by 'resource aggregation'. The resources required by each job are specified when its duration is calculated; then when the scheduled starting times have been decided, the requirements are totalled over concurrent jobs for each discrete time period. **1969** R. B. FULLER *Operating Man. Spaceship Earth* ii. 27 They had to control various resource-supplying mines, forests, and lands with which and upon which to build the ships and establish the industries. **1973** *Bulletin* (Sydney) 25 Aug. 50/2 Australia has no idea of proposing the formation of a consortium of resource-supplying countries. **1964** K. G. LOCKYER *Introd. Critical Path Anal.* viii. 72 It is necessary to know the amount of work available—that is, the capacity available. This, too, must be specified in resource-time. **1976** *Carn* Feb. 20/2 It is easy to demonstrate that a resource-wasteful and energy-intensive lifestyle is very damaging to the environment. **1965** *New Statesman* 12 Nov. 732/1 We should select important amenity 'resource zones' such as the Broads and, rather than trying to stop all industrial development in them, get all interested parties to hammer out ways of developing them for all-round use—with recreation taking a high priority.

Also as *v. trans.*, to supply (a person, etc.) with resources; hence **resou·rced** *ppl. a.*; **resou·rcing** *vbl. sb.*

1975 *Listener* 27 Mar. 398/1, I would have gone in for smaller [sc.ool] units..resourced from some central agency. **1975** *Library Assoc. Rec.* LXXVII. 258 The policy of the Association..must be against diminished resourcing of libraries. **1975** *Language for Life* (Dept. Educ. & Sci.) ix. 133 These and many other devices ensure that all the non-specialist teachers of English are fully resourced. **1979** *Observer* 23 Sept. 4/8 Social workers have been inadequately trained and inadequately resourced to meet the expectations upon them.

resp (resp), *a. nonce-wd.* Abbrev. of RESPECTABLE *a.* 4 a.

1922 JOYCE *Ulysses* 158 Resp girl (R.C.) wishes to hear of a post in fruit or pork shop.

respect, *sb.* Add: **I. 3. c.** Delete †*Obs.* and add later example.

1904 H. JAMES *Golden Bowl* I. i. 11 She had struck him, in respect to the beautiful world, as one of the beautiful, the most beautiful things.

IV. 16. e. *with* (*all due*) *respect* and varr.: a polite phr. expressing proper deference, freq. used before stating (with some insistence) disagreement with another person's views. Also const. *to*.

1826 M. R. MITFORD *Our Village* II. 207 [My greyhound] is sliding her snake-like head into my hand, at once to invite the caress which she likes so well, and to intimate with all due respect that it is time to go home. **1923** C. MACKENZIE *Parson's Progress* xviii. 257 With all respect to the gentleman at the back of the hall who passed that remark, I tell him that if you think you can do anything with your review.., you're mistaken. **1977** *Belfast Tel.* 28 Feb. 6/4 With respect, I feel that the views expressed by Lord MacDermott are unbalanced from the very isolated position of a judge who, by necessity of office, must lead a somewhat cloistered life. **1977** *Church Times* 22 July 10/1 It is, with the greatest respect to His Grace, very little use to say that the book has 'caused more hubbub than it is worth'. **1978** *Ibid.* 25 Aug. 11/3 With all due respect to your three correspondents, I do not think they have answered M. J. Feaver's question (August 11). **1978** *Times* 13 Mar. 19/4 With great respect, this well-intentioned suggestion seems almost wholly devoid of merit. **1980** J. FOLLETT *Churchill's Gold* II. i. 86 With respect, admiral, we should not be building boats for any other purpose than for sinking enemy shipping.

V. 18. *Comb.*, as *respect-worthy* adj.

1833 [in Dict., sense 16]. **1889** 'MARK TWAIN' *Connecticut Yankee* xiii. 154 They were about all that was useful, or worth saving, or respectworthy. **1915** A. QUILLER-COUCH *Nicky-Nan* xxiii. 296 A neighbours' quarrel, and between folks I know to be so respectworthy. **1973** *Times Lit. Suppl.* 7 Sept. 1024/4 Horatio Parker (who was none the less respect-worthy, never mean or petty).

respectability. Add: **1. a.** Also with a somewhat derogatory implication of affectation or spuriousness.

1907 G. B. SHAW *Major Barbara* III. 281 The seven deadly sins!—Yes, the deadly seven... Food, clothing, firing, rent, taxes, respectability and children. **1969** *Listener* 3 Apr. 443/1 The social cost was high: hypocrisy;..the substitute of 'respectability' (spurious morality) for thought-out morals. Now that couples who aren't ready to support them needn't have children at all, marriage is redundant. **1978** P. BAILEY *Leisure & Class in Victorian England* viii. 178 We approach respectability as a role rather than as an ideology... The myth of substantial working-class respectability.

c. *transf.* Of things.

1903 G. B. SHAW *Man & Superman* p. xxviii, He and I and Mr. Sidney Webb were sowing our political wild oats as a sort of Fabian Three Musketeers, without any prevision of the surprising respectability of the crop that followed. **1976** *Times* 21 May 1/4 The alleged smear campaign against the Liberal Party..gained new respectability yesterday when the Prime Minister confirmed..that attempts were indeed being made to discredit individual members of the Liberal Party.

respectabilize, *v.* Delete *rare* and add further examples.

1933 G. STEIN *Autobiogr. Alice B. Toklas* v. 119 Uhde wished to respectabilise himself and she wanted to come into possession of her inheritance, which she could only do upon marriage. **1940** C. P. SNOW *Strangers & Brothers* xliv. 318 But the café had been respectabilized since then. There were now two floors, and neat waitresses. **1977** *Daily Tel.* 30 Apr. 13/4 Journalists are urged not to 'respectabilize' racist organizations.

respectable, *a.* and *sb.* Add: **A.** *adj.* **4. c.** (Further examples.)

1974 A. PRICE *Other Paths to Glory* I. iv. 41 Make us all a lot of hot, strong coffee while I get myself respectable. **1978** K. ROYCE *Satan Touch* v. 77 Herb Stahm knocked on Ashley's bedroom door. 'It's me, Herb.' 'Come on in. That meant they were respectable.

6. *respectable-tawdry*.

1916 E. POUND *Lustra* 112 A quiet and respectable tawdry trio.

B. *sb.* (Further examples.)

1940 E. GILL *Autobiogr.* vii. 262 With the *young* snobs and the *young* sycophants, the *young* hangers-on of the academies and, above all the young *respectables* there is nothing to be done. **1966** *Guardian* 15 June 9/8 Middle aged respectables here tend to shy away from discussion. **1978** P. G. WINSLOW *Coppergold* 142 You think I'm rotten, don't you?.. You're like Daddy and all the old respectables.

respectfully, *adv.* Add: **b.** *Yours respectfully*: a conventional formula used in the subscription of letters.

1812 'L. IRVING' *Let.* 11 Jan. in C. Mackenzie *My Life & Times* (1963) I. 51 Your early reply through the post office will oblige. Yours respectfully, Leathley Irving. **1859** MRS. GASKELL *Let.* 10 Nov. (1966) 592 Believe me to remain, Yours respectfully E C Gaskell. **1885** R. D. BLACKMAN *Letter-Writer's Vade-Mecum* (ed. 9) 42 In addressing a stranger, the proper salutation is 'Sir',..and the letter may be concluded..'Yours respectfully'. **1907** F. CROCKER *Let.* July in T. H. Baynes *Early Hist. St. Andrew's Church, Oxford* (1973) 14 On behalf of my fellow-workmen I respectfully beg to solicit from you a donation towards the Fund. Yours respectfully. **1960** J. STROUD *Shorn Lamb* v. 56 'Dear mr mall,' it said '..if she says she don't want donald, we will have him, yours respectfully.'

respectively, *adv.* **2.** For †*Obs.* read *rare* and add later example.

1962 *John o' London's* 25 Jan. 82/2 The magistrate is the *Beak* or, less respectively, the *Old Tosser*.

respirate (re·spirēit), *v.* [Back-formation from RESPIRATION.] *trans.* To subject to artificial respiration.

1968 *Exper. Neurol.* XX. 416 The animals were artificially respirated by a Harvard 670 D pump. **1971** *Nature* 28 May 263/2 Trackers were cannulated and the animals were respirated artificially with a Harvard respirator.

respiration. Add: **1. a.** (Further *transf.* example.) *artificial respiration*: see *ARTIFICIAL *a.* 5.

1814 WORDSWORTH *Excursion* VIII. 365 Sails Which..Pass with the respirations of the tide.

d. *Biochem.* and *Biol.* The biochemical and cellular processes by which absorbed oxygen is combined with carbon in the organism to form carbon dioxide and generate energy; more widely, any metabolic process in which energy is produced by the net transfer of electrons from a substrate to an external oxidant (usu. called *anærobic respiration* when this is not free oxygen); also extended to include energy-producing metabolic processes (fermentations) not involving a separate oxidant.

1856 J. C. MORRIS tr. *Lehmann's Man. Chem. Physiol.* 272 This exchange of oxygen and carbonic acid, which we improperly call respiration, is not confined to any single spot of the organism. **1880** A. GAMGEE *Text-bk. Physiol. Chem. Animal Body* I. ix. 366 The respiration of muscles during contraction. **1900** A. J. EWART tr. *Pfeffer's Physiol. Plants* I. ix. 546 (*heading*) The relationship between aerobic and anaerobic respiration. *Ibid.* 549 The actual anaerobic respiration (or fermentation) continues unchecked. **1908** HALL & DEFREN tr. *Aberhalden's Text-bk. Physiol. Chem.* xviii. 412 It [*sc.* the blood] takes the oxygen from the lungs and gives it up to the tissues. The first gas-exchange is commonly spoken of as external respiration, and the latter as internal respiration. **1929** R. A. GORTNER *Outl. Biochem.* viii. 386 In the dark the respiration of cells is inhibited by carbon monoxide. **1934** A. T. HENRICI *Biol. of Bacteria* xi. 174 Those processes which yield energy for the organism are respiration. *Ibid.* 184 A great many, probably all, micro-örganisms capable of growing either as strict or facultative anaerobes..may obtain energy by this intermolecular respiration, by a simultaneous oxidation of one compound and reduction of another. **1949** KELLY & HITE *Microbiology* xiv. 175 The phenomenon of 'life without air' has since [Pasteur's time] been explained by the study of anaerobic respiration or fermentation. **1965** G. A. STRAFFORD *Essent. Plant Physiol.* viii. 131 Fermentation is the form of anaerobic respiration carried out by some fungi and bacteria. **1970** AMBROSE & EASTY *Cell Biol.* vi. 183 The pyruvic acid is used for respiration (oxidation of carbon to CO_2) in the mitochondria. **1974** *Nature* 13 Dec. 579/1 The enzymes responsible for the first step in nitrate assimilation and for nitrate respiration are the nitrate reductases, both these processes involving the conversion of nitrate to nitrite.

5. *attrib.*, as *respiration rate*.

1929 R. A. GORTNER *Outl. Biochem.* viii. 242 The sudden rise in respiration rate may be due to the fact that at, or below, 14·75 per cent moisture all, or practically all, of the moisture in the wheat kernel is in the form of bound water. **1970** AMBROSE & EASTY *Cell Biol.* ii. 89 An apparatus originally designed by Warburg can be used to measure respiration rates.

respirator. Add: **2.** Also, a gas mask, or any mask for providing protection against noxious substances in the air. (Further examples.)

1872 *Chem. News* 17 May 239/1 The charcoal respirators invented by me in 1854 are now coming into general use, especially in manufactories and laboratories. *Ibid.*, The respirator is suspended for ten or fifteen minutes over some strong solution of ammonia in a large beaker; in this way the charcoal absorbs a very large amount of ammoniacal gas... The wearer can remain for a considerable time in an atmosphere containing chlorine without suffering any inconvenience. **1915** MRS. BELLOC LOWNDES *Diary* 26 Apr. (1971) 61 The [German] soldiers..had their noses plugged with cotton wool and respirators over their mouths, but even so, 1000 perished by their own gas. **1915** *Sphere* 7 Aug. 146/1 Respirators for the use of the Russian soldiers, who have been again attacked by gas bombs. **1938** *Ann. Reg. 1937* 25 Not a little of the increase was due in reality to military reasons, the Home Office being assigned some £4,000,000 for the provision of respirators and the development of emergency fire brigade services for protecting the civil population in the event of war. **1971** *Brit. Med. Bull.* XXVII. 75/2 The use of..air-fed respirators..could greatly reduce the future incidence of asbestos-related cancers in this work. **1978** CADOGAN & CRAIG *Women & Children First* x. 219 Forbidden to play with his gas mask, he muses bitterly on the 'jolly good times' that should result from unrestricted access to respirators.

3. *Med.* An apparatus for maintaining artificial respiration.

1929 DRINKER & MCKHAN in *Jrnl. Amer. Med. Assoc.* 18 May 1659/2 (*caption*) The mechanical respirator, showing patient ready to be pushed into the tank. **1932, 1938** [see *IRON LUNG 1]. **1966** DUNLOP & ALSTEAD *Textbk. Med. Treatm.* (ed. 10) 941 The action of all positive pressure respirators (or ventilators, as they are more correctly described) is to produce inflation of the lungs at a rate of 14 to 20 cycles per minute, expiration being allowed to occur passively. **1967** SLONIM & CHAPIN *Respiratory Physiol.* iv. 48/2 Drinker-type 'iron lung' respirators are devices built to enclose all the body below the neck. **1977** C. STORR *Tales from Psychiatrist's Couch* xi. 118 His respiratory muscles were working again and he was out of the respirator for large parts of the day.

respiratory, *a.* Add: **2.** Special collocations: *respiratory centre*, a region of the brain which exercises control over respiration; *respiratory pigment*, a protein molecule with a pigmented prosthetic group, involved in the transfer of oxygen or electrons within living systems; *respiratory quotient*, the ratio of the volume of carbon dioxide evolved to that of oxygen consumed; *respiratory syncytial virus*, an RNA virus that causes disease of the respiratory tract; *respiratory therapy* (U.S.), the management of patients receiving artificial respiration or ventilation and of the apparatus involved; so *respiratory therapist*, one trained in this; *respiratory tract*, the passages through which air passes in respiration; *respiratory tree*, a branched system of respiratory passages.

1883 *Encycl. Brit.* XX. 480/1 The respiratory centre must be regarded as the seat of origin of the impulses which cause the muscular movements of inspiration and expiration. **1948** A. BRODAL *Neurol. Anat.* xi. 391 The existence of a respiratory centre is generally admitted in the medulla. **1896** *Q. Jrnl. Microsc. Sci.* XXXIX. 3 A respiratory pigment of a kind unusual in the Chætopoda. **1933** *Biol. Bull.* LXIV. 233 In several of the invertebrates also, respiratory pigments, either hemoglobin or hemerythrin, are found within special cells of the circulating blood or body fluids. **1968** PASSMORE & ROBSON *Compan. Med. Stud.* I. viii. 7/1 Cytochromes. These are respiratory pigments found in all cells. **1890** BILLINGS *Med. Dict.* II. 453/1 *Respiratory quotient*, the relation of the inspired free oxygen to that expired in the form of carbonic acid. **1900** W. S. HALL *Text-bk. Physiol.* iv. 218 The respiratory quotient varies considerably in different *species*, and in the same animal under different conditions. **1968** PASSMORE & ROBSON *Compan. Med. Stud.* I. xxxi. 7/1 Ketosis is associated with a low respiratory quotient (RQ), below 0·75, which reflects a predominance of fat metabolism. **1961** *New Eng. Jrnl. Med.* CCLXIV. 1174/1 The present report presents convincing evidence for the etiologic role of the respiratory syncytial virus in acute respiratory illnesses among children. **1968** *New Scientist* 15 Feb. 368/1 Respiratory syncytial virus (RSV) is now the most dangerous of the respiratory pathogens that affect young children. **1965** P. SAFAR (*title*) Respiratory therapy. **1973** D. A. HOLADAY in Caldwell & Moya *Adv. Respiratory Care & Physiol.* vi. 85 Their duties include..administration of other categories of respiratory therapy to patients in intensive care units, post-anesthesia recovery rooms, and other areas..where continuous ventilators are in use. **1974** TYLER & NETT in T. L. Petty *Intensive & Rehabilitative Respiratory Care* (ed. 2) vi. 79 The patient with acute respiratory failure is being cared for by both nursing and respiratory therapy staffs. *Ibid.* 98 The nurse, respiratory therapist, and, where available, the physical therapist have vital roles in the successful care of any patient from the onset of acute illness throughout the convalescent and rehabilitative phases. **1978** *Detroit Free Press* 16 Apr. (Parade Suppl.) 5/3 Alice Gaul, a former U.S. Navy nurse, and Mary Masal, a respiratory therapist, managed to keep that man alive through the eight-hour flight. **1936** *Discovery* July 206/2 It has not appeared to produce the usual irritative effects of ether on the respiratory tract. **1977** *Chest* LXXI. 346/1 Aerosol topical anesthesia for instrumentation of the respiratory tract was first described in 1949. **1932** BORRADAILE & POTTS *Invertebrata* xviii. 575 Finally there is a short, wide cloaca. Into the latter usually open two long, branched respiratory trees, whose ramifications end in thin-walled ampullae through which water, when pumped in by contractions of the cloaca, passes into the body cavity, carrying oxygen to the coelomic fluid, and so to the organs. **1970** W. H. PARKER *Health & Dis. in Farm Animals* iii. 35 The respiratory tree is an apt phrase used to describe the continuous subdivision of the bronchi.

respire, *v.* Add: **I. 2. d.** To carry out or exhibit the biochemical processes of respiration.

1927 M. BODANSKY *Introd. Physiol. Chem.* viii. 200 A tissue that has been washed until it no longer 'respires' will, upon the addition of glutathione, again take up oxygen and yield carbon dioxide. **1951** M. ABERCROMBIE et al. *Dict. Biol.* 197 Many organisms (or parts of them) respire anaerobically for some time when their supply of oxygen is insufficient for aerobic respiration. **1976** *Sci. Amer.* Feb. 68/2 When the crop is ensiled, the plant cells continue to respire for a period of time and the aerobic bacteria on the plant increase in number.

respirit, *v.* (Further example.)

1929 R. BRIDGES *Testament of Beauty* III. 84 All his Knights cleansed and respirited, reclothed as might be.

respirometer (respirǫ·mɪtəɹ). [f. L. *respirāre* to blow, breathe out + -OMETER.] **1.** *Med.* A device which measures the quantity of air expired, so that the condition of the lungs may be studied; = SPIROMETER.

1889 *Cent. Dict.* 5111/1 *Respirometer*, an instrument which is used to determine the condition of the respiration. **1907** *Arch. Otology* XXXVI. 401 A person afflicted with pulmonary tuberculosis..so weakened by his disease that he can raise the water only six inches in the 'Respirometer'. **1964** BATES & CHRISTIE *Respiratory Function in Disease* ii. 6 For bedside use, the Wright Respirometer, a pocket-sized instrument, is preferable.

2. *Physiol.* A device which measures the rate of consumption of oxygen by a living or organic system.

1915 *Jrnl. Physiol.* XLIX. p. xxiv, (*heading*) A quantitative respirometer. **1923** *Jrnl. Agric. Res.* XXIII. 101 (*heading*) A new and efficient respirometer for seeds and other small objects. **1948** *Jrnl. Exper. Med.* LXXXVII. 177 (*caption*) Details of respirometer chamber to hold twenty eggs. **1969** *Sci. Jrnl.* Nov. 19/2 A simple fully automatic respirometer which measures organic pollution in sewage has been developed.

Hence **respiro·metry,** the measurement of rates of oxygen consumption; **re:spirome·tric** *a.*, by means of or pertaining to the respirometer.

1932 *Jrnl. Gen. Physiol.* XVI. 5 (*heading*) Studies in respirometry. **1960** *Antibiotics Ann.* VII. 563 (*heading*) Respirometric assay of nystatin in animal tissue. **1972** *Ann. Rep. Freshwater Biol. Assoc.* XL. 41 Dr Jones has continued to use respirometric techniques..to study the factors that cause inhibition of bacterial and other biological activity. **1977** A. G. CALLELY et al. *Treatm. Industr. Effluents* xvii. 278 Biological test methods applied to this type of waste include respirometry, aeration flask tests, and continuous treatment tests.

responaut (re·spǫnǭt). [Irreg. f. RESPIRATOR + Gr. ναύτ-ης sailor, after *astronaut*, etc.] A patient dependent upon a mechanical respirator to maintain breathing.

1964 *Sunday Times* 12 Jan. 18/5 The story of the 'responaut' Ann Armstrong as told by Moira Keenan was an enthralling one. **1964** *Med. Officer* 7 Aug. 97/2 A new group of severely disabled people. This group, calling themselves Responauts (because they share similar problems with astronauts and oceanauts in establishing and maintaining communications and vital air supplies) are young people whose lives have been saved through the invention and use of the first artificial organ—an iron lung or mechanical respirator—following a severe attack of respiratory polio. **1965** *Daily Mirror* 30 Dec. 12/2 The Responaut..—a quarterly magazine Ann edits for people like herself who need a breathing machine to keep alive. People the doctors call 'responauts'. **1969** *Guardian* 22 Sept. 7/2 Responauts don't have to be chained to their apparatus in a hospital ward. **1979** *47th Ann. Rep. Pilgrim Trust 1977* 23 The Refresh Trust was formed in 1972..for the purpose of building and equipping a place where a few responauts could safely have a holiday.

respond, *v.* Add: **2. c.** *Bridge.* To make (a bid) in reply to a partner's opening (or subsequent) bid. Cf. *RESPONDER 1 b.

1958 *Listener* 16 Oct. 611/2 East should have responded Two Hearts. **1959** T. REESE *Bridge Player's Dict.* 158 If one diamond were opened and partner responded two clubs, there would not be a sound rebid. **1976** *Times* 1 May 12/6 To partner's opening One Heart it is dangerous to respond Two Clubs.

3. c. (Further examples.)

1876 *Jrnl. Soc. Telegr. Engin.* V. 514 We wish to construct a receiver to respond to a tone made by 128 vibrations per second. **1910** J. ERSKINE-MURRAY *Wireless Telephones* vii. 51 An electric circuit will respond most readily to impulses which come timed to its own natural rate of vibration. **1939** N. DE V. HART *Bridge Players' Bedside Bk.* xii. 57 North responded to the Forcing Two in Diamonds with Two Hearts. **1952** A. GRIMBLE *Pattern of Islands* 233 One or two cases had begun to respond to gynocardate injections. **1954** G. I. M. SWYER *Reproduction & Sex* xv. 203 More of the women..had responded until they were in their fifties or sixties. **1967** M. KENYON *Whole Hog* xx. 195 You could have responded to my three clubs, you could have said something.

‖**respondeat superior** (respǫ·ndeæt supe·riǭ°ɹ), *phr.* *Law.* [L., 'let the principal answer'.] A maxim embodying the rule of vicarious liability (see quots.).

[**1601** W. FULBECKE *Parallele or Conference Civill Law, Canon Law & Common Law* I. x. f. 76 A plea in Abatement of a writte bee not peremptorie, but a *respondeat vlterius*.] *a* **1634** E. COKE *Fourth Part Inst. Lawes Eng.* (1644) xi. 114 Now it is good to know, how the law commonly called *Respondeat superior*, holdeth in this Court and in other Courts. **1848** W. HARTON *Law Lexicon* 587/2 *Respondeat superior*, (let the superior answer). If a coroner of a county is insufficient, the county, as his superior, shall answer for him. **1903** *Automobile & Carriage Builders' Jrnl.* May 133/2 The maxim *respondeat superior*. **1965** *Mod. Law Rev.* XXVIII. 584 It is trite learning that civil responsibility could in general be imposed *respondeat superior*. **1976** *Billings* (Montana) *Gaz.* 6 July 3-A/2, I have no trouble with the concept of accepting responsibility for one's actions. It's called 'respondeat superior'. That is, the principal is liable for the actions of his agent which are authorized. It was what the Nuremberg trials were all about.

respondent, *sb.* Add: **1. a.** Also in recent use, one who supplies information for a survey. Chiefly *U.S.*

1955 *Bull. Atomic Sci.* 175/2 Almost one-third of the respondents said they were upset by witnessing the casualties. **1961** *Technology* Feb. 34/1 The respondents were not asked how often they had applied each technique, if more than once, since many would probably have no accurate record of this. **1968** *Amer. Jrnl. Public Health* LVIII. 327/1 The questionnaires were unsigned on their return and were thrown into a large pile at the front of the room in such a manner that no one from the military hierarchy could identify the responses of any respondent. **1970** D. GOLDRICH et al. in I. L. Horowitz *Masses in Lat. Amer.* v. 176 The interviewers selected respondents randomly from the household. **1975** *Amer. Speech 1973* XLVIII. 22 Recent behavioral tests..have shown that respondents are capable of identifying informants as black or white from auditory cues only.

4. *Psychol.* The result of, or response to, a specific stimulus (see quot. 1937).

1937 B. F. SKINNER in *Jrnl. Gen. Psychol.* XVI. 274 It is a necessary recognition of the fact that in the unconditioned organism two kinds of behavior may be distinguished. There is, first, the kind of response that is made to specific stimulation... I shall refer to such a reflex as a *respondent* and use the term also as an adjective in referring to the behavior as a whole. **1940** HILGARD & MARQUIS *Conditioning & Learning* iii. 66 Respondents..are movements elicited by recognized stimuli.

respondent, *a.* Add: **3. b.** *Psychol.* Responsive, or that occurs as a reflex, to some specific stimulus; esp. as *respondent conditioning*, the conditioning of an organism to a particular response through the controlled use of a stimulus.

1937 B. F. SKINNER in *Jrnl. Gen. Psychol.* XVI. 274 The distinction between operant and respondent behavior. **1940** HILGARD & MARQUIS *Conditioning & Learning* iii. 72/1 Respondent behavior, like ordinary spinal reflexes, is elicited by specified stimuli. **1966** I. G. SARASON *Personality* iv. 75 The actual role played by reinforcement has been studied intensively in..experimental situations, one involving classical or respondent conditioning. **1974** H. W. BERNARD *Personality* x. 241 Respondent conditioning can also be used as a therapeutic approach.

responder. Add: **1. b.** *Bridge.* The partner of the opening bidder.

1932 *Daily Tel.* 8 Oct. 15/5 The partner of the opening bidder is known as the responder. **1952** I. MACLEOD *Bridge* iv. 49 If the responder raises the opener's suit, it is a quantitative bid only. **1966** *Listener* 13 Jan. 78/2 What happens.. if the responder has only sufficient high card strength to raise One No Trump to Two No Trumps as an invitation to game? **1976** *Times* 1 May 12/7 When there is opposition bidding the responder's duty is to show the strength of his hand.

2. † **a.** An early, electrolytic, form of radio receiver. *Obs.*

1901 *Western Electrician* 27 July 49/1 To the new receiving device its inventors [*sc.* De Forest and Smythe] have applied the name 'responder', a term first suggested by the London Electrician as one suitable for any resistance device sensitive to electric radiations. **1904** *Electr. Rev.* 3 Sept. 330 The principle of this receiver or 'responder', is based upon the fact that the Hertzian oscillations produce sudden electrolytic action in a cell containing certain electrodes and solutions. **1915** W. H. ECCLES *Wireless Telegr.* 246 In one type of electrolytic detector, now merely of historic interest, the processes of electrolysis are employed to form fine threads of metal across the gap or gaps, and these threads are destroyed by the oscillations to be detected. The consequent alteration of resistance is observed by telephone or galvanometer. Of this type is..the 'responder' of de Forest and Smythe.

b. A device which automatically retransmits a pulse or signal on receiving one from an interrogator. Also *responder beacon*.

1945 *Nature* 15 Sept. 323/2 *A* and *B* are pulse-interrogator stations, the aircraft has a responder of constant and accurately known delay-time... 'G-H' and 'Babs'..utilize coded responses sent back by a ground responder-beacon in reply to pulses from an airborne or shipborne interrogator. **1945** *Electronic Engin.* Oct. 735 Vehicles could now carry small questioning transmitters ('interrogators') and obtain replies from 'responder' beacons on land or sea. **1957** *Encycl. Brit.* XVIII. 873Y/1 Beacons of the synchronous sort just described are variously called radar beacons, responder beacons, racons, or transponders, there being no important distinction among these terms. **1966** D. TAYLOR *Introd. Radar & Radar Techniques* iv. 45 A responder beacon..is a 'repeater', in the sense that it would retransmit after a very short time-interval any pulse signal received. **1977** *Offshore Engineer* Aug. 46/2 The remote vehicle's depth, horizontal range, slant range and relative bearing can all be determined by using this single responder in conjunction with Wesmar's new SS400TS sonar tracking system. The responder acts as a transmitter/receiver, receiving electronic signals through an umbilical.

3. *Biol.* and *Med.* An individual, structure, etc., that responds or reacts to some stimulus or treatment.

1963 *Jrnl. Exper. Med.* CXVIII. 954 Responders showed both Arthus and delayed allergic skin reactions to the immunizing conjugates. **1973** *Nature* 30 Nov. 245/1 Cells were classified as either β^+ or β^- (responders and non-responders to β-adrenergic stimulators) or P^+ or P^- (responders and non-responders to prostaglandin E_1). **1976** *Lancet* 30 Oct. 928/1 A further analysis was made into responders and nonresponders... The responders were defined as patients in whom the number of ulcers decreased by more than 50% over 2 or more months.

responding, *ppl. a.* (Further examples.)

1921 G. CECIL *Life Ld. Salisbury* II. vi. 219 His action was generally attributed to the decision to call out the Reserves..and Lord Beaconsfield implied as much in his responding speech. **1933** C. VANDYCK *Contract Contracted* 21 As the Playing Trick count is not used until the Trump Suit has been decided upon the Responding Hand is obviously the first one to use it. **1952** I. MACLEOD *Bridge* vi. 73 The responding hand..never assumes, at least until the second trial bid, that the opener is slamming. **1965** in J. Money *Sex Research* 107 Seldom does the responding female directly manipulate the clitoris through an entire sexual response cycle. **1968** R. KYLE *Love Lab.* (1969) xviii. 210 Data on several hundred responding individuals has more scientific validity than data on one.

responsa: see *RESPONSUM.

response. Add: **1. b.** Also *spec.* in *Psychol.* (freq. opposed to *stimulus*), an observable reaction to some specific stimulus or situation; the fact of such reaction.

1908 E. L. THORNDIKE in *Ess. Philos. & Psychol. in Honor W. James* 597 A situation arouses a response which brings an annoying state of affairs. The probability of a similar response in the future is lessened. **1919** J. B. WATSON *Psychol.* i. 16 Having now examined at some length into the general nature of both stimulus and response, we should be prepared to understand the object of a psychological experiment. **1934** H. DAVIS in C. Murchison *Handbk. Gen. Exper. Psychol.* 983 They..constitute an objective response of great value for analyzing the activities of the cortical tissue. **1948** A. C. KINSEY et al. *Sexual Behav. Human Male* v. 159 Evidence of minimal psychic components with good enough physical responses. **1952** FORD & BEACH *Patterns Sexual Behav.* xii. 239 Her capacity for complete response returned. **1957** B. F. SKINNER in Saporta & Bastian *Psycholinguistics* (1961) v. 228/1 Semantic theory is often confined to the relation between response and stimulus which prevails in the verbal operant called the tact. **1965** in J. Money *Sex Research* 101 Three women were able to achieve orgasmic response by breast manipulation alone. **1976** SENTER & DIMOND *Psychol.* vi. 102 Relaxation and anxiety are competing responses. You must behave in one way or the other.

c. The way in which an apparatus responds to a stimulus or range of stimuli.

1911 H. M. HOBART *Dict. Electr. Engin.* II. 630/1 The receiver must be sharply tuned so that the variations of frequency may be sufficient to make an appreciable difference in the strength of its response. **1915** W. H. ECCLES *Wireless Telegr.* 245 Fig. 176 shows the response of the detector (change of *I*) at various values of the intensity of magnetism *I* and the field *H*, for four different magnetic cycles. **1926**, etc. [see *frequency response* s.v. *FREQUENCY 6 a]. **1958** O. R. FRISCH *Nuclear Handbk.* xiv. 16 In designing a scintillation counter the spectrum of the fluorescent radiation must be marked as far as possible with the spectral response of the multiplier. **1961** G. MILLERSON *Television Production* iii. 41 Where the tube's response to red is excessive, this may be held back with an appropriate green or blue filter. **1970** J. EARL *Tuners & Amplifiers* iii. 68 The latest 'quality' amplifiers..boast a power response which is almost as good as the frequency response.

d. *Bridge.* A reply to a partner's opening (or subsequent) bid.

1939 N. DE V. HART *Bridge Players' Bedside Bk.* x. 52 South's response of Six Clubs showed first round control of clubs. **1947** S. HARRIS *Fund. Princ. Contract Bridge* I. iv. 35 It sometimes happens that South is able to make a positive response. **1958** *Listener* 16 Oct. 611/2 West makes her natural response of Three Diamonds. **1967** P. ANDERTON *Play Bridge* iii. 28 The negative response is 2 N.T. in which case the hand will probably be played in 3 H. **1976** *Times* 1 May 12/6 A minimum response can be shown only by a rebid of the suit.

6. *attrib.* and *Comb.*, as *response function, rate*; esp. in *Psychol.*, as *response bias, -movement, pattern, probability, set*; *response-contingent* adj.; **response time** *Electr.*, the time taken for a circuit or measuring device, when subjected to a change in input signal, to change its state by a specified fraction of its total response to that change.

1970 *Jrnl. Gen. Psychol.* LXXXII. 63 These findings appear to be incompatible with the notion that both scales measure 'response bias'. **1958** B. FLANAGAN et al. in Saporta & Bastian *Psycholinguistics* (1961) 415 (*title*) The control of stuttering through response-contingent consequences. **1972** *Sci. Amer.* May 97/1 The results of such analyses produce response functions..which can be plotted to show the mean responses of different species of trees to conditions of temperature, precipitation and prior growth. **1975** D. G. FINK *Electronics Engineers' Handbk.* xxv. 25 The antenna and receiver are configured to match a target signal at a particular angle, delay, and frequency. The radar will respond with reduced gain to targets at other angles, delays, and frequencies. This response function can be expressed as a surface in a four-dimensional coordinate system. **1892** VAN LIEW & BEYER tr. *Ziehen's Introd. Physiol. Psychol.* i. 14 Goltz has termed the automatic movements 'response-movements'. **1936** J. KANTOR *Objective Psychol. of Gram.* xx. 290 For objective psychology, moods are nothing but particular response-patterns or speech-community styles of utterance. **1965** *Brit. Jrnl. Psychol.* LVI. 217 (*heading*) Response patterns and strategies in the dynamics of concept attainment behaviour. **1960** W. N. DEMBER *Psychol. of Perception* (1970) viii. 287 All of the word-recognition experiments can be interpreted in terms of response-probability. **1946** *Jrnl. Amer. Statist. Assoc.* XLI. 522 The number of mail questionnaires and field interviews required to achieve a specified precision will vary with the response rate. **1966** *Rep. Comm. Inquiry Univ. Oxf.* II. 351 The figures for Great Britain are from a survey (with a response rate of 53 per cent.) of those university teachers who responded to the inquiry by the Robbins Committee in 1961–2 (in which the response rate was 86 per cent.). **1970** *Jrnl. Gen. Psychol.* LXXXII. 64 'Response set' is a generalized tendency to be agreeable. **1972** D. P. CAMPBELL in J. N. Butcher *Objective Personality Assessment* vi. 119 Response set, acquiescence and social desirability, are currently popular..even though the data in support of them [as concepts] are will-o'-the-wispy, at best. **1958** R. B. HURLEY *Junction Transistor Electronics* xix. 364 Output response times are reduced by a factor of 2·5. **1970** WILLARDSON & BEER *Semiconductors & Semimetals* V. i. 7 The response time is determined by the rate at which the [infrared detector] element warms and cools. **1975** D. G. FINK *Electronics Engineers' Handbk.* xvii. 49 Performance specifications usually include the response of the system to a step input, measured in terms of response time, rise time, delay time, settling time, and overshoot.

responsibility. Add: **1. a.** (Later examples.)

1973 *Black Panther* 17 Nov. 3/1 The attack was deliberate and appears to be associated with the letter from the 'Symbionese Liberation Army' claiming responsibility. **1974** *Oxford Mail* 26 Oct. 1/4 Four explosions shook New York's business and financial districts within a few minutes today and a self-styled Puerto Rican liberation movement claimed responsibility.

c. Const. *for*.

1903 G. B. SHAW *Man & Superman* I. 17 Mamma knows that she is not strong enough to bear the whole responsibility for me and Rhoda without some help and advice. **1928** E. O'NEILL *Strange Interlude* II. 66 Looks damnably upset... Wants to evade all responsibility for her, I suppose. **1971** G. K. ROBERTS *Dict. Polit. Analysis* 190 The individual responsibility of British ministers for the actions of their civil servants. **1974** *Black Panther* 19 Jan. 3/1 A communique signed by the Symbionese Liberation Army claimed responsibility for the attack. **1975** *Times* 20 Aug. 13/1 Just as one was wondering who would 'claim responsibility' for spoiling the Headingley wicket—Saór Eire, the Women's Liberation Army..—up popped the Campaign to Free George Davis. **1981** M. SPARK *Loitering with Intent* ii. 32, I can't take responsibility for your mother this afternoon.

2. c. A person to whom one is responsible; a person in authority. *Nonce-use.*

1893 E. DOWSON *Let.* c 22 Mar. (1967) 275, I have to let the responsibilities know exactly how many people I have invited.

responsible, *a.* Add: **3. c.** *responsible government*: (see quot. 1910). Also in extended use (esp. under influence of sense 5).

1839 LD. DURHAM *Rep. Affairs Brit. N. Amer.* 142 By creating high prizes in a general and responsible Government, we shall immediately afford the means of pacifying the turbulent ambitions, and of employing in worthy and noble occupations the talents which now are only exerted to foment disorder. **1865** EARL RUSSELL *Essay on Hist. Eng. Govt. & Constitution* p. lxviii, Others said, 'the grant of what is called "responsible government" [in Canada] is a grant of independence. It must be resisted.' **1906** W. S. CHURCHILL in R. S. Churchill *Winston S. Churchill* (1969) II. Compan. I. 506 We are not, of course, confined to any particular form of Responsible Government. **1910** *Colonial Office List* v. i. 633 The colonies possessing responsible government, in which the Crown has only reserved the power of disallowing legislation and the Secretary of State for the Colonies has no control over any public officer except the Governor. **1930** G. B. SHAW *Apple Cart* I. 33 The people have found out long ago that democracy is humbug, and that instead of establishing responsible government it has abolished it. **1957** *Encycl. Brit.* XII. 174/2 The device known as dyarchy, or double government,..was intended to train Indians for responsible government.

5. a. Comb., *responsible-minded* adj.

1960 *Times* 7 Mar. 13/5 Are publishers responsible-minded parents?

responsion. Add: **3.** Responsions was ended by statute in 1960.

responsive, *a.* and *sb.* Add: **A.** *adj.* **1. c.** *Bridge.* Of a double: used to invite a change to an unbid suit in response to a partner's take-out double.

1959 *Listener* 12 Mar. 489/3 His double would be 'responsive'. *Ibid.* 30 July 190/1 The responsive double is a double intended to give information when partner has already made an informatory double. **1964** *Official Encycl. Bridge* 456/2 The minimum strength for a responsive double varies slightly with the level of the auction. **1973** REESE & DORMER *Compl. Bk. Bridge* viii. 120 Responsive doubles are usually played up to the level of three spades.

Responsivist (rĭspǫ·nsivist), *sb.* (and *a.*). *Indian Hist.* [f. RESPONSIVE *a.* + -IST.] One who advocated working within the diarchical administrative system introduced in Indian provinces during British rule. Also *attrib.* or as *adj.*

1927 *Glasgow Herald* 1 Feb. 8 Agitation is afoot in Calcutta against the acceptance by Mr Ghuznavi of one of the two Bengal transferred portfolios with Mr Chakra Varti, the Responsivist leader, as his Hindu colleague. **1927** *Observer* 27 Mar. 11/2 The third general election, held at the end of last year, when the Swarajists, as distinct from the Responsivists, lost ground. **1936** J. NEHRU *Autobiogr.* xlviii. 387, I remember a frequent complaint of my father's that his Responsivist friends had no sense of humour. **1950** L. FISCHER *Life Mahatma Gandhi* xxvii. 234 A dissident group, headed by M. R. Jayakar and N. C. Kelkar, who believed in still more co-operation with the British, but less with the Moslems, split off from the Swarajists and formed the Responsivist party. **1951** D. G. TENDULKAR *Mahatma* II. 307 The Responsivists joined hands with the Independents in Bombay and formed the Indian National Party. **1969** D. DAS *India from Curzon to Nehru & After* xi. 122 The dissidents met at Akola and formed the Responsivist Party. **1970** J. S. SHARMA *India since Advent of British* iv. 271 The Conference of the Responsivists cooperation party which was held at Akola adopted its Manifesto.

responsivity. Delete *rare*$^{-1}$ and add further examples.

1901 *Ann. Rep. Board of Regents Smithsonian Inst. 1899–1900* 68 A principle of knowledge which may appropriately be styled the responsivity of mind. **1942** *Amer. Jrnl. Physiol.* CXXXV. 721 Suppression [of cortical response] is a decrease of responsivity of a certain area..caused by previous stimulation of specific cortical regions. **1976** *Lancet* 30 Oct. 953/2 Responsivity to change in America is remarkable.

responsor[2] (rĭspǫ·nsǭɹ). [f. RESPONSE + -OR.] A device that receives and processes the reply from a transponder, being usu. incorporated in the same unit as the interrogator.

1945 *Army & Navy Jrnl.* 18 Aug. 1534/4 *Responsor*, receiver used specifically in IFF system to receive the reply to a challenge. **1947** A. ROBERTS *Radar Beacons* xviii. 388 A simple responsor using a lamp to locate a beacon in range. **1976** P. HONOLD *Secondary Radar* v. 208 *Responsor*, the receiver of responses.

responso·rially, *adv. Eccl.* [f. RESPONSORIAL *a.* + -LY².] In a responsorial manner; with responses.

1901 PROCTER & FRERE *New Hist. Bk. Common Prayer* xi. 406 The Arians..went through the city..to their places of worship, singing responsorially all the way. **1929** E. C. THOMAS *Lay Folks' Hist. Liturgy* II. v. 186 On these occasions the processional chant, or Rogation psalm, was sung responsorially. **1978** P. G. COBB in C. Jones et al. *Study of Liturgy* II. iv. 186 It is not until the fourth century that there is clear evidence of psalms being sung responsorially.

‖ **responsum** (respǫ·nsǝm). Pl. **responsa**. [L., = answer, response.] **1.** A reply by a rabbi or Talmudic scholar to an inquiry on some matter of Jewish law.

1896 S. SCHECHTER *Studies in Judaism* 1st Ser. xi. 330 The greatest part of the literary activity of the Gaonim consists in their Responsa, in which they gave decisions on ritual questions, or explanations of difficult passages in the Talmud. *Ibid.* 331 The titles borne by the various collections of those Responsa belong to a period later than the author's. **1932** C. ROTH *Hist. Marranos* vii. 201 The responsa of the Levantine rabbis of the period are filled with discussions relating to the position of the Marranos in Jewish law. **1941** G. G. SCHOLEM *Major Trends in Jewish Mysticism* ix. 340 You find Zaddikim who write rabbinical responsa. **1962** B. ABRAHAMS tr. *Life Glückel of Hameln* iii. 66 Good Rabbi Asher..wrote a great responsum in my husband's favour. **1968** *New Scientist* 23 May 385/3 The team is investigating the responsa written by Jewish legal experts in Medieval Spain... A responsum..consists of a judgement usually revolving around a concrete problem in the commercial, social, moral or religious sphere. **1977** *N.Y. Times* 11 Jan. 30/1 Responsa..are fundamental in the evolution of Jewish law. Altogether, there are approximately 500,000 responsa written by 5,000 different authorities.

2. responsa prudentum (prude·ntǝm) *Law* [L., = the answers of the learned], in Roman Civil Law: the opinions and judgements of learned lawyers, variously forming part of this law (cf. Justinian *Inst.* I. ii. § 8). Also *transf.* in modern use.

In cl. Lat. *responsa prudentium* is the usual form, but most of the legal sources cited below have *prudentum* following the example of Blackstone (1765).

1681 LD. STAIR *Institutions Law Scotl.* I. i. 3 And so many times these *responsa prudentum*, have been received with as much Authority, and more heartiness for Laws, than the Dictates of Sovereigns. **1765** BLACKSTONE *Comm. Laws Eng.* I. 80 The *responsa prudentum* or opinions of learned lawyers. **1859** T. C. SANDARS *Institutes Justinian* 13 The *responsa prudentium* came to be enumerated among the direct sources of law. **1877** *Law Rep. Exchequer Division* II. 70 Of course the value of these *responsa prudentum* is affected by various circumstances. **1883** J. B. MOYLE *Imperatoris Iustiniani Institutionum* I. 30 The establishment of the empire brought with it a considerable change in respect of the responsa prudentum.

respray, *v.* [RE- 5 a.] *trans.* To spray (esp. with paint) again. Hence **respray·ed** *ppl. a.*; **respray·ing** *vbl. sb.* Also **re·spray** *sb.*, the action or fact of spraying again.

1934 *Webster*, Respray, *v.t.* **1959** *Listener* 16 Apr. 657/2 The car had been re-sprayed. **1962** *Law Rep.* 1961 26 July 35 The defendant appealed against so much of the decision as ordered him to pay for the respray. *Ibid.* 40 The defendant should be taken to have injured a motor-car that was already to contain respects [illegible] re-sprayed. **1968** 'D. RUTHERFORD' *Skin for Skin* ii. 16 The engine's nearly perfect but she does need a re-spray. **1975** G. SEYMOUR *Harry's Game* iii. 46 The stolen and resprayed Cortina. **1976** M. GILBERT *Night of Twelfth* vii. 65 A car which..was resprayed. First with a light grey undercoat, then with a dark grey finish.

Ressaldar: see *RISSALDAR.

‖ **ressentiment** (resa·ntimān). [Ger., a. F. *ressentiment*, f. *ressentir* to RESENT: cf. RESENTMENT.] = *RESENTMENT 2 d.

1945 G. A. MORGAN *What Nietzsche Means* vi. 151 Nietzsche illustrates the second type, which directs *ressentiment* outward, by anarchists, socialists and communists, who make 'society' to blame for their misery. **1949** R. K. MERTON *Social Theory* iv. 145 We must distinguish it [sc. rebellion] from a superficially similar but essentially different type, '*ressentiment*'. Introduced in a special technical sense, by Nietzsche, the concept of *ressentiment* was taken up and developed sociologically by Max Scheler. **1958** F. HEIDER *Psychol. of Interpersonal Relations* xi. 291 For Scheler, as for Nietzsche, '*ressentiment*' is an ever continued with a feeling of impotence to attain the value that another person has. **1968** *Internat. Encycl. Social Sci.* XII. 27/2 Scheler conceives of *ressentiment* as a complex syndrome involving conscious attitudes, feelings, and moral judgments as well as unconscious defenses and wishes. The *ressentiment*-laden person tends to devaluate authoritative persons and groups. **1975** W. S. SAHAKIAN *Hist. & Syst. Psychol.* i. 31 Nietzsche termed *ressentiment* that form of behaviour arising out of repressed hostility.

rest, *sb.*¹ Add: **I. 3. e.** A year's imprisonment. *Austral. slang* (*rare*).

1882 *Sydney Slang Dict.* 9/2 Dick went pulling down sawney for grub last week, when a cop pinched him. He's gone in the country for a *rest*. **1945** BAKER *Austral. Lang.* vii. 141 Here is a brief glossary of jail sentences: *lag*, three months...*rest*, twelve months [etc.].

f. In colloq. phr. *to give* (something or someone) *a rest*: to stop thinking or talking about.

[**1927** *Amer. Speech* II. 359/1 *Leave that rest a bit*..., let the matter alone for a while. 'Keep quiet, and leave that rest a bit.'] **1931** E. O'NEILL *Mourning Becomes Electra* (1932) 137 Give it a rest, Orin! It's over. Give yourself a chance to forget it. **1943** J. B. PRIESTLEY *Daylight on Saturday* xxiii. 226 I'm a bit tired of hearing about him today. So let's give him a rest. **1966** R. RENDELL *Vanity dies Hard* iii. 31 Could we give Nesta Drage a rest?..I was glad when she went away. **1971** — *One Across* ii. 9 'All right Mother,' said Vera. 'Let's give it a rest, shall we?'

II. 12. c. A support or hook for a telephone receiver when not in use, incorporating a switch that is automatically closed when the receiver is lifted.

1889 *Telephone* 15 Feb. 94/2 Improvements in telephone apparatus to be designated as a telephone-receiver rest. **1932** *Telegraph & Telephone Jrnl.* VIII. 97/2 Flash the Switchboard Telephonist into circuit by moving the Receiver rest up and down slowly until she answers. **1948** J. ATKINSON *Telephony* I. 291/1 The circuit is arranged automatically to switch the lines from the bell circuit to the speaking circuit when the receiver is lifted from its rest or hook. **1961** H. & N. SCHNEIDER *Your Telephone* iii. 26 In the exchange there is a huge electric battery. When you lift your telephone off its rest, a switch inside joins the two wires together.

d. A projecting part of a removable denture that gives it support by lying against a tooth.

1907 H. J. GOSLEE *Princ. & Pract. Crown & Bridgework* (ed. 2) xxv. 465 Providing a rest which will cause the piece to ride largely upon the supporting teeth instead of on the gum tissue. *Ibid.*, An occlusal rest consisting of a part of the clasp was early advocated by Dr. W. G. A. Bonwill. **1924** H. COWAN *Lectures Full Denture Prosthesis* II. 125 The insertion of occlusion rests or immediate dentures within thirty minutes after the teeth have been removed, contributes in a remarkable degree to the patient's confidence. **1942** J. R. SCHWARTZ *Mod. Methods Tooth Replacement* ix. 379 Unfavorable action that may be caused by rests not locked into an inlay occurs when the inclination of the tooth is such that the rest will slip or slide. **1976** TORRES & EHRLICH *Mod. Dental Assisting* xxvi. 815/2 A rest will not be strong unless it is designed to partially cover a tooth that needs to be built up to the height of the occlusion.

14. a. *rest area*, *billet*, *camp* (earlier and later examples), *centre*, *cure* (earlier and later examples), *day*, *home*, *pause*, *period*, *stop*, *therapy*; *rest-giving adj.*; **rest gown**, formerly, a gown used for evening wear at home; **rest room**, (*a*) a room (usu. in a public building) set aside for rest and quiet; (*b*) *U.S.*, a lavatory, a W.C.; **rest-tremor** *Path.*, a tremor in a part which is not being voluntarily moved.

1919 W. H. DOWNING *Digger Dial.* 41 *Rest area*, a district to which battalions, on leaving the danger zone, marched by long stages once a year for the purpose of polishing their brass work. **1944** K. DOUGLAS *Alamein to Zem Zem* (1946) 16 Here we halted, having left the heavy squadrons of Shermans and Grants still in our rest area. **1976** G. V. HIGGINS *Judgment D. Hunter* ix. 86 The youth..stopped in a rest area..so that the subject could relieve himself. **1978** W. GARNER *Möbius Trip* ix. 216 In the autoroute rest area at Chennevières..a chauffeured gray Mercedes. **1917** A. G. EMPEY *Over The Top* 306 *Rest billets*, shell shattered houses, generally barns, in which 'Tommy' 'rests', when relieved from the firing line. **1925** E. GRAVES *Welchman's Hose* 29 And back in rest-billets The Colonel congratulates 'B' Company on its kits. **1954** W. FAULKNER *Fable* 128 The..three others..had been spending their leaves and furloughs..among the combat-troop rest-billets. **1889** G. S. MACKENZIE *Let.* 11 Dec. in Ld. Lugard *Diaries* (1959) I. 50 Lay out rest camps and fortified posts at regular intervals. **1919** *Lit. Digest* 12 Apr. [illegible] The rest camp is so close..that [illegible] was going to a rest camp (cemetery). **1923** KIPLING *Irish Guards in Gt. War* I. 3 The city became almost a suburb to the vast rest-camps round it. **1971** *Rand Daily Mail* 27 Mar. 23/3 Incidentally, all Natal Parks Board rest camps are fully booked in the Easter holidays. **1940** *Economist* 5 Oct. 422/1 When their cool has gone the family seek refuge, and they find it in temporary rest centres run by the Public Assistance Department of The London County Council. **1976** *Liverpool Echo* 6 Dec. 8/4 Rest centres were made ready on the outskirts of all 'target' areas, like Merseyside, to house the virtual refugees. **1889** *Cent. Dict.*, Rest-cure. **1892** S. HALE *Let.* 28 Apr. (1919) 272 She is at a rest cure. [illegible] *War* 131 Then we went into the trenches which not long before had been so horrible..but now they..were voted 'a rest-cure sector'. **1959** T. S. ELIOT *Elder Statesman* ii. 57 Now's the time to take a long holiday, Let's say a rest cure. **1980** D. ADAMS *Restaurant at End of Universe* ix. 57, I only hope it's gone in for a rest cure... The way it's been living recently it must be on its last elbows. **1911** *Webster*, Rest day. **1959** *Encounter* Aug. 22/1 People were free to prepare their meals at home on 'rest days'. **1976** J. SNOW *Cricket Rebel* 90, I worked out that we had only one day free from cricket in those last six weeks in Australia apart from the rest days during the Test matches. **1888** BEERBOHM *Let. to R. Turner* (1964) 270 A solid and..rest-giving figure in the midst of that vile vortex. **1923** MRS. G. H. VAIZEY *College Girl* xviii. 257 Margaret herself, in a pink rest-gown. [illegible] 20 Nov. 326/1 [illegible] place being taken [illegible] and theatre [illegible] restful for home [illegible] *Stars* xvi. 286 [illegible] about her, Norma [illegible] 6 July 6/7 [illegible] workers may spend [illegible] **1976** B. BOVA *Multiple Man* (1977) [illegible] Why wouldn't the President simply..cart the old man off to a well-guarded rest home? **1926** *Encycl. Brit.* III. 454/1 An important innovation stressed by the industrial psychologist has been the introduction of short rests, in the middle of a working period, of about 10 or 15 min. duration. These regular breaks are technically known as rest pauses. **1954** J. A. C. BROWN *Social Psychol. of Industry* iii. 71 Two five-minute rest-pauses, morning and afternoon. **1922** *Encycl. Brit.* XXXI. 699/2 In spite of the considerable development of maximum hour regulation in the United States, not much attention has been paid to the question of legal rest periods. **1954** J. A. C. BROWN *Social Psychol. of Industry* iii. 74 The introduction of rest-periods which amounted to two ten-minute breaks in the morning and two in the afternoon. **1981** 'J. ROSS' *Dark Blue & Dangerous* xxii. 123 You will report off duty for a rest period. **1899** *Amer. Jrnl. Sociol.* May 729 Surely it would not be unreasonable to require that suitable rest-rooms be provided for the employés. **1918** A. BENNETT *Pretty Lady* xxvii. 182 Canteens, and rest-rooms, and libraries, and sanitation, and all this damned 'welfare'. **1942** 'M. INNES' *Daffodil Affair* iii. 103 His private research block..comprised a library, a museum, rest rooms and living quarters for the subjects. **1955** W. GADDIS *Recognitions* I. iii. 100 If you serve food you gotta have a rest room for ladies as well as men. **1975** *N.Y. Times* 8 Nov. 26/2 The totally unsubstantiated forecasts of effects ranging from the abolition of women's rest rooms to dire financial consequences may well have persuaded many women that it was safer to accept known evils than to risk unknown pitfalls. **1973** *Sunday Bull.* (Philadelphia) 14 Oct. (Parade Suppl.) 16/3 A truck driver napped at a rest stop. **1975** J. GRADY *Shadow of Condor* (1976) xiii. 214 It's another rest stop... Just a picnic table and some trash cans. **1949** *Radio Times* 15 July 24/1 Many people keep their vitality..by regularly practising Rest-therapy. **1925** S. A. K. WILSON in *Lancet* 8 Aug. 270/2 Some cases of disseminated sclerosis..definitely show what I may call a 'rest-tremor' as well as an 'action-tremor'. **1967** *Internat. Jrnl. Neuropharmacol.* VI. 122 There is a great similarity between rest tremor produced by tremorine and rest tremor in human pathology. *Ibid.*, It is likely that rest tremor in human pathology is due to the hyperactivity of these [alpha] cells [of the anterior horn].

c. In sense 8, as **rest energy** *Physics*, the energy inherent in a body by virtue of its possession of rest mass; **rest frame** *Physics*, a frame of reference relative to which a given body is at rest; **rest level** *Hydrology*, the natural level of water in an aquifer or on the ground surface; **rest mass** *Physics*, the mass of a body measured when it is at rest; **rest position** *Dentistry*, the relative position of the jaws when relaxed.

1938 *Forum* Feb. 95/2 From the Lorents transformation and the assumption of the impulse and energy principle for material particles..the equality of mass and rest-energy is derived. **1962** H. D. BUSH *Atomic & Nuclear Physics* ii. 54 Since the rest mass is m_0, the rest energy is considered to be m_0c^2. **1966** J. WERLE *Relativistic Theory of Reactions* v. 347 We are interested in the scattering angle of particle *d* in a rest frame of particle *b*. **1978** *Nature* 5 Oct. 411/1 The known general absence of Lyman continuum absorption in the rest frame of high redshift QSOs implies that < 0·1 of the continuum radiation is intercepted by optically thick clouds. **1908** W. COLES-FINCH *Water* xviii. 416 The rest-level in a well or boring is that level to which the water rises upon cessation of pumping. **1956** R. J. C. ATKINSON *Stonehenge* iii. 82 The level at which this pottery and the earliest fragments of bluestone occur is slightly above the rest-level which marks the end of the first phase of rapid silting in the ditch. **1914** L. SILBERSTEIN *Theory of Relativity* vii. 193 The coefficient *m* is called the *rest-mass* of the particle. **1938** R. W. LAWSON tr. *Hevesy & Paneth's Man. Radioactivity* (ed. 2) iii. 36 We differentiate between the 'rest mass' and the 'translational mass' of β-particles, i.e. the mass of the particle in motion. **1979** *Jrnl. R. Soc. Arts* CXXVII. 577/2 Nuclear fusion can convert 0·7 per cent of the rest mass of the hydrogen core into energy. [**1907** C. H. TURNER *Amer. Textbk. Prosthetic Dentistry* (ed. 3) iv. [illegible] This position is commonly called 'the resting bite'. **1921** G. H. [illegible] *Prosthetic Dentistry* iv. 76 The closed resting position of the mandible is one of more stable equilibrium than any other, and brings the greatest number of teeth into contact and the jaws closer together.] **1924** T. [illegible] *Art of Prosthetic Dentistry* iii. 55 Ask the patient to close, thus registering the proper rest position. **1962** BLAKE & TROTT *Periodontol.* xvi. 163 To bring the mandible into the rest position the patient should be seated comfortably in the chair, the head erect and away from the head rest, with the Frankfort plane horizontal. He is then asked to wet the lips with the tongue, swallow and remain quite still. The mandible should then assume the rest position.

rest, *sb.*² Add: **9.** (Further examples.)

In lawn tennis the usual term is *rally*.

1898 H. F. LAWFORD in W. A. Morgan '*House' on Sport* 428, I was told that one rest was eighty-one strokes. **1960** *Times* 18 July 14/4 Warburg..played six strokes in a fine rest. **1975** *Oxf. Compan. Sports & Games* 826/1 During a 'rest' (the real tennis term for a rally), while the player on the hazard side is trying to make a good chase, his opponent on the service side is in a stronger position.

10. *Med.* A small detached part of an organ, surrounded by tissue of another character; esp. as *adrenal rest*, a small displaced part of the adrenal cortex.

1892 R. BOYCE *Textbk. Morbid Histol.* viii. 160 In adult life 'rests' are frequently encountered in the lines of old incisions and punctures and are due to excess of reparative tissue. *Ibid.* x. 201 Another class of rests represents vestigial structures. **1898** I. N. KELYNACK *Renal Growths* xiv. 132 Growths arising in adrenal 'rests' appear to be sarcomatous in general characters rather than carcinomatous. **1912** *Q. Jrnl. Med.* V. 157 To present new reasons against the hypothesis that renal hypernephromata are derived from adrenal rests. **1928** J. F. BARNHILL *Nose, Throat & Ear* xi. 157 In sarcoma [of the accessory nasal sinuses] congenital 'rests' in the alveoli of the child..may be assigned as a cause. **1939** T. W. WIDDOWSON *Special or Dental Anat.*

(ed. 6) xii. 300 The epithelial cellular bodies, or rests,..are collections of epithelial cells sometimes seen in the inner portion of the periodontal membrane near to the cementum. **1963** C. L. DEMING in M. F. Campbell *Urology* (ed. 2) II. xxii. 912 Grawitz thought that these tumors came from adrenal rests and termed them hypernephromas.

rest, *v.*¹ Add: **I. 1. c.** orig. *N. Amer.* Of the body of a dead person: to remain at an undertaker's, a chapel, etc., before burial or cremation. (Usu. as pres. pple.)

1967 'CORIOLIS' *Death, Here is Thy Sting* iii. 54 Remains will be resting at the John Doe Funeral Home. **1968** *Globe & Mail* (Toronto) 13 Jan. 37/5 Resting at Bates and Maddocks Funeral Chapel..until 12 noon Monday. **1974** *Almonte* (Ontario) *Gaz.* 4 Apr. 4/3 Predeceased by a brother Harold... Rested at the Kerry Funeral Home. **1976** *Liverpool Echo* 22 Nov. 4/4 Funeral service at Anfield Crematorium... Resting at E. H. Roberts..where flowers may be sent.

2. f. *Theatr.* Of an actor: to be out of work (temporarily), to be unemployed. Also *transf.* (Usu. as pres. pple.)

1890 B. HALL *Turnover Club* 81 It would commend itself particularly to actors 'resting' for the summer. **1912** GALSWORTHY *Pigeon* I. 24 I am an interpreter... At present I am resting. **1923** A. CHRISTIE *Murder on Links* xxii. 250 They're in the provinces, somewhere, I believe—if they're not resting. **1938** G. HEYER *Blunt Instrument* x. 195 A very nice lady. Stage, but she's resting. **1947** J. SYMONS *Man called Jones* 109 She's in this new thing at the Splendid, and I'm resting. That's show business. **1958** *Times* 22 May 12/6 Ten shillings a week..used to go a long way with those who, in actors' parlance, were 'resting'. **1967** *Radio Times* 18 May 21 Except for the occasional coffee-bar job while 'resting', Dinsdale has always managed to earn a living from acting. **1976** R. HILL *Another Death in Venice* I. ii. 33 'Are you on holiday?'..'Resting, to use a theatrical term. Between jobs.'

g. *to rest up*: to recover one's strength by resting. orig. *U.S.*

1895 'MARK TWAIN' in *Harper's Mag.* Aug. 458 The other inquisitor could absent himself and rest up from his fatigues when he got worn out. **1911** H. S. HARRISON *Queed* x. 125 She had been remanded in bed for a day or two to rest up. **1922** Z. GREY *To Last Man* xiii. 284 Get rifle and ammunition, bake bread, and rest up before taking again the trail of the rustlers. **1936** J. CURTIS *Gilt Kid* xiii. 131 The best plan would be to..have something to eat and rest up. **1949** N. STREATFEILD *Painted Garden* xi. 127 She said..she was going to rest-up; I expect that means bed. **1965** M. SHADBOLT *Among Cinders* xxii. 210 Now it was just a place for shooters and trampers to rest up. **1972** J. BLACKBURN *For Fear of Little Men* xi. 116 Adder bites are unpleasant things... I advised the fool to rest up a bit longer. **1974** *Times Lit. Suppl.* 1 Feb. 97/1 Attila is resting up in Hungvar, exacting 'hospitality' from the German peoples.

III. 8. a. Also const. *up*. Cf. sense *2 g.

Quots. 1975 and 1976 are both *U.S. slang*.

1974 'J. LE CARRÉ' *Tinker, Tailor* xxviii. 241 They're resting you up for a season. **1975** L. DILLS *CB Slanguage Dict.* 50 *Rest 'em up place*, rest area (SE). **1976** LIEBERMAN & RHODES *Compl. CB Handbk.* vi. 157 Hey, we just spotted a smokey at that rest'em up area.

g. *U.S.* To bring the presentation of evidence pertinent to (a law case) to a close voluntarily.

1905 S. W. MITCHELL *Constance Trescot* xiv. 183 All the evidence for the plaintiffs was before the court, and Greyhurst sat down, stating that the plaintiff rested the case. **1950** *Chicago Tribune* 23 Jan. 1/8 Defense attorneys..elected to rest their case without calling a single witness. **1953** E. S. GARDNER *Case of Hesitant Hostess* xiv. 234 The prosecution objects. The prosecution has rested its case. **1972** *N.Y. Law Jrnl.* 14 Nov. 19/5 Plaintiff had not been cross-examined and certainly had not rested his case.

restart, *sb.* Add: Also stressed **re·start**. Also *attrib.* (Further examples.)

1971 *Computers & Humanities* V. 140 One of the most powerful features of this system is the restart procedure... The restart capability is especially useful under a system which limits the amount of time and records a user can run at any one time. **1977** *Gloss. Terms Data Processing* (*B.S.I.*) vii. 9/2 *Restart condition*, in the execution of a computer program, a condition that can be re-established and that permits a restart of the computer program. *Ibid.*, *Restart point*, a place in a computer program at which its execution may be restarted; in particular, the address of a restart instruction.

restatement. Add: Also in *Mus.* Cf. STATEMENT 1 b.

1944 W. APEL *Harvard Dict. Mus.* 638/2 *Restatement*, same as recapitulation in sonata-form. **1954** *Grove's Dict. Mus.* (ed. 5) III. 440/2 The restatement of the first subject is sufficient indication to the hearer as to what part of the movement he has arrived at. **1959** *Listener* 20 Aug. 297/2 The preparation for the restatement in his first movement is altogether classical in method. **1974** *Encycl. Brit. Macropædia* XVII. 5/2 The first theme is restated in the dominant key. This restatement could appear at first to be the second subject.

resta·ter. [f. RESTATE *v.*² + -ER¹.] A person who restates.

1925 G. O'BRIEN *Econ. Effects Reformation* ii. 125 Luther..regarded Christ and the Apostles as merely the restaters and expositors of the Decalogue.

restaurant. Add: **2.** *attrib.*, as *restaurant car*, *dinner*, *-keeper*, *lunch*, *manager*, *meal*, *proprietor*.

1875 Restaurant car [in Dict.]. **1913** KIPLING *Lett. of Travel* (1920) 220 That terrible restaurant-car dinner in the tunnel. **1967** 'T. WELLS' *What should you know of Dying?* x. 118 He's an elevator man at the Winchenden Arms... He's on the restaurant car. **1899** R. FRY *Let.* Oct. (1972) I. 174 The restaurant dinners all stop at 5.0. **1925** F. SCOTT FITZGERALD *Great Gatsby* iii. 69 A solitary restaurant dinner. **1876** W. WRIGHT *Hist. Big Bonanza* xlviii. 355 Some years ago a restaurant keeper had a number of these customers, who were eating him out of house and home. **1932** H. SIMPSON *Boomerang* vii. 141 The two young men..grinned up at him, and the first, the restaurant-keeper, answered: 'His money is as good as another's.' **1974** 'J. ROSS' *Burning of Billy Toober* ix. 91 Prosser..had had his restaurant lunch sent in on a linen-shrouded tray. **1976** *Liverpool Echo* 6 Dec. 6/6 Restaurant manager Jose Padilla..is always on the look-out for something new for his customers. **1938** *Cook's Continental Time-table* Jan. (Advt.), Comfort to the continent. Right through the winter, Imperial Airways will operate frequent services to Paris,..Full restaurant meals served during flight. **1960** M. SPARK *Bachelors* iii. 43 One can't afford two restaurant meals in one day. **1938** *Sun* (Baltimore) 11 June 9/2 Worley Carrico, middle-aged restaurant proprietor. **1974** *Encycl. Brit. Macropædia* XV. 778/1 The first restaurant proprietor is believed to have been one A. Boulanger, a soup vendor, who opened his business in Paris in 1765.

restauranter (re·stǫrǫntǝɹ). *U.S.* [f. RESTAURANT + -ER¹.] = RESTAURATEUR 1.

1887 *Ohio State Jrnl.* 20 July, The headquarters of Mr. Kiesewetter are at Diebold's, an opulent restauranter and general purveyor to the wants of delegates. **1938** *Sun* (Baltimore) 11 June 9/2 (*heading*) Restauranter jailed in fatal shooting.

restauranteur (re:stǫrāntō·ɹ). [f. RESTAURANT + -EUR as in RESTAURATEUR.] = RESTAURATEUR 1.

An erroneous form.

1949 E. HYAMS *Grape Vine in England* viii. 136 The second thinning may be done when the grapes are about the size of those very large and granite-like marrowfat peas which are beloved of English restauranteurs, or used to be. **1958** *Wall St. Jrnl.* 30 Sept. 7/3 Jack Dempsey, one-time heavyweight champion turned restauranteur. **1980** *Guardian Weekly* 20 Jan. 20/3 The growing number of hoteliers and restauranteurs.

re·staurantish, *a. rare.* [-ISH¹.] Resembling a restaurant; suggestive of a restaurant.

1896 A. BEARDSLEY *Let.* c 26 June (1970) 138 I have fallen on my feet here. Two palatial rooms and the additional comfort of being able to feed in a pretty little restaurantish dining-room.

restaurateur. Add: **1.** (Later examples.)

1969 *N.Y. Rev. Bks.* 2 Jan. 16/3 The editor of an intellectual journal has no less claim to being functional than a restaurateur. **1976** *Times* 21 May 4/8 Some had robbed a retired restaurateur of goods and money. **1980** *Amer. Speech* LV. 91 Books of advice to restaurateurs and menu designers suggest the use of foreign languages.

2. (Earlier example.)

1801 C. WILMOT *Let.* 13 Dec. in *Irish Peer* (1920) 13 The 'Palais Royal' is excessively new and entertaining... Libraries, Restaurateurs, Gambling Houses, Coffee Houses.

restauration. Add: **2. d.** (An) alteration or repair intended to restore a building, etc., to something like its original form or use. Cf. RESTORATION 4 b.

1949 E. POUND *Pisan Cantos* lxxx. 86 And I trust they have not destroyed the Old theatre by restaurations, and by late renaissance giribizzi. **1977** *It* May 29/3 Research into restauration of derelict urban or other areas.

restaurative (restǭ·rătiv), *a.*² *rare.* [ad. F. *restauratif*: see -IVE.] Having the function of a restaurant; providing restaurant facilities.

1875 in G. J. Holyoake *Hist. Co-operation in Eng.* I. ix. 214 In a short time the restaurative omnibuses will circulate through Paris... These vehicles will contain broth and sauce for the whole city.

rested, *ppl. a.* Add: Also const. *up* (see *REST *v.*¹ 2 g).

1922 E. O'NEILL *Anna Christie* I. 115, I was thinking maybe..he might be willing to stake me to a room and eats till I get rested up. **1929** G. ADE *Let.* 22 May (1973) 141, I am certainly glad to be home after a tedious tour and I am getting all rested up. **1935** M. M. ATWATER *Murder in Midsummer* xxi. 201, I guess we're all rested up. Come along, boys.

resteno·sis. *Med.* [RE-.] A recurrence of stenosis, esp. of a heart valve after surgery to correct it.

1954 *A.M.A. Arch. Internal Med.* XCIV. 777 Other examples of so-called restenosis following commissurotomy have been reported. **1977** *Proc. R. Soc. Med.* LXX. 816/2 Restenosis [of the ear canal] is very common.., especially in those cases where pneumatization is poor and the lateral sinus is abnormally anterior. **1980** *Brit. Med. Jrnl.* 12 Jan. 62/1 Many of these patients will have had a closed mitral valvotomy 10 to 30 years before, and the term 'mitral restenosis' is loosely used to describe the recurrence of their symptoms.

rester¹. Add: **1.** Also *spec.* an unemployed actor or actress.

1931 R. ALDINGTON *Colonel's Daughter* v. 300 The anteroom of a dramatic agency filled with restless resters.

2. A ledge for placing articles on in front of a balcony in a theatre.

1922 *Rep. Theatres & Music Halls Comm.* in *Minutes Proc. L.C.C.*, July–Dec. 1921 158 In order to prevent trays..being pushed off the resters in the front of balconies at theatres, etc...we have..decided to require..that the rester shall be sloped at, say, an angle of 30 deg.

rest-house. Add: **1. a.** Also, a building with a similar function in Malaysia or Africa. (Further examples.)

1954 [see *NIGERIAN *sb.* and *a.*]. **1964** (*title*) Malaysia: visitors guide. Hotel and rest house directory. **1966** D. FORBES *Heart of Malaya* iii. 43 He..drove to Belinggu and took Abigail to the rest house there. **1972** *Guardian* 22 Sept. 9/1 My night at the Rest House—a hang-over from the colonial days, where Government officials could stay when travelling across country—was interrupted by the distant sound of a gong. **1978** G. GREENE *Human Factor* v. ii. 241 He could even imagine himself in Africa, at some resthouse in the bush.

b. An establishment catering for persons requiring rest and recreation.

1928 GALSWORTHY *Swan Song* III. ii. 231, I feel I should be ever so much more interested if I ran a place of my own in the country—a sort of rest-house that I could make attractive for girls who wanted air and that.

2. *attrib.*, as *rest-house garden*, *keeper*.

1909 *Athenæum* 24 Apr. 492/1, I dislike the ramshackle rest-house, and its rude indifferent rest-house-keeper. **1973** 'B. MATHER' *Snowline* xv. 183, I withdrew into the resthouse garden again.

resti·mulate, *v.* [RE- 5 a.] *trans.* To stimulate again. Hence **restimula·tion**, **restimula·tory** *a.* [-ORY².]

1796 A. SEWARD *Let.* 11 Dec. (1811) IV. 282 Mr. B. will succeed in his design to re-stimulate the public mind to continue the war. **1822–34** [see RE- 5 a]. **1924** W. B. SELBIE *Psychol. Relig.* iv. 89 What are regarded as influences from the subconscious are due to a restimulation of brain tracts which have been influenced by previous experiences. **1925** T. DREISER *Amer. Trag.* (1926) II. xxxiii. 381 Even his obviously dwindling affection was restimulated by her quite visible need of help. **1962** *Economist* 27 Oct. 332/2 A Keynesian restimulation of internal demand. *Ibid.* 29 Dec. 1256/1 The orthodox thing..may be to take no further restimulatory action..and then to give big tax reliefs in his budget. **1968** *Ibid.* 20 July 18/2 In spurring the re-expansion made possible by a freeing of labour from uneconomic jobs, it will be sensible to rely on tax cuts as the main restimulatory weapon.

resting, *vbl. sb.*¹ Add: **1. c.** *Theatr.* Unemployment; being without an acting job.

1924 G. B. STERN *Tents of Israel* vi. 85 A young singer..[who] had sold all her things during her long period of enforced 'resting'. **1960** *Times* 28 Sept. 15/1 My theatrical colleagues who are only too familiar with the long periods of 'resting'—which being out of work is so politely called. **1973** J. BURROWS *Like an Evening Gone* i. 12 Though what..she did with herself in the great metropolis, in the frequent intervals of 'resting'—she didn't take typing jobs.

4. a. *resting period.*

1895 *Funk's Stand. Dict.*, Resting period. **1940** *Chambers's Techn. Dict.* 717/2 *Resting period*, any time in the life of a plant, or plant organ, when no growth or other activity appears to be in progress. **1952** [see *ASPECT *sb.* 14].

b. *resting ground.*

1921 *Daily Colonist* (Victoria, B.C.) 27 Oct. 10/3 Feeding and resting grounds scattered over the country for the use of birds during migration.

c. *resting stage* (Cytology) = *INTERPHASE *sb.* 1.

1896 E. B. WILSON *Cell* ii. 53 There are..some undoubted cases..in which the centrosome remains undivided during the resting stage and only divides as the process of mitosis begins. **1932** [see *INTERPHASE *sb.* 1]. **1957** C. P. SWANSON *Cytol. & Cytogenetics* iii. 48 Cells in interphase, or the resting stage, are characterized by a nucleus that shows little or no definable structure, except for the nucleoli and the prochromosomes.

resting, *ppl. a.* Add: **1. b.** *resting bud*; also in *Cytology*, as *resting* (= interphase) *cell*, *nucleus*; *Zool.*, *resting egg*, a fertilized egg which can survive the winter or other unfavourable period before hatching.

1895 *Sci. Progress* III. 333 Structural changes in the resting nucleus, which lead up to the formation of the reduced number of chromosomes, and which I have termed collectively the synapsis. **1896** M. HARTOG in *Cambr. Nat. Hist.* II. viii. 200 The fertilised eggs are of the kind termed 'winter' or 'resting' eggs. **1904** *Nature* 24 Nov. 76/2 Figures are given of resting-buds, twigs and their transverse sections [etc.]. **1934** *Biol. Rev.* IX. 160 Among the Cladocera..the same female may produce both females and males (by parthenogenetic reproduction) and resting eggs (by gametogenetic reproduction). **1953** H. L. EDLIN *Forester's Handbk.* ii. 22 The shoot increases in length only during the annual period of active growth. It grows..by actively dividing cells within the bud at its tip. When autumn comes, this suspends its activity, and becomes a resting bud, protected by bud scales, which remains dormant until the following spring. **1960** L. PICKEN *Organization of Cells* IV. ii. 103 The conflict between the genetic evidence, that the persistence of serial order of discrete genetic units in the linkage groups is of primary importance; and the light microscope evidence, that chromosomes are not visible in the resting nucleus in a majority of cells. **1975** J. B. JENKINS *Genetics* i. 27 In the resting, or interphase, cell the chromosomes are in an uncoiled condition and thus difficult to see. **1976** *Freshwater Biol.* VI. 408/1 Lack of food is probably the most important factor stimulating resting egg production.

c. *Theatr.* Between acting jobs; unemployed.

1958 A. WILSON *Middle Age of Mrs Eliot* ii. 143 The rich 'resting' stage stars. **1969** M. PUGH *Last Place Left* xxiii. 178 The waitress..looked like a resting actress who did her resting in daytime. **1971** M. BABSON *Cover-up Story* iv. 43 We still had the dubious privilege of representing two 'resting' actors.

3. Restful. *rare.*

1896 A. BEARDSLEY *Let.* 17 Nov. (1970) 205, I wish Mabel could have started from Southampton. I do hope her crossing may be quiet and resting after all her hard work.

restitute, *v.* Add: **3.** *intr. Genetics.* Of a break in a chromosome or chromatid: to be reversed by restitution (sense *8) of the two broken ends.

1945 *Jrnl. Genetics* XLVII. 13 Some of the breaks which restitute are lethals. **1971** LEVITAN & MONTAGU *Textbk. Human Genetics* iii. 135 Nonrearrangement breaks tend to restitute..quite readily.

‖ **restitutio in integrum** (restitiū·tio in inte·grǝm), *sb. phr. Law.* Also **ad integrum**. [L., restoration to the uninjured state, etc.] (See quot. 1909.) Also *transf.*

[**a 1633** T. HOPE *Major Practicks* (1938) II. 207 (*heading*) De Restitutione in Integrum.] **1704** T. WOOD *New Institute Imperial or Civil Law* I. ii. 55 A Restitution shall be adjudged in the behalf of the Minor if he is circumvented, tho' the Tutors or Curators were present and consenting, and tho' the Decree was made against the Minor judicially; and this is called *Restitutio in integrum*. **1845** T. THORNTON *Notes Cases in Eccles. & Maritime Courts* III. 77 The principle is, that the vessel should be placed *in statu quo* before the collision,—*a restitutio in integrum*. **1902** W. JAMES *Var. Relig. Exper.* vi. 156 When disillusionment has gone as far as this, there is seldom a *restitutio ad integrum*. **1909** *Halsbury's Laws of Eng.* X. 302 The underlying great principle by which the courts are guided in awarding damages is *restitutio in integrum*. By this is meant that the law will endeavour, so far as money can do it, to place the injured person in the same situation as if the contract had been performed, or in the position he occupied before the occurrence of the tort which adversely affects him. **1936** BUCKLAND & MCNAIR *Roman Law & Common Law* xii. 337 An action which had once reached the stage of *litis contestatio* could never be brought again, unless the facts brought the matter within the rather narrow range of cases in which a man could get *restitutio in integrum*. **1975** J. R. MURDOCH *Law Estate Agency & Auctions* xiii. 385 Since the principle underlying rescission is '*restitutio in integrum*' it will only be allowed where both parties can be substantially returned to their pre-contractual positions.

restitution. Add: **8.** *Genetics.* The coming together of the two parts of a broken chromosome or chromatid so as to re-form it; also *concr.*, the resulting chromosome or chromatid.

1941 *Cold Spring Harbor Symp. Quantitative Biol.* IX. 154/2 Even at ordinary doses there has been more than one break per gamete, but that break has usually been an invisible, 'restituted' one. How often may this 'restitution' have been imperfect or attended by some local alteration such as a 'gene mutation'? **1945** *Jrnl. Genetics* XLVII. 11 On this view it appears likely that lethals not apparently associated with any chromosome change are restitutions. **1980** R. P. WAGNER et al. *Introd. Mod. Genetics* x. 273/1 When a chromosome breaks, the two broken ends usually undergo restitution.

9. restitution nucleus *Cytology* [tr. G. *restitutions-kern* (O. Rosenberg 1927, in *Hereditas* VIII. 321)], a cell nucleus having twice the regular chromosome number, formed by an uncompleted mitotic or meiotic cell division.

1927 *Hereditas* VIII. 336 The semiheterotypic meta- and anaphase very often are not completed, but are interrupted by a premature homotypic division, whereby *Restitution-nuclei* are formed. **1950** *Adv. Genetics* III. 197 The zygoid chromosome number is restored through the fusion of two azygoid nuclei, the formation of a restitution nucleus or endomitosis. **1974** *Euphytica* XXIII. 631 The diploid chromosome numbers..may result either from first or second division restitution nuclei of meiosis.

restless, *a.* Add: **2. b.** (Further *Comb.* example.)

1914 'SAKI' *When William Came* vi. 101 You must remember that thousands and thousands of the more virile and restless-souled men have emigrated.

3. b. (Further *Comb.* example.)

1889 W. B. YEATS *Wanderings of Oisin* ii. 24 Where goes to gaze the restless-footed star Of twilight when he's weary.

re-stock, *v.*¹ Add: Also *absol.*

1906 E. DYSON *Fact'ry 'Ands* iv. 50 Ellis took pity on his emptiness, and Mumps was sent home to re-stock.

restoration. Add: **4. d.** *Dentistry.* Any structure provided to replace dental or oral tissue that has been removed or lost, such as a filling, crown, or bridge.

1934 F. W. FRAHM *Princ. & Technics Full Denture Constr.* xxvi. 475 The change from the natural to the artificial teeth is made so easily that the patient becomes accustomed to the supplied restoration in a few hours. **1962** BLAKE & TROTT *Periodontology* xvi. 170 Before the construction of such prosthetic or conservative restorations is begun the occlusion should be studied. **1976** TORRES & EHRLICH *Mod. Dental Assisting* xxvi. 802 To seat a restoration the casting is inserted into the tooth or onto the tooth crown.

6. *attrib.* and *Comb.*, as (sense 2 a) *Restoration comedy*, *drama*, *dramatist*, *pamphleteer*, *wing*; (sense 4 b) *restoration fund*.

1898 G. SAINTSBURY *Short Hist. Eng. Lit.* VIII. ii. 487 The cloven foot of Restoration comedy—the passionless and malevolent licentiousness of too much thereof. **1910** *Encycl. Brit.* IX. 630/2 Restoration comedy at first followed Jonson. **1925** B. DOBRÉE *Congreve's Comedies* I. p. xii, The Restoration comedy writers..keep us dancing along to a gay tune. **1955** N. MARSH *Scales of Justice* iii. 53 'Kettle,' Lady Lacklander said... Nurse Kettle did not resent being addressed in this restoration-comedy fashion by Lady Lacklander. **1976** *Amer. N. & Q.* XV. 35/1 In Restoration comedy, reference is often made to the 'canonical hours'. **1898** G. SAINTSBURY *Short Hist. Eng. Lit.* VIII. ii. 491 The glory..of this Restoration drama was not reached till long after the Restoration itself. **1923** A. NICOLL *Hist. Restoration Drama 1660–1700* i. 3 The study of Restoration drama demands a continual care. **1935** D. L. SAYERS *Gaudy Night* iv. 82 Calling people names that poor Miss Lydgate didn't know existed—the worst she knows being Restoration drama. **1977** J. AIKEN *Last Movement* vii. 130 I'd cast him as Rochester..or one of those fiends in Restoration Drama. **1912** E. GOSSE in *Restoration Plays* p. x, What all the Restoration dramatists suffer from is a tendency to produce common and inadequate poetry. **1923** A. NICOLL *Hist. Restoration Drama 1660–1700* iii. 184 Manners, in the mouths of the Restoration dramatists themselves, meant something quite apart from the modern meaning of the term. **1974** *Encycl. Brit. Macropædia* IV. 113/1 William Congreve, more than any other Restoration dramatist, shaped the English comedy of manners. **1898** G. B. SHAW *Mrs. Warren's Profession* III. 204 Gov'nor's ever so fond of it, because he got up a restoration fund and had it completely rebuilt. **1978** *Lancashire Life* Nov. 140/1 A restoration fund has been launched to restore the decaying 15th century tower of Mitton Church, near Whalley. **1875** Restoration pamphleteer [in Dict., sense 2 a]. **1920** 'O. DOUGLAS' *Penny Plain* xxv. 300 The austere Tudor front, the Restoration wing, the offices built under Queen Anne. **1936** J. BUCHAN *Island of Sheep* viii. 141 Nothing more modern than the Restoration wing built by Bruce of Kinross.

restorative, *a.* and *sb.* Add: **A.** *adj.* (Examples in *Dentistry*.)

1963 C. R. COWELL et al. *Inlays, Crowns, & Bridges* i. 1 Gold is stronger than other restorative materials and can be used in thin sections without danger of fracture. **1974** *News & Press* (Darlington, S. Carolina) 25 Apr. 17/3 It was first thought that the council might buy a mobile unit and employ a full time dentist who could..do restorative dentistry.

restore, *v.* Add: **3. a.** (Further examples.)

1868 A. J. MUNBY *Diary* 30 Oct. in D. Hudson *Munby* (1972) 258 The church..is being 'restored', as the phrase is. **1907** G. B. SHAW *Major Barbara* II. 244 He is one of the greatest of our public benefactors. He restored the cathedral at Hakington. **1920** W. B. YEATS *Michael Robartes & Dancer* 24, I, the poet William Yeats, With old mill boards and sea-green slates, And smithy work from the Gort Forge, Restored this tower for my wife George. **1931** J. BETJEMAN *Mt. Zion* 18 Look up! and how glorious He has restored the roof! **1959** N. PEVSNER *Yorkshire: W. Riding* 267 The church was restored in 1876 by Pearson.

d. (See quot. 1955.)

1943 *Amer. Speech* XVIII. 304/1 Restored cereal. **1955** M. REIFER *Dict. New Words* 177/2 *Restore*, *v.*, to give back to a processed food the nutritive value it originally had.

restorer. Add: **2.** That which restores.

1873, 1893 [see *hair-restorer* s.v. HAIR *sb.* 10 in Dict. and Suppl.]. **1911** G. B. SHAW *Doctor's Dilemma* I. 23 Its like a bald-headed man trying to sell a hair restorer.

resto·ringly, *adv.* [-LY².] In a restoring manner; restoratively.

1846 J. R. LOWELL *Let.* 18 Feb. (1894) I. 115 Falling gently and restoringly as dew on the withered youth-flowers of the oppressor.

restraint, *sb.* Add: **1. c.** Something which restrains or holds in check; esp. *head restraint*, an attachment to the seat of a motor vehicle to prevent the head from jerking back suddenly.

1968 *Wall St. Jrnl.* 5 Aug. 28/4 American Motors Corp. said it will make head restraints standard equipment on all 1969-model cars. **1972** *Times* 28 Nov. 5/6 (Advt.), Both front seats recline, have sockets for optional head restraints. **1973** *Times* 3 July 1/5 Technically, a head rest was purely for comfort, whereas a safety device for protecting the head was termed a head restraint. **1976** *Amer. Speech* 1973 XLVIII. 207 There, if he should become violent, he is placed in restraints 'straps' in his room. **1980** *Times* 29 Feb. 3 (Advt.), The seats are covered in crushed velour with head restraints at the rear as well as the front.

3. c. *restraint of trade.*

1890 *Statutes at Large U.S.A.* XXVI. 209 Every contract, combination in the form of trust or otherwise, or conspiracy, in restraint of trade or commerce among the several States..is hereby declared to be illegal. **1913** *Halsbury's Laws of Engl.* XXVII. 532 Under the head of restraint of trade by statute come all those cases in which certain trades have been absolutely forbidden by Parliament. **1933** *Sutton & Shannon on Contracts* xi. 164 A contract in restraint of trade is not contrary to public policy. **1941** *Economist* 5 Apr. 437/2 The greater danger lies in the growing influence of what the common law knows (and used to condemn) as 'agreements in restraint of trade'. **1973** *N.Y. Law Jrnl.* 31 Aug. 1/5 The two-count indictment charged conspiracy in restraint of trade in violation of the Sherman Antitrust Act.

restri·ctable, *a.* [f. RESTRICT *v.* + -ABLE.] Able to be restricted.

1973 I. ROBINSON *Survival of English* iii. 82 Economic management is management of human beings, which is just not restrictable to economics.

restricted, *ppl. a.* Add: **b.** In which a speed-limit is operative.

1933 E. CALDWELL *God's Little Acre* vii. 107 They passed through the other company towns slowing down in the restricted zones. **1939** *New Statesman* 29 July 196/1 The existing mild supervision of restricted areas is carried out by the so-called 'speed cops'. **1959** *Listener* 2 Apr. 603/1 There are cases where a driver is mistaken about a restricted area.

c. Of documents, information, etc.: for restricted circulation only (see also quot. 1975); not to be revealed to the general public for reasons of national security.

1944 [see *CLASSIFIED *ppl. a.* c]. **1950** [see *CLEAR *v.* 9 c]. **1957** *Ann. Reg.* 1956 345 A new model of the R.A.F.'s only fully supersonic fighter flying in Britain..was taken off the restricted list in August and was demonstrated at the aircraft industry's annual display at Farnborough. **1965** MRS. L. B. JOHNSON *White House Diary* 2 Sept. (1970) 313 Then President Truman gave us the tour [of the library], with Max asking all sorts of architectural questions—storage, humidity, traffic routing, vaults for documents that were still restricted. **1972** K. BENTON *Spy in Chancery* xv. 180 The Russians always start by asking you to hand over something quite harmless... And then they ask for something that's on the restricted list, but not really secret. **1972** P. RUELL *Red Christmas* ix. 83 What I'm going to tell you is restricted information. That means it's only known to the Prime Minister, [and] security top brass. **1975** *Times* 8 Feb. 2/1 'Restricted' papers ('restricted' is the lowest security classification) had been found on my doormat.

d. *U.S.* Limited to use by non-Jews; denying admission to Jews.

1947 *Cosmopolitan* Jan. 84/2 Is your inn restricted?.. You mean you do restrict your guests to Gentiles. **1953** P. FRANKAU *Winged Horse* I. ix. 41 Draw the slums. Draw the restricted hotels. **1973** W. P. MCGIVERN *Caprifoil* (1973) xii. 204 Anti-Semitism..is not only a matter of restricted clubs and colleges' quotas. **1979** *Listener* 16 Aug. 204/2 'I'm sorry, Mr Marx, but we can't let you use the pool, the country club is restricted.'..'Well, my daughter's only half-Jewish, could she go in up to her knees?'

e. *Biol.* Of a virus: unable to reproduce at its normal rate in certain hosts. Of DNA: subject to degradation by a restriction enzyme.

1957 *Virology* III. 500 A certain proportion of the T1 phage produced is now able to multiply on B(P1) or Sh(P1) ('unrestricted phage'). Single-burst experiments..showed that about 70% of the yielder cells liberate only normal, 'restricted' T1. The other yielder cells produce a mixture of restricted and unrestricted T1. **1965** *Ann. Rev. Microbiol.* XIX. 366 Phage λ variants can be classified with respect to their state of adaptation..by determination of the efficiency of plating (eop) on various hosts... λ·K has an eop of one on K12 and C, i.e., it is accepted, nonrestricted. The same phage λ·K, however, plates only exceptionally on B or on P1-lysogenic strains: it is said to be restricted. *Ibid.* 367 The whole population of restricted DNA molecules.

f. Of a language system: having a limited syntax and lexicon.

1962 B. BERNSTEIN in *Lang. & Speech* V. 32 Two general types of code can be distinguished: *elaborated* and *restricted*... In the case of an elaborated code, the speaker will select from a relatively extensive range of alternatives... In the case of a restricted code the number of these alternatives is often severely limited. **1964** M. A. K. HALLIDAY et al. *Linguistic Sci.* 96 Some registers are extremely restricted in purpose. They thus employ only a limited number of formal items and patterns... Such registers are known as *restricted languages*. **1968** E. W. GORDON in M. Deutsch et al. *Social Class, Race & Psychol. Devel.* xi. 390 Restricted language..develops as a product of unilateral decision making in the lower-class home. **1971** *Archivum Linguisticum* II. 67 Firth advocated what he called 'partial studies', e.g. the study of newspaper headlines *per se*, in which attention would be drawn to features of the 'restricted language' itself as well as to contrasts between it and other restricted languages. **1975** *Amer. Speech* 1973 XLVIII. 35 It has been further suggested that if speakers of a restricted code do not use such conjunctions, their language and probably their logical processes are somehow deficient.

g. Of a person: not allowed to move about freely; confined to a certain area or certain areas.

1971 *Sunday Express* (Johannesburg) 28 Mar. 9/5 'If a restricted person can satisfy me that he can obtain residence overseas and that he has a bona fide intention of not returning to South Africa, I cannot refuse his request for an exit permit,' Mr. Gerdener said. **1971** *Rand Daily Mail* (Johannesburg) 4 Dec. 13/1 Anglican bishops..are accused of defying the Government by wanting to offer help to people banned or restricted under the Terrorism Act. **1972** *Straits Times* (Malaysian ed.) 24 Nov. 21/2 Unemployed, and a restricted resident, Chua Ali Kow, 32, was sentenced to two years' jail.

restrictee (rīstrikti·). [f. RESTRICT *v.* + -EE¹, after *detainee*.] One whose freedom of movement is restricted, usu. for political reasons.

1959 *Observer* 31 May 6/8 This pay starts in the open camp—no guards, no wire, no warders—at 30s. monthly plus rations. When detainees thus become 'restrictees' they are trade tested and trained for various jobs as clerks or artisans. **1960** *Guardian* 24 Dec. 7/1 The Southern Rhodesian Government must now be keen to release the seven detainees and forty restrictees. **1965** *Spectator* 8 Jan. 35/2 The government can no longer prevent the restrictees from receiving visitors at will and today there is a constant stream on the trains. **1970** *Guardian Weekly* 21 Feb. 6/4 The main legal advantage that restrictees have over detainees is that their term of restriction is defined.

restriction. Add: **1. d.** Deliberate limitation of industrial output.

1888 W. E. NICHOLSON *Gloss. Terms Coal Trade* 71 *Restriction*, an arrangement or understanding among the hewers limiting their day's work to something less than a fair ordinary day's work. **1930** *Economist* 22 Mar. 652/1 At the same time, restriction is being maintained in Oklahoma, while in Texas, although the State Governor views all restriction schemes as a breach of the Anti-Trust laws, a certain amount of voluntary restriction is in force. **1931** *Brit. Jrnl. Psychol.* July 89 Restriction is practised by the non-union worker just as much as it is by the member of a trade union. **1961** *Problems of Progress in Industry* No. 11 p. 11 If their [sc. workers'] standards are lower than those considered as reasonable by managers, such behaviour is usually called 'restriction of output'.

4. *Math.* A function *f* whose domain is a subset of a given function *g*, whose codomain is the codomain of *g*, and for which $f(x) = g(x)$ for all x in the domain of *f*. Also *restriction mapping*.

1949 SPRINGER & POLLAK *Algebraic Topology* viii. 168 Since f is a restriction of g, we have that the map $\bar{g}$ of H(A) into H(B) which is reduced by g is exactly the same as the map f. **1963** D. BUSHAW *Elem. Gen. Topology* 147 If $f: X \to Y$ and $A \subset X$, the function $fj: A \to Y$, where j is the injection map from A to X, is called the restriction of f to A and is denoted by $f|A$. **1979** *Proc. London Math. Soc.* XXXVIII. 208 Recall that $D(G,H) = \mathrm{Ker}\ \rho_H$ where ρ_H: $\mathrm{Der}(G,ZG) \to \mathrm{Der}(H,ZG)$ is the restriction mapping.

5. *Biol.* Limitation of the rate of reproduction of a virus in certain hosts, owing to the destruction of viral DNA by a restriction enzyme.

1962 *Jrnl. Molecular Biol.* V. 47 Host-controlled modification is known to occur in many bacteriophage-host systems and is usually recognized by restriction in the efficiency of plating of the newly modified phage on its former host strain. **1968** *Proc. Nat. Acad. Sci.* LIX. 1305 The complementation studies suggest that restriction activity is conferred by at least two gene functions. **1979** *Nature* 1 Mar. 30/1 Bacteriophage T3 and T7 protect their DNA from restriction by producing, as the earliest detectable phage functions, anti-restriction proteins.

6. Special Comb.: **restriction endonuclease**, **enzyme** *Biochem.*, an enzyme that divides large molecules of DNA only if there is a specific sequence of several nucleotides (usu. four to six in number).

1977 *Sci. Amer.* Dec. 61/1 An important tool for plus-and-minus sequencing and for molecular biology in general was provided by the discovery several years ago of the enzymes called restriction endonucleases, which cleave large DNA molecules into discrete fragments. **1979** *Nature* 20 Sept. p. v, CP Laboratories Limited have in stock for immediate delivery more than 40 restriction endonucleases. **1965** W. ARBER in *Ann. Rev. Microbiol.* XIX. 368 One might like to assume that a highly specific 'restriction enzyme' only initiates the degradation, for example, by cleavage of DNA, and that these cleavage products are then subject to the action of less specific nucleases. **1977** *Time* 18 Apr. 48/1 Different plasmids, sometimes passed from one bacterium to another, can order up still another kind of chemical weapon, a so-called restriction enzyme, which can sever the DNA of an invading virus, say, at a predetermined point.

restri·ctionism. [f. RESTRICTION + -ISM.] A policy of restricting some practice, institution, etc.

1941 *Economist* 5 Apr. 436/2 It would be a disaster if wartime concentration were to leave any legacy of peacetime restrictionism. **1949** *Hansard Commons* 19 May 631, I do not want to say very much more about this matter of restrictionism, save to say that the body of Captain Ludd has long since mouldered in the grave but his ideas still march on in Britain. **1958** *Economist* 20 Dec. 1066/1 There was a prime example of this restrictionism earlier this week, when employers and building unions in the north together condemned the excellent practice whereby some building workers have entered into voluntary contracts to build some additional houses in certain developing areas after hours. **1968** *Physics Bull.* Nov. 394/2 Restrictionism relates to the absence of true competition due to restrictive trade associations etc. **a 1974** R. CROSSMAN *Diaries* (1975) I. 34 Office building was an area where, despite my dislike of restrictionism, I might be seriously prepared to think of physical controls to deal with the scandal.

restrictionist. Add: (Earlier and later examples.) Also *attrib.* or as *adj.*

1820 *Niles' Reg.* XVIII. 258/2 We undertake to say that there is not a single *confessed* restrictionist elected throughout the whole territory. **1941** *Economist* 29 Mar. 406/1 The preservation of the margins of, say, August, 1939, as maxima, may simply do the work of monopolists or restrictionists for them. **1951** *Sun* (Baltimore) 19 Mar. 1/2 A powerful bloc of 'restrictionists' is threatening to strike a successful blow at freedom of information in the name of the United Nations. **1962** *Listener* 8 Mar. 400/1 A restrictionist policy which the unions have denounced as the cause of most of the trouble. **1965** *New Statesman* 30 Apr. 673/1 What is needed is..a restrictionist philosophy in the field of reproduction..coupled with an expansionist philosophy in the field of production and distribution of all the other good things of life.

restrictive, *a.* Add: **2.** Also in *Gram.*, esp. of relative clauses.

1878 REED & KELLOGG *Higher Lessons in English* 98 The Adjective Clause, when not restrictive, is set off by the comma. **1895** *Funk's Stand. Dict.* s.v., A restrictive clause. **1924** O. JESPERSEN *Philos. Gram.* viii. 112 In English..only restrictive clauses can be introduced by *that* or without any pronoun. **1957** *Eng. Stud.* XXXVIII. 101 Restrictive clauses..are linked to their antecedents by close syntactic juncture. **1977** *Language* LIII. 70 The pragmatic distinction that Donnellan labeled the 'referential' and 'attributive' uses of definite descriptions is *not* reflected in the mood of restrictive relative clauses.

3. a. (Earlier *Econ.* example.)

1825 J. S. MILL in *Westm. Rev.* Apr. 412 If the landlords would attend a little to these, and some other effects of the restrictive system, we should no longer hear them clamouring..for a protecting duty of 20, 30, or 40 shillings.

b. *spec.* of a covenant.

1882 *Law Rep. Queen's Bench Div.* VIII. 410 With regard to the question of notice, *Tulk v. Moxhay* shews that a restrictive covenant will be enforced, and so do *Cox v. Bishop* and *Wilson v. Hart*. **1911** *Encycl. Brit.* XVI. 157/1 *Restrictive Covenants*:- These may be subdivided into two classes—covenants not to assign or underlet without the lessor's consent..; and covenants in restraint of trade. **1925** *Act 15 Geo. V* c. 22 § 10 A covenant or agreement (not being a covenant or agreement made between a lessor and lessee) restrictive of the user of land entered into after the commencement of this Act (in this Act referred to as 'a restrictive covenant'). **1935** *Discovery* Aug. 227/1 A new policy of preserving land by means of restrictive covenants has recently been adopted [by the National Trust], in addition to the older and more expensive method of purchase. **1953** [see *LILY-WHITE *a.* 2]. **1976** *Evening Post* (Nottingham) 15 Dec. 4/4 Mansfield District Council has decided to go ahead with its plan to close the market..despite the failure of their application to the Land Tribunal for the removal of a 100-year-old restrictive covenant on the site. **1979** *Internat. Jrnl. Sociol. of Law* VII. 339 In drafting the contract a planning authority condition that the future occupants do not erect fences has to be translated into a restrictive covenant.

c. *restrictive practice*: an arrangement in industry and trade which restricts or controls competition between firms; an arrangement by a group of workers to limit the output or restrict the entry of new workers: regarded by others as preventing labour or materials from being used in the most efficient way. Hence *restrictive practitioner*.

1928 *Britain's Industr. Future* (Liberal Industr. Inquiry) xiii. 146 The prevalence of these restrictive practices has varied very widely from trade to trade. **1946** *Sun* (Baltimore) 19 Feb. 10/3 The single argument on which all the restrictive practices have rested..was that the demand for housing was so limited that the various factors in the housing field in simple self-protection had to stretch the work out and keep costs high. **1948** *Act 11 & 12 Geo. VI* c. 66 § 1 For the purposes of this Act there shall be constituted a Commission, to be called the Monopolies and Restrictive Practices Commission... The Commission shall consist of not less than four nor more than ten members to be appointed by the Board of Trade. **1958** *Spectator* 31 Jan. 133/1 These two trades have been in the forefront as restrictive practitioners. **1964** *Mod. Law Rev.* XXVIII. 337 The House of Lords..handed down an important decision on the question of restrictive practices. **1966** *Economist* 12 Mar. 979/2 Mr Heath would apparently like to hand the supervision of labour relations over to the lawyers, the Tories' favourite restrictive practitioners. **1969** J. ARGENTI *Managem. Techniques* 207 Others blame labour for resisting changes to working methods and manning levels and for perpetuating restrictive practices. **1978** *Jrnl. R. Soc. Arts* CXXVI. 406/2 The moves made by successive Tory Presidents of the Board of Trade in the late '50s and early '60s to curb restrictive practices of various kinds.

restrictiveness. For *rare*⁻¹ in Dict. read *rare* and add later examples.

1978 *Language* LIV. 406 The second desideratum is the goal of restrictiveness: the notion that a theory of language should characterize just the class of possible human languages, and no more. **1980** A. N. WILSON *Healing Art* xviii. 223 Both of them found..emotional restrictiveness necessary.

restrictivist (rīstri·ktivist), *a.* [f. RESTRICTIVE *a.* + -IST.] Characterized by restriction; limiting.

1966 *New Statesman* 21 Oct. 574/1 Cost-inflation..forces the government to adopt restrictivist policies. **1977** P. JOHNSON *Enemies of Society* viii. 115 Part of the genetic coding of all children is knowledge of the highly restrictivist principles of universal grammar.

restrictor (rīstri·ktǫɹ). [f. RESTRICT *v.* + -OR.] **1.** One who restricts or advocates restriction.

1825 J. S. MILL in *Westm. Rev.* Apr. 291 It is a proposition which the restrictors themselves do not venture to dispute. **1952** *Chambers's Jrnl.* Apr. 230/1 Restrictive covenants in house or land deeds may be a curse to any buyer... It can happen that covenants are interpreted in a peculiar manner by the restrictors.

2. A device for restricting the flow of a fluid, e.g. by means of a porous medium or an orifice. Freq. *attrib.*

1940 E. MOLLOY *Aeroplane Maintenance & Operation* I. 105 *Restrictor jets*, these..are situated in two recesses opposite the lower edge of the throttle when in the closed position. **1961** R. V. WATTS in D. S. Carton et al. *Rocket Propulsion Technol.* I. 54 By opening the..restrictor, it became possible to set minimum thrust at the required angle. **1966** L. A. H. EASTMAN tr. *G. Schenkel's Plastics Extrusion Technol. & Theory* xiii. 368 The sheet die developed in the U.S.A. for the processing of impact-resistant polystyrene..has a flexible restrictor bar fitted in the upper part of the die, transverse to the flow direction. **1979** *Wear* LIV. 331 The stiffness of hydrostatic [journal] bearings is greatest when valve restrictors are used and decreases progressively with orifice and capillary restrictors.

restrike, *sb.* Add: Also, a reimpression of a print or medal.

1912 *Chambers's Jrnl.* May 327/2 Restrikes of these medals are not uncommon, as it is only within recent years that the Calcutta mint has been prohibited from issuing them. **1965** ZIGROSSER & GAEHDE *Guide to Collecting Orig. Prints* iv. 62 Prints, either late restrikes from the plate or genuine old uncolored impressions, have often been colored by a later hand. **1970** *New Yorker* 20 June 63/2 The reason for this low price..is that the print is a restrike. This means that it was made after the artist's death from the original plate. **1980** *Daily Tel.* 16 Feb. 27/6 In the case of badges, 'restrikes' are a common hazard.

2. *Electr. Engin.* The re-ignition of an arc.

1962 *Newnes Conc. Encycl. Electr. Engin.* 132/2 Suppression of the arc before natural current zero causes severe overvoltages and high frequency multiple restrikes.

restrike, *v.* Add: **b.** *intr. Electr. Engin.* Of an arc: to strike again (STRIKE *v.* 76 a). Also *trans.* (causatively).

1937 [implied in RESTRIKING *vbl. sb.* below]. **1942** A. ARNOLD *Switchgear Pract.* ix. 109 The arc will again restrike. **1955** E. MOLLOY *Electr. Engineer's Ref. Bk.* (ed. 8) vii. 24 It is the rate of rise of the restriking voltage that determines whether or not the arc will restrike. **1962** *Newnes Conc. Encycl. Electr. Engin.* 132/1 To achieve successful interruption [of alternating currents] it is necessary to prevent the arc restriking by separating the contacts to a distance which will withstand the restriking voltage. **1977** R. L. LITTLE *Metalworking Technol.* v. xxiv. 279/1 Restriking an arc... The welder strikes the arc approximately ¼ to 1 in. in front of the crater on the side opposite the weld bead.

Hence **restri·king** *vbl. sb.*, *esp.* in *Electr. Engin.*; *restriking voltage*, in a circuit-breaker, the value of the high-frequency voltage surge across the contacts following their separation and the extinction of the subsequent arc between them.

1937 TODD & THOMPSON *Outdoor High Voltage Switchgear* viii. 172 An important factor controlling restriking voltage is the type of machine..supplying the circuit. **1958** J. SHEPHERD et al. *Higher Electr. Engin.* xvi. 405 In oil circuit breakers, where dielectric strengths are normally higher than in air circuit breakers, re-striking may not take place.

restru·cture, *v.* [RE- 5 c.] *trans.* To give a new structure to; to organize into a new pattern; to rebuild, re-arrange. Hence **restru·ctured** *ppl. a.*; **restru·cturer**; **restru·cturing** *vbl. sb.*

1951 K. S. LASHLEY in L. A. Jeffress *Cerebral Mechanisms in Behavior* 119 Some children become very facile at such inversions of words, and re-structure new words without hesitation. **1951** G. HUMPHREY *Thinking* vi. 181 If every response has an element of newness,..why is not restructuring present all the time? **1958** W. J. H. SPROTT *Human Groups* iii. 48 The 'life-space' is constantly changed with 'locomotion', and also with changing awareness, in which case it is often said to be 'restructured'. **1962** *Economist* 22 Sept. 1123/2 The most comprehensive restructuring of international liquidity arrangements since the IMF was founded. **1967** *Observer* 19 Nov. 4/7 Wilson..thought that..his policy of controlled expansion, combined with the 'restructuring' of industry and a better balance between the regions, would in the end do the trick. **1969** *Listener* 17 July 65/3 He [sc. the Director-General] did not restructure the radio services of the BBC. **1970** *Times Lit. Suppl.* 23 Apr. 443/2 Lord Robbins is a tough-minded re-structurer. **1976** P. R. WHITE *Planning for Public Transport* x. 208 The question of restructuring private car costs to make them comparable in form is discussed below. **1977** *Belfast Tel.* 22 Feb. 6/2 Proposals for the re-structured council are being studied by the Secretary of State. **1978** W. F. BUCKLEY *Stained Glass* xxii. 212 He spent the first half hour with Overstreet, who pronounced the restructuring of the chapel's trussed roof complete.

‖ **Reststrahlen** (rest₁ʃtrā·lǝn), *sb. pl. Physics.* Also as sing. **Reststrahl** and with small initial. [Ger., residual rays, f. *rest* remainder + *strahlen* rays.] Electromagnetic radiation which is selectively reflected from the surface of a crystalline solid when the frequency of the incident radiation is nearly equal to the frequency of vibration of the ions constituting the solid.

1910 *Phil. Mag.* XIX. 761 For the investigation of the extreme infra-red portion of the spectrum the method of 'Reststrahlen' has proved most fruitful. **1942** *Rep. Progr. Physics* VIII. 28 To compare the high-frequency end of the spectrum..with that obtained from other data, e.g. from *Reststrahlen* in the case of ionic crystals. **1965** M. GARBUNY *Optical Physics* v. 285 (*caption*) Filter action of crystallite powders in their reststrahlen region. **1967** *Physical Rev. Lett.* XVIII. 601/2 The sample is opaque in the neighborhood of 23 meV due to *Reststrahl* reflection.

restu·dy, *sb.* [f. the vb.] The act of studying again.

1961 *Newark* (N.J.) *Even. News* 21 Nov. 24/2 (*heading*) N.J. fights restudy of WNTA sale stay. **1962** E. SNOW *Red China Today* (1963) liii. 402, I met a writer less well known than Hsiao Ch'ien, who had completed his 'restudy of the sources of art' in a commune. **1966** *Economist* 22 Jan. 309/1 When the commission sounded the retreat and sent the headache back to the engineers for 're-study', the state engineer suggested drily that the study would take just about a year. 1979 *Nature* 20–27 Dec. 832/1 A restudy of this genus has convinced us that it was based on part of the skull roof of a specialised placoderm.

resty, *a.*¹ **1. a.** (Further examples.)
1920 A. Huxley *Leda* 40 The machine is ready to start. The symbolic beasts grow resty, curveting where they stand. 1977 J. Aiken *Five-Minute Marriage* viii. 126 He guided his horses around the corner... The team appeared to be a trifle fresh and resty.

resty·le, *v.* [Re- 5 a.] *trans.* To style again; to give a new style to. So **resty·led** *ppl. a.*, **resty·ling** *vbl. sb.*
1934 in Webster. 1958 *Listener* 19 June 1006/2 The development of new equipment and the re-styling of the philosophy of design in this new-born industry are a fascinating branch of chemical engineering. 1958 *Times* 22 Oct. 5/5 Details like the front grille and the tail lights have been restyled. 1960 *Farmer & Stockbreeder* 8 Mar. 136/1 Entire pig herd replaced in re-styled buildings. 1965 D. Francis *Odds Against* xx. 249 Her hair had been re-styled... It.. curved in a bouncy curl. 1977 *Lancashire Life* Nov. 153/1 The Granadas have been completely restyled, and now look remarkably like the Audi 100. 1978 'M. Yorke' *Point of Murder* i. 15 She went..for restyling and emerged bouffant.

resuing (risiū·iŋ), *vbl. sb. Mining.* [Etym. unknown.] A method of stoping in which the rock wall adjacent to a narrow vein is removed before the vein itself, so that the ore can be extracted in a cleaner condition.
1909 H. C. Hoover *Princ. Mining* x. 101 (*heading*) Resuing. 1910 W. R. Crane *Ore Mining Methods* ii. 38 Resuing consists in opening up the stopes not in the vein but in the wall-rock, by whatever method of stoping seems best adapted to the existing conditions. *Ibid.* 43 Resuing is applicable to very narrow veins alone, *i.e.*, under 30 inches in width; its chief advantage being that a cleaner grade of ore can be mined than when both vein and walls are broken together. 1973 L. J. Thomas *Introd. Mining* vi. 188 In some cases narrow veins may be taken by resuing..[which] can be classed as either cut and fill, or as vein mining.

result, *sb.* Add: **3. d.** Usu. *pl.* The final marks, scores, and placings in (*a*) an examination, (*b*) a sports event.
1916 Joyce *Portrait of Artist* (1969) v. 210 Did you hear the results of the exams? 1937 Partridge *Dict. Slang* 695/1 *Results*, news of sports results. 1955 *Radio Times* 22 Apr. 42/2 Sport. Today's results and weekend preview. 1968 *Ibid.* 28 Nov. 8/5. 4.55 Racing Results. 1977 *Belfast Tel.* 28 Feb. 9/3 The following are the results of the November exams held by the Institute of Cost and Management accountants.

e. *pl.* Favourable or desired consequences. Also *sing.*, a good or favourable result against an opponent.
1922 E. O'Neill *Hairy Ape* vii. 73 Take some of those pamphlets with you to distribute aboard ship. They may bring results. 1927 —— *Marco Millions* III. i. 167, I kept my nose to the grindstone every minute... And I got results. 1931 *Punch* 18 May (*caption*) The charming young gold-digger who expected results of an Aberdonian. 1973 E. Dunphy *Only a Game?* (1977) ii. 52, I think we will get a result at Preston. 1976 *Observer* 21 Nov. 23/1 We needed a result... Perhaps we should have done better than win 1–0.

resultative (rizʌ·ltătiv), *a. Gram.* [f. Result *sb.* + -ative.] Expressing result. Also *absol.* as *sb.*
1926 H. Poutsma *Gram. Late Mod. Eng.* II. II. lviii. 545 The attributive past participle mostly has a momentaneous or resultative aspect. 1936 *Jrnl. Eng. & Germanic Philol.* XXXV. 368 The so-called Resultative Perfect requires a somewhat detailed examination. 1957 R. W. Zandvoort *Handbk. Eng. Gram.* I. iv. 61 English shares with other languages the use of the *resultative perfect*, which denotes a past action connected, through its result, with the present moment. I've bought a new car. 1963 F. T. Visser *Hist. Syntax Eng. Lang.* I. iv. 582 Examples of Old English verbs construed with a resultative predicative adjunct are not numerous. 1965 *Language* XLI. 109 In Indic, the perfect, whose function was originally stative,..developed..to a resultative. 1970 J. W. R. Lindemann *Old Eng. Preverbal Ge-* I. 4 *Ge-* may convert an intransitive verb into a resultative verb that is transitive. 1977 *Canad. Jrnl. Linguistics* Spring 51 Resultatives..represent a second category where *hə* occurs instead of a resonant plus shwa in reduplicative prefixes.

resume, *v.* Add: **I. 2. b.** Also with direct speech: to go on to say.
1765 H. Walpole *Castle of Otranto* i. 16 Yes, I sent for you on a matter of great moment, resumed he. 1789 J. Moore *Zeluco* I. xxii. 132 'Nay, my good friend,' resumed the Physician, 'it is a matter of indifference to me, what you do or do not believe.' 1850 F. E. Smedley *Frank Fairlegh* xliii. 376 'I have fancied that illness was beginning to sour your temper,' I replied. 'Illness of mind, not body,' he resumed. 1906 *Smart Set* May 9/2 'I guess, friend,' resumed the man with the pipe, 'she's been standin' out here coolin' off for some time, ain't she?' 1922 Joyce *Ulysses* 620 Mind you, I'm not saying that it's all a pure invention, he resumed. 1976 R. Hill *Another Death in Venice* I. iii. 63 'I'll say this, though,' resumed Wilf, 'I'm worried.'

résumé, *sb.* Add: Also with pronunc. (re-). Also **resume. b.** Chiefly *N. Amer.* = *curriculum vitæ* s.v. *curriculum. Also *fig.*
1961 Webster, *Résumé..specif.*: a brief account of one's education and professional experience. 1968 *Globe & Mail* (Toronto) 17 Feb. 51 If an interview is not convenient at this time, forward your resume, in confidence to Mr. Grossman. 1971 Golzen & Plumbley *Changing your Job after 35* viii. 86 The Résumé..will vary considerably with the type and level of job and can be the bare bones of a c.v. or a long, narrative account of your main achievements written up with a special bias. 1973 J. Ryder *Trevayne* (1974) iii. 384 There was an opening. What could look better on a résumé than the White House? 1976 *Glasgow Herald* 26 Nov. 25/1 Please submit detailed resume including personal data, educational background, and work experience. 1979 *Tucson Mag.* Feb. 88/2 She has added several fine credits to her resume since then, including a Washington D.C. debut this year.

resumptive, *a.* and *sb.* Add: **3.** *Gram.* **a.** In Jespersen's terminology: (see quot. 1917). **b.** Indicating resumption of a topic, etc.; having previous reference. So as *sb.* (see quot. 1954).
1917 O. Jespersen in *Historisk-Filologiske Meddelelser* I. v. 69 A second class comprises what may be termed *resumptive negation*, the characteristic of which is that after a negative sentence has been completed, something is added in a negative form with the obvious result that the negative effect is heightened. 1954 D. Bolinger in *Boletín de Filología Universidad de Chile* VIII. 48 Sometimes, for special effects, a presupposed element, even a lengthy one, is repeated though specifically known from the immediate context... We may call such a verbatim or near-verbatim presupposed element a 'resumptive'. *Ibid.* 49 An element which is explicitly resumptive comes after prosodic stress. 1957 *Publ. Amer. Dial. Soc.* xxviii. 145 A post-accentual resumptive is only a repetition, while a pre-accentual resumptive may be more. 1959 J. C. Catford in Quirk & Smith *Teaching of English* (1964) vii. 149, I know of no practical English grammar for foreign learners which describes the use of *oh*, *ah* or the introductory or resumptive *well*. 1970 M. Dahood *Anchor Bible: Psalms III* 337 M[assoretic] T[ext] *'asapperennāh* can be explained as employing the resumptive pronominal suffix. 1975 *Language* LI. 59, I regard Spanish resumptive intonation and beginning position as variant formal means of expressing syntagmatic complexity.

resupply, *v.* Add: Also *absol.*, to take on or acquire a fresh supply.
1977 'J. McVean' *Bloodspoor* xiii. 152 We'll have to resupply. We're mobile and there must be wells. *Ibid.* xviii. 218 They were down to the last few pints but..they'd be able to resupply from one of the tributary streams.

resu·rface, *v.* [Re- 5 a, c.] **1.** *trans.* To provide (a road, etc.) with a fresh surface.
1886 [implied in resurfacing *vbl. sb.* below]. 1894 *Westm. Gaz.* 23 Apr. 8/3 (Advt.), Old blocks bought up, sold, or resurfaced. 1901 *U.S. Dept. Agric. Yearbk.* 1900 352 When the road was resurfaced with limestone..it became excellent. 1929 *Daily Express* 11 Jan. 2/2 If these minor roads were..strengthened and resurfaced in accordance with modern road practice [etc.]. 1960 *Times* 4 July (Advt. Suppl.) 1/3 Safety razors had not resurfaced the New Man. 1973 'C. Aird' *His Burial Too* iii. 31 The Divisional Surveyor decided to resurface the road.

2. *intr.* To come to the surface again. Also *fig.*
1953 P. C. Berg *Dict. New Words in Eng.* 137/1 *Resurface..v.i.*, of a submarine: to come to the surface. 1962 *Times* 3 Apr. 14/6, I would re-surface before the water-splash fell back into the tank. 1968 J. Sangster *Foreign Exchange* ii. 57 It was midday when I resurfaced, too late for breakfast. 1971 *Flying* Apr. 73/2, I would resurface to find the airplane starting to turn back. 1973 L. Snelling *Heresy* II. i. 62 We lose sight of him... Killed by the Gestapo? In an air-raid? In any event he never resurfaced. 1978 J. McDowell in Hookway & Pettit *Action & Interpretation* 151 If it were to be so incorporated, all the difficulties of the relation between behaviour and action categories would presumably resurface.

So **resu·rfacing** *vbl. sb.*
1886 *Cyclist* 4 Aug. 1076/1 The re-surfacing of the Crystal Palace path. 1967 *Antiquaries Jrnl.* XLVII. 271 This structure had been..covered with a plaster surface, perhaps later than the surface of the atrium described above, but present over a sufficiently large area to indicate a general resurfacing. 1978 *Lancashire Life* Mar. 54/3 George's Lane and Belmont Road were selected for resurfacing.

‖ **resurgam** (resɔ̄·ɪgæm). [L.] 'I shall rise again', expressing Christian faith in resurrection at the Last Day. Usu. *transf.* and *fig.*
1662 J. Trapp *Annotations Old & New Testament* I. 142 Howbeit he had *hope in his death*, and might write *Resurgam* on his grave. 1847 Thackeray *Van. Fair* (1848) xiv. 126 Arms and Hatchments, Resurgam.—Here is an opportunity for moralizing! 1853 Geo. Eliot *Let.* 3 Jan. (1954) II. 79 *Now* nothing seems pleasant to me but—*Resurgam*. c 1859 E. Dickinson *Poems* (1955) I. 56, I slew a worm the other day—A 'Savant' passing by Murmured 'Resurgam'—'Centipede'! 'Oh Lord—how frail are we'! 1929 *Oxf. Poetry* 39 Arise..resurgam..for another day.

resurge, *v.* Add: (Further examples.) Hence also **resu·rged, resu·rging** *ppl. adjs.*
1912 R. Brooke *Let.* 23 Nov. (1968) 408, I shall be in Cambridge.., dispatching my resurged Dissertation. 1962 G. T. Warwick in C. Cullingford *Brit. Caving* (ed. 2) v. 184 The clear water from Legalough, which..may resurge at Hanging Rocks. 1965 *New Statesman* 30 Apr. 690/3 That both bomber fleets resurged so mightily in 1944 was due.. to the rapid conversion of the Mustang into a long-range escort. 1976 K. Royce *Bustillo* vii. 87 He fought off the resurging need for a drink. 1980 *Times* 1 Nov. 15/3 Mr Manley..leads a rump opposition smaller than that from which Mr Seaga has resurged.

resurgence. Add: **2.** The fissure through which a stream re-emerges at the end of an underground part of its course; the re-emergence of such a stream. [This sense results from the adoption of F. *résurgence* (cf. A. Vandel 1920, in *Bull. de la Soc. zool. de France* XLV. 46).]
1954 W. D. Thornbury *Princ. Geomorphol.* xiii. 327 The terms rise and resurgence have been applied to the reappearance of surface waters which have been diverted to underground routes. 1963 D. W. & E. E. Humphries tr. *Termier's Erosion & Sedimentation* xiv. 303 Sometimes the surface water plunges down into the underground system by way of a sink hole... The river, however, retains its individuality and may return to the surface through a resurgence or spring. 1965 *Geogr. Jrnl.* CXXXI. 37 This subterranean stream maintains a constant flow during all weather conditions, and is joined by a small seepage resurgence and streamlet within a large bedding plane cave.. through which it flows to the main resurgence. 1971 J. N. Jennings *Karst* v. 74 A useful distinction can be made between exsurgences fed entirely by seepage waters from the karst and resurgences supplied by the sinking of surface streams.

resurgent, *sb.* and *a.* Add: **A.** *sb.* **2.** = *resurgence 2. Also *attrib.*
1965 B. E. Freeman tr. *Vandel's Biospeleol.* i. 12 The outlets of large underground rivers are termed resurgents. 1972 Herak & Stringfield *Karst* xiii. 435 Resurgent caves..are associated with the uprising of water around the flanks of the Mendips.

B. *adj.* **2.** *Geol.* Applied to steam and other gases which after being absorbed by volcanic magma from groundwater and native rock are subsequently released into the atmosphere.
1908 R. A. Daly in *Amer. Jrnl. Sci.* CLXXVI. 48 These fluids were deposited and buried in the strata. They have been resurrected in their activity. They have 'risen again', both literally and figuratively; they may be called 'resurgent' emanations... All 'resurgent' emanations are of secondary origin. 1917 *Econ. Geol.* XII. 491 'Resurgent' was adapted by the writer..to signify the magmatic emanations of secondary origin, that is, those absorbed from country rock... Von Wolff..extends it to describe also certain pyroclastic deposits... Most authorities on the genesis of ore deposits appear to be opposed to the concept. 1932 F. F. Grout *Petrogr. & Petrol.* iii. 212 Magmas may acquire gases by assimilating or dissolving..some wall or roof rock that contained gas or water. This is 'resurgent water'.

resurrect, *v.* Add: **1. c.** (Earlier and later examples.)
1852 B. Young *Jrnl. of Discourses* (1854) I. 33/1 We shall not want to look upon our past actions; we shall say..I do not want that to be resurrected, but let it die in the grave. 1904 *Forum* July 132 The..offer made by General Reyes in behalf of the Bogota Government to resurrect and ratify the dead canal treaty. 1942 Z. N. Hurston in A. Dundes *Mother Wit* (1973) 31/1 They resurrected a joke or two and worried it like a bone.

2. (Further *fig.* example.)
1969 G. M. Brown *Orkney Tapestry* 52 The ribs of crag and tree Resurrecting with birds.

resurrected, *ppl. a.* Add: **1.** (Earlier example.)
1852 H. B. Kimball in B. Young *Jrnl. of Discourses* (1854) I. 355/2 You never will obtain your resurrected bodies, until you bring your spirits into subjection.

2. *Geomorphol.* Of a land form: exposed by erosion after having been covered by deposition.
1925 D. W. Johnson *New England–Acadian Shoreline* ii. 27 Resurrected peneplane shorelines appear to be fairly common along the coast of Acadia. 1954 W. D. Thornbury *Princ. Geomorphol.* ii. 25 Most resurrected features are of local extent and constitute a small portion of the present-day topography. 1970 R. J. Small *Study of Landforms* iii. 106 One of the most important types of escarpment associated with faulting is the 'resurrected' or 'exhumed' fault-scarp or fault-line scarp.

resurre·cting, *vbl. sb.* [f. Resurrect *v.* + -ing¹.] The action of the verb Resurrect.
1906 P. Lowell *Mars & its Canals* xii. 130 To call the lunar *maria* seas may..be..only a resurrecting in epitaph what was the truth in its day.

resurrection, *sb.* Add: **II. 5. a.** *resurrection appearance.*
1931 W. Temple *Thoughts on Some Probl. of Day* i. 19 The love revealed in Jesus Christ in His use of divine power..in His Death on the Cross, in His Resurrection-Appearances only to those whose love He had already won. 1977 G. W. H. Lampe *God as Spirit* vi. 151 Resurrection appearances and empty tomb cannot in themselves furnish first-hand evidence for God's vindication of Jesus.

c. resurrection plant: (*a*) (earlier example); (*b*) (earlier and later examples); (*c*) one of several other plants which stay quiescent during drought and revive when moistened.
(*a*) 1868 A. Gray *Field, Forest, & Garden Bot.* 374 'Resurrection-Plant'..is a nest-like ball when dry, but when moist it unfolds. (*b*) 1857 G. W. Johnson *Cottage Gardeners' Dict.* (ed. 2) 34/2 When it [*sc.* the rose of Jericho] alights in water, or on damp ground, the branches relax and open out, as if its life was renewed; hence its name Resurrection Plant. 1951 *Dict. Gardening* (R. Hort. Soc.) IV. 1753/1 The Rose of Jericho, *Anastatica Hierochuntica*, is sometimes called a resurrection plant. (*c*) 1902 L. H. Bailey *Cycl. Amer. Hort.* IV. 1507/1 Resurrection plants are great curiosities... The commonest ones are members of the mustard family and the club moss family. 1974 F. N. Howes *Dict. Useful & Everyday Plants* 217 Resurrection plant.

Several pl[ant]s that appear dead after drought or a dry season but revive again w[ith] rain or moisture have acquired this name.

d. resurrection pie (*transf.* example).
1903 G. B. Shaw *Let.* 11 Sept. (1972) II. 370 We have both got the same job..to strike out a line for the advanced guard that is..neither Manchester resurrection-pie on the one hand nor Protectionist resurrection-pie on the other.

e. resurrection fern, one of several ferns that survive drought, esp. the grey polypody, *Polypodium polypodioides*, of the southern United States.
1909 *Cent. Dict.* Suppl. 467/1 Resurrection-fern..contracts during drouth but revives in moist seasons. 1924 J. A. Thomson *Sci. Old & New* v. 30 The 'resurrection fern'.. curls up its fronds in drought, and uncurls them when the rains return. 1963 B. Cobb *Field Guide to Ferns* 56 Even though they wither in a drought they promptly become green again after getting moisture, and are therefore often referred to as Resurrection Ferns.

resuscitable, *a.* Delete *Obs. rare*—¹ and add further examples. Hence **resuscitabi·lity.**
1842 Carlyle *Let.* 12 Jan. (1904) I. 250 It lies buried under two centuries of quackeries, scepticisms, owleries,— *not* resuscitable. 1908 *Jrnl. Exper. Med.* X. 373 The resuscitability of the animal in such an instance clearly depends on the perfection of the technique employed to resuscitate it. 1919 W. De Morgan *Old Madhouse* xvii. 274 Flinder's mill-pool yielded when dredged a resuscitable corpse. 1971 H. Pacy *Road Accidents* iii. 73 As with resuscitability, 'life-and-death' decisions are required at the scene of accidents.

resuscitator. Add: **1.** (Further examples.)
1969 J. Secor *Patient Care in Respiratory Probl.* iv. 165 The nurse's role as a resuscitator in the event of cardio-pulmonary arrest has been an issue since the introduction of the procedure. 1977 *Lancet* 1 Jan. 9/1 It may well be that more intelligent parents are better resuscitators.

2. An apparatus used for resuscitation after asphyxia or arrest of respiration.
1929 *Jrnl. Amer. Med. Assoc.* 16 Nov. 1583/1 The E. and J. resuscitator is not a desirable apparatus for use by fire departments. 1938 *Surg., Gynecol. & Obstetr.* LXVI. 721/2 The ideal mechanical contraption for resuscitation is an apparatus which combines an inhalator and a resuscitator. 1943 *Science* 24 Dec. 548/1 The inventors of the pulmotor assumed, and the promoters of 'resuscitators' still claim, that by artificially forcing the lungs and chest through movements like those of breathing, a return of natural respiration should be induced. 1965 H. H. Bendixen et al. *Respiratory Care* xii. 122 Artificial ventilation is started by bag and mask. A self-inflating bag or a bellows resuscitator is suitable... A number of mechanical resuscitators are available.

ret, *sb.*¹ (Later examples.)
1949 *Publ. Amer. Dial. Soc.* xi. 62 *Ret*, *n.* and *v.t.*, a special form of *rot*. The process by which the stalk is prepared for separating the fibre—the rotting of the woody stalk. 1958 *New Biol.* XXVII. 15 In most countries the deseeded flax straw is now retted in warm water in concrete tanks. In the anaerobic ret largely practised in Belgium, the tanks are filled with air-dried straw which is then covered with water at a temperature of 18–27° C.

ret, *sb.*² (See quots.)
1874 Hotten *Slang Dict.* (new ed.) 268 *Ret*, an abbreviation of the word *reiteration*, used to denote the forme which, in a printing-office, backs or perfects paper already printed on one side. 1960 G. A. Glaister *Gloss. Bk.* 346/1 *Ret*, the second side of a sheet of paper.

ret., abbrev. of Retired *ppl. a.* 6. cf. *retd., ret'd.
a 1912 W. T. Rogers *Dict. Abbrev.* (1913) 166/1 *Ret.* (gen.), retired;—return. 1973 'D. Shannon' *No Holiday for Crime* v. 73 Lieutenant Colonel (ret.) Archer Pound. 1978 R. Condon *Bandicoot* i. 9 Captain Colin Huntington, R.N. (ret.).

retable. (Later examples.)
1965 C. D. Eby *Siege of Alcázar* (1966) vi. 122 Arab jewels, ceramics, Flemish retables. 1979 *Dædalus* Summer 136 He remembered having noticed at the foot of the retable 'a piece of a host about the size of a small coin'.

retablo. (In Dict. s.v. Retable.) Add: (Further examples.) Also, a votive picture displayed in a church.
1906 R. Fry *Let.* 13 Sept. (1972) I. 269 They [*sc.* tombs] are..in the middle of the choir with a huge gold and blue *retablo* behind them. 1930 *Mexican Arts* (Amer. Fed. Arts) 43 A type of Mexican painting that deserves special attention..is the retablo, or votive picture offering. 1939 *Mexican Art & Life* Apr., 'Retablos' in Mexico are little stories of religious character told in pictures and always represent a happening in which great misfortune was threatened but averted through the opportune intervention of some saint invoked by the person in distress. 1950 G. Brenan *Face of Spain* ii. 55 The whole of one end [of a church] was taken up by a vast gilt retablo, carved and scrolled and ornamented, in the centre of which..stood the miracle-working Virgin. 1965 C. Causley in *New Statesman* 12 Nov. 738/2 Christ hung down like a hawk-moth caterpillar,..As I walked through the glittering Precinct All the retablos burned like gold. 1976 *Arizona Republic* 16 May N-3/1 There is a companion exhibit of Mexican folk *retablos*, 70 images on tin plates as produced by various Mexican folk artists. 1980 *Times Lit. Suppl.* 9 May 535/3 Peeling, retablo-like posters of high-kicking chorines.

retail, *sb.*¹ (and *a.*). Add: **3.** (Earlier example.)
1851 D. Jerrold *Retired from Business* I. 6 And wholesales don't mix with retails? I think I see.

4. a. (Further examples.) Also *fig.*
1785 *Daily Universal Register* 1 Jan. 3/4 R. Croft, Taylor.. at his wholesale and retail warehouse..is now selling ladies' Italian Coats. 1848 Mill *Pol. Econ.* I. III. i. 520 The influence of those causes is ultimately felt in the retail markets. 1898 *Kansas City* (Missouri) *Star* 18 Dec. 2/1 Hatters, garment workers, shoe workers, retail clerks and textile workers. 1926 A. E. Taylor *Plato* xv. 379 This enables us to define the sophist again as a retail exporter of the knowledge of goodness.., though we must add that he sometimes retails his merchandise in the home market. 1926 *Times* 6 May 3/3 Coal is not being moved by rail, but retail distribution was being carried on in London yesterday. 1940 G. Crowther *Outl. Money* iii. 90 The second value of money which is usually distinguished is the value of money in buying the goods and services which the ordinary family consumes... This second variety can be called the retail value of money, or the cost of living. 1957 *Practical Wireless* XXXIII. 517/2 The British Radio Equipment Manufacturers' Association,..in their monthly retail survey. 1962 *Listener* 13 Sept. 376/2 Every famous jazz composer or retail-chain owner. 1967 G. Wills in Wills & Yearsley *Handbk. Managem. Technol.* x. 194 *Retail audit*,.. continuous research with a panel of retailers to study inventory levels and sales of products over the counter. 1970 *New Society* 5 Mar. 383/3 Retail margins (the difference between the price paid by the shopkeeper and the price paid by the consumer) had previously been gradually rising. 1973 *Times* 25 Apr. 33/5 The brief includes a supermarket of about 40,000 sq ft., other retail units, a night club. 1976 *Daily Tel.* 20 July 1/4 Retail trade showed a slight recovery last month. 1979 *Oil Majors in 1978* (Shell Internat. Petroleum Co.) 7 Gulf suffered reduced sales in Canada but increased them in Europe, partly owing to its acquisition of Mobil's retail outlets in Switzerland.

d. *Comb.*, as **retail price index,** an index of the variation in the prices of retail goods (see *index *sb.* 9 e); **retail price maintenance** = *resale price maintenance* s.v. *resale *sb.* 2.
1924 *University Jrnl. Business* June 263 In the construction of retail price and of cost-of-living indexes..significant developments have occurred. 1935 J. H. Cover *Retail Price Behavior* 86 Though it appears that the median may be more satisfactory for a retail price index, the mean is more logical for a cost-of-living index. 1974 *Times* 22 Mar. 17/3 Even by January, the retail price index had climbed half way to that seven per cent trigger point. 1938 S. Chase *Tyranny of Words* xiv. 175 The new laws for retail price maintenance. 1966 A. Battersby *Math. in Managem.* ix. 232 The appeal to elasticity..has given rise to the fierce arguments about Retail Price Maintenance and the dilemma of those supporters of a competitive economy who attempt to eliminate the effects of free competition.

retailed, *ppl. a.* (Later example.)
1811 Jane Austen *Sense & Sens.* III. xi. 225 It was neither in Elinor's power, nor in her wish, to rouse such feelings in another, by her retailed explanation, as had at first been called forth in herself.

retailing, *vbl. sb.* (Later example.)
1951 in M. McLuhan *Mech. Bride* (1967) 36 The specifications required to reach the goals now in sight for businesses of every description—manufacturing, distributing, retailing, servicing!

retain, *v.* Add: **I. 2. b.** (Later example.) *retain-and-transfer* attrib. phr., in *Association Football* (see quot. 1965).
1938 C. E. Sutcliffe et al. *Story of Football League* xiii. 120 A result which had completely vindicated the retain and transfer system and declared it to be legal. 1965 *Listener* 1 July 17/1 A 'retain and transfer' system... A player employed by a club could be 'retained' after the expiration of his contract of employment. No other club was then allowed to employ him, although his own club had no obligation to re-employ him. He could also be placed on a 'transfer' list, which signified that his club was willing to transfer him for a specified fee, only a small portion of which went to the player himself. 1974 *Scholarly Publishing* V. 235 The Times New Roman Type..was unacceptable, and consultants had to be retained to redesign it.

4. *absol.* (Later examples.)
1910 E. B. Titchener *Textbk. Psychol.* II. 405 The quick learner appears to retain as well as the slow. 1932 *New Yorker* 23 July 17/3 Prior to 1882, even a boy who didn't retain very well could make a kite out of two or three sticks.

retained, *ppl. a.* Add: **2.** (Later examples.) *retained object*, an object of a passive verb; *retained profit* (see quot. 1974).
1934 Webster, Retained object. 1941 Foerster & Steadman *Writing & Thinking* (rev. ed.) xvi. 153 A verb in the passive voice is transitive only if it has a retained object. 1961 R. B. Long *Sentence & its Parts* 504 Normal complements of passive-verb-form predicators are sometimes called 'retained objects'. An example is the *time* of *we weren't given time*. 1973 *Country Life* 29 Nov. (Suppl.) 9 (Advt.), Wanted for a retained client A period residence... Price is immaterial. 1974 *Terminol. Managem. & Financial Accountancy* (Inst. Cost and Managem. Accountants) 68 *Retained profits*, profits retained in the business, and not distributed to the owners/shareholders. 1977 *Cleethorpes News* 6 May 32/6 There are unlikely to be any surprises in Grimsby Town's retained list, which is due out today.

retainer¹. Add: **3. a.** (Earlier *transf.* example.)
1778 S. Foote *Trip to Calais* III. 78 As you gave me a handsome retainer, I have been in court and open'd the cause.

b. (Further examples.) Also in extended uses. Cf. *retaining fee* s.v. Retaining *ppl. a.* 2.
1975 O. Sela *Bengali Inheritance* vi. 46 'What are you saying? That we should all..take bribes?' 'We all do,' Chan said. 'We're all on retainers.' The retainer was the monthly share out, the market price for overlooking petty crimes [among Hong Kong police]. 1977 *Guardian* 13 Jan. 20/3 Richards has signed a contract with a Brisbane radio station which gives him a retainer. 1978 S. Brill *Teamsters* ix. 327 Milano is most probably getting similar retainers from several Teamster locals. 1978 *Lancashire Life* Nov. 76/2 The Clitheroe professional, Peter Geddes, for whom she works as assistant, earning a small retainer in exchange for free lessons, wise advice and a mixture of encouragement and constructive criticism. 1978 M. Puzo *Fools Die* xxvii. 310 Houlinan was Ugo Kellino's personal PR rep with a retainer of fifty grand a year.

retainer². Add: **1. b.** *Dentistry.* A structure cemented to a tooth and connected to a bridge to hold it in place.
1887 S. H. Guilford in W. F. Litch *Amer. Syst. Dentistry* II. II. 323 These little fixtures can be used as retainers with perfect success and to the exclusion of a more bulky plate. 1956, 1974 [see *pontic *a.*² B]. 1976 Torres & Ehrlich *Mod. Dental Assisting* xxvi. 797/1 The components of fixed bridgework are the pontics, the retainers and the connectors.

retake (rī·-), *sb.* [f. the vb.] **1. a.** The action of filming a scene, person, or object again; the picture or the scene obtained thus. Also *attrib.*
1918 H. Croy *How Motion Pictures are Made* v. 126 Directly on finishing the scene it is filmed again, the second exposure being called a 'retake'. 1919 H. L. Wilson *Ma Pettengill* ii. 67 Only one little retake, where she's happy over her boy's promotion in the factory. 1930 [see *can *sb.*¹ 3 c]. 1938 'E. Queen' *Four of Hearts* iv. 52 Corsi's the most finicky retake artist in pictures. 1941 G. Marx *Let.* 23 June (1967) 27 Here I am on Stage 18 waiting to shoot some retakes. 1960 *Guardian* 12 Dec. 6/2 There's so much to go wrong, the cameras jamming, the lens sticking, and no re-takes [in television]. 1972 D. Francis *Smokescreen* i. 9, I couldn't stand many more retakes of Scene 623... We had retaken it six times.

b. The action of recording music, etc., again.
1962 A. Nisbett *Technique Sound Studio* vi. 115 Music retakes should be recorded as soon as possible after the original. 1963 V. Gielgud in *Times* 22 Apr. 16/7 A certain perfectionism is possible in tape recording, but I feel that something is wrong if I personally have to go back and do retakes. 1973 'J. Marks' *Mick Jagger* (1974) 88 They suggested the use of their own studios for a retake of the same tune.

c. *fig.*
1937 *Sun* (Baltimore) 14 May 21/2 In motion-picture parlance, the Preakness at Pimlico Saturday will be a retake of the Kentucky Derby. 1959 *Punch* 20 May 686/2 She took a quick retake at the title, and..tried to stuff it.. inside her blouse.

2. *gen.* The action of taking something a second time.
1939 *Sun* (Baltimore) 17 Feb. 1/8 The purpose of repeating the testimony was to get a record which is to be made public... Senator Sheppard..announced, however, that the re-take of testimony was 'practically concluded'. 1977 C. Dexter *Silent World N. Quinn* vi. 56 The morning..had been fixed for the 'retake' of the Ordinary-level English Language papers. 1977 *Irish Press* 29 Sept. 18/6 McGhee took the spot-kick, but the 'keeper saved his shot, only to see the referee order a retake because the 'keeper had moved before the ball was kicked.

retake, *v.* Add: **1.** (Later examples in specialized senses.) Cf. *retake *sb.*
1929 H. L. Witwer *Yes Man's Land* xii. 304 This here's no quickie and I can't retake all that stuff and do business. 1972 [see *retake *sb.* 1 a]. 1973 'D. Halliday' *Dolly & Starry Bird* vi. 82 You have to retake all those pictures this morning. 1977 C. Dexter *Silent World N. Quinn* vii. 67 She's re-taking a few O-levels.

4. *absol.* To take a second time, take over again.
1962 A. Nisbett *Technique Sound Studio* vi. 115 Discreetly discourage phrases such as 'as I've just said', and retake if they are not cleanly editable. 1977 *Horse & Hound* 14 Jan. 43/2 (Advt.), Next intensive one month or week's preparation course for A1 (suitable for those retaking) in April.

retaliant (ritæ·liănt), *a. rare.* [ad. late L. *retāliānt-em*, pr. pple. of *retāliāre* to retaliate.] Retaliative.
1925 R. Bridges *S.P.E. Tract* xxi. 13 The universal attitude of Englishmen..naturally enough provoked a similar retaliant feeling in America.

retaliator. (Earlier U.S. and later examples.)
1788 W. Gordon *Hist. Independence U.S.A.* IV. 287 A set of vindictive rebels, known by the designation of *Monmouth retaliators*. 1946 G. B. Shaw *Geneva* 25 Women in every generation, look like vindictive retaliators, pugnacious sportsmen, and devout believers.

retama² (rĕtā·mă). Also **ratama.** [Amer. Sp. *retama*.] = *palo verde. Also *attrib.*
1891 J. M. Coulter *Bot. W. Texas* I. 94 P[*arkinsonia*] *aculeata*..[is] often cultivated for ornament and known as 'retama'. 1903 'O. Henry' in *Ainslee's* Dec. 139/2 One December in the Frio country there was a ratama tree in full bloom. 1909 B. Mackense *Trees & Shrubs San Antonio* 25 The retama is very elegant and is often planted for ornament. 1926 'O. Henry' in *Argosy* Sept. 63/2 How delicious was that morning breeze..fresh and sweet with the breath of the yellow ratama blooms! 1949 *Chicago Tribune* 20 Feb. 30/3 Cedar and mesquite alone are costing Texas ranchers 115 million dollars a year. Add the sage and cactus,..blackjack oak, retama and prickly pear and the toll is terrific.

retard, *sb.* Add: **1.** (Further example.)
1971 *Times* 14 Apr. 14/3 The Government, somewhat in retard of the fact, enacted the..eugenic protection law in 1949.

3. A device in a motor vehicle for retarding the ignition spark.
1932 *Motoring Encycl.* 10/3 The Bosch automatic advance and retard (Fig. 3) is a simple design for a stationary armature type of magneto. 1977 *Hot Car* Oct. 75/3 The old one is capped off still retaining the advance retard.

4. *U.S. slang.* A mentally retarded person.
1970 *Time* 23 Mar. 49 There are..heroin addicts, Air Force and CIA mental retards and Broadway Indians doing a Broadway Snake Dance. 1971 *New Yorker* 16 Jan. 76 The younger son, self-described as 'a hard-core retard', dreams of escaping to the wilds of Oregon to gambol with the bears and squirrels. 1979 *Observer* 21 Oct. 53/5 These are men who have been out of England for years on end... Social retards, they can still hold onto their given obsolete ideas and prejudices about women because of their geographical isolation, and their marooned intellects.

retardance. Delete *Obs. rare* and add: **2.** The action of retarding; also = *retardancy; usu. in *Comb.* with a preceding sb., as *fire, flame retardance*.
1948 *Industr. & Engin. Chem.* Mar. 400/1 Primary emphasis in fire retardance has been directed to coatings or impregnants which will protect a combustible substrate. 1954 *Adv. Chem. Series* IX. 1 Ordinary paints..possess a fair amount of fire retardance..during the first stages of a fire. 1971 *Nature* 4 June 335/1 The introductory chapter includes..smaller sections on the mechanisms of flame retardance..and effective concentrations of retardants.

retardancy (ritā·ɪdănsi.) [f. as prec. + -ancy.] The capability to retard; usu. used in *Comb.* with a preceding sb., as *fire, flame retardancy*.
1947 R. W. Little *Flameproofing Textile Fabrics* v. 172 The dehydration catalysis mechanism of flame retardancy. 1973 Kuryla & Papa (*title*) Flame retardancy of polymeric materials. *Ibid.* iii. 253 These compounds impart dimensional stability, abrasion resistance, water resistance, and flame retardancy to the backcoated fabric. 1973 *Sci. Amer.* Apr. 13/2 (Advt.), Are you trying to improve fire retardancy of materials or finished products?

retardant, *a.* Delete *rare* and add: Now usu. in *Comb.* with a preceding sb., as *fire-, *flame-retardant* adjs. (see also Retardent *a.*). (Further examples.)
1915 [see *fire-retardant* adj. s.v. *fire *sb.* B. 2]. 1947, 1966 [see *flame-retardant* adj. s.v. *flame *sb.* 10]. 1971 *Financial Mail* (Johannesburg) 26 Feb. 685/3 Flame-retardant paints. 1973 *Harrod's Xmas Catal.* 59/2 Christmas tree realistically reproduced. Fire retardant. 1976 *Horse & Hound* 3 Dec. 54/1 (Advt.), This specially shaped bed filled with Fire Retardant polystyrene beads takes the ache out of a tired wet dog.

B. *sb.* A substance that reduces or inhibits some phenomenon (usu. specified by a preceding sb., as in prec. sense).
1952 [see *fire-retardant* sb. s.v. *fire *sb.* B. 2]. 1959 *New Scientist* 8 Oct. 633/1 By the combined effects of daylength control and a growth retardant, the growth habit of petunia plants can be modified so that the plants flower earlier than usual yet remain compact and bushy. 1971 'R. Macdonald' *Underground Man* v. 31 The plane was lost in the smoke.. then climbed out trailing a pastel red cloud of fire retardant. 1975 *Nature* 3 July 4/3 Peel roughness could be controlled by spraying Alar (a growth retardant) on the citrus. 1976 *Shell in Industr. Chemicals* 8 Flame retardants for polyurethane foams.

‖ **retardataire** (rətardatę̄r'), *sb.* and *a.* Chiefly *Art.* [Fr., lit. '(one who is) late in arriving, acting, etc.'.] **A.** *sb.* A work of art executed in the style of an earlier period. **B.** *adj.* Behind the times; characterized by the style of an earlier period.
1903 R. Fry *Let.* 29 June (1972) I. 210, I have assumed that the man whom you called in your notes Lorenzo Bicci is meant for Bicci di Lorenzo. Lorenzo Bicci is much too early, b. 1333 (I'm sure), whereas Bicci di Lorenzo, d. 1452, would just suit this *retardataire*. I am speaking of the Quattrocento *Madonna* with two angels kneeling beneath. 1958 H.-R. Hitchcock *Archit.: 19th & 20th Cent.* viii. 148 Only in the design of public monuments..did a pompous and somewhat retardataire eclecticism rule. 1964 *Listener* 19 Mar. 488/2 The greatest Andrea [del Sarto] is *retardataire*, an artist who would have been more at home in the conditions of a quarter of a century earlier. 1966 *Ibid.* 1 Dec. 813/3 English art had always been retardataire. 1973 J. B. Trapp *Medieval Eng. Lit.* 7 In learning, too, England [by the end of the fifteenth century] was retardataire. 1977 *Times Lit. Suppl.* 14 Jan. 30/4 The *retardataire* appearance of much colonial architecture derives from the poor, often secondhand knowledge of contemporary architectural practice as well as from a conservatism in patrons' tastes.

retardate (ritā·ɪdēɪt), *sb.* [ad. L. *retardātus*, pa. pple. of *retardāre* to Retard.] One who is mentally or educationally retarded; also *attrib.*
1956 *Amer. Jrnl. Mental Deficiency* Jan. 531/1 That promotions to supervisory capacity are within reach of some retardates is demonstrated on Table VI. 1963 N. R. Ellis *Handbk. Mental Deficiency* v. 159 (*heading*) The role of attention in retardate discrimination learning. *Ibid.* xxi. 669 The achievement of regular-class and special-class retardates in single school systems. 1975 *New Society* 10 July 91/2 Perhaps they would care to visit this hostel for adult retardates..and see the emotionally deviant, socially inadequate, environmentally retarded adolescent. 1976 *Word 1971* XXVII. 521 One might speculate about what skills a reading retardate lacks. 1980 *Brit. Med. Jrnl.* 29 Mar. 930/2 The mentally deficient (call them mentally subnormal, retarded or 'retardates', if you must) have legitimate rights.

retardation. Add: **3. b.** (Earlier example.)
1834 [see *a tempo].

4. *Psychol.* Educational progress which is slower than average for the age-group; also, mental backwardness or subnormality in an adult. Cf. *retarded *ppl. a.* 1 b.
1907 *Psychol. Clinic* I. 98 The failure of many pupils to be promoted regularly from grade to grade—retardation—has been a subject for..serious consideration. 1914 W. B. Drummond tr. *Binet & Simon's Mentally Defective Children* ii. 16 According to a convention..we regard as defective in intelligence a child who shows a retardation of three years, when he himself is nine years of age or more. 1919 L. M. Terman *Measurement of Intelligence* i. 4 We can at least prevent the kind of retardation which involves failure and the repetition of a school grade. 1937 C. L. Burt *Backward Child* iv. 78 Thus, at the age of 10, the borderline for backwardness is a retardation of 1½ years (not, as is so commonly stated, of 2 years), or, in terms of the ratio, 15 per cent. 1963 N. R. Ellis *Handbk. Mental Deficiency* xxi. 678 Skill areas listed from most to least retardation were reading, arithmetic, writing, and spelling. 1970 Hinsie & Campbell *Psychiatric Dict.* (ed. 4) 666/1 Fashions in labeling this group change almost from year to year; in the 1960's, mental retardation was the favorite appellation, and justifiably so in that it does not imply that inheritance or constitutional defects are always the cause of mental retardation. 1975 Balthazar & Stevens *Emotionally Disturbed, Mentally Retarded* i. 3 The problems imposed by mental retardation and adaptive behavior.

retarded, *ppl. a.* Add: **1. b.** *Psychol.* Orig., applied to children whose mental or educational progress lags behind that of their contemporaries to a significant degree; later extended to anyone with a measured intelligence less than some value that is itself below the average, esp. when attributed to impaired learning or maturation in childhood and youth. Hence *ellipt.* as *sb.*
1895 G. E. Shuttleworth *Mentally-Deficient Children* ii. 19 Such children are also described as 'backward', or of 'retarded mental development'—terms corresponding to the 'Enfants arriérés' of French writers..and the 'Tardivi' of the Italians. 1910 J. D. Heilman (*title*) A clinical study of one thousand retarded children in the public schools of Camden, New Jersey. 1919 L. M. Terman *Measurement of Intelligence* i. 24 (*heading*) The intelligence of retarded children usually overestimated. 1937 C. L. Burt *Backward Child* iv. 79 In the following pages, the word 'retarded' will be used to mean any child whose educational ratio falls below 85 per cent. 1951 H. Loewy *Retarded Child* (rev. ed.) i. 14 Retarded children can be put into two broad classifications from the teaching point of view: (1) The child with a mental defect. (2) The mentally defective child. 1956 H. Michal-Smith *Mentally Retarded Patient* p. vii, The mentally retarded of all ages are subject to illness as are other people. 1960 Tansley & Gulliford *Educ. Slow Learning Children* ii. 44 If his reading age is two years below his mental age he is considered retarded and special efforts must be made to get him 'working to capacity'. 1967 Brussel & Cantzlaar *Chambers's Dict. Psychiatry* 144 A mentally retarded person is one whose intelligence quotient..is below 70. 1973 La Cruz & LaVeck (*title*) Human sexuality and the mentally retarded. 1975 Balthazar & Stevens *Emotionally Disturbed, Mentally Retarded* vi. 87 Critical statements of the value of psychotherapy for the retarded. 1979 *Books & Bookmen* Jan. 43/1 Of those who could be given reliable intelligence tests, thirty-two per cent were defective, with an IQ below fifty, and thirteen per cent were retarded, with an IQ between fifty and seventy. Fifty-five per cent were subnormal or normal, ie, with IQs above seventy.

2. *Physics.* Applied to parameters of an electromagnetic radiation field in which allowance is made for the finite speed of propagation of the radiation, so that the potential due to a distant source is expressed in terms of the state of the source at some time in the past.
1920 *Physical Rev.* XV. 312 (*heading*) Note on the retarded potentials. 1929 Mason & Weaver *Electromagn. Field* iv. 283, {ρ}, the so-called 'retarded value' of ρ, is given by {ρ} = ρ(x, y, z, t−r/c). 1941 C. A. Coulson *Waves* viii. 141 We call t−r/c the retarded time. 1962 Corson & Lorrain *Introd. Electromagn. Fields* xiv. 493 The retarded position [z] is less than z by a factor of 1/(1+(v/c)). 1974 *Encycl. Brit. Macropædia* VI. 660/2 The exact potentials corresponding to a point charge..are not obtained by substituting the total charge for the volume integral indicated in the retarded potentials. The problem arises because of the finite velocity of field propagation, so that the integral of the retarded charge density over space is not in general equal to the total charge. 1975 D. M. Cook *Theory Electromagn. Field* xiv. 402 A potential of this form is referred to as a retarded potential because the potential at time t is determined by the state of the charge at the so-called retarded time t'.

retardee·. *U.S.* [f. Retard *v.* + -ee.] A mentally retarded person.
1971 *Time* 5 Apr. 38 Almost half are geriatrics cases or mental retardees who receive only custodial care. 1973 *Rehabilitation* Jan.–Mar. 46/2 The Agency was established.. to market..products made by the mentally retarded and thus demonstrate employment possibilities of retardees. 1977 *Monitor* (McAllen, Texas) 28 June 8A/1 A one-day orientation on mental retardation is scheduled Thursday... Parents of retardees and others interested also are invited.

retarder. Add: **a.** (Further examples.)
1971 P. Gresswell *Environment* 264 Weed-killers or growth retarders are needed that keep rank growth in check without eliminating all wild flowers. 1980 *Jrnl. R. Soc. Arts* Mar. 175/1 We will find ourselves sooner or later seeking..to break through that asymptote determined by nature's built-in retarders.

b. A substance which slows down a reaction or process.
1878 [in Dict.]. 1946 *Sun* (Baltimore) 7 Oct. 2/5 The meat situation may prolong the housing problem because plaster cannot be made without 'retarder', which is obtained from horn and hoof meal. 1967 Margerison & East *Introd. Polymer Chem.* iv. 195 Those substances which affect the rate of the polymerization are termed inhibitors or retarders according to whether the rate is reduced to zero or to a finite value. 1974 *Encycl. Brit. Macropædia* IV. 1076/1 The characteristics of concrete often can be improved by including admixtures in the concrete mix. In addition to set-accelerators..there are set-retarders (usually of a sugar base) that are less commonly required.

c. (See quot. 1898.)
1898 W. S. Hutton *Steam-Boiler Constr.* (ed. 3) vi. 455 The smoke-tubes of multitubular boilers are sometimes fitted with either retarders or radiators, with the object of increasing the efficiency of the heating-surface. A retarder usually consists of a flat strip of sheet-metal twisted spirally, to compel the fuel-gases to travel through the tubes in a spiral-form. 1902 *Rep. Admiralty Comm. Naval Boilers* § 29. 15 The Committee..think it right to state that retarders will be found in many cases to render existing cylindrical boilers more efficient and economical than they are at present. 1903 *Sci. Amer. Suppl.* 24 Jan. 22625/1 The cylindrical boilers should be fitted with retarders in the tubes. 1953 J. N. Williams *Boiler House Practice* vii. 152 Refractory cores or retarders were employed but with only partial success as the refractory would not stand up to the temperatures which approximated to those of theoretical combustion.

d. *Railways.* An arrangement of rails placed inside and parallel to the running rails in a shunting yard which may be moved sideways so as to act as a brake on the flanges of wagon wheels.
1937 W. G. Raymond *Elem. Railroad Engin.* (ed. 5) xiv. 186 The car retarder is a device to control the speed of a moving freight car. 1966 G. F. Allen *B.R. after Beeching* viii. 248 Further B.R. applications of this ingenious device may be confined to the outer ends of reception sidings in existing yards already equipped with electronically-controlled, clamp-type retarders, to improve the speed control of shunted wagons after they have left the hump area. 1969 H. R. Broadbent *Introd. Railway Braking* ii. 11 The retarders used in marshalling yards do not come under the heading of train-operated brakes.

retarding, *ppl. a.* Add: *retarding field, potential.*
1947 F. G. Spreadbury *Electronics* xviii. 591 In order that no current shall flow it is..necessary to apply a retarding potential, i.e. the anode must be negative with respect to the cathode. 1950 *Jrnl. Physical Soc. Japan* V. 339/1 The assumption that electrons could enter the retarding field (grid-plate space) only a single time. 1953 *Ibid.* VIII. 182 The plate-current distribution in a retarding-field tube of concentric structure. 1963 B. Fozard *Instrumentation Nucl. Reactors* xi. 138 Such a circuit is provided by a thermionic diode working under what are known as retarding field conditions.

retardive, *a.* (Later example.)
1968 G. Jones *Hist. Vikings* II. i. 66 The retardive effect of high latitude, long winters, severing distance, and a barriered landscape upon the development of the northern kingdoms was considerable.

retd., ret'd, abbrevs. of Retired *ppl. a.* 6. Cf. *ret.
1942 Partridge *Dict. Abbrev.* 83/1 *Ret.*, retired. *Retd.*, returned. (2) A variant of *ret.* 1975 N. Luard *Travelling Horseman* i. 3 Colonel Stephen Wilmot, US Army (Ret'd), glanced at his watch.

rete. 1. a. Delete † *Obs.* and add later examples.
1905 *Sci. Amer.* 12 Aug. 120/3 Above the planisphere lies the neatly cut out and decorated 'rete' carrying upon its circular interior the constellations of the ecliptic. 1957 *Encycl. Brit.* II. 575/1 Having noted the sun's position for the day in the zodiac circle, rotate the *rete* until the sun's position coincides with a circle on the plate corresponding to the observed altitude. 1974 *Sci. Amer.* May 8/2 This modern descendant of the astrolabe comes packed in a leatherette case with a rete, nine climates, a planet-plotting device and instructions.

2. a. (Further examples.) Also *spec.* in *Zool.*, such a network that supplies the swim bladder of many fishes and releases gas from the circulation for secretion into the bladder so as to increase buoyancy. Also *rete* simply.
1896 [see *red body* s.v. Red *a.* 19 a]. 1897, 1949 [see *glomus 2]. 1961 *Jrnl. Gen. Physiol.* XLIV. 539 The gas gland consists of a cellular layer interposed between the gas gland capillaries emanating from the rete and the lumen of the swim-bladder. 1962 K. F. Lagler *Ichthyology* viii. 254 In the eel..over 100,000 arterioles and a slightly smaller number of venules give the rete a total surface of over 2 square meters where blood vessels are opposed to one another. 1973 *Sci. Amer.* Feb. 39/3 The tiny vessels intermingle to form slabs of vascular tissue, the *retia*, that lie close to the upper and lower surfaces of the dark muscle...

These *retia* are the heat exchangers that ensure the warmth of the dark muscle. 1974 [see *red gland* s.v. *red *a.* 19 a].

c. In full *rete testis*. A network of vessels through which spermatozoa pass before leaving the testicle for the epididymis.
1786 W. Cruickshank *Anat. Absorbing Vessels Human Body* II. 140 The absorbents which arise out of the rete testis are exceedingly large. 1821 J. Wilson *Lect. Struct. & Physiol. Male Urinary & Genital Organs* iv. 96 Each of the vessels forming the rete testis sends off a tube, which leaves the body of the testicle and enters the epididymis singly. 1849–52 R. B. Todd *Cycl. Anat. & Physiol.* IV. 977/2 The canals of the rete. 1906 *Practitioner* Nov. 663 Most of these recorded cases [of testicular tumour]..are characterised by *complexity* of growth; and this is explained, I believe, as regards my own, by its origin, which I have traced to the tubules of the rete testis. 1968 Passmore & Robson *Compan. Med. Stud.* I. xxxvii. 22/1 Each seminiferous tubule is highly coiled... They are looped in such a way that both ends join a series of about thirty short, so-called straight tubules which converge on a network of vessels lying in the mediastinum, the rete testis.

rete·ller. [f. Retell *v.* + -er¹.] One who tells or relates anew.
1929 *Cambr. Med. Hist.* VI. xxv. 827 It must be admitted that Chrétien himself does not claim to be an inventor, but rather a re-teller of tales.

retentate (rite·ntēɪt). [f. L. *retent-iō* a holding back (cf. Retention) + *-ate*, as in *distillate*, *filtrate*, etc.] That which fails to pass through a semi-permeable membrane, and so is retained on dialysis.
1959 Turner & Feinberg in *Nature* 10 Oct. 1139/1 We propose the term 'retentate' to designate those substances which are retained by semipermeable membranes in the course of dialysis. 1974 *Ibid.* 8 Nov. 176/1 The haemagglutinating activity in galactose..was found in the retentate and could be concentrated quantitatively and washed free of monosaccharide by ultrafiltration. 1977 *Ibid.* 6 Jan. 92/2 Retained molecules (the retentate) remain in the original sample container. Both the retentate and the filtrate fractions can be recovered.

retention. Add: **2. a.** Also *Psychol.*, the ability to retain specific previously learned mental, perceptual, or motor tasks; also *attrib.*, esp. as *retention curve*, the curve on a graph which shows the amount of learning retained over a period of time.
1902 J. M. Baldwin *Dict. Philos. & Psychol.* II. 470/2 The first [sound] must leave behind it some after-effect which so modifies the second as to determine the judgment. This may also be called retention. 1923 C. S. Spearman *Nature of Intelligence* xix. 304 Those who would trace memory back to retention have more particularly tried to depict it in the guise of associative reproduction. 1940 R. S. Woodworth *Psychol.* (ed. 12) x. 337 The retention curve, or 'curve of forgetting', was first obtained by the relearning method... The curve shows a gradual loss of retention with the lapse of time. 1949 Postman & Egan *Exper. Psychol.* xvi. 381 (*caption*) Retention curve showing the reminiscence phenomenon. 1952 McGeoch & Irion *Psychol. Human Learning* (ed. 2) x. 359 In general, the retention of perceptual-motor habits is quite high. 1963 L. J. Cronbach *Educ. Psychol.* (ed. 2) xi. 350 On a retention test several weeks later they [*sc.* pupils] did better than they had done at the end of the instruction. 1975 G. H. Bower in W. K. Estes *Handbk. Learning & Cognitive Processes* I. ii. 75 The 'fluctuation' mood of contextual alterations and their effect on retention.

b. (Later example.) Also *attrib.* as *retention rate*.
1972 *N.Y. Times* 3 Nov. 18/4 The retention rates for the addicts referred to the therapeutic communities—which typically hold less than 25 per cent of their patients—are high. 1974 *Amer. Jrnl. Epidemiol.* C. 104/2 In each cohort, the retention rate is similar for the ambulatory patients and for the patients inducted on an inpatient basis. 1977 D. Louria in M. M. Glatt *Drug Dependence* iv. 116 Initial efforts were directed to relating retention in a given treatment modality to nine demographic characteristics. *Ibid.*, Retention rates in the six programs studied ranged from over 60 to less than 5 per cent.

c. In Phenomenology, the continued consciousness of or existence in the present of a previous act or event. Cf. *protension 3 b.
1931 W. R. B. Gibson tr. *Husserl's Ideas* III. ii. 220 The absolute right of immanent retention, in respect of that in it of which we are conscious as 'still' living and having 'just' happened. 1943 M. Farber *Found. Phenomenology* xvi. 516 As long as the retention lasts the tone has its own time, it is the same, its duration is the same. 1962 Macquarrie & Robinson tr. *Heidegger's Being & Time* II. iv. 411 Circumspective making present, however,..is grounded in a *retention* of that context of equipment with which Dasein concerns itself in *awaiting* a possibility. 1966 A. Gurwitsch *Stud. Phenomenol. & Psychol.* vii. 137 The very reality of conscious life, when an act is an enduring one, is a phase of present actuality most intimately connected with a whole continuity of phases retained (in retentions of various degrees). 1974 R. Sokolowski *Husserlian Meditations* v. 134 The primitive elapsing of the now-consciousness into retention is an 'event' outside time.

3. d. Something that is kept back or retained.
1922 *Daily Mail* 15 Dec. 11 Major Doyle both rode and trained Ilderton, who looked a cheap retention at 100 gs. 1962 *Rep. Comm. Broadcasting 1960* 153 in *Parl. Papers 1961–2* (Cmnd. 1753) IX. 259 Gross licence revenue... Deduct retentions. 1970 P. Oliver *Savannah Syncopators* 63 There is a frequently expressed opinion that the use of the 'answering' guitar in some blues traditions is a retention from the custom of leader-and-chorus singing.

4. d. (Further example.)
c 1735 *Man of Manners* (ed. 2) 30 We ought to be plain and modest in our Discourse, so as he may take Notice of our Retention.

6. retention money (see quots.).
1911 W. Thomson *Dict. Banking* 452/2 *Retention money*, money which is retained for a certain time after completion of a contract; e.g. if a contract has been made for £5,000 it may be agreed that 10 per cent. of the money due to the contractor shall be retained till, say, six or twelve months after the completion of the contract. If an assignment of retention money is given as security, notice of the assignment must be given to, and an acknowledgment received from, the person who is liable to pay the money to the contractor. 1974 *Terminol. Managem. & Financial Accountancy* (Inst. Cost and Managem. Accountants) 17 *Retention money*, a sum of money representing an agreed proportion of a price for goods supplied or work completed, such proportion being withheld by the purchaser or contractee for an agreed period of time as security against failure by the supplier or contractor to fulfil his obligations under the terms of the contract.

retentional, *a.* [f. Retention + -al.] Of or pertaining to retention.
1931 [see *protension 3 b.] 1938 *Mind* XLVII. 517 But even on the 'retentional' interpretation, I am not sure that the doctrine is supported by the introspectible facts. 1966 A. Gurwitsch *Phenomenol. & Psychol.* xii. 305 The source and origin of these modifications is the 'strict present'; the modifications arise from the 'actual now' and are conveyed to the retentional phases.

retentionist. (In Dict. s.v. Retention.) Add: **2.** One who advocates the retention of capital (or occas. of corporal) punishment. Used also *attrib.*, esp. of countries which have retained capital punishment.
1956 [see *abolitionist c]. 1957 *Landfall* (N.Z.) Sept. 248 Those and only those who according to the Retentionists ought to be hanged. 1961 *Spectator* 7 Apr. 464 Retentionists do not..in fact base their desire to retain hanging on a belief that hanging reduces the number of murders. 1972 *Times* 28 Sept. 16/3 Retentionists [of caning] and agnostics can complain that the case for retention is made perfunctorily.

retentiveness. Add: **b.** *Physics.* The capability of retaining a residual magnetic field when a magnetizing field has been removed.
1886 J. Hopkinson in *Phil. Trans. R. Soc.* CLXXVI. 460 The ordinate OB is what is generally meant by the residual induction after great magnetising force, or the 'retentiveness'. 1902 *Encycl. Brit.* XXX. 430/2 The ratio of the residual magnetization to its previous maximum value measures the retentiveness of the metal. [*Note*] Hopkinson specified the retentiveness by the numerical value of the 'residual induction' (= 4πI). 1924 C. R. Underhill *Magnets* i. 21 The property whereby a magnetic material independently remains a magnet is called retentiveness. 1939 [see *retentivity].

retentivity. Add: **1.** Now usu. the strength of the magnetic field that remains in a sample after removal of a saturating inducing field. (Further examples.)
1912 Brooks & Poyser *Magnetism & Electricity* xxv. 414 Retentivity is measured by the 'residual' or 'remanent' magnetism, which persists when the magnetising force is removed. 1924 C. R. Underhill *Magnets* xvi. 291 The retentivity or the residual structural flux density..will be about 9,900 gausses, or about 66 per cent of the induction. 1931 S. R. Williams *Magnetic Phenomena* i. 52 The property of retaining to a greater or less degree a certain amount of magnetization is called the retentivity of the substance. The terminology of magnetism is rather confusing regarding some of these terms... The consensus of opinion among magneticians at present is to reserve the term *remanent magnetism* for the open-circuit residual magnetism as in the case of a U-shaped permanent magnet with the keeper off. 1939 L. F. Bates *Mod. Magnetism* ii. 58 The specimen is now no longer exposed to a magnetic field, but it still retains a considerable intensity of magnetisation, equal to *Ob*/4π, which is termed the retentivity of the material and is a measure of the ability to retain magnetism when not subjected to adverse treatment. Some authorities, e.g. Ewing, term the residual induction *Ob* the retentiveness. 1948 *Electronic Engin.* XX. 351/1 The principal requirements of steels for permanent magnets is that they shall have high remanence (retentivity) and coercive force. 1951 R. M. Bozorth *Ferromagnetism* xi. 499 In strong fields they [*sc.* the coercive force and the residual induction] approach limiting values called the coercivity and the retentivity, respectively. 1953 J. D. Kraus *Electromagnetics* v. 239 The retentivity of a substance is the maximum value which the residual flux density can attain. 1966 *McGraw-Hill Encycl. Sci. & Technol.* IV. 489/1 The difference between cores of an electromagnet and a permanent magnet is in the retentivity of the material used.

2. *Psychol.* The capacity or ability to retain learning or to remember.
1909 C. S. Myers *Exper. Psychol.* xiii. 173 Nor is the superior retentivity of the most distributed readings due to the involuntary revival of the syllables by the subject. 1923 C. S. Spearman *Nature of Intelligence* ix. 132 Our second quantitative principle may be called that of retentivity... It appears not to be restricted to cognition, but to extend to mental processes of almost all sorts. It even governs an immense number of purely physical events. 1938 *Times Lit. Suppl.* 19 Mar. 186/1 These are the laws of mental energy and its constancy, of retentivity. 1961 *Lancet* 12 Aug. 361/2 Spearman held that 'retentivity' was a factor not closely related to general intelligence.

3. *Geol.* The property of rocks and minerals of retaining gases, esp. radiogenic ones.
1960 *Amer. Jrnl. Sci.* CCLVIII. 600 The conditions imposed upon phlogopite..in order that its retentivity be greater than 95 per cent for 10⁸ years. 1968 Hamilton & Farquhar *Radiometric Dating for Geologists* i. 19 Chemical alteration which occurs penecontemporaneous with the event to be dated is not pertinent unless it affects the retentivity for argon of the mineral to be dated. *Ibid.* 22 The retentivities vary greatly within a mineral species. 1975 *Nature* 27 Feb. 704/2 The nature of the gas initially present in the heating apparatus markedly influences trace element retentivity at 1,000°C.

rete·xture (ri-), *v.* [Re- 5 a.] *trans.* To treat (material, a garment, etc.) so as to restore firmness to its texture. So **rete·xtured** *ppl. a.*, **rete·xturing** *vbl. sb.*
1953 *Laundries & Laundry Requisites* (ed. 12) 208 (Advt.), Re-texturing agent 'Esdefix' makes garments look newer—last longer. 1960 *Which?* May 110/1 Most dry cleaners..ask whether you want your clothes cleaned in the ordinary way or 'with retexturing'. *Ibid.* 111/1 A panel of 20 people was..asked to see if they could tell the difference between the retextured and other garments. *Ibid.* 111/2 There was no difference in appearance between retextured and simply dry cleaned clothes, but there was an indication that the retexturing process had put something back into the cloth, which might therefore last longer. 1976 B. Ball *Keegan* 9 Selecting a retexturing service for his lousy old jacket. 1977 *Punch* 31 Aug.–6 Sept. 349/1 Lord Kagan has had my raincoats retextured.

retgersite (re·tgəɪzəit). *Min.* [See quot. 1949 and -ite¹.] A polymorph of nickel sulphate hexahydrate, $NiSO_4.6H_2O$, a secondary mineral occurring as green crusts and tufts.
1949 Frondel & Palache in *Amer. Mineralogist* XXXIV. 188 A recent survey of the specimens of so-called morenosite in the Harvard collection..proved the natural existence of the tetragonal polymorph of nickel sulfate hexahydrate. The latter substance..is described beyond under the name retgersite. The name is given after the Dutch physical chemist and chemical crystallographer Jan Willem Retgers (1856–1896). 1972 *Mineral. Abstr.* XXIII. 314/1 X-ray powder data..and IR spectra are given for retgersite from sulphide-bearing quartz veins..from Chelmiec, Góry Kaczawskie Mts., Lower Silesia.

rethink, *v.* Add: Now usu. *spec.* with a view to changing intentions or attitudes. **a.** (Further examples.)
1930 E. Bullough tr. E. Gilson in T. F. Burns *Monument to St. Augustine* ix. 292 St. Augustine had but re-thought and deepened, from the point of view of a Christian, the essential elements of Platonism. a 1942 B. Malinowski *Sci. Theory of Culture* (1944) iii. 19 At times the thinker does nothing else but to re-think..what the primitive might have or ought to have thought or felt under certain conditions. 1944 J. S. Huxley *On Living in Revolution* 18 How can this disintegrating system be reintegrated on a new basis? One way of beginning to rethink our social framework is [etc.]. 1955 *Harper's Mag.* Jan. 12/1 A ranking Administration speaker..said that..the Republican party might 'have to rethink the power and resources policy'. The verb is from the advertising-agency jargon that the Administration has learned to speak so fluently but it is not likely to mean much. 1957 *Times Lit. Suppl.* 6 Dec. 729/2 A summons, in effect, to the younger German historians, which only a few of them have heeded, to rethink the whole of Germany's recent past. 1973 *Daily Tel.* 28 Sept. 7/1 Mrs Thatcher last night promised to rethink methods of awarding grants to married women students. 1977 F. Young in J. Hick *Myth of God Incarnate* ii. 30 In any attempt to rethink christological belief, the primacy of soteriology must be recognized.

b. (Further examples.)
1919 J. L. Garvin *Econ. Found. Peace* xviii. 439 Not to recognise this and not to re-think accordingly means either not being sincere in support of a real League of Nations or not being competently sincere. 1959 *Times* 16 Oct. 13/4 'Rethink'..has suddenly become modish... It means—it means—. Well... It certainly does not mean to think again, which is not only a rueful but an intransitive process. Think twice, perhaps? Never!.. You think twice before doing something silly; rethinking starts after you have done it. 1975 *Sunday Tel.* 22 June 32/4 (*heading*) Amin rethinks... As one of his conditions for the release of Mr. Hills, the President wants spare parts for Ferret patrol cars.

Hence **rethi·nker; rethi·nking** *vbl. sb.* (further examples); **rethou·ght** *ppl. a.*
1919 J. L. Garvin *Econ. Found. Peace* xviii. 440 The duty of rethinking is not only for some men, but more or less for all men, whatever their previous views. 1944 J. S. Huxley *On Living in Revolution* p. vii, Never, I suppose, has the process of re-thinking been so intense as in these past four years. There has been the re-thinking of old problems, the transvaluation of values. 1955 *Times* 25 Aug. 7/4 Much rethinking on the subject of disarmament has been going on in London and Washington. 1959 *Times* 16 Oct. 13/4 Repay, according to the dictionary, means to pay back, re-echo to echo back, regain to gain back; but although the rethinker may, while he is about his business, think back, one has the impression that he is mainly concerned with thinking forward. 1959 *Observer* 18 Oct. 24/7 She was ready with a dash of classless glamour for the new rethought brand image. 1963 *Times* 20 May 7/2 The need for rethinking and reorganization in art education is not questioned. 1971 *Farmer & Stockbreeder* 23 Feb. 25/3 To reverse the fall in the national breeding flock a great deal of rethinking was needed about production methods. 1977 *Meanjin* (Austral.) XXXVI. I. 64 Their language more clearly becomes that of inheritors of the shake-ups and rethinkings of the Fifties and Sixties.

re·think, *sb.* [f. the vb.] An act of rethinking; reappraisal; a product of rethinking.
1958 *Times Lit. Suppl.* 12 Sept. 511/2 Then came Mr. Khrushchev's speech at the Twentieth Party Congress and

close behind it the great Communist re-think. **1960** *Design* Feb. 29 The task of orientation towards a mass society required a rethink of..an ideal formula. **1968** *New Scientist* 8 Aug. 293/1 The need for a widespread rethink on attitudes in science education, particularly at university level. **1971** *Guardian* 1 Nov. 8/5 Industry must have a major rethink about the way it uses intelligent, qualified young people. **1976** *Jrnl. R. Soc. Arts* May 285/1 It is more difficult to apply the principles to famous modern buildings which look like a total rethink. **1977** *Listener* 3 Mar. 259/1 The whole area of prisoners' rights is long overdue for rethink.

rethrea·d (rī-), *v.* [Re- 5 a.] *trans.* To thread again. Also *absol.* and *transf.*

1904 *Westm. Gaz.* 16 Feb. 2/1 Should the thread break, it is immediately rethreaded by another device. **1906** *Ibid.* 12 Nov. 2/1 The boat rethreads the line of light. **1932** G. Heyer *Devil's Cub* iii. 35 Mary re-threaded her needle. **1963** *Times* 25 May 9/7 A woman nods in time as she rethreads. **1974** N. Freeling *Dressing of Diamond* 173 Just rethread and set it on automatic record. **1974** M. Babson *Stalking Lamb* iii. 28 The needle needed rethreading. She had come to the end of the length of silk.

retia, pl. Rete in Dict. and Suppl.

retiary, *a.* Add: **2. b.** Using a net to catch Lepidoptera.

1967 V. Nabokov *Speak, Memory* vi. 131 America has shown even more of this morbid interest in my retiary activities.

‖ **reticella** (retitʃe·lă). [It., dim. of *rete* net: see Rete.] A lace-like fabric produced esp. in Venice in the 15th, 16th, and 17th centuries. Also used *attrib.* to designate the type of geometric pattern characteristic of this fabric. Cf. next.

1865 F. B. Palliser *Hist. Lace* iv. 58 One Francesca Bulgarini also instructed the schools [at Siena] in the making of lace of every kind, especially the Venetian reticella. **1900** E. Jackson *Hist. Hand-Made Lace* 194 Reticellas, or Greek Point laces, were made chiefly from 1480 to 1620, the designs being always of the stiff geometrical type. **1932** D. C. Minter *Mod. Needlecraft* 54/2 Reticella is a style of work based on cut or drawn threads. **1960** B. Snook *Eng. Hist. Embroidery* 50 White work is much heavier in style in the late 16th and early 17th centuries... Drawn threadwork fillings and reticella motifs are combined in highly conventional leaves and flowers. **1977** Fleming & Honour *Penguin Dict. Decorative Arts* 658/1 *Reticella*, a decorative fabric made, like cutwork and drawn work, from panels of woven lines but with less use of the textile threads and much more free needlework.

‖ **reticello** (retitʃe·lo). [It., f. as prec.] **1. a.** As pl. **reticelli.** A network of fine glass threads embedded in some Venetian glass. **b.** Glass formed with this type of decoration. Also *reticello glass.*

[**1858** A. W. Franks *Vitreous Art* in J. B. Waring *Art Treasures of U.K.* 9 (*heading*) Lace and reticulated glass... The reticulated glass (*vasi a reticelli*) is a variety of the lace.] **1899** R. Glazier *Man. Historic Ornament* 95 The Venetians used with equal skill all the old methods of glass-making..*reticelli*, a network of white lines enclosing at the intersection a bubble of air. **1907** E. Dillon *Glass* xiii. 206 When two series of..rods are arranged to cross one another at an angle, we get a reticulated pattern, and within the *reticelli* thus formed a bubble of air may be caught up. **1926** W. Buckley *European Glass* 39 An example of 'Reticelli', or in German, 'Netz-glas'. **1977** Fleming & Honour *Penguin Dict. Decorative Arts* 658/2 *Reticello glass*, glass decorated with a mesh of opaque white threads beneath the surface. *Ibid.* 822/2 During the C16 they..began to make *reticello* or filigree glass.

2. = prec.

1953 M. Powys *Lace & Lace-Making* iii. 5 Soon the worker built up the lace decoration by throwing threads across these wide-open spaces in either direction, forming geometrical patterns and weaving in and out or button-holing over the threads that bridge the gap, Reticello.

reticle. Add: **3.** A disc or the like with a pattern of opaque and transparent portions which can be rotated in the path of a beam of light or other radiation so as to modulate it.

1959 *Proc. IRE* XLVII. 1566/2 The particular space-filtering properties associated with typical picket fence, checkerboard, and star reticles, as they operate in square or circular field stops, are shown in Figs. 5 and 6. **1961** *Jrnl. Optical Soc. Amer.* LI. 1011/1 In the detection of infrared radiation, it is common practice to interrupt the incident beam periodically to produce an alternating signal for electronic processing. The means used to accomplish this is usually a multisectored spinning aperture called a reticle. **1966** L. M. Biberman *Reticles in Electro-Optical Devices* iii. 30 A simple reticle providing 'two-level' phase modulation is shown... If a small image is focused upon such a rotating reticle, the reticle can provide an indication of the radial position of that image by means of the phase information in the pulses of transmitted radiation.

reticular, *a.* Add: **1. a.** *spec.* (i) *reticular tissue,* tissue of the reticuloendothelial system which helps to form the framework of lymphatic tissue, bone marrow, and the tissue of the spleen and liver.

1848 *Quain's Elem. Anat.* (ed. 5) I. p. cxiii, The substance known by the names of 'cellular', 'areolar', 'filamentous', and 'reticular' tissue. **1892** H. E. Clark *Wilson's Anatomist's Vade Mecum* (ed. 11) I. 16 Reticular tissue is found in all lymphatic glands, in the pharynx and tonsils, the solitary and agminate glands of the intestine, the thymus gland, and in the spleen... From its presence in lymphatic glands it has been named 'adenoid' and 'lymphoid' tissue. **1941** *Lancet* 11 Jan. 46/1 The relationship of the angioid proliferations to the reticular tissue is..complex.

(ii) Applied to a diffuse network of intermingled nerve fibres and nerve cell bodies in parts of the brain stem, one of the functions of which is to mediate changes in the degree of wakefulness.

1887 G. D. Thane *Ellis's Demonstr. Anat.* (ed. 10) ii. 197 In the dorsal portion of the medulla oblongata..the longitudinal fibres derived from the anterior and lateral columns of the cord..give rise to a structure that is known as the reticular formation of the medulla. **1949** Moruzzi & Magoun in *Electroencephalogr. & Clin. Neurophysiol.* I. 455/1 The following account..explores the relations of this reticular activating system to the arousal reaction to natural stimuli. **1962** A. Huxley *Island* xi. 171 Animal experiments indicated that it affected the reticular system. **1968** Passmore & Robson *Compan. Med. Stud.* I. xxiv. 20/1 It seems that consciousness is determined by the activity of the reticular formation and many anaesthetics act particularly upon it. **1975** D. & I. Jordan tr. M. Zimmermann in R. F. Schmidt *Fund. Neurophysiol.* vii. 221 There is a constant 'activating' afferent flow from the reticular system to the cerebrum that controls the state of consciousness. Therefore, the term 'reticular activating system' is used to denote this functional property of the Formatio reticularis.

(iii) *reticular cell,* a fibroblast or other unspecialized cell, esp. a phagocytic cell that helps to form the framework of the reticuloendothelial system and plays an essential role in blood formation; cf. *reticulum cell* s.v. *Reticulum 5.

1925 Strong & Elwyn *Bailey's Text-bk. Histol.* (ed. 7) iv. 75 Others maintain that the delicate fibers run in the peripheral cytoplasm (ectoplasm) of the reticular cells. **1927** *Amer. Jrnl. Path.* III. 523 Study of the so-called reticular cells of the spleen, lymph nodes and other organs show [*sic*] that they possess fibroglia fibrils and that they, therefore, are fibroblasts. *Ibid.* 525 There are no reticular cells other than fibroblasts. **1938** H. M. Carleton *Schafer's Essent. Histol.* (ed. 14) 48 The granular leucocytes, lymphocytes and monocytes are all derived from a columnar stem-cell called by Sabin the reticular cell. **1938** *Jrnl. Path. & Bacteriol.* XLVII. 461 The term 'reticular cell' was first introduced by Ribbert (1889) in describing the cells of lymphoid tissue to distinguish between the 'endothelial cells' of the lymph sinuses and the reticular cells proper. **1970** T. S. & C. R. Leeson *Histology* (ed. 2) vi. 105/1 Reticular cells may give rise to free macrophages, to early precursors of erythrocytes and leukocytes, and perhaps to other cell types. **1975** *Jrnl. Path.* CXVII. 119 The term reticular cell should be reserved for the dendritic reticular cell of Nossal *et al.* *Ibid.* 121 *Reticular cell*, a phrase with so many meanings as to be meaningless.

4. Of or pertaining to the reticulum of a ruminant.

1923 G. H. Wooldridge *Encycl. Vet. Med., Surg. & Obstetrics* II. 1025/1 This operation [*sc.* rumenotomy]..is sometimes performed for exploratory purposes in obscure cases of ruminal, reticular, or omasal indigestion. **1966** Dalling & Robertson *Internat. Encycl. Vet. Med.* V. 2633 The reticular contents are liquid and offer no resistance to a thorough examination.

reticulate, *a.* Add: **a.** (Further examples.)

1965 A. Holmes *Princ. Physical Geol.* (ed. 2) 1279/2 (Index), Reticulate (trellised) drainage. **1968** R. W. Fairbridge *Encycl. Geomorphol.* 90/1 A stream or river bed is said to have a braided pattern when the deeper channels form a lacy or reticulate network of divergent and convergent members. *Ibid.* 961/2 (*caption*) Braided pattern ('reticulate drainage') in semiarid environment.

b. *spec.* in *Bot.*, as an epithet of one kind of thickening of the walls of xylem elements.

1873 F. H. Hooker tr. *Le Maout & Decaisne's Gen. Syst. Bot.* 116 Cells may either be homogeneous, or punctate, or rayed, or reticulate, or spiral. **1907** W. C. Stevens *Plant Anat.* vii. 109 The tracheids are elongated cells especially adapted to be water carriers by numerous thin places in the walls in the form of bordered pits or associated with spiral, annular, or reticulate thickenings. **1976** Bell & Coombe tr. *Strasburger's Textbk. Bot.* (rev. ed.) II. ii. 304 Now begins the more extensive differentiation, particularly of the cell wall... The wall becomes thickened, usually by a process of opposition. In the conducting elements, for example, annular, spiral or reticulate thickenings are formed, and lignification sets in.

reticulate, *v.* **1.** Add: Also *fig.*

1971 *Nature* 17 Dec. 394/1 The central device employed in network thermodynamics is the conceptual separation of 'reversible' and 'irreversible' processes. That is, we mentally reticulate the system into subsystems, each of which either stores energy reversibly or dissipates energy without storage.

reticulated, *ppl. a.* Add: **1. c.** *spec.* of porcelain, etc. Cf. *pierced *ppl. a.* c.

1881 Audsley & Bowes *Keramic Art of Japan* 143 There are several specimens of pierced, or what is termed reticulated, porcelain. **1908** J. F. Blacker *Chats on Oriental China* xiii. 152 There is a white biscuit class, very rare, often having two walls or divisions, of which the outer one only is biscuit, reticulated or pierced with a fine network or trellis of various patterns, through which the interior wall can be seen. **1970** [see *pierced *ppl. a.* c]. **1974** Savage & Newman *Illustr. Dict. Ceramics* 245 (*caption*) Teapot with reticulated outer wall and handle with moulded terminals, creamware, Leeds, *c.* 1785. **1980** *Catal. Fine Chinese Ceramics* (Sotheby, Hong Kong) 166 Reserved on a reticulated florette and *wan* diaper-ground infilled in green.

reticulation. Add: **a.** *spec.* in *Photogr.*, (the

[...]

the peripheral blood..has..in the last decade..assumed clinical importance as an index of the activity of blood formation. I would suggest..that when the normal percentage of these cells in the peripheral blood is exceeded, the condition be designated 'reticulosis'... The word 'reticulocyte' might similarly be substituted for 'reticulated erythrocytes'. **1956** *Nature* 28 Jan. 190/1 Although reticulocytes have practically their full complement of haemoglobin, evidence from amino-acid incorporation studies suggests that these cells, unlike mature erythrocytes, still have protein-synthesizing capacity. **1968** H. Harris *Nucleus & Cytoplasm* i. 12 The mammalian reticulocyte continues to synthesize haemoglobin for some days after elimination of the cell nucleus.

b. *attrib.*, as *reticulocyte level*; **reticulocyte count,** the proportion or concentration of reticulocytes in the blood.

1922 E. B. Krumbhaar in *Jrnl. Lab. & Clin. Med.* VIII. 14 The temporary rise in the reticulocyte count immediately

[...]

staining methods, the growth is called a reticulosarcoma.

reticulosis (rītikiŭlōᵘ·sis). *Med.* Pl. **-oses** (-ōᵘ·siz). [ad. G. *retikulose* (E. Letterer 1924, in *Frankfurter Zeitschr. f. Path.* XX. 392): see Reticulo- and -osis.] Proliferative disease of reticuloendothelial cells.

Quot. 1922 s.v. *reticulocyte a illustrates a different sense.

1932 B. D. Pullinger in *Rose Res. on Lymphadenoma* 134 The term 'reticulo-endotheliosis' is not applicable to any member of the group. The term 'reticulose' or 'reticulosis' (Letterer) is more suitable. **1958** R. W. Raven *Cancer* II. xxiv. 453 The term *reticulosis* was first used in 1924, when Letterer described his original case of what is now often known as Letterer–Siwe disease: he interpreted the case as a variety of 'aleukaemia', which he named..*aleukämische Retikulose bzw. Retikuloendotheliose*... In 1934, Letterer stated that the term reticulosis should be used..only in

retie, *v.* Add: Also *absol.*

1954 A. G. L. Hellyer *Encycl. Garden Work* 106/2 If there is any doubt on this point, do not disturb the scion but retie at once.

retimber, *v.* Add: Also, to reforest.

1924 J. A. Hammerton *Countries of World* III. 1928/1 The state..is responsible for the systematic retimbering of the Alps in the upper valley of the Durance.

retinaculum. Add: **2. c.** In collembolans, a fused pair of appendages which hold back the furcula before releasing it for a spring.

1923 H. M. Lefroy *Man. Entomol.* 15 A curious appendage, the so-called 'catch' or retinaculum,..holds the furca in place when not in use. **1939** H. Womersley *Primitive Insects S. Austral.* 81 When the retinaculum releases the spring, the latter strikes the ground, forcing the insect to leap a considerable distance. **1969** R. F. Chapman *Insects*

[...]

which has been much studied, may be determined by at least five separate genes of which only one is clinically distinguishable from the rest. **1969** *Listener* 16 Jan. 66/1 One serious congenital abnormality—a form of blindness manifest in adults, *retinitis pigmentosa*—the early symptoms of which are likely to appear after a man is married and had children, who will then continue to carry the gene.

retino- (re·tinŏ-), comb. form of Retina, used in terms in *Med.*, as **re:tinoblasto·ma** (pl. **-omata**) [see Blasto-, *-oma], a malignant, familial tumour of the retina occurring chiefly in young children; **retino-ce·rebral** *a.*, of or pertaining to the retina and the brain; **re:tinochoroi·dal** *a.*, pertaining to the retina and to the choroid; **re:tinochoroidi·tis** = *choroido-retinitis; **retino·pathy** [*-pathy 2], non-inflammatory disease of the retina;

so **retinopa·thic** *a.*; **retino-te·ctal** *a.*, of or pertaining to the retina and the optic tectum; **retinoto·pic** *a.* [Gr. τοπικ-ός of or pertaining to place], (of a projection on the optic tectum) that preserves the spatial relations of the sensory receptors of the retina. Also Retinoscopy.

1924 *Trans. Amer. Ophthalm. Soc.* XXII. 26 We therefore recommend that the term glioma of the retina be not used, except temporarily as a synonym or explanation to one of the following, which may more properly be applied to this condition: Neuro-epithelioma..; Retino-blastoma, proposed by Mallory; or Retino-cytoma. **1940** S. Duke-Elder *Text-bk. Ophthalm.* III. xxxvi. 2832 Retino-blastomata..are common, forming the great majority of retinal 'gliomata'. **1966** Wright & Symmers *Systemic Path.* II. xl. 1637/1 A retinoblastoma is a highly malignant tumour that arises in the pars optica of the retina. It usually appears during the first two years of life. **1976** *Path. Ann.* XI. 319 Exfoliated cells of medulloblastoma, neuroblastoma, and retinoblastoma are characterized by nuclear molding and clustering of adjacent cells. **1891** Retino-cerebral [see *fatigue *sb.* 1 c]. **1930** *Jrnl. Physiol.* LXIX. 433 An interaction is occurring between the retino-cerebral apparatuses of the two eyes. **1962** H. C. Weston *Sight, Light & Work* (ed. 2) v. 168 The 'overpowering' of the retino-cerebral or visual sensory system does not occasion the painful feeling experienced when we are dazzled. **1895** *Arch. Ophthalm.* XXIV. 334 (*heading*) Three unusual cases of retino-choroidal degeneration. **1971** *Jrnl. Amer. Vet. Med. Assoc.* CLVIII. 740 (*heading*) Retinal and retinochoroidal lesions in early neuropathic canine distemper. **1881** G. Sigerson tr. *J. M. Charcot's Lect. Dis. Nerv. System* II. iii. 41 The lesion of the optic nerve which sometimes supervenes in glycosuria and syphilitic retino-choroiditis. **1975** *Ann. Ophthalm.* VII. 853/1 Toxoplasmosis is an important cause of focal exudative retinochoroiditis. **1933** *Amer. Jrnl. Ophthalm.* XVI. 612/1 From 1856..until the present, the question of a retinopathic entity due to diabetes has remained unsettled. **1976** *Lancet* 30 Oct. 961/2 The mean prolactin concentration in retinopathic patients was 11 : 5 ng/ml. **1932** *Amer. Jrnl. Med. Sci.* CLXXXII. 137 Retinal arteriosclerosis in association with hemorrhages and sharply defined white patches, so-called arteriosclerotic retinopathy. **1939** M. L. Hine *May & Worth's Man. Dis. Eye* (ed. 8) xviii. 288 To distinguish the non-inflammatory affections from the inflammatory, the now-accepted term 'retinopathy'..has been adopted. **1966** Wright & Symmers *Systemic Path.* II. xl. 1629/2 Formerly, it was supposed that the variety of ophthalmoscopical appearances associated with the vascular retinopathies merely represented different phases of the same disease: now, however, it is generally recognized that..three distinct forms can be differentiated—(i) arteriosclerotic retinopathy, (ii) hypertensive retinopathy, and (iii) diabetic retinopathy. **1978** *Jrnl. R. Soc. Med.* LXXI. 636/1 The impression is that encephalopathy and retinopathy are particularly related to the severity and rate of rise of blood pressure. **1962** *Nature* 1 Dec. 898/2 (*heading*) Retinotectal connexions after retinal regeneration. **1977** *Ibid.* 6 Jan. 52/1 The topography of the retino-tectal projection onto the optic tectum was found to be similar in the bullfrog and leopard frog. **1961** *Jrnl. Physiol.* CLVII. 27P In sixteen frogs the normal retinotopic projection on the optic tectum had been restored after optic nerve regeneration. **1979** *Nature* 12 Apr. 623/1 If these exchanges were cumulative, it is arguable that any nascent retinotopic order should become scrambled before axons reach the brain.

retinoid (re·tinoid), *sb. Biochem.* [f. *Retin(ol² + -oid.] Any substance displaying vitamin A activity.

1976 M. B. Sporn et al. in *Federation Proc.* XXXV. 1332/1 Natural forms of vitamin A and synthetic analogs of vitamin A; this entire set of molecules, both natural and synthetic, we shall call retinoids, in a manner analogous to the naming of carotenoids or steroids. **1976** *Lancet* 27 Nov. 1174/1 The value of using the synthetic retinoids in the treatment of these dermatoses lies not only in the excellent therapeutic response but also in the comparative lack of toxicity. **1980** *Nature* 17 Apr. 626/1 Retinoids reduce the saturation density and/or growth rate of many normal and tumorigenic cell lines.

retinol¹. Add: Now *Obs.*

retinol² (re·tinǫl). *Biochem.* [f. Retin(a + -ol.] Either of vitamins A_1 and A_2 (*spec.* the former), which are yellow carotenoid alcohols of formulæ $C_{20}H_{30}O$ and $C_{20}H_{28}O$ respectively; (sometimes followed by distinguishing numeral).

1960 *Jrnl. Amer. Chem. Soc.* LXXXII. 5581/1 The pure substance hitherto known as vitamin A_1 or axerophthol shall be designated retinol. **1968** A. White et al. *Princ. Biochem.* (ed. 4) l. 1048 Vitamin A activity in mammals is exhibited by α-, β-, and γ-carotenes, by retinol and retinol₂.. and by a few other carotenoids. **1976** *Nature* 4 Mar. 49/2 Vitamin A (retinol) is a nutrient essential for vision, growth, reproduction and proper differentiation of epithelial tissue.

Also **retino·ic** *a.*, in *retinoic acid*, the carboxylic acid obtained from retinol by oxidation; hence **retino·ate,** the anion of this; **re·tinyl** *attrib.*, denoting esters of retinol; **retiny·lidene,** [*-idene], the group in which form retinal exists in rhodopsin, i.e. a side chain linked to opsin by a double bond formed in a condensation reaction between the aldehyde group of retinal and an amino group of the opsin.

1960 *Jrnl. Amer. Chem. Soc.* LXXXII. 5581/1 The pure substance hitherto known as vitamin A acid shall be designated retinoic acid. **1968** A. White et al. *Princ. Biochem.* (ed. 4) l. 1049 Retinyl esters, the form present in ingested liver and fish-liver oils, are hydrolyzed in the intestine. *Ibid.*, Retinoic acid..readily replaces retinol in the rat diet. **1969** *Nature* 1 Feb. 435/1 There could perhaps be a direct interaction between the charged quanidinium ion and the molecular orbitals of the retinyledene [*sic*] chromophore. **1970** R. W. McGilvery *Biochem.* xxvi. 645 Since it won't save vision, it is apparent that retinoate is not readily reduced to retinal. *Ibid.*, Polar bear livers..contain as much as 30 micromoles of retinyl esters per gram—a 20-year supply for a human in each pound. **1973** *Nature* 16 Nov. 166/2 The only action of light in vision is the photoisomerisation of the 11-*cis* retinylidene (derived from vitamin A aldehyde) prosthetic or chromophoric group of rhodopsin, from the 11-*cis* to the all-*trans* configuration. **1976** *Ibid.* 4 Mar. 49/2 Some retinol is oxidised to retinoic acid (vitamin A acid) *in vivo*, but if an animal is provided with retinoic acid in place of dietary retinol, it can only partially substitute for the missing retinol.

retinula. Add: Also anglicized as **retinule** (re·tiniŭl). [Coined in Ger. by H. Grenacher in *Untersuchungen über das Arthropoden-Auge* (1877) ii. 17.] Also *attrib.* (*appositively*). (Further examples.)

1924 *Glasgow Herald* 31 May 4/2 The insect's compound eye..has its numerous lenses and retinules each wrapped up in a black mantle. **1978** *Nature* 29 June 772/1 The *Drosophila* compound eye consists of about 700 ommatidia, each containing six peripheral and two central retinula cells (photoreceptors).

retinyl: see s.v. *retinol².

re-ti·p, *v.* [Re- 5 c.] To supply with a new tip. Hence **reti·pping,** *vbl. sb.*

1839 Ure *Dict. Arts* 853 He had rendered entirely unserviceable 126 punches or borers, besides 26 others which had been re-tipped with steel. **1947** J. Steinbeck *Wayward Bus* i. 4 People stopped bringing..their ploughs for re-tipping.

retiracy. 1. (Earlier examples.)

1829 *Virginia Lit. Museum* 30 Dec. 460/1 *Retiracy*, 'solitude'. **1839** C. M. Kirkland *New Home* xi. 64 The important matter of supper being in some sort concluded, preparations were made for 'retiracy'.

retiral. Now chiefly *Sc.* **1.** Delete *rare* and add further examples.

1964 J. D. Mackie *Hist. Scotl.* ii. 22 The retiral which followed the departure of Agricola. **1976** *Scotsman* 24 Dec. 13/1 (Advt.), Retiral collection in aid of children's homes.

2. (Further examples.)

1939 *Daily Tel.* 18 Dec. 12/5 (Advt.), Owing retiral of Foreman Pattern Maker..a vacancy occurs for a first-class Man with organising ability. **1965** J. Potter *Death in Office* i. 10 You will recall my telling you of a verbal agreement between the late Chairman and myself that my retiral should be at my own discretion. **1973** *Stirling Observer* 25 July 11/6 A special retiral presentation is being made by the chairman. **1978** *Lochaber News* 31 Mar. 14/1 (Advt.), Young person required for civil engineering stores to fill vacancy due to retiral.

retire, *v.* Add: **I. 1. f.** Chiefly *Cricket.* (Earlier and later examples.) Also *to retire hurt*, of a batsman: to leave the field because of injury suffered at the crease; also *fig.*

1851 W. Clarke in W. Bolland *Cricket Notes* 128 You must..make the man play out... Perhaps before that is the case, you will have caused him to retire. **1863** *Lillywhite's Cricket Scores* III. 62 Wansell..was given out unfairly, and refused to retire. **1867** *Ball Players' Chron.* 6 June 2/1 His run, however, was the only one scored, as the next three strikers retired in succession. **1877** *Wisden's Cricketers' Almanack* 223 (*heading*) Lancashire v. Nottinghamshire. *Ibid.* 224 Nottinghamshire... F. Wyld, retired (*hurt*). **1892** *Ibid.* 209 Mr. E. C. Streatfield (Pembroke), not out 6—retired hurt. **1901** H. Bleackley *Tales of Stumps* iv. 105 Amidst..loud applause, he retired with thirty-eight runs to his credit. **1906** E. Dyson *Fact'ry 'Ands* xvii. 233 It..batted 'er in ther basket with 28lb. iv grey sugar paper done up tough, 'n' she retired 'urt. **1925** A. Christie *Secret of Chimneys* xv. 141 Poor little Michael didn't get it [*sc.* a disappointing answer] as straight from the shoulder as he might have done. But he retired hurt all the same. **1961** F. H. Burgess *Dict. Sailing* 170 *Retire*, withdraw from a race. **1977** *Arab Times* 3 Dec. 9/6 Kurian had to retire hurt after scoring 11 runs.

III. 11. a. Now also in *gen.* use.

1961 M. Spark *Prime of Miss Jean Brodie* iii. 71 She had been retired before time. **1980** *Times* 19 Feb. 1/1 One idea being strongly canvassed is that Sir Charles Villiers, British Steel's chairman, should be retired early.

b. (Earlier and later examples.)

1881 'Mark Twain' *Prince & Pauper* (1882) xix. 244 He was so awkward at this service that she retired him from it. **1974** *Sci. Amer.* Dec. 139/1 This material served for about a year before I retired it. **1980** D. Francis *Reflex* viii. 91 It's his last season... I'll have to retire him [*sc.* a racehorse].

c. *U.S. Baseball.* To cause (a batter or team) to retire; to put out.

1874 *Chicago Inter Ocean* 6 July 9/1 Schafer was retired on a fly caught by Meyerle in left field. **1889** N. F. Pfeffer *Scientific Ball* 33 Runners move up every time the ball is pitched;..the clever baseman will be guided by the action of the man he wants to retire. **1917** C. Mathewson *Second Base Sloan* 180 The first batsman was retired on an easy toss from Chase to Jim. **1949** *News-Herald* (Marshfield, Wisconsin) 19 July 9/4 Nowitzke gobbled up Bauer's grounder and threw him out to retire the side. **1972** *N.Y. Times* 4 June v. 1/7 Lyle retired the first 11 batters and wound up allowing just three singles.

retiree (rītaiə·ri·). *U.S.* [f. Retire *v.* + -ee¹.] One who has retired from a business or occupation; a pensioner.

1945 W. Pegler in *Times-Herald* (Washington) 3 Oct. 9/1 How many amputees were there, General Bradley wanted to know, resorting to a ghastly form that has given us, also, trainees, dischargees, and retirees. **1962** *Economist* 29 Dec. 1278/1 The 'retirees', people of modest means drawn from all parts of the United States. **1967** Mrs. L. B. Johnson *White House Diary* 10 May (1970) 516 One of the retirees of the ILGWU..came up to me with a big bouquet of yellow roses. **1972** C. Weston *Poor, Poor Ophelia* xvii. 102 The couple..were..retirees for whom living on a sailboat was apparently a life's dream. **1975** 'G. Black' *Big Wind* ii. 24 The retiree kept spraying my face with fury as he told me. **1979** *Tucson* (Arizona) *Citizen* 3 Oct. 18A/4 Waves of retirees have headed for country parts of Missouri.

retirement. Add: **6.** *attrib.* and *Comb.*, as (sense 2 b) *retirement age, city, community, home, pension, pensioner.*

1919 *Rep. Departmental Comm. Old Age Pensions* App. 157 in *Parl. Papers* (Cmd. 411) XXVII. 299 Your retirement age seems very low. **1976** A. White *Long Silence* xix. 164 The driver, one of the colonel's men, long past retirement age. **1964** V. Bartlett *Tuscan Retreat* i. 24 'Retirement cities' such as are growing up in the United States..lived in only by old people. **1976** *National Observer* (U.S.) 14 Feb. 9/1 Many elderly people trying to find a retirement community have heard stories of developments going broke. **1979** *Tucson* (Arizona) *Citizen* 20 Sept. 8c/1 At least 60 persons have put down deposits to buy security for their old age in the form of 'life contracts' with a proposed retirement community here. **1968** *Globe & Mail* (Toronto) 13 Feb. 31/4 (Advt.), Small retirement home, 2 bedrooms, 3 piece bath. **1977** E. Ambler *Send no more Roses* x. 248 Buy that nice retirement home you've always dreamed of owning. **1942** W. Beveridge *Soc. Insurance & Allied Services* III. 96 The pensions proposed in the Plan for Social Security are retirement pensions, not old age pensions. **1973** *Times* 13 Jan. 19/7 Basic retirement pensions are a flat amount..per week. **a 1974** R. Crossman *Diaries* (1976) II. 644 Should the exemptions be for retirement pensioners or for people over sixty-five?

retirer. (Later example.)

1978 *Oxf. Diocesan Mag.* July 5/3 The main problem of retirement has to be faced by the retirer himself.

retiring, *vbl. sb.* **2. b.** (Examples of *retiring age.*)

1919 *Rep. Departmental Comm. Old Age Pensions* App. 153/1 in *Parl. Papers* (Cmd. 411) XXVII. 299 The retiring age in this case is 50, but, providing the consent of the company is obtained, members are not bound to retire at that age. **1945** 'G. Orwell' *Animal Farm* iii. 26 There was a stormy debate over the correct retiring age for each class of animal. **1973** G. Greene *Honorary Consul* v. v. 329 If you were in the Service you would have passed the retiring age quite a while ago.

retoo·l, *v.* [Re- 5 a, c.] *trans.* **1.** To rework or shape again with a tool.

1866 [in Dict. s.v. Re- 5 a].

2. To furnish (a factory, etc.) with fresh tools; to provide new manufacturing equipment for.

1940 *Economist* 5 Oct. 432/2 In case of a change in the design or type of equipment to be produced, the manufacturer must be given ample time to plan its production and if necessary to re-tool his works. **1951** Cunningham & Sherman *Production of Motor Vehicles* iii. 76 It will be necessary to retool many existing machines already in the shop. *Ibid.*, To retool means to provide new jigs, fixtures, clamping devices, cutting tools..and other implementation that makes machines and machine tools effective for a particular operation. **1963** *Economist* 10 Aug. 510/2 The scheduled shut-downs this month to retool for the new 1964 models. **1976** Ld. Home *Way Wind Blows* iv. 58, I agreed with him that a home-based market was necessary if we were to re-tool our factories for export.

Hence **retoo·ling** *vbl. sb.*

1942 *Sun* (Baltimore) 8 Jan. 14/1 The publication goes on to say that 'retooling'..is not the only problem. **1951** G. Marx *Let.* 23 Jan. (1967) 34 Between retooling for the war effort and dueling with Wald and Krasna, I presume you are a fairly busy man. **1962** E. Snow *Red China Today* (1963) xxvii. 204 Plant in considerable confusion due to expansion and retooling. **1977** G. V. Higgins *Dreamland* xii. 146 Vulcan Forge Limited..[had] been a bit slow in retooling to resume production for its prewar markets.

retort, *sb.*¹ Add: **2. b.** *retort courteous*: in allusion to quot. 1600 in sense 2 a. Also *retort discourteous.*

1908 [see *oar *sb.* 5 a]. **1928** A. Huxley *Point Counter Point* iv. 64 The question..fairly invited the retort discourteous. **1977** H. L. McGuffie in Bond & McLeod *Newslett. to Newspapers* III. 197 The quarrel in print..ranged all the way from the Retort Courteous to the Lie Direct.

retort, *sb.*² Add: **5.** *retort furnace.*

1879 *Encycl. Brit.* X. 91/2 Retort furnaces are commonly fired or heated with a portion of the coke which forms one of the bye-products of the gas manufacture. **1958** A. D. Merriman *Dict. Metallurgy* 287/1 *Retort furnace*, a metallurgical furnace consisting of a fire-chamber, and frequently regenerative chambers, in which the retorts are placed for containing the materials for treatment. *Ibid.*, Cast-iron retorts are sometimes used in the retort furnace for treatment of mercury ores.

retort, *v.*¹ **I. 3. c.** (Earlier example.)

1811 Jane Austen *Sense & Sens.* III. i. 26 Marianne was going to retort, but she..forbore.

retort, *v.*² Substitute for def.: To heat in a retort in order to separate or purify substances. (Earlier and later examples.)

1850 N. Kingsley *Diary* 26 May (1914) 123 A warm [day;] the boys retorted the last weeks work. **1924** *Jrnl. Inst. Petroleum Technol.* X. 537 That refinery is supplied..with the crude oil and ammoniacal liquor derived from the shale retorted there. **1948** *Rep. Progr. Appl. Chem.* XXXIII. 40 The raw shale is retorted at four crude oil works and the crude products are refined at Pumpherston. **1964** J. E. Ransom *Range Guide to Mines & Minerals* iv. 58 There is a 200-ounce limit on the possession of retort sponge gold obtained by retorting the gold–mercury amalgam.

retortive, *a.* Add: **3.** Also, characterized by retorts.

1949 G. B. Shaw in *Strand Mag.* July 20/2 A trumpery farce may win an uproarious success by its retortive backchat.

retouch, *sb.* Add: **2.** *Archæol.* Secondary trimming or shaping applied to a stone implement at some period after initial manufacture; an instance of this.

[**1921** M. C. Burkitt *Prehist.* iv. 65 Having blocked out the implement in the rough it had then to be finished with what is known as secondary working or trimming (French, *retouche*).] **1926** D. A. E. Garrod *Upper Palæolithic Age in Britain* i. 38 The tang is well worked, with a steep retouch along both sides and enough of the upper part is left to give an idea of the form. **1932** *Jrnl. R. Anthrop. Inst.* LXII. 261 A fair proportion of lunates and other microliths showed a peculiar retouch which was found on three or four specimens, only at Shukba. **1957** V. G. Childe *Dawn European Civilization* (ed. 6) xi. 211 The technique of bifacial retouch on flint flakes and blades is more likely to have reached the north from the south-east than from the south-west. **1964** H. Hodges *Artifacts* vii. 103 The initial roughing out of a stone tool is usually referred to as *primary flaking*, while any later working..is called *secondary flaking* or *retouch*. **1977** *Antiquaries Jrnl.* LVII. 212 *Flint scraper*:..limited fine steep marginal retouch.

retouching, *vbl. sb.* Add: **3.** *retouching desk* (example), *varnish.*

1889 E. J. Wall *Dict. Photogr.* 163 Some sort of retouching desk is needed. **1895** *Montgomery Ward Catal.* Spring–Summer 253/2 French Retouching Varnish, for oil or water color paintings. **1934** H. Hiler *Notes Technique Painting* iii. 166 Retouching varnish is a quick-drying varnish used to bring out..parts of the picture which have gone flat or 'dead' in drying.

retract, *v.*¹ Add: **1. e.** *Phonetics.* To pronounce (a sound) with the tongue drawn back.

[**1889** A. J. Ellis *Early Eng. Pronunc.* v. 17 In D 6, 7, the tongue is often merely retracted.] **1890** H. Sweet *Primer of Phonetics* 73 The first element of the diphthong in *high* is retracted towards]. **1970** M. Swanton *Dream of Rood* 33 CGmc. *a* fronted to early OE *æ* and retracted instead of broken before an *l* or *l*-group.

retract, *v.*² Add: † **1. c.** *Chess.* In a chess problem: to take back or unmake (a move). *Obs.*

1871 *Dubuque Chess Jrnl.* 12 White retracts his last move and mates in 1 move. **1881** F. C. Collins *Sel. Chess Probl.* 111 Retract White's last move, then White to play, and Mate in One move. **1890** B. G. Laws *Two-Move Chess Probl.* iii. 16 Retract White's move by replacing White knight at K4, and Black pawn at QB4, then play P×P, *en passant*, discovering mate. **1907** [see *retractor 4].

retracting *vbl. sb.*² (later example).

1874 T. M. Brown *Bk. Chess Probl.* 20 The Dubuque Chess *Journal* Tourney..for 'retracting' problems.

retractable, *a.* Add: **3. a.** (Later examples.)

1920 *Flight* XII. 96/2 As a result of the unfortunate accident..Messrs. Vickers, Ltd., were unable to show their 'Viking' flying boat with retractable land undercarriage. **1933** *Jrnl. R. Aeronaut. Soc.* XXXVII. 446 Modern U.S.A. commercial aircraft attain higher speeds by clean design and use of retractable carriages. **1936** *Ibid.* XL. 715 Armament (defensive):..1 machine gun 12·7 calibre in retractable turret firing rear; [etc.]. **1968** Miller & Sawers *Techn. Devel. Mod. Aviation* III. ii. 54 One can find designs as early as Penaud's in 1871 that were well streamlined and incorporated a retractable undercarriage (first used by du Temple in 1858 on an airplane that did not fly).

b. Applied to an object which admits the retraction of a component part.

1961 *Lebende Sprachen* VI. 103/2 Retractable ball-point pen. **1962** A. Nisbett *Technique Sound Studio* ii. 39 Retractable ball-point pens can be a menace—people will fiddle with them, and produce a sharp, unidentified click every ten or fifteen seconds.

retracted, *ppl. a.* Add: **b.** *Phonetics.* Of a sound: pronounced with the tongue drawn back.

1874 A. J. Ellis *Early Eng. Pronunc.* iv. 1105/1 As he pronounced this *s* to me, it sounded like a retracted (ˌs) with a rattle of moisture. **1908** H. Sweet *Sounds of Eng.* 37 Quite distinct from the fully retracted back vowels. **1975** *Language* LI. 282 Less easy to pin down are the retracted sibilants.

retractile, *a.*¹ Add: **a.** *spec.* in *Med.*, applied to a testis in the inguinal region that can readily be manipulated into the scrotum or vice versa.

1938 Spence & Scowen in *Lancet* 29 Oct. 983/2 We have used the following terms:—(1) Retractile testes... (2) Retained testes. **1968** Turner & Bloodworth in J. M. B. Bloodworth *Endocrine Path.* xiii. 438/1 This has probably been due, in large part, to the failure of many observers to distinguish between the retractile testis and the true cryptorchid testis. **1977** *Lancet* 24–31 Dec. 1361/1 The 2 other boys, 1 aged 10 years with small inguinal testes and the other aged 13½ years with one impalpable and one retractile testis, showed no improvement.

retractile, *a.*² For *rare*—¹ read *rare* and add further example.

1920 H. G. Wells *Outl. Hist.* xxix. 327/2 Hadrian, his successor, was of a cautious and retractile disposition.

retraction. Add: **4. e.** *Phonetics.* The drawing back of the tongue in the articulation of speech sounds; articulation thus effected.

1890 H. Sweet *Primer of Spoken Eng.* 4 Each of the vowels formed by the different combinations of retraction and height is either *narrow* or *wide*. **1895** R. Morris *Hist. Outl. Eng. Accidence* (ed. 2) ii. 18 We distinguish three horizontal positions, or degrees of retraction of the tongue. **1927** *Year's Work Eng. Stud.* 1925 40 The third point deals chiefly with the phenomena of secondary retraction. **1977** *Archivum Linguisticum* VIII. 76 The only major objection to it must be that there are no other cases in Old English of non-velarized *l* preventing fronting or causing retraction.

retractive, *a.* and *sb.* Add: **A.** *adj.* † **3.** Of a chess problem: involving the retracting of a move or moves. *Obs.*

1890 J. Rayner *Chess Probl.* 9 A retractive problem is one in which some move..has to be retracted, and then mate or sui-mate in a given number of moves. **1890** B. G. Laws *Two-Move Chess Probl.* 16 These are called 'Retractive' problems.

retractor. Add: **4.** *Chess.* (See quot. 1907.)

1902 *Brit. Chess Mag.* 455 Two-move retractors. **1907** S. S. Blackburn *Terms & Themes Chess Probl.* 33 Retractors. Problems wherein the conditions require that the last move of one, or both, of the players shall be retracted, and that, when this is done, the problem shall be solved according to the usual conditions. **1937** T. R. Dawson *Caissa's Wild Roses in Clusters* 14/1 A simple changed-mate retractor theme.

retrai·n, *v.* [Re- 5 a.] *trans.* To train again; *spec.* to teach (a skilled or trained person) a new skill. Also **retrai·ning** *vbl. sb.*

1934 in Webster. **1937** *Jrnl. Compar. Psychol.* XXIV. 290 On the re-training trials of the following day, this behavior persisted. **1964** Mrs. L. B. Johnson *White House Diary* 11 Jan. (1970) 39 He had been out of work for two years and then he heard about the retraining program. **1977** P. Strevens *New Orientations in Teaching of Eng.* p. vii, Helping to train and re-train teachers. **1977** *Whitaker's Almanack 1978* 980/1 Under the rules of the Fund the Commission can assist in training and retraining schemes by reimbursing 50 per cent. of the cost financed by a public authority.

‖ **retraite** (rətrẹt). Also erron. **rétraite.** [Fr.: a mod. re-borrowing of *retraite* (see Retrait *sb.*¹).] **1.** = Retrait *sb.*¹ 3 b. Also *en retraite*, in retirement or seclusion.

1860 Mrs. Gaskell *Let.* 27 Aug. (1966) 631, I quite understand the wisdom of French ladies going into rétraite. **1958** L. Durrell *Mountolive* v. 102 Pombal is doing all this legitimately. He is on local leave. I am *en retraite*.

2. *Mil.* = Retrait *sb.*¹ 2.

1883 *Standard* 21 Sept. 3/1 A grand dinner..was followed by the performance of a *retraite* by the combined bands of the Eleventh Army Corps.

retranscri·be, *v.* [Re- 5 a.] *trans.* To transcribe again. So **retranscri·ption.**

1805 Retranscription [in Dict. s.v. Re- 5 a]. **1810** D. Wordsworth *Let.* 28 Feb. in W. Knight *Mem. Coleorton* (1887) II. 112 He [*sc.* Coleridge] has written a whole *Friend* more than once in two days. They are never re-transcribed, and he generally has dictated to Miss Hutchinson, who takes the words down from his mouth. **1841** I. D'Israeli *Amenities of Lit.* I. 175 Anthony Wood indignantly re-transcribed the whole of his English copy, and left the fair volumes to the care of the university. **1965** *Canad. Jrnl. Linguistics* Spring 85 Most forms are quoted from that source but are retranscribed. **1977** *Word* 1972 XXVIII. 256 The forms which Alonso has cited are not minimal pairs and should be retranscribed as follows.

retransfer, *sb.* Add: **2.** *Printing.* An impression taken from a lithographic image using special ink and paper for the purpose of transferring it to another lithographic surface.

1946 A. Kirk in H. Whetton *Pract. Printing & Binding* xviii. 207 Imperfect retransfers should be scrapped immediately. Good retransfers can be placed face down on a clean sheet. **1965** Zigrosser & Gaehde *Guide Coll. Orig. Prints* iii. 30 If the facts..are known, the authenticity might be questionable, since the artist did not actually work on each retransfer of the original master image. **1967** E. Chambers *Photolitho-Offset* xvii. 259 Before retransferring from litho plates or stones, a thin layer of retransfer ink should be applied with a nap roller to the image.

retranslate, *v.* **1.** (Earlier example.)

1860 G. H. Lewes *Let.* c 10 July in *Geo. Eliot Lett.* (1954) III. 319 There are paragraphs which read like a translation from a language into which one cannot retranslate them.

retransportation. (Earlier example.)

c **1751** S. Richardson *Let.* (1804) VI. 61 How I missed you, on my re-transportation!

retravirus (re·trăvəiᵊrŏs). *Biol.* [mod. L., f. initial letters of *re*verse *tra*nscriptase (see *transcriptase) + Virus.] = *retrovirus.

[**1974** *Intervirol.* IV. 202 Retraviridae are enveloped virions about 100 nm in diameter... They contain about 1.5% of 60–70S RNA... Some species that have been examined also contain a small amount of DNA... All viruses in the family contain antigenically specific reverse transcriptase (RNA-dependent DNA polymerase).] **1977** *Virology* LXXIX. 239 Xenotropic type C retraviruses were isolated from cell-free uterine extracts of normal adult NIH Swiss mice.

re·tread, *sb.* Chiefly *U.S.*, *Austral.*, and *N.Z.* [f. the vb.] **1.** A tyre supplied with a fresh tread; = *remould *sb.*

1914 *Auto-Motor Jrnl.* 4 Apr. 423 So exact is the work..that a retread is scarcely distinguishable from an original. **1921** *Daily Colonist* (Victoria, B.C.) 10 Apr. 10/6 Always carry a 'retread' as a spare. Don't throw away your old casings until you have had them retreaded at least once. **1943** *Sun* (Baltimore) 4 Feb. 7/2 Retread is new rubber over the same area as recap, but all the way down to the cord fabric. **1968** *Wanganui* (N.Z.) *Chron.* 15 Nov. 10/6 (Advt.), Insist on safe shoulder retreads. **1973** A. Cooke *Amer.* 12, I found myself rediscovering—on re-tread tires at the compulsory thirty-five miles an hour—the whole American landscape. **1976** *Drive* Sept.–Oct. 77/1 A retread is a reprocessed tyre made from a secondhand carcass welded to new tread rubber. **1977** *Bulletin* (Sydney) 22 Jan. 58/3 (Advt.), Because Bandag retreads run longer, you'll spend less time and labour changing worn tyres.

2. *transf.* and *fig.* **a.** *Mil. slang.* A retired soldier recalled for (temporary) service, a 'dug-out'. Also in extended use, esp. of retrained persons.

1941 *Salt* 22 Dec. 36/2 Characteristically the Australians call a small reconnaissance tank a 'dingo', and a 1914–1918 soldier enlisted a second time a 'retread'. **1945** Baker *Austral. Lang.* II. viii. 152 A soldier of the 1914–18 war who has joined up again is a *retread*. **1948** *Amer. Legion Mag.* Oct. 26 Retreads will reune: Retreads, men who served in both World Wars..will hold their first reunion..at Miami. **1953** *Economist* 17 Oct. 178/2 The pro-Eisenhower *Chicago Daily News* called his appointees 'Governor Stratton's team of Republican retreads'. **1962** *Listener* 16 Aug. 232/2 They also have shorter courses for older men, known rather depressingly as 're-tread' courses. **1965** C. Koch *Casual Company* I. i. 7 I've got retreads, and chickens bloated with fever, but they turn to at reveille. **1977** D. Beaty *Excellency* ii. 21 A diplomat with thirty years experience..not the re-tread given a job with other unwanted Civil Servants.

b. Of things.

1964 *Lebende Sprachen* IX. 35/3 Mr Kennedy's plans are retreads of older projects. **1976** *Times Lit. Suppl.* 1 Oct. 1232/1 An all-too-familiar exegesis that has nothing new to say about Eliot.., a critical retread representing the kind of discussion that came in the 1930s. **1979** *Arizona Daily Star* 22 July (Parade Suppl.) 8/3 It is hard to imagine that we shall face another 10 years of the same romantic nostalgia and retread pop culture as was dished out by the '70's.

retrea·d, *v.* [Re- 5 c.] *trans.* **1.** To furnish (a tyre) with a fresh tread.

1908 *Daily Report* 7 Feb. 11/4 (Advt.), 10–12 h.p. Wolseley, in excellent condition, front pneumatics, just been retreaded. **1912** *Motor Manual* (ed. 14) iii. 107 It is possible, in most cases, to have them retreaded by the makers.

2. *fig.* To retrain (a person) or provide with fresh employment, esp. after initial retirement. *U.S. slang.*

1963 *Lafayette Alumnus* Apr. 22/1 'Retreaded'..indicated that upon retirement, De Kay, Furness [etc.]..gave their learning, experience, and wisdom a capping. **1966** *Wall St. Jrnl.* 2 Dec. 1/4 To 'retread' many retired nurses and other skilled professionals through refresher courses.

Hence **retrea·der**; **retrea·ded** *ppl. a.*, **retrea·ding** *vbl. sb.*

1914 *Auto-Motor Jrnl.* 4 Apr. 423 During the retreading..it is very often found that the casing requires strengthening. **1921** *Dict. Occup. Terms* (1927) § 608 *Tyre retreader*, attaches new treads to worn tyre covers. **1937** *Sci. Amer.* Dec. 347/1 Retreading..involves nothing more nor less than replacing with new live rubber the tread which has become worn smooth under the abrasive action of the road. **1951** U. V. Troubridge tr. *Guareschi's Little World of Don Camillo* 207 The explosion had..been caused..by one of the retreaded tyres of the lorry. **1964** G. Mitchell *Death of Delft Blue* xxi. 228 Look over their used cars and retreaded tyres. **1976** *Globe & Mail* (Toronto) 3 July 6/1 A retreaded provincial politician who was fired from the Davis Cabinet. **1977** *Financial Times* 4 June 5/3 The retreading industry would welcome it, because casings are their raw material.

retreat, *sb.* **3. d.** Delete *rare* and add later examples.

1914 W. B. Wright *Quaternary Ice Age* vii. 155 They are named by Penck retreat-stadia ('Rückzugsstadien'), but it is considered probable that they were formed at the ends of periods of readvance. **1954** W. D. Thornbury *Princ. Geomorphol.* viii. 201 The factors influencing the retreat of slopes are far more complex than is generally realized. **1963** R. A. Daly *Changing World of Ice Age* i. 21 (*caption*) Two principal moraines of retreat in Finland. **1970** R. J. Small *Study of Landforms* xi. 388 If the ice is affected by episodes of retreat, separated by stillstands, a number of smaller, sub-parallel ridges ('recessional' or 'stadial' moraines) will be formed.

10. *Comb.*, as *retreat house.*
1920 J. F. Briscoe in *Rep. First Anglo-Catholic Congress* 182 There ought to be a retreat-house in every diocese. 1958 *Church Times* 14 Feb. 10/3 With its membership growing, part of the accommodation in the retreat house has had to be taken to house the brethren. 1979 *Country Life* 6 Dec. 2188/3 Rydal Hall..was leased to the diocese of Carlisle..as a retreat house and conference centre.

retreatal, *a.* (rītrī·tăl). [f. Retreat *v.* + -al.] Of or pertaining to the contraction and retreat of ice sheets and glaciers.
1896 *Amer. Geologist* XVIII. 156 The stages of retreatal deposits ending in Greenwich cove illustrate the shrinkage and final disappearance of a tongue of the ice. *Ibid.* 160 (*heading*) Retreatal formations on the central and eastern shores of the bay. 1937 *Geogr. Jrnl.* XC. 233 Most of this erosion occurred during the maximum stage of glaciation and very little during the retreatal stages. 1937 Wooldridge & Morgan *Physical Basis of Geogr.* xxii. 378 Retreatal stages of the ice in the valleys are marked by 'stadial moraines'. 1968 R. W. Fairbridge *Encycl. Geomorphol.* 327/1 Bergdahl (1963) has recently published a series of studies on the Närke plain where estuarine clays from the Baltic progressively overlap closely spaced ridges of retreatal moraines, each winter season being marked by another ridge, at 170–280 meter intervals.

retreating, *ppl. a.* **1.** (Further examples.)
1961 M. Levy *Studio Dict. Art Terms* 97 *Retreating colour*, a colour, such as blue, which in a painting appears to retreat into the distance. 1962 *Punch* 25 Apr. 655/2 'Retreating defence'—the rapid funnelling back and massing of defenders inside the penalty area. 1977 *Daily Express* 29 Mar. 32/4 The 'reserves' had an extra couple of players, in the form of the England coaching staff, to thicken a retreating defence still further.

retreatism (rītrī·tiz'm). [f. Retreat *sb.* + -ism.] **1.** A policy of retreat; advocacy of (military) withdrawal.
1951 *Times* 24 Feb. 7/3 General Eisenhower returned to Europe this week bringing assurance that his country has rejected the 'retreatism' advocated by Mr. Herbert Hoover and supported by Senator Taft. 1976 J. Rowan *Ordinary Ecstasy* ii. 25 If we see everything as perfect as it is, we may be inclined to quietism and retreatism in political terms.
2. *Sociol.* A state of passive withdrawal from society induced by a sense of inability to attain its norms or to offer resistance to them.
1957 R. K. Merton *Social Theory* (rev. ed.) iv. 153 Retreatism..is an expedient which arises from continued failure to near the goal by legitimate measures and from an inability to use the illegitimate route because of internalized prohibitions. 1963 T. & P. Morris *Pentonville* vii. 173 Retreatism in prison is comparatively rare, and can be identified with either an extreme manifestation of institutional neurosis..or with various stages of mental illness. 1969 in Lindzey & Aronson *Handbk. Social Psychol.* (ed. 2) IV. xxxiii. 352 With warfare no longer possible, there is a great deal of retreatism and social withdrawal. 1970 *New Society* 31 Dec. 1158/2 Thus, men like Roy..would still maintain that the only rational solution was retreatism.

retreatist (rītrī·tist), *sb.* (and *a.*) [f. as prec. + -ist.] **1.** One who advocates a policy of retreat; a supporter of (military) withdrawal.
a 1925 Curzon *Leaves from Viceroy's Notebk.* (1926) iii. 142 The Retreatists would not have these proposals at any price. 1951 *Times* 20 Feb. 4/5 Mr. Wherry and others like him..now dislike being called isolationists, but have been called 'retreatists' instead.
2. *Sociol.* One who has succumbed to retreatism (sense 2).
1957 R. K. Merton *Social Theory* (rev. ed.) v. 189 Retreatists are even more reluctant to enter into new social relations with others than are those described as 'alienated'. 1960 Cloward & Ohlin *Delinquency & Opportunity* i. 21 These terms..do not necessarily reflect the attitudes of members of the subcultures. Thus the term 'retreatist' does not necessarily reflect the attitude of the 'cat'. 1963 T. & P. Morris *Pentonville* vii. 173 The retreatist rejects both goals and means.
3. *attrib.* or as *adj.*
1957 R. K. Merton *Social Theory* (rev. ed.) v. 187 The retreatist pattern consists of the substantial abandoning both of the once-esteemed cultural goals and of institutionalized practices directed toward those goals. 1973 *Sociol. Rev.* XXI. 419 The attitudes and values of people whom they call retreatist, and the immediate conditions under which the response occurs.

retreative (rītrī·tiv), *a.* [f. Retreat *sb.* or *v.* + -ive.] Pertaining to or suggestive of retreat; tending to withdraw.
1899 B. Tarkington *Gentleman from Indiana* xix. 376 As they neared the brick house Harkless made out, through the trees, a retreative flutter of skirts on the porch. 1977 *Times Lit. Suppl.* 25 Mar. 332/3 A melancholic, self-retreative, self-distrustful constitution.

re-treatment. (Earlier example.)
1867 J. A. Phillips *Mining & Metallurgy Gold & Silver* x. 216 (*heading*) Re-treatment of first tailings.

retribalization (rītrəi:bălәizēi·ʃәn). [f. next.] The process of making or becoming retribalized.
1964 M. McLuhan *Understanding Media* ii. xxxii. 344 Today we appear to be poised between two ages—one of detribalization and one of retribalization. 1967 *Listener* 20 July 75/1 What Marshall McLuhan calls the tendency to retribalization in the multi-sensual experiences enjoyed by young people. 1970 *Internat. & Compar. Law Q.* XIX. i. 152 The general retribalisation designs of the present régime.

retribalize (rītrəi·bălәiz), *v.* [f. Re- + Tribal *a.* + -ize. Cf. *detribalize *v.*] *trans.* To restore (a person or society) to a tribal state; to encourage the tribal instincts and habits of. So **retri·balized** *ppl. a.*; **retri·balizing** *vbl. sb.*
1963 *Economist* 7 Sept. 805/2 A sprinkling of retribalised black statelets. 1964 M. McLuhan *Understanding Media* ii. xxiii. 229 It was easy for the retribalized Nazi to feel superior to the American consumer. *Ibid.* xxiv. 236 We have begun retribalizing with the same painful groping with which a preliterate society begins to read and write. *Ibid.* xxx. 304 The power of radio to retribalize mankind, its almost instant reversal of individualism into collectivism, Fascist or Marxist, has gone unnoticed. 1967 *Guardian* 9 Sept. 6/3 The effect of television as such is to retribalise and deliteratise mankind. 1969 A. Cohen *Custom & Politics in Urban Africa* i. 29 As the migrant becomes more settled, by being drawn into active participation in the social life of the Quarter—economically, politically, morally, and ritually—he becomes increasingly more '*retribalized*'.

retribute, *v.* **2.** (Later example.)
1933 W. H. W. Sabine *Guido & Girls* xi. 130 Those foul thoughts that lately have been mine, Thus justly retributed by the laws..that are divine.

retribution. 1. b. (Later example.)
1850 Thackeray *Pendennis* I. xxxvii. 355 She thought his retribution of the hundred pounds an act of angelic virtue.

retri·butivist, *sb.* (and *a.*) [f. retributive *a.* + -ist.] A believer in retributive justice. Also *attrib.* or as *adj.*
1939 *Mind* XLVIII. 157 Retributivists have been pushed into holding that pain *ipso facto* represses the worse self and frees the better, when this is contrary to the vast majority of observed cases. 1968 *Economist* 13 July 48/2 The current fashion..is to take it for granted that certain doctrines, such as the retributivist, are so discredited as to merit nothing but the most perfunctory and hostile attention. 1969 *Listener* 17 July 87/3 Why should we punish criminals? 'To ensure that they get their deserts,' says the Retributivist. *Ibid.* 88/1 Most of us have our retributivist moments ('Are you suggesting that Eichmann should have gone *scot free*?').

retri·ck, *v. rare.* [Re- 5 a.] *trans.* Of a heavenly light: to cause (a beam) to shine again. Also in fig. phr. *to retrick one's beams*, to restore one's mood; to regain one's happiness, shrug off despair.
Always with reference to Milton's line in *Lycidas*.
[1637 Milton *Lycidas* in *Poems* (1968) 253 The day-star..tricks his beams, and with new spangled ore, Flames in the forehead of the morning sky.] 1833 [implied in Retricked *ppl. a.*]. 1863 Trollope *Small House at Allington* (1864) II. iii. 28 We have retricked our beams in our own ways, and our lives have not been desolate. 1880 —— *Duke's Children* III. xxiv. 286 It is so that a man is stricken down... But it is given to him to retrick his beams.

retrievable, *a.* (Further examples.)
1967 *Times Rev. Industry* Oct. 84/1 Formalized techniques or procedures..for continuously carrying intelligence into retrievable form. 1974 C. Taylor *Fieldwork in Medieval Archaeol.* i. 18 The information..needs to be sorted and assembled in some form of retrievable system.

retrieval. Add: **1. b.** *spec.* = *information retrieval* s.v. *information 8. Freq. *attrib.*
1958 *Bull. Canad. Libr. Assoc.* Apr. 193/1 One of the reasons which makes these machines fundamentally uneconomical at present is that the frequency of demand for exactly the same mechanical retrieval process in reference work is rarely sufficient to justify the large initial cost of coding and mechanizing all that a machine incapable of judgment needs, to perform the process. 1966 *Jrnl. Amer. Med. Assoc.* 13 June 950/1 This technique allows medical narrative and numerical data to be collected efficiently in a form acceptable as computer input for subsequent storage, analysis, and retrieval. 1968 *Globe & Mail* (Toronto) 3 Feb. B7/5 The computer's instant retrieval mechanism can make available, within 24 hours, any information contained in major Canadian collective labor agreements. 1972 *Bookseller* 4 Mar. 1463/3 It is highly important..that new systems of cataloguing and retrieval shall be built into the new library. 1974 W. Garner *Big enough Wreath* xi. 138 How many men did you say you had on your..tapes?.. All tucked away in your retrieval system. 1979 J. E. Rowley *Mechanised In-House Information Syst.* i. 26 Retrieval keys, such as indexing terms are stored adjacent to the records to which they relate.

retrieve, *sb.* Add: **3. c.** A controlled exercise for a gun-dog simulating the retrieval of game; the object retrieved.
1932 L. Sprake *Art of Dog Training* v. 94 The pupil is taken to the regular training ground, and one or two retrieves of the usual dummy commence the proceedings. 1937 E. B. Moffit *Elias Vail trains Gun Dogs* ix. 134 Gallery critics at field trials are puzzled at the difficulty that many handlers experience in getting a dog to go out far enough to a retrieve. 1953 E. Stonex *Golden Retriever Handbk.* ix. 108 He must bring it right up to you—never let him run round you in circles with his retrieve. 1963 M. Brander *Gundogs* iv. 37 Only the very earliest retrieves of all should be made with the dummy or plaything thrown in full view in the open, so that the pup is encouraged to run in after it at once. 1976 *Field* 2 Sept. 474/1 He now performs the basic five retrieves as advocated by Maurice Hopper without any histrionics. 1979 *Country Life* 26 July 220/1 The gundog area has..a timed retrieve competition (the scurry).
d. *U.S. Sport.* The act of intercepting or otherwise regaining possession of the ball.
1961 in Webster. 1974 *State* (Columbia, S. Carolina) 15 Feb. 3-B/1 Barron is averaging 19.3 points a game and has been getting 11.5 retrieves per contest.

retrieve, *v.* Add: **I. 2. d.** To obtain again (stored information).
1962 *Communications Assoc. Computing Machinery* V. 12/2 Some kind of indexing scheme that can retrieve records ..within a short period of time. 1968 *Brit. Med. Bull.* XXIV. 195/1 By means of electronic pulses the data would be placed inside the computer system,..and be available for analysis at a future date as well as being able to be retrieved on demand. 1971 *Nature* 19 Mar. 155/2 Many short notes and letters contain the first 'rush' announcement of extremely important results—just what a current-awareness service should aim to retrieve. 1975 J. B. Harley *O.S. Maps* p. xiv, In the process of retrieving information the Survey's Librarian..has conjured otherwise elusive papers into my hands.

retrieveless, *a.* (Further example.)
c 1871 E. Dickinson *Poems* (1955) III. 836 Till some retrieveless Night Our Vigilance at waste The Garden gets the only shot That never could be traced.

retriever. Add: **1. b.** (Earlier example.)
1830 M. R. Mitford *Our Village* IV. 110 His noble Newfoundland dog, (a retriever is the sporting word).
2. (Later example.)
1977 *Tennis World* Sept. 17/3 'Baseliners', 'retrievers and 'counter-punchers' are players who stay back from the net.

retrieverish (rītrī·vәriʃ), *a.* [f. Retriever + -ish[1].] Resembling or suggestive of a retriever.
1909 H. G. Wells *Tono-Bungay* iii. ii. 243 There were two or three fox-terriers, a retrieverish mongrel, and an old, bloody-eyed, and very evil-smelling St. Bernard.

retrieving, *vbl. sb.* (Further examples.)
1962 *Communications Assoc. Computing Machinery* V. 11 (*heading*) Information structures for processing and retrieving.
attrib. 1972 J. S. Hall *Sayings from Old Smoky* 3 A computer-like retrieving process.

retrim, *v.* (Examples of *absol.* use.)
1966 D. Francis *Flying Finish* xviii. 217, I put on full flap, maximum drag..retrimmed..felt the plane get slower and heavier. 1977 'O. Jacks' *Autumn Heroes* v. 71 You drag 'em out, boss, while I re-trim.

retro-. Add: **3. a. retroana·lysis** *Chess*, analysis of a position so as to reconstruct the moves of the game leading to that position; also *transf.*; so **retroanaly·tical** *a.*; **retrocognition,** (*b*) *Psychol.*, paranormal cognition of events in someone or something else's past; **retrodispla·cement,** displacement rearwards; **re·trofocus** *a. Photogr.*, designating an optical system in which the distance of the rear surface from the image of an object at infinity exceeds the focal length, usu. achieved by placing a diverging group of lenses before a converging group.
[1933 H. Phillips *Week-End Problems Bk.* 182 Profound and puzzling retrograde analysis is needed to prove the legality of the key-move.] 1937 T. R. Dawson *Caissa's Wild Roses in Clusters* 13/1 Trio of retro-analyses. 1979 *Sci. Amer.* Dec. 20/2 Most chess problems deal with the future, such as how can White move and mate in three. Smullyan's problems belong to a field known as retrograde analysis (retro analysis for short), in which it is necessary to reconstruct the past. 1980 *Daily Tel.* 21 Apr. 13/7 Retro-analysis is..the root of much scientific thinking. It is as useful to the astronomer pondering the creating of the universe by observing space as it appears *now* as it is to the detective who solves a murder by deducing the series of events that led to the crime. 1966 *New Statesman* 10 June 858/3 It contains a good many highly complicated 'retro-analytical' problems. 1980 *Daily Tel.* 21 Apr. 13/7 The chess-board here is being used only as a tool for an exercise in retro-analytical deduction. 1962 C. D. Broad *Lect. Psychical Res.* 402 What I will call 'states of *direct* but not ostensibly *recollective retro-cognition*'. 1969 J. J. Macintosh in Macintosh & Coval *Business of Reason* 154 In the absence of a body there is no way of distinguishing between veridical memories and what might be called accurate retro-cognition. 1973 *Daily Tel.* (Colour suppl.) 30 Nov. 27/4 *Retrocognition*, as precognition, but of past events. 1903 *Med. Rec.* (N.Y.) 27 June 434/2 (*heading*) The treatment by anterior vaginal section of retrodisplacement of the uterus, complicated by adhesions. 1972 *Biol. Abstr.* LIII. 3376/2 Physical exertion of women employed in mechanical coal dressing does not affect the incidence of..retrodisplacement of the uterus. 1965 *Focal Encycl. Photogr.* (rev. ed.) II. 1293/2 Retrofocus lenses are almost invariably of the inverted telephoto type. 1977 J. Hedgecoe *Photographer's Handbk.* 323 In wide-angle, retro-focus constructions the back focus is much greater than the focal length which allows room for mirrors etc. within the camera construction. 1979 *Amat. Photographer* Feb. 95/2 The normal simple calculations for finding the effective f/number when engaged in close-up work with extension tubes or bellows do not always give the right answers when using a telephoto or retrofocus lens.
b. *retro-cæcal, -cardiac, -duodenal, -peritoneal, -pubic, -uterine* (earlier example);

retro-bu·lbar, situated or occurring behind the eyeball.
[1866 A. von Graefe in *Archiv für Ophthalm.* II. 147 Als solche erscheint mir die Annahme einer retrobulbären Neuritis.] 1879 *Archiv für Ophthalm.* VIII. 328 (*heading*) Three cases of retrobulbar, pulsating, vascular tumor. 1879 E. Nettleship *Student's Guide Dis. Eye* ii. xvii. 225 Neuritis behind the eye (retro-bulbar neuritis). 1961 *Lancet* 29 Apr. 908/2 Retrobulbar neuritis is a rare, though well-recognised complication of addisonian pernicious anæmia. 1964 S. Duke-Elder *Parsons' Dis. Eye* (ed. 14) 568 Retrobulbar injections should be preceded by an injection of procaine. 1903 *Amer. Med.* V. 836/1 Case 1. Retrocecal gangrenous appendicitis. 1961 *Lancet* 23 Sept. 671/2 *Staph[ylococcus] aureus*..was isolated from..the retrocæcal abscess. 1901 *Ibid.* 12 Jan. 118/2 This clear zone is somewhat triangular in shape. Dr. Mignon proposes to call it the retro-cardiac triangle. 1908 *Practitioner* Dec. 863 The 'retro-cardiac triangle', seen when the patient is in the lateral oblique position. This triangle is bounded by the heart in front, the spine behind, and the diaphragm below. *Ibid.* 827, I am inclined to think that transduodenal or retro-duodenal operations for this condition should be avoidable. 1893 *Trans. Path. Soc.* XLIV. 69 The case is..an example of what Astley Cooper termed the mesenteric variety of retro-peritoneal hernia. 1977 *Lancet* 28 May 1133/1 Retroperitoneal hæmorrhage occurred in 5 patients. 1967 Taylor & Cotton *Short Textbk. Surg.* xxvi. 397 Retropubic Prostatectomy. The prostate is approached..via a suprapubic incision through the cave of Retzius, the potential space between the back of the pubis and the bladder... A drain is inserted into the retropubic space. 1851 *Monthly Jrnl. Med. Sci.* XIII. 278 (*heading*) Retro-uterine sanguineous tumours.
c. Terms in *Astronautics* relating to retro-rockets and their use, as *retro-ignition, -impulse, -manœuvre, -propulsion, -system, -thrust.*
1967 *New Scientist* 21 Sept. 595/1 Retro-ignition was delayed for 12·5 seconds, coming at 150 000 feet instead of the normal 274 000 feet. 1961 H. H. Koelle *Handbk. Astronaut. Engin.* xxv. 29 The instantaneous retro impulses applied tangentially to the trajectory. 1976 *Sci. Amer.* June 7/3 In the Viking missions..the retromaneuver ..that will put the spacecraft in orbit around Mars will be based on commands sent from the earth at least a day earlier. *Ibid.* 59/1 Those commands will cause the spacecraft's retro-propulsion system to fire for 43 minutes, subtracting enough velocity to place the spacecraft in an elliptical orbit around Mars. 1962 S. Carpenter in *Into Orbit* 56 The retro-system is rigged so that the rockets will not fire..unless the capsule is in the correct attitude. 1962 Riley & Sailor *Space Syst. Engin.* iv. 86 The retro-thrust is directed at such an angle as to provide a desired range. Increases in retro-velocities result in small range angles.

retro (re·tro), *sb.* [Short for *retro-rocket.] = *retro-rocket b.
1961 *New Scientist* 21 Sept. 719/3 Whether because of the timing of the order, or the impulse of the retros, or drift of the wind.., the robot cosmonaut got away from its shepherds on the sea and fell into the water seventy miles from the nearest waiting ship. 1962 S. Carpenter in *Into Orbit* 56 The pilot must..have the capsule pointed exactly right when the retros fire. 1966 *Word Study* Dec. 1/2 Commander Alan B. Shepard, whose flight lasted only fifteen minutes, was precise and terse: 'Disarm.' 'Auto retro jettison circuit.'

retroaction. Add: **4. b.** *Psychol.* The effect of later learning on the memory of what was learnt previously. Also *attrib.*
1949 Postman & Egan *Exper. Psychol.* xvi. 378 Maximum retroaction occurs when original and interpolated learning are approximately equal in strength. 1953 C. E. Osgood *Methods & Theory in Exper. Psychol.* xiii. 562 The fact that so-called 'rest' control groups show loss in retention has already been noted in connection with the logic of transfer and retroaction experiments. 1971 L. Postman in J. W. Kling et al. *Woodworth & Schlosberg's Exper. Psychol.* xxi. 1110/2 The treatment corresponds to the rest condition in experiments on retroaction.

retroactive *a.* Add: **1. a.** (Later examples.) Also with *to*.
1952 *Sun* (Baltimore) 22 Mar. 6/4 A handsome wage increase..allows 12.5 cents an hour immediately and retroactive to January 1. 1978 J. Paxton *Dict. Europ. Econ. Community* (rev. ed.) 49 The United Kingdom..informed the Council of Ministers on 27 March that a subsidy would be granted on sugar and made retroactive to 1 Feb. 1979 *Arizona Daily Star* 5 Aug. c8/4 Placed Paul Hartzell, pitcher, on the 21-day disabled list, retroactive to last Thursday.
c. *retroactive infinitive* (Gram.), an active infinitive that has a preceding noun as its object.
1946 O. Jespersen *Mod. Eng. Gram.* V. xv. 233 Retro-active infinitives are found in connexion with the adverbs *yet* and *still*:..Rome and Naples—even Florence are yet to see (= we have not yet seen; *are yet to be seen* would mean 'can still be seen').
4. *Psychol.* That affects the remembering of what has been previously learned; esp. as *retroactive inhibition*, the inhibiting effect on recall that can be produced by attention given to new material after the original learning. Cf. *inhibition 4, *proactive *a.*
1909 C. S. Myers *Exper. Psychol.* xiii. 163 This 'retro-active inhibition' is yet another cause of the greater difficulty in learning longer than shorter studies. *Ibid.* 166 Similar experiments have been conducted with the object of proving retro-active, i.e. backward association. 1909 *Psychol. Monogr.* X. iv. 138 With G. retroactive inhibition was obvious. Each series was remembered fairly well until the next was given. 1915 *Ibid.* XIX. iv. (*title*) A study of retroactive inhibition. 1938 *Brit. Jrnl. Psychol.* Jan. 244 It could be argued that retroactive inhibition might influence the result by preventing the possibility of such recall. 1948 E. R. Hilgard *Theories of Learning* vi. 162 The natural conjecture on the assumption of continuous variation is that the amount of retroactive inhibition would increase gradually as dissimilarity was increased. 1963 Cofer & Musgrave *Verbal Behav. & Learning* 7 Many facts of acquisition are consistent with either formulation, but..others, especially the facts of proactive and retroactive interference, are not. 1965 J. M. Stephens *Psychol. of Classroom Learning* viii. 203 Your experience with the second task would work back to strengthen the rather tentative earlier association. This backward-working process is called retroactive facilitation. 1966 J. M. Brown et al. *Appl. Psychol.* 36 A common explanation for this loss of retention is that newly learned material inhibits that previously learned. A technical term for this is retroactive inhibition. 1975 G. H. Bower in W. K. Estes *Handbk. Learning & Cognitive Processes* I. ii. 75 An inherent restriction on retrieval times would then produce the observable phenomena of retroactive interference.

retrodi·ction. [f. Retro- + Diction, after Prediction *sb.*] The explanation or interpretation of past actions or events inferred from the laws that are assumed to have governed them. Cf. *postdiction.
1895 J. M. Robertson *Buckle & his Critics* x. 311 Let us first put a little order in our conception of prediction and 'retrodiction' as they indisputably take place in the settled sciences. *Ibid.* 316 The same reasoning applies to errors of interpretation, of what we have called 'retrodiction'. 1939 *Mind* XLVIII. 421 L-propositions are plainly useless save in so far as they assist prediction—or retro-diction—as to particular matters of fact. 1940 *Philosophy* XV. 22 It may be what Mr Ryle calls a *retrodiction*, as when I infer from marks seen in the snow that a cat has passed that way. 1956 J. N. Findlay in H. D. Lewis *Contemp. Brit. Philos.* 185 Prediction and retrodiction alike depend on the presence in our world of what have been called 'world-lines'. 1960 I. Berlin in W. H. Dray *Philos. Analysis & Hist.* (1966) 13 In the case of an historical study, retrodiction—filling in gaps in the past for which no direct testimony exists with the aid of extrapolation performed according to relevant rules or laws. 1975 J. W. Cornman *Perception, Common Sense & Sci.* viii. 302 The argument based on retrodiction might prompt another reply, namely, that the supposed fact of retrodiction violates Heisenberg's uncertainty principle. *Ibid.* 303 Epistemic and causal indeterminacy limit prediction but not retrodiction.
Hence (as a back-formation) **retrodi·ct** *v. trans.* and *absol.*, to infer by retrodiction (cf. *postdict *v.*); **retrodi·cting** *vbl. sb.* Also **retrodi·ctable, retrodi·ctive** *adjs.*, **retrodi·ctively** *adv.*
1932 H. H. Price *Perception* vii. 201 Any perceptual act is bound to be among other things a prediction,..and in the same way it must be 'retrodictive' as well. 1949 G. Ryle *Concept of Mind* v. 124 They are inference-tickets, which license us to predict, retrodict, explain and modify these actions, reactions and states. 1951 W. H. Walsh *Introd. Philos. Hist.* ii. 41 It has been said that whilst it is certainly not the business of historians to predict the future, it is very much their business to 'retrodict' the past. 1952 *Mind* LXI. 225 Attempts to predict or retrodict when no grounds for rational belief are obtainable. 1956 E. H. Hutten *Lang. Mod. Physics* vi. 222 We now interpret 'causality' as meaning that there can be formulated a universal sentence which ..allows us to derive a singular, descriptive, predictive or retrodictive, sentence. 1959 J. Blish *Clash of Cymbals* v. 104 Time in our experience is not retrodictable. *Ibid.* 105 Can we write a convergent retrodictive equation? 1966 C. G. Hempel *Philos. Nat. Sci.* vi. 72 The theory was used by Halley..to identify it retrodictively with comets whose appearances had been recorded on six previous occasions. 1975 J. W. Cornman *Perception, Common Sense & Sci.* viii. 303 Is there any reason to think that retrodicting the position and momentum of an entity more precisely than stated by the uncertainty principle violates the principle? 1976 *Times Lit. Suppl.* 27 Aug. 1057/4 What a historical model is meant to do: not to allow us to predict or retrodict..but to provide insight into and explanations of historical events.

retroduction. Delete *rare*—[0] and add later examples. *spec.* in *Philos.*, a type of logical reasoning that develops from some commonly accepted proposition until reasons are found that may alter the acceptance or understanding of the original proposition (see quot. *a* 1914). Also *transf.*
1786 J. Pinkerton *Anc. Scotish Poems* I. p. xxxiii, The poor man [*sc.* Walter Goodall] was writing a *retroduction* to Fordun. *a* 1914 C. S. Peirce *Coll. Papers* (1931) I. 28 There are in science three fundamentally different kinds of reasoning, Deduction.., Induction..and Retroduction (Aristotle's ἀπαγωγή, but misunderstood because of corrupt text, and as misunderstood usually translated *abduction*). *Ibid.* 29 Retroduction is the provisional adoption of a hypothesis, because every possible consequence of it is capable of experimental verification, so that the persevering application of the same method may be expected to reveal its disagreement with facts. 1939 *Mind* XLVIII. 378 In discussing 'the leap of the mind from data to hypothesis' he makes no reference to C. S. Peirce's *Retroduction*. 1958 N. R. Hanson *Patterns of Discovery* 217 Retroductions do not always lead to syntheses like those of Newton, Clerk Maxwell, Einstein and Dirac. They sometimes show the first chink in the old armour. 1965 P. Caws *Philos. of Sci.* xxxii. 243 This is the essential ingredient in what Peirce called retroduction, the intuitive jump from observed facts to hypotheses about them. 1976 C. Selltiz et al. *Res. Methods Social Relations* (ed. 3) ii. 32 In the process of determining explanations for observed events, social scientists often reason *from* conclusions to reasons *for* conclusions. We call this inference process *retroduction*, in contrast with deduction and induction. In retroduction, we try to think of plausible reasons why some event could have occurred in an attempt to construct an explanation of why the event did occur.
Also **retrodu·ctive** *a.*, pertaining to or characterized by retroduction; **retrodu·ctively** *adv.*
a 1914 C. S. Peirce *Coll. Papers* (1932) II. 491 Induction ..is manifestly adequate, with the aid of retroduction and of deductions from retroductive suggestions, to discovering any regularity. 1958 N. R. Hanson *Patterns of Discovery* iv. 86 *H* cannot be retroductively inferred until its content is present in 2. 1974 P. Achinstein in F. Suppe *Struct. Sci. Theories* 357 Retroductive or explanatory reasoning..is reasoning falling under the logic of discovery, whereas deductive reasoning from established theories is reasoning falling under the logic of justification. 1976 C. Selltiz et al. *Res. Methods Social Relations* (ed. 3) ii. 32 An example of retroductive reasoning appears in a study of the decline in trust in the national government during the last decade.

re·tro-engine. *Astronautics.* Also as one word. [f. Retro- + Engine *sb.*] = *retro-rocket b.
1967 *Britannica Bk. of Year* 804/1 *Retro-engine*, retro-rocket. 1971 *New Scientist* 6 May 305 Each Mariner..carries to Mars..about 1000 lbs of fuel for the 14-minute retroengine burn needed to slow it to the requisite orbital speed. 1977 *Sci. Amer.* Nov. 52/1 With the help of its retroengines and parachute it dropped to the surface of Mars.

retro-fire (re·trofəiᵊɹ), *v.* Also **retrofire.** [f. Retro- + Fire *v.*[1]] *trans.* To ignite or fire (a retro-rocket); to fire a rocket engine so as to give (a spacecraft) backward thrust. Also *absol.*
1961 *Ann. Rev. Med.* XII. 315 Periodically, a very light-weight disposable capsule may be retro-fired in such a way that the capsule together with all unwanted materials will be incinerated on re-entering the atmosphere. 1969 [see *free return* s.v. *free *a.* D. 2]. 1969 *Guardian* 14 Mar. 1/8 At 5 31 (BST) over Hawaii, Scott retrofired the main service propulsion engine..and Apollo 9 began its long arching descent.

retro-fire (re·trofəiᵊɹ), *sb.* Also **retrofire.** [f. Retro- + Fire *sb.*] The process or action of burning a retro-rocket. Also *attrib.*
1962 J. Glenn in *Into Orbit* 42 The clock is pre-set on the ground according to a timing for retro-fire which we have computed before the mission. 1962 *Flight Internat.* LXXXI. 263/1 Kauai Island acquires the spacecraft,..and checks the retro-fire clock for possible change if Mercury control so advises. Point Arguello..gives back-up command to the spacecraft after the clock initiates retro-fire. 1967 *New Scientist* 16 Nov. 424/2 It is not impossible to visualize a carefully controlled descent, the only limitations being the weight of propellants needed for continuous retro-fire. 1974 *Encycl. Brit. Macropædia* XVII. 369/2 Below the Command Module is the Service Module. It..contains the propulsion system for midcourse corrections, retrofire to achieve lunar orbit, and thrust to return from lunar orbit into Earth trajectory.
So **re·tro-fi:ring** *vbl. sb.*
1962 J. Glenn in *Into Orbit* 41 You superintend the retro-firing sequence from here with toggle switches. 1968 *Guardian* 24 Dec. 1/3, 3 21 p.m.: second retrofiring to bring spacecraft into a circular orbit 69 miles above the moon's surface. 1971 *Ibid.* 1 July 1/5 In order to carry out the descent to earth..retro-firing took place at 01.35.

retrofit (re·trofit), *sb.* orig. *U.S.* Also with hyphen. [f. Retro(active *a.* + Re)fit *sb.*] A modification made to a product, esp. an aircraft, to incorporate changes made in later products of the same type or model.
1956 in W. A. Heflin *U.S. Air Force Dict.* 441/1. 1962 *Flight Internat.* LXXXI. 292/1 Fig. 2 is a plot of the ratios of revenue lost versus maintenance and first costs for retro-fitted and 'designed-in' equipment. It points out.. that the economic cut-off point for retrofit is near the 100ft ceiling, ¼ mile point. 1963 T. Pynchon *V.* x. 286 An injury of the sexual organs could still be simulated by an attachable moulage, but then this blocked the cooling vent... A new retrofit, however, eliminated this difficulty, which was felt to be a basic design deficiency. 1967 *Times Rev. Industry* May 55/3 It is some indication of Avimo's position that it has been involved in three major retro-fits for aircraft—that is, the instruments already installed in the aircraft have been taken out and Avimo's put in instead. 1978 *Solar Energy* (Shell Internat. Petroleum Co.) 5 Thus the markets and products are likely to split into 'new' products incorporated, for example, into roof structures of new buildings, and 'retrofit' applications for existing housing stock.

retrofit (re·trofit), *v.* orig. *U.S.* Also with hyphen. [f. prec.] *trans.* To subject to a retrofit; to modify so as to incorporate changes made in later products of the same type or model. Also *absol.*
1956 in W. A. Heflin *U.S. Air Force Dict.* 441/1. 1971 *Sci. Amer.* June 2 The..passenger entertainment and service system..is now in service in the first class sections of a Boeing 747 which American Airlines retrofitted. 1973 *Guardian* 18 Jan. 4/4 American aircraft manufacturers are now researching a modification, known as 'retro-fitting' their engines to make them a good deal quieter... But Concorde ..cannot be retrofitted except at extreme cost. 1975 *Daily Colonist* (Victoria, B.C.) 3 May 14/4 You can't retrofit... That is, an existing furnace—coal or oil-fired—cannot feasibly be replaced with a solar system. 1979 *Nature* 5 Apr. p. xiii/1 The B305W can be supplied either as a module for retro-fitting to an existing pumping system or can be supplied installed as an integral package with the Ion Tech B500 high vacuum pumping system.

Hence **re·trofitted** *ppl. a.*, **re·trofitting** *vbl. sb.*
1960 *Aeroplane* XCIX. 145/2 Lately I've been collecting examples of dreadful Americanese. Such as unitized, retrofitted, heat treat, destruct button, [etc.]. 1962 [see *retrofit *sb.*]. 1975 *Times Lit. Suppl.* 23 May 563/1 The 'retrofitting' of jet aircraft to make them quieter. 1975 *Nature* 30 Oct. 727/1 New models and retrofitted older planes were, even in 1972, achieving 6 dB reductions on the figures quoted for their noisier brothers. 1977 *Blair & Ketchum's Country Jrnl.* (Brattleboro, Vermont) May 33/2 The Mount Washington hung on by virtue of its convention facilities, its retrofitted bathrooms and sprinkler system.

retroflex, *a.* Add: **2.** *Phonetics.* Pronounced with the tongue curled back; cacuminal.
1915 [see *cacuminal *a.*]. 1932 D. Jones *Outl. Eng. Phonetics* (ed. 3) xxv. 199 Retroflex sounds (also called 'cerebral', 'cacuminal' or 'inverted' sounds) are those in the formation of which the tip of the tongue is curled upwards towards the hard palate. 1942 *Amer. Speech: Reprints & Monogr.* No. 4. 41 This sound is generally clearly retroflex in the Great Smokies, as in most American speech. It is heard in such words as the following: Birch, bird, birth, Burchfield, burn. 1964 B. Honikman in D. Abercrombie et al. *Daniel Jones* 78 The frequent retroflex consonants in the languages of India and Pakistan are produced with the tongue curled back. 1973 J. C. Wells *Jamaican Pronunc. in London* 128 Dentals, alveolars, retroflex sounds, and palato-alveolars. 1977 *Publ. Amer. Dial. Soc.* 1974 LXI/LXII. 36 Again, the retroflex mid-central vowel comes from different speakers.

re·troflex, *v.* [Back-formation from Retroflexed *a.*] *trans.* and *intr.* To turn or fold back. So **re·troflexing** *ppl. a.*
1898 H. C. Porter tr. *Strasburger's Text-bk. Bot.* 396 The male branches give rise..to spherical stalked antheridia, which open at the apices by means of retroflexing valves. 1934 Webster, *Retroflex*, v.i., to turn or bend backward. 1954 S. Duke-Elder *Parsons' Dis. Eye* (ed. 12) xxiii. 405 A large corneal section is made as for cataract.., the cornea retroflexed by traction on the suture, and a triangular piece of iris more deliberately excised.

retroflexed, *a.* Add: **2.** *Phonetics.* = *retroflex *a.* 2.
1932 D. Jones *Outl. Eng. Phonetics* (ed. 3) xxv. 200 Retroflexed vowels may be represented in phonetic transcription. 1950 —— *Phoneme* p. xiii, ɹ fricative tongue-tip r; also the corresponding frictionless continuant, and a retroflexed variety of this. 1973 *Amer. Speech* 1969 XLIV. 263 The study consequently proposed to measure the degree to which the Midland dialect, in the form of heavily retroflexed postvocalic /r/, appears in Austin.

retroflexion. Add: **2.** *Phonetics.* Articulation of a sound with the tongue curled back.
1932 D. Jones *Outl. Eng. Phonetics* (ed. 3) ii. 11 In many parts..the effect of the *r* appears as a modification known as 'retroflexion' or 'inversion' of the preceding vowel. 1954 *Bull. School Oriental & Afr. Stud.* XVI. 558 It is well known as a feature of Sanskrit 'internal sandhi' that the coarticulation of retroflexion with constriction has more extensive syntagmatic implications than its coarticulation with occlusion. 1964 R. H. Robins *Gen. Linguistics* iii. 98 All vowel sounds may be characterized by retroflexion... This retroflexion is one of the characteristics of what is loosely called in Britain 'an American accent'. 1973 *Amer. Speech* 1969 XLIV. 263 A careful study of this feature was therefore planned and conducted to determine the degree of retroflexion of /r/.

retrogradation. Add: **3. b.** *Physical Geogr.* The landward retreat of a beach or coastline caused by wave-erosion.
1922 C. A. Cotton *Geomorphol. N.Z.* i. xxviii. 391 This process is termed progradation (as contrasted with retrogradation, the cutting-back of a coast by marine erosion). 1937 Wooldridge & Morgan *Physical Basis of Geogr.* xxi. 332 Retrogradation comprises not only beach recession but the general recession of the coastline under wave-attack. 1954 W. D. Thornbury *Princ. Geomorphol.* xvii. 442 Retrogradation of a shore line may go on so rapidly that small streams are unable to keep pace in downcutting with the rate of sea-cliff recession. As a result, these streams enter the sea from hanging valleys. 1968 R. W. Fairbridge *Encycl. Geomorphol.* 941/2 During retrogradation, a wide belt of beach ridges with their overlying dunes..may be rapidly removed.

retrograde, *a.* and *sb.* Add: **A.** *adj.* **3. e.** *Petrol.* Of a metamorphic change: resulting from a decrease in temperature or pressure. Opp. *prograde *a.* 1.
1932 A. Harker *Metamorphism* xx. 342 The changes which befall metamorphosed rocks subsequently to the culmination of metamorphism..are of the nature of degradation... This class of changes includes what Becke has styled 'diaphthoresis', implying ruin or corruption; but this rather cumbrous term has not been very widely adopted. It will be more convenient to speak of retrograde metamorphism. 1971 I. G. Gass et al. *Understanding Earth* i. 34/1 The metamorphism of many igneous rocks involves the replacement of a very high-temperature original mineral assemblage by a metamorphic assemblage at a lower temperature. This type of change, though not strictly referred to as retrograde since the starting material is not a metamorphic rock, nevertheless has all the essential characters of retrograde metamorphism. 1980 *Nature* 29 May 320/2 The Alpine uplift was accompanied by widespread retrograde metamorphism.
6. (Later examples.) Also *in retrograde* adv. phr.
1954 W. Faulkner *Fable* 5 For another instant, the cavalry held. And even then, it did not break. It just began to move in retrograde while still facing forward. 1959 A. G. Woodhead *Study of Greek Inscriptions* iii. 24 The assorted sherds of early date..show writing in both directions, but the majority of fragments..have their messages written retrograde. 1980 *Early Music* Jan. 111/2 Its slow movement incorporates the melody 'God Save the King', played first retrograde, later in inversion and finally in its normal form.
7. Of amnesia: pertaining to incidents preceding the causal event.
1935 *Lancet* 5 Oct. 763/1 The duration of this retrograde amnesia. 1960 *Jrnl. Compar. & Physiol. Psychol.* LIII. 524/1 Retrograde amnesia induced by electroconvulsive shock..or other trauma. 1969 *Times* 14 Apr. 6/8 Concussion, anaesthesia..and dosage with various drugs, are all known to impair the memory of the immediate past, a phenomenon known as retrograde amnesia. 1979 'S. Woods' *This Fatal Writ* 150 'He remembered too much..everything..up to and including the blow on the head. And you know..that's just not possible.' 'Retrograde amnesia.'

retrograde, *v.* Add: **1. b.** To cause to move backward.
1910 *Jrnl. Geol.* XVIII. 165 Headlands are cut back, or retrograded.

retrograding, *vbl. sb.* and *ppl. a.* (Further examples.)
1910 *Jrnl. Geol.* XVIII. 166 The retrograding of the shore ..due to active wave erosion. 1919 D. W. Johnson *Shore Processes & Shoreline Devel.* vi. 295 The phenomenon of a shifting fulcrum between a retrograding cliff and a prograding beach plain. 1968 R. W. Fairbridge *Encycl. Geomorphol.* 941/2 The retrograding shore line may cut back at an angle to previously formed ridges.

retrogressive, *a.* and *sb.* Add: **A.** *adj.* **3. c.** *Petrol.* = *retrograde *a.* 3 e.
1931 *Amer. Jrnl. Sci.* CCXXI. 8 No one criterion is a safe basis for the determination of retrogressive metamorphism. 1948 *Mem. Geol. Soc. Amer.* XXX. 299 Retrogressive metamorphism, or diaphthoresis, is the mineralogical adjustment of relatively high-grade metamorphic rocks to temperatures lower than those of their initial metamorphism. The process is thus a special case of polymetamorphism (repeated metamorphism).

retrolental (retro͵le·ntăl), *a. Path.* [f. Retro- + L. *lent-, lens* lentil (see Lens) + -al.] Situated or occurring behind the lens of the eye; *retrolental fibroplasia*, a disease caused by administration of high concentrations of oxygen to (esp. premature) infants, characterized by proliferation of retrolental tissue and sometimes leading to blindness.
1942 T. L. Terry in *Trans. Amer. Ophthalm. Soc.* XL. 262 Considerable thought has been given to the selection of an appropriate term for this disease condition. Early in the study..the term 'retrolental fibroplasia' was suggested by Dr. Harry K. Messenger. 1963 H. L. Birge in A. Sorsby *Mod. Ophthalm.* II. i. 6 (*heading*) Retinopathy of prematurity (retrolental fibroplasia). *Ibid.*, Retrolental fibroplasia is a primary retinal disease of non-inflammatory origin resulting from disordered retinal vascularization. 1975 *Daily Colonist* (Victoria, B.C.) 2 June 2/6 Born prematurely, he was blinded by retrolental fibroplasia, an eye disease brought on by the extra oxygen needed to keep him alive soon after birth.

retroreflective, *a.* (In Dict. s.v. Retro- 3 a.) Add: **2.** Also **retro-reflective.** Having or being the property of a retro-reflector.
1960 *Guide Civil Land Aerodrome Lighting* (*B.S.I.*) 40 In the case of taxiways which have occasional use, retro-reflective markers along the edges may be found to provide adequate guidance. 1961 *Space Res.* II. 287 Search light illumination of a retro-reflective satellite. 1970 *Sci. Amer.* Mar. 41/2 A billiard ball sent..into the corner of the table will, after two bounces, return along a path parallel to its incident direction. In the case of light three reflecting surfaces, all at right angles to one another, form a corner with the same retroreflective property.

re·tro-reflector. Also **retroreflector.** [f. Retro- + Reflector.] A device which reflects light back along the incident path, irrespective of its angle of incidence.
1961 *Space Res.* II. 290 A satellite carrying a radio beacon and an optical retro-reflector. 1969 *Daily Tel.* 21 July 17/3 The device which may foretell earthquakes is the laser ranging retro-reflector. It simply reflects..narrow beams of laser light sent up to it from earth. 1976 *Sci. Amer.* Feb. 51/1 The first is the lunar laser-ranging experiment, in which pulses of laser light are beamed through a telescope on the earth at one of the retroreflectors placed on the moon by the Apollo astronauts.

re·tro-rocket. [f. Retro- + Rocket *sb.*[3]] † **a.** (See quot.) *Obs.*
1948 W. Ley *Rockets & Space Travel* 347 *Retro-rocket*, anti-submarine weapon fired from planes backward with velocity matching plane speed so that the weapon fell vertically... Retro-rockets sank the last 'probable' German submarine in the war, on April 30, 1945.
b. *Astronautics.* An auxiliary rocket on a spacecraft that points in the forward direction, so as to provide thrust opposing forward motion when fired.
1957 *Times* 9 Nov. 6/6 To bring the satellite safely down from its orbit would require the use of what he called 'retro-rockets' to slow its speed. 1958 *New Statesman* 18 Oct. 512/1 It consisted of a three-stage rocket, plus a retro-rocket... The third stage was intended to circle the moon.. but it had attached a fourth stage—the retro-rocket—designed to discharge in the opposite direction and thus act as a brake. 1962 J. Glenn in *Into Orbit* 37 The engineers worked out three different ways of firing the small retro-rockets which would start us back towards the earth. 1963 *Guardian* 8 Apr. 1/3 It [*sc.* Lunik 4] will use retro-rockets to effect a soft landing on the lunar surface. 1974 *Encycl. Brit. Macropædia* XVII. 359/2 In the case of a spacecraft whose mission is to land softly on the Moon, on approach trajectory the attitude-control subsystem will rotate the spacecraft so that its retrorocket and landing radar are pointed toward the lunar surface.

retrospective, *a.* (and *sb.*) Add: **1. b.** Of an exhibition, programme of music, or the like: showing the development of the work produced, usu. by one artist, over a period. Freq. *ellipt.* as *sb.* (often const. *of* to introduce what is being exhibited).
1919 R. Fry *Let.* 22 Feb. (1972) II. 447 It's really a good show: a retrospective exhibition of Dudley..arranged all round the walls of the big room. 1931 —— *Let.* 3 Mar. (1972) II. 654 I've been having a retrospective show (forty years of work). 1932 [see *rayograph]. 1954 *Burlington Mag.* XCVI. 162/1 A retrospective exhibition..provides a much-needed opportunity to review Pasmore's..development.. from a Euston Road 'impressionism',..up to his most constructivist reliefs. 1964 *Listener* 5 Mar. 400/2 Once again the Marlborough has scooped all its rivals with a retrospective of the greatest of all the German Expressionists. 1969 *Vogue* Nov. 30/2 A true treasure at the Guggenheim, Constantin Brancusi—a complete retrospective of his work. 1972 *Village Voice* (N.Y.) 1 June 59/2 The Museum of Modern Art has been running a Will Rogers Retrospective. 1973 *Radio Times* 18 Jan. 49/1 As a prelude to tomorrow evening's major retrospective of music by Luciano Berio, Misha Donat introduces a work which represents a turning-point in his career. 1975 *Times Lit. Suppl.* 11 July 759/5 It is five years since Bill Brandt's retrospective of 125 photographs was shown at the Hayward Gallery. 1979 *Daily Tel.* 31 Dec. 12 The National Film Theatre gave him [*sc.* Alfred Hitchcock] a major retrospective. 1980 *Times* 8 Jan. 9/8 The retrospective does throw in a bonus in the shape of a room of very early work before..Dali became appreciably Dali.

re:trospecti·vity. *rare.* [f. Retrospective *a.* + -ity.] = Retrospectiveness.
1920 *Glasgow Herald* 23 Feb. 11/2 The adoption of the principle of the non-retrospectivity of financial law.

retrospectus (retrospe·ktŏs). [a. L. *retrospectus*, pa. pple. of *retrospicere* to look back, or f. Retro- + -*spectus* after Conspectus, Prospectus.] A retrospective review or summary.
1964 *Listener* 15 Oct. 603/1 Mr. Brooke's conspectus (or retrospectus) of the impact of the second world war on British writers. 1971 *Guardian* 22 May 9/3 The Nonesuch [Press]..has a Retrospectus rather than a Backlist.

‖ **retroussage** (rətrusāʒ). [Fr., f. *retrousser* to turn up, tie up.] In etching: (see quots.).
1959 P. & L. Murray *Dict. Art & Artists* 271 Retroussage is a term used in etching to describe the action of passing a ball of muslin lightly over an inked plate with the intention of dragging some of the ink out of the lines and smearing it across the plate. 1965 Zigrosser & Gaehde *Guide to Collecting Orig. Prints* iv. 45 For etchings a little ink is often left on the plate, giving the print a slight tone instead of a dead white, and with a cloth the ink is drawn up slightly out of the lines in an operation known as *retroussage*.

retroussé, *a.* Add: (Earlier and further examples.) Also *ellipt.* as *sb.*
1802 C. Wilmot *Let.* 25 Apr. in *Irish Peer* (1920) 57 General MacDonald..is tall, and thin, the nez retroussé and his eyes round and solemn. 1871 *Young Englishwoman* July 393/2 The Gabrielle dress..falling with puff or *retroussé* of any kind at the back, is very graceful in outline. 1885 [see *oyster-plant* s.v. *oyster 5 d]. 1930 *Daily Express* 6 Oct. 5/4 Whereas Yvonne's is retroussé, Jane's nose is—well, it's *not* snub. 1967 J. R. & P. H. Napier *Handbk. Living Primates* i. 16 In the gelada baboons..the face is moderately prognathic..but the nose is short and retroussé.

retrovirus (re·trovəiᵊrŏs). *Biol.* [mod. L., f. initial letters of *reverse transcriptase* (see *transcriptase) + -o + Virus.] An RNA virus of the family Retroviridæ, characterized by oncogenicity and the possession of reverse transcriptase. Cf. *retravirus.
[1976 *Virology* LXXI. 375 Family Retroviridae (RNA tumor virus group formerly genus *Leukovirus*).] 1977 *Nature* 8 Sept. 105/2 The successful transfection of a retrovirus by DNA extracted from infected cells was first achieved by Hill and Hillova. 1978 *Ibid.* 9 Feb. 543/1 There has been increasing interest in the isolation of mammalian retroviruses and their evolutionary relatedness to different species.
Hence **re·troviral** *a.*
1979 *Nature* 29 Mar. 420/1 Our results could also be seen as evidence of a marked propensity of retroviral DNA to undergo genetic gymnastics, including deletion.

retsina (retsī·nă). Also **retzina, rezina.** [ad. mod.Gr. ρετσίνα, f. ρετσίνι resin, f. Gr. ῥητίνη

pine resin (cf. L. *resina*).] A Greek resinated wine.
1940 H. J. Grossman *Guide Wines, Spirits, Beers* xvi. 160 Present-day Greeks still prefer a resinated to a natural wine... These wines are available in the United States. They are labeled *Retsina*. 1952 W. Plomer *Museum Pieces* xiii. 113 We drink retsina under the pine-trees. 1960 L. Durrell *Clea* i. i. 20 The promised farewell dinner of lamb on the spit and gold *rezina* wine. 1966 [see *moussaka]. 1973 'D. Jordan' *Nile Green* xxxii. 154 We went down on to the quay to eat fish and drink retsina. 1977 *New Yorker* 6 June 31/3 The three boys' choirs participating in the Berlioz presented him with a jeroboam of retsina.

re·tting, *ppl. a.* [f. Ret *v.*[2] + -ing[2].] That rets or rots.
1930 *Discovery* Dec. 408/1 The clean flax fibre obtained from the harvested flax straw is a silky lustrous material of a pale cream colour, not susceptible to the attack of retting bacteria. 1978 *Sci. Amer.* Aug. 75/3 The seeds then sink to the bottom of the retting liquid, and the residue is removed by flotation and screening.

retune, *v.* Add: **b.** To rephrase (the words of a song, etc.).
1959 I. & P. Opie *Lore & Lang. Schoolch.* vii. 107 In no great space of time the schoolchild ditty, When I was young and had no sense I bought a fiddle for eighteen pence,..was re-tuned in her honour: Lottie Collins, she had no sense, She bought a piano for eighteen pence.
c. To tune (electronic equipment) to a different frequency. Also *absol.*
1962 Simpson & Richards *Physical Princ. Junction Transistors* xiv. 357 It is desirable to retune the input and output circuits to satisfy this condition. 1970 R. Hill *Clubbable Woman* vii. 204 Jenny's portable radio began to play..brass-band music. This faded..but then returned louder than before as though the set had been re-tuned. 1977 *Sci. Amer.* June 77/2 He need only retune his receiver to a slightly lower frequency. 1978 *Daily Tel.* 27 Nov. 13/6 Does the BBC realise that the surest way to make you either switch off or re-tune is to make you unsure about what is actually going out?
d. To alter the tuning of (an engine).
1974 P. Cave *Mama* (new ed.) ii. 10 He was the unofficial repairman of the London Angels, the guy who could take a beaten-up old engine and retune it until it could fly.
Hence **retu·ning** *vbl. sb.*
1909 M. Greenwood in L. Hill *Further Adv. Physiol.* 410 'Retuning' with orange, red and green both act like orange itself in inducing a negative image. 1960 *Practical Wireless* XXXVI. 390/1 It provides a substantially constant power gain over the whole of the band 1·5Mc/s to 24Mc/s and therefore needs no retuning when the working frequency is changed. 1976 D. Munrow *Instruments Middle Ages & Renaissance* ix. 77/4 Mattheson's complaint that a lutenist spent most of his life tuning rather than actually playing reflects the inconvenient necessity of regular re-tunings, which upset the stability of the instrument.

retu·rfing, *vbl. sb.* [f. Returf *v.* + -ing[1].] The action or fact of covering with new turf.
1974 *Listener* 24 Jan. 119/1 She directed the returfing of a lawn overnight to please her husband. 1978 *Nature* 7 Sept. 51/2 The ground levels within the stones of Stonehenge have been altered during the last 25 yr by re-turfing, by the laying of gravel in 1963 and by its replacement with turf in 1978.

return, *sb.* Add: **I. 1. a.** *spec.* in *Theol.* = *Parousia.
1914 J. F. Silver (*title*) The Lord's return seen in history and in Scripture as premillennial and imminent. 1931 W. Montgomery tr. *A. Schweitzer's Mysticism of Paul* vi. 111 Since a time-interval has been interposed between the Resurrection and Return, the Resurrection of Jesus has become an independent event. 1970 J. L. Houlden *Paul's Lett. from Prison* 107 Saviour: Paul's only use of this as a title for Christ. It refers to him in his role at his final return. 1977 G. W. H. Lampe *God as Spirit* vi. 171 We can, perhaps, retain the idea of a visible *parousia*: not in the impossible sense of a personal return of Jesus from a heavenly throne, but in the form of the consummation of God's creation of mankind.
c. (Further examples.) Also as adj. phr.
1837 Dickens *Let.* 22 Feb. (1965) I. 237 The best plan therefore, will be for you to write me *by return*. 1889 E. Dowson *Let.* 18 Feb. (1967) 37 Tell me..by return of post if possible what will have happened when I am next to resume the story. 1905 Joyce *Let.* ?18 Aug. (1966) II. 105, I send you by this post *A Painful Case* which you are to copy and send back by *return of post*. 1949 N. Mitford *Love in Cold Climate* i. ii. 20 She always answered letters by return of post. 1957 *Practical Wireless* XXXIII. 509/1 A by-return service of all types and sizes. 1981 J. Stubbs *Ironmaster* i. 21 They say now that we shall have letters 'by return of Post', meaning that we write today and receive a reply..the day after tomorrow.
e. (Later examples.) Also *attrib.* (passing into *adj.*), and in *day-return* (= *day-ticket* s.v. Day *sb.* 24).
1905 E. M. Forster *Where Angels fear to Tread* vi. 164 It was an irritable couple who took tickets to Monteriano. 'Singles or returns?' said he. *a* 1911 D. G. Phillips *Susan Lenox* (1917) I. viii. 126 He had the return half of his own ticket. 1924 D. Moore *Fen's First Term* ix. 93 'Third single, please.' 'Return?' inquired the clerk. 1931 D. L. Sayers *Five Red Herrings* iii. 41 He had taken a first-class return to Glasgow. 1952 'J. Tey' *Singing Sands* iv. 63 'So he had a return ticket.' 'Yes. The return half was in his wallet.' 1973 D. Lang *Freaks* 41, I went down there for a visit on a cheap day-return. 1975 'A. York' *Dark Passage* (1976) vii. 88 I'll be back tomorrow, I should think. But make the return half of the ticket open, will you? 1977 S. Brett *Star Trap* xv. 167 The man didn't stop to buy a ticket. He must have a return.
g. *and return*, and back again.
1887 C. B. George *Forty Years on Rail* v. 88 The train.. ran from Waukegan to Chicago and return every day.
h. *return to nature*, the abandonment of urban life in favour of rustic simplicity.
1902 Chesterton *Twelve Types* 142 Some think that the return to nature consists in drinking no wine; some think that it consists in drinking a great deal more than is good for them. 1908 E. J. Banfield *Confessions of Beachcomber* i. ii. 54 The conviction that the career of the Beachcomber, the closest possible 'return to Nature' now popularly advocated, has charms none other possesses. *c* 1914 Wyndham Lewis *Let.* (1963) 65, I might devote two or three words..to scouring the 'banal' nakedness of various 'Return to Nature' shits. 1978 *Times* 3 Oct. 14/3 This is not a plea for the simple life, a return to nature, or a general rising against the machines.
2. b. See also Happy *a.* 3 in Dict. and Suppl.
II. 6. a. Also *pl.* (See quot. 1963.) *Canad.*
1809 A. Henry *Trav. & Adventures Canad.* i. iv. 40 Here, the returns, in furs, are collected, and embarked for Montréal. 1843 R. Campbell *Jrnl.* 10 June in C. Wilson *Campbell of Yukon* (1970) viii. 68 As soon as the river opened, Whitford left with a boat & crew and the returns for Fort Simpson. 1908 C. Mair *Through Mackenzie Basin* 183 There was still much work to be done in the way of transport of outfit and returns between Anderson and Fort Good Hope. 1963 G. S. McTavish *Behind Palisades* xx. 84 'Returns'..was the generic name for furs, and all trade results which were returned to the Old Country in exchange. 1971 *Alberta Hist. Rev.* Autumn 9 During the next few years he..visited the posts and carried the returns down-river to St. Louis.
b. *return on capital*, gain, profit, or income earned by capital (see also quot. 1970).
Various other phrases, e.g. *return for capital*, *return to capital*, and *return to invested capital* were used from the late nineteenth century onward.
1938 W. L. Crum in *Harvard Business Rev.* XVI. 336 The return on invested capital is one of the most significant.. among possible measures of corporate performance. No entirely satisfactory determination of the average rate of return on invested capital..can be made, but data accumulated over recent years..enable us to prepare a fairly close estimate. 1962 A. Battersby *Guide to Stock Control* x. 94 'Return on capital' has been used in this book as a concise description of the cost of holding stocks, for two principal reasons—it is often the predominant cost and the most readily variable. 1969 J. Argenti *Managem. Techniques* 102 Confusion over the term 'Return on Capital' is considerable since this can have either the traditional meaning or the DCF meaning and there is no fixed relationship between these two figures. 1970 M. Greener *Penguin Dict. Commerce* 285 *Return on capital*, a rather nebulous phrase. In the terminology of investment analysis and accounting it means the profit earned by *capital*... In company accounts return on capital is often the ratio that the profit bears to the total *equity* funds or shareholders' funds employed... The phrase 'return on capital' is frequently used when what is really meant is *dividend yield* or *earnings yield*.
III. 9. c. *spec.* in *Cricket*, a summary of bowling figures at the end of play.
1976 J. Snow *Cricket Rebel* 44 My return read nought for 117. 1977 J. Laker *One-Day Cricket* 137 Bob White.. must have surprised many Londoners with a number of fine returns since he left Lords.
11. e. *ellipt.* In various sports: a return match.
1958 F. C. Avis *Boxing Reference Dict.* 111 *Return*, a second contest with a boxer whom one has previously fought. 1964 *Guardian* 2 Mar. 7/6 Or we might arrange for a monkey's (monkey's paw—draw) so that we can have a return. 1977 *Daily Express* 29 Mar. 32/4 England..could only score four in the return at Highbury.
12. c. (Later examples.)
1974 *Country Life* 24 Jan. 152/3 West should have led the Heart Nine, to ask for a Diamond return. 1978 *Times* 22 July 9/4 East led his ♠7, in order to..invite a spade return for a ruff.
e. (Earlier and later examples.)
1833 J. Nyren *Young Cricketer's Tutor* 70 He had..such a rapid return..that I have seen many put out..in a single run. 1906 A. E. Knight *Compl. Cricketer* iv. 143 Inaccurate and wild returns not merely give away runs, they contribute to general slovenliness and slackness. 1972 J. Mosedale *Football* x. 139 McElhenny embellished the performance with 52 yards on two kickoff returns, 32 on the punt returns.
15. a. (Further examples.)
1892 I. Zangwill *Childr. Ghetto* III. 53 He let himself fall backwards, impinging noiselessly upon a heap of 'returns' of number one. 1902 *Chambers's Jrnl.* Feb. 114/1 The bags of tea, known as 'returns' which the samplers who come to the warehouse bring from the establishments they represent. 1934 T. Wood *Cobbers* xi. 132 'Any returns?' says the waitress, challengingly. 1971 *Publishers' Weekly* 20 Dec. 16/1 The average rate of returns (unsold books) in the industry as a whole is about 50%. 1977 *Private Eye* 13 May 6/3 As the evening's performance was a charity gala, there would be no cheap seats available. Returns were few and would start at £3.
b. = *recovery 5 f.
1909 *Brit. Birds* II. 364 The returns for a species so much shot as the Woodcock are shown to be scarcely more than 5 per cent. 1959 *Jrnl. du Conseil* XXV. 58 Haddock tagging..has shown that returns may come from the position of liberation after a very long period of freedom, or from a long way off after a comparatively short time. 1975 *Trans. Illinois State Acad. Sci.* LXVIII. iii. 282 Band returns from quail released September or earlier indicated a general dispersal of 0·3–0·5 miles from the release site.
IV. 16. a. *return chaise* (earlier example), *flight, load, mail, omnibus, post, ship*; **return address,** the address to which a postal item is to be returned in case of non-delivery; **return date** *U.S. Law*, the date on which a specified person is required to appear in court; **return envelope** *U.S.*, an addressed envelope enclosed with a letter for the recipient's reply; **return fare,** the fare for a return-ticket.
1928 *Publishers' Weekly* 30 June 2605 All envelopes must carry the name and return address of the sender in the upper left hand corner. 1798 G. Thompson *Sentimental Tour* 20 Falling in with a return-chaise, I agreed with the driver for a cast—So far, for so much. 1972 *N.Y. Law Jrnl.* 24 Oct. 18/7 A subpoena without a specific return date and a specific place of attendance is invalid. 1973 *Ibid.* 31 Aug. 1/1 There is a substantial probability that he will not appear in court on the return date. 1886 'Mark Twain' *Lett. to Publishers* (1967) 205 Enclose a stamped and printed return-envelop. 1974 *Spartanburg* (S. Carolina) *Herald* 19 Apr. A4/8 So send for my booklet 'How to Prevent Platonic Marriage', enclosing a long stamped return envelope, plus 25 cents. 1976 *Eastern Even. News* (Norwich) 22 Dec. 7/3 The clerk..should..see if it was possible to reintroduce the reduced rate return fare in off peak hours. 1979 *Homes & Gardens* June 26/4 The return air fare to Nice is a cracking £180. 1966 M. Woodhouse *Tree Frog* x. 74 I've booked my return flight. 1976 *New Yorker* 15 Nov. 83/1 The European booking agent who had arranged for the Belgrade and London appearances paid for their return flight. 1913 Jones & Wyatt *Motor Traction for Business Purposes* 15 This matter of light mileage naturally brings us to the question of return loads. 1977 'D. Rutherford' *Return Load* i. 25 We have a job for you... A return load. 1864 *Harper's Mag.* Jan. 205/2 Miss Amber answered the letter by return mail. 1975 P. Fussell *Gt. War & Mod. Memory* ii. 67 Geoffrey Keynes specialized in receiving antiquarian booksellers' catalogs and buying books by return mail. 1860 A. J. Munby *Diary* 21 Feb. in D. Hudson *Munby* (1972) 51 Waiting for the return omnibus, I discovered some pretty Gothic schools, new, on the green. 1885 *List of Subscribers, Classified* (United Telephone Co.) (ed. 6) 226 (Advt.), Estimates per return post. 1912 W. Owen *Let.* ?14 Nov. (1967) 168, I flatly disobeyed your Return-Post command. 1929 M. de la Roche *Whiteoaks* ix. 137 Alayne wrote by return post. 1977 *Private Eye* 13 May 22/2 Send small sample of urine and £3 for reliable and strictly confidential results by first class return post. *c* 1700 Evelyn *Diary* an. 1680 (1955) IV. 212, I went to visite a French Stranger,..who having been thrice at the *East Indias, Persia* & other remote Countries, came hither in our returne ships from those parts.
b. *return current* (further example), *flow.*
1957 G. E. Hutchinson *Treat. Limnol.* I. v. 268 We have a wind drift at the surface almost along the direction of the wind, and a region near the bottom of current flowing against the wind, the so-called return current. 1964 *Economist* 23 May 837/3 The only possible source is 'return-flow' from irrigation.
c. *return conductor.*
1847 *Phil. Mag.* XXX. 194 The earth has been made to act the part of the return conductor.
e. return room, a mezzanine room at the turn of a flight of stairs.
1914 Joyce *Dubliners* 82 He glanced up and saw Jack regarding him from the door of the return-room. 1922 —— *Ulysses* 296 The other boot which he had been looking for was at present under the commode in the return room. 1927 St. John Ervine *Wayward Man* i. ii. 35 The return-room at the top of the first flight of stairs could be converted into a kitchen.
17. *return crease* (earlier and later examples).
1775 *New Articles Game of Cricket* 2 The Bowling-Crease must be parallel with the Stumps, Three Feet in Length, with a Return-Crease. 1902 [see *call *v.* 4 m]. 1948 *Sporting Mirror* 21 May 6/1 Three creases are marked out at each end of a cricket pitch—the bowling crease, return crease and popping crease. 1963 *Times* 23 Apr. 4/7 He [*sc.* an umpire] is required..to see whether the bowler is no-balling on the return crease. 1976 J. Snow *Cricket Rebel* 99 Stackpole's bat was outside the return crease at the time the wicket was broken.
19. *return entertaining, leg, match* (later examples; also *fig.*), *thanks, tie.*
1899 'Mark Twain' in *Forum* (N.Y.) Mar. 29 You can judge..what sort of return-entertaining she has done. 1973 *Times* 31 Oct. 10/5 After Soviet objections to playing the return leg in the Santiago stadium..Fifa..sent a delegation to investigate. 1891 W. G. Grace *Cricket* iv. 108 The return match, at Lord's..was more encouraging to us. 1915 J. Buchan *Thirty-Nine Steps* i. 18 This is the return match for the *pogroms*. The Jew is everywhere. 1929 *Evening News* 18 Nov. 16/6 In a return ice hockey match yesterday Berlin beat London by four goals to two. 1971 *Nature* 23 July 213/1 This follows..a conference in Washington a month ago... A 'return match' meeting to clinch the project is scheduled for Madrid for early August. 1973 *Times* 21 Mar. 9/6 The suspense of the medical report and the outcome of the return match. 1977 *Times* 8 Dec. 17/4 Contrary to my prematch forecast and your diarists' ex post account, I did not win: nor has a return match been played or planned. 1907 *Yesterday's Shopping* (1969) 361/1 'Return thanks' cards... Return thanks for kind enquiries and sympathy. 1972 *V.A.T.: Scope & Coverage* (H.M. Customs) 33 Postcards..acceptance cards; 'thank-you' or 'return-thanks' cards. 1972 G. Green *Great Moments in Sport: Soccer* xi. 108 The return tie was played at Maine Road.

return, *v.* Add: **I. 4. b.** *to return to nature*: to abandon urban life in favour of rustic simplicity. Cf. *return *sb.* 1 h.
1902 G. K. Chesterton *Twelve Types* 141 This attempt to re-establish communication with the elemental, or, as it is sometimes more roughly and fallaciously expressed, to return to nature.
III. 16. d. *Cricket.* Of a bowler: to achieve (bowling figures) in an innings or other session of play.

1969 *Times* 29 July 9/2 The best bowling figures..were returned by Bore. **1976** J. Snow *Cricket Rebel* 112 Sussex..were rolled over for 104, Warwickshire left-arm spinner Jim Allan returning career-best figures of five for 11. **1977** J. Laker *One-Day Cricket* 69 Gilmour returned the staggering figures of 12–6–14–6.

returnabi·lity. [f. Returnable *a.* + -ity.] The fact or condition of being returnable; capacity to return or be returned.
1920 in Webster. **1973** *Daily Tel.* (Colour Suppl.) 23 Mar. 13/1 The main problem was one of 'returnability'. 'The British Government has to be sure some country will receive you when your visa for England has expired.' **1973** *Times* 29 Oct. 15/8 A device could be used..which the customer would..associate with returnability. The use of the device would clearly indicate..that the package is the property of the supplier. **1981** *Times* 19 Jan. 3/6 The commission's proposed directive on beverage containers..is now expected to appear without any of the references to mandatory recycling or returnability contained in earlier drafts.

returnable, *a.* **1. b.** (Later examples.)
1972 *V.A.T.: Gen. Guide* (H.M. Customs) 34 Containers (returnable). **1973** *Sat. Rev. Society* (U.S.) May 70/3 Buy returnable bottles..participate in your local recycling program when you must use nonreturnable containers. **1981** *Country Life* 7 May 1228/1 The price of a returnable bottle is shared by many users.

returned, *ppl. a.* Add: **2. b.** Designating a discharged serviceman who has returned home from a war. *Canad.*, *Austral.*, and *N.Z.*
1915 *Eye Opener* (Calgary, Alberta) 11 Dec. 2/3 An army of returned soldiers..who have been away fighting for the Empire are not going to..allow themselves to be told what is right and what is wrong by a bunch of stay-at-home aldermen or freak legislators. **1921** *Daily Colonist* (Victoria, B.C.) 24 Mar. 1/4 A body of 600 returned men today marched to the city hall and demanded..that no returned men be refused relief by the city council. **1930** W. K. Hancock *Australia* xiii. 274 The only notable non-political body in Australia (excluding the churches) is the Returned Soldiers' League. **1947** E. A. McCourt *Music at Close* I. vi. 65 Latimer stopped..to pick up..a returned man living on a settlement grant. *a* **1948** L. G. D. Acland *Early Canterbury Runs* (1951) xi. 314 After the 1914–18 War the Government resumed the Old Man Range country..and settled a returned soldier on it. **1950** *N.Z. Jrnl. Agric.* Jan. 26/2 Ten 10-acre sections have been allocated to returned servicemen. **1961** P. White *Riders in Chariot* viii. 236 You cannot tell me..that a home is not a home, with so many going roofless, and so many returned men.
3. *returned empty* (example in lit. sense, and later *transf.* examples).
1946 P. Bottome *Lifeline* xv. 137 A milk lorry..goes from here taking returned empties, and bringing back milk at dawn. **1961** B. Fergusson *Watery Maze* ix. 220 Several of these [ships] were veterans both of Ironclad and Torch, which meant that they had already spent months in unprofitable voyaging as returned empties, so to speak. **1977** C. Allen *Raj* i. 23/2 Early in the nineteenth century..unmarried women were regularly shipped out..to meet the demand for wives... Those who returned without husbands or fiancés were known as 'Returned Empties'.

returnee·. orig. *U.S.* [f. Return *v.* + -ee[1].] One who returns or is returned from abroad to his native land, *esp.* from war service or exile. Also *attrib.* and *transf.*
1944 *Newsweek* 17 July 62 The biography of a returnee fits into a regular pattern. **1945** *Christian Sci. Monitor* 17 Mar. 10/2 (*caption*) A launch filled with returnees and their wives sets out on exploration tour of the coves. **1955** *Sci. News Let.* 27 Aug. 142/3 They suggest a brainwashing in reverse as treatment for returnees showing symptoms of mental sickness. **1969** *Daily Tel.* 27 Dec. 5/2 The Communists counted 54 dead, nine taken prisoner and 31 'returnees' during the truce period. A 'returnee' is a Viet Cong guerilla who surrenders peacefully to the South Vietnamese Government, thereby securing all citizen's privileges. **1974** W. Hunt *North of 53 Degrees* x. 60 The Klondike gold had escaped the grasp of these returnees. **1977** *Guardian Weekly* 18 Sept. 15/3 They want to go back to Russia... Would-be Soviet returnees. **1981** *Times* 22 Jan. 8/3 The former hostages, already inelegantly dubbed 'returnees'.

return-ticket. Add: (Earlier and later examples.) Also *fig.*
c **1847** J. S. Coyne in M. R. Booth *Eng. Plays of 19th Cent.* (1973) IV. 195 There's poor Mary White gone..to that bourne from whence no traveller gets a return ticket. **1850** Thackeray *Pendennis* II. xxxv. 339 (*heading*) Chapter xxxv. Shows how Arthur had better have taken a return-ticket. **1922** [see *come-back *sb.*[1] 2]. **1946** *Happy Landings* July 8/3 If you take your parachute into the air you have a certain return ticket, and one that, should you need to use it, will bring you safely down to earth. **1977** *Times* 15 June 16/2 The cautious person travels from capitalism to socialism with a return ticket.

re-type, *v.* Add: Now usu. as one word without hyphen. **2.** (Further examples.)
Now the usual sense.
1898 H. G. Wells *Let.* 1 Jan. in *G. Gissing & H. G. Wells* (1961) 70 Parts I have reshaped, rewritten and retyped time after time. **1952** S. Kauffmann *Tightrope* iv. 58 Rose retyped Russell's memo for him in the usual five copies. **1961** 'E. Lathen' *Banking on Death* xiii. 105 The letter had to be retyped. **1976** H. MacInnes *Agent in Place* xiii. 136 Typed, retyped, torn up. The discarded pages filled the waste-basket. **1979** *Amer. N. & Q.* Apr. 126/1 Faulkner entrusted the job to Phil Stone, who in turn gave the typescript to his secretary, Grace Hudson, to retype.

Hence **rety·ped** *ppl. a.*, **rety·ping** *vbl. sb.* Also **re·type** *sb.*, a retyped copy.
1930 D. L. Sayers *Strong Poison* xiv. 183 She..threw it into the waste-paper basket together with the re-type which she had begun. **1943** J. Reith *Diary* 4 May (1975) vi. 303 The vice-controller had my retyped memo today. **1967** Karch & Buber *Offset Processes* iv. 74 The IBM Selectric Composer..allows automatic justification of the right-hand margin with a retyping. **1979** *Amer. N. & Q.* Apr. 126/1 William Faulkner sent the newly revised and retyped text to Liveright.

retzian (re·tsiăn). *Min.* [f. the name of Anders Jahan *Retzius* (1742–1821), Swedish naturalist.] A basic arsenate of manganese, calcium and rare earth elements, known as dark brown orthorhombic crystals from Sweden.
1895 H. Sjögren in *Bull. Geol. Inst. Univ. Upsala* II. 54 (*heading*) Retzian, a new arseniate from the Mossgrufva, Nordmark. *Ibid.* 59 The crystals of retzian are prismatic. **1921** *Bull. U.S. Geol. Survey* No. 679. 126 Retzian..strongly pleochroic. **1967** *Amer. Mineralogist* LII. 1610 The crystal structure of retzian is an interesting sheet structure.

retzina, var. *retsina.

Reub, Rube (rūb), abbrevs. *Reuben 1. Also *attrib.* Cf. *hey, Rube!* s.v. *hey *int.*
1896 Ade *Artie* i. 8 If I had time I'd go over to that church and make a lot o' them Reubs look like thirty-cent pieces. **1899** 'J. Flynt' *Tramping with Tramps* 396 *Rube*, a 'hoosier', or 'farmer'. **1904** [see *bark *v.*[1] 2 b]. **1915** *Dialect Notes* IV. 200 *Reub, rube*, local name for rustic... 'On the Fourth of July the Reubs all come to town.' **1919** [see *kike]. **1927** *Scots Observer* 26 Feb. 3/4 They know a Rube when they see him, or a guy, or a crook, or a bonehead. **1937** *Lit. Digest* 3 Apr. 21/3 Child labor laws..have forced children [of circus people] into schools for a certain number of months a year. This contact with 'rubes' and 'townies' has given rise to new interests. **1946** E. O'Neill *Iceman Cometh* iv. 253 The boys tell me the rubes are wasting all their money buying food and times never was so hard. **1949** F. Sargeson *I saw it in my Dream* 218 Mr Anderson said Cedric was just a rube. **1953** W. Moore *Bring Jubilee* ii. 20 Rube, huh? Much money you got?.. O.K., Reuben. Come along. **1956** W. R. Bird *Off-Trail in Nova Scotia* iii. 69 Don't try to talk down to someone you meet in those places no matter if they look like rubes for ten to one they've seen more of the world than you have. **1970** R. Price *Gt. Roob Revolution* 5 The word 'rube' (from Reuben) was originally used by carnival people to identify the farmers, yokels, and assorted rural types. **1973** J. Gores *Final Notice* xxiv. 158 The rube who wanders into the pool hall and loses a few games... Then the bets get bigger and he..starts clearing tables. **1975** *New Yorker* 29 Dec. 36/3 We had one girl who did a good rube act. **1976** *Time* 20 Dec. 17/1 They were an unsolemn pair, the young man who likes his rube image and the impeccably dressed man who looked more like a smooth character actor than a politician of enormous influence.

Reuben (rū·bən). *U.S.* and *Canad. colloq.*
1. The personal name *Reuben* applied to suggest the conventionally conceived figure of a farmer or rustic; a country bumpkin.
1804 in S. Larkin *Nightingale* 284 But she, tho' conscious of his worth, Had chose a youth more rare; a rustic Reuben was his name. **1890** B. Hall *Turnover Club* iv. 49, I overheard one of a knot of Reubens standing on a corner. **1901** (*song-title*) The wedding of the Reuben and the maid. *Ibid.*, You've heard about the Reuben and the time he came to town. **1905** 'H. McHugh' *You can search Me* 60 I've a couple of new card tricks..that will leave the Reubens gasping for air. **1911** H. Quick *Yellowstone Nights* xii. 313, I took a basket of eggs an went in among 'em, feelin' like a animal trainer in a circus parade as the Reubens gathered around the train. **1953** [see prec.].
2. In full, *Reuben sandwich.* A large sandwich containing cheese, meat, and sauerkraut, usu. made with rye bread and served hot.
Not obviously connected with sense 1.
1956 *Institutions* Oct. 44/2 'The Reuben'..is a grilled 3 decker sandwich of heroic proportions. Three slices of russian rye bread generously spread with thousand island dressing contain swiss cheese and corned beef slices interlaced with sauerkraut. **1967** *N.Y. Times* 10 Oct. 50 The Reuben sandwich..has become wildly popular in most areas of the United States. **1970** *Favorite Recipes of Lutheran Ladies* 15 Reuben Sandwich... Spread 6 slices bread with dressing, top with cheese, 1 tablespoon sauerkraut and corned beef. **1976** in K. M. Thomas *Winning Sandwiches for Menu Makers* 2/2 The recipes in this book represent the best of twenty years of sandwich competition, starting with the champion of the first contest in 1956, the now ubiquitous 'Reuben'.

reune (rĭ̦yū·n), *v.* *U.S. colloq.* [Back-formation from Reunion.] *intr.* To hold a reunion.
1901 *Princeton Alumni Weekly* 8 June 774/1 As the secretary of a class which has 'reuned' frequently..I wish to direct your attention to a breach of etiquette. **1929** *Amer. Speech* V. 175 Several years ago the late E. K. Graham, then president of the University of North Carolina, in speaking of class reunions, used somewhat facetiously the word *reune* instead of the locution *have a reunion*. It seems that the word has stuck. **1929** E. W. Springs *Above Bright Blue Sky* 237 She had seen in the paper where the 14th was going to reune at the banquet. **1948** *Amer. Legion Mag.* Oct. 26/1 (*heading*) Retreads Will Reune. **1949** Mencken in Kirby & Woolf *Philologica* 316 The embryologist of speech discerns several processes in the making of such novelties [*sc.* new verbs]. Some are simply nouns unchanged, e.g., *to contact*.. others are back-formations from nouns, e.g., *to locate, to enthuse, to reune.*

reunification. Add: (Further examples.) Hence **reunifica·tionist**, a supporter or advocate of reunification.
1955 *Times* 14 July 8/5 No policy is acceptable to Germany which does not have for its aim the reunification of the country in peace and freedom, and at the earliest possible moment. **1959** *Daily Tel.* 9 Apr. 10/3 Plans for control and inspection, and for a ceiling for forces, would be accepted only if progress towards German reunification went with them. **1967** *Listener* 6 July 20/2 For him [*sc.* Dr. Adenauer] German reunification was a necessary pre-condition of détente. **1970** W. Johnson *Cameroon Federation* v. 105 What incensed the reunificationists most, however, was the belief that the Territory had not received its just due even from the government of Nigeria. **1976** *Survey* Winter 18 One can imagine the Japanese reaction to the reunification of Korea under communist leadership.

reunify, *v.* (Later examples.)
1955 *Times* 8 July 8/3 The Assembly endorsed the main conclusions of M. de Menthon's report, which formed the basis of the debate—the need to establish a real system of security, to reunify Germany by means of free elections, and to construct a united Europe. *Ibid.* 13 July 6/4 French reactions to the suggestions, made in a recent leading article in *The Times*, that Germany might be reunified on the basis of a demilitarized eastern zone, are not *prima facie* favourable.

reunion. **3.** (Later *attrib.* examples.)
1929 E. W. Springs *Above Bright Blue Sky* 233 Are you going to the Aviators' Reunion Dinner to-night to celebrate the anniversary of the Armistice? **1952** M. Allingham *Tiger in Smoke* xv. 218 He was due to give an impersonation of his old officer at a reunion dinner. **1976** *Laurel* (Montana) *Outlook* 9 June 12/4 A two-hour program is being planned by a reunion committee and will be held from 2 to 4 p.m. in the Laurel Senior High School gymnasium.

re-up (rī·ˌʌp), *v.* *U.S. Services' slang.* [f. Re- 5 a + Up *v.*: see quots. 1930, 1942.] *intr.* To re-enlist. Also as *sb.*, one who re-enlists. Hence **re-u·pping** *vbl. sb.*
1906 *Soldier Slang* in C. M'Govern *Sarjint Larry an' Frinds, Re-up*, to re-enlist. **1913** [see *hitch *sb.* 8]. **1930** W. H. Waldron *Old Sergeant's Conferences* vii. 122 To 'Re-up' is to reenlist on the day following discharge... The 'Up' refers to holding up his hand to receive the oath of enlistment. **1942** E. Colby *Army Talk* 174 When enlisting and being sworn in, a man is said to 'hold up his right hand' for three years. So when he does it after being discharged, he 're-ups'. **1955** *Air Force Times* (U.S.) (Eastern ed.) 31 Dec. 13/1 Despite the surprising success of the re-up efforts, the problem of getting qualified replacements for the still critical 'hardcore' skills remains. **1958** *Ibid.* 2 Aug. 1/4 (*heading*) Re-Upping Quickly Pays Off. **1970** *Times* 28 May 7/7 The person they are likely to meet is the recruiting officer with his 're-up' quota to fill. **1972** J. Gores *Dead Skip* xi. 75 If he'd re-upped when his two years in the army were finished, he'd have been a sergeant by now. **1974** *Black Panther* 9 Feb. 22/1, I was told to talk to a recruiter on base about re-enlisting... He told me that if I re-up for the four-year reserve commitment he would fix it up so that I had a job waiting for me.

reupho·lster (rī-), *v.* [Re- 5 c.] *trans.* To upholster anew. So **reupho·lstery.**
1935 *Amer. Speech* X. 154/2 Entire [car] bodies rebuilt, reupholstered. **1943** F. Thompson *Candleford Green* vii. 116 The easy chair had been carried away to be re-upholstered. **1976** R. Rendell *Demon in my View* ii. 22 The sofa and the two armchairs had been re-upholstered. **1977** *Time* 4 July 6/3 Mortified French officials rushed the vehicle back to its manufacturer, where men on the assembly line worked frantically on reupholstery..and a new paint job. **1978** *N.Y. Times* 30 Mar. c13/3 (Advt.), Upholstery & reupholstery workroom. **1978** *P.O. Telephone Directory, Preston Area: Yellow Pages* 357 (Advt.), Specialists in upholstery & re-upholstery.

reusable (riyū·zăb'l), *a.* [f. Re-use *v.* + -able.] Capable of being re-used; suitable for a second or further use.
1959 *Sears, Roebuck Catal.* Spring & Summer 835/1 Paper plates..With re-usable plastic bag. **1962** *Guardian* 5 Dec. 6/3 Paper place mats..disposable, but certainly reusable several times. **1967** *Times Rev. Industry* Feb. 39/1 (Advt.), Dictamite takes a whole hour's dictation on one tiny reusable tape cartridge. **1970** *New Scientist* 26 Feb. 407/1 Those contractors who have plans for the airframe or engines of the reusable shuttle must submit them to NASA. **1971** *Engineering* Apr. 59/2 Some strippable coatings are reusable. **1975** *Nature* 17 Jan. 149/2 Out of the other end, or rather ends, come several potentially valuable by-products. Three grades of reuseable paper, two already pronounced as suitable for board making by commercial board mills, one suitable for high grade fuel.
Hence **reusabi·lity.**
1970 *Sci. Jrnl.* Aug. 32/2 'Reusability'—the introduction of launch vehicles and spacecraft that can be flown to space, returned and used again. **1973** *Daily Tel.* 24 Jan. 19 (Advt.), It had to meet rigid specifications. Uniform size and shape, chemical content, heat resistance, ventilation, reusability.

reu·sage (rĭ-). [Re- 5 a.] A second or further usage.
1956 *Discovery* July 230/2 Earlier examples of reusage of bricks by Saxon builders are numerous.

‖ **réussi** (reˌusi), *a.* Also fem. **réussie.** [Fr.] Fine, excellent, successful.
1948 F. R. Leavis *Great Tradition* ii. 37 Adam [Bede], we know, is a tribute to her father; but he is also the Ideal Craftsman, embodying the Dignity of Labour. He too is *réussi*. **1953** 'N. Blake' *Dreadful Hollow* vi. 75 She drew out of its wrappings an elaborate construction of shells enclosed by a glass dome. 'He must have made it himself. You see? It's a sort of mausoleum. Rather weird and *réussi*, isn't it?' **1958** B. Nichols *Sweet & Twenties* iii. 48 Here, I thought, is a woman who is completely *réussie*.

Reuter (roi·təɪ). The name of Baron Paul Julius von *Reuter* (1816–99), founder of a telegraphic and pigeon post bureau in Aachen in 1849, used *attrib.* and in the possessive to denote (the activities of) a news agency named after him, whose London headquarters were established in 1851. Also (in form **Reuters**) used *absol.*
[**1859** *Times* 31 Oct. 6/5 The following telegrams have been received at Mr. Reuter's office.] **1860** *Times* 1 Oct. 7/1 (*heading*) Reuter's Telegrams. **1913** Kipling *Lett. of Travel* (1920) 222 No newspapers come aboard, only clipped Reuter telegrams. **1964** M. McLuhan *Understanding Media* II. xxiv. 235 A Reuters dispatch for December 13, 1962, reported from Tokyo. **1967** *Listener* 28 Dec. 860/1 In the early days of BBC news when they were terrified of infringing the rights of newspapers they used to read out the Reuter tapes. **1969** *Observer* 5 Oct. 1/2 Reuter correspondent Anthony Grey was a free man in Peking tonight after being held hostage by the Chinese for 26 lonely months. **1977** *New Yorker* 3 Oct. 98/2 My guide..would be an Englishman.. John Peet, who had covered Germany for Reuters, the British news agency. **1980** M. Lutyens *Edwin Lutyens* xv. 232 Every morning Roderick's special Reuter messenger boy brought me a letter from him.

reutiliza·tion (rĭ-). [Re- 5 a.] A second or further utilization.
1936 *Discovery* July 230/2 Virtually all English brick-craft originated by the reutilisation of Roman bricks in mediaeval buildings. **1957** *New Biol.* XXIII. 81 The induced enzyme therefore appeared to be synthesized from amino acids present in the free amino acid pool and not by partial breakdown of other cell proteins, followed by the reutilization of the peptide moieties produced.

reu·tilize (rĭ-), *v.* [Re- 5 a.] *trans.* To utilize again. Hence **reu·tilized** *ppl. a.*
1889 *Lancet* 21 Sept. 585/2 After the white cells have lived their life and done their work, portions of their worn-out carcases may be reutilised in the body as nutriment. **1964** D. Nichols in *Oceanogr. & Marine Biol.* II. 408 Cornil, Mosinger and Calen (1935 a, b) were of the opinion that the matter eliminated by some amoebocytes in the epithelium was re-utilized by others. **1972** *Biblical Theol. Bull.* Feb. 30 Some of these reutilized traditional materials deserve a separate mention, namely the various formulas of prayer.

Rev., abbrev. of Reverend *a.* and *sb.* 2 c, d.
In U.S. usage freq. without preceding article.
1721 D. Wilkins *Leges Anglo-Saxonicæ* (list of subscribers), The Rev. Mr. Henry Briggs of Loose in Kent, [etc.]. **1785**, etc. [see Reverend *a.* and *sb.* 2 c]. **1876** W. Whitman *Daybks. & Notebks.* (1978) I. 37 Rev. R. P. Graves. **1884** [see Chasuble 1]. **1917** *Congress. Rec.* (Daily ed.) 14 Dec. 294/2 Rev. Joseph Burt Webster to be chaplain with rank from October 5, 1917. **1939** *Time* 27 Nov. 50/3 (*heading*) Mr. for Rev. **1943** R. Llewellyn *None but Lonely Heart* xxi. 118 Reverend Ernest Mott. Oxford, he is. Followed 'em on His bike as far as Hammersmith Bridge every year since '98. Ain't you, Rev, boy? **1969** C. Burke *God is Beautiful, Man* (1970) 74 The Revs. in the church got cut in on it too. **1972** J. Maryland in T. Kochman *Rappin' & Stylin' Out* 213 Say, Rev., tell them 'bout the little white lady. **1973** *Philadelphia Inquirer* (Today Suppl.) 7 Oct. 14/3 Rev. Velasquez recalls that before the first trial he heard Tony's defense attorney say [etc.]. **1978** *Listener* 9 Mar. 296/3 Religious broadcasting is not a job for a good man... Even the radio Revs and their staff are always set apart from other departments.

rev (rev), *sb.* Also **rev.** (with point). Abbrev. of Revolution *sb.* 4 c.
1901 *Catal. Mech. Engin. Coll.* (Victoria & Albert Mus.) I. 35 The example has 3 in. cylinders, with 4 in. stroke, and is intended to run at 300 revs. per min. **1918** *Chambers's Jrnl.* May 301/1 The revs dropped off. **1932** S. C. H. Davis *Motor Racing* vi. 87 The car went well, held its revs. with something in hand. **1942** *Tee Emm* (Air Ministry) II. 87 The revs. of the engine..can be controlled by the pilot in one of two ways. **1966** R. Maxwell in T. Wisdom *High-Performance Driving* v. 44, I stayed in each gear to maximum revs for best acceleration on the short, so-called straights. **1969** *Listener* 22 May 734/1, I admit to having enjoyed recently Kagel's *Improvisation Ajoutée* for organ.. at 45 r.p.m. In retrospect I am convinced that those extra 11⅔ revs per minute tightened up the form of this fascinating work to a slight but ideal degree. **1972** *Daily Tel.* 25 Feb. 2/7 As I got round the bend onto the main road I felt the revs begin to build up. When this happened I changed up a gear.
2. *Comb.* **rev-counter** (and varr.), † **rev-meter,** an instrument that measures and displays the rate of rotation, esp. of an engine, or the number of rotations.
1917 *Blackw. Mag.* May 803/2 The rev.-counter showed that the number of revolutions per minute had fallen off appreciably. **1948** M. Laski *Tory Heaven* iv. 58 The car.. had a searchlight and a wireless mast and outside gears and a rev counter. **1967** E. Chambers *Photolitho-Offset* viii. 103 It is possible to compute mathematically the required settings which are..mechanically applied in different ways. For example,..by using a rev-counter and vernier dials attached and synchronised with the movements of the lens board and copyboard carriages. **1976** *Glasgow Herald* 26 Nov. 19/5 The speedometer and rev. counter are..right in front of the driver. **1917** E. C. Middleton *Way of Air* vii. 49 The pilot is able to distinguish his instruments..the altimeter, which records the height, 'revmeter' which indicates the speed of the engine and the compass.

rev (rev), *v.* Pa. t. **revved,** pres. pple. **revving.** [f. the sb.] **1.** *trans.* To cause (an internal-combustion engine) to run quickly, esp. before bringing it into use; to speed *up*. Freq. *absol.* Also *fig.*
1920 *Blackw. Mag.* Oct. 449/1 A British Fighter whose pilot was revving up his 250-horse-power Rolls Royce Falcon. **1922** Joyce *Ulysses* 420 O get, rev on a gradient one in nine. **1934** [see *idle *v.* 4 a]. **1942** *Sat. Even. Post* 5 Sept. 22/1 Jimmy revved up and the engine burst into a deep-throated roar. **1944** A. Thirkell *Headmistress* i. 28 'Rev. her up, Copper.' The ginger-haired bicyclist..roared away up the street. **1956** *Sun* (Baltimore) 24 Apr. 21/1 Bill Hartack revved up his already torrid riding pace here today by winning four races out of seven tries to push his total victory for the year to 94. **1966** J. Betjeman *High & Low* 50 Loving relations Rev in the car park, changing gear at the bend. **1968** Mrs. L. B. Johnson *White House Diary* 31 Mar. (1970) 644 Lyndon came in with that jaunty step that I've seen him rev up under the most intense tension. **1973** A. Hunter *Gently French* iii. 23 The mech gently revved it, bringing in the supercharger. **1977** *Time* 13 June 47/2 He and a cousin revved up the company, branched into trucking and started hauling coal. **1978** J. Irving *World according to Garp* xii. 234 He revved his engine as if he were clearing his throat.
2. *intr.* Of an internal-combustion engine: to run (quickly), esp. with the clutch disengaged. Also said of the vehicle. Freq. with *up*.
1923 *Daily Mail* 9 Mar. 12 This little engine..cheats the taxation authorities by its long stroke and its capacity for 'revving' fast without vibration. **1930** *Times* 29 Mar. 17/6 (Advt.), The oil..is specially refined for the fast-revving engines of to-day. **1951** 'J. Wyndham' *Day of Triffids* i. 9 The released cross-traffic would rev. and roar as it started up the incline. **1955** G. Greene *Quiet American* II. ii. 111 All round me I could hear the cars of the soldiers and the diplomats revving up; the party was over for another year. **1960** J. Betjeman *Summoned by Bells* vii. 73 There's a Frazer–Nash. Gosh, what an engine! Did you hear her rev? **1965** *Motor* 17 July 8/1 The Anglia revs faster at any given speed in top. **1973** J. Pattinson *Search Warrant* vi. 90 He heard a motor engine revving up, dying away, revving again.
Hence **re·vved-up** *a.* (in quots., *fig.*); **re·vving** *ppl. a.* and *vbl. sb.*
1931 E. M. Brent-Dyer *Chalet School & Jo* xxi. 261 I'm so revved up,..I simply *can't* rest. Please, may I go for a walk instead? **1972** J. Aiken *Butterfly Picnic* vi. 107 My mind seemed revved up to operate at twice its normal pace. **1976** J. Carroll *Madonna Red* v. (1977) 174 The rush of his revved-up thinking. **1978** M. Harrison in *Islands* (N.Z.) Aug. 86, I didn't dream of standing by the revving plane, the wires of the fences caught the whisper of the landing-lights. **1979** K. Conlon *Move in Game* iii. 32 The man who drove his scarlet MG with much revving of the engine.

revaloriza·tion (rĭ-). [Re- 5 a.] The action or process of establishing a fresh price or value for something; revaluation. So **reva·lorize** *v. trans.*
1926 *Glasgow Herald* 6 Aug. 8 A preliminary revalorisation [of the franc] through a restoration of confidence should first be attempted. **1928** *Britain's Industr. Future* (Liberal Industr. Inquiry) v. xxviii. 410 The desire to 'revalorise' the currency (i.e. to raise its exchange value) before 'stabilising' it. This process of revalorisation..is infallibly accompanied..by severe trade depression. **1928** *Daily Tel.* 27 Mar. 14/3 The leading banks..replied that they would not advance a pfennig until the war debts had been revalorised and admitted. **1962** *Listener* 12 Apr. 624/2 If we are to have ministerial meetings they need not always be at the level of Heads of Government. Foreign Ministers must be revalorized. **1977** *Times* 27 Apr. 23/5 Mr Healey gets beta plus for revalorizing personal reliefs to counter about one year's fiscal drag. **1979** *Daily Tel.* 4 Apr. 10 These increases give effect to the re-valorisation of personal allowances provided for by Section 22 of the Finance Act 1977. **1979** *Maledicta* III. 71 In their attempts to enshrine and revalorize their own blackness and that of their intended audience, these poets of cultural nationalism frequently seem to assume [etc.].

reva·luate (rĭ-), *v.* [Back-formation f. Revaluation.] *trans.* To reassess, form a new valuation of. Also *absol.* Hence **reva·luating** *ppl. a.*
1949 Wellek & Warren *Theory of Lit.* iv. 34 The practicing critic..will revaluate the past. **1965** 'A. Burgess' in *Listener* 18 Feb. 275/1 'Another revaluation of Joyce would make one scream.' By the waters of Marylebone I sat down and wept, wondering whether to sink my revaluating galleys. **1977** *Times* 5 July 5/4 Even if a firm offer is made, however, both the British and French Governments will undoubtedly demand time to revaluate their respective commitments.

revaluation. Add: **a.** *spec.* in literary criticism.
1936 F. R. Leavis (*title*) Revaluation. **1976** *UCT Stud. in English* (Univ. Cape Town) Oct. 1 After the critical revaluations of the past twenty years, Wallace Stevens can no longer be thought of as an eccentric. **1978** *Times Lit. Suppl.* 1 Dec. 1393/2 What he is doing is practising the classic American trick of hurling effete European vice out of the window and then lugging it back through the door, wearing a coonskin cap and called Revaluation.
b. A second or further valuation of the financial or monetary worth of something, *spec.* a revision (usu. an increase) of the official value of a currency in relation to gold or another currency.
1925 *Times* 22 Sept. 22/6 The German Embassy has issued the following information regarding the German Law of Revaluation:—Mortgages, land and annuity debts..will on principle be revalued at 25 per cent. of their gold value. **1946** *Times* 13 July 7/1 Revaluation will tend to neutralize the impact [of disparity with U.S. prices] on the Swedish economy. **1955** *Times* 27 June 9/3 No ratepayer can properly gauge the effect of revaluation until he knows what rate his local authority intends to levy in the new dispensation. **1962** *Ann. Reg. 1961* 465 Sterling was under pressure as a result of general uncertainty following the revaluations of the Deutschemark and Dutch guilder. **1968** *Guardian* 20 Nov. 12/2 Revaluation means increasing the value of a currency in terms of other currencies. **1971** *Daily Tel.* 10 May 1/1 Dr Schmidt, president of the National Bank [of Austria] said losses to the reserves caused by the revaluation amounted to 1,500 million schillings. **1978** *Financial Times* 30 Jan. 2/6 The half per cent revaluation of the Dirham against the dollar..brought it into line with the currencies of Bahrain and Qatar.

revalue, *v.* Add: (Earlier and later examples.) *spec.* To adjust (usu. increase) the value of (a currency) in relation to gold or another currency.
1592 H. Wotton *Let.* in *Reliquiæ Wottonianæ* (1685) 668 The Sum should be revalued. **1925** [see *revaluation b]. **1946** *Times* 13 July 7/1 (*heading*), Swedish Krona Revalued. **1962** *Ann. Reg. 1961* 245 On 5 March the Deutschemark was revalued upwards by 4·75 per cent. **1968** *Guardian* 16 Nov. 1/2 The French..have been putting strong pressure on the Germans..to revalue. This would..make German exports less competitive. **1978** *Financial Times* 30 Jan. 2/5 The Bahrain dinar was revalued by 2 per cent against the U.S. dollar at the opening of business in Bahrain on Saturday.

revamp, *v.* Add: (Further examples.) Also, to rewrite in a new form; to renovate, remake, devise anew; to revise. Hence **revamped** *ppl. a.*; **revamping** *vbl. sb.* (further examples).
1850 D. G. Mitchell *Lorgnette* I. 141 Even the soberer subjects of History, he told me, must be re-vamped in some tasty way. **1878** 'Mark Twain' in *Atlantic Monthly* May 617/2 He had to keep on procuring magazine acceptances and then revamping the manuscripts to make them presentable. **1931** Ford & Crowther *Moving Forward* xi. 164 We decided to revamp our entire industry from top to bottom. **1934** J. A. & A. Lomax *Amer. Ballads & Folk Songs* i. 37 Sanders then changed the words 'Jimmie Jones' to 'Casey Jones'. Later it was picked up by some traveling vaudevillians and revamped to make the popularly known song, 'Casey Jones'. **1941** W. C. Handy *Father of Blues* xii. 169 The revamped outfit was acting strangely, whispering behind their hands. **1945** Nelson & Wright *Tomorrow's House* vii. 79 This is just the living-room function slightly revamped. **1947** *Sun* (Baltimore) 22 Dec. 2/1 Eastern and western states will revamp their foreign policies in the light of the situation in Germany. **1950** *Manch. Guardian Weekly* 7 Dec. 3 The British did some rapid discreet revamping of their agenda. **1956** W. H. Whyte *Organization Man* (1957) 99 Several of the technical schools have been revamping their curriculums. **1960** O. Skilbeck *ABC of Film & TV* 108 *Re-vamp*, to use an existing set for further shooting by altering it and adapting its practicals for another sequence. **1961** *Harper's Bazaar* Dec. 60 Brilliant revamping of the famous Helena Rubinstein flat in Paris. **1965** *Listener* 16 Dec. 984/2 Over the past three years they [*sc.* the East Germans] have done a lot of revamping, and are introducing what they rather ponderously call 'the new system of planning and guiding the national economy'. **1968** *Observer* 29 Dec. 32/3 On Wednesday, *Punch* appears revamped and under new editorship. **1979** *Jrnl. R. Soc. Arts* CXXVII. 122/1 The new or revamped public schools did not set out to equip their pupils to lead great industrial enterprises.
revamp *sb.* (later examples).
1943 *Yank* 10 Feb. 8 If a plane comes in badly shot up and has to go to basic engineering for a revamp, two crew members will go with her and personally inspect everything that's done. **1973** *Scotsman* 12 Jan. 3/1 Rockware Glass, the £16 million container group, are planning a major revamp of their factories. **1979** *Now!* 21–7 Sept. 59/3 The Navy IQ test was introduced in 1942, and given a 'revamp' in 1977.

‖ **revanchard** (rəvañʃāɪ), *a.* [Fr.] = *re-vanchist *a.*
1961 *Washington Post* 28 Aug. A-8/4 Placed before the fact..of a separate state which..immediately started to acquire a belligerent revanchard character, the population of East Germany in their turn created another German state. **1964** *Listener* 1 Oct. 523/2 In the eighteen-eighties, thanks partly to Boulanger, it was the Right [in France] which became militantly *revanchard*, and by the turn of the century it was denouncing the Left for being socialist, pacifist, semitic, and *dreyfusard*.

‖ **revanche** (rəvañʃ). [Fr.] Requital, revenge; the giving of like for like. *spec.* A nation's policy of securing the return of lost territory.
1858 Queen Victoria *Let.* 22 June in R. Fulford *Dearest Child* (1964) 117 She never allows a word to be said against Leopold who in revanche is much kinder to her than he was. **1870** G. Meredith *Let.* 9 Oct. (1970) I. 427 You greatmindedly took my criticism, and I long for the *revanche* of giving praise. **1889** M. S. Van de Velde *Cosmopolitan Recoll.* I. v. 162 He [*sc.* Prince Gortchakoff] has on his record the fate of Sleswig and of Denmark; Sadowa, the *revanche* of Sebastopol. **1894** G. du Maurier *Trilby* III. viii. 157 When they come back from *La Revanche*, may Madame Cantharidi ..welcome the returning heroes. **1895** M. Corelli *Sorrows of Satan* X. 108 Meanwhile I was surrounded by the rest of the men, all of them repeating the Viscount's suggestion of a 'revanche'. **1914** G. B. Shaw in *New Statesman* 14 Nov. (Suppl.) 20/1 France had given up hope of her Alsace-Lorraine *revanche*. **1919** J. M. Keynes *Econ. Consequences of Peace* (1920) iii. 32 Each guarantee that was taken, by increasing irritation and thus the probability of a subsequent *Revanche* by Germany, made necessary yet further provisions to crush. **1930** H. G. Wells *Autocracy of Mr. Parham* iv. ii. 285 An unhoped for *revanche* offered itself plainly and clearly to the German people. **1934** B. Russell *Freedom & Organization 1814–1914* xxxii. 495 He [*sc.* the Kaiser] caused England to join France and Russia, thereby encouraging Germany's enemies everywhere, awaking in France renewed hope of the *revanche*. **1939** A. J. Toynbee *Study of Hist.* IV. 118 The Justinianean *revanche*, in the sixth century, against the Vandals and the Ostrogoths. *a* **1943** R. G. Collingwood *Idea of Hist.* (1946) 169 History is regarded not only as a possible and legitimate form of knowledge but as the only genuine knowledge that exists or can exist. But this *revanche* not only fails in doing justice to natural science, it also misunderstands history. **1958** *Listener* 10 July 57/1 In 1914..France wanted Revanche.

revanchist (revǫ·nʃist), *sb.* and *a.* [f. prec. + -ist: cf. Fr. *revanchiste* (also used).] **A.** *sb.* One who seeks reprisal or revenge; *spec.* one who seeks to avenge the defeat of Germany in the war of 1939–45. **B.** *adj.* Also **revanchi·stic.** Pertaining to or characterized by a policy of reprisal or revenge.
1926 *Scots Observer* 13 Nov. 19/1 A Germany contented, satisfied, and prosperous would pay her debt and achieve a bloodless but a complete victory. It is France's policy that the sores be kept open even if they give a handle to Monarchist revanchists. **1953** *Ann. Reg. 1952* 195 The creation in Western Germany of hired forces of *revanchistes* under the command of Hitlerite generals. **1954** *Times* 5 Mar. 6/4 'Then their real aims will be revealed as well as their aggressive, revanchist calculations,' Mr. Molotov added... The Soviet Foreign Minister said the Adenauer Government was now 'the mainstay of the west German revanchists who want to clear the road to the resurgence of German militarism'. **1955** *N.Y. Times* 10 May 28/1 Far from being an instrument of either 'American aggressors' or 'German revanchists' (as Soviet propaganda would picture it), the North Atlantic alliance is entirely a defensive organization. **1959** *Times* 30 Mar. 5/4 Austria had undertaken to prohibit all Fascist and 'revanchist' organizations, and to wipe out all remaining traces of the Nazis. **1960** *Guardian* 25 Oct. 7/2 Revanchist speeches were being made..on the borders of Czechoslovakia. **1961** *Evening Bull.* (Philadelphia) 24 Aug. 1/8 Air corridors were set aside for the three Western Powers to insure the needs of their military garrisons, and not for subversive and revanchist (revenge-seeking) purposes. **1961** *Times Lit. Suppl.* 17 Nov. 815/4 Between the wars the Hungarians were only revisionists. No one labelled them as revenge-seekers and the communists had surely not yet invented that bastard word 'revanchist'. **1965** *New Statesman* 19 Mar. 432/1, I asked a prominent East German lawyer why, rather than sneer about 'revanchists', they did not support this attempt to convict Nazi criminals. **1965** *Listener* 16 Dec. 983/2 East German propaganda would say it [*sc.* living in West Germany] meant living under 'a state-monopolistic revanchistic militarist regime'. **1967** *Ibid.* 3 Aug. 155/3 When journalists and scholars..label Israeli revanchistes 'Jewish fascists', it's apparent that such tastelessness is even less bearable in non-fiction. **1968** *Daily Tel.* 1 Nov. 34/3 The statement referred also to the 'revanchist and expansionist forces in control in Bonn'. **1973** *Guardian* 19 Oct. 14/6 Not terrorists but revanchist states were the greater threat [to Israel]. **1976** T. Allbeury *Only Good German* xiii. 94 A sabotage team operating against the Warsaw Pact countries..supported by the CIA and West German revanchists. **1977** *Time* 7 Feb. 15/3 O'Brien supports a referendum to amend the Irish constitution to remove its revanchist claim on Ulster.
Hence **reva·nchism,** a policy of seeking reprisal or revenge.
1954 *Times* 5 Mar. 6/4 (*heading*) Mr. Molotov on German 'Revanchism'. **1957** *Wall Street Jrnl.* 6 Dec. 6/5 The struggle against West German militarism and revanchism (seeking for revenge), which are now threatening peace, is a vital task facing the peace-loving forces of the German people and all the nations of Europe. **1959** *Listener* 3 Sept. 344/2 The aim of this offer is to take the edge off accusations of revanchism. **1960** *Guardian* 7 Dec. 32/6 Developments in Germany do not go unscanned for signs of militarism and revanchism. **1971** *Listener* 7 Jan. 5/2 If imperialist aggression and German revanchism cease to be credible threats, what grounds are there for maintaining Soviet garrisons in satellite countries? **1974** *Times* 19 Feb. (European Defence Suppl.) p. ii/7 There is a body of opinion which holds that the Russians are looking more in the direction of the Bundeswehr, prompted by an undying fear of German revanchism.

Revd., abbrev. of Reverend *a.* and *sb.* 2 c.
1693 W. Kennett in W. Somner *Treat. Roman Ports & Forts* 1 The Life of Mr. Somner. To the Rev[d]. Mr. James Brome. **1811** Shelley *Let.* 17 Feb. (1964) I. 52 If any letter comes directed to the Revd. Charles Meyton, it is mine. **1926** Fowler *Mod. Eng. Usage* 502/1 *Reverend* is abbreviated *Revd* or now usually *Rev*. **1971** C. Colvin *M. Edgeworth's Lett. from Eng.* p. xxxvi, The Revd. Daniel Augustus Beaufort.., father-in-law of R. L. Edgeworth.

revealing, *ppl. a.* (Later example.)
1927 Belloc *Hist. Eng.* II. 210 The second rebellion against Henry II was a longer and much more revealing thing than the first.
revealingly *adv.* (further example).
1973 *Nature* 30 Mar. 306/2 There is the case of Graham Young..who, having been 'cured' at Broadmoor..of enjoying poisoning people, had an unfortunate relapse and poisoned several more. Revealingly, his last bout of evil.. was not attributed to mental disease: this time he was sent to prison.

revegetate, *v.* Add: **2.** *trans.* To produce the growth of new vegetation on (disturbed or barren ground). Of a plant: to colonize anew.
1955 *Sci. Amer.* June 50/3 Revegetating the area with plants more drought-resistant than the native flora and fertilizing the land are two of the most helpful measures. **1963** D. W. & E. E. Humphries tr. *Termier's Erosion & Sedimentation* i. 20 The tropical jungles spread out north and south, revegetating the sands and 'fixing' the dunes. **1971** *Nature* 4 June 287/1 Practical techniques were developed for revegetating bare, industrially derelict land near Swansea. **1975** *Sci. Amer.* Dec. 27/2 Halogeton, a toxic weed, is among the first and most tenacious plants to revegetate disturbed desert land. **1976** *Nature* 26 Aug. 733/2 All their contracts with coal suppliers contain a clause requiring that strip mined lands be restored and revegetated.
Hence **reve·getated** *ppl. a.*; **revegetation** (further examples).
1974 *Nature* 15 Feb. 428/1 Revegetation of toxic spoil from industrial processes. **1974** *Environmental Conservation* I. 60/1 Natural revegetation proceeds very slowly in arctic areas. **1976** *Billings* (Montana) *Gaz.* 2 July, Hodder's research involves comparing cattle weight gains on revegetated spoils that have been heavily fertilized to gains on unfertilized native range.

reveille. The pronunc. is now usu. (rivæ·li).

‖ **réveillon** (revē[i]yoṅ). [Fr.] A night-time feast or celebration, orig. one that took place after midnight on Christmas morning. Also *attrib.*
1803 E. Wynne *Diary* 25 Dec. (1940) III. iv. 98 Xmas Day. Sunday. We had midnight Prayers and a Reveillon. **1894** G. du Maurier *Trilby* I. iii. 240 The whole Quartier seemed alive with the *réveillon*. **1932** H. Nicolson *Diary* 1 Jan. (1966) 104 He had spent *réveillon* at the Fabre-Luces and had been kept up..till 8 am. **1965** M. Wallenstein *Merlin's Forest* xx. 205 A man was hanging out a large decorated menu for the Christmas Eve *réveillon*. **1967** C. Durrell tr. *Oliver's French at Table* ii. 69 Pope Julius I decreed that Christmas should be celebrated on 25 December, this providing an excuse for what soon became the traditional *réveillon*—an enormous midnight feast. **1967** *Guardian* 30 Dec. 7/8 French housewives..on Christmas Eve were hosts or guests at the traditional Réveillon supper. On New Year's Eve there is a Réveillon supper all over again. **1971** *Ibid.* 9 Dec. 11/5 A few loose segments [of garlic] to perfume the new year *Reveillon* soup. **1976** *Newmarket Jrnl.* 16 Dec., Called Reveillon, their elaborate suppers and festivities often continue until the early hours.

Revelation. Add: **4*.** A proprietary name for a make of leather goods, used esp. to denote an expanding suitcase.
1923 *Trade Marks Jrnl.* 2 May 876 Revelation... Bags, trunks, suitcases, attaché cases, card cases, cigar and cigarette cases, wallets and similar containers, all being goods made of leather or principally of leather... Jigger, Limited, ..London,..; merchants. **1935** G. Greene *Basement Room* 115 Grains of rice..fell on to his Revelation suitcase. **1938** D. du Maurier *Rebecca* vi. 60 My Revelation suit-case and the stout hold-all. **1964** J. Gardner *Liquidator* i. 15 The battered multi-labelled tan Revelation stood packed. **1967** 'F. Clifford' *All Men are Lonely Now* I. iv. 62 He had brought a Revelation from the flat and now he packed the twenty-five folders into it. **1975** *Listener* 9 Jan. 48/1 If one has a folding cycle, you can wrap it up..in a Revelation suitcase.

revelationism (revēlē[i]·ʃəniz'm). [f. Revelation + -ism.] The fact or process of making a revelation; advocacy of or belief in revelation.
1949 E. L. Mascall *Existence & Analogy* vii. 175 The theistic issue is..immediately raised..by a..discussion of.. the extreme 'revelationism' of Kierkegaard, Barth, Brunner and their disciples, and the Thomist *analogia entis*.

revelator. Add: **1.** (Earlier and later examples.)
1801 *Massachusetts Spy* 20 May 1/2 They shall have their part (saith John the Revelator) in the lake which burneth. **1840** *Latter-Day Saints Millenial Star* June 28/1 The prophet Daniel and the revelator John. **1980** *Times Lit. Suppl.* 25 July 842/3 Lawrence's rhapsodies on John the Revelator's horses.
2. A name given to the president of the Church of Jesus Christ of Latter-day Saints.
1867 W. H. Dixon *New Amer.* II. ii. 12 The Mormon will put his trust in Joseph, as a natural seer and revelator. **1895** *Denver Times* 5 Mar. 8/2 Joseph Smith, the son of the martyr Joseph Smith, is now prophet, seer and revelator. **1974** *Time* 7 Jan. 59/3 Lee was one of the youngest men ever to become 'prophet, seer, and revelator' of the Mormons.

revenant[2]**.** Add: Also fem. **revenante.** **1.** (Earlier and later examples.) Also *attrib.*, as *adj.*, and *fig.*
1827 T. J. Dibdin *Reminisc.* I. vi. 110 She will however frequently make her appearance in this narrative, not as a *revenante*, but prior to the period of her final departure. **1909** R. Brooke *Let.* 16 Apr. (1968) 166 It looks a little like Second Childhood, doesn't it? I think it is merely the first, revenant. **1910** J. C. Lawson *Mod. Greek Folklore & Anc. Greek Relig.* 407 If the devil in possession of the corpse chose to agitate it and drive it out of the grave, the dead demoniac was at once a *revenant*. **1942** 'M. Innes' *Daffodil Affair* III. iii. 91 The papers were full of strange elysiums, cigar-and-whisky empyreans, *revenants* who reported lawn-tennis tournaments on the pavements of paradise. **1955** [see *come-back *sb.*[1] 3]. **1958** *Times* 24 Nov. 12/2 Leonard Salzedo has written the concerto for this *revenant* among solo instruments. **1968** T. Kinsella *Nightwalker* 11 A revenant, A rain-worn, delicate Stone shape. **1969** P. Anderson in Cockburn & Blackburn *Student Power* 257 In the closed space of Gombrich's preoccupations, the psychology which was once exorcized is a revenant which necessarily returns to rule. **1970** R. Lowell *Notebk.* 179 (*title*) Revenants. **1972** *Daily Tel.* (Colour Suppl.) 12 May 61/1 In Eastern Europe..thousands of villagers still believe in the malignity of the *revenant* dead.

revenge, *sb.* Add: **7.** *attrib.*, as *revenge-killing*, *play, seeker, tragedy.*
1975 O. Sela *Bengali Inheritance* xvii. 147 That war was long over... A revenge killing after all this time was absurd. **1967** Revenge play [see *noia]. **1980** *Dædalus* Spring 133 He made *Hamlet* out of the tired revenge play tradition. **1961** *Guardian* 25 Nov. 1/4 It was essential..to 'tie the hands of West German militarists and revenge-seekers'. **1957** N. Frye *Anat. Crit.* 209 The revenge-tragedy is a simple tragic structure. **1977** S. Schoenbaum *William Shakespeare* x. 154 Chettle..is remembered, if at all, for his revenge tragedy of *Hoffman*.

‖ **revenons à nos moutons** (rəvənoṅz a no mutoṅ), *phr.* [Fr., lit. 'let us return to our sheep', with allusion to the confused court scene in the Old French *Farce de Maistre Pierre Pathelin* (*c* 1470).] 'Let us return to the subject': an exhortation to cease digressing.
[**1617** J. Chamberlain *Let.* 22 Feb. in T. Birch *Court & Times James I* (1848) I. 459 But, *pour retourner à nos moutons*, this feasting begins to grow to an excessive rate.] **1822** *Blackw. Mag.* May 610 *North.* On proceeds..Byron..the immoral, irreligious, and unpatriotic tendency of too many of his productions... *Omnes* (sing). Very good song, Very well sung, Jolly companions every one... *North.* Revenons *à nos moutons*! Childe Harold. **1839** Thackeray in *Fraser's Mag.* Aug. 228/2 But, *revenons à nos moutons*, let us return to that sweet lamb, Master Thomas. **1850** —— *Pendennis* I. xxxvii. 364 That brings me back to my point—revenons à nos moutons. Yes, begad! revenons à nos moutons. *c* **1909** D. H. Lawrence *Collier's Friday Night* (1934) II. 48 *Ernest.* I'll bet the dear boy enjoys that blush. *Beatrice.* Ra-ther! (*Artlessly revenant à son mouton.*) And he'll have the rose and all, to rejoice the cockles of his heart this time. **1926** J. S. Huxley *Ess. Pop. Sci.* 27 But *revenons à nos moutons*—to our chromosomes. **1932** A. Christie *Peril at End House* ii. 37 *Revenons à nos moutons*... I implore you to be serious. **1966** 'H. Carmichael' *Suicide Clause* xi. 124 You didn't come here to listen to irrelevant chatter. So, *revenons à nos moutons*. **1972** *Country Life* 9 Nov. 1243/2 *Revenons à nos moutons*. As soon as I saw my cards, the earlier hand flashed through my mind.

‖ **reventa** (reve·nta). [Sp., = resale.] (See quot. 1932.)
1932 E. Hemingway *Death in Afternoon* iv. 40 The re-venta are ticket brokers who take over all or most of the unsubscribed tickets from the bull ring management and sell them at a twenty per cent increase over their face value. **1960** *Times* 14 Nov. 16/7 If you want the best seats in Spain only the *reventa* has them.

revenue. Add: **6. c.** (Earlier example.)
1883 Zeigler & Grosscup *Heart of Alleghanies* 257 My pards mout tak' ye fer a revenoo, an' let a hole thro' ye.
7. a. *revenue account, act* (earlier example), *agent, boat, cutter* (earlier example), *department, expenditure, man* (further example), *stamp, tariff.*
1869 *Bradshaw's Railway Manual* XXI. 392 Confusion between capital and revenue accounts..has led to the undue increase of capital. **1970** R. W. Wallis *Accounting* x. 142 If the local authority charges similar fares and has similar expenses to those of the private undertaking, it too will make a profit or 'revenue account surplus'. *Ibid.* 147 The public authority 'provision for debt redemption' or 'loans redeemed' accounts tend to be equivalent to the 'depreciation provision' account..provided the revenue account is charged with an amount approximating to the amount of depreciation of the asset concerned. **1791** G. Washington *Diary* 4 June (1925) IV. 196 The discontents which it was supposed the last Revenue Act..would create subside as fast as the law is explained. **1864** *Statutes at Large U.S.A.* XIII. 224 Revenue agents..[shall] aid in the prevention, detection, and punishment of frauds upon the internal revenue. **1943** *Chicago Daily News* 24 Dec. 6/1 We would hate to be a revenue agent with that gal up in the cove. **1846** *Knickerbocker* XXVIII. 244 The revenue-boat from the *guardacosta* came on board before our sails were furled. **1790** *Deb. Congress U.S.* (1834) 1st Congress 2nd Sess., App. 2277 The officers of the revenue cutters hereinafter mentioned. *Ibid.* 2 July (1834) 1003, I likewise nominate the following persons to fill offices in the Revenue Department of the United States. **1964** S. Brittan *Treasury under Tories* iv. 101 The Inland Revenue and Customs and Excise (jointly known as 'the Revenue Departments'). **1961** Webster, Revenue expenditure. **1968** Johannsen & Robertson *Managem. Gloss.* 116 *Revenue expenditure*, expenditures of cash which are undertaken to maintain asset values (e.g. repairs) or to obtain current revenue (e.g. raw material purchases, factory payroll). **1970** R. W. Wallis *Accounting* x. 140 [In central government accounting] the distinction between capital and revenue expenditure is not recognized. **1979** *Daily Tel.* 6 Apr. 10/6 After deducting more than £3 million earmarked for capital projects, the council's new grant represents a rise in its revenue expenditure of about 19 per cent. **1895** *Century Mag.* July 378/2 I'm always skeered o' the revenue men bein' about. **1870** J. K. Medbery *Men & Mysteries Wall St.* 52 The acknowledgements are covered with revenue stamps. **1820** *Deb. Congress U.S.* 24 Apr. (1855) 1966 They enacted a treasury tariff, a revenue tariff, without the least regard to the situation of the country. **1887** *Courier-Jrnl.* (Louisville, Kentucky) 19 Feb. 4/1 They are the identical arguments which the Courier-Journal has been pounding into the understanding of the people in its fight for a revenue tariff.
b. *revenue-earner*; *revenue-paying, -sharing, -yielding.*
1963 *Times* 23 Mar. 11/1 Sport in general is a major revenue-earner. **1910** J. London *Let.* 9 Aug. (1966) 317 The idea that a judge of rectitude should..try a case between a non-resident like me and a revenue-paying, vote-swinging tenant like Muldowney. **1971** *New Yorker* 10 July 43 Brock's dominant theme was much the same as what President Nixon would later call 'revenue sharing'—that government had to be returned to local control by way of sending the taxpayers' money back to the states, counties, cities, and towns. **1973** *Black Panther* 17 Mar. A/3 Even the facade of Revenue Sharing as the promise of a 'new lease on life' for American cities is fast becoming a sad joke. **1976** *National Observer* (U.S.) 2 Oct. 10/1 Boston's Housing Improvement Program, an apparently successful effort to use Federal revenue-sharing money to upgrade city neighborhoods and preserve existing housing. **1898** E. Howard *To-morrow* v. 53 A considerable outlay would be incurred in respect of markets, water supply, lighting, tramways, and other revenue-yielding undertakings.

revenuer (re·vĕniuːəɪ). *U.S.* [f. Revenue + -er[1].] A revenue agent.
1880 *Daily Inter-Ocean* (Chicago) 1 June 12/1 His wife and daughter discharged their conjugal and filial duty by.. watching from their home for the approach of the 'Revenyors'. **1895** 'C. E. Craddock' *Mystery of Witch-Face Mountain* 15 The 'revenuers'..never rode alone. **1941** *Charlottesville* (Va.) *Daily Progress* 16 Aug. 1/5 The sharp rise in sugar and grain prices, coupled with the increased efforts of the 'revenooers' to ferret out the distilling outfits in the mountains have made it too perilous for most of the old-timers. **1949** *Américas* Aug. 10/1 The 'revenoo-ers' slowed the production of illegal whiskey. **1955** W. Faulkner *Fable* 173 The sheriff..and the city strangers in their city hats and neckties and shoes, smelling, stinking of excise officers, revenuers. **1963** *Times* 19 Mar. 10/7 The 'revenuer' keeps track of large sales of sugar. **1974** W. Garner *Big enough Wreath* x. 131 Moonshiner, made his own likker..scared stiff of being caught by the revenuers. **1974** *New Yorker* 3 June 43/2 Jon tried to persuade Anne and Ruth to take their bathing suits off... 'Listen, as long as we're back up here where the revenooers can't get at us, we might as well have a good time.'

reverb (rĭvɔ̄·ɪb), *sb.* Colloq. abbrev. of *reverberation 1 c.
1961 in Webster. **1968** *Blues Unlimited* Dec. 10 Muddy's new album Electric Mud is a morass of feedback,..reverb and every other trick. **1971** *Melody Maker* 13 Nov. 36/4 The features are the same as the bass amp, but with twice the channels, plus reverb and tremolo. **1976** Boot & Thomas *Jamaica* 34/1 They had about as much sheet reverb as the human body can stand.
So **reve·rbed** *a.*, adapted to produce a reverberating sound.
1977 *Rolling Stone* 30 June 102/2 Edmunds' treatment of Crudup-cum-Presley's 'My Baby Left Me'..captures the Sun sound from its string bass to its reverbed rhythm guitar.

reverberant, *a.* Add: Hence **reve·rberantly** *adv.*, in a reverberant manner.
1961 in Webster. **1976** *Gramophone* Oct. 644/1 He is made to stand out too reverberantly against the street-vendors' cries in Act 2. **1977** *Rolling Stone* 24 Mar., We've come to expect strong personalities in female C&W singers. Dolly Parton is the buxom, reverberantly mature earth mother.

reverberation. Add: **1. c.** *spec.* Temporary persistence or prolongation of sound without perceptible distinct echoes, produced either by repeated reflection from nearby surfaces or artificially. (Further examples.)
1828 E. Holmes *Ramble among Musicians of Germany* 194 The effect of an excellent orchestra is heightened by the structure of this edifice, which admits of a fine echo and reverberation. **1905** E. Atkinson tr. *Ganot's Nat. Philos.* (ed. 10) IV. i. 203 When a person is speaking in a large empty room, the reverberation from the walls and ceiling produces a confused effect. **1948** P. M. Morse *Vibration & Sound* (ed. 2) xxxii. 387 If the room is used chiefly for music, we can allow more reverberation without detriment. **1962** A. Nisbett *Technique Sound Studio* xi. 188 'Echo' is something of a misnomer for a studio technique which serves to extend reverberation without (it is hoped) introducing any actual echoes. *Ibid.* 191 Reverberation is added by an echo chamber. **1975** G. J. King *Audio Handbk.* i. 22 It is the aim of hi-fi reproduction..to reproduce the whole of the original, including the reverberation, which should not be unduly modified by the listening room.
3. c. *Plumbing.* = Water-hammer 2.
1967 *Gloss. Sanitation Terms (B.S.I.)* 63 *Reverberation*, a hammering sound caused by violent surges of pressure in water pipes. **1972** J. Hastings *Plumber's Compan.* 168 The working plumber is unlikely to use 'reverberation' or even 'concussion' when speaking of water hammer.
5. *attrib.*, as **reverberation chamber,** a room specially designed to reflect sounds produced within it; **reverberation time** *Acoustics*, the time taken, after cessation of production of a steady sound, for the average sound intensity at a given frequency in a room or enclosure to die away, *spec.* to decrease by 60 decibels.
1932 P. E. Sabine *Acoustics & Archit.* vi. 120 With a source of sound whose acoustic output can be varied in measured amounts the absorbing power of the reverberation chamber both in its standard condition and with the absorbent material present can be determined directly. **1966** *McGraw-Hill Encycl. Sci. & Technol.* I. 665/2 Two types of rooms especially constructed for research and development are..the random diffusion or reverberation chamber and the anechoic room. **1969** Reverberation chamber [see *sabin]. **1927** I. B. Crandall *Theory of Vibrating Syst. & Sound* v. 203 We define *T*, the reverberation time of the given enclosure as the time required for the energy density to sink from one prescribed level to another. *Ibid.* 211 A new concept of reverberation time more nearly in accordance with the actual conditions under which concert rooms are used. **1932** [see *acoustical *a.* b]. **1952** *Sci. News* XXIII. 83 Recent concert hall designs have included large quantities of polished wood as a surface finish in an endeavour to prevent the reverberation-time falling away at high frequencies. **1975** G. J. King *Audio Handbk.* i. 22 For the reproduction of music the reverberation time should not be too large otherwise the 'definition' will be impaired owing to the commencement of a new component of sound before those preceding have died away.

‖ **reverdie** (rəvęrdi). [Fr.] In medieval French lyric poetry: a song which celebrates the reappearance of spring.
1933 R. Tuve *Seasons & Months* i. 13 This description.. includes the same details as the opening stanza of a Middle English spring song or an Old French *reverdie*. **1976** *Scottish Rev.* Spring 19 In a song, 'The Dark Days are By', in March 1940, he uses the simplest and barest of words to convey the mood of a traditional *reverdie*.

reverend, *a.* and *sb.* Add: **2. b.** Also *absol.*, as a form of address.
1964 *Church of Eng. Newspaper* 11 Dec. 17/3 When I received the letter addressing me as 'Dear Reverend' I was so pleased..that I determined to subscribe to the journal. **1967** C. Armstrong in *Ellery Queen's Mystery Mag.* July 11/1 What you'd better do, Reverend, is go over to the motel and rest. **1974** J. Wainwright *Hard Hit* 8 'I dunno you, reverend. I think you're making some sorta..' 'Mistake?'
c. (Later examples.) Pl. *reverends.* Use without *the* is again current.
1939 R. E. Wolseley in *Ken* 9 Mar. 62/3 Ten devices.. invented by southern journalists to avoid using 'Mr.', 'Mrs.', and 'Miss', in front of the names of Negroes are: Mademoiselle Madame Professor Doctor Reverend [etc.]. **1961** R. B. Long *Sentence & its Parts* x. 230 There is a feeling that these [honorific modifiers] should be preceded by the article *the* and followed by given names and family names together, as in *the Reverend George Brewster*; but there is also a marked tendency to treat *reverend* exactly as the noun honorifics are treated, as in *Reverend Brewster will preach at the eleven-o'clock service*. **1973** *Publishers' Weekly* 7 May 61/2 (Advt.), The Reverends Rudolf Harvey and Lawrence Burke.
d. (Further examples of *sing.*)
1859 O. L. Jackson *Colonel's Diary* (1922) ii. 8, I..heard a very good sermon—from a Reverend from Pittsburgh. **1943** G. Greene *Ministry of Fear* III. ii. 191 That [car], sir—that's the reverend's... We thought it only right to let the vicar know. **1971** *Language* XLVII. 30 Whether or not the notorious Reverend or his students sat up nights inventing such errors, attested errors reveal the same kind of metathesis. **1976** J. McNeish *Glass Zoo* vii. 71 Funny, that cloak of yours. I know a Reverend in Leeds got about like that.

reverie, *v.* Add: (Later example.) Also *trans.*, to contemplate or recall with relish or enjoyment.
1961 B. Malamud *New Life* 315 He reveried accomplishment..foresaw an effective if..short career. **1980** P. Van Greenaway *Dissident* viii. 168 He reveried on.

reversal, *sb.* Add: **4.** *Photogr.* The process by which a positive image is produced on exposed photographic material without an intervening negative stage. Freq. *attrib.* (see sense *5).
1878 *Proc. R. Soc.* XXVII. 292 When the strength of the permanganate, or hydroxyl, was correct a reversed image of the least refrangible end of the spectrum was obtained,.. the reversal taking place somewhere near D, extending into the ultra-red. **1890** Hurter & Driffield in *Jrnl. Soc. Chem. Industry* 31 May 464/1 We accordingly distinguish four different periods of exposures... The last portion of the curve we term the period of 'reversal'. *Ibid.* 464/2 This period we have named the period of reversal because within this period happens that peculiar phenomenon, the transformation of the negative into the positive, the 'solarisation', 'reversal', &c. **1907** Sheppard & Mees *Invest. Theory Photogr. Process* II. vi. 212 Various forms of reversal in the photographic process are known, not all of equal interest or importance. **1927** C. B. Neblette *Photogr.* viii. 205 No photographic process is, strictly speaking, free from the effects of reversal. **1969** *Focal Encycl. Photogr.* 1294/1 The positive image produced by reversal is..intrinsically finer in grain than the negative from which it is derived.
5. *attrib.* and *Comb.* (chiefly in sense *4), as *reversal method, paper, stock, transparency*; **reversal film,** film that gives a positive image directly when it is processed; **reversal process** *sb.* = sense 4 above; so **reversal-processed** *ppl. a.*, **reversal processing** *vbl. sb.*; also **reversal-process** *v. trans.*; **reversal speed** *Aeronaut.*, the air speed above which the effect of a control surface is reversed, and at which it is nullified, by the aerodynamic effect of the elastic distortion of the aircraft caused by operation of the surface.
1931 J. H. Reyner *Cine-Photogr. for Amateurs* i. 10 Reversal film..will make excellent positives from under exposed film. **1962** *Which?* May 131/1 There are two types of colour film on the market—reversal film, which produces 'positive' transparencies (slides).., and negative film. **1970** *Amateur Photographer* 11 Mar. 60/3 While positive film is comparatively cheap and good results are possible,.. results cannot be nearly as pleasing as those obtained with conventional black-and-white reversal film. **1938** G. H. Sewell *Amateur Film-Making* ii. 13 It is also possible by applying different solutions [to an exposed film] to 'fix' the film, i.e. dissolve away the silver and leave the salts, and if this is done, the residue of silver salts can then be exposed to light and developed, and it will be found that a positive image will result. That is known as the 'reversal' method. **1969** M. J. Langford *Adv. Photogr.* xvi. 360 Using reversal paper to print direct from positive transparencies. **1920** W. Gamble *Photogr.* xi. 80 Development is as usual, and there is no reversal process as in the case of the Autochrome. **1944** T. A. Longmore *Med. Photogr.* 363 Development is carried out by the reversal process in order to produce a positive image to satisfy the requirements of the additive colour process. **1955** *Jrnl. Photogr. Sci.* III. 129/1 Deep-etch plates and most bi-metal processes require positives to produce a positive image by a reversal process during platemaking. **1973** Reversal-process *v.* [see *reversal processing* vbl. sb. below]. **1977** J. Hedgecoe *Photographer's Handbk.* 73 The paper prints must be reversal processed using 8 or 9 stages similar to Ektachrome processing. **1964** L. A. Mannheim tr. *Croy's Camera Copying & Reproduction* 150 Reversal-processed prints are liable to yield uneven image densities. **1938** D. A. Spencer *Colour Photogr.* x. 143 The correct exposure is normally that which will leave the highest highlight just transparent after reversal processing. **1973** W. Thomas *SPSE Handbk. Photogr. Sci. & Engin.* ix. 547 Films designed for reversal processing are available, and some black-and-white negative films can be reversal-processed. **1947** *Jrnl. R. Aeronaut. Soc.* LI. 5/2 For a stiff enough wing there will theoretically be one reversal speed in the subsonic range and two in the supersonic. **1966** *McGraw-Hill Encycl. Sci. & Technol.* I. 90/2 Elastic distortion of an airplane increases rapidly as its speed increases toward the divergence speed and reversal speed. **1969** J. Elliot *Duel* III. ii. 233 She went to..learn the mysteries of negatives and reversal stock. **1967** *Economist* 7 Jan. 67/3 Ciba's new process for making colour prints from reversal transparencies.

reverse, *sb.* Add: **1. b.** Also, in *Contract Bridge*, a rebid in a suit of higher rank than that which one has previously bid.
1936 E. Culbertson *Contract Bridge Complete* xi. 128 The bid of a higher ranking suit by a player who has bid one of lower rank is called a 'Reverse'. **1939** N. de V. Hart *Bridge Players' Bedside Bk.* 132 Logically..an exception occurs when the partner forces the original bidder to reverse. In that case, the reverse may not indicate additional values. **1951** I. Macleod *Bridge* ii. 23 The reverse primarily shows, or ought to show, the shape and not the strength of the holding. **1964** Frey & Truscott *Bridge Player's Dict.* 420/2 *Reverse*, an unforced rebid at the level of two or more, in a higher-ranking suit than that bid originally; a type of strength-showing bid... The English definition of a reverse by the opener is slightly wider in scope: a bid of a third suit in an uncontested auction which prevents responder returning to the original suit at the level of two... This allows for..a *high reverse*.
6. e. (See sense *12 a.)
12. a. Reverse gear; hence *in reverse.* Also *transf.*
1882 *Engineering* 10 Mar. 219/3 Locomotives having unbalanced slide valves are handled with the greatest ease by means of this steam reverse. **1900** W. W. Beaumont *Motor Vehicles & Motors* xiv. 246 The gear speeds with normal speed of motor are for 3½, 10 and 20 miles per hour, with reverse at 4 miles per hour. **1912** R. T. Nicholson *Bk. of Ford Van* x. 44 Use your reverse as a brake, in addition to the foot brake and the slow speed. **1925** B. Garbutt *Bk. of Austin 12* v. 29 You can imagine what would happen if, thinking you were going into..third speed, you really got into reverse in mistake. **1925** F. Scott Fitzgerald *Great Gatsby* iii. 67 'Back out,' he suggested... 'Put her in reverse.' **1937** *New Motoring Encycl.* 664/1 The ability to drive a car in reverse accurately and smoothly is essential to correct driving technique. **1958** *Spectator* 22 Aug. 245/1, I started to go into reverse immediately. I backed out of the dining-room. **1973** G. Moffat *Lady with Cool Eye* viii. 84 The car..was in neutral instead of bottom or reverse. **1976** *S. Wales Echo* 27 Nov. 5/7 He pleaded not guilty to..using a motor vehicle with a reversing light on when he was not travelling in reverse.
b. The name of the position of the gear lever or selector corresponding to reverse gear.
1963 [see *park *sb.* 5 c]. **1965** Priestley & Wisdom *Good Driving* v. 42 *Park* or *Reverse* must not be selected whilst the car is moving. **1970** F. Leigh *Bk. of Renault R16* ii. 8 On 1965–67 cars the reverse is found inwards down by second gear. **1974** D. Francis *Knock Down* xv. 191 The driver was shifting the gears from reverse to forward.

reverse, *a.* and *adv.* Add: **1. d.** *reverse fault* (Geol.), a fault in which the relative downward movement occurred in the strata situated on the underside of the fault plane (cf. *normal *a.* 2 d).
1878 [see *normal *a.* 2 d]. **1889** *Amer. Jrnl. Sci.* XXXVIII. 259 The explanation of the reverse faults seems obvious enough. They occur, as we have already said, mostly in strongly folded regions. Such folds can only be produced by lateral pressure. **1962** Read & Watson *Introd. Geol.* I. viii. 481 Reverse faults are those in which the hanging-wall rocks moved up the dip of the fault-plane relative to the rocks of the foot-wall. *Ibid.* 482 There is much more justification for referring to reverse faults, which involved a shortening of the crust, as compressional faults. **1976** S. Judson et al. *Physical Geol.* vii. 152/2 In a reverse fault, the least principal stress is vertical.
e. *reverse angle*: (*a*) *Cinemat.* and *Television.* The opposite angle from which the subject was seen in the preceding shot; freq. *attrib.*; also *reverse shot.* (*b*) *Squash.* Used with reference to a stroke played on to the side wall of the court opposite to the server; usu. *attrib.*
1934 Webster, *Reverse shot*, reverse-angle shot. **1939** B. C. Kiesling *Talking Pictures* 318 *Reverse shot*, the photograph of a scene from the opposite direction from which it was originally taken. **1948** [see *length *sb.* 10 b]. **1952** L. Anderson *Making a Film* 267 Shot in two set-ups: first

from the foot of the bed, then (for Penny's entrance) on the reverse angle, over Anselmo's shoulder. **1961** G. Millerson *Technique Television Production* 244 (*caption*) Location can be lost through reverse-angle cutting. **1968** *Squash Rackets* ('Know the Game' Ser.) 39/1 The reverse angle stroke is one played on to the opposite side wall. **1976** *Listener* 26 Feb. 230/3 The journalists were most patient with our retakes and reverse angles.

f. *reverse Polish*: see *Polish *a.* d.

g. *Contract Bridge. reverse bid* = *reverse *sb.* 1 b.

1939 N. de V. Hart *Bridge Players' Bedside Bk.* 132 A reverse bid logically shows considerable strength when it is made in such a way that the partner cannot put the bidder back to his first-bid suit without raising the bidding to a higher level than that at which the bidder could himself have returned to the suit. **1963** *Listener* 10 Jan. 102/3 He lacks the general strength for a 'reverse' bid of Two Hearts.

5. a. (Later examples.) *reverse dictionary*: a dictionary in which the words are arranged so that, read backwards, they are in alphabetical order; *reverse discrimination* = *positive discrimination* s.v. *positive *a.* 15; *reverse pass* (see quots. 1960, 1978).

1921 E. de Lissa in E. H. D. Sewell *Rugby Football up to Date* xv. 264 We are going to practise all the summer together to see if we can't bring off some of that 'reverse' passing, during next season. **1937** E. K. O'Brien in *Bridge World* Aug. 27 (*heading*) The reverse signal. **1951** *Sport* 7–13 Jan. 8/4 Ormondroyd..took a reverse-pass in fine style and went through the opposition like a bullet. **1954** *Newsweek* 26 Apr. 57 To help Scrabble fans, cross-word puzzle addicts, and other persons troubled for a word ending in 'X', 'Y' or 'Z', a 'reverse' dictionary has been compiled at the University of Massachusetts. **1954** *N.Y. Times* 30 May 34/5 Fisk officials cite her case as being possibly an example of 'reverse integration', a phrase enunciated in the light of the recent Supreme Court decision prohibiting segregation in the public schools. There are two others at Fisk who offer comparable examples. They are white students in the undergraduate school. **1959** *Listener* 27 Aug. 334/3 Of various other signalling methods, mainly continental, the one most interesting to note is the 'reverse peter', credited to the Austrian expert K. Schneider. **1960** E. S. & W. J. Higham *High Speed Rugby* i. 18 *The reverse pass.* This is a pass which is used to reverse the direction of the passing movement. As the ball is passed out to a player, he passes it back in again. **1961** *Times* 16 Jan. 3/5 Dodds was only inches wide with a reverse-stick shot. **1962** Y. Malkiel in Householder & Saporta *Probl. Lexicogr.* 17 Students of suffixes have derived considerable benefit from rhyming dictionaries and from their close congeners, the 'reverse' (*rückläufig*) dictionaries. **1962** T. Masters *Surfing made Easy* 65 *Reverse pullout*, kicking out with the body turning in the opposite direction. **1969** *Guardian* 11 Oct. 8/6 The Hunt report on Northern Ireland..sets its face firmly against the allocation of reserved places for Roman Catholics within the police—'reverse discrimination' it calls it. **1971** M. Lehnert (*title*) Reverse dictionary of present-day English. **1973** *Computers & Humanities* Mar. 221 There is a reverse concordance..and a key-word-in-context..index. **1976** *National Observer* (U.S.) 27 Nov. 3/1 If the Court rejects the concept of reverse discrimination, numerous educational and business programs across the country that give preferential treatment to minorities and women would be continued and probably expanded. **1978** *Sunday Times* (Colour Suppl.) 28 May 34/3 *Reverse pass*, when a player runs in one direction then passes in the other [in football].

b. *spec.* with reference to engines and vehicles; *reverse* (*idler*) *gear*, a gear wheel or mechanism which enables a vehicle or vessel to travel in reverse without reversing the rotation of its engine; also *fig.*; *reverse lever*, a lever by means of which the reverse gear of an engine may be brought into use; *reverse thrust*, thrust used to retard the forward motion of an aircraft, rocket, etc.; the condition of providing this; freq. *attrib.*

1882 *Engineering* 10 Mar. 219/3 An indicator R is..located upon the same pin as the reverse lever. **1900** G. D. Hiscox *Horseless Vehicles, Automobiles, Motor Cycles* xii. 242 Three speeds forward and a reverse slow speed. **1907** F. Strickland *Man. Petrol Motors & Motor Cars* xiii. 179 If the countershaft is driven faster than the engine, the reverse pinion would have to be driven very much faster. **1907** R. B. Whitman *Motor-Car Princ.* viii. 121 When the car is going forward, the square shaft and countershaft revolve in opposite directions; but when the reverse gear is introduced between them, the square shaft is revolved in the same direction as the counter-shaft, reversing the rotation of the driving wheels. **1907** W. F. M. Goss *Locomotive Performance* v. 105 An attempt is made to operate a locomotive with the reverse-lever in its extreme forward position at high speed. **1923** *Rep. Internat. Air Congr.* 594 (*caption*) The effect of reverse thrust on the gliding angle of the airplane. **1931** F. L. Allen *Only Yesterday* i. 7 He must remember to brake with the reverse pedal, or the low-speed pedal. **1933** *Discovery* July 221/1 The pilot has only to move the throttle and reverse lever, there being one position for forward and one for reverse. **1936** E. A. Phillipson *Steam Locomotive Design* x. 330 One of the most successful power reverse gears..is that operated by compressed air. **1947** *Shell Aviation News* No. 112. 22/1 The reduction of landing run by means of reverse-thrust braking will probably prove a very desirable feature for large aircraft. **1955** W. H. Crouse *Automotive Transmissions & Power Trains* i. 17 The reverse idler gear is always in mesh with the small gear on the end of the countershaft. **1959** *Listener* 26 Feb. 372/1 The October Moon probe carried eight small vernier rockets and one rather larger reverse-thrust rocket in addition to its main power plant. **1973** W. McCarthy *Detail* iii. 204 David..tried to accustom himself to the British car. He found the reverse gear and slowly backed out. **1976** B. Jackson *Flameout* (1977) i. 20 The catastrophe resulting from one of the engines being in reverse thrust at takeoff. **1979** *Guardian* 7 July 9/4 In the present era, everything has gone into reverse gear. New building is postponed, new hospitals..cannot open.

c. *reverse osmosis* (Physical Chem.), the process by which water or another medium tends to flow across a membrane in the direction opposite to that for natural osmosis when subjected to a hydrostatic pressure greater than the osmotic pressure.

1955 *Ann. Rep. Saline Water Comm. 1954* (U.S. Dept. Interior) 1. 10 Development of membranes and procedures for demineralization of saline water by reverse-osmosis methods are provided for in several contracts. **1970** *New Scientist* 14 May 337/3 Laboratory experiments have shown that cheddar cheese whey can be concentrated five-fold by reverse osmosis, or separated into a high-protein product. **1977** *Hongkong Standard* 14 Apr. 9/6 Other processes, notably reverse osmosis (whereby pure water is forced through membranes under pressure, leaving salts behind) show great longterm promise, but at present are suitable only for small plants or for purifying brackish water rather than sea-water.

d. *reverse transcriptase*: see *transcriptase.

5*. *reverse charge*: used *attrib.* to designate a telephone call for which the charge has been reversed (see *reverse *v.* 9 d).

1932 *Telegr. & Teleph. Jrnl.* XVIII. 118/1 Possible new types of service, such as..'reverse charge' calls. **1978** L. Heren *Growing up on The Times* i. 23, I..dialled the overseas operator and placed the reverse-charge call.

6. *adv.* **b.** *Comb.* with *ppl. adjs.*, as *reverse-acting, -applied, -biased.*

1957 E. B. Jones *Instrument Technol.* III. 117 Increase of the output pressure from the controller will cause the valve to close and so reduce the flow through the valve. Such a valve is described as 'reverse acting'. **1962** Simpson & Richards *Physical Princ. Junction Transistors* iv. 54 This condition applies to an even greater extent for reverse-applied voltages when the current flowing through the device is very small. *Ibid.* v. 75 The high resistance of the reverse-biased collector junction. **1975** D. G Fink *Electronics Engineers' Handbk.* vii. 32 When a *pn* junction is reverse-biased, the free electrons in the *n*-type material are attracted toward the positive terminal of the power supply and away from the junction.

reverse, *v.*[1] Add: **I. 8. a.** Also *refl.*

1944 *Sun* (Baltimore) 5 Apr. 10/2 The court has reversed itself, to be sure, but the reversal does not surely involve any real change in American attitudes on deeply controversial questions. **1978** *Guardian Weekly* 7 May 15/5 President Carter..has reversed himself, and now backs legislation to buy 225,000 tons. **1979** *Publishing News* 29 June 6/4 It..was turned down by the Literary Guild and *Reader's Digest*... A short time later, both book clubs reversed themselves and *Hotel* went on to spend a year on the best seller lists.

9. c. To put (a motor vehicle) into reverse gear; to drive (a motor vehicle) backwards. Also *absol.*

1902 W. Robinson *Gas & Petroleum Engines* (ed. 2) xvii. 782 The vehicle is reversed by a hand lever which moves sliding sleeves bolted to the bevel wheels of the reversing gear. **1902** *Motor-Car World* III. 106/2 When it is required to reverse, the car can instantly be driven forward again, without it being necessary to replace the reversing gear into its usual position. **1913** G. B. Shaw *Let.* 17 Sept. in *B. Shaw & Mrs. Campbell* (1952) 148, I..had to drive right up it..reversing at impossible hairpin corners. **1932** F. J. Camm *Bk. Motors* xiv. 127 It is dangerous and discourteous..to reverse unless you know that all is clear. **1939** J. Harrison *Motor-Cars To-day* ix. 111 When one wishes to reverse the car the bottom-speed wheel is moved towards the rear of the gearbox. **1941** L. D. Kitchin *Road Transport Law* 28/2 A vehicle must not be reversed for a greater distance or time than may be requisite for the safety or reasonable convenience of the occupants of that vehicle or of other traffic on the road. **1959** E. H. Clements *High Tension* viii. 134 His young step-cousin began to reverse her car. **1966** 'A. Hall' *9th Directive* xx. 187 He told the chauffeur to reverse the Lincoln. *a* **1976** A. Christie *Autobiogr.* (1977) vii. ii. 332, I don't think I can really reverse at all... The car never seems to go where I think it's going. **1977** D. MacKenzie *Raven & Kamikaze* iv. 49 Raven reversed and killed the motor.

d. Phr. *to reverse the charges*, to make the recipient of a telephone call responsible for paying for the call.

1927 *Sat. Even. Post* 30 Apr. 30/2 'Let's Call Up Sam and Tell Him What We think of Him.' 'Magnifique! And Don't Forget to Reverse ze Charges.' **1951** 'E. Crispin' *Long Divorce* xiv. 172 If you want to go on talking we must reverse the charges. **1972** 'W. Haggard' *Protectors* iii. 38 Phone me at once... Reverse the charges.

II. 10. b. (Earlier U.S. and later examples.)

1864 M. B. Chesnut *Diary* 6 July (1949) 418 A lame man can't reverse—as they call it in waltzing. **1912** V. B. Carter *Diary* 25 May in *Winston Churchill* (1965) xviii. 254 As he [*sc.* Ld. Fisher] never reverses I reel giddily in his arms and lurch against his heart of oak.

13. *Contract Bridge.* To rebid in a suit of higher rank than that which one has previously bid.

1939 N. de V. Hart *Bridge Players' Bedside Bk.* 132 An exception occurs when the partner forces the original bidder to reverse. In that case, the reverse may not indicate additional values. **1959** *Listener* 29 Jan. 229/1 He has already shown five hearts, inasmuch as he bid that suit and then reversed in spades.

reversed, *ppl. a.* Add: **b.** *reversed fault* = *reverse fault* s.v. *reverse *a.* and *adv.* d; **reversed-polarity** *Grammar*, used attributively of a complex interrogative construction in which positive and negative appear in each of two parts (e.g. *this is good, isn't it?*, and *this isn't any good, is it?*).

1876 Reversed fault [in Dict.]. **1882** A. H. Green *Geol.* xi. 492 The hade or slope is almost always towards the downthrow side: exceptions to this rule are called 'reversed faults'. **1929** M. H. Haddock *Disrupted Strata* iii. 36 In the common type of fault,..the striking angle θ will give a normal fault when obtuse and backthrust is present, or when acute and forethrust is present; the reversal of these remarks holds for reversed faults on the same fault plane. **1969** Bennison & Wright *Geol. Hist. Brit. Isles* x. 251 The Dent Line, a complex structure which has the effect of a reversed fault. **1957** D. L. Bolinger in *Publ. Amer. Dial. Soc.* xxviii. 113 The speaker courteously anticipates a possibly negative answer. This is perhaps related to the original sense of reversed-polarity conduciveness. **1975** *Language* LI. 26 What is the relation between constant polarity tags and reversed-polarity tags?.. The tags below both have reversed polarity:..a. Caterpillars have legs, don't they? b. Caterpillars don't have legs, do they?

reverser. Add: **1. b.** A control used to bring about reversal of the direction of motion of a vehicle; *esp.* a switch which changes the connections of electric traction motors to this end.

1905 Ashe & Keiley *Electric Railways* vi. 111 The function of the reverser..is to perform the proper connections of armature and field terminals, depending upon whether a forward or a backward motion of the car is desired. **1927** R. E. Dickinson *Electric Trains* iv. 62 The Reverser in multiple unit controls is usually in the form of a rocking arm carrying the contacts and making electrical connection with suitable fingers for reversing the direction of motion of a group of two motors. **1972** *Railway World* Oct. 449 Grif altered the regulator setting from time to time, but had no reverser with which to contend.

c. = *thrust reverser* s.v. *thrust *sb.* 7.

1967 D. P. Davies *Handling Big Jets* iv. 59 With a reverser efficiency of 50 % the net change will be from 1,000 lbs. forwards to 500 lb. reverse.

Reversi. **2.** (Later examples.)

1963 I. Gandy *Staying with Aunts* iii. 62 Another favourite game..was Reversi... Someone found the opportunity to change a whole row of counters to their own colour. **1976** *Time* 27 Dec. 2/3 The 'new' Japanese game Othello bears a remarkable resemblance to an English board game called Reversi.

reversible, *a.* and *sb.* For **b** read **B** and add: **A.** *adj.* **1. a.** (Further examples.)

1899 *T. Eaton & Co. Catal.* Spring & Summer 116/1 Men's reversible coats, made of napa tan leather, lined with heavy drab corduroy. **1910** F. Yeigh *Through Heart of Canada* ii. 36 The famous reversible falls of the St. John River..are best viewed from the suspension bridge. **1921** A. W. Grabau *Textbk. Geol.* I. xvii. 525 A reversible fall is produced, facing inward when the ocean is highest and outward when it is lowest. **1926** *Daily Colonist* (Victoria, B.C.) 9 Jan. 18/1 (Advt.), Reversible Rugs, shown in Oriental designs and colorings. They are made of hard-wearing jute and have the appearance of wool rugs. **1957** *Encycl. Brit.* XXIII. 459/1 Reversible fabrics may have..two series of differently coloured welts or warps to one of the other series, in which event they may be similarly figured on both sides by causing the threads of the double series to change places. **1974** *Times* 14 Apr. 7/2 The coats and capes ..are all reversible with soft knitted mohair on the outside and a sort of super deluxe dishcloth knitting inside.

b. Of a propeller: capable of providing reverse thrust, usu. by reversal of the pitch of the blades while the direction of rotation remains unchanged. So *reversible-pitch* attrib. phr.

1907 H. Chatley *Problem of Flight* vii. 86 Reversal of motion can be obtained by the use of change speed gearing, which is altered for the backward motion, so that its torque is reversed... Reversible wings to the helices can also be used for this purpose, but the rigidity would probably be impaired. **1922** *Flight* XIV. 657/2 An airscrew..has been produced by..one of the first airscrew making firms in America... It comprises a system of special blades, and a mechanism for varying the pitch of same..but the airscrew is reversible as well. **1923** *Rep. Internat. Air. Congr.* 588 This propeller is reversible as well as adjustable, and can be used as a brake in landing on rough terrain. **1930** F. E. Weick *Aircraft Propeller Design* xi. 188 The effectiveness of the reversible-pitch propeller as an air-brake..has been experimented with by both the Army and the Navy. **1935** J. W. Anderson *Diesel Engines* xv. 347 On reversible propellers the blades are twisted by suitable mechanism to give the proper angle for the desired direction of motion of the boat. **1951** P. Du Cane *High-Speed Small Craft* xi. 131 Variable (controllable) reversible-pitch propellers have so far been fitted in relatively few experimental craft built in Great Britain. **1965** C. N. Van Deventer *Introd. Gen. Aeronaut.* viii. 197/2 Towards the end of WW II the final refinement to propellers was perfected with the reversible pitch propeller. **1967** *Jane's Surface Skimmer Syst. 1967–68* 12/1 A 230 hp flat-six aero-engine drives a reversible-airscrew for propulsion.

2. *Physics* and *Chem.* **a.** *Thermodynamics.* Of a change or process: that is capable of reversal, completely and in detail; strictly, applicable to an ideal change in which the system is in equilibrium at all times. Also applied to a system undergoing such changes.

1856 *Phil. Mag.* XI. 216 The temperatures of two bodies are proportional to the quantities of heat respectively taken in and given out in localities at one temperature and at the

other, respectively, by a material system subjected to a complete cycle of perfectly reversible thermo-dynamic operations, and not allowed to part with or take in heat at any other temperature. **1878** [in Dict., sense 1 a]. **1922** Glazebrook *Dict. Appl. Physics* I. 929/1 The expansions and compressions and the transfers of heat that occur in a real engine are never strictly reversible, some of them indeed are far from being reversible. **1950** W. J. Moore *Physical Chem.* i. 22 Reversible processes are never realizable in actuality since they must be carried out infinitely slowly. **1958** Condon & Odishaw *Handbk. Physics* v. 5/1 All reversible engines working between the same two heat reservoirs must have the same efficiency. **1963** A. E. J. Hayes *Appl. Thermodynamics* v. 53 When a fluid undergoes a reversible process it passes through a series of equilibrium states, i.e. states in which the fluid properties are uniform throughout. The term reversible does not merely imply ability to operate in the reverse direction, but also that the process must be able to retrace its path through the same sequence of equilibrium states. **1978** P. W. Atkins *Physical Chem.* ii. 62 In thermodynamics a reversible change is one that can be reversed by an infinitesimal modification of a variable.

b. Applied to chemical reactions which in normal circumstances also take place in the opposite direction, and hence yield an equilibrium mixture of reactants and products.

1885 *Jrnl. Chem. Soc.* XLVIII. i. 117 Reversible chemical changes are thus brought into the class of reciprocal phenomena. **1899** J. McCrae tr. *A. Reychler's Outl. Physical Chem.* 241 The reversible reaction $BaSO_4 + K_2CO_3 \leftrightarrows K_2SO_4 + BaCO_3$. **1912** J. W. Mellor *Mod. Inorg. Chem.* vi. 99 Opposing reactions are also called incomplete or reversible reactions in contradistinction to irreversible or complete reactions typified by the action of zinc on sulphuric acid, where the reaction is completed in one direction and is not opposed by a counter reaction. **1950** N. V. Sidgwick *Chem. Elements* II. 1213 These acids..are formed by the action of water on the elementary halogen, which leads..to a reversible hydrolysis of the type $X_2 + H_2O \rightleftharpoons HX + HOX$. **1972** R. A. Jackson *Mechanism* iv. 85 When aliphatic fluorides are dissolved in antimony pentafluoride, a reversible reaction leads to the formation of carbonium ions.

c. Applied to an electrochemical cell in which the chemical reaction can be reversed by the application of a sufficiently large opposing e.m.f.; hence to quantities measured under, or pertaining to, conditions in which the e.m.f. of the cell is balanced by an applied e.m.f. and no current is flowing.

1895 C. S. Palmer tr. *Nernst's Theoret. Chem.* iv. vi. 599 Strictly speaking, all reversible batteries can be used as accumulators. **1936** R. W. Gurney *Ions in Solution* viii. 90 Under these circumstances it is found that the processes in each half-cell..are now simply taking place in the reverse direction. Such a cell is called a reversible cell—meaning reversible both in its mechanism and in the thermodynamic sense. **1957** G. E. Hutchinson *Treat. Limnol.* I. xi. 693 It is usual to term the actual potential measured the irreversible oxygen potential, while the theoretical value calculated may be termed the reversible oxygen potential. **1977** N. J. Selley *Exper. Approach to Electrochem.* iv. 63 The value of the applied p.d. for which the current is zero is the 'reversible cell potential'..for the back reaction. **1978** P. W. Atkins *Physical Chem.* xii. 346 A cell that is not truly reversible will respond to a small imbalance by supplying appreciable currents in one direction or the other, and it may be very difficult to detect its true balance point.

d. *Physical Chem.* Of a colloid or colloidal system: capable of being changed from a gelatinous state into a sol by a reversal of the treatment which turns the sol into a gel or gelatinous precipitate. Of a change of state: characterized by this property.

1899 *Jrnl. Physiol.* XXIV. 175 (*heading*) Heat-reversible colloidal mixtures. **1915** [see *irreversible *a.* 3]. **1930** *Engineering* 4 July 6/2 The power of absorption [of clay] being due to its containing a large proportion of matter in a state of fine division, forming a 'non-rigid reversible gel'.. composed of colloidal matter. **1955** J. Osborne *Dental Mech.* (ed. 4) i. 18 The cooling process is relatively slow, since all reversible hydrocolloids are bad conductors of heat. **1970** *Encycl. Polymer Sci. & Technol.* IV. 27 To redissolve reversible colloids, none of those drastic measures are required which are so characteristic of irreversible colloids.

B. *sb.* Also, a garment faced on both sides, so that it may be worn with either outside.

1863 *Cornh. Mag.* Jan. 89 The housebreakers' wives and children, maybe, take their turn during the day: at night, the men themselves watch. On such occasions they often wear 'reversibles', or coats which may be worn inside out; one side being of a bright, the other of a dark colour. **1905** *Eng. Dial. Dict.* V. 93/1 *Reversible*, a double cloth with a face, *gen.* made of better quality than the back. w. Yks. **1964** *McCall's Sewing* ii. 31/2 *Reversible*, fabric that has been finished so that either side may be used or a garment finished so that it may be turned and worn on either side.

reve·rsibly, *adv.* [f. Reversibl(e *a.* + -ly[2].] In a reversible manner.

1889 in *Cent. Dict.* **1944** [see *actin].

reversing, *vbl. sb.* Add: **b.** *spec.* The action of driving a motor vehicle backwards.

1900 W. W. Beaumont *Motor Vehicles & Motors* xiv. 246 The gearing..is for three speeds and reversing. *Ibid.* xxxi. 539 No reversing gear is provided, the reversing being effected by spur gearing. **1929** J. B. Priestley *Good Companions* i. ii. 67 Miss Trant discovered once again the terrors and dangers of reversing. **1959** *Motor Manual* (ed. 36) vi. 182 Reversing at night can be somewhat difficult.

2. Special Combs.: **reversing lamp, reversing light,** a light at the rear of a motor vehicle for illumination and to warn that the vehicle is reversing.

1960 R. Ireson *Penguin Car Handbk.* 125 Reversing lamps giving a white light to the rear are standard fittings on some higher priced cars. **1954** *Motor Manual* (ed. 35) xii. 250 A separate switch is provided to control the reversing light, and in some cases this is of normal type under the driver's direct control, but in others it is operated by the movement of the gear lever into the reverse position. **1968** *Radio Times* 28 Nov. 43/1 Not all cars have reversing lights fitted as standard. **1976** [see *reverse *sb.* 12 a].

reversing, *ppl. a.* Add: **1. a.** (Earlier example.) *reversing propeller*, a reversible propeller (see *reversible *a.* 1 b).

1804 M. Lewis in *Orig. Jrnls. Lewis & Clark Expedition* (1904) VI. 230 The reversing telescope when employed as the eye-piece gave me a more full..image. *a* **1817** Jane Austen *Northanger Abbey* (1818) II. xv. 320 No unworthy retraction.., no reversing decree of unjustifiable anger, could shake his fidelity. **1907** F. Strickland *Man. Petrol Motors & Motor Cars* x. 151 In the small sizes [of marine motor] this is done by having either a reversing propeller or a reversing gear worked with clutches, the engine being always kept running. **1935** J. W. Anderson *Diesel Engines* xv. 347 Small marine engines of 100 or 150 hp. are universally non-reversible. They usually run on the governor and reversing is obtained through reversing propellers in the very small sizes or reverse gears. **1973** D. Wright *Marine Engines & Boating Mech.* xiii. 184 A variable pitch and reversing propeller which will transmit and absorb ¾ hp per 100 rev/min giving..7½ hp at 1,000 rev/min.

b. *reversing layer* or *stratum*. Substitute for def.: A region of the solar atmosphere above the photosphere, formerly thought to be responsible for bringing about reversal of emission lines in the solar spectrum to absorption lines. Now *rare*. (Later examples.)

1926 H. N. Russell et al. *Astron.* I. vii. 196 The reversing layer, extending to a height of a few hundred miles above the photosphere and composed of the vapors of many of the familiar terrestrial elements. This merges gradually into the chromosphere. **1955** *Sci. Amer.* Sept. 194/2 The sun is entirely gaseous... From the outside in, the outer layers are the corona, the chromosphere, the reversing layer and the photosphere. **1974** *Encycl. Brit. Micropædia* II. 908/3 The lower chromosphere was formerly called the reversing layer because it was thought responsible for producing the dark lines of the solar spectrum that appear reversed against the bright continuous spectrum; actually the weak dark lines and bright continuum can be produced in essentially the same regions... The term reversing layer is now seldom used.

3. Special collocations: *reversing falls*, a waterfall or rapid in a narrow sea inlet in which the water flows in opposite directions when the tide is coming in and going out, because of the constricting effect of the narrow inlet; *reversing gear, lever* = *reverse gear, lever* s.v. *reverse *a.* and *adv.* 5 b; *reversing mill*, a rolling mill used in sheet metal production in which the metal is passed backwards and forwards between the same pair of rolls, which can have their direction of rotation reversed; *reversing thermometer*, a mercury-in-glass thermometer, normally used to obtain the temperature at depth in the ocean, which can be inverted at a depth and then retains its reading until its orientation is restored.

1910 W. O. Raymond *River Saint John* i. 3 Among the topographical features worthy of note are the remarkable 'reversing falls' at its mouth. **1961** E. Wahl *This Land* iv. 271 As the tide continues to rise the water begins to flow upstream and soon it is tumbling over the rock ledges in the opposite direction. This oddity of nature, called the Reversing Falls, has been a great tourist attraction for many years. **1974** *Encycl. Brit. Macropædia* VII. 780/2 The tide range is magnified by the narrowness and shape of the bay, a rise of 46 feet (14 metres) being common in Chignecto Bay and 53 feet (16 metres) in Minas Basin... The rising tide produces a 'reversing falls' at the mouth of the Saint John River. **1867** *Engineering* 29 Mar. 300 (*heading*) Screw reversing gear for locomotive engines. **1904** T. H. White *Petrol Motors & Motor Cars* ii. 108 When applied to an automobile, it is usual to make the reversing gear of a similar design to the slow-speed forward gear. **1972** S. H. Henshall *Medium & High Speed Diesel Engines for Marine Use* xv. 269 A single uni-directional engine with oil operated reversing gear could easily be combined on one lever on the bridge. **1887** *Engineering* 7 Oct. 393/1 A locomotive having an ordinary reversing lever to operate the links. **1925** *Marine Oil-Engine Handbk.* (ed. 6) ix. 76 The shaft (B) and with it the clutch (D) can be moved endways by the reversing lever through the usual double flange and collar. **1888** *Lockwood's Dict. Mech. Engin.* 285 Reversing mill. **1909** *Jrnl. Iron & Steel Inst.* LXXX. 15 The capital cost of an electrically driven reversing-mill is greater than that of a steam-driven mill. **1932** *Iron & Coal Trades Rev.* CXXV. 730/1 Our representative had an opportunity of inspecting an 18,000-h.p. (peak) reversing-mill motor... The motor is intended to drive a 3-stand reversing mill. **1967** A. H. Cottrell *Introd. Metall.* xxii. 439 The first rough rolling of such an ingot is done in a cogging mill between large cast iron rolls. This is usually a two-high reversing mill. **1912** Murray & Hjort *Depths of Ocean* v. 217 In 1878 Negretti and Zambra of London constructed a reversing thermometer, which has played a prominent part in physical oceanography. **1928** Russell & Yonge *Seas* xii. 260 A 'reversing thermometer' is used, that is one in which there is an S-shaped bend, so that if it be suddenly turned upside down the thread of mercury is broken and a permanent record of the temperature is obtained. The thermometer on reaching the required depth is reversed by means of a weight which slides down the wire and releases a spring catch. **1963** G. L. Pickard *Descriptive Physical Oceanogr.* vi. 89 After corrections..the reversing thermometer yields the temperature to an accuracy of about +0·02 C°.

reversion[1]. Add: **III. 8.** *attrib.*, as *reversion clause, duty.*

1933 P. Godfrey *Back-Stage* vi. 74 He should have a reversion clause in his contract, so that if in any year his play is not performed a certain number of times the rights will revert to him. **1909** *Westm. Gaz.* 28 May 8/1 Clause 7 deals with reversion duty. **1959** Jowitt *Dict. Eng. Law* II. 1554/1 *Reversion duty*, a duty formerly payable under the Finance (1909–10) Act, 1910, in certain cases on the determination of leases more than twenty-one years.

reversion[2]. For *rare*[−1] read *rare* and add: **b.** A drawing based upon an earlier sketch.

1848 D. G. Rossetti *Let.* Sept. (1965) I. 44 His last design is a re-version from Retzsch's outline of the same subject.

reversionary, *a.* (and *sb.*). Add: **2.** *spec.* in *reversionary bonus*, an increase in the amount of an insurance policy payable at the maturation of the policy or the death of the person insured.

1898 [see HM s.v. *H III.]. **1930** *Economist* 29 Mar. 702/2 The reversionary bonus for 'with profits' policies has been maintained at the rate to which it was raised. **1969** *Times* 30 Apr. 26 In the United Kingdom the rate of reversionary bonus for Ordinary branch assurances has been increased by 2s. per cent.

reversis. Add: **a.** (Earlier and later examples.)

1796 (*title*) Rules of Reversis, as played in the Fashionable Circles. By a Gentleman. **1977** *Jrnl. Playing-Card Soc.* May 23 Reversis is historically important as the earliest known negative complex trick-taking game.

b. (Later example.)

1977 *Jrnl. Playing-Card Soc.* May 28 To make reversis a player must win all eleven tricks.

revert, *sb.* Add: **2.** (Later and *transf.* examples.)

1927 Ld. Braye *Fewness of my Days* 159 A palace of antiquities, not only for the archæologist but also..for a Revert to the ancient faith. **1960** *Times Lit. Suppl.* 24 June 401/1 Professor Empson is anti-modern. Not a mere revert, of course,..he makes respectful acknowledgment to the views of Blake and Shelley.

revert, *v.* Add: **II. 8. c.** To cause to return *to* a former condition or practice.

1973 *Nature* 2 Mar. 16/1 It is possible to produce a crystalline area (memory state), to measure the change in optical transmission (read out), and to revert the same area to the amorphous state (erase). **1975** *Daily Tel.* 1 Aug. 2/5 Left-wingers in the Amalgamated Union of Engineering Workers tried to revert the union to its previous system of electing officials at branch meetings. **1977** *Evening Post* (Nottingham) 24 Jan. 2/7 The suggestion to revert a central site to agricultural use seems both practical and sensible.

revertant, *a.* Restrict *Her. rare*[−0] to sense in Dict. and add: **2.** *Biol.* Having reverted to the normal phenotype though of mutant or abnormal ancestry.

1955 *Genetics* XL. 893 Both revertant (adenine-independent, non-purple) and non-revertant (adenineless, purple) components could be separated by conidial platings. **1970** *Nature* 5 Dec. 908/1 Revertant cells selected from this line exhibit many of the properties of untransformed cells.. although they retain the SV40 genome. **1981** *New Scientist* 18 June 744/2 *E. coli* is so 'weak' that it can easily mutate back to its ordinary cousin... These 'revertant mutants' quickly take over.

B. *sb. Biol.* A revertant cell, organism, or strain.

1955 *Genetics* XL. 893 Tests of a number of these purple revertants revealed..that they were heterocaryons. **1971** *Nature* 25 June 488/1 By applying selection pressures to cultures of transformed malignant cells it is possible to select so-called revertants—cells which behave in culture as if they were normal. **1974** *Ibid.* 11 Oct. 483/1 Clones obtained on NFM-histidine were either prototrophic or revertants which grew on histidine as sole nitrogen source.

revertose (rĭvɜ·ɹtōᵘz). *Chem.* [f. Revert *v.* + -ose[2].] A substance obtained by the action of maltase on glucose, now regarded as a mixture of sugars.

1903 A. C. Hill in *Jrnl. Chem. Soc.* LXXXIII. 580 The larger portion..has optical and reducing properties between those of maltose and of glucose. It consists mainly of a biose..which I have called revertose. **1938, 1965** [see *gallisin].

revet (rīve·t), *v.*[2] [Re- 5 a.] *trans.* To 'vet' again; to recheck, to re-examine. Hence **reve·tting** *vbl. sb.*

1963 *Times* 26 Apr. 18/4 He asked whether Vassall had been vetted as necessary for his special post, and was told that he would be revetted before he went. This 'revetting' consisted simply of a check of his name against security service records to ascertain whether they contained any

information about him indicating that he held subversive beliefs. **1973** *Times Lit. Suppl.* 26 Jan. 96/4 Pictorial evidence..and..sculptural evidence..revetted by a wide choice of floor mosaics. **1979** P. Harcourt *Sleep of Spies* i. ii. 34 Everyone's being re-vetted, of course... They're doing some re-vetting there too.

review, *sb.* Add: **I. 3.** (Earlier *attrib.* examples.)

1740 *Life & Adv. Mrs. Christian Davies* 500 He allowed me exclusive of all others to sell beer in the Deer Park on a review day. **1781** R. F. Greville *Diary* 4 Aug. (1930) 9 We came on Ashford Common where the Blues were drawn up in Review Order.

5. d. Also *attrib.*

1965 *New Statesman* 19 Mar. 430/1 Only on one point does he reject the BMA case—he insists that the Review Body must work out the new pay-levels. **1968** *Panorama* (Austral.) May 2/4 They sat in 'review board' to use an advertising agency term. They were briefed on the new type of meal which the management had in mind. **1971** D. Bagley *Freedom Trap* ii. 53 'You have been classified as a high risk.. If you weren't so stupid you could get yourself out of this jam.' 'Out of this nick?' 'I'm afraid not... But the Review Board would look upon you very kindly if you co-operated with us.' **1977** *Belfast Tel.* 22 Feb. 12/1 The Industrial Relations (Northern Ireland) Order 1976 became law. It implements recommendations of the Review Body on Industrial Relations which reported in 1974.

7. a. Also, of music, drama, etc.

1929 [see *gag *sb.*[1] 3 c]. **1967** J. Philip et al. *Best of Granta* i. 18 Jazz features regularly in the record reviews.

c. (Later examples.) *review article*, an article that is a review; *review copy*, a copy of a new book sent for review to a periodical, writer, etc.

1836 J. S. Mill in *London & Westm. Rev.* Apr. 17 The attention cannot sustain itself on any serious subject, even for the space of a review-article. **1855** Bagehot *Coll. Wks.* (1965) I. 310 In truth review-writing but exemplifies the casual character of modern literature. *Ibid.* 312 The review-like essay and the essay-like review fill a large space. **1858** Carlyle *Fredk. Gt.* I. i. i. 16 The British Writer..images to himself a royal Dick Turpin, of the kind known in Review-Articles. **1903** G. B. Shaw *Let.* 15 July (1972) II. 338 To whom do you propose to send review copies? **1906** A. Bennett *Let.* 27 Feb. (1966) 69, I sent him a list of review-quotes but I have seen no result. **1907** F. M. Ford *Let.* ? Sept. (1965) 26, I had proposed to run to a column a week at, say, your ordinary review rates. **1933** Leavis & Thompson *Culture & Environment* 44 Examine the review pages of a Sunday newspaper. **1937** H. Nicolson *Diary* 11 July (1966) 307 Read my review books and spend a peaceful afternoon. **1965** *New Statesman* 19 Mar. 455/1 Judgments of this kind are proper to the review-article of modern length. **1966** C. Mackenzie *My Life & Times* V. 243, I asked..if we might count on review records coming regularly from H.M.V. **1976** *Guardian* 17 Apr. 11/6 Scarcely a day goes by without review copies reaching this office of books..on growing potato crops on the windowsills of Chelsea. **1978** *Amer. N. & Q.* XVI. 162/2 The front endsheets..are upside down in one review copy. **1979** *London Rev. Bks.* 25 Oct. 2/2 Any squeeze on review space squeezes fiction first.

review, *v.* Add: **7.** Also, to write criticism of (music, drama, etc.).

1873 Lytton *Kenelm Chillingly* II. iv. iv. 232 By the way, when we come by-and-by to review the exhibition at Burlington House, there is one painter whom we must try our best to crush. **1976** *Gramophone* Feb. 1404/3 The Denon PMA-500 amplifier has been a real pleasure to review.

reviewability (rĭviū·ăbi·liti). [f. Reviewable *a.* + -ity.] The quality or state of being judicially reviewable.

1975 *Columbia Law Rev.* LXXV. 132 Ultimate determinations as to the reviewability of particular decisions may take into account the nature of the agency's functions and the impact of review on its effectiveness and other legislative policies.

reviewable, *a.* Add: *spec.*, subject to judicial review.

1883 *Nation* (N.Y.) 20 Dec. 501/2 The proceedings in any criminal trial are reviewable by the full bench, whenever the judge who presides at the trial certifies that any point raised at it is doubtful. **1964** *Mod. Law Rev.* XXVII. 331 Nor can a decision reviewable on habeas corpus be considered a nullity which could be disregarded when attacked collaterally. **1974** *Sci. Amer.* June 22/1 We held that adequacy of treatment is reviewable in court.

reviewery (rĭviū·əri). *rare.* [f. Review *sb.* + -ery.] The genre of the review; reviews considered collectively.

? **1876** R. L. Stevenson *Lett.* (1911) I. 215, I was not a hundred miles from being miserably drowned, to the.. permanent impoverishment of British Essayism and Reviewery.

reviewing, *vbl. sb.* **1.** (Later *attrib.* examples.)

1951 M. McLuhan *Mech. Bride* (1967) 10/2 The editors stand at mock attention on the reviewing platform, thumb on nose. **1964** *Mod. Law Rev.* XXVII. 331 Inferior courts and tribunals are at the mercy of the court's reviewing power. **1977** *Time* 4 Apr. 53/1 At his Inaugural parade, Jimmy Carter's reviewing stand was warmed by a specially constructed solar heating unit.

‖ **revirement** (rəvirmaň). [Fr., f. *revirer*: see Revire *v.*[2]] An alteration of one's plans; a complete change of attitude or opinion.

1913 W. S. Churchill *Let.* 18 Dec. in *World Crisis 1911–14* (1923) viii. 173 If on a general *revirement* of Naval Policy the Cabinet decide to reduce the quota, it would be indispensable that a new exponent should be chosen. **1975** *Government & Opposition* X. 276 Contradictions are caused not so much by economic fluctuations—but rather by political *revirements*.

revise, *sb.* Add: **2. b.** (Earlier and later examples.)

1844 Dickens *Let.* 29 June (1977) IV. 153 Mr. Dickens will be glad if Mr. Newby will send a complete revise of the whole book..to Mr. Overs for his attentive perusal. **1875** A. Thackeray *Let.* in H. Ritchie *Lett. A. T. Ritchie* (1924) viii. 163 Mr. Payn writes sternly for the revise for my story and I must not write any more now. **1970** W. Garner *Puppet-Masters* xxxvi. 259 Here's the revise. We had it retyped with the new ending.

3. (Later *attrib.* example.)

1972 H. Evans *Newsman's English* i. 8 The copy is normally passed to a third executive, the 'revise editor'. On American dailies he is a 'slot man'.

revise, *v.* Add: **3. b.** (Later example.)

1939 N. Marsh *Overture to Death* v. 57 Then he gave himself four minutes to revise the conversation he had planned to have with Dinah.

c. To go over (a subject already learnt) in preparation for an examination. Also *absol.*

1946 'B. Truscot' *First Year at University* v. 66 Assuming the examination to be in May or June,..[the Easter] vacation should..be devoted to revision, and the work to be revised must be systematically divided up among the time available. **1977** C. Dexter *Silent World N. Quinn* xii. 109 You revise, I suppose?..I mustn't keep you from your revising.

revising, *vbl. sb.* (Later example.)

1969 C. Fremlin *Possession* xvii. 137 Janice said what about her revising?.. How could she ever get through her Mocks next term? **1977** [see *revise *v.* 3 c].

revising, *ppl. a.* (Earlier example.)

1757 Richardson *Let.* 19 Nov. (1964) 337 Your Ladiship's Correspondence & mine, will be ye safer, because of its length, & for its being only in our own Hands; & if it has yr. Ladiship's revising Eye, will be of ye less consequence.

revision. Add: **1.** Also *spec.*, the action of going over a subject that has already been learnt, esp. in preparation for an examination.

1916 A. Huxley *Let.* 1 May (1969) 98, I am busy with revision, doing papers for my tutor under examination conditions. **1937** Corbett & Eichele (*title*) Classified revision exercises in German. **1979** *Observer* 29 July 21/1, I thought the end of school term might be a good occasion for a little revision.

revisionism (rĭvi·ʒəniz'm). *Politics.* [f. Revision + -ism.] **1.** A policy first put forward in the 1890s by Edward Bernstein (1850–1932) advocating the introduction of socialism through evolution rather than revolution, in opposition to the orthodox view of Marxists; hence a term of abuse used within the communist world for an interpretation of Marxism which is felt to threaten the canonical policy.

1903 *Social-Democrat* VII. 84 (*heading*) Revisionism in Germany. **1909** E. C. Harvey tr. *Bernstein's Evolutionary Socialism* p. xxii, Subsequently the views put forward in the book have received the bye-name of Revisionism. **1934** P. & I. Petroff *Secret of Hitler's Victory* iii. 40 In the social democratic ranks extreme revisionism predominated. **1947** *Partisan Rev.* XIV. 396 Is it permissible, is it orthodoxy or revisionism, for a Marxist to take any significant ideas from the prevailing fashion of the 'neo-Kantian' revival? **1958** *Times* 4 Aug. 7/2 It is not only over Yugoslav 'revisionism' that China has lately taken a distinctive and uncompromising attitude. **1959** M. S. Levin tr. *Lenin's Against Revisionism* 122 The ideological struggle waged by revolutionary Marxism against revisionism at the end of the nineteenth century is but the prelude to the great revolutionary battles of the proletariat. **1962** *Listener* 8 Mar. 404/2 Russian Khrushchevian revisionism. **1965** *New Statesman* 18 June 945/2 China was going to fight Soviet revisionism to the bitter end. **1969** A. G. Frank *Latin Amer.* (1970) xiv. 221 There exists no dual society in the world today and all attempts to find one are attempts to justify and/or cover up imperialism and revisionism. **1975** J. De Bres tr. *Mandel's Late Capitalism* xvi. 517 Cheprakov's revisionism is here unequivocally spelt out.

2. A term used for a revised attitude to some previously accepted political situation, doctrine, or point of view; *concr.*, the name of the policy adopted by a right-wing Zionist group, active during the formative period of the State of Israel; mostly *U.S.*, a movement to revise the accepted versions of American history, esp. those relating to foreign affairs since the war of 1939–45.

1921 *Glasgow Herald* 4 Apr. 10 The British Foreign Office has got over its momentary lapse into revisionism. **1932** *Palestine Post* 25 Dec. 6/2 The leader of the Revisionism, Jabotinsky. **1939** G. F. Hudson *Turkey, Greece & E. Mediterranean* (1940) 28 Turkey, Greece, Rumania, and Jugoslavia therefore left out Bulgaria, and concluded between themselves a pact of mutual guarantee, the foundation of the so-called Balkan Entente (9 Feb. 1934). This was in effect an alliance against Bulgarian revisionism, just as the Little Entente was an alliance against Hungarian revisionism. **1940** *Economist* 6 Apr. 606/2 The Near Eastern States offer a happy hunting-ground to the roving ambition of predatory Great Powers. Once again, it is a question of revisionism *versus* the *status quo*. **1949** Koestler *Promise & Fulfilment* iii. 302 The conflict between Revisionism and official Zionism was mainly one of character and temperament. **1953** J. A. Lukacs *Great Powers & E. Europe* p. viii, The somewhat vague concept of historical revisionism is applicable only when there is an abundance of well-documented historical writing which, because of its unilateral emphasis or perspective, needs to be counter-balanced. *Ibid.* iii. 282 Ribbentrop..termed the Russian attitude [over Finland] as 'reasonable', stemming from an aim of 'modest revisionism'. **1959** *Encounter* Sept. 64/2 A humanistic revisionism can be secured only by revising the claims of science itself. **1960** S. G. Evans *Short Hist. Bulgaria* v. 175 The extent of German economic penetration and the degree of national chauvinism in the ruling class, leading to a clamant revisionism, plus the fact that Hitler appeared to be winning, made it certain that they would turn the way they did. **1965** *New Statesman* 1 Oct. 486/2 One linguistic difference between American and British historians lies in the frequency with which they use the word 'revisionism'. It is common currency in Transatlantic seminars and journals, but hardly ever heard in this country. *Ibid.*, 'Revisionism' goes on all the time because of disagreement about the moral and political significance of what happened. **1973** in R. Staar *Yearbk. on Internat. Communist Affairs* 80 The fact that the work was not condemned or suppressed, but indeed approved..by Georgian authorities, until Moscow intervened, indicated that historiographical 'revisionism' was rife in the republic. **1977** *Time* 15 Aug. 26/1 Revisionism is starting on Johnson as it has started on other Presidents.

revisionist, *sb.* and *a.* Add: **A.** *sb.* **1.** (Later examples in senses of *revisionism.)

1909 [see *opportunist 1 a]. **1915** L. B. Boudin *Theoret. Syst. K. Marx* i. 11 What they [*sc.* critics] claim is that they [*sc.* Marx's theories] were based on insufficient data, and that our present knowledge requires the revision of some of his tenets or the supplementing of them by some qualifying truths... Hence, the name *Revisionists*, under which most of the newer Marx-critics are known, and the term *Revisionism* applied to their writings and teachings. **1927** H. E. Fosdick *Pilgrimage to Palestine* (1928) xii. 288 Some [Zionists] are revisionists, trying to force Britain's hand and compel aggressive action towards making Jews the dominant political power. **1930** *Time & Tide* 6 Sept. 1125/2 Professor Barnes is what is known as a 'revisionist', in other words, an exponent of historical views on this subject, almost completely reversing those dominant in England and America eleven years ago. **1949** R. K. Merton *Social Theory* 115 The fallacious premise..found also in the writings of such Freudian revisionists as Fromm, that the structure of society primarily restrains the free expression of man's fixed native impulses. **1957** *Observer* 27 Oct. 6/2 The purge will be used to root out..the 'revisionists' who evince a negative attitude towards the Soviet Union. **1962** *Daily Tel.* 17 Nov. 16/3 The article is an important move in the Moscow–Peking dispute. It accused the 'revisionists' of retreating from the decisions taken jointly in Moscow. **1966** C. Potok *Chosen* (1967) xiii. 226 Every shade of Zionist thought was represented..from the Revisionists, who supported the Irgun, to the Neturai Karta, the Guardians of the City..Jerusalem. **1973** P. Hollander *Soviet & Amer. Society* i. 13 The revisionists simply cannot believe that if American foreign policy was wrong (criminal, bankrupt, or irresponsible) in regard to Vietnam, it could have been otherwise in different historical situations, especially if Communism was at issue. **1976** *Times* 9 June 16/2 It is not easy to strengthen contacts with Yugoslavian communists while keeping Spanish and French 'revisionists' at arm's length.

B. *adj.* That advocates or supports revision; pertaining to revisionism or revisionists.

1866, 1888 [in Dict., sense 1 *attrib.*] **1903** *Social-Democrat* VII. 97 Thus, the so-called democratising of industry through the company makes for the Social Revolution, and renders Revisionist Socialism impossible. **1925** *Zionist Rev.* IX. 48/1 The Radical and the Revisionist Opposition. **1934** *Sun* (Baltimore) 16 Oct. 10/3 The 'revisionist' historian of the war has transferred the responsibility for that great tragedy from the shoulders of the German military autocracy to those of M. Poincaré. **1949** Koestler *Promise of Fulfilment* ii. 259 A Revisionist doctor makes his living on Revisionist patients, goes to Revisionist cafés and frequents only Revisionist circles. **1961** *Listener* 30 Nov. 905/2 Much fruitful thinking is now going on within the framework of revisionist Marxism about the kind of relation which exists between basis and superstructure. **1969** *Amer. N. & Q.* Oct. 31/2 It is 'revisionist' history in the finest sense, a wholly new interpretation of the sources. **1971** *Human World* Nov. 18 In Chinese polemics of the same period, it was discernible that the primary object of Peking's venom was the 'revisionist social imperialism' of the 'new Tsars'. **1974** *Times* 18 Feb. 7/2 Their local Communist Party committees should carry out reforms aimed at preventing the growth of 'revisionist' tendencies. **1977** *Time* 30 May 4/3 The recent cluster of 'revisionist' books on Nazism, which would soften the frightening teachings of this maniacal movement.

revisit, *v.* Add: **2.** Also *fig.* (cf. sense 1). Always in pa. pple.

1971 *Language* XLVII. 503 (*heading*) The phoneme revisited. **1975** I. Robinson *New Grammarians' Funeral* ix. 169 (*heading*) 'The science of language' revisited. **1976** *Lingua* XL. 321 (*heading*) The *-ing* form revisited. **1977** *Language* LIII. 277 (*heading*) Conversational postulates revisited.

revitalization. Add: Also *attrib.*, as *revitalization movement*.

1956 A. F. C. Wallace in *Amer. Anthropol.* LVIII. 265 A revitalization movement is defined as a deliberate, organized, conscious effort by members of a society to construct a more satisfying culture. **1970** R. D. Abrahams *Positively Black* vi. 151 Perhaps the most important feature of resemblance between revitalization movements and the present separatist position is that both turn on the idea of social and cultural polarization within a pluralistic situation.

revival. Add: **1. b.** Also, the act of resuming a series of broadcast programmes.

1955 *Times* 13 May 16/1 The B.B.C. had left the door open... The minimum gap before such a revival would be six months. **1976** in *Amer. Speech* (1978) LIII. 58 We retain the right to edit our material which includes excessive emotionalism or statements which could be detrimental to S[tar] T[rek] fandom or revival.

3. b. (Earlier and later *attrib.* examples.)

1831 J. J. Strang *Jrnl.* 31 July in M. M. Quaife *Kingdom of St. James* (1930) 195 The revival meeting lasted a fortnight or more. **1891** *Atlantic Monthly* June 813/2 The old slaves are loath to give up the hysteric emotionalism of revival preaching. **1956** R. Macaulay *Towers of Trebizond* (1957) ix. 88 Then there had been the Billy Graham mission, which had passed through the other day, hired a room in the municipal buildings, and held a revival service through an interpreter. **1969** C. Burke *God is Beautiful, Man* (1970) 98 One day there was a guy named Peter in a revival tent on an empty lot. **1976** *Times* 13 Feb. 7/5 In 1876 he [*sc.* William Booth] set up as music publisher with Revival Music, a collection of hymns and gospel songs used by the Christian Mission, as his army was known until 1878.

revivalist. Add: **3.** *attrib.* or as *adj.*

1875 [in Dict., sense 2]. **1956** M. Stearns *Story of Jazz* (1957) iii. 30 The ceremonies became famous for their revivalist power and frenzy. **1965** G. Melly *Owning Up* xi. 128 Revivalist jazz was based on the Negro jazz of the 'twenties.

revive, *v.* Add: **II. 9. c.** Also *intr.* for *pass.* Also in modern use, to resume (a series of broadcast programmes).

1912 G. B. Shaw *Let.* 13 Feb. in *Lett. to Granville Barker* (1956) 179 The play [*sc. Hamlet*] revives sensationally every 15 years or so. **1955** *Times* 13 May 16/1 The B.B.C. had left the door open when its run ended. It would be too soon for Hughie Green to 'raise his eyebrows' in September, 1949, because he had not been told that it would be revived. **1977** *Times Lit. Suppl.* 25 Feb. 200/2 The plays..do not in my view revive successfully today.

revivication. Add: Now *rare.* (Further example.)

1929 V. Woolf *Let.* 11 May (1978) IV. 56 Last night we had a terrific revivication; the resurrection day was nothing to it.

revoker. Delete *rare* and add further examples.

1904 *Bridge & Progressive Bridge* 13 The penalty for a revoke is the addition of three tricks to the score of the opposing side, or the deduction of three tricks from the Revoker's score at the option of the opposing side. **1922** E. F. Benson *Miss Mapp* xi. 261 This way of taking a revoke was new to Tilling, for the right thing was for the revoker's partner to sulk and be sarcastic for at least twenty minutes.

‖ **revolera** (revolē·rā). *Bullfighting.* [Sp.] A movement in which the cape is fluttered above the matador's head (see also quot. 1957).

1957 A. MacNab *Bulls of Iberia* vi. 60 At times, good matadors prefer to finish off their series of passes in a gentler way with a *larga*,..in the form the *revolera* in which the cape swirls above the man's head... The term *revolera* is often loosely used for *serpentina* and *vice versa*. **1967** [see *mariposa].

‖ **révolté** (revǫlte). fem. **révoltée**; also unaccented and *erron.* **revolté.** [Fr.] One who revolts; a rebel or iconoclast; a nonconformist.

1890 E. Dowson *Let.* c 11 June (1967) 153 The *révolté*..I conceive as a violent & rather venomous person. **1910** G. B. Shaw in *Nation* 24 Dec. 544/1 Mary Fitton is quite modern, an *amoureuse* and a *révoltée*. **1934** R. Campbell *Broken Record* 21, I have never been a *révolté*. **1938** *Sat. Rev. Lit.* (U.S.) 19 Nov. 4/3 The book is worth reading for those few chapters dealing with the association of those young revoltees. **1963** 'Han Suyin' *Four Faces* 219 The British have always had the genius of stimulating revoltees. **1978** *Dædalus* Fall 98 Now they felt more competent to question the assumptions of the public school avant-garde revolté.

revo·ltingness. [f. Revolting *ppl. a.* + -ness.] The state or quality of being disgusting; repulsiveness.

1749 J. Cleland *Mem. Woman of Pleasure* II. 5 The revoltingness of gross, rank, and vulgar expressions. **1907** G. Murray *Rise Gk. Epic* v. 122 Homer has cut out these stories for their revoltingness. **1957** Wodehouse *Over Seventy* xvi. 153 He..smiled a smile of just the right degree of revoltingness.

revoltive (rĭvǫ·ltiv), *a. rare*[−1] [f. Revolt *sb.*[1] + -ive.] Characterized by revulsion; = Revolted *ppl. a.* 3.

1954 W. Faulkner *Fable* 241 A wounded man..was stared at with the same aghast distasteful revoltive pity and shock and outrage as a man in an epileptic seizure at high noon on a busy downtown corner.

revolute, *v.*[2] (Further examples.)

1921 [see *bolsh *v.*]. **1937** *Aeroplane* 16 June 744/2 Many of the supporters of the Red agitators are merely using Moscow money to ginger up their own Unions, without any intention of revoluting.

revolution, *sb.* Add: Now usu. with pronunc. (revǒlū·ʃən). **III. 6. c.** *Geol.* A major mountain-building episode, esp. one extending over a whole continent or occurring at the close of a geological era.

1802 J. Playfair *Illustrations Huttonian Theory* 2 The earth has been the theatre of many great revolutions, and..nothing on its surface has been exempted from their effects. To trace the series of these revolutions, to explain their causes,..is the proper object of a Theory of the Earth. **1845** C. Lyell *Trav. N. Amer.* I. iv. 99 The physical revolutions of the territory at present under consideration. **1863** J. D. Dana *Man. Geol.* iii. iv. 403 After the long ages of comparative quiet, when the successive Palaeozoic formations were in slow progress, and finally the rock-foundation of the continent east of the Rocky Mountains was nearly completed, a change of great magnitude began, which involved the Appalachian region with the continental border adjoining, and well merits the title of Appalachian revolution. **1915** C. Schuchert in Pirsson & Schuchert *Text-bk. Geol.* II. xli. 749 During the Appalachian Revolution much of eastern North America was again thrown into pronounced folds, faulted and widely thrust to the west and northwest. **1932** L. C. Snider *Earth Hist.* iii. 66 Environments of living things change very rapidly during such episodes, and as a result the faunas and floras before a revolution are quite different from those after it. **1978** W. L. Stokes et al. *Introd. Geol.* xvi. 379/2 There were no strong interactions between plates during the Cambrian Period, but there were extensive collisions during the Middle Ordovician to Devonian (Caledonian Revolution).

d. *the green revolution*, a great increase in the production of cereal crops in developing countries consequent on the introduction of high-yield varieties and the application of scientific methods to agriculture.

1970 *Financial Times* 23 Mar. 10/5 The primary factor behind the recovery has been the growth of agricultural output... Of course, the so-called Green Revolution is still very patchy. **1972** *Guardian* 30 Dec. 12/3 The green revolution took place only in wheat, not the coarse grains, pulses, and rice which the poor prefer. **1976** *Sci. Amer.* Sept. 31/1 The 'green revolution' of the late 1960's represented a significant improvement in grain-crop productivity, primarily in parts of Asia and Latin America. **1977** *Jrnl. R. Soc. Arts* CXXV. 226/1 The Green Revolution, in which new high yielding, short strawed varieties of wheat and rice were introduced into Third World countries.

7. c. In the Marxist doctrine of social evolution, the class struggle between the bourgeoisie and the proletariat leading in time to the downfall of capitalism and to its replacement by communism; also *continuing, continuous, uninterrupted revolution*, designating the concept of permanent revolution (cf. *permanent *a.* 1 d).

1850 H. Macfarlane tr. *Marx & Engels's Communist Manifesto* in *Red Republican* 16 Nov. 171/3 We have followed the more or less concealed civil war pervading existing society, to the point where it must break forth in an open revolution, and where the Proletarians arrive at the supremacy of their own class through the violent fall of the Bourgeoisie. **1920** B. Russell *Pract. & Theory Communism* ii. 32 The Third International is an organization which exists to promote the class-war and to hasten the advent of revolution everywhere. **1937** E. H. Carr *Internat. Relations* iii. 73 The duty of every good Communist was to spread throughout the world the same revolution which had been successful in Russia. **1964** Gould & Kolb *Dict. Social Sci.* 602/2 According to this [Marxist] view, great political and social revolutions are the instruments of the 'inevitable' progress of mankind towards a society in which freedom, self-government, social harmony, and equality will dominate. **1969** S. R. Schram *Pol. Thought Mao Tse-tung* (rev. ed.) iii. 203 The basic idea, that of a bourgeois-democratic revolution under the hegemony of the proletariat, was to be further developed by Mao Tse-tung. **1975** *Chinese Econ. Stud.* VIII. iv. 9 Lin Piao's, and other similar swindlers' advocacy of this fallacy was to vainly attempt to use the productivity-first viewpoint as a weapon to oppose the continuing revolution. **1975** J. Plamenatz *K. Marx's Philos. of Man* i. 26 There is still talk of 'revolution' as something that the 'contradictions' of capitalism or of 'bourgeois society' will bring about, but there is much more uncertainty as to who the makers of the revolution will be. **1977** 'S. Leys' *Chinese Shadows* (1978) viii. 169 The discrimination..between 'permanent revolution' (a Trotskyist heresy) and 'continuous revolution' (a genial and creative development brought to Marxist thought by Mao Tse-tung). **1978** *Fontana Dict. Mod. Thought* 464/2 Lenin..thought that the revolution would pass first through the 'bourgeois' and then through the 'socialist' stage in his scheme of 'uninterrupted' revolution.

11. *Russian Hist.* The revolutionary activity of different factions in Russia during 1917 which resulted in the overthrow of the existing regime and the establishment of a socialist state, the Union of Soviet Socialist Republics. Cf. *Russian Revolution* s.v. *Russian *a.* 2 d*.

1917 *Times* 13 Nov. 6/2 The Bolsheviks have dropped their offensive attitude, considering themselves now in the position of organizers of the defence of the Revolution against the armies of Kerensky, Korniloff, and Kaledin. **1932** R. H. B. Lockhart *Mem. Brit. Agent* iv. iii. 221 A Russian of good family, who long before the revolution had sacrificed a fortune for his Socialist convictions. **1948** B. D. Wolfe *Three who made Revol.* i. 38 The two decades of Russian history from the Revolutions of 1917 to the purges of 1937. **1959** P. Wiles tr. *Schakovskoy's Privilege was Mine* x. 107 Another young relative, born after the Revolution, became emotionally involved with a Soviet intellectual. **1978** J. Molyneux *Marxism & Party* iii. 58 During the rise of the revolution, the Mensheviks were in large part swept along by events.

12. *attrib.*, as *revolution counter* (= *rev-counter* s.v. *rev *sb.* 2), *indicator*. See also sense 8 b in Dict.

1961 Webster, Revolution counter. **1962** *Which? Car Suppl.* Oct. 139/1 Loud rattle from revolution counter cable. **1922** Revolution indicator [see *air sextant* s.v. *air *sb.*[1] B. III. 6].

revolutionary, *a.* and *sb.* Add: **1. a.** (Further examples.) Also *Comb.*

1919 G. B. Shaw *Heartbreak House* p. ix, Heartbreak House was quite familiar with revolutionary ideas on paper. **1930** [see *orgiastical *a.*]. **1937** 'C. Caudwell' v. 94 *Illusion & Reality* Revolutionary-puritan ideals of liberty. **1941** Koestler *Scum* 19 They had an explanation ready for every occasion... They called it 'revolutionary dialectics'. **1959** *Encounter* XIII. 78 Even after the revolutionary-existentialist mood in Poland faded. **1960** [see *luftmensch, luftmensh]. **1970** W. Klatt in D. J. Dwyer *China Now* (1974) xviii. 345 In 1967 and 1968 Chinese Press and radio reports revealed population data of a more serious type. These emanated from the Revolutionary Committees in China's chief administrative regions. **1975** C. P. Mackerras *Chinese Theatre in Mod. Times* x. 177 The well known Szechwanese Opera actress Ch'en Shu-fang..performed in a modern revolutionary opera in May 1971. **1977** *Undercurrents* June–July 10/1 Capitalism has developed (even if the revolutionary left's response to it has not).

Hence **revolutiona·rily** *adv.*

1864 R. Cobden *Let.* 3 May in J. Morley *Life R. Cobden* (1881) II. xviii. 448 It was feared..that if he went to the provinces he might be talking too revolutionarily. **1927** *Daily Express* 11 Feb. 1/1 Lisbon is traversed from north to south by wide avenues—splendid places for those who are revolutionarily inclined. **1956** W. S. Churchill *Hist. Eng.-Speaking Peoples* I. ii. vi. 181 He [*sc.* Richard I] made new and revolutionarily heavy demands for taxation. **1971** *Frendz* 21 May 8/2 Revolutionarily aware people can't be fooled by these kinds of people.

revolutioneering, *vbl. sb.* Add: Also, agitation for revolution. *rare.*

1941 *World-Telegram* (N.Y.) 8 Jan. 23/2 We had trouble with the Muscovites before and..made them promise..to leave off revolutioneering in our midst.

revolutionism. (Earlier and later examples.)

1841 C. Fox *Jrnl.* 20 Apr. (1972) 105 He [*sc.* John Sterling]..thinks 'The Misanthrope' his [*sc.* Molière's] best, and considers that all the din and stir of French Revolutionism is prefigured in it. **1959** C. Ogburn *Marauders* (1960) ii. 48 Equally he liked to inveigh against the English, for whom his animus seemed to compound Midwestern isolationism, an early American revolutionism, and an unexpected outcropping of Scottish nationalism. **1976** M. J. Lasky *Utopia & Revolution* (1977) v. 225 Before the emergence of a clear, simple concept of linear progress, both utopianism and revolutionism took their turns on the wheel of fortune.

revolutionizing, *vbl. sb.* and *ppl. a.* (Earlier examples.)

1799 J. Morse *Sermon Exhibiting Present Dangers* 17 The Clergy have been among the first victims to that sanguinary revolutionizing spirit which now convulses the world. **1854** Geo. Eliot tr. *Feuerbach's Essence of Christianity* vii. 77 Words possess a revolutionizing force; words govern mankind.

revolutionology (revǒlūʃənǫ·lŏdʒi). [f. Revolution *sb.* + -ology.] The science or study of revolution.

1905 D. M. Wallace *Russia* II. xxxv. 325 Such are a few characteristic extracts from a document which might fairly be called a treatise on revolutionology. **1932** *Sunday Times* 26 June 11/3 Trotsky is trying to work up a theory and science of revolutions, a sort of Revolutionology.

revolve, *sb.* **3. b.** A revolving stage. In full, *revolve stage.*

1938 *Times* 25 Aug. 8/4 A new electric revolve has been installed in place of the winch-controlled one. **1959** *Punch* 20 May 688/3 It is a big stage... In the middle is a revolve that goes round as easily as a bicycle wheel. **1972** T. Stoppard *Jumpers* 11 The National Theatre production was mounted on a revolve stage. **1975** *Country Life* 26 June 1689/2 A revolve has been added to the stage which, thanks to hydraulic machinery, can assume any shape the director pleases. **1980** *Daily Tel.* 29 Aug. 10/5 The technical difficulties bedevil the scenery hoists and drum revolve. *Ibid.*, The revolve had given the contractors and the board endless difficulties.

revolver. Add: **1. b.** *revolver grip, pistol, range, shooting, shot* (earlier example), *target*; *revolver-like* adj.

1971 I. Butykai tr. *Lukovich's Electric Foil Fencing* i. 20 Most foilists seem to prefer the revolver grip. **1925** T. Dreiser *Amer. Tragedy* (1926) II. iii. xix. 228 His voice..as opposed to Mason's revolver-like 'Excused!' **1866** J. Macgregor *Thousand Miles in Rob Roy Canoe* ii. 35 So my friend capped his revolver-pistol, and I acted as a pointer dog. ? **1891** W. S. Churchill *Let.* in R. S. Churchill *Winston S. Churchill* (1967) I. Compan. i. v. 262, I went to the 'Revolver Range' & shot 12 shots with a full sized weapon. **1918** W. Owen *Let.* 21 June (1967) 559 Today [I] have been Revolver Shooting. *a* **1861** T. Winthrop *John Brent* (1862) xix. 208 They were close to us, within easy revolver shot. **1918** W. Owen *Let.* 24 June (1967) 560 The hideous faces of the Advancing Revolver Targets I fired at last week.

revo·lvered *a.*, provided with a revolver or revolvers.

1901 *Pall Mall Gaz.* 6 Mar. 1 The revolvered footman..is not quite so grotesque as the revolvered Protestant lecturer.

1922 *Blackw. Mag.* Jan. 123/1 Whom should he see but his American friend the casket king, still picturesque and revolvered, on his way to the station bar. 1963 [see *LEATHER-JACKET 5].

revolving, *ppl. a.* Add: **c.** (Further examples.) Also *spec.* of an article of furniture or other simple mechanical construction, as *revolving door*, etc.

1866 Geo. Eliot *Let.* 4 Aug. (1956) IV. 294 Col. Hamley's volume..lies now on my revolving desk as one of the books I mean first to read. 1883 *Heal & Son Catal.: Bedsteads & Furnit.* 185/3 Mahogany, Walnut, or Oak Revolving Chair. 1892 G. Meredith *Let.* 25 Sept. (1970) II. 1101 Will..says that he and his Daisy had thoughts of a revolving book-case. 1907 *St. Nicholas* Oct. 1104/2 It was his stated duty to attend one of the big revolving doors that opened upon the main street. 1912 G. B. Shaw *Let.* 27 Oct. in *B. Shaw & Mrs. Campbell* (1952) 51 Revolving stages present no difficulty.. They are made in Germany. 1936 A. Thirkell *August Folly* vi. 175 We put the white wine in the revolving bookcase. 1966 J. Gloag *Social Hist. Furnit. Design* vii. 185 A late eighteenth century revolving bookcase in three tiers on a pillar and claw base. 1974 'E. Lathen' *Sweet & Low* ii. 23 Appeals for..a revolving stage at the repertory theater. 1977 *New Yorker* 9 May 35/1 They were not completely comfortable with escalators and revolving doors. 1980 L. Lewis *Private Life of Country House* ii. 19 A revolving bookcase which must not be twirled round at speed because the books fell out.

d. *Econ.* Applied to credit, etc., available in return for smaller regular repayments (see also quot. 1919). *revolving fund*, a fund that is continually replenished as it is drawn upon.

1919 W. Thomson *Dict. Banking* (ed. 2) 564/2 *Revolving credit*, a credit opened with a bank by an importer in order to enable an exporter to obtain payment for goods when ready for shipment. 1928 *Britain's Industr. Future* (Liberal Industr. Inquiry) II. ix. 101 The Local Loans Fund is a revolving fund vested in the National Debt Commissioners. 1962 *Economist* 16 June 1116/2 A 'revolving fund' is to aid the export industries. 1970 *Washington Post* 30 Sept. B13/4 (Advt.), Let Sears clean all your draperies—expert cleaning, prompt service! Charge it on your Sears revolving charge. 1972 *Daily Tel.* 15 Mar. 17 Midland Bank yesterday became the first of the big clearing banks to introduce a permanently revolving credit scheme for private customers. 1977 *Time* 28 Feb. 40/2 A department store, for example, must inform buyers that the interest charge of 1½ % a month on the unpaid balance in a revolving-credit account amounts to 18 % a year. 1980 *Jrnl. R. Soc. Arts* Apr. 303/2 *The Architectural Heritage Fund* describes the work of this revolving fund which makes low interest loans to building preservation trusts.

Revudeville (rĭviū·dəvil). [f. *REVU(E + VAU)DEVILLE.] A form of variety entertainment presented at the Windmill Theatre, London, between 1932 and 1964. Also *transf.*

1932 *Times* 3 Feb. 12/7 (Advt.), Windmill Theatre Revudeville 8.30. 1932 *Times* 4 Feb. 10/2 'Revudeville.'.. The Windmill Theatre turned last night from talking films to variety and revue turns. 1937 *Daily Express* 6 Feb. 15/1 Eric Barker, who holds forth at the Windmill Theatre, where they've just passed five years of non-stop revudeville. 1937 *Evening News* 19 Jan. 5/3 (*heading*) 'A Day at the B.B.C.'.. Skit in Revudeville. 1966 J. R. Taylor *Penguin Dict. Theatre* 228 Another phenomenon deserving of mention is *Revudeville* at the Windmill Theatre, a form of continuous variety show with comedy and nudes which ran in various editions from the 1930s... It closed in 1964.

revue (rĭviū·). [a. F. *revue* REVIEW *sb.*] **1.** A theatrical entertainment presenting a review (usu. satirical) of current events, plays, etc.; now also an elaborate musical show consisting of numerous unrelated scenes. Also without art., the genre of such entertainments.

1872 J. R. Planché *Recoll.* I. iv. 73 My theatrical labours in the year 1825 terminated with the production at the Adelphi..of a one-act piece on the 12th December, entitled 'Success; or a Hit if you like it', which I only mention because it was the first attempt in this country to introduce that class of entertainment so popular in Paris called 'Revue'. 1879 —— *Extravaganzas* III. 311 It is, in fact, 'a dramatic political allegory'. A *Revue*, not of theatrical and other popular novelties..but of the State of Europe at a critical period. *Ibid.* IV. 89 Jokes upon passing events..in a *Revue*..are, of course, indispensable. 1893 *Times* 27 Nov. 7/4 *Under the Clock* has been described as a *revue* of the sort which is popular in the French theatres towards the close of the year. 1899 *Times* 4 Apr. 5/2 It looks as if musical farces were declining in popularity when a specimen of this class of piece has to be called a *revue* and announced as an 'entirely new form of entertainment'. The *revue*..has never had the same favour here that it enjoys in Paris. 1908 M. Beerbohm in *Sat. Rev.* 18 Jan. 74/1 The attempts to popularise in London something like the Parisian 'revue' have been such dismal failures. 1913 *Punch* 9 July 41/3 The report that Mr. Lloyd George will shortly appear in a Revue at the National Liberal Club. 1920 A. G. Gardiner *Leaves in Wind* 71 One of those dismal things called revues, that are neither comedies nor farces, nor anything but shambling, huggermugger contraptions into which you fling anything that comes handy. 1923 R. Nevill *World of Fashion 1837–1922* xii. 269 Anything like a real *revue* is impossible in modern London... Any allusion to current politics such as are made in Parisian revues is at once denounced as being in bad taste. 1967 *Oxf. Compan. Theatre* (ed. 3) 798/1 During the Second World War the vogue for revue continued... Revue became a feature of the little theatres..but it was not until the advent in London of *Beyond the Fringe*,..which was seen in Edinburgh on 10 May, 1961, that satire entered the field. 1972 *Daily Tel.* 22 Nov. 15 The new revue at the Cambridge, 'Behind the Fridge', stars Peter Cooke, Dudley Moore, and no one else. 1981 *Country Life* 12 Feb. 409/4 Noël Coward's..*This Year of Grace*..was a revue and not a book or a play.

2. *fig.*

1899 A. E. W. Mason *Miranda of Balcony* iii. 35 After he had fallen asleep, a curtain was raised upon a fantastic *revue* of the past week. 1934 T. S. Eliot *Elizabethan Ess.* 167 The romantic comedy is a skilful concoction of inconsistent emotion, a revue of emotion... It consists in an internal incoherence of feelings, a concatenation of emotions.

3. *attrib.* and *Comb.*

1919 B. Tarkington *Let.* 30 Mar. in *On Plays* (1959) 10 What is really most valuable in it is the revelation, by an intimate, of the modern 'Revue' girl. 1933 M. Allingham *Sweet Danger* xvi. 197 She washed and changed with the speed of a revue star. 1933 P. Godfrey *Back-Stage* xiv. 174 The attempt to attract the smart, sophisticated section of the post-War public into patronizing the revue theatres. 1963 *New Yorker* 15 June 7 Ronny Graham's revue-eyed view of life..sometimes gets bug-eyed. 1975 *Listener* 9 Jan. 57/1 With all the resources of splendid new theatres..we..[are] dragging up old revue skits.

Hence **revue·tte** *nonce-wd.*, a short revue; **revue·-ish** *a.*, characteristic of a revue; **revu·ist**, a writer of revues.

1927 *Observer* 25 Sept. 11/1 Alfred Savoir, in collaboration with Rip, the revuist, has written 'Comme le temps passe'. 1930 *Daily Express* 6 Oct. 5/3 Jane Renouardt in Rip's revuette gives a marvellous imitation of Yvonne Printemps. 1969 *Listener* 20 Feb. 251/1 The authors of *Your Own Thing* can't handle this bold stuff... This is a revue-ish comment on rock music.

revusical (rĭviū·zikăl). orig. and chiefly *U.S. slang.* [f. *REVU(E + MU)SICAL *sb.*] A theatrical entertainment that combines elements of the revue and musical.

1931 *Amer. Speech* VII. 45 *Torso tossers* dance in *revusicals* which might even turn out *floperoos*. 1935 A. C. Baugh *Hist. Eng. Lang.* x. 378 Often such coinages [of portmanteau words] are formed with a playful or humorous intent... *Paradoxology, alcoholiday, anecdotage, revusical, yellowcution*, and the like, often reveal flashes of wit. 1936 *Daily Mirror* (N.Y.) 22 Apr. 10/4 Harriet Hoctor's class in 'The Ziegfeld Follies', the revusical. 1941 *Amer Speech* XVI. 16/1 Pardon Us Please, 'presenting thirty-five stars in person', advertises itself in newspapers as a *Revusical*. 1967 *New Idea* (Austral.) 23 Feb. 6/4 John McKellar's invention, a revusical, opening in Sydney..called Hail Gloria Fitzpatrick. The Phillip Theatre..coined the word revusical to describe a show that will cover every facet of theatre work.

revved, revving: see *REV *v.*

reward, *sb.*[1] Add: **II. 4. a. reward book,** a book given as a prize at school.

1820 [in Dict.]. 1865 C. M. Yonge *Clever Woman of Family* I. iii. 82 A summary pawning of all poor Lovedy's reward books. 1977 W. Feaver *When We were Young* 92/1 The cheap, spongy paper of reward books and annuals.

e. In phr. *to go* (*pass*, etc.) *to one's reward*, to die (and go to heaven). Also in ironic use. orig. *U.S.*

1883 'Mark Twain' *Life on Mississippi* li. 503 He went to his reward, whatever it was, two years ago. 1896 J. Curtin tr. *Sienkiewicz's Quo Vadis* lxx. 521, He went to his reward like a conqueror. 1942 E. Paul *Narrow St.* xxiii. 202 Vladimir de Pachmann..has recently died. I wept..when I read that the grand old poet of the piano..had gone to his reward at the age of eighty-five. 1949 G. Davenport *Family Fortunes* III. iv. 266 It was lucky for Mrs. Wilkens, Lou Belle thought, that her grandfather Aaron Toler had passed to his reward. 1975 D. Lodge *Changing Places* ii. 49 His beloved parent had passed to her reward from this very bed. 1977 D. Williams *Treasure by Degrees* ii. 19 When the old baggage was finally called to her reward..his own temporal benefit would be substantial.

f. *Psychol.* A recompense for a response which reinforces specific learning or behaviour. Freq. *attrib.*, esp. as *reward cell, centre, system*, with reference to areas of the brain in or near the hypothalamus which, when stimulated, reward the organism with sensations of pleasure. Cf. *PUNISHMENT 1 b.

1907 R. M. Yerkes *Dancing Mouse* vii. 100 In connection with the discussion of motives, it is an important fact that forms of reward are far harder to find than forms of punishment. 1912 *Jrnl. Animal Behavior* II. 50 It seems evident from this experiment that a combination of punishment and reward-motives is more effective in bringing about visual discrimination in the rat than is either punishment or reward used alone. 1929 *Brit. Jrnl. Psychol.* Oct. 173 Some experiments on the influence of the amount of food given as a reward on the rate at which young chickens learn. 1952 E. L. & R. E. Hartley *Fund. Social Psychol.* ix. 274 Without reinforcement or reward, no connection will be established between response and motivation. 1956 *Sci. Amer.* Oct. 108/3 At this point we assumed that the stimulus must provoke curiosity; we did not yet think of it as a reward. *Ibid.* 115/3 The main question..is to determine how the excited 'reward' cells act upon the specific sensory-motor systems. 1964 M. Argyle *Psychol. & Social Probl.* ii. 26 A most interesting series of attempts to demonstrate the acquisition of drives by reward learning has been carried out with monkeys. 1974 M. C. Gerald *Pharmacol.* x. 186 Drugs capable of causing the release of norepinephrine,.. are able to increase the rate of self-stimulation when electrodes are placed in the reward center. *Ibid.*, The reward system is located in the medial forebrain bundle. 1977 E. L. Deci in von H. Gilmer & Deci *Industrial & Organizational Psychol.* (ed. 4) viii. 208 Intrinsic rewards are ones which a person administers to himself or herself.

rewa·rdingness. [f. REWARDING *ppl. a.* + -NESS.] The quality or state of being rewarding.

1951 *Mind* LX. 124 The subject of modal logic is of course extremely difficult. These are two illustrations of the dangers of any sort of intuitive approach to it even in Aristotle, and at the same time of its rewardingness for anyone who forgoes that approach. 1972 M. Argyle *Social Psychol. of Work* iv. 60 Rewardingness may be an important part of this general ability—rewarding people are both more popular and influential. 1978 A. Ryan in Hookway & Pettit *Action & Interpretation* 66 If we want to know why someone does what he does, we must look to the goals he is seeking, and to his beliefs about their rewardingness to himself.

rewa-rewa. Also riwa-riwa. Substitute for first part of def.: A tall broadleaved forest tree, *Knightia excelsa*, of the family Proteaceæ, native to New Zealand and bearing long, narrow, toothed leaves and racemes of tomentose red-brown flowers followed by woody pods. (Earlier and later examples.)

1831 G. Bennett in *London Med. Gaz.* 5 Nov. 150/1 *Knightia excelsa*... This tree is abundant in the New Zealand forests, and is named Riwa-riwa by the natives. 1838 J. S. Polack *N.Z.* II. 396 The *Rewā rewā*..is a handsome grained wood, very serviceable to the builder and joiner... It grows to the height of sixty feet. 1966 *Encycl. N.Z.* III. 70/1 The rewarewa..grows to heights of up to 90 ft.

rewash, *v.* Add: Hence **rewa·shed** *ppl. a.*

1908 *Sears, Roebuck Catal.* 170/1 Inner tubes..made..of the very best quality of Upper River Para rewashed rubber. 1965 G. J. Williams *Econ. Geol. N.Z.* vii. 69/2 Detrital gold was mined in the Preservation Inlet area from rewashed glacial debris.

reweave, *v.* Add: (*absol.* example.) Hence **rewea·ving** *vbl. sb.*

1929 *Oxford Poetry* 6 These worn and tangled threads of day reweave Like memory's or music's threads. 1963 *Economist* 2 Feb. 392/2 It is this reweaving of America with Europe that General de Gaulle is now challenging. 1964 *McCall's Sewing* xv. 265 (*heading*) Reweaving/Invisible Patching. *Ibid.*, You can only use threads ravelled from the fabric for reweaving.

rewhisper, *v.* (Earlier example.)

1754 Richardson *Grandison* I. xxxvii. 271 I'll be hang'd if Miss Byron thinks so, re-whisper'd she.

re-wind, *v.* Add: (Further examples.) Also, to wind back. Also *absol.*

1967 H. Harrison *Technicolor Time Machine* xii. 118 'Reel is almost empty on the back projection.'.. 'Rewind it then.' 1972 D. E. Westlake *Bank Shot* iv. 23 He pushed the stop button on the cassette recorder, rewound, and played it back. 1977 D. Bagley *Enemy* ix. 65, I switched off the recorder and rewound the tape. 1977 J. Hedgecoe *Photographer's Handbk.* 12 After taking the last picture you disengage the film wind mechanism and rewind the exposed film into its cassette. 1980 C. Fitzgibbon *Rat Report* iii. 57 He re-wound the tape and played it over several times.

rewinding *vbl. sb.* (examples of *attrib.* use).

1930 *Daily Express* 6 Sept. 2/1 Fire broke out in the rewinding room of the Rialto Cinema..yesterday. 1958 *Newnes' Compl. Amat. Photogr.* 71 Nothing is more infuriating than to lose exposed film through a difficult rewinding system.

re·wind, *sb.* [f. the vb.] **1.** A mechanism for winding film, tape, or the like backwards.

1938 G. H. Sewell *Amateur Film-Making* v. 52 A rewind..consists essentially of two uprights each bearing a spindle, and one or both of them having a geared drive so that a spool of film placed on one can be wound off on to the other. 1958 *Newnes' Compl. Amat. Photogr.* 71 If possible, the rewind should be tested with an old film. 1973 C. Bonington *Next Horizon* xiii. 196, I got half a dozen pictures of him as I manipulated the rewind and trigger with frozen fingers.

2. The action or process of winding paper, recording tape, etc., backwards.

1964 *Gloss. Letterpress Rotary Printing Terms* (*B.S.I.*) 6 *Rewind*, the procedure for taking several butt ends of paper and winding them into a reel for use in the press. 1965 *Wireless World* July 34/1 (Advt.), Tape lifted during rewind from all heads or in contact with play-back head only if required. 1970 O. Dopping *Computers & Data Processing* xv. 236 In using long magnetic tape files comprising many reels of tape, the time for rewind and exchange of reels can become appreciable.

3. *attrib.*

1931 J. H. Reyner *Cine-Photogr. for Amateurs* x. 122 The film is then placed on the re-wind apparatus. 1964 *Gloss. Letterpress Rotary Printing Terms* (*B.S.I.*) 6 *Rewind join*, the adhesion between the butt ends of two webs which are wound into a reel in the printing works. 1972 Rewind starter [see *recoil starter* s.v. *RECOIL *sb.* 4 b]. 1972 F. Durbridge *Bat out of Hell* v. 182 He operated the rewind switch, took the tape back to zero. 1974 *News & Courier* (Charleston, S. Carolina) 19 Apr. (Wickes Lumber Advt. Suppl.) 8 Efficient and safe with a 3.5 h.p. Briggs & Stratton rewind-start engine. 1978 *SLR Camera* Nov. 69/1 The film speed setting is adjusted by the rotation of a collar surrounding the rewind knob.

rewi·re, *v.* [RE- 5 c.] **1.** *trans.* To provide with new wires, esp. for conducting electricity.

1903 *Motoring Ann.* 295 In cases of trouble from short circuits or broken wires, there is no remedy but to re-wire the car. 1910 *Installation News* IV. 64/2 They have decided

to rewire the whole building. 1973 E. Berckman *Victorian Album* 25 Would you object if we rewired the whole flat? 1981 *Times* 4 Mar. 23/3 The electrician who re-wires houses at the weekend.

2. To transmit (a telegraphic message) again. *rare.*

1907 *Westm. Gaz.* 2 Dec. 13/1 News..is telegraphed to Madrid, to be forthwith re-'wired' to the capitals.

Hence **rewi·ring** *vbl. sb.*; also **rewireable** *a.*

1930 A. Bennett *Imperial Palace* lxxi. 616 Obviously a lot of re-wiring would have to be done. 1934 Webster, Rewirable. 1976 *BSI News* Nov. 16/1 Three-pin rewirable or non-rewireable plugs will be available for the connection of Class I equipment (i.e. equipment requiring earthing). 1977 *Lancashire Life* Nov. 75/2 The amount raised will be put..to pay for some much needed re-wiring in the Institute.

re·wire, *sb.* [f. the vb.] A rewiring.

1934 in Webster. 1976 *Star* (Sheffield) 20 Nov. (Advt.), House rewires by qualified electrician.

reword, *v.* **3.** (Earlier example.)

1882 G. M. Hopkins *Note-bks. & Papers* (1937) 431, I..had to reword all passages speaking of God's kingdom as the elect.

rework, *v.* Add: **1. b.** *spec.* in *Geol.* Of a natural agent: to alter, esp. to remove and redeposit (rock or the like).

1870 C. F. Hartt *Thayer Exped.: Sci. Results Journey Brazil* xix. 573 Where the surface of the rock..had been covered by a thick layer of loose material, the glacier reworked this loose material, and when it disappeared left it as a paste. 1930 P. D. Trask *Recent Marine Sediments* III. 202 The .. 1962 Read & Watson *Introd. Geol.* I. vii. 381 Loose pyroclastic accumulations are always liable to be reworked by wind, streams and other surface-agents, and the reworked redeposits may be mixed with appropriate kinds of sedimentary rocks. 1964 *Oceanogr. & Marine Biol.* II. 122 Mud-feeding animals play a significant part in re-working the sediment even on the floor of the abyssal parts of the ocean. 1977 *Sci. Amer.* Mar. 94/3 One extreme hypothesis is that most if not all of it [*sc.* continental crust] was made early in the course of the earth's chemical differentiation and that ever since then it has been reworked, that is, heated, melted, recrystallized and deformed.

2. To change the variety of (a plant) by grafting.

1942 *Jrnl. Pomol.* XIX. 194 When reworking apple trees liable to attack by silver leaf after grafting, it is important to retain as much of the head of the tree as possible. 1950 *N.Z. Jrnl. Agric.* June 538/2 Auckland growers decided to rework thousands of Delicious trees to other varieties.

Hence **rewo·rked** *ppl. a.*; **rewo·rker,** one who works (something) again, *spec.* a reviser or redactor; **reworking** *vbl. sb.* (further examples).

1886 T. S. Hunt *Mineral Physiol. & Physiogr.* vii. 275 The layer of ground and 'reworked' decayed material resting on the gneiss, found by Hartt in Brazil. 1939 P. D. Trask *Recent Marine Sediments* III. 197 This type of bedding indicates frequent reworking of the sediments under the influence of currents of varying strength and direction. *Ibid.* 201 In these areas the sediments are reworked deposits, not new ones. 1955 J. G. Davis *Dict. Dairying* (ed. 2) 157 Reworking of butter invariably increases the count and also increases the size of the water globules. 1970 J. R. L. Allen *Physical Processes of Sedimentation* iv. 140 The beds preserved have properties denoting extensive sediment reworking. 1972 G. Jones *Kings, Beasts, & Heroes* II. i. 66 The story-tellers of the *Mabinogion* are to a high degree reworkers of old material, borrowers from an often remote antiquity. 1977 *Early Music* Oct. 581/1, I think it is a pity that the models of the two known reworkings are not identified more clearly. 1979 *Dædalus* Summer 94 The reader is not simply beckoned to catch an unwonted glimpse of the diverse states of a subtly reworked self-portrait.

rewrite, *v.* Add: **2. a.** Also *refl.*

1928 *Manch. Guardian Weekly* 5 Oct. 265/2 The events of the last four years in which the history of 1906 to 1914 was rewriting itself.

c. *Linguistics.* To write (an analysis of a phrase or sentence structure) in a different form, usu. by expansion. Also *absol.*

1955 N. Chomsky *Logical Struct. Linguistic Theory* vi. 235 We can produce derivations from this linear grammar by applying the conversions S_i (interpreted as the instruction 'rewrite X_i as Y_i') in sequence. 1957 —— *Syntactic Structures* v. 40 These rules can be dropped if we rewrite ..so that either *C* or *M*, but not both, can be selected. 1970 *Canad. Jrnl. Linguistics* XV. 95 Many contemporary linguists describe their data in terms of a system of rules which state that one string of symbols is to be rewritten as another string of symbols. *Ibid.* 97 We know that rule 1 must be the first rule executed because S is the initial symbol and rule 1 is the only rule that rewrites S. 1972 *Archivum Linguisticum* III. 14 A lexical redundancy rule could also be used to rewrite the semantic feature as a syntactic one.

Hence **rewri·ter,** one who revises or rewrites; **rewriting** *vbl. sb.* (earlier examples); also *attrib.*, as **rewriting rule** *Linguistics*, a rule that governs the rewriting of a structural analysis.

1838 J. S. Mill *Let.* 19 Oct. in M. Fawcett *Life Sir W. Molesworth* (1901) x. 204 This winter..will see me through the whole of it except the rewriting. 1854 C. M. Yonge *Heartsease* I. II. xii. 309 She has..given me some re-writing to do. 1912 *Collier's* 28 Sept. 17/2 By all the earmarks it was a 'three head' to be laconically telephoned into a 'rewriter' who would dutifully chronicle the event in 100 words. 1934 Blunden *Choice or Chance* 51 Rewriters, novelists and mirth-provokers. 1965 N. Chomsky *Aspects of Theory of Syntax* ii. 66 The natural mechanism for generating Phrase-markers..is a system of rewriting rules. A rewriting rule is a rule of the form..A→Z/X—Y where X and Y are (possibly null) strings of symbols, A is a single category symbol, and Z is a nonnull string of symbols. 1970 *Language* XLVI. 161 These PS rules generate the structure-sensitive $a^n b^n$ quite simply with one non-terminal category and two rewriting rules.

rewrite (ri·rait), *sb.* [f. the vb.] **1. a.** orig. *U.S. slang.* The act of revising a text; a revised text. (Chiefly in journalistic and publishing use.)

1930 E. B. White *Let.* Sept. (1976) 75 Carl..gravitated naturally down to Mr. Hearst's American and got a job on rewrite. 1933 Partridge *Slang To-day & Yesterday* iii. 181 To set a book is to revise it, whether for the author or his publisher; if the work entailed amounts to a virtual re-writing, the resulting typescript or manuscript is a re-write. 1938 E. Pound *Let.* 22 Apr. (1971) 307 How much revision do you propose to make in the proofs?..We're not out for collaboration and rewrite à la E.P. 1952 *Scrutiny* June 3/3 Dryden states the case nicely in the Preface to his rewrite of Shakespeare's *Antony and Cleopatra*. 1960 *News Chron.* 23 Feb. 6/3 ..a rewrite of Delilah's third act. 1972 H. Evans *Newsman's English* ii. 17 Mix both techniques—rewrite the lead into the story but edit the rest of it in copy. Or you can move nearer to a full-scale rewrite. 1976 G. McDonald *Confess, Fletch* (1977) xxv. 221, I haven't got the rewrite man I need... The guy I've got on rewrite..

b. *attrib.* and *Comb.*, esp. as **rewrite man,** one employed to rewrite newspaper copy for publication; also *transf.*

1901 *Munsey's Mag.* Nov. 222/1 Much of the copy that reaches the big newspaper offices is passed over to the 'rewrite man'. 1912 G. M. Hyde *Newspaper Reporting* ix. 125 The terms 'rewrite story' and 'follow-up, or follow story', are names which newspaper men apply to the rehashed or revised versions of other news stories. 1935 *Amer. Mercury* July 379/2, I have yet to discover an instance of real literary talent being brought to the surface by back-breaking chores on the rewrite desk. 1949 'G. Orwell' *Nineteen Eighty-Four* II. iii. 131 She could describe the whole process of composing a novel, from the general directive issued by the Planning Committee down to the final touching-up by the Rewrite Squad. 1955 H. Roth *Sleeper* xv. 126 She seemed..capable, especially of rewrite, and good rewrite people are hard to find. 1960 R. St. John *Foreign Correspondent* i. 12, I had been a rewrite man in Camden, New Jersey. 1973 *Times* 2 Mar. 14/4 He is..the GLC's chief strategic planner and 'rewrite man' on planning policy. 1977 *Detroit Free Press* 11 Dec. 3-B/3 He phoned the story in over a vodka and orange juice. 'It ain't much,' he told the rewrite man.

2. *Linguistics.* The act or process of writing an analysis of a phrase or sentence structure in a different form, usu. by expansion. Freq. *attrib.* and *Comb.*, esp. as **rewrite rule** = *rewriting rule* s.v. *REWRITING *vbl. sb.*

1960 R. B. Lees *Gram. Engl. Nominalizations* i. 2 The rewrite rules, which permit the conversion of given strings of symbols into more expanded strings. 1964 E. Bach *Introd. Transformational Gram.* ii. 17 In the grammar above the items to the left and right of the 'rewrite' arrow are strings of symbols. 1967 *Word* XXIII. 289 The equivalence of the different classes manifesting identical functions is implicit in the rewrite-rule formulations. 1971 *Amer. Speech 1968* XLIII. 132 If the universals generally posited by linguists for human language can be formulated as a set of re-write rules whose elements are unordered by nature, then the relatedness between languages is determinable. 1977 *Trans. Philol. Soc. 1975* 79 Many individual verbs will no doubt need extra specifications in terms of 'context-sensitive rewrite rules'.

rex[2]. Delete ‖ and *rare* and add: **1. b.** (With capital initial.) Used, during the reign of a king, esp. in law reports, to designate the prosecution in criminal proceedings. Cf. *REGINA.

1657 *Reports of Edward Bulstrode* I. 96 *Dominus Rex*, and Sir William Fitzwilliams against Ives. 1717 W. Salkeld *Reports of Cases Court of King's Bench* 324 Rex *versus* Bear. 1792 W. Boscawen *Treat. Convictions on Penal Statutes* 48 This was one of the points in the case of *Rex v. Hall*, last cited. 1848 E. W. Cox *Reports of Cases in Crim. Law* II. 423 *Rex* v. *Webb* is no authority for that distinction. 1976 *Law Reports Queen's Bench Div.* 373 Rex v. Bishop of Lincoln (1637).

2. A rabbit or mouse belonging to a variety so called, distinguished by a genetic mutation in which the guard hairs are reduced and wavy or completely lacking, giving the fur a soft, plushy texture; also, the characteristic mutation itself. Also *attrib.*

1929 *Jrnl. Heredity* XX. 192 Rex rabbits produce only rex offspring when bred with each other. 1935 *Nature* 14 Sept. 434/2 During the years 1927–29..a new class of rabbit, namely the 'Rex', with fur like the mole..was imported from France and Germany. 1939 *Ibid.* 23 Sept. 557/1 The rex mouse has a coat somewhat similar to that of the same character in the rabbit... Rex is an autosomal dominant to normal. 1956 C. Auerbach *Genetics in Atomic Age* iv. 16 The Rex rabbit..has a mole-like fur. 1977 *West Briton* 25 Aug. 34/2 (Advt.), Rabbits: Breeding pairs and young. Rex, dwarf, English and Dutch.

3. A cat belonging to the breed so called, developed in Cornwall and Devon during the 1950s and 1960s, and distinguished by short curly or wavy fur. Also *attrib.*

1960 *News Chron.* 28 July 6/3 There's going to be a Rex kitten. This is the curly-coated kitten—a new breed which experimental breeders are trying to establish. 1965 *Observer* (Colour Suppl.) 31 Jan. 4/3 A new sort of highly exclusive curly-coated cat called a Rex. 1972 M. Babson *Murder on Show* x. 125 The Devon Rex and the Cornish Rex. Strange, friendly, fantasy creatures, with curly astrakhan coats and great butterfly ears. 1977 D. S. Richards *Handbk. Pedigree Cat Breeding* vii. 107 No rex cats should be mixed with other breeds.

Rexine (re·ksīn). Also **rexine.** The proprietary name of a kind of imitation leather used in upholstery, book-binding, etc. Also *attrib.* and *Comb.*

1911 Coutts & Stephen *Man. Libr. Bookbinding* iv. 56 There are several textile fabrics, known as Rexine, ..and Pegamoid, etc., produced as substitutes for leather. 1915 *Trade Marks Jrnl.* 18 Aug. 880 Rexine... Artificial leather and goods manufactured wholly or principally thereof. British Leather Cloth Manufacturing Co., Limited, Hyde, near Manchester; leather cloth manufacturers. 1921 [see *PEGAMOID]. 1930 *Daily Tel.* 5 Apr. 11/3 (Advt.), Sale... Three-piece suites in hide, rexine and fabrics, etc. 1946 T. H. White *Mistress Masham's Repose* xix. 156 She carried a Rexine shopping bag. 1960 *Practical Wireless* XXXVI. 287/1 (Advt.), This contemporary cabinet in two-tone grey rexine is ideal for the modern home. 1963 J. Clements *Bookbinding* ix. 78 Rexine and some imitation cloths need to be cleaned with methylated spirit..before tooling. 1971 R. P. Jhabvala *Experience of India* 140 Their drawing-room was furnished with a blue rexine-covered sofa-set. 1975 E. Burdett *Craft of Bookbinding* vii. 561 Leathercloth..was the original waterproof cloth first made at the turn of the century, one of the best known being 'rexine', with a surface consisting of several coatings of nitro-cellulose. 1977 *Transatlantic Rev.* LX. 161 In the dining room, never used, there was a square table with six rexine-seated chairs around it.

Rexism (re·ksiz'm). [f. L. (*Christus*) *Rex*, (Christ) the King: see -ISM.] A right-wing Roman Catholic political movement established in 1935 in Belgium. So **Re·xist** *sb.* and *a.*

1936 *Times* 23 May 13/5 The Liberals represent the 'Rexists' as disguised Catholics. At all events, in spite of an entrance fee used for propaganda, the 'Rexist' meetings have drawn the biggest crowds. *Ibid.* 26 Oct. 13/3 M. Vandervelde, the Minister for Public Health, to-day addressed the Socialist Congress in Brussels on the attitude of the Socialist Party towards recent events, internal and external. His remarks on the relative merits of Rexism and Communism were loudly cheered by the Congress. 1937 'E. B. Ashton' *Fascist* viii. 245 M. Degrelle's 'Rexists'. 1954 B. & R. North tr. *Duverger's Pol. Parties* II. ii. 313 The violent surge of Rexism in Belgium in 1936 followed by just as violent a decline. 1966 'Han Suyin' *Mortal Flower* xvi. 344 Degrelle (who had organized the Rexists, something like the Nazi Brownshirts). 1971 M. Lyon *Belgium* v. 70 In 1935 a twenty-nine-year-old ex-student of Louvain University named Léon Degrelle started a Belgian movement in favour of his own brand of authoritarianism flavoured with Christianity, which he called Rexism from the name Christus Rex (Christ the King).

Reye (rəi, rēi). *Path.* [The name of Ralph Douglas Kenneth *Reye* (1912–78), Australian pædiatrician, who with others first described the syndrome in 1963 (*Lancet* 12 Oct. 749–52).] *Reye's* (or *Reye*) *syndrome*: an often fatal metabolic disease of young children.

1965 *Amer. Jrnl. Dis. Children* CX. 95/1 The possible occurrence of the Reye syndrome in the United States. 1969 *Ibid.* CXVII. 721/1 Thirty-three cases of Reye's syndrome (acute encephalopathy with fatty infiltration of the viscera). 1976 *Nature* 20 May 184/1 A number of children, mostly from areas that have been heavily sprayed in the past, have died from Reye's syndrome, a condition first described in 1963 following the US Air Force's aerial defoliation in Vietnam and Thailand. The disease causes increased susceptibility to viral infection, and it is the spray dispersal emulsifiers which are suspect; the sprays used are Fenitrothion and Phospamidon. 1977 *Daily Colonist* (Victoria, B.C.) 17 Dec. 12/1 There is no connection between the chemical Fenitrothion and the sometimes fatal children's disease called Reye's syndrome.

Reynolds (re·nǒldz). *Physics.* The name of Osborne *Reynolds* (1842–1912), Irish engineer and physicist, used *attrib.* (and in the possessive as *Reynolds'*, occas. *erron.* as *Reynold's*) to designate quantities discovered and concepts used by him, as **Reynolds number,** a dimensionless number used in fluid mechanics as a criterion to determine whether fluid flow past a body or in a duct is steady or turbulent, evaluated as $l v\rho/\nu$, where l is a characteristic length of the system, v is a typical speed, ρ is the mass-density, and ν is the kinematic viscosity of the fluid; *magnetic Reynolds number*, a number analogous in formation to the Reynolds number, used to describe the dynamic behaviour of a magnetized plasma; **Reynolds stress,** the net rate of transfer of momentum across a surface in a fluid resulting from turbulence in the fluid.

1910 *Sci. Abstr.* A. XIII. 406 The flow..is conditioned by the so-called Reynolds' number. 1930 Dougall & Deans tr. *Ewald's Physics of Solids & Fluids* vi. 273 This number, which gives the ratio of the forces of inertia and the forces of viscosity, is called Reynolds' number, in honour of Osborne Reynolds, the discoverer of this law of similarity. 1952 *Ark. f. Fysik* V. 322 The non-dimensional ratios are

analogous to Reynolds' number in ordinary hydrodynamics. In magneto-hydrodynamics there are many numbers of this kind entering in different combinations in different cases. 1958, etc. [see *NUSSELT]. 1962 W. B. Thompson *Introd. Plasma Physics* iv. 50 These conditions may be combined to give the condition for typical magneto-hydrodynamic behaviour... This condition was derived by Lundquist (1952) who called *M* the magnetic Reynold's number. 1974 *Encycl. Brit. Macropædia* XIV. 507/2 When the magnetic Reynolds number is much greater than one, resistance effects can be ignored and the magnetic lines of force are said to be frozen to (or to move with) the plasma. 1975 *Sci. Amer.* Nov. 83/1 In a large hummingbird.. the mean value of the Reynolds number for the flow past oscillating wings is 15,000; in a large wasp..it is 4,000; in the small fruit fly *Drosophila*..it is 200, and in the tiny parasitic wasp *Encarsia fomosa*..it is less than 20. 1977 J. L. Harper *Population Biol. of Plants* ii. 42 The flight Reynolds numbers for spores, pollen and the tiny dust seeds such as those of *Monotropa* and many orchids are usually so small that movement in air is dominated by viscous forces. 1943 *Q. Appl. Math.* I. 11 A new set of equations is obtained which differs from the first only in the presence of additional terms added to the mean values of the stresses due to viscosity. These additional terms are called the Reynolds stresses or eddy stresses. 1947 Hunsaker & Rightmire *Engin. Appl. Fluid Mech.* viii. 137 This fictitious [shear] stress is called either the turbulent shear stress..or the Reynolds stress, after Osborne Reynolds, who first pointed out the existence of turbulent momentum transfers. 1964 *Oceanogr. & Marine Biol.* II. 17 They concluded that the portion of the spectrum which they had measured contributed only about ¼ of the total turbulent energy and that the spectrum for $k > 0{\cdot}02\ \mathrm{cm}^{-1}$ could not make a major contribution to the Reynolds stresses.

rézbányite (re·zbānyəit). *Min.* Also **rezbanyite,** (sense 1) **retz-.** [f. *Rézbánya*, name of a village in Hungary (now Băiţa in Romania): see -ITE[1].] †**1.** (See quots.) *Obs.* [ad. G. *rezbanyit* (R. Hermann 1858, in *Jrnl. f. prakt. Chem.* LXXV. 450).]

1868 J. D. Dana *Syst. Min.* (ed. 5) 100 Retzbanyite... A lead-gray ore of bismuth. 1896 A. H. Chester *Dict. Names Minerals* 231 *Retzbanyite*, an obs. name for an impure var. of cosalite.

2. An orthorhombic lead and copper bismuth sulphide found as light grey masses. [ad. G. *rezbanyit* (A. Frenzel 1883, in *Mineral. und petrogr. Mittheil.* V. 178).]

1884 *Jrnl. Chem. Soc.* XLVI. 266 From the analyses, the formula of rezbanyite is calculated to be $4PbS,5Bi_2S_3$. 1968 *Mineral. Abstr.* XIX. 18/1 In the Fe–Cu deposits of l'Adrar Talіouine [Algeria], in addition to pyrite and chalcopyrite the rare minerals..rézbányite..and tetrahedrite occur.

‖ **rez-de-chaussée** (redəʃose). [Fr.] The ground floor of a building.

1802 C. Wilmot *Let.* 6 Oct. in *Irish Peer* (1920) 99 Its rez de chaussee is open'd by 60 arches through which one enters into the Amphitheatre. 1837 J. F. Cooper *Recoll. Europe* II. x. 245 Besides the *rez de chaussée*, which is but little above the ground, there are two good stories all round the building. 1842 Barham *Ingol. Leg.* 2nd Ser. 262 The *rez-de-chaussée*,—as some call the ground floor. 1871 C. Schreiber *Jrnl.* 3 June (1952) 124 All open spaces and cellars under the *rez-de-chaussée* of the houses. 1959 M. S. Briggs *Everyman's Conc. Encycl. Archit.* 272 *Rez-de-Chaussée* (Fr., literally 'level with the ground'), hence, the ground floor or ground storey of a building.

‖ **rézel, rezel** (rezel), **rézeuil** (rezöy), arch. varr. *RÉSEAU 1.

1865 F. B. Palliser *Hist. Lace* ii. 17 This plain netted ground was called réseau, rézel, rézeuil. *Ibid.* 19 Teachers of the art soon caused the various patterns to be reproduced in..samplars.., and young ladies worked at them diligently, as a proof of their competency in the arts of cut-work, lacis, and rézeuil. 1900 E. Jackson *Hist. Hand-Made Lace* 218 *Réseau*, (1) Identical with Rezel and Rezeuil.

re-ze·ro (rī-), *v.* [f. RE- 5 a + *ZERO *v.*] *intr.* and *trans.* To return to a zero position.

1971 *Amateur Photographer* 13 Jan. 51/2 This operates only when the camera is loaded and automatically re-zeros when the film cartridge is removed. 1978 *Nature* 17 Aug. 675/2 A further advantage is that the instrument can be re-zeroed simply by opening a valve which connects the two arms of the instrument allowing the pressure to equalise.

‖ **rezident** (rezide·nt). Pl. **rezidenty.** [Russ., in same sense.] = *RESIDENT *sb.* 4.

1968 W. Garner *Deep, Deep Freeze* xii. 132 The London *rezidenty*: 'illegal' agents who are members of Soviet *rezidentsii* in foreign countries. 1969 J. Fredman *Fourth Agency* vi. 47 Appointed 1st Secretary Soviet Embassy, Washington. This post believed to be official 'Rezident' for direction of all illegals under G.R.U. tutelage in U.S.A.

‖ **rezidentsia** (rezide·ntsiă). Pl. **rezidentsii.** [Russ., in same sense.] = *RESIDENTURA.

1968 W. Garner *Deep, Deep Freeze* xi. 130 A *rezidentsia* is a network of Soviet 'deep-cover' agents working in a foreign country. Its members..are known as Illegals. *Ibid.*, One country may have several independent and unrelated *rezidentsii*.

rezidentura, var. *RESIDENTURA. **rezina,** var. *RETSINA.

rezo·ne (rī-), *v.* [RE- 5 a.] *trans.* To assign (land, property, or people) to a new zone. Hence **rezo·ned** *ppl. a.*, **rezo·ning** *vbl. sb.*

1951 *Southern Reporter* 2nd Ser. LIII. 720 Chapter 17833..does not empower the Board of Adjustment to zone or rezone land located within Dade County but outside the boundaries of its municipalities. *Ibid.*, The action of the Board of Adjustment,..'was, in legal effect, the rezoning of said land by said Adjustment Board and is wholly void for want of jurisdiction in said Adjustment Board so to do.' 1961 *Guardian* 16 Jan. 6/3 On the opening of a new school our 8-year-old son was re-zoned to the new school. 1964 *Daily Tel.* 24 Jan. 23/7 (*heading*) Rezoned acres for houses. *Ibid.*, To curb office development London County Council is considering rezoning more than 54 acres, some already approved for offices. 1966 'D. Shannon' *With Vengeance* iii. 39 It had been a realtors' meeting: something to do with..a re-zoning matter. 1974 *Union* (S. Carolina) *Daily Times* 24 Apr. 1/3 The city council denied the request to rezone B-1 business districts to include the operation of retail liquor stores. 1976 *Time* 27 Sept. 55/2 In 1970 a rezoning brought black and white kids together in some school districts. 1978 *N.Y. Times* 30 Mar. B3/5 The Town Board in Harrison, N.Y., has voted unanimously to rezone 131 acres of land.

Rgveda, Rgvedic, varr. RIG-VEDA, *RIG-VEDIC *a.* (*sb.*)

1886–7 *Trans. Philol. Soc.* 658 Professor Max Müller says, on the metrical laws of the Rg Veda, 'The object of the Prātiçākhya is to register all the facts which possess a phonetic interest'. 1953 *Ibid. 1952* 125 In the Rgveda..the paradigms of *devī* and *vṛkīḥ* are essentially distinct. 1962 J. Gonda (*title*) The aspectual function of the Rgvedic present and aorist. 1971 *Archivum Linguisticum* II. 112 Circumstances have changed since the admirable study by Gonda on the present and aorist aspects of the Rgveda. 1974 *Encycl. Brit. Macropædia* IX. 440/2 The Rgveda, Atharvaveda, and Sāmaveda are purely metrical texts mainly used by priests in their ritual. *Ibid.* 442/2 Rgvedic has forms with affixes *ya* and *tva* functioning as future passive participles.

rhababe, var. *REBAB. **rhabd:** see *RHABDUS 2.

rhabditid (ræ·bditid), *a.* and *sb.* *Zool.* [f. RHABDIT(IS + -ID[3].] **A.** *adj.* Belonging to the family Rhabditidæ of nematodes or the order or suborder Rhabditida (or Rhabditata) containing it, the members of which are characterized by a rhabditiform œsophagus. **B.** *sb.* A rhabditid nematode.

1930 E. C. Faust *Human Helminthol.* 607/1 (Index), Rhabditid larvæ. 1933 *Jrnl. Washington Acad. Sci.* XXIII. 513 Male:..tail not extending beyond bursa; caudal papillae arranged as in the peloderan group of rhabditids. 1951 L. H. Hyman *Invertebrates* III. xiii. 292 Rhabditid juveniles have repeatedly been found in terrestrial snails. 1962 J. D. Smyth *Introd. Animal Parasitol.* xxv. 284 Subclass II. Phasmidia (= Secernentea)...Sub-order 3. Rhabditata (Rhabditids). Small transparent meromyarian worms. 1964 *Nematologica* X. 343 (*heading*) A xenic cultivation of two species of rhabditid nematodes on a commercial medium. 1980 *Parasitol.* LXXXI. II. p. lvii, Live rhabditid larvae were found contaminating the microfilarial cultures.

Rhabditis. Add: Also **rhabditis.** [Coined in Fr. by F. Dujardin in *Hist. nat. des Helminthes* (1845) 239.] Substitute for def.: A nematode worm of the genus so called (family Rhabditidæ), the members of which are found in soil, water, and decaying organic matter and in the larval stage are facultative parasites of mammals. (Further examples.)

1906 P. Falcke tr. *Braun's Animal Parasites of Man* 276 In the papules [on a boy's skin] the observer found one or more rhabditis. 1967 P. A. Meglitsch *Invertebr. Zool.* viii. 278 (*caption*) Anatomy of *Rhabditis*, a typical nematode. 1971 A. F. Bird *Structure of Nematodes* x. 235 Mapes..was unable to detect breakdown of food in the posterior bulb of *Rhabditis* and concluded that the principal role of the bulb flaps was a valvular rather than a crushing one.

b. *Comb.*, as **Rhabditis-form,** a form (of a nematode worm) characterized by a rhabditiform œsophagus; **Rhabditis-like** *a.* = RHABDITIFORM *a.* in Dict. and Suppl.

1886 W. E. Hoyle tr. *Leuckart's Parasites of Man* 96 Such is the case with..*Rhabdonema* (*Ascaris*) *nigrovenosum* .., whose *Rhabditis*-form, living in the excrement of frogs, differs very little from the animals related to it. 1897 Parker & Haswell *Text-bk. Zool.* I. VI. 287 The embryos [of the Nematode *Ascaris nigrovenosa*] pass from the lungs into the enteric canal of the host, are expelled with its fæces, and develop in water into a sexual Nematode, called the *Rhabditis*-form, in which the sexes are separate. 1906 P. Falcke tr. *Braun's Animal Parasites of Man* 276 It must have belonged to the rhabditis-like larva of a Nematode.

rhabditiform (also **rhabdiform**) *a.*, also *spec.* having or being a short, thick œsophagus with a bulb at the proximal end and another medially, as in the free-living, non-parasitic larval stages of certain nematodes; (further examples).

1924 *Jrnl. Helminthol.* II. 58 The latter [*sc.* the œsophagus of *Rhabdias fuscovenosa*] has completely lost its rhabditiform shape and is now very narrow for the greater part of its length. 1951 L. H. Hyman *Invertebrates* III. xiii. 304 The developing eggs pass into the host's buccal cavity and thence are swallowed into the digestive tract where they hatch into rhabdiform young, so called because of the rhabditoid form of their pharynx. 1969 A. M. Dunn *Vet. Helminthol.* I. 10/2 The oesophagus [in Nematoda]..differs in form according to the higher taxon to which the worm belongs... It may be filariform..; rhabditiform, with anterior and posterior swellings, in many free-living and plant-parasitic adult nematodes and..in the free pre-infective stages of the strongylates; bulb-shaped, [etc.]. 1973 T. C. Cheng *Gen. Parasitol.* xvii. 607/1 After molting, rhabditiform larvae become filariform larvae.

rhabditoid (ræ·bditoid), *a.* and *sb.* *Zool.* [f. RHABDIT(IS + -OID.] **A.** *adj.* **a.** = RHABDITIFORM *a.* in Dict. and Suppl. **b.** Belonging to the superfamily Rhabditoidea of nematodes. **B.** *sb.* A rhabditoid nematode.

1886 W. E. Hoyle tr. *Leuckart's Parasites of Man* 97 (*caption*) Rhabditoid form of *Rhabdonema* (*Ascaris*) *nigrovenosum*. 1932 Borradaile & Potts *Invertebrata* viii. 221 The classical life history of *Ancyclostoma*... First larval form (rhabditoid) with a buccal cavity like *Rhabditis*. 1951 [see *RHABDIFORM *a.*]. 1961 E. R. & G. A. Noble *Parasitol.* xv. 719 Members of all four rhabditoid families occur in intimate relationship with soil-dwelling insects. 1962 J. D. Smyth *Introd. Animal Parasitol.* xxv. 292 [Sub-order] Rhabditata. Usually possess the 'rhabditoid' type of pharynx. 1973 T. C. Cheng *Gen. Parasitol.* xvii. 593/1 In rhabditoids..the buccal capsule is divided into three sections.

rhabdo-. Add: **rhabdomere** (further examples); hence **rhabdome·ric** *a.*; **rhabdomy·o·lysis** *Path.*, the pathological lysis of skeletal muscle; **rha:bdomyosarco·ma** *Path.*, a malignant neoplasm of skeletal muscle (malignant rhabdomyoma), or (but see quots. 1958, 1976) of embryonic tissue; **rha·bdosome,** † **-soma** *Palæont.* [Gr. σῶμα body], a colony of conjoined graptolites; **rha·bdovirus** *Biol.*, any of the group of RNA viruses that includes the rabies virus.

1932 Borradaile & Potts *Invertebrata* x. 274 In the midst of each retinula is a vertical rod, known as the rhabdom, secreted by the cells of the sheaf in vertical sections which, when they are distinct, are known as rhabdomeres. 1974 *Nature* 29 Mar. 380/3 In the worker honey bee, ..eight photoreceptor cells..each contribute a wedge of microvilli (termed a rhabdomere). 1969 *Ibid.* 9 Aug. 641/1 No noticeable swelling of the rhabdomeric microvilli was observed during exposure to ultraviolet light. 1976 *Ibid.* 12 Aug. 626/2 The ultrastructure of invertebrate rhabdomeric visual cells have been studied. 1956 D. H. Bowden et al. in *Medicine* XXXV. 351 Biochemical and histological data have been obtained in three children with paroxysmal myohaemoglobinuria. These findings indicate that the recurrent symptom complex is the result of a pathological process in which striated muscle suddenly undergoes lysis. As a more satisfactory description of the disease, we propose that the condition be renamed acute recurrent rhabdomyolysis. 1963 *Amer. Jrnl. Med.* XXXIV. 554/1 Idiopathic recurrent rhabdomyolysis with myoglobinuria is a syndrome characterized by acute skeletal muscle weakness and pain associated with the passage of urine that is darkly pigmented owing to myoglobin. 1976 *Lancet* 18 Dec. 1343/1 Rhabdomyolysis is usually caused by serious muscle trauma (crush injuries), electric shock, arterial occlusion, or toxins (drugs, sea-snake bite). 1898 T. N. Kelynack *Renal Growths* xi. 104 (*caption*) Rhabdo-myo-sarcoma of left kidney. 1940 Pack & Anglem in H. W. Dargeon *Cancer in Childhood* 95 The recurrent rhabdomyosarcomas, the diffusely invasive sarcomas of striated muscle and those which are so deeply adherent that a local excision is not feasible, are best treated by amputation. 1958 *Trans. Ophthalm. Soc.* LXXVIII. 96 The exact cytological classification may eventually prove to be important from the prognostic and therapeutic point of view, so that the term rhabdomyosarcoma should not be applied unless striations have been demonstrated unequivocally. 1974 J. R. Wilbur et al. in *Neoplasia of Head & Neck* 281 Rhabdomyosarcoma is the most common soft tissue sarcoma in children. 1976 *Proc. R. Soc. Med.* LXIX. 897/1 It has been suggested (Ashton 1958) that all such tumours should be regarded as embryonal sarcomas because of their origin from undifferentiated foci of mesenchymal cells, and that the term 'rhabdomyosarcoma' should be applied only when there is evidence of longitudinal and cross-striations in the tumour cells, an indication perhaps of some degree of differentiation. 1893 *Lunds Univ. Årsskrift* XXIX. XII. 1, I have prepared.. a series of longitudinal and transverse sections through the rhabdosoma of several species. 1910 *Encycl. Brit.* XII. 366/2 It is the general practice of palæontologists to regard each graptolite polypary (rhabdosome) developed from a single sicula as an individual of the highest order. 1935 Twenhofel & Shrock *Invertebr. Paleontol.* iv. 90 The *Dendroidea* include graptolites with fan-shaped rhabdosomes composed of branches or stems joined together into a trellis-like framework by means of transverse bars. 1966 *McGraw-Hill Encycl. Sci. & Technol.* VI. 259/2 The Ordovician graptoloids have been subdivided into a large number of genera, based mainly on the branching and form of rhabdosome. [1965 *Ann. Inst. Pasteur* CIX. 633 The viruses possessing RNA, helical symmetry and a naked nucleocapsid, the order of *Rhabdovirales* is subdivided into two suborders according to the rigidity or flexibility of the virion.] 1966 *Progress Med. Virol.* VIII. 404 Rhabdovirus group. This group was recently proposed to include members whose structure is rod-shaped but is more like a bullet, flat at one end and rounded at the other. 1976 *Jrnl. Gen. Virol.* XXXII. 369 Rhabdoviruses infecting plant and vertebrate cells differ in their morphology.

rhabdom. (Further example).

1977 *Sci. Amer.* July 108/2 A compound eye is made up of ommatidia: tiny individual eyes that point in different directions. Each ommatidium consists of a lens that focuses light on several receptor cells sharing the common light-sensitive organ known as a rhabdom.

rhabdomancer. (Later examples.)

1944 S. Putnam tr. *E. da Cunha's Rebellion in Backlands* v. 237 Skilled rhabdomancers capable of indicating with

their mysterious wands the exact spot where a stream of water might be found. 1971 *Daily Tel.* 17 Aug. 8/1 Their predictions have no more validity..than the predictions of crystal-gazers, rhabdomancers, oömancers and observers of the flights of birds. 1976 *Worcester* (Mass.) *Telegram* 10 Sept. 7/1 This one is for *Rhabdomancers*. Among the other things dowsers are called [are] water witches, water diviners, water finders.

rhabdomancy. (Later example.)

1976 *Worcester* (Mass.) *Telegram* 10 Sept. 7/1 The formal term for dowsing is *rhabdomancy*. In Arkansas—but probably not in Danville, Vt.—..it is also known as witch-wiggling.

Rhabdopleura (ræbdoplū̆ᵊ·ră). *Zool.* [mod. L., f. RHABDO- + Gr. πλευρά rib, side.] A member of the genus of protochordates so called, belonging to the class Pterobranchia and having two tentacle-bearing arms.

[1869 G. J. Allman in *Q. Jrnl. Microsc. Sci.* IX. 58 The remarkable polyzoon for which the genus *Rhabdopleura* has been constituted is eminently distinguished by the presence within its cœnœcium of a rigid chitinous rod.] 1885 *Encycl. Brit.* XIX. 435/1 The dwelling of Rhabdopleura is a branched system of annulated tubes of a delicate membranous consistency, each tube corresponding to a single polypide. 1955 [see *CEPHALODISCUS].

rhabdus. Add: **2.** Also anglicized as **rhabd** (ræbd).

1940 [see *DIACTINE *a.*].

rhachi(o)-: see also *RACHI-, RACHIO- and related main entries in Suppl.

Rhætian, *a.* and *sb.* Add: Also **Rætian.** Substitute for def.: **A.** *sb.* **a.** A native or inhabitant of Rhætia or the Rhætian Alps in eastern Switzerland. **b.** The name of either of two languages (*a*) = RHÆTO-ROMANIC *sb.*; (*b*) = *RHÆTIC *sb.* (Earlier and further examples.)

1600 Holland tr. *Livy's Romane Hist.* v. 202 And doubtlesse, the nations about the Alpes, especially the Rhetians, had their beginning thus. 1607 Topsell *Four-f. Beasts* 33 The Italians call a Badger Tasso, the Rhetians, Tasch. *Ibid.* 40 The herbe Wolfbanie..is poison to Foxes.. as the Alpine Rhætians affirme. 1867 *Chambers's Encycl.* IX. 246/1 A third Teutonic people, the Goths, entered the country of the Rhætians, which nearly corresponded with the Grisons. 1933 L. Bloomfield *Language* iv. 64 The inscriptions in ancient Rhaetian show this language to have been an off-shoot of Etruscan. 1939 L. H. Gray *Foundations of Lang.* xii. 384 The Raetian of Italy has also been regarded as akin [to Etruscan]. 1970 *Times Lit. Suppl.* 8 Jan. 40/2 The five national tongues—French, Spanish, Portuguese, Italian and Rumanian—and..their five regional varieties—Occitanian or Provençal..Catalan..Dalmatian..Romansch or Rhaetian (now strictly West Rhaetian, for that alone has evolved a recognized written language), and Sardinian. 1972 W. B. Lockwood *Panorama Indo-European Lang.* iii. 23 The sphere of Latin..embraced the Alpine area, where it replaced Rhaetian, a language of unknown affinities attested in a handful of inscriptions.

B. *adj.* **1.** Of or pertaining to Rhætia or the Rhætian Alps.

1618 H. Wotton *Let.* in L. P. Smith *Life & Lett. Sir H. Wotton* (1907) II. 149 At the next general assembly of the Rhaetian communities. 1880–1 [in Dict.]. 1888 *Encycl. Brit.* XXIV. 45/2 Bormio and Valtellina were annexed..in 1815 (despite the remonstrances of the Rhætian leagues) to the kingdom of Lombardo-Venetia.

2. *Geol.* Of, pertaining to, or designating the highest (youngest) of the three stages constituting the Upper Triassic in Europe. Also *absol.*

1955 G. G. Woodford tr. *Gignoux's Stratigraphic Geol.* vii. 324 This Swabian facies of the Rhaetian is found everywhere in Europe outside of the Alpine domain. 1959 Wells & Kirkaldy *Outl. Hist. Geol.* (ed. 4) xv. 232 Ammonite specialists..recognise a Rhaetian Stage as the topmost stage of the Triassic System. 1966 D. T. Donovan *Stratigraphy* viii. 178 In 1864 E. Renevier proposed the Rhaetian Stage for the Rhaetic Formation. *Ibid.*, It is difficult to decide whether the Rhaetian fauna as a whole is more like that of the Trias or the Lias. 1969 *Proc. Geol. Soc.* Aug. 154 The use of *Rhaetian*, not Rhaetic, as the name for the latest Triassic age/standard stage is supported. It has been agreed (by the Stratigraphy Committee) to retire the word 'Rhaetic'.

Rhætic, *a.* Add: Also (as *sb.*), the epoch during which the strata were deposited. (Further examples.)

The Rhætic beds lie between the Keuper series of the Upper Triassic and the Liassic of the Lower Jurassic. They have been assigned variously to the Triassic and (esp. in Britain) to the Jurassic, and also to form a separate system between the two; they are now regarded as forming a series of the Triassic.

1904 [see *NON-SEQUENCE]. 1909 [see *OPHIOLITIC *a.*]. 1948 R. L. Sherlock *Permo-Triassic Formations* II. 244 Since most of the vulcanism took place in the Rhaetic, which we regard as Jurassic, it need not be described here, but the fact has an important bearing on the correlation of Gondwana Strata. 1963 [see next]. 1969 Bennison & Wright *Geol. Hist. Brit. Isles* xii. 279 The Rhaetic Beds, not exceeding 100 feet thick..are important as a datum of reference at the top of the Triassic and can be presumed to be approximately contemporaneous throughout the outcrop. 1971 B. W. Sparks *Rocks & Relief* x. 334 In the Alps of eastern Switzerland, ammonites found in the Rhaetic are more closely related to Triassic than to Jurassic forms. 1980 *Daily Tel.* 7 Apr. 8/8 Derived fossils from the Avon gravels are also among the gift with vertebrate remains from the Rhaetic Bone Bed at Aust Cliff including 34 teeth of the lung fish.

B. *sb.* Also **Rætic.** An ancient language of the Rhætian Alps and the area to the south of them. Also as *adj.*

1933 J. Whatmough in R. S. Conway et al. *Prae-Italic Dial. Italy* II. 4 Arrian (*Tact.* 44) distinguishes explicitly between Raetic spoken by the Raeti, and Keltic spoken by the Kelts, as late as the time of Hadrian. 1939 L. H. Gray *Foundations of Lang.* xi. 332 It is generally agreed that it [*sc.* Illyrian]..was connected with at least two ancient Italic dialects: Venetic..and Messapic..and to these we may perhaps add Raetic, with some sixty fragments. 1948 D. Diringer *Alphabet* II. ix. 501 The Raetic and other Alpine alphabets. 1952 *Archivum Linguisticum* IV. 87 Rhetic and Pyrenean scholars. 1974 R. A. Hall *External Hist. Romance Langs.* iii. 51 Among those [pre-Italic dialects of Italy] commonly considered to be Indo-European..are.. Rhaetic, in the Alps north of present-day Venetia, Lombardy, and Piedmont, and perhaps in the Po valley before the coming of the Etruscans..and of the Gauls. 1977 *Canad. Jrnl. Linguistics* Spring 31 The Early Italic dialects of Celtic, Ligurian, Lepontic, Raetic, and Venetic.

Rhæ:to-Etru·scan. [f. *Rhæto-* (see RHÆTO-ROMANIC *a.* and *sb.*) + ETRUSCAN *a.* and *sb.*] A hypothetical ancient language group.

1939 E. Prokosch *Compar. Gmc. Gram.* 55 Celtic, Finnic, Rhaeto-Etruscan.

Rhæ:to-Lia·ssic, *a.* *Geol.* [f. as prec. + LIASSIC *a.*] Of or pertaining to the Rhætic and Liassic series or the geological epochs during which they were deposited. Also **Rhæ:tic-Lia·ssic** *a.*

1963 D. W. & E. E. Humphries tr. *Termier's Erosion & Sedimentation* xviii. 357 The Rhaetic-Liassic Basin of Northwest Scania, Sweden, which is of paralic type, has been partly supplied by rivers, and partly by the sea. It provides an example of rhythmic sedimentation comprising at least twelve cycles in the Rhaetic and the Hettangian. 1965 *Senckenbergiana Lethaea* XLVI. 127 (*heading*) Rhaeto-Liassic plant microfossils from the Leigh Creek coal measures, South Australia. 1970 B. E. Balme in B. Kummel *Stratigraphic Boundary Probl.* 387/1 Species of *Falcisporites* ..are typically Triassic in Australia and are especially abundant in coals and other sediments of Middle to Late Triassic, and Rhaeto-Liassic age.

Rhæ:to-Ro·man, *a.* and *sb.* Also **Rhe-.** [Back-formation on RHÆTO-ROMANIC *a.* and *sb.* after *Roman*.] **A.** *adj.* = RHÆTO-ROMANIC *a.* **B.** *sb.* A speaker of Rhæto-Romanic.

1931 *Times Educ. Suppl.* 15 Aug. 320/4 The University of Geneva is creating a professorship in rheto-roman languages. 1933 C. D. Buck *Compar. Gram. Gk. & Lat.* 29 The Rhaeto-Roman dialects in parts of present Switzerland and northeastern Italy are a series of numerous dialects which cannot be reckoned as either French or Italian dialects, and of which some are used locally as written languages, as Romansh, Upper and Lower Engadine, Ladin, Friulian. 1942 K. W. Deutsch in J. A. Fishman *Readings Sociol. of Lang.* (1968) 601 Rheto-Romans found their ancient language introduced on a basis of full equality into the administration of Switzerland.

Rhæto-Romance. Also **Ræ-.** (In Dict. s.v. RHÆTO-ROMANIC *a.* and *sb.*) (Earlier and later examples.)

Now the more usual term.

1877 [see *LADIN]. 1934 Webster, Rhaeto-Romance languages. 1937 J. Orr tr. *Iordan's Introd. Romance Linguistics* i. 11 The dialects of the Rheto-Romance area. 1941 M. E. Maxfield in *Studies in Romance Lang. & Lit.* (Univ. N. Carolina) II. 1 (*title*) Raeto-Romance bibliography. 1952 *Archivum Linguisticum* IV. 87 Rheto-Romance *agör*. 1969 G. Price *Present Position Minority Lang. W. Europe* 9 Romansh—otherwise known as Raeto-Romance; it is no part of our concern to discuss whether or not the various Romansh dialects of the Swiss canton of the Grisons (Graubünden), the non-Italian Romance dialects of the Dolomite valleys (known as Ladin) and Friulan (spoken in north-east Italy) are rightly considered to be forms of the same language. 1976 *Language* LII. 703 Rohlfs' latest publication..represents an excellent introduction to the history of Rhaeto-Romance. 1977 C. F. & F. M. Voegelin *Classification & Index World's Lang.* 298 Rhaeto-Romance = Raeto-Romance = Rheto-Romance = Ladin = Rhaetian.

Rhages (rā·dʒiz). The name of a city in Persia (now Rayy in Iran, near Tehran), used *attrib.* to designate a type of pottery made there from the eleventh to the thirteenth centuries, characterized by polychrome enamelling.

1909 G. C. Pier *Pottery of Near East* v. 81 The Rhages lustre ware is well illustrated by the superb vase. 1925 B. Rackham tr. *E. Hannover's Pott. & Porc.* I. ii. 65 The greatest possible caution is..required in the case of the most expensive of all the Rhages wares, those with a wealth of figures and colours in the decoration. 1935 A. Christie *Death in Clouds* xi. 117 There was a good deal of Rhages ware and other Persian pottery. 1957 *Encycl. Brit.* XVIII. 371/2 The designs on this Rhages enamelled ware are pencilled with miniature-like fineness recalling..the beautiful workmanship in the manuscript illuminations of the early 13th century. 1974 Savage & Newman *Illustr. Dict. Ceramics* 240 *Rayy* (*Rhages*) ware, a type of Persian pottery made at Rayy, near Teheran, *c.* 1037–1256. The body is medium-hard and coarse. Many pieces have a blue ground, and the backs of some dishes have a deep-blue glaze.

rhamnose (ræ·mnǒᵘz). *Chem.* [a. G. *rhamnose* (Rayman & Kruis 1887: see *Chem. Centralblatt* (1888) XIX. 6): see RHAMNUS and -OSE[2].] A methyl pentose sugar with reducing properties which occurs widely in nature, esp. combined as glycosides in berries of the buckthorn and other shrubs of the genus *Rhamnus*.

1888 *Jrnl. Chem. Soc.* LIV. 667 As the analogy of the sugar $C_6H_{12}O_5$ to the dextrose series is brought into prominence, the authors propose to call it *rhamnose*, because it also appears to be identical with the sugar obtained by Liebermann from Rhamnus. 1888, 1890 [in Dict. s.v. RHAMNETIN]. 1931 E. C. Miller *Plant Physiol.* viii. 408 Rhamnose, $CH_3C_5H_9O_5$, is a pentose in which one atom of hydrogen has been replaced by a methyl group. 1970 R. W. McGilvery *Biochem.* xxiv. 585 Glycoproteins commonly contain residues of L-fucose..and some have residues of L-rhamnose (6-deoxy-L-mannose)... The mammalian nucleotide carrier of rhamnose is not known for certain.

Hence **rha·mnoside,** a glycoside in which rhamnose is the sugar.

1904 *Jrnl. Chem. Soc.* LXXXVI. I. 681 Rhamnosides. 1907 *Chem. Abstr.* I. 83 Violaquercetrin of *Viola tricolor* is identical with rhamnoside of buckwheat. 1959 *New Biol.* XXIX. 35 A substance having some characteristics of a sugar and provisionally called a 'rhamnoside', has been found in water from the Gulf of Mexico in concentrations of up to 50 milligrams per litre.

rhamphotheca (ræmfoþī·kă). *Ornith.* [mod. L. (J. C. W. Illiger *Prodromus Systematis Mammalium et Avium* (1811) 150), f. Gr. ῥάμφος beak + θήκη sheath.] The modified integument, usu. horny but in some birds leathery, that covers the jaws and forms the bill.

1870 Baird & Cooper *Ornithol.* I. 567 *Ramphotheca* [indexed as *rhamphotheca*], the horny covering, or sheath of the jaws. 1890 [see *gnathotheca* s.v. *GNATHO-]. 1972 P. Stettenheim in D. S. Farner et al. *Avian Biol.* II. i. 45 In ptarmigans (*Lagopus* spp.) and certain other grouse (*Tetrao* spp.), the rhamphotheca is shed intact after the breeding season.

rhapidosome (ræ·pidosǒᵘm). *Microbiology.* [f. Gr. ῥαπίς, ῥαπιδ- rod + -O + *-SOME[4].] A small rod-shaped body, frequent in bacteria of certain species.

1963 R. A. Lewin in *Nature* 6 Apr. 104/1 It is proposed that for the present these rod-shaped particles be called 'rhapidosomes'. 1967 *Jrnl. Bacteriol.* XCIV. 1755/1 It is not likely that the rhapidosomes of *Pseudomonas*, *Proteus*, and *Photobacterium* are either phage or parts of phage since similar structures have been seen in every strain examined. *Ibid.* 1755/2 The function of these rhapidosomes is unknown, but..they may be involved as anchorage for the nucleoplasm and may serve as an axis around which rotation occurs during replication. 1972 *Nature* 24 Mar. 144/3 Rhapidosomes are in essence two overlapping hollow cylinders... The two components are made of distinct protein subunits and there is no nucleic acid associated with the rhapidosome. 1977 *Canad. Jrnl. Microbiol.* XXIII. 1604/1 Rhapidosomes resemble bacteriophage tails.., but are often found alone, with no phage heads, coats, or other phage-like appendages.

rhapsodizing, *ppl. a.* (Earlier example.)

1814 Jane Austen *Mansf. Park* II. iv. 76 You will think me rhapsodizing.

rha·psody, *v.* [f. the sb.] **1.** = RHAPSODIZE *v.* 2.

1822 *Blackw. Mag.* Aug. 231/2 A conclusion, in which Sidney heartily joined, rhapsodying—'O Paris, fatal was the hour!' [etc.].

2. = RHAPSODIZE *v.* 3.

1899 *Westm. Gaz.* 13 June 2/3 Miss Jane H. Oakley.. rhapsodies this morning on 'Our Guns'. 1909 W. J. Locke *Septimus* xxi. 321 His face beamed as it had beamed in the days when he had rhapsodied over the vision of an earth one scab to be healed by Sypher's Cure.

rhasophore, var. *RASOPHORE.

rhe (rēi). *Physics.* [ad. Gr. ῥέ-ος stream.] A unit of dynamic fluidity, defined variously as the reciprocal poise or the reciprocal centipoise (see quots.); also, a unit of kinematic fluidity, equal to the reciprocal centistokes.

1928 Bingham & Thompson in *Jrnl. Amer. Chem. Soc.* L. 2879 The fluidity at 20° of a saturated solution of copper was 63·9 rhes. [*Note*] Reciprocal *poises*. 1933 *Physics* IV. 207/2 All the alcohols have a positive association as expected, and it is appreciably higher at a fluidity of 10 rhes than at 50 rhes. *Ibid.* 388 (*caption*) Fluidity data of organic compounds in rhes at regular intervals of temperature. 1958 *Van Nostrand's Sci. Encycl.* (ed. 3) 1417/2 *Rhe*, the absolute unit by which fluidity is measured or expressed. It is the reciprocal centipoise. 1961 *Handbk. Chem. & Physics* (ed. 43) 3151 *Fluidity*, the reciprocal of viscosity. The cgs unit is the rhe, the reciprocal of the poise. 1974 *McGraw-Hill Dict. Sci. & Techn. Terms* 1261/1 *Rhe*. 1. A unit of dynamical fluidity, equal to the dynamical fluidity of a fluid whose dynamic viscosity is 1 centipoise. 2. A unit of kinematic fluidity, equal to the kinematic fluidity of a fluid whose kinematic viscosity is 1 centistoke [*sic*].

RHEA

Rhea, Rea (rī·ă). Jocular aphetic form of DIARRHŒA, GONORRHŒA, etc.
1935 M. B. Carey *Mademoiselle from Armentieres* II. 65 Oh, the Rea sisters came to Brest, Parlez-vous, Leucor, Gonor and all the rest. 1971 *Coming, Dear!* 29 And now that famous rendition of *I got it bad and that ain't good* by the famous Rhea sisters Pya, Dia and Gonna.

rhe(e)bok, -buck. (Further examples.)
1834 A. Smith *Jrnl.* 14 Nov. (1939) I. 137 Saw the common rheebok. 1839 W. C. Harris *Wild Sports S. Afr.* 384 *Redunca Capreolus.* The Rhee Buck. Rhee-bok of the Cape Colonists. 1862 J. S. Dobie *Jrnl.* 10 Sept. in *S. Afr. Jrnl.* (1945) 23 Saw both oribi and rhebuck, the former very pretty animals. 1889 H. A. Bryden *Kloof & Karroo* xvi. 297 The Vaal or Grey Rhebock..is..plentiful on most of the mountain ranges. 1910 J. Buchan *Prester John* xiv. 230 There were droves of smaller game—rhebok and springbok and duikers. 1950 *Cape Argus* 17 July 4/4 Two dainty rhebuck stood and gazed at us. 1958 L. van der Post *Lost World of Kalahari* v. 91 Barely fifty feet from me five rhee-buck got up from their warm beds behind a ledge of rock. 1959 *Encounter* Sept. 53 These mountains of up-pointed spears Hold eland, oribi and rhebok.

rheid (rī·id), *sb.* and *a.* [f. Gr. ῥέ-ειν to flow + *-id*, after *liquid*.] **A.** *sb.* A substance which undergoes viscous flow when at a temperature below its melting point.
1954 S. W. Carey in *Jrnl. Geol. Soc. Australia* I. 70 The substance deforms as a fluid. In order to distinguish it from a liquid (seeing that it is below its melting point)..such a substance is here named a rheid. 1963 Turner & Weiss *Struct. Anal. Metamorphic Tectonites* viii. 321 We still have no knowledge of the threshold conditions beyond which rocks such as granite and schist..can be regarded as rheids.
B. *adj.* Characteristic of a rheid; that is a rheid.
1954 *Jrnl. Geol. Soc. Australia* I. 83 There are..numbers of empirical observations of solid intrusion of serpentine by rheid flow. 1962 L. C. King *Morphol. Earth* iv. 123 The long period over which the Jura folding has been accomplished, from Oligocene to late-Pliocene time, favours the idea of slow, rheid creep in the basement. 1965 A. Holmes *Princ. Physical Geol.* (ed. 2) ix. 204 While ice or any other solid body in a rheid state is slowly flowing, the deformation stress may temporarily increase to a point where the material is compelled to fracture, if only momentarily. 1972 J. G. Dennis *Structural Geol.* v. 102 A material is rheid by virtue of the time of observation. Rocks are not rheid in a quarry operation, but they are so during geological deformation.
Hence **rhei·dity**, the phenomenon of rheid behaviour; also, a measure of the rheid behaviour of a substance, equal to the time required for the viscous deformation to become one thousand times as great as the elastic deformation.
1954 S. W. Carey in *Jrnl. Geol. Soc. Australia* I. 70 Duration..to allow the viscous term to become a thousand times as great as the elastic term... This may be adopted arbitrarily as the threshold of wholly fluid behaviour and will be referred to as the rheidity of the substance for the given conditions. 1965 A. Holmes *Princ. Physical Geol.* (ed. 2) x. 238 The rheidity of salt is..about 10^8 to 10^9 seconds..i.e. between 3 and 30 years. *Ibid.* xi. 282 Fifty years ago little or nothing was known of the phenomena of rheidity and diapirism.

Rheingold (rəi·ngōuld). [Ger., lit. 'gold of the Rhine', name of an opera by Wagner.] An express train that runs between Amsterdam and Basle, along the course of the Rhine. Also *attrib.*
1935 C. Winchester *Railway Wonders of World* I. 492/3 The 'Rheingold' express is one of the most famous of European Pullman trains. It runs between the Hook of Holland and Basle. *Ibid.* 583/1 The 'Rheingold' is among the..most comfortable trains on the Continent. 1955 H. Nicolson *Diary* 25 Jan. (1968) 274 It is cold and dark when the steamer reaches Rotterdam... The Rheingold express is hot and comfortable. 1968 R. Sawkins *Snow along Border* ii. 6 You can catch the Rheingold—it's a marvellous train, one of the Trans-Europe expresses. 1969 *Sat. Rev.* (U.S.) 4 Jan. 48/2 To make a railroad voyage in like style today, an American must journey overseas. In Germany he rides the *Rheingold*; in France, the *Mistral*. 1973 *Guardian* 17 Mar. 14/5 The 'Rheingold' from Basle to Amsterdam..has a vista-dome observation car.

rhema: see *RHEME.

rhematic, *a.* Add: **2.** Of or pertaining to a *RHEME.
1959 J. Firbas in *Brno Studies in English* I. 39 The thematic elements are less important..than the rhematic elements. 1969 K. H. Wagner *Generative Gram. Studies Old Eng. Lang.* i. 49 Although there can be no definite proof, it is to be assumed that the rhematic constituent carries the primary stress. 1977 J. Lyons *Semantics* II. xii. 507 There is a very high correlation..between occupying initial position and being thematic, rather than rhematic.

rhe·matize, *v. Linguistics.* [f. *RHEMAT(IC *a.* + -IZE.] *trans.* To advance (a non-verbal term) to the status of a separate rheme. Hence **rhematiza·tion**; **rhe·matizing** *ppl. a.*
1959 J. Firbas in *Brno Studies in English* I. 43 Rhematic elements..communicatively strengthen basically thematic positions if they occur in them; they..'dynamize' them, rendering them transitional or 'rhematizing' them. *Ibid.* 51 Sentences that contain special rhematizing means of FSP [Functional Sentence Perspective]. 1969 K. H. Wagner *Generative Gram. Studies Old Eng. Lang.* i. 49 The rheme is placed immediately before the verb just in case there is a non-verbal constituent which can undergo rhematization. *Ibid.* 50 Generally speaking only previously mentioned constituents can be thematized, and only non-mentioned ones can be rhematized.

rheme (rīm). *Logic* and *Linguistics.* Also **rhema.** Pl. **rhemas, -ata** (*rare*), **rhemes.** [ad. Gr. ῥῆμα, -ατος that which is said, word, saying.] That part of a proposition or sentence which expresses a single idea. Specialized use in current linguistics: that part of a sentence giving new information about the theme. Cf. *MONORHEME, *RHETIC *a.*
1892 C. S. Peirce in *Open Court* VI. 3417/1 A rhema is closely analogous to a chemical atom or radicle with unsaturated bonds. A non-relative rhema is like a univalent radicle; it has but one unsaturated bond. A relative rhema is like a multivalent radicle. *Ibid.*, Two non-relative rhemas being joined give a complete proposition. *Ibid.* 3417/2 It follows that if we find three distinct and irreducible forms of rhemata, the ideas of these should be the three elementary conceptions of metaphysics. 1897 J. P. Postgate in *Fortn. Rev.* Sept. 428, I propose..to call the expression of a single idea or notion a *rheme* from ῥῆμα, 'a thing said' and to distinguish the expressions of qualifications and connections of such *rhemes* by calling them *epirrhemes* though, as a general term, *rheme* may serve for both. c 1903 C. S. Peirce in *Coll. Papers* (1932) II. II. ii. 144 A *Rheme* is a sign which, for its Interpretant, is a Sign of qualitative Possibility, that is, is understood as representing such and such a kind of possible Object. 1955 [see *PHEME]. 1959 J. Firbas in *Brno Studies in English* I. 39 Those sentence elements which convey something that is known, or may be inferred, from the verbal or from the situational context..are to be regarded as the communicative basis, as the theme of the sentence. On the other hand, those sentence elements which convey the new piece of information are to be regarded as the communicative nucleus, as the rheme of the sentence. 1972 J. Vachek in V. Fried *Prague Sch. Linguistics* i. 19 The other part, now usually called the rheme, contains the essential information transmitted by the given sentence-utterance and so substantially enriches the listener's knowledge. 1976 *Archivum Linguisticum* VII. 130 He points out that the distinction between old and new information has been recognized as the dichotomy of 'theme' and 'rheme' in the Prague school of linguistics, and under the terms 'topic' and 'comment' in recent times. 1976 Geach & Kenny *Prior's Doctrine of Propositions & Terms* 88 Predicates thus conceived as propositions with their subjects left out he called 'rhemes', from the Greek for 'verb'. More often he excepts proper names and especially demonstrative pronouns, i.e. the kinds of 'terms' which may fill the blanks in 'rhemes' but which are not 'rhemes' themselves.

† **Rhénan** (renañ), *a. Geol. Obs.* Also **Rhenan(e).** [a. F. *Rhénan* Rhenish, ad. L. *Rhēnanus*, f. *Rhēnus*, the river Rhine.] The name given by A. Dumont (*Bull. de l'Acad. R. de Belg.* (1848) XV. 684) to an independent *terrain* ('system') of the Palaeozoic found in the Rhineland (see quot. 1853[1]), identified by later writers with the Lower Devonian or the Siegenian and Emsian stages; of or pertaining to that series or those stages. Also *absol.*
1853 *Q. Jrnl. Geol. Soc.* IX. 24 [The *Terrain Rhénan* of M. Dumont comprises three divisions, of which the lowest or *Système gedinnien* appears to belong to the top of the Silurian system. The two upper divisions, called *Ahrien* and *Coblenzien*, which are consequently intermediate between the Old Red Sandstone and the Silurian system, have not yet been recognized in England.] *Ibid.* 25 The red sandstones are not naturally the base of the Eifel series, but rather the top of the Rhenane series. *Ibid.*, The 'Gedinnian system' must be entirely separated from the Rhenane series, with which..it has no organic remains in common. 1895 J. D. Dana *Man. Geol.* (ed. 4) 626 The Lower, Middle, and Upper divisions [of the Devonian] are named (1) the Rhénan, (2) the Eifelian, and (3) the Famennian. *Ibid.* 627 In the Eifel, the three divisions, the Rhénan, Eifelian and Famennian are well developed. 1931 Gregory & Barrett *Gen. Stratigraphy* 250 (*table*) Rhenan.

rhenate (rī·neᵗt). *Chem.* [f. *RHEN(IUM + -ATE[4].] A salt of the anion ReO_4^{-2}. So **rhe·nic acid,** H_2ReO_4, the parent acid of these salts, which is unstable and is known only in solution.
1929 *Chem. Abstr.* XXIII. 1833 A soln. of black ReO_2 in dil. HNO_3 gives a yellow-red coloration, yellow with NaOH, from which $Ca(OH)_2$ or $Ba(OH)_2$ ppts. bright yellow salts, rhenates, stable only in alk. soln. 1932 J. W. Mellor *Comprehensive Treat. Inorg. Chem.* XII. lxv. 478 The yellow product in all these cases is either an acidic soln. of rhenic acid, H_2ReO_4, or a salt of that acid, namely the rhenates, $R^1_2ReO_4$. 1950 N. V. Sidgwick *Chem. Elements* II. 1292 Rhenium trioxide is formally the anhydride of rhenic acid H_2ReO_4, the salts of which can be obtained in solution. 1962 J. E. Fergusson et al. in B. W. Gonser *Rhenium* 38 Of the complex compounds of Re^{VI}, the green rhenate $BaReO_4$ has been prepared as mixed crystals with $BaSO_4$. 1973 R. D. Peacock in J. C. Bailar et al. *Comprehensive Inorg. Chem.* xxxix. 940 Concentrated alkali induces dismutation to perrhenate and dioxide, so that it is not possible to prepare rhenates(VI) in aqueous solution—alkali fusion in the absence of air is necessary.

Rhenish, *a.* and *sb.* Add: **A.** *adj.* **1. a.** Also, in Archæology, used to designate a type of pottery made in the Rhineland in the Roman period.
1954 M. Wheeler *Rome beyond Imperial Frontiers* (1955) I. iv. 48 One was a black-glazed Rhenish beaker of familiar late type. 1955 I. A. Richmond *Roman Brit.* iv. 177 Production of Samian ware was brought to an end by the invasion of Gaul in the later third century... The gap was filled by..an increasing volume of imports of Rhenish glass-ware and Rhenish glazed pottery decorated with white slip. *Ibid.* 179 Rhenish pottery..comprises jugs, beakers, and vases. The fabric is thin, with a highly-polished dark colour coating, usually ornamented in simple running scrolls or medallions in white slip. 1957 *Archaeologia Æliana* XXXV. 187 (*heading*) Narrow-mouthed beakers in Rhenish ware. 1964 G. Webster *Romano-British Coarse Pottery* (Council for Brit. Archaeol. Research Rep. No. 6) 7 *Rhenish ware*, pottery from the Rhineland imported into Britain during the late second century and later.
2. *Rhenish wine* (later example).
1972 *Guardian* 30 Nov. 15/4 Of the rich Rhenish wines, the Alsatian Traminer and Gewurztraminer are becoming popular.
3. *Rhenish stoneware*, stoneware manufactured in the Rhineland, usually salt-glazed.
1925 B. Rackham tr. *E. Hannover's Pott. & Porc.* I. iii. 197 The manufacture of stoneware [in Germany] was confined to certain districts; its centre was for centuries the Rhineland from Coblenz to Cologne. Its present designation 'Rhenish stoneware' is, however, of comparatively recent date. 1974 Savage & Newman *Illustr. Dict. Ceramics* 245 *Rhenish stoneware*, saltglazed stoneware made in the Rhineland from the latter part of the Middle Ages, but principally in the 16th and 17th centuries, at a number of centres, the largest of which was Cologne. It was exported in great quantities at the time, and imitated extensively in the 19th century. It is especially noted for relief work.
4. *Rhenish fan*, a fan-shaped bundle of isoglosses in the Rhine valley, separating Low from High German.
1933 L. Bloomfield *Language* xix. 343 Some forty kilometers east of the Rhine the isoglosses of the great bundle that separates Low German and High German begin to separate and spread out northwestward and southwestward, so as to form what has been called the 'Rhenish fan'. 1961 R. E. Keller *German Dialects* 249 The Rhenish Fan..was seen as the result of a linguistic thrust from south to north along the River Rhine. 1977 *Word* 1972 XXVIII. 233 The 'Rhenish fan' has its counterpart in the isoglottic bundles separating North and South Wales.
B. *sb.* (Later examples.)
1922 Joyce *Ulysses* 331 What say you, good masters, to..a medlar tansy and a flagon of old Rhenish? 1970 M. Kelly *Spinifex* i. 24 Letting the chilled Rhenish dissolve a hard, nervy lump inside.

rhenium (rī·niəm). *Chem.* [mod.L., coined in Ger. (W. Noddack et al. 1925, in *Sitzungsber. d. Preuss. Akad. d. Wissensch.* 409), f. *Rhēn-us*, L. form of the name of the river Rhine: see -IUM.] A very dense, refractory, rare metallic element belonging to the manganese group, resembling platinum in appearance, and obtained in small quantities from molybdenite ores. Symbol Re; atomic number 75.
1925 [see *MASURIUM]. 1938 R. W. Lawson tr. *Hevesy & Paneth's Man. Radioactivity* (ed. 2) xvi. 160 In the case of the element 75 (rhenium), the regularities in the periodic classification..allowed definite predictions on the chemical nature of the missing substance to be made, so that the element could be looked for with success. 1950 N. V. Sidgwick *Chem. Elements* II. 1291 Such sulphides as molybdenite (MoS_2) may contain as much as 1, and in two cases as much as 10 and 21 parts of rhenium per million. 1961 *New Scientist* 30 Nov. 557/3 The resistance to corrosion offered by rhenium metal has for long been appreciated by industry and its special properties have frequently been put to good use. 1963 I. L. Finar *Org. Chem.* (ed. 4) I. xii. 276 Both rhenium heptaselenide and heptasulphide are good catalysts for hydrogenation. 1971 *World Petroleum* June 193/1 This catalyst, containing platinum and rhenium, exhibited such exceptional stability characteristics that it was possible to employ it at substantially lower operating pressures, so that marked increases in yields of motor fuel of a given octane rating, of hydrogen and of aromatics were obtainable. 1973 R. D. Peacock in J. C. Bailar et al. *Comprehensive Inorg. Chem.* III. xxxix. 905 Rhenium, the dvi-manganese of Mendeléev, and the least abundant of the natural chemical elements, was the last..to be identified by conventional chemical methods. *Ibid.* 938 Rhenium forms three anhydrous oxides: the dioxide, the trioxide and the heptaoxide.

rhenosterbos(ch), -bush (reno·stəɹbǫs, -bǫʃ, -buʃ). *S. Afr.* Also **renosterbos, -bush, rhinoster bos(ch), bush.** [Afrikaans, f. *renoster* rhinoceros + *bos* bush.] A shrub with greyish foliage, *Elytropappus rhinocerotis*, belonging to the family Compositæ, native to South Africa, and bearing small scale-like leaves and clusters of small purple flowers; = *rhinoceros-bush* s.v. RHINOCEROS 3 in Dict. and Suppl.
1812 A. Plumptre tr. *Lichtenstein's Trav. S. Afr.* I. 73 We found here in abundance a plant which..the colonists call..*rhinosterbosjes*, because, they say, that in the time when the rhinoceros was an inhabitant of the country, it used to feed very much upon this plant. 1822 W. J. Burchell *Trav. Interior S. Afr.* I. 101 A neat pale bushy shrub, of the height of three or four feet, called Rhinoster bosch (Rhinoceros bush). 1850 R. Gray *Jrnl.* 19 Dec. (1851) 199 More dreary than usual, with its unbroken monotonous rhinoster bush. 1896 R. Wallace *Farming Industries Cape Colony* 81 The rhenoster bush has spread more widely and more quickly

than exotic plants generally do. 1939 *Nature* 11 Mar. 412/2 The 'rhenosterbush'..represents the most arid kind of sclerophyll. 1948 *Cape Times* 6 Feb. 14/5 The rhenosterbos could be subdued by burning. *Ibid.* 22 Sept. 9/6 The stunted renoster bush has already appeared. 1955 V. M. Fitzroy *Dark Bright Land* 55 Papa means to clear the lower slope of a grey scrub they call rhenoster bush. 1966 E. Palmer *Plains of Camdeboo* xvii. 277 Tea made of the leaves of the rhenoster or rhinoceros bush to cure their indigestion. This low grey bush..sprawls over many of our Karoo mountains. 1973 Y. Burgess *Life to Live* 113 You can try some wildeals or renosterbos in a little brandy to make you strong.

rheo-. Add: **rhe·obase** *Physiol.* [ad. F. *rhéobase* (L. Lapicque 1909, in *Compt. Rend. Soc. de Biol.* LXVII. 283), f. *base* BASE *sb.*[1]], the minimum electrical stimulus which, applied continuously, can excite a nerve or muscle; cf. *CHRONAXIE, CHRONAXY; hence **rheoba·sic** *a.*; **rhe:ogonio·meter** *Physics*, a form of goniometer which can be used to measure shearing stresses in Newtonian and non-Newtonian fluids; hence **rhe:ogonio·metry**; **rhe·ogram** *Physics*, any diagram exhibiting experimental results pertaining to rheology (see quots.); **rheomo·rphism** *Geol.* [ad. G. *rheomorphose* (H. G. Backlund 1937, in *Bull. Geol. Inst. Univ. Uppsala* XXVII. 234), f. Gr. μορφ-ή form], the process by which a rock becomes mobile and partially or completely fused, usu. the result of heating by the addition of extraneous magmatic material; so **rheomo·rphic** *a.*; **rhe·ophil(e)** *a. Zool.* [-PHIL, -PHILE], tending to seek or inhabit an environment of flowing water; also as *sb.*, such an organism; hence **rheophi·lic, rheo·philous** (also stressed *-phi·lous*) *adjs.*; **rheopho·bic** *a. Zool.* [-PHOBIC], tending to avoid or not to inhabit an environment of flowing water; so **rhe·ophobe**, such an organism; **rhe·ophyte** *Bot.* [-PHYTE], a plant that is confined to flowing water; hence **rheophy·tic** *a.*; **rhe·oreceptor** *Zool.*, a sensory receptor that is sensitive to the flow of the surrounding water; hence **rhe·oreceptive** *a.*
1924 *Nature* 22 Mar. 427/1 The rheobase is the intensity in volts of a constant current closed instantaneously which will just excite if continued indefinitely. 1944 *Electronic Engin.* XVII. 27/2 In measuring chronaxie a current value is found which, when caused to flow for infinite time, will produce a minimal contraction in the muscle under observation. (For this purpose any period in excess of one second can be referred to as infinity.) This current value is called the Rheobase. 1952 *Ibid.* XXIV. 334/1 The ratio of this current to the rheobase depends on the rate of accommodation of the nerve. 1965 S. Ochs *Elements Neurophysiol.* ii. 21 A utilization or utilized time also occurs during the..shorter times of excitation found when using stimulating pulses stronger than rheobase. 1976 *Exper. Neurol.* L. 71 The C-fibers had a rheobase of 0·033 mA and a utilization time of 7·80 msec. 1942 *Chem. Abstr.* XXXVI. 6636 With faradization the rheobasic effect was decreased, the chronaxia increased. 1944 *Electronic Engin.* XVII. 27 Chronaxie is defined as that period of time for which a current, having twice the rheobasic value, must flow in order to produce the same minimal contraction. 1965 S. Ochs *Elements Neurophysiol.* ii. 20 When a rheobasic current is used, the nerve actually becomes excited a short time after the onset of the step pulse current. 1946 *Nature* 2 Nov. 614/2 The practical application of the Weissenberg rheogoniometer. 1949 K. Weissenberg in *Princ. Rheol. Measurement* (Brit. Rheologists' Club) iii. 39 It has therefore become necessary to design a new instrument, termed Rheogoniometer, which measures the macroscopically observable forces and displacements in a sufficiently comprehensive manner. 1974 [see *RHEOMETRIC *a.*]. 1974 *Physics Bull.* Jan. 20/2 These theories..led to the design of..the 'rheogoniometer' which allowed for the first time the movements and forces in flowing fluids to be measured as functions of time and in all three dimensions in space. 1976 *Nature* 5 Feb. 389/1 Rheogoniometry seems to measure T_1 as accurately as previous methods. 1933 Schofield & Blair in *Proc. R. Soc.* A. CXXXIX. 558 A study was made of the rate of elongation of cylinders of unyeasted dough hung vertically by their upper ends and allowed to extend under the action of gravity... It has been found convenient to mark on the dough cylinders a series of fine parallel lines accurately spaced 1 mm. apart. The marks were made by successive turns of a fine wire wrapped round a frame, which are wetted with enamel, the marks remaining wet long enough to be subsequently printed off on to a strip of duplicator paper... The print (which may be called a rheogram) is available for whatever analysis appears suitable. 1941 J. Stanley in *Industr. & Engin. Chem.* (Anal. Ed.) June 404/2 These facts are secured from the rheograms. 1944 G. W. S. Blair *Surv. Gen. & Appl. Rheol.* x. 123 Stanley..used the word 'rheogram' to describe a rate-of-shear stress curve. This is confusing, since such flow curves are quite different from the charts so described by Schofield and Scott Blair. 1974 P. L. Moore et al. *Drilling Practices Manual* vi. 224 One procedure suggested by Forbes includes construction of a regular rheogram chart of shear stress versus shear rate. 1937 *Proc. Geologists' Assoc.* XLVIII. 275 The quartz-felspar intergrowths of the rheomorphic veins are less perfectly micropegmatitic. 1954 *Amer. Jrnl. Sci.* CCLII. 602 The term 'rheomorphic' in accordance with its derivation means 'flow form'. It was introduced by Backlund (1937) to describe the flowage of rocks where there is no evidence to show that the rocks concerned were melted and capable of liquid flow. 1962 E. A. Vincent tr. *Rittmann's Volcanoes* vi. 199 Migmatitic granites thus frequently show very well marked rheomorphic structures on solidification. 1937 *Proc. Geologists' Assoc.* XLVIII. 260 These are clearly cases of rheomorphism. [*Note*] Backlund has introduced the term *rheomorphism* for the process whereby a pre-existing rock becomes partially or completely mobilised or fused as a result of the introduction of migrating materials (in great or small amount) with concomitant rise of temperature. 1964 L. U. De Sitter *Struct. Geol.* (ed. 2) xxix. 402 In some of these pockets of recrystallized nonoriented rock the random orientation of gneissic or schistose inclusions or xenoliths proves that this disturbance can develop into flow. This phenomenon has been called 'rheomorphism'. 1934 Webster, Rheophile, -phil *adj.* 1964 *Oceanogr. & Marine Biol.* II. 417 It appears as though *A. chiajei* is a slow-growing sediment feeder..whereas *A. filiformis* is a more rheophile suspension feeder with rapid growth. 1965 B. E. Freeman tr. *Vandel's Biospeleol.* xvii. 294 *Gammarus* are stream animals and are essentially rheophiles. 1974 *Nature* 8 Feb. 395/1 Rheophobic species may co-exist with the rheophiles. 1963 L. Birkett tr. *Nikolsky's Ecol. of Fishes* i. 76 Rheophilic fishes are also at the same time oxyphilic, i.e. they require plenty of oxygen. 1979 Rheophilic [see *rheophobic* adj. below]. 1951 L. H. Hyman *Invertebrates* II. x. 192 Some turbellarians are rheophilous, i.e., limited to flowing water. 1965 B. E. Freeman tr. *Vandel's Biospeleol.* xxiv. 391 According to Poulson (1961) *Amblyopsis spelaeus* is rheophilous. *Ibid.* xvii. 294 *Niphargus orcinus virei* is a 'rheophobe'. 1979 E. N. K. Clarkson *Invertebr. Palaeont. & Evol.* ix. 220/2 Some rheophobes live in deeper waters, so that a fair amount of detrital material has accumulated by the time it reaches them. 1965 B. E. Freeman tr. *Vandel's Biospeleol.* xxiv. 395 The fish, *Typhlichthys subterraneus*, is probably rheophobic. 1974 Rheophobic [see *rheophile* adj. above]. 1979 E. N. K. Clarkson *Invertebr. Palaeont. & Evol.* ix. 220/2 Rheophobic crinoids live in current-free waters though most deep water crinoids are rheophilic. 1932 *Bull. Jard. Bot. Buitenzorg* 3rd Ser. XII. 201, I draw the attention towards the remarkable fact that the leaves of *Neonauclea rheophila* and *Nauclea angustifolia* agree entirely with those of other well-known rheophytes. 1950 *Flora Malesiana* I. iv. p. lvii, Rheophytes are plants restricted to riverbeds. 1981 C. G. G. J. van Steenis (*title*) Rheophytes of the world. 1975 T. C. Whitmore *Trop. Rain-Forests Far East* 187 *Garcinia cataractalis*, a rheophytic shrub with narrow willow-like stenophyllous leaves and crimson fruits. 1951 L. H. Hyman *Invertebrates* II. x. 93 Rheoreceptive cells have been identified in *Mesostoma* and *Bothromesostoma*. 1948 *Nature* 25 Dec. 1000/2 Rheoreceptors [in Tunicata]: margins of the siphons—not only do the siphons bend to face a water-current, but also bend upwards against gravity. 1971 J. D. Carthy in J. E. Smith et al. *Invertebr. Panorama* x. 220 (*caption*) A section through the head of the flatworm *Mesostoma* to show four of the eight rheoreceptors, the two groups of chemoreceptors and four tactile receptors.

rheological (rīₒlǫ·dʒikăl), *a. Physics.* [f. as next + -ICAL.] Of or pertaining to the deformation and flow properties of matter.
1930 *Jrnl. Rheol.* I. 509 (*heading*) Teaching rheological technique. 1946 *Nature* 10 Aug. 199/2 With non-Newtonian liquids their peculiar rheological properties can be demonstrated by this method. 1956 *Sci. News* XL. 66 The rheological properties of clay dispersions are of particular importance in the ceramic and petroleum industries. 1959 *Times* 8 Jan. 2/6 (Advt.), The British Food Manufacturing Industries Research Association has vacancies for research staff to conduct investigations into..the rheological behaviour of chocolate, particularly at low rates of shear. 1974 *Sci. Amer.* Dec. 88/3 Solidification proceeds from spheroidal growth centers. The rheological (flow) properties of this structure are quite different from those of the dendritic structure.
Hence **rheolo·gically** *adv.*
1945 *Jrnl. Appl. Physics* XVI. 338/2 The asphalt industry has also been rationalized, rheologically speaking. 1963 *New Scientist* 22 Aug. 382/1 We may say that modern rheology began with J. L. M. Poiseuille's investigations of the flow of blood. He.., in 1842, published his law governing the flow through capillaries of rheologically simple liquids. 1973 *Nature* 14 Sept. 78/2 In a rheologically young planet, we expect that quakes will occur whenever the total elastic energy in a certain region exceeds some critical value.

rheologist (rīₒlǫ·lŏdʒist). *Physics.* [f. next + -IST.] One who studies or is expert in rheology.
1931 *Jrnl. Rheol.* II. 108 This number contains an index to the literature which should be useful to all rheologists. 1947 *Nature* 11 Jan. 70/1 The gap between the colloidal chemist and the physicist is being bridged by the rheologist. 1959 *Times* 5 Mar. 2/3 (Advt.), Dough rheologist. An experimental officer..to assist in rheological studies in a laboratory fully equipped with..dough testing machines. 1963 *New Scientist* 22 Aug. 382/1 A fundamental task of the experimental rheologist is to measure viscosity and other flow problems.

rheology (rīₒlǫ·lŏdʒi). *Physics.* [f. RHEO- + -LOGY.] **a.** The study of the deformation and flow of matter, esp. the non-Newtonian flow of liquids and the plastic flow of solids. **b.** The rheological properties *of* a substance.
1929 *Jrnl. Rheol.* I (*title-page*), Journal of Rheology published by the Society of Rheology. 1930 *Ibid.* I. 508 A large number of words have been suggested and John R. Crawford, late Professor of Latin at Lafayette College, inclined to the belief that 'rheology' is the most appropriate designation. 1947 *Nature* 11 Jan. 71/1 The rheology of suspensions is fundamentally dependent on the dispersion coefficient of the medium. 1968 B. S. Massey *Mechanics of Fluids* i. 23 The fluids with which engineers most often have to deal are Newtonian, that is, their viscosity is not dependent on either the rate of shear or its duration; and the term 'mechanics of fluids' generally refers only to Newtonian fluids (the study of non-Newtonian liquids is termed 'rheology'). 1968 M. Pyke *Food & Society* ix. 130 The firmness or toughness of foodstuffs is also a relevant measure of their quality; rheology is the branch of science which has been developed to measure and describe such characteristics. 1978 *Nature* 30 Mar. 478/2 During his eight years at M.I.T., Goetze and his associates became one of the world's leading groups in the study of the rheology of rocks.

rheometer. Add: **1.** (Earlier example.)
1839 *Ann. Electr., Magn. & Chem.* IV. 70 On this plan my Reometer (an instrument described in Gehler's Phil. Dictionary, new edition, vol. vi, sec. 3, p. 2494) is constructed and brought into action.
2. b. Any instrument for measuring the rheological properties of matter, esp. one which measures the viscous stress of a fluid.
1945 *Jrnl. Appl. Physics* XVI. 338 A modified Bingham-type rheometer. 1950 *Jrnl. Sci. Instruments* XXVII. 209 (*heading*) A continuous-shear rheometer for measuring total stress in rubber-like materials. 1967 Condon & Odishaw *Handbk. Physics* (ed. 2) III. iii. 48/1 Rheological properties are determined quantitatively in rheometers (plastometers), for which the viscometers are the prototype, either with their absolute value or relatively to the magnitude of the same property in a standard material. 1971 *Nature* 5 Nov. 50/1 The slime was diluted to various concentrations with clean water from the natural environment and then tested in a portable rheometer.
rheometric *a.* (earlier and further examples); **rheometry** (further example).
1843 C. Wheatstone in *Phil. Trans. R. Soc.* CXXXIII. 309 The voltaic element which I have employed in most of my rheometric researches. 1967 Condon & Odishaw *Handbk. Physics* (ed. 2) III. iii. 48/1 (*heading*) Rheometry. 1972 *Physics Bull.* Feb. 85/1 It has developed a rheometric device for measuring the reaction between oil, water and detergent, based on the rheometer at the School of Pharmacy at London University. 1974 *Nature* 22 Mar. 328/1 In order to characterize the rheological behaviour of the aqueous sugar solution, an extensive series of tests was done with a rheometric mechanical spectrometer and with a Weissenberg Rheogoniometer.

rheopexy (rī·ŏpeksi). *Physical Chem.* [f. RHEO- + *-PEXY.] The property, possessed by some sols, of undergoing accelerated gelation when subjected to gentle mechanical agitation. Cf. *THIXOTROPY. So **rheope·ctic** *a.*
1935 Freundlich & Juliusburger in *Trans. Faraday Soc.* XXXI. 921 It is too cumbrous to say 'thixotropic sols which can be solidified by orienting the particles'. Graham used the word 'pectous'..for sols which have turned to a jelly. We therefore propose to call a thixotropic sol, which may be solidified by gentle movement, 'rheopectic', and the phenomenon itself 'rheopexy'. 1958 *U.S. Patent* 2,846,394 1/1 The grease is a rheopectic lubricant composition which essentially comprises an oleaginous base thickened with..a lithium soap of a hydroxy stearic acid. 1963 *New Scientist* 22 Aug. 384/1 The viscosity in the return part of the programme is greater than the viscosity at the same rate of shear in the upward part. This behaviour is called rheopexy. 1967 *Ibid.* 20 July 143/2 The technique is to add to the fuel a high-molecular-weight polymer..which converts the fuel into a 'rheopectic gel'... Such a fluid undergoes a large change in viscosity and volatility when subjected to shear forces above a certain level. 1978 C. J. Geankoplis *Transport Processes & Unit Operations* ii. 105 In design procedures for thixotropic and rheopectic fluids for steady flow in pipes, the limiting flow-property values at a constant rate of shear are sometimes used.

rheostat. Add: Also, a resistor whose resistance can be varied by mechanical means, esp. a variable wire-wound resistor used for controlling large currents. (Further examples.)
1920 *Whittaker's Electr. Engineer's Pocket-bk.* (ed. 4) 561 In one form of rheostat, the grids are formed of rectangular wire in one continuous length. 1960 *Practical Wireless* XXXVI. 373/1 (Advt.), Variable rheostats, heavy duty slider resistor rated at 25 amps. 1972 P. Marks *Collector's Choice* i. 26 All the lights and rheostats. 1974 *Encycl. Brit. Micropædia* VIII. 548/2 The rheostat can adjust generator characteristics, dim lights, and start or control the speed of motors... A special type of rheostat is the potentiometer.
Hence **rhe·ostatted** *a.*, fitted with or controlled by a rheostat.
1978 J. Updike *Coup* (1979) v. 211 The symbolic riband of gold on his forehead took glitter from the rheostatted spotlight.

rheostatic, *a.* Add: **b.** *rheostatic brake*, a form of brake used in vehicles driven by electric traction motors, in which the motors are made to act as generators, the power so generated being dissipated in the starting rheostats; so *rheostatic braking.*
1920 *Whittaker's Electr. Engineer's Pocket-bk.* (ed. 4) 563 Brakes for tramcars may be classified as follows:..(2) Electric rheostatic brakes, in which the car motors are converted into generators and are loaded on rheostats, the brake being applied and regulated from the controller. 1931 R. A. Bishop *Electr. Trolley Bus* v. 118 The rheostatic brake so adjusted as not to overstress the transmission. *Ibid.* 192 (Index), Rheostatic braking. 1954 A. T. Dover *Electr. Traction* (ed. 3) viii. 130 For rheostatic braking either of the schemes..are employed. 1971 *Mod. Railways* May 197/2 As train speed is reduced and the rheostatic brake fades, proportional air brake is automatically re-established on the locomotive.

rheotaxis (rī₁ǫtæ·ksis). *Zool.* [a. G. *rheotaxis* (C. Herbst 1894, in *Biol. Centralbl.* XIV. 694): see RHEO- and TAXIS.] The orientation or movement of an animal (or, formerly, a plant) in response to a current of water
1900 in B. D. Jackson *Gloss. Bot. Terms* 318/1. 1903 *Amer. Naturalist* XXXVII. 201 The ear is held to be an organ directly concerned with the reactions of a fish to a current of water (rheotaxis). 1906 A. J. Ewart tr. *Pfeffer's Physiol. Plants* III. iv. 356 Roth..could detect no rheotaxis in *Euglena viridis*. 1951 L. H. Hyman *Invertebrates* II. x. 213 Planarians that habitually live in flowing water usually exhibit a positive rheotaxis, for obviously without such a reaction they could not maintain themselves in their habitat. 1970 *Nature* 24 Oct. 375/1 It is generally accepted that spermatozoa constrained to move in a fine tube swim preferentially upstream (they exhibit positive rheotaxis) when a current is established in the tube.
Hence **rheota·ctic** *a.*, exhibiting or pertaining to rheotaxis; **rheota·ctically** *adv.*
1900 B. D. Jackson *Gloss. Bot. Terms* 318/1 Rheotactic. 1903 *Science* 3 Apr. 531/1 They are..rheotactic, thigmotactic and phototactic in the highest degree. 1954 G. L. Clarke *Elem. Ecol.* ii. 57 The rheotactic reactions of most stream fish is such that they swim against the current and thus maintain their position in the stream. 1973 *Netherlands Jrnl. Sea Res.* VII. 180 No significant difference can be seen between the effect of bottom or surface currents on the rheotactic behaviour of the starfish. 1975 F. Creutzberg in O. Kinne *Marine Ecol.* II. II. viii. 599 *Palaemon serratus* and *P. xiphias* did not respond rheotactically in the experimental current channel.

rheotropism (ri₁ǫ·trǫpiz'm, ri₁ǫtrōᵘ·piz'm). *Zool.* and *Bot.* [ad. G. *rheotropismus* (B. Jönsson 1883, in *Ber. d. deutsch. bot. Ges.* I. 521): see RHEO- and TROPISM.] The orientation or movement of an animal or plant in response to a current of water.
1887 H. E. F. Garnsey tr. *A. de Bary's Compar. Morphol. & Biol. Fungi* ix. 449 Rheotropism. If a stream of water is made to flow slowly through a moistened porous substratum, ..the plasmodia which are vegetating on the moist surface wander in the direction of the stream. 1898 [see *CHEMOTROPISM]. 1965 B. E. Freeman tr. *Vandel's Biospeleol.* xxiv. 391 Gammarids show a very strong positive reaction towards running water and swim energetically against it... The name positive rheotropism has been given to this reaction.
So **rheotropic** (-trǫ·pik, -trōᵘ·pik) *a.*, exhibiting or pertaining to rheotropism.
1897 *Ann. Bot.* XI. 182 It [*sc.* the plasmodium] moved with the current, i.e. became negatively rheotropic. 1920 A. Arber *Water Plants* xxii. 282 Stems responded to the stimulus by their own activity and might thus be called positively rheotropic. 1931 G. K. Noble *Biol. Amphibia* xiv. 346 Steinmann..has reinvestigated the problem in tadpoles and newts and would attribute the rheotropic response to compensatory reflexes initiated by the labyrinth. 1968 F. R. H. Jones *Fish Migration* xii. 269 He studied the rheotropic responses of elvers in a circular flume.

rhesus. Add: [a. specific name of *Macacus rhesus*, formerly *Simia rhesus* (J. B. Audebert *Histoire naturelle des Singes* (1799) II. 5).] **1.** Also *rhesus macaque*. For *Macacus rhesus* substitute *Macaca mulatta*. (Earlier and later examples.)
1827 E. Griffith et al. tr. *Cuvier's Animal Kingdom* I. 289 The pig-tailed baboon, or Rhesus of Audebert, is the short-tailed Macaque of Buffon. 1906 E. Ingersoll *Life Animals: Mammals* 30 The widespread little yellowish Bengal or rhesus monkey..abounds in northern India. 1932 S. Zuckerman *Social Life Monkeys* vi. 88 More has been written about the reproductive mechanisms of the Rhesus macaque than about those of all the other primates together. 1948 *Psychol. Abstr.* XXII. 381/2 Large numbers of rhesus monkeys living in a semi-natural environment are shown with special emphasis on social interactions. 1967 J. R. & P. H. Napier *Handbk. Living Primates* III. 375 In spite of Fooden's proposition..that the crab-eating macaque, *M. fascicularis*, and the rhesus macaque, *M. mulatta*, should be considered conspecific, they are here retained as separate species. 1976 *Amer. Jrnl. Vet. Res.* XXXVII. 969/1 Rhesus macaques have been selected most for studying radiation sickness. 1978 D. Symons in E. O. Smith *Social Play in Primates* 209 Long term studies of free ranging rhesus monkeys indicate that the ranks of females and immature males are determined by the political milieu into which they are born.
2. *Med.* Used *attrib.* and in *Comb.* with reference to a major blood group consisting of three principal antigens to which naturally occurring antibodies are rare, and which are important because hæmolytic disease of the newborn is usually the result of antibodies produced in the blood of a rhesus-negative mother in response to the rhesus-positive blood of the fetus; (so called from being first discovered in the rhesus monkey); as *rhesus agglutinogen, antibody, antigen, incompatibility, system*; **rhesus baby**, an infant suffering from hæmolytic disease of the newborn owing to incompatibility between its own rhesus-positive blood and its mother's rhesus-negative blood; **rhesus factor**, any or all of the rhesus antigens, esp. the most important one; **rhesus-negative** *a.*, lacking the most important and common rhesus antigen, and therefore able to produce antibodies to it; so **rhesus-positive** *a.*, having this antigen.
1941 *Amer. Jrnl. Obstetr. & Gynecol.* XLII. 925 In the great majority of the cases the blood factor involved has been shown to be either identical with or related to the Rh (Rhesus) agglutinogen first described by Landsteiner and Wiener with the aid of rabbit sera prepared by injection of Rhesus blood. 1945 *Nature* 5 May 542/1 (*heading*) The new rhesus antibody. 1973 *Nursing Times* 24 May 668/2 Particular attention was paid to blood tests for Rhesus antibodies. 1947 G. F. Roberts *Rhesus Factor* i. 5 The rhesus antigen, being foreign to the recipient, provoked the formation of an antibody which would agglutinate rhesus positive red cells. 1950 *Sci. News* XV. 121 On the left is part of the surface of the red cell bearing a configuration which is responsible for the specificity of the rhesus antigen. 1967 *Spectator* 8 Dec. 706/1 A new serum to save rhesus babies was developed in Britain. 1968 *Sci. Amer.* Nov. 49/2 Among the 850,000 births each year in Britain,..the number of 'rhesus' babies is probably not more than about 5,000. 1942 *Jrnl. Amer. Med. Assoc.* 7 Mar. 843/1 Statistical data on the Rh (Rhesus) factor. 1958 *Listener* 17 July 87/1 In 1939 Levine..found a fourth system [of blood-groups], which has been named the Rhesus factor. 1969 *Times* 6 May 12/5 Incompatibility arises between the mother and her fœtus when the latter alone carries the Rhesus blood group factor. 1947 G. F. Roberts *Rhesus Factor* v. 32 The question of rhesus incompatibility must..be considered in terms of C, D and E. 1971 H. & M. Brant *Dict. Pregnancy, Childbirth & Contraception* 204 Rhesus incompatibility does not..cause any defect in development of the baby such as a hole in the heart or the blue baby condition. 1943 *Brit. Med. Jrnl.* 4 Sept. 293/2 Treat these infants by transfusion with rhesus-negative blood, free of agglutinins and thus test out the recommendation that rhesus-negative blood would produce better results than rhesus-positive blood. 1950 *Sci. News* XV. 126 The rhesus negative characteristic is almost unknown in the Far East. 1961 P. Mason *Common Sense about Race* i. iii. 52 In Britain five persons out of six are classed as Rhesus positive in the Rhesus blood-group system, while one in six is Rhesus negative. 1971 J. Z. Young *Introd. Study Man* xl. 587 In the majority of pregnancies Rhesus-negative women do not..become immunised by a Rhesus-positive foetus. 1943, etc. Rhesus-positive [see *rhesus-negative* above]. 1958 *Listener* 17 July 87/1 The Rhesus system has been of service not only in making the medical profession to some extent genotype-conscious but in demonstrating the genetic complexity of the regions of the germ-plasm responsible for blood-group polymorphisms. 1974 P. Svendsen *Introd. Animal Physiol.* x. 113 About twenty different blood group systems are known in man. Of these the ABO and the rhesus systems are the most important.

rhetic (rī·tik), *a.* [f. Gr. ῥητ-ός stated + -IC.] Designating or pertaining to an utterance that has the property of meaning (in its elements of sense and reference), as distinct from its identity as sound and words. Hence **rhe·tically** *adv.* Cf. *RHEME.
1955 J. L. Austin *How to do Things with Words* (1962) vii. 93 To say anything is..to perform the act of using that pheme or its constituents with a certain more or less definite 'sense' and a more or less definite 'reference' (which together are equivalent to 'meaning'). This act we may call a 'rhetic' act. *Ibid.* viii. 97 When different phemes are used with the same sense and reference, we might speak of rhetically equivalent acts. 1969 W. Cerf in K. T. Fann *Symposium on J. L. Austin* iv. 356 The suspicion arises that the phonetic act, the phatic act and the rhetic act are not subclasses, but parts of the locutionary act. 1971 M. Furberg *Saying & Meaning* (ed. 2) ii. 57 Reports of someone's words in indirect speech are reports of his rhetic act.

rhetoric, *sb.*[1] Add: **1. e.** Literary prose composition, esp. as a school exercise.
1828 R. Whateley *Elements Rhetoric* 4 Some writers have spoken of Rhetoric as the Art of Composition, universally; or, with the exclusion of Poetry alone, as embracing all Prose-composition. 1944 H. J. C. Grierson *Rhetoric & Eng. Composition* p. iii, Of University teaching in English I had enjoyed just fifty lectures at Aberdeen, of which twenty-five were devoted to Rhetoric or, as Rhetoric had come to mean under Dr. Alexander Bain and his successor William Minto, English Composition. 1953 T. S. Eliot *Amer. Lit. & Amer. Lang.* 5, I am happy to remember that in those days English composition was still called Rhetoric. 1972 *Lebende Sprachen* XVII. 35/2 *US* rhetoric—*BE/US* literary composition.
2. c. Delete † *Obs.* and add later examples.
1942 W. Stevens *Parts of World* 143 Midsummer love and softest silences, Weather of night creatures, whistling all day, too, And echoing rhetorics more than our own. 1949 Koestler *Promise & Fulfilment* II. v. 274 It was a disappointing speech—emotional rhetorics without a constructive programme. 1976 *Sunday Times* (Lagos) 3 Oct. 10/4 We cannot decide on the fundamental values and goals that will bind the present and future generations on the basis of vague ideas, irrelevant foreign slogans and rhetorics.
e. (*b*) (later examples); (*c*) of artistic style or technique.
1941 W. H. Auden in *Southern Rev.* VI. 729 Around them boomed the rhetoric of time. 1963 R. I. McDavid *Mencken's Amer. Lang.* 339 Among the neo-Aristotelian critics *rhetoric* is a current fashionable synonym for technique... The *Rhetoric* of Fiction. 1964 J. Summerson *Classical Lang. Archit.* iv. 33 Well, there are three buildings which, I believe, demonstrate..the 'rhetoric' of the Baroque. 1976 *Howard Jrnl.* XV. i. 52 The rhetoric of treatment will have to be replaced by the reality of treatment.

rhetorically, *adv.* Add: **a.** (Further examples.) Also in *Comb.*
1904 B. Russell in *Mind* XIII. 522 We may imagine a rhetorically minded soldier in battle saying to himself: 'To advance is to die, to retreat is dishonour; better death than dishonour.' 1927 E. O'Neill *Lazarus Laughed* iv. 177 He laughs with a wild triumphant madness and again rhetorically. a 1974 R. Crossman *Diaries* (1975) I. 210 By means of a conscious rhetorically- and demagogically-forced row with the Opposition I won back both their respect and my popularity among my own party. 1976 *Nature* 13 May 92/1 The choice between guns and butter was rhetorically offered by Goebbels to the German public.

rhetorician. Add: Also 8 **returrition. 3.** (Earlier and later examples.)
1676 in *Cath. Rec. Soc. Publ.* (1972) LXIII. 60 Rhetoricians. Jo. Brockholes. Jo. Townly. 1716 N. Blundell *Diary* 3 May (1952) ix. 164 Mr Carroll being to leave, gave his schoolfellows the Returritions a treat. 1773 H. T. Blount *Diary* 17 Apr. in *Cath. Rec. Soc. Publ.* (1972) LXIII. 354 Master John Bradshaw, Rhetorician, left the College, and returned to England. 1809 [see *PHILOSOPHER 1 c].

rhetoricize (retǫ·risəiz), *v.* [f. RHETORIC *sb.*[1] + -IZE.] **a.** *intr.* = RHETORIZE *v.* **b.** *trans.* (Chiefly in *ppl. a.*) To characterize with rhetoric; to make rhetorical. Also *fig.* Hence **rheto:riciza·tion**, the act or process of rhetoricizing.
1676 R. Meggott *Sermon on St. Paul's Day* 10 But we (as he very melancholily rhetoriciseth) are naked, impotent, and shiftless. 1932 C. S. Lewis in *Essays & Stud.* XVII. 62 A detailed study of the *Book of Troilus* would reveal this 'rhetoricization', if I may coin an ugly word, as the common quality of many of Chaucer's additions. 1934 T. Wood *Cobbers* xviii. 235 This youngster was born with a golden tongue in its mouth and was fed on rhetoricized milk. 1965 *New Statesman* 22 Oct. 604/1 All excellently put—though a tendency to rhetoricise is already apparent in the plea for 'the real impenetrable human person'. 1972 G. S. Fraser in Cox & Dyson *20th-Cent. Mind* II. xi. 395 The moral stance is strong just because, unlike Lawrence's changing and always dramatized or rhetoricized moral stances, it is not assertive.

rheumatic, *a.* and *sb.* Add: **A.** *adj.* **3. b.** Also *fig.*
1934 E. Sitwell *Aspects Mod. Poetry* i. 14 The cramped and rheumatic eight-syllable lines..are not suitable to his themes. 1958 J. Press *Chequer'd Shade* i. 14 It is obscure, less because of its vocabulary than because the entire movement of the syntax is rheumatic and muscle-bound.
4. b. *rheumatic fever* (later example).
1974 Passmore & Robson *Compan. Med. Stud.* III. xvi. 51/2 Rheumatic fever is an immunological disorder developing after an infection by a group A β-haemolytic streptococcus. Its major clinical features are fever, polyarthritis, carditis, chorea, erythema nodosum and annulare, and subcutaneous nodules.
d. Pertaining to rheumatism.
1886 H. Baumann *Londinismen* 155/2 *Rheumatic dodge*, List einer Bettler-klasse, welche Rheumatismus erheuchelt. 1976 *Lancet* 27 Nov. 1202/2 The hospital's closure was strenuously opposed by the consultant rheumatologists to whom it was obvious that to halve the number of beds in the regional centre would gravely impair the rheumatic service.
B. *sb.* **1.** Also *sing.*
1918 V. Woolf *Diary* 16 July (1977) I. 169 He is a ridiculous figure precisely like the false Mandrill, especially now that he's bent with the rheumatic, & can only creep and crawl.

rheumatism. 2. (Further examples.)
1910, 1915 [see *FIBROSITIS]. 1966 Wright & Symmers *Systemic Path.* I. iii. 113/2 The focal lesions of acute rheumatism are usually found in close relationship to arteries.

rheumatoid, *a.* Add: (Further examples.) Also, of or pertaining to rheumatoid arthritis; *rheumatoid factor*, any of a group of substances in the blood which are present in increased amounts in persons with rheumatoid arthritis.
1949 R. M. Pike et al. in *Jrnl. Immunol.* LXIII. 447 The property of agglutinating sensitized sheep cells to a marked degree appears to be associated almost exclusively with the serum in rheumatoid arthritis. The phenomenon is due to a factor which is present in the serum-globulins and its effect is not diminished on storage for considerable periods. Until this factor is more completely characterized, it may be conveniently termed the 'rheumatoid factor'. 1950 *Sci. News* XV. 134 The adrenal cortex may play an important role in the pathogenesis of rheumatic and rheumatoid conditions in man. 1956 *Proc. Soc. Exper. Biol. & Med.* XCII. 198/2 Sensitized cells which were strongly agglutinated by the rheumatoid serum. 1957 *Proc. R. Soc. Med.* L. 47 The appearance in the serum of rheumatoid arthritics of a factor similar to those induced in the rabbit sera..suggests that it, too, is an antibody to the non-specific groups of immune γ-globulin. 1957 *Ann. Rheumatic Dis.* XVI. 448/1 The term 'rheumatoid factor' (R.F.) has been applied to the substance or substances in the serum of most individuals having rheumatoid arthritis which combines with immune globulins or substances closely associated with immune globulins. 1970 Passmore & Robson *Compan. Med. Stud.* II. xxx. 9/2 This mode of origin was suggested for rheumatoid arthritis; in this disease numerous features..combine to support the idea that this is a 'disorder of the immunological mechanism' or even an autoimmune disease... Today there is more support for the alternative hypothesis that rheumatoid arthritis..may be caused by an exogenous agent like a virus or bacterium. 1980 *Sci. Amer.* July 42/2 Among the commonest autoantibodies in man are 'rheumatoid factors', a group of antibodies directed against gamma globulin.

rheumatology (rūmătǫ·lŏdʒi). [f. RHEUMAT(ISM + -OLOGY.] The study of rheumatism and rheumatic diseases.
1949 J. L. Hollander *Comroe's Arthritis* (ed. 4) i. 15 Rheumatology is the study of the Rheumatic Diseases, including arthritis, rheumatic fever, fibrositis, neuralgia, myositis, bursitis, gout, and other conditions producing somatic pain, stiffness and soreness. 1957 *Times* 20 Dec. 7/2 The Senate of London University have instituted a chair of rheumatology to be tenable at the Postgraduate Medical School, Hammersmith. 1973 *Daily Colonist* (Victoria, B.C.) 8 June 27/1 Dr. C. S. Man, a Taiwanese physician completing his post-graduate training in rheumatology and physical medicine in Canada. 1976 *Lancet* 25 Dec. 1419/1 He performed much fundamental research in rheumatology and was responsible for the first controlled trial of gold therapy in rheumatoid arthritis.
Hence **rheumato·logist**, a specialist or expert in rheumatology.
1949 W. Graham in J. L. Hollander *Comroe's Arthritis* (ed. 4) vii. 115 It is becoming more and more the conviction of rheumatologists that patients with arthritis respond best to treatment in a fully integrated rheumatism center. 1971 *Daily Tel.* 22 Mar. 3/7 The situation was noticed by a consultant rheumatologist who takes a clinic once a month at the cottage hospital. Every time he comes he has to deal with a fair number of cases of gout. 1977 *Lancet* 9 Apr. 793/1 The use of immuno-suppressive drugs to treat bulbous dermatoses was prompted by results obtained by rheumatologists treating autoimmune connective tissue diseases.

rhexia. In etym. insert after 'Linnæus': *Corollarium Generum Plantarum* (1737) 7. Substitute for def.: A perennial herb of the genus so called, belonging to the family Melastomaceæ, native to eastern North America and bearing white or purple flowers; also called meadow beauty or deer grass. (Earlier and later examples.)
1819 A. Rees *Cyclopædia* XXX. s.v., The species cultivated are; the Virginian rhexia..and the Maryland rhexia. 1824 H. E. Lloyd tr. *Spix & Martius's Travels in Brazil* II. iv. ii. 213 Large rhexias (*quaresima*) covered with purple flowers adorned the hill. 1887 *Cent. Mag.* July 327 Parts of New England have already a midsummer flower nearly as brilliant..in meadow beauty, or rhexia. 1951 *Dict. Gardening* (R. Hort. Soc.) IV. 1757/2 Rhexias do best in a bed of peat but may be grown in pots.

rhexigenous (reksi·dʒĕnəs), *a. Bot.* [f. Gr. ῥῆξις breaking (see next) + *-GENOUS.] Of an intercellular space in plant tissue: formed by the spontaneous mechanical rupture of cells. Hence **rhexi·genously** *adv.*, **rhexi·geny.**
1940 *Chambers's Techn. Dict.* 722/1 Rhexigenous. 1957 *Bot. Rev.* XXIII. 304 Rhexigenous lacunae would appear to be quite common in fleshy fungi, while lysigenous and schizogenous ones are more often found in higher, green plants. 1965 K. Esau *Plant Anat.* (ed. 2) viii. 188 Other large air spaces may arise lysigenously or rhexigenously. *Ibid.* xvii. 487 Large spaces may also result from a more or less extensive breakdown of cells by processes of lysigeny or rhexigeny. 1965 Bell & Coombe tr. *Strasburger's Textbk. Bot.* 103 (*caption*) Intercellular canal produced by the rupture of the annularly thickened tracheids of the protoxylem and the surrounding parenchyma (i.e. rhexigenously). 1978 *Acta Amazonica* VII. 484/2 Rhexigenous lacunae are present both in the petiole and midrib of the leaf blade.

rhexis (re·ksis). Also 5 **rixis.** [mod.L., ad. Gr. ῥῆξις, f. ῥηγνύναι to break.] † **1.** *Med.* Outpouring of blood through the ruptured wall of a blood vessel. *Obs.*
a 1425 Rixis [see SECONDARY *a.* 2 a]. 1684 tr. *Blancard's Physical Dict.*, *Rhexis*, the same that [*sic*] *Rhegma*. 1881 A. Flint *Treat. Princ. & Practice Med.* (ed. 5) I. i. 27 When the extravasation is through the ruptured wall of a vessel, it is called hemorrhage by rhexis. 1899 Allbutt *Syst. Med.* VI. 244 It cannot be doubted that the red corpuscles escape by diapedesis not by rhexis. 1914 T. W. Widdowson *Notes on Dental Surg. & Path.* iv. 53 In some acute inflammations the capillary walls may give way, resulting in an extravasation of blood (rhexis) into the surrounding tissue.
2. *Biol.* The fragmentation of a cell or cellular component.
1930 Maximow & Bloom *Text-bk. Histol.* iii. 66 In the granular leukocytes the nucleus may undergo fragmentation into separate parts (rhexis) or shrinkage (pyknosis). 1965 N. F. Rodman et al. in Sasahara & Stein *Pulmonary Embolic Dis.* 35 Platelet rhexis begins with disintegration of the central-most alpha granulomeres and is evidenced by an electron-lucent zone containing disintegrating alpha granulomeres. 1968 J. A. Serra *Mod. Genetics* III. xx. 130 To be efficient the polyploidizing chemicals must not cause rhexis or fragmentation of the chromosomes.

rhinarium. Add: **b.** (Further examples.)
1947 J. Stevenson-Hamilton *Wild Life S. Afr.* xiv. 102 The grey rhebuck..has woolly fur, a large rhinarium, and no gland patches under the ears. 1976 *Nature* 29 Jan. 339/1 These [*sc.* resemblances between *Tarsius* and the anthropoids] include the absence of a rhinarium, a retinal structure similar to the anthropoid fovea.
Hence **rhina·rial** *a.*
1929 F. W. Jones *Man's Place among Mammals* xiv. 86 The details of rhinarial structure. 1955 W. C. O. Hill *Primates* II. 3 Shortening of the face..is partly due to reduction in the size of the nasal apparatus, including under this head not only the olfactory organs but also the rhinarial complex.

Rhine[3]. Add: **b. Rhine daughter, maiden** [tr. G. *Rheintochter*], each of three water maidens, guardians of the Rheingold in Wagner's cycle of operas *Der Ring des Nibelungen* (1853–74); also used allusively of the fair blonde physique with which they are usu. portrayed.
1897 W. A. Ellis *Richard Wagner's Prose Works* VI. 108 The 'Rhine-daughters' were played by Frl. Lilli and Marie Lehmann and Frl. Lammert. 1899 G. Bell *Let.* 16 Aug. (1927) I. 51 We saw all the properties, and all the mechanism of the Rhine maidens. 1911 G. B. Shaw *Shewing-Up of Blanco Posnet* 356 From the Rhine maidens in Wagner's Trilogy, and the bathers in the second act of Les Huguenots, to the ballets of water nymphs in our Christmas pantomimes and at our variety theatres, the sound hygienic propaganda of the bath, and the charm of the undraped human figure, are exploited without offence on the stage to an extent never dreamt of by any novelist. 1922 T. S. Eliot *Waste Land* 59 The Song of the (three) Thames-daughters begins here... V. *Götterdämmerung*, III, i: the Rhinedaughters. 1940 W. Faulkner *Hamlet* iv. i. 276 To those below what Brunhilde, what Rhinemaiden on what spurious river-rock of papier mâché. 1950 A. Wilson *Such Darling Dodos* 82 With her enormous height and ample frame, her flaxen hair and bovine eyes, she was like a head prefect in the role of a Rhine maiden. 1965 V. Canning *Whip Hand* xiii. 148 The blond Siegfried type..and Katerina and Lottie, a couple of hand-picked Rhine-maidens. 1972 J. Symons *Players & Game* xviii. 136 Her hair is fine as that of a Rhine maiden.

rhinegrave. Add: **2.** [through F. *rhingrave.*] = *petticoat breeches* s.v. PETTICOAT *sb.* (*a.*) 9.
1885 H. A. Dillon *Fairholt's Costume in Eng.* (ed. 3) II. 349 *Ringrave* or *Rhingrave*, full breeches, with bunches of ribands at the knee. Molière, in the 'Bourgeois Gentilhomme', mentions them. 1934 *Times Lit. Suppl.* 25 Oct. 728/2 The word 'skirt' can then be kept for the rhinegrave or petticoat breeches of the Restoration. 1960 C. W. Cunnington et al. *Dict. Eng. Costume* 162/2 *Rhinegraves*,..immensely wide in the leg and pleated or gathered on to a waistband, falling like a divided skirt to the knees or just above.

Rhinelander (rəi·nlændəɹ). [f. RHINELAND + -ER[1]: cf. G. *Rheinländer*.] **1.** A native or inhabitant of the Rhineland.
1858 H. Mayhew *Upper Rhine* ii. 51 Chickens cost from 6*d.* to 1*s.* each..though it must not be imagined that for this price the birds are anything like our own in quality, it being the custom of the Rhinelanders to fatten no animal for food. 1861 *Times* 7 Oct. 6/3 What have the pious Belgians or the warm-hearted Rhinelanders..done in defence of the Pope's authority? 1866 J. Bryce *Holy Rom. Emp.* (ed. 2) xvii. 338 The fall in A.D. 1477 of the great principality..was seen with pleasure by the Rhinelanders. 1902 G. Meredith *Let.* ? 3 Mar. (1970) III. 1427 He is a descendant of those indomitable Lower Rhinelanders, who gave such trouble to the Romans. 1928 *Daily Tel.* 27 Mar. 14/2 A parallel case..is the veto placed by the Coblenz High Commission on the singing by the Rhinelanders of the 'Wacht am Rhein'. 1977 'R. West' *Celebration* 528 'From Frankfurt!' exclaimed my mother happily. 'You are a Rhinelander!'
2. (Freq. in G. form *Rheinländer*.) A variety of German polka.
1887 A. Zorn in P. J. S. Richardson *Social Dance of Nineteenth Cent. in England* (1960) ix. 102 In the year 1850 there appeared in all parts of Europe the 'Schottische', a round dance which had, as early as 1844, been executed in Bavaria under the name 'Rheinlaender', and in the Rhenish countries it was known as the 'Bavarian Polka'. 1912 A. Huxley *Let.* 13 May (1969) 41 There was a gingerbeer and other beer stall and a dancing floor where about fifty peasants in weird Hessian costume were dancing what I learnt was a Rhinelander, which looks like a mixture between a polka and a two-step. 1938 B. Schönberg tr. *Sachs's World Hist. of Dance* vii. 436 The *Rheinländer* or the Bavarian *polka*, in contrast to the *galop*, is a slow polka. 1953 J. Lawson *Europ. Folk Dance* xii. 116 Another polka-like step is called the *Rheinländer* and has an attractive zig-zag pattern.

rhinencephalon. Add: (Earlier and further examples.)
1846 R. Owen *Hunterian Lect.* II. viii. 184 The 'rhinencephalon' consists of two lobes of grey matter, which receive the prolongations of chiefly white fibres from the prosencephalon and its crura. 1890, etc. [see *PALLIUM 3 d]. 1901 *Jrnl. Anat. & Physiol.* XXXV. 438 After Turner's memoir of 1890, the old term 'rhinencephalon' attained a much greater vogue. *Ibid.* 441 The German Anatomical Nomenclature Commission adopted the teaching of His in this matter in its entirety, and called his 'lobus olfactorius' the 'rhinencephalon', under the mistaken idea that it represented the region so-called by Turner. 1908, 1921 [see *ARCHIPALLIUM]. 1978 C. Reid *Primer Human Neuroanat.* xv. 141 The rhinencephalon refers to the olfactory part of the brain and the use of this term at the present state of knowledge is fraught with difficulties.
rhinencephalic *a.* (earlier and later examples).
1846 R. Owen *Hunterian Lect.* II. viii. 181 A few of the medullary fibres..are continued forwards and outwards, as 'rhinencephalic crura'. 1956 *Sci. Amer.* Oct. 108/2 The test was performed on the animal with the electrode in the rhinencephalic nerve.

rhinestone. Add: **b.** (Earlier and later examples.) Also *fig.* and *attrib.* Hence **rhi·nestoned** *a.*, decorated with rhinestones.
1888 A. Randall-Diehl *Two Thousand Words* 179 *Rhinestone*, a stone which may be made to resemble the diamond. 'Rhinstones [*sic*] as well as paste, are taking the place of diamonds off as well as on the stage.' 1931 W. Faulkner *Sanctuary* xxviii. 283 The hat bore a rhinestone ornament. 1961 Webster, Rhinestoned adj. 1967 'La Meri' *Spanish Dancing* (ed. 2) vii. 96 Rhinestones covered boleros, *queso* hats and shoe-heels in a blinding flash. 1975 *New Yorker* 20 Oct. 154/1 This is an exasperating book... It has far too many rhinestones. 1976 L. St. Clair *Fortune in Death* xiii. 126 Blonde showgirls..in plumed headdresses and rhinestoned G-strings. 1977 *Time* 29 Aug. 24/2 The music got slicker and often sillier, turned from rock toward rhinestone country and spangled gospel. 1981 *Times* 7 July 11/1 The intricate rhinestone and bead embroidery.

rhinion (rəi·ni₁ǫn). *Anat.* [a. G. *rhinion* (A. von Török 1888, in *Monthly Internat. Jrnl. Anat. & Physiol.* V. 281), f. Gr. ῥῖν-ός, ῥίς nose: see *-ION[2].] The foremost point at which the two nasal bones meet.
1904 W. L. H. Duckworth *Morphol. & Anthrop.* x. 229 *Rhinion*, the most prominent point at which the nasal bones touch each other. 1937 *Amer. Jrnl. Physical Anthrop.* XXII. 486 *Rhinion*, the lower end of the internasal suture. 1968 H. de Villiers *Skull of S. African Negro* ix. 127 Nasion may be inset and thus lie at a greater or lesser distance from a line joining glabella to rhinion. *Ibid.* 128 Even moderately curved profiles ending abruptly at rhinion (category 1) or recurving slightly before rhinion (category 2) occur most commonly in the South African Negro.

rhino[2]. Add: **1.** (Further example.) Also *attrib.* and *fig.*
1890 F. D. Lugard *Diary* 22 Feb. (1959) I. ii. 111 Fresher rhino tracks, and elephant spoor everywhere. 1938 J. S. Huxley in *Times* 17 Aug. 11/5 Has not the time come when we can discard our etymological prejudices, accept the usage of the ordinary man, and frankly use 'rhinos'? 1973 M. Amis *Rachel Papers* 27 When I awoke that morning the bedroom was a rhino pen. 1979 *Monitor* (McAllen, Texas) 15 July 8E/5 Three of the surviving rhino species are Asian and two are native to Africa.
2. Special Combs.: **rhino beetle, bird, horn** = *rhinoceros beetle*, etc., s.v. RHINOCEROS 3 in Dict. and Suppl.
1958 J. Slimming *Temiar Jungle* i. 14 The bundle contained a large rhino beetle, still alive but half suffocated. 1951 R. Campbell *Light on Dark Horse* x. 131 The various types of rhino-birds and tick-birds..give warning to the game on whose parasites they feed. 1957 —— *Portugal* v. 87, I came across mermaids on the Rufigi River while trading in poached rhino-horns. 1974 O. Manning *Rain Forest* I. iv. 62 You could find his name on the dockets of some pretty strange things..Rhino horn, cantharides. 1979 *Monitor* (McAllen, Texas) 15 July 8E/5 Rhino horn has long been valued as an aphrodisiac by the oriental cultures.

rhino-. Add: **rhi:nolaryngi·tis** (see quot. 1900); **rhi:nolaryngo·logy** the study and pathology of the nose and larynx; hence **rhi:nolaryngolo·gic** *a.*; **rhi:nolaryngo·logist,** an expert or specialist in rhinolaryngology; **rhi:nopharyngi·tis**, inflammation of the nose and pharynx; **rhi:nopneumoni·tis** *Vet. Sci.*, any of three similar contagious diseases of horses, caused by herpes viruses and characterized by rhinitis; **rhinotheca**, for 'beak' read 'upper jaw'; [coined by J. C. W. Illiger in *Prodromus Systematis Mammalium et Avium* (1811) 150]; **rhino:trachei·tis** *Vet. Sci.*, a respiratory disease of cattle; also, a similar disease of cats.
1900 Dorland *Med. Dict.* 571/2 *Rhinolaryngitis*, inflammation of the mucous membrane of the nose and larynx. 1959 B. Wootton *Social Sci. & Social Path.* iv. 139 The incidence of, and the death rate from, rhinolaryngitis and infantile diarrhoea are abnormally high. 1933 *Jrnl. Amer. Med. Assoc.* 14 Oct. 1248/2 American physicians who studied in Vienna without having visited Professor Hajek's rhinolaryngologic clinic. 1938 Rhino-laryngologist [see *LEPROSARIUM]. 1959 *Jrnl. Laryngology & Otology* LXXIII. 65 Chordomata occur in the domain of the Rhino-laryngologist much more than anywhere else. 1900 Dorland *Med. Dict.* 571/2 Rhinolaryngology. 1905 *Amer. Jrnl. Med. Sci.* CXXX. 733 (*heading*) Local anæsthetics in rhinolaryngology. 1933 *Jrnl. Amer. Med. Assoc.* 18 Mar. 852/2 (*heading*) Substitute for cocaine and procaine in rhinolaryngology. 1896 *Amer. Yearbk. Surg.* 850 (*heading*) Neurasthenia due to ear-diseases and to rhinopharyngitides. 1951 M. Hynes *Med. Bacteriol.* (ed. 5) xi. 177 The meningococcus is spread from carrier to carrier by air-borne infection and settles in the nasopharynx, sometimes giving rise to rhinopharyngitis but usually causing no symptoms. 1957 E. R. Doll et al. in *Cornell Vet.* XLVII. 37 The most constant and severe lesions in the equine fetus also are in the respiratory tract. Uncomplicated natural infection by the virus causes a febrile reaction accompanied principally by a rhinitis. Accordingly, rhinopneumonitis, which embraces the nasal catarrh and pulmonary lesions, is designated as the name for the disease. 1978 *Jrnl. R. Soc. Med.* LXXI. 660 Foals aborted as a result of rhinopneumonitis..are also of expected normal weight and size. 1955 *Jrnl. Amer. Vet. Med. Assoc.* CXXVI. 463 (*heading*) Infectious necrotic rhinotracheitis of cattle. *Ibid.* 463/1 At first this condition was referred to simply as a virus disease... For the last two years it has been called necrotic rhinotracheitis or, for brevity, just 'necrotic rhinitis'. 1970 Jubb & Kennedy *Path. Domestic Animals* (ed. 2) iii. 166/2 Feline viral rhinotracheitis was undoubtedly the principal infection in 'cat distemper' or 'cat flu', etc., until Crandall and Maurer gave it pathological distinction. 1977 *Lancet* 13 Aug. 356/1 Hereford cattle may be susceptible to the carcinogenic action of an ocular herpesvirus (e.g., bovine rhinotracheitis virus) that has been inactivated by exposure to ultraviolet radiation.

rhinocerine, *a.* Add: **2.** Characteristic of or resembling a rhinoceros.
1958 *Listener* 16 Jan. 102/1 The white rhino, a shy dozy animal who likes to..lie in the mud and dream deep rhinocerine dreams. 1977 J. McClure *Sunday Hangman* iv. 34 The rhinocerine person of Sergeant Cecil Arnot.

rhinoceros. Add: **1. b.** (Earlier examples.)

1613 Marston *Insatiate Countess* I. sig.A4ᵛ, Mountebancke with thy Pedanticall action, Rimatrix, Buglors, Rhimocers [*sic*]. 1869 T. Taylor *Our Amer. Cousin* II. 29 There's that damned rhinoceros again.

3. *rhinoceros-run*; *rhinoceros-black, -like* adjs.; **rhinoceros auklet**, a puffin of the north Pacific, *Cerorhinca monocerata* (= *rhinoceros auk*); **rhinoceros bird**, (*b*) (earlier and later examples); **rhinoceros bush**, also = ***rhenosterbos**; (earlier and later examples); **rhinoceros puff-adder**, for def. read: = **rhinoceros viper*; **rhinoceros viper**, a venomous West African snake, *Bitis nasicornis*.

1887 R. Ridgway *Man. N. Amer. Birds* 12 Rhinoceros auklet. 1976 *Islander* (Victoria, B.C.) 19 Sept. 13/1 Common Murres and Rhinoceros Auklets were in abundance. 1822 J. Campbell *Travels S. Africa: Narr. Second Journey* I. xxiv. 282 There is a brown bird, about the size of a thrush, called the rhinoceros' bird, from its perching upon those animals and picking off the brush-lice which fix on him. 1947 J. Stevenson-Hamilton *Wild Life S. Afr.* v. 47 The black rhinoceros..is often accompanied by rhinoceros birds (*Buphaga*), which give the alarm on the approach of enemies. 1925 E. Sitwell *Troy Park* 82 Rhinoceros-black (a flowing sea!). 1731 G. Medley tr. *Kolb's Pres. State Cape Good-Hope* II. 104 The Delight of his [*sc.* the rhinoceros's] Tooth is a Shrub, not much unlike the Juniper... The Cape-Europeans call it the Rhinoceros-Bush. 1839 W. C. Harris *Wild Sports S. Afr.* 30 Barely sufficient quantity of fuel, from a shrub called the rhinoceros bush, could be obtained. 1915 E. R. Lankester *Diversions of Naturalist* xxviii. 260 The skulls and whole skeletons of great rhinoceros-like animals..are dug up in early tertiary sands. 1909 *Chambers's Jrnl.* Sept. 568/2 Elephant-tracks and rhinoceros-runs pierced the jungle here and there. 1910 R. L. Ditmars *Reptiles of World* iv. 327 The Rhinoceros Viper is..the most beautifully coloured of all poisonous snakes. 1965 R. & D. Morris *Men & Snakes* i. 16 The rhinoceros viper has a snout appendage of large pointed scales which looks like an erect horn.

rhinocerotid (rainosĕrǫ·tid). *Zool.* [f. mod.L. family name *Rhinocerotidæ*: see Rhinoceros and -id³.] A fossil mammal of the family Rhinocerotidæ.

1969 *Nature* 1 Feb. 452/1 Both forested and riverine environments are suggested by the Nagri tragulids, anthrocotheres, rhinocerotids and dinotheres. 1976 *Ibid.* 5 Aug. 464/1 The existence of two rhinocerotids has been established by the discovery of skulls of both *Ceratotherium* and *Diceros*.

rhinophyma (rainofai·mă). *Med.* Pl. **rhinophymata, -phymas.** [mod.L. (coined in Ger. by H. von Hebra 1881, in *Vierteljahresschrift f. Dermatol.* VIII. 603), f. Gr. ῥῑν-ός, ῥίς nose + L. *phȳma* swelling, tumour (ad. Gr. φῦμα growth, tumour).] Chronic enlargement and reddening of the nose with hypertrophy of its sebaceous glands.

1882 *London Med. Rec.* IX. 69/1 (*heading*) Hebra.—On rhinophyma. 1915 *Lancet* 18 Sept. 643/2 Rhinophymata give rise to much mental distress and worry, and a desire to shun all society on account of the unsightly appearance produced by their presence. 1920 *Jrnl. Amer. Med. Assoc.* 22 May 1450/1 Operative removal of rhinophymas has not offered much encouragement to surgeons. 1947 G. D. Wolf *Ear, Nose & Throat* xx. 407 Rhinophyma is the advanced and terminal stage of acne rosacea of the nose. 1973 *Sci. Amer.* Sept. 4/2 *An Old Man and His Grandson*, painted by Domenico Ghirlandaio late in the 15th century... The old man apparently suffers from rhinophyma, a condition in which the nose becomes knobby and enlarged as a result of rosacea, a chronic disease involving overactivity of the sebaceous glands. Usually the nose is also reddened, because small blood vessels become dilated.

rhinoplasty. (Earlier and later examples.)

1842 T. D. Mütter *Recent Improvem. Surg.* 33 Adopting the phraseology of Zeis, I include under the expression *plastic surgery*, all the specific terms..such as rhinoplasty when the nose is made, cheiloplasty when the lips, [etc.]. 1978 *Detroit Free Press* 16 Apr. (Detroit Suppl.) 14/4 If the same surgeon does a rhinoplasty he'll probably be paid $800 to $1,200.

rhinosporidiosis (rai:nospǭridi,ōᵘ·sis). *Path.* [f. mod.L. *Rhinosporidi-um*, name of a genus of fungi (Minchin & Fantham 1905, in *Q. Jrnl. Microsc. Sci.* XLIX. 521: see Rhino- and Sporidium) + -osis.] Chronic infection of mucous membrane and rarely of skin by the fungus *Rhinosporidium seeberi*.

1923 Ashworth & Turner in *Edin. Med. Jrnl.* XXX. 337 (*heading*) A case of rhinosporidiosis. 1935 *Indian Med. Gaz.* LXX. 76/1 Reports of rhinosporidiosis in females are rare. 1965 *New Scientist* 22 Apr. 243/3 Rhinosporidiosis is a widespread disease causing obstructive growths in the nasal passages of bathers and divers in southern Asia and S. America.

Also **rhinospori·dial** *a.*, pertaining to or caused by rhinosporidiosis.

1923 *Trans. R. Soc. Edin.* LIII. 332 The three series of Rhinosporidium, which I have studied..agree so closely with one another and with the examples described by Minchin and Fantham as *Rhinosporidium kinealyi*..that they obviously belong to the same species, and the other Rhinosporidial nasal polypi recorded in natives of India and Ceylon are also no doubt due to this species. 1954 R. D. G. P. Simons *Med. Mycol.* xxxiv. 369 Rhinosporidia granuloma has been found in many countries.

rhinoster bos(ch), bush, varr. *Rhenosterbos(ch), -bush.

rhinovirus (rai·novaiᵊrǫs). *Biol.* [mod.L., f. Rhino- + Virus.] Any of a group of picornaviruses including those which cause some forms of the common cold.

1961 C. H. Andrewes in *Yale Jrnl. Biol. Med.* XXXIV. 201 It was reported that some rather different viruses had been cultivated from colds in adults and that these had strong claims to be the agents which were being sought. These newly discovered agents, for which the name Rhinovirus is proposed, are the subject of this paper. 1970 *Daily Tel.* 1 Dec. 5/3 Symptomless rhinovirus infections occur. In addition, rhinoviruses and rhinovirus colds are most prevalent in the spring when sudden changes in temperature most commonly occur. The rhinoviruses are responsible for most colds. 1974 *Nature* 27 Sept. 276/1 Picornaviruses can be divided by physiochemical methods into five subgroups: enteroviruses (for example polio, swine vesicular disease), cardioviruses (such as EMC, Maus Elberfeld (ME)), foot and mouth disease viruses (FMDV) and human and equine rhinoviruses. 1978 G. A. Sheehan *Running & Being* x. 140 We become infected with one of the numerous rhinoviruses.

rhipidistian (ripidi·stiăn), *sb.* and *a.* [f. mod.L. *Rhipidistia* (E. D. Cope 1887, in *Amer. Naturalist* XXI. 1017), f. Gr. ῥιπίς, ῥιπιδ- fan + ἱστία pl. of ἱστίον sail + -an.] **A.** *sb.* A fossil fish belonging to the superorder Rhipidistia, which includes primitive crossopterygian fishes with paired fins that may have muscular lobes. **B.** *adj.* Of or pertaining to a fish of this kind or the group as a whole.

1916 *Mem. & Proc. Manch. Lit. & Philos. Soc.* LX. II. 23 Extensive fusions may have taken place between the bones of the head of Rhipidistians. 1931 J. R. Norman *Hist. Fishes* xvii. 358 One of the earliest members of the order is the interesting *Dipterus*, found in the Old Red Sandstone of Scotland along with the early Rhipidistians and Palaeoniscids. 1964 *Bull. Mus. Compar. Zool. Harvard* CXXXI. 313 (*heading*) The comparative anatomy of the snout in rhipidistian fishes. *Ibid.* 331 The etmoid endocranium has been described in the following rhipidistian genera: *Ectosteorhachis*.., *Eusthenopteron* [etc.]. 1978 *Sci. Amer.* Sept. 111/1 The earliest amphibians arose from a primitive group of lobe fins (the rhipidistians), and so all four-footed vertebrates (tetrapods) and their descendants also evolved from this vanished fish stock. 1979 *Nature* 18 Jan. 176/2 The mutual relations of salmon (an actinopterygian), lungfish (a dipnoan) and cow (a mammal, a class which can be traced back to rhipidistian crossopterygians by way of reptiles and amphibians) are certainly worthy of serious consideration.

rhipiphorid (ripi·fǫrid), *sb.* and *a.* *Ent.* Also **Rhip-.** [f. mod.L. family name *Rhipiphoridæ*, f. generic name *Rhipiphorus* (L. Bosc 1792, in *Jrnl. d'Hist. Nat.* II. 293, as *Ripiphorus*), f. Gr. ῥιπίς fan + mod.L. *-phorus* (see -phorous): see -id³.] **A.** *sb.* A beetle of the group Rhipiphoridæ, larvæ of which are endoparasites in insects, esp. wasps. **B.** *adj.* Of, pertaining to, or designating a beetle of this group.

1899 D. Sharp *Insects* II. v. 267 The elytra are, in several Rhipiphorids,..reduced to a very small size, and the wings are not folded. 1909 *Cent. Dict.* Suppl., *Rhipiphorid* adj. 1957 *Ann. Entom. Soc. Amer.* L. 96/1 The importance of genitalic structure in rhipiphorid taxonomy. 1968 R. H. Arnett *Beetles U.S.* 635 Hymenopteran larvae attacked by rhipiphorids generally complete their development up to pupation before they are killed. 1973 Evans & Eberhard *Wasps* vi. 236 (*caption*) *Macrosiagon flavipenne*, a rhipiphorid beetle that is a parasite of *Bembix* wasps.

rhipsalis (ripsēⁱ·lis). *Bot.* Also **Rhipsalis.** Pl. **rhipsalides** (ripsæ·lidīz), **rhipsalises.** [mod.L. (J. Gaertner *De Fructibus & Seminibus Plantarum* (1788) I. 137), f. Gr. ῥίψ wickerwork, mat + L. *-ālis* (see -al).] Any cactus of the genus so called, which is the only one to occur naturally in the Old World and comprises plants with branching stems, many of which are epiphytes with hanging branches; cf. *mistletoe cactus* s.v. *mistletoe 3.

1831 *Curtis's Bot. Mag.* LVIII. 3080 (*heading*) Rhipsalis Cassytha. Naked Rhipsalis. 1837 *Mag. Zool. & Bot.* I. 469 The fruit of Rhipsalis has been considered to possess a different structure from that of all other Cacteæ. 1884 L. Castle *Cactaceous Plants* 72 The chief distinguishing characters relied upon being the position of the flowers—namely, at the points of the branches, instead of being produced at the sides as in the other Rhipsalises. 1944 S. Putnam tr. *E. da Cunha's Rebellion in Backlands* i. 35 The sinuous rhipsalides, twining like green vipers through the tree branches. 1962 *Amateur Gardening* 24 Feb. 2/1 Several genera are epiphytes, growing..on the..branches of trees.. [including] the various species of mistletoe cactus, rhipsalis.

rhitidome, var. Rhytidome in Dict. and Suppl.

rhizina. Substitute for def.: A root-like structure on the underside of a lichen, consisting of one or several rhizoids. (Further examples.)

rhizine is the usual form of the word.

1914 M. Drummond tr. *Haberlandt's Physiol. Plant Anat.* v. 260 Among lichens the lower side of the thallus produces numbers of rhizoidal hyphae, which..act as root-hairs. In certain species they are united to form stout strands, the so-called rhizines. 1938 G. M. Smith *Cryptogamic Bot.* I. xv. 515 A rhizine may consist of a single simple to branched hypha or of a number of parallel hyphae that lie closely applied to one another. 1977 *Nature* 6 Jan. 46/1 In contrast to lichens and mosses, whose rhizines and rhizoids serve mainly for attachment, trees have root systems which transport sulphate upward.

rhizo-. Add: **rhizo·tomy** *Surg.*, section of a spinal nerve root.

1911 *Brit. Med. Jrnl.* 2 Sept. 523/2 Dorsal rhizotomy failed to give permanent and complete relief [from pain]. 1955 Naffziger & Adams in A. B. Baker *Clin. Neurol.* III. xxviii. 1445 The conversion of a spastic into a flaccid paraplegia by anterior rhizotomy is now well established. 1977 *Lancet* 17 Sept. 594/2 Before the advent of drug treatment for spasticity, the only remedy lay in irreversible operations such as tenotomies and rhizotomies.

rhizobium (raizōᵘ·biǫ̆m). Also **Rhizobium.** Pl. **-ia.** [mod.L. (coined in Ger. by B. Frank 1889, in *Ber. d. deutsch. bot. Ges.* VIII. 338), f. Rhizo- + Gr. βίος life.] A bacterium of the genus so called, which comprises heterotrophic aerobic individuals that form root nodules on leguminous plants and fix atmospheric nitrogen symbiotically with the plants.

1921 R. E. Buchanan *Agric. & Industr. Bacteriol.* iii. 32 Rhizobium.—These organisms are minute rods, motile when young, by means of flagellata. 1947 *Endeavour* VI. 130 When the seed of a legume develops in a soil containing *Rhizobia* the latter are attracted to the region of the developing root hairs. 1966 *McGraw-Hill Encycl. Sci. & Technol.* XI. 546/2 The treatment of leguminous seeds with bulk preparations of effective rhizobia is widely practiced.. for the purpose of..improving the yield and quality of leguminous plants. 1972 A. H. Gibson in Leigh & Noble *Plants for Sheep in Austral.* xi. 103 Considerable attention has been given to the development of techniques for estimating numbers of rhizobia in the soil. 1973 [see *nodulate *v.* a].

Hence **rhizo·bial** *a.*, of or pertaining to rhizobia; **rhizo·bially** *adv.*

1957 *Times* 2 Dec. (Agric. Suppl.) p. vi/5 Both were inoculated with a peat-bound rhizobial culture and the fortuitous treading of the grazing animal was the only way in which the seeds could be covered. 1971 *Nature* 16 July 175/1 The degree of rhizobial invasion after transfer of the cultured soybean root cells is related to the growth factor additions to the MS medium. 1977 *Whitaker's Almanack 1978* 1030/1 Much of the recent work has been concentrated on showing that the selection of the right legume with the appropriate rhizobial species depends on the lectin existing in the legume. 1977 J. L. Harper *Population Biol. Plants* xi. 365 Small amounts of ¹⁵N, a stable isotope, had been applied to the soil so that the nitrogen available to both species in the soil could be distinguished from the rhizobially fixed nitrogen which was unlabelled.

Rhizoctonia (raizǫktōᵘ·niă). [mod.L. (A. P. de Candolle *Flore française* (1815) VI. 110), f. Rhizo- + Gr. κτόν-ος murder + -ia¹.] A fungus of the form-genus so called, which comprises sterile fungi some of which cause disease in plants; also, a fungus formerly placed in this genus but reassigned following the discovery of a sexual state. Freq. *attrib.*

1897 W. G. Smith tr. *K. F. von Tubeuf's Diseases of Plants* II. i. 201 The spores open by a longitudinal slit, and a germ-tube emerging from each end branches into a mycelium which soon takes on the form of a rhizoctonia-strand. 1916 *Bull. Illinois Agric. Exper. Station* No. 189. 376 There are recognized in America two species of truly parasitic Rhizoctonias. 1927 *Ann. Appl. Biol.* XIV. 290 (*heading*) Rhizoctonia 'foot-rot' of the tomato. 1937 F. D. Heald *Introd. Plant Path.* xi. 226 The damping-off by Rhizoctonia is..serious in both garden and field. 1953 F. T. Brooks *Plant Diseases* (ed. 2) xv. 299 Other fungi producing small sclerotia also attack cereals and grasses, notably *Sclerotium rhizodes* Auersw., now considered to be a Rhizoctonia. 1965 Bell & Coombe tr. *Strasburger's Textbk. Bot.* 506 *Corticium solani* causes the *Rhizoctonia*-disease (a root rot) of potatoes. 1977 J. L. Harper *Population Biol. Plants* vi. 174 Density-dependent mortality is well known in forest nurseries where it is usually associated with pathogenic activity (e.g. Gibson (1956) showed that *Rhizoctonia* was responsible for killing a higher proportion of seedlings in overcrowded plots).

rhizoid, *sb.* Substitute for def.: A delicate root-like structure on the underside of many lower plants. (Further examples.)

1914 M. Drummond tr. *Haberlandt's Physiol. Plant Anat.* v. 226 The trichomes in question..are endowed with the properties of typical root-hairs, but in addition possess many of the capacities of roots; in particular, rhizoids agree with roots in being sensitive to the influence of light, gravity and moisture, whereas genuine root-hairs are quite free from these forms of irritability. 1952 J. Clegg *Freshwater Life* iv. 88 Stoneworts have no roots in the strict botanical sense, but root-like structures, or rhizoids, penetrate the mud and serve to anchor the plants. 1965 Bell & Coombe tr. *Strasburger's Textbk. Bot.* 70 Roots are lacking in the Bryophyta, but in their place most possess delicate rhizoids. 1977 [see *rhizine].

rhizomic (raizōᵘ·mik), *a.* *Bot. rare.* [f. Rhizom(e + -ic.] Pertaining to or of the nature of a rhizome.

1902 *Nature* 21 Aug. 399/2 The examination of rhizomic material of the unique fern *Matonia pectinata*.

rhizoplane (rai·zoplēⁱn). *Ecol.* [f. Rhizo- + Plane *sb.*³] (See quot. 1949.)

1949 F. E. Clark in *Adv. Agronomy* I. 246 It would be more logical to speak of this microflora in terms of the root surfaces than in terms of the soil adjacent to the roots of plants. This goal can be obtained by the use of the term *rhizoplane*. The rhizoplane is defined as the external surfaces of plant roots together with any closely adhering particles of soil or debris. 1961 *Canad. Jrnl. Bot.* XXXIX. 175 Infestation of the roots of tomato seedlings with naturally occurring rhizoplane fungi. 1975 *Sydowia Ann. Mycol.* XXVII. 312 (*heading*) Fungi isolated from rhizosphere, rhizoplane and soil.

rhizoplast (rai·zoplast). *Microbiology.* [f. Rhizo- + -plast.] A fibrous structure running from the nucleus of a protozoan to a blepharoplast (kinetosome).

1915 L. McCulloch in *Univ. Calif. Publ. Zool.* XVI. 3 In the central part of the hyaline body there is a large vesicular nucleus connected directly with the extranuclear organelles, the rhizoplast, 'kinetonucleus', flagellum, basal granule and the 'axostyle'. *Ibid.* 5 The rhizoplast is a faint line connecting the 'kinetonucleus' with the nucleus. 1947 *Ann. Rev. Microbiol.* I. 8 In the Polymastigina the anteriorly placed blepharoplast is the center of organization. From it originate the flagella, the rhizoplasts, the axostyles, [etc.]. 1976 Edington & Gilles *Path. in Tropics* (ed. 2) ii. 55 The eccentric spherical nucleus is about 1 μm in diameter and the rod-shaped rhizoplast may or may not be detectable.

rhizopod². Add: **2.** *Zool.* Usu. in L. form **rhizopodium** (pl. **-podia**). In Protozoa, a pseudopodium that branches and anastomoses to form a network.

1931 R. R. Kudo *Handbk. Protozool.* ii. 26 Rhizopodia. These are branching (reticulopodia) or anastomosing (syzopodia) temporary cytoplasmic projections. 1940 L. H. Hyman *Invertebrates* I. iii. 119 Thread-like pseudopodia that branch and anastomose into networks are termed reticulopods or rhizopods. 1958 G. A. Kerkut *Borradaile & Potts's Invertebrata* (ed. 3) ii. 17 Granules may often be seen to stream up and down the threads of the rhizopodia. 1972 M. S. Gardiner *Biol. Invertebrates* vi. 77/2 The rhizopodia of Foraminifera are characteristically filamentous and branching, with anastomoses of the branches forming a network.

Rhizopus (rai·zopəs). [mod.L. (C. G. Ehrenberg 1820, in *Nova Acta Physico-Med. Acad. Cæsareæ Leopoldino-Carolinæ Naturæ Curiosorum* X. 198), f. Rhizo- + Gr. πούς foot, in allusion to the form of its rhizoids.] A fungus of the genus so called (class Zygomycetes), the members of which are saprobic and include the bread-mould fungus and many that cause disease (*Rhizopus rot*) in fruit and vegetables.

1887 H. E. F. Garnsey tr. *A. de Bary's Compar. Morphol. & Biol. Fungi* v. 153 Branched gonidiophores..rise in a curve into the air from well-fed mycelia in a similar manner to the stolons of Rhizopus. 1912 *Jrnl. South-Eastern Agric. College, Wye* XXI. 187 On November 28th..the bruised pears showed signs of the *Rhizopus* rot and as other fungi began to appear the experiment was not continued. 1927 Gwynne-Vaughan & Barnes *Struct. & Devel. Fungi* 320 For rough cultures of *Rhizopus*, moist bread is used... 1953 F. T. Brooks *Plant Dis.* (ed. 2) vii. 115 The chief factors in preventing Rhizopus rot are temperature and humidity. 1957 [see *mucoraceous]. 1969 G. N. Agrios *Plant Path.* ix. 260 Attempts have been made to control *Rhizopus* rots by irradiating the fruit with gamma radiation. 1975 *Mycologia* LXVII. 194 Another problem was the luxuriantly tall, spreading growth of certain molds, such as *Rhizopus*, which obscured other fungi.

rhizosphere (rai·zosfīᵊr). *Bot.* [ad. G. *rhizosphäre* (L. Hiltner 1904, in *Arbeiten der deutsch. Landwirtschaftsges.* XCVIII. 69): see Rhizo- and Sphere *sb.*] The sphere of chemical and bacteriological influence of the roots of a plant.

1929 R. L. Starkey in *Soil Sci.* XXVII. 319 The plants withdraw considerable amounts of substances (principally inorganic) from the rhizosphere.., and eventually introduce large amounts of organic matter to the soil in the form of their dead tissues. ... A small epiphytic fern..common not only on trunks and branches, but also having a tendency to settle on leaves, to which it attaches itself by spreading roots and root hairs... Fungi living in the rhizosphere of the fern infect the tree-support and cause the die-back of the branches. 1972 J. G. Cruickshank *Soil Geogr.* vi. 175 At all sites, plant roots have a rhizosphere, created by their excreted organic substances and maintained by vast numbers of organisms. 1976 *Jrnl. R. Soc. Arts* CXXIV. 583/2 In some tropical species there is an association of nitrogen-fixing bacteria in the rhizosphere (not actually in the root itself, in nodules, but around the root) which are believed to exchange nitrogen and carbohydrates with the plant.

rho (rōᵘ). [Gr. ῥῶ.] **1.** The seventeenth letter of the Greek alphabet, P, ρ.

c1400 Mandeville *Trav.* (MS. Cotton Titus C. xvi, fol. 9ᵛ), An Abc of anoþer manere A. B. C...o..b..rho... 1611 [see *chi-rho]. 1746 T. Nugent tr. *De Port Royal's New Method of Learning Gk. Tongue* I. ii. 3 The Greeks have 24 letters, whose Figure, Name, and Power are as followeth:..Rho,..Sigma..Tau. 1883 I. Taylor *Alphabet* II. vii. 106 Even before the transmission of the alphabet to Italy the primitive Greek rho..which at Thera is hardly distinguishable from *delta*, began to be differentiated by a short tail. 1965 J. W. Wenham *Elements N.T. Gk.* i. 1 The alphabet:..ρ rho, r.

2. *Statistics.* Used to represent a coefficient of correlation calculated from the ranks of the data and answering to the product-moment coefficient.

[1910 C. Spearman in *Brit. Jrnl. Psychol.* III. 287 He [*sc.* K. Pearson] has worked out the relation between the coefficients of ranks..and those of measurements, assuming Gaussian distribution. $R = r = \sin(\pi/6)\rho$,..where ρ denotes the coefficient for ranks.] 1940 R. S. Woodworth *Psychol.* (ed. 12) iii. 89 'Rho' is a measure of correlation, known as the rank-difference measure. 1949 Q. McNemar *Psychol. Statistics* vi. 98 Rho is not as consistent from sample to sample as *r*, does not possess the mathematical advantages inherent in *r*, and therefore has been used only when the observations involve ranks, i.e. are not measures. 1970 *Jrnl. Gen. Psychol.* LXXXIII. 209 These two rankings were then correlated by use of Spearman's rho. 1978 ... II. 55 The computed rho of 0·9006 indicates that there is a high degree of systematic ordering between male and female rest room graffiti.

3. *Physics.* In full *rho meson*. A meson with zero hypercharge, unit isospin, unit spin, and negative parity that has a mass of about 770 MeV (1506 times that of the electron) and on decaying usu. produces two pions. Freq. represented by ρ.

1961 A. Bereia et al. in *Physical Rev. Lett.* VI. 370/1 The possibility of detecting experimentally the $T = 0$ unstable particle (which we shall refer to as ρ⁰).. 1970 *New Scientist* 16 Apr. 105/2 The rho meson is produced in this case from high energy photons incident on the complex nuclei. The rho is scattered in the nucleus before decaying. 1973 L. J. Tassie *Physics of Elementary Particles* x. 113 In subsequent experiments, the ρ* has also been observed. 1977 S. Weinberg *First Three Minutes* vii. 141 Rho mesons behave as if they consist of a quark and an antiquark.

rhochrematics (rōᵘkrĕmæ·tiks), *sb. pl.* [f. Gr. ῥοή or ῥόος stream, flow + χρηματ-, χρῆμα thing (in pl., goods) + -ic 2.] The science of the flow of materials and products from their sources to their final disposal. Hence **rhochremati·cian**, one who studies or makes use of this science.

1960 (*title*) Rhochrematics, a scientific approach to the management of material flows. 1960 *Aeroplane* XCIX. 405/1 The object of this discourse on 'rhochrematics' (a word which you won't find in your dictionary) was to show that the interaction among the separate functions of material flow and distribution and manufacturing costs must be studied and co-ordinated in order to arrive at an optimum product cost. 1962 *Flight International* LXXXI. 222/1 To be 'with' air freight you must be able to throw away words like rhochrematics, inventory management, regional warehousing, etc. 1977 *Containerisation Internat.* June 59/2 As for the Rhochrematicians, if the container had not existed, they would have had to invent it.

rhodamine (rōᵘ·dămīn). *Chem.* Also **Rhodamine.** [ad. G. *rhodamin* (see E. Weingärtner 1887, in *Chem.-Zeitung* 25 Dec. 1620): see Rhodo- and Amine.] Any of a class of synthetic xanthene dyestuffs, chiefly reds, which are obtained by condensation of phthalic anhydride with *m*-aminophenols.

1888 *Jrnl. Soc. Chem. Industry* 31 May 386/2 The Badische..have lately introduced a new colour Rhodamine. 1892 *Ibid.* 30 Apr. 345/2 Certain benzylated *m*-amidophenol derivatives can be condensed with phthalic anhydride to yield new basic rhodamine dyestuffs. 1903 C. Salter tr. *G. von Georgievics' Chem. Dyestuffs* 199 On eliminating alkyl groups from Rhodamines by means of [etc.]. 1930 C. L. Bird *Theory & Pract. Wool Dyeing* ix. 93 Rhodamine B is widely used, in spite of its very moderate fastness to light, on account of its brilliant shade, level-dyeing properties, fastness to washing and sulphur stoving, and cheapness. 1968 E. N. Abrahart *Dyes & Intermediates* x. 226 The use of Rhodamine B as a food colour has been discontinued for a number of years on account of its suspected carcinogenic nature. 1973 *Physics Bull.* July 419/3 The successful operation of the cw rhodamine 6G dye laser. 1976 *Sci. Amer.* Feb. 133/2 My method of detecting the location of colorless components of the sample is to stain them with a water solution of Rhodamine B.

rhodanizing (rōᵘ·dănaiziŋ). *Chem.* [f. Rhod(ium² + -an + -ize + -ing¹.] (See quot. 1971.)

1936 *Discovery* July 218/1 The recently-discovered 'rhodanizing' process for rendering silver untarnishable by the use of rhodium, is comparatively simple to operate, and can be applied to old as well as new silver. 1971 T. C. Collocott *Dict. Sci. & Technol.* 1007/1 *Rhodanizing*, the process of electroplating with rhodium, especially on silver, to prevent tarnishing.

rhodanthe (rodæ·nþi). [mod.L. (J. Lindley 1835, in *Edwards's Bot. Reg.* XX. 1703), f. Gr. ῥόδον rose + ἄνθος flower.] An annual herb, *Helipterum* (formerly *Rhodanthe*) *manglesii*, belonging to the family Compositæ, native to Australia, and bearing pink everlasting flowers.

1835 J. Lindley in *Edwards's Bot. Reg.* XX. 1703 (*heading*) Captain Mangles's Rhodanthe. 1927 *Jrnl. R. Hort. Soc.* LII. 271 They [*sc.* everlasting flowers] were represented in the trial by three stocks of Acroclinium, three of Rhodanthe, [etc.]. 1959 *Times* 17 Jan. 9/5 Helichrysums, acroclinium, rhodanthes: all annuals and easy to raise in a greenhouse or warmed frame. 1962 [see *helipterum].

Rhode Island (rōᵘd ai·lənd). The name of an eastern state of the United States, used *attrib.* or *absol.* to designate plants or animals associated with the region, as **Rhode Island bent** = Red-top 2 (in Dict. and Suppl.); **Rhode Island greening**, a variety of green-skinned apple, or the tree producing it; **Rhode Island Red**, a domestic fowl belonging to the breed so called, distinguished by brownish-red plumage; **Rhode Island White**, a chicken belonging to a variety of the Rhode Island Red, having white plumage.

1790 S. Deane *New-Eng. Farmer* 123/1 The Rhode Island bent... 1898 *U.S. Dept. Agric. Yearbk.* 1898 494 Creeping bent..and Rhode Island bent..are much prized for lawns. 1929 J. W. Bews *World's Grasses* v. 198 The former [*sc. Agrostis tenuis*] is known as 'Rhode Island Bent'. 1795 J. Jay *Let.* 12 Dec. in *Columbia Library Columns* (1970) May 43 Ten are of a kind called Rhode Island Greenings. 1817 W. Coxe *View Cultivation Fruit Trees* 129 Jersey, or Rhode-Island Greening. Sometimes called the Burlington Greening. 1884 E. P. Roe *Nature's Serial Story* 400 Those umbrella-shaped trees are Rhode Island greenings. 1896 *Country Gentleman* (Albany, N.Y.) 16 Nov. 911/1 The Rhode Island Red is a breed of fowls which is exciting considerable interest. 1904 *Ohio Farmer* 5 Aug. 96/1 The Rhode Island Red has originated in this country. A certain Mr. William Tripp, then of Little Compton, in the State of Rhode Island, bought from a neighbour..a rose-combed Brown Leghorn cockerel to cross upon his flock of buff Malay hens. 1931 E. Caldwell *Amer. Earth* 76 We lived on a small farm and had..a flock of Rhode Island Red chickens. 1937 Rhode Island Red [see *Kerry]. 1977 J. Hodgson *Invention of World* (1978) iii. 51 A dozen Rhode Island Reds..looked after for their eggs. 1926 *Bull. U.S. Dept. Agric.* No. 1506. 9 The Rhode Island White..is identical with the Rose-Comb Rhode Island Red, except that the plumage should be pure white. 1949 F. B. Hutt *Genetics of Fowl* 16 Color is the sole difference between the Rhode Island Whites and Rhode Island Reds.

Rhodes (rōᵘdz). [The name of Cecil John Rhodes (1853–1902), British financier and imperialist.] **1.** Applied *attrib.* to scholarships awarded annually since 1902 to students from the U.S., the British Commonwealth, South Africa, and Germany, for study at the University of Oxford; also to the scholars who receive these awards.

1902 G. Calderon *Adventure Downy V. Green* xvi. 97 He had done three years at Harvard, and had come up with a Rhodes scholarship to Oxford. *Ibid.* 101 This is Mr. Cheney, Rhodes Scholar of Pussy... 1911 M. Beerbohm *Zuleika Dobson* viii. 129 The American Rhodes Scholars..are a noble, rather than a comfortable, element in the social life of the University. 1914 C. Mackenzie *Sinister St.* II. iii. ix. 688 Do you really think these Rhodes Scholars from America and Australia and Germany are going to alter us? 1949 *Cavalier Daily* (Univ. Virginia) 23 Oct. 1/2 ..have been selected as the University's eligible candidates for this year's Rhodes Scholarship. 1955 [see *rock *n.*¹ 5 d]. ..Scholars cannot reasonably be expected, in all cases, to be paragons of intellect, literary and scholastic attainments, sporting ability, moral force, public spirit and leadership. ... 1977 *Oxf. Univ. Gaz.* 3 June 850/1 The generosity of the Rhodes Trustees enables Lady Margaret Hall to invite applications for a Rhodes Visiting Fellowship for Women tenable at the college for two years starting in October 1978 or January 1979.

2. Rhodes (or **Rhodes'**) **grass**, a perennial grass, *Chloris gayana*, native to Africa and widely cultivated elsewhere as a pasture grass.

1915 R. Marloth *Flora S. Afr.* IV. 20 In some districts *Chloris gayana* (Rhodes' grass) and *Phalaris coerulescens*..have also been found very valuable. ..cultivated in the S.W. States of N. America, in Hawaii, and in Australia. 1940 F. D. Davison ..Settlers were felling scrub..burning it off and sowing it to rhodes-grass..making fine grass paddocks. 1962 *New Scientist* 1 Mar. 487/3 Since the beginning of the century, when it was introduced into Australia from its native home in Africa, Rhodes grass (*Chloris gayana*) has become the dominant pasture species. 1977 J. L. Harper *Population Biol. Plants* xi. 364 Hill examined the nitrogen relations of *Chloris gayana* (Rhodes grass).

Rhodesian (rōᵘdī·ʃiăn, -ʒăn), *a.* and *sb.* [f. (sense A. 1) the name of C. J. *Rhodes* (see *Rhodes) + -ian; (other senses) *Rhodesia* (see note) + -an.

Rhodesia (so called in the late nineteenth century after Cecil Rhodes) was orig. the name of a large British territory in southern Africa comprising North-Eastern and North-Western (later Northern) Rhodesia, and Southern Rhodesia. Northern Rhodesia became known on independence (1964) as Zambia, and Southern Rhodesia (then Rhodesia) became Zimbabwe on independence (1980).]

A. *adj.* † **1.** With pronunc. (rōᵘ·dziăn). Of or pertaining to C. J. Rhodes. *Obs.*

1891 *Review of Reviews* Jan. 8/2 He maintains that the Portuguese..are most objectionable people, whose elimination from South Africa is one of the planks of the Rhodesian platform. 1893 *Work & Workers* Nov. 450/2 Lo Bengula's forces, bent on revenge, are slowly moving towards the Rhodesian settlements. 1897 [see *Krugerism].

2. Of, pertaining to, or belonging to Rhodesia.

1895 *Rhodesian Mining & Finance Co. Ltd. Prospectus* 5 (*heading*) Form of Application for Shares. To the Directors of the Rhodesian Mining & Finance Company, Limited. 1899 M. M. Verney *Mem.* IV. vii. 258 He might be in the 'moral meridian' of Rhodesian politics to-day. 1902 *Encycl. Brit.* XXXII. 232/1 In November 1897 the Rhodesian Railway..had reached Bulawayo. 1935 *Discovery* Nov. 315/2 Here is an argument that should be retained in the mind of the exploiters of the countryside, whether Rhodesian, American, or even British. 1956 *African Affairs* LV. 70 MacDonald..commanded a Rhodesian battalion in the second world war. 1974 *Standard Encycl. S. Afr.* X. 141/2 In the Second World War it was decided to use the Rhodesian military potential in a somewhat different fashion. Rhodesian youths..made good officers, especially in colonial units. 1977 *Whitaker's Almanack 1978* 596/2 The Rhodesian prime minister (Mr Ian Smith) announced major relaxations in the racial laws.

3. Special collocations. **a.** *Palæont.* *Rhodesian man*, a fossil hominid usually considered to be an early type of *Homo sapiens*, known from remains found at Broken Hill, Northern Rhodesia, in 1921; *Rhodesian skull*, the skull of Rhodesian man.

1921 A. S. Woodward in *Nature* 17 Nov. 372/2 The newly discovered Rhodesian man may therefore revive the idea that Neanderthal man is truly an ancestor of *Homo sapiens*. 1937 *Jrnl. R. Anthrop. Inst.* LXVIII. 51 *Javanthropus* appears to be more primitive than the European Neandertal man and has quite a resemblance to the Rhodesian Man. 1946 F. E. Zeuner *Dating Past* ix. 296 Another undated Neandertaloid is the Rhodesian Skull. 1952 *Proc. Prehist. Soc.* XVIII. 125 This fossil [*sc.* the Florisbad skull] is now generally accepted as being a specialised type not much older than Rhodesian man. 1973 B. J. Williams *Evol. & Human Origins* xi. 185/1 The Saldanha skull provided another specimen almost identical to that of Rhodesian Man. *Ibid.*, The Rhodesian skull differs in minor features from the classic Neandertal of Europe. 1978 *Sci. Amer.* Sept. 150/2 Rhodesian man..a form no longer considered a distinct species.

b. *Rhodesian* (*tick*) *fever, redwater, sleeping sickness* = *East Coast fever* s.v. *east D. 1 b.

1904 Rhodesian redwater, tick fever [see *east D. 1 b]. 1904 *Jrnl. Trop. Med.* VII. 322/1 We are dealing with true Texas fever and not the atypical South African or Rhodesian fever. 1962 Gordon & Lavoipierre *Entomol.* xxix. 184 The chief vector of Rhodesian sleeping sickness..is *G*[*lossina*] *morsitans*.

c. *Rhodesian lion dog* or *ridgeback*, a large, short-coated, light brown dog belonging to the breed so called, distinguished by a ridge of hair along the middle of the back, growing in the opposite direction to the rest of the coat.

1930 *Observer* 9 Feb. 13/2 Such rarities as..Rhodesian lion dogs, distinguished by the ridge of hair running along the back the reverse way of the rest of the coat. 1937 *Our Dogs* 10 Dec. 886/2 The belief that the Rhodesian Ridgeback is the direct result of crossing the old Cuban Bloodhound with the Hottentot Hunting Dog arose a considerable time ago. 1948 C. L. B. Hubbard *Dogs in Britain* xxi. 373 The Rhodesian Ridgeback obtains its name from this crest of hair which is present in all true specimens of the breed. *Ibid.* 375 It [*sc.* the Rhodesian Ridgeback] was for some years known in Britain as the Rhodesian Lion Dog. 1955 J. Packer *Valley of Vines* iv. 44 Their house stood empty save for the Native boy, Elias, and two Rhodesian ridgebacks. 1979 A. Price *Tomorrow's Ghost* xi. 182, I think we ought to have dogs... A pair of Rhodesian Ridgebacks—'Lion Dogs'.

B. *sb.* **1.** A native or inhabitant of Rhodesia. (Until independence usu. specifically a white inhabitant of Rhodesia.)

1897 *Rhodesia* 27 Nov. 62/2 To avow that the average Rhodesian is any better than other frail humanity would be untrue; but they are no worse. 1901 G. Grey in *Geogr. Jrnl.* XVIII. 64 No one could fail to be impressed by the magnificence of the mighty river, much less a Rhodesian. 1936 *Jrnl. R. African Soc.* XXXV. 47 It is well..that the present generation of Rhodesians should be occasionally reminded that their prosperous condition has not been attained without much endurance. 1956 *African Affairs* LV. 70 The attitude of most Rhodesians may be said to be that, although they see in advancement of the African a possible limitation of their own privileges, they would not wish that progress to be curbed. 1977 *Time* 8 Aug. 26/2, I [*sc.* Ian Smith] have the backing of the majority of Rhodesians. I believe Rhodesians are prepared to move, but not in the direction of a sellout.

† **2.** *pl.* *Stock Exchange.* The shares of Rhodesian mining companies. *Obs.*

1901 [see *Kaffir Circus* s.v. *Kaffir 4]. 1936 *Economist* 28 Mar. 726/1 Rhodesians also failed to share in the improvement in the Kaffir department, and prices tended to sag.

† **3.** A Rhodes Scholar. *Obs.*

Perhaps better regarded as formed on *Rhodes with pronunc. (rōᵘ·dziăn).

1903 *'Varsity* 3 Mar. 209 Herds of cattle..infest the streets on market-days... However, we may look for an improvement..when the stock whips of the Rhodesians have become common weapons. 1905 *Athenæum* 16 Dec. 837/2 The Rhodesians are well spread out amongst the Colleges. Of the present year Christ Church takes 7; Balliol and Oriel, 5;.. and the other Colleges 3 or less.

rhodesite (rōᵘ·dzəit). [f. the name of Cecil *Rhodes* (see *Rhodes) + -ite¹.] **1.** (With capital initial.) An admirer or supporter of Rhodes or his policies. *rare.*

1897 [see *Krugerism]. 1900 *Oldham Chron.* 15 Dec. 5/3 Sir Alfred Milner has displayed a singular lack of judgment. The High Commissioner has added one more to the many evidences on record of how completely he has been captured by the [Jameson] Raidites and Rhodesites.

2. *Min.* A hydrous silicate of calcium, sodium, and potassium, $(Ca,Na_2,K_2)_8Si_{16}O_{40}\cdot 11H_2O$, found as fibrous white orthorhombic crystals.

1957 E. D. Mountain in *Mineral. Mag.* XXXI. 607 A small specimen at first thought to be natrolite has proved to be a new mineral, for which the name rhodesite is proposed. 1969 *Amer. Mineralogist* LIV. 251 Rhodesite occurs as aggregates of radiating white, silky fibers associated with magadiite and trioctahedral montmorillonite in an altered silicic lava.

Rhodian, *a.* and *sb.* Add: **A.** *adj.* **c.** Designating a rhetorical style characteristic of the ancient Rhodian school of oratory; also, of the school itself.

1852 Disraeli *Ld. George Bentinck* xii. 203 The debate was opened by a dashing speech from Mr. McCarthy, worthy of the historical society in the most fervent hour of its Rhodian eloquence. 1876 —— in *Hansard Commons* 11 Aug. 1138 After the Rhodian eloquence to which we have just listened, it is rather difficult for the House to see clearly the point which is before it. 1954 H. Caplan in *Cicero's Ad Herennium* 193 The Rhodian school opposed the overloud delivery of the Asiatic orators. 1969 R. A. Lanham *Handlist Rhet. Terms* 90 *Rhodian style*, the middle style between Atticism and Asiatismus.

d. Designating a style of pottery characterized by brilliant red pigment, formerly thought to derive from Rhodes. Properly, a variety of later Isnik ware.

1899 R. Glazier *Man. Hist. Ornament* 80 In the Rhodian Ware..the purple is replaced by a fine opaque red of great body, called Rhodian red, produced from Armenian bole. 1900 F. Litchfield *Pott. & Porc.* vii. 244 We can readily recognise as a distinct class of decorative ware what is now known as Rhodian. 1957 [see *Isnik]. 1977 *Jrnl. R. Soc. Arts* CXXV. 482/2 A large Bichrome V ware crater, decorated on both sides between the handles with animals in what is certainly a perfect imitation of the Rhodian style of the seventh century BC.

rhodie, var. *rhody.

rhodinol (rōᵘ·dinǫl). *Chem.* [a. G. *rhodinol* (U. Eckart 1891, in *Arch. d. Pharm.* CCXXIX. 364), f. Gr. ῥόδιν-ος of or from roses (f. ῥόδον rose): see -ol.] An open-chain terpenoid primary alcohol, $C_{10}H_{20}O$, which is a red liquid first isolated from rose oil, and now known to be a mixture of two isomeric forms and to be identical with citronellol except in respect of the relative proportions of the isomers.

1892 *Jrnl. Chem. Soc.* LXII. I. 203 The author proposes the name rhodinol for the compound. 1894 *Ibid.* LXVI. I. 141 When oil of pelargonium is subjected to careful fractional distillation under reduced pressure, it yields rhodinol, ..identical..with the rhodinol obtained from oil of roses. 1929 [see *menthone]. 1953 D. H. R. Barton in E. H. Rodd *Chem. Carbon Compounds* IIb. xii. 496 The question of the constitution of citronellol and rhodinol is still a matter of controversy. 1956 I. L. Finar *Org. Chem.* II. viii. 265 Rhodinol is identical with citronellol, but the proportions of the two forms are different from those which occur in citronellol; the identity of citronellol and rhodinol is shown by the products of ozonolysis. 1959 P. de Mayo *Mono- & Sesquiterpenoids* i. 40 Rhodinol, isolated from rose oil, was for a long time considered to be distinct from citronellol, but it would appear that pure rhodinol and pure citronellol are identical.

Hence **rho·dinal** [a. G. *rhodinal* (op. cit., p. 373): see *-al³], the aldehyde, $C_{10}H_{18}O$, which is obtained by oxidation of rhodinol and is an analogous mixture of isomers resembling citronellal.

1900 *Jrnl. Chem. Soc.* LXXVIII. I. 452 Analyses are quoted in further proof of the isomeric change of rhodinal into menthone. 1929 [see *hydroxycitronellal* s.v. *hydroxy- 3]. 1951 A. W. Johnson et al. in E. H. Rodd *Chem. Carbon Compounds* Ia. viii. 504 Citronellal and its isomer rhodinal are olefinic aldehydes. 1956 I. L. Finar *Org. Chem.* II. viii. 265 A detailed study of rhodinal has shown that this compound is identical with citronellal, but consists of a mixture of the two forms in different proportions.

rhodium². Add: **b.** *rhodium plating, -plated* adj. (hence *rhodium-plate* vb. trans.).

1962 L. S. Sasieni *Optical Dispensing* i. 4 A very satisfactory alternative way of producing a frame with a white finish is to rhodium-plate a yellow gold filled frame. *Ibid.* vii. 170 A rhodium-plated or gold-filled supra. 1971 *Flying* Apr. 112/3 (Advt.), Pilots Sunglasses. Air Force style rhodium plated adjustable metal frame. 1978 *Neiman-Marcus Christmas Bk.* 29 The glitter of rectangular cut crystals, hand-set in rhodium-plated 2″ bar pins. 1950 *Thorpe's Dict. Appl. Chem.* (ed. 4) X. 539/1 Rhodium plating baths have come into extensive use during the past 20 years for producing a non-tarnishing finish on silverware and on electrical-contact materials. 1962 L. S. Sasieni *Optical Dispensing* i. 23 The effect of a white gold filled frame is often produced by rhodium-plating.

rhodo (rōᵘ·do). Colloq. abbrev. of Rhododendron 2. Cf. *rhody.

1920 *Punch* 19 May 399/2 *Jack's* passion for abbreviation ('rhodos' for rhododendrons) being the only ground of quarrel. 1939 *N. & Q.* CLXXVI. 412/1 *Narciss*... This is a common abbreviation among gardeners, florists, and others, like 'Chrysanths', 'Rhodos', and other curtailments of long botanical names. 1976 *Country Life* 5 Feb. 305/2 Budding rhodos and camellias could still be nipped.

rhodo-. Add: † **rho·dophane** [*-phane* after *diaphane*] = *rhodopsin; **rhodophycean** (-fəi·siăn) *a.* *Bot.* [ad. mod.L. *Rhodophyceæ*, f. Gr. φῦκος seaweed], of or pertaining to the class Rhodophyceæ, which comprises the red algæ and is coextensive with the division Rhodophyta; **rho·dophyte** *sb.* *Bot.* [ad. mod.L. *Rhodophyta*: see -phyte], a member of the division Rhodophyta, which comprises the red algæ and is coextensive with the class Rhodophyceæ; *a.*, of or pertaining to this division; **rho·doplast** *Bot.* [a. G. *rhodoplast* (A. F. W. Schimper 1885, in *Jahrb. f. wissensch. Bot.* XVI. 40): see -plast], a chromoplast containing a red pigment, phycoerythrin, found in red algæ.

1878–9 W. C. Ayres tr. W. Kühne in *Jrnl. Physiol.* I. 115 We have found it expedient to wash with slightly warmed alcohol, which though it removes a little Rhodophane, as we may call the third pigment, takes up without failure the last remains of xanthophane. 1951 Rhodophane [see *lipochrome* s.v. *lipo-]. 1935 J. E. Tilden *Algae* iii. 27 A long time must have elapsed between the Cyanophycean and Rhodophycean periods, in which the blue-green and the red algae in turn constituted the dominant flora of the sea. 1938 G. M. Smith *Cryptogamic Bot.* I. viii. 295 The gradual increase in temperature of surface waters as one passes from polar to tropical regions is correlated with a change in composition of the rhodophycean element in the flora. 1964 *Amer. Jrnl. Bot.* LI. 580/1 Rhodophycean algae which have similar external form..do not have a comparable internal organization. 1967 *Virginia Jrnl. Sci.* XVIII. 110 (*heading*) *Porphyridium sordidum* Geitler in America, with comments on the distribution of other North American rhodophytes. 1974 R. Y. Stanier in Carlile & Skehel *Evolution in Microbial World* 223 Both in terms of nucleo-cytoplasmic structure and in terms of the overall composition of their pigment systems, rhodophytes and cryptophytes show many differences. 1976 *Nature* 22 Jan. 176/3 Knoll and Barghoorn's observations..counsel caution in interpreting the Bitter Springs fossil biota as containing *bona fide* remains of chlorophyte or rhodophyte algae. 1886 *Jrnl. R. Microsc. Soc.* 640 For the chromoplasts of the Phæophyceæ the author proposes the term phæoplasts; for those of the Florideæ rhodoplasts. 1932 R. E. Torrey *Gen. Bot.* x. 184 In the brown and red algæ the pigment is brown or red and the plastids are called respectively phæoplasts and rhodoplasts. 1965 Bell & Coombe tr. *Strasburger's Textbk. Bot.* III. 466 The Rhodophyceae are usually red or violet, rarely dark purple or reddish-brown. Almost without exception the cells have a single nucleus and contain numerous simple chromatophores or rhodoplasts..in which chlorophyll..and the associated carotinoids in the lamellae of the plastids are masked by a red, strongly fluorescent, water-soluble pigment, phycoerythrin.

Rhodoid (rōᵘ·doid). Also **Rhodoïd, rhodoid.** [a. F. *Rhodoïd*, f. L. *Rhod*(*anus* the River Rhône + *-oïd*(*e* -oid.] A proprietary name for an incombustible thermoplastic derived from cellulose acetate.

1918 *Trade Marks Jrnl.* 5 June 590 *Rhodoid*... Celluloid in sheets, rods, tubes and similar forms. Société Chimique des usines du Rhone (anciennement Gilliard, P. Monnet et Cartier).., France; manufacturers.—18 Dec. 1917. 1934 *Archit. Rev.* LXXV. 20/2 The sparkling black rhodoid covers of this book. 1934 *Times Lit. Suppl.* 8 Feb. 93/1 Bound in black 'Rhodoid', a non-inflammable celluloid, the book opens easily. 1937 *Discovery* Jan. 25/2 The accompanying illustration of one of the artist's oil pictures on rhodoid. 1952 *Chem. Abstr.* XLVI. 2868 Studies on the effect of the base on crystn. of Au films were extended to: KCl,..Rhodoid, Plexiglas, and glass... It [*sc.* the resistivity] is higher on Rhodoid and Plexiglas than on glass. 1957 *Official Gaz.* (U.S. Patent Office) 14 May TM 49/2 *Rhodoid*... For raw or partly prepared plastic material and particularly plastic material having a cellulose acetate base in the form of powder, blocks, rods, sheets, leaves, plates, and plaques.

rhodologist (rōᵘdǫ·lŏdʒist). [f. Gr. ῥόδον rose + -ologist.] One who studies roses, esp. the taxonomic relationships of the members of the genus *Rosa*.

1911 *Jrnl. Bot.* XLIX. 298 Several important works on Scandinavian roses have been put forth by the distinguished Swedish rhodologist, Dr. S. Almquist. 1920 *New Phytol.* XIX. 156 *R*[*osa*] *hibernica*..is now regarded by all rhodologists as a hybrid. 1924 A. H. Wolley-Dod *Roses of Britain* 3 Continental rhodologists have expressed the opinion that the genus reaches its greatest degree of complication in Britain.

rhodopsin (rodǫ·psin). *Biochem.* [a. G. *rhodopsin* (Ewald & Kühne 1878, in *Untersuchungen aus d. physiol. Inst. d. Univ. Heidelberg* I. 181), f. *rhod-o* Rhodo- + Gr. ὄψ-ις sight, vision + *-in* -in¹.] The brilliant purplish-red light-sensitive pigment found in the retina of man and most other animal species, which consists of a protein (opsin) bonded to a prosthetic group (retinal) which is ultimately liberated by the action of light in the photochemical process of vision; = *visual purple* s.v. *visual *a.* and *sb.*

1886 [in Dict. s.v. Rhodo-]. 1937 *Nature* 12 June 1018/1 In every detail so far examined, the porphyropsin system faithfully reproduces the behaviour of the rhodopsin system, but with quite different components. 1959 S. Duke-Elder *Parsons' Dis. Eye* (ed. 13) iii. 26 Rhodopsin is a chromoprotein, the molecule of which consists of a reactive part, a chromophore, responsible for the preferential absorption of light, attached to a protein which acts essentially as a support. 1967 *New Scientist* 9 Feb. 330/2 Rhodopsin, the pigment of the retinal cones, has its absorption maximum at a wavelength of 5550 angstroms. 1968 M. Pyke *Food & Society* vii. 93 One of the chemical functions of vitamin A is known to be the re-conversion of the retinal pigment, rhodopsin [*printed* rhodospin], which becomes bleached in bright light, to its receptive state as visual purple. 1976 *Sci. Amer.* July 108/1 In all animals, invertebrates and vertebrates alike, the visual pigment rhodopsin is present in the photoreceptor membrane of the visual cells.

rhodora (rodōᵊ·ră). [mod.L. (Linnæus *Species Plantarum* (ed. 2, 1763) 561), a. L. *rhodōra* the plant *Filipendula ulmaria*, f. Gr. ῥόδον rose.] A deciduous shrub, *Rhododendron canadense*, formerly the only species of the genus so called, belonging to the family Ericaceæ, native to north-eastern North America, and bearing rose-purple or white flowers early in spring. Also *attrib.*

1786 J. Abercrombie *Gard. Daily Assistant* 32/2 Rhodora or Canada rosebay. 1838 [see *kalmia]. *c* 1859 E. Dickinson *Poems* (1955) I. 101 The Crocus stirs her lids—Rhodora's cheek is crimson. 1869 J. G. Fuller *Uncle John's Flower Gatherers* 59 The Azaleas..are beautiful cousins of the Rhodora. 1942 W. R. van Dersal *Ornamental Amer. Shrubs* vi. 84 Valued for its early, bright rose-purple flowers and rich in its associations is the Rhodora, *Rhododendron canadense*. 1951 E. Graham *My Window looks down East* xx. 176 The lavender-pink Rhodora is in bloom. 1958 G. A. Petrides *Field Guide Trees & Shrubs* 363 Rhodora Azalea... A low shrub with blunt-tipped somewhat hairy leaves.

rhodusite (rōᵘ·dŭsəit). *Min.* [ad. G. *rhodusit* (H. von Foullon 1891, in *Sitzungsber. d. k. Akad. d. Wissensch.* (Math.-Nat. Classe) C. 176), f. *Rhodus* L. name of the Mediterranean island Rhodes: see -ite¹.] A fibrous variety of riebeckite in which some of the iron is replaced by magnesium.

1894 *Jrnl. Chem. Soc.* LXVI. II. 461 He [*sc.* von Foullon] describes an asbestos-like variety of glaucophane... For this he proposes the name of rhodusite. 1965 E. W. Heinrich *Microsc. Identification of Minerals* x. 271 There are.. varieties of riebeckite (rhodusite) in which Mg replaces Fe^{2+} to a variable extent.

rhody (rōᵘ·di). *colloq.* Also 9 **roddy**; **rhodie.** [f. Rhod(odendron + -y⁶.] = Rhododendron 2. Cf. *rhodo.

1851 H. Mayhew *London Labour* I. 133/2 I've bought roddies, as I calls them (rhododendrons), at 4*s.* a dozen. 1937 Partridge *Dict. Slang* 696/1 *Rhody, -ie*, a rhododendron: 'nursery' and familiar coll.: C. 20. 1943 C. S. Churchill *Let.* 17 May in M. Soames *Clementine Churchill* (1979) xxii. 336 Your favourite rhodie was in full bloom. 1951 W. Sansom *Face of Innocence* viii. 106 Among the rhodies a polishing and a stoking were abroad. 1973 *Country Life* 25 Jan. 231/3 I'm overrun by..*Rhododendron ponticum*... The pontic rhodie is drear indeed.

rhombencephalon (rǫ:mbense·fălǫn). *Anat.* [a. G. *rhombencephalon*: see Rhombo-, Encephalon.] The hind-brain, comprising the metencephalon together with the myelencephalon.

1897 C. L. Dana *Textbk. Nerv. Dis.* (ed. 4) i. 3 The posterior vesicle develops into the rhombencephalon or posterior brain. 1900 A. Hill tr. *H. Obersteiner's Anat. Central Nerv. Syst.* (ed. 2) ii. 48 His has divided the brain somewhat differently; into (1) Rhombencephalon, which includes the after-brain, hind-brain, and isthmus..(2) Mesencephalon..(3) Prosencephalon. 1934 R. R. Grinker *Neurol.* i. 12 Division of the rhombencephalon into metencephalon (cerebellum), and myelencephalon (medulla). 1962 *Gray's Anat.* (ed. 33) 976 The rhombencephalon comprises the medulla oblongata, pons and cerebellum; its cavity is the fourth ventricle. 1966 *McGraw-Hill Encycl. Sci. & Technol.* IX. 58/2 In dorsal view the ventricle of the myelencephalon, together with the metencephalon, has a rhomboid appearance, which is why this brain section is sometimes called the rhombencephalon.

Hence **rho:mbencepha·lic** *a.*

1954 Penfield & Jasper *Epilepsy & Functional Anat. of Human Brain* vii. 285 (*heading*) Cerebellar vertigo (rhombencephalic).

rhombic, *a.* Add: **1.** *spec.* applied to a radio aerial with a horizontal rhombic shape. Also *ellipt.* as *sb.*

1935 *Proc. IRE* XXIII. 24 The structure was descriptively termed the 'diamond-shaped' antenna.., but it has since become known as the 'rhombic' antenna and will be so designated here... Checking measurements were made on small-scale rhombic antennas operating at correspondingly short wavelengths. 1950 K. Henney *Radio Engin. Handbk.* (ed. 4) xiv. 653 When two rhombics are used in broadside as a single system, one common support can be used without balance problems, when fed symmetrically. 1968 *Radio Communication Handbk.* (ed. 4) xiii. 62/1 The average gain of a rhombic is 3db greater than that of a *V* aerial with the same legs and this is to be expected because the rhombic is twice as large. *Ibid.* 62/2 The design of *V* and rhombic aerials is quite flexible and both types will work over a 2 : 1 frequency range. 1975 D. G. Fink *Electronics Engineers' Handbk.* xxii. 64 The rhombic antenna is in wide use.

rhombiferan (rǫmbi·fĕrăn), *a.* and *sb.* *Palæont.* Also **Rhombiferan.** [f. as Rhombifer + -an.] **A.** *adj.* Of or pertaining to a group of cystoid echinoderms whose calyx plates are penetrated by groups of tubes forming a rhombic pattern. **B.** *sb.* A rhombiferan echinoderm.

1920 H. L. Hawkins *Invertebr. Palæont.* II. iii. 160 Late forms of Rhombiferan Cystids, and perhaps of Amphoridea, have been reported from the Devonian, but they are inconspicuous. 1962 D. Nichols *Echinoderms* xi. 136 A section of a pore-rhomb..shows that in some rhombiferans the depressions between the pores of each pair were deep grooves. 1966 *McGraw-Hill Encycl. Sci. & Technol.* IV. 366/1 (*caption*) Rhombiferan cystoid (*Echinoencrinus*, Ordovician), showing pore rhomb and anus on side of calyx. 1978 *Jrnl. Paleont.* LII. 717/2 The elongate, cylindrical stems of *Caryocrinites ornatus*..resemble those of many crinoids and differ from the short conical stalks typical of rhombiferans.

rhombochasm (rǫ·mbokæz'm). *Geol.* [f. Rhombo- + Gr. χάσμ-α yawning hollow.] (See quot. 1958.)

1958 S. W. Carey in *Continental Drift* (Geol. Dept., Univ. of Tasmania, Hobart) 192 Rhombochasm..will be used for a parallel-sided gap in the sialic crust occupied by simatic crust, and interpreted as a dilatation. 1976 *Nature* 15 Apr. 590/1 Figure 3 *a* shows a rhombochasm, like the Gulf of California, forming behind the arc and marginal basin fragment.

rhomboclase (rǫ·mboklēⁱz, -s). *Min.* [ad. Hungarian *rhomboklas* (J. S. Krenner 1891, in *Magyar tudományos Akad. Értes.* II. 96), f. *rhombo-* Rhombo- + Gr. κλάσ-ις breaking, cleavage (cf. Orthoclase).] A hydrated ferric acid sulphate, $HFe(SO_4)_2\cdot 4H_2O$, found as colourless or pale-coloured thin tabular crystals.

1910 *Mineral. Mag.* XV. 429 (*heading*) Rhomboclase. 1938 *Amer. Mineralogist* XXIII. 742 Rhomboclase occurs at Alcaparrosa, near Cerritos Bayos [N. Chile], intimately associated with szomolnokite and roemerite. 1968 I. Kostov *Mineralogy* 499 Rhomboclase is orthorhombic-dipyramidal, occurring as colourless to white or greenish thin tabular crystals with perfect {001} cleavage.

rhombogen. Add: [ad. F. *rhombogène* (E. van Beneden 1876, in *Bull. de l'Acad. R. de Belgique* XLI. 1195).] (Earlier and later examples.)

1883 [see *infusorigen]. 1899, 1972 [see *nematogen* s.v. *nemato-].

Rhone (rōᵘn). Also **Rhône.** The name of a river which runs from Switzerland, through southeastern France, to the Mediterranean Sea, used *attrib.* and *absol.* to designate wines made from grapes growing in the Rhone valley (esp. between Lyons and Avignon).

[1833 C. Redding *Hist. Mod. Wines* v. 124 The red wines Glun, Châteaubourg, Soyons, Tournon, St. Jean de Musois, Vion, and others, are Rhone flavoured wines.] 1852 E. Twisleton *Let.* 9 Oct. (1928) iv. 56, I have now fallen peaceably into liking Bordeaux, Burgundy, and the Rhone wines in all their variety. 1935 A. E. Housman *Let.* 21 Sept. (1971) 376 There are also many good restaurants and some excellent Rhone wine. 1963 L. Charteris *Saint in Sun* (1964) 184 You and I..can wash down our *cailles* with a red Rhône. 1965 A. Sichel *Penguin Bk. Wines* III. 165 The general name 'Rhône wines' can be misleading, since some, grown in the *département* of the Rhône, belong to other appellations, and some wines with the appellation 'Rhone' come from the valley of the Rhône in other *départements*. 1976 *Times* 3 Apr. 13/4 The curious white Rhônes that are made from the Viognier grape.

rhoorkee, var. *Roorkee.

Rhoosian (rū·ʃiăn), *sb.* and *a.* Also **Rooshan, Rooshian, Roosian.** Joc. var. of Russian *sb.* and *a.*

1843 [see *Prooshan *a.* and *sb.*]. 1864 *Daily Tel.* 26 July 5/3, I never did 'room' with a Rooshian before, and I'd like to know them stript. 1878 W. S. Gilbert *H.M.S. Pinafore* II. 28 For he might have been a Roosian, A French, or Turk, or Proosian, Or perhaps Itali-an! 1898 J. D. Brayshaw *Slum Silhouettes* 151 I'm off to the Cattle Market to buy one o' those little Rooshian ponies. 1914 W. de la Mare in *Sat. Westm. Gaz.* 22 Aug. 2/3 With Dreadnoughts here, two kinds of French, the Rooshans at your heel. 1940 E. Pound *Cantos* lix. 86 And the rhoosians (Orosians) served a sort of lunch To the chinese ambassadors. 1974 F. Selwyn *Cracksman on Velvet* II. 70 The Rhoosians couldn't stop him at Inkerman and nor will you.

rhopalic, *a.* Add: (Further examples.) Also as *sb.*

1943 J. T. Shipley *Dict. World Lit.* 482/1 *Rhopalic verse*, ..wedge verse, in which each word is one syllable longer than the preceding one. Occasionally, of a stanza in which each line is a foot longer than the preceding line, as Crashaw's *Wishes to His Supposed Mistress*. 1956 [see *epanaleptic* adj. s.v. *epana-]. 1975 W. R. Espy *Almanac of Words at Play* p. xxii, *Rhopalic*, a snowballing line or passage in which each successive word has one more syllable (or letter) than the last.

rhopalium (rōᵘpēⁱ·liǫ̆m). *Zool.* Pl. **-ia.** [mod.L., f. Gr. ῥόπαλον club.] Each of a number of marginal sense-organs in some jellyfish. Hence **rhopa·lial** *a.*

1888 Rolleston & Jackson *Forms Animal Life* (ed. 2) 782 These tentacles shorten, and their basal portions are converted into the sensory bodies or rhopalia of the Medusa. 1979 *Nature* 13 Sept. 141/2 Swimming in scyphozoans is under control of rhopalial pacemakers. *Ibid.*, Jellyfish stopped swimming when we removed the rhopalia. *Ibid.*, Motor impulses originating in the rhopalium and conducted into the subumbrellar nerve-net were recorded.

rhopheocytosis (rōᵘfi:osəitōᵘ·sis). *Biol.* [ad. F. *rhophéocytose* (Policard & Bessis 1958, in *Compt. Rend.* CCXLVI. 3196), f. Gr. ῥοφεῖν to gulp down: cf. Phagocytosis and *pinocytosis.] The process whereby a cell absorbs material into small vacuoles without first sending out projections.

1962 *Blood* XIX. 645 The cell does not extend any veils as is characteristic of micropinocytosis, but appears to 'aspirate' the external material. This process, therefore, appeared to deserve a special designation, and coined [*sic*] for it the term 'rhopheocytosis'. 1964 D. W. Allen in Bishop & Surgenor *Red Blood Cell* viii. 324 The authors assume that the ferritin is transferred from reticulum cell to erythroblast by a form of micropinocytosis which the authors call 'rhopheocytosis'. 1974 *Nature* 15 Feb. 462/1 A dense and continuous dark labelling was seen in contact with the membrane of all normoblasts, surrounding the whole perimeter of the cell, including rhopheocytosis invaginations.

rhotacism. Add: Also as mod.L. **rhotaci·smus.**

1897 *Syd. Soc. Lex.*, Rhotacismus. 1959 'A. Burgess' *Beds in East* ii. 49 'But I'm hungwy.' A pathetic little-girl's rhotacismus. 1977 *Times Lit. Suppl.* 11 Feb. 147/3 The privileged families, productive of good shots with..pronounced rhotacismus, are still there to help kill pheasants.

rhotacization (rōᵘ:tăsəizēⁱ·ʃən). *Philol.* [f. Rhotacize *v.* + -ation.] **a.** = Rhotacism 2. **b.** = *r-colouring* s.v. *R I. 1 c.

1973 J. M. Anderson *Struct. Aspects Lang. Change* vi. 119 (*heading*) Rhotacization. 1975 *Amer. Speech 1973* XLVIII. 111 The symbol [ɜ] is used for the stressed mid-central vowel whether with or without rhotacization.

rhotacize, *v.* Add: **2.** *trans.* To convert (a sound, esp. *s*) into *r*.

1965 *Language* XLI. 64 The prothesis rule must apply before *b* is deleted and..*b* must be deleted before the *s* can be rhotacized. 1977 *Ibid.* LIII. 51 The traditional form **esom* was rhotacized into **erom*.

Hence **rho·tacized** *ppl. a.*; **rho·tacizing** *ppl. a.* and *vbl. sb.*

1887 Rhotacizing *ppl. a.* [in Dict.]. 1893 J. Clark *Man. Linguistics* vii. 181 Wherever, medially, in Italic, an *s* between two vowels followed an unaccented syllable, the final result gave *z* in the non-rhotacising dialects, such as Oscan, and *r* (through *z*) in the rhotacising dialects, such as Latin and Umbrian. 1976 *Language* LII. 342 The nasal vowel would be found in a completely oral environment—and thus subject to denasalization, giving the standard examples of rhotacized forms. 1977 *Ibid.* LIII. 53 The form **som* was not sensitive to rhotacism, even when placed after a word ending in a vowel.., because the word boundary suspended the rhotacizing effect.

rho-theta (rōᵘ,þī·tă). [The Greek letters ρ (*rho*) and θ (*theta*), used symbolically in radar to denote distance and bearing respectively.] Used *attrib.* to denote radar navigational systems in which the position of aircraft is measured as range and bearing from a single ground station. Also *absol.*, denoting a position measurement.

1957 *Electronics* 1 June 156 (*heading*) Vortac beacons for rho-theta navigation. 1961 *Engineering* 7 Apr. 479/3 Any type of fix designation, fix identifier, longitude-latitude, relative direction, or rho-theta is accepted by the direct processing routine. 1975 D. G. Fink *Electronics Engineers' Handbk.* xxv. 94 In ICAO practice, the DME is nearly always associated with a VOR, the two systems forming the basis for a rho-theta area navigation system.

Rhovyl (rōᵘ·vil). Also **rhovyl.** A proprietary name for a type of polyvinyl chloride fibre.
1949 *Official Gaz.* (U.S. Patent Office) 7 June 45/1 Société Rhovyl, Paris, France... *Rhovyl*... For threads and yarns containing filaments of polyvinyl chloride. **1949** *Trade Marks Jrnl.* 21 Dec. 1148/2 *Rhovyl*... All goods included in Class 23. Société Anonyme Rhovyl..21, Rue Jean-Goujon, Paris, France; manufacturers. **1963** A. J. Hall *Student's Textbk. Textile Sci.* ii. 97 Since Rhovyl fibres are very inert chemically and are not inflammable..they can be used in the production of fabrics having industrial rather than textile uses. **1972** *Guardian* 31 Oct. 11/5 Underwear; Heavy-weight tights in nylon or rhovyl or wool/Helanca.

rhubarb. Add: **3*. a.** The word 'rhubarb' as repeated by actors to give the impression of murmurous hubbub or conversation. Hence allusively.
1934 A. P. Herbert *Holy Deadlock* 194 The chorus excitedly rushed about and muttered 'Rhubarb!' **1952** *Radio Times* 17 Oct. 11/3 The unemployed actors had a wonderful time. We'd huddle together in a corner and repeat 'Rhubarb, rhubarb, rhubarb' or 'My fiddle, my fiddle, my fiddle' —and it sounded like a big scene from some mammoth production. **1958** *Observer* 7 Dec. 18/5 Actors, who shout 'rhubarb—rhubarb—rhubarb' to give the impression of a distant riot. **1960** J. Betjeman *Summoned by Bells* ix. 105 And in the next-door room is heard the tramp And 'rhubarb, rhubarb' as the crowd rehearse A one-act play in verse. **1972** P. Dickinson *Lizard in Cup* xi. 174 The conversation.. was meaningless; they might just as well have been muttering 'rhubarb, rhubarb'. **1976** *Gramophone* June 71/2, I wondered if the chorus would have made a better effect had the words been less clearly articulated (like actors in a crowd scene muttering 'rhubarb').

b. *Mil. slang.* A low-level flight for opportune strafing.
1943 *Time* 22 Mar. 51/1 When a fighter pilot flies low over France, strafing whatever he finds—trains, troops, airdromes—he is 'on a rhubarb'. **1945** C. H. Ward-Jackson *Piece of Cake* (ed. 2) 51 *Rhubarb*.., a ground-strafing, go-for-anything-you-see worthwhile sortie. **1956** J. E. Johnson *Wing Leader* vi. 79 We began to carry out low-level flights over France. These operations were known by the code name *Rhubarb*. *Ibid.* 80 Usually our *Rhubarb* efforts yielded little more than a staff car. *Ibid.*, I loathed those *Rhubarbs* with a deep, dark hatred.

c. *U.S. slang.* A heated dispute, a row; *spec.* a disturbance or argument on the field of play at a sporting (orig. *Baseball*) event.
1943 *N.Y. Herald Tribune* 13 July 22/3 Mr 'Red' Barber,..who has been announcing the games of the Brooklyn Dodgers, has used the term 'rhubarb' to describe an argument, or a mix-up, on the field of play. **1947** *Time* 22 Sept. 70/1 Next inning, at the plate, there was a face-to-face exchange of hot words..—the kind of rough passage that fans appreciatively call a 'rhubarb'. **1949** *Richmond* (Va.) *Times-Dispatch* 17 Jan. 3/2 The citizen waiting for a streetcar yesterday was of several minds about the 'rhubarb' between the Virginia Transit Company and its drivers. **1950** *Sun* (Baltimore) 20 June 21/2 Such talk is not publicized for obvious reasons. But this has reached type, and one of those so-called 'rhubarbs' is in the pot stewing away. **1959** *Washington Post* 11 Feb. c5/1 Among those who had bets on Dorothy's Best in Monday's false start rhubarb,..most were back to racing's cold war. **1973** *Times* 15 Aug. 7/3 'Rhubarbs', the name used for noisy arguments that break out on the field, started when a Yankee batter, after missing a Perry special, yelled 'spitter' at him. **1975** F. Kennedy *Alberta was my Beat* xii. 147 It was at the Montreal Stampede that the big 'rhubarb' occurred, involving..Canada's top bucking horse rider. **1976** *National Observer* (U.S.) 3 July 15/3 To be conned? Now that always starts a rhubarb, and in this week's hand it started when a defender fell for a *pseudo* end-play.

d. *slang.* Nonsense, worthless stuff.
1963 *Radio Times* 3 Oct. 17/1 Dig this Rhubarb..a new kind of television entertainment... Which (we hope) will prove that there is no shortage of writers for television so long as you are not particular about whether they are still alive or not. **1976** *Telegraph* (Brisbane) 6 July 12/2 They gave me some rhubarb about violating the firework zone. **1977** *Times Lit. Suppl.* 3 June 673/1 Peking opera..employed..a huge repertoire consisting almost entirely of rhubarb. **1979** *Times* 22 Feb. 5/7 We should look at the individual... Whether he or she went to the right school.. that's rhubarb.

4. a. *attrib.* and *Comb.*, as (senses 1–3) *rhubarb crumble, fritters, jam, juice, powder* (earlier example), *pudding, tart* (earlier example); (sense *3* a) *rhubarb noise*; **rhubarb disease** = *crown rot* s.v. *crown *sb.* 35.
1958 Rhubarb crumble [see *crumble *sb.* 2 b]. **1976** *Cumberland & Westmorland Herald* 1 Aug. 1/6 For their sweet course, pupils could make a choice from fresh fruit.. hot mince pies and custard and rhubarb crumble and custard. **1924** Rhubarb disease [see *crown rot* s.v. *crown *sb.* 35]. **1855** E. Acton *Mod. Cookery* (rev. ed.) xix. 383 (*heading*) Rhubarb fritters. **1861** Mrs. Beeton *Bk. Househ. Managem.* 797 (*heading*) Rhubarb jam. **1873** *Young Englishwoman* Nov. 571/1 A receipt for making rhubarb jam. The rhubarb must..be well boiled with..preserving sugar and.. bitter almonds. **1977** K. O'Hara *Ghost of T. Penry* viii. 67 Harriet ate a lot of bread-and-butter with the rhubarb jam. *c* **1863** T. Taylor in M. R. Booth *Eng. Plays of 19th Cent.* (1969) II. 84, You will make it champagne?.. None of your home-brewed; I buy my rhubarb-juice at the greengrocer's. **1971** *Guardian* 15 Feb. 6/5 His father-in-law drinks rhubarb juice by the glassful. **1958** D. Wallace *Forty Years On* v. 63 There was an uproar. From the general rhubarb noise the Dean could be heard. **1977** *Gramophone* Aug. 340/1 There is almost no theatrical production, sound-effects or 'rhubarb' noises. **1784** J. Woodforde *Diary* 9 Mar. (1926) II. 121 Mr. Thorne left..a Rhubarb Powder to take to Morrow. **1861** Mrs. Beeton *Bk. Househ. Managem.* 672 (*heading*) Boiled rhubarb pudding. **1946** *Farmhouse Fare* (Farmers Weekly) 135 Rhubarb pudding..flour.. dates..syrup..milk..cocoa..margarine..rhubarb... Bake in hot oven. **1793** J. Woodforde *Diary* 11 May (1929) IV. 28 We had for Dinner to day, hashed Calfs Head..& a Rhubarb Tart.

rhu·barb, *v. slang* (orig. *Theatr.*). [f. the sb.] *intr.* Of an actor: to repeat 'rhubarb' (*rhubarb 3* a); to mumble indistinctly in order to represent the noise of a crowd. Freq. *transf.* in gen. use. Also redupl. Occas. *trans.* with direct speech as obj.
1958 *Spectator* 11 July 47/1 'Hear, hear,' they rhubarb-rhubarbed. **1965** *Observer* 20 June 25/5 The barons, mildly rhubarbing in some awkwardly symmetrical pieces of stage grouping, rightly suspect the King of double-dealing. **1966** I. Jefferies *House-Surgeon* iv. 79 We rhubarbed till he had gone. **1967** D. Skirrow *I was following this Girl* xxxv. 210, I listened hard and rhubarbed my way through, trying to make any sort of sense of what I was hearing. **1976** *Daily Tel.* 21 Sept. 11/2 Livia faced the Roman mob, all seven of them, rhubarbing at the Palace back door. 'You wait,' she cried imperiously, 'until my husband gets home!'
So **rhu·barber,** an actor who repeats 'rhubarb'; also *transf.*; **rhu·barbing** *vbl. sb.* and *ppl. a.*
1953 A. McQueen *Let.* 31 Aug. in Partridge *Dict. Slang Suppl.* (1961) 1247/2 When a few actors gathered backstage and represented 'noise without' made by a mob, they intoned the sonorous word 'rhubarb'. The action was called 'rhubarbing', the actors 'rhubarbers'. **1965** *New Statesman* 20 Aug. 266/2 A floodlit market place at night with hucksters arriving to tempt the rhubarbers. **1970** *Times* 27 Feb. 13/5 The few attempts at pageantry generally come to grief against painted backdrops, watched by a handful of rhubarbing commoners. **1978** *Daily Tel.* 10 May 36/6 Mr Jenkin showed himself to be a formidable rhubarber in his own right. He nodded and moved his lips with the utmost vigour.

rhumba, var. *rumba *sb.* and *v.*

rhumbaba, var. *rum baba* s.v. *baba².

rhumbatron (rʊ·mbătrǫn). *Electronics.* [Irreg. f. Gr. ῥύμβ-ος, ῥόμβ- whirling motion (f. ῥομβεῖν to turn around) + *-tron.] A cavity in which bunching of an electron beam is produced by an applied radio-frequency field, or conversely a pulsed beam produces a radio-frequency voltage across a gap in the cavity.
1939 R. H. & S. F. Varian in *Jrnl. Appl. Physics* X. 322/1 The resonant properties of practically closed metal vessels... In our laboratory, these have been called 'rhumbatrons', from the Greek word 'rhumba', meaning rhythmic oscillation. **1939** *Jrnl. R. Aeronaut. Soc.* XLIII. 388 A beam of electrons representing a constant current is sent through two metal containers acting as resonators and known as Rhumbatrons. In the first Rhumbatron is an oscillating electric field, parallel to the stream and of such strength as to change the speeds of the electrons by appreciable fraction. .. In the second Rhumbatron..a considerable fraction of the power of the arriving electron bunches is converted into power of high frequency oscillations. **1947** *Electronic Engin.* XIX. 149/1 Examination of the 'rhumbatron' type of circuit used by the Varians introduces us to an interesting new phenomenon..the existence of an isolated electric field within each rhumbatron, in spite of the fact that the whole of the exterior of each rhumbatron and the drift-tube itself are constantly at R.F. earth potential. **1968** *Radio Communication Handbk.* (ed. 4) ii. 23/1 A klystron is a valve containing an electron gun..from which a narrow beam of electrons is..focused through small apertures across which one or more oscillatory circuits are connected; these circuits are in the form of hollow toroidal chambers known as rhumbatrons.

rhydectomy, var. *rhytidectomy.

rhyme, *sb.* Add: **4.** (senses 2 and 3) *rhyme-analogy, -compulsion, -form, -type, -word* (further examples); *rhyme-like* adj.; **rhyme scheme,** the ordered patterning of end-rhymes in metrical composition; **rhyme sheet,** a broadsheet containing verses for display.
1930 A. W. Aron in *Curme Vol. Linguistic Studies* 19 By *rhyme-analogy* we mean the associational process by which the gender of a word is influenced by other words with similar suffixes or other similar endings. **1923** R. Graves *Feather Bed* 17 Sacred Carnivals trundle through my mind, With Rhyme-compulsion mottoing each waggon. **1927** W. E. Collinson *Contemp. Eng.* 7 Rhyme-forms which have aroused H. G. Wells' anger, like *roly-poly*. **1934** Rhyme-form [see *letter-name* s.v. *letter *sb.*¹ 7 a]. **1957** N. Frye *Sound & Poetry* 125 Having rejected formal rhyme, he by no means avoids subdued rhymelike effects. *a* **1931** E. Pound *Make it New* (1934) vii. 400 Where both Rossetti and I went off the rails was in taking an English sonnet as the equivalent for a sonnet in Italian. I don't mean in overlooking the mild difference in the rhyme scheme. **1962** W. Nowottny *Lang. Poets Use* viii. 192 This..is very emphatically signalled as some kind of new departure by a change in the rhyme-scheme. **1981** J. Brabazon *Dorothy L. Sayers* iii. 27 Her mind constantly absorbed in..a language, a rhyme-scheme, an astronomical phenomenon. **1920** (*title*) Rhyme sheet (Poetry Bookshop). **1943** N. Marsh *Colour Scheme* i. 16 The archly-reproachful rhyme-sheets in bathrooms and lavatories. **1968** D. Hopkinson *Incense-Tree* ii. 22 On the walls of our nursery, my mother had pinned rhyme sheets published by the Poetry Bookshop. **1945** C. L. Wrenn in *Slavonic & East Europ. Rev.* XXIII. 123 Both have sought..to preserve alike the rhythm and the rhyme-types of their originals. **1943** —— in *Trans. Philol. Soc.* 32 A glance at the apparatus, for instance, of any well-edited Middle English text will show how a study of the orthography of the rhyme-words in a poem..may point the way to an original reading. **1960** A. Clarke *Horse-Eaters* 27 The nasal syllable in Houyhnhnm Brings rhyme-word.

rhyme, *v.* Add: **3. d.** *U.S. Blacks.* Const. *up.* To improvise (a blues composition).
1968 P. Oliver *Screening Blues* 16 Blues singers pride themselves on their ability to 'rhyme up a song' but they do not consider this an essential requirement of their music. **1968** *Blues Unlimited* Sept. 13, I rhymed up a song and called it 'The Auction Day blues'.

rhymelet (rəi·mlėt). [f. Rhyme *sb.* + -let.] A short piece of rhyme or poetry.
c **1870** M. Collins in F. Collins *Lett. & Friendships* (1877) I. v. 66, I meant to say your rhymelets smell More of the country than Pall Mall. **1917** J. Adams *Student's Guide* 76 Many of us..have been grateful to the author of certain flagrant rhymelets.

rhyming, *vbl. sb.* Add: **b. rhyming dictionary** (earlier and later examples); **rhyming slang,** a variety of (orig. Cockney) slang in which a word is replaced by a phrase which rhymes with it (see quot. 1933); also *rhyming slang(st)er.*
1775 J. Walker *Dict. Eng. Lang.* p. v, A rhyming dictionary in a living language, for the purposes of poetry, seems no very unnatural or useless production. *a* **1846** B. R. Haydon *Autobiogr.* (1927) i. 4, I remember him with his rhyming dictionary, composing his verses and scanning with his fingers. **1964** 'E. McBain' *Ax* x. 191 I'd like to buy a rhyming dictionary... I promised somebody I'd find a rhyme. [*c* **1850** in R. Pearsall *Worm in Bud* (1969) I. ii. 49 The new style of cadgers' cant..is done all on the rhyming principle.] **1859** Hotten *Dict. Slang* 134 The cant, which has nothing to do with that spoken by the costermongers, is known in Seven Dials and elsewhere as the Rhyming Slang, or the substitution of words and sentences which rhyme with other words intended to be kept secret. *Ibid.* 135, I learn that the rhyming slang was introduced about twelve or fifteen years ago. **1911** J. W. Horsley *I Remember* xi. 252 The more modern rhyming slang..invented and chiefly used by costermongers, to whom 'daisy roots' is a substitute for the word boots. **1933** 'G. Orwell' *Down & Out in Paris & London* xxxii. 238 Twenty-five or thirty years ago..the 'rhyming slang' was all the rage in London. In the 'rhyming slang' everything was named by something rhyming with it —a 'hit or miss' for a kiss, 'plates of meat' for feet, etc. **1973** B. Aylwin *Loads of Cockney Cobblers* p. vi, Of course the Cockney—a born linguist—will adapt his rhyming slang specifically *for* the Common Market. **1977** *Listener* 5 May 588/3 Any speaker of Anglo-American knows that 'bread' as a synonym for 'money' is the work of those dreary cockney rhyming-slangers. 'Bread and honey', money. **1948** Partridge *Dict. Forces' Slang* 155 Before the First World War, it was in common use in the Forces, even in Scottish regiments, such was the influence of the Cockney, who is the expert rhyming slangster.

rhy·mingly, *adv.* [f. Rhyming *ppl. a.* + -ly².] In a rhyming manner.
1880 Ruskin in *19th Cent.* Aug. 198 The necessity..of completing the nomenclature rhythmically and rhymingly.

rhymsterette (rəi:mstəre·t). *nonce-wd.* [f. Rhymester + -ette.] A female rhymer; an inferior poetess.
1957 V. Nabokov *Pnin* vii. 180 A fair sample of her production is the kind of stuff that émigré rhymsterettes wrote after Akhmatova: lackadaisical little lyrics.

rhynchocœle, *a.* and *sb.* For **a, b** read **A, B** and add: **B.** *sb.* **2.** *Zool.* Also **-cœl, † -cœlum.** A body cavity in nemertean worms that contains the introverted proboscis.
1902 *Encycl. Brit.* XXXI. 121 (*caption*) Rhynchocœl. *Ibid.*, The cavity into which the proboscis is retracted, the rhynchocœlum, is formed by a split which appears in the mesobiast surrounding the epiblastic pit. **1932** Borradaile & Potts *Invertebrata* vii. 205 When the muscles of the proboscis sheath contract and press upon the fluid in the rhynchocoel the proboscis is everted. **1972** M. S. Gardiner *Biol. Invertebrates* ii. 51/1 Except for this rhynchocoele, the body is solid in the sense that the digestive tract and other organs are surrounded by parenchyma cells.
Hence **rhynchocœ·lic** *a.*, of or pertaining to the rhynchocœle.
1902 *Encycl. Brit.* XXXI. 121 (*caption*) Rhynchocoelic blood-vessel.

rhynchodæum (riŋkodī·ǒm). *Zool.* Pl. **-æa** (-ī·ă). [mod.L. *rhynchodæum* (coined in Ger. by A. Oswald 1893, in *Vierteljahrsschr. der Naturf. Gesell. in Zürich* XXXVIII. 347), f. Gr. ῥύγχ-ος snout + ὁδαῖον, neut. of ὁδαῖος that is on or by the way (f. ὁδός way).] The cavity anterior to and partially containing the proboscis of certain invertebrates, esp. ribbon-worms of the phylum Nemertea and gastropods of the genus *Buccinum*. Hence **rhynchodæ·al** *a.*
1894 *Jrnl. R. Microsc. Soc.* 328 Oswald proposes to call the space between the introvert or proboscis [of certain gastropods] and its sheath the rhynchodæum, its anterior opening the rhynchostome, the opening at the apex of the proboscis the pharyngostome, and the oral aperture proper, where ectoderm joins the endoderm, the gastrostome. **1912** *Mem. Liverpool Marine Biol. Comm.* XX. 30 The cavity (part of the external world) between the proboscis and its sheath..is known as the Rhynchodaeum, and the opening of the latter, or the false mouth, is the Rhynchostome. **1963** R. D. Barnes *Invertebr. Zool.* viii. 119/2 The lumen of the rhynchodaeum is continuous with that of the proboscis proper. **1972** R. Gibson *Nemerteans* i. 38 The proboscis pore opens immediately into the rhynchodaeum, a rather short tubular chamber that in many pelagic hoplonemerteans is very short. *Ibid.*, Rhynchodaeal musculature, if present, is usually weak and predominantly composed of circular fibres. **1975** *Zeitschr. f. Morphol. der Tiere* LXXX. 67 In *O[enopota] levidensis* the rhynchodaeal walls have a series of annular folds, whose musculature appears to allow dilation and contraction. The dilation could enlarge the rhynchodaeal space, conceivably of assistance in prey capture and engulfment.

rhynchokinesis (riŋkǒkəinī·sis, -kin-). *Zool.* [ad. G. *rhynchokinetik* (H. Hofer 1949, in *Acta Zool.* XXX. 213), f. Gr. ῥύγχο-ς beak, snout + κινητικ-ός moving: see Kinetic *a.* (*sb.*) and *kinesis.] A process, found in some birds and lizards, by which the upper bill or jaw may be raised relative to the cranium by extensive bending of nasal and premaxillary bones.
1963 W. J. Bock in C. G. Sibley *Proc. XIII Internat. Ornithol. Congr.* I. 46 Hofer showed that rhynchokinesis is correlated with a schizorhinal nostril. **1964** *Jrnl. Morphol.* CXIV. 15/1 Some birds have a form of kinesis that cannot be easily assigned to either prokinesis or rhynchokinesis. **1964, 1973** [see *prokinesis]. **1974** P. J. K. Burton *Feeding & Feeding Apparatus in Waders* ii. 33 Two principal modifications are found in present day birds. These are prokinesis, in which the upper jaw moves about a pivot formed by a narrow strip of flexible bone at the fronto-nasal hinge; and rhynchokinesis, in which bending occurs over a wider zone, some distance along the bill. Rhynchokinesis involves a change in bill shape.
So **rhynchokinetic** (-kəine·tik) *a.*
1960 [see *prokinetic *a.*]. **1964** *Jrnl. Morphol.* CXIV. 14/1 The rhynchokinetic skull is characterized by a region of bending located somewhere along the dorsal bar of the upper jaw. **1974** P. J. K. Burton *Feeding & Feeding Apparatus in Waders* ii. 33 It is..convenient to distinguish between moderately rhynchokinetic species in which bending occurs along the whole length of the upper jaw anterior to the internasal septum, and highly rhynchokinetic species in which reinforcement of the dorsal bar anterior to the nasal partition has moved the zone of bending much further forward.

rhynchonellid (riŋkone·lid). *Zool.* and *Palæont.* [ad. mod.L. family name *Rhynchonellidæ*, f. generic name *Rhynchonella* (G. Fischer von Waldheim 1809, in *Notice des Fossiles du Gouvernement de Moscou* 35), f. *rhynchon-* (irreg. f. Gr. ῥύγχος snout) + *-ella*, dim. suffix (see -el²): see -id³.] An articulate brachiopod of the family Rhynchonellidæ, whose members are found as both fossils and living animals.
1913 C. R. Eastman tr. *Zittel's Text-bk. Zool.* (ed. 2) I. v. 399 Few Rhynchonellids agree with the genotype, which is *Rhynchonella loxia* Fischer..from the Upper Jura of Russia. **1920** A. M. Davies *Introd. Palæont.* i. 37 Young specimens of this group are the best on which to study the deltidial plates.., which in so many rhynchonellids become obscured by the curvature of the beaks. **1936** *Geogr. Jrnl.* LXXXVII. 431 Other fossils include cephalopods, gastropods, rhynchonellids and crinoid stems. **1970** R. M. Black *Elements Palaeont.* xi. 158 A number of different stocks of brachiopods survived into the Mesozoic... The most commonly occurring forms..are rhynchonellids and terebratulids. The rhynchonellids are in general less important after the Jurassic.

Rhynchosporium (riŋkospǭᵊ·riǒm). Also rhyncho-. [mod.L. (E. Heinsen 1900, in *Jahrb. d. Hamburg. Wissensch. Anstalten* (3 Beiheft) XVIII. 43), f. Gr. ῥύγχο-ς snout + σπορ-ά sowing + L. *-ium*, neut. of *-ius*, adj. suffix.] A fungus of the genus so called, members of which cause leaf blotch in cereals and other grasses; also, the disease caused by this fungus. Also *attrib.*
1920 *Phytopathology* X. 54 (*heading*) Evidence of disease resistance in barley to attacks of Rhynchosporium. **1937** *Jrnl. Agric. Res.* LV. 175 (*heading*) *Rhynchosporium* scald of barley, rye and other grasses. *Ibid.* 180 In the early stages of the development of *Rhynchosporium* in the host tissue the mycelium is wholly subcuticular. **1967** *Punch* 26 July 140/3 Then there is Vulcan—an interesting new six-row winter barley, which, it is thought, will eventually be granted rights. It has, in effect, double resistance, being almost immune to Mildew and the dreaded Rhynchosporium. **1971** *Country Life* 7 Oct. 936/1 Another important fungoid disease..is rhynchosporium..which makes black blotches on the leaves of barley.

rhynchostome (ri·ŋkostōᵘm). *Zool.* [ad. G. *rhynchostom* (A. Oswald 1893, in *Vierteljahrsschr. der Naturf. Gesell. in Zürich* XXXVIII. 347). f. Gr. ῥύγχος snout + στόμα mouth.] The anterior opening of the rhynchodæum in proboscis-bearing gastropods.
1894, 1912 [see *rhynchodæum]. **1975** *Zeitschr. f. Morphol. der Tiere* LXXX. 71 The prey could be engulfed by the rhynchostome and rhynchodaeal sphincter functioning as gripping organs and working in ratchet fashion to pass the prey in.

rhynchotal (riŋkōᵘ·tăl), *a. Ent.* [f. as Rhynchotous *a.* + -al.] = Rhynchotous *a.*
1903 *Nature* 23 Oct. 616/1 Additions to the rhynchotal fauna of Central America.

rhyodacite (rəi,ŏdēⁱ·səit). *Geol.* [f. Rhyo(lite + Dacite.] A type of extrusive volcanic rock having a porphyritic texture and intermediate in composition between rhyolite and dacite.
1913 A. N. Winchell in *Jrnl. Geol.* XXI. 214 The volcanic equivalent of granodiorite, which may be appropriately called rhyodacite, forms another intermediate group between the alkalcic and the alkaline rocks. **1932** A. Johannsen *Descriptive Petrogr. Igneous Rocks* II. 356 Rhyodacites are the extrusive equivalents of granodiorites. They are rocks of porphyritic habit, usually white, yellowish, reddish, or grayish in colour, and showing phenocrysts of quartz, plagioclase, and some ferromagnesian constituent, in a microcrystalline, cryptocrystalline, or glassy groundmass. **1970** Macdonald & Abbott *Volcanoes in Sea* vii. 143/2 The rhyodacite of Mauna Kuwale, in the Waianae Valley on Oahu, appears to be an unusual, extreme type derived by differentiation of the tholeiitic basalt magma in the same way that trachyte is derived from alkalic basalt. **1977** A. Hallam *Planet Earth* 164/1 Most provinces of rhyolitic volcanics contain substantial amounts of rhyodacite as well as true rhyolite.

rhyolite. Add: In mod. use, a type of acidic extrusive volcanic rock, usu. pale in colour and porphyritic in texture, having phenocrysts esp. of quartz or potassium-feldspar in a fine-grained or glassy groundmass which commonly shows flow structure. (Earlier and later examples.) [First as G. *rhyolith* (not *-lit*) (*Jahrb. d. K. K. Geol. Reichs-Anstalt* (Wien) (1860) XI. 156).]
1868 *Mem. Calif. Acad. Sci.* I. 50 The name 'rhyolite' was proposed, early in 1860, for certain rocks frequently occurring on the southern slope of the Carpathians. **1932** A. Johannsen *Descriptive Petrogr. Igneous Rocks* II. 268 Most rhyolites are light in colour: white, yellow, brown, or pink, but yellowish green and even deep red rocks are found. **1944** A. Holmes *Princ. Physical Geol.* xx. 455 During the late Pleistocene there were paroxysmal eruptions of rhyolite-tuffs, followed later by flows of rhyolite. **1955** *Sci. News Let.* 9 July 20/2 The ancient implements are made chiefly of slate and shale but some are of jasper, quartz and rhyolite porphyry (rock of lava origin). **1971** I. G. Gass et al. *Understanding Earth* i. 22/1 The rock types granite, diorite and gabbro are all coarse-grained and usually found in intrusions. They are matched compositionally by fine-grained rocks, respectively rhyolite, andesite and basalt, which are characteristically found as surface lava flows (extrusions). **1977** A. Hallam *Planet Earth* 164/1 Rhyolite eruptions are characteristic of areas of active mountain-building such as island arcs (e.g., Sumatra, Japan, West Indies), and mountain chains such as the Andes.

Rhyssa (ri·să). *Ent.* [mod.L. (I. L. C. Gravenhorst *Ichneumonologia Europæa* (1829) III. 260), f. Gr. ῥυσ(σ)ός shrivelled, wrinkled.] An ichneumon of the genus so called, the members of which prey upon the larvæ of wood-boring insects, esp. in conifers, and are usu. black with white face and markings.
1908 C. Morley *Ichneumons Gt. Brit.* III. 26 *Rhyssae* are found flying in pine woods, where the larvae of Sirices attack the solid timber. **1931** *Discovery* Dec. 386/1 As a popular study of the woodwasp and its major parasite, Rhyssa, the film is in its essentials admirable. **1969** R. F. Chapman *Insects* xxx. 629 The parasite *Rhyssa* (Hymenoptera) is able to detect the larva of its host *Sirex* (Hymenoptera) through several inches of wood as a result of olfactory stimulation.

rhythm, *sb.* Add: The pronunc. (ri·ð'm) is now standard. **II. 5. c.** *ellipt.* A rhythm instrument or musician (also *sing.* for *pl.*); a rhythm section. orig. *U.S.*
1938 D. Baker *Young Man with Horn* iii. 145 Every fourth dance..turned out to be a trumpet solo by Rick Martin, flanked by some rhythm. **1947** R. de Toledano *Frontiers of Jazz* 68 The instrumental blend and precision of each section—reed, brass, rhythm. **1956** M. Stearns *Story of Jazz* (1957) xv. 170 The standard number of musicians in a dance band was nine: two trumpets, two saxes, one trombone, and four rhythm (banjo, piano, drums, and tuba). **1970** *New Yorker* 23 May 80/3 Sy Oliver..is holding a retrospective of his work..with the help of a nine-piece group that includes two trumpets,..and three rhythms.

8. *Geol.* and *Physical Geogr.* Regularity in the way something is repeated in space; also, a feature that is repeated at regular intervals of space.
1914 C. B. Crampton *Geol. of Caithness* ix. 89 Sedimentary rhythm in marine deposits usually depends on the principle of the development of calcareous and terrigenous strata from opposite directions to overlap one another. **1924** *Proc. Yorks. Geol. Soc.* XX. 125 There is a rhythm in the succession of the beds, occasionally interrupted, but always discernible. **1937** *Bull. Geol. Soc. Amer.* XLVIII. 1935 Many mud-sand rhythms include strata too thick for seasonal accumulation. **1951** *Geol. Mag.* LXXXVIII. 166 The major rhythmic units attain a thickness of 100 or so feet but minor or partial rhythms also occur suggesting small changes in the intensity of the flow movement. **1976** P. D. Komar *Beach Processes & Sedimentation* x. 278 For many reported occurrences it is not possible to determine whether crescentic bars or some other shoreline rhythm are being described.

9. a. *attrib.* and *Comb.*, as (sense 1) † *rhythm prose*; (sense 4) *rhythm-foot, -word*; *rhythm-deaf, -drunk* adjs.; (sense 5) *rhythm-accent, dancer, group, musician, -pattern, singing*; **rhythm and blues** orig. *U.S.*, blues music with a strong rhythm; **rhythm club** *Hist.*, a club which specialized in the presentation of jazz; **rhythm guitar,** a guitar upon which the chord sequences of a melody are played; so *rhythm guitarist*; **rhythm instrument,** an instrument that constitutes part of the rhythm section of a musical band; **rhythm man,** one who plays a rhythm instrument; **rhythm section,** that part of a musical (orig. jazz) band whose main function is to supply the rhythm, often consisting of a piano, double-bass, and drums, sometimes with a guitar or other instruments.
1948 *Penguin Music Mag.* Oct. 36 For instance, Russian composers rely on rhythm-accent for their effect more than on harmony or melody. **1949** *Billboard* 25 June 30/2 Records listed are rhythm and blues records that sold best in stores according to the Billboard's special weekly survey. **1955** A. J. McCarthy *Jazzbook 1955* 84 Rhythm and blues is a Negro music and it was already in full flower when the jazz fans became aware of its existence. The actual musical content of the average rhythm and blues record is very limited. **1960** *Melody Maker* 31 Dec. 5/2 When we got to 'The Preacher', they both objected violently. 'No, no, no, no! We can't record that! That's rhythm and blues!' **1967** J. Wilson in L. Deighton *London Dossier* 25 The Stones.. had their first central London club engagement at the old Marquee. Rhythm 'n Blues in England began there with the late Cyril Davis, who died a few years ago. **1977** McKnight & Tobler *Bob Marley* 10 Unlike soul, jazz or rhythm and blues, reggae remains the inviolable preserve of black musicians. **1933** *Melody Maker* 2 Dec. 2/1 (*heading*) All the rhythm clubs doing well. **1959** 'F. Newton' *Jazz Scene* xiii. 246 The student will observe the absence of Rhythm Clubs in Hampstead, Kensington or Chelsea. **1963** *Listener* 14 Mar. 458/1 People of my generation had to rely on their gramophones and the rhythm clubs to carry them through this period of deprivation. **1942** Rhythm dancer [see *jitterbug *sb.* 2]. **1926** Fowler *Mod. Eng. Usage* 505/2 This writer has produced a single sentence seventeen lines long without a single slip in grammar. That so expert a syntactician should be rhythm-deaf is amazing. **1916** W. B. Yeats *Reveries over Childhood* 194 Verse spoken by a man almost rhythm-drunk at some moment of intensity. **1883** G. M. Hopkins 7 Nov. *Let.* (1956) 329 Music is..the very place where the difference of time-feet and rhythm-feet recognised in Greek poetry is still in force. **1959** G. Freeman *Jack would be Gentleman* vi. 129 'Give us a song first.'.. 'Okay; but you'll have to do without the rhythm group.' **1973** *Guitar* Sept. 5/2 Plenty of familiar faces in the band:.. Charlie McCracken (ex-Taste, bass) and Spencer on rhythm and slide guitar. **1977** McKnight & Tobler *Bob Marley* vi. 79 The instrumental chores of the band were shared like this: Marley—vocals, rhythm guitar; Family Man—bass. **1967** *Listener* 21 Dec. 802/1 If you encounter a little riot of colour ambling along Charing Cross Road,..it is the rhythm guitarist of The Who. **1977** *Zigzag* Mar. 17/5 Bill Cheatham ..switched from being their roadie to their rhythm guitarist. **1927** *Melody Maker* Sept. 926/3 The bass, being a rhythm instrument, must conform to the rhythm set by the rhythm section. **1952** B. Ulanov *Hist. Jazz in Amer.* (1958) vi. 51 The rhythm instruments—drums, bass, and guitar—made up the engine that powered the jazz machine: their function was to keep the syncopated beat going in regular almost inflexible alternations of weak and strong accents. **1949** L. Feather *Inside Be-Bop* iii. 96 Seldom takes solos, but fine rhythm man. **1952** B. Ulanov *Hist. Jazz in Amer.* vi. 66 St. Cyr is the redoubtable rhythm man who kept such a fine beat going for Jelly Roll Morton's Red Hot Peppers and Louis Armstrong's Hot Five and Seven. **1977** *Rolling Stone* 19 May 83/1 Problems related to studio overdubbing (not Benson, but rhythm musicians). **1946** R. Blesh *Shining Trumpets* i. 18 Rhythm-pattern of dancers' feet. **1970** Rhythm pattern [see *rhythmic *a.* 1 a]. **1599** Rhythm prose [in Dict., sense 1 a]. **1927** Rhythm section [see *rhythm instrument* above]. **1938** D. Baker *Young Man with Horn* iii. 117 Rick.. started setting chairs together..in threes: reed section, brass section, rhythm section. **1955** A. Morgan in A. J. McCarthy *Jazzbook 1955* 18 With a coloured front line.. and a mixed white and coloured rhythm section (Dodo Marmorosa, Barney Kessell, Red Callender and Don Lamond) these recordings are amongst the best examples of the use of a guitar in a bop rhythm section. **1977** *Times* 15 Nov. 17/6 An enthusiastic rhythm section whose drummer.. sometimes allowed his effervescence to occlude his marvellous sense of swing. **1934** S. R. Nelson *All about Jazz* iv. 73 Whiteman..started the fashion for organized rhythm singing with the Rhythm Boys. **1977** *Belfast Tel.* 22 Feb. 8/2 The American group's breed of close harmony and tight rhythm singing. *c* **1874** G. M. Hopkins *Jrnls. & Papers* (1959) 279 In Greek the scanning is by..rhythmic beat, that is beat belonging only to the rhythm-words, not to the sense-words.

b. Used *attrib.* with reference to the periodic variation of fertility in women, esp. in **rhythm method,** a method of birth control depending upon continence during the period of ovulation.
[**1920** W. M. Gallichan *Critical Age of Woman* iii. 52 (*heading*) The sexual rhythm in women.] **1934** *New Republic* 5 Sept. 98/2 The rhythm theory of fertility and sterility in women was briefly explained to the Committee. **1940** N. E. Himes *Pract. Birth-Control Meth.* viii. 125 Most authorities..believe the rhythm method unreliable. *Ibid.* 126 If a Catholic woman will not use a scientific method, it is perhaps better that she use the rhythm method than none at all. **1955** P. S. Henshaw *Adaptive Human Fertility* xi. 208 Couples who wish to use the rhythm method for preventing conception have been advised to avoid coitus during the period of the nineteenth through the ninth day before the beginning of the next expected menstrual period. **1961** *Guardian* 3 Mar. 13/3 Even those most hostile to the use of contraceptives are in favour of the rhythm method. **1971** *Sunday Mirror* 25 July 11 We follow the rhythm system. But we don't use it to avoid procreation. We use it to *have* children. That's the way it was meant to be. **1972** *Human World* May 23 You use the rhythm method not just by having intercourse now, but by not having it next week, say.

rhythmal (ri·ðmăl), *a. rare.* [f. Rhythm *sb.* + -al.] = Rhythmical *a.* 3.
1908 Hardy *Dynasts* III. i. vi. 346 Whose emissaries knock at every door, In rhythmal rote.

rhythmic, *a.* and *sb.* Add: **A.** *adj.* **1. a.** Also appositively, as *rhythmic-melodic* adj.
1946 R. Blesh *Shining Trumpets* ii. 36 The..principle of free rhythmic-melodic improvisation. **1970** P. Oliver *Savannah Syncopators* 55 Repeated rhythm patterns and rhythmic-melodic patterns..such as are characteristic of boogie-woogie and blues piano.

3. *Geol.* and *Physical Geogr.* Exhibiting or characterized by a spatial rhythm or periodicity.
1914 C. B. Crampton et al. *Geol. of Caithness* ix. 89 In the Helman Head Beds the rhythmic sequence is confined to continental, alluvial, and lacustrine deposits. **1960** Turner & Verhoogen *Igneous & Metamorphic Petrol.* (ed. 2) xi. 292 Gravitational settling of heavy dark minerals within the layer of mush carpeting the floor..is thought to be responsible for the rhythmic layering so widely prevalent in the lower levels. **1976** P. D. Komar *Beach Processes & Sedimentation* x. 265 Attempts at classifying rhythmic shoreline features generally stress their spacings. Beach cusps are considered to have smaller spacings.., while sand waves.. and giant cusps have larger spacings.

rhythmicity (riðmi·siti). [f. Rhythmic *a.* + -ity.] Rhythmical quality or character; the capacity for maintaining a rhythm.
1901 *Buck's Handbk. Med. Sci.* (ed. 2) III. 109/2 The pulse rate presents more or less regular and extensive variations in the course of a day... They are hardly the expression of an inherent rhythmicity. **1910** *Heart* I. 217 The ventricles are freed of the influence of their pace-maker, and their rate of beat is consequently slow until the dormant rhythmicity becomes fully awakened. **1944** *Mind* LIII. 271 Variability in size, number, direction, space, distance and rhythmicity is probably the most important characteristic in the primitive visual field. **1969** *Nature* 23 Aug. 782/1 Cutting the nerves..which connect the corpora cardiaca to the sub-oesophageal ganglion..caused a gradual loss of rhythmicity in otherwise intact cockroaches. **1971** Chin-Wu Kim *Survey Linguistic Sci.* 63 It definitely has a tendency to maintain iambic rhythmicity.

rhythmicize (ri·ðmisəiz), *v. rare.* [f. Rhythmic *a.* + -ize.] *trans.* To make (a song) rhythmical; to endow with rhythm. Hence **rhy:thmiciza·tion.**
1907 R. Boughton *Bach* iii. 47 We must have before us the melody of Luther's noble hymn, rhythmicized according to Bach. **1979** *Early Music* Apr. 283/1 It is true that we felt at liberty to experiment with rhythmicization and other aspects of performance style.

rhythmite (ri·ðməit). *Geol.* [ad. G. *rhythmit* (B. Sander 1936, in *Mineral. und petrogr. Mitt.* XLVIII. 190) f. *rhythm(us* Rhythm *sb.* + *-it* -ite¹.] Each of the repeated units in a sedimentary formation exhibiting a rhythmic structure; *spec.* a varve.
1946 *Prof. Paper U.S. Geol. Survey* No. 212. 30/2 The word 'rhythmite', used by Sander for the individual units of rhythmic beds, will be adopted here as a brief term to designate the couplet of distinct sedimentary types of rock, or the graded sequence of sediments, that form a unit bed or lamina in rhythmically bedded deposits. **1951** E. B. Knopf tr. *Sander's Contributions to Study of Depositional Fabrics* 135 We have the same need for names to denote rocks of this sort (possibly rhythmites) as we had for tectonites and similar names. [*Translator's note*] It has been suggested by A. Knopf that 'laminites' might be a better term than 'rhythmites' in order to avoid the positive implication of perfect periodicity in the recurrence of laminae. **1969** Bennison & Wright *Geol. Hist. Brit. Isles* xvi. 361 In lake deposits counting of seasonally banded deposits, varves (or rhythmites)..can be used for dating such sediments up to 10,000 years old. **1976** *Nature* 18 Mar. 235/1 The rhythmites make up a regular series of plane beds with only occasional discontinuities caused by local penecontemporaneous erosion and minor faulting.

rhythmize, *v.* Delete *rare* in Dict. and add: (Further example.) Also *absol.*, to establish a steady rhythm.
1933 *Brit. Jrnl. Psychol.* Jan. 263 In some simple factory tasks it seems impossible not to rhythmize. **1946** *Scrutiny* Dec. 94 A number of impressions that would otherwise be merely a sequence can, if rhythmized, be perceived as an organized whole.
Hence **rhy:thmiza·tion**; **rhy·thmized** *ppl. a.*; **rhy·thmizing** *vbl. sb.*
1901 E. B. Titchener *Exper. Psychol.* I. ii. x. 339 The object of this experiment is to bring out the fact of subjective rhythmisation, or (as it is also termed) subjective accentuation. **1933** *Brit. Jrnl. Psychol.* Jan. 263 The rhythmized impressions are differentiated according to their position within the rhythm unit. *Ibid.*, In typewriting,.. individual differences in rhythmizing are far more marked. **1946** *Scrutiny* Dec. 94 The experience of rhythm is..an active process, the process of rhythmization. **1955** G. Abraham in H. Van Thal *Fanfare for E. Newman* 21 Similarly rhythmised string chords replace the 'scrub' tremolo.

rhytidectomy (rəitide·ktǒmi). *Surg.* Also in contracted form **rhydectomy.** [f. Gr. ῥυτιδ-, ῥυτίς wrinkle + *-ectomy.] The surgical removal of wrinkles, esp. from the face.
1931 R. J. E. Scott *Gould's Med. Dict.* (ed. 3) 1217/1 Rhytidectomy. **1939** S. Fomon *Surg. of Injury & Plastic Repair* xix. 1348 (*caption*) Ehrenfeld rhydectomy. **1954** Zubek & Solberg *Human Devel.* v. 106 Once character lines are fully developed, the only recourse is rhytidectomy—that is, plastic surgery through which wrinkles are removed by traction on the skin—and even such drastic measures are only temporary. **1960** S. Fomon *Cosmetic Surg.* viii. 455/2 The benefits that accrue from a facial rhydectomy are manifold. **1971** *Daily Colonist* (Victoria, B.C.) 15 Aug. 29/4 The face lift or rhytidectomy takes care of sagging jowls and heavy expression lines—seldom the fine ones. **1977** *Chicago Tribune Mag.* 2 Oct. 23/1 Whereas the price of a complete rhytidectomy (face-lift) in the United States..averages between $2,500 and $5,000..Lotter's maximum charge is $2,500.

rhytidome. Add: Also (*rare*) **rhitidome.** Substitute for def.: The outer part of the bark of a woody plant, composed of layers of dead phloem and cork. (Further examples.)
1896 W. R. Fisher *Schlich's Man. Forestry* V. 486 The Turkey-oak is used here and there in Austria for the production of bark, but on account of its forming, at an early age, a deeply-cracked rhitidome, or dead bark..it is of little value. **1925** Eames & MacDaniels *Introd. Plant Anat.* ix. 215 Hypodermal layers..may be combined with periderm.. to form rhytidome, as in *Dracaena*. **1953** K. Esau *Plant Anat.* xiv. 332 In some rhytidomes parenchyma and soft cork cells predominate; others contain large amounts of fibers. **1976** *Flora* CLXV. B. 326 Rhytidome composition.. is considered also to be important for fungal bark infections.

‖ **ri** (rī). Pl. ri. [Jap.] **1.** A traditional Japanese unit of length equal to 36 *cho*; in modern use equivalent to approximately 2·44 miles (3·93 kilometres).
1845 *Encycl. Metrop.* XX. 486/2, 1 long ri = 3 standard miles nearly. 1 short ri = 2 ditto ditto. **1874** *Trans. Asiatic Soc. Japan 1873–4* 33 After a somewhat disagreeable walk of 3 *ri* Kusatsu is reached. **1876** H. W. Bates *Illustr. Trav.* 248/1 The *ri* is little short of two and a half miles. **1890** B. H. Chamberlain *Things Japanese* 352 Tokyo.. covers an immense area, popularly estimated at four *ri* in every direction, in other words, a hundred square miles. **1897** *Japan Times* 10 Apr. 3/4 How many *ri* of the inundated tracts of land along the rivers could possibly be polluted by the poisonous matter from the mine? **1964** *Japan* (Unesco) (rev. ed.) 847 *Ri* (36 chō), 3·92727 kilometre, 2·44033 miles.
2. In Japan (in ancient times) and Korea, the smallest subdivision of rural administration.
1959 R. K. Beardsley et al. *Village Japan* iii. 40 We have already mentioned the division of old Kibi into provinces (*kuni*). These were further divided into districts (*kōri*, later *gun*), and still further into villages (*gō*, today 'mura') and hamlets (*ri*). **1966** J. W. Hall *Govt. & Local Power in Japan, 500 to 1700* iii. 82 Groups of fifty or fewer *ko* formed the administrative village, called *ri*. *Ibid.* 83 In 715 the name *ri* was abandoned, and the fifty-*ko* units were renamed *gō*. These newly named administrative villages were then divided into two or three smaller units to which the term *ri* was applied... For some reason the *ri* was discontinued as a unit of official organization in 740. **1971** *Korean Folklore & Classics* III. 7 Mr. Kwak lived in Sa Dong Ri. **1972** P. M. Bartz *South Korea* 54/1 A *myon* may be divided into one or more *ri*... It is usually translated into English as 'village', but a *ri* does not *ipso facto* contain a village. It is merely the smallest area of rural subdivision.

ria. Delete ‖ and substitute for def.: **1.** A long narrow inlet of the sea formed by the partial submergence of an unglaciated river valley. [Adopted as a technical term in Ger. by F. von Richthofen *Führer für Forschungsreisende* (1886) ix. 309.]
The wider use represented by quots. 1899, 1915 has not been generally adopted.
1899 *Proc. Amer. Acad. Arts & Sci.* XXXIV. 220 The term *ria*, from the Spanish, may be advantageously used to cover all types of subaerially carved troughs, including von Richthofen's fjord, ria, dalmatian, and limian types. **1915** *Ann. Assoc. Amer. Geographers* V. 82 The Spanish word, ria, introduced into German by Richthofen, and now widely adopted, means any broad or estuarine river mouth, and not necessarily an embayment produced by the partial submergence of an open valley in a mountainous coast, in the sense that Richthofen originally proposed. **1939** *Geogr. Jrnl.* XCIII. 235 Many of the rivers enter the heads of rias which extend many miles inland. **1946** L. D. Stamp *Britain's Struct. & Scenery* v. 49 The best example of a coastline of drowned valleys or ria coast is the south-west of Ireland. Soundings show that the floor of the ria..slopes steadily seawards and there is no 'lip' as there is in the case of a glaciated valley. **1974** *Encycl. Brit. Micropædia* IX. 560/2 Generally occurring along a rugged coast perpendicular to a mountain chain, many rias were formed by the rise in sea level after the melting of the vast continental glaciers. **1977** Millward & Robinson *Landscape of Brit.* 46 In the South-West Peninsula, the deeply set valleys give rise to long branching arms of the sea known as 'rias'.
2. *attrib.*, as *ria shoreline*; **ria coast** (also † **rias coast,** after G. *riasküste* (von Richthofen, loc. cit.)), a coast marked by numerous rias.
1902 *Encycl. Brit.* XXVIII. 623/1 A further subdivision depends on the character of the inter-relation of land and sea along the shore producing such types as a fjord-coast, ria-coast, or lagoon-coast. **1924** tr. *Wegener's Orig. Continents & Oceans* 53 These strongly planed-down chains.. stretch..westwards so as to form a wild jagged type of coast (the so-called rias-coast) in south-western Ireland and Brittany. **1946** [see above]. **1977** A. Hallam *Planet Earth* 94 For example, drowned river valleys of the ria coasts, such as those of southwest Ireland, can be differentiated from glacially eroded drowned coasts. **1919** D. W. Johnson *Shore Processes & Shoreline Devel.* iv. 173 Shorelines formed by the partial submergence of a land mass dissected by normal river valleys, which may be called ria shorelines, after the ria coast of northwestern Spain..; thus used, the term ria is not restricted to the narrow meaning assigned to it by von Richthofen.

rial, *sb.*¹ Restrict 'Now only *Hist.*' to senses in Dict. and add: **5.** Also **riyal. a.** The monetary unit of Iran, introduced in 1930 and equal to one hundred dinars.
1932 A. T. Wilson *Persia* x. 278 The Gold Standard Act [of 28 Mar. 1930] establishes as the legal unit of currency the gold rial, to be represented in coinage by the *pahlevi* of 20 rials and the half pahlevi of 10 rials. **1936** *Encycl. Islam* III. 1162/3 The modern Persian riyāl is a money of account: originally (1930) 20 riyāls = £1 stg. but by the system finally adopted in 1933, 100 dinārs = 1 riyāl = 1 pahlavi = £1 stg. **1948** D. N. Wilber *Iran* 213 The rial is also popularly referred to as the kran. **1953** A. Smith *Blind White Fish in Persia* vii. 115 Eggs, 11 rials; mast, 4 rials; grapes, 18 rials. **1972** *Times* 20 Oct. 9/4 Smuggled opium is sold in Iran at a price below the 17·5 rials (about 10p) a gram the Government charges registered addicts. **1973** E. Hyams *Final Agenda* v. 61 'Can I have that map?' 'Forty-five *rials*, sir.' **1976** R. Moore *Dubai* i. 6 Fitz removed a thousand-rial note, about fifteen dollars, and placed it in the envelope.
b. The monetary unit of Saudi Arabia.
1939 H. St. J. B. Philby *Sheba's Daughters* ii. 21 It is perhaps sufficient to mention that the camels hired by him for our use cost no more than 15 Riyals (about £1 sterling) apiece for a journey that occupied two and a half months. **1951** *Chambers's Jrnl.* Nov. 689/2 In the 19th century the dollar was banned in the Ottoman Empire, including Egypt, and became obsolete in Tunis and Algeria. Rather later, Ibn Saud of Arabia substituted, not without difficulty, his own rials for it. **1964** *Ann. Reg. 1963* 308 The sum of 244 million riyals was allocated to provide free education for everyone. **1970** *Times* 3 Apr. (Arab League Suppl.) p. x/5 The Government has already spent more than 4,000m. rials in five years developing communications. **1977** *Arab Times* 13 Dec. 4/5 Recently Shahwa found at Ahmad Al Jaben Street a bag containing KD 1,300 and Saudi Rial 2,594 and an expensive watch.
c. Any of various monetary units of other countries of the Middle East (see quots.).
1959 *Statesman's Year-bk.* 1383 The monetary unit is the Sudanese pound (£S), divided into 10 Riyals. **1968** *Listener* 4 Jan. 7/1 A guerrilla attack on the outskirts of Sanaa for a few Yemeni riyals. **1970** *Whitaker's Almanack 1971* 917/2 Oman... On May 7, 1970, a new currency was brought into circulation. The main unit is the *Rial Saidi* = £1. Each *rial* is divided into 1,000 new *Baiza*. **1972** *Times* 15 May (Qatar Suppl.) p. iv/6 Banknotes of one, five, 10, 25, 50 and 100 Riyals. **1977** *Times* 18 Feb. (Banking Suppl.) p. v/5 In ..1976 the Qatari rial joined the Bahrain dinar and the United Arab Emirates dirham in the travellers' reciprocity scheme.

rialto. Add: (Earlier examples.) Also *fig.*
1596 Shakes. *Merch. V.* I. iii. 108 Signior Anthonio, many a time and oft In the Ryalto you haue rated me About my monies and my vsances. **1611** Coryat *Crudities* 169 The Rialto which is at the farther end of the bridge as you come from Sᵗ. *Marks*, is a most stately building, being the Exchange of Venice, where the Venetian Gentlemen and the Merchants doe meete twice a day, betwixt eleuen and twelue of the clocke in the morning, and betwixt fiue and sixe of the clocke in the afternoone. **1869** S. R. Hole *Bk. about Roses* i. 11 There the poor Rose-trees stand..in the spring a *rialto, rendezvous*, common-room, and tap for all the riff-raff of the insect world.

riata. Add: Also **reata.** (Earlier and later examples.) Also in Comb., as *riata-man.*
1846 *Californian* 12 Sept. 1/1 A riata (rope) was made fast to the broken bone and the jaw dragged out. **1853** *Harper's Mag.* Aug. 308/2 Each mule being secured by a long *reata*..was permitted to graze until sunset. **1924** F. R. Bechdolt *Tales of Old-Timers* 103 They had failed to take into account..a skill with the reata. **1934** *Amer. Ballads & Folk Songs* 409 Old Sandy Bob was a riata-man. **1948** F. Blake *Johnny Christmas* i. 41 Johnny loosened the horsehair knots, then..took down his rawhide riata. **1964** F. O'Rourke *Mule for Marquesa* 126 He sliced rope lengths from the reatas tied to the saddles. **1970** *Arizona Highways* Oct. 48/1 Now Buster Jig was a *riata* man, With his gut-line coiled up neat.

rib, *sb.*¹ Add: **I. 1. d.** A joke; a teasing or joking remark. Chiefly *U.S.*
1929 J. P. McEvoy *Hollywood Girl* iii. 39 Of course it was just a rib to see him, as the only thing I know about newspapers is that they smell fresh. **1952** *Herald-Tribune* (N.Y.) 25 Jan. 18/3 Wilt said, 'He is a parasite.' Maybe this was a rib. **1958** J. Wain *Contenders* 8 The mere idea of Robert as a schoolmaster is, of course, a screaming rib. **1964** 'E. Queen' in *Cavalier* Aug. 15/2 'Dad, is this a rib?' 'I wish I could joke about it.'

e. Slang phr. *to get into* (someone's) *ribs*: to borrow or otherwise obtain money from (someone). (Only in Wodehouse.)
1939 Wodehouse *Uncle Fred in Springtime* iii. 47 Leave it to me. I will get into his ribs for you. **1951** —— *Old Reliable* xviii. 200 It was..the worst possible moment anyone could have selected to approach him with the idea of getting into his ribs for twenty thousand dollars. *a* **1975** —— *Sunset at Blandings* (1977) viii. 59 Did he discover that Jeff was the son of the man who got into his ribs for that substantial sum, there would be no question of engaging him as his secretary.

III. 11. e. *Aeronaut.* A structural member in an aerofoil, positioned more or less fore-and-aft and serving to define the contour of the aerofoil and sometimes also as part of the load-bearing structure.
1888 [see *propeller 3 c]. **1919** H. Shaw *Text-bk. Aeronaut.* ix. 109 The chief function of the ribs is to give the wing its correct shape, while they also serve as compression members between the two spars, and as a framework for the attachment of the fabric. **1941** N. H. Anderson *Aircraft Layout & Detail Design* ii. 24 The term 'skewed rib' is applied to a rib that is bent inboard or outboard along its line of intersection on a spar. **1965** C. N. Van Deventer *Introd. Gen. Aeronaut.* vii. 141/2 The ribs are secured to the spars, and since they are curved they give the wing its familiar aerodynamic shape when it is covered with a 'skin' of aluminium alloy. **1966** D. Stinton *Anat. Aeroplane* xi. 206 Ribs may be built up like frames, be light as formers, or be made like bulkheads.

IV. 13. a. *rib-chop, -end, -steak*; **rib-cage,** the chamber formed by the ribs and their connecting tissues, which contains the lungs, heart, etc.; also *fig.*; **rib-eye** *N. Amer.*, a cut (cf. *eye *sb.*¹ 16 e) of meat that lies along the outer side of the rib (of beef-cattle); usu. *attrib.*, as *rib-eye muscle, steak.*
1909 *Daily Chron.* 9 July 9/4 By any outward pressure about the waist, the diaphragm is hampered in its movement, and breathing becomes costal; that is, a woman then breathes only within her chest or rib-cage. **1932** W. Faulkner *Light in August* x. 212 Feeling..his white chest arch deeper and deeper within his ribcage. **1959** *Times* 17 Aug. 10/6 The builder showed me some of the points of a racing pigeon... The full rib-cage inside which the heart and lungs generated the energy for sustained flight. **1978** J. Irving *World according to Garp* i. 25 There were two waist gunners tucked into the rib cage of the plane. **1936** Rib chop [see *best *a.* 5]. **1979** *Lore & Lang.* Jan. 29 Rib chops. **1851** H. Melville *Moby Dick* II. xxxix. 259 So did this old whale leave his aged bulk, and now and then partly turning over on his cumbrous rib-ends, expose..the unnatural stump of his starboard fin. **1979** *Lore & Lang.* Jan. 27 Rib end of sirloin. **1926** *Nat. Provisioner* 3 Feb. 10/1 Roast-ready rib is prepared..starting at a fixed point determined by measuring off 3 in. from extreme outer tip of rib-eye muscle at 12th rib. **1943** P. T. Ziegler *Meat we Eat* xvi. 251 The quartering is done by inserting the knife at the desired spot below the rib eye muscle. *Ibid.* 255 A recent practice is to cut out the rib eye of the better grades of ribs and serve them as *club steak*. **1966** A. Hawkins *Steak Book* (end-paper), Club steak. Country Club steak. Delmonico steak. Ribeye steak. **1974** *Greenville* (S. Carolina) *News Piedmont* 20 Apr. 5/1 (Advt.), 4 oz. Ribeye steak... 59c. **1978** *Chicago* June 229/1 Recommended: the special rib-eye steak and the Lithuanian roast duck. **1922** Joyce *Ulysses* 659 On the middle shelf..a small dish containing a slice of fresh ribsteak. **1976** *Columbus* (Montana) *News* (Joliet Suppl.) 27 May 3/4 (Advt.), Rib steak..lb. $1.69.

b. rib-bender (earlier and later examples); **rib-digger,** a person given to light-hearted banter; so **rib-digging** *ppl. a.*; **rib-tickler,** that which amuses; a joke or aphorism; hence **rib-tickling** *ppl. a.*
1861 H. Rhys *Theatr. Trip through Canada & U.S.* xii. 108 He was licked in five minutes by as many of Morrissey's rib-benders. **1901** G. B. Shaw *Admirable Bashville* ii. 311 My calling hath apprenticed me to pangs. This is a rib-bender; but I can bear it. **1929** Rib-digger [see *back-slapping *ppl. a.* and *vbl. sb.*]. **1925** H. J. Laski *Let.* 28 Apr. in *Holmes–Laski Lett.* (1953) I. 737 In a very different genus, *Love* by the Countess von Arnim which seemed to me devilish clever, with a sly rib-digging quality not unworthy of Jane Austen. **1933** P. Godfrey *Back-Stage* vii. 98 By the introduction of 'stage business'..the skilled comedian can transform very thin stuff into a real 'rib-tickler'. **1963** *Times Lit. Suppl.* 24 May 379/2 He does include some real rib-ticklers. **1976** M. Millar *Ask for me Tomorrow* viii. 66 'Isn't that a funny idea?' 'A real rib-tickler.' **1936** J. L. Hodson *Our Two Englands* vi. 108 This centre for providing violent or exquisite or rib-tickling emotions has a Fun-house resembling a super-modern factory. **1977** *It* June 26/4 Your old pal has written this rib-tickling tale specially for jubilee freaks all.

14. b. *rib-knit, -weave.*
1968 J. Ironside *Fashion Alphabet* 247 A rib knit (i.e. purl and plain) fabric is much more elastic than a plain knit. **1973** *Guardian* 10 Apr. 13/2 Cape faced with chunky ribknit. **1964** *McCall's Sewing* iv. 51/2 The Rib Weave which has ribbed or corded effects in either the warp or filling direction is another variation of the plain weave. **1968** J. Ironside *Fashion Alphabet* 261 *Rib weave*: the ribbed effect is obtained by using thicker or doubled yarn in one direction.

c. *rib-cloud*; **rib-joint** *U.S. slang*, a brothel; **rib-randing** *Basketry* (see quot. 1953); also (as back-formation) *rib-rand* vb. trans.; **rib-roll** *sb.*, a farm roller with a ribbed or hooped surface; hence as *vb. trans.*; **rib-stall,** a set of wall-bars for physical exercises.
1868 G. M. Hopkins *Jrnls. & Papers* (1959) 170 At sunset featherbed sky with a fluffy and jointed rib-cloud. **1943** M. Shulman *Barefoot Boy with Cheek* xx. 192 One night Scott became involved in a fracas in a Minneapolis rib joint. **1965** C. Colter in A. Chapman *New Black Voices* (1972) 70 Forty-third Street,..the street of rib joints and taverns. **1975** *Amer. Speech 1969* XLIV. 91 The great majority represent middle-class restaurants. There are a few exceptions: a few drugstores, 'rib joints', snack bars, and a couple of 'soul food' places in the black ghetto. **1912** T. Okey *Introd. Art of Basket-Making* ix. 92 *Rib-randing*..is used..where it is desirable to break the monotony of a deep space of simple randing. **1953** A. G. Knock *Willow Basket-Work* (ed. 5) 17 *Rib-Randing*. This is produced by carrying the randing rod in front of two stakes instead of one, as in ordinary randing. **1959** D. Wright *Baskets & Basketry* ii. 36 *Rib-randing*..must be worked over a number of stakes not divisible by three. The close weave is useful on lids. *Ibid.* iv. 125 Rib-rand the ball..finishing with a round of pairing. **1969** G. E. Evans *Farm & Village* 176 *Rib-roll*, land roller with corrugated surface. **1971** *Arable Farmer* Feb. 29/2 The crop is not rib rolled until the peas have chitted and are just coming through. **1908** *Mosquito* Dec. 2 The heart of the Games' Mistress still yearns after some more ribstalls, a Norwegian ladder, and a bench for remedial work.

rib, *v.*¹ Add: **4. b.** To discredit; to incriminate. *U.S. slang. rare.*
1926 *Clues* Nov. 162/1 *Rib*, to frame up. **1929** Hostetter & Beesley *It's a Racket!* 236 *Rib*, to talk about anyone, to talk slander about anyone. **1935** *Amer. Speech* X. 52/2 To rib (to discredit).
c. To annoy or threaten; to put pressure on (someone). *U.S. slang. rare.*
1929 D. Hammett *Red Harvest* iv. 49 If Max Thaler means anything to you, you ought to pass him the word that Noonan's trying to rib him.
d. To fool or dupe; to tease; to make fun of (someone or something). *slang* (orig. *U.S.*). Also *absol.*
1930 *Amer. Mercury* Dec. 457/2 *Rib*, *v.*, to beguile. 'We rib the sap that it's McCoy and he goes for it.' **1934** J. O'Hara *Appointment in Samarra* (1935) 98 Mr. Robert Hermann is in his best form, ribbing me about last night. **1938** *New York Panorama* xi. 262 Armstrong..could take a pop tune and reinterpret it, often *ribbing* (satirizing) it. **1941** K. Tennant *Battlers* xiii. 143 The busker pricked up his ears when he heard the merry-go-round owner 'ribbing' Fosdick on the number of lodgers he was taking in nightly. **1942** *Sun* (Baltimore) 2 Feb. 8/4 I've been here about a year and every once in a while I get 'ribbed' about Baltimore. **1955** L. P. Hartley *Perfect Woman* viii. 78 When the chaps rib her she doesn't quite know how to act up. **1958** J. Wain *Contenders* 30 Baxter just thought I was ribbing him. **1960** 'N. Shute' *Trustee from Toolroom* 294 Quit ribbing, Sol. **1967** H. Storey in *Coast to Coast 1965–6* 203 One of the neighbours had ribbed him about 'being on the bottle'. **1972** M. Williams *Inside Number 10* xii. 303 Jim Callaghan has always been very charming to me, though I won't say that once I began to work for Harold I was not often ribbed and left with the odd, slightly acid remark ringing in my ears. **1976** *Times Lit. Suppl.* 31 Dec. 1640/3 [Ben] Jonson was ribbed on more than one occasion for daring to put forth his collected works even before he could have been called middle-aged.

Ribaga (rība·gă). *Ent.* The name of Constantino *Ribaga* (*fl.* 1897–1929), Italian zoologist, used *attrib.*, in the possessive, and with *of* to designate a pouch on the abdomen of female insects, chiefly of the family Cimicidæ, into which spermatozoa are deposited by the male during copulation.
Orig. restricted in application to a part of the organ now so called (see quot. 1920).
1920 F. W. Cragg in *Indian Jrnl. Med. Res.* VIII. 46 Its only attachment is to the ventral wall of the abdomen, by means of a structure to be described as the organ of Ribaga. *Ibid.* 51 Berlese considered that these granules are waste products, produced during the digestion of the spermatozoa, and excreted through the organ of Ribaga. **1957** Richards & Davies *Imms's Textbk. Entomol.* (ed. 9) I. 189 In the Cimicidae and in *Xylocoris* the sperms are deposited in Ribaga's organ..and from there pass through the haemocoele to the ovaries. **1970** *Entomol. Abstr.* II. 38/1, 5829 ♀ C[imex] lectularius from a variety of sources showed 185 specimens with an abnormality of the Ribaga organ.

riband, *sb.* Add: **5. b.** *riband-back* adj.
1955 R. Fastnedge *Eng. Furnit. Styles* vii. 166 Chippendale had good reason for satisfaction. The riband-back chair..notably expressed the style of his *Director* period. **1960** H. Hayward *Antique Coll.* 238/1 '*Ribband-back' chair*, a mahogany chair, the splat carved in the form of knotted ribbons and bows in a manner highly expressive of rococo taste.
c. riband cane, var. *ribbon cane* s.v. *ribbon *sb.* 10 c.

ribbed, *ppl. a.* **2. b.** (Earlier examples.)
1756 in *Essex Inst. Hist. Coll.* (1907) XLIII. 277 Smith wore when he went away..blue ribb'd Stockings. **1782** in *Ibid.* (1859) I. 13/1 A patton [*sc.* pattern] of White Ribed Stuff for a Wescoat & Briches.

ribbing, *vbl. sb.* Add: **1. b.** (The action of) teasing or joking at another's expense (see *rib *v.*¹ 4 d). *slang* (orig. *U.S.*).
1935 Wodehouse *Luck of Bodkins* xxiv. 301, I was just kidding. Sure! Ribbing, we call it over here. **1935** *Sun* (Baltimore) 9 Mar. 6/2 A dozen fat rainbow trout were in the [swimming] pool—another case of Hollywood 'ribbing'. **1939** I. Baird *Waste Heritage* xv. 192 There was a brief outbreak of good-natured ribbing. **1951** *Sport* 30 Mar.–5 Apr. 3/2 Stan had to contend with a good deal of ribbing from his team-mates. **1956** C. P. Snow *Homecomings* 139 Orthodox opinions, collective gibes, a bit of ribbing—that was enough to keep them zestful. **1968** *Globe & Mail Mag.* (Toronto) 13 Jan. 12/3 Unlike..rustic yokels..they can even take a ribbing. **1976** M. & G. Gordon *Ordeal* (1977) iii. 15 He was..the only man at the party to wear a tie. He took a ribbing about that.
2. (Further examples in *Knitting*.)
1948 'P. Wentworth' *Traveller Returns* xxi. 134 She had begun Johnny's second stocking and almost finished the ribbing at the top. **1958** C. Fremlin *Hours before Dawn* xxi. 173 Her niece..was frowning dreadfully over the spacing of the button holes in the front ribbing. **1962** 'E. Peters' *Funeral of Figaro* iii. 69 Two inches of neat ribbing stood stiff on the bright blue plastic needles. **1973** D. Biggs *Knitting for Fun* 41/2 If you can knit a square you can make this gay poncho... Join corners of neck ribbing.

ribbok (ri·bǫk), var. Reebok.
1947 H. C. Bosman *Mafeking Road* xii. 67 He remembered having had that same feeling once before when he had shot a ribbok. **1953** *Cape Argus* 25 Apr. 8/4 The animals include ribbok, grysbok,..and one blue wildebeest. **1966** E. Palmer *Plains of Camdeboo* vii. 124 'The skull is ribbok,' he said, 'and that's the tibia of a buffalo.'

ribbon, *sb.* Add: **I. 2.** Also *pl.*, prizes or decorations awarded to the winners of a competition or show; chiefly in phr. *in the ribbons*, among the prize-winners. Also *fig.*
1932 *New Yorker* 23 July 42/3 He [*sc.* a horse] went off like the stock market on Saturdays and wasn't in the ribbons. **1974** *News & Press* (Darlington, S. Carolina) 25 Apr. 11/1 Bobby Gurley's efforts in the 50-yd. dash and softball throw won first-place ribbons. **1977** *Horse & Hound* 10 June 24/2 (*caption*) Julia Watts on her Song of the Sea,..again in the ribbons among the hacks at the Hertfordshire Show.
5. d. Also *spec.* a strip of land, esp. a path or road.
1923 *Times* 26 May 7/4 Pedestrians should not walk in the road..but should keep to the footpaths or ribbons on the Downs specially prepared for them. **1948** 'J. Tey' *Franchise Affair* ix. 90 The London–Larborough road was a black straight ribbon in the sunshine. **1958** *Spectator* 6 June 721/3 The factories are in other parts of the borough, along the ribbon of Western Avenue. **1966** *Guardian* 10 Aug. 14/2 The consultants put great emphasis on the need for landscaping and the creation of ribbons of open spaces in and around Warrington. **1976** M. Maguire *Scratchproof* xii. 182 Staring out at the ribbon of road ahead.
e. A long, narrow strip of thin impregnated fabric or coated plastic wound on a spool, used as the inking agent in a typewriter or printer (in earlier use, a similar strip used in hand-stamping devices). Also *attrib.*
a **1877** Knight *Dict. Mech.* III. 1936/2 Ribbons for hand-stamps are tapes saturated with an oily pigment. **1883** J. G. Petrie *Man. Type-Writer* 16 For ordinary writing a black ribbon is best. *Ibid.*, Ribbons will last a long time if they are taken care of. **1907** *Yesterday's Shopping* (1969) 357/2 The..Typewriter Desk..has drawers for holding Carbon, Ribbons, Stationery. **1951** A. C. Clarke *Sands of Mars* iv. 33, I should have warned you to use a black ribbon. Contrast isn't so good with blue. **1962** *Which?* Dec. 358/2 All the models had automatic ribbon reverse. **1976** N. Meyer *West End Horror* viii. 77 Address typewritten—on a Remington in need of a new ribbon.
f. *ellipt.* = *ribbon microphone* (sense 9 c below).
1937 *Printers' Ink Monthly* May 40/3 *Ribbon*, a velocity microphone. **1962** A. Nisbett *Technique Sound Studio* ii. 44 The two microphones can be so placed that the only move necessary to get from one to the other is a slight turn. The ribbon will have a smooth, realistic quality.
II. 9. a. *ribbon-manufacturer* (earlier and later examples), *threader.*
1799 *Times* 1 June 2/1 The Bill for Regulating the Wages of Ribbon Manufacturers, in Coventry. **1883** M. Blind *George Eliot* iii. 26 Mr. Bray was a wealthy ribbon manufacturer. **1973** M. Laski *Geo. Eliot* 23 Mr Pears was a ribbon manufacturer. **1931** Ribbon threader [see *mushroom *sb.* 3 e]. **1969** E. H. Pinto *Treen* 318 The ribbon threader, a round-ended, tapering rod, nearly 1 ft. long, with a large eye at the narrow end, was once a commonplace implement.
c. *ribbon-knit, shop, work*; **ribbon cartridge,** (*a*) a pick-up cartridge that works on the same principle as the ribbon microphone; (*b*) a cartridge containing a spooled typewriter ribbon for easy and clean insertion into and removal from a typewriter; **ribbon chute** = *ribbon parachute* below; **ribbon clerk** *U.S. slang* (see quot. 1953); also fig.; **ribbon-copy** *U.S.*, the top copy of a (typed) letter or document; **ribbon microphone** or (*colloq.*) **mike,** a microphone whose electrical output results from the motion of a thin metal ribbon mounted between the poles of a permanent magnet; **ribbon parachute,** a parachute having a canopy consisting of an arrangement of closely spaced tapes.
1975 G. J. King *Audio Handbk.* viii. 172 Instead of a coil of wire a single conductor in the form of a 'ribbon' may be used, the result being a ribbon cartridge. **1977** *New York Rev. Bks.* 12 May 17/2 (Advt.), The Lexikon portables use Ribbon Cartridges that snap in. **1946** *Sun* (Baltimore) 18 July 13/2 The parachute itself is of revolutionary design. Known as a 'ribbon chute' it is composed of 150 individual ribbons approximately two inches wide. **1927** E. Pound *Let.* 29 Dec. (1971) 215 Am I expected to respect either

myself or anyone else because some graduated ribbon-clerk offers me 75 bucks for writing blah in a false-pearl and undies monthly? **1953** BERREY & VAN DEN BARK *Amer. Thes. Slang* (1954) §563/1 Stock Market... *Ribbon clerks, lambs*, amateur, small-fry traders who take occasional flings in the market. **1977** *Time* 11 Apr. 24/1 Flying has become so routine that the notably pragmatic insurance companies charge pilots no more for policies than they do ribbon clerks. **1968** R. LOCKRIDGE *Plate of Red Herrings* xi. 147 Always they send us carbons... They send the ribbon copies to book publishers. **1976** M. MILLAR *Ask for me Tomorrow* 28, I can type this letter up for you... How many copies do you need?.. You'll want to give her the ribbon copy. **1969** C. ARMSTRONG *Seven Seats to Moon* xii. 109 Her hands kept busy with a ribbon-knit costume she had been working on for weeks. **1974** *Country Life* 3/10 Jan. 54/2 Wool bouclé suits, ribbon knits and embroidered sweaters. **1931** *Jrnl. Acoustical Soc. Amer.* III. 59 The ribbon microphone consists of a lightly corrugated metallic ribbon suspended in a magnetic field and freely accessible to air vibrating from both sides. **1944** *Electronic Engin.* XVI. 328/3 A simple detector for supersonics in the region of 20 kc/s can be made from an old ribbon microphone. **1975** G. J. KING *Audio Handbk.* ix. 200 The ribbon microphone..tends to respond to the particle velocity of a sound wave. **1957** *Practical Wireless* XXXIII. 714/2 Most ribbon mikes have a transformer built in at the bottom of the case. **1948** *Jrnl. Brit. Interplanetary Soc.* VII. 95 The containers were lowered successfully for the first time by ribbon parachutes. **1956** W. A. HEFLIN *U.S. Air Force Dict.* 442/2 *Ribbon parachute*, a type of parachute consisting of numbers of ribbons held in place by equally-spaced tapes, with spacing between the ribbons to give the required porosity. The ribbon parachute was developed in Germany and used during WW II. **1699** M. LISTER *Journey to Paris* 176 Many of the great Ribban Shops remove out of the Palais hither. **1857** GEO. ELIOT in *Blackw. Mag.* July 62/1 The Dissent..had let off half its chapel area as a ribbon-shop. **1856** O. JONES *Gram. Ornament* xv. 3 Various patterns, mostly geometrical, consisting of interlaced ribbon-work,..and strange monstrous animals and birds. **1960** H. HAYWARD *Antique Coll.* 238/2 *Ribbon work*, embroidery in fine, narrow silk ribbons often combined with chenille thread and aerophane.

10. a. *ribbon bed, decoration, flame, handle, ornament, pattern, road*; *ribbon-back(ed)* adj.; **ribbon-building** = next; **ribbon development** [*DEVELOPMENT 3 d], the building of houses in a single line along a main road, usu. one leading out of a town or village; hence **ribbon-developed** *a.*

1920 *Burlington Mag.* Oct. 195/1 Thomas Chippendale is usually credited with being the originator of the ribbon-back chair. *Ibid.* 196/1 Nearly all the known variations of the ribbon-back have the ball-and-claw foot in addition. **1980** *Times* 2 June 4/7 A set of six ribbon-back Chippendale chairs. **1966** A. W. LEWIS *Gloss. Woodworking Terms* 14 Ribbon-backed chairs with cabriole legs are a typical feature [of the Chippendale period]. **1969** *Sears Catal.* Spring/Summer 26 Cardigan Sweater... Matching color buttons on a ribbon-backed border. **1879** C. M. YONGE *Magnum Bonum* III. xxxi. 642 Cutting scarlet geraniums in the ribbon beds. **1928** *Daily Express* 27 Sept. 10/6 Ribbon-building should be abolished. **1935** D. L. SAYERS *Gaudy Night* xiv. 299, I was born there, and I shall be sorry if I live to see the land sold for ribbon-building. **1939** V. G. CHILDE *Dawn European Civilization* (ed. 3) ii. 16 Punctured ribbon decoration and pedestalled goblets have analogies also in the Balkans. **1931** A. L. ROWSE *Politics & Younger Generation* viii. 199 New colonies of semi-detached villas, and..ribbon-developed roads. **1927** *Garden Cities & Town Planning* XVII. 172/2 The writers are well aware of the disadvantages of 'ribbon' development. **1929** *Times* 19 Jan. 8/1 Your condemnation of ribbon development building along arterial roads is especially applicable to the new arterial Reigate to Dorking road. **1934** J. B. PRIESTLEY *Eng. Journey* x. 336 We passed through several villages that looked hardly more than slums that had been scattered along the road. This odd ribbon development is fairly common in colliery areas. **1954** M. BERESFORD *Lost Villages* viii. 275 With ribbon-development, petrol-stations and roadhouses it may find itself no longer the deserted village. **1979** J. E. MORPURGO *Allen Lane* iv. 138 The desert of ribbon development and light industrial estates..had been allowed to sprawl all over Middlesex. **1958** *Chambers's Techn. Dict.* Add. 1010/1 *Ribbon-flame burner*, a tubular gas-burner on which a ribbon of flame is produced by means of alternating corrugated and plain steel strips inserted in a milled slot, thus forming honeycombed flame ports, or by the tube being drilled with lines of very fine holes in close formation. **1973** *Times* 30 July 11/2 The cooker embodies another step forward in technology—a ribbon flame instead of the old jets, inherently more stable and less likely to blow out in a draught. **1957** V. G. CHILDE *Dawn European Civilization* (ed. 6) ii. 17 The vases may be provided with..flanged ribbon handles. **1972** *Trans. Oriental Ceramics Soc.* XXXVIII. 54 A grey earthenware tripod vessel, three-lobed, with a ribbon handle. **1941** *Oxoniensia* VI. 32 There were a fair number of vaulting ribs found *passim* at the east end of both aisles... A few ribs were found decorated with ribbon ornament. **1940** *Burlington Mag.* Dec. 180/2 The traditional Anglo-Saxon Ribbon Pattern themes. **1954** M. RICKERT *Painting in Brit.: Middle Ages* 231 *Ribbon pattern*, sometimes used to describe plait work or strap work; any flat ribbon-like ornament. **1956** G. TAYLOR *Silver* ii. 24 During the ninth and tenth centuries Scandinavian influence was strong.. mostly characterized by asymmetrical ribbon patterns and animal forms. **1929** *Times* 29 Aug. 13/6 (*heading*) Park ways and ribbon roads. **1934** T. S. ELIOT *Rock* i. 21 And now you live dispersed on ribbon roads, And no man knows or cares who is his neighbour. **1969** *Daily Tel.* 18 Jan. 9 A ribbon-road framed with cream-painted filling stations..and bright blue and white bus signs every few miles.

b. ribbon figure, grain, a striped pattern of grain seen in some quarter-sawn hardwoods (see quots.).

1940 BROWN & PANSHIN *Commercial Timbers U.S.* x. 242 The quarter-sawn surface of interlocked-grain wood exhibits a characteristic striped or ribbon figure. **1964** P. KOCH *Wood Machining Processes* ii. 16 The interlocked and alternately inclined grain, which forms a ribbon figure, presents a machining problem. **1934** *Archit. Rev.* LXXVI. 64/3 Very handsomely figured timber is exported... Material possessing 'ram's horn' figure, is also met with, and 'stripe' or 'ribbon-grain' is relatively common. **1948** *New Biol.* IV. 87 Mahogany is also best converted in this way [*sc.* by quarter-sawing], in this case to show the 'ribbon-grain' figure due to the inclination of the fibres to the vertical axis differing in successive growth rings; this results in differential light reflection from the boards comparable to that shown by a freshly rolled lawn.

c. ribbon (also † **rib(b)and**) **cane,** a variety of sugar cane whose mature stalks have red or purplish longitudinal stripes.

1811 G. MATHISON *Notes Jamaica* 65 The riband or striped cane is no longer cultivated by judicious planters. **1833** B. SILLIMAN *Man. Sugar Cane* 10 The varieties of Cane cultivated in the United States, are the Creole,..the Ribbon Cane [etc.]. **1834** *Visit to Texas* x. 92 The ribband cane requires to be planted every three years. **1929** L. F. CARR *America Challenged* 252 Some extra fine ribbon-cane molasses. **1938** M. K. RAWLINGS *Yearling* xviii. 215 The red ribbon cane had made a fair stand. **1961** F. G. CASSIDY *Jamaica Talk* xvi. 350 Ribbon and stripe or striped cane..is still being grown.

ribbon, *v.* Add: **3.** Of a road, track, etc.: to continue in the manner of a ribbon; to stretch *out* like a ribbon. Also *transf.* Hence **ri·bboning** *vbl. sb.* and *ppl. a.*

1926 J. B. AMES *Valley of Missing Men* iv. 29 The trail ribboned endlessly through a rough, hill country that seemed utterly deserted. **1926** *Hutchinson's Best Story Mag.* Nov. 88/2 He led it [*sc.* a horse] silently down to the river bank, where the great white road ribbons out eastward to the sea. **1959** *Observer* 30 Aug. 7/1 The same ferro-concrete and glass blocks are going up in Zürich and in New Delhi, accompanied..by the same ribboning-out of metropolis into suburbia. **1968** R. V. BESTE *Repeat Instructions* xix. 197 The Bestyet Fish Saloon was frying and a queue of customers ribboned out through the door on to the pavement. **1969** 'J. MORRIS' *Fever Grass* xxi. 185 The road ribboned in three lazy twists. **1976** H. TRACY *Death in Reserve* xxi. 168 He drove back..looking at the ribboning road with unblinking bright eyes.

ribbonism. (Earlier example.)

1828 ANGLESEY *Let.* 31 Mar. in R. Peel *Mem.* (1856) I. 34 Ribbonism is extending.

ribby, *a.* **1.** Delete *rare*[-1] and add later examples. Also *fig.*, suggestive of or resembling ribs.

1851 H. MELVILLE *Moby Dick* II. xv. 129 In bony, ribby regions of the earth. **1924** C. E. MONTAGUE *Right Place* ix. 122 All sorts of ribby ridges and intercostal hollows dropping down from that spine to the water-line on each side. **1934** T. WOOD *Cobbers* viii. 101 The homesteads, iron roofed and set about with ribby water-butts. **1970** *Daily Tel.* 7 May 17/1 The fortunes of Courtaulds were founded in Victorian times on black crêpe,..a ribby material widows and mourners wore. **1977** 'H. OSBORNE' *White Poppy* xv. 118 The horses..were miserable creatures, ribby and pathetic.

2. *slang.* Dirty, shabby; seedy, run-down; unpleasant, nasty.

1936 J. CURTIS *Gilt Kid* 33 Nearby was a little café... Ribby kind of a gaff, but I might as well go in. **1976** P. ALEXANDER *Death of Thin-Skinned Animal* xx. 207 She lived at the ribby end of Maida Vale. **1977** M. RUSSELL *Dial Death* i. iv. 28 'How are—things?' 'Ribby'.

ribible. (Further examples in *poet.* use.)

a **1770** CHATTERTON *Poems* (1777) 3 The swote ribible dynning yn the dell. **1928** BLUNDEN *Retreat* 19 And Chatterton's ribibles dinned in the dell.

ribitol (rəi·bitǫl, ri·bitǫl). *Chem.* [f. *RIB(OSE + *-ITOL.] A colourless crystalline pentahydric alcohol, $HOCH_2{\cdot}(CHOH)_3{\cdot}CH_2OH$, which is obtained by reduction of ribose and occurs in the leaves of pheasant's eye, *Adonis vernalis*; also known as *adonitol.*

1946 *Adv. Carbohydrate Chem.* II. 117 Upon recrystallization from ethanol the melting point and mixed melting point with an authentic sample of ribitol was 101·5–102°. **1948** *Jrnl. Amer. Chem. Soc.* LXX. 2809/1 All of the acetates of the pentitols have been described in crystalline form save that of ribitol (synonym, adonitol). **1954** [see *PENTITOL]. **1968** A. WHITE et al. *Princ. Biochem.* (ed. 4) xli. 913 When ribitol is present, the sugar substituent appears always to be linked to the 2-position of this alditol. **1970** R. W. MCGILVERY *Biochem.* x. 188 Ribitol is the alcohol corresponding to ribose rather than the sugar itself, so riboflavin is not a nucleoside.

Hence **ri·bityl** [-YL], the univalent radical derived from ribitol by the loss of a terminal hydroxyl group.

1946 *Jrnl. Org. Chem.* XI. 83 Hydrogenation of these triacyl arylamine ribosides yields triacyl ribityl amines. **1970** R. W. MCGILVERY *Biochem.* x. 189 The ribityl group is a part of the flavin, not something extra.

riblet. (Further examples.)

1949 M. MEAD *Male & Female* xviii. 375 In ancient Samoa, the women made lovely bark-cloth, pressing out the fluctuating, beautifully soft lines against mats on which the pattern was sewed in coconut-leaf riblets. **1962** *Aeroplane* 23 Aug. 24/3 To this metal structure are attached the glass-fibre leading- and trailing-edge units. The former is in one piece from tip to engine nacelle and contains 11 glass-fibre riblets. **1975** *Nature* 16 Oct. 577/2 It [*sc.* a fossil bivalve of genus *Neoschizodus*] is closely related to the widely distributed *N. laevigatus* (Ziethen) of Lower to Middle Triassic age, but differs in having a radial riblet on the posterior area, and in possessing striations on some teeth.

ribo- (rəi·bo). *Biochem.* Comb. form of *RIBOSE: **ri:bohomopo·lymer** [*homopolymer* s.v. *HOMO-], a polymer of any one of the ribonucleosides; **ribonu·clease** [*NUCLEASE], any enzyme which catalyses the hydrolysis of RNA into oligonucleotides and smaller molecules; **ri:bonucleopro·tein** [*NUCLEOPROTEIN], a combination of a protein with RNA; **ribonu·cleoside,** a nucleoside containing ribose; **ribonu·cleotide,** a nucleotide containing ribose. Also *RIBOFLAVIN, *RIBONUCLEIC *a.*

1971 *Nature* 23 July 254/1 It seems probable that both the ribohomopolymers and the natural ribonucleic acids do catalyse a reaction *in vitro.* **1975** *Ibid.* 25 Sept. 327/2 Enzymatic activities for synthesising four ribohomopolymers, including poly(G), have been found in rat liver, and all are stimulated by RNA. **1938** DUBOS & THOMPSON in *Jrnl. Biol. Chem.* CXXIV. 502 The ribonuclease used in the following experiments was extracted from a commercial preparation of dried pancreatin. **1946** *Nature* 27 July 129/2 Many enzymes are not specific as to substrate; but ribonuclease, a depolymerase, is highly specific for ribonucleic acid. **1971** J. Z. YOUNG *Introd. Study Man* xxvi. 372 Haldane noted that ribonuclease is the smallest known enzyme, containing 124 amino-acid residues. **1940** *Science* 18 Oct. 361/1 The virus of equine encephalomyelitis (Eastern strain) is a complex of high molecular weight consisting of phospholipids, cholesterol, fatty acid and ribonucleoprotein. **1968** A. WHITE et al. *Princ. Biochem.* (ed. 4) ix. 204 Since that time, many additional plant viruses have been isolated as crystalline substances, and they are all ribonucleoproteins. **1974** D. & M. WEBSTER *Compar. Vertebr. Morphol.* ix. 182 The perikaryon [of a neuron] contains..the same organelles found in most cells. However, it has a far greater concentration of densely folded, rough endoplasmic reticulum (ribonucleoprotein). **1931** LEVENE & BASS *Nucleic Acids* vi. 129 A fifth ribonucleoside was discovered by Benedict and his coworkers. **1968** A. WHITE et al. *Princ. Biochem.* (ed. 4) xxix. 667 Synthesis of RNA occurs only when all four ribonucleoside triphosphates are present. **1929** *Jrnl. Biol. Chem.* LXXXI. 215 There occur in nature ribonucleotides of two types. **1959** *Ibid.* CCXXXIV. 2351/1 Studies designed to elucidate the mechanism by which ribonucleotides are converted into deoxynucleotides in extracts of *Escherichia coli.* **1970** R. W. MCGILVERY *Biochem.* xx. 473 The deoxyguanosine and deoxyadenosine nucleotides required for synthesis of DNA are formed by reduction of the corresponding ribonucleotides.

riboflavin (rəiboflē[i]·vin). *Biochem.* Also **-ine.** [a. G. *riboflavin* (P. Karrer et al. 1935, in *Helv. Chim. Acta* XVIII. 429): see *RIBO- and *FLAVIN 2.] Vitamin B_2; a yellow pigment that is a flavin having a ribityl side-chain and is present in many foods (esp. milk, liver, and green vegetables), deficiency of which leads to poor growth and deterioration of the skin.

1935 *Brit. Chem. Abstr.* A. 1286/2 Synthetic *d*-riboflavin.. has a growth-promoting power equal to that of the purest samples of natural lactoflavin, with which it appears in other respects to be identical. **1945** [see *ANEURIN, -INE]. **1954** *New Biol.* XVII. 117 In rats, riboflavin deficiency produces cleft palate and shortening of the limbs. **1967** M. KENYON *Whole Hog* xvi. 166 Thirty grams of niacin and about ten each of pantothenic acid and riboflavin. **1970** R. W. MCGILVERY *Biochem.* xxvii. 676 Riboflavin not only occurs in most foods but is also synthesized by the intestinal flora. As a result, a primary deficiency of riboflavin is very rare.

ribonucleic (rəiboniū·klī,ik, -niukli·ik, -ē[i]·ik), *a. Biochem.* [f. *RIBO- + *NUCLEIC *a.*] *ribonucleic acid*: a generic term for any of the nucleic acids yielding ribose on hydrolysis; they occur chiefly in the cytoplasm of cells, where they direct the synthesis of proteins, and in some viruses, where they also store the genetic information. Abbrev. RNA.

1931 LEVENE & BASS *Nucleic Acids* ix. 299 Yeast nucleic acid..is the most readily available starting material for chemical work on ribonucleic acids. **1944** *Ann. Reg.* 1943 354 Ribonucleic acids..were found to be widely distributed in high concentrations in mammalian tissues. **1958** *Times* 26 Sept. 15/2 This slightly damaged ribonucleic acid may be the key to the action of interferon. **1967** *Daily Tel.* 8 June 14/6 High concentration in the brain of a chemical known as ribonucleic acid greatly increases the capacity for learning. **1971** J. Z. YOUNG *Introd. Study Man* xi. 147 Nerve cells are characterized by a larger amount of ribonucleic acid in their cytoplasm.

ribose (rəi·bō[u]z, -s). *Biochem.* [a. G. *ribose*, f. *rib-onsäure* (both E. Fischer 1891, in *Ber. d. Deut. Chem. Ges.* XXIV. 4215), f. *ribon* (formed arbitrarily by rearrangement of some of the letters of *arabinose*, name of the related sugar from which Fischer prepared ribose) + *säure* acid: see -OSE[2].] **1.** An aldopentose sugar, the lævorotatory (D-) isomer of which occurs widely in nature as a constituent of many nucleosides and several vitamins and enzymes.

1892 *Jrnl. Chem. Soc.* LXII. I. 439 The precipitate..contains the chief portion of the ribose. **1927** [see *PENTOSE]. **1937** *Nature* 30 Oct. 745/2 Phosphopyridine nucleotides.. consist of a pyridine derivative, namely, nicotine acid amide, adenine, two molecules of a sugar, ribose, together with..phosphoric acid. **1938** *Thorpe's Dict. Appl. Chem.* (ed. 4) II. 287/1 Ribose is also a component of lactoflavin. **1948** W. W. PIGMAN *Chem. of Carbohydrates* iii. 104 The universal occurrence of D-ribose in all living cells should make this sugar of the greatest interest to biochemists and biologists. **1960** *New Biol.* XXXI. 40 Another type of nucleic acid containing a different sugar, ribose, and named ribonucleic acid or RNA. **1970** AMBROSE & EASTY *Cell Biol.* vii. 247 The pentose sugar ribose is formed from glucose by an oxidative pathway known as the hexose monophosphate shunt.

2. *attrib.* or as prefix = *RIBO-; **ribose nucleic, ribosenu·cleic acid** = **ribonucleic acid*; also *ribose nucleoprotein, nucleotide*, etc.

1920 *Jrnl. Biol. Chem.* XLIII. 329 Referring only to the ribose nucleotides, the observations are as follows. **1929** *Ibid.* LXXXI. 712 The sugar was obtained in the same manner as in the case of the ribose nucleoside. **1942** *Ibid.* CXLIV. 589, 5 to 6 per cent of the virus is a ribosenucleic acid. **1943** *Nature* 11 Dec. 693/1 Ribose nucleic acid is not restricted to plant tissues. **1944** *Ibid.* 10 June 711/2 The chloroplasts..contain plasmonucleoproteins (ribose nucleoproteins). **1957** *New Biol.* XXIII. 84 Ribonuclease, an enzyme that destroys ribose nucleic acid. **1960** *Ibid.* XXXI. 16 A different kind of nucleic acid (RNA, ribosenucleic acid, which differs from DNA in having a slightly different pentose sugar). *Ibid.* 17 RNA is found in living material closely combined with protein, as ribosenucleoprotein.

ribosome (rəi·bosō[u]m). *Cytology.* [f. *RIBO-(NUCLEIC *a.* + *-SOME[4].] Each of the particles of ribonucleic acid and associated proteins found in the cytoplasm of living cells, which bind to messenger RNA and synthesize polypeptides.

1958 R. B. ROBERTS *Microsomal Particles & Protein Synthesis* p. viii, To some of the participants, microsomes mean the ribonucleoprotein particles of the microsomal fraction contaminated by other protein and lipid material; to others, the microsomes consist of protein and lipid contaminated by particles. The phrase 'microsomal particles' does not seem adequate, and 'ribonucleoprotein particles of the microsome fraction' is much too awkward. During the meeting the word 'ribosome' was suggested... The present confusion would be eliminated if 'ribosome' were adopted to designate ribonucleoprotein particles in the size range 20 to 100 S. **1961** *Times* 18 Aug. 12/4 Protein synthesis seems in fact to be localized in small bodies, known as ribosomes. **1967** *New Scientist* 28 Sept. 661/1 The ribosomes of mitochondria are probably of the 70S (bacterial) variety. **1974** M. C. GERALD *Pharmacol.* iii. 52 Some of the reticulum appears 'rough', because of the presence of dense bodies termed ribosomes.

Hence **riboso·mal** *a.*, of or pertaining to a ribosome; *ribosomal RNA*, the RNA of a ribosome.

1960 *Jrnl. Molecular Biol.* II. 109 It appears likely..that ribosomal protein, like histone.., consists largely of α-helical protein. **1961** *Nature* 13 May 581/2 It was thought most likely that ribosomal RNA was genetically specific. **1970** AMBROSE & EASTY *Cell Biol.* iii. 114 It now seems that one of the chief metabolic functions of the nucleolus is the synthesis of ribosomal RNA, and possibly even the formation of complete ribosomal particles..from RNA and ribosomal proteins.

ribosyl (rəi·bosəil, -il). *Biochem.* [f. *RIBOS(E + -YL.] A univalent radical derived from ribose by loss of a hydroxyl group.

1947 *Jrnl. Biol. Chem.* CLXXI. 377 The preparation from the deacylated free nucleosides with a 5 to 10 per cent excess of chlorine, as described for the synthesis of 1-D-ribosyl-5-chlorouracil, is undoubtedly the method of choice. **1957** *Biochem. Jrnl.* LXV. 609 (*heading*) Enzymatic transfer of the ribosyl group from inosine to adenine. **1970** R. W. MCGILVERY *Biochem.* xx. 465 The rate of the reaction is probably always limited by the supply of the ribosyl compound.

ribuck, var. *RYEBUCK.

ribulose (rəi·biŭlō[u]z, -s). *Biochem.* [f. *RIB(OSE + *-ULOSE[2].] A ketopentose sugar which in the form of phosphate esters is an important intermediate in carbohydrate metabolism and photosynthesis.

1936 *Jrnl. Biol. Chem.* CXV. 731 The formation of *d*- and *l*-ribulose from *d*- and *l*-arabinose respectively has been demonstrated. *Ibid.* 747 Five portions (75 gm. each) of *l*-arabinose, treated with pyridine.., gave a total of some 100 gm. of crude ribulose. **1959** I. L. FINAR *Org. Chem.* (ed. 3) I. xviii. 433 Two monosaccharides, ribulose and sedoheptulose, play an essential part in the photosynthesis of carbohydrates. **1959** A. WHITE et al. *Princ. Biochem.* (ed. 2) xviii. 417 Were all the 3-phosphoglyceric acid converted to hexose by reversal of glycolysis,..no ribulose diphosphate would be available to serve as acceptor for CO_2 in subsequent fixation reactions. **1976** *Nature* 27 May 344/2 It is easy to formulate schemes for the synthesis of 2'-amino-2'-deoxyribose from ribose or ribulose and ammonia..but we have been unable to obtain appreciable yields of the amino sugars in this way.

ribwork. (Earlier example.)

1848 E. BRYANT *What I saw in California* xxi. 271 These *rancherias* consist of a number of huts constructed of a rib-work or frame of small poles.

Ricard (ri·kaɹ). [Name of the manufacturers.] The proprietary name of an aniseed-flavoured aperitif; a drink of this.

1965 *Trade Marks Jrnl.* 21 July 977/1 Ricard... Aperitif wines containing aniseed for sale in England, Scotland and Wales. Ricard (a Société anonyme organised under the laws of France),..Paris, France; distillers and merchants. **1967** G. GREENE *May we borrow your Husband?* 11, I sat with a Ricard on the terrace. **1970** T. KENRICK *Only Good Body* 8 Sit down at your table, order a Ricard. **1971** *Guardian* 22 May 7/4 The local aperitif is Ricard, a dry *pastis* taken with plenty of ice and water which makes it change colour in much the same manner as Dettol. **1976** 'A. YORK' *Dark Passage* i. 6 Harrington..ordered a Ricard and another beer.

ricasso (rīka·so). [It.] The part of the blade of a sword that is next to the hilt.

1884 R. F. BURTON *Bk. of Sword* vii. 125 In the Italian foil, which preserves the plate, the section of the blade between that and the grip is called the *Ricasso.* **1885** E. CASTLE *Schools & Masters of Fence* xv. 235 We have adopted the French word pas d'âne, and we may as well likewise, for want of a better, adopt the Italian word 'ricasso', used to designate that part of the blade between the cup guard and the quillons of the Italian foils and duelling swords. **1948** *Proc. Prehist. Soc.* XIV. 155 Only those swords with parallel-sided blades, ogival tips and deep rectangular ricassos have been included. **1959** G. SAVAGE *Antique Collector's Handbk.* 23 Swords are often decorated in a variety of ways... A common place for decoration is the *ricasso*—the flat, rectangular part of the blade immediately below the hilt. **1970** F. WILKINSON *Edged Weapons* iii. 38/2 The use of the *ricasso* to obtain a better grip on the sword had led to the introduction of loops, rings and bars attached to the hilt. **1978** N. K. SANDARS *Sea Peoples* 100 In neither area did the parallel-edged flange-hilted sword develop a leaf-edged blade for strong cutting action, nor a milled ricasso at the grip.

Ricci (ri·tʃi). *Math.* [Name of C. G. *Ricci* (1853–1925), Italian mathematician.] *Ricci tensor*: a symmetric second-order tensor obtained by contracting the Riemann–Christoffel tensor.

1923 VEBLEN & THOMAS in *Trans. Amer. Math. Soc.* XXV. 554 This we shall call the Ricci tensor because it reduces to the tensor studied by Ricci for the case of the Riemann geometry. **1926** L. P. EISENHART *Riemannian Geom.* i. 22 The Ricci tensor..was first considered by Ricci who gave it a geometrical interpretation in case g_{ij} is the fundamental tensor of a Riemann space. **1967** CONDON & ODISHAW *Handbk. Physics* (ed. 2) II. vi. 50/1 The vanishing of the Ricci tensor does not imply the vanishing of the Riemann-Christoffel tensor and consequently does not insure the Minkowski character of Einstein space-time in the large. **1974** *Nature* 5 Apr. 472/1 The precise mathematical quantity which describes the curvature of space-time is known as the Ricci tensor R_{jk}, which is a 4 × 4 array of numbers defined at each point of space-time.

rice, *sb.*[1] **4.** Delete † *Obs.* and add later examples.

1895 R. MARSDEN *Cotton Weaving* viii. 272 The hanks are placed upon light, collapsible hexagon reels termed rices. **1953** A. JOBSON *Househ. & Country Crafts* xi. 124 The winder worked in conjunction with a wrap wheel, or an adjustable wool winder, which was a stand to which were attached rices or runners. **1957** SIMPSON & WEIR *Weaver's Craft* (ed. 8) viii. 94 It is a great saving of time if a 'rice' or 'swift' is available, in which case the wool may be taken directly from the skeins to the warping board.

5. rice creel, a frame for holding rices (sense 4).

1895 R. MARSDEN *Cotton Weaving* viii. 272 They are very light, and easily revolve with the pull of the thread. This is termed the rice creel.

rice, *sb.*[2] **4. a.** Substitute for def.: In full, *wild rice.* An aquatic North American grass, *Zizania aquatica*, or its seeds. (Earlier and later examples.)

1775 A. HENRY *Trav. & Adventures Canada* (1901) 242 The women brought me a further and very valuable present, of twenty bags of rice. **1950** *Chicago Tribune* 18 Mar. 10/3 New rice will stir in the lake's dark bed.

5. a. *rice ball, bran, brandy, bun, cracker, -flake, flour* (further examples), *gin, meal* (earlier example), *mould, noodle, oil, polish, pudding* (earlier and later examples), *-salad, -slop, soup, spirit.*

1850 *New England Farmer* II. 322/1 Rice Balls.—Pour upon half a pound of rice three pints of boiling milk [etc.]. **1974** G. JENKINS *Bridge of Magpies* ix. 141 His contribution consisted of some rice balls and bright sticky sweets. **1909** *Cent. Dict.* Suppl., Rice-bran. **1970** SIMON & HOWE *Dict. Gastron.* 327/1 Rice bran is the first major by-product removed from the rice kernel during the milling process. **1937** M. COVARRUBIAS *Island of Bali* v. 109 *Arak*, distilled rice brandy. **1971** R. PURVIS *Treasure Hunting in Brit. Columbia* ix. 114 Old rice-brandy flasks are considered..rare collectors' items. **1889** J. WHITEHEAD *Steward's Handbk.* 420/1 Ground Rice Buns. **1963** A. L. SIMON *Guide Good Food & Wines* 222/1 Rice Buns. **1970** J. KIRKUP *Japan behind Fan* 3 Peppery rice crackers wrapped in seaweed. **1973** *Sunday Bull.* (Philadelphia) 7 Oct. (Parade Suppl.) 28/3 Fredericks is a six-meals-a-day man, and so paused two hours after breakfast for an ounce of Gruyere cheese, some brown rice crackers and a glass of milk. **1947** J. STEINBECK *Wayward Bus* i. 5 Boxes of cornflakes, riceflakes, grapenuts, and other tortured cereals. **1970** SIMON & HOWE *Dict. Gastron.* 327/1 There are rice flakes..that may be eaten cold. **1878** *Amer. Home Cook Bk.* 15 Mix the rice-flour smoothly with the water. **1959** A. K. LANG in H. Q. Masur *Murder most Foul* (1973) 71 His pushers would..cut the strong brown heroin with lactose or rice flour. **1971** *Fashion Panorama* (Ceylon) Apr.–June 30 The pounding of rice flour continued. **1969** *Listener* 12 June 814/1 The village chief himself asked us to a dinner of dried deer and shrimp crackers, chicken and lettuce, sweet potato and duck and rice gin. **1789** G. WHITE *Selborne* II. lviii. 280 The dogs..are fed for the table with rice-meal and other farinaceous food. **1906** *Mrs. Beeton's Bk. Househ. Managem.* 1044 (*heading*) Whole rice mould. **1922** 'R. CROMPTON' *More William* ii. 31 Rice-mould! Every single day. I *hate* it, don't you? **1965** H. BURKE *Chinese Cooking for Pleasure* 17/2 Rice noodles with prawn and chicken. **1980** *Times* 6 Dec. 11/3 Chicken.. and rice noodle Army-style as a change from rice. **1970** SIMON & HOWE *Dict. Gastron.* 327/1 Rice oil is the oil extracted from rice bran and polish. **1972** *Guardian* 25 Sept. 1/6 Rice oil—a widely used foodstuff. **1909** *Cent. Dict. Suppl.*, Rice-polish. **1970** SIMON & HOWE *Dict. Gastron.* 327/1 Rice polish is the final layers removed from the rice kernel during the polishing process. **1709** W. KING *Useful Trans. in Philos.* III. 53 Of these I shall discourse at large, when I treat of butter'd Wheat,..Rice-Pudding, and Oatcakes. **1948** *Good Housek. Cookery Bk.* 427 *Fruit Rice Pudding.* As for Rice Pudding, adding a cupful of seeded raisins..to the milk and rice. **1966** F. SHAW et al. *Lern Yerself Scouse* 75 *Yer cudden knock de skin off a rice puddn*, you are a weakling. **1973** 'I. DRUMMOND' *Jaws of Watchdog* ix. 122 Melons were sliced into, prawns shelled, rice-salads scooped out. **1922** JOYCE *Ulysses* 391 She queasy for a bowl of riceslop. **1747** H. GLASSE *Art of Cookery* ix. 78 A Rice Soop. **1797** J. WOODFORDE *Diary* 8 May (1931) V. 34 We had for Dinner, some Rice Soup..rost Beef &c. **1827** M. WILMOT *Jrnl.* 27 Aug. in *More Lett.* (1935) 297 The supper for all begun with a cup of rice soup, which was handed round to each. **1909** J. JOYCE *Let.* 20 Dec. (1966) II. 277, I would like roast beef[,] rice-soup, [etc.]. **1963** A. L. SIMON *Guide Good Food & Wines* 220/2 Rice Soup. **1967** O. WYND *Walk Softly* i. 9 Crack a couple of bottles of rice spirit over two bows on the same day. **1978** *Nagel's Encycl.-Guide: China* 1447 Rice spirit, known as yellow wine, is drunk warm, from little china bowls.

b. *rice-clearing, -country* (later example), *-flat, lake, paddy, plantation* (earlier and later examples), *swamp* (earlier and later examples), *terrace.*

1895 CONRAD *Almayer's Folly* vi. 116 Finding shelter under that man's roof in the modest rice-clearing. **1975** *Country Life* 2 Jan. 9/2 Georgia [USA] proved good rice country. **1882** *Trans. Asiatic Soc. Japan* X. 44 The river, after making a bend round some rice flats, turns in and sweeps right under the high bank. **1905** 'L. HOPE' *Indian Love* 57 The velvet rice-flats lie so emerald green. **1820** J. D. DOTY *Let.* 27 Sept. in *Wisconsin Hist. Coll.* (1876) VII. 199 The Indians around Sandy Lake, in the month of September, repair to Rice Lake to gather their rice. **1831** J. M. PECK *Guide for Emigrants* 14 A third [boat] may start from the rice lakes at the head of the Mississippi. **1870** J. W. MCCLUNG *Minnesota* 235 It is finely watered by Rum River passing out of Mille Lac through numerous small marshy rice lakes. **1933** N. WALN *House of Exile* i. 18 Rice paddies roughened by dead stubble. **1945** *Newsweek* 24 Sept. 50 Calling themselves members of 'The Rice Paddy Navy', American naval personnel—disguised as Chinese—set up more than fifty weather stations in unoccupied China. **1977** J. CLEARY *High Road to China* v. 186 The road that ran through the rice paddies. **1787** in *Documentary Hist. Amer. Industr. Society* (1910) I. vii. 323 Great Encouragement will be given to an Overseer of a sober industrious Character, to manage a Rice and Lumber Plantation, about Thirty Miles from Charles-Town. **1957** P. WORSLEY *Trumpet shall Sound* iv. 81 Another cult-leader..was interested in government schemes for establishing rice-plantations. **1975** *Guardian* 2 Jan. 38/1 In Colonial days each [estate] was a prosperous rice plantation. **1731** *Pennsylvania Gaz.* 29 Apr.–6 May 2/2 These Rice Swamps are flat low Grounds, by the Sides of Rivers or Runs. **1957** M. BANTON *W. Afr. City* i. 21 The first clearing of rice swamps in the Scarcies involved much arduous and dangerous work. **1952** *Oxf. Jun. Encycl.* VI. 378/1 (*caption*) Rice terraces in Bali, Indonesia. **1973** D. MAY *Laughter in Djakarta* iii. 52 They were driving right past the rice terraces... Women were..thinning out the young rice on several of the terraces.

c. *rice-barn, -crater, -crop* (later examples), *cultivation, diet, husk, kernel, kettle, -measure, -port, -pot, sack, shoot; rice-white* adj.

1901 *Chambers's Jrnl.* Aug. 538/1 Nearer the water a few small houses..and a higher two-storied rice-barn. **1926** T. E. LAWRENCE *Seven Pillars* (1935) IV. xlvi. 266 Two raised each smaller cauldron and tilted it, letting the liquid splash down upon the meal till the rice-crater was full, and the loose grains at the edge swam in the abundance. **1965** 'LAUCHMONEN' *Old Thom's Harvest* ii. 24 We aint gwine get no water to plant ricecrop. **1973** 'B. MATHER' *Snowline* iii. 34 A tidal wave in the Bay of Bengal last year that.. ruined the rice crop. **1841** M. EDGEWORTH *Let.* 14 Mar. (1971) 584 Entertaining accounts by master of agricultural-different processes in different parts of the world—Rice cultivation—Irrigation—India. **1960** I. A. STANTON *Dict. for Med. Secretaries* 44/2 *Rice diet*, a diet consisting of rice, sugar, fruit and fruit juice. **1965** H. BURKE *Chinese Cooking for Pleasure* 12/2 Chinese pickled eggs..[are] bought ready to eat. One simply cracks and removes their outer casing of rice husks and clay. **1975** *Country Life* 16 Jan. 153/3 A pillow of rice-husks. **1970** SIMON & HOWE *Dict. Gastron.* 326/1 The edible rice kernel is found in a hard shell-like hull surrounded by several layers of bran. **1895** CONRAD *Almayer's Folly* v. 86 The fire was burning in the cooking shed, with the rice kettle swinging over it. **1936** *Burlington Mag.* Jan. 40/2 The great Imperial rice-measure from the Metropolitan Museum. **1903** CONRAD *Typhoon* i. 6 Outlandish names of lumber-ports, of rice-ports, of cotton-ports. **1862** MRS. MASON *Civilizing Mountain Men* I. ii. 24 Holding in his hand a wicker rice-pot, which shone in the dimness like a great bowl of gold. **1895** CONRAD *Almayer's Folly* iv. 81 He eyed..the aged statesman sitting..by his domestic rice-pot. **1978** L. DEE tr. *Hsia Chih-Yen's Coldest Winter in Peking* vii. 133 They took these scraps, even the burnt crusts in the rice pots, and made them into 'nutrition supplements'. **1970** 'D. HALLIDAY' *Dolly & Cookie Bird* iii. 33 You all meet over the trolleys with your long-grain rice sacks at the mainline Cash & Carry. **1895** CONRAD *Almayer's Folly* xi. 216 In the middle of a shadowless square of moonlight, shining on a smooth and level expanse of young rice-shoots, a little shelter-hut perched on high posts,..seemed very small. **1966** 'A. HALL' *9th Directive* xxv. 229 The flooded earth where the tender rice-shoots stood. **1973** 'B. MATHER' *Snowline* xiv. 167 Village women.. were dibbling rice shoots into the monsoon-flooded paddy. **1871** G. M. HOPKINS *Jrnls. & Papers* (1959) 216 In returning the sky in the west was in a great wide winged or shelved rack of rice-white fine pelleted fretting.

6. a. *rice-cooker, -farmer, -husker, -planter* (earlier and later examples).

1974 *Daily News* (Tanzania) 27 Sept. 8/4 (Advt.), Complete Akai stereo system,..rice-cooker, crockery, cutlery, imported suits. **1977** *South China Morning Post* (Hong Kong) 22 July 1/5 My daughter was taking a broken rice-cooker to a repair shop. **1961** *Nat. Geogr. Mag.* Aug. 242/1 Practically all the people of Laos,..about two million of them—are rice farmers. **1977** P. THEROUX *Consul's File* 127 The Malay rice-farmers met and decided to bring their problem to a *bomoh*—a medicine-man. **1901** KIPLING *Kim* iv. 106 They could hear the old lady's tongue clack as steadily as a rice-husker. **1775** *Amer. Husbandry* I. 66 It concerns only those who have dealings with London, these are the tobacco and rice planters. **1838** J. J. AUDUBON *Ornith. Biogr.* IV. 318 In South Carolina the Golden-eye is abundant during winter, when it at times frequents the reserves of the rice-planters. **1949** C. S. MURRAY *This our Land* iv. 74 With favorable prices, and slave labor comparatively cheap, quite a few rice planters piled up large fortunes.

b. *rice-dressing, -growing* (further examples), *-planting.*

1901 *Chambers's Jrnl.* Feb. 125/1 An English firm.. erected a rice-dressing mill on the shores of the Caspian. **1946** *Nature* 5 Oct. 462/1 A rice-growing village neither very rich nor very poor. **1972** *Country Life* 28 Dec. 1796/2 Rice growing is another highly photogenic occupation. **1852** W. G. SIMMS *Sword & Distaff* xxx. 210 It's lucky I do know something of rice planting. **1903** 'P. PENNINGTON' *Woman Rice Planter* (1913) i. 1 You have asked me to tell of my rice-planting experience. **1962** E. SNOW *Red China Today* (1963) xxviii. 212 Of numerous rice-planting machines invented by peasants the most popular is operated by two people. **1965** 'LAUCHMONEN' *Old Thom's Harvest* i. 8 He.. dream about them blasted ricefields and riceplanting.

7. rice and peas = *peas and rice* s.v. *PEA[1] 1 d; **rice bowl,** (*a*) a dish out of which rice is eaten; (*b*) an area in which abundant quantities of rice are grown; also *attrib.*; **rice-Christian,** substitute for def.: one, esp. an Asian, who adopts Christianity for material benefits; (further examples); **rice crispies** = *Rice Krispies* below; also *sing.*; **rice-convert** = *rice-Christian* above; **rice-grain,** (*a*) *pl.* (in Dict.); (*b*) used *attrib.* to describe a type of decoration on porcelain in which perforations are made and allowed to fill with melted glaze; **Rice Krispies,** proprietary name of a breakfast cereal made from rice; **rice powder,** a face powder with a pulverized rice base; hence *rice-powdered* adj.; **rice-shell** (earlier example); **rice-stitch,** (*b*) a type of cross-stitch (see quots.); **rice table, tafel** = *RIJSTTAFEL.

1947 E. N. BURKE *Stories told by Uncle Newton* II. 8 Some people call the midday meal 'lunch' but we always called it dinner on Sundays. As usual it consisted of a half moon of rice and peas; a few blocks of yams, [etc.]. **1958** B. HAMILTON *Too Much of Water* iv. 74 He swore that all three must come to a real Barbadian breakfast..at Fair Hope. 'Maan,' he said, 'I give you flying fish an' pepper-pot, an' pudding and souse, an' rice and peas.' **1969** *Daily Tel.* 11 Jan. 14/1 To bring tears to ex-patriate Jamaican eyes just mention..rice an' peas. **1922** W. S. MAUGHAM *On Chinese Screen* li. 206 The coolie's rice bowl has its rough but not inelegant adornment. **1943** *Sun* (Baltimore) 10 June 12/2 Hunan and Hupeh, 'rice bowl' provinces of East Central China. **1950** P. BOTTOME *Under Skin* xx. 166 On my tray there was a little porcelain bowl... She said, 'This is my rice bowl—my mother gave us each one.' **1950** *Times* 25 Apr. 7/2 In Cambodia and Laos, and even in Cochin-China, important as a rice-bowl area, the situation is better. **1969** *Commerce* (Bombay) 26 July 153/1 Agrarian unrest..has been a hardy annual for more than two decades now in Thanjavur district—Tamil Nadu's 'rice bowl' which has 1·4 million acres under paddy cultivation. **1972** 'M. HEBDEN' *Killer for Chairman* I. iv. 52 She moved an empty rice bowl on the table. **1975** *Times* 4 Mar. 1/6 According to reports from Cambodia, once the rice bowl of Indo-China, civilian deaths from starvation are increasing. **1979** *Times of India* 17 Aug. 14/3 Waters in Kuttanad, the rice bowl of Kerala, have been severely polluted by massive application of pesticides in paddy fields. **1883** *Encycl. Brit.* XVI. 518/1 The Propagation Society is now proclaiming the gospel in nearly six hundred and fifty villages in the Tinnevelly district, amongst not merely food-seeking 'rice Christians' but those who have had the courage to face severe persecution for joining the Christian church. **1941** A. J. CRONIN *Keys of Kingdom* (1942) IV. vii. 179 The wisdom of Father Chisholm's determination to have no rice-Christians in his flock was now apparent. **1959** P. FLEMING *Siege at Peking* iii. 40 The missions, whose less spiritually minded adherents were known as 'rice-Christians', thus tended to become centres of privilege. **1973** *Listener* 31 May 721/2 'Rice Christians', people who were outcasts in China..willing to come into the hands of the Church and the reformers. **1956** 'C. BLACKSTOCK' *Dewey Death* ii. 36 The fourteen-year-old at the breakfast table..can devour the Black Mass with her rice crispies. **1963** *Guardian* 5 Oct. 5/6 The mouse..eats a Rice Crispy like a sandwich. **1977** P. B. & J. S. MEDAWAR *Life Sci.* i. 14 The sound—as of innumerable mice eating Rice Crispies—that sometimes accompanies long-distance telephone calls. **1926** T. E. LAWRENCE *Seven Pillars* (1935) IX. xcix. 548 We wanted no rice-converts. Persistently we did refuse to let our abundant and famous gold bring over those not spiritually convinced. **1910** *Encycl. Brit.* V. 746/1 Sometimes the perforations are left clear, but in the rice-grain pattern the incisions are generally filled up with the melted glaze so that they become like so many windows in the walls of the piece. **1960** H. HAYWARD *Antique Coll.* 238/2 '*Rice-grain' decoration*, a decoration used on Chinese porcelain in which small perforations in the body are filled with transparent glaze; a technique adopted from Persian pottery, popular during the 18th cent. **1971** L. A. BOGER *Dict. World Pott. & Porc.* 285/2 The rice grain pattern is characteristic of the Ch'ien Lung period, 1736–1795. **1974** SAVAGE & NEWMAN *Illustr. Dict. Ceramics* 139 (*caption*) Persian pottery bowl with rice-grain piercing, 17th century. **1936** *Trade Marks Jrnl.* 30 Sept. 1212/1 Kellogg's Rice Krispies... A food made of rice, for human consumption. Kellogg Company of Great Britain Limited,..London,.. merchants. **1961** *Esquire* Aug. 57 She'd..pour the milk over the Rice Krispies, to wake up the fellow. **1967** *Trade Marks Jrnl.* 24 May 685/1 Rice Krispies... Cereal preparations principally of rice, being breakfast foods. Kellogg Company of Great Britain Limited,..Manchester; manufacturers. **1971** R. K. SMITH *Ransom* (1972) I. 22 A bowl of Rice Krispies with sliced bananas. **1975** *New Yorker* 7 July 32/1 Harry prepared his breakfast in the kitchen—Rice Krispies with skimmed milk, and Sanka sweetened with saccharin. **1874** M. T. NASH tr. *Cazenave's Female Beauty* i. 28 Various means are proposed as protection against them [*sc.* alterations produced by tan and sunburn]... That which I prefer most is a little pure rice powder. **1892** C. CAMPBELL tr. *Baroness Staffe's Lady's Dressing-Room* III. 335 If you buy your rice-powder, be careful not to choose it perfumed with orris-root. **1934** M. VERNI *Mod. Beauty Culture* xlii. 416 A simple and useful face powder is made in the following way—2 oz. of pure rice powder. 2 dr. of french chalk. ½ dr. of oxide of zinc. **1972** R. CORSON *Fashions in Makeup* xv. 355 Rice powder and powdered arrowroot starch were commonly used to remove shine from the skin. **1922** JOYCE *Ulysses* 523 As they are now, so will you be, wigged, singed,..ricepowdered. **1955** T. STERLING *Evil of Day* iii. 33 Mrs. Sheridan's rice-powdered hand made a familiar gesture. **1838** J. J. AUDUBON *Ornith. Biogr.* IV. 33 Those beautiful shells, which, on account of their resemblance to grains of rice, are commonly named rice-shells. **1934** M. THOMAS *Dict. Embroidery Stitches* 169 *Rice stitch*.., a Canvas Stitch usually worked in a thick thread for the large crosses and a fine thread for the smaller stitches. **1960** G. LEWIS *Handbk. Crafts* 87 Rice stitch has as its foundation a large cross, with the two arms double the size of the simple cross stitch, and when it has been made 4 short stitches are worked across all four corners of the main cross. **1976** P. CLABBURN *Needleworker's Dict.* 227/1 *Rice stitch*.., variation of cross stitch much used in modern canvas work and needlepoint. **1909** WEBSTER, Rice table. **1926** A. HUXLEY *Jesting Pilate* II. 184 The only truly Rabelaisian feature of Javanese diet is the Rice Table. **1943** D. WELCH *Maiden Voy.* xiv. 114 Mr MacDonald decreed we should have rice tafel.

rice, *v. Cookery.* [f. RICE[2].] *trans.* To press (food) through a coarse sieve to produce granular shapes. So **riced** *ppl. a.*, **ri·cing** *vbl. sb.*

1923 J. CONRAD *Handbk. Cookery* 16 If the potatoes are not to be used at once..it is a good idea either to rice them in a ricer or to mash them. **1926** I. C. SMITH *Blue Bk. Cookery* xxiii. 331 Riced Potatoes. Boil the potatoes. While still hot, press them through a potato-ricer. **1933** *Sun* (Baltimore) 27 Feb. 3/7 Boiled stuffed capon, with riced potatoes, cauliflower, [etc.]. **1947** M. GIVEN *Mod. Encycl. Cooking* II. 1502 Put hot, freshly boiled, peeled potatoes into a ricer... Rice the potatoes directly into a hot serving dish. **1957** M. MCCARTHY *Memories Catholic Girlhood* 224 A salad.. sprinkled with riced egg yolk. **1966** *Vogue* Nov. 148/2 Mashed, riced, boiled or baked potatoes. **1969** R. & D. DE SOLA *Dict. Cooking* 191/2 *Ricer*, utensil for ricing cooked vegetables and fruits by forcing them through a perforated container.

rice-bird. 3. Substitute for def.: *U.S.* One of several small birds found in rice-fields, esp. the bobolink, *Dolichonyx oryzivorus.* (Earlier and later examples.)

1731 M. CATESBY *Nat. Hist. Carolina* I. 14 The Rice-bird ..[is] esteemed in Carolina the greatest delicacy of all other Birds. **1957** O. BRELAND *Animal Friends & Foes* II. 70 Robins, meadowlarks, bobolinks, and even flickers were often served in restaurants as 'rice birds'. **1958** S. A. GRAU *Hard Blue Sky* 17 A yellow and black ricebird whizzed over his head.

rice grass (in Dict. s.v. RICE[2] 7). **a.** Either of two grasses of the genus *Leersia*, *L. oryzoides* (= CUT-GRASS) and *L. hexandra*, which are tall perennial grasses having rhizomatous roots and growing in wet ground.

1857 C. JOHNSON *Grasses Gt. Brit.* 4 *Leersia oryzoides.* Rice Grass... First noticed as a native of Britain in September 1844..at Henfield, Sussex. **1889** [in Dict. s.v. RICE[2] 7]. **1973** TOTHILL & HACKER *Grasses S.E. Queensland* I. 17 The coastal region provides conditions for numerous fresh-water streams, ponds, bogs, etc... Common grasses of these sites are water couch.., common reed.., swamp rice grass (*Leersia hexandra*).

b. *Austral.* and *N.Z.* In full *meadow rice grass.* A perennial grass, *Microlæna stipoides*, that grows in tufts from creeping rhizomes in semi-shaded ground. Also called *weeping grass.*

1889 [in Dict. s.v. RICE[2] 7]. **1973** ATKINSON & BELL in G. R. WILLIAMS *Nat. Hist. N.Z.* xv. 375/1 On Macauley Island..a combination of fire..and goats..had by 1966 reduced the vegetation to a short grassland dominated by rice grass, *Microlaena stipoides*, [etc.].

c. A grass of the genus *Spartina*, esp. *S. townsendii*, a tall, rhizomatous, perennial grass that grows on salt marshes and estuaries and has useful soil-binding properties. Also called *cord grass, marsh grass.*

The event referred to in quot. 1907[1] is said to have occurred in 1870.

1907 *1st Rep. R. Comm. Coast Erosion* 367/2 in *Parl. Papers* (Cd. 3683) XXXIV. 1 Many years ago a ship from the River Plate is said to have come in with a kind of rice grass. It was quite accidental: it came over in a wheat cargo. It sprang up near Southampton. These little plants, which are estuarine plants from Argentina, gradually spread out over Southampton Water, and the whole of Southampton Water from Carlshot Castle right up to Redbridge is entirely covered with what some people call 'Sea Rice' and others 'Rice Grass'. **1907** *Bull. Misc. Information R. Bot. Gardens, Kew* 196 In response to a letter asking whether any light could be thrown on the name 'rice-grass', which has now for the first time been quoted in connection with *Spartina*, the vernacular name hitherto given being 'cord-grass', Lord Montagu of Beaulieu, writes on May 2, 1907,:—'Mr. Rankin..told me he thought the plant growing on my foreshore was *Spartina stricta*, and not the *alterniflora.* He seemed to be quite aware of the term "rice-grass"... Personally I have never heard this word.' **1917** *Chambers's Jrnl.* Mar. 207/1 One of these schemes relates to a certain form of seaweed, locally known as rice-grass, which is rapidly covering Poole Harbour. **1944** J. S. HUXLEY *On living in Revolution* 72 During the last half-century a new type of rice-grass appeared in Western Europe, and has been so successful that the Dutch have used it to reclaim land from the sea. **1961** M. ASHBY *Plant Ecol.* x. 193 Wherever the vigorous rice-grass..is present it replaces most of the stages in the succession.., and may form almost a pure stand from pioneer mud-binding to the high marsh, which can be used for grazing.

d. *Austral.* and *N.Z.* Any of various grasses of the genus *Tetrarrhena* (see quots.).

1930 A. J. EWART *Flora of Victoria* 137 *T[etrarrhena] accuminata* R. Br., Pointed Rice Grass. Stems long and slender. *Ibid.* 138 *T. distichophylla* R. Br., Hairy Rice Grass. A tufted, branched perennial grass, creeping at the base, ascending to a height of 1 foot or more. **1962** N. C. W. BEADLE et al. *Handbk. Vascular Plants Sydney District & Blue Mts.* 529 *Tetrarrhena*... Wiry Rice Grass. **1962** J. H. WILLIS *Handbk. Plants Victoria* I. 91 *T. juncea*... Forest Wire Grass (Wiry Rice Grass, Tangle Grass.) **1973** TOTHILL & HACKER *Grasses S.E. Queensland* III. 261 *Tetrarrhena* sp. Wiry Rice Grass. A scrambling, wiry, but fairly slender perennial.

e. *U.S.* In full *Indian rice grass.* A perennial grass, *Oryzopsis hymenoides*, growing in clumps in semi-arid regions of the western U.S.

1935 A. S. HITCHCOCK *Man. Grasses U.S.* 415 Nearly all the species are highly palatable to stock, but are usually not in sufficient abundance to be of importance, except *O. hymenoides* (Indian ricegrass), which..furnishes much feed. **1968** F. W. GOULD *Grass Systematics* v. 195 *Oryzopsis hymenoides* (Roem. & Schult.) Ricker, Indian ricegrass.., is an important forage species.

f. *U.S.* In full *pinyon rice grass.* A perennial grass, *Piptochætium fimbriatum*, with stiffly erect stems and leaves that is common in open pine forests in the mountains of northern Mexico and the southwestern U.S.

1951 KEARNEY & PEEBLES *Arizona Flora* 115 (*heading*) *Piptochaetium.* Pinyon rice grass. **1973** R. H. WAUER *Birds of Big Bend National Park* 26 Grasses that include Pinyon Ricegrass.

ricer (rəi·səɹ). *Cookery.* [f. *RICE *v.* + -ER[1].] A utensil for ricing food.

1896 *Columbus* (Ohio) *Dispatch* 21 Nov. 11 Cook one quart of blanched chestnuts in boiling stock till tender, press them through a ricer [etc.]. **1936** *Sears Catal.* 655/1 Potato Ricer Removable heavy retinned seamless bowl... 26c. **1963** M. MCCARTHY *Group* xiv. 325 Girls today were.. opening a tin of baby food, instead of..pressing beef in the ricer for beef juice. **1969** [see *RICE *v.*]. **1970** SIMON & HOWE *Dict. Gastron.* 327/2 The ricer is usually built in the shape of a metal basket with a snugly fitting metal plunger which exerts pressure on the vegetables within as they are pushed through.

‖ **ricercar, ricercare** (ritʃərkā·r, -kā·re). *Mus.* Pl. **ricercare, ricercares, ricercari, ricercars.** [It., f. *ricercare* to search out.] Any of a variety of musical forms, characteristically in a fugue style, found between the 16th and 18th centuries (see quots.). Cf. RESEARCH *sb.*[1] 4.

1789 C. BURNEY *Gen. Hist. Music* III. ii. 176 Short instructions for composing masses, motets, psalms, madrigals, and *ricercari.* *Ibid.*, This term [*sc. ricercari*], which implied any work of fancy, and original invention, was succeeded by *fantasia*, as *fantasia* was by *sonata.* **1881** GROVE *Dict. Mus.* III. 126/2 *Ricercare* or *Ricercata*.., an Italian term of the 17th century, signifying a fugue of the closest and most learned description. Frescobaldi's Ricercari (1615), which are copied out in one of Dr. Burney's note-books.., are full of augmentations, diminutions, inversions, and other contrivances. **1944** W. APEL *Harvard Dict. Mus.* 643/1 These ricercares could be sung. *Ibid.* 644/1 In the later literature for the lute..the lute ricercar more and more approaches the style of the imitative ricercar. **1945** *Musical Q.* XXXI. 459 These works..fall into three types: the one monothematic ricercare, a few ricercari with one main subject..and the commonest type—the polythematic ricercari. **1947** A. EINSTEIN *Mus. Romantic Era* xvii. 294 The 'schools' of the ricercar, whether called fugue as in Germany, *tiento* as in Spain, or fancy as in England, were different, but the spirit and form were the same. **1954** *Grove's Dict. Mus.* (ed. 5) VII. 154/2 The distinction which has been drawn between the *ricercare* as a more learned and the *fantasia* as a more fanciful variety is not really supported by the facts. **1959** *Collins Mus. Encycl.* 547/2 Contrapuntal ricercars were written for instrumental ensembles in the 16th cent. **1961** I. HORSLEY in *Acta Musicologica* XXXIII. 29 (*title*) The solo ricercar in diminution manuals: new light on early wind and string techniques. **1961** *Times* 9 Sept. 11/3 The three-part ricercare for solo harpsichord starts the work. **1968** *New Oxf. Hist. Music* IV. xi. 557 The *ricercari* and fantasias up to the end of the sixteenth century all preserve the principal features of the motet. **1969** *Daily Tel.* 25 Jan. 12 Palestrina wrote few madrigals, and the ricercare is his only known purely instrumental compositions. **1976** D. MUNROW *Instruments Middle Ages & Renaissance* 72/1 Like keyboard instruments, the strings developed a vast solo repertory with their own idiomatic forms of the prelude, ricercar, and *tastar de corde.* **1977** *Folio* Winter 30 It seems very possible that the keyboard ricercare could be held up as the one and only time that J. S. Bach actually wrote with the piano rather than the harpsichord, clavichord or organ in mind. **1980** *Early Music* Apr. 248/1 The first-named turn out to be descendants of the lute preludes so described two centuries before, not ricercars in the fugal sense at all.

‖ **ricercata** (ritʃərkā·tă). *Mus.* Pl. **ricercate.** [It., pa. pple. (fem.) of *ricercare* to search out.] = *RICERCAR, RICERCARE.

The more usual term is now *ricercar(e.*

1740 J. GRASSINEAU *Mus. Dict.* 198 *Research*, or *Ricercata*, a kind of prelude or voluntary play'd on an Organ, Harpsichord, Theorbo, etc., wherein the composer seems to look out or search for strains and touches of harmony, which he is to use in the regular piece to be play'd afterwards. **1819** REES *Cycl.* XXX, *Ricercata*, a research, a flourish, a prelude, an impromptu, a voluntary. **1881** [see *RICERCAR, RICERCARE]. **1958** A. JACOBS *New Dict. Mus.* 310 Ricercata. **1977** *Early Music* July 415/1 Practising Virgiliano's ricercate without slurs..will also encourage properly relaxed double tonguing.

rice-water. Add: **1.** (Earlier example.)

1789 G. PARKER *Life's Painter* iv. 31 Genuine Hollands, or right Jamaica, qualified with rice-water, spirited lemons, and a proper dash of fragrant oranges.

2. (Earlier example.)

1833 *Lancet* 12 Oct. 116/2 The rice-water evacuations which notoriously result from the disease.

rich, *a.*, *adv.*, and *sb.* Add: **I. 7. d.** (Later examples.)

1936 N. COWARD *To-night at 8.30* II. 40 Me, grumble! I like that, I'm sure. That's rich, that is. **1977** J. ANDERSON *Appearance of Evil* i. 6 'You have experienced a spontaneous demonstration of disapproval..at your last recital.' 'Spontaneous! That's rich.'

8. d. Of the mixture in an internal-combustion engine: containing an amount of fuel greater than that required for complete combustion.

1915 G. A. BURLS *Aero Engines* vi. 106 The rich mixture is diluted by additional air supplied through the light adjustable spring-loaded automatic air-valve. **1917** WODEHOUSE *Uneasy Money* x. 117 Your chauffeur, having examined the carburettor, turns to you and explains the phenomenon in these words: 'The mixture is too rich.' **1920** R. B. WHITMAN *Tractor Princ.* iv. 55 The most serious result of a rich mixture..is in the production of carbon, and the carbonization of the engine. **1971** P. J. MCMAHON *Aircraft Propulsion* vi. 196 During an acceleration..the fuel/air ratio will already be richer than for normal steady running.

II. *adv.* and *Comb.* **9. b.** (Later example.)

1864 G. M. HOPKINS *Poems* (1967) 15 From easy runnels the rich-pieced land I water with my foot.

c. (Later example.)

1925 BLUNDEN *Eng. Poems* 93 Rich-tongued anew The foreign birds are come.

10. b. (Later examples.)

1922 JOYCE *Ulysses* 726 Those richlooking..drinks those stagedoor johnnies drink with the opera hats. **1930** E. WAUGH *Vile Bodies* x. 195 A really good story my second day on the paper. This ought to do me good with the *Excess*—*very* rich-making.

10*. In reduplicated form.

a **1725** E. TAYLOR *Dialogue between Justice & Mercy* in *Poet. Wks.* (1939) 40, I to him will hold out in my hand The golden scepter of my Rich-Rich Grace. **1963** P. H. JOHNSON *Night & Silence* x. 58 Living with the rich-rich suited him right now. **1974** 'J. MELVILLE' *Nun's Castle* viii. 176 Ted probably was rich. Not superlatively rich (rich rich as the slang goes) but solidly, comfortably endowed. **1977** *South China Morning Post* (Hong Kong) 13 Apr. 14/3 Karl Lagerfeld always puts together a mouth-watering collection for Chloe (aiming towards those rich-rich women who buy from Valentino).

III. *absol.* or as *sb.* **11. a.** Also in phr. *the rich get richer* (*and the poor get poorer*).

1921 KAHN & EGAN *Ain't we got Fun* (song) 5 There's nothing surer The rich get rich and the poor get children. **1936** C. SANDBURG *People, Yes* 164 The rich get richer and the poor get children. The rich have baby napkins, the poor have diapers. **1972** *Times* 8 May 14/1 It is the old story of the rich getting richer and the poor getting poorer, this time by courtesy of Dr Borlaug's miracle grains. **1973** *Black Panther* 31 Mar. C/1 County underassessment too, perpetuates the 'rich get richer' cycle in Oakland.

12. Also in phr. *new rich*: cf. *NEW *a.* 8 d.

1909 *Daily Chron.* 3 Nov. 3/5 It is concerned with the intrusion of Saul Dene, a 'new rich', into a Yorkshire country set.

rich, *v.*[1] Restrict † *Obs.* to sense 2 in Dict. and add: **1.** (Later examples.)

1912 J. MASEFIELD *Widow in Bye Street* 86 Sunwarm gorses rich the air with scent. **1925** E. O'NEILL *Desire under Elms* I. iv. 50 Blood an' bone an' sweat—rotted away—fertilizin' ye—richin' yer soul—prime manure, by God, that's what I been t' ye! **1955** E. POUND *Classic Anthol.* II. 126 Falls snow, fine sleet Plus drizzle and soak Riching, by mulch, full favour the grain Of all our folk.

Richard. Restrict † *Obs. rare* to sense in Dict. and add: **2.** [More formal equivalent of *DICK *sb.*[6]] A detective.

1914 JACKSON & HELLYER *Vocab. Criminal Slang* 70 *Richard, noun.* General currency. A detective. Derived from the process of nicknaming, but in reverse of the usual custom. Thus from the term '*detective*', 'dick' was suggested and hence '*Richard*' was derived. Or, following the corruption of the English 'Robert' to 'Bob' and 'Bobby', the American parallel was suggested. **1952** in Wentworth & Flexner *Dict. Amer. Slang* (1960) 426/2 Mickey Spillane, who turns out epics about a private richard named Mike Hammer. **1964** H. KANE *Snatch an Eye* xxiv. 147 The richard tails his own client... If the private detective shows up, it means that the client did not shake him. **1967** E. MCGIRR *Hearse with Horses* i. 7 Private eye, poor Richard, or shamus. **1974** —— *Murderous Journey* 12 A surprisingly high proportion of well-to-do murderers hire private richards to delve into the demise of the victim.

3. [Shortening of *Richard the Third*, rhyming slang for *BIRD *sb.* 1 d.] A girl, woman.

1950 P. TEMPEST *Lag's Lexicon* 180 *Richard.* A girl. The girl friend. **1960** 'A. BURGESS' *Doctor is Sick* xiv. 107 Judged by some of vem glamorous richards on ve telly. **1964** *Listener* 31 Dec. 1053/1 What Richard's gonna look at them in them 'airy outfits and them big boots. **1970** G. F. NEWMAN *Sir, You Bastard* viii. 232, I was just sleeping at this Richard's place during the day... I didn't know she was brassing.

4. Prefixed to another word, so as to form a name or nickname, or used in a phrase with specific sense: **poor Richard** (see quot. 1970); **Richard Roe** *Law*, the name formerly given to a fictitious defendant in actions of ejectment; *U.S.*, also an unidentified defendant in criminal proceedings; **Richard's himself again** (orig. in quot. 1700) (see quot. 1911).

1700 CIBBER *Richard III* v. 52 Conscience avant; Richard's himself again. Hark! the shrill Trumpet sounds, to Horse: away! My Soul's in Arms, and eager for the Fray. **1768**, etc. [see *John Doe* s.v. JOHN 4]. **1779** 'R. SAUNDERS' *Poor Richard's Prophesy* 4, I have frequently heard one or other of my adages repeated, with, as *poor Richard says*, at the end on't. **1870** *Brewer's Dict. Phr. & Fable* 747/1 *Richard Roe*, a mere nominal defendant in actions of ejectment. The name used to be coupled with *John Doe*, but these airy nothings are no longer the lawyer's tools. **1911** *Conc. Oxf. Dict.* s.v. *Richard*,..*Poor R.*'s *sayings*, maxims from almanacs issued by Benjamin Franklin with *Poor R.* as pseudonym; *R.'s himself again*.., said by or of person recovered from despondency, fear, illness, &c. **1928, 1957** [see *John Doe* s.v. *JOHN 4]. **1970** *Brewer's Dict. Phr. & Fable* (rev. ed.) 851/1 *Poor Richard*, the assumed name of Benjamin Franklin in a series of *almanacs* from 1732 to 1757. They contained maxims and precepts on temperance, economy, cleanliness, chastity, and other virtues; and several ended with the words 'as poor Richard says'.

richardia (ritʃā·ɹdiă). [mod.L. (C. Kunth 1818, in *Mém. Mus. Hist. Nat. Paris* IV. 433), f. the name of Louis Claude Marie *Richard* (1754–1821), French botanist + -IA[1].] = *CALLA 2.

1859 *Curtis's Bot. Mag.* LXXXV. 5140 (*heading*) Spotted-leaved Richardia. **1914** W. F. ROWLES *Garden under Glass* vi. 100 Any good potting soil will suit Richardias. **1951** [see *CALLA 2].

Richardson[1] (ri·tʃăɹdsǫn). The name of Sir John *Richardson* (1787–1865), Scottish naturalist and explorer, used *attrib.* or in the possessive in the names of birds or other animals first collected by him or named in his honour, as **Richardson('s) (ground) squirrel,** a spermophile, *Citellus richardsoni*, found in central North America; cf. *picket-pin gopher* s.v. *PICKET *sb.*[1] 7; **Richardson('s) grouse,** the North American spruce grouse, *Canachites canadensis*; **Richardson's jager** = *Richardson's skua*; **Richardson's owl,** the Arctic owl, *Ægolius funerea richardsoni*; **Richardson's skua,** the Arctic skua, *Stercorarius parasiticus.*

1831 W. SWAINSON in R. Jameson *Wilson & Bonaparte's Amer. Ornithol.* IV. 334 The next species in size and importance is Richardson's grouse (*Tetrao Richardsonii*), so named in honour of Dr. Richardson, the distinguished traveller. *Ibid.* 354 Richardson's jager, whole plumage, brown. **1835** L. JENYNS *Man. Brit. Vertebr. Animals* 282 Richardson's Skua. **1856** J. CASSIN *Illustr. Birds Amer.* 185 Richardson's Owl... The largest of this genus. **1868** *Amer. Naturalist* II. 529 Richardson's squirrel. **1896** R. RIDGWAY *Man. N. Amer. Birds* (ed. 2) 260 Northern North America; south, in winter, to northern border of United States.. Richardson's Owl. **1897** R. B. SHARPE in A. H. Miles *Conc. Knowl. Nat. Hist.* 264 Richardson's skua..nests in the Orkneys and Shetland Isles, as well as in some of the Hebrides and on the north-west of Scotland. **1901** *Daily Colonist* (Victoria, B.C.) 30 Oct. 1/6 The Richardson grouse, or fool hen, a smaller species..abound everywhere at lower levels. **1927** *Daily Express* 30 Nov. 9/5 At Regent's Park.. a 'Richardson's Owl'. **1940** E. T. SETON *Trail of Artist-Naturalist* 189 The Richardson or yellow ground squirrel.. nests in colonies like those of the prairie dog. **1947** R. T. PETERSON *Field Guide to Birds* (ed. 2) 137 The facial discs of Richardson's Owl are framed with black. **1963** D. A. BANNERMAN *Birds Brit. Isles* XII. 17 As a summer visitor to nest in the British Isles, Richardson's skua is much more numerous and more widely spread than the very local great skua.

Richardson[2] (ri·tʃăɹdsǫn). *Electronics.* [Name of Sir Owen W. *Richardson* (1879–1959), English physicist.] *Richardson('s) equation*: an equation giving the maximum current density of electrons emitted by a hot metal surface in terms of its temperature and work function. Also called *Richardson–Dushman equation* [Saul *Dushman* (1883–1954), U.S. physicist].

1925 [see *OUTGASSING *vbl. sb.*]. **1939** H. J. REICH *Theory & Appl. Electron Tubes* ii. 17 Richardson's equation holds only for the saturation current. **1950** P. G. ANDRES, *Surv. Mod. Electronics* ii. 41 At the turn of the century O. W. Richardson, using the analogy of evaporation at the surface of a liquid, developed an equation for electron emission. Later S. Dushman modified Richardson's equation. **1958** *Chambers's Techn. Dict.* 1010/1 Richardson-Dushmann equation. **1967** CONDON & ODISHAW *Handbk. Physics* (ed. 2) VIII. vi. 77/1 (*heading*) Statistical derivation of the Richardson equation. **1975** D. G. FINK *Electronics Engineers' Handbk.* I. 32 If the electrons are removed from the emitting surface as rapidly as they are released, the emission is temperature-limited and the density of current of emitted electrons obeys the Richardson-Dushman equation $J = (1-R)A_0T^2\epsilon^{-\phi/kT}$. *Ibid.* XXVII. 12 If $V_c < \phi_c$,..the saturation current density j_s is as given by Richardson's equation.

Richardson[3] (ri·tʃăɹdsǫn). *Physics.* The name of Lewis Fry *Richardson* (1881–1953), English physicist, used *attrib.* and in the possessive to designate quantities and concepts discovered by him, as **Richardson('s) criterion,** a criterion, depending on the value of the Richardson number, used to determine whether flow in a stratified fluid will be turbulent or laminar; **Richardson('s) number,** a dimensionless number given, essentially, by the ratio of the fluid density gradient to the square of the velocity gradient.

1934 D. BRUNT *Physical & Dynamic Meteorol.* xiii. 255 Richardson's criterion agrees with that given by Prandtl except for a factor ½. **1948** *Jrnl. Marine Res.* VII. 280 One type of nondimensional number, the Richardson number.. $r = gE/(V')^2$. **1956** *Nature* 3 Mar. 435/2 A regime of effectively free convection of heat begins to operate in the surface layers of the atmosphere at a negative Richardson number as low as 0·02 or 0·03. *Ibid.*, This transforms the Rayleigh criterion into a Richardson criterion. **1957** G. E. HUTCHINSON *Treat. Limnol.* I. v. 255 Richardson's number. **1964** *Oceanogr. & Marine Biol.* II. 19 If, as is usually the case in the ocean, the density of the water increases with depth, stability effects will tend to reduce the intensity of the turbulence... The stability effect is usually considered in terms of the Richardson number. **1974** *Encycl. Brit. Micropædia* VIII. 570/1 The Richardson number, or one of several variants, is of practical importance in weather forecasting and in investigating density and turbidity currents in oceans, lakes, and reservoirs.

Richardsonian, *a.* and *sb.* Add: **A.** *adj.* (Earlier and later examples.)

1786 A. SEWARD *Let.* 29 Mar. (1811) I. 135 Miss Reeves' reply to my Stricture on her Richardsonian absurdity, is at once weak and artful. **1844** THACKERAY in *Fraser's Mag.* XXIX. 709/1 It is entirely unnatural, theatrical, of the Davidgian, nay, Richardsonian drama. **1930** A. BIRRELL *Et Cetera* ix. 161 Doddridge's love affairs are narrated by him at Richardsonian length. **1952** *Essays in Crit.* II. 388 Fielding ridiculed the Richardsonian use of the present tense in *Shamela.* **1975** M. BUTLER *J. Austen & War of Ideas* xiii. 295 She [*sc.* Jane Austen] is the Fielding of her period, reacting against the Richardsonian individualism of the sentimental genre.

B. *sb.* (Later examples.)

1892 A. BIRRELL *Res Judicatæ* i. 32 The great Napoleon was a true Richardsonian. **1932** J. M. S. TOMPKINS *Pop. Novel in England 1770–1800* iv. 120 Such arrant Richardsonians as..the inevitable book-sellers' hacks.

Richard's pipit (ri·tʃăɹdz). [tr. F. *pipi Richard* (L. J. P. Vieillot in *Nouveau Dict. d'Hist. Nat.* (1818) XXVI. 491), named after Monsieur *Richard*, an amateur ornithologist of Lunéville, who first made it known.] A large pipit, *Anthus novæseelandiæ* (which includes the former *A. richardi*), found in large areas of the Palæarctic region.

1833 P. J. SELBY *Illustr. Brit. Ornith.* (ed. 2) I. 264 (*heading*) Richard's pipit. **1870** *Zoologist* V. 1984 Richard's Pipit... Some of them are doing well in confinement. **1921** *Ibis* III. 653 La Touche..appears to recognize this southern Chinese race of Richard's Pipit. **1953** D. A. BANNERMAN *Birds Brit. Isles* II. 51 Richard's pipit has been found over a hundred times in the British Isles since the first established record in 1812 when one was netted near London. **1971** *Country Life* 9 Sept. 616/2 It is the Scarborough bird-watchers who have made Filey Brigg famous..searching for and identifying rare visitors..such as Richard's pipit.

Richebourg (riʃbūr). Name of a wine-growing district of the Côte de Nuits, France, used to designate the red wine produced there.

1833 C. Redding *Hist. Mod. Wines* v. 97 Further on is the Vosnes... The most celebrated of these wines are the Romanée St. Vivant, Romanée-Conti, Richebourg, and la Tache. 1878 G. Meredith *Lett.* (1970) II. 559 Hockheimer or dry, still, red Boury, Richebourg and your friend to wash all down. 1926 [see *Melba*]. 1932 E. Hemingway *Death in Afternoon* i. 12 A liver that will not allow me to drink Richebourg, Corton, or Chambertin. 1973 H. McCloy *Change of Heart* v. 49 What would you like with the *tournedos?* Richebourg?

Richelieu (rī·ʃlyö). [Of obscure origin. Possibly the name of Cardinal *Richelieu* (1585–1642).] Used *attrib.* to designate a form of cut-work in which the spaces are connected by picoted bars, used in clothing and household accessories.

1880 *Myra's Jrnl. Dress & Fashion* Sept. 437/2 (*heading*) Edging in Richelieu guipure... The scallop is worked in the same way, and the material cut away underneath. 1882 [see *Strassburg*]. 1895 *Montgomery Ward Catal.* 301/2 Ladies' Extra Fine Lisle Thread Hose, Richelieu, ribbed. 1923 *Daily Mail* 26 June 15 Another lovely set is in Richelieu embroidery on linen edged with Point de Venise. 1967 E. Short *Embroidery & Fabric Collage* ii. 29 Another type of white work used for accessories and children's clothes was Richelieu work, an elaborate form of cut work. 1975 P. Clabburn *Needleworker's Dict.* 227/1 *Richelieu embroidery*, type of whitework where, as in Renaissance embroidery, the design is outlined in buttonhole stitch, the fabric cut away, and the spaces joined with bars or brides. The difference between the two types is that here the joining bars are decorated with picots.

richellite (rīfe·lait). *Min.* [a. F. *richellite* (Césaro and Despret 1883, in *Ann. de la Soc. géol. de Belgique* X. 36), f. *Richelle*, name of a locality near Visé in Belgium, where it was discovered; see -ite¹.] A hydrated basic phosphate of calcium and ferric iron, usu. also containing fluorine, which occurs as amorphous reddish- or yellowish-brown masses.

1884 *Jrnl. Chem. Soc.* XLVI. 1102 (*heading*) Richellite, a new mineral species. 1935 J. W. Mellor *Compr. Treat. Inorg. & Theoret. Chem.* XIV. lxvi. 412 G. Césaro and G. Despret [*sic*] described a mineral from Richelle, Belgium, which they called richellite. The analysis corresponds with calcium ferric fluophosphate, although G. Cesaro gave $8FePO_4.Fe_2OF_2(OH)_3.36H_2O$ for the formula. 1963 *Amer. Mineralogist* XLVIII. 300 Richellite, a calcium-iron hydrated phosphate, appears to be virtually amorphous.

ri·chening, *ppl. a. rare.* [f. Richen *v.* + -ing².] That is becoming richer.

1930 G. K.'s *Weekly* 26 July 318/2 Each generation of mankind may be said to inherit an ever-richening treasure of synthetic wisdom from the total past.

Richi, obs. var. Rishi in Dict. and Suppl.

richish, *a.* For *rare*⁻¹ read *rare* and add further example.

1955 Priestley & Hawkes *Journey down Rainbow* xii. 179 Yet another poet (wisely married to a richish wife).

Richter (ri·χtəɹ). *Geophysics.* The name of Charles Francis *Richter* (b. 1900), American seismologist, used *attrib.* with reference to a logarithmic scale he devised for expressing the magnitude of an earthquake (see quot. 1935 s.v. *magnitude 2 c) as calculated from the oscillations of a seismograph trace, and ranging from negative values for micro-earthquakes to about 8·9 for the most catastrophic.

1938 L. D. Leet *Pract. Seismol. & Seismic Prospecting* ix. 304 Since the Richter magnitude scale is logarithmic in amplitudes, doubling the magnitude gives a scale logarithmic in energies and log E = 12 log E_0 + 18 where M is the magnitude on the Richter scale. 1957 *Bull. Seismol. Soc. Amer.* XLVII. 187 'Magnitude' refers to an estimate of the Richter magnitude by the Seismological laboratory at Pasadena. 1969 E. L. Wiegel *Earthquake Engin.* (1970) ii. 50 Richter has correlated Modified Mercalli intensity with the earthquake's Richter magnitude as follows. 1971 [see *magnitude 2 c*]. 1976 *National Observer* (U.S.) 8 May 19/3 It is when the magnitude starts reaching 5·5 on the Richter Scale and higher that destruction is wrought. 1977 A. Hallam *Planet Earth* 63/3 In general, the larger events—those with Richter magnitudes greater than 5 or 6—do not occur in regions of plate creation but are associated only with the more violent process of plate underthrusting.

b. *fig.*

1967 *Boston Sunday Herald* 7 May (Show Guide) 17/1 This novel about the Nation's Capital in the period from Pearl Harbor to Senator McCarthy falls somewhere between the sententiousness of Allen Drury and the political astuteness of Anthony Trollope—say about 4 on a Richter scale that measures the quakes of getting and holding office. 1975 *New Yorker* 6 Oct. 140/2 We correspondents who remained in the city tried to assess the rumours as best we could on a sort of Richter scale of our own: if we heard enough reports from different sources about a particular occurrence, we figured that the story was true and gave it a high rating. 1977 H. Greene *FSO-1* xix. 171 In an arm of government as seismologically sensitive as the Foreign Service, the news..was a shock on the level of six or seven on the diplomatic Richter scale.

richterite (ri·χtərait). *Min.* [ad. G. *richterit* (A. Breithaupt 1865, in *Berg- und Hüttenmännische Zeitung* XXIV. 364), f. the name of H. Theodor *Richter* (1824–98), German metallurgical chemist: see -ite¹.] **a.** A manganiferous mineral of the amphibole group, $(Na,K)_2(Mg,Mn,Ca)_6Si_8O_{22}(OH)_2$, found as elongated brown, yellow or rose-red monoclinic crystals. **b.** The end member of the series of amphiboles containing this, which lacks manganese and has the ideal formula $Na_2CaMg_5Si_8O_{22}(OH)_2$.

1868 J. D. Dana *Syst. Mineral.* (ed. 5) 235 The Richterite of Breithaupt[...] is near schefferite in composition. 1946 *Chem. Abstr.* XL. 7085 Richterite is found to form a special type of Mg-rich amphiboles distinguished by the replacement of Ca by Na (K) as compared with tremolite. 1966 E. W. Heinrich *Microsc. Identification of Minerals* x. 273 Richterite,..for a rare manganiferous amphibole, is closely allied to soda tremolite. 1978 *Mineral. Mag.* XLII. 543 [Report on Nomenclature of Amphiboles as approved by International Mineralogical Association]. Sodic-calcic Amphiboles. This group is defined as monoclinic amphiboles in which (Ca + Na_B) ≥ 1·34 and 0·67 < Na_B < 1·34... End Members: Richterite (etc.).

Richter's hernia (ri·χtəɹz). *Med.* [Named after August Gottlieb *Richter* (1742–1812), German surgeon, who described the condition in *Abhandlung von den Brüchen* (1778) ch. xxiv.] A hernia which does not involve the whole width of the gut, so that the lumen remains open.

1887 F. Treves in *Medico-Chirurg. Trans.* LII. 153, I would venture to suggest the title 'Richter's hernia'... to make still more clear the lines that should separate this hernia from that known as Littre's, and partly because with Richter must rest the main credit of establishing the individuality of this lesion. 1940 *Jrnl. Surg.* VII. 864/1 A strangulation of part only of the circumference of the intestinal wall is a Richter's hernia. *Ibid.* 864/2 One should be suspicious of a Richter's hernia when the present illness was ushered in by the incarceration of a long-standing hernia, and where the illness persists. 1974 R. M. Kirk et al. *Surgery* viii. 139 Part of the bowel wall may be strangulated, leaving the lumen unobstructed, in Richter's hernia.

ri·cing². [f. Rice³.] (See quot. 1937.)

1937 E. J. Labarre *Dict. Paper* 201/1 Ricing, a term applied, chiefly in esparto mills, to the mottled or grainy look through due to 'crowding' or crushing... In Scotland also referred to as 'rawns' (Sc. rawns, the roe of fish). It is generally due when the paper shows characteristic small knots due to the stuff being too wet, between ..and too much rubbing in the machine chests. 1969 R. R. A. Higham *Handbk. Papermaking* ii. 42 This type of beater (sc. Taylor beater) is suitable for short-fibred stock such as esparto, where only light treatment is called for and which at the same time tends to prevent *ricing* of the fibres, produced by excess whipping of stock.

rick, *sb.*¹ Add: **2. a.** *rick-builder, -building, -burner* (earlier example), *-burning* (later example), *-lifter, -shifter*.

1909 *Eng. Dial. Dict.* V. 96/1 *Rick-builder*, a man who builds ricks. 1936 *Discovery* Nov. 363/2 The old lacemakers and the hurdle maker, the shepherd and the rick builder, the last being an artist in hay, inhabit the old-world village. 1960 *Farmer & Stockbreeder* 23 Feb. 55/3 Many of us are so steeped in the old rick-building traditions of hay-making to imagine that we still do it with silage. 1844 *Punch* VII. 17 (*caption*) The home of the rick-burner. 1939 D. Cecil *Young Melbourne* vii. 206 The riotings and rick-burnings...roused his fear of revolution. 1910 *Encycl. Brit.* XIII. 108/1 Various forms of rick-lifters are in use, the characteristic feature of which is a tipping platform on wheels to which a horse is attached between shafts. 1924 *Glasgow Herald* 14 Dec. 6/7 Skiffs have been transported across the island on rick-lifters to augment the temporary fishing fleet. 1957 E. E. Evans *Irish Folk Ways* xi. 155 Carried at last to the haggard or rick-shifters or slipes... the hay is built into small circular stacks (pikes).

rick, *sb.*⁴ *slang.* [Origin unknown.] = *gee *sb.*⁴ 3. Also *attrib.* or as *adj.*, fictitious, sham.

1928 [see *gee *sb.*⁴]. 1937 P. Allingham *Cheapjack* vii. 63 On..occasions the worker has a rick, that is to say, a confederate planted in the crowd, whom he could always choose as the first bidder. 1969 *Sunday Tel.* 7 May 5/5 If you are standing near a bookie's joint, somebody, and a merchant dashes in and places a bet, such as 'Seventy pounds to forty. On top', don't take a blind bit of notice. It's a rick bet.. It don't even go in the book. His sole object is to push or goad you into making your bet.

rick, 'rick (rik), abbrev. Rickshaw, ricksha.

1886 Kipling in *Pioneer Mail* 16 June 743/1 All the sahibs hailed 'ricks'—they call them 'ricks' here [*sc.* in Hong Kong]. 1962 *Coast to Coast 1961–62* 81 Hell, wouldn't they get 'em! There was a 'rick outside. He was in a 'rick at the taxi rank. 1970 ..Wouldn't credit it, his going for a Jap sheila in a bloody rick.

rick, *v.*¹ (Later N. Amer. examples.)

1914 *Dialect Notes* IV. 78 *Rick up, v. phr.*, to pile up (brush). 1951 H. E. Giles *Harbin's Ridge* xi. 102 When I got there Faleecy John was ricking up cook wood in the corner of the yard. I pitched in to help. I like to rick wood. 1962 M. E. Murie *Two in Far North* I. ii. 23 Thank heaven for the nine cords of good spruce wood ricked up in the back yard.

rickardite (ri·kādɔit). *Min.* [f. the name of Thomas A. *Richard* (1864–1953), U.S. mining engineer + -ite¹.] A tetragonal telluride of copper found as brittle, metallic, purplish-red masses in a number of copper mines in Colorado and elsewhere.

1903 W. E. Ford in *Amer. Jrnl. Sci.* CLXV. 69 Rickardite occurs at Vulcan, Col., in the Good Hope mine... Rickardite itself occurs in small lens-shaped masses, generally rather closely associated with native tellurium. 1943 R. D. George *Minerals & Rocks* viii. 223 Weissite, Cu_2Te_3..occurs associated with rickardite. 1949 *Amer. Mineralogist* XXXIV. 342 Rickardite is a mineral of striking and unmistakable appearance. 1963 *Structure Rep.* 1956 XX. 101 Cu_7Te_5 is rickardite, and was formerly described as $Cu_{4-x}Te_2$.

ricker. For *Naut.* read *orig. Naut.* and add later examples in various *spec.* senses.

1905 W. B. *Where White Man Treads* 20 Posts..driven into the ground at such an angle over the fire that when the rickers, on which the birds were hung, were laid into them, the fat dripping from the upper tier missed the one below. 1940 *Chambers's Techn. Dict.* 724/2 *Rickers*, round timber of less than 5 in. diameter in the middle. 1958 *N.Z. Timber Jrnl.* May 56/1 *Ricker*: (a) A long pole or sapling used for handling floating timber at the docks. (b) A kauri tree in the pole stage, N.Z.

ri·cket, *sb.*² *Criminals' slang.* [Origin unknown.] A blunder, mistake.

1958 F. Norman *Bang to Rights* i. 36 If you are chopping all day..you must at some time make a ricket and chop your toe. 1963 *Oxford Mail* 23 Aug. 1/6 You obviously know a lot. I have made a ricket (mistake) somewhere.

ri·cket, *v. rare.* [Back-formation from Rickety *a.*] *intr.* To move in a rickety manner; to lurch.

1903 T. De Leon *Bachelor's Box* v. 44 Van muttered to himself, as the cab rocked and ricketed down the street. 1955 L. A. G. Strong *Deliverance* ii. 65 One of the back legs of his chair was short, so that..it ricketed to and fro.

ricketic (rike·tik), *a. Med.* [f. Rickets + -ic.] = Rickety *a.* 1, 3.

1972 *Lancet* 20 May 1113/2 A biochemist friend...is campaigning for the use of 'ricketic' as the proper adjective for rickets. 1973 *Nature* 28 Sept. 180/1 Both parent compounds have long been known to possess similar anti-ricketic activity in man.

ricketiness. (In Dict. s.v. Ricketily *adv.*) (Earlier example.)

1867 G. M. Hopkins *Further Lett.* (1956) 48 The more frankly you confess the 'ricketiness of' yr. position, do you see, the less excuse you leave yourself for staying in it?

rickets. Add: Also 7 rackets, rekets. **1.** (Earlier and further examples.) Now known to be a vitamin D deficiency disease.

1634 *Parish Clerks' Company Bill of Mortality* (Guildhall Libr. MS. 5424/3), Diseases and Casualties this year... Rickets 14. 1653 J. Bulwer *Anthropometamorphosis* xx. 332 This new disease, commonly called the Rickets, or more properly the Rachitis. 1661 J. Bisse *Osteola Caroliniana* 55 The disease is not expresst by a word of the singular number *reket*, but plurally *rekets. Ibid.*, The calling of it, as we do, *rekets* pronouncing of it as if it were written with (*i*) in the first syllable..the Greek may give light, that it would then be written with (*ë*) or (*ı*). 1883 *Medico-Chirurg. Trans.* LXVI. 204 The characteristic symptoms of the so-called acute rickets..are not due to rickets at all but are truly scorbutic. 1919 *Lancet* 15 Mar. 408/2 An examination of the results obtained suggests that rickets is a deficiency disease which develops in consequence of the absence of some accessory food factor or factors. 1974 Passmore & Robson *Compan. Med. Stud.* III. xxiv. 20/1 Rickets..is caused by deficiency of vitamin D, due to a combination of a dietary lack and inadequate exposure to sunlight, necessary for the synthesis of the vitamin in the skin. In the absence of vitamin D the absorption of calcium from the gut is impaired and there is a defective calcium and phosphorus metabolism.

rickettsia (rike·tsiä). Also Rickettsia. Pl. -iæ, ias. [mod.L. *Rickettsia* (coined in Ger. as the name of a genus by H. da Rocha-Lima 1916, in *Berliner klin. Wochenschr.* 22 May 567/2), f. the name of H. T. *Ricketts* (1871–1910), U.S. pathologist, who first described such organisms in 1909 (*Jrnl. Amer. Med. Assoc.* 30 Jan. 379–80) and died of typhus contracted as a result of his research on them: see -ia².] Any of a group of very small rod-shaped or coccoid micro-organisms that are mostly intracellular parasites in vertebrates and include the causative agents of several febrile diseases in man. Freq. *attrib.*

1919 *Jrnl. Med. Res.* XLI. 87 The name 'Rickettsia' has been applied by da Rocha-Lima to minute bacillary forms found by Hegler and von Prowazek in typhus fever, and regarded as identical with bodies described by Ricketts in Mexican typhus. 1922 Hiss & Zinsser *Textbk. Bacteriol.* (ed. 5) xlviii. 944 Rickettsia-like microorganisms may be present in lice fed upon healthy people... It is not at all conclusively definite that the Rickettsia bodies are micro-organisms. 1935 *Discovery* Dec. 375/2 Very significant is the distinction between the various infection-chains of the different forms of Rickettsia disease. 1947 *Ann. Rev. Microbiol.* I. 337 This preparation contains rickettsiae and

tissue particles in suspension. 1951 Whitby & Hynes *Med. Bacteriol.* (ed. 5) 360 The rickettsiae are primarily intestinal parasites of arthropod blood-sucking insects...but some half-dozen species have become adapted to invade the animal body and cause disease. 1969 *Times* 3 June 6/2 The trachoma agent is one of a group of tiny organisms called the Rickettsiae. 1976 *National Observer* (U.S.) 21 Aug. 8/4 The disease is caused by rickettsiae—micro-organisms called rickettsiae, which resemble very small bacteria but, like a virus, grow only in susceptible cells. 1976 A. L. Smith *Microbiol. & Path.* (ed. 11) xxvii. 271/2 The rickettsias of the genus *Rochalimaea* can be cultured in host cell-free media. 1979 *Sci. Amer.* Oct. 88/2 Although..we had isolated the causative agent of Legionnaires' disease, for several weeks we did not know whether its organism was a rickettsia or a bacterium.

Hence **ricke·ttsial** *a.*, of, pertaining to, or caused by rickettsiae.

1940 W. H. Holmes *Bacillary & Rickettsial Infections* v. 76 They discovered that trench fever is a true rickettsial disease transmitted by the body louse. 1947 *Ann. Rev. Microbiol.* I. 338 A method for the preparation of rickettsial antigens from yolk sacs without the use of ether has also been developed. 1977 *Time* 31 Jan. 56/2 The discovery was made not by a man who hunts ordinary bacteria, but by a specialist in leprosy and rickettsial diseases like typhus and spotted fever.

rickettsialpox (rike·tsiälpǫks). Also rickettsial pox. [f. *rickettsial *a.* + Pox *sb.*] A mild rickettsial disease transmitted by mites.

1946 R. J. Huebner et al. in *Public Health Rep.* LXI. 1605 Because of a clinical resemblance to chickenpox and because the organism isolated from one patient has the morphological and cultural characteristics of rickettsiae, the name 'rickettsialpox' is proposed. 1955 *Sci. News Let.* 11 June 373/2 New diseases, Q fever and rickettsialpox, have appeared and some old ones, polio, Coxsackie virus infections, and infectious hepatitis, or jaundice, have increased in the period covered by Dr. Dauer's report. 1970 *New Scientist* 27 Aug. 407/2 Mice probably play a greater part than rats in the transmission of..rickettsial pox (which closely resembles chickenpox).

rickety, *a.* Add: **3. b.** *rickety rosary*, a line of swellings on either side of the chest, reminiscent of strings of beads and symptomatic of rickets.

[1887 Vickery & Knapp tr. *Strümpell's Text-bk. Med.* 868 There is a swelling at the junction of the cartilages with the ribs, which can be felt and seen through the skin, and produces what is called the 'rosary of rickets'.] 1907 G. F. Still in W. Osler *Mod. Med.* I. xxxiii. 876 The most frequent [osseous] manifestation [of rickets] is the so-called 'rickety rosary', or beading of the ribs, a thickening at the costochondral junction which in a thin child can be seen and in others easily felt. 1970 W. H. Parker *Health & Dis. in Farm Animals* i. 7 In a pup or child suffering from rickets they can actually be seen because of the enlargement of the joints which occur with this disease, giving rise to the grim phrase 'a ricketty rosary'. 1974 Passmore & Robson *Compan. Med. Stud.* III. xxiv. 21/2 In the chest an early sign [of rickets] is bleeding of the costochondral junctions. This is called rickety rosary as the beads extend in chain fashion down both sides of the thorax.

ri·ckety-ra·ckety, *a.* [redupl. f. Rickety *a.*] Unsteady; shaky; tottering.

1895 *Punch* 21 Sept. 135 (*caption*) We're a rare old—fair old—rickety, rackety crew. 1931 J. R. Macdonald in *Times* 5 Nov. 14/4 The whole world has got into what we in this corner of it would call a ricketty-racketty state. 1976 D. Heffron *Crusty Crossed* v. 39 We were given some rickety-rackety set of train coaches with old wooden bench seats.

rickey (ri·ki). orig. *U.S.* Also **Rickey, ricky.** [prob. f. the surname *Rickey.*] An iced drink consisting of gin, whiskey, or the like, mixed with lime or lemon juice and carbonated water. Freq. with defining word prefixed, esp. *gin rickey*.

1895 G. J. Kappeler *Mod. Amer. Drinks* 97 *Brandy rickey.* In a thin medium-sized glass put one lump of ice, the juice of half a lime, one jigger brandy, fill the glass with siphon carbonic water; drink while effervescent. *Ibid.*, Canadian rickey. *Ibid.*, Gin rickey. 1906 *Mrs. Beeton's Bk. Househ. Managem.* xlix. 1511 *Gin rickey. Ingredients.*—1 wineglassful of gin, 1 dessertspoonful of lemon or lime-juice, seltzer water, ice... *Note.*—Any other spirit may be used..and would..give its name to the compound. 1908 'O. Henry' *Gentle Grafter* i. 13 Why, in Bryan's second campaign,..they used to give me three gin rickeys and I'd speak two hours longer than Billy himself could on the silver question. 1925 F. Scott Fitzgerald *Great Gatsby* vii. 140 Tom came back, preceding four gin rickeys. 1928 [see *gimlet *sb.*¹ 1 c]. 1930 G. R. Brown *Washington* xi. 369 In course of time, the 'rickey' came to be referred to as the 'gin rickey', whereas it had always been a 'whiskey rickey'. 1945 P. Cheyney *I'll say she Does!* iii. 72 The lyin' woman is like a gin ricky without any gin. 1975 M. Amis *Dead Babies* xiii. 67 By 12.30, Giles had consumed five gin-rickies, four gin-and-tonics, three gin-and-its, two gin-and-bitters, and one gin. 1975 *Washington Post* 27 Jan. A 19/1 A certain Col. Joseph Rickey used to frequent a drinking place in E Street known as Shoomaker's... The original Rickey, a la Shoo's, consisted of Bourbon, soda and lime juice. In time gin was substituted for the Bourbon and..[the Colonel's] name was forever attached.

rickle, *sb.*² Add: Also *Anglo-Ir.* **1. a.** (Later examples.)

1922 J. Buchan *Huntingtower* iii. 57 Huntingtower was the auld rickle o' stanes at the sea-end. 1934 E. Linklater *Magnus Merriman* 197 There was a smug trim smooth little minister, and a rickle of rural inanity behind. 1963 *Field Archaeol.* (Ordnance Survey) (ed. 4) 53 A careful search will often show associated hut sites and even field boundaries in the form of long rickles of stones.

2. (Later example.)

1871 G. M. Hopkins *Jrnls. & Papers* (1959) 213 [Lancs.] Roberts says the first grass from the scythe is the *swathe*, then comes the *strow* (tedding), then *rowing*, then the foot-cocks, then *breaking*, then the *kubrows*, which are gathered into *bats*, then sometimes another break and *turning*, then *rickles*, the biggest of all the cocks, which are run together into *placks*, the shapeless heaps from which the hay is carted.

3. (Earlier and later examples.)

1700 *Black Bk. Kincardineshire* (1843) 130 He hid the said web among a rickle of turffs. 1957 E. E. Evans *Irish Folk Ways* xiv. 184 The peats are turned and built into larger and larger piles, turf-foots, castles, rickles, hungs and clamps. 1979 *Country Life* 27 Sept. 964/1 Turf..will be built into..piles with varying names such as rickles, castles, lumps and clamps.

rick-rack: see *ric rac, rick-rack.

rickshaw, ricksha. (Later *attrib.* examples.)

1915 G. Frankau *Tid'apa* ii. 10 'Come here. Come up here, dear,' they cry To the drunkly-waving seaman as his rickshaw-wheels roll by. 1923 D. H. Lawrence *Birds, Beasts & Flowers* 173 The rickshaw boys begin to understand. 1933 N. Lowry *Ultramarine* iii. 144 The rickshaw-wallah ran away with us, his sandals padding. 1951 R. Campbell *Light on Dark Horse* ii. 34 Zulu Rickshaw-pullers..emptied quarts of heady 'tchwala'. 1969 *Observer* 29 Oct. 29/2 A posse of rickshaw-cycles and cyclists come tearing round the corner. 1973 *Country Life* 1 Nov. 1321/2, I found myself face to face with a hundred rickshaw-tricycle boys. 1977 'S. Leys' *Chinese Shadows* (1978) 196 His father, a rickshaw 'boy', always beaten up by capitalists, was covered with welts and scars. 1977 *N.Y. Rev. Bks.* 21 Apr. 3/2 They were privileged to support servants and clerks. Chinese teachers,..horse boys, amahs, cooks, guides, rickshaw men. 1978 *Chinese Lit.* xii. 13 He decides to become a rickshaw-puller, the lowest of the low.

ri·cky, *a. rare.* [f. Rick *sb.*² + -y¹.] = Rickety *a.* 2 c.

1922 D. H. Lawrence *Aaron's Rod* viii. 87 A maudlin crying to be loved, which makes your knees all go ricky.

ricky, var. *rickey.

ricky-tick (ri·ki,ti·k), *sb.* and *a. slang* (chiefly *U.S.*). [Imitative.] **A.** *sb.* An even, repetitive, or monotonous rhythm, as in early jazz; old-fashioned 'straight' jazz or ragtime. **B.** *adj.* Of musical rhythm or tempo: even, repetitive, monotonous; of music: trite, old-fashioned, 'corny'. Also *transf.* and as *adv.* Cf. *rinky-dink *sb.* and *a.*, *rinky-tink *a.*

1938 *Brit. Empire Mod. Eng. Illustr. Dict.* 1257/1 Ricky-tick (*Am.*): Old-fashioned jazz. 1942 Berrey & Van den Bark *Amer. Thes. Slang* § 578/31 Of tempo and tone... *clubhouse, hotel, rickty-tick* [*Index*, ricky-tick], gently softly. *Ibid.* § 579/2 'Straight jazz.' (Old-fashioned jazz, which reproduces the score faithfully, as distinguished from 'swing'.)..*ricky-tick. Ibid.* § 579/14 Outmoded; non-'swing', *ricky-tick.* 1958 *Gramophone* Dec. 328/2 A rather ricky-tick rhythm. 1967 *Time* 22 Sept. 61/1 It sizzles with musical montage, tricky electronics and sleight-of-hand lyrics that range between 1920s ricky-tick and 1960s raga. 1968 *N.Y. Times* 5 Feb. 29/1 To the ricky-tick of Guy Lombardo's 'I don't want to get well, I'm in love with a beautiful nurse', three maimed Army veterans fumble through a dance routine. 1968 *New Yorker* 18 May 6/2 A big red fire engine, peanut shells on the floor, and ricky-tick banjo music summon up the plinking eighteen-nineties, or whatever. 1969 *N.Y. Rev. Bks.* 21 Aug. 8/3 The rickytick, ingroup details sprinkled through the text. 1970 *Atlantic Monthly* June 104/1 Ricky-tick Conductor jointed like a walking-stick Insect, with silver glasses, silver punch, And silver seat in his sere serge. 1970 *Wall St. Jrnl.* 15 June 1/1 The critics..have dismissed the Welk brand of music as hopelessly square, bland ricky-tick rhythms aimed at a fading generation.

Hence **ri·cky-ti·cky** *a.* = *ricky-tick *a.*

1952 B. Ulanov *Hist. Jazz in Amer.* xiv. 157 They rejected his ricky-ticky beat with distaste. 1970 *Time* 12 Jan. 44/1 White music has always been very ricky-ticky, steppity-step, plunkety-plunk-banjo. 1976 *Times Lit. Suppl.* 23 Apr. 486/2 Weill's errant and loudly stated bass-line throwing up the odd chord that violently subverts the triteness of the ricky-ticky melody.

ricochee (ri·kǒʃi), repr. colloq. pronunc. of Ricochet *sb.*

1903 Kipling *Five Nations* 188 An' waits till the sights come on. For 'Im an' 'Er an' a hit (Direct or ricochee).

ricochet, *sb.* Add: **2. b.** *ricochet mound*; ricochet word (see quots.).

1902 *Kynoch Jrnl.* Apr.–May 80/2 At various points we have placed ricochet mounds, the first of which will catch any low or accidentally fired shot. 1881 *Brewer's Dict. Phr. & Fable* (ed. 12) 742/2 Reduplicated or Ricochet words, of intensifying force. Chit-chat, click-clack. 1967 W. & M. Morris *Dict. Word & Phr. Origins* II. 228 His hobby is collecting what he calls 'Siamese-twin words'..from *argy-bargy* to *zoot suit*. The word experts call these 'reduplicated words' or, in lighter moments, '*ricochet words*'.

ricochet, *v.* **2.** (Earlier example.)

1758 J. Amherst *Jrnl. Siege of Louisbourg* in *Gentl. Mag.* XXVIII. 387/2 [Guns] to ricochet the works and the town.

|| **ricordo** (rikǫ·rdo). Pl. **ricordi.** [It., lit. 'memory'.] A token of remembrance, souvenir; in *Art*, a copy made by a painter of a composition by another painter.

1926 *Contemp. Rev.* Dec. 770 Such is the little Play acted in Arzua, and for me a beautiful last *ricordo* of a place I had learned to love. 1974 *Times Lit. Suppl.* 25 Jan. 73/1 This much disputed, and admittedly unusual, drawing may in fact be Rembrandt's 'ricordo' of a composition by another, earlier, artist. 1975 *Ibid.* 21 Mar. 300/3 The authors have analysed La Tour's paintings one by one, paying attention to, and discussing, even the poorest and most miserable copies or *ricordi* of lost compositions.

|| **ricotta** (rikǫ·tā). [It.: see Ricocт.] A kind of Italian cottage cheese. Also attrib. in *ricotta cheese*.

1877 *Encycl. Brit.* VI. 772/2, 524,700 lb of 'ricotta', a fresh common cheese made after the butter and cream have been for the most part removed from the milk. 1923 [see *milk-sapped s.v.* *milk *sb.* 6]. 1966 T. Pynchon *Crying of Lot 49* i. 10 Her trip to the market..to buy ricotta and listen to the Muzak. 1971 *Sunday Times* (Colour Suppl.) 28 Mar. 39/1 Ricotta.. Made of ewe's milk (or rather the whey left over from making Pecorino) it is soft, unsalted and short-lived. Ricotta is particularly popular in Rome..where it is eaten with salt and black pepper or with sugar and cinnamon. Used extensively in Italian cookery... It is also the basis of two popular puddings. 1976 *Gourmet* June 23/1 For dessert, try cassata, delicate ravioli like dumplings filled with ricotta and sprinkled with ground cinnamon. 1980 *Times* 9 Apr. 9/7 Fritters of soft ricotta cheese... For Easter ricotta fritters are favoured.

ric-rac, rick-rack (ri·kræk). *Fashion.* Also as one word. [Origin unknown: perh. redupl. f. Rack *v.*³ or Rick *v.*²] A decorative zigzag braid used as a trimming for garments. Also *attrib.*

1884 L. M. Reyner in *Jrnl.* 25 Aug. in R. L. Stout *Maud* (1939) 326 Splendid sempstress—clothes nearly all done—Rickrack and feather-edge on the moon. 1911 C. Harris *Eve's Second Husband* 279 A wife who wore rick-rack braid on her petticoats. 1926 *Daily Colonist* (Victoria, B.C.) 7 Jan. 11/1 (Advt.), House Dresses..trimmed with attractive colored chintz and rick-rack braid. 1941 *Amer. Speech* XVI. 99 A *wee-waisted* model may have alliterative *ric-rac-rows*. 1956 J. Potts *Diehard* i. 5 Rick-rack trimming around the collar and long sleeves. 1960 *Harper's Bazaar* Apr. 76 Loosely sashed sheath..trimmed with black ric-rac. 1972 *Vogue* Feb. 51/1 White cotton piqué decorated here by waves of ricrac. 1974 *Sumter* (S. Carolina) *Daily Item* 24 Apr. (Belk Stroman Advts. Suppl.) 13 'Our Miss B' flare skirt, shirt-look bodice with white collar or ric-rac. 1976 *Daily Tel.* 29 Oct. 17/3 Yves St Laurent..resolutely translated autumn's long wool peasant skirts into spring's long cotton ones. He threw rick-rack and red curtain bobbles on them.

Hence **ri·c-rac, ri·ck-rack** *v. trans.*

1972 *Vogue* Feb. 46/2 *Poppy dress*..pinked and ric-racked. 1974 *State* (Columbia, S. Carolina) 3 & 4 Mar. G12 (*caption*) A green tablecloth rickracked with pink, blue and lemon completes the look.

rictus. Add: **2. a.** (Later examples.) Also *transf.* and *fig.*

1912 R. B. C. Graham *Charity* 12 A skin like parchment, which gave her face, when worked upon by a slight rictus in the nose she suffered from, a look, as of a horse about to kick. 1957 L. Durrell *Justine* ii. 107 This ghastly *rictus* gouged out in his taut cheeks. *a* 1963 S. Plath *Crossing Water* (1971) 21 Under the eyes of the stars and the moon's rictus He [*sc.* an insomniac] suffers his desert pillow. 1969 *Listener* 2 Jan. 27/3 To the comedians I like I find I go more than half-way. I find myself sitting with a rictus of amused incredulity and surprise. 1974 R. Adams *Shardik* xlix. 397 A child with a continuous twitching of the head gazed up wide-eyed, his mouth gaping in a kind of rictus of startled alarm. 1977 'M. Innes' *Honeybath's Haven* iv. 38 The muscular effort required..had the effect of contorting his features into a ferocious *rictus*.

rid, *v.* Add: **I. 1. c.** (Later dial. and U.S. examples.)

1891 Hardy *Tess* III. lii. 199 Are you house-ridding today? 1919 T. K. Holmes *Man from Tall Timber* xii. 144 I'll rid up the place and get our dinner. 1939 F. Thompson *Lark Rise* vi. 104 Fireplaces were 'ridded up', and tables and floors were scrubbed.

riddance. Add: **5. b.** (Further examples.) Also in phr. *good riddance of* (or *to*) *bad rubbish*.

1847 Dickens *Dombey* (1848) xliv. 438 A good riddance of bad rubbish!.. Get along with you, or I'll have you carried out! 1863 [in Dict.]. 1924 M. Irwin *Still she wished for Company* xviii. 220 If all they say downstairs is true..it's good riddance to bad foreign rubbish. 1928 S. Sassoon *Mem. Fox-Hunting Man* viii. ii. 287 It may well be wondered how the Hunt had survived the despotism of the old-world grandee, with whom previous Masters had been obliged to cooperate (as 'best Master we've ever had' while they reigned, and 'good riddance of bad rubbish' when they resigned). 1975 *Times* 4 July 5/3 The American War of Independence..can be seen, as George III consoled himself by looking at it, as good riddance to bad rubbish. 1977 H. Fast *Immigrants* ii. 110 A dead Chinese was good riddance to bad rubbish.

ridden, *ppl. a.* **2.** (Later example.)

1976 *Leicester Advertiser* 26 Nov., Best ridden Dartmoor awards have been anonymously donated at the NPS and Bath and West shows, and there is also the Allendale Award for the most versatile pony under saddle—for ridden classes have for a long time been a poor relation in all pony breed classes.

riddle, *sb.*¹ Add: **4.** *riddle-rid* adj.

1918 W. de la Mare *Motley* 44 Old and alone, sit we, caged, riddle-rid men.

5. *riddle-ballad, -book* (earlier example), *-game, -song*; **riddle canon** (see quot. 1889).

1816 Jane Austen *Emma* I. x. 176 There go you and your riddle-book one of these days. 1889 *Cent. Dict.* s.v. *canon*¹, *Enigmatical canon*,..*riddle canon*, in old music, a canon in which one part was written out in full and the number of parts was given; the remaining parts were to be written out by the student in accordance with the requirements of an enigmatical inscription written upon the music. 1934 Webster, Riddle ballad. 1950 M. J. C. Hodgart *Ballads* i. 14 The 'riddle' ballads are at the beginning of Child's collection... The basic theme of these is that of a mortal outwitting a supernatural being by quickness of wit, and of the magic power of the Word expounded in riddles. 1954 J. R. R. Tolkien *Fellowship of Ring* 21 Gollum challenged Bilbo to the Riddle-game. 1965 M. Spark *Mandelbaum Gate* ii. 32 She heard the familiar lilt of the riddle song, 'One Kid', from the lips of her lolling cousins.

riddle, *sb.*² Add: **4. riddle-board, -land** (see quots.).

1818 *Massachusetts Spy* 14 Oct. 4/3 'And what..is riddle land?' That which is of so open and loose a texture as to let the rain falling on it pass through it. 1969 E. H. Pinto *Treen* 141 Riddleboards were formerly used in the North of England and Scotland for making oatcake or oatbread.

riddle, *v.*¹ **2. c.** (Later example.)

1941 T. S. Eliot *Dry Salvages* v. 14 Release omens By sortilege, or tea leaves, riddle the inevitable With playing cards.

riddle, *v.*² Add: **2. a.** (Earlier *fig.* and later examples.)

1872 'Mark Twain' *Curious Dream* 87, I dosed him with bad jokes, and riddled him with good ones. 1916 'Boyd Cable' *Action Front* 209 You machine-gunners riddling holes in a target or a row of posts.

b. (Later *fig.* example.)

1928 *Daily Mail* 3 Aug. 16/7 London is riddled through and through with receivers of textile goods.

riddle-bread, -cake. (Later examples.) Also *U.S. dial.*

1915 *Dialect Notes* IV. 240 *Riddle cakes*, griddle cakes. 1969 E. H. Pinto *Treen* 141 In Scotland and North-east England, oatbread was often known as haver or riddle cake and in North-west England as haverbread or riddlebread.

riddled, *ppl. a.*¹ For *rare*⁻¹ read *rare* and add later example.

1918 G. Frankau *One of Them* x. 76 '*À vous le plein, Madame.*' Keen brains unravel The riddled sum. '*Deux louis — soixante-dix* [sic].'

riddler¹. (Later example.)

1978 *Jrnl. Lancs. Dialect Soc.* xxvii. 24 The earliest recorded English riddles, in which the riddler looks at familiar objects in an unfamiliar light.

riddling, *vbl. sb.*¹ (Later *attrib.* example.)

1969 G. M. Brown *Orkney Tapestry* 60 Into the riddling region where gods and men negotiated only an elite could trespass—men who had made a long study of the black arts.

ride, *sb.*¹ Add: **1. a, b.** (Earlier example.) Also esp. in colloq. phr. *to have* (or *give*) *a rough* (*easy*, etc.) *ride*.

1759 J. Wesley *Jrnl.* 13 Aug. (1764) 74 Monday, 13. I took a little ride to Croydon. 1969 S. Hyland *Top Bloody Secret* ii. 234 Judging by what he said on the phone, he won't give you an easy ride. 1969 *Times* 11 July 8/8 The new 'town managers' seem to be having a rough ride in London. 1974 *Times* 5 Nov. 15/1 President Giscard d'Estaing has had a fairly quiet ride until now. 1976 *Eastern Even. News* (Norwich) 9 Dec. 10/4 It seems to me that he is being given an unnecessarily rough ride. 1978 *Times* 17 Jan. 19/5 Union officials..met the new Leyland chief... He was given a rough ride.

f. *to take for a ride* and varr. (orig. *U.S.*): (*a*) *colloq.* to tease; to mislead deliberately, to hoax, to cheat; (*b*) *slang* to take on a car journey with the intention of murdering or kidnapping.

(*a*) 1925 *Dialect Notes* V. 344 *Take for a ride*, jolly; josh. 1929 J. P. McEvoy *Hollywood Girl* vii. 109 What was the name of that girl he was crazy about?.. Dugan. But Jack isn't bragging about it. She certainly took him for a ride. 1931 E. Linklater *Juan in Amer.* iv. ii. 283 'Do you mean that?' asked Mr. Adelaide nervously. 'Nonsense,' said Mr. Boles. 'He's taking you for a ride.' 1956 A. Wilson *Anglo-Saxon Att.* ii. i. 203 But for Vin, there were winks and the tongue stuck in the cheek, the wide boy who wasn't to be taken for a ride by anyone. 1962 *John o' London's* 25 Jan. 94/3 A young American teacher who goes to Persia and gets taken for quite a few rides. 1973 'D. Jordan' *Nile Green* ix. 43 She said, quickly, '..Are you an expert?'.. I said, 'Not at all... You could take me for a ride any day you chose.' 1976 J. I. M. Stewart *Memorial Service* xvi. 271 The provost had taken me on a ride to Otby in more senses than one.

(*b*) 1927 *Vanity Fair* (N.Y.) Nov. 132/3 'Taking him for a ride' is underworld for enticing a person to death. 1929 *Sun* (Baltimore) 15 Mar. 2/6 Possibility that Joseph Drell, wealthy cigar store owner, may have been 'taken for a ride' by underworld enemies, tonight was engaging the attention of detectives investigating his kidnapping. 1931 F. L. Allen *Only Yesterday* x. 261 Another favourite method was to take the victim 'for a ride': in other words, to lure him into a supposedly friendly car, shoot him at leisure, [etc.]. 1933 Wodehouse *Heavy Weather* i. 12 Perhaps some Duke who doesn't want to see himself in the 'Peers I Have Been Thrown Out Of Public Houses With' chapter has been threatening to take him for a ride. 1935 'J. Guthrie' *Little Country* ix. 166 When they shoot a man [in America] they call it sending him for a ride. 1941 *Strand Mag.* Aug. 332/1 You're not only being taken for a sail, Mrs. de Courcy, but also for a ride. Your passage-money back to Singapore is just five thousand pounds. 1944 E. S. Gardner *D.A. calls Turn* vii. 64 These persons whispered that some day Carr would mysteriously disappear, and no one would ever know whether he had quietly faded into voluntary oblivion or had been 'taken for a ride'. 1979 T. Barling *Olympic Sleeper* ix. 103 Not Costas driving... They must be dumping the cab. Or taking Costas for a ride.

g. *slang.* An act of sexual intercourse. Cf. Ride *v.* 3 and 16.

1937 Partridge *Dict. Slang* 696/2 *Ride*,..an act of coition. 1977 J. B. Hilton *Dead-Nettle* xviii. 155 I'd slipped in once or twice while Frank was out. She deserved a slow ride in-between times with that big old oaf. 1981 'J. Sturrock' *Suicide most Foul* i. 25 He reckons to have a ride on her..even if he has to marry her to get it.

h. *for the ride*: for fun; as an observer only. *colloq.*

1960 Wentworth & Flexner *Dict. Amer. Slang* 427/1 *Ride, go along for the*, to join in passively, usually for the fun of it, without making an active contribution. 1963 'J. Melville' *Burning is Substitute for Loving* ii. 41 Everyone went to the supermarket for the ride. It was fun. 1968 *Economist* 12 Oct. 99/1 The pension funds could keep buying equities for the ride. 1971 M. Babson *Cover-up Story* xi. 122 They're a little nervous about getting a courtroom full of fans just along for the ride. 1977 J. Wainwright *Nest of Rats* ii. iv. 167, I wouldn't know—I am along strictly for the ride.

i. *spec.* in *Surfing*.

1968 *Surfer Mag.* Jan. 48/2 I'd try and get long rides. 1971 *Times* 9 Aug. 5/2 You compulsively turn and start paddling your board out to sea for 'just one more ride'. 1975 *Oxf. Compan. Sports & Games* 1005/2 For a straight-wave ride, the surfer stands with flexed joints in a relaxed stance with one foot either side of the lateral and longitudinal centre line.

4. a. (Further examples.)

1949 'J. Tey' *Brat Farrar* xiv. 123 He's a lovely ride. And he can jump anything you can see the sky over. 1963 A. Duggan *Story of Crusades* iv. 39 The warhorse was usually a stallion, which made it a nuisance in camp and an uncomfortable ride. 1977 *Horse & Hound* 10 June 36/1 (Advt.), Brown gelding... A great ride, would be ideal for Master.

b. A motor vehicle. *N. Amer. slang.*

1930 *Amer. Speech* VI. 134 *Ride*, n., automobile. 1972 C. Milner *Black Players* v. 136 With his unspectacular conservative suits and modest 'ride' (a Toyota station wagon).

c. A roundabout or other device on which one rides at an amusement park or fair. Chiefly *U.S.*

1934 in Webster. 1953 *Amer. Speech* XXVIII. 117 *Kiddie rides, n.*, general term for the merry-go-round and other rides which attract children's patronage. 1966 T. Pynchon *Crying of Lot 49* v. 121 Down at the city beach, long after the pizza stands and rides had closed, she walked unmolested. 1968 D. Braithwaite *Fairground Archit.* 103 In 1927 Joseph and Robert Lusse of Philadelphia produced an accomplished design. Since that time, 'Dodg'ems' have been the most consistently popular ride on the fairground. 1977 H. Fast *Immigrants* ii. 138 He'd go on all the rides with me.

5. b. *Jazz slang.* A swinging rhythm; also, an improvised passage in such a rhythm.

1936 *Delineator* Nov. 49/2 *Ride*, easy-going rhythm. 1952 B. Ulanov *Hist. Jazz in Amer.* (1958) xi. 122 In that famous record he gave the chords of the tonic and the dominant a noisy ride. 1956 E. Hunter *Second Ending* i. 75 You give him all the hot rides. 1970 P. Oliver *Savannah Syncopators* 37 The 'ride' of a New Orleans jazz band, the 'slow and easy' slow-drag of a country blues band, have no counterpart in the forceful thrust of the multilineal drum rhythms. 1973 J. Wainwright *Pride of Pigs* 175 The washboard player tapped the off-beats..lifting the rhythm and giving it ride.

6. a. The characteristic motion of a motor vehicle or other means of transport in respect of passenger comfort, smoothness, etc.

1937 [see *ride-control* below]. 1947 [see *ride comfort* below]. 1962 *Which? Car Suppl.* Oct. 126/2 The Sunbeam Rapier's ride was choppy. 1974 *Country Life* 21 Feb. 378/3 The ride is very comfortable, with normal road shocks eliminated before reaching the occupants. 1980 *Times* 29 Feb. 23 Ride is on the firm side, but acceptably so, and the same can be said of the seats.

b. The quality of a horse's gait when being ridden.

1955 H. Smith *Horseman through Six Reigns* xix. 189 Most of the Irish hunters are shown fairly green... Their ride can either improve or deteriorate..after two or three years in England. 1955 *Times* 10 June 4/2 Mr. Deptford's grey Valeta..stood second, and Miss Wainwright's chestnut Lovely Boy third. With hacks nearly everything is in the ride.

7. *attrib.* and *Comb.*, as (sense 1 a) *ride-sharing*; (sense *1 f) *ride job, killing*; (sense 2 a) *ride-gate, -path, -side*; (sense *5 b) *ride solo, tempo*; (sense *6 a) *ride characteristic, comfort, control, height, luxury*; **ride cymbal,** a cymbal used by jazz drummers for keeping up a continuous rhythm (opp. to a *crash cymbal* used mainly for single 'crashes').

1961 *Which? Reports on Cars* 14 *Ride characteristics* for driver and passenger are assessed subjectively and, where possible, measured by instruments. 1947 Ride comfort [see *gust effect* s.v. *gust *sb.*¹ 3]. 1977 *Lancashire Life* Mar. 118/3, I..was impressed with its ride comfort over rough mountain roads. 1937 *Times* 13 Apr. (Brit. Motor Suppl.) p. xxxviii/2 The chassis specification includes a 21-h.p. 6-cylinder engine..; ride-control hydraulic shock-absorbers front and rear. 1956 M. Stearns *Story of Jazz* (1957) xviii. 234 Clarke made the single right-hand 'ride' or 'top' or 'front' cymbal the rhythmic center. 1965 *Crescendo* Oct. 10/1 The ride cymbal with snare/bass drum independence.. is a free lesson for any student of jazz drumming. 1977 *New Yorker* 9 May 51/1 He had a ride cymbal, and he played it.. four beats to the bar. 1919 J. Masefield *Reynard* i. 53 A jam of horses in the spinney, Close to the ride-gate. 1971 *E. Afr. Standard* (Nairobi) 10 Apr. 8/6 Suspension is fully independent all round, incorporating automatic ride-height control. 1978 *Hot Car* June 91/2 The ride-height of cars used for towing caravans and trailers. 1955 *People* (Austral.) 5 Oct. 5/2 No policeman would have to risk being mistaken for a mobster doing a gangland ride job. 1935 A. J. Pollock *Underworld Speaks* 96/2 *Ride killing*, when assassins take the victim for a ride in an automobile and slay him. 1979 *Jrnl. R. Soc. Arts* Dec. 38/2, I shall always remember the impression of ride luxury when we first interconnected that Alvis. 1892 B. Potter *Jrnl.* 25 Oct. (1966) 293 There were rights-of-way, ridepaths over the hills, notably by the Spittal. 1973 *Sunday Bull.* (Philadelphia) 7 Oct. (Parade Suppl.) 14/3 The Minnesota Highway Department has set up a state employees' ride-sharing operation. 1977 *Chicago Tribune* 2 Oct. xiii. 3/2 For a considerable number of motorists polled, it would take $2 gasoline to push them into sacrificing pleasure driving or switching to..ride-sharing to get to their jobs. 1943 J. W. Day *Farming Adventure* xvii. 187 For two miles we rode through half-grown timber, the ride-sides choked with thistles, blue- and red-headed, tangled with long grasses and tall flowering weeds. 1949 *Ebony* June 41 When Willie plays a ride solo he is better received than anyone else in the band. 1940 *Swing* Jan. 25 The other side is *Bugle Call Rag* at ride tempo.

ride, *v.* Add: **I. 1. d.** *to ride for a fall* (earlier and later examples).

1884 E. W. Hamilton *Diary* 16 Jan. (1972) II. 544 He [*sc.* Goschen] believes that C. [*sc.* Chamberlain] is 'riding for a fall' and has doubts as to his loyalty towards Mr. G. 1900 G. B. Shaw *Let.* 31 Dec. (1972) II. 479, I conclude that on turning it over you have concluded that you had better ride for a fall than face the economies that would be needed to allow the shop to clear itself. 1951 J. D. Salinger *Catcher in Rye* xxiv. 242, I have a feeling that you're riding for some kind of a terrible, terrible fall. 1963 *Review & Herald* 31 Jan. 7/3 People who spread their physical favors around over all the available members of the opposite sex are riding for a fall in the realm of final, lasting happiness with any one person.

h. *to ride herd on* (rarely, *over*): to guard and control (a herd of cattle) by riding on its perimeter; also *transf.*, to keep guard over, be in charge of, keep in check; to boss, subject to discipline. *N. Amer.*

1897 A. H. Lewis *Wolfville* xviii. 235 The way them pore darkened drunkards rides herd on each other..is as good as sermons. 1902 —— *Wolfville Nights* xviii. 266 I'm romancin' leesurly along the street when I encounters a party who's ridin' herd on one of these yere telescopes. 1906 H. Green *At Actors' Boarding House* 367 Buck was riding herd on all the Dutch ovens in camp, filled with baking bannocks. 1940 *Variety* 3 Apr. 39 The name bands are come on for the record jockeys who ride herd over not only Decca records but all the others. 1944 *Sun* (Baltimore) 17 Nov. 8/6 Her mother-in-law..and a cousin of her husband..arrived in Albany to-day to ride herd on the Dewey small fry. 1955 R. P. Hobson *Nothing too Good for Cowboy* vi. 51 Some of the horses would be night horses for riding herd on the cattle. 1973 *Whig-Standard* (Kingston, Ontario) 11 Aug. 7/1, I was riding herd on a hundred head of beeves. 1977 'E. McBain' *Long Time no See* xi. 172 Two men who should be taking care of people getting robbed or mugged, go to waste our time instead riding herd on a bunch of street hoodlums.

i. *Jazz slang.* To play with an easy or easily flowing rhythm.

1929 T. Waller (record title) Ridin' but walkin'. 1933 *Melody Maker* 23 Dec. 7/1 Ellington..never played the sort of music to cause Sonny Green to 'ride' in the Sidney Mansie manner. 1938 *Metronome* Feb. 25 When they ride you can't help getting a lift. 1977 J. Wainwright *Do Nothin'* xi. 184 When Ellington opens on an eight-bar piano intro..you know that..when the full outfit starts leaning back and riding, you are going to be lifted cloud-high.

j. —— *rides again*: someone or something makes a reappearance, usu. under different or unexpected circumstances or in modified form.

1939 (*film title*) Destry rides again. 1941 *Pleasures of Publishing* (Columbia Univ. Press) 3 Feb., Our good friend Helen Bower of the Detroit *Free Press* sends us a circular which..is headed, 'Blackstone Rides Again', and is an announcement for a new edition of Blackstone's Commentaries. 1961 C. Willock *Death in Covert* xii. 203 One headline said: *Regency Rakes Ride Again.* 1972 J. Wainwright *Requiem for Loser* iv. 72 The publisher said: '..sequel?'.. He [*sc.* the author] said: 'What d'you think we should call it? *Drover Rides Again*, or *Son of Drover?*' 1977 I. Shaw *Beggarman, Thief* iii. iv. 240 Willie Abbott rides again.

k. *to ride work*: to exercise a racehorse.

1950 *Landfall* IV. 19 Gordon, did you know I ride work now in the mornings? 1959 M. Gee in C. K. Stead *N.Z. Short Stories* (1966) 271 He'd ridden work on the horse and knew her well. 1979 D. Francis *Whip Hand* vi. 74 A work jockey is a lad who rides work on the gallops. *Ibid.* xiv. 174 My girl assistant says she saw you riding work on the Heath.

l. *spec.* in *Surfing*.

1963 *Observer* 13 Oct. 15/4 Riding a wave gives me a feeling of control. I know when I'm riding well and when I'm not. 1975 *Oxf. Compan. Sports & Games* 1005/2 A surfer who prefers to ride with his right foot forward is called a 'goofy footer'.

m. *to ride shotgun*: to travel as an (armed) guard in the seat next to the driver of a vehicle. Hence *transf.* and *fig.*, to act as a protector; to ride in the passenger seat of a motor vehicle. Chiefly *U.S.*

1963 B. S. Johnson *Travelling People* v. 107 And if you want anyone to ride shotgun for you, just you let Henry know. 1966 *National Observer* (U.S.) 26 Dec. 1/2 The gunships 'ride shotgun' on the highly vulnerable, more lightly armed transports. 1971 M. Tak *Truck Talk* 131 *Ride shotgun*, to ride in the passenger seat of the tractor. 1972 *National Observer* (U.S.) 27 May 1/1 Bob Earle was riding shotgun when suddenly my car skidded hard, its rear whipping out to the right. 1972 J. Wambaugh *Blue Knight* (1973) xv. 284 On nightwatch it's comforting sometimes to have someone riding shotgun or walking beside you. 1976 *Wymondham & Attleborough Express* 3 Dec. 2 He offered to ride shotgun for me. 1980 *Times* 21 Jan. 5/2 It was quite by chance that *The Times* found itself riding shotgun for the Red Army.

6. c. (Further examples.)

1931 *Daily Express* 21 Sept. 15/4 The course rode dead, and was not in favour of weight-carrying. 1974 *Country Life* 3 Oct. 925/3 While there were refusals in plenty, most of them due to rider-failure, the course rode well.

d. Of a motor vehicle: to admit of being driven, in respect of passenger comfort, etc. Cf. *ride *sb.*¹ 6 a.

1973 *Milestones* Winter 27/3 The car rode quite well, handling bad surfaces with aplomb.

II. 9. c. To pursue a course without intervention; to rest without further elaboration; to pass without comment; chiefly in phr. *to let* (something) *ride*, to leave alone; to allow to take its natural course.

1921 S. Ford *Inez & Trilby May* iii. 52 If I'd been brought up in a pawn shop I might describe 'em better, but, being no gem expert, I'll have to let it ride at that. 1933 H. L. Ickes *Diary* 14 Nov. (1953) I. 121, I was assured through Marx that the thing would be allowed to ride. 1938 'R. Hyde' *Nor Years Condemn* 171 Ah, let her ride and see what happens. 1944 R. Chandler *Lady in Lake* xi. 63 'Kind of smelly work, to my notion.' I let that ride. 1959 H. Holt *Wreath for Lady* xi. 68 'We'll let it ride like that for the present,' said the detective. 1961 J. Wade *Back to Life* vi. 56, I let it ride. I couldn't be bothered to reply. 1975 J. Symons *Three Pipe Problem* xvi. 158 You think you can just let it ride?

10. a. Also *fig.*

1956 B. Holiday *Lady sings Blues* (1973) viii. 76, I knew that night I had the future of the whole band riding on me. 1964 G. B. Schaller *Year of Gorilla* (1965) iv. 87 They had tried to raise pigs and failed, and their luck now rode on a crop of turnips. 1972 *Village Voice* (N.Y.) 1 June 70/4 Basically, however, the election will ride on Lowenstein. 1976 *National Observer* (U.S.) 25 Sept. 10/2 A lot was riding on grapes, but the quality of the white wine made from the postwar harvest was poor.

11. b. (Further examples.)

1890 *Dialect Notes* I. 19 Your collar rides up behind. 1922 Joyce *Ulysses* 528 The scanty, daringly short skirt, riding up at the knee to show a peep of white pantelette, is a potent weapon. 1951 *Sunday Times* 28 Oct. 11/4 'Ski pyjamas' with deeply-ribbed ankles and wrist-bands that won't ride up. 1971 *Guardian* 24 Aug. 9/1 The pantie..holds the blouse from riding up.

III. 14. c. (Later *transf.* examples.)

1974 *State* (Columbia, S. Carolina) 3 Mar. 1-D/3 Clemson pulled ahead midway the second half and rode the fine play of Terrell Suit to a 71-58 triumph over Georgia Tech. 1976 *Southern Even. Echo* (Southampton) 15 Nov. 13/2 Vinazzani appeared to ride a tackle from Earles and then fall over his own feet.

IV. 15. c. *to ride a hobby*: to pursue a favourite occupation or subject to an excessive degree. Cf. Hobby *sb.*¹ 5.

1823, 1874 [see Hobby *sb.*¹ 5]. 1875 *N. Amer. Rev.* CXX. 189 He must of course be naturally of a rather attitudinizing turn, fond of brooding and spouting and riding a theological hobby. 1876 J. Fergusson *Hist. Indian & Eastern Archit.* vi. ii. 425 It may look like riding a hobby to death, but I cannot help suspecting a wooden origin for it.

d. *to ride* (*the*) *gain* (Broadcasting): to reduce or increase the gain when the input signal becomes too large or too small, in order to keep the output within the limits of succeeding equipment. *colloq.*

1937 *Printers' Ink Monthly* May 40/3 *Ride gain*, to compress the volume range of a program electrically in order to transmit it over lines and equipment within proper limits. 1962 A. Nisbett *Technique Sound Studio* v. 100 Overmodulation on music must be avoided not by 'riding the gain' but by careful preparation—gradually pulling back on the gain control for half a minute or so before the peak is due.

e. *to ride the clutch*: in driving a motor vehicle, to keep one's foot too long on the clutch pedal, depressing it slightly.

1965 Priestley & Wisdom *Good Driving* iii. 30 The second common fault is 'riding the clutch'. 1968 *Practical Motorist* Feb. 622/4 Unless you are moving off, changing gear, or braking keep your foot away from the clutch pedal. 'Riding' the clutch is the fastest possible way of wearing out the friction lining.

16. (Later examples.)

1922 Joyce *Ulysses* 762 That blackguard-looking fellow with the fine eyes peeling a switch attack me in the dark and ride me up against the wall without a word. 1978 S. Allan *Inside Job* iii. 41 She mounted him and rode him.. until they climaxed together.

17. c. To annoy, worry, rile. *colloq.* (orig. *U.S.*).

1918 H. C. Witwer *From Baseball to Boches* 359 Well, Joe, the mob begins to ride me. 1926 S. Lewis *Mantrap* viii. 92, I guess I *have* been riding you pretty hard... I'm sorry. 1938 G. Heyer *Blunt Instrument* viii. 155 They won't arrest you. You needn't let that bugbear ride you. 1940 R. Chandler *Farewell, my Lovely* iii. 22 Go ahead and ride me. Everybody else does. 1948 F. Brown *Murder can be Fun* (1951) vii. 104 Dineen didn't like that so he rode Jerry every chance he had. 1959 'J. R. Macdonald' *Galton Case* (1960) xvii. 140 Don't let it ride you. You're a willing man, but you can't take on responsibility for all the trouble in the world. 1973 N. Graham *Murder in Dark Room* x. 65 'It's still murder.' 'That's what's riding me,' he said. 'I feel I almost had a hand in it.'

18. c. *spec.* in *Surfing*. Also *fig.*

1953 'S. Rattray' *Bishop in Check* 101 One-half per cent of them play tennis—or swim, or surf-ride. 1968 W. Warwick *Surfriding in N.Z.* 2 The Duke..caught a wave and rode it until a short distance from the shore. 1969 *Times* 14 Apr. 6/7 Powerful enough to accelerate atomic particles, the particles 'riding' on the waves and gaining energy from them. 1970 *Studies in English* (Univ. Cape Town) I. 26 Most surfers leave the wave before they reach the soup, as it is tricky to ride. 1975 *Oxf. Compan. Sports & Games* 1005/1 A surfer can ride the wave in a straight line..to the beach or, by weight transference, turn his board to the right or left.

d. To travel in or on (a train, public transport vehicle, etc.), to be a passenger on. Chiefly *U.S.* Also in various phrases, chiefly connected with tramp life, as *to ride the blind(s)* [*blind *sb.* 10], to travel on a blind baggage car without paying one's fare; *to ride* (*the*) *cushions*, to travel luxuriously; to travel as a paying passenger; *to ride the rails*, to travel by rail (esp. without a ticket); *to ride the rods* [*rod *sb.*¹ 9 g], to travel on a bar underneath a railway carriage or wagon.

1906, etc. [see below]. 1926 J. Black *You can't Win* vii. 80 'Let's ride the passenger trains,' I said. 1934 D. Runyon in *Collier's* 3 Mar. 9/1 He is a great hand for riding the tubs back and forth between here and Europe. 1941 B. Schulberg *What makes Sammy Run?* vi. 122 He spent three weeks riding the street cars and the buses and trying to get by the studio reception desks. 1962 A. Shepard in *Into Orbit* 95, I honestly never felt that I would be the first man to ride the Mercury capsule. 1964 Mrs. L. B. Johnson *White House Diary* 11 Sept. (1970) 195 They said they would be delighted to ride the train with me through North Carolina. 1972 *Publishers' Weekly* 24 Jan. 25 Bill is too young to remember when you could ride the subway for a nickel. 1978 *Detroit Free Press* 16 Apr. 1A/2 Willie Ramsey leaves his house on the west side of Detroit and rides buses for nearly two hours to reach his factory job.

1906 *Dialect Notes* III. 153 *Ride* (*the*) *blind, v. phr.*, to steal a ride on a blind baggage car or on a railway train. 'I've spent all my money; I'll have to *ride the blind back*.' 1950 A. Lomax *Mr. Jelly Roll* 172, I..was ashamed to wire for money. So I decided to hobo..and, when the train pulled out the Denver station, I was riding the blinds.

1918 *Dialect Notes* V. 29 *To ride the cushions*, vb. phr...to travel first class. 1924 'Digit' *Confessions 20th Cent. Hobo* 12 *Ride the cushions*, travel by passenger train in the orthodox manner. 1929 H. W. Odum in A. Dundes *Mother Wit* (1973) 195 Thought I would ride cushions till my money give out. 1946 Mezzrow & Wolfe *Really Blues* i. 17 Riding the cushions on my way home made me think of another train ride I once took. 1972 *Amer. Speech* 1968 XLIII. 289 *Ride the cushions*, to deadhead.

1946 Mezzrow & Wolfe *Really Blues* i. 17 We hit Cape Girardeau, Mo., dirty from riding the rails. 1977 *It* June 4/4 According to the Railways Act of 1889, Section 538.., it is an offence to ride the rails 'without having previously paid the proper fare'.

1907 J. London *Road* 24 The tramp, snugly ensconced inside the truck, with the four wheels and all the framework around him, has the 'cinch' on the crew—or so he thinks, until someday he rides the rods on a bad road. 1925 G. H. Mullin *Adventures of Scholar-Tramp* xi. 157 He wanted to get into Philly late that very afternoon. That would probably mean riding the rods of a fast mail on the Pennsylvania, he said, adding the casual hope that we would have no trouble finding rods fit to ride. 1935 J. T. Farrell *Studs Lonigan* iii. xvi. 377 Or maybe lose a leg or get killed under a train, or freeze to death riding the rods in winter. 1946 Mezzrow & Wolfe *Really Blues* i. 5 He spent years.. digging the riffs the wheels were knocking out when he rode the rods.

e. *to ride the lightning*: to suffer execution on the electric chair. *U.S. slang.*

1935 *Jrnl. Abnormal Psychol.* XXX. 364 *Ride the lightning*, to be electrocuted. 1965 *Daily Progress* (Charlottesville, Va.) 22 Mar., He pleaded for an opportunity to 'ride the lightning' of the electric chair. 1968 *Sunday Truth* (Brisbane) 15 Sept. 10/3 Four hours before I was 'to ride the lightning' in the chair a man came to the jail and confessed.

20. d. *transf.* To lead (a person) *off* a subject or away from a desired end; to sidetrack.

1908 A. Chamberlain *Politics from Inside* 128 We got several of our discussions in before them and 'rode them off', to use a racing phrase. 1928 'Sapper' *Female of Species* x. 161 Look here, Peter—we've got to try and ride them off.

21. a. *to ride on a rail* (earlier and later examples).

1837 Hawthorne *Twice-Told Tales* 161 The millmen.. [hesitated whether to] ride him on a rail, or refresh him with an ablution at the town-pump. 1843 Dickens *Mart. Chuz.* (1844) xxi. 267 'If I can realise your meaning, ride me on a rail!' returned the General. 1854 S. Smith *'Way Down East* x. 251 Others..proposed..giving him a good coat of tar-and-feathers and riding him out of town on a rail. 1929 E. Linklater *Poet's Pub.* xxiii. 253, I was ridden on a rail..in nothing but a torn shirt. 1949 *Sat. Even. Post* 2 July 67/1, I feel somewhat like the man who was being ridden out of town on a rail.

e. To bring *in* or introduce (a cinematographic picture) with an accompaniment of music.

1927 *Observer* 17 Apr. 3/3 The orchestral prelude is usually quite elaborate, and the picture is what is called 'ridden in'.

rideable, *a.* Add: **3.** Suitable for being hunted on horseback. *rare.*

1910 *Blackw. Mag.* Apr. 557/1 We were a long time before we found a rideable pig.

Rideal–Walker (ridī·l wǭ·kəɹ). [f. the names of Samuel *Rideal* (1863–1929) and J. T. Ainslie *Walker* (1868–1930), English chemists, who described the test in 1903 (*Jrnl. R. Sanitary Inst.* XXIV. 424).] *Rideal–Walker method* or *test*, a procedure for determining the germicidal efficiencies of disinfectants relative to phenol as a standard; hence *Rideal–Walker coefficient*, a measure of germicidal efficiency obtained by this method.

1907 *Jrnl. R. Sanitary Inst.* XXVIII. 379 He wished to draw attention to a few points..in connection with the practical details of the Rideal-Walker method of standardisation of disinfectants. 1919 *Jrnl. State Med.* XXVII. 62 The Rideal-Walker test..is difficult in execution on account of the attempt to include in one test all the variable factors. 1940 *Thorpe's Dict. Appl. Chem.* (ed. 4) IV. 32/1 The Rideal-Walker coefficient is obtained by dividing that dilution of the disinfectant which shows life in 2½ and 5 minutes, but no life thereafter, by that dilution of carbolic acid..which shows life in 2½ and 5 minutes, but no life thereafter. 1965 Turk & Porter *Short Textbk. Microbiol.* v. 27 The Rideal-Walker coefficient was at one time widely used as a measure of the potency of an antiseptic. 1971 H. Chick et al. *War on Dis.* ix. 89 The errors in the Rideal-Walker test were avoided by the modifications introduced in the Chick-Martin technique.

rided, *a.* (Further example.)

1938 J. W. Day *Dog in Sport* v. 73 The bracken on his downland slopes was rided into broad squares.

rideman. Restrict † *Obs.* to sense in Dict. and add: Also **ride man, ride-man. 2.** *U.S. slang.* An operator of a roundabout or similar device at an amusement park or fair.

1926 *Amer. Mercury* Dec. 464/2 Every reader along Broadway knows that 'Lemon Stands Don't Interest Ride Men' means that poor-paying amusement parks fail to attract operators of roller-coasters, chute-the-chutes and carousels. 1942 Berrey & Van den Bark *Amer. Thes. Slang* § 524/9 *Ride man*, a roller-coaster operator.

3. *Jazz slang.* A musician who improvises in a pronounced rhythm (see also quot. 1970). *U.S.*

1935 *Vanity Fair* (N.Y.) Nov. 38/2 *Ride-men* is the term applied to the improvisers of these licks. 1942 Berrey & Van den Bark *Amer. Thes. Slang* § 576/4 '*Swing' musician*, ..rideman. 1945 *Band Leaders & Record Rev.* Mar. 20 Within a horn blast of Hollywood and Vine, the crossroads of Glamour-town, can be found many lairs of the hepcats—haunts of gates and ride men. 1970 C. Major *Dict. Afro-Amer. Slang* 97 *Ride man*,..in jazz, the leading soloist, who establishes the pace.

ride-off, rideoff (rai·dǫ̈f). [f. Ride *v.* + Off *adv.*, after *play-off, etc.] In a competitive event involving horsemanship: a round or phase of competition to resolve a tie or determine qualifiers for a later stage.

1973 *Country Life* 19 July 148/2 Mrs Lorna Johnstone winning on El Farruco, the horse with whom she reached the ride-off in the Olympic dressage. 1976 *Billings* (Montana) *Gaz.* 17 June 1-H/3 About 300 riders..qualified at regional rideoffs nationwide.

ride-on (rai·dǫn), *a.* [f. Ride *v.* + On *prep.*] Of a power-driven lawn mower, etc.: on which the operator rides.

1969 *Times* 13 Dec. p. vii/5 He would more than recoup the cost of a 24 in. ride-on type of motor mower. 1972 *Country Life* 23 Mar. 703/3 A number of ride-on rotaries are also available. 1977 *Sunday Tel.* (Colour Suppl.) 5 June 11 (Advt.), The Simplicity range of ride-on tractors is the most competitive of its kind on the market.

rideout. Restrict † *Obs.*⁻¹ to sense in Dict. and add: **2.** *Jazz slang.* Also **ride-out.** A final chorus. Also *attrib.*

1939 *Metronome* May 19 *Pussy Willow* has a great ride-out. 1946 R. Blesh *Shining Trumpets* xii. 286 The Hot Seven's..ride-out ensemble sounds like a free-blowing parade band. 1952 B. Ulanov *Hist. Jazz in Amer.* vi. 52 Whether they are playing a solo.., or pushing a ride-out chorus to its obstreperous end, their harmonic thinking is vertical. 1962 *Jazz Monthly* Oct. 24 *Folk Forms* is reduced to a short bass solo, a short drum solo and a ride-out. 1977 *New Yorker* 8 Aug. 68/2 'On the Other Side of the Tracks'.. has an ebullient and remarkable rideout section.

rider. Add: **I. 2. e.** A passenger, esp. one using public transport. Now chiefly *U.S.*

1886 H. Baumann *Londinismen* 156/1 *Rider*.., Droschkenkutscher-Slang: Passagier, Fahrgast. 1930 [see *job *sb.*² 4 f].

1966 *Economist* 26 Feb. 798/1 The bankrupt suburban railway services in the East..want to get out of the money-losing passenger business entirely, letting their hundreds of thousands of daily riders fend for themselves as best they can. 1972 J. GORES *Dead Skip* (1973) xiii. 89 Kearny opened the rider's door for her: the tall blonde slid in. 1977 *New Yorker* 1 Aug. 28/1 Over the decade 1965 to 1975, the city's transit system, according to the Temporary Commission, lost twenty per cent of its riders.

f. *easy rider*: see *EASY *a.* 14 c.

g. *spec.* in *Surfing*.

1963 *Observer* 13 Oct. 15/4 The wave traps and dumps the rider, burying him for half a minute or longer. 1968 W. WARWICK *Surfriding in N.Z.* 13/1 Quite often a rider can be so locked in on a fast breaking wave that an ordinary pullout is impossible. 1975 *Oxf. Compan. Sports & Games* 1005/1 The belly-board rider slots in by paddling ahead of the oncoming wave.

6. b. Short for *circuit-rider* s.v. *CIRCUIT *sb.* 10.

1884 'C. E. CRADDOCK' *In Tennessee Mts.* I. 15 The rider says ther's some holp in prayer. *Ibid.* III. 143 All them Peels, the whole lay-out, war gone down ter the Settlemint ter hear the rider preach.

II. 14*. *Ophthalm.* Each of a set of linear opacities extending radially outward from the main opacity in some kinds of cataract.

1892 A. DUANE tr. *Fuchs's Text-bk. Ophthalm.* viii. 371 The riders on the periphery of the lamellar cataract originate from the fact that a second layer, peripherally situated with regard to the first, is beginning to become opaque, doing so first only at isolated spots corresponding to the equator of this first layer... These partial opacities embrace the equator of the inner opacity in front and behind; they ride upon it, as it were, whence the name riders. 1937 E. WOLFF *Dis. Eye* vi. 65 When viewed with the ophthalmoscope mirror..the disc and riders appear black or grey, and the clear lens around shows the normal red reflex. 1964 S. DUKE-ELDER *Parsons' Dis. Eye* (ed. 14) xix. 266 The opacity is usually sharply demarcated and the area of the lens within and around the opaque zone is clear, although linear opacities like spokes of a wheel (called riders) may run outwards towards the equator.

ri·der, *v.* *U.S.* [f. the sb.] *trans.* To strengthen (a fence) with riders.

1760 G. WASHINGTON *Diary* 15 Apr. (1925) I. 155 Good part of my new fencing that was not Riderd was leveld. 1787 *Ibid.* 27 Apr. III. 208 Women staking and ridering fence of the said field. 1858 J. A. WARDER *Hedges & Evergreens* 151 In Delaware..worm-fences, not ridered, were to be five feet high.

ridered, *a.* Add: **2.** (Earlier and later examples.) Also, of a building: having a ridered fence (*rare*). Cf. STAKE *sb.*¹ 2 c.

1852 *Trans. Mich. Agric. Soc.* III. 333 The staked and ridered domicil. 1855 *Chicago Weekly Times* 17 May 3/5 A whirlwind..scattered in every direction a strong 'staked and ridered fence'. 1885 'C. E. CRADDOCK' *Prophet Gt. Smoky Mts.* xii. 231 The corn that Dordain had ploughed on the steep slope was high, and waved above the staked and ridered fence. 1946 G. FOREMAN *Last Trek of Indians* 169 All their farms were inclosed with good rail fences sufficiently high to secure their crops, many of them 'staked and ridered'.

ridership. Restrict *rare* to sense in Dict. and add: **2.** orig. *N. Amer.* The number of passengers (using a particular form of public transport). Also *attrib.*

1972 W. G. DAVIS *Urban Transportation Policy for Ontario* 5 This emphasis on the needs of the passenger and the improvement of service has enlarged ridership considerably. 1975 *Nature* 20 Feb. 580/2 This move has already been reflected in increased public transport 'ridership' figures in most communities. 1979 *Monitor* (McAllen, Texas) 9 July 8B/8 Allegheny [Airlines] operates no DC-10's, so it has picked up part of the ridership while those planes are grounded. 1981 *Daily Tel.* 7 Mar. 18 Costs are more or less fixed whatever the level of ridership and the only way of improving revenue is to attract more off-peak passengers.

ridge, *sb.*¹ Add: **4. a.** (Examples referring to submarine features.)

1944 A. HOLMES *Princ. Physical Geol.* xv. 319 In the shallower depths, over sub-tropical and tropical submarine banks and ridges, the shells of pteropods become abundant. 1954 W. D. THORNBURY *Princ. Geomorphol.* xviii. 462 Off the coast of California..there is a series of submarine basins and ridges similar in origin to the faulted structures landward from them. 1961, etc. [see *MID-OCEAN *a.* b, *MID-OCEANIC *a.*]. 1977 A. HALLAM *Planet Earth* 99/3 Topographic forms can greatly influence currents, which may for example be..diverted round large hills and ridges.

c. For def. read: an elongated region of high barometric pressure. (Further examples.)

1914 *Seaman's Handbk. Meteorol.* (Meteorol. Office) vii. 81 An area of considerably higher barometric pressure.., either as a ridge..or in the more extensive form of an anticyclonic system. 1968 G. M. B. DOBSON *Exploring Atmosphere* (ed. 2) vi. 136 Troughs and ridges tend to circulate round the pole from west to east, but the general westerly wind at these heights has a much greater speed, and the air actually flows through these troughs and ridges. As the air blows into a low pressure trough it descends, while as it approaches a ridge it ascends. 1977 *Hongkong Standard* 14 Apr. 16/2 A ridge of high pressure covers the northern part of the south China Sea.

5. a. *transf.* (Later examples.)

1815 WORDSWORTH *Spanish Guerillas* in *Poems* II. 255 They have learnt to open and to close The ridges of grim War. 1895 W. B. YEATS *Death of Cuhoollin* in *Poems* 203 My father dwells among the sea-worn bands, And breaks the ridge of battle with his hands.

Comb. (Further examples.)

1919 J. MASEFIELD *Reynard* 51 Meadows ridge-and-furrow ploughed. 1958 *New Biol.* XXVI. 40 This is particularly well shown in grasslands in which there are marked variations in the height of the water table, such as the characteristic ridge and furrow grasslands of Britain. 1967 *Listener* 6 July 10/3 The head gardener to the Duke of Devonshire, Joseph Paxton, then invented ridge and furrow roofing, without rafters. 1974 C. TAYLOR *Fieldwork in Medieval Archaeol.* iii. 57 The ridge and furrow ends on a well-marked terrace which was both a trackway through the fields and a headland on which the plough was turned.

6. d. One of the many raised lines on the skin that are esp. noticeable on the fingers and palms of the hand and the sole of the foot.

1842 *Penny Cycl.* XXII. 86/2 Each such ridge shows on its summit a little furrow dotted with minute apertures. 1866 *Chambers's Encycl.* VIII. 756/1 The cross grooves that intersect the ridges and papillæ on the hands and feet. 1892 F. GALTON *Finger Prints* i. 1 Let no one despise the ridges on account of their smallness, for they are in some respects the most important of all anthropological data. 1920 E. WALLACE *Daffodil Mystery* xxviii. 220 Compare them!.. Line for line, ridge for ridge,..it is Milburgh's thumb-print. 1940 R. MORRISH *Police & Crime-Detection* x. 89 The ridges ('papillary' ridges as they are called) are formed by the mouths of the ducts of the sweat-glands. 1966 T. S. & C. R. LEESON *Histology* xiii. 250/1 Ridges are absent on the forehead, external ear, perineum, and scrotum. 1980 A. SILVERSTEIN *Human Anat. & Physiol.* vii. 97/2 The patterns of grooves and ridges we see are in the epidermis, but they do not originate there. They are produced by variations of folds and ridges in the underlying dermis.

7. a. *ridge-cap* (later example), *rope*; *ridge-roofed* adj.

1975 *Whig-Standard* (Kingston, Ontario) 4 Sept. 23/6 [Her] specialty is to make ridge cap shakes which are joined and overlapped alternately to make a leak-proof seal for the roof ridge. 1963 H. N. SAVORY in Foster & Alcock *Culture & Environment* iii. 34 A small rectangular, ridge-roofed house. 1933 Ridge rope [see *A TENT].

b*. In sense 4, as *ridge crest, prairie* (U.S.), *road* (U.S.), *system, -top, walk*.

1963 L. F. CHITTY in Foster & Alcock *Culture & Environment* vii. 179 From Onibury, the general trend of the way is clear, but its actual line is partly problematical: there is no longer an extended ridge-crest to give it definition. 1971 I. G. GASS et al. *Understanding Earth* xxi. 303/1 This relationship..poses the question of whether these volcanoes in the flanking basins originated on ridge crests and remained active while they were carried away upon the spreading sea-floor. 1977 A. HALLAM *Planet Earth* 97/3 The mean depth below the water surface of the world's ridge crests is 2700 m (8775 ft). 1882 *Econ. Geol. Illinois* II. 73 The prairies are..of two classes—those that are a little elevated and rather level near the lower course of the streams, and more elevated and rolling prairies on the higher ridges. The latter are the so-called 'ridge prairies'. 1817 *N. Amer. Rev.* IV. 185, I have returned by the ridge road. 1871 *Harper's Mag.* Dec. 46/2 These 'ridge-roads'..form a system of ready-made highways. 1961 Ridge system [see *MID-OCEAN *a.* b]. 1971 I. G. GASS et al. *Understanding Earth* xxi. 302/2 Such volcanoes are mostly found along the ridge systems, particularly in the Atlantic. 1877 Ridge-top [in Dict., sense 7 c]. 1977 G. SCOTT *Hot Pursuit* viii. 74 The ground began to climb..and we were on a ridge-top... The mountain ranges stretched away. 1940 W. A. POUCHER *Lakeland through Lens* 43 Their proximity to the Buttermere Valley makes the western end of these ridges equally approachable, and incidentally very fine ridge walks either way. 1976 *Lancs. Even. Post* 7 Dec. 8/4 The switchback skyline of the Troutbeck fells..provides one of the best ridge walks in eastern Lakeland.

c. *ridge-bank, -chain, -line* (later example, in sense *6 d); (sense *6 d) *ridge characteristics, -count, system*; also *ridge-hop* vb.

1945 C. MANN in B. James *Austral. Short Stories* (1963) 72 The house they had on the ridge-bank was near the middle of the river bend. 1968 J. ARNOLD *Shell Bk. Country Crafts* 164 This..left the wagoner free to throw the ridge-chain over the back-pad. 1954 F. CHERRILL *Cherrill of Yard* vii. 75 The incriminating impressions of the ridge characteristics of a fragment of his palm. 1970 P. LAURIE *Scotland Yard* ix. 200 These ridge-counts go in as well. 1976 Ridge-count [see *phenylthiocarbamide*]. 1973 C. BONINGTON *Next Horizon* xviii. 247 The twin-engined Fokker Friendship ridge-hops over the tree-covered tentacles of the great peaks. 1940 R. MORRISH *Police & Crime-Detection* x. 91 Ridge-lines on the print may represent furrows which have become filled with blood on the finger. 1954 F. CHERRILL *Cherrill of Yard* vii. 69 The patterns on the ends of the fingers are simply a culmination of the ridge system which covers the whole of the palmar surface of the hands.

8. ridgeback = *Rhodesian ridgeback* s.v. *RHODESIAN *a.* 3 c; **ridge barrow** *Archæol.* [BARROW *sb.*¹ 3], a type of long, earthen grave-mound; **ridge cucumber,** a cucumber of a variety suitable for growing outside in a temperate climate, freq. grown on ridges of soil (cf. RIDGE *sb.*¹ 5 c); **ridge runner** *U.S. slang*, a southern mountain farmer, a hill-billy; also in Blacks' use, any white person; **ridge tent** (see quot. 1963); **ridgewise** *adv.*, in the manner of a ridge.

1937 *Our Dogs* 10 Dec. 886/3 The ridge in the breed..is present in practically every Ridgeback puppy. 1945 L. G. GREEN *Where Men Still Dream* 167 The finest type of Bushman hunting dog, a light brown ridgeback mongrel with dark stripes and a trace of the greyhound in its appearance, is now verging on extinction. 1977 P. C. VENTER *Soweto* 51 A ridgeback yawned and got up from the polished door step. 1951 *Field Archæol.* (Ordnance Survey Prof. Papers No. 13) (ed. 3) 15 The whole affair, which belonged to Neolithic times, was..interpreted as an eccentric form of long barrow to which the term 'ridge barrow' has been applied. 1963 *Ibid.* (ed. 4) 28 A variant of the earthen long barrow which seems to be confined at present to Dorset is the so-called 'ridge-barrow' which was found during the excavation of Maiden Castle near Dorchester. 1851 Ridge cucumber [in Dict., s.v. RIDGE *v.* 4 a]. 1933 H. H. THOMAS *Pop. Encycl. Gardening* 257/2 The plants must be grown out in a greenhouse or frame, with the exception of the Ridge Cucumber and the Gherkin, which can be grown out of doors in summer. 1962 *Listener* 25 Oct. 698/1, 2 small ridge cucumbers. 1933 *Amer. Speech* VIII. III. 31/1 *Ridge runner*, originally an Arkansas, rather than a Kentucky, hill billy. Any uncouth, stupid fellow. 1947 A. M. TROUT *Greetings from Old Kentucky* 9 While strolling through the woods one day with my friend, Bill Curry, a ridge runner from London, in Laurel County, we came upon a large bunch of hogs. 1966 *Publ. Amer. Dial. Soc.* 1964 XLII. 31 The most characteristic feature of Negro terms for Caucasian is the tendency to expand particularized designations for Caucasians to include all members of the race, e.g., terms for poor Southerners:..*ridgerunner* [etc.]. 1913 J. F. M. H. STONE *Caravanning & Camping-Out* xiv. 125 The tent I have the most liking for..is the type known as 'ridge tent', 'patrol tent', or 'emigrant tent'. 1926 E. E. REYNOLDS *Camping for All* iii. 13 Stanley..states that one of these double-roof ridge tents withstood three hundred days of rain. 1963 *Camping* ('Know the Game' Series) 3/2 *Ridge Tents.* These have long sides to the roof supported by upright poles at each end... A ridge tent with roof running down to the ground is called an A-tent. 1977 *Grimsby Even. Tel.* 5 May 3/6 (Advt.), Wanted, small ridge tent with fly-sheet. 1725 Ridge-wise [in Dict., sense 7 b]. 1743 W. ELLIS *Mod. Husbandman* Nov. vii. 316 They..cover the Turneps and Foss with Earth, Ridge-wise. 1902 J. BUCHAN *No-Man's-Land* i. 5 A sort of plateau, benty and rock-strewn, running ridge-wise above a chain of little peaty lochs.

ridge, *sb.*² For † *Obs.* read '*Obs.* exc. *U.S.*' and add: (Later example.) Also, any metal coin.

1931 *Writer's Digest* Oct. 29 *Ridge*, a gold coin of any denomination. 1935 *Amer. Speech* X. 13/2 *Chink*, metal money; loose change. (Obs.) Modern *ridge*, *coin*. 1955 *Publ. Amer. Dial. Soc.* XXIV. 78 Pockets were actually picked for metal coins—*ridge* or *smash*.

ridge (ridʒ), *a.* *Austral. slang.* [f. RIDGE *sb.*²] Good, all right, genuine.

1938 PARTRIDGE *Dict. Slang* (ed. 2) 1026/2 *Ridge*, adj., good; valuable: Australian. 1953 K. TENNANT *Joyful Condemned* i. 4 'It's ridge, Hec,' she assured him. 'He won't come here again.' 1971 D. IRELAND *Unknown Industrial Prisoner* vii. 130, I convinced her the whole thing was ridge!

ridgeling. (Further example.)

1974 E. C. STACEY *Peace Country Heritage* ii. 103 Reddon studied the occurrence of ridglings in boars and sought to determine if it was a hereditary trait, which it proved to be.

ridge-pole. Add: **2.** (Earlier example.) Also *fig.*

1814 *Niles' Weekly Reg.* V. 322/2 At the time I left the boat the waters were about midway on the roofs of the houses generally, and quite to the ridge poles of several. 1955 E. POUND *Classic Anthol.* III. 184 High, pine-covered peak full of echos, Proud ridge-pole of Heaven.

ridger. Add: **1.** (Earlier example.)

1733 W. ELLIS *Chiltern & Vale Farming* 322 Two Ropes or Chains..are held by the Ridger of the Cart-saddle.

2. (Further examples.)

1947 T. HENNELL *Countryman at Work* 70 The 'free ends' of the thatch (where the building ends vertically, and the thatch cannot be continued round the corners) must be made up to the proper thickness within long rods or ridgers, which are first bent upwards to a right angle, pinned in the edges of thatch, finally bent again and pinned down across it. 1977 *Grimsby Even. Tel.* 5 May 3/6 (Advt.), BMB Iron Horse, in perfect working order, with two pair of wheels,..plough, drag, skerry, potato lifter and ridger.

ridgey-dite (ri:dʒi dəi·t), *a.* *Austral. rhyming slang.* [After *RIDGY-DIDGE *a.*] All right; = *RIDGE *a.*

1953 K. TENNANT *Joyful Condemned* xxx. 295 He'd tell you himself I'm ridgey-dite. I worked for him.

ridgie-didge, var. *RIDGY-DIDGE *a.*

Ridgway (ri·dʒwē¹). The name of the *Ridgway* family, used *attrib.* and *absol.* to designate pottery and porcelain produced from 1792 by the brothers George (*c* 1758–1823) and Job Ridgway (1759–1814), or by their descendants.

1911 J. F. BLACKER *Nineteenth-Cent. Eng. Ceramic Art* ix. 207 To say that there will be little old Ridgway ware and china amongst future collections is to assume that fashion will never fail in its pursuit of fine specimens. That is probably true; useful articles stand in another class, and though there is fine old Ridgway it is not easy to get. 1960 H. HAYWARD *Antique Coll.* 239/1 Ridgway earthenware & porcelain. 1972 G. A. GODDEN *Illustr. Guide Ridgway Porcelains* ii. 10 It is extremely unlikely that we shall ever be able to identify positively the eighteenth-century Ridgway earthenwares. *Ibid.* 14 The Ridgway porcelain dessert and tea-wares do not at this pre-1830 period bear any factory trade-mark. 1974 *Encycl. Brit. Micropædia* VIII. 577/3 (*caption*) Ridgway porcelain inkpot, a caricature of Job Ridgway's wife..*c* 1810.

ridgy-didge (ri:dʒidi·dʒ), *a.* *Austral. slang.* Also **ridgie-didge, ridgy-dig, rigi-dig.** [Elaboration of *ridgy*, f. *RIDGE *a.* + -Y⁶, -IE.] = *RIDGE *a.*

1953 BAKER *Australia Speaks* 102 *Ridgy-didge* or *ridgy-dig*.., honest, genuine, okay. 1953 R. BRADDON in I. Bevan *Sunburnt Country* 130 The phrase is used invariably either as a simple question 'Ridgy Didge?' or as an unequivocal assurance: 'Ridgy Didge!' 1963 L. GLASSOP *Rats in New Guinea* xii. 153 'It's ridgie-didge,' said Eddie. 'Spit me death.' 1968 *Courier-Mail* (Brisbane) 27 July 11 Strike me handsome! What a fair dinkum, dinki-di, rigi-dig, bonzer, curl-the-mo, nob of coots these O'Grady blokes are, mate! 1976 *Sunday Sun* (Brisbane) 1 Aug. (Fun Holiday Guide) 1/1 Just to prove I'm ridgie-didge, I talked to my friends at Sunday Sun.

ridiculosity. (Further examples.)

1773 J. HOADLY *Let.* 16 Nov. in D. Garrick *Private Corr.* (1831) I. 583 You seem now to give into Dr Goldsmith's ridiculosity in opposition to all sentimentality. 1960 C. ACHEBE *No Longer at Ease* xii. 119 Pat said it was all silly and ridiculous. She actually used the word *ridiculosity*, which made Obi smile internally.

ridiculous, *a.* Add: **1. c.** *slang.* Outstanding, excellent.

1959 *Jazz* Summer 209 His technique is ridiculous! 1960 D. CERULLI et al. *Jazz Word* 95 To a jazzman..*ridiculous* is wonderful. 1968 *Scottish Daily Mail* 3 Jan. 6 Superlatives ..gradually increased with the years into 'out-of-sight', 'ridiculous' and 'unbelievable'.

2. b. *adv.* Also *colloq.* (Further examples.)

1834 C. F. HOFFMAN *Winter in West* (1835) I. 270 Those Indians behaved most *ridiculous*. They dashed children's brains against the door-posts. 1976 *Daily Mirror* 11 Mar. 24/2 Don't talk ridiculous!

riding, *sb.* Add: **1.** (This ceased to be an official designation after Local Government reorganization outside Greater London on 1 Apr. 1974.)

2. b. *spec.* An administrative or electoral district in Canada. Also *transf.*

1792 in *Rep. Bureau Archives Ontario* (1906) IV. 180 The said county of Glengarry, bounded as afore said, shall be divided into two ridings. 1853 *Elora* (Ontario) *Backwoodsman* 21 Apr. 2/5 When I do seek the votes of the electors of the north riding, I shall fearlessly submit my qualifications and character to the judgment of all who can cast aside their personal feelings and look only to the public good. 1867 [in Dict.]. 1890 *Grip* (Toronto) 29 Mar. 213/1 In that riding the New Party had only 50 pledged members, but pulled nearly 800 votes. 1921 [see *BÂTONNIER]. 1957 *Maclean's Mag.* 6 July 36/2 At one time Spadina regularly returned J. B. Salsberg as an LPP member to the Provincial Legislature, a habit that inspired a waggish reporter to tag it the 'Little Red Riding'. 1970 D. WATERFIELD *Continental Waterboy* ii. 10 H. W. Herridge, C.C.F. member for West Kootenay, our M.P., gave a talk at Nakusp, warning of the considerable changes that might be expected in the economics of his riding. 1975 *Times Lit. Suppl.* 10 Oct. 1189/3 A communist member once sat briefly in parliament for a Montreal riding. 1978 *Toronto Star* 26 Feb. A4/3 The allegations are directed against Georges Marchais, 58, France's outspoken Communist leader who is standing for re-election by the 80,000 voters in the Kremlin riding of Val-de-Marne.

riding, *vbl. sb.* Add: **I. 1. e.** Provoking, teasing, annoying. *U.S. colloq.*

1927 *Amer. Speech* Dec. 167/2 *Riding*, being sarcastic. 1930 D. HAMMETT *Maltese Falcon* xviii. 220 The boy said: 'You bastard, get up and shoot it out if you've got the guts. I've taken all the riding from you I'm going to take.'

II. 5. a. *riding-blanket, -boot* (later examples), *-breeks, cap* (later example), *costume, gauntlet, hat* (later examples), *mac*.

1935 H. L. DAVIS *Honey in Horn* xi. 165 She pulled her riding-blanket down on her bare shoulders with a temperish jerk. 1851 HAWTHORNE *House of Seven Gables* i. 18 With such a tramp of his ponderous riding-boots as might of itself have been audible in the remotest of the seven gables, he advanced to the door. 1952 E. O'NEILL *Moon for Misbegotten* I. 56 He is dressed in..immaculately polished English riding boots with spurs. 1980 G. M. FRASER *Mr American* xxvi. 533 A field officer in Sam Browne, red tabs, and gleaming riding boots. 1916 W. OWEN *Let.* 10 Feb. (1967) 380 Brown has riding-breeks. 1976 *Alyn & Deeside Observer* 10 Dec. 21/6 (Advt.), Riding macs, riding caps with new collapsible peaks, hunting bowlers. Bargain price. 1944 A. CLARKE *Viscount of Blarney* 42 A Gallant in eighteenth-century riding costume appears at doorway. 1855 F. DUBERLY *Let.* 29 Jan. in E. E. P. Tisdall *Mrs. Duberly's Campaigns* (1963) iv. 124, 6 pairs of doeskin riding gauntlets, 7½. 1782 J. WOODFORDE *Diary* 24 Apr. (1926) II. 19 For a riding Hat for Nancy..pd. 1. 13. 6. 1811 JANE AUSTEN *Let.* 18 Apr. (1932) II. 69, I must have a straw hat, of the riding hat shape. 1978 'F. PARRISH' *Sting of Honeybee* iv. 51 The other old lady..fitted them out with riding hats, and helped them on to small, fat, elderly ponies. 1961 M. KELLY *Spoilt Kill* II. 91 She was..wearing an old riding mac. 1974 L. DEIGHTON *Spy Story* xvi. 163 A figure in a dirty white riding mac.

b. *riding-quirt.*

1951 R. CAMPBELL *Light on Dark Horse* xxiii. 337 The dried penis of a bull which we carry in those parts in place of a riding-quirt.

c. *riding boat, carriage* (earlier example), *chair* (later example), *horse* (earlier and later examples), *-pony, stallion*; also of machinery on which the operator rides (cf. *RIDE-ON *a.*), as *riding mower, plough*.

1908 G. SANGER *70 Yrs. Showman* xvi. 57 My father was able to add 'riding' or 'over and over' boats, as they were called, to his peep-show and roundabouts. 1785 T. JEFFERSON *Notes Virginia* VIII. 158, 5,126 wheels of riding carriages. 1971 *Country Life* 11 Mar. 528/2 It is impossible to judge how many of these chamber horses, also called dandy horses and riding chairs, were made. 1641 *Rec. Early Hist. Boston* (1877) II. 61 Its ordered that all dry cattle shall be driven of the necke, and not be suffered to abide there, except Riding horses. 1714 *Essex Inst. Hist. Coll.* (1883) XX. 179 Voated that the Neck of Land..be granted and reserved for the use of the Town of Salem for a pasture for Milch Cows and Rideing Horses. 1940 W. FAULKNER *Hamlet* IV. i. 255 The tethered wagons and riding horses and mules. 1977 *Horse & Hound* 10 June 5/3 Our present efforts to breed riding horses. 1969 *Sears Catal.* Spring/Summer 5 A compact yet powerful Riding Mower that's truly easy to start. 1976 *Billings* (Montana) *Gaz.* 17 June 6-F/1 (Advt.), Bolens 5 HP riding mower. 1911 *Encycl. Brit.* XXI. 851/2 The 'sulky' or *riding plough* is little known in the United Kingdom. 1960 DAVIES & VAUGHAN *Beyond Old Bone Trail* xi. 74 We bought..a riding plough—the 'walking' model was right out of date. 1954 J. R. R. TOLKIEN *Fellowship of Ring* I. xi. 190 The two or three riding-ponies. 1940 W. FAULKNER *Hamlet* I. i. 15 Colonel John Sartoris his self shot Ab for trying to steal his claybank riding stallion during the war.

d. *riding code, instructor, lesson, mistress, -muscle, tournament*; **riding ballad** (earlier example).

1800 W. SCOTT *Let.* 18 Oct. (1932) I. 105, I do not mean entirely to limit my collection to the Riding Ballads, as they are called in our country, those namely which relate to Border feuds and forays. 1971 *Riding for Recreation* (Brit. Horse Soc.) ii. 3/1 The British Horse Society in consultation with other organisations has compiled a Riding Code. 1946 M. C. SELF *Horseman's Encycl.* 342 The proper choice of a riding instructor is of utmost importance. 1977 N. MARSH *Last Ditch* iii. 63 A riding instructor or some such in the army. 1886 F. H. BURNETT *Little Lord Fauntleroy* viii. 159 Fauntleroy took his first riding lesson. 1975 J. MCCLURE *Snake* viii. 110 Business so bad..and yet his kid goes to riding lessons. 1895 *Funk's Stand. Dict.*, Riding mistress. 1926 GALSWORTHY *Silver Spoon* III. vii. 275 A riding mistress was teaching a small boy to trot. 1914 D. H. LAWRENCE *Prussian Officer* 2 His orderly, having to rub him down, admired the amazing riding-muscles of his loins. 1934 *Sun* (Baltimore) 15 Aug. 9/3 The amateur riding tournament today. 1941 *Ibid.* 20 Aug. 11/7 The mountain folk..also will have their wood-chopping and sawing contests, riding tournament and rifle shoot.

ridley (ri·dli). [Etym. unknown.] Either of two species of marine turtle, *Lepidochelys kempii* of the Atlantic or *L. olivacea* of the Pacific and Indian Oceans.

1942 A. F. CARR in *Proc. New England Zool. Club* XXI. 8, I believe that a change in the non-technical designation of *kempii* is indicated. Some time ago..Stewart Springer..told me of a species of sea turtle, known locally as the 'ridley', which was recognized as distinct by the natives of the Keys. *Ibid.* 9 Henceforth I shall use ridley. 1952 — *Handbk. Turtles* 402 A ridley will die, apparently of rage and frustration, if placed on its carapace out of water. 1966 R. C. STEBBINS *Field Guide Western Reptiles & Amphibians* 89 Pacific ridley... A relatively small sea turtle with uniformly olive-colored heart-shaped carapace... More a bottom dweller than other marine turtles. 1974 *Encycl. Brit. Micropædia* IX. 12/2 The Atlantic ridley is widely known as the 'bastard turtle', the name stemming from the mistaken belief that it is the offspring of a green turtle and a loggerhead. 1980 *Economist* 16 Aug. 26/1 The Kemp's ridley is being wiped out.

ridotto, *sb.* (Earlier *attrib.* example.)

1826 M. KELLY *Reminisc.* I. 204 The ridotto rooms, where the masquerades took place, were in the palace.

riebeckite (ri·bekəit). *Min.* [ad. G. *riebeckit* (A. Sauer 1888, in *Zeitschr. d. Deut. geol. Ges.* XL. 138), f. the name of Emil *Riebeck* (d. 1885), German explorer: see -ITE¹.] A mineral of the monoclinic amphibole group, $Na_2Fe^{2+}{}_3Fe^{3+}{}_2Si_8O_{22}(OH)_2$, found as dark blue or black prismatic crystals and often containing magnesium.

1889 *Jrnl. Chem. Soc.* LVI. 109 This hornblende, named riebeckite, has the same composition as arfvedsonite, and is the analogue of ägirin of the augite series. 1941 *Proc. Prehist. Soc.* VII. 65 This hornblende..recalls the habit and mode of occurrence of riebeckite in the paisanites. 1965 G. J. WILLIAMS *Econ. Geol. N.Z.* ix. 141/1 It should not pass unnoticed that the Taipo Valley thorite occurrence also contains rutile..and that the presence of riebeckite suggests an alkali environment. 1977 A. HALLAM *Planet Earth* 136/2 Riebeckite is a member of the amphibole group.

Riedel (ri·dĕl). *Med.* The name of B. M. K. L. *Riedel* (1846–1916), German physician, used in the possessive to designate (*a*) an unusually long lobe of the liver; (*b*) a rare condition of uncertain status in which the thyroid gland is largely replaced by dense fibrous tissue.

1905 KINNICUTT & POTTER tr. *Sahli's Diagnostic Methods* 311 (*caption*) Riedel's projection of the liver in cholelithiasis. 1932 *Practitioners Libr. Med. & Surg.* I. vi. 567 Riedel's lobe is an occasional linguiform process of liver substance projecting caudally from the caudal border of the right lobe. If detectable by palpation or percussion it may be mistaken for an enlarged gallbladder; of course the two may coexist. 1978 *Jrnl. R. Soc. Med.* LXXI. 200 The homogeneous echogenic mass of the lower border of the liver is either a Riedel's lobe, if of the same echogenicity as the liver; or a tumour, if of differing echogenicity. 1917 H. FRENCH *Index Differential Diagnosis Main Symptoms* (ed. 2) 158 Progressive and extreme fibrosis of the organ such as is seen in ligneous thyroiditis, or Riedel's disease. 1966 WRIGHT & SYMMERS *Systemic Path.* II. 1111/2 (*heading*) Riedel's thyroiditis... This disorder, formerly known as Riedel's struma, 'woody thyroid', or invasive fibrous thyroiditis, is not unanimously accepted as a distinct disease entity. Up to twenty years ago the term was widely used synonymously with Hashimoto's disease.

‖ **riegel** (ri·gĕl). *Physical Geogr.* Also **Riegel.** Pl. **riegels, riegeln.** [Ger., a. MHG. *rigel* crossbar for fastening, OHG *rigil* bar; see RAIL *sb.*²] A low, transverse ridge of resistant bedrock on the floor of a glacial valley; = *rock bar* s.v. *ROCK *sb.*¹ 9 a.

1910 *Geogr. Jrnl.* XXXV. 284 Where a barrier of more resistant rock has hemmed in a portion of the valley (*Riegel*), narrow picturesque gorges have been cut. 1916 G. TAYLOR *With Scott* iii. 136 Leaving the glacier and the upper lake, I proceeded east to the Riegel... This bar across a great glacial gorge was paralleled by many in the Swiss Alps... In my opinion this bar (or riegel), and the more important one we discovered..are relics of 'steps' in the original topography. 1954 W. D. THORNBURY *Princ. Geomorphol.* xv. 370 Steps..have been attributed to the effects of varying rock hardness, the idea being that risers and associated riegels mark hard rock 'bars' across a glacial trough. 1957 G. E. HUTCHINSON *Treat. Limnol.* I. i. 71 The basins are divided by transverse ridges, presumably *Riegeln* or rock bars. 1968 R. W. FAIRBRIDGE *Encycl. Geomorphol.* 745/1 Partly worn-down cross-valley bars of resistant rock ('riegels') remain, impounding lakes, though these have commonly been drained by rivers that have cut gorges through the riegels since the melting of the glaciers.

‖ **riel** (ri·əl). [Khmer.] The basic monetary unit of Kampuchea, equal to 100 sen.

1956 *Whitaker's Almanack* 1957 886/1 The official rate of exchange (1956) was 97.7 *riels* = £1. 1964 *New Statesman* 17 Jan. 79/1 The black market value of the pound sterling rose to 300 riel after the quarrel with America. 1977 'J. LE CARRÉ' *Hon. Schoolboy* xv. 341 Going rate is three million riels to the Minister... Less if you pay gold.

riem. Add: Also with pronunc. (rim). (Earlier and later examples.)

1822 W. J. BURCHELL *Trav. S. Afr.* I. 151 The *riem* (or halter), is a leathern thong about twelve feet in length, with a noose at one end, by which it is fixed round the ox's horns. 1840 [see *BRAY *v.*⁴]. 1932 C. FULLER *Louis Trigardt's Trek* 100 We loosened the *riem* on one hind wheel. 1952 *Cape Argus* 15 Nov. 7/9 She then tied a riem around my neck and hit me with a leather belt. 1966 E. PALMER *Plains of Camdeboo* x. 184 Finally he tied a riem round the ant-bear's hindquarters.

Riemann (ri·mæn). *Math.* The name of G. F. Bernhard *Riemann* (1826–66), German mathematician, used *attrib.* and in the possessive to designate various concepts of his, as **Riemann geometry,** Riemannian geometry; **Riemann('s) hypothesis,** the hypothesis (unproved by 1981) that all the zeros of the Riemann zeta function, except those on the real line, have a real part equal to ½; **Riemann integral,** a definite integral obtained by subdividing the interval of integration, multiplying the width of each subdivision by the greatest or the least value of the integrand within it, summing the products so obtained, and taking the limit of the sum as the width of the subdivisions tends to zero; so **Riemann integrable** *adj. phr.*, **Riemann integration; Riemann('s) surface,** a surface which covers a plane more than once, and so could be used to plot a function that is not single-valued; **Riemann tensor,** the Riemann–Christoffel tensor; **Riemann zeta** (or ζ) **function,** an analytic function ζ of the complex variable *s*, equal almost everywhere to $\{1^{-s}+2^{-s}+3^{-s}+\ldots\}$.

1922 *Proc. Nat. Acad. Sci.* VIII. 23 The functions $\lambda_{g_{ij}}$ satisfy (2.4) and give a Riemann geometry. 1974 R. M. PIRSIG *Zen & Art of Motorcycle Maintenance* (1976) III. xxii. 257 He turned to the question, Is Euclidian geometry true or is Riemann geometry true? He answered, The question has no meaning. 1924 *Proc. Cambr. Philos. Soc.* XXII. 296 We assume Riemann's hypothesis. *Ibid.*, We assume the truth of the Riemann hypothesis. 1959 *Listener* 23 Apr. 715/2 This conjecture, now known as the Riemann hypothesis, has never been either proved or disproved. 1957 K. S. MILLER *Advanced Real Calculus* iv. 49 If $I' = I''$ we shall say that $f(x)$ is Riemann integrable or simply integrable on $[a,b]$ and write $I = \int_a^b f(x)dx$. 1972 A. G. HOWSON *Handbk. Terms Algebra & Anal.* xxvii. 136 If f is bounded on I then it is Riemann integrable..on I if and only if it is continuous almost everywhere..on I. 1914 *Proc. London Math. Soc.* XIII. 133 Corresponding to Riemann's extension of the notion of an integrable function, we now have a certain class of functions which may be said to possess a 'Riemann' integral with respect to the monotone increasing function $g(x)$, that is to say a function such that the summation..has a unique and finite limit, however the points x are chosen in their corresponding intervals, and however those intervals are constructed, provided only the length of the greatest of them approaches zero as $n \to \infty$. 1970 S. KOTZ tr. *Pesin's Classical & Mod. Integration*

Theories vii. 112 The Riemann integral can be defined in two ways: by the Riemann process as the limit of Riemann's sums and by the Darboux process as the common value of the lower and upper integrals. 1939 I. S. SOKOLNIKOFF *Advanced Calculus* iv. 99 It seems desirable to begin the study of Riemann integration by presenting a reasonably careful definition of the definite integral based on the intuitive concept of the area under the curve. 1893 A. R. FORSYTH *Theory Functions Complex Variable* xv. 336 The region, in which the variable z exists, no longer consists of a single plane but of a number of planes.., often called sheets... The aggregate of all the sheets is a surface, often called a Riemann's Surface. 1893 HARKNESS & MORLEY *Treat. Theory Functions* vi. 205 We shall show how to form a Riemann surface in some simple special cases. 1932 A. HUXLEY *Brave New World* iv. 73 Two thousand Beta-Minus mixed doubles were playing Riemann-surface tennis. 1974 *Encycl. Brit. Macropædia* I. 726/2 A compact Riemann surface is homeomorphic to the (topological) surface obtained from a sphere by cutting g pairs of holes in it and attaching to each pair of holes a handle. 1922 *Proc. Nat. Acad. Sci.* VIII. 24 Where $R_{pq,rs}$ is the Riemann tensor of the first kind. 1957 *Physical Rev.* CV. 1089/1 It is the Riemann tensor which characterizes the presence of radiation. Physically, this is because the Riemann tensor describes the variations in the gravitational field from event to event in space-time. 1973 HAWKING & ELLIS *Large Scale Structure of Space-Time* ii. 42 Having split the Riemann tensor into a part represented by the Ricci tensor and a part represented by the Weyl tensor, one can use the Bianchi identities..to obtain differential relations between the Ricci tensor and the Weyl tensor. 1899 *Messenger of Math.* XXIX. 114 If $\zeta(s, a, \omega)$ be the extended Riemann ζ function. 1931 *Q. Jrnl. Math.* II. 161 The theory of the Riemann zeta-function. 1966 OGILVY & ANDERSON *Excursions in Number Theory* iii. 35 We needed a value of the Riemann Zeta-function, the technical name for the series that converged to $\pi^2/6$.

Riemann–Christoffel (ri:mæn,kri·stǫ̈fĕl). *Math.* [f. prec. + the name of E. B. *Christoffel* (1829–1900), German mathematician.] Used to designate a tensor of the fourth order whose components are functions of a covariant co-ordinate system and the corresponding contravariant system, and which occurs in the mathematical description of curved space-time.

1918 A. S. EDDINGTON *Rep. Relativity Theory of Gravitation* iii. 40 The required equations of the law of gravitation must, therefore, include the vanishing of the Riemann–Christoffel tensor as a special case. 1956 G. C. MCVITTIE *Gen. Relativity & Cosmol.* ii. 30 In relativity theory applications of the tensor calculus a very important part is played by a symmetrical tensor of rank two, called the Ricci tensor, which is obtained by contraction from the Riemann–Christoffel tensor. 1974 *Encycl. Brit. Micropædia* VIII. 580/2 The Riemannian curvature is obtained by first contracting the Riemann–Christoffel curvature tensor, applying it to two vectors spanning the subspace in question, and then dividing by the area of the parallelogram formed from the vectors.

Riemannian (rimæ·niăn), *a.* *Math.* [f. as prec. + -IAN.] Used to designate a non-Euclidean geometry which is everywhere positively curved, and various associated concepts.

1920 A. S. EDDINGTON *Space, Time & Gravitation* xii. 183 The world became non-Euclidean; a new geometry called Riemannian geometry was adopted. 1923 *Ann. Math.* XXIV. 367 A generalization of Levi-Civita's concept of infinitesimal parallelism in a Riemannian manifold. 1926 L. P. EISENHART *Riemannian Geom.* ii. 35 The metric defined..is called the Riemannian metric and a geometry based upon such a metric is called a Riemannian geometry. Also we say that the space whose geometry is based upon such a metric is called a Riemannian space. 1965 H. EVES *Survey of Geom.* II. xiv. 341 One can describe Riemannian geometry as the mathematical study, couched in geometrical terminology, of an arbitrary quadratic differential form. 1974 *Encycl. Brit. Micropædia* VIII. 580/2 In Riemannian geometry, a straight line of finite length can be extended continuously without bounds, but all straight lines are of the same length.

‖ **riempie** (ri·mpi). *S. Afr.* Also **riempje.** Var *rheimpy* s.v. RHEIM. Freq. used for the seats of chairs or stools. Also *attrib.* and in *Comb.* Cf. RIMPI in Dict. and Suppl.

1913 D. FAIRBRIDGE *That which hath Been* xx. 229 Heavy teak chairs with *riempje* seats were ranged round the wainscot with mathematical precision. 1920 R. Y. STORMBERG *Mrs. Pieter de Bruyn* xxiii. 76 Ouma du Preez on the riempie settee devoutly clasped her cotton-mittened hands. 1937 S. CLOETE *Turning Wheels* vi. 100 While dozing in her wide riempie-bottomed chair, she had seen one girl and two men leave the camp. 1938 E. A. WALKER *Great Trek* 41 A couple of deal tables and *riempie* chairs and stools..as seats along the stoep. 1939 'D. RAME' *Wine of Good Hope* I. iv. 44 Even Lowell, perched on a little riempie stool beside Grim..smiled at the torrent of high-pitched abuse. 1943 'B. KNIGHT' *Covenant* (1944) i. 34 Jannion saw her mother hook her foot under the top of the *riempie*-strung stool. 1949 M. MASSON *Narrowing Lust* xix. 180 Calmly she seated herself on a stool of riempje. 1953 M. MURRAY *Fire-Raisers* x. 101 He saw that Sarel was lying on the riempje sofa, his feet on one arm. 1971 *Cape Times* 13 Feb. 21/3 (Advt.), 2 riempie seated chairs.

‖ **riempiestoel** (ri·mpistu:l). *S. Afr.* Also **riempies-stoel.** [Afrikaans, f. prec. + *stoel* stool.] A *riempie*-seated chair or stool.

1933 W. MACDONALD *Romance of Golden Rand* 72 Louw Geldenhuys, sitting on a *riempiestoel* in the little office. 1953 U. KRIGE *Dream & Desert* i. 75 Marta, sitting on her *riempiestoel* in front of the door leading to the kitchen. 1963 H. C. BOSMAN *Unto Dust* 122 A newly-appointed veldkornet, looking important, seated on a riempies-stoel.

riemschoen, var. *REMSKOEN.

Riesling (ri·sliŋ, -zliŋ; *also* rəi·zliŋ). [Ger.] The name of a variety of vine and grape widely grown in Germany, Austria, Alsace, and elsewhere; the dry white wine produced from this grape. Also *attrib.*

1833 C. REDDING *Mod. Wines* V. 168 The white wines are ranked in quality as follows: Riesling, muscadine, Kléber, or Klebner, and common. Riesling wine is distinguished by a particular bouquet, by strength, and durability. 1888 *Encycl. Brit.* XXIV. 609/1 The grape from which it [*sc.* Sercial] is produced is of the Riesling variety. *Ibid.* 611/1 Amongst the leading descriptions of vine plants in German vineyards the Riesling stands out pre-eminent. 1902 E. R. EMERSON *Story of Vine* vii. 127 The Steinberg is a hill about three miles from the Rhine... Only Riesling vines are grown, but several grades of wine are made. 1954 P. HIGHSMITH *Blunderer* (1956) ii. 18 Walter ordered a broiled fish and a bottle of Riesling. 1959 *Times* 21 Sept. 13/3 *Johannisberg*.., as it is made from the *riesling* grape, has some of the fuller fruitiness of German wines. 1961 *New Statesman* 15 Dec. 921 (Advt.), *Riesling*. Per crate of 12 at 7/9 each. 1977 *Time* 21 Nov. 59/1 Riesling, the small yellow grape from which come the classic wines of Moselle, Alsace and the Rhine. 1980 D. BLOODWORTH *Trapdoor* x. 56 A chilled bottle of Jugoslav Riesling.

Rifa'ee, var. *RUFAI.

rifampicin (rifæ·mpisin). *Pharm.* [f. *RIFAM(Y)CIN with inserted *pi-* f. *PIPERAZINE.] A reddish-brown crystalline powder, a member of the rifamycin group of antibiotics, which is used to treat a wide range of diseases, esp. pulmonary tuberculosis.

1966 N. MAGGI et al. in *Chemotherapia* XI. 285 The hydrazone..3-(4-methyl-piperazinyl-iminomethyl)-rifamycin SV, named by us 'rifampicin'. 1969 *Times* 12 May 16/3 The antibiotic called rifampicin could interfere with the production of R.N.A. molecules. 1970 *Nature* 25 July 382/1 Rifampicin has also been found to be active against some viruses. 1975 *Sci. Amer.* Mar. 126/2 The fast-acting antibiotic rifampicin, found in a soil organism..in 1957, was proved against leprosy in the 1960's. 1980 *Brit. Med. Jrnl.* 29 Mar. 900/2 In three [families] rifampicin had been given to contacts and family members because the initial presentation of the child suggested meningococcal disease.

rifampin (rifæ·mpin). *Pharm.* [f. as *RIFAMPICIN.] The equivalent in the U.S. Pharmacopeia of *RIFAMPICIN.

1968 *Antimicrobial Agents & Chemotherapy* 519/1 The in vitro and in vivo activity of rifampin against gram-positive bacteria. 1972 *Evening Telegram* (St. John's, Newfoundland) 27 June 4/8 The great danger in rifampin is that tuberculosis bacteria, which cause the disease, can readily become resistant to the drug. 1978 *Pediatr. Res.* XII. 491 Haemophilus influenzae type b infection in a day care center: eradication of carrier state by rifampin.

rifamycin (rifămai·sin). *Pharm.* orig. **rifomycin.** [Prob. f. It. *rifo-rmare* to reform + *-MYCIN.] Any of a class of natural and semi-synthetic antibiotics of which the first examples were isolated from the fungus *Streptomyces mediterranei*.

1959 P. SENSI et al. in *Farmaco* (Ed. Sci.) XIV. 146 The antibiotic, named by us rifomycin, shows a high activity against gram-positive bacteria and mycobacteria. 1960 *Antibiotics Ann.* 1959–1960 262 Rifomycin has been isolated in our laboratories from the fermentation broths of a strain of *Streptomyces mediterranei*. 1963 *Antibiotics & Chemotherapy* XI. 61 The Rifamycins constitute a group of antibiotics, produced by a new Streptomyces. 1970 *New Scientist* 24 Dec. 546/1 Rifamycin B *per se* is not active against bacteria but slowly degrades in aqueous solution to yield an active product, rifamycin SV. 1976 *Lancet* 11 Dec. 1304/2 The rifamycin antibiotics have been used in the treatment of leprosy since 1963.

riff, *sb.*³ (Later example.)

1952 M. TRIPP *Faith is Windsock* iv. 67, I believe I am right in diagnosing his dog's trouble as riff. This is invariably picked up in damp places and frequently appears in the clefts between the toes.

Riff (rif), *sb.*⁶ and *a.* Also **Rif.** [f. *Rîf*, the name of a district in Morocco.] **A.** *sb.* A Berber of the Rîf district of Morocco. **B.** *adj.* Of or belonging to the Riffs.

1923 *Contemp. Rev.* Nov. 616 Representatives of the Riffs now demand the formal recognition of the Riff republic. 1925 *Glasgow Herald* 6 July 9/2 Several Riff attacks in the neighbourhood of the River Liben have been repulsed. 1926 *Chambers's Jrnl.* Mar. 158/1 Hussein was a good Riff, and the killing lust was upon him. 1964 C. MACKENZIE *Life & Times* III. 76 There might be other Riffs about with rifles who would shoot us in the hope of plunder.

riff, *sb.*⁷ [Abbrev. of RIFFLE *sb.*] **1.** A simple musical phrase repeated over and over, often with a strong or syncopated rhythm, and freq. used as background to a solo improvisation. Also *attrib.* and in *Comb.*

1935 [see *DIG *v.* 6 c]. 1940 [see *BACKING *vbl. sb.* 7 b]. 1946 R. BLESH *Shining Trumpets* xii. 277 Henderson..established the current trend of riff-swing. 1951 *Landfall* V. 286 She never could manage riff rhythms or any jazz rhythms. 1955 *Jazzbook 1955* 8 When played with the passion and abandon of a great orchestra like Basie's the riff is an exciting revelation; as a trick phrase in the hands of a mediocre band it becomes the most banal of musical devices. 1962 J. BALDWIN *Another Country* I. i. 16 They might swap stories of..gigs they'd played, riffs they remembered. *Ibid.* 20 He..beat a riff with his fingers. 1972 *Blues & Jazz* Sept. 11/3 The saxes come in with an unoriginal but beautifully swinging riff when Fats makes his vocal entry. 1976 A. MURRAY *Stomping Blues* v. 60 The actual mood of the performance..is determined by the up-tempo beat plus the riff-style call-and-response choruses. 1976 *Zigzag* Apr. 28/1 The band shine best when they are tackling unison riffs and counter riffs with customary ease. 1978 *Oxford Times* 31 Mar. 19/1 The harmonica introduction is unforgettable and the tune has a very catchy backing riff. 1980 M. BOOTH *Bad Track* iii. 68 The sax cut in with Mel's violin scrawking against the drums and the low riff of the electronic piano.

2. In extended use: a repeated phrase, idea, or situation.

1970 *New Yorker* 14 Nov. 166/2 He has an opportunity for some lovely comic riffs. 1972 *Times Lit. Suppl.* 29 Dec. 1570/4 These schizoid riffs are not the half-hearted impersonations of madness..but characterizations which exhibit the mania and pain of true suffering. 1974 *Black World* Dec. 52/2 There are some new riffs played out in this novel, riffs which are significant when measured against Baldwin's earlier novels. 1976 *Times Lit. Suppl.* 19 Mar. 318/4 Mr Morphet has not sufficiently discussed Hodgkin's community of ideas with painters of his generation, for he uses riffs, as it were, from many of his contemporaries.

riff, *sb.*⁸ = *RIFFLE *sb.* 3 b.

1960 *20th Cent.* Mar. 256 The most cursory riff through the..daily and weekly press.

riff, *v.*¹ Restrict † *Obs.* to sense in Dict. and add: **2.** With *through* = *RIFFLE *v.* 3 c.

1956 W. H. WHYTE *Organization Man* (1957) xxv. 347 Riff through these maps quickly, and in the few seconds..you can see in crude animation the fissures begin to widen.

riff, *v.*² [f. *RIFF *sb.*⁷] *intr.* To play riffs in popular music. Also *fig.* (in context *trans.*). Hence **ri·ffing** *ppl. a.*

1955 *Jazzbook 1955* 8 Few things were more satisfying than Jimmy Rushing shouting the blues with the band riffing away behind him. 1959 C. MACINNES *Absolute Beginners* 158 He said life was a junction: the junction, he said, of composite opposites (he liked that group, and riffed it several times). 1966 *Melody Maker* 7 May 13/1 The Faces storm into this new one with powerful guitar, and riffing organ. 1968 *Blues Unlimited* Sept. 23 Horns coming in to riff nicely after the build-up. 1977 *Gramophone* Sept. 525/2 What comes across most forcefully is the band's ability to riff fearlessly and hypnotically.

Riffian (ri·fiăn), *sb.* and *a.* Also **Rifian.** [f. *RIFF *sb.*⁶ and *a.* + -IAN.] = *RIFF *sb.*⁶ and *a.*

1867 'MARK TWAIN' *Innoc. Abroad* (1869) viii. 77 There are stalwart Bedouins of the desert here, and stately Moors ..and Jews, whose fathers fled hither centuries..ago; and swarthy Riffians from the mountains. 1899 A. E. W. MASON *Miranda of Balcony* ii. 23 A Riffian sauntered by..a great coarse tail of hair swinging between his shoulders. 1909 *Westm. Gaz.* 13 Aug. 5/4 The guns are now trained against the town, and are evidently being handled by Riffian merchants familiar with the main buildings. 1924 *Glasgow Herald* 30 July 13 The Riffians planned to surprise the convoy sent to Xauen. 1931 *Times Lit. Suppl.* 10 Dec. 1008/4 A plot is being hatched for the organization of a fresh Riffian rebellion against the French in Morocco. 1965 C. D. EBY *Siege of Alcázar* (1966) 21 The Riffian tribes of Spanish Morocco. 1977 P. RAYMOND *Matter of Assassination* viii. 88 The Hamitic accent of Rifian Berber.

ri·ffing, *vbl. sb.* [f. *RIFF *v.*² + -ING¹.] The action or fact of playing riffs.

1936 *Harper's Mag.* Apr. 570/2 'Jamming', 'cat-time', 'swing', 'riffing', 'getting off', 'going to town', 'ragging', 'gut-bucketing', and all the rest are names for the *hot* performance, which is the heart and soul of jazz. 1949 L. FEATHER *Inside Be-bop* vi. 42 Jo Stafford's arrangement of *The Gentleman is a Dope* began with four bars of unmistakably bop riffing. 1958 G. BOATFIELD in P. Gammond *Decca Bk. of Jazz* xxiv. 310 These are extraordinary and unique tracks, with Dodds making soaring and at times agonizing music against the sombre riffing of the brassmen. 1976 A. MURRAY *Stomping Blues* vii. 104 The process of riffing (from the verb, *to riff*) refers not only to making riff phrases..but also to improvisation in general.

riffle, *sb.* Add: **I. 3. a.** Also *riffle-shuffle*.

1970 K. ROOS *What did Hattie See?* xi. 105 He gave the deck a riffle shuffle, carefully not disturbing the four bottom cards. 1972 *Sci. Amer.* July 104/1 With your back still turned, ask him to cut the deck several times, then give it one thorough riffle-shuffle.

b. A quick skim or leafing through (of pages of a book, papers, etc.).

1960 *Guardian* 28 Oct. 6/4, I skimmed the book in a first riffle. 1967 *Adv. Inorg. Chem. & Radiochem.* X. 373 A fast riffle through the thallium volume of Gmelin's *Handbuch*.. uncovers the following mixed valence compounds. 1979 J. GARDNER *Nostradamus Traitor* vii. 22 Just a riffle through the books.

II. For *U.S.* read 'orig. and chiefly *U.S.*'

4. a. Also *Canad.* (Earlier and later examples.)

1785 R. BUTLER *Jrnl.* in *Olden Time* (1847) Oct. 440 Met riffles, all very shallow, struck with the barge several times. 1948 *Sat. Even. Post* 23 Oct. 36/2 In the lingo of our Rogue River guides, a riffle is anything between a foaming cataract and white-water rapids heavily sprinkled with boulders.

1962 M. E. MURIE *Two in Far North* II. vi. 153 In the course of the morning we fought our way over six riffles. 1976 *Nature* 5 Aug. 483/1 Other research workers have noted that the spacing of riffles in straight and meandering channels is between five and seven channel widths and that skew shoals in flumes have been found to be 6.5 channel widths apart. 1980 *Birds* Summer 22/2 Dipper and common sandpiper are associated with rapid flowing, relatively well aerated water with riffles (shallow sections)—features generally only found on upland rivers.

b. *to make the riffle*: to succeed in crossing a rapid; hence *fig.*, to be successful in an attempt or undertaking.

1853 F. A. BUCK *Let.* 31 Dec. in *Yankee Trader in Gold Rush* (1930) 130 'Madam La Marquise'..built a splendid saloon, opened and flourished for about two months but couldn't make the riffle. 1873 'MARK TWAIN' & WARNER *Gilded Age* xxxi. 279 There's old Balaam, was in the Interior—..he's made the riffle on the Injun; great Injun pacificator and land-dealer. 1887 [see sense 4 a in Dict.]. 1911 R. D. SAUNDERS *Col. Todhunter* i. 19, I aint got no business doin' that, but I'll try if I can make the riffle. 1950 *Daily Ardmoreite* (Ardmore, Okla.) 14 Feb. 8/1 [The] Rexroat girls [were] doing their best trying to win first in the basketball tournament but they couldn't quite make the riffle.

c. A ripple or ruffle.

1938 M. K. RAWLINGS *Yearling* x. 92 A slight breath of air rippled across the marsh and the water rippled under it... 'Jest enough of a riffle,' Penny said, 'and the moon jest right.' 1960 J. STROUD *Shorn Lamb* ii. 17, I had one sensational glimpse of a long white leg and a riffle of white underwear. 1975 *New Yorker* 27 Oct. 111/3 We stood and listened to the silence—to the voices of birds, to the stir of a breeze in the trees, to the riffle of moving water.

5. a, b. More widely, a transverse bar or channel, usu. one of a series, for breaking a flow of fluid or sorting or separating a stream of particles. (Earlier and later examples.)

1850 N. KINGSLEY *Diary* 7 May (1914) 120 Finished the riffles to the machine to day. 1915 *Trans. Amer. Inst. Mining Engineers* LI. 405 During the past three years there has been developed at Morenci a new type or arrangement of riffles. 1939 A. M. GAUDIN *Princ. Mineral Dressing* xiii. 298 Riffles are of great importance to tabling. They are responsible for the increased capacity of riffled decks over smooth ones. 1951 KIRK & OTHMER *Encycl. Chem. Technol.* VII. 307 A sluice is essentially an inclined trough or launder, the bottom of which is provided with transverse strips or riffles, with variable spacing. 1955 H. R. COX *Gas Turbine Princ. & Pract.* xviii. 9 The primary air and fuel stream is divided on leaving the pump by a simple riffle arrangement. 1964 F. CHICHESTER *Lonely Sea & Sky* v. 48 We shovelled the gravel into long boxes with 'riffles', small pieces of wood across them, which stopped the heavy gold when the pay dirt was washed through the box by the river. 1968 COULSON & RICHARDSON *Chem. Engin.* (ed. 2) II. xvii. 691 A series of slats, or riffles as they are termed, about ¼ in. in height... The large particles and the less dense material are carried downwards, and the remainder is carried parallel to the riffles.

riffle, *v.* Restrict *rare* to senses 1–2 c and add: **2. d.** *U.S.* To ruffle in a slight or rippling manner. Also *fig.*

1901 S. E. WHITE in *Century Mag.* LXII. 466/1 The breeze and the sun played with the prairie grasses, the breeze riffling them over. 1926 *Ladies' Home Jrnl.* Nov. 228 Even the wail of music from the Palace of Dance barely riffled his preoccupation. 1977 D. HARSENT *Dreams of Dead* 24 A wind riffling the orchard.

3. a. (Later examples.)

1939 *Reader's Digest* May 29/1 Pick up and riffle a deck of cards. All you see is a blur of card ends. 1973 'J. ASHFORD' *Double Run* iv. 30 Comyns collected up the cards and riffled them into a pack with slow and deliberate movements.

b. To flick through (papers, books, etc.); to thumb (a block of paper, a book, etc.), releasing the leaves in (usu. rapid) succession.

1922 H. TITUS *Timber* viii. 77 He riffled the pages slowly. 1938 R. FRANKEN *Gold Pennies* xv. 159 Mrs. Miller glanced at the opening paragraph, and then riffled the pages tentatively. 1948 *Electronic Engin.* XX. 367/3 The material to be counted must be riffled so that a step-like edge is presented, though the riffling need not be perfect. 1957 T. STURGEON in D. Knight *100 Yrs. Sci. Fiction* (1969) 125 Slim lifted the paper out of the box,..riffled a thumbful of the sheets at the top. 1958 J. BALDWIN in *Mademoiselle* Mar. 149/1 He quickly riffled some papers on his desk, putting on a business air as rakishly as she had seen him put on his hat. 1977 G. DURRELL *Golden Bats & Pink Pigeons* iv. 87 He took my passport and riffled it..like an expert card sharper.

c. *intr.* To thumb or leaf *through*. Also *transf.* and *fig.*

1931 *Sun* (Baltimore) 19 Feb. 10/7 Yesterday, while riffling through a new book of memories and character sketches,..I saw it said that Mr. Bertrand Russell..reads detective fiction. 1959 *Times* 5 Nov. 14/6 A blonde girl.. propped a music-stand in front of her, and riffled through a script. 1960 V. NABOKOV *Invitation to Beheading* vi. 64 The draught riffled through the papers on the table. *Ibid.* vii. 74 He produced..a thick batch of home snapshots of the smallest size..Riffling through them as through a deck of cards, he began placing them one by one on the table. 1962 *Listener* 22 Nov. 845/2, I was riffling through these morbid thoughts. 1967 *Boston Sunday Globe* 23 Apr. B41/5 Riffling through a few recent editions of morning and evening Globes and New York, Washington, and Philadelphia newspapers we come across lines that read: Mortgage cut near [etc.]. 1979 A. HAILEY *Overload* I. x. 60 At his desk he riffled through the messages.

riffled (ri·f'ld), *a.* [f. RIFFLE *sb.* + -ED².] Provided with riffles.

1915 *Trans. Amer. Inst. Mining Engineers* LI. 416 The effect of the uniform feed and water distribution over this completely riffled table is to produce uniform velocities in the water. 1927 *Chambers's Jrnl.* Feb. 109/2 In action the human dry-blower shovels some sand on to the top perforated riffled sheet: and then he shakes the entire machine. 1951 KIRK & OTHMER *Encycl. Chem. Technol.* VII. 310 The modern bumping table is a deck concentrator of the riffled or strake type. 1968 COULSON & RICHARDSON *Chem. Engin.* (ed. 2) II. xvii. 691 Riffled tables can be used for separating materials down to about 300-mesh in size, provided that the difference in the densities is large.

riffler². Restrict *dial.* to sense in Dict. and add: **2.** A trough or sluice containing one or more upright boards to divide or retard the flow through it.

1850 N. KINGSLEY *Diary* 17 May (1914) 122 Stoped down to day and made a panning trough to pour quicksilver from the riffler into. 1924 E. C. TUCKER et al. *Manuf. Pulp & Paper* IV. vi. 24 Rifflers, or sand traps, are wood troughs through which the stock flows from the regulating box to the screens. *Ibid.* 25 The riffler is divided into two runs by the central dividing board. 1963 R. R. A. HIGHAM *Handbk. Papermaking* ii. 67 With rifflers and sand tables the stock is passed at approximately 0·5% consistency along narrow channels in the base of which are felt-covered boards which are set at right angles to the flow of stock.

riffling (ri·fliŋ), *vbl. sb.* [f. RIFFLE *sb.* and *v.* + -ING¹.] **1.** An arrangement or system of riffles (sense 5).

1915 *Trans. Amer. Inst. Mining Engineers* LI. 408 It was found impossible with this system of riffling to handle successfully the range of sizes..coming from the jigs. 1951 KIRK & OTHMER *Encycl. Chem. Technol.* VII. 307 There are endless variations of riffling.

2. The action of RIFFLE *v.* 3.

1948 [see *RIFFLE *v.* 3 b]. 1964 A. WYKES *Gambling* i. 15 A quick riffling through the replies..confirmed my suspicion.

riffling (ri·fliŋ), *ppl. a.* [f. RIFFLE *v.* + -ING².] Of water: moving in riffles; agitated.

1754 J. PREBLE in *New-England Hist. & Geneal. Reg.* (1868) XXII. 408 The navigation to Norridgewalk is considerably difficult by the reason of the rapidity of the stream, and riffling falls. 1911 J. F. WILSON *Land Claimers* ix. 123 She..heaved her catch up out of the grip of the riffling water.

riff-raffy (ri·f,ræfi), *a.* [f. RIFF-RAFF + -Y¹.] Having the character of riff-raff; disreputable.

1928 D. H. LAWRENCE *Lady Chatterley* 309 She was lost, lost utterly in this world of riff-raffy expensive people. 1962 *New Yorker* 10 Mar. 29/2 Brunswick..kept improving the tone of bowling alleys..hiring architects and decorators to make sure the places wouldn't look riff-raffy.

riffy (ri·fi), *a.* *Jazz.* [f. *RIFF *sb.*⁷ + -Y¹.] Full of riffs; repetitive.

1952 B. ULANOV *Hist. Jazz in Amer.* xvi. 194 Artie recorded some small band jazz, riffy but fresh. 1964 *Listener* 27 Aug. 319/3 Lyttleton played the sort of music (riffy swing) he first achieved fame eschewing. 1976 *Gramophone* June 101/3, I didn't go too much on *Mop mop*, which got a bit 'riffy'.

Rifian, var. *RIFFIAN *sb.* and *a.*

rifle, *sb.*³ Add: **1. b.** *rifle-calibre, cannon, -grooved, -pistol*.

1940 *War Illustr.* 5 Jan. 554/2 The armament is believed to consist of two shell-firing guns as well as four rifle-calibre machine-guns. 1861 E. COWELL *Jrnl.* 14 May in M. W. Disher *Cowells in Amer.* (1934) 347 The steamer, Pawnee was moored in front of the city, yesterday, with her guns (rifle cannon) and mortars so commanding it. 1977 'J. GASH' *Judas Pair* ii. 27 [Duelling pistols] should have walnut stocks, and usually be rifle-grooved. 1883 'MARK TWAIN' *Life on Mississippi* 317, I arose and drew an elegant rifle pistol on him.

2. a. (Earlier examples.)

1770 G. WASHINGTON *Diary* 5 Mar. (1925) I. 368 By John Jost for my Rifle in full £6.10. 1772 D. TAITT *Jrnl.* 5 Apr. in N. D. Mereness *Trav. Amer. Colonies* (1916) 537 Others took the Cock off his riffle and Sixteen Carrots of Tobacco.

b. (Earlier and later examples.)

1843 W. C. MACREADY *Diary* 18 Sept. (1912) II. 222 He came back to introduce Mr Webster of the Rifles to me. 1952 E. WAUGH *Men at Arms* I. ii. 77, I got seconded to the African Rifles. 1971 S. HILL *Strange Meeting* iii. 200 The Rifles were still advancing.

3. a. *rifle-butt* (hence occas. as *v. trans.*), *-oil, -rag, -sight, -thong*.

1968 *Listener* 27 June 826/3 But strip away the abstract terminology and watch tough heavily-armoured cops beating up boys and girls with rifle-butts, and the spectacle becomes embarrassing. 1977 *Rolling Stone* 16 June 36/1 There's just this part of him that's seven feet tall, big as a house, and wants to rifle-butt everybody around. 1923 KIPLING *Irish Guards in Gt. War* I. 46 It is very hard to get any sort of rifle-oil. 1933 F. RICHARDS *Old Soldiers never Die* 43 We had..no rifle-oil or rifle-rag to clean our rifles with. *a* 1914 in *Penguin Bk. Austral. Ballads* (1964) 221 They had sought him over their rifle-sights. 1932 AUDEN *Orators* III. 112 That rifle-sight you're designing: is it ready yet? 1917 W. OWEN *Coll. Poems* (1963) 39 Knit in the webbing of the rifle-thong.

b. *rifle-gallery, -match*.

1880 E. LEATHES *Actor Abroad* xiv. 165 Beneath the building, scooped out of the cliff, are a bowling-alley and a rifle-gallery. 1880 'MARK TWAIN' *Tramp Abroad* 626 No information about prize fights.., rifle-matches, or other sporting matters.

c. *rifle artillery, battalion, company* (earlier example), *corps* (earlier example).

1861 LD. PALMERSTON *Let.* 30 Dec. in Ld. Cowley *Paris Embassy during Second Empire* (1928) xi. 234 They still attack in columns which would be blown to pieces by the French rifle artillery long before they could come into contact with their opponents. 1775 *Jrnls. Continental Congress U.S.* (1905) III. 305 To John Biddle,..Commissary of the riffle Battalion. *Ibid.* (1905) II. 250 The expences incurred for raising and arming the riffle companies. 1777 J. THACHER *Military Jrnl.* (1823) 121 The gallant Colonel Morgan, at the head of his famous rifle corps..commenced the action.

d. *rifle-shooting* (later example), *twirling*.

1974 *Ridge Citizen* (Johnston, S. Carolina) 18 Apr. 1/6 Her special training while in school consists of several years of rifle twirling while a member of the Thurmond Color Guard. 1981 E. NORTH *Dames* xiv. 268 There was a rifle range nearby and Major Frimley taught rifle shooting to the girls.

e. rifle-coat = *rifle-frock*; **rifle-frock** (earlier examples); **rifle-grenade,** a grenade discharged from a rifle; **rifle microphone** (or *colloq.* **mike**) *U.S.* = *gun microphone* s.v. *GUN *sb.* 15; **riflescope** *U.S.*, a telescopic rifle sight.

1877 *Rep. Indian Affairs* (U.S.) 5 The coats to be in shape like the old fringed rifle-coat or blouse. 1782 J. TRUMBULL *M'Fingal* IV. 78 While rifle-frocks sent Gen'rals cap'ring. 1811 *Niles' Weekly Reg.* I. 45/2 In this valuable class of cotton goods are included rifle-frocks. 1915 Rifle grenade [see *drain-pipe* s.v. *DRAIN *sb.* 5]. 1928 C. F. S. GAMBLE *Story of N. Sea Air Station* v. 87 It was also intended that the Marten Hale rifle-grenade should be issued for use from aircraft. 1962 A. NISBETT *Technique Sound Studio* 255 *Gun microphone* (Am.: *Rifle Microphone*), moving coil microphone with a battery of tubes of different lengths leading out along the axis. 1974 *Some Technical Terms & Slang* (Granada Television), *Rifle mike*, a specialised form of microphone used in filming. 1978 *Broadcast* 17 July 9/3 He..feels better off handing in his rifle mike. 1961 WEBSTER, Riflescope. 1978 *Detroit Free Press* 5 Mar. c 9/4 (Advt.), A spring sale... All riflescopes..½ off.

rifle, *v.*¹ Add: **1. e.** *intr.* To make a vigorous search *through*.

1966 D. F. GALOUYE *Lost Perception* xiv. 147 He turned to see Weldon Radcliff sitting at a polished desk and rifling through a file holder. 1977 *Woman & Home* Nov. 154/2 Grace started to rifle through the contents of her bag. 1978 *Vogue* Feb. 88/2 Visitors from all over the world rifle through the tweeds and tartans.

rifle, *v.*³ Add: **2. b.** *transf.* To hit or kick (a ball) hard and straight. Hence **ri·fling** *ppl. a.*²

1948 [see *OVERRUN *v.* 10 a]. 1973 *Times* 13 Apr. 13/3 In the end it was the low rifling backhand drive from Gomozkov that defeated him. 1975 *New Yorker* 7 Apr. 91/1 The pivot play, in which the ball was rifled in to the center positioned at the foul line with his back to the basket. 1977 *Daily Mirror* 16 Mar. 30/5 Referee James Brimmell stepped in to save the 22-year-old Scot after the challenger had pinned him on the ropes for fully a minute, rifling in a furious stream of punches. 1978 *Guardian* 27 Feb. 16/4 Robson.. seized on a rebound..and joyfully rifled the equaliser past Shilton.

rifle-gun. For Now *arch.* read Now *arch.* exc. *U.S. dial.* and add *U.S. dial.* examples. Also *attrib.*

1913 H. KEPHART *Our Southern Highlanders* xiii. 286 Everywhere in the mountains we hear of biscuit-bread,.. rifle-gun, rock-clift, [etc.]. 1949 'J. NELSON' *Backwoods Teacher* x. 105 A generation was growing up which..knew only by hearsay..of 'rifle-gun' (a proper distinction from shotgun) balls run in molds on the 'chimley hurth'. 1951 L. CRAIG *Singing Hills* 13 Mr. Givens and his son had taken their rifle-guns and their dog and had gone off hunting.

rifle range. 2. (Earlier and later examples.)

1862 *St. Andrews Gaz.* 3 Oct. 3/4 At the rifle range, the corps was divided into two squads, the one party firing against the other. 1978 J. IRVING *World according to Garp* xii. 236 We would not allow an outdoor rifle range in the suburbs!

riflery (rəi·f'l,ri). Chiefly *U.S.* [f. RIFLE *sb.*³ + -RY.] Rifles collectively; firing from rifles; shooting with rifles. Also *attrib.*

1846 R. B. SAGE *Scenes Rocky Mts.* xxiv. 198 A volley of riflery was discharged among the promiscuous throng, with fatal effect. 1869 *Punch* 27 Mar. 119/1 Once before Shakespeare's Cliff reverberated with the roar of riflery. 1962 *Transylvania Times* (Brevard, N. Carolina) 9 Aug. 5/4 The 1962 Riflery season at Camp Illahee went off with a 'bang' as 196 girls started their National Rifle Association qualifications. 1976 *National Observer* (U.S.) 17 Apr. 13/5 (Advt.), Counselor instruction in..riflery, hiking, arts and crafts.

rifle-shot. Add: **1.** (Earlier examples.)

1803 A. ELLICOTT *Jrnl.* 249 We anchored about rifle shot from the sloop. 1816 *Niles' Reg.* IX. Suppl. 190/2 The schooner now thought it prudent to claw off, and had just escaped out of rifle shot.

3. (Earlier examples.)

1837 W. IRVING *Capt. Bonneville* I. viii. 160 There were the remains of the rude fortress in the swamp, shattered by rifle shot. 1846 R. B. SAGE *Scenes Rocky Mts.* xvii. 145 The sharp crack of a rifle-shot.

rift, *sb.*² Add: **2. f.** *Quarrying* (chiefly *U.S.*). Any of a series of parallel planes along which (normally igneous) rock may most easily be split, freq. distinct from the natural bedding planes. Also, the property by which such rocks tend to split most easily in one direction.

1886 T. S. HUNT *Min. Physiol. & Physiogr.* vii. 274 A phenomenon..apparently due to superficial alternations of temperature on certain crystalline rocks, which have resulted in establishing in them..a series of rifts or divisional planes parallel to the present surface, which are well known to quarrymen. **1907** *Bull. U.S. Geol. Survey* No. 113. 27 Another peculiarity of rift is that the angle of its inclination may at some places be modified by gravity. **1912** H. RIES *Building Stones & Clay-Products* iii. 96 The rift is an obscure foliation, either vertical (or nearly so) or horizontal, along which the granite splits more readily than in any other direction. **1934** O. BOWLES *Stone Industries* vii. 81 In quarrying, a good rift assists greatly as it facilitates bed lifting where open bed planes are absent. *Ibid.*, Rift is due chiefly to orientation of grains. **1937** *Mem. Geol. Soc. Amer.* V. 19 The term, rift, denotes the peculiar property of granitic rocks to split relatively easily in a direction other than the 'bedding' (which is parallel to the earth's surface). **1960** O. BOWLES in J. L. Gillson et al. *Industr. Min. & Rocks* (ed. 3) xv. 327/1 [Paving stones] are shaped by hand processes involving expert knowledge of the 'rift' and 'run' directions of easy splitting.

g. A large fault running parallel to the major regional relief, esp. one bounding a rift valley; also, a rift valley.

1921 E. J. WAYLAND in *Geogr. Jrnl.* LVIII. 345 The term 'Western Rift' is used in this paper to denote that part of the Great Rift Valley system which forms the western boundary of Uganda... The two terms 'Eastern Rift' and 'Western Rift' are here introduced for the sake of convenience; they are not proposed as geographical names, or intended for other than local use. **1924** J. G. A. SKERL tr. *Wegener's Orig. Continents & Oceans* xi. 166 The East African rift-valleys form the most beautiful example of such rifts. **1936** B. WILLIS *E. Afr. Plateaus* 55 The great San Andreas rift of California, 500 miles long, is a fault. **1957** G. E. HUTCHINSON *Treat. Limnol.* I. i. 12 There has been some uptilting all around the margins of the plateau bounded by the Albert–Edward–Tanganyika system of rifts. **1970** *Kenya Farmer* Feb. 12/3 Looking at the rainfall figures for 1969 in the Lower Rift we see that..only a few lucky farmers got enough rain to begin planting.

4. *rift basin, fault, faulting, -system, tectonics*; **rift block** *Physical Geogr.*, a horst or a graben; **rift saw** (see quot. 1958); **rift sawing** *vbl. sb.* = *quarter sawing* vbl. sb. s.v. *QUARTER *sb.* 30; so **rift-sawn** *ppl. a.*; **rift valley,** substitute for def.: a large, elongated valley with steep walls formed by the relative depression of a block of the earth's surface and bounded by nearly parallel faults or fault-zones; cf. *GRABEN; (further examples); **Rift Valley fever,** a disease of animals of the Rift Valley of East Africa which can affect humans as a mild fever and is caused by an arborvirus.

1978 *Nature* 13 July 133/1 For most ancient rift basins, it is very difficult to demonstrate whether rifting was preceded or accompanied by doming. **1929** D. JOHNSON in *Compt. Rend. 15th Internat. Geol. Congr.* II. 361, I shall consider only four major types of blocks included between bounding faults:..II. Rift blocks; or those relatively raised or lowered between normal faults. **1944** A. HOLMES *Princ. Physical Geol.* xix. 439 The boundary faults are then regarded as steep upthrusts and the rift blocks as wedges (widening in depth) held down by pressure from the upriding sides. **1954** W. D. THORNBURY *Princ. Geomorphol.* x. 266 The relatively raised blocks are commonly called horsts and the lowered blocks grabens, when they are the direct effects of faulting. Johnson (1929) preferred the name rift blocks for them, but there are objections to its use because rift is often applied to earthquake rifts.., along which movement has been largely horizontal. Johnson would call an uplifted block a rift block mountain or horst and a lowered block a rift block basin or graben. **1977** A. HALLAM *Planet Earth* 78/3 Although the Cenozoic phases of rift faulting are well documented, geological mapping in older rocks has shown that the rift faults frequently coincide with ancient tectonic dislocations. **1909** WEBSTER, Rift saw. **1958** *N.Z. Timber Jrnl.* May 56/2 *Rift saw*, a special type of circular saw, usually with inserted teeth, for converting slabbed logs into flooring strips. **1909** WEBSTER, Rift sawing. **1920** F. T. HILL *Pract. Aeroplane Construction* iv. 97 Rift sawing..is to be preferred [for aircraft work]. **1968** F. HILTON *Craft Technol. for Carpenters & Joiners* i. 17 (*heading*) Quarter or rift sawing. **1920** F. T. HILL *Pract. Aeroplane Construction* iv. 98 Rift-sawn spruce can now be obtained in this country. **1965** W. H. BROWN *Introd. to Seasoning Timber* ii. 20 The boards so cut [*sc.* in radial plane] are known as quarter sawn, edge grain, or rift sawn. **1972** M. VERNEY *Boat Repairs & Conversions* v. 93 All pine timber used for decking needs to be rift-sawn. **1908** *Geogr. Jrnl.* XXXI. 217 (*heading*) The East African rift-system. **1978** *Nature* 13 July 133/1 The East Greenland Triassic–Jurassic rift seems to be part of a three-armed rift system where the two other arms are represented by a little known basin along the coast of Southeast Greenland, and by parts of the northern North Sea graben complex. **1976** *Ibid.* 9 Sept. 119/1 Rift tectonics are amply demonstrated by lateral inconsistence and temporal heterogeneity in Upper Palaeozoic and Mesozoic stratigraphy. **1946** L. D. STAMP *Britain's Structure & Scenery* xxiii. 228 The area is structurally a rift valley let down between parallel faults but is not otherwise a valley in the ordinarily accepted meaning of the word. **1954** W. D. THORNBURY *Princ. Geomorphol.* xix. 508 The volcanoes along the African rift valleys and the Rhine graben, as well as the *puys* of central France, display a..linear arrangement. **1968** R. W. FAIRBRIDGE *Encycl. Geomorphol.* 947/2 Detailed surveys of rift valleys in Africa, the Middle East and elsewhere show that idealized symmetric graben profiles are rarely seen, that the opposite sides are frequently asymmetric, and that there is often only a single fault. **1976** K. THACKERAY *Crownbird* ii. 33 He found himself looking at the sides of the rift valley, recognizing different strata. **1931** DAUBNEY & HUDSON in *Jrnl. Path. & Bacteriol.* XXXIV. 578 We have proposed the name *Rift Valley fever* as a popular alternative to our first suggestion *enzootic hepatitis*, which was originally applied to the disease in sheep. The latter is hardly a suitable name for the human disease, since we have as yet no evidence that the liver is involved in man. **1962** GORDON & LAVOIPIERRE *Entomol. for Students of Med.* xix. 121 The demonstration in 1949 that a species of *Eretmapodites* (*E. chrysogaster*) was capable of acting as a vector of Rift Valley fever under natural conditions, has increased the interest of the epidemiologist in this group of mosquitoes. **1978** *Nature* 26 Jan. 308/2 Rift Valley fever or enzootic hepatitis is a severe viral infection which primarily affects sheep and cattle causing many deaths in pregnant and newborn animals. The disease occurs naturally only in Africa and by 1912 it was recognised in the Rift Valley in Kenya.

rift, *sb.*^{4} **1.** (Earlier and later examples.)

1727 in *Documents Colonial Hist. New-York* (1855) V. 826 The French..have no way but to come up from Montreal to the Lake against a Violent Stream, all full of Rifts and Falls and Shallows. **1845** J. F. COOPER *Chainbearer* II. vi. 80 The most that can be done with it [*sc.* the lumber]..will be to float it down to the next rift. **1879** *Scribner's Monthly* Nov. 21/1 In one hanging rift close by the bank..I took at five casts fifteen fish.

rift, *v.*^{1} **1. a.** (Later examples.)

a **1861** T. WINTHROP *John Brent* (1862) xix. 209 A little pathway in the sage-bushes suddenly opened before me, as a lane rifts in the press of hurrying legions 'mid the crush of a city thoroughfare. **1898** H. S. CANFIELD *Maid of Frontier* 75 The mass of vapor overhead rifted for a moment. **1979** *Nature* 22 Nov. 378/2 The block rifted from mainland South-east Asia and met with the northwards drifting Australian continent during the Neo-gene.

ri·fting, *vbl. sb.*^{2} [f. RIFT *v.*^{1} + -ING^1.] **1.** Chiefly *U.S.* **a.** *Quarrying.* The occurrence or development of rifts in rock. Cf. *RIFT *sb.*^{2} 2 f.

1889 *9th Ann. Rep. U.S. Geol. Survey* 602 (*heading*) Rifting of the quarried rocks. **1891** *Amer. Jrnl. Sci.* CXLI. 269 Rifting is..dependent upon the thousands of minute dislocations which occur in every cubic inch of rock. **1937** *Mem. Geol. Soc. Amer.* V. 20 Rifting in granites has been known for a long time.

b. The action of splitting mica into thick sheets.

1939 G. A. ROUSH *Strategic Min. Supplies* xii. 345 The books [of mica] are first split into slabs about ¼ inch thick (rifting). **1954** S. J. JOHNSTONE *Minerals for Chem. & Allied Industries* 331 The preparation of mica for the market usually involves..rifting into sheets sufficiently thin to be cut by hand with a blade. **1960** S. A. MONTAGUE in J. L. Gillson et al. *Industr. Min. & Rocks* (ed. 3) xxvii. 559/1 The operations of rough splitting and trimming mica are termed 'rifting' or 'sheeting'.

2. *Physical Geogr.* The process of severing of areas of the earth's crust into distinct blocks or plates by formation of rifts. Cf. *RIFT *sb.*^{2} 2 g.

1924 J. G. A. SKERL tr. *Wegener's Orig. Continents & Oceans* xi. 166 Folding and rifting are only different effects of one and the same process. **1969** *Times* 20 May 12/7 Major rifting of the Earth's surface. **1969** *Nature* 11 Oct. 140/1 Although rifting was completed during the Lower Cretaceous, significant continental drift did not occur until the Lower Turonian. **1978** *Ibid.* 13 July 127/1 The eruptions were accompanied by extensive rifting..which led to formation of the Sveinagjá graben.

fig. **1975** W. SAFIRE *Before the Fall* x. iv. 667 There was no rift..between the President and his national security adviser. There was plenty of rifting going on between Henry and the White House staff.

riftless, *a.* (Earlier example.)

1797 *Spirit of Farmers' Museum* (1801) 71 From the gnarl'd oak..To the green vine, which twines its riftless sides.

rig, *sb.*^{4} Add: **1.** Also [infl. by Da. *rig*, Norw., Sw. *rigg*], = RIGGING (*vbl.*) *sb.*^{2} 2.

1888 L. A. SMITH *Music of Waters* 222 Out in Pamerent in Holland, There lay a brig so old; Worn out was her hull, And worn out was her rig. **1889** *Musical Times* 1 July 394/2 On page 222 she uses the word 'rig' as 'rigging', whereas even a landsman knows the difference. **1903** *Rudder* Feb. 52/1 Mr. Crowninshield has changed the overhangs slightly..and has put on a slightly larger rig. **1934** *Yachting Monthly* LVI. 463/1 The weight and center of gravity of the rig must be considered.

2. (Earlier examples.) *rig of the day* (Naut.), the uniform to be worn on any particular day.

1843 T. C. HALIBURTON *Attaché* I. xii. 218 Congregations are rigged out in their..bran new clothes, silks, satins.., and all sorts of rigs. **1853** 'P. PAXTON' *Stray Yankee in Texas* xxvii. 267 Here was a rig for a July day in Texas, with the thermometer at 105° in the shade! **1914** [see *NUMBER *sb.* 5 c]. **1948** PARTRIDGE *Dict. Forces' Slang* 155 *The rig-of-the-day.* The dress that is to be worn for the day is 'piped' with 'Hands to breakfast and clean; rig-of-the-day, number—whatever it is'. **1961** F. H. BURGESS *Dict. Sailing* 154 *Out of the rig of the day*, incorrectly dressed.

3. Delete *U.S.* and add: **a.** (Further examples.) *spec.* = *oil rig* s.v. *OIL *sb.*^{1} 6 e.

1885 *Encycl. Brit.* XVIII. 716/2 When the location of a well has been determined, a derrick or 'rig' is built, which consists of the derrick itself and a small house for an engine. **1946** E. HODGINS *Mr. Blandings builds his Dream House* i. vi. 89 Mr Blandings' well was being drilled... The rig's motor roared and stank. **1957** *Economist* 21 Dec. 1080/3 During 1958 test wells will be drilled in the Benin area in Western Nigeria..; five rigs are now drilling. **1974** E. AMBLER *Dr. Frigo* i. 46 It's not the same as ordinary off-shore drilling. The rigs have to be different. **1975** *Daily Tel.* 24 May 11/4 Some gas production rigs built up to 10 years ago for the southern North Sea had designs which underestimated the full vigour of the waves.

b. orig. and chiefly *N. Amer.* (Earlier and later examples.) Also, any vehicle; *spec.* a lorry, a truck.

1831 A. STODDARD *Diary* 30 Nov. in *Michigan Hist. Mag.* (1927) XI. 472 Breakfast swallowed we stepped into our next rig, which was a lumber wagon. **1872** 'MARK TWAIN' *Roughing It* xlvi. 325 I've heard tell of carriages all my life... I mean to have the nobbiest rig that's going. **1931** H. F. PRINGLE *Theodore Roosevelt* I. i. 4 A rig had been..summoned from a near-by stable. **1938** *Amer. Speech* XIII. 307 *Rig*, also a designation for any bus. **1938** *Commercial Car Jrnl.* June 60/3 *Rig*, a truck. **1942** BERREY & VAN DEN BARK *Amer. Thes. Slang* § 766/2 (*Bus, taxicab, truck.*)..oilcan, rig, tub. **1957** J. KEROUAC *On Road* (1958) 16 A great big tough truckdriver..got his rig under way. **1963** *Amer. Speech* XXXVIII. 45 Rig.., a tractor and semitrailer; sometimes just a tractor. **1969** *Publ. Amer. Dial. Soc.* LII. 35 *Rig*, any truck on the fire fighting force. 'Back the rig in.' **1970** J. H. GRAY *Boy from Winnipeg* 63 The [fire brigade] horses were kept in box stalls behind the rigs which were all parked facing the street. **1972** J. MINIFIE *Homesteader* x. 85 When the last rig had pulled out my father went down to the barn. **1974** *Times* 11 Feb. 6/7 Mr Nixon came on the air..to urge the drivers to get their 'big rigs' back on the road. **1976** *Daily Tel.* (Colour Suppl.) 16 July 10/1 No CBer ever drives. He cooks or pushes a rig. **1976** L. DEIGHTON *Twinkle, Twinkle Little Spy* xxiii. 225 The gargantuan trailer-trucks rolled south to Timbuktu in convoy, enough drivers in each rig to eat and sleep in relays. **1977** *Hot Car* Oct. 76/2 Pulled into Steeple Claydon early Sunday morning and looked out across nearly 250 vans, trucks, rods and rigs.

c. Any apparatus or device.

1868 *Rep. Iowa Agric. Soc. 1867* 174, I consider the Victor mill & Cook's evaporator the best rig for making sirup profitably from cane. **1950** *Engineering* 10 Mar. 265/2 Regrinding is extremely simple; the drill steel is placed on a rig and the tip is ground either by a pedestal grinder or by a hand grinder. **1973** *Sci. Amer.* Nov. 1/3 (Advt.), Before a new seat enters production, its suspension must survive 300,000 load changes on a vibrating rig. **1977** *Amer. Speech* *1975* L. 65 *Rig n*, hypodermic needle used to inject drugs into a vein. 'A rusty rig will give you hepatitis.' **1979** *Sci. Amer.* Dec. 52/1 Intravenous infusion, which is normally done in a hospital with an elaborate rig involving bottles of fluid, connecting tubes and cannulas.

d. An amateur's radio transmitter and receiver; also, a telegraph, a radar set, or the like.

1942 BERREY & VAN DEN BARK *Amer. Thes. Slang* § 810/2 *Telegraph*, ethegraph, kid, Old Betsy, rig, she. **1960** *Daily Progress* (Charlottesville, Va.) 29 Sept. 39/3 Hams have their own slang terms. 'Rig' is what they call their sending and receiving equipment and 'shack' is the name of the room or structure housing their sets. **1966** M. WOODHOUSE *Tree Frog* xxi. 155 This [radar] rig is operational two per cent of the time. **1976** A. HOPE *Hi-Fi Handbk.* 10 After the purchase of a house and a car, an impressive hi-fi rig may well be the next most expensive item ever bought by a householder. **1977** *Rolling Stone* 30 June 118/4 But by the time you added a booster amplifier, and perhaps a separate cassette unit—not to mention a CB rig—you ended up with no leg room. **1978** *Observer* 18 June 4/2 The normal range of a regular watt rig is between five and ten miles.

e. The penis. *coarse slang.*

1964 *Amer. Folk Music Occasional* i. 12 There is a certain latitude in the approach to Negro blues where women are sweet food..and sex is hard, virile labor (I got a big tall rig, it drills way down deep). **1973** M. AMIS *Rachel Papers* 92 All weekend I cried,..thought of ways of committing suicide,..considered lopping off my rig with a razor-blade.

4. *attrib.*, as (sense 3 a) *rig crew, medic* (colloq.), *operator.*

1972 L. M. HARRIS *Introd. Deepwater Floating Drilling Operations* xviii. 191 The shortage of personnel is not limited only to rig crews. **1977** *Observer* 24 Apr. 1/6 The rig crew was about to install a safety valve on the top of a production pipe. **1977** *Lancet* 8 Oct. 751/1 Resuscitation during this period is the responsibility of a sick-bay attendant ('rig medic') or, if life-supporting measures are required in a pressure chamber, of a suitably trained diver. **1975** *Offshore* Aug. 51/1 A rig operator, on the average, will call on the divers two or three times a month for inspection, repairs or retrieval of lost equipment.

rig (rig), *sb.*^{7} *dial.* Also **rigg.** [Perh. f. WRIG *v.*, as the fish is remarkable for the way it twists itself round the line on which it is caught.] The tope, *Galeorhinus galeus*, a shallow-water shark found in the eastern Atlantic and in the Mediterranean; in quot. 1963 = *HUSS *sb.*

[**1547** *Act 1 Edw. VI* VII. m. 5 pisces regal, viz' sturgions balenas catas porpec' delphinos rigges graspes. **1549** *Act 3 Edw. VI* II. m. 40 pisces regal videl' sturgiones balenas cetas porpec' delphinos rigges graspes.] *c* **1700** in *Newnes Encycl. Angling* (1963) 90/2 Royal fish, viz. sturgeons, grampuses, whales, porpoises, dolphins, riggs and graspes, and generally whatsoever other fish having in themselves great and immense size or fat. **1887** PARISH & SHAW *Dict. Kentish Dial.* 128 *Rig*, the common tope. *Galeus vulgaris*.—Folkestone. **1927** *Glasgow Herald* 7 Sept. 12/7 On the south-east coasts of England a species of small shark, known locally as 'rigg', is caught on strong lines baited with mackerel. **1963** [see *HUSS *sb.*].

rig, *v.*^{2} Add: **1. d.** (See quot. 1956.)

1909 *Aero Man.* 104 By rigging diagonal wires across top and bottom and across both sides, the entire frame of the glider will be rendered quite rigid. **1916** *Aëronaut. Jrnl.* XX. p. ii (Advt.), Steel wire strands & cords of galvanised or plated steel wire (for rigging aircraft). **1956** W. A. HEFLIN *U.S. Air Force Dict.* 443/1 *Rig.* 1. To assemble, adjust, and

align the major components of an *aircraft*; *specif.*, to assemble and align the airfoils or other surfaces of an *aircraft*. 2. To fit out an *aircraft* with control cables, bracing cables, pulleys, turnbuckles, and the like. **1978** *Sci. Amer.* Nov. 135/1 The Wrights also flew the machine as a glider, experimenting by rigging it with a dihedral angle.

rig, *v.*^{6} Add: **1. b.** To take to task; to rag or tease. *U.S.*

1892 *Dialect Notes* I. 231 *Rig*, to tell a joke on. 'He rigged him good.' **1899** A. H. QUINN *Pennsylvania Stories* 100, I rigged him about it once and he said he'd reform.

2. (Further examples.)

1933 *Lit. Digest* 1 July 40/3 'It's in the bag' is an expression..to designate that prize-fights or horse-races have been rigged. **1935** *Sun* (Baltimore) 30 Oct. 1/2 The way the 'election' was rigged made an opposing farmer seem a sap. **1955** *Times* 23 June 10/2 Mr. J. McShane..claimed that that meeting had been 'rigged' and that rank and file London dockers would not resume work on Monday. **1966** *Wall St. Jrnl.* 25 July 1 'Top 40' record lists that allegedly are rigged by gifts of free records by the carload. **1967** W. SOYINKA *Kongi's Harvest* 72, I did my best to rig the results in favour of the state co-operatives.

3. *to rig the market*, also *fig.*

1978 A. RYAN in Hookway & Pettit *Action & Interpretation* 76 The only question at issue here is whether we should treat his account of social interaction as a story about how we rig the market, or as a story about how we engage in putting on a good show.

riga^2 (rī·gă). [Hausa.] A man's loose-fitting robe, worn in West Africa.

1923 *Daily Mail* 14 June 7 The two Sudanese hunters..saw London... Wearing long white cloaks, or rigas,..they began a search for Captain Buchanan. **1928** *Daily Express* 6 Oct. 9 There appears on the verandah steps a dignified Mohammedan in flowing 'riga'. **1932** J. CARY *Aissa Saved* liii. 287 God, in a white riga and a new indigo turban. **1973** J. J. MCKELVEY *Man against Tsetse* iii. 171 A flowing yellow *riga* Nigerian style over a brown business suit. **1977** P. NEWMAN et al. *Mod. Hausa-Eng. Dict.* 100/1 The *riga*, worn over a shirt and trousers, is the standard attire in much of West Africa.

rigaree (rigărī·), *a.* Also **rig-a-ree.** *Glass-Blowing.* [Perh. ad. It. *rigare*: see next.] Applied to a pattern of raised bands upon a glass vessel, or the method of producing such bands.

1923 H. J. POWELL *Glass-Making in Eng.* iii. 22 (*caption*) Base of tumbler, showing rig-a-ree decoration. *Ibid.* 44 *Rigaree marks*, parallel lines on a raised band, tear or collar of glass, produced by ridges on the edge of a small metal wheel. **1946** W. B. HONEY *Victoria & Albert Mus. Glass* i. 8 Softened glass is readily drawn out into threads, and these may be wound upon ('trailed') and attached to a vessel and melted in to give a more or less smooth surface; such threads may be drawn together with tools into a network or be impressed with a patterned wheel, making the so-called 'rigaree' trail. **1960** H. HAYWARD *Antique Coll.* 240/1 *Rigaree trail*, a process of glass-blowing in which applied threads of glass are attached to a vessel. **1965** [see *PASSGLAS].

‖ **rigatoni** (rigatō^u·ni), *sb. pl.* (Also used as *sing.*) [It. f. *rigato* pa. pple. of *rigare* to draw a line, to make fluting.] Short hollow tubes of pasta in fluted form; a dish of this pasta.

1930 H. BURKE *Cookery Bk.* 100 'Rigattoni' is the Italian name for a special kind of macaroni which comes in short thick tubes. **1958** *Catal. County Stores, Taunton* June 17 *Naples Macaroni*... Cut—Millenghe, Rigatoni, Cannelloni—a lb. ½. **1962** *Punch* 25 July 129/1 *Capellini*,..*rigatoni*, *tagliatini*, noodles and a hundred other traditional shapes [of pasta]. **1967** A. BAILEY in L. Deighton *London Dossier* 56 A plate of really excellent spaghetti, rigatoni, tagliatelli or ravioli. **1976** *Times* 10 July 10/7 Home-made cannelloni or rigatoni with mussels, are other first courses worth trying.

‖ **rigaudon** (rigodoṅ). [Fr., of obscure origin: see RIGADOON *sb.*] = RIGADOON *sb.*

1706 [see note in small type below RIGADOON *sb.*]. **1923** E. SITWELL *Bucolic Comedies* 26 Dancing with angels all in a round, Hornpipe and rigaudon on the Fair's ground. **1938** *Oxf. Compan. Mus.* 801/2 The rigaudon was gay and lively until its promotion from the village green to the aristocratic ball-room and the court, when it took on dignity. **1959** *Collins Mus. Encycl.* 549/2 *Rigaudon*.., a Provençal dance in lively 2/2 time which was adopted into the suite and into the ballet of French opera in the late 17th cent. **1970** W. APEL *Harvard Dict. Mus.* (ed. 2) 734/1 The *rigaudon* also occurs in the suites of modern composers (Grieg, *From Holberg's Time*; Ravel, *le Tombeau de Couperin*). **1979** *Early Music* Jan. 40/1 The French use this (2-)metre in various dances such as..rigaudons, gavottes, rondeaus, [etc.].

rigescent, *a.* Delete *Bot.* and add examples.

1966 *Punch* 20 July (Advt. facing p. 107), The tough, rigescent and unyielding bar of iron is heated to the required temperature. **1974** *Nature* 5 Apr. 512/1 Sodium dodecyl sulphate..left an intact rigescent integument inside which was the intact starch sheath.

rigged, *ppl. a.*^{1} Add: **3.** Fitted *up*, esp. as an expedient or makeshift.

1932 C. ISHERWOOD *Memorial* i. iii. 36 The rigged-up stage was backed by sackcloth curtains.

4. Fixed *up*, equipped.

1935 A. J. POLLOCK *Underworld Speaks* 97/2 *Rigged up*, all ready with auto, tools, pistol, machine gun, etc., for a criminal job.

rigger^1. Add: **1. b.** One who attends to the rigging of aircraft.

1912 *Flight* 20 Apr. 348/2 N.C.O.'s and air mechanics will be required as engine drivers, fitters,..riggers, &c. **1920** [see *RIGGING (*vbl.*) *sb.*^{2} 1 b]. **1930** *Daily Express* 6 Oct. 2/5 Norcott, A. W. J., aged twenty-nine, rigger... No previous flying experience before joining R 101 in October 1929. **1942** *R.A.F. Jrnl.* 30 May 30 By what, then, is the R.A.F. kept in the air?.. Perhaps by the riggers..or the Observer Corps?

4. (Earlier and further examples.)

1893 [see *IN-RIGGER]. **1935** L. LUARD *Conquering Seas* i. 9 They drove those full riggers in the good old days.

6. a. One who works on an oil rig (orig., on a rotary one). **b.** One who is employed in the construction of oil rigs.

1949 *Amer. Speech* XXIV. 35 From this conflict emerged the titles..*rigger*..and *swivel neck* for rotary drillers. **1974** *BP Shield Internat.* Oct. 8/3 Highlands Fabricators created their own skilled local workforce..by training 610 former farm workers..as welders, riggers and fabricators. **1977** *Off-shore Engineer* May 10/2 (Advt.), Manpower—experienced Fitters, Riggers and Welders.

rigger^3 (ri·gəɹ). *Austral.* and *N.Z. slang.* [Origin unknown.] (See quot. 1941.)

1941 BAKER *Dict. Austral. Slang* 60 *Rigger*, a quart bottle of beer, esp. a quart of draught beer in a square-faced gin bottle. **1943** F. SARGESON in *Penguin New Writing* XVII. 66 I'd meet him every day, and I'd always bring a couple of riggers.

rigging, (*vbl.*) *sb.*^{2} Add: **1. b.** The work of adjusting the wires, control surfaces, etc., of an aircraft.

1914 *Flight* 3 July 700/2 The list of lectures..are as follow..Rigging, and Training of Riggers. **1920** G. C. BAILEY *Compl. Airman* xxii. 173 Rigging is the art of erecting the machine and so adjusting the various surfaces, controls, etc., that it is in a fit condition for flying. The upkeep of the machine is also part of the rigger's work. **1937** *Discovery* Sept. 293/1 A very elementary account of the rigging and truing-up of an aircraft. **1941** NORCROSS & QUINN *Aviat. Mechanic* viii. 345 (*caption*) Cadet class in rigging at Randolph Field, Tex. A biplane involves all the fundamentals of rigging.

2. *to climb the rigging*: to become angry, lose one's temper. *Naval slang.*

1916 'TAFFRAIL' *Carry On!* 27 To get angry is to 'get dizzy' or 'climb the rigging'.

b. The system of ropes and wires used to support the structure of and distribute the load of an airship or aircraft.

1843 *Mechanics' Mag.* 8 Apr. 276/1 The direct resistance of the car, masts, and rigging, in the construction of aerial vehicles, will..probably put a limit to their velocity not much exceeding 24 to 30 miles per hour. **1900** J. M. BACON *By Land & Sky* i. 12 The balloon on rising fouled a big elm, and for a moment remained caught high up among the boughs. Then it tore off a large branch entangled in the rigging. **1919** H. SHAW *Text-bk. Aeronaut.* xvii. 198 In the Parseval type [of airship], the car is suspended from a large elliptical rigging band just below the centre of the envelope. **1969** *Gloss. Aeronaut. & Astronaut. Terms* (*B.S.I.*) VII. 6 *Rigging*, the system of wires or cords and their attachments, by which the dead weight, or the main cable tension, is distributed over the hull or envelope. **1977** D. BEATY *Excellency* i. 7 The weekends he'd spent..servicing Cirrus engines and checking the rigging of Austers in return for free flying lessons.

c. In full *rigging lines.* In a parachute, the system of cords which join the canopy to the harness.

1929 *Aeronautics* (H.M.S.O.) 109 The rigging is so parcelled that it is pulled out gradually, the part already released remaining taut and straight, the whole of the rigging emerging before the commencement of the withdrawal of the body. **1935** C. G. BURGE *Encycl. Aviation* 489 *Rigging Lines*, the cords which transmit the load from the harness (or life lines) to the body of the parachute. **1952** *Chambers's Jrnl.* May 262/2 A bad exit can cause the rigging-lines to tangle and prevent the 'chute from opening. **1958** *Listener* 7 Aug. 202/2 There were casualties..due to tangling of the rigging lines, or faulty exit. **1972** *Daily Tel.* 16 May 3 The Prince of Wales was turned upside down when his feet caught in the rigging lines of his parachute as he hurtled seawards from 1,200ft during his controversial jump from an RAF plane.

ri·gging, (*vbl.*) *sb.*^{3} [f. RIG *v.*^{6} + -ING^1.] The action of the verb RIG^6; trickery, swindling; ragging or teasing; manipulation of prices; electoral fraud.

1839, etc. [see THIMBLERIG *v.*]. **1912** H. CROLY *Marcus A. Hanna* 460 He was constantly on the lookout for a chance to joke about the peccadilloes of his friends. There were few of them who escaped this kind of rigging. **1932** *Sun* (Baltimore) 27 Apr. 15/7 His general chronicle of stock-price 'rigging'. **1946** *Ibid.* 27 Dec. 1/8 One of the sharpest breaks came in New York and Chicago wholesale butter markets... Much of the mystery surrounding the break in New York, where charges of 'rigging' had been heard, was removed. **1959, 1961** [see *ballot-rigging* s.v. *BALLOT *sb.*^{1} 4]. **1980** G. M. FRASER *Mr American* xxii. 430 It isn't rigging, you see. You couldn't rig a British judge and jury nowadays.

riggish, *a.* Add: Occas. in wider (i.e. non-dialectal) use. (Later examples.)

1963 T. PYNCHON *V.* i. 15 Which after forty-five years was nothing for any riggish Pappy Hod to be finding out. **1973** *Daily Tel.* 25 Apr. 15/3 Both have a sweetness that disinfects Wycherley's basic cynicism, whose spirit is admirably personified by Frances Cuka's unashamedly riggish Lady Fidget.

Riggs (rigz). *Path.* [Name of John Mankey Riggs (1810–85), U.S. dentist, who described the condition in 1876 (*Pennsylvania Jrnl. Dental Sci.* III. 99–104).] *Riggs'(s)* (also erron. *Rigg's*) *disease*: a disease of the tissues adjacent to the teeth, characterized by suppuration and inflammation of the gums.

1877 *Dental Cosmos* XIX. 70 What do I know about Riggs's disease (so-called).—How does it happen that this title has come into existence? *Ibid.* 185 The trouble which has come to be called Riggs's disease has not hitherto responded to..treatment. **1887** C. S. TOMES *Syst. Dental Surg.* (ed. 3) 695 The discharge, which is a prominent symptom in the more acute cases, suggested the name pyorrhoea alveolaris, which has passed into general currency, while the warm advocacy of a particular form of treatment, and the consequent attention drawn to its advocate in connection with the disease, has led to the frequent use of the term Rigg's disease. **1900** 'MARK TWAIN' in *McClure's Mag.* Jan. 290 If you don't know what Riggs's Disease of the Teeth is, the dentist will tell you. **1923** K. H. THOMA *Teeth, Diet & Health* vi. 78 Other names for pyorrhea are Rigg's disease, receding gums, and pericementoclasia, the latter being the latest scientific term.

right, *sb.*^{1} Add: **II. 9. b.** (Further examples.) *right-to-life* adj. phr., designating persons, etc., opposed to the abortion of the unborn fœtus or concerned with this issue; hence **right-to-lifer.**

1916 G. B. SHAW *Androcles & Lion* p. lxx, We must begin by holding the right to an income as sacred and equal, just as we now begin by holding the right to life as sacred and equal. **1972** *Times* 22 Jan. 18/7 English law does not yet recognize a right to privacy. **1973** A. E. WILKERSON *Rights of Children* 312 It is doubtful that in the right-to-life controversy the rights of the unborn child will be inviolate. **1973** *Austral. Humanist* XXVI. 1/1 The victory of the Roman Catholic-dominated Right to Life Association over the McKenzie–Lamb private members Bill to provide abortions on request..will be seen by social historians of the future as pyrrhic. **1977** *Time* 25 July 2/3 If the Government provided its citizens with alternative birth control methods (free of charge), there would be fewer abortions. Even the right-to-lifers would like that.

c. (Further examples.) *right-to-work*, phr. used *attrib.* with reference to a worker's right not to be required to join a trade union. *U.S.*

1958 *Economist* 15 Nov. 599/2 So far the only change that Mr Meany has committed himself to seek is one to remove the provision which allows the states to adopt 'right-to-work' laws. These are laws forbidding employers and unions to enter into contracts that require workers to join the union within a specified period. **1967** *Harper's Mag.* Mar. 8 You might have then seen the virtue of the right-to-work policy... What would have been gained..if you sanctioned a system by which this honest and qualified ex-con were compelled to join a union against his will as the price of holding his new-found job? **1976** *Globe & Mail* (Toronto) 22 Mar. 6/1 In 1967 the great bulk of federal civil servants were given the right to strike. **1979** *Tucson Mag.* Mar. 31/3 Both [cities] are situated in a right-to-work state.

f. In *pl.* A title or authority to perform, publish, film, or televise a particular work, event, etc. *serial rights*: see SERIAL *adj.* b in Dict. and Suppl.

1890 KIPLING *Let.* in C. E. Carrington *Rudyard Kipling* (1955) vii. 162 Harper & Co. bought the serial rights for American and paid me. **1913**, etc. [see *film rights* s.v. *FILM *sb.* 7 c]. **1935** *Discovery* Sept. 277/1 The cost of the televising rights for the fight would have to be such as to compensate for the loss of cinema rights. **1939** D. L. SAYERS *In Teeth of Evidence* 208 There were the touring rights..and film rights..and probably radio rights. **1953** [see *ASTRONOMICAL *a.* 1 b]. **1959** *Bookseller* 17 Jan. 124/1 American publishers..claim..that the financial rewards of book-publishing come not from mere publishing but from the sale of film, paperback, serial and other rights. **1974** I. PARSONS in A. Briggs *Ess. Hist. of Publishing* 49 Richardson had made binding agreements with a succession of Dublin booksellers under which he was to receive certain sums in return for exclusive rights. **1981** *Bookseller* 27 June 2226/1 We've sold the US and UK rights in books on the South Island of New Zealand. **1981** M. SPARK *Loitering with Intent* xii. 220 The Triad [Press] sold the American rights, the paperback rights, the film rights, and most of the foreign rights.

g. *rights issue.* An issue of shares offered at a special price by a company to its existing shareholders. Also *ellipt.*

1955 *Times* 20 Aug. 11/2 The Commercial Bank of Australia's 'rights' issue of 2,105,868 Ordinary shares of 10s. (Australian currency) at 15s. each has been over-subscribed without recourse to the underwriters. **1960** *Economist* 8 Oct. 187/1 Successive mergers..reduced its proportionate holding. So, too, did the rights issues of ordinary shares, for as an American investor Bendix could not subscribe for the new shares but had to sell its rights. **1968** *Sun* 25 Oct. 10/5 With last night's price for the ordinary 18s. 9d., the 'rights' are worth a little over 2d. per share, which is not very much. **1970** *Money Which?* Mar. 61/2 Under a rights issue, a company in which you already own shares offers you the chance to buy new shares at a special price. **1976** *Birmingham Post* 16 Dec. 9/5 Lazards are to discuss the intricate and difficult problems of the conflicting timing of its offer for Dunford and Elliott and Dunford's rights issue with the takeover Panel. **1981** *Times* 24 Apr. 15/4 Rowntree Mackintosh..is to raise £42m after expenses from shareholders with a rights issue... Terms of the issue are one new ordinary share at 160p for every four held.

10. b. (Further examples of phr. *in one's own right.*) Now freq. used in general senses,

without reference to a particular title or claim.

1863 *Times* 11 Mar. 5/2 The Crown Princess of Prussia ..has always been popular in her own right. **1939** G. B. SHAW *In Good King Charles's Golden Days* i. 47 When I am King—as I shall be, in my own right, and not by the leave of any Protestant parliamentary gang. **1965** *Listener* 2 Sept. 331/1, I shall try to say something of the fundamental problems of science which are of the deepest significance in their own right. **1966** *Ibid.* 20 Oct. 560/1 It is a pity that Tvardovsky is so little known in the west. He is a very perceptive critic, and a considerable poet in his own right. **1971** *Guardian* 12 Nov. 9/4 The Domaine de la Rayre..is a wine of character..in its own right. **1978** I. B. SINGER *Shosha* iii. 50 He has a nasty wife and estranged children who are rich in their own right.

IV. 17. d. (Earlier and later examples.) Usu. applied to any political group holding conservative principles.

1825 *Ann. Reg. 1824* 152/1 M. de la Bourdonnaye (leader of the extreme right). **1830** *Ann. Reg. 1829* 157/2 All the new ministers belonged to the extreme right. **1848** MRS. GASKELL *Let.* 2 Nov. (1966) 60, I never can ascertain what I am in politics; and veer about from extreme Right,—no, I don't think I ever go as far as the extreme Left. **1940** W. TEMPLE *Thoughts in War-Time* iii. 24 The Right tends to have a fuller sense of historical continuity than the Left. **1954** KOESTLER *Invisible Writing* xxxvi. 385 The trial of the so-called 'Anti-Soviet Block of Rights and Trotskyists' took place in Moscow. **1955** *Times* 23 May 11/7 There is much division among the delegates of free countries on how to deal with the Left and Right dictatorship delegations. **1960** O. MANNING *Great Fortune* xviii. 220 We did nothing to establish a liberal policy that could save the country from either extremity—Left or Right. **1969** A. G. FRANK *Latin Amer.* xix. 316 The current wave of government repression against the Left need not mean a permanent move to the Right. **1974** J. WHITE tr. *Poulantzas's Fascism & Dictatorship* iv. 224 In the struggle against the Left Opposition..the Comintern took a turn to the 'right'.

e. In various sports, the right side or wing of the field of play; a player occupying this position (cf. *RIGHT WING 2).

1867 *Ball Players' Chron.* 8 Aug. 6/3 The nine will be as follows:..Peters, right. **1892** *College Index* (Auburn, Alabama) Nov. 27 Mr. G. O. Shackleford, the Athens left guard, who gave our right so much trouble at the match game in Atlanta last February has entered college here. **1934** in B. James *England v Scotland* (1969) 152 Attack after attack on the English goal. Superb work by that sprite of a player, Cook, on the extreme right. **1949** *Telephone-Reg.* (McMinnville, Oregon) 4 Aug. 2/1 Jimmy 'Whiskers' Beard then drove both in with a single to right.

18. c. (Further examples.)

1898 [see *LAND *v.* 2 e]. **1930** *Daily Express* 6 Oct. 11/5 The blow with which he dropped Compere for the full count was a right to the jaw. **1937** C. HIMES *Black on Black* (1973) 130 He'd go out fast with the garrote in his left hand, throw a hard right at the bastard's face. **1972** J. MOSEDALE *Football* iv. 49 Bob Snyder..threw a roundhouse right that knocked Matheson out the door.

d. A shot fired at game with the right barrel of a double-barrelled shotgun; a bird or beast hit by such a shot. Cf. RIGHT AND LEFT, RIGHT-AND-LEFT *sb.* 2.

1893, etc. [see *LEFT *sb.* 4].

e. *Surfing.* The (use of the) right foot.

1968 *Surfer Mag.* Jan. 73/1 Eamonn Matthews..caught some nice rights. **1970** *Surf '70* (N.Z.) 44/2 There were good lefts and occasional rights with Ted Spencer carving turns people thought were impossible.

f. A right-hand turn. *U.S.*

1961 WEBSTER, s.v., Take the right at the fork. **1969** D. E. WESTLAKE *Up your Banners* v. 35 The light turned green and she made a right. **1977** R. E. HARRINGTON *Quintain* vi. 49 A blue Rover had made a right off a side street to pull in ahead of the cab. **1981** G. V. HIGGINS *Rat on Fire* xiii. 96 Leo Proctor took a right in Dorchester Avenue and drove the van south.

V. 21. *Comb.* **right-left** *attrib. phr.*, of or between right and left.

1968 M. S. LIVINGSTON *Particle Physics* vii. 132 The parity conservation expresses this symmetry between events in the real world and their mirror images, this right-left symmetry of form. **1970** S. ROKKAN *Citizens, Elections, Parties* x. 335 He sees in this circumstance a possible explanation for the absence of a clear-cut tradition of right-left polarization in the United States. **1978** *Science* 24 Feb. 852 (*heading*) Right-left asymmetries in the brain.

right, *a.* Add: **I. 3. a.** *right triangle*, a right-angled triangle.

1903 J. MCMAHON *Elem. Geom.* i. 62 In a right triangle the side opposite the right angle is called the hypotenuse. **1970** J. R. BYRNE *Mod. Elem. Geom.* v. 153 Suppose that we had a 'Book of Standard Right Triangles' for angles from measure 1 to 89°.

II. 7. e. *that's right*, used to express affirmation or agreement; *is that right?* phr. inviting confirmation of a statement or proposal; *to get* (something) *right*, to be accurate or correct (in a certain matter), to have (something) clear in one's mind.

1905 *N.Y. Even. Post* 7 Apr. 2 The President's address was frequently interrupted with applause and cries of 'That's right'. **1922** 'K. MANSFIELD' *Voyage* in *Garden Party* 186 'And you've got your cabin tickets safe?' 'Yes, dear.'.. Grandma felt for them inside her glove and showed him the tips. 'That's right.' **1925** A. CHRISTIE *Secret of Chimneys* x. 96 'You recognized the body as that of one of your guests?' 'That's right, inspector.' **1930** G. B. SHAW *Apple Cart* i. 12, I say to them 'You are supreme: exercise your power.' They say, 'That's right: tell us what to do'; and I tell them. **1932** *N. & Q.* June 415/2 We not are even simple-minded enough to say 'yes'; we prefer to say 'That's right'. **1933** *Punch* 16 Aug. 195 (*caption*) Tourist (*after two miles of it*). 'Look here, Alfonso, let's get this right. Have you adopted us or have we adopted you?' **1938** G. GREENE *Brighton Rock* VII. vii. 335 'So he was at your school.' 'That's right.' **1951** C. P. SNOW *Masters* III. xxxvii. 301, I expect Eliot has got everything he said right. **1958** *Spectator* 4 July 26/2 He *does* contrast the Spencers and the Churchills (Sir Winston is a Churchill not a Spencer, or have I got it right?). **1976** L. DEIGHTON *Twinkle, twinkle, Little Spy* x. 99 'They would take it real bad.' 'Is that right,' said Mann.

8. c. *right sort, stuff*: an alcoholic drink. *slang.*

1820 J. H. REYNOLDS *Fancy* 110 *Right sort*, gin. **1927** WODEHOUSE *Meet Mr Mulliner* vi. 198 A sharp spasm had reminded him how much of the right stuff he had in him at that moment.

d. Of persons and things: regarded with approval; socially acceptable; potentially influential.

1842, 1883 [see SORT *sb.*^{2} 11 b]. **1900** H. JAMES *Little Tour in France* (ed. 2) vii. 76 The middle of the eighteenth century..was surely, in France at least, the age of good society, the period when the 'right people' made every haste to be born in time. **1901** J. VAIZEY *About Peggy Saville* x. 76 Travelling is good fun if you..provide yourself with introductions to the right people. **1928** KIPLING *Bk. of Words* xi. 86 It may be a snobbish way of putting it, but a man should know 'the right people' in the great world of books. **1931** S. JAMESON *Richer Dust* xv. 445 She went regularly to their houses, and with them to the right restaurants. **1936** A. CHRISTIE *Murder in Mesopotamia* xx. 178 She's young and she's crude, but she's the right sort. **1946** G. MILLAR *Horned Pigeon* i. 6 All of these officers were 'the right sort'. Which meant that their parents had all had sufficient money to send them to the more expensive schools. **1956** I. BROMIGE *Enchanted Garden* i. iii. 28 The importance of knowing the right people, of being seen with the right people. **1963** L. PETERS *Tarakian* vii. 98 Firth..had been to one of the 'right' schools, and one of the 'right' universities, and.. belonged to one or two of the 'right' clubs. **1973** 'D. HALLIDAY' *Dolly & Starry Bird* xi. 161 He had been to all the right schools and belonged to all the right clubs. **1981** J. CAREY *John Donne* ii. 77 Goodyer knew the right people... He mingled with the dispensers of power and lobbied them on Donne's behalf.

e. *Criminals' slang.* Reliable, trustworthy (from the criminal's point of view); friendly or sympathetic to criminals. Freq. in phr. *right guy*. Also *right croaker*, a doctor who will treat criminals without informing the police, or give prescriptions for drugs.

1856 in G. L. Chesterton *Revelations of Prison Life* I. ix. 137 They [*sc.* the swell mobsmen] frequent those public-houses the landlords of which they know to be what they term right (*i.e.* a thief's friend). **1886** A. PINKERTON *Thirty Years a Detective* 324 You will find him game, a good workman an a ded rite man. **1891** J. BENT *Criminal Life* 272 Will you go and tell Dutch Doll to come up to try and get me right twirl (good warder). **1891** 'F. W. CAREW' *No. 747* i. 4 Warder—who, when accessible to a bribe, is termed a 'right-screw'. **1906** H. GREEN *At Actors' Boarding House* 61 Sammy explained that..the remainder had dwindled rapidly, what with treating the gang and being a right guy generally. **1911** [see *BOOB *sb.* 1]. **1914** JACKSON & HELLYER *Vocab. Criminal Slang* 70 *Right*,..sympathetic in a criminal sense; fixed; squared. **1926** [see *IN AND IN *adv.* b]. **1929** W. R. BURNETT *Little Caesar* i. 8 Some day he'll turn yellow. Hear what I say. He's not right. **1929** HOSTETTER & BEESLEY *It's a Racket!* vii. 93 To him [*sc.* the racketeer], a physician or surgeon is only a 'croaker', a 'right croaker' if he is the sort who will treat a fugitive criminal's wounds or injuries without notifying the police. **1938** J. PHELAN *Lifer* viii. 70 'Friend of yours?' inquired Mansell. 'Sure. Mine and Art's and Bill Weldon and—and anyone that's *right*.' **1951** *Evening Sun* (Baltimore) 27 Mar. 4/1 He [*sc.* a dope addict] may have found he could acquire prescriptions..from a doctor who had his price... (The doctor was a 'right croaker'). **1953** W. BURROUGHS *Junkie* vii. 71 When I told Gains what the hotel clerk said to me and how lucky we were he happened to be a right guy, he said, 'We've got to pack in. We can't last with this crowd.' **1969** *Publ. Amer. Dial. Soc.* LI. 23 We shall use this vocabulary..to indicate how both races equally—but separately—participate in such aspects of a common 'inmate culture' as those of *right guys* and *stoolpigeons*.

10. d. (Earlier example with reference to a person's age.)

1777 P. THICKNESSE *Year's Journey* II. 228 Get it [*sc.* information] from a French officer, or a priest, provided they are on the *wrong* side of forty... Avoid all acquaintance with either, on the right side of thirty.

13. b. (Further examples.) Now chiefly *Austral.* and *N.Z. colloq.* (influenced by *all right*: see sense 15 c below). Also *right as rain* (see sense 15 a below).

1873 C. M. YONGE *Pillars of House* IV. xlvii. 320 With the shout 'As right as a trivet', Charles Audley the younger ..rode on. **1909** BEERBOHM *Yet Again* 229 He looked,..'fit as a fiddle', or 'right as rain'. **1914** W. W. JACOBS *Night Watches* i. 14 'It was done in the collision,' said Mr. Scutts... 'I was as right as rain before then.' **1959** *Listener* 15 Jan. 115/2, I gave him a drink..and he was right. **1960** G. SANDERS *Mem. Professional Cad* i. iii. 35 It had severed some ligaments or what-not that caused him to have a slight limp afterwards, but apart from that he was as right as rain. **1963** A. LUBBOCK *Austral. Roundabout* 48 She poked her head out and cried: 'Are yer right?' I answered that I was far from right. **1944** P. W. COWAN in *Coast to Coast 1943* 166 He gave Mac the lantern. 'Think you'll be all right? Anything you want?' 'I'll be right,' Mac said. **1967** N. FREELING *Strike Out* 30 'Was there anything wrong with the horse?' 'Not a thing... Horse was right as rain.' **1977** J. AIKEN *Last Movement* ii. 48 He will be right as rain,.. don't you worry. **1980** J. MELVILLE *Chrysanthemum Chain* 169 He'll surface as right as ninepence in due course.

e. Colloq. phr. *I'm all right, Jack*: see *JACK *sb.*^{1} 3 c.

f. *to see* (someone) *right*, to look after (someone), to protect (that person's) interests.

1974 D. FRANCIS *Knock Down* xii. 147 Get me ten good two-year-olds and I'll see you right. **1975** *Times* 4 Jan. 12/6 No one's ever come to me before a fight and said 'Do this or do that and we'll see you right.'

14. b. (Further examples.)

1856 C. M. YONGE *Daisy Chain* II. xvi. 515 'I must be away from it all.'.. 'Forgetting yourself,' said Ethel. 'Right. I want to have no leisure to think about myself,' said Norman. **1877** A. J. MUNBY *Diary* 23 Jan. in D. Hudson *Munby* (1972) 386 'I think I should go lengthwise down the boards,' said Massa, gravely. 'Right!' said the servant. **1883** H. SMART *At Fault* xviii. 213 Everybody will come to you; they will say Mr. Marlinson knows all about it. And you can reply, 'Right you are, I do... But..my lips are sealed.' **1912** E. R. BURROUGHS in *All-Story Mag.* Oct. 246/2 'If the mutineers are victorious our one slim hope lies in not having attempted to thwart or antagonize them.' 'Right you are, Alice. We'll keep in the middle of the road.' **1923** D. H. LAWRENCE *Kangaroo* xiv. 310 The same good-humoured, right-you-are approach from everybody to everybody. 'Right-you-are! Right-O!' Somers had been told so many hundreds of times, Right-he-was, Right-O!, that he almost had dropped into the way of it. **1935** 'L. FORD' *Burn Forever* xiii. 120 'I reckon you'd jes' as soon drive him over yonder to th' Crossroads? Hit ain't but a little piece.' 'Right,' Ben said... He opened the car door and turned back. **1970** R. GADNEY *Drawn Blanc* iv. 28 'Have you ever thought of working for the Foreign Service?' 'The British one?' 'Right.'

c. *too right* (*Austral.* and *N.Z. colloq.*), expressing agreement or approval.

1919 W. H. DOWNING *Digger Dial.* 51 *Two eyes right* or *too right*, certainly. **1926** K. S. PRICHARD *Working Bullocks* iv. 41 'That's to say you're as big a fool fruit-growing as Bill Graeme is catching wild horses.' Wally Burke's eyes took the fire-light as he smiled. 'Too right,' he said. **1936** *Punch* 22 Apr. 456/2 For she is crumbling, the poor old boat... 'Take care you don't put the brush through her.' 'Too right,' as the girls of Melbourne used to say. 'Too painfully right.' **1943** D. STEWART *Ned Kelly* I. i. 23 Devine's still in the lock-up?.. Too right he is. **1963** D. BALLANTYNE in C. K. Stead *N.Z. Short Stories* (1966) 162 'Dad, you're not going out at this time of night, are you?' my mother said. 'Too right I am,' Grandad said. **1977** *Zigzag* Aug. 31/1 Something better change—too right mate!

d. Phr. *how right you are* (and variants) used to express strong agreement or affirmation.

1927 H. T. LOWE-PORTER tr. *Mann's Magic Mountain* iv. 183 He saw how right Joachim had been in saying that it was hard to get acquainted here. **1935** D. L. SAYERS *Gaudy Night* viii. 167 'She's awfully kind. But I'm always having to be grateful... It makes me want to bite.' 'How right you are,' said Harriet. **1942** E. WAUGH *Put out More Flags* ii. 98 'I expect they thought that if we had time we should try and stop them coming.' 'How right they were.' **1957** M. KENNEDY *Heroes of Clone* i. 4. 8 'Don't say exclusive when you mean expensive.'.. 'How right you are!' **1967** O. NORTON *Now lying Dead* i. 1 Not that anything they decided could embarrass the police, as the coroner gently pointed out. And how right he was. **1970** G. BUTLER *Coffin from Past* xiv. 147 'I always said I'd move away from here... I never liked the district.'.. 'How right you were,' he said.

e. Appended as an interrogative to the end of a statement, inviting agreement or approval, = *am I right?* orig. *U.S.*

1961 'E. FENWICK' *Friend of Mary Rose* (1962) x. 126, I gave her fifty dollars..fifty bucks, that's not raisins, right? **1968** *New Yorker* 4 May 49 We'll always groove with one another, right? **1969** 'P. KAVANAGH' *Such Men are Dangerous* i. 27 He was just doing his job, right? **1971** *Melody Maker* 13 Nov. 36/6 You're getting paid, right? Let's have a little co-operation around here. **1977** *Time* 26 Sept. 43/2, I was living alone on the West Side, in a one-room apartment with the bathroom out in the hallway and the bathtub in the kitchen, right? **1981** G. V. HIGGINS *Rat on Fire* xxvi. 159 You been here before, right? You can probably read the sign.

15. a. Colloq. phr. *right as rain* (see also sense 13 b above).

1894 W. RAYMOND *Love & Quiet Life* x. 108 ''Tes so right as rain, Zir,' zes I. **1894** SOMERVILLE & 'ROSS' *Real Charlotte* II. xx. 78 If only this infernal Fitzpatrick girl would have stayed with her cads in Dublin everything would have been as right as rain. **1908** A. S. M. HUTCHINSON *Once aboard Lugger* VI. vi. 332 We'll pull through right as rain. **1929** W. P. RIDGE *Affect. Regards* 110 If your husband hadn't put a spoke in, it would have been as right as rain. **1930** G. B. SHAW *Apple Cart* i. 15 *Proteus.* How did you get on with the King? *Boanerges.* Right as rain, Joe. You leave the king to me. **1973** *Times* 15 Dec. 7/5, I wished to renew my membership in the club. He licked his lips with relish as he lifted my familiar card from his file cabinet. 'Right as rain, sir.'

c. (Further examples.) Also in predicative use, in the sense 'satisfactory, acceptable', and as *adj. phr.* (see *ALL RIGHT 2). *all right by* or *with* (someone): acceptable to (that person).

1898 G. B. SHAW *Candida* II. 124 It will be all right: he will forgive me. **1914** —— *Fanny's First Play* III. 198 Is your aunt all right? **1939** R. CHANDLER *Let.* 19 Feb. (1966) 195 *The Big Sleep* is very unequally written. There are scenes that are all right. **1950** W. HAMMOND *Cricketers' School* i. 20 That was all right by Jim! **1958** *Listener* 20 Nov. 835/1 If ever he wants to use what I'm going to say, it's perfectly all right with me. **1959** *Ibid.* 22 Jan. 153/1 Joe said: 'That's all right by me.' **1962** L. DEIGHTON *Ipcress File* iii. 25 'You want a meal?' 'Yes,' I said quickly. 'That's all right, then, sit down.'

d. *she's* (or *she'll be*) *right*: all is well, that is fine. *Austral.* and *N.Z. colloq.*

1947 D. M. DAVIN *For Rest of our Lives* xvi. 81 She'll be right as soon as we get back with the boys again. **1950** *N.Z. Listener* 3 Mar. 12 'She's right!' Miss Cooper said, with a good Pig Islander's inflexion. **1958** *Ibid.* 9 May 6/4 'They'll be very upset to think that they couldn't have met you gentlemen here, but I'll convey your regards to them, and thank you very much.' Ike said, 'She's right, ma'am, she's right.' **1959** *Times* 27 June 7/7 He calls in the appropriate tradesman, who fixes the thing so that it works again, but not very well. He protests and is told: 'She'll be right.' **1961** B. CRUMP *Hang on a Minute* 33 Thanks for the beer, added Jack. She's right, said the barman. Hope you get the job. **1974** A. BUZO *Coralie Lansdowne says No* 65 'There's more in the car. I'll go and get it.' 'Do you want a wheelbarrow?' 'She'll be right.'

III. 17. e. As an intensifying word in derogatory and ironical contexts. Phr. *a right one*: a fool; an extremely stupid or awkward person.

1960 *Daily Tel.* 21 Jan. 15/7 Two young soldiers of the R.A.S.C. were described as 'right mugs' in written plans that apparently referred..to Pte. John Terence Bush's attempted escape. **1962** L. DEIGHTON *Ipcress File* xxx. 190 We'd been a couple of right ninnies. Followed all the way! **1965** *Oxford Mail* 17 Nov. 11/5 Mr. Weir said that when student teacher Ian Brodie first broached his dream of a monster sandwich he thought: 'I've got a right one 'ere.' **1968** B. TURNER *Sex Trap* xiv. 131 'I got a right one today,' she said. 'Thought I'd had them all, you know, but this one wanted me to hold his John Thomas.' **1973** *Observer* 14 Jan. 1/4 'The Government did not know that there was no settlement in writing, and how could an order apply to something which did not exist,' he said. 'The Government made a right mess of it.' **1976** *Southern Even. Echo* (Southampton) 13 Nov. 9/4 He said the Conservative party would inherit a 'right old mess'. **1978** I. MURDOCH *Sea* 104 'You look a right clown,' I said.

18. b. *Math.* Used to denote an entity whose definition involves a pair of elements in a conventionally defined order (see quots.).

[**1905** *Trans. Amer. Math. Soc.* VI. 203 So that 6× and the right-hand distributive law fail. **1933** *Ann. Math.* XXXIV. 483 When $F(x) = D_1(x)D_2(x)$ we shall call $D_2(x)$ a right-hand and $D_1(x)$ a left-hand divisor of $F(x)$.] **1937** [see *IDEAL *a.* and *sb.* B. 3]. **1938** F. D. MURNAGHAN *Theory of Group Representations* iv. 91 The set Hs^{-1} (which consists of the inverses of the elements of the left coset sH) has no element in common with H. We term it the right coset of H determined by s^{-1}. **1962** CURTIS & REINER *Representation Theory Finite Groups* ii. 50 The group M is called a right R-module if there is a product mr defined such that $mr \in M$ and $(m_1+m_2)r = m_1r+m_2r$, $m(r_1+r_2) = mr_1+mr_2$, $m(r_1r_2) = (mr_1)r_2$, $m.1 = m$, for all $r \in R$, $m \in M$... The right regular module R_R is defined similarly, and its submodules are the right ideals of R. *Ibid.* viii. 395 Let S be a subset of A... The right annihilator $r(S)$ is given by $r(S) = \{a \in A: Sa = 0\}$. **1971** E. C. DADE in Powell & Higman *Finite Simple Groups* viii. 280 We regard the R-endomorphisms as right operators on M. **1972** A. G. HOWSON *Handbk. Terms Algebra & Anal.* v. 28 Similarly, we can form the set of all elements of the form zx, where $z \in S$ and, denoting this by Sx, refer to it as a right coset of S. *Ibid.*, There is no universal convention regarding which set of cosets should be termed 'right' and which 'left'.

21. Special collocations in sport: *right back, centre, corner, end, forward, guard, half.* Also *right field* (Baseball): the part of the outfield to the right of the batter as he faces the pitcher; also, a fielder in this position; *right fielder*: a fielder in the right field. See also *RIGHT WING 2, *RIGHT-WINGER (*b*).

1897 *Encycl. Sport* I. 419/2 The right back and the right half-back look after the opposing left wing forwards. **1956** *Granta* 18 Feb. 5/1, I had a very easy game at right back. **1951** *Sport* 6–12 Apr. 18/4 It was only justice when their right-centre Hunt opened their account with a try. **1969** EAGLESON & MCKIE *Terminol. Austral. Nat. Football* III. 9 *Right centre*, the player occupying the right centre position in a team. **1955** R. SMITH *Doyle's Lifetime in Hurling* iv. 34 Harney roving down to the right corner angled a grand point. **1975** *Irish Independent* 27 May 12/1 Kilkenny will be without..right corner back Fan Larkin and right corner forward Mick Brennan in next Sunday's Wembley [hurling] tournament. **1896** CAMP & DELAND *Football* 344 Instructions to Right End... You should help the right tackle block his man. **1970** *Washington Post* 30 Sept. D3/5 Brown, the 6-foot-5, 265-pound defensive right end, charged through Baltimore's offensive line. **1974** *Cleveland* (Ohio) *Plain Dealer* 13 Oct. C12/7 The Jackets locked up the game with 10:09 remaining as quarterback Jim Tressel jaunted around right end for a 16-yard touchdown. **1857** *Spirit of Times* 29 Aug. 404/3 Enterprise Club. Maxfield, catcher;.. Davis, right field; Knight, second base. **1867** [see *left field* s.v. *LEFT *a.* 3 c]. **1949** *Marshfield* (Wisconsin) *News-Herald* 19 July 9/4 Corbett hit a change of pace pitch on a line into right field for the second Tomahawk hit. **1970** *Globe & Mail* (Toronto) 25 Sept. 31/3 With one out, Bob Robertson lined a double down the rightfield Line. **1974** *Evening Herald* (Rock Hill, S. Carolina) 18 Apr. 6/3 Mike Williams then unloaded a triple down the rightfield line... Roberson and Williams came around to score on a Mussman balk. **1867** *Ball Players' Chron.* 25 July 1/4 The right fielder was active at times, once making a very handsome stop. **1889** 'MARK TWAIN' *Connecticut Yankee* xiii. 533 What a handy right-fielder he was! **1912** C. MATHEWSON *Pitching in a Pinch* 27 Kane, the little rightfielder on the Cincinnati club, was the first man up. **1957** [see *left fielder* s.v. *LEFT *a.* 3 c]. **1977** *Time* 12 Dec. 57/1 In 1927, with Leftfielder Meusel hitting .337, Centerfielder Earle Combs .356 and Rightfielder Babe Ruth .356 plus 60 home runs, the Yankees had what many students of the game consider the best outfield in baseball history. **1911** P. H. DAVIS *Football* xviii. 446 *Yale*... Right Guard, R. C. Tripp, '06. **1970** *Washington Post* 30 Sept. D1/3 Right guard Vince Promuto was trying to loosen up a bruised knee and rookie Paul Laaveg of Iowa filled in for him. **1905** P. WALKER *How to play Assoc. Football* 36 In this and following diagrams.. L.B., R.H...represent respectively..left-back, right-half. **1928** in B. James *England v Scotland* (1969) vi. 134 Right-half Edwards worked hard and at times effectively. **1947** *Sporting Mirror* 7 Nov. 9/1 Tunstall, United forward, raced after a long pass from his right half. **1951** *Sport* 7–13 Jan. 13/4 Perseverance has repaid Billy Stroud, right-half-back of Newport County. **1969** EAGLESON & MCKIE *Terminol. Austral. Nat. Football* III. 10 *Right half forward*, the player occupying the right half-forward position in a team. **1975** *Irish Times* 24 May 3/1 Their right-half forward, Francis Loughnane, a man who..has won many games for the county.

IV. 22. a. Conservative, reactionary; applied *spec.* to (members or supporters of) that part of a political party or grouping especially noted for its conservatism (see RIGHT *sb.* 17 d in Dict. and Suppl.); *right of centre* [CENTRE *sb.* 15]: tending to hold conservative political views.

The use originates with the seating positions of the 1789 French National Assembly: see quot. 1837 and the note in Dict. s.v. CENTRE *sb.* 15.

1794 tr. *C. Desmoulins's Hist. Brissotins* 40 La Source, the least corrupted of those who voted with the left, and dined with the right side of the Convention, but whose pride was excited against Robespierre. **1829** *Ann. Reg. 1828* 162/2 The result of this election proved that..the union—scarcely a natural union—of a fraction of the extreme right or royalist side, with the whole of the left, or liberal side, which, at the general election had driven him from office, was still continued to keep him out. **1837** CARLYLE *French Revol.* I. vi. 307 There is a Right Side (Coté Droit), a Left Side (Coté Gauche); sitting on M. le President's right hand, or on his left: the Coté Droit conservative; the Coté Gauche destructive. **1933** *Labour Monthly* July 419 In order to defend 'democracy' it is necessary to maintain a Left Cartel Government in office so as to defeat the menace of a Right Bloc Government. **1938** D. M. PICKLES *French Polit. Scene* iii. 60 A group labelled 'Democratic' is invariably Right, as (with one exception) are all groups labelled 'Republican'. **1958** P. JOHNSON in N. Mackenzie *Conviction* 207, I had a job on..a glossy Paris magazine. It was vaguely right-of-centre, superficially progressive. **1962** *Listener* 18 Jan. 115/1 The party has in fact swung round so fast that some worthy stalwarts who think of themselves as right-of-centre are still surprised suddenly to find themselves stranded way out on the left of the official party line. **1964** GOULD & KOLB *Dict. Social Sci.* 383/2 The term *right* is most commonly (though not exclusively) used as a pejorative term, by those who believe themselves to be on the left. **1973** R. E. M. IRVING *Christian Democracy in France* viii. 260 Although boosted by extreme right votes, the CD did even worse than the Centre in 1962. **1974** T. ALLBEURY *Snowball* xiii. 69 A right-of-centre Trades Union delegate and a Midlands MP passed notes. **1976** *National Observer* (U.S.) 28 Aug. 13/2 They survive by being cleaner than clean, righter than right, and purging everybody else except their own carefully selected proteges. **1977** *Times* 18 Feb. 12/4 Constituency pressure against a number of right-of-centre MPs..has undermined the party's position in the country. **1978** *Times* 26 July 15/4 We were dazzled..whether we were politically right, left or centre.

b. *right deviationism*: in a Communist party or society, (advocacy of) departure or divergence from orthodox principles or policies towards more conservative principles. So *right deviationist*, an advocate or proponent of such principles.

1930 [see *DEVIATION 3 e]. **1957** R. N. C. HUNT *Guide to Communist Jargon* xxix. 103 The Party and its leaders..will be accused of that form of opportunism which goes by the name of right-deviationism. **1958** G. M. CARTER *Politics of Inequality* ii. 61 In September, 1931, Bunting, W. H. Andrews, Solly Sachs and other prominent Communists were expelled from the Party for 'right deviationism'. **1958** P. KEMP *No Colours or Crest* vii. 141 This attitude branded him as a Right deviationist in the eyes of his fellow Communists.

V. 23. *Comb.* **a.** Parasynthetic, as *right-eyed, -footed, -twisted.* **b.** *right-counter, -cross*; **right bower,** the jack of trumps in Euchre; also *fig.*; **right-to-left** *attrib. phr.*, designating movement from the right to the left.

1839 *Spirit of Times* 24 Aug. 294/2 The *right* and *left bower* in the game of *Euchre.* **1863** in *Ann. Army of Cumberland* 501 Smith..said he would..furnish him with a letter of introduction to his 'right bower' in Nashville. The right bower proved to be Dr. Hudson. **1872** G. P. BURNHAM *Mem. U.S. Secret Service* p. vii, *Right bower*, the second-best card in *euchre*; next to the white 'joker'. **1946** MOREHEAD & MOTT-SMITH *Penguin Hoyle* 29 The second-highest is the jack of the trump suit, called the *right bower*. *a* **1964** H. P. TRITTON in *Penguin Bk. Austral. Ballads* (1964) 228 Gently turn the sheep around so the right bower couldn't see. **1950** J. DEMPSEY *Championship Fighting* 10 He landed a right-counter to the head. **1965** G. MCINNES *Road to Gundagai* iii. 52, I..followed it up with a..right-cross with the school cap. **1904** *Science* 8 Apr. 592/1 If right-eyed, the engineer can see the track and signals better from the right side of the boiler than from the left. *Ibid.* 593/2 It may be better for the oculist to leave a person right-eyed rather than to give such lenses as suddenly compel left-eyedness. **1932** *Amer. Jrnl. Ophthalm.* XV. 321/1 Most right handed people are right eyed. **1934** *Times Lit. Suppl.* 4 Jan. 4/2 All the animals that set themselves to grip their prey, are right-footed. **1960** V. JENKINS *Lions Down Under* 174 He came up into an orthodox right-to-left threequarter movement. **1964** E. A. NIDA *Toward Sci. Transl.* vi. 136 In many instances a very long sentence, even though having right-to-left attribution and with several potential terminal points, must be broken up into smaller segments to avoid formal overloading of the communication. **1969** *Listener* 24 July 119/3 Clark acknowledges the assistance of Carlo Pedretti, with his unrivalled knowledge of Leonardo's hand (especially of that maddening right-to-left script). **1934** *Times Lit. Suppl.* 4 Jan. 4/2 A right-twisted shell lies conveniently for the waves breaking upon it so that they may press it more closely to the rock. **1934** 'H. MACDIARMID' *Stony Limits* 120 A right-twisted shell lies apt for the waves..to press it more closely to the rock.

right, *adv.* Add: **2. a.** (Further examples of *right along.*) Also in sense 'all along' (chiefly *U.S.*).

1880 'MARK TWAIN' *Tramp Abroad* xlii. 497 We tore right along, over rocks, rubbish, gullies, open fields. **1965** *Listener* 21 Oct. 608/1 Is the American Government prepared to take up this kind of initiative..? *Harriman.* Yes, we have been doing that right along.

3. b. (*b*) (Further examples of *right off*; also used *absol.*)

1931 [see *LEVEL *v.*^{1} 8]. **1952** M. LOWRY *Let.* 11 Jan. (1967) 285 Also, right off, I want to thank yourself..most sincerely for the superlative Christmas present. **1976** *National Observer* (U.S.) 14 Aug. 8/1 Right off, I'm overwhelmed by the vegetables inside: mounds of crimson, vine-ripened tomatoes [etc.].

(*c*) (Further examples of *right away.*)

1903 G. B. SHAW *Man & Superman* IV. 151 *Violet.* Go away until I have finished speaking to your father. *Hector.* No, Violet: I mean to have this thing out, right away. **1909** R. E. KNOWLES *Attic Guest* ix. 121, I told him one day how Charlie was still urging me to consent that it should be soon, right away soon. **1942** *Short Guide Gt. Brit.* (U.S. War Dept.) 5 You will find out right away that England is a small country. **1957** *Practical Wireless* XXXIII. 727/2 (Advt.), Post this off right away. **1978** D. QUINN *Fear of God* 121 'It's very important—see what you can turn up.' 'Right away,' Helen said.

c. For *U.S.* read orig. *U.S.* and add further examples.

1955 'J. CHRISTOPHER' *Year of Comet* iii. 108 Find yourself a drink. I'll be right up. **1969** R. WILLIAMS in D. Knight *100 Yrs. Sci. Fiction* 286 A few minutes later, however, Pete Martens in Appliances called me up... 'Thank you, Pete,' I said. 'I'll be right down.' **1972** 'T. COE' *Don't lie to Me* vii. 70 'We're all set, Dan.' He looked at her in surprise, as though he'd never seen her before, and then said, 'Fine, be right with you.'

6. a. *right now, then* (further examples). Now also in sense 'immediately', 'without delay' (chiefly *U.S.*).

1932 KIPLING *Limits & Renewals* 84 Mackworth: 'Swear you will. At once.' Haylock: 'I swear I will. Right now.' Me (and it's not my fault that I love English): 'None of your Transatlantic slang here. Say "at once".' **1948** 'N. SHUTE' *No Highway* iv. 82 He's not the only passenger that's in a nervous state right now. **1953** *Manch. Guardian Weekly* 19 Mar. 7 There never was a time when respect for law and legal process was more important to our people than right now. **1957** *New Yorker* 2 Nov. 68/3 Explaining that she had to go back to the hotel herself right then to get dressed.. she returned to the car. **1959** C. FREMLIN *Uncle Paul* iii. 30 *I'll* find her a hotel. Right now. **1959** *Listener* 19 Nov. 895/1, I hardly think readers will be reaching it off the shelves a quarter of a century hence as eagerly as, right now, I am reaching for *The Small Years.* **1975** R. L. SIMON *Wild Turkey* (1976) i. 2 We'd better start right now.

7. a. *right on!* Used as an expression of enthusiastic agreement, approval, or encouragement. Also as *attrib. phr.* *U.S. slang.*

Freq. in Black English in the U.S.

1925 in Odum & Johnson *Negro & his Songs* vii. 202 Railroad Bill was a mighty sport, Shot all buttons off high sheriff's coat Den hollered, 'Right on, Desperado Bill!' **1968** B. SEALE in *Ramparts* 20 Oct. 32/1, I said, 'Right on.' I was right behind him. This brother..knew what to do, when to do and how to do it. **1970** *New Yorker* 20 June 50/2 Miss Williams said, 'You have not adequately answered those questions,' and sat down, to cheers, whistles, and a cry of 'Right on!'—conceivably the first time the chairman had ever heard the phrase live. **1970** *Melody Maker* 12 Sept. 34 'Only in a capitalist society could art be turned to profit. 'Right on.' **1970** *Time* 19 Oct. 45 In Boston, Homans is known as a 'right-on lawyer'—he defends blacks, war protesters and poor people. **1971** J. KILLENS in A. Chapman *New Black Voices* (1972) 59 But now, on with the story. And Black Blessings to you all. Right on. **1972** *Listener* 27 Jan. 123/2 A whole albumful of 'right on' soul music. **1973** *Black World* May 32/2 If Marx were alive he could see his way clear to say to this observation, 'Right on, Brother!' **1974** K. MILLETT *Flying* II. 198 Right on, Vita, so you must have waged your woman's war for years. **1976** *Spare Rib* Oct. 32/1, I had just read *Sappho Was a Right-On Woman* by Sidney Abbott and Barbara Love. **1979** *Daily Tel.* 3 Sept. 4/1 A correspondent remarked: 'You don't portray any crisis feeling.' The President replied: 'Right on.'

b. (Later example of *right there.*)

1948 AUDEN *Age of Anxiety* v. 116 He'll be right there With His Eye upon me.

9. a. (Further examples.)

1891 W. B. YEATS *Let.* Apr. (1954) 167 The 'proofs' of the Blake book are coming in.... The illustrations look right well. **1954** *Manch. Guardian Weekly* 5 Aug. 12/3 Sometimes ..there is a fight. If the incident takes place in a public-house it is often worse. 'It's broken bottles right fast; and, brother, I get out.' **1981** P. MALLORY *Killing Matter* ix. 97 Cale was doing right well for himself.

b. (Further examples.) Now chiefly *U.S.* *right smart*: see SMART *a.* 7 b in Dict. and Suppl.

1869 'MARK TWAIN' *Innoc. Abr.* 134, I did not feel right comfortable for some time afterward. **1881** —— *Prince & Pauper* iii. 40 They were waved aside with a right royal gesture. **1884** C. H. SMITH *Bill Arp's Scrap Bk.* vi. 75 Go get me a switch, right straight. **1936** M. MITCHELL *Gone with Wind* xlii. 755 Miz Wilkes is right sensible, for a woman.

1952 *Manch. Guardian Weekly* 20 Mar. 4/3 Yes, sir, as far as this state's concerned, he looks right nice where he is. 1977 *Washington Post* 7 Sept. c5/2 It turned out that some of them were women who would like to be in my place—who hate me and would like to see me gone. I slipped into a right unpopular place when I married George Wallace.

13. c. (Further examples of *all right.*) Now usu. in weakened senses, 'indeed', 'certainly'.

1911 *Maclean's Mag.* Mar. 96/2 He's a high flier, all right. 1915 J. Buchan *39 Steps* iv. 81 It was no question of preventing a war. That was coming, as sure as Christmas... Karolides was going to be the occasion. He was booked all right, and was to hand in his checks on June 14th. 1930 J. B. Priestley *Angel Pavement* v. 249 Yes, I saw you all right. You looked very annoyed, too. 1933 M. Lowry *Ultramarine* vi. 245 Boy, but it was hot all right. 1943 *Illustr. London News* 9 Jan. 30/1 Coffee was being burnt in Brazil because there were not enough purchasers for it in Europe. But this was solely because the people of Europe lacked the currency with which to buy it. They wanted the coffee all right. 1944 B. Hutchison *Hollow Men* viii. 107 I'm afraid, all right.

15. b. *right, left, and centre*: everywhere; in all directions.

1956 H. & M. Williams *Plaintiff in Pretty Hat* ii. 61 Pontificating..and expressing your damned opinions and judgements right, left and centre as if you're Solomon in all his glory? 1958 *Times* 19 Sept. 6/7 Mr. Eric Longsworth.. suggested that the duty of the committee should be to take action whenever a theatre was threatened, lobbying 'right, left, and centre' local and national bodies. 1963 [see *bundobust]. 1967 'G. North' *Sgt. Cluff & Day of Reckoning* x. 91 Do any of us want Barker asking folk questions, right, left, and centre? 1970 J. Porter *Rather Common sort of Crime* x. 115 A boy like Rodney..couldn't possibly be earning enough money to go around buying whisky right, left and centre? 1977 *Times* 25 Jan. 5/7 People are doing it [sc. committing suicide] right, left and centre all the time. There is nothing to it nowadays.

16. a. *right-hearted, -justified* (sense 15).

1848 J. R. Lowell *Fable for Critics* 41 All honour and praise to the right-hearted bard Who was true to The Voice when such service was hard. 1908 M. & J. Findlater *Crossriggs* xlvii. 346 Alex gave it the reverent attention that details of the kind will always command from right-hearted people. 1970 *Computers & Humanities* IV. 167 References were to be right-justified.

b. *right-acting, -feeling* (examples), *-judging, -seeing*; **right-reading**, such as can be read without first being reversed by a mirror.

1973 E. Bullins *Theme is Blackness* 160 If I was a right-actin' sister I wouldn't go into this, you understand..but you know I ain't right-actin'. 1869 Mill *Subj. Women* iv. 149 Among right-feeling and well-bred people, the inequality is kept as much as possible out of sight. 1883 'Mark Twain' *Life on Mississippi* 432 Any right-feeling reptile would do that. 1868 Mill *Let.* 9 June (1910) II. 113 The detestation of the right-judging among his fellow-creatures. 1955 J. Ashworth *Operation & Mech. of Linotype & Intertype* II. xxxiii. 329 The film magazine is a holder containing the film and a mechanism for advancing it between lines... Production models will be able to produce either wrong- or right-reading positives as desired. 1967 Karch & Buber *Offset Processes* iv. 87 Images can be right reading or wrong (as seen in a mirror) reading to suit the process used in offset-lithography. 1855 Bagehot *Coll. Wks.* (1965) I. 284 A particularly rational and right-seeing man.

c. Delete † and add: *right-justify* [back-formation from *right-justified* in sense *16 a].

1970 *Computers & Humanities* IV. 169 The point at which the line is to be padded with blanks to right-justify the reference. 1980 Glass & 'De Nim' *Second Coming* 58 He could right justify and multiple column per page and footnote with only a little more work.

right about, *sb.*, *adv.* (and *a.*). Add: **A.** *sb.* **3.** A change of mind; the adoption of an opinion or policy contrary to that previously held.

1936 O. Nash in *Bad Parents' Garden of Verse* 59 You ask, and properly ask, no doubt, Whence this astonishing right-about? 1956 *Sun* (Baltimore) 28 Sept. 12/2 The resolution completes the policy rightabout which the American comrades..began after Khruschev denounced Stalin and Stalinism.

right and left, right-and-left, *adv.* (*v.*), *a.*, and *sb.* Add: **A.** *adv.* **1. b.** (Earlier and later examples.) Also, from every quarter; indiscriminately.

1839 F. Hastings *Let.* 8 Mar. in R. B. Martin *Enter Rumour* 51, I wished you to know the truth..and you are welcome to tell it right and left. 1863 Mrs. Gaskell in *All Year Round* VIII. 560/1 Mr. Wilkins has written everywhere, right and left. 1873 'Mark Twain' *Gilded Age* 319 You still begin to squander a fortune right and left. 1876 C. Schreiber *Jrnl.* 26 July (1911) I. 472 Froeschel, of Hamburg, was there buying right and left. 1884 G. Meredith *Let.* 13 May (1970) II. 735, I am now rejecting kind invitations right and left, for I see I shall not be released before the first or second week of next month. 1912 T. Dreiser *Financier* xxx. 341 The stocks had been unloaded freely right and left. 1927 J. N. McIlwraith *Kinsmen at War* xix. 194 Stephen struggled and slashed right and left. 1962 P. Gregory *Like Tigress at Bay* xiv. 142 Rudy was there now, smirking, bowing to the customers, giving orders right and left.

C. *sb.* **2.** Also *transf.* and *fig.*

1935 A. P. Herbert *What a Word!* vii. 221 This is the best right-and-left I have found. Issue is a 'conclusion' in one sentence and 'question' in the next. 1978 R. V. Jones *Most Secret War* xli. 390, I felt that we had achieved one of the best 'right and lefts' of all time by picking out the 14th and 15th Companies.

right angle. Add: *right-angle fold* (Printing), a fold at right angles to a previous fold; so *right-angle folding*.

1931 H. Jahn *Hand Composition* xvi. 264 Models..which operate at high speed on both parallel and right-angle fold work. 1946 A. Monkman in H. Whetton *Pract. Printing & Binding* v. 59 Machines are available to fold large and small sheets and for parallel or right-angle folding. 1956 H. Williamson *Methods of Book Design* xix. 312 Some machines are equipped with turn-tables, so that..successive parallel folds can be converted at will into right-angle folds. 1972 A. G. Martin *Finishing Processes in Printing* ii. 19 When a number of single sheets are collectively folded once, the technique is known as 'lump folding'. This method is used for make-up of account-book sections and no right-angle folds are employed.

right-branching, *a. Linguistics.* [Right *adv.*] Of a grammatical construction, etc.: having the majority of its constituents on the right of its tree diagram. Also as *vbl. sb.*

1961 N. Chomsky in *Proc. Symposia Appl. Math.* XII. 13 From the fact that memory is finite, the fact that a sentence is heard or spoken from 'left-to-right'.., or the fact that the rules of a generative grammar..may be partially ordered, nothing whatsoever can be concluded about left- and right-branching in P-markers. *Ibid.* 14 We would predict that a hearer..would tend to treat right-branching units..as successive and disjointed, rather than integrated segments, on first hearing. 1965 —— *Aspects of Theory of Syntax* i. 13 Right-branching structures are those with the opposite property [to left-branching], for example,..[this is [the cat that caught [the rat that stole the cheese]]]. 1965 [see *left-branching *a.*]. 1978 *Language* LIV. 84 In English, sentential complement constructions are right-branching.

righteous, *a.*, *adv.*, and *sb.* Add: **2*.** *U.S. slang.* Fine, excellent; of good quality. Freq. in collocations (see quots.), esp. *righteous moss* [see *moss *sb.*[1] 5 e], hair of good texture or characteristic of a white person.

Freq. in Black English in the U.S.

1942 Z. N. Hurston in *Amer. Mercury* July. 84 It [sc. his hair] looked just like that righteous moss, and had so many waves you got seasick from looking. 1944 C. Calloway *Hepsters Dict.*, *Righteous*, splendid, okay. Ex., 'That was a righteous queen I dug you with last black.' 1944 D. Burley in A. Dundes *Mother Wit* (1973) 212 *Righteous*, pleasing to the senses, glorious, pretty, beautiful, mighty. 1945 L. Shelly *Jive Talk Dict.* 31/2 *Righteous riff*, interesting conversation. 1946 Mezzrow & Wolfe *Really Blues* (1957) 377 *Righteous bush*, marihuana. 1963 T. Pynchon *V.* x. 281 'She's up there' Matilda said, with a smile for everybody, even musicians with a headful of righteous moss, who were making money and drove sports cars. 1967 [see *lame *sb.*[2] 2]. 1968–70 *Current Slang* (Univ. S. Dakota) III–IV. 103 *Righteous egg, n.* a good guy. 1970 C. Major *Dict. Afro-Amer. Slang* 97 *Righteous Moss*, white folks' hair. 1976 *Daily Mirror* 2 Apr. 20/2 *Righteous*, can mean same as dread; also good or excellent.

righter[2]. Add: Also in *equal righter*, an advocate of equal rights for women.

1896 *Truth* 22 Oct. 1029/2 Some ladies—evidently Equal Righters—reproach me with having last week devoted far more space to the Czar than to his consort.

right hand. Add: **4.** *right-hand drive*, a steering system in a motor vehicle in which the steering wheel and other controls are fitted on the right side instead of on the left. Also *attrib.* So *right-hand driving* vbl. sb.

1912 *Horseless Age* 21 Feb. 375/2 In right hand drive cars it [sc. the steering column] is located directly in front of the right hand front passenger. 1936 M. Kennedy *Together & Apart* iii. 178 'But don't you,' she asked, 'find the right-hand driving rather confusing?' 'I don't mind it. But in Sweden it's left-hand.' 1940 R. Chandler *Farewell, my Lovely* viii. 55 The car had a right-hand drive. 1956 [see *left hand 3]. 1965 *Listener* 10 June 878/2 The two latest French saloons, the Renault 16 and the Peugeot 204, did not appear at Silverstone last Wednesday because right-hand drive versions are not yet available. 1969 J. Leasor *They don't make them like that any More* vi. 185, I thought that all Mercedes exported to England before the war had right-hand drive? 1974 A. Price *Other Paths to Glory* ii. viii. 208 Got 'GB' in the back..the Jag had—that's English. An' right-hand drive—that's English too.

right-handed, *a.* (*adv.*) **6.** Delete *rare* and add further examples as *adv.*

1910 *Blackw. Mag.* July 24/1 A single stag has swung right-handed. 1929 *Morning Post* 30 Dec. 13/1 He [sc. a fox] ..left at the Toft end going straight..over the railway, leaving Kingston village right-handed. 1936 P. Fleming *News from Tartary* 194 We turned right-handed and slanted down to the foothills. 1976 *Horse & Hound* 3 Dec. 22/2 He ran right-handed over Clawson Lodge, Clawson Hill Farm.. where he tried the earths.

right-hander. Add: **2.** (Further examples.)

1968 *Listener* 19 Sept. 357/1 If to right-handers this heart-cry sounds wildly exaggerated, it can only be because they are unaware of the inconvenience..they perpetually inflict on those of us who live the half-life of the submerged left. 1972 *Sci. Amer.* Apr. 83/3 It is an interesting fact that right-handers with a strong family history of left-handedness show better speech recovery than people without left-handed inheritance. 1978 *Amer. Speech* LIII. 285 Stone tools surviving from that epoch were mainly chipped by right-handers.

3. A bend to the right in a road, track, etc.

1970 J. Miles *Expert Driving Police Way* xv. 139 On a right-hand bend I usually do not dip so early, as the lights are pointing away from oncoming traffic, but on certain right-handers I find it an advantage to have my lights dipped. 1973 'J. Ashford' *Double Run* xiv. 113 The first right-hander with slightly banked curve was dead ahead.

right-ho, var. *righto *int.*

rightie, var. *righty. **rightie-ho,** var. *righty-ho *int.*

rightism (rəi·tiz'm). Also with capital initial. [f. Right *sb.*[1] 17 d + -ism.] The political view or principles of 'the right'.

1939 H. G. Wells *Holy Terror* ii. i. 101 The formulae of the gangsters varied from extreme leftism to extreme rightism. *Ibid.* iii. ii. 270 Two great systems of ideas faced each other in the world, 'Leftism' and 'Rightism'. 1940 W. Stevens *Let.* 12 Jan. (1967) 351 There are a lot of things that the workers are doing that I do not believe in... I think this explains my rightism. 1951 Koestler *Age of Longing* i. 27, I take it for granted that one must fight in self-defence... I merely wish to point out that this has nothing to do with..leftism, rightism, capitalism, and socialism or any other idea or ism. 1959 B. & R. North tr. *M. Duverger's Pol. Parties* (ed. 2) ii. ii. 346 The 'rightism' of governmental alliances is consequently opposed to the leftism of public opinion. 1960 *Times Lit. Suppl.* 22 Apr. 260/3 The anti-rightism which Mr. Clark saw was the tail-end of the campaign against the intellectuals begun in 1957. 1961 *Spectator* 20 Oct. 535/2 Some comparisons and conclusions may now be drawn as a basis for gauging the strength and prospects of American rightism. 1966 *Punch* 28 Dec. 971/1 He has set down the worst excesses of radical rightism of the past two hundred years. 1974 *Time Out* 27 Sept. 23/1 Walker's own naivete, not content with equating criticism of the Kremlin line with rightism, extends to a total misunderstanding of political activity.

rightist (rəi·tist), *sb.* and *a.* Also with capital initial. [f. as prec. + -ist.] **A.** *sb.* A member or adherent of 'the right' in politics.

1937 *Times* 22 Sept. 8/2 The assassinations of the past six years [in Japan] were due to the discontent of younger officers and Rightists with depression or threats of depression. 1945 C. Burney *Dungeon Democracy* iii. 70 He was, in Austrian politics, known to be a Rightist and a Catholic. 1958 *Times Lit. Suppl.* 26 Sept. 546/4 The Chinese intellectual of to-day, pilloried as a 'bourgeois rightist' might well feel that the old cycle is at work once again. 1959 *Listener* 10 Sept. 379/1 There have been a good many people, so-called rightists, who thought that the authorities were trying to drive the country [sc. China] rather too fast. 1960 G. Mikes *How to be Inimitable* 53 The Labour Party has a few real leftists and the Tories a few real rightists (and vice versa). 1965 *New Statesman* 2 July 18/2 You are told that President Clark Kerr has created the myth that there are people worse than himself—rightists baying in the legislature. 1970 [see *law *sb.*[1] 3 a]. 1974 J. Mitchell *Death & Bright Water* v. 45 Greek fighting Greek, Right against Left... By that time Michael was a rightist too.

B. *adj.* Of, pertaining to, or characteristic of 'the right' in politics; tending to conservatism.

1938 *Time* 30 May 34/3 By daybreak enraged Rightist students found their university had been taken over by the United Socialist Youth of Mexico. 1940 C. S. Lewis *Let.* 17 Jan. (1966) 176 You will presently see both a Leftist and a Rightist pseudo-theology developing. 1957 *Economist* 28 Sept. 1032/1 In Hongkong the continuing reiteration of official reports from the Chinese Communist press of unrest, crime, 'rightist' plotting, food black-marketing and counter-revolutionary activity..gives an impression of smouldering crisis or impending explosion. 1959 B. & R. North tr. *M. Duverger's Pol. Parties* (ed. 2) ii. ii. 347 Within the Third Force the influence of the M.R.P. continually declined between 1947 and 1951, to the advantage of the Radicals and the Right wing, in accordance with the 'rightist' trend of alliances. 1966 *Listener* 3 Nov. 660/3 He identified himself wholeheartedly with the 'moderate' and 'rightist' aspects of Stalinism. 1969 F. Halliday in Cockburn & Blackburn *Student Power* 309 It was decided..to reactivate an old system whereby rightist students were formed into 'University Defence Groups'. 1970 R. A. H. Robinson *Origins Franco's Spain* ii. 67 The exception in the two Rightist minorities was Royo Villanova. 1974 *Times* 4 Nov. 14/6 The [Durham University] debating society is regarded as a Rightist organization. 1974 T. P. Whitney tr. *Solzhenitsyn's Gulag Archipelago* I. i. ii. 52 The nonexistent 'rightist opposition' would come later. 1977 *Time* 3 Jan. 9/1 In the Middle East, Syrian President Hafez Assad gained new stature by forcibly bringing to a halt the civil war in Lebanon involving rightist Christians, left-wing Moslems, and their Palestinian allies.

rightly, *adv.* Add: **3. d.** Phr. *I don't rightly know*, expressing reserve: 'I am not sure.'

1861 T. Hughes *Tom Brown at Oxf.* III. iii. 48, I don't rightly know, sir. 1954 R. Bissell *High Water* viii. 93, I spose so, but I don't rightly know.

right-minded, *a.* Add: **1.** (Further examples.) Also *transf.*

1933 *Granta* 19 Apr. 358/1 These appointments have been virtually closed to Jews, Socialists, and indeed any but right-minded nationalists for many years. 1961 *Times* 28 Dec. 11/4 For those who enjoy books set in Oxford, there were at least two detective stories, a right-minded farce by Mr. Dacre Balsdon [etc.].

rightmost (rəi·tmōᵘst), *a.* [f. Right *a.* + -most.] Situated or occurring farthest to the right. Also *absol.*

1963 *New Yorker* 15 June 28/2 The three casement windows on my left (the rightmost of which is missing a hook

latch). 1964 *IBM Systems Jrnl.* III. 125 Decimal digits.. appear in fields of variable length..and are accompanied by a sign in the rightmost four bits of the low-order byte. 1971 *Language* XLVII. 263 All the cases discussed involve the movement or deletion of verbal objects rather than subjects. The reason is that since the NSR [nuclear stress rule] assigns primary stress to the rightmost element, only cases in which the underlying rightmost element has been affected by transformations can provide crucial evidence. 1973 [see *packed *ppl. a.*[1] 1 c]. 1978 *Language* LIV. 411 Complex NP Shift applies in 9, where the complex NP does not substitute for the rightmost constituent in S.

rightness. Add: **5.** (Earlier example.)

1884 [see *bilaterality].

righto (rəi·tōᵘ), *int.* Also **right-ho, right-o, right-oh.** [f. Right *a.* + H)o *int.*[1] 5.] An exclamation expressing agreement with or acquiescence in an opinion, proposal, etc., or compliance with a request. Also as *sb.*, an acceptable person, and as *v. intr.*, to acquiesce, agree. Cf. *righty-ho *int.*

1896 E. Turner *Little Larrikin* i. 12 'Hurry up now and be a good kid.' 'Right-O!' said Lol cheerfully. 1899 *Punch* 7 June 274/3 *Dear Clarence*, —Yours to hand. Righto! Announcement sent to *Morning Post*. Thine, *Mabel*. 1902 C. J. C. Hyne *Mr. Horrocks, Purser* 52 Do it how you like, only anyway do it. 'Right-O,' said the fourth officer. 1905 *Pall Mall Mag.* Oct. 456/2 'We will leave Mrs Bergman's name out of the question.' 'Right oh', assented Leonard. 1912 *Collier's* 26 Oct. 16/3 'Right-oh,' said the young man. 1915 T. Burke *Nights in Town* 191 Righto, laddie, righto. I get you. 1916 'Boyd Cable' *Action Front* 134 'Then if the shells pitch too near we can slip the cable and run for it.' 'Right-oh!' said the captain. 1920 Wodehouse *Damsel in Distress* viii. 110 'Give this note to Lady Maud.' 'Right-ho.' 1921 Galsworthy *To Let* i. v. 53 Stable secret! Righto! 1922 Joyce *Ulysses* 420 We two, she said, will seek the kids where shady Mary is. Righto, any old time. 1924 F. M. Ford *Some do Not* i. ii. 44 On Monday I shall telegraph: 'Righto' and nothing else. 1930 [see *go *v.* 58 d]. 1930 Kipling *Limits & Renewals* (1932) 267 We'll expect her at nine, then... Righto! 1936 Wodehouse *Laughing Gas* iv. 41, I had met her when she was taking a holiday at Cannes. We became chummy. I asked her to marry me. She right-hoed. So far, so good. 1937 G. Frankau *More of Us* xi. 120 Thought she like that? He wouldn't put it past her. Righto. Then let her learn which was the master. 1949 N. Marsh *Swing, Brother, Swing* ix. 212, I never say 'right-ho'. He'll think I'm demented. 1952 Wodehouse *Pigs have Wings* ii. 56 All set. She right-hoed like a lamb. 1974 J. Thomson *Long Revenge* ix. 106 'You're Mr Finch..?' Finch replied that he was and Ken replied, 'Right-oh.'

right of way. Add: **1. a.** (Further example.)

1925 A. J. Toynbee in *Survey Internat. Affairs 1920–23* 41 The part played by the League in the problems of the right of way through the Kiel Canal..is noticed in Part III.

b. The right to build and operate a railway line on land belonging to another; the land on which a railway line is built. *U.S.*

1839 *Jrnl. Indiana Ho. Representatives 1838* 23 Sess. 101 Mr. Blair introduced bill no. 32, to grant the right of way to Illinois. 1883 *Rep. Indian Affairs* (U.S.) p. xxii, I had the honor to submit to the Department..the draft of a bill.. to grant a right of way to the Carson and Colorado Railroad Company. 1902 S. E. White *Blazed Trail* 407 On either side of the right-of-way lay mystery in the shape of thickets. 1919 T. K. Holmes *Man from Tall Timber* xxix. 365 The right-of-way of the railroad was not wide enough to halt the conflagration in any case. 1931 H. F. Pringle *Theodore Roosevelt* i. viii. 94 The locomotives of the Northern Pacific were still a novelty to those who lived along the right of way. 1949 *Boston Sunday Globe* 1 May (Fiction Mag.) 13/5 My suggestion is that you put some guards out there patrolling the right-of-way.

2. (Earlier example.)

1805 *Times* 7 Nov. 4/1 In passing through the inclosures belonging to the Defendant, he arrived at a gate which was fastened, and, considering it a public way, he forced the lock... [He] wrote... I always considered it a public right of way.

3. a. The legal right of a pedestrian or user of a (motor) vehicle to proceed with precedence over other vehicles and road-users at a particular point where their paths cross or converge. Also of sea-traffic.

1913 *Evidence Sel. Comm. Motor Traffic* I. 343/2 in *Parl. Papers* VIII. 93 One of the objections raised here by other witnesses to it was that vehicles going along the main thoroughfares would not slow up at all for the side streets if they thought they had the right of way? 1920 *Statutes of Manitoba* 10 Geo. V. c. 81 s. 10 p. 268 When a person operating a motor vehicle meets another person operating a motor vehicle or driving any draft animal at a crossroad or intersection of roads or streets, the person to the right hand of the other shall have the right-of-way. 1925 *Motor* 29 Dec. 1100/1 The existing rule of the road being 'keep to the left', surely so long as a driver can drive on the left of the centre of the road he has right of way. 1934 *Glasgow Herald* 16 Apr. 11/5 There must be crossing-places where the pedestrians' right of way is inviolable and the motorist must pass over at his peril. 1954 *Highway Code* ii. 6 Give way to pedestrians on uncontrolled zebra crossings. They have the right of way. 1954 *Lloyd's List Law Rep.* I. 144/2 That does not, of course, mean to say that the upcoming ship has anything in the nature of an absolute right of way, because, as has been frequently laid down, there must be some give and take between the two vessels. 1958 *U.S. & Canad. Aviation Reports* 453 As a seaplane was coming down to a lake for a landing..another seaplane started its take-off run..its pontoons damaged the wing and rudder as it passed over. *Held*: The take-off airplane pilot was solely at fault for failing to yield right-of-way to the descending seaplane. 1975 *N.Y. Times* 31 Oct. 38/2 Syvester Lachut was cited by a Castle Shannon patrolman for failing to yield the right-of-way at a street crossing.

b. *transf.* in *Sport.*

1963 *Amer. Speech* XXXVIII. 204 Reference is made to.. *trail hogs* who do not give the *right of way* to slower skiers. 1967 J. Severson *Great Surfing* Gloss., *Right of way*, the surfer who is already on the wave and riding has the right of way and others coming into the wave should respect this right.

4. *attrib.*, as (in sense *1 b) **right-of-way man**, one who surveys a right of way for a railway.

1891 *Harper's Mag.* Nov. 886/2 The first men to follow the engineers..are 'the right-of-way men'. 1904 F. Lynde *Grafters* xxviii. 361 Our right-of-way man has just sent a telegram to all agents.

right-thinking, *a.* [Right *adv.* 16 b.] Thinking rightly; holding sound or acceptable views. Also *transf.* Hence **right-thinker** *sb.*

1829, 1850 [see Right *adv.* 16 b]. 1890 Kipling in *Pioneer Mail* (Allahabad) 19 Mar. 394/1 And verily it is not a good thing to live in the East for any length of time. Your ideas grow to clash with those held by every right-thinking white man. 1920 A. Huxley *Limbo* 137 Jacobsen himself rarely volunteered a remark about the war; it was taken for granted that he thought about it in the same way as all other right-thinking folk. 1922 S. Lewis *Babbitt* iii. 34 It was the very best of water-coolers, up-to-date, scientific, and right-thinking. 1930 E. Rice *Voyage to Purilia* xix. 293 Was marriage..regarded by all right-thinking Purilians as the summit of happiness? 1931 A. Huxley *Music at Night* 172 Granted the preliminary assumption that concupiscence is wicked, right-thinkers are justified in specially discriminating against the representations of this sin... Among the right-thinking the doctrine of the inherent wickedness of concupiscence is still held with an extraordinary intensity. 1934 B. Russell *Freedom & Organization 1814–1914* xxviii. 394 The area must not be in Asia or Africa; this was clear to all right-thinking people. 1939 *Ann. Reg. 1938* i. 12 Lord Snell..gave eloquent expression to the grief and indignation felt by all right-thinking men at what was happening in Austria. 1961 *Times* 11 July 12/6 Now no right-thinking suburbanite would be seen on his lawn without wearing Bermudas. 1961 *Times Lit. Suppl.* 1 Dec. 861/2 *Midcentury* is one prolonged grouse..about those 'right-thinkers' dedicated to 'the word with the welfare label'. 1974 'J. Le Carré' *Tinker, Tailor* xxviii. 335 He ought to censure him on behalf of right-thinking men.

right-turn. [f. Right *a.* + Turn *sb.*] A movement to the right, *spec.* one made by a motor vehicle from one road to another. Also *attrib.* and as *v. intr.* So **right-turning** *ppl. a.*

1960 *Guardian* 24 June 20/4 Much of the present congestion in the area is due to right-turning traffic. 1967 *Gloss. Highway Engin. Terms* (*B.S.I.*) 11 *Right-turn lane*, an auxiliary deceleration and waiting lane constructed between an overtaking lane and a central reserve or traffic island so that vehicles approaching a junction can slow down and wait clear of other traffic. 1968 *Highway Code* 17 When it is safe, give a right turn signal and..take up position just left of the middle of the road, or in the space marked for right-turning traffic. 1972 *Guardian* 26 June 1 The ring system has been evolved to do away with the conflicts of right-turning traffic. 1973 J. Gores *Final Notice* xvi. 102 They *should* right-turn toward the freeway on-ramp.

rightward, *adv.* and *a.* Add: **B.** *adj.* **a.** (Further examples.) Also in political sense (see Right *sb.*[1] 17 d in Dict. and Suppl.).

1941 'G. Orwell' in *Partisan Rev.* July–Aug. 319 The recent rightward swing means that we are being regimented by wealthy men and aristocrats rather than by representatives of the common people. 1972 *Sci. Amer.* July 86/3 Usually a reader sees a page of text and moves his eyes over it, ordinarily in a rightward and downward direction. 1977 *Times* 4 Oct. 14/1 A rightward trend towards the centre on the part of M. Mitterand.

b. *rightward welding* (see quots.).

1941 A. C. Davies *Sci. & Pract. Welding* iii. 146 Rightward welding..was introduced some years ago to compete with electric arc welding in the welding of plate over ⅜ in. thick... In this method the weld progresses along the seam from left to right, the rod following the blowpipe. 1967 M. D. Jackson *Welding* iv. 203 There are two methods of making downhand welds with the oxy-acetylene process and these are generally known as leftward welding or rightward welding, the terms being descriptive of the direction of welding. 1975 Bram & Downs *Manuf. Technol.* ii. 48 In rightward welding the flame is directed towards the completed part of the joint and welding proceeds from left to right.

rightwards *adv.* (later examples, in political sense).

1968 *Listener* 7 Nov. 610/3 The novelist Kingsley Amis is regarded, by himself among others, as having moved 'rightwards' with the years. 1971 P. Worsthorne *Socialist Myth* iii. 33 Any Labour Government has to lean over Rightwards to avoid being suspected of lack of patriotism.

right wing. [f. Right *a.* + Wing *sb.*] **1.** *Mil.* See Right *a.* 18, Wing *sb.* 7 a.

2. In football and similar games: the position of a player on the right side of the centre(s); a player occupying this position; the part of the field in which the right wing normally plays. Cf. Wing *sb.* 7 b.

1882 in Charles-Edwards & Richardson *They saw it Happen* (1958) 299 Of the Rovers forwards the right wing pair first became prominent. 1900 in B. James *England v Scotland* (1969) iii. 72 For England, Crabtree was cleverly holding the Scottish right wing in check and showing grand form. 1901 *Encycl. Sport* II. 419/2 The left back and left half-back attend to the opposing right wing forwards. 1947 *Sporting Mirror* 7 Nov. 3/2 Appearing in their league side on the right wing and, later, at centre forward he became very popular with the Shepherds Bush spectators. 1969 Eagleson & McKie *Terminol. Austral. Nat. Football* iii. 10 *Right wing*, a variant for right centre. 1974 *Observer* 1 Sept. 18/3 Hall improvised brilliantly up and down Liverpool's right wing interpassing with Heighway.

3. That section of a political party, assembly, or other body most tending to hold conservative or reactionary views (see Right *sb.*[1] 17 d in Dict. and Suppl.). Freq. (with hyphen) *attrib.* or as *adj.*

1905 W. James *Meaning of Truth* (1909) v. 124 If the formula ever became canonical, it would certainly develop both right-wing and left-wing interpreters. 1927 U. Sinclair *Oil!* xiv. 348 Jacob Menzies had been clubbed almost insensible on the picket-line. Jacob was the 'right wing' brother..who had been earning his education by pressing students' pants. 1928 E. & C. Paul tr. *Stalin's Leninism* 322 Is our Party..making any concessions to the right-wing deviation of the Communist International? 1933 *Labour Monthly* Apr. 237 We are not yet at the point where the right wing advocates dictatorship of the proletariat. 1937 'G. Orwell' *Road to Wigan Pier* xii. 243 The thinking person, by intellect usually left-wing but by temperament often right-wing, hovers at the gate of the Socialist fold. 1940 W. Temple *Thoughts in War-Time* iii. 23 The Right Wing tends to emphasise..the distinction between the Government and the community and apply to the Government the ethics of trusteeship. 1941 J. S. Huxley *Uniqueness of Man* ii. 80 It is not eugenics but right-wing politics if we merely talk of favouring the breeding of the upper classes of our present social system at the expense of the lower. 1955 *Times* 7 July 8/4 The right-wing monarchists and neo-Fascists [in Italy] are expected to 'wait and see'. 1957 *Economist* 26 Oct. 299/2 President Castillo's strong-arm methods steered the country between communism and right-wing nationalism. 1958 M. Argyle *Relig. Behaviour* viii. 83 It seems likely that Catholics have an intrinsically right-wing attitude. *Ibid.*, Protestants support right-wing parties, Jews support the left and have radical attitudes. 1964 *Ann. Reg. 1963* 5 The job of foreign affairs spokesman went to Mr Patrick Gordon Walker..and the remainder of the 'shadow' portfolios were distributed among other right-wing personalities. 1973 M. Amis *Rachel Papers* 57 Had he poked a switchblade into an enlightened student (for Norman was passionately right-wing)? 1974 J. White tr. *Poulantzas's Fascism & Dictatorship* iii. 126 The Liberal Party itself was in an increasing state of decomposition, with Salandra's right wing now dominant. 1981 *Times* 15 June 2/7 A new right-wing pressure group publication circulating among moderates.

Hence **right-winger**, (*a*) a member of a political right wing; (*b*) in *Football*, a player on the right wing. Also **right-wingery, right-wingism**; **right-wingy** *a.*

1928 E. & C. Paul tr. *Stalin's Leninism* 322 They hesitate between embracing the tenets of Leninism and throwing in their lot with the right-wingers. 1929 Right-winger [see *cross-pass* s.v. *cross- B.] 1937 Wyndham Lewis *Blasting & Bombardiering* 306 Mr. Douglas Jerrold stands out from the hardboiled crowd: he is the brains of the Right. I think he is the Sotelo of English right-wingery. 1951 *Sport* 16–22 Mar. 13/3 In 20-year-old Brian Siddall..Stoke appear to have found a right-winger of unbounded promise. 1960 *News Chron.* 1 July 4/3 The back-bench Right-wingers want a strong-man Chancellor. 1962 *Tuscaloosa* (Alabama) *News* 12 Feb. 4/6 A meeting of the American Jewish Congress which had been discussing right-wingism in the USA. 1973 *Scotsman* 13 Feb. 1/1 Almost all students, including moderates and Right-wingers, are convinced that they must not go on, and that serious disturbances will result if they do. 1974 'D. Kyle' *Suvarov Adv.* iv. 36 'Why do I represent the Russians?' he would ask. 'Because they want me to. Though why they pick a right-winger like me I can't imagine.' 1977 *Listener* 13 Jan. 60/4 Some people have thought that your heroes are rather snooty and right-wingy and so forth. 1978 *Listener* 13 July 59/4 An..irrational right-wingism, which eschews all 19th-century liberalism.

ri·ghtwise, *adv.* [f. Right *a.* + Wise *sb.*[1] II.] In a right-hand direction.

1908 Hardy *Dynasts* iii. ii. ii. 369 Henceforward, masses of the foe Withdraw, and, firing as they go, Pass rightwise from the cockpit out of sight.

righty (rəi·ti). Also **rightie.** [-y[6].] **1.** *U.S.* A right-handed person; *spec.* in *Baseball*, a right-handed pitcher; a batter who stands on the left side of the plate and swings the bat from right to left. Also quasi-*adv.*

1949 *Sun* (Baltimore) 3 June 18/7 Pie—a rightie..compiled an all-time average of 300-plus against all kinds of pitching. 1963 R. I. McDavid *Mencken's Amer. Lang.* 564 In baseball, switch hitters speak of their relative prowess batting *lefty* and *righty*. 1974 *Anderson* (S. Carolina) *Independent* 18 Apr. 8c/4 Smith, a righty with an assortment of pitches, struck out 15 Mets and yielded only three walks as he overpowered the visitors. 1975 *New Yorker* 15 Sept. 27/3 The Sox put in..Cleveland, a righty who hadn't seen action in sixteen days. *Ibid.* 22 Sept. 102/3 He never hit righty in a game.

2. A right-winger in politics. Also *transf.* and *attrib.*

1967 *Listener* 10 Aug. 164/1 'The lefties are almost completely in control of the nation's communications.'.. The righties have, then, a case. *Ibid.* 164/3 He might just as easily be a militant atheist or Trotskyite as a rightie.

1969 *Time* 31 Oct. 53 England's thin red line of intellectual royalists is being overrun by 'progressive' reformers who deliberately sabotage old-fashioned academic virtues... The Manchester Guardian called them a 'tightly knit bunch of righties'. 1970 *Guardian* 19 Nov. 9/4 This sounds a bit like his Righty chum, Bernard Levin, but not like any Lefty I know. 1976 *Sunday Tel.* 29 Aug. 14/4 This ecclesiastical equivalent of Enoch Powell has done and said much more than demand the old Latin Mass, and this is where the majority of Catholics part company from the Rightie of Trent. 1978 *Spectator* 21 Oct. 17/2 Such rightie code-phrases as 'the need for discipline in our schools'.

righty-ho (rəi·tihōᵘ), *int.* Also **rightie-ho, righty-oh,** etc. [f. Right *a.* + -y[6] + Ho *int.*[1] 5.] = *righto *int.*

1927 D. L. Sayers *Unnatural Death* xv. 170 'You will remember who you are, won't you?' 'Righty-ho.' 1938 [see *auntie, aunty b]. 1942 A. Christie *Body in Library* v. 51 She said she had a headache..so I said righty ho, and that was that. 1945 E. Waugh *Brideshead Revisited* 13 For Christ's sake don't say 'rightyoh'. 1959 T. B. Morris *Death among Orchids* xix. 170 Righty-ho, darling! That's the lot for now. 1960 B. Cobb *Don't lie to Police* iii. 43 Oh, righty-oh. Come right in. 1960 I. Cross *Backward Sex* 65 O.K. kid and righty ho old top. 1973 J. Porter *It's Murder with Dover* ii. 20 'I should make it now, Prissy.'.. 'Rightie-ho!' Lady Priscilla set off..for the kitchen.

rigid, *a.* and *sb.* Add: **A.** *adj.* **1. a.** (Further *Comb.* example.)

1932 W. Faulkner *Light in August* vii. 140 He walked stiffly past her, rigidfaced.

c. *spec.* of an airship: belonging to the type whose shape is maintained by a framework and not (chiefly) by the pressure of gas in the envelope.

[1902 *Encycl. Brit.* XXV. 101/2 He [sc. Count Zeppelin] has gained an advantage by attaching his propellers to the balloon,..but this requires a rigid framework and a great increase of weight.] 1909 A. Berget *Conquest of Air* ii. 26 There is obviously another way..; it is to make the balloon rigid. 1910 C. C. Turner *Aerial Navig. To-day* iv. 62 One of the most famous airships of the rigid type was Zeppelin No. 4. 1930 *Daily Express* 6 Oct. 2/4 He..was engaged on research into problems connected with rigid airship construction. 1977 *It* June 16 And so began the era of the rigid airship. 1980 *Nature* 20 Mar. 288/1 In retrospect, this was the most successful rigid airship designed in this country but the Air Staff decided that they had no use for it and it flew for the last time on 20 September 1921.

d. Phr. *to bore, scare*, etc., (someone) *rigid*: to be excessively boring, frightening, etc. *colloq.*

1943 [see *bind *v.* 22]. 1952 M. Tripp *Faith is Windsock* xiii. 200 Dick's bloody unruffled ways and his bloody reasonable talk bind me rigid. 1970 O. Norton *Dead on Prediction* vii. 136 *You* shook him rigid, producing that photograph of her. He was afraid he'd be suspected. 1972 K. Campbell *Thunder on Sunday* 58 It's no tourist place, I assure you... You'd be bored rigid. 1976 R. Barnard *Little Local Murder* viii. 102 They'll have made the connection: anonymous letters—murder. And..they'll be scared rigid.

3. b. (Further examples.) Also of personality, and of traits and mental processes. Cf. *rigidity 2 b.

1949 *Jrnl. Personality* XVII. 322 A man with rigid delusions may readily bring his delusions in and out of conscious processes. 1960 M. Rokeach *Open & Closed Mind* ix. 183 Rigid thinking should be expected to lead to difficulties in thinking analytically. 1964 Gould & Kolb *Dict. Social Sci.* 494/1 Personalities which are highly integrated may be spoken of as 'rigid'. 1972 *Jrnl. Social Psychol.* LXXXVII. 66 Of the three tests used, only the inventory gave meaningful differences, with Communists and Fascists being more rigid than normals. 1972 A. Storr *Dynamics of Creation* viii. 92 The obsessional [personality] is controlled, inhibited, and rigid in his ideas.

5. b. *Logic.* (See quots.)

1972 S. A. Kripke in Davidson & Harman *Semantics of Natural Lang.* 269 Let's use some terms quasi-technically. Let's call something a *rigid designator* if in any possible world it designates the same object, a *non rigid* or *accidental designator* if that is not the case. 1975 N. Rescher *Theory of Possibility* iv. 86 A *rigid designator*—i.e., one that inevitably specifies precisely the same individual in the context of *every* possible world. (This concept coincides pretty much with what is traditionally referred to..as a 'logically proper name' for an individual.) 1976 *Times Lit. Suppl.* 12 Nov. 1434/3 Ruth Marcus's suggested criterion for rigidity of predicates; that *F* is rigid if necessarily anything that is *F* is necessarily *F*.

B. *sb.* **2.** A rigid airship. Cf. sense 1 c of the adj.

1919 *Sphere* 19 July 54/1 Reports to the Admiralty.. urging them to build rigids without delay. 1920 *Glasgow Herald* 3 Dec. 5 The lecturer said that after comparing non-rigids, semi-rigids, and rigids it was obvious that it would be the rigid airships which would be developed for commercial work. 1928 *Daily Tel.* 12 June 17/2 The Mayfly, the first naval rigid, was being built. 1963 A. Smith *Throw out Two Hands* ii. 31 Lighter-than-air flying suddenly took on a sterner aspect with the development of the dirigible... The word blimp comes from this period for, so the generally accepted story goes, the dirigibles were classified into A—rigid and B—limp. Be that as it may, the rigids were the most exciting.

rigidification (ridʒiːdĭfikē̆ⁱ·ʃən). [f. Rigidify *v.*: see -fication.] The action of making or becoming rigid. Also *fig.*

1947 *Sun* (Baltimore) 31 Oct. 10/2 Would not the proposed rigidification of the whole approach..offer hazards of its own? 1965 *Jrnl. Immunol.* XCIV. 130/1 Rigidification of a portion of a normally motile cell membrane may lead to its rupture. 1973 B. R. Wilson *Magic & Millennium* xii. 384 An over-institutionalization, a rigidification, not only of religious practices, beliefs, and procedures, but also of the entire pattern of life of a new community. 1980 P. Greenhalgh *Pompey* i. 208 Cato's faction displayed that inflexibility of purpose which reacts to increasing isolation by rigidification rather than relaxation.

rigidify, *v.* Delete *rare* and add: **a.** (Further examples.)

1920 A. S. Eddington *Space, Time & Gravitation* ii. 39 If a substantial aether analogous to a material ocean exists, it must rigidify, as it were, a definite space. 1942 *Nature* 14 Nov. 564/1 If we conceptually rigidify such a system into a definite formal scheme, we can think of it as a set of alternative canalized paths. 1961 B. R. Wilson *Sects & Society* 1 The sect..seeks itself to rigidify a pattern of behaviour. 1973 P. Evans *Bodyguard Man* ii. 23 Rigor mortis had already begun to rigidify the cells.

b. (Further examples.)

1979 *Nature* 16 Aug. 538/3 The slightly enhanced 'mobility' of the substrate binding area is intriguing, in particular, if it were to be found that these segments 'rigidify' on substrate binding. 1981 *Times Lit. Suppl.* 20 Feb. 202/1 Now he [sc. A. Goehr] is assumed, as a Cambridge professor, to be rigidifying in the academy.

Hence **rigi·dified** *ppl. a.* (chiefly in *pred.* use); **rigi·difier.**

1912 J. Galsworthy in *Daily Mail* 27 May 4/4 Which of us does not know the deflecting power of trusteeship, rigidified, as it is, by law? 1956 J. S. Bruner et al. *Study of Thinking* v. 128 One could begin either way—adopting either a part or a whole hypothesis—and arrive at the same conclusion provided one did not become rigidified before the process of proof was completed. 1968 *Listener* 21 Mar. 365/2 In literature, some artists are naturally rigidifiers and others naturally moulders. 1977 T. Berger *Who is Teddy Villanova?* i. 6, I am, with expanded chest and rigidified tendons, just..five feet ten.

rigi-dig, var. *ridgy-didge *a.*

rigidity. Add: **1. b.** *modulus of rigidity* (also *rigidity modulus*).

1877 *Encycl. Brit.* VII. 805/2 The 'modulus of rigidity' of an isotropic solid is the amount of tangential stress divided by the deformation it produces. 1930 *Engineering* 11 Apr. 465/2 In this problem, accurate values of the bulk modulus..and the rigidity modulus..were required. 1966 *McGraw-Hill Encycl. Sci. & Technol.* IV. 419/2 The shear or rigidity modulus..measures the resistance of the material to change in shape without change in volume.

2. b. *Psychol.* Inflexibility and unadaptability in a person's outlook and responses.

1943 K. F. Walker et al. in *Character & Personality* XII. 35 This additional source of interference might well be called *disposition-rigidity*, and consists of the influence of an habitual activity on a less well-established activity. 1946 P. M. Symonds *Dynamics Human Adjustm.* vii. 171 Forms of personal rigidity either in gesture or posture..or thoughts again represent methods of defense against active impulses and hence anxiety. 1949 *Jrnl. Personality* XVII. 321 We hope to demonstrate experimentally what principal forms of rigidity exist. 1960 M. Rokeach *Open & Closed Mind* ix. 184 High rigidity should lead to difficulties in the analytic phase of thinking. 1973 D. L. Schaeffer tr. *Cohen's Patterns of Personality Judgment* iv. 277 These findings would lead us to anticipate positive correlations between judgmental variance and both the neuroticism and the rigidity scores on the personality questionnaire.

rigidize (ri·dʒidəiz), *v.* [f. Rigid *a.* + -ize.] *trans.* To make rigid. Chiefly as **ri·gidized** *ppl. a.*; also **ri·gidizing** *vbl. sb.*

1948 *Rev. Sci. Instruments* XIX. 725/2 The case of the.. meter is made of rigidized metal. *Ibid.* Rigidizing, or texturing, of the sheet from which the case is formed contributes greater impact strength..than could be provided by plain metal. 1961 *Aeroplane* CI. 379/1 Both are made in 'rigidized' stainless steel to enable them to withstand the severe handling to which this type of equipment is subjected. 1962 *Flight Internat.* LXXXI. 104/1 The 'rigidizing' is accomplished mechanically by stretching the aluminium foil beyond its elastic limit when the structure is fully inflated. 1964 *New Scientist* 23 Jan. 201/1 A balloon that was supposed to become 'rigidised' by the stretching of its plastic skin as it inflated. 1977 *Film & Television Technician* Mar. 10/1 (Advt.), Reward! £200 for information leading to the recovery of a rigidised case containing spare parts for Steenbeck Editing Machines.

rigmarole, *sb.* (and *a.*). Add: **1. c.** *transf.* A succession of tiresome duties; a lengthy procedure; a fuss, a 'palaver'.

1955 *Times* 24 June 12/5 The Government set up..the whole rigmarole of scheduling, listing, and building preservation orders. 1965 M. Spark *Mandelbaum Gate* i. vi. 178 The whole rigmarole about the film unit visiting Qumran was plainly an occasion for citing the case of the producer. 1977 *Grimsby Even. Tel.* 14 May 10/5 The Wakefield Trophy series is a good way to introduce lads or girls in the area to cycle racing without having to go to the expense and rigmarole of training.

rigmarole *v.* (further example).

1922 Joyce *Ulysses* 756 O beau pays de la Touraine that I never even sang once explaining and rigmaroling about religion and persecution.

rigmo. Also **rig-mo.** A slang shortening of *rigor mortis* (s.v. Rigor 2).

1966 *New Statesman* 8 Apr. 512/3 'The foot curls a bit, you know,' confides Liberace as a coffin-salesman, 'when rigmo sets in.' 1975 *Observer* 6 Jan. 17/2 Embalmers' aids like the Natural Expression Former (a plastic device which, inserted into the mouth after rig-mo—as we call it in the trade—sets in, can produce a seraphic smile on the deceased face).

rigol, *sb.* Restrict '*Obs.* exc. *dial.*' to senses in Dict. and add: **1.** (Later examples with reference or allusion to Shakes. 2 *Hen. IV* iv. v. 36.)

1733 L. Theobald *Wks. Shakes.* III. 517 Hence a Rigolet, or Rigol, may, I presume, stand in English for a Circle, any Thing round. 1826 Hazlitt *Plain Speaker* II. ix. 263 Here love's golden rigol bound his brows. 1883 G. Macdonald *Princess & Curdie* xix. 145 His crown..lay in front of him, his long, thin old hands folded round the rigol, and the ends of his beard straying among the lovely stones.

2. c. *Sailing.* (See quots.)

1961 F. H. Burgess *Dict. Sailing* 172 *Rigol*, the outboard semicircular gutterway over a porthole. 1976 *Oxf. Compan. Ships & Sea* 711/2 *Rigol*, a curved, semicircular steel strip riveted to a ship's side over a scuttle with the object of deflecting any water which runs down the side of the ship and preventing it from entering the scuttle when it is open.

‖ **rigolette** (rigolet). *Obs. U.S.* [Fr., f. the name of a character in *Mystères de Paris* (1842–3) by Marie-Joseph ('Eugène') Sue.] A variety of woollen head-scarf worn by women.

1863 L. M. Alcott in *Commonwealth* 29 May 1/3 All night hovering like a massive cherubim in a red rigolette over the slumbering sons of man. 1866 'E. Foxton' *Herman* I. ii. 40 At last his sister appeared, in her blue evening cloak and *rigolette*, with the flowers in her hair peeping out between its strings of tiny pearl balls. 1867 *Student & Schoolmate* Mar. 90 Kitty wore a Rob Roy cloak and a dainty white hood, called in those days a rigolette (though how the fashion-maker spelt the word I don't know).

rigor. Add: **2. b.** *fig.*

1929 W. Faulkner *Sanctuary* (1981) v. 59 A tradition.. in the throes of its own rigor-mortis. 1934 C. Lambert *Music Ho!* ii. 131 *Socrate*..has something of the rigor mortis always associated with overtheoreized music. 1977 *Rolling Stone* 16 June 39/2 He showed up at the producer's black-tie party, an occasion to normally give him rigor mortis.

rigoristic (rigŏri·stik), *a.* [f. Rigorist + -ic.] Of, pertaining to, or characteristic of a rigorist; austere, stringent.

1907 W. James *Pragmatism* iv. 160 Pluralism..has no need of this dogmatic rigoristic temper. 1934 *Times Lit. Suppl.* 7 June 398/2 Carlyle's failure to impose his narrow, rigoristic, moralistic, joyless Annandale view of the world upon the world, added an element of tragedy to his deeply tragic sense of life. 1955 L. Rosenfeld in W. Pauli *Niels Bohr* 92 One might well regard any rigoristic conception of this problem as superfluous.

‖ **Rigsdag** (ri·gzdag). [Da., f. gen. of *rige* realm + *dag* Day; cf. *Reichstag, *Riksdag.] The name of the Parliament of Denmark, formerly comprising two Houses, the Landsting and the Folketing.

In 1953 the upper chamber and the name Rigsdag were abolished. The remaining single chamber is called the Folketing.

[1645 J. Howell *Lett.* vi. i. 1 He was at Rensburgh..at a Richsdagh an Assembly that corresponds our Parliament.] 1883 *Students' Encycl.* II. 368/1 D[enmark] is governed by a hereditary monarchy, and by a National Assembly or *Rigsdag*. 1938 J. H. S. Birch *Denmark in Hist.* xx. 337 The Constitution provided for two Assemblies, the Folketing and the Landsting. The Rigsdag, which comprised the two Assemblies and was to meet annually, exercised jointly with the King legislative power and the right of imposing taxes. 1956 W. L. Shirer *Challenge of Scandinavia* iv. 278 There was a national election on April 21, 1953, for the Lower House (Folketing) of the Rigsdag. 1972 S. Oakley *Story of Denmark* xiii. 178 The new ministry put forward a very liberal project, largely the work of Lehmann and Monrad, which envisaged a *Rigsdag* of two houses, both elected by universal male suffrage.

‖ **Rigsmaal** (ri·gzmǭl). Also **rigsmaal.** [Da., f. as prec. + *mål* (formerly *maal*) language; cf. *Riksmål.] = *Dano-Norwegian *sb.* Cf. *Landsmål.

1902 *Encycl. Brit.* XXXI. 275/1 The close of 1899 and the beginning of 1900 were occupied by a discussion, which drowned all other interests, and in which every Norwegian author took part, as to the adoption of the *landsmaal*, or composite dialect of the peasants, as the national language in place of the *rigsmaal* or Dano-Norwegian. 1932 W. L. Graff *Lang.* 387 A form of Danish with Norwegian peculiarities ('rigsmaal') was adopted as the official and literary language [of Norway]. 1959 A. C. Baugh *Hist. Eng. Lang.* (ed. 2) 36 This regeneration of the national speech has not succeeded in displacing Dano-Norwegian (*Rigsmaal*).

Rig-vedic (rig,vē̆ⁱ·dik), *a.* (*sb.*) Also **Rig Vedic, Rigvedic.** [f. Rig-veda + -ic; cf. *Rgvedic *a.* (*sb.*).] Of or pertaining to the Rig-veda. Also as *sb.*, Rig-vedic Sanskrit.

1881 *Encycl. Brit.* XII. 781/1 The conquest of the vast new tracts thus included seems not to have commenced till the close of the Rig-Vedic era. 1895 E. W. Hopkins

Relig. India iii. 74 The end of the Rig Vedic period. 1923 H. D. Griswold *Relig. of Rigveda* ii. 28 The primary source for the Rigvedic age is, of course, the *Rigveda*. 1968 *Language* XLIV. 8 The dual and the present indicate that the full grade (whether developed in common or separately in Rigvedic and Avestan) is secondary. 1975 *Language* LI. 115 The most intriguing bit of evidence presented in EIP involves glide formation and glide vocalization in Rigvedic Sanskrit.

rijsttafel (rəi·stāfĕl). Also **reistafel, rijstafel.** [Du., f. *rijst* Rice[2] + *tafel* Table *sb.*] A South-East Asian rice dish (see quots.).

1889 S. J. Hickson *Naturalist in N. Celebes* ix. 224 As an article of food these small prawns are much prized by the natives, and the Dutchman is very glad to add the *kleine garnalen* to his list of comestibles at the *rijsttafel*. 1932 W. S. Maugham *Narrow Corner* xvi. 124 *Reistafel* was served. They piled their plates with rice and curry, fried eggs, bananas and a dozen strange concoctions. 1933 L. Deeley *Internat. Cookery* 12 Travellers to the Far East all know the famous Rijstafel, to be met on board and ashore, from Ceylon to Shanghai. To describe every dish served is almost impossible. It includes curried fish, rabbit, chicken, served with rice and the famous Kroepoek Oedang which is a special kind of dried shrimp. 1937 C. Connolly in L. Russell *Press Gang!* 89 Some stewards from the P. & O. worked away at the punkahs, or at distributing reistafel and planters' punch. 1962 P. Helm *Death has Thousand Entrances* iv. 86 The meal, an Indonesian rijsttafel, was over. 1965 B. Sweet-Escott *Baker St. Irregular* viii. 243 The problem was discussed by us with the Dutch over many a *reistafel* in Colombo. 1969 *Daily Tel.* 14 Jan. 15/7 The rijsttafel, served in a multitude of dishes, is delicious—with rare strange spices, sometimes hot, but always delicate and more subtle than much Chinese food. 1977 J. Didion *Bk. Common Prayer* ii. ii. 65 She knew the Indonesian couple who did the rijsttafel.

‖ **rikka** (ri·ka). Also **rikkwa.** [Jap., lit. standing flowers.] A traditional form of Japanese flower-arrangement.

1889 J. Conder *Theory Jap. Flower Arrangem.* (1935) 25 The use of many different kinds of flowers in one composition though followed in the earlier styles of *Rikkwa* and *Shin-no-hana* is opposed to the principles of the purer styles. 1914 M. Averill *Jap. Flower Arrangem.* i. 28 Rikkwa and Nageire are the two branches into which Ike-bana has been divided. 1934 A. Koehn *Art Jap. Flower Arrangem.* xvii. 131 A huge well-shaped Pine branch or Bamboo spray formed the centre of these 'rikkwa' designs. 1965 W. Swaan *Jap. Lantern* iv. 49 There are two main styles [of ikebana]: the formal *rikka* arrangements used to decorate the altars of Buddhist temples..and the *nageire* or 'thrown-in' style. 1976 *San Antonio* (Texas) *Express* 14 Oct. 3-G/1 *Rikka* is the oldest established form [of flower arrangement], dating back five centuries. It is a complex style with 11 main branches presenting a stylized representation of the landscape of Buddhist paradise.

‖ **Riksdag** (ri·ksdag). Also **riksdag.** [Sw., f. gen. of *rike* realm + *dag* Day; cf. *Rigsdag.] The name of the Swedish Parliament.

Since 1971 a one-chamber Parliament.

1887 *Encycl. Brit.* XXII. 743/1 The riksdag meets every year on January 15, and consists of two houses. 1893 W. W. Thomas *Sweden & Swedes* xvii. 217 The Riksdag, or Diet, of Sweden is opened with great pomp and ceremony... Both houses of the Riksdag leave the parliament building and come to the palace to have their sessions opened by the King. 1955 *Ann. Reg. 1954* 234 These elections had, possibly, important effects on the Riksdag elections. 1976 S. Lloyd *Mr Speaker, Sir* iii. 93 The Speaker..cannot adopt the Swedish practice under which, when the votes are equal, the Talsmand of the Riksdag has to draw lots.

‖ **Riksmål** (ri·ksmǭl). Also **riks-, -maal.** [Norw., f. gen. of *rike* realm + *mål* language; cf. *Rigsmaal.] = *Dano-Norwegian *sb.* Cf. *Landsmål.

The usual term now for this language is *Bokmål*.

1926 *Encycl. Brit.* II. 1090/1 Simultaneously with this lyric flourishing in the *riksmaal*, no adequate revival has taken place in the *landsmaal*. 1939 L. H. Gray *Foundations of Lang.* 346 Norwegian died out as a literary language in the Middle Ages, when it gave place to Danish, which, under the appellation of *Riksmaal*, served as the standard speech until the nineteenth century and, indeed, still so functions to a considerable extent. The *Riksmaal* is now yielding place to a truly national *Landsmaal*, based upon indigenous Norwegian dialects. 1952, 1966 [see *Nynorsk]. 1972 W. B. Lockwood *Panorama Indo-Europ. Lang.* 125 *Landsmål* 'National Language'..in 1885 won official recognition beside Dano-Norwegian, now known as *Riksmål* 'Language of the Realm'.

rile, *v.* Add: **2. a.** Also const. *up*.

1857 J. G. Holland *Bay-Path* iii. 32 It only raises the devil in me, and riles me all up. 1872 W. A. Hickman *Brigham's Destroying Angel* iii. 72 Some of the boys began to get terribly riled up. 1929 W. Faulkner *Sound & Fury* 67 'Getting Quentin all riled up.' Dilsey said. 'Why can't you keep him away from her.' 1966 T. Pynchon *Crying of Lot 49* iii. 68 Calculating this ought to rile up public opinion some.

riled *ppl. a.* (further example); also **riled-up.**

1856 S. Smith in *National Intelligencer* 26 Jan. 3/1, I found the President..looking kind of riled and very resolute. 1978 G. A. Sheehan *Running & Being* x. 137 If we go to bed mad we are likely to wake up with a riled-up stomach.

Riley (rəi·li). orig. *U.S.* Also **Reilly.** [A common Irish surname.] In colloq. phr. *the life of Riley*, a comfortable, enjoyable, and carefree existence.

The phrase is freq. said to owe its origin to one of a number of late nineteenth-century songs (as, for example, in W. & M. Morris, *Dict. Word & Phr. Origins* (1962) 215, and S. J. Raff in *Amer. Speech* (1976) LI. 944 ff.) but has not been traced earlier than the song of 1919, which gave it currency.

1919 H. Pease *My Name is Kelly* (song) 3 Faith and my name is Kelly Michael Kelly, But I'm living the life of Reilly just the same. 1923 E. Rice *Adding Machine* vi. 85 This is the life of Riley all right. 1936 H. Miller *Black Spring* 142 The old man's leading the life of Reilly. 1949 J. B. Priestley *Delight* 7 The life of Reilly, which some people imagine me to lead, has been further away than a fading dream. 1960 *20th Cent.* May 387 The executive is..living the life of Riley by tax evasion and expense account. 1965 J. Porter *Dover Two* vi. 70 The native cunning which had enabled him to live the life of Riley with no visible means of support..failed him dismally now. 1978 *Daily Tel.* 8 Feb. 6/7 It is simply not true that we don't pay tax and are living the life of Riley.

rill, *sb.*[1] Add: **1. a.** *spec.* A small trickle of water formed temporarily in soil or sand after rain or tidal ebb.

1883 T. C. Chamberlin *Geol. Wisconsin* I. 43 Rills, especially those following outgoing tides furrow the sand or mud, particularly in flowing over a pebble, shell or other obstruction. Such grooves where preserved constitute rill marks. 1908 *Jrnl. Geol.* XVI. 748 The word 'rill' will be used to indicate such a streamlet in an overloaded condition, that is previous to the degree of concentration necessary to cut a gully. 1925 *Water-Supply Papers U.S. Geol. Survey* No. 499. 96 As the supply of débris is small these rills are not fully loaded and are effective eroding agents. 1939 *U.S. Dept. Agric. Yearbk. 1938* 1167 Frequently in sheet erosion the eroding surface consists of numerous very small rills. 1966 *McGraw-Hill Encycl. Sci. & Technol.* XII. 430/2 All traces of the rills are removed after the land is tilled. 1968 R. W. Fairbridge *Encycl. Geomorphol.* 1102/1 In calm weather, the backwash of the returning water [on a beach] has little energy, creating a diamond pattern of small rills. 1975 R. V. Ruhe *Geomorphol.* vi. 108/1 During intense runoff, a divide between two rills is broken by caving, by undercutting by a deeper rill, or by overflow.

c. *rill action, channel, -cutting, erosion, -mark* (further example), *wash*; *rill-threaded* adj.

1960 B. W. Sparks *Geomorphol.* iv. 69 In the wet season the middle section of the slope may be under the influence of concentrated rill action. 1962 L. C. King *Morphol. of Earth* v. 137 Following the cutting of a steep hillside by rill and gully action.., surface water requires to be discharged across a relatively flat terrain to an adjacent stream channel. *Ibid.* 138 The water..may be insufficient to form sheets and then passes across the pediment in rills only. Where this occurs frequently, pediments are scored by rill channels. 1968 R. W. Fairbridge *Encycl. Geomorphol.* 819/2 Frequently the cutting of rill channels and gully heads is by far the most active process operating upon the steeper slopes. 1925 *Water-Supply Papers U.S. Geol. Survey* No. 499. 96 Rill cutting at the foot of mountain slopes. 1962 L. C. King *Morphol. of Earth* v. 138 The zone of laminar flow..is often elided, and rill-cutting and gullying appear extensively upon many pediments. 1939 *U.S. Dept. Agric. Yearbk. 1938* 1167 Rill [*erosion*],.. accelerated erosion by water which produces small channels that can be obliterated by tillage. 1966 *McGraw-Hill Encycl. Sci. & Technol.* XII 431/1 (*caption*) Rill erosion showing how water has followed the old corn rows. 1963 D. W. & E. E. Humphries tr. *Termier's Erosion & Sedimentation* x. 211 Tidal currents which occur during the retreat of the sea from a beach form a pattern of fine channels, particularly where the water is retarded by obstacles, pebbles or shells. These channels or rill-marks formed on the surface of moist, soft sand can be preserved by fossilization. 1933 R. Campbell *Flowering Reeds* 28 The shimmering beams of a morning that sinewed The lowlands with silver, and trawled to the plains, Rill-threaded, the sweep of its glittering seines. 1908 *Jrnl. Geol.* XVI. 751 (*heading*) Hypothesis of rill-wash applied to the St. Louis region. 1937 *Bull. Geol. Soc. Amer.* XLVIII. 343 The interstream areas..are today being lowered chiefly by tributary streams and rill wash. 1972 A. Young *Slopes* vii. 62 Surface flow may be divided into sheetwash, where the ground is entirely or largely covered by a moving layer of water, and rillwash, when the water flows mainly as micro-channels.

3. (Earlier and later examples.)

1876 E. Nelson *Moon* iii. 71 There is one class of formations..which, from their unknown nature, cannot well be classified. These are the rills or clefts. 1954 *Proc. Nat. Acad. Sci.* XL. 1103 The rills are cracks in the surface, about 1" (2 km.) in width and narrower, occurring almost anywhere on the moon. 1963 A. N. Strahler *Earth Sci.* v. 83/2 Rills may be on the order of a mile wide and as long as 150 miles.

4. *Phonetics.* [ad. Da. *rille* (O. Jespersen *Fonetik*, 1899).] Used *attrib.* to designate a fricative produced by forcing air through a groove-like aperture between the tongue and the roof of the mouth (see quots.).

1912 E. Prokosch in *Amer. Jrnl. Philol.* XXXIII. 197 Spirants of these places of articulation can be formed in two ways: either, the surface of the tongue is convex, so that the breath passes through a narrow slit, as with *þ, χ*; or, the tongue forms a more or less distinct rill in its median line, as with *s, sh*. The former may be called slit sounds, the latter rill sounds. 1918 A. W. Aron in C. F. Hockett *Leonard Bloomfield Anthol.* (1970) 56 They are synonymous with what we call in Jespersen's terminology 'rill spirants', namely unvoiced *s*, voiced *z*, and the sibilants in *shall* and *azure*. 1958 C. F. Hockett *Course in Mod. Linguistics* viii. 72 Both English /sz/ and English /θð/ are normally apico-alveolar, but the former are *rill* spirants, the latter *slit* spirants.

rille. (Further examples.)

1957 *New Scientist* 7 Nov. 16 Criss-crossing the lunar surface there are alignments of mountains, valleys and trench-like formations called rilles. 1971 *Daily Colonist* (Victoria, B.C.) 25 July 2/3 A gorge called Hadley Rille cuts an 80-mile long 1,200-feet-deep gash across the..face of the moon. 1977 A. Hallam *Planet Earth* 15/3 These features, known as rilles, are shallow, often sinuous, valleys that cut across the maria.

rilled (rild), *a.* [f. Rill *sb.*[1], Rille + -ed[2].] Marked by or possessing rills or rilles; having narrow grooves.

1924 R. Newstead *Guide to the Study of Tsetse-Flies* 24 Wings with the membrane slightly rilled. 1935 O. Davies *Roman Mines in Europe* i. 39 In prehistoric times ore was pounded with rilled hammers. 1957 N. K. Sandars *Bronze Age Cultures in France* iii. 130 Rilled [G. *gerillt*] ware pots are of various shapes... Rilled decoration appears quite suddenly in France with graves and bronzes. 1970 R. E. Horton in G. H. Drury *Rivers & River Terraces* v. 154 In the latter case..a rilled surface may not develop. 1974 *Nature* 12 Apr. 573/2 The two large craters Oppenheimer and Schrödinger were also rilled. Small rilled craters are again found within larger flat-floored craters.

rillett(e)s. Add: The form ‖ **rillettes** (riyet) is now usual. (Earlier and later examples.) Also **rillettes de Tours,** a finely-minced pork pâté from the Tours region of France.

1889 J. Whitehead *Steward's Handbk.* iv. 420/1 *Rillettes de Tours*, cold cakes of meat of the head-cheese order. At the Paris ham fairs the rillette makers build up fancy pyramids of small rillettes and decorate them. 1894 G. du Maurier *Trilby* I. iii. 238 Tongues, hams, *rillettes de Tours*..for the supper. 1931 A. de Croze *What to eat & drink in France* xix. 160 *Rillettes de Tours* (same preparations as for 'Rillons', but the dices of meat are chopped and warmed again before being potted). 1960 E. David *French Provincial Cooking* 215 Another speciality..was the pâté made largely with *grattons*, the little browned scraps which are the residue after the pork fat has been melted down, and which were also the original ingredient of the renowned *rillettes de Tours*. 1973 *Times* 26 May 12/5 One local resident..writes, properly conscience-stricken, in praise of its [sc. a restaurant's] rillettes and ratatouille.

rilling (ri·liŋ), *sb.*[2] *Archæol.* [f. Rill *sb.*[1] + -ing[1].] Pottery decoration or marking of a rilled nature. Cf. *rilled *a.*

1940 C. F. C. Hawkes *Prehist. Found. Europe* iv. 98 There are straight or curving bands of rilling. 1957 N. K. Sandars *Bronze Age Cultures in France* iii. 130 The rilling is close-set, lightly drawn with a blunt implement on the moist clay, leaving a noticeable ridge between the rills. 1964 H. Hodges *Artifacts* i. 29 The presence on the inner and outer surfaces of pottery of fine horizontal grooves and ridges, called *rilling*, caused by the coarser particles in the clay body and irregularities of the potter's hands, is commonly held to be a certain indication that pottery was wheel-thrown.

rilling, *vbl. sb.* Delete † and add: **1.** (Later examples.)

1975 R. V. Ruhe *Geomorphol.* vi. 107/2 Rilling, a more effective erosion agent than sheet wash.., is a dominant process in hillslope reduction. 1978 W. W. Emmett in M. J. Kirkby *Hillslope Hydrol.* v. 171 Rilling is generally considered to be evidence of more accelerated erosion than sheet erosion.

2. Flow in rills.

1929 R. Bridges *Testament of Beauty* i. 12 Valleys vocal with angelic rilling of rocky streams.

Rilsan (ri·lsæn). Also **rilsan.** [Invented word: *ri-* f. F. *ricin* castor-oil plant (see Ricinus).] A proprietary name (orig. used in France) for a kind of nylon used esp. as a fibre.

1951 R. Dumon in *Rev. Gén. Caoutchouc* XXVIII. 348 Rilsan, a new plastic of the polyamide type..made out of castor oil, may be spun in order to get textile threads. 1952 *Sun* (Baltimore) (B ed.) 28 Feb. 4/3 Two plants to produce rilsan already have been built and a third is under construction. 1956 *Official Gaz.* (U.S. Patent Office) 27 Mar. TM 185/1 Rilsan Corporation, Boston, Mass. Filed May 17, 1955. *Rilsan*... For synthetic yarns. Use since July 27, 1953. 1963 *Trade Marks Jrnl.* 2 Oct. 1393/2 Rilsan... All goods included in Class 22 made of plastics. Organico.., Paris... Manufacturers. 1963 A. J. Hall *Textile Sci.* ii. 84 Rilsan fabrics can be permanently pleated under appropriate high temperature heat-setting conditions. 1969 —— *Stand. Handbk. Textiles* (ed. 7) i. 60 At present Rilsan fibres are used in articles such as brushes.

rim, *sb.*[1] Add: **1. b.** (Further examples of a drum.)

1934 [see *rim-shot*, sense 6 below]. 1976 *New Yorker* 8 Mar. 108/3 He would hit the snare directly, or hit the snare-head and the rim (a rim shot).

d. *pl. spec.* that part of the frame of a pair of spectacles which surrounds the lens.

1865 Dickens *Mut. Fr.* II. iii. vi. 48 Mr Wegg, in fitting on his spectacles, opened his eyes wide, over their rims, and tapped the side of his nose. 1903 G. B. Shaw *Man & Superman* iv. 147 He pauses quietly to take out and put on his spectacles, which have gold rims. 1923 A. Huxley *Antic Hay* x. 156 For semi-evening dress, shell rims with gold ear-pieces—and gold nose-bridge. 1937 J. Squire *Honeysuckle & Bee* i. 24 They certainly wouldn't swallow my yarn at a 'Spike' if I turned up in tortoiseshell rims.

3. c. *U.S. slang.* The outer edge of the semi-circular or horseshoe-shaped desk at which a newspaper's sub-editors work.

1923 W. G. Bleyer *Newspaper Writing & Editing* (rev. ed.) i. 10 The copy-desk is semicircular in form, and the head copy-reader sits in the 'slot', or inside of the desk, while the copy-readers occupy places around the outside, or 'rim'. **1933** Garst & Bernstein *Headlines & Deadlines* ii. 26 The copy desk usually is semicircular or horseshoe-shaped...The copy editors sit 'on the rim'. **1976** *National Observer* (U.S.) 2 Oct. 4/2 At 9:05 p.m., Crimson President Jim Kramer hurries over to the rim, where Managing Editor Margaret Shapiro is marking copy.

4. d. *ellipt.* = *rim-rock*, sense 6 in Dict. and Suppl. *N. Amer.*

1869 J. Anderson *Rough but Honest Miner* in *Sawney's Lett.* (ed. 2), He hammers at the rock, Believin' it's a rim, When ten to ane 'tis naething But his fancy's whim. **1946** *Notes Placer-Mining* (Brit. Columbia Dept. Mines Bull. No. 21) 21 The first shallow diggings gave way to.. sniping operations along the rims or on weathered bed-rock. **1976** *Billings* (Montana) *Gaz.* 4 July 8-D/4 (Advt.), Between the rims, comfortable, pleasant and quiet 2 and 3 bedroom duplex.

6. rim band, a driving belt or rope passing around a rim wheel; **rim-brake** (earlier and later examples); **rim drive,** a method of driving a gramophone turntable by means of frictional contact between the motor shaft and the inner rim of the turntable, often with an intermediate wheel between the two; so **rim-driven** *a.*; **rim-fire** (examples in latter sense); **rim light** *Photogr.* and *Cinemat.*, a lamp placed behind the subject in order to produce the appearance of a halo of light; also, the light produced by a lamp in such a position; so **rim lighting**; **rim man** *U.S. slang*, a newspaper sub-editor (cf. sense 3 c above); **rim-rack** *v. trans. U.S. dial.* (chiefly *Naut.*), to injure or damage (something) (see also quot. 1929); also *fig.*; hence **rim-racked**; **rim-rock,** for *U.S.* read *N. Amer.* and add: earlier and later examples; hence as *v. trans.*, to drive (sheep) over a cliff (see quot. 1944); also **rim-rocker,** (*a*) one who rim-rocks sheep; (*b*) (see quot. 1968); **rim-shot,** a drum-stroke in which the stick strikes the rim and the head of the drum simultaneously; **rimstone** *Geol.*, a thin layer of calcite deposited round the rim of an overflowing basin or in an evaporating pool of water, characteristic of karst topography; **rim wheel,** substitute for def.: on a spinning mule, a large pulley which transmits drive to the spindles and serves as a fly-wheel; (later examples).

1890 J. Nasmith *Mod. Cotton Spinning Machinery* xi. 184 Over this the endless cord or band driving the spindles is passed—being known as the 'rim band'. **1970** H. Catling *Spinning Mule* ix. 163 Minders naturally spared neither trouble nor expense to keep their rim bands in the best possible condition. **1896** A. Sharp *Bicycles & Tricycles* xxxii. 528 Tyre and Rim Brakes.—The brake is usually applied to the tyre of the front wheel. **1917** *Cycling Man.* 18 Rim brakes are the most common in use. **1974** *Listener* 14 Feb. 212/2 The effectiveness of rim brakes is reduced..in wet weather. **1961** E. N. Bradley *Records & Gramophone Equipment* iii. 69 Wow could also occur as a result of damage to the turntable rim (in a rim-drive system). **1962** A. Nisbett *Technique Sound Studio* viii. 147 Quick-start techniques fall into two categories, depending on the type of drive employed by the turntable. (i) Rim drive. **1976** A. Hope *Hi-Fi Handbk.* iv. 43 In an idler wheel or rim drive design the motor drives a sequence of rollers... Belt drive tends to be quieter, and more expensive, than rim drive. **1956** G. Slot *Hi-Fi from Microphone to Ear* vii. 67 The rim of the turntable (if it is rim-driven) [must be] perfectly circular. **1900–1901** *Kynoch Jrnl.* Dec.–Jan. 41/2 The first real rifle I possessed was an antiquated rim-fire rook rifle. **1977** D. Seaman *Committee* 14 Sedgwick assembled the ·22 Marlin rimfire repeater. **1940** W. Nurnberg *Lighting for Photogr.* iv. 105 The greatest luminosity in a rim-light picture is obviously always the main light produced by the basic illumination. **1948** —— *Lighting for Portraiture* ii. 84 A combination of basic cross lighting..supplemented by a rim light on the near side, is bad. **1961** P. Snow *Electronic Flashlight Photogr.* xiv. 115 We are already using two backlights for 'rim-light' effects on the subject. **1977** J. Hedgecoe *Photographer's Handbk.* 94 Generally it is advisable to reveal some detail in the rest of the face by using frontal light. This should be soft, shadowless and less intense than the rim light. **1940** W. Nurnberg *Lighting for Photogr.* iii. 61 If we now move the light source horizontally we obtain side-lighting, and when continuing this movement rim-lighting. **1971** *Listener* 11 Nov. 671/2 The Fonda film has everything:..minimal rim-lighting, fancy focus-pulls. **1933** R. M. Neal *Newspaper Desk Work* ii. 7 Copyreaders..are designated as 'rim men'. **1970** R. K. Kent *Lang. Journalism* 113 Rim men are the copy-readers who sit at the rim to do their work. **1914** *Dialect Notes* IV. 78 *Rimrack*, to injure, damage. **1929** F. C. Bowen *Sea Slang* 112 *Rim-Rack, to*, to strain or damage a vessel, particularly by driving her too hard in a sea. Very frequently used on the Grand Banks. **1952** *Amer. Legion Mag.* July 5/2, I am going to tell you right now my husband is not joining the American Legion another year if you keep up this rim-racking the Administration. **1957** *Maine Coast Fisherman* July 21/1 If nets have been rimracked it means working on deck under flood repairing nets and gear. **1974** J. Dowell *Look-Off Bear* 32 Faded denim workshirt, rim-racked old felt hat..and calf-high lumberman's rubbers. **1860** H. Greeley *Overland Journey* 350 It is one of the arts of the miner to know just where to tunnel through the 'rim rock' so as to strike what was the bottom of the lake. **1942** Berrey & Van den Bark *Amer. Thes. Slang.* §917/9 *Rimrock*, to drive sheep over a precipice. **1944** R. F. Adams *Western Words* 128/2 *Rim rockin' sheep*, running sheep over a cliff to destruction. This was often done during the wars between the cattle and the sheep factions. **1949** *World-Herald Mag.* (Omaha, Nebraska) 18 Sept. 18/5 Cattle raisers destroyed the flocks [of sheep] by clubbing, shooting, dynamiting,..poisoning, and stampeding them over cliffs—a practice sometimes called rim-rocking. **1958** 'W. Henry' *Seven Men at Mimbres Springs* ix. 101 They're..crawling the rimrock to smell out what took Major Mobry off so suddenlike. **1976** *Billings* (Montana) *Gaz.* 27 June 5-D/8 (Advt.), Canyon setting surrounded by the rimrocks, nestled in Spring Valley. **1934** *Nat. Geogr. Mag.* Feb. 213/2 My first companion, John Mullens, of Homestead, a 'rim-rocker'.. steered me through in six days. **1950** *Amer. Speech* XXV. 305 The term *rimrockers* goes back to the days of the cattle-sheep wars. **1968** R. F. Adams *Western Words* (rev. ed.) 251/2 *Rim rocker*, a cowboy's name for a horse agile enough to climb steep hills and travel over rocks and rough country. **1934** E. Little *Mod. Rhythmic Drumming* 25 A characteristic feature of rhythmic drumming is the rim-shot. This is the effect obtained by striking the snare drum head and the edge of the counter hoop simultaneously. **1968** *Crescendo* May 30/3 Bellson's rhythm is tremendous. His sympathetic fills, from 'cute' rimshots to make-'em-'ave-it roar-ups kicking the phrases are worth the price of the record alone. **1977** J. Wainwright *Do Nothin'* viii. 133 The right stick is used for the fireworks: for the rimshots, for the off-beats, for the roll-work. **1930** W. M. Davis in *Bull. Geol. Soc. Amer.* XLI. 485 Rimstone has been added to the list to name calcareous deposits formed around the rims of overflowing basins. **1968** R. W. Fairbridge *Encycl. Geomorphol.* 1051/1 Rimstone dams greater than 40 feet in height are known while micro-forms only a few millimetres high abound on stalagmites and flowstone. **1971** *Islander* (Victoria, B.C.) 24 Jan. 5/3 Rimstones are the layers of calcite left by receding pools. **1894** J. Lister *Cotton Manuf.* vi. 54 The main shaft drives the spindle by the rim wheel. **1950** J. W. Radcliffe *Woollen & Worsted Yarn Manuf.* xxv. 351 (*caption*) Fast and loose pulleys, rim wheels for driving spindles, and the backing-off friction.

rim, *sb.*⁴ (Earlier example.)

1817 G. Barker *Diary* 15 May in *Voorloper* (1976) 674 Oxen rims were also cut.

rim, *v.*¹ Add: **2.** *intr.* Of a steel ingot: to form an outer skin of relatively pure steel. Also *trans.* (causatively).

1958 A. D. Merriman *Dict. Metallurgy* 290/2 Steels which contain not more than 0·15% C can be made to 'rim'. **1967** A. H. Cottrell *Introd. Metallurgy* xi. 138 Only low-carbon steels contain enough oxygen to rim satisfactorily. **1974** *Encycl. Brit. Macropædia* XVII. 651/2 Plain carbon steels containing between 0·04 and 0·25 percent carbon can be rimmed successfully.

rim, *v.*³ [Perh. var. of Ream *v.*³ in Dict. and Suppl.; cf. Rim *v.*¹ and *rimmer³.]

1. a. *intr.* (See quot. 1923.) *U.S. dial.* **b.** *trans.* (See quot. 1972.) *coarse slang* (orig. *U.S.*).

1923 *Dialect Notes* V. 219 *Rim*, v., to be desirous of sexual intercourse, wanting to be bred. Said especially of sows. **1959** W. Burroughs *Naked Lunch* 90 'Darling, I want to rim you,' she whispers... 'All right. I'll go wash my ass.' 'No, I'll wash it.' **1972** B. Rodgers *Queens' Vernacular* 172 *Rim*, 1. To lick or suck anus. 2. to lubricate the anus with saliva, usually as a prelude to fucking. 'Tense? Try getting rimmed.' **1975** M. Amis *Dead Babies* xv. 74 Skip'd rim a snake so long as someone held its head.

2. *N. Amer. slang.* To cheat, to swindle (a person). Cf. *Ream *v.*³ 4.

1945 V. J. Monteleone *Criminal Slang* 84 *Rim*,..to cheat; to swindle: to defraud. **1973** D. Hughes *Along Side Road* vii. 54 Ten bucks? For that old thing? I'd be rimming you, Charles.

‖ **rimaye** (rimāy). [Fr., f. L. *rima* fissure + F. *-aye*, collective suffix.] = *bergschrund.

1920 H. Raeburn *Mountaineering Art* viii. 107 At the foot of the snow-slope, just where it eases into the glacier, runs a long chasm in the ice, the *bergschrund*, *randkluft*, or *rimaye*. **1949** A. Roch *Climbs of my Youth* xiv. 118 Taking to a couloir of easy broken rock, we soon reached the rimaye, where one last rappel had to be placed. **1957** E. A. Wrangham *Sel. Climbs Range Mont Blanc* III. 67 Go up the Dôme glacier,..to the rimaye below the Col des Aiguilles. **1974** G. Sutton tr. *Mazeaud's Naked before Mountain* i. 22 Slide he did, but into the rimaye, and we had to look for him and pull him out. *Ibid.*, The rimaye..is the large crevasse formed around the head of a glacier where it detaches itself from the mountain.

Rimbaldian (rimbæ·ldiăn), *a.* Also (ræmbō·diăn) **Rimbaudian**; ‖ **Rimbaldien, Rimbaudien.** [ad. F. *Rimbaldien*, f. the name of Arthur *Rimbaud* (1854–91), French poet + *-ien* -ian.] Of, pertaining to, or characteristic of Rimbaud or his poetry.

1924 R. Macaulay *Orphan Island* xxii. 296 A hundred birds unballed themselves and sang... 'Very Rimbaudien,' said Charles dizzily. **1952** *Essays in Crit.* II. 21 Her capacity for dissatisfaction..might have led to Rimbaldian adventures or Baudelairean visions. **1957** L. Varèse in *Rimbaud's Illuminations* p. xiv, The cardinal tenet of the *Rimbaldien* cult has been that *Une Saison en Enfer* was the nineteen-year-old boy's farewell to literature, his *mea culpa*. **1965** R. G. Cohn *Toward Poems of Mallarmé* 5 *Le Sonneur* indicates he [*sc.* Mallarmé] could have trodden further along the earthy Rimbaldian path. **1976** *Listener* 25 Nov. 691/3 The roared Rimbaudian lyrics of Peter Hammill. **1978** *Amer. Poetry Rev.* Nov./Dec. 30/1 Alcohol is the means but it is really fate at work—something a little Faustian, a little Coleridgian or Rimbaldian.

rimbellisher (rimbe·lifəɹ). [f. Rim *sb.*¹ + (Em)bellisher.] An ornamental chromium-plated rim placed round the wheel-hub of a motor vehicle; a wheel-trim. (No longer current.)

1949 *Motor* 10 Aug. 2 We announce the 'Rimbellisher' as an important addition to the already extensive range of ACE roadwheel disc equipment. This handsome accessory is a superimposed chromium rim which adds the finishing touch and makes wheel cleaning easier. **1955** *Times* 14 June 10/1 The wheels are fitted with rimbellishers as standard equipment. **1958** *Times* 12 Dec. 1/5 (Advt.), Jaguar 3·4 litre Saloon..fitted with overdrive, radio, rimbellishers, wing mirrors.

Rimbochay, Rimboché, obs. varr. *Rinpoche.

‖ **rimbombo** (rimbǫ·mbo). *rare*⁻¹. [It.; cf. Rimbomb *v.*] A booming roar; a resounding or reverberation.

1873 Geo. Eliot *Let.* 20 June (1956) V. 422 Shallow undiscriminating scorn, made to seem profound by a 'rimbombo' of rhetoric (like the singing into big jars to make demon-music in an opera).

rime, *sb.*¹ Add: **2.** ‖ **e.** *rime couée* = *tailed rime* s.v. Tailed *a.* 1 d.

c **1330** [see Couwee *a.*]. **1775** T. Tyrwhitt *Canterbury Tales of Chaucer* IV. 72 Though Robert of Brunne in his Prologue professes not to attempt these elegancies of composition, yet he has intermixed several passages in Rime Couwée. **1893** [see Tailed *a.* 1 d]. **1935** *Essays & Stud.* XX. 97 The rime couée or romance stanza of six lines (double eight and six). **1945** E. K. Chambers *Eng. Lit. at Close of Middle Ages* i. 25 The metre of the Chester plays..is a Romance metre of the type known as *rime couée* or tail-rhyme.

‖ **f.** *rime riche* = *rich rhyme* s.v. Rich *a.* 7 c.

[**1903** H. J. Chaytor *Compan. French Verse* iv. 23 Rimes are also distinguished as rich and sufficient (riche, suffisante).] **1904** Brandin & Hartog *Bk. French Prosody* iv. 53 Victor Hugo uses the *rime riche* more than any of his predecessors. **1930** A. Huxley *Vulgarity in Lit.* 35 When Laforgue wrote of that '*roi de Thulé, Immaculé*' his *rime riche* was entirely above suspicion. **1961** A. Clarke *Later Poems* 92 With the exception of the sonnet and the little experiment in *rime riche*, these pieces came to me quite unexpectedly.

5. *rime-making.*

1935 C. S. Lewis in *Lysistrata* May 22 Rude rime-making wrongs her beauty, Whose breasts and brow.. Bewitch the worlds.

rime, *sb.*² Add: In scientific use now distinguished from hoar-frost (see *hoar-frost). (Further examples.)

1895 [see *hoar-frost]. **1912** W. de la Mare *Listeners* 23 At midnight 'neath a maze of stars I flame with glittering rime. **1921** A. E. M. Geddes *Meteorol.* vi. 182 Hoar frost must not be confused with rime, which is an accumulation of frozen moisture on trees, &c., and is formed only during fog. **1947** *Jrnl. R. Aeronaut. Soc.* LI. 274/1 With a smaller rate of catch of water, at a lower temperature of the air, the water will freeze in the area of catch producing ice which has a porous structure, and a mat surface. This type is known as rime ice. **1978** *Sci. Amer.* Apr. 144/3 When the saturation is higher than 140 percent, the growth of crystals is so rapid that rime (an amorphous deposit of frozen droplets) grows on the crystals and destroys their optical faces.

Comb. **1891** Kipling *Barrack-Room Ballads* (1892) 204 It's North you may run to the rime-ringed sun. **1910** W. de la Mare *Three Mulla-Mulgars* iii. 46 The rime-laden branches of the trees.

rime, *v.*¹ **3.** (Later examples.)

1915 *Encycl. Relig. & Ethics* VIII. 258/1 'Riming people to death—a practice used by the *filid* as well as by the druids—was connected with the power of the spoken word. **1969** J. Wainwright *Big Tickle* 166 He'll rime you—you lousy git—he'll rime you to ribbons.

rime, *v.*² Add: (Further examples.) In scientific use now restricted to mean: to cover with rime (see *rime *sb.*²). Also *intr.*, to become rimed.

1907 N. Munro *Daft Days* xxxii. 266 Oh, London, London!.. The multitudinous monuments rimed by years. **1966** *Jrnl. Faculty Sci. Hokkaidō Univ.* 7th Ser. II. 331 In Japan, most of snow crystals are more or less rimed. Therefore the grade of riming is very important. **1973** *Sci. Amer.* Jan. 105/1 When a crystal rimes, material is added mostly on its underside, thus increasing its weight without greatly increasing its air resistance. *Ibid.*, The more a crystal is rimed, the faster it rimes. **1978** *Nature* 24 Aug. 791/2 (*caption*) Ice fragments collected downwind of an ice-coated sphere riming at −7°C.

Hence **ri·ming** *vbl. sb.*²

1966 [see above]. **1969** *Jrnl. Atmospheric Sci.* XXVI. 138/1 This apparent multiplication of ice crystals may be associated with the process of riming of the crystals. **1973** *Sci. Amer.* Jan. 105/1 When a snow crystal evolves within a cloud of supercooled water drops, it can grow not only by stealing vapor from around the drops but also by

actually colliding with individual drops. The process is called riming, and it is an important complication in all types of snowflakes and snow crystals. **1978** *Nature* 24 Aug. 791/1 (*heading*) A possible mechanism of ice splinter production during riming.

rimed, *ppl. a.*² Delete *rare*⁻¹ and add further examples. Cf. *rime *sb.*²

1954 U. Nakaya *Snow Crystals* ii. 87 Snow crystals with numerous water droplets attached are very frequently observed in our climate... This type is better called a rimed crystal. **1973** *Nature* 26 Oct. 451/2 Mossop *et al.* report small ice crystals..but always in association with rimed ice pellets. **1975** *Ibid.* 22 May 317/1 Unrimed and lightly rimed planar and spatial crystals were sampled.

rime-frost. For ? *Obs.* read *rare* and add later examples.

1904 in *Eng. Dial. Dict.* **1957** Blunden *Poems of Many Years* 298 Their poor limbs shook With the wind's or the rimefrost's blue stroke.

rimer, *sb.*² Add: Now *rare.* Also (*U.S.*) rimmer (ri·məɹ). (Further examples.)

1865 J. H. A. Bone *Petroleum & Petroleum Wells* 22 But the hole must be as nearly round as possible, and therefore the tools are taken out, and a 'rimmer', or 'reamer', sent down, which cuts down the irregularities of the hole. **1894** W. J. Lineham *Text-bk. Mech. Engin.* vi. 209 Round holes are cleaned by the Parallel Rimer..and taper holes by means of a Taper Rimer. **1907** *Yesterday's Shopping* (1969) 701/2 Rimers, square or half round.

rimland (ri·mlænd). [f. Rim *sb.*¹ + Land *sb.*] A peripheral area of land of political or strategic significance.

1944 N. J. Spykman *Geogr. of Peace* iv. 38/1 The central continental plain can continue to be called the heartland but..it is..to be equated with the political extent of the Union of Soviet Socialist Republics. Beyond the mountain barrier, the coastland region..may..be referred to as the rimland, a name which defines its character accurately. *Ibid.* 43/2 In the three great world wars of the nineteenth and twentieth centuries,..the British and Russian empires have lined up together against an intervening rimland power as led by Napoleon, Wilhelm II, and Hitler. **1973** J. C. Plano et al. *Polit. Sci. Dict.* 331 *Rimland theory*, the theory that emphasizes the rimlands of Europe, the Middle East, Africa, South Asia, and the Far East as the keys to the security of the United States. **1979** *Daily Tel.* 23 June 17/3 Commanders of the small forces of Western Europe are extremely disturbed by the strong hold the Soviet Union and its satellites have acquired in what are dubbed 'the rimlands' of the world's greatest concentration of crude oil in the Persian Gulf.

rimless, *a.* Delete *rare* and add further examples of the word used of spectacles.

1908 *Sears, Roebuck & Co. Catal.* 187/2 Rimless spectacles are the very latest and most stylish spectacles made. **1923** A. Huxley *Antic Hay* x. 156 And for full dress, gold-mounted rimless pince-nez are refinement itself. **1934** J. B. Priestley *Eng. Journey* i. 5 He had a sharp nose, a neat moustache, rimless eyeglasses. **1953** K. Tennant *Joyful Condemned* xviii. 162 The magistrate looked over the top of his rimless glasses. **1977** D. James *Spy at Evening* vii. 38 He..balanced a pair of rimless spectacles on the tip of his nose.

rimmed, *a.* Add: **3.** *rimmed steel* = *rimming steel* s.v. *rimming (*ppl.*) *a.*

1926 *Iron Age* CXVII. 1778 (*heading*) Rimmed steel and how it is made. **1926** *Jrnl. Iron & Steel Inst.* CXIV. 579 The particular physical features of an ingot of good rimmed steel are a solid outer skin and certain gas-holes. **1959** *Ibid.* CXCI. 231 A statistical investigation into the existence of a functional relationship between tapping slag iron and carbon and manganese losses in rimmed steels. **1963** W. H. Dennis *Metallurgy of Ferrous Metals* xiii. 199 Rimmed steels possess a high degree of cleanliness.

rimmer³. U.S. var. Rimer *sb.*² in Dict. and Suppl.

rimming, (*ppl.*) *a.* [f. Rim *sb.*¹ or *v.*¹] *rimming steel*: a low-carbon steel in which deoxidation has been controlled and limited to produce ingots having an outer rim or skin relatively free from carbon and impurities.

1926 *Iron Age* CXVII. 1778 Rimming steel is also called by some 'open steel', because the top of the ingot does not freeze over as rimming in proceeds, but the central metal continues fluid and in active motion for some minutes after teeming. **1930** *Engineering* 14 Mar. 357/2 Exceptions are metals such as steel of the 'rimming' type, from which very large quantities of gas are rapidly evolved during solidification. **1956** J. Dearden *Iron & Steel To-Day* (ed. 2) ix. 148 Low carbon steel which has been only partially killed is known as 'rimming steel' because of the rim of almost pure iron which forms the outer portion of the ingot. **1967** A. H. Cottrell *Introd. Metallurgy* xi. 138 The alternative is to prevent the metal from shrinking by allowing a small amount of CO to form as bubbles, when it freezes. This is done in rimming steels. These are low-carbon (< 0·15 per cent C) steels usually used for sheet steel pressings.

‖ **rimon** (rimō·n). Pl. rim(m)onim. [Heb., lit. pomegranate.] A pomegranate-shaped ornament for a Jewish Law-scroll. Cf. Pomegranate 2.

[**1845** J. Kitto *Cycl. Bibl. Lit.* II. 635/1 Rimmon..is mentioned in numerous places in the Old Testament, and is universally acknowledged to denote the Pomegranate-tree and fruit... We find frequent mention of it as an ornament..in the temple.] **1946** *Hebrew Union Coll. Ann.* XIX. 363 Three Torah scrolls,..in vertical positions, are visible, dressed in beautiful mantles and crowned with golden *Rimmonim*. **1962** 'E. McBain' *Empty Hours* 112 'I was putting the *rimonim* back onto the handles of the scroll.' 'Putting the what, sir?' Carella asked. 'Listen to the big Talmudic scholar,' Meyer said, grinning. 'Doesn't even know what *rimonim* are. They're those decorative silver covers, Steve, shaped like pomegranates.' **1976** Y. L. Bialer *Jewish Life* 188 *Rimon*, an adornment for the *sefer Torah*. *Ibid.* 189 *Rimonim* are made of gold, silver, crystal,..shaped like upright crowns, arched towers, [etc.].

rimpi. Add: Also rimpey, rimpje. (Further examples.)

1887 A. A. Anderson *25 Yrs in Waggon* I. xii. 280 A rope of beads of sufficient length to go round the loins twice and fastened in front with a piece of rimpey. **1914** Kipling in *Geogr. Jrnl.* XLIII. 372 The smell of home-made rimpje on a Dutch farm at the other side of the world.

Rimpoche, var. *Rinpoche.

rim-schoen, var. *remskoen.

‖ **rin** (rin). [Jap.] A Japanese monetary unit, equal to $\frac{1}{10}$ sen; also, a coin of this value. Also *collect.* as *pl.*

1875 [see Yen]. **1875** [see Sen]. **1891** A. M. Bacon *Japanese Girls & Women* vi. 160 There is something picturesque about these sen and rin. **1931** *Economist* 2 May 947/1 [Japanese] savings banks followed suit..reducing the rate [on deposits] by 1 rin per diem or 0·365 per cent per annum. **1962** R. A. G. Carson *Coins* 548 The bronze sen and 5 rin pieces of this reign had as obverse type the kiri-flower crest which had appeared on the gold obans and kobans of the shogunate.

‖ **rinceau** (ræ̃nso). *Art.* Also 8 **rainçeau.** [Fr., in the same sense.] (See quot. 1962.)

1778 R. & J. Adam *Works in Archit.* I. 5 We..have added grace and beauty to the whole, by a mixture of grotesque (E) stucco, and painted ornaments, together with the flowing rainçeau (F) with its fanciful figures and winding foliage. **1917** A. D. F. Hamlin *Hist. Ornament* vii. 98 The *rinceau*..which is a combination of the S-line, the spiral, the vine-motive and the acanthus leaf, was developed during the Alexandrian age into an ornament which has contributed a most important element into the splendor of Roman, mediæval and modern art. **1941** *Burlington Mag.* July 25/1 There is an eagle at each corner with *rinceaux*, and above are half-compartments with garlands. **1962** R. G. Haggar *Dict. Art Terms* 290/2 *Rinceau*, an elaborate foliated spiral or scroll pattern.

rind, *sb.*¹ Add: **1. d.** *Bot.* A hard outer layer on a fungus.

1887 H. E. F. Garnsey tr. *A. de Bary's Compar. Morphol. & Biol. Fungi* iii. 58 In other forms the rind is distinguished from the medulla by gelatinous cell-walls, as in the pileus and stipe of Agaricus (Mycena) vulgaris. **1927** Gwynne-Vaughan & Barnes *Struct. & Devel. Fungi* 1 The hyphae..may give rise to root-like strands known as rhizomorphs, or to a compact resting body, the sclerotium, the outer cells of which are modified to form a rind, protecting the inner regions from desiccation. **1951** J. A. MacDonald *Introd. Mycol.* ii. 14 In a few cases the mycelial cord is surrounded with a dark rind similar to that which surrounds a sclerotium. **1974** *Canad. Jrnl. Bot.* LII. 1128/2 About 5½ days after inoculation, a definite organization of mycelia to form the rind was observed.

4. (Further examples in *Anat.*)

1950 [see *adrenal *a.*]. **1974** *Nature* 4 Oct. 428/2 The ganglia have a cell rind formed by nerve cell perikarya.

6. c. *slang.* Impudence, effrontery, 'cheek'. Cf. *crust *sb.* 7 b.

1903 A. M. Binstead *Pitcher in Paradise* iii. 79 With that preface they had the immortal rind to pull out a fifth document for me to sign, guaranteeing them the starting-price as returned nightly in *The Evening Standard*. **1915** Wodehouse *Something Fresh* v. 154 You have the immortal rind to suppose that I will stand being nagged and bullied. **1977** *Times Lit. Suppl.* 1 July 791/1 *The Björn Borg Story* (I'm glad they didn't have the rind to use the word 'Life').

7. rind grafting, grafting in which the scion is inserted between the bark and the wood of a stump; = *crown-grafting* s.v. Crown *sb.* 35; so **rind graft.**

1947 R. J. Garner *Grafter's Handbk.* viii. 181 Unlike the oblique cleft the rind graft must be tied firmly with soft string. **1956** *Dict. Gardening* (R. Hort. Soc.) (ed. 2) II. 918/1 All those [branches and spurs] left are grafted by whip-and-tongue or rind-grafts according to size. **1726** R. Bradley *Improvements Planting & Gardening* (ed. 5) 558 The first Sort of Graffing which I shall mention, is that Sort which we call Whip Graffing, or Rind Graffing. **1881** *Encycl. Brit.* XII. 236/2 Crown-grafting or rind-grafting..is preferable to cleft-grafting, inasmuch as it leaves no open spaces in the wood. **1882** [in Dict.]. **1969** *E. Afr. Agric. & Forestry Jrnl.* XXXV. 144/2 An attempt was made to try 'rind grafting' on the species at breast height..under field conditions.

rind, *v.*¹ Add: **b.** To rub or remove skin from (a person or animal) or from (an item of food, *esp.* bacon: see sense 5 of the *sb.*).

1893 *Eng. Illustr. Mag.* Sept. 872/1 Tom Walker used to rub his bleeding fingers in the dust after being rinded by David Harris. **1962** *Sunday Times* 14 Jan. 30/3 Rind the bacon rashers and peel the bananas.

rinforzando (rinfǫɹtsæ·ndo). *Mus.* [It., gerund of *rinforzare* to strengthen.] A sudden stress or crescendo made on a short phrase; a direction to make this. Also *fig.*

1801 Busby *Dict. Mus.*, *Rinforzando*, the same as *Forzando*. **1812** [see Crescendo 1 a]. **1858** Geo. Eliot *Scenes Clerical Life* I. 48 Mr. Spratt was boxing the boys' ears with a constant *rinforzando*. **1944** W. J. Finn *Conductor raises his Baton* (1946) iv. 132 *Rinforzando* is only akin to *sforzando* but is often treated as a *subito* by conductors. The term *rinforzando* is not found in many modern scores. **1965** E. Kahn *Conducting* v. 58 *Rinf.* or *rfz* (rinforzando) indicate an emphasis on a short phrase.

ring, *sb.*¹ Add: **I. 1. e.** Phr. *to get the ring*, to become engaged to be married (usu. said of a woman).

1914 Joyce *Dubliners* 122 Lizzie Fleming said Maria was sure to get the ring and..Maria had to laugh and say she didn't want any ring or man either. **1951** in M. McLuhan *Mech. Bride* (1967) 95 The girls who get the rings. **1979** D. Cook *Winter Doves* II. iii. 64 Got the ring yet, Walter? Has she proposed to you yet?

3. g. = *curtain ring* s.v. Curtain *sb.*¹ 8.

1814 Jane Austen *Mansf. Park* I. xv. 296 The curtain will be a good job... We shall be able to send back some dozens of the rings. **1847** C. Brontë *Jane Eyre* II. i. 2 A woman..sewing rings to new curtains. **1926–7** *Army & Navy Stores Catal.* 1106/1 Rod for curtains... Plain Ball Ends... Rings, per dozen (⅛ in. larger than pole).

h. A metal or plastic band placed round the leg of a bird, usu. when a nestling, so that it may be uniquely identified when caught on a later occasion; a *leg-ring* (see *leg *sb.* 17 a); also, a similar marker placed on a limb of a bat.

1907 *Brit. Birds* I. 58 The plan of marking birds by an aluminium ring round the leg has often been tried, but never in a really systematic fashion... To place rings on the legs of young birds just before they fledge would not be a great difficulty. **1909** *Ibid.* III. 5 We have had prepared a number of rings, of which we will send a supply..to any reader of the Magazine who will undertake to mark birds. **1925** Turner & Gurney *Bk. about Birds* vii. 71 Much has been discovered about the movements of birds by fixing small numbered rings on the legs of young birds in the nest. **1958** *Listener* 30 Oct. 684/1 The 'ring' is generally a thin strip of aluminium, shaped like the letter 'C' and is marked with a number and some sort of address. On a bird, it is clamped round the leg; but a bat's leg is too delicate for this, so the ring is put round the fore-arm. **1973** *Guardian* 5 Feb. 11/5 The British Bird Fancy Council..is now introducing a system of coded rings..and encouraging breeders to keep full records.

i. A bottomless vessel used in **ring culture.*

1962 H. G. W. Fogg *Chrysanthemum Growing* ix. 62 The roots in the rings will not need any watering between feeds. **1964** *Times Lit. Suppl.* 21 May 442/2 Much of the early work on ring culture, a system of growing plants in bottomless pots or 'rings', was carried out at the Lenton Research Station under the direction of the author [*sc.* A. W. Billitt]. Ring culture has become now firmly established both as a commercial and amateur method of growing chrysanthemums, carnations and tomatoes. **1976** *Observer* (Colour Suppl.) 9 May 12/2 The best way to grow tomatoes is by 'ring culture'. The soil in the greenhouse..is replaced with aggregate, such as clinker, and the tomatoes planted in rings, or bottomless pots.

4. b. *pl.* A competitive game in which rings are thrown on to hooks.

1906 B. Kennedy *Wander Pict.* 245 Over yonder on the [inn] wall is the target with hooks at which they play the game of rings. They stand off and pitch rubber rings on to the hooks.

II. 7. f. *slang.* The anus. Phr. *to spew one's ring* (and similar phrases), to vomit violently.

1949 Partridge *Dict. Slang* (ed. 3) 1154/1 *Ring*,.. anus (also *ring-piece*): low: late C. 19–20. **1952** T. A. G. Hungerford *Ridge & River* 130 'I'd get shot in the ring, that's what I'd get,' said Wallace. **1965** R. Stow *Merry-go-Round in Sea* 174, I bet I would have booted him in the ring if he hadn't run. **1966** K. Amis *Anti-Death League* I. 32 Then the technique is to slip him a glass of Scotch or whatever he's hooked on about half a minute before the emetine makes him spew his ring. **1971** B. W. Aldiss *Soldier Erect* 105, I can't take this sodding shitting Wog beer... Makes me spew my ring every time! **1978** R. Busby *Garvey's Code* iii. 28 We just left the husband and he's bringing his ring up.

8. a. (Earlier and later examples of *rings round* (or *under*) *the eyes.*)

1850 Thackeray *Pendennis* II. vii. 70 The rings round his eyes were of the colour of bistre. **1856** C. M. Yonge *Daisy Chain* II. xii. 472 He looked very wan, with the dark rings round his eyes, a deeper purple than ever. **1902** A. Bennett *Anna of Five Towns* x. 258 It's a shame to send you home with those rings round your eyes. **1911** O. Onions *Widdershins* 270 The bistred rings that weeks of nursing had put under her dark eyes. **1973** R. Thomas *If you can't be Good* xiii. 111, I thought there were some new lines in his face. I wasn't imagining the dark rings under his eyes. **1981** T. Heald *Murder at Moose Jaw* xi. 133 There were dark rings under her eyes and her face was pinched.

e. A gold-coloured band worn on the sleeve to designate rank in the armed services. Also *transf.*

1942 *R.A.F. Jrnl.* 13 June 32 They were all of superior rank to myself... My solitary ring did not allow me to voice my suspicions. **1943** Hunt & Pringle *Service Slang* 55 *Rings*, abbreviated reference to an Officer's rank, denoted in the Navy and R.A.F. by the number of rings on his sleeve. **1950** 'D. Divine' *King of Fassarai* xv. 116 Bull's

got more rings than I have. Why shouldn't he have his headaches too? **1952** M. Tripp *Faith is Windsock* i. 14 Now that Bergen has got his ring there doesn't seem so much point in staying N.C.O. with the others. **1953** 'N. Shute' *In Wet* v. 164 Go into the R.A.F. and try to make a go of it, and get the rings because you've earned them, not because you're heir to the Throne. **1970** D. Francis *Rat Race* iv. 55 No such thing was possible in one of his aircraft, he had told me stiffly, and I could take my four gold rings away... I hadn't worn my captain's jacket for nearly two years. **1976** 'A. Hall' *Kobra Manifesto* xvi. 218 The pilot stood there, a tall mahogany-faced type with four gold rings on his sleeve.

9. i. (Earlier and later examples.)

1835 A. Smith *Diary* 27 July (1940) II. 131 The old ring-kop was speaking... This ring-kop was the person in charge of the party... None of the others had rings. **1866** [see *head-ring* s.v. *head *sb.* 66]. **1925** D. Kidd *Essential Kaffir* 33 Only married men are allowed to wear this ring. **1952** [see *head-ring* s.v. *head *sb.* 66].

III. 10. d. *Chem.* A number of atoms bonded together to form a closed chain.

1869 *Jrnl. Chem. Soc.* XXII. 361 The hydrocarbon benzole is of so much interest from its derivatives, that it has attracted a good deal of attention, and to explain its molecular constitution, the six atoms of carbon have been represented as arranged in a ring. **1889** G. M'Gowan tr. *Bernthsen's Text-bk. Org. Chem.* 461 Phthalic acid or its derivatives ensue on the breaking up of the compound, not only from one but from both of the six-cornered rings. **1927** N. V. Sidgwick *Electronic Theory of Valency* xiv. 251 Only a very few chelate rings of more than six atoms have been observed. **1950** *Thorpe's Dict. Appl. Chem.* (ed. 4) X. 339/1 Treatment of pyrrole with hydroxylamine causes a smooth opening of the ring. **1964** N. G. Clark *Mod. Org. Chem.* ii. 21 Pyridine..has a structure very similar to that of benzene; the six-membered ring, however, contains a trivalent nitrogen atom. **1974** D. M. Adams *Inorg. Solids* vi. 183 Sulphur has quite a complex allotropy: the thermodynamically stable form consists of S_8 crown-shaped rings in close-packed layers.

e. *Cytology.* A chromosome, group of chromosomes, or part of a chromosome in the form of a loop, without free ends.

1929 *Jrnl. Genetics* XXI. 44 A single ring of four may be formed at diakinesis, the other chromosomes assorting themselves in separate pairs. **1949** Darlington & Mather *Elem. Genetics* vi. 129 From the exchange system of pairing a ring of four, chain of four, two chains of two, or a chain of three with a univalent, can arise at metaphase. *Ibid.* xii. 263 *Oenothera lamarckiana*, whose chromosomes normally form a ring of 12 and one pair. **1962** *Lancet* 29 Dec. 1384/1 Monosomic chromosomes may be subject to hazards in meiosis—from autosynapsis, leading to centric and acentric rings and fragments, to centric aberrations leading to iso-chromosomes.

11. a. (Further examples.) *spec.* a combination of dealers, contractors, or the like, who cooperate in buying or selling at agreed price-levels, in order to increase their profits. Also *attrib.*

1870 W. W. Fowler *Ten Years in Wall St.* i. 28 The bulls often unite to raise the value of particular stocks, and form those combinations known in the 'Street' as 'cliques', 'rings' or 'pools',—terms nearly synonymous. **1909** [see *conference 4 e]. **1929** *Times* 31 Oct. 14/4 In order to safeguard the home consumer against exploitation by the producers' 'rings', which the coal-marketing schemes will establish, the Government will set up in every district a special committee to keep a watch on prices. **1936** *Sun* (Baltimore) 2 Dec. 3/1 In addition Thorp advocates.. more stringent regulations of auction and sales rings. **1949** J. Symons *Bland Beginning* 22 That's Foskiss. Buys everything for the ring. Doesn't give the small men a chance. **1955** *Times* 13 May 9/1 Wellington City Council, which recently protested strongly against the submission of equal tenders by a number of British firms, has now decided to accept a tender for electric cable which is £3,000 below the 'ring price'. **1961** R. Godden *China Court* vii. 274 She comes back from sale after sale... She gets to know their 'tricks' as she calls them, a childish word for the ring. **1964** [see *monopsony]. **1968** R. H. R. Smithies *Shoplifter* viii. 175 The Ring..was an organization of English dealers who attempted to take over the entire antiques business—rigging auctions, intimidating retailers, fixing prices. **1968** *Sunday Times* 10 Nov. 1/2 Ring dealing is when dealers agree not to force the price up by bidding against each other at an auction. **1971** *Nature* 6 Aug. 365/1 The overt reason for the anti-trust laws is to prevent the formation of cartels and similar price-fixing rings. **1972** 'M. Delving' *Shadow of Himself* ii. 24 'I don't doubt there'll be a ring,' he added, bitterly... He was referring to the system by which several dealers agree to let one of their number bid for all, thus cutting down the competition and squeezing out opposition.

c. An organization or network of people engaged in espionage.

1943 D. Powell *Time to be Born* vi. 128 He had no secret mission to investigate the spy rings and unmask the Nazi agents. **1961** R. Seth *Anat. Spying* iii. 45 In a very short time counter-espionage knew the names and whereabouts of all twenty-six members of the ring. **1972** A. Morice *Murder on French Leave* viii. 104 The cinema had been the meeting-place with another operator in the spy ring. **1981** R. Airth *Once a Spy* xii. 139 When Franklyn's ring broke up..I went back to work for Bonn.

11*. *Math.* [a. G. *ring*, used in this sense.] A set of elements with two binary operations ('addition' and 'multiplication') which is a group under addition and closed under multiplication, and with the property that multiplication is distributive over addition and associative.

Some writers also require a ring to have an identity element for multiplication.

1935 *Ann. Math.* XXXVI. 406 It is only in integrally closed rings like in the theory of algebraic numbers that the decomposition theorems naturally take on a multiplicative form. **1967** [see *field theory]. **1968** D. G. Northcott *Lessons on Rings, Modules & Multiplicities* i. 1 From this point onwards, when we speak of a ring it is to be understood that we always mean a ring with an identity element. **1971** J. H. Conway in Powell & Higman *Finite Simple Groups* vii. 244 The ring of 24 × 24 matrices.

IV. 13. a. Phr. *to keep* (or *hold*) *the ring*, to be an onlooker at a fight; to stand by while others quarrel. Chiefly *fig.*

1829 P. Egan *Boxiana* 2nd Ser. II. 678 At least 500 Irishmen, armed *à la shilelah*, kept the ring. **1905** *Spectator* 21 Jan. 79/1 There is a cynicism which nothing but conscious impotence could excuse in the thought of 'keeping a ring' while the Bulgarians of the Principality..are drawn into a life-and-death struggle with the Turks. **1924** Galsworthy *White Monkey* I. vii. 54 Keep clear and keep the ring!.. Good friendly terms with..all the outlying countries that we can get at by sea. And let the others dree their weirds. **1928** *Britain's Industr. Future* (Liberal Industr. Inquiry) III. xv. 166 There are still many people who hold that the State ought not to meddle with industry,..but should confine itself to holding the ring while the disputants fight out their differences. **1970** *Times* 8 Apr. 1 Is there a limit to the amount of time that the Army can reasonably be expected to stay here [*sc.* in Ulster] holding the ring, if the humans involved here are not going to solve their own problems? **1970** *Atlantic Monthly* Nov. 6 Does the phrase 'Asians helping Asians', or, as critics put it, 'Asians fighting Asians', mean that the United States will provide everything but the front-line manpower, or that it will step back and help hold the ring, or that it will become a mere spectator? **1978** *Jrnl. R. Soc. Arts* CXXVI. 400/1 Its [*sc.* Government's] task in the economic field is to hold the ring between the many popular pressures, frequently irreconcilable, which assert themselves.

b. (Earlier examples.)

1775 G. Selwyn *Let.* 8 Dec. in *15th Rep. R. Comm. Hist. Manuscripts* App. vi. 306 in *Parl. Papers 1897* (C. 8551) LI. 1 The devil a bit will he ever part with, but by putting it into the Ring, where he is nicked, and the money gone. **1822** *Sporting Mag.* X. 4/2 Mr. Bayzand was well known in the ring as a betting man. *Ibid.* 192/2 He never opened his mouth in the ring *under five hundred*. **1845** Disraeli *Sibyl* I. i. ii. 12 'Will any one do anything about Hybiscus?' sang out a gentleman in the ring at Epsom.

d. A prison-yard; a fence or wall surrounding this.

1898 O. Wilde *Ballad of Reading Gaol* 7 And I and all the souls in pain, Who tramped the other ring. **1900** 'Odysseus' *Turkey in Europe* i. 28 They [*sc.* the Avars] were celebrated for their 'rings', enormous circular fences with which they surrounded the prisoners and plunder they had taken.

e. *Austral.* (See quot. 1941.)

1941 Baker *Dict. Austral. Slang* 60 *Ring*, the scene of operations of a two-up school or the school itself. **1948** [see *double-header e].

14. a. Also *ring-around.*

1907 Joyce *Chamber Music* p. ix, Winds of May, that dance on the sea, Dancing a ring-around in glee From furrow to furrow.

d. (Further examples.) Also with *make*, and in extended uses.

1907 Wodehouse *White Feather* viii. 88 Dexter's had taken thirty points off the School House just after half-time. 'Mopped them up,' said the terse and epigrammatic Painter. 'Made rings round them.' **1917** 'Contact' *Airman's Outings* 139 Snatches of familiar flying-talk..made rings round the Hun. *c* **1928** T. E. Lawrence *Lett.* (1938) 572 It riles me unbearably to lose my scalp to a lot of fellows round whom I can make rings. **1939** *War Illustr.* 4 Nov. 252/1 We saw a large black aeroplane travelling at a high speed. It was being pursued by two British fighters and they made rings round it. **1947** *Sporting Mirror* 7 Nov. 10/3 The return of Dodds revitalised the Everton attack which ran rings round Sheffield United. **1950** 'S. Ransome' *Deadly Miss Ashley* iii. 39, I can't help feeling we're having rings run around us. **1961** E. Waugh *Unconditional Surrender* III. iv. 292 Tito has..gone to join the Russians... Our chaps are rather annoyed about it... I bet Winston is. I told you he'd make rings round the old boy. **1973** 'D. Jordan' *Nile Green* xxxiv. 167 The deal's been bust... The Russians ran rings round us. **1980** *Times Lit. Suppl.* 27 June 724/2 Balfour was an undeniable flop and Joseph Chamberlain made rings round him.

V. 15. a. *ring-bearer, -bright, -finder, -game, -having, -lore, -plait.*

1932 W. Faulkner *Light in August* xi. 237 He was twelve then, and they wanted him to be the ringbearer. **1954** J. R. R. Tolkien *Fellowship of Ring* II. viii. 393 'And you, Ring-bearer,' she said, turning to Frodo. **1976** *Laurel* (Montana) *Outlook* 30 June 3/5 Gary Martin of Laurel, brother of the groom, was ring bearer. **1949** Blunden *After Bombing* 14 A child's eye drooped, so gleamed the ring-bright shell. **1954** J. R. R. Tolkien *Fellowship of Ring* II. i. 237 And you are the heir of Bilbo, the Ring-finder. **1886** *Lippincott's Monthly Mag.* Mar. 239 The ring-games, or 'carols', are great favorites, as they were among the English court ladies. **1916** A. S. Neill *Dominie Dismissed* vii. 98 The ring games down at the school there nearly all deal with love and matrimony. **1972** *Times* 7 Aug. (Jamaica Suppl.) p. ii/4 She has recorded Kumina music, ring games and Pocomania meetings. **1912** E. Pound *Ripostes* 27 He hath not heart for harping, nor in ring-having Nor winsomeness to wife. **1890** W. Jones *Finger-Ring Lore* p. vii, In thus contributing to the extension of knowledge, the subject of ring-lore has a close affinity to that of numismatics, but it possesses the supreme advantage of appealing to our sympathies and affections. **1954** J. R. R. Tolkien *Fellowship of Ring* II. ii. 57 All that he would reveal to us of his ring-lore. **1908** W. G. Collingwood *Scandinavian Brit.* 245 The ornament with ring-plaits.. cannot be earlier than the tenth century.

c. *ringwise* adj.

1944 *Sun* (Baltimore) 29 Mar. 15/2 Salica, ringwise and cunning, was unwilling to trade punches. **1958** F. C. Avis *Boxing Ref. Dict.* 95 *Ringwise*, gifted in ring tactics.

e. *Chem.* In sense 10 d, as *ring-closure, -compound, -formation, -opening, -structure*, etc.

1946 *Nature* 28 Dec. 930/1 The problem of effecting ring-closure through the meta- and para-positions of the benzene nucleus. **1964** N. G. Clark *Mod. Org. Chem.* xv. 294 On heating..these compounds readily undergo ring-closure through loss of a molecule of ethanol. **1932** *Discovery* Aug. 246/1 The changes during the coalification process proceed only to a half-way stage between the most stable kind of carbon chain compound, and the most stable kind of ring compound. **1961** *New Scientist* 16 Mar. 668/1 Almost any organic compound gets decomposed in time, even stable ring compounds like phenol, naphthalene and toluene. **1913** J. B. Cohen *Org. Chem. Adv. Students* II. ii. 111 Nearly all the above reactions may become intra-molecular if the necessary grouping is present, and in such cases ring formation follows. **1967** Margerison & East *Introd. Polymer Chem.* iii. 129 These formulations of the reactions of difunctional monomers have ignored the possibility of ring formation. **1959** Ring-opening [see *decarboxylate *v.*]. **1967** Margerison & East *Introd. Polymer Chem.* v. 261 Other ring-opening polymerizations in this class include the conversion of caprolactam to nylon 6. **1930** *Engineering* 18 Apr. 525/2 These investigations indicated that the cellulose molecule had a ring structure.

16. *ring-base, -brooch, -ditch, -foot, -gasket, -handle, -hook, -scissors, -weight.*

1957 V. G. Childe *Dawn European Civilization* (ed. 6) v. 60 Ring bases and genuine handles betoken an unusual degree of sophistication. **1960** T. Burton-Brown *Early Mediterranean Migrations* i. 27 The comparative frequency of ring-bases in the red-polished..wares..is a detail of some significance. **1883** C. T. Gatty *Catal. Mediæval & Later Antiquities Mayer Museum* 37 *Ring brooch*, in silver; inscribed on one side, in niello work, +AVE: MARIA: GRACIA: PLENA: D, and on the other +AGLA+NO. A+S: BLASIV. **1931** J. Evans *Eng. Posies & Posy Rings* p. xii, Such inscriptions were commonly engraved in the thirteenth and fourteenth centuries on the ring-brooches that were used to fasten the dress at the neck. **1977** *Antiquaries Jrnl.* LVII. 457/1 Note on the significance of the Londesborough ring brooch. **1936** *Oxoniensia* I. 8 In the following pages some account is given of the investigation of barrows and ring-ditches in the Oxford area. **1954** S. Piggott *Neolithic Cultures* ii. 32 These cultures are sometimes associated with causewayed ring-ditches of a funerary or ceremonial nature. **1927** Peake & Fleure *Priests & Kings* x. 162 Deep cups have been found with rounded bottoms, also bottles with lugs instead of handles and sometimes with hollow ring-feet. **1980** *Catal. Fine Chinese Ceramics* (Sotheby, Hong Kong) 76 The well-finished ring foot decorated with incised key-fret. **1972** L. M. Harris *Introd. Deepwater Floating Drilling Operations* ix. 93 An AX ring gasket provides a metal-to-metal seal between the connector and the wellhead. **1936** *Burlington Mag.* July 26/1 The pair of..candlesticks with ring-handles has the Britannia standard marks with the date-letter for 1736–37. **1972** *Trans. Oriental Ceramics Soc.* XXXVIII. 58 A grey earthenware jar and cover, with ring-handle masks in relief. **1913** J. Masefield *Daffodil Fields* 72 Its open door, With old wrought bridle ring-hooks at each flank. **1961** *B.S.I. News* Nov. 25 (*title*) Dimensions of ring-hook automatic hitch. **1908** *Practitioner* June 769 Later Mr. Jessop introduced his 'ring-scissors', which made this piecemeal removal of the gland an easier matter. **1922** Joyce *Ulysses* 665 Though ringweight lifting had been beyond his strength..he had excelled in his stable and protracted execution of the half lever movement on the parallel bars.

18. a. ring armature *Electr.*, an armature having a ring winding; **ring beam,** a ring-shaped beam of yarn; **ring binder,** a loose-leaf binder having clasps that pass through holes in the paper and can be closed to form rings; similarly **ring binding; ring book,** a notebook having the form of a **ring binder*; **ring-building** *Archæol.*, the forming of vessels by adding successive layers of ring-shaped pieces of clay; hence **ring-built** *a.*; **ring circuit,** (*a*) *Electronics* = *ring counter* below; (*b*) *Electr.*, a wiring arrangement for power distribution in domestic or similar premises in which sockets and fixed appliances are connected to a single loop of cable which starts from and returns to a fuse-box; **ring complex** *Geol.*, an association of igneous intrusions arranged in an arcuate or ring-like plan; **ring counter** *Electronics*, a counting circuit consisting of a number of flip-flops or other bistable devices wired in a closed loop; **ring-craft** (further examples); also *transf.* of other sports; **ring culture,** the technique of growing plants in a bottomless cylinder containing nutrients and resting on an inert bed through which water is provided; **ring current,** (*a*) *Geophysics*, a belt of charged particles which orbit the earth, trapped by the magnetic field in its ionosphere; (*b*) *Chem.*, a circulation of electrons in an annular molecular orbital (e.g. in aromatic molecules) under the influence of a magnetic field; **ring dike, dyke** *Geol.*, a dike that is arcuate or roughly circular in plan, formed by upwelling of magma along ring fractures following cauldron subsidence of a circular block; **ring flash** *Photogr.*, a circular electronic flash tube that fits round the camera lens to give shadowless lighting of a subject near the lens; **ring-fort** *Archæol.*, a fort or other position defended by ringed entrenchments; **ring fracture** *Geol.* [tr. G. *kreisbrüche*], a conical or nearly cylindrical fault associated with cauldron subsidence; **ring gland** *Zool.*, a gland in dipteran larvæ which secretes ecdysone; **ring-junction,** a road junction at which traffic is channelled in two directions round a central island, entering and leaving by smaller islands; **ring-keeper,** (*b*) (examples); **ring-kop** *S. Afr.* [Afrikaans *kop* head], an African tribesman or warrior entitled to wear a head-ring (see Head *sb.* 66 in Dict. and Suppl.); **ring-lock** (examples); **ring main,** (*a*) an electric main that starts from and returns to a particular power station or sub-station, so that each consumer has an alternative path for supply in the event of a failure; also = *ring circuit* (*b*) above; (*b*) *Plumbing*, an arrangement of pipes forming a closed loop into which steam, water, or sewage may be fed and whose points of draw-off are supplied by flow from two directions; **ring modulator** *Electronics*, a circuit that incorporates a closed loop of four diodes and can be used for balanced mixing and modulation of signals; **ring oiling,** a method of automatic lubrication of bearings in which a ring rests upon and turns with the journal and also dips into a reservoir containing the lubricant; so **ring-oiled** *ppl. a.*; **ring oiler; ring-opener,** a seal on a tin container which is broken by pulling a ring attached to it; **ring-play** *U.S.*, a circular dance movement accompanied by song; **ring-pored, -porous** *adjs. Forestry*, applied to woods in which the large pores produced in spring form partial or complete rings; cf. *diffuse-porous* adj. s.v. *diffuse *a.* 2 f; hence **ring porosity, -porousness; ring-pull** *a.*, designating a tin container fitted with a **ring-opener*; **ring-rope,** (*b*) usu. *pl.*, = Rope *sb.*¹ 2 c; **ring rot** [tr. G. *bakterienringfäule* (A. Spieckermann 1914, in *Landwirtsch. Jahrb.* XLVI v. 660)], a bacterial disease of the potato, affecting the tubers, caused by *Corynebacterium sepedonicum*; **ring scaler** *Electronics* = *ring counter* above; **ring-seat** = *ringside seat* s.v. *ringside b; **ring shake** *Forestry*, a partial or complete separation of two or more consecutive growth rings in a tree; = Cup-shake; so **ring-shaken** *a.*; **ring-shout** *U.S.*, a religious dance consisting of loud singing and circular movement; **ring-sight** *Mil.* (see quot. 1973); **ring spanner,** a spanner in which the jaws are in the form of a ring with internal serrations, which fit completely around the nut and put pressure on all its faces; **ringspot, ring spot,** (*a*) any of several plant diseases characterized by annular spots or marks on the leaves; (*b*) an annular mark on a plant or animal; so **ring-spotted** *a.*; **ring-toss** (earlier examples); **ring velvet,** velvet so fine that a width of it can be drawn through a ring; **ring-watch** (see quot. 1962); **ring winding** *Electr.*, a form of armature winding in which each turn of the winding passes through the centre of the hollow armature core (cf. *Gramme²); **ring-work,** (*a*) (later examples); **ring-worm** *U.S. slang* (see quot. 1929); **ring-yarn,** yarn produced by ring-spinning.

1893 Ring armature [see *Gramme²]. **1974** *Encycl. Brit. Macropædia* VI. 610/1 The ring armature..enormously improved the efficiency of early electric generators. **1924** *Times Trade & Engin. Suppl.* 29 Nov. 247/1 Ring beams are now worse off in margin by 1d. per lb. compared with a month or six weeks ago. **1929** A. J. Vaughan *Mod. Bookbinding* II. 136 *The ring binders.* These employ rings which may be opened or closed, the paper being pierced with round holes. **1977** *New Yorker* 27 June 30/1 Forced into proximity as we were by ledgers, ring binders, and jars of mucilage. **1977** P. D. James *Death of Expert Witness* II. v. 80 A quarto-sized loose-leaf notebook with a ring binding. **1923** H. A. Maddox *Dict. Stationery Terms* 67 *Ring books*, loose-leaf books arranged on the principle of split or hinged rings, which by a finger lever motion open to receive the leaves punched to fit the rings. **1942** H. A. Maddox *Dict. Stationery* 94 Ring book. **1965** P. Wylie

They both were Naked I. iv. 149 My second address was neatly arranged in a ring-book. **1957** V. G. Childe *Dawn European Civilization* (ed. 6) xi. 204 From Sweden to Siberia..all pots were manufactured by the same technique of ring-building. **1964** H. Hodges *Artifacts* i. 26 *Ring-building*... The walls of the vessel are formed by the addition of sausage-shaped rolls of clay of various lengths. **1963** E. M. Jope in Foster & Alcock *Culture & Environment* xiii. 337 In the tenth to twelfth centuries, the distinctive bar-lip style of pottery, flat-based and coil- or ring-built, rooted firmly in the simplest techniques of prehistoric pottery-making. **1964** H. Hodges *Artifacts* i. 29 Much early pottery that has been described as wheel-thrown was probably initially ring-built and only finally given its form on a rotating wheel. **1931** *Proc. R. Soc.* A. CXXXII. 306 The simplified 'chain' arrangement of thyratrons, from which the more useful 'ring' circuits..were developed. **1946** *Rev. Sci. Instruments* XVII. 185/2 The two pentodes of each trigger pair form opposite partners in the ring circuit. **1961** B. Pym *No Fond Return of Love* xxiv. 236 Her loud clear tones were addressed to the London Electricity Board, and the conversation seemed to be about power plugs and something called a 'ring circuit'. **1962** N. D. Watts in G. A. T. Burdett *Automatic Control Handbk.* xviii. 10 Fig. 11 shows a ring circuit in which the primer electrodes are omitted for simplicity. **1963** *Times* 6 May (Electric Power Suppl.) p. vii/3 Complete interchangeability at multiple points throughout the house is achieved by the 'ring circuit'. **1974** A. Douglas *Noah's Ark Murders* viii. 75, I thought lighting these days was on ring circuits—surely one burned-out fuse shouldn't mean total failure? **1916** E. B. Bailey et al. *Geol. Ben Nevis & Glen Coe* viii. 109 The fault serves as the inner boundary of the ring-complex known as the Fault-Intrusion. **1965** A. Holmes *Princ. Physical Geol.* (ed. 2) xi. 261 Intrusions in the form of concentric arcs or rings are of two distinct types, forming ring complexes such as are exceptionally well developed around the Tertiary volcanic centres of NW Britain. **1976** *Nature* 23 Sept. 307/1 The duration of about 5 Myr for the igneous activities in this ring complex is especially informative in terms of the formation of one complete ring complex. **1942** W. B. Lewis *Electr. Counting* viii. 90 A thyratron ring counter which could have any number of thyratrons arranged in a ring with an arc in one of them. **1969** J. J. Sparkes *Transistor Switching* viii. 195 If the outputs of the last flip-flop..are connected back to the inputs of the first flip-flop then a Ring Counter is formed. **1971** J. H. Smith *Digital Logic* vi. 121 Ring counters, which were used for many years using gas filled trigger valves, will give a count related to the number of stages. If there are *n* stages the counter will count to 2*n*. **1900** A. Conan Doyle *Green Flag* iv. 125 A hard veteran, full of cool valour and ring-craft, could give ten or fifteen years and a beating to most striplings. **1922** Joyce *Ulysses* 313 Handicapped as he was by lack of poundage, Dublin's pet lamb made up for it by superlative skill in ringcraft. **1957** A. MacNab *Bulls of Iberia* xii. 136 As for the *lidia*, the general ringcraft, he [*sc.* Belmonte] admits he knew nothing of it and cared less. **1976** *Daily Record* (Glasgow) 29 Nov., Len Harvey, one of the greatest exponents of ringcraft British boxing has produced, died at his London home yesterday. **1961** *Amateur Gardening* 4 Nov. Suppl. 42/2 Ring culture. In recent years the growing of certain plants, chiefly tomatoes, in bottomless containers has come into vogue. **1962** H. G. W. Fogg *Chrysanthemum Growing* ix. 60 Ring culture, a term which was, I believe, first used as a result of experiments, at Tilgate Horticultural Research Station. **1964, 1976** [see sense 3 i above]. [**1933** *Terrestrial Magn.* XVIII. 82 Whether the latter [current] will close upon itself so as to become an isolated current-ring.] **1941** *Ibid.* XLVI. 1 In our theory of geomagnetic storms, we attribute the main phase to a hypothetical electric ring-current. **1956** *Proc. R. Soc.* A. CCXXXVI. 522 To find the effect of the ring current on the proton resonance line we have to average over orientations. **1962** F. I. Ordway et al. *Basic Astronautics* iv. 164 The ring current system probably consists of low energy particles and is able to effectively disturb the terrestrial magnetic field... It is estimated that a current flow of 1,000,000 amp exists. **1966** Williams & Fleming *Spectrosc. Methods Org. Chem.* iv. 88 Protons attached to systems which can sustain a ring current suffer a paramagnetic shift relative to olefinic protons of isolated double bonds. **1976** *Jrnl. Geophysical Res.* LXXXI. 2701/1 The decay of the proton ring current by charge exchange loss. **1915** *Summary of Progr. Geol. Survey 1914* The Loch Bà felsite is the most perfect example of a ring-boss, or ring-dyke, anywhere known. **1931** *Amer. Jrnl. Sci.* CCXXII. 145 The ring-dike is found between the granite frame and the volcanics of the mountains wherever bed rock is exposed. **1976** *Nature* 23 Sept. 307/1 The 7·6 ± 0·2 Myr isochron..combines various samples from the three inner distinct ring dykes..made up of alkaline granites and quartz syenites. **1969** L. Gaunt *Commonsense Photogr.* xiii. 218 There are special ring flash units which fit round the lens and give even, almost shadow-free lighting. **1975** G. Skoglund *Colour in your Camera* (ed. 6) 144 Shadowless 'multi-directed' lighting, can be produced by a piece of equipment known as a 'ring-flash'. **1978** *SLR Camera* Aug. 45/2 To my knowledge it is the only 100mm lens which has a built-in ring flash that will continuously focus from 1/15 down to life size. **1935** *Discovery* Apr. 102/1 Ring-forts of earth or stone represent one of the most common kinds of monuments of antiquity throughout Ireland. **1937** *Proc. Prehist. Soc.* III. 407 It is hardly justifiable to class ring-forts as fortified sites; they were ordinary farmsteads with a natural measure of protection, and assumed the character of a fortified settlement only if the site was very large. **1960** S. Cruden *Scottish Abbeys* 21 The massive circular stone wall of the 'ring-fort', of late Iron Age or Dark Age date. **1976** *Country Life* 6 May 1163/1 Ring forts..are common all over Ireland. **1919** *Geol. Mag.* LVI. 469 In some ways the most interesting feature of Old Iceland is the ring-fracture system (*Kreisbrüche*) of the north-west peninsula. **1924** E. B. Bailey et al. *Tertiary & Post-Tertiary Geol. Mull, Loch Aline, & Oban* i. 7 It was in Iceland that the word *ring-fracture* (*Kreisbrüche*) was introduced by Thoroddsen. **1965** A. Holmes *Princ. Physical Geol.* (ed. 2) xi. 261 The ring dyke..represents the case where the weight of the keystone..has produced the ring fractures within which subsidence—often referred to as cauldron subsidence—has taken place. **1937** E. Hadorn in *Proc. Nat. Acad. Sci.* XXIII. 481 Since it has been shown that the ring has a glandular function, the term 'ring-gland' may be used so long as no homology to other structures in other insects has been established. *Ibid.* 484 Puparium-formation can be accelerated by transplantation of a normal ring-gland to lethal larvae. **1978** *Molecular & Gen. Genetics* CLXIV. 79/2 The ring gland donors were at the late larval stage by which ecdysone production was likely to have begun. **1972** *Guardian* 26 June 1/3 In this new ring junction traffic turning right goes between the island in the centre and the off-side mini-roundabout. **1973** *Daily Tel.* 3 Aug. 17/1 The experimental 'ring junction' to ease traffic congestion at the Plough roundabout..may be made permanent by winter. **1912** *Chambers's Jrnl.* June 394/1 He organized a body of 'ring-keepers' to preserve order as far as possible. **1922** Joyce *Ulysses* 510 The virgins..burst through the ring-keepers and the ropes and mob him with open arms. **1835** Ring-kop [see sense 9 i above]. **1910** J. Buchan *Prester John* viii. 149 In such a man one would have looked for a *ring-kop*, but instead he had a mass of hair..long and curled like some popular musician. **1935** *Brit. S. Afr. Ann.* 35 Our principal native warrior, an old 'ringkop', promptly took up his quarters in a friendly tree. **1856** G. Price *Treat. Fire & Thief-Proof Depositories & Locks & Keys* xiv. 205 Two centuries ago the puzzle-lock attracted far more attention than any other... The chief among them are *ring-locks*. **1868** A. C. Hobbs *Constr. Locks* iii. 17 According to the kind of handle employed, it [*sc.* the room-lock] may be a knob lock or a ring lock. **1965** G. McInnes *Road to Gundagai* ix. 137, I..ambled gently down the drive toward the ringlock gate. **1892** J. A. Fleming *Alternate Current Transformer* II. ii. 207 (*caption*) The Metropolitan Company's system of ring mains. **1901** F. B. Crocker *Electric Lighting* II. 505 (Index), Ring mains. **1904** *Electr. World & Engineer* 27 Feb. 396/2 The four batteries of boilers constituting each section of the boiler plant are interconnected by a 10-in. main and a 4-in. auxiliary line, both in the form of a ring main... With the location of the valves adopted in the event of rupture of any section of a ring main the trouble may be localised by shutting off the disabled section, the remaining half of the main being kept in service. **1930** *Engineering* 17 Jan. 92/1 In the near future it would probably be necessary to connect these lines by ring mains. **1945** *Jrnl. R. Aeronaut. Soc.* XLIX. 529/1 It is avoided in the larger warships by the spread of the ring main (all below armour) which is admitted as not feasible in aircraft. **1959** Goodin & Downing *Domestic Sanitation* iv. 97 The use of ring mains tends to reduce the size of pipes which must be used and are [*sic*] of great value in maintaining supplies when bursts occur. **1962** *Which?* Mar. 82/1 You will also need a convenient electric power point. Having a new one installed—especially in an older house where there is no ring main—may mean a lot of new wiring. **1962** *Newnes Conc. Encycl. Electr. Engin.* 202 Ring mains offer the possibility of a firm supply at all times, but the cost of the complicated protective gear.. makes such schemes justifiable only in cases where no loss of supply can be tolerated. **1976** G. Moffat *Short Time to Live* vi. 53 He'd..modernise the place: put in a ring main, dig drains, build a septic tank. **1938** *Nippon Electr. Commun. Engin.* Apr. 118/2 Although there are many considerations when it comes to the connecting method of rectifiers in the various carrier suppressed modulators, one of the most effective is that of the so-called Ring modulator. **1974** Harvey & Bohlman *Stereo F. M. Radio Handbk.* ii. 17 Another way of generating sidebands without the carrier appearing in the output is to employ a ring modulator. **1974** *Down Beat* 18 July 42/2 The ring modulator adds both upper and lower sidebands of sound to the original tone. These modulated outputs will be the sum of and the difference between the frequencies of the original tone and an internal sine wave or external source. **1920** J. R. Battle *Handbk. Industr. Oil Engin.* vii. 421 There have been cases where hot running, ring-oiled bearings have been made to run cool. **1930** *Engineering* 11 July 39/3 The outer end of the crankshaft is supported by a ring-oiled outboard bearing and pedestal. **1968** J. J. O'Connor *Stand. Handbk. Lubric. Engin.* xxxv. 11 Lead babbitt.. is frequently used as the bearing material for ring-oiled motors. **1962** G. A. T. Burdett *Automatic Control Handbk.* xii. 11 Ring oiler. **1970** B. Pugh *Pract. Lubric.* viii. 164 Small [steam] turbines usually have the simplest arrangements for bearing lubrication, i.e. ring oilers or standard type lubricators. **1972** R. C. Gunther *Lubrication* xi. 304 Ring oiler consists of a free metal ring that rides on the journal and carries oil from a reservoir located below the bearing. **1904** *Electr. Rev.* 10 Sept. 410 The journal bearings are of the ring-oiling, self-aligning type. **1920** T. C. Thomsen *Pract. Lubric.* ix. 158 Ring oiling is employed largely on modern high-speed shafting bearings. **1974** P. Cave *Mama* (new ed.) vii 46 Mama took a can of beer which was thrust towards her and ripped off the ring opener. **1975** G. V. Higgins *City on Hill* iii. 74 He.. brought out two cans of beer... They stripped off the ring openers and drank. **1977** H. Innes *Big Footprints* II. iii. 187 Abe snapped the ring-opener of his [beer] can..and drank. **1935** *Bull. Folk-Song Soc. Northeast* ix. 11 The movement, with upraised hands, is not dissimilar to that in certain types of *ring-play*. **1942** L. Parrish *Slave Songs Georgia Sea Islands* iv. 99 This ring-play varies in action wherever I see it done... The tune, however, always remains the same. **1972** R. D. Abrahams in T. Kochman *Rappin' & Stylin' Out* 222 The numerous songs, ring-play, and verbal routines in Tobagonian Bongo (wake), as performed by adults in the community. **1909** P. T. Maw *Pract. Forestry* viii. 164 The so-called 'ring-pored' trees—Oak, Ash, Elm, Spanish Chestnut, and Acacia. **1978** A. Bernatzky *Tree Ecol. & Preservation* iv. 53 *Fraxinus*, *Castanea*, *Quercus*, and *Robinia* are ring-pored. **1940** *Bot. Gaz.* CII. 115 Ring porosity is restricted to..the North Temperate zone. **1950** Metcalfe & Chalk *Anat. Dicotyledons* I. p. xlvii, Ring-porousness, or the development of a marked zone of larger vessels at the beginning of a growth ring, appears to be accompanied by an increase in the length of the complete vessels in the pore zone. **1902** F. Roth *First Bk. Forestry* III. 222 The ring-porous woods, like oak, ash, chestnut. **1928** [see *diffuse *a.* 2 f]. **1956** F. W. Jane *Struct. Wood* xi. 250 Very approximately,..the timber of a ring porous hardwood possesses maximum strength when its growth rings number between 6 and 10 to the inch. 1968 G. Tsoumis *Wood as Raw Material* ii. 14 Growth rings are generally more distinct in ring-porous than in diffuse-porous woods. 1970 *Times* 16 Feb. (Food in Britain Suppl.) p. iii/3 Easy opening devices are undergoing considerable development—and ring-pull and zip-top cans are already available. 1973 'D. Halliday' *Dolly & Starry Bird* vii. 89 Poor Jacko, who treated birds and ring-pull cans as one problem. 1922 Joyce *Ulysses* 313 The Santry boy was declared victor to the frenzied cheers of the public who broke through the ringropes and fairly mobbed him with delight. 1942 J. Masefield *Generation Risen* 30 The seconds' faces Watch through the ring-ropes. 1920 E. F. Smith *Introd. Bacterial Dis. Plants* iv. i. 474 Spieckermann's ring rot of Potato. 1946 *Sun* (Baltimore) 30 May 7/5 Even though potatoes 'look nice' they may harbor the dangerous ring-rot disease. 1974 Z. Király et al. *Methods Plant Path.* viii. 180 (*caption*) Processed plant extract of a ring-rot infected tuber. 1960 *Gloss. Atomic Terms* (U.K. Atomic Energy Authority) 46 Ring scaler. 1963 B. Fozard *Instrumentation Nucl. Reactors* viii. 79 If the scheme of Fig. 8.4 is extended to form a ring scaler, all the indicating cathodes may be connected together. 1920 Ring-seat [see *goods *sb. pl.* 2 b]. 1929 Wodehouse *Mr. Mulliner Speaking* iv. 134 The sportsman in him whispered that he was missing something good, for ring-seats to view which many men would have paid large sums. 1905 *Sci. Amer. Suppl.* 25 Mar. 24433/1 Ringshake..consists in a partial or entire separation of two consecutive annual rings, and appears on a cross section as one or more splits running concentrically around the log. 1938 H. E. Desch *Timber* xi. 133 Serious splits are often called 'shakes'... These are of several types, *e.g.*, ring-shake where the separation follows a growth ring, star-shake where the ruptures radiate outward from the pith. 1968 J. Arnold *Shell Bk. Country Crafts* 84 They used to 'split the heart'..as this obviated what were known as 'ring-shakes'. 1851 J. Brown *Forester* (ed. 2) ii. 199, I have seldom seen one [chestnut] which had arrived at the age of fifty or sixty years, which was not ring shaken in the heart-wood. 1931 R. W. Gordon in A. T. Smythe et al. *Carolina Low Country* 199 One of the simplest forms [of shouts], known as the 'ring shout', is apparently widespread. In this, the shouters form a circle and proceed around and around in a sort of slow processional, facing always in one direction. 1942 L. Parrish *Slave Songs Georgia Sea Islands* iii. 54 Shouting appears to be of two types: Along the coast of Georgia and South Carolina the most popular form is the ring-shout. 1970 P. Oliver *Savannah Syncopators* 56 A 'ring-shout'—a shuffling dance in counter-clockwise direction performed by a circle of worshippers which gradually intensified in tempo and collective excitement. 1940 N. Monks *Squadrons Up!* viii. 213 The circular (ring) sight for his eight machine-guns. 1942 *Tee Emm* (Air Ministry) II. 133 Before he had judged range with his ring sight, now he had to estimate without it. 1973 J. Quick *Dict. Weapons* 374/2 *Ring sight*, a sight, especially a gunsight, in the shape of a ring or concentric rings, through which aim is taken and range is estimated. 1930 *Buck & Hickman Ltd. Gen. Catal. Tools & Supplies* 285 Ring spanners. 1970 K. Ball *Fiat 600, 600D Autobook* x. 121/2 (*caption*) Adjusting brake shoe cams. Movement of the ring spanner..on the adjusting nut. 1973 J. Leasor *Host of Extras* ix. 166, I unlocked the..tool box..and took out a roll of ring spanners. [1906 M. C. Cooke *Fungoid Pests Cultivated Plants* 50 The ringed brown spot (*Septoria Chrysanthemi*)..has apparently been confined to Italy.] 1923 Salmon & Wormald in *Jrnl. Min. Agric.* XXX. 148 The diseased areas on the leaf first appear as brown spots... As the spots reach a diameter of 1/8 in. or so, the central dead portions drop out, leaving the characteristic perforations... The name 'ring-spot' disease appears suitable to designate this form of injury which strikes the eye of the grower as he walks over the field. 1927 *Phytopathology* XVII. 325 (*caption*) Ringspots in an early stage of development with margins of alternating zones of chlorotic and normal tissue. 1939 *Jrnl. Pomol.* XVII. 27 Ring Spot, the common name accepted in this country for the disease [of lettuces] caused by the fungus mentioned [sc. *Marssonina Panattoniana*], has been widespread in temperate regions from an early date. 1964 *Phytopathology* LIV. 702/1 No ringspot virus was detected in uninoculated primary leaves through the sixteenth day. 1971 *Country Life* 18 Feb. 381/1 Ringspot, Mottle and Vein Mottle are only three of the viruses afflicting carnations. 1974 *Ibid.* 28 Nov. 1648/1 In Scotland..members of this race [*sc.* large heath butterflies] differ from English specimens in having fewer ring-spots on the under-surface of the wings. 1980 *Daily Tel.* 16 Jan. 8/3 There are some Cornish cauliflower types which have not been available for about 30 years which we think were resistant to a disease called ringspot. 1961 R. W. Butcher *Brit. Flora* II. 762 Leaves broad, lanceolate, grey-green, ring-spotted, flat and broadened from the base. 1871 *Sports & Games* July 121 A new Game, *ring toss*,..affords an attractive out-door sport, and furnishes a degree and kind of physical exercise that improves and develops the general health and strength. 1874 *St. Nicholas* Jan. 171/2 There are many other games to be found in the shops.. such as 'Ring-toss', 'Magic Hoops', and 'Parlor Croquet'. 1927 *Times* 20 Oct. 17/4 The bride..wore a picture gown of white ring velvet. 1931 *Daily Tel.* 21 May 6/3 A green satin beauté gown..lined with green ring velvet. 1932 G. Greene *Stamboul Train* II. ii. 90 Could you get me five yards of ring velvet? 1952 C. W. Cunnington *Eng. Women's Clothing* vii. 232 Evening gown in mulberry ring velvet. 1939 F. Scott Fitzgerald *Let.* Mar. (1964) 53 As it is a lavish gesture it should be a simple present..on the other angle from a ring-watch. 1962 E. Bruton *Dict. Clocks & Watches* 147 *Ring watch*, a watch mounted in a finger ring. 1892 S. P. Thompson *Dynamo-Electr. Machinery* (ed. 4) xii. 309 (*caption*) Development of ring winding for 4-pole machine. 1893 [see *drum-winding* s.v. *drum *sb.*[1] 13]. 1922 A. H. Avery *Dynamo Design & Constr.* ix. 122 The winding is electrically continuous, and progresses steadily forward round the armature just as with the ring winding. 1963 L. F. Chitty in Foster & Alcock *Culture & Environment* vii. 177 A ringwork beside the Knighton road may be medieval. 1975 J. G. Evans *Environment Early Man Brit. Isles* vii. 164 Earthworks such as motte-and-bailey castles and ring-works preserve a buried soil and ditch sequence which can be used to extract environmental evidence. 1929 Hostetter & Beesley *It's a Racket!* 236 *Ringworm*, one who is a regular attendant upon prize fights and boxing matches. 1930 *Forum* Dec. 373/2 Many of these words..are employed daily in our own sports pages. Even the most casual American *ringworm* will recognize these: *knob, mugg, the one-two punch*, [etc.]. 1954 *Sun* (Baltimore) 30 Mar. 18/7 'Ring worms', as some are in the habit of referring to fight fans, have a hot one coming up Friday night when Kid Gavilan takes a shot at Bobo Olson and his middleweight title. 1892 J. Nasmith *Students' Cotton Spinning* ix. 348 A method of winding ring yarn on a cylindrical surface. 1909 *Westm. Gaz.* 13 Aug. 8/4 A determined effort is being made by cotton-spinners in Lancashire who produce ring yarn to form an association to keep up prices.

b. ringbarker *Austral.*, a stick-insect, *Podacanthus wilkinsoni*, which in swarms devours the leaves of eucalypts; **ring-billed** († **mew**) **gull**, a New World gull, *Larus delawarensis*; **ring-eye**, (*b*) (examples); = *silver-eye* s.v. Silver *sb.* and *a.* 21 c; cf. *white-eye* s.v. White *a.* 12 e; **ring perch** (examples); = *ringed perch* s.v. Ringed *ppl. a.* 5.

1935 K. C. McKeown *Insect Wonders Austral.* xviii. 149 Another fine stick-insect is found in the Walcha district, where it is popularly known as the 'Ringbarker', because of the dying brown appearance of the trees after the insects have stripped them of their foliage. 1965 *Austral. Encycl.* VIII. 295/2 Most phasmids are solitary in habit. One of the few gregarious forms, the ringbarker (*Podacanthus wilkinsoni*), sometimes appears in countless numbers in various districts of New South Wales, where it defoliates the eucalypts. 1831 Ring-billed mew-gull [in Dict., sense 17]. 1834 T. Nuttall *Man. Ornithol. U.S. & Canada* II. 300 Ring-billed Mew Gull... Sp. Charact.—Commissure of the stout ringed bill rather longer than the tarsus. 1844 J. E. DeKay *Zool. N.Y.* II. 309 The Common gull.., although called the Ring-billed Gull in the books, has received no other popular name than Brown Winter Gull. 1917 T. G. Pearson *Birds of Amer.* I. 47/1 The California and Ring-billed Gulls generally nest together in big colonies on the inland lakes. 1975 *Behaviour* LII. 143 Parent Ring-Billed Gulls are able to recognize their own chicks after about 7 to 9 days posthatching. 1951 J. Frame *Lagoon* 47 She was big and warm and knew about cats and little ring-eyes. 1953 *Landfall* VII. 21 They caught a blackbird and two ring-eyes. 1877 C. Hallock *Sportsman's Gazetteer* 272 Yellow Perch; or Ring Perch.—*Perca flavescens*. 1947 J. H. Brown *Outdoors Unlimited* 233 'Ring' perch, or yellow perch as they are more widely called, offer the earliest fishing in this part of the country.

ring, *sb.*[2] Add: **2. c.** (Later *transf.* examples.)

1957 G. Ryle in C. A. Mace *Brit. Philos. in Mid-Cent.* 264 The word 'analysis' has..a good laboratory or Scotland Yard ring about it. 1973 *Times Lit. Suppl.* 21 Sept. 1091/3 The 'special relationship' with Japan, using a phrase which will have an ironic and melancholy ring in British ears.

d. *Electronics.* A sequence of damped oscillations at the resonant frequency of a circuit; an individual oscillation in such a sequence.

1949 *Electronic Engin.* XXI. 207/3 Thus the deliberate introduction of a ring into the response can be a means of improving the sharpness of the transition edges of the observed picture. 1971 J. Earl *How to choose & use Pickups & Loudspeakers* iii. 87 (*caption*) This pulsed-tone shows only slight 'ringing'... An insufficiently low value load could incite worse rings than this. 1975 G. J. King *Audio Handbk.* ii. 43 No amplifier worthy of the hi-fi label should exhibit rings or overshoot into a load of pure resistance. *Ibid.* v. 122 Since the ring is very quickly damped, giving virtually an overshoot effect, this sort of performance is perfectly acceptable.

3. c. orig. with *up*. A call on a telephone. Also, each of a series of ringing sounds produced by a telephone receiving a call; *to give* (someone) *a ring*, to call by telephone. Cf. *ring *v.*[2] 10 b.

1900 [see *hullo *int.*]. 1910 *Daily Chron.* 26 Feb. 6/2 It is only that most modern of human summonings, a telephone 'ring up'. 1930 J. B. Priestley *Angel Pavement* iv. 157 I'll just give the City Transport a ring to see if they've heard anything about that lot we sent to Norwich. 1934 T. E. Lawrence *Let.* 14 Sept. (1938) 819 Give me a ring at Hythe and let us meet. 1948 [see *central *sb.*]. 1951 T. Sterling *House without Door* x. 109 The telephone rang... It went on ringing... She listened, anticipating each ring to the second. c 1952 A. Huxley *Lett.* (1969) 660 If you ever come into town, why don't you give me a ring? 1960 [see *book *sb.* 4 d]. 1963 V. Nabokov *Gift* ii. 137 There's no guarantee the room is not already disposed of, but still I would advise you to give her a ring. 1973 'M. Underwood' *Reward for Defector* vi. 44 Give me a ring later this evening and let me know. 1976 H. Nielsen *Brink of Murder* i. 9 He dialled his own number... Kevin.. answered on the fourth ring.

d. Phr. *the dead ring* (*of*): see quot. 1916. Cf. *ringer[2] 5. *Austral.* and *N.Z. slang.*

Possibly belongs under some other sense of the noun.

1916 C. J. Dennis *Songs of Sentimental Bloke* 124 *The dead ring*: a remarkable likeness. 1948 D. W. Ballantyne *Cunninghams* I. xv. 81 They [*sc.* the sons] were the dead ring of Gil. 1951 —— in *Landfall* V. 166 A fine little chap. Dead ring of his old man, eh?

5. *ring-in*: a fraudulent substitution; the action of 'ringing in' (see Ring *v.*[2] 13 b). *Austral. slang.*

[1924 *Truth* (Sydney) 27 Apr. 6 *Ring in*, false.] 1941 Baker *Dict. Austral. Slang* 60 *Ring-in*, a horse or dog that is fraudulently entered in a contest under an assumed name and/or disguised. 1969 C. Drummond *Odds on Death* vii. 152 The elderly book-makers..were his hosts... Past losses were debated... 'A ring-in after all these years,' had said a ruined giant of a man disgustedly. 1971 *Telegraph* (Brisbane) 26 June 5/1 All these are checked against the dog, and the chances of a 'ring-in' are completely eliminated.

ring, *v.*[1] Add: **1. d.** (Earlier and later examples.) Also *S. Afr.*

1868 [implied in *ringing *vbl. sb.*[1] 4]. 1890 'R. Boldrewood' *Colonial Reformer* II. xviii. 111 A desultory entry into the receiving yard then takes place... The 'ragers' observing this movement keep wildly and excitedly 'ringing', like a first class Maelstrom. 1928 'Brent of Bin Bin' *Up Country* x. 172 'Well, are you going to stay?' 'Can't, thank you. Our mob was ringing a bit when I left.' 1941 I. L. Idriess *Great Boomerang* vii. 56 The cattle began to ring, the centre beasts edged outwards, then turned inward and began to sniff, to paw the earth. 1947 J. Stevenson-Hamilton *Wild Life S. Afr.* x. 72 The [buffalo] herds ring to protect themselves against marauders in much the same way as the herds of domestic cattle do. 1959 H. P. Tritton *Time means Tucker* 71 When they [*sc.* cattle] drift the only thing that can be done is try to turn them and get them circling, or ringing as the drover terms it.

6. c. To put a numbered ring on the leg of (a bird) so that it may be identified subsequently; to treat (a bat) similarly.

1908 *Brit. Birds* I. 327 A large number of birds of various species are ringed each year at Rossitten on the Baltic. 1925 Turner & Gurney *Bk. about Birds* vii. 72 When he escaped from the Sudan long after, he was able to tell those who ringed the bird that it had been found in Omdurman. 1958 *Listener* 30 Oct. 684/2 We have now ringed nearly 3,000 bats. 1971 *Daily Tel.* 22 June 8/4 A tufted duck, ringed in Essex two years ago by the British Trust for Ornithology, has been found in West Pakistan, where it was shot by a hunter in Rawalpindi. 1978 P. Conder *RSPB Guide to Birdwatching* 102 Many ringers, particularly those working on a population of a particular area, ring nestlings.

11. *Austral. slang.* To beat (a shedful of men) at sheep-shearing. Also *transf.* Cf. Ringer[1] 5 a in Dict. and Suppl.

1895 A. B. Paterson *Man from Snowy River* (1896) 136 The man that 'rung' the Tubbo shed is not the ringer here, That stripling from the Cooma side can teach him how to shear. 1899 'S. Rudd' *On our Selection* 84 He shore..at Welltown, and rung the shed by half a sheep. 1905 in A. B. Paterson *Old Bush Songs* 27 And once I rung Cudjingie shed, and blued it in a week. 1957 D. Niland *Call me when Cross turns Over* v. 132 He would take on anything, wheat-lumping, tree-felling, shearing—always ringing the shed—droving, anything at all that suited him. 1967 *Telegraph* (Brisbane) 25 Mar. 2/5 To 'ring the shed' a shearer's cook has to earn more money than the top shearer.

12. To draw a circle round (something printed) so as to focus attention on it.

1970 R. K. Kent *Lang. Journalism* 113 *Ring*, to draw a circle around; encircle, as to signify various directions in copy editing. 1981 *Times* 23 June 2/2 She [*sc.* the Prime Minister] has probably already ringed a date in the autumn of 1983 for the election.

ring, *v.*[2] Add: **B. I. 1. d.** Of an electric circuit or a solid body: to undergo damped oscillation at its resonant frequency.

1952 G. C. Smith in Molloy & Poole *Television Engineers' Servicing Man.* 12 The flyback 'overshoots' and 'rings', but it is frequently damped out by a capacitor and a resistance in series across the coils. 1973 *Newnes Colour Television Servicing Manual* I. ii. 56/2 The amplified signal delivered from its collector causes T1 to 'ring'. 1975 *Nature* 24 Jan. 233/1 It has been widely accepted..that a [nuclear] test fired while the Earth was ringing from a really major earthquake..would be impossible to detect.

2. d. Of a telephone (bell): to produce the ringing sound which indicates that there is a caller on the line (Line *sb.*[2] 1 e).

1924 J. Reith *Broadcast over Britain* III. v. 168 If people are moving about the room, or the telephone rings..the [radio] play has simply no chance. 1951 M. Kennedy *Lucy Carmichael* I. iv. 27 The telephone rang at intervals all the evening. 1979 T. Wiseman *Game of Secrets* iv. 48 He spent an hour getting himself reconnected, and after that he waited for the telephone to ring.

II. 6. a. (Later examples with reference to the summoning of servants.)

1856 Dickens *Dorrit* (1857) I. x. 81 'I must refer you,' returned Mr. Barnacle, ringing the bell, 'to the Department.' 1914 L. Woolf *Wise Virgins* ii. 31 They [*sc.* servants] won't stay because you ring the bell for them while they're at dinner!

d. *to ring the bell*: to win recognition; to be a complete success (see also *bell *sb.*[1] 7 c).

1915 *Munsey's Mag.* Apr. 561/2, I am reading your 'Barry Newton' yarn. It scores a bulls' eye, it rings the bell, it brings a coconut to earth. 1925 E. Wallace *Strange Countess* x. 93 'You've certainly rung the bell this time, Lois.' 'It seems too good to be true, doesn't it?' 1945 *Daily Mirror* 15 Aug. 3/3 Leeds Corporation has fifty 'retired' trams to sell... They think a tram would 'ring the bell' as a home, week-end bungalow, or greenhouse. 1976 *Church Times* 30 July 7/5 The wise sight-seer knows that, however alert and receptive he is, even the treasures of Florence aren't going to ring the bell every time.

e. *to ring a bell* (colloq.): see *bell *sb.*[1] 7 d.

7. a. (Further examples.) Also *spec.* to summon or send *for* a servant or required object by this means.

1847 E. Brontë *Wuthering Heights* I. x. 202 Why not have Mrs. Dean up to finish her tale?.. I'll ring: she'll be delighted to find me capable of talking cheerfully. 1864 H. Cullwick *Diary* June in D. Hudson *Munby* (1972) 195 Me by myself in that kitchen..ready to do any thing for 'em whenever they rang for me. 1926 D. L. Sayers *Clouds of Witness* ii. 49 Ring for anything you want. 1980 N. Marsh *Photo-Finish* vii. 188 Alleyn..put his thumb on the bell.. and Marco came in... He said: 'You rang, sir?'

10. a. (Later *fig.* examples.) *to ring in*, also, to include, take into consideration; to bring (someone) into an operation, activity, etc.

1900 Ade *Fables in Slang* 74 The Pew-Holders didn't even admit..that the Preacher had rung in some New Ones [*sc.* names]. 1922 D. H. Lawrence *England, my England* 45 The clanging pain in his head rang out the rest of his consciousness. 1925 T. Dreiser *Amer. Tragedy* II. III. xvi. 202 She can't be kept out of the case... We'll have to ring her in, I'm afraid. 1954 Wodehouse *Jeeves & Feudal Spirit* viii. 72 I've got the whole family here... I only wanted Trotter, but Mrs. T. and Percy rang themselves in. 1973 *New Yorker* 17 Feb. 88/2 With that one stroke, the union could ring in a lot of public figures. 1974 *Publishers Weekly* 30 Dec. 90/1 Mr. Brooke is summoned from far away Lima, and an old suitor of Philippa's, Lord Tancred, is rung in to help.

b. *spec.* to call (someone) by telephone. Freq. with *up*. Also *absol.*, and with *round* (= to call a succession of people by telephone), *through*.

1880 *Punch* 17 July 13/2 For you upon them both may frown, And say that you are shocked, or May knock the Secretary down, And then ring up the Doctor. 1882 T. D. Lockwood *Pract. Information for Telephonists* 130 Ask the office operator to ring up the complaining person and await results. 1889 [in Dict.]. 1906 S. Ford *Shorty McCabe* vi. 150 He was goin' to ring up the police reserves. 1913 G. B. Shaw *Let.* 14 July in *B. Shaw & Mrs. Campbell* (1952) 132, I shall ring up tomorrow in spite of my dread of being unwelcome. I rang a second time today; but the answer was buzz, buzz. 1930 J. B. Priestley *Angel Pavement* viii. 398 He rang me up last night, at home, to say he'd just arrived and would be down this morning. 1934 T. E. Lawrence *Let.* 14 Sept. (1938) 819 It lies on my conscience that you so often ring up vainly. 1934 N. Marsh *Man lay Dead* xii. 207 I'll ring through at about one o'clock. 1940 H. G. Wells *Babes in Darkling Wood* II. i. 134 There were one or two people he might ring up, but probably they would be holiday-making now and out of town. 1948 'N. Shute' *No Highway* vi. 169 While I was down there, Miss Learoyd rang through. 1955 *Times* 22 July 10/6 Two young friends rang me up rather excitedly the other evening and asked me if I would go round and give them my advice. 1958 L. A. G. Strong *Treason in Egg* vii. 127 You'd better ring the police. 1960 J. Stroud *Shorn Lamb* xv. 170 If he does [turn up], I'll ring round for a hostel. 1969 *Listener* 6 Feb. 187/1 When Jelly Roll Morton..played his compositions to the Harlem team, eyebrows were raised no further than the pejorative 'don't ring *us*' level. 1974 A. Morice *Killing with Kindness* ii. 14 He was going to ring round in the morning and fix up for us all to go and see them. 1974 *Times* 7 Feb. 14/6 By ringing and writing to every MP, Service and the bill's sponsors have so far secured 74 pledges to be present. 1977 'M. Underwood' *Murder with Malice* x. 91 'Thanks for ringing, sir. I appreciate it.'.. Nick dropped the receiver back. 1980 A. Auswaks *Trick of Diamonds* iii. 80 'Don't ring us, we'll ring you,' grunted Bob Jones sarcastically. 1981 J. Wainwright *All on Summer's Day* 66 Ring round the other divisions. I want some C.I.D. men.

c. (Earlier and later examples.) Also *fig.* and *intr.* for passive (with the curtain as subject).

1772 D. Garrick *Peep behind Curtain* II. i. 30 Pray be so good as to ring down the curtain, that we may rehearse in form. 1807 *Monthly Mirror* Aug. 133 The prompter rings the lofty curtain down. 1913 F. H. Burnett *T. Tembarom* xv. 186 'Now,' he said, 'we can ring up for the first act.' She filled the teapot. 1916 S. Kaye-Smith *John Galsworthy* 63 Thus the curtain rings down on Irene Forsyte, crushed under the heel of prosperity. 1950 H. F. Maltby (*title*) Ring up the curtain.

d. *to ring off* (earlier and further examples). Now more usu., to discontinue a telephone conversation by replacing the receiver, = *hang up* s.v. *hang *v.* 28 a. Also *fig.*

1882 T. D. Lockwood *Pract. Information for Telephonists* 85 Frequently an annunciator between two circuits when connected to allow the subscriber, if he please, to ring off. 1899 *Electrician* 1 Dec. 181/2 Ringing off is avoided, as this is performed automatically by replacing the receiver on the hook. 1900 [see Phone *sb.*[1] and *v.*]. 1906 A. Bennett *Whom God hath Joined* iv. 158 He rang off, curtly, without another word. 1920 R. Macaulay *Potterism* III. i. 104 You mustn't ring off yet, indeed you mustn't. Hold on while I tell daddy. 1935 *Punch* 21 Aug. 223/1 'I'm coming round to wring your wretched little neck!' shouted Mr. Applestalk as he rang off. 1938 E. Bowen *Death of Heart* III. i. 328 'So then you rang off?' 'No, he did. It was his tea-time, no doubt.' 1967 M. Kenyon *Whole Hog* i. 12 I'll ring him anyway... 'Bye now—Yes, I'm going to ring off. 1973 S. Dobyns *Man of Little Evils* xii. 127 The operator came back on the line. 'I'm afraid your party has rung off.'

fig. 1895 *Inlander* Dec. 114 *Ring off*, stop talking. 1906 E. Dyson *Fact'ry 'Ands* vi. 71 'Shut up! D'yeh 'ear?.. Arr-r-r ring off, cant yeh!' The girl..opened a startled eye. 1938 F. M. Ford *Let.* 26 Nov. (1965) 305, I will ring off. Let us know from time to time how things go with you. 1940 F. Sargeson *Man & his Wife* 34 Wouldn't you like to stay out here for good? Fred said. Ring off, Ken said. I got a bite.

e. *to ring back*, to reply to (a previous caller) by telephone. Also *absol.* and as *sb.*

Quots. 1971 and 1972 represent technical senses in Telephony.

1942 N. Balchin *Darkness falls from Air* iii. 57, I hung up. 'Pearce is going to ring you back,' I said. 1944 H. McCloy *Panic* 118 This is Jim, testing. I'm going to hang up and then I want you to ring me back. 1960 I. Jefferies *Dignity & Purity* xi. 181 I'll ring you back Gobbo. Couple of minutes. 1969 P. N. Walker *Carnaby & Conspirators* vii. 68 'Make a check on Henry Pritchard too, sir.'.. 'Will do. Will you ring me back?' 1971 *Gloss. Electrotechnical, Power Terms* (B.S.I.) III. ii. 28 *Ring-back signal*, a backward signal to recall a calling subscriber held by the operator. 1972 *Sci. Amer.* Sept. 120/1 When the path is found, the signal-distributor sends electrical signals to ring the telephone of the called subscriber and sends ringback tones to the caller. 1974 M. Birmingham *You can help Me* iii. 47 There was one caller from a public call-box who..didn't ring back. 1977 W. Marshall *Thin Air* i. 11 He said quickly, 'I'll ring you back.'

f. *to ring in*, to report by telephone. Also *trans.*, to transmit (a verbal message) by telephone.

1949 N. Marsh *Swing, Brother, Swing* xi. 254 I'll ring in then and get something to eat. 1956 *New Statesman* 18 Aug. 180/3 We had to think about finding a telephone booth from which to ring in a preliminary story. 1964 M. Banton *Policeman in Community* iii. 83 The beat officer has to 'ring in' to headquarters every hour from automatic boxes mounted on standards at the kerbside. 1971 B. Graham *Spy Trap* i. 7 He..drove to the secondary rendezvous point. .. Maybe Hannifin had rung in. 1975 *Listener* 16 Oct. 505/1 People ring in, wanting help.

11. c. *to ring up*: to make a record of; *spec.* to record (a sale) on a cash register or similar device. Also fig.

1937 J. T. Farrell *Fellow Countrymen* 180 He paid Kitty fifteen cents, which she rang up. 1939 *Sun* (Baltimore) 18 Jan. 5/5 Asked if the ship's speeds in any of its previous trips through the canal had been 'rung up', Leonard Nieberline, first officer, answered in the negative. 1948 C. Himes *Black on Black* (1973) 267. When she stopped at the cash register across from him to ring up a sale he said, 'Baby, I really love you.' 1956 A. Huxley *Adonis & Alphabet* 167 Energies which, if canalized and directed, can be made to do useful work and ring up handsome profits. 1957 *Economist* 21 Dec. 1051/1 Last Saturday, when the strike was crumbling, the shops stayed open until nine; many rang up record sales. 1962 *Times* 15 Mar. 9/7 The items in connexion with which he was accused did not appear upon the cash receipt slip. The cashier must have omitted to ring these up. 1968 [see *driver 2 b]. 1970 *Daily Tel.* 14 Nov. 15/3 One technique is to bully the check-out girl and get her sufficiently confused to miss the fact that half the goods have been pushed through without having been rung up. 1976 'E. McBain' *Guns* vii. 148 The cashier rings up the check, money comes tumbling down the cash register chute.

13. a. *spec.* to effect a fraudulent change in the identity of a motor vehicle.

1967 N. Lucas *CID* vi. 80 The two cars are..rebuilt into one 'bastard' car... The process is known as 'ringing' cars. 1971 *Drive* Summer 22/1 Like any commercial venture, the business of car ringing—changing a vehicle's identity—has to be cost-effective. 1977 A. Hunter *Gently Instrumental* ii. 19 The Parry brothers..copped three apiece for ringing cars.

ring-a-ring (riŋări·ŋ). Also **ringaring**, etc. [Fanciful extension of Ring *sb.*[1]] A circle or circular movement. *Ring-a-ring o' roses* (and variants), a game played by children holding hands in a circle (also *transf.*).

1881 K. Greenaway *Mother Goose* 48 Ring-a-ring-a-roses, A pocket full of posies. 1922 Joyce *Ulysses* 506 I'm a tiny tiny thing Ever flying in the spring Round and round a ringaring. 1927 W. E. Collinson *Contemp. Eng.* 11, I can be brief in mentioning the various games we played during childhood. I will mention..ring a ring o' roses. 1945 C. S. Forester *Commodore* xxiii. 259 The Governor..tried to dance a sort of ring-a-ring-of-roses with the two Englishmen. 1953 A. Clarke *Moment Next to Nothing* II. 32 He Was near that oak, speaking of Marravaun—A ring-a-ring o' birds around him. 1957 J. Masters *Far, Far the Mountain Peak* i. 5 Why don't I suggest a game of ring-a-ring-a-roses, or kiss-in-the-ring? 1963 *Times* 30 May 4/2 England now were playing arrogant football and stroking the ball from man to man as though they were playing ring-a-ring-o'roses. 1972 G. Green *Great Moments in Sport: Soccer* ii. 38 While the opposition was being enticed into these closely woven webs of ring-a-roses, two..of the front runners would be streaking ahead into the unguarded places. 1974 *Times* 1 Apr. 1/8 Strong men blenched and broke into a sweat of embarrassment when made to dance 'Ring-a-ring o' roses' in public outside Guildhall.

ring-bark, *v.* Add: **a.** Also, to remove a narrow or incomplete ring in order to check rapid growth. (Further examples.) **ring-barking** *vbl. sb.* (further examples).

1938 C. P. Ackers *Pract. Brit. Forestry* vi. 215 This is the time to ring-bark and let these rough poles, after they have done their job, topple over as branchless rotten logs. *Ibid.* xi. 341 Partial ring barking is practised in some orchards so as to check the sap of the tree and so induce more free fruiting. 1961 *Observer* 18 June 35/5 Successfully greedy roots can mean fruitless apples and pears, and 'ring barking' is the remedy. 1975 H. F. Heady *Rangeland Managem.* iv. 57 Ring-barking by voles killed as much as 84 percent of the *Artemisia tridentata* in some Montana stands.

ring-bolt. For *Naut.* read 'orig. *Naut.*' and add further examples.

1840 *Crockett Almanac 1841* 19 A ring bolt in the barn floor. 1958 L. Durrell *Balthazar* iv. 90 Firm as a figure held by ringbolts. 1968 E. R. Buckler *Ox Bells & Fireflies* xv. 215 He just grabbed the sonofabitch by the scruff o' the neck and lifted him about two feet offa the floor—and, by God, by the time he was through with him, he brought him to the ringbolt and don't you think he didn't. 1974 R. Adams *Shardik* xxvi. 210 The tie-bars..were secured by chains to ring-bolts set..in the walls and floor.

ring-dance. (Further examples.)

1910 [see *polska]. 1933 E. K. Chambers *Eng. Folk-Play* 202 There is a general resemblance between Sword Dances proper and..Ring, Hoop, or Garland Dances. 1946 R. Blesh *Shining Trumpets* v. 98 He had the work-songs, the spiritual, the ring-dance. 1960 Auden *Homage to Clio* 28 Rustics in a ring-dance pantomime.

ringed, *ppl. a.* Add: **2. c.** Of a bird: bearing a ring or rings on one or both legs.

1908 *Brit. Birds* I. 298 Should anyone come across any of Herr Mortensen's ringed birds at any time, it is hoped they will send the ring, foot, and data of capture either to him or to me. 1948 *Ibid.* XLI 233 Recoveries of ringed Mallard are in the vast majority of cases obtained from birds shot. 1978 P. Conder *RSPB Guide to Birdwatching* 104 The table of longevity shows that ringed birds are able to survive to a good age.

5. ringed carpet, a pale grey moth, *Cleora cinctaria*, with dark patches on the forewings; **ringed penguin**, the chinstrap penguin, *Pygoscelis antarctica*.

1866 Ringed carpet [see Carpet *sb.* 4]. 1896 J. W. Tutt *Brit. Moths* xi. 300 The Ringed Carpet..is abundant at Lyndhurst in May and early June, on the dwarf firs among the heather. 1948 W. J. Stokoe *Caterpillars Brit. Moths* II. 197 The Ringed Carpet... The New Forest in Hampshire is said to be the district *par excellence* for this species. 1919 Ringed penguin [see *Adélie]. 1964 A. L. Thomson *New Dict. Birds* 611/1 The Chinstrap (or Ringed) Penguin..is most abundant in the Antarctic Peninsula.

Ringelmann (ri·ŋgĕlmæn). [Of uncertain attribution; perh. the name of Maximilien *Ringelmann* (1861–1931), French scientist.] Used *attrib.* and † in the possessive with reference to a means of estimating the darkness and density of smoke by visual comparison with a chart bearing different shades of grey (formed by lines ruled with different spacings on a white card); as *Ringelmann card, chart, scale*.

1898 S. B. Donkin *Heat Efficiency of Steam Boilers* ix. 184 Ringelmann's smoke scale.—Professor Ringelmann has conceived the idea of representing different intensities of smoke, from light through grey to black, not by small sample tints, which are apt to be misleading, but by an arrangement of lines on white paper, drawn in a particular way. 1905 W. Nicholson *Smoke Abatement* vii. 144 Copies of the Ringelmann Smoke Scales..are obtainable. 1923 H. G. Clinch *Smoke Inspector's Handbk.* vii. 74 (*heading*) Measurement of smoke—Ringelmann's shade cards. 1954 [see *dark smoke* s.v. *dark *a.* 13 c]. 1958 *New Scientist* 5 June 101/1 Discharge of smoke darker than Shade 2 on the Ringelmann chart..for a total of more than ten minutes within eight hours is forbidden. 1962 *B.S.I. News* May 20 The B.S.I. has..undertaken..the preparation of standard Ringelmann charts and the measurement of solids emitted from chimneys.

ringer[1]. Add: **5. a.** (Earlier and later examples.) Also *N.Z.*

1871 'R. Boldrewood' in *Cornh. Mag.* XXIII. 85 The 'Ringer', or fastest shearer of the whole assembly. 1888 —— *Robbery under Arms* I. ix. 110 Jim..was trying to shear sheep and sheep with the 'ringer' of the shed. 1910 C. E. W. Bean *On Wool Track* 196 The man who shears most sheep is the 'ringer'. 1927 M. M. Bennett *Christison* xxii. 193 With the new shearing machines a hundred sheep a day were shorn easily, while ringers scored over two hundred. 1934 T. Wood *Cobbers* 196 He can shear a hundred a day: a hundred and twenty, a hundred and fifty; two hundred—even three hundred and twenty, at times, if he is a Ringer—that is the quickest of the team. 1952 J. Cleary *Sundowners* iii. 138 By the end of the day he wanted to be the 'ringer' shearer. 1963 N. Hilliard *Piece of Land* 90 The ringer of the Maori shearing gang that year had been Keko. 1965 *N.Z. Listener* 26 Feb. 15/2 In shed shearing the position of 'Ringer' is sought after and competition among shearers has always been keen.

b. An expert. *Austral. slang.*

1918 C. J. Dennis *Digger Smith* 112 *Ringer*, expert. 1923 'B. L. Standish' *Lego Lamb, Southpaw* iii. 26 'That guy's a ringer,' declared Shultz. c 1926 'Mixer' *Transport Workers' Song Bk.* 24 For I'm classed among the 'ringers', And from others stand apart. 1943 *Amer. Speech* XVIII. 256 With Americans a *ringer* is a double; here [*sc.* in Australia] he is an expert. 1965 *Telegraph* (Brisbane) 5 July 8 Ringer (the best—old shearing-shed term later adopted by townies).

6. One who rings birds.

1909 *Brit. Birds* III. 5 Any finder of a ring so marked should realise that communication with the 'ringer' is intended. 1946 *Ibid.* XXXIX. 260 The ringer probably keeps a sharper look-out for dead birds than the ordinary person. 1966 *Punch* 14 Dec. 897/3 Ringers try to recapture each bird in colonies of sea birds for an annual ring check, and quite often they catch the same bird each year for six years or more in their gardens. 1978 P. Conder *RSPB Guide to Birdwatching* 103 In the early days when there were few ringers, most of the birds were recovered dead but as more ringers took part in the scheme..so the numbers of live recoveries increased.

7. *Austral.* A stockman; a station hand.

1909 J. X. A. Cameron *Spell of Bush* (1910) 48 Damsinkers, fencers, scrub-cutters, ringers, and other men doing contract work in the vicinity. 1942 C. Barrett *On Wallaby* i. 14 'Jim the Ringer' came in every month for 'a bender'. 1953 'N. Shute' *In Wet* x. 348 His camp consisted of..a humpy shelter made of gum tree boughs for his white ringer, Phil Fleming. 1954 B. Miles *Stars my Blanket* xxiii. 204 The stockman—or 'ringer' as he is called—rides into the yard with a lassoo and 'rings' his bullock in true wild-west style. 1964 *Sunday Mail Mag.* (Brisbane)

27 Sept. 3/1 A ringer..is a Queensland stockman who holds his cattle during a muster by 'ringing' them on horseback. 1977 *Telegraph* (Brisbane) 12 Jan. 3/1 The pub, local waterhole for stockmen, ringers and station hands.

8. *colloq.* An officer in an air force; a member of an air-crew. Also with preceding numeral, indicating rank.

The use is derived from the rings indicating rank worn on the sleeve of an officer.

1943 C. H. Ward-Jackson *Piece of Cake* 51 Pilot Officer is a 'Half-ringer', and Squadron Leader a 'Two and a half ringer'. 1945 Baker *Austral. Lang.* 163 *Ringer*, an officer. 1976 'A. Hall' *Kobra Manifesto* v. 65 One of the air-crew, a two-ringer.

ringer[2]. Add: **4.** *U.S. slang.* A horse or other competitor fraudulently substituted for another in a race or other sporting activity; one who engages in a fraud of this kind.

1890 *Stock Grower & Farmer* 9 Aug. 8/2 At the same time 'Andy Croker' is the most notorious 'ringer' on the turf. 1914 'High Jinks, Jr.' *Choice Slang* 17 *Ringer*, a name applied to a man or horse dishonestly entered in an event with others far below his class and a class in which he could not be entered legitimately. 1928 Foy & Harlow *Clowning through Life* 188 We had scarcely made the match when we were given a secret tip that Bennett was a 'ringer'. 1935 A. J. Pollock *Underworld Speaks* 98/1 *Ringer*, a race horse who has been substituted under the name of another horse in a race. 1938 M. Lane *Edgar Wallace* III. v. 304 He was an attractive young man, known in his own profession as 'Ringer' Barrie for his ability to ring the changes of disguise on race-horses. 1944 *Fortune* Sept. 140/2 The chance that 'ringers'—good horses masquerading as poor ones under assumed names—can be sneaked into races. 1958 *Sun* (Baltimore) 15 Aug. 20/2 Evidence.. tends to show that it was not a horse called Bye Bye Will that won that race but a 'ringer'. 1966 *Listener* 27 Oct. 613/3 He rode third in a regimental steeplechase: the winning horse was later found to be a ringer. 1973 B. Broadfoot *Ten Lost Years* xxi. 240 Some teams used to bring in ringers, a Yankee, or a guy from the East. 1980 *Times* 11 Mar. 6 The Crown claimed that the horse had been switched and that the winner was in fact a 'ringer', a more successful stablemate called Cobblers March.

5. *to be a* (*dead*) *ringer for* (or *of*): to resemble closely; to be an exact counterpart of. *slang* (orig. *U.S.*).

1891 *Sporting Times* (N.Y.) 4 July 10/4 Homan is a 'dead-ringer' for Anson. 1900 Ade *More Fables* 162 Bob.. was a Ringer for a United States Senator, all except the White Coat. 1903 'O. Henry' in *Ainslee's* Mar. 129/2 The man was a ringer for the pictures of the fat Weary Willie in the funny papers. 1916 J. Buchan *Greenmantle* xiii. 174 Now you're in these pretty clothes you're the dead ringer of the brightest kind of American engineer. 1946 *New Yorker* 16 Mar. 22/1 The Nissen hut..which were [*sic*] dead ringers for the council lodges the Iroquois Indians used to build. 1954 Wodehouse in *Encounter* Nov. 43/2 We also felt like minor infusoria at the bottom of a well. 'Wodehouse,' I remember saying to myself more than once, 'Alter your appearance very slightly, and you would be a ringer for a waterbeetle.' 1959 *Punch* 21 Oct. 251/1 He [*sc.* a shark] has life pretty easy and apart from the gill-rakers is a ringer for Patrick Joseph. 1960 B. Keaton *My Wonderful World of Slapstick* 122 Going through the casting book we found a man and woman who were dead ringers for the Belgian rulers. 1970 'T. Coe' *Wax Apple* xii. 89 Doctor Fredric Cameron..was an almost dead ringer for J. Roger Urbermann. 1973 C. Sagan *Cosmic Connection* xiii. 92 There is little doubt that the average person's view of Hell—sizzling, choking, sulfurous, and red—is a dead ringer for the surface of Venus.

6. *U.S. slang.* An outsider or intruder; an imposter, *spec.* one who attaches himself to a political or other group to which he does not belong.

1896 Ade *Artie* xi. 100 About a dozen ringers followed us in and stood around rubberin'. 1904 *N.Y. Tribune* 18 Oct. 1 The members of the Manhattan & Democratic clubs occupied front seats. The press seats were largely occupied by ringers. *Ibid.* 8 Nov. 3 The Democratic leaders to-day started to send a lot of alleged 'ringers' across the line into West Virginia to vote to-morrow. 1926 *Clues* Nov. 162/1 *Ringer*, one who butts in on another's racket. 1928 *Manch. Guardian Weekly* 26 Oct. 335/2 Perhaps seventy-five were really newspaper men and women, the others being what the American language calls 'ringers', 'gate-crashers', or 'dead-heads'. 1940 Wodehouse *Eggs, Beans & Crumpets* 59 Too often, when you introduce a ringer into a gaggle of Pekes, there ensues a scrap like New Year's Eve in Madrid. 1963 S. Greer *Metropolitics* v. 104 We have omitted the 'ringer'; none of our respondents claimed to recognize the spurious name. 1965 M. Bradbury *Stepping Westward* vii. 357 This is quite a party. I'm going to feel a real ringer. 1978 *Detroit Free Press* 2 Apr. (Detroit Suppl.) 8/1 Inside the lobby of the dilapidated building, Blow Dry scans the inhabitants, hoping..that no one will spot her as the ringer. 1981 C. R. Lajeunesse *Dead Man Running* xi. 35 A ringer for you will be leaving your place, same car, same registration.

7. *slang.* A false registration plate attached to a stolen motor vehicle; a thief who uses these.

1962 *New Statesman* 21 Dec. 899/1 The driver stays with the car regardless, and the car is equipped with ringers (false number-plates). 1964 E. Parr *Grafters All* ii. 25 The car is now driven to a hideaway, where ringers (false number-plates) are substituted. 1970 P. Laurie *Scotland Yard* iii. 69 All the ringer has to do is buy a [car] key, come along as innocent as pie, open the door and drive off to wherever he does his ringing. 1971 *Drive* Summer 21/2 When the professionals—the car 'ringers'—get to work, the profit on a skilfully doctored vehicle can be more than £500. 1971 *Road Ahead* (Brisbane) Sept. 18 In Britain, 'ringers' produce very special cars. 'Ringers' are experts in modifying stolen cars, giving them a new identity.

8. *ringer-up*. A person making a telephone call; = *caller *sb.* 1 f. Cf. *ring *v.*[2] 10 b.

1963 N. Marsh *Dead Water* (1964) v. 117 The ringer-up was Miss Cost. 1968 P. Dickinson *Skin Deep* v. 104, I don't take on casual ringers-up. 1970 Y. Carter *Mr. Campion's Falcon* xxii. 161 The lunatic fringe—the compulsive ringers-up.

Ringer[3] (ri·ŋəɹ). *Biol.* The name of Sydney *Ringer* (1834–1910), English physician, used *attrib.*, *absol.*, and in the possessive to denote physiological saline solutions of a type which he introduced and which usu. contain (in addition to sodium chloride) salts of potassium and calcium.

1893 *Jrnl. Physiol.* XIV. 200 Ringer's solution kept the heart beating vigorously for more than thirty hours. 1913 *Rep. Brit. Assoc. Adv. Sci. 1912* 660 A heart perfused with a Ringer solution without lime stops much earlier than when perfused with a Ringer solution without lime and without potassium. 1915 *Jrnl. Physiol.* L. 138 Isotonic Ringer or sodium chloride is not an indifferent fluid. 1932 W. Burridge *Excitability* xxi. 173 We..washed out the muscle with saline or Ringer. 1956 *Nature* 18 Feb. 340/1 Ringer solution was injected through the catheter into the sinus in order to show that no larger anastomoses still existed between the sinus and other veins. 1964 W. G. Smith *Allergy & Tissue Metabolism* ii. 19 Schacter has demonstrated that Ringer perfused rabbit tissues release histamine in anaphylaxis. 1967 K. M. Smith *Insect Virol.* v. 105 The sucrose is then removed by dialysis in a cellophane bag immersed in cold Ringer's solution. 1975 *Nature* 10 Jan. 99/2 When the Ringer was made sufficiently hypertonic so that twitch movement was essentially eliminated, the second component propagated throughout the fibre.

Comb. 1958 [see *centrifugation]. 1977 *Proc. R. Soc. Med.* LXX. 160/1 His blood pressure fell from 130/80 mm Hg before anaesthesia to 50 mm Hg during the operation, despite rapid injection of 2 litres of Ringer-lactate solution.

Ringerike (ri·ŋərikə). The name of a district centred on Honefoss north of Oslo in Norway used *attrib.* (after H. Shetelig *Norske Aarsberetning* (1909), 96–107) to designate a style of late Viking art, characterized by abundant use of plant motifs as ornament.

1924 A. F. Major tr. *J. Brøndsted's Early Eng. Ornament* iii. 293 Shetelig has more recently dealt with these runestones; he christens them after the reddish sandstone, which is quarried at Ringerike in the south of Norway, the 'Ringerike Group' and their plant ornament the 'Ringerike style'. 1936 A. W. Clapham *Romanesque Archit.* viii. 193 The decoration also includes a free use of interlacement, chevron-ornament, and a foliage of Ringerike or late Viking type. 1937 E. V. Gordon tr. *Shetelig & Falk's Scandinavian Archæol.* xvii. 303 In the Ringerike style the ancient animal ornament was thrust firmly into the background. 1952 D. T. Rice *Eng. Art 871–1100* v. 127 The new manner is known as the Ringerike style, and it is distinguished by the employment of a characteristic leaf ornament. 1970 Foote & Wilson *Viking Achievement* ix. 307 The style was directly succeeded by the Ringerike style, which first appeared early in the first quarter of the eleventh century. *Ibid.* 310 Ringerike elements were still appearing on Irish objects in the 1120s. 1973 *Times* 1 Nov. 4/5 A unique discovery is a wooden panel with decoration in the Ringerike style of the twelfth-century, with an animal design. 1976 D. Verey *Cotswold Churches* ii. 18 At Bibury, the capital..on the north [of the chancel arch] is in Ringerike style. Many motifs are Scandinavian.

ring-fence, *sb.* Add: **b.** (Later *fig.* examples.)

1950 *Times* 8 Feb. 6/3 The decline is understood to be due mostly to the removal of the 'ring-fence' round the industry, that is, the provision of the Control of Engagement Order requiring miners to stay in coal-mining employment. 1965 *Economist* 6 Nov. 623/2 Imports into ECSC have not been held out by the tariff 'ring fence' erected after the last recession.

ring-fence, *v.* Add: (Further examples.) Also *fig.*

1903 R. Fry *Let.* 30 Jan. (1972) I. 203 B.B. should not have it said that he is capable of political scheming to ring-fence Italian art. 1974 *Northern Times* (Golspie, Sutherland) 2 Aug. 3/4 The proposal to ringfence Embo village was not acceptable for grant purposes.

ringhals (ri·ŋhals). Also **rinkhals.** [ad. Afrikaans *rinkhals*, f. *ring* ring + *hals* neck.] A large venomous spitting cobra, *Hemachatus hæmachatus*, of the family Elapidæ, found in southern Africa, and distinguished by a white ring or two across the neck of an otherwise brown or black skin. Also *attrib.*

1793 tr. *C. P. Thunberg's Trav. Europe, Afr. & Asia* I. 208 A colonist had been bitten in the foot some time before by a serpent of the species called Ringhals (or Ringneck). 1835 J. W. D. Moodie *Ten Years S. Afr.* I. xv. 316 The puff-adder, the ring-hals, and the berg-adder, are very poisonous and very numerous. 1864 T. Baines *Explor. S.-W. Afr.* xiv. 449, I think the species is called 'ring hals' (or ringed throat) in the Colony. 1906 *Westm. Gaz.* 16 Jan. 4/1 A Spurred Chameleon, a small Monitor, and a couple of Ring-hals snakes. 1925 *Other Lands* July 44/2 They pointed to the half-open door, where she saw uncoiling itself a large ringhals. 1931, etc. [see *cobra]. 1939 S. Cloete *Watch for Dawn* v. 62 With the bite of a mamba or a ringhals the heart sometimes beat even after life had gone. 1947 *Cape Argus* 29 Nov. (Mag. Section) 1/3 Cobras, like the ringhals and indeed all other snakes make for cover when they feel the approach of some intruder. 1956 A. G. McRae *Hill called Grazing* vi. 47 The six-foot-long Rinkhals cobra, reared above its coils, hood flattened and tiny, evil head weaving. 1972 L. van der Post *Story like Wind* ix. 292 She..saw, sitting upright on its tail, black as ebony and shining as with oil, a seven-foot rinkhals cobra.

ringie (ri·ŋi). *Austral. slang.* [f. Ring *sb.*[1] + -ie.] The keeper of the ring in a game of *two-up.

1941 Baker *Dict. Austral. Slang.* 65 *Ringie*, the keeper of a two-up school. 1949 *Strand Mag.* Dec. 100/1 The ringie takes care of operations inside the ring of side betters which gathers round the two protagonists. 1951 F. Hardy *Power without Glory* 323 Red Ted was 'Ringie'. He supervised the game in the ring itself, seeing that the pennies were spun fairly, and calling the results.

ringing, *vbl. sb.*[1] Add: **1. a.** Also, the putting of a numbered ring on a bird or a bat. (Further examples.)

1910 *British Birds* III. p. iii, There is every indication that facts of the utmost interest and importance will be brought to light by the ringing of birds. 1953 Lockley & Russell *Bird-Ringing* i. 5 It was in 1899 that Herr Christian C. Mortensen laid the foundations of scientific bird-ringing when he placed his first aluminium rings, stamped with numbers, on the legs of young starlings. 1958 *Listener* 30 Oct. 684/2 For our first attempt at bat-ringing..my wife and I and a friend climbed up into the roof of a barn.. where we knew there was a colony of bats. 1978 P. Conder *RSPB Guide to Birdwatching* 101 The Protection of Birds Act 1967, prohibits ringing except under licence issued by the Nature Conservancy Council through the British Trust for Ornithology.

4. The action of cattle in forming a ring. Cf. Ring *v.*[1] 1 d in Dict. and Suppl.

1868 C. W. Browne *Overlanding in Australia* 77 After an hour's amusement of this sort, they stop of their own accord. This evolution is termed 'ringing'. 1941 Baker *Dict. Austral. Slang* 60 Ringing, the milling of cattle.

ringing, *vbl. sb.*[2] Add: **1. c.** *ringing-up*, in senses 10 b, c of the verb.

1835 Dickens *Sk. Boz* (1836) 1st Ser. II. 205 Let us take a peep 'behind' previous to the ringing up. 1924 Galsworthy *White Monkey* III. xiv. 316 He..closeted himself in the telephone booth... Ringing-up was quicker. 1949 N. Mitford *Love in Cold Climate* I. v. 51 This ringing-up of Paris seemed to me a most dashing extravagance. Aunt Sadie..only made trunk calls in times of crisis.

2. b. *Electronics.* The phenomenon of transient damped oscillation occurring in a circuit at its resonant frequency as a result of a sudden change in voltage level; also, in *Television*, the occurrence on the screen of black lines to the right of a white object, caused by transient oscillation in the video amplifier of the receiver.

1949 *Electronic Engin.* XXI. 207/2 If the attenuation of the high frequencies takes place too suddenly (i.e. sharp cut-off) then we get ringing. 1953 H. A. Chinn *Television Broadcasting* xvi. 640 A spurious peak or a sharp cutoff in the response-frequency characteristic..discloses itself in the resulting picture display as..repetitions of the original signal. This manifestation of 'ringing' of the video circuits can be detected by observation of vertical lines in the picture. 1961 *Times* 20 Nov. (Television Suppl.) p. xii/1 Gross defects are obvious to any viewer either by poor contrast, lack of definition, 'ringing', or unwanted interference. 1969 J. J. Sparkes *Transistor Switching* iv. 113 In practice the inductance in the emitter wire through which the drive current has to flow is likely to limit switching speed or cause 'ringing' in the output voltage. 1978 *Gramophone* Jan. 1336/2 There is no suggestion of overshoot or ringing, showing that the amplifier is extremely stable and has a good damping factor.

4. ringing tone *Teleph.*, the sound produced in a caller's telephone to indicate that connection has been made to another telephone and it is ringing.

1924 W. Aitken *Automatic Telephone Syst.* III. lvi. 272 A portion of the ringing current also passes over the upper condenser,..and the caller receives the ringing tone. 1943 G. Greene *Ministry of Fear* I. v. 80 He dialled the number... He was almost afraid to hear the ringing tone. 1970 T. Lewis *Jack's Return Home* 181 The ringing tone only went once before someone lifted the receiver.

ringle-ji·ngle, *v.* *nonce-wd.* [f. Ringle *v.*[2] + Jingle *v.*] *intr.* To write in verse.

1913 G. B. Shaw *Let.* 7 Feb. in *B. Shaw & Mrs. Campbell* (1952) 83, I never have to think of how to say anything in prose... Yet, when I want frightfully to ringle-jingle with words they wont come that way.

ring-master. Add: (Further examples.) Also as *v. trans.*

1943 C. H. Ward-Jackson *Piece of Cake* 51 *Ringmaster*, squadron commander. 1952 L. Bell *Inside Fight Game* iv. 73 A match maker can be, and often is, the promoter of a boxing show. His license also permits him to officiate as ringmaster. 1958 F. C. Avis *Boxing Ref. Dict.* 95 *Ringmaster*, the official in charge of the arrangements at a boxing venue. 1964 *New Statesman* 17 Apr. 616/1 To bring off this effect, *Caligula* needs producing as a kind of play-within-itself: a fantastic circus of cruelty ringmastered by Caligula for a circle of witnesses as sane as he is. 1969 P. West *Words for Deaf Daughter* iii. 72 Speeding up, you ringmaster him [*sc.* a budgerigar] through his full repertoire of forward rolls and swift, thudding vaults.

1972 *N.Y. Times* 4 June v. 9/6 James Fallon, the manager, and C. L. (Honey) Craven, the ringmaster, both expressed their pleasure with the new surface. 'The new ring makes things much easier for the rider,' said Fallon.

Hence **ri·ngmastership,** the art or status of a ringmaster; also *fig.*

1966 *Economist* 22 Oct. 378/2 That they did not do so during last week's conference was due in considerable measure to some masterly ringmastership on the part of the party chairman, Sir Dan Mason. 1969 *Daily Tel.* 6 Oct. 15/7 Most circuses get by without script-writers but this one had five, doubling up as clowns performing what they had written under the ringmastership of Ian McNaughton.

ring-neck, *sb.* Add: **1. a.** Also, a ring-necked pheasant, etc.

1921, 1965 [see *Mongolian *a.* 4].

ring-net, *sb.* Add: **2. b.** A form of long seine-net which is supported at the ends by separate boats, one of which moves in a circular path towards the other in order to trap the fish within the net, used esp. in the Scottish fishing grounds to catch herring.

1949 Mitchison & Macintosh *Men & Herring* 21 A ring net is about a hundred fathoms long and twenty fathoms deep and mostly set so that the back rope with the corks is almost on the top of the water. 1950 P. F. Anson *Scots Fisherfolk* vii. 80 Ring-nets have always been the favourite gear employed on the Firth of Clyde and Loch Fyne. 1969 D. B. Thomson *Seine Net* vii. 131 Most of these [boats] are dual-purpose seine net and ring-net boats... They have a low rail and large fish hold hatch to suit the ring net operation. 1978 M. Gray *Fishing Industries of Scotland, 1790–1914* vi. 122 With fish from ring-net boats the transports could be in Glasgow by early in the day although the supply from the drift-net boats, slower to unload, would not arrive till later in the day.

So (in sense *2 b) **ri·ng-net** *v. trans.*; hence **ri·ng-netter,** a fishing boat intended for use with ring-nets; **ri·ng-netting** *vbl. sb.*

1936 *Discovery* Dec. 388/1 Scottish West Coast fishermen in Loch Fyne..were engaged in ring-netting herring. 1952 G. Maxwell *Harpoon at Venture* ii. 47 The Mansons.. brought the first new-ring-netter to Mallaig after the war. 1960 Williamson & Boyd *St. Kilda Summer* i. 23 Cunningham had sent his ring-netter, *A' Mhaighdean Hearrach*, to take us off. 1969 D. B. Thomson *Seine Net* vii. 131 The mast is stepped so it can be lowered when the vessel is engaged in ring netting. 1978 M. Gray *Fishing Industries of Scotland, 1790–1914* vi. 120 In the 1830s the drift-net fishermen found, or felt, themselves threatened by a rival body—the fishermen who adopted the technique variously described as 'trawl-', 'seine-' or 'ring-netting'. As a description, 'ring-netting' is probably the most accurate term. In this new form of fishing..two boats operated together, holding the different ends of a net some 150 yards in length; the net was cast as one boat moved from a starting-point near its partner, to form a wide circle as it returned to the original point, and the haul was completed by gradually constricting the circle till the fish were in a concentrated mass ready to be hauled aboard. *Ibid.* 121 It was possible..for ring-netters also to carry drift-nets which they would shoot on occasion.

ringocandy. *nonce-wd.* [? f. Ringo var. Eryngo or Ring *sb.*[1] + *o'* (= of) + Candy *sb.*[1]] Some kind of confection.

1922 Joyce *Ulysses* 199 Hot herringpies, green mugs of sack, honeysauces, sugar of roses, marchpane, gooseberried pigeons, ringocandies.

ri·ng-road. [f. Ring *sb.*[1] + Road *sb.*] A by-pass road encircling a town or urban area.

1928 *Daily Express* 27 Aug. 8/6 London has no form, no symmetry. I suggest that we could give her this by cutting a broad ring-road through the old nineteenth century suburbs. 1933 L. P. Abercrombie *Town & Country Planning* iv. 144 The external Ring road has been frequently made to avoid the destructive widening of an old village. 1942 *Country Life* 9 Oct. 692/3 Among recommendations are..removal of markets from the central areas to positions on the ring road. 1943 Forshaw & Abercrombie *County of London Plan* iv. 53 We are strongly of the opinion that a relatively fast traffic ring-road is essential. 1952 *Ann. Reg. 1951* 402 The system of arterial ring-roads was abandoned, largely on grounds of cost. 1956 *Sun* (Baltimore) 17 July 12/2 Mr. McVoy establishes a priority system in which the first need is the ring road around the inner city. 1963 *Times* 22 Feb. 5/2 A ring road will surround the new centre and a recently built open market is to be doubled. 1971 *Country Life* 3 June 1377/1, I knew a priest in Tirana who fought the City Council over a plan to drive a ring-road through his church. 1981 B. Hines *Looks & Smiles* 31 They..caught a bus out to the Ring Road where a Trading Estate was being developed.

ringrou·ndabout, *v.* *nonce-wd.* [Ring *v.*[1]] *trans.* To surround.

1922 Joyce *Ulysses* 189 The faithful hermetists..ringroundabout him.

ringside (ri·ŋsəid). Also **ring side, ring-side.** [f. Ring *sb.*[1] + Side *sb.*[1]] **a.** The area immediately surrounding a boxing ring or other sports arena; more generally, the area which accommodates spectators; the scene of a sporting activity. Also *transf.*

1866, 1896 [in Dict. s.v. Ring *sb.*[1] 15 c]. 1926 [see *barrera]. 1930 [see *hook-up]. 1956 B. Holiday *Lady sings Blues* (1973) xix. 159 There at the ring-side was Mrs. Helen Hironimus, the Alderson warden. 1965 *Universe* 15 Oct. 12/4 (*heading*) My conversations at the ringside. 1976 [see *opening *ppl. a.* 1 b].

b. *attrib.* and *Comb.*, as *ringside judge, table*; **ringside seat,** a seat immediately adjacent to a boxing contest or other sporting activity; also *transf.* and *fig.*

1976 *Scotsman* 15 Dec., The ringside judges may score it a clear victory for Kohl, leader of the Christian Democrats. 1932 *Daily Express* 20 Sept. 19/5, I trust my health will be good when he makes his debut. I want to be in a ringside seat. 1934 Wodehouse *Right Ho, Jeeves* xvii. 212 From the fact that he spoke as if he had a hot potato in his mouth without getting the raspberry from the lads in the ringside seats, I deduced that he must be the head master. 1940 *War Illustr.* 19 Jan. 638/3 From our 'ringside' seats in the air, we saw a spectacle which made us heartily glad it was not our duty to attack a British warship! 1947 G. Greene *19 Stories* 75 Like a bull he was on show, sitting there mournfully in the plaza with his dog, a magnificent spectacle for which we all had ring-side seats. 1975 V. Canning *Kingsford Mark* viii. 140 Carlo found the rifle... Unseen, he would have a ringside seat. 1929 D. Runyon in *Cosmopolitan* July 59/1, I see Waldo Winchester, the scribe, sitting at a ringside table all by himself.

Hence **ri·ngsider,** one who occupies a position at a ringside; a spectator.

1898 A. M. Binstead *Pink 'Un & Pelican* iv. 87 Old Jack Baldock, always the noisiest of ringsiders, was howling at the Fulham lad. 1954 C. L. B. Hubbard *Compl. Dog Breeders' Man.* 202 A woman exhibitor can handle her dog well, feel comfortable, and please the ringsiders..by wearing a simple but practical outfit. 1960 *Times* 28 Sept. 16/6 With money at stake among some ringsiders, not everyone was thinking calmly. 1976 J. Snow *Cricket Rebel* 113 He spooned a simple catch from a stroke ringsiders described as a 'protective jab'.

ringster. Add: **1.** (Earlier and later examples.)

1875 *Chicago Tribune* 15 Dec. 4/1 The support secured for Mayor Cobb was sufficient..to defeat the unholy alliance of ringsters and politicians by which Boardman's nomination was first obtained. 1908 *Nation* 16 Apr. 344/3 Hereafter the word [grafter] cannot be lightly used as a synonym for any malefactor at the head of a corporation, or any political ringster.

2. *U.S.* A member of a price-ring. Cf. *ring *sb.*[1] 11 a.

1878 *Congress. Rec.* 20 Mar. 1915/1 As the honest contractor will not go into a business where he has to evade the law, the ringster has it all his own way. 1879 *Harper's Mag.* Oct. 717 The inopportune arrival of several cargoes of Texan beef broke the ring and ruined the ringsters. 1904 I. M. Tarbell *Hist. Standard Oil Co.* I. 107 'Deserters', 'ringsters', 'monopolists' were the terms applied.

3. A boxer.

1926 G. Carpentier *Art of Boxing* 5 Some 'ringsters'—I use this word in a particular sense, as you will see—have severely battered countenances after a career of a few years as pugilists. 1965 *Eng. Stud.* XLVI. 465 A boxer, among scores of other appellations, may be..a *ringster*.

ringtail, ring-tail. Add: **3. a.** (Later examples.)

1873 'Vanderdecken' *Yachts & Yachting* 185 A racing cutter will be fitted with four gaff-topsails, viz., a jib-headed or ring-tail topsail that is set without a yard. 1901 S. H. King *Dog-Watches* 59 The Victoria, had a throat and peak mainsail instead of the mutton-leg mainsail and ringtail gaff topsail of the Excelsior. 1934 [see *Jamie Green].

4. (Written ring tail.) A dog's tail which is curled so as to form nearly a complete circle.

1872 'Idstone' *Dog* x. 87 The tail should be of a moderate length,..not curved over the back, not carried low, nor curved at the end like what in Bulldogs is called a 'ring tail'. 1961 J. Lanning *Great Danes* viii. 73 Ring tails are a hereditary fault and were common at one time.

5. *U.S. slang.* A worthless person, a hobo.

1926 *Amer. Speech* I. 652/2 *Ring-tail*, a Hobo who is carrying a 'grouch'. 1931 'D. Stiff' *Milk & Honey Route* 205 He [*sc.* an unpopular fellow] is also a *ring tail*. Such hobos are often under suspicion. 1935 A. J. Pollock *Underworld Speaks* 98/1 *Ring tail*, an ignorant, loud mouthed, vulgar person. 1947 *Amer. Speech* XXII. 214 In the Pacific Theater..the sobriquets applied to the Japanese were particularly hateful, as *ringtails, yellow bastards*, and a host of unprintables.

6. *Austral. slang.* (See quots.)

1941 Baker *Dict. Austral. Slang* 60 *Ringtail*, a coward. 1943 *Amer. Speech* XVIII. 256 With Americans a *ringtail* is a grouchy person; with Australians he is a coward.

7. *attrib.*, as *ringtail roarer, snorter*, etc., = Ring-tailed *a.* 4 in Dict. and Suppl.

1832 J. K. Paulding *Westward Ho!* I. xiv. 124, I got tired of making fun of the ringtail roarer. 1859 *Oregon Argus* 10 Dec. 1/1 Here lies James D. Porter, Who lived as he hadn't orter, But as a Methodist exhorter Was a regular ring-tail snorter. 1862 J. R. Lowell *Biglow Papers* 2nd Ser. 25 My eldes' boy's so took up, wut with the Ringtail Rangers An' settin' in the Jestice-Court for welcomin' o' strangers.

ring-tailed, *a.* Add: **3.** (Further example.) Cf. *ringtail, ring-tail 4.

1894 R. B. Lee *Hist. & Descr. Mod. Dogs* (*Non-Sporting Division*) ix. 239 The tail should be..not curved upwards at the end, called 'ring tailed'.

4. *ring-tailed roarer* (earlier and later examples applied to persons). Also in other fanciful collocations, applied to persons and things, as *ring-tailed snorter, squealer*, etc.

1830 *Painesville* (Ohio) *Tel.* 15 June 1/5 *Ringtailed Roarers*, a most violent fellow, a Crockett. 1836 *Crockett's Yaller Flower Almanac* 9, I am a raal ringtailed roarer. 1837 R. M. Bird *Nick of Woods* I. iii. 56 Stranger, my name's Ralph Stackpole, and I'm a ring-tailed squealer! 1944 B. A. Botkin *Treas. Amer. Folklore* I. 175 The ring-tailed roarer is a comic version of the frontiersman who wrestles single-handed with the wilderness. 1947 *Chicago Tribune* 2 Nov. IV. 9/2 The 'ring-tailed Roarers' of the pioneer days shout their raucous delight, unsubtle, earthy, outlandish, direct. 1950 *Ithaca* (N.Y.) *Jrnl.* 1 Aug. 6/2 You'll have to hand it to this..secretary of agriculture... He's a ring-tailed snorter. 1972 J. Mosher *Adultery* IV. xxiii. 190 Listen to that wind coming up, would you? She's a ring-tailed snorter.

ring-wall. Add: Also **ringwall, ring wall.** **1. a.** (Later examples.) Also *fig.*

1944 *Cape Times* 25 Oct., In trade they did not want the ring-wall of the British Empire round them. 1950 H. L. Lorimer *Homer & Monuments* i. 27 A ring-wall of slabs was erected round the graves on the new level. 1963 L. F. Chitty in Foster & Alcock *Culture & Environment* vii. 175 When its site was first 'bull-dozed' in 1947, the ruins of a broad ring-wall were revealed, constructed of local tilestones. 1970 I. Petite *Meander to Alaska* II. xi. 111 He was in the process of repairing the foundation [of a house] with a concrete ring wall.

b. A roughly circular eminence surrounding a crater or mare on the moon or a similar formation on the earth, freq. of volcanic origin.

1950 W. Ley *Conquest of Space* 86 Copernicus is probably the most beautiful of all lunar craters. Its ringwall is 12,000 feet high at the highest point; its diameter is 56 miles. 1962 E. A. Vincent tr. *Rittmann's Volcanoes* iii. 122 Ring-walls (ramparts) with outflows of lava are common. 1966 *Earth-Sci. Rev.* I. 231 The continents [on the Moon] are very mountainous and the mountains and lesser eminences generally form parts of the 'ringwalls' of maria and craters. 1968 R. W. Fairbridge *Encycl. Geomorphol.* 682/2 The island [*sc.* Makatea] rises on all sides from very deep water with a fringing reef.., bounded in the inner side by an abrupt or overhanging cliff of ancient coral limestone... To the inside again, there is a second cliff.., and a second rocky terrace. Within this ring-wall is a moat-like depression.

ringway (ri·ŋwēɪ). [f. Ring *sb.*[1] + Way *sb.*[1]] A circular system of major roads round a town or urban area.

1969 *Daily Tel.* 16 Apr. 17/6 The council's plans for ringways were absolutely essential if London was to remain a major world city. 1970 [see *orbital *a.* 3]. 1971 *New Scientist* 8 July 102/2 Terence Bendison describes the traffic-swept gulf between Shepherds Bush and Holland Park, where part of the inner ringway has been imposed on streets and houses. 1973 R. Busby *Pattern of Violence* vi. 96 The familiar streets..down the east ringway to the Queen's underpass.

ri·ngwise, *adv.* [f. Ring *sb.*[1] + Wise *sb.*[1] II.] In the form of a ring or rings; so as to produce a ring-shape.

1889 *Lancet* 3 Aug. 244/1 Their foreheads are tattooed ringwise, with singularly shaped cuttings in the skin. 1915 E. R. Lankester *Diversions of Naturalist* 101 Backbone-pieces and the body muscles attached ring-wise to them. 1922 Joyce *Ulysses* 62 Pepper. He sprinkled it through his fingers, ringwise, from the chipped eggcup.

ringy, *a.*[1] Add: Also **ringey.** **2.** *N. Amer. slang.* Irritable, contentious, angry.

1932 *South of Market Tribune* (San Francisco) 23 Dec. 22 Nothing will make the boss more ringey. 1934 M. C. Boatright *Tall Tales from Texas* 30 He's a good-natured bird and don't git ringy about it. 1942 Berrey & Van den Bark *Amer. Thes. Slang* § 284/6 *Ill-tempered*..ringy. 1955 R. P. Hobson *Nothing too Good for Cowboy* xviii. 187 To take some of the snoose out of the ringy Batnuni bunch [of cattle]. 1962 [see *onery, onnery, o'n'ry]. 1977 *N.Y. Times* 11 Jan. 20/1 He thought Miss Longet had been 'ringy' the day she shot and killed Mr. Sabich.

rink, *sb.*[2] Add: **2. b.** (Later examples.)

1968 *Globe & Mail* (Toronto) 3 Feb. 35/5 Webb, whose rink is composed of Jean Dye, Bill and Helen Ferguson, scored 39 points for its three wins. 1976 *S. Wales Echo* 26 Nov., The only Merthyr rink which returned a winning card was that skipped by Noel Tippett.

3. c. *N. Amer.* A frozen surface on which ice hockey is played. Also, a hall or stadium containing this.

1896 *Times* (Niagara-on-the-Lake, Ont.) 27 Feb. 1/4 The Niagaras..know just how to toss the puck around from one end of the rink to the other in order to score. 1945 W. H. Pugsley *Saints, Sinners & Ordinary Seamen* 90 In the early months of the war..ratings lived in a converted hockey rink. 1953 *Canad. Geogr. Jrnl.* XLVI 138/2 The children maintain their own open air hockey rink on the ice. 1974 *Plain Dealer* (Cleveland, Ohio) 26 Oct. 5-D/1 The Elysium had been a little rink which could accommodate almost 2,000.

4. A measured strip of bowling-green on which a match is played. Also, the players allotted to a rink.

1864 W. W. Mitchell *Man. Bowl-Playing* 21 When..any number of players, not exceeding eight, form sides and commence a game, they make what is called a rink. *Ibid.*, The space or division of the Green is also commonly called a rink. 1897 *Encycl. Sport* I. 129/2 *Rink*, (1) a narrow section of a bowling-green, some twenty feet in breadth taken by one party for their game. (2) All the players upon the two sides. (Both terms are more common in Scotland than England.)

1906 *Canadian Mag.* Sept. 475/2 Like curling.., [bowling] permits of an adjournment occasionally in order that..the opposing rinks may 'join' each other and have 'something'. 1975 *Oxf. Compan. Sports & Games* 94/2 The green is divided by boundaries of fine string into six 'rinks' the length of the green and 19 to 21 ft. (5.8–6.4 m.) wide. 1976 *Laws of Game* (Eng. Bowling Assoc.) (ed. 3) 4 The green shall be divided into spaces called rinks, each not more than 19 feet nor less than 18 feet wide.

5. *attrib.* and *Comb.*, as *rink boot, -side*; **rink rat** *N. Amer.*, a youth who seeks casual work at an ice-hockey rink in return for free admission, etc. (see also quot. 1945); **rink string**, a length of string which marks the boundary of a rink on a bowling-green.

c 1885 in M. Johnson *Amer. Advertising* (1960), Childs's Cash Shoe Store. 'Ladies' rink boots' a specialty. 1945 L. Shelly *Jive Talk Dict.* 31 *Rink rat*, skating rink enthusiast. 1965 *Victoria* (B.C.) *Daily Times* 20 July 10/6, I was a rink rat at the Forum. 1970 J. H. Gray *Boy from Winnipeg* 51 Two Fort Rouge schoolboys from the Kennedy rink rats eventually made good as professionals. 1916 A. Bridle *Sons of Canada* 26 The genial boss..sits in his fur-lined greatcoat at the rinkside. 1972 'E. Lathen' *Murder without Icing* (1973) ii. 17 He was at rinkside. On the ice, the New York Huskies were having a workout. 1960 R. Williams *Border Country* iv. 113 Harry went up to the bowling green to..set out the rink-strings and the mats.

rink, *v.* Add: (Earlier and later examples.) Also *trans.* (in quot. *fig.*).

1877 H. Sidgwick in A. & E. M. Sidgwick *Henry Sidgwick* (1907) v. 326, I 'rinked' or 'runk' (I do not know how the verb is conjugated)... It is not half as amusing as real skating. 1909 'W. N. P. Barbellion' *Jrnl.* 25 Dec. (1919) 24, I..idly scan magazines in the Library and occasionally rink—with palpitation of the heart as a consequence. 1946 R. Campbell *Talking Bronco* 25 The zephyr from the blue nevadas, Stirruped with kestrels, smoothly rinking The level wave where halcyons drowse.

rinker (ri·ŋkəɹ). *dial. rare.* [f. *rink*, dial. var. Ring *sb.*[1] + -er[1]: cf. *ring* a circle into which marbles are thrown (*Sc. Nat. Dict.*) and Ringer[1] 1.] A marble (see quot.[1]); a game of marbles.

1910 A. Bennett *Clayhanger* i. i. 6 They were not the paltry marble of today, plaything of infants, but the majestic 'rinker', black with white spots, the king of marbles in an era when whole populations practised the game. *Ibid.* 9 The open gates of a manufactory disclosed six men playing the noble game of rinkers on a smooth patch of ground... They were celebrated marble-players, and..they shot the rinkers from their stubby thumbs with a cannon-like force and precision.

rinkite (ri·ŋkəit). *Min.* [ad. Sw. *rinkit* (J. Lorenzen 1884, in *Öfversigt af k. vetensk. Förh.* 111), f. the name of Henrik Johannes *Rink* (1819–93), Danish geologist and explorer: see -ite[1].] A silicate mineral similar to (or identical with) mosandrite, found as reddish- or yellowish-brown crystals from Greenland.

1886 *Jrnl. Chem. Soc.* L. 676 Rinkite. 1887 *Mineral. Mag.* VII. 234 Fluorite is fairly abundant, and also two of the rarer minerals—laavenite and rinkite. 1937 *Mineral. Abstr.* VI. 179 Rinkite from Greenland gave sp. gr. 3.458 and an orthorhombic cell..containing two molecules $[SiO_4]_2[(Ti, Ce)F]Ca_2Na$. 1971 *Ann. Rep. Delegates Sci. Area Univ. Oxford 1970* 131 Accessions... Rinkite..from south-west Greenland. 1971 *Acta Crystallogr.* B. XXVII. 1277 Rinkite, $(Ti,Nb,Al,Zr)(Na,Ca)_3(Ca,Ce)_4(Si_2O_7)_2(O,F)_4$ approximately, is a silicate crystallizing in the monoclinic system.

rinktum (ri·ŋktəm). *rare. Southern U.S. dial.* alteration of Rectum.

1929 W. Faulkner *Sound & Fury* 86 You know what I'll do. I'll skin your rinktum.

rinky-dink (ri·ŋkidiŋk), *sb.* and *a. slang* (chiefly *U.S.*). Also **rinkey-dink, rinkydink, rinky-dinky**. [Orig. unknown: cf. *ricky-tick *sb.* and *a.*]

A. *sb.* Something that is worn out or antiquated; a worthless object. *spec.* a cheap place of entertainment. Also in phr. *to give* (someone) *the rinky-dink* and varr., to cheat or swindle (someone).

1912 A. H. Lewis *Apaches N.Y.* xii. 265 They was lyin'..an' givin' each other th' rinkey-dink in th' old days same as now. 1922 J. A. Dunn *Man Trap* i. 8 Jimmy abhorred mining corporations with a lot of stockholders and a few of those liable at any moment to hand you the rinky-dink and freeze you out by due process of legal indifference toward small-fry claimants. 1942 *Harper's Bazaar* July 21/2 Don't give me the rinkydink. 1951 *Atlantic Monthly* Mar. 80/1, I think of Sweet Mama Stringbean as she was called when she played the Rinky-dinks for $25 a week. 1956 S. Longstreet *Real Jazz* 147 *Rinky-dink* is broken-down stuff. 1969 *New Yorker* 1 Nov. 6/2 *Red Garter*..eighteen-nineties rinky-dink, complete with fire engine, but the banjo band is above average. 1977 *Amer. Speech 1975* L. 65 *Rinky-dink*..*n*, something that is cheap or worn out. 'His car is a real rinky-dink.'

B. *adj.* Worthless, worn out, trivial; old-fashioned, outmoded.

1913 *Wells Fargo Messenger* I. 105/3 She did not care to ruin her life as a Sunday supplement feature to some rinky-dinky foreign count. 1942 Berrey & Van den Bark *Amer. Thes. Slang* § 675/12 Baseball... *Rinky-dink ball*, inferior playing. 1946 Mezzrow & Wolfe *Really the Blues* vii. 87 My struggle-buggy was getting to look like a rinky-dink old tin can on wheels. 1951 E. Waters *His Eye is on Sparrow* vi. 77 Of all those rinky-dink dumps I played, nothing was worse than the Monogram Theatre in Chicago. 1973 *Globe & Mail* (Toronto) 24 Sept. 5/3 He thinks it hasn't been legalized only because of 'rinkydink politics' but he decries other aspects of the drug scene and says he thinks he's kept some of his young patients from getting involved. 1979 *Fortune* 15 Jan. 5 Facet [was]..a rinky-dink outfit with no real resources and not much to offer.

rinky-tink (ri·ŋkitiŋk), *a. slang* (chiefly *U.S.*). [Imitative: cf. prec. and Tink *int.* and *sb.*] Designating a jazz or ragtime piano on which simple, repetitive tunes are played; tinkling, jangling. Cf. *ricky-tick *a.*

1962 E. Lucia *Klondike Kate* ii. 34 A rinky-tink piano or a scratchy gramophone had its intoxicating effect upon her. 1974 *News & Courier* (Charleston, S. Carolina) 25 Apr. 5-a/1 Scott Joplin played his toe-tappers on a rinky-tink piano. 1975 J. Gores *Hammett* (1976) i. 12 A rinky-tink piano was bashing out 'Ja-Da'.

Rinne (ri·nə). *Med.* Also (*erron.*) Rinné. The name of Heinrich Adolf *Rinne* (1819–68), German otologist, who described a form of the test in 1855 (*Vierteljahrschr. f. d. prakt. Heilkunde* XLV. 72), used in the possessive, *attrib.*, and *absol.* to designate a diagnostic test for deafness (see quot. 1883).

1883 tr. *Politzer's Text-Bk. Dis. Ear* 693 Rinne's experiment..consists in setting a tuning-fork..on the vertex or on the mastoid process, and allowing it to vibrate till the note is no longer heard; the prongs of the fork are then brought close to the ear, and in normal circumstances the tone will be heard again (positive experiment). 1899 G. Bacon *Man. Otology* ii. 81 Rinné's test... If heard again, which is the case in the normal ear, this is called the positive Rinné test. 1902 tr. *Brühl & Politzer's Atlas & Epitomy Otol.* 87 Positive result of Rinne's experiment, or positive Rinne. 1959 H. A. Newby *Audiol.* iv. 61 If the patient replies affirmatively, the result of the test is said to be a Rinné negative... A Rinné test on a normal ear will yield a positive result. 1974 Passmore & Robson *Compan. Med. Stud.* III. xxxii. 3/2 In conductive deafness, bone conduction is better heard than air conduction and Rinne's test is negative.

rinneite (ri·nə,ait). *Min.* [a. G. *rinneit* (H. E. Boeke 1908, in *Chem. Zeitung* XXXII. 1228), f. the name of Friedrich *Rinne* (1863–1933), German mineralogist: see -ite[1].] A rhombohedral chloride of iron, potassium, and sodium, $K_3FeNaCl_6$, which is known as colourless, pale rose, violet, or yellow granular masses from saline deposits, and can be prepared artificially.

1909 *Jrnl. Chem. Soc.* XCVII. ii. 153 Rinneite... An anhydrous mineral..has been found in considerable quantities at the Nordhausen Works. 1953 *Q. Jrnl. Geol. Soc.* CVIII. 289 At Rockhead the marl is much richer in rinneite and carnallite than at the other boreholes. Several coarse, colourless to yellow, rinneite inclusions..were found. 1970 *Rocks & Minerals* XLV. 376/1 Rinneite has been reported as a secondary mineral in the salt mine at Winnfield Dome.

‖ **Rinpoche** (ri·npǫtʃe). Also Rimpoche, etc. [Tibetan, lit. 'precious (jewel)'.] An honorific title given to a chief priest among Tibetan Buddhists. Cf. *panchen.

1774 G. Bogle *Narr. Mission to Tibet* (1876) iii. 26 A tower, about five or six stories high,..is appropriated to Lama-Rimboché. 1784 S. Turner *Let.* 2 Mar. in *Acct. of Embassy to Court of Teshoo Lama in Tibet* (1800) iii. 364 The death of Gesub Rimbochay, offered a new prospect of opening..communication. 1800, etc. [see *panchen.] 1863 tr. *E. Schlagintweit's Buddhism in Tibet* xii. 153 The *Panchen Rinpoche* is considered to be an incarnation of Chenresi's celestial father, Amitābha. 1882 *Encycl. Brit.* XIV. 230/1 The Dalai Lāma..is actually called the *Gyalpo Rinpotshe*, 'the glorious king',—his companion being content with the title *Pantshen Rinpotshe*, 'the glorious teacher'. 1889 M. Monier-Williams *Buddhism* xi. 284 The other Grand Lāma who resides in the monastery of Tashi Lunpo..has the Tibetan title of Panchen Rinpoche (Pañ-ćen Rin-po-će), 'the great Paṇḍit Jewel'. 1929 D. Macdonald *Land of Lama* ii. 43 This cleric..was now allowed to assume the incarnation of Buddha Amitabha, with the title of *Panchen Rimpoche* or 'Precious Gem of Learning'. 1980 D. Hart-Davis *Heights of Rimring* xvi. 174 Access to the monastery could be gained only by means of a contraption which the *rinpoches*, the lamas, let down on a rope.

rinse, *sb.* Add: **2. c.** A solution (or cream) which temporarily tints or conditions the hair. Also, an application of this.

1928 *Daily Mail* 25 July 3/6 Though the price of Icilma Shampoos remains at 3d., each packet now contains a wonderful Toning Rinse suitable for every shade of hair, which removes *all trace of lather*, and leaves the hair in a state of exquisite burnished beauty. 1942 M. Dickens *One Pair of Feet* ix. 189, I think I shall go and have a platinum rinse next payday. 1944 M. Laski *Love on Supertax* x. 93, I think I'll have a blue rinse to-day. 1948 M. Sturges-Jones *In Wedlock Wake* 137 Why don't you try a blond rinse? It wouldn't make you look bleached. 1949 N. Mitford *Love in Cold Climate* 221 A blue rinse for her grey hair. 1958 J. Cannan *And be a Villain* vii. 155 Age must be disguised, hushed up with dyes and rinses. 1959 *Spectator* 24 Sept. 394/3 Women..with bitter-sweet smiles and blue-rinse bobs. 1962 D. Lessing *Golden Notebk.* iii. 345 'I did try a rinse,' he remarked, 'but the grey shows through.' 1977 B. Pym *Quartet in Autumn* i. 2 Letty knew that there were white hairs interspersed with the brown and that most people would have had a brightening 'rinse' anyway.

3. In *Comb.*, as **rinse-aid** (see quot. 1963).

1963 *Which?* 6 Feb. 50/1 Five [dishwashing] machines..supplied rinse aids. These are liquids added to the final rinse water to make it flow more easily and prevent it from remaining as drops on the surface. 1970 *Ibid.* Oct. 294/1 Some detergents—or rinse aids—leave a white deposit on everything.

rinse, *v.* Add: **4. c.** Const. *out.* Also *absol.*

1953 [see *drip-dry *v.*]. 1976 M. Millar *Ask for Me Tomorrow* (1977) xvi. 132 You're not helpless. Can't you rinse out your own socks?

d. To treat (hair) with a rinse. Cf. *rinse *sb.* 2 c. Also *absol.*

1959 N. Lofts *Heaven in your Hand* 125 His mother, of course, used make-up too, and had her hair 'rinsed' and waved. 1962 J. Braine *Life at Top* xiii. 170 Tom was..very proud of his mane of white hair... There were some people in Dufton who said that he blue rinsed it a little. 1971 M. Kelly *25th Hour* i. 53 Louise's hair seemed to go an elegant grey all at once (she rinses a bit of course).

Rinyo–Clacton (ri·nyo,klæ·ktən), *a. Archæol.* [f. the place-names *Rinyo*, Orkney + *Clacton*, Essex.] Of, pertaining to, or designating the culture represented by Late Neolithic grooved ware first discovered at Rinyo and Clacton.

1954 S. Piggott *Neolithic Cultures Brit. Isles* x. 280 *The Rinyo-Clacton culture.* Defined originally by 'Grooved ware'..the culture has a curious distribution, mainly concentrated in two separate areas, one in south-east England and the other in the Orkneys, though intermediate sites are now being discovered. 1963 E. S. Wood *Collins Field Guide Archaeol.* ii. ii. 105 The northern branch of the Rinyo-Clacton culture..have left in Orkney remains from which a fairly complete picture of their economy can be built up: at Skara Brae and at Rinyo are groups of squarish stone huts. 1967 *Antiquaries Jrnl.* XLVII. 182 Rustication as a decorative motif can occur on three styles of British Late Neolithic pottery—on Beakers, Fengate Ware, and Grooved or Rinyo-Clacton ware. 1970 Bray & Trump *Dict. Archæol.* 256/1 To the Late Neolithic belong the highly decorated wares of Peterborough and Rinyo-Clacton styles.

rio, var. *ryo.

Rioja (rio·ha). Also rioja. [The name of a district of northern Spain.] A wine produced in this district.

1907 *Yesterday's Shopping* (1969) 96/1 *Claret*..Rioja (Spanish), per dozen flagons 21/9. 1920 G. Saintsbury *Notes on Cellar-bk.* ii. 18 Other light Spanish wines of this class (of the Riojas, etc., we may speak separately) are excellent. *Ibid.* vi. 89 White Rioja (a capital beverage liquor). 1926 E. Hemingway *Sun also Rises* xix. 257 We had roast young suckling pig and drank *rioja alta*. 1934 R. Macaulay *Going Abroad* xxiii. 182 Wicker flasks of rioja. 1951 R. Postgate *Plain Man's Guide to Wine* vii. 108 Of table wines which a foreigner may be offered, Rioja ('J' pronounced as 'H') is..a full, strong wine produced in the north, mostly red and with a heavy plush-like flavour. 1965 G. Household *Olura* 157 Allarte's capacity for red Rioja, anchovies and bread was astonishing. 1973 *Country Life* 26 Apr. 1162/1 Rioja..comes from the north of the country, about 100 miles south of the western Pyrenees. *Ibid.* 1162/3 There is both red and white Rioja..and the red is certainly the more distinguished.

riometer (ri,ǫ·mitəɹ, rəi·omitəɹ). *Geophysics.* [f. the initial letters of *relative ionospheric opacity* + -meter.] An instrument which permits continuous measurement of the absorption of cosmic radio waves by the ionosphere.

1959 Little & Leinbach in *Proc. IRE* XLVII. 315/2 Thirteen of the commercially built units, now called 'riometers' (Relative Ionospheric Opacity Meters) are currently in use... The riometer is a self-balancing receiving system in which a local noise source is continuously made equal to the noise power from the antenna. 1968 G. M. B. Dobson *Explor. Atmos.* (ed. 2) viii. 151 An entirely different method which is now being much used is to measure the intensity of cosmic radio waves which are constantly reaching the earth from the galaxy. These radio waves will be absorbed as they come down through the D region, so that low intensity of cosmic radio waves received at the ground means strong absorption by the D region. These instruments are generally known as riometers. 1972 *Nature* 28 Jan. 215/1 Ground based observations using a 27 MHz riometer and an H magnetometer indicated almost quiet ionospheric conditions. 1980 *Ibid.* 17 Jan. 278/2 The Siple riometer operates at 30 MHz and has a response time of 0.25 s to increases or decreases in signal intensity.

‖ **rione** (rio·ne). Pl. rioni. [It.] A district or administrative division of Rome. Cf. Region 5.

1927 *Daily Express* 23 Nov. 12 Nine new fountains have been inaugurated in Rome, each corresponding to a rione, or section of the city. 1936 G. F.-H. & J. Berkeley *Italy in Making* II. vii. 115 The orders were sent out to the heads of each of the rioni (fourteen in number) in Rome. Each head of a rione sent his orders to his capi-squadra (heads of squadrons). 1965 C. Hibbert *Garibaldi & his Enemies* i. iv. 50 In every *rione* deputies were appointed to take command of the citizens when the bells of the Capitol and Montecitorio summoned them to arms. 1979 *Jrnl. R. Soc. Arts* Nov. 775/1 The triangular shed..is a new social centre meant to revive and support a local form of civic organization, the *rione*, an alternative to the centralized bureaucracy, church and *municipio*.

riot, *sb.* Add: **2. d.** (Further examples.)

1969 *Morning Star* 9 July 4/3 The trees are flourishing better than ever, and the borders have been a riot of colour. 1974 'S. Woods' *Done to Death* 184 The garden was a riot of colour.

4. c. *to read the Riot Act* (also with small initials): in *transf.* use, to announce or declare that (unruly) action or conduct must cease; to reprimand or caution sternly.

The Riot Act was repealed in 1973.

1819 W. Bradford *Let.* 17 Dec. in M. Wilmot *More Lett.* (1935) 39 She has just run out to read the riot act in the Nursery. 1842 C. Fox *Jrnl.* 18 Apr. (1972) 123 Sydney Smith said, 'Lady Holland is not one woman, but a multitude; just read the Riot Act and you'll presently see them disperse!' 1906 J. London *Let.* 20 Oct. (1966) 211 You might have found out, before you read me the riot act. 1946 D. Hamson *We fell among Greeks* xvii. 187, I met the E.D.E.S. envoys and read them the riot act, so to speak. 1976 P. Hill *Hunters* x. 131 Read her the riot act, tell her to be a good girl and take her home.

d. *colloq.* (orig. *Theatr.*). Something extremely successful or amusing; *spec.* an uproariously successful performance or show, a 'smash hit'. Also of persons.

1909 P. G. Williams in *Sat. Even. Post* 5 June 17/2 *A riot*, great success. 1919 F. Hurst *Humoresque* 195 If you think that is a riot..you wait until you see the way they're going to eat me up in the court scene. 1929 J. B. Priestley *Good Companions* ii. i. 249 There isn't a more promising little show anywhere... It could have been an absolute riot. 1933 N. Streatfeild *Tops & Bottoms* xxii. 287 He was a riot. 1936 P. Quentin *Puzzle for Fools* xxvi. 253 He'd be a riot in a mental hospital. 1943 J. B. Priestley *Daylight on Saturday* xv. 101 These shows—they're a riot. 1959 E. H. Clements *High Tension* viii. 134 Get that word-perfect..and you'll be a riot tomorrow. 1976 J. Snow *Cricket Rebel* 110 His rendering of 'Barnacle Bill the Sailor' was a riot and became his party piece.

e. In full, *riot sale*. A sale. *U.S. slang.*

1952 *N.Y. Post* 26 Sept. 73/1 (Advt.), Auto 'riot' sale. 1969 *Punch* 5 Feb. 193/1 Some of New York's stores are having a shoe riot.

5. *attrib.* and *Comb.* **a.** General attrib. uses (sense 4), as *riot area, call, control, zone*; **b.** designating equipment worn or carried (esp. by peace-keeping forces) in a riot, as *riot equipment, gear, gun, helmet, shield, stick*; also (parasynthetically) *riot-helmeted* adj.; also of persons or vehicles equipped to quell riots, as *riot police, squad, tank, van, wagon*; **c.** in instrumental, etc., uses, as *riot-battered, -prone, -ripe, -scarred, -torn* adjs.; **d. riot gas**, an irritant gas fired in capsules into a mob to quell rioting, tear-gas.

1973 *Freedom* 2 June 3/1 No one who lives in the riot areas of Belfast needs any reminders of what violence can look like. 1976 *Daily Times* (Lagos) 27 Aug. 9/2 Unofficial reports said 20 bodies were found in the riot battered township on Wednesday. 1905 *N.Y. Even. Post* 7 Nov. 2 Charges of illegal voting resulted in a disturbance which police were unable to subdue, and a riot call was sent in. 1964 Kirk & Othmer *Encycl. Chem. Technol.* (ed. 2) IV. 877 In the last few years, an agent, CS, has been developed for riot control. 1974 *North Myrtle Beach* (S. Carolina) *Times* 17 Apr. 1/5 The riot control unit then had to make its way to Hillside Drive and make a similar sweep to quell trouble spots on that street. 1955 H. Kurnitz *Invasion of Privacy* (1956) xii. 79 The old man is handing out riot equipment and orders are shoot to kill. 1968 *Punch* 4 Dec. 804/1 Jelly-crazed five-year-olds can now be quietened with a discreet blast of MACE lavender-perfumed riot gas. 1969 *Guardian* 22 Jan. 1/3 A powerful fragmentation grenade which will scatter CS riot gas among demonstrators is being developed by the Ministry of Technology. 1978 *N.Y. Times* 30 Mar. A5/6 Thousands of policemen..in riot gear. 1930 *Morning Post* 9 Apr. 11 Detectives in motor-cars equipped with 'riot guns' toured through the districts where violence was anticipated. 1976 'B. Shelby' *Great Pebble Affair* 49 Officer Hodgson with his riot gun. 1969 S. Greenlee *Spook who sat by Door* xiii. 112 Watching the police in riot helmets and the angry faces of the crowd. 1973 'S. Harvester' *Corner of Playground* iii. v. 212 Tight-lipped young officers, whites of eyes gleaming under rim of riot-helmet. 1970 *Daily Tel.* 18 Apr. 5/3 Three young women were wounded by shotgun pellets when 200 riot-helmeted, sheriff's deputies dispersed a crowd of about 700 smashing windows at a bank. 1958 *Daily Sketch* 2 June 2/5 Riot police armed with rifles, machine-guns and tear-gas tensed for an attack by Communist demonstrators. 1977 *Times* 18 Feb. 7/5 Riot police took control of Rome University tonight after using tear gas to disperse angry left-wing students. 1967 *Economist* 7 Oct. 45/2 Property-owners in riot-scarred (and riot-prone) neighbourhoods. 1968 *Ibid.* 20 July 40/3 Many state capitals are now humming with reports of 'mass' cancellations by insurers of policies covering property in riot-prone areas. 1910 Kipling *Rewards & Fairies* 103 Tom Dunch an' some of his kidney was drinkin' themselves riot-ripe. 1965 *Jet* 16 Sept. 50 Martin Luther King, Jr. was forced to invoke the name of Elijah Muhammed to gain a hearing in the riot-scarred community of Watts. 1967 Riot-scarred [see *riot-prone* above]. 1967 *Economist* 7 Jan. 31/1 (*caption*) Calcutta police behind riot shields. 1968 *Guardian* 21 Oct. 18/7 Mr Callaghan has issued no instructions to the Commissioner of the Metropolitan Police,..about the use of tear gas, water cannon, riot shields, troops, or other weapons familiar to demonstrators in Paris. 1978 *Peace News* 25 Aug. 4/1 The police included a large contingent of the Special Patrol Group with riot shields. 1981 *Daily Tel.* 3 Mar. 2/1 Police, some with riot shields and crowbars, stormed the building. 1948 *New Yorker* 1 May 75/1 A trio of jeeps, bringing the *celere*, or riot squad. 1955 *Times* 25 Aug. 9/2 The requirements for this are good intelligence arrangements and a mobile riot squad such as the strategic reserve in Cyprus will provide. 1977 *New Yorker* 24 Oct. 132/3 The prison's three riot squads..officially called..correctional emergency-response teams. 1930 J. Dos Passos *42nd Parallel* v. 404 He'd been halfstunned by a riotstick. 1972 R. Perry *Fall Guy* i. 28 The police arrived..with their riot sticks in evidence..lead-weighted batons. 1978 D. Francis *Trial Run* xi. 156 They were armed..with riot sticks. Nasty hard things like baseball bats, swinging from a loop of leather round the wrist. 1966 L. Cohen *Beautiful Losers* i. 134 It is not enough that she and Prince Philip will be greeted by police cordons, riot tanks, and the proud backs of hostile crowds. 1969 C. Himes *Black on Black* (1973) 286 The riot tank didn't know where to look for him. 1968 *Economist* 12 Oct. 3/2 The..riot-torn country of Northern Ireland. 1976 *Ulverston* (Cumbria) *News* 3 Dec. (Suppl.) p. i/3 *Shoot!*..Containing articles by well-known footballers, the 'lowdown' on the riot-torn 1975 European Cup Final. 1981 *Yorks. Post* 9 July 1/3 Prince Charles wants to help youngsters in Liverpool's riot-torn Toxteth area. 1973 J. Drummond *Bang! Bang! You're Dead!* xxxiv. 116 There were extra police on duty, and several riot vans under the trees. 1976 J. McClure *Rogue Eagle* xiv. 234 There were a couple of riot vans double-parked outside. 1969 *Guardian* 5 Aug. 1/7 Finally riot wagons were moved in to block the oblique junction. 1973 *Black Panther* 28 July 3/2 Police arrived in buses and riot wagons. 1975 R. H. Rimmer *Premar Experiments* (1976) i. 56 According to my informant, Bren and Merle were living in a rundown tenement house in the 'riot zone'.

riot, *v.* Add: **II. 6. b.** To engage in a riot or violent disturbance.

1755 Johnson, *To Riot*..4. To raise a sedition or uproar. *a* 1832, etc. [implied in Rioting *vbl. sb.* 2]. 1981 W. Ebersohn *Divide the Night* v. 71 When they [*sc.* blacks] rioted they did it with greater anger here [*sc.* in Johannesburg] than anywhere else. 1981 *Yorks. Post* 9 July 4/5 As thoroughly decent a group of people as you would wish to meet, they did not riot in the streets.

7. *Hunting.* = *to run riot* s.v. Riot *sb.* 3 a. Also const. *after, on.*

1954 J. I. Lloyd *Beagling* 143 *Riot*, to hunt anything other than their legitimate quarry. 1971 *Country Life* 7 Oct. 897/1 Hounds will riot more readily after roe deer than any other species. 1976 *Horse & Hound* 3 Dec. 38/1 A great deal of time would be spent, however, correcting the pack rioting on Scotch sheep.

rioty (rəi·əti), *a. nonce-wd.* [f. Riot *sb.* + -y[1].] Riotous; noisy, rackety.

1819 Keats *Let.* 20 Sept. (1958) II. 206 Bless the child, how rioty she is!

rip, *sb.*[4] **3.** (Earlier and further examples.) Also *U.S.*, esp. in *Music.* Also *transf.*, a burst (of laughter).

1855 *Knickerbocker* XLV. 129 List to the rip and the roar of the song. 1867 'T. Lackland' *Homespun* ii. 271 Sometimes he could not help giving a rip of laughter that drew the eyes of the whole school round to him in an instant. 1933 *Metronome* Mar. 34 The rip is produced by short and quick glissando up to the tone, attacked sforzando and cut off quickly. 1961 A. Berkman *Singers' Gloss. Show Business* 75 *Rip*, an effect in which the entire band plays a fast glissando up to a heavily accented note to emphasize or punctuate a violent action or thought.

4. *U.S. police slang.* (See quots.)

1939 *Fortune* July 101/3 An inspector's lieutenant..found the patrolman lounging with his gloves off, smoking a cigarette. Probable penalty: one day's 'rip' (fine of a day's pay). 1958 *N.Y. Times Mag.* 16 Mar. 88/3 Rip—A fine imposed for infraction of police regulations: e.g., 'I got a five-day rip' (fined five days' pay).

5. *attrib.* **rip cord**, (*a*) *Aeronaut.* = **ripping cord*; (*b*) *Aeronaut.*, a cord which holds a parachute pack closed and which, when pulled, opens the pack and allows the parachute to unfold and inflate; (*c*) *fig.*; **rip line, panel, valve** *Aeronaut.* = *ripping line, panel, valve* s.v. *ripping *vbl. sb.* 2 b.

1909 V. Lougheed *Vehicles of Air* 108 Practically a valve is the 'rip cord', by means of which a seam running along the side of a balloon can be laid open. 1911 *Sci. Amer.* 25 Mar. 300/1 In case of accident the aviator, by pulling a rip cord, can open the parachute. 1925 [see *pilot chute* s.v. *pilot *sb.* 6]. 1946 W. F. Burbidge *From Balloon to Bomber* 40 In 1908, an American parachutist, invented a 'free parachute'... The parachute was packed into a container worn on the airman's chest and released by a whale-bone spring operated by a rip-cord. 1969 *Daily Colonist* (Victoria, B.C.) 7 Feb. 36/1 'Then you'd better put on a parachute if you're afraid,' Beachel said. 'How does the chute work?' Till asked. 'Where is the rip-cord?' 1974 [see *ripping line* s.v. *ripping *vbl. sb.* 2 b]. 1975 'D. Jordan' *Black Account* i. xix. 100 He was big..and wide but ripcord lean. 1981 W. Winward *Ball Bearing Run* ix. 111 It would be necessary to fall clear of the bomber stream before pulling the rip-cord. 1963 A. Smith *Throw out Two Hands* viii. 95 Above it [*sc.* the basket of a balloon]..were the valve line (for gentle release of gas) and the rip line (for a total release of gas). 1933 *Sun* (Baltimore) 22 Nov. 2/7 Rather than take a chance on crossing the bay with his diminishing gas, he pulled the rip panel and down they came. 1963 A. Smith *Throw out Two Hands* v. 63 You lose the rope. Then you pull it again, and this time you can feel it jerking open the rip panel. 1978 A. Welch *Bk. of Airsports* v. 81/1 For this purpose a rip panel is built into the balloon near the top and is kept closed with either a parachute rip or Velcro. 1907 *Jrnl. Soc. Arts* 19 Apr. 602/2 By means of the rip valve they were able to come down pleasantly and easily.

rip, *sb.*[5] Add: **1. b.** *ellipt.* = **rip current*.

1941 *Jrnl. Geol.* XLIX. 338 The term 'rip' might also be used as an abbreviation, which removes the unfortunate tidal connotation of the popular term 'rip tide'. 1968 R. W. Fairbridge *Encycl. Geomorphol.* 950/2 The width of the central 'trunk' of the rip can be quite narrow, perhaps only a few tens of feet, but the effect of the current can sometimes be detected up to a mile or more from the shore. 1968 W. Warwick *Surfriding in N.Z.* 20/2 Remember to dive beneath broken waves and do not swim against rips, they will only take you a short distance out to sea before fading out. 1977 *Herald* (Melbourne) 17 Jan. 6/9 He was helping to rescue four people caught in a rip when the accident happened.

2. (Earlier and later examples). Chiefly *U.S.*

a 1828 in B. James *Jrnl.* (1896) ii. 195 We passed several very dangerous places, which they there [on the Kennebec R.] termed 'rips', which was [*sic*] a confused number of rocks and large stones in the direct way we were obliged to pass, and which generally had a fall of some few feet. 1839 E. Holmes *Explor. Aroostook River* 7 The existing obstacles which present themselves to the present navigation of this river, are, the 'rips', which are occasioned principally by loose boulders of rock. 1941 B. A. Williams *Strange Woman* iii. i. 128 After that we'd skin 'em [*sc.* scurry ducks] out and tie a string to 'em and let 'em hang in rips of the quickest water we could find. 1977 *New Yorker* 9 May 106/3 A couple of tributaries came into the river,..and they deepened the pools and improved the rips.

3. Special Comb.: **rip current**, an intermittent, strong, narrow current on or near the surface of the sea, flowing directly out from the shore and acting to remove water which has been brought to the shore by waves and wind; **rip tide**, (*a*) = *rip current*; (*b*) = sense 1; (*c*) *fig.*

1936 F. P. Shepard in *Science* 21 Aug. 181/2 The name 'rip tide' is certainly not appropriate, since the current described has nothing to do with the tide... The name 'rip current' is suggested, since it is close to the other name and describes the way in which the current rips through the oncoming breakers. 1941 *Jrnl. Geol.* XLIX. 339 The chief seaward return of water moved in by the waves seems to be in the form of rip currents. 1957 G. E. Hutchinson *Treat. Limnol.* I. v. 358 The usual pattern..is the development of longshore currents which return to the free water at particular points..as rip currents. 1973 *Daily Tel.* 15 Aug. 6/3 He believed a rip current may have been responsible for carrying the 20 bathers out to sea. 1862 Rip tide [in Dict., sense 1]. 1931 *Daily Progress* (Charlottesville, Va.) 26 Aug. 1/2 Surf bathing as a means of avoiding the heat was made unattractive to many because of the possibility of the recurrence of dangerous 'rip tides'. 1936 [see **rip current*]. *a* 1963 S. Plath *Ariel* (1965) 45 Your stooges..Riding the rip tide to the nearest point of departure. 1970 I. Petite *Meander to Alaska* i. v. 47 Riptides, eddies,..complicate a cruiser's traffic patterns. 1973 *Sunday Bull.* (Philadelphia) 14 Oct. (Parade Suppl.) 14/4 Probably the toughest rescue I had was a Mexican family unfamiliar with the riptides. They wandered out about 400 yards into the surf, then started yelling. 1976 *New Yorker* 22 Mar. 106/2 Their echoings of Futurism and Expressionism—for example, the riptides of black diagonals in Dove's 'Field of Grain Seen from Train'. 1977 *Time* 28 Mar. 45/3 In Colombia, surging coffee revenues have been accompanied by a riptide of 26% inflation. 1978 J. A. Knauss *Introd. Physical Oceanogr.* x. 219 Rip tides can be dangerous to the unwary swimmer.

rip, *sb.*[6] Add: **2. a.** (Later examples in the milder sense.)

1918 Galsworthy *Five Tales* 77 My grandfather lived to be a hundred; my father ninety-six—both of them rips. 1935 S. Desmond *Afr. Log* li. 264 A humble repentant sinner—once perhaps 'a bit of a rip'—but very appealing. 1951 *Chambers's Jrnl.* Nov. 656/2 Would you believe it, the old rip had a flutter in Norland Deeps himself?

b. (Further examples.)

1893 G. B. Shaw *Let.* 4 Sept. (1965) I. 404 The mother a most deplorable old rip. 1910 P. W. Joyce *English as we speak it in Ireland* xiii. 313 *Rip*, a coarse ill-conditioned woman with a bad tongue.

rip, *v.*[2] Add: **I. 1. a.** (Further examples.)

1861 *Once a Week* 10 Aug. 180/1 The joy of ripping out the middle stump of a good batter surpasses even that of wiping a man's eye at an overhead cock-pheasant. 1977 P. Hill *Fanatics* 33 They've ripped out the phone.

b. *slang.* To steal.

1904 'No. 1500' *Life in Sing Sing* 252/1 *Rip*, to steal with impunity. 1970 *Time* 22 June 52/3 For extra, unanticipated personal needs, he 'rips off'—or steals... Some of those who take jobs in department stores or markets steal what they can... Some who work in restaurants or drugstores let their friends in to eat or rip what they need. 1976 *Telegraph* (Brisbane) 20 Apr. 1/3 They believe some have ripped millions of dollars from Medibank since it began. 1977 *Guardian* 23 June 3/4 (Advt.), While the fluff saps the mark, the dip rips the wad.

c. *Baseball.* To score (a hit) in a vigorous manner.

1970 *Washington Post* 30 Sept. D2/5 Renko ripped his run-scoring hit in the second. 1974 *Anderson* (S. Carolina) *Independent* 19 Apr. 5B/4 Designated-hitter Duke Sims..ripped a run-scoring double over first base.

d. Of a competitor or team: to defeat overwhelmingly. *U.S.*

1974 *State* (Columbia, S. Carolina) 3 Mar. 1-D/6 South Carolina..ripped the Cougars, 104–86, behind a sparkling 37-point performance by Alex English. 1976 *Springfield* (Mass.) *Daily News* 22 Apr. 40/4 Three runaways featured action in the Bi-County League. St. Mary's romped over Westfield Voke, 16–3; Gateway ripped Belchertown, 12–1; Smith School blitzed Holyoke Trade, 12–0. 1978 *Detroit Free Press* 2 Apr. 5E/11 (*heading*) Borg rips Smith, gains WCT finals.

2. a. (Later examples.) Also with *through.*

1930 [*best *a.* 7 b]. 1973 'J. Patrick' *Glasgow Gang Observed* iii. 30 He had..fifty-nine stitches on the side of his face; he had been 'ripped' only a few weeks ago. 1976 *Honolulu Star-Bull.* 21 Dec. H-6/1 Meanwhile, the Davis Cup nations are ripped by political squabbling.

e. To open or release, or to deflate, by the use of a rip cord. Also *absol.*

1902 J. Alexander *Conquest of Air* iii. 54 There was an arrangement for ripping the cover when ready to descend. *Ibid.* 55 They immediately ripped the balloon and commenced their descent. 1907 G. Bacon *Record of Aeronaut* xiv. 263 Mr. Spencer was in favour of ripping open the valve. 1920 G. C. Bailey *Compl. Airman* xxxi. 242 The ripping panel is a specially sewn section of the fabric, a cord lead to which enables the balloonist to rip it at will. 1963 A. Smith *Throw out Two Hands* xxi. 219, I..remember fumbling for the rip-panel cord... Had I ripped?

3. a. Also *fig.*

1976 *National Observer* (U.S.) 14 Aug. 12/1 What rips you up is the craziness... I felt like, well, since I loved them both, they should love each other. They don't.

d. *Cricket.* Of a ball: to hit (a stump or stumps) on delivery at speed, so as to knock it (or them) back or out of the ground. Also with bowler as subj.

1832 P. Egan *Bk. Sports* 348/2 She [*sc.* the ball]..Ripp'd up the *off* and *centre*! 1887 F. Gale *Game of Cricket* xiv. 244 Nothing would be better for cricket itself than for a young unknown cricketer..to rip up the wickets of some of the county cracks.

4*. *Austral. slang.* To annoy intensely; usu. in phr. *wouldn't it rip you*, used as an expression of exasperation.

1941 *Argus* (Melbourne) *Week-End Mag.* 15 Nov. 1/3 Another universal favourite is still the famous 'Wouldn't it—!' Never given the final words (the completed sentence has several variations on 'Wouldn't it rock you!' or 'Wouldn't it rip you!') the explanation depends upon inflexion as to whether it conveys disgust, amazement, or pleasure. 1944 L. Glassop *We were Rats* xiii. 74 'I can't do it,' he said again. 'There are no partitions between the places... It's disgusting..'. Everyone gaped at him. 'What's wrong with this galah?' asked somebody and another said, 'Well wouldn't it rip you? What do you expect him to throw?' *Ibid.* xxviii. 162, I had the idea that if you joined the A.I.F. you had to fight in the front line. I know now how many men it takes to keep one in those trenches. Do you know our divisions have even got a mobile laundry, decontamination unit and mobile bath unit? Wouldn't it rip you?

4.** *to rip off. slang* (orig. *U.S.*). Cf. also *rip-off *sb.* (i) To steal; to embezzle.

1967 *Trans-Action* Apr. 7 The hustler 'burns' people for money, but he also 'rips off' goods for money; he thieves, and petty thieving is always a familiar hustle. 1971 *It* 4–18 Nov. 3/5 An analysis of 800 documents ripped off from the Pennsylvania FBI office. 1972 *National Observer* (U.S.) 27 May 12/2 Bank robbery? It's only Establishment money that's being ripped off. 1974 *Black Panther* 16 Mar. 6/2 Spiro got caught ripping off tax money. 1977 *New Yorker* 9 May 34/2 First he owned an Atala, but it got ripped off, so he bought a Peugeot. 1981 A. Cross *Death in Faculty* viii. 90 Soldiers are always ripping things off, from their own outfit, from the enemy, everything.

(ii) To exploit financially; to cheat or defraud; to rob; to deceive.

1971 *Frendz* 21 May 16/4 The young people are well aware that they are being ripped off by these parasites, and, quite naturally, think that the visiting musicians are on the side of the promoters. 1973 *Black World* Jan. 33/1 Individuals within the group felt that there were too many instances of their singly being 'ripped off' and exploited as Black artists. 1973 *N.Y. Law Jrnl.* 2 Aug. 13/5 He, Harris, and Sydnor had 'ripped' off patrons and the owner of the bar. 1974 S. Ellin *Stronghold* 55 Experimental group therapy sessions which we all attended..partly because ripping off the amiable idiot who conducted them was better than another game of checkers. 1976 *Kingston* (Ontario) *Whig-Standard* 19 Jan. 19/3 Mrs. Baird and Mrs. LaMarche were in complete agreement as to who were the worst offenders at ripping the consumer off: Television and auto repairs. 1976 *Observer* 22 Feb. 6/3 Many women think all garages consider they can 'rip off' women drivers. 1977 *Spare Rib* July 35/2 A police guard formed in front of Mothercare in Oxford St—afraid we'd attack it for the way it rips off motherhood. 1978 *Detroit Free Press* 5 Mar. A 23/1 Sid Luft's 14-year-old lawsuit charging that his late wife Judy Garland was ripped off..by Hollywood executive David Begelman. 1981 *Times* 23 Apr. 4/8 Martin was not ripping me off.

(iii) To have sexual intercourse with; *esp.*, to rape.

1971 *Black Scholar* Sept. 32/1 If she had been any other broad he would have ripped her off..that night. 1973 *Black World* Sept. 53, I done shot dope, been to jail, swilled wine, ripped off sisters, passed bad checks. 1974 *Guidelines to Volunteer Services* (N.Y. State Dept. Correctional Services) 42 *Rip off*, rape, pull a job.

(iv) To burgle, to steal from (a store, etc.).

1972 'E. McBain' *Sadie when she Died* iii. 30 Q. Why did you go into the apartment? A. To rip it off. Q. To burglarize it? A. Yes. 1973 *Black World* Jan. 54/2 They were ripping off a furniture store in a few hours. 1977 *Rolling Stone* 24 Mar., Not when young blacks have ripped off bookstores across the country to get illegally what the lack of a job prevents them from getting legally.

(v) To copy; to plagiarize.

1975 *Radio Times* 12–18 July 9/4 Just about everyone (including the Immaculate Jean-Luc Godard) ripped off Dick Lester's cool style. 1977 *Undercurrents* June–July 11/4 I've never yet refused a request to reproduce anything of mine but I've lost count of the times I've been ripped off. 1978 *Sci. Amer.* June 26/1 Two books, one an instruction manual for a geometrical instrument.., the other a witty polemic against a Padovan student who had sought to rip off that very instruction book!

II. 7. a. (Examples of the simple use.) *let her rip* (earlier and later examples); *to let* (someone or something) *rip*: to allow (that person or thing) to go, to continue unchecked, etc.; *to let rip*: to let fly, to let oneself go.

1853 *Daily Morning Herald* (St. Louis) 19 Jan. (Th.), We've got 'em on the hip, Letter Rip! Letter Rip! 1863 *Harper's Mag.* Oct. 716/1 We cannot raise a tip To pay our board and laundry bill, And have to 'let 'em rip'. 1869 H. Phillips *Jrnl.* 14 Dec. (typescript) 203 All hands tailing sheep let them rip at night. 1881 A. Bathgate *Waitaruna* ix. 139 Most of the [diggers], when they found I would not buy, would throw their picks down and say if I would not buy them I could take them or let them rip as I pleased. 1888 'R. Boldrewood' *Robbery under Arms* II. xiii. 211 He rips over to Daly's mob, borrows a horse, saddle, and bridle, and leads him straight down to our camp. 1894 F. A. Barkly *Among Boers & Basutos* (ed. 2) xiv. 186, I galloped round the Kopje with my police and half-a-dozen volunteers..and we 'letrip' to use the Africander expression. 1909 C. Owen *Philip Loveluck* xii. 175, I can rub along somehow..by letting the pressing rip. 1915 Wodehouse *Psmith Journalist* xxv. 203 And now..let her rip. What can I do for you? 1916 'Boyd Cable' *Action Front* 255 A shell cracked overhead, and the shrapnel ripped down along the trench behind them. 1926 Galsworthy *Silver Spoon* iii. iv. 246 Alec would know where he was when it was over, and so would she!.. Let it rip! 1930 A. Bennett *Imperial Palace* ix. 45 He let them rip..because he enjoyed the grand spectacle of their passion. 1947 H. Read *Grass Roots of Art* iii. 71 We cannot..oppose the machine. We must let it rip, and with confidence. 1965 *Listener* 22 July 140/1 Should one try to make the action clear and comprehensible..or let rip with the poetry? 1966 *Ibid.* 17 Nov. 718/3 What would you do about all these many wage increases... Would you let them all rip? 1971 C. Bonington *Annapurna South Face* viii. 95 Almost as soon as I had let rip, however, I realized the injustice of my complaint. 1977 McKnight & Tobler *Bob Marley* x. 134 The other view, which was expressed by the minority, was 'let her rip!'. 1977 *Sounds* 9 July 8/1 A frantic 'live' sounding version of The Stones' 'The Last Time' which rips along grandly, seven minutes and forty-four seconds of unadulterated pure fire. 1978 *Dumfries Courier* 20 Oct. 28/1 The present difficult decisions..will be thrust into insignificance if inflation lets rip again.

c. *to rip and tear*: to rage, to rave; to go raging *about.*

1873 'Mark Twain' & Warner *Gilded Age* xxvii. 249 A man wants rest, a man wants peace—a man don't want to rip and tear around *all* the time. 1884 'Mark Twain' *Huck. Finn* xxi. 207 It was perfectly lovely the way he would rip and tear. 1886 Baumann *Londinismen* 157/1 Ripping and tearing about. 1917 *Dialect Notes* IV. 342 *Rip and tear*, to rave.

d. In quasi-adverbial use.

1884 [see Sockdolager 3].

III. 9. *Comb.* as **rip-and-read**, used *attrib.* to designate material supplied by teletype which is read on radio or television; also of an organization supplying such material; **rip-and-tear** *U.S.*, used *attrib.* to designate crude and violent methods in crime; also *transf.*; cf. sense 7 c above; **rip-off**, used *attrib.* to designate an opening device that has to be torn off; **rip-stop**, used *attrib.* and *absol.* of nylon clothing or equipment woven so that a tear will not spread; **rip track** *N. Amer.*, a section of railway line used as a site for repairs to carriages.

1973 *New Journalist* (Austral.) July–Aug. 6 The 'rip-and-read' news service of Sydney's labour [radio] station, 2KY. 1974 Hawkey & Bingham *Wild Card* ii. 26 The newscaster was reading..rip-and-read copy—a story that had just come up on an agency teletype machine that the news editor rated too big to hold while it was rewritten. 1937 E. H. Sutherland *Professional Thief* 241 *Rip-and-tear*, adj., without caution; same as 'raw-jaw', or 'murder grift'. 1955 D. W. Maurer in *Publ. Amer. Dial. Soc.* xxiv. 93 They do not constitute the upper echelons of the profession. They are also known as *clout and lam mobs, hijackers*, or *rip and tear mobs.* 1965 G. Jackson *Let.* 12 Mar. in *Soledad Brother* (1971) 66 Understand though that you do not live in the real rip-and-tear world. 1973 *Nation Rev.* (Melbourne) 31 Aug. 1436/6 The knife edged ripoff tag on the top of some cans. 1971 C. Bonington *Annapurna South Face* App. B. 249, 2-man [tent], in ripstop nylon. 1976 *National Observer* (U.S.) 14 Feb. 9/3 (Advt.), Prime Duck Down socks... Covered with blue ripstop nylon. 1978 *Sci. Amer.* Feb. 158/3 Other covering materials include sailcloth..and nylon rip-stop. 1960 *Glossaria Interpretum: Chemins de Fer* 1882 Voie de réparations,..Repair track, rip track *Am.* 1973 *Amer. Speech 1969* XLIV. 246 Various parts of the yard have names..such as *riptrack* (a long section of track—possibly several tracks if the yard is large—which is used for car repair).

ripe, *a.* (*sb.*[3] and *adv.*). Add: **2. a.** Also, *ripe peeler* (Peeler[1] 2 b).

1952 *Sun* (Baltimore) (B ed.) 23 June 12/5 Language peculiar only to soft-crabbing... *Ripe peeler*—Has the same characteristics as the 'green peeler' but is more advanced in the shedding process.

b. (Later example.)

1949 G. Davenport *Family Fortunes* i. iv. 54, I swear, Martha, if I'd of met you when you was still ripe, I'd have left Hattie's mother, kids and all, to follow you clear to California, I would.

3. a. (Later example.) Also *absol.*

1930 Wodehouse *Very Good, Jeeves!* iv. 102 Having got me in sporting mood with a bottle of the ripest.

b. (Later examples.)

1810 E. Weeton *Let.* 25 Feb. (1969) I. 240, I have had another boil on my face... I neither lanced, nor poulticed it, but when ripe, let the matter out with a needle. 1909 *Dialect Notes* III. 363 *Ripe*, said of a boil when it is ready to be lanced.

d. (See quot. 1949.)

1949 A. R. Daniel *Bakers' Dict.*, *Ripe dough*, technical term for a dough ready for scaling having received a period of fermentation sufficiently protracted to enable the gluten to reach its most extensible condition. 1962 *Listener* 22 Mar. 511/1 There is a stage in breadmaking when the dough is said to be 'ripe'.

8*. In various slang senses. **a.** Drunk (cf. sense 7 d in Dict.). **b.** Fine, excellent; thoroughgoing (also used ironically); hence, beyond reasonable bounds, excessive. **c.** Angry.

1823 'J. Bee' *Slang* 149 *Ripe*—drunk. First cousin to mellow. 1923 Wodehouse *Inimitable Jeeves* ix. 89, I liked the place, and was having quite a ripe time there. 1925 *Flynn's* 14 Mar. 281/1 *Ripe*, drunk. 1932 A. J. Worrall *Eng. Idioms* 33 He was shooting at cats with darts. I told him it was a bit ripe and asked him to stop. 1948 Partridge *Dict. Forces' Slang* 156 *Ripe*, complete, thoroughgoing. Usually allied with 'bastard'. 1959 I. & P. Opie *Lore & Lang. Schoolch.* iii. 53 They come down like a ton of bricks on people who tell a stale joke. 'Do you know where Smudger takes his girl?' gags the would-be comic, 'He takes her behind a bush because it's very *privet.*' Whereupon the 'ripe one' is complimented: 'Oh lor, last time I heard that the tears rolled down my bib.' 1964 *Australasian Post* 21 May 13 Even a ripe shiner isn't just a black eye to the man in the white coat. It is a peri-optic ecchymosis. 1966 R. Jeffries *Death in Coverts* iii. 93 We all joked about it and Bill got really ripe. No sense of humour. 1967 R. Campbell in *Coast to Coast 1965–66* 20 Jack'll be ripe pickings by the time that old buzzard comes around from the police station to close the pub. 1969 'J. Fraser' *Cock-pit of Roses* xvi. 127 'What the bloody hell are you playing at?' 'That's ripe considering you just near broke my arm!'

10. a. *ripe-aged, -bearded, -meated.*

1922 Joyce *Ulysses* 59 Slapping a palm on a ripemeated hindquarter. 1934 Webster, *Ripe-aged.* 1944 E. Sitwell *Green Song* 11 We heard in the dawn the first ripe-bearded fire Of wheat. 1952 C. Day Lewis tr. *Virgil's Aeneid* v. 95 Ripe-aged Acestes.

ripely, *adv.* **3.** (Later examples.)

1939 Joyce *Finnegans Wake* (1964) 474 His locks of a lucan tinge, quickrich, ripely rippling. 1973 *Daily Tel.* 13 Aug. 12/3 Stravinsky's 'Firebird'... Ozawa encourages a ripely romantic performance.

ripener. Add: **3.** A device in which honey is allowed to stand until it is fit to be put in jars.

[1883 *Brit. Bee Jrnl.* XI. 209/2 The above simple arrangement..will combine that of a honey-extractor and a honey-ripener in one compact piece of apparatus.] 1905 *Instruction in Bee-Keeping* (Dept. Agric. & Techn. Instruction for Ireland) 20 The ripener..is a tinned iron cylinder about 19 inches in depth by about 8½ inches in diameter, and fitted with a treacle tap at the base. 1930 W. Herrod-Hempsall *Bee-Keeping* I. ix. 525 The ripener was so named because at one time both the unsealed and sealed honey used to be extracted together, it was then run into the ripener and allowed to stand for some time in a warm room. 1971 *Country Life* 18 Nov. 1347/1 The ripener, into which the extracted honey is put, so that air bubbles may escape in due time, was so heavy that it was all both of us could do to lift it.

ripening, *vbl. sb.* Add: **3.** In various industrial processes, applied to a stage in which a material is left to stand until desired properties are attained; *spec.* in rayon manufacture (see quot. 1957).

1919 *Jrnl. Soc. Chem. Industry* 31 Oct. 373T/1 There result, upon application of partial hydration ('ripening'), esters of high viscosity and great strength, suppleness, and wearing qualities. 1932 *Discovery* Sept. 289/1 After ripening has been completed the soluble salts are removed. 1937 *Ibid.* Aug. 247/1 The actual process is a long one.., the polymerisation process alone occupying from 90 to 120 hours, after which a 'ripening' process for the synthetic rubber of three to eight days is necessary. 1950 R. W. Moncrieff *Man-Made Fibres* viii. 96 Ripening is an essential part of the viscose process; 'young' viscose cannot be spun satisfactorily. 1957 *Textile Terms & Definitions* (Textile Inst.) (ed. 3) 82 *Ripening*,..(1) A process in the production of cellulose acetate consisting of the splitting off of some of the acetic acid and most of the combined catalyst present in the primary cellulose acetate. (2) A process in the manufacture of viscose rayon in which the viscose is matured prior to spinning. The rate of ripening is controlled by the time and temperature at which the spinning fluid is maintained. The process is sometimes called maturing or ageing. 1973 *Materials & Technol.* VI. iv. 307 This 'ripening'..results in a solution which will later give easier coagulation, and it is allowed to proceed until a required amount of hydrolysis has taken place.

4. *attrib.*, as *ripening-time.*

1910 W. de la Mare *Three Mulla-Mulgars* i. 7 The great Ukka-tree, which he had climbed at ripening-time.

ripicolous, *a.* (Examples.)

1906 J. B. Smith *Explanation Terms Entomol.* 117 Ripicolous: dwelling on river banks: riparian. 1965 B. E. Freeman tr. *Vandel's Biospeleol.* xiii. 196 The Bembiinae are essentially ripicolous.

ripienist. (Later examples.)

1935 P. A. Scholes *Radio Times Music Handbk.* 18 A 'full' body of strings and a smaller body of two or three solo string players—the 'ripienists' and the 'concertinists', as we may call them. 1944 W. Apel *Harvard Dict. Mus.* 646/2 *Ripienista* (Ripienist) is an orchestral player.

ripieno, *a.* and *sb.* Add: Also ripiano. **1. a.** (Further examples.)

1740 J. Grassineau *Mus. Dict.* 203 *Ripiano*, or *Ripiéno*, signifies *full*, and is used in pieces of music in parts, to distinguish those parts that play now and then to fill up, from those that play throughout the piece. 1933 *Radio Times* 14 Apr. 82/3 Miners, cotton spinners and the like who get a real kick from the ripiano cornet and the bombardon. 1954 *Grove's Dict. Mus.* (ed. 5), I. 914/1 In Britain the instrumentation ultimately became as follows for the purpose of contests: 1 Soprano Cornet in E♭... 1 Repiano Cornet in B♭. [*Note*] 'Repiano' is a spelling peculiar to the brass band. *Ibid.* 914/2 Nowadays only one [*sc.* flugelhorn] is used, playng a separate part or combining with the repiano cornet. 1962 [see *concertino 2]. 1978 *Daily Tel.* 13 Mar. 11/1 The Choir—with the youthful ripieno Chorus—responded with clear tone and vitality.

b. Occas. with It. pl. ripieni. (Earlier and later examples.) Also *collect.*, the group of accompanying instruments that form the main orchestral body in a concerto, as distinct from the concertino (*concertino 2).

1740 J. Grassineau *Mus. Dict.* 203 There are..two kinds of *Ripiénos*, one whereof plays the part of the little chorus exactly... The other sort is much better, because they play a different part. 1930 *Radio Times Dict. Mus. Terms* 50 *Ripieno*, the instruments which form the accompaniment, as opposed to those which have solo parts. 1935 P. A. Scholes *Radio Times Music Handbk.* 18 In the Concerto Grosso..the idea is..that *two bodies* of instruments are..responding to one another antiphonally... The larger body is called *Ripieno*.., and the smaller one '*Concertino*'. 1944 W. Apel *Harvard Dict. Mus.* 646/2 'Ripieni' indicates the full orchestra.., as distinguished from the soloists... The term 'senza ripieni', however, is not identical with 'orchestra silent', but calls for the leading members only of the orchestra, i.e., for a smaller ensemble used for the accompaniment of the soloists. 1960 *Times* 2 Nov. 16/6 A *ripieno* of six players (five strings and harpsichord). 1961 [see *concertino 2]. 1976 *Gramophone* Dec. 1028/3 The balance and interplay between voices and ripieno are two of the several outstandingly successful features of the present performances.

riplet: see *ripplet.

rip-off (ri·pǫf), *sb. slang* (orig. *U.S.*). Also rip off, ripoff. [f. *to rip off* s.v. *rip *v.*[2] 4**.]

1. One who steals, a thief.

1970 *Manch. Guardian Weekly* 2 May 16/4 'Who do you have on Haight Street today?' he [*sc.* a San Francisco drug peddler] said disgustedly... 'You have burn artists (fraudulent dope peddlers), rip-offs (thieves), and snitchers (police spies).' 1971 *Rolling Stone* 24 June 8/3, I call them rip-offs, and they are, nothing but pirates and vultures.

2. A fraud, a swindle; a racket; an instance of exploitation, esp. financial.

1970 *Melody Maker* 12 Sept. 29 *Rip off*, capitalist exploitation. 1970 *Time* 21 Dec. 4/1 This is what, in contemporary parlance, is called a rip-off. 1971 *It* 9–23 Sept. 12 Fun Caterers of Battersea..had the main catering concession (the biggest rip-off there.) 1973 *Houston* (Texas) *Chron.* 21 Oct. 7/3 Dunlop said the increased spring markups had been 'inflationary', a polite word in the context for 'ripoff'. 1974 *Sunday Sun* (Brisbane) 28 July 24/2 The great snackbar rip-off that had city workers weeping into their salad rolls. 1975 *N.Y. Times* 14 Apr. 30/4 A five-day week, with ten paid holidays, plus a ten-week paid vacation yearly. Such a contract is a 'rip-off'. 1977 *Time* 4 July 21/1 They [*sc.* French soldiers and civil servants] get rich and Djibouti gets nothing. That's not enlightened colonialism. It's a bloody rip-off. 1980 *Times* 31 May 2/3 Britain's 41 motorway service areas..have attracted such accolades as 'poor', 'appalling' and 'a rip-off'.

3. An imitation or plagiarism, usu. one made in order to exploit public taste.

1971 *Newsweek* 18 Oct. 38/3 Most of the architecture is Inspired Bastard, most of the historical re-creations are Shameless Ripoff. 1974 *Publishers Weekly* 4 Mar. 72/2 This kaleidoscopic fantasy, a ripoff on everything from spy novels to the Oedipus complex. 1976 *Time* (Canada) 19 Jan. 16/3 Flynt runs three Hustler Clubs in Ohio, tacky rip-offs of the Playboy Clubs, offering expensive drinks and leggy 'hostesses'. 1977 *Private Eye* 1 Apr. 4/1 *Blue Belle* [*sc.* a film], yet another of the seemingly endless *Emmanuelle* rip-offs. 1980 *Jewish Chron.* 29 Feb. 30/2 We were treated to a kaleidoscopic mess of fifties rip-offs, sixties platitudes and seventies mistakes; shirtwaisters, minis, halter-necks, op art, sloppy joes, bermudas and, latest ubiquity, the flying suit.

4. a. *attrib.* passing into *adj.*

1971 *National Times* (Austral.) 15–20 Feb. 1/3 In Sydney comics and books have been appearing from the 'rip-off' press—the underground printers and publishers who are printing editions of banned books sneaked singly through Customs. 1973 *Nation Rev.* (Melbourne) 24–30 Aug. 1399/6 The poor unfortunate buyer getting lumbered..with the cost of the device (at ripoff prices). 1973 *National Observer* (U.S.) 6 Oct. 23/3 The 'rip-off' blues, the blues that musicians get when they write songs that make other people rich and leave them poor as before. 1975 *Time* 12 May 17/1 The rip-off capital of the world [*sc.* Saigon]. 1976 *New Yorker* 5 Apr. 31/2 Cargo leaving New York for places like South America is often a kind of object lesson in rip-off economics. 1976 *Times* 11 June 8/1 The trade in old books is an incongruous mixture of fine art almost beyond price and the rascally hustle and rip-off hugger-mugger of a flea market.

b. *Comb.*, as **rip-off artist, merchant**, one who carries out a rip-off; a thief, fraud, or racketeer.

1971 *Frendz* 21 May 11/2 Rip-off artists are only occasionally armed or violent; more usual is..the traditional con-man. 1971 J. Mandelkau *Buttons* xiii. 149 From now on my club was going to have nought to do with the Alternative society and its rip-off merchants. 1974 *Amer. Speech 1970* XLV. 210 Bring your own food. There won't be any ripoff merchants there. 1977 *It* May 5/2, I am not suggesting that the Pink Floyd are rip-off artists, but it is undeniable that much contemporary music is a response to alienation. 1977 C. McFadden *Serial* xxxix. 84/2 He checked out the chain lock that secured his Motobecane against rip-off artists.

Ripolin (ripolæ̃n). [Fr.] The proprietary name of a make of paint.

[1889 *Trade Marks Jrnl.* 25 Dec. 1/1 Ripolin 92,910. Chemical substances used in manufactures or philosophical research, and anti-corrosives... 2nd October 1889.] 1907 *Yesterday's Shopping* (1969) 157/1 Ripolin enamel. White, black, sulphur, [etc.]. Tin... 1/0. 1922 C. K. S. Moncrieff tr. *Proust's Swann's Way* II. 29 My room in the Grand Hôtel de la Plage..the walls of which, washed with ripolin, contained..a finer air. 1934 A. Huxley *Beyond Mexique Bay* 32 We prefer the lighter woods, we prefer metal and glass and ripolin.

riposte, *v.* **2.** (Later examples.)

1898 J. M. Cobban *Angel of Covenant* xiii. 147 'Had I not taken pains with the foolish old man,' riposted Maudlin, blushing high, 'we should yet be sitting at mumchance.' 1958 *Observer* 6 Dec. 864/1 The western powers might logically riposte by offering to discuss a free and neutral status..for all Berlin. 1972 *Daily Tel.* 27 May 12 'But I'm David Broome,' says the show jumper. 'I don't care if you are Basil Brush,' ripostes the sergeant-major. 1977 T. Heald *Just Desserts* 191 'But what do we have to show for it?' 'We have unlimited access to the former Scoff network,' riposted Bognor.

ripped (ript), *ppl. a.* (and *pa. pple.*). [f. Rip *v.*[2] + -ed[1].] **1.** *dial.* Cut, slit.

1823 E. Moor *Suffolk Words & Phr.* 129 *Fleeches.* The portions into which a tree or piece of timber is cut by the saw, in its first position over the saw-pit... When turned and *ripp'd* or *rippen'd*, that is, cut into smaller portions, such portions are called *Scantlins*, in Suffolk and..elsewhere. 1880 J. H. Maxwell *Sheep-Marks* 15 Topped on both ears, ripped on near.

2. With *up.* **a.** Angry. **b.** Torn up.

1941 *Amer. Speech* XVI. 190 *Ripped up*, angry. 1973 'E. McBain' *Hail to Chief* i. 3 The ripped-up sections of planking.

3. (As *pa. pple.*) Under the influence of a drug. *U.S. slang.*

1971 *Rolling Stone* 24 June 16/1 It comes on so fast all with one toke. By the third you're ripped to the gills for several hours. 1973 D. Lang *Freaks* 17 Oh, wow, boy, are you ripped! 1975 C. James *Fate of Felicity Fark* vi. 55 On he gabbled as if ripped on Speed. 1981 *Listener* 1 Jan. 23/3 The story is of a threatened pop singer who, though he gets ripped at need, is really very moral.

4. *slang.* With *off.* Of a person: robbed; exploited; of a thing: stolen.

1971 *New Yorker* 22 May 40 (*title*) Ripped off. 1975 *Maclean's Mag.* Sept. 8/2, I can 'hate men' but many of them are in the same spot as I am: co-opted, ripped-off and mad. 1976 Scott & Koski *Walk-In* iv. 23 The garages where you can buy hot tires and ripped-off hub caps.

ripper. Add: **1. a.** (Further examples.)

1909 *Daily Chron.* 20 Aug. 1/1 The theory most generally entertained is that the canister..was left in a coal train in the mine by a ripper or repairer. 1967 *Gloss. Mining Terms* (*B.S.I.*) viii. 22 *Ripper*, a man who rips. 1979 B. Hines *Price of Coal* 31 Sid and the other rippers stayed on the paddy until it reached the end of the track.

b. A criminal who rips the bodies of his victims; *spec.* = *Jack the Ripper* s.v. *Jack *sb.*[1] 36. Also *transf.* and *Comb.* Hence (*nonce-wds.*) **rippero·logist**, a student of the crimes of Jack the Ripper; **rippero·logy**.

1890, etc. [see *Jack the Ripper* s.v. *Jack *sb.*[1] 36]. 1909 J. R. Ware *Passing Eng.* 209/2 *Ripper*, daring murderer of women. 1935 *Amer. Speech* X. 20/1 *Ripper.* 1. A degenerate who molests, rapes, or mutilates women in parks or other secluded spots; probably from Jack the Ripper. 2. A shrewd or lucky fellow who 'gets away with murder'. 1970 C. Major *Dict. Afro-Amer. Slang* 98 *Ripper*, one who has a reputation for cutting others with a knife. 1974 A. Douglas *Noah's Ark Murders* iv. 40 The series of Ripper-like murders that had all but ended his own life. 1974 G. Moffat *Corpse Road* iii. 41 'The Ripper stopped at six [murders].' 'Perhaps this one won't stop.' 1975 D. Rumbelow *Compl. Jack the Ripper* 5 To..'Ripperologists' everywhere—not forgetting *Jack* who brought us all together. 1977 *Guardian Weekly* 27 Feb. 20/3 This file-work gives a little more definition to the blurred (and in places bizarrely touched-up) picture of the dark side of late-Victorian high life that has been emerging during our seventies, mostly through the addition of new bits of information to that arcane old science known as Ripperology. 1979 W. J. Fishman *Streets of E. London* 102/2 It is still debated whether Turner was the first victim [of Jack the Ripper]. Some 'Ripperologists' suggest that Polly Nichols was the first. 1979 *Guardian* 1 June 4/1 The 'Yorkshire Ripper'..has murdered 11 women in less than four years. 1981 *Yorks. Post* 8 July 7/1, I don't want people to cash in on what the Ripper did to my daughter.

2. a. (*c*) *Criminals' slang.* A tool for opening safes, etc.

1889 Farmer *Americanisms* 460/1 *Ripper* or *Mason Ripper*, (1) a new and ingenious implement of burglary, used in opening safes or vaults with iron surfaces. 1925 *Flynn's* 14 Mar. 281/1 *Ripper*, a can-opener (a burglar's tool). 1963 R. I. McDavid *Mencken's Amer. Lang.* 717 The use of *stew* is declining, modern *heavy gees* preferring to use a *stick, ripper* or *can opener* on laminated safes.

b. An implement that is attached to a tractor to break up concrete or hard soil.

1955 'N. Shute' *Requiem for Wren* i. 20 This war [against the rabbits] went on continuously with tractor-drawn rippers to destroy the warrens. 1963 Oglesby & Hewes *Highway Engin.* (ed. 2) xiv. 476 In recent years large rippers mounted on huge crawler tractors..have been used successfully to break up loose or fractured rock. 1976 *Billings* (Montana) *Gaz.* 2 July 11-c/8 (Advt.), Power Tilt Straight dozer with Ateco swinging swank ripper. 1979 *Arizona Daily Star* 5 Aug. (Advt. Section) 9/2 Teeth for backhoes, loaders, ripper points, corner bits, cutting edges for all makes.

3. a. (Earlier and later examples.) Also, of a person, *spec.* an attractive young woman. In recent use, chiefly *Austral. slang.* Also *attrib.*

1838 *Bell's Life* 26 Aug. 4/1 One of Mr. Mynn's best balls, technically a 'ripper', took the top of the middle stump. 1846 *Swell's Night Guide* 40 In conjunction with the above is Miss Emma Watling, a regular ripper. 1848 [see *creeper 10**]. 1848 J. Mitchell in *Amer. Speech* (1935) X. 41/1 *Ripper.* anything very large of its kind. [A Nantucketism.] 1905 *Pall Mall Mag.* July 111/1 He had found her a ripper as to looks. 1935 Auden & Isherwood *Dog beneath Skin* i. ii. 37 What do you think of her? Isn't she a ripper? 1951 E. Lambert *Twenty Thousand Thieves* x. 182. 'Good letter, Chips?' A gurgle. 'It's a ripper!' 1969 A. O'Toole *Racing Game* xviii. 200 'Not a bad run,' I observed... 'A ripper,' Badger agreed. 1970 *Sunday Truth* (Brisbane) 10 May 64 If I'm ever asked what Australia is like, I'll say, 'She's a bloody ripper, mate!' 1973 *Australian* 7 July 16, I love this ripper country. 1976 *Courier-Mail* (Brisbane) 13 Mar. 19/11 Nagle has a fine ear for Australian dialect. The books' a 'ripper', as his characters might say.

4. *U.S. Pol.* (See quot. 1937.) Also *attrib.*

1895 *Columbus* (Ohio) *Dispatch* 1 Apr. 4/2 The Merryman ripper bill looks very much as if the Republicans of this city were going to the legislature for offices. 1937 J. R. Schultz in *Amer. Speech* XII. 319 The word 'ripper' is commonly used in Pennsylvania political parlance to describe a bill that abolishes an office or commission of state or city. Such an act is said usually to have as its purpose the elimination of an officer or member of a commission who is unfriendly to the party in power.

5. *Sc.* A simple fishing tackle consisting of 'a line having attached to it a heavy metal bar fitted with hooks' (*S.N.D.*).

1925 *Glasgow Herald* 17 Aug. 7 He..caught a 20 lb. ling with a ripper. *c* 1930 in *Scot. Nat. Dict.* (1968) VII. 455/3, I vrocht aa efterneen at the ripper, bit I hid nae luck. 1946 *Aberdeen Press & Jrnl.* 2 July, The primitive method of fishing with dandy lines, consisting of ripper and hooks.

ripping, *vbl. sb.* Add: **1. a.** (Later examples.)

1900 in Farmer *Public School Word-bk.* 167. 1911 R. Nevill *Floreat Etona* vii. 224 In the days when such a close connection existed between Eton and King's, a Colleger leaving to go to Cambridge used to go through the old form known as 'Ripping'... The two folds of the Colleger's serge gown were sewn together in front, and the Provost 'ripped' them asunder. 1973 'J. Patrick' *Glasgow Gang Observed* iv. 41 The case for the defence collapsed..when Big Dick boasted..that the attack was 'the biggest rippin' Ah've done'. 1979 *Jrnl. R. Soc. Arts* Jan. 93/1, I was very interested in your ideas about shifting the rippings from the coal face and spreading it into the sea.

2. b. *Aeronaut.* Used *attrib.* with reference to a strip of fabric sown into and forming part of the skin of a balloon and to the cord which, when pulled, tears this strip away to bring about rapid deflation, as *ripping cord, line, panel, rope, valve.*

1907 *Strand Mag.* Feb. 149/1 The utility of the ripping-cord was brought home to me..in a recent ascent. 1908 A. Hildebrandt *Airships Past & Present* xvi. 184 The ripping-cord was the invention of the American aeronaut, Wise, in 1844; Godard introduced it in France in 1855. 1910 Ripping cord [see *ripping panel* below]. 1907 *Strand Mag.* Feb. 151/2 Faure..is the first aeronaut who had the courage to experiment with the ripping-line in mid air. Till he made his first attempt about two years ago, all balloonists thought that to pull the ripping-line anywhere but on the ground meant suicide. 1974 *Oxford Jun. Encycl.* (rev. ed.) IV. 35/2 Among other important equipment is the ripping-line or rip-cord, which is painted red so that it shall not be pulled in mistake for the valve line. 1908 A. Hildebrandt *Airships Past & Present* xvi. 184 The ripping-panel is placed on that side of the covering to which the guide-rope is attached. 1910 C. C. Turner *Aerial Navigation* iii. 45 On coming within a few yards of the ground he pulls the ripping panel open, the cord from which comes down through a hole near the neck of the balloon. In some balloons,..the ripping cord passes through the neck. 1919 H. Shaw *Text-bk. Aeronaut.* xvii. 195 For emergency cases a 'ripping panel' is fitted, for use when a quick descent is necessary. 192- *Encycl. Brit.* XXX. 89/1 One complete series of balloons came down with unexpected suddenness, all being deflated by the rupture of their ripping panels. On examination, it was found that moisture had condensed on the ripping ropes and frozen there, until each cord was about as thick as a man's forearm. 1907 G. Bacon *Record of Aeronaut* xiv. 255 Aeronautical experts..advised..that, as the balloon might have to remain inflated for a long while before starting, one of large size should be employed, and a 'solid' or 'ripping' valve substituted for the usual 'Butterfly' variety. 1912 C. B. Hayward *Pract. Aeronaut.* 51 The usual 'ripping valve' is also provided in the form of a narrow strip of balloon fabric glued over a long cut in the envelope.

ripping, *ppl. a.* Add: **1.** (Further *fig.* examples.)

1896 G. F. Northall *Warwickshire Word-bk.* 192 *Ripping, adj.* Sharp, cutting, as applied to cold weather; e.g. 'a ripping frost'. Midlands. 1932 H. J. Massingham *World without End* 297 Words like..'ripping' or 'sniping', the adjectives of a very sharp frost..are playlets in themselves. 1978 J. Wambaugh *Black Marble* x. 237, I have a ripping headache.

2. a. Also advb., and as a complement. (Further examples.) *Obsolescent.*

1846 *Swell's Night Guide* 74 One calls for the 'lanciers'; another, the 'caledonians'; when the Boshman, ripping innocent of either..strikes up the college hornpipe. **1894** 'A. Hope' Dolly Dialogues ix. 51 She did look ripping in that white frock. **1898** [see *knock-out sb. 4]. **1911** D. H. Lawrence White Peacock ii. ii. 229 She was very fine and frank and unconventional—ripping, I thought her. **1921** [see *dug-up a.] **1921** H. Williamson Beautiful Years 137 Jack was excited about his friend going to school with him. 'I say, how ripping, man!' he cried. **1944** [see *decent a. 5 b]. **1978** Palin & Jones (title) Ripping yarns.

†**b.** Very fast or rapid. Obs.

1826 [in Dict., sense 2]. **1828** Bell's Life 27 July 4/1 The ripping bowling of the Captain. **1846** W. Denison Cricket: Sk. Players 22 Mr. Osbaldeston, and Brown of Brighton, afterwards launched forth as 'fast ripping bowlers'. **1868** H. Woodruff Trotting Horse xi. 116 There is no occasion for the ripping spurts which intervene in the other training. **1877** London Society June 537/1 Hinkly's bowling was ripping indeed.

ri·ppingness. Obsolescent. [f. Ripping ppl. a. + -ness.] Splendid quality; excellence.

1910 A. Bennett Clayhanger iv. xii. 566 'She's a ripping woman.'.. His preoccupation with the rippingness of Mrs. Chris Hamson. **1927** C. E. Montague Right off Map xviii. 174 It's the doing the thing that..makes you half drunk with the rippingness of it.

rippit, var. Rippet. Sc. and U.S. dial.

1851 W. Anderson Rhymes 195 Sic a rage an' a rippit I seldom hae seen. **1870** J. C. Duval Adventures Big-Foot Wallace xliii. 270 At last the manager threw his hat among 'em and called out, 'Stampede all', and the 'rippit' commenced. **1887** [see Rippet]. **1913** H. Kephart Our Southern Highlanders xiii. 294 If they quarrel, it is a ruction, a rippit, a jower, or an up-scuddle—so be it there are no fatalities which would amount to a real fray. **1913** J. Service Memorables 67 Some dreidfu' nicht rippit there had been amang the cairters. **1928** M. Chapman Happy Mountain 313 Degrees of feeling among unfriendly neighbors... A rippit, fight with fists. **1958** Huntly Express 19 Dec. 6 For fear ony rippits brook out at the dance.

ripple, sb.³ Add: **1. a.** (Later examples.)

1941 L. D. Baldwin Keelboat Age 71 The breaking of the water over the bars and chains was known to the boatmen as ripples, or riffles. **1964** F. O'Rourke Mule for Marquesa ii. 50 'How do we cross?'..'Quicksand to the right. You move out the bar, angle downstream to that big ripple, then straight on across.'

2. a. spec. in Physics, a wave on the surface of a fluid the restoring force for which is provided by surface tension rather than by gravity, and which consequently has a wavelength shorter than that corresponding to the minimum speed of propagation.

1871 W. Thomson in Phil. Mag. XLII. 374 The 'Capillary waves'..referred to by Russell are what I, in ignorance of his observations.., had called 'ripples'. **1887** Amer. Jrnl. Math. IX. 67 Sir W. Thomson proposes to distinguish by the name of ripples those waves whose length is less than the above critical value of λ [wavelength]. **1938** L. M. Milne-Thomson Theoret. Hydrodynamics xiv. 377 Ripples are waves on [ed. 2: in] whose propagation capillarity plays the predominating part. **1966** McGraw-Hill Encycl. Sci. & Technol. XIV. 417/1 Ripples generated by wind at the interface between air and water on oceans and lakes are of importance to the friction of air flowing over water, and to the reflection and scattering of electromagnetic and sound waves.

b. (Further examples.)

1968 J. Winearls Mod. Dance (ed. 2) ii. 38 Raising and lowering can follow each other in uninterrupted snake-like ripples. **1972** Courier-Mail (Brisbane) 3 July 2/6 Laurence (Lord) Olivier, on every official opening night at the Old Vic, was reserving front-blocks of seats, 'the Ripples' for students whose natural reaction might..introduce 'ripples' to break down the first-night social-or-coterie stuffiness. **1977** Design Engin. July 19/3 High number of commutator segments giving very smooth operation (low torque ripple) over a speed range from zero to several thousand rev/min. **1979** Time 2 Apr. 20/3 As Finland bottomed out of recession, pre-election polls showed a ripple to the right.

d. A name for an ice cream manufactured with an admixture of coloured syrup that gives it a rippled appearance.

Registered as a proprietary name in the U.S.

1939 Ice Cream Trade Jrnl. Nov. 19 (Advt.), Fudge ripple ice cream. Ibid., This delicious fudge syrup is incorporated in your own vanilla ice cream as you fill the can from a batch freezer or incorporated in your vanilla ice cream by a special fudge ripple attachment as it flows from your continuous freezer. **1942** Official Gaz. (U.S. Patent Office) 16 June 511/1 Gerald G. Balch, doing business as Balch Flavor Co., Pittsburgh, Pa... Ripple... For ice cream and flavoring concentrate for use in the manufacture of variegated type of ice cream and for use in the manufacture of a flavoring adapted to be applied on or to ice cream, sold in bulk. Claims use since Oct. 23, 1939. **1973** Hyde & Rothwell Ice Cream vii. 141 (caption) Diagram of Hoyer ripple machine. **1976** Milton Keynes Express 25 June 10 (Advt.), Ice Cream... Raspberry Ripple, 4 litre £1.10. **1977** 'J. Bell' Such Nice Client xvi. 156 A pork pie..and an ice ripple from the freezer..brought her back into fighting shape. **1981** Times 2 May 7/4 Dessert..angel food cake, blueberry cheese cake, fudge, ripple ice cream.

3. Also fig., a rumour.

1977 N. Adam Triplehip Crackman vii. 71 There's a ripple he's walking a bit heavy these days.

4. (Earlier and later examples.)

1853 Alta California (San Francisco) 31 May 2/1 Repeated instances have come to our knowledge when the amount of gold saved has been doubled by a little alteration or improvement made in the ripple-box. **1902** Chambers's Jrnl. Mar. 176/1 This process consists of passing the sand through an arrangement of inclined sieves, which by reason of 'ripples' or bars of wood fastened transversely across their surfaces, discard all light material that cannot pass the various meshes. **1938** D. Forbes My Life in S. Afr. vi. 87 One had to keep the earth moving all the time. If it once settled to the bottom of the box it formed a solid mass, and did not settle into what are called the ripples on the bottom of the box made to catch the gold.

5. Electr. Small periodic variations in voltage superposed on a direct voltage or on an alternating voltage of lower frequency. Freq. attrib.

1920 Whittaker's Electr. Engineer's Pocket-bk. (ed. 4) 152 When the rotor is slotted, as in turbo-alternators with cylindrical rotors, a ripple is induced in the conductor-pressure. **1928** Observer 17 June 26/3 Manufacturers should publish a curve of inductance against direct current component, measured at ripple frequency. **1947** R. Lee Electronic Transformers & Circuits iii. 66 Reactors are used in electronic power equipment to smooth out ripple voltage in d-c supplies, so they carry direct current in the coils. **1967** Electronics 6 Mar. 308/2 Quadrature rejection is 40 db minimum, while ripple is 0·5% peak-to-peak of full scale output. **1972** Physics Bull. Aug. 491/2 An automatically adjusting transformer ensures high stability; ripple is 1 mV peak to peak.

6. Applied to a method of firing or discharging in succession or at intervals.

1944 Shorter Oxf. Eng. Dict. (ed. 3) Add., Ripple.., applied to a method of firing torpedoes in succession. **1951** Life 18 June 54/3 (caption) Using six rocket launchers with 24 tubes each, the platoon in a matter of seconds blasted Red positions with 144 rounds—a barrage which rocketmen call a 'ripple'. **1968** Daily Express 12 Feb. 2/5 Ripple, an attacker's technique of sending off his missiles in waves, so that a defender fires his defensive missiles at the first waves and has nothing left to deflect the final blow.

7. attrib. and Comb., as **ripple burnish,** (poet.) **-dripple, ware; ripple black, fresh, -warped** adjs.; **ripple (pony) cloth,** cloth having a rippled appearance; also attrib.; **ripple control** Electr., a method of performing simple control operations, such as switching of street lights, by superposition of a high-frequency switching signal on the mains supply; **ripple(-through) counter** Electronics, a type of binary counter consisting of a number of bistable circuits wired in cascade, so that each changes its state only after all the preceding ones have changed state; **ripple effect,** the continuous and spreading results of an event or action; **ripple-fired** adj. [cf. sense 6 above], (of missiles) fired in rapid succession or at intervals; so **ripple-firing** a.; **ripple-flaking** Archæol., a method of flaking flint; so **ripple-flaked** a.; **ripple sole,** a kind of rubber sole having thick ripple-shaped ridges; hence **ripple-soled** ppl. a.; **ripple stitch,** a drawn fabric stitch (see also sense 2 in Dict.).

1965 Wireless World Aug. 28 (Advt.), Dial escutcheon measures 6" long by 4¼" wide, finished ripple black. **1964** H. Hodges Artifacts i. 31 The effect of this ripple burnish can be highly decorative. **1922** Ripplecloth [see house-jacket s.v. *house sb.¹ 19]. **1930** Daily Express 6 Oct. 5/2 (Advt.), Pure wool ripple cloth, for cosy dressing gowns. **1957** M. B. Picken Fashion Dict. 276/1 Ripple cloth, woolen dress fabric with long silky hairs on right side. Also called zibeline. **1977** B. Pym Quartet in Autumn xvi. 139 Her old blue ripple-cloth dressing gown. **1938** Jrnl. Inst. Electr. Engin. LXXXIII. 827/1 The fullest advantage can only be taken of the ripple control principle if a central ripple transmitter ..can be designed to cover the entire network fed at high tension from that point. **1952** J. M. Waldram Street Lighting xx. 381 A ripple control system operating about 500 lamps in an urban area, pre-war. **1974** Times 4 Feb. 16/6 It seems..opportune to raise again the..question of ripple control. For the uninitiated, it is a means of transmitting coded signals from a central control point over the existing mains network to operate switching contactors at any point in the network, e.g., domestic consumers' premises. **1967** Electronics 6 Mar. 160/1 It cannot be used in ripple counters or other circuits requiring the toggle function. **1973** Sci. Amer. May 110/3 Combinations of flip-flops and NAND gates interconnected as described are known as ripple counters. The reason is that input pulses 'ripple' through the string of flip-flops sequentially, each device triggering the next one in the series. **1916** Blunden Harbingers 49 The ripple-dripple of the brooks. **1966** Wall St. Jrnl. 14 Feb. 10/3 Price-boosting already is producing a 'ripple effect' in which companies pass on increased costs in higher price tags on their own products. **1973** Times 20 Dec. 12/4 Industrial expansion is now halted by the three day week, not to say by the ripple effect of the public expenditure cuts. **1977** Detroit Free Press 11 Dec. 7-B/4 The ripple effect from the 'Star Wars' film has taken on tidal wave proportions. **1954** Sun (Baltimore) (B ed.) 25 June 10/3 Proximity-fused bombs or photo flash bombs..high explosive rockets, ripple fired rockets, smoke rockets and guided missiles. **1949** A. R. Weyl Guided Missiles 31 Control of the launching manœuvre can be electric, for example, by closing a circuit which ignites a propelling charge, or electronic. The former is preferred for small missiles, with 'ripple-firing' control for the multiple and successive launching of A.A. missiles. **1960** Oxf. Univ. Gaz. 4 Mar. 805/2 A fine ripple-flaked flint knife with gold leaf attached to the blunt end. **1899** Ripple-flaking [in Dict., sense 2 b]. **1921** Chambers's Jrnl. 5 Feb. 145/1 Brierly Stretton and I were..discussing the intensive examination of ripple-flaking. **1944** Blunden Shells by Stream 44 While some freed fountain of delight Played beauty ripple-fresh and bright. **1952** C. W. Cunnington Eng. Women's Clothing 296 Ripple pony cloth..Resembles a finely ribbed miroir velvet with a bright finish. **1963** N. Marsh Dead Water (1964) viii. 201 Gentleman's country shoes, size nine-and-a-half ripple soles. **1970** Globe & Mail (Toronto) 26 Sept. 52/2 (Advt.), Women's Slippers... Some ripple rubber soles. **1977** C. McFadden Serial xli. 88/2 Kate had already gone springing off in her ripple-sole Famolares. **1977** Austral. Furnishing Buyers Guide Spring/Summer 58/4 With heels, steel tips, crepe and ripple soled shoes, carpets are getting much more wear and tear. **1933** K. S. Lofthouse Compl. Guide Drawn Fabric 22 Ripple stitch... Worked from right to left... Three pairs are worked with six threads between. **1971** J. H. Smith Digital Logic vi. 103 In the asynchronous counter..the trigger input of each binary is from the preceding output... This mode of operation gives an inherent delay in the ripple-through counter. **1931** Ripple ware [see *Badarian a.]. **1962** V. Nabokov Pale Fire 143 The ripple-warped reflection of a ledge that jutted high above his present position.

ripple, v.³ **1. b.** (Later transf. and fig. examples.)

1924 R. Campbell Flaming Terrapin iii. 45 In spangled pride A python ripples from his shrivelled hide. **1929** D. H. Lawrence Pansies 39 We ripple with life through the days.

d. To pass quickly through each of a series in turn.

1967 Electronics 6 Mar. 47/3 It will ripple through a truth table in maybe 100 microseconds,..moving through all possible logic combinations. **1973** Sci. Amer. May 110/3 Input pulses 'ripple' through the string of flip-flops sequentially, each device triggering the next one in the series.

2. a. (Further examples.)

1922 T. S. Eliot Waste Land iii. 36 The brisk swell Rippled both shores. **1974** Sat. Rev. World (U.S.) 2 Nov. 32/3 The lake..was rippled by a sailboat.

c. (Later example.) Also, to let out or utter with a rippling sound.

1901 A. E. W. Mason Clementina xiii. 155 The girl reading it drew a breath and rippled out a laugh of gladness. **1913** C. Mackenzie Sinister St. I. ii. x. 313 'Why, you silly old thing..,' rippled Stella.

rippled, ppl. a. **a.** (Later examples.)

1903 H. G. Wells War in Air v. 154 A rippled veil of still, thin sunlit cirrus. **1927** Peake & Fleure Peasants & Potters vii. 100 Plain, incised, and rippled ware. **1930** R. Campbell Poems 2 The rippled silver of her breasts. **1950** Fransden & Nelson Ice Cream xvii. 177 Rippled ice cream: as the plain vanilla ice cream is drawn from the freezer into the package, specially prepared syrups..are added by means of a special nozzle so as to produce a marbled effect.

ri·ppleless, a. [-less.] Without causing ripples, unmarked by ripples.

1923 J. S. Huxley Ess. Biologist iii. 113 The Crested Grebe... Its brilliant white belly, protective grey-brown back, rippleless and effortless diving, [etc.]. **1941** Steinbeck & Ricketts Sea of Cortez xiv. 131 The lagoon [was] rippleless.

ripplet. (Earlier example in form riplet.)

1820 Shelley Orpheus in Compl. Poet. Wks. (1904) 699 Each riplet makes A many-sided mirror for the sun.

rippling, vbl. sb.³ Add: **2.** (Earlier and later examples.)

1745 W. Pote Jrnl. Captivity French & Indian War (1896) 55 This Day was foul weather the Greatest part of the Day, and likewise verey bad Paddling, on account of Ripplings and falls. **1832** W. D. Williamson Hist. Maine I. 57 Here are ripplings, to avoid which, a canal was cut twenty rods in length.

rip-rap, sb. Add: **1. b.** (Later examples.) Also, the sound of fireworks detonating. Also transf.

1909 Hall Caine White Prophet I. i. 8 Once more the words rang like a rip-rap down the line. **1930** Sea Breezes Dec. 72 The glare and rip-rap of the fireworks ashore..told that the Old Year 1882 had expired. **1942** Sun (Baltimore) 13 Mar. 13/3 Sea-coast guns on Fort Wool, known to civilians as the Rip Raps, will be fired Saturday between 9 A.M. and noon. **1974** Country Life 28 Nov. 1682 A firework party..rockets sped into the sky..followed by..sparklers and even 'rip-raps' round our feet.

c. (See quot.)

1959 I. & P. Opie Lore & Lang. Schoolch. xviii. 381 There are more than sixty established names for the pursuit of illegally knocking at doors... Rip Rap. Derby.

3. For U.S. read orig. U.S. and add: More widely, loose stone used for revetments, embankments, or the like; also, a structure made of this. (Earlier and further examples.)

[**1822** Niles' Reg. 15 June 252/1 The expense of getting stone, and delivering it at the Rip Raps and Point Comfort.] **1833** H. Barnard Let. 4 Mar. in Maryland Hist. Mag. (1918) XIII. 314 We passed in our way..Old Point Comfort, upon which you know Fortress Monroe is situated—the ripraps directly opposite, which two will effectually secure the Bay. **1838** J. Childs Western Railroad (1839) 25 To guard the embankments by rip-rap walls. **1880** News & Press (Cimarron, New Mexico) 9 Sept. 3/3 It is the intention of the company to raise the grade of the approaches on both sides, protecting the exposed surfaces by what is technically known as 'rip rap', a kind of loose rock work. **1892** Outing July 254/1 At the worst places ripraps of brush and stones have been built to confine the river to its bed. **1926** Daily Colonist (Victoria, B.C.) 11 July 36/3 It was necessary to blast out a way at the cliffside, with occasional banks of rip-rap, and stretches of trestle. **1931** Sun (Baltimore) 1 Sept. 9/6 A railroad section foreman, blasting for rip rap stone, found the first sample of the ore in western Pike county [Arkansas]. **1975** Offshore Aug. 36/2 After placement the base was grouted into position and a wall of rip-rap placed right round the base to protect it against scour. **1976** Billings (Montana) Gaz. 16 June 7-A/2 The time required to obtain a riprap permit for Montana rivers and streams.

rip-rap v., **rip-rapping** vbl. sb. (Further examples.)

1884 Harper's Mag. Sept. 504/1 Cliff ledges..[are] connected one terrace above by the other, by..a natural riprapping of fallen fragments. **1897** M. B. Kerr in E. L. Wilson et al. Mountain Climbing vi. 286 We riprapped the bottom of the slope to prevent slipping down the hill. **1904** Dialect Notes II. 420 Highland Avenue is rip-rapped on one side. **1938** Sun (Baltimore) 8 Nov. 5/3 Another [project] is for rip-rapping the river bank. **1943** B. A. De Voto Year of Decision 332 They had to..pry boulders out of their course, riprap swamp patches, sometimes bridge brooks that could not be crossed otherwise. **1976** Columbus (Montana) News 27 May 8/3 The resolution was signed for the riprapping of property by the Absarokee lagoon.

‖ **ripresa** (riprē·sa). Mus. [It.] A repeat; a refrain (see quot. 1947).

1740 J. Grassineau Mus. Dict. 204 Ripresa. See Repeat. **1876** Stainer & Barrett Dict. Mus. Terms 379/2 Ripresa... (1) A reprise or burden. (2) A repeat. **1947** E. Blom Everyman's Dict. Mus. 501/1 Ripresa,..a refrain, especially in the 14th–15th cent. It. ballata and frottola. **1977** Early Music July 326/2 The ballata consists of a refrain or ripresa of one to four poetic lines, followed by two piedi..and a volta.

ri·pripple v., nonce semi-reduplication of Ripple v.³

1922 Joyce Ulysses 85 He saw his trunk and limbs riprippled over.

ri·p-roaring, riproaring, a. orig. U.S. [Cf. Riproarious a.] Full of vigour, spirit, or excellence; first-rate; boisterous; full-blooded.

1834 W. A. Caruthers Kentuckian in N.Y. I. 62 There was a rip-roaring sight of slight o' hand and tumbling work there. **1845** J. J. Hooper Some Adventures Simon Suggs x. 127 And I seed the biggest, longest, rip-roarenest, blackest, scaliest..allegator. **1884** E. W. Nye Baled Hay 231 He thought..Kirke was there..to give Laramie the grandest, riproaringest tempest of mirth that she had ever experienced. **1905** Dialect Notes III. 64 We had a rip-roaring time. **1906** N.Y. Even. Post (Sat. Suppl.) 8 Sept. 1 When he was called upon to address the conference he got a rip-roaring welcome. **1923** Daily Mail 28 Feb. 10 (Advt.), It's a rip-roaring, red-blooded yarn that no man or woman will be able to read unmoved. **1948** Sunday Pictorial 18 July 11/3 A rip-roaring performance by Oscar Homolka. **1950** C. Fry Venus Observed ii. ii. 70 Well, here's a rip-roaring gauntlet to be run By a couple of God's children. **1958** Times 3 Nov. 3/1 It was a match in which little skill was shown, but for all that a rip-roaring, babel-like affair of incessant movement and high good spirits. **1979** Guardian 14 Apr. 9/7 Rip-roaring commercial [film] successes.

Hence **ri:p-roa·ringly** adv.

1951 Sport 27 Jan.–2 Feb. 12/2 Their young side..is having a rip-roaringly successful cup season.

riproarious, a. For U.S. read orig. U.S. (Earlier and later examples.)

1830 N.Y. Constellation 11 Sept. 2/5 The English traveller had put up at a little log tavern on the banks of the Savannah, where the riproarious conduct took place. **1840** Congress Globe 2 Apr. 376/1 Here and there a gentleman from both political parties, who had been drawn out by curiosity to witness their riproarious proceedings [at the Whig 'powwows']. **1948** R. W. Chapman Lexicography 6 The Dictionary of American English..stopped at 1900, before the trickle of that rip-roarious idiom became a flood. **1975** J. I. M. Stewart Gaudy iii. 57 It was one of those rare books which, while enjoying riproarious popular success, at the same time owns sufficient intrinsic merit to achieve among the critical a kind of classic status straight away.

riproa·riously, adv. [f. Riproarious a. + -ly².] In a riproarious or boisterous manner.

1834 D. Crockett Narr. Life xi. 78 The next day it rained rip-roriously.

ri·psnorter. orig. U.S. Also **rip-snorter.** [f. Rip v.² + Snorter¹ 2: cf. Riproarious a.] Someone or something exceptionally remarkable in appearance, quality, strength, or the like; spec. a storm, a gale. Cf. Snorter¹ 2 b.

1840 Crockett Almanac 20/1 Of all the ripsnorters I ever tutched upon, thar never war one that could pull her boat alongside of Grace Peabody. **1885** Santa Fé Weekly New Mexican 20 Aug. 2/6 Any galoot who wants the Ripsnorter for a year can have it left at his bar-room on payment of three red chips in advance. **1889** K. Munroe Dory Mates (1890) vi. 84 Boys, we are in for a regular 'rip-snorter'. I never saw a nastier night. **1924** R. Clements Gipsy of Horn v. 87 It came on to blow in a way that the packet-rats called a 'rip-snorter'. **1931** A. J. Cronin Hatter's Castle ii. ix. 368 Did you see that shot of mine, cocky?.. It was a—a regular nor'easter—a pickled ripsnorter. **1941** Baker Dict. Austral. Slang 60 Ripsnorter, something particularly good. An amusing or eccentric person. **1951** New Yorker 3 Mar. 28/1 The villain is a real ripsnorter. **1972** Last Whole Earth Catalog (Portola Inst.) 445/3 This is Gurney Norman the author speaking, bringing you the end of this folk tale, and it's a rip-snorter.

ri·psnorting, a. orig. U.S. Also **rip-snorting.** [f. prec.] = *rip-roaring, riproaring a.

1846 Yale Lit. Mag. June 336 What a rip-snorting red head you have got! **1904** Topeka (Kansas) Capital 2 June 4 It is now stated that Bryan will make a rip-snorting speech at the St. Louis convention. **1926** Spectator 10 July 44/1 It's a ripsnorting, red-blooded show—a wow. **1926** [see *hotsy-totsy a.]. **1956** N. Coward South Sea Bubble i. i. 12 You're a rip-snorting old careerist, darling. **1959** Daily Mail 17 Feb. 4/5 The Count Basie riff-number Alright, Okay, You Win, with ripsnorting backing. **1976** P. Henissart Winter Quarry xxvii. 276 These ripsnorting professional anti-communists. **1978** Detroit Free Press 14 Apr. 15B/4 A ripsnorting cockroach race, with the men on their hands and knees, urging their bugs down makeshift lanes.

Hence (as a back-formation) **ri·psnort** v., to go boisterously, to rollick; **ri:psno·rtingly** adv.

1974 Publishers Weekly 24 June 59/2 It's a definitive book, ripsnortingly adult. **1975** C. A. Haddad Moroccan vii. 83 'Come to Daddy,' I said..she clambered on my back and together we ripsnorted our way to Ilanit. **1979** 'A. Hailey' Overload ii. x. 159 Water which promptly became high-pressure steam and ripsnorted to a separate superheater section.

Ripuarian, a. and sb. Add: **A.** adj. **3.** Designating a northern dialect of Middle Franconian German.

1910 Encycl. Brit. XI. 779/2 The boundary-line between Low and High German..may roughly be indicated by the..place-names, on the understanding..that the Ripuarian dialect..is to be classed with High German. **1934** Priebsch & Collinson German Lang. ii. vii. 323 Within Middle Franconian the Ripuarian dialect..keeps unshifted rp.

B. sb. **2.** The name of a dialect of Middle Franconian German.

1910 Encycl. Brit. XI. 780/1 Middle Franconian.., which ..represents a kind of transition dialect to Low German, is itself divided into (α) Ripuarian or Low Rhenish with Cologne and Aachen..as centres, and (β) Moselle Franconian with Trier..as principal town. **1934** Priebsch & Collinson German Lang. ii. i. 88 In Ripuarian—the Northern dialect of Middle Franconian—the unshifted form ūp (up) 'auf' is found. **1939** L. H. Gray Foundations of Lang. xi. 349 Middle Franconian appears in two sub-forms: Ripuarian in the north, and Moselle in the south. **1961** R. E. Keller German Dial. vii. 249 Luxemburgish appears ..as a relic area..separated..from Ripuarian by the Dorp/Dorf isogloss in the north. **1970** Chambers & Wilkie Short Hist. German Lang. iii. 21 Middle Franconian, divided into Ripuarian (in the Cologne area) and Moselle Franconian.

Rip Van Winkle (rip væn wi·ŋk'l). The name of a character in Washington Irving's Sketch Book (1819–20), a good-for-nothing who falls asleep for twenty years, applied: **a.** transf. to a person unfamiliar with prevailing conditions.

1833 Advocate (Shelbyville, Kentucky) 28 Sept. 2/4 Wm. C. Preston, of South Carolina, in one of his furious tirades, applied to the State of North Carolina, the somewhat degrading epithet of 'the Rip Van Winkle of the South'. **1852** Dickens Bleak Ho. (1853) ii. 5 Both the world of fashion and the Court of Chancery are things of precedent and usage; over-sleeping Rip Van Winkles, who have played at strange games through a deal of thundery weather. **1892** G. B. Shaw Fabian Soc. 20 There are some Rip Van Winkles in our movement who are only now waking up. **1939** C. Day Lewis Child of Misfortune iii. i. 262 A Rip Van Winkle's self-pity would seize him. **1945** R. Hargreaves Enemy at Gate 187 That somewhat shop-soiled military Rip van Winkle, Giuseppe Garibaldi. **1974** H. McCloy Sleepwalker ii. 17 What was she? A girl Rip Van Winkle who had been asleep since 1940?

b. attrib. to something characteristic of or resembling Rip Van Winkle or (an aspect of) his experience.

1849 Picayune (New Orleans) 21 July 1/6 A person absent for three weeks, on returning, almost fancies that he has been taking a Rip Van Winkle slumber. **1893** I. Zangwill Ghetto Tragedies 133 Is it possible that I can get into touch again with my youth..after a sort of Rip Van Winkle sleep? **1959** Times 6 May 15/7 Time, in Česky Krummau, seems to have stopped about the year 1800. In this Rip-van-Winkle atmosphere, we were taken to the tiny castle chapel, where there is a baroque organ in which every single pipe is in its original state. **1977** W. H. S. Smith Young Man's Country iii. 93 On the wall, as if to convince me that I was not in a Rip Van Winkle dream, was a list of S.D.O.s.

So **Rip Van Wi·nkleish** a., characteristic of or resembling Rip Van Winkle, ignorant of present conditions; hence **Rip Van Wi·nkledom,** (a) the Catskill Mountains in the state of New York, the site of Rip Van Winkle's sleep; (b) a state of prolonged sleep; **Rip Van Wi·nkleism,** an outmoded custom or opinion.

1829 Mechanic's Press (Utica, N.Y.) 5 Dec. 28/1 His Rip Van Winkleish habits asked no more than to pursue 'the even tenor of their way'. **1842** C. M. Kirkland Forest Life II. 228 [Reading an old-fashioned book] was counted among my Rip-Van-Winkle-isms. **1852** Harper's Mag. Aug. 420/2 A Pilgrim from the backwoods..had just been awakened from a Rip-Van-Winkleish existence of a quarter of a century by the steam-whistle of the Erie Railroad. **1888** G. B. Shaw Let. 9 Feb. (1965) I. 185 He persists in his 18th century Rip-van-Winkleism. **1892** Outing Apr. 48 (title) A cyclist's visit to Rip Van Winkledom. Ibid. 50/2 We are already in the confines of Rip Van Winkledom. **1911** Beerbohm Lett. to R. Turner (1964) 195 It made me feel very Rip-Van-Winkleish to find no Alfred Douglas. **1956** M. Lowry Let. 13 Nov. (1967) 391 She..has also for the latter part of this time been mostly asleep... She emerged from this Ripvanwinkledom, feeling and sounding better than she has in ten years.

Riquewihr (ri·kviəɹ). The name of a town in Alsace, applied to white wines produced there.

1947 T. E. Carling Wine-Wise vi. 34 The latter [sc. Gentil, Riesling and Traminer] are usually sold under the name of the grape, coupled with that of the village—as Riesling-Riquewihr, and Traminer-Ribeauvillé, etc. **1951** E. Paul Springtime in Paris x. 184 Was it Chablis, Pouilly or Riquewihr with the oysters? **1963** I. Fleming On H.M. Secret Service xxiv. 264 You are just in time for some good Strasbourg sausage and a passable Riquewihr. **1974** C. Wilson Bk. of Booze iii. 124 Alsace wines..are usually sold under the name of the grape—Riesling, Traminer, Sylvaner, etc.—and sometimes the town—Riquewihr being perhaps the best known.

riroriro (rĭ·roriro). [Maori.] Also (with hyphen) riro-riro. The New Zealand grey warbler, Gerygone igata, a small wren-like bird belonging to the subfamily Malurinæ of the family Muscicapidæ.

1835 W. Yate Acct. N.Z. (ed. 2) ii. 58 Riroriro, a very small brown bird, with white feathers under the wings and tail. **1860** [see *fan-tail sb. 3]. **1884** M. A. Martin Our Maoris viii. 125 There is a little bird called Riro-riro in New Zealand from its note. **1939** D. Cresswell Present without Leave 32 The riro-riro..has a small wistful song. **1955** W. R. B. Oliver N.Z. Birds (ed. 2) 477/1 Mention of the Riroriro occurs in many stories, songs and proverbs of the Maori.

‖ **ris de veau** (ridəvō). Also erron. 9 riz de veau. [Fr.] A dish of sweetbread of veal. Freq. in Comb. Also fig.

1820 M. Edgeworth Let. 4 June (1979) 144, I give you one dinner by which you may judge all the rest—Bouilli de boeuf—..ris de veau piqué—maquereau, [etc.]. **1861** Mrs. Beeton Bk. Househ. Managem. 909 Entrées. Riz de Veau aux Tomates. **1877** E. S. Dallas Kettner's Bk. of Table 452 Roasted Sweetbreads. This is what the French set down as the Ris de veau à l'Anglaise. **1927** N. Wainwright tr. Dekobra's Madonna of Sleeping Cars xiv. 187 There are no great men except the inventors and the developers of science. All the rest amount to more or less ris de veau surrounding an Adam's apple. **1938** L. MacNeice I crossed Minch vii. 101 I'll have a little ris de veau, I think. **1964** L. Deighton Funeral in Berlin xxiii. 126 There is entrecôte or ris de veau. **1975** R. Stout Family Affair viii. 188 Sweetbreads poached in white wine, dipped in crumbs and eggs, sautéed, and doused with almonds in brown butter... They call it ris de veau amandine.

rise, sb. Add: **I. 4. c.** Cricket. The upward movement of a ball after pitching.

1851 J. Pycroft Cricket Field viii. 165 Slow balls can be pitched nearer to the bat, affording a shorter sight of the rise. **1897** P. Norman West Kent Cricket Club 40 There was always a good spin on his ball, and he could..give it that 'abrupt rise', as it has somewhere been called, which is so fatal to many a good batsman.

d. Theatr. The raising of the curtain at the beginning of a scene. In phr. at rise, whereby the playwright introduces the description of the opening situation.

1905 [see curtain-rise s.v. *curtain sb. 8]. **1933** S. Kingsley Men in White ii. i. 77 Three months later... At rise: Mr Houghton, short, stodgy, aggressive..the economist, has just finished reading a report. **1961** Bowman & Ball Theatre Lang. 17 At rise, at the moment when the rising curtain first discloses a scene; said often of the relative positions of actors at such a moment. Ibid. 301 Rise,..the going up of a curtain (also as the curtain rise or the rising). **1962** [see curtain-fall s.v. *curtain sb. 8].

e. slang. = Erection 4. Usu. in phr. to get a rise.

1949 Partridge Dict. Slang (ed. 3) 1154/2 Rise, get a, to experience an erection. **1973** M. Amis Rachel Papers 55 'Have you fucked Sue?'.. 'What was it like?'.. 'It was okay, except I couldn't get a proper rise.'

6. c. Also of other precious metals and stones.

1940 I. L. Idriess Lightning Ridge xv. 99 Andy sank five hundred shafts, toiled for years and years, and never made a rise.

d. slang. A fit of anger.

1877 'Eton Boy' Day of my Life i. 11, I told him for his good; he needn't get in such a rise about it. **1895** H. W. Nevinson Neighbours of Ours i. 21 Mrs. Sullivan was in a fair rise about it, sayin' as 'e'd took us all in.

8. For Obs.—¹ read Obs. and add later examples.

1848 D. Cox Let. 29 July in F. G. Roe David Cox (1924) ii. 59 It was expect[ed] there would be a rise among the Irish. **1853** Dickens Child's Hist. Eng. in Househ. Words 17 Sept. 71/1 He had some wild hope of gaining the Irish people over to his side by favoring a rise among them.

II. 9. b. A long, broad, gently sloping elevation rising from the sea bed, esp. that at the edge of a continental shelf.

1903 Geogr. Jrnl. XXII. 193 The elevation is either entirely surrounded by depressions or is a prolongation of the continental border. The rise is an elevation which rises gradually with an angle of only a few minutes of arc, irrespective of whether it is wide or narrow... Rises carry the chief features of suboceanic relief, so that if the ocean floor was changed into dry land they would act as the main watersheds. **1954** W. D. Thornbury Princ. Geomorphol. xviii. 477 A good example of a rise or swell is that in the Pacific Ocean known as the Hawaiian swell or rise. It is a comparatively gentle rise some 600 miles wide and about 1900 miles long. **1974** Nature 30 Aug. 694/2 States like Argentina, Australia, Canada..will prefer to go beyond 200 miles to the outer edge of the continental rise if this is larger than 200 miles.

III. 14. Also in Phonetics. Cf. rise-fall below.

1911 Encycl. Brit. XXI. 465/2 A high rise, which begins high, and consequently can only rise a little higher, expresses simple question, while..a low rise..expresses

various degrees of surprise or indignation. **1932** [see fall-rise s.v. *fall sb.¹ 29]. **1965** Language XLI. 210 This scale maintained its neatness only when the nuclear exponent was a simple 'fall' or 'rise'. When the nucleus was 'fall-rise', 'rise-fall', or 'fall-plus-rise', two phenomena were observed.

15. b. Delete colloq. and add further examples.

1915 W. S. Maugham Of Human Bondage cvii. 564 If they were not worth a rise it was better to sack them at once. **1921** [see *raise sb.¹ 5 c]. **1957** E. H. Shepard Drawn from Memory vii. 131 She..said that her 'intended' was going to receive a rise and work on the passenger trains. **1978** Verbatim Sept. 12/1 The influence of British usage in America is more and more evident: The New York Times, 1 April 1978, in referring to the increased settlement for transit workers in New York City, called it a '6% Rise'.

c. For '(See quot.)' read: the rise of (an amount or period of time), more than, above (that quantity); and the rise, and more. (Earlier and later examples.) Now rare. Cf. Rising pr. pple. 3.

1834 in J. S. Bassett Southern Plantation Overseer (1925) 66, I muste plante the rise of a hundred aceres in coten. **1839** Southern Lit. Messenger V. 379/1 It is the rise of a week since I last shifted. **1845** J. J. Hooper Some Adventures Simon Suggs xii. 141 Bill..has been ded the rise of twenty year. **1858** in N. E. Eliason Tarheel Talk (1956) 290 At Mr Collins thar has bin de rise of A hundred down with the measles. **1905** 'O. Henry' in Everybody's Mag. Dec. 820/1, I will undertake for to say that I've seen the rise of $50,000 at a time in that tin grub box that my adopted father calls his safe.

IV. 20. Comb., as **rise-and-fall** adj.; **rise-fall** Phonetics, a rise and subsequent fall of pitch compressed into one syllable (cf. *fall-rise s.v. *fall sb.¹ 29); also attrib.; **rise time** Electronics, the time required for a pulse to rise from 10% to 90% of its steady value.

1926 Gloss. Terms Electr. Engin. (Brit. Engin. Stand. Assoc.) 159 Rise-and-fall pendant, a pendant the height of which can be regulated by means of a pulley and counterweight or similar device. **1950** Engineering 21 July 59/2 Sellers include a 'rise and fall clause' in their contracts. **1974** tr. Wertheim's Evolution & Revolution i. 64 Oswald Spengler ..elaborated the rise-and-fall concept as a world-wide cyclical movement from which no human civilization could escape. **1977** Grimsby Even. Tel. 14 May 6/5 (Advt.), 3-phase saw bench with rise and fall table, £80. **1964** Crystal & Quirk Syst. Prosodic & Paralinguistic Features in Eng. iv. 50 Such nuclei are of the following seven types: fall, rise, level, fall-rise, rise-fall, fall-plus-rise, and rise-plus-fall. **1966** G. N. Leech Eng. in Advertising v. 49 The..advertisement ..contains three examples of the rise-fall tone... A contour line above each example indicates the position of the rise-fall. **1973** Archivum Linguisticum IV. 25 In paratone I..the 'low rise-fall'..is likely to be followed by the 'wide fall'. **1947** Rev. Sci. Instruments XVIII. 643/1 To obtain best possible rise times for the pulses, care is required in laying out the components. **1952** Proc. IRE XL. 962/1 Because of limited bandwidth, the pulse takes a certain length of time to build up its amplitude, i.e., it requires a certain 'rise time'. **1969** J. J. Sparkes Transistor Switching i. 22 The turn-on time is divided into two parts, the delay time..and the rise time. **1975** G. J. King Audio Handbk. ii. 41 Extended high-frequency response is..required to ensure that the rise time of the amplifier is not less than that of transient-type programme signal components.

rise, v. Add: **B. I. 1. h.** Welsh dial. Of a funeral party: to depart from the home of the deceased or bereaved before the interment.

1959 Western Mail (Cardiff) 18 Feb. 3/1 The funeral on Friday, Feb. 20, rising at 3 p.m., for interment at Gwaelod-y-Brithdir Cemetery. **1976** Ibid. 8 Jan. 10/2 The funeral..will be rising at 2 p.m. for New Bethel Chapel, from her daughter's residence.

3. c. In imp. phr. rise and shine, a command to wake up and leave one's bed. orig. Armed Forces'.

1916 Recruiters' Bull. (U.S. Marine Corps) Apr. 11/2 He rapped at the door and in stentorian tones cried, 'Rise and shine... Wiggle a toe.' **1917** Kipling Diversity of Creatures 237 A high sun over Asia shouting: 'Rise and shine!' **1927** P. Riley Mem. Blue-Jacket 89 Hands were called at 5:30 a.m., the bo'sun's mates going around the deck shouting 'All hands, rise and shine.' **1946** J. Irving Royal Navalese 146 Rise and shine!, the boatswains' mates' call to The Hands to roust them out in the morning. **1953** G. Bell Black Marigolds xix. 189 Wakey, wakey, rise and shine, or have you fainted? **1973** H. Nielsen Severed Key x. 105 'Rise and shine, lovebirds!' he shouted. 'The honeymoon is over!'

II. 14. b. (Further fig. examples.)

1932 E. M. Brent-Dyer Chalet Girls in Camp xii. 193 'I said it to make Juliet rise—and she hasn't risen!' cried Jo. **1966** Listener 6 Oct. 507/2, I should perhaps apologise for having risen to the bait of Mr Wilkinson's provocative letter. **1974** 'J. Le Carré' Tinker, Tailor i. 17 He knew they were teasing him but he was unable not to rise.

III. 15. b. to rise to the occasion (further examples).

1889 E. Dowson Let. 12 Apr. (1967) 66 Limehouse won't produce a dinner but if you can stand it 'possibly my old Sonia's larder may rise'—not 'to the occasion' but to a scratch tea. **1906** G. B. Shaw Let. 7 May (1972) II. 622 The tenor..rose to the occasion and was bully. **1922** H. Crane Let. 7 Dec. (1965) 107 This parodist and facile assessor could so gracefully rise to the occasion of a new attitude. **1952** M. Laski Village ii. 38 Sheila had risen splendidly to the occasion and offered..to wash up. **1975** Harper's & Queen May 138/2, I told my parents... It would be a mistake to say they were enthusiastic, but they do rise to the occasion.

V. 29. b. U.S. To exceed in number or amount.

1838 'B. Smith' Motley Bk. 177 Brother George counted the strokes of his arm upon the cushion, and he thinks he rose a hundred in the course of the sermon. **1877** S. O. Jewett Deephaven 133, I like well enough to see a hog that'll weigh six hundred,..but for my eatin' give me one that'll just rise three.

riser. Add: **II. 7. b.** Geomorphol. The steeply sloping part of each of the step-like parts of a glacial stairway or similar landform.

1930 F. E. Matthes Geol. Hist. Yosemite Valley 95/2 All the other steps in the upper Merced Canyon have risers and sills composed of very sparingly jointed or wholly massive rock. Ibid. 98/1 These steps..have conspicuously sheer, smooth fronts, or risers. **1954** W. D. Thornbury Princ. Geomorphol. xv. 369 Each step [in a glacial stairway] typically has three component parts: a riser, a riegel, and a tread. **1974** Encycl. Brit. Macropædia XIX. 641/1 The establishment of runoff after wastage of the ice has occurred will lead to a series of waterfalls or cascades at the site of each riser in the stairway. **1975** C. Taylor Fields in Eng. Landscape iv. 90 The most characteristic features of strip lynchets are usually their steep risers, sometimes of considerable height.

9. (Further examples.)

1950 Times Rev. Industry Sept. 16/2 (Advt.), Risers act as reservoirs of molten metal to compensate for liquid shrinkage. **1967** A. H. Cottrell Introd. Metallurgy xiii. 184 The metal is poured into a sand mould through a gate, down a sprue and along a runner. A riser has also to be provided, leading upwards from the top of the casting, to provide a pool of molten metal to feed the casting as it freezes and shrinks. **1973** J. G. Tweeddale Materials Technol. II. ii. 36 Risers are vertical channels rising to the top of the mould, which help to control flow by trapping slag and in which the rise of metal in the mould and the progress of solidification can be followed.

10. (Earlier examples.)

1821 J. Fergusson Brit. Pat. No. 4594. 18 Oct., In the process of printing from stereotype plates the plates are put upon and fastened to certain materials or apparatus called by different names, such as blocks, matrix plates, risers, &c., which are made either of iron, brass, type metal, ..gypsum, wood [etc.]. **1841** W. Savage Dict. Art of Printing 702 Risers, the material upon which stereotype plates are fixed, in order to be printed.

12. a. = rising main (b) s.v. *rising ppl. a. 6; **riser diagram,** a diagram of the risers in a building.

1898 Engin. Rec. 26 Nov. 566/1 The grouping of risers has been made..so that the average current demand on each pair of lighting bars will be about equal. **1924** T. Croft Conduit Wiring iii. 68 (caption) Riser diagram of conduit wiring in a hotel building. **1930** Moyer & Wostrel Industr. Electr. & Wiring v. 146 Wires and cables serving as main risers or feeders in buildings of fire-resistive construction, may be run bare..under special conditions. **1967** G. A. T. Burdett Electr. Installations 36 Especially suitable for short runs and for individual risers from a main distribution board on the ground floor.

b. A vertical pipe for the upward flow of a fluid; spec. (a) one carrying water or steam from one floor to another in a central heating system; (b) one extending from an offshore drilling or production platform to the sea-bed, through which drilling may be done or oil or gas may flow. Also **riser pipe(line).**

1908 A. G. King Pract. Steam & Hot Water Heating xi. 111 The riser or risers..rise directly to the top floor or attic ..and here branch in the several directions necessary to feed the various drop risers supplying the radiators. **1941** Nature 15 Mar. 315/1 The tall 'risers', that is, vertical 5-in pipes which convey the water up to the projector nozzles, are spaced throughout the plantation as required. **1961** B.S.I. News Nov. 17/1 The code will recommend the installation of fire lifts and internal fire mains or 'risers' to assist firemen in applying water to a fire as early as possible. **1969** T. Stanley More Small Bore Heating ii. 77/3 The pipework is being run under..the first floor with drops and risers to radiators on the ground floor. **1972** L. M. Harris Introd. Deepwater Floating Drilling Operations xii. 133 Selection of the riser-pipe steel is critical for a long-life, trouble-free operation. **1975** Offshore Progress: Technol. & Costs (Shell Internat. Petroleum Co.) 6 As the rig rises and falls with the heaving surface of the sea, the riser must be held in vertical tension in order to prevent it from buckling. **1977** New Yorker 9 May 38/1 Behind the riser pipe in the bathroom Puttermesser kept weeks' worth of Sunday Times crossword puzzles.

13. Cricket. A ball that rises sharply on pitching.

1955 Times 5 July 4/4 Shackleton could not draw away quickly enough from a riser from James and was acrobatically caught in the slips.

14. = lift-web s.v. *lift sb.² 18.

1927 C. A. Lindbergh We—Pilot & Plane viii. 140, I left the ship head first and was falling in this position when the risers whipped me around into an upright position and the chute opened. **1975** tr. Melchior's Sleeper Agent (1976) iii. 230 When the [parachute] canopy is fully open you will swing under it... You will check the oscillation by tugging on the two risers in the direction of your swing... He grabbed hold of two of the webbed risers. **1976** L. Sanders Hamlet Warning (1977) xxix. 277 He pulled the ring, saw the silk deploy... He fought the risers, stopping his oscillation.

rishi. Add: Also 8 Richi. (Earlier and later examples.)

1766 J. Cleland Way to Things by Words 91 The language of the antient Richi, or of the Vedams, is often hardly intelligible, even by the most skilful, who know only the Sanscort, fixed by the grammars. **1916** [see *jnana]. **1934** A. D. Waley Way & its Power App. ii. 114, I see no reason to doubt that the 'holy mountain-men' (shêng-hsien) described by Lieh Tzŭ are Indian rishi... It is at least a possibility that some knowledge of the yoga-technique which these rishi used had also drifted into China. **1939** Antiquity XIII. 15 The Aryan rishis sang their Vedic hymns. **1971** Shankar's Weekly (Delhi) 11 Apr. 22/2 Mythological stories of irate Rishis hurling curses at trespassers came to mind and every moment one feared our Rishi's wrath.

rishon (ri·ʃɒn). Particle Physics. [a. Heb. ri'šôn first, primary; cf. *-on¹.] A hypothetical particle postulated as a constituent of quarks and leptons.

1979 H. Harari in Physics Lett. LXXXVIb. 84/2 The most economical set of building blocks consists of two $J = 1/2$ objects: one charged ($Q = 1/3$) and one neutral..We denote the charged particle by T and the neutral one by V. We name these particles 'rishons'. [Note] 'Rishon' means first, primary (in Hebrew). **1979** New Scientist 5 July 22/3 The charged leptons can consist of the three charged rishons (TTT) and the neutrinos of three neutral rishons (VVV), but the fractionally-charged quarks are made up of different combinations such as TTV or VVT. **1981** Sunday Times 1 Mar. 15/2 The latest theory reduces all the matter in the universe to just two kinds of basic particles. Its originator—Haim Harari of the Weizmann Institute in Israel—has named the particles rishons, from the Hebrew for primary.

‖ **rishta** (riʃta·). Also **rishtu, reshta.** [Tadzhik.] A local name in part of the Soviet Union for the guinea-worm (Dracunculus medinensis), a common parasite of man and other mammals, and the disease which it produces.

1834 A. Burnes Trav. Bokhara II. 180 Among the diseases of Bokhara, the most distressing is the guinea-worm, or Dracunculus, here called 'rishtu': it is confined to the city. The inhabitants believe that the disease arises from drinking the water of the cisterns in summer, when they become fetid and infested with animalculæ. **1885** H. Lansdell Russ. Central Asia II. liii. 146 What I had seen and been warned of in Samarkand, made me specially fearful of the rishta, a well-known disease in Bokhara. **1923** Chambers's Jrnl. July 472/2 The barber-surgeon trying to operate on a young man for 'rishta', that troublesome parasite which burrows into the skin. **1966** E. Schuyler Turkistan iv. 82 The Sarts are not only attacked by the usual maladies.. but they have besides two or three which are peculiar to the country... One of these is the reshta, or 'Guinea-worm'.

rising, vbl. sb. Add: **II. 8.** Also const. up.

1942 W. S. Churchill End of Beginning (1943) 145 People very often fall by the very means which they have used and built their hopes upon for their rising-up!

11. b. (Further U.S. examples.)

1834 W. Sewall Diary 7 Dec. (1930) 160/1 Laid up with a bad rising on my hand. **1867** A. D. Richardson Beyond Mississippi xi. 133 He spoke of a swelling upon his knee as a 'rising'. **1938** M. K. Rawlings Yearling xix. 236 None of us ain't got risin's. **1949** T. Capote Other Voices v. 104, I had me a rising on my butt big as a baseball. **1972** E. Wigginton Foxfire Bk. 244 Scrape the white of an Irish potato and place the scrapings on the bump. Bind them on with a clean cloth. This will draw the risin' (boil) to a head.

15. a. Also Comb., as (salt-)rising bread (N. Amer.).

1865 Mrs. Stowe House & Home Papers 236 Salt-rising bread. **1882** G. M. Barbour Florida iii. 56 The feast of hog, hominy, beef..and likely a few villainous compounds of flour, cheapest brown sugar, or sirup, and called cake or 'risin'-bread'. **1933** Sun (Baltimore) 3 Feb. 10/7 The Western correspondent..is talking about a foodstuff that resembles salt-rising bread..about as much as lady fingers resemble Russian black bread... Only a slight quantity of corn meal is used in the preparation of salt-rising bread. **1960** J. J. Rowlands Spindrift 172 The meat..was flanked by plates of moist and closely knit salt-rising bread. **1973** L. Russell Everyday Life Colonial Canada viii. 96 'Salt-rising' bread was made without benefit of yeast.

rising, ppl. a. Add: **4. c.** (Further examples.) Also, characterized by increase in vocal stress or a rise in pitch. Also Comb., as rising-falling.

1881 G. M. Hopkins Lett. to R. Bridges (1955) 40, I call rising rhythm that in which the slack comes first, as in iambs and anapests, falling that in which the stress comes first, as in trochees and dactyls. **1894** H. Sweet Anglo-Saxon Reader (ed. 7) p. xciv, There is a tendency to combine different types in a line, the falling types A and D being most frequent in I, while in II the rising types B and C are preferred. **1931** G. Noël-Armfield Gen. Phonetics (ed. 4) xiii. 69 These [signs] may be combined to showing falling-rising, rising-falling, and so forth. **1955** Archivum Linguisticum VII. 155 The Greek circumflex is not, essentially, a rising-falling accent. **1964** R. H. Robins Gen. Linguistics iii. 111 Tones may..rise or fall, or rise and fall, or fall and rise (rising, falling, rising-falling, falling-rising tones, respectively), and be distinguished by the actual direction in which the pitch moves. **1973** Archivum Linguisticum IV. 19 Typical sequences of tones..in which a final falling tone is preceded by a rising tone.

6. rising damp, moisture absorbed from the ground into a wall; **rising diphthong** Phonetics, a diphthong in which the final vowel is more prominent; **rising main,** (b) an electricity main passing from one floor of a building to another; **rising sun:** see Sun sb. 2 a.

1956 W. A. G. Bradman Taking Care of Your Home iv. 61 Rising damp..is invariably characterized by a line of dampness appearing above the skirtings. **1975** Times 30

Oct. 6/5 The walls had been sodden with rising damp for years. **1888** H. Sweet Hist. Eng. Sounds (ed. 2) 9 A 'rising'..diphthong. **1892** J. Wright Primer Gothic Lang. viii. 43 A diphthong may be defined as the combination of a sonantal with a consonantal vowel. And it is called a falling or rising diphthong according as the stress is upon the first or second element. **1960** P. H. Reaney Orig. Eng. Place-Names 45 In Devon, OE ēa frequently became a rising diphthong in ME and survives with initial y: Yalland, Yelland. **1940** Chambers's Techn. Dict. 727/2 Rising mains, in an electrical installation, a mains circuit which runs from one floor of a building to another. **1967** G. A. T. Burdett Electr. Installations 37 Where conditions allow there are advantages in using purpose-made rising mains.

rising, pr. pple. Add: **3. a.** (Earlier example.)

1837 W. Jenkins Ohio Gazetteer 64 It enjoys a yearly income of rising $4,500.

risk, sb. Add: **1. d.** at (or † in) risk, at high (etc.) risk: in danger, subject to hazard. Also as adj. (See also sense *2 c.)

1901 'L. Malet' Counsel of Perfection xi. 243 Whether the capital owned by his better nature was not in risk of being exhausted—whether the drafts made on it might not eventually be dishonoured. **1965** New Statesman 10 Dec. 951/2 (Advt.), The appointment should be of interest to those who are prepared to assist in training child care officers and actively supervising casework of 'at risk' families. **1966** Listener 10 Feb. 199/1 It is necessary to know both the number of legitimate children born to women in this age-group and the number of married women at risk. **1972** Daily Tel. 5 May 6/8 Eight thousand historic churches in England are at risk through damage and decay. **1973** Sci. Amer. July 20/2 Women who were at high risk of bearing retarded infants. **1977** National Trust Spring 9/1 Soon nearly half our elms will be dead and the remainder all at risk. **1977** Lancet 23 July 203/1 The baby was considered to be at high risk.

e. A person who is considered a liability or danger; one who is exposed to hazard. (Freq. with qualifying word.)

1948 [see *clearance 5 c]. **1954** Manch. Guardian Weekly 22 Apr. 3/1 A loyalty risk is a man whose paramount allegiance to the United States is in doubt. A security risk is one who may be consciously the most adamant patriot but whose judgment or tactlessness may cause him to make decisions or disclose information that could harm the national security. **1961** Lancet 12 Aug. 328/2 That patients classified as 'poor risks' according to Russek's criteria..show a higher mortality-rate is no cause for wonder. **1976** W. Greatorex Crossover 32 He was frozen out... He was treated as a security risk..but it didn't bother him.

2. a. Also (freq. without article), the chance that is accepted in economic enterprise and considered the source of (an entrepreneur's) profit. all risks: see *all a. E. 13. Cf. *uncertainty.

1776 Adam Smith Wealth of Nations I. i. x. 136 The ordinary rate of profit always rises more or less with the risk. **1848** Mill Pol. Econ. I. ii. xv. 479 The difference between the interest and the gross profit remunerates the exertions and risks of the undertaker. **1921** F. H. Knight Risk, Uncertainty, & Profit ii. 41 The doctrine that profit is to be explained exclusively in terms of risk has been vigorously upheld. **1944** A. Cairncross Introd. Econ. vi. 76 The more fickle the demand, either from one season to another, or from year to year, the stronger will be the tendency to spread risks and steady production by diversifying output. **1977** B. Benjamin Gen. Insurance xi. 271 The mathematics of risk theory and of model building do not at present cover these kinds of business risks other than by incorporating past investment experience.

c. Law. In phr. at (one's, etc.) risk, of merchandise, etc.: at the liability of a stated party. Also of persons, liable to repay loss or damage.

1798 [in Dict., sense 2 a.] **1887** Law Rep. Queen's Bench Div. XVIII. 65 The expression 'at ship's risk' cannot be strictly correct, because the ship has no risk. **1970** New Society 5 Feb. 209/1 He therefore should be at risk where the car causes damage.

3. attrib. and Comb., as **risk aversion, -bearing, category, factor, level, management, -taker, -taking; risk-free** adj.; **risk analysis,** the systematic investigation and forecasting of risks in business and commerce; similarly **risk–benefit analysis; risk capital,** money that is put up for speculative business investment; **risk money,** (b) = risk capital above; **risk profile,** a forecast of the probable range of hazards in an enterprise; **risk-rate,** a rate of interest related to a degree of hazard in an enterprise.

1964 Harvard Business Rev. Jan.–Feb. 95 (heading) Risk analysis in capital investment. **1977** R. E. Megill Introd. Risk Analysis xvi. 173 In the search for new oil and gas fields, risk analysis takes the judgments of explorationists and engineers and translates them into the language of probability. Risk analysis, thus, helps a manager make reasonable decisions. **1964** W. S. Vickrey Metastatics & Macroecon. v. 88 The differential between long- and short-term interest rates thus requires both liquidity preference and risk aversion to sustain it. **1972** Accountant 21 Sept. 349/2 Any application of probability theories to decision-making must have regard to the susceptibilities of the decision-maker, his own attitudes and those of his corporation—in particular, their 'risk aversion factor'. **1931** Economist 21 Nov. 957/1 The establishment of confidence, which would make possible a reduction of the premium on risk-bearing. **1958** Times Lit. Suppl. 25 July 426/4 In discussing the reward of risk-bearing the author refers to the special information that some may have and to the riches of others, thus finding that 'risk-taking surpluses' accrue. **1975** Physics Bull. May 203/2 One fears that until recently no such risk-benefit analysis would have been attempted. **1976** Conservation News Nov.–Dec. 3/1 On risk/benefit analysis, where many people hold that value judgements are involved, common ground seems impossible to find. **1948** Sun (Baltimore) 7 Apr. 19/1 He contended that newly saved risk capital in 1946 and 1947 supplied only $700,000,000 of the $50,500,000,000 of new money required by American industry. **1962** Economist 19 May 693/1 'Incentives' have not been dulled and 'risk capital' has not dried up. **1976** F. Zweig New Acquisitive Society i. vii. 69 The old acquisitiveness provided long-term risk-capital for industrial development. **1973** Sci. Amer. Sept. 65/1 Women were assigned to one of four risk categories. **1971** Brit. Med. Bull. XXVII. 23/2 Does the prevalence of individual risk factors..differ between soft-water and hard-water areas? **1950** Mind LIX. 126 The 'reactionaries' are those who believe that scientific enquiries can proceed from risk-free observational records immune from statistical tests. **1980** Sci. Amer. Mar. 33/3 People do not seek a risk-free society, but they do find it hard to manage risks that are not fully understood. **1962** A. Battersby Guide to Stock Control iii. 28 We already have two possible figures... Why are these figures different from each other? Because they correspond to different risk levels. **1970** New Scientist 15 Jan. 93/2 The risk-level of the former can be detected at interview. **1963** Mehr & Hedges (title) Risk management in the business enterprise. **1978** Financial Rev. (Austral.) 27 July 25/3 Risk Management, whatever you call it, is part and parcel of a big Insurance Broker's business. **1944** H. A. Wallace Century of Common Man xiv. 70 A business man ought to be able to get his 'risk money' back before he has to pay too much in the way of taxation. **1969** Daily Tel. 29 Dec. 12/2 The oilmen are also aware that if the Gas Council is pushed further into exploration the Government will have to provide 'risk money' on a large scale. **1969** J. Argenti Managem. Techniques 233 When the forecast is made, an estimate of the probability of the range of errors is also made and this is used to calculate the 'risk profile' of the project. **1928** Britain's Industr. Future (Liberal Industr. Inquiry) iii. xvi. 187 Capital will not be forthcoming for any enterprise unless it can expect (a) a normal rate of interest..and (b) in addition to that a 'risk-rate' corresponding to the chance of loss in the particular business; and this risk-rate must vary according to the conditions of every industry and of every concern. **1944** R. Lehmann Ballad & Source iii. 148 That is just a phrase the petty-cautious use against the fiery ones, the risk-takers. **1957** A. C. L. Day Outl. Monetary Econ. xxxviii. 491 Willing risk-takers (e.g. settlers in new countries). **1979** N.Y. Rev. Bks. 25 Oct. 49/2 McCagg's attempt is evidently to reconcile the sober war-time Stalin..with the postwar risk-taker. **1921** F. H. Knight Risk, Uncertainty, & Profit ii. 46 If risk were exclusively of the nature of a known chance or mathematical probability, there could be no reward in risk-taking. **1936** J. M. Keynes Gen. Theory Employment xxiv. 372 Diminishing unduly the motive towards risk-taking. **1948** Sun (Baltimore) 7 Apr. 19/1 Mr. Hooper..explained that eager risk taking by individual investors is essential to the smooth operation of the free enterprise system. **1975** 'E. Lathen' By Hook or by Crook xiv. 132 'He's always taken big chances.'.. 'That's the risk-taking his children are alarmed about.'

riskiness. Add: **2.** The quality of being risqué. Cf. Risky a. 3.

1877 Argus (Melbourne) 5 Dec. 6/1 Mr. Albery..has so far brought it [sc. a play] to the level of English requirements that we have it now with its original humour, but freed from its accompanying riskiness. **1938** H. Granville-Barker Quality 5 What rule would he [sc. Gilbert Murray] make for distinguishing a play written to exploit the mere riskiness of a subject and such a one as Ibsen's Ghosts.

risky, a. Add: **1. b.** Social Psychol. Phr. risky shift: in decision-making, the shift of opinion towards an option involving greater risk that may take place when responsibility for the decision rests with a group rather than an individual.

1964 M. Wallach et al. in Jrnl. Abnormal & Social Psychol. LXVIII. 272/1 Group responsibility in the presence of group decision lead to a strong risky shift. **1967** Kogan & Wallach in New Directions in Psychol. III. 240 Risky shifts took place with high regularity for groups of both sexes. **1972** M. Argyle Social Psychol. of Work vi. 133 Whatever its causes the risky shift is clearly a source of unwise decisions in groups. **1978** Lamm & Myers in Adv. in Exper. Social Psychol. XI. 149 Some researchers have used a rating scale (degree of preference for the risky vs. the cautious course of action). By and large, risky shift is obtained on items which elicit risky individual responses.

Risley (ri·zli). The name of Richard Risley Carlisle (d. 1874), U.S. gymnast and circus performer, used attrib. (and absol.) to designate an act in which a supine acrobat juggles another with his feet, as Risley act, business, etc. Also transf.

[**1843** N.Y. Herald 4 May 2/2 One of the chief attractions ..was Prof. Risley and his boy. **1846** Illustr. London News 7 Feb. 101/1 The very clever performances of Mr. Risley and his two sons continue to be nightly received with loudest acclamations.] **1861** H. Mayhew London Labour III. 94/2 (heading) The Street Risley. Ibid., There is but one person in London who goes about the street doing what is termed 'The Risley performance'. Ibid. 97/1 We've been continuing ever since at this Risley business. I lay down on a carpet, and throw then summersets from feet to feet. Ibid. 98/1 I've done the Risley in the streets of London. **1901** Cassell's Mag. Sept. 389/1 (caption) A Risley Pose. Ibid. 389/2 There will be a day when a Japanese is an underman and a young Westerner a top-mounter, or vice versâ; and we shall see a brilliant show of Risley act or juggling. **1912** M. B. Leavitt 50 Yrs. Theatr. Managem. xxv. 381 The second European 'hit' was made by Risley and his two sons, presenting the tossing and tumbling of the youngsters, to this day called the 'Risley Act'. **1931** Amer. Mercury Nov. 353/2 Risley act, one in which three acrobats lie on their backs and toss a fourth from one to the other. **1938** N. Streatfeild Circus is Coming vii. 117 He told them about the first Risley who had the idea of juggling with a real boy. Of how the idea caught on, and that kind of performance was always known as a 'Risley act'. **1957** J. & A. Durant Pict. Hist. Amer. Circus ii. 20 The 'Risley Act' (balancing with the feet while lying on the back) was performed by the Aztecs.

‖ **risoluto** (rizolū·to), a. and adv. Mus. [It.] † **A.** adj. (See quots.) Cf. Resolved ppl. a. 7. Obs.

1740 J. Grassineau Mus. Dict. 19 Canone partito, or risoluto,..is when all the parts of a perpetual fugue are writ either in partitions, or different lines, or in separate parts, with the proper pauses that each is to observe, and therein differs from Canone Chiuso. Ibid. 204 Risoluto, resolved: thus we say a syncoped discord is resolved.

B. adv. With resolution or emphasis. (Used as a direction on a musical score.)

1837 J. A. Hamilton Dict. Mus. Terms (ed. 4), Risoluto (Italian), with boldness and resolution. **1876** Stainer & Barrett Dict. Mus. Terms 379/2 Risoluto,..with resolution. **1976** Gramophone Feb. 1337/3 Harrell enters with a confidence and firmness that may contradict the overall marking quasi improvisando, but which faithfully reflects the more specific risoluto.

Risorgimento (risǭːɹdʒime·nto). Also **risorgimento.** [It., = renewal, renaissance.] **1.** The movement which led to the unification of Italy as an independent state with its capital at Rome in 1870.

1889 J. A. R. Marriott Makers Mod. Italy ii. 38 In 1847 Cavour, in conjunction with Santa Rosa, Cesare Balbo, and others, founded a new journal, named the Risorgimento, for the purpose of disseminating constitutional ideas of government. **1902** Encycl. Brit. XXIX. 628/2 Few dates in modern European history equal in significance that of 20th September 1870, when the Italian troops under General Cadorna took possession of Rome in the name of the Italian nation, and completed at one stroke..the work of the Risorgimento. **1902** G. Meredith Let. 15 Apr. (1970) III. 1436 Mazzini.. never wavered in the faith he had that their sacrifices would lead to the Risorgimento. **1905** Mrs. H. Ward Marriage of W. Ashe iv. xviii. 363 He had sat late with his hosts,—men prominent in the Risorgimento, and in the politics of the new Kingdom,—discussing the latest intricacies of the Roman situation and the prospects of Italian finance. **1910** W. H. Griffin Life Robert Browning x. 158 The events of the risorgimento. **1933** N. & Q. 4 Mar. 161/2 The Risorgimento was..kept free from that violence of popular fury.. which touches both the French and the Russian re-making of the nation with a sort of devilishness. **1937** A. Huxley Ends & Means x. 155 Before the Risorgimento the Austrians governed Italy by means of gendarmes, spies and agents provocateurs. **1955** Times 6 June 8/7 Count Alessandro Casati, one of Italy's elder statesmen and a true representative of the Italian liberal tradition of the Risorgimento, died on Saturday night. **1957** Sunday Times 8 Dec. 7/6 The Risorgimento heroes. **1961** Listener 19 Oct. 611/1 The partisan movement [in Italy] and the second Risorgimento which followed. **1977** New Yorker 2 May 101/1 Prosperity united Italy in ways the risorgimento never had.

2. transf. A revitalization or renewal of activity in any sphere.

1957 R. Chase in Partisan Rev. Summer 369 The poetic risorgimento of Ezra Pound and his group. **1959** Listener 17 Dec. 1062/2 This has been more than an economic and industrial risorgimento. **1978** Ld. Birkenhead R. Kipling xv. 215 The lethargy and frivolity of his own countrymen, upon whom he now turned..his indignation in a passionate but vain attempt to inspire a risorgimento.

risorius (ris-, rizǭ·riǫs). Anat. [ellipt. for mod.L. musculus risorius (J. D. Santorini Observationes Anatomicæ (1724) i. 33), f. L. musculus muscle + risor laugher (f. ridēre to laugh) + -ius, adj. suffix.] A muscle of facial expression running from the parotid fascia to the corner of the mouth, variable in form and sometimes lacking.

1829 J. & C. Bell Anat. & Physiol. Human Body (ed. 7) I. 279 [The Platysma myoides] terminates on the face and jaw. Some of its fibres, mounting over the bone of the jaw, are inserted near the depressor anguli oris; and others, a little higher on the face, are called risorius santorini. **1867** W. Sharpey et al. Quain's Elem. Anat. (ed. 7) I. 176 The risorius or smiling muscle (Santorini), consisting of some very thin fasciculi,..joins the orbicularis and depressor anguli oris at the angle of the mouth. **1902** D. J. Cunningham Text-Bk. Anat. 377 The risorius is a thin flat muscle which forms partly a continuation of the platysma myoides on the face, partly a separate muscle, with an origin from the masseteric fascia. **1936** A. Huxley Eyeless in Gaza xviii. 233 The whole mechanism of the excruciating grimace, the upward and outward pull of the zygomaticus major, the sideways tug of the risorius. **1961** L. F. Brosnahan Sounds of Language iv. 80 This suggests again a cline in gene frequency, the decrease in the mean height running in the same direction through the African-European-Asiatic population as the increase in the frequency of the occurrence of the risorius muscle in the face.

risotto. Delete ‖ and add: Now usu. with pronunc. (rizǫ·to). (Earlier and later examples.) Also Comb.

1855 E. Acton *Mod. Cookery* (rev. ed.) xxxii. 615 (*heading*) Risotto à la Milanaise. 1950 E. Linklater *Mr. Byculla* vi. 68 A Risotto Bolognese for one and threepence. 1960 *Housewife* May 91/1 My deep freeze usually contains..a risotto or two. 1981 E. Dewhurst *Trio in Three Flats* iii. 19 Cathy's asked me to supper..risotto and a complicated salad.

‖ **risqué** (ri·ske, ri·ske), *a.* Also **risque.** [Fr., pa. pple. of *risquer* Risk *v.*] = Risky *a.* 3.
1867 'Ouida' *Under Two Flags* II. iv. 121 She..sang..the most wicked and *risqué* of her slang songs. *Ibid.* viii. 207 It was..too simple, too little *risqué*. A child might do it. 1894 A. Beardsley *Let.* c 3 Jan. (1970) 61 Our idea [in starting *The Yellow Book*] is that many brilliant story painters and picture writers cannot get their best stuff accepted in the conventional magazine, either because they are not topical or perhaps a little risqué. 1899 J. London *Let.* 30 Mar. (1966) 26 Our magazines are so goody-goody, that I wonder they would permit a thing as risque and as good as that. 1913 E. Pound *Let.* 30 Mar. (1971) 18 Again to your note: 'Risqué.' Now really!!! Do you apply that term to all nude statuary? I admit the verse 'To Another Man on his Wife' might deserve it, but you're not including that. Surely you don't regard the Elizabethans as 'risqué'? 1924 *Brit. Weekly* 18 Dec. 301/3 In remote corners others are reading *risqué* novels with a sex interest. 1952 *Scrutiny* XVIII. 317 Not infrequently Mr. Auden seems to be trying to atone for this by indulging in slightly *risqués* side-glances. 1962 V. Connaught *Secret Heart of Princess Alexandra* i. 15 She enjoys a good joke, even if it is a little risqué, provided it is well told. 1975 *Country Life* 25 Dec. 1804/1 Directors were turning out sophisticated, risqué social comedies.

Riss (ris). *Geol.* The name of a tributary of the Isar in Austria and Germany, adopted by A. Penck (in Penck & Brückner *Die Alpen im Eiszeitalter* (1909) I. I. 110) and used *attrib.* to designate the third (penultimate) Pleistocene glaciation in the Alps, and in conjunction with *Würm to designate the following interglacial period. Also *absol.*
1910 *Zeitschr. f. Gletscherkunde* IV. 244 The interval between the Riss and the Würm.., the Riss–Würm interglacial stage. *Ibid.*, The moderate erosion..which the Riss drift has experienced. 1927 [see *Mousterian, Moustierian *a.* and *sb.*]. 1931 *Discovery* Sept. 282/2 The fluvio-lacustrine deposits, rich in volcanic elements, of the Tiber Valley, which Roman geologists hold to be contemporary with the warm interglacial period intervening between the Riss and the Würm glaciations of the European Ice Age. 1944 A. Holmes *Princ. Physical Geol.* xii. 247 The depth reached by weathering during Riss-Würm time is found to be about three times that achieved on similar but later deposits exposed during post-Würm time. Since the latter is about 25,000 years, it follows that the Riss-Würm interval cannot have been less than 75,000 years. 1968 R. W. Fairbridge *Encycl. Geomorphol.* 335/1 There appears to be unanimity that during the Last Interglacial (Riss-Würm or Sangamon) there were sea-levels of the order of 7·5 meters..and 3·5 meters..higher than the present. 1971 *Nature* 22 Jan. 253/2 It is now recognized that the species *Homo sapiens* is of at least Riss age. 1974 *Sci. Amer.* June 96/1 The next to last major advance, known to scholars as the Penultimate Glacial (Riss II in the Alpine sequence), marked the end of the Middle Pleistocene some 125,000 years ago.

rissaldar (risāldā·ɹ). *Indian.* Also **risal(a)dar** and with capital initial. [Hind. *risāldār*, *risālādār*, f. Pers. *risāla* troop of horse.] Now the more usual form of Ressaldar. Also *Comb.*, as *rissaldar-major*.
1842 W. Miles tr. *Meer Hussein Ali Khan Kirmani's Hist. Hydur Naik* xxiii. 327 The Nawaub now gave orders to the Risaladárs of the regular and irregular infantry, to encircle the fort, and then commence the attack with their artillery and musketry. 1848 J. H. Stocqueler *Oriental Interpreter* 198/1 *Rissaldar*, an officer of the Irregular India cavalry, whose rank corresponds with that of a captain of a troop. 1851 J. B. Fraser *Mil. Mem. Lieut.-Col. J. Skinner* I. ix. 274 The rissaldars finding so much money in their hands, began to quarrel about the division of it. 1927 Rissaldar-major [see *en grande tenue* s.v. *en *prep.*] 1951 J. Masters *Nightrunners of Bengal* 337/2 There were three grades of Native Officers..: infantry—subadar-major, subadar, jemadar; cavalry—rissaldar-major, rissaldar, jemadar. 1964 A. Swinson *Six Minutes to Sunset* ii. 30 A Risaldar-Major with the impressive name of Khan Bahardur Fazal Dad Khan turned up. 1981 V. Powell *Flora Annie Steel* xiii. 109 Roshan has risen to the rank of *rissaldar* (Indian officer).

Risso (ri·so). The name of Giovanni Antonio *Risso* (1777–1845), Italian naturalist, used in the possessive in **Risso's dolphin** to designate the grampus, *Grampus griseus*, first described as *Delphinus rissoanus* by A. G. Desmarest in 1822 (*Mammalogie* II. 579).
1871 *Proc. Zool. Soc.* 506 Prof. Flower, F.R.S., read a paper on the so-called Risso's Dolphin. 1924 [see *caa'ing whale]. 1927 *Daily Express* 5 Sept. 2/5 Pelorus Jack, the famous white Risso's dolphin, which for years has piloted ships into Wellington Harbour, New Zealand,..is missing. 1960 G. Maxwell *Ring of Bright Water* v. 62 Contrary to information contained in the majority of text-books, in which Risso's dolphin is described as a rarity, it is in fact the commonest of all the lesser whales to visit the Hebrides in summer. 1971 M. & R. Moffett *Dolphins* 18 The grampus, or Risso's dolphin, lives far out at sea.

ristocetin (ristosi·tin). *Pharm.* [Arbitrarily formed; *-cetin* f. *actinomy)cet(es + -in¹.] An antibiotic substance (now known to be a mixture) obtained from the actinomycete *Nocardia lurida* and formerly used to treat staphylococcal infections.
1957 W. E. Grundy et al. in *Antibiotics Ann. 1956–7* 687 A new antibiotic that has been given the generic name ristocetin was isolated from the fermentation beer of an actinomycete. 1960 M. E. Florey *Clin. Appl. Antibiotics* IV. iii. 77 Isolated from the fermentation liquor of an unidentified actinomycete called by the authors *Nocardia lurida* and obtained from soil at Colorado Springs, ristocetin gives promise of being of value in the treatment of staphylococcal and other infections. 1968 J. H. Burn *Lect. Notes Pharmacol.* (ed. 9) 111 Vancomycin and ristocetin do not appear to induce the emergence of resistant staphylococci. 1980 *Brit. Med. Jrnl.* 18 Oct. 1039/1 The coagulation studies were diagnostic of Glanzmann's thrombasthenia with diminished platelet aggregation in response to ristocetin.

‖ **ristorante** (ristŏra·nte). [It.] An Italian restaurant in Italy or elsewhere.
1925 W. J. Locke *Great Pandolfo* iv. 40 Washer-up in a relative's *Ristorante* in a mildewed corner of Soho. 1967 *Observer* 24 Sept. 31/5 Stopping at the first *ristorante* for a drink. 1968 R. Sawkins *Snow along Border* xviii. 148 The Villa Messina was near a *ristorante* called 'Sud-Est'. 1973 'S. Harvester' *Corner of Playground* II. iv. 102 A table outside a *ristorante* on a Rome side-street. 1981 'M. Yorke' *Hand of Death* xii. 102 The Ristorante Sorrento was in a narrow street that led to Fletcham Abbey from the market square.

† **rit** (rit), *sb.*² Slang. abbrev. of Ritualist 2. *Obs.*
1878 *Oxf. Times* 23 Mar. 8/1 *On dit* that five notoriously ritualistic undergraduates were received into the Church of Rome... Most of these young gentlemen have distinguished themselves in Oxford as Roaring Rits. 1898 A. Cavalier *Let.* in C. Mackenzie *My Life & Times* (1963) II. 243 My mater calls me a 'dirty Rit'. 'So we've got a Rit in the house, have we.' 1909 J. R. Ware *Passing Eng.* 209/2 *Rit*,.. a ritualistic clergyman.

rit, *sb.*³ *dial.* [Shortened form of Ritling: see Reckling.] The smallest and weakest pig of a litter; a ritling. Also *transf.* of a person.
1885 R. Holland *Gloss. County of Chester* (1886) 288 *Rit*.., the smallest pig of a litter. Also applied to a puny child. 1940 *Manch. Guardian Weekly* 15 Mar. 216 We gave special food to the rit of one brood (the little one known in other parts [than Cheshire] as the runt or reckling, and by other names). 1962 Orton & Halliday *Survey Eng. Dial.* I. 1. 279 Q[uestion] What do you call the smallest and weakest pig of the litter?.. La[ncashire]..ɹit. 1969 Orton & Barry *Ibid.* II. 1. 288 Ch[eshire] ɹit... St[affordshire].. ɹit.

rit., *sb.*⁴ *Mus.* Abbrev. of *ritardando.
1886 [see *a tempo]. 1959 *Collins Mus. Encycl.* 552/1 *Ritardando*,..commonly abbreviated *rit.*

Ritalin (ri·tălin). *Pharm.* Also **ritalin.** A proprietary name (orig. used in Switzerland) for the drug methylphenidate hydrochloride, a central nervous system stimulant related to amphetamine; methyl-α-phenyl-α-piperid-2-ylacetate hydrochloride, $C_{14}H_{19}NO_2 \cdot HCl$.
The appearance of the word in trade-mark literature predated the introduction of the drug by a number of years.
1949 *Trade Marks Jrnl.* 1 June 477/1 *Ritalin*... Pharmaceutical preparations for use in the treatment of the nervous system. Ciba Limited..manufacturers and merchants.—28th May 1948. 1949 *Official Gaz.* (U.S. Patent Office) 16 Aug. 594/2 Ciba Limited, Basel, Switzerland. Filed July 9, 1948. *Ritalin.* Applicant claims ownership of Swiss Registration No. 107,093, dated Mar. 28, 1944. 1954 *Chem. Abstr.* XLVIII. 8945 Ritalin..has a psychomotor excitatory effect, producing psychic stimulation and a coördinated increase in motility in exptl. animals. 1955 *Ann. N.Y. Acad. Sci.* LXI. 101 This new drug, phenidylate—phenyl-(α-piperidyl)-acetic acid methyl ester (Ritalin)—is a synthetic preparation which could be described as a psycho-analeptic, a mental and physical stimulator which..seems to counterbalance the reserpine-depressing activity. 1970 *Daily Tel.* 30 June 3/1 The United States Food and Drug Administration has urged doctors to exercise extreme caution in prescribing Ritalin because of the danger of addiction. *Ibid.* 3 Dec. 13 More than 200 youngsters regularly visited his surgery and home and were given National Health Service prescriptions for the 'soft' drugs drynamyl, ritalin and mandrax. 1976 H. Ferguson *Confessions Long Distance Acid Head* 7 Apart from cannabis, I have used barbiturates,..procaine, ritalin, even apomorphine once.

‖ **ritardando** (ritaɹdæ·ndo). *Mus.* Pl. **ritardandi, -os.** [It., gerund of *ritardare* to slow down.] A musical direction indicating a gradual reduction of speed; as *sb.* = Retardation 3 b; a passage where this occurs.
1811 Busby *Dict. Mus.* (ed. 3) Ritardando. (Ital.) An expression implying a slackening of the time. c 1865 E. Dickinson *Poems* (1955) II. 724 Dying at my music!.. Hold me till the Octave's run! Quick! Burst the Windows! Ritardando! 1889, etc. [see *accelerando]. 1893, etc. [see *allargando]. 1958 *Times* 22 Nov. 9/3 He is inclined to begin his ritornelli faster than is comfortable to the soloists and at the end to draw out an absurd *ritardando*. 1966 *Listener* 25 Aug. 286/1 This constant variation can be heard on many levels; in the brilliant interplay of pure unmixed tone-colours, in the fluctuations of tempo between fast and slow, accelerando and ritardando, and, in particular, in the rhythm. 1978 *Jrnl. R. Soc. Arts* CXXVI. 356/1 There must always be a reason and purpose for these slight accelerandos and ritardandos which we call rubato.

rite. Add: **1. a.** *the last rites* = *the last sacraments* s.v. *Sacrament *sb.* 2 e; *rite A, B*: the two classes of Eucharistic rite in the Church of England's *Alternative Service Book 1980*, distinguished by being in present-day English and traditional liturgical English, respectively.
[1786 Burns *Poems Sc. Dial.* 190 The last, sad, mournful rites bestow!] 1922 C. Kerr *Cecil Marchioness of Lothian* xv. 228 Dr Talbot said Mass in her room and she was given the last rites of the Church. 1927 *Times* 11 July 14/4 Canon Breen, the local parish priest, was hurriedly brought and administered the last rites. 1961 P. J. Hepburne-Scott tr. J. C. Didier (*title*) The last rites. 1975 *Times* 8 Nov. 1/7 The cardinal..received the sacrament of the sick (previously called 'last rites'). 1977 *Belfast Tel.* 27 Jan. 9/5 Their call is being backed up by local priest, the Rev. Peter Burns, who gave Mr. Moyna the Last Rites. 1980 *Daily Tel.* 24 Oct. 3/1 (*heading*) Queen [to be] at new Synod Rite A service. *Ibid.*, The form of service will be Rite A from the new Alternative Service Book... The service is a revision of that known hitherto as Series 3. 1980 *Alternative Service Bk. 1980* 5 The Order for Holy Communion Rite A. *Ibid.* 6 The Order for Holy Communion Rite B. 1980 *Alternative Service Bk. 1980: Commentary* 74 The main further change made from Series 3 to Rite A is the separation of the commemoration of the departed from the summary sentences commending all the worshippers to God. 1980 *Times Lit. Suppl.* 14 Nov. 1281/2 Even the Lord's Prayer is now on sale in three versions—that of the Book of Common Prayer..; that of Rite A *et passim* and that of Rite B *et passim*.
d. *pl.* Used as a journalistic term for any ceremony (*U.S.*).
1950 *Richmond* (Va.) *Times-Dispatch* 1 Apr. 1/2 (*heading*) Airport rites set for 2 p.m. today at Byrd. Planes, personalities will mark ceremony. 1957 *Sun* (Baltimore) 21 Jan. B-1/6 (*heading*) Rites at White House performed before 80. Press is barred from ceremony in East Room.
e. Anthrop. *rite of intensification*: a rite marking a special event affecting a social group and tending towards strengthening the bonds uniting its members; usu. *pl.*; *rite of passage* = *Rite de passage.
1909 *Folk-Lore* XX. 510 What M. van Gennep has here done is to enforce his contention by considering..a number of the sequences of rites to which he has given the title of Rites of Passage. 1947 Chapple & Coon *Princ. Anthropol.* xxi. 507 A Rite of Intensification..restores equilibrium for the group after a disturbance affecting all or most of its members. 1959 W. Goldschmidt *Understanding Human Society* v. 178 Rituals involving the whole community, called rites of intensification, re-enforce the initiate's sense of belonging and serve to strengthen group ties. 1960 Vizedom & Caffee *Van Gennep's Rites of Passage* p. vii, Passage might more appropriately have been translated as 'transition', but in deference to van Gennep and general usage of the term 'rites of passage', this form of the translation has been preserved. *Ibid.* p. ix, Ceremonies which accompany and assure the changes of the year, season, or month are rites of passage. 1970 P. Spencer in P. Mayer *Socialization* 148 The second type, performed when misbehaviour was expected from the moran and when their corporate unity and morale were low, were rites of intensification pure and simple. 1971 K. Thomas *Relig. & Decline of Magic* iii. 57 The subsequent raising of the age at which children are expected to undergo it [*sc.* confirmation] to fourteen or so has given it a more pronounced role as a rite of passage marking the arrival of 'social' puberty. 1978 W. A. Haviland *Cultural Anthropol.* (ed. 2) xiii. 346/1 Rites of intensification..are particularly common among horticultural and agricultural people, with their planting, first fruit, and harvest ceremonies. 1978 *Times Educ. Suppl.* 13 Jan. 15/1 The transition from fifth-year courses to A level involves a rite of passage to 'real' history 1978 *Chatelaine* Dec. 17/1 And hockey is not just a sport. It's a rite of passage.

‖ **rite de passage** (rit də pasāʒ). *Anthrop.* Pl. **rites de passage.** [Fr., lit. 'rite of passage', a term coined by Arnold van Gennep: see quot. 1909.] Any of the rites of separation, transition, and incorporation that mark an individual's social existence from birth to death as he passes from one stage of life to another; ritual that marks the end of one phase and the start of another. Cf. *Rite 1 e.
[1909 A. van Gennep (*title*) Les rites de passage.] 1911 *Man* XI. 30 Should we be right in including many of the cases of *rites de passage* in a general category of rites de *première fois*? 1934 R. Benedict *Patterns of Culture* ii. 25 In order to understand puberty institutions, we do not most need analyses of the necessary nature of *rites de passage*. 1949 G. Bateson in M. Fortes *Social Structure* 45 A poor man was about to undergo one of the important and expensive *rites de passage* which are necessary for persons as they approach the top of the Council hierarchy. 1957 M. Banton *W. Afr. City* xi. 210 Both native and Aku *rites de passage* appear to have been influenced by Creole practices. 1964 W. McCord in I. L. Horowitz *New Sociol.* xxv. 435 Most..practiced traditional *rites de passage*. 1972 M. Argyle *Social Psychol. of Work* iv. 67 The transition to a new job involves some degree of re-socialization, and the shift is sometimes assisted by a public ceremony..known to sociologists as a rite de passage. 1977 *Times* 22 Mar. 12/2 The [Newfoundland] seal hunt is..a necessary *rite de passage* for all young men.

‖ **ritenuto** (ritenū·to), *a.*, *adv.*, and *sb.* *Mus.* [It., pa. pple. of *ritenere*, f. L. *retinēre* to hold back.] **A.** *adj.* and *adv.* Of musical movement: restrained, held back in tempo. Used adverbially as a direction indicating immediate

reduction of speed. **B.** *sb.* (*pl.* **ritenuti** or **ritenutos**). A phrase or passage thus indicated.
1828 Busby *Mus. Manual* 148 Ritenuto. (Ital.) Movements to which this term is prefixed, are to be performed in a gentle, delicate, and restrained manner. 1888 L. A. Smith *Music of Waters* 15 Chorus. *Ritenuto molto.* Low-lands, Low-lands, Hur-rah, my John! 1952 *Conc. Oxf. Dict. Mus.* 500/2 *Ritenuto*, 'Held back', i.e. 'Slower' (immediately, not gradually as with *Ritardando* and *Rallentando*; but it may be that some composers have not observed this distinction). 1955 G. Abraham in H. Van Thal *Fanfare for E. Newman* 14 A boisterous tutti fortissimo ending by ten quiet bars, *un poco ritenuto*. 1955 [see *diminuendo]. 1959 *Times* 12 Jan. 12/6 Their Bach was of the sewing-machine school, their Brahms full of ugly *ritenuti*. 1975 *Daily Tel.* 4 Feb. 11/1, I must protest against end ritenutos that came as an anti-climax in a good many of the variations. 1976 *Guardian* 13 Apr. 10/5 He plays strongly and straightforwardly; the rubatos and ritenutos are there.

rithe. (Further examples.)
1925 A. Moore *Last Days of Mast & Sail* vii. 216 [The Bosham boats] are most dangerous-looking and would not live long in a seaway, but in the channels and rithes of Chichester Harbour the water is generally smooth and they continue in use generation after generation. 1931 Belloc *Cranmer* ii. 21 There stood on the Eastern edge of the town of Cambridge, just beyond the King's Ditch, as it was called (a runnel of water, the Long Rithe, which drained that flooded land and led from a mill above), a little place already known in this year, 1503, as 'Jesus' College.

ritodrine (ri·todrīn). *Pharm.* [Invented word.] A sympathomimetic agent used esp. as a uterine relaxant in cases of premature labour, when it is administered as an intravenous infusion of the hydrochloride; 2-*p*-hydroxyphenethylamino-1-*p*-hydroxyphenyl-propanol, $C_{17}H_{21}NO_3$.
1971 *Amer. Jrnl. Obstetr. & Gynecol.* CX. 111/1 Ritodrine hydrochloride, previously identified as Du-21220, is a sympathomimetic amine with beta-adrenergic-inhibitory properties. 1977 *Lancet* 8 Oct. 777/2 In Britain two [drugs] have been licensed for use as agents which inhibit labour—namely, salbutamol and ritodrine.

ritornello. Add: Also *fig.*
1977 *Times Lit. Suppl.* 1 July 796/4 The Marxist theme is a ritornello appearing at the end of each chapter rather than at the conclusion of the book. 1978 *Ibid.* 13 Oct. 1153/3 What is the correlation between schools and literacy?.. The irrelevance of schools is a persistent *ritornello* here.

Ritschlian (ri·tʃliăn), *a.* and *sb.* [f. the name of Albrecht *Ritschl*, German theologian (1822–89).] **A.** *adj.* Of or pertaining to Ritschl or his doctrines. **B.** *sb.* A follower of Ritschl or a student of Ritschlianism.
1891 *Chambers's Encycl.* VIII. 733/2 The distinguishing feature of the Ritschlian theology is perhaps the eminence it gives to the practical, ethical, social side of Christianity. *Ibid.* 734/1 The Ritschlians now form a large and important school in Germany. 1938 *Times Lit. Suppl.* 30 Apr. p. x/2 For a generation past the influence of the Ritschlian tradition has been far more potent in Scottish theology. 1952 *Hibbert Jrnl.* 1951 L. 12 The Ritschlian contempt for 'speculative Theism'. *Ibid.* 15 The *Fourth Gospel*, with such New Testament epistles as *Romans* or *Ephesians*, neglected if not deliberately excluded by Ritschlians in favour of the Synoptists' parables of a 'Kingdom of God' to be established here and now. 1957 *Oxf. Dict. Chr. Ch.* 1168/2 The so-called 'Ritschlian School' was characterized by its stress on ethics and on the 'community', and by its repudiation of metaphysics and religious experience. 1970 *Evangelical Q.* XLII. 95 The last four words tend to nullify the import of the preceding, which as such is the usual Ritschlian explanation; the doctrine of the subjective origin of the thing. *Ibid.* 99 The Ritschlians were, therefore, strongly opposed to the separation which had been made in traditional theology between the person and work of Christ.
Hence **Ri·tschlianism**, the theological or philosophical doctrines of Ritschl.
1892 J. Orr in *Thinker* Aug. 148 Ritschlianism has a metaphysic, and a specially dangerous one. 1917 Bartlet & Carlyle *Christianity in Hist.* v. v. 596 Ritschlianism and Catholic Modernism are the most marked movements in this direction. 1969 D. L. Mueller *Introd. Theol. A. Ritschl* iii. 105 This idea so crucial for comprehending Ritschl's conception of faith became a kind of watchword of later Ritschlianism. 1970 *Evangelical Q.* XLII. 97 All possibility of speaking of an absolute nature in the Deity as the ground of His historical manifestations is in Ritschlianism swept aside.

Ritsu (ri·tsu), *sb.* Also **Risshu** (ri·ʃu). [f. Jap. *ritsu* law, moral law.] A Buddhist sect of the early Tang period, introduced in the 8th century to Japan where it flourished, concerned primarily with the study of monastic discipline and ordination rites.
[1727 J. G. Scheuchzer tr. *Kæmpfer's Hist. Japan* I. II. v. 199 In the 1850 streets of this city, there were 1050 of the *Ten Dai's* religion,..9912 of *Rit.*] 1880 E. J. Reed *Japan* I. iv. 91 The Ritsu, introduced by the Chinese priest Kanshin, under the empress Koken. 1917 A. K. Reischauer *Stud. Jap. Buddhism* iii. 86 Last of these older sects to reach Japan was the Ritsu (Vinaya Sect), introduced in 754, though it would seem that its doctrines..were among the first teachings to be introduced into Japan. 1931 G. B. Sansom *Japan* II. vi. 121 The Ritsu sect did not trouble much about doctrinal questions, but paid special attention to discipline and correct spiritual succession. 1935 C. Eliot *Jap. Buddhism* viii. 232 After the establishment of this new Kaidan, the Risshū seems to have declined though it somewhat revived in the twelfth century.

ritter, var. *rutter².

Ritter's disease (ri·təɹz). *Med.* [Named after Gottfried *Ritter* von Rittershain (1820–83), Bohemian physician, who first described it as a distinct entity in 1878 (*Central-Zeitung f. Kinderheilkunde* II. 3–23).] A sometimes fatal form of dermatitis affecting newborn infants.
1888 *Amer. Jrnl. Med. Sci.* XCV. 11 The pemphigus simplex acutus, like Ritter's disease, belongs to the first weeks of life, but appears earlier, sometimes in the first few days after birth. 1931 B. Williamson *Handbk. Dis. Children* xvii. 143 Pemphigus neonatorum (Ritter's disease)..occasionally is luetic in origin, but more often it is a septic infection associated with unclean midwifery. 1966 *Amer. Jrnl. Dis. Children* CXI. 391/2 The exact etiology of Ritter's disease has never been finally established, many investigators having reported an association with staphylococcal infection.

ritual, *a.* and *sb.* Add: **A.** *adj.* **1.** *spec.* in *Archæol.*, applied to objects or constructions.
1901 A. J. Evans *Mycenaean Tree & Pillar Cult* 9 At the foot of the handle of axe, namely, appears in each case that distinctive piece of Mycenaean ritual furniture..described as 'the horns of consecration'. 1934 *Burlington Mag.* Mar. 139/1 Other ritual bronzes said to have been found with the *huang*. 1941 *Antiquity* XV. 142 (*title*) A datable 'ritual barrow' in Glamorganshire. 1951 *Field Archæol.* (Ordnance Survey Prof. Papers No. 13) (ed. 3) 42 Another curious type of site about which our knowledge is imperfect is the so-called 'ritual' well, or Belgic burial shaft. 1954 S. Piggott *Neolithic Cultures* ii. 56 It is possible that some of the 'ritual holes' frequently recorded from near the burials in the Wessex barrows may have been post-holes, but the majority seem to have served some other unexplained purpose connected with the funerary rites. 1963 W. F. Grimes in Foster & Alcock *Culture & Environment* v. 105 No doubt in due course something like 'sacred circle' or 'ritual circle' frankly avowing ideas about their use will become permissible for a majority of them. 1975 P. Warren *Aegean Civilization* ii. 43 The individual who made the so-called Harvester Vase..created a masterpiece. Here are 27 figures marching around a ritual vase a few inches high.
2. *ritual murder*, murder carried out as a rite; also *fig.* and *attrib.*; similarly *ritual killing*.
1896 [in Dict.]. 1936 C. Day Lewis *We're not going to do Nothing* 14 Organised mass-murder—as apart from ritual-killings, blood-feuds and the like—can admit of only one satisfactory explanation. 1950 M. Hay *Foot of Pride* v. 119 In Germany, the ritual-murder legend was the chief..excuse for a series of riots all over the country which threatened the Jews with complete extermination. 1962 L. Deighton *Ipcress File* i. 9 The man..was now using knife and fork to commit ritual murder on a cream pastry. 1964 M. Suther et al. tr. *Maritain's Moral Philos.* 142 (*heading*) The ritual murder of Realities that are elevated to the skies. 1966 *New Statesman* 15 July 78/3 The sudden revival in Russia of the ritual-murder myth after half a century of anti-religious mass education. 1972 J. McClure *Caterpillar Cop* ii. 24 Back marked by long cuts... Those wounds suggest a ritual killing.
3. In extended and trivial use: pertaining to or constituting a social or psychological ritual (see sense B. 2 below); used, occurring, etc., as a social convention or habit.
1947 *Atlantic Monthly* July 114/2 Many political speeches ..are delivered in a language which is above people's heads and is in fact a 'ritual survival' from an age in which electors were few and literate. 1953 H. S. Sullivan *Interpersonal Theory of Psychiatry* xviii. 307 All these ritual avoidances and preoccupations give one a feeling that one is making some sense in an important area of living. Actually one is not making any sense at all, because one is completely inaccessible to any data. 1972 W. Labov *Language in Inner City* viii. 305 Those who have some knowledge of urban ghetto culture will recognize Rel's remark *Your mother's a duck* as a ritual insult. 1975 R. Collins *Conflict Sociol.* iii. 115 We can tell the difference by the increase in ritual elements as the talk becomes more sociable, the orienting gestures are toward each other rather than toward the topic. 1977 B. Pym *Quartet in Autumn* i. 5 He offered her the bag of jelly babies, but this was only a ritual gesture and he knew that she would refuse. *Ibid.* v. 40 Most of the inhabitants of the village were retired married couples with the ritual grandchildren.
B. *sb.* **2. a.** (Later examples.) Now freq. in *sing.* Also in extended and trivial uses.
1906 J. G. Frazer *Adonis, Attis, Osiris* 325 When the Bechuanas are about to found a new town, they observe an elaborate ritual. 1911 Beerbohm *Zuleika D.* iii. 29 He cared for his wardrobe and his toilet-table..merely as..a ritual in which to express and realise, his own idolatry. 1914 J. S. Huxley in *Proc. Zool. Soc.* I. 506 We must now go on to consider a very different question... I mean the gradual change of a useful action into a symbol and then into a ritual. 1958 J. O'Conor *Iron Harp* in J. C. Trewin *Plays of Year* XVI. 179 He's had no lunch... Ah, now, that won't do. A man should never miss the ritual of a good meal. 1975 R. Collins *Conflict Sociol.* iii. 97 All that animal language lacks in comparison to human rituals, is a symbolic significance or *naming* quality... We usually assume that animal rituals are innate but that human rituals are learned. 1975 L. Lee *I can't stay Long* (1977) 203 The ritual of bargaining was long and elaborate.
b. *spec.* in *Psychol.* A series of actions compulsively performed under certain circumstances, the non-performance of which results in tension and anxiety.
1932 M. Gabain tr. *Piaget's Moral Judgment of Child* iv. 359 The rituals attached to eating, going to bed, etc., show the hold which habit has over his [*sc.* the child's] nature. 1946 O. Fenichel *Psychoanal. Theory of Neuroses* xiv. 268 Touching rituals replace taboos; washing compulsions, fears of dirt; social rituals, social fears. 1956 H. S. Lippman *Treatm. Child in Emotional Conflict* ix. 113 The diagnosis of obsessional neurosis in a child depends primarily on the presence of ceremonials or rituals which he cannot control. 1968 L. Eidelberg *Encycl. Psychoanal.* 382/2 Although similar in pathogenesis, it is useful to distinguish between rituals and other obsessive-compulsive manifestations. 1972 A. Storr *Dynamics of Creation* viii. 98 The ritual had started originally as an attempt to purge himself of the guilt surrounding masturbation.
3. (Later examples.)
1923 J. S. Huxley in *Jrnl. Linnean Soc.* XXXV. 255 The effect as of tension, of emotional ritual, so familiar to all those who have watched birds during courtship, is marked. 1947 Chapple & Coon *Princ. Anthropol.* xix. 481 This question of the conditioned nature of symbols is a basic requisite to the understanding of the whole subject of ritual. 1961 L. Thompson *Toward Sci. of Mankind* xi. 182 Hopi ritual may be viewed as a complex, but logical and ordered, whole, which expresses symbolically the Hopi conception of the universe, the law, and the life process. 1969 in Halpert & Story *Christmas Mumming in Newfoundland* 112 'Ritual' here is broadly defined as largely symbolic activity, aimed towards controlling social relations. 1971 L. Neal in A. Gayle *Black Aesthetic* 285 Like all good ritual, its purpose is to make the audience stronger, more sensitive to the historical realities that have shaped our lives and the lives of our ancestors.
Comb.
1868 J. G. Whittier in *Atlantic Monthly* Feb. 221 Nor ritual-bound nor templeward Walks the free spirit of the Lord! 1930 D. H. Lawrence *A Propos Lady Chatterley* 36 The strange priest-controlled, ritual-fulfilled condition of the earlier Egyptians. 1937 M. Covarrubias *Island of Bali* viii. 216 The ritual-magic dances characteristic of primitive peoples.

‖ **ritualia** (ritiuē¹·liă). *nonce-wd.* [L., pl. of *rituāle* relating to rites or ceremonies.] Objects used in or connected with religious rites and ceremonies.
1931 *Times Lit. Suppl.* 5 Nov. 864/2 The records of this synagogue may contain papers of great interest, and its *ritualia* certainly include objects of considerable beauty.

ritualism. Add: **1.** (Later examples.)
1935 B. Malinowski *Found. Faith & Morals* i. 6 It is rather the recognition of his practical and intellectual limitations, and not the illusion of the 'omnipotence of thought', which leads man into ritualism. 1952 Gerth & Martindale tr. *Weber's Anc. Judaism* xiii. 336 Prophecy together with traditional ritualism of Israel, brought forth the elements that gave to Jewry its pariah place in the world. 1975 M. Douglas *Implicit Meanings* v. 79 So the tension between ritualism of established authority and enthusiasm from the outlying borders of society, the dynamic of reform in European history, must have its counterpart in the unwritten history of any primitive tribe.
2. *Sociol.* (See quot. 1957.)
1949 R. K. Merton *Social Theory* iv. 141 The socialization patterns of the lower middle class thus promote the very character structure most predisposed toward ritualism. 1955 P. M. Blau *Dynamics of Bureaucracy* xii. 193 Three variations of displacement of goals were observed, all of which differed from ritualism. 1957 R. K. Merton *Social Theory* (rev. ed.) v. 184 Ritualism refers to a pattern of response in which culturally defined aspirations are abandoned while 'one continues to abide almost compulsively by institutional norms'. 1961 O. J. Harvey et al. *Conceptual Syst.* ii. 43 (*heading*) Concreteness disposes toward ritualism. 1963 T. & P. Morris *Pentonville* vii. 172 Two kinds of ritualism may be distinguished—the *ritualism of identification* and the *ritualism of dependency*. 1974 H. R. Bobbitt et al. *Organizational Behav.* iii. 61 Ritualism appeared to result more from lack of security in important social relationships ..than from overidentification with rules or strong habituation.

ritualist. Add: **2.** *spec.* in *Anthrop.* In a tribal society, one who performs a ritual.
1969 M. Douglas *Natural Symbols* ii. 35 The primitive ritualist, in his ascribed social system, expresses cosmic orientations and moral directives in condensed symbols. 1974 B. & R. Hill *Spirit in Stone* iii. 35 A man whom we will call the ritualist and several assistants are fishing for salmon with a reef net. *Ibid.*, The ritualist carefully hands the children the fish he has caught. 1977 L. J. Bean in Fogelson & Adams *Anthropol. of Power* 123 Individuals may ..obtain it [*sc.* power] by inheriting or purchasing ritual equipment, and the knowledge that goes with it, from a shaman and/or ritualist.
3. Someone whose behaviour is characterized by ritualism (esp. in sense *2). Also *attrib.*
1949 R. K. Merton *Social Theory* iv. 140 The syndrome of the social ritualist is both familiar and instructive. 1957 *Ibid.* (rev. ed.) v. 185 Situations patterned by the social structure which invite the ritualist response of overconformity to normative expectations. 1963 T. & P. Morris *Pentonville* vii. 172 The ritualist has largely rejected the socially approved goals altogether, and fallen back upon punctilious conformity. 1969 M. Douglas *Natural Symbols* i. 2 The ritualist becomes one who performs external gestures which imply commitment to a particular set of values, but he is inwardly withdrawn, dried out and uncommitted.

ritualistic, *a.* Add: (Later examples.) Also, characteristic of ritual actions or behaviour.
1949 R. K. Merton *Social Theory* iv. 140 The ritualistic type of adaptation can be readily identified. 1952 Gerth & Martindale tr. *Weber's Anc. Judaism* xiii. 336 In Israel,

originally, ritualistic segregation from strangers was totally absent. 1962 I. Sarnoff *Personality Dynamics & Devel.* xii. 350 An individual may be led to develop ritualistic acts symbolizing both aspects of his conflict. 1971 P. Greenacre *Emotional Growth* I. xi. 169 The nature of the early shattering experience could sometimes be deciphered from the ritualistic behavior accompanying the use of the fetish. 1973 *Black World* June 4/1 One of the most salient characteristics of the New Black Theater..is its ritualistic aspect. By ritualistic, I mean the strong presence of *symbols, characterizations, themes* and *language styles* which are frequently repeated from play to play. 1977 *Lancashire Life* Nov. 136/3 Very many people..do not attend church because they are bored by ritualistic services.

rituality. Restrict † *Obs.* to sense 1 in Dict. and add: **2.** (Later example.) *rare.*
1974 *Times* 16 Apr. 7/6 What Solzhenitsyn writes about ideological rituality, about the harmful waste of millions of people's time and efforts on this chatter that inculcates twaddle and hypocrisy, is indisputable.

ritualization (ritiuăləizē¹·ʃən). [f. Ritualiz(e *v.* + -ation.] **1.** *Zool.* The evolutionary process by which an action or behaviour pattern in an animal loses its ostensible function and changes into an effective social signal for other members of the species.
[1914 J. S. Huxley in *Proc. Zool. Soc.* I. 506 A very different question..is also well brought out in the pairing-habits of the Great Crested Grebe: I mean the gradual change of a useful action into a symbol and then into a ritual. 1923 —— in *Jrnl. Linnean Soc.* (*Zool.*) XXXV. 278 The 'ritual' use of non-sexual actions during courtship.] 1942 N. Tinbergen in *Bibliotheca Biotheoretica* I. 90 Substitute movements often have signal functions, and, doubtless in connection with these functions, may be extremely ritualised, so that they become unrecognisable when no comparative study is made. The degree of ritualisation actually reached is impressively illustrated by Lorenz' careful study of the epigamic movements of Anatinas. 1952 —— in *Q. Rev. Biol.* XXVII. 23/2 This new function [of releasing responses in other individuals] must have started a new evolutionary development during which the displacement activities became increasingly better adapted to it. This evolutionary displacement I have called 'ritualization', following Huxley (1923). 1965 *New Scientist* 17 June 768/2 The chief survival value of ritualization is obviously in the recanalization or the discharging of aggression. 1966 W. H. Thorpe in Thorpe & Zangwill *Current Probl. Animal Behaviour* II. 94 In the process of ritualisation, certain elements are exaggerated whilst other elements tend to disappear entirely. 1974 *Nature* 3 May 8/3 Another feature of mutual courtship [in birds] is the extent of ritualisation; but this also applies to many male unilateral displays where the sexes have different plumage.
2. *Psychol.* The formalization of certain actions that serve to express a particular emotion or state of mind which is either innate or acquired as part of a social code.
1932 M. Gabain tr. *Piaget's Moral Judgment of Child* i. 42 This phenomenon..is the counterpart of that sort of ritualization of behaviour which can be observed in any baby before it can speak or have experienced any specifically moral adult pressure. 1934 Gesell & Thompson *Infant Behav.* iii. 190 Ritualization is a reinstatement of the situation, a method of defining and perhaps improving new abilities; but it is itself a general ability, an intrinsic product of growth. 1951 J. Holloway *Lang. & Intelligence* x. 189 The systematization of language is exactly parallel to the ritualization of behaviour. 1961 C. & W. M. S. Russell *Human Behav.* ii. 93 Ritualization serves to make the movement provide simple key stimuli, by simplifying it and exaggerating it. 1970 E. Klinghammer tr. *Eibl-Eibesfeldt's Ethology* vi. 101 The cultural ritualizations of man follow the pattern of phylogenetic ritualization. 1978 E. Erikson *Toys & Reasons* ii. 69, I will chart only one use of playfulness throughout life which has received little attention, namely what I call ritualization in everyday life.
3. The action of forming a social or religious ritual.
1952 Gerth & Martindale tr. *Weber's Anc. Judaism* xiii. 353 This ritualization of dietary habits made commensalism very difficult. 1952 T. Parsons *Social System* 414 The particularism, traditionalism, and 'ritualization' of traditional Chinese society. 1962 M. Gluckman *Ess. Ritual of Social Relations* 24, I propose to use the phrase 'ritualization of social relationships'—and for brevity 'ritualization'—to define this tendency... I can only plead that my readers should..take 'ritualization' as referring to a stylized ceremonial. 1966 *Phil. Trans. R. Soc.* B. CCLI. 413 If we consider the ritual of coronation from the point of view of its manifest subject matter, we describe it as the ritualization of eminent office. 1978 J. D. Crichton in C. Jones et al. *Study of Liturgy* I. 5 One of the most constant features of human history is the ritualization of the great events of human life, birth, marriage, and death.

ritualize, *v.* Add: **2.** (Further examples.)
1967 V. W. Turner *Forest of Symbols* iv. 94 A number of critical moments of transition which all societies ritualize and publicly mark. 1975 M. Douglas *Implicit Meanings* ix. 142 To provide such detailed analyses of life crises, afflictions and of how the Ndembu ritualise them, creates thorny problems of interpretation. 1979 *Dædalus* Summer 130 Strong emotion can be ritualized.
3. *Zool.* To cause (an action or behaviour pattern) to become ritualized.
1961 *Centennial Rev. Arts & Sc.* Fall 406 Because the pattern [of behaviour] was empirically determined..one is never quite certain which behaviour elements are effective, and the whole pattern becomes ritualized. 1965 *New Scientist* 17 June 768/1 Animals tend to ritualize their aggression, rearing up, roaring, showing their teeth, or erecting their ruffs,..in such a way that an adversary recognizes the intention and reacts accordingly. 1968 R. F. Ewer *Ethology of Mammals* viii. 192 In the spotted hyaena, smelling of the ano-genital region has been ritualised into a greeting ceremony which is regularly performed when members of a group meet. 1972 *Sci. Amer.* Sept. 59/1 Birds intending to fly..typically crouch, raise their tail and spread their wings slightly just before taking off. Many species have ritualized these movements into effective signals... The signals serve to coordinate the movement of flock members, and also may warn of approaching predators.

ritualized, *ppl. a.* Add: (Further examples.)
1932 M. Gabain tr. *Piaget's Moral Judgment of Child* i. 16 The child handles the marbles at the dictation of his desires and motor habits. This leads to the formation of more or less ritualized schemas. 1976 R. M. Keesing *Cultural Anthropol.* xvii. 369/1 Ritualized combat on the traditional 'battlefield' gives a vivid picture of one side of tribal warfare. 1978 E. Erikson *Toys & Reasons* ii. 73 Any ontogenetic reconstruction of the relation of play to politics would have to begin with the meaning of ritualized interplay for the development of the individual ego.
2. *Zool.* Of an animal's action or behaviour pattern: not having its ostensible function but serving as a signal to other members of the species.
Some of the examples are better interpreted as *pa. pple.*
1942 [see *ritualization 1]. 1964 G. B. Schaller *Year of Gorilla* (1965) ix. 239 Natural selection may act on such displacement activities by enhancing their effectiveness as communicatory signals; they may become stereotyped and be incorporated into a definite display—they become ritualised. The fact that the gorilla often places a leaf between its lips suggests that this curious gesture may be a ritualised act of displacement feeding. 1966 K. Z. Lorenz *On Aggression* xiv. 241 All the culturally evolved norms of 'fair fighting', from primitive chivalry to the Geneva Convention, are functionally analogous to phylogenetically ritualized combat in animals. 1971 *Nature* 18 June 469/1 The rutting conflicts of most species [of cervid] are largely ritualized, antlers serving the function of display more than as jousting weapons. 1978 D. Symons in E. O. Smith *Social Play in Primates* 208 Because fighting entails serious risk, ritualized signals have evolved by which animals communicate aggressive intent and, thereby, avoid fighting.

ritually, *adv.* (Further examples.)
1914 J. S. Huxley in *Proc. Zool. Soc.* I. 506 The Grebe is interesting as showing all three stages of the process at one time—the passive attitude employed sometimes directly, sometimes symbolically, and sometimes ritually. 1966 K. Z. Lorenz *On Aggression* v. 48 In our European common shelduck for example, the whole process..contains no ritually fixed parts. 1972 T. R. Williams *Introd. Socialization* viii. 183 Some American Indian cultures such as the Hopi and Zuni, ritually observe puberty for one or for both sexes. 1978 G. Wainwright in C. Jones et al. *Study of Liturgy* II. 1. 37 Ritually, a novel feature of the Protestant worship was the giving of independent value to hymns in the structure of the service.

Ritz, *sb.* (*a.*) [The name of the Swiss-born hotelier César *Ritz* (1850–1918), given to his luxury hotels in Paris, London, New York, and elsewhere.] **a.** Used allusively of a large and luxurious hotel, *esp.* in negative phrases. Also *attrib.* and as *adj.*
1910 R. Fry *Lett.* (1972) I. 336, I will not pretend that my cuisine rivals the Ritz. 1922 F. Scott Fitzgerald (*title*) The diamond as big as the Ritz. 1926 E. Hemingway *Sun also Rises* xix. 238 We drove in to Biarritz and left the car outside a very Ritz place. 1928 W. S. Maugham *Ashenden* vi. 93 They came to a tavern in a blind alley, noisome and evil... 'It's not the Ritz,' he said, 'but at this hour of the night it's only in a place like this that we stand a chance of getting something to eat.' 1942 M. Dickens *One Pair of Feet* vii. 111 A phrase that often sprang, unvoiced, to my lips, was 'This ain't the ruddy Ritz'. 1960 R. Kirkbride *Innocent Abroad* xii. 87 Lousy as the room was, I was damn' glad to have it... 'It isn't the Ritz,' I said. 'But we got nowhere else to go.' 1973 W. H. Canaway *Harry doing Good* I. i. 14 The outhouse..was warm and fusty..but..for fifty pence who would expect the Ritz? 1978 *Vogue* Feb. 8/2 Creating charming country house suites with prints, quilting, Roman blinds, pretty colours, real Ritz comfort.
b. Colloq. phr. *to put on the ritz*, to assume an air of superiority. *U.S.*
1926 R. Lardner in *Hearst's Internat.* Jan. 33/2 If you mention some really worth while novel like, say, 'Black Oxen', they think you're trying to put on the Ritz. 1929 I. Berlin *Puttin' on the Ritz* (song) 3 If you're blue and you don't know what to do Why don't you go where Harlem sits Puttin' On The Ritz. 1945 L. Saxon et al. *Gumbo Ya-Ya* i. 11 You had to put on the ritz downtown, which some of the gals didn't like. 1980 H. Luce *In Midst of Death* iii. 34 We'll have to decide how long we can go on putting on the Ritz in this house. Personally, I'd much rather live in a cottage.

ritz, *v.* *U.S. colloq.* [f. prec.] *trans.* To behave haughtily towards (someone); to snub. (See also quot. 1962.) Also *refl.*, to give oneself airs.
1911 G. Ade in *Chicago Daily News* 16 Sept. 28/2 They went abroad and began to Ritz themselves. 1924 H. C. Witwer in *Cosmopolitan* Nov. 42/2 We graciously presented Bertha with permission to bring him up to dinner at our flat one night, and he Ritzed and four-flushed us all till me and Hazel had to either dash out into the great outdoors or else give this big blah a sofa pillow shower! 1939 R. Chandler *Big Sleep* iii. 30 You sent for me. I don't mind your ritzing me. 1941 [see *high-hat *v.*]. 1962 S. Strand *Marketing Dict.* 636 *Ritzed it*, in the fashion field, adding glamour to a fabric or dress so that it will have public appeal. One way would be to advertise it in a class magazine. 1978 *Vogue* Feb. 8/2 (*heading*) Ritzing the Ritz.

Ri·tzian, *a.* [f. as prec. + -ian.] Worthy or typical of the Ritz.
1908 F. Hopwood *Let.* 11 Jan. in R. S. Churchill *Winston S. Churchill* (1969) II. Compan. II. 742 A banquet of Ritzian splendour. 1918 G. Frankau in *Poetical Wks.* (1923) II. 179 Veil from these eyes their last too vivid canto:—Those Ritzian chambers on the Place Vendôme. 1940 'Gun Buster' *Return via Dunkirk* II. vii. 142 Veal cutlets, tinned peas, cheddar cheese and vin rouge awaited me at the Command Post. In the circumstances, a Ritzian repast.

ritzily, ritziness: see *ritzy *a.*

ritzy (ri·tsi), *a.* *colloq.* (orig. *U.S.*). Also **Ritzy.** [f. *Ritz *sb.* + -y¹.] **a.** (In a complimentary sense.) Having class, poise, or polish; smart, stylish, glamorous, 'classy'. **b.** (In a derogatory sense.) Of persons: haughty, pretentious, ostentatious; of things: flashy, pretentious-looking.
1920 Wodehouse *Jill the Reckless* (1922) xvi. 240 The Duchess, abandoning that aristocratic manner criticized by some of her colleagues as 'up-stage' and by others as 'Ritz-y', [etc.]. 1923 *Variety* 15 Nov. 19/2 The upstaginess and ritzy attitude of the movie stars has become a matter of common talk. 1926 S. Lewis *Mantrap* i. 20 Now there's some real Ritzy dancing-pumps... Had 'em done specially. 1930 E. Wallace *White Face* v. 49 She's got the only respectable apartment... All Ritzy. 1932 S. Gibbons *Cold Comfort Farm* xvii. 237, I want a new Gary Cooper..only more ritzy. Someone who can look good in a tuxedo. 1938 E. Bowen *Death of Heart* II. iv. 243 Friend of your sister-in-law?.. He'll be a bit ritzy for us, then, won't he? 1943 *Archit. Rev.* XCIII. 80/4 There is no disguising the vulgarity of his Tokio hotel, the ritzy streamlining of his Johnson house, the mere ugliness of the Jones house of 1929. 1947 *People* 22 June 5/1 That glamour gal of British trains, the ritzy, resplendent Golden Arrow. 1949 'G. Orwell' *Let.* 5 Sept. in B. Crick *G. Orwell* (1980) xvii. 399, I feel ghastly and can't write much, but we had a wonderful journey down yesterday in the most ritzy ambulance you can imagine. 1951 M. McLuhan *Mech. Bride* (1967) 80/2 'Ritzy dames' who are provided with custom-built allure. 1958 *Woman's Own* 10 Sept. 33/1 For that ritzy touch, get octopus or ham in champagne. 1959 W. D. Pereira *North Flight* iv. 54 You should see 'is eldest kid... Bone idle. Goes to one of them ritzy schools, but it won't 'elp 'im none. 1963 *Observer* 1 Dec. 8/3 A stunning report, with its ritzy format and private jokes. 1970 R. Freeth *Lighting* iv. 32/2 If you want to be really 'Ritzy' install a filament reflector bulb as well over the cooker and sink. 1976 *Evening Advertiser* (Swindon) 31 Dec. 10/7 When I buy skin-food it's the product I'm interested in, not a ritzy bottle to decorate the dressing table. 1979 *Daily Tel.* 4 June 17/4 (Advt.), It looks most glamorous..for lounging in a ritzy beachside cafe.
Hence **ri·tzily** *adv.*, **ri·tziness.**
1928 *Motion Picture Classic* Jan. 27 They whirl ritzily out to Beverly Hills. 1929 M. Lief *Hangover* 234 'Oh that?' she said sort of ritzily, 'why, that's only her filling-station!' 1965 *Harper's Bazaar* Nov. 70 Lids can now be coated ritzily and memorably with Max Factor's Iridescent Gold shadow. 1967 N. Tomalin in L. Deighton *London Dossier* 285 Neighbourhoods of unbelievably stuffy ritzyness. 1970 'O. John' *Diamond Dress* xiii. 145 The smooth, spacious ritziness of this large store.

‖ **riva** (ri·va). [It.] In Italy, a river-bank, sea-shore, or quay; *spec.* the *Riva degli Schiavoni* in Venice.
1880 F. J. Sitwell tr. *Yriarte's Venice* p. xii, The Riva and its Denizens. 1888 H. James *Aspern Papers* I. i. 13 The old palace..overlooked a..canal, which had a narrow *riva* or convenient footway on either side. 1909 J. Joyce *Let.* 7 Sept. (1966) II. 249, I long to see the lights twinkling along the *riva* as the train passes Miramar. 1928 Beerbohm *Variety of Things* 186 As I passed along that Riva, I would try to imagine Venice as she was. 1965 H. Honour *Compan. Guide to Venice* v. 77 The Fondamenta dell'Arsenale leads down to the *Riva degli Schiavoni*... To your left the Riva sweeps up to the public gardens.

rivalrous, *a.* Add: **2.** Given to rivalry; acting as a rival. orig. *U.S.*
1920 in Webster. 1961 F. H. Allport in Webster, s.v., Ascendant, expansive, and rivalrous students. 1963 *Observer* 21 Apr. 29/4 In the three- to six-year-old stage boys become rivalrous with their fathers, girls with their mothers. 1965 *Amer. N. & Q.* Sept. 14/2 A rivalrous and divided Italy had little more to offer. 1972 M. Mead *Blackberry Winter* (1973) vi. 70 Sisters, while they are growing up, tend to be very rivalrous and as young mothers they are given to continual rivalrous comparisons of their several children. 1980 *N.Y. Times* 28 Oct. C1/1 Even geographical separation cannot sever their closeness or quell their rivalrous strivings.

rivalry. Add: **1. b.** *Psychol.* Lack of fusion of the visual fields presented separately but simultaneously to each eye when these are sufficiently different, there being instead an alternation of perceived images.
1878 A. Gamgee tr. *L. Hermann's Elem. Human Physiol.* (ed. 2) x. 427 (*heading*) Rivalry of the fields of vision. 1950 K. N. Ogle *Res. Binocular Vision* vi. 61 Depending on the particular characteristics of the image patterns falling on the retinas of the two eyes, fusion will be complete or partial, or as an antithesis of fusion, the patterns may actually exhibit a rivalry or resistance to fusion. 1974 L. Kaufman *Sight & Mind* viii. 306 Even when the inner square is seen floating above the background lines in the stereoscope, all the lines are in a constant state of binocular rivalry.

rived, *ppl. a.* (Later examples.)
1853 *Trans. Mich. Agric. Soc.* IV. 156 Either the rived or sawed bolt may be used. 1887 *Century Mag.* Apr. 901/1 The

earliest houses of worship in America belonged to the make-shift order of architecture,—four walls of..rived clapboards with earth filled in between.

‖ **rive gauche** (rīv gōʃ). [Fr.] = *left bank* s.v. *Left *a.* 3.
1862 Mrs. Gaskell *Jrnl.* Feb. in *Fraser's Mag.* (1864) Apr. 435/1 We went to-day along the Boulevard Sévastopol, Rive Gauche, to pay a call. 1894 G. du Maurier *Trilby* I. II. 51 Then back again to the quays on the *rive gauche* by the Pont Neuf. 1928 R. Hall *Well of Loneliness* xxxi. 289 Of course you'll have to live on this side, the Rive Gauche is the only possible Paris. 1948 A. Waugh *Unclouded Summer* ii. 16 It was his belief that the Americans who were creating an *emigré Rive gauche* colony had sold their birthright. 1959 *Listener* 28 May 941/2 All those romantic stories..which have been going the rounds of the *rive gauche* ever since. 1977 *Times* 19 Feb. 9/1 Those whose shelves are filled with rarities, generally secured for 'next to nothing' on the Rive Gauche.

rivelling, *ppl. a.* (Later example.)
1878 G. M. Hopkins *Poems* (1967) 74 But his eye no cliff, no coast or Mark makes in the rivelling snowstorm.

river, *sb.*¹ Add: **I. 1. f.** *Printing.* (See quot. 1948.) Also *river of white*.
1898 G. B. Shaw *Let.* 5 Jan. in *Ellen Terry & Shaw* (1931) 287 Oh those proofs, those proofs! Imagine..sticking in words to make the printing look decent—to get the rivers of white out of it! 1927 —— *Let.* 7 Mar. in *To Young Actress* (1960) 114 They avoided white patches and rivers in the rich black block of letterpress. 1929 *S. P. E. Tract* XXXIII. 437 In careful book-printing the possibility of manipulating spaces is limited, because evenness is desirable, and rivers of white on the page must be avoided. 1948 M. E. Skillin et al. *Words into Type* 546 *River*, a streak of white space in printed matter caused by the spaces between words in several lines happening to fall one almost below another. 1967 *Guardian* 13 Oct. 5/3 Morison holding up a book, inspecting the printed page for rivers of white caused by bad printing.
g. The finest grade of diamond. Cf. *river stone* in sense 4d below and Water *sb.* 20.
1934 in Webster. 1946 G. Stimpson *Bk. about Thousand Things* 267 *River* and *extra river* are now used to denote diamonds of the finer qualities. 1965 J. Y. Dickinson *Bk. of Diamonds* viii. 219 *River*,..the finest color grade in diamonds; an extraordinarily transparent stone may be called 'an extra river stone'. 1973 *Times* 25 Aug. 17/3 The (more or less) accepted English classes run thus in descending order: (1) finest fine white or river *alias* blue-white.
3*. Phrases. **a.** *to sell down the river*: to sell (a troublesome slave) to the owner of a sugar-cane plantation on the lower Mississippi, where conditions were harsher than in the northern slave States; hence *fig.*, to deliver (one) over to slavery (*rare*); to let down, betray. *colloq.* (orig. *U.S.*).
1851 Mrs. Stowe in *National Era* 14 Aug. 1/2 I've had one or two of these fellers, and I jest sold 'em down river. 1894 'Mark Twain' *Pudd'nhead Wilson* ix. 113 Ole Marse Driscoll 'll sell you down de river. 1927 Wodehouse *Small Bachelor* i. 21 When Sigsbee Waddington married for the second time, he to all intents and purposes sold himself down the river. 1941 Auden *New Year Let.* II. 44 'I'll fix you something for your liver'; And thus he sells us down the river. 1943 K. Tennant *Ride on Stranger* ix. 98 'Perhaps we could persuade Mrs. Brewster to abandon that part of the pageant?' 'Oh don't!.. She'd like to sell me down the river as it is, cheap.' 1955 E. Pound *Section: Rock-Drill* (1957) lxxxvi. 24 England not yet sold for the Suez—That would have been 20 years later, or was it '74? At any rate, sold down the river, passed over Parliament. 1958 Hayward & Harari tr. *Pasternak's Dr. Zhivago* vi. 155 It's my considered opinion, Yurochka, we've been sold down the river. 1976 *Southern Even. Echo* (Southampton) 16 Nov. 3/3 Some aspects of Britain's education system needed to be put right but 'we should not sell it down the river' Education Secretary Mrs. Shirley Williams said last night.
b. *up the river*: (orig.) to Sing Sing prison, situated up the Hudson River from the city of New York; hence *fig.*, to or in prison. *colloq.* (orig. *U.S.*).
1891 in H. Campbell *Darkness & Daylight* (1892) ii. 75 Lager-beer had come up since I went up the river. 1905 C. H. Day *Actress & Clerk* v. 53, I didn't go up the river for several stretches for nothing, I didn't. I've got a record. 1946 *Chicago Daily News* 5 Mar. 8/3, I done it. Send me up the river. Give me the hot seat. 1951 Wodehouse *Old Reliable* i. 18 A member of the jury which three years before had sent him up the river for what the Press of New York was unanimous in describing as a well-earned sentence. 1963 J. N. Harris *Weird World Wes Beattie* (1964) iii. 24 But I *still* want to talk to Mrs. Leduc and find out why she sent the boy up the river.
c. *down the river*, used in various senses, as: into slavery (cf. sense 3* a above); finished, past, over and done with; to prison (cf. sense 3* b above). *colloq.* (orig. *U.S.*).
1893 'Mark Twain' in *Century Mag.* Dec. 238/1 Percy Driscoll slept well the night he saved his house-minions from going down the river. 1930 J. B. Priestley *Angel Pavement* ii. 80 And up to eighteen months ago, I'd have told you that Claridge and Molton was one of the soundest concerns in the business. And look at 'em now. Properly in Queer Street. Absolutely down the river. 1931 *Sun* (Baltimore) 31 Jan. 1/5 True enough, I used to hustle a little beer in the old days —but that's all down the river. 1939 'E. Queen' in *Blue Bk. Mag.* Oct. 18/1 'Mike's car's gone down the river.' 'I thought the champion was wealthy,' said Mr. Queen. 'Not any more.' 1974 *Times* 31 Jan. 4/5 He had overheard Miss Jones threatening Mr Dee 'to send him down the river for life'.

II. 4. a. *river-bridge, -coast* (later example), *flat* (earlier and later examples), *-front, -glade* (earlier and later examples), *-grove, hill, -island* (later examples), *-isle, -lane, -marsh, -meadow, road, -shore* (earlier example), *state, terrace* (later example), *town, -trail, -walk* (later examples).
1915 E. Pound *Cathay* 20, I had to be off to So, far away over the waters, You back to your river-bridge. 1940 W. Faulkner *Hamlet* I. iii. 77 His destination was not far: a little under a mile to the river bridge, a little more than a mile beyond it. 1960 R. Campbell tr. *A. Rimbaud's Drunken Boat* in *Coll. Poems* III. 17, I felt no more the guidance of my tow-men As I came down by listless river-coasts. a 1816 B. Hawkins *Sk. Creek Country* (1848) 47 On the right side, off from the river flats, the land is waving. 1977 *Weekly Times* (Melbourne) 19 Jan. 62 (Advt.), A very scenic property rising from irrigated river flats to undulating and hilly terrain, with magnificent outlook. 1855 *Chicago Weekly Times* 16 Jan. 1/1 To lease for a term of years. 200 feet river front, ready docked. 1978 J. A. Michener *Chesapeake* 237 He had been discussing her with young men of the region, offering them..even a stretch of river-front, if they would marry his eldest daughter. 1848 B. Smith in *Rep. to authorize Draining of Ever Glades* (U.S. Congress Comm. Publ. Lands) 19 The name Ever Glades is doubtless of English gift, and probably was originally 'River Glades'. 1957 Blunden *Poems of Many Years* 284 You see Old Night Begin to shade the river-glade. 1930 —— *Poems* 290 The secret paths of river-groves. 1793 J. Filson in G. Imlay *Topogr. Descr. W. Terr. N. Amer.* (ed. 2) II. 118 After passing the Miami River hills..the country in places is broken. 1948 *Clarke County Democrat* (Grove Hill, Alabama) 29 Apr. 4/2 The river hill, while not yet quite subdued, is nothing like the formidable barrier that it once was. 1913 J. London *Valley of Moon* xvii. 479, I wouldn't trade a square mile of this kind of country for the whole Sacramento Valley, with the river islands thrown in and Middle River for good measure. 1939 Auden & Isherwood *Journey to War* i. 31 The British Consulate is in the foreign concession, on the river-island of Shameen. 1900 J. A. Joyce in *Fortn. Rev.* Apr. 577 Through the trees can be seen the town harbour, and the fjord,..as it stretches past headland and river-isle out to the sea. 1947 C. S. Lewis in *Punch* 21 May 434/1 He held at the finish but a small river-isle. 1968 G. Jones *Hist. Vikings* III. iii. 224 The river-lanes of France and the Low Countries. 1978 C. Tomlinson *Shaft* 43 They..narrow out into A now-smooth riverlane. 1930 E. Pound *XXX Cantos* vi. 24 By river-marsh, by galleried church-porch. c 1847 Thoreau in J. L. Shanley *Making of Walden* (1957) 198 Men who frequent the river meadows and solitary ponds in the horizon-connecting links between towns. 1912 P. S. Allen *Let.* 11 Apr. (1939) 101 We even found fritillaries growing..flowers which we have hitherto always associated with the moisture of Thames' river-meadows. 1776 G. Washington *Let.* 27 Dec. in *Boston Gaz.* (1777) 20 Jan. 2/1, I formed my detachment into two divisions, one to march up the lower or river road, the other by the upper or Pennington road. 1829 J. MacTaggart *Three Yrs. Canada* II. 202 When the snow falls deep, before the ice has had time to freeze to any considerable thickness, the *river roads* remain dangerous all the season. 1955 E. A. Collard *Canad. Yesterdays* 302 Some experienced travellers on the river-roads even carried 'choke-ropes'. 1770 G. Washington *Tour to Ohio* in *Olden Time* (1846) I. 423 At the lower end of the Long Reach..is a large bottom, but low, and covered with beach near the river shore. 1845 *Southern Lit. Messenger* XI. 578/1 There, too, should be present..all the river States, to deliberate upon the present condition of those great arteries of commerce among them. 1976 *Daily Times* (Lagos) 4 Sept. 2/2 Divisional administration in the River state has been abolished with immediate effect. 1969 Bennison & Wright *Geol. Hist. Brit. Isles* VI. xvi. 366 River terraces..are remnants of former floodplains dissected by the rejuvenation of rivers consequent upon uplift. a 1850 G. G. Foster *New York Naked* (c 1855) vii. 74 Here he fell in, accidentally, with a rich banker and capitalist, from one of the river towns. 1938 H. Asbury *Sucker's Progress* ix. 212 When they came ashore they demanded women and whisky, and the river towns provided both in great abundance. 1977 B. F. Chamberlin in Bond & McLeod *Newslett. to Newspapers* IV. 248 The legislature had been established as the 'supreme power' of the Commonwealth in an agreement among the early Seventeenth-Century river towns. 1902 S. E. White *Blazed Trail* III. xxx. 211 The little procession..took its way up the river-trail. 1923 L. Y. Erskine *River Trail* viii. 58 In the morning Geoffrian..set out for the river trail. 1914 Kipling in *Nash's & Pall Mall Mag.* Nov. 181/1 A river front, a narrow terraced river-walk in front of semi-oriental houses. 1976 *State Jrnl.* (Lansing, Michigan) 11 July B-7/1 Ducks are a long-time attraction to visitors... Now residents can enjoy them while strolling along the new 'Riverwalk'.
b. *river-bend, -edge, -line, mouth* (earlier example), *ravine, -reach* (earlier example), *-system* (later examples).
1898 W. H. Ogilvie *Fair Girls & Gray Horses* 33 By stock-routes brown and burnt and bare, by flood-wrapped river-bends, They've hunted them from gate to gate—the drover has no friends! 1972 R. G. Kazmann *Mod. Hydrol.* (ed. 2) iv. 115 On the Mississippi River..a number of river-bend cutoffs have been constructed. 1883 'Mark Twain' *Life on Mississippi* xxii. 256 St. Louis is a great city; but the river-edge of it seems dead. 1945 *Finito! Po Valley Campaign* (15th Army Group) 7 Behind these riverlines were the machine gun nests. 1958 N. Levine *Canada made Me* v. 127 Old bits of dead grass, like tufts of hair, stuck out of the mud. Near the riverline the snow had not melted. 1979 R. Cox *Auction* ii. 41 'The Yanks must be going flat out,' said Horst. 'Thank God we're not defending that riverline.' 1790 J. Backus *Diary* 6 Dec. in W. W. Backus *Geneal. Mem. Backus Family* (1889) 93 Came down to the river mouth of a large run. 1788 J. May *Jrnl. & Lett.* (1873) 75 We contemplated in our plans a grand bridge over the river ravine. 1879 *Encycl. Brit.* X. 276/2 The river-ravine likewise crept backward, but at a more rapid rate. 1849 Thoreau *Week Concord Riv.* 370 There is a pleasant tract on the bank of the Concord, which I have in mind;..the open wood, the river-reach. 1879 *Encycl. Brit.* X. 272/1 In a vast river system like that of the Mississippi, the area of drainage is..extensive. 1962 H. R. Loyn *Anglo-Saxon Eng.* i. 9 For the main part the river-systems drain west in this area.
c. *river-craft* (earlier and later examples), *steamboat, steamer, traffic*.
1837 De Quincey in *Blackw. Mag.* July 94/1 From the want of bridges, or sufficient river craft for transporting so vast a body of men. 1963 *Times* 18 May 16/5 (*caption*) Various kinds of rivercraft on the Yangtze Kiang in China. 1979 A. Morice *Murder in Outline* xiv. 115 A boat-yard belonging to..an old-established family firm, who hired out river craft. 1857 M. H. Stacey *Jrnl.* 24 May in *Uncle Sam's Camels* (1929) II. 28 What an immense difference we find between the quiet Sundays at home and the bustling ones on board these river steamboats. 1902 Conrad *Youth* 67, I was going to take charge of a two-penny-halfpenny river-steamboat with a penny whistle attached! 1833 E. T. Coke *Subaltern's Furlough* v. 70 The American river steamers are noble vessels. 1903 Joyce *Let.* 8 Feb. (1966) II. 26, I..came back to Paris in one of the little river-steamers. 1936 *Discovery* Dec. 379/2 The ordinary river-steamer services. 1879 *Rep. Comm. Navig. River Thames* p. xxx in *Parl. Papers 1878–79* (C. 2338) XLI. 245 As to the hour at which the ordinary river traffic or daylight excursions should end, there is more difference of opinion. 1968 W. Warwick *Surfriding in N.Z.* 40/3 Dangerous currents and river traffic.
d. *river board, -crossing, -cult, -damp* (later example), *-debris, -dream, -fancy, -flow, -glimpse, -link, lot, -mist, police, trip, -voyage, week*; **river blindness,** (blindness due to) onchocerciasis; **river capture** *Physical Geogr.*, the natural diversion of the headwaters of one stream into the channel of another, freq. resulting from rapid headward erosion of the latter stream; **river engineering,** the branch of civil engineering concerned with the improvement and control of rivers; **river gravel,** gravel that was formed on the bed of a river; **river ooze, River Ouse,** rhyming slang for 'booze'; **river-pay** (later example); **river stone,** a diamond found during river-digging.
1955 *Times* 8 July 9/7 Some types of blindness are more intractable. Such, for example, is the notorious 'river blindness' of the Gold Coast, where in the northern territories there are estimated to be 40,000 blind people. 1972 *Daily Tel.* 22 Nov. 4/4 One French project is to eradicate 'river blindness', an insect-born disease which has ravaged and depopulated the valleys of the Volta rivers. 1975 *Sci. Amer.* Oct. 53/2 There are villages in tropical Africa and Central America where as many as 15 per cent of the people are blind. They are victims of 'river blindness', a frequent complication of the parasitic disease onchocerciasis which.. has recently been recognized as a major public-health problem throughout the tropical world. 1948 *Act* 11 & 12 *Geo. VI* c. 32 § 1 The Ministers shall..by order establish boards (to be known as 'river boards') for the areas so defined, who shall have the functions conferred on or transferred to them by or under the following provisions of this Act, being functions relating to land drainage, fisheries and river pollution and certain other functions relating to rivers, streams and inland waters. 1963 *Times* 23 Jan. 6/3 The information they had gathered in the past fortnight from river boards throughout the country made it seem likely that when the thaw came it would reveal an urgent need for many land drainage and flood control schemes. 1901 *Geogr. Jrnl.* XVIII. 227 Examples taken from various parts of Italy of alteration in the direction of valleys due to river-capture, etc. 1937 Wooldridge & Morgan *Physical Basis Geogr.* xv. 211 The river-captures of the first cycle will still be legible in the pattern of the drainage, but there will now be no direct evidence of the former continuity of consequent drainage lines. 1960 B. W. Sparks *Geomorphol.* vi. 112 The type of river capture described is largely explained by differences in rock resistance, but another factor becomes of importance in areas of permeable rocks: the possibility of underground diversion preceding and aiding surface capture. 1977 A. Hallam *Planet Earth* 76/1 Possible signs of river capture that can often be detected in the landscape include windgaps and elbows of capture, incision of the capturing stream below the capture, and the evident misfit nature of the beheaded stream. a 1951 E. Hill in Murdoch & Drake-Brockman *Austral. Short Stories* (1951) 292 The blacks ran with him for four or five miles, as far as the river-crossing. 1965 Auden *About House* (1966) 29 Shrines where a subarctic fire-cult could meet and marry A river-cult from torrid Greece. 1963 *Landfall* Mar. 23 River-damp softened her hair: her skin smelled of soap. 1951 R. Campbell *Light on Dark Horse* 121 The ridge of river-debris after the flood, ran along the base of these strandlooper-dunes. 1936 Auden *Look, Stranger!* 15 Whose river-dreams long hid the size And vigours of the sea. [1882 L. F. Vernon-Harcourt *Rivers & Canals* I. p. v, In preparing a course of lectures on 'River and Canal Engineering'..it appeared to me that a book might be useful.] 1886 *Encycl. Brit.* XX. 571/2 River Engineering. The improvement of rivers may be considered under two aspects, [etc.]. 1966 *McGraw-Hill Encycl. Sci. & Technol.* XI. 585/1 Technical knowledge is inadequate to explain fully the relationship between stream form and valley slope, but it is necessary in river engineering to recognize it. 1934 Blunden *Mind's Eye* III. 170 Spenser, in his *Faerie Queene*..marries the Medway to the Thames with a great display of river-fancy. 1960 *Times* 25 July 11/6 A rise in river-flow. 1964 *Oceanogr. & Marine Biol.* II. 34 When they have a sufficiently high sediment concentration to give a density of the riverflow exceeding that of the salt water, rivers entering the ocean may similarly sometimes continue as underflows. 1875 'Mark Twain' in *Atlantic Monthly* Jan. 70/1 The 'point' above the town, and the 'point' below, bounding the river-glimpse and turning it into a sort of sea. 1874 *Q. Jrnl. Geol. Soc.* XXX. 229 A careful examination of this very interesting deposit convinces me that we have here preserved portions of an old river-gravel. 1975 J. G. Evans *Environment Early Man Brit. Isles* iii. 61 River gravel is a major

economic concern. It is used extensively for road and building foundations... It is also much needed as land for building on. *a* **1862** THOREAU *Maine Woods* (1864) 251 The Allegash..here consists principally of a chain of large and stagnant lakes, whose thoroughfares, or river-links, have been made nearly equally stagnant by damming. **1704** *Public Rec. Colony of Connecticut* (1868) IV. 493 Part of a lot called the River lot, purchased of the said Nathan[ll] Holt. **1968** E. RUSSENHOLT *Heart of Continent* II. v. 76 Families already living along the Assiniboine, exercise 'squatter's rights', and lay claim to the newly-surveyed River Lots. **1926** KIPLING *Debits & Credits* 233 And the river-mist runs silver round their knees! **1931** BROPHY & PARTRIDGE *Songs & Slang 1914–1918* (ed. 3) 350 *River Ouse*, a booze, a drink(ing). **1962** R. COOK *Crust on its Uppers* ix. 76 The place still bulging with smoke and river ooze. **1825** G. F. LYON *Brief Narr. Unsuccessful Attempt to reach Repulse Bay* 2 On the 16th Commissioner Cunningham arrived from Chatham, and the ship's company received their river pay, with three months' advance. **1800** P. COLQUHOUN *Treat. Commerce & Police River Thames* i. 22 The nature of the several articles of Trade and Manufacture..cannot fail to produce a conviction of the indispensable necessity of a well-planned and energetic System of *River-Police*; to regulate and control the economy of so vast a machine, and to protect such an astonishing mass and variety of Property. **1974** *Times* 15 Apr. 2/1 It sank before river police could note its registration marks. **1887** J. MACKENZIE *Austral Africa* II. IV. iv. 87 The 'river stones', as they are called, are usually more valuable than those found in 'dry diggings' or mines. **1904** L. J. SPENCER tr. *Bauer's Precious Stones* I. II. 186 The higher quality of the river stones as compared with those from the dry diggings does not militate against the truth of this theory as to their origin. **1898** J. S. WEBB in *Century Mag.* Mar. 672 (*title*) The river trip to the Klondike. **1912** W. OWEN *Let.* 23 June (1967) 142 Mrs. Lott's River Trip is to be next Tuesday. **1954** J. R. R. TOLKIEN *Fellowship of Ring* 400, I hoped the river-voyage would beat him, but he is too clever a waterman. **1932** D. H. LAWRENCE *Last Poems* 30 Come then!.. In a week! The ancient river week, the old one.

e. *river-bailiff, family, Indian, pilot, pirate, -thief* (earlier and later examples); *river-rat* (earlier and later examples); **River Brethren** *pl.*, members of a Christian sect originating (*c* 1770) among settlers on the Susquehanna river, characterized esp. by the performance of baptisms only in rivers; **river hog, pig** *N. Amer. slang* = RIVER-DRIVER.

1905 W. OWEN *Let.* 7 Aug. (1967) 25 He was fishing this morning when a river bailiff came up. **1854** J. BELCHER *Relig. Denominations U.S.* 919 Others were organized into a body called, *The River Brethren*, partly from the locality in which they were first found, near the Susquehanna, and Conestoga, and chiefly from their baptisms being celebrated only in rivers. **1951** H. E. GILES *Harbin's Ridge* xxiii. 202 And they baptized different, too. Face forward in the water, three times. In the early days, back in Pennsylvania, they'd been named the River Brethren on account of it, I'd heard. **1937** A. HUXLEY *Let.* 15 Dec. (1969) 429 This last is an appendage on one of the numerous vast estates of what are called 'The River Families', who have been living here in a feudal sort of way, in some cases, for two or more centuries. **1902** S. E. WHITE *Blazed Trail* lvi. 384 And now we've gone and bust, just because that infernal river-hog had to fall off a boom. **1964** *Outdoorsman* (Campbellford, Ontario) Dec. 1/2 One may see a visitor with a misty look in his eye; an old blacksmith, top loader, barn boss, teamster, cookie or river hog who has returned to a fleeting glimpse of an era long gone by. **1680** W. HUBBARD *Gen. Hist. New Eng.* in *Mass. Hist. Soc. Coll.* (1815) 2nd Ser. V. 33 The River Indians, such who had seated themselves in seuerall commodious plantations up higher upon Connecticutt river. **1785** T. JEFFERSON *Notes Virginia* 388 The Mohawks carried on a furious war down the Hudson against the Mohiccons and river indians. **1907** F. W. HODGE *Handbk. Amer. Indians* I. 786/2 Mahican ('wolf'). An Algonquian tribe that occupied both banks of the upper Hudson r... To the Dutch they were known as River Indians. **1921** *Dialect Notes* V. 113 *Riverpig*,..a lumberman who follows the drive in low water and dislodges logs from bars, mud, etc. **1947** *Sat. Even. Post* 8 Mar. 20/1 River pigs bristled all around him, men who hadn't seen a town or a saloon for nine months. **1883** *Harper's Mag.* Oct. 799/2 Mr. Clemens..in his character, first, as an apprentice to the occupation of a river pilot. *c* **1849** 'N. BUNTLINE' *B'hoys of N.Y.* iv. 30 Alvorado began to see how well his friend and rival River Pirate was situated. **1962** S. E. FINER *Man on Horseback* xii. 230 The force of river pirates known as the Binh Xuyen. **1883** J. GREENWOOD *Tag, Rag, & Co.* 35 With enough of 'river rats' to occupy my thoughts during my overland journey home, I paid my old waterman his due. **1905** *Bull. U.S. Forest Service* LXI. 44 *River rat*, a log driver whose work is chiefly on the river; contrasted with Laker. **1976** *Kingston* (Ontario) *Whig-Standard* 4 June 28/1 Tom Harrison, a 'river rat' since 16, has purchased the Gananoque Water Taxi. **1853** DICKENS *Down with Tide* in *Househ. Words* VI. 481/2 River thieves can always get rid of stolen property..by dropping it overboard. **1882** J. D. MCCABE *New York* xxxiv. 518 Another dangerous class of criminals are the river thieves, or 'River Pirates'.

f. *river-crosser, -inspector.*

1936 M. FRANKLIN *All that Swagger* xviii. 167 He saddled the river-crosser—a tall old grey. **1875** 'MARK TWAIN' in *Atlantic Monthly* Feb. 221/2 We had a fine company of these river-inspectors along, this trip.

g. *river-borne, -caught, -cut, -encircled, -fed, -formed* (later example), *-rounded*; also with pres. pples., as *river-winding*.

1928 *Daily Tel.* 4 Dec. 12/4 Splitting the market into two, for river-borne and rail-borne supplies respectively. **1924** A. J. SMALL *Frozen Gold* xii. 248 Others sat round the braziers and held great slabs of river-caught salmon against the red-hot grids. **1957** G. E. HUTCHINSON *Treat. Limnol.* I. i. 81 Other authors have believed the lakes to occupy river-cut valleys. **1951** River-encircled [see *mountain-cresting* adj. s.v. *MOUNTAIN 7 b]. **1913** E. F. BENSON *Thorley Weir* i. 21 A strip of river-fed grasses and herbs of the waterside. **1977** A. HALLAM *Planet Earth* 50 Glacial erosion modifies river-formed valleys into U-shapes. **1879** G. M. HOPKINS *Poems* (1967) 79 Cuckoo-echoing, bell-swarmèd, lark-charmèd, rook-racked, river-rounded. **1951** S. SPENDER *World within World* ii. 39, I used to go for long walks and bicycle rides into the hilly, tree-scattered, river-winding countryside.

h. With adjs., as *river-dark, -thick, -wise*.

1925 E. SITWELL *Troy Park* 100 She swims across the river-dark vast floors. **1924** —— *Sleeping Beauty* xv. 53 How river-thick flow your fleeced locks. **1934** WEBSTER *Riverwise*, adj. **1940** *Sun* (Baltimore) 22 Apr. 3/6 River-wise city officials..expressed belief the inundation would prove more annoying than damaging.

5. *river-bird, -fly*; **river bass,** substitute for def.: one of several freshwater fishes of the family Centrarchidæ, esp. the black bass, *Micropterus salmoides*; (earlier and later examples); **river chub** (earlier example); **river herring** (earlier and later examples); also, formerly, the mooneye, *Hiodon tergisus*; **river pearl,** a pearl from a freshwater mussel, esp. *Margaritifera margaritifera*; **river trout** (earlier examples).

1857 *Spirit of Times* 11 Apr. 86/2 The Oswego (sometimes known as the 'river bass') is the heavier fish, often attaining to eight pounds weight. **1890** W. D. HOWELLS *Boy's Town* 30 There were men who were reputed to catch at will, as it were, silvercats and river bass. **1910** W. DE LA MARE *Three Mulla-Mulgars* iv. 52 They heard the trump-billed river-birds calling their secrets one to another. **1882** River chub [see *horny-head* s.v. *HORNY *a.* (*sb.*) 7]. **1958** J. CAREW *Black Midas* vi. 128 A river fly 'lighted on the tip of Belle's nose. **1842** J. E. DEKAY *Zool. N.Y.* IV. 266 [The river mooneye] is known under the popular names of Herring, River Herring, and Toothed Herring. **1977** *Hongkong Standard* 12 Apr. 2/8 Officials of the National Marine Fishery Services found illegal amounts of river herring in the trawler's hold 240 miles southeast of Boston. **1885** *Encycl. Brit.* XVIII. 447/2 River-pearls are produced by the fresh-water mussels inhabiting the mountain-streams of temperate climates in the northern hemisphere. **1963** P. MOYES *Murder à la Mode* iii. 55 Get me lots of gold bracelets and some river pearls. **1975** *Times* 6 Mar. 5/1 The borders..are adorned with sapphires, rubies, emeralds, and river pearls. **1834** *Chambers's Edin. Jrnl.* 6 Dec. 357/3 Fish (we speak of river trout) spawn seldom in such [slow, muddy] waters. **1867** *Harper's Mag.* Dec. 48/1 He has already achieved unequalled success in breeding river-trout.

6. *river-cress, palm, willow*; **river birch** (examples); **river (red) gum,** substitute for def.: the most widespread of the red gum-trees, *Eucalyptus camaldulensis*; cf. *RED GUM[2] 2; (later examples); **river white gum,** a gum-tree, *E. andreana*, with smooth white bark; (earlier and later examples).

1853 W. DARLINGTON *Flora Cestrica* (ed. 3) 275 Black Betula. Black Birch. Red Birch. River Birch. **1884** C. S. SARGENT *Rep. Forests N. Amer.* 161 Red Birch. River Birch... Used in the manufacture of furniture. **1969** T. H. EVERETT *Living Trees of World* xiii. 104/1 The river birch grows in lowlands from Massachusetts to Florida. **1953** A. CLARKE *Moment Next to Nothing* I. i. 18 I've little to offer a guest... But it is yours, a round of bread, a pick Of river-cress and goat-cheese. **1911** C. E. W. BEAN *'Dreadnought' of Darling* ii. 17 A single line of railway runs straight out into the back country..and stops within sight of the river gums. **1963** W. E. HARNEY *To Ayers Rock & Beyond* iii. 29 The gaunt river-gums..grow along the bank and bed of this one-time mighty river. **1964** D. VARADAY *Gara-Yaka* ix. 74 In the lush valleys among the rock forts..there stand magnificent River Palms. **1930** A. J. EWART *Flora Victoria* 821 River Red Gum. A fairly tall tree, up to 80 or 150 feet high, with a greyish-white bark shedding in thin leaves or flakes. **1973** G. M. CHIPPENDALE *Eucalypts W. Austral. Goldfields* 183/2 The river red gum is the most widespread eucalypt in Australia. **1884** A. NILSON *Timber Trees New South Wales* 58 River White Gum.—Trunk smooth and nearly white. **1961** PENFOLD & WILLIS *Eucalypts* xii. 249 The tree [sc. *E. andreana*] is known as 'White Top', or 'River White Gum', and occurs fairly plentifully on the river banks and mountain ranges of eastern New South Wales. **1963** M. SHADBOLT *Summer Fires & Winter Country* 233 In summer we swam down under the river-willows.

river, *v.* Add: **2.** *intr.* To follow a river-like course.

1921 A. CLARKE *Sword of West* 23 Far below me lay A deep green valley rivering through grey mist.

river-bottom. (Earlier examples.)

1752 C. GIST *Jrnl.* 20 Feb. (1893) 75 Then continued our course..the last 5 [miles] thro the river bottoms, which were a mile wide and very rich. **1793** G. IMLAY *Topogr. Descr. W. Terr. N. Amer.* (ed. 2) 411 At the edge of the wood lands, and before your descend [*sic*] into the river bottoms, one of the most charming prospects..displays itself.

ri·ver-di:gging. orig. *U.S.* [f. RIVER *sb.*[1] + DIGGING *vbl. sb.*] **a.** *pl.* Gold or diamond diggings in the neighbourhood of a river or stream, or in a dried-up river-bed. **b.** The action of digging at such a place. Hence **river-digger.** Cf. DIGGING *vbl. sb.* 4.

1851 D. B. WOODS *Sixteen Months at Gold Diggings* i. 13 The 'river diggings' include the bars and auriferous portions of the channels of the tributaries of the Sacramento and San Joaquin, during their passage through the foot-hills. **1862** [see *dry diggings* s.v. *DRY *a.* C. 3]. **1881** E. E. FREWER tr. *Holub's Seven Yrs. S. Afr.* I. iii. 60 The settlement at the river-diggings sprang up with a rapidity as marvellous as those of California. **1904** L. J. SPENCER tr. *Bauer's Precious Stones* I. II. 185 The amalgamation of the 'dry diggings' to form the De Beers Consolidated Mines, has had the effect of increasing the number of river diggers. **1920** F. C. CORNELL *Glamour of Prospecting* i. 10 A modicum of genuine men of past experience—principally ex-'river-diggers'—men whose small capital was running away like water for bare necessities in this miserable dust-hole of creation. **1947** L. HASTINGS *Dragons are Extra* i. 10 Dabbling in river-digging for diamonds. **1947** *E. Afr. Ann. 1946–7* 122/1 The 'river diggings' of South Africa where workings extend..along present or ancient river beds.

river-driver. For *U.S.* read *N. Amer.* and add later examples. Hence **ri·ver-drive,** a drive of logs down a river; **ri·ver-driving,** the action of driving logs down a river.

1854 F. J. BULLARD *Now-a-Days* 65 River drivin' is the pootiest part of loggin', I think. **1908** S. E. WHITE *Riverman* v. 50 How does river-driving strike you? **1920** *Rod & Gun in Canada* Nov. 646/1 We were enjoying the sights we saw along the line: the lakes, rivers, the river-drives. **1937** C. M. WILSON *Aroostook* 105 The 'river-drives' were the consummation of turning out the timber. **1963** *Canada Month* Nov. 22/1 The lumber trade furnished employment for thousands of lumber jacks, river drivers, and sailors. **1972** [see *pick-pole* s.v. *PICK *sb.*[1] 8]. **1974** D. SEARS *Lark in Clear Air* ii. 27 He..got a job as boss on a river-drive. *Ibid.* xiv. 178 Tommy bought a new pair of river-driving boots.

riverine, *a.* and *sb.* Add: **A.** *adj.* **2.** Also, resembling a river.

1884 E. JENKINS *Week of Passion* II. iv. 156 His face,..deeply rutted, here and there, with expressive valleys and riverine lines of wrinkle.

rivering (ri·vəriŋ), *ppl. a. nonce-wd.* [Cf. RIVER *v.* in Dict. and Suppl.] Flowing in river form.

1939 JOYCE *Finnegans Wake* (1964) I. 216 Beside the rivering waters of, hitherandthithering waters of. Night!

riverrun[1] (ri·vəɹ,rʌn). *nonce-wd.* [Cf. RUN *sb.*[1] 29 a.] The course which a river shapes and follows through the landscape.

1939 JOYCE *Finnegans Wake* (1964) I. 3 Riverrun, past Eve and Adam's, from swerve of shore to bend of bay, brings us by a commodius vicus of recirculation back to Howth Castle and Environs.

river runner. *N. Amer.* [f. RIVER *sb.*[1] + RUNNER.] **a.** One who drives a river-vessel. **b.** One who engages in the leisure activity of running, or travelling down, a river in a small craft (as a rubber dinghy, etc.). Hence **river run**[2], **river running.**

1913 O. A. ROTHERT *Hist. Muhlenberg Co.* xxxi. 393 The coal barges were taken up and down the river by men known as 'river runners'. **1962** *Nat. Geographic* Apr. 561/1 (*caption*) River runners stop for a swim in the Little Colorado's Blue. **1965** MRS. L. B. JOHNSON *White House Diary* 28 Oct. (1970) 330 I'll never be satisfied after having been down the Snake River until I've become a river runner. **1968** *Sunset* Mar. 34/2 On your first river-running excursion..you'll..welcome all the advice you can get. *Ibid.* (*heading*) What to take on a River Run. **1974** C. F. MARTIN *Sierra Whitewater* 8 The drive from San Francisco Bay to good river runs is thus about an hour shorter than the drive to good skiing. **1976** *National Observer* (U.S.) 24 Jan. 12/2 River running in an inner-tube, skeet shooting, and trout fishing are speedily arranged at minimal cost.

riverscape (ri·vəɹskēɪp). [f. RIVER *sb.*[1] + SCAPE *sb.*[3]; formed in imitation of LANDSCAPE.] **1.** A picturesque view or prospect of a river.

1903 C. S. SMITH *Barbizon Days* 17 Moret has noble turreted gateways and Grez a church more picturesque than that of Montigny, riverscapes more alluring, and a ruined château. **1927** H. V. MORTON *In Search of England* xii. 213 All the beauty and peace of the Warwickshire countryside have been packed into one riverscape. **1966** *New Statesman* 25 Nov. 793/1 Nobody listened to his proposal to replace Rennie's Waterloo Bridge with a tunnel to save spoiling a fine riverscape.

2. A painting of a river or riverside scene.

1930 *Time & Tide* 9 May 606 Mr. Lamorna Birch and Sir H. Hughes Stanton show very capable riverscapes. **1964** N. FREELING *Double-Barrel* iii. 91 The arrangement opposite: a wholly bluey-greeny Monet riverscape over a little writing table. **1975** *Gramophone* Dec. 1065/1 The record is very well produced with a fine autumnal riverscape by Monet on the front.

riverside. Add: **b.** *riverside bathing, sand, village.*

1760 G. WASHINGTON *Diary* 14 Apr. (1925) I. 153 Has 2 Pecks of sd. Earth and 1 of Riverside Sand. **1914** W. OWEN *Let.* 1 June (1967) 257 They went to visit some friends who lived in a riverside village. **1946** *Nature* 31 Aug. 290/1 At the age of fourteen, he contracted severe rheumatic trouble, as a result of river-side bathing.

riverward, *adv.* and *a.* Add: **B.** *adj.* (Later example.)

1966 D. VARADAY *Gara-Yaka's Domain* xi. 121 In the riverward spoor a few drops of blood showed up.

ri·verwise, *adv.* [f. RIVER *sb.*[1] + WISE *sb.*[1]] In the manner of a river; in relation to a river.

1927 JOYCE *Lett.* (1966) III. 164 She has grown—riverwise—since the night you heard her under the sign of Ursa Minor. **1946** R. GRAVES *Poems 1938–45* 20 They carry Time looped so river-wise about their house There's no way in by history's road To name or number them. **1946** R. BLESH *Shining Trumpets* vii. 151 The French laid out the city, eleven squares riverwise and six squares deep.

rivery, *a.* Delete *rare* and † *Obs.*[—1] and add: **1.** (Later examples.)

1916 A. HUXLEY *Burning Wheel* 13 Lorelei, Combing the silken mystery, The glaucous gold of her rivery tresses. **1977** *Rolling Stone* 19 May 86/3 'The Wheel' is more Western bop than swing, a brilliant, rivery thing that makes it easy to understand why Charlie Parker liked jamming with Ray Price's band.

2. (Later examples.)

1889 W. B. YEATS *Wanderings of Oisin* 78 From rolling valley and rivery glen, With horsemen hurrying near and far, I drew at evening my mailed men. **1973** P. LIVELY *Ghost of Thomas Kempe* ii. 17 Green, rivery, elm-scattered Oxfordshire.

Rivesaltes (rivsalt). The name of a town near Perpignan in southern France, used *attrib.* and *absol.* of a sweet wine produced there.

1824 A. HENDERSON *Hist. Anc. & Mod. Wines* II. ii. 178 At Salces..a white wine is grown, which..is thought to resemble Tokay; but, in point of richness, it is inferior to the Rivesaltes. **1836** C. REDDING *Mod. Wines* (ed. 2) v. 131 The quantity of Rivesaltes muscadine made is about sixty-five hogsheads per annum. *c* **1870** in H. W. Allen *No. 3 St. James's St.* (1950) 185/1 Rivesaltes, pints—42/-. **1920** [see *JURANÇON]. **1971** A. DURKAN *Vendange* ix. 85 A speciality from this district are the dessert wines called Vin Doux Naturel... The best-known names are Banyuls, Muscat de Frontignan, Muscat de Lunel, Rivesaltes, Maury and Grand Roussillon.

rivet, *sb.*[1] Add: **1. e.** *pl.* Money, coins. *slang.*

1846 *Swell's Night Guide* 130/1 Rivets, money. **1848** *Sinks of London laid Open* 121 Rivits, money. **1937** 'J. CURTIS' *You're in Racket* xviii. 190 'So you got a bit of rivets to speculate?' 'I ain't said so. All I said as I could put up a bit.'

2. *rivet bar, head, hearth, tail*; **rivet gun,** a hand-held tool for inserting rivets.

1890 D. K. CLARK *Steam Engine* I. 657 The diameter of the ⅜-inch rivet-bars was reduced to ·03 inch. **1950** *Nat. Geogr. Mag.* Sept. 297 (*caption*) Chattering rivet guns attach the door to a barrel section of a Constellation. **1902** *Encycl. Brit.* XXXII. 597 The riveters also work in squads, ..with sometimes a catcher, *i.e.*, a boy to pass on the heated rivets when the distance from the rivet-hearth is great. **1978** *Jrnl. R. Soc. Arts* CXXVI. 690/2 On long-term creep tests at operating temperatures, it was found that the rivet heads and tails cracked.

riveter. Add: **2.** (Further examples.) (See also quot. 1963.)

1928 [see *BOATSWAIN 3]. **1963** JONES & SCHUBERT *Engin. Encycl.* (ed. 3) 1068 'Riveting machines', according to common usage, differ from 'riveters' in that the riveting operation with a machine is effected by a succession of blows or by a compressive rotating action, whereas a riveter merely subjects the rivet to compression.

riveting, *ppl. a.* (In Dict. s.v. RIVET *v.*) (Further examples.)

1854 *Rambler* Feb. 198 Riveting as were these narratives when we first read them..those who now read them for the first time are generally disappointed. **1967** N. MARSH *Death at Dolphin* v. 129 Her smart friends..said things like: 'Absolutely riveting' and 'Loved your play'. **1975** *Evening Standard* 24 June 20/1 Landladies..are far more genned up on local history and things to do (plus some riveting gossip) than the managers of more impressive emporia. **1979** E. H. GOMBRICH *Sense of Order* x. 271 'Some 2500 years ago'—Kurz sums up his riveting article—'a Greek artist conceived the strange idea of putting a movable ring into the mouth of a lion.'

ri·vetingly, *adv.* [f. RIVETING *ppl. a.*] In a riveting manner.

1971 *Guardian* 21 June 11/1 'Yesterday's Men' was an extreme example: it provided rivettingly viewable entertainment. **1973** *Daily Tel.* 22 Nov. 10/3 His animation of wartime London is uncannily detailed, rivetingly atmospheric.

Riviera (rivyē͡ə·rǎ). Also **riviera.** [It., lit. 'coast, shore'.] **1.** The name of the Italian sea-board about Genoa, applied also to the Mediterranean coast from Marseilles in France to La Spezia in Italy, a fashionable winter resort in the 19th century and more recently popular for summer holidays; usu. with *the*. Also *attrib.* Cf. RIVER *sb.*[1] 3.

[**1632**: see ROOT *sb.*[1] 4 c]. **1766** J. NORTHALL *Trav. through Italy* VI. ii. 471 The dominions of this State consist of the countries extending along the sea-coast, on both sides, from the city of Genoa, which are called the eastern and western Rivieras. This word in Italian signifies a strand. **1797** *Encycl. Brit.* VII. 426/2 The people of Genoa revolted..and reduced a great part of the Riviera. **1852** DICKENS in *Keepsake* 120 He had hired an old place on the Riviera, at an easy distance from my city, Genoa. **1863** GEO. ELIOT *Let.* Oct. (1956) IV. 111, I shall imagine you winding along the Riviera, and then settling in sight of beautiful things. **1892** I. ZANGWILL *Childr. Ghetto* II. II. i. 270, I had better take a hansom to the Riviera at once. **1909** C. F. G. MASTERMAN *Condition of England* ii. 57 In Biarritz, Pau, Dinard—he might have said in the whole *côte d'azur* of the Riviera—'the English have conquered us,' he declares **1939** S. DE MADARIAGA *Christopher Columbus* iii. 25 Eating and drinking..in that Genoese riviera so sunny and full of the joy of existence. **1941** KOESTLER *Scum of Earth* 28 There was an elderly Riviera-Englishman on the platform. **1978** *Times* 18 Mar. 11/1 The Riviera or Cote d'Azur has a special affinity with the English who apart from the Romans were its first holidaymakers.

2. *transf.* Applied to other coastal regions considered to resemble the Mediterranean Riviera. Also with *a* and *pl.*, and *attrib.*

1891 M. F. SWEETSER *King's Handbk. U.S.* 175 The old convents and churches..and the yachting in the adjacent waters, furnish a great variety of interest for visitors to the American Riviera. **1904** *Railway Mag.* Sept. 258 'The Riviera Express' is the title chosen by Mr. J. C. Inglis, the General Manager of the Great Western Railway, as the most apposite name for the Plymouth–Paddington non-stop express. **1910** *Bradshaw's Railway Guide* Apr. 1126 Lyme Regis, Dorset. Hotel Alexandra... The only hotel in its own grounds in the English Riviera. **1911** HEATH & HASLEHURST (*title*) The Cornish Riviera. **1922** WODEHOUSE *Girl on Boat* xvii. 307 'Why not Cornwall?' said Sam. 'The Riviera of England!' **1951** W. SANSOM *Face of Innocence* iii. 36 He did not know the rivieras, I knew them well. **1974** *Sat. Rev. World* (U.S.) 19 Oct. 43/1 Every properly equipped nation must have a Riviera.

Rivieran (rivyē͡ə·rǎn), *a. rare.* [f. prec. + -AN.] Of, pertaining to, or characteristic of the Riviera.

1897 A. BEARDSLEY *Let.* 1 Apr. (1970) 290, I have been getting into quite a glow over a packet of Rivieran photographs. **1909** *Westm. Gaz.* 21 Sept. 4/2 'The climate,' she writes, 'is of quite Rivieran mildness during the winter months.'

rivière. Add: **1.** (Earlier and later examples.)

1879 M. E. BRADDON *Vixen* I. xvi. 304 The special presents which stood out..were—a *rivière* of diamonds.., a pair of priceless crackle jars, a Sèvres dinner-service of the old *bleu-du-roi*. **1958** M. KELLY *Christmas Egg* II. 61 Rivière—what was that?—of large brilliant diamonds, with ear-rings en suite, silver.

2. *needlework.* A row of open-work.

a **1855** C. BRONTË *Professor* (1857) II. xviii. 9 A 'rivière', or open-work hem round a cambric handkerchief. **1886** in A. Adburgham *Shops & Shopping* (1964) xviii. 209 In Blonde de Seville, it [*sc.* a lace fichu] falls into cascades and rivières.

riwa-riwa, var. *REWA-REWA.

rix-dollar. Add: **2.** A unit of currency introduced into certain former colonies, as by the Dutch in Cape Province and the English in Ceylon. (See also quot. 1962.)

1785 G. FORSTER tr. *Sparrman's Voy. Cape Good Hope* I. 19 Board and lodging are paid for here as at the Cape, from one rix-dollar to one and a half a day. **1790** E. HELME tr. *Le Vaillant's Trav. Afr.* I. ii. 21 While I was there [*sc.* at Capetown] butchers meat was very cheap; I have seen thirteen pounds of mutton bought for an escalin; (elevenpence English) an ox for 12 or 15 rix dollars; (at six shillings and nine-pence English, the rix dollar) ten quarters of corn for 14 or 15 Rix-dollars, and other things in proportion. **1827** G. THOMPSON *Trav. & Adventures Southern Afr.* I. I. iv. 79 He provides a salary of 400 rix-dollars to encourage a day-school for females. **1836** *Penny Cycl.* VI. 453/2 In the district of Putlam they were faced boldly in the open forest, and ensnared singly, for a reward varying according to the size and description of animal, from 11 to 352 rix-dollars. **1866** J. LEYLAND *Adventures Far Interior S. Afr.* III. 233 The charge for crossing in the barge was fifteen shillings, or ten rix dollars. **1900** A. H. KEANE *Boer States* p. xviii, *Rixdollar*, a coin current in the Cape in colonial times. **1962** R. A. G. CARSON *Coins* 533 The Dutch monetary system of a rix-dollar or rijksdaalder of 48 stuiver was continued [in Cape Province] by the British in the early nineteenth century.

riyal: see *RIAL *sb.*[1] 5. **riyo,** var. *RYO.

riza (ri·zǎ). [Russ., f. OSlav. *riza* garment.] A metal shield or plaque framing the painted face and other features of a Russian icon, and engraved with the lines of the completed picture.

1927 E. H. MINNS tr. *Kondakov's Russ. Icon* ii. 37 As long ago as the fourteenth century, under Greek influence, the Russians began to cover..the figures with plates of silver showing in more or less relief the outlines and folds of the clothes and vestments: such a plate is called a *riza*, properly speaking a garment, especially a chasuble:.. The parts of the figures left unclothed, faces, hands, and the like, all the flesh tints, show through holes in the *riza*. **1931** P. ROMANOV tr. *Zarine's Three Pairs of Silk Stockings* ii. 9 One hall was full of pictures... Another was covered with ikons with old rizas of pearl. **1963** T. T. RICE *Russian Icons* 25 It became usual to encase an icon in a costly metal cover or *riza*, in which openings were cut to show the essential sections of the painting. **1970** *Guardian* 1 Apr. 16/5 Sometimes elaborate silverwork was added, covering much of the surface and leaving only the faces and figures visible. This metalwork, the riza, was occasionally ornamented with precious or semi-precious stones. **1978** *Daily Tel.* 24 Aug. 12/5 Among the collection is a 17th century icon of the Virgin of Kazan, with embossed silver-gilt riza..dating from the end of the 19th century.

rizalite (rizā·ləit). *Geol.* [f. *Rizal*, the name of a province of central Luzon in the Philippines (named after Dr. José *Rizal* (1861–96), Philippine patriot) + -ITE[1].] Any tektite of the type characteristically found in the neighbourhood of Manila.

1928 H. O. BEYER *Philippine Tektites* (1961) I. 21 Since.. our Luzon material cannot be properly classed as either Billitonite or Australite..I am proposing the name *rizalite* as a temporary designation for it. **1937** *Mineral. Abstr.* VI. 403 A popular account is given of the tektites ('rizalites'), mostly from Rizal province, which are spherical or cylindrical. **1940** *Pop. Astron.* XLVIII. 44 Rizalites (pitted spheroids, ovals, and cylindrical forms being most characteristic). *Ibid.* 45 The most typical Rizalites occur only in Luzon, although a few similar, pitted specimens are known from Borneo and Java.

r-less: see R I. 1 c.

RNA. *Biochem.* Also (*rare*) **R.N.A. 1.** Abbrev. of *ribonucleic acid*.

See also *messenger RNA, *ribosomal RNA, *transfer RNA.

1948 *Jrnl. Biol. Chem.* CLXXV. 989 Protein, phospholipides, and ribonucleic acid (RNA) are lost from the liver. **1959** *Sunday Times* 14 June 24/3 RNA carries the genetic information in the small viruses only, while DNA does this in all other organisms. **1968** H. HARRIS *Nucleus & Cytoplasm* iii. 57 No precise mechanism for this RNA-mediated regulation [of genetic activity] has yet been put forward. **1968** *Times* 19 Oct. 4/8 Further decipherment of the genetic code depended on synthesizing artificial RNAs containing all 64 possible triplet combinations. **1969** R. B. FULLER *Operating Man. Spaceship Earth* iv. 52 The genes and the R.N.A. and D.N.A. and other fundamental principles governing the fundamental design controls of life systems. **1977** *Nature* 27 Oct. 834/1 Enzymatic rather than chemical methods were used to degrade the RNA.

2. Special Combs. **a. RNA virus,** a virus in which the genetic material is RNA.

1963 *Nature* 17 Aug. 664/1 The use of specific inhibitors has led to the conclusion that the replication of small RNA viruses is not dependent on the integrity of host DNA. **1975** *Ibid.* 23 Oct. 634/3 Many RNA viruses are involved in animal cancers and it is believed that some human cancers may be caused in the same way.

b. In names of enzymes acting on RNA, as *RNA polymerase, replicase, synthetase.*

1962 *Biochem. & Biophysical Res. Communications* VII. 30 The stimulation by the phage DNA is dependent on the addition of purified RNA polymerase with polymerase-poor fractions. **1973** R. G. KRUEGER et al. *Introd. Microbiol.* xi. 331/1 RNA polymerase is a complex enzyme whose affinity for different promoters..depends upon protein factors which are not tightly bound to the core enzyme. **1965** *Biochem. & Biophysical Res. Communications* XVIII. 283 (*heading*) Isolation of turnip yellow mosaic virus RNA replicase. **1969** *New Scientist* 18 Dec. 590/1 These small spherical viruses contain RNA, which functions directly as messenger, coding for three proteins, coat protein, 'maturation factor' and RNA replicase. **1962** F. H. BERGMANN in *Methods in Enzymol.* V. 708 Since the term 'amino acid-activating enzyme' thus represents only a partial description of the catalytic activity of such enzymes, the term 'amino acyl RNA synthetase' is more appropriate. **1968** *Times* 29 Oct. 7/4 The third [gene] specifies an enzyme, called RNA synthetase, which produces copies of the RNA strand.

RNase (ārenē͡ɪ·z). *Biochem.* Also **RNAase** (ārenē͡ɪ·ē͡ɪz). [f. *RN(A + *-ASE.] = *ribonuclease* s.v. *RIBO-.

1957 *Jrnl. Biochem.* XLIV. 761 This RNase was inhibited by various metal ions. **1961** *Biochimica & Biophysica Acta* XLVII. 145 When yeast protoplasts are incubated with RNase, protein synthesis is inhibited. **1961** *Ibid.* LI. 190 The gel-filtration method is recommended as a suitable method to obtain the highly active fraction from RNAase I core. **1968** A. WHITE et al. *Princ. Biochem.* (ed. 4) xii. 258 The histidine residues of RNase are uniquely reactive with iodoacetate or bromoacetate at pH 5·5. **1977** *Nature* 22/29 Dec. 760/2 The ability of cassava latent virus nucleic acid to infect *N*[*icotiana*] *clevelandii* was destroyed by DNase but not RNase.

roach, *sb.*[1] Add: **2. b.** *roach fisherman, swim*; **roach pole,** a type of rod used in fishing for roach.

1974 *Country Life* 30 May 1333/2, I didn't have much success as a roach fisherman. **1897** *Encycl. Sport* I. 24/2 On the Thames, the bank-angler commonly uses a long bamboo roach-pole and tight line. **1944** 'N. SHUTE' *Pastoral* i. 3 With his long greenheart roach-pole, his bag of ground bait, and his gentles. **1971** *Country Life* 16 Dec. 1730/4, I believe the Lea roach-pole anglers were then regarded as supreme masters of that particular art. **1902** *Chambers's Jrnl.* Nov. 699/1, I was preparing to fish a not unpromising roach-swim, and was trying the depth.

roach, *sb.*[3] Add: [App. a *transf.* use of ROACH *sb.*[1]] **1.** (Further *transf.* example.) Cf. *roach-back(ed)* s.v. ROACH *sb.*[1] 2 a.

1955 W. W. DENLINGER *Compl. Boston* I. 105 Many Boston terriers with level backs will show roach on a cold day.

2. a. *U.S.* A roll of hair brushed upwards and back from the face; a topknot (see quot. 1959[2]). Also *attrib.*, as *roach cut*.

1881 J. C. HARRIS in *Century Mag.* June 244/2 Den he take en walk upter de little Gal, Brer Rabbit did, en pull he roach, en bow, en scrape he foot, en talk mighty nice en slick. **1929** W. FAULKNER *Sound & Fury* 177 A man with a fierce roach of iron grey hair. **1959** E. TUNIS *Indians* 47/2 Most of the head was plucked or singed bare for a roach cut; only a strip of hair was left from front to back. *Ibid.* 48/1 The most usual ceremonial headdress in the East was the artificial roach.

b. *attrib.* in sense of *ROACHED *a.* 2.

1781 *R. Georgia Gaz.* 8 Mar. 4/2 (Advt.), A Black Horse, about 13 and an half hands high, half roach main [etc.]. **1835** J. T. IRVING *Indian Sk.* II. 4 She was mounted upon a little wall-eyed, cream-coloured pony, with a roach mane and a bobtail.

roach, *sb.*[4] Add: Chiefly *U.S.* **1.** (Further examples.)

1942 E. PAUL *Narrow St.* x. 75 Her failure to get results kept her hopping like a roach in a skillet. **1950** *Harlem Q.* Fall–Winter 21 A fat roach sluggishly made its way ceilingwards. **1976** *National Observer* (U.S.) 29 May 15A/3 The Postal Service is the kind of problem most members think shouldn't exist. It's like roaches in the kitchen.

2. *slang.* A policeman.

1932 *Evening Sun* (Baltimore) 9 Dec. 31/5 *Roach*, policeman. **1968** *Word Study* Dec. 5/1 Not only is a policeman a *bull*; he may also be a *roach*.

3. *slang.* The butt of a cigarette, *spec.* a marijuana cigarette. (Perh. a different word.)

1938 *New Yorker* 12 Mar. 47/2 A pinched-off smoke, or stub, is a roach. **1953** W. BURROUGHS *Junkie* ii. 29 'Would you like to get high?' Mary asked. 'There may be a roach around here somewhere.' **1966** T. PYNCHON *Crying of Lot 49* iii. 64 Holding up the glowing roaches of their cigarettes..to spell out alternate S's and O's. **1971** *It* 2–16 June 2/1 Freaks from several other pads saved all our roaches for a week and posted them to Det. Sgt. Boothe who is the chief of Bournemouth drug squad. **1972** M. J. BOSSE *Incident at Naha* i. 46, I..took out my pot pouch and cigarette paper... I..rolled myself a joint... I had finished the roach down to my fingernails.

4. *attrib.* and *Comb.*, as (sense 1) *roach killer, poison, -powder*; *roach-crawling* adj.; (sense *3) *roach holder*; **roach clip** = *roach holder* above.

1968–70 *Current Slang* (Univ. S. Dakota) III–IV. 103 *Roach clip*, a small metal clip used to smoke marijuana, resembling a paper-clip or tweezers. **1979** *Christian Science Monitor* (Eastern ed.) 21 Nov. B1/1 Clearly visible through the shop's windows are..'roach clips' for holding the butt of a marijuana cigarette. **1964** S. LEAKS in J. H. Clarke *Harlem* 23 *Harlem*, a dingy-dirty cluster of roach crawling rat-infested brownstones and tenement flats. **1967** *Evening Standard* 26 June 7/3 In Haight-Ashbury [in San Francisco] there are now over 30 psychedelic boutiques..selling at inflated prices such items as..roach holders (cigarette holders specially designed for reefer smoking). **1973** *Sunday Express* (Trinidad & Tobago) 8 Apr. 13/2 I'm on the floor, surrounded by pots and pans etc. shaking the can of roach killer. **1975** *New Yorker* 23 June 29/2 St. Vincent's emergency room is one of my favorite emergency rooms in the whole world. I know it well, from the time..my daughter ate the roach poison. **1930** J. DOS PASSOS *42nd Parallel* 60 A smell of ham and coffee and roachpowder. **1931** E. POUND *Let.* 6 Oct. (1971) 236 In reply to yr. last: I am not interested in roach-powder but if the janitors and swabbers can't keep the place clean, I take it *somebody* has got to provide insecticide or even squash the individual cockroach.

roach, *v.* Add: **2. a.** (Earlier example.)

1818 *Missouri Gaz.* 25 Dec. 4/5 His mane has been divided..and that part that laid on the left side, cut off as if to roach him.

b. Of persons: to brush or cut (the hair) in a roach. Also with *up*.

1833 *Sk. & Eccentr. D. Crockett* ii. 38 His hair was roached, and he wore an air of much dignity. **1853** J. G. BALDWIN *Flush Times Alabama* 108 His hair was roached up, and stood as erect and upright as his body. **1900** ADE *More Fables* 62 He would go to School with his face scrubbed to a shiny pink and his Hair roached up on one side. **1919** H. L. WILSON *Ma Pettengill* iii. 84 She was..a kind of a slaty blonde with bobbed hair—she'd been roached fore and aft. **1932** L. C. DOUGLAS *Forgive us our Trespasses* (1937) iv. 79 His hair roached high to show he had an intellectual forehead. **1950** L. HUGHES *Simple speaks his Mind* xvi. 86 Her head was all done fresh and shining with a hair-rocker roached up high in front.

roached, *a.* For *U.S.* in Dict. read 'Chiefly *U.S.*' and add: **1.** (Earlier and later examples.) Also *Comb.*, as *roached-backed* adj.

1776 *New England Chron.* 25 Jan. 3/3 (Advt.), Strayed or stolen..a sorrel horse..roach'd back, 3 white feet, [etc.]. **1844** E. C. WATMOUGH *Scribblings & Sk.* (ed. 2) 176, The two [horses] with roatched backs, and ears glued to their necks, were scrambling. **1894** 'MARK TWAIN' in *St. Nicholas* Feb. 355/1 Roached-backed ones that he said was hyenas. **1945** C. L. B. HUBBARD *Observer's Bk. Dogs* 56 Body [of Dandie Dinmont terrier] low-to-ground, roached-backed. **1955** W. W. DENLINGER *Compl. Boston* I. 172 For a dog with a naturally roached spine, nothing can be done. **1979** T. GRAY *Chihuahua* (ed. 5) vii. 98 Another fault, a slightly roached back, will often level out by the time the puppy is three-parts grown.

2. Of hair (on a horse or person): brushed or cut in a roach. Hence, having hair dressed in this way. Also *roached-up*.

1790 *Augusta* (Georgia) *Chron.* 13 Mar. 3/1 A Bay Horse, roached mane and a small switched tail. **1836** *Southern Lit. Messenger* II. 303/1 The..fat, impudent pony, with roached main and bobtail. **1856** P. CARTWRIGHT *Autobiogr.* xii. 141 This young man had a mighty bushy roached head of hair. **1891** *Appeal-Avalanche* (Memphis) 26 Apr. 7/2 Strayed... one dark bay colt, roached mane and end of tail cut off. **1944** DUNCAN & NICKOLS *M. Graham* 91 There were twenty-seven big, little, and middlin'-sized boys and girls, giggling and whispering about the master's roached, curly red hair. **1949** 'J. NELSON' *Backwoods Teacher* 268 If she wants to go 'round lookin' like an old roached-up mule, *we* don't care.

roaching, *vbl. sb.*[2] *U.S.* [f. ROACH *v.* + -ING[1].] The action of brushing or cutting the hair in a roach. Also, the process or fact of this.

a **1883** G. W. BAGBY *Sel. Misc. Writings* (1885) II. 27 You see it [*sc.* his individuality] in the tie of his cravat, the cut of his coat,..the roaching of his hair. **1903** *N.Y. Even. Post* 24 Oct. (Saturday Suppl.) 2/1 When brought to market he [*sc.* the mule] undergoes the process of 'roaching', which consists of removing all the hair of poor quality and scanty growth.

roachy (rōu·tʃi), *a. U.S.* [f. ROACH *sb.*[4] + -Y[1].] Infested with cockroaches; resembling cockroaches (see also quot. 1900).

1900 *Dialect Notes* II. 54 *Roachy, adj.*, pertaining to poor work or preparation. **1937** F. SCOTT FITZGERALD *Let.* 8 Oct. (1964) 19 You are right that romantic things really happen in roachy kitchens and back yards. **1979** P. L. G. BATEMAN *Household Pests* II. 93 They contaminate more than they consume, polluting everything with a foul 'roachy' odour which is persistent.

road, *sb.* Add: **I. 4. b.** (Later example.)

1978 *Lancashire Life* July 63/1 He took young Sam down the pit and showed him the seam, eighteen inches high, which he had to work in a road about six feet wide.

c. For *U.S.* read 'Chiefly *U.S.*' and add further examples.

1898 H. E. HAMBLEN *Gen. Manager's Story* 68, I was passed along from one road to another, my transportation costing me nothing. **1921** *Daily Colonist* (Victoria, B.C.) 11 Mar. 2/3 The railway official quoted said he could not state whether the Canadian roads would follow the lead of the United States roads in cutting wages. **1932** *Atlantic Monthly* Mar. 318/2 Those railways..were once so prosperous that men..thought the roads would own the country unless curbed. **1942** *R.A.F. Jrnl.* 3 Oct. 3 (*caption*) Blast area with damage to roads and wagons, and a group of derailed wagons. **1950** O. S. NOCK *Brit. Locomotives* 4 The regulator opening was varied a little to suit the rise and fall of the road. **1963** *Wall St. Jrnl.* 24 Jan. 29/3 The road..operates in Guatemala and El Salvador. **1967** *Listener* 6 Apr. 461/2 At Edenbridge sidings they..told us to stand in number three road to get our breakfast.

e. *spec.* with qualifying word, a common trade-route (now freq. for illicit goods).

1931, etc. [see *silk-road*, s.v. *SILK *sb.* 10 a]. **1977** *Listener* 1 Dec. 733/1 The lost city of Zufar, the port which marked the beginning of the incense road, where ships unloaded spices from Asia for the classical Roman world. **1977** H. OSBORNE *White Poppy* xliii. 272 The gendarmes would be..watching..the opium roads.

5. a. *on the road*: also *spec.* of a person travelling as (*a*) a salesman, (*b*) a tramp; (earlier and later examples in sense 'on tour'). Also *N. Amer.*, = *AWAY *adv.* 11 (cf. sense 9 d below).

1860 DICKENS *Uncomm. Trav.* in *All Year Round* 28 Jan. 321/1, I am both a town traveller and a country traveller, and am always on the road. **1870** O. LOGAN *Before Footlights* xxviii. 367 The organ of the circus people..gives many curious details of circus-life Behind the Scenes, and 'on the road'. **1897** *Forum* Feb. 735 It is the man who wilfully and knowingly makes a business of crime..that I have found in largest numbers 'on the road'. **1907** J. LONDON *Road* 194 As a sample of life on The Road, I make the following quotation from my diary. **1908** A. BENNETT *Old Wives' Tale* I. iv. 70 He was a traveller for the most renowned and gigantic of all Manchester wholesale firms... He had been on the road for Birkinshaws for several years. **1920** WODEHOUSE *Jill the Reckless* (1922) xiv. 210 You've got to stick around with this show after it opens on the road. **1931** M. ALLINGHAM *Police at Funeral* xxi. 277, I know a 'busy' when I see one. I 'aven't been on the road for thirty years without gettin' inside once or twice. **1937** 'G. ORWELL' *Road to Wigan Pier* ix. 182, I would find out about tramps..and then, when I..knew the ropes well enough, I would go on the road myself. **1956** B. HOLIDAY *Lady sings Blues* (1973) i. 2 When he went on the road with that band it was the beginning of the end of our life as a family. **1967** J. B. PRIESTLEY *It's an Old Country* xv. 162 He was drinking hard, always a dam' silly thing to do on the road, except with a few old customers. **1968** *Globe & Mail* (Toronto) 15 Jan. 17/6 A team with the experienced potential of the defending cup champions is letting its fans down badly when it wins only four of 19 games on the road. **1977** *Daily Express* 29 Mar. 20/2 We start touring America in May, move on to Europe and England by September, and we are taking the 70 musicians on the road.

f. In extended uses based on *all roads lead to Rome* (*ROME 1 b (*d*)).

1917 E. THOMAS in *Ann. New Poetry* 55 Now all roads lead to France. **1942** E. PAUL *Narrow St.* xxiii. 209 All roads lead straight to me, as you have so often remarked. **1974** D. G. COMPTON, *Continuous Katherine Mortenhoe* vi. 165 'Where to now?'..I gestured widely... 'All roads lead out of town.'

g. *one for the road*: see *ONE *numeral a.* 1 d. Also with other numerals.

1955 J. P. DONLEAVY *Ginger Man* (1957) xix. 184 'You've had a few.' 'Five for the road. Never let it be said that I took to the highway or even byway without fuel for me little heart.'

6. c. (Further examples.)

The expression stems from a saying attributed to Euclid by Proclus (*Comm. on Euclid* Prol.): μὴ εἶναι βασιλικὴν ἀτραπὸν ἐπὶ γεωμετρίαν there is no royal short cut to geometry.

1857 TROLLOPE *Barchester T.* II. i. 2 There is no royal road to learning; no short cut to the acquirement of any valuable art. **1918** A. S. EDDINGTON *Rep. Relativity Theory Gravitation* ii. 27 Some readers will find the next two chapters difficult,..but I doubt if there is any royal road to relativity. **1954** 'N. SHUTE' *Slide Rule* vii. 158 There is no royal road to risk capital, no tap that can be turned on. **1966** P. GREEN tr. *Escarpit's Novel Computer* xiv. 173, I had to obtain that last official stamp of approval which could open up the royal road towards an easy and fruitful career. **1971** *Daily Tel.* 19 Apr. 23 The royal road for the research scientist in industry is sign-posted by positive answers to some critical questions: [etc.].

g. *capitalist road* [tr. Chinese *zīběn zhǔyì dàolù*], esp. during the Cultural Revolution in China, an observable tendency to adopt political ideals and practices leading towards capitalism. Cf. *ROADER[1] 7.

1966 *Peking Rev.* 12 Aug. 8/2 The main target of the present movement is those within the Party who are in authority and are taking the capitalist road. **1966** *N.Y. Times* 11 Dec. 3/3 Peking newspapers..attacked Hsia Yen, a playwright and former Deputy Minister of Cultural Affairs, as one of those in authority in the Communist party who had taken 'the capitalist road'. **1971** W. F. DORRILL in T. W. Robinson *Cultural Revol. in China* ii. 72 Numerous accusations were raised against the 'handful' in authority in the Party who were following the 'capitalist road'. **1973** T. R. TREGEAR *Chinese* iii. 58 The reliance on the profit motive and bonuses—Economism and 'taking the capitalist road', as he [*sc.* Mao] dubbed it. **1978** *China Reconstructs* Nov. 2/2 Anyone who pushed production or technical research was labeled as taking the capitalist road and ignoring politics.

7. b. (U.S. and Austral. examples.)

1826 A. ROYALL *Sk. Hist., Life, & Manners U.S.* 58 Put them cheers, (chairs) out of the road. **1924** E. O'NEILL *Desire under Elms* III. i, in *Compl. Wks.* II. 200 Git out o' my road! Give me room! I'll show ye dancin'. **1943** K. TENNANT *Ride on Stranger* xxiv. 269 Yes, I was cowardly enough to wait until you were out of the road. **1953** *Amer. Speech* XXVIII. 253 You are in the road... Get out of my road.

8. d. *any road, anyroad* = ANYWAY *adv.* and *conj.* 3. (Chiefly *north. dial.*)

1896 F. M. T. PALGRAVE *Hetton-le-Hole Words* 38 'Any road' (anyhow). **1932** P. MACDONALD *Rope to Spare* xi. 156 Anyroad, sir, to cut a long story brief, I gets down to the mill-'ouse. **1964** O. E. MIDDLETON *Walk on Beach* ii. 28 And how is the arm today, anyroad? **1968** M. WOODHOUSE *Rock Baby* vii. 67 We knew we'd have to expect one or two failures... Any road, we crossed it off. **1976** 'J. CHARLTON' *Remington Set* i. 5 Any road, what's it to you?

II. 9. a. *road-car* (further examples), *engine, tanker, -wagon* (earlier and later examples).

1906 *Chambers's Jrnl.* 24 Feb. 205/2 In appearance it is a very different thing from the road-car which may perhaps be regarded as its parent. **1955** A. BUDRYS in D. Knight *100 Yrs. Sci. Fiction* (1969) 251 Just before he reached the Boonesboro town line, he saw the locked and weathered cottage standing for sale... He had pulled his roadcar up to a gentle stop, swung sideways in his seat, and looked at it. **1976** *Times* 20 Mar. 14/6 Many off-track activities including a concours d'elegance of classic road cars. **1875** *Encycl. Brit.* I. 323/2 (*heading*) Road-Engines. **1886** *Walla Walla* (Washington) *Union* 24 Nov. 3/4 The 'hog' will haul nine loaded cars up the heavy Alto grade, while the ordinary road engine had a hard tussel to haul four or five. **1925** S. O'CASEY *Let.* 11 Sept. (1975) I. 147 A lumbering road-engine, with its monstrous, monotonous rumble. **1971** J. TERRELL *Bunkhouse Papers* xii. 155 A road engine purring over beyond the depot, waiting to hook on to the Limited. **1968** *Guardian* 1 Oct. 5/2 Road tanker drivers..are protesting against the proposal..to install a tachograph in lorries. **1979** *Jrnl. R. Soc. Arts* CXXVII. 406/2 Oil..would be taken by road tanker to Furzebrook. **1743** W. ELLIS *Mod. Husbandman* June iv. 37, I sent thirty-four Bushels at one Time..by a common Road Waggon. **1880** HARDY *Trumpet-Major* II. xvi. 14 This gentleman..suggested that Bob should wait till three or four that afternoon, when the road-waggon would arrive. **1968** J. ARNOLD *Shell Bk. Country Crafts* 150 Road-wagons could not be run during the months of winter.

b. *road-bank* (earlier example), *-bed* (further examples), *-bend, -cut, -cutting, -dust* (earlier example), *frontage, island, -rail, sign, -stone* (further examples), *stud, surface.*

1863 B. TAYLOR *Hannah Thurston* III. ii. 54 As they drove homewards through the cool evening air, through.. the golden-rods on the road-banks. **1902** *Chambers's Jrnl.* Jan. 61/1 In order that the oiling may be confined to the road-bed only, the rails are kept free from spraying by guards on the sprinkling-car. **1938** L. MUMFORD *Culture of Cities* 316 Small wonder that the Nile and the Euphrates ..were the roadbeds of their civilization. **1979** *Arizona Daily Star* 5 Aug. I. 3/1 In addition to some rocky road-beds and a pretty rotten record for staying on schedule, it is true that there are a number of well-aged cars. **1911** J. MASEFIELD *Jim Davis* iii. 36 The watcher at the road-bend came running back. **1978** *Nature* 8 June 459/1, I have collected unweathered samples from new exposures in quarries and roadcuts in the type area. **1936** *Discovery* Jan. 21/2 The Aculeate Hymenoptera, many of which take advantage of banks in road-cuttings and well-trodden paths, all made by man. **1854** DICKENS *Hard T.* II. vi. 195 So strange to have the road-dust on his feet instead of the coal-grit. **1942** *London Replanned* 5/2 Better building sites on important road frontages. **1976** *Evening Post* (Nottingham) 15 Dec. 15/4 (Advt.), Pleasant semi-detached house with half an acre of land. Road frontage 39′ 6″. **1932** L. GOLDING *Magnolia St.* III. iii. 511 She..took up her stand just off the pavement... Tram-drivers, chauffeurs, cyclists..accepted her as part of the landscape, like a road-island. **1903** J. MASEFIELD *Cargoes* in *Broad Sheet* No. 17, With a cargo of Tyne coal, Road-rails, pig-lead. **1904** *Car* VII. 240/2 Road signs... The conference held..to consider the desirability of uniformity of action with regard to signs and notice boards. **1914** *Autocar* XXXIII. 574/2 Owing to the fact that the mutilation of road signs by sportsmen and others has caused considerable confusion..the California State Automobile Association is reported to have lately placed a bull's eye target on each post. **1949** N. MARSH *Swing, Brother, Swing* ix. 211 There's one thing.. that's sticking out of this mess like a road-sign and I can't read it. **1976** P. LIVELY *Stitch in Time* i. 3 Maria saw this place announce itself with a road-sign. Lyme Regis. She had been studying road-signs throughout the journey. **1886** *Encycl. Brit.* XX. 583/2 The qualities required in a good road stone are hardness, toughness, and ability to resist the action of the weather. **1958** [see *AGGREGATE *ppl. a.* and *sb.* B 5]. **1970** *Railway Mag.* Oct. 577/2 The roadstone is transferred to road vehicles for transport to the motorway site. **1935** *Economist* 11 May 1112/1 Several rather special branches of the local steel industry have experienced some increase of activity.., notably armaments, aircraft steel..and stainless steel road studs. **1959** E. K. WENLOCK *Kitchin's Road Transport Law* (ed. 12) 112/2 No vehicle, except a solo bicycle.., may stop on a road between the road studs and the crossing. **1886** *Encycl. Brit.* XX. 583/1 The road surface should have just enough convexity to throw the wet off freely. **1976** 'J. ROSS' *I know what it's like to Die* xxiv. 151 Seeing the road surface slipping sideways as he toppled.

c. *road accident, atlas, casualty, construction, death, expenses, haulage, junction, kill, maintenance, manners, map, -march, marching, -marker, -meet, -meeting, -name, noise, project, race, -railway, rumble, safety, signing, space, speed, system, toll* (later examples), *-tour, traffic, transport, -web, wheel.*

1935 'OWNER-DRIVER' (*title*) Road accidents and speed limits. **1976** P. DRISCOLL *Barboza Credentials* I. i. 16 A policeman..had seen the bodies of enough road-accident victims to know what to expect. **1905** (*title*) Pratt's road atlas of England and Wales for motorists. **1963** *Which?* July 196/1 When maps are bound together in book form, they are called road atlases. **1971** *Guardian* 7 July 1 The Minister for Transport..announced..new measures designed to reduce road casualties. **1961** *Suspense* Mar. 119 One of his officers found him a job with a road construction company. **1977** *Borneo Bull.* 7 May 17/1 (Advt.), Sakai..of Japan have been making road construction machines since 1918. **1936** A. CHRISTIE *ABC Murders* xvii. 124 There are, what is it—about 120—road deaths every week? **1966** *Listener* 22 Sept. 412/1 The natural life of man..may still be short for many of us in organized society today, if road deaths continue at their present rate. **1839** DICKENS *Let.* 1 Mar. (1965) I. 515 The money for the coach-fares and road expences will be paid by you and Mitton. **1938** *Act 1 & 2 Geo. VI* c. 44 §1 For the purpose of regulating the remuneration of workers employed upon road haulage work..there shall be established by the Minister of Labour..a board..to be called the Road Haulage Central Wages Board. **1977** 'D. RUTHERFORD' *Return Load* ii. 30 The road haulage world was a friendly one. **1936** *Discovery* Oct. 317/2 The first busy road junction in the country to be equipped with invisible ray apparatus, to enable pedestrians to cross the roads in safety. **1972** R. & R. WRIGHT *Cariboo Mileposts* 40 They [*sc.* magpies]..usually feed on carrion or road-kills. **1976** *Islander* (Victoria, B.C.) 28 Mar. 7/2 Road kills have taken a few sheep of breeding age annually. **1961** *Atlanta Constitution* 6 Mar. 4 It [*sc.* DeKalb's budget] includes..increased expenditures for essential services such as..sanitation and road maintenance. **1942** *Ann. Reg. 1941* 98 A serious deterioration of road-manners. **1963** BIRD & HUTTON-STOTT *Veteran Motor Car* 159 The result may be a hybrid but it is undeniably magnificent with better-than-100 m.p.h. performance and perfect road manners. **1883** *Wheelman* (Boston, Mass.) I. 315 The preparation of road maps and posting of guide-boards are to be important features in next season's work. **1959** *Times* 25 June 12/6 My best aid was a road-map, which showed all dry sandy areas in Holland. **1972** 'H. BUCKMASTER' *Walking Trip* 112 She..stopped at a magazine kiosk and..bought a road map of Scotland. **1960** *Times* 17 Sept. 7/7 A steelband may have any number of instrumentalists, from the basic four up to a 'carnival road-march side' of over 100. **1977** *R.A.F. News* 8–21 June 7/1 Although very little is heard of road marching in the UK..on the Continent it is almost a national pastime. **1970** J. MCN. DODGSON *Place-Names of Cheshire* I. 166 ME *clywe, cle(o)we*..might be used here..to denote a mound serving as a landmark and road-marker on the wild moors. **1976** *Billings* (Montana) *Gaz.* 20 June 1-B/4 Mocabee said Ferguson's motorcycle struck a road marker. **1924** J. MASEFIELD *Sard Harker* III. 114 The tracks at the road-meet led away to the left. **1954** J. R. R. TOLKIEN *Fellowship of Ring* 18 As still passed to and fro through that ancient road-meeting. **1914** *Sat. Even. Post* 4 Apr. 12/1 Beside the monakers or road names of a hundred hoboes were scratched such messages as: 'Beware of dog'. **1965** *Eng. Stud.* XLVI. 266 The variant..can be confirmed from the Windsor road-name. **1970** J. MCN. DODGSON *Place-Names of Cheshire* I. 49 The road-name *Lymestrete*. **1963** *Times* 4 June 7/7 Above 65–70 m.p.h...road noise was high. **1973** *Times* 24 May 35/1 The engine is remarkably quiet and there is almost no road or wind noise. **1976** *Southern Even. Echo* (Southampton) 18 Nov. 32/5 Six Filipinos working on a road project in the southern island of Mindanao were killed in an ambush. *a* **1904** W. J. FISHER *Let.* in S. Weintraub *London Yankees* vi. 201, I am anxious to do nothing to discourage motoring, and I do not at all object to this road race. **1926** E. HEMINGWAY *Sun also Rises* xix. 247 Organizing the road races had made him know France... All spring and all summer and all fall he spent on the road with bicycle road-racers. **1954** *Sun* (Baltimore) 24 Nov. 17/2 The world's longest road-race test of car stamina and driving skill. **1976** *Cumberland News* 3 Dec. 19/3 The Border mens' team came home in eleventh position in the Brampton to Carlisle ten mile road-race. **1960** E. BOWEN *Time in Rome* i. 18 Hilarious buses, electric road-railways zooming into the hills. **1952** *Jrnl. Acoustical Soc. Amer.* XXIV. 661/1 In playback of a monaural recording of road rumble through a speaker, the observer

can assign a direction to the source. **1976** *Honolulu Star-Bull.* 21 Dec. F-1/1 Undercoating insulates you from hot roads, reduces road rumble, and protects against stone chip damage. **1920** *Sci. Amer.* 6 Nov. 467 Automobile Signals for Danger Spots... New illustrations of old ideas for street comfort and road safety. **1937** M. BORDEN *Black Virgin* i. 4 Road-safety instruction for school children. **1977** C. WATSON *One Man's Meat* x. 95 The chief constable..muttered 'Good gracious me, road safety committee.' **1968** *Autocar* 14 Mar. 24/1 British road signing is often the best in Europe. **1979** *Internat. Jrnl. Sociol. of Law* Feb. 68 There are various ways by which the police could prevent and control traffic disorders and road accidents:..proper and adequate road-signing, [etc.]. **1911** *Encycl. Brit.* XXIII. 393/1 The remainder of the road space is formed as an earthen track. **1975** *Times* 14 Mar. (Small Car Suppl.) p. ii/3 If you took 3 ft off the average car, you would have another six million feet of road space [in London]. **1964** L. DEIGHTON *Funeral in Berlin* xxiv. 135 It's a good road... There was no need to burn up any road-speed records. **1977** 'E. TREVOR' *Theta Syndrome* vi. 83 The TR-2 had collided with another vehicle..at a much higher road speed. **1845** *Chambers's Edin. Jrnl.* 19 Apr. 242/1 The whole road system of Great Britain..is..the most awkward and absurd institution on the face of the earth. **1904** W. M. RAMSAY *Lett. Seven Churches* xxix. 416 Laodicea was a knot on the road-system. **1932** F. L. WRIGHT *Autobiogr.* III. 321 The United States everywhere already affords increasingly great road-systems. **1976** G. SEYMOUR *Glory Boys* i. 7 The maps..showed..the road system of northern France. **1966** B. CASTLE in *Highway Code* 1 The road toll is a tragic waste—a waste of lives, a squandering of resources. **1977** *Borneo Bull.* 7 May 36/1 Razali's death pushed the road toll to 17. **1920** WODEHOUSE *Jill the Reckless* (1922) xviii. 269, I sold it half-way through the road-tour. **1864** *Great Western Mag.* Jan. 36 Some idea of the startling effect which it had upon the road traffic may be formed from the fact that the Commissioners of the Metropolitan Roads.. ceased to light the roads near Kensington. **1909** *Chambers's Jrnl.* June 340/1 A successful attempt was made to conduct road-traffic without the use of animal-power. **1977** *Whitaker's Almanack 1978* 140/1 Lighting-up Times, ..under the Road Traffic Act, 1956, are from half an hour after sunset to half an hour before sunrise. **1913** H. E. WIMPERIS (*title*) The principles of the application of power to road transport. **1969** *Jane's Freight Containers 1968–69* 231/1 Road-transport-weigh-bridge at the entrance of the Uberseehafen. **1925** W. DEEPING *Sorrell & Son* vi. 58 Then take the road-web for the ordinary tourist. London some hundred miles. Salisbury thirty or so. **1939** H. HODGE *Cab, Sir?* 266 The meter records the fare..switching from miles to minutes automatically as soon as the road-wheels stop turning. **1975** *Country Life* 2 Jan. 32/3 Sports road wheels with radial-ply tyres.

d. N. Amer. *attrib.* or as *adj.* with reference to sporting fixtures played away from home, as *road game, trip*, etc. Cf. *on the road* above (sense 5 a) and *AWAY *adv.* 11.

1961 *Newsweek* 14 Aug. 44/3 He broadened teammates' minds by reading sensitive passages aloud during road trips. **1961** *Dallas Morning News* 10 Oct. II. 3 The Texans have two more road games—at Buffalo and Houston—before they play for the old folks again. **1968** *Globe & Mail* (Toronto) 13 Jan. 42/5 The Leafs..have won only four of 19 road games. **1973** *Weekend Mag.* (Montreal) 27 Jan. 12/2 Working with ropes and on stools in stretching exercises in hotels on long road trips. **1976** *Billings* (Montana) *Gaz.* 5 July 1-C/3 Hoff also said the Mustangs two road victories at Lethbridge helped settle the club down and give it some confidence. **1979** *Arizona Daily Star* 1 Apr. C2/4 On the road games, Bill wanted the home team to take the court first because he loved to come out and kick them off the end of the court he wanted.

10. a. *road board, -contractor, haulier, -master* (earlier and later examples), *-party* (earlier and later examples), *scout.*

1865 *Geelong* (Austral.) *Advertiser* 27 Feb. 123/8 (*heading*) Meredith Road Board. **1915** *Political Q.* May 180 The Road Board..has restricted the grants to completing advances promised before the war. **1885** *List of Subscribers, Classified* (United Telephone Co.) (ed. 6) 74 (*heading*) Contractors—road and sewer. **1936** *Discovery* Feb. 55/1 The diversion of the roadway..did not suit the plans of the road-contractors. **1937** *Daily Tel.* 16 Feb. 7/2 (*heading*) Road hauliers win test case. **1977** *Modern Railways* Dec. 473/1 Rail movement cannot yet match the norm of around 60,000 miles a year which properly run road hauliers get from their vehicles. **1825** *Kingston* (Ontario) *Chron.* 7 Jan. 3/3 Bulls and Oxen to run at large—Fences 5 feet high. Road Masters to be Judges of Fences. **1856** *N.Y. Herald* 12 Jan. 1/4 James Flood is road master of his section; any obstruction being on the track it is the duty of the flagman to exhibit his red flag. **1905** KIPLING *Actions & Reactions* (1909) 21 On my uncle's farm, in Connecticut. He was what they call road-master there. **1966** *Kingston* (Ontario) *Whig-Standard* 1 Sept. 3/2 The roadmaster came down on his track speeder and gave us a fatherly talking to. **1840** Road party [see *iron gang* s.v. *IRON *sb.*[1] 12]. **1945** BAKER *Austral. Lang.* ix. 182 Heavy boots were called *road party boots*. **1931** *Star* 8 May 16/1 While being chased by a road scout on a motor-cycle..a car collided with a lamp post.

b. *road-repairer, -upper, -user* (further examples).

1921 *Dict. Occup. Terms* (1927) §44 *Road repairer*;.. keeps roadways in repair below ground; [etc.]. **1932** AUDEN *Orators* II. 49 Acting suspiciously as road-repairers. **1928** *Daily Express* 12 June 3/4 The 'road-uppers' are busy in London again. **1922** *Daily Mail* 25 May 4/4 Always show..courtesy to all other road users. **1959** *Radio Times* 23 Oct. 3/1 Certain categories of road-users are barred..; these include pedal cyclists, invalid carriages and 'L' drivers. **1976** *Oadby & Wigston* (Leics.) *Advertiser* 26 Nov. 6/5 As an ordinary citizen and road user he said he had been concerned enough about the road to see the police and Harborough District Council about it.

11. a. *road-building, -burning, -hugging, -patching, -pricing, -surfacing.*

1910 W. JAMES in *McClure's Mag.* Aug. 467/2 To coal and iron mines,..to road-building and tunnel-making,.. would our gilded youths be drafted off. **1980** *Times* 29 Feb. 18 Civil engineers claim £200m debt backlog on road-building contracts. **1931** T. E. LAWRENCE *Let.* 11 Mar. (1938) 716 After that some road-burning [i.e. fast travelling on the roads], I hope. **1963** *Times* 29 Jan. 3/7 The lightness of steering and smooth road-hugging feel of the..suspension give the car..a steady gait. **1977** *Custom Car* Nov. 70/1 (Advt.), An incredibly wide tyre for road-hugging traction-action! **1974** *Evening Herald* (Rock Hill, S. Carolina) 18 Apr. 4/1 These costs—for paving, road-patching materials, garbage containers, county employe salaries, all keep climbing each year. **1964** *Punch* 17 June 878/3 A panel research statement on 'road pricing'. **1976** P. R. WHITE *Planning for Public Transport* x. 210 If road pricing were introduced in urban areas to indicate scarcity of road space, then some reduction in national fuel-tax rates would be appropriate. **1912** KIPLING *Diversity of Creatures* (1917) 10 The sputter and crackle of road-surfacing machines. **1959** *Chambers's Encycl.* XI. 725/1 Many other methods of road surfacing have been experimented with.

b. *road-shy.*

1914 'SAKI' *Beasts & Super-Beasts* 32 He [*sc.* a horse] was not really road-shy, but there were one or two objects of dislike that brought on sudden attacks of what Toby called the swerving sickness.

c. With ppl. adjs., as *road-hauled, -killed, -stained.*

1969 *Jane's Freight Containers 1968–69* 2/2 European road-hauled transport. **1980** *Topeka* (Kansas) *Capital* 23 Feb., I once unsuspectingly ate and enjoyed a turkey vulture (which my funloving host passed off as a road-killed wild turkey). **1964** F. WARNER *Early Poems* 67 Gathering her road-stained dress She lay within a rock recess.

12. road allowance *Canada*, (*a*) a strip of land retained by government authorities for the construction of a road; (*b*) an area at either side of a road which remains a public right-of-way; **road apples** *pl. N. Amer. slang*, horse droppings; **road band,** a touring group of musicians; **road-borne** *a.* (further examples); **road-bound** *a.*, dependent on roads; restricted to using roads; **road brand** *N. Amer.*, a temporary brand given to cattle in transit; hence as *vb. trans.*; **road breaker,** (*a*) one employed to break up the road surface prior to repair, etc.; (*b*) a mechanical tool used for this; **road bridge,** a bridge that carries a road; **road company** *U.S.*, a travelling theatrical company; **road control,** a station for checking travellers' credentials, etc.; a group of people making such checks; **road-craft,** (*a*) knowledge of or skill in matters pertaining to the use of the road; road sense; (*b*) *collect.* = *road traffic; **road crew,** the group of 'roadies' which accompanies a touring band of musicians; **road drill,** (*a*) a mechanical drill used for breaking up road surfaces; (*b*) the routine for crossing a road safely; **road driver** *U.S.*, (*a*) one who drives animals on the road; (*b*) a long-distance lorry-driver; **road-farer,** one who travels by road; also **road-faring** *sb.* and *a.*; **road-ferry,** a ferry serving a road; **road fund,** a fund, esp. that established by the Roads Act of 1920, to meet provisions for roads; **road hand** (earlier U.S. example); **road-head,** (*a*) *Mining*, the part of a roadway between the last support and the face; so **road-heading,** mining at a road-head; (*b*) = *road-end*; **road hockey** *Canada*, a type of hockey played in the road; **road-holding,** the ability of a car to retain its stability; **road-house,** also, any roadside establishment providing refreshment or entertainment; (further examples); **road hunter,** a hound which is adept at following a scent on the road; so **road-hunting** *a.*; **road kid** *slang*, a boy tramp or hobo; **road life,** the life of those who are 'on the road'; **road manager,** an organizer of tour details and supervisor of equipment, etc., for musicians 'on the road'; **road-mark,** (*a*) a road sign (in quot. *fig.*); (*b*) *U.S.* = *road brand*; **road-mobile** *a.*, suitable for transporting by road; **road oil** *N. Amer.*, oil sprinkled on the roads to lay dust; **road patrol,** (*a*) a person or group of people patrolling the roads; (*b*) *Canada*, a machine used in the maintenance of unpaved roads; **road racer,** (*a*) a vehicle used in road-racing; (*b*) a contestant in a road race; **road–rail,** used *attrib.* with the sense 'suitable for use on both road and railway', or 'accommodating a road and a railway'; **road-railer,** a goods vehicle that can run on both road and rail (see quot. 1964); a container which can be transported by both road and rail; cf. *Ro-

Railer; **road rash** *slang*, grazing caused by falling from a skateboard; **road roller**, a heavy mechanical roller used for flattening road surfaces; **road-runner** (earlier and later examples); **road-running** *vbl. sb.*, running on the roads for sport or exercise; also as *ppl. a.*; **road sense**, capacity for intelligent handling of vehicles or coping with traffic on the road; **road show**, a show given by touring actors or musicians, usu. with the minimum of equipment and preparation; also *transf.* and *attrib.*; hence as *vb. intr.*; **road sweeper**, (*a*) a person who sweeps roads; (*b*) a device for sweeping roads; **road train**, a large lorry pulling one or more trailers; **road tunnel**, a tunnel through which a road passes.

1844 *Niagara* (Ontario) *Chron.* 29 May 2/2 A bill was introduced..entitled 'An act to close up the Road Allowance between Lots Nos. 42 and 43..in the township of Cayuga'. **1947** E. A. McCourt *Music at Close* 43 He..rode west along the road-allowance until he reached a part of the country which was new to him. **1958** J. G. MacGregor *North-West of 16* ii. 27 He carefully stepped off this distance, which was the 'road allowance', and came to another great spruce at the exact corner of his land. **1973** *Kingston* (Ontario) *Whig-Standard* 14 Mar. 3/1 In the 1783 survey of the lakefront townships, the provision had been made for a 60-foot road allowance across the front. **1942** Berrey & Van den Bark *Amer. Thes. Slang* § 124/2 *Road apples*, horse dung. **1951** M. Spillane *One Lonely Night* v. 112 Smart? Sure, just like road apples that happen behind horses. **1970** J. H. Gray *Boy from Winnipeg* 53 The best pucks were always those supplied by passing horses, 'road apples' we called them. **1937** *Amer. Speech* XII. 48/1 *Road band*, a traveling unit. **1976** *Casper* (Wyoming) *Star-Tribune* 29 June 17/6 (Advt.), Wanted: vocalist for road band. **1977** *Zigzag* Mar. 2/1 Those guys had been my road band anyway. **1914** Kipling *Years Between* (1919) 78 That I may sing of Crowd or King or road-borne company. **1973** *Daily Tel.* 10 Aug. 14 In 1971 road-borne freight amounted to 52,000 million ton-miles. **1937** L. Hart *Europe in Arms* x. 120 The limitations of the large road-bound coaches or lorries which compose such bus columns were made manifest. **1941** *Washington Post* 4 Sept. 12/5 Members of both divisions pointed today to powerful motor vehicles roadbound because of continuous rains. **1874** J. G. McCoy *Hist. Sk. Cattle Trade* i. 7 The slight brand put on the stock at that time [when the herd is started to market over the trail] is called a road brand, in contradistinction to the ranch brand, which is usually put on the animal when young. **1933** J. V. Allen *Cowboy Lore* ii. 44 When cattle were driven to market, it was easy for them to get mixed up with others, and this accounts for the *road brand*, used for rapid identification. **1955** W. Foster-Harris *Look of Old West* viii. 229 Unless a trail herd was under one brand, which ordinarily it wasn't, it was customary to road-brand the animals—that is, give them an extra insignia to identify them on their journey. **1928** *Daily Mail* 31 July 13/3 One side of Kensington-road..is also in the hands of the road-breakers. **1967** *Gloss. Highway Engin. Terms* (*B.S.I.*) 39 *Road breaker*, a power driven tool for breaking up road pavements by impact. **1976** *Southern Even. Echo* (Southampton) 6 Nov. (Advt. Suppl.) 8/5 Road breakers, angle and straight girders, rock drills. **1819** *Massachusetts Spy* 3 Nov. 2/3 A salute was fired from a road-bridge by a detachment..of artillery. **1870** E. G. E. Ward *Jrnl.* 13 Sept. in D. P. Carew *Many Years, Many Girls* (1967) i. 16 He had heard the Road bridge blown up, and feared the railway would follow, and that I might not get out of Paris! **1935** *Discovery* Oct. 300/2 The new road-bridge over the Severn. **1976** *Liverpool Echo* 7 Dec. 17/5 The swim is downstream of the new road bridge over the River Dee at Eccleston. **1900** *Everybody's Mag.* II. 583/2 In the years of association which I have had..with 'road companies' I have become familiar with the types. **1959** W. Faulkner *Mansion* ix. 205 The old road-company drammer reversed in gender: the frantic child clinging this time to the prospective groom's coattail. **1977** *New Yorker* 3 Oct. 129/1 The carryings-on of Gavino's mother, a road-company Anna Magnani in an Italian version of 'Sons and Lovers'. **1946** R. Capell *Simiomata* ii. 69 Evert himself drove him through the German road-controls to..the east coast of Attica. **1966** M. R. D. Foot *SOE in France* x. 326 They were arrested by a road control that for once searched the greengrocer's lorry they were hidden in. **1868** T. Wright *Great Unwashed* 265 The old tramp..has a beneficial knowledge of what may be called road-craft. **1897** 'H. S. Merriman' *In Kedar's Tents* xii. 130 Conyngham learnt much of that road-craft which had raised Concepçion Vara to such a proud eminence among the rascals of Andalusia. **1917** 'Contact' *Airman's Outings* 9 Mirrors of polished steel, as used on the handlebars of motor-cycles, to give warning of roadcraft at the rear. **1934** Webster, *Roadcraft*..., skill or dexterity in driving on a road. **1963** *Times* 19 Feb. 11/3 What is wanted is a more radical reform in the driving test calculated to lift the standard of roadcraft as quickly as possible to the much higher level demanded by the scrum of the modern highway. **1974** *Country Life* 17 Oct. 1108 Apart from the roadcraft side, the mechanical side can pay dividends if one learns to use the car sympathetically. **1976** *Evening Post* (Nottingham) 15 Dec. 5/6 The threat is thought to follow Stewart's visit to Newcastle two years ago, when there was an incident at a city nightclub involving some of his road crew. **1977** *Sounds* 9 July 19/2 We've got the best road crew we could find. **1934** S. Spender *Poems* (ed. 2) 41 At corners of day Road drills explore new areas of pain. **1973** *Scottish Sunday Express* 5 Aug. 4/2 Clever dog to obey the road drill! **1976** 'J. Fraser' *Who steals my Name?* viii. 99 You've got the road drills outside your office... I can't hear myself think. **1897** *Boston Morning Jrnl.* 4 Jan. 5/6 The half-mile track is convenient of access to road drivers from the city. *Ibid.*, The road driver frequently drives his own horse a trial mile. **1929** *Sat. Even. Post* 16 Nov. 41/3 R is for Road Driver, the name long-distance haulers give the lad that knows his cucumbers. **1973** *Amer. Speech* 1969 XLIV. 207 *Road driver*, driver who drives long distances. **1923** *Daily Mail* 22 May 4 Hotels,..which set out to cater efficiently for the growing army of roadfarers. **1961** *Times* 6 Sept. 13/4 There has even been a proposal to collar 'roadfarer' for drivers of private motor cars. **1915** R. Wellbye (*title*) The roadfaring handbook to inexpensive motor touring. **1925** *Chambers's Jrnl.* June 379/2 She would probably not average over 2 m.p.h., which seems almost incredible to a different section of the road-faring fraternity. **1920** *Act* 10 & 11 *Geo. V* c. 72 Sched. 1, The definition of 'roads' shall be extended so as to include road-ferries and footways. **1793** *Jrnl. House of Keys* (I.O.M.) 18 Dec. (MS.), Several matters, which I would mention as worthy your Serious and frequent Consideration...1st. The Road Fund. **1845** *Chambers's Edin. Jrnl.* 19 Apr. 242/1 The whole together along with 69 steelyards, or cast-weighing machines, having cost the road fund not less than L. 10,000. **1920** *Act* 10 & 11 *Geo. V* c. 72 §3 There shall be established for the purposes of this Act,..a fund to be called the Road Fund. **1975** M. Simpson *Chrome Connection* vi. 143 Could I see your road fund licence, sir? **1873** J. E. Lester *Atlantic to Pacific* v. 28 At this station ..we shall see the 'John Chinamen' as road-hands. **1883** W. S. Gresley *Gloss. Coal-Mining* 205 *Road-head* (S[cottish]), see *Gate-end*. **1934** *Webster*, *Roadhead*, the end of a road (*dial.*). **1950** E. Mason *Pract. Coal Mining* I. viii. 121 (*caption*) Arrangement at the roadhead of a double unit face. **1958** J. C. F. Statham *Coal Mining Pract.* IV. iv. 244 About 30 per cent. of fatal and serious non-fatal accidents from falls occur at roadheads, i.e. the short length of roadway within 10 yd. of the working face. **1958** A. J. Toynbee *East to West* xlviii. 144 We rounded a corner and saw our car waiting for us at the road-head. **1976** *Nature* 12 Aug. 532/3, I..walked..back to the road-head. **1969** *New Scientist* 27 Feb. 444/1 These four tunnels are being driven by the four roadheading machines. *Ibid.*, This method reduces the manpower needed for road-heading by about half. **1965** *Kingston* (Ontario) *Whig-Standard* 28 Dec. 24 (*caption*) Road hockey was back in style Monday as these youngsters proved in a rough game played in Portsmouth during the afternoon. **1976** *Ibid.* 13 Feb. 22/1 A nearby road hockey game played with the numerous readily available road apples. **1932** *New Yorker* 14 May 32/2 Great attention has been devoted to suspension and road-holding. **1959** G. Freeman *Jack would be Gentleman* iii. 50 Sports cars are..better than the average family saloon—steering, road-holding, everything that adds up to real safety. **1975** *Times* 14 Mar. (Small Car Suppl.) p. ii/9 The Mini's greatest asset is probably its road-holding. **1936** O. Lancaster *Progress at Pelvis Bay* 63 Many..who..motor down..by way of the new Flushbrook By-pass, must be familiar with the..'Hearts Are Trumps' roadhouse. **1944** 'N. Shute' *Pastoral* i. 8 In peace-time it had been something of a road-house, with a snack-bar. **1957** J. Braine *Room at Top* x. 92 Four months in Warley had given me a fixed taste for either the roadhouse or the authentic country pub. **1972** D. Anthony *Blood on Harvest Moon* ii. 18 Across the highway was a roadhouse where, according to the sign, you could dine, dance, and drink. **1897** *Encycl. Sport* I. 560/1 The only thing he [*sc.* the huntsman] *can* do is to..try every gateway and likely-looking spot where she [*sc.* the hare] may have turned off or lain down, but it is a mere matter of luck unless he has the real road hunter. **1977** *Horse & Hound* 14 Jan. 20/1 In the old days they [*sc.* foxes] probably ran the roads much more because they were quieter. The minor roads were not even tarmacked, with the result that you had your 'road hunting hounds' much more frequently. **1909** J. London *Road* 173 A boy on the Road is never a gay-cat; he is a road-kid or a 'punk'. **1937** 'D. Boyle' *Keeping in Trouble* iii. 38 He was a 'road-kid', that is to say he found his company, within the great confederation of tramps, with youngsters of his own age or less. **1902** *N.Y. Times Mag.* 28 Dec. 12/1 A comedian who has seen so much of the unlovely side of 'road' life that he might well be the writer of tragedies. **1956** B. Holiday *Lady sings Blues* viii. 93 He always wanted me to ride up with him.. and sometimes Benny, our road manager. **1978** *Detroit Free Press* 16 Apr. 23A/3 Police..accused the road manager, who wasn't identified, of giving alcohol to a 14-year-old and an 18-year-old girl. **1879** G. M. Hopkins *Lett. to R. Bridges* (1955) 77 The island was so Marian that the very Milky Way we made a roadmark to that person's shrine. **1881** *Lippincott's Mag.* XXVII. 570/1 Every animal, besides the regular brand of the owner, has his tail bobbed and a 'road-mark' put upon him during the drive. **1908** Murray & Miller *Round-Up* 268 Having cut out the stock for the drive, a road mark, a supplementary brand for identification, is burned into the hides. **1922** *Encycl. Brit.* XXX. 249/1 The French guns up to the 6 in., and howitzers up to the 9.45 in. inclusive will be road-mobile... All heavier natures will be on railway mountings. **1921** *Daily Colonist* (Victoria, B.C.) 8 Apr. 6/3 The Saanich works committee last night authorized the purchase of $7,000 worth of road oil and one hundred tons of asphalt. **1976** *Billings* (Montana) *Gaz.* 1 July 1-c/3 Finally, the commission approved the purchase of 25,000 gallons of road oil from the state. **1899** Kipling *Stalky & Co.* 252, I engineering myself..into command of a road-patrol—no shovellin', only marching up and down. **1958** *Cut Knife* (Saskatchewan) *Grinder* 3 Apr. 1/6 In the Rural Municipality of Cut Knife, the burgesses were asked to vote upon Bylaw 12 authorizing the Council to purchase a new road patrol. **1963** *Times* 8 Jan. 11/1 Their road patrols say that many motorists are unaware that a dry cloth, or some paper handkerchiefs used to soak up the condensation or melted snow thrown over the sparking plugs, leads and distributor, could often save them from being stranded. **1970** *R.A.C. Guide & Handbk.* 39 Road Patrols..equipped with vans, are on daily duty. **1908** H. G. Wells *War in Air* i. 14 Even a road-racer, geared to a hundred and twenty, failed to satisfy him. **1954** *Sun* (Baltimore) 24 Nov. 17/2 (*heading*) Road racers top 3 marks. **1976** *Norwich Mercury* 19 Nov. 12/1 For road racers and scramblers the financial drain is even worse. **1927** *Times* 20 Dec. 11/3 The 'road-rail' truck represents in a modified form the upper part of a railway wagon which can be exchanged between road and rail vehicles. **1963** *Times* 8 June 14/3 Two-day talks between British and French Government officials on whether there should be a Channel rail bridge or a road-rail tunnel ended in London yesterday. **1977** *Modern Railways* Dec. 494/1 The works include..the rebuilding of 20 stations to create fully-equipped road-rail interchanges. **1960** *London Midland Region Staff News* (Brit. Railways) Feb./Mar., The Roadrailer..is similar in appearance to the normal tractor drawn semi-trailer seen on the roads. It differs in two respects, however. At the rear, there is a special device which retracts the road wheels and lowers a pair of rail wheels. At the front, a special coupling fits into the rear end of another similar trailer. **1964** *Economist* 15 Aug. 663/3 British Railways' latest ploy to attract traffic from the roads involves..freight wagons which ride the rails for most of the journey and then take to the roads as trailers to deliver goods directly to the customers' doorstep. British Roadrailer Services is being set up as a joint company by British Railways and the Transport Holding Company to develop the use of these roadrailers. **1965** *Ibid.* 26 June 1559/2 The lorry people might make use of liner trains, or better, the new 'roadrailers' (of which they own half with the railways), for convenient loads of odd packages on the longer runs. **1976** *Daily Tel.* (Colour Suppl.) 30 July 10/3 It is hard on both rider and wardrobe. Young skateboarders proudly show off their 'road rashes'. **1978** *Skatcat's Quiz Bk.* (R. Soc. Prevention of Accidents) 5 'Road rash' isn't clever. As well as hurting for over a week, bad grazes or cuts are the signs of bad skateboarding. **1886** *Encycl. Brit.* XX. 583/1 In Great Britain horse-rollers have to a great extent been superseded by steam road rollers. **1971** *Daily Nation* (Nairobi) 10 Apr. 25/2 Tenders are invited for one three-wheel road roller. **1856** *Hutching's Mag.* Nov. 201/2 The Road-Runner is seldom seen in trees, unless pursued very closely. **1930** R. Macaulay *Staying with Relations* xix. 275 Not a thing to look at, on this so-called road, only cactus and chaparral and road-runners and those darned flowering aloes. **1972** G. Durrell *Catch me a Colobus* ix. 188 A Road-runner—a strange little bird with a crest and a long tail and enormous flat feet. **1908** A. Shrubb *Running* x. 66 Of all forms of pedestrianism..there can be nothing superior to cross-country running... Track or road running is apt to grow monotonous, however exciting it may be; but there is nothing monotonous in an open country run. **1934** V. Woolf *Writer's Diary* (1953) 216, I cannot without more labour than my roadrunning mind can compass describe the queer impression of sunny impersonality. **1962** Lydiard & Gilmour *Run to Top* viii. 68 Ordinary tennis shoes..don't cushion against the constant jarring of road running which can damage leg joints. **1976** *Cumberland News* 26 Nov., Go into the road running strongholds. **1923** *Daily Mail* 10 Aug. 6/3 The good driver uses care instinctively because he has the imagination or 'road-sense' which tells him instantly what he can and what he cannot do. **1947** J. Dean *Murder Most Foul* v. 98 Before the war the B.B.C. conceived the..idea of engaging racing drivers to give broadcasts on 'road sense'. **1975** 'M. Yorke' *Small Hours* ii. 20 She was..bad at parking..though her road sense was good. **1908** *Variety* 16 May 1 De Dio, a foreigner,..has been engaged by Martin Beck for his Orpheum Road show next season. **1939** J. B. Priestley *Let People Sing* xiv. 412 I'm booking acts for a road-show. **1942** Berrey & Van den Bark *Amer. Thes. Slang* § 596/2 *Go on tour*,..road-show. **1959** *Time* 28 Sept. 32/3 Road-showing in Cole Porter's *Can-Can* in Toronto, the French singer-comedienne..had to negotiate a ramp leading out of the tent-theater. **1961** *Wall St. Jrnl.* 8 Nov. 3/2 The Kennedy Administration opened its road show to inform citizens across the country what's going on in Washington. **1976** J. McClure *Rogue Eagle* ii. 28 The sprained ankle which forced her to drop out of an American roadshow. **1977** *Time* 25 Apr. 52/1 Some were impressed more by the viewpoint espoused in the road-show tactics of Phyllis Schlafly, an Alton, Ill., housewife and an active Republican. **1909** Road-sweeper [used in def. of Roader[1] 4]. **1937** *Times British Motor Number* 13 Apr. p. xxxviii/2 The Karrier road sweeper.. is also popular abroad. **1939** G. B. Shaw *Geneva* i. 16 The president and parliament are elected by adult suffrage every two years. So are all the judges and all the officials, even the road sweepers. **1973** *Times* 18 June 3/2 Teachers at a local school have been parking outside and thereby frustrating the mechanical road sweeper. **1959** Road train [see *pedal-radio* s.v. *Pedal sb.* 7]. **1964** L. Deighton *Funeral in Berlin* xxxi. 164 One of those heavy trucks with two trailers that they call 'road trains'. **1977** 'D. Rutherford' *Return Load* i. 12 The driver of the Scania road train drew level with the cab of the British vehicle. **1939** *Encycl. Brit. Bk. of Year* 684/1 A project of a new road tunnel through the St. Gothard. **1976** J. Lund *Ultimate* x. 91 The long road tunnels through the mountain got them there quickly.

roadability (rōudăbi·lĭti). [f. Road *sb.* + Ability.] Suitability for being driven on the road; roadworthiness; road-holding ability.

1925 C. Morley *Safety Pins* 116 When the car has a 126-inch wheelbase, it makes it very easy riding and gives it charming 'roadability'. **1928** *Sunday Dispatch* 16 Sept. 5 (Advt.), Wider track, lower centre of gravity, improved roadability. **1973** *Sci. Amer.* Dec. 2/1 (Advt.), Put these characteristics of engine performance and roadability together, and it's not surprising that BMW owners will tell you no other car comes close for combining brilliant responsiveness with precise control.

roadable (rōu·dăb'l), *a.* [f. Road *sb.* + -able; see prec.] Suited to being driven on roads.

1929 *Bookman* (N.Y.) May 301/1 Motor car advertising of the past decade has brought forth the remarkable word 'roadability'... A car that has 'roadability' is, presumably, 'roadable'; that is, it can be roaded—whatever that might mean. **1935** A. P. Herbert *What a Word!* ii. 51 What sort of a car, I wonder, is a car which is not 'roadable'? **1972** *Daily Tel.* 22 Nov. 4/5 A 'roadable' aircraft has been shown at the Detroit Motor Show... It can be driven on the road or flown from an area smaller than a football field.

roadblock (rōu·dblǫk). Also **road block, road-block**. [f. Road *sb.* 4 + Block *sb.* 19 a.]

a. A barrier or obstruction on a road, usu. one set up by the army or police.

1940 *Hutchinson's Pict. Hist. War* 7 Aug.–1 Oct. 4 Demonstrations of how to build road-blocks, how to deal with tanks,..how to deal with refugees on the road, are all part of the Home Guard training. **1943** *Ann. Reg. 1942* 63 The road blocks for obstructing tanks were also found to be much more formidable than had been anticipated. **1954** X. Fielding *Hide & Seek* v. 67 Pedestrians..were often searched by the sentries manning the road-blocks. **1958** 'A. Gilbert' *Death against Clock* xii. 167 When you put up your road blocks, then I have to take the diversion. **1971** *Daily Tel.* 23 July 1/5 A soldier fired at a car which broke through a roadblock, hitting the back window. **1977** *Centuryan* (Office Cleaning Services) Christmas 15/4 They had not gone far before they came to a roadblock caused by an accident. **1978** *N.Y. Times* 30 Mar. A5/2 The complex is now guarded by 14,000 riot policemen who man roadblocks and monitor highway traffic from camouflaged roadside positions.

b. *fig.*

1945 *Tuscaloosa* (Alabama) *News* 19 June 4/5 The French general is probably the only remaining road-block to Communism in France. **1952** *Sun* (Baltimore) 29 Feb. 19/3 This bill has been subjected to roadblock after roadblock by those attempting to delay its passage during these closing days of 1952 General Assembly session. **1957** *Economist* 7 Sept. 770/1 In every direction the logic of the Korean economy runs headlong into a political roadblock. **1963** J. Mitford *Amer. Way of Death* v. 177 An English contributor to the American *Professional Embalmer* describes some of the roadblocks he has encountered. **1977** Miller & Swift *Words & Women* v. 71 Nowhere are the semantic roadblocks to sexual equality more apparent—or significant—than in the language of the dominant organized religions.

Hence as *v. trans.*

1954 in Webster Add. **1955** *Newsweek* 10 Jan. 20/1 The debate that followed roadblocked Mr. Eisenhower's legislative program for weeks. **1965** *Canad. Jrnl. Linguistics* X. 153 The analysis of texts, opening the way to ethnographic and historical studies that otherwise would still be roadblocked. **1972** *Listener* 10 Feb. 167/2 The Army road-blocked Newry.

road-book. Add: **1.** Also *transf.*

1931 *Times Lit. Suppl.* 29 Jan. 80/2 The civil aviation section [of *Jane's All the World's Aircraft*] continues to act as a world-wide aviation 'road-book'.

2. Also *spec.*, a log-book kept by the driver of a commercial vehicle.

1939 'N. Blake' *Smiler with Knife* xvii. 246 You sign my road-book, cock, or I'm not opening this van.

roadeo (rōu·di,o). *U.S.* [f. Road *sb.*, after Rodeo.] A gathering of lorry drivers for competitive events in and exhibitions of driving skill.

1948 *Sun* (Baltimore) 23 Aug. 11/1 To be eligible to enter the roadeo drivers must have completed one year of no-accident driving for the same employer. **1952** *N.Y. Times* 12 Oct. ii. 18/2 The American Trucking Associations, Inc., was holding its annual Roadeo, a competition for champion, professional truckdrivers. *Ibid.* 18/2 The Roadeo course is a hard test of man's ability at the wheel. **1955** *Amer. Speech* XXX. 150 The roadeo or competition of truck drivers is now a familiar institution. **1971** M. Tak *Truck Talk* 131 *Roadeo*, the National Truck Roadeo, an exhibition and competition of driving skills in various classifications of motor vehicles for expert truckers.

roader[1]. Add: **6.** *Taxi-drivers' slang.* A long-distance taxi fare or journey.

1939 H. Hodge *Cab, Sir?* ii. 28 It may be a long job—a 'roader' as we call it—out to Richmond or Highgate. **1978** *London-Wide Radio Taxis* (Licensed Taxi Drivers Assoc. Ltd.) [Publicity leaflet] p. iii/2 Roaders are an everyday event on radio. Put yourselves into the shoes of a director of a company who requires a taxi for a long distance haul. Does he go out into the street and hail a cab or send his secretary to find one? Of course he doesn't. He rings for a cab.

7. *capitalist roader* [tr. Chinese *zǒuzīpài*, short for *zǒu zīběn zhǔyì dàolù dāngquánpài*], esp. during the Cultural Revolution in China, a term for Party officials, e.g. the secretary of a provincial Party committee or the chairman of a people's commune, who were alleged to have capitalist tendencies. Cf. *capitalist road* s.v. *Road *sb.* 6g.

Used of people of various views who are out of favour with the Chinese leadership.

1967 *Economist* 7 Oct. 26/1 The unsurprising absence of Liu Shao-chi, Teng Hsiao-ping and their fellow 'capitalist-roaders' brings the politburo membership down from the 24 who were elected in the first flush of the cultural revolution in August 1966 to the 15 who are still appearing today. **1970** E. Snow *Red China Today* (rev. ed.) xlix. 389 Lau Shaw himself committed suicide in 1966 when he was attacked by Red Guards as a 'revisionist' and 'capitalist roader'. **1973** R. Taylor *Educ. & Univ. Enrolment Policies in China, 1949–1971* 15 The so-called capitalist-roaders refused to admit any of them on the grounds that worker-peasant children were not of high enough scholastic calibre. **1976** *Financial Times* 24 Nov. 6/1 The official Hsinhua news agency, writing about Mme. Mao and the so-called 'gang of four' or the 'new capitalist roaders', was saying that she was a 'big careerist' who 'prostrated herself in admiration before Western bourgeois things'. **1978** Hua Kuo-Feng in *Peking Rev.* 10 Mar. 11/1 The 'gang of four'..openly dished up a counter-revolutionary political programme equating veteran cadres with 'democrats' and 'democrats' with 'capitalist-roaders' and agitated for rounding up 'capitalist-roaders' at all levels from the central down to the local.

road hog, *sb.* (See Road *sb.* 12.)

1891 *Outing* Dec. 238/2 The 'road hog' curses him and the wayside brute calls out the dog. **1898** [see Road *sb.* 12]. **1909** *Q. Rev.* Jan. 143 The habitually reckless motorist..commonly known as the 'road-hog'. **1925** *Public Opinion* 14 Aug. 151/3 Road-hogs who run down pedestrians. **1932** D. L. Sayers *Have his Carcase* ii. 29 A fast saloon car..overtook them... 'The beastly road-hog!' said Mr Perkins. **1970** 'D. Halliday' *Dolly & Cookie Bird* iv. 52 You're a road-hog... What do you drive at home? **1972** K. Bonfiglioli *Don't point that Thing at Me* xiv. 122 'Lost my temper... Bloody road-hog.' 'He might easily have done us a mischief,' I agreed.

Hence **roa·d-hog** *v. intr.*, to be or act like a road hog; so **roa·d-hogging** *vbl. sb.* and *ppl. a.* Also **roa·d-hoggery**, behaviour characteristic of a road hog; **roa·d-hoggish** *a.*, having the driving habits of a road hog; hence **roa·d-hoggishness**; **roa·d-hoggism**, a road-hoggish act.

1914 'I. Hay' *Knight on Wheels* xviii. 174, I wouldn't go road-hogging if I were you... Road-hogging is rotten bad form. **1923** *Daily Mail* 28 May 5 In four days, without road-hogging, we have covered 646 miles of Britain. **1926** *Glasgow Herald* 12 July 6/3 The perfect Sunday—to hide at home while the rest of the world road-hogged it out of town. **1927** *Scots Observer* 16 July 10/4 Avoiding excess of speed and other road-hoggisms. **1928** *Daily Express* 24 Apr. 10/2 The road-hogging motor-coach. **1930** *Time & Tide* 7 Feb. 172 No motorist, however road-hoggish he may be, deliberately slays a child or any other person. **1933** A. Morris *Digging in Southwest* 80 Road-hogging is one of the most anti-social characteristics in the world. **1963** *Guardian* 21 Jan. 6/3 Chief vice of bad drivers.—Men: Sheer selfishness and road-hoggery. **1965** *Punch* 28 July 138/3 The film started with the customary pop music and some sinister hints of teenage drug-taking, road-hoggishness, violence and debauchery. *Ibid.* 20 Oct. 569/2 Some road-hoggery was due to the imperfect construction of the motor car. **1974** D. Francis *Knock Down* xiv. 174 Letting loose that road-hogging two-year-old.

roadie (rōu·di), *sb.* Also **roady**. [f. Road *sb.* + -y[6], -ie.] = *road manager*; an assistant employed by a touring musical band whose duties include the erection and maintenance of equipment. Hence as *v. intr.*

1969 Fabian & Byrne *Groupie* ix. 72 Bill, the roadie, buys me a drink. **1972** *Daily Tel.* (Colour Suppl.) 17 Nov. 26/3 Each individual musician has his own personal roady who is an expert in the instrument that his governor plays. **1976** *New Musical Express* 17 Apr. 24/5 Even allowing for roadies and so on, aren't there far too many people being supported by far too few? **1976** *Star* (Sheffield) 20 Nov., The author once roadied for the band. **1980** *Times Lit. Suppl.* 28 Mar. 365 He even gets paid as a roadie on tour.

roading, *vbl. sb.*[1] Add: **2. b.** *concr.* A road surface. *rare.*

1857 J. E. Ritchie *Night Side of London* 5 The cost of this paved roading was 14 millions.

roadman. Add: **2.** A person using the roads for any purpose. An itinerant canvasser or seller of goods, a travelling salesman; a tramp or vagrant; a road-racer.

1906 S. E. Sparling *Introd. Business Organiz.* 206 Another method of direct selling is found in the system of canvassers and road-men sent out by factories. **1912** A. S. M. Hutchinson *Happy Warrior* v. vi. 318 There cried to them 'Away! away!' all the instinct that, since first law came on the land, has bade roadmen, gipsies, outlaws, take immediate flight from trouble. **1949** A. Miller *Death of Salesman* ii. 80 You're a road man, Willy... We've only got a half-dozen salesmen on the floor here. **1951** *Sport* 27 Apr.–3 May 11/4 Sutherland..is a sprinter turned roadman, and it was in a sprint finish that he won his Empiad title in New Zealand. **1955** *Publ. Amer. Dial. Soc.* xxiv. 84 One can discern two different kinds of thief..: the *road man* is the more venturesome, the more restless. **1962** J. D. MacDonald *Key to Suite* (1968) ii. 16 Once in Atlanta..one of Federal's road men.. goosed the wife of the executive vice-president. **1976** *Star* (Sheffield) 29 Oct. 27/1 The Horseshoe climb..is noted as a 'roadman's' course where riders such as Waugh can beat the hill specialists at their own game.

roadmanship (rōu·dmănʃip). [f. prec. + -ship.] Ability to drive on the roads; skill in using the roads.

1953 P. C. Berg *Dict. New Words* 137/2 *Roadmanship*, skill in using public roads; e.g. the Highway Code tells you the principles of good roadmanship. (Perhaps after horsemanship.) **1958** *Archit. Rev.* CXXIII. 300/2 Skill in driving and roadmanship, allied with the glamour of the superior vehicle, enable the ordinary person to express his innate sense of craftsmanship. **1959** *Economist* 16 May 9/2 An Israeli can give the impression that no other country has ever suffered from juvenile delinquency or bad roadmanship. **1973** *Daily Tel.* (Colour Suppl.) 7 Dec. 25/2 Drivers with high standards of car control and roadmanship.

† **roadometer** (rōudǫ·mĭtəɹ). *Obs.* [f. Road *sb.* + -ometer.] **a.** A device for measuring distance travelled. **b.** (See quot. 1926.)

1848 W. Clayton *Latter-day Saints' Emigrants' Guide* 3 The distances from point to point are shown as near as a *Roadometer* can measure. **1859** B. D. Williams *Let.* 9 May in L. R. Hafen *Overland Routes to Gold Fields* 238 The road..from Leavenworth city to Denver city, is 689 miles in length by the roadometer. **1926** *Glasgow Herald* 14 Oct. 6 There is a considerable range of motor car attachments, including a roadometer which automatically indicates the position of the car on the road by a scale which shows the camber.

roadscape (rōu·dskēip). [Road *sb.*] A view or prospect of a road; a picture of a road. Also, landscaping of a road.

1942 Berrey & Van den Bark *Amer. Thes. Slang* § 137/4 *Roadscape*, a view of a road. **1959** *Archit. Rev.* CXXV. 245/3 In turning over pictures of German autobahnen, so competent in detail, it comes almost as a relief to find a piece of roadscape that doesn't work,—an example of how not to do it (the only one in this article). **1968** *Radio Times* 28 Nov. 21/1 Roadscape with rusting rails... The second of four talks on Los Angeles.

roadside. 2. (Further *attrib.* examples.)

1925 F. Scott Fitzgerald *Great Gatsby* v. 98 Light, which fell unreal on the shrubbery and made thin elongating glints upon the roadside wires. **1939** Joyce *Finnegans Wake* (1964) i. 31 The roadside tree the lady Holmpatrick planted. **1936** *Discovery* Apr. 125/1 A pair of roadside thieves in Nevada. **1949** *Sun* (Baltimore) 14 July 8/2 The Pennsylvania Highway Department is receiving appreciative comment from the motorists on its roadside rests. **1961** *A.A. Handbk.* 12 A.A. Roadside Telephone Boxes are placed at carefully selected points along main roads. **1976** *Deeside Advertiser* 9 Dec. 24/3 His companion was thrown against roadside signs.

roadster. Add: **2. a.** (Further *attrib.* example.)

1974 *Greenville* (S. Carolina) *News* 23 Apr. 11/2 Thursday's performances have two Roadster pony classes.

b. (Further *attrib.* example.)

1922 [see *basket-car* s.v. *Basket *sb.* B. 1 b].

c. *U.S.* A light carriage.

1892 *Hist. Rev. Industr. & Commerc. Growth York County* 68 The former [repository and office] carries a fine line of..everything in light and heavy work from the most substantial farm truck to the lightest finished roadster. **1901** *Dialect Notes* II. 146, I went to a farmer near and hired a young horse and a roadster.

d. orig. *U.S.* A type of motor car, esp. an open two-seater. Also *attrib.*

1908 *Sci. Amer.* 8 Feb. 104 Cadillac... Model G—Roadster, $2000. **1922** H. Titus *Timber* xxix. 255 Rowe stood..a long interval,..watching her roadster disappear into the jack pines. **1928** F. N. Hart *Bellamy Trial* ii. 30 We drove out from New York in my roadster. **1938** G. Greene *Brighton Rock* vi. ii. 249 Where did you bring a swell blonde to if not the Cosmopolitan,..driving over the down in a scarlet roadster. **1948** W. Sansom *South* 122 Round the narrow corner of old grey walls nosed the chromium grill, the long bonnet, and then all the pale gleaming length of a torpedo-shaped roadster. **1962** *Punch* 17 Oct. 560/2 Drag entries come in a profusion of classes: dragsters, modified roadsters, roadsters, [etc.]. **1973** D. Lees *Rape of Quiet Town* ii. 33 It was lucky I left my purse in your roadster. **1977** *Custom Car* Nov. 5/4 As for handicap racing in the Roadster and Production classes, the spectators seem to enjoy watching it.

3. b. For *U.S.* read 'orig. *U.S.*' and add earlier and later examples.

1890 N. P. Langford *Vigilante Days* II. vi. 92 Henry Plummer was chief of the band;..Cyrus Skinner, fence, spy, and roadster. **1925** G. H. Mullin *Adventures Scholar Tramp* iv. 56 The roadsters, or hobos who travel, are seldom without smoking or chewing tobacco. **1970** *Oxf. Mail* 27 Apr. 1/3 When we first saw the man we thought nothing about it. Roadsters are a common sight in these parts. **1974** 'J. Ross' *Burning of Billy Toober* iv. 35 The body was probably that of a roadster using the shed to sleep in.

roa·d test. [Road *sb.*] A test of the performance of a vehicle on the road. Hence (usu. with hyphen) as *v. trans.*, to test (a vehicle) on the road; also *transf.* and *fig.*; so **roa·d-tester**; **roa·d-testing** *vbl. sb.*

1906 S. Krausz *ABC of Motoring* 134 This completes the chassis, and..the automobile is ready for examination and road test. **1921** *Dict. Occup. Terms* (1927) 115/2 *Tester, car*;..*road tester* (*motor*); drives finished chassis to test it under road conditions. **1937** *Times British Motor Number* 13 Apr. p. xxiii/2 It is the practice of Vauxhall Motors to road test every Vauxhall car and Bedford truck which is made. *Ibid.*, A certain proportion of the cars still undergo extended road tests. **1946** *Time* 29 July 80 This 60-ft. aluminium-magnesium bus was road-tested in California last week. **1949** *Life* 4 Apr. 129/1 In the U.S. many of the best brains go into business and the public secures their services only after they have been seasoned and road-tested on this severe and impartial proving ground. **1953** *Sun* (Baltimore) 10 July 4/6 Road tests are planned at the Army's Aberdeen (Md.) Proving Ground. **1961** *Times* 14 Feb. 17/2 We cannot feel sorry that these cars are coming off the road. We road-test the best of them and 50 to 60 per cent are definitely death traps. **1968** *Guardian* 10 Sept. 3/7 Maiden voyagers on the Cunarder QE2 will have..to road-test the ship. **1971** *Daily Tel.* 16 July 7 (Advt.), Howard Hunt is a road-tester. Every XJ6 is roadtested twice. **1977** *Custom Car* Nov. 80/2 (Advt.), Koni road testing is real tough, and includes a full international rally and racing programme.

roadwork. Also **road work, road-work.** [Road *sb.*] **1.** Work done in building or repairing roads. Also *pl.*, repairs to roads.

1869 *Bradshaw's Railway Manual* XXI. 100 To effect a diversion of road and other works. **1895** *Funk's Stand. Dict.* s.v. *road, n.*, *road-work, n.*, labor expended in making or repairing roads. **1951** R. Firth *Elem. Social Organiz.* iii. 118 There is one frame of organization.. which regards a farm as a place for a man and his family ..to use as a base from which to operate..in getting cash from road-work. **1958** J. G. MacGregor *North-West of 16* vii. 95 Up to this time all roadwork had been voluntary and consisted of a settler doing enough along the road each way from his place to remove the trees that fell across it. **1965** F. Sargeson *Mem. Peon* ii. 21 His regular hours of road-work labour. **1966** D. Francis *Flying Finish* x. 126 The horsebox drivers..had to make a detour because of roadworks.

2. (See Road *sb.* 12.)

3. The work of an itinerant thief. *Criminals' slang.*

1925 H. Leverage in *Flynn's* 14 Mar. 281/1 *Road work*,..pocket picking, etc., done while traveling. **1955** *Publ. Amer. Dial. Soc.* xxiv. 84 Because of the stresses and strains of road work, he [*sc.* the road man] is usually a sharp, alert thief.

4. Work done on the roads, esp. exercise and training by athletes, sportsmen, etc.

1903 Somerville & 'Ross' *All on Irish Shore* 159 The five couple and Carnage were..on a scent that was a real comfort to them after nearly five miserable months of kennels and road-work. **1928** *Funk's Stand. Dict.* s.v. *road, n.*, *road-work, n.*, exercise taken on a road, as by athletes in training. **1950** J. Dempsey *Championship Fighting* xxiv. 183 *Roadwork* means running on the road. **1964** D. Francis *Nerve* xvi. 190 A..little used secondary road..served only two farms and one private house, and because of its quietness it was a regular route for the Axminster horses on roadwork days. **1969** G. E. Evans *Farm & Village* x. 113 The most trouble we had with Suffolks on road-work was *splinters* and *side-bone*. **1971** A. Burgess *MF* vi. 69 They trotted along, as in road-work, in their orange-and-cream jerseys. **1977** *Time* 25 July 39/2 He was doing roadwork and punching bags in preparation for his role in Martin Scorsese's *Raging Bull*, a film about Fighter Jake La Motta.

roa·dworthiness. [f. Roadworthy *a.* + -ness.] Roadworthy character; reliability on the road.

1923 *Daily Mail* 6 Aug. 4/4 One object was to demonstrate the roadworthiness or otherwise of these..family cars. **1928** *Daily Tel.* 16 Oct. 17 The low centre of gravity gives to the new Humber a road-worthiness unequalled. **1955** *Times* 1 July 6/6 The Government have dropped the proposals for compulsory tests of vehicles for road-worthiness. **1980** N. Freeling *Castang's City* xvi. 105 Polish those boots, boy, and examine them for road-worthiness.

roady, var. *Roadie.

roaf, var. *Rouf.

roak. Add: *roke* is the usual form. (Further examples.)

1914 [see *Lap *sb.*[2] 2 e]. **1923** Glazebrook *Dict. Appl. Physics* V. 363/2 Rokes are formed from ingot cracks, blowholes at or near the surface, and certain kinds of surface defect of the ingot, and in the case of rolled bar they are usually radial when examined on a cross-section. **1945** Greaves & Wrighton *Pract. Microsc. Metallogr.* x. 173 Rokes..consist of fissures..with their surfaces separated by a thin film of scale or other impurity. **1951** G. R. Bashforth *Manuf. Iron & Steel* II. x. 320 Subcutaneous blowholes, occurring very near the skin of the ingot, may become oxidized during reheating, resulting in the formation of 'roaks' and seams in the finished bars or blooms. **1967** A. K. Osborne *Encycl. Iron & Steel Industry* (ed. 2) 354/2 Rokes. (Roaks.) **1974** P. Wright *Lang. Brit. Industry* xix. 184 Ingot defects have various names, for instance the *roke*, into which a surface blowhole rolls out.

roam, *v.* Add: **4.** *trans.* To cause (the eyes) to look over a scene. *rare.*

1900 J. Bloundelle-Burton *Seafarers* xii. 118 As he spoke he roamed his eye around the tranquil, glassy sea.

roan, *a.* and *sb.*[1] Add: **A.** *adj.* **b.** *Hippotragus equinus.* (Earlier and further examples.)

1839 W. C. Harris *Wild Sports S. Africa* xxii. 194 We descended into a valley, bent upon the destruction of a roan antelope. **1970** *Daily Nation* (Nairobi) 16 Jan. 13/1 Going south along the lake is the Lambwe Valley Game Reserve, small and still undeveloped, but possessing the rare Roan Antelope.

B. *sb.*[1] **1. c.** (Further examples.)

1958 L. van der Post *Lost World of Kalahari* vi. 111 It's unbelievable! They're there in thousands! Zebra, wildebeest, roan! **1978** *Times* 23 Nov. 13/9 Extensive softwood planting in the [Kenyan] plains is displacing much of its game, such as roan and sable.

3. A fashion shade: cf. sense A. c.

1960 [see *Grège *a.* and *sb.*].

roar, *sb.*[1] Add: **2. b.** *to go with a roar*, to make uninterrupted progress or be a conspicuous success. *colloq.*

1845 Dickens *Let.* 6 Aug. (1977) IV. 347 It was a most prodigious success; and went, with a roar, all through. **1903** G. B. Shaw *Let.* 12 June (1972) II. 331 'The Admirable Bashville'..went with a roar from beginning to end. **1907** *Punch* 1 May 308/2 Everything went with a roar.

roar, *v.* Add: **1. c.** (Earlier examples.)

1815 B. Wynne *Diary* 28 July (1940) III. xii. 378 The Girls, who *roared* the whole way, laughing at the odd vehicle. **1828** T. Creevey *Let.* 3 Mar. in J. Gore *Creevey's Life & Times* (1934) xii. 260 Brougham's letter is..in folly and insanity by no means inferior to his former effusions. We both *roared* at it.

3. e. To travel on a vehicle which is making a loud noise; to motor rapidly. Also *fig.*

1923 *Motor Cycling* 26 Sept. 658/3 Marsden roared through on his last lap. **1951** *Amer. Speech* XXVI. 230/2 Wesleyan roars to victory. **1958** B. Nichols *Sweet & Twenties* x. 128 They were all roaring off to Ascot. **1963** *New Yorker* 15 June 58 George Rotan..roared back to win eleven of the next twelve. **1970** P. Laurie *Scotland Yard* iii. 69 The one getting in slams the door and roars off, nearly running my mate over. **1973** *Times* 22 Jan. 9/8 The closest he came to betraying anxiety last evening was when he suddenly started roaring ahead.

4. c. Const. *up.* To abuse, to reprimand. *slang* (chiefly *Austral.*).

1919 W. H. Downing *Digger Dial.* 42 *Roar up*, upbraid; abuse. **1925** Fraser & Gibbons *Soldier & Sailor Words* 244 *Roar up*, to abuse. **1944** W. E. Harney *Taboo* (ed. 2) 63, I roared him up, but it was no good. **1947** N. Lindsay *Halfway to Anywhere* 69 Bill was able to roar him up, anyway, for having the blinkin' cheek to come shoving his nose into Bill's affairs. **1962** [see *Roaring forties 2].

roarer[1]. Add: **1. a.** (Further example.)

1903 W. S. Blunt *Seven Golden Odes* 33 Fled to the land of the lions, roarers importunate.

4. (Earlier example.)

1827 *Massachusetts Spy* 10 Jan. 1/4 The Albany beau.. drinks brandy and talks politics, swears at the servants, and quarrels with his landlord and is in fact what he styles himself, 'a real roarer'.

5. *U.S.* An oil-well from which the oil pours rapidly and noisily.

a **1885** B. J. Crew *Pract. Treat. Petroleum* (1887) vii. 227 We have no right, perhaps, to expect a continuance of the 'roarers', or 'gushers' as they are termed.

roaring, *ppl. a.* Add: **4.** (Further examples.) *roaring days* (Austral.), the time of the Australian gold-rush; also *transf.*, hey-day; *the roaring twenties*, the third decade of the twentieth century (with reference to the post-war buoyancy of that period).

1897 H. Lawson *Coll. Verse* (1967) I. 339 But these seem dull and slow to me compared with Roaring Days. **1921** M. E. Fullerton *Bark House Days* (1931) xiv. 144 We loved the stories of the 'roaring fifties'. **1930** *Sat. Rev.* 15 Mar. 328/1 The giants of the roaring 'twenties ought to be able to achieve glory of some sort in half as many years. **1936** 'W. Hatfield' *Australia through Wind Screen* 53 In its roaring days 'The Duchess' was better than many a goldmine. **1973** *Times* 2 Mar. 14/2 The theme [of the ball] will be the roaring twenties. **1978** *Dædalus* Fall 30 For those belonging to the classes of the immediate post-World War I period, the massacre of the young officers..meant that countless positions..had become vacant in all spheres of society; this led to an ephemeral but marked shift to a more youthful establishment; hence, the Roaring Twenties.

7. (Earlier and later examples.)

1755 C. Charke *Life* 153 But was..fully convinced, that I should carry on a roaring Trade. **1976** *Milton Keynes Express* 25 June 7/2 These attractions did a roaring trade round the perimeter of the sports hall.

8. Also as a general intensive: full-blooded, whole-hearted; unqualified, out-and-out.

1963 D. Lessing *A Man & Two Women* 302 If you are going to make love, what does it matter who with? Why shouldn't she simply walk into the street, pick up a man and have a roaring sexual affair with him? **1965** *Listener* 18 Nov. 806/1 Psychiatric treatment has not proved a roaring success. **1970** Mrs. L. B. Johnson *White House Diary* p. xiv, I feel..a deep, roaring faith in and love for this country.

roaring forties. 1. See Forty *a.* and *sb.* B. 4. Now usu. restricted to those parts of the oceans between latitudes 40° and 50° south, where strong westerly winds blow; occas. also applied to the winds themselves.

1883, 1893 [see Forty *a.* and *sb.* B. 4]. **1906** W. Marriott *Hints to Meteorol. Observers* (ed. 6) 68/1 *Roaring forties*, the regions between lat. 40° and 50°S., where the 'brave West winds' blow. **1924** R. Clements *Gipsy of Horn* vii. 126 Right 'roaring forties' weather came down on us with a swoop. **1953** A. A. Miller *Climatology* (ed. 8) xi. 199 In the southern hemisphere the disturbance of the planetary winds is much less; 'Roaring Forties' and the 'Brave West Winds' blow all the year round with considerable force. **1966** F. Chichester in *Sunday Times* 30 Oct. 3/6 Twice I have entered the Roaring Forties and been driven out by gales and squalls.

2. *fig.* **a.** The fifth decade of life. **b.** *Naut. slang.* (See quot. 1948.)

1867 *Harper's Mag.* Sept. 509/2 A very pretty woman, whose bark of life had not as yet drifted into the 'roaring forties'. **1930** H. K. Pasma *Close-Hauled* ii. 211, I am in my roaring forties now. **1948** Partridge *Dict. Forces' Slang* 156 *Roaring forties*, Lieutenant Commanders between 40 and 50 years of age. **1962** Granville *Dict. Sailors' Slang* 96/2 *Roaring forties*, rough seas in 40–50 degrees south latitude, hence, a slang name for certain taut-handed lieut.-commanders in their forties, who are always roaring up the hands.

roaringly, *adv.* (In Dict. s.v. Roaring *ppl. a.*) (Further examples.)

1947 Dylan Thomas *Let.* 1 Mar. (1966) 298, I was roaringly well, then, some minutes after, a little mewling ruin. **1980** *Daily Tel.* 21 July 10/3 This festival built its name in the 'fifties under Jean Vilar's direction, with Gérard Philipe as star, by staging French classics with a zest and a roaringly romantic appeal to basic theatrical values which gave birth to the rightly named Théâtre National Populaire.

roast, *sb.* Add: **4.** Also, an instance of this. (Further examples.) (See also quot. 1900.) Now chiefly *N. Amer.*

1900 *Dialect Notes* II. 54 *Roast, n.* 1. Unfair treatment, as hard marking in a course. 2. A partial decision, as from an umpire. 3. A severe criticism. 4. A reproof. **1903** *Booklovers Mag.* Dec. 663/1 This national love for a good 'roast', this spirit of mockery, this national habit of joking, is the one great thing about us that foreigners can't understand. **1976** *Globe & Mail* (Toronto) 16 Feb. 16/1 (*caption*) It was billed as a roast to mark Mr. Sniderman's 25th year in the music business, but in reality it was a heart-warming evening because the roasters had only kind words for this beloved couple, who've done so much for Canada.

roast, *v.* Add: **4. b.** (Earlier and later examples.) Also, to criticize, to denounce.

1710 *Let. to Noble Lord occasion'd by Proc. against Dr. Henry Sacheverell* 16 As for Dr. Sacheverell, nothing will serve some of 'em but Roasting him; using the Expression of a Furious Zealot against him, who is since Dead. **1890** in Barrère & Leland *Dict. Slang* II. 183/1 Another letter received from one W. T. Nelson, of Cleveland, severely roasts both. **1895** W. C. Gore in *Inlander* Dec. 114 *Roast, v.* 1. To censure. 2. To ridicule. **1905** 'H. McHugh' *You can search Me* iii. 50 If he were to roast our Skinski it might hurt our business. **1912** J. Sandilands *Western Canad. Dict.*, *Roast*, to expose, to abuse, to rate, to tell a person off. A roasting, a severe rating or castigation in a speech. **1920** Wodehouse *Jill the Reckless* (1922) xviii. 267 I've an idea..that the critics will roast it. **1966** *Listener* 27 Oct. 613/3 Their methods caused a scandal and they were roasted in the press by Labouchère. **1976** F. Trueman *Ball of Fire* ii. 39 They made me twelfth man and I was roasted for falling asleep in a deck-chair during play. **1977** *Times* 28 Oct. 8/5 During the evening the Prince was 'roasted' by Martin—a friendly American custom of insulting a person as a sign of favour.

roast beef. Add: **b.** *roast-beef-of-old-England-man, sandwich*; **roast beef coat** = *roast beef dress.*

1818 'A. Burton' *Adventures J. Newcome* ii. 117 His ship-washed linen out he laid, And roast beef coat in smart parade. **1831** M. Edgeworth *Let.* 29 Mar. (1971) 507 Her husband is one of the thin dried old race of true hunter and shooter men and roast beef of old England-men. **1967** 'D. Shannon' *Chance to Kill* (1968) xiii. 189, I even remember what she had... It was the hot roast-beef sandwich. **1971** D. Enefer *Screaming Orchid* xii. 103, I had stopped..for roast beef sandwiches and bitter beer.

roasting, *vbl. sb.* Add: **1. b.** (Further examples.)

1895 *Wales* May 222/1 Your father will make short work of giving you a roasting. **1900** 'Flynt' & 'Walton' *Powers that Prey* 122 Nettie was emboldened to continue her 'roasting'. **1942** J. B. Priestley in *R.A.F. Jrnl.* 3 Oct. 2, I..have taken and dished out uproarious insults. .. My friendly hosts have been anxious about the way in which I would take this elaborate 'roasting'. **1963** *Times* 25 Feb. 3/6 The crowd, which cheered wildly all the time for the popular Mormon elder from neighbouring Utah, gave Tiger a 'roasting' several times and they booed lustily when the decision was announced as a draw. **1977** *Time* 28 Mar. 8/3 In their exchanges with the Russians, members of the U.S. delegation anticipate a bit of a roasting.

2. a. *roasting device, fork, kitchen, machine, tin.*

1923 H. Crane *Let.* 5 Dec. (1965) 159 The ten pound bird was put into a wonderful roasting machine... You put the bird on a long spit that had a crank and catches... You must have seen one of these roasting devices. **1950** W. Bird *Nova Scotia* ii. 37 On the wall are such items as an otter head, and feet,..and brass roasting forks. **1960** E. David *French Provincial Cookery* 66 A shallow rectangular baking or roasting tin. **1965** E. Tunis *Colonial Craftsmen* vi. 67 Sometime near the beginning of the eighteenth century, perhaps earlier, somebody invented the 'roasting kitchen', a reflecting oven built as an arch-topped box on legs, with one open side to face the fire.

rob, *v.* Add: **1. a.** (Further *transf.* and *fig.* examples.)

1926 *Publishers' Weekly* 19 June 1966/1 You may improve your golf game... Why not get rid of that disconcerting slice which robs your drive? **1948** R. M. Ayres *Missing Tide* i. 44 The food's quite good, and they don't rob you, anyway.

d. *Association Football.* To deprive (an opposing player) of the ball.

1882 *Blackburn Times* 1 Apr. 6/3 Goodhart started the ball from the centre, but he was instantly robbed by Strachan. **1970** *Times* 30 Sept. 15/4 Novak held on too long in midfield and was robbed by Graham. **1976** *Morecambe Guardian* 7 Dec. 8/2 Towers and Thomas forced the defence into some confusion when a backpass went astray. Finch had to move quickly to rob Thomas who was charging through.

5. (Further examples.)

1919 G. B. Shaw *Heartbreak House* ii. 77, I should rob all the money back from Mangan. **1939** Joyce *Finnegans Wake* (1964) iii. 453 Robbing leaves out of my taletold book. **1953** [see *robber trench* s.v. *Robber 2 b]. **1977** *Irish Press* 29 Sept. 5/5 Vincent Walker..was found guilty of robbing the sum of £8,798.

robber. Add: **2. a.** *robber-book, -haunt.*

1884 'Mark Twain' *Huck. Finn* ii. 13 The rest [of the oath] was out of pirate books, and robber books. **1937** J. W. Day *Sporting Adventure* 91 The magpies will go off to their robber-haunts in lonely carrs of willows down on the marshes.

b. robber baron [Baron 1], a feudal lord who engaged in plundering; also *transf.*, *spec.* [*Baron 2 b] in *U.S.*, a financial or industrial magnate of the late nineteenth century who behaved with ruthless and irresponsible acquisitiveness; also *attrib.*; **robber trench** *Archæol.*, a trench representing the foundations of a wall, the stones of which have been partially or entirely removed.

1878 C. F. Adams *Railroads* 145 The commissioner has not hesitated to give his opinion of the foreign owner as a 'robber baron'. **1882** C. Schurz in *Boston Herald Suppl.* 30 June 1/3 It will not be surprising at all to see some day a movement set on foot to put an end to the operations of the modern robber barons, who, by corporate rascality, supplemented with tricks of the stock exchange, manage to plunder at will not only their fellow-gamblers, but the innocent bona fide investors in corporate enterprises. **1930** J. S. Huxley *Bird-Watching* ii. 32 Predaceous glaucous gulls, the robber barons of the Arctic bird-world. **1934** M. Josephson (*title*) The robber barons. **1949** *Jrnl. Econ. Hist.* Nov. 187 In studying the so-called 'robber barons', Destler was impelled to consider also a few early 'career men'. **1957** *Times Lit. Suppl.* 8 Nov. 670/3 Next she builds up an immensely lucrative cosmetic business, backed by a robber-baron tycoon named Jim Seymour. **1962** J. Braine *Life at Top* x. 131 A robber baron of the Middle Ages. **1976** M. J. Lasky *Utopia & Revolution* (1977) ii. 74 Bakunin joined the call for a crusade of destruction, and he, too, became a robber baron in a holy cause. **1979** *Time* 2 Apr. 45/1 For years psychiatrists have also been regarded as medicine's robber barons. **1953** R. J. C. Atkinson *Field Archaeol.* (ed. 2) ii. 72 On many Roman and later sites where ancient buildings have once stood, the stone will have been partially or completely robbed from the walls and foundations for re-use elsewhere. In such cases the walls can be traced only as 'robber-trenches'. **1967** *Antiquaries Jrnl.* XLVII. 196 The outer edge of the wall and robber trenches has been found along most of the edge of the north aisle and around the west end. **1978** *Ibid.* LVIII. 106 A late Roman beaded and corrugated pin similar to one found at Lydney was found in robber trenches of the medieval cloister.

c. *robber gull*; **robber-fly** (earlier and later examples).

1871 *Amer. Naturalist* IV. 686 A robber-fly..burrows in the sand. **1970** *Age* (Melbourne) 22 June, Another [family] comprising the predatory robber-flies. **1946** J. W. Day *Harvest Adventure* vi. 86 The big robber gull dropped like a sack of wheat, without a kick, at seventy yards.

Hence **ro·bberish** [-ish[1]], *a.*, suggestive of robbers; **ro·bberism** [-ism], control by or the business of robbers; robbery; **ro·bberling** [-ling[1]], a little or puny robber.

1855 Swinburne *Let.* 4 Aug. (1959) I. 6, I longed for you all to be there,..for it [*sc.* a cave] was admirably robberish. **1913** D. H. Lawrence *Love Poems & Others* 8 Under the glistening cherries... Three dead birds lie: Pale-breasted throstles and a blackbird, robberlings Stained with red dye. **1921** *Glasgow Herald* 18 Jan. 6 Communism in Russia is robberism.

robbery. Add: **3.** *fig.* An excessive financial demand; a proposal which wholly or chiefly benefits the proposer; an outrageous injustice; esp. in *daylight robbery*, *highway robbery* (s.v. *Highway 4).

c **1863** T. Taylor *Ticket-of-Leave Man* i. 10 *Dalton*: I won't go higher than fifteen bob for a fiver. *Moss*... Only fifteen—it's robbery. **1874** E. P. Roe *Barriers burned Away* v. 38 'I want five dollars out of you before you take that trunk off.' 'Why, this is sheer robbery,' exclaimed Dennis. **1886**, etc. [see *highway robbery* s.v. *Highway 4]. **1949** D. M. Davin *Roads from Home* i. i. 8 'I can never afford it,' said his sister. 'It's daylight robbery.' **1976** *Springfield* (Mass.) *Daily News* 23 Apr. 39/1 Though the Celtics are well known for their game-long verbal abuse of officials, Wednesday night they got away with robbery. **1977** *Times* 28 Feb. 8/5 It was, in fact, a bit of daylight robbery. As Jimmy Andrews, the disappointed Cardiff manager, said later: 'Everton had all the big names and the luck.'

robbo (rǫ·bo). *Austral. local slang.* [f. the name *Rob*(*inson* (see quot. 1897) + *-o[2].] A horse and trap; a sulky; a poor horse; the driver of a horse and trap. Also, anything or anyone not up to standard, and in other extended uses.

1897 *Bulletin* (Sydney) 23 Jan. 11/3 In answer to a correspondent's question as to the derivation of 'robbo' (Sydney slang for the vehicle ordinarily called a 'sulky') 'J.P.' writes as follows..:—'Four Bob Robbo'—four shillings Robinson. Robinson, who lived in the classic suburb of Waterloo, Sydney.., came into a bit of money and bought a horse and trap. The money was spent, and Robinson tired of feeding the horse, which got poor; so he then sometimes let out the horse and trap (both somewhat worse for wear) for 4s. per half-day. There was a run on the cheap hire, and Rob. bought two other horses and traps, which he let out at the same price. A neighbouring livery-stable keeper and his employés resented Rob's cutting-down prices; and, when any of the rival's equipages passed, used to cry out, in derision, 'Four Bob Robbo!' The cry was taken up by the kids, and has now become a Waterloo classic. *Ibid.*, 'Robbo' has in an extensive Sydney circle come to mean anything unsatisfactory. For instance, a girl enters a jeweller's shop with: 'Watcher been givin' us? Look at the clasp of this 'ere bracelet I bought of yer last week. It's gone bung already. It's a fair robbo.' *Ibid.*, Also 'robbo' has come to mean amateur. **1906** A. J. Tompkins *With Swag & Billy* 51 Right out of the haunts of the motor, the bike and the Robbo. **1939** K. Tennant *Foveaux* iv. v. 430 There was old Bert Robinson... 'E kept a livery stable down at the Foot. I s'pose you've 'eard of the Four-bob Robbos, then? The chaps used to go an' hire a cart for four bob and take it round loaded with vegetables. The kids used to call after 'em, 'Four Bob Robbo, Four Bob Robbo.' Old Bob Noblett, 'e's an old man now, but I can remember when Bob Noblett was a four-bob robbo. **1956** *Collins New Eng. Dict.* (Austral. & N.Z. Suppl.) 1279/2 *Robbo, Four-Bob-Robbo*, a horse and sulky... Now used only for a decrepit horse.

robe, *sb.*[1] Add: **1. c.** A dressing-gown. See also *bath robe* s.v. *Bath *sb.*[1] VI.

1854 Dickens *Hard T.* ii. viii. 223 She arose, put on a loose robe, and went out of her room in the dark. **1931** J. B. Fagan *Improper Duchess* i. 25, I puta on my silk robe, I go down to his rooms. **1938** M. Allingham *Fashion in Shrouds* vi. 73 'Robe's' the new name for dressing-gown. **1945** 'L. Lewis' *Birthday Murder* (1951) x. 152 She..put on the white terry-cloth robe. **1955** T. Sterling *Evil of Day* xiii. 134 A man in a robe and slippers. **1957** F. & R. Lockridge *Practise to Deceive* (1959) xiii. 181 Susan wore a white towelling robe. **1966** *Wall St. Jrnl.* 7 Jan. 2/2 Penney officials noted heavy sales in women's robes and sleepwear. **1970** G. F. Newman *Sir, You Bastard* viii. 246 Tying his robe, he stepped out and along to the kitchen. **1976** *New Yorker* 26 Jan. 50/3 At lunch, Mrs Fox, still in pajamas, slippers, and robe, nearly drops a tray on Mrs Tompkins' head.

7. *robe-cloak, -coat.*

1908 G. B. Shaw *Lett. to Granville Barker* (1956) 139, I have persuaded her to be discovered next time in a robe-cloak. **1746** in R. Chambers *Traditions of Edinburgh* (1846) 47 No misses in skirts and jackets, robe-coats, nor stay-bodied gowns, to be allowed to dance in country-dances. **1911** C. Mackenzie *Passionate Elopement* xviii. 170 Swansdown misses..put into corsets almost as soon as they were out of robe-coats. **1964** *New Shetlander* No. 70. 27 She wuir a hap, rob cott an bratt.

robe (rōub), *sb.*[2] Also **'robe**. Abbrev. of Wardrobe.

1935 *Spectator* 7 June 972/1 Mr. Toop, a wholesale furniture-maker in the Curtain Road with whom I once had dealings, introduced me to some pretty examples of what grammarians, I believe, call aphaeresis. 'If you want a 'Board,' he would say, 'I'd choose wawnut every time: but when it comes to a 'Robe, there's nothing to touch m'yogany.' **1969** *Sydney Morning Herald* 24 May 43/9 (Advt.), Built-in robes. **1974** T. R. Dennis in J. Burnett *Useful Toil* iii. 354, I did a bedroom suite for £21; it had a six-foot robe..dressing-table and tall-boy and bed to match. **1977** *Evening Gaz.* (Middlesbrough) 11 Jan. 11/7 (Advt.), Two double bedrooms, one with fitted unit and 'robe.

robe, *v.* Add: **1.** Also, to apparel (oneself) in a dressing-gown.

1969 *New Yorker* 31 May 32/1 If I am resolute, I will arise and robe myself.

‖ **robe de nuit** (rōb də nwī). [Fr.] A night-dress.

1855 Trollope *Warden* ii. 21 He has exchanged..those shining black habiliments for his accustomed *robe de nuit*. **1897** G. Stevens *Let.* 2 Nov. in *Lett. W. Stevens* (1967) 16 Your Mother is making up some sort of..a Robe-de-Nuit something to cover your abused anatomy as you wander ..to the toilet. **1911** E. M. Clowes *On Wallaby* v. 119 A lady whose husband had seen another lady going to the bath in her *robe de nuit* alone. **1968** J. Ironside *Fashion Alphabet* 71 *Robe de Nuit*, nightdress.

‖ **robe de style** (rōb də stīl). [Fr., lit. 'robe of style'.] (See quot. 1969.) Also *fig.* and *attrib.*

1928 [see *eau-de-Nil* s.v. *Eau]. **1931** *Times Lit. Suppl.* 25 June p. i/4 Dignincont's..etchings and Coulouma's clear setting in Baskerville lend an expensive *robe de style* to a novel which seems more at home in a yellow jacket. **1963** *Times* 24 Jan. 12/4 John Cavanagh's bridal model, a soft vision in white chiffon, cut on *robe de style* lines, with chiffon veil falling from a flowered chignon, has given rise to much speculation whether this studied simplicity will be reflected in the gown that Princess Alexandra will choose for her wedding. **1969** R. T. Wilcox *Dict. Costume* (1970) 293/1 *Robe de style*, the twentieth century infanta style, an evening fashion for which Lanvin of Paris became famous; its vogue was in the nineteen twenties and thirties. It had a tight bodice with a bouffant skirt, ankle or floor length.

Robert. Add: **3.** (Further examples.)

1899 'J. Flynt' *Tramping with Tramps* ii. 231 But look out for the Robert and the Dee (the policeman and the detective). **1929** T. L. Davidson *Murder in Laboratory* xiv. 108, I stopped and asked a Robert the time. **1968** J. Lock *Lady Policeman* iv. 34 Believe it or not PCs are still occasionally wished, 'Good morning, Robert!'

5. *Robert sauce, Sauce Robert*: see Sauce *sb.* 1 in Dict. and Suppl.

6. *Naut. slang.* A spell off duty; a sleep, a 'nap'.

1935 'L. Luard' *Conquering Seas* xii. 140 I'll get head down for a proper robert.

Robertian (rǫbə̄·ɹtiăn), *a.* [f. prec. + -ian.] Of or pertaining to Robert the Strong (*d.* 866), count of Anjou and of Blois, or his descendants, who became kings of France. Also as *sb.*, a follower or successor of Robert the Strong.

1903 D. C. Munro *Hist. Middle Ages* vii. 66 Charles the Simple..had little power, and the kingdom was wrested from him in 923 by a member of the Robertian house. **1942** Strayer & Munro *Middle Ages* vi. 147 Otto had the advantage of being the brother-in-law both of the Carolingian king and of the head of the rival Robertian family. **1957** *Encycl. Brit.* IX. 589 Henceforth there ensued a long duel between the Robertians and the Carolingians in which three times the Robertians were chosen and might have taken the crown.

Robertine (rǫ·bəɹtīn, -əin), *sb.* and *a.* [f. as prec. + -ine[1].] **A.** *sb.* A follower of Robert of Melun (d. 1167), English-born scholastic theologian.

1846 T. Wright *Biogr. Brit. Lit.: Anglo-Norman Period* iv. 201 His disciples formed a sect which was long known by the name of Robertines. **1906** W. H. Schofield *Eng. Lit. to Chaucer* ii. 52 At Mont St. Geneviève the Robertines' long continued to discuss their leader's great work..the *Summa Theologiae*, which above all gave warrant for his repute as a metaphysician.

B. *adj.* = *Robertian *a.*

1938 Z. N. Brooke *Hist. Europe* iv. 96 The ambition of the Robertine house dictated the fortunes of the French kingdom in the tenth century.

Hence **Roberti·nian** *sb.*, a follower or successor of Robert the Strong.

1910 *Encycl. Brit.* X. 813/1 The struggle against the Robertinians went on relentlessly. *Ibid.* 813/2 There was a kind of *entente cordiale* between the Carolingians and the Robertinians and Otto.

Robertonian (rǫ:bəɹtōu·niăn). [f. as prec. + *-onian* as in *Caledonian*, *Patagonian*, etc.] A follower of Robert the Strong (see *Robertian *a.*).

1936 H. A. L. Fisher *Hist. Europe* i. xvii. 208 Robert the Strong, Count of Paris, fought for ten years against the Northmen... The Robertonians were as distinguished for caution as for courage.

Robertsonian (rǫ:bəɹtsōu·niăn), *a.* *Cytology.* [f. the name of William R. B. *Robertson* (1881–1941), U.S. biologist, who first described such translocations in 1916 (*Jrnl. Morphol.* XXVII. 220) + -ian.] Applied to the formation of a metacentric chromosome from two heterologous acrocentric chromosomes by the fusion of their centromeres or by a translocation with the loss of a small fragment; and to karyotypic changes brought about by this process.

1954 M. J. D. White *Animal Cytol. & Evol.* (ed. 2) x. 192 In certain groups such as the Acrididae, 'Robertsonian' rearrangements or whole-arm transpositions account for a large part of the obvious differences in karyotypes. **1955** *Nature* 2 Apr. 601/1 The wide variation in the mitotic numbers must be attributed to Robertsonian changes. **1960** *Jrnl. Nat. Cancer Inst.* XXIV. 1187 A large mediocentric chromosome and a heterochromatic minute were formed, apparently at the expense of two acrocentric chromosomes, providing a classic example of a Robertsonian relationship, manifesting itself within the neoplastic cell population of a transplantable tumor. **1973** *Nature* 3 Aug. 262/1 The most common chromosomal changes seen in vertebrate evolution are Robertsonian fusions which create one metacentric from two acrocentrics and inversions which, if pericentric in nature, change the position of a centromere. **1974** *Ibid.* 10 May 164/1 These consisted of thirty-eight Robertsonian translocations, forty-seven reciprocal translocations and nine pericentric inversions.

Robertson's law (rǫ·bəɹtsənz). *Cytology.* [f. as prec. + Law *sb.*[1]] The law that states that the number of chromosome arms of a population or species tends to remain constant, although the number of chromosomes may vary. Cf. prec.

1945 M. J. D. White *Animal Cytol. & Evol.* viii. 170 In certain groups 'Robertson's law' explains many of the more obvious changes in chromosome shape. **1956** *Jrnl. Morphol.* XCIX. 265 Well-documented instances of chromosomal phylogeny conforming to Robertson's law have been reported from several groups of animals.

Robespierrist (rōu·bspiēərist), *sb.* and *a.* [f. the name of *Robespierre* (see below) + -ist.] **A.** *sb.* A follower of Maximilien François Marie Isidore de Robespierre (1758–94), one of the leaders in the French revolution; a Jacobin (sense 2). **B.** *adj.* Associated with, or adhering to, Robespierre.

1834 [see *Orleanist]. **1904** J. R. M. Macdonald in *Cambr. Mod. Hist.* (1907) VIII. xii. 338 The Commune, whose conception of the ultimate ends of the *coup d'état* differed *toto cœlo* from those of both Dantonists and Robespierrists. **1929** L. R. Gottschalk *Era of French Revol.* i. iii. iii. 263 (*heading*) The Robespierrists destroyed. **1937** *Downside Rev.* Oct. 519 It is certain that the laws [of Ventôse] contributed to the fall of the

Robespierrists. *Ibid.*, The majority of the Committee were unsympathetic and obstructive, which increased the tension between them and the Robespierrist minority. **1975** G. RUDÉ *Robespierre* 9 It marks a welcome addition to Robespierrist studies. *Ibid.* 43 While the Jacobins and the Convention—even the Robespierrists among them—were prepared to tolerate controls and State-direction of the nation's economy merely as exceptional and temporary measures, [etc.].

robiboo, var. *RUBBABOO.

robin¹. Add: **II. 3.** (Earlier and later examples.) For *U.S.* read *N. Amer.*
1703 S. SEWALL *Diary* 16 Mar. (1879) II. 75 The Robbins cheerfully utter their Notes this morn. **1944** S. BELLOW *Dangling Man* 172 A few large birds, robins and grackles, appeared in the trees. **1966** *Vancouver Province* 19 Nov. 1/5 The robin had been sitting in a mountain ash tree in his front yard.

4. b. *blue robin*, add: *Sialia sialis*; (earlier example).
1844 J. E. DEKAY *Zool. N.Y.* II. 65 The Bluebird, or Blue Robin as it is called in the western counties.

c. robin-chat, one of several African thrush-like birds belonging to the genus *Cossypha* of the family Turdidæ.
1901 A. C. STARK *Birds S. Afr.* II. 209 (*heading*) Noisy Robin-Chat. **1931** *Discovery* May 138/2 The robin chat..is smart in appearance, with..blue shoulder patches and bright orange-rufous underparts. **1960** *Times* 29 Sept. (Nigeria Suppl.) p. xxi/7 The colourful white-headed robin chat, that richest and most versatile of Nigerian songsters.

5. *robin-anthem*; **robin's egg** (further examples); usu., **robin's egg blue**; **robin-snow** (earlier example in a more general sense).
1853 THOREAU *Jrnl.* 11–12 Jan. in *Writings* (1906) X. 462 He says that the most snow we have had this winter (it has not been more than one inch deep) has been only a 'robin snow' as it is called, *i.e.* a snow which does not drive off the robins. **1881** Robin's-egg blue [see *PEACOCK *sb.* 4*]. **1887** M. E. WILKINS *Humble Romance* 15 A dress-pattern of robin's-egg blue silk. **1910** *Busy Man's Mag.* Dec. 65/1 On either side were swift hills mottled with green and gold, ahead a curdle of snow-capped mountains, above a sky of robin's-egg blue. **1933** N. WALN *House of Exile* I. iii. 43 Two lovely robin's-egg-blue bowls and two pairs of ivory chopsticks. **1951** E. PAUL *Springtime in Paris* v. 91 Cloud battalions retreating, and stragglers streaked with red—geranium, salmon, vermilion, magenta. Between them, their complements of robin's egg, turquoise, and faint bottle green. **1951** AUDEN *Nones* (1952) 15 A robin with no Christian name ran through The Robin-Anthem which was all it knew. **1970** R. LOWELL *Notebk.* 27 The boys.. Crawling the swimming pool's robin's-egg sky. **1978** M. PUZO *Fools Die* xi. 118, I gave up all thoughts of buying a Cadillac and settled for the robin's-egg blue Dodge.

III. 6. a. (Later example.)
1913 D. H. LAWRENCE in *New Statesman* 16 Aug. 595/2 We called the purple primroses 'robins', for no reason, unless that they bloomed in winter.

7. b. (Earlier example.)
1853 J. RICHARDSON *Let.* 24 May in N. E. Eliason *Tarheel Talk* (1956) 290 We caught 19 brim & robbins.

robing, *vbl. sb.* Add: **4.** *robing-table*.
1927 T. WILDER *Bridge San Luis Rey* ii. 19, I slipped into the sacristy, climbed the robing-table..and walked in.

Robin Hood, *sb.* Add: **1.** (Further examples of allusive use.) Also, more widely, any person who acts irregularly for the benefit of the poor.
1931 J. BUCHAN *Blanket of Dark* v. 96 This Catti.. robbed especially rich men..but spared the Church and the poor—a shabby Robin Hood. **1948** G. V. GALWEY *Lift & Drop* iv. 70 Strip cartoons relating the adventures of Hugh Stinton the Robin Hood of private enquiry agents. **1967** *Listener* 30 Mar. 421/2 The challenge of an Asian-style Robin Hood telling the poor that they will be fattened with good food for which the rich and corrupt..will be forced to pay. **1973** P. B. AUSTIN tr. *Sjöwall & Wahlöö's Locked Room* xxiv. 203 She thought of him as a Robin Hood who stole from the rich to give to the poor. **1976** *Oadby & Wigston* (Leics.) *Advertiser* 26 Nov. 2/4 The plot involves five main characters, all budding Robbin Hoods who realise there is money in fur coats. **1978** M. PUZO *Fools Die* xi. 119, I still had a little bit of the Robin Hood in me.

attrib. and *Comb.* (Further examples.)
1951 KOESTLER *Age of Longing* II. iii. 235 Pierre.. practised a kind of Robin Hood democracy. **1963** A. LUBBOCK *Austral. Roundabout* 161 The bushrangers also had a number of allies..for their Robin Hood attitude to their victims. **1975** *Times* 18 Apr. 4/7 A home loan fraud with a Robin Hood quality..to obtain mortgages for Asian immigrant families. **1977** *It* May 29/3 Ideally these would be assessed on a Robin Hood basis.

5. Robin Hood's barn, used as the type of an out-of-the-way place; esp. in phr. *(a)round Robin Hood's barn*, by a circuitous route (lit. and fig.).
a **1854** J. F. KELLEY *Humors of Falconbridge* (1856) 220 The way some folks have of going round 'Robin Hood's barn' to come at a thing. **1878** *N. & Q.* 22 June 486/2 'Where have you been today?' 'All round Robin Hood's barn! I have been all about the country, first here and then there.' **1913** E. M. WRIGHT *Rustic Speech* xi. 189 To go round by Robin Hood's barn (Cmb. w. Midl.) is to go a roundabout way, to go the farthest way. **1928** S. LEWIS *Man who knew Coolidge* 17 When it came to *talking*, why say, he wandered all round Robin Hood's barn! **1934** E. M. RHODES *Beyond Desert* 201 Wagon-road goes all around Robin Hood's barn to get to my place. **1951** H. WOUK *Caine Mutiny* xxxix. 464, I have gone all the way around Robin Hood's barn to arrive at the old platitudes, which I guess is the process of growing up. **1977** *Time* 31 Jan. 1/3 Your article on birth control goes around Robin Hood's barn for an answer to the birth control problem.

6. Used *attrib.* and *absol.* to designate a type of high-crowned hat with the brim turned up at the back and down at the front, trimmed with a feather.
1894 C. G. HARPER *Revolted Woman* ii. 43 Rational Dress..is only Bloomerism with a difference... A 'Robin Hood' hat, even as in the bygone years, crowns this confection. **1939** R. CHANDLER *Big Sleep* xi. 79 Her black hair was glossy under a brown Robin Hood hat. **1944** A. THIRKELL *Headmistress* x. 220 What interested him was her hat; a kind of Robin Hood hat of green felt with a long quill stuck jauntily through the crown. **1960** *News Chron.* 11 Apr. 8/4 Gone are the heavy-looking trilbies... In their place have come the delta and the Robin Hood. **1966** 'A. YORK' *Eliminator* iv. 66 Wilde..dressed in flannel bags and a sports coat, added an old Robin Hood. **1975** W. HILDICK *Bracknell's Law* 36, I was accosted by a little old woman in black: black coat, black stockings, black hat—and all a throwback to the forties, with the hat of the Robin Hood type.

Hence **Robin-Hoo·dish** *a.*, characteristic of a Robin Hood; beneficent to or benefiting the poor.
1974 *Listener* 18 July 86/3 The morally respectable, Robin-Hoodish bank robbery.

Robinocracy (rǫbinǫ·krăsi). [f. the name *Robin* (ROBIN¹) + -OCRACY.] The régime of Sir Robert Walpole (1676–1745), the predominant figure in British politics between 1721 and 1742; the clique led by Walpole; the period of Walpole's supremacy.
1727 *Craftsman* 22 July 71 This week was publish'd Robin's Panegyrick on Himself and his Friends at Westminster; modestly proving that they were all very honest Fellows and deserving Patriots; with a full Confutation of the charge of Bribery and Corruption Offered to the consideration of the Freeholders; Citizens, Burgesses and Freemen of Great-Britain. *Populus me sibilat, at mihi plaudo.* Hor. Printed for S.B. W.W. and T.W. Printers to the Robinocracy. *a* **1902** ACTON *Lect. Mod. Hist.* (1906) xvi. 274 After the fall of Walpole it was observed..that the country felt itself superior to the government. This was the natural result of the time known as the Robinocracy; not because he devised liberal measures, but because he was careful to be neither wiser nor more liberal than the public. **1974** J. B. OWEN *Eighteenth Cent.* i. 23 On 19 April 1722 Sunderland died of pleurisy, and the way was open for Walpole to assert his supremacy. The Robinocracy had begun. **1977** W. A. SPECK *Stability & Strife* x. 222 Bolingbroke could be highly persuasive and his essays were the most substantial contemporary critiques of the Robinocracy.

robin redbreast. Add: **1. d.** *U.S.* = ROBIN *sb.*¹ 3 in Dict. and Suppl.
1696 S. SEWALL *Diary* 4 Jan. (1878) I. 242 Some say they saw a Robin-Redbrest to-day. **1865** *Atlantic Monthly* May 517/1 Shortly after Robin-Redbreast.. [arrives] the Golden-Winged Woodpecker. **1949** *Hobbies* Oct. 155/1 Robin Redbreast—most familiar of North American birds—has thrived as man's close neighbor.

robinre·dbreasted, *a. nonce-wd.* [f. ROBIN REDBREAST + -ED².] Clad in a red waistcoat.
1922 JOYCE *Ulysses* 465 Tom Rochford, robinred-breasted, in cap and breeches.

Robinsonade (rǫ:binsənē¹·d, ‖ -ādə). Also **Robinsonnade** and with lower-case initial. Pl. **Robinsonades,** ‖ **-aden**. [ad. G. *Robinsonade* (coined by J. G. Schnabel, *Die Insel Felsenburg* (1731), Preface): see next and -ADE.] A novel with a subject similar to that of *Robinson Crusoe*; a story about shipwreck on a desert island.
1847 *Blackw. Mag.* Sept. 330/2 These outcasts from civilisation, the adventures of most of whom would furnish abundant materials for a Robinsonade. **1941** P. B. GOVE *Imaginary Voy. Prose Fiction* p. ix, The late Hermann Ullrich, whose knowledge of the influence of *Robinson Crusoe* has probably never been equaled, put on the title page of his bibliography of robinsonades in 1898 'Teil I'. *Ibid.* I. v. 125 Imitations of *Robinson* have been known usually as *Robinsonaden* or *robinsonades* (only rarely as robinsoniads), and similar works published before 1719 as *prerobinsonades*. **1967** B. W. ALDERSON tr. *B. Hürlimann's Three Centuries Children's Bks. in Europe* xvii. 252 Robinsonnades, like history, geography, and travel books, have always had a big attraction for the Swiss. **1974** *Encycl. Brit. Micropædia* VIII. 618/3 *Robinsonade*, novel written in imitation of *Robinson Crusoe*.. dealing with the problem of the castaway's survival on a desert island. **1978** D. WAGGONER *Hills of Faraway* 16 The Robinsonade is, of course, named for *Robinson Crusoe*, and is the story of a castaway—a voyage cut short —in an isolated setting, which the author can use to describe his ideas of the basic elements separating man from beast.

Robinson Crusoe (rǫ·binsən krū·so). The name of the eponymous hero of Daniel Defoe's fictional narrative (1719), who survives shipwreck on a desert island, used allusively. Also *attrib.* and (rare) *ellipt.* as *Robinson*. Cf. *CRUSOE. So **Ro·binson Cru·soe** *v. trans.*, to maroon on a desert island; **Ro·binson Cru·soic** *a.*
1768 *London Mag.* Oct. 543/1, I am of late from a sprightly fellow become a peevish mal-content; and am misadventure had Robinson-crusoed me, by throwing me into a desert-isle, where I could have nothing but seals and wild goats for my companions. **1849** L. HUNT *Bk. for Corner* I. 14 There are Robinson Crusoes in the moral as well as physical world,..—men, cast on desert islands of thought and speculation; without companionship; without worldly resources; forced to arm and clothe themselves out of the remains of shipwrecked hopes, and to make a home for their solitary hearts in the nooks and corners of imagination and reading. **1856** E. K. KANE *Arctic Explorations* I. xxvi. 348 A host of expedients were to be resorted to, and much Robinson Crusoe labor ahead. **1878** TROLLOPE *How 'Mastiffs' went to Iceland* iii. 6 Though the life of a Robinson Crusoe or a few Robinson Crusoes may be very charming, humanity always desires to restore a Robinson Crusoe back to the community of the world. **1929** G. B. SHAW *Matter with Ireland* (1962) 213 His Robinson Crusoic independence of his neighbors. **1930** R. CAMPBELL *Adamastor* 50 Of all the ocean-gods and nagas That the last surviving Robinson. **1941** L. MACNEICE *Poetry W. B. Yeats* x. 218 Eliot..in *After Strange Gods* has grouped Lawrence and Yeats..as writers who have suffered from the lack of an established religion and a traditional moral code and who have invented for these things Robinson Crusoe substitutes. **1974** H. MACINNES *Climb to Lost World* iv. 53 The army compound..had a Robinson Crusoe atmosphere with the palm and pau trees. **1979** 'G. BLACK' *Night Run from Java* viii. 76 One [island] where Robinson Crusoe the marine accident victims.

robinsonite (rǫ·binsǫnait). *Min.* [f. the name of S. C. *Robinson* (b. 1911), Canadian geologist + -ITE¹.] A bluish or grey lead antimony sulphide occurring as slender prismatic crystals and fibrous or compact masses.
1952 L. G. BERRY et al. in *Amer. Mineralogist* XXXVII. 438 The powder pattern obtained by Professor Peacock was found to be identical with one of several unidentified patterns obtained by Dr. S. C. Robinson at Queen's University during his investigation of the synthesis of lead antimony sulphides... The name robinsonite is given to this new mineral in honor of Dr. Robinson, whose synthesis made its identification possible. *Ibid.*, Robinsonite occurs as a primary mineral with pyrite, sphalerite, stibnite, and boulangerite in small pieces in oxidized ore bodies at the Red Bird mercury mine, Pershing County, Nevada. **1973** *Canad. Mineralogist* XII. 199/1 The Pb-Sb-S system..has been studied between 300° and 700°C. Five phases have been synthesized: Phase I ($3PbS{:}Sb_2S_3$)..., Phase II ($5PbS{:}2Sb_2S_3$)... [illegible] robinsonite ($PbS{:}Sb_2S_3$)... of robinsonite, which appears to be stable down to below 318°C, as a mineral is a perplexing problem; natural robinsonite may be stabilized by small amounts of impurities.

robomb (rǭ·bǫm). *temporary.* [f. *ROB(OT + BOMB *sb.*] = *robot bomb, *flying bomb.
1944 *Saturday Night* (Toronto) 22 July 17 (*heading*) Germany's robombs another case of 'too little and too late'. **1945** H. S. ZIM *Rockets & Jets* xix. 266 Like the V-1 robomb it must be given a high initial speed before the engine begins to operate.

robot (rǭ·bǫt). [Czech, f. *robota* forced labour; used by Karel Čapek (1890–1938) in his play *R.U.R.* ('Rossum's Universal Robots') (1920).] **1. a.** One of the mechanical men and women in Čapek's play; hence, a machine (sometimes resembling a human being in appearance) designed to function in place of a living agent, esp. one which carries out a variety of tasks automatically or with a minimum of external instruction.
1923 P. SELVER tr. *Čapek's R.U.R.* 28 You see..the Robots have no interest in life. They have no enjoyments. **1923** *Times* 9 June 11/1 Almighty God has populated the world with Robots, legislation of this sort would have been reasonable. **1928** *Daily News & Westm. Gaz.* 20 Apr. 11/4 The latest..Rotary Press, a veritable Robot in the complicated work it performs night after night without hitch. **1937** *Spectator* 23 Apr. 758/1 Men who will go to their doom with the unswerving directness of robots. **1942**, etc. [see *ROBOTIC *sb.* 1]. **1945** *Sun* (Baltimore) 9 Feb. 6-0/2 A robot, which never forgets, will do the job. **1958** [see *ANDROID]. **1969** I. & P. OPIE *Children's Games* xii. 340 'They' pretend to be robots goes mad,' reports a headmaster. **1978** *Sci. Amer.* Feb. 77 (*caption*) Spot-welding robots..are used in assembling the underbodies of Chevrolet Novas. **1979** *Daily Tel.* 7 Nov. 6/8 The British Robot Association believes between 6,000 and 7,000 robots were in use world-wide in industry last year. **1980** *Times* 1 July 19/5 A *real* robot is programmable; it can be programmed to perform different, and changing tasks. In 1978 Japan put 1,100 playback or programmable robots into its factories.

b. A person whose work or activities are entirely mechanical; an automaton.
1923 *Westm. Gaz.* 22 June 7/5 Mr. G. Bernard Shaw defined Robots as persons all of whose activities were imposed on them. **1926** C. E. M. JOAD *Babbitt Warren* 82 Robots live by standardization. **1929** C. CONNOLLY *Let.* in *Romantic Friendship* (1975) 325 America is..a great

youthful boisterous robot. **1943** J. B. PRIESTLEY *Daylight on Saturday* ix. 55, I thought it would be better having a fairly intelligent..girl instead of one of these little office robots. **1977** G. W. H. LAMPE *God as Spirit* ii. 51 The person who is 'seized' by the Spirit is thought of as a passive object, temporarily reduced to the status of a robot.

c. Chiefly *S. Afr.* An automatic traffic-signal.
1931 *Even. Standard* 5 Aug. 2/1 (*heading*) Traffic 'Robots' in the City. **1939** *Forum* (Johannesburg) 1 Feb. 31/1 The Daily Dispatch, East London, is critical of a proposal to fix robots in the town's streets. **1948** H. V. MORTON *In Search of S. Afr.* 17 Another word used in South Africa, but long discontinued in England, is robot for traffic lights. **1958** *Johannesburg Star* 16 Dec. 6/7 Johannesburg drivers..want to turn right or left while pedestrians, with the robot in their favour, are crossing. **1966** A. FUGARD *Boesman & Lena* ii. 38 When the robot said 'Go' there at Berry's Corner I was nearly bang in my broek. **1974** *Eastern Province Herald* 1 Oct. 9 Vandals removed the lamps from seven traffic robots and the flashing head from a warning pole.

d. A robot bomb. *temporary.*
1944 *Daily Tel.* 11 July 1/5 Many of the robots launched against England on Sunday night fell into the sea. **1944** J. LEES-MILNE *Prophesying Peace* (1977) 86 From here [Ashes] saw his first robot.

2. *attrib.* and *Comb.*, as *robot army, astronaut, -brain, clerk, -land, -maker, masses, (petrol) station, -pilot, satellite, system, type, -worker; robot-controlled, -like* (also adj.), *-run* adjs.; **robot bomb** = *flying bomb* s.v. *FLYING *vbl. sb.* 3; **robot plane,** (*a*) = *queen bee* s.v. *QUEEN *sb.* 13; (*b*) = *robot bomb*; **robot roost,** a place for the storage of robot bombs; **robot teacher,** an electronic teaching aid; **robot train,** a robot-controlled underground train.
1927 *Morning Post* 20 Aug. 9 (*heading*) Robot army 'gassed'. **1961** *Daily Tel.* 14 Sept. 1/4 Technicians at Cape Canaveral, Florida, successfully sent a Project Mercury space capsule carrying a robot astronaut and recovered it from the Atlantic. **1944** *Sun* (Baltimore) 20 June 9/1 Most military authorities here are generally agreed that the robot bomb or plane is of..little military value. **1944** *N.Y. Times* 25 June 4E/1 (*heading*) German robot bomb is a potential menace. **1945** G. MILLAR *Maquis* xiv. 294 A false report that a certain factory there was making parts for the robot bombs that the Germans had begun to send to England. **1951** KOESTLER *Age of Longing* I. viii. 140 The ancient Neanderthaler with a modern robot-brain. **1954** *Britannica Bk. of Year* 637/1 *Radar-Brain*, a device used to guide supersonic missiles from the ground, and *Robot-Brain*, a similar apparatus built into the missile. **1928** *Daily Express* 8 June 3/2 A new automatic selling machine, described as the 'Robot clerk', which will say 'Thank you' and give change, will replace the present automatic machines. **1964** *Ann. Reg.* 1963 304 London transport had also developed a robot-controlled underground train. **1960** KOESTLER *Lotus & Robot* II. vi. 173 The robotland reflected in the mirror makes us shudder. **1927** *N.Y. Times* 7 Mar. 16 An iron Robotlike woman Rotwang had meanwhile prepared. **1928** *Daily Express* 11 Aug. 3/7 The romance of past centuries and under-land Dickens-like notions then meet at Sandwich. **1972** T. MCHUGH *Time Pawns* xix. 93 Of the robot-like buffalo skull that pursued and devoured people. *Ibid.* 133 Marching robotlike after the coyote, the skull eventually devoured him. **1976** B. BOVA *Multiple Man* (1977) xiv. 147 That same robot-like Oriental butler served us steaks. **1946** J. T. SHIPLEY in W. S. Knickerbocker *20th Cent. Eng.* 313 Despite robot-makers..human nature changes, if at all, but slowly. **1946** J. S. HUXLEY *Unesco* ii. 43 The robot masses and class-types of ancient Mesopotamia and Egypt. **1973** *Times* 9 Nov. 33/1 (*heading*) Robot petrol stations. *Ibid.*, The two trends are being combined to produce what BP..calls robot stations, namely self-service and automatic money acceptance. **1940** *Aberdeen Press & Jrnl.* 31 Mar. 7/3 One of these robot-pilots has been fitted to a big twin-engined Supermarine Napier flying boat. **1952** A. T. BRAMBLE *Air-Plane Flight* xv. 247 Automatic control or 'robot-pilot' is really a piece of electronic gadgetry rather than an instrument as generally understood. **1938** Robot plane [see *queen bee*]. **1944** J. LEES-MILNE *Prophesying Peace* (1977) 84 Diane made stately preparations to dive under the table at the first sound of a robot plane. **1944** *Sun* (Baltimore) 9 Jan. 1/3 United States heavy bombers struck twice today at the robot coast around Pas de Calais. **1958** I. ASIMOV *Naked Sun* i. 11 Fear of open spaces that barred them from the robot-run farming and mining areas of their birth. **1958** C. C. ADAMS *Space Flight* 142 A manned satellite will be a formidable project... Unlike the robot satellite, it cannot be built in the laboratory and then installed in or on a rocket for delivery to orbit. **1976** *Sci. Amer.* Feb. 77 (*heading*) Robot systems. **1961** *Daily Tel.* 5 Sept. 21/5 Two robot teachers were on show in the Psychology Section. One, like a portable television set, had nine black knobs and a red button on the front. The pupil presses the black knobs to give his answer and the red button to obtain the correct solution. **1963** *Ibid.* 9 Apr. 17/4 (*heading*) Robot train tested with passengers. **1959** H. BARNES *Oceanogr. & Marine Biol.* 177 It is convenient to mount a Robot-type camera in a watertight case, usually fastened to a pole. **1935** H. G. WELLS *Things to Come* 13 All the balderdash..about 'robot workers' and ultra skyscrapers, etc., etc., should be cleared out of your minds.

Hence **robotee·r,** an expert in the making of robots; **robote·sque** *a.*, resembling or suggestive of a robot; **robo·tian** *a.*, of or belonging to a robot or robots; **ro·botism,** mechanical behaviour or character; **robo·tnik** [*-NIK], a person behaving with mindless obedience to authority; **ro·botry,** the condition or behaviour of robots; **ro·boty** *a.*, robot-like.
1924 *Observer* 6 Jan. 12/2 When we reach the gloomy depths of 'commercial English'..we are dealing with a mere thing of use, the very pith and genius of Robotry. **1927** *Daily Express* 30 Aug. 3/4 The Girl in the Lift must on some occasions drop her magnificent Robotry. *Ibid.* 5 Sept. 9/1 There are times when they seem to be purely robotesque, automata driven by impulses of destruction beyond their control. [illegible] **1946** *Amer. Jrnl. Psychol.* LIX. 190, I wish to define the role of robotism in psychology, to show what sense there is in talking about robots. [illegible]

robotic (rǭbǫ·tik), *a.* and *sb.* [f. *ROBOT + -IC.] **A.** *adj.* Of or pertaining to robots; characteristic of or resembling a robot.
1941 I. ASIMOV in *Astounding Sci. Fiction* May 99 [illegible] **1973** M. AMIS *Rachel Papers* 49, I said in a robotic voice. **1974** L. DEIGHTON *Twinkle, Twinkle Little Spy* viii. 84 The kind of phlegmatic robotic bastard that communism breeds.

B. *sb.* **1.** *pl.* The art or science of the design, construction, operation, and application of robots and the like; the study of robots; *laws of robotics*, a set of rules devised to govern the actions of robots, enunciated in the science fiction stories of Isaac Asimov (see quot. 1968¹).
At first a science-fiction term but now more generally used of automatic processes in industry.
1941 I. ASIMOV in *Astounding Sci. Fiction* May 53 There's irony in three of the greatest experts in robotics in the world falling into the same elementary trap, isn't there? **1942** —— in *Ibid.* Mar. 100/1 We have: One, a robot may not injure a human being... The three fundamental Rules of Robotics—the three rules that are built most deeply into a robot's positronic brain. **1957** —— *Naked Sun* (1958) ii. 21 The robot showed no overt response. It couldn't, of course. Its responses were limited and controlled by the Three Laws of Robotics. **1968** —— in *Sci. Jrnl.* Oct. 116/2 Eventually, I formulated these safeguards in the shape of 'The Three Laws of Robotics'. 1. A robot may not injure a human being or, through inaction, allow a human being to come to harm. 2. A robot must obey the orders given it by human beings, except where such orders would conflict with the First Law. 3. A robot must protect its own existence, except where such protection would conflict with the First or Second Law. **1968** *Times* 1 Nov. 23/3 Significant technological advances in the field of 'robotics'—the science of using robots in industrial automation—were announced today. **1970** A. BUTLER *Coffin for Canary* viii. 100 Robotics... To tend towards the human. **1978** *Observer* (Colour Suppl.) 22 Oct. 15/2 In dealing with the many aspects of robotics—historical, philosophical, mythical, actual and projected—Jasia Reichardt defends the machine against the reproachful excessive fears about this artificial image of ourselves. **1979** *Topic* (Imperial Coll., London) 22 Jan. 9/1 Support is planned for..new computer applications (e.g., robotics).

2. *sing.* A robot. *rare.*
1951 C. SIMAK *Time & Again* (1956) v. 20 The robotic clicked and chuckled. It moved a pawn... A human simply can't beat a robotic expert. **1981** *Times* 10 Mar. 4/2 Will he consider direct grants for the purchase of such robotics?

So **robo·tical** *a.*; **robo·tically** *adv.*; **robo·ticist,** an expert in making and operating robots; **robo·ticized** *a.*, robotized.
1940 I. ASIMOV in *Super Sci. Stories* Sept. 70/2 Johnson is an expert Roboticist. **1942** —— in *Amazing Stories* Feb. 227/1 Austin Wilde, Robotical Engineer, turned to Sam Tobe and said, 'Did you get anything out of the robot?' **1947** —— in 'E. Crispin' *Best SF Two* (1956) 127 The government cruiser was making ready to carry the two roboticists back to Earth. **1957** —— *Naked Sun* iii. 35 A thoroughly roboticized economy. **1960** M. SCRIVEN in S. Hook *Dimensions of Mind* xiii. 120 The roboticist in his task of duplicating the brain functions of higher vertebrates. **1972** *Internat. Jrnl. Man-Machine Stud.* IV. 444 The most obvious computer solution would be to simple search serially through each region to find which was the largest. For a roboticist using a serial computer, that may well be the best approach. **1976** K. BONFIGLIOLI *Something Nasty in Woodshed* v. 53 Sam got up in a robotical sort of way. **1979** C. THOMAS *Snow Falcon* 24 Asked to rehearse once more lines he knew by heart... Robotically, he began.

robotize (rǭ·bǫtaiz) *v.* [f. *ROBOT + -IZE.] *trans.* **a.** = *AUTOMATE *v.* 1. **b.** *fig.* To render mechanical or lifeless, to cause to act as if lacking will or consciousness. So **ro·botized** *ppl. a.*; also **robotiza·tion.**
1927 C. M. GRIEVE *Albyn* 46 Dostoevsky's mistake was to imagine that Russia alone could prevent the robotization of Europe. **1927** *Daily Express* 7 Nov. 10/2 Lacking a skilled class of artisans, it is only by Robotising industry that she can hope to fight her way back to prosperity. **1928** *Ibid.* 20 Apr. 13/3 Sir William Joynson-Hicks..protested that he had not seen any sign during the last few months that the House [of Commons] had become robotised. **1928** *Observer* 15 Jan. 11/2 These robotised people..are only employed and allowed to exist because no one has yet been sufficiently energetic to invent a machine to replace them. **1930** *Ibid.* 16 Feb. 17/6 He adds..that in machine mass production lies the foundation of the evil, saying: 'We must not robotise America.' **1952** R. WOLFF *Lines* xx. 236 Even when I was a kid the big plants had been pretty completely robotized. **1967** L. VON BERTALANFFY *Robots, Men & Minds* vi. 64 The robotization of the human individual. **1969** *New Yorker* 4 Jan. 13/4 The masses..victory chants, book burning, robotized phalanxes of soldiers, devour their enemies. **1975** *New Yorker* 21 Apr. 24/2 Katharine Ross plays the young New Yorker who moves to Stepford and discovers that the wives have been robotized by their husbands. **1976** *Sci. Amer.* Feb. 77/1 During the 1930's and 1940's petroleum refineries and petrochemical plants were essentially 'robotized' by inserting rather simple analogue control instruments in the feedback loops that regulated the pressure, temperature and flow rates in distillation columns, catalytic crackers and other equipment designed to process continuously flowing materials.

robotology (rǭbǫtǫ·lǫdʒi). [f. *ROBOT + -OLOGY.] The study of robots; robotics. So **roboto·logist.**
1946 *Amer. Jrnl. Psychol.* LIX. 190 The second robot is beginning the progress in his word which, if generalized, would make him into a scientist,..or at least a robotologist. *Ibid.* 192 When the physiological picture is complete it will be found that physiology is not necessarily identical with robotology. **1970** A. TOFFLER *Future Shock* (1971) ix. 210 In a quite different field of robotology there is progress, too. Technicians at Disneyland have created extremely life-like computer-controlled humanoids capable of moving their arms and legs, grimacing, smiling, [etc.]. **1972** *Computers & Humanities* VI. 135 Such a theory will..be part of a general performance theory, certain aspects of which are covered in what one might call 'general robotology'..such as questions pertaining to the interaction between robot and man.

robotomorphic (rǭ:bǫtǫmǭ·ɹfik), *a.* [f. *ROBOT + -o- + -MORPHIC, after ANTHROPOMORPHIC *a.*] Designating or pertaining to a view of man as a robot or an automaton.
1969 KOESTLER in Koestler & Smythies *Beyond Reductionism* 2 The common target of these 'holy discontents'..seems to be what von Bertalanffy called the robotomorphic view of man. **1970** *Times* 17 Dec. 15/4 You say I overestimate the dangers of the Robotomorphic or Ratomorphic view of man. **1974** *Nature* 20 Aug. 765/1 The 'robotomorphic' mechanistic view of man implied in behaviouristic psychology.

Rob Roy. Add: **1.** Also *ellipt.*
1938 J. BETJEMAN *Oxf. Univ. Chest* v. 97 At the ferry you board a punt or canoe or rob roy and paddle down the stream of the Cherwell. **1976** *Country Life* 8 Apr. 870 The Rob Roys were essentially all-purpose boats.

2. A cocktail made of Scotch whisky and vermouth.
1960 'P. QUENTIN' *Green-Eyed Monster* i. 5 The two of them were..drinking Rob Roys. **1968** H. KANE *Killer's Kiss* xi. 75 'A drink, Mr. Chambers?' 'Double Rob Roy, not too sweet.' **1975** M. H. CLARK *Where are Children?* vi. 44 Jonathan's favorite drink—a Rob Roy with a twist.

robust, *a.* Add: **1. d.** (Further examples.) Also *Anthrop.* Opp. *gracile*.
1948 S. B. KRAUS *Basis Human Evol.* vii. 224 Those [individuals] of Swartkrans and Kromdraai were considerably taller and more robust, perhaps attaining the stature and weight of modern Man. **1977** *Times Educ. Suppl.* 21 Oct. 21/2 The first gracile australopithecines to be found was the skull that Dart found at Sterkfontein half a century ago, but more recently robust forms have also been identified in the South African sites.

4. Applied to a statistical test that yields approximately correct results despite the falsity of certain of the assumptions on which it is based; also, to a calculation, process, or result if the result is largely independent of certain aspects of the input.
1955 BOX & ANDERSEN in *Jrnl. R. Statistical Soc.* B. XVII. 1 To fulfil the needs of the experimenter, statistical criteria should (1) be sensitive to change in the specific factors tested, (2) be insensitive to changes, of a magnitude likely to occur in practice, in extraneous factors. A test which satisfies the first requirement is said to be powerful and we shall typify a test which satisfies the second by calling it 'robust'. **1966** S. BEER *Decision & Control* x. 232 What is important is the recognition of common features in the set of outcomes; these are the inductive inferences which may be classed as forecasts. We say that the system is robust in respect to a particular

set of outcomes. **1972** *Jrnl. Social Psychol.* LXXXVIII. 204 The tests are robust regarding the assumptions of normality and equality of variances, but only when sample sizes are equal. **1978** *Nature* 16 Nov. 264/1 The ANOVA assumes equality of variances, a condition not satisfied here, however the test is robust to small deviations in homoscedasticity. **1979** *Sci. Amer.* Apr. 69/2 This conclusion, they point out, 'is robust, in that we have derived it from the global geochemical distribution of uranium, and we have also derived it from the U.S. uranium-mining history and from a wide variety of subsets of the U.S. uranium-mining history'.

robusta (robɒ·stă). [fem. of L. *rōbusta* ROBUST, specific (now varietal) epithet (L. Linden *Catal. Plantes économiques de l'Horticole coloniale* (1900) 64).] An evergreen variety of coffee, *Coffea canephora* var. *robusta*, native to Africa and widely cultivated elsewhere for its heavy crops of small beans; also, the beans produced by a tree of this kind. Also *attrib.*
1909 *Philippine Agric. Rev.* II. 590 A new variety of coffee known as 'robusta'..was discovered some years ago growing wild on the estates in Africa. *Ibid.*, The robusta coffee planted in east Java yields after three years. **1922** [see *ARABICA]. **1944** *Empire Jrnl. Exper. Agric.* XII. 191 Robusta coffee..grows wild in many of the wetter forests of Uganda. **1959** [see *ARABICA]. **1961** F. L. WELLMAN *Coffee* v. 81 The most common variety is named Robusta, and this type has large, dark-green leaves... Trees of the Robusta variety tend to have a flattened top. **1976** *Times* 7 May 22/5 Fears of a shortage ..sent robusta coffee prices to new all-time highs on the London markets. *Ibid.*, Shortages will increase the demand for African robustas.

robusticity. For *rare* read 'Chiefly *Anthrop.* and *Zool.*' (Further examples.)
1946 F. E. ZEUNER *Dating Past* ix. 299 They show nothing of the robusticity and exuberance of bodily growth of Crô-Magnon Man, whose contemporaries they were. **1959** *Chambers's Encycl.* I. 460/2 Other features than size, e.g. shape and robusticity, are expressed in anthropometry by indices. **1971** *Nature* 5 Feb. 407/1 In overall size and robusticity the fossil closely resembles the pygmy chimpanzee.

robustness. Add further examples in senses 1d and *4 of the adj.
1953 G. E. P. BOX in *Biometrika* XL. 318 This remarkable property of 'robustness' to non-normality which these tests for comparing means possess, and without which they would be much less appropriate to the needs of the experimenter, is not necessarily shared by other statistical tests. **1973** J. BUETTNER-JANUSCH *Physical Anthropol.* viii. 240 [Modern man] may walk bipedally. Among the specializations that permit this are the shape of the arch and the position and robustness of the big toe. **1974** ADBY & DEMPSTER *Introd. Optimization Methods* iv. 78 A concept both more vague and much more difficult to ensure is termed robustness. A robust algorithm is one which in practice usually yields the global minimum or a good local minimum of any function of even a large number of variables from a poor initial approximation.

roc. β. (Further examples.)
1924 *Nature* 19 Apr. 564/2 Purely fabulous species, such as the phœnix and the rukh, are likewise dealt with. **1974** *Encycl. Brit. Micropædia* VIII. 619/2 The Kublai Khan inquired..about the *rukh* and was brought what was claimed to be a *rukh*'s feather, which may really have been a frond of the *Raphia* palm.

‖ **rocaille** (rokāʸ). Also **rocail** and with capital initial. [a. F. *rocaille* rock-work, rococo.] An artistic or architectural style of decoration characterized by ornate rock- and shell-work; a rococo style. Also *attrib.* or as *adj.*
1856 M. DIGBY WYATT in O. Jones *Grammar of Ornament* I. xix. 109 The twisted and foliated scrolls and shells grew into the 'rocaille' and grotto-work of [baroque]..; degenerating at last into..'Chinoiserie'. **1905** *Scribner's Mag.* July 47 Rocaille differs from our rockwork in that it does not attempt to imitate the natural formation of rocks, but rather seeks to create architectural forms by combinations of pebbles and shells, such as conventionalized figures of sea-gods, and dolphins. **1936** *Burlington Mag.* Oct. 187/1 Louis XVI, who did not care for the 'rocaille' style like his grandfather. **1944** J. LEES-MILNE *Prophesying Peace* (1977) 69 A cliff-like structure hung with reliefs, and encrusted with shells, sea urchins and rocaille ornaments. **1958** *Listener* 2 Oct. 530/1 The staggering rocaille on Bena Lulua figures. **1960** *Times* 14 Jan. 14/5 Sauceboats of 1737 having shell shaped bowls or *rocaille* bases. **1975** J. GORES *Hammett* (1976) xi. 82 The ornate rocaille pier glass. **1979** E. H. GOMBRICH *Sense of Order* vii. 189 To what extent can Riegl's method be used for the explanation and analysis of the Rocaille? Are these playful shells..just another metamorphosis of the acanthus?

‖ **rocambolesque** (rǫkæ:mbǫle·sk), *a.* [a. F. *rocambolesque* fantastic, f. *Rocambole* the name of a character in the novels of Ponson du Terrail (1829–71), French author, the subject of improbable and fantastic adventures + -ESQUE.] Of or resembling Rocambole (see etym.); incredible, fantastic.
1949 KOESTLER *Promise & Fulfilment* I. viii. 91 It was the first anti-British terror act of the Irgun and it displayed already all the features of rocambolesque etiquette. **1960** B. MARSHALL *Divided Lady* I. xxi. 75 How amused the General would have been by this rocambolesque religiosity. **1976** *New Society* 13 May 370/3 An exemplary surrealistic life..which included a *rocambolesque* episode.. in which he kidnapped his Bulgarian mistress from her husband.

‖ **roche,** *sb.*² Delete entry: see *ROCHE MOUTONNÉE.

Roche (rǭʃ). *Astron.* The name of Edouard Albert *Roche* (1820–83), French mathematician, used *attrib.* and in the possessive to denote concepts arising out of his work, as **Roche('s) limit,** (*a*) the closest distance to which a self-gravitating body (strictly a fluid body: see quot. 1900) can approach a more massive body without being pulled apart by the gravitational field of the latter body; (*b*) the smallest continuous equipotential surface (having the form of two lobes meeting at a point) which can exist around both members of a system of two gravitating bodies, *spec.* a binary star system; **Roche lobe,** either of the two volumes of space (meeting at a point) that are bounded by Roche's limit (*b*) in a binary system; **Roche zone,** the region of space within Roche's limit (sense (*a*)).
1889 G. H. DARWIN in *Harper's Mag.* June 73/1 The distance of..2·44 of a planet's radius I call Roche's limit for that planet. **1900** *Astrophysical Jrnl.* XI. 122 In the derivation of Roche's limit the assumption was made that the satellite was a perfectly homogeneous incompressible fluid, and that its rotation and revolution were performed in the same period. **1930** R. H. BAKER *Astron.* v. 212 All parts of Saturn's rings lie within Roche's limit. **1959** Z. KOPAL *Close Binary Syst.* iii. 133 Such configurations represent the largest closed equipotentials capable of containing the whole mass of the respective components, and will hereafter be referred to as their Roche limits. **1969** *Times* 10 July 12/8 If the moon had ever come within a critical distance of earth, known as the Roche limit, the tidal forces raised by the earth would have disrupted it. **1972** W. STROHMEIER *Variable Stars* vii. 182 Expansion of the components in close binaries towards their Roche limits, in a time scale of 10^6 years, can also give rise to an exchange between the rotational and orbital momenta. **1974** *Sci. Amer.* Feb. 53/1 Only a body with more than gravitational cohesion can withstand the tidal effects within Roche's limit. [**1960** *Astrophysical Jrnl.* CXXXII. 149 (*caption*) The radii of the Roche-limit lobes for a mass ratio of unity.] **1969** *Ibid.* CLVIII. 571 Morton concluded that on such a time scale the contact component would be stable, shrinking within the Roche lobe after initial mass loss. **1975** *Sci. Amer.* Mar. 30/3 In the evolution of a typical binary, as soon as one of the components expands to a volume larger than that of its Roche lobe the matter outside the lobe will begin to flow toward the companion star. **1971** I. G. GASS et al. *Understanding Earth* vii. 112/2 What would happen if the Moon were to enter the Roche zone? **1978** *New Scientist* 23 Nov. 607/2 One or more former moons of Uranus spiralled into the planet's Roche zone where they broke up because of tidal forces, producing the parent fragments of the rings.

rochea (rǭu·ʃiă). [mod.L. (A. P. de Candolle *Plantarum Historia Succulentarum* (1803?) 103), f. the name of François de la *Roche* (d. 1813), French botanist + -A 2.] A succulent plant of the genus so called, belonging to the family Crassulaceæ, native to South Africa, and bearing leathery leaves and clusters of white, pink, or red flowers.
1932 A. J. MACSELF *Amateur's Greenhouse* xi. 253 Hybrid Kalosanthes or Rocheas in white, rose, etc., are similar in habit. **1955** V. HIGGINS tr. *Bertrand's Indoor Plants* 84 Rochea... For some years now the florists have offered this attractive plant on Mother's Day. **1979** A. HUXLEY *Reader's Digest Success with House Plants* 340/3 Rocheas are small shrubs grown primarily for their clusters of flowers.

‖ **roche moutonnée** (rǫʃ mutǫne). *Physical Geogr.* [Fr., f. *roche* rock, ROCHE *sb.*¹ + *moutonnée* MOUTONNÉE.] A bare rock outcrop which has been shaped by glacial erosion, characteristically smoothed and rounded by abrasion but often also displaying one side (the 'downstream' side) which is rougher and steeper because of plucking. Hence **ro:che mouto·nnéed** *a.*, abounding in *roches moutonnées*.
De Saussure (see quot. 1786), to whom the term is frequently attributed, applied the adj. *moutonnée* to small rounded hillocks (usu. covered with vegetation) which suggested, *en masse*, a fleece or a wig of a style termed *moutonnée*. These features do not correspond to the meaning of *roches moutonnées* which later became accepted, and were not associated by de Saussure with glaciers. (See also s.v. in *Gloss. Geol.* (Amer. Geol. Inst., 1972) 613.)
[**1786** H.-B. DE SAUSSURE *Voyages dans Alpes* II. xlviii. 512–3 Plus loin, derriere le village de *Juviana* ou *Envionne* on voit des rochers qui ont une forme que je nomme *moutonnée*... Les montagnes que je désigne par cette expression sont composées d'un assemblage de têtes arrondies, couvertes quelquefois de bois, mais plus souvent d'herbes, ou tout au plus de broussailles. Ces rondeurs contigues & répétées forment en grand l'effet d'une toison bien fournie, ou de ces perruques que l'on nomme aussi *moutonnées*.] **1843** J. D. FORBES *Trav. through Alps* iii. 53 The surface of rock..is even and rounded, often dome-shaped or spheroidal, showing the structure of the rock in section... Such surfaces were called Roches Moutonnées by De Saussure. **1862** *Q. Jrnl. Geol. Soc.* XVIII. 187 For many miles in the Alb Valley, both above and below St. Blasien, *roches moutonnées* stand like islands through the alluvium. **1865**, etc. [see ROCHE *sb.*²]. **1905** *Bull. Geol. Soc. Amer.* XVI. 51 The northern slopes..have been considerably smoothed by ice action... The whole surface is rochemoutonnéed, especially on the north, where nearly all rocks are absolutely fresh. **1935** *Discovery* Mar. 79/2 Dome-like rocks are exposed which in appearance recall the ice-formed *roches moutonnées*. **1957** J. K. CHARLESWORTH *Quaternary Era* I. xi. 251 De Saussure gave the name roche moutonnée to the distinctive, rounded forms which abound in glaciated terrain (he himself failed to associate them with ice) and give the effect of a thick fleece or the wavy wigs styled *moutonnées* in his day (they were slicked down with mutton tallow). **1977** A. HALLAM *Planet Earth* 86/3 Many valleys are very deeply incised, with U-shaped cross-profiles and floors composed of smoothed, striated and streamlined rock hummocks (called roches moutonnées).

Rōchū, var. *RŌJŪ.

rock, *sb.*¹ Add: **I. 1. b.** For '*U.S.* and *Austr.*' read 'orig. *U.S.*', and add: Also freq., a stone used as a projectile. (Further examples.)
1890 BARRÈRE & LELAND *Dict. Slang* II. 183/2 *Rocks* (American), small stones or pebbles... The term is used in some parts of England. **1939** J. STUART in *Esquire* May 55/1, I pull a round rock from my pocket. I let th' rock go. I holler: 'Rocks! Watch out!' **1968** *New Society* 29 Aug. 304/1 It is now frequent for British newspapers to record that during some riot or disturbance the crowd has thrown 'rocks' (= 'stones'). **1969** *West Australian* 5 July 1/1 Several policemen fell to the ground after they were hit with rocks. **1976** *Billings* (Montana) *Gaz.* 17 June 1-F/5 Ambulance services were suspended when mobs hurled rocks at the vehicles, injuring drivers. **1979** *Observer* 16 Sept. 1/1 The Belfast house of Mr Gerry Fitt, Social Democratic and Labour MP for East Belfast, was besieged by about 200 youths armed with rocks yesterday.

d. *Canad.* = CURLING-STONE.
1911 R. E. KNOWLES *Singer of Kootenay* 296 Every man of them held his breath as the flying rock came to the port. **1963** *Times* 25 Feb. (Canada Suppl.) p. xvi/1 The Scots melted cannon balls to fashion their 'rocks' and played the game on the frozen St. Lawrence river and ponds in the area. It is interesting that rocks made of iron were still being used in parts of Ontario and Quebec as late as 20 years ago. **1974** *Globe & Mail* (Toronto) 20 Feb. 33/7 Dr. Will McTavish of Winnipeg and Ralph Smith of Noranda excelled at getting their draws into the centre of the house and knocking out opposition rocks.

2. d. *on the rocks*: also (esp. of a marriage, etc.), on the point of dissolution; finished.
1958 E. WILSON in *N.Y. Post* 1 June 2/3 [Roberto Rossellini's] headlined romance with Sonali Das Gupta is now reported on the rocks. **1975** *Globe & Mail* (Toronto) 12 Sept. 12/6 Simply adding more ice to a marriage that's already on the rocks won't save the partnership, the Law Reform Commission of Canada says in a paper on divorce.

e. (Earlier and later examples.) Also *spec.* a dollar. In phr. *a pocketful of rocks*, a large amount of money.
1840 *Picayune* (New Orleans) 31 July 2/2 He was just on the eve of leaving town with his 'pockets full of rocks'. **1846** in D. Corcoran *Pickings from New Orleans Picayune* 143 Here I am in town without a rock in my pocket. **1847** J. S. ROBB *Streaks of Squatter Life* 165 If I had a 'pocket full of *rocks*', you should share them. **1858** J. R. LOWELL *Poet. Wks.* II. 284 A pocket-full of rocks 'twould take To build a house of free-stone. **1905** *Dialect Notes* III. 17 *Rocks*, dollars. **1942** Z. N. HURSTON in A. Dundes *Mother Wit* (1973) 224/1, I don't bet, but I'll doubt you. Five rocks! **1949** *Cavalier Daily* (Univ. of Va.) 22 Oct. 4/1 They got a campaign goin' around here to try to stick us students six rocks just to go..and listen to some old bag yell her fool head off.

f. *slang* (orig. *U.S.*). A precious stone, *spec.* a diamond. Cf. *rock-diamond* (sense 6 d) in Dict.
1908 H. GREEN *Maison de Shine* 83 'So that's his new wife, eh?' said Goldie later. 'Did you pipe the rocks she had on?' **1926** J. BLACK *You can't Win* xiii. 178 I'll unharness these 'rocks'. **1929** [see *LOOGAN]. **1936** J. CURTIS *Gilt Kid* v. 57 Some of the women present, he saw, were wearing goodish rocks. **1953** 'S. RANSOME' *Drag Dark* (1954) vi. 60 Goodlee wrote his check..then walked out with the rock. **1968** A. DIMENT *Bang Bang Birds* vii. 106 He..listened to my vague replies like my advice was worth its weight in sparkling rocks. **1973** 'I. DRUMMOND' *Jaws of Watchdog* i. 12 'We will see some of the most beautiful jewellery in the world... The emeralds.'.. 'Personally,' said Jenny, 'I call it vulgar, having all those rocks on a yacht.'

g. In *U.S.* colloq. phr. *between a rock and a hard place*: without a satisfactory alternative, in difficulty (see also quot. 1921).
1921 *Dialect Notes* V. 113 *To be between a rock and a hard place*,..to be bankrupt. Common in Arizona in recent panics; sporadic in California. **1959** L. ROBERTS *Up Cutshin & Down Greasy* v. 82 That was one time dad was between a rock and a hard place. **1963** D. OGILVY *Confessions Advertising Man* xi. 160 As a private person, I would gladly pay for the privilege of watching it without commercial interruptions. Morally, I find myself between the rock and the hard place. **1976** T. WOLFE *Mauve Gloves & Madmen* 37 The dive brings you down so low,

you are now down into the skeet range of that insidiously well-aimed flak! This, as they say, puts you between a rock and a hard place.

h. Usu. *pl.* An ice-cube or crushed ice for use in a drink. In phr. *on the rocks*, (of a spirituous liquor) served with ice. *slang* (orig. *U.S.*).
1946 *Amer. Speech* XXI. 35 *Rocks*, ice. **1948** F. BROWN *Murder can be Fun* (1951) iii. 44 A slug or two of rock and rye won't hurt you. **1949** *Life* 14 Nov. 63 Ordering a Scotch on the rocks at the bar. **1952** N. SPAIN in C. Asquith *Second Ghost Bk.* 36, I..went in and fixed myself a Scotch on rocks, neat. **1955** J. B. PRIESTLEY *Journey down Rainbow* 220 They all drank a lot of whisky-on-the-rocks. **1959** 'J. CHRISTOPHER' *Scent of White Poppies* vi. 82 Rocks in your Scotch, Cam? I can get some from the fridge. **1966** *Listener* 20 Oct. 573/2 For some reason, no one knows quite why, Americans insist on having ice, or 'rocks' as they call it, always in easy reach. **1978** R. LUDLUM *Holcroft Covenant* iii. 39 That was scotch on the rocks, wasn't it?

i. pl. *slang.* The testicles; = STONE *sb.* 11a. In coarse phr. *to get one's rocks off*, to achieve sexual satisfaction, to ejaculate; also, in weakened sense, to obtain enjoyment.
1948 *Amer. Speech* XXIII. 249/1 *Get your rocks off*, an expression used to denote extreme enjoyment. **1961** *Ibid.* XXXVI. 150 Expressions using *rocks* and *stones* to mean testes are at least as old as the Renaissance, but in the mouths of today's teen-agers, *hot rocks* seems to imply only a warm romantic interest by a teen-ager of either sex in one of the opposite gender. **1971** *Frendz* 5 Aug. 22/2 Get yer rocks off Seymour. OK. But there are limits. Surely. **1972** *Show* Sept. 55/2 Unrelenting sequences of unsympathetic characters getting their rocks off. **1975** J. BRAINE *Pious Agent* vi. 23 I'd get a swift kick in the rocks. **1975** G. V. HIGGINS *City on Hill* vii. 195 I've been reduced to dressing up in order to get my rocks off. **1978** *Chicago* June 90/3 This is a good film for getting your rocks off. but not the sort you will remember much about two weeks later. **1978** J. IRVING *World according to Garp* xi. 205, I don't get my rocks off by humiliating myself, you know.

j. *U.S. Baseball slang.* An error. In phr. *to pull a rock*, to make a mistake.
1939 E. J. NICHOLS *Hist. Dict. Baseball Terminol.* (thesis, Pa. State College) 57 *Pull a rock*, see 'boner'. *Ibid.* 9 *Boner*, an error in judgment. **1951** *Birmingham* (Alabama) *News* 31 July 16/3 How does a guy who has been labeled 'the perfect player' feel after pulling his first 'rock' in a long and brilliant baseball career? **1952** *Philadelphia Even. Bull.* 4 Oct. 13/2 Who deserved the rap for the 'rock' that cost the Yankees yesterday's World Series game? **1955** *Daily Progress* (Charlottesville, Va.) 5 May 10/7 'Luckily, it didn't hurt us but I pulled a rock.' Durocher then went on to explain his 'rock', which didn't prevent the Giants from winning. .. 'Good strategy, my foot,' mocked Durocher after the game. 'It was a real rock.' **1956** *Sun* (Baltimore) 26 Apr. B 25/3 Bill changed his mind and lifted Rhodes out of the lineup... The criticism was that he had 'pulled a rock'.

3. g. Mineral ore. *U.S.*
1830 *Workingman's Gaz.* (Woodstock, Vermont) 28 Oct. 38/1 The surface is almost covered with rock, all which contains gold..which is obtained by breaking or pounding the rock. **1896** C. H. SHINN *Story of Mine* 78 The quartz prospector..only pans out a few ounces of powdered rock. **1902** O. WISTER *Virginian* xv. 172 Are they taking much mineral out? Have yu' seen any of the rock? **1948** *Los Angeles Times* 12 Jan. II. 8/3 (*heading*) Ruby mine runs into rich rock.

5. a. Delete *U.S.* and add later example.
1977 *Grimsby Even. Tel.* 5 May 18/2 Principal sorts were: Cod 1,712 kits, haddock 1,059,..rocks 23, skate 58,..monk 16.

II. 6. a. *rock-arch, -barrier, -bluff, -chamber, -chimney, -cliff, -crust, -drift, -flat, -floor, -fortress, -hill, ledge, -point, -pool* (further examples), *-rampart, -shelter* (further examples), *-shrine, -stack, -terrace, -wall*.
1936 H. NICOLSON *Let.* 28 Sept. (1966) 274 The precipices,..the rock-arches..roared back at us. **1940** C. DAY LEWIS tr. *Virgil's Georgics* IV. 90 Proteus shelters within behind a huge rock-barrier. **1886** A. WINCHELL *Walks & Talks in Geol. Field* 53 We have seen..the rock-bluffs bounding..the basins of the great lakes. **1954** J. R. R. TOLKIEN *Fellowship of Ring* 217 There was a cave or rock-chamber behind. *Ibid.* 401 There were many birds about the cliffs and the rock-chimneys. **1972** *Shooting Times & Country Mag.* 1 July 17/3 A rock 'chimney' up which I was none too happy in ascending. **1952** S. SPENDER *Learning Laughter* 9 From the ship we saw houses on a green shelf above a rock-cliff. *a* **1963** C. S. LEWIS *Poems* (1964) 45 Down far under his rock-crust. **1951** W. DE LA MARE *Winged Chariot* 16 Sweet salt-tanged air, birds, rock-drift. **1959** BLUNDEN *Hong Kong House* 2 It was no garden—so adust, red-dry the rock-drift soil was. **1967** *Oceanogr. & Marine Biol.* V. 483 In the most subtropical areas,..the fauna living between the algae covering these rock-flats may be greatly impoverished. **1905** *Jrnl. Geol.* XIII. 393 The desert plain may be reduced to a lower level than that of the deepest initial basin; and then a rock-floor,..unrelated to normal baselevel, will prevail throughout. **1946** F. E. ZEUNER *Dating Past* vii. 223 Resting on an irregular rock-floor at about 7·5 metres above low sea-level, a beach conglomerate is found. **1934** W. S. CHURCHILL *Marlborough* II. xv. 331 Coblenz..stands opposite the majestic rock-fortress of Ehrenbreitstein. **1946** R. CAMPBELL *Talking Bronco* 41 Rock-fortress of your sex and gender! *a* **1862** THOREAU *Maine Woods* (1864) 262 Being struck with the perfect parallelism of these singular rock-hills,..I took out my compass. **1963** D. W. & E. E. HUMPHRIES tr. *Termier's Erosion & Sedimentation* xiii. 260 This type of coastal cornice, or rock ledge, is thus a phenomenon of marine abrasion. **1849** D. G. ROSSETTI *Let.* 18 Oct. (1965) I. 74 One rock-point standing buffeted alone, vexed at its base with a foul beast unknown. **1948** L. MACNEICE *Holes in Sky* 31 Foam-quoits on rock-points. **1907** E. GOSSE *Father & Son* vi. 156 The antiquity of these rock-pools..used to occupy my Father's fancy. **1955** V. PALMER in B. James *Austral. Short Stories* (1963) 32 He stared into rock-pools. **1924** R. CAMPBELL *Flaming Terrapin* vi. 94 The brink of the abyss, Where into space the sharp rock-rampart drops. **1927** PEAKE & FLEURE *Hunters & Artists* 40 A small rock-shelter, now quarried away. **1971** *World Archaeol.* III. 147 Puntutjarpa is a minor sacred site about 150 ft west of the rockshelter. **1933** *Burlington Mag.* June 290/2 A small seated Buddha-image,..cut out of a rock-shrine at Yiin-Kang. **1969** *Tanzania Notes & Rec.* July 3 The great rock-shrine of Tita, in southern Turu near Puma, is in the eyes of the Turu themselves less powerful than similar shrines in the mountains of Sandawe country and in Isanzu. **1969** G. M. BROWN *Orkney Tapestry* i. 17 There among them, standing out to sea a little, is the rock-stack called The Old Man of Hoy. **1892** *Bull. Geol. Soc. Amer.* III. 65 Raised marine deposits with an arctic fauna occur over the latest moraines in Scandinavia... Bravais, half a century ago came to the conclusion that two elevated rock-terraces in northern Norway examined by him are not horizontal but descend toward the north. **1968** R. W. FAIRBRIDGE *Encycl. Geomorphol.* 1184/2 [River] terraces may be cut into the solid rock or consist of a rock bench veneered with a comparatively small thickness of alluvium (rock terrace). **1904** W. M. RAMSAY *Lett. Seven Churches* xxv. 360 At times an oblique crack develops in the rock-wall. **1954** J. R. R. TOLKIEN *Two Towers* 73 At the far end the rock-wall was sheer.

b. *rock art, -carving* (examples), *-drawing* (examples), *-engraving, -painting, -picture*.
1959 J. D. CLARK *Prehist. S. Afr.* ix. 248 The rock art tells us little for certain about marriage customs. **1974** B. & R. HILL *Spirit in Stone* 11 With very few exceptions most rock art sites are located near villages. **1907** H. M. CHADWICK *Origin Eng. Nation* xii. 306 The rock-carvings at Tegneby. **1950** H. L. LORIMER *Homes & Monuments* vi. 354 The well-known rock-carving of Ivriz on which a King appears before the god of vegetation. **1938** H. A. WINKLER *Rock-Drawings of Southern Upper Egypt* I. 26 The discovery of rock-drawings showing boats of a type foreign to Egypt. **1977** H. INNES *Big Footprints* I. 33 The location of the rock drawings. **1920** H. G. WELLS *Outl. Hist.* I. xvii. 126/1 From rock engravings we may deduce the theory that the desert was crossed from oasis to oasis. **1959** J. D. CLARK *Prehist. S. Afr.* ii. 29 The first European in Rhodesia to see rock engravings was probably Thomas Leask who saw those near Wankie in 1869 when on a hunting trip to the Zambezi. **1908** *Encycl. Relig. & Ethics* I. 822/2 The rock-paintings..are either stencilled..or painted in outline. **1977** *Times* 23 Apr. 14/1, I read of a Bushman woman wearing a circle of beads..'exactly like that of her ancient prototype' in an early rock painting. **1939** *Man* No. 119. 178/2 On one of the stalactite pillars..was found a big round stone with.. traces of red paint on its surface, as used in the rock-pictures. **1952** V. G. CHILDE *New Light Most Anc. East* (ed. 4) ii. 17 The rock-pictures..demonstrate the survival of the 'Rhodesian fauna'.

c. *rock-movement, porosity, -type*.
1907 *Bull. Geol. Survey N.Z.* No. 3. 95 Differential rock-movement is recorded by the well-slickensided faces and the plastic finely comminuted rock-material occurring in the plane of contact. **1946** *Nature* 6 July 31/1 G. A. Maximovich.., after making a compilation of several thousands of determination[s] of rock-porosity,..has calculated the average porosities of different types of rocks. *Ibid.* 3 Aug. 172/1 The commonest rock-type [on Heard Island] is trachybasalt.

d. *rock-cut, -cutting, -land, -top, -vine, -wilderness*.
1873 J. H. BEADLE *Undevel. West* 139 A long rock-cut. **1965** G. MCINNES *Road to Gundagai* ix. 134 Each rattling rock-cut, each looping embankment and low trestle bridge carried us further into an unbelievable land. **1873** 'MARK TWAIN' & WARNER *Gilded Age* 420 There is Newark,..then marshes, then long rock cuttings, devoted to the advertisements of patent medicines. **1891** M. E. RYAN *Pagan of Alleghanies* 96 The rest of that rock-land is going to break away sometime. **1946** W. DE LA MARE *Traveller* 17 He..had wakened to the rock-land. **1960** *Wall St. Jrnl.* 30 Nov. 7/3 The technique of 'rockland' farming was developed a few years ago in south Florida... Growers use the rockland in preference to more fertile soil partly because..rocklands are on higher ground and are less subject to flooding. **1927** D. H. LAWRENCE *Mornings in Mexico* 135 High on a narrow rock-top. **1927** JOYCE *Pomes Penyeach*, Goldbrown upon the sated flood The rockvine clusters lift and sway. **1927** D. H. LAWRENCE *Mornings in Mexico* 162 The great, hollow, rock-wilderness space of that part of Arizona.

7. a. *rock-crushing, -infesting, -loving* (earlier example), *-rolling*.
1966 A. BATTERSBY *Math. in Managem.* i. 16 This rock-crushing argument may bell be used to suppress a bright boy. **1940** A. H. GARDINER *Theory of Proper Names* i. 7 The rock-infesting monsters. **1847** EMERSON *Poems* 230 A wild-rose, or rock-loving columbine, Salve my worst wounds. **1957** R. CAMPBELL *Coll. Poems* II. 106 But now the longed-for sound, As of rock-rolling torrents underground, Approaches.

b. *rock-climbing* (further examples), *-folding, -painting*.
1923 G. D. ABRAHAM *First Steps to Climbing* iv. 45 It is a well-considered opinion that rock climbing is the most important branch of mountaineering. **1977** *Times* 19 Jan. 14/1 Wasdale Head proclaims itself as the birthplace of rock climbing. **1965** G. J. WILLIAMS *Econ. Geol. N.Z.* iii. 20/1 It is very clear that there is a general parallelism between rock-folding and the trend of the lodes. **1965** R. & D. MORRIS *Men & Snakes* i. 17 Australia is the only continent where rock painting is still practised regularly today.

c. *rock beater, climber* (further examples), *-hopper, -hunter, -painter*.
1935 *Discovery* July 203/2 The rock is..crushed in rock beaters. **1940** F. SMYTHE *Adventures of Mountaineer* vii. 75, I knew my companion to be a magnificent rock climber, as agile and as active as a cat. **1977** *R.A.F. News* 27 Apr.–10 May 20/6 St Athan's mountain rescue team.. were asked to help rescue a rock climber who had fallen in the Brecon Beacons. **1959** S. J. BAKER *Drum* 140 *Rock-hopper*, a person who fishes from rocks on a sea-coast. **1969** *Man* (Austral.) Mar. 12/2 Many rock-hoppers are experienced rock climbers, of a breed who, for sport, crawl like flies over the granite. **1971** R. PURVIS *Treasure Hunting in Brit. Columbia* ii. 47 The first rockhunter to emerge in the early stone age wasn't interested in the beauty or gem quality of stones. **1919** H. G. WELLS *Outl. Hist.* xii. 77/2 The simplicity, directness, and detachment of a later Palæolithic rock-painter appeal more to modern sympathies than does the state of mind of these Neolithic men. **1961** L. VAN DER POST *Heart of Hunter* 9, I gave a brief account of the tragic extermination of this little hunter and rock-painter by the Black and the White invaders of his ancient country.

8. a. *rock-born, -bound* (further examples), *-bred* (further examples), *-bristled, -cut* (further examples), *-girt* (earlier example), *-guarded, -living, -nurtured, -perched, -rooted* (further examples), *-staked*.
1849 J. R. LOWELL in *National Anti-Slavery Standard* 23 Aug. 50/6 Taghkanic's rockborn child Dares gloriously the dangerous leap. **1913** W. B. YEATS in *Brit. Rev.* Apr. 89, I have kept my faith though faith was tried To that rock-born, rock-wandering foot. **1937** DE LA MARE & JONES *This year: Next Year* 12/2 The lovely sirens sing.. in their rock-bound solitude. **1978** *Amer. Poetry Rev.* Nov./Dec. 20/3 Along the stormy, rockbound Ligurian coast. **1920** W. B. YEATS in *Dial* Nov. 462 She seemed to have grown clean and sweet Like any rock-bred, sea borne bird. **1941** L. B. LYON *Tomorrow is Revealing* 22 Encounter The rock-bred wolf or risk the valley road. **1847** J. R. LOWELL *Summer Storm* in *Poems* 2nd Ser. 66 Like the toothless sea mumbling A rock-bristled shore. **1933** *Burlington Mag.* Nov. 237/1 The paintings in the rock-cut Chapels of Cappadocia. **1979** *London Rev. Bks.* 25 Oct. 14/2 (Advt.), Two hundred years ago.. Buddhist rock-cut shrines, the mosques of Moslems,..were all but unknown to Europeans. **1845** E. A. POE in *Graham's Mag.* Dec. 251/1 No billow breaking into foam Upon the rock-girt shore of Time. **1929** C. E. ROBINSON *Hist. Greece* vi. 63 One great advantage indeed the Attic folk possessed in the admirable rock-guarded harbours adjacent to their capital. **1923** D. H. LAWRENCE *Birds, Beasts & Flowers* 41 A rock-living, sweet-fleshed sea-anemone. **1913** W. B. YEATS in *Brit. Rev.* Apr. 87 Rock-nurtured Aoife took a pin. **1913** W. DE LA MARE *Peacock Pie* 64, I long to watch the sea-mew wheel Back to her rock-perched mate. **1890** *Congress. Rec.* 7 June 5802/1, Every rock-rooted advocate of the gold standard is in favor of [this provision]. **1930** W. B. YEATS *Wild Apples* 1 Unsheltered by..steading, Rock-rooted and grown, A great tree of Erin, It stands up alone. **1894** KIPLING *Seven Seas* (1896) 131 Thou hast not toiled at the fishing..Nor worked the war-boats outward through the rush of the rock-staked seas.

b. *rock-arched, -browed, -chested, -floored, -walled, -wombed* (later example).
1833 J. G. WHITTIER *Poet. Wks.* (1898) 559/2 Through rock-arched Winooski the salmon leaps free. **1944** BLUNDEN *Shells by Stream* 19 Above the rock-browed shag-haired weir. **1939** DYLAN THOMAS *Map of Love* 4 If the dead starve, their stomachs turn to tumble An upright man in the antipodes Or spray-based and rock-chested sea. **1905** W. M. DAVIS in *Jrnl. Geol.* XIII. 388 The initial relief will be extinguished even under the slow processes of desert erosion, and there will appear instead large, rock-floored plains sloping toward large waste-floored plains. **1879** J. G. WHITTIER *Poet. Wks.* (1898) 257/1 Church that..Saw within the rock-walled bay Treville's lilied pennons play. **1954** W. FAULKNER *Fable* 260 The rock-wombed powder magazines under the Gates of Hercules.

c. *rock-footed, -hard, -solid, -steady, -still, -white*; also *rock-blackness*.
1968 R. P. WARREN *Incarnations* (1970) 11 The moon, eastward and over The ridge and rock-blackness, rears. **1911** BEERBOHM *Zuleika D.* xxii. 317 Sole and splendid survivor he stood, rock-footed, before her. **1935** 'L. LUARD' *Conquering Seas* v. 69 Plenty of rock-hard, shelf cod. **1978** S. SHELDON *Bloodline* xxii. 250 The man was fully aroused now, rock-hard. **1972** *Ulster* (Sunday Times Insight Team) xvi. 273 They had seen their support in the area—once rock-solid—steadily and severely eroded. **1976** A. PRICE *War Game* I. ix. 175 He'll never sit for this seat... It's rock-solid Conservative. **1928** *Outlook* 26 May 650/1 Consols were rock-steady at 112. **1976** J. WAINWRIGHT *Who goes Next?* 151 The killer held the rifle rock-steady. **1976** J. B. HILTON *Gamekeeper's Gallows* ii. 20 He cocked his eye up to the pressure-dial. The needle was rock-still, not even trembling. **1916** BLUNDEN *Pastorals* 21 Through the bindweed's rock-white mesh.

9. a. rock bar *Physical Geogr.* = *RIEGEL; **rock biscuit** (earlier example); **rock bit** *Oil Industry*, a drilling bit for use in hard formations; **rock bolt** *Mining*, a tensioned rod passing through a bed of rock and anchoring it to the body of rock behind; so **rock bolting,** the practice or technique of using rock bolts; **rock-bottom,** substitute for def.: bed-rock; also *fig.*, the fundamental or lowest possible level, nadir (see also quot. 1866); also *attrib.* or as *adj.*, lowest possible, unbeatable (of prices, etc.); fundamental, firmly grounded, honest, sound; (earlier and later examples); **rock-bun** (earlier example); **rock cake** (earlier example); **rock candy** (earlier and later examples); also in *Big Rock Candy Mountain(s)*, a song about a mythical earthly paradise, used allusively in sense 'utopia'; **rock climb**, the ascent of a rock-face; also as *v. intr.*; **rock coal** *U.S.*, anthracite; **rock creep**, the creep (*CREEP *sb.* 7a) of rock, boulders, etc.; **rock-crusher**, (*a*) a machine used to break down rocks; (*b*) *fig.* in *Bridge*, a superlative hand; also *attrib.*; **rock-dust** *N. Amer.*, pulverized stone used to prevent explosions in coal mines; so **rock-dusting** *vbl. sb.*; hence **rock-dust** *v. trans.*, to treat (a mine) with pulverized stone; **rock-duster** (see quot. 1975); **rock-face**, a vertical expanse of natural rock; also *transf.* and *fig.*; also **rock-faced** *a.*; **rockfall**, the descent of loose rocks; a mass of fallen rock; **rock fan** *Physical Geogr.*, an eroded rock surface similar in shape to an alluvial fan, with a convex profile in transverse section; **rock fence** chiefly *Southern U.S.*, a stone wall; **rock-fill** *Engin.*, large rock fragments used to form the bulk of the material of a dam; freq. *attrib.*; **rock-flour**, substitute for def.: finely powdered rock, esp. that formed as a result of glacial erosion; (further examples); **rock-garden** (later example); so **rock-gardener**; **rock glacier**, a large mass of rock debris, in some cases mingled with ice, which moves gradually downhill in the manner of a glacier; **rock gong** *Archæol.* (see quots.); **rock happy** *U.S. Mil. slang*, mentally disturbed through serving too long on a (Pacific) island; **rock-hog**, a labourer engaged in tunnelling through rock; **rock hole**, (*a*) a tunnel; (*b*) *Austral.*, a natural depression in a rock that catches water; **rock hound** *colloq.* (orig. *U.S.*), (*a*) a geologist; (*b*) an amateur mineralogist; hence **rock-hounding** *vbl. sb.*, the hobby or activity of an amateur mineralogist; **rock-house**, (*a*) a house built of stone or quarried rock; (*b*) a shady place under overhanging rocks providing a suitable habitat for ferns; **rock mechanics**, the branch of science and engineering concerned with the mechanical properties and behaviour of rock; **rock of ages** *Rhyming slang*, wages; **rock-peg** *Mountaineering*, a nail-like device hammered into rock to assist climbing; **rock phosphate**, a sedimentary rock containing phosphates in high proportion; phosphorite; **rock pile** *U.S. slang*, (*a*) a heap of stones; (*b*) a jail or prison, in allusion to the convict's task of breaking stones; also *transf.* and *fig.*; **rock pitch** *Mountaineering*, an expanse of rock between belay points; **rock piton** *Mountaineering*, a piton used to assist climbing of rock; **rock river** = *rock glacier* above; **rock scorpion** (earlier and later examples); **rock-slide** orig. *U.S.*, a slippage of rock; a rough mass of rock that has subsided thus; also *fig.*; **rocksman** *Sc.* = ROCKMAN 1; **rock stream** = *rock glacier* above; **rock sugar** (later example); **rock waste**, fragments of rock produced by weathering; **rock well**, an oil well drilled through superficial deposits of clay, sand, or the like into underlying rock; **rock wool**, a material such as limestone, slag, or the like, made into the form of a fine, matted fibre, esp. for use in thermal insulation or soundproofing.

1912 W. H. Hobbs *Earth Features* xxvi. 377 When the backward grades upon the valley floor are especially steep, the rock step becomes a rock bar, or riegel, of which nearly every Alpine valley has its example. **1954, 1957** [see *RIEGEL]. **1963** D. W. & E. E. Humphries tr. *Termier's Erosion & Sedimentation* v. 121 Special characteristics such as cirques..and rock bars can always be recognized. **1861** Mrs. Beeton *Bk. Househ. Managem.* 852 (*heading*) Rock biscuits. **1920** *Engin. & Mining Jrnl.* 7 Feb. 404/1 The invention and development of rotary rock bits lagged behind the introduction and successful application of the rotary method of drilling. **1974** R. D. Grace in P. L. Moore et al. *Drilling Practices Man.* iv. 66 Rock bits should be economical in the shale. **1955** L. A. Panek in *Rep. Investigations U.S. Bureau of Mines* (1956) No. 5154. 1 The practice of roof bolting or rock bolting to stabilize rock surrounding underground excavations has increased..within a few years. **1957** *Q. Colorado School of Mines* July 235 Because of their increasingly extensive use in non-coal mines, we prefer to call these devices 'rock' rather than 'roof' bolts and will refer to them as such. **1973** L. J. Thomas *Introd. Mining* viii. 310 Rock bolts, sometimes called roof bolts, are the first line of defence in many mining and civil engineering applications. [**1856** 'Old Colonist' *How to Farm & Settle in Austral.* 56 This lowest bottom, 'the rock' as it is emphatically termed, in reference to its character as a bar to further digging for gold.] **1866** *Oregon State Jrnl.* 24 Nov. 2/2 A sound democrat, or 'rock bottom', never shrinks from the requirements of his master. **1890** in Barrère & Leland *Dict. Slang* II. 183/1 Other freight wars, covering much less territory than the present, have gone to rock bottom before any attempt has been made to restore rates. **1902** W. N. Harben *Abner Daniel* 273 See here, I've got a rock-bottom proposal to make to your people. **1904** —— *Georgians* 200 Now cool off, an' let's git down to rock-bottom. **1923** D. L. Sayers *Whose Body?* vii. 167 There aren't many men who wouldn't be nice—to her, and even then, if they aren't rock-bottom she can see through them. **1930** *Sat. Even. Post* 26 July 14/1 'Pay you?.. How much, Angelo?' 'The rock bottom is half a million.' **1935** H. Edib *Clown & his Daughter* xlv. 258 By the time she had touched the rock-bottom of misery she had also reached a decision. **1955** D. Davie *Articulate Energy* vii. 69 We are sobered and shocked when the mood reaches rock-bottom. **1977** *Belfast Tel.* 19 Jan. 7/8 Builders engaged in this work were rapidly reaching 'rock bottom'. **1980** *Daily Tel.* 16 Jan. 23/4 In this way, the service can be offered at rock-bottom prices. **1889** J. Whitehead *Steward's Handbk.* iv. 420/2 *Rock buns*, rough rocky looking cakes made of ½ lb. each butter, sugar and currants, [etc.]. **1868** M. Jewry *Warne's Model Cookery* 613/1 *Rock cakes*,..butter..flour..sugar..lemon..eggs..brandy. **1723** J. Nott *Cook's & Confectioner's Dict.* sig. U6, *To candy Nutmegs*... Pour your Candy to them,..set them in a warm Place for about three Weeks, and they will be of a Rock Candy. **1906** M. P. W. Locke (*song-title*) Big Rock Candy Mountains. *Ibid.*, Come to the Big Rock Candy Mountains, And I'll show you the bees and the cigarette trees And the soda water fountains. **1930** G. Milburn *Hobo's Hornbk.* 61 To homeguards 'The Big Rock Candy Mountains' may appear a nonsense song, but to all pied pipers in on the know it is an amusing exaggeration of the ghost stories used [by jockers] in recruiting kids. *Ibid.*, Said the bum to the son, 'O, will you come To the Big Rock Candy Mountains.' **1949** C. Himes *Black on Black* (1973) 278 He set up all the boys in the neighbourhood to peanut brittle and icecream and rock candy. **1961** *Life Treas. Amer. Folklore* 294 The Hobo Special climbs Big Rock Candy Mountain. **1975** *Daily Colonist* (Victoria, B.C.) 16 Apr. 4/4 The goal is their 20th century version of the big rock candy mountain. **1895** W. M. Conway *Alps from End to End* 402/2 (Index), Rock Climbs, Where to find. **1929** F. S. Smythe *Climbs & Ski Runs* iii. 14 My first rock climb was the Little Gully. **1934** Webster, Rock climb *v.* **1960** *Guardian* 11 July 1/4 He had intended to rock-climb with a companion. **1976** G. Moffat *Over Sea to Death* v. 52 She would be even happier were she to reach the top by way of a rock climb. **1858** *Southern Lit. Messenger* XXVI. 189/2 Ef thar had bin..a fier-plais instid uv a great to burn rock cole, the thing would uv bin kumpleat. **1913** O. A. Rothert *Hist. Muhlenberg County* 389 The early blacksmiths called this fuel 'rock coal', thus distinguishing it from charcoal. **1938** C. F. S. Sharpe *Landslides* iii. 31 Rock-creep.—It is sometimes observed that although creeping masses of rock have moved many feet their original relation to the bedrock can still be recognized. **1960** [see *CREEP *sb.* 7a]. **1968** R. W. Fairbridge *Encycl. Geomorphol.* 275/2 Such movement of rock debris induced by gravity as talus creep, rock creep, and debris slides all transport rock fragments to lower elevations. **1897** Rock-crusher [see sense 7c in Dict.]. **1952** I. Macleod *Bridge* iv. 39 The Two Club bid..has the double advantage of freeing the other bids of Two of a suit for specialized use, and coping with the rockcrusher hands which do not qualify for a Strong Two opening. **1965** *Times* 9 Jan. 9/7 Her bidding was cautious to a degree, requiring a positive rock crusher for anything above the level of one. **1973** *Country Life* 24 May 1503/1 A first-class collection of hands... 5 [Clubs]..shows a rock-crusher, asking partner to choose the suit. **1938** *Richmond* (Va.) *Times-Dispatch* 7 Nov. 6/2 Sprinkling of the rock-dust through the mines allays the highly explosive coal dust. *Ibid.*, The Bureau of Mines has a new argument in support of its plea that all coal mines be rock-dusted to prevent explosions. **1947** *Sun* (Baltimore) 27 Mar. 1/2 Reports of the State Inspection said the mine was 'not adequately rock-dusted'. **1977** *Transatlantic Rev.* LX. 80 Inside, [there is] the rockdust rumble of grinding teeth, molar on molar. **1975** *Publ. Amer. Dial. Soc. 1973* LIX. 48 *Rock duster*,..1, a mechanical blower, often caterpillar track or train wheel mounted, which forces rock dust against the dangerously dusty areas of the mine... 2, a worker in charge of distributing rock dust throughout the mine, by hand or machine. **1926** *Daily Colonist* (Victoria, B.C.) 20 July 12/4 By rock dusting is meant the spreading of incombustible dust throughout a mine in sufficient amount to cool and extinguish the flame of an incipient explosion. **1932** *Durant* (Okla.) *Daily Democrat* 10 Mar. 4/5 By rock dusting, a practice made mandatory by the 1929 legislature, this coal dust is mixed with an equal amount of rock dust, the latter lowering the ignition point of the mixture. **1938** *Richmond* (Va.) *Times-Dispatch* 7 Nov. 6/2 Rock-dusting also is advocated as a means of increasing visibility in the mines and preventing minor accidents. **1855** Rock-face [see sense 6a in Dict.]. **1931** C. Day Lewis *Coll. Poems* (1935) 76 As one who wanders into old workings Dazed by the noonday, desiring coolness, Has found retreat barred by fall of rockface. **1940** *Chambers's Techn. Dict.* 728/2 *Rockface*, (*Masonry*) the form of face given to a building-stone which has been quarry-faced. **1968** *Amer. Speech* 1967 XLII. 295 *Rock-face stone*,..slabs of stone sawed on the top and bottom surface (up to five surfaces), which are then placed in a machine exerting pressure and cracking the stone. This leaves a pleasing rough surface toward the outside. **1972** *Times* 29 Nov. 28/8 (Advt.), The..appeal is to the younger person who really wants a vital and interesting job as a change agent at the 'rock face'. **1944** K. Levis in Murdoch & Drake-Brockman *Austral. Short Stories* (1951) 427 The men were separated in bunches cut off by rock-faced water-beds. **1970** H. Braun *Parish Churches* xviii. 216 The face of the stone is left 'rock-faced' and not worked at all except along its margins. **1930** *Times Educ. Suppl.* 24 May p. i/2 Crossing the débris of a huge rockfall which apparently came down recently. **1967** M. J. Coe *Ecol. Alpine Zone Mt. Kenya* 87 The Tarn..is enclosed at its lower edge by the rock fall. **1971** *World Archaeol.* III. 150 The rockfall layers at Puntutjarpa were of considerable archaeological interest. **1900** W. M. Davis in *Bull. Geol. Soc. Amer.* XI. 210 Near the base of the mountain front nearly all of the ravines broaden and their floors become distinctly convex, thus imitating the form well known in alluvial fans, though rarely matched in an eroded surface of solid rock. These convex floors will be called rock fans. **1932** *Amer. Jrnl. Sci.* CCXXIII. 392 The alluvial fan is the expression of that form where deposition alone has occurred, or where considerable deposition has accompanied erosion of bedrock. The 'rock fan'..is the same form..where erosion has exceeded deposition. *Ibid.* 393 Rock fans must be carved by streams, and cannot be produced by simple weathering back of the mountain front. **1968** R. W. Fairbridge *Encycl. Geomorphol.* 965/1 The rock fans described range in area from several acres when steep, 20–26°, to one or more square miles when gentle, ½–7°. **1970** R. J. Small *Study of Landforms* ix. 310 The existence of rock fans.. has been disputed by many geomorphologists, who have claimed that such fans are in reality no more than the alluvial fans that they are supposed to resemble so closely. **1896** *Dial. Notes* I. 423 *Rock fence*, a stone wall. **1949** H. Kurath *Word Geogr. Eastern U.S.* 31/2 For a fence built of loose stone the North Midland uses *stone fence* as against the Northern *stone wall* and the Southern *rock fence*. **1974** *Amer. Speech* 1971 XLVI. 60 *Rock fence* appears in southern Illinois (a South Midland settlement area) as expected. **1911** *Sci. Amer.* 17 June 592 (*caption*) Characteristic rockfill across a creek. **1960** *Times* 7 Mar. 8/3 It is hoped..to save the Temples from inundation by means of an earth and rock-fill dam. **1969** E. W. Morse *Fur Trade Canoe Routes* ii. v. 55 The dam and its rockfill now obscure the upper portion of the portage. **1976** *National Observer* (U.S.) 19 June 4/1 The Interior Department's Bureau of Reclamation has been building dams, including 250 out of rock-fill and earth. **1937** *Geogr. Jrnl.* LXXXIX. 43 Great angular blocks of rock are embedded in a jumble of fragments from the size of dust upwards, it is, in fact, a loose breccia of large pieces associated with smaller ones grading down to the finest rock-flour. **1963** G. L. Pickard *Descriptive Physical Oceanogr.* iii. 23 In fjords fed by rivers from glaciers, the surface low-salinity layer may be a milky white from the finely divided 'rock flour'. **1975** J. G. Evans *Environment Early Man Brit. Isles* ii. 44 A prerequisite of the formation of wind-blown deposits is a dry land surface from which frost-shattered rock flour can be whipped up and transported. **1962** R. Page *Education of Gardener* viii. 231 By the end of the nineteenth century rock-gardens had become a lasting feature of British gardens. **1942** E. Waugh *Put out More Flags* iii. 172 The word 'Colonel' for Basil had connoted an elderly rock-gardener. **1905** W. Cross *Silverton Folio* (U.S. Geol. Survey. Geol. Atlas of U.S. No. 120) 25/2 All the accumulations..just described impress one with the sense of motion.. So noticeable was this that in the field they were spoken of as 'rock-glaciers' and upon the map receive the name 'rock streams'. **1910** S. R. Capps in *Jrnl. Geol.* XVIII. 360 The special agents of degradation with which I wish to deal at present..I have called rock glaciers... In material the rock glaciers are composed of angular talus. **1954** W. D. Thornbury *Princ. Geomorphol.* iv. 85 Kesseli concluded that the rock glaciers of the Sierra Nevada were essentially fossil glaciers. **1968** R. W. Fairbridge *Encycl. Geomorphol.* 711/1 If the surface of the ice is densely covered with blocks, one may speak of a rock glacier, but, in North America, 'rock glacier' does not necessarily involve ice. **1955** B. E. B. Fagg in *3rd Pan-Afr. Congr. Prehist.* (1957) xlvii. 310 Very extensive exploration of the granite hills revealed the existence of large numbers of these hammered rocks, which I think can best be described as rock gongs. They consist of huge natural spalls or exfoliations of rock which happen to rest or be wedged in a position favourable to the production of musical notes. **1959** *S. Afr. Archaeol. Bull.* XIV. 112/2 Rock gongs should be described as 'ringing rocks' or 'sounding stones'. **1961** K. P. Wachsmann in A. Baines *Mus. Instruments through Ages* i. 30 Recent studies have revealed many instances of slabs of rock being used as if they were drums. These 'rock gongs', as their discoverers called them, occur in Africa north of the equator, in Europe, and in Asia. **1945** *Yank* 15 June 2/2 The set routine can drive a man nuts wherever he is... Out here [*sc.* the Marianas] the expression is rock happy... In the Pacific there is no escape from places like Kwajalein. **1946** *Richmond* (Va.) *Times-Dispatch* 9 May 12/1 GI's..were growing rock happy from too long internment on a coral island. **1909** *Chambers's Jrnl.* Dec. 828/2 The rock-hogs had not proceeded far before they pierced a large pocket. **1954** V. Lysenko *Yellow Boots* 190 They spoke of dynamite and flying rock responsible for the death of many a 'rock-hog'. **1895** M. Pemberton *Impregnable City* ii. xiv. 285 Darkness of the rock-hole. **1936** I. L. Idriess *Cattle King* iv. 30 He learned probabilities and signs by means of which waterholes may be located in apparently dry creeks, and in rock-holes in valley or gorge. **1944** *Living off Land* iii. 50 Locating water in the form of soaks, springs and rock-holes. **1922** *Daily Ardmoreite* (Ardmore, Okla.) 10 Jan. 6/2 (*heading*) Interesting tale of work in Africa told by Texas rock hound. **1940** *Fortune* Mar. 83 Drillers consider themselves a superior breed, look with scorn upon 'rockhounds' (geologists), 'chemicos,' pipemen, roughnecks, etc. **1949** *Natural Hist.* LVIII. 220/1 There are numerous semiprecious stones to interest the 'rock hound'. **1962** E. Lucia *Klondike Kate* viii. 175 Kate was central Oregon's first serious rock hound, of which there are thousands today. **1970** *Nature* 4 Apr. 45/2 Its bias is naturally towards the United States, where 'rock hounds' and geological societies are much more common than in the United Kingdom. **1979** *N.Y. Times Mag.* 30 Sept. 88/2 (Advt.), Exclusive metal ratchet device permits shovel to be locked into any position... A necessity, too, for sportsmen,..fishermen, hikers, rockhounds. **1949** *Desert Mag.* June 31/1 In all my rockhounding I have never seen sand fly so fast. **1973** *Daily Tel.* 25 Aug. 16/1 The objects of his search might be coins, lost jewellery, Victorian ceramics, or if he feels like a change a spot of rockhounding—searching for semi-precious stones—or gold-panning in Scotland. **1976** *Globe & Mail* (Toronto) 3 July 35 (*heading*) Go rockhounding or trail riding, for everything goes in Ontario. **1818** E. P. Fordham *Jrnl.* 26 Jan. in *Personal Narr.*

Trav. (1906) 154 They had a strong rock house among the hills. **1883** E. A. Smith *Rep. Geol. Survey Alabama 1881–82* 438 Underneath the overhanging cliffs, or 'rock houses', as they are termed, grow abundantly some of our rarest and most beautiful ferns. **1889** *Harper's Mag.* Dec. 120/1 Thet thar rock house o' his'n, which he hev quayried the rock an' put up hisse'f, I 'low it's the beatenes' house in creation. **1901** C. Mohr *Plant Life Alabama* 17 The..fern, *Trichomanes petersii*,..with others like it hidden in the dark recesses of rocky defiles and the so-called 'rock houses'. **1948** E. N. Dick *Dixie Frontier* 26 Along the rivers in certain places the rocks projected out over the banks. Hunters and early settlers sometimes lived in the shelter of these for months. They were known as rock houses. **1956** *Q. Colorado School of Mines* LI. iii. (*title-page*) Symposium on rock mechanics. Papers and discussion from the first annual symposium on rock mechanics. **1966** *McGraw-Hill Encycl. Sci. & Technol.* XI. 599/2 An understanding of rock mechanics is essential to elucidate the processes which mold the face of the earth. **1977** A. Hallam *Planet Earth* 104/2 The engineering geologist works with experts in the related fields of soil mechanics and rock mechanics. **1937** Partridge *Dict. Slang* 702/2 *Rock of ages*, wages. **1974** P. Wright *Lang. Brit. Industry* x. 89 If there's no *rock of ages* (wages), there may well be a *bull an' cow* (row). **1971** C. Bonington *Annapurna South Face* x. 118 He put in a couple of ice-screws, then, having run out of these, hammered in ordinary rock-pegs, which are much shorter than ice-screws and not nearly as secure. **1868** *Chem. News* 13 Nov. 238/2 (*heading*) Rock phosphates. **1900** [see *phosphatizing* vbl. sb. s.v. *PHOSPHATIZE *v.*]. **1936** [see *PHOSPHATE 1b]. **1949** *Thorpe's Dict. Appl. Chem.* (ed. 4) IX. 482 Sedimentary rock phosphate, or phosphorite, occurs in two forms: in thick beds, usually of high phosphatic content; and in layers of nodules, commonly of lower phosphatic grade. **1965** G. J. Williams *Econ. Geol. N.Z.* xvi. 250/2 Free went on to suggest the possibility of fusing rock-phosphate with greensands and dolomite to give a mixed fertilizer. **1888** *Congress. Rec.* 1 May 3571/1 If this were a police court, the Senator from Indiana would be sent to the rock-pile for being drunk and disorderly. **1927** K. Eubank *Horse & Buggy Days* 127 We were..given 30 days on the rock pile or the privilege of leaving town on the first rattler out, which took us into Memphis. **1945** *Seafarers' Log* 13 Apr. 6/3 Had one of the Bull Line rock piles [*i.e.* a ship on board which work is hard] in. **1945** L. Shelly *Jive Talk Dict.* 16/2 *Rockpile*, a very tall building. **1947** *Sat. Even. Post* 23 Oct. 132/3 Everybody was dead-pan and silent. But disciplined—like convicts on a rock pile. **1949** W. Stevens *Let.* 12 Dec. (1967) 659 We call the office the rockpile, yet so large a rockpile is a good deal more than that. **1970** C. Major *Dict. Afro-Amer. Slang* 98 *Rock pile*, any tall building. **1929** Rock pitch [see *ice-ridge* s.v. *ICE *sb.* 7b]. **1955** [see *ETRIER]. **1934** *Canad. Alpine Jrnl.* XXII. 128 These analogs of rock pitons..have now definitely passed their test for usefulness. **1972** D. Haston *In High Places* i. 12 It should be noted that ice overhangs can be tackled on ice-pitons in the same way as rock overhangs on rock-pitons. **1920** *Natural Hist.* XX. 172/1 In rate of flow these rock rivers are probably slower than the ice rivers, or glaciers. **1954** W. D. Thornbury *Princ. Geomorphol.* iv. 85 Striking examples of rock glaciers, rock streams, or rock rivers have been described..in the Sierra Nevada of California. **1818** 'A. Burton' *Adventures J. Newcome* iv. 239 Fagged he was in every limb, And the Rock Scorpions laughed at him. **1916** 'Taffrail' *Pincher Martin* viii. 145 They arrived at Gibraltar, where the ships went alongside the Mole..to take in coal. But here..the fuel was carried on board in small baskets on the backs of nondescript, garlic-scented aliens known as 'rock scorpions'. **1976** *Daily Colonist* (Victoria, B.C.) 10 June 4/1 The 29,000 Gibraltarians—affectionately known to generations of British seafarers as Rockscorpions—have been eyeing Madrid hopefully. **1851** H. Melville *Moby Dick* II. xiv. 123 Some mossy rock-slide from the Patagonian cliffs. **1921** *Daily Colonist* (Victoria, B.C.) 29 Mar. 13/5 Owing to a big rockslide west of Terrace, B.C., the G.T.P. passenger train from Prince Rupert is now twenty-four hours late. **1959** R. Campbell *I would do It Again* xviii. 127 We took saddle horses across the flat as far as the mountain slope, which was covered by a rockslide. **1970** R. Lowell *Notebk.* 203 Is it my imagination or..Pound's Cantos lost in the rockslide of history? **1971** *Islander* (Victoria, B.C.) 13 June 2/4 The pica and ground squirrel may be seen around the rock slides on the edge of the meadow. **1980** *Beautiful Brit. Columbia* Summer 33 The main trail to Eva Lake..across rock slides. **1852** W. Macgillivray *Hist. Brit. Birds* V. 434 The dexterity of these rocksmen is truly astonishing. **1905** Rock stream [see *rock glacier* above]. **1909** *Prof. Papers U.S. Geol. Survey* No. 67. 31 The name 'rock stream', which has been found a convenient descriptive term, was suggested by the peculiar streamlike appearance of the deposits, which look as if they had moved down the cirques or valleys after the manner of glaciers. **1964** W. C. Putnam *Geol.* x. 238/1 (*caption*) A rock stream or rock glacier is composed of frost- and ice-shattered rock filled with interstitial ice which slowly moves downslope. **1970** R. J. Small *Study of Landforms* x. 333 Fossil rock streams exist in many temperate lands today. **1889** J. Whitehead *Steward's Handbk.* iv. 420/2 *Rock sugar*,.. candy rock work used to build up ornamental pieces of confectionery and to sell as sponge candy. **1907** *Bull. Geol. Soc. Amer.* XVIII. 358 If the moist epoch last long, the mountains of arid countries, such as Persia,..must lose their naked character and become well shrouded with rock waste. **1946** F. E. Zeuner *Dating Past* vii. 220 This cave, the Grotta Gualtari..was completely sealed by rock-waste. **1865** A. Gesner *Pract. Treat. Coal, Petroleum, & Other Distilled Oils* (ed. 2) ii. 40 The rock wells, as they are termed, are those deeper borings which resemble those of Pennsylvania. **1867** *Ure's Dict. Arts* (ed. 6) III. 404 The rock-wells are of two characters, namely, 'pumping' and 'flowing'. **1928** E. R. Powell *U.S. Pat.* 1,656,828 2/1 It is to be understood that so-called rock wool is made directly from the rock which contains only the slight trace of sulphur; while so-called mineral wool is made from the slag which contains the higher percentages of sulphur. **1936** [see *PLASTERBOARD]. **1959** *House & Garden* Dec.–Jan. 57 The outer walls of the house are constructed of two brick skins with a cavity between filled with rock wool for insulation. **1975** G. J. King *Audio Handbk.* vi. 136 A material..which possesses a high value of acoustical absorbitivity; some commonly used materials being rockwool..and polyurethane foam.

b. rock borer, for 'the family *Petricolidæ*' substitute 'the superfamily Saxicavacea'; (later examples); **rock chuck,** the North American yellow-bellied marmot, *Marmota flaviventris*; **rock crab** (earlier example); **rock goat** (earlier and later examples); **rock hyrax** = *rock rabbit* (*a*), *DASSIE 1; **rock lizard,** an African or Australian dragon lizard belonging to the family Agamidæ; (see also sense 9a in Dict.); **rock lobster,** esp. the marine crayfish, *Palinurus vulgaris*; (later examples); **rock python,** one of several large snakes of the family Boidæ, esp. the African *Python sebæ*; **rock rabbit,** substitute for def.: a hyrax belonging to the genus *Procavia* or *Dendrohyrax*, esp. the African *P. capensis*; (earlier and later examples); (*b*) = *PIKA; **rock rat,** (*c*) a South American rodent, *Aconæmys fuscus*; (*d*) an Australian thick-tailed rat belonging to the genus *Zyzomys*; **rock scorpion,** a southern African scorpion, *Hadogenes lawrencei*; (see also sense 9a); **rock worm,** add def.: a marine polychæte worm belonging to the family Eunicidæ; (later examples).

1928 Russell & Yonge *Seas* vi. 148 The largest and most efficient rock borers are bivalve Molluscs. **1971** *Oxf. Bk. Invertebr.* 86/1 *Hiatella* is a rock-borer (if the rock is soft enough). **1913** *Outing* Jan. 451 (*caption*) Not a woodchuck, but a 'rockchuck'. **1947** B. A. De Voto *Across Wide Missouri* 162 Robes..were made from..beaver,.. wolf, or even rockchuck. **1968** *National Observer* (U.S.) 22 July 6/5 Bones of such small mammals as the rock chuck, rock squirrel, northern pocket gopher, and pygmy rabbit, were found in the Altithermal strata. **1837** J. L. Williams *Territory of Florida* 105 The Rock Crab is common on the Atlantic coast. **1607** Topsell *Foure-f. Beasts* 244 (*heading*) The Helvetian Alpian wilde or Rocke-Goat. **1820** J. Campbell *Trav. S. Afr.* I. ii. 29 The rock-goat..had found its way to a place, which no human foot had ever yet trod. **1954** G. Durrell *Bafut Beagles* iii. 59 'Rock hyrax.'.. 'Yes. How you de call um for Bafut?' 'Here we call um N'eer.' **1966** C. Sweeney *Scurrying Bush* iii. 36 Although the rock hyrax is only about sixteen inches long and weighs little more than eight pounds when adult, it is not an animal to trifle with. **1937** *Discovery* May 137/2 The Rock Lizard can be met with all over the [Nullarbor] plains. **1947** J. Stevenson-Hamilton *Wild Life S. Afr.* xxxv. 318 There are nine species of so-called Rock Lizards (*Agama*) known as koggelmannetjies in South Africa. **1928** Russell & Yonge *Seas* xiv. 316 The handsome Spiny or Rock Lobster or Crawfish..differs from the lobster in its larger size. **1953** *Sun* (Baltimore) 9 Sept. 10/7 The name of the South African crawfish was changed by law to 'rock lobster'. **1961** [see *CRAWFISH *sb.* 1b]. **1969** *N.Z. News* 17 Dec. 5/3 Under the [new] regulations crayfish are referred to as rock lobster. The change is necessary to promote the labelling of crayfish as rock lobster in marketing in overseas countries. **1974** *Times* 9 Dec. 12/4 A notice outside advertised baby rock lobster tails with two veg. **1910** R. L. Ditmars *Reptiles of World* iv. 227 Another big serpent is the African Rock Python. **1934** A. Russell *Tramp-Royal in Wild Austral.* xxxviii. 251 We rode almost on top of a rock python one day... He was ten feet long. **1965** R. & D. Morris *Men & Snakes* i. 16 Bushmen will eat snakes when available and especially prize the large rock python. **1840** B. Shaw *Memorials S. Afr.* xii. 147 There were numerous traces of rock rabbits. **1846** [see *DASSY]. **1878** J. H. Beadle *Western Wilds* 457 The rock rabbits..ran from covert to covert with a peculiar low moaning cry. **1927** A. Philip *Painted Cliff* 69 Rock-rabbits shrilled, darting amongst the rubble. **1931, 1952** [see *DASSIE 1]. **1962** *Field, Horse & Rodeo* (Calgary) Nov. 15/3 The Pika (or Rock Rabbit) spends most of the daylight hours cutting and gathering vegetation. **1972** J. McClure *Caterpillar Cop* iv. 46 Danny was going up and down those stairs like a rock rabbit. **1976** T. Walker *Spatsizi* vii. 68 A rock rabbit had been busy storing dried leaves and grasses for winter. **1964** E. P. Walker et al. *Mammals of World* II. 1045/2 Rock rats are active beneath the snow in winter. **1970** W. D. L. Ride *Guide Native Mammals Austral.* ix. 148 Adults of these rock-rats are easily recognized because they have thick tails. **1971** L. H. Matthews *Life Mammals* II. vii. 216 The rock rat, *Aconæmys fuscus*, the only species of the genus, with short, hairy, but untufted tail is about the size of a common rat. **1976** *Nature* 24 June 639/2 Many surveys produce novelties; animals considered very rare or extinct are discovered in some numbers (for example..the rock rat, *Zyzomys woodwardi*). **1789** W. Paterson *Narr. Four Journeys Country of Hottentots* 165 The Black, or Rock Scorpion, is nearly as venomous as any of the serpent tribe. **1973** *Stand. Encycl. S. Afr.* IX. 544/2 The rock-scorpion, *Hadogenes*, is extraordinarily flattened, being adapted for living in narrow fissures between rocks. **1963** R. P. Dales *Annelids* 14 Some relatives of the eunicid rockworms..are surprisingly like earthworms. **1971** *Oxf. Bk. Invertebr.* 98 The eunicids (rock worms and palolos) form a large and varied family of rock-dwellers and mud-burrowers.

c. rock duck, add: *Histrionicus histrionicus*; (later example); **rock sandpiper,** substitute for def.: the purple sandpiper, *Erolia maritima*, or a similar bird, *E. ptilocnemis*, of the Pacific coast of North America; (later examples); **rock snipe** (earlier and later examples); **rock wren,** substitute for 'see quots.': (*a*) one of several wrens belonging to the genus *Salpinctes*, found in parts of western North America; (*b*) a New Zealand wren, *Xenicus gilviventris*; (earlier and later examples).

1965 E. Richardson *Living Island* 185 The handsomest duck I have seen..the male harlequin or rock-duck. **1903** E. Coues *Key to N. Amer. Birds* (ed. 5) II. 817 Feather-leg Sandpipers. Rock Sandpipers. **1961** R. T. Peterson *Field Guide Western Birds* (ed. 2) 112 Rock Sandpiper... In winter, very similar to Purple Sandpiper of Atlantic. **1835** J. J. Audubon *Ornith. Biogr.* III. 558 Their marked predilection for rocky shores has caused them to be named 'Rock Snipes' by the gunners of our eastern coast. **1917** T. G. Pearson *Birds Amer.* I. 232 Purple Sandpiper... Other Names.—Rock Sandpiper; Rock Snipe. **1858** S. F. Baird *Birds Pacific Railway Routes* 357 Rock Wren... High central plains through the Rocky mountains. **1869** [see *cactus wren*]. **1946** D. C. Peattie *Road of Naturalist* iii. 39 The rock wrens and the canyon wrens..watered the air with rapture. **1966** R. A. Falla et al. *Field Guide Birds N.Z.* 194 Rock Wren... Habitat is distinctive, mainly open screes, moraines and fell-fields of mountains above the bushline. **1973** R. D. Symons *Where Wagon Led* I. v. 77 All around us rose the bubbling songs of the rock wrens.

d. rock beauty, substitute for def.: a small, dark brown and yellow, Caribbean reef fish, *Holacanthus tricolor*; (examples); **rock eel,** (*b*) = **rock salmon* (*c*); **rock salmon,** (*c*) a commercial name for the catfish, *Anarhichas lupus*, or a dogfish, *Scyliorhinus stellaris* or *S. caniculus*; cf. *HUSS; **rock skipper,** a small marine fish belonging to the family Gobiidæ, able to survive out of water for a limited time; **rock sole,** a flatfish, *Lepidopsetta bilineata*, found in the Pacific Ocean off the western coast of North America; **rock sturgeon,** add to def.: *Acipenser fulvescens*; (examples).

1892 T. D. A. Cockerell in *Bull. Inst. Jamaica* I. 9/1 Rock Beauty. Head, anterior part of the trunk, caudal and margins of the soft dorsal and anal fins yellow: the remainder brownish-black. **1959** R. P. L. Straughan *Salt-Water Aquarium in Home* iv. 85 Rock Beauties.. need plenty of aeration. **1965** Mrs. L. B. Johnson *White House Diary* 3 June (1970) 281 Rock beauties, yellow about half-way back and yellow tails, and the rest of them brown. **1977** D. J. Coffey *Encycl. Aquarium Fish* 141/1 Rock Beauty... From the Caribbean, this yellow fish has a black patch that expands with age. **1969** A. Wheeler *Fishes Brit. Isles & N.-W. Europe* 46/1 The dogfish..is sold as rock eel and rock salmon. **1931** J. R. Norman *Hist. Fishes* xix. 385 It has been found convenient to market this perfectly wholesome fish [*sc.* cat-fish] under a more pleasing name, and..it is sold as 'Rock Salmon'. **1957** R. Campbell *Portugal* iv. 65 In the [London] fried-fish shops..'rock salmon'..is the trade name for shark. **1958** *Times* 18 July 7/1 Rock salmon is in fact usually catfish. **1963** [see *HUSS *sb.*]. **1969** A. Wheeler *Fishes Brit. Isles & N.-W. Europe* 452/2 It [*sc.* the catfish] is sold with the related species under the names rock salmon and rockfish. **1977** *Times* 8 Feb. 9/4 Rock salmon has had its name changed..by the trades description act, and is now called Huss. **1905** D. S. Jordan *Guide to Study of Fishes* II. xxix. 510 The rock-skippers..are herbivorous, with serrated teeth set loosely in their jaws. These live in rock-pools of the tropics and leap from rock to rock. **1966** C. Sweeney *Scurrying Bush* xi. 154 There are perhaps a hundred or more kinds of rock skippers in tropical oceans, all able to shuffle and wriggle along on spray-soaked rocks. The kind that I found was as goggle-eyed as any and..resembled a small seal with a frog's head, the flipper-like pectoral fins supporting the front of the body. **1965** A. J. McClane *Stand. Fishing Encycl.* 730/2 Rock sole are only occasionally taken by hook and line. **1971** *Islander* (Victoria, B.C.) 21 Mar. 2/3, I have taken in recent years..rock sole and sand sole. **1877** C. Hallock *Sportsman's Gazetteer* i. 329 Rock Sturgeon.—*Acipenser rubicundus*. This is the sturgeon of the great lakes. **1961** E. S. Herald *Living Fishes of World* 67/2 The eastern American lake or rock sturgeon has been known to reach a weight of 300 pounds.

e. rock beauty, also called *Petrocallis pyrenaica*; (examples); **rock chestnut-oak** (examples); **rock elm,** substitute for def.: one of several North American elms, esp. *Ulmus thomasii* or its timber; (earlier and later examples); **rock maple,** substitute for def.: the sugar maple, *Acer saccharum*, or its timber; (earlier and later examples); **rock melon** = CANTALOUP; **rock oak** (earlier and later examples); **rock tripe** (earlier and later examples).

1870 W. Robinson *Alpine Flowers for Eng. Gardens* ii. 272 *Petrocallis Pyrenaica*—Beauty of the Rocks... A 'rock beauty!' as everybody must confess who sees its fresh light-green tufts, not more than an inch high, and cushioned snugly amidst the broken rocks. **1930** H. Correvon *Rock Garden & Alpine Plants* xii. 249 D[raba] (*Petrocallis*) *pyrenaica*. Rock Beauty. **1964** A. N. Griffith *Collins Guide to Alpines* 205 This very small genus [*sc. Petrocallis*] provides us with one charming little plant which well deserves its name of Rock Beauty. **1810** F. A. Michaux *Hist. Arbres Forestiers de l'Amérique Septentrionale* I. 23 Rock chestnut oak..seul nom donné à cette espèce dans les Etats de New-York et de Vermont. **1832** D. J. Browne *Sylva Amer.* 285 The rock chestnut oak is sometimes 3 feet in diameter, and more than 60 feet high. **1897** G. B. Sudworth *Nomencl. Arborescent Flora U.S.* 156 Chestnut Oak..[also called] Rock Chestnut Oak. **1830** *Trans. Lit. & Hist. Soc. Quebec* III. 84 The

timber of this variety..is known by the name of Rock Elm. **1955** *Bush News* (Port Arthur, Ontario) Feb. 3/5 Southern Ontario..is sending..rock elm timbers to Britain. **1972** *Handbk. Hardwoods* (Forest Prod. Res. Lab.) (ed. 2) 73 Rock elm may be distinguished from other commercial elms by the small size and sparse distribution of its early wood pores. **1775** S. Thayer *Jrnl.* 30 Sept. in *Rhode Island Hist. Soc. Coll.* (1867) VI. 4 The timber [is] large and of various kinds, such as Pine, Oak, Hemlock and Rock Maple. **1949** 'J. Nelson' *Backwoods Teacher* xxvi. 265 Hi Slocum..had tapped a few rock maples he knew about—though this is not really maple syrup country. **1977** *New Yorker* 27 June 30/3, I..fell over a rock-maple chair and skinned my knee. **1980** *Early Music Gaz.* July 12/1 (Advt.), Made of impregnated rock maple and finished in dark color. **1871** V. Lush *Jrnl.* 16 Mar. (1975) 105 Blanche bought 8 fine rock melons for 4/o... Blanche has reduced 6 of them into jam. **1882** [see **pie-melon*]. **1929** [see *CANTALOUP]. **1972** J. S. Gunn in G. W. Turner *Good Austral. Eng.* iii. 60 Is there any difference at all between..rock melon/cantaloup and many other pairs? **1699** *Public Rec. Colony Connecticut* (1868) IV. 304 Running eastward three hundred rod to a rock-oak tree markt. **1949** Collingwood & Brush *Knowing your Trees* 224 Sometimes this tree is called rock oak or mountain oak because it grows on high, rocky slopes. **1854** Mayne Reid *Young Voyageurs* 384 Rock-tripe..was a black, hard crumply substance that nearly covered the surface of the rock, and was evidently of a vegetable nature. **1907** *St. Nicholas* July 847/1 'Rock-tripe', another lichen, has been eaten in the arctic regions in times of famine. **1952** F. Mowat *People of Deer* 37 Sometimes she scrabbled through the drifts on hilltops and found..a handfull of rock-tripe, a kind of moss.

rock, *sb.*[3] Add: **1. b.** Phr. *rock of eye* = *rack of (the) eye* s.v. Rack *sb.*[1] 4f.

1890 Barrère & Leland *Dict. Slang* II. 183/2 *Rock of eye and rule of thumb* (tailors), refers to doing anything which requires scientific treatment by guesswork. **1957** N. Squire *Theory of Bidding* xiii. 216 Honour-tricks will be counted at their normal value as in the Table of Honour-tricks, but with additions found by rock-of-eye.

2. orig. *U.S.* **a.** Musical rhythm characterized by a strong beat.

1946 Mezzrow & Wolfe *Really Blues* vii. 90 The Cotton Pickers..came on with a steady rock that was really groovy. **1952** H. Sinclair *Music out of Dixie* vi. 245 He played eight bars of a new introduction he had thought up and..[said] 'I want that steady rock.' **1952** R. A. Waterman in S. Tax *Acculturation in Americas* 217 Musical terms like 'rock' and 'swing' express ideas of rhythm foreign to European folk tradition, and stem from African concepts. **1970** P. Oliver *Savannah Syncopators* 36 Jazz developed a different kind of rhythmic feeling with a lifting movement between adjacent beats which the jazz musician identifies as 'rock' or 'swing'.

b. = *ROCK AND ROLL. Now freq. used to encompass most modern popular music with a rocking or swinging beat. Also the last element in *Combs.*, as *acid rock*, *folk rock*, etc.; see *hard rock* (*c*) s.v. *HARD *a.* 22a, *PUNK ROCK, *raga rock* s.v. *RAGA 2.

1957 *Beat* Sept. 7/1 'It's the answer to Rock,' said one and all... But a new sound package of diluted Rock, Hill-Billy tunes and ersatz Blues assails our ears. **1959** *Daily Mail* 17 Feb. 4/4 *Yellow Dog Blues* played in basic style by Joe Darensbourg's band..unexpectedly popped up among the rock. **1960** M. Spark *Ballad of Peckham Rye* (1964) iv. 58 Findlater's rooms were not given to rowdy rock but concentrated instead upon a more cultivated jive, cha-cha, and variants. **1963** J. T. Story *Something for Nothing* v. 166 'It's only folk singing,' Albert told him. 'Well, it makes a change from all this old rock,' said Sid. **1965** *Time* 17 Sept. 102/2 Folk rock owes its origins to Bob Dylan, 24, folk music's most celebrated contemporary composer. **1968** *National Observer* (U.S.) 3 Nov. 24 It has been clear for some time that 'rock' is getting longer, more sophisticated, more ambitious, restless with chordal limitations and the three-minute format. **1969** *Britannica Bk. of Year* (U.S.) 799/1 *Acid rock*, rock 'n' roll songs whose titles or lyrics make cryptic reference to drugs. **1969** *Rolling Stone* 28 June 38/4 (Advt.), Two guitarists needed immediately... Booked for TV show in a few mos. Have material, underground & acidrock. **1972** *Saturday Night* (Toronto) Sept. 42/2 Like light shows, psychedelic posters and acid rock, it seems to have emerged first in California. **1976** *New Statesman* 17 Dec. 884/1 The whole of rock..had grown away from its roots, absorbing the influences of poetry, folk and protest music, and in the Sixties becoming central to the internal communications of a whole generation.

c. *attrib.* and *Comb.*, as *rock album*, *artist*, *band*, *beat*, *club*, *critic*, *criticism*, *culture*, *fan*, *festival*, *group*, *guitarist*, *history*, *idiom*, *lyric*, *movement*, *movie*, *music*, *musical*, *musician*, *number*, *opera*, *press*, *record*, *show*, *singer*, *singing*, *song*, *star*, *thing*; also *rock-dominated*, *-tinged* adjs.

1979 *Yale Alumni Mag.* Apr. 30/3 Many sociable Soviets turned out to be dealers on the thriving Soviet black market..interested primarily in acquiring American blue jeans..., rock albums, dollars, or chewing gum. **1973** *Black World* Nov. 45/2 Many rock and soul artists have retained their interest in..gospel music. **1968** *Listener* 13 June 774/1 There was a rock band that whooped it up all the louder, to drown the inevitable news. **1978** G. Vidal *Kalki* vi. 154 A rock band deafened us. **1969** *Listener* 20 Feb. 251/1 They..claim to have brought 'the rock beat, the *now* sound, to the American Musical Theatre'. **1972** *Jazz & Blues* Sept. 4/2 A slashing rock beat. **1965** M. Morse *Unattached* v. 177 A 'rock' club was started for younger teenagers. **1977** *Rolling Stone* 13 Jan. 8/3 If rock critics recognize and understand this as a problem, why don't they do something about it? **1977** *N.Y. Rev. Bks.* 14 Apr. 40/4 Mark Miller's 'review'.. of recent rock criticism seriously distorts its subject. **1967** *Economist* 8 Apr. 144/1 This is politics fashioned for the young: 'the rock culture', it is being called. **1977** *Rolling Stone* 24 Mar., His full-blown, upper-register style is ingeniously contrasted with Walden's simple, melodic, rock-dominated charts. **1961** *Times* 12 Aug. 7/6 More intelligent than the majority of 'rock' fans. **1968** *Rolling Stone* 12 Oct. 1/3 'The best freaking scene ever,' said one musician. The Sky River Rock Festival and Lighter Than Air Show was not dampened by the rain that fell over Labor Day weekend. **1971** M. Smith *Gypsy in Amber* xix. 144 I've never seen a rock festival. **1967** *Listener* 14 Sept. 350/2 Rock groups..concentrated on achieving the authentic and personal expression. **1977** *It* May 10/1 Perhaps there are also rock groups who would be prepared to perform at benefit concerts. **1977** *Gay News* 24 Mar. 28/2 His cohorts perform well too especially Ray Russell, even if he is inclined to go in for circular solos, just like a rock guitarist. **1976** *New Statesman* 17 Dec. 884/2 It is this concern with rock history which distinguishes them from others who have called for a return to the basic virtues of good ole rock-'n'roll. **1976** *Gramophone* Dec. 952/2 The Amazing Rhythm Aces, a band from Tennessee..successfully combine country, rockabilly, swing and nostalgia into the rock idiom. **1976** *Listener* 18 Nov. 645/2 Most rock lyrics are straight melodramas. **1975** *Ibid.* 18 Sept. 370/2 The rock movement saw our present crisis coming and died of shock. **1971** *It* 4–18 Nov. 19/1 The financial success of cheap rock movies. **1967** *Listener* 23 Nov. 681/3 Is there an analogy between films and rock music? **1978** *Hi-Fi News* Sept. 7 Popular and rock music benefit from this performance. **1969** L. Roxon *Rock Encycl.* 420 By the end of 1968, in spite of all the talk about rock and the new music, the big rock musical had still to be done. **1977** F. Weldon in *Winter's Tales 23* 192 Brian offers..Hugo a part in a new rock musical going on in the West End. **1968** *Listener* 4 July 18/1 Rock musicians can now sing anything that can be said by traditional forms of creative expression, and more besides. **1969** L. Roxon *Rock Encycl.* 42 There is not a rock musician working today who has not consciously or unconsciously borrowed from his [*sc.* Chuck Berry's] sound. **1957** *Sat. Rev.* (U.S.) 5 Oct. 6 You feel it in a beat—in jazz—real cool jazz or a good gutty rock number. **1969** *Newsweek* 9 June 95 It was almost inevitable that the British group The Who should write the first rock opera. **1979** *Newsday* 31 Dec. 26 Francis Coppola's dazzlingly beautiful, nightmarish Vietnam combat adventure, staged like a psychedelic rock opera, is a provocative drama flawed by a murky ending. **1977** *Zigzag* Aug. 6/2 He's one of the only rock press geezers worth reading. **1977** *Chainsaw* Sept./Oct. 7/1 The national weekly rock press do have articles on new-wave groups. **1971** B. Malamud *Tenants* 45 They danced to some rock records Willie had brought along in a paper bag. **1960** *New Left Rev.* May–June 33/1 He met Mr. Parnes at a Liverpool rock show. **1959** *Punch* 10 June 788/2 Richard, like most rock singers, dances from the knees in a style borrowed from African warriors. **1973** J. Jones *Touch of Danger* xxvii. 164, I met this boy and dropped out with him. He wanted to be a rock singer. **1977** *Rolling Stone* 30 June 25/1 One good reason Elliman wants to remain with Clapton is that his band serves as a fine outlet for her rock singing. **1960** *Times* 26 Feb. 16/4 This song conforms to the pattern of the teenagers' acceptance to-day. It is a rock song with a rock gimmick. **1976** *New Yorker* 17 May 125/2 A rock star with a limp feather boa draped around her shoulders. **1978** G. Vidal *Kalki* vi. 153 Deafening was what H.V.W. would call the din from the rock stars' dressing rooms where electric guitars whined. **1959** C. MacInnes *Absolute Beginners* 56 The days when the Rock thing first broke. **1977** *Rolling Stone* 7 Apr. 26/2 His first solo album, *Solid*, a fine mixture of love ballads with jazz- and rock-tinged soul, has been selling short of hit status.

rock, *v.*[1] Add: **5. c.** *colloq.* (orig. *U.S.*). To cause to move with musical rhythm, esp. with the beat of *ROCK *sb.*[3] 2b. Occas. (esp. in early use) with sexual connotations (see also sense *7b).

1922 T. Smith in Godrich & Dixon *Blues & Gospel Records 1902–1942* (1969) 648 (*song-title*) My man rocks me (with one steady roll). **1938** C. Calloway *Hi de Ho* in R. S. Gold *Jazz Lexicon* (1964) 256 *Rock me*, send me, kill me, move me with rhythm. **1939** W. Hobson *Amer. Jazz Music* (1940) iii. 54 Albert Ammons's *Boogie Woogie Stomp*.., in jazz slang, might be said to 'rock the joint'. *Ibid.* iv. 87 Simple jazz-rhythmic phrases..may be blasted out by players with enough lip and lung strength in a way that will 'rock' the crowd. **1951** Davis & Hunter *Rock Little Baby* (song) 3 Some girls like men who are big and strong, You'll be my man, Just as long as you Rock little daddy, Work little daddy, Rock little daddy, Rock me all night long. **1956** B. Holiday *Lady sings Blues* (1973) xi. 103 We used to rock that joint. **1961** *Jazz Notes* Feb.–Mar. 39, I don't remember anyone who could 'rock' a Kenilworth audience before! **1972** *Even. Telegram* (St. John's, Newfoundland) 24 June 10/4 Joan Morrissey and a group known as The Commanders Showband were 'really rockin' 'er' at the Staff Club. **1977** *Rolling Stone* 7 Apr. 3/2 (Advt.), Boston, man. They really rock and roll. They rocked the place apart.

d. *fig.* To distress, perturb, upset; to surprise, startle, dumbfound. *to rock the boat*: see *BOAT *sb.* 1d. *colloq.*

1940 E. Pound *Cantos* lii. 11 Gold brokers made profit Rocked the exchange against gold. **1941** *Argus Weekend Mag.* (Melbourne) 15 Nov. 1/3 Another universal favourite [in Australia] is still the famous 'Wouldn't it ——!' Never given the final words (the completed sentence has several variants on 'Wouldn't it rock you!' or 'Wouldn't it rip you!'), the exclamation depends upon inflexion as to whether it conveys disgust, amazement, or pleasure. **1947** N. Marsh *Final Curtain* ix. 139 Has Troy seen about the Will?... It'll rock them considerably. **1951** *Sun* (Baltimore) 9 June B1/1 His diplomatic phrasing wrapped the punch in polite words, but Grady was nonetheless rocked. **1955** 'N. Shute' *Requiem for Wren* vii. 197 It turned out you were a Rhodes scholar, which rocked her a bit. **1960** *Sunday Express* 24 July 1/3 It is not only from the Opposition that Mr. Macmillan can expect criticism. His decision will rock the Tory Party too. **1966** [see *OFF-BROADWAY *sb.*]. **1981** *Observer* 22 Mar. 7/6 (*heading*) New sex scandals rock Washington.

6. d. (Further example.)

1937 I. Baird *John* ix. 103 Tiber rocked back on his heels.

e. *to rock along*, to continue in typical fashion. *U.S. colloq.*

1906 J. W. Carr in *Dial. Notes* III. 153 *Rock along*, to continue unsettled... 'So the matter rocked along and nothing was done.' **1946** *Sun* (Baltimore) 10 Oct. 2/1 The creation of a new board or administrator..would permit the program to rock along much as it is now. **1972** J. S. Hall *Sayings from Old Smoky* 115 *Rockin' along*, going along as usual. 'Everything is rockin' along just like when Lena was here.'

f. In Mountaineering: to work one's way *up* a chimney by a rocking movement.

1920 G. W. Young *Mountain Craft* 168 The body is kept upright in the middle on the spring of the bent knees and supported by the pressure of the hands, placed like the feet one against each wall. In this fashion we can 'rock' up satisfactorily.

7. a. Of popular music: to possess a rocking rhythm (see *ROCKING *ppl. a.* 1c); to exhibit the characteristics of rock music.

1938 *Metronome* July 21 *Harry James' Lullaby in Rhythm* really rocks. **1946** R. Blesh *Shining Trumpets* xiii. 309 The music..jumps rather than rocks. **1977** *Rolling Stone* 24 Mar., Waters has written six new tunes for the album, marking the end of a long dry spell, but his standards and one old Willie Dixon tune rock the hardest.

b. To perform, or dance vigorously and in an improvised way to, popular music with a strong beat (*ROCK *sb.*[3] 2a); hence, to play or dance to rock music (*ROCK *sb.*[3] 2b), to rock and roll. Occas. (esp. in early use) with sexual connotations (see sense *5c).

[**1934**: see *ROCK AND ROLL.] **1948** Moore & Reig (*song-title*) We're gonna rock. **1951** Davis & Hunter (*song-title*) Rock little baby. **1953** Freedman & De Knight (*song-title*) We're gonna rock around the clock. **1956** *Look* 26 June 45 Elvis Presley. The hottest thing rockin', sings throbbing lyrics that sound almost unintelligible. *c* **1956** 'L. Slim' *Rooster Blues* in P. Oliver *Screening Blues* (1968) vi. 193 We got to rock tonight baby, yes, we got to rock tonight. **1974** *Down Beat* 18 July 38/2 The band now isn't together enough to play all that... I mean they try to rock and they don't. **1977** *Time* 1 Aug. 16/2 In a Salisbury discothèque last week..'troopies' (soldiers) and their birds were rocking to a song about the country's bad news.

c. *to rock out*, to enjoy oneself enthusiastically, *esp.* by playing or dancing to rock music. Also as *attrib. phr. colloq.* (chiefly *U.S.*).

1968 *Surfer Mag.* Jan. 47/3 Maria likes rock-out dancing and surfing. **1972** B. Rodgers *Queens' Vernacular* 173 *Rock out*,..to enjoy oneself to the fullest. **1977** *Rolling Stone* 5 May 74/3 Even on the Stones' 'Happy', Lofgren changes the rock-out showpiece of hero Keith into a more subdued and funky shuffle. **1977** C. McFadden *Serial* (1978) lii. 110/2 Kate..went to find the Reverend Thurston on the dance floor, where he was rocking out with Marlene.

rock, *v.*[2] Add: **2.** (Earlier example.)

1836 *Public Ledger* (Philadelphia) 30 Aug. 1/4 Rock him! rock him! cried the boys, rock him round the corner... The wearer was 'rocked' till he turned his cloak inside out.

3. *W. Country dial.* To remove the calcareous deposit or 'fur' from the inside of (a kettle).

1880 Hardy *Trumpet-Major* II. i. 4 The broken clock-line was mended, the kettles rocked, the creeper nailed up, and a new handle put to the warming-pan. **1905** in *Eng. Dial. Dict.* V. 138/1 Kettle wants rocking.

ro·ckabi:lly. orig. *U.S.* Also **rock-a-billy.** [Blend of *ROCK (AND ROLL and *HILL)-BILLY.]

1. A type of popular music, originating in the southeastern U.S., combining elements of rock and roll and hill-billy music. Also *attrib.*

1956 *Billboard* 8 Dec. 22/3 Johnny Burnette is on hand to inject a touch of rockabilly in 'Lonesome Train'. **1957** *Variety* 23 Oct. 18/1 [The film] 'Rockabilly Baby' will hit a responsive chord among teenagers. **1959** *Times* 9 Nov. 9/6 Rockabilly and hula hoops came and went. **1962** *Globe & Mail* (Toronto) 19 Nov. 6/6, I suggest that as a public service the rock-a-billy radio stations join together to present an informative, unsponsored, prime time, two-hour program. **1971** R. A. Carter *Manhattan Primitive* (1972) xx. 190 A drummer, a guitarist, and a trumpet player pounded out rockabilly. **1976** *Time* 27 Sept. 90/3 Honky-tonk songs..came out of Texas in the late 1930s and early '40s. Elvis and Jerry Lee Lewis adapted the style to rock 'n' roll in the '50s. Sometimes called rockabilly, it celebrates booze, gambling, fighting, steppin' out, temptation and, like all country music, love. **1980** *Daily Mirror* 10 Apr. 12/2 Rockabilly is Eighties style. Special shops are opening to cater for the revived demand.

2. A person who performs this music.

1958 *Britannica Bk. of Year* 519/2 *Rockabilly*, a word ingeniously compounded from the expressions *Rock 'n' roll* and *Hillbilly*, and meaning a country singer. **1968** *Rolling Stone* 25 May 1/3 They called Elvis the 'rock-

abilly'. **1969** B. C. Malone *Country Music U.S.A.* 246 A hybrid specimen developed: an individual who possessed characteristics of both the rock-and-roll and country singer, the rockabilly.

ro·ckaboo:gie. Also **rock-a-boogie.** [Blend of *ROCK (AND ROLL and *BOOGIE(-WOOGIE.] A type of popular music, combining elements of rock and roll and boogie-woogie. Also *attrib.*

1956 B. Darnell et al. (*song-title*) Rock-a-boogie baby. *Ibid.*, He's got a little rattle He shakes around, I don't know where that rhythm he ever found, Oh he's a Rock-a-boogie Baby, Rock-a-boogie Baby of mine. **1956** Benson & Jones (*song-title*) Rock-a-boogie swing. *Ibid.*, You can see the kids a-dancin', Hear the music ring, It's a brand new beat, It's the rock-a-boogie swing... You got-ta rock. You got-ta roll. **1975** *Listener* 25 Dec. 889/2 That noted rockaboogie guide, the *Daily Express*. **1977** *Zigzag* Apr. 14/2 We were more refined, tasteful if you like, as opposed to the old good time rock-a-boogie.

rock-a-bye (rǫ·kăbəi), *v.* Also **rock-a-by.** [f. Rock *v.*[1] + *bye* (see Bye-bye[1]).] *imp. phr. rock-a-bye, baby*: a traditional phrase (esp. in a nursery rhyme) to induce an infant to fall asleep, used as an accompaniment to the rocking of a cradle. Also with joc. var. (see quot. 1954). Cf. Hushaby *int.*

1805 *Songs for Nursery* 36 Rock-a-bye, baby, thy cradle is green, Father's a nobleman, mother's a queen. **1812** *Mother's Gift* 5 Rocka-by baby bunting, My father's gone a hunting. *c* **1820** in I. & P. Opie *Oxf. Dict. Nursery Rhymes* (1951) 58 Rock a bye baby—puss is a lady..So hush a bye babe lie still. **1954** Dylan Thomas *Under Milk Wood* 81 Rockabye, grandpa, in the tree top..When the bough breaks the cradle will fall, Down will come grandpa, whiskers and all. **1975** *101 Favourite Nursery Rhymes* 58 Rock-a-bye, baby, on the tree top, When the wind blows the cradle will rock; When the bough breaks the cradle will fall, Down will come baby, cradle and all.

rock and roll. Also **rock-and-roll, rock 'n' roll.** [f. vbl. phr. *to rock and roll*: see Rock *v.*[1] and Roll *v.*[2] in Dict. and Suppl.] **1.** A type of popular dance-music characterized by a heavy beat and simple melodies, often with elements of the 'blues'. Cf. *rhythm and blues* s.v. *RHYTHM *sb.* 9a.

[**1934** S. Clarke (*song-title*) Rock and roll. **1951** Davis & Hunter *Rock Little Baby* (song) 1 Rock little daddy, Work little daddy, Rock little daddy, Send me with a rock and roll. **1954** *Billboard* 27 Nov. 29 (Advt.), Rockin' rollin' rhythmic! Ella Mae Morse with Big Dave and his orchestra.] *Ibid.* 25 Dec. 18/4 Alan Freed..will sponsor his first 'Rock and Roll Jubilee Ball' at the St. Nicolas Arena here on January 14 and 15. **1955** *N.Y. Times* 26 Mar. 17 According to William E. Kelsey Jr., a business man who has organized such parties, 'Rock 'n Roll' is less to blame for the situation than are the alcoholic beverages taken straight by boys and girls from 16 to 19 years old. **1956** *Observer* 30 Dec. 8/8 What else happened in 1956? Elvis Presley happened. So did Rock 'n' Roll. **1957** D. Hague in S. Traill *Concerning Jazz* 113 The only blot on the ledger of good productive jazz is that monstrosity at first referred to as 'Rhythm and Blues' and now more popularly called 'Rock and Roll'. **1959** *Times* 27 June 7/3 'Rock 'n' Roll' was so closely followed by Skiffle that the uninitiated were apt to confuse the two. **1962** *Listener* 1 Nov. 703/2 The bulletins last night kept breaking into the rock-'n'-roll. **1968** A. Lipson *Russian Course* 1 Hooliganism is defined [in the Soviet Union] as 'behavior exhibiting disrespect for the social order'. Includes: scoffing at authority, playing rock and roll, wearing loud clothes, as well as rowdyism and petty criminal acts. **1969** N. Cohn *AWopBopaLooBop* (1970) i. 15 In 1951, a DJ called Alan Freed launched a series of rhythm reviews at the Cleveland Arena... These shows featured coloured acts but were aimed at predominantly white audiences and, to avoid what he called 'the racial stigma of the old classification', Freed dropped the term R&B and invented the phrase Rock'n'Roll instead. **1974** A. Shaw *Rockin' '50s* xii. 106 In a liner note for an End album *Alan Freed's Top 15*, Freed claimed that he began using the term 'rock' and 'roll' in 1951. **1977** *Rolling Stone* 21 Apr. 74/1 You know her life was saved by rock and roll. **1977** *Zigzag* Aug. 5/1 Y'see I write about punk and you just want to read about something that'll scare your mum, I'm sure that elsewhere someone will let ya know it IS only rock'n'roll.

2. A dance to this music.

1958 [see *Charleston]. **1958** *Times Lit. Suppl.* 21 Nov. p. xxxii/5 Alex Moore leads beginners of any age through the mysteries of ballroom dancing, from the waltz..to the Mambo and 'Rock 'n' Roll (Jive)'. **1960** *Master Detective* July 83/2 Rock and Roll, that's what I'm good at. I got a terrific collection of platters.

3. *attrib.*, as *rock and roll ball*, *band*, *craze*, *dance*, *dancer*, *group*, *lyric*, *music*, *number*, *party*, *record*, *revivalism*, *revivalist*, *riot*, *road show*, *singer*, *song*, *star*, *station*, *stuff*, *tune*.

1954 *Billboard* 25 Dec. 18 (*heading*) Freed to sponsor 'Rock & Roll' ball. **1972** *Guardian* 28 June 16/1 It [*sc.* the Rolling Stones] is the raunchiest, flashiest, most exciting rock 'n' roll band in existence. **1980** *Oxford Times* 1 Feb. 23/3 On tracks like 'Rock Music' they seem to want to be a hard rock 'n' roll band. **1977** *Times* 18 Apr. (Gramophone Suppl.) p. iv/3 By the 1950s..popular music was to make another revolution for the record business... The rock'n'roll craze symbolized in the figure of Elvis Presley. **1955** *Life* 2 May 19, I am a teen-ager and I see no future in the rock 'n roll dance craze. **1966** A. Young in *Spero* I. ii. 21 New things he'd worked out on drums for a Rock & Roll dance. **1957** E. Paul *That Crazy Amer. Music* 243 The contemporary crop of rock and roll dancers include the young folks near the head or toward the foot of each class. **1968** *Crescendo* Feb. 16/3 This is a rock and roll group! **1955** *Life* 2 May 19, I think you need a teen-ager's view on rock 'n roll lyrics. **1956** *Look* 26 June 47 A record spins, and the boys and girls react to rock 'n' roll lyrics with laughter, not involvement. **1956** *Newsweek* 23 Apr. 32 Rock-and-roll music, he [*sc.* Asa Carter] said, 'is the basic, heavy-beat music of Negroes. It appeals to the base in man, brings out animalism and vulgarity.' **1958** *Publ. Amer. Dial. Soc.* xxx. 39 Now that that age group [*sc.* teenagers] is concerned with 'rock and roll' music, the jazz audience still remains composed..of young people of college age. **1955** *Life* 18 Apr. 168/2 *Cash Box*..challenged anybody to find smut in the top rock 'n roll numbers. **1969** *Listener* 3 Apr. 470/3 He can get them swinging with a rock-and-roll number. **1955** *N.Y. Times* 26 Mar. 17/3 A month ago Mr. Kelsey organized a 'Rock'n Roll' party for 2,000 at an armory here. **1957** *Gramophone Popular Record Catal.* III. 157 Rock and roll party. **1955** *Life* 18 Apr. 168 But parents and police were startled by other rock 'n roll records' words which were frequently suggestive and occasionally lewd. **1974** *Guardian* 27 Mar. 12/6 Bill Haley's return to London for yet another bout of rock and roll revivalism. **1972** *Jazz & Blues* Sept. 5/1 The Rock 'N' Roll revivalists. **1958** *Punch* 19 Nov. 665/1 If in the fullness of time it sparks off a rock-'n'-roll riot in Hamburg or Tokyo they receive the news with a grunt. **1977** *Sounds* 1 Jan 2/4 The Glitter Band will no longer back him up and the old rock 'n' roll road show has been wound up. **1956** *Look* 26 June 42 Lillian Briggs is a rock 'n' roll singer who also plays at the trombone. **1964** *Amer. Folk Music Occasional* i. 16, I think a lot of Rock and Roll singers and blue singers, they really have the feeling exactly as we about gospel. **1955** *Life* 18 Apr. 168 *Variety*..cranked out indignant stories about 'leerics' in the rock 'n roll or rhythm and blues songs. *Ibid.* (*caption*) Contingent of bounding dance fans at Easter show in Brooklyn's Paramount Theater greet roster of rock 'n roll stars performing latest songs on stage. **1972** J. L. Dillard *Black English* vi. 261 A title from one of the rock and roll star's songs. **1973** E. Bullins *Theme is Blackness* 46 Steve fumbles with the radio dial and finds a rock 'n' roll station. **1957** *New Yorker* 26 Oct. 35/2, I do enjoy a good opera production... I don't enjoy this rock-'n'-roll stuff. **1972** D. Haston *In High Places* iii. 45 Robin..was whistling a rock-'n-roll tune.

Hence as *v. intr.* (freq. considered as comprising two separate verbal units for the purpose of forming derivatives, etc.: see also *ROCKING *vbl. sb.*[1] 4), to play or dance to rock-and-roll music; **rock and roller; rock and rolling** *vbl. sb.* and *ppl. a.*

[**1951** Davis & Hunter *Rock Little Baby* (song) 1 Rock little baby, Rock and roll, Rock me in rhythm, Satisfy my soul.] **1956** *Time* (Overseas/Atlantic ed.) 18 June 37 Pop Record Maker Mitch Miller, no rock 'n' roller, sums up for the defense: 'You can't call any music immoral.' **1956** *Time* 24 Sept. 48 *My Boy Elvis*..is a real rock 'n' roller. **1956** *N.Y. Times Mag.* 4 Nov. 44/3 (*heading*) Europe rocks 'n' rolls. **1957** *Economist* 21 Sept. 946 (Advt.), Fibreglass..baffles the chatter of rock-'n-rollers—and keeps the place warm as neatly as it keeps the peace. **1957** *Observer* 15 Dec. 7/4 Gay shirts for the boys who skiffle and rock 'n' roll are equally hard to discover. **1958** H. Maxwell *Railway Mag. Miscellany* 6 We can..for an hour or so Rock and Roll to a more thunderous and more significant beat than that which passes for Music with us today. **1959** A. Wesker *Roots* i. 19 There's nothing wrong with rock 'n rolling, only God preserve me from the girl that can do nothing else! **1960** *News Chron.* 29 Mar. 3/2 Rock 'n' roller Cliff Richard. **1960** P. Hastings *Sandals for my Feet* i. vi. 61 Those rock and rolling Romeos. **1966** *Listener* 20 Oct. 568/1 There were already a dozen or so young people rocking and rolling wildly to some vintage discs of Elvis Presley. **1976** *Weekend Echo* (Liverpool) 4/5 Dec. 1/2 A concert by veteran rock-and-roller Bill Haley. **1977** [see *ROCK *v.*[1] 5c]. **1977** *Sounds* 9 July 18/1 Silly really that the American rock 'n' rolling public..should take Starz to its heart so readily.

Rockaway. Substitute for etym.: [f. the place name of *Rockaway*, New Jersey.] and add: (Earlier and later examples.) Also *attrib.*

1845 M. M. Noah *Gleanings from Gathered Harvest* 174, I keep a little Rockaway wagon. **1846** *Spirit of Times* 9 May 121/1 The price of a 'Rockaway' carriage which will carry eight persons depends very much on its finish. **1944** T. D. Clark *Pills, Petticoats & Plows* 292 Everywhere carriage makers turned out fancy..'cutunders', 'rockaways',..and 'heavy duties'. **1948** J. D. Rittenhouse *Amer. Horse-Drawn Vehicles* 19 Rockaway or depot wagon.

rockbridgeite (rǫ·kbridʒəit). *Min.* [f. the name of *Rockbridge* County, Virginia, where the first specimens were recognized + -ITE[1].] A basic phosphate of iron and manganese which is found as dark green or black masses and crusts (turning brown in air owing to oxidation) in limonite and pegmatite deposits.

1949 C. Frondel in *Amer. Mineralogist* XXXIV. 513 The specific name dufrenite is here re-defined to apply to a particular member of the dufrenite-complex, and the new name rockbridgeite is proposed for another common member of this complex. **1951** C. Palache et al. *Dana's Syst. Min.* (ed. 7) II. 868 Rockbridgeite occurs in the United States in a limonite deposit on South Mountain near Midvale, Rockbridge County, Virginia. **1970** *Amer. Mineralogist* LV. 166 A remarkable occurrence at Fodderstack Mountain, Montgomery Co., Arkansas, shows fibrous masses of rockbridgeite with mammillary surfaces and color banding.

rockburst (rǫ·kbɜ̄ɪst). Also **rock burst.** [f. Rock *sb.*[1] + Burst *sb.*] A sudden, violent rupture or collapse of highly stressed rock in a mine.

1928 *Daily Mail* 3 Aug. 18/1 The rockburst at a depth of nearly 5,000ft. in the City Deep mine that has..caused the deaths of two Europeans and eleven natives, is described as one of the biggest pressure bursts experienced on the Rand goldfield. **1942** *Mine & Quarry Engin.* VII. 233/1 One of the major problems of mining at depth lies in the occurrence of rockbursts. **1946** C. B. Jeppe *Gold Mining on Witwatersrand* I. x. 790 'Pressure bursts' or 'rock bursts'..major effects of excessive differential stresses. **1967** *New Scientist* 14 Dec. 678/1 The success of US and Soviet workers in predicting rockbursts underground. **1978** *Mining Equipment Internat.* June 17/1 Rockbursts currently account for about 80% of South African underground mine fatalities.

Rockefeller (rǫ·kəfelər). The name of John D. *Rockefeller* (1839–1937), Amer. financier and philanthropist, used as the type of an immensely rich man. Also *attrib.* Hence **Ro:ckefelle·rian** *a.*, designating that which only a rich man could afford.

1938 I. Goldberg *Wonder of Words* vii. 136 Anciently, men looked up to Crœsus..as a man of immense wealth... Thousands of years later men still say, 'He is a Rockefeller,' or 'He is a Rothschild.' **1939** 'F. O'Brien' *At Swim-Two-Birds* i. 63, I rejoined..that I..was no Rockefeller, thus utilizing a figure of speech to convey the poverty of my circumstances. **1941** B. Schulberg *What makes Sammy Run?* i. 14 I'm not exactly Rockefeller, but I'm always good for a little touch. **1975** P. Lorraine *Ask Rattlesnake* i. v. 88 'You don't remember anything?.. But, James, you could be—' 'A Rockefeller?' **1976** *National Observer* (U.S.) 24 Apr. 14/2 The daily rates are Rockefellerian,.. beginning at about $183 for two in high season from Dec. 20 through April 21. **1976** R. Condon *Whisper of Axe* i. xiii. 77 Nobody, not even the Rockefeller boys, can shake up the cash for a hunnert keyes [of heroin]. **1979** *N.Y. Times Mag.* 30 Sept. 14/4 That uncle, whom I still think of as a Mellon, a Rockefeller,..would collapse in awe of my annual grocery bill today.

rocker[1]. Add: **1. c.** A popular song that rocks (see *ROCK *v.*[1] 5c and 7a); a rock song.

1954 *Billboard* 6 Nov. 52/4 The deep-voiced chanter hands the rocker a good performance... The boys bow on the label with a so-so reading of a new rocker here. **1970** *New Yorker* 12 Dec. 187/1 'One More Weekend', a honky-tonk rocker..is about getting away. **1974** *Guardian* 22 Mar. 14/5 'Raised on Robbery' is a successful all-out rocker (with witty but still bleak lyrics). **1977** *Sounds* 1 Jan. 4/2 The songs they write alternate between brittle rockers, melodramatic movies and clever pop songs.

d. One who performs, dances to, or enjoys rock music (see *ROCK *sb.*[3] 2b); *spec.* a teenager or young adult of a type characterized by liking rock and roll, typically wearing long hair and a leather jacket, and riding a motorcycle (freq. contrasted with *MOD *sb.*[3]). Also *transf.*

1963 [see *MOD *sb.*[3] and *a.*]. **1963** *Economist* 28 Dec. 1332/2 Teenagers want..motor bikes and leather jackets to show that they are 'rockers'. **1964** *Spectator* 17 Apr. 503/1 Brighton cancelled its proposed 'beat festival' next month on hearing that the Mods and Rockers were coming in force. **1965** *New Statesman* 19 Nov. 801/1 A couple of literary Rockers rang up anonymously: 'We'll bomb the gallery if [Ezra] Pound turns up.' **1966** C. Mackenzie *Paper Lives* iv. 54, I cannot think that *Romeo and Juliet* is made more accessible to the imagination of young people by making the Capulets what I believe are called Mods and the Montagues what I believe are called Rockers. **1972** *Listener* 10 Aug. 187/2 Chuck Berry is the rocker's rocker and the real man. **1973** J. Wainwright *Pride of Pigs* 82 He was..a nineteen-year-old who had once identified himself as a greaser and, before that, as a Rocker, but who now led a provincial chapter of Hell's Angels. **1977** *Time* 3 Jan. 56/2 The debuting Barbra brings a hostile rocker audience to their feet with the wonder of her funkiness.

2. c. (Further examples.)

1923 Wodehouse *Inimitable Jeeves* viii. 78 The Duke is off his rocker. **1932** E. Waugh *Black Mischief* v. 195 It's going to be awkward for us if the Emperor goes off his rocker. **1943** 'C. Dickson' *She died a Lady* vii. 58 We're wondering if there was anybody who cared enough about Mrs. Wainright to go off his rocker and kill both of 'em when she fell for somebody else. **1953** 'M. Innes' *Christmas at Candleshoe* xxi. 221 'His behaviour is certainly very aberrant. Would it, one wonders, be occasioned by a sudden abnegation of the ratiocinative faculty?' 'Off his rocker—eh?' **1961** [see *FLIP *sb.*[3] 5]. **1976** T. Sharpe *Wilt* xiv. 144 'To put the record straight, what I said was that some of them were...' 'Off their rockers?' suggested the reporter.

4. b. For *U.S.* read 'orig. *U.S.*' and add earlier and later examples.

1852 in Mrs. Stowe *Key to Uncle Tom's Cabin* (1853) 136/1 Will be sold,..Haircat Chairs, Sofas and Rockers. **1855** S. Robinson *Jrnl.* 18 Nov. in *Kansas* (1856) viii. 98 [He was] seated in the nice large rocker drawn up before [the fire]. **1905** *Delineator* May 829/1 The Windsor rockers are not so common as the side chairs. **1911** *Daily Colonist* (Victoria, B.C.) 30 Apr. 4/5 (Advt.), Bedroom suites in white enamel..rockers, in white enamel to match. **1978** *Lancashire Life* Apr. 42/3 Ah sit theer i' mi rocker, Just startin' to nod off.

c. (Earlier examples.)

1830 *Boston Transcript* 15 Dec. 2/3 The surface mines, which are of very inferior importance, require no other labour than that necessary in washing the earth in *rockers*, or large inclined troughs with mercury. **1833** H.

BARNARD *Let.* 18 Apr. in *Maryland Hist. Mag.* (1918) XIII. 346 The top soil is removed—then the gravel is washed, by being thrown into what is called a rocker, or cradle, which is in fact a little more than a large cradle. **1852** [see *LONG TOM 2].

f. A tanning vat in which hides are rocked to and fro on a pivoted frame. Freq. *attrib.*

1885 C. T. DAVIS *Manuf. Leather* xviii. 353 Another form of handler in use is known as the rocker handler, and it consists of a frame constructed of wood, and hung by pivots in the centre of the top of the vat so as to give a dipping movement of 7 or 8 in. to each end of the frame. **1897** *Ibid* (ed. 2) xxiv. 382 The hides are next suspended in 'rockers'... They remain in the 'rockers' from seven to ten days. **1969** T. C. THORSTENSEN *Pract. Leather Technol.* v. 69 The hides are tied and hung in rocker racks and tanned by the vegetable tanning process. Prior to vegetable tanning, the hides in the lime condition are put in rockers containing deliming and bating materials.

g. A rock-shaft, rocker arm, or any similar rocking device forming part of a mechanism; *esp.* (*a*) a device for controlling the positions of brushes in a dynamo, and (*b*) a rocker arm in an internal-combustion engine.

1888 S. P. THOMPSON *Dynamo-Electric Machinery* (ed. 3) iii. 63 The rockers which support the brush-holders should admit of sufficient angular displacement being given to the brushes. **1893** R. GRIMSHAW *Locomotive Catechism* 178 Lengthen the rocker, so as to lower the entire motion. **1915** G. A. BURLS *Aero Engines* vi. 108 When the 'plus' part of the cam comes into contact with the roller the upper end of the rocker U is pushed upwards and the lower-end depressed, thus opening the valve. **1921** *Motor Electr. Manual* iii. 50 A magneto that has been standing out of use..is very liable to have a sticking rocker. **1928** *Evening News* 18 Aug. 1/3 Gallop then had to go into the pits with rocker trouble. **1935** T. E. LAWRENCE *Lett.* (1938) 855 The front rocker ran dry... So we pulled down the rocker assembly, and found more bits of Harry M's. string in the rocker-fulcrum pipe! **1961** *Carbon Brushes & Electr. Machines* xiv. 221 The correct location of the brush rocker is usually indicated by the machine constructor by a pair of marks, one on the rocker and one on the frame of the machine. **1970** K. BALL *Fiat 600, 600D Autobk.* i. 9/1 The design is conventional, incorporating..overhead valves operated by pushrods through rockers. **1975** F. PORGES *Design of Electr. Services for Buildings* i. 3 The switch opens when the bottom of the rocker is pressed and shuts when the top is pressed.

5. b. (Earlier example.)

1854 B. F. TAYLOR *Jan. & June* II. 155 The boys sha'n't skate? Who grudges them the 'rockers'?

c. *Ice-skating.* = *rocking turn* s.v. *ROCKING *vbl. sb.*^{1} 3.

1892, 1902 [see *COUNTER *sb.*^{4} 6]. **1936** *Sun* (Baltimore) 15 Feb. 13/2 There are 72 different school figures the skater must learn... Counters, rockers, brackets, loops, threes. **1973** *Times* 7 Feb. 15/8 He looked ill at ease on the first figure, the forward outside rocker.

d. 'One of the curved stripes under the three chevrons that indicate the grade of a sergeant (as in the U.S. Army and Marine Corps)' (Webster 1961). Also *transf.*, any similar badge bearing a slogan or device.

1944 *Yank* 6 Oct. 15/1 'Woddya want, fellah?' said the sergeant. He was thin for a top kick and his blouse was much too big. The bottom rocker reached down to his elbow. **1948** *Christian Sci. Monitor* (Mag. Sect.) 6 Nov. 8/1 Sergeant second class—three stripes, two 'rockers'—has been changed to sergeant first class. **1967** E. E. KERRIGAN *Amer. Badges & Insignia* i. 27 Next come first, second, and third grade sergeants, which were separated into both line and staff grades. The arcs, or 'rockers', indicated line grade, and straight lines below indicated grade. **1971** J. MANDELKAU *Buttons* vii. 81 The patch consisted of the top rocker—three inches wide, red on white—*Hell's Angels*. The bottom rocker said, *England* and between them I carried the small death's head. **1976** *New Yorker* 15 Mar. 102/2 Wetsel, a staff sergeant (E-6) with three stripes and one rocker, arrived in the company.

e. The upward curve on a surfboard.

1963 *Surfing Yearbk.* 43/1 *Rocker*, the slight upward slope in a surfboard. **1968** W. WARWICK *Surfriding in N.Z.* 3/2 The nose, was rounded with a slight uplift or rocker. **1970** *Studies in Eng.* (Univ. of Cape Town) I. 27 Less familiar words include *rocker*, or *banana*, which indicates the curvature of the surfboard along its length.

6. For *attrib.* in Dict. read '*attrib.* and *Comb.*' and add: (also in sense 2 b) *rocker bearing, box, gear, shaft* (earlier and further examples), *sieve, tailing*; *rocker-less* adj.; **rocker arm,** a rocking lever in an engine; *esp.* one in an internal-combustion engine which serves to work a valve and is operated by a push-rod from the camshaft; **rocker(-bottom) foot** *Med.*, a foot with the sole curved downwards; **rocker panel,** in a motor vehicle, a panel forming part of the bodywork below the level of the passenger door; **rocker switch,** an electrical switch having a mechanism incorporating a spring-loaded rocker.

1860 CLARK & COLBURN *Rec. Pract. Locomotive Engine* 67/1 The block..is carried upon the upper end of an arm, attached to, and vibrating upon the lower end of the rocker-arm. **1874** *Railroad Gaz.* 9 May 170/2 This block is attached to the lower rocker-arm by a pin, *c*, which works freely in the block. **1928** *Evening News* 18 Aug. 1/3 Baron d'Erlanger's Lagonda broke a valve rocker arm, and had to retire. **1970** K. BALL *Fiat 600, 600D Autobk.* i. 12/2 Take off cylinder head cover, rocker arm and shaft assembly. **1930** *Engineering* 30 May 696/3 The Aintree end [of the bridge] was treated..by means of four 100-ton jacks, and fixed rocker bearings weighing 5 tons each were placed in position. **1975** *New Yorker* 17 Feb. 26/2 He [*sc.* a robot] has rocker bearings for hands, shock absorbers for forearms,..hubcaps for shoulders. **1950** *Jrnl. Bone & Joint Surg.* XXXIIA. 344 Sonnenburg saw 688 cases of congenital club-foot, 42 cases of flat-foot, and 15 cases of congenital convex pes valgus or 'rocker-bottom' foot... In two of the cases club-foot was present originally; due to mistreatment, rocker-foot developed. **1956** *Clin. Orthopaedics* VIII. 94/2 If applied properly to give pressure under the cuboid, the packed felt will prevent the disastrous deformity of rocker foot. **1977** N. E. SHAW in *Bone & Joint Dis.* (Brit. Med. Assoc.) 114 (*caption*) Calcaneus is in equinus and is tucked up behind ankle joint. Rocker-bottom deformity has developed. **1892** J. G. A. MEYER *Mod. Locomotive Constr.* 199 The lifting-shaft bearing and rocker-box..are bolted to the front splice. **1965** *Motor* 17 July 3/2 Condensation in the rocker box of his B.M.C. 1100. **1902** F. J. A. MATTHEWS *Electr. Motor Installations* iii. 51 No rocker-gear is fitted to the machine, so that it is impossible..to alter the position of the brushes. **1950** NEWTON & STEEDS *Motor Vehicle* (ed. 4) vi. 118 (*caption*) Rocker gear of refined design for overhead camshaft. **1922** D. H. LAWRENCE *Aaron's Rod* i. 9 A baby was cooing in a rocker-less wicker cradle. **1921** C. W. TERRY *Pract. Motor Body Building* xxxviii. 255 A standard pattern taxi-cab with recessed rocker panels at the back. **1952** T. A. WOHLFEIL et al. *Automobile Body Reconditioning* vii. 88/1 Rocker panels are boxlike sections consisting of inner and outer panels welded to the edge of the floor pan. **1978** *N.Y. Times* 29 Mar. A25/4 (Advt.), Porsche '77... Sport wheel, fogs, alloys, rocker panel. **1842** R. & G. L. SCHUYLER in *Q. Papers Engin.* (1844) I. IV. 2 The valves..are of our own contrivance, and peculiar to this ship; they are worked by a separate eccentric and rocker shaft, which is set as to follow the motion of the steam valve. **1939** *Automobile Engineer* XXIX. 300/2 Rubber gland packing effectively seals the rocker shaft against leakage. **1950** *Engineering* 4 Aug. 104/1 A number of parts such as valves and valve gear..rocker shaft components, and bearings, are interchangeable with the six- and eight-cylinder engines. **1869** *Overland Monthly* III. 301/2 The united crash of pebbles on hundreds of quickly agitated rocker sieves, sounded in his ear like the roar of a cotton factory. **1884** *Bull. U.S. Nat. Museum* No. 27. 575 Cradle or Rocker Sieve, for washing the contents of the dredges. **1964** *Electr. Engineer's Ref. Bk.* (ed. 11) xxx. 79 (*heading*) Rocker switch. **1971** *Daily Tel.* 24 Nov. 11/4 The facia has been tidied up and rocker switches provided. **1975** F. PORGES *Design of Electr. Services for Buildings* i. 3 The advantages of the rocker switch are that it is easier to operate and that it is almost impossible to hold half open, even deliberately. **1906** C. DE L. CANFIELD *Diary of Forty-Niner* ii. 18 Worked out the claim and before I moved the Tom, tried some of the rocker tailings.

rockeried (rǫ·kərid), *a.* [f. ROCKERY + -ED^{2}.] Furnished with a rockery or rockeries.

1966 J. BETJEMAN *High & Low* 67 But strew the roads with tin signs 'Keep Left', 'M4', 'Keep Out!' Command, instruction, warning, Repetitive adorning The rockeried roundabout.

rocket, *sb.*^{3} Add: **1. b.** In proverbial phr. *to rise like a rocket and fall like a stick* (cf. STICK *sb.*^{1} 4 h) and varr., describing a sudden, meteoric rise and subsequent fall, as of fortune, etc.

1792 T. PAINE *Let. to Addressers* 4 As he [*sc.* Burke] rose like a rocket, he fell like the stick. **1838** R. H. BARHAM *Let.* 7 Mar. (1870) II. vii. 48 Poor man, he has gone up like a rocket and is coming down like the stick. **1909** *Brit. Weekly* 7 Jan. 405/3 We know the talk about a man going up like a rocket and coming down like a stick... It is generally the man's own fault. **1922** JOYCE *Ulysses* 364 My fireworks. Up like a rocket, down like a stick. **1950** G. B. SHAW *Farfetched Fables* 83 Political adventurers and 'tin Jesuses' rose like rockets to dictatorships and fell to earth like sticks.

c. Any elongated device or craft (as a flying bomb, a missile, a spacecraft) in which a rocket engine is the means of propulsion.

1919 R. H. GODDARD *Method of reaching Extreme Altitudes* (Smithsonian Misc. Coll. LXXI. No. 2) 1 The problem was to determine the minimum initial mass of an ideal rocket necessary, in order that on continuous loss of mass, a final mass of one pound would remain, at any desired altitude. **1920** *Photo Play* 7 Sept. 1/1 The theory of a Professor Goddard that a rocket could be sent to the moon. *Ibid.*, The propulsive power of the rocket..is derived from a series of multiple charges. **1929** *Amazing Stories* May 151 In the meantime, Dr. Mueller busied himself with making the rocket shipshape, for in spite of every precaution the supplies were in chaos. **1944** *Times* 11 Nov. 2/1 For the last few weeks the enemy has been using his new weapon, the long-range rocket, and a number have landed at widely scattered points in this country. **1949** W. LEY *Conquest of Space* (1950) i. 21 The rocket is a mile high and the spectators realize that there is still a deafening sound beating upon their ears. **1964** *Yearbk. Astron. 1965* 160 The rocket plummeted down near Guericke in the Mare Nubium, within a few miles of its intended position. **1970** N. ARMSTRONG et al. *First on Moon* iv. 76 At the time of Apollo 11 there was no doubt that the Saturn V was the most powerful operational rocket on earth. **1977** *Whitaker's Almanack 1978* 595/1 Mozambique troops fired rockets into the centre of Rhodesia's border city of Umtali but damage was stated to be minimal.

d. In full *rocket engine* or *motor*. An engine operating on the principle of the pyrotechnic rocket, providing thrust by the same method as a jet engine but without depending on the surrounding air for combustion (see also quot. 1971).

1919 R. H. GODDARD *Method of reaching Extreme Altitudes* (Smithsonian Misc. Coll. LXXI. No. 2) 6 By application of the above principles, it is possible to convert the rocket from a very inefficient heat engine into the most efficient heat engine that ever has been devised. **1929** *Sci. Wonder Q.* Fall 7 Prof. Hermann Oberth, a German of Mediarch, and Prof. Robert H. Goddard, an American of Worcester, Massachusetts..have solved it, though..only theoretically, by means of the rocket motor. **1931** *Jrnl. R. Aeronaut. Soc.* XXXV. 34 The fuel loading for rocket engines is a different matter from that of an engine of the explosion type. **1939** *Astounding Sci. Fiction* May 61/1 Each man in the crew tensed himself, gathering his abdominal muscles to resist the enormous acceleration developed by the launching catapult and the ship's own rockets acting in conjunction. **1960** C. H. GIBBS-SMITH *Aeroplane* xv. 122 The Messerschmitt Me 163 *Komet* fighter..had swept-back wings, a Walter liquid-fuel rocket motor, and a speed of 590 m.p.h. **1965** W. R. CORLISS *Space Probes & Planetary Exploration* x. 204 Because they will be used for delicate maneuvers, the on-board rockets have to be precisely controlled not only in thrust level but also in thrust duration and direction. **1971** P. J. MCMAHON *Aircraft Propulsion* x. 294 The convention of speaking of liquid fuel rocket *engines* but of solid fuel rocket *motors* is established in Britain. **1972** *Guinness Bk. Records* 128/2 The car was powered by a liquid natural gas/hydrogen peroxide rocket engine delivering 22,000 lb. s.t. maximum and thus theoretically capable of 900 m.p.h. **1977** *Engin. Materials & Design* Aug. 25/1 In rocket motors extremely high temperatures are developed (up to 3500°C). **1977** I. RIDPATH *Signs of Life* viii. 153 In its simplest form, the nuclear rocket uses as a propellant liquid hydrogen, which is heated to a gas by the reactor and expelled at high speed.

e. *off one's rocket*, mad. Cf. ROCKER^{1} 2c. *slang*.

1925 FRASER & GIBBONS *Soldier & Sailor Words* 244 *Rocket, off one's*, mad. **1959** I. & P. OPIE *Lore & Lang. Schoolch.* x. 179 He is cracked, he's cuckoo... He's off his rocket ('Off your rocket' is a development of 'off your rocker').

f. A severe reprimand. Freq. *to give* (or *get*) *a rocket. slang* (orig. *Mil.*).

1941 *New Statesman* 30 Aug. 218/3 [War-time slang.] *To stop a rocket*, receive a reprimand. **1942** E. WAUGH *Put out More Flags* ii. 153 The C.O. led Captain Brown away. 'He's getting a rocket,' said the anti-tank man. *a* **1944** K. DOUGLAS *Alamein to Zem Zem* (1946) xii. 77, I contended [*sic*] myself with giving him a rocket, and told them to hurry up and mend the tank. **1949** 'N. BLAKE' *Head of Traveller* III. xiv. 231 Your Superintendent gave me a rocket yesterday about 'harbouring her', as he put it. **1957** I. MURDOCH *Sandcastle* vii. 104 Demoyte had pondered the outrage..made a mental note to give Mor a rocket when he next saw him,..and felt immensely better. **1961** A. WILSON *Old Men at Zoo* i. 36 If Beard's to blame, then he should get the rocket. **1975** J. I. M. STEWART *Young Pattullo* vii. 155 Fish was sent to the Provost and given a rocket.

3. *rocket aeroplane, age, airplane, base, battalion, boat* (earlier and later examples), *engineer, flight, flyer* (later examples), *flying, frame* (earlier example), *fuel, installation, jet, pilot, projectile, propellant, propulsion, research, scientist*; *rocket-launching* vbl. sb. and ppl. adj., *-shooting* vbl. sb.; *rocket-assisted, -boosted, -borne, -carrying, -driven, -firing, -like* (later example), *-powered, -propelled, -tracking* adjs.; **rocket astronomy,** the branch of astronomy in which measurements are made by instruments carried by rockets above the atmosphere; **rocket-bomb,** (*a*) (see quot. 1895); (*b*) = *flying bomb* s.v. *FLYING *vbl. sb.* 3; (see also quot. 1973); **rocket car,** a car powered by a rocket engine; **rocket chamber,** the combustion chamber of a rocket engine; **rocket gun,** a gun firing rockets; **rocket launcher,** a device or structure for launching rockets; **rocket-man,** (*a*) a soldier responsible for firing rockets (? *obs.*); (*b*) *colloq.* an astronaut; **rocket net** *sb.*, a net having small rockets attached, which is laid on the ground and then propelled so as to envelope a group of feeding birds for ringing; hence **rocket-net** *v. trans.*, to trap in this way; **rocket netting** *vbl. sb.*; **rocket pad,** a launching pad for a rocket; **rocket plane,** (*a*) an aircraft powered by a rocket motor; (*b*) an aeroplane armed with rockets; **rocket projector** = *rocket launcher* above; **rocket range,** (*a*) a rocket-launching range (cf. *RANGE *sb.*^{1} 11c); (*b*) the area within range of a rocket; **rocket-rattling** *vbl. sb.* and *ppl. a. colloq.*, threatening with the military use of rockets or nuclear weapons (after *sabre-rattling* s.v. *SABRE *sb.* 4a); **rocket ship,** (*a*) a spaceship powered by rockets; (*b*) a warship armed with rockets; **rocket sled**: see *SLED *sb.*^{1} 2b; **rocket tube,** (*a*) a tube out of which a rocket is fired; † (*b*) a rocket motor.

1932 H. NICOLSON *Public Faces* I. 16 With this explosion chamber the problem of the rocket aeroplane was finally solved. **1959** *Globe & Mail* (Toronto) 15 July 1/3 Scout laws created in the horse and buggy days don't

always fit into today's rocket age. **1959** *Listener* 10 Dec. 1024/1 The strange and striking contrasts that must exist between the buildings of the Victorian and Edwardian periods and those that will be put up in the Rocket Age. **1928** *Sci. Amer.* Sept. 260/1 (heading) Can there be a Rocket Airplane? **1929** *Amazing Stories* May 148 The series of experiments were given their first impetus by the German rocket airplanes, successfully designed for the Berlin-to-New York air service. **1941** *Flight* 23 Jan. p. b/1 It may be expected that rocket-assisted take-off can be made more effective if not very efficient. **1950** *Sci. News* XV. 82 Rocket assisted take-offs by heavy bombers are quite common. **1959** *Economist* 17 Jan. 221/1 The RAT (rocket-assisted torpedo), a complicated but highly praised anti-submarine device. **1960** *McGraw-Hill Encycl. Sci. & Technol.* XI. 600/2 Rocket astronomy, first used in 1945 in the United States with German V-2 rockets, has been especially fruitful in studies of solar phenomena. **1971** *New Scientist* 18 Mar. 636/2 The emphasis is on the more modern approach which has grown up over the past 10 years as balloon and rocket astronomy have aided observations. **1954** M. CAIDIN *Worlds in Space* 177 In the opinion of many, the combination of the moon-launched rocket with an atomic bomb war head merited a thorough investigation of the value of the lunar rocket base. **1958** *New Statesman* 4 Jan. 1/1 The government seems determined to go ahead and establish American rocket-bases in Britain. **1976** *New Yorker* 15 Mar. 79/1 He explained that Intelligence had come to suspect that a North Vietnamese Army rocket-battalion command group had moved into the Song Quan Valley. *c* **1829** D. JERROLD in M. R. Booth *Eng. Plays of 19th Cent.* (1969) I. 168 I'd fight yard-arm to yard-arm for you.., or fight in a rocket boat. **1948** W. LEY *Rockets & Space Travel* 197 They..were massed on the decks of special 'rocketboats', rack after rack of self-propelled projectiles, fired electrically from below deck... One such rocket boat could, in the space of a few minutes, throw as much steel and high explosive as the turrets of three battleships. **1895** *Funk's Stand Dict.*, *Rocket-bomb*, a harpoon-rocket. **1943** MRS. BELLOC LOWNDES *Let.* 20 Dec. (1971) 247 A good many people believe the rocket-bomb is coming, but a famous airman laughed at the idea of its being a real danger to London. **1949** 'G. ORWELL' *Nineteen Eighty-Four* I. viii. 85 'Steamer' was a nickname which..the proles applied to rocket bombs. **1973** J. QUICK *Dict. Weapons & Mil. Terms* 375/3 *Rocket bomb*, an aerial bomb equipped with a rocket to give it added velocity and penetrating power after being dropped from an aircraft. **1958** *Technology* Mar. 25/1 Scientists and the services have hurried into print with space plans.., among them rocket-boosted.. engines to fire a payload to the moon. **1962** W. B. THOMPSON *Introd. Plasma Physics* i. 4 Recently, rocket- and satellite-borne counters have detected belts of energetic radiation..high above the earth's atmosphere. **1971** *Sci. Amer.* July 74/2 X-ray observations with rocket-borne instruments have shown that the remnant of Tycho's nova is also a strong source of X rays. **1930** *Times Educ. Suppl.* 25 Oct. p. iv/1 The rocket-car experiments of the late Herr Max Valier. **1976** *Star* (Sheffield) 30 Nov. 12/9 Gabelich strapped himself into a rocket car named Blue Flame and covered two measured miles..at an average speed of 630 mph. **1961** *Guardian* 25 Oct. 11/2 Rocket-carrying submarines. **1936** *Smithsonian Misc. Coll.* XCV. No. 3. 2 In these experiments it was shown that a rocket chamber and nozzle, since termed a 'rocket motor', could use liquid oxygen together with a liquid fuel, and could exert a lifting force without danger of explosion. **1939** *Astounding Sci. Fiction* May 59/1 Injecting excess charges of fuel into the rocket chambers. **1928** Rocket-driven [see *rocket-propelled* below]. **1978** R. V. JONES *Most Secret War* viii. 68 The German Navy was said to have developed remote-controlled rocket-driven gliders of about three metres span and three metres long. **1951** *Jrnl. R. Aeronaut. Soc.* LV. 92/1 Rocket propellants must have certain undesirable features, and it is the task of the rocket engineer to minimise the consequences of these. **1970** H. TREVELYAN *Middle East in Revolution* 149 The Iraqi Air Force.. attacked the rebel headquarters with rocket-firing aircraft. **1978** R. V. JONES *Most Secret War* xliii. 403 The attack took place on 16th March with rocket-firing Typhoons of No. 198 Squadron. **1930** H. CHATLEY *Princ. Rocket Propulsion* 3 This is the basis of the dreams of rocket flight to the moon. **1934** H. G. WELLS *Exper. Autobiogr.* I. vi. 328 They did not so much climb to success; they were rather caught by success and blown sky high... Only one item in this rocket flight is significant here. **1959** *Times Lit. Suppl.* 20 Mar. 167/2 The book is, in the main, a really excellent elementary account of rocket flight and space travel. **1927** *Amazing Stories* Nov. 725 Many schemes have been proposed for space flying, and some of the more recent ones, notably the Goddard Rocket Flyer, seem to come closest toward a strictly scientific solution of the problem. **1939** *Ibid.* Sept. 112 She had attached herself to him, demanding that he teach her how to pilot a rocket flier. **1929** *Sci. Wonder Q.* Fall 7 While writing the story, the author has had the collaboration of practically all the German scientists who have of late come into prominence in their researches into not only rocket flying, but space flying and astro-physics. **1931** *Wonder Stories* Jan. 900 We have succeeded in securing near Berlin a suitable rocket flying field, a large field on which the starting supports for the different rockets were set up. **1835** J. E. ALEXANDER *Sketches in Portugal* v. 126 Saldanha's artillery consisted of four 5½-inch howitzers, six 9-pounders, six of 6, two of 3, and three rocket frames. **1931** *Amazing Stories* Dec. 804 A battleship has destroyed Albertville, Ontario, site of the Canadian rocketfuel factory and magazine. **1937** *Discovery* Sept. 277/2 Equipping this type of aeroplane with rocket engines, complete with rocket fuel. **1977** *Time* 18 July 35/3 It is known as phencyclidine hydrochloride, but youngsters on this latest and fastest-spreading high know it as 'angel dust', 'rocket fuel' and 'goon'. **1884** *Bull U.S. Nat. Museum* No. 27. 281 The rocket-gun..throws a large rocket and explosive lance weighing eighteen or twenty pounds, which acts in the capacity of a harpoon and bomb, and is used mainly in coast whaling. **1935** *Jrnl. R. Aeronaut. Soc.* XXXIX. 410 The main characteristic of..rocket guns..is the increase of their efficiency when the ratio of weight of the rocket..to the weight of the charge decreases. **1944** *Jane's All World's Aircraft* p. iii/2 The rocket-guns with which some..fighters were equipped.. enabled them to attack. **1959** E. H. CLEMENTS *High Tension* ii. 33 The Hebridean rocket-installations. **1944** *Aviation* Jan. 149/3 The mass of a rocket jet can be readily varied by feeding more or less fuel..into the rocket motor. **1944** Rocket launcher [see *LAUNCHER 2]. **1977** *N.Y. Rev. Bks.* 23 June 3/2 He learned to slaughter people with rifles and knives and explosives or to blast them to pieces with rocket launchers. **1956** A. H. COMPTON *Atomic Quest* 223 The great installations along the coast..turned out to be rocket launching platforms. **1968** *Times* 16 Dec. 7/3 The systematic recording of disturbances like these could be used to provide remote observers with information about rocket launchings. **1973** D. KYLE *Raft of Swords* (1974) iii. 19 Our force of rocket-launching submarines came into service. **1952** S. SPENDER *Learning Laughter* 38 There was a screen of cypress trees with the column of one palm tree in their midst, bursting rocket-like at the apex. **1821** G. R. GLEIG *Narr. Campaigns Brit. Army* XIX. 270 Attached to this corps of infantry, were a party of rocket-men, and two light three-pounder guns. **1938** *Sci. Amer.* May 270 (*heading*) Number one rocket man. **1964** *Galaxy Mag.* Oct. 181/1, I was a Rocketman 3/c on the Moon, guarding the Aristarchus base against invaders from outer space. **1972** *Melody Maker* 20 May 16/2 Elton John remains..a writer (with Bernie Taupin) of songs of lasting merit... The success of his beautiful 'Rocket Man' single proves that he has survived all the flack. **1948** *Severn Wildfowl Trust Ann. Rep.* 43 Although this report is primarily concerned with the activities of the Trust during the year 1947, it seems that such an important development as the first attempt with the Trust's new rocket nets for ringing the wild geese should be included although it took place early in 1948. **1952** *Blackw. Mag.* Feb. 106/1 When they want to tell t'other from which, they rocket-net them and paint their sterns. **1954** *Brit. Birds* XLVII. 316 By rocket nets, it has been possible to make an intensive study of the population of the Pinkfoot. **1973** *Wildfowl* XXIV. 164/2 A lot of effort went into attempting to rocket net Barnacle Geese. **1979** *Ibid.* XXX. 165/2 A single catch of 372 Barnacle Geese at Caerlaverock in October (one of the largest catches made with rocket nets) provided much valuable data. **1953** *Severn Wildfowl Trust 5th Ann Rep.* 22 The rocket-netting technique has undergone considerable modification during the four years since the first experiments were made. **1969** *Wildfowl* XX. 86/1 The Wildfowl Trust's rocket-netting programme had concentrated on the Pink-footed Goose since this proved the easiest to trap in the large numbers needed. **1965** *Time* 23 July 36 For those pictures, JPL's boss..and his crew had sweated out Mariner's launch from a Cape Kennedy rocket pad. **1977** *Jersey Even. Post* 26 July 14/3 It was vandalized by the German rocket-pad crews. **1949** R. A. HEINLEIN *Red Planet* i. 6, I still think I'd like to be a rocket pilot. **1958** C. C. ADAMS *Space Flight* p. vii, There have been space books for children—our present space cadets and future rocket pilots. **1928** *Pop. Mechanics* Nov. 718/2 Valier has calculated that a rocket plane could be shot from Berlin to New York in ninety-three minutes. **1929** *Mech. Engin.* Nov. 865/1 The rocket plane with its possibility of moving at speeds..would seem to be the answer to the problem of quick transportation. **1932** A. HUXLEY *Brave New World* iv. 70 The deeper drone of the rocket-planes hastening, invisible, through the bright sky five or six miles overhead. **1945** *Daily Tel.* 7 Aug. 1/6 R.A.F. shattered panzer counter-attack in Normandy. Rocket planes knocked out 35 tanks. **1949** *Ann. Reg. 1948* IV. 416 Earlier in the year (in May) it was reported from America that the rocket-plane Bell XS-1 had been flown faster than sound. **1965** W. R. CORLISS *Space Probes & Planetary Exploration* x. 209 Rocket planes and helicopters are possible prime movers for unmanned landers, but surface locomotion is much more likely. **1936** *Pop. Science* May 16/2 An experimental rocket-powered glider..carried the cargo to its intended destination. **1948** *Electronics* June 93/1 Rocket-powered engines of one particular type employ two kinds of fuel. **1959** *Daily Tel.* 23 Feb. 11/6 This year two test pilots are expected to make the first flights in the rocket-powered North American X-15. **1943** *Fortune* June 92/2 A strange gun called the bazooka that fires a rocket projectile. **1944** T. N. DALTON *Jet Propulsion* 44 The Encyclopaedia Britannica says that rocket projectiles were used by the Chinese. **1945** L. E. O. CHARLTON *R.A.F. & U.S.A.A.F. July 1943—Sept. 1944* 292 (*caption*) Thunderbolt showing rocket projectors fitted to one of its wings. **1961** B. FERGUSSON *Watery Maze* ix. 235 The mass of rocket projectors pointing into the air from an LCT 2. **1932** *Bull. Amer. Interplanetary Soc.* Feb. 8 How best can we utilize each of these as a rocket propellent? **1944** C. P. LENT *Rocket Res.* 67/1 After using the rocket propellants the flying weight is only 1780 Kg and during the period of ascent the total weight averages less than 2000 Kg. **1928** *Sci. Amer.* Sept. 260/1 The recent German experiments with rocket propelled cars and gliders have attracted much attention, and it is now asked whether rocket-driven airplanes are not possible, navigating at fantastic speeds. **1951** *Mind* LX. 119 We have the idea now of a rocket-propelled missile capable of flying from Moscow to New York. **1978** R. V. JONES *Most Secret War* xxxix. 371, I already knew of two liquids used by the Germans in the rocket-propelled glider bombs they had been using against our ships. **1928** *Explosives Engineer* VI. 457 (*heading*) Motoring by rocket propulsion. **1929** *Sci. Wonder Stories* Aug. 265 Aeronautical authorities have stated recently that the future development of the airplane will be along the lines of rocket-propulsion. **1942** *Aeronautics* Aug. 49/2 The greatest reason against rocket propulsion of aeroplanes is the question of oxygen, its weight and storage in an aeroplane. **1963** *Listener* 23 May 866/1 Even before Verne's death the idea of using rocket propulsion for space-travel had been put forward by..Konstantin Eduardovich Tsiolkovskii. **1948** *Hansard Commons* 15 Mar. 1805 We have joint research stations; for instance, the one about which there has been considerable publicity, the rocket range in Australia. **1971** 'D. HALLIDAY' *Dolly & Doctor Bird* ii. 24 The tracking of moonshots and other missiles from the American rocket range is done by the electronic brains in these stations. **1976** *New Yorker* 15 Mar. 79/1 A..command group had moved into the Song Quan Valley, ten miles to the west and almost within rocket range of the division headquarters. **1960** *News Chron.* 21 July 4/5 The..retaliation threats, the rocket-rattling over Cuba. **1961** *Sunday Express* 29 Jan. 1/4 President Kennedy has put a sharp curb on rocket-rattling, anti-Russian speeches. **1969** *Guardian* 31 Mar. 10/1 Rocket-rattling by any large Power over a weaker neighbour is deplorable. **1937** *Discovery* Sept. 269/1 Fundamental problems of rocket research. **1977** *Whitaker's Almanack 1978* 165/1 The progress of rocket research during the last war led to the development by the Germans in 1944 of the V.2 rocket. **1952** *Sun* (Baltimore) 5 Sept. 2/6 Take it from the rocket scientists who expect to fly to Mars some day—flying saucers are not space ships from another planet. **1959** *Listener* 5 Mar. 410/1 Rocket-scientists are not unaware of this. **1927** *Literary Digest* 25 June 20/1 He [*sc.* Max Valier] is even now building a rocket-ship. **1928** *Amazing Stories* Aug. 427 Not to mention the rocketships that might be in the air. **1936** *Forum & Century* July 38/2 But the question of whether rocket ships will ever reach the planets can be even approximately answered only when intensive research has been carried on over many years. **1951** W. LEY *Rockets, Missiles & Space Travel* p. viii (*caption*) One of the Navy's 'rocket ships' bombarding Peleliu Island on the same morning that was D-Day in Europe. **1969** *New Yorker* 12 Apr. 53/1 Anybody on earth with a pair of binoculars can see that setting a rocket ship down there would be a tricky operation. **1981** *Daily Tel.* 15 Apr. 1 The American space shuttle landed on a dry lake bed in California's Mojave Desert yesterday to complete the maiden flight of the first re-usable rocketship. **1925** R. GRAVES *Welchman's Hose* 35 And watched the nightly rocket-shooting, varied With red and green, and livened with gun-fire. **1971** 'D. HALLIDAY' *Dolly & Doctor Bird* ii. 24 There are four main rocket-tracking stations in the Bahamas. **1881** W. D. HAY *300 Years Hence* iv. 70 The largest [projectiles] requiring apparatus like the old rocket-tubes and howitzers, and dealing certain death to every living thing within a mile of the place of explosion. **1898** D. BEATTY *Diary* 8 Apr. in W. S. Chalmers *Life & Lett. David, Earl Beatty* (1951) ii. 33, I with the Rocket tube first occupied a position on the left of the Artillery. **1932** *Flight* XXIV. 1023/1 The rocket tube or rocket motor, as it is called in Germany..is filled with powder of special composition.

rocket, *v.* Add: **1. a.** (Later examples.)

1967 [see *NAPALM *v.*]. **1973** *Houston* (Texas) *Chron.* 21 Oct. 1/1 The 22-year-old officer, whose tank had been rocketed, said the Egyptian soldiers..threw what they apparently thought was his body out of the ditch. **1978** *Guardian Weekly* 12 Feb. 6/1 Six vessels..are rocketing besieging guerrilla forces around the port of Massawa.

b. To propel (someone) at speed, as by a rocket; to send by rocket. Also *fig.*

1837 J. COTTLE *Killcrop* in *Early Recoll.* II. 316 From yon tall rock I'll hurl him to perdition... I'll rocket him. **1958** *Listener* 16 Oct. 603/2 Probably he [*sc.* an astronaut] will come down in a large sphere..because the retardation he will experience in this way will expose him to no worse strains than those he suffers in any case as he is rocketed upwards. **1959** *Times* 11 June 3/6 A boundary rocketed his score to a dozen. **1966** I. ASIMOV *Fantastic Voyage* i. 10 We would pile him into an X-52 and rocket him through inner space.

c. To reprimand severely. Cf. *ROCKET *sb.*^{3} 1f. *slang* (orig. *Mil.*).

1948 PARTRIDGE *Dict. Forces' Slang* 156 He rocketed me like hell. **1971** J. WAINWRIGHT *Dig Grave* 96 The assistant chief constable was still rocketing Sergeant Sykes.

2. c. In general use: to move like a rocket, to speed; (of prices, etc.) to increase substantially, to soar. Also const. with advbs.

1881 *Baily's Mag.* Oct. 37 He played too forward and a little too quick to a very quick, straight ball, and she 'rocketted' between the wickets. **1924** W. J. LOCKE *Coming of Amos* xxiv. 312 A flash of lightning rocketed across the black gap of the open window. **1924** H. G. WELLS *Exper. Autobiogr.* I. v. 219 The more brilliant investigators rocket off into mathematical pyrotechnics and return to common speech with statements that are.. nonsensical. **1937** J. BETJEMAN *Coll. Poems* (1958) 44 The heart of Thomas Hardy flew out of Stinsford churchyard... It rocketed over the elm trees. **1947** *Evening News* 5 Nov. 1/5 A hectic day's trading..sent the shares rocketing on Monday from 13s. 3d. to 23s. 9d. **1952** DYLAN THOMAS *Coll. Poems* 132 Up through the lubber crust of Wales I rocketed to astonish The flashing needle rock of squatters. **1957** *Economist* 2 Nov. 375 Manufacturer's exports rocket 23 times in 7 years! **1972** D. HASTON *In High Places* ii. 29 The rope rocketed out. This was really high-quality ice climbing in action. **1976** *Times* 17 Mar. 2/8 Mr Benn rocketed to prominence as a potential future party leader..in the early 1970s. **1979** *Jrnl. R. Soc. Arts* CXXVII. 431/2 It goes along relatively steadily till 1972, then rockets up and comes back down and has just rocketed up again.

rocketeer. Delete *rare* $^{-1}$ and add: **1.** (Further examples.)

1945 F. H. M. LLOYD *Hurricane* i. 12 Fighter, night-fighter, fighter-bomber, tank-buster, 'rocketeer', ship-fighter, merchant-ship protector...in thirty different forms and on thirty-seven different fronts, the Hurricane went into action. **1974** D. SEAMAN *Bomb that could Lip-Read* ix. 70 The flash..won't affect the man firing the rocket-launcher... I want every rocketeer to fire his two rounds.

2. One who experiments or works with rockets; a rocket expert or enthusiast.

1929 *Review of Reviews* Sept. 91 (*heading*) The new race of rocketeers. **1935** *Jrnl. Brit. Interplanetary Soc.* Oct. 13 *Rocketeer*, one who experiments with rockets. **1953** J. N. LEONARD *Flight into Space* 25 One philosophical rocketeer pointed to the fact that man's body stands midway in size between the atoms it contains and the great galaxies that float beyond the stars. **1957** P. MOORE *Sci.*

& Fiction 18 Lucian's seamen are the logical ancestors of the rocketeers and space-cadets of to-day. **1960** *New Statesman* 2 Jan. 7/3 While the rocketeers burst into space, the advances that touched the man in the street were mostly of the kind of wide-screen movies and stereophonic records. **1971** *Nature* 23 Apr. 494/2 The group of astronomers at the University of Leicester, who have been among the most successful X-ray rocketeers, are still producing valuable data. **1972** *Sci. Amer.* Dec. 118/3 Model rocketeers fly light refiyable rockets they make themselves or from kits.

rocketee·ring, *vbl. sb.* [f. prec. + -ING^{1}.] = *ROCKETRY.

1932 D. LASSER *Conquest of Space* vii. 116 The support of a financier of world-wide experience, and of men of science..comes as a revelation to those who viewed the field of 'rocketeering' as a visionary dream. **1938** *Forum* Feb. 96/2 Rocketeering may, in some distant future, take us to the moon or elsewhere in space. **1962** *Times Lit. Suppl.* 2 Feb. 68/5 The race is therefore on, for a greater prize..than any that can be won by rocketeering.

rocketer. Add: † **b.** *Cricket.* = SKYER. *Obs.*

1886 *Cricket* 25 Feb. 18/1 A good man..is the man to go for a 'pocketer' [*sic*] between the wickets. *Ibid.* 25 Mar. 35/2 In my last letter of course 'Pocketer' ought to read 'Rocketer'. **1887** F. GALE *Game of Cricket* iv. 70 He hit [the ball] too quick, and instead of going out of the field, as it might have done, it went up a tremendous rocketer. **1900** *Badminton Mag.* Oct. 380 The great smiter..hit a ball very high straight to the young blacksmith... But for some reason—whether he was thinking too much of the style of the thing, or was unused to rocketers—..the ball fell with a hollow clank on the top of his head.

ro·cketing, *ppl. a.* (Further examples.)

1952 DYLAN THOMAS *Coll. Poems* 172 He..prays, Who knows the rocketing wind will blow The bones out of the hills. **1959** *Economist* 25 Apr. 318/2 Behind the increase in imports there is cited a 'rocketing' increase in Dutch production. **1964** M. ARGYLE *Psychol. & Social Probl.* xvi. 199 We are astounded however by the rocketing crime rate, [etc.]. **1970** *Daily Tel.* 8 May 17 Doctors, headmasters and health educationists are disturbed by the rocketing numbers of young smokers.

ro·cketing, *vbl. sb.* [f. ROCKET *v.* + -ING^{1}.] The action or practice of the vb. in various senses.

1928 P. F. NOWLAN in *Amazing Stories* Aug. 427 The favorite American method of propulsion was known as 'rocketing'. **1972** *Guardian* 4 May 15/8 Rocketing of Saigon, plus ground probes near the capital, could shake the politicians..out of their current isolation. **1975** *Church Times* 14 Mar. 1/5 Rocketing and shelling of the airfield was going on as the plane left.

rocketry (rǫ·kėtri). [f. ROCKET *sb.*^{3} + -RY.] The science or use of rockets and rocket propulsion. Also *fig.*

1930 G. E. PENDRAY in *Bull. Amer. Interplanetary Soc.* Nov.–Dec. 4 The practical work of getting a liquid-fuel rocket actually into the air was a contribution of America's, as were the three most fundamental achievements of modern rocketry. **1934** *Jrnl. Brit. Interplanetary Soc.* I. I. 3 There you have the nucleus of the British movement in rocketry. **1934** *Astronautics* Mar. 7 'Rocketry' itself is a coined word, first suggested at a meeting of the American Interplanetary Society in 1930 and since widely adopted. **1943** C. S. LEWIS *Perelandra* vi. 91 He was a man obsessed with the idea which is.. circulating all over our planet in obscure works of 'scientifiction', in little Interplanetary Societies and Rocketry Clubs. **1951** 'J. WYNDHAM' *Day of Triffids* i. 30 Sustained research in rocketry had at last succeeded in attaining one of its objectives. It had sent up a missile which stayed up. **1957** *Times* 10 Oct. 10/1 The American programme..called for the launching of small test spheres this year..to check rocketry instrumentation. **1958** *Sunday Times* 14 Sept. 7/4 His [*sc.* Thomas Wolfe's] absurd rocketry about great America, decadent Europe and so on. **1962** F. I. ORDWAY et al. *Basic Astronautics* i. 21 A scientific analysis of rocketry applied to high altitude meteorological research. **1968** A. DIMENT *Bang Bang Birds* v. 70 A collection of scientists..who weren't quite bright enough to get into the rocketry racket. **1977** *Daily Tel.* 28 July 1/6 The small spaceport at Kagoshima ..looked more like a station for amateur rocketry than a serious rival to Cape Canaveral.

rocketsonde (rǫ·kėtsǫnd). Also **rocket-sonde, rocket sonde.** [f. ROCKET *sb.*^{3} + *-sonde*, after *RADIOSONDE.] A package of meteorological or other scientific instruments which is carried aloft by a rocket, released in the upper atmosphere, and floats down by parachute, transmitting measurements automatically by radio.

1949 E. DURAND in G. P. Kuiper *Atmospheres Earth & Planets* iv. 134 (*heading*) Rocket sonde research at the Naval Research Laboratory. **1951** *Jrnl. Brit. Interplanetary Soc.* X. 18 Direct measurements have been made by means of rocket-sondes to altitudes of over 100 miles. **1963** *Proc. 1st Internat. Symp. Rocket & Satellite Meteorol.* 3 The rocketsonde, at the present time, measures the temperature only; the wind is determined by radar tracking of the metallized parachute. **1969** MCINTOSH & THOM *Essent. Meteorol.* vii. 112 Above the ceiling of balloons, information is provided by rocket sonde or by satellite. **1976** *Nature* 13 May 124/1 The phase of the annual oscillation in the zonal winds at 82 km is consistent with the phase of this oscillation in the lower mesosphere determined from rocketsonde data.

rock-fish. Add: **1.** Also = *rock salmon* (*c*) s.v. *ROCK *sb.*^{1} 9d. (Earlier and later examples.)

1598 FLORIO *Worlde of Wordes* 279/1 Piota,..a kind of rock fish. **1969** [see *rock-salmon* s.v. *ROCK *sb.*^{1} 9 d].

3. = *KLIPFISH 1.

1731, 1806 [see *KLIPFISH 1].

rockfoil (rǫ·kfoil). [f. ROCK *sb.*^{1} + FOIL *sb.*^{1}] = SAXIFRAGE.

1879 RUSKIN *Proserpina* I. viii. 160 Their names..can be pleasantly said..in this order..Roof-foil, Rock-foil, Primrose. **1887** G. NICHOLSON *Illustr. Dict. Gardening* III. 311/1 Rockfoil. A name, suggested by Ruskin, for the genus *Saxifraga*. **1914** IRVING & MALBY *Saxifrages* i. 1 There is no other genus of rock plants that is so extensive as that of the Rockfoils. **1963** R. D. MEIKLE *Garden Flowers* 159 (*heading*) Saxifrage, Rockfoil.

rockie, var. *ROCKY *sb.*

rocking, *vbl. sb.*^{1} Add: **2. b.** The action of using a rocker (*ROCKER^{1} 4c) in gold-mining.

1850 J. W. AUDUBON *Western Jrnl.* (1906) 202 The men began 'rocking' yesterday, one cradle, and get about a dollar an hour. **1859** *Brit. Colonist* (Victoria, B.C.) 3 Apr. 2/1 The lowest sum named by any miner as the product of a day's rocking is three to five dollars. **1896** C. H. SHINN *Story of Mine* 42 We started to rocking with my water.

3. rocking-turn, a movement or figure in skating (see quot.). Cf. *ROCKER^{1} 5c.

1869 VANDERVELL & WITHAM *Syst. Figure-Skating* 219 After having exhausted the Q, I began to consider the feasibility of making the change direct from the inside forwards to inside backwards and *vice versâ*..by the employment of a kind of turn, for which..I can find no more simple..name than the 'Rocking Turn'.

4. The action or practice of playing or dancing to popular music with a strong beat and rocking rhythm, *esp.* rock and roll.

1948 R. BROWN (*song-title*) Good rockin' tonight. **1956** *Newsweek* 18 June 42/3 (*heading*) Rocking and rolling. **1974** *Down Beat* 18 July 38/2 The lyricism is as fervent as ever, but the rocking isn't... I don't mean..that the band ought to be rocking.

rocking, *ppl. a.* Add: **1. b.** *Prosody.* Designating a metre in which each foot consists of a stressed syllable standing between two unstressed syllables.

c **1883** G. M. HOPKINS *Poems* (1918) Pref. 1 If the stress is between two slacks there will be Rocking Feet and Rhythms. **1932** F. R. LEAVIS *New Bearings in Engl. Poetry* v. 167 Rocking Feet and Outriders will help no one to read his [*sc.* Hopkins's] verse. **1957** B. DEUTSCH *Poetry Handbk.* (1958) 130 When the stress falls between two unstressed syllables, as in this line of Swinburne's 'Far oút to the shállows and stráits of the fúture, by róugh ways or pléasant', the rhythm is a rocking one. **1965** A. F. SCOTT *Current Lit. Terms* 249 When the metrical stress falls between two unstressed syllables, the rhythm is called rocking.

c. Of popular music: characterized by a strong beat and rocking rhythm (cf. *ROCK *sb.*^{3} 2); that is performed in the style of rock music.

1949 *Billboard* 3 Dec. 108/2 Combo drives thru a rocking riffer, with a punching bary sax leading the way. **1954** *Ibid.* 13 Nov. 98 Another spirited rhythmic side in which the lead singer soars out wild and free over the rocking beat provided by the rest of the group. **1959** G. AVAKIAN in M. T. Williams *Art of Jazz* (1960) ix. 88 One of the fastest rocking blues ever made. **1968** *Melody Maker* 30 Nov. 6/6 A rocking version of B. B. King's 'Sweet Sixteen'. **1976** *Leicester Trader* 24 Nov. 4/7 One rocking track called Winnebago even reminds me of..the opening track from Argent's second album.

rocking-chair. Add: **1.** (Earlier Amer. example.)

1766 in *Hobbies* (1949) Sept. 50/2 1st Mo. 1766, to a rocking Chair for andrew hunter 3/.

2. Used *attrib.* to designate something considered as being conducted or obtained at home, without first-hand experience of normal difficulties; spec. *rocking-chair money*, unemployment benefit. Cf. *ARM-CHAIR, ARMCHAIR. Chiefly *U.S.*

1933 *Sun* (Baltimore) 27 Feb. 6/3 A question which delights every sewing circle and rocking-chair parade in the country. **1944** *Amer. Speech* XIX. 156/2 *Rockin' chair money*, unemployment compensation. **1946** *Richmond* (Va.) *Times-Dispatch* 16 Jan. 5/2 An increase in 'rocking chair money' for the State's unemployed. **1959** *Globe & Mail* (Toronto) 22 Apr. 7/1 The somewhat less sophisticated people of the Maritimes have a happier name for Unemployment Insurance... 'Rockin' Chair Money'. **1962** *Daily Tel.* 10 Sept. 10/2 A third U2 incident over Communist territory..will renew groans at the President's rocking-chair diplomacy.

Rockingham (rǫ·kiŋăm). The title of Charles Watson-Wentworth, second Marquis of *Rockingham* (1730–82), applied *attrib.* to earthenware, china, a variety of glaze, etc., produced on his estate at the Old Works, Swinton, Yorks., from *c*1745 to 1842. Also applied loosely to similar products. Now usu. designating pieces of a tea-service. Also *ellipt.*

[**1832** G. RICHARDSON in *Cabinet Cycl.* No. 26. 22 At the Rockingham works, which have been so named in compliment to their early patron, the celebrated marquess of Rockingham, porcelain is now produced which vies successfully in every kind of excellence with that of older English establishments.] **1857** J. MARRYAT *Hist. Pott. & Porc.* (ed. 2) xii. 291 They [*sc.* the works] also manufactured the brown or chocolate-coloured ware, which obtained the name of 'Rockingham ware'. **1863** W. CHAFFERS *Marks Pott. & Porc.* 134 A sort of brown or chocolate-coloured ware, made in the beginning of the present Century, obtained the name of Rockingham ware. **1869** C. SCHREIBER *Jrnl.* 4 Nov. (1911) I. 60, 3 Rockingham cups and saucers and 3 plates. **1870** *Ibid.* 11 Feb. (1911) I. 66 A few good bits of Chelsea,..one of Rockingham (very good). **1881** C. C. HARRISON *Woman's Handiwork* II. 110 The temperature at which Rockingham ware is fired is suitable for this [underglaze] work. **1900** F. LITCHFIELD *Pott. & Porc.* vii. 94 The ware is highly glazed, some of it being not unlike the brown Rockingham ware. **1935** G. GREENE *England made Me* ii. 57 The Minister stood above his Rockingham china. **1957** MANKOWITZ & HAGGAR *Conc. Encycl. Eng. Pott. & Porc.* 194/1 '*Rockingham' glaze*, a lustrous, purple-brown lead glaze (stained with manganese) made at the Rockingham factory, Swinton, early nineteenth century, and in Staffordshire. **1960** R. COLLIER *House called Memory* xii. 171 Croquet on the lawn at four o'clock with the Rockingham tea-service as background. **1963** *Times* 23 Feb. 11/7 There were many bone china potters in Staffordshire making cottages: their productions, without any justification, are classed arbitrarily as Rockingham and thus acquire the market value of that establishment's brilliant reputation. The flowers on Rockingham cottages are usually less disproportionately large than those ornamenting cottages made elsewhere. **1965** [see *DAVENPORT^{4}]. **1973** L. COOPER *Tea on Sunday* 11 The dark blue and gold Rockingham cup.

rocking-horse. Add: (Earlier and later examples.) Also *fig.*

1724 in *N. & Q.* (1942) 7 Feb. 76/1 Rocking Horse. William Bird, turner, just without Newgate. **1922** JOYCE *Ulysses* 495 Result of the rockinghorse races. **1936** F. R. LEAVIS *Revaluation* iv. 112 Prior takes happily to those anapaestic, rocking-horse rhythms. **1964** D. VARADAY *Gara-Yaka* vii. 64 Then she bounded back in rocking-horse cheetah gait.

Rockite. Add: Now only *Hist.* (Further example.) Also **Ro·ckism.**

1830 *Times* 6 Nov. 3/6 There is little doubt that distress and desperation form the mystery, and that a kind of English *Rockism* is in operation. **1848** MILL *Pol. Econ.* I. II. ix. 375 Rockism and Whiteboyism are the determination of a people, who have nothing that can be called theirs but a daily meal.., not to submit to being deprived of that for other people's convenience. **1949** C. GRAVES *Ireland Revisited* viii. 101 The Rockites, hidden among the cliffs and intending to trap the Bantry party, revealed their presence.

rockless, *a.* (Later example.)

1977 *Lancashire Life* Nov. 82/1 What, on its almost rockless coast, has the County Palatine got?

rock-like, *a.* Substitute for def.: Resembling a rock; possessing the qualities of rock; hard or firm as rock. (Further examples in *fig.* use.)

1959 P. BULL *I know Face* ii. 29 This line had to be delivered in a steady rock-like voice. **1963** *Times* 5 June 16/3 His endearing character with its rocklike qualities won him innumerable friends from many generations. **1975** R. BARCLAY *Ernest Bevin & Foreign Office* iv. 79 Just as he himself had always been rock-like in his loyalty to Churchill, so from now on he remained absolutely firm in his support of Attlee.

rock 'n' roll: see *ROCK AND ROLL.

rockoon^{2} (rǫkū·n). [Blend of ROCK(ET *sb.*^{3} and BALL)OON *sb.*^{1}] A rocket fired from a balloon; a balloon carrying a rocket to be fired in the upper atmosphere.

1953 *Sci. News Let.* 8 Aug. 89/2 The balloon-rocket technique, commonly referred to as Balloon Assisted Take-Off or Rockoon, was developed by Dr. James A. Van Allen, head of the State University of Iowa physics department. **1955** *Sun* (Baltimore) (B ed.) 23 Nov. 14/6 The release of rockoons will be synchronized with similar releases in the Northern Hemisphere. **1959** *New Scientist* 20 Aug. 244/2 Analysis of magnetic measurements by balloon-borne rockets (rockoons)..has revealed a second 'electrojet' current..around the Earth. **1963** A. MACLEAN *Ice Station Zebra* iv. 59 Drift Stations habitually carried rockets..and radio-sondes and rockoons. The sondes were radio-carrying balloons..: the rockoons, radio rockets fired from balloons.

rock-ribbed, *a.* Add: **2.** *fig.* Resolute, uncompromising, staunch; esp. of political allegiance. orig. *U.S.*

1887 *Courier-Jrnl.* (Louisville, Kentucky) 3 May 4/4 Mr. Straus is a rock-ribbed Democrat. **1911** H. S. HARRISON *Queed* 292 Various feelings had gradually stiffened an early general approval into a rock-ribbed resolve. **1925** T. DREISER *Amer. Trag.* (1926) I. I. xvi. 122 Clyde always struck her as one who was not any too.. rock-ribbed morally or mentally. **1950** *Manch. Guardian* 20 Feb. 6/6 The dyed-in-the-wool Democrat can be fanatical in devotion to his party's creed and traditions. So can the rock-ribbed Republican. **1961** *Economist* 28 Oct. 341/2 He is a man of such rock-ribbed integrity. **1969** *Daily Tel.* 11 Oct. 12 A Massachusetts seat that has always been held by rockribbed Republicans. **1976** *Pub-*

lishers Weekly 16 Apr. 88/1 Goldwater, rockribbed in his sincerity, speaks for many Americans currently disenchanted with Washington's government-by-bureaucracy.

rock-rose. Add: **4.** *N. Amer.* The bitter root, *Lewisia rediviva*, a small perennial herb belonging to the family Portulacaceæ, native to western North America, and bearing solitary pink or white flowers.

1906 *Contrib. U.S. Nat. Herbarium* XI. 49 The rock-rose or bitterroot..occurs abundantly in crevices of 'scab', making a brave show with its beautiful rose-colored flowers. **1963** *Beaver* Autumn 53/1 The rolling hills, cactus and rock roses..flooded the dry land with character and colour. **1973** R. D. SYMONS *Where Wagon Led* I. v. 77 We should find some rock roses here.

5. An aggregate of tabular crystals of a mineral suggestive of the petals of a rose; = *ROSE *sb.* 16e, *ROSETTE *sb.* 5e.

1933 *Amer. Mineralogist* XVIII. 261 The barite occurs as sand barites or barite rosettes (locally called 'rock roses') and barite accretions. **1962** W. A. DEER et al. *Rock-Forming Minerals* V. 193 Concretions of barytes in sandstone sometimes take the form of rosettes known as 'sand barites', 'rock roses' or 'desert roses'. **1977** A. HALLAM *Planet Earth* 85 (*caption*) Desert roses (or rock roses) display one of the more unusual modes of occurrence of evaporitic minerals... Found..only in arid areas, these clusters of platy crystals are typically of barite or gypsum.

rock-staff^{2}. Substitute for entry:

rock-staff^{2}. *E. Anglian dial.* [ROCK *sb.*^{2}] A distaff. Also *fig.*, a superstition; a fancy, crotchet; esp. in phr. *an old woman's rock-staff*.

1765 *Compl. Maltster & Brewer* p. xxiii, The notion of pease bloom, and weeds being up in the water, is but a meer old woman's rockstaff. *a* **1825** R. FORBY *Vocab. E. Anglia* (1830) II. 279 *Rock-staff*,..a distaff; from which.. the wool *was* spun 'by twirling a ball below'... 'An old woman's *rockstaff*,' is a contemptuous expression for a silly superstitious fancy. **1867** *N. & Q.* 3rd Ser. XI. 215 She is so full of her old woman's rock-staffs. **1895** P. H. EMERSON *Birds, Beasts, & Fishes Norfolk Broad-land* II. xix. 396 There is a curious rockstaff in the marshlands that a viper's slough will draw thorns from your flesh.

rocksteady (rǫ·kste:di). Also **rock-steady, rock steady.** [f. *ROCK *sb.*^{3} 2 + STEADY *a.* 4.] A style of popular music, originating in Jamaica, characterized by slow tempo and stressed off-beat. Also, a dance to such music. Also *attrib.* Cf. *REGGAE.

1969 *Observer* 12 Jan. 3/8 West Indian teenagers..nowadays..danced to music called 'rocksteady'. *Ibid.* 23 Nov. 25/8 Aspiring Kingston..dancing the Reggae, Jamaica's successor to the Ska and the Rocksteady. **1971** A. SALKEY in *One Love* 7 We have been quick to recognise the excellence and the appeal in the musical alternative of the..*Rock Steady* and *Reggae*. **1971** *Daily Tel.* (Colour Suppl.) 30 July 7/4 As ska, rocksteady or blue-beat, music like this has been around in Britain for more than a decade, hidden amidst the West Indian subculture in London, Birmingham and elsewhere. **1973** *Telegraph-Jrnl.* (St. John, New Brunswick) 28 July 5/2 A West Indian rock-steady band was playing. **1977** MCKNIGHT & TOBLER *Bob Marley* iii. 41 Ska mutated into 'rock steady'... Rock steady was slower in tempo than ska—again to assist the singers in their unenviable task of shouting louder than the volcanic eruptions produced by a bass player with..amplification... Rock steady is further distinguished from ska by the extra syncopation involved. *Ibid.* vi. 72 It was 1968 and Johnny Nash, a black American singer..had achieved several hits with rock steady type songs. **1980** *Melody Maker* 19 Jan. 25/4 Saxa played with some of the early rock-steady acts that toured Britain, like Desmond Dekker and Laurel Aitken.

rock-water. (Later example.)

1948 F. BLAKE *Johnny Christmas* I. 40 Yet in this utter lack of sound, except for the sputter of flames and the musical seep of rock-water, Johnny came alert, strangely, as if awakened by a thin cry of warning.

Rockwell (rǫ·kwel). The name of Stanley P. *Rockwell*, 20th-cent. U.S. metallurgist, used *attrib.* with reference to a hardness test which he introduced, in which the depth of penetration of the material (usu. a metal) by a steel ball or a diamond cone is measured under specified conditions; hence also used to denote values of relative hardness determined by such methods.

1920 *Foundry Trade Jrnl.* XXII. 778/2 A new hardness testing apparatus, called the Rockwell hardness tester, is now available. **1922** S. P. ROCKWELL in *Trans. Amer Soc. Steel Treating* II. 1013 The Rockwell hardness tester is at present made in three sizes. **1922** *Chem. Abstr.* XVI. 3296 Formulas..are given for conversion from the Rockwell value to the Brinell value. **1930** *Engineering* 19 Sept. 358/1 Rockwell hardness measurements and X-ray diffraction patterns had shown that lattice distortion could be accompanied by appreciable softening. **1945** A. T. BIRKBY *Phenolic Plastics* v. 56 The V.P.N. and Rockwell 'C' test equivalents of this are 700–750 and 58–60 respectively. **1976** *Islander* (Victoria, B.C.) 16 May 15/3 Get a blade with a Rockwell hardness factor of 57 to 59... Steel less than Rockwell 57–59 is too soft and won't hold a cutting edge.

rocky, *a.*^{1} Add: **1. b.** (Earlier example.) *Rocky Mountain Indian* (earlier example); **Rocky Mountain bee plant,** an annual herb, *Cleome serrulata*, belonging to the family Capparidaceæ and bearing racemes of pink flowers; **Rocky Mountain canary,** a burro or jack-ass; **Rocky Mountain feathers,** wood shavings; **Rocky Mountain (spotted) fever,** a sometimes fatal rickettsial disease transmitted by ticks; **Rocky Mountain goat,** the North American mountain goat, *Oreamnos americanus*; = MAZAME 2, *mountain goat* s.v. MOUNTAIN 9c in Dict. and Suppl.; **Rocky Mountain grasshopper** = *Rocky Mountain locust*; **Rocky Mountain iris,** a blue-flowered iris, *Iris missouriensis*, found in south-western North America; **Rocky Mountain juniper,** a small conifer, *Juniperus scopulorum*, found in the south-western United States; **Rocky Mountain locust,** a migratory North American grasshopper, *Melanoplus spretus*; **Rocky Mountain oyster,** lamb's fry; **Rocky Mountain sheep,** the bighorn sheep, *Ovis canadensis*; = *big-horn* s.v. BIG *a.* B. 2 in Dict. and Suppl.; **Rocky Mountain spotted fever** = *Rocky Mountain fever*; **Rocky Mountain spotted (fever) tick,** a brown or grey tick, *Dermacentor andersoni*, found in parts of western North America, where it is the vector of Rocky Mountain fever; **Rocky Mountain wood tick** = prec.

1802 in *Med. Repository 1803* 238 In the fall of 1800 I was on an excursion, on horseback, through the plains that are situated between the Sascatchievan and Missisourie Rivers, along the rocky mountains. *attrib.* **1900** B. B. SMYTH *Plants & Flowers of Kansas* i. 14 Such temperatures..are generally supposed to be destructive to plant life; but the following plants live through them and continue to thrive: Pincushion cactus, prickly pear,..Rocky Mountain bee plant. **1939** *Nat. Geogr. Mag.* Aug. 227/2 Bees are attracted in such great numbers to the nectar secreted abundantly by these dainty blossoms..that the species is often called 'Rocky Mountain bee plant'. **1963** J. J. CRAIGHEAD et al. *Field Guide Rocky Mt. Wildflowers* 69 Rocky Mountain Bee-plant..is a much-branched annual 2–5 ft. tall. **1905** *Outing* Apr. 47/2 His varied outfit he packs on the back of the 'Rocky Mountain canary'. **1929** *Amer. Speech* V. 147 The burro or jackass of the early days is still called a *Rocky Mountain canary*, because of its tuneful voice. **1962** *Maclean's Mag.* 18 July 44/2, I started a fire with a handful of 'Rocky Mountain feathers'—dry shavings—made that morning before we left our previous camp. **1886** *Buck's Handbk. Med. Sci.* III. 85/2 The 'Rocky Mountain Fever', so called by practitioners on the slope of that great mountain chain, exhibits frequent divergences from the true clinical features of typhoid fever, and may show a continued remittent type, but the pathology observed in not a few of these cases links them to typhoid fever. **1917** *Indian Med. Gaz.* LII. 16/1 Most observers place the incubation of the Rocky Mountain fever at three to seven days. **1939** *Brit. Encycl. Med. Pract.* XII. 340 The other types [of tick-borne typhus] can be most easily understood by considering the respects in which they differ from Rocky Mountain fever. **1842** J. E. DEKAY *Zool. N.Y.* I. 112 Rocky Mountain Goat..larger than the common goat. **1884–5** [see GOAT 1 b]. **1901** *Daily Colonist* (Victoria, B.C.) 11 Oct. 1/6 The Rocky Mountain goat captured..last spring..is to be sent to the London Zoo this week. **1949** *Canad. Alpine Jrnl.* May 55 We had seen elk, moose, Rocky Mountain goats, and bighorn. **1909** WEBSTER 944/1 The allied migratory Rocky Mountain grasshopper..sometimes travels in vast hordes in the region west of the Mississippi. **1966** DAVIDSON & PEAIRS *Insect Pests* (ed. 6) viii. 129 The Rocky Mountain grasshopper is considered the most important migratory species in the United States and Canada. **1806** P. WAKEFIELD *Excursions N. Amer.* xliii. 380 We saw some straggling parties of Rocky Mountain Indians. **1880** T. MEEHAN *Native Flowers & Ferns U.S.* 2nd Ser. I. 103 As it is the only species of *Iris* found there the common name of 'Rocky Mountain Iris' has suggested itself to us. **1963** J. J. CRAIGHEAD *Field Guide Rocky Mt. Wildflowers* 34 Rocky Mountain Iris..usually bears from 1 to 4 variegated violet-blue flowers. **1898** G. B. SUDWORTH *Check List Forest Trees U.S.* 35 *Juniperus scopulorum* Sargent. Rocky Mountain Juniper. **1949** COLLINGWOOD & BRUSH *Knowing your Trees* 135 The twigs of the Rocky Mountain juniper..are four-sided, with leaves arranged alternately in pairs. **1972** L. HANCOCK *There's a Seal in my Sleeping Bag* viii. 182 Old dried whitened Rocky Mountain juniper trees sprout the artistic bulky stick nests of the double-crested cormorants. **1878** *Rep. Comm. Agric. 1877* (U.S. Dept. Agric.) 264 The Rocky Mountain Locust, or Grasshopper of the West. **1930** S. HENRY *Conquering our Great Amer. Plains* 319 Coming home late one afternoon for supper I stepped back surprised to see what became known as Rocky Mountain locusts covering the side of the house. **1972** V. A. LITTLE *Gen. & Applied Entomol.* (ed. 3) vii. 94 Although the Rocky Mountain locust is found throughout most of the United States, it is considered a pest of importance only in the Great Plains region. **1889** J. WHITEHEAD *Steward's Handbk.* IV. 420/2 *Rocky Mountain oysters*, Lambs' fries. **1940** C. L. BROWN et al. *Amer. Cooks* 71 (*heading*) Fried lamb's fries, or Rocky Mountain oysters. **1804, 1818** Rocky Mountain sheep [in Dict.]. **1904** [see *ARGAL]. **1936** D. MCCOWAN *Animals Canad. Rockies* v. 45 Rocky Mountain sheep..are almost entirely guided by what we call instinct. **1977** D. ANTHONY *Stud Game* xxviii. 188 They hunted Rocky Mountain sheep with bow-and-arrow. You have to be good to do that. [**1903** Rocky Mountain spotted fever: see *spotted fever* s.v. *SPOTTED *a.* and *ppl. a.* 3 a.] **1905** *U.S. Hygienic Lab. Bull.* XX. 8 Rocky Mountain 'spotted fever' is reported for Idaho, Montana, Nevada, Oregon, Wyoming, ?Washington State, and possibly Utah and Alaska. **1906** *Jrnl. Amer. Med. Assoc.* 7 July 33/1, I arrived in Missoula, Montana, April 21, 1906, equipped for the bacteriologic and hematologic study of the so-called Rocky Mountain spotted fever. **1947** *Ann. Rev. Microbiol.* I. 346 The fatality rate in Rocky Mountain spotted fever is greatly influenced by age. **1974** *Greenville* (S. Carolina) *News* 23 Apr. 3/2 The little ticks that carry Rocky Mountain spotted fever didn't have to find a warm spot under a log this winter. **1937** *Jrnl. Econ. Entomol.* XXX. 52 The first, known popularly as the Rocky Mountain spotted fever tick, or Rocky Mountain wood tick, is our most versatile species as a disease vector. **1976** *Islander* (Victoria, B.C.) 7 June 14/2 The most serious type of infection, called tick fever, is transmitted by the Rocky Mountain spotted tick. **1937** Rocky Mountain wood tick [see *Rocky Mountain spotted fever tick*]. **1951** METCALF & FLINT *Destructive & Useful Insects* (ed. 3) xxiii. 983 The Rocky Mountain wood tick is the most important tick in the United States. **1976** *National Observer* (U.S.) 21 Aug. 8/4 The Rocky Mountain wood tick carries the illness in the West.

c. (Earlier examples.)

1827 J. SMITH *Let.* 12 July in *Dict. Americanisms* (1951) II. 1409/2, I allude to the country of the Great Salt Lake, West of the Rockies. *a* **1861** T. WINTHROP *John Brent* (1862) vi. 60 At the foot of those bare, bulky mounds of mountain by which the Wasatch range tones off into the great plains between it and the Rockys, we overtook the Salt Lake mail.

Hence **ro·ckily** *adv.*

1972 D. HASTON *In High Places* ix. 100 Almost an ideal mountain panorama. Fitzroy, Poincenot and satellites rockily bounding the right, rounding off towards Pere Giorgio at the head of the valley. **1981** *Times Lit. Suppl.* 6 Feb. 147/5 The difficulties of absorbing women into a group so rockily traditional as the police.

rocky, *a.*^{2} Add: **a.** (Earlier and later examples.) Also, in recent use, tipsy, drunken.

1737 *Pennsylvania Gaz.* 6–13 Jan. 2 He's Rocky, Raddled,..Lost his Rudder. **1912** A. BENNETT *Matador of Five Towns* 157 'What's up with that wheel?' 'It's rocky, that's what that wheel is.' **1938** 'N. SHUTE' *Ruined City* x. 204 The whole thing was a pretty rocky deal. **1941** *Direction* IV. v. 15/2 Stack had been..gettin leapin drunk... One morning in April, 1906, after he had had a rocky night and had a headache built for a hippopotamus he was out..to cool the burnin thirst in his throat. **1947** *Partisan Rev.* XIV. 493, I was drinking scotch on an empty stomach, and beginning to feel a trifle rocky myself. **1951** A. BARON *Rosie Hogarth* I. vi. 62 A chap always felt a bit rocky after he'd shown what he was made of. **1977** D. BEATY *Excellency* vii. 90 The régime's rocky. The future of the country's uncertain.

b. For *slang* read *colloq.* and add to def.: Now usu. in sense 'difficult, hard'. (Earlier and later examples.)

Some cases may be equally well interpreted as examples of a *fig.* use of ROCKY *a.*^{1} 1 a.

1873 J. MILLER *Life amongst Modocs* 71 We may have a rocky time down there, my boy. **1960** B. KEATON *My Wonderful World of Slapstick* 13, I am by no means overlooking the rough and rocky years I've lived through. **1976** E. DUNPHY *Only a Game?* iii. 90, I had played well in the first half at Swindon when things were rocky.

c. *colloq.* In poor health; ill, unwell.

The sense in quot. 1792 is unclear; it may be ROCKY *a.*^{1} 2 b.

1792 F. BURNEY *Let.* 28 Jan. (1972) I. 106 A former Patient is often alarmed..but she is very *Rocky*..& she will be glad when the alarm passes over. **1926** E. HEMINGWAY *Sun also Rises* I. vii. 56 What's the matter, darling? Do you feel rocky? **1929** M. DE LA ROCHE *Whiteoaks* iii. 56 Is Wake feeling rocky to-night? **1932** G. GREENE *Stamboul Train* II. 62, I guess you're a bit rocky. You haven't escaped from anywhere, have you? **1954** G. DURRELL *Bafut Beagles* viii. 151 'Made you feel a bit rocky?' inquired the doctor cheerfully, feeling my pulse.

ro·cky, *sb.* *Naval slang.* Also **rockie.** [f. ROCKY *a.*^{2}] (See quots.) Also *attrib.* or as *adj.*

1919 W. LANG *Sea Lawyer's Log* 29, I have seen an officer who presides directly over our naval destinies fix the eye of a basilisk upon a luckless 'rockie' who incautiously spoke of a parade. **1927** 'GIRALDUS' *Musings of Merry Matloe* (ed. 2) 186 *Rocky*, a Royal Naval or Fleet Reserve man. Also a R.N. Reserve officer who once was more commonly known as a 'Cargo Shifter'. **1929** F. C. BOWEN *Sea Slang* 112 *Rockies*, R.N.V.R. ratings training in H.M. ships in peace time and very highly valued as worked ratings. Before the War it was also applied to R.N.R. ratings, seldom since. **1948** PARTRIDGE *Dict. Forces' Slang* 156 *Rockies*, officers of the Naval Reserves. **1957** KERR & GRANVILLE *Royal Naval Volunteer Reserve* vi. 91 The active-service men labelled them [*sc.* RNVR ratings] a 'rocky' lot—'rocky' being an oblique reference to unstable sea-legs and the waved tapes in their blue jean collars.

rococo, *a.* and *sb.* Add: **A.** *adj.* **1.** (Further examples.)

1870 M. ARNOLD in *Pall Mall Gaz.* 29 Nov. 3/2 We heard the honest German soldiers *Hoch-ing*, hurrahing, and God-blessing in their true-hearted but somewhat *rococo* manner. **1902** H. L. WILSON *Spenders* ix. 92 She is rather a beauty, you'll find;..a bit rococo in manner, I suspect.

2. a. (Earlier and later examples.) Also of interior decoration.

1841 C'TESS BLESSINGTON *Idler in France* I. 21 The whole [of the terraces near La Tour-Magne at Nîmes] offering a curious mixture of military and *rococo* taste.

1918 Heal & Son Catal.: Cottage Furnit. 1 The 'new art' overmantel smothered in rococo photograph frames. 1967 N. Freeling Strike Out 40 Presentation silver..in every conceivable pattern from curliest rococo to bleakest Swedish. 1972 Country Life 7 Dec. 1574/1 All these rooms have delicate rococo plaster ceilings picked out in pale pastel shades. 1980 Early Music Apr. 173/2 The organ sonatas of C. P. E. Bach from the 1750s are a good match for the rococo organ in Midwolda (1772).

transf. (Further examples.)

1931 N. & Q. 15 Aug. 109/2 It is further planned to give Goethe plays and rococo concerts on an open-air stage. 1938 W. S. Maugham Summing Up 28 Dryden flourished at a happy moment... He was the first of the rococo artists. 1941 Jazz Information Nov. 21/2 James P. [Johnson] made his first player piano rolls..as a 'race' feature alongside the rococo but immensely popular efforts of Phil Ohman. 1955 Times 21 July 7/7 Stravinsky's choice of a more or less definite rococo pastiche is a highly appropriate musical idiom. 1959 Listener 26 Nov. 952/1 Haydn's symphonic music began as rococo entertainment. 1967 G. Steiner Lang. & Silence 28 This would..lead one to ask whether..the rococo virtuosity of Salinger is arguing an absurdly diminished and enervating view of human existence. 1970 Oxf. Compan. Art 987/1 The painter to whom the epithet 'Rococo' has most often been loosely applied is, perhaps, Watteau, and in his rejection of the grand sujet and his fanciful and curvacious rhythms he does..fit into the movement.

b. *Embroidery.* (See quots.)

1882 Caulfield & Saward Dict. Needlework 426/1 Rococo Embroidery is used for table borders, fire screens, and cushion covers, and is made with écru linen foundations, ornamented with filoselles. 1934 M. Thomas Dict. Embroidery Stitches 171 Rococo stitch,..must be worked on a very wide-meshed canvas of even weave and the little groups or bundles of stitches are set in alternate squares, leaving the others open and rather 'pulled' in effect. 1960 G. Lewis Handbk. Crafts 38 Frequently these two are the only stitches used to the neglect of the many others which would greatly enrich many pieces of work, such as.. French stitch..and Rococo stitch. 1960 B. Snook Eng. Hist. Embroidery 82 Rococo stitch is particularly effective. Ibid. 98 Designs of flowers, bouquets, ferns, ribbons and tassels..were either in varied silk stitchery on satin, or in tent stitch on canvas if enclosed in a rococo border.

B. *sb.* **1.** (Further examples.)

1935 W. S. Maugham Don Fernando x. 224 Decoration ..degenerated with time to the frivolous ornament of rococo. 1947 A. Einstein Mus. Romantic Era iii. 20 The 18th-century stylistic period that preceded the Empire, the Rococo, had been a last tremulous echo of the grandeur of the Baroque. 1954 [see *baroque a. (sb.)]. 1965 Listener 3 June 830/1 The drawing in nearly all Monticelli's pictures is reminiscent of the rococo.

rocococity. (Further example.)

1916 A. Huxley Let. 29 Dec. (1969) 118 My monocle is very grandiose, but gives me rather a Greco-Roman air of rococDocity.

rod, *sb.*[1] Add: **I. 5. b.** (Further examples.)

1935 B. Perry And gladly Teach viii. 181 After showing me how often he had been 'high rod' on his stretch of the river, he would 'O.K.' all of my estimates with a smile. 1975 Oxf. Univ. Gaz. 16 Jan. 428/1 (Advt.), Prospective rods may apply for descriptive booklet.

c. Used in association with Gun *sb.* to designate the twin pursuits of fishing and shooting.

1840 J. Wilson (title) The rod and the gun, being two treatises on angling and shooting. c 1860 in A. Adburgham Shops & Shopping (1964) vii. 74 Their [sc. the fabrics'] close imitation of the colour of the natural objects surrounding the Sportsman have rendered them an absolute necessity for the pursuits of the rod or gun. 1912 E. D. Cuming (title) With rod and gun. 1966 Times 28 Feb. (Canada Suppl.) p. ii, Canada is a..catch-your-limit rod-and-gun of a country.

d. *transf.* The right to fish a length of river.

1932 G. Cornwallis-West Edwardians go Fishing v. 60 A friend of mine was invited to fish..on one of the upper reaches of the Test owned by Colonel Sneyd, who had kept a rod himself but had let two other rods to men who had little if any knowledge of the art of dry fly fishing. 1958 Angling Times 28 Feb. 11/3 The Board offers 14 rods to let during the 1958 season for its fishery on the River Piddle. 1970 Daily Tel. 9 Nov. 8/6 Fishing fees range from £250 to £400 a rod annually.

III. 9. a. *spec.* = *control rod* s.v. *control *sb.* 5; also (in full *fuel rod*), a long, slender piece of fuel for a nuclear reactor.

1956 Ann. Rev. Nucl. Sci. VI. 329 The fuel loading consists of roughly five tons of natural UO_2 or uranium metal as round rods clad with Zircaloy. Ibid. 334 Two automatic regulating rods ordinarily hold the power level within 3 per cent of the desired value. 1959 C. Hodder-Williams Chain Reaction ix. 113 Did he leave it so late that the 'X holes' had warped out of alignment and the rods wouldn't drop? 1964 Jrnl. Brit. Nucl. Energy Soc. III. 298/1 By utilizing a high energy, high strain rate deformation process, the fuel rod is not only completely encapsulated but..the can wall is pressurized into the anti-rachetting grooves. 1975 J. R. Lamarsh Introd. Nucl. Engin. vii. 262 The ordinary movement of the rods in most power reactors is controlled..by an on-line computer. 1979 K. Follett Triple iv. 77 The reactor has three thousand fuel channels, each channel containing eight fuel rods.

d. One of the elongated light-sensitive cells in the retina responsible primarily for vision in poor light. Cf. Cone *sb.*[1] 10.

1866 [in Dict., sense 9 b]. 1905 A. Flint Handbk. Physiol. xxvi. 658 The rods are regular cylinders, their length corresponding to the thickness of the layer, terminating above in truncated extremities, and below in points which probably are continuous with the filaments of connection with the nerve-cells. 1937 Carlson & Johnson Machinery of Body xi. 447 At the point where the optic nerve enters the retina there are no rods and cones. 1958 Brocklehurst & Ward Gen. School Biol. xiv. 185 Rods are more numerous near the periphery of the retina, and cones..near its centre.

e. *slang.* The penis; the erect member.

1902 in Farmer & Henley Slang s.v. Penis. 1934 E. Pound Eleven New Cantos xxxix. 46 His rod hath made god in my belly. 1960 A. West Trend is Up x. 454 'I want you to love me and cherish me all the days of my life.' 'You want the rod, you silly bitch, you fouled-up boarding-house bitch,' he said, 'that's what you want.' 1975 B. Meggs Matter of Paradise vi. iii. 142 He was seventeen..rod cocked and aimed at every passing female object.

f. *slang* (chiefly *U.S.*). A gun; a pistol or revolver.

1903 H. Hapgood Autobiogr. of Thief xii. 289 The dago dropped the smoke-wagon and the bartender threatened to put him in prison for pulling a rod on respectable people. 1926 J. Black You can't Win xi. 145, I think I shall put a small 'rod' in my coat pocket hereafter. 1929 D. Runyon in Hearst's Internat. Oct. 201/1 Dave the Dude takes personal charge of Wild William and removes a rod from his pants pocket. 1934 A. Merritt Burn Witch Burn! v. 65 'Pass your rods, Paul.' Without a word the chauffeur dipped into his pockets and handed him a pair of automatics. 1942 Wodehouse Money in Bank (1946) xxv. 224 If I've got to stick up an eat-'em-alive baby like her with nothing but a finger in my pocket, I want an extra cut... Either I have a rod, or it's seventy-five–twenty-five. 1953 K. Tennant Joyful Condemned iv. 34 What's wrong with you, waving that rod about like a bloody half-wit. 1965 [see *Betsy]. 1978 J. Carroll Mortal Friends ii. iv. 179, I ain't getting my ass blown off because you're stupid. You won't get near Zorelli with a rod anyways.

g. *N. Amer. slang.* The draw-rod of a railway carriage or truck. Cf. *ride *v.* 18d.

1904 Outing July 486/2 Thousands of miles in the United States and Canada he has wandered on rods and blind baggages. 1924 J. Tully Beggars of Life 56, I beat it through De Kalb last night on the rods of a mannerfest meat train. 1931 'D. Stiff' Milk & Honey Route 192 We beat it on the run... Some rode the rods on passengers, While some blew out on freights. 1959 Punch 17 June 799/2 One does not picture train-robbers lurking at the top of the bank, nor even hoboes riding the rods.

h. *slang* (chiefly *U.S.*). = *hot rod.

1945 [see *hot rod]. 1948 Hot Rod Mag. July 4/3 With Carson upholstery And all the fine gear Of a more beautiful rod You never will hear. 1957 J. Kerouac On Road i. xii. 79 A burly blond kid in a souped-up rod. 1972 J. Gores Dead Skip vii. 45 A two-bit Mission District auto and accessory dealer who specialized in old cars for conversion to dune buggies, drag cars, rods, and the like. 1978 Hot Car June 91/1 With just a beach buggy to his credit at that stage, Brian saw some US rods in Chicago and planned to build himself a C Cab on his return to this country.

IV. 11. b. *rod-bender*; *rod-case* (earlier and later examples), *licence*, *-ring*, *-stand*.

1956 People 13 May 12/6 He reckons that morning and evening are the best times for rod-benders, using bread flake and paste, on a 16- or 18-hook, for roach. 1976 S. Wales Echo 26 Nov., Steve..thought he had hooked a good cod when he struck into a real rod-bender at The Leys. 1852 C. M. Yonge Two Guardians i. 2 His numerous equipments, consisting of a long rod-case, a fishing-basket and landing-net. 1918 Kipling in Story-Teller Dec. 221/2 'Did you? Good!' he replied heartily over the rod-case on his shoulders. 1969 Sears, Roebuck & Co. Catal. Spring–Summer 713/1 Ted Williams Rod Case... Foam-lined compartments at both ends for reels, accessories. 1976 Eastern Daily Press (Norwich) 19 Nov. 21/4 The proposed new charges..will be in addition to, and completely separate from, the rod licence charge. 1885 J. W. Martin Float Fishing & Spinning (ed. 2) 181/2 Rod-rings. 1972 Shooting Times & Country Mag. 1 July 12/1 Take care to ensure that the line can still run through the rod rings. 1902 Chambers's Jrnl. July 425/1 Any trout-rod, even after a brief life spent in such a manner, might have accepted its pensioner peg on the rod-stand.

c. In sense *9d as *rod cell*, *pigment*, *vision*; *rod-free* adj.

1940 Chambers's Techn. Dict. 728/2 Rod-cell. 1970 Ambrose & Easty Cell Biol. i. 26 In the complex light-sensitive rod cells of the vertebrate retina the membrane and fibre structure of cilia have been modified to receive light and convert it into an electrochemical stimulus. 1915 J. H. Parsons Introd. Study of Colour Vision iii. 204 Beyond the rod-free central area the cones diminish continuously in number. 1921 E. B. Titchener Text-bk. Psychol. xxii. 89 Animals whose eyes lack this rod-pigment—fowls, pigeons—are strictly diurnal in their habits. Ibid., Whether the visual purple is essentially concerned in rod-vision. 1921 R. S. Woodworth Psychol. (1922) x. 226 Dim-light vision, or twilight vision as it is sometimes called, is rod vision and not cone vision.

12. rod puppet, a puppet operated and supported by rods; **rod-rider** *U.S.*, one who rides the rods (see sense 9 g above); hence **rod-riding** *ppl. a.*

1930 Puppetry i. 64 Stick or Rod Puppets. 1949 P. McPharlin Puppet Theatre in Amer. xx. 347 Rod-puppets have stirred up interest for their novelty and adaptability. 1960 Guardian 19 Oct. 9/5 The Chinese theatre, with its impressive traditional rod puppets. 1976 Jrnl. R. Soc. Arts Apr. 254/1 A larger form of glove puppet can be produced by placing the figure on a rod; this is called a rod puppet. 1952 L. Hughes Laughing to keep from Crying 60 The rod-riders got off nowhere near the station. 1967 B. J. Banfill Pioneer Nurse iv. 43 Many readers may not have heard about the Rod Riders, who formed a part of history in opening the west. During the 'Awful Thirties', this name was given to wandering embittered men. Ibid. 44 Rod Riders..planned to steal rides on trains. 1953 W. Burroughs Junkie x. 95 This type cop could just as well be an oldtime rod-riding thug.

rod, *v.* Add: **4.** *trans.* To push a rod through (a drain or pipe) in order to clear it. Hence **ro·ddable** *a.*, capable of being rodded.

1924 E. G. Blake Plumbing I. xvii. 149 A manhole should be provided at each alteration of the direction of the drain, and at intervals of not more than from 80 yds. to 100 yds. in all straight runs. This will enable any obstruction to be cleared by rodding the drain. 1949 Escritt & Rich Work of Sanitary Engineer (ed. 2) xxii. 275 The disconnecting trap..is provided with a rodding arm which permits the outgoing line of pipe to be rodded. Ibid. xxiv. 283 The outlet should..be arranged so that it is easily roddable. 1971 B. Linden Home Owner's Maintenance Guide ii. 28 The drain pipes will probably have to be rodded to clear the blockage.

5. *intr.* Const. *up.* To arm oneself with a gun or guns. Cf. *rod *sb.*[1] 9f. *U.S. slang.*

1929 [implied in *rodded ppl. a. 4]. 1935 N. Ersine Underworld & Prison Slang 63 Rod up and we'll blow. 1950 Harper's Mag. Feb. 75/2 They do not rod up, or arm themselves.

6. *trans.* To 'soup *up*' or convert (a car) into a hot rod. Cf. *rod *sb.*[1] 9h. *U.S. slang.*

1972 J. Gores Dead Skip v. 32 A young man with an old car (hence, probably rodding it up, hence, probably, a car-lover).

rodded, *ppl. a.* Add: **4.** Const. *up.* Armed with a gun or guns. Cf. *rod *v.* 5. *U.S. slang.*

1929 D. Runyon in Hearst's Internat. Oct. 201/1 None of the guests are supposed to come rodded up, this being strictly a social matter. 1930 Flynn's 12 Apr. 402/2, I was rodded up an' I could 'a' give him the works..but it wasn't worth it. 1938 D. Runyon Furthermore ix. 171 It is very much against the law for guys to go around rodded up this way in New York City.

rodden: see *rodham.

rodder (rǫ·dəɹ). *slang* (chiefly *U.S.*). [f. Rod *sb.*[1]] = *hot rodder* s.v. *hot rod; one who converts cars into hot rods. Cf. *rod *sb.*[1] 9h.

1949 Hot Rod Mag. Feb. 18/2 The rod news is rather short this month. I guess the California rodders now know why we keep our cars parked in the winter months. 1953 Berrey & Van den Bark Amer. Thes. Slang (1954) § 81a/2 'Hot rod' owner,..hot rodder, rod jockey, rodder. 1972 World of Wild Wheels (Custom Car) 58/1 When the British rodder glances through American Street Rod magazines he is faced with 99 per cent American cars. 1977 Hot Car Oct. 15/2 Brighter news for the rodder is another kit available from G. A. Stanley Palmer Ltd.

rodding, (*vbl.*) *sb.*[1] Add: **2. b.** The action of *rod *v.* 4.

a 1890 [in Dict., sense 2]. 1945 W. J. Woolgar Pract. Plumber & Sanitary Engin. 313/1 Rodding eyes are useful at changes of direction and at the top end of a long branch drain. 1953 L. B. Escritt Building Sanitation (ed. 3) viii. 94 A light manhole cover is provided in a central position to facilitate rodding. 1973 BSI News Apr. 5 (caption) Members of the code drafting committee..watch a demonstration of rodding to remove blockages in drainage pipes.

4. *Geol.* A linear structure in metamorphic rocks characterized by the arrangement of grains of a constituent mineral, esp. quartz, in parallel rods. Also *rodding structure.*

1907 B. N. Peach et al. Geol. Structure of N.-W. Highlands of Scotland (Mem. Geol. Survey) xii. 211 In the exposures of these dykes, which happen to cross the direction of rodding almost at right angles, no planefoliation is observable. 1923 Q. Jrnl. Geol. Soc. LXXIX. 439 In the hornblende-schists there is actual rodding or elongation of the crystals. 1953 Proc. Geologists' Assoc. LXIV. 119 Rodding is developed from quartz that has been introduced into, or has segregated in, the rocks. 1970 K. C. Jackson Textbk. Lithology vii. 426 Associated with the larger-scale folding may be the segregation of more mobile constituents such as quartz, resulting in a small-scale rodding structure.

roddon: see *rodham.

roddy, obs. var. *rhody.

rode, *sb.*[2] For *U.S.* read *N. Amer.* and add further examples.

1792 G. Cartwright Jrnl. I. p. xiii, Rode, a small tow-line, of four inches and an half; made use of by shallops, by way of a cable. 1950 R. Moore Candlemas Bay i. 45 His anchor and rode were stowed down under the stern. 1963 J. T. Rowland North to Adventure x. 147 With both hooks down and a long scope of rode she should be able to ride out anything.

rode, *v.*[2] Add: Hence **ro·ding** *vbl. sb.* (freq. *attrib.*) and *ppl. a.*

1888 [in Dict.]. 1927 E. Sandars Bird Bk. for Pocket 156 Cock's nuptial display ('Roding'),..plumage fluffed out and uttering call. 1955 Times 10 May 12/5 These two sounds, so useful as an aid to seeing woodcock, are constantly uttered during its 'roding' flights, which start at sunset, are continued for perhaps an hour, and are resumed at dawn. 1973 Ibis CXV. 135 Many workers have described the roding behaviour of the Woodcock and speculated on its biological meaning.

rodent, *a.* and *sb.* Add: **B.** *sb.* **2.** *attrib.* and *Comb.*, as *rodent controller*, *officer*, *operative*, *operator*; *rodent-carried*, *infested*, *-like*, *proof* adjs.; **rodent-run** *Ornith.*, a run made by some birds when disturbed in which they resemble a running rodent.

1959 New Biol. XXIX. 96 A rodent-carried bacterium caused the Black Death. 1970 Daily Tel. 14 May 2/8 A senior rodent controller..was yesterday presented with the British Empire Medal for his 18 years' work destroying rats. 1979 Dædalus Summer 137 They all confirm that the chapel was rodent infested. 1978 P. Porter Cost of Seriousness 3 Just as at seven the teeth stick out which Later slope in, rodent-like. 1944 Liverpool Echo 31 Jan. 2/2 Westminster City Council's rat-catcher is in future to be called Rodent Officer. 1973 in Fremdsprachen (1976) XX. 212/1 Fourteen visits from a rodent officer..had not stemmed the infestation. 1944 Sunday Times 5 Nov. 6/2 When it comes to official jargon, can you beat turning our old friend the rat-catcher into a 'Rodent Operative'? 1972 Daily Tel. 1 May 3/4 The Ministry also advises people to consult their local authority's rodent operative on the best way of applying whatever measures they decide on. 1979 J. Gardner Nostradamus Traitor li. 248 'Are you a rat-catcher, Mr. Kruger?'..'They are called rodent operatives nowadays.' 1946 Word Study May 2/2 Euphemisms..rodent operator for rat-catcher. 1972 C. Drummond Death at Bar vii. 179 The lunatic..is now a traveller for a firm of rodent operators. 1977 Field 13 Jan. 68/1 (Advt.), It's horse sense to buy your food storage bins direct from the manufacturer. A quality product. Rodent proof. 1950 Ibis XCII. 28 We observed that a husky dog would immediately chase the Purple Sandpiper that made a 'rodent-run' and would be completely fooled. 1961 D. A. Bannerman Birds Brit. Isles X. 197 Birds of this northern race [of golden plover] will sit until almost trodden upon, and then shuffle away looking much like a teal in the process, though some would prefer—without too much regard for the truth—to speak of its 'rodent-run'. 1976 Van Tyne & Berger Fund. Ornith. (ed. 2) iv. 209 The 'rodent-run' display.. has been reported for a number of shorebirds.., for some tundra species, and for the Green-Tailed Towhee.

rodentian, *a.* For *rare*[−1] read *rare* and substitute for def.: Of, pertaining to, or consisting of rodents.

1974 Amer. N. & Q. XIII. 25/1 The latent comedy in Chaucer's further gathering of words with obvious rodentian associations.

rodenticide (rode·ntisəid). [f. Rodent *sb.* + -cide 1, after *insecticide*, etc.] A poison used to kill rodents.

1938 Bull. Calif. Dept. Agric. XXVII. 172 This material ..might serve both as a rodenticide and an insecticide. 1961 Times 18 July 6/6 The use of toxic chemicals is governed by a voluntary notification scheme agreed between the Government departments and the industrial association concerned, covering insecticides, fungicides, herbicides and rodenticides. 1975 Nature 24 Jan. 275/2 Field populations of common rats..resistant to anticoagulant rodenticides..have appeared in Scotland, Denmark, [etc.].

Hence **rodentici·dal** *a.*, of or pertaining to a rodenticide; poisonous to rodents.

1939 Jrnl. Amer. Vet. Med. Assoc. XCV. 486/1 Human poisoning has resulted from..accidental ingestion of the metal used for entomological and rodenticidal purposes. 1979 P. L. G. Bateman Household Pests i. 25 Pest control chemicals are selected for their insecticidal or rodenticidal efficiency.

rodeo. Delete ‖, for *Amer.* read 'orig. *U.S.*', and add: Now also with pronunc. (rōu·dɪ,o). **3. a.** A public exhibition of skill, often in the form of a competition, in the riding of unbroken horses, the roping of calves, wrestling with steers, etc.

1914 B. M. Bower Flying U Ranch 16 They have them rodeos on a Sunday, mostly, and they invite everybody to it, like it was a picnic. 1925 Annual Rodeo Program (Tucson, Arizona) 3 We extend a cordial invitation to you to come to Tucson for our Annual Rodeo. 1938 D. Coolidge Arizona Cowboys ii. 27 The round-up had just begun. They call it rodéo, in Spanish, but the cactus cowboys pronounced it rodér. The contest riders of today have given it another twist and call it ró-deo. 1940 Arizona (Arizona Work Projects Administration Writers' Project) 72 That distinctively western entertainment, the rodeo, was originally an exhibition of cowboy skill in the regular activities of cattle ranch and range. But today it is largely commercialized and many of its features are of the circus type, remote from the cowpuncher's everyday life. 1950 Manch. Guardian Weekly 5 Oct. 6/2 Madison Square Garden is presently dedicated to Gene Autry and the annual rodeo. 1976 Columbus (Montana) News 27 May 1/4 One of those injured was a prime mover and instigator of this rodeo, Ed Miller, who is currently in St. Vincent's with a broken leg.

b. *transf.* A similar exhibition of competitive skill in the riding of motor-cycles, fishing, etc.; also used more generally of other types of competition. Also *fig.* Cf. *roadeo.

1927 My Oklahoma July 23/1 Oklahoma is going to have a state-wide baby rodeo next year. 1928 Daily Express 7 May 15/3 On Wednesday..a motor-cycle rodeo in the afternoon will be followed by a carnival procession through the town. 1940 Sun (Baltimore) 11 Sept. 1/7 Nazi bombers smashed at London with increasing violence early today in their fourth consecutive dusk-to-dawn rodeo of destruction. 1949 Daily Progress (Charlottesville, Va.) 22 Aug. 9/1 Entries for the fishing rodeo for youngsters here must be in by Thursday.

4. *attrib.* and *Comb.*, as *rodeo circuit*, *clown*, *cowboy*, *ground*, *parade*, *queen*, *rider*, *riding* (*sb.* and *a.*).

1961 M. S. Robertson Rodeo 101/1 The California Rodeo..is one of the Big Four, the rodeos whose pioneering and consistent greatness bridged the years from the inception of the 'cowboy tournaments' to the modern rodeo circuit. 1980 Country Life 13 Nov. 1819/1 He.. started bronc riding in the rodeos... Demobbed, he returned to the rodeo circuit. 1927 Progressive Arizona IV. ii. 7 The arena presents a scene of animation with the judges, time-keepers, contestants, performers, event clerks, and Jolly, 'the funniest rodeo clown of them all' milling about. 1941 L. B. Chaffin Sons of West xv. 222 This trick, in almost every identical move, is practiced by modern rodeo clowns of today. 1958 E. H. Peplow Hist. Arizona II. xx. 405 The competitive life of a rodeo cowboy is shorter than that of an athlete in almost any other sport. 1892 Rodeo ground [in Dict., sense 1]. 1979 Tucson Mag. Apr. 68/3 Admission is charged for this event at the rodeo grounds. 1941 La Fiesta de los Vaqueros Program (Tucson, Arizona) 9 Tucson and its guests spend two hours..standing on each other's toes in order to see the rodeo parade. 1976 Billings (Montana) Gaz. 10 June 1-A/5 In Hardin, activities include a week-long carnival beginning Monday followed by a rodeo Friday and a rodeo parade at noon Sunday. 1945 Pueblo (Colorado) Star-Jrnl. 3 June 7/3 Nine girls at the Pueblo ordnance depot will don cowboy hats, bright shirts, and jeans to vie for the honor of rodeo queen. 1975 R. Hoban Turtle Diary xxv. 115 She'd..been a rodeo rider, done roller derbies, wrestled, had three husbands and all kinds of troubles. 1979 'G. Black' Night Run from Java i. 7 A rodeo rider thrown by a bronco. 1974 Times 7 Jan. 8/3 We did hunter trials, show jumping, rodeo riding, and so on. 1976 Evening Standard 29 Dec., The Hard Breed. Rodeo-riding Cain slain by younger brother.

Hence **rode·o** *v. intr.*, to compete in a rodeo. Also **rode·oer**; **rode·oing** *vbl. sb.* (All also with stress on first syllable.)

1959 Rodeo Sports News 1 Nov. 2/1 I've wondered.. what the contestants do when they quit rodeoing. Ibid. 15 Nov. 2/4 A cowboy who would rather hunt than rodeo —we've got everything in the northwest! 1970 Ibid. 15 Nov. 2/2 A top bull rider who rodeoed up through the mid-sixties stopped by and said hello the other day. 1976 Oregonian (Portland, Oregon) 14 June C3/2 Because it is not a sport sanctioned by the National Collegiate Athletic Association, rodeoers are free to compete in professional rodeo while they are still on the college circuit. 1977 New Yorker 6 June 74/2 They loved making cowhands of their frisky little girls—they took them riding and roping and rodeoing.

rodgersia (rǫdʒə̄·ɪziā). Also **Rodgersia.** [mod.L. (A. Gray 1859, in Mem. Amer. Acad. New Ser. VI. 389), f. the name of John *Rodgers* (1812–82), American admiral + -ia[1].] A large perennial herb of the genus so called, belonging to the family Saxifragaceæ, native to eastern Asia, and bearing compound leaves and terminal panicles of small white flowers.

[1902 Gardeners' Chron. 23 Aug. 131/2 At present in the rock garden at Kew there is in flower, for the first time in Europe, a new species of Rodgersia.] 1908 R. Farrer Alpines & Bog-Plants ix. 185 Of foliage plants for the lake-side..there is nothing to surpass the Rodgersias. 1962 Amateur Gardening 24 Mar. 4/2 Among foliage plants for the waterside there are none to surpass the rodgersias. 1976 Country Life 6 May 1172/3 There is a wealth of plants..including hardy plants such as gunneras, rodgersias, hostas and hellebores.

rodham (rǫ·dām). *E. Anglia.* Also **rodden, roddon.** [Of obscure formation (see note below).] A raised bank in the Fen district of East Anglia, consisting of the bed and levees of a dry river-course which have been raised above the level of the adjacent land by deposition of silt, usu. by the incoming tide, and by compaction and lowering of the adjacent peat soil; occas. used to signify only a levee bounding such a river-course.

The spelling *roddon* was adopted and popularized by G. Fowler (see quot. 1932), whilst the older form in -(*h*)*am* remains dominant in local use. Any connection with *roddin* (cf. E.D.D. and Rod *sb.*[2]) is unlikely.

[1857 T. Wright Dict. Obsolete & Provincial Engl. II. 806 Roddam, a bed of sand resting on the clay beneath the peat, in the fens of Cambridgeshire.] 1932 G. Fowler in Geogr. Jrnl. LXXIX. 210 There are numerous raised banks of laminated silt or shell maul meandering through the Fens. Neither historians nor geologists appear to have noticed them... Fen dwellers however have noted these banks but generally without realizing their origin. They call them roddons. This word appears to be allied to roddin or rodden, which Wright in his 'English Dialect Dictionary'..gives as meaning a narrow road, path or sheep track. I spell the word roddon as it sounds; and I prefer it as a spelling to rodham, as used in the name Rodham Farm, as the latter appears corrupted in the second syllable. 1945 B. E. Dorman Story of Ely i. 3 These raised river beds are known as roddens... One fine example..can be seen..alongside..the road from Littleport to Shippea Hill. The few houses along this road are nearly all built on the rodden, for it provides a firmer foundation than peat. 1957 A. K. Astbury Black Fens v. 26 Levees formed as parallel ridges one on either side of the main channel... Where subsequent cultivation has been long and constant the two levees may tend to merge into one general bank of silt... But all such levees, whether separate or merged, are in the black fens known as rodhams. Ibid. 27 Fowler..used the form roddon—influenced by the fact that in the north of England the word roddin or rodden means a narrow road... But although later writers have adopted Fowler's spelling, the fact remains that fenmen themselves call these things rodhams. 1957 G. E. Hutchinson Treat. Limnol. I. i. 119 The tidal water from this shallow arm of the sea, the southward continuation of the modern Wash, was evidently very turbid and deposited levees at the sides of the channels, known as roddons, along which it ran. 1963 E. S. Wood Collins Field Guide to Archaeol. ii. 199 Silt Banks, otherwise known as roddons, or rodhams, are caused by tidal action depositing silt up slow rivers. 1971 Norfolk Fair Feb. 36/3 Old extinct watercourses can be traced by the rodhams of silt and the slades of chalky material. 1974 J. R. Ravensdale Liable to Floods i. 21 Gordon Fowler..noticed a roddon in the north-east corner of Cottenham parish.

Rodinesque (rodæne·sk), *a.* [f. the name of Auguste *Rodin* (1840–1917), French Romantic School sculptor + -esque.] Of, pertaining to, or reminiscent of Rodin or his work, marked by masterly realism and love of movement.

1905 G. B. Shaw Let. 13 Mar. (1972) II. 521 It is a bad case of helpless genius in the first blaze of youth; and the drawings are queer and Rodinesque enough to be presentable at this particular moment. 1934 Sunday Times 11 Feb. 6/1 Mr. O'Casey's eye-appeal is the old business of Rodinesque, Volga-Boatmannish stage-grouping. 1934 F. Scott Fitzgerald Tender is Night i. iv, As if the features..had been molded with a Rodinesque intention. 1962 Times 14 Feb. 15/1 Little Rodinesque bronze figures..are on show in the front gallery.

rodingite (rōu·diŋəit). *Petrogr.* [f. the name of the River *Roding*, S. of Nelson, New Zealand + -ite[1].] A crystalline rock consisting of diallage and grossularite (or hydrogrossularite), often with prehnite and chlorite, and typically found in or adjacent to serpentinite masses, having been formed by the calcium metasomatism of basic or ultrabasic igneous rocks.

1911 J. M. Bell et al. in Bull. N.Z. Geol. Survey No. 12. 31 Dykes of a coarse-grained gabbro-like rock penetrate the serpentines in many places... The writers have applied to the rock the name 'rodingite', owing to the typical exposure occurring on the River Roding. 1954 Mineral. Mag. XXX. 525 (heading) Rodingite from the Girvan–Ballantrae complex, Ayrshire. 1976 Neues Jahrb. für Mineral. (Monatshefte) 188 The rodingites described here show an absolute prevalence of garnet, which has a variable composition ranging from an almost pure grossularite..to an intermediate member of the grossularite-andradite series.

Hence **rodingi·tic** *a.*; also **ro·dingitized** *ppl. a.*, converted into rodingite; **ro·dingiti:zing** *ppl. a.*, **ro:dingitiza·tion.**

1953 Jrnl. Faculty of Sci., Hokkaido Univ. 4th Ser. VIII. 419 (heading) On the rodingitic rocks within the serpentinite masses of Hokkaido. 1971 Canad. Jrnl. Earth Sci. VIII. 642/2 This subdivision..improved the permeability for rodingitizing fluids. Ibid. 643/1 (caption) Serpentinized and partially rodingitized peridotite. Ibid. 644/1 Development of garnets, hydrogarnets (hibschite), in a process somewhat similar to rodingitization. 1975 Contrib. Mineral. & Petrol. XLIX. 233 The latter described rodingitic material in their study of serpentinized ultramafic rocks dredged from the Mid-Atlantic Ridge at 45°N. Ibid., Partially rodingitized olivine gabbros were also found. Ibid. 253 The rodingitization of the gabbroids and the serpentinization of the ultramafics appear to be two concomitant and complementary metasomatic processes.

rodman. Add: **2.** (Earlier and later examples.) Chiefly *U.S.*

1853 A. W. Whipple in Rep. Explorations for Railroad to Pacific Ocean (U.S. War Dept.) (1856) i. i. 5 The chainmen and rodmen being ignorant of their duties, little more than teaching them could this day be accomplished. 1972 Publishers' Weekly 7 Aug. 48/2 In Chicago in 1925, when he was a rodman with a Cook County surveying crew.

3. *slang.* One who handles a gun; a gunman. Cf. *rod *sb.*[1] 9f.

1924 G. C. Henderson Keys to Crookdom 396 Assaulter, rough guy, hard bird, rod man, rod toter. 1931 Amer. Speech VII. 113 He used to be a rodman on the convoy, but he didn't have the guts for that job. 1940 W. R. Burnett High Sierra i. 12 We need a rodman... You're it. a 1953 Dylan Thomas Quite Early one Morning (1954) 39 A raid by the vice-squad on a clip-joint for retired rod-men. 1962 John o' London's 4 Oct. 325/2 Robert is victim number two of this assassination, the only witness who could identify the rod-man.

rodney. Add: **3.** [Perh. a different word.] A small fishing boat or punt. *Canad.*

1895 Christmas Rev. (Newfoundland) 18 Jim Leary, whose handiwork, whether displayed in the construction of baitskiff, smack, skiff, punt or rodney, was always superior to what any other man in the settlement could turn out. c 1900 in Regional Lang. Stud.—Newfoundland (1978) viii. 24 Rodney, small, single crosshand[ed] punt. 1908 N. Duncan Every Man for Himself ix. 260 'Launch that rodney,' Wull directed, 'an' put me on shore.' 1923 Sunday at Home Dec. 153/1 The punt..was one of the small, light type of boat called a 'rodney', and it was used mainly for shooting about the harbour, or on sealing trips. 1931 J. R. Smallwood New Newfoundland 266 Often he'd take us to sail in his rodney, Out over the water. 1966 A. R. Scammell My Newfoundland 36 Sid, go over to Blanchard's and keep an eye on what they're doing. Take the rodney.

Rodriguan (rǫdrī·găn), *a.* and *sb.* [f. *Rodrigues* (see below) + -an.] **A.** *adj.* Of, pertaining to, or characteristic of the island of Rodrigues, a dependency of Mauritius in the western Indian Ocean, or its people. **B.** *sb.* A native or inhabitant of Rodrigues.

1973 Times 5 Mar. (Mauritius Suppl.) p. v/1 The islanders are very Rodriguan-minded. Ibid., An elected island council..would give Rodriguans a chance to put their views. 1974 Islander (Victoria, B.C.) 25 Aug. 2/4 We found the Rodriguans interesting. 1977 G. Durrell Golden Bats & Pink Pigeons iv. 86 A large, chocolate-coloured Rodriguan in a handsome, khaki uniform.

roe[1]. Add: **2. roe ring,** a track worn by roe deer running in circles prior to mating; **roe-stalking,** the hunting of roe-deer on foot; so **roe-stalker.**

1951 H. Tegner Roe Deer iii. 27 (caption) Shape of roe rings: small circles denote small trees or bushes as axes around which roe form these runs. 1960 M. Burton Wild Animals 128 A feature of the rutting season [of roe deer] which has attracted a good deal of attention in recent years has to do with the use of 'roe rings', in which a form of courtship takes place. 1974 F. Holmes Following Roe i. 10 Roe rings, well-trodden runs in the shape of a circle or a figure-of-eight, are evidence of roe residence if they have been recently used. 1927 Edwards & Wallace Hunting & Stalking Deer xlii. 237, I have never yet met a roe-stalker who did not love the roe. 1906 J. G. Millais Mammals Gt. Brit. & Ireland III. 178 There are of course hundreds of estates in the North where Roe-stalking is not possible. 1973 Country Life 26 July 254/3 A week's roe-stalking in Britain is..one of the most sought-after privileges among European devotees.

roe[3]. Substitute for etym. 'Perh. a transf. use of Roe[2]' and add further examples.

1920 A. L. Howard Man. Timbers of World 144 Mahogany, Cuba... Many of the logs are beautifully figured or marked with wavy and curly grain, which is variously termed splash mottle, roe and mottle, fiddle-back, plum, snail, blister and cross-bar. 1938 B. J. Rendle Commerc. Mahoganies 5 Mahogany is remarkably stable and does not shrink and swell so much as most woods. Irregularities of the grain produce a variety of figure—fiddleback, blister, stripe or roe, curl, mottle, etc. 1952 J. Gloag Short Dict. Furnit. 396 Roe, a name given to the regular appearance of dark figures and spots in figured mahogany, which give a mottled effect, like a fish roe. A form of roe figure occurs in flowered, or East Indian, satinwood. 1968 Canad. Antiques Coll. Aug. 24/2 Roe figure, this is alternate bands of twisted grain which produce stripes parallel to the length of the tree. When viewed in certain lights from either end the light and dark stripes are reversed.

rœblingite (rə̄·bliŋəit). *Min.* [f. the name of W. A. *Rœbling* (1837–1926), U.S. civil engineer + -ite[1].] A rare, monoclinic, basic sulphate-silicate of lead, calcium, and other elements, occurring as compact white masses of minute crystals.

1897 Penfield & Foote in Amer. Jrnl. Sci. CLIII. 415 At the request of Mr. Nason the authors take pleasure in naming this mineral roeblingite in honor of the celebrated engineer, Mr. W. A. Rœbling of Trenton, N.J. 1966 Amer. Mineralogist LI. 507 Only a tentative formula, $(Pb_2S_2O_{14})R_{14}Si_{12}O_{44}H_{20}$, where R = Ca, Mn, Sr, Na, and K,..can be assigned to roeblingite.

Roedean (rōu·dīn). The name of an independent public school for girls (founded 1885) in the borough of Brighton, applied (usu. *attrib.*) to refined speech or behaviour in (young) women, such as is popularly associated with the girls of this school. (Freq. in derogatory use.)

1948 C. Day Lewis Otterbury Incident 83 Now don't go all Roedean with me, beautiful. 1958 J. Cannan And be a Villain iv. 94 Dropping the high clear Roedean voice she affected and speaking naturally. 1963 M. Frayn in Sissons & French Age of Austerity 336 The orange-girls.. articulating 'Come, gentle people, buy', in sub-Roedean accents. 1969 S. Hyland Top Bloody Secret ii. 114 The switchboard girl spoke English in the best Roedean manner. 1972 J. Rossiter Rope for General Dietz iii. 31 Her accent was a creamy 1969-vintage Roedean. 1977 F. Branston Up & Coming Man x. 95 A Roedean accent which could have flayed the skin off a waiter.

Roederer (rō·dərəɹ). Also **Rœderer.** The proprietary name of a champagne produced by the firm of Roederer in Rheims.

1872 B. Jerrold London vii. 68 The would-be aristocrat flashing his silver mug of foaming Rœderer in the eyes of the Vulgar. 1876 Trade Marks Jrnl. 15 Nov. 803/1 Theophile Roederer & Co. Reims. Maison fondée en 1864. ..Gustave Bousigues, dit Bley, of and on behalf of the firm of Theophile Roederer and Co., Reims, France; champagne wine merchants... Champagne wine. 1907 [see *Mumm]. 1920 G. Saintsbury Notes on Cellar-bk. v. 74, I had some good wines of it—Pommery and Krug and Roederer among them. 1974 H. R. F. Keating Underside xvi. 156 There was Roederer to drink, a small case of it packed in ice.

‖ **roemer** (rō·məɹ). [a. Du. *roemer*, G. *römer*; cf. Rummer.] A type of decorated German or Dutch wine-glass with a knobbed or 'prunted' stem.

1897 A. Hartshorne Old Eng. Glasses 47 Germans were making roemers in the Low Countries before the middle of the seventeenth century, and there also the quaint vessel had a long course, being the glass par excellence, with its delicate shades of blue, yellow, or green, that the painters never tired of painting. 1926 W. Buckley European Glass 55 Particular mention should be made of the 'roemer', a form that is usually acknowledged not only to be the most beautiful product of the German industry but one of the most beautiful forms that has been made in glass in any country. 1942 [see *pass-glas]. 1972 Country Life 23 Mar. 700/2 No one who is familiar with German and Dutch glass..can fail to notice ..the prunts on the stems of the popular wine glass of those days—the roemer.

roemerite (rə̄·mərəit). *Min.* Also **römerite.** [ad. G. *römerit* (J. Grailich 1858, in Sitzungsber. d. K. Akad. d. Wissensch. in Wien XXVIII. 272), f. the name of Friedrich Adolph *Römer* (1809–69), German geologist: see -ite[1].] A hydrated sulphate of ferrous and ferric iron, often containing zinc, which occurs as rust-brown to yellow triclinic crystals, usu. as an oxidation product of pyrite.

1877 E. S. Dana Text-bk. Mineral. iii. 373 Alum and Halotrichite Groups. Here belong: Tschermigite, ammonium alum. Kalinite, potassium alum... Also Roemerite, and Voltaite. 1903 Jrnl. Chem. Soc. LXXXIV. ii. 555 Römerite was prepared by allowing a mixture of powdered ferrous sulphate and acid ferric sulphate to remain in contact with moist air for several months. 1927 Amer. Mineralogist XII. 282 Chemically, roemerite is a double sulphate of ferrous and ferric iron, the ferrous iron of which may sometimes be replaced by zinc and magnesium, and the ferric iron by aluminum. 1970 Ibid. LV. 78 Roemerite is generally the result of an oxidation of iron sulfides.

Roentgen, roentgen (rō·ntχyĕn; now usu. anglicized, as rɒ·ntyĕn; also rə̄·nt-, rǫ·nt-; -gĕn, -ʒĕn). Also **Röntgen, röntgen.** [The name of Wilhelm Conrad *Röntgen* (1845–1923), German physicist, who discovered X-rays in 1895 (Sitzungsber. d. Phys.-Med. Ges. z. Würzburg 132).] **1.** *attrib.* († or in the possessive), as *Roentgen rays*, X-rays. Hence *Roentgen photograph*, *therapy*, etc. Occas. written as a prefix (cf. *roentgen-, roentgeno-). Now chiefly *U.S.*

1896 [in Dict. s.v. Röntgen]. 1896 Lancet 1 Feb. 326/2 Do Roentgen's rays possess germicidal properties? Ibid. 22 Feb. 477/1 Two preliminary short exposures to Roentgen rays indicated that the metal..was probably embedded among the bones of the wrist. 1910 Arch. Roentgen Ray XV. 85 In Roentgentherapy the filter has brought the treatment of hypertrichosis once more into the domain of practical politics. 1911 Encycl. Brit. XXIII. 695/1 The radiation..from tin is as penetrating as that given out by a fairly efficient Röntgen tube. Ibid., The incidence of Röntgen rays on matter causes the matter to emit cathodic rays. 1933 U. V. Portmann in O. Glasser Sci. of Radiol. xii. 221 In the beginning of roentgen therapy only skin diseases and some superficial malignant conditions were treated. 1940 H. K. Pancoast et al. (title) The head and neck in roentgen diagnosis. Ibid. xi. 773 The chest should always be included as a part of any roentgen examination of the neck, particularly in infants and young children. 1953 A. J. Delario Roentgen, Radium & Radioisotope Therapy iii. 17 Because roentgen rays have such short wave-lengths, they cannot be diffracted by various diffraction gratings, as can visible light. 1956 L. A. Hadley Spine iv. 95 By this method it is possible to furnish roentgen evidence of ligamentous or soft tissue injury. 1959 W. T. Moss Therapeutic Radiol. ii. 35 With few exceptions all roentgentherapy techniques, by necessity, entail skin irradiation. 1959 W. T. Murphy Radiation Therapy xxxvi. 770 Roentgen sickness is not frequently seen after pelvic irradiation. 1972 H. L. Kundel in E. J. Potchen Current Concepts in Radiol. I. i. 1 (heading) Factors limiting roentgen interpretation—physical and psychologic. 1978 S. Sheldon Bloodline ii. 33 A Roentgen desk in the library.

2. (Usu. in the form **roentgen.**) In full **roentgen unit.** The unit of exposure to X or gamma radiation, equal to the quantity of radiation that gives rise to ions carrying a total charge of 2·58 coulombs (regardless of sign) per kilogramme of air. Abbrev. *r.* [Proposed (in Fr.) by I. Solomon 1921, in Arch. d'Électr. méd. expér. & clin. XXIX. 362.]

The precise definition of this unit has been altered several times. Cf. *rem *sb.*[1], *rep[3].

1922 [see *R II. 4]. 1932 Radiology XVIII. 95/2 At the second International Congress of Radiology in 1928..the measurement of air ionization was accepted as the basis of international dosage measurement and a definition was given of the unit of dosage designating a roentgen unit and written in abbreviated form as 'r'. 1938 R. W. Lawson tr. Hevesy & Paneth's Man. Radioactivity (ed. 2) xxiv. 258 The maximum daily dosage of γ-rays that a human being can tolerate without apparent harm is 0·1 röntgen unit. 1950 Radiology LV. 744/1 As our exciting energies increased, we were placed in a position of having continually to modify the definition of the roentgen in order to cope with the new properties of the higher-energy radiations. Because of this situation, minor modifications in the definition of the roentgen were made in 1931, 1934 and 1937. 1955 Bull. Atomic Scientists Sept. 257/2 It is known that radiation dosages in the levels of 400 roentgen units..are lethal to about half the individuals exposed. 1956 Brit. Jrnl. Radiol. XXIX. 355/2 The radiation dose in röntgens within any volume element is determined by the number and energy of the photons passing through that element during the exposure, irrespective of the local distribution of matter. Ibid. 356/1 Difficulties have only arisen when we have tried to make the röntgen do service for a unit of absorbed dose. 1959 Listener 26 Nov. 929/2 The fall-out from testing bombs gives a thirty-year dose of ·1 roentgen. ..The dose from natural radiation is about 3–5 roentgens. 1970 Passmore & Robson Compan. Med. Stud. II. xxxiii. 4/2 The same exposure (in roentgens) may result in different absorbed doses (in rads) in different tissues. 1977 [see *protocol sb. 3].

Hence † **Roe·ntgenized** *ppl.a.*, subjected to the action of X-rays; also † **Roe·ntgenize** *v. trans.* (*rare*: in quot. *fig.*); † **Roe:ntgeniza·tion.**

1897 Phil. Mag. XLIII. 243 The effect can in no way be due to conduction through the Röntgenized air. 1899 Röntgenized [see Röntgenize v.]. 1907 Med. Rec. (N.Y.) 9 Nov. 760/2 In the use of radium, if enclosed in aluminium or mica receptacles, we utilize the beta or cathode ray. This we cannot do in Roentgenisation. 1909 E. Reich Nights with Gods 17 It [sc. jealousy] has Röntgenised the most hidden interiors. 1920 Arch. Radiol. & Electrotherapy XXIV. 270/1 Röntgenisation of the lymphatic glands should always supplement radium therapy.

roentgen-, roentgeno-. Chiefly *U.S.* Also **röntgen(o)-.** Comb. forms of *Roentgen, as in **roentgenky·mogram** [ad. G. *röntgenkymogramm* (Gött & Rosenthal 1912, in München med. Wochenschr. 17 Sept. 2033)], a recording made with a kymograph (sense *2); **roentgenky·mograph** = *kymograph 2; **roe:ntgenkymogra·phic** *a.*, of, pertaining to, or involving the roentgenkymograph; hence **roe:ntgenkymogra·phically** *adv.*; **roe:ntgenkymo·graphy** [ad. G. *röntgenkymographie* (Gött & Rosenthal 1912, loc. cit.)], the process or technique of using a kymograph (sense *2); kymography; **roe·ntgenogram,** an X-ray photograph; cf. Radiograph *sb.* 2 in Dict. and Suppl.; **roe·ntgenograph** *sb.* = prec.; also as *v. trans.*, to take an X-ray picture of (an organ, etc.); = *radiograph *v.*; **roentgenogra·phic** *a.*, pertaining to or involving roentgenography; hence **roentgenogra·phically** *adv.*; **roentgeno·graphy,** radiography carried out by means of X-rays; **roentgenolo·gic, -lo·gical** *adjs.*, of, pertaining to, or involving roentgenology; hence **roentgenolo·gically** *adv.*; **roentgeno·logist,** one who practises roentgenology; **roentgeno·logy,** † (*a*) (see quot. 1905); (*b*) (the field of science concerned with) the medical use of X-rays, esp. as a diagnostic tool; cf. *radiology; **roe·ntgenoscope** *sb.* = *fluoroscope; hence as *v. trans.*, to examine by means of a fluoroscope; **roentgenosco·pic** *a.*, fluoroscopic; hence **roentgenosco·pically** *adv.*; **roentgeno·scopy,** fluoroscopy; **roent-genothe·rapy,** radiotherapy carried out by means of X-rays.

1913 Arch. Roentgen Ray XVII. 379 (caption) Roentgenkymogram of the left ventricle. 1930 Arch. Internal Med. XLV. 63 The slit in the lead sheet was placed over this point and the roentgen-kymogram taken. 1942 A. M. Master Electrocardiogram (ed. 2) 226 The character of the ventricular pulsations, as seen..in the roentgenkymogram, may..lead to the suspicion of cardiac aneurysm. 1913 Jrnl. Amer. Med. Assoc. 4 Apr. 1127/2 (heading) Analysis of electrocardiogram by means of the Roentgenkymograph. 1968 Luisada & Sainani Primer of Cardiac Diagnosis xviii. 117 The waves revealed by the electrokymograph are similar to those of the roentgenkymograph. 1930 Arch. Internal Med. XLV. 71 Variability of stroke volume was surmised because of..the different shape of the roentgen-kymographic curves produced by ventricular systole. 1970 G. H. Alexander Heart i. 10/2 The roentgenkymographic studies of the heart which have been done at St. Francis by the author since 1957. 1940 Amer. Heart Jrnl. XIX. 462 Characteristic abnormalities in left ventricular pulsation as recorded roentgenkymographically in 200 cases of myocardial infarction. 1913 Arch. Roentgen Ray XVII. 378 (heading) Roentgenkymography: a roentgenographic method of demonstrating the movement of the heart. 1971 Amer. Jrnl. Roentgenol. CXI. 868/1 The authors have studied by analytic roentgen kymography the pulsations of the thoracic aorta in 5 patients. 1904 F. P. Foster Appleton's Med. Dict. 1707/1 Röntgenogram, a Röntgen ray picture. 1907 Med. Rec. (N.Y.) 10 Aug. 246/1 He remembered a case where a Roentgenogram demonstrated an object in the right ureter, and a diagnosis of a stone was made; but on incision no stone was found. 1977 Jrnl. Bone & Joint Surg. LIX. 575/1 The fracture was demonstrated.. by a roentgenogram of the carpal tunnel. 1905 Jrnl. Amer. Med. Assoc. 23 Dec. 1971/1 The film is then inserted, held in place by an assistant or the patient himself, the point of pressure being..on the particular tooth to be Roentgenographed. 1909 Cent. Dict. Suppl., Roentgenograph,..a radiograph; a Röntgenogram. 1940 Amer. Jrnl. Roentgenol. XLIV. 944/1 The side of a bone cavity opposite the focus is to be roentgenographed. 1965 Ibid. XCV. 135/1

When roentgenographed, the involved areas show widened diploic space and radial striation of bone spicules. 1977 Environmental Res. XIII. 47 Serial roentgenographs of histologically confirmed massive fibrotic lesions in 14 deceased gold miners were retrospectively studied. 1909 Amer. Jrnl. Med. Sci. CXXXVII. 377 (heading) A Röntgenographic study of peristalsis. 1961 R. D. Baker Essent. Path. ix. 169 It is the calcification in the primary tuberculous complex which makes possible the recognition of the condition by roentgenographic study. 1977 Amer. Jrnl. Med. LXII. 366/2 Eight of the 14 atypical cases were diagnostic problems because of atypical roentgenographic and clinical features. 1909 Amer. Jrnl. Med. Sci. CXXXVII. 420 The width of the apices of the lungs has not yet been sufficiently studied roentgenographically to find employment diagnostically. 1965 Arch. Internal Med. CXV. 580/2 Her hands were normal and roentgenographically both hands and feet were normal. 1905 Nature 27 July 301/1 Among the results of the recent Röntgen congress at Berlin has been the authoritative adoption by a special committee of the following terminology:— Röntgenology = the study of Röntgen rays, Röntgenoscopy = observation by Röntgen rays, Röntgenography = photography by the rays... Röntgenotherapy and the verb to Röntgenise in their obvious meanings. 1958 Optima Sept. 130/2 The double platinum salts, such as platinum lithium cyanide or platinum thorium cyanide, are brilliantly fluoroscopic, and are used in roentgenography and fluoroscopy. 1912 Amer. Jrnl. Med. Sci. CXLIII. 754 (heading) A Röntgenologic study of spastic obstipation. 1967 Amer. Jrnl. Roentgenol. CI. 457/2 (caption) Roentgenologic signs of hypertrophic pyloric muscle in a 66 year old man. 1911 Arch. Roentgen Ray XV. 328, I propose to give my impressions from the Roentgenological point of view. 1925 S. Lewis Martin Arrowsmith xxv. 292 The clinic did, perhaps, give overmany roentgenological examinations to socially dislocated women who needed children and floor-scrubbing more than pretty little skiagraphs. 1977 Surgery LXXXII. 848 (heading) An aggressive roentgenological and surgical approach to acute mesenteric ischemia. 1915 Amer. Jrnl. Roentgenol. II. 795/1 The cases to be examined roentgenologically are selected from those which have gastric symptoms. 1968 Jackman & Beahrs Tumors of Large Bowel iv. 50 This method enabled them to find additional polyps, other than those detected roentgenologically. 1905 Jrnl. Amer. Med. Assoc. 23 Dec. 1971/1 Permit me to make a few remarks about the technic employed, which, I think, will be of value to every Roentgenologist. 1961 R. D. Baker Essent. Path. xix. 533 Granules of calcium often form in the pineal after puberty and are helpful in locating the midline of the brain for the roentgenologist. 1905 Roentgenology [see roentgenography above]. 1914 Jrnl. Amer. Med. Assoc. 22 Aug. 651/2 The utility of some insight into Roentgen procedures on the part of the general practitioner does not nullify the advantages of specialization in roentgenology. 1977 Surg., Gynecol. & Obstetr. CXLIV. 563/2 Conventional roentgenology confirmed the presence of a tumor at the gastroesophageal junction. 1923 Amer. Jrnl. Roentgenol. X. 722 (heading) A plea for the use of the roentgenoscope in the diagnosis of urinary calculi. 1924 Ibid. XI. 93/2 With most..types of the tilt-table roentgenoscope, it will be necessary to remove the screen ordinarily used. 1926 Jrnl. Amer. Med. Assoc. 19 June 1904/2 An arrangement should be made by which the patient can be roentgenoscoped at right angles without change of position. 1955 Amer. Jrnl. Roentgenol. LXXIV. 812/1 Anyone who has roentgenoscoped an infant's chest realizes that an infant can squeeze virtually all the macroscopic air out of his lungs..with crying. 1909 Amer. Jrnl. Med. Sci. CXXXVII. 418 When both apices are equally or only slightly darkened, I have grown cautious with röntgenoscopic diagnosis. 1965 Biol. Abstr. XLVI. 8241/2 (heading) Roentgenoscopic investigation of oil-bearing seeds. 1909 Amer. Jrnl. Med. Sci. CXXXVII. 420 The process, röntgenoscopically at least, slowly advances, and after weeks is recognized as tuberculosis. 1945 Amer. Jrnl. Roentgenol. LIII. 608/1 The alimentary canal as observed roentgenoscopically. 1904 F. P. Foster Appleton's Med. Dict. 1707/1 Röntgenoscopy, examination with the aid of the Röntgen rays. 1914 Jrnl. Amer. Med. Assoc. 21 Nov. 1828/1 Roentgenoscopy has proved that ossification may ..occur in the epiglottis. 1971 Biol. Abstr. LII. 5081/1 The essential methods of roentgenological examination ..are roentgenoscopy and roentgenography. 1903 Med. Rec. (N.Y.) 31 Jan. 168/2 Extravagant promises will discredit the new and delicate field of Röntgenotherapy. 1925 Jrnl. Amer. Med. Assoc. 28 Feb. 671/1 Roentgenotherapy effects a complete disappearance of palpable lymph nodes and reduces the spleen to its normal size. 1960 Biol. Abstr. XXXV. 4048/1 Giving 2–3 drops each of validol..on sugar to patients with carcinoma of various sites (50) after a session of roentgenotherapy..usually eased the general reaction to the irradiation.

roe·ntgenite. *Min.* Also **röntgenite.** [f. *Roentgen, roentgen + -ite[1].] A rhombohedral fluorocarbonate of cerium, lanthanum, and calcium, $Ca_2(Ce,La)_3(CO_3)_5F_3$, found as small yellow or brown crystals in association with other rare-earth minerals at Narsarsuk, Greenland.

1953 G. Donnay in Amer. Mineralogist XXXVIII. 868 (heading) Roentgenite, $3CeFCO_3.2CaCO_3$, a new mineral from Greenland. 1966 Z. Lerman tr. K. A. Vlasow's Geochem. & Mineral. of Rare Elements II. viii. 272 Röntgenite $Ce_3Ca_2(CO_3)_5F_3$. Identity established by Donnay in 1953 during X-ray analysis of rare-earth fluocarbonate specimens from Greenland. 1975 Amer. Mineralogist LX. 351 Intimate syntaxy between parisite, synchisite, roentgenite, and bastnaesite was..observed even on a very fine scale.

roepperite (rə̄·pərəit). *Min.* [f. the name of William T. *Roepper* (1810–80), German-born U.S. mineralogist + -ite[1].] A black mineral of the olivine group containing iron, manganese, and zinc.

The name *röpperit* (G.) was also proposed in 1872 by A. Kenngott (in Neues Jahrb. f. Mineral. 188) for a species of manganiferous dolomite described by Roepper.

1872 G. J. Brush in Dana's Syst. Min. (ed. 5) App. i. 13 Roepperite. Iron, manganese, zinc, chrysolite... Roepperite, G. J. Brush. 1875 E. S. Dana Ibid. App. ii. 49 Kenngott..has proposed to give the name rœpperite to the manganese dolomite, analysed by Rœpper.... Almost simultaneously..Brush gave the name rœpperite to the new chrysolite of Rœpper, and there is no question but that this name should be received. 1961 Amer. Mineralogist XLVI. 549 Roepperite is black... It has been described as a variety of tephroite containing notably high amounts of FeO and ZnO. 1972 Ibid. LVII. 977 The infrared spectrum of the zincian olivine roepperite..is comparable to those of Fe-Mn olivines and bears no resemblances to the spectrum of willemite containing tetrahedral Zn^{2+}.

roesslerite (rə̄·slĕrəit). *Min.* Also **rösslerite.** [ad. G. *rösslerit* (R. Blum 1861, in Jahresber. der Wetterauischen Ges. für die ges. Naturkunde 33), f. the name of Carl *Rössler*, 19th-c. German scientist: see -ite[1].] A hydrated acid arsenate of magnesium, $MgHAsO_4.7H_2O$, which occurs as small colourless plates forming an oxidized crust on some arsenical deposits, and has been prepared artificially.

1868 J. D. Dana Syst. Min. (ed. 5) 556 A mineral in monoclinic crystals occurs at Joachimsthal and Kremnitz. ..which is probably rœsslerite. 1903 Jrnl. Chem. Soc. LXXXIV. ii. 656 Crystals..of the arsenic compound, $(NH_4)MgAsO_4.6H_2O$, isomorphous with struvite were obtained, and at the same time crystals of rösslerite ($MgHAsO_4.7H_2O$). 1951 C. Palache et al. Dana's Syst. Min. (ed. 7) II. 712 Artificial roesslerite is precipitated together with $MgNH_4AsO_4.6H_2O$..from an acid solution of disodium arsenate and ammonium sulfate by a solution of magnesium sulfate. 1973 Acta Crystallographica B. XXIX. 287/1 Roesslerite is biaxial negative.

‖ **roesti** (rō·sti). Also **rosti, rösti.** [Swiss Ger.] A Swiss style of fried potatoes. (Variously taken as *sing.* and *pl.*)

1952 H. Sutton Footloose in Switzerland ii. 55 The most typical dish of Zurich is something known as g'schnetzeltes... It comes served with noodles or roesti, which are home fries. 1953 New Horizons (Pan Amer. World Airways) (new ed.) 122/1 Each region [of Switzerland] has its specialties: Fondue and Raclette..in the French section..sausages, roasts and fried potatoes (Rösti) in the German section. 1961 P. Cannon Eating European 223 The Swiss Rosti, which is close to being their national home dish, is made with cooked potatoes. 1961 N., S. Hazelton Continental Flavor 304 Roesti is really a version of home-fried potatoes, and certainly the best. 1973 M. Waldo Compl. Round-World Cookbk. 145 Fried Potatoes... Rösti.

rog (rǫdʒ), *int.* Abbrev. of *Roger[2] 6.

1955 R. J. Schwartz Compl. Dict. Abbrev. 155/3 Rog, roger. 1969 Guardian 21 July 1/3 Back came the single syllable answer from the spacecraft... 'Rog.' 1970 N. Armstrong et al. First on Moon xiv. 354 'You're cleared for landing.'.. 'Rog. Gear is down and locked.'

Rogallo (rogæ·lo). Also **rogallo.** The name of Francis M. *Rogallo*, 20th-c. U.S. engineer, used *attrib.* and *absol.* to designate a light, flexible, triangular wing deployed by means of tension lines or rigid tubes and used on spacecraft and for hang-gliding.

1961 Rogallo wing [see *paraglider]. 1968 McGraw-Hill Encycl. Space i. 127 The tanks..are specially arranged to accommodate the housing of the recovery wing. The latter, known by the names of Paraglider, Flex-wing or Rogallo Wing (from the name of its inventor, Francis M. Rogallo, director of the large wind-tunnel at NASA's Langley Research Center) is a flexible wing which is deployed in flight and enables the rocket to glide. 1974 Observer (Colour Suppl.) 7 Apr. 68/1 The rogallo, a triangular kite-shape wing made of aluminium tubing with a nylon sail. 1974 Sci. Amer. Dec. 138/1 Most sky surfers first learn to fly on a Rogallo kite. 1978 A. Welch Bk. of Airsports i. 9/2 They come in an increasing variety of shapes and sizes, ranging from the basic rogallo for club and school use to almost aeroplane planform hang gliders.

rogan (rōu·găn). *Canad.* Also 8 **roggan,** 9 **roggin.** [ad. Canad. Fr. (*h*)*ouragan*, f. Algonkian (see W. S. Avis Dict. Canadianisms).] A water-tight container made of birch-bark.

1743 J. Isham in Publ. Hudson's Bay Rec. Soc. (1949) XII. 188 A Roggan. Slawee. 1791 P. Fidler Jrnl. 10 Nov. in Publ. Champlain Soc. (1934) XXI. 523 We are obliged to roast all & make water by immersing red hot stones into a roggan of Snow. 1820 J. Clouston Jrnl. 10 July in K. G. Davies Northern Quebec & Labrador Jrnls. & Corr. 1819–35 (1963) 57 He had a wooden roggin which would hold about five gallons. 1894 Outing Nov. 127/1, I saw..the 'rogans', or water-tight vessels of birch-bark, beautifully stitched with roots, and trimmed around the opening with colored porcupine-quills. 1922 Beaver June 7/1 Ornamented work baskets, plain baskets or 'rogans', for holding fish, game, berries, or canoe pitch are also made of the bark. 1968 E. S. Russenholt Heart of Continent i. i. 1 These hunters heat stones in their open fires; and drop them into birch bark rogans filled with water and meat—until the water boils and cooks the meat.

Roger[2]. Add: **3. a.** (Further examples.)

Quot. 1653 in Dict. seems to be a ghost.

1720 D'Urfey Pills VI. 201 Here's a Health to the Queen, let's Bumpers take in hand, And may Prince G——'s Roger grow stiff again and stand. c 1800 Burns in Merry Muses of Caledonia (1959) 147 Bonie lassie, braw lassie, Will ye hae a soger? Then she took up her duddie sark, An' he shot in his Roger. c 1863 'Philo Cunnus' Festival of Passions ii. 25 With my right hand, I grasped my flaming Roger.

6. Also with small initial. As *int.* Used to represent the letter *r* (= received) in radio transmission (see quot. 1947). Also *transf.* in general use, an expression of affirmation.

1941 Amer. Speech XVI. 168/1 Roger! Expression used instead of okay or right. (Air Corps). 1943 Signal Training (All Arms): Signal Procedure I. 9 Roger, used to mean 'message received'. 1945 Sun (Baltimore) 25 Jan. 9 Sometimes a voice called, 'Flak.' Once I heard one pilot say, 'Are you hit?' The reply was, 'Roger, I am hit. Going home.' 1947 Amer. Speech XXII. 110 In radio procedure the letter R, or roger, possesses the code designations 'received', or 'I have received your message', when signalled by the station addressed... Nevertheless, since radio operators or pilots signalling roger are receipting for a message, it has also come to mean unofficially 'O.K.' or 'I understand'. 1954 J. Masters Bhowani Junction xix. 170, I heard the duty officer on the R/T: 'Dogfish Six speaking... Roger, over.' 1960 Sunday Express 18 Sept. 1/3 If he wanted to speak to you he would ask you to go in. Roger? 1963 D. Irving Destruction of Dresden iii. ii. 132 'Tell the aircraft in top height band to come down below the medium cloud.' 'Roger.' 1971 D. Haston in C. Bonington Annapurna South Face xvii. 210 At first it had been a chore to use them [sc. radios] and the jargon of Roger, Over, etcetera, had seemed artificial, but sitting in the specific loneliness of Camp VI it was a good feeling to communicate with others.

roger (rǫ·dʒəɹ), *v.*[1] *slang.* Also **rodger.** [f. Roger[2].] *trans.* To copulate with (a woman); to have sexual intercourse with. Also *absol.* Hence **ro·gering** *vbl. sb.* and *ppl. a.*

1711 W. Byrd Secret Diary 26 Dec. (1941) 459, I rogered my wife. c 1750 A. Robertson Poems 98 Dear sweet Mr. Wright..Go rodger to-night Your Wife, for ye want her. 1763 Boswell Jrnl. 4 June in London Jrnl. (1950) 273, I picked up a little profligate wretch and gave her sixpence... 'Should not a half-pay officer r-g-r for sixpence?' 1771 [see Ragmatical a.]. 1870 Cythera's Hymnal 81 He rogered the National School. 1884 tr. Abishag in Old Man Young Again (1898) I. 36, I gave Mrs. P—. a really good rogering, and sent her to sleep perfectly contented. 1919 E. Pound Sel. Lett. (1971) 150 If I were, however, a professor of Latin in Chicago, I should probably have to resign on divulging the fact that Propertius occasionally copulavit, i.e. rogered the lady to whom he was not legally wedded. 1931 E. Waugh Diary 14 Jan. (1976) 347 He got very drunk and brought a sluttish girl back to the house. He woke me up later in night to tell me he had rogered her and her mama too. 1942 E. Paul Narrow St. xvi. 116 When Rudolph Valentino died... 'Hey, American,' yelled Madame Absalom... 'What did that type have that other men have not? He must have rogered half the women in your country?' 1953 Landfall Sept. 179 You black-mouth, you night bird, you rogering swine. 1953 Dylan Thomas Let. 22 June (1966) 409, I..sulked all morning over my warm beer as they..rolled rodgering down. 1961 A. Wilson Old Men at Zoo i. 54 I'm not at all sure about the Empress Theodora. I fancy she was rogered by an ape more than once in her circus acts. 1967 D. Pinner Ritual xvii. 167 He singed the rogering labourers... It took minutes before fornication subsided. 1972 'R. Gordon' Doctor on Brain xxiii. 168 'Who is the father of the child?' 'The man who rodgered her, of course.' 1976 K. Bonfiglioli Something Nasty in Woodshed iii. 32 You won't catch him... The bloke who rogered Mrs Breakspear.

roger (rǫ·dʒəɹ), *v.*[2] *U.S.* [f. *Roger[2] 6.] *trans.* To acknowledge (a message, etc.) as received.

1962 J. Glenn in Into Orbit 195 Both of these readings were within limits and I rogered the message. 1977 J. Wambaugh Black Marble (1978) vi. 83 'We just got a call,' he said. 'Roger it, please.'

Roger de Coverley. (Earlier examples with *Sir* and in abbreviated form.)

1804 H. Wynne Diary 19 Dec. (1940) III. v. 147 We danced Sir Roger de Coverly. 1875 L. Troubridge Life amongst Troubridges (1966) x. 101, I danced every dance except Sir Roger, at the end.

Rogerene (rǫ·dʒərīn). *U.S.* [f. the name *Rogers* (see below) + *-ene* as in Nazarene *a.* and *sb.*] A member of a small religious sect founded by John *Rogers* (1648–1721) in Connecticut, opposed to some of the formal practices of churches and participation in military service. Also *attrib.* in *Rogerene Quaker.*

1754 J. Hempstead Diary 17 Mar. in Coll. New London Co. Hist. Soc. (1901) I, A Company of the Rogerens..held their meeting after our meeting was over. 1820 Niles' Reg. 22 July 366/1 A contagious disorder is now raging among the sect known by the name of Rogereen Quakers in Grotan. 1865 Harper's Mag. May 812/2 In the year 1720 a sect arose in New London, Connecticut, called, from their leader, 'Rogerenes'. 1865 Massachusetts Hist. Soc. Coll. VII. 584 John, the third son of James Rogers, of New London, and the founder of the sect of Rogerenes, of whom a small number still remain in that vicinity. 1931 Times Lit. Suppl. 6 Aug. 602/3 The Shakers, the Christadelphians and the Rogerenes are other sects. 1943 New England Q. Mar. 3 On a wooded hill above Mystic,

Connecticut, live the remnants of a little-known religious sect called the Rogerenes, or sometimes Rogerene Quakers.

Roget (rǫ·ʒēi). The name of Peter Mark *Roget* (1779–1869), English physician and philologist, used *absol.* with reference to his *Thesaurus of English Words and Phrases*, a catalogue of synonyms first published in 1852. Also in *Comb.*

1940 *Times* 19 Apr. 7/4 To journalists and other writers, weary of racking their brains or raking the well-thumbed pages of Roget in search of alternatives, the word 'Quisling' is a gift from the gods. **1955** E. Blishen *Roaring Boys* III. 152 Charles was like some oral Roget, uttering long lists of horrid synonyms. **1962** L. Deighton *Ipcress File* xiii. 75 A few books remained on the shelves, a *Roget*, a business directory..and a *Chambers's Dictionary*. **1970** D. L. Emblen *P. M. Roget* xv. 276 Again and again, letters to *The Times* and other papers call upon other writers..to consult their 'Roget' before making such execrable use of the language. **1973** M. Amis *Rachel Papers* 113 So ended my short, derivative, *Roget*-roughaged essay, complete with stage-directions.

‖ **rognon** (rǫnʸoṅ). [Fr.] **1.** Chiefly *pl.* In Gastronomy, (a dish of) kidneys. Also *attrib.* and *Comb.*

1828 Lytton *Pelham* I. xii. 79 What cook *can* possibly respect men who..eat *rognons* at dinner instead of at breakfast. *c* **1864** S. O. Beeton in N. Spain *Mrs. Beeton* (1948) II. vi. 212 Everybody had just well breakfasted upon cotelletes, omelettes, Rognons. **1877** C. Reade *Woman-Hater* I. v. 97 After the *rognons à la brochette*, and a bottle of champagne, he let out. **1923** A. Huxley *Antic Hay* iv. 61 'And where are my *rognons sautés*?' he shouted at the waiter. **1967** A. Wilson *No Laughing Matter* iii. 337 She..cooked specially for him as she had not done for ages, rognons Bercy and omelette confiture. **1972** *Guardian* 11 Mar. 15/3 [The] Brasserie du Nord..is noted..for its *saucisson* and *rognon* dishes. **1979** *Times* 15 Dec. 6/6 My mother followed the sun... She lay down.. darkening like *rognons* on a spit.

2. *Mountaineering.* A rounded outcrop of rock or stones surrounded by a glacier or an ice-field.

1935 S. Spencer *Mountaineering* 364 *Rognon*, rounded rock in the centre of a glacier. **1957** R. G. Collomb *Dict. Mountaineering* 127 In Victorian days some *rognons* were used as sleeping places.., e.g. the Stöckje on the Scheonbuhl [*sic*] glacier near Zermatt. **1958** *Jrnl. Glaciology* III. 264 On the upper parts of the Glaciar Universidad there had been little change. After 1945 the surface sank slightly around a *rognon* (rounded nunatak) at 3930 m. **1963** *Oxford Mountaineering* 11 At the top of the rognon the snow steepened and we had to traverse up and round to the foot of the..couloir which leads on to the face. **1973** C. Bonington *Next Horizon* xxi. 279 Our way lay across the glacier and up a rognon, a sort of island of rocks round which the glacier flowed on either side.

rogue, *sb.* Add: **2. d.** *rogue and villain*: rhyming slang for 'shilling'.

1859 Hotten *Dict. Slang* 145 *Rogue and villain*, a shillin,—common pronunciation of shilling. **1887** J. W. Horsley *Jottings from Jail* i. 3 Come, cows and kisses, put the battle of the Nile on your Barnet Fair, and a rogue and villain in your sky-rocket. **1965** *Australasian Post* 4 Mar. 46 *Shilling*... Sometimes known in rhyming slang as a 'rogue and villain'. **1973** B. Aylwin *Load of Cockney Cobblers* xiv. 62 *Rogue & villain*, shilling.

5. a. Also *fig.* (*attrib.* or as *adj.* in quots.).

1963 *Times Lit. Suppl.* 18 Jan. 44/3 His role is that of the *advocatus diaboli* rather than the rogue elephant. **1978** K. Gregory *First Cuckoo* 21 The nation's rogue elephants rampage, shattering complacency and compelling many to an agonizing reappraisal.

b. Also *fig.*

1926 J. Masefield *Odtaa* x. 171 He roused up as a big, elderly rogue-bull of a man..came in. **1939** G. Household (*title*) Rogue male. **1977** N. Adam *Triplehip Cracksman* v. 56 I wasn't sure I liked myself..rogue male acting instinctively, obeying the territorial imperative.

6*. *attrib.* or as *adj.* in general use, denoting: **a.** An inexplicably aberrant result or phenomenon; an extra or misplaced item in a list, table, etc.

1952 *Analyst* LXXVII. 171 With the exception of one rogue result, the present estimates are as concordant as can reasonably be expected. **1964** C. Dent *Quantity Surveying by Computer* iii. 30 A device enabling you to switch the machine to manual and continue using it as an ordinary typewriter is very useful if, for instance, you wish to insert a 'spot' item, or other 'rogue' item in the bill of quantities, at the last moment. **1972** *Physics Bull.* Oct. 611/1 The tables have been well produced and very few errors were detected. In the body of the table, only one rogue point was noticed. **1979** *Personal Computer World* Nov. 73/2 When the program detects the rogue value, this is an indication that the input list is complete and further processing can continue.

b. Something that is inexplicably faulty or defective.

1962 *Daily Tel.* 18 Jan. 12/7 Manufacturers are aware that 'rogue' and sub-standard cars are sold to the public. **1965** *New Statesman* 30 Apr. 695/2 A group of American bombers..go rogue through a mechanical foul-up. **1971** *Atom* Apr. 99/1 Such differences are readily detectable and allow experimental fuel elements to be checked for rogue fuel pellets. **1974** *Guardian* 14 Mar. 9/2 His counsel ..told Mr Justice Phillips... 'You are familiar with the expression 'rogue car'. Well, this was rather like a rogue house.'

c. That which lacks appropriate control; something which is irresponsible or undisciplined.

1964 *Daily Tel.* 22 Feb. 14/6 Frequent complaints were made about a very small number of 'rogue' firms which belonged to neither the Association of British Travel Agents nor the Travel Trade Association. **1972** *Accountant* 19 Oct. 496/1 How is the ordinary man in the street to tell whether it has been calculated on the 'fair' basis as laid down by the legislation, or the unfair basis which will no doubt be perpetrated by a considerable number of rogue traders? **1979** *Daily Tel.* 4 Apr. 3 A housewife's game of patience came to an abrupt end when a 20-ton 'rogue' mechanical shovel begun crunching its way through the walls of her semi-detached home. **1981** *New Scientist* 29 Jan. 278/3 (*caption*) Gamma-ray bursts may come from collisions between rogue asteroids and neutron stars.

7. *rogue-word*; *rogue-eyed* adj.

1867 Meredith *Vittoria* I. ix. 133 She had, in tripping down the Piazza with her rogue-eyed cousin from Milan, looked away [etc.]. **1922** Joyce *Ulysses* 48 Roguewords, tough nuggets patter in their pockets.

8. rogue's gallery: also **rogues' gallery**; (earlier and later examples); also *transf.* and *fig.*

1859 *Amer. Jrnl. Photogr.* II. 75 The Rogues' Gallery is located at the police head quarters... The photographer is a regularly appointed policeman. **1904** [see *Pink *sb.**]. **1945** 'E. Queen' (*title*) Rogues' gallery. **1955** *Publ. Amer. Dial. Soc.* xxiv. 41 His [*sc.* a pickpocket's] face appears more frequently than any other type of criminal in 'rogue's galleries' and police files. **1959** *Listener* 26 Nov. 946/1 Mr. Klein's second collection of arch-deceivers, his new rogues' gallery, is fascinating. **1973** 'I. Drummond' *Jaws of Watchdog* x. 133 His face was not recognised in the rogues' gallery, nor did his description tally with any known criminal. **1977** McKnight & Tobler *Bob Marley* 10 Bob Dylan, Arthur Lee, Keith Richard, Bob Marley—the rogue's gallery of rebel input that forms the hard stuff at the centre of rock.

rogue, *v.* Add: **4.** Also, to take *out* (inferior plants) from a crop. (Further examples.)

1965 *Sunday Mail Mag.* (Brisbane) 26 Sept. 15 Sometimes we speak of 'rogueing' a crop, which means taking out the plants which aren't typical of the variety or which have become diseased. **1967** *Sunday Times* 19 Feb. 31/4 Small flowered plants should be rogued out or not allowed to seed. **1978** *Country Life* 20 July 184/3 White foxgloves..once established will seed themselves. If you keep roguing out any coloured throwbacks they could become a permanent feature.

roguer (rōu·gəɹ). [f. Rogue *v.* + -er¹.] A person employed to identify and eliminate inferior plants in a crop, esp. of potatoes.

1945 T. Whitehead et al. *Potato in Health & Dis.* (ed. 2) vii. 77 The roguer should remove or mark all plants which in any character differ substantially from the variety under consideration. **1960** *Times* 29 July 12/6 The roguer's job is to sample the crop for wrong varieties or disease. **1967** A. E. Cox *Potato* vii. 157 The roguers working in pairs—one marking plants and the other lifting and carrying them off—should deal with only two drills at a time.

roguing, *vbl. sb.* **3.** (Later examples.)

1968 *Punch* 18 Sept. 410/3 Any plants not true to type are removed by systematic 'roguing', which may account for a further ton per acre. **1978** Hide & Lapwood in P. M. Harris *Potato Crop* xi. 432 Roguing (negative selection) and later multiplication from disease-free plants (positive selection) were used to improve the health of seed.

Rohilla (rōᵘhi-lā), *sb.* and *a.* Also † **Rohella, Rohila.** [Pashto 'inhabitant of Roh', f. place-name *Rōh*, a district of Afghanistan: see also quot. 1885.] **A.** *sb.* A member of a people of Afghan origin inhabiting the Bareilly district of Northern India. **B.** *adj.* Of or pertaining to this people.

1773 W. Hastings *Diary* 21 Aug. (1948) 6 The Vizier added that the Abdallee maintained a Correspondence with the Rohellas. *Ibid.*, Money..due to the said Nabob by virtue of any engagement between him and the Rohilla chiefs. **1786** Burke *Wks.* (1868) IV. 221 That the said Warren Hastings..did, in September, 1773, enter into a private engagement with the said Nabob of Oude..to furnish him..with a body of troops for the declared purpose of 'thoroughly extirpating the nation of the Rohillas'. **1829** J. Tod *Rajast'han* I. xxv. 672 The fragments were..placed in position to receive the flesh-pots of the sons of Ishmaël, the mercenary Rohilla Afghan. **1885** G. C. Whitworth *Anglo-Indian Dict.* 269/2 *Rohillā*, [Pashto, from *roh*, a mountain.] The name of a highland clan of Paṭhāns who early in the eighteenth century took possession of the district, now called after them, Rohilkhand. **1892** Kipling *Barrack-Room Ballads* 104 We drove the black Rohillas back. **1921** G. A. Grierson *Linguistic Survey of India* X. 9 After the death of Aurangzēb, in 1707, the dissensions among the Hindūs of Bareilly enabled 'Ali Muḥammed Khān, the leader of the Rōhilā Paṭhāns, to obtain possession of the country which is now called, after the name of the tribe, Rohilkhand... It is hardly necessary to point out the connexion between Rōh and Rōhilā. The latter word means literally an inhabitant of the Rōh. **1960** J. S. Watson *Reign of George III* xii. 309 These Afghan soldiers of fortune, called the Rohillas, were suspected of co-operation with the Marathas. **1971** R. Russell tr. *Ahmad's Shore & Wave* vi. 52 Some Rohillas had recently been arrested and charged with a series of burglaries and armed robberies.

‖ **rohrflöte** (rō·rflōtə). *Mus.* Also **rohr flute**; *pl.* -**n.** [G., f. *rohr* tube + *flöte* flue-stop.] An organ stop having its pipes partly closed, the stopper at the top of each pipe being pierced by a thin tube.

1773 C. Burney *Present State of Mus. Germany* II. 305 Catalogue of stops in the Great Organ at Haarlem built by Müller 1738... 5. *Roer fluit*, 8 ft. with a funnel or small pipe upon the top. Eng. equivalent, Diap. half stopt. **1855** [see *gedackt, gedact]. **1898** Stainer & Barrett *Dict. Mus. Terms* 387/2 Rohrflöte, (Ger.) Reed-flute. An organ stop consisting of closed pipes, the tone of which is slightly reedy in quality, but very sweet. **1911** W. & T. Lewis *Mod. Organ Building* v. 93 Many stops.. occupy an intermediate position between two classes; as, for instance, a Rohr Flute. This stop is made to produce an upper harmonic which renders its timbre rather flutey in character, but at the same time, the fundamental tone is sufficiently obvious to link it with a Stopped Diapason. **1938** *Oxf. Compan. Mus.* 668/1 *Rohrflöte* or *Rohr Flute* (literally 'Reed Flute', but 'reed' here means a tube), of metal stopped pipes, with a slender tube through the stopper (hence the name). **1959** *Collins Mus. Encycl.* 554/1 *Rohrflöte*,..an organ stop of the flue type. The pipe is stopped at one end, but the stopper is pierced by a hole, in which is inserted a metal tube or chimney. **1966** P. Williams *European Organ 1450–1850* vii. 238 Organ-builders evidently brought with them new stops like *Rohrflöten*.

roibek, var. *rooibekkie. **roibok,** var. *rooibok.

‖ **roi fainéant** (rwa fęneaṅ). [Fr., lit. 'sluggard king': see Fainéant *sb.* and *a.*] One of the later Merovingian kings of France, whose power was merely nominal. Also *transf.* and *fig.*

1879 *Encycl. Brit.* IX. 530/2 Children were kings in both Austrasia and Neustria; we reach the days of the 'do-naught' princes, the *rois fainéants*, and of the struggle between the mayors of Austrasia and Neustria. **1898** L. Sergeant *Franks* xiv. 199 Dagobert's death.. marked the beginning of a series of Merovingian *rois fainéants*. **1929** W. R. Inge *Assessments & Anticipations* ii. 35, I have acquiesced in the undignified rôle of a *roi fainéant*. **1935** *Chambers's Encycl.* IV. 810/1 Charles Martel, *maire du palais* to the 'Rois Fainéants', defeated Arab invaders at Poitiers. **1966** *Economist* 1 Jan. 23/1 The launching of the Sputnik in 1957, in the reign of the *roi fainéant*, President Eisenhower, seemed to justify Khrushchevian boasts that America's days of..supremacy were numbered. **1975** J. H. M. Salmon *Society in Crisis: France in 16th Cent.* viii. 193 The last years of Charles IX were those of a *roi fainéant*. The king ordered a new offensive against the Protestants of the south and La Noue in the west, but there were no resources to finance his armies.

roil, *v.*³ Add: **2.** (Later example.)

1907 *Springfield* (Mass.) *Weekly Republ.* 17 Jan. 6 The publication of such a work naturally roiled the publishers of Webster's international dictionary.

3. *intr.* To move in a confused or turbulent manner; to billow.

1939 W. Faulkner *Wild Palms* 26 As something recognisable roils momentarily into view from beneath stagnant and opaque water, then sinks again. **1963** T. Pynchon *V.* i. 22 Engine exhaust roiled in clouds around him. **1964** D. F. Galouye *Counterfeit World* xiii. 113 The waters roiled with the restless presence of thousands of—. **1977** *Time* 6 June 46/2 Strange currents flow for years in the deeps of the American society, then for reasons unclear suddenly roil to the surface.

Hence **roi·ling** *ppl. a.*

1967 C. O. Skinner *Madame Sarah* viii. 171 Sarah glanced down at the roiling flood water. **1976** U. Curtiss *Birthday Gift* xiv. 132 One thing stood clearly out of the whole roiling mess.

roiled, *ppl. a.* For 'of the passions' read 'esp. of the passions' and add further *fig.* examples. Also with *up*.

1929 *Sun* (Baltimore) 12 June 1/1 It will be seen when the Senator gets roiled he can go the paces with the next one among the friends of the people. **1939** J. Steinbeck *Grapes of Wrath* x. 123 Your Pa's pa, he quoted Scripture all the time. He got it all roiled up, too. **1975** in W. Viereck *Lexikalische Ergebnisse des Lowman-Survey* I. iv. 279 If he lost his temper, you say he got... roiled.

roily, *a.* Add: *roily oil*, petroleum containing much emulsified water. Hence **roi·liness.**

1912 *Mem. Geol. Surv. India* XL. 121 [The well] being shut down at first on account of the 'roiliness' or emulsification of the oil. *Ibid.*, Two wells; one of these was a 'dry hole', but the other flowed during the first 24 hours 18,000 gallons of an emulsion of oil and water known in America as roily oil. **1915** Redwood & Eastlake *Petroleum Technol. Pocket-bk.* iv. 214 On recommencing pumping the well gave nothing but 'roily oil' for more than a month. **1920** E. H. C. Craig *Oil-Finding* (ed. 2) iii. 71 In a porous rock...oil and water may be inextricably intermingled... Such a rock struck in a well will probably yield 'roily oil', an emulsion very difficult to separate into its constituents, oil and water.

roineck, var. *rooinek.

‖ **roi soleil** (rwa sǫlęy). [Fr., lit. 'sun king'.] A title commonly used to designate Louis XIV of France, derived from a heraldic device used by him; applied *transf.* to any similarly pre-eminent individual, ruler, or divinity. Also *attrib.*

1890 G. Birdwood *Rep. Misc. Old Rec. India Office* 222 The earliest coins minted by the English in India were of copper, stamped with a figure of an irradiated *lingam*, the phallic 'Roi Soleil'. **1943** E. M. W. Tillyard *Elizabethan World Picture* vii. 83 The *roi soleil* is indeed one of the most persistent of all Elizabethan commonplaces. **1958** *Spectator* 6 June 721/2 Her rule is no longer the *roi-soleil* variety. **1961** *Listener* 31 Aug. 319/2 Popular books on Picasso in which the artistic *roi soleil* of the post-war years is still pictured as on some barricade or other. **1966** *Guardian* 31 Dec. 5/3 Cecil Harmsworth King, the *roi soleil* of Long Acre. **1978** *Times* 27 May 7/1 The palmy days of the Roi Soleil.

‖ **Rōjū** (rō·dʒū). Also **Rōchū, rōjiu, rōjū,** etc. [Jap.] The senior councillors or ministers of state in Japan under the Tokugawa government (1603–1867).

1874 F. O. Adams *Hist. Japan* I. I. x. 70 The successors of Iyéyasû..were mostly *fainéants*, as were their almost hereditary ministers, the rôjiu. **1893** F. Brinkley tr. *Hist. Empire Japan* viii. 329 In the event of the Shogun himself taking the field, he had to be accompanied by all the feudal barons, the Ministers of State (*Rochu*) becoming generals and the *Wakatoshiyori* holding chief command over the bannerets. **1912** E. Lee tr. *Saito's Hist. Japan* 147 The board of the 5 Rōchū, the treasurers... controlled the imperial court officials and Daimiō. *a* **1922** J. Murdoch *Hist. Japan* (1926) III. i. 4 These five constituted the Great Council, which was presently to become known as the Rōjū. **1970** J. W. Hall *Japan* x. 175 The *Rōjū* were given authority over matters of national scope. **1974** *Encycl. Brit. Macropædia* X. 71/2 By reorganizations in 1633–42 the executive..was almost completed, as represented by the offices of senior councillors (rojū), [etc.].

roky (rōᵘ·ki), *a.*² *Founding.* [f. *roke*, var. Roak (in Dict. and Suppl.) + -y¹.] Possessing or characterized by rokes.

1932 E. Gregory *Metallurgy* ii. 58 Some of these cracks may escape detection in the forge, and unsound roky billets are thus produced. **1940** Simons & Gregory *Steel Manuf.* xix. 141 It is not uncommon for these cracks to be unnoticed in advance of forging or rolling, and they then elongate and 'open-out', producing 'roky' billets.

rolag (rōᵘ·læg). *Spinning.* [a. Gael. *rolag*, dim. of *rola* a roll.] A roll of carded wool ready for spinning.

1932 Simpson & Weir *Weaver's Craft* v. 30 The wool is now lifted lightly on the right-hand card, and placed on the back of the left, where it is rolled between the two card backs into a neat roll, or 'rolag'. **1964** H. Hodges *Artifacts* ix. 128 The roll of carded or combed fibres, the *rolag* or *sliver*, may be wound round a second rod. **1977** Y. Deutch *Weaving & Spinning* 129 The wool must first be teased and carded to separate the fibres and prepared for spinning by forming a roll of wool or rolag. *Ibid.*, Before spinning you will need to prepare by hand about 45 cm (18 in) of woollen thread from a rolag.

Rolandic (rolæ·ndik), *a. Anat.* Also **rolandic.** [f. next + -ic.] Used to designate various features of the central nervous system associated with Rolando (see next), as (*a*) the motor region or area of the cerebral cortex; (*b*) the fissure or sulcus of Rolando; (*c*) the angle at which the fissure of Rolando meets the median plane of the brain.

1881 C. G. Comegys tr. *J. M. Charcot's Lect. Dis. Spinal Cord* v. 49 We have enclosed within these pyramidal lines, what is known as the Rolandic region of the cerebral cortex, and which represents..a region endowed with special physiological properties. **1883** W. B. Hadden tr. *J. M. Charcot's Lect. Localisation of Cerebral & Spinal Dis.* v. 193 The central, median, or, if you will so term them, the Rolandic convolutions. **1890** *Jrnl. Anat. & Physiol.* XXV. 3 In most hemispheres a small variable tertiary furrow may be detected below the lower end of the Rolandic fissure. *Ibid.* 19 By the 'Rolandic angle' I mean the angle which is formed by the meeting of the upper end of the sulcus with the mesial plane. **1908** H. E. Santee *Anat. Brain & Spinal Cord* (ed. 4) iii. 56 The average Rolandic angle is 71° 7'. **1910** G. G. Davis *Appl. Anat.* 36 The Rolandic area embraces the ascending frontal, or precentral, and posterior portion of the three frontal convolutions, the former being in front of the fissure of Rolando, or central fissure. **1921** Tilney & Riley *Form & Functions Central Nerv. Syst.* xxxvi. 643 The entire length of the Rolandic fissure is 8 cm. *Ibid.*, The *gyrus Rolandicus*... extended the entire length of the two Rolandic fissures. **1962** M. C. H. Dodgson *Growing Brain* vii. 55 The determination of Cunningham's Rolandic Index may be useful in doubtful instances. **1972** M. L. Barr *Human Nerv. Syst.* xiii. 207 The rolandic sulcus indents the superior border of the hemisphere about 1 cm behind the midpoint between the frontal and occipital poles. **1974** L. F. Sies *Aphasia Theory & Therapy* i. 54 The third area..is located within the mid-saggital fissure, just anterior to the Rolandic motor foot area.

Rolando (rolæ·ndo). *Anat.* The name of Luigi *Rolando* (1773–1831), Italian anatomist, used with *of* and *attrib.* to designate: **a.** A fissure or sulcus of the brain separating the frontal lobe from the parietal lobe, described by him in 1825 (*Mem. d. R. Accad. d. Sci. di Torino* XXIX. 163). [tr. F. *sillon de Rolando* (F. Leuret 1839, in *Anat. Comparée du Syst. Nerv.* (1839–57) I. vi. 398).]

1839–47 R. B. Todd *Cycl. Anat. & Physiol.* III. 696/2 Two superior convolutions are met with above the fissure of Sylvius, between which is placed..the fissure of Rolando. **1861** *Proc. Zool. Soc.* 248 Fig. 1 was drawn from an almost fresh brain, fig. 2 represents a brain which had been for several months in spirit. The roundness of outline of the latter as compared with the former, and the more transverse direction of the fissure of Rolando, are very remarkable. **1890** *Jrnl. Anat. & Physiol.* XXV. 139 It is a question if the fissure of Rolando is present in any other brains than those of Apes and Man. **1921** Tilney & Riley *Form & Functions Central Nerv. Syst.* xxxvi. 643 The fissure of Rolando has been found interrupted near its middle in the brains of several distinguished men. **1974** *Encycl. Brit. Macropædia* XII. 982/2 The two major grooves on the lateral surface of the brain are the lateral fissure (fissure of Sylvius), which starts at the base of the brain and extends upward and backward on the lateral surface, and the central sulcus (sulcus of Rolando), which runs from the middle of the dorsal border of the hemisphere downward almost to the lateral fissure.

b. The translucent gelatinous substance which fills the ends of the posterior grey horns of the spinal medulla.

1853 Busk & Huxley tr. *A. Kölliker's Man. Human Histol.* I. 408 The posterior, longer and thinner [horns].. at the front are covered with a more transparent layer, consisting of a preponderance of smaller nerve-cells —the substantia gelatinosa of Rolando. **1872** H. Power tr. J. Gerlach in S. Stricker *Man. Human & Compar. Histol.* II. xx. 361 The posterior cornua are divisible into two portions..an anterior and a posterior, which last, owing to its peculiar translucency when examined with the naked eye, has long been termed the substantia gelatinosa of Rolando. **1929** Hewer & Sandes *Introd. Study Nerv. Syst.* vi. 50 Fibres from the collaterals anastomosing round cells of the substantia gelatinosa of Rolando. **1976** *Expo. Brain Res.* XXVI. 77 Peripheral neurotomy induced a series of peculiar, *sui generis* alterations, both in the Rolando substance and in the dorsal column.

rôle. Delete ‖ and add: Now usu. spelt **role.** **1. a.** For 'Chiefly *fig.*' read 'freq. *fig.*' and add: Also *spec.*, a part in a play, opera, film, or broadcast drama; = Part *sb.* 9. See also *title-rôle* s.v. Title *sb.* 11.

1886, 1900 [see *title-role* s.v. Title *sb.* 11]. **1912** M. B. Leavitt *50 Yrs. Theatr. Managem.* xv. 184 Jennie Winston, an Australian, was likewise famous as a male impersonator and was also a favorite in leading operatic rôles. **1937** D. Frohman *Encore* xv. 199 In the course of his subsequent long career on the stage, he included in his repertoire more than a hundred and thirty difficult rôles. **1973** R. Rolph in P. Noble *Favorite Movies* I. 103 In *Citizen Kane*... Welles does indeed play a role in his film. **1980** D. Garfield *Player's Place* iv. 157 Studio actors have been found wanting in the performance of 'classical' roles.

For *transf.* read: **b.** The typical or characteristic function performed by someone or something; freq. in phr. *to play a role*. (Later examples.)

1944 J. S. Huxley *On Living in Revolution* 73 He [*sc.* Darwin] was aware that isolation might play a role in the production of new species. **1957** T. Lehmann tr. *N. A. Gorchakov's Theater in Soviet Russia* x. 108 Did the Communist Party have any ideas of its own about the role of the theater before the Revolution of October, 1917? **1961** [illegible] One of the maternal grandmother's chief roles..is being steadily taken over by the doctor, the midwife and the health visitor. **1973** A. R. Prest in Crick & Robson *Taxation Policy* ix. 129 A more recent study does seem to suggest a more positive role for these devices. **1981** *Newsweek* 4 May 74/3 The so-called hypothalamic-pituitary axis is the control center for hormones throughout the body and also plays an important role in emotions.

2. *Social Psychol.* The behaviour that an individual feels it appropriate to assume in adapting to any form of social interaction; the behaviour considered appropriate to the interaction demanded by a particular kind of work or social position.

1913 G. H. Mead in *Jrnl. Philos.* X. 377 This response to the social conduct of the self may be in the rôle of another—we present his arguments in imagination and do it with his intonations and gestures... In this way we play the rôles of all our group; indeed, it is only as we do this that they become part of our social environment. **1936** R. Linton *Study of Man* viii. 114 Every individual has a series of rôles deriving from the various patterns in which he participates and at the same time a *rôle*, general, which represents the sum total of these rôles. **1949** R. K. Merton *Social Theory* iii. 110 A conception basic to sociology holds that individuals have multiple social roles and tend to organize their behavior in terms of the structurally defined expectations assigned to each role. **1950** T. M. Newcomb *Social Psychol.* viii. 280 A position has no meaning without its accompanying role. **1961** E. Goffman *Encounters* 85 In sociology there are few concepts more commonly used than 'role', few that are accorded more importance, and few that waver so much when looked at closely. **1967** M. Argyle *Psychol. Interpersonal Behaviour* iv. 73 By a 'role' is meant a pattern of behaviour which is shared by most occupants of a position, and which comes to be expected of them. The role usually includes a series of distinct relationships with people in other positions. **1977** R. Holland *Self & Social Context* v. 82 There is no attempt to explore the possibility that psychologists' and sociologists' own roles may influence their definitions and uses of the concept of role.

3. An expression, usu. in the form of a symbol or series of symbols, of the function or signification of a term appearing in an index or thesaurus, used esp. as a means of linking its possible relevance to other terms with which it may be associated. Usu. *attrib.*, as *role indicator*, *operator*.

1961 *Amer. Documentation* XII. 98 (*heading*) Notes on the use of roles and links in coordinate indexing. **1963** *Aslib Proc.* XV. 207 With 'roles' each keyword is classified by function. **1970** A. Chandor et al. *Dict. Computers* 332 *Role indicator*, a code associated with a keyword to identify it as a noun, verb, or adjective, etc. **1976** *Program* X. 18 [illegible] **1979** A. C. Foskett *Subject Approach to Information* (ed. 3) vi. 81 [illegible] **1979** J. E. Rowley *Mechanised In-House Information Syst.* i. 46 Roles or role indicators are appended to an index term at the indexing stage to indicate the use of the term in that context.

4. *attrib.* and *Comb.* (sense 2), as *role absorption*, *-assumption*, *-creating*, *-differentiation*, *-expectation*, *-structure*, *-structuring*, *theory*, *theorist*; *role-assuming*, *-determined*, *-determining* adjs.; **role conflict**, the difficulties encountered when one role makes conflicting demands on an individual or when an individual has several roles whose demands are conflicting; **role distance**, detachment from one's role; also (with hyphen) as vb.; **role model**, someone who, in the performance of a role, is taken as a model by others; **role-play**, the performance of a role, esp. the deliberate rehearsal or acting of a particular role, freq. used as a technique in training or psychotherapy; so **role-play** *v. intr.* and *trans.*, **role-player, role-playing** *vbl. sb.*; **role relation, relationship** (see quot. 1957); **role reversal**, the assumption of a role which is the reverse of that normally performed; **role-set** (see quot. 1957); **role-taking**, the imaginary assumption, leading to understanding, of another's role; hence (as back-formation) **role-take** *v. intr.*

1937 J. L. Moreno in *Sociometry* I. 53 The weaker the role absorption by the ego, the more often can the ego substitute [illegible] adopt affected forms of speech as a means of 'role-distancing' from the lower classes, from whom they had hitherto been differentiated by speaking French. **1978** A. Ryan in Hookway & Pettit *Action & Interpretation* 68 The question whether the crucial element in the dramaturgical picture is that cluster of insights which goes under the general heading of 'role distance'. **1979** *Internat. Jrnl. Sociol. of Law* VII. 289 Not that the average performer seems conscious of any evidence on his part of role-distance; on the contrary, such ritual commitment furnishes the core of his identity. **1951** Parsons & Shils *Toward Gen. Theory of Action* iv. 190 Role-expectations are patterns of evaluation. **1969** in Halpert & Story *Christmas Mumming in Newfoundland* 142 Cat Harbour..is normally characterized by rather rigid and formal role expectations. **1957** W. Thielens in R. K. Merton *Student-Physician* II. 138 By the time students enter law or medical school, those whose decisions were made earliest are most likely to have a

role model. **1977** *N.Y. Times Mag.* 26 June 10/2 If the teacher was a 'role model', parents were obviously unaware of it. **1961** R. J. Corsini et al. *Roleplaying in Business & Industry* i. 9 If they stopped now and then and discussed, evaluated, and practiced alternative ways of reacting, they roleplayed. **1964** M. Argyle *Psychol. & Social Probl.* x. 133 Students role-play some of the situations they will meet on the job. **1970** *Peace News* 2 Oct. 3/4 During a strategy game, a situation may arise which is so interesting that the group may want to roleplay it. When the roleplay is over, people can return to the strategy game. **1979** *Lore & Lang.* Jan. 4 Dylan (5:3) is taking part in a 'spiderman' role-play, and another participant tries to drag him away. **1943** J. L. Moreno in *Sociometry* VI. 438 Role-player is a literary translation of the German word 'Rollenspieler' which I have used. **1978** *Dædalus* Summer 137 He [*sc.* the bourgeois] is the man who, when dealing with others, thinks only of himself, and, in his understanding of himself, thinks only of others. He is a role-player. **1980** *Times Lit. Suppl.* 31 Oct. 1221/3 The hero of Anne Tyler's new novel, Morgan Gower, is an inveterate role-player. **1943** J. L. Moreno in *Sociometry* VI. 438 It may be useful to differentiate between *role-taking*..—*role-playing*—which permits the individual some degree of freedom—and *role-creating*. **1951** *Amer. Sociol. Rev.* XVI. 181/2 In *role-playing* one does not pretend anything. A policeman arresting a person is..performing or *playing* a role expected of one holding the position of public protector. **1960** W. H. Whyte *Organization Man* v. 56 The role of slugger—not just a role-playing role, either—was assigned in advance. **1980** *Times Lit. Suppl.* 23 May 575/5 Role-playing is perhaps the true subject of the modern novel. **1940** *Sociometry* III. 20 The pattern of rôle relations around an individual as their focus, is called his cultural atom. **1950** T. M. Newcomb *Social Psychol.* xiii. 453 These four kinds of role relationships call for quite different sets of activities on his part. **1957** E. Bott *Family & Social Network* i. 3 A *role-relationship* is defined as those aspects of a relationship that consist of reciprocal role expectations of each person concerning the other. **1977** R. Holland *Self & Social Context* v. 97 The person with a strong ego can integrate experience of his past role-relationships and put it to the use of role-performance. **1951** *Occupational Psychol.* XXV. 65 The method of role-reversal is designed to change the cognitive structure of disputants so that their social perception changes from divergence to convergence. **1967** M. Argyle *Psychol. Interpersonal Behaviour* x. 188 Role-reversal: here a trainee takes the reverse of the role he would take in real life, e.g. a foreman takes the role of a shop-steward. **1975** W. A. Haviland *Cultural Anthropol.* xiii. 319/2 During the installation rites of a chief among the Ndembu.., a different type of role reversal is manifest; the chief must sit in silent humility while he is reviled.. by anyone who feels so inclined. **1957** R. K. Merton in *Brit. Jrnl. Sociol.* VIII. 110 Unlike Linton, I begin with the premise that each social status involves not a single associated role, but an array of roles. This basic feature of social structure can be registered by the..term, role-set... By role-set I mean that complement of role-relationships in which persons are involved by virtue of occupying a particular social status. **1968** P. K. Bock in J. A. Fishman *Readings Sociol. of Lang.* 215 Radically different behavioral expectations *are* attached to the role of 'teacher' in connection with various members of the corresponding role-set. **1977** Warren & Ponse in Douglas & Johnson *Existential Sociol.* x. 274 Instead, they have been concerned with social roles, role sets, and so on. **1940** *Sociometry* III. 21 This often produces a typical conflict in the rôle-structures of two marriage partners. **1978** A. Ryan in Hookway & Pettit *Action & Interpretation* 67 The sociologist may, perhaps, rest content with giving a structural description of a society's role structure. **1967** D. Cooper *Psychiatry & Antipsychiatry* v. 84 There seemed an obvious need for a separate unit with less ritual and less rigid role-structuring. **1972** *Jrnl. Social Psychol.* Dec. 247 The ability to role-take accurately, or empathize, is the ability to see, feel, respond, and understand as if one were the other person. **1934** G. H. Mead *Mind, Self & Society* IV. 254 The immediate effect of such rôle-taking lies in the control which the individual is able to exercise over his own response. **1951** *Amer. Sociol. Rev.* XVI. 180/2 The term role-taking meant, for Mead, a strictly mental or cognitive or empathic activity, not overt behavior or conduct. **1964** M. Argyle *Psychol. & Social Probl.* x. 136 It [*sc.* indoctrination induced by Chinese 'thought reform'] can perhaps best be described as a piece of ego-involved role-taking, produced by extreme coercion, together with the temporary adoption of a new frame of reference. **1972** *Jrnl. Social Psychol.* Dec. 247 Role taking refers to the imaginative reconstruction by ego of alter's role. **1954** G. Lindzey *Handbk. Social Psychol.* I. 238/1 More than any other single group, the role theorists have developed and used the conception of the self as an intervening variable. **1977** R. Holland *Self & Social Context* v. 91 Up to this point I have dealt with the *structural* role theorists represented in Coser and Rosenberg's book of readings. **1954** G. Lindzey *Handbk. Social Psychol.* I. 238/1 The trend in role theory is in the study of the interactions of self and role as coordinates and not as parallels. **1972** W. C. Coe *Challenges of Personal Adjustment* viii. 215 Role theory bridges the gap between the constructs of sociology and of psychology.

Hence **role** *v. trans.*, to provide (a term) with a role indicator; **ro·ling** *vbl. sb.*

1976 *Program* X. 14 (*heading*) A minicomputer retrieval system with automatic root finding and roling facilities. *Ibid.* 24 The presence of the connective merely ensures that the first word in the multi-word phrase is roled and stored.

roleo (rōᵘ·lio). *U.S.* Also **rolleo.** [f. Roll *v.*² + Rod)eo.] (See quot. 1942.) Also *attrib.*

1933 *Nat. Geographic* Feb. 166 (*caption*) A floating log affords precarious footing; yet this expert woodsman.. rides it standing... Contests in this sport are a part of the 'Rolleo' celebration held annually. **1942** Berrey & Van den Bark *Amer. Thes. Slang* §513 *Rolleo*,..a log-rolling contest. **1948** *Chicago Tribune* 6 July 1. 1/4 The others learned that the roleo is like a rodeo—except that instead of riding ornery bronchos, the contestants ride on floating logs that spin so fast the water churns up like a lawn sprinkler. **1949** *Boston Globe Mag.* 9 Oct. 8/1, I only wish your dad could see you roll at the Roleo. **1954** *Ocean Press* 24 Aug. 7/3 Running first won the world's championship at a 'roleo' in 1942 and has held the title ever since. **1956** *Sun* (Baltimore) 29 June 11/1 Lawrence Bergeron,..president of the National Roleo Association, drowned today while competing in the 'old-timer' finals of the world championship log-rolling championships.

Rolex (rōᵘ·leks). The proprietary name of a make of watch. Also *attrib.*

1912 *Trade Marks Jrnl.* 14 Aug. 1242 'Rolex'... Watches. Wilsdorf & Davis,..London,..; watch manufacturers. **1957** J. Braine *Room at Top* iv. 40 A young man with sleek black hair, a shiny red face and a gold Rolex Oyster. **1970** W. Wager *Sledgehammer* (1971) xvi. 99 He glanced down at the face of his gold Rolex. **1973** J. Rossiter *Manipulators* xiv. 138 The watch on his wrist [was] an upper-bracket Rolex. **1977** B. Freemantle *Charlie Muffin* xvii. 170 He looked at the heavy Rolex watch that had been part of the élite snobbism of the Green Berets in Vietnam.

Rolf (rǫlf). Also **rolf.** The name of Ida P. *Rolf* (1897–1979), U.S. physiotherapist, used *attrib.* to designate her technique of deep massage (also known as 'structural integration') aimed at reducing muscular, and consequently psychic, tension. Hence as *v. trans.* Also **Rolfed** (rǫlft) *ppl. a.*; **Ro·lfer,** a practitioner of this technique; **Ro·lfing** *vbl. sb.*, the Rolf technique.

1958 D. Lawson-Wood *Psycho-Logics & Posture* 11 The author is a fully qualified and registered masseur and physiotherapist, and is engaged..in further intensive study and research in the Rolf Technique. **1970** *Psychol. Today* IV. 58/1 Rolf and the older rolf practitioners recognize the importance of this emotional component. *Ibid.* 88/3 In the case of the rolfed subjects there was no training. **1971** W. C. Schutz *Here comes Everybody* 176 One man has been Rolfed many times and one area of difficulty is a rounded back. **1972** G. Downing *Massage Bk.* 155 The pain..stops immediately when the Rolfer's hand is taken away. **1972** *New Yorker* 21 Oct. 34/2 Everyone under twenty-five discussed Rolfing... Rolfing is a system of deep massage that stretches and rearranges the tissue surrounding the muscles. **1977** *N.Y. Times* (City ed.) 15 July c. 22/2 We already spend far too much time practicing artificial modes of sociability, such as group encounters, sensitivity training, 'T' groups, Rolfing and the like. **1979** *Brit. Med. Jrnl.* 24 Mar. 796/2 There are also those..who heal with rolfing, shiatsu, polarity treatment, aeriontherapy, or psionic medicine.

roll, *sb.*¹ Add: **II. 6. e.** A quantity of bills or notes rolled together; hence, the money a person possesses. *U.S.* and *Austral.* Also phr. *a roll Jack Rice couldn't jump over*, a large quantity of money (*Austral. slang*).

1846 *Dollar Newspaper* (Philadelphia) 22 Apr. 4/6 He also had a roll which he said contained $600. **1904** *N.Y. Times* 16 May 5 It was as easy to be separated from one's 'roll' at a shell game there a quarter of a century ago as it was ten years ago. **1907** 'O. Henry' *Trimmed Lamp* 171 He drew out his 'roll' and slapped five tens upon the bar. **1912** J. Sandilands *Western Canad. Dict.*, *Roll*, or *Wad*, a person's present supply of dollar bills or paper money. Roll him is to rob him of his money. **1919** H. L. Wilson *Ma Pettengill* ii. 62 [He] asked her how big her roll was, saying that he lived out there and it cost something to make a home. *a* **1925** [see *cut *v.* 56 q]. **1926** J. Black *You can't Win* iv. 35 No Missouri dip would take his roll, extract two fifty dollar bills, and put the rest back in his pocket. **1945** Baker *Austral. Lang.* v. 107 A man.. may even be fortunate enough to have a roll Jack Rice couldn't jump over. Jack Rice was a racehorse noted for his performances over hurdles. **1954** T. Ronan *Vision Splendid* II. 119 'I've got a roll Jack Rice couldn't jump over.' Marty produced one of those wads of currency Mr. Tappingham had seen only in the cruder American films and started peeling off ten-pound notes. **1960** 'N. Culotta' *Cop this Lot* v. 82 Man walks around with a roll in 'is kick Jack Rice couldn' jump over, an' 'e's not worth a zac.

f. A quantity of photographic or cinematographic film supplied rolled up; a spool of film.

1890 [see sense 13 c]. **1925** *Kodak Mag.* July 109 It is quite a good idea to develop just one or two rolls, to make sure that you are giving correct exposures. **1960** O. Skilbeck *ABC of Film & TV Working Terms* 110 Some Magazines hold..only two hundred feet of Stock and the Rolls are smaller than normal. **1973** C. McCarry *Miernik Dossier* (1974) 147 I'm sending you a roll of snaps to keep for me... You can have them developed. **1976** K. Thackeray *Crownbird* v. 82 Priest was loading a roll of Tri-X into a black Nikon.

g. *spec.* = *music roll* (*b*) s.v. *Music *sb.* 12 d.

1902 *Encycl. Brit.* XXXI. 767/1 The use of the perforated roll acts by means of the ingenious and indeed faultless application of pneumatic leverage to the ordinary piano. **1906, 1913** [see **music roll*]. **1921** A. Huxley *Crome Yellow* x. 94 The music stopped... He.. turned to the cabinet where the rolls were kept. He trod off the old roll and trod on the new. **1926,** etc. [see **piano roll*]. **1928** *Melody Maker* Feb. 161/3 Holding back the licenses of the 'Mechanical' reproductions on records and rolls. **1956** S. Longstreet *Real Jazz* 129 James P. Johnson was a great man on the rolls. Till 1920 he punched a lot of rolls. After that he recorded sides. **1972** *Jazz & Blues* Oct. 6/3 Changing the playing speed of the roll does not alter the pitch. **1977** *Times* 25 June 26/9 (Advt.), Pianola piano..100 rolls..£700.

h. A rolled-up quantity of a prohibited drug.

1962 'K. Orvis' *Damned & Destroyed* v. 36 Loaded. Full of heroin. Carrying a roll, too. **1976** *Whig-Standard* (Kingston, Ontario) 21 Jan. 45/3 Bruce denied any knowledge of the roll, claiming his suitcase had been left unopened in the motel.

10. c. An item of food (other than bread) that is rolled up or doubled over before being cooked; chiefly with defining words, as *fig-*, *meat*, *potato roll*. See also *jelly roll* s.v. *Jelly *sb.*¹ 4, *pancake roll* s.v. *Pancake *sb.* 3, *sausage roll* s.v. Sausage 4 d, *Swiss roll* s.v. Swiss *a.* 2.

1845 E. Acton *Mod. Cookery* xvi. 420 Excellent meat rolls. Pound..veal, chicken, or turkey... Form it into small rolls..fold them in good puff-paste, and bake them. **1922** Joyce *Ulysses* 25 A bag of figrolls lay snugly in Armstrong's satchel. *a* **1944** K. Douglas *Alamein to Zem Zem* (1946) 62 Meat roll and excellent ersatz coffee graced our menu. **1950** *Mrs. Beeton's Bk. Househ. Managem.* 1181 *Potato rolls*... Cut the potatoes into small pieces... Roll out the paste to the thickness of ¼ of an inch, cut in rounds or squares 4 inches across, fill each with the vegetables, fold it over like a turnover, and bake.

11. b. (Examples of *hollow roll*.)

1904 Goodchild & Tweney *Technol. & Sci. Dict.* 288/2 *Hollow roll*, a lead roll made by bending over the edges of sheet lead, and so forming a tube. **1960** *B.S.I. News* May 23 Guidance on the use of lead sheet used as a covering for roofs... Design methods for both the wood-roll and hollow-roll systems.

12*. *Geol.* An ore body in sedimentary rock that has a C- or S-shaped vertical cross-section cutting across strata. Freq. *attrib.*

1942 *Bull. U.S. Geol. Survey* No. 936. 363 The vanadium-bearing hydrous mica is..in part concentrated..in thin zones that cut across bedding. As the zones..are curved or wavy, they are called rolls by the miners. **1955** *Prof. Papers U.S. Geol. Survey* No. 300. 239/1 Similarities between roll ore bodies and the more prevalent tabular ore bodies in sedimentary rocks of the Colorado Plateau. *Ibid.* 239/2 In cross section, rolls commonly show C, S, and 'socket' shapes.., but in plan are linear. **1976** R. I. Rackley in K. H. Wolf *Hand-bk. Strata-Bound & Stratiform Ore Deposits* VII. iii. 116 The uranium 'roll' has long been known to uranium producers.

13. a. *roll film, -shutter, -tobacco* (later examples).

1895 *Montgomery Ward Catal.* Spring & Summer 217/1, 1 Roll Film, for 25 exposures—.20 1 Box of 5 Rolls of Film (for 25 exposures each) 1.00. **1902** *Encycl. Brit.* XXXI. 690/2 In many ways the most convenient and compact hand cameras are those made specially for use with the roll-film cartridges in many different sizes. **1933** *Discovery* Feb. 59/2 Roll films are used, each roll containing one hundred exposures. **1951** Yarsley & Kitchen in H. M. Langton *Synthetic Resins* (ed. 3) ii. 116 The bulk of the many millions of feet of cine film used throughout the world to-day is still celluloid, and 3½ mil celluloid is the base for amateur roll film. **1977** J. Hedgecoe *Photographer's Handbk.* 10 By the 1890s George Eastman's rollfilm camera allowed many pictures to be taken at one loading. **1911** *Chambers's Jrnl.* Feb. 141/2 At the kerb end the front is provided with a roll-shutter. **1922** Rollshutter [see *down-coming *ppl. a.*]. **1913** I. Cowie *Company of Adventurers* 311 Accordingly Whitford placed on the dressed buffalo skin which they had placed on the ground before them, two pint measures of tea and a yard of thick Canadian roll tobacco. **1929** Moberly & Cameron *When Fur was King* 35 One and a half feet of Canadian roll tobacco sold for one..made-beaver.

c. *roll-end*; *roll-munching* adj.

1970 *Toronto Daily Star* 24 Sept. 27/1 (Advt.), Roll ends at cost. **1976** *Bridgwater Mercury* 21 Dec. 5/1 (Advt.), Room-size remnants. We must clear dozens of roll-ends to make room for new stocks. **1970** G. F. Newman *Sir, You Bastard* viii. 210 The bar was packed with fat roll-munching office workers.

III. 17. *roll-produced* adj.; **roll feed,** a feed mechanism supplying paper, strip metal, etc., by means of rollers; so **roll-feeding** *vbl. sb.*; **roll-fed** *ppl. a.*; **roll-forming** *vbl. sb.*, cold forming of metal by repeated passing between rollers; so **roll-form** *v. trans.*; **roll-formed** *ppl. a.*; **roll mark,** a mark produced on sheet metal in flattening it with an imperfect set of rollers.

1967 Karch & Buber *Offset Processes* ii. 31 *Flexographic*. This process involves the use of rotary (web or roll fed) printing from rubber plates. **1968** *Gloss. Terms Mechanized & Hand Sheet Metal Work* (*B.S.I.*) 19 *Roll feed*, a feed mechanism that imparts continuous or intermittent motion to strip by means of rollers in contact with both surfaces. **1967** V. Strauss *Printing Industry* vi. 362/2 Roll feeding was originally developed for the production of metropolitan newspapers by relief printing. *Ibid.* 363/1 Designers of roll-feeding machinery have devised a number of different roll-feeding methods. **1949** *Tool Engineers Handbk.* (Amer. Soc. Tool Engineers) 989 Most sheet and strip metals can be successfully roll-formed. **1958** *Times Rev. Industry* June 20/1 The cylindrical body sections are made from two plates, roll-formed cold to shape and welded together. **1971** *Engineering* Apr. 59/1 Some permanent plastic coatings..will endure the metal to which they are applied being drawn, roll-formed, bent, and pressed, without cracks developing or the coating peeling. **1949** *Tool Engineers Handbk.* (Amer. Soc. Tool Engineers) 989 Generally speaking, the sharpest corner practicable to maintain on a roll-formed section would be one having an outside radius equal to

the metal thickness. **1977** *Engin. Materials & Design* Aug. 50/3 Because the rivets are roll-formed, they are straighter than extruded rivets. **1932** E. V. Crane *Plastic Working of Metals* v. 91 Bending Operations.— Bar-folders, brakes, drawbenches, roll-forming machines and bending dies in presses share the field. **1954** J. F. Young *Materials & Processes* (ed. 2) xix. 805 Roll forming consists of passing strip stock between sets of shaped driven rollers. **1967** S. Kalpakjian *Mech. Processing of Materials* vi. 202 A further development of roll forming is the production of welded tubing, starting with a flat strip. **1923** Glazebrook *Dict. Appl. Physics* V. 364/1 The effect of alternate heating and cooling is to cause small cracks in the surface of the rolls, which lead to slight ridges, or 'roll marks'. **1962** G. R. Bashforth *Manuf. Iron & Steel* (ed. 2) IV. iv. 138 Sections are liable to develop certain defects, such as.. 'roll marks' due to defective or badly worn rolls. **1952** J. B. Oldham *Eng. Blind-Stamped Bindings* i. 4 The use of a roll-produced decoration.

roll, *sb.*² Add: **1. b.** In *go and have a roll*: go away, 'get lost'. *slang.*

1941 Baker *Dict. Austral. Slang* 34 *Have a roll!, go and*, go to the devil! **1959** I. & P. Opie *Lore & Lang. Schoolch.* x. 192 Juvenile language is well stocked..with expressions inviting a person's departure, for instance:..go and have a roll.

c. Esp. in phr. *to have a roll on* and varr.: to have a conceited bearing, to give oneself airs. *Eng. Public School slang.*

1881 C. E. Pascoe *Everyday Life in our Public Schools* 160 Anything approaching 'swagger' is severely rebuked; there is no more objectionable quality than that understood by the expression, 'He's got such a horrid roll on'. **1908** D. Coke *House Prefect* i. 11 Brereton, they decided, had a bit of a roll on. **1913** A. Lunn *Harrovians* iii. 53 Ewen was an ugly lout, and was beginning to put on roll...after the game, his tendency to 'put on roll' was duly checked in the approved fashion.

d. (An act of) rotation of a vehicle or craft about an axis parallel to its direction of motion.

In the case of ships the movement consists of a partial rotation, immediately reversed, caused by wind or waves; with aircraft it is either a similar unintended movement or a deliberate manœuvre consisting of a complete turn through 360°; with motor vehicles and helicopters it is a tipping (outwards and inwards respectively) in cornering.

1862 W. Froude *Rolling of Ships* 75 All ships having the same 'periodic time', or period of natural roll, when artificially put in motion in still water, will go through the same series of movements. **1907** J. Masefield *Tarpaulin Muster* xvi. 161 At the last of her rolls there comes a clattering of tins, as the galley gear and whack pots slither across to leeward, followed by cursing seamen. **1912** *Techn. Rep. Advisory Comm. Aeronaut. 1911–12* 102 The pendulum movement from side to side..misleads the pilot into operating his wing flaps to recover the vertical position of his body. By this he may aggravate the roll. **1918** W. G. McMinnies *Pract. Flying* x. 194 The roll, which consists of making the machine loop sideways and continue in the same direction as it was travelling before the manœuvre, is done with the engine on or off. **1920** *Nature* 11 Mar. 47/2 For use on board ship the compass must be mounted..so that the rolls..shall have but small effect on the compass. **1942** N. MacMillan *How to pilot Aeroplane* xv. 100 The full roll, the half roll, and the double half roll can all be made on the glide or during a zoom as well as on the level. **1945** J. M. Labberton *Marine Engineers' Handbk.* ix. 1389 The maximum velocity will occur at the vertical position and diminish to zero at the extremities of the roll... The maximum dynamic effect will occur at the maximum angle of roll. **1953** M. Rauscher *Introd. Aeronaut. Dynamics* 660/2 (Index), Roll or Bank, angle of. **1957** J. Shapiro *Helicopter* iii. 52 The sideways attitude of the helicopter against the horizon is known as its 'roll'. A more frequently used term for the same condition is 'bank'. **1961** *Times* 28 Mar. 4/6 There is a good deal of roll when cornering fast. **1967, 1974** [see *pitch *sb.*² 2 b]. **1974** *Physics Bull.* Jan. 11/1 The six component wind tunnel balance..will be able to measure three forces (lift, drag and side force) and three moments (pitch, yaw and roll) on any aircraft model it supports. **1978** *Lancashire Life* Apr. 141/1 The Fiat 132-2000 rides very well indeed on all kinds of road surface and corners capably with a minimum of roll.

e. *Gymnastics.* An exercise in which the body is rolled into a tuck position and turned in a forward or backward circle.

1898 F. Graf et al. *Hints to Gymnasts* III. 176 Before attempting any kind of..rolls..or handstands, the pupil should have mastered thoroughly all kinds of straight arm swinging exercises. **1920,** etc. [see *neck-roll* s.v. *Neck *sb.*¹ 16]. **1935** *Encycl. Sports* 331/1 When half the roll has been accomplished the hands are changed from behind the head to a position in front, so that the body is then pivoted on the inside of the arms. **1955** *Simple Gymnastics* ('Know the Game' Series) (ed. 2) 25 *Forward Roll* —Bend forward and take the weight of the body on the hands. Tuck the head well under and roll with the knees on the chest. **1956** Kunzle & Thomas *Freestanding* ii. 32 From a forward roll to stand, to a cartwheel sideways down the same line as the roll.

f. A throw (at dice).

1926 G. Ade *Let.* 26 Oct. (1973) 114 This kind of party [*sc.* a 'Monte Carlo' party] is the wildest and most hilarious thing you ever heard. Before we got through Sunday evening the crap shooters were rolling for a hundred thousand a roll. **1966** O. Norton *School of Liars* iv. 62 Ben rattled the dice-box. 'Now..we'll have a quick roll before Hank the Bank comes, and I think Scott's in the chair.' **1969** R. C. Bell *Board & Table Games* II. v. 91 The first caster throws all five dice on his first roll. **1974** *Times* 20 Feb. 19/2 We are still schooling craps on this and we think we can make it on the next roll.

g. *colloq.* An act of sexual intercourse. *a roll in the hay*: see *Hay *sb.*¹ 3.

1942 Berrey & Van den Bark *Amer. Thes. Slang* §362/1 *Copulation*,..roll. **1962** P. Green tr. *S. de Beauvoir's Prime of Life* I. ii. 80, I had several unpleasant incidents with truck drivers, not to mention a commercial traveler who wanted me to have a roll with him in the ditch, and left me flat in the middle of the road when I refused. **1976** P. Ferris *Detective* viii. 146 It involves State Security. Your Rosemary has been having a roll with a Cabinet Minister.

3. b. *Phonetics.* = Trill *sb.*² 3. Cf. Roll *v.*² 4 c, *rolled *ppl. a.* 4.

1950 D. Jones *Pronunc. of Eng.* (ed. 3) I. 95 Another variety of the 'burr' is a uvular fricative sound (without roll)... One may also hear a uvular roll with accompanying friction. **1973** J. D. O'Connor *Phonetics* ii. 47 Rolls consist of several rapidly repeated closures and openings of the air passage, as in the rolled *r*-sounds of Scottish or Italian... [The] uvular roll is common in Dutch for *r* and may be heard in French and German too—the sound is reminiscent of a gargling noise.

8. *attrib.* and *Comb.*, as *roll angle*, *plane*; **roll axis,** the axis about which a vehicle or craft rolls; **roll bar,** an overhead metal bar to protect the occupants of a motor vehicle in the event of its overturning; **roll cage,** in a motor vehicle, a centre box section designed to protect the occupants if the vehicle overturns; also *attrib.*; **roll cast** *Angling*, (see quot. 1960¹); hence as *vb. trans.* and *intr.*; *roll-casting* vbl. sb.; **roll rate,** the angular velocity of a vehicle or craft about its roll axis.

1961 *Which? Reports on Cars* 14 Published reports so far have been based on subjective assessment of roll angle. **1970** *Motoring Which?* July 99/2 All three had low roll angles. **1950** Newton & Steeds *Motor Vehicle* (ed. 4) xxxi. 566 The roll axis for a car having axles at front and back will be some distance above ground level while that having independent suspensions at front and back will have a roll axis lying at ground level. **1962** Roll axis [see *pitch axis* s.v. *Pitch *sb.*² 26]. **1954** *Amer. Speech* XXIX. 101 *Roll bar, n.*, a curved bar welded or bolted to the frame rails extending upward in back of the driver's seat to protect him in case he 'flips' over. **1957** *Life* 29 Apr. 133 In sanctioned meets cars must have roll bars over driver's seat. **1969** H. Nielsen *Darkest Hour* xiii. 143 Goddard.. went over the embankment... The car has a roll bar, but you can see what happened. The door sprung open and he went out of it head first. **1979** *Tucson Mag.* Mar. 25/1 A removable forward hardtop and a convertible softtop rear window are separated by a Targa-style rollbar. **1972** *Sci. Amer.* Apr. 9/3 (Advt.), And 'roll cage' construction. The kind that soon, by law, may be required on all cars. **1973** *Times* 18 Oct. 35/3 The body comprises a one-piece glass-fibre outer shell on a steel monocoque centre section, with built-in rollcage. **1976** *Good Motoring* May 12/1 The roll-cage passenger compartment and anti-intrusion bars in the doors to help in side impacts. **1934** R. Kelly *Fishing* 9 (*heading*) The Roll Cast. This cast is used where trees and brush overhang the banks of the stream. **1947** R. Bergman *With Fly, Plug & Bait* II. vi. 113 A skillful and long roll cast is essential. *Ibid.*, When roll casting I grease my line carefully. *Ibid.*, My torpedo head tapered line also has rather a stiff finish and roll-casts well. **1960** Edwards & Turner *Angler's Cast* x. 101 The Spey cast..is the simplest roll cast. *Ibid.* 105 If the angler is fishing the left bank of the river, with obstructions behind him, his only method of getting the line out is to roll cast. **1960** C. Willock *Anglers' Encycl.* 158/2 *Roll-cast*, a fly cast in which the line is picked off the water without being thrown behind. **1972** *Trout & Salmon* June 58/3 Then make a roll-cast, but instead of roll-casting the line on to the water, roll it into the air. **1947** R. Bergman *With Fly, Plug & Bait* II. vi. 113, I have a special level line with a rather stiff finish which is especially fine for roll casting. **1960** Edwards & Turner *Angler's Cast* x. 101 The average angler..thinks nothing of roll casting, with constant changes of direction, for half-a-day on his own trout stream. **1971** *Aeronaut. Jrnl.* LXXV. 295/2 Some selected type of manoeuvre, such as pitch attitude, or pitch rate, or normal acceleration, and corresponding quantities in the roll plane. **1961** W. R. Kolk *Mod. Flight Dynamics* viii. 146 An airplane's ability to roll is properly a characteristic of its maneuverability, but is also a cornerstone of its flying qualities by reason of the unstable pitch-yaw resonance encountered at roll rates exceeding the natural frequencies in either pitch or yaw. **1975** G. H. Saunders *Dynamics of Helicopter Flight* v. 178 When the roll rate builds up to the point where the damping moment equals the control moment, no further increase in roll rate is achieved.

roll, *v.*² Add: **I. 1. a.** *to roll the bones*, to play dice. *U.S. slang.*

1929 H. W. Odum in *Amer. Mercury* Sept. 49/2 So we sets 'round in circle an' starts rollin' them bones. *Ibid.* 58/1 Gonna roll them bones. Gonna git some money an' play bad. **1945** L. Saxon et al. *Gumbo Ya-Ya* vii. 127 Today in the colored sections of the city there are always circles of men 'rollin' the bones' playing *Indian Dice*, which is any game of Craps unsupervised by a syndicate and without a player for the 'house'.

e. Also *trans.* To bowl (a game making a specified score, a number of strikes). *U.S.*

1974 *Cleveland* (Ohio) *Plain Dealer* 13 Oct. c.8/3 Marge Dimario, bowling in the Top Ten League at Ambassador Brookpark Lanes, rolled a 275 game. **1979** *Arizona Daily Star* 1 Apr. c2/6 Earl Anthony..rolled nine strikes in the championship match yesterday.

f. Computers. *to roll out*, (*a*) (see quots. 1954, 1962¹); (*b*) to transfer (data held in a main memory) to an auxiliary store when a program of greater priority requires the former; similarly *to roll in* (in two senses).

1954 *Computers & Automation* Dec. 20/2 *Roll out*, to read out of a register or counter by the following process: add to the digits in each column simultaneously; do this 10 times (for decimal numbers); when the result in each column changes from 9 to 0, issue a signal. **1962** *Gloss. Terms Automatic Data Processing* (*B.S.I.*) 86 *Roll out* (*to*), for a counter which counts modulo *n*, to read its content by causing it to count a sequence of *n* pulses, determining at what stage in the sequence the content passes through zero. *Ibid.* 87 *Roll in* (*to*), to increase the content of a counter by causing it to count a sequence of pulses. **1969** P. B. Jordain *Condensed Computer Encycl.* 435 When main memory is released by any program, or a task terminates and its space becomes available, a task that had been rolled out can be rolled in and restarted. **1970** O. Dopping *Computers & Data Processing* ix. 123 The programs are..often stored in secondary memories, and the necessary program parts are rolled in to the primary storage when needed. When another program needs the memory space, some program parts may have to be rolled out to secondary storage again.

2. a. (Earlier and later U.S. *fig.* examples.)

1859 *La Crosse* (Wisconsin) *Union* 24 Oct. 2 He ought.. to pitch in and help roll up a big majority for Randall. **1900** *Congress. Rec.* 23 Jan. 1103/2 They answered them by rolling up a plurality of 5,665 votes for the member from Utah out of a total of 67,805. **1951** *Sun* (Baltimore) 19 June 7/3 In the state elections of Lower Saxony..the SRP rolled up nearly 400,000 votes. **1976** *Billings* (Montana) *Gaz.* 4 July 11-A/1 The powerful PRI has always rolled up massive victories in every election during the past half century.

5. f. To rob (esp. someone drunk, drugged or sleeping). *slang.*

1873 A. S. Evans *À la California* xii. 298 When one of these fellows makes a raise by 'rolling a drunk' (i.e., taking the valuables from the pockets of a drunken man on the sidewalk), he will take a single bed at 37½ cents. **1892** C. C. Jenkyns *Hard Life in Colonies* 165 To 'roll drunks' was to frequent drinking saloons, to follow any man who left drunk, roll him into the gutter and rob him. **1912** [see *roll *sb.*¹ 6 e]. **1923** A. Price *Dreams* 3 My money, I kept in my cutter shoes, And I wasn't rolled the endurin' trip, So the whole ten days I hit the booze, With a downhill haul, and I let her rip. **1935** *Sun* (Baltimore) 2 July 1/1 We decided to get him drunk in his room the next night..and roll him. **1939** R. Chandler *Big Sleep* xx. 167 Here we are with a guy who..has fifteen grand in his pants... Somebody rolls him for it and rolls him too hard, so they have to take him out in the desert and plant him among the cactuses. **1949** *Life* 24 Oct. 23 She heard her new friends kidding about rolling guys. **1955** W. Gaddis *Recognitions* III. v. 940 She paid all the bills at the George Sank and gave him a terrific time for a couple of days and then rolled him. **1958** G. Greene *Our Man in Havana* v. v. 245 In some of these places they try to roll you. **1960** *Times* 21 Sept. 3/7 We walked through a few back streets and Lutt suggested 'rolling' (robbing) someone. **1962** E. Lucia *Klondike Kate* iv. 107 The dames seldom rolled the miners or slipped them a Mickey. **1968** *Globe & Mail Mag.* (Toronto) 13 Jan. 7/4 If a hustler is not himself homosexual,..he is called 'trade'. Rough trade refers to hustlers who are liable to beat up or roll the homosexual, either after or instead of sexual relations. **1974** in W. R. Hunt *North of 53 Degrees* vii. 42 If you don't get drunk, you don't get rolled. **1978** *Courier-Mail* (Brisbane) 22 Apr. 10/4 He had given much thought before sentencing two aboriginal women.. for 'rolling' a man in a hotel.

g. To start moving; *spec.* (esp. in command *roll 'em*) to start (cameras) filming. *slang.*

1939 J. Dell *Nobody ordered Wolves* iv. 51 Someone shouted, 'Roll 'em,' and, someone else, 'Quiet there.' **1949** R. Chandler *Little Sister* xix. 131 He went back beside the camera. The assistant shouted 'roll 'em' and the scene went through. **1959** *Elizabethan* June 26/1 The director call out 'Action' to the actors, then 'Roll 'em' and the cameraman starts the camera. **1973** J. Drummond *Bang! Bang! You're Dead!* xxxi. 107 We may need the trucks at any time... I'll phone if I want you to roll 'em. **1977** *Rolling Stone* 21 Apr. 63/6 'Roll 'em'! crackled over the radio.

h. *fig.* To reduce, cut *back* (esp. prices). *U.S.*

1943 *Funk & Wagnalls New Stand. Encycl. Yearbk. 1942* 81/1 In many instances, wholesale or manufacturers' prices were 'rolled back' to an earlier date. **1943** *Sun* (Baltimore) 29 May 1/3 We are, therefore, confronted with the choice of rolling back the cost of living..or permitting an adjustment of wages and other income in line with the increase in the cost of living. **1944** *Ann. Reg. 1943* I. 287 Four 'pressure groups'..decisively vetoed the President's plan to 'roll back' farm prices to the level of the previous September. **1951** *Manch. Guardian Weekly* 15 Mar. 10/4 The Tampa *Tribune* in Florida..prominently reported,..the Government's promise to roll back meat prices. **1972** *Daily Tel.* 25 Apr. 4/7 From first reports, he thought perhaps 10 per cent. of America's large businesses would be required to 'roll back' prices because of excessive profits. **1975** *Washington Post* 26 Dec. A 22/1 The focus of this..bill is its attempt to roll back oil prices. **1977** *Time* 25 July 5/2 Fully 1.15 million workers were jobless in June... Unless the Giscard regime can roll back that figure, it could become a lethal weapon in the hands of the left.

i. *Econ.* With *over*. To finance the repayment of (maturing stock, or the debt it represents) by the issue of new stock.

1957 *Jrnl. Finance* Mar. 52 Since the success of a refunding offer is measured in terms of the percentage of the maturing obligation which is 'rolled over' into the new issue, it is required that the Treasury tailor its terms to the needs of the market. **1959** *Wall St. Jrnl.* 27 Jan. 17/3 Government bond dealers said that they expect the Treasury to announce late this week its plans for refunding nearly $15 billion of Federal debt maturing next month. How the Treasury will roll over these securities is anybody's guess. **1973** *Daily Tel.* 15 Sept. 23/3 Existing maturities are normally 'rolled over' (refinanced on their redemption date by the issue of further bonds at whatever the going rate of interest happens to be), thus giving the municipal treasurer a virtually perpetual access to the money market. **1976** *Economist* 16 Oct. 105/2 Even without any increase in interest rates since early April, 1976, the cost of servicing the national debt was bound to increase..from the need to roll over £2.8 billion of gilt-edged stock due to mature during the year.

8. a. *spec.* To make (a cigarette) by rolling paper round loose tobacco. Freq. in phr. *to roll one's own*: to make one's own cigarettes; also *fig.* Hence *roll-your-own* attrib. and ellipt.

1885 Dickens *Dorrit* (1857) I. i. 6 He was now rolling his tobacco into cigarettes by the aid of little squares of paper. **1892** H. G. Parker *Pierre & his People* 128 He slowly rolled a cigarette. **1893** [in Dict.]. **1903** A. Bennett *Leonora* iii. 69 He had extraordinary aptitudes for drawing corks..and rolling cigarettes. **1930** *Amer. Speech* VI. 92 The following expressions belong to colloquialisms and slang, including movie and radio neologisms... *Rolls its own.* [Etc.]. **1932** J. D. Carr *Poison in Jest* xi. 157 He produced papers and tobacco... 'Good American,' he announced. 'I roll my own.' **1934** Webster, *Roll one's own*,..to do things without outside aid. **1936** L. Hellman *Days to Come* I. 18 He has started to roll a cigarette. Quickly Julie offers him a box from the table. **1940** *Amer. Speech* XV. 335/2 Cigarettes..may be *home-made, rolls*, or *roll-your-owns.* **1941** *N.Y. Times* 25 July 14/5 'Ghosting' is routine in public papers in the United States, and has been since our history began... Mr. Roosevelt proved again today that he can roll his own whenever he has the time and the inclination. **1952** *Arena* (N.Z.) xxxi. 2 But then Charlie would have rolled himself one, and looked up at the hills and pretended he didn't hear. **1960** J. McNamee *Florencia Bay* 59 Looks sixty. Thin face. Dark. Looks a little Indian but not our kind of Indian. Rolls his own. **1975** R. L. Simon *Wild Turkey* (1976) xx. 149, I..took out some papers and started to roll a joint. **1977** *Daily Mirror* 30 Mar., Roll-your-own cigarette tobacco will also go up—but pipe tobacco and cigars escape. **1980** *Forest Products News* (Wellington, N.Z.) XVII. I. 6 He had come straight out of the bush with his roll-your-own.

fig. (Further examples of phr. 'rolled into one'.)

1879 T. Hardy *Let.* 26 Mar. (1978) I. 64 It is possible that he & the ancestor of your relative were two different persons who were in India at the same time, & so got rolled into one. **1907** G. B. Shaw *Major Barbara* 167 My methods..would be no use if I were Voltaire, Rousseau, Bentham, Mill, Dickens, Carlyle, Ruskin, George, Butler, and Morris all rolled into one. **1951** M. McLuhan *Mech. Bride* (1967) 135/2 He is the Supreme Court and human fate rolled into one. **1978** *N. & Q.* Feb. 94/1 Social and political historian, literary critic and man of the theatre rolled into one.

b. *fig.* (Further examples.)

1906 Kipling *Puck of Pook's Hill* 221 Then the Winged Hats began to roll us up from each end of the wall. **1949** *Sun* (Baltimore) 25 Nov. 1/3 Capture of these critical defenses..placed the Americans in position to roll up the whole Yamashita defense front. **1963** 'J. Le Carré' *Spy who came in from Cold* iii. 24 He had made a mistake in Berlin, and..his network had been rolled up.

10. c. To make or form by passing a material between rollers.

1967 A. H. Cottrell *Introd. Metall.* xxii. 442 This principle has been particularly developed in Rohn and Sendzimir mills for rolling thin foil. **1972** *Daily Tel.* 14 Apr. 21/2 This hid a 6 p.c. rise for the billets and nil rise for the reinforcing bars which are rolled from them.

10*. *to roll off*, to cause (the frequency response of audio apparatus) to decrease smoothly at the end of its range; also *to roll in* or *on*, to cause a similar increase. Cf. sense 24 below.

1970 J. Earl *Tuners & Amplifiers* v. 105 The receiver must incorporate a network which rolls the treble response off at the same rate as the transmitter rolls it on. **1975** *Hi-Fi Answers* Feb. 69/3 In a three-way speaker you've got to get the mid-range to cover the whole of the speech band all in one go,..and you've got to get it down to at least two octaves below the frequency at which you want to roll it in. **1975** G. J. King *Audio Handbk.* ii. 38 Some designers prefer deliberately to roll-off the bass around 30 Hz. **1976** *Gramophone* Aug. 359/3 It also rolls off the curve sharply from 12·5 Hz to 6 Hz.

II. 11. c. Also (gen.), to start moving. (Further examples.) Also *transf.* and *fig.*

1944 L. Lariar *Man with Lumpy Nose* viii. 75 'Do me a favor and go home and write it!' McEmons stood over the reporter menacingly. 'Get rolling!' The reporter shuffled out of the room. **1952** S. Kauffmann *Philanderer* (1953) iv. 61 'Let's roll, dreamer', said Perry. **1956** A. H. Compton *Atomic Quest* i. 55 To help get the atomic program rolling. **1959** I. Jefferies *Thirteen Days* viii. 108, I..waved the drivers on. As they rolled I gave them one last treat..by taking my hat off. **1965** *New Statesman* 14 May 753/1 The private train is ready to roll. **1970** W. Smith *Gold Mine* xii. 31 Wake up. Time to roll. **1977** *Observer* 3 Apr. 11/3 The PanAm captain then shouted: 'The bastard's not been given permission to roll. We're on the runway. We're on the runway.'

d. Of the foot: to slip on or upon an object.

1878 R. L. Stevenson in *Temple Bar* LII. 55 His foot rolled upon a pebble. **1904** L. Tracy *King of Diamonds* ix. 123 Philip..almost fell too, for his left foot rolled on the constable's staff.

e. *roll on* ——: expressing a wish that time may pass quickly until a particular event; may (something) come soon.

1885 M. Davitt *Leaves from Prison Diary* I. 150 'A burst in the City. Copped while boning the swag. 7 Stretch, 1869. Roll on 1876. Cheer up, pals.' Another—'Hook, 7 ys. Roll on time.' **1917** F. T. Nettleinghame *Tommy's Tunes* 21 When this ruddy war is over, Oh! how happy we shall be!.. Roll on, when we go on furlough; Roll on, when we go on leave. **1936** J. Curtis *Gilt Kid* ii. 19 'Well,' said the Gilt Kid, 'this is a whole lot better than making scrubbing brushes back in the old Monastery Garden.' 'Yes, and saying to yourself, "Roll on Cocoa".' **1958** M. K. Joseph *I'll soldier no More* xiii. 237 'What's your new gaffer like, Tom?' 'Like a barber's cat... I should worry—roll on my ticket.' **1962** *Sunday Express* 21 Jan. 15/5 Roll on the mid-twentieth century Venus. And the best of synthetic luck to her. **1970** M. Tripp *Man without Friends* i. 15 He wakes at seven..saying 'Roll on my retirement.' **1978** K. Amis *Jake's Thing* x. 98 Roll on wrist-watch television.

f. To taxi in an aircraft. *Obs.* exc. *Hist.*

1910 *Flight* 24 Sept. 776/1 Messrs B. H. Barrington Kennett.., A. Aitken, and St. Croix Johnstone..are 'rolling' whenever the weather is suitable. **1915** Kipling *Diversity of Creatures* (1917) 423 Wynn..had finished 'rolling'..and had gone on from a 'taxi' to a machine more or less his own. **1961** C. B. Smith *Testing Time* iii. 48 It was still quite an event to leave the ground, and many would-be fliers spent their whole time 'rolling' (as taxying was then called).

g. *heads will roll* and varr.: there will be executions; also *fig.*, some will be ousted from power or position; also in extended and weakened uses.

1930 *Daily Herald* 26 Sept. 1/1 Giving evidence, Hitler declared..'If our movement is victorious there will be a revolutionary tribunal which will punish the crimes of November 1918. Then decapitated heads will roll in the sand.' **1940** *Time* 5 Aug. 22/1 Echoes of 'Heads will roll' Hitlerism were heard from Paris to Marseille. **1961** *Time* 1 Dec. 77/3 A.M.C. made it clear, too, that more heads would roll if the workers still failed to get the message. **1963** A. Howard in Sissons & French *Age of Austerity* 17 Mr Macmillan's head rolled at Stockton-on-Tees at 10.25 a.m. **1966** P. O'Donnell *Sabre-Tooth* xvi. 225 'Suppose this improbable thing happens?'..'Then no doubt my head will roll.' **1972** *National Observer* (U.S.) 27 May 7/2 President Nixon decreed 'heads will roll' if 'petty bureaucrats' hinder Jaffe's war on narcotics. **1978** *Rugby World* Apr. 45/1 Wales lost, and heads rolled.

h. Of a movie camera or cameraman, etc.: to be in action; to start filming.

1938 'E. Queen' *Four of Hearts* iv. 53 'Then it's okay to shoot the works now, Butch?'.. 'We're rolling, Sam.' **1938** F. Scott Fitzgerald *Let.* 18 Apr. (1964) 28 It may come right at the crucial point of this picture (due to roll in June, but perhaps not starting till the fifteenth). **1958** *Punch* 17 Sept. 382/3, I can imagine the whole cast falling about with hysterical laughter the moment the cameras stop rolling. **1971** D. E. Westlake *I gave at the Office* (1972) 178 'Okay, Jay,' Joe finally said, from behind the camera and lights and sound equipment. 'Let her roll.' **1978** G. Vidal *Kalki* iii. 75 A man with a clapboard stood between the camera and the door. 'Start rolling,' said the director.

12. b. *to roll up.* (Earlier and later examples.) Also, to arrive; to appear on the scene.

1861 *Goulburn* (New South Wales) *Herald* 18 Sept. 2/2 It is not by accident that flags are unfurled with mottoes upon them, as 'roll up', 'no Chinese'. **1920** G. Bell *Let.* 24 Oct. (1927) II. xix. 567 When the Mayor of Bagdad rolled up at 9 or the Naqib sent his son Saiyid Mahmud I was obliged to 'endosser' dressing-gown and go out to see them. **1929** 'Sapper' in *Legion Bk.* 214 The man hasn't rolled up yet, but he won't be long. **1955** W. Gaddis *Recognitions* III. v. 863 The sight of a soiled limousine parked up the street..clouded his face with the memory of the girls from the American Embassy in Madrid who had rolled up the day before. **1968** M. Woodhouse *Rock Baby* xxiv. 232 They had to wait for me to roll up because I had the D.F. set, which meant I was the only one who could pin it down precisely. **1976** J. Wainwright *Who goes Next?* 161 A townie. A bit overdressed..he once rolled up in a velvet jacket. **1977** *Water Sport* (Austral.) Jan. 56/2 So please roll up and bring some of your friends.

e. *to roll along* = sense 12b in Dict. and Suppl.

1928 A. Waugh *Last Chukka* 82 She entertains whoever there may be that chooses to roll along.

16. c. (Later examples.)

1978 *Chicago* June 124/2 With money rolling in from the rest of the family empire..he began buying, parcel by parcel, the farmland around his family's estate. **1979** D. Lowden *Boudapesti 3* ii. 16 No, it's not money... We were quite well off... It was rolling in.

18. a. Also of motor vehicles. *rolling in the aisles*: see *aisle 5b.

1954 *Amer. Speech* XXIX. 101 *Roll, v.i.*, to overturn. **1968** *Sun* 12 Nov. 8/4 While the world sleeps, they [*sc.* rally drivers] 'yump' and 'wrong slot' and sometimes have the misfortune to 'roll'. **1976** *Billings* (Montana) *Gaz.* 1 July 2-A/4 The patrol said American Horse's vehicle rolled three times after the collision.

e. *to roll out*: to get out of bed, to get up. *U.S. colloq.*

1884 W. Shepherd *Prairie Experiences* 237 The cook's voice shouts 'Roll out'... Before you have time to dress.. it is 'Breakfast!' **1930** L. Hughes *Not without Laughter* xv. 183 When his mother rolled out at six o'clock to go to work, he woke up again. **1942** Z. N. Hurston in A. Dundes *Mother Wit* (1973) 223/1 All you did by rolling out early was to stir your stomach up. **1963** *Amer. Speech* XXXVIII. 271 The term *roll out* means 'to get out of bed in the morning'.

19. a. Hence *rolling-in-money* absol. as *sb.*

1960 Auden *Homage to Clio* 74 The rolling-in-money, The screamingly-funny.

21. a. Also with *down.*

1898 *Forest & Stream* 19 Feb. 156/2 Before the wind reached us the schooner rolled down at such an angle that her crew commenced to shorten sail. **1916** F. W. Wallace *Shack Locker* (1922) 166 She rolled down an' came up with a dory a-hangin' on her fore-cross-trees.

c. Of an aeroplane: to turn about its longitudinal axis.

1909 'Aero-Amateur' *Flying* ix. 55 If the wings of a soaring, or gliding machine are curved upwards in the form of a bow the machine certainly has a tendency to travel in a straight line, but will have also a tendency to roll badly. **1918** J. M. Grider *War Birds* (1927) 69 He was looping and rolling between the church spires. **1976** *Times* 17 July 12/3 The Pitts, a small and exceptionally manoeuvrable biplane..can roll through more than 360° in a second.

f. *to roll with the punches* (and varr.), of a boxer: to move the body away from the opponent's blows in order to lessen their impact; *fig.*, to adapt oneself to difficult circumstances, take troubles in one's stride.

[**1941** F. Gilmore *Push Yourself* iv. 27 In boxing it is called 'rolling the punch' when a boxer, not having time to avoid being hit, deliberately moves with the punch when it hits him.] **1951** J. J. Walsh *Boxing Simplified* viii. 32 In an actual bout he will not have so much time to roll with the blow. **1956** H. Kurnitz *Invasion of Privacy* ii. 15 He had mastered the trick of rolling with the punches, rendering himself invisible when a crisis darkened the neighbouring skies. **1963** J. Crosby *With Love & Loathing* 48 Madison Avenue rolls with the blow; it watches carefully which direction the cookie crumbles. **1979** *Now!* 21–27 Sept. 74/1 It would be possible to roll with such punches were it not for the fact that the 1980 election season has already begun and a seemingly invincible Democratic contender has suddenly launched himself into the fight.

24. *to roll off*, (of audio apparatus) to exhibit a response decreasing smoothly to zero with increasing signal frequency; so *to roll in*, to exhibit a response increasing similarly from zero. Cf. sense *10*.

1959 *Consumer Reports* Sept. 452/2 The newer *Jansen* [tweeter] also rolled off slightly in the extreme high-frequencies. **1962** A. Nisbett *Technique Sound Studio* 253 The simplest form of filter (one resistor and one condenser) rolls off at 6 dB/octave above or below a certain frequency. **1970** J. Earl *Tuners & Amplifiers* i. 25 Many [loudspeaker] systems employ two units, one.. going from about 30 or 40Hz and rolling-off due to the action of the crossover around 1 or 2 kHz and the other for treble rolling-in at about 1 or 2 kHz and responding up to 16 kHz or higher. *Ibid.* iii. 69 The majority of amplifiers have in-built high-pass filtering, rolling off around the 20 to 30Hz mark. **1976** *Gramophone* July 235/1 Further tonal correction is provided by push-buttons, to provide separate filters rolling off at 7 and 10kHz respectively.

ro·ll-around, *a.* [f. *to roll around.*] That can be moved around on wheels or castors.

1973 *Sunday Bull.* (Philadelphia) 7 Oct. (Parade Suppl.) F12 (Advt.), The GE Countertop Oven is versatile in other ways, too. You can use it on a roll-around cart, or build it in. **1976** *Amer. Speech 1974* XLIX. 116 *Roll-around wet bar*, portable counter equipped with a sink.

ro·ll-away, *a.* Also **rollaway.** [f. *to roll away.*] That may be removed on wheels or castors. Also *absol.* as *sb.*, a roll-away bed.

1938 *Sun* (Baltimore) 13 Oct. 11/3 Mr. Latham invented the 'roll-away' stage over the footlights which is used in many motion-picture houses. **1941–2** *Sears, Roebuck Catal.* 689/4 Odora Roll-Away Chest... 4 easy-rolling ball bearing casters. *Ibid.* 1274/7 (Index), Roll-a-way Chests 689 Roll-a-way Cots 904C. **1958** *Daily Progress* (Charlottesville, Va.) 4 June 14/2 Cots and roll-aways donated by townspeople after a radio appeal. **1960** *Farmer & Stockbreeder* 29 Mar. 119/1 There are no litter problems, eggs are laid in rollaway nests, the floor area per bird is only 1 sq ft. **1960** *Washington Post* 27 Nov. E3 Three rollaway beds and a cot are stacked during the day on the back porch. **1966** A. Cavanaugh *Children are Gone* II. iii. 30 A musty old building... Three rooms with a rollaway bed, which she and two other girls had taken turns occupying. **1969** *Islander* (Victoria, B.C.) 16 Nov. 2/2 The charming young hostess..offered to get a roll-away cot for me. **1971** A. A. Michele *You don't have to Ache* ix. 188 Frequently the 'rollaways' that motels keep for the use of children are harder and better for your back than the regular bed mattress. **1974** *Country Life* 21 Nov. 1580/2 You get another big double bed downstairs and a roll-away bunk.

roll back, roll-back, rollback. [f. *to roll back.*]

1. The action or fact of rolling backwards.

1937 *Times British Motor Number* 13 Apr. p. xxii/4 More than one method is available for automatically preventing roll back on an incline. **1949** Shurr & Yocom *Mod. Dance* v. 181 The swift roll-back carries body weight onto the shoulders. **1978** *Daily Tel.* 4 Mar. 1/8 They died as nine Gnats swept across the airfield to practise a 'roll back'—a manoeuvre performed by the Red Arrows for several years without incident.

2. [*roll *v.*² 5 h.] *U.S.* A reduction or decrease; *spec.*, a return (of commodity prices, etc.) to a lower level. Also *attrib.*

1942 *Time* 11 May 80 OPA have denied their [*sc.* retailers'] plea for a 'roll-back' of ceiling dates that would

recognize the lag between rising wholesale and retail prices. **1943** *Funk & Wagnalls New Stand. Encycl. Year Bk. 1942* 373/1 For Price Roll Back and Price Squeezes, see *Business Review* under *Commodity Prices.* **1943** *Sun* (Baltimore) 19 May 7 Promised 'roll backs' on cabbage and lettuce will still leave prices of these foods, at the farm, three times as high as they were a year ago. *Ibid.* 1 July 14/3 Congress had been equally unkind to his eager plan to detour around the thornier difficulties of farm-price control by a rollback food-subsidy scheme. **1945** *Richmond* (Va.) *News Leader* 20 Aug. 10 The OPA is standing pat on its prediction of a clothing price rollback—eventually. **1959** *Time* 27 Apr. 15/2 The cold war's boundaries in 1959 were much as they had been in 1953—the rollback had been in men's minds, not real estate. **1972** *Fortune* Jan. 101/1 Even the 3.9 percent price increase the company posted for its 1972 models last July, before the price freeze forced a rollback, would have recovered only 65 percent of the unit cost increases anticipated for the current model year. **1973** *Black Panther* 21 Apr. 2/2 This decision provides a legal basis to begin a roll-back of that power. **1974** *Financial Times* 15 Aug. 17/6 We do need the 'roll back' which ex-President Nixon promised but (predictably) did not deliver. **1976** *Lebende Sprachen* XXI. 153/2 The pilot noticed an inflight engine roll back from 90 to 80% rpm.

roll-call. Add: **1.** (Earlier examples.) Also (*U.S.*), a calling over of a list of members of a legislative or similar body in order to ascertain how each wishes to vote on a particular measure.

1775 *Essex Inst. Hist. Coll.* (1912) XLVIII. 61 This morning we went to rol col & then got our Brefust. **1777** *Ibid.* (1877) XIII. 118 Ordered that the Hour for Roll Call be altered to Nine o'clock in the morning. **1899** [in Dict., attrib.]. **1902** *Ann. Rep. Amer. Hist. Assoc. 1901* I. 323 Except for the provision in the constitution there would have been no roll call on these votes. **1947** *Economist* 27 Dec. 1047/1 The sponsors of ERP avoided a roll-call on the first vote in order to take one on the agreed Bill. **1955** *Times* 4 Aug. 6/7 In foreign policy the President secured bipartisan support on nearly every occasion; in 32 Senate roll-calls the Democratic majority failed to agree with the Republican majority only once. **1972** *Computer & Humanities* VI. 184 The data are placed on cards with one record holding the yea..or nay..votes of one congressman on every roll-call.

attrib. *roll-call analysis, vote* (further examples).

1950 *N.Y. Times* 20 Apr. 1/6 The Senate adopted today, by a roll-call vote of 66 to 0, a resolution directing the Secretary of the Navy to confer appropriate posthumous decorations on the crew of the Privateer that presumably was shot down..over the Baltic sea. **1963** *Midwest Jrnl. Polit. Sci.* VII. 156 (*heading*) A second look at the validity of roll-call analysis. **1970** *Computers & Humanities* V. 8 Several embarked on similar scalogram excursions into roll-call analysis of collegial bodies, both with and without computers. **1970** *Internat. & Compar. Law Q.* XIX. I. 68 On a vote by show of hands, the required two-thirds majority was not obtained, but a second (roll-call) vote was taken, and the retention of the reference..was confirmed. **1979** *Tucson* (Arizona) *Citizen* 20 Sept. 2A/4 Rep. Robert Bauman, R-Md., said the House voted against the bill because it came on a roll call vote, which puts each member on record as either supporting or opposing the measure to which the pay raise was attached.

ro·ll-call, *v.* [f. the sb.] *trans.* To call the roll for (a group or body of persons). Also *fig.*

1928 *Daily Express* 19 Mar. 12/2 The German officers were counted or 'roll-called' in their rooms to save them the trouble of having to assemble or fall in with the other prisoners. **1962** V. Nabokov *Pale Fire* 55 But who can teach the thoughts we should roll-call When morning finds us marching to the wall?

roll-collar. Delete '(Now *rolled collar*)' and add: (Later examples.) Also *attrib.*

1907 E. P. Oppenheim *Secret* ii. 15 He wore..a made-up white tie, with the ends tucked in under a roll collar. **1922** Joyce *Ulysses* 72 Stylish kind of coat with that roll collar. **1929** *Even. News* 18 Nov. 6/5 (Advt.), Below we show the D. B. Chester with long roll collar. **1963** *Times* 23 Jan. 12/4 The ubiquitous, exquisitely soft, reversible woollens were used for perfectly tailored, gently precise coats with tiny roll-collars. **1973** S. B. Jackman *Guns covered with Flowers* vi. 98 He dressed quickly—pants, T-shirt, roll-collar shirt, dark slacks.

ro·ll down, roll-down. [f. *to roll down.*] A game in which balls are rolled down a board into numbered holes, slots, or the like; a table or stall where this is played. Also *attrib.*

1926 *Variety* 29 Dec. 7/4 The outdoor show game with its 'rag front',..'roll downs', [etc.]. **1942** Berrey & Van den Bark *Amer. Thes. Slang* § 626/21 *Roll-down*, a gambling device using small balls rolling down between pins to holes in the bottom of the table. **1943** K. Tennant *Ride on Stranger* vii. 61 But the Roll-Down Table was not doing too badly despite the rain... The players could shelter while they rolled the billiard balls down the green felt into shallow cups.

rolled, *ppl. a.* Add: **1. a.** (Further examples.) Also with sbs. used *attrib.*

1819 M. Edgeworth *Let* ?10 Mar. (1971) 181 All the fashionable trimmings are of that rolled sort of flounces. **1866** Geo. Eliot *Felix Holt* III. xlvi. 221 The grandeur of barbaric forms—when rolled collars were not yet conceived. **1928** *Daily Mail* 3 Aug. 10/4 Members of the audience looked twice before they could..appreciate the fact that she had rolled stockings. **1962** L. Deighton *Ipcress File* xxv. 164 The British man..put on his rolled blue roll-neck sweater. **1955** 'D. Cory' *Phoenix Sings* v. 89 A big fellow in a rollneck pullover. **1968** *Daily Mirror* 20 Aug. 9/1 And I got a couple of bright flowery roll-neck tops which will also go with the slacks. **1968** *Guardian* 19 Sept. 8/3 John Cranko came on stage after the première of his latest ballet in a white silk roll-neck. **1970** T. Lewis *Jack's Return Home* 148 He had on a white silk rollneck and a bright red cardigan. **1977** *Time Out* 17–23 June 80/1 (Advt.), Former male model—but more the jeans and rollneck type.

rollock (rǫ·lək). *slang.* [Alteration of Ballock, *bollock.] **a.** *pl.* As *int.* = *bollock 3. **b.** Comb., *rollock naked* adj. = **ballock-naked* adj.

1961 'B. Wells' *Day Earth caught Fire* ii. 31 'Rollocks!' said Maguire and his voice was deliberately gruff to hide his embarrassment. **1962** A. Wesker *Chips with Everything* I. i. 11 Even if you're stark rollock naked, you'll all spring to attention.

rollocking, var. *rollicking *vbl. sb.* 2.

rolloff, roll-off (rōu·lǭf). [f. *to roll off.*] **1.** *Ten-Pin Bowling.* A game to resolve a tie or determine the qualifier for a later round of competition.

1947 *Richmond* (Va.) *News-Leader* 2 May 22 (*heading*) Scher wins Major Men's Pin Championship after rolloff. **1975** *Cleveland* (Ohio) *Plain Dealer* 6 Apr. 7-c/3 When they bowl in the rolloffs each contestant will be presented a Brunswick bowling bag as the prize for gaining the finals. **1976** *Eastern Even. News* (Norwich) 27 Aug., He won the roll-off competition with an eight game total of 1474 for a tournament average of 185. **1979** *Arizona Daily Star* 22 July c2/4 He won in the two-frame roll-off.

2. The smooth fall of response with frequency of a piece of audio equipment or the like at an end of its range. Cf. *roll *v.*[2] 10* and 24.

1950 *Audio Engin.* Aug. 28/2 It is due to irregularities in the groove walls, and if heard through a flat system, sounds smooth or satiny, corresponding to initially white (flat spectrum) noise with perhaps some roll-off of the highs. **1956** *IRE Trans. Audio* IV. 35/2 To obtain the smooth transition between the woofer and tweeter, the two units should be designed to complement each other, both as to level balance and for obtaining the required rolloff in the low-frequency unit at the crossover frequency. **1959** Kuh & Pederson *Princ. Circuit Synthesis* xiii. 212 In practice, the roll-off of the magnitude of the transfer function of an *m*-derived delay line is primarily due to the dissipation of the inductances. **1975** G. J. King *Audio Handbk.* ii. 39 From the practical point of view, it is commonly necessary to introduce infrabass roll-off at least to attenuate rumble and other infrasonic noises which are often present on the programme signal.

3. First throw at dice.

1966 O. Norton *School of Liars* vi. 94 [We] flipped for roll-off. Wally's ace took the box.

roll-on (rōu·lǫn), *sb.* and *a.* [f. *to roll on.*] **A.** *sb.* **a.** Also *pl.* A type of elasticated corset designed to be stepped into and rolled up on to the body. **b.** A deodorant, etc., applied by means of a rolling stopper at the mouth of the container.

a. 1941 *Amer. Speech* XVI. 96 Do you like these *Roll-ons?* **1945** *Richmond* (Va.) *Times-Dispatch* 29 Mar. 3/2 A start could be made by changing the corsets—reminiscent of cumbersome bone and lacing styles—and attracting young women with the idea of..roll-ons. **1960** M. Cecil *Something in Common* 40 Holding the base of her roll-ons firmly, she wriggled inside them optimistically. **1963** *Times* 7 June 8/1 He was searched and underneath a woman's roll-on was found to have 28 bars of gold weighing 61lb. in a belt. **1972** *Times* 26 June 13/5, I..found myself in a bedroom in which was a very surprised lady struggling into her roll-on.
b. 1960 *Which?* Feb. 35/2 Body mist, the only roll-on to contain hexachlorophane, weakened nylon and viscose rayon.

B. *adj.* That rolls on; involving rolling on.

1950 B. Pym *Some Tame Gazelle* i. 9 She liked her clothes to fit tightly and always wore an elastic roll-on corset. **1960** *Which?* Feb. 35/1 Most of those who commented preferred the roll-on method to either spray or stick. **1962** *Fuller Brush Products* Feb. 2/2 'Roll-On' anti-perspirant and deodorant. **1975** N. Freeling *What are Bugles blowing For?* xx. 118 Her mind was furnished with moisture cream and roll-on deodorants. **1981** *Radio Times* 16–22 May 26/2 (Advt.), Two-way stretch roll-on girdle.

ro·ll-on, ro·ll-off. [f. *to roll on, to roll off.*] Used *attrib.* with reference to a method of transportation of vehicles by ship in which they are simply driven on to the vessel at the beginning of the voyage and off it at the end. Cf. *lift-on, lift-off* s.v. *lift *v.* 14.

1955 *Sun* (Baltimore) 3 Nov. 10/2 At Palm Beach, he said, two major facilities of immediate value and interest are available to roll-on, roll-off shippers. **1958** *Engineering* 21 Mar. 354/3 Improvements are being made, including.. the introduction of special containers, palletisation and of roll-on/roll-off methods. **1963** *Times Rev. Industry* Dec. 31 Collect a combined load from a number of British factories. Utilize the rollon-rolloff ferry services and deliver the goods direct to the continental buyer without using any expensive foreign warehouse space. **1967** *Sunday Times* 1 Jan. 24/6 Car-carrying liners with roll-on, roll-off facilities. **1969** *Guardian* 23 July 16/8 A ferry terminal for roll-on, roll-off passenger traffic. **1970** *Times* 2 June (Container Suppl.) p. ii/2 By the mid-1970s it is estimated that about 160 ocean-going purpose-built cellular container ships and roll-on roll-off ships with container capacity will be operating on the world's trade routes. **1972** *Guardian* 14 Aug. 20/2 Plans for a roll-on, roll-off ferry to Shetland from Aberdeen. **1976** *Southern Even. Echo* (Southampton) 16 Nov. 1/5 Dockers have threatened to close Portsmouth Docks in a move to stop Brittany Ferries from opening a freight service. They argue that yet another roll-on roll-off service would affect the jobs of dockers at ports like Poole.

ro·ll-out. Also **rollout.** [f. *to roll out.*] **1.** An act of moving or wheeling out; *spec.* the official rolling out of a new aeroplane or spacecraft. Also *attrib.*

1957 *Britannica Bk. of Year* 512/1 *Rollout*, the rolling of an aeroplane from the production line. **1967** *Economist* 16 Dec. 1158/2 The roll-out ceremony and the preparations now being made for Concorde's first flight not only mark the half-way stage in the aircraft's development, but also the point at which costs begin to rise really sharply. **1973** *Nature* 9 Feb. 360/2 The landing of Luna-21 and roll-out of Lunokhod-2 into the Mare Serenitatis area ..is already being hailed by the Soviet press as a precursor of cooperation in space between the United States and the USSR. **1976** *Daily Tel.* 18 Sept. 15 A milestone in the development of the next era of space travel was reached yesterday with the roll-out of America's first shuttle spacecraft.

2. The part of a landing during which an aircraft travels along the runway losing speed.

1959 *IRE Trans. Aeronaut. & Navig. Electronics* VI. 59/1 The objective..was to develop a full instrument-landing system which included touchdown and rollout. *Ibid.* 69/2 The roll out was accomplished by using the large hand-brake lever to keep the directional gyro centered until a full stop was completed. **1964** *Sci. Amer.* Mar. 33/1 If landings in Category III are adopted, however, the approach, flare decrab, touchdown and possibly rollout will be performed automatically. **1970** *Graphic* 12 Nov. 3 Two weeks following the first multiple flame-out, that same plane was loaded with 128 passengers headed for Mexico City. It landed safely. But no sooner was it on the ground—still on rollout—than all four engines quit again!

3. *Amer. Football.* A play in which a quarterback moves away from his protective blockers before attempting to pass. Also *attrib.*

1959 *Washington Post* 8 Nov. c6/4 A series of quarter-back roll-outs. **1969** *Eugene* (Oregon) *Register-Guard* 3 Dec. 2D/1 The Spartans would run what looked like a rollout and then fire the ball from one side of the field to the other. **1970** *Globe & Mail* (Toronto) 28 Sept. 18/7 A rollout quarterback, with outstanding running ability, Gabler has had difficulty connecting with receivers. **1977** *Chicago Tribune* 2 Oct. III. 4/3 We switched from our regular 5-2 defense into a 4-3 in the second half to try to keep their quarterback from getting around the corner on rollouts. **1969** D. Tallman *Directory of Football Defences* iii. 58 The roll-out pass is identical to the sprint-out, except that the quarterback, when clearing from the center will execute a roll-out or reverse-pivot technique.

ro·ll-over. orig. *U.S.* Also **rollover, roll over.** [f. *to roll over.*] **1.** An overturning, a turning upside down; a complete revolution.

1945 *Richmond* (Va.) *Times-Dispatch* 19 Mar. 2/4 (*heading*) Yank describes B-29 roll over in Osaka raid. **1950** *Dance Mag.* Dec. 35/2 Her 'dance' included back-bends, cartwheels, splits and 'four successive rollovers'. **1955** *Sun* (Baltimore) 19 July 17/6 Crash rollovers, head on collisions and T-bone crashes. **1962** *Amer. Speech* XXXVII. 272 All sorts of strapping and cushioning devices to insure survival of the driver in case of a high-speed collision or roll-over. *c* **1973** J. Cholerton *Adv. Acrobatic Tricks & Dances* (Assoc. Amer. Tap Dancing) 11 Two backward rollovers, two nip-ups (to left), to centre. **1973** *Sci. Amer.* Feb. 80/1 Although frontal collisions are the most frequent accidents, side collisions and roll-overs are by far the most dangerous. **1977** *Time* 24 Jan. 2/3 (Advt.), The ESV's have proved their life-saving value in head-on and rear-end collisions, side-swipes and roll-overs.

2. *Econ.* Extension or transfer of a debt or other financial relationship; *spec.* reinvestment of money realized on the maturing of stocks, bonds, etc.; an issue of stocks or bonds replacing one which matures.

1958 *Fortnightly Rev.* (Anderson & Strudwick, Richmond, Va.) 27 June 4 *Roll-over*, a stock or bond issue which takes the place of one that is maturing (and into which the money realized on the maturing issue may be put). **1958** *Washington Post* 6 Nov. A18/1 We have..a roll-over, something in the order of $23 billion worth of short-term notes, I think four times, certainly three. **1972** *Times* 27 June (Tokyo Suppl.) p. v/4 The Japanese phrase 'circular-buying' covers the roll-over of speculative funds from one sector to another. **1976** *National Observer* (U.S.) 22 May 9/5 (Advt.), If you've been buying notes, bills or certificates of deposit for yourself or your organization, you'll find it's a lot easier to buy Dreyfus Liquid Assets. There's no paperwork, no worrying about maturity dates, roll-overs, safe-keeping or delivery—and there's no sales charge.

3. *attrib.* and *Comb.*, as (sense 1) *roll-over accident, bar, protection*; (sense 2) *roll-over contract, contribution, credit, facility, provision, relief.*

1970 *Motoring Which?* July 107/2 Door release buttons stuck out a little—might allow doors to open in a roll-over accident. **1970** *Daily Tel.* 15 June 6/5 Roll-over bars have long been standard fittings on open racing cars. **1973** P. Einzig *Roll-Over Credits* xv. 102 At the time of writing there is a complete lack of standardisation in the terms of roll-over contracts. **1976** *National Observer* (U.S.) 3 July 9/1 In such a situation the transfer of funds is treated as a 'roll-over contribution'. **1973** P. Einzig *Roll-Over Credits* i. 3 One of the most important changes has been the evolution..of the system of roll-over credits—medium term credits with variable interest rates which are adjusted at fixed intervals to changes in the current market rates for short-term credits. **1975** *Daily Tel.* (Colour Suppl.) 25 July 30/3 If a creditor will only lend his money at short-term, but a debtor cannot hope to pay his debts for a long time, he is offered what is called a 'rollover facility'—a debt which is nominally short-term, but is automatically relent when it falls due for payment. **1972** *Times* 28 Dec. 21/1 Meeting all the foreseeable regulations on crash and roll-over protection. **1974** *Daily Tel.* 27 Apr. 21/2 Some developers had been hoping that the 'roll over' provisions would have been extended to the new tax. **1973** *Scotsman* 12 Jan. 6/2 Changes in capital gains tax should give either tapering relief on gifts or, if this is rejected, roll-over relief provisions on gifts. **1976** *Incorporated Linguist* XV. 72/2 The 'roll-over relief' where the gain on the disposal of an asset is transferred into the asset which replaces it, so that although the [capital gains] tax is payable, it is so only on the disposal of the final asset (eg on retirement). **1977** D. W. Healey in *Times* 30 Mar. 4/8, I propose an improved rollover relief for what is often called 'domestication', that is the transfer of an overseas branch to a separate non-resident company.

rolloway, var. *roloway.

Rolls (rōulz), colloq. abbrev. *Rolls-Royce. Also *attrib.* and *fig.*

1928 E. Wallace *Double* i. 9 Dick knew the gentleman very well by name; indeed, he had recognised his big yellow Rolls standing outside the hotel. **1932** Auden *Orators* III. 101 In Rolls or on bicycle they bolt for mama. *a* **1963** 'Beachcomber' *Best of Beachcomber* (1963) ix. 102 She has a Rolls body and a Balham mind. **1965** *New Statesman* 14 May 753/3 He is essentially a working.. journalist who sees the paper 'off the stone' in the composing room every night before driving by Rolls to his Greenwich home. **1977** C. McCullough *Thorn Birds* x. 224 Do you think Bob might let me borrow the old Rolls, if not the new one?

Rolls-Royce (rōulz roi·s). [Name of the manufacturing company.] **1.** A Rolls-Royce motor car.

1908 *Trade Marks Jrnl.* 26 Feb. 300 *Rolls Royce*... Motor cars and chassis included in this class. Rolls Royce, Limited, Cooke Street, Hulme, Manchester; Motor car manufacturers. **1915** 'I. Hay' *First Hundred Thousand* xiii. 173 Not long ago he was..driving a Rolls-Royce for a Duke. **1932** D. H. Lawrence *Last Poems* 256 Do you hear my Rolls Royce purr, as it glides away? **1936** J. B. Priestley *Walk in City* i. 5 So he popped into a long black Rolls-Royce. **1958** [see *bish]. **1975** *Sunday Times* 16 Nov. 44/3 The electric oven..had a dark glass front like a pop star's Rolls-Royce.

2. *fig.* **a.** Any product considered to be of the highest quality.

1916 W. A. Robson *Aircraft in War & Peace* xi. 161 None of the different machines made for these specialised purposes..will compare with the best pleasure aeroplane, the Rolls-Royce of the air. **1923** A. Bennett *Things that have interested Me* II. 107 The Row was flanked by processions of nun-like nursemaids pushing single prams and double prams—the Rolls-Royces of the pram-world. **1957** A. MacNab *Bulls of Iberia* ii. 28 The famous Murubes of old. Principal herd of main-line Vistahermosa. The 'Rolls-Royce' of taurine breeds. **1974** *Daily Mail* 24 Aug. 12/6 A lustrous Isphahan—the finest are the Rolls-Royces of rugs—size 6ft. by 4ft. might take two women two years to complete, and the value would be £1,800. **1977** *New Yorker* 19 Sept. 92/2 There was a nine-foot Bechstein—which many people feel is the Rolls-Royce of pianos—but when she opened the lid she found a mouse inside, eating the felt.

b. *attrib.* passing into *adj.*

1951 H. Hastings *Seagulls over Sorrento* in *Plays of Year 1950* IV. I. 45 Wot the 'ell's a bloke like 'im with a Rolls Royce accent..want to get mixed up in this mob for? **1960** *Sunday Express* 6 Nov. 16/6 Stanley Baker's rugged style has put him up in the Rolls-Royce class of actors. **1961** Partridge *Dict. Slang* Suppl. 1251/2 A bit Rolls Royce in his ideas. **1974** *Daily Tel.* 7 June 16 Vintage port—the Rolls-Royce end of the trade—accounts for only about one per cent. of port production. **1977** *Times* 23 Dec. 14/1 Norman Royce..disclaimed any pretensions to a 'Rolls-Royce performance', as a speaker.

Hence (*nonce-wds.*) **Rolls-Roy·ced** *a.* travelled in a Rolls-Royce; **Rolls-Roy·celess** *a.*, without a Rolls-Royce, devoid of Rolls-Royces; **Rolls-Roy·cey** *a.*, suggestive of a Rolls Royce; exceedingly wealthy.

1918 G. Frankau *One of Them* iii. 29 And the shrill cycle-bell's first tintillations Resounded from the dawning to the dark In a Rolls-Royceless, Peter Pan-less Park. *Ibid.* xxi. 162 For scarce a score of Rolls-Royced miles away..Miss Parker sat. **1926** D. H. Lawrence *Let.* 7 July (1932) 664 He was very nice; and apparently rich, too rich: Rolls-Roycey.

roll stone. *U.S.* A stone rounded by friction or attrition on a beach or in the bed of a river.

1845 J. C. Frémont *Rep. Exploring Expedition* 124 A swift current, over a bed composed entirely of boulders or roll stones. **1872** *Rep. Vermont Board Agric.* I. 688 A fine specimen of gold from a rollstone he found, while digging a well.

roll-top, *a.* and *sb.* Add: **A.** *adj.* **1.** (Earlier and later examples.)

1887 *Trial H. K. Goodwin* (Massachusetts Supreme Judicial Court) 15 That shows the position of the roll-top desk which was in the front office. **1923** R. Herrick *Homely Lilla* xi. 173 A young woman looked up from the roll-top desk where she was running over a typed list of names. **1933** H. Nicolson *Jrnl.* 27 Jan. (1980) I. 47 Two brown-wood roll-top desks are pushed against the wall. **1977** C. McCullough *Thorn Birds* vii. 130 The roll-top desk stood alongside one of the big windows.

2. Applied to other items having a roll-over top or a top with a rolled shape.

1977 *Wandsworth Borough News* 7 Oct. 22/1 (Advt.), Kitchen/breakfast room..with Ascot sink water heater, solid fuel boiler, built-in larder, glazed china cupboard and small roll-top bath. **1977** *Time* 19 Dec. 43/1 A trendy new kitchen..may..include a Fasar range,..chopping-block islands with separate vegetable sinks, a rolltop condiment 'garage', [etc.].

B. *sb.* **1.** A roll-top desk.

1895 in *Funk's Stand. Dict.* **1912** W. Owen *Let.* 23 Mar. (1967) 125, I have suffered in being informed that the 'Roll Top' is sold. **1932** E. Bowen *To North* xii. 115 It was a relief not having her sprawling to telephone over Emmeline's roll top. **1980** *Family Handyman* Sept. 28/1 He'd priced rolltòps in a downtown department store and knew he'd cut the cost in half.

2. The flexible top of a roll-top desk.

1913 in Webster. **1978** M. Kenyon *Deep Pocket* viii. 91 Peckover sat at the desk, slid the roll-top up.

roll-up, *sb.* Add: **1. b.** Also applied to salad, cooked food, etc., that is rolled up to form (part of) a dish. *U.S.*

1949 *New Yorker* 19 Nov. 94/3 Lunch box salad roll-ups. Roll up individual servings of finely shredded vegetable salad or coleslaw in a cabbage or lettuce leaf, fasten with a toothpick, [etc.]. **1952** *Sun* (Baltimore) 17 Jan. (B ed.) 12/1 (*heading*) Apple-ham rollups. **1977** *Chicago Tribune* 2 Oct. XI. 3/1 Place 2 tablespoons of mixture on each slice of roast beef; roll up. Place green beans on bottom of buttered shallow 10-by-6-by-2-inch casserole. Top with roll-ups.

c. (Earlier example.)

1831 M. Edgeworth *Let.* 6 May (1971) 536 A dressing-case—leather-roll-up which he preferred to a box.

d. A hand-rolled cigarette. *slang* (orig. *Prisoners'*).

1950 P. Tempest *Lag's Lexicon* 181 A 'good' roll-up is one that has a reasonable amount of tobacco in it. **1958** F. Norman *Bang to Rights* 55 A real snout, not even a roll up. **1963** T. Parker *Unknown Citizen* i. 23 The cigarette..was tasteless after the strong roll-ups in prison. **1967** *Daily Tel.* 21 Feb. 15/7 They smoked what one girl described as a 'roll-up', a loosely-rolled cigarette containing hemp. **1977** *New Society* 6 Oct. 24/2 An old man with one leg coughed in an agonising manner as he inhaled a very thin and slightly sad looking roll-up.

B. *adj.* That can be rolled up; suitable for rolling up; made by rolling up.

1908 *Sears, Roebuck Catal.* 97/1 Two roll-up straps. **1923** *Daily Mail* 12 Feb. 2 (Advt.), Roll-up felts at half price. **1939–40** [see *fold *v.*[1] 1 e]. **1948** *Sunday Pictorial* 29 Aug. 6/4 A prisoner will do anything for tobacco. He will sell his dinner for two thin 'roll-up' cigarettes which have less tobacco in them than there is in a respectable cigarette-end. **1964** *McCall's Sewing* 168/1 The true roll-up sleeve..must be made of fabric with no right and wrong side. **1966** J. Gardner *Amber Nine* x. 148 Back down the cat-walk to the main roll-up garage-type door. **1974** *Camping & Caravanning* Sept. 12/3 The awning has a nylon zip arched 'stable door' type doorway and a complete roll-up front.

roll-uppable (rōulʊ·păb'l), *a.* *nonce-wd.* [f. *to roll up*, Roll *v.*[2] 8b + -able.] Able to be rolled up, suitable for rolling up.

1961 T. Hughes *Meet my Folks!* 33 Or the roll-uppable rubber ladder.

rolly, *a.* Add: Also *Comb.*, as *rolly-eyed* adj.

1965 G. McInnes *Road to Gundagai* xi. 197 Another.. known to us as 'The Rolly-Eyed Duke'.

Rolly (rǫ·li). Representation of a popular pronunc. of *Rollei.

1961 Partridge *Dict. Slang* Suppl. 1251/2 *Rolly*, a Rolliecord or Rollieflex [*sic*] camera. **1971** 'A. Hall' *Warsaw Document* xxi. 271 Let everyone know you're the press, take plenty of Rollies. **1973** K. Benton *Craig & Jaguar* iii. 28 There was nothing of value..only his Rolly, and that was broken.

roloway. Also **rolloway.** Delete † *Obs.* Substitute for etym. and def.: [a. the specific name of *Simia roloway* (J. C. D. von Schreber *Säugthiere* (1774) I. 186), prob. f. the animal's native name in Ghana.] A large black and white guenon, *Cercopithecus diana roloway*, found in parts of tropical West Africa and distinguished from the Diana monkey by a longer, white beard. Also *attrib.* (Later examples.)

1894 R. Lydekker *Royal Nat. Hist.* I. iv. 102 The real name of the diana monkey in its native districts is said to be Roloway on the Gold Coast. **1910** W. P. Westell *Bk. Animal Kingdom: Mammals* ix. 163 The Roloway Monkey is often mistaken for the handsome Diana, to which it is very similar. **1966** W. C. O. Hill *Primates* VI. 531 This form has been much confused with the preceding; many authors..having treated the Roloway as the Diana. **1966** R. & D. Morris *Men & Apes* viii. 234 Diana and rolloway monkeys have been considerably reduced in numbers.

rolwagen (rōu·lwægə̆n). Also **rollwagon, rolwaggon, row-waggon,** etc. [a. Du. *rolwagen*, lit. 'roll-wagon'.] A kind of Chinese cylindrical porcelain vase, or a Dutch imitation of this (see quot. 1960).

1675 [see Roll-wagon]. **1761** H. Walpole *Let.* 13 June (1928) II. 119 Don't trouble yourself about Delft—nay, I am now afraid you should get any, lest you should pack it up in an old china jar, and really find a meaning for that strange auctioneers word, a *rowwaggon*. **1786** *Catal. Portland Museum* 19 A 2-handled jar, and 2 row waggons. **1895** Rimbault & Clinch *Soho & its Associations* ii. 35, 1 pair of blew china rowlwaggons. **1954** T. Volker *Porcelain & E. India Co.* v. 20 Two carrack flasks, one broken, 15, two small *rolwagens*. *Ibid* [Note] *Rolwagen* is a name still in use in the porcelain trade in Holland, denoting a cylindrical vase, usually with a flat lip. **1957** *Apollo* June 251/1 'Rollwaggon' (or however you care to spell it) is still used of the cylindrical-bodied vases of the type frequently found in Transitional and K'ang Hsi blue-and-white. **1960** R. G. Haggar *Conc. Encycl. Continental Pott. & Porc.* 383/2 *Rol-wagens*, corrupted in English to *row-waggons, roll-wagons*, the name given to cylindrical vases found in K'ang Hsi blue-and-white.

roly-poly, *sb., a.,* and *adv.* **5.** *attrib.* For 1848 quot. substitute:

1841 Thackeray *Gt. Hoggarty Diamond* (1849) xii. 168 You said I make the best rolly-polly puddings in the world.

Rom. Add: Also pl. Rom.

1910 *Encycl. Brit.* XII 38/1 Only those who starting from the ancient Byzantine empire have travelled westwards..call themselves by the name of Rom. **1973** *Guardian* 26 Mar. 7/1 The continued historical prejudice against the Rom—as the gipsies call themselves. **1976** *Sci. Amer.* Jan. 131/2 The Rom always try to cooperate.

b. *attrib.*

1973 *Guardian* 26 Mar. 7/2 The non-conforming way of life of the Rom people seems to attract increasing intolerance. **1976** *Sci. Amer.* Jan. 131/2 The Rom families that are studied here..spent almost half of their time away from home, travelling.

rom., abbrev. of Roman *sb.*[1] 4, used esp. as a proof-correctors' mark.

1824 J. Johnson *Typographia* II. 216 Rom. **1902** A. E. Housman *Let.* 30 Nov. (1971) 62 It ought to stand upright, not to slant (I have written 'rom.' in the margin). **1954** T. W. Chaundy et al. *Printing of Math.* ii. 53 Italic is restored to normal (i.e. roman) type by 'rom.' in the margin. **1973** S. Beale *Collins's Authors' & Printers' Dict.* (ed. 11) 379/1 *Rom.*, roman type.

Romagnol, Romagnole (rōumānʸǫ·l, -ōu·l), *sb.* and *a.* Also in It. form **Romagnolo** (fem. **-ola**, pl. **-oli**). [ad. It. *Romagnolo*, f. *Romagna* (see below).] **A.** *sb.* A native or inhabitant of the Romagna, a district of northern Italy (now part of the region of Emilia-Romagna). **B.** *adj.* Of or pertaining to the Romagna or its inhabitants.

1821 Byron *Don Juan* IV. 118 Juan's companion was a Romagnole. **1841** *Penny Cycl.* XX. 70/2 The Romagnoli are lively and quick, but they have the character of being hasty and violent. **1860** E. B. Browning *Napoleon III in Italy* in *Poems before Congress* 11 Piedmontese, Neapolitan, Lombard, Tuscan, Romagnole, Each man's body having a soul. **1901** M. Carmichael *In Tuscany* iii. 115 A hot-headed Romagnol, Alfonso Cerquetti,..had the hardihood to publish a pamphlet pointing out errors in the new Vocabolario of the Academy. **1926** *Contemp. Rev.* Mar. 297 General Pangalos might imitate the ancient Roman, rather than the modern Romagnole Dictator,..and retire to his Eleusinian farm. **1934** E. Pound *Eleven New Cantos* xxxv. 25 The Romagnols wd. come here to Mantua. **1966** S. Mann *Collecting Playing Cards* i. 31 The Romagnole pack resembles the Piacentine in many ways. **1973** M. West *Salamander* i. 42 Ask that fellow over there, the street-cleaner, what he is. He will answer, 'I am a Sard, a Calabrese, a Neapolitan, a Romagnolo.' **1975** *Times Lit. Suppl.* 31 Oct. 1311/1 Even the title *Amarcord*, Romagnolo dialect for 'I remember', hints..at the identity of the director and the subject of the memories.

So **Roma·gnan** [-an] *a.*; **Romagne·se** [-ese] *sb.* and *a.*

1845 *Encycl. Metrop.* XII. 385/2 Demoralization has never, perhaps, sunk human nature lower than we find it among the Romagnese [*c* 1500]. **1931** M. Yeo *St. Francis Xavier* vi. 62 Fogs from the Romagnan marshes enveloped the city. **1933** *Times Lit. Suppl.* 8 June 385/3 Already in Dante's 'Purgatorio' a Romagnese gentleman laments the days when 'ladies and knights,..toils and ease, inspired love and courtesy'. **1972** *Sansoni-Harrap Stand. Italian & Eng. Dict.* I. II. 1120/2 Romagnese dialect.

romaine (romēi·n). [a. Fr. fem. of *romain* Roman.] **1.** *U.S.* = Cos. Also *attrib.*

[**1885** W. Miller tr. *Vilmorin-Andrieux's Veget. Garden* 309 The Paris market gardeners grow, under the name of *Romaine Plate*, a variety which appears to be intermediate between the Green and the Gray Paris Cos Lettuces.] **1907** H. W. Wiley *Foods & their Adulteration* VI. 284 Among the varieties which are most highly prized for this purpose [*sc.* salad] are the cabbage lettuce and the variety known as Romaine. *Ibid.*, The Romaine lettuce is more highly prized by most connoisseurs. **1942** E. Paul *Narrow St.* vi. 50 Cabbages, cauliflowers,.. potatoes, lettuce, romaine, *chicorée* and other salad leaves. **1966** T. Pynchon *Crying of Lot 49* i. 11 The..garlicking of a bread, tearing up of romaine leaves. **1972** *New Yorker* 22 July 22/2 Fifteen romaine lettuces for four-fifty. **1978** *N.Y. Times* 30 Mar. c7/1 Cos or romaine..grows upright and forms tight sheaths of rich green leaves.

2. Any of various crêpe fabrics. *romaine crêpe* (see quot. 1968 and cf. quots. 1923 s.v. *crêpe). Also *attrib.*

1922 *Glasgow Herald* 26 Apr. 10 The bride..wore a gown of white romaine. *Ibid.*, The two bridesmaids.. wore white crepe romaine dresses. **1932** *Daily Tel.* 25 Apr. (Advt.), Coat of wool romaine. **1932** *Daily Express* 25 June 9/5 Lady Haslam..was in love-in-a-mist romaine armure. **1939** M. B. Picken *Lang. Fashion* 122/1 *Romaine*, sheer silk fabric in basket weave, made of fine yarns, having smooth surface and slightly more body than triple sheer. **1952** C. W. Cunnington *Eng. Women's Clothing* 296 *Romaine*.., a lining fabric of French make, in warp satin weave. A sheer silk fabric in basket weave. Also 'a light woollen with dull surface and flat square weave'. **1957** M. B. Picken *Fashion Dict.* 277/1 *Romaine crepe*, heavy sheer crepe. **1968** J. Ironside *Fashion Alphabet* 223 *Romaine crêpe*, a heavy, semi-sheer crêpe usually in a dull finish rayon, sometimes silk, made to resemble wool crêpe.

‖ **romaji** (rōu·madʒi). Also **romazi**, and with capital initial. [Jap., f. *roma* Roman + *ji* letter(s).] A system of Romanized spelling for the Japanese language.

[**1888** B. H. Chamberlain *Handbk. Colloq. Japanese* i. 9 There is a party in favour of the adoption of the Roman alphabet. Its organ, the '*Rōmaji Zasshi*', gives articles..romanised according to a simple phonetic system.] **1903** R. Lange *Text-bk. Colloq. Japanese* p. xviii, *Romaji* is designed to represent phonetically the standard pronunciation of the present day. **1935** *Amer. Speech* X. 274/2 Several movements for Romaji, Japanese written phonetically in Roman letters, are under way. [**1939** *Jrnl. Amer. Oriental Soc.* LIX. 99 The Japanese had long ago worked out a diaphonic spelling called *Nipponsiki* (Japanese-Style) *no Rômazi* and propagated it in competition with the reigning bastard English-Italian system named after Hepburn.] *Ibid.* 102 The present use of rômazi during an extended period is a hindrance rather than a help in the mastery of Japanese as it is normally written. **1950** D. Jones *Phoneme* 102 The name of the mountain which used to be written Fuji in the old Rōmaji system is now written Huzi. *Ibid.* 105 In the new Rōmazi orthography hi and si are distinguished. **1966** P. S. Buck *People of Japan* (1968) x. 130 The Japanese have to master *romaji*, Japanese spelled in Roman letters.

‖ **Romalis** (romā·lis). [Sp.] A Spanish gypsy dance (see quots.). Also, the music of this dance.

1841 Borrow *Zincali* I. vi. 317 Chicharona danced the Romalis (Gypsy dance) before her. **1846** [see *olé[1]]. **1889** L. A. Smith *Through Romany Songland* 54 The famous Romalis, the dance which Tiberius may have seen, and which no one but a gipsy dances in Spain. It is danced to the ancient Oriental music of hand-clapping, and to an old religious Eastern tune, low and melancholy, diatonic, not chromatic, and full of sudden pauses which are strange and startling. It is sung in unison, and has a chorus in which every one joins. **1967** 'La Meri' *Spanish Dancing* (ed. 2) v. 74 This Romalis is part of a wedding ceremonial. The melody is Eastern; diatonic, low, melancholy and with sudden breathtaking pauses. Some..even identify it with the dance which Salome did before Herod.

Roman, *sb.*[1] Add: **I. 1. c.** (Further examples.)

1950 T. Williams *Roman Spring of Mrs. Stone* i. 34 Patience, said the Contessa. Rome was not built in a day! I am a Roman, said Paolo, but I am not Rome. **1967** C. Seton-Watson *Italy from Liberalism to Fascism* ix. 334 He was an intelligent man, but vain and irresolute, with a Roman's liking for *combinazioni*.

3. c. The dialect of the modern Romans.

1598 [see *Neapolitan *sb.* d]. **1642** J. Howell *Instructions Forreine Trav.* xi. 138 There is in Italy the Toscan, the Roman, the Venetian, the Neapolitan,..and others..and all these have severall Dialects and Idiomes of Speech. **1973** *Daily Tel.* (Colour Suppl.) 22 Feb. 65/3 She..spoke such a flowery Roman that I wondered if this wasn't a sort of cultural or social dust thrown into one's eyes.

5. b. A Roman hyacinth.

1925 *Glasgow Herald* 26 Aug. 8/7 Early Romans are in, but at a very high price. **1934** 'E. M. Delafield' *Provincial Lady in Amer.* 126 Early Romans should certainly be well above ground now.

II. 6. (Further examples.)

1936 S. Dark *Manning* iii. 101 Tory Romans were henceforth allowed to wear primroses in their buttonholes on the anniversary of the death of Benjamin Disraeli. **1956** R. Macaulay *Towers of Trebizond* xxii. 255, I decided that it should stick to Anglican churches, eschewing both Knox and the Romans. **1962** V. J. K. Brook *Life Abp. Parker* xix. 343 He had constantly to entertain those given into his charge by the Council—Romans or others—that he might reason with them. **1965** M. Spark *Mandelbaum Gate* ii. 45 Latest bulletin from the Holy Romans..they'll take at least another month to decide. **1975** Byfield & Tedeschi *Solemn High Murder* (1976) i. 2 The Romans might have a nice large new.. church right in town..but the little Anglican wooden shack..had wine.

roman, *sb.*[3] *S. Afr.* Also **roo(i)man.** [Afrikaans, f. *rooi* red + *man* man.] **1.** A marine fish, *Chrysoblephus laticeps*, belonging to the family Sparidæ and having reddish skin. Also *attrib.*

1790 E. Helme tr. *Le Vaillant's Trav. Afr.* I. ii. 22 Among those [fish] in greatest estimation, they distinguish the *rooman*, a red fish [etc.]. **1801** [see Steenbrass]. **1804** R. Percival *Acct. Cape of Good Hope* 43 The most common is the Roman fish... It is of a deep rose colour and of the perch kind. **1893** [see *kingklip]. **1957** S. Schoeman *Strike!* iii. 32 It is universally known as 'roman', although some anglers and fishermen call it 'rooi roman' (red roman). **1971** *Cape Argus* 14 May 14 John Hughes shot a roman of 4,1 kg—which is equal to the South African spearfishing record.

2. A large nocturnal sun-spider belonging to the order Solifuga (or Solpuga) and having a sandy-coloured body.

1905 F. Purcell in Flint & Gilchrist *Sci. in S. Afr.* III. iii. 178 The large nocturnal..species of *Solpuga*..are variously known by the name of Romans, Jagd-spinnekoppen (Hunting Spiders) or Haar-sheerders (Hair cutters). **1966** E. Palmer *Plains of Camdeboo* xiv. 233 On the farm we know..the nocturnal species [of spiders] as Rooimans or Red Men, and of these latter I can neither think nor speak except in capitals.

‖ **roman** (romañ), *sb.*[4] [Fr.: see Romaunt *sb.* and *a.*] A romance; a novel. Esp. in phrases: **roman à clef**, a novel in which actual persons are introduced under fictitious names; **roman à thèse**, a novel that seeks to further a viewpoint or expound a theory; **roman d'aventure** = Romance *sb.* 2; **roman de geste** = **chanson de geste*; **roman expérimental**, a realistic novel based upon deterministic theories of human nature of an alleged scientific character; also *fig.*; **roman fleuve**, a sequence of self-contained novels; **roman noir**, a Gothic novel, a shocker, a thriller; **roman policier**, a story of police detection.

1765 [see Romaunt *sb.* and *a.* 1]. **1868** Roman de geste [see *chanson 2]. **1884** W. James *Will to Believe* (1897) 173 Like the friends of M. Zola, we pique ourselves on our 'scientific' and 'analytic' character, and prefer to be cynical, and call the world a 'roman expérimental' on an infinite scale. **1889** E. Dowson *Let.* 5 May (1967) 75 We.. may..evolve a brilliant *roman*. **1893** H. James *Let.* 23 Jan. in P. Gunn *Vernon Lee* (1964) x. 138 Her books of fiction are a tissue of personalities of this hideous roman-à-clef kind. **1905** Roman d'aventure [see *lai[1]]. **1913** G. Turquet-Milnes *Influence of Baudelaire* v. v. 250 In his [*sc.* Arthur Machen's] works we again meet the distrust of nature from the documentary point of view—the distrust of 'Romans à Clef'. **1928** A. Christie *Mystery of Blue Train* x. 81, I see, Madame, that you have a *roman policier*. **1931** *Times Lit. Suppl.* 26 Feb. 151/4 The 'Radcliffian' novel, or *roman noir*, as the French call it. *Ibid.* 31 Dec. 1054/4 The study aims at giving a detailed analysis of the German criticism of Zola's Rougon-Macquart cycle and his theory of the *roman expérimental*. **1934** Webster, Roman à thèse. **1936** *Times Lit. Suppl.* 15 Feb. 121/3 Those great *romans-fleuves* whose unnumbered volumes have no other purpose than to show us to ourselves as we appear. **1940** H. G. Wells *Babes in Darkling Wood* 5 They pass at last..into more or less honest fact telling, into 'historical reconstruction', the *roman à clef*, biography, history and autobiography. **1954** K. Tillotson *Novels of Eighteen-Forties* I. 3 The *roman à thèse* is already establishing itself. **1955** *Times* 4 Aug. 10/3 *The Typewriter* is not great Cocteau, but it is fine melodramatic fare and strangely compelling emotionally. It has wit and good dialogue, but this is essentially a tragedy set in the frame of a *roman policier* and the answer to the mystery remains unanswered and unguessable until the very last scene. **1957** *Encycl. Brit.* I. 945/1 The inter-influence of French and English literature can be studied in the Breton romances and the *romans d'aventure* even better than in the epic poetry of the period. **1959** *Listener* 3 Dec. 1007/3 The re-creation of the medieval *roman*..in *The Story of Reynard*. **1965** *Ibid.* 27 May 799/1 *David in Silence*..is also in its way a *roman à thèse*. **1965** *Observer* 5 Sept. 21/5 The film is a *roman policier*. **1966** J. Carter in Glover & Greene *Victorian Detective Fiction* p. xiv, The early *roman policier* writers. **1971** J. Pope-Hennessy *A. Trollope* xvii. 364 We might claim..that *An Eye for an Eye* initiated the series of Trollope's *romans noirs*. **1972** V. Gielgud *Black Sambo Affair* xxvii. 208 A fine collection of *romans policiers*. **1974** *Bookseller* 15 June 2696/3 He [*sc.* Anthony Powell] obviously feels reasonably protective towards the maestro of the *roman fleuve*. **1977** *New Yorker* 24 Oct. 184/2 A roman à clef whose skeleton key would seem to be the unsavory case of Alice Crimmins and her two murdered children. **1978** *Times Lit. Suppl.* 1 Dec. 1405/5 Success takes him to Hollywood, and there you feel a roman à clef, based on Puzo's profitable frustrations in movieland, is intended.

Roman, *a.*[1] Add: **I. 3.** Also *Comb.*, as *Roman-looking* adj.

1869 'Mark Twain' *Innoc. Abr.* xlviii. 505 We came to a..Roman-looking ruin.

4. a. spec. *Roman father*, a dominating head of a family.

1750 W. Whitehead *Roman Father* v. ii. 74 Has not a Roman father power to take The lives of all his children? **1906** Kipling *Puck of Pook's Hill* 148 There can't be much of the Roman Father about you! **1922** T. E. Lawrence *Let.* 26 Aug. (1938) 361 Perhaps I'm playing the Roman father trick, and it's not as bad as I think. **1940** H. G. Wells *Babes in Darkling Wood* I. iv. 89 My Roman father! The Cadi of Clarges Street! **1962** *Listener* 25 Oct. 694/1 Dr Borosdin, the almost Roman father. **1977** P. G. Winslow *Ditch Hill Murder* II. 153 Richard had taken to playing the Roman father to Lerida.

d. *Roman holiday*, an occasion on which entertainment or profit is derived from injury or death; a scene of suffering considered as an object of amusement; a pitiable spectacle.

Orig. a holiday for a gladiatorial combat: see quot. 1818 s.v. Holiday *sb.* 2.

1886 'S. Coolidge' *What Katy did Next* ix. 223 (*heading*) A Roman holiday. **1931** R. Ferguson *Brontës went to Woolworth's* xxv. 220 There. I've made a Roman holiday of my dear little acquaintance, and I only hope I'm right. **1951** G. Greene *Lost Childhood* 47 The critics..were perhaps influenced by horror at the Roman holiday. **1957** 'H. Carmichael' *Put out that Star* x. 103 All you people ever think about is how you can turn any damn' thing at all into money: anything to make a Roman holiday. **1966** P. O'Donnell *Sabre-Tooth* xv. 205 She had watched impassively, caring nothing for the man's death but loathing the Roman-holiday manner of it. **1967** A. Wilson *No Laughing Matter* III. 329 Was this what Picasso's wonderful Guernica stood for, this Roman holiday? **1972** A. Hunter *Vivienne* x. 131 A Roman holiday was in the making, and the number of reporters had risen to five.

5. c. *Roman uncial* = Semi-uncial *a.* (*sb.*)

1897 [see *half-uncial* s.v. *half- II. n]. **1906** E. Johnston *Writing & Illuminating* i. 38 Roman uncials were fully developed by the fourth century.

8. Also, *Roman-Doric, -Dutch* (later examples).

1928 R. Nevill *Romantic London* ii. 39 The lower order is Roman-Doric. **1957** Ld. Hailey *Afr. Survey 1956* xxii. 1520 In South Africa ownership of the land is, in accordance with the principles of Roman-Dutch law, held to comprise all values in the land including mineral rights. **1964** J. Summerson *Classical Lang. Archit.* 49/1 The Greek order has no base, nor is a base prescribed by Vitruvius, though in practice the Roman Doric always has a base, the Greek never. **1972** *Mod. Law Rev.* XXXV. I. 46 There is no warrant in Roman-Dutch law for a discretion as wide as that enunciated by the Appellate Division.

II. 10. *Roman fever* [transf. use of 13c], a fondness for the Church of Rome, a desire to be converted to Rome.

1877 O. Wilde *Lett.* (1962) 45 Poor Dunskie: I know he looks on me as a renegade; still I have suffered very much for my Roman fever in mind and *pocket* and happiness. *a* **1884** M. Pattison *Mem.* (1885) vi. 226 The daughter had got the Roman fever in her veins. **1929** S. Leslie *Anglo-Catholic* xii. 171 Edward..assured him he had not left the Anglican Church, though the Anglican Church, he thought, had probably left him. The Canon only said, 'For Roman fever there is no cure and for Rome there is no leechdom.' **1952** R. Macaulay *Let.* 12 Jan. (1961) 248, I am glad Dom Gregory Rees thinks 'Roman fever' abated; I don't notice it myself, anyhow among the laity.

III. 13. d. Applied to a bidding system in Bridge orig. used by certain Italian players, or to various conventions and signals within this system.

1959 Belladonna & Avarelli *Roman Club Syst. Distrib. Bidding* 2 In this fashion the person who is already playing bridge can learn one phase of the Roman Club expertly before going on to the next formula. **1959** Reese & Dormer *Bridge Player's Dict.* 190 *Roman system*... Opening bids of one diamond, one heart, and one spade, show a genuine suit and are forcing for one round. **1964** *Listener* 21 May 851/1 'Roman leads'..is a method whereby the lower of two touching honours, rather than the higher, is the normal lead. **1970** S. Hughes *Art of Coarse Bridge* iv. 93 Patiently South explained that the Roman Club..meant that he had either a minimum balanced hand or a very strong one. **1975** *Times* 27 Sept. 10/7 Opening bids with double meanings which we now associate with the Roman and other artificial systems.

IV. 14. b. *Roman chamomile* (earlier example).

1815 F. P. Chaumeton *Flore Médicale* II. 123 Latin Anthemis nobilis... *Anglais* Chamomile; Roman Chamomile; Sweet-scented chamomile.

15. b. (Earlier example.)

1607 Topsell *Foure-f. Beasts* 655 The hornes..are so lively expressed by Pliny... They are..long, about two Roman feet and three palmes... They are in breadth where they ioyne to the head, three Roman fingers and a half.

16. c. (Further examples.) = *Parker's cement.

1889, 1917 [see *Parker's cement]. **1919** A. T. Bassett *S. Barnabas', Oxford* vi. 67 Some amusement was caused at the time by a box bearing the words 'Roman Cement', in large letters, being delivered at the Church for use in the connection with repairs to the campanile. **1970** H. Braun *Parish Churches* xix. 223 Roman cement is difficult to procure, but an admixture of lime with the Portland cement will help to improve the colour to some small extent. **1977** *Sci. Amer.* July 82/2 The high quality of Roman cements, which is evident in the number and solidity of the Roman structures still standing, was due in large measure to the added discovery that lime mixed with reactive siliceous material (in the form of crushed tiles or volcanic ash) gave a cement that developed superior strength and water resistance. 'Roman cement' made in this way enjoyed wide prestige and retained its popularity with little improvement or development until the end of the 18th century.

f. *Roman ring* = *flying ring* s.v. *flying *ppl. a.* 3.

1911 *Daily Colonist* (Victoria, B.C.) 26 Apr. 5/4 They [*sc.* vaudeville athletes] begin by drawing their body from the floor on Roman rings with snail-like slowness. **1965** F. Sargeson *Mem. Peon* vi. 165 The creak of parallel bars and Roman rings.

17. *Roman sandal, satin* (further examples), *scarf*; Roman bath = **Roman tub*; **Roman tub** *U.S.*, a large sunken bathtub.

1855 E. Twisleton *Let.* 6 Apr. (1928) xiv. 266 Mrs. Carlyle was sumptuous, in a black velvet and Roman scarf. **1914** C. Mackenzie *Sinister St.* II. III. iv. 559 Stella..was rushing from window to window, trying patterns of chintz and damask and Roman satin. **1934** Webster, Roman sandal. **1939** M. B. Picken *Lang. Fashion* 130/2 *Roman sandal*, sandal with front composed entirely of straps, equally spaced. **1961** *Harper's Bazaar* Feb. 69 The splendid Roman striped satin of a Heppelwhite chair. **1971** *Sunday Nation* (Nairobi) 11 Apr. 20/1 (Advt.), Boots, Roman Sandals. **1972** *Fortune* Jan. 140c (Advt.), You'll have a cocktail lounge and restaurant on the premises. An outside elevator joining the opulent pool deck area and the beach. A Roman tub in your master bath. **1976** *Bathroom Ideas* 58/2 Blue, blue is this Roman bath set apart from the rest of the room. **1979** *Arizona Daily Star* 5 Aug. (Advt. Section) 16/9 Features formal dining, atrium, 2-way fireplace, roman tub in master bedroom.

Romanaccio (romanā·tʃio). [It.] A modern dialect spoken in the city of Rome.

1963 *Guardian* 6 June 11/2 Cardinal Agagianian.. speaks 11 languages including Romanaccio (Roman cockney). **1973** M. West *Salamander* ii. 163 A stable-boy laughed and I flew at him, clawing and punching and screaming in Romanaccio.

Roman candle. [Roman *a.*[1]]

1. See Roman *a.*[1] 16d.

2. A parachute jump on which the parachute fails to open; a parachute which fails to open. Also (in full *Roman candle landing*) an unsatisfactory landing by an aircraft. *slang.*

1943 Hunt & Pringle *Service Slang* 56 *Roman candles.* 'When a parachute simply fails to open. (Of course, on landing, you dash to the stores and get another.)' **1943** C. H. Ward-Jackson *Piece of Cake* 52 *Roman candle landing*, a bad landing. **1952** *Chambers's Jrnl.* May 261/2 It is not so very long ago since parachute-jumping was a stunt indulged in by steel-nerved men of boundless courage performing in air circuses. In those days, the sight of some hapless individual streaming to earth with a 'Roman candle' (an undeveloped 'chute) was not exceptional. **1959** *Chambers's 20th Cent. Dict.* Add. 1389/1 *Roman candle*, a bad landing by aeroplane: a landing by parachute when the parachute fails to open.—v.i. to make such a landing. **1961** E. Waugh *Unconditional Surrender* II. v. 141 The first thing the commandant asked when I reported Crouchback's accident. 'A Roman Candle?' he asked. **1976** A. White *Long Silence* vii. 59, I experienced the sense of relief that says, 'This time, no roman candle!'

3. A Roman Catholic. *slang.*

1941 G. Kersh *They die with their Boots Clean* II. 57 There's services for C. of E-ers and Roman Candles. **1959** I. & P. Opie *Lore & Lang. Schoolch.* xvi. 344 In Staines, Catholics are 'Roman Candles', and R.C. children call the Protestants 'Old Proddy Dogs'. **1974** P. Haines *Tea at Gunter's* ii. 18 She said: 'I've noticed you lots—you're a Roman candle, aren't you?' 'What?'..'*R.C.*, silly.'

Hence **Roman-ca·ndle** *v. intr.*, to make a parachute jump with a parachute that fails to open.

1959 [see sense 2 above]. **1975** tr. *Melchior's Sleeper Agent* (1976) III. 230 *He had roman-candled!* The chute had not opened! He was plunging toward oblivion.

Ro·man Catholi·city. [f. Roman Catholic *sb.* and *a.* + -ity.] Roman Catholicism.

1806 M. B. Pembridge *R.C. Church Vindicated* i. 40 These edifices still bear the external signature of Roman Catholicity. **1965** E. O'Brien *August is Wicked Month* xvii. 218 'It's your Roman Catholicity,' he said.

Romance, *sb.* and *a.* Add: **I. 1.** *Comb.* (Earlier and later examples.)

1882 E. A. Freeman *Lect. Amer. Audiences* I. v. 155 Did not the Norman Conquest..bring with it a settlement of strangers, of Romance-speaking strangers, enough to destroy all pretence on the part of the English nation to pure Teutonic descent? **1964** Romance-based [see *Latin-based* s.v. *Latin *sb.* 5]. **1964** E. Palmer tr. *Martinet's Elem. Gen. Linguistics* v. 150 The Romance-speaking clerks of the eighteenth century..used.. variously a local Romance language and another language, Latin.

II. 3. b. (Earlier example.)

1816 [see *historical *a.* 3].

5. a, b. *spec.* (also with pronunc. rōu·mæns) a love affair; idealistic character or quality in a love affair; a love story; that class of literature which consists of love stories.

1916 G. B. Shaw *Overruled* 81, I felt my youth slipping away without ever having had a romance in my life; for marriage is all very well; but it isnt romance. Theres nothing wrong in it, you see. **1921** Mencken *Amer. Lang.* (rev. ed.) vii. 209 The chief movement in American.. would seem to be toward throwing the accent upon the first syllable... I might add..*defect, excess, address, magazine, decoy* and *romance*. **1922** Joyce *Ulysses* 280 Chorusgirl's romance. Letters read out for breach of promise. **1936** 'G. Orwell' *Keep Aspidistra Flying* x. 264 When a customer demanded a book of this category or that,.. 'Sex' or 'Crime' or 'Wild West' or 'Romance' (always with the accent on the *o*), Gordon was ready with expert advice. **1939** N. Monsarrat *This is Schoolroom* xvii. 385 The dance-band world..has given us a new pronunciation—'bokay' for bouquet, 'rómance' thus accented. **1942** T. Rattigan *Flare Path* I. 26 He was on a week's leave, and we were married before he went back to his Squadron. What the papers would call a whirlwind wartime romance. **1951** in M. McLuhan *Mech. Bride* (1967) 24/2 She loved him with another woman's body.. one of the tensest, most passionate romances you have ever experienced. **1954** [see *fantasy, phantasy *sb.* 4 f]. **1960** *Times* 28 Sept. 15/4 Harry, undaunted by a succession of parties at which he has done nothing whatever, always attends in the hope of finding romance. **1966** C. Mackenzie *My Life & Times* V. 193 The cinema audience wants rómance. We must give them rómance. **1971** J. Fleming *Grim Death & Barrow Boys* vii. 87 It's the end of ro-mance, is marriage.

7. a. *romance-writer* (further example); *romance-reading* (later example), *-weaving, -writing* (further examples); *romance-wards* adv.

1876 *Westm. Rev.* XLIX. 361 The novelist proper studies to represent his little world as the great world is; whereas the romance-writer..builds an ideal world. **1887** *Contemp. Rev.* LI. 172 Really good romance writing is the most difficult art practised by the sons of men. **1904** 'Mark Twain' in *Harper's Weekly* 10 Dec. 11/1 There was no romance-reading that night. **1904** W. H. Hudson *Green Mansions* 2 Let us hope that now, at last, the romance-weaving will come to an end. **1920** R. Macaulay *Potterism* VI. iii. 228 He was also leaning romancewards and departing from the realm of pure truth. **1979** *N. & Q.* Feb. 90/2 Hanning tends to brush aside these technical difficulties of romance-writing.

b. *romance-like* adj.

1971 K. Millett *Sexual Politics* (1972) I. i. 5 That Ida has dressed herself in a collapsible bathrobe and silk stockings is not only accommodating but almost romance-like.

8. *romance-literature, thriller.*

1905 *N. Amer. Rev.* CLXXX. 5 You have made the American home..beautiful with your..noble romance-literature. **1961** *Times Lit. Suppl.* 8 Dec. 2/1 The growing success of the romance-thriller, where the basic plot of virgin-marries-older-man is sharpened..by often well-devised and dramatic villainy. **1975** *Listener* 20 Nov. 685/1 The many competent women writers of 'romance thrillers'.

romance, *v.* Add: **5.** *trans.* To have a romance or affair with, to court.

1942 Berrey & Van den Bark *Amer. Thes. Slang.* § 354/4 *Court*,..race, romance, run with, run *or* chase after, rush. **1956** B. Holiday *Lady sings Blues* (1973) vi. 59, I was accused of romancing everyone in the band. **1970** M. Butterworth *Vanishing Act* xi. 125 A good-looking chap..could do a bit of counter-jumping and romance the lady customers..if he had the cheek. **1976** T. Gifford *Cavanaugh Quest* v. 79 They were working on my kind of music... I'd romanced Anne to old stuff like that and we'd made love to Claude Thornhill recordings. **1980** N. Dempster in *Daily Mail* 10 Apr. 19/3 He has been romancing Antonia for a year.

‖ **romancé** (romañse), *f.pl. a.* Fem. **romancée.** [Fr., f. *romancer* (see Romance *v.*).] Fictionalized, rendered in the form of a novel; *spec.* of a biography.

The fem. *romancée* is used in quot. 1962 because of Fr. *la biographie*.

1938 *Times Lit. Suppl.* 10 Sept. p. v/2 Signor Falta carefully avoids whatever may appear *romancé*. **1962** *Times* 5 July 17/4 The result is a solid, readable biography, slightly *romancée*.

romancing, *vbl. sb.* Add: Also occas. with pronunc. (rōu·mænsiŋ).

1879 G. Meredith *Egoist* II. vii. 156 Oh! Mrs. Montague, that is what the country people call roemancing...

roma·ncingly, *adv.* *rare.* [f. Romancing *ppl. a.* + -ly[2].] In a romancing manner.

1908 H. James *Spoils of Poynton* p. xii, By just so much would the muse of 'dialogue', most usurping influence of all the romancingly invoked, be routed without ceremony.

Romanée (romane). The name of a vineyard in the commune of Vosne-Romanée in the Côte d'Or department of France used *absol.* to designate the red wine produced there. Also, **Romanée-Conti, Romanée St. Vivant,** similar wines of this commune.

1833 [see *Musigny]. **1845** *Encycl. Metrop.* XXV. 1279/1 The Romanée St. Vivant comes from a vineyard called by that name, at a monastery so styled, where it was brought to perfection by the sons of the church. It is little, if any thing, inferior to Romanée-Cônti. **1858** Thackeray *Virginians* I. xxix. 226 He..could distinguish between Clos Vougeot and Romanée with remarkable skill. **1904** A. Bennett *Great Man* xvi. 174 He had gathered..that the greatest of all burgundies was Romanée-Conti. **1920** G. Saintsbury *Notes on Cellar-bk.* iv. 55 A Romanée of '87 which was good. **1936** Bentley & Allen *Trent's Own Case* xii. 143 The Romanée St. Vivant 1904 which was to follow the Meursault. **1976** G. Moffat *Over Sea to Death* xv. 175 Maynard..ordered a second bottle of Romanée-Conti.

Romanesco (rōumāne·sko), *a.* and *sb.* [It.] (Of or pertaining to) a modern dialect spoken in the city of Rome.

1967 P. E. H. Durston *Mortissimo* (1968) xvi. 135 The Romanesco accent had been perfect. **1973** M. West *Salamander* i. 56, I had friends there: Castiglione, who used to be a great locksmith.., Giuffredi, the poet, who wrote satires in Romanesco which nobody read any more.

romanesque, *a.* (and *sb.*) Add: **4.** Romantic. ? *Obs.*

1799 Malthus *Diary* 24 June (1966) 87 He spoke of him [*sc.* Gustavus III] as..a little too romanesque and bizarre. **1850** C. M. Yonge *Kenneth* xx. 237, 'I know he thinks your point of honour rather romanesque,' said Effie, in her French-English. **1869** K. H. Digby *Little Low Bushes* 260 All fair things, lovely,..wild, or romanesque.

Romani: see Romany[3] *sb.* and *a.*

1899 F. H. Groome *Gypsy Folk-Tales* p. lxxx, *Bakht*, the Rómani word for 'luck' or 'fortune'. **1972** *Guardian* 28 Nov. 14/3 The Gipsy Council is..printing readers..in Romani and English. **1973** *New Society* 6 Dec. 595/1 Joint general secretary of the World Romani Congress. **1976** *Word 1971* XXVII. 357 Romani-English is distinctive because it has developed largely within a closed community.

Romanian, *a.*[1] (Earlier example.)

1841 Borrow *Zincali* II. III. ii. 104 The curiosity of some learned individuals..induced them to collect many words of the Romanian language, as spoken in Germany, Hungary, and England.

Romanian (romēi·niăn), *sb.* and *a.*[3] [f. *Romănia*, f. the native name *Romăn*: see Rouman *sb.* and *a.*, and -an, -ian. This is now the officially preferred form: cf. Roumanian *sb.* and *a.*, *Rumanian *sb.* and *a.*]

A. *sb.* **1.** A native or inhabitant of Romania (now the Socialist Republic of Romania); = Roumanian *sb.* 1, *Rumanian *sb.* 1.

1956 S. Fischer-Galati *Romania* I. i. 2 One [theory], proposed by the Romanians, is that most of the inhabitants 'took to the mountains' of Transylvania. **1967** P. Latham *Romania* IV. 216 The Romanian from ancient times has embellished his surroundings with painting, carving, pottery and weaving. **1969** W. Forwood *Romanian Invitation* i. 18 Of their wines, Romanians are rightly boastful. **1971** O. Manning *Romanian Short Stories* p. viii, The more the Romanians change, the more they are themselves. **1973** *Ann. Reg. 1972* 123 Possibly the replacement of Ulbricht by Honecker and the normalization of relations betwen the two Germanies made the move more acceptable to the Romanians. **1974** *Encycl. Brit. Macropædia* XV. 1053/2 The widespread rich folk costumes and the ancient folklore of Romanians.. provide a reminder of the country's long traditions.

2. The language of Romania, a Romance language which has been exposed to many foreign, esp. Slavonic and Greek, influences; = Roumanian *sb.* 2, *Rumanian *sb.* 2.

1954 Pei & Gaynor *Dict. Linguistics* 187 *Romanian*, a Romance language, the national tongue of Romania, and the native language of approximately 15,000,000 people. **1956** S. Fischer-Galati *Romania* I. i. 2 The Balkan Kutso-Vlach people speak a language akin to Romanian. **1969** W. Forwood *Romanian Invitation* ii. 25 The name Bucharest,..in Romanian Bucureşti, evokes the..days of leisurely railway journey across the Ruritanian map of Europe. **1972** M. L. Samuels *Linguistic Evol.* vi. 95 Romanian, Bulgarian and Albanian..share a number of features. **1974** K. Katzner *Lang. of World* (1977) II. 95 Rumanian, more correctly spelled Romanian, is, as its name suggests, one of the Romance languages. **1976** 'D. Fletcher' *Don't whistle 'Macbeth'* 48 None of us spoke Romanian.

B. *adj.* Of or pertaining to Romania, its inhabitants, or their language; = Roumanian *a.*, *Rumanian *a.*

1956 S. Fischer-Galati *Romania* I. i. 2 The Romanian nation was formed through the union of the Romans and the native population. **1967** P. Latham *Romania* I. 18 The Romanian Black Sea coast faces east. **1969** W. Forwood *Romanian Invitation* i. 17 The pleasures of the Romanian kitchen are very special. **1971** O. Manning *Romanian Short Stories* p. ix, Romanian music is either gypsy music or peasant music. **1973** *Ann. Reg. 1972* 122 It was a relatively inward-looking year for the Romanian Communist Government and its leader. **1976** P. Clabburn *Needleworker's Dict.* 228/1 Romanian stitch has one cross at a slight angle, romanian couching has several crosses at the same angle. **1979** *Records & Recording* Aug. 61/2 The distinguished Romanian soprano sings the part quite beautifully.

Romanichal: see *Romany[3] 3 b.

Romanicist (romæ·nisist). [f. Romanic *a.* (*sb.*) + -ist.] A student of Romance (sense 1); a scholar versed in Romance languages or literature.

1930 K. Malone in *Studies in Honor of H. Collitz* 328 *Romanicist*..seems to be the only word that fits the case [of the Romance philologist]. **1937** J. Orr tr. *Iordan's Introd. Romance Linguistics* iv. 279 Ferdinand de Saussure..was an Indo-Europeanist, not a Romanicist. **1957** *Archivum Linguisticum* IX. 90 From the Romanicist's viewpoint.

‖ **romanità** (romanitā·). Also with capital initial. [It., f. next.] **a.** = next. **b.** The spirit or influence of the central Roman authorities of the Roman Catholic Church; acceptance of papal policy.

1927 *Observer* 19 June 13/1 He [*sc.* Machiavelli] was too great a realist for his intellect to suffer imprisonment by humanist admiration for Rome, or, as we call it today, *romanità*. **1963** *Economist* 7 Dec. 1007/2 The assertion of collegial status for the bishops has been advanced jointly with a demand that the Curia should be internationalised... Yet internationalisation by itself is secondary. Indeed, it might sharpen the existing reasons for complaint: foreigners have been known to succumb to *Romanita*; they lack the engaging and emollient Italian supposition that laws are, on the whole, unlikely to be obeyed.

‖ **romanitas** (romā·nitas, romæ·nitās). Also with capital initial. [late L.] The spirit or ideals of ancient Rome; Romanism.

1947 *Horizon* Feb. 84 The *romanitas* upon which Europe was founded. **1961** *Listener* 16 Nov. 814/1 The natural pride in *Romanitas* which has passed into his [*sc.* Pope Leo I's] thinking from secular tradition. **1975** *Times Lit. Suppl.* 25 Apr. 464/4 Amid the decline and fall of a civilization to which he owed both his religion and his *romanitas*, St Augustine inscribed..the charter of an enduring Christian culture. **1977** *History* LXII. 174 One would wish to know..the extent to which there remained a concept of *romanitas* in Celtic Britain.

romanization. 4. (Further examples.)

1925 C. H. Brewitt-Taylor *San Kuo* I. p. ii, The Wade system of romanisation, in which the vowels are Italian, has been used. **1934** *Bull. Inst. Hist. & Philol. Acad. Sinica* IV. IV. 387 The French system of romanization of Chinese. **1961** *Amer. Speech* XXXVI. III. 177 Chang Ker Chiu gives both characters and the romanization *in—'yan* for 'craving for tobacco or opium'. **1963** [see *Pinyin]. **1973** *Lancet* 14 July 78/1 Even this difficulty is compounded by the fact that there are about a dozen systems of romanisation (e.g., the same point may be described as Ho-ku, He-gu, or Ro-Kou). **1975** *Daily Colonist* (Victoria, B.C.) 22 Aug. 5/1 China is preparing to move into..romanization of the written Chinese language.

Ro·manness. [-ness.] The quality of being influenced by Rome or by Roman Catholicism.

1959 *Catholic Times* 20 Mar. 5/4 His theory that the Welsh were profoundly conscious of their Roman-ness.

Roman-nosed, *a.* Add: (Further example.) Hence **Ro·man-no·sedness** (*nonce*).

1912 J. S. Huxley *Individual in Animal Kingdom* iii. 80 In all Metazoa there is, before and during the sexual process, a shuffling and recombination of the chromosomes of the nucleus—those bodies which taken together appear to determine the characteristics of the offspring, or at least those which mark it off from others of the same species,—whether it shall be tall or short, fair or dark, chubby or lanky, tip-tilted or Roman-nosed. **1912** D. H. Lawrence *Let.* 24 Dec. (1932) 87 If it's destined to have a snub nose, it's sheer waste of time to harass the poor brat into Roman-nosedness.

Romano (romā·no). [It., = Roman.] In full *Romano cheese*. A strong-tasting hard cheese, orig. made in Italy.

1908 Doane & Lawson *Varieties of Cheese* 39 The Formaggio Pecorini are the sheep's-milk cheeses made in Italy... The most common cheese of this sort is the one designated Cacio Pecorino Romano, or merely Romano. *Ibid.*, In making Romano cheese the milk is heated to 100° F. and coagulated by rennet in fifteen minutes. **1918** J. L. Sammis *Decker's Cheese Making* (ed. 6) xxx. 218 *Romano*. This is usually made from skim milk [in America]. **1949** N. Standen *Art of Cheese Cooking* 30 Romano, in Italy, used to be made from sheep's milk which gave it a pungent flavor. In this country, though, it's made from cow's milk and is correspondingly milder. **1955** R. C. Brown *Compl. Bk. Cheeses* iii. 26 Romano is not as expensive as Parmesan. **1966** Marquis & Haskell *Cheese Bk.* II. 64 The black-rinded and..grainy-looking Romano..is much stronger than Parmesan. **1976** N. Thornburg *Cutter & Bone* viii. 199, I have..some ridiculously expensive Romano cheese compliments of George.

Romano-. Add: Now usu. with pronunc. (romā·no). Also, *Romano-British* (earlier and later examples), *-Briton, -canonical, -cosmopolitan, -Egyptian, -Germanic* (later example), *-Hellenistic, -Saxon, -Visigothic.*

1847 J. Y. Akerman (*title*) An archaeological index to remains of antiquities of the Celtic, Romano-British and Anglo-Saxon periods. **1963** *Times* 21 Feb. 5/7 Caerwent was the only fully developed Romano-British town in Wales. **1975** J. G. Evans *Environment Early Man Brit. Isles* vi. 157 There was some colonization by the Belgae and Romano-British people of areas not previously taken up. **1896** A. H. Keane *Ethnol.* II. xiv. 398 The Teutons merged everywhere in diverse proportions with the Romano-Britons. **1956** Auden *Old Man's Road*, So thought (I think) the last Romano-Briton. **1909** Webster, Romano-canonical. **1974** A. Watson *Legal Transplants* vii. 45 Some Roman law was creeping in primarily as a result of the acceptance of rules of romano-canonical procedure. **1923** R. G. Collingwood *Roman Britain* iii. 68 Houses..never losing their Celtic stamp or becoming Romano-cosmopolitan instead of Romano-British. **1934** Webster, Romano-Egyptian. **1964** W. L. Goodman *Hist. Woodworking Tools* 33 Most of the adzes..appear to be of a similar pattern to the Romano-Egyptian adze described earlier. **1980** *Jrnl. R. Soc. Arts* July 534/2 In the Romano-Germanic tomb of a little girl, was found a decorated goose egg. **1972** D. Dakin *Unification of Greece* i. 5 Many Byzantines began to see themselves not as heirs to the Romano-Hellenistic traditions of the West but as the successors of ancient Hellas. **1956** J. N. L. Myres in D. B. Harden *Dark-Age Brit.* 16 (*heading*) Romano-Saxon pottery. **1970** J. L. Shneidman *Rise of Aragonese-Catalan Empire* I. v. 154 In the city of Urgel..during the period of Arab domination the bishop had..governed the city under the Romano-Visigothic law called the *Fuero Juzgo*.

ROMANOWSKY

Romanowsky (rǭᵘmänǫ·fski). *Histology.* Also **-ofsky, -ovski, -ovsky.** The name of Dmitriy Leonidovitch *Romanowsky* (1861–1921), Russian physician, used *attrib.*, in *Comb.*, and in the possessive to designate a stain and staining technique devised by him, and a class of derived stains and techniques, used for the detection of parasites in blood.

1903 *Brit. Med. Jrnl.* 30 May 1253/1, I was struck by the curious appearance..of small round or oval bodies... On staining them by Romanowsky's method, they were found to possess a quantity of chromatin, of a very definite and regular shape, which clearly differentiated them from blood plates or possible nuclear detritus. *Ibid.* 28 Nov. 1401/1 The deep red of the Romanowsky-stained chromatin of the bodies is represented by black in the drawings. **1906** *Boston Med. & Surg. Jrnl.* CLIV. 643/1 A staining fluid, devised by me for use in the staining of blood films according to the method of Leishman, which gives the so-called Romanofsky polychrome staining. **1930** [see *ORTHOCHROMATIC *a.* 2]. **1947** *Ann. Rev. Microbiol.* I. 48 They compared their findings made in this way with parallel studies employing the ordinary technics of smears stained with Romanowsky's stain. **1960** E. GURR *Encycl. Microsc. Stains* i. 267 The azurs I, A, B and C, and methylene violet..are present in varying degrees in the Romanowsky type of stains. The latter consists of methylene blue and its oxidation products in combination with eosin. **1970** J. C. SWARTZWELDER et al. in J. E. Blair et al. *Man. Clin. Microbiol.* xlix. 440/1 Stained with Giemsa or some other Romanowski dye. **1978** *Nature* 22 June 595/1 The cytoplasm stains a pale blue with Romanovsky stains, and the single nucleus..a reddish-purple.

Romansh, *sb.* and *a.* Add: Also *α.* **Romantsch.** *β.* **Rumansch.** (Further examples.) Also used of other Rhæto-Romance dialects and of this group of dialects as a whole.

1946 *Archit. Rev.* C. 58/1 It is for us a matter of course that the Alemanic part of the country would speak German, the French Swiss part French,..and the Rhaetian districts Romansch, that, for instance, the Grisons, which comprise districts speaking German, Italian and Romansch, should publish their decrees in all three languages. **1969** *Language* XLV. 185 To counter Italian nationalist claims, Romansh was officially established as the fourth national language of Switzerland in 1939. **1970** *Times Lit. Suppl.* 8 Jan. 40/2 He is fully conversant with the five national tongues—French, Spanish, Portuguese, Italian and Rumanian—and with their five regional varieties—Occitanian or Provencal..Catalan..Dalmatian..Rumansch or Rhaetian (now strictly West Rhaetian, for that alone has evolved a recognized written language), and Sardinian. **1971** *Language* XLVII. 797 (*heading*) Targets and paradigmatic borrowing in Romantsch.

b. (Further examples.)

In quot. 1920 referring to an ethnic group.

1880 [see *LADIN]. **1920** *Q. Rev.* Apr. 443 Its population is not of German, but of Allemanic race, the only exception being that part which is of Romantsch origin. **1969** [see *RHÆTO-ROMANCE].

romantic, *a.* and *sb.* Add: **A.** *adj.* **1. c.** Of a work of modern literature, etc.: having romance as its subject; treating of a love affair.

1960 R. REES *For Love or Money* ii. 30 The doctrine of D. H. Lawrence's *Fantasia of the Unconscious*: that sexual passion, unrelated to the religious impulse..leads to sterility and death—as in *Anna Karenina*, in *Carmen*, and in the greater part of European 'romantic' literature. **1977** B. PYM *Quartet in Autumn* i. 3 Unable to find what she needed in 'romantic' novels, Letty had turned to biographies of which there was no dearth. **1981** S. RADLEY *Chief Inspector's Daughter* i. 15, I get depressed because I write romantic fiction instead of straight novels.

4. b. (Earlier and further examples.) Also of ballet (see quot. 1957). Also, of, pertaining to, or characteristic of romanticism in literature, etc.

1812 H. C. ROBINSON *Jrnl.* 19 May in E. J. Morley *Henry Crabb Robinson on Bks.* (1938) I. 84 We proceeded to Coleridge's first lecture... He spoke of religion, the spirit of chivalry,..and a classification of poetry into ancient and romantic. **1813** *Edin. Rev.* Oct. 206 The poetry of the Spanish peninsula seems to have been more romantic and less subject to classical bondage than that of any other part of Europe. **1814** W. TAYLOR in *Monthly Rev.* Apr. 364 The eleventh [chapter] divides European poetry into two schools, the classical, and the romantic. **1833** W. MAGINN in *Fraser's Mag.* VIII. 64 'The noticeable man [*sc.* Coleridge] with large grey eyes'—the worthy old Platonist—the founder of the romantic school of poetry. **1908** P. E. MORE *Shelburne Ess.* 5th Ser. 119 Like Friedrich Schlegel, he indulges in the romantic irony of smiling down upon himself and walking through life like a *Doppelgänger*. **1928** [see *CLASSICAL *a.* 6 d]. **1930** W. EMPSON *Seven Types of Ambiguity* i. 27 Before the Romantic Revival the possibilities of not growing up had never been exploited so far as to become a subject for popular anxiety. **1937** D. BUSH *Mythology & Romantic Trad. in Eng. Poetry* p. xiii, The effect of both the romantic and the industrial movements was to make the artist, if not an anti-social figure, at any rate an isolated one. **1938** *Oxf. Compan. Mus.* 810/1 By the 'Romantic School' in music is meant the group of active spirits in that movement which began in Germany with Weber (born 1786)... Or it can be carried back as far as Schubert (born 1797) and Beethoven (born 1770). **1951** F. KERMODE *Romantic Image* vii. 132 The next step forward in Romantic aesthetic depended upon a new theory of language. **1957** G. B. L. WILSON *Penguin Dict. Ballet* 230 *Romantic ballet*, used, somewhat narrowly, to describe the ballets produced during the period of the Romantic revival in literature in the early nineteenth century, or roughly from 1830–1850, taking as their theme the odyssey of mortal man in love with some female spirit of the air or water or with some maiden risen from her tomb... The dividing line is a slender one, i.e. in the romantic ballet the accent is on colour or mood rather than form and design which is predominant in the classical ballet. **1959** F. GADAN et al. *Dict. Mod. Ballet* 329/1 Several other great Romantic dancers appeared as La Sylphide. **1960** BECKSON & GANZ *Reader's Guide Lit. Terms* (1961) 108 Romantic irony occurs when a writer builds up a serious emotional tone and then deliberately breaks it and laughs at his own solemnity. **1977** J. A. CUDDON *Dict. Lit. Terms* 573 *Romantic revival*, a term loosely applied to a movement in European literature (and other arts) during the last quarter of the 18th c. and the first twenty or thirty years of the 19th c.

5. (Further examples.)

The examples given here, illustrating the collocation of the adjective with *love, lover, friendship*, and the like, provide evidence of the emergence of its common present-day use to convey the idealistic character or quality of a love affair. Cf. *ROMANCE *sb.* 5 a, b.

1728 F. HUTCHESON *Ess. Passions* I. iv. 94 A Romantick Lover has..no Notion of Life without his Mistress, all Virtue and Merit are summed up in his inviolable Fidelity. **1754** R. BERENGER in *World* 4 July 474, I know several unmarried ladies, who in all probability had been ..good wives and..mothers, if their imaginations had not been early perverted with the chimerical ideas of romantic love,..upon which principle, a footman may as well be the hero as his master. **1769** J. USHER *Clio* (ed. 2) 82 Innocent and virtuous love..inspires us with heroic sentiments,..a contempt of life, a boldness for enterprize, chastity, and purity of sentiment... People whose breasts are dulled with vice, or stupified by nature, call this passion romantic love; but when it was the mode, it was the diagnostic of a virtuous age. **1778** S. TIGHE *Let.* 2 Apr. in G. H. Bell *Hamwood Papers* (1930) 27 There were no gentlemen concerned, nor does it appear to be anything more than a scheme of Romantic Friendship. **1806** BYRON *Fugitive Pieces* 23 And friendships were form'd, too romantic to last. **1858** LYTTON *What will he do with It?* (1859) III. VII. xiv. 135 (*heading*) Romantic Love pathologically regarded by Frank Vance and Alban Morley. **1866** C. M. YONGE *Dove in Eagle's Nest* II. ii. 41 Good substantial wedded affection was not lacking, but romantic love was thought an unnecessary preliminary, and found a vent in extravagant adoration not always in reputable quarters. **1942** T. BAILEY *Pink Camellia* vii. 50 The lovemaking was of the purely romantic kind, for Cecily would have no other. **1945** *New Statesman* 23 June 408/3 The book opens with a tale of romantic friendship at Oxford in the years following the first great war. **1966** *Listener* 7 Apr. 509/3 Nowadays, however, educated young West Africans have discovered the alleged virtues of romantic love. They stress the idea of marriage being a true union of husband and wife as well as an economic partnership. Love will be the most important thing when they marry. **1971** E. MAVOR *Ladies of Llangollen* v. 96 The strange ambivalence of the pre-Freudian romantic friendships. **1975** J. PLAMENATZ *Karl Marx's Philos. of Man* xiv. 400 The idea of romantic love has flourished in the same kind of society as the small family. Indeed, this family is quite often seen as the creature of romantic love: it is set up by a man and a woman who come to love one another and who choose each other as life partners. **1978** *Morecambe Guardian* 14 Mar. 17/2 Partnerships flourish. A romantic attachment is possible, but do not take it too seriously.

B. *sb.* **2.** (Earlier and further examples.) Also, a composer of romantic music.

1827 CARLYLE in C. E. Norton *Two Notebks. of T. Carlyle* (1898) 111 Grossi..has written a new Epic... Grossi is a Romantic. **1927** R. H. WILENSKI *Mod. Movement in Art* 30 Nineteenth-century romantics deliberately left out all the features which the admirers of classical painting were accustomed to regard as indispensable to art. **1932** W. B. YEATS *Words for Music* 11 We were the last romantics, chose for theme Traditional sanctity and loveliness. **1933** A. DAVIDSON tr. *Praz's Romantic Agony* 4 The thirst for the infinite..animates the lines of the Romantics. **1938** *Oxf. Compan. Mus.* 113/1 Despite their sheer musical beauty, his [*sc.* Brahms's] compositions are strongly charged with what may be called an extra-musical emotion; hence the classification of their composer as a romantic. **1960** A. O. LOVEJOY in M. H. Abrams *Eng. Romantic Poets* 15 To be unsophisticated, to revert to the mental state of 'simple Indian swains', was the least of the ambitions of a German Romantic... The greatness of Shakespeare, in the eyes of *these* Romantics, lay in his Universalität. **1961** C. CLUTTON in A. Baines *Mus. Instruments* ii. 66 The [organ] works of Liszt and Franck,..and of such late romantics as Reger, Jongen, and Elgar, rely upon a very large instrument. **1966** H. G. SCHENK *Mind of European Romantics* i. 6 Rationalism was attacked by the Romantics not on the grounds that the intellectual results yielded by it were false, but rather on the grounds that they were inadequate. **1977** *Times* 18 Oct. 24/9 White tuxedos are occasionally supplied to shipboard romantics.

roma·ntic, *v.* [f. the adj.] *trans.* = ROMANTICIZE *v.* 1. Also with *up*.

1969 G. LYALL *Venus with Pistol* xxii. 137 It was a fairly flat scene of somewhere in Venice, a bit romanticked up. **1972** *Guardian* 8 June 2/1 'Elizabeth R' starts a new run if you like your history romanticked.

roma·nticalism. *rare.* [f. ROMANTICAL *a.* + -ISM.] = ROMANTICALITY.

1922 W. J. LOCKE *Tale of Triona* xiii. 142 She..was driven by she knew not what idiot romanticalism into the grey worries of wifehood and motherhood.

romantically, *adv.* Add: **3.** *Comb.*, as *romantically-minded* adj.

1952 'M. COST' *Hour Awaits* 227 It appears that this Professor..is romantically minded. **1965** HOUSE & STOREY *Lett. C. Dickens* I. p. xxi, The kind suggested here by the romantically-minded Kate.

romanticism. 3. (Earlier and further examples.)

1823 *New Monthly Mag.* IX. 175/2 The French Academy ..has determined never to receive within its bosom any one polluted by the dramatic heresy of romanticism. **1830** [see *CLASSICISM 1]. **1934** C. LAMBERT *Music Ho!* II. 118 A title like the *Pathetic Symphony* is looked on as an example of decadent romanticism. **1937** D. BUSH *Mythol. & Romantic Trad. in Eng. Poetry* xii. 398 In various ways and for various reasons the broad deep stream of romanticism had run thin. **1941** P. H. LANG *Music in Western Civilization* xv. 746 In Beethoven classicism became romantic, and in Schubert romanticism became classic. **1957** F. KERMODE *Romantic Image* viii. 145 Romanticism is just the new disease at the stage of mania. **1960** A. O. LOVEJOY in M. H. Abrams *Eng. Romantic Poets* 5 The offspring with which Romanticism is credited are as strangely assorted as its attributes and its ancestors. **1978** *Times Lit. Suppl.* 25 Aug. 944/5 This..biography has its interest..as a portrait of a very unhappy man whom Romanticism destroyed.

romanticist. Add: **1.** (Earlier example.) Also in music.

1827 CARLYLE in *Edin. Rev.* XLVI. 325 Their grand controversy, so hotly urged, between the Classicists and Romanticists..shows us sufficiently what spirit is at work in that long stagnant literature. **1883** GROVE *Dict. Mus.* III. 152/2 We cannot acquit the younger romanticists of the charge of an excessive realism. **1938** *Oxf. Compan. Mus.* 810/1 It is..sometimes considered that the classical element..in the work of those two [*sc.* Schubert and Beethoven] was strong enough to rank them as the last of the Classicists rather than as the first of the Romanticists. **1941** P. H. LANG *Music in Western Civilization* xv. 746 If Weber, Chopin, and Schumann are accepted as full-blooded romanticists, we..cannot enrol the composer of the *Unfinished Symphony*..in their company. **1970** W. APEL *Harvard Dict. Mus.* (ed. 2) 738/1 These traits need not imply that nonromantic music lacks emotional appeal. Nor does it mean that the romanticists were not form-conscious.

romanticize, *v.* Add: Hence **roma:nticiza·tion; roma·nticizing** *vbl. sb.*

1899 *Speaker* 14 Apr. 424/2 Enlivened by champagne and some grotesque romanticising on the part of the amorous Duchess. **1935** *Mind* XLIV. 95 His [*sc.* Nietzsche's] 'Dionysus philosophy' is a typically Germanic brutalisation, exaggeration, romanticisation of something borrowed. **1968** G. ASHE *Quest for Arthur's Britain* i. 28 Leland's romanticisation of Henry VIII was elaborately transferred to Elizabeth by Edmund Spenser.

roma·nticky, *a. rare.* [f. ROMANTIC *a.* + -Y[1].] Of a romantic character.

1912 D. H. LAWRENCE *Let.* ?5 Nov. (1962) I. 154, I want to read something romanticky—feel like it.

romanticness. (Later example.)

1968 H. KONINGSBERGER *Revolutionary* v. 13 The romanticness of..tears shed by women in Turgenev.

Romantsch, var. ROMANSH *sb.* and *a.* in Dict. and Suppl.

Romany[3]. Add: **3. b.** Special Combs., as **Romany chal, Romanichal** [CHAL], a (male) gypsy; **Romany chi** (tʃəi) [Romany *chai* girl], a gypsy girl; **Romany rye** [*RYE *sb.*[3]], a man, not a gypsy, who associates with gypsies.

1843 BORROW *Zincali* (ed. 2) I. 32 Those were brave times for the Rommany chals. **1851** [see GORGIO]. **1953** J. DE B. LEVY *As Gypsies Wander* i. 38 His pleasure was extreme when he first heard that non-Gypsy people had written poems in praise of *Romanichals*. **1960** G. E. C. WEBB *Gypsies* i. 19 Whoever heard of a *gorgio* coming up to a *Romanichal* and greeting him with words of the old language? **1857** Romany chi [see GORGIO]. **1876** [see CHAL]. **1933** K. BERCOVICI (*title*) The Romany chai. **1851** BORROW *Lavengro* II. xxvi. 236 Here the Gypsy gemman bee, With his Roman jib and his rome and dree—Rome and dree, rum and dry Rally round the Rommany Rye. **1857** —— *Romany Rye* II. ix. 113 'I'll bet a crown,' said the jockey, 'that you be the young chap what certain folks call "The Romany Rye".' **1915** F. CUTTRISS *Romany Life* xi. 242 He introduced me as a Romany Rye. **1929** K. BERCOVICI *Story of Gypsies* x. 231 What is most astounding..is the mention made of the natives of England of good families who were found in the company of these Gypsies... Who could they be, these gentlemen, these first Romany ryes? **1973** *Cassell's Encycl. World Lit.* (rev. ed.) I. 489/1 The Gypsy Lore Society in Liverpool, which was founded by the American Romany Rye Charles Godfrey Leland.

romanza. Restrict † *Obs.* to sense in Dict. and add: ‖ **2.** *Mus.* [a. It., Sp. *romanza.*] Pronunc. (roma·nzä); in Sp. contexts also (romã·nþä). A romantic song or melody; a lyrical piece of music; = ROMANCE *sb.* 4 b.

1834 *Chambers's Edin. Jrnl.* III. 110/2 Another youth.. begins singing a Spanish romanza. **1938** *Oxf. Compan. Mus.* 810/2 *Romanza*.., a song or song-like instrumental composition. **1970** J. BLADES *Percussion Instruments* x. 196 Alexander Goehr scores for the lion's roar in his *Romanza for 'cello and orchestra* (1968). **1975** *New Yorker* 16 June 97/3 Then the romanza begins.

‖ **romanze** (roma·ntsə). *Mus.* Pl. **romanzen.** [Ger., = romance.] A composition of a tender or lyrical character; *spec.* a slow, romantic instrumental piece or movement. Cf. prec. and ROMANCE *sb.* 4 b.

1883 GROVE *Dict Mus.* III. 147/2 The Romanze in Mozart's D minor PF. Concerto differs..from the slow movements of his other Concertos in the extremely tender and delicate character of its expression. **1947** A. EINSTEIN *Mus. Romantic Era* xi. 130 It joins together five movements—Introduction, Allegro, Romanze, Scherzo, and Finale—into an uninterrupted whole. **1970** W. APEL *Harvard Dict. Mus.* (ed. 2) 736/2 The German *Romanze* is primarily an instrumental composition of a lyrical character. *Ibid.*, Vocal *Romanzen* occur mostly in operas.

Romary (rǭᵘ·mări). [f. the name of the manufacturer.] The proprietary name of a brand of biscuits.

1926–7 *Army & Navy Stores Catal.* 8/2 Biscuits.. Romary Ginger Nuts. **1929** *Trade Marks Jrnl.* 4 Dec. 2029 Romary's... Biscuits. A. Romary & Company, Limited,..Tunbridge Wells, Kent; manufacturers. **1934** E. BOWEN *Cat Jumps* 192 The Romary biscuits. **1977** P. HARCOURT *At High Risk* I. viii. 101 His secretary.. placed beside me a plate of Romary biscuits.

Romayne, obs. f. ROMAN. For 'obs.' read 'obs. exc. as applied to carving, etc. with a motif of heads in medallions'. Pronunc. (romē̆i·n). (Later examples.)

1904 P. MACQUOID *Hist. Eng. Furnit.* iii. 50 Chair.. decorated..with medallioned heads surmounted by conventional ornament in the Italian manner, and which in this century obtained the name 'Romayne Work'. **1955** R. FASTNEDGE *Eng. Furnit. Styles* 287 *Romayne carving*, decorative *motifs* taking the form of small profile heads in medallions, introduced in the early sixteenth century. **1961** *Times* 2 Dec. 11/7 Small objects, carved with Romayne heads. **1969** E. H. PINTO *Treen* 196/1 The first two snuff boxes of this type, carved with 'Romayne heads'..were very worn and I thought that they were genuinely mid or late 16th-century. **1975** *Oxf. Compan. Decorative Arts* 672/2 *Romayne work*, contemporary term for a decorative motif consisting of small profile-heads in medallions carved on furniture and panelling. This form of decoration was introduced into England from Italy in the time of Henry VIII and was often combined with Tudor roses and traditional Gothic tracery or linenfold.

romazi, var. *ROMAJI.

Romberg (rǫ·mbəɹg). *Med.* The name of Moritz Heinrich *Romberg* (1795–1873), German physician, used in the possessive, *attrib.*, and *absol.* to designate (*a*) the test of requiring a patient to stand with feet together and eyes closed, and (*b*) the sign or symptom, diagnostic of ataxia, shown by a patient who then sways or falls (described by Romberg in *Lehrb. d. Nervenkrankheiten des Menschen* (1846) I. 795).

1885 *Jrnl. Nervous & Mental Dis.* XII. 354 'Romberg's symptom'—*i.e.* inability to stand with the eyes shut and the feet together—is not always present. *Ibid.* 355 'The Romberg symptom'. **1932** *Practitioners Libr. Med. & Surg.* II. xxiii. 809 The Romberg test for static ataxia is carried out by having the patient stand with heels and toes together with open eyes and then with closed eyes. The tendency to sway appreciably with the eyes shut constitutes a positive Romberg. **1961** *Lancet* 9 Sept. 569/1 Both legs were slightly weak, and Romberg's sign was positive. **1977** *Ibid.* 10 Dec. 1228/2 Neurological examination revealed a vertical gaze palsy,..and a tendency to fall backwards on Romberg's test.

Rome, *sb.* Add: **1. b.**

(*a*) (Earlier and later examples.)

1545 R. TAVERNER tr. *Erasmus's Adages* sig. D 1ᵛ Ye may use this prouerbe when ye wyll signyfye that one daye..is not ynoughe for..acheuynge..a greate matter ..Rome was not buylt in one daye. **1822** SCOTT *Fortunes of Nigel* II. x. 237 Rome was not built in a day—you cannot become used to your court-suit in a month's time. **1849** C. BRONTË *Shirley* I. v. 123 'As Rome,' it was suggested, 'had not been built in a day, so neither had mademoiselle Gérard Moore's education been completed in a week.' **1873** 'F. FERN' *Memorial Vol.* 347 Rome wasn't built in a day;—cooks can't be manufactured in a minute. **1901** S. LANE-POOLE *Sir H. Parkes* xvii. 316 The Japanese..went too fast and fell into grave commercial, monetary, and administrative troubles. Neither Rome nor New Japan could be built in a day. **1941** P. CHEYNEY *Trap for Bellamy* iv. 58 Bellamy said: 'Life is what you make it. Rome wasn't built in a day.' **1950** T. WILLIAMS *Roman Spring of Mrs. Stone* I. 34 Patience, said the Contessa. Rome was not built in a day!

(*b*) (Later examples.)

1976 N. THORNBURG *Cutter & Bone* i. 22 'When in Rome,' he said finally, shuddering. **1977** *Rolling Stone* 21 Apr. 73/3 He had a point—when in Rome and all that—but it was a point he was not exactly loath to make.

(*d*) *c* **1380** CHAUCER *Troilus & Criseyde* (1894) II. 36 For every wight which that to Rome went, Halt nat o path, or alwey o manere. *c* **1391** —— *Astrolabe* (1872) Prol. 29 Ryht as diuerse pathes leden diuerse folk the rihte wey to Rome. **1806** R. THOMSON tr. *La Fontaine's Fables* IV. XII. xxiv. 110 Three pious men, having one end in view, Their way to heaven with equal zeal pursue.— Three diff'rent roads the three concurrents chose, All roads alike conduct to Rome...—So those Thought they might part, and yet get on secure. **1861** C. READE *Cloister & Hearth* I. xxiv. 270 All roads take to Rome. **1872** W. BLACK *Strange Adv. Phaeton* vi. 111 'Surely the road to Oxford is easy to find.' 'It is,' I say to her. 'For you know all roads lead to Rome, and they say that Oxford is half-way to Rome—*argal*—.' But knowing what effect this reference to her theological sympathies was likely to have on Tita, I thought it prudent to send the horses on. **1911** J. A. THOMSON *Introd. Sci.* iii. 63 All roads lead to Rome, and he must be a bold man who will declare any of Nature's beckonings to be unworthy of attention. **1912** J. S. HUXLEY *Individual in Animal Kingdom* vi. 154 All roads lead to Rome: and even animal individuality throws a ray on human problems.

3. c. Special Comb.: **Rome-Berlin axis** [*AXIS[1] 4 b], the association formed in 1936 between Fascist Italy and National Socialist Germany.

1936 [see *AXIS[1] 4 b]. **1938** E. AMBLER *Cause for Alarm* viii. 128 The Rome-Berlin axis is one of the most effective principles of European power-politics that has ever been stated. **1939** 'G. ORWELL' *Coming up for Air* III. i. 182 Rubber truncheons, Rome-Berlin axis, Popular Front. **1976** S. HYNES *Auden Generation* vii. 193, 1936 is the peripeteia, the point where the action turned:..the Rome-Berlin Axis was formed [etc.].

roméite (rǭᵘ·me,əit). *Min.* Also **romeite.** [f. ROME(INE + -ITE[1].] An antimonate of calcium, $Ca_2Sb_2O_7$ (usu. also containing other elements esp. iron or titanium), which occurs as yellow or yellow-brown octahedral crystals. Formerly known as *roméine*.

[**1850** J. D. DANA *Syst. Min.* (ed. 3) 416 (*heading*) Romeine, *Damour*,..Romeit.] **1868** *Ibid.* (ed. 5) 547 Romeite was found by B. de Lom at St. Marcel in Piedmont. **1916** *Bull. U.S. Geol. Survey* No. 610. 96 The Brazilian so-called atopite is..identical with the romeite from Italy. **1953** *Mineral Mag.* XXX. 101 The definition of stibiconite..includes the mineral roméite, which has long been recognized as a calcium antimonate. **1968** I. KOSTOV *Mineral.* II. 265 Roméite and hydrocervantite.. have adsorbed water.

Romeo (rǭᵘ·mio). [Name of the hero of Shakespeare's tragedy *Romeo and Juliet.*] **1.** A lover, a passionate admirer; a seducer, a habitual pursuer of women. Also *attrib.*

1766 C. ANSTEY *New Bath Guide* ix. 59 May I oft my Romeo meet, Oft enjoy his Converse sweet. **1867** TROLLOPE *Claverings* I. iii. 35 He has come out so strong in the Romeo line... We shall have him under your bedroom window with a guitar. **1917** E. O'NEILL *Long Voyage Home* in *Smart Set* Oct. 56/2 *Driscoll*..Shut up, ye Rooshan baboon! A foine Romeo you'd make in your condishun. **1942** *Sun* (Baltimore) 26 Mar. 10/2, I think from the way I so often see his eyes cast far up and around through the open spaces that he is also playing a bit of Romeo. **1974** G. MITCHELL *Javelin for Jonah* iii. 48 Henry..locked up the mansion to keep out any prospective Romeos who might fancy a visit to the women students' rooms. **1976** *Leicester Chron.* 26 Nov. 2/4 He's known as the studio's resident Romeo, with a social life and a string of girlfriends which must be exhausting rather than exhilarating.

2. (With small initial.) Also *romeo slipper*. A type of high slipper, now only for men, usu. made of felt and with elasticated gores. *U.S.*

1895 *Montgomery Ward Catal.* Spring & Summer 514/2 Men's leather sole Romeo... This slipper is made of one piece of black felt. *Ibid.* 514/3 Ladies' Romeo... Made of toilet felt, fur trimmed. **1898** *Morse & Rogers Money Saver* Oct. 17 (Advt.), Men's romeo slippers. Don't be out of slippers. Romeos are becoming more and more popular. **1924** E. FERBER *Show Boat* i. 5 Peeping..around this, the baffled eye could just glimpse oddments and elegancies such as..a pair of scuffed tan kid bedroom slippers (men's) of the sort known as romeos. **1952** R. BISSELL *Monongahela* xix. 205 While sitting on the bunk pulling on your romeos you wonder what side of the family this madness comes from that makes you live like this.

3. *Romeo and Juliet*: Anglicization of Sp. *Romeo y Julieta*, the proprietary name of a Havana cigar. (Also semi-Anglicized as *Romeo and Julieta, Romeo y Juliet.*) Also *attrib.*

1904 *Trade Marks Jrnl.* 7 Sept. 1105 Romeo y Julieta... Tobacco, cigars, cigarettes and snuff. Rodriguez Argüelles & Co.,..Havana, Cuba; cigar manufacturers. **1907** *Yesterday's Shopping* (1969) 64/2 *Havana cigars*... 'Romeo and Julieta'. **1945** A. HUXLEY *Time must have a Stop* v. 51 His Uncle Eustace lighted the massive Romeo and Juliet. **1957** J. OSBORNE *Entertainer* II. 43 Gave me a box of Romeo and Juliet cigars. **1966** *Guardian* 19 Mar. 6/6 The host had his mouth full of Romeo y Juliet cigar. *Ibid.* 12 Oct. 5/5 One could always finish with a cigar; say Romeo y Julieta, coronas, 66os the hundred, 5½ inches long.

romer (rǭᵘ·məɹ). Also **Romer.** [Named after Carrol *Romer* (1883–1951), British barrister, who invented it.] A small piece of plastic or card with scales along two edges meeting at a right angle, or (if transparent) bearing a grid, used for measuring the map reference of a point within the grid printed on a map.

1933 *Geogr. Jrnl.* LXXXII. 47 This Romer, as it was called after the gentleman who invented it. **1943** F. F. CROSSLEY *Map Reading* iii. 15 In order to facilitate the estimation of the last figure of the reference it is useful to construct a Romer... The sides of the Romer are subdivided into tenths. **1960** S. TURNER *Rallying* 38 The cross-roads might thus be given as 25401464. For plotting of this accuracy a romer is necessary. **1963** P. DRACKETT *Motor Rallying* iii. 41 Lining up the Romer so that one straight edge of the rectangle is in line with 386 and the straight edge meeting it is in line with 443, the point of bisection gives the place referred to. **1969** G. C. DICKINSON *Maps & Air Photographs* viii. 114 Of course a different romer is needed for each scale. **1975** J. B. HARLEY *O.S. Maps* ii. 24 A point..can be indicated still more closely by estimating the tenths of the appropriate grid square either by eye or by means of a romer.

rōmerite, var. *ROEMERITE.

Romeward, *adv.* and *a.* Add: **3. b.** Directed towards or facing the city of Rome.

1850 J. MILEY *Hist. Papal States* I. 6 Not less so on the Adriatic side of the Apennines than on the Romeward side. **1979** *London Rev. Bks.* 25 Oct. 5/3 Salim's flight to London can be compared..to the Romeward journey in Virgil.

Ro·mewardness. [-NESS.] Tendency towards Roman Catholicism.

1901 *Daily Chron.* 27 Aug. 5/1 The young Duchess was forgiven by her relatives for her Romewardness.

Romish, *a.* **3.** For † *Obs.* read *arch.* (Further examples.)

1816 T. J. HOWELL *Stranger in Shrewsbury* 182 During its subjection to the Romish yoke, this country formed part of the province of Flavia Cæsariensis. **1917** W. OWEN *Let.* 14 Feb. (1967) 435 Do you need a Brooch? I saw an Egyptian one, rather huge,..and an Assyrian. No they wouldn't do since you no longer wear your Ancient British frocks. All the others in this town are either Romish, or nondescript.

‖ **rommelpot** (rǫ·məlpǫt). Also **rommel pot, rommel-pot.** [Du., = rumble pot.] **a.** A type of drum used in southern Africa (see quot. 1840). **b.** A type of drum used in the Low Countries (see quots. 1964, 1976).

1840 B. SHAW *Memorials S. Afr.* iii. 44 The rommel pot is a bamboo over which a piece of skin is tightly stretched, and is used as a drum at their [*sc.* the Namaquas'] dances. **1881** *Encycl. Brit.* XII. 311/1 The 'rommel-pot' was a kind of drum. **1964** S. MARCUSE *Mus. Instruments* 446/2 *Rommelpot*.., Dutch and Flemish friction drum with friction stick, made of an earthenware pot sometimes containing water, closed at the top by an animal bladder. The friction stick penetrates the center of the bladder and is rotated or pushed up and down. **1970** J. BLADES *Percussion Instruments* x. 196 In Flanders the rommelpot is particularly associated with Christmas. **1976** D. MUNROW *Instruments Middle Ages & Renaissance* v. 34/1 The rommelpot..is a type of friction drum... The action required is not scraping the stick to and fro but rubbing it gently with moistened fingers.

Romney (Marsh) (rɒ·mni, rǫ·mni māɹʃ). The name of an area of rich grazing land on the coast of Kent, used *absol.* and *attrib.* to designate a stocky, long-woolled sheep of the breed so called.

1837 W. YOUATT *Sheep* vii. 239 In some parts of the east of the county a polled breed of middle size, a cross between the Romney Marsh and the South Down, is found. **1861** I. BEETON *Bk. Househ. Managem.* 327 (*caption*) Romney-Marsh Ewe. *Ibid.*, The Romney Marsh..is a highly useful..variety..of the English domestic sheep. **1891, 1894** [see *KENT *sb.*[3]]. **1922** V. SACKVILLE-WEST *Heir* iv. 42 It's like sheep..Oxfordshire don't do on Romney Marsh, and Romney Marsh don't do in Oxfordshire. **1923** [see *LEICESTER]. **1926** *Daily Colonist* (Victoria, B.C.) 16 Jan. 15/5 Two handsome Romney rams, proud and bellicose,..were landed today. **1950** *N.Z. Jrnl. Agric.* Jan. 15/3 The Romney sheep was by then [*sc.* about 1890] the dominant breed. **1957** [see *KENT *sb.*[3]]. **1974** T. ALLBEURY *Snowball* xiv. 72 It's been said that Romney Marsh sheep are bred only for survival. **1976** *Leicester Advertiser* 26 Nov. 11/5 The Animal Breeding Research Organisation is importing four hairy Romney rams from New Zealand. **1978** *Jrnl. R. Soc. Arts* CXXVI. 590/2 Factors limiting lamb production in New Zealand Romneys.

romp, *sb.* Add: **2. b.** Phr. *in a romp*, with the greatest ease.

1901 J. RALPH *War's Brighter Side* xv. 249 One said to me, as he pointed at Maghersfontain Kopje, 'Set a brewery upon top of that and my regiment will take the place in a romp.' **1904** 'O. HENRY' in *Everybody's Mag.* Feb. 192/1 Rompiro will win in a romp... We'll carry the country by 10,000.

3. *attrib.*, as *romp-suit* = *ROMPER 2.

1961 W. SANSOM *Last Hours of Sandra Lee* iv. 70 A fresh-faced girl in a romp-suit.

romp, *v.* Add: **2. a.** Also *transf.*

1928 *Sunday Express* 22 July 1/1 The child of 1928 simply romps through papers which were 'teasers' for the child of 1914. **1951** *People* 17 June 2 Petula Clark, who romps away with her first grown-up part with all-star honours. **1960** *Times* 16 May 19/1 He and Davies romped to a 5–1 lead. **1964** *Amer. Folk Music Occasional* I. 40 Those..lists of Broadway..romp up past the million mark in a few months. **1968** J. SANGSTER *Touchfeather* ii. 8, I romped through the training, passing out eventually with the highest marks anyone could remember. **1976** *Southwest Times Record* (Fort Smith, Arkansas) 20 Sept. 1 B/1 The Dallas Cowboys overcame a rash of early errors and romped to a 24–6 National Football League victory over the New Orleans Saints.

b. Also *transf.* (Earlier and later examples.)

1881 E. W. HAMILTON *Diary* 28 Oct. (1972) I. 178 The Liberal candidate, though a Roman Catholic and not supposed to be a good candidate, simply 'romped in'. **1910** A. BENNETT *Clayhanger* III. xvii. 444 A demy poster ..to inform the public that the true friend of the public was 'romping in'. **1927** *Observer* 18 Sept. 17/2 It is a bad blow to official Labour that Mr. Larkin should have romped home in north Dublin. **1950** *Sport* 22–28 Sept. 4/1 On Saturday the Forest 'stiffs' romped home to a 5–1 victory over Halifax. **1974** *Times* 2 Mar. 4/7 Mr Thorpe..romped home in his own constituency while Liberals elsewhere were generally less successful. **1977** *West Briton* 25 Aug. 18/1 Troon were handsome winners on Saturday when, facing a Hayle score of 200 for seven, they romped home by seven wickets.

romper. Add: **2.** Usu. *pl.* Also *romper suit*. A one-piece garment for a child to wear at play; a casual one-piece garment worn esp. by young women. (See also quots. 1941, 1943.)

1909 *Dialect Notes* III. 364 *Rompers, n. pl.* A one-piece garment for children to play in. **1915** R. W. LARDNER *Bib Ballads* 3 Hark! A voice from the easy chair: 'He hasn't a romper that's fit to wear.' **1922** *Moving Picture Stories* 23 June 24/1 The dainty bit of femininity, by the way, wore a suit of gingham rompers. **1922** *Westm. Gaz.* 20 Oct. 9 (Advt.), An attractive romper suit for a small child is made of white washing material. **1928** L. NORTH *Parasites* 18 Many of them wore sweaters that would have put Joseph's coat to shame. And very long, very baggy knickers, Hollywood rompers. **1941** *Amer. Speech* XVI. 186/2 [British Army slang] *Rompers*, battle dress. **1943** 'T. DUDLEY-GORDON' *Coastal Command* 85 Sipping hot coffee as he took off his rompers (combined parachute harness and Mae West life-jacket) he told us of his first night raid. **1964** C. WILLOCK *Enormous Zoo* vii. 128 He wore his one-piece romper suit and his three-foot-wide straw hat. **1970** *Women's Wear Daily* 23 Nov. 31/2 We see little rompers..as a possible replacement. **1974** A. GODDARD *Vienna Pursuit* II. 77 A toddler in pale blue rompers.

rompish, *a.* (Earlier and later examples.)

1709 W. KING *Useful Trans. in Philos.* I. 37 The Dance was something Rompish. **1977** *Listener* 5 May 592/1 *Albert Herring* is altogether an awkward, disconcerting affair—so rompish on the surface.

romulea (rǫmiū·liă).. Also **Romulea.** [mod. L. (J. F. Maratti *Plantarum Romuleæ et Saturniæ in Agro Romano* (1772) 13), f. *Romul-us*, name of the mythical founder of Rome.] A small bulbous plant of the genus so called, belonging to the family Iridaceæ, native to coastal regions of southern Europe and South Africa, and bearing yellow, red, or purple flowers resembling a crocus.

1876 J. G. BAKER in *Jrnl. Bot.* XIV. 236 There are specimens in the herbaria either of Kew or the British Museum, with the exception of three of the *Romuleæ*. **1887** *Gardeners' Chron.* 5 Feb. 184/2 The hardier section of Romuleas belonging to the Mediterranean regions are also worthy our attention. **1909** R. FARRER *In Yorkshire Garden* viii. 148, I was quite terrified at the aspect of the Romulea clumps that my kind Cornish friend sent me the other day, so wild, so long and wiry-haired was their aspect. **1928** R. MACAULAY *Keeping up Appearances* ii. 14 Back from the beach stretched grassy slopes, purple and pink with romulea and silene. **1964** A. N. GRIFFITH *Collins Guide to Alpines* 243 Other romuleas, including the less hardy species from S. Africa, will be found described in detail.

‖ **rondavel** (rǫndā·vel). *S. Afr.* Also † **rondabel, ronddawel.** [a. Afrikaans *rondawel.*] A round tribal hut of primitive construction, usu. with a thatched, conical roof. Also *transf.*, a similar simple building used esp. as a holiday cottage; also as an outbuilding on a farm, etc.

1891 J. WIDDICOMBE *Fourteen Yrs. in Basutoland* 84 Mr. Charles Bell had very kindly engaged a Mosuto..to build us a round hut, or *rondavel*, as the whites usually call it. **1900** A. H. KEANE *Boer States* p. xviii, Rondabel, ronddawel, a round hut..; is now an outhouse detached from the dwelling, and used as a kitchen. **1904** A. WILMOT *Life & Times Sir R. Southey* iii. 69 At present the Landdrost lies in a 'Rondavel' of reeds and mud. **1910** J. BUCHAN *Prester John* iii. 53 There were some twenty native huts, higher up the slope, which the Dutch call *rondavels*. **1924** *Chambers's Jrnl.* Jan. 53/1 At the scene of operations rondavels have been constructed to accommodate the workers. **1936** 'THE IDLER' *Rolling Home* xxxi. 385 It consisted of a dozen rondavels grouped round a central thatched dining-room... A rondavel is a circular room built of brick or mud with a door and windows and is roofed with stout thatch. **1951** R. CAMPBELL *Light on Dark Horse* ii. 51 Even our Governor-Generals sleep, in the hot weather, in thatched rondavels walled with a mixture of cow-dung and mud. **1958** M. SPARK *Go-Away Bird* 102 She had used to squat with old Makata..outside his large rondavel. **1965** *Spectator* 8 Jan. 35/3, I slept in one of the rondavels..vacated for me for the occasion. There are twenty of these—tin huts twenty feet in diameter and partitioned to form two tiny semi-circular rooms. **1973** G. DURRELL *Beasts in my Belfry* vi. 121 The [bears'] dens, which were scattered among the bramble bushes that filled the enclosure, were beehive-like rondavels of stone covered with earth and turf. **1976** *Vogue* Jan. 114/3 Antigua..the Anchorage Hotel.. accommodation units, ranging from rondavels to air-conditioned rooms with patios.

‖ **rond de cuir** (roṅ də küir). [Fr., lit. 'circle of leather'.] A round leather cushion, commonly used on office chairs in France; hence *transf.*, a bureaucrat.

[**1893** G. COURTELINE (*title*) Messieurs les Ronds-de-Cuir.] **1915** W. J. LOCKE *Jaffery* viii. 101 Do you think a leather seat for that hard wooden chair—what the French call a *rond-de-cuir*—would very greatly impair the poor fellow's imagination? **1938** *Times Lit. Suppl.* 28 May 368/3 Into the next twelve years he crowded all his life's work, his volumes of stories and novels..his good-bye to a *rond de cuir*. **1963** I. FLEMING *On H.M. Secret Service* xxiv. 259, I am just a pilot. I am not a 'rond de cuir', a chairborne flyer. **1969** *Punch* 5 Mar. 363/2 The island in this Octave is Barra, where he is in charge of the Home Guard and conducts a running fight on its behalf with the *ronds de cuir* of Whitehall. **1975** *Listener* 4 Sept. 314/4 How many *ronds-de-cuir* in peripheral *mairies*.. must have lived through Robespierre!

‖ **rond de jambe** (roṅ də ʒaṁb). *Ballet.* Pl. **rond(s) de jambes, ronds de jambe.** [Fr., lit. 'circle of the leg'.] A circular movement of the leg in dancing. Freq. in *Comb.* (see quots.).

1828 R. BARTON tr. *Blasis's Code of Terpsichore* II. 101 Suppose it is the left leg that stands on the ground whilst the right, in the second position, is prepared for the movement, make it describe a semicircle backwards, which brings your legs to the first position, and then continue on the sweep till it completes the whole circle, ending at the place from whence it started. This is what we technically term *ronds-de-jambe*. *Ibid.* 102 The practice of *grands et petits battemens*, the *rond-de-jambes* on the ground and in the air,..&c. **1889** G. B. SHAW in *Star* 4 Oct. 2/4 The *entrechats, battements, ronds de jambes, arabesques, élévations*, and what's-his-names of the art of theatrical dancing. **1922** BEAUMONT & IDZIKOWSKI *Man. Classical Theatr. Dancing* II. i. 34 *Ronds de Jambe à Terre* serve to enable you to turn your leg well outwards. *Ibid.* 53 The celebrated dancers Gardel and Vestris are credited with the invention of the *rond de jambe en l'air*. **1930** CRASKE & BEAUMONT *Theory & Pract. Allegro in Classical Ballet* II. 70 Execute a *Double Rond de Jambe en dedans, sauté*, that is:- Spring upwards off the *right* foot. While the body is in the air—Execute with the *left* foot a *Double Rond de Jambe en l'air en dedans*. **1975** *New Yorker* 26 May 91/1 His passé leg in a multiple pirouette sweeps through rond de jambe en l'air into the opening battement of a series of grands jetés en tournant.

ronde. Restrict *Typog.* to sense in Dict. and add: **2.** A round dance; a dance in which the participants move in a circle or ring.

[**1931** G. L. NUTE *Voyageur* 41 The call for *la ronde* was issued. This dance was another customary part of the journey, and it was entered into heartily despite the moralizing tone of the verses.] **1950** MARCEL-DUBOIS & ANDRAL *Dances of France* I. 21 In the Bourbonnais a peasant wedding is the great occasion for traditional Rondes, ring dances, by all the guests after the banquet. One such Ronde is..round a fire. **1974** *Islander* (Victoria, B.C.) 30 June 16/3 Wives and daughters of the traders joined in, and Warre partook of a Canadian ronde, a dance in which, he wrote, 'your admiration of your partner is expressed by kissing her openly as often as she will permit'. **1977** *Early Music* July 431/3 There are no descriptions of dances specifically called 'Ronde'. The name implies a round dance and rhythmically they fit either a Branle..or an Almaine.

3. A round or course of talk, activity, etc.; a treadmill. Cf. *ROUND *sb.*[1] 13c.

1957 *Economist* 19 Oct. 196/1 The subject has been completely submerged in the economic discussions which dominate the current *ronde*. **1977** *Times Lit. Suppl.* 1 Apr. 401/4 Heinz already represented the first step away from what was ultimately unbearable about the homosexual *ronde*.

rondelier (rǫndēli°·ɹ). *nonce-wd.* [f. RONDEL + -IER.] A poet who composes rondels.

1878 G. M. HOPKINS *Lett. to R. Bridges* (1955) 49, I am very glad to hear the Rondeliers have come to see the beauty of your poetry.

‖ **rondeña** (rǫnde·nyă). [Sp.] A variety of song or dance native to Ronda in Andalusia.

1883 GROVE *Dict. Mus.* III. 599/2 Songs and dances often derive their names from the provinces or towns in which they are indigenous; thus *rondeña* from Ronda. **1954** *Ibid.* (ed. 5) III. 372 Most forms [of Andalusian song]..have four lines of eight syllables, and these include forms such as *granadinas, rondeñas*,..descended directly or indirectly from the *fandango*. **1967** 'LA MERI' *Spanish Dancing* (ed. 2) vi. 82 The Rondeñas originated as a lover's serenade under the window of his sweetheart, as did the Tarantos of the Levant.

‖ **rondeur** (roṅdör). [Fr.] *pl.* Roundness, rounded forms or lines; *spec.* the curves of the female body.

1938 H. G. WELLS *Apropos of Dolores* iii. 113 A vast majolica plaque insisting upon the Rape of the Sabines, but always recalling to my mind, I don't know why—the rondeurs I suppose—that bustling cheese market at Alkmaar. **1966** *Guardian* 5 Aug. 8/4 These wide coats and dresses..will be able to take our rondeurs in their swing.

rondine (rǫ·ndīn), *a. nonce-wd.* [f. F. *rond* ROUND + -INE[4], after *blondine*.] Made round, rounded.

1923 E. SITWELL *Bucolic Comedies* 70 Fat blondine pearls Rondine curls Seem.

‖ **rond-point** (roṅ,pwaṅ). [Fr., f. *rond* ROUND + *point* centre.] **a.** In a garden: a circular space or centre whence paths radiate. **b.** In a town or city: a circus or roundabout where roads converge.

1884 H. JAMES in *Atlantic Monthly* May 631/2 A *jardin français*..with little blue-green perspectives and alleys and *rond-points*. **1903** A. H. BEAVAN *Tube, Train, Tram, & Car* x. 120 Sloane Street, where anyone approaching town by way of Kensington, meets the first of the numerous metropolitan 'rond-points'. *Ibid.* 121 A few doors from the 'rond-point' in Brompton Road. **1948** *Archit. Rev.* CIII. 158/2 Foremost in his mind he must have had such questions as where to place his rond-points and their radiating streets, and how to co-ordinate and integrate the various units of the plan. **1964** *Shell Gardens Bk.* 169 *Rond-Point*, a circular space or clearing from which avenues and alleys diverge or upon which they converge and from which one may get vistas of various parts of a garden or park. **1967** C. HUSSEY *Eng. Gardens & Landscapes 1700–1750* v. 41 The enclosing shrubberies were threaded by walks, straight for the most part but irregularly aligned, connecting sundry *rond-points* and mounds, to debouch at unexpected angles into the glade.

Roneo (rǭᵘ·nio), *sb.* Also **roneo.** [f. the initial letters of ROTARY *a.* and *Neostyle* (see quot. 1901 below).] The proprietary name of various kinds of office equipment, esp. a duplicating machine. Freq. *attrib.* and *Comb.*

1901 *Trade Marks Jrnl.* 27 Nov. 1182 Roneo... Paper.., stationery and bookbinding. Neostyle Manufacturing Company, Limited. **1914** D. FRASER *Winning a Primitive People* III. xxvi. 272 A native clerk is there.. turning on the Roneo hundreds of copies of some circular to the teachers. **1919** *Trade Marks Jrnl.* 13 Aug. 1138 Roneo... Type printing machines, type setting machines, and embossing machines. Roneo, Limited,.. London. **1920** *Glasgow Herald* 3 Dec. 10 The staff of typists and Roneo operators required is very numerous and well paid. **1926** *Socialist Rev.* June 10 No printed newspapers (a few roneo bulletins), no trains. **1934** *Times Lit. Suppl.* 8 Mar. 162/3 *Roneo*.—As one must have it, why not have the derivation—Rotary Neostyle? **1941** E. R. EDDISON *Fish Dinner* xiii. 229 I'll go on for a bit: get my covering memorandum into shape... You've got the annexes all off the roneo now, have you? **1947** *Trade Marks Jrnl.* 5 Mar. 120/1 Roneo... Duplicating apparatus, duplicating machines..Roneo Limited,.. Romford, Essex, and.. London,..; manufacturers and merchants. **1950** *Official Gaz.* (U.S. Patent Office) 12 Dec. 377/1 *Roneo*. Applicant claims ownership of British Registration No. 241,483 dated Oct. 23, 1901, and United States Registration No. 182,682. For duplicating machines and their parts, duplicating apparatus and their parts. **1958** S. HYLAND *Who goes Hang?* xxxi. 136 The complexities of the Roneo strip-index on which he was working. **1973** J. WAINWRIGHT *High-Class Kill* 119 An official statement; typed-out, Roneo-copied. **1977** *Gay News* 24 Mar. 22/4 One available in bookstores is Alain Huet's *Agence Tasse*, a roneotyped newssheet.

Hence as *v. trans.*, to copy or reproduce with a Roneo duplicating machine (usu. *pass.* with pa. pple. *roneo'd*); **ro-neo'd** (also **roneo-(-)ed**) *ppl. a.*, **ro-neo-ing** *vbl. sb.*

1921 *Spectator* 7 May 584/1, I have had the memorandum 'Roneo'd' for circulation among near acquaintances. **1926** *Contemp. Rev.* June 682 Newspapers were reappearing in weird type-written or roneo-ed form. **1928** *Bull. Internat. Fed. League of Nations Societies* May–June 5 Among the documents which have been translated, roneoed and distributed by the Secretariat may be cited the following important Memoranda. **1934** *Planning* I. XXIII. 8 In addition to this broadsheet a roneoed bulletin goes round fortnightly to the hundred working members telling them what each group is doing. **1935** *Punch* 6 Mar. 262/2 He said he liked the little poem and was having some copies 'roneo'd'. **1940** W. S. CHURCHILL *Second World War* (1949) II. 631 The officials concerned in roneo-ing the various circulations. **1956** *Eng. Stud.* XXXVII. 146 The preface to the present (roneo'd, not printed) book. **1966** *Listener* 23 June 915/1 They studied the roneoed exam papers. **1974** D. SEAMAN *Bomb that could Lip-Read* xv. 144 Back would come a roneoed letter.

Rong (roŋ), *sb.* and *a.* [Native name.] = *LEPCHA *sb.* and *a.*

1840 A. CAMPBELL in *Jrnl. Asiatic Soc. Bengal* IX. I. 379 The real Lepcha, or Rong proper, has no tradition whatever..connected with the advent of his tribe into this part of the world. **1854** J. D. HOOKER *Himalayan Jrnls.* I. v. 127 They, or at least some of their tribes, call themselves Rong, and Arratt, and their country Dijong. **1876** G. B. MAINWARING *Gram. Róng (Lepcha) Lang.* p. vii, The proper name of the Lepchas, as they call themselves, is—Róng. *Ibid.* I. 1 The Róng (Lepcha) Alphabet may be divided into two parts. **1909** G. A. GRIERSON *Linguistic Survey of India* III. I. 233 The Lepchas are considered as the oldest inhabitants of Sikkim... They call themselves Róng... The number of speakers of Róng in Sikkim and Darjeeling were..estimated. *Ibid.*, Róng literature comprises Buddhistic and other religious books. **1938** G. GORER *Himalayan Village* i. 35 The Lepchas do not call themselves Lepchas; they call themselves *Rong*.

rongeur (roṅʒö·ɹ). *Surg.* [a. F. *rongeur* gnawing, a rodent, f. *ronger* to gnaw.] A strong surgical forceps with a biting action, used for removing small pieces from bone. Also *rongeur forceps*.

a **1884** KNIGHT *Dict. Mech.* Suppl. 764/1 Post's rongeur is specifically for the mastoid bone. **1888** *Buck's Handbk. Med. Sci.* VI. 176/2 The gnawing, or rongeur, forceps are necessary for the removal of the edges of bone and of

diseased parts not otherwise accessible. **1908** J. W. SLUSS *Emergency Surg.* II. iv. 401 Provide, besides the ordinary instruments, Rongeur forceps, a mallet and chisel, or a trephine. *Ibid.* 405 The dura is now exposed, and if the opening..needs to be enlarged, the dura should be detached from the edge of bone and the chisel or rongeur employed. **1927** J. B. MACALPINE *Cystoscopy* xi. 160 The cystoscopic rongeur..may be used to break up stones which are very soft and friable. **1938** D. MUNRO *Cranio-Cerebral Injuries* vii. 90 At least 1 large and 1 small biting rongeur. **1966** G. L. HOWE *Minor Oral Surg.* xi. 221 The ease and speed..are due to the use of the side-cutting rongeurs (alveolotomy shears).

‖ **ronggeng** (rǫ·ŋgeŋ). [Malay.] **a.** A dancing-girl in Malaysia. **b.** A form of Malaysian popular dancing, often accompanied by singing.

1817 T. S. RAFFLES *Hist. Java* vii. 342 The common dancing girls of the country..are called *róng'geng*, and are generally of easy virtue... The *róng'gengs* accompany the dance with singing. **1849** ONG-TAE-HAE *Glance at Interior of China* 57 Native actresses are called ronggengs ..they flourish a paper fan, sing native songs, and perform savage dances. **1910** R. J. WILKINSON *Papers on Malay Subjects: Life & Customs* III. 28 A *ronggeng* sings and acts. **1927** R. J. H. SIDNEY *In Brit. Malaya Today* xxiv. 281 We were treated to a pukkah Malay *ronggeng*. **1966** G. BLACK *You want to die, Johnny?* vi. 112 We got radio Binton, with *ronggeng* music. **1972** M. SHEPPARD *Taman Indera* 89 The most popular of these [dances] was called *Ronggeng*—a word which now means a professional dancing girl who dances and sings. **1976** LD. HOME *Way Wind Blows* vii. 116 Tunku Abdul Rahman..taught me —or rather tried to teach me—the ronggeng, the Malaysian dance in which the male, as far as I could see, plays a subordinate role to the female with hilarious results.

rongo-rongo (rǫ·ŋgo,rǫ·ŋgo). *Archæol.* [Native name.] Hieroglyphic signs or script found on wooden tablets on Easter Island, a Chilean dependency in the eastern Pacific Ocean; the art of incising these. Also *attrib.* and *ellipt.*

1919 K. ROUTLEDGE *Mystery of Easter Island* xvi. 243 The tablets, known as 'kohâu-rongo-rongo', were an integral part of life on the island. *Ibid.* 244 Every clan had professors in the art who were known as rongo-rongo men ('tangata-rongo-rongo'). *Ibid.* 249 Kaara was servant to the Ariki, and had been taught rongo-rongo by him. ..The matters with which..the rongo-rongo would deal, such as genealogies, lists of ariki, or the wanderings of the people. **1947** D. DIRINGER *Alphabet* viii. 137 The script *rongo-rongo* was the monopoly of organized teachers; every clan had its own 'writing professors', that is, experts in the art who were known as *tangata-rongo-rongo*, 'rongo-rongo-men'. *Ibid.*, A less elaborate kind of rongo-rongo was called *tau*. **1957** M. BULLOCK tr. *Métraux's Easter Island* xii. 188 The spear carried by the leader..recalls the staff sometimes borne by the Easter Island *rongorongo*. *Ibid.*, The name given to these [hieroglyphic] tablets, *kohau rongorongo*. **1958** T. HEYERDAHL *Aku-Aku* vi. 164 The cave contained every possible thing except *rongo-rongo*.

‖ **ronin** (rǭᵘ·nin). Also with capital initial. [Jap.] In feudal Japan, a lordless wandering samurai; an outlaw. Also *transf.* in recent use, a Japanese student who has failed and is permitted to retake a university (entrance) examination.

1871 A. B. MITFORD *Tales of Old Japan* I. 4 The word *Rônin*..is used to designate persons of gentle blood, entitled to bear arms, who have become separated from their feudal lords. *Ibid.* 18 Then the Rônins lost patience. **1876** W. E. GRIFFIS *Mikado's Empire* I. xxvii. 278 When too deeply in debt, or having committed a crime, they left their homes and the service of their masters, and roamed at large. Such men were called *rōnins*, or 'wave-men'. **1899** KIPLING *From Sea to Sea* I. xxi. 415 And now let us go to the tomb of the Forty-Seven Ronins. **1947** R. BENEDICT *Chrysanthemum & Sword* vii. 138 The huge invincible *ronin* (a lordless samurai who lives by his own wits), the hero Benkei. **1967** D. & E. T. RIESMAN *Conversations in Japan* 17 Many had been *ronin* (the name given those who try again and again over a period of years to pass the exams), and finally when they made it were exhausted. **1970** *Observer* (Colour Suppl.) 8 Feb. 32/2 High school students who fail the university exam and are waiting to try again are called *ronin*, a reference to the landless samurai of old Japan which clearly describes their unhappy displaced position in a chronically status-sensitive society. **1974** *Encycl. Brit. Micropædia* VIII. 663/1 *Ronin*, in Japan, masterless samurai (warrior aristocrats) of the Kamakura (1192–1333) and Tokugawa (1603–1867) periods who were often vagrant and disruptive and sometimes actively rebellious.

ronk (rǫŋk), *a. dial.* [var. RANK *a.*] Unmanageable, refractory, unruly; depraved, libidinous; cunning. Hence **ro·nkness.**

1877–1905 in *Eng. Dial. Dict.* **1908** E. M. SNEYD-KYNNERSLEY *H.M.I.* xvii. 191 'Well, sir, he is not a bad sort of boy, but he is—er—er—' I broke in to his relief: 'His mother says he is *ronk*.' The master jumped at it: 'That's exactly what he is, sir: he's ronk.' *Ibid.*, 192 Choir-boys have an established reputation for ronkness. **1972** *Times* 31 Oct. 14/7 'Ronk' means what the yokels of London would describe as 'randy'.

Ronson (rǫ·nsən). The proprietary name of a brand of cigarette lighter. Also *attrib.*

1929 *Trade Marks Jrnl.* 11 Sept. 1516/1 Ronson... Pyrophoric lighters. The Ronson Art Metal Company, Limited. **1957** C. MACINNES *City of Spades* I. vii. 44, I..held out my Ronson...when, hey presto! the lighter was flicked from my hand. **1961** C. WILLOCK *Death in Covert* xi. 199 Mr Goss took his Ronson lighter from his pocket and flicked it into flame. **1972** O. SELA *Bearer Plot* ii. 18, I leaned forward with my Ronson. The flame whooshed up.

Röntgen. Add: see *ROENTGEN (now the more usual spelling).

röntgenite, var. *ROENTGENITE.

Ronuk (rǫ·nɒk), *sb.* Also **ronuk.** The proprietary name of a make of polish; *spec.* a brand of floor polish. Hence as *v. trans.*, to polish with Ronuk. Also **ro·nuked** *ppl. a.*; **ro·nuker,** one who uses Ronuk.

1896 *Trade Marks Jrnl.* 8 Apr. 325 Ronuk... Polishing preparations. Thomas Mottley Fowler and Thomas Horace Fowler, trading as T. M. Fowler,.. Brighton; manufacturers. **1912** *Daily Chron.* 5 Mar. 4/5 'Ronuk' imparts a brilliant polish. *a* **1913** 'BARON CORVO' *Desire & Pursuit of Whole* (1934) i. 3 The profane vulgar who want..brown boots properly ronuked, and linen stiffly starched. **1916** *Yorkshire Post* 29 June 10/3 In one hall three or four shovel-fuls of dirt are taken up by the ronukers as against one in any other hall, and twice as long is taken to ronuk this hall than any other hall, and three times the amount of ronuk is needed. **1918** KIPLING in *Story-Teller* Dec. 227/1 You'll often find half a dozen Brethren..polishing and ronuking and sweeping everything they can get at. **1927** W. E. COLLINSON *Contemp. Eng.* 66 One curious development of some nouns is their power to form verbs: to zog a stain off (with a scouring powder), to ronuk [rɔnʌk] a floor (polish). **1929** *Trade Marks Jrnl.* 6 Mar. 404/2 Ronuk... Floor polish. Ronuk, Limited,.. Portslade,.. London,.. Manchester; and.. Cape Town, South Africa. **1955** N. BALCHIN *Fall of Sparrow* ii. 62 He was a big, bulky man, completely bald, and it was believed that the house matron polished his head every day with Ronuk.

'roo, roo *sb.*[2] (rū). *Austral. colloq.* [Shortened form of *kangaroo.*] **a.** = KANGAROO *sb.* 1.

1904 'S. RUDD' *Sandy's Selection* 11 Dead 'roos were common enough, but seldom was a live one thrown in Sandy's way. [**1926**: see *KANGA[2].] **1933** *Bulletin* (Sydney) 30 Aug. 21/3 The whites have a kindly feeling for the 'roo. **1945** *Coast to Coast 1944* 80 First time he's seen a blasted roo. **1966** *Southerly* XXVI. 110 Possums and roos to trap. **1977** *Caravan World* (Austral.) Jan. 59/1 The river had brought emus and 'roos close to the road. **1979** *Daily Tel.* 23 Aug. 11/7 The baby roo tries to get herself adopted by other animals but they prove to be unsatisfactory means of transport.

b. *attrib.* and *Comb.*, as *roo bus, meat, shooter, steak*; **roo-bar** (see quot.); **'roo rat** = KANGAROO-RAT 1.

1976 *Car Facts & Feats* (ed. 2) III. 158 (*caption*) The cage in front of the head lamps is affectionately known as a 'roo-bar'. **1968** K. WEATHERLY *Roo Shooter* 12 The roo bus swept round a corner, into full view about twenty yards away; the shooter hit the brakes, depressed the clutch and grabbed the ·22 all in the same instant. **1977** *Weekly Times* (Melbourne) 19 Jan. 33/5 Not bad this 'ere veal. Not a touch on roo meat though. **1947** I. L. IDRIESS *Isles of Despair* xxxii. 212 Hungrily she.. satisfied her craving for land flesh. Kangaroo, wallaby.. flying fox, 'roo rat, porcupine. **1980** *Age* (Melbourne) 1 Apr. 11/1 The national controversy between conservationists and those who want to increase the commercial exploitation of the kangaroo, notably the professional 'roo shooters and the farmers, [etc.]. **1926** K. S. PRICHARD *Working Bullocks* 21 When he had cooked and eaten the piece of 'roo steak he had been carrying.

Hence **roo** *v.*[2] *intr.*, to hunt kangaroos.

1932 K. S. PRICHARD *Kiss on Lips* 82 Rooin' this week, Colonel? ... Cripes, that's something like!

roocoocoo (rūkūkū·), *v.* Also **roocooroo** (rūkūrū·). [Imitative.] *intr.* Of a pigeon or a dove: to coo.

1922 JOYCE *Ulysses* 225 The..porch..where pigeons roocoocooed. **1960** C. DAY LEWIS *Buried Day* ii. 31 The roo-coo-rooing of doves.

rood, *sb.* Add: **6.** **Rood-fair** (further examples).

1931 J. BUCHAN *Blanket of Dark* 86 Old John Naps was at the Rood Fair on Barton Heath. **1957** *Dumfries & Galloway Standard* 26 Jan. 3/2 The 'Reed' Fair, as we pronounced it in our Dumfries dialect—'Reed' was a corruption of Rude or Rood or Cross.

roode bec etc. (varr. of words of Afrikaans origin normally spelt *rooi-*): see *ROOIBEKKIE etc.

roof, *sb.* Add: **1. a.** (Later examples of pl. *rooves.*)

1903 *Dialect Notes* II. 352 *Roof, n. pl.* rooves. Common plural in Mass. **1938** C. HIMES *Black on Black* (1973) 165 W'en de panic cum an' de Lawd tek yo' food an' yo' clothes an' de rooves off'n yo' haids, den laff. **1939** [see *council (housing) estate* s.v. *COUNCIL 17].

c. Also (not usu. *poet.*), a home, a household; a dwelling-place.

1853 C. BRONTË *Villette* I. iii. 29 The evening, by restoring Graham to the maternal roof (his days were passed at school), brought us an accession of animation. **1922** D. L. SAYERS *Let.* 18 Dec. in J. Brabazon *Dorothy L. Sayers* (1981) ix. 96 He simply has not a red cent or a roof. **1979** J. RATHBONE *Euro-Killers* iii. 34 She had been happy to give them a free roof until they got work.

e. In phrases (chiefly *colloq.*). (*a*) *to raise* (or *lift*) *the roof*: to create an uproar, to make a resounding noise; (*b*) *the roof falls in*: something disastrous occurs, everything goes wrong; (*c*) *come off the roof*: don't put on airs; (*d*) *to hit the roof* = *to hit the ceiling* (*CEILING *vbl. sb.* 5b); (*e*) *to go through the roof*: to become very angry (see also sense *2a).

(*a*) **1860** M. J. HOLMES *Cousin Maude* 57 Ole master'll raise de ruff, case he put 'em away to sell. **1894** 'MARK TWAIN' in *Century Mag.* June 233/1 She was here to watch the trial now, and was going to lift up just one 'hooraw' over it... 'When dat verdic' comes, I's gwine to lif' dat *roof*, now, I *tell* you.' **1905** *Eng. Dial. Dict.* V. 147/2 *Oxf.* Do be quiet, or you'll raise the roof (G.O.). **1922** WODEHOUSE *Girl on Boat* xvi. 253, I couldn't get within ten feet of that dog without its lifting the roof off. **1959** *Times* 19 Jan. 3/3 A good song to raise the roof. **1972** J. W. THOMPSON in W. King *Black Short Story Anthol.* 255 She flew from the kitchen like a startled sparrow, her hands perched nervously upon her hips—all set to raise the roof!

(*b*) **1866** D. BOUCICAULT *Flying Scud* in Nicoll & Cloak *Forbidden Fruit & Other Plays* (1940) 172 She undertook with all her weight to sit upon my knee... Fourteen stun six, I thought the roof had fell in. **1958** J. MORGAN *Expense Account* ii. 26 And it all worked out exactly right—up to the moment he walked into his office. Then the roof fell in. **1976** H. MACINNES *Agent in Place* xvii. 191 Georges said, 'I think the roof just fell in.' Tony had no reply. For once he was quite speechless.

(*c*) **1895** W. P. RIDGE *Minor Dialogues* ix. 86 She took up such a 'igh and mighty attitude..so I says to her, I says, 'Come off the roof.'

(*d*) **1925** FRASER & GIBBONS *Soldier & Sailor Words* 245 *Roof, to hit the*, to get into a temper. **1928** J. P. MCEVOY *Show Girl* ix. 133 Milton gave me a couple of drinks early in the evening out of his flask and Jimmy hit the roof. **1971** V. CANNING *Firecrest* x. 149 The P.M. and his cabinet..would hit the roof if they knew half of the things that went on.

(*e*) **1958** *Spectator* 25 July 133/1 Would it have hurt if someone had done it to you before? You'd have gone through the roof? **1975** J. SYMONS *Three Pipe Problem* xviii. 179 The company are simply wild. They have gone through the roof.

2. a. Also, = *CEILING *vbl. sb.* 6d. *to go through the roof* (and varr.), of bids, prices, sales, etc.: to surpass the expected limit, to reach extreme heights.

1939 *Richmond* (Va.) *Times-Dispatch* 16 Aug. 17/8 Spokesmen for the shellers contended that since the price pegging program put a 'roof' on the price they must pay for peanuts they were entitled to a 'floor' against possible losses from 'innocent' over-buying. **1946** E. HODGINS *Mr. Blandings builds his Dream House* viii. 118 The Knapp sales curves were going through the roof. **1947** *Forum* (Johannesburg) 24 May 15/3 The Labour Party continues to snipe at the Government for refusing to take the roof off the maize price. **1962** *Listener* 28 June 1113/2 (Advt.), Starting salary from £11.10.0 per week to £12.16.0 per week according to qualifications, rising to roof of £14.2.0 per week. **1965** *New Statesman* 16 July 101/4 (Advt.), Starting salary £2,185–£2,445 according to relevant experience and qualifications rising to a roof of £2,835. **1972** *Times* 24 Oct. 10/3 Only a few special treasures were bid through the roof. **1973** *Times* 30 Oct. 19/6 On lots that were rare and undamaged they [*sc.* prices] went through the roof.

c. (Later examples.) Also, *spec.* in *Mountaineering* (see quot. 1963[2]).

1963 A. GREENBANK *Instructions in Rock Climbing* ix. 98 On the lip of an overhang which has no footholds immediately below... You jockey one boot up the rock, pressing on the flat face, then throw a knee over the edge of the 'roof'. *Ibid.* 151 *Roof*, the underside of an overhang. **1972** D. HASTON *In High Places* viii. 94 After an easy first pitch there was a series of overlapping roofs leading to a big ledge, followed by a line of cracks and chimneys to the top. **1976** *Newmarket Jrnl.* 16 Dec., Left-back Mickey Fordham latched on to a pass from Eley to fire home a third into the roof of the net.

d. Also (*Austral.*), the stratum lying immediately over material that contains opal.

1931 M. S. BUCHANAN *Prospecting for Opal in Australia* 8 Almost all the sheet of potch containing opal lies within two ft. from the roof. **1960** *People* (Broadway, New South Wales) 27 Apr. 51 Pipe opal..is mostly found in soft, white clay between one and six inches below the overlying sandstone 'roof'.

e. A high mountain range or plateau; the highest part (*of* a region); *the roof of the world* [tr. Wakhani *bam-i-dunya*], orig. applied to the Pamirs, later also to Tibet or the Himalayas.

1842 *Chinese Repository* XI. 143 The Wakhanis name this plain Bam-i-Duniah, or 'Roof of the World', and it would indeed appear to be the highest table-land in Asia. **1876** T. E. GORDON (*title*) The roof of the world, being the narrative of a journey over the high plateau of Tibet to the Russian frontier and the Oxus sources on Pamir. *Ibid.* ix. 121 We were now about to cross the famous 'Bam-i-dunya', 'The Roof of the World', under which name the elevated region of the hitherto comparatively unknown Pamir tracts had long appeared in our maps. **1889** G. N. CURZON *Russia in Central Asia* v. 144 Descending from the hidden 'Roof of the world', its waters tell of forgotten peoples and whisper secrets of unknown lands. **1902** D. G. HOGARTH *Nearer East* 31 The course of this ridge in the Anatolian roof..determines the parting of all the waters. **1953** J. MASTERS *Lotus & Wind* xix. 235 Few travellers used this route that arched over the roof of the world to link India with Chinese Sinkiang. **1956** G. N. PATTERSON *God's Fool* i. 13, I sat there in that remote valley in Tibet where no white man had ever been, fifteen thousand feet above sea-level on the roof of the world. **1959** *Listener* 15 Jan. 140/3 If you want to give yourself a test of stamina and skill in map reading you can walk the Roof of Wales. **1968** N. Tranter *Cable from Kabul* ii. 29 Look—this area's called the Roof of the World. It's no place for aircraft. **1973** *Guardian* 20 Oct. 13/3 Identification of Church and State in Tibetan Buddhism dates from the 1640s. But the *feel* of an independent way of life on the battlemented roof of Asia is immensely older. **1979** *Yale Alumni Mag.* Apr. 2/3 (Advt.), Snow-capped peaks of the Himalayas along the roof of the world.

f. *Aeronaut.* = *CEILING *vbl. sb.* 6 b. ? *Obs.*
1917 [see *CEILING *vbl. sb.* 6 b]. **1940** *S. P. E. Tract* lv. 193 *Roof* is the zenith of a plane's ascent.

6. a. (Later examples.)
1926 Maines & Grant *Wise-Crack Dict.* 7/1 *Dropping one's roof*, losing one's hat. **1949** R. M. Howe *H. Gross's Criminal Investigation* (ed. 4) viii. 162 *Titfa, roof, bonnet or tile*, hat.

c. An umbrella. ? *Obs.*
1844 E. Hall *Diary* in O. A. Sherrard *Two Victorian Girls* (1966) xi. 106 [A] family roof [umbrella] and a great blanket shawl.

7. a. Also, *roof-board, -capping, comb, -deck, -decking, -glass, -outlet, -pane, -ridge* (earlier and later examples), *-roller, -screen, -shelter, -slab, -space* (later examples), *-terrace, -thatch, -thatching, -truss* (later example).
1848 O. S. Fowler *Home for All* (1851) 90 Since the roof boards cross these rafters, so as to form an arch the other way, surely no greater solidity or strength can be required. **1940** *Chambers's Techn. Dict.* 730/1 *Roof boards*, boards laid on a roof to provide a foundation and an undercovering to the covering materials proper, such as slates, tiles, etc. **1968** O. S. Nock *Railway Enthusiast's Encycl.* 246 (*caption*) G.W.R.: an early example of dining-car service, with rather flamboyant roof-boards, c. 1900. **1977** *36 Home Handyman Projects* (Austral. Home Jrnl.) 74/1 Roof capping is usually put on with cement. After a few years of sun, wind and rain the cement cracks and falls out. **1908** *Encycl. Relig. & Ethics* I. 687/1 On the roof was a roof comb—one of the most distinguishing features of Maya architecture. **1971** *Country Life* 4 Nov. 1219/2 A three-room temple surmounted by an enormous stone roof-comb originally carved with an impressive seated figure. **1947** *Archit. Rev.* CII. 117 The whole of the area on which this house is built is utilized for outdoor functions; above the living-room-kitchen part is an open roof-deck, below the bedrooms a car-port. **1979** *Arizona Daily Star* 5 Aug. (Advt. Section) 20/1 The roof deck affords fantastic views. **1960** *Farmer & Stockbreeder* 16 Feb. (Suppl.) 40/1 When used as the roof-decking to your new buildings or as a suspended ceiling to your existing ones, Stramit keeps temperatures constant. **1978** C. Tomlinson *Shaft* 39 Leaves might fall On to the roof-glass. **1967** *Gloss. Sanitation Terms* (*B.S.I.*) 41 *Roof outlet*, a rainwater fitting, normally provided with a grating, for building into a flat roof to receive rainwater for discharge into a rain-water pipe. **1922** Joyce *Ulysses* 265 Roll of Bensoulbenjamin rolled to the quivery loveshivery roofpanes. **1874** B. F. Taylor *World on Wheels* 218 There is singing everywhere:..from the second rail of the fence, a gust of melody; from the roof-ridge, a solo. **1917** Conrad *Shadow-Line* ii. 79 Here and there in the distance, above the crowded mob of low, brown roof ridges, towered great piles of masonry. **1936** *Discovery* Aug. 251/2 A roof roller [excavated at Tell Duwein, near Jerusalem] was identical in form with that in use today in Palestine. **1971** *Gloss. Terms Fire* (*B.S.I.*) ii. 7 *Roof screen*, a vertical screen fitted internally to the roof of a building to divide the roof into bays, so that smoke and hot gases from a fire are contained within the bay of origin. **1928** D. H. Lawrence *Lady Chatterley* x. 145 Only one or two [chicks]..still dibbed about in the dryness under the straw roof-shelter. **1963** *Gloss. Gen. Building Terms* (*B.S.I.*) 20 *Roof slab*, a slab forming the continuous loadbearing structure of a roof and spanning between supports. **1960** *Farmer & Stockbreeder* 16 Feb. 49/2 (Advt.) Agrecon buildings give..a maximum roof-space. **1970** J. Earl *Tuners & Amplifiers* i. 12 A simple indoor or roofspace aerial would be suitable for the reception of local stations. **1937** *Archit. Rev.* LXXXII. 119 (*caption*) The roof-terrace is paved with 'Paropa' patent slabs. **1912** 'Q.' *Hocken & Hunken* xix. 239 With a rampant climber such as Rosa Devoniensis it is advisable to cut out each autumn, and clean remove some of the old wood; and this is no easy job when early neglect has allowed the plant to riot up and over the roof-thatch. **1968** J. Arnold *Shell Bk. Country Crafts* 329 Hazel rods have many other uses, such as for salmon-traps on the Severn.. and pegs for roof-thatching. **1964** W. L. Goodman *Hist. Woodworking Tools* 197 It may be, however, that the original French word *fermoir* has something to do with the *ferme* or roof-truss.

8. a. *roof-walker*; *roof-levelling* adj.
1863 G. M. Hopkins *Let.* 4 May (1956) 78 Leaving a candle burning, which I thought would keep the roof-walker in..suspense. **1920** W. B. Yeats *Michael Robartes & Dancer* 20 And one bare hill Whereby the haystack and roof-levelling wind..can be stayed.

c. Locative, instrumental, etc., as *roof-clustered, -mired, -rent, -wrecked* adjs.; *roof-ward(s)* adv.
a **1879** W. Howitt in M. Howitt *Autobiogr.* (1889) I. vii. 227 Clouds of smoke..burst from the windows and streamed up roofwards. **1880** 'Mark Twain' *Tramp Abroad* xix. 171 A hill..with..its..roof-clustered cap of architecture. **1922** Hardy *Late Lyrics & Earlier* 283 The bower we shrined to Tennyson, Gentlemen, Is roof-wrecked. **1932** Auden in *Rev. Eng. Stud.* (1978) Aug. 302 A sleepy liftboy swirled us roofward. **1933** C. Day Lewis *Magnetic Mountain* 50 Yet passing derelict mills and barns roof-rent. **1955** A. Clarke *Later Poems* (1961) 48 A cage-bird came among sparrows..Plucked, roof-mired, all in mad bits.

9. roof bolt *Mining*, a tensioned rod anchoring the roof of a working to the strata above; so **roof bolting** *vbl. sb.*, the practice of using roof bolts; **roof-brain**, the cerebral cortex; **roof-climb** *v. intr.*, to climb over the roofs of buildings; so *roof-climber, roof-climbing* vbl. sb.; **roof-drip**, a drip or dripping of water from a roof; **roof-garden** (*a*) (earlier and further examples); freq. applied to a place for eating or entertainment situated on the roof of a building; also *attrib.*; (*b*) (see quot. 1932); **roof-jack**, (*a*) *Canad.*, a pole supporting the roof of a tent; (*b*) *Canad.*, a smoke vent of a chimney; (*c*) *U.S.*, a support for a house painter engaged in painting a roof; **roof-man** = *gutter-man* (*c*) s.v. *GUTTER *sb.*[1] 8; **roof organization** [tr. G. *dachsorganisation*], a parent organization; **roof pendant** *Geol.*, a mass of country rock projecting downwards into an intrusive body such as a batholith; **roof-rack**, a framework upon the roof of a motor vehicle to which luggage is attached; **roof-rail** (see quots.); **roof rat**, *Rattus rattus alexandrinus*, a climbing rat which has a brownish back and greyish underparts; (examples); **roof-scraper** (see quot.); **roof-spotter**, an observer posted at the top of a building to give warning of hostile aircraft; so **roof-spotting**; **roof-top** (in Dict., sense 7 a), used *attrib.* of something situated on top of a building; **roof-watcher** = **roof-spotter*; **roof-water**, rain-water collected from or falling from the roof of a building.
1955 *Trans. Inst. Mining Engineers* CXIV. 849 Roof bolts cannot be used to replace normal supports at the face, but they have been used to advantage to bolt weak immediate beds together or to stronger beds above. **1973** L. J. Thomas *Introd. Mining* vii. 276 It is unwise to rely on roof bolts to hold up the roof bar and the lip of the entry. **1954** *Jrnl. Chem., Metall. & Mining Soc. S. Afr.* LIV 285/1 It was necessary to resort to roof and side bolting and pig netting in order to protect the personnel. **1958** I. C. F. Statham *Coal Mining Pract.* II. ii. 144 An inversion of roof bolting, so-called floor bolting, in which bolts are inserted in holes drilled into the floor has proved successful in reducing floor lift. **1940** C. S. Sherrington *Man on his Nature* vii. 222 Observation indicates rather a roof-brain which overseers subordinate mechanisms. **1960** *20th Cent.* Dec. 549 At the beck and call of those more primitive regions..the roof-brain wakes or sleeps. **1951** 'M. Innes' *Operation Pax* vi. vi. 286 If you roof-climb,..then you just can't..sit in libraries too. **1932** *Daily Mirror* 28 May 6/4 An appeal to undergraduate roof-climbers is made in the 'Cambridge Review'. **1932** G. Greene *Stamboul Train* iii. i. 123 He intended to do no more roof-climbing that night. **1970** R. Lowell *Notebk.* 238 Thud of roofdrip. **1893** M. Holley *Samantha at World's Fair* 286 Why, the very elevator you rode up to the ruff garden on wuz made by a woman. **1895** *N.Y. Dramatic News* 6 July 2/1 The growth of the roof garden idea has undoubtedly tended toward the obliteration of the regular forms of theatrical amusement during the heated term. **1911** *Chambers's Jrnl.* Feb. 113/1 Thus the director..has an opportunity to go to any part of the hotel, from the kitchen to the roof-garden, if he wants to look into matters. **1932** *Santa Fe Mag.* Jan. 34/1 A mallet type or a helper engine on a mountain job is a *sacred ox* or a *roof garden*. **1959** P. Oliver in M. T. Williams *Art of Jazz* (1960) xii. 110 The musicians and singers who had recently enjoyed a booming success at the..dance-halls and roof-gardens, were now finding themselves 'laid off'. **1958** J. G. MacGregor *North-West of 16* iii. 42 Beside the roof-jack it [*sc.* some animal] lay all night, and there..was a bulge it made in the roof of the tent. **1970** R. & J. Paterson *Cranberry Portage* xiv. 88 Cranberry settlement squatted beneath a shifting smoke blanket, upheld by blue-grey columns spinning upwards from a hundred galvanized roof jacks. **1975** *Amer. Speech* 1969 XLIV. 23 *Roofjack, n.* 1. A 12' to 14' plank with cleats affixed to shingles or embedded into roof material and which supports the painter; it serves as a platform from which steeply pitched roofs are painted. 2. A platform made for the pitch of a roof, flexible and made of wood; it is raised level against the pitch of the roof and thus allows the roof to be worked. **1921** Roof-man [see *gutter-man* s.v. *GUTTER *sb.*[1] 8]. **1948** W. R. Benét *Reader's Encycl.* 541/2 s.v. *Institute of France*, It is a roof organization and embraces these five academies. **1906** R. A. Daly in *Bull. Geol. Soc. Amer.* XVII. 336 The whole forms a huge irregular block of roof rock almost completely surrounded and probably underlain by..granite. Such a block, once a downwardly projecting part of a roof in stock or batholith, may be named a 'roof pendant'; it is analogous to the pendant of Gothic architecture. **1934** C. R. Longwell et al. *Outl. Physical Geol.* ix. 173 Batholith: partly uncovered by erosion; P and P′ are masses of invaded country rock projecting deeply downward into the batholith. They are termed roof-pendants. **1961** *Amer. Mineralogist* XLVI. 249 Erosion has lowered the surface to the point where only patches of the metamorphics have been preserved, embedded in a matrix of the quartz diorite as roof pendants or 'curtains'. **1960** *News Chron.* 29 Apr. 10/5 Anyone..can..have his car fitted with..a roof-rack. **1976** P. Cave *High Flying Birds* i. 13 Just load the kites on to my roof-rack, drive down to the nearest Channel ferry service and go looking for the perfect hill. **1930** *Motor Body Building* LI. 105/1 'Coachwork Nomenclature' *Front roof rail*, the cross bar joining the front ends of the cant rail. *Rear roof rail*, the cross bar at the back of the top of the body joining the rear ends of the cant rail. **1969** *Jane's Freight Containers* *1968–69* 18/3 6. Freight container components... 6.8. *Roof rails*, longitudinal structural member situated at the top edge on either side of the freight container. **1882** D. C. Beard *Amer. Boys Handy Bk.* xxiii. 210 The roof rat in the Southern States came originally from Egypt. **1926** *Genetics* XI. 456 The roof rat..is common in the southern states, especially along the seaboard. **1957** C. Breland *Animal Friends & Foes* i. 17 The two more important kinds of house rats are the brown, or Norwegian, rat and the black, or roof, rat. **1971** *New Scientist* 15 Apr. 178/2 The term 'rat' could refer to..the black rat..and its colour varieties such as the roof rat. **1909** J. R. Ware *Passing Eng.* 210/2 *Roof scrapers* (*Theatrical*), gallery boys—especially those standing behind the highest row of seats—and therefore nearer the roof. **1940** *Manch. Guardian Weekly* 18 Oct. 288 If we are appointed roof-spotters to our office, then we must have sharp eyes, good ears, shrewd judgment, and a knowledge of aeroplane types. *Ibid.* 1 Nov. 322 The reports to the Ministry show that the roof-spotting system is welcomed by the workers. *Ibid.*, In the aggregate many thousands of man-hours have been saved by efficient roof-spotting. **1935** *Discovery* Apr. 94/2 He was cautious as to the advent of roof-top landings and city aerodromes. **1961** Conyn & Marten *Bali Ballet Murder* xxi. 220 One of those roof-top nightclubs [in New York]. **1963** *House & Garden* May 55/2 A rooftop garden in the Palazzo Wolkov. **1972** F. Forsyth *Odessa File* i. 6 The underground car park..200 yards from the house where he had his roof-top flat. **1979** *Tucson Mag.* Apr. 65 Distinguished by its..rooftop solar collectors, the system has 621 square feet of flat plate collectors. **1941** Roof watcher [see *JIM CROW[4]]. **1879** *Harper's Mag.* June 134/1 During storms the roof water increases this action. **1910** W. De Morgan *Affair of Dishonour* iv. 55 To him who drinks no water, roof-water and well-water are welcome alike.

roof, *v.* Add: **2. a.** (Further examples.) Also with a person as object and with *over*.
a **1820** Blake *Jerusalem* xix, in *Compl. Writings* (1972) 642 And Los was roof'd in from Eternity in Albion's Cliffs. **1891** E. Arnold *Light of World* 10 The impartial skies Roof one race in. **1935** A. J. Cronin *Stars look Down* i. xxii. 211 The inrush had roofed in the Swelly: for fifty yards a barrier of water blocked the ropeway. **1972** R. Adams *Watership Down* ix. 36 Far around.. stood the orderly rows of beans,..roofing them over.

roofage. (Later *transf.* example.)
1950 M. Peake *Gormenghast* xviii. 108 A posse of professors in a whirl of gowns and a shuffling roofage of mortar-boards.

roofed, *ppl. a.* **1.** (Later examples.)
1909 C. F. G. Masterman *Condition of England* viii. 254 The roofed-in labyrinthine airless ant-heaps of Mr. Wells's nightmare. **1923** D. H. Lawrence *Birds, Beasts & Flowers* 27 All your ponderous roofed-in erection of right and wrong. **1931** [see *BREEZE-WAY, BREEZEWAY]. **1934** L. B. Lyon *White Hare* 11 The roofed glade's a sieve That lets drip through sweet water. **1946** F. Sargeson *That Summer* 107 We all went under a little roofed-in part. **1976** 'G. Black' *Moon for Killers* vi. 83 A roofed-over area that looked almost big enough to be a bus depot.

roofer. Add: **2.** A hat. *slang.* Cf. Roof *sb.* 6 a.
1859 G. W. Matsell *Vocabulum* 74 *Roofer*, a hat. **1941** Baker *Dict. Austral. Slang* 61 *Roofer*, a hat.

3. = *Collins[1]. In full, *hospitable roofer*. Also *attrib.*, as *roofer letter*.
1914 *Sphere* 7 Mar 312/3, I learn from *The Evening News* that that which some call..a 'roofer'—that is to say, a letter of thanks for hospitality—is also known as a Collins. **1935** F. M. Ford *Let.* 27 Sept. (1965) 243 They [*sc.* references] will at least serve as a hospitable-roofer. **1937** G. Frankau *More of Us* vii. 75 That bashful Hebe, too, received fair tip..Sophie, red roses and a roofer letter, With 'kind regards to her young ladyship', And, 'hoping the old boy's catarrh is better'. **1971** C. Williams-Ellis *Architect Errant* vii. 103, I felt that I really ought to write to the owner, Lord Townshend, at least some sort of a roofer or an apology. *Ibid.* xv. 217 In my roofer letter to Vita, I, of course, paid due homage to her so passionately loved Knole.

roofing, (*vbl.*) *sb.* Add: **1. b.** *roofing felt.*
1894 *Country Gentlemen's Catal.* 269 *Roofing felt.*—The best known weather-resisting material yet introduced for roofing purposes. **1929** *Morning Post* 2 Oct. 6/5 These industries include electrical engineering, paint and varnish making, roofing-felts,..and the manufacture of linoleum. **1954** *Paper Terminol.* (Spalding & Hodge, Ltd.) 51 *Roofing felt*, a very porous, soft and thick paper made from low-grade materials and used as a base for impregnation with bitumen, tar, etc.

roofless, *a.* Add: **3.** Applied to poker played with no limit to the raise.
1912 R. W. Service *Rhymes of Rolling Stone* (1913) 104 Your trouble was a roofless game of poker now and then.

roof light. Also roof-light, rooflight. [Roof *sb.*] **1. a.** A flashing warning light that projects upwards from the roof of a motor vehicle. **b.** A small interior light attached to the underside of the roof of a motor vehicle.
1958 'Castle' & 'Hailey' *Flight into Danger* viii. 107 At the turn-off..a police cruiser stood..its roof-light blinking a constant warning. **1961** *Times* 11 July 3/6 Coathooks, rooflights, clocks, and mirrors. **1970** A. Ross *Manchester Thing* 58 The automatic roof light gave me a quick glimpse of two men, then the driver reached up to switch it off. **1977** *Daily Tel.* 13 Jan. 17/1 An RAF helicopter crew..were guided up a valley in thick fog and darkness by the roof-light of a Land-Rover.

2. A panel in or on the roof of a building or vehicle made of a material which admits light.
1961 *Engineering* 27 Oct. 552 A new series of rooflights have been designed. **1967** *Gloss. Caravan Terms* (*B.S.I.*) 2 *Rooflight*, for the purpose of expressing rooflight dimensions for catalogues, the roof aperture through which light and air pass. **1978** A. Fenton *Island Blackhouse* 18 All other windows..are fixed roof lights in the thickness of the thatch.

roo·fline, roof-line. [Roof *sb.* 7 a.] **1.** The outline or silhouette of a roof or a collection of roofs.
1857 C. Vaux *Villas & Cottages* 54 Some degree of picturesqueness can always be obtained by the treatment of the roof-lines. **1886** [see Roof *sb.* 7 a]. **1936** W. Faulkner *Absalom, Absalom!* ix. 366 It loomed, bulked, square and enormous, with jagged half-toppled chimneys, its roofline sagging a little. **1955** E. Bowen *World of Love* i. 9 The long low roofline framed by too much sky. **1976** 'Trevanian' *Main* i. 14 Above the roofline, defused city-light glows in the damp, sooty air.

2. The outline of the roof of a car, usu. as seen in side elevation.
1967 [see *notch-back* s.v. *NOTCH *sb.* 6]. **1971** *Daily Tel.* 20 Oct. 7/2 The styling competition winner is the Cirrus, a dart-shaped 2 + 2 GT car with a roofline only 49in high.

roofscape (rū·fskēip). [f. Roof *sb.* + Scape *sb.*[3]] A scene or view of roofs.
1928 *Observer* 22 July 9/3 It appears I do not mention certain roofs and arcades that Mr. Gwynn has seen. Let me inform Mr. Gwynn that every week..I pass twenty wonderful roofscapes which are probably quite unknown to Mr. Gwynn. **1949** *Archit. Rev.* CV. 277/2 (*caption*) Roofscape: a view from the roof of almost any St. John's Wood house reveals at once the factor which contributes most to the character of the place—trees. **1954** L. P. Hartley *White Wand* 23 From my bedroom windows I enjoyed a roofscape. Domes and towers gave it grandeur and formal beauty, but what chiefly fascinated me was the roofs themselves. **1965** N. Freeling *Criminal Conversation* ii. iv. 112 He was high..and possessed several roof- and streetscapes. **1975** C. N. Manlove *Mod. Fantasy* vi. 215 The actual shape and relief of Sark are throughout close to what we are told of those of Gormenghast, with its long wings, its broken tower and its huge flat roofscapes.
Also **roo·fscaping** *vbl. sb.* [after *LANDSCAPING *vbl. sb.*], (see quot. 1967).
1962 *Spectator* 30 Mar. 426 One characteristic of roofscaping has all but vanished over the past decade or so—that of building rose-covered archways or pergolas. **1967** *Britannica Bk. of Year* (U.S.) 804/2 *Roofscaping*, the landscaping of rooftops (as of apartment houses and office buildings).

roof-tile. Delete ? *Obs.* and add: In mod. use, a tile used as a roofing material; a roofing tile. Also *fig.* (Later examples.)
1793 J. Woodforde *Diary* 3 Mar. (1929) IV. 12 It blowed off many Tiles from the old part of the House, some Roof Tiles. **1936** *Discovery* May 142/2 The dangerous sprinkling of snow over the 'roof tile' slabs [of Mount Everest]. **1940** 'Gun Buster' *Return via Dunkirk* i. i. 11 The house was its old self even to the last brick, the last roof-tile. **1957** *Jrnl. Geol.* LXV. 239/1 The deeply trenched roof-tile shapes of figure 8 and the delicate fern-frond pattern..shown in figure 9 are two of several different shapes exposed in a single quarry of that area.
So **roof tiler.**
1921 *Dict. Occup. Terms* (1927) § 570 Roof tiler. **1973** *Times* 16 Oct. 4/8 Mr Walsh, aged 24, a roof tiler. **1976** *Star* (Sheffield) 29 Nov. 10/2 (Advt.), Roof tiler required in the very near future.

roof-tree. Add: Also **rooftree. 1.** (Later examples.) Also *fig.*
1923 T. S. Eliot *Waste Land* v. 23 Only a cock stood on the roof-tree. **1950** D. Gascoyne *Vagrant* 38 Entangled in the thicket of World Roof-Tree's dense leaves. **1955** E. Pound *Classic Anthol.* III. 184 High, pine-covered peak full of echos, Proud ridge-pole of Heaven, roof-tree Whence descended the whirl of spirits. **1969** *New Scientist* 13 Mar. 554/1 An enormous rooftree 558ft long has just been inched up from ground level to serve as the chief single member of the new hangar which is being built by BOAC.

rooi-aas (roi·ās). *S. Afr.* [Afrikaans, f. *rooi* red + *aas* bait.] = *RED-BAIT.
1895, **1905** [see *RED-BAIT]. **1913** W. Thompson *Sea Fisheries Cape Colony* ii. 48 Eastwards of Cape Point the place of this crustacean is filled by red-bait (*rooi-as*). **1930** [see *RED-BAIT].

rooibaadjie (roi·baiki, -baikyi). *S. Afr.* Also **Roed Vatje, rooiba(a)tje, -baaitje.** [Afrikaans, f. *rooi* red + *baadjie* jacket.] **1.** A British regular soldier, a redcoat. Now chiefly *Hist.*
1848 H. Ward *Five Yrs. in Kaffirland* i. v. 164 And how Umhala would laugh at the *Roed Vatjes*! **1858** [see sense 2]. **1885** J. Nixon *Compl. Story Transvaal* x. 183 First of all the officers, regular and irregular, should be fired at, and then the men with the puggarees round their hats (that is, the volunteers); and as for the rovi-baatjes [*sic*] (red-backs, i.e. regulars), it didn't matter about them—they would be sure to run when their officers were killed. *Ibid.* xi. 202 The moment the Boers rushed out to attack, after they saw their friends coming down the hill-side, the 'bastards' naturally took to flight, and sixty of his [*sc.* the Boer commandant's] men followed them, and thereupon the 'rooibatjes' fled also. **1897** E. Glanville *Tales from Veld* xxvi. 200 Sonny, them *rooibaaitjes* can fight, but they're foolish. **1941** S. Cloete *Hill of Doves* (1942) viii. 116 Why, our men were soldiers, veterans of wars, when these Rooibaadjies were but children. **1971** *Daily Dispatch* (East London, Cape Province) 18 Dec. 9 A young British rooibaadjie lurched towards him from the shadows!

2. A red larval form of the South African brown locust, *Locustana pardalina*.
1858 H. Calderwood *Caffres & Caffre Missions* xii. 157 The youngest locusts..are then partly red and partly black... Sometimes they are called *roibatjes*—that is, red-coats, in allusion to the soldiers. **1875** C. B. Bisset *Sport & War in Afr.* 170 You see the very earth become alive with diminutive insects,.. increasing in size and becoming the colour of the brightest red. At this stage they are called the Rooi baatyes or red soldiers. **1902** *Trans. S. Afr. Philos. Soc.* XI. p. xlv, The young of the migratory one [*sc.* a locust]..are so gaily coloured as to have earned for them the local name of '*rooi-batjes*', or redcoats. **1972** *Stand. Encycl. S. Afr.* VII. 21/1 Young crowded hoppers would develop into typical phase *gregaria* 'rooibaadjies'.

rooibekkie (roi·beki). *S. Afr.* Also **roibek, rood(e)bec, -bekje, rooibe(c)k(ie), rooibekje.** [Afrikaans, f. *rooi* red + *bek* beak + *-ie* dim. suff.] Either of two birds with red beaks, the waxbill, *Estrilda astrild*, or the pin-tailed whydah, *Vidua macroura*.
1793 tr. *C. P. Thunberg's Trav. Europe, Afr., & Asia* I. 312 The *Loxia Astrild*, on account of its red beak, was called Rood-beckje, or *Red-beak*, and was found in great numbers in the farmers gardens. **1822** W. J. Burchell *Trav. Interior S. Afr.* I. xii. 266 The *Roode-bekje* (Red beak), a small finch,..is a very common bird. **1861** Lady Duff Gordon *Let.* 10 Dec. in *Lett. from Cape* (1921) 60, I will try to bring home some cages of birds—Cape canaries and 'roode bekjes' (red bills), darling little things. **1868** J. Chapman *Trav. Interior S. Afr.* II. i. 17 We shot and skinned some birds, among which was a long-tailed finch (king rooi bekkie). **1890** A. Martin *Home Life on Ostrich Farm.* i. 18 Another soft-voiced little singer is the *rooibeck*, or red-beak. **1899** R. B. & J. D. S. Woodward *Natal Birds* 66 This bird [*sc.* the pin-tailed widow bird] derives its name from its pretty wax-like red bill, which resembles that of the estrilda, and in common with them it is called *roibek*. **1900** A. C. Stark *Birds S. Afr.* I. 98 Common Waxbill..'Roodebec'..and 'Rooibeckie' of Dutch Colonists. **1913** [see *KING *sb.* 7 a]. **1936** E. L. Gill *First Guide S. Afr. Birds* 32 Pin-tailed Widow-bird, King Rooibekkie,..is also parasitic; the eggs are laid in nests of the common Rooibekkie. **1952** *Cape Times* 15 Jan. 9/8 Farmers are demanding the extermination of the common rooibekkie, or waxbill. **1963** S. Cloete *Rags of Glory* 548 The rooibekkies were still chattering in the trees.

rooibok (roi·bǫk). *S. Afr.* Also **roibok, roodebok, rooibuck, rooye bok.** [Afrikaans, f. *rooi* red + *bok* buck.] = *IMPALA.
1824 W. J. Burchell *Trav. Interior S. Afr.* II. xi. 215 One [antelope] is called *Paala* (Parla) by the Bichuanas, and is known by the name of *Roodebok* (Redbuck). **1835** A. Smith *Diary* 27 Aug. (1940) II. 182 They had much difficulty in getting them to..carry the flesh of a rooye bok. **1866** T. Leask *Diary* 12 May (1954) 44 Saw some roibok, but felt too done up to go after them. **1875** [see *IMPALA]. **1896** H. A. Bryden *Tales S. Afr.* 65 Smallfield ..had shot a good rooibok. **1926** *Glasgow Herald* 31 Aug. 2/6 He had got a rooibuck (or impala) ram. **1932** Fuller & Fouche *Louis Trigardt's Trek* vi. 65 He told..about a *roobok* and two geese they had shot. **1947** [see *IMPALA]. **1968** L. G. Green *Full Many a Glorious Morning* 189 Palapye Road, named by an extinct tribe after the rooibok, was once the site of the 'post office tree'.

rooibos (roi·bǫs). *S. Afr.* Also **rooibosch, rooibostee.** [Afrikaans, f. *rooi* red + *bos* bush.]
1. An evergreen South African shrub of the genus *Aspalathus* (formerly *Borbonia*), belonging to the family Leguminosæ, and cultivated for its leaves which are used to make a kind of tea; also, the beverage made from the leaves. Also *attrib.*
1911 *S. Afr. Jrnl. Sci.* VII. 374 The author described.. a *Borbonia*, the source of Cape 'rooibos' tea. **1932** Watt & Breyer-Brandwijk *Medicinal & Poisonous Plants S. Afr.* 70 *Borbonia pinifolia* Marl.,..Rooibostee..and *Borbonia cordata*..are also used as teas. **1946** [see *bush tea* s.v. *BUSH *sb.*[1] 11]. **1948** *Cape Argus* 18 Dec. 7/9 Dr. Pieter le Fras Nortier..established a flourishing rooibos industry. **1949** L. G. Green *In Land of Afternoon* iii. 52 Bush tea is..an entirely different plant—a legume. You hear it called rooibos. *Ibid.* 53 The ants of the district collect rooibos seeds and store them underground. **1951** *Cape Times* 17 Oct. 2/6 Mr. Riordan was..a well-known rooibosch tea farmer. **1977** *Daily Tel.* 5 Mar. 18 Rooibosch is almost entirely free from tannin and contains no caffein.

2. A shrub or small tree, *Combretum apiculatum*, belonging to the family Combretaceæ, native to central and southern Africa, and bearing red or yellow foliage in winter, and spikes of scented yellow flowers.
1932 Watt & Breyer-Brandwijk *Medicinal & Poisonous Plants S. Afr.* 128 The Zulus regard *Combretum erythrophyllum* Sond., Bush willow,..Rooibos,..as poisonous. **1972** Palmer & Pitman *Trees S. Afr.* III. 1633 The rooibos is a valuable fodder tree.

rooi-els (roi·els). *S. Afr.* Also **rood(e) els, elze.** [Afrikaans, f. *rooi* red + *els* alder.] An evergreen tree, *Cunonia capensis*, belonging to the family Cunoniaceæ, native to southern Africa, and bearing compound leaves and racemes of fragrant cream flowers; also, the reddish wood of this tree. Cf. *red alder, red els* s.v. Red *a.* 17 d in Dict. and Suppl.
1801 J. Barrow *Acct. Trav. Interior S. Afr.* I. v. 339 Roode els..stands water well. **1822** W. J. Burchell *Trav. Interior S. Afr.* I. vii. 143 Its colonial name is *Rood Elss* (Red Alder), although the tree has not..the least resemblance to the Alder of Europe. **1948** *Cape Times* 5 Aug. 8/7 If trees must be planted, let them rather be the ..rooi els and other local forest trees. **1972** Palmer & Pitman *Trees S. Afr.* I. 665 The rooi-els grows in moist mountain forests.

rooigras (roi·χras). *S. Afr.* Also **rooigrass.** [Afrikaans, f. *rooi* red + *gras* grass.] A southern African grass, *Themeda triandra*, which goes a reddish colour in winter. Also *attrib.*
1889 H. A. Bryden *Kloof & Karroo* iv. 88 Much of the lower parts of these hills is clothed with rooi-grass. **1907** T. R. Sim *Forests & Forest Flora Cape of Good Hope* iv. 37 The rushes have given place to rooi-grass. **1929** J. W. Bews *World's Grasses* vi. 253 'Rooi gras'..is a valuable forage grass. **1950** *Cape Times* 30 Oct. 9/6 A. C. Erasmus scattered the first *rooigras* seeds in the..bare land recently cleared. **1966** C. A. W. Guggisberg *S.O.S. Rhino* iv. 92 The dominant grass throughout the..range of the southern white rhino is *Themeda triandra*, which stands quite high and is popularly known as 'rooigrass'. **1972** *Stand. Encycl. S. Afr.* V. 320/2 A large number of species, e.g. Guinea grass.., rooigras.., the finger-grasses ..are important pasture grasses.

rooihout (roi·hǝ̄ᵘt). *S. Afr.* Also **roodehout, roye-houtt.** [Afrikaans, f. *rooi* red + *hout* wood.] One of several trees with reddish wood, esp. the Cape plane, *Ochna arborea*, or its wood. Also *attrib.*
1790 E. Helme tr. *Le Vaillant's Trav. Afr.* II. xiv. 288 Another tree, called *Roye-houtt*, (red wood) so named from its deep red colour, grows very thick. **1896** E. Clairmonte *Africander* i. 2 A flock of long-tailed house birds..would dash past to settle in a *rooihout* tree. **1907** T. R. Sim *Forests & Forest Flora Cape of Good Hope* xiv. 163 Cape Plane..Roodehout. **1973** *Eastern Province Herald* (Port Elizabeth) 28 May 13/2 No self-respecting woodcutter would have the handle of his axe made from any timber other than 'rooihout' (Cape plane) a reddish close grained wood.

rooikat (roi·kat). *S. Afr.* Also **roode-kat.** [Afrikaans, f. *rooi* red + *kat* cat.] = Caracal.
1785 G. Forster tr. *Sparrman's Voy. Cape of Good Hope* I. 150 Another kind of cat, as it is called, or the *roode-kat*, is..supposed to possess a great medicinal power in its skin. **1880** J. Nixon *Life among Boers* vi. 142 There is another kind, known as the rooikat, whose skin is highly prized for making karrosses. **1939** S. Cloete *Watch for Dawn* v. 70, I am as slim as a rooikat and this is my country. **1948** *Cape Times* 4 Dec. (Mag. Section) 3/4 A lithe red shape trotting through the bush ahead.. was my first sight of the African lynx or rooikat. **1966** E. Palmer *Plains of Camdeboo* ix. facing p. 160 (*caption*) The brilliant coloured lynx or rooikat—rich red with jet-black ear tufts.

rooikrans (roi·krans). *S. Afr.* Also **rooikran(t)z.** [Afrikaans, f. *rooi* red + *krans* wreath, in allusion to the red aril of the seed.] A yellow-flowered shrub, *Acacia cyclops*, of the family Leguminosæ, native to Western Australia, and naturalized in southern Africa, where it is also called the golden willow. Also *attrib.*
1917 R. Marloth *Flora S. Afr. Suppl.: Common Names* 71 Rooikrans... From Australia..one of the Golden willows. **1920** *S. Afr. Smallholders' & Fruit Growers' Year Bk. 1920–1921* 175 Rooikrantz—A shrubby wattle useful for firewood. **1950** [see *Port Jackson willow* s.v. *Port Jackson]. **1953** *Cape Times* 9 July 14/2 Silver trees are being choked by young self-seeded cluster pines and rooikrans. **1957** *Cape Times Mag.* 20 July 10/4 He wielded the chopper as easily as a man—severing the tough branches of the rooikrantz trees. *Ibid.* 10/6 On the outskirts of the rooikranz trees they began to dig.

rooiman, var. *ROMAN *sb.*[3]

rooinek (roi·nek). *S. Afr.* Also **roineck** and with capital initial. *Pl.* **rooineks, rooinekke.** [Afrikaans, f. *rooi* red + *nek* neck.] A term applied by Afrikaans-speaking South Africans to the British or to English-speaking South Africans.
1890 'S. Erasmus' *Prinsloo of Prinsloosdorp* ii. 14 One morning he was on the market with his waggon when two men—English Rooineks—came and said: 'Piet, do you want to make £15?' **1896** H. A. Bryden *Tales S. Afr.* 210 Cornelis would open up, and yarn to me in a way that, until you know him well, the Boer seldom manifests to the *rooi-nek*. **1900** *Captain* III. 121/1 *Rooinek* scout. **1900** Kipling in *Daily Express* 13 June 4/5 And you will see how we can shoot rooineks. **1921** *Chambers's Jrnl.* Jan. 32/1, I was thinking of the efforts that that infernal rooinek (red-neck) of a son of yours is making to deprive me of my only child. **1937** C. R. Prance *Tante Rebella's Saga* 47 A rascally Irish 'rooinek' whose real name had been Pat Murphy till he changed it to Piet van der Merwe when he turned Afrikander. **1947** H. C. Bosman *Mafeking*

Road 117 But of course no rooinek can make a living out of farming, unless they send him money every month from England. **1962** [see *Limey a]. **1963** S. Cloete *Rags of Glory* xxxix. 316 The Englishmen were sunburned, red as lobsters. They did not go brown like the Boers. That's why we call them rooineks—rednecks—Renata thought. **1969** Visct. Buckmaster *Roundabout* xviii. 279 An English taxi-driver told me that he had lived for twenty years in Cape Town, only still to be called 'A bloody roineck', the name given to our troops in the Boer war. **1972** *Daily Dispatch* (East London, S. Afr.) 2 Feb. 6 Nasty little racist jibes, which we South Africans have been listening to for the past 10 years, about Van der Merwe and the Rooinekke and 'a bantu'. **1975** 'D. Jordan' *Black Account* v. 32 The Afrikaner industrialists ..had emerged since 1948 to challenge the English *rooineks* and their dominance of South African finance.

rooirhebok (roirē·bǫk). *S. Afr.* Also **roode, rooye rheebok.** [Afrikaans, f. *rooi* red + *RHE(E)BOK.] The mountain reedbuck, *Redunca fulvorufula*. Also *attrib.*
1835 A. Smith *Diary* 11 July (1940) II. 107 During this season the male and female of the rooye rheebok are generally apart. **1835** A. Steedman *Wanderings S. Afr.* I. ii. iv. 176 We observed, at some little distance, several antelopes of the description called *roode-rheebok*, which were running up the mountain. **1850** [see Rheeboc]. **1904** [see Nagor]. **1912** J. Stevenson-Hamilton *Animal Life Afr.* x. 157 Rooi rhebok favour the lower slopes of hills. **1957** *Cape Argus* 13 July 7/5, I have seen a rooi-rhebok ewe battling to save her kid from an eagle.

rook, *sb.*[1] Add: **4. a.** *rook-babble, -roost, -scarer*; *rook-scaring* vbl. sb.; *rook-crowded, -delighting, -racked, -routed* adjs.
1948 C. Day Lewis *Poems 1943–47* 21 The rook-babble of bathers. **1964** *Listener* 24 Dec. 1009/2 Goal-posts, a public-house, a rook-crowded birch. **1914** W. B. Yeats *Responsibilities* 37 Suddenly I saw the cold and rook-delighting Heaven. **1879** G. M. Hopkins *Poems* (1967) 79 Lark-charmèd, rook-racked, river-rounded. **1937** J. W. Day *Sporting Adventure* 88 The great rook-roosts of winter, the annual nightly gatherings of thousands of these birds, are breaking up. **1923** Blunden *To Nature* 46 In the rook-routed vale. **1910** Rook-scarer [used in Dict. in def. of *roof-starver*]. **1946** J. W. Day *Harvest Adventure* xvi. 266 A rat-catcher is a Pest Control Officer and a rook-scarer a Corvine Operator! **1895** 'Rosemary' *Under Chilterns* i. 20 The poor child ain't fit for sech work as that there rewk-scarin' this weather. **1910** [used in Dict. in def. of *roof-starving*]. **1969** G. E. Evans *Farm & Village* v. 55 Some called it *bird-keeping* or *bird-tending*—keeping the birds off the newly sown land—while others referred to it simply as *rook-scaring*.

b. rook-drive, an expedition to shoot rooks; **rook rifle** (further examples); **rook-worm**, add: esp. the larva of the cockchafer, *Melolontha melolontha*; (later examples).
1969 R. Blythe *Akenfield* 20 'Did you kill men, Davie?' 'I got several'—the same answer to a question on how he did on a rook-drive. **1900** E. Glyn *Visits of Elizabeth* 50 She amused herself..by shooting at rabbits..with a rook rifle. **1907** [see *EJECTOR 2]. **1921** 'K. Mansfield' *Let.* 3 Feb. (1977) 215 My grandpa said a man could travel all over the world with a clean pair of socks and a rook rifle. **1972** *Shooting Times & Country Mag.* 4 Mar. 11/1 Somehow we got hold of a .300 rook rifle cartridge. **1976** *Ibid.* 16–22 Dec. 46/4 (Advt.), Holland and Holland .410 converted rook rifle, £70. **1959** E. F. Linssen *Beetles Brit. Isles* II. 124 The larvae of the Cockchafer—sometimes called by farmers White Grubs or Rookworms—are exceedingly destructive. **1973** J. M. Chinery *Field Guide Insects Brit. & N. Europe* 303 It is said that rooks are particularly fond of both adult and larval cockchafers and the larvae are often called rookworms.

rook (ruk), *sb.*[4] U.S. shortening of *ROOKIE.
1905 *Bluejacket* Mar. 190/1 The sailors there said we were 'rooks'. **1927** *Amer. Speech* II. 278/1 *Rook*,.. novice. **1935** *Our Army* (U.S.) June 12 A life-long profession from club-footed 'Rook' to Top Soak. **1941** G. Kersh *They die with their Boots Clean* ii. 85 This here Spencer drops weight...millions of stones that rook lost. **1942** *Yank* 23 Sept. 17 In the horse cavalry, recruits do not complain as loudly about kitchen police as do the rooks in other branches.

rook, *v.*[1] **1.** (Further examples.)
1938 *Sun* (Baltimore) 11 Oct. 24/2 There have been numerous complaints that the growers have been 'rooked'. **1969** *Listener* 10 Apr. 482/3 Because we had been rooked at the door, none of us ever thought of boycotting the desk where another seated veteran..was selling postcards. **1977** *Capital Times* (Madison, Wisconsin) 27 Jan. 10/3 The Federal Trade Commission thinks that a lot of people have been rooked by these buying clubs.

rookery. Add: **2. a.** (Earlier example.)
1832 A. Earle *Narr. of Residence on Tristan d'Acunha* 357 We visited what they call a 'penguin rookery'.

b. (Earlier and later examples.)
1832 B. Morrell *Four Voyages* p. xxiv, The word *rookery*..has been applied by all our South Sea navigators to the breeding encampments of various oceanic animals, such as seal, penguins, &c. **1846** *McLean Papers* 22 Dec. 42 (typescript), A boat goes out in search of a 'rookery' as they term the collected seals on any spot. **1932** S. Zuckerman *Social Life Monkeys & Apes* v. 69 Bull seals fight each other..for territory in the rookery or mating ground. **1972** L. Hancock *There's a Seal in my Sleeping Bag* vii. 145 The rocky shores of Triangle Island itself are used by the Steller sea-lions as hauling-out grounds, while those of the large islets immediately off the main island are breeding grounds and rookeries.

3. (Earlier and later examples.)
1792 G. Galloway *Poems on Various Subjects* 74 Then I begin my follies to repent, With naked elbows and a coat thread bare... So for to hide my gold I need no bags, While like to rookry dogs I'm cloth'd with rags. **1971** *Daily Tel.* 2 Dec. 6/4 The artisans, a potent radical force, were very different from the a-political desperate starvelings of the London rookeries. **1973** *N.Y. Law Jrnl.* 4 Sept. 5/3 Look at the city's unrepairable slums housing miserably over a million people... These rookeries are beyond repair. **1976** M. Butterworth *Festival!* vii. 106 Arnold..lay in a crude shelter..in the heart of a close-packed slum of some of the worst habitations at the festival; a rookery so vile that it had been totally rejected by all the rest.

4. (Earlier and later examples.)
c **1820** *Oh, What a Row!* (song) People toiling, roasting, boiling, bless us! such a rookery. **1925** *Dialect Notes* V. 340 *Rookery*, confusion, ruckus. **1942** Berrey & Van den Bark *Amer. Thes. Slang* § 5/1 *Disorder n.*,..riz-raz, rookery, [etc.].

rookie (ru·ki). *slang.* Also **rookey, rooky.** [Origin uncertain; perh. corruption of Recruit *sb.*] **1.** A raw recruit, *spec.* (*a*) an army or police recruit; (*b*) a novice at a sport, etc., esp. a first-year player in a particular team (chiefly *N. Amer.*).
1892 Kipling *Barrack-Room Ballads* 68 So 'ark an' 'eed, you rookies, which is always grumblin' sore. **1893** [in Dict. s.v. Rooky *sb.*]. **1900** I. L. Reeves *Bamboo Tales* 100 One of his men, a green 'rookie'. **1909** R. A. Watson *Happy Hawkins* 142 'Why, you blame rookie,' sez I, 'You don't really think I was mad do you?' **1913** *Chicago Record-Herald* 1 Mar. 12/2 Cal tried out Lefty Delano, a New Brunswick southpaw rookie. **1913** H. A. Franck *Zone Policeman 88* ix. 275 From the lieutenant to the newest uniformless 'rookie' every member of the police was swarming in and out of the building. **1918** I. Crump *Conscript 2989* 6 Oh, you rookey. **1929** *Daily Express* 15 Jan. 9/7 In 1915..he was a member of the Baltimore baseball team as a gawky 'rookie'. **1930** *Punch* 16 Apr. 443/1 The keen young rooky, just fresh from his public school. **1939** *Airman's Gaz.* Dec., Who was the Rookie here who thought..Blanco was a Spanish General. **1956** 'E. McBain' *Cop Hater* (1958) ix. 81 A squad car driven by a young rookie. **1961** F. H. Burgess *Dict. Sailing* 173 *Rooky*, a brand new hand. **1966** *Time* 8 July 60/3 The new enlarged S. & S. will be in a more powerful position to bid for blockbusting authors, whose contracts have been escalating as rapidly as those of prize pro football rookies. **1975** *Listener* 16 Jan. 66/2 [He] flew with Hunt to Hong Kong, as a police rookie. **1976** *Billings* (Montana) *Gaz.* 16 June 5-c/1 Rookies Chris Evert..and Rod Laver..headed the list of players named Tuesday to the World Team Tennis Western Division All-Star team. **1979** *Tucson* (Arizona) *Citizen* 20 Sept. 1D/1 He was rookie of the year on the Professional Golfers Association tour in 1965.

2. *attrib.* passing into *adj.*
1930 E. H. Lavine *Third Degree* (1931) 102 The shooting of 'rookie' Patrolman James A. Broderick. **1944** *Chicago Daily News* 21 Oct. 11/1 They expressed themselves..as willing to trade..for the rookie outfielder. **1954** L. Armstrong *Satchmo* 37 They gave me the rooky greeting saying, 'Welcome, Newcomer'. **1963** *Listener* 4 Apr. 585/1 To the eye of the rookie policeman they were as unobtrusive as the pavement stones. **1968** [see *Charley-horse, charley horse]. **1972** M. Williams *Inside Number 10* xii. 322, I recall the times he was censorious with us about security and times when he treated us like the latest members of an awkward squad in a rookie army. **1974** *Spartanburg* (S. Carolina) *Herald* 18 Apr. C1/7 The Dallas Cowboys lost a fifth player to the World Football League Wednesday when rookie tight end John Kelsey of Missouri signed with the Honolulu Hawaiians. **1977** *Daily Mirror* 30 Mar., Rookie police constable Paul Weaver, 19, saved a 74-year-old widow from her blazing home. **1977** D. L. Altheide in Douglas & Johnson *Existential Sociol.* iv. 137 Rookie reporters..learn the ropes and eventually become bored with the questions of newcomers. **1980** *Washington Star* 18 July A6/3 Reagan's nomination is the penultimate step in a steady 12-year drive to the White House that began with a feeble tentative pass at the office in 1968 when he was derided as a Grade B Hollywood movie actor and a rookie right-wing fringe governor of California.

rooking, *vbl. sb.*[1] (Later examples.)
1936 *Sat. Even. Post* 19 Dec. 11/1 What else could we do but quit? Mel wasn't treatin' you square. Can you deny it? I won't speak for myself, though I was gettin' a fancy rookin', too, singin' and all. **1955** *Sun* (Baltimore) 10 Sept. 6/4 They submit to a rooking involving thousands of dollars.

rooking, *ppl. a.* (Later example.)
1934 Dylan Thomas *18 Poems* 20 A rooking girl who stole me for her side.

rookus (ru·kǝs), var. *RUCKUS.
1893 H. A. Shands *Some Peculiarities of Speech in Mississippi* 53 *Rookus*, a word signifying a quarrel or row. **1902** 'O. Henry' in *Ainslee's Mag.* Mar. 130/1 He talks all spraddled out..'bout the rookuses he's been in. He claims to have saw the elephant and heard the owl. **1924** W. M. Raine *Troubled Waters* vi. 60 Course there'll be a rookus between him and Joe Tait.

b. rookus-juice *U.S. slang*, liquor.
1929 *Amer. Speech* IV. 385 Such terms as *rookus juice, third rail*,..and *bust-head* are evidently references to the potency or the effect of the liquor designated. **1942** Berrey & Van den Bark *Amer. Thes. Slang* §99/1 *Liquor*..rinse, rookus-juice, [etc.]. **1951** *Western Folklore* X. 80 Give me a shot of:..rookus juice.

rooky, *sb.* Substitute for entry:
rooky, var. *ROOKIE.

room, *sb.*[1] Add: **I. 2. a.** Also in phr. *room at the top*.
1900 W. James *Let.* 2 Apr. (1920) II. 121 Verily there is room at the top. S— seems to be the only Britisher worth thinking of. **1914** A. Bennett *Price of Love* vii. 143 The Imperial had set out to be the most gorgeous cinema in the Five Towns; and it simply was. Its advertisements read: 'There is always room at the top.' **1929** *Times* 11 Jan. 13/4 When successful men give schoolboys their prizes they generally throw in a little advice. They recommend industry, neatness, punctuality, and other virtues, but they also dwell on the saying that there is always room at the top. **1933** W. S. Maugham *Sheppey* iii. 89 You have to be pretty smart with all the competition there is nowadays... There's always room at the top. **1947** 'G. Orwell' *Eng. People* 22 The masses..know it is not true that 'there's plenty of room at the top'. **1957** J. Braine *Room at Top* xxviii. 230 You're the sort of young man we want. There's always room at the top. **1960** *20th Cent.* July 79 Academically speaking, the room at the top in sociology is lessening. **1980** *Times* 14 Jan. 5/5 In that last crisis, McEnroe suddenly looked young and vulnerable and Borg's tennis told him bluntly that, for the time being, there was no room at the top.

b. *no room to swing a cat in* and varr.: see Swing *v.*[1] 7 a.

II. 6. c. (Earlier and later examples.) Cf. *fishing-room* s.v. Fishing *vbl. sb.*[1] 5 b in Dict. and Suppl.
1620 R. Whitbourne *Discourse & Discovery of New-found-land* 30 [They] doe cut downe many of the best trees they can finde, to build their stages and roomes..; hewing..and destroying many others that grow within a mile of the Sea, where they use to fish. **1937** P. K. Devine *Folk-Lore Newfoundland* (Gloss.), *Room*, a fishing premises: stage, flakes and store. **1948** *Canad. Geogr. Jrnl.* Mar. 110/1 Fishing off Labrador is carried on by fishermen who leave Newfoundland in May or June and reside at 'rooms' (buildings used by the fishermen) on various parts of the Labrador coast until the close of the season. **1954** F. Briffett *Story of Newfoundland & Labrador* 32 A man's fishing property—flakes, stages and stores—was known as his room. **1963** J. T. Rowland *North to Adventure* iv. 54 Most of the schooner men had permanent stations, or 'rooms',..with storehouses and fish stages. **1975** *Canad. Antiques Collector* Mar.–Apr. 10/2 Of a crew of 40, there would be 24 to man eight small boats and 16 to work on the room.

8. a. Also *spec.* (chiefly *pl.*) a room or rooms for public gatherings, an assembly room, auction room, gambling room, etc.; at Lloyd's of London, the area where insurance business is carried out.
1766 C. Anstey *New Bath Guide* vii. 45 The Captain is come, And so kind as to go with us all to the Room. *Ibid.* viii. 48 (*heading*) Mr. B-n-r-d goes to the Rooms. His opinion of Gaming. **1771** Smollett *Humph. Cl.* I. 115 In the forenoon, they crawl out to the Rooms or the coffee-house. **1779** F. Burney *Diary* Oct. (1842) I. 254 In the evening we all went to the rooms. The rooms, as they are called, consisted, for this evening, of only one apartment, as there was not company enough to make more necessary. **1822** W. Hazlitt in *New Monthly Mag.* IV. 112 An old gentleman..who looked as if he had played many a rubber at the Bath rooms. **1876** Geo. Eliot *Dan. Der.* I. ii. xv. 291 They moved off together to saunter through the rooms, Sir Hugo saying as they entered the large *saal*—'Did you play much at Baden, Grandcourt?' **1904** A. E. W. Mason *Truants* xxiii. 217 She..bought a visitor's list at the kiosk in front of the rooms. **1928** A. Christie *Mystery of Blue Train* xxvii. 214 He found [him] in the Rooms, jauntily placing the minimum stake on the even numbers. **1931** *N. & Q.* 29 Aug. 155/2 Book auctions. —May I voice a long overdue protest against the pernicious and iniquitous custom prevailing in the 'Rooms' of doing up parcels of books with string. **1933** D. C. Peel *Life's Enchanted Cup* x. 121 The Rooms were so crowded that I could not get near enough to play at my chosen table. **1946** G. Stimpson *Bk. about Thousand Things* 51 This bell, which hangs in a clock-topped tower in 'Lloyd's Rooms' was salvaged from the frigate Lutine. **1962** H. O. Beecheno *Introd. Business Stud.* xvi. 153 He [*sc.* the broker] will then take this in to 'the Room' at Lloyd's and approach one or more leading underwriters. **1972** [see *OLD BOY].

10. a. *room air, -fellow, maid, number, paper, rent* (earlier and later examples), *ticket*; *room-keeper* (later example); *room-breaking* vbl. sb.; **room clerk**, in a hotel, guest-house, etc., a clerk who assigns rooms to patrons; **room divider**: see *DIVIDER 9; **room service**, the provision of food or drink for a hotel guest in his room, or the department providing this; **room temperature**, the temperature of a, or the, room, esp. that which is comfortable for occupants, conventionally taken as about 20°C; also *attrib.* and *fig.*; **room-to-room** *attrib.*, (of a telephone) connecting rooms within the same building.
1957 *Encycl. Brit.* XI. 353/1 Louis Savot..developed for the Louvre a fireplace in which room air was drawn through passages under the hearth and behind the fire grate. **1975** *New Yorker* 22 Dec. 78/3 He decided to discontinue Berger's oxygen therapy. He wanted him breathing room air again. **1951** S. Spender *World within World* 36 After the room-breaking episode the attitude of my fellow freshmen towards me altered. **1916** W. A. Du Puy *Uncle Sam, Detective* 49 The room clerk had suggested that it was the custom of the hotel that guests without baggage should pay in advance. **1978** S. Sheldon *Bloodline* xxxiii. 315 Before Max left Chamonix, he stopped at the desk of the Kleine Scheidegg hotel and talked to the room clerk. **1930** R. Macaulay *Staying with*

Relations ix. 122 Snakes might be her companions, wild cats her room-fellows, but she..abandoned herself to these. **1922** Joyce *Ulysses* 304 The meal should be divided..among the members of the sick and indigent roomkeepers' association. **1955** A. Ross *Australia 55* xv. 212 The room-maid says the world will end, not with an atom bomb, but with a flood. **1976** *Evening Standard* 14 June 26/3 (Advt.), Room maids, m/f, required by London Penta Hotel. **1959** A. Christie *Cat among Pigeons* iv. 54 Must have given me the wrong room number. **1968** 'M. Carroll' *Dead Trouble* ii. 22 The receptionist..asked what room number. **1976** H. Nielsen *Brink of Murder* xii. 94 Send up a fifth of Buchanan's. ..You know my room number. **1870** P. Fitzgerald in *All Year Round* V. 112/1 The decorations of the Jericho Theatre are rather of a homely cast, room paper garnished with bead mouldings. **1973** *Canad. Antiques Collector* Jan.–Feb. 20/1 A few scattered stories of the early elegance of room paper survive. **1818** *N. Amer. Rev.* Mar. 427 The room rent and wood are estimated upon the condition that two students live in a College room. **1851** C. Cist *Sk. Cincinnati in 1851* 65 The annual term bill for room-rent..and incidental expenses is ten dollars. **1942** *Amer. Mercury* July 90 That meant..maybe room rent and a reefer or two. **1930** A. Bennett *Imperial Palace* vii. 32 The head floors-waiter did not conceal his belief that the room-service was the basis of prosperity. **1935** Wodehouse *Luck of Bodkins* xxiv. 311 If you go to that phone and call Room Service, you can get all the champagne you want. **1949** O. Nash *Versus* 99 (*title*) Mrs. Purvis dreads room service. **1965** I. Fleming *Man with Golden Gun* vii. 94 Order what you want from Room Service. **1971** R. Thomas *Backup Men* xviii. 156 The room service waiter..brought the hamburgers and coffee. **1974** P. Gore-Booth *With Great Truth & Respect* 155 The only way I found of getting any relaxation at all was to be extravagant and have 'room service' breakfast. **1978** *Time* 3 July 47/2 Half-eaten room-service sandwiches. **1924** J. G. A. Skerl tr. *Wegener's Orig. Continents & Oceans* 128 They can prove that the earth is about two or three times as rigid at room-temperature as steel. **1945** R. T. Rolfe *Dict. Metallogr.* 4 It may consist..of a period of standing at room temperature. **1959** J. Braine *Vodi* xiv. 193 Of course the red wine should be at room temperature. **1962** Simpson & Richards *Physical Princ. Junction Transistors* ix. 211 The effect of these two factors is to make it desirable to place the operating point at a lower point than would be decided by room-temperature conditions. **1974** *Times* 13 Nov. 12/5 Put the mixture..overnight in the refrigerator. Allow to come up to room temperature again before baking. **1976** I. Levin *Boys from Brazil* v. 143 He wasn't accorded a warm or even room-temperature welcome. **1977** *Nature* 17 Feb. 660/2 The bK$_{610}$ intermediate..has a lifetime of 2 μs at room temperature. **1905** A. Bennett *Tales of Five Towns* ii. 264 She pushed his room-ticket across the page of the big book. **1938** Room-to-room [see *BUZZ *sb.*[1] 1 d].

b. *room-sealed* ppl. adj.
1963 *B.S.I. News* May 15/2 B.S. 3561 refers to fan-assisted air heaters,..and room-sealed heaters, giving requirements for their construction and performance. **1967** *Gloss. Terms Gas Industry* (*B.S.I.*) 78 *Room-sealed appliance*, an appliance which, when in operation, has the combustion air inlet and the combustion products outlet isolated from the room in which the appliance is installed.

III. 13. d. (Further example.)
1861 *Macm. Mag.* Dec. 141/2 Missis would still keep going on with her parties and company, o' rum o' minding her farm and dairy.

room, *adv.* For *Obs.* read *arch.* or *dial.* and add: **2.** (Later example.)
1969 G. M. Brown *Orkney Tapestry* 134 Guidman, go to your bacon And cut us doon a daggon Cut it lucky, cut it room, Look 'at you dunno cut your toom.

3. a. (Later example.)
1902 J. M. Barrie *Little White Bird* xv. 163 He was drifted towards the far shore, where are black shadows he knew not the dangers of, but suspected them, and so.. went roomer of the shadows until he caught a favouring wind.

room, *v.*[2] **2.** For *U.S.* read 'chiefly U.S.' and add: **a.** (Further examples.) Also *to room it*.
1912 F. M. Hueffer *Panel* i. i. 19 She and me were on the old North Circuit. Roomed it and ate off the same old herring together. **1937** *Observer* 22 Aug. 7/2 He dressed like a hobo, hitch-hiked from San Francisco to Los Angeles, and roomed on the town's Main Street as a plain British seaman. **1969** L. Michaels *Going Places* 167 Slotsky helped him with chemistry and French—Finn's reason for rooming with him in the first place. **1973** *Time* 25 June 11/1 John..roomed with Barry Goldwater Jr., who now is his neighbor. **1977** *Western Mail* (Cardiff) 5 Mar. (Rugby Suppl.) 4/2 It was on that tour that I developed a close friendship with many of the Welsh players... TGR and JPR I roomed with on many occasions, waiting on them hand and foot. **1979** *Yale Alumni Mag.* Apr. (Suppl.) CN12/1 Charlie, with whom I roomed, stayed through freshman year.

b. (Earlier example.)
1860 *Blackw. Mag.* Jan. 112/2 A miserable public house, where I was 'roomed', or in other words, put into the same room with, a rising medical practitioner.

rooman, var. *ROMAN *sb.*[3]

roomer, *sb.* For *U.S.* read 'orig. *U.S.*' and add: **1.** (Earlier and later examples.)
1871 L. H. Bagg *4 Years at Yale* 46 *Roomer*, a word used by landladies to designate a lodger or occupant of a room who takes his meals elsewhere. **1905** 'O. Henry' in *N.Y. World Mag.* 20 Aug. 4/2 There was rejoicing among the gentlemen roomers whenever Miss Leeson had time to sit on the steps. **1912** J. Sandilands *Western Canad. Dict.*, *Roomer*, a lodger who has living accommodation in a house and gets his food elsewhere. **1919** *Studies* VIII. 304 There is no 'board' provided, but simply a room... A 'roomer' has all the perfect liberty of a latchkey... She can entertain what visitors she likes in her own room. **1939** *Sun* (Baltimore) 6 Jan. 1/2 Held with Joseph Malone, 27, as a material witness was Elizabeth Gelula, 23,..a third-floor roomer. **1959** M. Chamberlin *Dear Friends & Darling Romans* v. 50 There were, besides me, two other roomers in the apartment. *a* **1968** M. Richler in R. Weaver *Canad. Short Stories* (1968) 2nd Ser. 151 'Meet your new roomer,' Mervyn said. **1973** *Kingston* (Ontario) *Whig-Standard* 5 Mar. 28 (Advt.), Roomers, $15 weekly with kitchen facilities. **1976** *National Observer* (U.S.) 25 Sept. 8/2 A roomer who had been helping to install some additional bathrooms was heating wax on a hot plate in his room.

2. With a numeral prefixed: a house with that number of rooms, as *six-roomer*.
1853 Dickens *Bleak Ho.* lxiv. 612 'It's a six roomer, exclusive of kitchens,' said Mr. Guppy, 'and in the opinion of my friends, a commodious tenement.' **1972** *Daily Tel.* 8 Mar. 22 Some high figures for Chelsea houses—£29,000 for a six-roomer in First Street.

roome·tte. *N. Amer.* [f. Room *sb.*[1] + -ette.] A small sleeping-compartment on a train; also, a small bedroom for letting. Also *attrib.*
1938 *Sun* (Baltimore) 14 June 20/2 The roomettes are small, completely inclosed rooms with accommodations for one traveler. Washstands fold into the walls after use. **1945** *Sci. Amer.* Mar. 170 Two-Story Pullmans—..the cost margin is reduced to a narrower point in the new duplex roomette car... Each roomette has individual control of heat, light, and air conditioning. **1957** *New Yorker* 26 Oct. 68/3 Seated in his roomette, he opened the paper nervously. **1960** *Times* 19 Jan. 16/5 Travellers seeking rest between planes will rent roomettes, containing bed and bath, by the hour. **1971** *Guardian* 15 Sept. 11/1 Wandering from my roomette to the observation car as Canadian National transported me through the Rockies. **1974** *News & Courier* (Charleston, S. Carolina) 17 Feb. 1-A/8 The 9:40 arrives at 9:55 and everyone piles on—except me. My reserved 'roomette' is gone because a car was taken off the train somewhere between Jacksonville and Savannah. **1976** *Billings* (Montana) *Gaz.* 30 June 5-D/6 (Advt.), Chalet Roomette, private bath, refrigerator, washing facilities, $75 + dep. **1980** *Times* 4 Feb. 12/5, I was in a single sleeper, which Amtrak [*sc.* an American railway corporation] call a 'roomette'—probably the ugliest composite word ever invented. **1980** *Daily Tel.* 7 Nov. 15/1 He settled into his cramped, double-glazed Australian train cabin ('Roomette' in Strine).

roo·mie, *sb.* *U.S. colloq.* Also **roomy.** [f. Room *sb.*[1] + -ie.] A room-mate.
1918 *Sat. Even. Post* 6 July 5/3, I wouldn't make no holler at that if they had of left us pick our own roomys. **1945** *Richmond* (Va.) *Times-Dispatch* 26 July 14/7 Shortstop Skeeter Newsome, 'Inky's' roomie on the road, went home because of illness in his family. **1967** 'E. Queen' *Face to Face* xvi. 76 The Temple girl is given an alibi by her roomie. **1973** J. Wambaugh *Blue Knight* viii. 122, I called Craz my old roomie because..twenty years ago, I moved into this big house with him. **1976** 'B. Shelby' *Great Pebble Affair* 77 One of the screws came and rattled the bars on my cage. 'Hey... We finally got you a roomie.'

roo·ming, *vbl. sb.* [f. Room *v.*[2]] **1. a.** The letting of rooms to lodgers. **b.** The occupying or sharing of rooms. Chiefly *attrib.* See also Rooming-house in Dict. and Suppl.
1959 *Listener* 11 June 1012/1 Houses that are now usually converted into rooming apartments of inconvenient character. **1967** *Economist* 21 Oct. 264/2 The crowded rooming areas of central London. **1968** *Globe & Mail* (Toronto) 3 Feb. 41/1 (Advt.), Large family home 8 rooms huge kitchen 2 washrooms garage, walk-out basement to backyard. Perfect area for rooming. **1970** *Cape Times* 28 Oct. 20/2 (Advt.), Fine 7-roomed residence. Excellent rooming proposition. Fully furnished. **1977** *Rolling Stone* 5 May 50/1, I already have the rooming list.

2. rooming-in orig. *U.S.* (see quot. 1947); also *transf.* (see quot. 1978) and *attrib.*
1946 *Mod. Hospital* Dec. 44/2 Three Detroit hospitals.. have been..experimenting in certain selected cases with a 'rooming-in' plan which permits the mother to have her baby at her bedside. **1947** Jackson & Thoms in *Connecticut State Med. Jrnl.* XI. 175/1 Rooming-in is a term applied to that form of hospitalization where mother and newborn baby room together and in which the mother takes as much care of the baby as possible. **1950** *N.Y. Times* 30 Dec. 16/2 About 75 per cent of expectant mothers who plan to have their babies at Grace-New Haven Community Hospital..now are requesting the rooming-in plan whereby they may keep their babies with them instead of having them cared for in the hospital nursery... Rooming-in is preferred by more women with higher education and by women with husbands in upper occupational classifications. **1960** F. W. Goodrich *Maternity* vii. 103 Most ideal rooming-in set-ups make some provision for the baby to be returned to the nursery when the mother so desires. **1965** *Nursing Times* 5 Feb. 182/1 The siting of a nursery far distant from the wards added to the difficulties of the staff in spite of the fact that rooming-in was practised. **1974** G. B. Lipkin *Psychosocial Aspects of Maternal-Child Nursing* x. 56/1 The nurses in the rooming-in unit stress their function as instructors in child care. **1978** *Who's Who* 1934/1 A 'Rooming-In' hospital for mother nursing of infants and small children with congenital defects requiring plastic surgery.

rooming-house. For *U.S.* read orig. *U.S.* and add further examples.
1909 *Washington Times* 2 Mar. 1 Hundreds of persons who never slept in any but first-class hotels when away from home will tonight get their rest in rooming houses. **1911** *Daily Colonist* (Victoria, B.C.) 1 Apr. 13/1 (Advt.), 14 roomed house, just overhauled, repainted and in first class condition. One of the best rooming house propositions in the city. **1923** E. F. Wyatt *Invis. Gods* iii. ii. 105 Fairfax Avenue was now a nondescript street of rooming houses, apartments and carpet-cleaning establishments. **1932** *New Yorker* 9 Apr. 36/2 Miss Cedarholm..succeeded ..to the ownership of the brick dwelling at 338 Schermerhorn Street. This she continued to operate as a rooming-house. **1957** V. Nabokov *Pnin* iii. 64 There had been—in yet another rooming house—a still cozier bedroom-study. **1958** 'N. Shute' *Rainbow & Rose* iii. 76 Ma went there for the movies, and then when her contract ended she kept a rooming house. **1961** *Daily Tel.* 25 Mar. 16/5 The body of a woman.., was found by police officers last night in a rooming house in Lorenzo Street, King's Cross. **1973** *Sun-Herald* (Sydney) 26 Aug. 3/3 Police said another grandchild witnessed the stabbings in a rooming house. **1976** *Billings* (Montana) *Gaz.* 7 July 10-A/6 She moved to Washington D.C., where she was a proprietor of a rooming house for 20 years.

roomless, *a.* (Later example.)
1971 *Daily Tel.* 27 Sept. 6 (*heading*) Sleeping bags and camp beds for roomless students.

room-mate. For *U.S.* read orig. *U.S.* and add earlier and further examples.
1789 W. Dunlap *Father* iv. 42 We were room mates at Halifax. **1873** C. M. Yonge *Pillars of House* III. xxx. 170 The room and the room-mate that had seemed so disgusting to home-bred Felix. **1912** A. Brazil *New Girl at St. Chad's* i. 19 One of my room-mates snored atrociously. **1923** *Jrnl. Exper. Psychol.* VI. 436 She admitted having sold..clothing taken from her roommate. **1949** A. Huxley *Let.* 11 May (1969) 598 It would be very unwise to try to pay back two thousand in the first year—particularly at the price of having a room mate. **1951** *Sport* 27 Apr.–3 May 8/3 Stan and Jack are fellow-Geordies who were England room-mates last summer. **1954** W. K. Hancock *Country & Calling* ii. 66 My room-mate, Percy Dicker, is now my brother-in-law and a man of renown in the Wangaratta district of Victoria. **1975** D. Lodge *Changing Places* ii. 101 His roommate freaked out on LSD. **1978** H. Wouk *War & Remembrance* xxviii. 287 The third roommate, the squadron exec, was writing in the ready room.

roomy, var. *ROOMIE.

roorback. Substitute for etym. [The name of the fictitious author Baron von *Roorback* (see quot. 1864)] and add: Also **roarback.** (Earlier and later examples.)
[**1844** *Republican Sentinel* (Richmond, Va.) 4 Oct. 3/3 The rapid succession of events in the 'Roorback' line, has satisfied us, that the whole matter is a *quiz* or a *forgery*.] **1855** I. C. Pray *Mem. J. G. Bennett* 368 Among the efforts made to defeat the election of Mr. Polk was one to which allusion is frequently made in political discussion, politicians speaking of a political lie as a 'Roorback'. **1870** L. Baugh *To Voters of Washington County* (broadside), Beware of 'Roarbacks' on the eve of the election. **1913** A. B. Reeve *Poisoned Pen* xii. 367 Billy McLoughlin knows how to make the best use of such a roorback on the eve of an election. **1947** *Chicago Daily News* 27 Mar. 6/1 The roorback stage of the closing days of the campaign broke wide open today with appeals to racial and religious prejudice coming to the surface in many sections of the city. **1963** R. I. McDavid *Mencken's Amer. Lang.* 180 Since the exposure of the fraud actually helped Polk, it is sometimes spelled *roarback*.

Roorkee (rūᵊ·ɹkī). Also **Roorkhee, rhoorkee, Roorkie.** The name of a town, northeast of Delhi, in Uttar Pradesh, India, used *attrib.* in **roorkee chair**, a type of collapsible chair, with wooden frame and canvas back and seat, originally produced there; also *ellipt.*; **roorkee work**, a kind of canvas work associated with Roorkee.
1905 *Army & Navy Co-op. Soc. Rules & Price List* 15 Mar. 261/2 Roorkhee folding chair. Each 18/9 (Supply uncertain) Do. do., English make, of ash throughout, better finished. Each 21/6. **1907** *Yesterday's Shopping* (1969) 281/2 Roorkhee Folding Chair. English make, of ash throughout, well finished... Do., in green canvas with back... An improved form of Roorkee chair, it having a leg rest. **1936** J. Cary *Afr. Witch* viii. 158 Four chairs, including the Resident's well-known *rhoorkee*, taller than the rest, were empty. **1937** G. Frankau *More of Us* xiv. 147 Splendid sank Circe to a canvas throne Of rhoorkee work. **1953** J. Masters *Lotus & Wind* v. 69 There was no furniture left except a folding table and a Roorkie chair on which the colonel sat. **1973** 'B. Mather' *Snowline* xvi. 189, I lay back in a long Roorkhee chair. **1975** C. Allen *Plain Tales from Raj* v. 65 Perhaps some Roorkee chairs, made of canvas stretched on wood.

Roosevelt (rōᵘ·z(ǝ)velt, rū-·). The name of Theodore *Roosevelt* (1858–1919), President of the United States 1901–9, used *attrib.* or in the possessive in **Roosevelt('s) elk, wapiti** to designate a large, dark-coloured elk, *Cervus canadensis roosevelti*, found in coastal forests of north-western North America, and named in his honour by C. H. Merriam in 1897 (*Proc. Biol. Soc. Washington* XI. 271).
1897 *Proc. Biol. Soc. Washington* XI. 272 Roosevelt's Wapiti... Size large; head and legs black (probably only in winter pelage); skull and antlers massive. **1902** Stone

& Cram *Amer. Animals* 34 Roosevelt's Elk... Larger and darker coloured, with heavier horns. **1923** *Outing* Apr. 3/1 The Olympic peninsula..contains vast, unmapped forests..teeming with the lordly Roosevelt elk. **1975** *Islander* (Victoria, B.C.) 26 Jan. 16/4 The Olympic, or Roosevelt elk, are common in Olympic Park.

Rooseveltian (rōu·z(ə)veltiăn, rū·-), *a.* [f. the family name *Roosevelt* + -IAN]. Of, pertaining to, or characteristic of Theodore Roosevelt (see prec.), or Franklin Delano Roosevelt (1882–1945), President of the U.S. 1933–45, or the Roosevelt family in general. Hence **Roo·seveltism.**

1908 *Sci. Amer.* 25 Jan. 59/3 In this advanced twentieth century we had fondly hoped that the 'nature faker' at least was a product of the Rooseveltian age of literature. **1909** *Weekly Ardmoreite* (Ardmore, Okla.) 24 Feb. 8/1 Quarantine lifts Mar. 4... Get rid of Rooseveltism. **1915** *Fatherland* (N.Y.) 20 Dec. 346 We do not believe that the German Americans will have to choose between the devil of Rooseveltism and the deep sea of Wilson. **1923** A. Huxley *On Margin* 164 In true Rooseveltian style, he admired energy for its own sake. **1940** *Economist* 16 Nov. 601/2 He can make the official policy of the United States Government more completely Rooseveltian than it has been. **1948** *Chicago Tribune* 1 Feb. I. 37/3 It is a typical example of Rooseveltian democracy where charity is invited where it is not needed. **1953** *News* (Lynchburg, Va.) 6 May 6/1 That is not Rooseveltism. It is not New Dealism... It is good old-time Jeffersonian democracy. **1953** R. Niebuhr *Christian Realism & Polit. Probl.* (1954) v. 59 The *status quo*, until the Rooseveltian era, permitted a degree of non-interference by the state.. which must make America a paradise for all true devotees of *laissez-faire*. **1965** *Economist* 16 Jan. 221/2 The profile ..bears a distinctly Rooseveltian chin. **1974** *Listener* 31 Jan. 148/1, I recall being the only child in my class who did not come from the Rooseveltian atmosphere of the homes of the Thirties. **1977** *National Observer* (U.S.) 22 Jan. 17/1 And Eleanor was an accomplished smiler in her own right, adding weight to the theory that smiling was a major Rooseveltian impulse. **1979** *N.Y. Rev. Bks.* 17 May 30/4 Schaller thus finds in early 1941, long before Pearl Harbor, the origin of the Rooseveltian program which wound up five years later in General Marshall's unavailing postwar mediation.

rooseveltite (rōu·z(ə)veltait, rū·-). *Min.* [ad. Sp. *rooseveltita* (R. Herzenberg 1946, in *Bol. Técnico* (Facultad Nacional Ingeniería, Univ. Técnica, Oruro, Bolivia) No. I. 10), f. the name of Franklin D. *Roosevelt* (see prec.): see -ITE[1].] An arsenate of bismuth, $BiAsO_4$, which is found as a white or grey crust in veinlets of wood-tin in Bolivia and Argentina.

1947 *Amer. Mineralogist* XXXII. 372 (*heading*) Rooseveltite. **1949** *Mineral. Abstr.* X. 9 It is named rooseveltite, and is perhaps isomorphous with pucherite. **1972** *Tschermaks Mineral. und Petrogr. Mitt.* XVII. 65 Rooseveltite occurs in the weathering zone of the San Francisco de los Andes and Cerro Negro de la Aquadita mines, located in the San Juan Province, Argentina.

Rooshan, Rooshian, Roosian, varr. *RHOOSIAN *sb.* and *a.*

Roosky (ru·ski), var. *RUSKY *a.* and *sb.*[2] in Dict. and Suppl.

roost, *sb.*[1] Add: **1. c.** (Later example.) Also without const.

1966 D. Lack *Population Stud. Birds* ix. 156 At night they assemble in immense roosts, often in thorns.

d. (Earlier and later examples.)

1818 *London Guide* p. xii, *Roost*, bed. *Ibid.* 225 Coming from roost one morning,..I met old acquaintance, B—e, in Barbican. **1930** R. Campbell *Adamastor* 72, I.. Who now am but a roost for empty words. **1944** [see *robot roost* s.v. *ROBOT 2]. **1946** Mezzrow & Wolfe *Really Blues* xii. 219, I know I'm gonna call some hogs soon as I hit my roost.

e. *to rule the roost*, now the more usual form of *to rule the roast* s.v. ROAST *sb.* 1 b.

1769 in *William & Mary Coll. Q. Mag.* (1908) Jan. 175 They say she rules the Roost, it is a pity, I like her Husband vastly. **1828** A. N. Royall *Black Bk.* II. 315 These priests will rule the roost. **1893** *Boston Jrnl.* 20 Apr. 5/3 England rules the roost. Her ships at Hampton Roads admittedly the finest. **1926** Fowler *Mod. Eng. Usage* 509/1 *Rule the roast* (*roost*). The OED gives no countenance to *roost*, it does not even recognize that the phrase ever takes that form; but most unliterary persons say *roost* & not *roast*; I have just inquired of three such, & have been informed that they never heard of *rule the roast*, & that the reference is to a cock keeping his hens in order. **1931** R. Campbell *Georgiad* ii. 29 The great Tu Quoque rules the golden roost. **1938** A. Christie *Appointment with Death* v. 37 Her husband thought a lot of her and adopted her judgment on almost every point. He was an invalid for some years before he died, and she practically ruled the roost. **1955** 'A. Gilbert' *Is she Dead Too?* vi. 114 'Miss Bannerman was very jealous, and she didn't care for children or animals.' 'Then she could have found herself some other lodgings. You'd scarcely let her rule your roost.' **1963** *New Yorker* 15 June 16/3 The London underworld, where Peter Sellers rules the roost. **1974** S. Ellin *Stronghold* (1975) 36 My grandfather..ruled the roost, and he was a firm ruler for all his mild manner.

rooster. Add: **1. a.** (Earlier and further examples.)

1772 A. G. Winslow *Diary* 14 Mar. (1894) 45 Their other dish..contain'd a number of roast fowls—half a dozen, we suppose, & all roosters at this season no doubt. **1806** *Balance* (Hudson, N.Y.) 22 July 227 (Th.), The New York Rooster—may he continue to crow! **1870** J. H. B. Nowland *Early Reminisc. Indianapolis* 149 It was during this canvass [in 1840] that Tom gave to the Democratic party their emblem, which they have claimed ever since, the chicken cock, or rooster. **1922** Joyce *Ulysses* 646 Chalk a circle for a rooster. **1923** E. W. Benjamin *Marketing Poultry Products* iv. 120 Cock, or Rooster.—These are the mature males. **1951** M. A. Jull *Successful Poultry Managem.* (ed. 2) xi. 348 A cock or old rooster is a mature male chicken with coarse skin, toughened and darkened meat.

b. (Earlier and further examples.)

1785 Grose *Dict. Vulgar T.*, *Queer rooster*, an informer that pretends to be sleeping, and thereby overhears the conversation of thieves in night cellars. **1821** P. Egan *Life in London* II. v. 276 Roosters and the 'peep-o'-day boys' were out on a prowl for a spree. **1840** *Log Cabin* 5 Sept. 3/2 *Chapman*, the great Rooster of the Loco-Foco party,..was formerly one of the editers [*sic*] of an Infidel paper, the Boston Investigator. **1855** *N. Amer. Rev.* CXLI. 434 The toughest set of roosters that ever shook the dust of any town. **1883** *Bird o' Freedom* 7 Mar. 3/1 Whether the returned member be a rooster or not time will tell. **1923** R. D. Paine *Comrades of Rolling Ocean* xiv. 252 What was that rooster's name?

c. *U.S.* A wild violet as used in a children's game.

1884 *Harper's Mag.* June 94/1 Purple violets..were slaughtered by hundreds, for the projecting spur under the curved stem at the base of the flower enabled the boys to hook them together and 'fight roosters', as they termed it. **1946** C. Richter *Fields* 231 In April they played Hens and Roosters, yoking their wild white and blue violets to see which would get its head pulled off.

d. A bird that is roosting or about to roost.

1949 *Brit. Birds* XLII. 323 The more leisured flight of the roosters [*sc.* starlings] was in contrast to the steady procession of the migrants.

3. rooster comb *U.S.* = *rooster head; **rooster('s) head** *U.S.*, the American cowslip, *Dodecatheon meadia*, or a wild violet of the genus *Viola*; also *attrib.*; **rooster tail** *N. Amer.*, the curved plume of water thrown up by a speedboat or surfboard.

1964 Mrs L. B. Johnson *White House Diary* 21 May (1970) 142 One little girl..offered me a bunch of red and yellow wildflowers—'snake tongue' and 'rooster comb'. **1894** *Jrnl. Amer. Folk-Lore* VII. 94 *Dodecatheon Meadia*, var., shooting stars, roosters' heads. **1934** H. Vines *This Green Thicket World* 171 One not thicketed might have felt sorry for the blue daisies, white daisies, roosterheads. **1947** *Atlantic Monthly* July 41/2 Spring not only brought tadpoles but..big bunches of rooster-head violets that the children picked in the woods. **1953** *Marine Digest* 19 Sept. 29/1 She was boxed in on the first turn by Gale and the two Such Crusts and their combined rooster tails just about sank her. **1956** *Sun* (Baltimore) 26 Oct. 34/5 It ruled..that it was 'highly probable' the wake or fall of water from the 'rooster tail' of the boat travelling at high speed had caused the damage. **1963** *Pix* 28 Sept. 62/4 *Rooster-tail*, wake of a board. **1976** *Telegraph-Jrnl.* (St. John, New Brunswick) 7 Aug. 3/3 The small craft skim the river at incredible speeds. Their giant rooster tails sometimes reach a height of 90 feet.

Hence **roo·sterish** *a.*

1898 'Mark Twain' in *Harper's Mag.* Mar. 536/2 He stands vast and conspicuous..self-satisfied and roosterish.

roosting, *vbl. sb.*[1] Add: **2. a.** *roosting area, behaviour, habit, site, -stick, -time* (later examples), *-tree* (earlier example).

1976 T. Soper *Everyday Birds* v. 59 When birds reach the vicinity of the roosting area there will often be a spectacular flying display. **1953** *Brit. Jrnl. Animal Behaviour* I. 91 (*heading*) The winter roosting and awakening behaviour of captive Great Tits. **1964** A. L. Thomson *New Dict. Birds* 709/1 Roosting behaviour varies from species to species, and to some extent within a species. **1976** H. M. Dobinson *Bird Count* ix. 150 (*heading*) Roosting habits of our most common species. **1953** *Brit. Jrnl. Animal Behaviour* I. 91/2 Aggressive behaviour over roosting sites was seen on a few evenings. **1927** W. de la Mare *Told Again* 287 The hens on their roosting-sticks. **1743** W. Ellis *Mod. Husbandman* July xvi. 77, I..took the Hen and her Young at Roosting Time. **1825** C. Waterton *Wanderings in S. Amer.* 119 All the Toucanets feed on the same trees... You will find it has only been a dinner party, which breaks up and disperses towards roosting time. **1953** *Brit. Jrnl. Animal Behaviour* I. 91/1 In England the changes in the roosting times [of Great Tits] are similar to those in the north but less marked. **1834** J. J. Audubon *Ornith. Biogr.* II. 41 These roosting-trees of the Buzzards are generally in deep swamps, and mostly in high dead cypress trees.

root, *sb.*[1] Add: **I. 1. d.** In phr. *on* (*its*) *own roots*, used to describe a plant whose tissues all developed from the same embryo; not grafted or budded.

[**1822** J. C. Loudon *Encycl. Gardening* II. 436 The scion is a part of the living vegetable, which, united or inserted in a stock or other vegetable of the same nature, identifies with it, and grows there as on its natural stem and roots.] **1869** S. R. Hole *Bk. about Roses* viii. 112 The two trees ..are on their own roots, but the Rose thrives stoutly on the Brier and the Manetti, budded and grafted, wherever roses grow. **1914** H. H. Thomas *Gardening for Amateurs* 696/1 Plants may grow rampantly on their own roots to the material disadvantage of any useful products. **1944** Kains & McQuesten *Propagation of Plants* (rev. ed.) xiv. 334 Why do not nurserymen sell us plants on their own roots? The answer is that in no other way [than grafting] can fruit trees true to name be propagated so rapidly. **1968** *Horticultural Abstr.* XXXVIII. 630/2 Gialla Precoce Morettini on its own roots flowered earlier on a sandy soil than on a clay soil.

3. c. *U.S. dial.* A spell effected by the supposedly magical properties of certain roots. Cf. *root doctor, worker*, sense 22 below.

1935 Z. N. Hurston *Mules & Men* 340 Nearly all of the conjure doctors practice 'roots'. **1962** *Jrnl. Amer. Folklore* LXXV. 313 Local synonyms for the spell are 'curse', 'trick', 'fix', 'conjure', 'root', and 'hoodoo'.

4. a. (Later examples used of hair.)

1940 W. Faulkner *Hamlet* III. ii. 198 The bleached hair darkening again at the roots since it had been a year now since there had been any money to buy more dye. **1970** G. F. Newman *Sir, You Bastard* viii. 213 Brown roots growing through her split blonde hair.

c. (Further examples.)

1910 *Aëronaut. Jrnl.* XIV. 115 The angle of incidence of each wing gradually decreases from the root to the tip. **1948** H. Constant *Gas Turbines* v. 77 The blades stall at the root and tip. **1978** D. Küchemann *Aerodynamic Design of Aircraft* vi. 429 The upwash generated by that part of the body ahead of the root of the gross wing should also be close to that generated by the portion of the gross wing ahead of the root and should again be small.

d. *slang.* The penis.

1846 *Swell's Night Guide* 119/1 *Flash*, to sport, to expose, he flashed his root. **1902** Farmer & Henley *Dict. Slang* V. 289/2. **1970** K. Millett *Sexual Politics* III. vii. 329 It measures intelligence as 'masculinity of mind', condemns mediocre authors for 'dead-stick prose', praises good writers for setting 'virile example' and notes that since 'style is root' (penis), the best writing naturally requires 'huge loins'.

5. b. The bottom of the groove of a screw thread.

1892 *Screws & Screw-Making* (Britannia Co., Colchester) iii. 39 The diameter at the root of the thread. **1920** F. J. Camm *Screw Cutting* i. 6 In some instances American screws are measured at the bottom of the thread B; this portion is often called the root. **1964** S. Crawford *Basic Engin. Processes* (1969) xiv. 299 The root is the bottom portion of the groove between the flanking surfaces of the thread.

II. 11. d. *pl.* Established ties with a locality or region; one's social, cultural, or ethnic origins or 'background'. Also in colloq. phr. *to put down roots*, to become established in a place, to settle down.

1921 R. Macaulay *Dangerous Ages* iv. 82 The..infinitely loved Barry, who was going to give her roots. **1949** G. B. Shaw *Buoyant Billions* II. 21 Plenty of money and no roots. No traditions. **1969** A. G. Thomas in L. Durrell *Spirit of Place* 117 On three occasions, when he has bought a house and put down roots, the whole collection has been posted out to him. **1977** *Gay News* 7–20 Apr. 10/4 In Scots and Welsh schools children are taught about their national roots, culture and history. **1977** P. Theroux *Consul's File* 184 Is it possible to put down roots here?.. The Chinese won't, the Tamils can't, the Malays pretend they have them.

III. 14. a. (Further examples.)

1876 *Rep. Brit. Assoc. Adv. Sci. 1875* II. 13 (*heading*) Theorems on the *n*th roots of unity. **1941** Courant & Robbins *What is Math.?* ii. 100 The complex cube roots of 1..are the roots of the equation $x^2 + x + 1 = 0$. **1966** *Math. Rev.* XXXI. 29/1 If α is an algebraic integer, $\alpha \neq 0$, α not a root of unity, then at least one of the conjugates of α has absolute value greater than 1.

c. A unique node or vertex of a graph from which every other node can be reached. Also *root node*.

1857 A. Cayley in *Phil. Mag.* XIII. 172 The inspection of these figures will show at once what is meant by the term in question, and by the terms *root, branches*,..and *knots* (which may be either the root itself, or proper knots, or the extremities of the free branches). **1881** *Amer. Jrnl. Math.* IV 266 In a tree of N knots, selecting any knot at pleasure as a root, the tree may be regarded as springing from this root, and it is then called a root-tree. **1965** *Proc. Cambr. Philos. Soc.* LXI. 499 A tree is a connected topological graph without circuits. The vertices will also be called nodes or individuals. There is often one distinguished node called..the root. **1973** C. W. Gear *Introd. Computer Sci.* vii. 282 Formally, a tree is a set of nodes connected by branches such that there is one and only one way of going from one node to another via branch connections, and which has a distinguished node called the root node. **1973** S. Even *Algorithmic Combinatorics* vi. 109 A vertex *v* is called a root..of the graph if every vertex of the graph is reachable from *v*. **1976** E. Minieka tr. *Berge's Graphs & Hypergraphs* (ed. 2) iii. 32 A graph does not always have a root. **1977** *Sci. Amer.* Apr. 70/1 The location of the first key to be examined in a binary tree is traditionally known as the root; in the 31-word example the root is 16.

d. *digital root*: the digit obtained when all the digits of a number are added and the process is repeated on successive results until the outcome is a single digit.

1956 G. A. Montgomerie *Digital Calculating Machines* vii. 140 One such check number is the 'digital root' obtained by adding all the digits of the number. **1973** *Sci. Amer.* Dec. 120/1 One way to do it makes use of the old accountant's trick for checking addition by digital roots.

16*. Miscellaneous senses of uncertain affinity. Cf. *ROOT *v.*[1] 9.

a. *slang.* (orig. *Schoolboys'*). A forceful kick. Also **root about** (see quot. 1900).

1900 Farmer *Public School Word-Bk.* 169 *Root-about*.. (The Leys), promiscuous football practice. **1934** N. Scanlan *Winds of Heaven* 46 Matt gave him 'a root in the gear' and told him not to talk like a stable boy. **1961** in Webster, Caught him a great root with his boot on the backside.

b. *Austral. coarse slang.* An act of sexual intercourse. Also, a (female) sexual partner.

1959 in R. Chamberlain *Stuart Affair* (1973) xi. 111 Did you have a root? **1961** F. Hardy *Hard Way* iii. 77 The conversation led inevitably to women. Our shabby criminal struck a match revealing..a sign scrawled on the wall: 'Best American root—ring such and such a number.' **1969** *Private Eye* 4 July 14/3, I hear tell these *artists* in London don't exactly have to chase the odd *root*. **1973** A. Buzo *Rooted* I. 43 Hey, do you remember the time he got pissed out of his mind and fronted up to this old duck and asked her for a root? **1974** P. Kenna *Hard God* I. 33 Have you ever gone all the way with a girl?.. You know what I mean. Have you ever had a real root? **1976** D. Ireland *Glass Canoe* 147 Johnny Bickel..thought she'd be an easy root and began to take notice of her.

IV. 17. a. *root-bud* (later example), *-system, -thread, -tip, -zone*.

1977 J. L. Harper *Population Biol. of Plants* 290 Both sexes spread clonally by means of root buds. **1902** *Encycl. Brit.* XXV. 439/1 The presence of a feeble absorptive root-system and an extended surface of the shoot for transpiration and transudation are the outstanding points [of hydrophytes]. **1969** P. Thrower *Every Day Gardening* iv. 85/2 Budding is really a form of grafting and enables the grower to unite a garden rose with a root system or 'stock' obtained from a wild or vigorous rose. **1954** J. R. R. Tolkien *Fellowship of Ring* I. vii. 141 His grey thirsty spirit drew power out of the earth and spread like fine root-threads in the ground. **1954** —— *Two Towers* 66 Something between root-tip and leaf-tip. **1967** L. Picken *Organization of Cells* iv. 127 In the presence of low concentrations of phenolic compounds growing root tips also showed a temporarily increased frequency of fragmentation [of chromosomes]. **1953** J. Ramsbottom *Mushrooms & Toadstools* xviii. 206 The microflora is greater in the region of actively growing roots than in the soil generally: this is perhaps particularly true of bacteria, but also holds for fungi. This root-zone of increased population is known as a rhizosphere.

b. *root-canopy*.

1930 Blunden *Poems* 318 Thus the sacred well Is passed, and now the far root-canopy Issues its people, swift and slippery.

c. *root-forming, -room, -sort*; *root-eaten, -filled, -fringed, -pale, -weary* adjs.

1915 *Proc. Soc. Antiquaries London* XXVII. 149, I have often picked up on the surface of the camp pieces of old root-eaten human bones. **1918** *Atlantic Monthly* CXXII. 122 The Place d'Etoile was perhaps first adumbrated by wild boars concentrating on a root-filled marsh. **1946** *Nature* 19 Oct. 555/1 The root-forming capacity of penicillins G and X almost certainly resided in these substances themselves. **1944** Blunden *Shells by Stream* 5 Upon the root-fringed dais. **1960** S. Plath *Colossus* 63 Root-pale her meagre frame. *c* **1887** G. M. Hopkins *Poems* (1967) 103, I do advise You, jaded, let be; call off thoughts awhile Elsewhere; leave comfort root-room. **1977** *Interim* IV. iv. 4 Strips of unripened green, retarded from maturity by the moisture and greater root-room in the ditch's silts below. **1960** T. Hughes *Lupercal* 33 Worm-sort, root-sort, going where it is profitable. **1931** A. Huxley *Cicadas* 51 Never a tortured flower Shudders, root-weary, on the verge of flight.

18. *root-cellar* (later examples), *-crop* (later examples), *-cutting board, -field, -puller, -pulper, vegetable* (later examples); *root-loving, -pulping* adjs.

1872 Root-cellar [see *grain-box* s.v. *GRAIN *sb.*[1] 18 a]. **1965** E. L. Myles *Emperor of Peace River* II. ii. 184 After that we collected the potatoes and put them in the root cellar. **1901** L. H. Bailey *Princ. Vegetable-Gardening* ix. 271 Root crops require a cool season and a deep soil. **1969** *Oxf. Bk. Food Plants* 172 (*heading*) Crucifer and composite root crops. **1969** E. H. Pinto *Treen* 95 The introduction and gradual increase throughout the 18th century, in the growing of root crops for animal winter feed, led to the importance of the well worn root cutting board. **1932** Blunden *Fall in, Ghosts* 9 The crucifix surmounting the steps of granite in the middle of the rootfields. **1977** F. Parrish *Fire in Barley* ii. 18 Dan heard the bloodhounds..race across the root field towards the farm. **1947** C. S. Lewis in *Punch* 1 Oct. 324/1 Fruit-loving, root-loving gods. **1856** *Trans. Mich. Agric. Soc.* VII. 54 D. O. & W. S. Penfield, Detroit, one iron root puller. **1952** S. Selvon *Brighter Sun* ix. 161 With a root-puller attached the tractor would move up to a tree and the arms would reach down into the earth and wrest the tree out. **1940** *Chambers's Techn. Dict.* 730/1 *Root pulper*, a machine comprising a rotating steel disc furnished with cutters, to which roots that have been cleared of soil are fed. **1978** *Morecambe Guardian* 14 Mar. 22/3 (Advt.), Bale Sledge, Buck Rake, Root Pulper. **1910** *Chambers's Jrnl.* Mar. 207/2 The electricity furnished by this means serves to light the house.., and drives a chaff-cutter, a circular saw, and a root-pulping machine. **1957** P. Worsley *Trumpet shall Sound* 15 The people live by cultivating..root-vegetables. **1976** *Southern Even. Echo* (Southampton) 1 Nov. 4/3 The sandy soil there, he reckons, suits root vegetables just fine.

19. *root-treatment*; *root-filling, -planing, -rising*; *root-filled* adj.

1963 C. R. Cowell et al. *Inlays, Crowns, & Bridges* viii. 84 A post-retained crown is commonly indicated for a root-filled anterior tooth the natural crown of which has become discoloured. **1977** *Proc. R. Soc. Med.* LXX. 439/1 Teeth root-filled or crowned before operation were excluded from these results. **1963** C. R. Cowell et al. *Inlays, Crowns, & Bridges* viii. 85 The root filling should be well condensed **1969** *Gloss. Terms Dentistry* (*B.S.I.*) 23 *Root filling*, the permanent filling and sealing of the root canal of a tooth to avoid the accumulation within the root canal of fluids or micro-organisms. **1962** Blake & Trott *Periodontology* x. 105 For pockets under 3mm, only removal of calculus and root planing and polishing are necessary. **1922** D. H. Lawrence in *Poetry* XXI. 65 Until your veiled head almost touches backward To the root-rising of your erected tail. **1927** W. E. Collinson *Contemp. Eng.* 60 If a tooth is decaying or hollow we have it stopped or filled..or we have root-treatment (sterilization and withdrawal of nerve).

20. *root-cause, -conception* (further example), *-confusion, -divergence*, etc. (Now passing into adj.)

1915 E. Carpenter *Healing of Nations* i. 12 One might be on safer ground by trying to get at the root-causes of this war. **1977** J. Wainwright *Day of Peppercorn Kill* 191 He didn't trust us, Dick—that's the root cause. **1934** *Downside Rev.* LII. 223 As to the second root-conception of Cistercianism, the mere enunciation of an opinion..cannot be allowed. **1940** W. Empson *Gathering Storm* 25 The mind..now less easily decides On a good root-confusion to amass Much safety from irrelevant despair. **1927** Auden & Day-Lewis *Oxf. Poetry* p. vi, The logical conflict, between the denotatory and connotatory sense of words, which is the root-divergence of classic and romantic. **1933** E. Partridge *Words, Words, Words!* I. 88 The root-idea of blood as something vivid or distressing or both still colours the use of the adjective. **1923** D. H. Lawrence *Kangaroo* vii. 141 Hardly sympathy at all, but an ancient sort of root-knowledge. **1960** *Spectator* 7 Oct. 518/2 Mr. Kimche is arguing against the consistent record, and against the very root-logic of Zionism. **1933** H. Read *Art Now* I. 47 This brings us down to the root-problem of aesthetics. **1957** M. Swan *Brit. Guiana* iv. 68 It is one of the root problems of the country. **1924** R. Hichens *After Verdict* III. xiv. 491, I hated her Then because I loved you. That was the root reason. *a* **1957** R. Campbell tr. A. de Campos in *Coll. Poems* (1960) III. 138 Which, once constructed, announce themselves As Real-Things, Spirit-Things, or Entities of the Stone-Soul, Made ours at certain moments by root-sensations. **1976** S. Hynes *Auden Generation* ii. 56 As the decade moved on, these images took on heavier symbolic meanings..but the root-sense of the images remained the same.

21. b. (Further examples.) *root-accent, -class, determinative, -element, -enlargement, -expansion, form, -language, -morpheme, -noun, -play, -stem, stress, -syllable* (later examples), *-word* (later examples); *root-accented, -final, -initial, -forming, stressed* adjs.

1935 G. K. Zipf *Psycho-Biol. of Lang.* 133 The explanation offered by Jespersen for extensive root-accent. **1975** *Language* LI. 140 The more commonly occurring root-accented forms *trắmane, dắmane, dhắrmane, bhắrmane*. **1879** W. D. Whitney *Sanskrit Gram.* ix. 208 The root-class [of verbs]..its present-stem is coincident with the root itself. **1965** G. Y. Shevelov *Prehist. of Slavic* xxiv. 367 It is to be assumed that in these words the varying consonants had not been originally a part of the root but were the so-called root determinatives, a kind of suffixes whose function is no longer discoverable. **1935** G. K. Zipf *Psycho-Biol. of Lang.* 145 When the accent.. was not on the endings, it was always on the stem-formative (suffix or infix) and not on the root-element. **1976** *Archivum Linguisticum* VII. 63 The discrepancy between the consonants..is easily accounted for by the assumption of different root-enlargements. **1895** P. Giles *Short Man. Compar. Philol.* xxv. 370 The details of the theory of root-expansion are..as yet too little worked out. **1965** H. M. Hoenigswald in W. Winter *Evidence for Laryngeals* 93 Such extra-Indoiranian etymologies as have been advanced with any promise mostly involve root-final position for the voiceless aspirates. **1973** *Trans. Philol. Soc.* 1971 68 *sil-* is not a permissible Indo-European root form. **1933** L. Bloomfield *Language* 275 Even our root-forming morphemes..have some flexibility. **1956** *Language* XXXII. 453 The root-initial verb aspect markers are most aptly described in terms of simulfixation. **1972** *Ibid.* XLVIII. 477 The alternations in the non-nasal prefixes are conditioned by the voicing of the root-initial consonant. **1885** *Encycl. Brit.* XVIII. 774/1 A Chinese monosyllable or an Egyptian or Polynesian dissyllable is radical, unless there can be demonstrated in some part of it a formative value; and a language wholly composed of such words is a root-language. **1935** G. K. Zipf *Psycho-Biol. of Lang.* 177 The total magnitude of complexity of the root-morpheme *fac*, a typical example, was diminished. **1950** *Lingua* II. 241 He makes only a few isolated remarks about the morphemes that occur most frequently, i.e. the root-morphemes. **1972** *Language* XLVIII. 477 Such a proto-initial is very poorly attested by the comparative data in root-morphemes. **1879** W. D. Whitney *Sanskrit Gram.* xiii. 314 The root-noun used as infinitive has the same form, and the same accent,..as in its other uses. **1962** C. Watkins *Indo-European Origins of Celtic Verb* I. 185 The verbal root *med- being identical with the athematic root noun *med-. **1970** M. Dahood *Psalms* III. 109 The rootplay evident in *yilbʿšu* and *boštām*..is of a piece with the wordplays that wryly characterize many biblical and Canaanite laments. **1879** W. D. Whitney *Sanskrit Gram.* v. 129 Root-stems, having in them no demonstrable element added to a root. **1979** T. Burrow *Problem of Schwa in Sanskrit* 66 The root stem *nās-* f. 'nostril, nose' inflects with long vowel in the only strong case which occurs in the Veda. **1965** G. Y. Shevelov *Prehist. of Slavic* iv. 68 In all these cases the Li[thuanian] F[alling]P[itch] type has root stress. *Ibid.* 69 Analogy with the root stressed instr[umental] and loc[ative] pl[ural]. **1900** H. Sweet *Hist. Lang.* vi. 103 The place of the accent [in Aryan] was not restricted by any considerations of quantity or distance from the end of the word,..nor was it restricted to the root-syllable of a word, as was afterwards the tendency in the Germanic language. **1972** *Language* XLVIII. 477 Arranging the words in the alphabetical order of their root-syllables. **1918** R. Bridges in G. M. Hopkins *Poems* 100 Passages where, in a jungle of root-words, emphasis seems to oust euphony. **1954** H. Read *Anarchy & Order* 196 The root-word *vir* [in *virtue*] has the implication of masculinity.

22. root-aorist *Philol.*, in certain Indo-European languages, an aorist formed by adding personal endings directly to the root-syllable of the verb; **root-ball**, (*a*) = NIGGER-HEAD 1 a in Dict. and Suppl.; (*b*) the mass formed by the roots of a plant and the soil between and around them; hence **root-balled** *a.*; **root beer** (earlier and later examples); **root-bound** *a.*, † (*a*) bound or held by roots; (*b*) = POT-BOUND *a.*; also *fig.*; †**root bread** *U.S.*, the bulbs of *Camassia quamash* (cf. CAMAS, QUAMASH), formerly baked and eaten in western North America; **root canal,** the pulp-filled cavity within the root of a tooth; **root cutter,** (*a*) an implement for cutting edible roots; (*b*) one for cutting tree roots underground; **root cutting,** a cutting taken from the root of a plant; **root digger,** (*b*) (earlier and later examples); **root doctor** *U.S. dial.*, one who treats ailments by means of roots, a herb-doctor; also = *root worker* below; **root gall** (see quot. 1902); **root-graft** *sb.*, (*a*) a graft of a scion on to a root; (*b*) a naturally occurring graft between the roots of neighbouring trees; hence **root-graft** *v. trans.*, to graft by means of a root-graft; **root-grafted** *ppl. a.*, **root-grafting** *vbl. sb.*; **root-knot,** a disease of many crop and other plants, caused by infestation of the roots with the nematode *Heterodera marioni* producing characteristic swellings or nodules; freq. *attrib.*; **root-mean-square** *Physics*, a mean calculated as the square root of the arithmetic mean of the squares of a set of values; freq. *attrib.*; **root nodule,** a swelling on a root of a legume or other higher plant containing symbiotic micro-organisms which fix nitrogen; **root pressure** *Bot.* [tr. G. *wurzelkraft* (J. von Sachs *Handb. der Exper.-Physiol. der Pflanzen* (1865) IV. vii. 199)], the hydrostatic pressure generated in the roots of a plant, which helps the sap to rise in the xylem; **root rot,** a disease of plants, attacking the roots; **roots reggae,** a style of reggae music considered as an expression of the black Jamaicans' cultural identity; **root swell(ing),** an outgrowth of a tree above a root, forming a natural buttress; † **root tubercle** = *root nodule* above; **root worker** *U.S. dial.*, one who uses roots to work spells, a conjurer (cf. sense 3 c above); so **root work.**

1879 W. D. Whitney *Sanskrit Gram.* xi. 276 Imperative forms of the root-aorist are not rare in the early language. **1955** H. G. Lunt *Old Church Slavonic Gram.* iv. 89 The most wide-spread type of the older aorists was the so-called 'root-aorist', attested by over 650 examples with some 27 verbs. **1976** *Archivum Linguisticum* VII. 62 In Oscan-Umbrian *-e-* is the sign of a secondary thematization of the Indo-European root-aorist. **1930** *Sat. Even. Post* 13 Dec. 11/2 Bogs of black muck dotted with devilish, rotating root-balls that throw a man waist-deep. **1956** X. Field *Housewife Bk. House Plants* I. 31 If the outside of the root ball is a network of roots re-potting is called for. **1973** J. L. Faust *Bk. House Plants* 37 If the root ball of the plant is very tightly packed and hard, it can be squeezed a bit to break it apart. **1966** *Gloss. Landscape Work* (*B.S.I.*) IV. 19 *Root-balled*, with roots contained within a well-defined mass of soil (in practice usually wrapped with protective material). **1843** *Knickerbocker* XXII. 85 Let..the temperance halls and the root-beer perambulatories make answer. **1921** [see *COCA-COLA]. **1974** E. Brawley *Rap* II. xix. 250 Sucking on his root beer freeze through a red plastic straw. **1634** Root-bound [in Dict., sense 17 c]. **1885** R. T. Cooke *Root-Bound* 11 My plants do blossom well.. and I don't know why unless it is because they are root-bound. *Ibid.* 12 It's good for folks and flowers too to be root-bound..sometimes; especially, if we want to bring forth good fruit. **1946** *Nature* 23 Nov. 762/2 Further experiments show the importance of..the feeding of root-bound plants with a balanced fertilizer prior to transplanting. **1976** *S. Wales Echo* 27 Nov. 4/1 He informed the schoolmaster that finding a square root meant looking for limp leaves in seedlings or pot plants, this condition being caused by them being root-bound in box or pot. **1805** W. Clark in *Orig. Jrnls. Lewis & Clark Expedition* (1905) III. 85 Traded for some root Bread & skins to make shirts. **1806** J. Ordway in Lewis & Ordway *Jrnls. Western Explor.* (1916) 352 We bought a little dark couloured root bread which is not good but will Support nature. *c* **1840** D. Thompson *Narr. Explorations W. Amer. 1784–1812* (1916) II. iv. 413 An old Man made a short speech, and made a present of two cakes of root bread (not moss). **1893** *Dental Rec.* XIII. 523 (*heading*) Filling root canals with coal wadding. **1923** *Ibid.* XLIII. 269 The root-canals afford excellent hold for posts. *Ibid.* 682 The first requisite for root-canal filling is the complete sterilisation of the root-canal and tubuli. **1978** S. Sheldon *Bloodline* xli. 356 A dental bill for root-canal work for Charles Martel. **1807** Root cutter [in Dict., sense 18].

a **1877** Knight *Dict. Mech.* II. 1975/2 To bring the roots to a convenient size for the stock and to remove the danger of choking, root-cutters were introduced. **1943** J. Stuart *Taps for Private Tussie* xxii. 226 You could follow the mule behind a locust-beamed plow with a sharp root cutter in it and hear the roots pop. **1969** E. H. Pinto *Treen* 18 Root Cutters. The traditional type illustrated..with pivoted knife at one end, was made in a considerable range of hardwoods.., and was used generally..by apothecaries. **1954** A. G. L. Hellyer *Encycl. Garden Work* 68/2 As a rule root cuttings are taken while the plant is dormant, which means generally, in winter. **1969** P. Thrower *Every Day Gardening* v. 108/1 Propagation by root cuttings is the best way of increasing many thick-rooted perennials like verbascums, Oriental poppies, phlox, anchusa and *Limonium latifolium*. **1831** W. Gordon *Let.* 3 Oct. in A. H. Abel *Chardon's Jrnl. at Fort Clark* (1932) 346 Many of these [Snake Indians] go by the name which signifies Root digger, because they live by digging roots. **1837** W. Irving *Capt. Bonneville* II. xii. 204 These are of that branch of the great Snake tribe called Shoshokoes, or Root Diggers, from their subsisting, in a great measure, on the roots of the earth. **1947** B. A. De Voto *Across Wide Missouri* 432 'Root-digger'..describes all the tribes, most of them superior tribes, that lived in localities where there were staple crops of edible roots and bulbs. **1821** J. Howison *Sk. Upper Canada* xii. 195 'Oh!' said the woman, 'if I had but the *root* doctor that used to attend our family at Connecticut; he was a dreadful *skeelful* man.' **1890** *N.Y. Age* 19 Apr. 1/1 Carmier was what people call down here a root doctor... He only rode around the county..and made his living curing the sick and selling his medicine. **1900** *Jrnl. Amer. Folklore* XIII. 228 People git conjur from the root-doctors and one root-doctor often works against another, the one that has the most power does the work. **1934** [see *MOJO[1]]. **1962** *Jrnl. Amer. Folklore* LXXV. 315 She finally went to a root doctor and was informed that her husband and three women had placed a spell upon her. **1902** L. H. Bailey *Cycl. Amer. Hort.* IV. 1545/2 The term root-gall is usually applied to the abnormal enlargement of roots due to insects and other animal organisms. **1933** *Jrnl. R. Hort. Soc.* LVIII. 233 The absence of detailed information regarding the infective stage of the root-gall nematode and its life history has been remedied. **1977** J. L. Harper *Population Biol. Plants* xvi. 484 (*caption*) *Biorrhiza pallida* forms root galls and meristem galls on the oak at different seasons. **1824** J. C. Loudon *Encycl. Gardening* (ed. 2) II. 396 Such root-grafts grow with uncommon vigour. **1900** L. H. Bailey *Cycl. Amer. Hort.* II. 661/2 In the West apples at least are usually root-grafted. **1951** F. J. Chittenden *Dict. Gardening* II. 919/2 Rhododendrons..are..frequently root-grafted, using roots of common species of their genus as stocks. **1956** *New Biol.* XX. 101 There is evidence that the fungus can infect trees only through wounds that penetrate the bark. The disease spreads locally by means of natural root grafts. **1900** L. H. Bailey *Cycl. Amer. Hort.* II. 663/2 In the East..budded apple trees are preferable to root-grafted trees. **1942** Kains & McQuesten *Propagation of Plants* (rev. ed.) xii. 294 Ten Walldow root-grafted trees were all dead but one limb on one tree. **1707**, *c* **1820** Root grafting [see sense 17 a in Dict.]. **1886** G. Nicholson *Illustr. Dict. Gardening* II. 91/2 Plants largely propagated by Root-grafting are Bignonias, Clematis, Hollyhocks, and Wistarias. **1977** J. L. Harper *Population Biol. of Plants* 235 It is doubtful whether any careful search has ever been made to detect the extent of root grafting in other communities of herbs. **1889** *Bull. U.S. Dept. Agric. Div. Entomol.* No. 20 (*title*) The root-knot disease of the peach, orange, and other plants in Florida, due to the work of Anguillula. *Ibid.*, 9, I..can find no mention of the root-knot..prior to the year 1857. That year Hon. P. J. Berckmans..found this disease prevalent. *Ibid.*, In 1876 I found the root-knot prevalent over Florida, and learned from old residents that as far back as 1805 it had been known. **1912** E. W. Swanton *Brit. Plant-Galls* viii. 107 Miss Ormerod first reported the occurrence of this pest, known as the 'root-knot' eelworm, in Britain. **1954** *New Biol.* XVI. 113 The Root Knot Eelworm..is a tropical or sub-tropical species which in Britain infests the soil of heated glasshouses... Its host-range includes almost all the decorative plants grown in conservatories. **1976** *Daily Times* (Lagos) 8 June 2/2 The workshop is discussing Integrated Crop Protection System with emphasis on root-knot diseases affecting economic crops. **1895** *Electrician* 27 Sept. 721/1 A short time ago Dr. Fleming published a new and ingenious method of plotting wave forms with polar co-ordinates, and of directly obtaining therefrom the root mean-square value. **1927** S. H. Long *Navigational Wireless* i. 7 Thus the effective value I equals the square root of the mean value of the squares of all the instantaneous values. This is often called the root-mean-square value, or R.M.S. value or virtual value. **1956** A. A. Townsend *Struct. Turbulent Shear Flow* iii. 51 The rate of increase of the decay scale is proportional to the root-mean-square turbulent velocity. **1978** *Nature* 9 Mar. 143/2 We note here that sound pressures as well as displacement are expressed as root-mean-squares. [**1899** J. B. Farmer *Botany* ix. 44 Plants which have not these nodules on their roots are unable to utilize the free atmospheric nitrogen.] **1907** F. Cavers *Plant Biol.* iii. 119 The root-nodules of leguminous plants contain a micro-organism which fixes free atmospheric nitrogen. **1949** A. Nelson *Introd. Bot.* xxv. 391 The root nodule, so typical of this bacterial association with a legume, commences when the bacterium enters the root hair of the legume. **1976** Bell & Coombe tr. *Strasburger's Textbk. Bot.* (rev. ed.) 293 In the root nodules of alder, *Hippophae, Eleagnus*, and also probably of *Myrica* and *Casuarina*, the organisms concerned are symbiotic actinomycetes, also capable of fixing atmospheric nitrogen. **1875** Bennett & Dyer tr. *Sachs's Text-bk. Bot.* III. 600 (*caption*) Apparatus for observing the force with which water escapes under root-pressure from the transverse section of a stem. **1896** *Phil. Trans. R. Soc.* B. CLXXXVI. 572 An important function of root-pressure, *i.e.*, to dissolve up and clear out the gaseous contents of such conduits as are occupied by bubbles. **1931** E. C. Miller *Plant Physiol.* iv. 168 Under conditions of low transpiration or in the spring before the leaves are unfolded, water is forced into the conducting vessels of the root and up through the stem under pressures varying from a fraction of an atmosphere to several atmospheres. This pressure is evidenced by the bleeding of cut vines and branches of certain species of plants and is apparently connected with the exudation of water from the leaves of plants, which occurs under certain conditions. This pressure, which is set up in the fibrovascular bundles of the stem and root due to the water which is being forced in them is known as 'root pressure'. **1976** *Sci. Amer.* May 104/3 Although Hales had discovered the existence of root pressure, he concluded that the roots are not solely responsible for the pressure of the sap in the branches. **1883** Root rot [in Dict., sense 17 a]. **1933** *Jrnl. R. Hort. Soc.* LVIII. 280 The occurrence of root-rot of Sweet Peas ..is described as one of the causes possibly associated with the streak disease of Sweet Peas. **1978** Evans & Kumm *Woman's Own Pot Plant Doctor* 21/1 The commonest reason for all house plants dying off is root rot. **1977** McKnight & Tobler *Bob Marley* x. 127 What reaches our ears is no longer roots reggae. **1978** *Oxford Times* (City ed.) 24 Feb. 15 This is a good example of roots reggae complete with chunky rhythm and 'dub' echoes. **1932** *Sun* (Baltimore) 6 Sept. 6/11 The famous Wye oak..is reported to be 27 feet 8 inches in circumference four and a half feet above ground, but the measurement taken at this point is said to include large root swells. **1902** *Forestry Q.* I. 56 The influence of the enlarged base of the bole (root-swelling) is appreciable at the breast-high point, and gives the stem a neiloid form. **1954** W. E. Hiley *Woodland Managem.* ix. 134 By girthing at 6 feet instead of 5 feet it may be possible to get away from the root swelling which usually occurs at the base of a large tree and often gives rise to inaccurate estimating. **1887** H. Marshall-Ward in *Phil. Trans. R. Soc.* B. CLXXVIII. 539 The first close investigation of these root-tubercles (as they may be shortly termed) is due to Woronin. **1894** *Knowledge* 1 Mar. 68/1 (*heading*) The root-tubercles of peas, beans, and vetches. **1897** W. G. Smith tr. *Tubeuf's Dis. Plants* ix. 101 First-year alders without tubercles do not thrive in soil free from nitrogen..; when, however, provided with root-tubercles they assimilate nitrogen. **1967** D. C. Tinling in *Psychosomatic Med.* XXIX. 483 (*heading*) Voodoo, root work, and medicine. **1970** M. Walker *Prophets for New Day* 29, I run down to Sis Areny's And told her what I seen 'Root-worker's out to git me What you reckon that there mean?'

root, *v.*[1] Add: **II. 3. b.** *Austral. slang.* (See quot. 1959.)

The placing of this sense is uncertain; it may be, or be apprehended as, a *fig.* use of sense 9 b below.

1945 Baker *Austral. Lang.* viii. 152 The authentic digger form is *Wouldn't it root you!* A regimental paper 'Wiry' (1941) took its name from the first letters of the words in this phrase. **1951** D. Stivens *Jimmy Brockett* 244 'It looks as though we're rooted, smacker,' I told Herb. **1959** Baker *Drum* II. 140 *Root*,..to outwit, baffle, exhaust, utterly confound (someone). Whence, *to be rooted*, to be exhausted or confounded; *get rooted!* Go to blazes! **1961** M. Calthorpe *Dyehouse* (1962) xl. 186 'He can get rooted, for all I care,' Collins said bitterly. **1973** *Telegraph* (Brisbane) 15 Nov. 3/1 Mr. Whitlam later admitted having said in an aside: 'It is what he put in his guts that rooted him.' **1974** J. Powers *Last of Knucklemen* III. ii. 93 'What the hell's goin' *on* here?' 'The Hun's rooted—that's what!' 'Done like a dinner!'

3*. To cause (a cutting) to grow roots.

1824 J. C. Loudon *Encycl. Gardening* (ed. 2) II. 400 All plants which are difficult to root..will be found in the first instance..to throw out roots only, from the ring of herbaceous matter. **1884** D. T. Fish *Pop. Gardening* I. 212/1 One strong argument in favour of rooting roses at that season [*sc.* spring] consists in the fact that they have all the summer before them to grow into plants. **1925** W. Watson *Gardener's Assistant* VI. 82/1 We root a Cactus by drying it in the sun. **1969** P. Thrower *Every Day Gardening* iii. 45/1 Cuttings which have been rooted under mist, or in a heated propagator, must be hardened off..before planting them in the open ground.

4. c. (Further examples.)

1941 *Sun* (Baltimore) 25 Nov. 14/3 The trouble into which he intervened roots in a controversy over whether welding is a separate 'art' or not. **1955** E. Pound *Section: Rock-Drill* lxxxix. 56 The Civil War rooted in tariff.

III. 9. Miscellaneous senses of uncertain affinity. (Perh. properly developments of *v.*[2])

a. *trans.* and *intr.* To kick, esp. in the backside. *slang* (chiefly *Schoolboys'*).

1890 Barrère & Leland *Dict. Slang* II. 186/1 *Root, to* (schools and London), to give one a kick behind. **1914** 'I. Hay' *Lighter Side School Life* ii. 52 We rooted Sowerby afterwards for grinning. **1934** *Bulletin* (Sydney) 31 Jan. 32/2 Give the horse that can root a bit to the horse-breaker or the head stockman is the general rule, or, better still, to the blacks. **1946** B. Marshall *George Brown's Schooldays* xxxvii. 145 Rooting them [*sc.* new pupils] up the backside is the only way of dealing with them.

b. *Austral. coarse slang. trans.* (usu. with a male subject). To copulate with; *intr.*, to copulate, to engage in sexual intercourse. Also in phr. *to root like a rattlesnake*, to copulate vigorously.

1958 R. M. Stuart in R. Chamberlain *Stuart Affair* (1973) ii. 12, I took her bathers off. Then I raped her. She was hard to root. **1966** P. White *Solid Mandala* 185 We'll root together so good you'll shoot out the other side of Christmas. **1969** *Private Eye* 1 Aug. 14 The Pope's a Jew if that jam tart doesn't root like a rattlesnake. **1974** K. Cook *Bloodhouse* 110 We found this bloody little poofter down on the beach fiddling with a bird... Couldn't even root her.

root, *v.*[2] Add: **1. c.** (Further examples.) Also const. *about, around*. Now also *colloq.*

1904 in *Eng. Dial. Dict.* V. 151/2 They like to rute about the house. **1916** Joyce *Portrait of Artist* v. 203 He allowed his mother to..root into the folds of his ears. **1920** R. Macaulay *Potterism* I. ii. 20 Watching Tane's.. hand with its short square fingers rooting in the sand for shells. **1943** V. Palmer in *Coast to Coast 1942* 29 Charlie rooted about in the nose of the dinghy drawn up above the tide. **1977** C. Rocks *Winter's Tales 23* 132, I rooted around till I found the kettle.

d. For *U.S. slang* in Dict. read *colloq.* (orig. *U.S. slang*) and substitute for def.: To cheer *for* a (baseball, etc.) team. Also *transf.*, to be active *for* a person or thing by giving support, encouragement, or applause. Also without const. (Earlier and later examples.)

1889 *N.Y. Semi-Weekly Tribune* 5 Nov. 5/4 Murphy has done little but 'root' for the Giants this year. **1895** J. S. Wood *Yale Yarns* 152 We rooted hard, too, and did a lot of shouting and yelling. **1922** S. Lewis *Babbitt* v. 66 Zilla keeps rooting for a nice expensive vacation. **1943** *Crisis* July 201/3 The papers of Los Angeles crowed... They rooted and cheered. **1951** *Sport* 30 Mar.–5 Apr. 3/1 If the rules of the tournament made it possible for Stan to be transferred to Newcastle tomorrow, then the whole country would be rooting for the 'Magpies' on April 28th. **1951** in M. McLuhan *Mech. Bride* (1967) 8/1 He rooted fiercely for the underdog, perhaps because he was so much the underdog himself. **1959** N. Mailer *Advts. for Myself* (1961) 400 If he dares not to castrate his hatred of society..then I would have to root for him because he may have been born to write a great novel. **1967** *Boston Sunday Herald Mag.* 9 Apr. 4/3 You'll find it becomes a whole different game from just sitting in your armchair, rooting blindly. **1971** A. Burgess *MF* xii. 140 A popcorn-eating audience roots for two youths fighting a huge engulfing python. **1976** A. Miller *Inside Outside* vii. 81, I..wound up in front of the Visiting Committee with the Governor rooting for me.

e. *root hog or die*, used of or addressed to persons, implying the necessity of labour or exertion to maintain life or prosperity. Also as *attrib. phr. N. Amer.*

1834 D. Crockett *Narr. Life* viii. 60 We therefore determined to go on the old saying, root hog or die. **1843** *Amer. Pioneer* II. 419 This letter exhibits his as well as my own case in that day; for it was 'root hog or die', and hard times have come back again! **1879** A. W. Tourgée *Fool's Errand* xxv. 150 The 'root-hog-or-die' policy. **1904** *N.Y. Even. Post* 20 Aug. 4 'The school and college', explained plains President Eliot, 'cannot use the method of Nature—root, hog, or die.' **1931** J. T. Adams *Epic of Amer.* i. 37 At the beginning of most settlements it was 'root, hog, or die' for all. **1976** *Globe & Mail* (Toronto) 9 June 41/6 Many of that generation, however, no longer put up with that root-hog-or-die kind of motivation.

rootage. Add: **3.** A system of roots; a root-stock.

1895 in *Funk's Stand. Dict.* **1927** H. E. Fosdick *Pilgrimage to Palestine* 60 The very rootage from which came Aaron's rod that budded..the solemn monks still show to the visitor.

rooted, *ppl. a.* Add: **1. a.** (Further examples.) *spec.* Having been made to grow roots.

1852 G. W. Johnson *Cottage Gardener's Dict.* 304/2 A rooted cutting is not a new plant, it is only an extension of the parent. **1969** P. Thrower *Every Day Gardening* viii. 196/2 Give the rooted cuttings as much light as possible to prevent them from becoming drawn.

rooter[1]. Add: **1. b.** A machine for loosening the surface of the ground.

1950 *N.Z. Jrnl. Agric.* Oct. 333/2 At first a small rooter was used which was designed for scarifying pavements and roads and which could be drawn by a 40 h.p. tractor. **1965** G. J. Williams *Econ. Geol. N.Z.* Plate xxxix, This very pure calcareous material is soft enough to be excavated by rooters, and in consequence can be produced at low cost.

rooter[2]. **2.** For *rare* in Dict. read '*colloq.* (chiefly *U.S.*)' and substitute for def.: One who cheers or 'roots' for a (baseball, etc.) team. Also *transf.*, one who supports or encourages another; a warm advocate, a partisan. (Earlier and later examples.)

1890 *N.Y. Press* 8 July 6/1 At this juncture the New York rooters began to 'pull' for the home team, but the effort was useless, not a man..succeeded in reaching first base. **1901** *Daily Colonist* (Victoria, B.C.) 31 Oct. 4/3 'What makes him look so very white?' inquired the fairy maid. 'He's had the starch knocked out of him,' the woolly rooter said. **1931** L. Steffens *Autobiogr.* II. III. xxxiii. 593 They..don't ask about, they don't hear of, the always existing few quiet students with concealed gifts in the rooters at a football game. **1939** G. Ade *Let.* 7 July (1973) 212, I attended the [1912 Republican] convention as a spectator and also as a rooter for Theodore Roosevelt. **1952** *Manch. Guardian Weekly* 12 June 3/3 General of the Army Eisenhower..came..to hold his first political press conference before the New York reporters... There was also present a motley and vociferous band of rooters. **1959** *Times Lit. Suppl.* 6 Nov. p. xix/1 The exclusive audience that goes to Twickenham or Lord's is replaced by millions of rooters. **1963** D. Ogilvy *Confessions Advert. Man* i. 14 This will give us 49,700 rooters for Ogilvy, Benson and Mather. **1978** *N.Y. Times* 29 Mar. B 5/1 The Wildcats' coach..had walked back into the stands to embrace his family and shake hands with some rooters with blue and white buttons.

rootfast, *a.* (Later *poet.* example.)

1953 C. Day Lewis *Italian Visit* III. 36 A character root-fast Like a man's in the deposit of all his acts.

root-house. 2. (Earlier and later examples.)

1790 *Pennsylvanian Packet* 30 Mar. 4/2 On the premises are..two arched stone root-houses. **1961** W. O. Mitchell *Jake & Kid* 28 'Bin a real fine summer fer vegetables,' he said then. 'Too bad yer ma don't have no root house.' **1970** *Islander* (Victoria, B.C.) 22 Nov. 13/2 Into the root house went the potatoes, carrots and other root vegetables.

rootiness. Add: Also *fig.* Cf. RACINESS.

1937 G. M. Young *Daylight & Champaign* 193 Here is exactly the harshness and rootiness, the integrity and objectivity that our poetry needed.

rooting, *vbl. sb.*[1] Add: **1. b.** *rooting medium.*

1935 A. F. Hort *Garden Variety* IV. 229 The rooting medium is about six inches of ordinary builders' sand.

5. *coarse slang.* Of a male: the action or process of copulating. (Now chiefly *Austral.*: cf. *ROOT *v.*[1] 9 b.)

1922 Joyce *Ulysses* 754 All the poking and rooting and ploughing he had up in me. **1970** G. Greer *Female Eunuch* 41 All the vulgar linguistic emphasis is placed upon the *poking* element; *fucking, screwing, rooting, shagging* are all acts performed upon the passive female.

rooting, *vbl. sb.*[2] Add: **2.** *slang* (chiefly *U.S.*). Cheering, encouraging, or otherwise supporting. Also in *Comb.*, as *rooting interest*. Cf. ROOT *v.*[2] 1 d.

1937 D. Runyon in *Collier's* 21 Aug. 32/4 No talking and no rooting from the spectators is permitted. **1971** L. Koppett *N.Y. Times Guide Spectator Sports* viii. 155 There is the team element as a rooting interest. **1977** *Time* 25 July 51/2 *One on One* is a picture that..transcends its humble conception and develops what movie people used to call a 'rooting interest' in its characters.

rooting, *ppl. a.*[2] Add: **2.** In redupl. Comb., as **rootin' tootin'**, (*a*) *dial. rare*, inquisitive, meddlesome; (*b*) *slang* (chiefly *N. Amer.*), noisy, rumbustious, boisterous; rip-roaring, lively. Cf. *ROOTY-TOOT.

1875 Nodal & Milner *Gloss. Lancs. Dial.* 228 He's a *rootin'* tootin' sort of a chap. **1924** L. B. Kozlowski in *Catal. Copyright Entries* (U.S. Copyright Off.) III. 5698 (*song-title*) Roottin-toottin-Lou; from Kalamazoo. **1937** *Film Daily* 1 May 4/3 (*heading*) Gene Autry in Rootin' Tootin' Rhythm. **1948** *Sun* (Baltimore) 7 Jan. 13/1 Basketball games today have developed into rootin', tootin' contests, with more of the emphasis on the tootin'. **1949** *N.Y. Times Bk. Rev.* 27 Mar. 32 'Smoke up the Valley' is actually a rootin'-tootin' romance of blazing six-shooters and gore. **1963** *New Statesman* 24 May 786/3, I also..expressed mild surprise about a rootin' tootin' night club that advertised 'Girls! Girls! Girls!'

rootle, *v.* Add: **1.** Also const. *about, round*. (Further examples.)

1917 Kipling *Diversity of Creatures* 261 There's a tin of cocoa in my study somewhere... Rootle round till you find it. **1929** V. Woolf *Room of One's Own* i. 14 The chapel itself was marsh too, where the grasses waved and the swine rootled. **1936** A. Christie *Cards on Table* ix. 85 I'll leave you my keys and..you can rootle to your heart's content. **1943** *Theology* xlvi. 159 It is coming to be seen that he [*sc.* Nietzsche] rootled about in the subsoil of the modern mind to the profit of few things so much as the Christian Faith. **1959** *Elizabethan* Apr. 10/2 We rootled among the debris for something to eat. **1964** P. White *Burnt Ones* 203 On the way, as she rootled after the lovely little lighter, I was relieved to see her bag was still stuffed with notes. **1977** *Zigzag* Apr. 43/3 He rootled about under the stairs and found an unopened Christmas present bottle of Glenfiddich.

2. (Further examples.) Also *transf.*

1945 D. Rees *Cambridge Murders* xiii. 135 He set one or two members of his staff to rootle out the past histories of all the people whose names had been mentioned. **1955** M. Banks *Commando Climber* x. 189 Their [*sc.* the reindeers'] disappearance has been attributed to a late autumn thaw..which covered their winter pasture with a crust of ice that prevented them rootling out their fodder. **1978** *New Scientist* 20 July 171/2 Pigs which rootle out the eggs and eat the vulnerable young.

rootless, *a.* Add: **1. b.** (Further examples.)

1934 A. Woollcott *While Rome Burns* 93 Those rootless widows who wear buttoned shoes. **1977** R. Barnard *Blood Brotherhood* viii. 79 Rootless young men, without families.

rootlessness (further examples); also **roo·tlessly** *adv.*

1927 E. Bowen *Hotel* xiv. 167, I haven't had time for a feeling of rootlessness. **1929** A. Huxley *Do what you Will* 157 Nature-worship is..so refined, so rootlessly high-class. **1958** *Times Lit. Suppl.* 3 Jan. 4/4 Mr Mellers characterizes the nineteenth century Russian Westernizers' music as 'rootlessly European' and believes that Stravinsky has merely made this rootlessness symbolic of the modern artist in general. **1978** P. Moore *Man, Woman, & Priesthood* i. 4 In an age of rootlessness which is searching for significant tradition, we have much to offer if we have the courage to dig deep enough. **1980** *Church Times* 1 Aug. 2/4 Drinking problems are, more often than not, connected with other social problems such as homelessness and rootlessness.

root-stock. Add: **3.** A stock on to which another variety has been grafted or budded.

1933 H. H. Thomas *Pop. Encycl. Gardening* 392/1 The shoot of the required variety is united with a suitable rootstock or with a branch of an established tree of the same kind. **1954** A. G. L. Hellyer *Encycl. Garden Work* 208/2 An apple may be described as grafted upon paradise rootstock. **1969** P. Thrower *Every Day Gardening* iv. 84/2, I keep a close lookout throughout the season for suckers growing from the roots below the union of the rose and the rootstock.

rooty, *sb.* Add: Also **rootey. a.** (Further examples.) Cf. *ROTI[2].

1900 Kipling in J. Ralph *War's Brighter Side* (1901) xv. 253 And the 'umble loaf of 'rootey' Costs a tanner, or a bob. **1900** 'M. Thyme' in *Ibid.* xx. 316 Bully beef and rooty, and Something's give me a pain. **1957** [see *JILDI]. **1959** *Listener* 5 Mar. 406/1 Eight ounces of 'rooty'—that is bread.

b. rooty gong (*GONG[2] 2 a), a medal formerly awarded to members of the British Army in India (see quots.).

1925 Fraser & Gibbons *Soldier & Sailor Words* 245 *Rooty gong*, long service Medal. **1936** F. Richards *Old-Soldier Sahib* vi. 108 The Good Conduct medal or 'Rooty Gong'..was so called because it was a regular ration-issue, like bread or meat or boots. **1948** Partridge *Dict. Forces' Slang* 157 *Rooty gong*,.. *Rooti* is the Indian Army word for bread, the implication being that the wearer has eaten a tremendous aggregate of Service loaves and therefore deserves it.

rooty, *a.* Add: Also, belonging to or suggestive of roots.

1905 E. F. Benson *Image in Sand* xviii. 292 The warm wind bore with it..the good, moist, rooty smell of the dusky heather.

rooty-toot (rū·ti,tūt). *slang* (chiefly *U.S.*). Also **root-a-toot**, etc. [A redupl. form, ult. of echoic origin, usu. representing the sound of a trumpet; cf. *rootin' tootin'* s.v. *ROOTING *ppl. a.*[2] and TOOT *v.*[2]] Something noisy, riotous, or lively; *spec.* an early style of jazz music. Also as *adj.* and (in various nonce-uses) *vb. intr.*

1887 T. Darlington *Folk-Speech of S. Cheshire* 319 There was a *rooty-tooty* at Cholmondeley last Setterday, an' everybody from raïnd about went bu' mey. **1907** G. B. Shaw in *Neolith* Nov. 9 The trumpet angel..root-a-tooted at the sky. **1931** O. Nash *Hard Lines* 24 Oh rooti-ti-toot for Smoot of Ut. **1936** *Amer. Mercury* XXXVIII. p. x/2 *Rooty-toot*,..razz-ma-tazz. **1937** G. Frankau *More of Us* iii. 38 So Izzy Cohen (y sus Boys) root-tooting Moved Innocent to choric rhapsody. **1938** *Brit. Empire Mod. Eng. Illustr. Dict.* 1257/2 *Rooty-toot* (Am.), old-fashioned jazz. **1951** W. Morum *Gabriel* I. iii. 39 He knew hambone and joanna meant trombone and piano, But what could be a rooty-toot, a gobstick, skins and skeletons? **1976** *Listener* 29 July 120/3 It's all done to the most cheerful, rooty-tooty music imaginable. **1977** *Time Out* 28 Jan.–3 Feb. 17/3 Ma, some of them songs are rooty-toot-toot but the whole damn show is as real as George Wallace fairy and as sassy as a pile of rocking horse sheet.

rope, *sb.*[1] Add: **I. 2. c.** (Earlier and later examples.) Phr. *on the ropes*: see quot. 1958; also *fig.* Also, the ropes marking the boundary of a cricket ground.

1829 P. Egan *Boxiana* 2nd Ser. II. 158 Lenney found himself hanging on the ropes, where he was milled down. **1888** R. H. Lyttelton in Steel & Lyttelton *Cricket* xvi. 439 There is a strong cord running all round the ground, every decently hard hit is certain to reach the ropes if the ball once passes the fieldsman. **1892** *Sporting Life* 31 May 3/4 Among his hits were three drives over the ropes for 6. **1901** G. B. Shaw *Admirable Bashville* II. i. 302 The Australian Champion and his challenger..fought to a finish... The bold Ned Skene revisited the ropes to hold the battle for his quondam novice. **1904** A. A. Milne in *Punch* 18 May 358/1 Time was I cared for cricket,..Cutting a ball to the ropes for four. **1924** 'W. Fabian' *Sailors' Wives* xv. 175 You've got him on the ropes. They tell me he shows signs of matrimony. **1958** F. C. Avis *Boxing Reference Dict.* 78 *On the ropes*, said of a boxer who is forced back on to the ropes by his opponent, or is lying helpless on them. **1971** *Times* 27 Sept. 9/8 Griffith was defenceless against the ropes and his own corner as Monzon unleashed a string of straight rights and lefts. **1972** *Times* 16 May (Wall Street Suppl.) p. iv/2 A good section of the industry was on the ropes and there were times when I wondered if it would survive. **1976** J. Snow *Cricket Rebel* 168 (*caption*) Ray Illingworth hooks..in the England v West Indies Test at the Oval, 1973. The ball, arrowed, is on its way to the ropes. **1977** J. Laker *One-Day Cricket* 88 Three further perfectly timed shots had cleared the boundary ropes. **1977** *New Yorker* 25 July 70/3 Miss Wade was on the ropes several times in the first set, but she stayed in there and managed to win it, 7-5. **1980** *Tablet* 26 Jan. 81/3 There is talk that the Kennedy campaign is not just 'on the ropes', but that it is plain dead.

f. *transf.* A type of lodging-house (see quot. 1836).

1836 Dickens *Pickw.* (1837) xvi. 160 The twopenny rope..is just a cheap lodgin'house, vere the beds is twopence a night... They has two ropes, 'bout six foot apart, and three from the floor, which goes right down the room; and the beds are made of slips of coarse sacking, stretched across 'em... At six o'clock every mornin', they lets go the ropes at one end, and down falls all the lodgers. **1973** L. Heren *Growing up Poor in London* i. 10 One of the ropes, or lodging houses, was home for Indian pedlars... The rope was next to a pub.

g. A skipping-rope. Cf. *to jump rope* s.v. *JUMP *v.* 1 f.

1874 R. L. Stevenson in *Portfolio* V. 116 A mistress of the art of skipping..the rope passed over her black head and under her scarlet-stockinged legs with a precision and regularity that was like machinery. **1927** C. V. Goddard in *Word Lore* II. 128 Never leave the rope empty Go to church on Ash Wednesday. **1959** I. & P. Opie *Lore & Lang. Schoolch.* xii. 239 People from the surrounding villages bring great lengths of clothes-line with them, and skip ten and even fifteen abreast in each rope. **1978** J. Irving *World According to Garp* vii. 133 Jumping rope for half an hour in a corner of the gymnasium.

h. *Mountaineering.* A climbing-rope. So *transf.*, a group of climbers, *esp.* one that is roped together. Also *attrib.* and *fig.*

1892 C. T. Dent *Mountaineering* ii. 71 There is no part of the Alpine equipment for those who intend to go above the snow line..more important than the rope. **1919** G. D. Abraham *On Alpine Heights & Brit. Crags* i. 10 The legs of both were steadied by the second climber, who secured the rope around a projecting knob of rock. **1935** R. L. G. Irving *Romance of Mountaineering* vii. 114 Tom de Lépiney runs out fifty metres of rope in crossing, held by the rope from as high as possible. **1935** D. Pilley *Climbing Days* iv. 84 And for those who did not lead, but still desired to take the share of responsibility which falls to any genuine member of a *rope*, a climb would help. **1941** C. Kirkus *Let's go Climbing* iii. 46 When a rope travelling south meets a rope travelling west the result is apt to be rather like a Maypole dance. **1955** M. Banks *Commando Climber* v. 89 Lower down we passed under some tottering, unstable-looking sêracs, in company with an Italian and a French rope. **1965** A. Blackshaw *Mountaineering* vii. 195 A wet rope should not be left coiled, as this will delay drying and encourage deterioration. **1968** P. Crew *Encycl. Dict. Mountaineering* 100/2 In artificial climbing rope management can become very complicated. **1972** D. Haston *In High Places* iv. 52 We reckoned on teaming up two ropes of two, Eley with me and Geoff with Dennis English. **1979** D. Clark *Dread & Water* ii. 33 It's up to you and your pals on the same rope to make your own decisions as the need crops up.

i. A rope suspended vertically in a gymnasium for climbing and other exercises.

1903 *Handbk. Physical Training* (Admiralty) I. 30 For rope climbing the class will be formed up about 4 paces from the ropes. **1940** McClow & Anderson *Play Gymnastics* 87 Small boys who are free in the gymnasium show a great interest in equipment upon which they can climb or from which they can hang. They never seem to tire of swinging on the ropes. **1965** D. R. Casady et al. *Handbk. Physical Fitness Activities* xii. 96/2 When climbing the rope, one must climb down as well as up.

3. (Later examples.) So *transf.*, capital punishment. Also in phr. *to take a rope*, to hang oneself.

1934 H. N. Rose *Thes. Slang* 18/1 Jim got a rope this morning. **1935** A. J. Pollock *Underworld Speaks* 98/2 *Rope*, hanging. **1935** J. Hargan *Gloss. Prison Lang.* 7 *Rope, take a*, to hang oneself, to commit suicide. **1950** H. E. Goldin *Dict. Amer. Underworld Lingo* 180/2 *Rope*, capital punishment by hanging; (loosely) capital punishment by any means. **1976** *Leicester Mercury* 14 Oct. 4/4 The complete disregard for law and order which is so prevalent today is the direct result of the policies of himself and most members of the Labour Government which resulted in the cane being abolished for disobedient schoolboys, the birch for thugs and the rope for murderers.

4. b. *to come to the end of one's rope*, (*b*) to come to the end of one's resources, to be at the end of one's tether. So *at the end of one's rope*, etc.

1931 F. L. Allen *Only Yesterday* ii. 32 Physically the President was almost at the end of his rope. **1943** M. Carpenter *Experiment Perilous* 214 I've come, I think, to the end of my rope. **1954** N. Coward *Future Indefinite* v. vi. 321 What I had been dreading for a long time happened. I collapsed finally and knew that I had come to the end of my rope. **1971** *Ink* 12 June 7/4 On Monday, 24 May, the Mans strikers—now at the end of their rope financially—voted to accept the compromise proposals. **1977** *Transatlantic Rev.* LX. 79 Judy was at the end of her rope.

c. (Further examples.) Also *to show one, understand, the ropes.*

1850 'J. Timon' *Sketch* 18 Aug. in *Opera Goer* (1852) II. 186 The belle of two weeks standing, who has 'learned the ropes'. **1854** *Congress. Globe* 33rd Congress I Sess. App. 893/2 They are familiar with all the dodges of the season, understand the ropes about town [etc.]. **1860** T. C. Haliburton *Season-Ticket* viii. 226 Tell me..about Canada, and show me the ropes. *a* **1911** D. G. Phillips *Susan Lenox* (1917) II. ii. 20 I'll show you the ropes... You'll find the job dead easy. **1937** 'G. Orwell' *Road to Wigan Pier* ix. 182, I would find out about tramps and how you got in touch with them..and then, when I.. knew the ropes well enough, I would go on the road myself. **1949** E. Waugh *Loved One* 133 Mr Schultz had found a young man to take Dennis's place and Dennis was spending his last week at the Happier Hunting Ground in showing him the ropes. **1973** G. Greene *Honorary Consul* I. i. 26 Fortnum knew the local ropes. He saved the Ambassador a lot of trouble. **1976** J. I. M. Stewart *Young Pattullo* ii. 43, I was being made aware.. that I didn't quite know the ropes.

e. *to pull the ropes*, to direct or influence events. ? *Obs.*

1876 W. G. Nash *Century of Gossip* iv. 70, I cum purty near..tellin' 'em that Elton wouldn't pull a rope for him, if he got the nominashun. **1900** G. N. Boothby *Maker of Nations* i. 19 You *do* require to know the ropes. And what is more, you require to be very careful how you pull those ropes when you are familiar with them.

f. *money for old rope*: see *MONEY *sb.* 6 h.

II. 5. c. (See quot. 1950.) Also *attrib.*, as *rope silk*.

1880 L. Higgin *Handbk. Embroidery* i. 4 'Embroidery', or Bobbin Silk..is manufactured in what is technically called 'rope', that is, with about twelve strands in each thread. When not 'rope' silk, it is in single strands, and is then called 'fine' silk. **1910** *Art Needlework* 2/2 Arden's 'Hazel' Embroidery No. 3... As thick as (and closely resembling) those silks called 'Rope' and 'Cable', it can be used for merely outlining with long and short stitch. **1950** *Mercury Dict. Textile Terms* 430/1 *Rope silk*, an embroidery silk thread consisting of singles doubled into threads and these doubles again doubled to form a strong thread.

d. *U.S. slang.* A cigar.

1934 H. McLellan in *Detective Fiction Weekly* 10 Nov. 29/2 He jerked a cigar out of her mouth... 'It burns my stomach to see a dame smoking a rope'. **1940** *Amer. Speech* XV. 335/2 A cigar is *rope*. **1960** Wentworth & Flexner *Dict. Amer. Slang* 433/2 *Rope*, a cigar... Occasional use in comic papers and by would-be wits. **1978** H. Wouk *War & Remembrance* vii. 66 Carter Aster was smoking a long brown Havana tonight. That meant his spirits were high; otherwise he consumed vile gray Philippine ropes.

e. *Anthrop.* A system of descent or inheritance in which the link is formed from father or mother to the children of the opposite sex (see quot. 1935).

1935 M. Mead *Sex & Temperament* x. 176 Instead..of organizing people into patrilineal groups or matrilineal groups..the Mundugumor have a form of organization that they call a *rope*. A rope is composed of a man, his daughters, his daughters' sons, his daughters' sons' daughters; or if the count is begun from a woman..her sons, her sons' daughters..[etc.]. **1953** A. K. C. Ottaway *Educ. & Society* ii. 25 Inheritance [among Mundugumor] passes from father to daughter, and then to her son. This is known as a 'rope'. **1968** *Internat. Encycl. Social Sci.* VIII. 405/2 Men may be linked cross-sexually to their mothers, and women to their fathers, to produce the alternating or cross-sexual system of the 'rope'. **1976** H. Reading *Dict. Social Sci.* 181 *Rope*, descent group resulting from alternating descent.

f. *slang.* Marijuana.

1944 D. Burley in A. Dundes *Mother Wit* (1973) 211 Marijuana-Weed—..*rope*. **1945** L. Shelly *Jive Talk Dict.* 16/2 *Rope*... Marijuana cigarette. **1972** [see *muggle[3]].

g. *Astr.* A group of magnetic lines of force twisted together.

1961 H. W. Babcock in *Astrophysical Jrnl.* CXXXIII. 577 The fluid shear will be affected by the increased magnetic viscosity of local field concentrations, and these will be twisted into more or less discrete flux strands or 'ropes'... The flux ropes may be visualized as roller bearings. **1977** *Nature* 21 Apr. 686/1 More than 90% of the total magnetic flux, outside pores and sunspots, that emerges from the sun is confined to ropes that are only a few hundred kilometres across.

6. b. (Earlier and later examples.)

1617 T. Roe *Jrnl.* 6 Oct. in Purchas *Pilgrimes* (1625) I. iv. xvi. 571, I told him I had a rich Pearle, and some other ropes faire. **1931** *Amer. Speech* VII. 113 Get this rope to the fence before we fall for receiving. **1966** A. Loos *Girl like I* vii. 145 Gaby Deslys..wore 'ropes' of pearls, as they were then called.

7. b. A bacterial condition of bread and the like in which it may be drawn into strands.

Cf. quot. 1850 s.v. Ropiness.

1899 J. Blandy *Baker's Guide* (ed. 4) iii. 169 (*heading*) Rope in cakes. **1921** W. & W. C. Jago *Technol. Bread-Making* xvii. 345 During hot weather bread is liable to an outbreak of the disease called 'rope'. *Ibid.*, Modern writers agree in ascribing rope to bacterial activity. **1972** *Sci. Amer.* Mar. 18/1 Baked goods, for example, go stale rapidly. Once made, they are often exposed to mold spores that become active in warm weather or high humidity. In bread the spores produce a condition called 'rope'.

III. 8. a. *rope bed, bedstead, -bit, -bridge* (later examples; also *fig.*), *sling, sole, tow.*

1925 H. Crane *Let.* 17 June (1965) 208 A lot of wonderful old rope beds and furniture came right along with it. **1972** E. Wigginton *Foxfire Bk.* 140 The rope bed was once the only bed to be found in this area. **1971** *Canad. Antiques Collector* Sept.–Oct. 15/1 Another early..bed is the low poster rope bedstead. **1940** C. Day Lewis tr. *Virgil's Georgics* iii. 61 Try a rope-bit In his mouth now and then. **1923** Rope-bridge [see *jhula]. **1961** L. van der Post *Heart of Hunter* 10, I was possibly the only person who could start this kind of interpretation; who could be this kind of improvised little ropebridge over the deep abyss between the modern man and the first person of Africa. **1965** A. Nicol *Truly Married Woman* 39 It used to take them about an hour to make a detour to cross ..on the swinging rope bridge. **1901** Merwin & Webster *Calumet 'K'* i. 5 'Slack away!' he called to the engineers, and he cast off the rope sling. **1957** Clark & Pyatt *Mountaineering in Brit.* xvi. 239 Rope-slings were used thus as early as 1931. **1971** C. Bonington *Annapurna South Face* xi. 127 Standing in a rope sling, suspended from a peg, he was able to reach up to another crack above the overhang and hammer in a further peg, clipped in another sling and pulled himself up. **1894** *T. Eaton & Co. Catal.* Spring & Summer 31/1 White canvas bathing shoes, rope soles. **1964** Rope sole [see *mess-boy* s.v. *mess *sb.* 7]. **1965** *Economist* 25 Dec. 1416/3 In the [U.S.] National Forests there are 199 developed winter sports sites equipped with 164 chair lifts..312 rope tows and 48 ski jumps. **1968** *Globe & Mail* (Toronto) 13 Jan. 35/7 London's Ski Club with seven rope tows. **1978** W. F. Buckley *Stained Glass* xv. 147 He found it irresponsible that his thoughts should turn to skiing, which he longed to attempt in the lofty Alps after several winters of rope tows in Vermont during hectic weekends away from Yale.

b. *rope-boy, -knout, skipping, socket.*

1952 *Landfall* Sept. 206 Ropeboys just standing can feel cocky pride in shouting. **1970** *Guardian* 26 Nov. 13/2 A rope boy, in climbing diction, is a second man who spends patient hours securely belayed as he holds or pays out the rope for a leader. *a* **1918** W. Owen *Mental Cases* in *Poems* (1920) 8 Thus their hands are plucking at each other; Picking at the rope-knouts of their scourging. **1969** R. D. Abrahams *Jump-Rope Rhymes* p. xv, Rope skipping..with men..is now part of the training program for some athletic activity..rather than a game. **1889** *Cent. Dict.*, Rope-socket. **1935** *Discovery* Apr. 118/2 Actual drilling is done by a 'string' of tools... At the top of the string is the connecting rope socket, which permits the tools to turn freely, ensuring a round hole.

d. *rope-climbing, spinning* (later examples).

1903 [see sense 2 i above]. **1965** D. R. Casady et al. *Handbk. Physical Fitness Activities* xii. 96/2 Rope climbing promotes the development of a strong grip. **1926** *Daily Colonist* (Victoria, B.C.) 17 July 10/2 Mr. Ash..had plenty of thrills among the Mexican bandits and cattle thieves, during which time he became expert with the revolver, the lasso, and rope spinning. **1969** G. E. Evans *Farm & Village* xi. 126 This saddler's shop, with 'a rope-spinning ground' behind it was sold by auction in July 1875 at the Lion Inn, Debenham. **1975** F. Kennedy *Alberta was my Beat* viii. 92 Guy Weadick ..when he found that he could not successfully ride outlaw horses, turned to rope spinning.

e. *rope-soled, -swung.*

1920 *Blackw. Mag.* Apr. 507/2 He was dressed quaintly in well-washed dungarees,..a gaudy waist-cloth, rope-soled shoes [etc.]. **1955** M. Allingham *Beckoning Lady* iii. 39 She was wearing a bright blue dress..and rope-soled shoes. **1973** G. Mitchell *Murder of Busy Lizzie* iv. 47 She pulled on a pair of rope-soled shoes. **1957** A. Clarke *Too Great a Vine* 23 Rope-swung victims ring that bell.

9. rope border (esp. in *Basketry*), a border resembling the twisted strands of rope; **rope-boring**, the boring of wells with a drill suspended and worked by means of a rope; **rope brown**, a type of strong brown paper orig. made from old rope; **rope burn**, a burn caused by the friction of a rope; hence as *v. trans.*; **rope embroidery silk** = sense 5 c above; **rope horse**, a horse ridden by one roping an animal; **rope race**, the compartment or passage through which a driving-rope passes; **rope rider** (see quot.); **rope-sight**, in bell-ringing, facility in judging when to pull a rope, from the position and movement of others; **rope silk**: see sense 5 c above; **rope stitch** (earlier and later examples); **rope-trick**, restrict † to sense in Dict. and add: (*b*) a juggling trick or sleight-of-hand involving a rope or ropes; freq. in *Indian rope-trick*; also *fig.*; **rope-walker** (later examples); hence *rope-walking* vbl. sb.; **rope-way**, (*a*) (examples); (*b*) a rope used as a means of transport; **rope wrapping** = *rope brown* above.

1897 *Private Life of Queen* xxiv. 201 A very simple cornice..composed of the conventional 'egg and dart' and 'rope' borders. **1912** T. Okey *Introd. Art of Basket-Making* ix. 100 *The Rope Border*—This, a modification of the plaited border, may be carried out by numbering six stakes in succession and doubling the first two. **1953** A. G. Knock *Willow Basket-Work* 26 The simplest and smallest rope border was used on the oval buff shopping basket. **1888** *Chambers's Encycl.* II. 331/2 The rope-boring machinery of Mather and Platt of Salford..is in extensive use. **1902** *Encycl. Brit.* XXX. 763/2 In Europe rods, either of iron or wood, seem to be preferred, though rope boring is by no means unknown. **1908** R. W. Sindall *Manuf. of Paper* vi. 27 *Rope browns* are common papers made of fairly strong material of a miscellaneous character, this name having been derived from the fact that rope and similar fibre were at one time used exclusively. **1914** E. A. Dawe *Paper & its Uses* xvii. 115 Brown wrapping papers are made of various materials and in many qualities and substances. Rope browns, air-dried, cylinder-dried are three kinds. **1926** *Paper Terminol.* (Spalding & Hodge, Ltd.) 23 *Rope brown*, a quality of brown paper manufactured from old rope. **1955** S. C. Gilmour *Paper* xxii. 251 The thickness of a quality such as Rope Brown would appear to the touch to be much in excess of the same substance in an M. G. Pure Kraft. **1905** *Outing* July 415/1 Before we left that camp Rodney and Sue were sleek and fat, and my bruises and rope-burns were healed. **1944** B. A. Botkin *Treas. Amer. Folklore* i. iv. 132 The red rope-burn that he wore about his permanently stiff neck, usually hidden by a bandana, was his only diploma. **1948** Faulkner *Intruder in Dust* vii. 159 A big saddleless black mule with a rope-burn on its neck. **1965** A. Blackshaw *Mountaineering* viii. 217 (*caption*) He is wearing gloves to protect his hands from rope burns should the leader fall. **1966** M. & O. Murie *Wapiti Wilderness* v. 54 There was such friction from the mule's wild lunges that my palms were rather badly rope-burned. **1895** *Montgomery Ward Catal.* Spring & Summer 90/3 Rope Embroidery Silk..very coarse. **1897** *Sears, Roebuck Catal.* 321/2 Corticelli Rope Embroidery Silk..A course [*sic*] silk, for bold designs..when rapid work is required. **1944** R. F. Adams *Western Words* (1945) 131/2 When running an animal to be roped, the educated rope horse knows when the cowboy takes down his rope and what is expected of him. **1961** R. P. Hobson *Rancher takes Wife* i. 19 Rhino was a good rope horse. **1892** J. Nasmith *Students' Cotton Spinning* xii. 400 In arranging the blowing rooms it is now customary to separate them from the main building by the rope race. **1903** *Sci. Amer.* 23 May 392/2 In soft-coal mines the man in charge of the cable train is called a rope rider. In bringing his cars out of the mine he sits upon the ring which connects the cable with the train. **1902** *Encycl. Brit.* XXVI. 521/1 He [*sc.* the bellringer] has to bear in mind,..what bell or bells are striking immediately before or after him—this being ascertained chiefly by 'rope-sight' *i.e.*, the knack..of seeing which rope is being pulled immediately before and after his own. **1956** G. E. Evans *Ask Fellows who cut Hay* xviii. 143 The science of change-ringing is something of a mystery to the layman... 'It's all right once you get *rope-sight*,' one old ringer confided. **1975** *Islander* (Victoria, B.C.) 16 Mar. 12/3 He must start pulling his bell before the bell that he is following has sounded, so he must be able to recognize from the movement of the ropes (without hearing the sound) when he should start to pull his own rope, and this art of recognition is called 'ropesight'. **1977** *Church Times* 20 May 8/5 The ability to see one's path in this dancing maze is called ropesight, and is an essential attribute for a change ringer. **1880** L. Higgin *Handbk. Embroidery* iii. 28 Rope stitch..should..have the appearance of a twisted rope. **1932** D. C. Minter *Mod. Needlecraft* 55/2 Chain, pekinese, appliqué, Portuguese border and rope stitch..are useful for working this type of letter. **1887** *Encycl. Dict.* VI. i. 182/1 *Rope-trick*.., a juggling feat, introduced into England from America by the Brothers Davenport, in 1864. The performer was bound with ropes in a cabinet, or to a chair; the lights were then lowered, and on their being raised he was discovered at liberty, having been released, it was said, by spiritual agency. **1894** A. Lang *Cock Lane* 106 Thus, when Ibn Batuta, the old Arabian traveller, tells us that he saw the famous rope-trick performed in India—men climbing a rope thrown into the air, and cutting each other up, while the bodies revive and reunite—he very candidly adds that his companion, standing by, saw nothing out of the way, and declared that nothing occurred. **1907** Maskelyne & 'Devant' in 'D. Devant' *My Magic Life* (1931) xii. 131 We are prepared to pay a salary at the rate of £5,000 a year to any man who can perform the Rope Trick as described in the legend... He is to stand out in the open air... He is to throw one end of a rope into the air, and the other end is to be on the ground. The rope is to become stiffened; a boy is to climb up it and disappear into space. **1922** L. H. Branson *Indian Conjuring* ix. 76 (*heading*) The Indian rope trick. **1953** —— *Lifetime of Deception* xxxviii. 206 The Indian '*Jadoo-wallah*' is a much over-rated performer, particularly as the world-famous Indian rope trick has never been performed. This is a statement of fact. **1958** *Times Lit. Suppl.* 17 Jan. 26/5 These are no ugly moral questions, no probings of primitivism: Sir Maurice blandly marshals the material..and the miracle (or rope trick, according to one's viewpoint) is duly performed. **1977** *Private Eye* 4 Mar. 17/3 For what such massive buying operations did was to ensure that the Slater share price resembled the Indian rope trick, defying gravity. **1862** E. A. Hall *Diary* 2 Jan. in O. A. Sherrard *Two Victorian Girls* (1966) ii. 289 Spent an hour at the Crystal Palace and saw the rope-walker, Blondin. **1942** E. Sitwell *Street Songs* 11 We watched the sonambulists, rope-walkers, argonauts. **1881** *Gen. Statutes State of Michigan* (1882) I. 539 Any person..who shall apprentice, give away, let out or otherwise dispose of any such child to any person in or for the vocation, service or occupation of rope or wire walking..shall be deemed guilty of a misdemeanor. **1890** [see *act *sb.* 7 c]. **1957** *Encycl. Brit.* XIX. 547/1 *Rope-walking*, the art of walking, dancing and performing tricks on a rope or wire stretched between two supports. **1889** *Engineer* LXVIII. 454/1 Rope railways, as they were called, or ropeways, for transmitting minerals and goods, seem to be rapidly growing in favour, especially for mining purposes. **1928** *Daily Mail* 7 Aug. 8/5 Next week's programme includes instruction in the use of heavy derricks and aerial ropeways. **1941** 'R. West' *Black Lamb & Grey Falcon* II. 925 If you have to have a rope-way, you have to have Germans... All the decent funiculars in the world are made by a German company. **1950** tr. *Mountaineering Handbk.* (Assoc. Brit. Members Swiss Alpine Club) x. 116 To transport loads, injured people or materials over precipices, ravines, large crevasses or torrents, where possible fix a rope over the obstacle..the anchorage at the ends of a ropeway should be firm enough to meet all eventualities. **1963** *Economist* 30 Nov. 911/1 Aerial ropeways and chairlifts can be pretty profitable. **1937** Rope wrapping [see *acid-proof* adj. s.v. *acid *sb.* 4].

rope, *v.*[1] Add: **1. b.** (Further examples.)

1974 *Times* 18 Feb. 2/7 They had set out yesterday morning to climb Zero Gulley. About half way up Mr Beattey slipped and fell 100ft, landing on a ledge. Mr Thomas was roped to him. **1976** D. Clark *Dread & Water* i. 8 Redruth was climbing solo on a pretty easy pitch... Silk was roped to a partner.

absol. (Further examples.) Also with *up*.

1922 E. R. Eddison *Worm Ouroboros* xii. 177 They roped at the foot of the glacier that came down from the saddle, some five thousand feet above them. **1925** *Climbers' Club Jrnl.* XVII. 41 We roped up at the foot of the rocks at ten o'clock and serious climbing began at once. **1950** T. Longstaff *This my Voyage* ii. 16 We struck the arête at seven fifteen and after a bite, roped up. **1952** Morin & Smith tr. *Herzog's Annapurna* ix. 138 We roped up in the same order in which we had camped. **1965** A. Blackshaw *Mountaineering* vii. 198 The party should rope at the bottom of the first pitch of the climb.

d. (Further examples.)

1925 E. F. Norton *Fight for Everest: 1924* 115 It was one of our rules that any party of porters..must be met at the Col and escorted and roped over the intricate route into camp. **1976** A. White *Long Silence* ii. 18 It had been a difficult climb... He..roped me most of the way.

e. *to rope down* (intr. and trans.), to descend by means of a double rope fixed above; to make an abseil.

1931 *Climbers' Club Jrnl.* XVII. 204 The next little excitement was when we came to the top of..the Grand Diable... One has to rope down it. **1935** D. Pilley *Climbing Days* vi. 122 This roping down..is a trick one gets used to. **1943** E. Shipton *Upon that Mountain* iv. 78 We reached a gap about 30 feet deep, and roped down into it. **1945** G. W. Young *Mountain Craft* (ed. 4) iv. 152 Climbers, shy still of claiming it as a national practice, still struggle alternatively with 'rappel' and 'Abseilung', so as to put a wrapper..of dark foreign distinction about new methods of roping down. **1955** P. Bauer *Kanchenjunga Challenge* i. i. 22 We roped down with flashes of lightning as our only illumination. **1965** A. Blackshaw *Mountaineering* viii. 239 (*heading*) Roping down (abseiling or rappelling).

2. (Examples with *off*.)

1889 *Cent. Dict.* s.v. rope[1] v., A space in front of the pictures was roped off to prevent injury to them. **1921** A. Huxley *Crome Yellow* xxviii. 298 It was the hour for the dancing..a space had been roped off. **1976** *S. Wales Echo* 23 Nov., A section of the centre had to be roped off yesterday to enable schools to use the sports facilities.

4. b. *to rope in* (further examples); also, to arrest (*rare*).

1916 'Boyd Cable' *Action Front* 10 They..roped in my captain to identify me. **1925** D. G. Mackail in *Strand Mag.* Sept. 254/2 I'm sorry for you, my man, but ..another twenty-four hours, and we might have been roping you in, too. **1929** —— *How Amusing!* 108 He remembered now; they'd roped him in as a godfather. **1970** *Nature* 2 May 395/1 Despite its ability to attract private funds, the zoo has been less successful at roping in the public. **1973** E. Page *Fortnight by Sea* xi. 120 I've roped in the Pagets for a game, it seems she plays golf too. **1978** *Lancashire Life* Mar. 96/2 Much of it can be a do-it-yourself operation, in which the whole family can be roped in to help. **1981** N. Freeling *One Damn Thing* iv. 30 The gendarmes..sent the urban police to rope in the rest of the band.

5. b. (Earlier and later examples.)

1882 *Sydney Slang Dict.* 7/2 *Rope*, to lose a race purposely: to swindle one's backers or the public by a 'cross' or prearranged race, in which the best man or best horse is made to 'rope' or run behind. **1904** R. Thomas *Swimming* ii. 44 A racer is said to rope when he does not exert himself to the utmost, in order to make out that he is not so good a swimmer as he really is, that he may thus get an advantage in the next handicap for which he enters.

ropeable, *a.* Add: Also *N.Z.* Also **ropable.** (Earlier and later examples.)

1874 C. de Boos *Congewoi Corr.* 195, I don't know a nastier smell than the smeller new togs just fresh from the tailor's goose, and the thoughter that amost made me ropable. **1919** H. Lawson *Coll. Verse* (1969) III. 385 Don't get ropable, or moony—and, above all, don't get spoony. **1955** P. White *Tree of Man* (1956) 278, I often remember how you broke that washstand at Yuruga. Mother was ropeable. **1957** D. Niland *Call me when Cross turns Over* (1958) 216 God, she thought, as she sat down, he's ropable. **1958** *N.Z. Listener* 16 May 21/3 There was—with a walking stick, his leg in plaster. And was he ropeable! He came down and ripped into them: 'Who do you think you're going to play—a kindergarten? You're playing New Zealand. Now get cracking.' **1963** J. Cantwell *No Stranger to Flame* viii. 125 She was going to have my kid, but she dropped it when another bloke put the acid on. I got ropeable and did her.

roped, *ppl. a.* Add: **2. a.** (Earlier and further examples.) Also *fig.* Also *roped-off*.

1829 P. Egan *Boxiana* 2nd Ser. II. 350 Harry Jones and Bob Simmonds entered a twenty-four-foot roped ring at one o'clock. **1876** G. M. Hopkins *Wreck of Deutschland* iv, in *Poems* (1967) 52, I steady as a water in a well, to a poise, to a pane, But roped with, always, all the way down from the tall Fells or flanks of the voel, a vein Of the gospel proffer. **1921** A. Huxley *Crome Yellow* xxvii. 295 In a roped-off space beyond, Mary was directing the children's sports. **1927** Wodehouse *Meet Mr. Mulliner* iii. 70 He seemed to be always on the point of introducing into debates on parish matters the methods which had made him so successful in the roped ring. **1932** Auden *Orators* iii. 109 To stand with the wine-dark conquerors in the roped-off pews. **1943** J. B. Priestley *Daylight on Saturday* iv. 23 They went..to the enormous canteen. There was a roped-off space in the middle. **1955** E. Hillary *High Adventure* 10 George Band with Sherpas ..on a roped cliff. **1965** F. Sargeson *Mem. Peon* vi. 163 His meagre, roped-off end of the..area. **1976** A. Price *War Game* ii. 217 Posters directing motorists to roped-off fields. **1977** *Custom Car* Nov. 58/1, I can't help thinking a roped-off enclosure and a bit of creative parking for the various classes would have made for more interest.

b. (Further example.)

1935 *Encycl. Sports* 431/2 No one should ever attempt a roped climb without at least one experienced mountaineer in the party.

rope-like, *adv.* and *a.* (Earlier example.)

1840 Emerson *Woodnotes* i, in *Dial* Oct. 244 The rope-like pine roots crosswise grown Composed the network of his throne.

rope-maker. Add: **b.** *rope-maker's eye*: a special eye made on a rope.

1883 *Man. Seamanship for Boys' Training Ships R. Navy* (Admiralty) (1886) 127 A Rope-Maker's Eye is generally made in the end of a jibstay when fitted with a slip at the jib-boom end, and has a thimble in it to receive the slip. **1911** *Encycl. Brit.* XV. 874/1 Ropemakers' Eye.. is formed by taking out of a rope one strand longer by 6 in. or a foot than the required eye, [etc.].

rope-over, *a.* *rare*[−1]. [Rope *sb.*[1]] ? With muscles like twisted strands of rope.

1887 G. M. Hopkins *Poems* (1967) 104 Lank Rope-over thigh; knee-nave; and barrelled shank.

roper. 4. For *U.S.* read 'chiefly *U.S.*' and add further examples.

1964 C. Willock *Enormous Zoo* v. 92 At last her head was coming within range of the ropers. **1976** *Billings* (Montana) *Gaz.* 5 July 1-c/1 At Sunday's performance,

cowboys in the calf roping event were the stars as each of the last five ropers managed to conquer Mike Cervi's rowdy black calfs to place in the money.

5. (Earlier examples.)

1840 *Picayune* (New Orleans) 31 Oct. 2/3 He had not well landed on the Levee, so famous for cotton bags, sugar,..'ropers in', and other 'dry goods'. **1844** J. H. Greene *Expos. Arts Gambling* (ed. 2) 158 Those secret partners, by gamblers, are termed *ropers*, or *stool-pigeons*: their business is to delude the inexperienced into their dens of iniquity.

rope's end, *sb.* Add: **1.** Also in phr. *not to care a rope's end for*.

1858 Trollope *Three Clerks* II. viii. 178 Uncle Bat.. did not care a rope's end for Undy Scott.

rope's-ending, *vbl. sb.* (Earlier example.)

1840 H. Cockton *Life Valentine Vox* xii. 88 You shall catch, my dear, the blessedest rope's-ending you ever had any notion on yet.

rope-walk. Add: (Earlier and later examples.) Also *attrib.*

1672 *Rec. Early Hist. Boston* (1881) VII. 72 John Harrisons rope walke. **1886** *Encycl. Brit.* XX. 844/1 *Ropewalk Spinning*.—The sequence of operations in this ancient but still greatly used method of working is—(1) heckling the fibre; [etc.]. **1963** G. Blake *Gourock* ii. 22 It is interesting to note that the site on which was built the original Gourock, Ropework Company's rope-walk in 1777 was already marked 'Ropework' in 1721. *Ibid.* 24 The original Gourock ropewalk, 200 fathoms long and slated for half its length ran down to a rocky spit. **1971** *Daily Progress* (Charlottesville, Va.) 21 July 23/4 Born in an era of sailing vessels with their need for miles and miles of rope for riggings, the Ropewalk flourished in both world wars and declined after each.

rope-work. Add: **3.** Use of ropes in climbing.

1901 G. Bell *Let.* 8 Sept. (1927) I. vii. 127 After a great deal of complicated rope work we reached the Gemse Sattel. **1979** D. Kyle *Green River High* xvi. 211 All *I* had to do was shin a hundred feet up a rope!.. I've never been a one for rope-work, even in the gym.

4. Decoration with a rope motif.

1952 J. B. Oldham *Eng. Blind-Stamped Bindings* i. 11 Italian and Spanish blind bindings are so unlike English as to be easily distinguishable, with the former's characteristic leafy or rope-work frame, and the latter's rather unusual and fantastic stamps.

ropey, var. Ropy *a.* in Dict. and Suppl.

rope-yarn. Add: **3.** Used *attrib.* to designate a day given as a holiday or, more usu., a half-holiday (see quots.). Chiefly *Naval slang*.

1886 H. Baumann *Londinismen* 159/1 *Rope-yarn.. Sunday*, freie(r) Sonntag. **1914** *Dialect Notes* IV. 151 *Rope-yarn holiday*, *n. phr.*, a half holiday. **1929** F. C. Bowen *Sea Slang* 113 *Rope Yarn Sunday*. In the British Navy, '*Make and Mend time*' q.v., on Thursday afternoon. The Americans use the term for Saturday afternoon when there are no drills or inspections, but ship's work is done. Their make and mend is on Wednesday, known as *Rope Yarn Holiday*. **1952** J. V. Noel *Naval Terms* 184 *Rope Yarn Sunday*, any afternoon, except a week-end, that is free of work. Usually Wednesday afternoon is called *rope yarn Sunday* since liberty is often granted early on that day aboard ships in port. **1956** E. N. Rogers *Queenie's Brood* 79 Rope-yarn Sunday is the seaman's Monday. Actually, it is a half day off and comes on a Wednesday afternoon. **1962** Granville *Dict. Sailors' Slang* 97/2 *Rope-yarn Thursday*,..the original Naval..half holiday.

roping, *vbl. sb.* Add: **1.** (Further examples.)

1883 *Man. Seamanship for Boys' Training Ships R. Navy* (Admiralty) (1886) 50 Q. What is the roping? A. The bolt rope round the edges of the sail to prevent it from rending. **1977** D. James *Spy at Evening* vi. 29, I jumped. Into the black spray, clawing wildly for the rope ladder..my outstretched fingers touched the stiff, wet roping.

4. For '*U.S.* and *Austral.*' read 'chiefly *U.S.* and *Austral.*' (Further examples.) Also *roping-up*.

1907 S. E. White *Arizona Nights* iii. i. 241 The roping and throwing and branding..filled our days with..the unusual. **1932** Blunden *Face of England* 102 The last shocks are on the wagon, the roping-up is done. **1973** *Times* 2 Oct. 15/2 Giraffes are among the most difficult animals to capture. Straight-forward roping is unsuitable because a fall is more likely to break the neck of a giraffe than of a more compact animal. **1976** *Billings* (Montana) *Gaz.* 16 June 1-c/2 Cooper, who was being pressed in the second go-round of steer wrestling, held the lead in team roping with teammate Phil Longacre at 7.34 seconds. **1979** *Tucson Mag.* Apr. 68/3 The charreada is a fast-paced program of riding, roping, music, and dancing.

7. *roping arena, horse, stick*.

1979 *Farmington* (New Mexico) *Daily Times* 27 May 10 c/5 (Advt.), 40 acres with..iron fenced corral, roping arena, etc. **1949** *Los Angeles Times* 12 July ii. 2/5 The final resting place of Soap Suds, once the greatest humorist's favorite roping horse. **1976** *Billings* (Montana) *Gaz.* 27 June 8-D/1 (Advt.), Saddle horses of all kinds & roping horses for sale. **1878** E. S. Elwell *Boy Colonists* 190 Ernest..had to get behind the animal, and by dint of prodding with the roping stick..to force it to run up towards the corner. **1888** 'R. Boldrewood' *Robbery under Arms* III. xiv. 201 He stuck to his roping-stick—good, heavy-ended gum sapling, six or seven feet long.

ropy, *a.* Add: Also **ropey. 1. a.** (Further examples.)

1935 J. Lawrance *Painting from A to Z* xi. 103 Thick, ropey coats of paint look unsightly. **1940** H. L. Hind *Brewing* ii. xxxvii. 917 The general consensus of practical brewing opinion in Great Britain seems to be that a species of coccus or 'sarcina' is nearly always to be found in samples of ropy beer. **1950** *N.Z. Jrnl. Agric.* Feb. 165/3 Stock should not be allowed to drink from stagnant pools, as these are a source of the organism responsible for ropey milk.

c. *fig.* Bad, unsatisfactory, unreliable, unwell. *slang* and *colloq.*

1942 *R.A.F. Jrnl.* 18 Apr. 10, I then commanded a scratch squadron of rather ropey machines. **1942** *Tee Emm* (Air Ministry) II. 131 It will probably show how ropy your judgement is on modern types. **1944** S. Gibbons *Bachelor* xvi. 144, I think you must have a ropey time. Worrying about the poor, and giving all your money away and not eating much. **1945** *Gen* 30 June 51/1 He's feeling a bit ropey now, and he's going to have a bad night. **1953** E. Hyams *Gentian Violet* iii. 42 Their aircraft are pretty ropey. **1957** *Daily Mail* 3 Dec. 14/1 It is, of course, very difficult to get waiters on New Year's Eve. If you hire them outside, you may get a few ropey types. **1959** W. D. Pereira *North Flight* ii. 33 You look a bit ropey, Dad, why don't you go home..and relax? **1961** *Sunday Express* 15 Jan. 4/6, I feel a bit ropey... I think I've picked up some sort of virus. **1963** *Times* 20 May 6/5 Some of the acting is a bit ropy. **1971** J. Aiken *Nightly Deadshade* iii. 22 Mother..was crippled and Father left her... Everything was one hundred per cent ropy except that she had a little money. **1978** F. Mann *Acupuncture* (ed. 3) ix. 138 Sometimes a ropy pulse occurs in elderly people who have had many illnesses affecting several bodily systems, all of which have been only partly cured.

d. Of a cow: producing ropy milk.

1960 *Farmer & Stockbreeder* 19 Jan. 60/3 But what did farmers do? They sent all their ropy cows to F[atstock] M[arketing] C[orporation] and flogged the best in the auction mart.

2. (Further examples.)

1973 M. Amis *Rachel Papers* 132 Otherwise it was mostly long whitened grass, frizzled bushes, and hundreds of ropey little trees, fifteen feet high. **1976** *Nature* 26 Feb. 650/1 Sample ARP 74-14-31, with a ropey surface showing flow direction, has a preserved glassy crust and a palagonite layer overlain by a thin manganese coating.

roque (rōᵘk). [An arbitrary alteration of Croquet *sb.*; cf. Roquet *sb.*] A form of croquet played in the U.S., differing from croquet chiefly in the use of a hard-surfaced, embanked court, ten hoops, and short-handled mallets. Also *attrib.*

1899 *Boston Even. Transcript* 15 Sept. 6/5 The players of the new croquet, having developed a new and scientific game, have adopted a new name, and call it roque. **1906** *Springfield* (Mass.) *Weekly Republ.* 30 Aug. 16 A 16-years-old lad who never before had played in a big-tournament won the national championship at roque. **1909** *Chicago Daily News* 12 Aug. 8/1 They are holding a roque tournament at Norwich, Conn. **1924** R. Lardner in *Current Opinion* July 128/2 'Good gracious!' I said. 'Imagine being married to a woman that plays five hundred like she does and drops her teeth on the roque court!' **1954** J. Steinbeck *Sweet Thursday* viii. 55 Roque is a complicated kind of croquet. **1968** *Punch* 9 Oct. 495 Archery and golf and rackets and roque and lawn tennis have all been in and out again. **1976** *Webster's Sports Dict.* 356/2 *Roque*, a variation of croquet played with short-handled mallets on a hard-surfaced court.

Roquefort. Add: **a.** (Further examples.) (Now a proprietary name in the U.K.)

[**1766** W. C. Croker *Compl. Dict. Arts & Sci.* I. s.v. *Cheese*, At Rochfort, in Languedoc, they make a cheese of ewe's milk.] **1885** [see *Edam]. **1927** M. A. Hulbert *Treasures of Hundred Cooks* ix. 212 Use equal parts of Roquefort and Philadelphia cream cheese. **1933** *Gourmet's Bk. Food & Drink* ix. 136, I do not think a creamy or soft cheese is the best..nor, on the other hand, would I advise starting the meal with the brutal pungency of Roquefort. **1955** *Times* 10 May 12/4 Roquefort, a French blue veined cheese, is perhaps the best example of a cheese that is still to-day definitely 'regional'. **1963** *Trade Marks Jrnl.* 26 June 885 Roquefort... Roquefort cheese. Société Anonyme Des Caves & Des Producteurs Réunis de Roquefort (Aveyron). **1976** *National Observer* (U.S.) 23 Oct. 10/3 Other favorite menu frauds, Reeves reports, include substituting..blue cheese for Roquefort.

b. In full *Roquefort dressing*. A variety of salad dressing made with Roquefort cheese. Also used *attrib.* of a salad served with this. Chiefly *U.S.*

1943 D. Powell *Time to be Born* ix. 210 Cheever must ..eat oysters Rockefeller and Roquefort salad dressing. **1961** J. Heller *Catch-22* (1962) ix. 98 I've got some live Maine lobsters hidden away that I can serve you tonight with an excellent Roquefort salad and two frozen éclairs. **1962** *Listener* 11 Jan. 90/3 It is worth trying to order one of those marvellous green salads unaccompanied by the demand: 'Roquefort, thousand-isle or French?' **1968** C. Drummond *Death & Leaping Ladies* i. 21 Plain salad, with a little touch of Roquefort dressing. **1973** M. & G. Gordon *Informant* xl. 152 A hamburger dinner with lettuce salad and Roquefort dressing. **1977** *Transatlantic Rev.* lx. 87 Andrea, in bad humor, portions out breakfast—leg of armadillo.., imitation roquefort dressing, a half-ration of water and sour mix.

roquesite (rǫ-kesəit). *Min.* [ad. F. *roquésite* (Picot & Pierrot 1963, in *Bull. de la Soc. franç. de Min. et de Crist.* LXXXVI. 7/2), f. the name of Maurice *Roques*, 20th-c. French geo-

logist: see -ite[1].] A sulphide of copper and indium, $CuInS_2$, occurring as small greyish blue crystals.

1963 *Mineral. Abstr.* XVI. 372/1 A new mineral determined by electron microprobe technique is named roquésite. It occurs as inclusions in bornite in the Charrier Cu, Sn, and Fe mine, Allier, central France. **1970** *Doklady Earth Sci.* CXCI. 138/1 During field investigation of ores from depth at a pyrite deposit in central Kazakhstan, we detected roquesite, a mineral reported in such ores for the first time. **1974** *Mineral. Abstr.* XXV. 329/1 Roquesite..occurs in grains up to 1 mm in length in bornite associated with chalcopyrite, wittichenite,.. etc. in veins cutting a Devonian volcanic-sedimentary succession.

Ro-Railer (rōᵘ,rēi-ləɹ). Also **ro-railer.** [f. Ro(ad *sb.* + Rail *sb.*[2] + -er[1].] The name of an experimental vehicle, introduced by the London, Midland, and Scottish Railway, which could be adapted to run on either road or railway. (No longer current.) Cf. *road-railer* s.v. *road *sb.* 12.

1931 *Times Educ. Suppl.* 31 Jan. p. iv/1 Trials of the 'ro-railer', a vehicle which..can be driven either on the railway or on the road. **1934** *Discovery* Nov. 317/1 The L.M.S. have experimented with various types [of railcar], including a 'Ro-Railer' which can run equally well on road and on rail, and more conventional Diesel-engined car. **1959** H. Ellis *Brit. Railway Hist.* II. iii. i. 334 The L.M.S. tried out, on the Stratford upon Avon and Midland Junction line, a motor bus called a Ro-Railer, with two sets of wheels which could be very quickly substituted, one for another, according to whether it were to go on road or railway.

‖ **Roriz** (rori·ʃ). The name of a wine-growing estate in the Douro valley of Portugal, used *absol.* to designate a variety of port produced there.

1873 C. M. Yonge *Pillars of House* I. ii. 30 What right had you to know that I knew the taste of Cape from Roriz? **1907** *Yesterday's Shopping* (1969) 96/1 *Ports for laying down*..Quinta Roriz. **1917** *Harrods Catal.* 1286/1 *Old bottled vintage ports*..Kopke Roriz. **1920** G. Saintsbury *Notes on Cellar-Bk.* iii. 38 Kopke's famous 'Roriz' did not..appeal to me.

ro-ro (rōᵘ·rōᵘ), *a.* Abbrev. of *roll-on, roll-off.

1969 *Jane's Freight Containers 1968–69* p. iii/2 Ro-Ro berths were also included. **1969** *Australian* 30 Oct. 13/1 First the Scandias and now it's to be an even larger series of roll-on roll-off (Ro-Ro) ships for service in the Australia-Europe trade. **1974** *Times* 23 Jan. 16/7 Colonel Frank Bustard, one of the great shipping pioneers..was the father of the 'ro-ro' revolution. **1975** *Globe & Mail* (Toronto) 2 Oct. B14/1 It is adapted for roll-on, roll-off operation and will be served by two..ro-ro ships. **1978** *N.Y. Times* 30 Mar. D12 (Advt.), The new Farrell fleet incorporates the most sophisticated and efficient methods of cargo handling. We have breakbulk, container, LASH, RoRo and giant container vessels.

Rorschach (rǭ·ɹʃāχ). The name of Hermann *Rorschach* (1884–1922), Swiss psychiatrist, used *attrib.* and *absol.* to designate a type of projective personality test first devised by him, in which a standard set of ink blots of different shapes and colours is presented one at a time to a subject with the request that he should describe what they suggest or resemble. Also *Rorschach (ink) blot, method*, etc. Also *fig.*

1927 Moir & Gundlach in *Jrnl. Exper. Psychol.* Apr. 151 Each subject was given the Rorschach test. *Ibid.*, In a classification of the individual records according to the Rorschach diagnostic tables. **1935** *Amer. Jrnl. Psychiatry* July 109 The Rorschach test has recently received considerable attention in psychiatric and psychological circles. **1942** [see *insecurity 1]. **1948** *Personnel Psychol.* I. 357 (*heading*) Can the Rorschach pick sales clerks? **1951** Koestler *Age of Longing* i. 40 The brandy..expanded slowly into a Rorschach blotch on the marble surface. **1953** A. K. C. Ottaway *Educ. & Society* viii. 147 The Rorschach method of giving a verbal interpretation of ink-blots. **1956** *Publ. Amer. Dial. Soc.* xxvi. 121 The social or clinical psychologist will want to go still more deeply into the individual's imaginative response to the Rorschach ink blots. **1958** *Listener* 17 July 93/2 Rorschach blots and tapestries. **1960** *Commentary* June 486/2 Rorschach (inkblot) ratings. **1966** T. Pynchon *Crying of Lot 49* i. 18 His [*sc.* a psychotherapist's] theory being that a face is symmetrical like a Rorschach blot. **1971** *Jrnl. Gen. Psychol.* LXXXV. 295 Free associative responses and responses to ambiguous stimuli—a kind of auditory Rorschach. **1974** B. M. & D. D. Braginsky *Mainstream Psychol.* vi. 118 The two most widely used psychological tests, the Rorschach inkblot test and the Draw-a-Person test (both projective tests), have also been widely researched. **1980** *Dædalus* Spring 136 The Rorschach test is effective..because it forces people to be creative.

Rörstrand (rǭ·rstrand). Also *erron.* **Rostrandt.** The name of a building in Stockholm in which a ceramics factory was opened in 1725, used *attrib.* and *absol.* to designate the varieties of pottery and porcelain manufactured there.

1881 C. Schreiber *Jrnl.* 7 Oct. (1911) II. 354 Stockholm, Rostrandt, and Marienberg pottery and porcelain. **1901** W. P. Rix tr. *E. Bourry's Treat. Ceramic Industries* i. i. 753 Rörstrand crockery ware. **1906** W. Burton *Porcelain* xvi. 188 The body and the glaze of the Rörstrand porcelain are as beautiful as those of Copenhagen. **1925** B. Rackham tr. *E. Hannover's Pott. & Porc.* I. iv. iv. 482 Rörstrand faience has not infrequently..an even faint violet tone. **1961** *Guardian* 3 Mar. 10/5 Foreign buyers do not want Wedgwood or Doulton or Spode or Minton to look like Arabia, Rörstrand, or Rosenthal. **1974** *Encycl. Brit. Micropædia* VIII. 669/3 *Rörstrand faience*, first faience (tin-glazed earthenware) produced in Sweden.

rort (rǭɹt), *sb.* *Austral. slang.* [Back-formation f. Rorty *a.*] **1.** A trick, a 'dodge'; a fraud or dishonest practice. Now freq. with qualifying word.

1936 J. Devanny *Sugar Heaven* 20 The cockies are supposed to pay this retention money into the bank..but normally they don't pay it in... It's the greatest rort ever. **1958** *Sunday Mail Mag.* (Brisbane) 24 Aug. 4/4 'If they don't it will be a rort.' 'But why should it be a rort?' asks the man. **1973** *Nation Rev.* (Melbourne) 31 Aug. 1441/3 Items such as the health scheme have yet to be introduced and others—such as the removal of many of the more outrageous tax rorts—could still be frustrated were there to be an early election. **1979** *Sunday Sun* (Brisbane) 7 Jan. 20/3 Many professional people.. previously were denied access to the typical expense account rort.

2. A crowd; a wild party.

1941 Baker *Dict. Austral. Slang* 61 *Rort*,..a crowd. **1952** T. A. G. Hungerford *Ridge & River* 81 Out we go on another bloody rort, so what's the use of saving a day? **1969** G. Johnston *Clean Straw for Nothing* 78, I am not, strictly, a true devotee of the wild Australian 'rort' and always remorseful in my hangovers. **1972** *Sydney Morning Herald* 26 Aug. 20/3 One of her annual St Teresa's Day parties—a decorous..underworld rort in honour of St Teresa.

rort (rǭɹt), *v. slang.* [Back-formation from Rorty *a.*] **1.** *intr.* To shout, complain loudly; to shout abuse. Also, to call the odds at a race-meeting. Also with *at*.

1931 T. H. Dey *Leaves from Bookmaker's Book* ii. 35 How he could 'rort', and keep his customers on the racecourse in a perpetual roar of laughter with his witty remarks. **1935** M. Harrison *Spring in Tartarus* iii. 327 It isn't you..that I'm rorting at. **1962** Granville *Dict. Sailor's Slang* 97/2 *Rort*, to shout in argument or act truculently when charged with indiscipline... In Cockney Slang *to rort* is to 'shout the odds'.

2. (Examples both exemplify the *vbl. sb.*) (See quots.) *Austral. slang.*

1941 Baker *Dict. Austral. Slang* 61 *Rorting*, shrewd practices, confidence trickery. **1980** *Sunday Mail* (Brisbane) 15 June 6 (*heading*) Overseas tax havens and 'rorting' claimed. $3000m a year in tax dodges.

rorter (rǭ·ɹtəɹ). *Austral. slang.* [f. *rort *sb.* or *v.* + -er[1].] One who engages in dishonest practices; a professional sharper or trickster.

1941 Baker *Dict. Austral. Slang* 61 *Rorter*, a professional sharper: a hawker of worthless goods: one who practises sly dodges to obtain money. **1945, 1961** [see *poofter rorter* s.v. *poofter]. **1962** A. Marshall *This is Grass* 159 Rorters like Flogger prepared to fleece any man who stood staring around him.

rorty, *a.* Add: (Further examples.) Also in extended senses: (of persons and things) boisterous, rowdy, noisy; (of drinks) intoxicating; (of behaviour, speech, etc.) coarse, earthy, of dubious propriety; crudely comic. Also as quasi-*adv.*

1888 ''Arry' in *Punch* 6 Oct. 156/1, I like to feel rorty and free. **1898** R. Hichens *Londoners* xvi. 280 'Tell us a good story, Rodney—one of your rorty ones.' Mr. Rodney shrivelled. 'I fear', he murmured—'I fear I am scarcely in the—the—er—rorty vein to-night.' **1899** R. Whiteing *No. 5 John St.* ix. 95 She is Boadicea..no 'British warrior queen' of nursery recitation, but a right-down 'raughty gal', leading her alley [*sic*] to battle against the Roman 'slops'. **1904** G. B. Lancaster *Sons o' Men* 190 If Sandy or Towse get into a row we must back 'em up, of course. But it's been a rorty piece of work. **1914** 'Bartimeus' *Naval Occasions* xvii. 144 I've heard her talking like a Mother to a rorty Midshipman. **1923** —— *Seaways* vii. 96 Isn't he a little man?.. Bettin' with bookies and actin' rorty. **1932** S. Gibbons *Cold Comfort Farm* xvi. 217 Compared with the heavy, muffling darkness of the night in which the countryside was sunk, the lights looked positively rorty. **1950** 'D. Divine' *King of Fassarai* xxix. 255 It [*sc.* coconut milk] comes rortier if you leave it..to ferment, but it's got a kick now. **1969** W. Tute *Matter of Diplomacy* ii. 16 The rorty brigadier must have a taste for lean stringy meat, though of course she had been a baronet's daughter and that made up for a lot. **1971** *Sunday Nation* (Nairobi) 11 Apr. 29/3 The once-called 'power roar' from the engine compartment.. never became obtrusive or 'rorty'. **1973** *Daily Tel.* 23 Aug. 6/8 He has a wide and rorty selection of illustrations. **1974** *Good Motoring* Mar. 22/1 It is..odd that such a comfortable car should have such brilliant performance ..which is characterised by a 'rorty' exhaust note. **1978** C. Beaton *Parting Years* 160 Anne Tree is likewise an oversize personality and character—rorty, Hogarthian and with exquisite understanding of character.

Rory O'More (rǭə·ri ōᵘmōə·ɹ). *Rhyming slang.* Also **Rory O'Moore** and *ellipt.* as **Rory, rory.** [The name of a legendary Irish rebel, the eponymous hero of a popular ballad (1826) and novel (1837) by S. Lover (1797–1868), Irish writer.] **a.** The floor; also *on the Rory*, poor, penniless. **b.** A door.

1857 'Ducange Anglicus' *Vulgar Tongue* 17 Rory-O'More, *n.* floor. **1859** Hotten *Dict. Slang* 145 *Rory O'Moore*, the floor. **1892** 'Doss Chiderdoss' in *Sporting Times* 29 Oct. 1/2, I fired him out of the Rory quick. **1935** A. J. Pollock *Underworld Speaks* 98/2 *Rory O'More*, the door. **1936** J. Curtis *Gilt Kid* xviii. 178 Some lousy berk must have been snooping around the place and found that rory open. **1938** F. D. Sharpe *Sharpe of Flying Squad* 332 *On the Rory*, penniless. **1953** Berrey & Van den Bark *Amer. Thes. Slang* (1954) § 466/3 *Rory O'Moore*, a door. **1973** B. Aylwin *Load of Cockney Cobblers* x. 49 *Rory O'More*, floor or door, *Rory*.

Rosalia[2] (rozā·liă). *Mus.* Also with small initial. Pl. **Rosalias, Rosalie.** [Prob. the personal name *Rosalia*, occurring in the title of an Italian popular song, *Rosalia, mia cara*, the melody of which employs this device.] The repetition of a phrase or passage one note higher, with the retention of the same intervals and a consequent change of key.

[**1773** C. Burney *Present State of Music in Germany* II. 327 The French have a term for this tediousness, which is wanting in other languages, they call it *Rosalie*.] **1801** Busby *Dict. Mus.* s.v., *Rosalia*, a term applied by the Italians to the repetition of a passage one note higher. **1883** Grove *Dict. Mus.* III. 160/1 *Rosalia*,..a form of Melody, Vocal or Instrumental, in which a Figure is repeated several times in succession, transposed a note higher at each reiteration. *Ibid.* 160/2 Schumann has been recently accused of writing Rosalie, *usque ad nauseam*. **1937** G. B. Shaw *London Music in 1888–89* 30 To give the orchestra symphonic work instead of rosalias and rum-tum. **1944** W. Apel *Harvard Dict. Mus.* 652/2 *Rosalia*, a disparaging term denoting the schematic and unimaginative application of sequential treatment... The word applies in particular to sequences which, owing to the exact repetition of the intervals, involve modulation of the key to the higher second. **1954** *Grove's Dict. Mus.* (ed. 5) VII. 230/2 *Rosalia*, the name given to the identical repetition of a melody a tone higher, keeping the exact intervals of the notes.

rosaline (rōᵘ·zălīn). Also with capital initial. [App. of Fr. origin (see etym. note).] In full, *rosaline point*. A variety of fine needlepoint or pillow lace. Used esp. of Venetian or Belgian rose-point lace. Also *Comb.*

The word is not found in standard Fr. dicts., but occurs in specialist works (cf. C. Mague *Dentelles Anciennes* (1930)). Webster (1961) prefers to regard it as irreg. f. Rose *sb.*, but its meaning and use in Eng. suggest a Fr. origin.

1900 E. Jackson *Hist. Hand-Made Lace* vii. 66 Richness of workmanship distinguished the early eighteenth century specimens, and the firm yet delicate laces such as Rosaline Point..were especially suitable for the purpose. **1921** Galsworthy *To Let* i. ii. 25 'There's a bit of rosaline point in here,' he said, stopping before a shop, 'that I thought you might like.' **1953** M. Powys *Lace & Lace-Making* vi. 47 In the Treasury of St. Marks in Venice there is a Bas d'Aube of the finest quality with smaller flowers and detail, a type of Point de Venise which is called 'Rosaline', the whirls and picots sometimes rising up to the height of half an inch and giving a moss-like effect. *Ibid.* iv. 23 Rosaline is made with the same technique as the Point d'Angleterre and Duchesse, the only difference being that it has a winky pin at the border of the braid instead of the usual edge. **1966** 'E. Kyle' *Love is for Living* iii. 27 They hadn't any rosaline perlé... Their best lacemaker is ill. *Ibid.* vi. 51 This is rosaline perlé. Here is the real rose-point. You see that the first has these small knots like pearls to diversify the pattern? **1971** *Country Life* 4 Nov. 1197/3 This *rosaline* or *point de neige* is the summit of virtuosity.

rosary. 6. a. (Further *fig.* examples.)

1951 N. Marsh *Opening Night* v. 106 Trying to cheer herself up by telling over her rosary of romantic memories. **1960** S. Becker tr. *Schwarz-Bart's Last of Just* (1961) vi. 291 One day Fräulein Blumenthal arrived to visit, leading her rosary of tiny Levys.

rosasite (rōᵘ·zăsəit). *Min.* [a. It. *rosasite* (D. Lovisato 1908, in *Atti d. R. Accad. d. Lincei* XVII. ii. 726), f. *Rosas*, name of a mine at Sulcis, Sardinia: see -ite[1].] A carbonate-hydroxide of copper and zinc, $(Cu,Zn)_2(OH)_2CO_3$, a secondary mineral found as a bluish-green deposit on copper and zinc ores.

1909 *Jrnl. Chem. Soc.* XCVI. ii. 246 (*heading*) Rosasite, a new mineral from the mines of Rosas (Sulcis, Sardinia). **1958** *Mineral. Mag.* XXXI. 501 Rosasite has not so far been reported in Britain, but we are now able to record no less than seven occurrences. *Ibid.*, The rosasite generally forms small bluish-green wart-like aggregates.

‖ **rosbif** (rǫzbīf). [Fr., repr. Roast beef.] **a.** In *Gastronomy*, beef (and occas. other types of meat) roasted in the English manner. Also *Comb.* and *transf.*

1846 R. Ford *Gatherings from Spain* xi. 120 Our true love for the *ros-bif* of old England. **1877** E. S. Dallas *Kettner's Bk. of Table* 7 In the most popular cookery books of France..roast mutton and lamb are designated Rosbif de Mouton, and Rosbif d'agneau. **1897** A. Beardsley *Let.* 15 Apr. (1970) 303 She..has been

lecturing me about diet. Hot water and rosbif make up her programme. **1923** Joyce *Let.* 12 July (1966) III. 78 My complexion is now cinnabar and rosbif a l'anglaise. **1972** *Country Life* 20 Jan. 159/1 The Frenchman, if given the opportunity, prefers *le rosbif anglais*. **1972** J. Aiken *Butterfly Picnic* i. 19 How about a nice grilled steak—chicken—rosbif?

b. A French pejorative term for an Englishman. *rare.*

1858 Thackeray *Virginians* (1859) II. iii. 23 Only my white cockade and coat had saved me from the fate which the other *canaille* of Rosbifs had deservedly met with.

roscherite (rǫ·ʃĕrəit). *Min.* [ad. G. *roscherit* (F. Slavík 1914, in *Bull. internat. de l'Acad. tchèque des Sciences, Prague* XIX. 109), f. the name of Walter *Roscher* (fl. 1914), German apothecary and mineral collector: see -ite[1].] A hydrated basic phosphate of beryllium, calcium, iron, and manganese, $(Ca,Mn,Fe)_3Be_3(PO_4)_3(OH)_3.2H_2O$, sometimes also containing magnesium, found as yellowish-green to brown crystals in granite and pegmatite.

The presence of beryllium in roscherite was not discovered until 1958, the mineral previously having been believed to contain aluminium.

1916 *Chem. Abstr.* X. 31 Roscherite is a new monoclinic mineral occurring as short thick plates and also as slender prismatic and thin tabular crystals. **1958** *Amer. Mineralogist* XLIII. 824 Roscherite from the Sapucaia pegmatite mine occurs as single crystals, crystal aggregates, and granular crusts in vugs in muscovite. **1975** *Tschermaks Mineral. und Petrogr. Mitteilungen* XXII. 266 Roscherite from Lavra da Ilha, Taquaral, Minas Gerais is a magnesian roscherite.

Roscius (rǫ·siŏs, rǫ·sk-, rǫ·ʃiŏs). Also **Roshus, Rossius.** The name of Quintus *Roscius* Gallus (see Roscian *a.*), used to designate an actor, usu. one of outstanding ability, success, or fame (now chiefly *Hist.*, with reference to David Garrick). Also *fig.*

1647 Herrick *Noble Numbers* 74 Thou art that Roscius, and that markt-out man, That must this day act the Tragedian. *a* **1661** Fuller *Worthies* (1662) London 224 Edward Allin..was the Roscius of our age. *a* **1706** Evelyn *Diary* an. 1662 (1955) III. 338 His best [painting] in my opinion is Lacy the famous Rossius or comedian, whom he has painted in three dresses. **1749** W. R. Chetwood *Gen. Hist. of Stage* 155 Mr. George Powel, a reputable Actor, with many Excellencies, gave out that he would perform the Part of Sir John Falstaff in the Manner of that very excellent English Roscius, Mr. Betterton. **1763** Boswell *Jrnl.* 21 Jan. (1950) 163, I was sitting with the great Roscius of the age [*sc.* Garrick]. **1793** W. B. Stevens *Jrnl.* 13 May (1965) 82 The little Roscius of a Baronet tortures his *Crura Podilla* into Harlequin Agility. **1804** *Times* 27 Nov. 3/1 The Young Roscius was at Covent-Garden Theatre last night. **1826** Hazlitt in *New Monthly Mag.* Jan. 38 Of our party only two persons present had seen the British Roscius [*sc.* Garrick]. **1888** Kipling *Soldiers Three* 58 Captain dear,.. the gallery have enjoyed the performinces av a Roshus. **1958** C. Oman *David Garrick* xiv. 372 The Garricks set out for home next day... John O'Keefe, the Dublin playwright, saw Roscius for the last time 'walking very quick (his way)' up and down the Adelphi terrace. **1973** C. Price *Theatre in Age of Garrick* ii. 6 To the eighteenth century, Garrick was the outstanding actor of modern times, and to call him 'Roscius' as was so often done was merely to indicate that in one respect at least England could rival ancient Rome.

roscoe (rǫ·sko). *U.S. slang.* Also **rosco** and with capital initial. The surname *Roscoe* used as a term for a gun, usu. a pistol or revolver. See also *John Roscoe* s.v. *John 4.

1914 Jackson & Hellyer *Vocab. Criminal Slang* 72 *Roscoe*,.. a revolver... 'Stash your roscoe before you come back to the kip.' **1927** *Amer. Speech.* II. 387/2 During the harvest season, individuals and gangs harvested the harvesters at the point of the *rod* or *Roscoe* (gun). **1930** *Sat. Even. Post* 28 June 161 Slide back the shutters in the steel windows, jam your roscoes through and blast hell out of everything in sight. **1930** *Amer. Mercury* Dec. 457/2 They settle him on a rosco rap. **1941** *Sun* (Baltimore) 15 Oct. 10/16 Favorite roscoes and tommy guns of gangland's paladins. **1958** *Sat. Even. Post* 20 Sept. 86/1 Mickey and a sinister young hood..waved a couple of loaded roscoes at the cashier. **1965** E. Lacy *Moment of Untruth* vii. 111 'I know you're a dick, but..' ..'What makes you think that?' 'At the bull fight, when I stood up to let you pass—felt the roscoe on your hip.' **1979** E. Newman *Sunday Punch* viii. 62 'You'll shoot me if I don't sell?'.. His hand went to the bulge again. 'Is that what they call a "roscoe"?'

roscoelite (rǫ·skolǝit). *Min.* [f. the name of Sir Henry *Roscoe* (1833–1915), English chemist + -ite[1].] A vanadium ore that is a basic silicate of potassium, vanadium, and aluminium belonging to the mica family and occurring as minute green or brown scales.

1876 J. Blake in *Amer. Jrnl. Sci.* CXII. 31 The mineral, to which I have given the name of Roscoelite,—in honor of Professor Roscoe, of Manchester, who has done so much for the chemical history of vanadium,—is a well marked species of mica, containing quite a large percentage of vanadium. **1943** R. D. George *Minerals & Rocks* iii. 92 As an ore of vanadium, roscoelite has been found chiefly in western Colorado. **1966** *Mineral. Abstr.* XVII. 670/1 Haradaite ($SrVOSi_4O_{10}$), goldmanite (calcium vanadic garnet $Ca_3V_2Si_3O_{12}$), and roscoelite ($KV_2AlSi_3O_{10}(OH)_2$) have been hydrothermally synthesized..at moderate *PT* conditions. **1974** *Encycl. Brit. Micropædia* X. 346/3 Vanadium is the 22nd most abundant element in the Earth's crust. Some commercial sources are the minerals carnotite, vanadinite, and roscoelite.

rose, *sb.* and *a.* Add: **A. I. 1. c.** *sugar of roses* (earlier and later examples).

1686 W. Denton *Let.* in M. M. Verney *Mem.* (1899) IV. ix. 359, I could wish you would take sugar of roses with yr. asses' milke. **1922** [see *ringocandy].

3. b. *rose of the sun* = Rosa solis 1.

1910 Kipling *Rewards & Fairies* 275 Excellent herbs had our fathers of old—..Cowslip, Melilot, Rose of the Sun.

II. 4. e. *to pluck a rose*: see *pluck *v.* 9.

f. *not to be the rose but to be near it* (and variants), phr. expressing a person's proximity to some admired person, ideal, or the like.

[**1808** F. Gladwin tr. *Sâdy's Gûlistân* p. x, I was a worthless piece of clay, but having for a season associated with the rose, the virtue of my companion was communicated to me.] **1818** C. R. Maturin *Women* I. x. 191 'I am not the rose,' said he, 'but I have been near the rose.' **1825** H. Wilson *Memoirs* (ed. 2) I. 234, I considered her with respect and admiration, unmixed with jealousy. This was not the rose; but she had dwelled with it. **1848** Thackeray *Pendennis* (1849) I. ii. 11 If they were not the roses, they lived near the roses, as it were, and had a good deal of the odour of genteel life. **1866** Mrs. Gaskell *Wives & Daughters* II. xviii. 181 The great reason why she did not hear of the gossip against Molly as early as anyone, was that, although she was not the rose, she lived near the rose. **1872** S. R. Hole *Six of Spades* xii. 106 It seems to say, with the perfumed earth in the Persian fable, 'I am not the rose; but cherish me, for we have dwelt together.' **1899** H. James *Awkward Age* ii. viii. 84 Mrs. Grendon, though not perhaps herself quite the rose, is decidedly, in these days, too near it. **1907** E. Gosse *Father & Son* iv. 91, I was not permitted to go forth and trade with this old person, but sometimes our servant-maid did, thereby making me feel that if I did not hold the rose of merchandise, I was very near it. **1917** 'O. Douglas' *Setons* xiii. 151 It was not the rose but it was someone who at times was near the rose—and he went and sat down beside Jessie. **1979** *Country Life* 7 June 1863/4 Laura moves to Candleford Green, which, if not the rose of Candleford itself, is still nearer the rose than was Lark Rise.

g. *pl.*, expressing favourable circumstances, ease, success, etc., in various phrases, as *roses, (roses,) all the way, not all roses, everything's roses, come up roses* (U.S.).

[**1629**: see sense 4 c in Dict.] **1855** Browning *Patriot* in *Men & Women* I. 191 It was roses, roses, all the way. **1899** [see sense 4 c in Dict.]. **1930** A. P. Herbert *Water Gipsies* xiii. 173 The tunnel was too much for you, eh? Well, I told you it weren't all roses on the 'Cut'. **1938** G. Greene *Brighton Rock* vii. i. 283 'Sometimes he's bad to me. Oh, I can tell you,' she urged, 'it's not all roses.' **1948** Wodehouse *Uncle Dynamite* vi. 84, I should have thought you would be so glad to get back from a ghastly country like Brazil that life would have been roses, roses all the way. **1969** *Times* 12 Dec. 24 If some disaster hit us..we would have to soldier on, pretending that everything in the column was coming up roses. **1971** 'E. Lathen' *Longer the Thread* (1972) vi. 60 'We don't have to worry about where the next thunderbolt will hit.' 'So everything's roses,' Eric Marten growled derisively. **1974** *New Yorker* 1 Apr. 95/2 (Advt.), There is a splendid hotel on a marvelous corner of Park Avenue where everything's coming up roses and crystal and gilt. **1976** C. Weston *Rouse Demon* (1977) xviii. 89 This kid's from a good solid home. Parents are okay. Everything's roses. **1977** *Time* 7 Feb. 59/1 Aired over eight consecutive nights, *Roots* came up roses for ABC. **1977** *World of Cricket Monthly* June 42/2 Although Australia lost the Ashes, it was roses, roses, all the way for him.

h. *the last rose* (with allusion to quot. 1820), the last flowering of an era, an art form, or the like, before its end.

[**1820** T. Moore *Irish Melodies* 119 'Tis the last rose of summer, Left blooming alone.] **1965** C. Mackenzie *My Life & Times* IV. 147 The summer of 1912 blows in my memory like a flower of time that was; it is for me the last rose of a London that vanished during the First World War. **1978** *Times* 5 Aug. 14/6 The 'Pervigilium Veneris' is one of the most haunting incantations of love ever written. This last rose of pagan poetry is also appropriately mysterious. **1981** *Sunday Tel.* 14 June 12/8 This book is a literary curiosity. It is the last rose of a pre-Vatican II summer.

i. *to come out smelling of roses* (and variants): to emerge with an (apparently) unblemished record.

1968 'E. Lathen' *Come to Dust* xvii. 167 No matter how you sliced it, the old grads..were not going to come out of this smelling like roses. **1976** J. Porter *Dover & Claret Tappers* xii. 146, I intend to emerge from this business smelling of roses. If, to achieve this, I have to wash my hands in your blood, that's perfectly OK by me.

j. *roses round the door*, phr. used to denote marital (or rural) domestic happiness.

1934 L. Golding *Five Silver Daughters* xiii. 315 Talking about my mother and her pearls—it all sort of reminded me of the roses round the door. **1977** B. Pym *Quartet in Autumn* iv. 38 'Roses round the door and all that', as Norman used to say when Letty's retirement plans were mentioned.

5. *English rose*: see *English *a.* 2 e.

6. a. *Wars of the Roses* (examples).

1835 M. Graham *Little Arthur's Hist. Eng.* II. xxxii. 3 For more than thirty years afterwards, the civil wars in England were called the wars of the Roses. **1841** S. Bamford *Passages Life Radical* (ed. 2) II. xxvi. 132, I passed the Obelisk at Barnet, where the famous battle was fought in the wars of the roses. **1879** Trollope *Eye for an Eye* I. ii. 30 They have held the same property since the wars of the roses. **1939** W. S. Maugham in *Hearst's Internat.* Feb. 30/2 The barony held by the first earl dated from the Wars of the Roses. **1966** A. L. Rowse (*title*) Bosworth Field and the Wars of the Roses.

IV. 12. b. (Later examples.) Also printed on fabric, woven in a carpet, etc.

1897 *Sears, Roebuck Catal.* 290/1 Imported Paris Organdies..with colored roses, buds and leaves. **1955** R. P. Jhabvala *To whom she Will* xv. 102 She was very fine now in a pink silk kamiz with blue roses on it. **1964** C. Mackenzie *My Life & Times* III. 13 The Surgeon's lessons [in putting] were given along the corridor.., the hole being one of the roses in the Brussels carpet. **1972** *Country Life* 6 Jan. 25/3 Many quilts [were] named after roses..Cactus Rose, Desert Rose, Rose of Sharon, Rambling Rose (whose other name was Old Maid's Ramble).

c. Also as an emblem of the rival sporting teams of Yorkshire and Lancashire.

1907 F. Thompson in *Athenæum* 23 Nov. 654/3 It is little I repair to the matches of the Southron folk, Though the red roses crest the caps, I know. **1954** A. W. Ledbroke *Lancashire County Cricket* xxv. 244 The bank holiday Battle of the Roses provides..the nearest approach to the atmosphere of Sydney or Melbourne—when the crowds are orderly. But there is more than tenseness to a match between Lancashire and Yorkshire.

14. c. For (? *obs.*) read (now usu. *compass rose*) and add: More generally, a circular pattern showing the points of the compass. (Later examples.) Cf. Wind-rose 2.

1919 S. F. Card *Air Navigation* ii. 11 When the chart or map contains the true and magnetic 'roses'..the conversion can be done by putting a straight edge from the centre to the true direction. **1937** M. Covarrubias *Island of Bali* iv. 76 The Nawa Sangga, the magic rose of the winds, the Balinese cardinal directions. *Ibid.* ix. 280 A chicken with feathers of five colours was placed in the centre [of an offering], next to a small circular Rose of the Winds made of rice dyed in the eight different colours of the cardinal directions. **1943** [see *pelorus]. **1951** N. Monsarrat *Cruel Sea* i. 13 The ship was his: he was to commission and to command H.M.S. *Compass Rose*... Compass Rose was nothing out of the ordinary; it had to be a flower name because she was one of the new Flower Class corvettes. **1960** E. L. Delmar-Morgan *Cruising Yacht Equipment & Navigation* ii. 30 The dial or, as we call it these days, Pelorus or dummy compass rose, was the navigator's instrument and was in use for many centuries long before it was given magnets and mounted on a pivot.

f. *Building.* A circular, sometimes ornamental mounting through which the shaft of a door-handle may pass.

1857 *Commissioners of Patents' Jrnl.* 16 Jan. 41 Patent 37, January 5, 1857, Andrew Brundish: for mounting knobs, and in constructing and mounting roses for locks, latches and other such like fastenings. **1902** J. T. Rea *How to Estimate* xi. 217, 2-in cast brass knobs with solid necks, cast rose and escutcheon. **1945** N. W. Kay in R. Greenhalgh *Building Repairs* iv. 110/1 A knob may be held by its rose and be free to swivel in it. **1957** M. T. Telling in *Pract. Building & Decorating* ii. iii. 154 To cover holes for keys and spindles, escutcheon plates and roses are fixed with small brass nails or screws.

g. A circular mounting on a ceiling through which the wiring of an electric light passes; = *ceiling rose* s.v. *ceiling *vbl. sb.* 7.

1889 *Illustr. Official Jrnl.* (Patent Office) 24 July 616/1 Improvements in roses for supporting electric lamps. **1944** A. C. Greenwood *Pract. Electr. Wiring & Contracting* v. 152/2 Covers of roses should screw down with an easy motion. **1967** *Times Rev. Industry* June 74/3 Electric fittings (e.g. switches, roses, lampholders, fuseboxes, junction boxes). **1977** L. R. Wakelin *Home Electr. Repairs* 24/2 A neutral conductor must be taken to the light rose.

h. A figure in *Sword-dancing* (see quots.).

1913 C. J. Sharp *Sword Dances N. England* iii. ii. 106 The leader should call 'Nut', a bar or two before the end of a strain, so that the Rose may be begun at the commencement of the next strain... The dancers leave the Nut in the hands of No. 1 and fall back into line facing the audience, returning to the original Rose position. **1933** E. K. Chambers *Eng. Folk-Play* 129 There is a persistent figure..in which each dancer presses the hilt of his sword under the point of his neighbour's so as to mesh the swords together..in a form which may be anything from a pentagon to an octagon... This is called the Lock or Nut, which probably means Knot, and at Whitby the Rose. **1971** D. Kennedy *North Skelton Sword-Dance* 9 *The Rose.* The leader raises the Lock in his right hand and all dance round clockwise.

i. An award (differentiated as the *Golden, Silver*, and *Bronze Rose*) presented at the International Television Festival at Montreux for successful light entertainment programmes.

1961 *Times* 27 May 7/5 The B.B.C.'s 'Black and White Minstrel Show' won the main prize tonight in the Montreux international television festival's Golden Rose contest. The jury awarded the first prize of a 'golden rose' and 10,000 Swiss francs..to the B.B.C. show. **1964** *Ann. Reg. 1963* 457 The C.B.S. spectacular, *Julie and Carol at Carnegie Hall*, won the top prize for light entertainment—the Golden Rose—at the Montreux International Television Festival in May. **1972** *Times* 5 May 5/3 Britain carried off both the Golden Rose and the

ROSE

Silver Rose television awards here today for the best television light entertainment shows. **1975** *Times* 5 May 4 *The Goodies*, the BBC entry, has won the Silver Rose award at the television festival at Montreux... Italy won the contest and the Golden Rose... The Bronze Rose.. went to Austrian television.

16. d. The rounded end of a potato, esp. one being used for sprouting.

1851 H. Stephens *Bk. of Farm* (ed. 2) I. 630/2 The sets should be cut with a sharp knife, be pretty large in size, and taken from the rose or crown end of the tubers. **1976** *Country Life* 5 Feb. 325/4 Seed tubers of [illegible] [sc. potatoes] will be stood 'rose' or blunt end uppermost.. to sprout.

e. *Geol.* = *ROCK-ROSE 5, *ROSETTE 5 e.

1911 *Proc. U.S. Nat. Museum* XXXVIII. 19 In Rockenberg occur well-developed rosettes or 'roses', often uniting in extensive groups. **1954** R. L. Parker tr. *Niggli's Rocks & Mineral Deposits* vii. 274 (*caption*) Rosette-like arrangement of tabular crystals of hematite, known as iron roses. **1955** F. H. Pough *Field Guide Rocks & Minerals* (ed. 2) iii. 182 [Barite] is found in perfect imitative 'roses' of a red-brown color and sandy texture near Norman, Oklahoma. **1973** A. J. L. Deeson et al. *Collector's Encycl. Rocks & Minerals* 122/1 Gypsum 'roses' occur in many areas where gypsum has been dissolved in percolating waters which are drawn to the surface by capillary action and evaporate.

V. 19. a. *rose-bloom* (later example), *-blossom, -bough, -breath, -culture, -dust, -flake, -flower* (later example), *-fruit, -grower* (earlier and later examples), *-petal* (later example), *-prickle, -scent, -stem, -time* (later examples), *-tribe*.

1929 C. Day Lewis *Transitional Poem* ii. 34 Heedless if truth maintain On the rose-bloom her station? **1878** Swinburne *Forsaken Garden* in *Poems & Ballads* [Ser. 2] 29 The foam-flowers endure when the rose-blossoms wither. **1927** E. Sitwell *Rustic Elegies* 37 Beneath the twisted rose-boughs of the heat. **1895** W. B. Yeats *Countess Kathleen* 93 Ah, leave me still A little space for the rose-breath to fill. **1846** T. Rivers *Rose Amateur's Guide* (ed. 4) ii. 131 Modern gardening has made rapid strides in rose culture. **1924** E. Sitwell *Sleeping Beauty* xiii. 44, I shall be but thin rose-dust, He will be cold, unkind. **1951** L. MacNeice tr. *Goethe's Faust* 294 [illegible] rose-dust in my eyes. **1876** G. M. Hopkins *Wreck of Deutschland* xiii, in *Poems* (1967) 58 Sigma, signal, cinquefoil token For lettering of the lamb's fleece, ruddying of the rose-flake. **1917** D. H. Lawrence *Look! We have come through!* 60 To me it seems the seed is just left over From the red rose-flower's fiery transience. *Ibid.*, How will you have it?—the rose is all in all, Or the ripe rose-fruits of the luscious fall? **1963** S. Plath *Uncoll. Poems* (1965) 9 First frost, and I walk among the rose-fruit. **1857** Rose-grower (see *rose show*, sense 23 a below). **1930** G. Saintsbury *Notes on Cellar-Bk.* iv. 57 A friend of mine.. had some official business with one of the great rose-growers in the neighbourhood of London. **1966** R. Campbell tr. [illegible] in *Coll. Poems* III. 83 [illegible] Showers with a thousand rose-petals of light. **1944** E. Sitwell *Green Song* 1 Remember the rose-prickles of bright paws Though we shall mate no more. **1859** Geo. Eliot *Adam Bede* I. xv. 279 The delicate rose-scent of his hair. **1885** A. Edwardes *Girton Girl* I. v. 111 [illegible] its rose scents. **1960** S. Plath *Colossus* 56 Thorns on the bloody rose-stem. **1859** Mrs. Gaskell *Let.* 26 Apr. (1966) 121 The Shaens begged me to come in rose-time to them. **1913** E. Pound *Ripostes* 33 [illegible] Till the rose-time will be over. **1924** E. Sitwell *Sleeping Beauty* v. 16 If none of the rose-tribe can survive The snow, then how can our poppet live?

b. *rose-alley, -arbour* (later example), *-bower* (later examples), *-farm, -garden* (further examples), *-hedge, -land*.

1934 E. M. Wright *Story of Joseph Wright* viii. 232 We formed a torchlight procession down the rose-alley, and buried Mary's playfellow in the new part of the garden. **1977** *Belfast Tel.* 22 Feb. 9/6 The rose arbor in Belfast's Botanical Gardens provides little shelter from the rain. **1876** O. Wilde [illegible] Roses are white in the rose-bower. **1975** L. O'Faolain *Women in Wall* iii. 46 'Please... Let me do it.' 'Tell,' he went to sit down in a rose bower, 'about the Call.' **1970** T. Hughes *Crow* 13 The woodpecker drummed clear of the rotovator and the rose-farm. **1920** *Granta* 5 Feb. 201 It is said.. that he keeps poultry and a cow, plays tennis.. on a pan pipe, bathes every evening at sunset, and takes all his meals in a rose garden. **1936** T. S. Eliot *Burnt Norton* in *Coll. Poems 1909–35* 185 The door we never opened Into the rose-garden. **1867** A. J. Evans *St. Elmo* xxi. 291 A Cherokee rose-hedge is not more thickly set with thorns than a literary career with grievous, vexatious, tormenting disappointments. **1929** R. Bridges *Testament of Beauty* III. 88 In the New World far Pasadena's roseland.

c. *rose knot* (later example).

1947 A. Ransome *Great Northern?* i. 20 A hand.. took hold of the rose knot worked in the end of the bit of rope that dangled from the clapper of a small ship's bell.

20. a. (Further examples.)

1858 W. Bagehot in *National Rev.* Apr. 455 The harsh outlines of poverty will not bear the artificial rose-tint. **1870** F. Kilvert *Diary* 13 Mar. (1938) I. 56 The mountain clad in deep snow and tinged with rose colour... As the sun set a lovely rose tint stole over the snowy mountains. **1906** W. de la Mare *Poems* 124 From the day, The rose-light ebbed away. **1916** Joyce *Portrait of Artist* (1969) v. 218 The dull white light spread itself east and west, covering the world, covering the roselight in his heart. **1922** C. Mackenzie *Altar Steps* xxvii. 315 A set [of vestments] in old rose damask for mid-Lent. **1929** Blunden *Near & Far* 47 And sounding works whose smoke lifts proud Through towers of force to you rose-cloud. **1930** E. Pound *XXX Cantos* ii. 13 The coral face under wave-tinge, Rose-paleness under water-shift. **1932** Blunden *Face of England* 85 And that far rose-reflection burns On the dusk water far too red. **1949** Dylan Thomas *Let.* 13 Oct. (1966) 328, I have the hot and cold rose-flush coming and going after elderberry wine last night. **1962** [illegible] I held her hand to examine its dusky rose-bloom.

c. (Later examples.)

1916 E. & O. Sitwell *20th-Cent. Harlequinade* 23 Rose-silver base. **1928** T. Eaton & Co. *Catal.* Spring & Summer 11 Felt hat.. Colors Sand; Rose-beige; Gobelin Blue. **1930** E. Pound *XXX Cantos* xvii. 78 Stone trees, white and rose-white in the darkness. **1932** H. Crane *Let.* 15 Feb. (1965) 390 Wine glasses of a smoky rose-purple transparency that set one dreaming. **1935** E. Pound [illegible] And within the crystal, went up swift as Thetis In colour rose-blue before sunset And carmine and amber. **1953** W. de la Mare *O Lovely England* 33 Rose-green the light where a hermit knelt praying, His solitude verdurous, vision-like, still. **1966** [illegible] Their rose-brown flesh burnt by the sun. **1977** [illegible] (Advt.), 14 hands in outstanding quality rose grey gelding, 4 yrs.

21. a. *rose-fringed, -flecked, -flushed, -footed, -impearled, -lit, -shadowed, -shot, -spotted, -stained, -veiled*.

1920 Blunden *Waggoner* 20 The rose-fringed roach and bluish bream. **1965** E. Bishop *Questions of Travel* 25 Hastily, all alone, a glistening armadillo left the scene, rose-flecked, head down, tail down. **1913** C. Mackenzie *Sinister St.* I. iv. 80 Over a red wall hung down the branch of a plum tree, loaded with creamy ovals of fruit, already rose-flushed with summer. **1944** E. Sitwell *Street Songs* 18 Then, who knows Rose-footed swan from snow, or girl from rose. **1917** [illegible] Rose-impearled o'er a wonder-world Glowed the last of the sunset-gleams. **1920** A. Bennett *Clayhanger* [illegible] He left the crowded and rose-lit dining-room early. **1867** G. Meredith *Vittoria* I. i. 11 The gleam of the distant rose-shadowed Alpine snows. **1957** R. Campbell *Coll. Poems* II. 100 Freckled like rose-shot apricots. **1953** A. G. L. Hellyer *Sanders' Encycl. Gardening* (ed. 22) 75 Sepals with a white, rose-spotted, ribbon-like appendage. **1977** J. R. R. Tolkien *Silmarillion* [illegible] rose-stained in the sunset. [illegible] R. Campbell tr. *Baudelaire's Poems* 46 The balcony beneath a rose-veiled sky.

b. *rose-shaped*.

1933 E. Sitwell *Five Variations* 4 And many a rose-shaped heart must lie beneath The maps on strawberry leaves. **1945** C. L. B. Hubbard *Observer's Bk. Dogs* 87 Ears rose-shaped.

22. a. *rose-embroidered, -frail, -full, -heavy, -hot, -pale, -soft, -solemn, -towering*.

[illegible] Alas, no rose-embroidered work men share [illegible]. **1932** Joyce *Pomes Penyeach* 4 Frail the white rose and frail are Her hands that gave.. [illegible]. **1932** E. Sitwell [illegible] The rose-full, rose-soft, hooped dresses are wet with dew. **1899** W. B. Yeats in *Sat. Rev.* [illegible] The rose-heavy twilight. **1922** D. H. Lawrence in *English Rev.* [illegible] In rose-hot tips, and flakes of rose-pale snow. [illegible] **1860** J. R. Lowell [illegible] Rose-soft, and feeling in her bosom stir The rose-soft palms of that first murderer. **1916** Joyce *Portrait of Artist* (1969) [illegible] within rosesoft stuffs that he dared to conceive of the soul or body of a woman moving with tender life. **1932** [see *rose-full* above]. **1935** W. Empson *Poems* 22 Snow puppy curves, rose-solemn dado band. **1949** S. Spender *Edge of Being* 16 [illegible] the rose-towering heads of silence grow.

b. *rose-circled, -clustered, -crowned* (later fig. example), *-embowered, -entangled, -festooned, -garlanded, -lamped, -wreathed* (earlier and later examples).

1975 G. Ewart *Be my Guest!* i. 32 Or I see Gertrude waving from a cottage with a very attractive rose-circled door. [illegible] rose-clustered, floating island. **1910** R. Brooke *Hill* in *Camb. Rev.* XXXII. 181 We shall go down with unreluctant tread Rose-crowned into the darkness! **1918** G. Frankau *One of Them* xiv. 251 [illegible] That rose-embowered private sitting-room. **1962** I. Murdoch *Unofficial Rose* [illegible] He wandered towards her through the rose-entangled forest. [illegible] merely a rose-festooned [illegible]. **1917** A. Waugh *Loom of Youth* [illegible] he would step rose-garlanded out of the lighted room, in the flush of his success. **1925** C. Day Lewis *Beechen Vigil* 24 Like a queen of fable In rose-lamped gardens. **1847** H. W. Longfellow *Evangeline* II. iii. 108 Rose-wreathed, vine-encircled, a broad and spacious veranda. **1924** R. Graves *Mock Beggar Hall* 61 While incense burns beside the rose-wreathed couch.

23. a. *rose-berry* (later example); **rose bit**, a countersink bit having a conical head with a number of radial cutting teeth that meet at the tip; † **rose blanket** *U.S.*, a blanket decorated with a rose motif; **rose box**, (*b*) *Naut.* (see quot. 1976); **rose diagram**, a diagram in which values of a quantity in various directions are shown graphically according to compass bearing, in the manner of a wind-rose; **rose-fever** (earlier example); **rose gold**, delete † and (?) and add def.: an alloy of gold with a little copper, having a reddish tinge; (later examples); **rose head**, (*c*) = ROSE *sb.* 17; (*d*) a spreading top on an upright rain-pipe; **rose-hip**, also = HIP *sb.*¹; (examples); **rose-hip syrup**, a syrup containing extract of rose hips, taken as a source of vitamin C; **rose-hip tea**, a beverage made from rose-hips and hot water; **rose-jar**, a jar for holding dried rose petals; **rose-petal**, used *attrib.* of various preserves, wine, etc., made from rose petals; **rose show**, an exhibition mainly or entirely of roses; **Rose Sunday** orig. *Hist.*, the fourth Sunday in Lent; **rose-temple**, a belvedere over which climbing roses may be trained; **rose-wine** = ROSOLIO.

c **1921** D. H. Lawrence *Mr. Noon* iv, in *Mod. Lover* (1934) 224 Gilbert helped her to pick scarlet rose-berries, [illegible]. **1858** *Min. Proc. Inst. Civil Engineers* XVII. 178 A 'rose-bit' is.. employed to remove the intervening metal. **1962** A. W. Lewis *Gloss. Woodworking Terms* 7 The chief type of [countersink bit] is the 'rose' bit which has radial flutes [illegible]. [illegible] best Rose Blankets. **1850** *Columbian Centinel* [illegible] Rose Blankets. [illegible] **1972** [illegible] The rose boxes and strainer plates are clear, clean and sound. **1976** *Oxf. Compan. Ships & Sea* 723/1 *Rose Box*, the name given to the strainer at the end of the suction pipe of a bilge pump... It is also widely known, particularly in yachts, as a strum box. **1938** *Bull. Geol. Soc. Amer.* XLIX. 1887 The orientation of the long axes are [*sic*] plotted as a conventional 'rose' diagram. **1956** *Q. Jrnl. Geol. Soc.* CXII. 71 A rose diagram is made showing the directions, in 10-degree classes, of the long axes of the stones. **1971** I. G. Gass et al. *Understanding Earth* xiii. 175 (*caption*) Rose diagram representing readings of dip directions of the cross laminae of either linguoid ripples or barchan dunes. **1851** [illegible] This complaint [sc. hay-asthma] is known in the United States, and is called there, rose-fever. **1869** [illegible] (see *KARAT). **1948** A. Selwyn *Retail Jewellers' Handbk.* (ed. 3) 137 Red gold had a revival when Paris jewellers re-introduced it in jewellery in 1937–38, and other countries followed. Pale shades are known as rose gold. **1742** W. Ellis *Mod. Husbandman* [illegible] with a Rose-head, that the Water may run on their Roots. **1883** [illegible] Four-inch rain-pipes are to be provided, with proper rose-heads. **1857** Rose-hip [in Dict., sense 19 a]. **1915** *Chambers's Jrnl.* 20 Mar. 271/2 These [sc. plants for feeding pigs and fowls] were supplemented in the autumn by non-fattening foods, such as acorns.. and rose-hips. **1976** [illegible] rosehips. **1940** *Q. Jrnl. Pharmacy* [illegible] During the winter of 1941–42 the Ministry of Health experimented with two new large-scale sources of the vitamin, black-currant purée and rose hip syrup. **1947** [illegible] Gripe water, rose hip syrup and everything else a baby could conceivably need. **1966** [illegible] The pink rose hip tea.. is the great favorite at the famous Bircher-Brenner Sanatorium in Switzerland. **1973** C. Bonington *Next Horizon* xii. 175 John and Dougal settled down to their meal of dried meat and nuts, followed by rose-hip tea. **1894** *Harper's Mag.* Jan. 310/1 A rose-jar stood on one [table] in the corner. [illegible] I've brought stuffed vine-leaves and rose-petal jam, Tewfik. **1963** M. McCarthy *Group* i. 18 An Armenian restaurant in the twenties, where you got rose-petal jelly for dessert. **1968** J. Rathbone *Hand Out* [illegible] white cheese, rose-petal syrup and tea. **1970** *Rose Ann.* 93 [illegible] Home-made rose-petal wine is a pleasant and unusual wine. **1975** P. Somerville-Large *Couch of Earth* vii. 115 The waiter.. came over.. the napkin over his arm stained with rosepetal jam. **1977** *Times* 25 Nov. [illegible] Experimenting with rose petal oils to soften the skin. **1857** *Florist* Apr. 122 A suggestion to Rose growers—amateurs and professionals:—why should we not have, near some central station (such as Rugby) A Grand National Rose Show? **1978** *Lancashire Life* Sept. 40 The Lakeland Rose Show this year cost £25,000 to stage. **1860** [illegible] It is not known when the ceremony of consecrating the rose was introduced... The day is always the fourth Sunday in Lent, which is consequently known as 'Rose-Sunday'. **1891** [illegible] Golden roses were blessed each year on Laetare Sunday, hence called Rose Sunday. **1864** [illegible] *Rose Bk.* [illegible] To form a simple rose temple is a matter of no great difficulty. **1894** *Country Gentlemen's Catal.* 295/2 Rose Temple.. Price—£5. With Openings filled in to form a Summer House. Price—£7 5/-. **1852** in *Venerabile* (1930) Apr. 357 A good dinner and caffe after with beautiful rose-wine.

b. rose-briar (later example); **rose geranium**, add: or *P. graveolens*, or one of several varieties of them; also, a perfume resembling the scent of these flowers; (earlier and later examples); **rose gum**, a large gum-tree, *Eucalyptus grandis*, found in eastern Australia; **rose mahogany**, an eastern Australian timber tree, *Dysoxylum fraserianum*, of the family Meliaceæ, or its fragrant reddish wood; **rose pea**, add '17th and' before '18th century'; (earlier examples).

1932 D. H. Lawrence *Last Poems* 178 Rose-leaves to bewilder the clever fools And rose-briars to strangle the machine. **1832** *Chambers's Edin. Jrnl.* 7 Apr. 76/2 [illegible] rose geranium, intending to ornament his shop. **1890–1** T. Eaton & Co. *Catal.* Fall & Winter 43/2 Perfumes. Italian violet, rose geranium, white heliotrope. **1939** L. MacNeice *Autumn Jrnl.* 69 Clouding the cooling water with rose geranium soap. **1964** C. Lowenfeld *Herb Gardening* ii. 171 Rose geranium is a shrubby plant with deeply cut and divided leaves and clusters of pink and lavender flowers. **1977** *Vogue* 15 Sept. 83/1 Bubble Bath in four fragrances—Lemon Verbena, English Fern, Rose Geranium and Lavender Blue. **1947** H. H. Anderson *Trees New South Wales* (ed. 2) 158 Flooded Gum or Rose Gum (*Eucalyptus grandis*)... A tall, frequently majestic tree. **1967** A. Rule *Forests Austral.* iii. 35 Species such as tallowwood and rose gum, common in the humid coastal forests of eastern Australia. **1929** W. D. Francis *Austral. Rain-Forest Trees* 165 *Dysoxylum fraserianum* Benth. Rosewood, Rose Mahogany. **1958** *N.Z. Timber Jrnl.* June 59/1 Rose mahogany.. resembles Honduras mahogany. **1966** *Austral. Encycl.* III. 319/2 One of the best-known and most abundant of these [trees] is the rose mahogany, which is often called rosewood. [illegible] **1629** J. Parkinson *Parad.* [illegible] The Scottish or tufted Pease, which some call the Rose Pease. [illegible]

c. rose-bug (earlier and later examples); **rose-hopper** = *rose leaf-hopper; **rose leaf-hopper**, a greenish-yellow sucking insect, *Typhlocyba* (or *Edwardsiana*) *rosae*, of the family Cicadellidæ, which attacks the foliage of roses, making the leaves pale and mottled.

1806 *Massachusetts Spy* 1 Oct. 3/4 He suggests that the Rose-bug is the pre-existing state of those worms. [illegible] **1852** T. W. Harris *Treat. Insects New England Injurious to Vegetation* (ed. 2) 190 There is another little leaf-hopper that.. lives upon the rose bushes. [illegible] **1890** *Insect Life* II. 340 Original figures are given of the.. Rose Leaf-hopper. **1950** Metcalf & Flint *Destructive & Useful Insects* (ed. 3) xvii. 585 The rose leafhopper and another common apple leafhopper, pass the winter in the egg stage. **1961** J. Hollis *Roses* x. 106 Rose leaf-hopper.. sucks up the sap and causes mottling to appear on the leaves.

24. a. rose-comb, substitute for second part of def.: also, a bird bearing a comb of this kind; also *attrib.*; (earlier and later examples); **rose-fish** (earlier and later examples); **rose-madder**, the rose colour produced by madder dye or pigment; **rose-mole**, a mark or mole of a reddish colour.

1850 D. J. Browne *Amer. Poultry Yard* 52 The Bushy rose comb of the golden Hamburgh terminating in a sharp point behind.. is seen in no other variety of fowl. **1927** Haldane & Huxley *Animal Biol.* ii. 66 The original pure-bred rose-comb stock gives nothing but rose-combs. *Ibid.* 69 The offspring will clearly be blue Andalusian rose-combs. **1973** *Country Life* 15 Nov. 1265/1 Of the 21 [bantams] two were cockerels of rosecomb blood... One of the two rosecombs was supposed to be a hen. [illegible] & Supposing it to be a Rose fish, ran forward to see it. [illegible] **1866** H. C. Standage *Artists' Man. Pigments* v. 52 Madder Lakes (Madder Carmine.. Lake or Rose Madder). **1890** Montgomery Ward *Catal.* Spring & Summer [illegible] Pink Madder, Rose Madder. **1902** *Encycl. Brit.* XXXI. 773/1 Amongst the former [sc. natural colouring matters] may be named.. rose-madder.. and the madder-lakes from the alizarin and allied bodies derived from the root of the ordinary madder plant *Rubia tinctorum*. [illegible] For rose-moles all in stipple upon trout that swim.

b. rose-finch, a small European or Asian finch belonging to the genus *Carpodacus*, the males of which have red or pink plumage.

1863 T. C. Jerdon *Birds India* II. 399 The Rose-finch is found as a cold weather visitant throughout the greater part of India. **1890** E. W. Oates *Fauna Brit. India Birds* II. 212 The genus *Propasser* belongs to the Rose-finches, the males of which are characterized by.. rose-coloured plumage. **1953** D. A. Bannerman *Birds Brit. Isles* II. 175 The eastern races of the rosefinch also winter in India.

rose, *v.*¹ Add: **2. b.** *intr.* To become rosy; to blush. *rare*⁻¹.

1922 Hardy *Late Lyrics* 22 You grew elate, And rosed, as maidens can, For a brief span.

rosé (rōᵘ·ze, ‖ roze), *sb.* (*a.*) [*ellipt.* for F. *vin rosé* pink wine.] **1.** A wine that is light red or pink in colour.

1897 A. Beardsley *Let.* 17 May (1970) 320, I have just vomited up the meals of the last two or three days. You see your Saint Marceau (royal rosé) has been avenged. **1932** E. Hemingway *Death in Afternoon* 491 Valdepenas is.. excellent in both white and rosée [*sic*]. **1951** E. David *French Country Cooking* 26 Rosés of Anjou and Tavel. **1959** *News Chron.* 9 Dec. 6/6 It says of a Portuguese Rosé: '.. Pale pink but sparkling'. **1960** *House & Garden* June 118/3 A Vin Rosé from the Côte du Rhône will cost you about 9s. 6d. and a Rosé from Anjou a little more. **1974** *Guardian* 24 Jan. 13/5 True rosé is made from black grapes when the skins are left in the vat for the first two or three days of fermentation. [illegible] **1960** *Harper's Bazaar* Apr. 115/2 Add 1 bottle dry rosé wine. **1967** M. Lichine *Encycl. Wines* 417/2 Not to be forgotten are the rosé wines of Portugal. **1974** *Harrods Christmas Catal.* 14 Ostrich feather boa.. shades of mandarin/rosé. **1977** *Times* 12 Feb. 7/2 [illegible] rosé meat, rough-cut pâté and lambent juices.

rose-apple. 2. (Earlier and later examples.)

1790 W. Beckford [illegible] The orange, the rose-apple, the paw, and plantain. **1975** J. R. & A. Mancinelli tr. *Bianchini & Corbetta's Fruits of Earth* 168 The rose apple is about the size and colour of an apricot, with one to three seeds inside.

rose bowl. [f. ROSE *sb.* + BOWL *sb.*¹] **1.** A bowl designed to hold cut roses; *spec.* such a bowl offered as a prize in a competition.

1883 Montgomery Ward *Catal.* [illegible] Rose Bowl, [illegible] imitation of heavy cut glass. **1916** *Daily Colonist* (Victoria, B.C.) 25 July 7/1 (Advt.), Brilliant Beauty Glassware Assortment.. Rose Bowls—Clear crystal, beautiful design. Special, each 50c. **1958** L. Durrell *Mountolive* xi. 232 The room was full of the scent of the pastels burning in the great rose-bowl by the telephone. [illegible] **1977** *Daily Tel.* 13 Oct. 13 Today's young people hardly know what a rose-bowl is. [illegible] The winner will receive a silver rose-bowl which is competed for annually.

2. *U.S.* (With capital initials.) The name of a football stadium in Pasadena, California; used *attrib.* and *absol.* of a football match played between rival college teams annually on New Year's Day at the conclusion of the local Tournament of Roses.

1930 *Los Angeles Times* 2 Jan. (Sports section) 1/1 Southern California's smashing victory gives that institution two wins in as many Rose Bowl games.. The other triumph was scored over Penn State in 1923. **1947** *Collier's* 29 Nov. 89/2 Getting to the Rose Bowl is even simpler this year because Illinois, as the winner last year, isn't eligible to go [illegible]. [illegible] Michigan's football team began working out without equipment Tuesday in preparation for the trip to Pasadena, Calif., for the Rose Bowl game against Southern California New Year's Day. [illegible]

rosebud. Add: **2. b.** (Later non-*attrib.* example.)

1973 *Times Lit. Suppl.* 1 June 608/1 He married.. a college beauty queen (a 'Rosebud' of 1922).

†c. A member of the junior section of the Girl Guides Association, now called a 'Brownie' or 'Brownie Guide' (see *BROWNIE¹ 2). *Obs.* exc. *Hist.*

1914 A. Baden-Powell in *Girl Guides' Gaz.* June 2/1 The age at which a Rose Bud may join the Baden-Powell Girl Guides is eight years. [illegible] **1973** *New Society* 27 Sept. 755/2 Brownies were started in 1914 by Baden-Powell's sister, under the name of Rosebuds.

rose-colour, *sb.* Add: **1.** (Earlier *transf.* example.)

1865 Mrs. Gaskell *Wives & Daughters* (1866) I. xxv. 261 Such were the facts, but rose-colour was the medium through which they were seen.

2. (Earlier example.)

1857 Trollope *Barchester T.* III. ix. 177 It was not all rose colour with Mr. Slope, although his hopes ran high.

rose-colour, *v.* (Later example in *transf.* or *fig.* sense.)

1974 M. C. Gerald *Pharmacol.* xv. 281 Stimulation of RAS results in an enhanced state of arousal to environmental stimuli,.. 'rose-coloring' the individual's subjective appraisal of the world around him.

rose-coloured, *a.* Add: **3.** (Earlier examples.) *rose-coloured spectacles*, used in *fig.* phrases to indicate that a person's view of something is unduly favourable, optimistic, or idealistic.

1854 C. M. Yonge *Castle Builders* iv. 56 The first rose-coloured lights in which they had viewed everything, was wearing off. **1856** Dickens *Dorrit* (1857) I. xxxiv. 299, I don't like to dispel your generous visions, and I would give any money.. to live in such a rose-coloured mist. **1861** Rose-coloured spectacles [in Dict.]. **1867** A. J. Evans *St. Elmo* [illegible]. **1921** Princess P. Metternich *Days that are no More* iii. 104, I was young and favoured by fortune, no troubles had yet befallen me, and I saw everything through rose-coloured spectacles. **1981** *Oxford Jrnl.* 15 May 8/1 Mrs M— must be viewing Castleton crossroads through rose-coloured spectacles. Far from being vastly improved, it is becoming a ghastly eyesore.

rose du Barry (rōᵘz du ba·ri). Also **rose du Barri**. [f. the name of the Comtesse du Barry (1746–93), a patron of the Sèvres porcelain factory.] A soft shade of pink developed *c* 1757 for use as a ground colour on Sèvres porcelain. Also *attrib.* or as *adj.* Cf. *ROSE POMPADOUR.

1856 [see *CARAT¹ 3]. **1879** C. Schreiber *Jrnl.* 13 Oct. (1911) II. 227 Some very fine Rose du Barry Sèvres vases. **1902** *Encycl. Brit.* XXIX. 730/2 The experts of Tōkyō and Nagoya have produced very beautiful specimens of monochrome enamels—yellow (canary or straw), rose du Barry, liquid-dawn red, [etc.]. **1912** C. Mackenzie *Carnival* [illegible]. [illegible] Du Barry Red is sometimes wrongly called Rose du Barri. **1960** H. Hayward *Antique Coll.* 243/2 *Rose du Barry*, a popular name for a crimson ground used on Sèvres porcelain and properly known as rose Pompadour. **1974** *Country Life* 3 Oct. 939/3 Vases of delicate pink which we know now as rose Pompadour but which used to be called rose du Barry.

rose-leaf. Add: (Further *attrib.* and *Comb.* examples.) Hence **ro·se-lea:fy** *adj.*

1854 C. M. Yonge *Heartsease* II. xviii. 50 A little pair of socks, in delicate rose-leaf knitting. [illegible] **1884** 'Mark Twain' *Huck. Finn* xxii. 222 Every lady's rose-leafy dress flapping soft and silky. [illegible] **1913** C. Mackenzie *Sinister St.* I. ii. 167 As for Dora's face, Michael found it beautiful with the long-lashed blue eyes and rose-leaf complexion. [illegible] On rose-leaf heights we too were born, But spirit falters, or the proud foot slips.

rosella¹. 1. Substitute for def.: A brightly coloured seed-eating Australian parakeet belonging to the genus *Platycercus*. (Earlier and later examples.)

1829 *Sydney Gaz.* 21 July 4/1 The doleful dying quails, And rosellas [*sic*] golden. **1838** J. Hawdon *Jrnl.* 15 Mar. in *Journey New South Wales to Adelaide* (1952) 51 Parrots now appeared more numerous, Rosellas and others. **1941** *Coast to Coast 1941* 86 Thither were transferred the assortment of rosellas and grass parrots [illegible]. **1966** [illegible] Pale-headed and Northern Rosellas.

2. *Austral.* and *N.Z.* A sheep whose wool is beginning to fall off naturally, and which is therefore easy to shear.

1849 D. McLeod in *Stephen's Adelaide Miscell.* III. 81 At shearing he.. pick[s] all the 'Rosellas' [clean-bellied sheep]. [illegible] **1954** R. C. Studholme *Te Waimate* (ed. 2) [illegible] rosella [illegible]. **1972** J. S. Gunn in G. W. Turner *Good Austral. Eng.* 61 The shearer has recognised *flyer*, *cop*, *gunbarrel*, as terms for what is usually called a *barebelly* or *rosella*.

‖ **rosemaling** (rō·səmɑːlɪŋ, -mɔ̄lɪŋ). [Norw., f. *rose* rose + *maling* painting.] The art of painting (wooden implements, furniture, etc.) with decorative flower motifs. Hence **ro·semaled, -malt** (-mald, mālt) *ppl. a.* [Norw. *-malt* pa. ppl. of *male* to paint], decorated with rosemaling; **ro·semaler**, one who practises rosemaling.

1948 *School Arts* Mar. 223/1 The Rosemaling (flower painting) of Norway has been brought to this country [sc. the U.S.A.] and skillfully adapted to American living by Per Lysne of Stoughton, Wisconsin. **1950** H. Major *Norwegian Holiday* 85 There will be a bowl or two decorated in the colorful rose-painting (*rosemaling*) design of the district. *Ibid.* 93 (*caption*) A student must first design his rosemaling. **1953** J. Stewart *Folk Arts of Norway* iv. 87 Rosemaling.. has a counterpart in the Swedish *blomstermålning* (flower painting), in the decorative painting of Russia, and in Pennsylvania Dutch art. **1956** T. Bøhn in *Norwegian-Amer. Stud. & Rec.* XIX. 120 The multitude of Norwegian immigrant items grew, including such things as *rosemalt* chests, cupboards, boxes. **1972** Dean & Smith *Wisconsin* 145/1 Rosemaling, the colorful art of floral painting that grew up in Norway, is enjoying a booming revival in Wisconsin. **1975** Lovoll & Bjork *Norwegian-Amer. Hist. Assoc., 1925–1975* v. 21 The museum has also conducted workshops in *rosemaling* (rose painting) and wood carving, with teachers from Norway. **1975** J. Langland in *Massa-*

chusetts Rev. (Univ. of Mass.) Summer 568 They stuffed their childhood into rosemaler trunks, clamped them with iron bands locked once and for all on the eastern hemispheres. **1976** P. Virch *Rosemaling in Round* i. 1/2 The largest public collection of rosemaled chests, bowls, furniture and boxes can be seen in Vesterheim, the Norwegian-American museum in Decorah, Iowa. *Ibid.*, (*caption*) This collection of bowls owned by the author was painted for her by.. rosemalers of Norway.

rosenbuschite (rōᵘ·zěnbuʃəit, rōᵘzěnbu·ʃəit). *Min.* [ad. Norw. *rosenbuschit* (W. C. Brögger 1887, in *Geol. För. i Stockholm Förh.* IX. 254), f. the name of K. H. F. *Rosenbusch* (1836–1914), German mineralogist and geologist: see -ITE¹.] A fluorine-containing aluminosilicate of calcium, sodium, zirconium, and titanium occurring as radiating groups of slender triclinic crystals of an orange or grey colour.

1890 *Jrnl. Chem. Soc.* LVIII. 1079 Rosenbuschite.. presents sufficient analogies to pectolite for it to be described as a zircon-pectolite. **1966** W. A. Deer et al. *Introd. Rock-Forming Minerals* 60 Rosenbuschite occurs as an accessory mineral in nepheline-syenite, but is a more frequent constituent of nepheline-syenite pegmatites.

rosenhahnite (rōᵘ·zěnhā-nəit). *Min.* [f. the name of Leo *Rosenhahn*, U.S. amateur mineralogist, who first found it in 1962 + -ITE¹.] A hydrous calcium silicate, $(CaSiO_3)_3.H_2O$, occurring as buff to white, tabular or lath-like, triclinic crystals.

1967 A. Pabst et al. in *Amer. Mineralogist* LII. 336 (*heading*) Rosenhahnite, a new hydrous silicate from Mendocino County, California. **1973** *Nature* 5 Jan. 42/1 Thermal dehydration data showed that the molecule of water is given off very slowly at temperatures between 400 and 500°C and a single crystal of rosenhahnite transforms into an almost perfect single crystal of wollastonite, $CaSiO_3$.

Rosenkreuzian, var. ROSICRUCIAN *sb.* and *a.* in Dict. and Suppl.

Rosenthal (rōᵘ·zěntāl). The name of Philip *Rosenthal*, founder of a porcelain factory at Selb in Bavaria *c* 1880, used *attrib.* of pottery made there.

1947 M. Penkala *European Porcelain* 64 Rosenthal porcelain is carefully modelled and decorated. **1962** P. Purser *Peregrination 22* xxx. 136 The tray was silver and the tea service Rosenthal china. **1970** E. Pace *Saberlegs* xvi. 152 Oranges heaped in an old Rosenthal bowl. **1977** C. McFadden *Serial* (1978) i. 8/2 They spent it rapidly on .. Rosenthal china.

rose of Sharon. Add: **1. b.** (Later examples.)

1876 E. G. White *Testimonies for Church* (1948) I. 19 There was a beautiful pink flower in the garden called the rose of Sharon. **1974** *Daily Colonist* (Victoria, B.C.) 23 July 20/6 The other Rose of Sharon is a real beauty whose true name is *Hibiscus syriacus*.

c. Delete *dial.* and add later examples.

1938 A. T. Johnson *Garden To-day* xl. 141 The old 'Rose of Sharon'.., a lowly carpeting shrub, is fairly well known. **1979** C. E. L. Phillips *New Small Garden* xv. 243 There are 'carpeting' plants for covering rough places or growing under trees. The Rose-of-Sharon is one of the most useful of these.

3. Chiefly *U.S.* The name of a pattern used in quilting. Also *attrib.*

1894 *Scribner's Mag.* Sept. 363/1 Other floral designs, the sunflower, double peony, rose of Sharon hint at flower-borders lovingly tended by the over-taxed hands of a busy housewife. **1915** M. D. Webster *Quilts* caption facing p. 75 Rose of Sharon. Made in Indiana about 65 years ago. It has a wool interlining instead of the usual cotton. **1929** R. E. Finley *Old Patchwork Quilts* xi. 126 The quilt shown.. presents the original somewhat crude but very old type of 'Rose of Sharon' block. **1964** D. Brightbill *Quilting as Hobby* 88/1 The most popular pattern [of Bride's Quilts].. was the Rose of Sharon... Its name was probably derived from the Song of Solomon: 'I am the Rose of Sharon And the Lily-of-the-Valley.' **1966** D. A. Hinson *Quilting Man.* iii. 45 The Rose of Sharon quilt patterns are legion. **1974** *Times* 11 Jan. 9/5 Quilting is a special craft of [the Mennonites].. and they love the traditional patterns like Rose of Sharon.

ro·sepath. [f. ROSE *sb.* + PATH *sb.*] A pattern used in weaving.

1932 Simpson & Weir *Weaver's Craft* x. 92 Other suitable 'threading drafts' will be given; but the 'Rosepath' entering offers plenty of scope for experiment and a great variety of patterns. **1960** A. Gibbs in G. Lewis *Handbk. Crafts* 103 Let us assume that we are going to use a simple pattern, Rosepath (this is a universal pattern, being found in peasant weaving throughout the world; it is capable of a very large number of variations).

rose-pink, *sb.* and *a.* Add: **A.** *sb.* **1.** (Later example.)

1877 E. S. Dallas *Kettner's Bk. of Table* 338 The rose-pink, which is sometimes used for colouring.. is obtained from the peach tree.

2. Also *fig.*, sentimentality, sentimental writing.

1872 G. Meredith *Let.* 3 Dec. (1970) I. 473 Read the first chapter for a specimen of modern rose-pink.

rose Pompadour (rōᵘ·z pǫ·mpădūᵊɹ). [f. ROSE *sb.* + POMPADOUR.] = *ROSE DU BARRY. Cf. POMPADOUR 2.

Considered by some authorities to be the more correct term.

1884 P. Villars tr. *Gasnault & Garnier's French Pottery* ix. 123 We must mention, among the most remarkable productions of the manufactory of Sèvres during the first period whose history we have briefly sketched, the beautiful ornamental vases with.. the pink ground termed *rose carné* or *rose Pompadour* (sometimes erroneously called *rose Dubarry*). **1905** W. Burton tr. *Auscher's Hist & Descr. French Porcelain* viii. 65 In 1757 the painter Xrowet invented that most famous ground-colour, rose-Pompadour. **1935** *Burlington Mag.* May 249/2 The 'claret-colour', which must have started as an attempt to ape the lovely rose Pompadour of Sèvres. **1936** *Ibid.* Dec. p. xix/2 The Sèvres porcelain especially the apple-green and *rose Pompadour* pieces. **1949** *Dict. Colours Interior Decoration* (Brit. Colour Council) III. 23/1 *Rose Pompadour.* This colour (matched to porcelain in the Wallace Collection, London) is named after Madame de Pompadour who took a great interest in the porcelain manufactory, first at Vincennes and later at Sèvres. **1960** R. G. Haggar *Conc. Encycl. Continental Pott. & Porc.* 386/2 *Rose Pompadour*.. was invented according to Garnier by Jean Hellot, or, according to Auscher by Xhrouet... It takes its name from the Marquise de Pompadour (died 1764). **1960** H. Hayward *Antique Coll.* 243/2 Coalport.. produced a rose Pompadour in the 19th cent. **1975** *Oxf. Compan. Decorative Arts* 135/2 An outstanding result of this [development of Sèvres, fostered by royal privileges] was the development of a series of very splendid ground-colours: dark blue in 1749,.. pink ('rose Pompadour') in 1757 and so on.

ro·se-ti:nted, *a.* [f. ROSE *sb.*] = ROSE-COLOURED *a.* spec. *rose-tinted spectacles* = *rose-coloured spectacles* s.v. *ROSE-COLOURED *a.* 3.

1855 [see ROSE *sb.* 21 a]. **1885** 'Mark Twain' *Let.* 2 Mar. (1917) II. 452 He sold me $10,000 worth of another rose-tinted stock. **1956** C. Wilson *Outsider* viii. 245 The good-natured, eupeptic vulgarian who sees life through rose-tinted spectacles. **1966** *Listener* 20 Oct. 559/2 The official propaganda machine, whose task is to present every aspect of Soviet life through rose-tinted spectacles. **1977** *Church Times* 25 Feb. 7/2 There are no rose-tinted specs shading the very clear eyes of Elizabeth West as she describes life in her *Hovel in the Hills*.

Rosetta stone (rōᵘze·tă stōᵘ·n). [The name of a celebrated stone, bearing a trilingual inscription dating from the 2nd c. B.C., found in 1799 near Rosetta in Egypt.

The stone, now in the British Museum, bears an inscription in Greek, demotic, and hieroglyphics; the decipherment of the latter two parts of the inscription by Jean-François Champollion in 1822 led to the interpretation of all the other early records of the Egyptian civilization.]

Used *transf.* and *fig.* of something that resembles the Rosetta stone, usu. by acting as a key to some previously unattainable understanding (in quot. 1969, something indecipherable).

1902 *Encycl. Brit.* XXVI. 721/2 Although not the Rosetta stone which enabled him to decipher the minute structure of glucose and its congeners, this compound made possible for the first time the separation and identification of such compounds. **1933** H. G. Wells *Shape of Things to Come* 24 About a third of the shorthand stuff was already represented by longhand or typescript copy in the folders. That was my Rosetta Stone. **1969** R. Lowell *Notebk. 1967–68* 102 The typescript looked like a Rosetta Stone. **1975** *New Yorker* 29 Sept. 29/1 A knowledge of accounting is a kind of Rosetta stone. **1979** *Sci. Amer.* Mar. 72/1 The spectrum of the hydrogen atoms has proved to be the Rosetta stone of modern physics: once this pattern of lines had been deciphered much else could also be understood.

rosette, *sb.* Add: **1. a.** (Earlier and later examples.) Also, applied to such a decoration awarded to prize-winners at horse shows and similar events.

1790 *Pennsylvania Packet* 11 Dec. 3/2 Imported.. Ladies.. elegant.. beaded rosettees, for shoes. **1951** J. Pullein-Thompson *Radney Riding Club* ii. 31 When the ponies were fed and settled for the night Eric nailed the new rosettes beside the others in the saddle-room. **1957** R. Ferguson *Rosettes for Jill* v. 52, I fastened my two red rosettes and the blue one on Black Boy's browband, and collected my prizes. **1965** T. Gunn in *New Statesman* 14 May 768/1 To enter Jerusalem on an ass.. or wear a rosette for Arsenal. **1967** *Listener* 26 Oct. 552/3, I walked up to one of the chaps and said.., pointing at the two Welsh rosettes upon his chest: 'Excuse me, what are these for?'

d. *transf.* A rose- or star-shaped symbol used in guides to hotels and restaurants to indicate the standard of service or cuisine provided.

1966 P. V. Price *France* 318 The famous 'stars' of Michelin, which are indicated by rosettes in the guide book, refer *solely* to the standard of the food and drink. **1974** *Country Life* 24 Jan. 167/3 (Advt.), Our AA Rosette reflects.. the high standards of country house hotel-keeping of which we are justly proud. **1976** *Times* 2 Oct. 10/1 In Avignon.. a truly Lucullan dinner at *Hidly-Lucullus*.. the food worthy of its two rosettes in Michelin.

4. b. Also, a similar formation as a symptom of plant disease, the leaves on a stem being clustered owing to its greatly reduced internodal growth. (Further examples.)

1891 *Jrnl. Mycol.* VI. 143 The lower leaves on these tufts or rosettes roll and curl, turn yellow,.. and fall early. **1937** F. D. Heald *Introd. Plant Path.* ii. 20 Rosettes, or closely grouped clusters of leaves caused by the failure of axes to make a normal elongation. This should not be confused with the normal rosette habit of certain plants. **1952** [see next sense]. **1980** *Amateur Gardening* 18 Oct. 21 The lovely wide-faced flowers.. are held over the hairy rosettes of leaves in May and June.

c. Delete *U.S.* and substitute for def.: Any of various plant diseases in which there are rosette-like malformations of leaves. Also *rosette disease*. (Examples.)

1891 E. F. Smith in *Jrnl. Mycol.* VI. 143 It seems best, therefore, to call it [sc. the disease] 'the peach rosette' until it can be determined whether it is identical with yellows. *Ibid.* 146 This rosette disease resembles yellows very closely. **1923** *Phytopathology* XIII. 41 The symptoms of the rosette disease of wheat bear certain resemblances to the symptoms of corn mosaic as described by Kunkel. **1946** *Ann. Reg. 1945* 349 K. M. Smith.. has shown that the two separable complexes of tobacco rosette disease may be transmitted.. by Aphis. **1950** *Times* 2 Feb. 9/2 This new discovery brings within reach a means of prevention of virus diseases which are transmitted by aphids, such as yellow virus on sugar beet, strawberry virus and rosette disease in groundnuts. **1952** tr. *Gram & Weber's Plant Dis.* 489/2 Rosette is a disease of *Lilium longiflorum* and its varieties... Infected plants have yellow leaves which remain in a basal rosette. **1972** J. T. Slykhuis in Kado & Agrawal *Princ. & Techniques Plant Virol.* vii. 208 Evidence that pigeon pea sterility and rose rosette are caused by viruses rests on transmission of the disease by grafting.

d. *Med.* A group of red cells bearing one factor adhering to one red cell bearing another factor, produced in tests for antigens, antibodies, and related substances on the cell surfaces.

[**1958** *Jrnl. Clin. Invest.* XXXVII. 1216/2 The antibody-sensitized red cells were seen to cluster around individual leukocytes, often producing rosette-patterns.] **1964** *Immunol.* VII. 477 Many of the peritoneal cells were seen to be coated with sheep red cells, giving, in many instances, a characteristic 'rosette' appearance. **1966** *Jrnl. Exper. Med.* CXXIII. 144 (*caption*) Rosettes of sheep red cells adsorbed onto the surface of guinea pig lung macrophages. **1971** I. M. Roitt *Essent. Immunol.* iii. 50 When lymphocytes are incubated with, say, sheep red cells, those with surface receptors for the erythrocytes will bind them to form a rosette. **1976** *Nature* 15 July 216/1 This heat treatment does not destroy the ability of the macrophages to form rosettes with sheep red blood cells.

5. a. (Earlier and later examples.)

1767 J. Wedgwood *Let.* 23 May (1965) 53 At Birmingham I saw a Lathe executed upon the plan of that which is full of Rosetts, and every Rosett had a projection from the edge. **1931** A. U. Dilley *Oriental Rugs & Carpets* (caption to plate 33) India Rug of Persian star, palinette, rosette, and leaf design with border of realistic flowering plants.

d. = *ROSE *sb.* 14 g.

1896 R. Robb *Electr. Wiring* v. 152 The holes in the socket bushing and in the rosette are little larger than enough to allow the cord to slide through them. **1904** *Electr. World & Engin.* 7 May 887/2 The finish and smoothness of the rosette are in every way admirable. **1961** C. C. Carr *Craft's Amer. Electricians' Handbk.* (ed. 8) iv. 126 The drop cord passes through the hole in the center and is attached to connections inside the body of the rosette.

e. *Geol.* = *ROCK-ROSE 5, *ROSE *sb.* 16 e.

[**1902** H. A. Miers *Mineral.* iv. i. 249 Irregular conjunctions are distinguished as—... Rosette-shaped, when they overlap round a centre like the petals of a rose.] **1905** *Bull. U.S. Geol. Survey* No. 239. 59 The rosettes are sometimes a foot in diameter, while there is every gradation from this to submicroscopic size. **1923** *Proc. Oklahoma Acad. Sci.* III. 102 Barite and especially the form known as 'sand barite rosettes', has long attracted attention as one of the most widely disseminated of Oklahoma minerals. **1971** Fejer & Walker tr. *H. Baegel's Collector's Guide Minerals & Gemstones* i. 31 Tabular minerals (gypsum, barite, hematite) may occasionally form rosettes ('desert roses').

f. *Engin.* An arrangement about a point of three or more coplanar lines that represent the axes of strain gauges used to determine the strain existing in a structure or material at that point.

1931 W. Hovgaard in *Trans. Soc. Naval Architects & Marine Engineers* XXXIX. 26/1 We arrive thus at what may be called a 'rosette' of strain measurements consisting of one horizontal (longitudinal) strain, one vertical (transverse), and two at 45 degrees inclination. **1946** G. Murphy *Advanced Mechanics of Materials* iii. 64 Valuable information concerning the stresses can be obtained by measuring the strains developed in a model or a trial design of the prototype and converting the strains into stresses. The usual procedure involves the measurement of the normal strains on a rosette of three or four intersecting gage lines at the point. **1950** M. I. Hetényi *Handbk. Exper. Stress Anal.* ix. 400 The four-gage 45° rosette combines all the advantages enumerated.. for the equiangular and the rectangular rosettes. **1969** H. N. Norton *Handbk. Transducers for Electronic Measuring Syst.* xiii. 559 Gages with multiple grids (rosettes) were developed for simultaneous measurement of strain in different directions.

6. *rosette bud, habit, symptom, virus*; *rosette-forming* adj.; **rosette gauge** *Engin.*, an assembly of strain gauges whose axes correspond to the lines of a rosette (see sense *5 f); **rosette plant** (see quot. 1934).

1977 J. L. Harper *Population Biol. of Plants* xviii. 543 Digitalis purpurea regenerate from rosette buds if the inflorescence is damaged before the seed is required. *Ibid.* xiv. 437 The flora of Port Meadow is composed of perennial grasses.. plus laterally spreading clonal dicots.. and some rosette-forming species. **1943** *Exper. Stress Anal.* I. i. 13/2 In connection with the use of wire resistance rosette gages.. small corrections must sometimes be applied to the initial strain observations. **1969** H. N. Norton *Handbk. Transducers for Electronic Measuring Syst.* xiii. 561 When using rosette gages, it is necessary to operate upon the output readings using Poisson's ratio to convert strain rates to stress. **1937** Rosette habit [see sense 4 b above]. **1903** W. R. Fisher tr. *Schimper's Plant-Geogr.* III. iv. 706 Perennial rosette-plants play a leading part, especially on alpine meadows. **1934** H. Gilbert-Carter tr. *Raunkiaer's Life Forms of Plants* ii. 47 A transition is formed from the rosette plants in which.. the foliage leaves are all gathered into a rosette at the base. **1965** *Austral. Encycl.* VII. 123/2 It [sc. the Western Australian pitcher-plant] is a small rosette plant. **1928** C. E. Owens *Princ. Plant Path.* xix. 423 Potato plants which are attacked by the Rhizoctonia fungus on the underground parts sometimes show leaf-roll and rosette symptoms. **1937** K. M. Smith *Textbk. Plant Virus Dis.* ii. 186 Numerous other [groundnut] plants in the Gambia have been observed showing typical rosette symptoms. **1960** J. E. van der Plank *Plant Path.* III. vii. 262 Rosette virus ordinarily spreads slowly from peach to peach. **1977** J. L. Harper *Population Biol. of Plants* xvi. 488 The rosette virus of peach.. kills the host rapidly.

rosette (roze·t), *v.* [f. prec. sb.] **1.** *intr. Med.* Of a cell: to form a rosette.

1969 [implied in ROSETTING *ppl. a.* below]. **1973** *Jrnl. Immunol.* CXI. 1834 Lymphocytes with binding sites for complement.. do not rosette with R[abbit] R[ed] B[lood] C[ells]. **1977** *Lancet* 5 Nov. 988/1, 2%.. of the lymphocytes rosetted with sheep red blood-cells (T cells).

2. *trans.* To award a rosette-like symbol to, as a mark of excellence.

1974 *Guardian* 20 Mar. 1/3 Three restaurants much rosetted by English guides—the Ritz, the Savoy, and the Mirabelle.

So **rose·tting** *ppl. a.*

1969 *Internat. Arch. Allergy* XXXV. 220 Vicarious loss of potentially rosetting cells would distort the results.

rosetted, *a.* Add: **1. a.** (Further examples.)

1901 *Bull. Geol. Soc. Amer.* XII. 166 Even the loose and less coherent matrix reveals, under the action of wind and rain, an ill-defined, though unmistakably radiate, or rosetted structure. **1955** *Mineral Abstr.* XII. 573 Rosetted crusts of silver-white semseyite on galena. **1966** D. Varaday *Gara-Yaka's Domain* iv. 47 There they [sc. cheetahs].. glared up at their rosetted relative. The humourless leopard glared back at them. **1969** *Internat. Arch. Allergy* XXXV. 214 (*caption*) Typical appearance of rosetted cells obtained with the suspension-centrifugation technique. **1975** *Daily Tel.* 8 Sept. 13 Triumphantly rosetted and bristling with the familiar red-and-white battle insignia of Manchester United.

b. Of skin or pelt: marked with rosette-like blemishes.

1905 W. E. Castle *Heredity of Coat Characters in Guinea-Pigs & Rabbits* 75 A rosetted or rough coat is unknown in rabbits. **1960** O. Manning *Great Fortune* I. 20 The skin was mottled purple and rosetted with yellow scabs.

c. Having been awarded a rosette.

1972 *Times* 6 May 9/3 As delicious as any rosetted *specialité*.

2. Affected with rosette disease.

1891 *Bull. U.S. Dept. Agric. Div. Veg. Physiol. & Path.* I. 48 Many of the roofs of rosetted trees were honey-combed by gum-pockets. **1937** K. M. Smith *Textbk. Plant Virus Dis.* ii. 186 The rosetted plant may flower, but few of the pegs make any growth. **1949** Butler & Jones *Plant Path.* viii. 289 'Rosetted' peaches nearly always die the following autumn or winter.

rosetting (roze·tiŋ), *vbl. sb.* [f. ROSETTE *sb.* and **v.* + -ING¹.] The occurrence or development of rosettes.

1948 Melhus & Kent *Elem. Plant Path.* x. 246 These plants nearly always remain dwarfed and often show a 'rosetting' caused by the increase in the number of branches and the shortening of the internodes. **1970** *Internat. Arch. Allergy* XXXIX. 658 It is possible to obtain a rosetting reaction with antigen-coated erythrocytes around receptor-bearing dead lymphoid cells. **1978** T. T. Kozlowski in Horsfall & Cowling *Plant Dis.* III. ii. 31 Symptoms of mineral deficiencies include necrosis, dieback of shoots, rosetting, [etc.]. **1978** *Nature* 13 Apr. 619/2 (*caption*) Cells were tested for rosetting with IgG-OE immediately or after incubation overnight at 37 °C, and with or without trypsin treatment.

rose-water. Add: **1. c.** *rose-water bowl, dish* (later examples), *ewer* (later example); **rose-water pear** (earlier example); **rose-water pipe**, an oriental tobacco-pipe in which the smoke passes through rose water before reaching the mouth.

1629 J. Parkinson *Parad.* III. xxi. 592 The Rosewater peare is a goodly faire peare, and of a delicate taste. **1835** N. P. Willis *Pencillings by Way* II. xxi. 234 A string of beads in one hand, and a splendid narghilé, or rose-water pipe, in the other. **1956** G. Taylor *Silver* v. 97 A rose-water dish of 1672 belonging to St John's College, Oxford. **1960** H. Hayward *Antique Coll.* 243/2 *Rose-water ewer and dish or basin*, used for finger-washing at table. **1968** *Canad. Antiques Collector* June 9/3 The rose water bowl or basin was like an enormous soup plate, 12 to 20 inches in diameter. It had an extra wide rim, two inches or more, around the slightly depressed center and was usually ornately decorated.

Hence **ro·se-wa:tery**, *a.*

1902 G. B. Shaw *Let.* 20 June (1972) II. 277 The comparatively rose-watery part of it [sc. a situation in *Mrs. Warren's Profession*].

rosewood. Add: **5*.** A shade or tint of the colour of rosewood.

1853 *Heal & Son Catal.: Bedsteads* 33/1 Parisian Bedstead, with handsome cast iron side and ends, sheet iron head and foot board, japanned rosewood. **1897** *Sears, Roebuck Catal.* 22/3 Wood stains.. Perfect imitations of natural woods, Cherry, Rosewood, Mahogany, Walnut [etc.]. **1907** *Yesterday's Shopping* (1969) 145/1 Stains.. as used by the Working Ladies Guild, colours:—Rosewood, Satinwood, Oak, [etc.]. **1927** *Daily Tel.* 21 Feb. 14 (Advt.), Two-Piece Suits, Coats, &c. Available in shades of Light Grey, Rosewood, Fawn, Sand, &c. **1930** *Daily Express* 6 Oct. 13/1 (Advt.), Shades, beaver, brown, rosewood, and drab. **1971** [see *MOSS *sb.*¹ 5 d].

Rosh Chodesh, Rosh Hodesh (roʃ χō·deʃ, rōᵘʃ χōᵘ·děʃ). [Heb., lit. 'head of the month'.] A Jewish half-holiday observed at the appearance of the New Moon, the beginning of the Jewish month.

1879 C. E. Sachau tr. *Albîrûnî's Chronol. Anc. Nations* xiv. 274 (*heading*) Nisan... has only one Rosh-Ḥodesh and 30 days. **1934** Webster, Rosh Ho'desh.. or Cho'desh. **1963** *Times* 24 Apr. 16/1 He initiated the delightful presidential custom of inviting to his residence a representative gathering of citizens from a different ethnic group each 'Rosh Chodesh', the start of a Hebrew calendar month. **1976** G. Jessup *No Strange God* xi. 75 The first day of every month is a semi-festival, called *Rosh Ḥodesh*, when mourning and fasting are forbidden.

Rosh Hashana (roʃ haʃanā·, rōᵘʃ hăʃōᵘ·nə). Also **Rosh Hashanah, Rosh Hashonoh**, etc. [Heb., lit. 'head of the year'.] The Jewish New Year, celebrated on the first and second day of the month Tishri.

1846 *Jewish Chron.* 21 Aug. 199/1 Rosh Hashana (New Year), and Yom Kippur (Day of Atonement), were kept by most of our co-religionists. **1862** *Chambers's Encycl.* IV. 304/1 The most exalted of new-moon festivals was that of the first day of the seventh month, 'the day of remembrance of the sounding' or 'of trumpets' (Lev. xxiii. 24), to which in later times.. the name of Rosh hashana (New Year) was given. **1907** I. Zangwill *Ghetto Comedies* 18 We lived somehow till *Rosh Hashanah* (New Year), hoping it would indeed be a New Year. **1957** L. Stern *Midas Touch* III. xx. 152 So this year you will come to *shul* maybe for *Rosh Hashona*? **1960** F. Raphael *Limits of Love* II. i. 165 It's *Rosh Hashana*... I'm going to read some prayers. **1970** *Challenge* (Lubavitch Foundation) 281 *Rosh Hashonoh* is Coronation Day for G-d as our King and for Israel as His people. **1973** *Jewish Chron.* 19 Jan. 5/4 After the Munich massacre of Israeli Olympic sportsmen.. it was decided that memorial prayers would be recited at the Rosh Hashana service in Copenhagen's main synagogue. **1973** *Synagogue Light* Sept. 26 Jewish residents of New York State, will join on September 26, with others of their faith throughout the world, in the observance of Rosh Hashanah. **1978** *Detroit Free Press* 16 Apr. (Record) 9/5 On September 16, 1939, the eve of Rosh Hashana.. the *Luftwaffe*.. bombed the Jewish quarter of Warsaw.

Roshi (rōᵘ·ʃi). [Jap.] The spiritual leader of a community of Zen Buddhist monks.

1934 D. T. Suzuki *Training of Zen Buddhist Monk* iv. 31 Yinji-ryō which attends on the master known as Rōshi. **1949** C. Humphreys *Zen Buddhism* vii. 142 Laymen, accepted for teaching by the Roshi, may come for a period... The whole of the monastery is locked in full meditation with a queue of.. monks waiting their turn for the Roshi to confirm, reject or make further suggestions for their inward labours. **1959** *Encounter* Jan. 21/1 The Zen Master, or *Roshi*, the spiritual head of the monastery. **1972** *Last Whole Earth Catalog* (Portola Inst.) 50/2 She's sort of like an elderly Zen priest, an old roshi who after years of work and study has distilled a large burden of 'knowledge' into a single gem of wisdom which he renders in a single haiku. **1974** R. C. Zaehner in *Times Lit. Suppl.* 6 Dec. 1389/4 Anglican priest turned Zen Rōshi, reconciled to his church shortly before his death, but belly-laughing, genial, infuriating Rōshi to the end. **1978** C. Humphreys *Both Sides Circle* xv. 165 There is no likelihood of Zen roshis, in the full meaning of the term as used in the Rinzai School, arriving in Europe in sufficient numbers to give us expert training.

Rosicrucian, *sb.* and *a.* Add: Also (after G. *Rosenkreuz*) **Rosenkreuzian**. **A.** *sb.* Also, a member of a present-day Rosicrucian society (see sense B below).

1957 *Encycl. Brit.* XIX. 560/1 That it is not always possible to prove the existence of the order in a given country at any particular moment does not disturb the Rosicrucians, for it seems to be recognized that there occur periods when the order is deliberately 'in sleep'.

B. *adj.* (Further examples.) Now also applied to various societies that claim to continue the Rosicrucian tradition.

1837 K. H. Digby *Mores Catholici* VIII. vi. 193 The cabalistic learning, expressed in the unintelligible language of Theosophy, which was in the seventeenth century to be the foundation of the Rosenkreuzian society. **1959** *Chambers's Encycl.* XI. 840/2 In the mid-19th century there came into being a Rosicrucian Society in England, an offshoot of masonry, the leading figure being R. W. Little. The name has also been appropriated by American theosophy. None of these later developments can claim historical continuity with any original group. **1961** *Listener* 21 Sept. 443/2 Nothing of his marriage or his dealings with Rudolf Steiner, the Rosicrucian Order, or the Catholic Church. **1977** *Daily Times* (Lagos) 5 Jan. 2 (Advt.), Members of the Rosicrucian Order Amorc (Nigeria) have applied to be registered under the Land Perpetual (Succession) Act.

Rosicrucianism. (Later examples.)

1955 C. S. Lewis *Surprised by Joy* iv. 62 She was.. floundering in the mazes of Theosophy, Rosicrucianism, Spiritualism; the whole Anglo-American Occultist tradition. **1961** *Listener* 21 Dec. 1089/1 For a time Satie dabbled in Rosicrucianism.

rosied, *a.* (Later example.)

1910 W. de la Mare *Three Mulla-Mulgars* xxii. 291 The faintly-rosied starlight.

Rosie Lee, var. *ROSY LEE.

rosier. (Later *poet.* example.)

1925 E. Sitwell *Troy Park* 88 Under a rosier Stood the Bishop Walked with a crozier.

rosin, *sb.* Add: **1. c.** *slang.* (*a*) Alcoholic drink. Cf. ROSIN *v.* 2. (*b*) A fiddler, a violinist; also, *rosin-the-bow*.

1734 *Select Trials* I. 227/1 Says I to the Gentleman, I hope, Sir, you won't forget your Coachman—a little Rozzam wou'd do very well. *Ibid.*, Rosin, strong Drink: A Metaphor first used among Fidlers. **1864** Hotten *Slang Dict.* 215 *Rosin*, beer or other drink given to musicians at a dancing party. *Ibid.*, *Rosin-the-bow*, a fiddler. **1901** F. E. Taylor *Folk-Speech S. Lancashire*, *Rozzin*, a jocular term for musician's drink. **1904** S. Watson *Wops the Waif* (1924) iii. 9/1 A short, lame man,.. with a violin beneath his arm, suggesting the identity with the 'rosin' announced.

2. *rosin-distiller, -flux*; **rosin-back** *Circus slang*, (*a*) a horse used by a bareback rider or acrobat; (*b*) a bareback rider.

1923 C. R. Cooper *Under Big Top* 170 She is trained to the 'rosinback', as the ring horse is called. **1931** *Amer. Mercury* Nov. 353/2 *Rosinbacks*, bareback riders. **1933** P. Godfrey *Back-Stage* xvii. 213 One of the least spectacular, yet most difficult, tasks is to train the trick-rider's horse, or 'rosin-back'. These horses are massive Flemish animals, capable of supporting on their broad backs several performers at a time. **1945** C. B. Cochran *Showman looks On* iii. 33 A 'rosin-back' is a ring-horse used by bareback riders... Rosin is rubbed into the horse's back to help the rider to get a firm footing as he jumps from the ring on to the horse. **1974** V. Canning *Painted Tent* iii. 51 There were a few horses in the stable, a couple of rosin backs and a small black pony. **1885** *List of Subscribers, Classified* (United Telephone Co.) (ed. 6) 207 Tar, rosin, benzole and naphtha distillers. **1960** Cooke & Marcus *Electronics & Nucleonics Dict.* 410/2 *Rosin-core solder*, solder made up in tubular or other hollow form, with the inner space filled with rosin flux to serve as a noncorrosive flux for soldering joints. **1976** *S9* (N.Y.) May/June 101/3 For electronics work only rosin flux is used as it is non-corrosive.

Rosinante. Add: (Earlier example.)

1745 *Life & Adventures B.-M. Carew* 65 Who, enraged by their several Losses, began to curse the Doctor and his Rosinante.

Hence **Rosina·ntine** *a.*, lean, worn-out. *nonce-wd.*

1936 P. Fleming *News from Tartary* ix. 235 Plump and naked.. they [sc. camels] were a great contrast to our shaggy, Rosinantine beasts.

rosined, *ppl. a.* Add: Also *fig.*

1774 Goldsmith in *London Chron.* 28–30 Apr. 416/3 And shall I mix in this unhallow'd crew? May rosin'd lightning blast me if I do.

rosiner (rǫ·zinəɹ). *Ir.* and *Austral. slang.* Also **rosner, rossiner, rozener.** [f. ROSIN *sb.* or *v.* + -ER¹.] A drink of spirits; a stiff drink. Also *transf.*

1932 D. Johnston *Moon in Yellow River* I. 35 Well, you must step across to the store some time and we'll give you a rosner. **1933** *Bulletin* (Sydney) 10 May 20/1 Fill up the cup, a rozener, a hummer! **1934** S. Beckett *More Pricks than Kicks* 119 'And the rosiners' said Mrs Tough, 'will you have that in the lav too?' Reader, a rosiner is a drop of the hard. **1947** H. D. Brockman *Fatal Days* 114 I've not had a solitary spot since four. I need a rosiner. **1954** T. Ronan *Vision Splendid* 345 Two nips that old Block had and the one I poured into Peter. They were rozeners I'll admit, but still the three I've had out of this second bottle haven't exactly been small ones. a **1966** 'M. na Gopaleen' *Best of Myles* (1977) 310 A rossiner wouldn't be bad, have a double one after this. **1973** D. Stuart *Morning Star, Evening Star* 53 There's no harm in a bit of rosiner after a hard day's travel, just once in a while.

Rosminian. (Earlier examples.)

1837 L. D. Shrewsbury *Let.* 16 Apr. in D. Gwynn *Fr. Luigi Gentili* (1951) ix. 128, I thought the Rosminians would bring their means with them. **1843** M. Pattison *Diary* 1 Oct. in W. Meynell *John Henry Newman* (1890) iii. 32 Talk of some Rosminian Nuns coming to England.

rosner, var. *ROSINER.

rosolio. Add: Also **rossolio.** (Earlier and further examples.)
1818 'A. Burton' *Adventures J. Newcome* IV. 238 At each Jew Agent's did he stop, Each Wine-house, and Rosolio-shop. 1850 [see *liqueur-glass* s.v. *LIQUEUR *sb.* 3]. 1973 *Times Lit. Suppl.* 14 Sept. 1064/2 He plied the abbot with sweet pink rosoglio and was then admitted to a closet in the cellar.

Ross (rɒs), *sb.*[4] The name of Sir James Clark *Ross* (1800–62), Scottish explorer, used *attrib.* and in the possessive in **Ross('s) gull,** to designate a pinkish-white Arctic gull, *Rhodostethia rosea,* formerly named *Larus rossii* in his honour by J. Richardson in 1825 (*App. W. E. Parry's Jrnl. Second Voy. N.-W. Passage 1821–23* 359).
[1872: see Rosy *a.* 5.] 1902 *N. Amer. Fauna* XXII. 80 Ross Gull. The first known specimen of this beautiful species was killed at Alagnak, Melville Peninsula, by James Clark Ross, in June, 1823, during Parry's second voyage. 1926 A. Thorburn *Brit. Birds* IV. 70 The Wedge-tailed Gull... A specimen of this small and very beautiful species, known also as Ross's Gull, is said to have been obtained at Tadcaster, Yorkshire, in December, 1846. 1957 L. L. Snyder *Arctic Birds Canada* 222 The rather fragmentary information pertaining to Ross's Gull has come largely from the Old World. 1971 *Country Life* 23 Sept. 751/1 The announcement of two rare arctic gulls sighted in the north east of England last December: a Ross's gull and an ivory gull. 1976 *New Yorker* 26 Jan. 25/3 Last year around this time, a Ross's gull—the only species of gull with a conspicuously pink breast—caused a considerable stir when it arrived at Salisbury, Massachusetts, thousands of miles from its normal home, in Siberia.

Ross (rɒs), *sb.*[5] The name of Bernard R. *Ross* (1827–74), factor of the Hudson's Bay Company, used in the possessive in **Ross'(s) goose** to designate a small Arctic goose, *Chen rossii,* formerly *Anser rossii,* named in his honour in 1861 by John Cassin (*Proc. Acad. Nat. Sci. Philadelphia* 72).
1874 E. Coues *Birds of Northwest* 553 Horned Wavy; Ross' Goose. 1908 C. Mair *Through Mackenzie Basin* 320 At Fort Chipewyan.. Ross's goose is the last to arrive in the spring. 1947 C. E. Gillham *Raw North* 175 He kindly interpreted for me while I interviewed his natives regarding the whereabouts of the Ross's goose. 1966 W. E. Godfrey *Birds of Canada* 52/2 White or greyish-white geese with black wing tips, they are likely to be confused with Ross's Goose.

Ross (rɒs), *sb.*[6] The name of Sir Charles A. F. L. *Ross,* Scottish-born engineer and soldier, used to designate a type of rifle used by the Canadian Army, esp. in the war of 1914–18.
1906 *Canadian Mag.* Dec. 66 (Advt.), Ross Rifles... The best in the World... 303 Calibre. 1917 *Grit* (Toronto) 7 Dec. 4/5 Think of the Ross Rifle, the lame horses, the sham shoes, the Allison rake-off. 1963 *Military Arms of Canada* 43 The Ross Rifle was a straight pull bolt action rifle that was made in calibre .303 British and .280 Ross for the Canadian Government. 1972 J. Minifie *Homesteader* xix. 169 We had been issued Ross rifles, a heavy weapon with a straight-draw action which jammed. 1973 J. Quick *Dict. Weapons* 377/1 *Ross .303 rifle*... Under actual trench-warfare conditions it was found unsuitable. A serious problem with this weapon is the bolt. If reassembled wrong, it will permit firing in an unlocked position, resulting in serious injury or death to the shooter.

ross, *v.* (Earlier and later examples.)
1853 S. Strickland *27 Yrs. Canada West* II. 230 As soon as the tree is felled, a person, called a liner, rosses and lines the tree on each side. 1958 *N.Z. Timber Jrnl.* June 59/2 *Rossing,* removing the bark from logs.

Rossby wave (rɒ·sbi). *Physics* and *Meteorol.* [f. the name of Carl-Gustaf Arvid *Rossby* (1898–1957), Swedish meteorologist + Wave *sb.*] A long wavelength fluctuation of a current in a fluid system having no divergence and subject to Coriolis force; *esp.* a lateral fluctuation of a jet stream, with wavelength comparable with the radius of the earth.
[1951 *Jrnl. Meteorol.* VIII. 264/2 The velocity..of Rossby long waves relative to a basic current.] 1963 *Deep-Sea Res.* X. 735 Damped, stationary Rossby waves can occur in the ocean superimposed on a steady west to east flow. 1974 *Earth-Sci. Rev.* X. 203 Planetary or Rossby waves, though probably unimportant in the fluid interior of the Earth, are of interest to earth scientists in general, because of their pervasive role in the general circulation of oceans and atmospheres. 1974 *Nature* 5 Apr. 539/1 The intense Kuroshio current may generate a series of Rossby waves, which can propagate across the entire Pacific Basin. 1974 *Encycl. Brit. Macropædia* X. 163/1 If floor conditions are neither divergent nor convergent.., the absolute vorticity should not change with time... This explains the reason for the formation of long planetary waves, the so-called Rossby waves, in the upper-tropospheric flow patterns.

Rossettian (rǫze·tiăn), *a.* [f. the name of D. G. *Rossetti* (see below) + -IAN]. Pertaining to or characteristic of Dante Gabriel Rossetti (1828–82), English poet and Pre-Raphaelite artist, or his work.
1881 'V. Lee' *Let.* in P. Gunn *Vernon Lee* (1964) vii. 79 All the Rossettian poeticules. 1905 G. B. Shaw *Let.* 11 Sept. (1972) II. 555 His [*sc.* W. Morris's] old Rossettian associates called him Topsy. 1908 D. H. Lawrence *Let.* 9 Oct. (1962) I. 30 What's the M.R.? Mary Rose? It sounds Rossettian. 1914 G. B. Shaw *Pygmalion* (1916) III. 144 One of the beautiful Rossettian costumes which.. led to the absurdities of popular estheticism. 1927 E. Sitwell in *Daily Mail* 30 June 10/4 In provincial cities we may still find relics of Rossettian heads of fox-coloured hair. 1974 K. Clark *Another Part of Wood* v. 179 An invitation to dine was in Rossettian language 'Come and grub with me', but the dinner was more Beardsleyan.
So **Rossettia·na** [Ana *suff.*], relics of, or information about, D. G. Rossetti.
1928 R. L. Megroz in *Daily Express* 28 Sept. 10/3 Sir Hall Caine has now made explicit what every careful student of Rossettiana has realised, that Lizzie Siddal committed suicide.

rossie (rɒ·si), *sb. Anglo-Ir.* [ad. Ir. *rdsaidhe, rdsai.*] A wandering woman, a jilt; used as a disparaging term for a woman.
1922 Joyce *Ulysses* 359 If they could run like rossies she could sit so she said she could see from where she was. 1927 P. S. Dinneen *Irish-Eng. Dict.* (ed. 2) 879/1 *Rásach, -aighe, -acha, f.,* a rambling woman, a gipsy, a jilt; *cf.* rossie (*Dublin*); *rásaidhe, id.* 1939 Joyce *Finnegans Wake* (1964) 327 All the prim rossies are out dressparading. 1961 'F. O'Brien' *Hard Life* v. 71 She told us all about her dear friend, Emmeline Pankhurst. Now there is a bold rossie for you if you like.

Rossi–Forel (rɒ:si fore·l). Also **Rossi Forel.** [The names of Michele Stefano Conte de *Rossi* (1834–98), Italian geologist, and François-Alphonse *Forel* (1841–1912), Swiss physician and limnologist, who in 1883 collaborated in proposing the scale (a modification of Rossi's scale of 1873).] *Rossi–Forel scale*: a ten-point scale used to measure the local intensity of an earthquake.
1885 *Science* 6 Mar. 197/1 The intensity of shocks is measured on the Rossi-Forel scale. 1900 *Jrnl. Geol.* VIII. 304 The continuous curves represent the isoseismal lines of intensities 4 and 3 of the Rossi-Forel scale. 1946 *Nature* 13 July 65/1 The shock on May 8 was apparently the most pronounced felt in Dunedin for many years. The intensity recorded there was IV–V on the Modified Mercalli Scale (equivalent to 5 on the Rossi–Forel Scale). 1976 *Daily Colonist* (Victoria, B.C.) 17 Aug. 1/1 The quake, measuring 7 on the Rossi Forel scale of 10, struck shortly after midnight.

rossiner, var. *ROSINER.

Rossinian (rǫsī·niăn), *a.* [f. the name of G. A. *Rossini* (see below) + -IAN.] Pertaining to or characteristic of Gioacchino Antonio Rossini (1792–1868), Italian operatic composer, or his music.
1869 H. S. Edwards *Life of Rossini* v. 55 The melody, as it now exists, is eminently Rossinian in form and style. 1897 G. B. Shaw in *Sat. Rev.* 20 Mar. 290/1 Passages which are Rossinian in their reliance on symmetry of melody and impressiveness of march to redeem poverty of meaning. 1927 *Daily Tel.* 14 June 12/6 The performance..was as full of spirit and élan as the delightful Rossinian 'snippets' that help to make this ballet [*sc.* La Boutique Fantasque] perhaps the most exhilarating thing in the whole of M. Diaghileff's repertory. 1955 E. Dent in H. Van Thal *Fanfare for E. Newman* 92 We sometimes find the most tragic situation set to the most cheerful music merely because the Rossinian style demands a big *finale* and the impression of a happy end. 1977 *New Yorker* 11 July 82/1 'Un Giorno di Regno', Verdi's first comic opera, which is a bit old-fashioned for its period and full of Rossinian idioms.

rossite (rɒ·səit). *Min.* [f. the name of Clarence Samuel *Ross* (1880–1953), U.S. geologist + -ITE[1].] A hydrated calcium vanadate, $CaV_2O_6.4H_2O$, found as yellow triclinic crystals occurring in glassy masses in sandstone.
1926 Foshag & Hess in *Amer. Mineralogist* XI. 66 (*heading*) Rossite, a new calcium vanadate from Utah. *Ibid.,* This mineral has been called rossite, in honor of Dr. C. S. Ross of the U.S. Geological Survey. 1963 *Canad. Mineralogist* VII. 713 The crystal structure of rossite, $Ca(VO_3)_2.4H_2O$, has been determined. 1968 I. Kostov *Mineral.* 470 Rossite group. The group comprises exclusively vanadates of aluminium, iron, sodium, and calcium, most of which are closely related to the montroseite group of oxides.

rösslerite, var. *ROESSLERITE.

‖ **rosso antico** (rɒ·so a·ntiko). [It., lit. 'ancient red'.] **1.** The name given by Josiah Wedgwood (see Wedgwood) to the red stoneware produced at his Staffordshire factories.
1776 J. Wedgwood *Let.* 3 Mar. (1903) II. 163, I am afraid we shall never be able to make the *Rosso Antico* otherwise than to put you in mind of a red-Pot-Teapot. 1875 E. Meteyard *Wedgwood Handbk.* 28 Wedgwood made much red ware from the same Bradwell-wood clay as that used by Elers, only he glazed the insides of his vessels. Some of Wedgwood's earliest portrait medallions and bas-reliefs were in rosso antico, but the results were not satisfactory. 1976 *Times* 7 Dec. 16/3 The same American bidder paid £1,000..for a Wedgwood *rosso antico* pot-pourri vase of 1805.
2. A rich red marble found in Italy, and employed as a decoration. Also *attrib.* or as *adj.*
1816 J. Dallaway *Of Stat. & Sculpt.* 248 That [*sc.* the marble] of Lybia, is called, by the present antiquaries, 'rosso antico': of this marble there is no known quarry. 1848 Mill *Pol. Econ.* I. III. iv. 552 The materials of many of the ornamental articles manufactured in Italy are the substances called rosso, giallo, and verde antico. 1863 Lytton *Caxtoniana* II. 15 The columns of its lofty portico were of the *rosso antico* marble. 1882 *Athenæum* 30 Dec. 906/1 The material altogether Tuscan, the white marble having been brought from Serravezza, the red (like a fine *rosso antico*) from the neighbourhood of Siena. 1969 *Listener* 16 Jan. 79/1 Of marbles I have found *cipollino, pavonazzetto, giallo* and *rosso antico,* but no harder materials such as porphyry or serpentine.

rossolio, var. Rosolio.

ros solis. 2. (Later example.)
1877 E. S. Dallas *Kettner's Bk. of Table* 21 The Italians..brought with them into France at least two liqueurs—this acqua d'oro with a predominant flavour of rosemary; and rossolis, with a predominant flavour of sundew.

roster, *sb.* Add: Also with pronunc. (rōᵘ·stəɹ).
2. (Later examples.) Also in extended uses.
1930 *New Statesman* 3 May p. iii/1 One particular point is worth noting as to the names in this amazing roster of public enemies. 1942 E. Paul *Narrow St.* xxxix. 337 Practically the entire roster of the Cagoulards was in the new Vichy Government. 1955 *Railway Mag.* May 365/2 This involves over 300 miles daily and is normally a 4-6-0 steam locomotive roster. 1967 W. W. Newcomb *Rock Art of Texas Indians* iv. 38/1 The roster of game animals which could figure prominently in human subsistence is not large. 1971 M. Tak *Truck Talk* 133 *Roster.* When the dispatch department has no immediate assignment for a driver, the driver's name is entered on the dispatch roster. 1978 *Detroit Free Press* 16 Apr. (Detroit Suppl.) 25/3 There he burned up American Association and made Tigers' roster in Spring Training, 1977.
3. *attrib.* and *Comb.,* as *roster-board, game, sheet, system.*
1963 J. Lusby in B. James *Austral. Short Stories* 233 Thwaites..walked to the roster-board. 1977 *Weekly Times* (Melbourne) 19 Jan. 71/2 Feature of the roster games in the Kentish Cricket Association was the 6–15, including the hat-trick, by Railton's David Castles against Kimberley. 1977 R. Ludlum *Chancellor MS.* xvii. 184 He kept mementoes... Photographs, roster sheets. 1976 B. Jackson *Flameout* vii. 122 The roster system played unfair tricks on investigators: you could.. find yourself involved in two [air crash] investigations in three months.

roster (rɒ·stəɹ, rōᵘ·stəɹ), *v.* [f. the sb.] *trans.* To place (someone or something) on a roster.
1922 *Glasgow Herald* 26 Jan. 8 The men can be rostered up to nine hours, with overtime paid after eight hours. 1962 *Mod. Railways* May 350/1 The day Birmingham–Glasgow trains in each direction are now rostered for Type 4 diesel haulage. 1967 *Times Rev. Industry* July 65/2 Even opening on Saturday morning is a vexed question among bank staffs. The board's suggestion that staffs might be rostered..ignores the factors of personal convenience and prestige. 1970 *Railway Mag.* Oct. 561/2 The stud of 'Deltics' is regularly being rostered right up to the limit of locomotives in traffic. 1973 C. Mason *Hostage* ii. 34 Eighteen men..had been rostered on for the two shifts of the guard. 1975 *New Yorker* 13 Oct. 152/3 One of McInally's replacements, Saxon, is rostered at five feet ten and a hundred and forty pounds. 1977 *Daily Tel.* 18 Feb. 2/5, I also found examples where more staff were rostered to work on Sundays than on week-days.
Hence **ro·stered** *ppl. a.,* placed on a roster; assigned in accordance with a roster.
1973 *Daily Tel.* 13 Dec. 2/8 They [*sc.* train drivers] are continuing to do rostered overtime as agreed locally, and are only banning voluntary additional overtime and rest-day working. 1977 *N.Z. Herald* 8 Jan. 2-9/6 (Advt.), Fitter-turners will be required to work alternating shifts or a rostered day work scheme.

rösti, var. *ROESTI.

rostral, *a.* (*sb.*) Add: **4.** *Anat.* (See quot. 1975.)
1894 *Amer. Naturalist* XXVIII. 374 The two ends of the principal axis are respectively 'rostral' instead of 'cephalic' or 'oral' or 'proral'..and..'caudal' [according to Schulze]. *Ibid.* 375 Among Ascidia..there is perhaps a rostral extremity, but there is no caudal extremity in adults. 1953 *Brit. Jrnl. Psychol.* XLIV. 184 The procedures employed were..transorbital leucotomy and thermocoagulation of the cerebral cortex in the rostral portion of the frontal lobes. 1954 T. L. Peele *Neuroanat. Basis Clin. Neurol.* iv. 39/2 Running transversely across the rostral end of the cerebral peduncles are the optic tracts. 1975 E. Gardner et al. *Anatomy* (ed. 4) i. 5/2 Rostral means nearer the 'front end', which is taken to be the hypophysical area in the early embryo and the region of the nose and mouth in post-embryonic life.
Hence **ro·strally** *adv.,* towards the rostral part.

1936 *Jrnl. Anat.* LXX. 208 Sagittal serial sections show that rostrally it runs into continuity with the medial preoptic nucleus. 1954 T. L. Peele *Neuroanat. Basis Clin. Neurol.* iv. 38/1 Beginning at the caudal point of transection and proceeding rostrally, the following visible divisions can be made: medulla, pons,..and cerebral hemispheres. 1970 *Brain* XCIII. 42 Most of the degeneration is a little more rostrally situated than in the preceding experiment. 1978 C. Reid *Primer Human Neuroanat.* xvi. 149 A larger area of the reticular formation facilitates or augments reflexes at lower levels and this area extends rostrally through the pons and mid-brain and into the diencephalon.

rostrifacture (rɒ·strifæ:ktiŭɹ). *rare*[−1]. [f. L. Rostrum beak: after Manufacture *sb.*] A structure made by a bird with its beak.
1884 E. Coues *Key to N. Amer. Birds* (ed. 2) 408 Distinguished as the orioles are for the dexterity and assiduity they display in their elaborate textile rostrifactures [etc.].

rostro-. Add: *rostrocaudally* adv.; **rostro-ca·rinate** *a. Archæol.,* of or pertaining to stone implements of a keeled and beaked shape, esp. those characteristic of the Oldowan and Sangoan cultures of the African Pleistocene, and to flint objects from the Red Crag deposits of East Anglia, formerly thought to be hand tools of late Pliocene date, but now believed to be natural formations; also *ellipt.* as *sb.*
1912 R. Lankester in *Phil. Trans. R. Soc.* B. CCII. 295 We distinguish..an anterior surface, narrowed to the form of a keel and ending in a beak (hence we call the implement 'rostro-carinate') as a consequence of the oblique direction and convergence of the lateral surfaces, which approach one another so as to leave only a narrow keel-like ridge between them. 1934 *Jrnl. R. Anthrop. Inst.* LXIV. 337 Among these large tools (which were afterwards called Sangoan), a number of well-made rostro-carinate forms is to be distinguished. 1952 *Mem. Geol. Survey Uganda* VI. II. 64 The most finely finished product is somewhat canoe-like in shape—sharp prow, blunt stern..; the less finished or those not elaborately shaped, rather like a flat bottomed boat or rostro-carinate. 1957 J. K. Charlesworth *Quaternary Era* II. xxxviii. 1016 The Cromerian implements..are ochreous or orange-brown artefacts, often striated as at East Runton. The tools are usually made from heavy flakes but include rostrocarinates and crude Abbevillean forms. 1964 K. P. Oakley *Frameworks for dating Fossil Man* iv. 176 Some [Oldowan flakes] were beak-shaped. [*Note,* p. 263] 'Rostro-carinate', a term which is better avoided since it suggests identification with the flaked flints well known under that name from the Crags of East Anglia which are now regarded to be of natural origin. 1960 *Jrnl. Compar. Neurol.* CXV. 166/2 In the medial nucleus a topographic organization is suggested in which the nucleus has effectively made a 180° rotation rostrocaudally. 1975 *Nature* 17 Apr. 617/2 There is also a gradient, though less steep, rostrocaudally along the eminentia.

rostrum. Add: **1. a.** (Earlier example.)
1542 N. Udall tr. *Erasmus's Apophthegmes* II. f.315ᵛ, The place called *Rostra* (where oracions wer made to the people).
2. a. (Later example.)
1974 R. Adams *Shardik* xxiv. 186 The rostra, barracoons and blocks of the slave market.
b. Also *transf.*
a 1964 G. Underwood *Pattern of Past* (1968) ix. 96 *Rostra* are sloping recumbent stones, sometimes projecting from a hillside, and sometimes half buried in a slanting position on level ground. At their upper points they mark small terminal blind springs, and their situation and appearance suggests that they may have been intended as pulpits. I have named them accordingly.
d. A platform for a policeman when superintending the traffic at a crossing.
1930 *Morning Post* 16 July 12/5 First and foremost.. there is the constable on the rostrum.
e. *Theatr.* (See quot. 1951.)
1930 W. G. Fay *Short Gloss. Theatr. Terms* 17 When it is necessary to use a rostrum to get elevation on the stage it is generally concealed behind a ground row. *Ibid.* 24 *Ramp,* a slope made of planks from a rostrum to the stage where steps are not used. 1951 *Oxf. Compan. Theatre* 678/1 *Rostrum,* any platform, from a small dais for a throne to a vast battlement, placed on the stage. It is usually made with a removable top and hinged sides, to fold flat for packing. It is reached by steps or a ramp, and quitted off-stage by 'lead-off' steps. A rostrum-front is a canvas-covered flat placed to conceal the front of the platform.
f. *Cinemat.* and *Television.* A platform used to support a camera employed in the filming of animated sequences and the like. Also *attrib.*
1951 Halas & Privett *How to Cartoon for Amateur Films* 105 Let's leave the camera on one side for the moment and consider the *rostrum.* That is the frame on which the camera and the board which holds the animation drawings are mounted. 1959 Halas & Manvell *Technique Film Animation* xix. 235 Such scenes as night bombing, wrecked aircraft, submarines under water and flying through cloud were done with one or two drawings, a little wood-carving, cotton-wool and the full use of single and multiplane shooting on the animation rostrum. *Ibid., Rostrum camera,* apparatus for producing an image on cine-film. Its minimum requirements for animation work are that it must be capable of exposing one frame or film at a time when required. 1975 *Gloss. Terms Motion-Picture Industry* (*B.S.I.*) 9 *Rostrum,* adjustable but rigid support for the camera and the animation table, so constructed that they do not alter position relative to each other in an uncontrolled way. 1976 *Oxf. Compan. Film* 19/2 In the diagram, the camera (a) is shown mounted on the rostrum (US term 'animation stand'). 1977 *Broadcast* 18 Apr. 43/2 Vacancy for aerial image rostrum cameraman.
3. b. *transf.*
1782 H. Walpole *Let.* 18 May (1904) XII. 251 To-day we hear that Sir George Rodney has defeated—ay, and taken—Monsieur de Grasse in his own ship... These naval rostra arrived very opportunely to stay our impatience for a victory over the Dutch.
4. d. (Earlier example.)
1740 J. Grassineau *Mus. Dict.* 205 *Rostrum,* is the name of an instrument wherewith they rule paper for musical compositions.

rosy, *a.* (*sb.*) Add: **1. f.** *slang.* Drunk; tipsy.
1905 *Dialect Notes* III. 17 *Rosy, adj.* Slang. Drunk. 1931 *Princeton Alumni Weekly* 22 May 798/1 When 'the lid is off' one gets 'rosy',..and maybe 'passes out'. 1975 D. Bagley *Snow Tiger* xii. 104 Sure, there was drinking. Some of the boys..got pretty smashed... I was a bit rosy myself.
5. a. **rosy apple,** (*a*) used in skipping formulas; (*b*) (see quot. 1959[2]); **rosy-bill,** a South American pochard, *Netta peposaca,* which has a pink bill.
1916 N. Douglas *London Street Games* 64 Rosy apples lemon and a pear A bunch of roses shall she wear. 1959 I. & P. Opie *Lore & Lang. Schoolch.* xv. 339 The skipping formula usually begins.. Rosy apple, lemon tart, Tell me the name of your sweetheart. A, B, C, D, [etc.]. *Ibid.* xviii. 381 There are more than sixty established names for the pursuit of illegally knocking at doors... *Rosy Apple.* Derby. 1964 *Western Folklore* XXIII. 258 Rosy apples, Mama's little tart, Tell the initials Of your sweetheart! 1888 R. Hubbard *Ornamental Waterfowl* II. v. 162 The Rosy-bill is a native of South America. 1956 G. Durrell *Drunken Forest* i. 16 Rosybills, immaculate in their gleaming black-and-grey plumage, their beaks looking as though they had been freshly dipped in blood. 1976 *Eastern Daily Press* (Norwich) 16 Dec. (Advt.), Sale, pairs black swans, Emperor, Barnacle, Egyptian, Tested, Gadwell, Shelduck, Rosybill, Carolinas, Pintails.
6. *rosy-blue, -gilt, -golden, -mauve, -red* (later example); also, *rosy-pale.*
1862 G. M. Hopkins *Vision of Mermaids* (1929), Which, lightening o'er the body rosy-pale, Like shiver'd rubies dance. 1925 V. Woolf *Common Reader* 115 The apples rosy-gilt in the afternoon sun. 1926 D. H. Lawrence *Sun* iii. 11 The child and she were now both tanned with a rosy-golden tan, all over. 1952 A. G. L. Hellyer *Sanders' Encycl. Gardening* (ed. 22) 94 [*Cattleya*] *Harrisoniana,* light rosy-mauve, variable, summer, autumn. 1956 H. Gold *Man who was not with It* (1965) xi. 89 His hungry mouth, rosy-red with fever. 1976 I. Murdoch *Henry & Cato* I. 3 Leaving New York in daylight, his plane had soon risen into a sort of radiant rosy-blue stratospheric gloom. 1978 *New York* 3 Apr. 94/3 The terrine de poisson, a rosy-pale slice of fish pâté.
7. *rosy-billed, -fleeced, -flowered, -footed* (later example), *-lipped, -muzzled, -petalled, -rayed;* also, *rosy-glistening, -rising, -torturing.*
1876 *Proc. Zool. Soc.* 399 The Rosy-billed Duck has been successfully introduced into Europe. 1936 R. Campbell *Mithraic Emblems* 57 The rosy-fleeced Arrival of the Moon. 1927 V. Woolf *To Lighthouse* I. vii. 63 A rosy-flowered fruit tree. 1928 O. Gogarty *Wild Apples* 8 Fair skin and smooth as the rosy-footed dove's wing! a 1918 W. Owen *Poems* (1963) 127 The crunch of boots on blue snow rosy-glistening. 1862 G. M. Hopkins *Vision of Mermaids* (1929), The waves were rosy-lipp'd. 1923 D. H. Lawrence *Birds, Beasts & Flowers* 63 Violets, Pagan, rosy-muzzled violets. 1928 Blunden *Retreat* 44 This retinue Of rosy-petalled sauntering joys. 1925 —— *Eng. Poems* 90 Others like opals rosy-rayed convene. 1916 —— *Pastorals* 19 Nothing Eastern come to us Save the rosy-rising sun. 1929 —— *Near & Far* 19 No rosy-torturing desert.
B. *ellipt.* or as *sb.* **1.** *the rosy* (*a*) (see Dict., sense 1 e); (*b*) blood; (*c*) a good time; phr. *to do the rosy,* to have a good time. *slang.*
1891 *Sporting Life* 25 Mar. 7/3 Goddard was smothered in the rosy as he went to his chair, and Choynski bled at the mouth. 1892 E. J. Milliken *'Arry Ballads* 69/2 A doin' the rorty and rosy as lively as 'Opkins's lot. *Ibid.* 77 Not *my* idea of the rosy.
2. *Naut. slang.* A ship's rubbish-bin.
1937 D. Marlowe *Coming, Sir!* ii. 46, I struggled with the heavy garbage bins, called 'rosies'. 1962 Granville *Dict. Sailors' Slang* 97/2 *The rosy,* the Merchant Navy's *gash bucket*: a 'rose by any other name'. 1966 'L. Lane' *ABZ of Scouse* 86 *Put a crust on ther Rosy fer 'im,* he has a good appetite. *Rosy,* a ship's swill bin.

Rosy Lee, Rosie Lee (rōᵘ·zi lī·). Also with small initials. Rhyming slang for 'tea'. Also *ellipt.* as Rosie.
1925 Fraser & Gibbons *Soldier & Sailor Words* 246 *Rosy Lee,* tea. (Rhyming slang.) 1929 J. B. Priestley *Good Companions* I. iv. 133 'Ow about a drop o' Rosie Lee? *Ibid.* 134 We'll 'ave the Rosie now, George. 1964 A. Prior *Z Cars Again* iv. 35 This is the best cup of rosy I get all day, Janey. 1968 J. Boland *Breakdown* i. 4 Want a drop of rosie, do yer, Dad? 1970 A. Draper *Swansong for Rare Bird* ix. 90 We were having a cup of rosy lee.

rot, *sb.*[1] Add: **5.** (Further examples.) Also used of activities, objects, etc. Also as *int.* Cf. *tommy-rot* s.v. Tommy 6.
1879 M. E. Braddon *Cloven Foot* iv. 96, I thought he despised ballet-dancing. Yet this is the third time I have seen him looking on at this rot. 1880 Henley & Stevenson *Deacon Brodie* IV. 79 Portrait of George as a gay hironmonger... O rot! Hand it over, and keep yourself out of that there thundering moonlight. 1894 G. Moore *Esther Waters* xxxix. 302 All bloody rot; who says I'm drunk? 1905 H. James *Golden Bowl* vi. 74 He had not many things, none of the redundancy of 'rot' they had elsewhere seen. 1914 G. B. Shaw *Fanny's First Play* 158, I quite agree that harlequinades are rot. a 1953 E. O'Neill *Long Day's Journey* (1956) I. 35 It's damned rot! I'd like to see anyone influence Edmund more than he wants to be. 1977 C. McCullough *Thorn Birds* ii. 36 'What if it isn't the Eyetie girl?'..'Rot!' said Paddy scornfully.
6. (Earlier and later examples.) Usu. in phrases *the rot set in, to stop the rot.* Now chiefly in extended uses: a decline (in resources, standards, behaviour, etc.).
1868 *J. Lillywhite's Cricketers' Compan.* 61 A terrible 'rot' set in at the commencement of their second 'venture'. 1882 *Australians in Eng.* 71 After the fall of Leslie's wicket, however, a complete 'rot' set in. 1912 P. F. Warner *Eng. v. Austral.* ix. 100 Ransford..had rendered great service to his side by helping to 'stop the rot'. 1926 G. M. Trevelyan *Hist. Eng.* VI. ii. 642 By these all too drastic measures the rot of pauperism was stopped. 1930 J. B. Priestley *Angel Pavement* i. 38 He could not pretend to himself now that such pitiful economies as these could stop the rot. 1938 R. Warner *Professor* v. 113, I really don't know how the rot set in, but the process may have been something like this. 1951 C. P. Snow *Masters* xix. 157 We must take care that a rot doesn't set in. 1955 *Times* 6 June 3/1 The rot began when Appleyard came into the attack. 1958 *Spectator* 22 Aug. 251/1, I have a feeling that, recently, airlines have been allowing this precious asset to depreciate. The rot set in with the introduction of bus fares. 1969 *Listener* 17 July 68/1 The rot set in, I think, with the President's speech to the Air Force Academy at Colorado Springs in June. 1973 M. Woodhouse *Blue Bone* ii. 12, I went up to London..and that, as the saying goes, is where the rot set in.

rot, *v.* Add: **1. e.** *N. Amer.* Of sea or river ice: to melt or thaw. Cf. *ROTTEN *a.* 4 c.
1892 [implied in Rotting *vbl. sb.*[1]]. 1977 *New Yorker* 20 June 86/2 Ice was beginning to rot.
2. b. *slang. to rot about,* to fool about, waste time. Now *rare.*
1902 E. Nesbit *Five Children & It* viii. 198 When we're all rotting about in the usual way heaps of things keep cropping up. 1909 J. R. Ware *Passing Eng.* 211/1 *Rotting about..,* wasting time from place to place. 1927 W. E. Collinson *Contemp. Eng.* 116 'To play the fool' is to rag about, rot about, fool about, play the [giddy] goat, bucket around.
3. a. Also *fig.,* to languish (*in* a place).
1927 *Scribner's Mag.* Feb. 168/1 A man must do something. It's better than rotting in the saloons in Casper. 1975 T. Allbeury *Special Collection* ii. 10 The Moscow Centre has just left them to rot. 1978 I. B. Singer *Shosha* ii. 39, I asked for Dora and he replied 'Rotting in Siberia'.
4. b. Also with *down.*
1912 Galsworthy *Inn of Tranquility* 79 'They don't do a stroke more than they're obliged,' he ended;..'Yes,' he muttered, 'the nation is being rotted down.'
c. (Earlier example.)
1811 *Weekly Reg.* 5 Oct. 86/1 (*heading*) Process for rotting hemp.
d. *slang.* To spoil, interfere with; to ruin. Also const. *up.*
1908 A. S. M. Hutchinson *Once aboard Lugger* VI. viii. 344 You rotted my show all right. 1908 D. Coke *House Prefect* viii. 104 You can see Bob's off you, and we don't want to rot the whole thing up, just when he's begun to be decent again. 1932 'A. Bridge' *Peking Picnic* xxv. 323 I've got a complex about the whole business, and you know why. Well, that might rot it all up, at any moment. 1973 N. W. Schur *British Self-Taught* 335 To *rot* a plan is to *spoil* it. 1978 *Sunday Times* 15 Jan. 42/7 A turquoise velvet top (detested since I rotted up a quiz programme in it).
7. a. (Further examples.) Also, to abuse, denigrate. Also (in *absol.* use), to joke.
1890 W. E. Henley *Let.* 6 June in J. Connell *W. E. Henley* (1949) vii. 182 He'd have given much to hear you rotting the Alien. 1905 H. A. Vachell *Hill* vii. 155 Has anybody been rotting you? 1914 'I. Hay' *Lighter Side School Life* vii. 181 We don't do any *work*: we just rot Duck-face. We simply rag his soul out. 1914 G. B. Shaw *Fanny's First Play* III. 200 But I'm serious: I'm not rotting. Really and truly—. 1922 S. Leslie *Oppidan* iii. 38 A sport taking the mysterious form of 'rotting the Flea'. 1934 R. Macaulay *Going Abroad* xxx. 264 There are things one simply mustn't rot about, I feel.

rota. Add: ‖ **4.** *Mus.* A musical composition which has the form of a round; this form itself. Used esp. of medieval English songs (as 'Sumer is icumen in', where this designation appears in the original manuscript). Cf. Round *sb.*[1] 19 b.
1876 Stainer & Barrett *Dict. Mus. Terms* 381/1 *Rota*.., a Round, but the word is sometimes applied to anything with frequent repeats, as for instance a Hymn tune. 1883 Grove *Dict. Mus.* III. 180/1 It..is written for six voices, four of which sing the round proper or 'rota' (as it is termed in the Latin directions for singing it). 1944 W. Apel *Harvard Dict. Mus.* 652/2 *Rota,*.. medieval name for a round, particularly the Sumer-canon, probably with reference to the 'turnover' of the melody in the different parts. 1955 *New Oxf. Hist. Mus.* (rev. ed.) II. xi. 462 Its form, which is described in the manuscript itself as a *rota,* is that of an infinite canon. 1979 *Early Music* July 391/2 The piece in question is the famous *rota* 'Sumer is icumen in'.

5. *attrib. and Comb.,* as *rota committee, system;* **rota cut,** an interruption or reduction of power or water supplies which is imposed on different areas by rotation in time of shortage. See also sense 1 in Dict.
1935 *Planning* II. XLIV. 13 As much as possible of the actual assessment of need should be left to local rota committees, using the regional scale with fairly wide discretion. 1974 *Times* 15 Feb. 15/2 The third course..is to impose rota cuts designed..to avoid the working days of industry, and..essential services such as hospitals. 1977 *Times* 20 Apr. 5/2 Rota-cuts (when water supplies are cut off for a certain number of hours a day) might appear a less painful alternative than standpipes (involving total cut-off of domestic supplies). 1955 *Times* 25 Aug. 9/6 Only if a six-day shopping week is adopted, with a rota system to give staff a five-day week, will distribution costs be materially increased.

rota-, var. *ROTO-.

rotal, *a.* **3.** (Further example.)
1960 *Tablet* 18 June 586/2 The Rotal sentence referred to by Dr. McReavy relies on the pronouncement of Benedict XIV.

rotamer (rōᵘ·tăməɹ). *Chem.* [f. Rotational *a.* + *-MER.] Any of a number of distinct conformations of a molecule which can be interconverted by rotation of part of the molecule about a particular bond; a rotational isomer.
1963 *Chem. & Industry* 29 June 1086/1 The single sharp N-methyl signal indicates magnetic averaging of signals and hence very rapid rotamer interconversion even at low temperatures. 1969 *Jrnl. Chem. Soc.* B. 1019/2 The near temperature-invariance of the n.m.r. spectrum over the range studied could be the consequence either of one minimum-energy rotomer being strongly preferred, or of insufficient difference in magnetic shielding at *N*-methyl for different rotamers. 1976 *Nature* 26 Aug. 780/1 The analysis of the side-chain rotamer population..shows that the Met, Phe, and Tyr side chains favour one of the two possible *trans-gauche* rotamers, but for each of these residues all three rotamers have significant populations.

Rotameter (rotæ·mītəɹ, rōᵘ·tămītəɹ). Also **rotameter.** [partial tr. G. *rotamesser* (*Chem. Rev. über die Fett- u. Harzind.* (1911) XVIII. 55), f. *rota*(*tion* Rotation, etc.: see -METER.]
1. A proprietary name for a device with a transparent wall that is fitted into a pipe or tube and indicates the rate of flow of fluid through it.
1911 *Chem. Abstr.* V. 1695 (*heading*) The rotameter. 1914 G. Lunge *Techn. Gas-Analysis* 50 The 'Rotameter' of the Deutsche Rotawerke, Aachen, allows of directly reading off the quality of gas (or liquid) passing through per hour. 1925 *Industr. Chem.* I. 474/2 In oxy-acetylene welding, if rotameters are inserted in the oxygen and acetylene tubes, it is possible to obtain perfect uniformity. 1949 E. Chain in H. W. Florey et al. *Antibiotics* II. xvii. 701 From the air-filter the air passed through a rotameter..and a check-valve..into the fermenter through the sparger. 1952 *Trade Marks Jrnl.* 18 June 555/1 *Rotameter*... Apparatus for measuring, indicating, or recording the rate of flow of gases or liquids. Rotameter Manufacturing Co. Ltd., Derwent Works, Purley Way, Croydon, Surrey; Manufacturers—29th Nov. 1948. 1979 A. L. Lydersen *Fluid Flow & Heat Transfer* iii. 61 Standard rotameters are delivered for connection to pipes with diameters from 3 to 150 mm.
2. Var. *ROTOMETER.

rotang. The form **rotan** is now freq. used, as being closer to the Malay. Also *attrib.* of objects made of rotan.
1821 J. Leyden tr. *Malay Annals* 124 The whole Siamese army retreated; and, as they took their departure, they threw down large quantities of their baggage rotans in the district of Moar, where they took root. 1927 H. M. Tomlinson *Gallions Reach* xxxi. 242 The climbing palms, the rotans, flourished about it. 1939 *Geogr. Jrnl.* XCIV. 419 It [*sc.* the nutmeg fruit] falls to the ground with a heavy 'plop', or is gathered in a rotan cage at the end of a long stick. 1954 R. H. Holttum *Plant Life in Malaya* xiii. 186 The most important group of climbing monocotyledons are the Rotans or Rattan canes. 1959 *New Biol.* XXX. 51 The rotan lashings and ladders are renewed annually. 1963 J. Kirkup *Tropic Temper* 23 The..hall of the hotel was full of bamboo and rotan furniture. 1972 *Straits Times* (Singapore) 28 Nov. 1/2 The Government is bringing in the rotan as a new weapon to fight the drug menace in Singapore.

Rotarian (rotēᵊ·riăn), *a.* and *sb.* Also occas. with small initial. [f. Rotary *a.* and *sb.* + -AN.] **A.** *adj.* Of, pertaining to, or characteristic of the Rotary organization, a Rotary Club, or Rotarians. **B.** *sb.* A member of a Rotary Club. Cf. *ROTARY *a.* 4.
1911 (*title of periodical*) The Rotarian. 1915 *Chicago Herald* 9 Nov. 10/5 The Rotarians will observe 'Moving Picture day' at a luncheon in the crystal room of the Hotel Sherman. 1921 *Glasgow Herald* 10 Feb. 9/4 The President..said the Prince lived out consistently the motto of Rotarians, 'Service, not self'. *Ibid.* 15 June 11/1 An effective programme on education must produce a more intense study of Rotarian literature, a more liberal use of the Rotarian speakers available [etc.]. 1923 R. Herrick *Homely Lilla* xi. 181 Lilla, on opening the newspapers, often found his name and a brief report of his remarks at a Rotarian lunch. 1928 L. North *Parasites* 270 Rotarians and women's clubs wrote her letters applauding her patriotic stand for home-products. 1931 [see *Main Street, main street b]. 1935 G. Greene *Basement Room* 141 He wouldn't have recognized himself among the rotarians. 1939 W. Fortescue *There's Rosemary* lv. 313, I quietly repeated the *rondeau* he had written for the Rotarian garden-party we had so lately given at Admiral's House. 1947 *Britannica Bk. of Year* (U.S.) 704/1 Rotarians throughout the world devoted their programs during the month of November to the United Nations Educational, Scientific and Cultural Organization. 1955 *Times* 9 May 3/2 The father, a pillar of rotarian society, expects the worst and is won over. 1968 M. Bragg *Without City Wall* xxiv. 225 There, too, was the herd, just as herd-like as the Institutes, Rotarians, Churches, and Social Activators. 1976 *Time* 27 Sept. 30/1 When a Southerner calls his territory 'God's country,' he is less Rotarian than religious. 1978 G. Vidal *Kalki* iii. 53 Dr. Ashok detached himself from a group of American secret agents (or Rotarians or salesmen).
Hence **Rota·rianism,** the Rotarian system; the way of life held to be characteristic of Rotarians.
1922 *Nation* (N.Y.) 19 Apr. p.v (Advt.), Do you know your state? How it stands in intelligence, rotarianism, bootlegging, evangelism, crime? 1928 *Daily Express* 20 June 2/5 A pantomime symbolising the story of the spread of Rotarianism throughout the world. 1942 Berrey & Van den Bark *Amer. Thes. Slang* §231/1 *Rotarianism,* middle-class propriety.

rotary, *a.* and *sb.* Add: **A.** *adj.* **2. a.** (Further examples.)
1906 *Daily Colonist* (Victoria, B.C.) 6 Jan. 7/3 Big rotary snowplows and gangs of men have been unable to cope with the conditions. 1939 *Archit. Rev.* LXXXV. 76/3 The laundry is all electric, and is equipped with a Rotary Washer and Spin Dryer, and Rotary Ironer. 1960 *Which?* Mar. 48/1 The rotary mower differs from the side-wheel and roller in the way it cuts, which is a chopping action, like a scythe, rather than a shearing action, like a pair of scissors. 1963 R. R. A. Higham *Handbk. Papermaking* ii. 68 Rotary screens may be divided into inward and outward flow types which may be either oscillating or stationary. 1970 *Which?* Mar. 84/2 If you have got long, rough, grass, you will still need a rotary mower. 1977 *Evening Gaz.* (Middlesbrough) 11 Jan. 10/4 (Advt.), Kenwood rotary ironer, excellent condition, £20. 1979 *SLR Camera* June 39/3 Don't put the prints in a conventional flat-bed or rotary dryer designed for drying fibre-based papers.
b. *Printing.* Designating a press in which a cylindrical printing surface is rotated continuously in contact with moving paper, usu. from a web, and the resulting method of printing.
1880 F. J. Wilson *Typogr. Printing Machines* IV. 135 Owing to the speed at which rotary machines are driven, slight difficulties frequently arise. 1899 J. Southward *Mod. Printing* III. xvi. 148 Rotary web printing was in England first rendered practicable by engineers employed in the office of *The Times,* who produced the 'Walter Press,' which was completed in 1866. 1926 R. W. Polk *Pract. Printing* xv. 114 There are rotary presses (called sheet-fed rotaries) which print sheets of paper previously cut to size, but most of them print from large rolls of paper which feed a continuous web through the machine at a high rate of speed. 1962 *Penrose Ann.* LVI. 103 It is still too early to predict what the future will hold for the letterpress rotary machine utilizing wrapround plates. 1968 J. R. Biggs *Basic Typogr.* 80/2 With a rotary press, in which an impression is made at every revolution of the cylinder, very high speeds are possible. 1979 P. G. New *Bk. Production* vi. 81 Letterpress..is taking a leaf from litho's book by adopting some of the same techniques, such as rotary printing.
c. Designating or pertaining to a system of drilling, used esp. in drilling for oil, in which the drilling column with the bit attached to it is rotated; *rotary table,* in rotary drilling, a power-driven steel turntable which is attached to the top of the drilling column and serves to rotate it.
1906 B. Redwood *Treat. Petroleum* (ed. 2) I. 287 The rotary system, which is in general use in the oil-fields of the coastal plain of Texas, is a modification of that invented by Fauvelle in 1845. *Ibid.* 288 There are three styles of rotary rigs in use. *Ibid.* 289 The lower end of the drilling-rod or casing with the bit attached is passed through the rotary table. 1912 E. H. C. Craig *Oil-Finding* viii. 150 Thus through a thick soft argillaceous group it may be found most profitable to use a rotary rig, while drop drills and under-reamers may suit a variable series containing hard calcareous bands. 1939 D. Hager *Fund. Petroleum Industry* ix. 200 The drill is turned by means of the rotary table. 1944 B. A. Botkin *Treas. Amer. Folklore* IV. 493 Rivalries take such subtle forms as the feud between cable-tool drillers and rotary workers in the oil fields. 1974 *Petroleum Rev.* XXVIII. 724/3 Samples may be retrieved by conventional rotary coring. 1974 *BP Shield Internat.* Oct. 18/4 Once it's drilled all the way down to the 'rotary-table', we pull the kelly back, unscrew it, and then make it up to another 30 feet section of pipe.
4. (With capital initial.) Of or pertaining to a world-wide organization of clubs for business and professional men (of which the first, formed in Chicago in 1905, met at each member's premises in rotation) which have the aim of promoting unselfish service and international goodwill. *Rotary Club,* a local branch of this organization.
1910 *Chicago Record-Herald* 10 June 2/4 'The National Association of Rotary Clubs will be one of the most powerful factors in the civic life of the nation,' declared Paul P. Harris... 'Its membership, limited to one man in each line of business,..fighting together in the seventeen largest cities of the country, will be able to win on about any proposition they undertake.' 1921 *Glasgow Herald* 10 Feb. 9/4 Sir Harry Lauder was the guest of honour at a Rotary Club luncheon at the Hotel Cecil, London, yesterday afternoon. *Ibid.* 15 June 11/1 The subjects under consideration included Rotary education, publicity, and business methods. 1930 G. O. Thomas *Calm Weather* 61 The Rotary Club of which I am a member is very luckily composed of such persons as are engaged in different ways of life, and deputed as it were out of the most conspicuous classes of mankind. 1945 *Business Week* (U.S.) 30 June 44/3 But final decision is to be made..at the next fully attended Rotary convention. 1951 *Britannica Bk. of Year* 558/1 Eleven young men and one young woman nominated by Rotary Clubs in Great Britain and Ireland were awarded Rotary foundation fellowships. 1963 *Sat. Even. Post* (U.S.) 9 Feb. 60/2 A new club adopts the standard Rotary constitution in its native tongue. *Ibid.* 62/3 Rotary volunteers are setting up small-business clinics in backward areas. 1972 T. P. McMahon *Issue of Bishop's Blood* xii. 143 He was pudgy, five-seven—the type that would fit at any Rotary luncheon in the country. 1976 *S. Wales Echo* 25 Nov. 20/4 The Rotary Club of Cardiff run a Christmas bargain shop from Tuesday next to Saturday.
5. Special collocations: *rotary camera,* a type of automatic camera, used to photograph documents, in which the subject material is moved automatically past the lens in synchronization with the film; *rotary clothes-drier* or *-line,* an approximately circular clothes line supported by spokes from a central pole and capable of rotation; *rotary converter,* an electric motor adapted for use with either alternating or direct current and capable of converting one to the other; *rotary cutting* or *cut,* a method of making veneer by rotating a log longitudinally against a knife-edge so that a layer of wood is peeled off; hence *rotary-cut* adj.; *rotary cutter,* the apparatus used in this method; (see also sense 2 in Dict.); *rotary engine,* any engine which produces rotary motion or of which the action depends upon the rotation of some part or parts; *spec.* (*a*) an aircraft engine with a fixed crankshaft around which cylinders and propeller rotate; (*b*) a Wankel engine; hence *rotary-engined* adj.; *rotary table*: see sense 2 c above; *rotary-wing,* used *attrib.* to denote any aircraft deriving its lift from aerofoils that rotate, usu. in an approximately horizontal plane.
1955 H. Ten Eyck *Gloss. Terms Microreproduction* 68 *Rotary camera,* any microfilm camera which photographs documents while they are being moved by some form of transport mechanism. 1962 A. Günther *Microphotogr. in Library* (Unesco) 18 For the production of roll microfilm..there are rotary cameras in which separate original pages and the film move synchronously. 1974 G. G. Baker et al. *Guide to Production of Microforms* iii. 15 Some rotary cameras have been specially designed to accept continuous line-printout stationery. 1971 *Guardian* 10 Apr. 4/1 (Advt.), Rotary clothes drier with 100 foot line. 1971 *Country Gentlemen's Mag.* May 222/2 Rotary Clothes Lines..offer more line space than the conventional fixed line. 1978 P. Porter *Cost of Seriousness* 3 A camera, an eye Of memory is recounting inches along from the pea-trellis, The cement-block fence, the rotary clothes-line. 1899 Franklin & Williamson *Alternating Currents* xiii. 166 An ordinary direct current dynamo may be made into an alternator by providing it with collecting rings..in addition to its commutator. Such a machine is called a rotary converter. *Ibid.* 167 The rotary converter may be used as an ordinary direct-current dynamo or motor. 1934 *Discovery* Nov. 324/2 Their products include D.C. to A.C. rotary converters,.. constant current changing dynamos and an entirely portable petrol-driven alternator. 1950 *Times Rev. Industry* Sept. 25/1 One of the first three locomotives is to have a rotary converter set and d.c. motors. 1927 Knight & Wulpi *Veneers & Plywood* xvii. 151 The modern methods of manufacturing veneer are practically three-fold, the oldest being sawn, the later, sliced, and the recent, rotary cut. 1974 *Encycl. Brit. Macropædia* XIX. 922/2 More than 90 percent of all veneer is rotary cut... Logs of hard woods, intended for rotary cut or sliced veneer, are softened by submersion in hot water or steam. 1936 *Archit. Rev.* LXXX. 180/3 If you produce your veneers in any required size by the rotary cutter (invented about 1892)..you superimpose something on your materials which is not natural to them according to established standards. 1973 *Materials & Technol.* VI. i. 85 The handling of veneers as they are produced by the rotary cutters varies from factory to factory. 1927 Knight & Wulpi *Veneers & Plywood* xvii. 148 There are four ways of converting logs into veneer... If these are to be arranged according to volume of production, rotary cutting will easily stand at the head of the list. 1957 *Encycl. Brit.* XXIII. 42/2 Veneers are also produced by means of the rotary cutting process as a raw material for plywood. 1838 Rotary engine [in Dict., sense 2]. 1887 *Encycl. Brit.* XXII. 516/2 In all rotary engines, with the exception of steam turbines,—where work is done by

the kinetic impulse of steam,—there are steam chambers which alternately expand and contract in volume. 1909 *Flying: the Why & Wherefore* x. 91 The recent successes of the seven-cylinder rotary Gnome engine. 1928 C. F. S. Gamble *Story of N. Sea Air Station* xiii. 216 Although rotary engines were falling gradually into disfavour owing to their heavy lubricating-oil consumption, lack of reliability, and large head resistance, one engine was designed during this year—the Bentley Rotary..which gave excellent service in single-seater machines in 1918. 1960 C. H. Gibbs-Smith *Aeroplane* I. viii. 30 In 1887 he [*sc.* Lawrence Hargraves] invented the rotary engine (driven by compressed air) in which the cylinders and propeller revolved about a stationary crankshaft. 1968 S. E. Ellacott *Everyday Things in Eng. 1914–68* xii. 182 The Wankel rotary engine..was on show at Earl's Court in October 1967. 1969 J. D. Storer *Simple Hist. Steam Engine* i. 16 If the water or wind could be replaced by a man-made stream of steam, or hot gases, an ideal rotary engine would result. This type of heat engine is known as a turbine. 1973 H. Jones *Steam Engines* iv. 47 Between 1785 and 1800, Boulton and Watt supplied 110 rotary engines fitted with sun and planet gear to textile mills. 1909 *Westm. Gaz.* 23 Oct. 9/1 Delagrange brought out his rotary-engined Blériot. 1973 *Times* 28 June 31/2 The Mazda RX3, now the cheapest rotary-engined car at £1,615 in Britain. 1908 Rotary-wing [see *gyropter* s.v. *GYRO-]. 1935 *Jrnl. R. Aeronaut. Soc.* XXXIX. 53 The main objects of this invention is [*sic*] to increase the stability and manœuvrability of helicopter machines, to improve the controllability of rotary wing aircraft to reduce the drag of such aircraft. 1958 *Times* 1 Mar. 7/3 A company engaged in manufacturing rotary wing aircraft is seeking permission to operate a base which would consist of a small platform built over the Thames, connected to an aircraft parking area on the river bank.
B. *sb.* **1.** *spec.* A rotary printing machine or press. (Further examples.)
1926 *Penrose's Ann.* XXVIII. 135 A battery of reel-fed litho. offset rotaries are running most efficiently. 1978 R. Clay in J. Moran *Clays of Bungay* xiii. 145 In 1938 the Company had purchased two old Cottrell sheet-fed rotaries.
2. (With capital initial.) The Rotary organization or its ideals; an individual Rotary Club. *Rotary International,* the official title (since 1922) of the world-wide organization of Rotary Clubs.
1921 *Glasgow Herald* 10 Feb. 9/4 Sir Harry Lauder.. said Rotary was like the lamplighter who came into a dark street. 1922 *Rotarian* May 234/1 No more important question can ever come before a Rotary convention than one which will be discussed at the convention in Los Angeles—a Constitution for International Rotary. 1935 D. Fahey *Mystical Body of Christ in Mod. World* vi. 112 Let us see what attitude the Catholic Church has adopted towards Rotary. 1944 B. Johnson *As Much as I Dare* 275 When the members of the Denver Rotary attend Eastern conventions they wear ten-gallon hats. 1963 *Sat. Even. Post* 9 Feb. 58 Once a mutual-aid society for Midwestern businessmen, Rotary busily promotes peace and good brotherhood on a global scale. 1977 H. Fast *Immigrants* IV. 243 I'm due to speak to Rotary in thirty-five minutes.
3. *U.S.* = *ROUNDABOUT *sb.* 4 d.
1940 N. Bel Geddes *Magic Motorways* v. 91 Progress around the rotary is slow, for all cars have to weave from lane to lane and are slowed down by the cars feeding in ahead. 1955 *New Yorker* 12 Mar. 38/2 At eight the next morning we came to the first traffic rotary outside New York, in New Jersey. 1966 *PMLA* LXXXI. II. 11/1 In my lifetime I have seen the *traffic circle* of the Middle Atlantic States become the *rotary* of New England. 1976 A. Cross *Question of Max* i. 15 She executed..several rotaries which seemed specifically designed to enable cars going in opposite directions to meet head-on.

rotatably (rotē[i]·tăbli), *adv.* [f. Rotatable *a.* + -LY[2].] In a manner that allows rotation.
1918 H. Seymour *Reproduction of Sound* 263 The stylus 9 is mounted in a lever 10 rotatably supported on a pivot 11.

rotate, *v.* Add: **3.** (Earlier and later examples.)
1861 *Trans Illinois Agric. Soc.* IV. 318 We must rotate crops. 1950 *N.Z. Jrnl. Agric.* Jan. 4/1 By rotating the calves through the paddocks ahead of the cows at intervals of 3 or 4 days. 1980 W. Safire in *N.Y. Times Mag.* 20 Jan. 10/3 She explained, 'We're going to rotate the house and we even rotate the cars. We've been separated for four months, and it's a growing experience.'
5. *U.S. Mil.* (See quot. 1973[2]: chiefly *pass.*) Also *intr.* for *pass.*
1944 *Yank* 4 Feb. 6 The policy on leaves and furloughs includes provision that individuals who have had two years Alaskan service and who do not desire to be rotated may volunteer for an additional Alaskan tour. 1951 *Sun* (Baltimore) 3 Nov. (B ed.) 6/2 The assurances to G.I.'s in Korea that they would be rotated home were regarded with unmodulated disbelief. 1954 *Britannica Bk. of Year 1953* 354/2 Each side might rotate up to 35,000 men a month on a man-for-man basis. 1973 *Washington Post* 13 Jan. A3/3 You look at an NLF..soldier, who can't..get R and R to Hong Kong, time off in Vungtau, and then rotate in a year back to the States. 1973 J. Quick *Dict. Weapons* 377/2 *Rotate,* to remove a person, crew, unit, or the like from service in an overseas area, from combat service, or from service in a hardship environment and to return such person, crew, or unit to service in the zone of the interior or other less exacting environment. 1976 'B. Shelby' *Great Pebble Affair* 12 Donnely and I rotated back to the States together.

rotating, *ppl. a.* **2.** (Further examples in sense 'rotative'.)
1975 *Saturday Night* (Toronto) July–Aug. 18/2 It has been the unions in the public service, not surprisingly, who have perfected the kind of random sniper-fire known as 'selective' strikes or 'rotating' strikes or 'twenty-four-hour' strikes. 1976 *National Observer* (U.S.) 21 Aug. 15/1 Asolo now offers nine plays in rotating repertory from mid-February to Labor Day.

rotation. Add: **1. b.** *Cryst., Math., Physics.* The conceptual operation of turning a system about an axis.
1899 W. J. Lewis *Treat. Crystallogr.* iii. 19 When the least angle which gives interchangeability is 90°, the rotation can be effected four times before the crystal returns to its original position. 1965, etc. [see *REFLECTION 9*].
c. *Math.* = *CURL *sb.* 3 e.
1909 J. G. Coffin *Vector Analysis* v. 117 The operator ∇ × applied to **F** or curl **F** (read del cross **F** or curl of **F**), also sometimes written in German books, rot **F** (read rotation of **F**), is a new vector derived from **F**. 1911 [see *CURL *sb.* 3 e]. 1923 H. Levy tr. *Runge's Vector Analysis* ii. 111 The vector field **f**, when its rotation is not zero, that is, when **f** is not the gradient of a scalar function, leads to a second vector field **g** = ∇ × **f**. 1972 A. G. Howson *Handbk. Terms Algebra & Anal.* xxxv. 175 In physical applications, curl represents some measure of rotation (older texts often describe curl **f** as the rotation of **f** and denote it by rot **f**).
d. *Statistics.* The mathematical rearrangement of a body of data, regarded as representing a set of points in a space, so that the axes of the space come to lie in directions of particular relevance.
1935 L. L. Thurstone *Vectors of Mind* ix. 222 Each independent rotation may be regarded as a disturbance of a pair of columns in the factorial matrix. 1952 R. B. Cattell *Factor Analysis* xxi. 411 A great advance in speed in rotation processes is available now through the I.B.M. multiplier (a rotation of a 15 × 18 factor matrix can be done in a day). 1972 *Jrnl. Social Psychol.* LXXXVII. 69 Rotation to simple structure was made for five factors in each case, though not all five could be interpreted.
2. d. *Forestry.* The cycle of planting, felling, and replanting; the period of this, the (actual or planned) time between the formation or regeneration of a crop and its felling.
1888 E. E. Fernandez *Man. Indian Sylviculture* i. 6 The existence of a rotation implies the more or less simultaneous appearance of the old generation, and a similarly more or less simultaneous removal of that generation and the appearance of the new one. There can hence be no rotation in the case of selection-worked forests. 1889 W. Schlich *Man. Forestry* I. II. ii. 170 The selection of the rotation..should be so fixed under the method of natural regeneration as to admit of a proper regeneration of the wood, whether by seed or coppice shoots. 1890 [see sense 2 a in Dict.]. 1927 *Forestry* I. 101 A more detailed investigation of the returns from quality class V forests showed that at 100 years (the financial rotation) the yield was 1.06 per cent., and at 120 years (the more usual rotation) it was 1.03 per cent. 1938 G. S. Cansdale et al. *Black Poplars* 9 The poplar is essentially a tree to be grown on short rotation. 1977 M. Clawson *Decision Making in Timber Production* 26 For many sites, the earnings from the next rotation of timber growing are highly speculative or dubious, partly because it is so far in the future.
3. *rotation axis, group; rotation-like* adj.
1903 Rotation axis [see *holoaxial* s.v. *HOLO-]. 1971 I. G. Gass et al. *Understanding Earth* i. 19/1 The stereogram of zircon..shows a four-fold rotation axis in the centre and also shows a number of reflection planes. 1974 G. Reece tr. *Hund's Hist. Quantum Theory* xiii. 177 These two-valued representations of the rotation group had been discovered by Hermann Weyl. 1952 Dylan Thomas *Let.* 11 Dec. (1966) 388 It was, I think, originally a little 'organ'..full of rotation-like gossip.

rotational, *a.* Add: **2.** *Physics.* Of, pertaining to, or designating the (quantized) energy possessed by molecules, etc., by virtue of their rotation.
1914 *Chem. Abstr.* VIII. 859 According to the quantum hypothesis the rotational energy of a mol[ecule] varies discontinuously, from which it follows that a band is made up of a series of lines whose vibration difference is a measure of the moment of inertia of a mol[ecule]. 1939 J. W. T. Spinks tr. *G. Herzberg's Molecular Spectra* I. iii. 72 The rotational quantum number *J*..gives approximately the angular momentum in units *h*/2π. 1950 W. J. Moore *Physical Chem.* xi. 327 A set of closely packed rotational levels is associated with each of these vibrational levels. 1973 C. Sagan *Cosmic Connection* (1974) iv. 27 Molecules undergo rotational transitions, due to the free rotation of the molecule. 1978 P. W. Atkins *Physical Chem.* xvii. 561 According to the gross selection rule, methane cannot give a rotational Raman spectrum, whereas the hydrogen molecule, and any other diatomic..can.
3. *Agric.* Applied to methods of land management in which animals are grazed on successive areas of land in turn, so that each area is empty for a time after having been grazed.
1931 H. E. Woodman et al. in *Jrnl. Agric. Sci.* XXI. 267 It may be inferred that a similar result would follow from a system of rotational grazing, where the pasture enclosures, after being closely grazed, are permitted a 3-weeks' interval of unchecked growth before being grazed again. 1950 *N.Z. Jrnl. Agric.* Apr. 307/1 Rotational cropping is now an excellent fertility builder. 1967 C. D. Blake et al. *Fundamentals Mod. Agric.* ix. 208/1 In the broad sense, rotational grazing is any system of handling animals which involves holding them for short periods on small sections of the total area of an available pasture and then regularly moving them, as a group, to other sections of the whole area. There is an infinite number of rotational grazing systems available.

rota·tionally, *adv.* [f. prec. + -LY[2].] In a rotational manner; by or with respect to rotation.
1894 *Phil. Trans. R. Soc.* A. CLXXXV. 817 A simple theory of free electrons in a rotationally fluid æther. 1946 *Nature* 3 Aug. 176/1 These velocities are so high that rotationally stable stars cannot be formed unless one of two conditions is fulfilled. 1950 *N.Z. Jrnl. Agric.* Jan. 4/2 Calves which are rotationally grazed. 1976 *Physics Bull.* July 292/2 Eddies of rotationally dominated turbulence. 1978 *Sci. Amer.* Jan. 122/3 The four letters p, b, d and q are rotationally similar.

rotativism (rōᵘ·tătiviz'm, rotē[i]·t-). [f. Rotative *a.* + -ISM.] A system whereby different political parties hold office in turn according to a pre-arranged plan.
1908 *Sun* (N.Y.) 3 Feb. 2/1 Each party held office by arrangement alternately. This arrangement was known as rotativism. 1921 *Edin. Rev.* Jan. 158 The country must in effect sink back into the slough of 'rotativism'. 1960 W. C. Atkinson *Hist. Spain & Portugal* xiii. 313 The system in Portugal was known as rotativism. 1969 I. Gilmour *Body Politic* I. i. 55 The party struggle was also a stage fight in Portugal, where the parties agreed under a system called rotativism to alternate in office.

rotativist (rōᵘ·tătivist, rotē[i]·t-), *a.* (and *sb.*) [f. as prec. + -IST.] **1.** Of, pertaining to, or characterized by rotativism in politics. Also *ellipt.* as *sb.* Also *transf.*
1909 *Spectator* 12 June 918/2 The late King [of Portugal] yielded to the intrigues of the 'rotativist' parties. 1917 G. Young *Portugal* vi. 217 The collapse of the Portuguese colonial empire must come with the continuance of the struggle between rotativist Royalism and revolutionary Republicanism. 1926 *Glasgow Herald* 24 Dec. 4/7 They recalled the indifference of the electorate on the frequent occasions of general elections by order of the Rotativists. 1929 *Camb. Anc. Hist.* (ed. 2) III. 253 A regular 'rotativist' arrangement made by the son of Smendes with the Thebans by which he was to be succeeded by the Theban high-priest..and he again by a Tanite.
2. That relates to rotative movement. *rare.*
1939 A. J. Toynbee *Study of Hist.* VI. 173 Stoic and Epicurean philosophers who apparently were not put out by the incongruity between their rotativist conception of the nature of Reality and their ethical aim of Detachment.

rotator. Add: **2. a.** (Further examples.)
1930 *Jrnl. Sci. Instruments* VII. 22 It is necessary, for the purposes of wool examination, to be able..to examine the filament completely at any point by rotating it through 360°, and for this purpose a fibre rotator has been designed. 1967 *Stain Technol.* XLII. 107 The rotator consists of a 12 inch disc to which 8 glass jars.. are held firmly... It is rotated by means of a rubber belt driven by a small electric motor. 1971 *Sci. Amer.* July 85/2 Perhaps a much more massive rotating object in the core of a galaxy could account for the existence of quasars. In these models the rotator must have a mass of about a billion solar masses. 1977 *Lancet* 28 May 1150/2 This allows rapid separation of antibody-bound from free T_3/T_4 by simple inversion of the assay tubes in a rotator.
b. A device for rotating an aerial.
1959 *Sears, Roebuck Catal.* Spring/Summer 884/2 TV Antenna Rotators... Turns antenna 360° per min. 1970 *Globe & Mail* (Toronto) 28 Sept. 30/4 (Advt.), Al's TV, towers, color, rotator, U.H.F., Channel 17 installed. 1974 Harvey & Bohlman *Stereo F.M. Radio Handbk.* vii. 163 For long-distance (DX) reception an aerial amplifier and/or aerial rotator are useful accessories to a high gain aerial in extracting the largest possible signal to drive the receiver into full amplitude limiting. 1977 *Gramophone* Nov. 965/1 Feeding the tuner/amplifier from a four-element J-Beam FM aerial mounted on a rotator, it was possible to receive several continental FM stations in mid-Surrey.

rotavate, rotovate (rōᵘ·tăvē[i]t, rōᵘ·to-), *v.* [Back-formation from next.] *trans.* To prepare (a field, garden, etc.) with a Rotavator; to work (a substance) into the soil by means of a Rotavator. Hence **ro·ta-, ro·tovating** *vbl. sb.*; **rota-, rotova·tion.**
1959 *Birmingham Post* 10 June 1/1 A new allotment.. had been 'adequately fenced, roadways constructed and the site ploughed and rotavated'. 1960 *Farmer & Stockbreeder* 26 Jan. 78/3 It is not claimed that Rotavation is suitable for all cultivations, all the time, in all soils. 1962 *Times* 11 Apr. 24/3 Garden rotovated. 1962 *Times* 3 Dec. (Agric. Suppl.) p.vii/5 The usual procedure..is to rotavate the dead herbage. 1971 *Nature* 13 Aug. 446/2 The British Ministry of Agriculture suffered some criticism for suggesting that the best way of disposing of surplus and unwanted DDT..was to rotovate it into the soil. 1977 *Jersey Even. Post* 26 July 24/6 (Advt.), Sub-soiling and rotovating undertaken at competitive rates. 1978 *Rescue News* Summer 4/2 Methods of working peat include rotovation.

Rotavator, Rotovator (rōᵘ·tăvē[i]təɹ, rōᵘ·to-). Also with small initials. [f. Rot(ary *a.* + Culti)vator; see *ROTO-.] Proprietary names of a machine with rotating blades designed to break up or till soil.

1936 *Trade Marks Jrnl.* 1 July 808 Rotavator... Ploughs, cultivators, diggers, harrows and hoes, all being agricultural machines. Rotary Cultivators Limited. 1951 *Official Gaz.* (U.S. Patent Office) 13 Mar. 362/2 Rotary Hoes Limited, East Horndon, England. Filed Dec. 27, 1949. *Rotavator*... For ploughs, cultivators, diggers, harrows, and hoes. 1954 *Encounter* Dec. 31/2 Nobody said..that it was a handicap to have a holding so large you could *choose between* a rotavator and a proper tractor. 1959 *Times* 13 Mar. 6/6 Mr. Merricks then bid £5 and the Rotovator was knocked down to him. 1963 *Times* 17 Jan. 3/7 The machine, a 'Rotavator', breaks up the packed snow and ice into powder, so that it can be swept away. The machine has been used on Watford's football ground and there are plans to use it at race-courses. 1970 T. Hughes *Crow* 31 The woodpecker drummed clear of the rotovator. 1971 'S. Smith' *Grave Affair* vii. 102 They went back to the farm for the rotavator to break up the soil. 1977 *Trade Marks Jrnl.* 21 Sept. 1895/1 *Rotovator*... Agricultural machinery and power operated agricultural implements. 1978 *Morecambe Guardian* 14 Mar. 22/3 (Advt.), Agrotiller and Landmaster Rotovators.

rotavirus (rōᵘ·tăvəiᵊrɒs). *Biol.* [mod. L., f. L. *rota* wheel + Virus.] Any one of a genus of wheel-shaped double-stranded RNA viruses.

1974 T. H. Flewitt et al. in *Lancet* 13 July 61/1 Since these viruses differ morphologically both from reoviruses and orbiviruses, the name 'rotavirus' is suggested for them. 1977 *Lancet* 11 June 1263/2 Rotaviruses cause acute enteritis in man and animals. 1977 *Rec. Adv. Clin. Virol.* I. 158 Rotaviruses differ both in size and shape from the reoviruses... They resemble orbiviruses in the appearance of rings..seen on some particles..; but the [rotavirus] diarrhoea viruses also differ from the orbiviruses in their smooth outline. 1979 *Brit. Med. Jrnl.* 15 Dec. 1551/1 Rotavirus infection is the commonest cause of acute non-bacterial gastroenteritis in infancy and childhood.

R.O.T.C., ROTC (rǫ·tsi; *also* ɑ̄ɹ,ōᵘ,tī,sī). *U.S.* [Acronym f. the initials of *R*eserve *O*fficers' *T*raining *C*orps.] A military division with units established at civilian educational centres to qualify students for appointment as reserve officers.

1916 *N.Y. Times* 27 Dec. 6/1 The cap ornament for members of the Reserve Officers' Training Corps is to consist of a wreath inclosing the letters R.O.T.C. 1919 [see *drip *sb.* 3 b]. 1925 *Scribner's Mag.* July 15/2 He goes to the R.O.T.C. and prepares himself for a berth in the adjutant-general's office. 1936 *N.Y. Herald Tribune* 1 June 17/3 (*heading*) Flint defends the R.O.T.C. in Syracuse talks. 1959 N. Mailer *Advts. for Myself* (1961) I. 35 He had been allowed to go to this university only on the agreement..that he..was to join the R.O.T.C. and to remain in it until after graduation. 1974 *Hartsville* (S. Carolina) *Messenger* 22 Apr. 2-A/8 Early in the school year he went to Myrtle Beach Air Force Base for a military physical that was sent to all the academics and to the ROTC headquarters. 1975 *Publishers Weekly* 26 May 57/1 Out of his campus experiences he examines ROTC from the academic viewpoint. He ponders the fairness of grading prospective draftees.

rote, *sb.*[2] Add: **3.** *rote-like* adj.; **rote learning** (further examples); also *spec.* in *Psychol.*, the learning by rote of meaningless material designed to be free of associations, as a technique in the study of learning.

1914 *Brit. Jrnl. Psychol.* VII. 253 (*title*) The value of distributed repetitions in rote learning. 1940 G. Katona *Organizing & Memorizing* vii. 164 We shall study the classic material used in investigating the memory, that is, rote learning of nonsense syllables. 1970 *Jrnl. Gen. Psychol.* LXXXII. 54 The similarity of the present results to those in the rote literature argues against the sharp distinction drawn between meaningful and rote learning by Ausubel. 1954 W. Faulkner *Fable* 35 He said, repeated, rote-like, cold, unemphasised, almost telegraphic: 'Comité des Forges'.

rote, *sb.*[6] (Further examples.)

1909 *Newfoundland Q.* Dec. 9/1 The fishermen are accustomed, in foggy weather, to find their bearings by carefully listening to the rout of the sea on the shore, which they (very correctly) call rote, or rut. 1941 T. S. Eliot *Dry Salvages* i. 8 The menace and caress of wave that breaks on water, The distant rote in the granite teeth. 1965 S. E. Morison in *Amer. Neptune* Oct. 236 Often have I heard a Maine man say, 'Sea's making up. Hear that rote!' *Ibid.*, T. S. Eliot doubtless listened to the rote from his parents' house, during the windless calm after a storm, or on a 'weather-breeder' day when swells from the eastward begin crashing on the 'granite teeth' of Cape Ann before a storm breaks.

rotenone (rōᵘ·tĕnōᵘn). *Chem.* Orig. †**-on.** [ad. Jap. *rotenon* (K. Nagai 1902, in *Jrnl. Tokyo Chem. Soc.* XXIII. 753), f. *roten* derris: see -one.] A toxic crystalline polycyclic ketone, $C_{23}H_{22}O_6$, obtained from the roots of several species of plant (notably derris, cubé, and timbo), which is widely employed as an insecticide in the form of a powder or an emulsified spray. Also as *v. trans.*, to treat with rotenone.

1924 *Chem. Abstr.* XVIII. 408 From air-dried roots, 0·93% crude rotenon is obtained which is mixed with waxy impurities. 1925 *Ibid.* XIX. 1708 Further analysis of rotenone, an active insecticidal principle of the root of *Derris elliptica* Benth. 1962 Gordon & Lavoipierre *Entomol.* ix. 57 Fine dusts containing pyrethrum or rotenone are highly successful when employed against various insects such as fleas and lice. 1975 *New Yorker* 19 May 45/3 We poison out a small reef by squirting in emulsified rotenone, a chemical poison derived from the root of a South American plant called cubé, originally used by Indians for fishing. *Ibid.* 46/1 You rotenone a reef, and for the next hour or two you pick the samplings up. 1977 Lewis & Elvin-Lewis *Med. Bot.* ii. 45 *Tephrosia cinerea*, yielding rotenone and the toxic principle tephrosin, is used in Venezuela and Africa as a fish poison.

rot-gut, rotgut. Add: **1.** (Further examples.)

1911 E. M. Clowes *On Wallaby* vi. 164 The cattle-men, shearers, and shepherds get their internal machinery completely ruined in time by the quantity of inferior boiled sugar and fruit that they consume, and which they have inelegantly christened 'rot-gut'. 1923 C. E. Mulford *Black Buttes* xiv. 220 Yes, even a drink of rot-gut would 'a' bought you! 1939 Joyce *Finnegans Wake* (1964) 381 And suck up..whatever surplus rotgut, sorra much, was left by the lazy lousers of maltknights and beerchurls. 1946 *Time* 7 Oct. 10/2 For 50 years we have been hearing how the drought-smitten Jayhawkers were poisoning themselves on bootleg rotgut because we couldn't get decent liquor. 1952 E. O'Neill *Moon for Misbegotten* iv. 173 That isn't Phil's rotgut. That's real, honest-to-God bonded Bourbon. 1969 *Private Eye* 4 July 14/3 But don't drink that rotgut. Here warm your gizzard with a tot of rum from my flask. 1976 *Times* 8 July 16/4 It was being killed mercilessly by the whisky posts with their rotgut.

2. (Further examples.) Also *transf.* and *fig.*

1877 H. Ruede *Let.* 24 Apr. in *Sod-House Days* (1937) 57 They have a brand called 'Old Style', some of Catlin's (St. Louis) cheap rotgut tobacco, and from that price up. 1927 L. Bromfield *Good Woman* xiii. 140 A glass filled many times with the rot-gut whisky that Hennessy sold. 1948 F. Blake *Johnny Christmas* I. 5 Not a man in that line but hated Santa Ana and his Mexicans, hated their talk, the way they killed, their rot-gut laughter. 1970 J. Howard *Please Touch* 6 Many kinds of wine: sweet, dry, nutty, fruity, insouciant, rotgut, presumptuous and noble. 1978 *Sunday Times* (Colour Suppl.) 18 June 42/3 Traders stack their boats with liquor, rotgut whisky and cachaca, cheap spirit.

rother-beast. For *Obs.* read '*Obs.* exc. *arch.*' and add later examples.

1836 in A. R. Stedman *Marlborough & Upper Kennet Country* (1960) xxvii. 270 No burgess shall keep on the common more than two rother beasts... that is to say, kine or bullocks. 1933 *Catholic Bull.* Mar. 215 Let us read over again the overwhelmingly impressive reasonings of this great among the greatest pontiffs, then contrast the fruits of imperial connection, of Masonry, of that trade in pasturing and larding the sides of rother-beasts, to champion which the Knight of the Ranchers..has couched his lance.

Rothschild (rǫ·þstʃəild). [Name of Mayer Amschel *Rothschild* (1744–1812) of Frankfurt, and his descendants, proprietors of an international banking firm.] **1.** One who resembles a member of the Rothschild family in being exceptionally rich; a millionaire. Also in colloq. phr. *to come the Rothschild*: to pretend to be rich (see Come *v.* 28 c).

1833 Carlyle *Sart. Res.* I. v, in *Fraser's Mag.* VIII. 670/2 All miracles have been out-miracled for there are Rothschilds and English National Debts. 1850 E. Ruskin *Let.* 18 Apr. in M. Lutyens *Effie in Venice* (1965) I. 167 We..called on Madame Chabrillan who is married to the Rothschild of the family. 1863 Geo. Eliot *Romola* I. I. 74 The Bardi..[were] standing in the very front of European commerce—the Christian Rothschilds of that time. 1885 H. James *Little Tour in France* xii. 92 Jacques Coeur..was a Vanderbilt or Rothschild of the fifteenth century. 1893 W. S. Gilbert *Utopia (Limited)* I. 32 Though a Rothschild you may be..As a company you've come to utter sorrow. 1905 Farmer & Henley *Dict. Slang* 111/2 *To come the Rothschild*, to pretend to be rich. 1910 W. J. Locke *Simon* xvi. 221, I had wealth—not a Rothschild or Vanderbilt fortune but enough to assure me ease and luxury. 1938 [see *Rockefeller]. 1974 J. Gardner *Return of Moriarty* 57 A relatively young whore, Mary Jane Kelly, who sometimes came the Rothschild about her past, calling herself Marie Jeannette Kelly.

2. as *adj.* See *Mouton Rothschild.

‖ **rôti**[1] (roti). [Fr.] In *Gastronomy*, a main course consisting of roasted meat; (a dish of) roasted meat. Also as *adj.* (with preceding sb.) and *Comb.*

1771 Smollett *Humph. Cl.* III. 143 The rotis were scorched and stinking, for the honour of the fumet. 1806 J. Pinkerton *Recoll. Paris* II. vi. 102 Upon the appearance of the *roti*, the ordinary wine is changed for the richer kinds of Burgundy or Bordeaux. 1841 Thackeray *Mem. Gourmandizing* in *Fraser's Mag.* XXIII. 714/1 Saddle of mutton rôti. 1864 Mrs. Gaskell *French Life* in *Fraser's Mag.* LXIX. 440/2 The rôti and the salad follow. 1906 Mrs. Beeton *Bk. Househ. Managem.* lxii. 1669 *Rôti* (Fr.), the roast..the course of a meal which is served before the entremets. 1951 *Good Househ. Home Encycl.* 639/1 *Rôti*, the meat, poultry or game course in a dinner. 1980 G. Greene *Dr. Fischer* xvi. 114 With the Mouton Rothschild there was a *rôti de boeuf*.

‖ **roti**[2] (roti). [a. Hindi, Urdu *roṭī* bread, Rooty.] A cake of unleavened bread of a type originating in India. (Now also current in the W. Indies.)

1920 *Chambers's Jrnl.* 29 May 407/1 What are two hours to an Indian peasant? They had had their *roti* (bread), and, the stomach being full, having to wait was a small matter. 1952 [see *Johnny-cake a]. 1958 J. Carew *Black Midas* vi. 98 We bought fruit and roti from peddlers on the stelling. 1971 *Leader* (Durban) 7 May 9/1 (Advt.), Only the best is good enough... That's why I use only Bakers Homo Flour for my roti. 1971 *Advocate-News* (Barbados) 17 Sept. 6/4 Among the specialties to be presented to Barbadians by the Roti Shop are chicken, beef and shrimp roti, dholl puri and potato roti. 1974 *Socialist Worker* 9 Nov. 8/3 The starving stream in from all directions to receive a roti (a thin flat piece of bread) or perhaps two if they are lucky. 1976 *Sunday Standard* (Bombay) 11 July 4/3 Roti, the unleavened bread, is the stuff of our lives.

rotisserie (rōᵘtī·səri). orig. *U.S.* Also **rôtisserie.** [a. F. *rôtisserie*, f. *rôtiss-*, stem of *rôtir* to roast + *-erie* -ery.] **1.** A restaurant where meat is roasted or barbecued, freq. at a grill in the front window.

1868 *Overland Monthly* Nov. 470/1 At some of these French houses, especially designated as *rotisseriés* [*sic*], the kitchen is nominally open to inspection. 1914 S. Lewis *Our Mr. Wrenn* i. 15 A rôtisserie, before whose upright fender of scarlet coals whole ducks were happily roasting to a shiny brown. 1925 *Restaurant News & Managem.* Dec. 10 (*caption*) An instance of successful catering to business and professional people The Rotisserie Inn, Salt Lake City. 1936 Mencken *Amer. Lang.* (ed. 4) 215 *Rôtisserie*, with the accent omitted, seems to be an Americanism. It signifies an eating-house wherein chickens and butcher's meat are roasted at a charcoal-grill, usually in the show-window of the establishment.

2. A cooking appliance which has a rotating spit for roasting and barbecuing meat. Also *attrib.* and *Comb.*

1953 *Home Beautiful* Apr. 133 Cooking on a rotating spit or rotisserie is high gourmet cooking. 1953 J. & M. Robertson *Compl. Small Appliance Cookbk.* ii. 37 Rotisserie heat is *beside* or *above* the revolving food. *Ibid.* ii. 42 Serve with rotisserie-browned potatoes. 1960 *Guardian* 17 Mar. 9/4 All the glittering machines, the washers, the electric rotisseries. 1969 *Daily Colonist* (Victoria, B.C.) 6 Dec. 40/1 They borrowed a commercial rotisserie, got the charcoal white hot, and loaded the apparatus with 200 pounds of wild boar. 1973 *Times* 30 July 11/1 Rôtisserie spits, continuous cleaning ovens, the use of colour..these and other innovations..are maintaining..the popularity of the gas cooker. 1978 *Lancashire Life* Apr. 125/3 There are, in fact, three variations of this cooker, the most expensive one having a built-in rotisserie and kebab attachments.

Hence (as a back-formation) **roti·ss(e)** *v. trans.* and *intr.*, to cook meat on a rotisserie; **roti·ssed** *ppl. a.*, **roti·ssing** *vbl. sb.*

1958 *Word Study* Dec. 5/1 The manufacturer has created the verb 'rotiss'. *Ibid.* 5/2 The housewife is advised to set her pointer according to what meat is being 'rotissed', and is informed that she needn't preheat when 'broiling or rotissing'. *a* 1963 P. Bracken *I hate to housekeep Bk.* (1969) viii. 72 She is a little scared of the rotisserie in her new double oven, so she continues to buy her chickens ready-rotissed. 1978 *Verbatim* Feb. 1/2 One San Francisco appliance dealer boasts of a stove which will not only roast and broil, it will also *rotisse*!

Rotissomat (rōᵘtī·somæt). [f. *rotiss(erie + -o + *-mat.] The proprietary name of a commercial automatic cooking appliance with rotating spits for roasting meat.

1947 *Official Gaz.* (U.S. Patent Office) 8 July 187/1 Rotiss-o-mat Corporation, Astoria, Long Island, N.Y. *Rotiss-o-mat for electric rotisseries.* Claims use since January 1946. 1958 *Trade Marks Jrnl.* 9 Apr. 372/2 *Rotiss-o-mat*... Electric installations for cooking poultry and parts of such installations included in Class II. Harley Manufacturing Corporation..New York. 1960 *Observer* 13 Nov. 3/3 Chicken restaurants depend on a shiny piece of plant called a 'rotissomat', invented by the American, Sol Leder. 1961 *Times* 24 July 13/5 An Oxford supermarket has a rotissomat. 1963 M. Beadle *These Ruins are Inhabited* xi. 160 The one near me on the London Road has installed a rotissomat.

roto (rōᵘ·to). *N. Amer.* Abbrev. of *rotogravure 2, an illustrated or pictorial (section of a) newspaper or magazine.

1932 G. A. Chappell *Evil through Ages* v. 70 To-day our Sunday illustrated sections are taken up by wives and mothers..who sign testimonials, exhibit dogs, dress up as colonial dames, anything to get into the rotos. 1942 Berrey & Van den Bark *Amer. Thes. Slang* §522/17 *Rotogravure section*, roto, rotogravure, roto section. 1975 *Boston Globe* 22 Feb. 7/3 (Advt.), The Bride: a special roto magazine devoted to the newlyweds and soon-to-be-weds of '75 this Sunday in The Boston Globe. 1978 *Amer. Poetry Rev.* Nov./Dec. 30/1 He was writing for the roto section of a Sunday newspaper. 1979 *Globe & Mail* (Toronto) 25 Jan. 7/3 A few publications such as Chatelaine and the rotos—Weekend and the Canadian—slept right through it, as far as ad revenues were concerned.

roto- (rōᵘ·to-), in some words also *rota-*, comb. form of L. *rōta* wheel, roller and Eng. Rotary *a.*, Rotation, etc., as in Rotograph, *rotochute, etc.

rotochute (rōᵘ·toʃūt). [f. *roto- + Para)-chute *sb.*] A mechanical device with rotating blades which can be attached to objects dropped from a great height so as to slow their fall.

1949 *Jrnl. Brit. Interplanetary Soc.* VIII. 139 A new type 'high-speed' parachute, intended for use in retrieving recording equipment from high-altitude rockets, was tested recently... Ordinary parachutes were ripped to shreds after falling free into the denser atmosphere... Expelled from A-4 rockets at altitudes up to 100 miles.., the rotochute attains supersonic velocity before atmospheric density builds up and the blades begin to revolve, gradually being forced out against air pressure to assume a horizontal position. 1955 *Sci. News Let.* 5 Mar. 160/3 Supplies can now be dropped to troops from low altitudes by a 'rotochute', a bomb-shaped device with rotor blades to slow the descent. 1962 *Aeroplane* CII. 45/2 In the recovery of space vehicles, the 'rotochute' is capable—according to this company—of satisfying three basic requirements not easily met in a single recovery system.

Rotodyne (rōᵘ·todəin). Also **Rotadyne** (rōᵘ·tă-). [f. *roto- + *-dyne.] A proprietary name of an aircraft equipped with rotors, capable of vertical take-off and rapid flight.

1949 *Trade Marks Jrnl.* 17 Aug. 727/1 *Rotadyne* 676,644. Rotary wing aircraft. The Fairey Aviation Company Limited,..Hayes, Middlesex. *Ibid.*, *Rotodyne* 676,645... To be associated with No. 676,644. 1958 *Times* 9 Nov. 4/4 The..Rotodyne, the world's first vertical take-off and landing airliner. 1959 [see *assault *sb.* 8]. 1959 [see *convertiplane].

rotogravure (rōᵘ:to,grăviūᵊ·ɹ). *Printing.* Also ‖**rotogravur,** †**rotagravure,** and with capital initial. [orig. the name of the *Rotogravur* Deutsche Tiefdrück Gesellschaft (Berlin), said to be f. the names of two other companies, *Roto*phot (Berlin) and Deutsche Photo*gravur* AG (Siegburg), adopted in Eng. with assimilation of the ending to that of Photogravure. The form *rotagravure* (in sense 1) is an etymologizing re-formation f. L. *rota* wheel, roller + Photo)gravure or F. *gravure* engraving.] **1.** A method of printing by means of a rotary press with intaglio cylinders, usu. used at high speed for long print runs.

1913 *Photography* 7 Jan. 2/1 The half-tone block..has advantages for certain purposes which it does not share..with Rotogravure. 1913 *Illustr. London News* 8 Feb. (Suppl.) p. iii/1 The rotogravur method is that more generally called the carbon. 1914 *N.Y. Times* 29 Mar. 11/1 Advance copies of the rotogravure section of *The Times* of next Sunday..awakened enthusiasm. This is the first rotogravure section to be printed upon the new rotogravure presses of *The Times*, and it contains thirty-eight additional famous paintings from the Altman collection. 1919 S. H. Morgan in *Inland Printer* July 407/1 The proper name for the process and its product is 'rotary photogravure', and it is quite natural that in these busy times there would be an effort to abbreviate these two words. So why not use..'rota', meaning a wheel or roll, and 'gravure',..and by combining the two call it 'rotagravure' hereafter? 1940 *Chambers's Techn. Dict.* 731/1 Rotagravure. 1942 J. Steinbeck *Moon is Down* ii. 29 Lieutenant Prackle took from his pocket a folded rotogravure page and he unfolded it and held it up and looked at it. It was a picture of a girl. 1957 *Gravure* Mar. 38/3 The first use of rotogravure in a periodical occurred in 1897, when a gravure illustration was included with an article by W. Burger describing the Castle Kreuzenstein, and which appeared in the monthly bulletin of the Imperial Austrian Museum of Art and Industry. 1972 *Physics Bull.* Sept. 532/2 For high quality colour work with long runs (one million or more) rotogravure printing is universally used.

2. A sheet or other object, or a section of a newspaper or magazine, that has been printed by this process.

1914 *N.Y. Times* 29 Mar. 11/3 The rotogravures are superior to any group of reproductions I have ever seen issued in this way, except for the occasional photogravure that some publication has put forth. 1943 D. Powell *Time to be Born* iv. 94, I suppose business experience never can quite make up for your picture in the Sunday rotogravure. 1968 L. J. Braun *Cat who turned on & Off* vii. 64 His pleasurable dreams were always in colour; others were in sepia, like old-time rotogravure. 1978 J. Updike *Coup* (1979) v. 197 The American press loved this artful clown; in their rotogravures he looked like a negative print of Santa Claus.

rotometer (rotǫ·mitəɹ, rōᵘ-tomītəɹ). Also **rota-.** [f. *roto- + -meter.] A hand-held measuring device incorporating a small wheel whose revolutions are registered in terms of distance travelled, e.g. on a map or plan.

1901 F. W. Taylor *Art of Cutting Metals* (ed. 3) 91 An instrument called a 'rotameter', which we have found the best appliance for practical use in measuring the cutting speed. *Ibid.*, The small wheel..which projects beyond the rim of the rotameter is firmly pressed directly against the surface of the rotating forging. 1913 T. W. Croft *Amer. Electricians' Handbk.* 134 The rotometer (Fig. 141), is a convenient tool for scaling distances. The little wheel is run over the course of the circuit. The pointer indicates feet direct for drawings of certain scales. 1949 R. Ashley *Electr. Estimating* i. 14/1 The map measure, commonly known in the profession as the 'rotometer', is considered by most electrical estimators engaged in large construction work as the most valuable of tools. 1957 G. E. Hutchinson *Treat. Limnol.* I. ii. 166 The shore line may be measured on the map by means of a rotometer. 1971 W. N. Alerich *Electr. Construction Wiring* xvii. 456/2 The measuring can be done with the aid of a rotameter or an architect's scale.

roton (rōᵘ·tǫn). *Physics.* [f. Rot(ation + *-on[1].] A quantum or quasiparticle associated with vortical motion in a liquid, esp. in liquid helium.

1941 L. Landau in *Jrnl. Physics U.S.S.R.* V. 75/2 An 'elementary excitation' of the vortex spectrum might be called a 'roton'. [*Note*] This name was suggested by I. E. Tamm. 1947 *Physical Rev.* LXXII. 852/1 In Landau's theory, the excited atoms of the Bose-Einstein theory are replaced by 'rotons'. 1968 *New Scientist* 25 July 198/3 The evaporation of helium atoms due to the decay of 'rotons', the elementary rotational excitations of liquid helium. 1973 *Nature* 9 Nov. 66/1 A new type of elementary excitation, christened the ^{3}He roton, may exist in liquid ^{3}He–^{4}He mixtures. 1977 *New Scientist* 3 Mar. 507/1 The physical nature of the roton still remains a mystery some thirty years after their existence was first postulated by Landau, although they have been detected in numerous experiments.

rotor. Add: **3.** A rotor arm.

1920 V. W. Pagé *Useful Hints for Motorists* iii. 89 The distributor head and rotor are made of bakelite. 1959 *Motor Man.* (ed. 36) viii. 215 Before replacing the rotor, which should also be cleaned with a petrol-damped rag and dried with a soft cloth, apply a few drops of engine oil to lubricate the cam bearing. 1975 tr. *Melchior's Sleeper Agent* III. 212 Someone's been monkeying with the jeep... The rotor! Someone's pinched the distributor rotor.

4. A cylinder mounted vertically on a ship and designed to be rotated on its axis, so that the Magnus effect will provide a forward propulsive force in a cross-wind.

1924 *Public Opinion* 14 Nov. 483/3 Very little electric power is required to work the rotors. 1924 *Glasgow Herald* 17 Nov. 4 The navigational importance of Herr Anton Flettner's 'rotor' as an auxiliary. 1925 *Ibid.* 6 Feb. 9 These towers—technically called rotors—are supported on strong internal masts, about which they are revolved by small electric motors. 1943 [see *Magnus effect]. 1957 *Encycl. Brit.* XIX. 577/2 The inventor states that it is not intended to drive ships solely by wind rotors, but that they shall serve as an auxiliary power upon steam and motor vessels.

5. A hub with a number of radiating arms that is rotated in an approximately horizontal plane to provide the lift for a helicopter or other rotary-wing aircraft.

1930 *Jrnl. R. Aeronaut. Soc.* XXXIV. 915 The wings of the aeroplane and the rotor of the autogiro. 1945 *Tee Emm* (Air Ministry) V. 56 Note the main rotor, that is the big propeller affair on top... Get rid of any assumption that because the rotor is *above* the aircraft it is also above *you*. 1973 R. Lewis *Blood Money* viii. 107 The helicopter..dropped lower towards the surface of the tarn until finally the downthrust of air from the rotors churned the water into a maelstrom.

6. The rotating vessel in a centrifuge.

1939 *Industr. & Engin. Chem.* Sept. 1073 The rotor or bowl must be removed for cleaning when it becomes filled with bowl cake. 1958 M. G. Larian *Fund. Chem. Engin. Operations* (1959) xiv. 566 The centrifuge is shown in Fig. 24. It consists of a tubular bowl rotor enclosed in a stationary casing. 1978 *Nature* 14 Sept. 147/1 Brain homogenates..are centrifuged at 100,000*g* for 18 h in a Beckman 42.1 rotor.

7. A part of an encoding or decoding machine, rotation of which changes numerous electrical circuits and thereby the code.

1946 *U.S. Patent* 2,402,182 3 The selected ratchet and pawl mechanism..then rotates the rotors one step while the contacts carried by the rotors are disengaged. 1973 H. Gruppe *Truxton Cipher* xviii. 189 That moment in '42 when he had handed over the Truxton Cipher rotors to the Russians. 1979 *Books & Bookmen* Jan. 31/1 The object was to enable the recipient of the message to set the rotors of his own machine for deciphering that particular message.

8. A large eddy in which the air circulates about a horizontal axis. [a. G. *rotor*, introduced in this sense by J. Küttner 1938, in *Beiträge zur Physik der freien Atmosphäre* XXV. 108.]

1949 *Q. Jrnl. R. Meteorol. Soc.* LXXV. 54 The crests of waves may be capped by clouds under suitable conditions of humidity. Beautiful examples are the Helm Bars.., and the 'Rotors' associated with the Moazagotl. 1955 *Tellus* VII. 367 The stationary lee-waves produced by a big mountain often break up into turbulent whirls or 'rotors' in the lower layers of the air flow. 1960 *Aeroplane* XCVIII. 390/3 They moved downwind into the downdraught and at 2,500 ft. dropped into the rotor and its turbulence, which became especially violent from 1,500 ft. down to the ground. 1979 *Courier-Mail* (Brisbane) 15 June 6/10 It was believed that wind conditions against the cliff face caused what was known in the sport as a 'rotor'. 'The wind spins across from the hill face and grabs you,' he said.

9. *attrib.* and *Comb.*, as **rotor arm,** the part of the distributor of an internal-combustion engine which, by its rotation, successively makes and breaks electrical contacts so that each sparking plug fires in turn; **rotor blade,** each of the radiating arms of the rotor of a helicopter or other rotary-wing aircraft; **rotor cloud,** a turbulent cloud in a rotor (sense *8) in the lee of a mountain; **rotor disc,** (*a*) the space swept out by rotor blades as they rotate; (*b*) the rotor head; **rotor head, hub,** the structure at the upper end of a shaft of a rotorcraft, to which the rotor blades are attached; **rotor ship,** a ship whose motive power is derived from cylindrical rotors. See also *rotorcraft.

1919 Fraser & Jones *Motor Vehicles* xviii. 195 The rotor arms are placed at right angles to each other and project from both sides of the shaft. 1964 [see *distributor b (ii)]. 1968 *Listener* 25 July 109/1, I stopped the jeep in the middle of the crowd while Kim automatically removed the rotor arm and padlocked the gears. 1931 De la Cierva & Rose *Wings of Tomorrow* viii. 118 This was the only Autogiro that ever broke a rotor blade in flight. 1947 *Times* 16 Apr. 2/1 An airscrew which can be extended to the dimensions of a rotor blade such as that used in helicopters has long been sought. 1973 R. Lewis *Blood Money* viii. 106 Crow heard the chatter of rotor-blades and saw the helicopter coming in. 1959 *Gloss. Meteorol.* (Amer. Meteorol. Soc.) 487 *Rotor cloud*, a turbulent, altocumulus-type cloud formation found in the lee of some large mountain barriers. 1967 R. W. Fairbridge *Encycl. Atmospheric Sci.* 609/1 The moazagotl condition is set up by a standing wave established when the warm chinook or foehn-type air flows down the lee slope of the mountain range, initiating a series of undulations with eddies, the rising air of which develop cumuloform rotor clouds. 1974 T. Beer *Atmospheric Waves* iv. 182 The base of the rotor cloud is near the level of the crest while the top may be several thousand feet higher. 1944 H. F. Gregory *Anything Horse can Do* xiv. 145 The direct-control Autogiro is controlled by tilting the rotor disk in the desired direction. Actually it is accomplished by rocking the hub. 1976 B. Jackson *Flameout* (1977) iv. 54 Klein bent to stare at the forward stages of the compressor rotor. All the wing-like blades were rooted in the rotor disk. 1931 De la Cierva & Rose *Wings of Tomorrow* vi. 93, I propose to overcome the inequality of lift by building into the rotor head a device for changing the angle of incidence of the blades as they made their circle. 1958 Lambermont & Pirie *Helicopters & Autogyros of World* 32 The rotor head..was controlled by feathering and impressed flapping. 1931 De la Cierva & Rose *Wings of Tomorrow* viii. 118 The rotor hub is almost entirely a machine shop product. 1949 *Electronic Engin.* XXI. 292/2 The testing tower..was erected to provide a means of testing rotor hubs and blades independently of helicopter aircraft. 1924 *Glasgow Herald* 17 Nov. 4 (*heading*) The rotor ship. *Ibid.* 12 Dec. 8 The Rotor ship is apparently a thing devoid of beauty. It reminds one of a match-box with two cigarettes placed vertically on top. 1926 *Ibid.* 27 Apr. 11 The big new rotorship Barbara will be launched at Bremen tomorrow. Her tonnage is 3000, and she has three rotors, each measuring nearly 100ft. in height and 13ft. in circumference. 1949 O. G. Sutton *Sci. of Flight* iv. 84 The idea of a rotor-ship does not seem to have appealed to ship-builders in general and it is now regarded as a scientific curiosity and no more. 1957 *Encycl. Brit.* XIX. 578/1 An ordinary sailing vessel requires to take down all her canvas in a hurricane, but the rotor ship could continue sailing, with more stability for manoeuvring.

rotorcraft (rōᵘ·tǫɹkrɑ̄ft). [f. Rotor + Craft *sb.*] A rotary-wing aircraft.

1940 *Jrnl. Aeronaut. Sci.* VII. 444/1 Theoretical studies of rotorcraft vibration..were initiated about two years ago by the Kellett Autogiro Corporation. 1955 Liptrot & Woods *Rotorcraft* ii. 12 Rotorcraft, which is the generic name for rotary wing aircraft, may be classified under six categories... Helicopter... Gyroplane... Cyclogyro... Gyrodyne... Compound helicopter... Convertible aircraft. 1969 *New Scientist* 28 Aug. 421/1 At least 60 per cent of the loaded weight of a tactical monoplane is normally disposable... The corresponding figure for rotorcraft rarely exceeds 45. 1979 *Daily Tel.* 14 Sept. 10/5 Britain's only helicopter museum, the British Rotorcraft Museum at Weston-super-Mare, Avon, said yesterday that it had been forced to close because the local airfield lease has run out.

Rotoscythe (rōᵘ·tosəið). Also **roto-scythe, rotoscythe.** [f. *roto- + Scythe *sb.*] The proprietary name of a machine with rotating blades, designed to cut rough grass or vegetation.

1948 *Times* 20 Mar. 6/7 The roto-scythe..has come to take the place of the brushing-hook for the removal of grass and herbage too tall for the mowing machine. 1949 *Trade Marks Jrnl.* 19 Jan. 48 *Rotoscythe*... Lawn Mowers. Power Specialities Limited,..Slough, Buckinghamshire; manufacturers. 1955 *Radio Times* 22 Apr. 46/1 Rotoscythe cuts long or short, neater–quicker. 1966 'J. Berrisford' *Wild Garden* i. 16 Rough grass..must be cut (by scythe or rotoscythe).

Rototiller (rōᵘ·totiləɹ). Chiefly *N. Amer.* Also **rototiller, roto-tiller.** [f. *roto- + Tiller *sb.*[1]] A machine with rotating blades or prongs designed to break up or till soil (registered in the U.S. as a proprietary name). Hence **ro·totilling,** the preparation of soil with a Rototiller.

1923 *Sci. Amer.* Dec. 411/1 From England comes the description of one which differs radically from the American variety in that the soil is worked by a revolving member called a miller... The rototiller is driven by a two-cycle, 8 to 10 horse-power engine. 1932 *Official Gaz.* (U.S. Patent Office) 25 Oct. 949/1 *Rototiller* for farm machine for soil cultivation. Claims use since Feb. 9, 1929. 1938 C. Culpin *Farm Machinery* vii. 106 One of the most successful types of rotary cultivator is the small 'Rototiller' type, of which the best-known example is the

Simar Rototiller... The loose tilth produced by the Rototiller is very suitable for much horticultural work. 1959 *Times* 13 Mar. 6/6 Another man bid £5 for a rototiller. 1969 *Daily Colonist* (Victoria, B.C.) 15 July 22/5 In cultivated ground, hoeing or rototilling doesn't help much—in fact, in can spread the pest. 1974 *Globe & Mail* (Toronto) 5 Mar. 29/5 A friend of mine who deals in rototillers tells me he has more orders and enquiries this year than he's ever had. 1976 *Casper* (Wyoming) *Star-Tribune* 29 June 17/7 (Advt.) Custom plowing and rototilling, dozer work. 1978 *Sunday Star* (Toronto) 21 May A3/6 Rental agencies reported that most..of their garden tools, and especially roto-tillers, were rented for the day.

Rotovator, var. *Rotavator.

‖ **rotta** (rǫ·ta). *Hist.* Also **rota.** [med. Lat.: see Rote *sb.*[1]] = Rote *sb.*[1]

1883 Grove *Dict. Mus.* III. 179/1 *Rota*, or *Rotta*.., not, as might be supposed from its name, a species of vielle or hurdy-gurdy, but a species of psaltery or dulcimer, or primitive zither, employed in the middle ages in church music. It was played with the hand, guitar-fashion, and had seven strings mounted in a solid wooden frame. 1942 E. Blom *Music in England* i. 12 A miniature in an eighth-century codex in the British Museum..shows King David playing on a rotta (a generic link between the lyre and the harp). 1964 S. Marcuse *Musical Instruments* (1966) 102/2 In the 8th c., Cuthbert had mentioned the cithara that 'we' call rotta. 1977 *Early Music* July 300/1 When the musicians and minstrels adopted it [*sc.* the ancient psaltery] for their own purposes, ..they made its shape and form suitable to their convenience, applying additional strings and calling it by the vernacular name *rotta*.

rotten, *a.* Add: **I. 4. c.** orig. *N. Amer.* Of ice: weak; melting, disintegrating. (Cf. *rot *v.* 1 e.)

c 1665 P. E. Radisson *Voyages* (1885) 133 We cutt the ice wth hattchetts & we found places where [it] was rotten, so we hazarded ourselves often to sinke downe to our necks. 1746 T. Walker *Diary* 20 Mar. (1889) 9 Went over yᵉ River upon yᵉ ice. It grew very rotten. 1795 E. P. Simcoe *Diary* 7 Feb. (1911) 266 At Jacques Cartier the ice was so rotten I was obliged to go a league higher to cross the river with safety. 1849 J.-E. Alexander *L'Acadie* II. ii. 31 Thence we proceeded to Montreal, which we reached after four days and three nights of most unpleasant travel, and even dangerous, on account of exposure and the rotten ice. 1916 N. Duncan *Billy Topsail* xvi. 120 [The ice] had yielded somewhat—it must have gone rotten—in the weather of that day. 1935 *Monthly Weather Rev.* (Washington) LXII. 133/1 The boatman, fisherman, and lots of others ..swear that at this season [*sc.* spring] surface ice becomes rotten, or honeycombed, and sinks. 1966 T. Armstrong et al. *Illustr. Gloss. Snow & Ice* Fig. 7 (*caption*) Rotten ice. The puddles on the surface have mostly joined together and in places have melted right through the ice.

II. 8. b. (Earlier and later examples.) Also in weakened sense in *rotten luck, shame*, etc. Also as intensifying word and as quasi-*adv.*

1880 'Mark Twain' *Tramp Abroad* xxiii. 226 I'm most rotten certain 'bout that. 1911 G. B. Shaw *Blanco Posnet* 405 You that always talk as if He never did anything without asking your rotten leave first. 1914 —— *Fanny's First Play* I. 177, I was copped in the Dock Road myself: rotten luck, wasn't it? 1922 Joyce *Ulysses* 748 It was rotten cold too that winter. 1930 [see *electric blanket* s.v. *electric *a.* 2 b]. 1943 [see *due *sb.* 4 c]. 1952 E. O'Neill *Moon for Misbegotten* II. 107 You rotten bastard! 1959 I. & P. Opie *Lore & Lang. Schoolch.* ix. 161 Juvenile repugnance continues to be expressed by the old standbys..rotten, rotten shame, rotten swiz. 1964 *Daily Mail* 14 Dec. 1/3 The other girls sent me up rotten when they heard about my date. 1976 M. Machlin *Pipeline* xli. 446, I would not say that they are your friendliest people. And sometimes they are downright rotten. 1977 *Listener* 12 May 626/2 Mr Wood is not only brave enough to send himself up rotten, but also to make a hilarious series out of the whole literary game. 1980 *Jewish Chron.* 26 Dec. (Lit. Suppl.) p. vii/2, I was the only girl among 50 reporters and of course I was spoilt rotten.

d. *to knock rotten*, to kill or stun. *Austral. slang.*

1919 W. H. Downing *Digger Dial.* 31 *Knocked rotten*, killed or stunned. 1941 *Coast to Coast* 179 'He pulled it down on top of him,' continued Jo... 'It knocked him rotten.' 1945 Baker *Austral. Lang.* vi. 120 The development of an extensive vocabulary of fighting terms... *knock rotten*.

e. *Austral. slang.* Drunk. Also in phr. *to get rotten.*

1941 Baker *Dict. Austral. Slang* 61 *Rotten, to get*, to become exceedingly drunk. 1953 T. A. G. Hungerford *Riverslake* 135 Monday to-morrow—blasted work again. God, could I get rotten! 1971 J. Famechon *Fammo* 145 A reporter from one of the Sydney papers—he was the last to leave, rotten.

10. a. *rotten-boned, -chested, -fleshed, -livered* adjs.

1912 D. H. Lawrence *Let.* 3 July (1962) I. 134 My cursed, rotten-boned, pappy hearted countrymen, *why* was I sent to *them*. 1927 R. Graves *Poems* (*1914–26*) 210 As counterbalance in my mind To being rotten-boned and blind. 1969 L. Michaels *Going Places* 63, I..coughed again, a rasping, rotten-chested hack. 1908 Hardy *Dynasts* III. iv. vi. 417 We kings? Kings of the underground country, then, by this time, if we hadn't been too rotten-fleshed to follow the drum. 1929 R. Graves *Poems* 20 Lame, rotten-livered, this and that canaille.

b. *rotten-sweet*; also *rotten-woven*.

1947 M. Morris in B. James *Austral. Short Stories* (1963) 348 She stood over the bin inhaling the queer rotten-sweet smell of the blossoms. 1868 G. M. Hopkins *Jrnls. & Papers* (1959) 184 In the train I was noticing that strange rotten-woven cloud.

c. (Further example.)

1936 W. Greene *Death in Deep South* 69 You are rotten-egged out of a jerkwater town—rotten egged!—by a handful of hoodlums.

rottenly, *adv.* Delete *rare*⁻⁰ and add examples in sense 8 b of adj.

1905 G. B. Shaw *Let.* 29 Nov (1972) II. 584, I see..that the papers all say that..Major Barbara [is] a rottenly undramatic play. 1927 H. Walpole *Jeremy at Crale* iii. 43 'They weren't half pleased at your playing so rottenly.' 'I didn't play rottenly.' 1934 D. L. Sayers *Nine Tailors* II. 79 I'm afraid—if I go west this time—I'll be leaving you rottenly badly off, old girl.

Rotten Row. 2. (Further example.)

1975 *Listener* 30 Oct. 581/3 The majority of our line-of-battle ships had been rotting in reserve [by 1778]... Many of the battleships laid up in 'Rotten Row' were mere stacks of decayed timber.

rotting, *vbl. sb.* Add: **1.** Also *rotting-down* (in quot. *fig.*).

1916 Galsworthy *Sheaf* 269 Economically..such rotting-down of the boys is grievously short-sighted.

rotto (rǫ·to), *a.* *nonce-wd.* [f. Rott(en *a.* + *-o[2].] A jocular var. of Rotten *a.* 8 in Dict. and Suppl.

1922 Joyce *Ulysses* 23 The father is rotto with money. *Ibid.* 630 There was the case of O'Callaghan..among whose other gay doings when rotto and making himself a nuisance to everybody all round he was in the habit of ostentatiously sporting in public a suit of brown paper.

Rottweiler (rǫ·twəiləɹ, -vəiləɹ). Also **Rott-weiler.** [a. Ger., f. *Rottweil*, the name of a town in Württemberg, West Germany + *-er* -er[1].] A large black-and-tan dog belonging to the breed so called, having a short, coarse coat, docked tail, and a broad head with pendent ears. Also *attrib.*

1907 R. Leighton *New Bk. Dog* 521/2 The Rottweil Dog, usually called the *Rottweiler Metzgerhund*, or butcher's dog of the town of Rottweil in South Germany. 1917 *Policeman's Monthly* Jan. 5/3 Nowadays four breeds of dogs are being used for police purposes: the Continental Sheepdog,..the Airedale Terrier, the Doberman Pinscher and the Rottweilers. 1939 Kinney & Honeycutt *How to raise Dog* ii. 73 Very large (Breeds such as Great Danes, Newfoundlands,..mastiffs, Rottweilers, Italian bulls, and Pyrenean mountain dogs). 1948 C. L. B. Hubbard *Dogs in Brit.* xvi. 163 Although a comparatively new arrival to Britain the Rottweiler is a very well-known dog on the Continent. 1962 *Times* 24 Aug. 1/3 (Advt.), Very special homes required by 3 Rottweiler puppies. 1963 *Guardian* 5 Jan. 5/3 Among the dogs successfully trained for police work are the Rottweiler..and the Bouvier. 1971 'L. Egan' *Malicious Mischief* (1972) viii. 127 Katharine thought of Labrador retrievers,..Newfoundlands, and Rottweilers. 1978 *Daily Tel.* 11 Apr. 3/6 Dulwich College staff are alleged to have been attacked by Rottweilers—German hunting dogs—belonging to a neighbour.

rotunda. Add: **2. a.** (Earlier examples.)

1780 A. Young *Tour in Ireland* I. 2 In the evening to the Rotunda, a circular room, 90 feet diameter, an imitation of Ranelagh, provided with a band of music. 1808 M. Wilmot *Russ. Jrnls.* (1934) III. 316 From the Gallery one enters a Green House..in the Center of which is a rotunda for Entertaining Company.

b. *N. Amer.* The main hall of a public building; a lobby, a concourse.

1905 *Eye Opener* (Calgary) 28 Jan. 3/1 The hotels have no drinking-water tanks in their rotundas for the use of guests or local patrons—for obvious reasons. 1912 J. Sandlands *Western Canad. Dict.*, *Rotunda*, the hall or main entrance and waiting-room of a railway depot or an hotel. 1924 J. F. Dorrance *Never Fire First* xiii. 134 The scene in the rotunda of Montreal's impressive Windsor Station was as lively as it was metropolitan. 1958 *Edmonton Jrnl.* 19 June 33/3 Place and Date of Sale [is] Rotunda, Fifth Floor, Natural Resources Building, Edmonton, Alberta. 1973 H. Kemelman *Tuesday Rabbi saw Red* vi. 41 Inside the enclosed area of the Marble, the marble-tiled rotunda, students were swarming about.

2*. *Typogr.* A type-face of gothic inspiration used in some early printed books, based on a rounded script developed in the 13th century and popularized by the Bolognese law-school; also, the manuscript hand on which this type-face was based.

1929 A. F. Johnson in *Library* IX. 364 This is the rounded gothic of the Italians... In typography the Germans call it Rotunda. *Ibid.*, Jenson's Rotundas had a much wider vogue than his roman. 1954 R. Stokes *Esdaile's Student's Man. Bibliogr.* (ed. 3) iv. 141 The third class of *rotunda* types..is largely an Italian style of gothic and shows a much more open quality. 1969 H. Carter *View of Early Typogr.* iii. 50 This Italian rotunda was a somewhat cramped letter with short ascending and descending strokes... It was ideally suited to printing and printing was ideally suited to it. 1976 *Times Lit. Suppl.* 22 Oct. 1328/3 Caxton's mainstay bâtardes 2, 4, and 6 are surely best understood as belonging to the great calligraphic types of the earliest period, in fourth place after the Gutenbergian or Mainz texturas and rotundas and the Venetian romans. 1978 *Jrnl. R. Soc. Arts* CXXVI. 378/1 The widely practised 'Rotunda' hand of the Marmion, Soane, Serristori and other manuscript Hours.

‖ **Rotwelsch** (rō·tvelʃ). Also † **Rothwelsch.** [Ger., f. MHG. *rot* beggar or *rôt* red + *welsch* Welsh.] A form of slang or cant used by vagrants and criminals in Germany and Austria.

1841 Borrow *Zincali* II. III. ii. 130 The name of this [robber] jargon varies... In Germany [it is called] 'Rothwelsch' or red Italian. 1892 [see Yiddish *a.*]. 1916 H. Bradley *Shakespeare's English* in *Shakespeare's England* II. xxx. 567 The Hebrew words that are so conspicuous in the contemporary *Rotwelsch* of Germany. 1934 Priebsch & Collinson *German Lang.* 260 The lowest type of speech is the thieves' and beggars' cant known in German as *Rotwelsch* or *Gaunersprache*. 1961 *John o' London's* 30 Nov. 610 The *argot* of the French underworld, the *Rotwelsch* of Germany, is paralleled by the *Cant* of English rogues and vagabonds. 1973 *Reader's Digest* Feb. 122/2 'Please, put away that firecracker,' said the Major, lapsing into *Rotwelsch*—the slang of Vienna's underworld.

rouble. Add: **1, 2.** The rouble is now available primarily in paper form.

‖ **roucoulement** (rūkūlmaṅ). *rare.* [Fr.] The soft cooing sound made by doves. Also *transf.*

1863 J. A. Symonds *Let.* 12 Aug. (1967) I. 413 This *roucoulement* (cooing) in the throat is different from the yogel & seems to be peculiar to Swiss singing. 1926 E. Sitwell *Elegy on Dead Fashion* 2 Roucoulement of doves and veilèd belles.

Rouen. Add: **a.** Also used to designate earthenware of a type made at Rouen (esp. in the sixteenth and seventeenth centuries), as *Rouen faience, plate, ware.*

1863 W. Chaffers *Marks Pott. & Porc.* 94 From this period [*sc.* the early sixteenth century] until the middle of the XVIIth Century no notices of the Rouen Fayence have been discovered. 1869 C. Schreiber *Jrnl.* 11 Oct. (1911) I. 51 In all these shops there was a profusion of Rouen ware, or what pretended to be such, which proves how common it is. 1971 J. R. Bernasconi *Collectors' Gloss.* 191 *Rouen ware*, an enamelled faience ware produced at Rouen in the 17th century. 1974 N. Freeling *Dressing of Diamond* 26 His valuable pieces of Nevers and Rouen faience. 1977 *Western Mail* (Cardiff) 5 Mar. 14/1 Elwyn D. Thomas & Co., F.S.V.A. will Sell by Auction.. 'Dresden', 'Mason', 'Delft', 'Rouen' plates and plaques.

rouf (rōᵘf). Also **roaf, rofe, roof.** Backslang, esp. among costermongers and criminals, for 'four'; *spec.* four shillings, four pounds; a four-year prison sentence.

1851 H. Mayhew *London Labour* I. 23/1 *Rouf-yenep*, Fourpence. 1882 *Sydney Slang Dict.* 11/2 *Roaf Yanneps*, four pence. 1950 P. Tempest *Lag's Lexicon* 212 All [prison] sentences are referred to in slang... 4 years, a 'lagging' or a 'rofe' (pron. 'roaf'). 1957 *Evening News* 12 Nov. 6/6 Newcomers [to Cockney slang] are a 'rouf' (4s), 'a deuce' (£2), and 'anarf' (10s). 1958 F. Norman *Bang to Rights* III. 138, I tried to tell them that it had been a business deal, but you know what it's like talking to a moronic coszer, so that was it I got a rouf. 1972 K. Royce *Miniatures Frame* v. 64 From under a pottery sugar jar..protruded two jacks... I found a roof under them.

rouge, *a.* and *sb.*[1] Add: **B.** *sb.*[1] **1. a.** (Further *fig.* example.)

1882 'F. Anstey' *Vice Versâ* xvii. 312, I saw through his rouge with half an eye.

c. *rouge compact.*

1931 F. L. Allen *Only Yesterday* v. 107 For every adult woman in the country there were being sold annually over a pound of face powder and no less than eight rouge compacts.

2. a. Also (usu. with qualifying adj.) applied to polishing powders other than ferric oxide (see quot. 1937). (Further examples.)

1937 *Industr. Minerals & Rocks* (Amer. Inst. Mining & Metall. Engineers) i. 55 Briefly, they [*sc.* metallic oxide buffing materials] consist of various iron oxides such as crocus (red-brown), rouge (red), black rouge (magnetic iron oxide) mainly for glass, green rouge (chromium oxide) mainly for platinum and stainless steels; satin rouge (lampblack) for celluloid and bone; [etc.]. 1962 R. Webster *Gems* I. xx. 366 The polishing powder..may be either rouge (iron oxide), green rouge (chromium oxide), putty powder..or rotten-stone.

b. *rouge flambé* (rūʒ flaṅbe), a brilliant red glaze for porcelain, orig. Chinese, made from copper oxide.

1902 *Encycl. Brit.* XXXI. 875/2 Even the long-sought secret of the Chinese *sang de boeuf* and *rouge flambé* glazes has been worked out in Europe. 1912 A. Bennett *Matador of Five Towns* 4 A few specimens of modern *rouge flambé* ware made at Knype. 1960 R. G. Haggar *Conc. Encycl. Continental Pott. & Porc.* 124/1 He succeeded in producing a fine *rouge flambé* and an admirable turquoise blue glaze. 1967 M. Chandler *Ceramics in Mod. World* iii. 95 Copper oxide..can—under suitable conditions in a reducing atmosphere—produce the extremely brilliant red of rouge flambé. *a* 1977 *Harrison Mayer Ltd. Catal.* 14/2 *Rouge flambe, sang de boeuf*, a red glaze originating in China, its rich colour being due to a copper glaze fired under reducing conditions.

c. *rouge de fer* (rūʒ də fĕr), an orange-red enamel colour made from a base of ferric oxide and used on Chinese porcelain.

1922 *Daily Tel.* 12 June 20/5 (Advt.), A pair of famille verte baluster vases enamelled in birds and flowering prunus, in green, rouge de fer, yellow, and aubergine, 17¼ in. high. 1939 *Burlington Mag.* Apr. p.xv/1 A pair of famille-verte jars and covers,..with *rouge-de-fer* borders. 1959 *Times* 3 Mar. 7/2 A pair of eighteenth century models of dogs enamelled in *rouge de fer* realized 170 guineas. 1980 *Catal. Fine Chinese Ceramics* (Sotheby, Hong Kong) 166 All in tomato-red 'rouge-de-fer' and gilding.

4. a. (Further example.) Also *fig.*

1886 Hardy *Mayor Casterbr.* I. x. 123 The rich *rouge-et-noir* of his countenance underwent a slight change. 1920 H. Crane *Let.* 18 Aug. (1965) 41, I am sure there is more of a 'rouge et noir' cast to your surrender. 1958 L. Durrell *Balthazar* iii. 61 You are not the sort of man to stake everything on a single throw at *rouge et noir*.

5. a. (Earlier and later examples.)

1805 *New Pocket Hoyle* 117 Another parcel is then dealt for rouge in a similar manner. 1867 [see *couleur 2]. 1928, 1964 [see *noir 2 b].

b. The red numbers in the game of roulette.

[1850 *Bohn's Hand-bk. Games* 348 The other chances are also designated on the green cloth,..on one side 'l'impair, la manque, et le rouge'.] 1923, etc. [see *manque]. 1928, etc. [see *noir 2 a].

6. French red wine; = Red *sb.*[1] 3 b. Also in *Comb.*

1957 L. Durrell *Spirit of Place* (1969) 143 You should see..the care with which they select a good bottle of champagne..or even an ordinary rouge at a shilling. 1976 N. Roberts *Face of France* xv. 153 The accompanying glass of wine..is only *rouge supérieure*.

rouge, *sb.*[2] Add: **2. b.** (Further examples.) (See also quot. 1954.)

1954 *Sun* (Baltimore) 11 Dec. 11/7 The 'rouge' is a point scored when a kick into the end zone isn't run out or when the kick goes clear through the 25-yard-wide zone. 1959 *Times* 30 Nov. (Canada Suppl.) p.xx/1 In Canada we also have the rouge, which is a kicked single point. 1966 *Weekend Mag.* (Montreal) 27 Aug. 20/2 If only our rouge were added to the American game, football..wouldn't leave a thing to be desired as a game. 1976 *Webster's Sports Dict.* 357/1 *Rouge*, Canadian football, a score of one point awarded to the kicking team when a member of the receiving team fails to run a kickoff or a punt out of his own end zone.

rouge, *v.*[1] **2. b.** (Later example.)

1954 H. Gold in *New World Writing* VI. 13 You should have seen me rouge all over.

rougeless (rū·ʒlĕs), *a.* *rare.* [f. Rouge *sb.*[1] + -less.] Lacking rouge (in quot. *fig.*).

1857 [see *intime].

‖ **rouget** (rūʒe). [Fr.] = *red mullet* s.v. Red *a.* 17 c.

1885 A. Edwardes *Girton Girl* III. v. 83 He invited me to eat red mullet with him..Rougets en papillottes, accompanied by fine old graves. 1960 E. David *French Provincial Cooking* 286 *Rouget* proper..is red mullet. 1967 G. Greene *May we borrow your Husband?* 59, I ordered a small *rouget* and a half bottle of Pouilly. 1975 *Harpers & Queen* June 110/3 We ate *rougets* next; such *rougets*, simply grilled. 1977 *Times* 28 Jan. 15/7 Loup, daurade and rouget are among the best local fish [in Nice].

Rouget cell (rūʒe). *Histology.* Also **Rouget's cell.** [tr. G. *Rougetsche zelle* (B. J. Vimtrup 1922, in *Zeitschr. f. Anat. u. Entwicklungsges.* LXV. 178), f. the name of C. M. B. *Rouget* (1824–1904), French physiologist, who described such cells in 1873 (*Arch. de Physiol.* V. 603).] = *pericyte.

1922 A. Krogh *Anat. & Physiol. Capillaries* iii. 54 As there can be no doubt that the richly ramified muscle cells on the capillary wall are the same as those originally found by Rouget in the hyaloid membrane, Vimtrup has named them after the first discoverer, and we shall speak of them henceforth as Rouget cells. 1928 [see *pericyte]. 1939 W. E. Le Gros Clark *Tissues of Body* vii. 158 In appearance, Rouget cells (or pericytes, as they have been called), are quite similar to connective-tissue cells. *Ibid.* 159 Vimtrup..reported that, in Amphibia, local contraction of capillaries always started at the site of one of the Rouget cells. 1961 G. Bevelander *Essent. Histol.* (ed. 4) x. 106 According to some authors the tubule is clasped at intervals by Rouget's cells. These are branching cells which are said to be contractile and to cause the constriction of the capillaries. 1970 T. S. & C. R. Leeson *Histol.* (ed. 2) xi. 217/1 Among the pericapillary elements, peculiar cells (Rouget cells) with long branching processes which surround the capillary wall have been described. Early studies indicated that these cells were contractile and were responsible for the contractility of capillaries. More recent work suggests that true capillaries in mammals do not possess Rouget cells and that capillary contractility is independent of them.

rough, *sb.*[1] Add: **I. 2. c.** (Further examples.)

1955 [see *bunker *sb.*[1] 4]. 1971 'D. Halliday' *Dolly & Doctor Bird* iii. 29, I played well that morning, and the two balls I shot into the rough I recovered. 1977 *Cork Examiner* 6 June 7/2 At the 13th, Higgins was in the rough off the tee. 1980 *Guardian* 10 June 25/3 A spectator found another in the left rough and Cisco found the other in the right rough.

II. 6. c. The heavier, rougher part of housework; freq. in phr. *to do the rough*.

1946 M. Dickens *Happy Prisoner* vii. 114 Cosy discussions on clothes and curtains and women to do the rough. 1950 J. Cannan *Murder Included* iii. 39 He..suggested having a woman for the rough. 1959 *Times* 21 Nov. 1/3 No cooking or rough. 1974 'A. Gilbert' *Nice Little Killing* v. 70 The woman who came to do the rough twice a week.

d. Sharp, acid, or harsh drink; *spec.* (*a*) *slang*, draught bitter beer; (*b*) rough cider.

1946 J. Irving *Royal Navalese* 147 'Rough', draught bitter beer. 1960 'R. East' *Kingston Black* xiii. 129 He was selling the rough at three shillings a gallon.

III. 9. c. Applied to alum used as an adulterant in bread.

1855 *2nd Rep. Comm. on Adulteration of Food* 47 in *Parl. Papers* 1854–5 VIII. 373 There are several trade names for alum; one of them, being very characteristic of its effects on the mucous surface, is 'roughs', and another is 'seasoning'. 1909 *Practitioner* Feb. 263 All the samples of bread contained alum, and an instance was mentioned of flour, with which as much as ten per cent. had been mixed... In the trade, the adulterant received the name of 'roughs'.

10. a. Delete *rare*⁻¹ and add further examples. Also, a rough sketch, layout, etc.

1710 in *Publ. Colonial Soc. Mass.* (1925) XV. 395 A rough of sundry Articles wᵉ drawn up. 1796 J. Steele *Papers* (1924) I. 144 A rough of a letter which may at some future period compose part of a circular. 1936 *Punch* 12 Aug. 170/2, I don't suggest for a moment that these are *finished* ideas. They are no more than artists' roughs. 1961 Webster, [4]Rough,..4d rough proof. 1970 R. K. Kent *Lang. Journalism* 114 *Rough*, a preliminary layout or drawing, without details. 1975 J. Butcher *Copy-Editing* iv. 48 Alterations to artwork are caused as often by authors' inadequate or incorrect roughs..as they are by draughtsmen's mistakes. 1976 *Vogue* 15 Mar. 24/1 The roughs of my column are completed.

d. Uncut precious stone; an uncut gem, esp. a diamond.

1920 in Webster. 1961 in Webster s.v., A huge piece of rough was cut to a superb gem of 128 carats. 1974 L. St. Clair *Emerald Trap* (1975) i. 6, I want lots of big roughs. Finsch, Top Wesselton, maybe some Jaeger. 1976 W. Greatorex *Crossover* 162 No thefts of rough have been reported, so I suppose they're clean?

rough, *a.* Add: **I. 1. b.** Also, having a long nap.

1848 J. Ruskin *Let.* 29 June in M. Lutyens *Ruskins & Grays* (1972) xiii. 123, I beg very *particular* thanks for the *Rough* towels. 1939 *Army & Navy Stores Catal.* 623/1 'Christy' bath towels rough brown linen pile... 'Christy' bath towels in mixed linen and cotton..a semi-rough towel for hard wear.

d. Applied to the surface of a tennis- or squash-racket on which the loops formed by the string(s) looped around others project; freq. in context of spinning a racket to decide the choice of service or ends. Opp. *smooth *a.* 1 d.

1890 [see *smooth *a.* 1 d]. 1911 C. H. B. Quennell in L. Weaver *House & its Equipment* 204 It spoils the game if, as a result of guessing 'rough' or 'smooth' [etc.]. 1973 M. Russell *Double Hit* xxv. 186 Nevil spun his racket. 'Smooth,' said Colleano. 'Rough. I'll serve.'

e. *Bacteriol.* Applied to a bacterial phenotype characterized by corrugated and irregular colonies, and by cells lacking polysaccharide capsules.

[1920: see *R. II. 2 a.] 1921 J. A. Arkwright in *Jrnl. Path. & Bacteriol.* XXIV. 38 The irregularity of the surface has led to this variant being called the Rough ('R') form in distinction from the Smooth ('S') form. 1949 L. R. Thompson *Introd. Microorganisms* viii. 106 Rough (R-type) colonies are characterized by a dull appearance, and a folded or uneven surface. 1974 Q. N. Myrvik et al. *Fund. Med. Bacteriol. & Mycol.* ii. 25 When freshly isolated gram-negative pathogens are cultivated in the laboratory, they often undergo a smooth to rough (S→R) colony mutation.

II. 5. c. (Earlier example.) Also, unfortunate, unreasonable, unfair.

c 1856 W. Whitman *Daybks. & Notebks.* (1978) III. 670 That's rough. 1867 'Mark Twain' *Let.* 5 Dec. in C. Clemens *Mark Twain* (1932) 16 Another devilish thing is that the Alta [California] copyrighted the letters—that was rough. 1889 A. Lang *Letters on Lit.* 183 As we had also lots of..boomerangs..the poultry used to have rather a rough time of it. 1941 Baker *Dict. Austral. Slang* 61 *A bit rough*, unreasonable, unfair. 1942 *Yank* 23 Sept. 14 At best the going's very rough. 1944 *Yank* 4 Aug. 5 'We were 66 days on the beach at Anzio,' said Egan. 'It was rough.'

10. Also in wider use. (Further examples.) Also, miserable, dejected, in a bad way.

1883 'Mark Twain' *Life on Mississippi* lii. 513, I spent my last 10 cts for..cheese & i felt pretty rough. 1961 M. Dickens *Heart of London* II. 204 He looks rough. Someone ought to do something. Take im to ospital. 1971 C. Bonington *Annapurna South Face* xvi. 196 'I'll never make it to Camp VI,' said Nick. 'I feel dead rough.' 1972 *Times* 22 June 4/1, I felt really rough..before I was admitted to hospital.

III. 11. c. Of the sound of an internal-combustion engine: irregular, excessively noisy.

In quot. 1945 with a pun on sense 10.

1930 *Engineering* 24 Oct. 534/3 A state of affairs which would cause the engine to be 'rough' in its running. 1945 C. H. Ward-Jackson *Piece of Cake* (ed. 2) 53 When an engine sounds rough it is not well.

13. b. Of language, expression, etc.: coarse, vulgar, indelicate.

1958 *Spectator* 1 Aug. 176/2 It badly needs its rough jokes. 1961 in Webster s.v., A rough anecdote for such an audience. 1976 *Honolulu Star-Bull.* 21 Dec. E-1/4 You learn to live with the rough language so it doesn't bother you.

14. b. *rough and tough* (earlier example, used as a nickname).

c 1813 I. Pocock in M. R. Booth *Eng. Plays of 19th Cent.* (1969) I. 67, I suppose old rough-and-tough, master Grindoff, will be here presently.

d. In slang phrases *rough as bags, guts*, etc., uncouth, coarse. Chiefly *Austral.* and *N.Z.*

1919 W. H. Downing *Digger Dial.* 42 Rough as bags. 1925 Fraser & Gibbons *Soldier & Sailor Words* 246 *Rough as a sandbag*,..a term for a person who behaves unpleasantly. Uncouth. Objectionable. 1929 K. S. Prichard *Coonardoo* ii. 22 Ted was as rough as bags..a good-looking, good-natured bloke who could neither read nor write. 1938 E. Lowe *Salute to Freedom* 318 Rough as bags. Cleared his throat..and spat, just missing a pile of ribbons. 1941 Baker *Dict. Austral. Slang* 61 *Rough as a sandbag*, as for next. *Rough as bags, rough as a bag*, unpolished, crude, coarse. Esp. applied to persons. 'Rough as a pig's breakfast' is an equivalent. 1941 —— *N.Z. Slang* 53 [20th cent. N.Z. slang includes] rough as a bag (the Australians also have rough as bags), and rough as a pig's breakfast. 1946 E. G. Webber *Johnny Enzed in Middle East* 23 Smarten 'em up... Rough as bags. 1948 P. White *Aunt's Story* 34 Tom Wilcocks was as rough as bags. His neck was red and strong. The pollard had caked hard on his hard hands. 1966 G. W. Turner *Eng. Lang. Austral. & N.Z.* vi. 115 There is simile:..'rough as bags' (which I know better in the variant 'rough as sacks'). 1966 B. Beaver *You can't come Back* 118 I'm shy all right, but I'm not smooth... I'm rough as guts. 1968 F. Hardy *Unlucky Australians* 11 The old Territorian is a good bloke, rough as guts but his heart's in the right place. 1970 *Guardian* 25 July 6/1 Behan was most obviously gross and cussed and tragic and rough as ould bags. 1977 C. McCullough *Thorn Birds* x. 235 Even Dot MacPherson, the Bingelly heiress, ..was rough as bags, no posh Sydney boarding school and all that crap.

IV. 17. d. Of stationery, etc.: for use in writing rough notes or exercises; in which preliminary records are written.

1867, etc. [see *rough book*, sense 21, in Dict. and Suppl.]. 1884, etc. [see *rough log(-book)*, sense *21]. 1928 E. Scott *War among Ladies* I. iv. 44 Blotting-paper, foolscap, 'rough' paper..were laid out. 1960 *Sc. Nat. Dict.* V. 343/2 *Jot-book*, a rough note-book. *Ibid.*, A pupil's rough exercise book. 1977 P. D. James *Death of Expert Witness* II. 101 His rough notebook?... Anything of importance was noted in that book, and subsequently transferred to the files.

e. Applied to a vacuum of the lowest degree of evacuation.

1927 G. W. C. Kaye *High Vacua* vi. 74 For industrial purposes, such as exhausting rough vacuum mains, furnaces, or ovens, the so-called 'dry air pump' of the engineer is normally employed. 1949 S. Dushman *Vacuum Technique* iii. 141 With a 'rough' vacuum of about 10 mm mercury, such a pump could reduce the pressure to about 1 micron. 1969 *Gloss. Terms Vacuum Technol.* (*B.S.I.*) I. 7 *Rough vacuum*, 10^5 N/m^2 to 10^2 N/m^2. 760 torr to 1 torr. [*Note*] Not intended to be precise definitions, but to provide convenient and practical subdivisions of the vacuum range. 1976 *Physics Bull.* Apr. 161/1 Medium vacuum is used extensively for freeze drying and rough vacuum is also used for specimen handling and sample transfer.

V. 21. a. **rough book,** (*a*) also = *rough log(-book)*; (*b*) a book in which rough notes are written, a jotter; **rough calf** (see quot. 1952); **rough cut** *Cinematogr.*, the first edited version of a film, the state of a film after preliminary editing; also *attrib.*; **rough grazing,** uncultivated land used for grazing; an area of such land; **rough log(-book)** *Naut.*, a book in which the particulars of a ship's voyage are first entered, to be written up later in the main log-book; **rough mix,** a preliminary blend of separately recorded parts of a piece of music; **rough pâté,** pâté made with coarsely-chopped or -minced meat; **rough scruff** *U.S.* = *rough-scuff*; **rough spin** *Austral. slang*, a misfortune; **rough stuff,** (*c*) unruliness, violent behaviour; **rough-tonguing,** rude speech; verbal abuse, disparaging; a scolding; **rough trade** *slang*, a tough or sadistic element among male homosexuals, esp. prostitutes; the activities of homosexual prostitutes; (see also quots. 1935, 1973).

1902 Conrad *Typhoon* v. 47 He copied neatly out of the rough-book the number of miles, the course of the ship. 1969 A. Laski *Dominant Fifth* ii. 43 She had been drawing on her rough book. 1912 Monk & Lawrence *Text Bk. Stationery Binding* 85 Rough calf or its substitutes require the surface well cleaned before tooling. 1952 A. W. Lewis *Basic Bookbinding* ii. 17 *Rough calf*, calf skin finished on the flesh side and used on books with the flesh side outermost. 1963 B. C. Middleton *Hist Eng. Craft Bookbinding Technique* 286 Rough calf was much used in the seventeenth and eighteenth centuries. 1975 *Sotheby & Co. (Hodgson's Rooms) Catal.* 31 July–1 Aug. 45 Abelard (Peter) and Heloise. Opera,..a few leaves slightly soiled, eighteenth century rough calf, slightly rubbed.

1939 N.Y. Times 2 Apr. x. 4/4 The only demands we have made on the producers as a Guild were to have two weeks' preparation time for 'A' pictures, one week preparation time for 'B' pictures and to have supervision of just the first rough cut of the picture. 1952 L. Ross Picture (1953) iii. 108 Actually, every director should make the rough cut—the film as assembled from start to finish for the first time—himself. 1957 Manvell & Huntley Film Music iii. 59 Functional music may be composed after the film has been shot and assembled in rough-cut. 1970 Daily Tel. 23 Sept. 12/3 There was no censorship apart from the cutting of a single frame at the request of an East German Government representative who saw the pictures at rough-cut stage. 1978 P. J. Kavanagh People & Weather vi. p. ciii, He returned to the studio with his film, triumphant... But when he put together the rough-cut he was appalled. 1932 Jrnl. Min. Agric. XXXIX. 37 White clover can be successfully established on certain types of rough grazings without mechanical cultivation. 1953 E. Smith Guide Eng. Traditions 1 'Rough grazing', wild open land over which various owners of livestock have grazing rights. 1966 I. Moore Grass & Grasslands iv. 30 The transition from ley to permanent pasture or meadows, thence to rough grazing and scrub, and finally to forest, is an orderly, gradual process. 1970 Sruth (Inverness) 16 Apr. 3/1 The reconditioning of regenerated areas of heath land and rough grazings. 1884 Naval Encycl. 701/2 Rough log, the book in which the journal of the ship is originally written. A smooth copy, signed by the watch-officers, is inspected by the commanding officer, and forwarded to the Navy Department. 1917 D. Wilson-Barker Man. Elementary Seamanship (ed. 7) vii. 225 Every officer keeps an account of the work..during his watch. This record he enters on a log slate, scrap, deck, or rough log, as it may be called. 1922 F. Riesenberg Stand. Seamanship for Merchant Service xviii. 761 The smooth log is a copy of the rough log. The latter is the original and valuable record. 1948 R. de Kerchove Internat. Maritime Dict. 667/2 Ship's log book, a nautical record compiled from entries taken from the rough log book. 1961 F. H. Burgess Dict. Sailing 68 Deck log, a ship's rough log book, in which is recorded all information about working the ship, and other events as they occur. Ibid. 174 Rough log, the deck log. 1962 G. Danton Theory & Pract. Seamanship xiii. 288 The Chief Officer's logbook.. is virtually a diary of the ship's activities. The information contained therein is derived from the rough logbook, which is kept by the individual Officers-of-the-watch. 1970 D. M. Henderson Seamanship xxvi. 464 The rough, original or chart room log-book is written up by the Officer of the Watch or Officer of the Deck. 1977 Rolling Stone 24 Mar., I'd sneak back in and listen to the rough mix. 1977 Zigzag Apr. 6/2 The way Stevie likes to work is to record something, then take a rough mix down to the country where he's got a little demo studio, and work out what he's going to put down within what's already there. 1961 G. Smith Business of Loving xi. 222, I settled on a rough paté, some pheasant with game chips. 1974 Times 4 Nov. 14/8 We chose rough country pâté and Vichyssoise to start with. 1977 P. Harcourt At High Risk i. 19 We had a rough pâté de la campagne. 1831 Boston Even. Transcript 1 Oct. 1/2 The roughscruf of St Louis called my deliverer a Watchenago. 1865 'Mark Twain' in Californian 18 Mar. 8/2 The ruff-scruff and rag-tag-and-bob-tail of noble old Calaveras. 1924 Truth (Sydney) 27 Apr. 6 Rough spin, bad luck. 1940 F. D. Davison Woman at Mill 150, I had a rough spin. 1913 J. London Valley of Moon i. iv. 32 There's goin' to be rough stuff down there in a minute. 1915 H. L. Wilson Ruggles of Red Gap ii. 30 But you'll have to be firm, because he's full of tricks. And if he starts any rough stuff, just come to me. 1919 W. H. Downing Digger Dial. 42 Rough stuff, an undisciplined, reckless, indecent, disorderly or disrespectful person or thing. 1925 Chesterton Everlasting Man i. i. 24 His chief occupation..was.. treating women in general with what is, I believe, known in the world of the film as 'rough stuff'. 1940 Wodehouse Eggs, Beans & Crumpets 238 Your aunt..has a right to early information about any rough stuff that is being pulled on the premises. 1959 'M. M. Kaye' House of Shade xx. 275 I'd have got that pro-Red nancy-boy before he started any rough stuff. 1978 Lancashire Life Apr. 73/2 The presence of a girl in a group of tipsy young men keeps them in check, however: the laughs are there but the rough stuff isn't. 1916 'Boyd Cable' Action Front 98 An' I admit I felt easier after that rough-tonguin'... That slobberin' an' kissin' business..may be all right for a lot o' bloomin' Frenchies. 1919 J. Buchan Mr. Standfast xii. 217 He would relish the rough-tonguing of non-coms. 1956 N. Marsh Off with his Head (1957) ix. 191 Maids up to castle heard his great-auntie giving him a terrible rough-tonguing. 1935 A. J. Pollock Underworld Speaks 99/1 Rough trade, a person picked up on the street by a sexual pervert. 1965 Playboy Aug. 124/2 The gay boys call us 'rough trade'! We're the ones they date.... We're the ones they buy presents for. 1967 Evening Standard 11 July 10/3 The gradual destruction..of.. Boyde Ashlar..as he gets involved with what I believe is called the Rough Trade. 1973 Amer. Speech 1970 XLV. 58 Rough trade n, pick-up from one of the occupations typical of tough men, such as truck drivers or dock workers. 1976 M. Machlin Pipeline xxxviii. 412 There were no gay bars or hangouts, and very few gays dared walk the streets in the more extravagant, deviant-type-wardrobes. Any gay activity in Fairbanks was probably confined to rough trade. 1978 C. Beaton Parting Years vi. 159 He made friendships too easily with the 'rough trade'. 1980 Times Lit. Suppl. 7 Mar. 253/3 Auden was a homosexual who..seems to have had a greater craving for a settled relationship and being loved than for rough trade or other casual excitement.

b. **rough collie**, a long-coated black and white, or black, tan, and white collie; **Rough Fell**, a large long-woolled sheep of the breed so called, found in parts of the Pennine area; **rough greyhound** = Deer-hound.

1806 Rough collie [see Collie]. 1872 'Stonehenge' Dogs Brit. Islands (ed. 2) ii. viii. 175 The rough or shaggy-coated colley..has a fine foxlike muzzle. 1931 A. C. Smith About our Dogs xvii. 275 The Smooth Collie should be identical in all features with the Rough, except in coat. 1977 Grimsby Even. Tel. 5 May 3/5 (Advt.), Cairns, Westies, Rough Collies, Old English Sheepdogs. 1916 W. J. Malden Brit. Sheep & Shepherding vi. 58 The Rough Fell sheep of the moors and hills of North-west Yorkshire..and adjoining districts are clearly allied to the Scotch Black-face. 1945 J. F. H. Thomas Sheep ii. 30 The Rough Fell. Again a breed not numerically strong. 1960 Rough Fell [see *Exmoor]. 1888 Rough greyhound [see fleet-hound s.v. *fleet a.1 4]. 1948 C. L. B. Hubbard Dogs in Brit. xv. 122 The Deerhound, or Rough Greyhound as it was then called, was a prized possession of the Scottish chieftain.

22. *rough-water*, *-weather* (earlier and later examples), *-wood*.

1833 Chambers's Edin. Jrnl. 1 June 141/1 Those who are tough, keep the deck in their rough-weather cloaks. 1898 H. E. A. Coate Realities of Sea Life xiv. 124 All hands very busy in unbending rough-weather sails and bending fine-weather ones. 1921 W. de la Mare Crossings 67 A garden chair beside a roughwood table. 1967 Gloss. Terms Air-Cushion Vehicles (B.S.I.) 7 Rough water drag, the increment in the drag during operation in rough water over the drag, under otherwise identical conditions, in calm water. 1971 Flying Apr. 27/1 A deep V bottom provides lower impact loads on rough-water landings. 1978 J. A. Michener Chesapeake iv. 184 The entire group of Quakers went..to the rough-wood house of James Lamb.

23. a. *rough-barked* (earlier example), *-edged* (further examples), *-faced* (earlier example), *-grained* (further example), *-mouthed*, *-surfaced*.

1836 J. G. Whittier Mogg Megone 11 The gnarled trunk of the rough-barked oak. 1932 D. Gascoyne Roman Balcony 36 Glittering, rough-edged shadows on the dull lawn. 1970 Daily Tel. 23 Jan. 6 Rough-edged men who cannot complete a sentence without a four-letter word. 1978 J. Carroll Mortal Friends ii. vii. 216 He threw himself into the fray with a fierceness that was rough-edged and merciless even for him. 1812 E. Weeton Let. 25 May (1969) II. 15 A rough-faced fellow, a journeyman saddler. 1962 in E. E. Evans-Pritchard Ess. Social Anthropol. v. 115 He represented a sultan who excels the ordinary people in body and spirit, and one gained the impression that one was dealing with a rough-grained, able and cunning man. 1899 A. Bennett Jrnl. 29 Oct. (1932) I. 96 The actual coarse, ignorant, crude-thinking, rough-mouthed maiden of past times. 1926 Rough-surfaced [see *offset sb. 10 b]. 1962 Science Survey XI. 166 There are smooth-surfaced vesicles, vacuoles and tubules; flattened sacs whose limiting membranes are encrusted with particles and therefore 'rough'-surfaced.

b. (Further example.)

1901 Nature 19 Sept. 523/2 Seven rough-keeled snakes (Dasypeltis scabra).

c. *rough-stalked meadow-grass* (later examples).

1901 H. M. Ward Grasses iii. 42 Poa trivialis. (Rough-stalked Meadow-grass.) Conspicuous in deep rich pastures. 1960 Farmer & Stockbreeder 8 Mar. 117/1 Rough-stalked meadow grass, bent, and wild white clover together with a few so-called 'weeds'.

rough, *adv.* Add: **1. a.** (Examples corresponding to sense *11 c of the adj.)

1954 L. Klemantaski tr. Fraichard's Le Mans Story v. 52 'The engine is running rough!' he cried. 1978 J. Gardner Dancing Dodo xxxviii. 308 The port engine faltered... She had started to run rough.

b. (Further examples.) Also, *to live rough*.

1960 Guardian 7 Dec. 1/4 He had been sleeping rough with the others on a haystack. 1974 Whig-Standard (Kingston, Ontario) 11 Jan. 7/1 A 'dosser'... Sleeps on a bench, wrapped in a newspaper, living 'rough'. 1974 J. I. M. Stewart Gaudy ix. 172 We neither of us had a bean, you see, and I was just going to sleep rough. 1977 Jrnl. R. Soc. Arts CXXV. 148/2 There are going to be 2000 single people in London without homes sleeping rough this Christmas.

2. a. *rough-dig*, *-edit*, *-land*, *-school*, *-sketch*, *-sort*.

1909 Country Life 23 Oct. 577/1 One could see him rough-schooling younger brothers and companions. 1910 W. J. Locke Simon xxiv. 315 The story of his marriage is a little lunatic drama all to itself and I will tell it some day. But now I can only rough-sketch the facts. 1950 Partridge Here, There & Everywhere 166, I should like to rough-sketch the position occupied by him and Lewis Carroll in the chronology of the subject. 1960 Aeroplane XCIX. 541/2 Turning to lunar and interplanetary research, Mr. Stoller said that in 1962 three Ranger vehicles were planned to rough-land payloads on the surface of the Moon. These will be followed by the soft-landing mission. 1962 A. Nisbett Technique Sound Studio vii. 124 It will often be possible to rough edit without bothering to mark the tape. 1969 W. Rutherford Gallows Set iv. 55 This film has already been rough-edited... That means that the editor will have done all the obvious things, taken out false starts, put in the cutaway questions. 1972 H. Evans Newsman's Eng. i. 1 The international news..has been checked, rough-edited. 1976 Norwich Mercury 19 Nov. 11/3 It is advisable to rough-dig all uncropped land. 1978 Cahiers de Lexicologie XXXII. 31 Assembling and rough-sorting a citation collection.

b. *rough-bound*, *-built*, *-cut* (later example), *-dug*, *-hurled*, *-plucked*, *-scored*, *-split*, *-trimmed*.

1865 G. M. Hopkins Poems (1967) 20 Those crookèd rough-scored chequers may be pieced To crosses meant for Jesu's. 1870 D. G. Rossetti Let. 21 Apr. (1965) II. 851, I suppose the inscription at the back of the rough-bound copy sent is from the real block. 1882 W. D. Hay Brighter Britain! I. v. 120 Rough-split sections of the great logs. a 1892 J. G. Whittier in S. T. Pickard Life & Lett. Whittier (1894) I. i. 13 And lo! in the midst of a clearing stood The rough-built farmhouse, low and lone. 1909 Daily Chron. 18 Oct. 4/5 Fowls are sold both dead and rough-plucked, and alive for fattening. 1925 Blunden Eng. Poems 31 As wave-wise Rough-hurled they rose, With a sweet sureness. 1965 G. J. Williams Econ. Geol. N.Z. xiv. 221/1 During the second world war about 1½ tons of rough-trimmed mica were won from this area. 1967 E. Short Embroidery & Fabric Collage ii. 38 Woven silks were embellished with rich all-over embroidery which incorporated pearls and rough-cut gem stones. 1973 R. Adams Watership Down xx. 125 The Honeycomb was still rough-dug and half-finished. 1976 Cumberland News 3 Dec. 35/3 (Advt.), Oven ready and rough plucked birds.

d. *rough-editing*, *-landing*, *-schooling*.

1909 Spectator 30 Oct. 678/2 Sir Percy Fitzpatrick is certainly right in picking out..the 'rough-schooling of younger boys and companions'. 1959 Fortune July 157/1 A somewhat more difficult trip..will be the rough landing of a fifty-pound payload on the moon. 1962 A. Nisbett Technique Sound Studio vii. 117 Rough editing is assembling the main body of the programme in the right order and taking out the longer stretches of unwanted material.

rough, *v.1* Add: **I. 2. c.** (Earlier and later examples.) Also with inanimate object. See also sense *6 g.

1845 W. G. Simms Wigwam & Cabin 1st Ser. 58 She [sc. a bear] roughed me once or twice more with her paws. 1904 Baltimore American 1 Aug. 2 (heading) Badman roughs a train, but is shot in the hand by a plucky trainman. 1928 Daily Mail 25 July 12/4 Tunney knows he will be roughed and bustled around for the first few rounds. 1957 D. Niland Call me when Cross turns Over vi. 153 They grabbed Shelton and roughed him outside into the rising wind. 1971 Frendz 21 May 2/4 Cant put the rest down because the tin hats will rough my kin. 1978 N.Y. Times 29 Mar. B 6/5 The Mets..roughed Pete Falcone with a pair of runs apiece in the fourth, fifth and sixth innings.

II. 4. a. (Earlier example.)

c 1771 M. Suckling Let. in Southey Life of Nelson (1813) I. i. 5 What..has poor Horatio done,..that he.. should be sent to rough it out at sea?

5. c. *Austral.* and *N.Z.* To shear (a sheep) badly.

1878 'Ironbark' Southerly Busters 180, I allus roughs 'em when the boss Ain't on the shearin' floor. 1897 D. McK. Wright Station Ballads 37 But he wouldn't shear at Maimai, started in to rough them through. 1956 G. Bowen Wool Away! (ed. 2) 156 Rough 'em, the opposite to 'pink 'em', and meaning rough shearing and a bad job of the sheep.

III. 6. b. (Earlier and later examples.)

1770 C. Carroll Let. 25 Apr. in Maryland Hist. Mag. (1917) XII. 352, I think you are wrong to Have the Capitalls &c. finished there, they may be defaced in the Carriage, wh Danger would be avoided if only roughed out there. 1955 Times 4 Aug. 7/6 The first act has been already roughed out. a 1974 R. Crossman Diaries (1975) I. 370 By the early hours I had roughed out a reasonably intelligent script.

g. Const. *up*. To deal roughly with, assault, damage, upset, intimidate; = sense 2 c.

1942 Berrey & Van den Bark Amer. Thes. Slang §341/2 Treat roughly; 'manhandle'... rough (up), rough-house, strong-arm, treat 'em rough. 1943 R. Chandler Lady in Lake (1944) xxxvi. 192 You know how to rough up a bum that hasn't any money. 1959 'M. M. Kaye' House of Shade iii. 29 When am I supposed to have roughed up your room? 1963 Times 14 Jan. 3/1 They had roughed-up France's pack a year ago. 1970 M. Braithwaite Never sleep Three in Bed xi. 135 They began to rough us up and we kicked and pulled and yelled about what our dads would do if they didn't leave us alone. 1973 Time Out 2–8 Mar. 15/1 A lot of teachers got roughed up, but that's not to say beaten up. 1977 M. Goulder in J. Hick Myth of God Incarnate iii. 58 Not only must he be prepared to be roughed up by southern policemen; he must also risk assassination. 1978 N.Y. Times 30 Mar. D 19/4 But at that point, the Phillies dealt Bruhert a cruel blow. They roughed him up with six rapid-fire singles. 1978 J. A. Michener Chesapeake xiv. 864 Amos Turlock..led an expedition to Caveny's home, which had been roughed up but not destroyed.

h. *to rough down*, to give (wood) a rough, preliminary planing. Cf. Roughing *vbl. sb.* 2 b.

1960 McGraw-Hill Encycl. Sci. & Technol. XIV. 543/1 Flat or uniformly contoured surfaces of wood are roughed down, smoothed, or made level by the shaving and cutting action of a wide-edged blade or blades.

7. a. (Earlier and later examples.)

1770 C. Carroll Let. 20 Apr. in Maryland Hist. Mag. (1917) XII. 351 The stone cutters wish to have a draft of the Bases & Capitalls, they could rough the stones to that draft, & save a great deal of carriage. 1937 Times 13 Apr. (British Motor No.) p.xii/2 Machines of particular interest are the Gleason completing machine for producing differential pinions, roughed and finished at a speed of 65 seconds each, and the lapping machines.

d. To subject to a partial or preliminary evacuation. Also with *down*, *out*.

1948 [implied in roughing down s.v. *roughing vbl. sb. 2 b]. 1971 Physics Bull. July 423/2 This consists of a large ion pump and liquid N_2 cooled titanium sublimation pump combination, 'roughed' by two high capacity sorption pumps and an oil free mechanical roughing pump. 1976 A. Roth Vacuum Technol. v. 200 The removal of the atmospheric air from the system to some acceptable operating pressure is referred to as roughing out the system... Mechanical rotary pumps, and ejectors are the typical roughing and backing pumps.

roughage. For '*dial.* and *U.S.*' read 'orig. *dial.* and *U.S.*' and add: **1.** Also *N.Z.* (Further examples.)

1940 E. C. Studholme Te Waimate x. 80 The first work of the settlers was to burn off the roughage on large areas of country. 1950 N.Z. Jrnl. Agric. Feb. 122/3 On much of the country a fairly high proportion of cattle to sheep is carried, the cattle being used largely to clean up roughage left by the sheep.

2. The indigestible fibrous matter or cellulose in vegetable foodstuffs. Also *attrib.*

1927 Lancet 16 July 106/1, I suspect that the advocacy of this bread was begun on the ground that the extra cellulose which it contains, forming what has been termed 'roughage', is of advantage as a mechanical irritant to the mucous membrane of the colon. 1932 Times Lit. Suppl. 7 Jan. 1/3 French art has had in the past its extra-artistic responsibilities, its burdens and entanglements, leading to the presence in the work of art of what corresponds to 'roughage' in food. 1948 Good Housek. Cookery Bk. 13 Much constipation is due to insufficient quantities of roughage in the diet. 1963 Times 4 Feb. 4/7 Roughage diets. 1976 Which? Nov. 244/3 Wholemeal bread is a simple and effective source of roughage.

b. *fig.*

1931 Musical Times Jan. 74/1 This was another of the oddities that we have to accept as roughage to the Hallé [orchestra] fare. 1948 C. S. Lewis Lit. & Life 58 There is no roughage in a Kipling story—it is all unrelieved vitamins from the first word to the last. 1962 [see *funk sb.2 2]. 1963 Punch 23 Jan. 141/3 Twelve months in jail provided the roughage for his first book. 1974 Times 28 Feb. 15/3 The electorate..finally gagged on their traditional roughage of internecine strife.

rough-and-ready, *a.* Add: **2.** (Earlier examples, as a nickname.)

1843 in Amer. Speech (1965) XL. 132 But Rough and Ready made dem smell Gunpowder a la Poker. 1846 Congress. Globe 24 May 865 Col. [Zachary] Taylor..had won for himself by his gallant conduct in the field the soubriquet of 'Old Rough and Ready'.

Hence **rough-and-readiness.**

1956 Essays in Crit. VI. 185 Beliefs that allow Pound to be satisfied..with a surprising emotional rough-and-readiness. 1960 Guardian 30 Nov. 7/1 The rough-and-readiness of such Italian neo-realists as Rossellini.

rough-and-tumble, *a.*, *sb.*, and *adv.* Add: **A.** *adj.* **4.** Roughly constructed or improvised; makeshift. *rare.*

1912 Kipling Land & Sea Tales (1923) 70 They heaved up their rough-and-tumble anchor, and made after a.. sailing-ship.

C. *adv.* (Further U.S. example.)

1935 Z. N. Hurston Mules & Men (1970) i. viii. 178 Mr. Allen might have eaten by the rules but Cliffert and I went at it rough-and-tumble with no holds barred.

rough-and-tumbling. For *nonce-wd.* read *rare* and add further example.

1832 Chambers's Edin. Jrnl. I. 130/2 The scene can only be compared to a rough-and-tumbling in the back woods of America.

roughback (rʌ·fbæk). [f. Rough *a.* + Back *sb.1*] One of several flatfishes with rough skins, esp. the long rough dab, *Hippoglossoides platessoides*. Also *attrib.*

1795 J. Sinclair Stat. Acct. Scotland XVI. 548 Flounders of all kinds, roughback, plaise. 1815 J. Arbuthnot Hist. Acct. Peterhead 15 Dab, vulgarly called Rough Back Fluke, Rochie. 1903 G. Sim Vertebr. Fauna of Dee 244 Long Rough Dab. 'Rochie'..'Rough-back Fluke'..is not looked upon with much favour as an article of food. 1935 Fisheries Notice (Min. Agric.) No.23.4 Suggested Trade Name. Rough-back... General English Equivalent. Long Rough Dab. 1973 J. Grigson Fish Cookery 281 Roughback (i American plaice) (ii rock sole; flat-fish).

rough-cast, roughcast, *ppl. a.* and *sb.* Add: **I. 1. c.** Of glass: cast in a particular manner (see next).

1939 Archit. Rev. LXXXV. 99 (caption) Bent dome of 'rough-cast' glass used, at the Saint-Gobain pavilion, Paris 1937, as a simple and elegant alternative to the ordinary opaque basin. 1973 Technical Translation Bull. XIX. 103 Plate glass..is expensive to produce owing to the need for grinding and polishing of the 'rough cast plate' produced as a first stage... Indeterminate patterns such as 'rough cast' or 'cathedral' are normally called 'cast.'

2. d. A type of glass (see quot. 1962).

1962 Gloss. Terms Glass Industry (B.S.I.) 31 Rough cast, rolled translucent glass, one surface of which has a definite texture, made by rolling molten glass either on a table or between rollers. 1973 [see prec. sense].

rough-casting, *vbl. sb.* **1.** (Later example.)

1977 S. Wales Echo 18 Jan. (Advt.), Pebble Dashing, Spar Dashing, Rough-casting, all types of Plastering carried out.

rough-draft, *v.* (Further example.)

1975 I. K. Martin Regan & Manhattan File 110 Regan wrote his own report and rough-drafted Cassidy's report as a favour.

rough-dry, *v.* Add: (Further examples.) Now more generally, to dry roughly or imperfectly. Hence **rou·gh-dried** *ppl. a.*

1900 Kipling in Daily Express 12 June 4/5 Sweating men, rough-dried sweating horses with wisps of precious forage. 1942 Z. N. Hurston in A. Dundes Mother Wit (1973) 222/1 He was born with this rough-dried hair. 1952 'J. Tey' Singing Sands i. 8 Service..had lost its starch and its high glaze. It had become what housewives call rough-dried. 1978 P. Harcourt Agents of Influence xii. 146 She..continued to rough-dry her hair on the towel.

rough-dry, *a.* (Earlier examples.)

1856 Mrs. Stowe Dred I. 181 Clothes look rough-dry, as if they had been pulled out of a bag. 1865 M. Eyre Lady's Walks S. of France i. 8 The articles..are neither starched nor ironed, but simply sent home rough dry.

rough-grind, *v.* Add: So **rou·gh-ground** *ppl. a.*

1901 Kipling Kim xi. 289 A clothful of atta—grayish, rough-ground native flour,—twists of down-country tobacco.

rou·gh-hew, *sb. rare.* [f. the vb.] Something lacking the finishing process; a preliminary version.

1889 G. M. Hopkins Lett. to R. Bridges (1935) 301 It was only a sketch, a rough-hew of a song.

rough house, rough-house, *sb. slang* (orig. *U.S.*). [Rough *a.* 21.] An uproar, a disturbance, a row; horseplay, boisterous behaviour; a fight, a struggle.

1887 M. Roberts Western Avernus 54 He called the bridgeman a very opprobrious name, and for a moment there was great danger of a 'rough house' out of hand. 1895 [see Rough *a.* 21]. 1900 Dialect Notes II. 55 Rough-house, n. 1. A disorderly class. 2. Rough play. 1902 G. H. Lorimer Lett. Merchant xvi. 238 [He] said he liked things lively, but there was altogether too much rough house on Beacon Hill for him. 1906 Dialect Notes III. 154 Rough-house, n. Scuffle (in a room). 'The room looked like they had had a roughhouse.' Common slang. 1908 A. J. Dawson Finn xix. 292 Seems to me you've been havin' a pretty rough house with somebody. 1911 R. D. Saunders Col. Todhunter iii. 49 But an ominous cry rose from his front. 'Rush 'em, boys! Make a rough-house.' 1913 R. H. Barbour Around the End xxi. 258 The audience.. indulged in wild 'rough-house'. 1919 'Etienne' Strange Tales from Fleet 136, I need not enlarge on the subsequent entertainments, which consisted of a sing-song followed by a 'rough house', in which a certain amount of furniture was broken. 1923 H. L. Foster Beachcomber in Orient xi. 240, I..watched Singapore fade into the distance with its memories of vice, iniquity, and general rough-house. 1933 E. O'Neill Ah, Wilderness! i. 25 That's enough, now. No more roughhouse. You sit down here, Richard. 1941 Penguin New Writing II. 87 Someone being funny and turnin' off your juice and you keep striking at your electrode and wondering why the hell it doesn't strike and flash, and losing your temper, and then seeing the rest laughing and having a rough house for a while. 1952 [see *Boche]. 1973 'B. Graeme' Two & Two make Five xiii. 123 He's smaller and lighter than me; not nearly so useful in a rough house.

b. *attrib.* or as *adj.* (Also *roughhouse*.)

1898 F. P. Dunne Mr. Dooley in Peace & War 25 Other gin'rals iv th' r-rough-house kind, like Napoleon Bonypart, th' impror iv th' Frinch, Gin'ral Ulis S. Grant, an' Cousin George Dooley, hired coarse, rude men. 1901 Official Basket Ball Rules (Spalding's Athletic Library) 49 Mr. Naismith never invented the game for 'rough house' work. 1906 N.Y. Globe 22 Aug. 6 On that rough-house occasion more force than was necessary was used to eject the Bryanite faction. 1919 H. L. Wilson Ma Pettengill ii. 74 Two or three other directors..had put him into rough-house funny plays where he got thrown downstairs or had bricks fall on him. 1932 Amer. Speech VII. 241 Jazz is free, Jazz is roughouse. 1938 E. Bowen Death of Heart ii. iv. 240 Mr. Bursely was shoved against the bookcase by Wallace Parker shoving, that rude way. .. I didn't like him to see us so rough house. 1938 R. Graves Coll. Poems 121 Time and Space Do but amuse us with their rough-house turn. 1946 Casper (Wyoming) Tribune-Herald 29 Mar. 9/3 Rocky, with his striking black hair and roughhouse tactics in the ring, has become a gallery idol.

rou·gh-house, *v. slang* (orig. *U.S.*). Also **rough house, roughhouse.** [f. the sb.] **1.** *intr.* To make a disturbance or row; to behave or act boisterously or violently; to fight or engage in horse-play *with*. Also quasi-*trans.* with *it*.

1900 Dialect Notes II. 55 Rough-house, v.i. To put a room in disorder. 1904 R. L. McCardell Show Girl & her Friends 107 When they teased him about having to keep him out of the place if he did not stop rough-housing it, poor Dopey smiled..and asked if he had hurt anybody very bad. 1908 U. B. Sinclair Metropolis 57 She's always wanting to rough-house it. 1920 'Sapper' Bull-Dog Drummond ix. 239 Somebody has been rough-housing by the look of things. 1928 Chambers's Jrnl. Apr. 211/2 He had a veteran in combat to deal with, a man who had 'rough-housed' it all over the world. 1929 E. L. Rice Street Scene ii. 167 Rough-housing with your kid brother. 1971 Daily Colonist (Victoria, B.C.) 5 May 43/2 Police spokesmen said the boys were 'rough-housing' on the grass. 1977 Time 10 Jan. 13/1 Twice, a broken leg set him back—once when he was caught in an avalanche while skiing, later while roughhousing with friends.

2. *trans.* To handle (a person) violently; to assail roughly; to maltreat by rough usage.

1902 H. L. Wilson Spenders xxxvi. 436 You rough-housed the boy considerable yesterday. 1903 N.Y. Times 6 Oct. 1 After the rush the classes adjourned to Massachusetts Avenue and began to 'rough house' the passing street cars. 1925 H. L. Foster Trop. Tramp with Tourists 188 The S[ocial] M[anager] must not risk the loss of their future patronage by mauling or roughhousing such as might not enjoy it, or at least tolerate it. 1928 Daily Express 4 Sept. 9/5 Harvey as a boxer of exceptional skill should not have allowed West to rush into close quarters and 'rough-house' him. 1938 X. Herbert Capricornia (1939) xxvi. 389 He explained that when he had tried to keep order at the station he had been rough-housed. 1952 in Wentworth & Flexner Dict. Amer. Slang (1960) 434/1 The question of whether..gun-toting bodyguards rough-housed Swedish citizens.

Hence **rou·gh-houser** (also *attrib.* and *fig.*); **rou·gh-housing** *vbl. sb.*

1904 N.Y. Even. Post 2 Jan., In fiction, whether it is historic, society, or the work of literary rough-housers. 1927 Blackw. Mag. June 843/2 It is more than mere rough-housing. 1933 D. L. Sayers Murder must Advertise v. 90 Stand by..in case there's any rough-housing. 1949 Koestler Promise & Fulfilment i. xii. 131 Apart from some rough-housing..the troops found no occasion for any martial activity. 1951 Sun (Baltimore) 2 Mar. 19/2 In some sections, wrestling is outdrawing boxing... Lord Carlton, who poses as a titled Englishman; the Golden Terror, a roughhouser type..have contributed to the success. 1974 H. L. Foster Ribbin' vi. 244 Many male teachers relate physically to male students through playful roughhousing where some form of physical body contact is made.

roughie (rʌ·fi). *dial.* and *slang*. Also **roughy.** [f. Rough *sb.1* + -ie, -y6.] **1.** A rough or rowdy; a brawler; a hooligan.

1905 Eng. Dial. Dict. V. 158/1 Roughy, sb. Sc. Irel. 1 A coarsely made, bullying fellow. Ant. Grose (1790) MS. add. (C.) 1933 Bulletin (Sydney) 15 Nov. 33/1 With such a lot of roughies in the hall. 1953 Amer. Speech XXVIII. 118 Rough, roughy, n. A carnival roustabout, a manual labourer. 1966 W. S. Ramson Austral. Eng. iv. 62 Roughie, meaning 'a coarsely made, bullying fellow' was used in both Scotland and Ireland. 1971 P. Driscoll White Lie Assignment vii. 60, I know a roughie when I see one... He's just one of those blokes who can't stay away from trouble.

2. *Austral.* In dog- and horse-racing: an outsider.

1934 'S. Rudd' Green Grey Homestead 155 Those who had lost a wager or two will turn to Bell and say: 'You knew something about the roughie!' 1951 Cusack & James Come in Spinner 40 He's a roughie so 'e'll go out at long odds. 1958 F. Hardy Four-Legged Lottery 14, I might just have a shilling on a roughie. 1973 Sun-Herald (Sydney) 26 Aug. 58/2 Punters were reluctant to support him... Consequently Pepper Moss went out as a 12–1 'roughie'.

3. *Austral.* A trick, an unfair practice; esp. in phr. *to put a roughie over*.

1939 K. Tennant Foveaux ii. i. 122 Kelly put a roughie over Charlie to-day. Ibid. iii. ii. 151 'They're putting over a roughie at Central,' the secretary..mentioned casually. 1945 Baker Austral. Lang. xv. 265 A roughie, toughie, hottie, crookie, swiftie, smartie will all be heard in male conversation to describe a joke or trick that is either agreeable or disagreeable. 1970 R. Beilby No Medals for Aphrodite 269, I bluffed him, put a roughie over him.

roughing, *vbl. sb.* Add: **2. b.** (Further examples.) Also with *off*, *out* (cf. Rough *v.1* 6 b in Dict. and Suppl.; also *attrib.*).

1846 Roughing out [in Dict., sense 5]. 1883 Proc. Inst. Mech. Engineers 226 They are used in different machine tools principally for 'roughing out', or..rapidly reducing castings, forgings, &c., from their rough state to nearly their finished forms. 1901 Jrnl. Inst. Electr. Engineers XXXI. 312 Two drills should in all cases be provided, one a roughing-out drill, and the other a finishing drill. 1947 J. C. Rich Materials & Methods of Sculpture ix. 252 The bushhammer is a very useful and fairly rapid tool, which may be employed from the coarse, initial roughing-out stages almost to the very delicate or final stages of the carving. 1947 Dylan Thomas Let. Jan. (1966) 292 And I also worked upon the preliminary roughing-out of the script with Taylor. 1948 Rev. Sci. Instruments XIX. 15/1 The holding pump serves as a fore-pump for the diffusion pump during the roughing-down portion of the cycle. 1969 E. H. Pinto Treen 388/1 A mid-19th-century roughing off plane. 1970 R. J. Small Study of Landforms iv. 128 Many of these joints are evidently post-denudational, having appeared since the 'roughing-out' of the main elements in the landscape by Tertiary erosion.

3. (Earlier examples.)

c 1823 Byron Don Juan xii. lxiii. variant line 6 in Hagelman & Barnes Concordance to Byron's Don Juan (1967) 682 But those who have been a little used to roughing. 1836 T. Power Impressions Amer. II. 211 This little city [sc. Mobile] was to me one of the most attractive spots I visited south of the Potomac. I came upon it..after a severe roughing, and found a fine climate and old friends.

4. (Further example.) Also *N.Amer.*, in *Football*, *Ice Hockey*, and *Lacrosse*: foul tackling, punching, or pushing. Also, *roughing-the-kicker* attrib.

1938 Sun (Baltimore) 27 Oct. 12/6 Brooks, of Yale, broke through and rushed the kicker... Being unable to stop himself in midair, he naturally crashed into the kicker, and Michigan was given the ball for roughing the punter. 1958 F. C. Avis Boxing Reference Dict. 96 Roughing, questionable tactics in a boxing contest, and likely to involve disqualification of the offender. 1958 Herald-Tribune (Grande Prairie, Alberta) 28 Feb. 5/6 Bryan McCurdy..and Bill Oakford..went off together for roughing and slashing at the 18:55 mark. 1961 J. S. Salak Dict. Amer. Sports 371 Roughing the kicker (football), making unnecessary bodily contact with the punter, which is illegal. Roughing the passer (football), unnecessary roughness to a player who has thrown a forward pass. A penalty is involved for this infraction. 1968 Globe & Mail (Toronto) 15 Jan. 20/2 The skirmish provoked a pushing duel that netted every player on the ice, with the exception of the goalkeepers, minors for roughing. 1973 Houston (Texas) Chron. (Suppl.) 14 Oct. 5/1, I was unaware of just how specialized pro football had become until..discovering that one team's kicker had not punted the entire evening. He had, however, drawn eight roughing-the-kicker penalties... It's 15 yards and an automatic first (five yards and an automatic first for the less flagrant 'running into the kicker'). 1976 Webster's Sports Dict. 357/1 Roughing the kicker (football), a personal foul that results when a defensive player runs into or knocks down the kicker on a scrimmage kick without first touching or deflecting the ball... When it is called, the penalty is 15 yards from the previous spot. Roughing the passer (football), a personal foul in professional play that results when a defensive player runs into or tackles a passer after a forward pass has been thrown... When it is called, the penalty is 15 yards from the previous spot.

b. With *up*: see *rough *v.1* 6 g.

1960 Sunday Express 23 Oct. 17/4 His roughing-up of George Bernard Shaw can't compare for butchery with what Mr. Mankowitz did last week to Robert Louis Stevenson. 1973 J. Pattinson Search Warrant iv. 67 He could already feel the stiffness in his limbs that was the result of the roughing-up. 1977 R. Barnard Blood Brotherhood xvi. 181 Their gang knifings and roughings-up.

5. *roughing filter*, *plane*, *shop*; **roughing pump**, a pump for evacuating a system from atmospheric pressure to a lower pressure at which a second pump can operate.

1904 Rep. Brit. Assoc. Adv. Sci. 1903 761 The other two original tanks were converted into six roughing filters containing 3 feet in depth of fine gravel, to intercept particles which have escaped the precipitation process. 1960 McGraw-Hill Encycl. Sci. & Technol. XIV. 543/1 The scrub or roughing plane..has heavy, rounded blades making it suitable for cleaning up rough boards. 1958 Rev. Sci. Instruments XXIX. 368/1 With the roughing pump pinched off..and the system then thoroughly baked out while pumping with the electronic pump, very low pressures can be achieved. 1971, 1976 Roughing pump [see *rough *v.1* 7 d]. 1910 Daily Chron. 15 Jan. 7/2 The cause of the fire was the fusing of an electric motor in the 'roughing' shop [of an opticians' factory].

rough lock, rough-lock. *N.Amer.* [Lock *sb.2* 4.] A device, as a chain, for slowing the passage down a slope of a vehicle or of logs. So **rou·gh-lock, rou·ghlock** *v. trans.*, to slow a vehicle by means of a rough lock, to attach chains to a vehicle so as to slow it; **rou·gh-locking** *vbl. sb.*

1859 Rough-locking [see Rough *adv.* 2 d]. 1884 W. Shepherd Prairie Experiences 197 The hind wheels were rough-locked, that is, a large linked chain was tied round the rim of the wheel in such a way that the wheel rides upon the chain, which drags along and cuts into the ground. 1913 E. MacLennan in MacLennan & Snow Songs of Neukluk 15 She had a skookum load of logs, but I couldn't understand With the rough-lock and the gee-pole how it scaped from her command. 1962 J. Onslow Bowler-Hatted Cowboy v. 46 The rough-lock bit deep into the softening ice and mud. 1973 R. D. Symons Where Wagon Led i. viii. 128 The stout sleigh groaning in protest against the logging chain rough-locked around a hind runner.

rough neck, rough-neck, roughneck. *colloq.* (orig. *U.S.*). [Rough *a.*] **1. a.** A rough or rowdy; a person of rough habits or quarrelsome disposition; an uncultivated or ignorant person.

1836 Col. Crockett's Exploits & Adventures Texas iv. 58 You may be called a drunken dog by some of the clean shirt and silk stocking gentry; but the real rough necks will style you a jovial fellow. 1903 Sun (N.Y.) 25 Nov. 2 The police were kept on the jump chasing away gangs of 'rough necks' (the pet name for the rowdies in Sam Park's late union) who went from building to building trying to intimidate members of the new union. 1903 N.Y. Evening Post 17 Aug. 7/7 His [sc. Sam Park's] stated income amounts to union wages from his union of 'rough-necks', as the iron-workers call themselves, as walking delegate. 1917 J. M. Grider War Birds (1927) 30 But there are a few rough-necks in every outfit that will cause trouble and get the whole bunch in wrong. 1918 [see non-academic s.v. *non- 3]. 1929 J. Buchan Courts of Morning i. iii. 51 The water-front was a perfect rat-hole for every criminal in the Pacific—every brand of rough-neck and dope-smuggler and crook. 1940 E. N. Teall Putting Words to Work i. xxi. 147 The business man will say that if a university can afford to write such letters there is no need for a roughneck like him to bother. 1959 'J. Christopher' Scent of White Poppies vi. 91 She has me tabbed for a roughneck... She has enough on with civilizing you, without having me to cope with as well. 1972 D. Haston In High Places i. 14 Jimmy was twenty-eight, and already a qualified architect; we were seventeen-year-old roughnecks. Basically I think..he was at heart a roughneck himself. 1979 Time 13 Aug. 28/3 Like Lewis, countless other managers and entrepreneurs are coming to Denver to live amid its comfort and culture while their hired roughnecks and miners squeeze the energy from the rural out-posts.

b. *transf.* and *fig.*

Some examples are hardly distinguishable from sense 1 a.

1916 Rio Grande Rattler 13 Sept. 1 Ten buck [private] packers, known in the army as 'rough necks'—a title that usually fits the situation nicely. 1916 H. L. Wilson Somewhere in Red Gap v. 208 And so the party moved on for an hour or two, with the roguish young rough-necks cutting up merrily at all times, pretending to be cowboys coming to town on pay day. 1918 Dialect Notes V. 27 Rough-neck, n. 1. A rowdy. 2. A woman or girl of easy morals but not a prostitute. 3. A dance, open to anyone who 'has the price', at which 'anything goes'. General. 1926 Maines & Grant Wise-Crack Dict. 13/1 Razor back, roughneck or stake driver in a circus. 1926 F. Scott Fitzgerald Great Gatsby iii. 59 I'm Gatsby, he said... I was looking at an elegant young rough-neck. 1941 E. P. O'Donnell Great Big Doorstep iv. 59 'Are you a rough-neck?' 'Yes. I'm in the bull-gang so far. I'm trying to get in the office.' 1960 [see *mauvais coucheur]. 1978 Amer. Poetry Rev. July/Aug. 36/2 He was also an intellectual roughneck.

c. A worker on an oil-rig, esp. a labourer on the floor of a rig.

1917 Dialect Notes IV. 421 Roughneck, n. A man who works about an oil derrick. 1932 Amer. Speech VII. 270 Roughneck, n., the regular term for a member of a driller's crew on a rotary rig; not applied to the driller. 1948 Chicago Tribune 5 Dec. i. 14/3 Among today's roughnecks you'll find college men—petroleum engineers and geologists. 1958 Times 15 May 14/6 Any such rig [for oil drilling] is known throughout the industry as a wildcat, and unskilled members of drilling crews are technically classified as 'roughnecks'. 1972 Guardian 11 Feb. 12/1 A Dutch oilman endorsed this. 'When the exploration is over, the 'roughnecks' (local labour) will go.' 1976 M. Machlin Pipeline xi. 135 He..had worked as a Rough-neck in the Louisiana area and in East Texas on the oil rigs. 1977 Time 14 Mar. 37/1 The centre of the rig's activities is the mud-slicked drill floor, where half a dozen roughnecks struggle day and night with heavy chains and power-driven winches to shove 90-ft.-long pieces of drill pipe into the narrow hole.

2. *attrib.* Rough; rowdy; uncultivated; characteristic of a rough-neck.

1916 H. L. Wilson Somewhere in Red Gap vii. 288 He really wanted..to study insect life and botany and geography and arithmetic,..instead of being killed off in a sudden manner by his rough-neck parent. 1920 C. Sandburg Smoke & Steel 7 The others were rough-neck singers a long ways from home. 1931 'R. West' in Time & Tide 19 Sept. 1091, I commend to every reader the essay on 'Foreheads Villainous Low', with its entertaining satire on the new 'roughneck' movement among the intellectuals. 1973 A. Hunter Gently French xiv. 128 Those risks..would be part of the fun for a roughneck Romeo. 1976 R. Sanders in D. Villiers Next Year in Jerusalem 209 The roughneck genius of a Walt Whitman.

So **rou·gh-neck** *v. intr.*, to work as a rough-neck on an oil-rig; **rou·gh-necking** *vbl. sb.*

1932 Amer. Speech VII. 270 Roughneck, intr. v., to work as a member of a rotary driller's crew. 1976 Globe & Mail (Toronto) 16 Feb. 3/1 About 200 a year are beginners ready to try roughnecking, the industry's term for the beginners' job. 1977 New Yorker 6 June 47/2 One.. roughnecked in the oil fields near Houston.

roughness. 1. d. For *U.S.* read *local* (chiefly *U.S.*) and substitute for def.: Fodder, hay, corn-husks, etc., as used to feed cattle or horses, as opp. grain. (Earlier and later examples.) Also *transf.*

1813 J. Hartsell Jrnl. 29 Oct. in East Tennessee Hist. Soc. Publ. (1939) XI. 99 Did not draw any rufness for our teeme. 1846 Knickerbocker XXVIII. 313 The truck's all soaked, and there can't nobody stay here to save souls without some kind of roughness to keep up natur'. 1859 W. Dickinson Gloss. Dial. Cumberland 93 Roughness,..grass left for winterage. 1888 C. D. Warner On Horseback iv. 142 'Roughness', we found out at the other house, meant hay in this region. 1938 J. Stuart Beyond Dark Hills iv. 88 We don't feed the cattle anything but roughness. 1949 Publ. Amer. Dial. Soc. xi. 10 Roughness, ..fodder; roughage. 1966 Ibid. 1964 xlii. 22 Roughness, roughage: fodder, corntops, coarse hay.

e. *Bacteriol.* The quality of being rough in sense *1 e of the adj.

1929 Topley & Wilson Princ. Bacteriol. & Immunity vii. 191 The property of colonial roughness is associated.. with a characteristic change in the method of cell division. 1934 Jrnl. Bacteriol. XXVII. 559 Roughness is a relative term when applied to colonial form for many rough colonies may have a smooth appearance on ordinary media. 1960 L. Picken Organization of Cells iii. 63 The roughness expresses itself in the formation of dry, membranous, or brittle colonies, with irregular margins and corrugated surface, and a granular appearance under the microscope; in contrast to the colonies of the Smooth type which are creamy or butter-like in consistency, with even margins and homogeneous in texture.

roughometer (rʌfǫ·mɪtəɹ). *U.S.* [f. Rough *a.* + -o + -meter.] = *profilometer 2 (*b*).

1926 Public Roads VII. 144/2 The roughometer consists of a rack which is attached in a vertical position to the front axle of the vehicle. 1947 Daily Progress (Charlottesville, Va.) 20 Mar. 8 (caption) State Highway Department engineers have developed the 'rough-o-meter' pictured above to detect some of the tiniest irregularities in the surface of highways.

rough-out, roughout (rʌ·faut). [Rough *v.1*] **1.** *Archæol.* A prototype of an artefact. Cf. Rough *v.1* 6 b.

1936 Proc. Prehist. Soc. II. 214 Some sixty odd specimens were found most of which were implements gone wrong in the manufacture or rough-outs never proceeded with. 1959 J. D. Clark Prehist. S. Afr. v. 120 Unfinished roughouts are found associated with factory debris and many finely-made developed Acheulian and even later transitional forms. 1980 Rescue News Sept. 4/3 Many large fragments of stone roof slates in various stages of rough-out were found in the fill.

2. Used *attrib.* to designate informal outdoor clothing. *U.S.*

1963 New Yorker 29 June 75 Chinos, shirts, roughout jackets. 1976 National Observer (U.S.) 10 Apr. 21/5 One scuffed cowboy boot sits poised atop the other; the baggy, rough-out trousers are now slung low beneath the prodigious belly.

rough-rider. Add: **1.** (Earlier example.)

1733 J. Bramston Man of Taste 17, I would with Jockies from Newmarket dine, And to Rough-riders give my choicest wine.

2. Also *fig.*

1977 T. Allbeury Man with President's Mind vi. 62 At the Pentagon end Langham's going to need a man who grinds away diplomatically. Not a rough-rider.

rough-riding, *vbl. sb.* (Earlier and later examples.)

1776 G. Colman Let. 21 July in Private Corr. D. Garrick (1831) I. 231 After a great deal of rough riding, I have got him to accept bills of exchange payable in two and four months. 1969 Telegraph (Brisbane) 17 May 3/2 Thrill-seeking boys..have discovered a dangerous new pastime—riding in tall buildings on the top of a lift cage... The boys call it 'rough-riding'.

rough shoot. [Shoot *sb.1*] An act of shooting game without beaters; an area in which one has a right to shoot in this manner. So **rou·gh-shoot** *v. intr.*; **rough-shooter; rough-shooting.**

1900 [see Shoot *sb.1* 1 g]. 1934 F. Ellis Summers of Yesterday ii. 78 It was a wild bit we had taken, sea-trout fishing, brown trout, and rough shooting, that was all. 1937 L. Durrell Panic Spring vii. 113 They used to ..rough-shoot across the parklands. 1943 G. Greene Ministry of Fear ii. i. 122 The Home..had its own hens and pigs and a good many acres of rough shooting. 1972 'M. Innes' Open House v. 42 Manage you a bit of rough shooting, too. Brought your gun? 1976 Shooting Times & Country Mag. 18–24 Nov. 28/2 (Advt.), A perfect fowler's or roughshooter's gun. Ibid. 16–22 Dec. 48/2 (Advt.), Yellow dog, 2 years, trained for rough shoot. 1976 Evening Post (Nottingham) 16 Dec. 21/8 (Advt.), Rough shoot wanted, 30 miles radius of Nottingham.

rou·gh-towel, *v. rare.* [Rough *a.*] *trans.* To rub or dry with a towel of long-napped material.

1889 E. Sampson Tales of Fancy 23 He sponged his men..and rough-towelled them.

rough-up. Add: **a.** (Further examples.)

1933 Bulletin (Sydney) 6 Dec. 24/4 [Bridge] The council, while signifying approval, wants to make it clear that in future all inter-State rough-ups will be its pigeon. 1951 E. Rickman Come racing with Me x. 85 This one may be fit enough in a week or two to be given a 'rough-up' (a good gallop with companions but not a formal trial).

c. A fight; a brawl.

1896 Sessions Papers Cent. Criminal Court 22 June, There was a little rough up, and I found myself stabbed in my arm. 1950 K. S. Prichard Winged Seeds 26 There'd 've been a rough-up in no time, and only half a dozen of us with Paddy against forty or fifty men.

d. (See quot.)

1919 V. Marshall World of Living Dead 69 The 'donkey-dipper' is another kind of pick-pocket. He works alone, and his methods are to grip, to rip, and to run. 'A dead rough-up'—thus the more scientific of the fraternity designate him in their scorn.

roughy, var. *roughie.

‖ **rouille** (rūy). [Fr., lit. 'rust'.] Mayonnaise flavoured with pimento or the like.

1951 R. Campbell Light on Dark Horse xix. 275 No decent fisherman will eat bouillabaisse without the rouille. 1976 N. Roberts Face of France ix. 101 A fish soup..a bourride, with a spoonful of rouille stirred into it.

roulade. Add: **2.** *Cookery.* A dish prepared by rolling up a slice of meat or a sponge or similar base, esp. with a filling (see quots. 1969, 1975[1]). Also *attrib.*

1885 Tasty Dishes (James Clarke & Co.) 151 Roulades of Beef. 1958 Catal. County Stores, Taunton June 3 Pâté de foie gras..Roulade for slicing—a tin 14/9, 22/3. 1965 A. R. Daniel Up-to-Date Confectionery (ed. 4) xxviii. 407/2 Roulade Slices. Ibid. 408/1 Spread the sheet of roulade with pink-coloured kirsch-flavoured butter icing. Ibid. 409/1 The Swiss..make a special type of butter-sponge sheet from which to make roulade and dresden slices. 1969 R. & D. De Sola Dict. Cooking 194/2 Roulade,..slice of meat, covered with forcemeat or other savoury filling, rolled up, and cooked. 1975 J. Child From Julia Child's Kitchen iii. 109 A roulade is a flat soufflé baked in a rectangular shape, then rolled up with a filling. 1975 Times 18 Dec. 7/5 The roulade of avocado is..delicious. 1980 Times 24 May 24/3 Puddings include chocolate roulade.

rouleau. Add: **3.** (Earlier example.) Also in extended use (see quot. 1976). Also *attrib.* and *Comb.*

1820 M. Edgeworth Let. 8 June (1979) 160 Muslin gowns each trimmed with rouleaux of sattin. 1968 J. Ironside Fashion Alphabet 105 Rouleau, a rounded padded belt which may be finished with a knot or bow. 1970 Kay & Co. (Worcester) Catal. 1970–71 Autumn/Winter 173/1 Deep back opening with rouleau tie fastening. 1972 Country Life 23 Mar. 737/3 A tweed suit..has a rouleau belt. 1974 Janet Frazer Catal. Spring & Summer 184/3 Nylon slip with lace trimmed rouleau straps. 1975 Times 14 Oct. 9/5 The hand made rouleau belts and the gold and ivory buttons. 1976 P. Clabburn Needleworker's Dict. 228/3 Rouleau, any trimming or part of an article which is rounded or rolled. It may be in the form of piping or may mean a turned hollow tube as in the 'shoelace' shoulder straps of the 1930s and 1940s. 1977 Daily Tel. 4 Apr. 15/1 She wears a charming little rouleau-edged cap tilted over her brow. Ibid. 15/3 There were..dresses of black cotton.., and the academic robe, with high, padded rouleau neck, shone out..in most unacademic tangerines.

4. Used *attrib.* to designate a type of vase with a cylindrical body and narrow neck, made in China from the late seventeenth century, or an imitation of such a vase.

1915 R. L. Hobson Chinese Pott. & Porc. II. x. 165 (caption) Club-shaped (rouleau) Vase finely painted in famille verte enamels. 1936 Burlington Mag. Nov. p. xix/1 Famille-verte vases..a large rouleau specimen. 1937 Ibid. June p. xxv/2 Other specimens of K'ang-hsi in brilliance and quality are a pair of large rouleau vases. 1964 M. Medley Handbk. Chinese Art 81/2 Rouleau vase is a vase with cylindrical body, short rather flat shoulders, a short thick neck, also cylindrical, and a slightly spreading mouth, which sometimes turns up a little at the rim. The term applies to a type of vase produced from the late 17th century onward. 1977 Times 17 May 16/4 Two large (17½ inches) famille verte rouleau vases.

‖ **roulement** (rulmañ). [Fr., lit. 'roll, roster'.] A movement of members or equipment of the armed services; rotation of units, relief of troops. Also *attrib.*

1918 W. S. Churchill in M. Gilbert Winston S. Churchill (1977) IV. Compan. i. 290 Although roulement can proceed on both sides, this is a formidable preponderance and it tells more and more as reserves are used up. 1930 Times Lit. Suppl. 10 July 567/1 After the Lys offensive he was in what was known as the roulement, the transfer of troops to the Rheims area. 1941 W. S. Churchill Second World War (1950) III. 735 Reserves of pilots and machines should be disposed in squadrons, and thus allow roulement to be extended in the event of protracted fighting. 1971 H. Wilson Labour Govt. (1974) xix. 467 Moreover, there was a problem of rotating troops, what is known in the West as roulement. 1977 Guardian Weekly 14 Aug. 4/2 Their individual battle tours on Irish soil..Roger spent four months here once with his roulement unit.

roulette. Add: **2. b.** Also *roulette ball*, *system*, *-wheel* (earlier example).

1844 Rep. Sel. Comm. Gaming 210 in Parl. Papers VI. 1 Seized..2 roulette balls, 2 dice-boxes, 2 bags containing 366 counters, [etc.]. Ibid. 211, I seized a roulette-wheel and a quantity of gambling apparatus. 1926 A. Christie Murder of R. Ackroyd iii. 26 Caroline visibly wavered.. much as a roulette ball might coyly hover between two numbers. 1976 'J. Fraser' Who steals my Name? xi. 134 A roulette ball has no memory... In the South of France a ball went into the same slot seven times running. 1976 P. Cave High Flying Birds iv. 47, I once sold a fool-proof roulette system to a professional gambler for 500 francs.

d. *Russian roulette*: see *Russian *a.* 2 d*.

5. (Earlier example.)

1835 Brit. Cycl. Arts & Sci. I. 508/1 A more expeditious way of multiplying the dots has been contrived in the instrument called a roulette, a toothed wheel, fixed to a handle which, by being rolled forcibly along the copper, produces a row of indentations.

rouletted, *pa. pple.* Add: **b.** Of archaeological objects: impressed with lines or dots by means of a cogged wheel or a comb. **c.** *gen.*

1938 Oxoniensia III. 28 Sherd with shell-grit; fine stabs and rouletted horizontal lines. 1939 V. G. Childe Dawn European Civilization (ed. 3) xii. 214 The 'rouletted' decoration is executed with a comb with very short teeth, separated by extremely narrow interstices, and probably with a curved edge. It yields a practically continuous line of round or, more often, rectangular dots, separated by low septa. 1969 G. Bibby Looking for Dilmun xv. 323 They were Attic ware, imports from Greece itself. Some of them were even rouletted, decorated with a close pattern of semi-circles made with a toothed wheel, a characteristic which proved their Greek origin beyond a doubt. 1975 J. B. Harley O.S. Maps v. 72 It..was characterized by National Grid lines rouletted in black. [Note] That is, the line consists of small, closely spaced dots. 1977 Antiquaries Jrnl. LVII. 381 Fine wares include bowls..with a foot-stamp within rouletted circles (second century A.D.), and lead-glazed wares.

rouletting, *vbl. sb.* Add: **b.** Decorating pottery, etc., with dotted lines by means of a cogged wheel or comb; ornamentation produced in this way.

1937 Oxoniensia II. 19 It contained a great quantity of Roman pottery, mostly early grey ware jars but including part of a butt-beaker in fine buff ware with two zones of rouletting, and a bit of a buff amphora handle. 1973 A. H. Whiteford N. Amer. Indian Arts 148 Coins and ingots were hammered into sheets and decorated by stamping and rouletting. 1979 Archaeology July–Aug. 31/1 The carinated bowls were stamped with palmettes inside concentric spirals of rouletting, a technique of producing hatch lines around the central design.

Roumanian, *sb.* and *a.* Add: See also *Romanian *sb.* and *a.3*, *Rumanian *sb.* and *a.* **A.** *sb.* **1.** (Further examples.)

1925 N. Iorga Hist. Roumania iii. 35 Anthropology and ethnography do not find the Slav type amongst the Roumanians. 1934 R. W. Seton-Watson Hist. Roumanians i. 9 The Roumanians claim that they are the true descendants of Trajan's colonists. 1964 Whitaker's Almanack 1965 913/1 By the Treaty of Berlin..the Principality was recognized as an independent State, and part of the Dobrudja (which had been occupied by the Roumanians) was incorporated. 1974 M. B. Brown Econ. of Imperialism xii. 296 What worries the Roumanians and other underdeveloped countries in the Soviet bloc is precisely the results of the artificial world division of labour in which they have for so long been held.

2. (Further examples.)

1925 N. Iorga Hist. Roumania viii. 145 Coresi.. began to publish religious works in Roumanian, or in Roumanian and Slavonic. 1933 L. Bloomfield Language xviii. 314 A feature common to both Roumanian and the western Romance languages is presumably guaranteed as Latin. 1964 Whitaker's Almanack 1965 913/2 Roumanian is a Romance language with many archaic forms and with admixtures of Slavonic, Turkish, Magyar and French words.

B. *adj.* (Earlier and later examples.) *Roumanian stitch* = *Oriental stitch* s.v. Oriental *a.* 3 c.

1860 Universe 8 Dec. 1/2 His projects extend to the formation of a great Roumanian State. 1883 R. Torceanu Simplified Gram. Roumanian Lang. p. vii, The Roumanian tongue can claim..attention on more grounds than one. 1925 N. Iorga Hist. Roumania viii. 145 Religious manuscripts..in which the Roumanian text in red letters follows the Slavonic text in ink. 1932 D. C. Minter Mod. Needlecraft 10/1 Intergradating one stitch and colour with another, as is possible with Irish stitch..and Roumanian..or satin stitch. 1934 M. Thomas Dict. Embroidery Stitches 177 Roumanian Stitch consists of a long stitch across the shape tied down with a shorter slanting one in the centre. 1959 R. N. C. Hunt Books on Communism 37 An exhaustive study by a group of experts on aspects of Roumanian life under Communism. 1974 M. B. Brown Econ. of Imperialism xii. 295 In the Roumanian dispute with Comecon, the basis of the Roumanian argument has been one that we have met before.

roumanite, var. *rumänite.

Roumelian (rume·liăn, -ī·liăn), *a.* (*sb.*) Also **Rumelian.** [f. as Roumeliote: cf. Turk. *rum* Byzantine Greek (of Turkish nationality), *il* province.] Of or pertaining to Roumelia (see Roumeliote in Dict. and Suppl.), with particular reference to Ottoman territories of the southern Balkans inhabited by Greeks and now forming parts of northern Greece and Bulgaria; of or pertaining to the form of Greek spoken there. Also as *sb.*, a Greek inhabitant of Roumelia.

1859 J. F. Maguire Rome (ed. 2) xxviii. 323 The different languages in which poetical compositions were recited at the Polyglot Academy, for Epiphany, 1858 [were]..Danish, Roumelian, Albanian, Polish, [etc.]. 1880 E. W. Hamilton Diary 29 Aug. (1972) I. 41 Colonel Wilson who is reporting on the state of the Bulgarian and Eastern Roumelian provinces finds..a better state of affairs than he expected. 1888 Encycl. Brit. XXIII. 655/2 This is due partly to the Christian communities, notably the Maronites and others in Syria, the Anatolian and Roumelian Greeks, and the Armenians. 1902 D. G. Hogarth Nearer East 155 Most thinly peopled are the mountainous districts between the Vardar and the Rumelian plains. 1935 H. Edib Clown & his Daughter xxi. 114 Why should they insult him and spit at him because he was a Rumelian?

Roumeliote. Add: Also **-iot.** (Earlier and later examples.)

Normally used of the area corresponding to Aetolia and Acarnania. Cf. *Roumelian *a.* (*sb.*).

1835 N. P. Willis Pencillings II. xxii. 248 At the Adrianople gate, we found a large troop of horsemen.. who had accompanied a Roumeliote chief from the mountains. 1886 Times 5 Apr. 5/1 Instinctively the Roumeliots prefer to blame Turkey rather than Russia. 1939 A. Toynbee Study of Hist. VI. 331 The Rumeliot Greek Armatole and Klephtic ballads. 1964 A. A. Pallis Greek Miscellany 144 He took the side of the Roumeliots against the party of the primates of the Morea. Ibid., His hero was that greatest of Roumeliot captains, George Karaiskakis. 1969 C. M. Woodhouse Philhellenes iv. 118 The same feelings were expressed by an old Rumeliot whom James Emerson met. 1973 —— Capodistria xv. 347 Two Roumeliote kapetánioi, Grivas and Stratos, were bombarding each other..to settle a private feud. Ibid. xvi. 382 The Roumeliotes..were eager for action.

‖ **Roumi** (ru·mi). Also 6 **Rumi**, 9 **Roumy.** Fem. **roumia.** [ad. Arab. *rūmī* Byzantine, Pers. *rūmī* Turk, Greek.] Among Arabs, a term for a European.

1576 R. Eden tr. Vertomannus's Navigation & Vyages vi. xiv. 401 They affyrmed also, that there are certayne Christian kynges (whiche they call Rumi) of great power, confynyng or borderyng on the dominions of the great Turke. 1819 J. L. Burckhardt Trav. Nubia 542 Such a misfortune had never been heard of in the time of the Islam, and before the Roumy had ever come into these parts. 1867 'Ouida' Under Two Flags II. iii. 74 Not but what our Roumis are brave fellows enough; better comrades no man could want. 1924 Public Opinion 27 June 619/3 The Roumis had got round the flanks and were attacking the Arab Camp. 1927 Daily Express 17 June 9 An intelligent and up-to-date caid..in the Sahara ..answered..'most of my people are unaccustomed to the ways of the roumia (foreign woman).' 1958 Times Lit. Suppl. 11 July 393/4, I did happen to be the first roumia allowed to a remote branch of the sect who live off the beaten track.

rounce, *sb.2* Add: [Perh. ad. G. *ramsch* a variety of Skat.] (Earlier and later examples.) Also, a similar domino game. Hence **rounce** *v.2* (see quots. 1864, 1868).

1855 in Calif. Hist. Soc. Q. (1929) VIII. 352 Had a great rounce game, a little noise but no fun. 1857 Hoyle's Games 99 The Game of Rounce is played by each player taking five pieces, (after having turned for the trump, the highest piece turned deciding that point); the trump is then turned up for the trump-holder by his right hand adversary, the highest end being trump. 1864 W. B. Dick Amer. Hoyle 397 Rounce. This is a pleasant game [of dominoes], and from two to four may participate in it. Ibid., The player who fails to take a trick with his hand is 'Rounced', i.e., sent up five points. 1868 —— Mod. Pocket Hoyle 196 The game of Rounce, as played in the United States, is derived from the German game of Ramsch, and in its principal features resembles Division Loo. Ibid. 197 Each trick taken in play counts one point, and if a player fail to take a trick after entering to play his hand, he is Rounced, that is, sent up five points, which adds a X to his score. 1975 Way to Play 66/1 Forty-two, or domino rounce, is an adaptation of a card game for play with dominoes. The object is to score points by winning tricks.

round, *sb.1* Add: **I. 2. c.** (Earlier examples.)

1660 W. Denton Let. 29 Feb. in M. M. Verney Mem. (1894) III. xiii. 469 The Beef the best that ever was eat, I eat a whole Round last night my self. 1771 J. Woodforde Diary 5 Jan. in Parson Woodforde Soc. Q. Jrnl. (1970) III. i. 24, I gave them for Dinner..a Round of Beef boiled.

attrib. 1934 Webster Round steak. 1972 'L. Egan' Paper Chase (1973) x. 161 Athelstane was..condescending to eat the best round steak cut into bite-size pieces. 1975 Evening Herald (Dublin) 8 May 6/7 Round steak.. dropped by 10p per pound.

e. pl. *Comm.* Articles that are naturally or artificially produced in round shapes.

1911 Chambers's Jrnl. 8 Apr. 297/1 Formerly 'flats' and 'rounds' used to be spoken of to distinguish the imports of this drug [sc. rhubarb]. 1928 Daily Mail 25 July 19/3 Potatoes.—Spitalfields: English Kidneys 6s to 7s. rounds 5s to 5s 6d per cwt.

5. a. Also *fig.*, a condition which displays a given subject from all aspects; three-dimensionality. Usu. in phr. *in the round*.

1931 Times Lit. Suppl. 31 Dec. 1052/3 One cannot tamper with a screen character who speaks like a human being..and has his being 'in the round'. 1933 Punch 12 July 51/2 It is not an easy part, seeing that it is the only character in the whole cast to be drawn in the round. 1948 'M. Westmacott' Rose & Yew Tree ix. 72 Up to now Lord St. Loo had been a name, an abstraction... Now he came into the round—a living entity. 1959 Spectator 7 Aug. 164/3 The camera also gives an impression in the round of the man who seems one-dimensional in print.

d. *Theatr.* In phr. *in the round*, alluding to performance on a stage or arena surrounded by the auditorium, as distinguished from a proscenium; esp. in *theatre-in-the-round*. Cf. *arena 5.

1944 Bull. National Theatre Conf. (U.S.) Apr. 19 In this country, Glen Hughes out in Seattle has operated his Studio and Penthouse theatres, playing sophisticated comedies to small audiences 'in the round'. 1948 Sat. Rev. 3 Apr. 24/1 'Theatre-in-the-round' is the way it is described by John Rosenfeld, who is not a czar but the czar in matters dramatic and musical in the Southwest. 1950 Sun (Baltimore) 8 June 16/1 The New York debut of theatre-in-the-round was off to a rousing start last week. 1958 New Statesman 22 Feb. 228/3 At the Mahatma Gandhi Hall, Fitzroy Square, Miss Margaret Rawlings is giving Phèdre in English In-the-Round. 1963 Listener 28 Mar. 559/2, I do object to playing to them in the round, because it gives them a chance to get at the actor physically. 1963 'E. McBain' Ten Plus One (1964) vii. 73 We did the play in the round... we banked rows of rented bleachers on the stage, and the performers worked in the centre. 1967 Oxf. Compan. Theatre (ed. 3) 941/2 Modern theatre-in-the-round first came into prominence in Russia, where in the 1930s Okhlopkov in his Realistic Theatre produced a number of Soviet plays on stages set up in the central area with the audience pressing close on all sides. 1980 Times Lit. Suppl. 5 Sept. 973/1 Audience involvement was not new to Tudor Drama: medieval theatre-in-the-round had already thrived on it.

e. *out-of-round* sb., the extent to which an object departs from being circular in section; also as *adj.* Hence *out-of-roundness*.

1951 C. W. Kennedy Inspection & Gaging iv. 67 Standards for allowable taper, out-of-round or eccentricity should be established in every shop. 1955 W. H. Crouse Automotive Engines xiv. 403 Some bearing failures may result..from a tapered or out-of-round crankpin. Ibid. 412 Bearings working against out-of-roundness or taper of more than 0·0015 inch will not last long. 1962 Mod. Petroleum Technol. (Inst. Petroleum) (ed. 3) xxvi. 848 'Out-of-roundness'..must always be expected [in a pipe] as a result either of poor manufacture or of damage in transit. 1970 K. Ball Fiat 600, 600D Autobk. xi. 130/2 The out-of-round must not exceed ·0004 inch. 1975 Bram & Downs Manuf. Technol. iv. 110 Disadvantages of self-centring chucks are that they cannot clamp blank or out-of-round items to maintain accuracy. 1979 B.S.I. News Jan. 4/1 Yielding in stiffeners due to out-of-roundness and buckling.

III. 13. c. *spec.* A recurring succession or series *of* meetings for discussion or negotiation; one stage in such a process. Also without *const.*

1964 *Ann. Reg. 1963* 252 They disagreed on what should be the approach of the Six in preparation for the forthcoming 'Kennedy round' of negotiations. **1977** *Economist* 22 Oct. 89/1 There is still no sign (two months into the current wage round) that wages are about to go through the roof. **1978** *Internat. Relations Dict.* (U.S. Dept. State Library) 42/2 The talks, which opened in Geneva in October 1973, were called the 'Tokyo Round' because they were initiated by a declaration signed in Tokyo.

14. c. *pl. Naut.* Inspection.

1902 L. Delbos *Naut. Terms* (ed. 4) 140/1 *Rounds*, inspection. **1914** 'Bartimeus' *Naval Occasions* xviii. 158 The Sub-Lieutenant, 'standing the rounds' in the doorway. **1916** 'Taffrail' *Stand By!* 13 Except on Sundays, when the latter is specially tidied up for the 'rounds', it will not bear close investigation. **1961** F. H. Burgess *Dict. Sailing* 174 *Rounds*, inspection.

15. Also *spec.* a visit to each of the in-patients in a ward or under the care of a particular doctor or nurse.

1904 *Sci. & Art of Nursing* I. iv. 99 There is always the danger that in the haste and pressure to have all in order for the rounds of the medical staff, the minor requests of patients may be postponed. **1928** A. T. Schofield *Behind Brass Plate* xiii. 94 Samuel Fenwick, in his rounds, was very droll... After the usual examination of a new patient he performed his well-known trick. **1954** A. Huxley *Let.* 9 May (1969) 706 He [*sc.* a physician] takes foreign pupils—mostly doctors..—young men who live near by and go the rounds with him and learn by listening, answering questions and doing. **1965** Spencer & Tait *Introd. Nursing* vii. 31 It is generally considered that visitors can be in a main ward outside main meal times, sanitary rounds, rest times and doctors' rounds. **1974** G. B. Mair *Confessions of Surgeon* v. 58 When added to routine clerking, ward rounds, night rounds, dealing with emergencies,..no day had enough hours.

16. c. (Earlier examples.)

1775 C. B. Clapcott *Rules of Golf* (1935) 24 No member of this Society pay the Cadies more than one penny per round. **1834** P. Buchan *Peterhead Smugglers* 63 To gang wi' you to the links ilka morning at five o'clock to a round o' the golf. **1866** *Golfer's Year Bk.* 65 The order of play was the reverse of the wonted 'round' over Bruntsfield, in order that strangers might cope on equal footing with players who were up to the green. Each round consisted of 7 holes, and four rounds were fixed on for the decision of the Tournament.

17. b. Now usu. *pl.* Also *to make the rounds.*

1837 *Jamestown* (N.Y.) *Jrnl.* 22 Mar. 3/2 There is a story going the rounds in relation to the president-elect. **1862** O. W. Norton *Army Lett.* (1903) 55 Everything of the kind has to go the rounds, you know. **1934** H. L. Ickes *Secret Diary* (1953) I. 254 He expressed the fear that in some way I connected him with all of these stories that are going the rounds about me. **1977** *Rolling Stone* 13 Jan. 39/2 The rumor that the FBI started about her being a Soviet spy is still making the rounds at parties she no longer attends.

IV. 20. a. (Further examples.)

1928 C. Mackenzie *Extraordinary Women* x. 176 Two rounds of stingers brought the evening to a close. **1960** M. Spark *Ballad of Peckham Rye* (1964) vii. 107 Dixie, at first under the impression that Humphrey was buying the round, asked for a ginger ale. **1970** G. F. Newman *Sir, You Bastard* viii. 255 Just a slag avoiding his round. **1978** J. Porter *Dead Easy for Dover* xii. 125 The local chap had proved himself more than willing to stand his round, and Dover didn't ask more than that of anyone.

b. (Further examples.) Also, a sandwich or sandwiches made of two slices cut from a loaf of bread.

a **1902** S. Butler *Way of All Flesh* (1903) lxxii. 330 She..had made him a round of toast. **1947** A. Ransome *Great Northern?* xix. 232 Peggy was cutting rounds of bread to make potted meat sandwiches. **1974** L. Deighton *Spy Story* xvi. 162 'Have you come for your sandwiches?'..'Last night's pork, and one round of cheese.'

23. b. Also *fig.* and in attrib. phr. *round-by-round.*

1937 'M. Innes' *Hamlet, Revenge!* II. iii. 137 Gott sighed. 'You certainly know the habits of your friends. Round Two to you.' **1955** T. H. Pear *Eng. Social Differences* 246 Championship fight..with a leading article and a back-page 'round-by-round' report. **1959** *Listener* 22 Oct. 681/2 He began round two by making a fresh application for *habeas corpus.* **1961** *Times* 25 May 15/4 It is a remarkable round-by-round study in the art of politics. **1967** *Listener* 3 Aug. 147/3 By quoting, selectively, two rounds of a three-round controversy.. [you] gave an inaccurate picture of the course of the argument.

d. (Further examples.)

1921 [see **end sb.* 20 b]. **1951** *Sport* 30 Mar.–5 Apr. 2/4 In the Amateur Cup they reached the second round, losing to Pegasus.

24. a. (Earlier examples.)

1794 C. Mathews *Let.* 28 Dec. in A. Mathews *Mem. Charles Mathews* (1838) I. vi. 129 He came forward at the end of the play.., and he had six successive rounds of applause. **1808** *Monthly Mirror* Mar. 268 The audience.. with not three, but six rounds of applause, greeted his return.

25. *ellipt.* = *round-the-houses* s.v. *round *prep.* 1 a. *slang.*

1893 P. H. Emerson *Signor Lippo* xiv. 55 One day he walked straight into this kitchen clobbered in a black pair of rounds, tight to his legs.

round, *sb.*[2] (Further example.)

1943 T. Harsley *Find, fire, & Strike* 38 The 'round down' at the stern where the aircraft ready to take off are ranged.

round, *a.* Add: **I. 3. b.** Also *fig.* of character.

1927 E. M. Forster *Aspects of Novel* iv. 106 The test of a round character is whether it is capable of surprising in a convincing way. If it never surprises, it is flat.

c. (Further examples.) See also quot. 1960.

1788 E. Sheridan *Jrnl.* 22 Dec. (1960) vi. 138 As to gowns all kinds—Chemises—Round gowns with flounce or not. **1815** *La Belle Assemblée* June 274/1 *Morning-Dress.* Round dress of jacconet muslin. **1836** Dickens *Sk. Boz* (1837) 2nd Ser. 100 They were decent people, but not over-burdened with riches, or he would not have so outgrown the suit when he passed into those corduroys with the round jacket. **1960** C. W. Cunnington et al. *Dict. Eng. Costume* 184/2 *Round dress* or *gown*,..a term indicating a dress with joined bodice and skirt, the latter closed all round... 18th c. Occasionally made with a slight train. 19th c. No train, the term now meaning a dress without a train.

e. Designating any of several styles of circular, conical, or pill-box hat.

1795 tr. *C. P. Moritz's Trav.* 141 A fellow in a brown frock and round hat. **1804** [in Dict., sense 4]. *c* **1806** D. Wordsworth *Jrnl.* (1941) I. 303 A fine fellow..in tight clean clothes and a nice round hat. **1825** H. Wilson *Mem.* II. 175 Down came Colonel Palmer..his laced jacket covered with an old, short, brown great coat, and a shabby round hat. **1828** D. Wordsworth *Jrnl.* (1941) II. 403 Women often with round hats, like the Welsh. **1890** C. M. Yonge *More Bywords* 137 Those foolish girls thought me too fine a lady to like to be seen with her in her round hat on a Sunday. **1968** [see *pill-box b].

5. a. *round dance,* (*a*) (further examples); also *round dancer, dancing*; (*b*) [tr. G. *rundtanz* (K. von Frisch 1923, in *Zool. Jahrb., Abt. f. Allgemeine Zool. u. Physiol. der Tiere* XL. 31], a circular movement performed by bees at their hive or nest, believed to indicate a source of food to other bees.

1919 *Ladies' Home Jrnl.* May 31/1 My dear lady, are you going to give up round dancing? **1947** A. Einstein *Mus. Romantic Era* x. 110 The sharp rhythm of round dances and torch dances. **1950** K. von Frisch *Bees* iii. 71 The round dance and the waggling dance are two different terms in the language of bees, the former meaning a source of food near the hive and the latter a source at 100 metres or more. **1952** C. R. Ribbands *Behaviour & Social Life of Bees* xix. 153 The characteristic of the 'round dance' is that the bee performs a complete circle, whereas the 'waggle dance' is a figure-of-eight. **1973** R. A. Morse *Compl. Guide Beekeeping* xiii. 203 Certain races of bees use a dance intermediate between the round dance and the wag-tail dance incorporating parts of both. **1976** *Columbus* (Montana) *News* 3 June 1/2 Mr. and Mrs. Robert Shanks, Columbus, are ardent and avid square and round dancers.

II. 7. a. (Further example.)

c **1829** D. Jerrold in M. R. Booth *Eng. Plays of 19th Cent.* (1969) I. 175, I deserve a round dozen [*sc.* thirteen lashes] for the question.

IV. 15. a. round ball, (*b*) also *spec.* an early alternative name for Base-ball (earlier examples); **round barrow** *Archæol.*, a Bronze Age burial mound of circular form; **round bilge,** a curved, as distinct from an angular or stepped, hull; also *attrib.*; hence **round-bilged** *a.*; **round cell** *Path.*, used *attrib.* = next; **round-celled** *a. Path.*, (of a neoplasm) characterized by round, undifferentiated cells; **roundeye** *slang*, a European, as distinguished from a *slant-eye* (*slant *a.* 3); **round heels** chiefly *U.S.*, rounded heels that allow the wearer to rock backwards easily; usu. *transf.* and *fig.* (*slang*) implying the inability to remain upright, as in an incompetent boxer or sexually compliant woman; hence **round-heeled** *a.*; **round-heeler; round log** *U.S.*, a log that has been felled but not hewn; also *attrib.*; **round lot** *U.S.*, a unit of trade (see quot. 1962); **round timber** *U.S.*, timber that has been felled but not hewn; also *transf.*; **round towel** (earlier and later examples); **round turn,** also in colloq. (orig. *Naut.*) phr. *to bring* (*fetch*) *up with a round turn*, to check or stop suddenly; **round wood,** (*a*) = **round timber*; (*b*) short logs of small diameter from the tops of pine and spruce trees, used for box-making.

1834 R. Carver *Bk. of Sports*, This game is known under a variety of names. It is sometimes called 'round ball', but I believe that 'base', or 'goal ball' are the names generally adopted in our country. **1856** *Porter's Spirit of Times* 27 Dec. 276/3, I have thought..a statement of my experience as to the Yankee method of playing 'Base', or 'Round' ball, as we used to call it, may not prove uninteresting. **1869** J. Thurnam in *Archaeologia* XLII. 168, I propose to classify the barrows of this part of England according to the following scheme:.. I. Long Barrows. (*Stone period*)... II. Round Barrows. (*Bronze period*). *Ibid.*, In none of the..long barrows.. have objects of metal..been found... In the..round barrows, not only are there objects of stone, but..chiefly, those of bronze, and..iron... They may be regarded..as belonging to the Bronze period. **1926** M. C. Burkitt *Our Early Ancestors* vi. 151 In England we have..passage graves (generally called 'Long Barrows'..) in many places.., and stone kists (generally called Round Barrows from the circular shape of the tumuli). **1975** J. G. Evans *Environment Early Man Brit. Isles* vi. 130 Two Bronze Age round barrows known as the Burton Howes. **1980** *Encounter* May 59/1 Long barrows and causewayed camps signal the territories of the early and middle Neolithic groupings, with the henges and round barrow cemeteries appearing in the late Neolithic/early Bronze Age. **1951, 1961** Round bilge [see *hard chine* s.v. *hard *a.* (*sb.*) 22]. **1977** *Austral. Sailing* Jan. 38/2 The round-bilged 'mouldie' hull has virtually disappeared. **1889** Round cell [see sense 17 in Dict.]. **1961** R. D. Baker *Essent. Path.* xix. 533 Malignant tumors of the thymus may arise from epithelial structures (carcinoma) or from a fibrous component (fibrosarcoma) but are perhaps most often round-cell tumors and presumably lymphomas. **1873** Round-celled [in Dict., sense 16]. **1907** J. H. Parsons *Dis. Eye* xxix. 605 Sarcoma is rare; it may be round or spindle-celled, pigmented or non-pigmented. **1961** R. D. Baker *Essent. Path.* xiii. 302 The undifferentiated sarcomas must be classified according to how they look microscopically, and may be described as spindle-celled or round-celled. **1967** *Guardian* 16 Aug. 6/5 Many Europeans have been assaulted simply because they were 'roundeyes'. **1977** 'J. Le Carré' *Hon. Schoolboy* vi. 125 In the East a roundeye could live all his life in the same block and never have the smallest notion of the secret tic-tac on his doorstep. **1957** J. Kerr *Please don't eat Daisies* 118, I know I'm just a broad, Mike. I'm a round-heeled babe. **1975** 'R. Rostand' *D'Artagnan Signature* (1976) xiv. 83 You said that as if I'm some round-heeled little chippie who dragged you to the floor. **1927** *Vanity Fair* (N.Y.) Nov. 67/2 Others contend that 'a round-heeler' was applied to street-walkers many years ago. **1926** Abbott & Weaver *Love 'em & leave 'Em* III. 109 You want people to say you got round heels. Why don't you go on the streets and be done with it? **1926** *Variety* 29 Dec. 7/4 A push-over, which means a fighter with round heels along cauliflower alley, was, by the same token, a dame on rockers. **1929** E. Wilson *I thought of Daisy* i. 16 Myra Busch is a push-over!..She's got round heels! **1944** R. Chandler *Lady in Lake* v. 35 You'd think..I'd..pick me a change in types at least. But little roundheels over there ain't even that. **1963** 'G. Bagby' *Murder's Little Helper* (1964) viii. 84 Little Miss Roundheels..specialized in gentlemen who were otherwise committed. **1975** P. de Vries *Glory of Humming Bird* xiii. 192 Her famous round heels did not seem to rule out a stern morality on other counts. **1869** S. Haycraft *Hist. Elizabethtown, Kentucky* (1921) ii. 15 In the winter time they met in the round log cabins with dirt floors. **1871** E. Eggleston *Hoosier Schoolmaster* 95 He came upon a queer little cabin built of round logs. **1884** 'Mark Twain' *Huck. Finn* xxxii. 329 Phelps's was one of these little one-horse cotton plantations..round-log kitchen. **1942** *Sun* (Baltimore) 27 Feb. 17/5 The rates charged by brokers for 'round lots' (units of 100 shares) range from 3 cents a share..to 13 cents. **1962** S. Strand *Marketing Dict.* 638 *Round lot*, a trading unit. 1) On the New York Stock Exchange, 100 shares. 2) On the Chicago Board of Trade, 5000 bushels. **1874** J. F. Rusling *Across Amer.* xxvii. 429 Snow galleries consumed in all nearly forty-five million feet, board measure, of sawed timber, and over a million and a quarter feet of round timber. **1905** *Bull. Bureau of Forestry* (U.S.) No. 61. 45 *Round timber*, pine trees which have not been turpentined. **1964** *Times Rev. Industry* Mar. 56/1 The firm has arranged with Boys and Boden to reopen the sawmill section of British Sawmills at Welshpool from March 1 for the conversion of round timber. **1972** *Gloss. Terms Timber* (*B.S.I.*) 8 *Round timber*, felled trees, logs or poles. **1845** *Ainsworth's Mag.* VIII. 71, I at last became quite tired of him and his string of repetitions, or round towel speaking. **1908** G. Jekyll *Children & Gardens* ii. 12 If it can have a small pantry containing a water supply and a sink,..and a round towel handy, it will be better than if these necessaries were in the kitchen itself. *a* **1910** in *Amer. Speech* (1979) LIV. 99 *Round turn*,..'fetched up with a round turn'. Suddenly. **1920** Galsworthy *In Chancery* II. vi. 175 The end came swiftly on the 20th of January with a telegram... It brought him up with a round turn. **1961** F. H. Burgess *Dict. Sailing* 174 *Round turn, Bring up with a*, stop someone or something abruptly. **1910** *Timber Trades Jrnl.* 8 Jan. 37/1 The wood shipped from Archangel is the now well-known roundwood. **1930** *Aberdeen Press & Jrnl.* 3 Apr. 8 So far as the 'round wood' or 'pulp wood' is concerned, most of the Aberdeen contracts for the season are now fixed-up. **1971** *Country Life* 25 Nov. 1450/1 The cash value..from the sales of roundwood and timber..in East Anglia.

16. a. *round-barred, -bellied* (later example), *-bodied, -bottomed* (later examples), *-browed, -budded, -cheeked, -cornered* (later example), *-ended, -eyed* (further example), *-hatted, -lipped, -necked, -paned, -pollened, -soled, -spectacled, -sterned, -walled*; also *round-looking.*

1923 D. H. Lawrence *Birds, Beasts & Flowers* 143 The upstart of your round-barred, sun-round tail! **1919** J. Masefield *Reynard the Fox* 9 Round-bellied like a drinking-cup. **1963** R. P. Dales *Annelids* ii. 42 *Sternapsis* is a small, round-bodied burrower in which the septa have mostly broken down. **1909** B. Lubbock *Deep Sea Warriors* 37 Three men came.., each shouldering a 'round-bottomed chest', as the sailor's bag is called. **1964** V. J. Chapman *Coastal Veg.* iii. 73 Round-bottomed flasks are completely filled with sea water. **1921** Round-browed [see *firm-lipped* s.v. *firm *a.* C. 1 b]. **1925** W. de la Mare *Two Tales* 95 Minute plants, their round-budded clusters showing. **1605** Marston *Dutch Courtezan* I. i. sig. B1, A softe plumpe round cheekt froe. **1871** Geo. Eliot *Let.* 17 June (1956) V. 153, I hope she will be round-cheeked and strong. **1967** Karch & Buber *Offset Processes* xii. 503 Round-cornered cards are usually purchased already round-cornered, die cut and absolutely rectangular. **1951** Whitby & Hynes *Med. Bacteriol.* (ed. 5) xx. 314 The *Clostridia* in their most typical form are straight or slightly curved round-ended bacilli 0·4μ–1·2μ × 3μ–8μ. **1960** *Farmer & Stockbreeder* 2 Feb. (Suppl.) 8/3 Square or round-ended sticks are also useful..

in building up a design. **1923** D. H. Lawrence *Birds, Beasts & Flowers* 160 Yellow eyes incomprehensible with thin slits To round-eyed us. **1962** *Times* 21 Dec. 10/7 A round-hatted drummer. **1906** Hardy *Dynasts* II. iv. i. 230 The Archduchess, a fair, blue-eyed, full-figured, round-lipped maiden. **1958** S. Spender *Fool & Princess* 161 Round-looking lips. **1962** L. Deighton *Ipcress File* xxxi. 198 Jean wore a new round-necked, sleeveless.. dress in tangerine linen. **1974** *Country Life* 17 Jan. 107/1 A round-necked, sleeveless top. **1937** de la Mare & Jones *This Year, Next Year* 39/1 Through its round-paned window. **1909** W. Bateson *Mendel's Princ. Heredity* I. v. 93 It was..more usual to find whites exclusively produced by the cross of two extracted F_2 whites, long-pollened and round-pollened respectively. **1964** W. L. Goodman *Hist. Woodworking Tools* 73 It is a round-soled plane, 14in. long, with a type (b) mouth carving enclosing the date 1706. **1945** W. de la Mare *Burning-Glass* 53 Round-spectacled Chardin's Passion for life. **1897** J. L. Allen *Choir Invisible* xiii. 195 Where some round-sterned packet from New England or New Amsterdam was unloading its cargo. **1931** G. O. Russell *Speech & Voice* 67 A..round-walled organ pipe.

b. *round-mouthed, -winged.*

1945 Step & Wells *Shell Life* 228 Such a form as *Littorina rudis*..was probably the ancestor of the Round-mouthed Snail (*Cyclostoma elegans*), which is clearly a marine snail that has been so modified that it lives far inland on the dry chalk-downs. **1907** R. South *Moths Brit. Isles* 1st Ser. 175 (*heading*) The round-winged muslin. **1908** *Ibid.* 2nd Ser. 257 In most of such aberrations the tips of the fore wings are rather more rounded than in typical specimens, and these are referable to ab. *rotundaria*, Haworth (Round-winged Wave).

c. *round-seeded* (later example).

1970 *Daily Tel.* 10 Jan. 7/3 Round-seeded peas lack the flavour of the wrinkled varieties.

17. (Further examples.)

1856 Round-bend [see *Limerick 2 b]. **1936** J. Steinbeck *In Dubious Battle* iv. 64 A roundwick Rochester lamp. **1940** E. Molloy *Electric Wiring* vi. 156 The British Standard Specification for domestic plugs and sockets is confined to the round-pin type. **1941** H. I. Chapelle *Boatbuilding* 44 The round-bottom model is considered by most amateur builders too difficult to construct. **1946** *Fortune* Apr. 142/1 [He] obviously doesn't want the job or he wouldn't have put in any round-figure bid that size. **1956** 'J. Wyndham' *Seeds of Time* 231 The doctor's round-figure price made him frown. **1967** Karch & Buber *Offset Processes* xii. 504 Perforating machines allow round-hole perforating, like that found on postage stamps and grocery store stamps. **1968** J. Arnold *Shell Bk. Country Crafts* 160 Later wagons, built after 1850 or so, had round-section ironwork. **1970** *Which?* Sept. 280/2 We have criticised this plug before since it will fit into a 5-amp round-pin socket which leaves the appliance unearthed. **1976** *Woman's Day* (U.S.) Nov. 158 Following construction detail, assemble legs and stretcher with 3" roundhead stove bolts and nuts. **1979** *Nature* 7 June 537/1 (*caption*) The 5-doxylstearic acid..was dried down from chloroform/methanol (2:1 v/v) solution in a round-bottom flask. **1979** *Jrnl. R. Soc. Arts* Nov. 746/1 The round-hole capsules resemble round-hole tea ceremony rooms.

round, *adv.* and *prep.* Add: **A.** *adv.* **I. 1. a.** Also in phr. *round and round* (further examples).

1898 A. B. Gomme *Tradit. Games* II. 143 Round and round went the gallant, gallant ship. **1936** *Billboard* 11 Jan. 12/4 That latest contagion, *Music Goes Round and Round*,..is selling at the rate of 16,000 copies daily. **1977** *Washington Post* 26 Dec. c8/2 The music sells the movie. The movie sells the albums. The TV and radio and newspapers sell both. Round and round we go.

e. *all the year round* (further *attrib.* examples). Also as sb. phr. (see quot. 1883).

1883 Hardy in *Graphic* Summer 4/2 One of those curious summer shelters sometimes erected on exposed points of view, called an all-the-year-round. **1910** *Busy Man's Mag.* Feb. 58/2 Vancouver is becoming an all-the-year-round resort. **1939** G. Greene *Confidential Agent* IV. i. 269 We want to make it an all-the-year-round resort. **1963** *Times* 5 Feb. 7/5 Equipped for all-the-year-round motoring.

g. (Earlier example.)

1857–8 in W. Whitman *Daybks. & Notebks.* (1978) III. 676 The Doctor has evidently been 'round some'.

III. 11. *round-turning* (later example); *round-girdled* adj.

1878 O. Wilde *Ravenna* 5 A moon of fire Round-girdled with a purple marriage-ring. **1923** D. H. Lawrence *Birds, Beasts, & Flowers* 25 Am I not blind, at the round-turning mill?

B. *prep.* **1. a.** Also in phr. *round and round* (further examples); *round-the-world* (further examples); **round-the-** (also †**me**) **houses,** (*a*) *Rhyming slang*, trousers (see also *round *sb.*[1] 25); (*b*) *attrib. phr.* applied to a motor race or circuit following the streets of a city.

1898 A. B. Gomme *Tradit. Games* II. 122 Round and round the village, As we have done before. **1951** in *Oxf. Dict. Nursery Rhymes* 184 Round and round the garden like a teddy bear; One step, two step, Tickle you under there! **1977** *Washington Post* 26 Dec. c8/2 Travolta himself..is going round and round the country with a cordon of publicists.

Comb. **1857** 'Ducange Anglicus' *Vulgar Tongue* 17 *Round me houses*,..trousers, pronounced trouses [ed. 2, 1859, trousies]. **1858** A. Mayhew *Paved with Gold* II. x. 169 Philip intimating that, as soon as he had put on his *trousers*, he would blacken Bill's *eyes*, roared out, 'Wait till I've togged my 'round-the-houses', and then I'll cook your 'mince-pies' for you'. **1898** J. D. Brayshaw *Slum Silhouettes* 220 An' as fer 'is rahnd-the-'ouses, they 'ad a crease right dahn 'em. **1906** E. Dyson *Fact'ry 'Ands* xiii. 164 No man that wore 'ome-made round-th'-'ouses ever done wonders in this world. **1932** A. Christie *Peril at End House* xi. 131 It was he who financed..the expenses of the round-the-world flight. **1935** Eyston & Lyndon *Motor Racing* iv. 38 The introduction of events run over short circuits planned within the confines of a town. These have become known as 'round-the-houses' races, the first of which was the Grand Prix of Monaco, inaugurated in 1929. **1957** S. Moss *In Track of Speed* i. 14 The Manx Cup race in the Isle of Man, which was a sort of 'round-the-houses' contest in the environs of Douglas. **1970** *N.Z. News* 21 Jan. 16/1 Champion of the American circuit, British-born Ron Grant revelled in conditions he had not experienced before in the Wanganui annual round the houses motorcycle race. **1974** *Times* 10 Jan. 12 The West German round-the-world racing yacht, Peter von Danzig, is putting into Bluff, on New Zealand's South Island. **1974** P. Wright *Lang. British Industry* x. 87 Some of it [*sc.* rhyming slang] apparently doesn't even rhyme properly; e.g. *round the houses* (trousers).

c. *round the clock, the clock round*: see *clock *sb.*[1] 4. Also *Comb.*, as *round-the-year.*

1959 *News Chron.* 28 Nov. 3/1 Round-the-year sea bathing.

2. c. *colloq.* Of time: About; approximately. Cf. *around *prep.* 4 b.

1928 F. N. Hart *Bellamy Trial* iii. 92 It must have been round quarter to nine. **1942** Partridge *Usage & Abusage* 277/1 *Round* for *on* or *about* is a characteristic of Cockney speech: e.g. 'Meet me round seven o'clock'.

5. a. *round-the-corner* (further examples).

1881 [see *lazy-tongs]. **1915** A. Conan Doyle *Valley of Fear* I. vii. 121 You get to your point, I admit, but you have such a deuced round-the-corner way of doing it.

c. *round the wicket*: see *bowl *v.*[1] 4 b. (Further examples.)

1867 G. H. Selkirk *Guide to Cricket Ground* iv. 61 The discretion allowed to the bowler to deliver the ball either over or round the wicket. **1894** N. Gale *Cricket Songs* 26 If round the wicket, medium pace, Won't make the batsman budge,.. Sling him a grub. **1956** N. Cardus *Close of Play* 14 Macaulay bowled off-spin from round the wicket. **1966** B. Johnston *Armchair Cricket* 109 *Round the wicket*, a method of delivery where the bowler has his back to the stumps at the bowling end as he delivers the ball, i.e. a right-arm bowler bowls on the right-hand side of the stumps, a left-arm bowler on the left-side. **1974** *Sunday Tel.* 9 June 34/6 Titmus, fancying his chances, went round the wicket and induced the predictable catch to short leg.

d. Phr. *round the bend*: see *bend *sb.*[4] 10 c.

round, *v.*[1] Add: **I. 4. d.** To approximate (a number) by expressing it in fewer significant figures (the rightmost digit(s) being replaced by 0 and the last unaltered digit being increased by 1 when the digit that followed is 5 (or 6) or more); to express (a number) in a less exact but more convenient form. Also with *down, off, up* (see senses 5 h, 6 e, 8 d below).

1934 in Webster. **1935** Shuster & Bedford *Field Work in Math.* iv. 14 Round the following numbers to three significant figures. *Ibid.* 15 Multiply 2·87 ft. (*a*) by 3·14, (*b*) by 3·142... Round in each case to three figures. **1956** G. A. Montgomerie *Digital Calculating Machines* vii. 129 The usual rule is to ignore a digit less than five and to add one in the next place for five or more. 3·54 would be rounded to 3·5, 3·55 to 3·6, 3·56 to 3·6. **1962** *B.S.I. News* Jan. 25/1 The results are either exact or have been rounded by the accepted convention to the number of significant figures given. **1966** *Rep. Comm. Inquiry Univ. Oxf.* II. p. xxxii, In the tables each figure is rounded separately. **1971** *Jrnl. Gen. Psychol.* LXXXV. 72 The loadings have been rounded to two figures.

5. *round up.* **d.** (Further example.)

1947 A. Ransome *Great Northern?* viii. 111 The *Sea Bear* slipped on in silence towards the big white motor yacht... She rounded up perhaps forty yards away.

e. (Later *lit.* and earlier and later *transf.* examples.) Also *absol.*

1869 *Overland Monthly* III. 126 At night they 'round up' or 'corral'. **1907** C. E. Mulford *Bar-20* 15 They shore outer be here now. They rounded up last week. **1925** E. F. Norton *Fight for Everest 1924* 26 Kingston and I amused ourselves by trying to round up some kiang [*sc.* wild donkeys]. **1949** *Sky Line Trail* Oct. 18/1, I met some cowboys rounding up strayed horses.

transf. **1885** *Weekly New Mexican Rev.* 15 Jan. 2/5 Mr. Twitchell went down to 'round up' the gang and was so far successful as to spot the leader. **1910** *Chambers's Jrnl.* June 384/1, I have seen groups of these unfortunates 'rounded up' and marched off to the nearest police-station. **1931** *Daily Express* 15 Oct. 6/3 The star-traders of the talkies have been out rounding up fresh material from which to carve the box-office idols of the future. **1944** M. Laski *Love on Supertax* xii. 118 They delay.. arrests in the futile hope of 'rounding up the whole gang'. **1975** P. G. Winslow *Death of Angel* vi. 136, I heard about your difficulty and immediately rounded up Cecil.

h. To increase (a number) when rounding it (cf. sense 4 d above) by adding 1 to its rightmost remaining digit, or by expressing it as the next higher round number.

1956 G. A. Montgomerie *Digital Calculating Machines* vii. 129 In a long calculation, all these increases may accumulate, and it is better to round some of them up and some of them down. **1963** *Rep. Comm. Inquiry Decimal Currency* iv. 30, in *Parl. Papers 1962–3* (Cmnd. 2145) XI. 195 The custom with some of these goods is to round up, sometimes down, to the nearest halfpenny. **1969** *Guardian* 30 July 16/1 The Shell-Mex and BP group..will not be advising the 17,000 stations it supplies whether to 'round-up' or 'round-down' petrol prices when the halfpenny ceases to be legal tender. **1975** *Language for Life* (Dept. Educ. & Sci.) xxi. 310 Some Authorities with schools of under 100 pupils round up the number on roll to the nearest 50 and calculate their *per capita* allowance on that basis. **1976** [see *rounding *vbl. sb.* 1 c].

6. *round off.* **e.** = sense 4 d above. Also *absol.*

1935 Shuster & Bedford *Field Work in Math.* iv. 14 The product given above, 20·671728 ft., should be 'rounded off' to 20.7 ft. **1945** J. P. Eckert et al. *Description of ENIAC* (PB 86242) (Moore School of Electr. Engin., Univ. Pennsylvania) B-5 The products c_i are rounded-off to the same number of places. **1977** K. M. E. Murray *Caught in Web of Words* xi. 211 James had rounded off sums downwards rather than upwards—writing £900 for an actual £975 for example. **1978** Green & Lewis *Sci. with Pocket Calculators* ii. 21 Many calculators..round off automatically when displaying results.

8. a. (Further examples.) Also *fig.* and *refl.*

1926 *Publishers' Weekly* 29 May 1789/2 Presently we came away. The inquiry was rounding itself out. **1937** A. L. Rowse *Sir Richard Grenville* 10 New discoveries.. helping to round out and present at length a fairly full portrait of the man. **1947** J. C. Rich *Materials & Methods of Sculpture* Pl. 35/4 (*caption*) Rounding out the forms from the front of the sheet. **1966** *Listener* 24 Nov. 763/1 Now, with three full-length plays behind him..it is possible to round out a little that first impression. **1972** *Daily Tel.* 30 Nov. 21 Lloyds Bank's new merchant bank, set up yesterday to round out the bank's services, has no name. **1979** *Arizona Daily Star* 5 Aug. I. 1/4 The third act very satisfactorily rounds out what has long been a frustrating, partially finished production.

d. *round down*, to decrease (a number) when rounding it (cf. sense 4 d above) by making no alteration to its remaining digits, or by expressing it as the next lower round number.

1956, etc. [see sense 5 h above]. **1970** *Guardian* 19 Feb. 13/6 The new conversion table would enable prices sometimes to be rounded down, although some may be rounded up. **1971** *Daily Tel.* 9 Nov. 14 It is Post Office practice for telephone bill totals ending in ½p to be rounded down to the nearest whole penny. **1976** [see *rounding *vbl. sb.* 1 c].

II. 11. c. Delete *rare*⁻¹ and add later example.

1911 W. James *Some Probl. Philos.* vi. 99 Rationalistic philosophy has always aspired to a rounded-in view of the whole of things, a closed system of kinds.

III. 13. a. (Later example.) Now *rare.*

1941 *Penguin New Writing* II. 14 Early as it is women and old men are hunting for scraps of coal on the side of the incline. They have to be away before the police start to round.

b. Also const. *in.*

1924 Galsworthy *White Monkey* I. xiii. 109 He rounded-in from the Embankment towards home.

f. To turn round *on*; to assail, assault, esp. with words; to abuse, berate.

1882 *Sydney Slang Dict.* 7/2 *Round* (*on a man*),..to abuse. **1909** *Blackw. Mag.* Sept. 413/1 On one occasion.. she had rounded on him and scolded him for a full half-hour. **1932** E. Bowen *To North* xx. 211 Cecilia did not round on Julian. **1966** *Listener* 24 Nov. 764/2 It may be possible to find a parallel in the work of other writers whose first impulse, as young men, was to round on society. **1973** *Times* 16 Nov. 4 Professor Peters also rounds on the Inner London Education Authority for exceeding its brief.

14. a. Also with *out.*

1912 *Red Mag.* Apr. 510/2, I guess she didn't know how she had rounded out in the mountain air.

d. (Earlier example.)

1879 *Missouri Republican* 22 Oct. 3/7 Are you going to 'round up' at Maj. B.'s tonight?

round about, *adv.* and *prep.* Add: **B.** *prep.* **3.** Of time, amount, etc.: about; approximately; = *around *prep.* 4 b.

1913 P. Reeves (*title*) Round about a pound a week. **1926** W. R. Inge *Lay Thoughts* 182 In the Middle Ages the births and deaths in the undrained towns were both round about 50 per thousand in each year. **1961** N. Cardus *Sir T. Beecham* 64 It was round about 1931 that he told me he was about to form a new orchestra in London.

roundabout, *sb.* and *a.* Add: **A.** *sb.* **2. b.** (Earlier and later examples.)

1823 J. F. Cooper *Pilot* I. vi. 66 The young sailor.. slipped his arms into the sleeves of a morning roundabout, covered with the trappings of his profession. **1825** in *Trans Illinois State Hist. Soc. 1910* (1912) 177, I have twelve shirts six pair Pantaloons 6 vests..two roundabouts. **1904** *N.Y. Even. Post* 7 Jan. 7 Only yesterday this young man was playing about the streets of Washington, a schoolboy in roundabouts.

c. (Earlier example.)

1844 *Lowell Offering* IV. 175 [He sat] in a large flat-bottomed 'roundabout' on the opposite side of the fire-place.

†**d.** *U.S.* A loose dressing-gown worn by women. **e.** *N.Z.* (See quot. 1861.) *Obs.*

1841 *Southern Lit. Messenger* VII. 525/1 The garment is a long, loose roundabout, connecting in front with strings, and is much worn, even at the present time. **1856** V. Lush *Jrnl.* 17 Jan. (1971) 176 The whole lot stood quietly looking at us, clothed from top to toe in their long full roundabouts. **1861** R. B. Paul *N.Z.* 17 [The Maori women's] usual dress is..a shapeless sack of printed calico, called a 'roundabout', tied round the neck but loose at the waist. **1874** W. M. Baines *Narr. E. Crewe* 118, I also gave [the Maori girl] 8 yards of Navy blue print (which everyone knows is enough for a 'roundabout'). **1890** P. A. Philips *Reminisc. Early Days* 7 The hostess did not dress for dinner..her usual attire being a Maori roundabout. **1895** K. D. Wiggin *Village Watch-Tower* 103 Mother had let her slip on her new green roundabout over her nightgown.

4. b. *to gain on the swings and lose on the roundabouts*: see *swing *sb.*[2]

d. A junction at which traffic moves one way round a central island. Cf. *rond-point b, *rotary *sb.* 3.

1927 *Glasgow Herald* 3 Jan. 7/2 There is only one drawback to the roundabout, and that is the inconvenience caused to pedestrians. **1937** *Times* 13 Apr. (British Motor No.) p. viii/1 Roundabouts..have the advantage of keeping vehicles on the move. **1947** *Daily Mail* 22 May 3/4 Removal of the Mansion House to make room for a big round-about. **1955** *Times* 2 Aug. 9/7 Makeshift tactics are particularly evident in the proposed treatment at Hyde Park Corner which includes an extremely complicated roundabout. **1967** *Listener* 28 Sept. 398/1 People make only occasional use of their speedometer.. on such critical occasions as the approach to roundabouts. **1977** *Belfast Tel.* 14 Feb. 5/9, 12 shots were fired at an armoured police vehicle near the roundabout at Narrow-water Castle.

B. *adj.* **1. d.** (Earlier example.)

1737 *S. Carolina Gaz.* 30 Apr.–7 May 1/1 [Drunkenness] bears no kind of Similitude with any sort of Virtue, from which it might..borrow a Name; and is therefore reduc'd to the wretched Necessity of being express'd by distant round-about Phrases.

f. Of or pertaining to a junction at which traffic moves one way around a central island. Cf. sense A. 4 d above.

1927 *Rep. Commissioner Police Metropolis, 1926* 18 During the past year round-about systems of traffic have been put into operation at Parliament Square [etc.]. **1939** *War Illustr.* 7 Oct. 127 This car was found abandoned the morning after colliding with the posts of a 'roundabout' island. **1976** *Alyn & Deeside Observer* 10 Dec. 1/6 From there it runs to Broughton in the Welsh county of Clwyd and ends at a roundabout junction.

3. b. Designating a type of chair with a rounded seat or back (see quots.). Cf. sense A. 2 c.

1741 in J. S. Moore *Goods & Chattels of our Forefathers* (1976) 286 Six India Back Chairs and a Round about Ditto with Leather seats £2 0s od. **1840** *Knickerbocker* XVI. 115, I sat in my roundabout chair the other evening. **1936** F. C. Morse *Furniture* 170 'Roundabout' chairs are met with in inventories from 1738 under various names,—'three-cornered chair', 'half round chair', 'round about chair'. **1952** J. Gloag *Short Dict. Furnit.* 398 The round-about chair has a circular seat, either upholstered or caned, a semi-circular back, and six legs. **1960** H. Hayward *Antique Coll.* 84/1 Corner chairs were known in the 17th cent. and were sometimes called elbow chairs or roundabout chairs. **1966** M. M. Pegler *Dict. Interior Design* (1967) 380 *Roundabout chair.* This chair is usually designed to fit into a corner, the square seat diagonally set and the back extending across two adjoining sides.

5. a. (Earlier and later examples.)

1802 C. Findlater *Gen. View Agric. Peebles* 40 The round-about fire side..was universally in use in the kitchen. **1978** T. Henderson *Shetland* 142 (*caption*) A round-about fire in Walls about 1910.

round-arm, *a.* and *adv.* Add: **1.** Also, of a bowler who delivers the ball thus.

[**1833** J. Mitford in *Gentl. Mag.* Sept. 238/2 Ashby.. introduced the round bowling, by throwing the arm in a sweeping circular position.] **1836** *New Sporting Mag.* Oct. 358 Mr. Lowth is a round-arm, left handed bowler.

round-armed *a.* (earlier examples); hence **round-armer,** a round-armed delivery.

1854 Round-armed [see *bowl *v.*[1] 4 b]. **1863** *Lillywhite's Cricket Scores* III. 43 A medium-paced round-armed bowler. **1951** R. Robinson *From Boundary* ii. 39 He rings in a leg-break or a round-armer now and again. **1954** J. Fingleton *Ashes crown Year* 255 Then came Miller's..appeal..as he rapped May's pads with a round-armer.

rounded, *ppl. a.* Add: **II. 6. c.** Of a number: having been approximated by rounding; expressed in fewer significant figures. Also with advb.

1947 *Math. Tables & Other Aids Computation* II. 286 He had taken 10-figure logarithms of rounded-off quantities containing only five or six significant figures. **1953** *Proc. IRE* XLI. 1271/1 *Long Right* places the bits to be dropped into the *MQ* register; *Round* then leaves a rounded number in the accumulator. **1956** G. A. Montgomerie *Digital Calculating Machines* vii. 129 A convenient rule to ensure this is to make the rounded digit even rather than odd in case of doubt. **1973** Phillips & Taylor *Theory & Applic. Numerical Anal.* 359 If the amount neglected is exactly $\frac{1}{2}10^{-t}$ we can avoid statistical bias by forcing the last digit in the rounded number to be even.

roundel. Add: **I. 2. c.** Now only *Hist.* (Further example.)

1971 R. Howe *Mrs. Groundes-Peace's Old Cookery Notebk.* 56 Wooden trenchers were also known as treen roundels.

5. d. (Further examples.)

1885 [see *bullion[3] 2]. **1898** W. Gandy *Romance of Glass-Making* ix. 145 Now and then one comes across an old window—generally a cellar window—where the panes have been filled with bull's-eyes, 'roundels', or the waste centres left from the discs of crown glass after cutting. **1908** A. L. Duthie *Decorative Glass Processes* i. 28 Circular in form..are *roundels*, which have always been largely used in leaded lights and are characteristic of German and Italian windows. They are made in an infinite variety of colour and size. **1933** R. Mollet *Leaded Glass Work* ii. 13 Bullions are very popular... The smaller sizes (2 or 3 in. in diameter) are sometimes called 'roundels'.

e. An identification disc painted on an aeroplane; *spec.* that of the Royal Air Force and Royal Naval Air Command, comprising a design of concentric red, white, and blue circles.

1948 *Daily Tel.* 5 July 1/1 The R.A.F. plane—I could see the roundel—was spiralling down without a tail. **1963** J. Lusby in B. James *Austral. Short Stories* 225 Wingtip clear of the next man's, able to move forward or back and level with his roundel. **1975** T. Allbeury *Palomino Blonde* xxiii. 142 A helicopter came..across the bay. The RAF roundels looked fresh and clean.

rounder. Add: **I. 1. d.** For *U.S.* read *N.Amer.* and add earlier and later examples. Also *transf.*

1854 *Congress Globe* 33rd Congress 1st Sess. App. 1220/3 I have always found him a very kind and agreeable man—what the 'rounders' in New York would term a 'glover'. **1879** A. Daly *Let.* 20 Oct. in J. F. Daly *Life A. Daly* (1917) xxi. 330 [We] are old 'rounders' and familiar with the voice, gait and peculiarities of most of the actors and actresses on the American stage. **1920** C. Sandburg *Smoke & Steel* 51 A rounder leered confidential. **1935** Z. N. Hurston *Mules & Men* (1970) I. iv. 93 'What make de rooster crow every morning at sun-up?' 'Dat's to let de pimps and rounders know de workin' man is on his way.' **1943** W. H. Chase *Sourdough Pot* v. 24 [She] possessed all the earmarks of a 'rounder'—the evidence of much dissipation was remarkably developed. **1962** 'K. Orvis' *Damned & Destroyed* iv. 29 An all-night dive patronized by cheap women and rounders and drunks. **1975** *Globe & Mail* (Toronto) 7 Oct. 4/5 He agreed that rounders—a term used to describe criminals whose haunts include hotels in the Jarvis Street–Dundas Street area of Toronto —have a great fear they will be seen talking to the police.

e. *U.S. slang.* (See quot. 1903.)

1881 *Bradstreet's* 29 Jan. 51/4 The 'rounder' in almstaking is headed off. **1903** *Charities* 3 Oct. 283 The class of persons known as 'rounders', people who go from one hospital to another seeking advice and treatment, a species of medical mendicants.

f. *U.S. slang.* A transient railway worker.

1908 *Casey Jones* (song) in *Railroad Man's Mag.* May 764/1 Come all you rounders, for I want you to hear The story told of an engineer, Casey Jones was the rounder's name, A heavy right-wheeler of a mighty fame. **1939** F. J. Lee *Casey Jones* 287 The word 'rounder' as applied to Casey must be taken as a light, affectionate appellation. **1961** *Listener* 24 Aug. 270/2 His was a six-pipe job whose moans sent every coloured 'rounder' from Chicago to New Orleans into ecstasies.

2. a. (Earlier and later examples.)

1828 W. Clarke *Boy's Own Book* (ed. 2, London) 20 *Rounders.* In the west of England this is one of the most favourite sports with bat and ball... In Rounders, the players divide into two equal parties, and chance decides which shall have the first innings. **1854** Dickens *Let.* 12 July (1938) II. 566 The keeping up of a 'home' at rounders. **1939** *Bull. N.Y. Public Libr.* Apr. 303 Is baseball an offshoot of rounders? **1969** I. & P. Opie *Children's Games* 5 At Sedgley Park School in Staffordshire, about 1805, the boys were content with Kites, Marloes (marbles),..Rounders,..and even with 'playing horses'. **1977** *Cleethorpes News* 27 May 18/1 In addition to exploring the mines, they played rounders among the sheep on the hillside.

II. 5. (Earlier example.)

1774 T. Pennant *Tour in Scotl. & Voy. Hebrides 1772* I. 99 On each side the gateway..are two rounders.

III. 11. *Newfoundland.* Small unsplit cod, freq. eaten as a delicacy.

1908 C. W. Townsend *Along Labrador Coast* v. 132 The very small cod are not boned, but are salted whole. These are called 'leggies' or 'rounders'. **1966** A. R. Scammell *My Newfoundland* 32 School fees could not be put on the account and the schoolmaster wouldn't accept fish, tomcods or rounders. *Ibid.* 91 When the last rounder was aboard he gave a quizzical glance at the sun. **1974** *National Geogr. Mag.* Jan. 129/2 We dined sumptuously on native dishes..'rounders' (baby cod, salted and dried whole like kippers, and boiled for breakfast).

round-faced, *a.* Add: **2.** Also, *round-faced monkey.* For *Macaques* substitute *Macaca.*

1872 *Proc. Zool. Soc.* 777 It would seem that our Round-faced Monkey, in the proportionate dimensions of fore limb to spine's length, presents closest agreement with man.

Roundhead, round-head. Add: **1. a.** (Later *transf.* and *attrib.* examples.)

1963 *Times* 11 Feb. 6/7 Now that industry is 'dishoarding' labour and achieving results more effectively than Roundhead policy at the Treasury. **1973** D. Aaron *Unwritten War* 345 Southern magazines featured articles contrasting invidiously planter 'Cavaliers' and Yankee 'Roundheads'. **1976** *Listener* 5 Feb. 140/3 Under the Cromwellian leadership of Peter Hall, the roundheads of the new professionalism drove the cavalier dilettanti largely from the scene.

b. *N.Amer. slang.* An immigrant from northern Europe, *spec.* a Swede.

1895 *Dialect Notes* I. 393 *Roundhead*, a Swede. **1902** S. Clapin *New Dict. Americanisms* 341 *Roundhead*, in the North-West, frequently said of a Swede. **1931** 'D. Stiff' *Milk & Honey Route* iii. 38 Swedes are 'roundheads' or 'salve eaters'. **1976** 'Trevanian' *Main* (1977) iii. 57 'He's not a bad type, for a Roundhead,' Gaspard says.

c. *Ethnol.* One of a race or type of man characterized by roundness of the head. Cf. *round-headed *a.* 1 b. *rare.*

1896 A. H. Keane *Ethnol.* I. v. 106 Mounds differing in type from those of the round-heads.

4. (Further example.)

1908 A. W. Tilby *Eng. People Overseas* I. ii. 72 The former was strongly cavalier and episcopal; the latter was as strongly roundhead and puritan.

Hence **rou·ndheader** = sense 1 b above.

1934 J. O'Hara *Appointment in Samarra* iii. 80 The schwackies, the roundheaders..—regional names for non-Latin foreigners—probably were inside getting drunk.

round-headed, *a.* Add: **1. b.** *Ethnol.* Designating a race or type of man characterized by possessing a skull of rounded shape, usu. distinguished from a Long-head (sense 2). *rare.*

1896 A. H. Keane *Ethnol.* I. v. 106 Mr. W. K. Moorehead..recognises two distinct mound-building races, the old long-headed, and the later round-headed intruders.

Hence **round-hea·dedness,** the state of being round-headed; the condition of having a round head.

1935 [see *high frequency 1 b].

round-house, *sb.* Add: **2. b.** (Earlier example.)

1808 J. Davis *Post Captain* (ed. 3) i. 5, I..was obliged to get up in the night to go to the roundhouse.

3. a. (Further example.)

1971 *Country Life* 11 Nov. 1325/1 An example..is a farm building specifically termed the wheelhouse, but more popularly known as roundhouse or gin gan.

4. For *U.S.* read 'orig. *U.S.*' and add earlier and later examples. Also *fig.* and *attrib.*

1856 W. Ferguson *Amer. by River & Rail* 249 The engine-house..is open in the centre; and this arrangement ..is much less expensive, than the 'round house', where all is covered in. **1895** *Rep. Chicago Strike 1894* (U.S. Strike Comm.) 214 A number of switch tenders, yard clerks, flagmen, tower men, and roundhouse men left their work. **1945** F. H. Hubbard *Railroad Avenue* ii. 10 Many runners considered it smart to keep roundhouse work on an engine down to a minimum—the fewer the defects they reported, the better standing they had at the roundhouse. **1953** *Manch. Guardian Weekly* 5 Nov. 15/1 This restoration of power [over the money supply] was hailed by a writer in 'Harper's Magazine' as a..guarantee that 'the engine of inflation has been stowed firmly in the roundhouse'. **1966** M. R. D. Foot *SOE in France* ix. 269 Six large engines in the Troyes locomotive roundhouse. **1980** *Dædalus* Spring 121 Myths provide a conceptual system through which we may understand..a roundhouse where we can move from the track of one person's reality to another's.

5. a. *U.S. Baseball.* A pitch made with a sweeping side-arm motion. Also *attrib.*

1910 *Amer. Mag.* June 224/2 The first curves discovered were of the variety now known as the 'barrel hoop' or 'round house'. **1912** C. Mathewson *Pitching* 19 When I first joined the Giants, I had what is known as the 'old round-house curve', which is no more than a big, slow outdrop. **1926** *Amer. Speech* I. 369/2 Pitched balls are designated by obvious terms. A 'spitter', a 'hook',..a 'round-house'.

b. *slang* (orig. *U.S.*). A blow delivered with a wide sweep of the arm. Also *fig.* Freq. *attrib.*, esp. as *roundhouse left, right.*

1920 *Collier's Mag.* 3 July 34/4 He swung a roundhouse left, square to the Kid's unprotected face. **1927** *Daily Express* 16 Dec. 3 It is necessary to take the [golf] ball cleanly, and with something akin to a 'round-house' swing from this position. **1932** J. T. Farrell *Young Lonigan* iii. 133 They fought, slugging, socking away, rushing, swinging with haymakers and wild swishing roundhouses. **1945** *Tee Emm* (Air Ministry) V. 51 Discourage him by jabbing his snout or gills with an oar,.. and don't take round-house swings that may upset you. **1946** *Sun* (Baltimore) 14 Dec. 2/6 Mr. Collins leaped to his feet and swung a round-house right at the witness. **1948** *Ibid.* 12 May 17/6 Trainer Jimmy Jones obviously was disappointed at failure to have Coaltown [*sc.* a horse] on hand to deliver Columet's [*sic*] winning jab and round-house... Jimmy will saddle the Calumet starter. **1958** J. Kerouac *On Road* iv. 126 Damion's girl suddenly socked Damion on the jaw with a roundhouse right. **1967** *Boston Herald* 1 Apr. 17/7 Harris built up an early lead over the baffled 29-year-old veteran with his roundhouse blows. **1976** M. Machlin *Pipeline* lvii. 573 Doheny's arm drew back, as though to launch a roundhouse upper-cut.

round-house *v.*, (*b*) *slang*, to hit (a person) with a round-house blow.

1974 W. Garner *Big Enough Wreath* xi. 137 She roundhoused Smith with a white plastic handbag that must have had a brick in it.

rounding, *vbl. sb.*[1] Add: **1. a.** (Further examples.) Also with *off, out, up.*

1876 M. Whilldin *Descr. Western Texas* 16 It soon became evident that a place near us had been selected for 'rounding up'. **1916** *Daily News* 6 Sept. 4 There can rarely..have been a better example of the insolence of Zabernism than the 'rounding up' of crowds of unoffending people at the stations. **1932** W. C. Holden *Rollie Burns* xiii. 172 Our rounding-up outfit was camped about eight miles from the Yellow House Canyon. **1936** *Trans. Philol. Soc.* 78 The rounding of a to [ɔ] is in Middle English generally held to be a West Midland feature. **1947** A. Einstein *Mus. Romantic Era* xi. 127 The perfect rounding-out of the form, which from the musical standpoint is entirely self-contained. **1949** *Jrnl. R. Aeronaut.*

Soc. LIII. 957/1 The cabin floor angle in the steeper types, such as Dakotas and Lancastrians, is changed as slowly as possible by slow rounding out and by landing with the tail just off the ground. **1961** E. A. Powdrill *Vocab. Land Planning* iii. 38 The private developer refers to most forms of peripheral development as 'rounding-off', whilst the planning authority merely contends that it is not, usually without saying what rounding-off really means. **1971** P. Gresswell *Environment* 132 Development will be severely restricted except for a reasonable amount of infilling and/or rounding off. **1977** *Canad. Jrnl. Linguistics 1976* XXI. 176 At this level we want to state, for example, that a language has vowels which are opposed in rounding.

c. The action of *round *v.*[1] 4 d. Also with *down, off, up* (cf. *round *v.*[1] 5 h, 6 e, 8 d).

1935 Shuster & Bedford *Field Work in Math.* iv. 14 In computation with approximate numbers, rounding off should be done by these rules. **1953** *Proc. IRE* XLI. 1270/1 Different calculations may require different methods of rounding in order to reduce the residual rounding error to an acceptably low level. **1963** *Rep. Comm. Inquiry Decimal Currency* iii. 18 in *Parl. Papers 1962–3* (Cmnd. 2145) XI. 195 Outstanding balances would convert exactly on the changeover date, with no discrepancies which might or might not be explained by decimalisation roundings. **1973** C. W. Gear *Introd. Computer Sci.* vi. 249 The errors introduced by rounding and truncation are initially small, but sometimes their effect is amplified by subsequent operations. **1976** C. Birtwistle *Electronic Calculator* iii. 35, 23 is rounded down to 20 and 28 is rounded up to 30. Where is the critical point at which rounding down changes to rounding up? Obviously it is the half-way mark, 25.

3. a. *rounding plane*; **rounding error** = *round-off error* s.v. *round-off *sb.* 1.

1948 *Math. Tables & Other Aids to Computation* III. 79 The operator is concerned with numerous questions of digital accuracy and the accumulation of rounding errors. **1962** A. Battersby *Guide to Stock Control* 115 Allowing for rounding errors, the stockholding cost at ¼ per cent would be £15. **1973** C. W. Gear *Introd. Computer Sci.* vi. 249 The first source is called rounding error, and is due to the fact that only a finite set of all of the real numbers can be represented in the computer as floating-point numbers. **1940** *Chambers's Techn. Dict.* 732/2 Rounding plane. **1969** E. H. Pinto *Treen* 389 Different versions of the same tool, known as a stail-engine, witchet, or rounding and tapering plane.

b. rounding-off error = **round-off error.*

1945 J. von Neumann in B. Randell *Origins Digital Computers* (1973) 362 A reasonable precision for many differential equation problems is given..by keeping the relative rounding-off errors below 10^{-8}. **1974** W. T. Welford *Aberrations Symmetrical Optical Syst.* vi. 84 The difference between them is to be calculated to a fraction of a wavelength; this may make a heavy demand on the computer when rounding-off errors are allowed for, since the computer word length may correspond to only about eight decimal digits.

roundness. Add: **1. a.** (Further *fig.* example.)

1927 E. M. Forster *Aspects of Novel* iv. 98 Dickens's people are nearly all flat (Pip and David Copperfield attempt roundness, but so diffidently that they seem more like bubbles than solids).

d. *out-of-roundness*: see *round *sb.*[1] 5 e.

round-off, *a.* Add: **B.** *sb.* **1.** = *rounding *vbl. sb.*[1] 1 c. Freq. *attrib.* as *round-off error*, the error introduced thereby.

1946 Goldstine & von Neumann in J. von Neumann *Coll. Wks.* (1963) V. 17 A very complicated calculation in which the accumulation and amplification of the round-off errors threatens to prevent the obtaining of results of the desired precision. **1947** *Bull. Amer. Math. Soc.* LIII. 1026 The transition from the true operations to their pseudo-operations is effected by any one of the familiar methods of round off. **1962** *Times Lit. Suppl.* 20 Apr. 268/1 Whenever approximations due to round-offs or other reasons are indicated, the degree of approximation is perfectly well determined. **1973** C. W. Gear *Introd. Computer Sci.* vi. 258 Sometimes a bad choice of method will make the solution very sensitive to round-off or truncation errors.

2. The act of rounding off or completing an operation appropriately.

1964 *Trampolining* ('Know the Game' Ser.) 40/1 It is similar in action to the round-off in ground work tumbling.

Round Robin. Add: Now usu. with small initials. **3. a.** (Earlier and later examples.) Now used loosely of any such document signed by many persons, freq. in alphabetical order to indicate that responsibility is shared.

(*a*) **1730** *Weekly Jrnl.* 3 Jan. 3/4 A Round Robin is a Name given by Seamen, to an Instrument on which they sign their Names round a Circle, to prevent the Ring-leader being discover'd by it, if found.

(*b*) **1978** B. Levin in K. Gregory *First Cuckoo* 13 Writers of round robins ('We, the undersigned, each in his or her personal capacity..') also choose *The Times* for preference, the second elevens being accommodated elsewhere.

transf. **1977** *N.Y. Rev. Bks.* 4 Aug. 7/1 As if to point up the homosexual theme, rather than to offer a round robin of sexuality.

b. orig. *U.S.* A tournament in which every player or team competes once with each of the others. Freq. *attrib.*

1895 *Official Lawn Tennis Bull.* 3 Jan. 1/2 The so-called round-robin tournament, where each man plays every other, furnishes the best possible test of tennis skill. *Ibid.* 3/1 No one would..argue that a man of that rating could win in a round-robin. **1904** J. P. Paret *Lawn Tennis* iii. 24 Invitation tournaments are of American origin, and the matches are generally played on what is called the 'round robin' system, each of the players meeting all of the others in turn. *Ibid.* iv. 65 The British visitors next played a round-robin at Chicago. **1943** M. Kraitchik *Math. Recreations* ix. 231 In a round-robin tournament among teams of four or two we must arrange a schedule by which every team meets every other just once. **1952** E. Lasker *Chess Secrets* 379 Arrange the players in groups, and have the winners, or the first two or three of each group, play a final round robin. **1974** *Times* 20 Apr. 11/1 The 'Aces' of America [*sc.* a bridge team] held their own against the Italians in the preliminary round-robin to decide who should compete in the final. **1978** *Time* 3 July 50/1 In June, the 14 survivors and the West German team..moved to Argentina to join the host country's team in an exhausting series of round-robin matches.

c. (See quot. 1976.)

1972 J. Wambaugh *Blue Knight* (1973) x. 171 Would you care for a round robin or a three-horse parley to-day? **1976** *Daily Tel.* (Colour Suppl.) 26 Mar. 31/4 *Round robin*, three horses linked in Up and Down bets on each pair, plus three doubles and a treble (10 bets).

Hence **round-robinee·r,** a subscriber to a round robin (sense 3 a), an importuner; **round-ro·bining,** the act of subscribing to a round robin (sense 3 a).

1933 H. P. Long *Every Man a King* xii. 179 If ever again you fifteen round-robineers find me drowning, for Heaven's sake, let me drown! **1968** *Guardian* 13 July 8/6 On Tuesday—after much round-robining—hundreds of BBC secretaries and studio staff intend boycotting the corporation's canteen.

round-shouldered, *a.* Add: (Further *transf.* example.)

1950 H. L. Lorimer *Homer & Monuments* v. 262 Here the round-shouldered short-tanged blade [of a sword] which alone was found at Arkalochon is in the majority, but side by side with it a more serviceable type is developed.

Hence **round-shou·lderedness,** the state or quality of having round shoulders.

1940 S. Spender *Backward Son* 14 References to..his round-shoulderedness.

roundsman. Add: **2.** (Earlier and later examples.)

1868 *N.Y. Herald* 31 July 6/5 Patrolman Jas. Mee..is hereby appointed roundsman on the force. **1870** 'Mark Twain' in *Galaxy* Sept. 430/1 The rank of constable or even roundsman. **1902** *Chambers's Jrnl.* Oct. 674/1 The first grade of promotion is to roundsman... The roundsman is an important man, for on him the discipline of the patrolmen largely depends. **1937** *Sun* (Baltimore) 1 Dec. 3/1 John McAdams, a former customs roundsman, who was dismissed recently, also was accused of being a member of the smuggling ring.

3. (Further examples.)

1935 *E. Anglia Daily Times* 18 Dec. 4/2 A Lowestoft milk roundsman, who persuaded his customers to buy more than 12 per cent. more milk in three months, has won the £25 and cup awarded for salesmanship. **1977** 'M. Underwood' *Fatal Trip* xix. 106 It did not take long to discover the name of the dairy [and]..that Frey Chaytor was the roundsman.

round table, *sb.* Add: **4.** *spec.* an assembly of people for a conference or discussions at which all participants are accorded equal status (in this sense freq. *attrib.*). Also *transf.*, a collection of opinions or remarks on a particular subject.

1889 [in Dict.]. **1892** *Review of Reviews* Feb. 148/1 The subject of the 'Round Table Conference'..is what part churches should take in labour problems. **1901** H. W. Paul *Life Gladstone* xxi. 245 The year 1887 opened with an attempt to reconcile the conflicting elements of the Liberal Party, which came to be known as the Round Table Conference. **1910** (*title of periodical*) The Round Table: a quarterly review of the politics of the British Empire. **1928** *Daily Express* 3 July 2/4 The Archbishop of Canterbury made a striking proposal yesterday for a 'round table' to discuss the future relations of Church and State. **1929** *Times* 31 Oct. 14/3 Mr. Benn himself proposed to visit India forthwith for.. a round-table conference. **1943** M. McCarthy in *Partisan Rev.* May–June 280 The problems..are..opened for discussion in an atmosphere reminiscent of the Chicago Round Table. **1943** *Times* 8 July 5/5 In certain instances we stayed overnight to enjoy the free and easy of a 'round table conference'. **1947** *Radio Times* 14 Mar. 1/2 Round-table controversial political discussions, which the BBC will continue to originate. **1952** D. Riesman in *Antioch Rev.* Dec. 418 A roundtable..of which he was chairman at the Corning Conference. **1955** *Times* 30 July 5/1 Round-table talks may be held in London about a year hence to consider a new constitution for a self-governing Singapore. **1973** *Word 1970* XXVI. 120 He took an active part in the linguistic and anthropological life of Mexico, attending conferences, round tables, and the like. **1976** *National Observer* (U.S.) 10 Apr. 1/3 The Observer convened an informal round-table talk involving six young people with differing perspectives. **1978** *Jrnl. R. Soc. Arts* CXXVI. 768/1 In 1930–31 he was Adviser to the Indian States Delegation to the Round Table Conference in London.

5. (With capital initials.) A formal association whose members meet regularly for discussion, *spec.* an organization (or a branch of it) founded in 1927, in which professional people between the ages of 18 and 40 hold discussions, debates, and similar activities, and undertake community service and the promotion of international understanding. Also allusively as *adj.*, designating the qualities or characteristics associated with the Round Table or its members.

1917 L. Curtis *Let. to People of India* 13 The Round Table organisation..is merely a system for enabling people to unite for the study of their duties as citizens of this Commonwealth, as a guide to their own individual action. **1928** *Review of Reviews* Mar.–Apr. 253/2 At Norwich..an Association of Young Men has established a Club for young business men, meeting once a week in the evening for the reading of papers, discussions, and debates, under the title of 'The Round Table'. **1955** A. Huxley *Let.* 25 Mar. (1969) 739 Possibly also to go up ..and stay a few days near Puharich's Round Table Foundation outfit. **1968** *Guardian* 10 Sept. 2/7 Mr Powell was speaking to the Rowley Regis Round Table at Cradley Heath, Staffordshire. **1972** J. Burmeister *Running Scared* xvii. 213 The woman was..wearing carefully bleached out jeans... Round Table, thought Ginny, with a touch of country. 'I'm terribly sorry.' The voice was Round Table too. **1973** *Stand. Encycl. S. Afr.* IX. 412/2 The Round Table movement in South Africa was founded in East London in 1948. **1977** *Times of Zambia* 7 Sept. 2/4 The committee is made up of the Rotary, Round Table and Lions clubs members.

round table *v.*, **round-tabler.** Delete *nonce-uses* and add later examples.

The use in quot. 1923 is with allusion to the periodical *The Round Table* (see quot. 1910 under sense 4 above).

1923 T. E. Lawrence *Let.* 27 Mar. (1938) 413 You [*sc.* Lionel Curtis] have tried (Round Tabling and by mouth) to tell all whom you can reach. **1976** *National Observer* (U.S.) 1 May 19/3 They might squabble bitterly from time to time, but in the main the Round Tablers quoted each other, promoted each other, wrote books and plays and articles about each other with an incestuous zeal. **1976** *Norwich Mercury* 10 Dec., Children of Round Tablers sing for the 250 old folk from Wymondham and Attleborough who attended a Christmas party in Wymondham Central Hall on Sunday.

round-top. 2. (Further examples.)

1897 *Sears, Roebuck Catal.* 251/2 Black enameled iron, round top trunk. **1962** L. S. Sasieni *Princ. & Pract. Optical Dispensing* xii. 312 Round Top Fused [Trifocals]. **1966** J. S. Cox *Illustr. Dict. Hairdressing* 129/1 *Round-top butch*, a hair style for men in which the hair is cut very short to a round contour.

round trip. orig. *U.S.* Also **round-trip.** [f. Round *a.* 15.] **a.** A circular tour or trip; an outward and return journey.

1860 *Railroad Guide* (Dinsmore & Co.) Sept. 142 Round trip tickets. **1868** *Putnam's Mag.* Mar. 351/1 Time for the round trip..44 hours. **1892** [in Dict. s.v. Round *a.* 15 a]. **1923** R. D. Paine *Comrades of Rolling Ocean* xiii. 223 We signed for the round trip in the *Liberty Chimes*, but we don't feel like taking her all the way home. **1956** *People* 13 May 9/4 The 30-mile round trip to the lonely isle took seven hours. **1976** Scott & Koski *Walk-In* (1977) iii. 21 'Round trip?' the ticket seller asked... 'You want to go and come back?'

b. *transf.* and *fig.*

1932 *Amer. Speech* VII. 270 *Round-trip*.., one *pull-out* and the subsequent *run-in* of the drill pipe in rotary drilling. **1935** *Econ. Geol.* XXX. 739 Very deep wells may require as much as eight hours to make the 'round trip'. **1938** *Amer. Speech* XIII. 220/2 Round-trip words, i.e. from O[ld] F[rench] to Eng[lish], thence to French. **1945** L. Shelly *Jive Talk Dict.* 33/1 *Round trip*, anything unusually good or outstanding. **1963** *Gloss. Mining Terms* (*B.S.I.*) III. 12 *Round trip*, the operation of withdrawing the drill rods and bit, etc., from the hole, of extracting core, replacing rods and bit and resuming drilling. **1973** *N.Y. Law Jrnl.* 20 July 5/1 The..Stock Exchange said yesterday it would like to attract small investors back to the stock market with a 'round trip' rate.

c. *attrib.*

1860 [see sense a above]. **1890** *Brighton* (Colorado) *Reg.* 25 Jan. 1/4 The railway company has rehashed its round-trip rate, from suburban towns to Denver. **1939** G. Greene *Another Mexico* ii. 31 He had got a round-trip ticket to Mexico City. **1976** 'M. Barak' *Secret List H. Roehm* v. 58 Enclosed was an Air France round-trip ticket to Montevideo. **1978** *Detroit Free Press* 5 Mar. D21/3 (Advt.), Commodore's unique Fly Free/Cruise Easy program..pays half of the lowest applicable round-trip air fare..to Miami.

Hence **round-tri·pper,** (*a*) a traveller who makes a round trip; (*b*) in *Baseball*, a home run; a batter who hits a home run; **round-tri·pping** *Econ.*, the practice of earning profit by borrowing on overdraft and relending in money markets.

1944 D. Burley in A. Dundes *Mother Wit* (1973) 219 That's why I'm out here..instead of being a round-tripper. **1962** P. Purser *Peregrination 22* xii. 60 He was a tourist.. the only genuine round-tripper in the first-class. **1974** *Saturday* (Charleston, S. Carolina) 20 Apr. 5–8/4 Jeff Grantz..powered four round-trippers in one inning. **1974** *Daily Tel.* 25 May 20/3 Interest arbitrage operations, otherwise known as round tripping. **1977** *Times* 23 Nov. 29/8 It is believed there has been a good deal of 'round tripping'—borrowing in the money markets to finance purchases of CTDs at a profit. **1978** J. A. Michener *Chesapeake* 649 Home Run Baker..would hit in one year the unheard-of total of twelve round-trippers. **1980** *Boston Globe* 30 Mar. 76 Home runs were circuit clouts. Then they became round trippers, until George Scott renamed them taters. **1980** *Times* 5 Sept. 15/2 It becomes attractive for blue-chip corporate borrowers to take loan from their banks and re-lend the funds to the short-term money markets. This is known as 'round-tripping'.

round-up. Add: **1. b.** *transf.* (in quot. *attrib.*)

1926 J. Masefield *Odtaa* 214 He took one of the big round-up stew cauldrons which lay against a wall.

2. a. For *U.S.* (and *Austr.*) read orig. *U.S.* (Earlier and later examples.) Also *fig.*

1873 in *Ann. Wyoming* (1927) V. 74 The herders of this Co. start a Round-up tomorrow... Each man picks out his stock and drives them in. **1878** J. H. Beadle *Western Wilds* xxviii. 437 These cattle, having run wild upon the plains of western Texas, are collected by a grand 'round-up'. **1879** *Tinsley's Mag.* XXIV. 353 One's companions, when camping out on the 'round up' are often anything but desirable. **1907** S. E. White *Arizona Nights* iii. 60 We had our first round-up, found the natural increase much in excess of the loss by Indians. **1909** [see *pay *v.*^{1} 10]. **1951** E. Paul *Springtime in Paris* xv. 287 Busse, rattled as a rabbit in a roundup, bounced back to the pavement just in time to bump into the burly Chestnut Man. **1976** *Billings* (Montana) *Gaz.* 10 June 1-c/1 At the Spanagel ranch west of Forsyth all hands are in the middle of roundup and putting up hay.

b. (Earlier and later examples.)

1880 *Harper's Mag.* Feb. 380/2 We old fellows have a round up 'most every year in Denver. **1936** L. C. Douglas *White Banners* iv. 70 An unexpected invitaton to read a paper at the first monthly round-up of the University Club.

c. The group of men and horses engaged in a round-up.

1878 in *Colorado Mag.* (1939) XVI. 152 Most of the round-up gone; a few still lingered at the bar. **1903** 'O. Henry' in *Everybody's Mag.* June 519/1 The round-up had ridden on but a few moments before. *a* **1918** G. Stuart *40 Yrs. on Frontier* (1925) II. 178 It was a novel sight to witness the big spring roundup pull out.

d. A survey of opinion; a résumé of facts or events; *spec.* in *Broadcasting*, a summary of newsworthy items.

1886 *Philadelphia Times* 3 May 1/1 That exception.. will probably be included in the general round-up by tomorrow. **1892** *Boston Jrnl.* 29 Nov. 3/1 Round-Up of the Boston Aldermanic Districts. **1904** F. Crissey *Tattlings of Retired Policeman* ii. 42 A hatchet-faced lawyer..made a quick round-up of the representatives of the corporate interests and vested rights of the state. **1932** *Sun* (Baltimore) 21 Dec. 8/2 The 'round-ups' of Congressional opinion on the war debts issue. **1949** *Lincoln County News* (Oceanlake, Oregon) 4 Aug. 3/4 Another periodic survey of censorship conditions..shows no major barriers have come down since the last roundup of the situation. **1958** *Spectator* 1 Aug. 159/1 The BBC's Middle East round-up on Saturday night was a notable project. **1962** *Listener* 22 Mar. 528/1 He presented a brilliant newsreel round-up 'Cease-fire in Algeria'. **1967** *Economist* 2 Dec. 915/1 A round-up of how America has reacted to the first rush into gold. **1974** *Radio Times* 21 Feb. 46/2 Commentary from Trinidad on the final day's play in the Second Test, together with a round-up of the day's sport.

e. The systematic rounding-up of people or of objects; *spec.* the arrest of people suspected of crime.

1899 *Chicago Rec.* 17 Jan. 12/1 A 'round-up' of all suspicious characters was begun. **1927** A. Christie *Big Four* v. 54 A short time ago a round-up was made of certain crooks and gunmen. **1943** *Sun* (Baltimore) 17 Nov. 6/1 The collection today will be of all tins accumulated since June 9, when the last such roundup took place. **1966** M. R. D. Foot *SOE in France* ix. 264 Contacts in the Rouen police..provided reliable warnings of impending round-ups. **1978** *Detroit Free Press* 16 Apr. (Record) 9/3 The round-up began Aug. 26 at the command of the Germans, who at the same time began arresting 20,000 foreign Jews in the occupied zone.

f. = *rodeo 3 b.

1914 *World's Work* Feb. 444/2 During the three days of The Round-Up, a constant stream of humanity pours into Pendleton. **1948** *Great Falls* (Montana) *Tribune* 18 Sept. 5/4 Malta is preparing to welcome at least 5,000 people this weekend when the two-day fall roundups will be staged.

g. In *fig.* phr. *the last round-up*, death, resurrection, or the Last Judgement.

1932 G. Brown (*song-title*) The last round up. **1940** *Hoofs & Horns* Dec. 11/1 Tom Mix..has laid down his honors..and taken the sunset trail that leads to the Last Roundup.

5. *attrib.*, as (sense 2 a) *round-up boss, camp, captain, outfit, party, wagon*; (sense 2 d) *round-up article, programme, review*; (sense 2 f) *round-up pennant, week.*

1960 V. Jenkins *Lions down Under* 14 'Potentially the greatest team ever to tour New Zealand,' was how Graeme Jenkins..described them in a round-up article. **1977** *Irish Times* 8 June 4/7 A round-up article on the celebrations written by the agency's court correspondent was delivered by hand to some London offices last night. **1920** J. M. Hunter *Trail Drivers of Texas* 313 The round-up boss would let no one drive through the herd. **1923** R. Pocock in *Outward Bound* Mar. 410/2 Seventeen miles across the Mesa la Sal, in Utah, brought me to a round-up camp. **1907** Round-up captain [see *cut *sb.*^{2} 23 c]. **1890** *Stock Grower & Farmer* 14 June 5/3 The herder was found by the Long S roundup outfit, about six miles west of Sulphur draw. **1885** *Weekly New Mexican Rev.* 26 Mar. 1/6 Round-up parties have already been started for that section. **1891** *Fur, Fin & Feather* Mar. 188 Wolves naturally follow in the wake of round-up parties. **1931** Y. Winters *Coll. Wks.* (1952) 66, I remembered.. The sprawling streets,.. The Round-up pennants.

1979 *Jrnl. R. Soc. Arts* CXXVII. 358/1 What is reported on the news, and current affairs, something very different indeed from..election round-up programmes. **1978** *Amer. Poetry Rev.* Nov./Dec. 32/3 A round-up review by George Dillon in *Poetry*. **1893** T. Roosevelt *Wilderness Hunter* ii. 23 Close beyond the trees on the farther bank stood the two round-up wagons. **1973** R. D. Symons *Where Wagon Led* I. i. 10 Two of the three men.. were out in the south country with the roundup wagon. **1924** W. M. Raine *Troubled Waters* iv. 40 A poster.. announced Round-up Week,..roping, and other Western sports.

roundward, *a.* and *adv.* Delete *nonce-word* in Dict. and add: **A.** *adj.* (Further example.)

1927 D. H. Lawrence *Mornings in Mexico* 80 The reeling, roundward motion of tree-tips in a wind.

roundwise, *adv.* and *a.* Add: **2.** *adj.* **b.** = Roundward *a.* *rare*$^{-1}$.

a **1930** D. H. Lawrence *Apocalypse* (1932) 42 The roundwise moving of the cosmos.

roundy, *a.* **1.** (Later example.)

1882 G. M. Hopkins *Poems* (1918) 54 As tumbled over rim in roundy wells Stones ring.

Rous (raus). *Biol.* The name of Francis Peyton *Rous* (1879–1970), U.S. physician, used *attrib.* to designate (*a*) a type of virus-induced sarcoma which afflicts birds, described by him in 1910 (*Jrnl. Exper. Med.* XII. 696); (*b*) an RNA virus which causes such sarcomata (its existence was suggested by Rous et al. in 1912 (*Jrnl. Amer. Med. Assoc.* 20 Nov. 1794/1)). So **rou·svirus.**

1911 *Jrnl. Exper. Med.* XIII. 389 We succeeded..in cultivating malignant tissues such as the Rous chicken sarcoma. **1925** W. E. Gye in *Lancet* 18 July 109/2 It has now been shown that Rous tumour No. 1 is caused by a virus which has been cultivated. **1931** *Brit. Jrnl. Exper. Path.* XII. 127 To decide whether the fragility of the Rous virus were due to its inability to resist oxidation, or whether a proteolysis were the inactivating cause. **1945** Duran-Reynals & Shrigley in *Res. Conf. on Cancer* (Amer. Assoc. Adv. Sci.) 13/1 When the Rous tumor virus is inoculated into ducks two sets of lesions occur. **1945** H. S. N. Greene in *Ibid.* 19/1 Dr Shrigley has succeeded in transplanting the Rous sarcoma in the anterior chambers of guinea pigs' eyes, and this is the first instance in which a virus-induced tumor has been proved to be a cancer. **1961** R. D. Baker *Essent. Path.* xiii. 311 Carcinogenic viruses are known to produce the Rous sarcoma in chickens and the Shope papilloma of rabbits. **1961** *Lancet* 5 Aug. 301/1 The ability of cells infected with Rous-sarcoma virus to multiply has been demonstrated, but such chronically infected tumour cells cannot be considered to have recovered. **1972** *Sci. Amer.* Jan. 26/1 The rousviruses do not transfer information from RNA to RNA, as other RNA viruses do. **1977** *Nature* 15 Dec. 631 The genome of Rous sarcoma virus (RSV) is a 30–40S RNA of 10,000 nucleotides.

rousable (rau·zăb'l), *a.* [f. Rouse *v.*^{1} + -able. Cf. *arousable *a.*] Capable or admitting of being roused.

1910 *Daily Chron.* 7 Mar. 8/1, I thought her endowed with a temper that might be very violent when roused, though not easily rousable. **1961** *Lancet* 5 Aug. 323/1 She was semicomatose but rousable.

rouse, *sb.*^{2} Add: **3.** (Later examples.) Also *rouse-out.*

1881 C. A. Stephens *Knockabout Club in Woods* (1882) xi. 122 The result was a most unwelcome rouse-out shortly after ten o'clock. **1916** C. Sandburg *Chicago Poems* 125 The silk and flare of it [*sc.* a red scarf] is a great soprano leading a chorus Carried along in a rouse of voices reaching for the heart of the world.

4. *attrib.*, as (sense 2) *rouse-parade.*

1937 D. Jones *In Parenthesis* I. 4 We've got too many buns—and all those wads—you knew they were going—why did you order them—they won't be in after rouse-parade even—they've gone.

rouse (raus), *v.*^{4} *Austral.* and *N.Z. colloq.* Also **rous.** [Cf. Roust *v.*^{1}] *intr.* To scold. Freq. const. *at, on, onto*: to upbraid (someone). Hence **rou·sing** *vbl. sb.*^{3}

c **1910** in G. A. Wilkes *Dict. Austral. Colloquialisms* (1978) 279/2 *Rouse*, abuse or vilify. **1911** L. Stone *Jonah* v. 47 It's gittin' late; 'ow'll yer ole woman rous w'en yer git 'ome? *Ibid.* xi. 126 'E niver rouses on me. W'en 'e gits shirty, I just laugh, an' 'e can't keep it up. **1915** C. J. Dennis *Songs of Sentimental Bloke* 88 If she 'ad only roused I might 'a' smiled. She jist seems 'urt an' crushed; not even riled. **1934** V. Palmer *Sea & Spinifex* 182 Combo's one of those sulky devils that forget nothing. .. Can't take a bit of rousing as part of the day's work. **1940** F. Sargeson *Man & his Wife* 27 Then Mrs Bowman roused on to me for putting too much sugar in her tea. **1951** D. Cusack *Say No to Death* 30 Auntie used to rouse on me frightfully because I spent so much time on the beach. **1961** R. Lawler *Piccadilly Bushman* 31 Don't rouse at me, Alec.

rouseabout. Add: **2. a.** (Earlier and later examples.) Also *N.Z.*

1861 [see *fleece-picker* s.v. *fleece *sb.* 6]. **1901** M. Franklin *My Brilliant Career* xvii. 150 Joe Slocombe, the man who acted as groom and rouseabout, was waiting for me. **1909** 'S. Rudd' *From Selection to City* ix. 76 Numbers of shearers and rouseabouts had arrived on the scene before us, and the station was all astir. **1917** A. B. Paterson *Saltbush Bill* 8 Come all you little rouseabouts and climb upon my knee. **1936** A. Russell *Gone Nomad* iii. 19, I..was able to throw, skirt, and roll a fleece with the crack rouseabout of the shed. **1947** D. M. Davin *Gorse blooms Pale* 81 I'm not letting Joe get buggered about by any bloody fly-by-night rouseabout. **1961** *N.Z. Listener* 26 May 8/1 The rouseabout and presser had just finished clearing up. **1966** 'J. Hackston' *Father clears Out* 181 As a boy milker, manager, rouseabout, stud-master, and jackeroo on our place I became aware of the prevailing bad manners among some cows.

b. *transf.* An odd-job man or general worker; a casual labourer. *Austral.* and *N.Z.*

1906 E. Dyson *Fact'ry 'Ands* ii. 15 Billy the Boy, the juvenile rouseabout from the printer's flat. **1911** E. M. Clowes *On Wallaby* iii. 64 It is to land so won and so held that every casual 'rouseabout' or street loafer feels that he has a perfect right. **1933** *Bulletin* (Sydney) 7 June 20/1 The rouseabout at Casey's pub was sacked. **1951** Cusack & James *Come in Spinner* 39 They'll probably stick you in as rouseabouts in a lunatic asylum, seeing the experience you've 'ad 'ere.

c. *attrib.*, as *rouseabout swiper, work.*

1887 J. Farrell *How he Died* 19 It may be that the rouseabout swiper who rode for the doctor that night Is in Heaven with the Hosts of the Blest, robed and sceptred, and splendid with light. **1906** E. Dyson *Fact'ry 'Ands* xii. 148 A man was engaged to assist at the guillotines and do the rouseabout work of the factory. **1934** J. Lilico *Sheep Dog Mem.* 27 [The dogs] would head, lead, huntaway, force and back, though..they were best at rouseabout work.

rouseabout (rau·zăbaut), *v.* *Austral.* [f. Rouseabout 2.] *intr.* To work as a rouseabout. Hence **rou·seabouting** *vbl. sb.*

1914 *Bulletin* (Sydney) 17 Dec. 44/2, I never done no shearin'; but I rouseabouted one year in a shed near Muttaburra. **1945** Baker *Austral. Lang.* II. iii. 61 We.. now speak of *jackerooing* (just as we speak of *rouseabouting*), for work as a station-hand. **1967** *Southerly* XXVII. 205 Writing, he said, came easier than..shearing, station rouseabouting.

rousedness (rau·zĕdnĕs, -zd-). *rare.* [f. Roused *ppl. a.*^{1} + -ness.] The state of being aroused; a roused condition, alertness.

1915 D. H. Lawrence *Rainbow* iv. 94 And there was a kind of bristling rousedness in the room. *Ibid.* x. 261 Everywhere was a sense of mystery and rousedness.

rouser. Add: **1. b.** (Earlier example.)

1765 H. Jackson *Ess. on Brit. Isinglass* 56 Previous to cleansing, conquassate the whole Aggregate with a Rouser.

2. a. (Earlier and later examples.)

1839 C. F. Briggs *Adventures H. Franco* I. xiv. 127 We never exchanged another word until we reached the fire, and then, says he to me, I tell you what, Smith, it is going to be a rouser. **1977** *Time* 24 Oct. 8/3 It cleared the way for a rouser of a speech by Thatcher.

4. (Earlier and later examples.)

1897 H. Lawson *While Billy Boils* 85 They are all shearers, or at least they say they are. Some might be only 'rousers'. **1900** —— *Verses, Popular & Humorous* 168 The 'rouser' has no soul to save. Condemn the rouseabout! *a* **1964** E. Harrington in *Penguin Bk. Austral. Ballads* (1964) 262 The rousers gave a billycan and brand new tucker bag.

rousie (rau·zi). *Austral.* and *N.Z. colloq.* Also **roussie, rousy.** [Abbrev. of Rouseabout.] = Rouseabout 2.

1933 *Bulletin* (Sydney) 29 Nov. 20/1 The rousie entered the office to be paid off. **1952** [see *dagger *sb.*^{1} b]. **1956** F. B. Vickers *First Place to Stranger* ix. 135 He tried to run all the rousies till Ivor stopped him. *a* **1964** H. P. Tritton in *Penguin Bk. Austral. Ballads* (1964) 228 Then try to catch the rousy's eye, and softly whisper, 'Tar'. **1966** G. W. Turner *Eng. Lang. Austral. & N.Z.* vii. 147 There are names for woolshed workers,..the *rousie* or rouseabout and the *sheepo* who fills the catching pens.

Rousseauan, *a.* Add: (Later examples). Also as *sb.* **Rousseauesque** *a.*^{1} (examples), **Rousseauian, Rousseauish** *adjs.* (later examples); similarly **Rousseaui·stic, Rousseau·vian** *adjs.*; **Rousseauism** (later examples); **Rousseauist** (examples) (also as *adj.*), **Rousseauite** (later example); similarly **Rousseaua·rian.**

1775 H. Walpole *Let.* 3 Apr. (1904) IX. 174 The Rousseaurians [*sic*] will imagine that I interpolated the condemnation of his *Eloïse*. **1806** J. Mackintosh *Let.* 24 Dec. in R. J. Mackintosh *Mem. Life Sir J. Mackintosh* (1835) I. vi. 306 It is certainly a most ingenious, and the only reasonable, modification of the Lockian and Rousseauvian principle. **1889** *Cent. Dict.*, *Rousseauist*.., a follower or an admirer of J. J. Rousseau. **1905** H. G. Wells *Mod. Utopia* v. 171 The sweetish, faintly nasty slops of Rousseauism. **1914** *Blast* 20 June 18 Blast.. Rousseauisms (wild Nature cranks). **1928** C. Hollis *Dr. Johnson* iii. 67 The Rousseauan argument was that man was good. *Ibid.* 68 Rousseauans had been responsible for the generalisation that man was good. **1930** H. Read *Wordsworth* iii. 131 They had taken a ward to educate.. but they did not take their duties any more seriously than a pair of Rousseauites should. **1935** D. Fahey *Mystical Body of Christ in Mod. World* iv. 30 The Rousseauist dogma of the natural goodness of man. *a* **1937** J. L. Stocks *Reason & Intuition* (1939) vi. 84 In modern times the Rousseauistic strain still survives. **1937** 'C. Caudwell' *Illusion & Reality* v. 98 Wordsworth, like

Shelley profoundly influenced by French Rousseauism. **1938** R. Graves *Coll. Poems* 96 And being no Rousseauist, Nor artists-of-the-world-unite. **1947** A. Einstein *Mus. Romantic Era* xviii. 339 This Rousseauesque theory of the linguistic origin of music, basically wrong as it is, still reappears even in professional psychological literature of the 19th century. **1962** Gregor & Nicholas *Moral & Story* v. 144 This somewhat Rousseauistic view of nature contrasts strangely with the determinist one, which Hardy runs alongside it. **1962** C. Walsh *From Utopia to Nightmare* ix. 126 Americans are Rousseauans by temperament. **1965** M. Hodgart *Faber Bk. Ballads* 12 We are all Rousseauists or Wordsworthians to the degree that we are discontented with the artificiality of our culture. **1967** B. W. Alderson tr. *Hürlimann's Three Cent. Children's Bks. in Europe* ix. 120 The Romanticism and Rousseau-ish idealism in books about America and the Indians. **1969** J. Mander *Static Society* 170 The 'good' Indian myth of Rousseauvian Europe. **1971** Rousseauian [see *neo-primitivism* s.v. *neo- 1 a]. **1971** G. Steiner *Bluebeard's Castle* i. 24 What needs close attention is the extent to which critiques of urban society tend to become indictments of all formal, complex civilization as such ('civilization', of course, has in it the word for city). Rousseauist naturalism has an obvious destructive edge. **1974** *Listener* 21 Mar. 372/3 The editor of his [*sc.* Aaron Burr's] journal calls it 'Rousseauistic'... Couldn't one just call it unselfconscious? **1977** *Times Lit. Suppl.* 11 Feb. 148/2 The Machiavellian and Rousseauesque hints of subordination of church to state. **1977** in *Private Eye* 10 June 5/2 In the name of directness, of authenticity, of courage, of any number of Rousseauian virtues that belong exclusively to the noble savage. **1977** *Listener* 16 June 790/3 Throngs of Indonesians who.. appear to conform to the most elevated models of Rousseauvian *noblesse*. **1978** *Dædalus* Summer 2 The Rousseauan longing for a pastoral idyll is not yet dead.

Rousseauesque (rusō^u,e·sk), *a.*^{2} [f. the name of Henri 'le Douanier' *Rousseau* (1844–1910), French primitive painter + -esque.] Characteristic of the style of Rousseau.

1962 *Guardian* 9 July 5/3 She is also extremely successful in depicting Rousseau-esque foliage. **1978** A. Huxley *Illustr. Hist. Gardening* iii. 76 (*caption*) This Rousseauesque painting from the Jodhpur [*sic*] school depicts a lush private garden.

Roussette. Add: **3.** A white wine produced primarily in the French departments of Savoy and Jura.

1926 P. M. Shand *Bk. of Wine* v. 171 Frangy and Digny-Musièges belong to the more or less immediate neighbourhood of Seyssel, as does also a good, rusty-coloured wine called Roussette, named after the informing grape of the Savoy vineyards. **1946** G. Millar *Horned Pigeon* xx. 309 With the *fondue* we drank *Roussette*, an excellent local white wine which was then new to me. **1967** A. Lichine *Encycl. Wines* 459/1 *Roussette*, the principal wine of Seyssel, in Haute-Savoie, France. It is white, flinty-dry, and made from Roussette grapes. **1968** *Vogue* 15 Apr. 121/1 To the South of the Jura lie the vineyards of the Savoy. Both red and white wines are grown, but it is the latter that are worth looking for, especially seyssel and..crépy and roussette.

roussie, var. *rousie.

roust, *v.*^{2} For '*dial.* and *U.S.*' read 'orig. *dial.* and *U.S.*' and add: **1. a.** (Further examples.) Also, to rouse or stir *up*, to raise or arouse (*from* one's bed, etc.).

1850 W. Colton *Deck & Port* 299 We rousted our anchors this afternoon from the bed in which they have slumbered for the last six weeks. **1871** J. Hay *Little Breeches* 12 But we rousted up some torches, And sarched for 'em far and near. **1883** *Peterson Mag.* June 469/2 Awhile ago you was all rousted-up about goin' to New York village to see Mrs. Larne. **1905** J. C. Lincoln *Partners of Tide* xi. 221 'Now, then,' he added, 'while we're waitin' for the tide to turn we might's well roust out a little more of the cargo.' **1939** J. Steinbeck *Grapes of Wrath* 123 Don't roust your faith bird-high an' you won't do no crawlin' with the worms. **1972** E. Wigginton *Foxfire Bk.* 169 You get up and roust up your fire. **1978** R. Ludlum *Holcroft Covenant* xliii. 496 I've been rousted from my bed to take additional scrapings from the dead man's room.

b. *intr.* To get *up*, turn *out*; to rummage *around*.

1884 J. C. Harris *Mingo & Other Sk.* 162 It twon't never do in the roun' worl' for to be a-makin' faces at 'im frum the groun'. Roust up, roust up. **1900** C. C. Munn *Uncle Terry* 172, I ginerally roust out by daylight. **1912** R. A. Wason *Friar Tuck* 67, I knew it was my duty to roust out an' keep Horace from gettin' more sleep'n my treatment for his nerves called for. **1941** J. Street *In my Father's House* i. 19 Suppose you were a dominecker rooster—I mean hen... And you were in a coop and just outside the coop was a heap of grass that you wanted to roust around in. **1977** 'O. Jacks' *Autumn Heroes* v. 67 He made them go over the check lists with him... He rousted around.

2. To jostle (see quot. 1942); (esp. of police) to harass, rough up. Hence as *sb.*^{2} *N.Amer.* (orig. *Criminals'*) *slang.*

1904 'No. 1500' *Life in Sing Sing* 252/1 *Roust*, to jostle. **1942** Berrey & Van den Bark *Amer. Thes. Slang* §490/5 *Roust, rousting, rowdy-dowdy*, the act of causing a crush in a crowd or jostling the victim in order to pick his pockets. *Ibid.* §490/12 *Roust*,..to cause a crush in a crowd or jostle the victim in order to pick his pockets. **1961** Rigney & Smith *Real Bohemia* p. xvi, *Roust, a*, a bust, an arrest, a raid. **1972** J. Wambaugh *Blue Knight* (1973) v. 68, I can't take this kind of heat. I can't take being rousted and hurt. **1976** N. Thornburg *Cutter & Bone* viii. 193 He ran into Sergeant Verdugo, one of the detectives who had rousted him the night of the murder. **1978** R. Thomas *Chinaman's Chance* xxiii. 234 'I'm..your friendly, conscientious chief of police.' Wu nodded slowly. 'And this is a roust, huh?' **1978** *Globe & Mail* (Toronto) 11 Jan. 8/1 Like one afternoon last spring,..when plain-clothes cops are rousting us when we're trying to do business. At the same time the bank at Church and Carleton gets robbed and the robber got away on foot.

roust (raust), *v.*^{3} *Austral. colloq.* [var. of *rouse *v.*^{4}; cf. Roust *v.*^{1}] *intr.* = *rouse *v.*^{4} Also with quasi-obj. in phr. *to roust hell out of.* Hence **rou·sting** *vbl. sb.*^{1}

1916 C. J. Dennis *Songs of Sentimental Bloke* 124 *Roust* or *rouse*, to upbraid with many words. **1918** —— *Digger Smith* 82 All me roustin' leaves 'em both serene. **1938** X. Herbert *Capricornia* xviii. 250 'All dem sister proper humbug.' 'How's that?' 'All time roustin'. All time tink we go out wid boys. We no can talk boys. But dem sister proper mad long boys demself.' **1941** S. Campion *Mo Burdekin* 139 And 'avin 'im roust hell outa me for it. **1970** P. White *Vivisector* i. 11 He hung around Mumma, waiting for her to settle, and she didn't roust on him.

roustabout. 2. a. For *Austr.* read 'orig. *U.S.*' and add earlier and further examples. Also, a casual or unskilled labourer; a vagrant or layabout.

1877 *Harper's Weekly* 17 Mar. 3/3 The vagabonds, the roustabouts, the criminals, and all the dregs of society seem to be Democrats. **1880** A. A. Hayes *New Colorado* v. 77 He was a kind of rostabout [*sic*], or dish-washer, to a camping outfit. **1883** 'Mark Twain' *Life on Mississippi* li. 454 Do you mean the Roman army?—those six sandalled roustabouts in nightshirts? **1896** J. McDougall *Saddle, Sled & Snowshoe* xv. 187, I have been Mr. Woolsey's interpreter, guide, and general 'roust-about', his confidante and friend, for the past two years. **1911** H. S. Harrison *Queed* 35 It takes a Whitney to invent the cotton gin, but the dullest negro roustabout can operate it. *a* **1918** G. Stuart *40 Yrs. on Frontier* (1925) II. 179 Every man, whether owner of the largest herd or a humble roustabout, takes his orders from the captain. **1942** E. H. Paul *Narrow St.* xix. 152 Butchers, roustabouts and helpers..toiled steadily in the lamplight. **1960** H. Miller *Nexus* (1964) xiv. 237 I've got a good wife, only we're temperamentally unsuited to one another. I'm too common for her. Too much of a roustabout.

b. *spec.* A workman in a circus. *N. Amer.*

1931 *Amer. Mercury* Nov. 353/2 *Razorbacks*,.. Workmen who load and unload the circus train; never called *roustabouts* or *flunkeys*. **1949** *Los Angeles Times* 9 Apr. 2/3 Roustabouts from the Clyde Beatty circus appeared to offer any manual labor needed. **1957** *Harper's Bazaar* Feb. 175 He may earn his living as a petty criminal, a hobo, a carnival roustabout or a free-lance moving man in Greenwich Village. **1976** *Telegraph-Jrnl.* (St. John, New Brunswick) 4 Sept. 32/1 From a carnival *roustabout* to owner of the show in 25 years.

3. A general or manual labourer on an oil installation.

1948 H. L. Mencken *Amer. Lang.* Suppl. II. 763 *Roustabout*, a laborer on an oil lease, not a member of the rig crew. **1959** Larson & Porter *Hist. Humble Oil & Refining Co.* xii. 291 The next annual crop of new engineers was..put to work for a year as roughnecks or roustabouts. **1971** C. Simpson *New Australia* 518 The average young oilfield worker, called a 'roustabout', needed to have more than muscles. Technical competence was also called for. **1972** L. M. Harris *Introd. Deepwater Floating Drilling Operations* iv. 35 Drilling and roustabout crew requirements differ little from rig to rig. **1975** *Offshore Engineer* Dec. 54/1 (Advt.), The clothing was tested on the rig Sedco 700, operating close to the 62nd parallel, by supervisors and roustabouts on the nightshift.

roustabout (rau·stăbaut), *v.* [f. prec.] *intr.* To be, or work as, a roustabout.

1907 'O. Henry' in *Everybody's Mag.* Nov. 593/1, I hurried the rest of the way up the river, roustabouting on a lower coast packet that made a landing for every fisherman that wanted a plug of tobacco. **1934** in *Amer. Ballads & Folk Songs* 494 When Jack is old and weather-beat, Too old to roustabout.

rouster. Add: **1.** (Further examples.)

1929 B. L. Burman *Mississippi* iii. 47 A rouster, with two coffee sacks tied around his body..lay on the boiler deck strumming a guitar. **1938** —— *Blow for Landing* xiv. 236 Barrels began to roll swiftly down the gang-plank, like a bass-drum accompaniment for the feet of the negro rousters. **1941** [see *coonjine *sb.* and *v.*].

2. *U.S.* and *Austral.* = Roustabout 2 in Dict. and Suppl.

1890 in Barrère & Leland *Dict. Slang.* **1911** C. E. W. Bean *'Dreadnought' of Darling* xxxviii. 338 There tumbled out of it all the sweepings of Sydney, all the old cripples, and beggars, and rousters in Christendom.

rousting (rau·stiŋ), *vbl. sb.*^{2} *U.S. colloq.* [f. Roust *v.*^{2} + -ing^1.] (An act of) police harassment, a police raid (see also quot. 1942).

1942 Berrey & Van den Bark *Amer. Thes. Slang* §490/5 *Rousting*,..the act of causing a crush in a crowd or jostling the victim in order to pick his pockets. **1960** *Washington Post* 25 Jan. 1 There's more vice on Pacific Heights and Nob Hill and there's no rousting (police raiding) up there. **1968** S. Ellin *Valentine Estate* II. vi. 51 Rousting was the word for it. Keep pushing a man until he either left the territory or did something he could be nailed for. **1972** B. Garfield *Line of Succession* I. 75 The prisoners each morning complained to their lawyers of the nightly roustings. **1975** *High Times* Dec. 31/1 So far, however, the little hungo town has been spared the midnight roustings.

rousy, var. *rousie.

rout, *sb.*^{10} (Later example.)

1912 C. Mackenzie *Carnival* xxxiv. 358 'I'm going to have a rare old rout-out this morning,' Jenny announced.

rout, *v.*^{8} Add: **2. b.** (Further examples.) Also with *away*; *spec.* (also without const.) to cut a groove in (a wooden or metal surface), to machine or work with a router.

1818 *Trans. R. Soc. Arts* XXXV. 123 In the old way of routing the wood the grooves are torn and uneven at the bottom. **1934** *Woodworker* XXXVIII. 158/3 He first routs out his templet as suggested to the..shape he desires. **1946** C. H. Hayward *Light Machines for Woodwork* xii. 155 It is of special value for routing the ends of pieces..of odd shape. **1948** H. Missingham *Student's Guide Commerc. Art* ii. 100 The line block is finished by first routing away all unwanted metal from the work. **1958** *New Scientist* 17 July 441/2 (*caption*) Routing an aircraft bulkhead from a solid billet of high-tensile aluminium alloy. **1960** 'N. Shute' *Trustee from Toolroom* vi. 131 You routed each plank all along its length to fit the next one?

3. a. (Earlier example.) Also to fetch (a person) out of a house, etc.

c **1776** H. Newdigate *Let.* in A. E. Newdigate-Newdegate *Cheverels* (1898) i. 9 My Lord routed us out ye moment we had breakfasted to pass sentence upon some trees that are to be fell'd.

b. (Earlier and later examples with *out.*) Also, to turn *out* (a room, etc.).

1814 Jane Austen *Let.* Aug. (1952) 397 As soon as my Trunk & Basket could be routed out from all the other Trunks & Baskets in the World, we were on our way. **1929** J. Masefield *Hawbucks* 101 He went home to his cellar and routed out a bottle of port. **1938** M. K. Rawlings *Yearling* iv. 31 He dashed in to his room and routed out his heavy cowhide brogans. **1950** R. Moore *Candlemas Bay* iii. 142 Ordinarily, he would have enjoyed routing out the fishhouse. **1973** A. Christie *Postern of Fate* I. v. 39, I shall go up and rout him out.

routing *vbl. sb.* (further examples, in sense 2 b of the vb.); also with *out.*

1935 *Times* 9 Nov. 4/4 Blind men..use, unaided, the circular saw and the routing machine (a speed of 15,000 revolutions a minute) to carve the animals. **1946** C. H. Hayward *Light Machines for Woodwork* xii. 147 For such work as the free-hand routing out of..the groundwork of a piece of carving it is essential that the wood is laid flat..and the machine passed over it. **1953** E. G. Hamilton *Power Tools for Home Craftsman* vii. 231 Routing with a pattern is a fast and simple method of doing production work. **1958** *New Scientist* 17 July 441/1 The chief applications of the intricately shaped parts produced from solid metal by routing are in the aircraft industry. **1976** C. H. Groneman *Gen. Woodworking* (ed. 5) xlvi. 224 (*caption*) Freehand routing of a penciled design.

route, *sb.* Add: The pronunc. (raut) is used by many speakers in the U.S. and Canada.

1. a. Also, used in various countries, esp. the U.S. and France, with a following numeral to designate a particular highway. Also *fig.*

1924 *N.Y. Times* 21 Dec. viii. 9/7 Route 2 is the highway from Scranton Pa...to Montreal. **1933** Kipling *Souvenirs of France* i. 18 That was the Rhone Road, Route 7. **1938** *Travel* June 37/1 From New York there are three delightful motor routes..all picking up Route 6. **1962** 'K. Orvis' *Damned & Destroyed* v. 41 You haven't fooled me. You're on Route Zero. **1970** *Washington Post* 30 Sept. B4/1 Fredericksburg location is just off route 95.

c. *U.S.* An established stage by which post is conveyed prior to delivery.

1792 *Deb. Congress U.S.* 10 Jan. (1849) 58 The route by which the mails are at present conveyed shall in no case be altered. **1821** *Ibid.* 31 Dec. (1855) 47 Praying that the route of the mail from Savannah to Augusta..may not be altered. **1874** *Ann. Rep. Postmaster-General* (U.S. Post-Office Dept.) 209 Each railway post-office clerk..is required to attach to each package of letters he makes up a facing or label-slip bearing the address of the package, the office or route upon which it was made up.

d. *N.Amer.* A round travelled regularly by someone collecting, delivering, or selling goods, such as newspapers or milk.

1841 *Jamestown* (N.Y.) *Jrnl.* 5 May 2/4 He succeeded in obtaining possession of a route for a morning penny paper. **1849** C. Mathews *Moneypenny* xiii. 119 Go upstairs, and tell Wages to give you the St. John's Park route. He'll fix your pay. **1868**, etc. [see *paper route* s.v. *paper *sb.* 12]. **1874**, etc. [see *milk-route* s.v. *milk *sb.* 10]. **1939** J. P. Marquand *Wickford Point* x. 108 Her father ran a milk route and drank hard cider. **1976** *Washington Post* 19 Apr. c14/4 (Advt.), Routes are available in the metropolitan Washington area to aggressive persons who are experienced Route Sales people.

e. Phr. *to go the route*, in *Baseball*, to pitch for an entire game; also *transf.* in *Boxing*; *fig.*, to go the full distance, to go all the way. *U.S.*

1913 *Chicago Record-Herald* 16 Mar. viii. 1/5 This was the first complete battle Cicotte has pitched, and he was watched closely to see if he could go the route. **1926** J. Black *You can't Win* xvi. 230 If a Chinese doesn't like you he will keep away from you; if he does like you he will go the route. **1933** *Amer. Speech* Oct. 36/1 He went the route without being kayoed. **1948** *Chicago Tribune*

8 May ii. 3/3 Bill Voiselle went the route for the Braves. **1963** I. Fleming *On H.M. Secret Service* iv. 45 She made love with the fervour and expertness of a girl who, in the American phrase, had 'gone the route'. **1974** *Index-Jrnl.* (Greenwood, S. Carolina) 18 Apr. 11/3 Steve Rogers went the route, giving up six hits.

3. a. (Earlier example.)

1751 Fielding *Amelia* I. i. ix. 68 This Letter was from his Captain, to acquaint him, that the Rout, as they call it, was arrived, and that they were to march within two Days.

c. (Earlier and later examples.)

1802 C. James *New Mil. Dict.*, s.v. *March*, His next care must be the arrangement of all its different component parts, with which he will form his column of route. **1976** *Broadcast* Dec. 17/2 There can be few occupations so completely degrading as marching in column-of-route.

4. (sense 1) *route book, card, check, marker, -mile, -mileage, number, planning, proficiency*; *route-proving* adj.; (sense 3) *route column, march* (hence as *vb. intr.*), *-marching* (earlier example); **route-goer** *Baseball*, one who goes the route (see sense 1 e above); hence **route-going** *a.*; **routeman, route man** *N.Amer.* = Roundsman 3; also, a salesman who works a particular route (see sense 1 d above); hence **routemanship; route salesman** *N.Amer.*, a salesman who works a particular route (cf. **routeman*); so **route sales, route salespeople; route sheet** *N.Amer.*, an engagement itinerary for a touring company or artist.

1910 (*title*) Route book for the British Isles. **1975** *Oxf. Compan. Sports & Games* 702/2 The correct route, which the organizers convey to the crew in route-books. **1931** *Amer. Speech* VI. 335 *Route-card*,..a table or schedule issued to show people giving the 'stands' for about ten days in advance. **1963** P. Drackett *Motor Rallying* iii. 39 British rallies require only the accurate plotting of six-figure map references plus the ability to..read from a simple route-card. *Ibid.* 37 Route checks, or passage controls, are also a feature of the majority of rallies. **1954** W. Faulkner *Fable* 6 It was a whole battalion.. emerging from the *Place de Ville* in close route column. **1967** *Boston Sunday Herald* 26 Mar. ii. 3/1 The Philadelphia 76ers had five route-goers..while Hal Greer missed only one game. **1976** *Billings* (Montana) *Gaz.* 17 June 2-h/4 Doug DeCinces and Lee May each hit three-run homers to support 39-year old Mike Cuellar's first route-going performance of the season. **1918** *Nat. Laundry Jrnl.* 1 May 56/1 It really matters little whether he be known as a route man, salesman, or representative. .. Good route men are scarce. **1943** *Daily Progress* (Charlottesville, Va.) 29 May 8 (Advt.), In the face of a 40% reduction in gasoline mileage your routeman will call on you three times each week as in the past. **1976** *Billings* (Montana) *Gaz.* 2 July 9-c/1 (Advt.), Excellent opportunity for a Route Man to take over present route & expand. **1945** *U.S. Armed Forces Educ. Man.* em 991. xiii. 136 This chapter explains the elements of routemanship. **1895** W. S. Churchill *Let.* 27 Feb. in R. S. Churchill *Winston S. Churchill* (1967) I. Compan. I. viii. 559, I went out with the regiment on Friday to a route march—which was very fine. **1909** *Blackw. Mag.* Sept. 396/1 They have all been called out to some absurd inspection, or route march, or manœuvres, or something. **1934** Webster, Route-march, v.i. **1939** *Airman's Gaz.* Dec., You route march into the local swedeville. **1977** D. Bagley *Enemy* xxix. 231 Benson had a cushy billet for a soldier in wartime. Not for him route marches in the pouring rain. **1789** *Rules & Regulations Field Exercises & Movements Army in Ireland* I. 54 In Common Route marching the same regularity of step cannot be required, as is necessary in the operation of manœuvre. **1925** *N.Y. Times* 5 Aug. 8/3 The shield of the United States was adopted today as a model for the outline of route markers for the system of national highways. **1968** M. Woodhouse *Rock Baby* xv. 148 Plenty of people cross various borders..when they can't see the route-markers. **1911** *Encycl. Brit.* XXII. 824/2 In Europe the average route-mile capital is £27,036. **1962** *Observer* 25 Mar. 1/5 Route-mileage is the length of routes, as distinct from the length of individual sets of tracks. British Railways cover about 18,500 route-miles. **1967** *Listener* 26 Jan. 123/1 A maximum of 100 m.p.h. is now possible over about 360 route-miles of British Rail. **1924** *N.Y. Times* 21 Dec. viii. 9/7 The route number is painted in figures five inches high. **1973** D. Westheimer *Going Public* iv. 64 He memorized the route numbers of the buses. **1967** *Jane's Surface Skimmer Syst.* 1967–68 51/1 The Central Electricity Generating Board..is constantly faced with route-planning problems. **1959** Wallis & Blair *Thunder Above* iii. 15 His first trip as captain would be a route-proficiency check. **1957** *Times* 21 Dec. 5/3 The flight..completed their route-proving programme before the beginning of commercial operations. **1937** DeArmond & Graf *Route Sales Managem.* 4 Another factor in route sales distribution, the accurate anticipation of customer demand. **1937** *Job Descriptions of Laundry Industry* 253 Route-Man, Route Driver, Route-Salesman.. Drives a Delivery Truck over an established route to collect washing from and deliver it to customers' houses. **1968** *Globe & Mail* (Toronto) 13 Feb. 33/7 (Advt.), Route Salesman..required by supplier of industrial garments. **1976** *Washington Post* 19 Apr. c14/4 (Advt.), *Route Sales.* Routes are available..to aggressive persons who are experienced Route Sales people. **1916** *Variety* 27 Oct. 12/2 Sam Sidman's Own Show is on the Columbia route sheet to play there. **1941** W. C. Handy *Father of Blues* xiv. 195 Each one..had been copying other pluggers' borrowed route-sheets and submitting them to me as evidence of work done by themselves.

5. (Later examples.) Also *en-route*, (N.Amer.) *enroute.*

1955 *Times* 10 May 10/3 In the course of his journey to Leeds the Prime Minister made several speeches *en route*. **1956** R. Braddon *Nancy Wake* ix. 93 He would attempt to escape somewhere en route to Gerona. **1967** *Boston Sunday Herald* 26 Mar. vi. 4/1 (Advt.), Enroute to Miami with overnight stays in Wash., D.C., [etc.]. **1976** *National Observer* (U.S.) 7 Feb. 11/3 You can take advantage of Sitmar's 'Cruise Plus' feature which allows stopovers enroute home. **1978** *Nature* 5 Oct. 363/2 US spacecraft are now en-route to Venus, Jupiter, Saturn and Uranus.

route, *v.* (in Dict. s.v. Route *sb.*). Delete '(Chiefly in railway use.)' and add further examples.

For the pronunc., see the sb. in Dict. and Suppl.

The pres. pple. is spelt *routeing* (the better form: cf. note s.v. **routing vbl. sb.*) or *routing.*

1893 M. H. Cushing *Story of our Post Office* 235 Here are the carriers themselves, engaged in 'routing' the mail. **1926** J. Black *You can't Win* ix. 113 The papers were carefully read at night, and the next morning 'routed' through the prison. **1926** *N.Y. Times Mag.* 15 Aug. 6 Complaints were routed past the complaint department to the President's office. **1952** *Oxf. Mag.* 24 Jan. 142 An attempt is made to govern, by routing it through the Proctors, the growing spate of information. **1959** *Daily Tel.* 17 Mar. 13/2 A minute later he routed the Liverpool Street–Norwich train through. **1960** *Washington Post* 16 Nov. a16 West Germany's share of a greater European effort in NATO would have to be routed through NATO organs. **1961** L. Mumford *City in Hist.* (1966) xvi. 567 Major through-traffic streams must be routed around residential areas. **1971** P. Gresswell *Environment* 105 Others [*sc.* footpaths] might be better routed round field edges than through the middle of fields. **1971** D. Potter *Brit. Eliz. Stamps* xiii. 142 Very large postings in bulk attract substantial discounts. They are routed by second-class mail. **1977** *Daily Tel.* 20 Jan. 17/4 The organising committee intends to prevent a repetition by routeing the procession through wide streets.

b. To schedule or bill.

1916 *Variety* 22 Oct. 12/2 Rud Hynicka's show..will not play the Star and Garter next week as routed. **1932** L. C. Douglas *Forgive us our Trespasses* (1937) xiii. 253 Deducing from time-tables, Dinney hypothetically routed Joan to arrive at six-thirty on Thursday evening.

c. To direct (an electrical signal or transmission of any kind, as a telephone call) over a particular circuit or path, or *to* a particular location.

1948 J. Atkinson *Telephony* I. xii. 234/2 The group centre extends the call to the zone centre where it is routed to the distant zone centre exchange. **1956** *B.B.C. Handbk. 1957* 50 The sound components of the various contributions are routed and switched simultaneously with the vision. **1962** *Listener* 3 May 770/2 If one operator on the lunar surface wished to communicate with another operator a dozen miles away, his only method would be to route his signal by way of the Earth. **1964** F. L. Westwater *Electronic Computers* i. 6 By means of electronic switching devices a word is routed to the correct address in the store. **1973** *Daily Tel.* 22 Jan. 2/1 The dialling code for Rome is 010 39 6. The digits 010 route the call to the international automatic exchange in London, 39 routes to Italy, and the final 6 is the code number for Rome. **1973** *Physics Bull.* Feb. 109/1 The reference channel plug-in..accepts a reference signal derived from a chopper, which enables the plug-in to control and route the signal pulses to two counting channels.

‖ **route nationale** (rūt nasyonal). Pl. **routes nationales.** [Fr., = national highway.] In France, a main or trunk road constructed and maintained by the central government.

1896 A. P. Rockwell *Roads & Pavements in France* 26 The 22,000 miles of *Routes Nationales* are periodically examined with great care in order to ascertain the actual thickness of the stone layer. **1906** C. Neville *Round France in Motor* vi. 39 The next morning we ought to have continued along the Route Nationale, which goes to Macon *via* Châlon. **1924** Kipling in *N.Y. World* 23 Apr. 15/2 Route Nationale No. 20 conducts from Paris to the Spanish frontier at Bourg-Madame. **1949** M. Laski *Little Boy Lost* iii. 49 This isn't Paris—it's some shabby village away from all the *routes nationales*. **1973** G. Sims *Hunters Point* xix. 177 The frontier town of Menton.. where the French Route Nationale meets the Italian Via Aurelia.

router, *sb.*^{5} **1.** Substitute for def.: A cutter that removes wood from a groove or recess, as in a router plane. (Earlier example.)

Quot. 1875 in Dict. belongs to next sense.

1818 *Trans R. Soc. Arts* XXXV. 123 With my plane, as fast as the cutters pierce the wood, the router follows after, and clears the wood out of the groove.

b. A router plane.

1846 C. Holtzapffel *Turning & Mech. Manipulation* II. 979 Mr. Wm. Lund has constructed the router..with a screw adjustment to the cutter. **1923** R. Greenhalgh *Pract. Joinery & Carpentry* xix. 245 A number of grooves are first run round the wreath [of a handrail] in suitable places, a useful tool for this purpose being the router. **1954** W. E. Kelsey *Carpentry, Joinery & Woodcutting Machinery* i. 14 Router or old woman's tooth... This is a tool for cleaning out and levelling the bottoms of trenches. **1974** G. Blackburn *Illustr. Encycl. Woodworking Handtools* 169 The Pattern Maker's Router is similar to the Router Plane, but with a machined, larger sole.

c. (See quot. *a* 1877.)

a **1877** Knight *Dict. Mech.* I. 288/2 The center-bit consists of three parts: a center point or pin..; a thin cutting point or nicker that..circumscribes the hole; and a broad chisel-edge or router, placed obliquely, and tearing up the wood within the circle marked out by the point. **1947** H. E. King *School Cert. Woodwork* vi. 63 Boring Bits... The router and nicker are sharpened on the inside only. **1955** M. M. Waters *Woodwork* 107 The nicker extends lower than the router and so engages the wood slightly ahead of it.

d. A woodworking machine similar to a spindle moulder but using a much higher speed of rotation and able to produce finished work; also, a portable hand-held version of this.

1946 W. B. McKay *Joinery* i. 24 Another form of vertical boring machine is known as a router or recessing machine or overhead spindle moulder. **1954** W. E. Kelsey *Carpentry, Joinery & Woodcutting Machinery* xvii. 517 The router has taken over a great deal of the lighter work up to 1 in. or 1¼ in. thick which was formerly done on the spindle-moulder. *Ibid.* 546 Portable electric router... This machine works on the same principle as the overhead-router. **1958** *Wall St. Jrnl.* 30 Sept. 7/4 An official..enthuses over a new power wood-working tool called a 'router'. **1976** *Arizona Republic* (Phoenix) 9 May k2/6 Harman uses a router (similar to an electric drill..) to make a hole in the center of the slice for the hand shaft to go through. **1976** *S. Wales Echo* 23 Nov. 11/5 (Advt.), One overhead router £200. Various other items for woodworking shop.

3. *attrib.*, as **router bit, cutter** = sense 1 c above; **router plane,** a plane with a cutter projecting below the sole so that the bottom of a groove or recess can be planed.

1953 E. G. Hamilton *Power Tools for Home Craftsman* vii. 230 Small router bits are usually of the single-flute type. **1976** C. H. Groneman *Gen. Woodworking* (ed. 5) xlvi. 220 Most routers use ¼- or ⅜-in.-shank (6·35- or 9·52-mm-shank) router bits. *a* **1877** Knight *Dict. Mech.* III. 1995/2 *Routing-machine*, a shaping-machine which works by means of a router-cutter..revolving above a bed with universal horizontal adjustment. **1846** Router plane [in Dict., sense 1]. **1934** *Planecraft* (C. & J. Hampton Ltd.) xiv. 105 The Record Router Plane..is made both with an open and with a closed mouth. **1966** A. T. Collins *Newnes Compl. Pract. Woodworking* 30 Router planes are used for levelling and smoothing the surface on the bottom of a groove, slot or cavity which is inaccessible to an ordinary grooving plane. **1974** [see sense 1 b above].

routier^1. (Later *Hist.* examples.)

1962 J. Needham *Sci. & Civilisation in China* IV. i. 285 Huang Shêng-Tsêng named as one of his sources a *Chen Wei Pien* (Collection of Needle Positions), which may or may not have been a specific printed book. If it was, it must have been a 'routier' or 'rutter' like the *Yüeh Yang Chen Lu Chi* (Record of Courses Set by the Needle in the Cantonese Seas), which is known to have still existed in the 18th century. **1971** S. E. Morison *European Discovery Amer.: Northern Voy.* xiv. 465 Alfonce attributes the discovery of this river to the Portuguese, and the rhymed *routier* of 1547 by Jean Mallart agrees.

‖ **routier**^{2} (rutye). [Fr., f. *route* Route *sb.*]

1. *Hist.* A member of any of numerous companies of mercenary soldiers that were active in France during the later Middle Ages.

1845 *Encycl. Metrop.* XI. 620 They [*sc.* the mercenary adventurers] were named also..*Routiers*, for numerous reasons too unsatisfactory to deserve quotation. **1910** *Encycl. Brit.* II. 683/1 Arthur now resumed the war against the English, and at the same time took vigorous measures against the plundering bands of soldiers and peasants known as *routiers* or *écorcheurs*. **1924** K. Norgate *Richard the Lion Heart* ii. 53 If these Routiers could have been controlled by their employers, Henry and Richard might probably have been easily surrounded and captured. **1951** W. B. Wells tr. *Perroy's Hundred Years War* iv. i. 149 When he entered the king's service, ..Duguesclin was no more than a captain of *routiers*, fond of pillage and raids. **1961** P. Green tr. *Oldenbourg's Massacre at Montsegur* iv. 105 The *routiers*, or mercenary companies, who formed a large part of the infantry. **1965** Auden *About House* (1966) 17 Conventional Blunderbuss war and its routiers. **1970** M. Jones *Ducal Brittany* vi. 167 The payments made by John IV to the *routier* companies in 1368–9 were..rather of protection money.

2. In France, a long-distance lorry driver. Also *attrib.*

In the *attrib.* examples the reference is usually to the *Guide des Relais Routiers*, a guide-book originally designed for lorry-drivers in France.

1961 L. Durrell in *Holiday* Feb. 114 We planned to stop somewhere on the road to Béziers and have a bite of supper—Raoul knew a little place patronized by the *routiers*. **1971** *Guardian* 18 Aug. 10/5 One hotel (Routiers) stop in each direction including dinner and breakfast.. [£]9.00. **1975** *Ibid.* 27 Jan. 7/3 Setting out with your Green Card and your Routier Guide to storm the Alpine passes. **1976** *Times* 14 Feb. 13/4 A cheap and cheerful *routier* halt where..the *café au lait* came in something more like a *pot de chambre* than a cup.

routinary, *a.* Delete *rare* and add later examples. Also, in wider senses: that acts according to routine; occurring, performed, etc., routinely.

1963 V. Nabokov *Gift* ii. 105 Some sort of routinary hallucination, like a harmless domestic ghost that sits down..every evening by the fireside. **1967** D. Flakoll tr. *Asturias's Cyclone* iii. 33 'The second ball is routine... Perhaps you can tell us where..routine commences.' 'It ends the instant the ball leaves your hand. You are no longer a rutinary [*sic*] bowler. The adventure begins.' **1976** *Word 1971* XXVII. 61 The exceptionality, for today, of prenatal assessment of foetal neurophysiology will be the routinary procedure of the future.

routine, *sb.* (*a.*) Add: **1. c.** *Theatr.* A carefully rehearsed act or sequence of actions (in dancing, singing, dialogue, etc.); a sketch, turn, or 'number'; the manner in which an act is performed. Similarly in *Gymnastics*, a performance comprising a sequence of exercises carried out either on the floor or on apparatus, usu. in competition. Also *transf.* and *fig.*

1926 *Dance Mag.* June 25/3 No one ever taught him a routine. When he hummed a tune, dance steps just came to him. **1930** *Dancing Times* July 354/2 If a student goes through the same routine of steps (I am not talking of exercises, but of combined steps constituting a dance) [etc.]. **1932** N. Coward *Words & Music* I. 9 Don't do a pratfall in your first routine. **1949** N. Marsh *Swing, Brother, Swing* xii. 280 He wasn't meant to fall. They'd altered the routine. **1956** H. Kurnitz *Invasion of Privacy* xiv. 92 Do you know the blackmail routine that Jarrold gave me tonight? **1959** Loken & Willoughby *Compl. Bk. Gymnastics* xvii. 196/1 For example, a fast, snappy mass tumbling act would be good following a slow, precise doubles balancing routine. **1963** 'E. McBain' *Ten plus One* vi. 78 What the hell is this?.. A vaudeville routine at the Palace? **1975** *Oxf. Compan. Sports & Games* 452/1 In C-difficulty routines he may perform movements such as going from a handstand between the bars and then resting again in another handstand. **1977** *Time* 22 Aug. 43/1 A teacher in Peoria had encouraged him to become a performer, and when he returned from Germany he started a routine there at a little club.

d. *Computers.* A set of instructions which performs a specific task and is stored so that it may be executed many times; now esp. one which may be part of a longer, self-contained program.

1945 J. P. Eckert et al. *Description of ENIAC* (PB 86242) (Moore School of Electr. Engin., Univ. Pennsylvania) b-3 Suppose it is desired to..carry out a computational routine of m line steps, print the final results, and then perform the same routine n times. **1948** Goldstine & von Neumann in J. von Neumann *Coll. Wks.* (1963) V. 217 We call the coded sequence of a problem a routine, and one which is formed for the purpose of possible substitution into other routines a subroutine. **1948** *Proc. IRE* XXXVI. 1453/1 The iterative methods of numerical analysis involve the repeated performance of computing routines. **1951** M. V. Wilkes et al. *Preparation of Programs for Electronic Digital Computer* iii. 22 A 'closed' subroutine is one which is called into use by a special group of orders incorporated in the master routine or main program. **1967** *Technology Week* 23 Jan. 11/2 (Advt.), Software for Sigma 5 includes..a library of mathematical, business and utility routines. **1971** Dudrap & Emery in R. A. Wisbey *Computer in Lit. & Linguistic Res.* iii. 90 It is often better to provide a few assembly-code routines than to try doing character editing in 'raw' fortran. **1980** K. D. Wise *Microcomputers* v. 102 Transfers of data or control between routines should occur only when the programmer specifically requests them and only as called for in the specification of the routines.

3. a. (Later examples.) Now passing into *adj.* Also, in wider senses: of a customary or standard kind; usual, typical, standard.

1940 H. Spencer *Art & Life W. Shakespeare* v. 197 No routine braggart-soldier he. **1960** 'E. McBain' *Give Boys Great Big Hand* vi. 59 'Maybe you can find some of Karl's skull on it. Isn't that what you'd like to find?' 'This is just a routine investigation, Mrs. Androvich.' **1961** W. Sargeant in Webster s.v. *routine* adj., The level of artistry..was altogether routine and uninspired. **1964** L. Deighton *Funeral in Berlin* xxvii. 146, I shouldn't worry about it. It's just a routine check. **1979** *Sci. Amer.* Dec. 112 Recently it has become routine in many laboratories and hospitals to record evoked potentials from the brain stem.

b. *Comb.*, as *routine-chained, -ridden, -sodden* adjs.

1920 *Chambers's Jrnl.* 19 June 453/2 Routine-chained staffs worked on into the night. **1929** A. Huxley *Holy Face* 64 Our routine-ridden, mechanized world of flabbily sub-human sentimentalists. **1964** M. McLuhan *Understanding Media* x. 103 The need for advanced knowledge presses on the spirits of the most routine-ridden minds. **1920** *Contemp. Rev.* June 866 The Soviet authority had to destroy everything in this department—the laws themselves, the routine-sodden institutions.

routine (rutī·n), *v.* [f. prec.] *trans.* To apply a routine to; to organize according to a routine. Hence **routi·ning** *vbl. sb.*

1897 G. B. Shaw in *Sat. Rev.* 18 Dec. 712/1 No actor can possibly play leading parts of the first order six nights a week all the year round unless he underplays them, or routines them mechanically in the old stock manner. **1941** W. C. Handy *Father of Blues* iv. 39, I was consulted by Whalen and Martelle relative to routining their shows. **1948** 'La Meri' *Spanish Dancing* iv. 43 There are a variety of typical steps which can be routined at the will of the dancer. **1959** R. Condon *Manchurian Candidate* ii. 31 Yen Lo got three implantation teams started on them, staying with each team through the originating processes until he had assured himself that all had been routined with smoothness. **1976** W. Goldman *Magic* ii. 65 He spent the intervening days working out his routining. Start with the flashy stuff or save those for the end?

routined (rutī·nd), *ppl. a.* [f. Routine *v.* + -ed^1.] Subjected to or regulated by (a) routine.

1913 E. F. Benson *Thorley Weir* i. 22 The gleaming romance and glory that lie so close below the surface of the most routined and rutted life. **1928** *Manch. Guardian Weekly* 23 Nov. 407/2 Criticism will do well not to base on this routined output a judgment which leaves out of account the Schubert of the year of his death. **1948** 'J. Tey' *Franchise Affair* xiii. 138 One result of stepping out of a routined life was..that you couldn't..stroll home at four o'clock of an afternoon. **1964** F. Bowers *Bibliogr. & Textual Crit.* vi. iii. 180 An inexperienced.. compositor might be supposed to feel the influence of copy more strongly than a thoroughly routined workman.

routineer. (Later examples.)

1928 G. B. Shaw *Intelligent Woman's Guide Socialism* lxx. 340 The civil servant, the judge, the navy captain, the field marshal, the archbishop, however extraordinary able, gets no more than any routineer of his rank and seniority. **1956** 'H. MacDiarmid' *Stony Limits & Scots Unbound* 129 The routineer Haig, Whose lack of imagination carried him through. **1977** *National Observer* (U.S.) 22 Jan. 6/1 The innovator, in both business and the arts, is always to be contrasted..with the routineer.

routinely (rutī·nli), *adv.* orig. *U.S.* [f. *routine *sb.* (*a.*) 3 a + -ly^2.] As a matter of course or of routine; according to (a) routine; by rote, mechanically.

1924 *Scribner's Mag.* Aug. 216/1 Even now I think we take policemen, professors, conductors, etc., much too routinely. **1948** A. Cooke in *Manch. Guardian Weekly* 29 Apr. 13/2 It has been routinely filed away and incorporated in the body of American dogma. **1956** *Nature* 25 Feb. 383/1 The inhibition of the enzyme by the osmium tetroxide fixative routinely used for electron microscopical preparations. **1965** *Economist* 10 July 137/3 Such infractions are routinely settled with a small bribe for the policeman. **1968** J. D. Watson *Double Helix* xviii. 129 He was only routinely enthusiastic as he went over Griffith's quantum-mechanical arguments. **1971** D. Lambert in C. Bonington *Annapurna South Face* 289 Appendicectomy is done routinely on members of expeditions to the Antarctic. **1974** J. Heller *Something Happened* 101 'What did you do today?' I ask routinely (before she can ask me).

routiner (rutī·nəɹ). [f. Routin(e + -er^1.] †**1.** (see Routineer.) *Obs. rare*$^{-1}$.

2. *Teleph.* A set of equipment for testing circuits and switching apparatus in an exchange.

1929 *P.O. Electr. Engineers Jrnl.* XXII. 24 Routiners are composed of two main parts: the test and control apparatus, and the access equipment. **1948** J. Atkinson *Telephony* I. xxi. 453/1 Specially designed test boxes or routiners..are arranged to apply test conditions somewhat more onerous than the conditions normally encountered in practice. **1973** *P.O. Electr. Engineers Jrnl.* LXVI. 44/2 Trunk circuit routiners..are used to verify that a call can be established to a distant answering relay-set over each outgoing trunk circuit in turn and that the transmission loss..is within maintenance tolerance.

routing, *vbl. sb.* (in Dict. s.v. Route *sb.*). Add: Also **routeing.** Also, the action of the vb.: direction along, or allocation to, particular routes. (Further examples.)

Routeing is the better form to distinguish it from Routing *vbl. sb.* and *ppl. a.* (pronounced rau·tiŋ).

1903 *Electr. World & Engineer* 23 May 856/1 The facility which such combination lends to through routing of cars. **1930** M. Clark *Home Trade* xxiv. 198 All waste of time in the passage of any piece of work through the factory, scientific management methods endeavour to eliminate by 'routeing'. **1947** A. Harris *Bomber Offensive* 188 Against this we devised a new kind of routeing. **1964** F. L. Westwater *Electronic Computers* iii. 28 The routing and control of a sequence of pulses throughout a computer depends on an appropriate assembly of switching circuits. **1975** *Daily Tel.* 12 Feb. 14 Postal addresses are in effect routing directions. **1976** P. R. White *Planning for Public Transport* vii. 139 The weakness is that journey times between major towns are often far too long, especially in relation to potential direct routeings.

‖ **routinier** (rutinye), *a.* and *sb.* *Mus.* [Fr.: cf. Routineer.] **A.** *adj.* Of a piece of music: composed in a routine or orthodox manner. *rare.* **B.** *sb.* A conductor who performs in a mechanically correct, but uninspiring, way.

1934 C. Lambert *Music Ho!* v. 279 Walton's..mature but regrettably consonant *Belshazzar's Feast* was dismissed, particularly by the older critics, as 'routinier', conventional, and unworthy of its place in so selectly revolutionary a festival. **1970** *Guardian* 1 Jan. 8/2 Everything was first-rate with the exception of Boris himself..and the conductor Boris Khaikin, a tired routinier. **1970** *New Yorker* 26 Sept. 114/3 A new conductor..didn't seem to arouse much tension. He proved to be what is usually described as a useful *routinier*.

routinization (rutī:nəizē^i·ʃən). [f. *routinize *v.* + -ation.] The being or becoming routine in character or operation; the action of superimposing a routine upon that which was previously less systematized or controlled.

1934 in Webster. **1946** Gerth & Mills *From Max Weber* (1947) iii. 54 By tracing out the routinization of charisma, Weber is able to assign a heavy causal weight to institutional routines. **1954** *Encounter* June 12/1 Coal mining..now—with the mechanisation of cutting and conveying—takes on much of the routinisation of factory work. **1960** *Guardian* 29 Apr. 11/1 Beatnik Tanya.. excoriates the routinisation of sex in marriage. **1965** H. Kahn *On Escalation* xiii. 258 It is a major purpose of current command-and-control efforts to facilitate the 'routinization'..of the various aspects of crisis management. **1972** A. Giddens *Politics & Sociol. in Thought of Max Weber* iii. 39 The 'routinisation' of politics—that is to say, the transformation of political decisions into decisions of administrative routine..—is specifically foreign to the demands which are most basic to political action. **1976** *National Observer* (U.S.) 26 June 17/3 We are all equally liable to the lumpish routinization of compassion both maddening and necessary.

routinize (rutī·nəiz), *v.* [f. Routine + -ize.] *trans.* To subject to (a) routine; to make into a (matter of) routine.

1928 *Amer. Speech* III. 434 An investigator of nurses' training asks whether nurses are to become 'machinized and routinized'. **1937** J. Dollard *Caste & Class in Southern Town* xv. 344 All such behavior patterns have emotional value, even when they seem most routinized. **1960** W. H. Whyte *Organization Man* xxvi. 364 They know how to routinize crisis. **1965** *Listener* 24 June 925/2 Ours is a time in which man..has been mechanized and routinized. **1973** J. S. Bruner *Beyond Information Given* (1974) xvii. 300 When..the child has routinized the task of holding two objects, one in each hand, there then occurs a first storage activity. **1978** *Dædalus* Summer 70 Duty emerges..as an attempt to codify, systematize, and routinize behavior which springs from desire.

routi·nized, *ppl. a.* [f. prec.] Subject to (a) routine; made into a (matter of) routine.

1938 *Sun* (Baltimore) 21 July 18/1 There was nothing to indicate an approaching disaster in the routinized handling of the dynamite before the blast. **1945** G. Williams *Women & Work* ii. 41 Women..were mostly confined to the lower-grade, more routinised categories. **1949** M. Mead *Male & Female* xiii. 269 The baby would not eat, having very doubtfully enjoyed the routinized, cloth-enveloped experience. **1952** B. Ulanov *Hist. Jazz in Amer.* xiii. 151 The fairly tight, routinized Nichols sessions set the style. **1978** *Jrnl. R. Soc. Arts* CXXVI. 412/2 Administration is the conduct of affairs in a routinized fashion. **1981** *Times Lit. Suppl.* 27 Feb. 216/1 The twentieth century may be the century of..routinized labour.

roux. (Earlier and later examples.)

1813 L. E. Ude *French Cook* vi. 140 Cut your chops.., then fry them in a little butter, of a nice brown colour, drain this butter, and make a *roux* very *blond*. *Ibid.* xiii. 297 Put a lump of butter into a stew-pan... Then make a *roux* by mixing a little flour. When your *roux* begins to get brown, put in two large onions cut [etc.]. *Ibid.* xiv. 361 After having stewed your oysters.., you make a *roux blanc* into which you put a few small onions, [etc.]. **1845** E. Acton *Mod. Cookery* (ed. 2) iv. 97 Sauce tournée is..rich pale gravy..thickened with delicate white roux. **1861** Mrs. Beeton *Bk. Househ. Managem.* 251 *White roux*, for thickening White Sauces... Allow the same proportions of butter and flour..and proceed in the same manner as for brown roux. **1945** *ABC of Cookery* (Ministry of Food) xiii. 51 This mixture of fat and flour is called a roux. **1965** *House & Garden* Dec. 84/4 Cook for a further few minutes to dispel the raw flour taste, then stop, if you want the *roux* white, carry on a little longer for the blonde, and longer still for a brown. **1976** *National Observer* (U.S.) 6 Nov., The roux should be brown but not burned. Do not burn or you will ruin the gravy.

rov, var. *rav.

rover^1. Add: **3. a.** Also *spec.* (see quot. 1944).

1933 H. G. Wells *Bulpington of Blup* v. 177 They were to go as 'Rovers' to the Russian Ballet. **1944** G. B. Shaw *Everybody's Political What's What?* xxxi. 279 Complaisant critics were welcomed in the theatre even when all the stalls were sold out and they had to be content as 'rovers' without allotted seats, sitting or standing about wherever they could.

c. *Australian Rules Football* (see quot. 1969^1). *Rugby Football*, formerly, an extra forward performing some of the functions of the scrum half and fly half. *Amer. Football*, a defensive linebacker who is assigned to move about to anticipate opponents' plays.

1894 A. Sutherland in M. Shearman *Athletics & Football* (ed. 4) ii. vii. 422 The rover is an individual chosen for his quickness and readiness to go wherever he is wanted. He observes the turn of the game, and follows when he sees his own followers being overweighted by their adversaries [in Australian Rules Football]. **1909** E. G. Nicholls *Mod. Rugby Game* iv. 54 A fifth three-quarter..as the 'rover' or flying half is frequently styled. **1916** *Colliers* 30 Dec. 30/3 He is all over the field as a rover, diagnosing the play quickly and with unfailing accuracy. *a* **1917** J. E. Raphael *Mod. Rugby Football* (1918) xvii. 225, I played 'rover' for England on a memorable day at the Crystal Palace. **1927** Wakefield & Marshall *Rugger* ii. vi. 268 The formation used in New Zealand... In this formation..the outsides consist of the extra forward, who may be described either as a rover or a half-back, a scrum-half, two five-eighths, a centre, [etc.]... The rover puts the ball into the scrum, while the scrum-half stands at the base of the scrum to take it out. **1954** J. B. G. Thomas *On Tour* 26 They [*sc.* the All Blacks] packed 2-3-2, with Gallaher acting as a 'rover', whose duty it was to put the ball into the scrum while the scrummage half back waited behind the scrum. **1969** Eagleson & McKie *Terminol. Austral. Nat. Football* iii. 10 *Rover*, a member of the ruck, usually smaller than the other two members (*followers*), and selected for agility in capturing the ball. **1969** *Australian* 24 May 39/4 Essendon has Barry Davis back as a ruck-rover, to help captain Don McKenzie, and this should strengthen the side's following division and provide more opportunities for rovers Bob Greenwood and Don

Gross. **1970** *Univ. of Alabama Football Press Guide* 17 The rover slot was very similar to linebacking. **1975** *Oxf. Compan. Sports & Games* 385/1 A team [in Australian Rules Football] is made up of three full-forwards [etc., and]..the 'ruck'. The ruck consists of two followers and a rover, who moves with the flow of play.

d. Formerly, a member of a senior branch of the Scout Association (see SCOUT *sb.*[4] 2 c). Also *rover scout.*

In 1967 this branch of the Scout Association was replaced by the venture scout branch (see quot. 1966).

1922 R. BADEN-POWELL *Rovering* 210 Rovers are a Brotherhood of the *Open Air* and *Service*. **1933** A. G. MACDONELL *England, their England* xiii. 235 Lots of the young chaps are Rovers and don't drink so as to be an example to the Scouts and Cubs. **1959** *Chambers's Encycl.* II. 481/2 The movement in Great Britain is divided into four groups: rover scouts, i.e. young men of 18 and over; [etc.]. **1966** *Times* 10 June 11/1 The Scout Association will have three main sections—cub scouts.., aged 8 to 11; scouts, aged 11 to 16; and venture scouts (replacing senior scouts and rovers), aged 16 to 20. **1972** K. BONFIGLIOLI *Don't point that Thing at Me* i. 7 His bedroom is..full of fresh air; just what you would wish your Rover Scout son's room to be.

e. The name given to an R.A.F. reconnaissance patrol flown in 1940 and 1941. Also *attrib.*

1942 *R.A.F. Jrnl.* 3 Oct. 18 'Strike' and 'rover' patrols were on the board every day. **1957** R. BARKER *Ship-Busters* ii. 35 The Rover was a roving commission, an armed reconnaissance against enemy shipping.. carried out by a small number of aircraft working independently.

f. Also **Rover, 'Rover.** *ellipt.* A Land-Rover (see *LAND *sb.*[1] 12).

1961 A. WILSON *Old Men at Zoo* iv. 207 You hop into the rover, Carter. You're frozen. **1973** G. MOFFAT *Lady with Cool Eye* vi. 66 Slade was taking the spare wheel off the 'Rover's bonnet. **1975** *Country Life* 13 Feb. 373/1 Some elderly sportsmen have high seats constructed on their 'rovers'..for shooting.

5. A remote-controlled surface vehicle for extraterrestrial exploration.

[**1967** *Jrnl Spacecraft & Rockets* IV. 209/1 A dynamic analysis in preliminary design of a lunar roving vehicle should have at least two basic aspects.] **1970** *Science Jrnl.* Jan. 16 The first rover is scheduled to take four trips of up to 32 km each with travel limited to 4·8 km radius from the landing site. **1971** [see *moon buggy* s.v. *MOON *sb.* 16]. **1971** *Nature* 19 Nov. 125/3 The rover had a responsive steering, and..climbed slopes where the dust layer was deep enough to make walking difficult. **1972** [see *LUNAR *a.* and *sb.* A. 1 c]. **1978** *Sci. Amer.* Mar. 89/3 Another possibility is an unmanned surface rover capable of traversing hundreds of kilometers over a period of several years, which would be able to analyze the regolith in more detail than any satellite in orbit could.

roving, *vbl. sb.*[1] Add: **1. b.** *roving course.*

1939 P. H. GORDON *New Archery* iii. 21 Royal edict set aside places for shooting in the towns and provided long roving courses over the distances between towns.

2. b. roving commission (see quot. 1846); also *transf.* and *gen.*, (a body given) authority to pursue any inquiry or investigation in whatever quarters it may be considered necessary.

1846 [in Dict.]. **1867** *Congress. Globe* 22 Mar. 273/2, I think it would be safer to leave this matter [of certain state claims] to the direct inspection of the War Department, than to send out a roving commission. We have had enough of these roving commissions. **1894** *Congress. Rec.* 25 Apr. 4098/1 Is it a legitimate expenditure of the public money to send up consuls with roving commissions to hunt out commerce for a certain class of our people? **1930** W. S. CHURCHILL (*title*) My early life: a roving commission. **1936** A. CHRISTIE *ABC Murders* xviii. 130, I had a kind of roving commission to purchase things for my brother. **1954** 'N. BLAKE' *Whisper in Gloom* I. ii. 31 Who'd you put him on to? Or was it a roving commission? **1959** *Ann. Reg. 1958* 179 The Russians opposed the Western proposal that they should be mobile and on a permanent footing on the ground that such 'roving commissions' would engage in espionage. **1981** *Listener* 1 Jan. 12/1 For several years as West Africa correspondent he had what was in effect a roving commission.

roving, *vbl. sb.*[3] Add: **2. a.** (Further examples.)

1960 *New Scientist* 10 Mar. 597/1 The quantity of glass fibre applied..is governed by the rate of operation of the rollers that feed the multi-ply coil of glass fibre rovings to the cutter block. **1964** H. HODGES *Artifacts* ix. 128 Sometimes the rolag may be drawn out to a thickness approaching that of the required thread, and even given a slight twist before winding on to the distaff. Prepared fibres in this state are usually called rovings. **1972** *Physics Bull.* Nov. 663/3 The glasses are produced in continuous strands, consisting of 204, or multiples of 204, filaments which are subsequently processed into rovings or into yarns for weaving purposes. **1977** *Austral. Sailing* Jan. 51/2 The construction..sounds strong, including a hand-laid layer of woven rovings and additional strengthening in stress areas.

b. (Further examples.)

1946 A. J. HALL *Stand. Handbk. Textiles* iii. 105 The roving at this stage is about as thick as coarse string. **1972** *Physics Bull.* Nov. 663/3 S-Glass is most commonly employed in 'roving' or other unwoven forms. **1972** *Sci. Amer.* Dec. 47/2 Roving, on bobbins, is put into spinning frames, where it receives a final drawing out and the twisting necessary to make it into yarn.

roving, *ppl. a.* Add: **2. d.** Of an ambassador, journalist, etc., required to travel to various locations to deal with events as they occur.

1938 E. WAUGH *Scoop* III. i. 258 Will you accept five year contract five thousand year roving correspondent. **1946** *R.A.F. Jrnl.* May 147 This month our roving reporter..went back to Germany. **1958** *Listener* 26 June 1043/2 He was in the recent past Mr. Kishi's roving trade ambassador in south-east Asia. **1965** B. SWEET-ESCOTT *Baker St. Irreg.* iii. 98 Staying in the Middle East as 'a kind of roving ambassador'. **1967** *Boston Sunday Globe* 23 Apr. 29/2 Boston's roving inspectors—who check on city services for the mayor. **1968** J. DRUMMOND *Gantry Episode* ix. 72 It can't be done. Not by ordinary methods. That's why Purnell wants you as his roving reporter. **1970** A. SINCLAIR *Guevara* vi. 71 From 1960 onwards, Che had often served as a roving ambassador for Fidel Castro. **1972** D. BLOODWORTH *Any Number can Play* xii. 111 She would return..as a roving correspondent for the Worldover Syndication Service.

3. a. (Later examples.)

1951 N. MITFORD *Blessing* I. ii. 17, I'm afraid she's deeply romantic, and Valhubert has a roving eye. **1968** D. GRAY *Died in Red* xiv. 73 The man with the roving eye comes along, and he tells you you're wonderful. **1970** V. GIELGUD *Candle-Holders* xi. 98 Angela Baynes had caught Tarzan's permanently roving eye.

row, *sb.*[1] Add: **I. 1. a.** *spec.* A line in a chorus.

1849 THACKERAY *Pendennis* I. xiv. 125 Who's that gal in the second row, with blue ribbons, third from the stage. **1932** D. L. SAYERS *Have his Carcase* xxiii. 303 O.K. darling. 'Aeroplane Girl', first row, song and dance. **1967** A. WILSON *No Laughing Matter* II. 85 The thousands of silly boys who join the back row of the chorus every year.

2. d. *Mus.* = *tone-row* s.v. *TONE *sb.* 11. Also *Comb.*, as *row-note.*

1936 *Musical Q.* XXII. 14 The chief contribution towards the organization of the twelve-tone system is that peculiarly Schoenbergian concept—part abstract theory and part pure inspiration—the 'row': a semi-arbitrary arrangement of the twelve chromatic tones into a horizontal motival structure. *Ibid.* 31 Examples could be multiplied indefinitely, but,..most of them would boil down to some similar types of random distribution of the row-notes. **1958** A. JACOBS *New Dict. Mus.* 390 This method works through the 'note-row' (or 'series'), in which all the twelve notes are placed in a particular order as the basis of a work. No note is repeated within a row, which accordingly consists of twelve different notes and no others. **1965** *Listener* 20 May 757/3 An important aspect of the work is the extraction from the note-series of innumerable motivic elements and of the great variety of ways in which the row is itself presented. **1971** *Times Lit. Suppl.* 1 Oct. 1180/2 He [*sc.* Webern] was particularly partial to rows whose second half is a mirror inversion of the first.

4. a. Also *Comb.*, as **row house** *N. Amer.*, a terraced house; also (with hyphen) *attrib.*; hence **row housing.**

1936 H. HAGEDORN *Brookings* i. 10 So the Brookings children moved to Baltimore..and went to live in a pleasant brick row-house with the canonical white stoop which Baltimoreans cherished. **1940** *Sun* (Baltimore) 16 Feb. 24/1 Mr. Pagon, in proposing that the entire area be rezoned from a row-house status, pointed out that [etc.]. **1949** *Ibid.* 29 Nov. 12/1 The Housing Authority's plans call for row housing. **1952** *Ibid.* 11 Jan. 12/3 The narrowest row house will be 16 feet across the front..and there will be only seven to a group. **1957** W. H. WHYTE *Organization Man* xxiii. 305 A study of several new Philadelphia row-house neighborhoods. **1968** *Globe & Mail* (Toronto) 13 Feb. B2/4 There is no doubt that municipalities..would be agreeable to more realistic zoning in respect to smaller lots, row housing, etc. **1979** *Kingston* (Ontario) *Whig-Standard* 29 Mar. 21/4 The township has called for ten feet of yard between the end of a line of row houses and the next building.

b. Also *attrib.*

1871 J. M. LANGFORD *Let.* 2 Dec. in *Geo. Eliot Lett.* (1956) V. 223 Some of the Row Houses whose subscription was partially delivered have been in for more.

e. Chiefly *U.S.* A line of cells in a prison; esp. in phr. *death row*, the part of a prison where condemned prisoners are kept.

1950 in M. McLuhan *Mech. Bride* (1967) 4/2 The doomed men..were filmed in death row yesterday afternoon. **1968** *Listener* 15 Feb. 210/2 Until he arrived, the ten prisoners in death row never left their cells, not even for exercise. **1971** *Black Scholar* Apr.–May 19/1 He is now waiting on Death Row in the Ohio Penitentiary. **1973** *Philadelphia Inquirer* 7 Oct. (Today Suppl.) 26/3 After all that time, you'd think we would all be exhilarated to be off the row. But everyone of us wanted, on some level, to go back. We didn't want to face the responsibilities of being out and having to fend for ourselves. **1973** *Publishers Weekly* 27 Aug. 231/2 Sentenced to death, he cut off his penis and has spent 23 years in a cage on a row reserved for lunatics. **1980** *Sci. Amer.* Apr. 63/3 Texas, which already has such a capital-punishment law on the books, currently ranks second in the nation, with 119 on death row.

6. a. Also *Comb.*, as **row boss** *U.S.* (see quot. 1937); **row crop** (see quot. 1930).

1930 *Amer. Speech* VI. 11 Irrigation farmers call beets, potatoes, and beans *rowcrops* in distinction from alfalfa and the grains, the *flood* crops, for in the former case the water is directed down rows instead of being allowed to flow over the whole field. **1937** *Sun Mag.* (Baltimore) 11 July 9/3 'The young ones aren't as good as their parents,' said the row boss. *Ibid.*, The term 'row boss' comes..from his being in charge of the pickers when they are working along the rows of vegetables. **1943** J. S. HUXLEY *TVA* 45 Indian corn and other row crops. *Ibid.* 58 Row crops are confined to the more level spots. **1950** *Engineering* 5 May 505/3 Light, medium and heavy tractors were all represented, and each group was subdivided into rowcrop and general-purpose tractors. **1960** *Farmer & Stockbreeder* 1 Mar. 125/1 We pioneered pneumatic tyred row-crop wheels. **1971** *Arable Farmer* Feb. 29/1 Wheel tractors on the farm being used solely for haulage and row-crop work. **1977** *New Yorker* 29 Aug. 48/1 So much for row-boss supervision, so much for harvest labor, so much for trucking, so much for tractor overhead, so much for fertilizer and pesticides.

b. For *U.S.* read 'orig. *U.S.*' and add later examples.

1912 J. MASEFIELD *Widow in Bye St.* IV. 56 Bessie, the gipsy, got with child by Ern... 'I hear the gipsy has a row to hoe.' **1955** *Times* 2 Aug. 4/6 The lecturer then set himself a hard row to hoe; the scholarly correction of everything his audience may have been taught at school about King John, Runnymede, and Magna Carta. **1961** B. FERGUSON *Watery Maze* v. 119 Mountbatten had therefore no easy row to hoe; but he had a definite course to steer: the invasion of France. **1969** *Listener* 26 June 894/2, I recognise full well that there are many people who always find life a pretty difficult row to hoe and our society must be a compassionate society. **1976** *New Yorker* 26 Apr. 62/3 Women have a God-damned hard row to hoe.

7*. In knitting, one line of stitches.

1800 M. EDGEWORTH *Parent's Assistant* (ed. 3) II. 79 Her mother's unfinished knitting lay upon a table near the bed, and Susan sat down in her wicker arm chair and went on with the row, in the middle of which her hand stopped the preceding evening. **1872** GEO. ELIOT *Middlem.* II. III. xxvi. 69 Mrs Taft who was always counting stitches and gathered her information in misleading fragments caught between the rows of her knitting. **1909** W. J. LOCKE *Septimus* i. 4 She counted the rows of her knitting. **1932** [see *KNIT *sb.* 1 a]. **1970** M. HAMILTON-HUNT *Knitting Dict.* 19 Cast on in usual way, work a few rows of st st, the depth of the hem required.

II. 12. Also *transf.* of occurrences: in succession, consecutively. *colloq.*

a **1961** in WEBSTER s.v. *row*, Won the state tourney for four years in a row. **1969** 'E. LATHEN' *When in Greece* xiii. 139 Acute gastric distress..kept him awake..for a second night in a row. **1979** M. BABSON *So soon done For* i. 10, I burnt the clothes he'd been wearing yesterday, and I shampooed his hair three times in a row.

14. Special Combs.: **row matrix** *Math.*, a matrix consisting of a single row of elements; **row vector** *Math.*, a vector represented by a row matrix.

1941 BIRKHOFF & MACLANE *Survey Mod. Algebra* viii. 203 The coordinates of a vector ξ relative to a given basis in an *n*-space *V* form a one-rowed array $X = (x_1, \ldots, x_n)$, *X* a 'row matrix'. This may be considered as a vector in $V_n(F)$ or as a $1 \times n$ 'row matrix'. **1954** BEAUMONT & BALL *Introd. Mod. Algebra & Matrix Theory* i. 23 Any *m* by *n* matrix..may be thought of as an *m* by 1 column matrix with elements which are 1 by *n* row matrices. **1928** H. W. TURNBULL *Theory of Determinants* iii. 36 There are two distinct types of vector, the row vector, and the column vector. **1978** *Nature* 13 Apr. 605/2 A row vector..may be derived which has elements representing the magnitude of growth response to each climatic variable.

row, *sb.*[2] **1. a.** (Earlier and later examples.)

1746 S. BARRY *Let.* 6 June in D. Garrick *Private Corr.* (1831) I. 41 This occasioned a pleasant scene, for immediately, a terrible 'Row' ensued, between the few who paid ready money, and those who brought in his benefit-tickets. **1753** J. POULTER *Discoveries* (ed. 2) 13 He would prick again for thirty [Guineas]; we were afraid he would make too big a Row when he lost that, that is, a great Noise. **1955** *Times* 11 May 14/3 The Barons Court contest gains piquancy from a big local row which both sides expect to favour the Conservatives.

row, *sb.*[3] (Earlier example.)

1832 F. TROLLOPE *Dom. Manners Amer.* (ed. 2) I. xvi. 249 A row upon the Ohio was another of our favourite amusements.

row, *v.*[1] Add: **I. 2. c.** Also *to row in*, to conspire. *slang.*

1897 *Daily Tel.* 12 Feb. 5/7 It's very likely the sellers and the general public concerned in auction sales are anything but satisfied with the results of sales by auction where a 'knock-out' is arranged, and especially where the auctioneer 'rows in' with the crew. **1909** J. R. WARE *Passing Eng.* 211/2 *Row in*,..unfair conspiracy. From Thames life through centuries. A man 'rowed in' in a river robbery, or even a murder. **1934** P. ALLINGHAM *Cheapjack* xvi. 202, I think these boys had better row in with us...We may as well stick together. **1970** G. F. NEWMAN *Sir, You Bastard* vii. 194 What if they try to row in? **1977** P. MOYES *To kill Coconut* vii. 99 'Rowing in' is slang for implicating somebody in a crime.

II. 8. a. (Earlier examples.) Cf. *SALT RIVER 2 b.

1828 *Reg. Deb. Congress U.S.* 2 Feb. 1341 But, sir, I will venture to say this, that, in playing this game, if the Secretary of State is not influenced by the same courtesy which governed the courtiers of the great Frederick, never to beat the monarch at chess, that he could give the President twenty-nine, and as they say in Kentucky, 'row him up salt river'. **1832** *Spirit of Times* 28 Apr. 3/1 He 'rowed' Stanberry 'up a salt creek', and is now *being* tried by the House of Representatives for his unlucky propensity. **1832** J. K. PAULDING *Westward Ho!* I. ix. 77 See if I don't row you up Salt River before you are many days older.

9. e. *to row out*, to exhaust by rowing.

1928 *Daily Express* 7 Aug. 12/6 Both pairs finished in a distressed condition, Boardman being completely rowed-out.

row, *v.*[3] Add: **1. a.** (Earlier example.)

1789 *Loiterer* 14 Nov. 10 We..looked into every coach, *rowed* the waggons, examined both the boxes, the roofs, and the baskets.

b. For quot. substitute:

1789 *Loiterer* 21 Feb. 11 Racket *rowed* me up at seven o'clock—sleepy and queer but forced to get up to make breakfast for him.

2. a. Also const. *out, out of.*

1908 *Smart Set* June 143/1 Most fathers would have rowed me out of the house. **1976** *New Mus. Express* 31 July 6/4 But you get these weird, insecure feelings that they might be trying to row you out, which wasn't the case. No one was talking about sacking me.

3. Now usu. in the more limited sense 'to have a row, to quarrel noisily or heatedly'.

1890 T. A. JANVIER *Aztec Treasure-House* xvi. 195 Some of these Indians are friendly, and we don't want to start a row with them if they are willing not to row with us. **1914** S. LEWIS *Our Mr. Wrenn* xvii. 227 Why, Mouse! I thought you'd be glad to see me. I've never rowed with you, have I? **1970** G. F. NEWMAN *Sir, You Bastard* viii. 209 He couldn't remember rowing. Rows were usually loud demonstrative things. **1978** R. RENDELL *Sleeping Life* viii. 73 We row, of course we do, that's healthy in a marriage, but we love each other.

row-de-dow. (Earlier examples.)

1790 R. TYLER *Contrast* III. i. 42 There was a soldier fellow, who talked about his row de dow, dow, and courted a young woman. **1832** *Deb. Congress U.S.* 13 Mar. (1833) 2128 The rub-a-dub and row-de-dow excitement. **1885** *Referee* 8 Mar. 5/1 With regard to the Prince and Princess's visit to Ireland, the 'row-de-dow'—that is, we believe, the Hibernian term for it—which took place [etc.].

rowdy, *sb.*[1] **a.** (Earlier example.)

1808 W. LITTELL *Festoons of Fancy* (1814) 62 But it seems to this court that the loss to him would be the same, as if he had lost it among those, whom his gentlemanship is pleased to call *rowdies.*

rowdy, *v.* Add: **b.** *trans.* To treat in a rowdy manner.

1825 J. K. PAULDING *John Bull in Amer.* xii. 209 Notwithstanding..their being regulated and rowdied, and obliged to cut down trees as big round as a hogshead.

rowdy-dow, *sb.* [Cf. ROW-DE-DOW.] Boisterous noise; uproar. Also *attrib.* passing into *adj.* and quasi-*adv.* So **rowdy-dowy** *a.*

1852 J. LABERN *Popular Comic Song Bk.* 75 While Spifflicating Charlie Coker and Jane of the Hatchet-face divine, Just did the Rowdydowy Poker. **1935** W. STEVENS in *Poetry* XLV. 245 The heavy bells are tolling rowdy-dow. **1946** *Time* 22 July 40 This rowdy-dow roundup is the wild-cow milking contest. **1950** *N.Y. Times* 9 July II. 1/1 To restore the old rowdy-dow of burlesque, Mr. Mike Todd and Mr. Clark have gathered a handful of authentic drolls.

rowdy-dow, *v.* [f. prec.] *intr.* To be noisy or boisterous.

1966 T. PYNCHON *Crying of Lot 49* v. 110 She collided with a gang of guided tourists come rowdy-dowing out of a Volkswagen bus.

rowdy-dowdy, *a.* (Earlier and later examples.)

1854 M. S. CUMMINS *Lamplighter* 260 To offer herself as a champion for that rowdy-dowdy child. **1928** [see *NIGHTMAN 2].

rowdying (rau·di,iŋ), *vbl. sb.* [f. ROWDY *v.* + -ING[1].] Causing a disturbance.

1839 *Picayune* (New Orleans) 26 Feb. 2/4 There *is* more quiet and less rowdying..here than in Boston, with all its anti-drinking, anti-bellringing and other anti-noise making laws. **1887** *Courier-Jrnl.* (Louisville, Kentucky) 18 Feb. 1/3 There was a good deal of noise and 'rowdying'. **1913** D. H. LAWRENCE *Sons & Lovers* xiii. 430 'We've heard that song before,' snapped the old man. 'Now you get off, and don't be long about it. Comin' *here* with your rowdying.'

rowdyism. (Earlier and later examples.)

1842 S. LONGFELLOW *Let.* 8 Feb. in *Lett. Charles Dickens* (1974) III. 40/1 [Charles Dickens was] very animated and talkative,..with..the slightest tincture of rowdyism in his appearance. **1936** I. L. IDRIESS *Cattle King* xx. 189 Perhaps the hostility and rowdyism that the Salvation Army had to suffer..aroused his sympathy. **1955** *Times* 20 Aug. 6/1 The seriousness with which the commission regards the situation..from the relatively minor demonstration in Saarbrücken last Saturday night to the rowdyism in Neunkirchen on Wednesday night. **1976** *Southern Even. Echo* (Southampton) 15 Nov. 2/8 The problems presented by late night rowdyism.

roweite (rōu·ait). *Min.* [f. the name of George *Rowe*, 20th-cent. U.S. mine official and mineralogist + -ITE[1].] A basic borate of calcium and manganese, usu. also containing magnesium and zinc, first found as light brown elongated orthorhombic crystals in zinc ore at Franklin, New Jersey; $(Mn,Mg,Zn)_2Ca_2B_4O_7(OH)_6$.

1937 BERMAN & GONYER in *Amer. Mineralogist* XXII. 301 The crystals of roweite are light brown in color, lath shaped and without measurable terminations. **1975** *Soviet Physics: Doklady* XX. 244/1 The isostructural nature of the (Mn, Mg) and Mn roweites and the $[B_4O_5(OH)_4]$ tetraradical..met in sodium borate.

rowel, *sb.* Add: **I. 1. d.** *rowel-deep* adv. (earlier example).

1832 W. C. BRYANT *Poems* 45 His spurs are buried rowel-deep, he rides with loosened rein.

II. 8. For *Obs.*—[1] read *Obs.* and add later example.

1836 *Col. Crockett's Exploits & Adventures Texas* i. 1 Though they start at the lowest rowel of the ladder.

rowel, *v.*[1] Add: **2. c.** *fig.*

1918 F. HACKETT *Ireland* xii. 331 The indecency and indignity of personal subjection rowelled Parnell like a spur with teeth in it. **1931** E. LINKLATER *Juan in Amer.* II. xii. 137 Now the staccato ear-splitting *rafale* of cheering rowels them afresh. **1967** S. BECKETT *Stories & Texts for Nothing* 42 Between the caressing voice and the fingers rowelling my neck the contrast was striking. **1975** E. BERCKMAN *Indecent Exposure* viii. 94 Her visit..was strong enough to rowel and disturb her.

rowelled, *ppl. a.* For *rare*—[1] read *rare* and add: **b.** Pricked by rowels (example is *fig.*).

1924 R. CAMPBELL *Flaming Terrapin* ii. 29 Rowelled by that sharp prow to hissing hate, The waves washed round her.

rowing, *vbl. sb.*[1] Add: **2. b. rowing machine,** an appliance in which exercises may be done that simulate rowing; **rowing stick** *poet.*, an oar; **rowing tank** (see quot. 1976).

1848 J. DE C. LOCKE tr. *Montolieu's Swiss Family Robinson* 2nd Ser. I. x. 95 (*heading*) The rowing-machine. **1894** *Outing* Mar. 458/1 The exercises consist of hard work on rowing-machines or in the tank, vigorous dumb-bell exercise, and a run of two miles per day. **1935** C. ISHERWOOD *Mr. Norris changes Trains* iv. 68 The Baron made a hobby of his figure. He tortured himself daily on an electric horse, a rowing-machine and a rotating massage belt. **1944** T. RATTIGAN *While Sun Shines* II. 58 If you want exercise I've got a rowing machine in the bathroom. **1977** 'E. CRISPIN' *Glimpses of Moon* i. 20 Exercising on a rowing-machine. **1923** E. POUND *XXX Cantos* xx. 93 Their names are not written in bronze Nor their rowing sticks set with Elpenor's. **1892** *Outing* Jan. 277/2 In 1887 the rowing tank was first put into practical use in the Yale gymnasium. **1939** NICKALLS & MALLAM *Rowing* iv. 87 The object of the rowing tank is to allow oarsmen to indulge in..rowing without going out on the river... Tank rowing originated in America, where ice prevents any outdoor rowing for a considerable part of the year. **1976** *Webster's Sports Dict.* 360/1 *Rowing tank*, a large tank of water containing a mock-up of a shell in which an oarsman or sculler can practice his stroke and work on technique. The effectiveness of a stroke is indicated by a meter which measures the turbulence of the water.

c. *rowing-song.*

1888 L. A. SMITH *Music of Waters* p. xxvii, Rowing-songs should..also be included in this class. **1956** M. W. STEARNS *Story of Jazz* ix. 96 Whereas early travelers heard rowing songs and sea shanties, later specialists found work songs.

rowing, *vbl. sb.*[3] (Earlier example.)

1832 E. GROSVENOR *Let.* 15 Dec. in G. Huxley *Lady Elizabeth & Grosvenors* (1965) iv. 103 To some magistrates who behaved shabbily B. gave what was called 'a proper rowing'.

rowing, *vbl. sb.*[5] (Later example.)

1964 H. HODGES *Artifacts* x. 145 Finally, the clean felted cloth was often brushed with teazles (*teazling, rowing*) to raise a nap of fine hairs on the surface.

rowing, *ppl. a.*[2] Add: **b.** Quarrelling; disposed to quarrel.

1961 *Guardian* 20 Oct. 7/6 The grey Depression background, the rowing parents.

Rowland (rōu·lănd). *Physics.* The name of H. A. *Rowland* (1848–1901), U.S. physicist, used *attrib.* and in the possessive to designate certain devices and concepts associated with his work, as **Rowland('s) circle,** a circle on which must lie the entrance slit, (curved) grating, and photographic plate of a spectrograph if all the spectral lines are to be brought to a focus on the plate; **Rowland ghost,** a spurious spectral line produced by a periodic error in the spacing of the lines of a diffraction grating; **Rowland grating,** a diffraction grating ruled on a machine built by Rowland; **Rowland('s) mounting** (see quot. 1966); **Rowland ring,** a torus made of a magnetic material whose properties it is wished to investigate and linked with a coil of current-carrying wire.

1932 *Jrnl. Optical Soc. Amer.* XXII. 245 Symmetric adjustment of the grating about the point of tangency to the Rowland circle. **1952** R. W. DITCHBURN *Light* vi. 196 If then a point source of light is placed at a point Q on the circle whose diameter is equal to the radius of the grating, and which touches the grating at its centre, the spectra will be focused along the circle. This circle is known as the 'Rowland circle'. **1967** G. W. STROKE in S. Flugge *Handbuch der Physik* XXIX. 477 Eq. (25.27) is clearly satisfied on Rowland's circle. *Ibid.*, The best foci are obtained on the Rowland circle when the source is also placed on that circle. **1922** *Jrnl. Optical Soc. Amer.* VI. 419 The separations of the Rowland ghosts from the parent lines are readily deduced from the characteristics of the ruling engine. **1969** D. RICHARDSON in R. Kingslake *Appl. Optics & Optical Engin.* V. ii. 27 In contradistinction to Rowland ghosts, which usually arise from errors extending over large areas of the grating, each satellite usually originates from a small number of misplaced grooves in a localized part of the grating. **1910** *Phil. Mag.* XX. 773, I had a polished flat plate of speculum metal, such as is used for making Rowland gratings, silver-plated and polished. **1926** R. W. LAWSON tr. *Hevesy & Paneth's Man. Radioactivity* IV. i. 44 The grating space of a Rowland grating is about 10^{-4} cm. **1974** *Encycl. Brit. Macropædia* II. 235/1 Some Rowland gratings are still in use. **1901** *Physical Rev.* XII. 10 The second grating..was of 21 ft. radius, 14,438 lines to the inch. It was arranged on Rowland's mounting. **1914** *Astrophysical Jrnl.* XL. 205 It provides for a grating of 15 ft. (4·57 m) radius, and, optically considered, is the Rowland mounting with the plane of the focal circle vertical. **1966** *McGraw-Hill Encycl. Sci. & Technol.* IV. 141/1 In the Rowland mounting, camera and grating are connected by a bar forming a diameter of the Rowland circle. **1953** J. D. KRAUS *Electromagnetics* v. 232 (*caption*) Rowland-ring method of obtaining magnetization curve. **1966** *McGraw-Hill Encycl. Sci. & Technol.* VIII. 51/1 When the core of the Rowland ring is initially demagnetized, B = 0 and H = 0.

rowlock. Add: Now freq. with pronunc. (rǫ·lǫk).

rownsepyked (rau·nspəikėd), *ppl. a. rare*—[1]. [f. ROUNSEPIKE.] Of a tree, having branches stripped of leaves.

1937 D. JONES *In Parenthesis* III. 39 More leper-trees pitted, rownsepykėd out of nature, cut off in their sap-rising.

row-off. [f. ROW *v.*[1] + OFF *adv.*] In rowing, a race giving the losers in previous heats a second chance to qualify for the final.

1928 *Daily Tel.* 7 Aug. 12/1 Under the repechage system of rows-off between previous losers, J. Wright.. and T. D. A. Collet..had another chance in the sculling event.

row-over. [f. ROW *v.*[1] + OVER *adv.*] An instance of rowing over. Cf. ROW *v.*[1] 1 f.

1868 W. BROUGH *Field of Cloth of Gold* v. 41 Here I am you see, Coming to trial, should the plaintiff halt, Defendant claims a judgement by default. So you are mine; and I my rival crow over. It's what they call in boat-racing a row over.

Rowton (rau·tǝn). The name of Montague William Lowry-Corry, 1st Lord *Rowton* (1838–1903), used *attrib.* in **Rowton (lodging-) house,** a type of cheap lodging-house intended to provide better conditions than a common lodging-house.

1892 *Times* 16 Dec. 8/1 Yesterday a large model lodging-house which has been erected by Lord Rowton at Bond-street, Vauxhall, for the accommodation of working men, was opened for the inspection of visitors. The building, which has been named 'Rowton-house', stands upon a site within a few yards of Vauxhall-cross. **1911** *Encycl. Brit.* XXIII. 789/1 In 1894 a company, Rowton Houses (Limited), was incorporated to extend the scheme, a main characteristic of which was that the houses should not be charitable institutions but should be on a paying commercial basis. **1932** KIPLING *Limits & Renewals* 388 In what they call a Rowton lodging-house. **1937** H. G. WELLS *Brynhild* vi. 74 One man wrote from a Rowton lodging-house on ruled paper torn from an exercise book. **1956** A. WILSON *Anglo-Saxon Att.* II. ii. 355 He had drifted from lodgings to Salvation Army hostels and Rowton Houses. **1960** C. WILSON *Ritual in Dark* I. ii. 41 It would have destroyed his appetite, like a meal in a Rowton House. **1968** *Listener* 28 Nov. 735/2 Eventually we drove him round Camden Town looking for a night's lodging. We went first to the local Rowton House. **1972** *Guardian* 19 Feb. 9/3 He ended up in a hostel, like Rowton House, for the down and out. **1977** *Vole* No. 3. 23/2 Gone is the once normal category of 'the lodger' and gone are such institutions as Rowton House, providing decent short-term accommodation for single people.

row-waggon, var. *ROLWAGEN.

rowy (rau·i), *a.* [f. ROW *sb.*[2] + -Y[1].] Noisy; characterized by quarrelling.

1922 JOYCE *Ulysses* 763 Hes running wild now out at night away from his books and studies and not living at home on account of the usual rowy house.

Roxbury (rǫ·ksbĕri). The name of a town in Massachusetts, used *attrib.* in **Roxbury russet** to designate a variety of green-skinned apple with russet markings, originally grown in New England.

1822 J. THACHER *Amer. Orchardist* 136 Roxbury russeting..is one of the best known, and most valuable fruits in Massachusetts. **1834** *N.Y. Sun* 23 Sept. 4/1 The sweet side of the apple is of a bright yellow colour, and the sour side of the same colour as the Roxbury Russet. **1861** [see *BALDWIN]. **1880** *Harper's Mag.* Mar. 573/2 She set right to a-parin' them Roxbury russets. **1949** *Amer. Forests* Sept. 20/1 Some of the apples sound familiar: Smoke House, Roxbury Russet, Jonathan, Baldwin. **1975** *New Yorker* 11 Aug. 35/1 The first American apple of which there is any record is the now all but forgotten Roxbury Russet.

Roxy (rǫ·ksi), *sb.* The nickname of Samuel Lionel Rothafel (1882–1936), U.S. radio and film entrepreneur, used *attrib.* of persons and things connected with the chain of cinemas built by him.

1940 F. SCOTT FITZGERALD *Let.* 12 July (1964) 84 It's very modern to be taking dramatic criticism although it reminds me vaguely of the school for Roxy ushers. **1957** *Encycl. Brit.* XV. 868/1 The Roxy theatre,..which opened in 1927 in New York city, with a 6,250 seating capacity, cost $8,000,000 and grossed in one week $144,267. **1961** A. BERKMAN *Singers' Gloss. Show Business* 76 *Roxy ending*.., the fanfare ending of a song, first used at the Roxy Theatre in New York. The Roxy Ending is sometimes played at the end of a production number, or where majestic fullness is required. (Also *Paramount ending, Publix ending*).

Roy (roi), *sb.*[3] *Austral.* [f. the personal name *Roy.*] A smart, fashionable, or 'smooth' person. Also *attrib.*

1960 *Encounter* May 28 The Australian business-man or big land-owner, the button-down shirt, lightweight suit type of smoothie from the North Shore line in Sydney or the Toorak Road in Melbourne, with his spurious 'taste' and 'culture'... In current Australian terminology, this is the 'Roy' type. **1965** *Nation* (Austral.) 27 Nov. 21 Middle-class 'Roys' in sports cars and yachting jackets. **1971** F. HARDY *Outcasts of Foolgarah* xi. 143 The young executives, the in-people, call them what you like, the Roys, the jet set, the status symbol seekers from Perisher Valley to Palm Beach, and none of them worth a pinch of shit if it comes to doing an honest day's work.

royal, *a.* and *sb.* Add: **A.** *adj.* **I. 3. a.** (Further examples.)

1815 J. MAYNE *Jrnl.* 3 Feb. (1909) xi. 270 The Princess of Wales was present, and towards the end of the opera she went round into the royal box. **1874** W. P. LENNOX *My Recoll.* I. iii. 77, I..upon reaching the theatre dismounted, and followed the royal party into the ante-room of the royal box. **1930** E. WALLACE *Lady of Ascot* x. 105 Julian had a Royal Enclosure badge, and was the only member of the party possessed of this privilege. **1958** *Spectator* 8 Aug. 201/1 The physicists, who sometimes exercise their sovereignty by barring psychology from the Royal Enclosure of the physical sciences. **1963, 1968** [see *ENCLOSURE 4 a]. **1971** H. TREVELYAN *Worlds Apart* xxiv. 282 My wife did her part, showing great endurance and invariable good humour, finding herself, while I was in England, on Khruschev's right at a lively dinner in the 'Royal Box' in the Bolshoi Theatre during the celebration of Shakespeare's four hundredth anniversary. **1974** 'G. BLACK' *Golden Cockatrice* vii. 113 The kind of people who might one day get him into the Royal Enclosure at Ascot.

4. (Earlier and further examples of special designations.)

1592, 1769 [see NAVY[1] 3]. **1786** [see ARTILLERY *sb.* 5]. *a* **1865** SMYTH *Sailor's Word-bk.* (1867) 583 Royal naval reserve. **1889** [see ENGINEER *sb.* 2 c]. **1911** *Shipping World* 15 Mar. 276/1 The Royal Fleet Auxiliary *Burma* is a vessel with considerable claims to notice. **1913** [see *flying officer* s.v. *FLYING *vbl. sb.* 3]. **1918** *Times* 16 Mar. 10/2 It is Our Will and Pleasure that the Air Force to be established pursuant to the said Act shall be styled the 'Royal Air Force'. *Ibid.* 28 Nov. 9/3 The three Corps will henceforth be known as the 'Royal Army Service Corps', the 'Royal Army Veterinary Corps', and the 'Royal Army Ordnance Corps'. **1922** JOYCE *Ulysses* 48 Her fancyman is treating two Royal Dublins in O'Loughlin's of Blackpitts. **1935** *Ann. Reg. 1934* 23 The Minister announced that a new branch of the Territorial Army, under the name of the Royal Defence Corps, was to be formed. **1937** *Ann. Reg. 1936* 65 The King had approved the creation of a new Reserve called the Royal Air Force Reserve, which would be open only to men in civil life. **1943** [see R.E.M.E., REME]. **1946** *Times* 10 Dec. 4/4 The King has approved that the following regiments and corps shall in future enjoy the distinction of 'royal', their new titles being..Royal Army Educational Corps. Royal Army Dental Corps. Corps of Royal Military Police. Royal Pioneer Corps. **1950** *Jrnl. R. United Service Inst.* XCV. 289 In the same Army Order it was also stated that on transfer to the R.A.C. the R.T.C. would be re-designated Royal Tank Regiment. **1955** *Times* 20 July 8/7 The Queen has approved the promotion of Prince Bernhard of the Netherlands to the honorary rank of Air Vice-Marshal Royal Air Force Volunteer Reserve, with effect from July 8. **1965** *Oxford Mail* 29 June 1/4 A new Army corps will come into being next month. It is the Royal Corps of Transport made up of the Royal Army Service Corps..the Royal Engineers' transport units and the Movement Control Service.

5. b. *Royal Borough*, part of the title of three English boroughs (Kensington (and Chelsea), Kingston-upon-Thames, and Windsor) that have a royal connection.

1897 *Private Life of Queen* xxvii. 226 Our Queen..gave the plot of land..to the people of the 'Royal Borough' [of Windsor] for a recreation ground. **1901** *London Gaz.* 19 Nov. 7472/2 The King has been pleased to direct Letters Patent to be passed..granting the title 'Royal' to the Metropolitan Borough of Kensington, and ordaining and declaring that the said Borough shall henceforth be called and styled the 'Royal Borough of Kensington'. **1923** *Victoria Hist. Co. Berkshire* III. 56/2 The borough of Windsor..was from the first, as it has since remained, a royal borough, owning no overlord but the King. **1930** G. B. SHAW *Apple Cart* II. 75 It is my intention to offer myself to the Royal Borough of Windsor as a candidate at the forthcoming General Election. **1975** G. EVANS *Kensington* p. xvi, Although the granting of the title 'Royal' does not carry with it any special precedence or privilege, there are only three English Boroughs—Kensington, Kingston-upon-Thames..aud Windsor—on which the Sovereign has conferred the title. **1976** *Equals* Dec. 8/1 She is one of four Conservative councillors for the St. Mary's ward of the Royal Borough of Windsor and Maidenhead.

6. (Further examples.)

1759 in Hodges & Hughes *Sel. Naval Documents* (1927) 135 Whereas Mr. Nathaniel Peacock has been educated in the Royal Academy at Portsmouth, and is well qualified to serve His Majesty at sea. **1834** *Times* 25 Jan. 5/3 The mayor and other members were willing to show proper deference to the Royal commission. **1871** [see COMMISSION *sb.*[1] 6]. **1873** *London Gaz.* 21 Jan. 265/1 We do therefore beg leave to recommend that your Majesty will be graciously pleased, by your Order in Council, to approve of the closing of the Royal Naval College at Portsmouth, and the founding of a College at Greenwich, to be styled 'The Royal Naval College'. **1894** *Times* 19 May 7/3 A report by Mr. R. Hunter Pringle, Assistant-Commissioner to the Royal Commission on Agriculture, has been laid before Parliament. **1907** *Times* 9 Mar. 12/2 The ninth annual dinner of the Automobile Club was held..last evening... The chairman announced that a letter had been received from the Home Office stating that his Majesty had been pleased to command that the club should be henceforth known as the Royal Automobile Club. **1926** *Daily Chron.* 13 May 1/7 The proposals in this direction tentatively made in the report of the Royal Commission should be pressed and the powers of the proposed board enlarged. **1926** *Encycl. Brit.* II. 1020/1 The constant vigilance and activity of the Royal Society for the Protection of Birds and the Selborne Society have secured legislative and administrative protection for birds. **1927** T. M. LONGSTRETH *Silent Force* 344 We are members of the Royal Canadian Mounted Police..and I'm going to seize your ship and cargo. **1928** *Times* 24 May 11/4 The Royal Colonial Institute, which celebrates its diamond jubilee this year, has decided to change its name to 'The Royal Empire Society'. **1930** [see *National Trust* s.v. *NATIONAL *a.* 5]. **1955** *Times* 15 June 8/7 The executive committee of the Royal Automobile Club, which is responsible for the conduct of motor races in the United Kingdom, is to meet to-day. **1958** *Times* 9 May 7/7 The Queen has approved, by Order in Council, that henceforth the Royal Empire Society shall be known as the Royal Commonwealth Society. **1965** *Listener* 17 June 892/6 Three royal commissions are at this moment examining the general parts of the local government body. **1971** *Whitaker's Almanack 1972* 1098/2 Royal British Legion, Headquarters, Pall Mall, S.W.1. **1976** *Fundy Tourist* (St. Stephen, New Brunswick) 1 July 1/1 The Royal Canadian Mounted Police—the words conjure up visions of red knights on slick black steeds. **1977** *R.A.F. News* 11–24 May 5/6 The collection..in aid of the Warboys Branch of the Royal British Legion Honorary Association. **1977** *Western Morning News* 1 Sept. 6/1 The following.. adhere to a code of conduct laid down by the Royal Institute of Chartered Surveyors. **1980** *Brit. Med. Jrnl.* 29 Mar. 925/1 The obvious and only course of action was a Royal Commission.

7. a. (Further examples.)

1780 T. DAVIES *Life David Garrick* I. xvi. 180 The king was prevailed upon to give a kind of sanction to this entertainment, by a royal command, on the first night of representation. **1863** [see ASSENT *sb.* 2]. **1869** *Bradshaw's Railway Manual* XXI. 298 The Fusion Bill ..was..carried through both Houses of Parliament and received the Royal assent on the 31st of July. **1976** *Times* 1 Sept. (Fashion Suppl.) p. ii /5 Norman Hartnell and Hardy Amies, both royal warrant holders.

II. 8. a. Also applied to the use of the plural pronoun 'we' by a single person to denote himself. Cf. WE *pron.* 2 a.

1835 [see WE *pron.* 2 a]. **1931** *N. & Q.* 6 June 414/1 The writer uses 'we' throughout—rather unfortunately, as one is sometimes in doubt whether it is a sort of 'royal' plural, indicating only himself, or denotes himself and companions. **1960** J. RAE *Custard Boys* II. xv. 175 'In the absence of the accused we will continue with the trial.' ..He used the royal 'we', but he spoke for us all. **1964** R. H. ROBINS *Gen. Linguistics* vii. 287 Somewhat similar is the use of the 'royal *we*'; in strictly ceremonial circumstances reigning sovereigns in some countries (of which Great Britain is one) use what are otherwise first person plural pronouns in reference to themselves in their official or constitutional capacity. **1966** J. CLEARY *High Commissioner* ii. 26 'May we ask whom you wish to see?' Monarchs and butlers, Malone thought: who else has the right to speak in the royal plural? **1975** M. BABSON *There must be Some Mistake* i. 1 '*We* simply can't take it in,' Lydia drawled, her '*we*' not only royal, but universal.

d. (Later examples.) Also (chiefly *U.S. colloq.*) used as an intensifier, freq. with ironic force.

1938 G. GREENE *Brighton Rock* III. i. 99 She remembered: a face in a bar. She had a royal memory. **1951** J. D. SALINGER *Catcher in Rye* iv. 27 He gave out a big yawn while he said that. Which is something that gives me a royal pain in the ass. I mean if someone *yawns* right while they're asking you to do them a goddam favor. **1960** WENTWORTH & FLEXNER *Dict. Amer. Slang* 435/1 *Royal*... Used as a term of emphasis, esp. before taboo words and expressions, most freq. in 'a royal screwing'. **1972** *Dict. Contemp. & Colloq. Usage* 24/2 *Royal screw* (*fuck*),..an ultimate or complete put-down; total failure where success was expected; an unmitigated defeat or deception. **1976** *Times Lit. Suppl.* 9 July 841/2 Life principle Elspeth may be, but she is also a royal pain in the neck. **1977** C. MCFADDEN *Serial* x. 26/2 Kate had been Harvey's idea of a royal Bengal pain in the ass for the last year.

13. b. royal antelope, a tiny antelope, *Neotragus pygmæus*, found in forested areas of West Africa; **royal Bengal (tiger),** an Indian variety of the tiger, *Panthera tigris*, distinguished by unbroken stripes.

1771 T. PENNANT *Synopsis Quadrupeds* 28 Antelope.. Royal..with very short strait horns. **1871** E. G. E. WARD *Jrnl.* 3 June in D. P. Carew *Many Years, Many Girls* (1967) i. 52 One lady..had arrayed herself in a complete suit of tiger-stripes..so that she looked like a Royal Bengal. **1872** *Proc. Zool. Soc.* 639 The type of the former I believe to have been a genuine specimen of the Royal Antelope. **1964** L. S. CRANDALL *Managem. Wild Mammals in Captivity* 675 The royal antelope..was represented in the Zoological Gardens of London in 1914. **1964** R. PERRY *World of Tiger* xv. 233 The..very rare Royal Bengal tiger is distinguished by unbroken black stripes.

14. b. *Royal Ann(e)* (U.S.) a variety of bigarreau cherry, having red skin and white flesh, or a tree bearing fruit of this kind; *Royal Sovereign*, a variety of strawberry or its large, early-ripening fruits.

1892 *Proc. R. Hort. Soc.* XV. p. lxvii, First Class Certificate. To Strawberry Royal Sovereign (votes, unanimous), from Mr. T. Laxton, Bedford. **1897** S. T. WRIGHT *Fruit-Culture* xviii. 116 For early forcing, Royal Sovereign is a grand acquisition, as it is remarkably early and prolific, with large fruit of excellent quality. **1900** L. H. BAILEY *Cycl. Amer. Hort.* I. 293/2 The Napoleon Bigarreau (locally known as Royal Ann) is the ideal for a white cherry. **1930** V. R. GARDNER *Cherry* xiii. 111 The outstanding light-fleshed sweet cherry is the Napoleon,—also known as Napoleon Bigarreau and Royal Ann. Indeed it is the *one* white-fleshed sweet cherry of real commercial importance in the United States. **1941** M. L. SMITH *Going to God's Country* iv. 175 It was very beautifull with all the groves of tall fir trees and the Royal Ann cherys. **1946** *Nature* 6 July 24/2 The virus..produces faint chlorotic spots on the leaves of Royal Sovereign strawberry. **1957** M. MCCARTHY *Memories Catholic Girlhood* viii. 202 Two kinds of cherry trees, black and Royal Anne. **1960** B. K. WILSON *Lovely Summer* i. 11 Slade is going to pick the Royal Sovereigns this afternoon.

15. royal binding (see quot. 1952); **royal flush** (earlier and later examples); formerly also = *straight flush* s.v. STRAIGHT *a.* 9 b; **royal icing,** a hard, shiny icing, the ingredients of which include egg whites; **royal jelly:** see *JELLY *sb.*[1] 2; **Royal Scot,** a familiar name for the London to Glasgow express; also, the name of the class of locomotive designed to haul this train; **Royal Stewart (tartan):** (see quot. 1975); also known as *dress Stewart* or *Royal tartan*; **royal straight (flush),** in Poker, = *royal flush* in Dict.; also *fig.*; **royal tennis** = *real tennis* s.v. *REAL *a.*[2] 4 e.

1929 G. D. HOBSON *Bindings in Cambr. Libraries* 141 More royal bindings were turned out at this bindery [*sc.* Samuel Mearne's] than at any other. **1939** *Guide Exhib. in King's Library* (Brit. Mus.) 119 The later royal bindings do not, perhaps, maintain the same high level of excellence. **1952** J. CARTER *ABC for Book-Collectors* 157 A book described as being in a 'royal binding' may be expected to have a sovereign's arms on one or both covers; but it must not necessarily be supposed that it therefore has a royal provenance. **1868** W. B. DICK *Amer. Hoyle* (ed. 5) 177 *Royal Flush*, a Straight or Sequence, all of the same suit. *Ibid.* 178 *Straight Flush*, the same as Royal Flush. **1888** B. MATTHEWS *Pen & Ink* 197 The Straight Flush (called a Royal Flush when it begins with the ace and ends with the ten). *Ibid.* 198 The Royal Flush is not often seen; like other exalted monarchs it does not make itself common in men's eyes. **1922** [see *FULL HOUSE 2]. **1845** E. ACTON *Mod. Cookery* xvi. 423 (*heading*) Tourte meringuée, or, tart with royal icing. **1974** *Times* 13 Nov. 12/5 Royal icing or glacé icing must be made with proper icing sugar. **1927** *Times* 27 Sept. 16/2 With the beginning of the winter train service yesterday the London, Midland and Scottish Railway introduced a non-stop run of 299½ miles... This run will be made daily by the 10 a.m. Royal Scot from London to Carlisle, which proceeds to Edinburgh and Glasgow. **1933** *Triumph of Royal Scot* 16 The locomotive which hauled the train throughout the tour was LMS Number 6100 Royal Scot, the first of 50 locomotives of the Royal Scot class to be constructed in 1927. **1942** *Model Railway News* Jan. 9/1 The exhaust steam injector on the 'Royal Scot' was omitted on the model. **1953** *Manch. Guardian* 15 Aug. 3/3 Good fortune and the fact that all the coaches were of the new all-steel type reduced casualties when the Royal Scot was derailed near Abington last Saturday. **1977** *Times* 30 Apr. 12/3 On Monday..the Royal Scot is making a celebration run from Euston to Glasgow—50 years..after the first train of that name chugged its way out of London. **1842** J. S. STUART in *Vestiarum Scoticum* Pl. III (*caption*) The Royal Stuart. **1855, 1969** [see *HUNTING *vbl. sb.* 3 b]. **1975** J. SCARLETT *Scotland's Clans & Tartans* 94 The origin of the Royal Stewart tartan is unknown... There is no record of the Royal Stewart sett having borne that name before the nineteenth century—little that it even existed. ..General Stewart of Garth..stage-managed George IV's visit in 1822 and costumed him in 'Royal Stewart'. **1895** W. STEVENS *Let.* 4 Aug. (1967) 7 Girls charming lots of money but am always open to engagements in finance where I hold a royal straight. **1907** J. C. HARRIS in *Uncle Remus's Mag.* Oct. 28/3 The hand I've dealt to you is known as a royal straight flush, an' it sweeps ever' thing before it. **1962** R. COOK *Crust on its Uppers* iii. 40 They're all *diamonds*, ace, king, queen... Suppose he makes royal straight flush? *Ibid.* 41 He hasn't made royal straight. **1902** E. MILES *Racquets, Tennis, & Squash* v. xi. 270 *Note on the name 'Tennis'*.—In Tasmania the game is called Royal Tennis; in England it is occasionally called Real Tennis; and in America it is always called Court Tennis. **1912** G. INGLIS *Sport & Pastime in Australia* xii. 175 Tennis—or Royal Tennis as it is often called in Australia—was first introduced into the Commonwealth by Mr. S. S. Travers about 1875... The Hobart court was originally built by Mr Travers as a

private court. In 1882 it was taken over by the 'Royal Tennis' Club. **1965** *New Statesman* 6 Aug. 185/1 The royal tennis court, which is enclosed. **1977** *Times* 19 Aug. 12/8 Playing royal tennis at Hampton Court.

c. *royal blue* (earlier example), *green, purple* (later examples).

1789 J. WOODFORDE *Diary* 9 Sept. (1927) III. 139, I took 2 Inside Places in the Royal Blue Coach..to London. **1881** C. C. HARRISON *Woman's Handiwork* I. 61 The cushion-cover..has a ground of royal purple velvet. **1902** *Recipes for Colour, Paint, Varnish, Oil, Soap & Drysaltery Trades* i. 36 Royal Green. Prussiate of potash..Sulphate of iron, [etc.]. **1913** C. L. UEBELE *Paintmaking & Color Grinding* x. 171 Foremost in the line of greens is what we [*sc.* Americans] call chrome green, which, however, is known on the other side as Brunswick or royal green, an intimate mixture of chrome yellow and Prussian blue. **1951** R. MAYER *Artist's Hand-bk. Materials & Techniques* ii. 60 *Royal green*, chrome green. **1956** G. DURRELL *My Family & Other Animals* 18 The endless, meticulous curves of the sea flamed for an instant and then changed to a deep royal purple flecked with green.

16. *royal-sized.*

1974 *Publishers Weekly* 26 Aug. 302/1 A royal-sized volume containing a spectacular gallery of 306 illustrations in full colour.

B. *sb.* **2. e.** A name projected, but not adopted, in Great Britain and Australia, for a decimal unit of currency.

1920 *Rep. R. Comm. Decimal Coinage* 11 in *Parl. Papers* (Cmd. 628) XIII. 467 The second scheme (Lord Leverhulme's) proposes the creation of a new unit of 100 halfpennies to be called a Royal. *Ibid.* 12 We must regard the halfpenny and Royal scheme as impracticable. **1962** A. C. AITKEN in *Listener* 26 Jan. 159/1, I would rectify this..by simply having a pound of a dozen shillings. I will call it a 'royal'—for that has the proper sound and connotation, and, besides, a stag of twelve points is a 'royal'. **1963** *Guardian* 6 June 11/2 The Cabinet decided today that Australia's main currency units will be the Royal and the Crown... The royal, equal to 10 of the present Australian shillings, will be subdivided into 100 cents.

3. c. (Earlier example.)

1848 QUEEN VICTORIA *Jrnl.* 18 Sept. (1980) 64 A magnificent stag, 'a royal', which had dropped, soon after Albert had hit him.

5. a. Also, a royal mast.

1937 C. S. FORESTER *Happy Return* I. i. 10 He had.. noted that the wind was from the west, and just strong enough to give the ship steerage way, with all sail set to the royals. **1970** *Parade* (Austral.) June 26/2 The ship must have sunk almost instantly because when she was found two days later, only the top of one of her royals was just visible out of the water.

b. (Earlier and later examples.)

1839 *Knickerbocker* XIII. 42 Send him some ratlin-stuff, so that he can set up brace-backstays abaft, and cross his royal yards, and call all hands up anchor. **1927** G. BRADFORD *Gloss. Sea Terms* 146/1 *Royal yard*, the next above the topgallant yard.

7. b. (Later example.)

1977 *Navy News* June 4/1 Next opportunity for the Hermes and 845 Squadron to work with the Royals was during Exercise Dawn patrol in the first weeks of May.

8. Short for *royal blue.*

1885 *Queen* 24 Oct. (Advt.), *Ladies' gloves*... Shades, Tan, Golden, and Royal. *c* **1900** in *American Mail Order Fashions* (1961) 30 *Sweater*, made with alternate stripes.. of..red and royal, or royal with red stripes. **1922** *Daily Mail* 11 Dec. 1 (Advt.), Beautifully made in Duvetyn or Monchon,..Black, Grey, or Royal. **1939** J. B. PRIESTLEY *Let People Sing* iv. 82 It [*sc.* a van] had been generously rather than neatly painted, in a manly scheme of crimson and royal. **1974** *Harper's & Queen* Sept. 35/1 Crêpe dress. Black, sand, mink, red, emerald or royal.

9. *ellipt.* for: **a.** The Royal Society. Also *attrib.* **b.** The Royal Show (of the Royal Agricultural Society).

1951 C. P. SNOW *Masters* v. 45 There's not been a day pass in the last three years when he hasn't reminded me that he is a Fellow of the Royal, and that I am not. *Ibid.* 49 He would not get into the Royal Society now. But as March came round each year, he waited for the announcement of the Royal elections. **1958** *Spectator* 27 June 829/2 The Royal is the shop-window of British farming. **1966** 'W. COOPER' *Mem. New Man* II. vi. 172, I was not agreeing off the cuff that Bill's getting into the Royal would necessarily enhance his prospects. **1975** *Country Life* 26 June 1676/1 Beneath the surface of a highly professional modern 'Royal', one may sense the old-time garden-party atmosphere... Verona and the Royal Show are each unique in their own way because they have evolved.

‖ **royale** (rwayāl). [Fr., lit. 'royal', in same sense.] = IMPERIAL *sb.* 8.

1842 W. C. MACREADY *Diary* 26 May (1912) II. 171 A middle-aged man..with moustache and royale oiled to points which curled up at considerable distance from his face. **1877** E. CREER *Lessons in Hairdressing* 95/2 *Royale* (or *Imperiale*), tuft of beard just below the under lip.

Royalist. Add: **1. a.** *spec.* in Canada, a United Empire Loyalist (see LOYALIST in Dict.).

1785 R. HUNTER *Quebec to Carolina* (1943) 65 The Royalists have settlements along Lake St. Francis and up to Cataraqui. **1972** J. MOSHER *Some would call it Adultery* IV. xxi. 176 'Oh, a Royalist, eh?' said the admiral, using the Yankee term for United Empire Loyalist, as they were known in Canada.

royalty. Add: **6. c.** Also, a payment made, or a portion of the production given, by a producer of minerals, oil, or natural gas to the owner of the site or of the mineral rights over it. Also *attrib.*

1896 B. REDWOOD *Petroleum* I. v. 250 Such leases are often transferred at a larger royalty, especially after the territory has been proved productive. **1949** *Our Industry* (Anglo-Iranian Oil Co.) (ed. 2) i. 8 Oil companies wishing to prospect in a foreign country have first to make an agreement with the Government of that country... This agreement determines the royalties payable. **1971** WILLIAMS & MEYERS *Oil & Gas Terms* (ed. 3) 390 The landowner's royalty is typically 1/8th of production. **1977** *Time* 5 Dec. 59/1 What had really blown was a giant natural-gas well that probably will make Lucy and her husband, Walter Parlange, royalty rich.

e. (Earlier and later examples.)

1857 MRS. GASKELL *Let.* 26 Nov. (1966) 484 He was to have the sale of them for three years..paying me a royalty of 3d on each copy sold. **1867** J. SPEDDING *Publishers & Authors* 25 In order to translate the substance of the bargain into a percentage upon the sale, (or a 'royalty', as we call it,) it is only necessary to divide the total estimated profit by the number of copies through the sale of which it is to be made. **1875** HARDY *Let.* 4 Nov. (1978) I. 40 Name of book. Copies sold in the half year. Retail price of same. 10 per cent royalty. **1974** R. RENDELL *Face of Trespass* iv. 43 He began worrying about his royalty statement. *a* **1976** A. CHRISTIE *Autobiogr.* (1977) VI. iv. 318, I had not kept any of the royalty statements sent me.

f. A periodic payment for the right or privilege of using another person's know-how under a know-how or trade secrets agreement.

1962 *Conveyancer* XXVI. 368 Some sort of lump-sum payment on the signing of the agreement will usually be appropriate, since it must be recognised that the seller runs a risk when he hands over the initial batch of information and documents. Beyond that, it is common to provide for some sort of royalty on turnover. *Ibid.* 369 The seller..will be willing to accept a reasonably widely drawn royalty clause on the basis that if the seller's methods turn out to be usable..they are certain to incorporate a good deal of indispensable information. **1973** J. P. CUNNINGHAM *Competition Law of E.E.C.* viii. 172 The know-how agreements between Happich and Gallino and between Happich and Maglum contained provisions requiring the licensees to pay royalties to Happich.

7. d. *Ir.* (See quot.)

App. a nonce-use based on Ir. *Ráth na Ríoghradh* the rath of the royalty, a name for the rath at Tara (A. J. Bliss).

1893 W. B. YEATS *Celtic Twilight* 104 They came to a royalty (a name for the little circular ditches, commonly called raths or forts, with which Ireland is covered since Pagan times).

royster, var. ROISTER.

roz (rǫz), abbrev. *ROZZER.

1971 J. WAINWRIGHT *Dig Grave* 79 Not that he gave a fart about the roz crowd. **1977** —— *Do Nothin'* v. 83 The roz has removed his helmet.

rozener, var. *ROSINER.

rozzer (rǫ·zəɹ). *slang.* [Origin unknown.] A policeman, a detective.

1893 P. H. EMERSON *Signor Lippo* xviii. 87 If the rozzers was to see him in bona clobber they'd take him for a gun. **1903** A. M. BINSTEAD *Pitcher in Paradise* iii. 75 He..nearly knocked down the rozzer in the mackintosh suit who was regulating the traffic from the middle of the road. **1936** M. ALLINGHAM *Flowers for Judge* xiii. 193 Aven't seen 'er since she went off with a rozzer. **1942** WODEHOUSE *Money in Bank* (1946) xiv. 126 You mean a rozzer? A detective? **1958** E. HYAMS *Taking it Easy* III. iii. 298 Then some nosy rozzer come up on a motorbike. **1962** R. GORDON *Doctor in Swim* v. 36 You can always try a bit of give and take with an English rozzer, and no hard feelings. **1977** 'E. CRISPIN' *Glimpses of Moon* xii. 236 The rozzers were after him, not a doubt about that.

r-process: see *R III. 7.

-rrhaphy, formative element [ad. Gr. -ῥαφία, f. ῥάπτειν to sew: see -Y[3]], used to form words denoting surgical suturing of a wound or part, as *gastrorrhaphy, hysterorrhaphy.*

-rrhoea, -rrhea, formative element [ad. Gr. -ῥοια (as in διάρροια DIARRHŒA, γονόρροια GONORRHŒA), f. ῥοία flux, flow], used in various medical terms, as *LOGORRHŒA, MUCORRHŒA.

Rualla (ru,a·lă), *sb.* and *a.* Also **Ruála, Ruwalla,** etc. **A.** *sb.* A Bedouin people; a member of this people. **B.** *adj.* Of or pertaining to this people.

1831 J. L. BURCKHARDT *Notes on Bedouins & Wahábys* I. 6 *El Rualla*..generally occupy the desert from Djebel Shammar towards the Djof, and thence towards the southern vicinity of the Hauran; but they frequently encamp between the Tigris and Euphrates. **1875** *Encycl. Brit.* II. 247/2 First, the Anezah clan, whose pasture-grounds extend from Syria southward to the limits of Jebel Shomer... Their principal subdivisions are the Sebaá on the north, the Woold-Alee on the west, and the Ruála on the south. **1888** C. M. DOUGHTY *Trav. Arabia Deserta* I. vii. 194 The poor man had been maimed thus by a Ruwàlla lance-thrust in the mouth, when riding in the North. **1917** [see *HOWEITAT]. **1918** T. E. LAWRENCE *Lett.* (1938) 248 We decided to carry out a flying attack.. with our regular troops, the Rualla horse.., and such Hauran peasants as should be brave enough to declare for us. **1926** —— *Seven Pillars* (1935) III. xxx. 174 His was the chief family of the Rualla, but Nuri had no precedence among them at birth, nor was he loved. *Ibid.*, One of the chief men of the Ruwalla. **1959** W. THESIGER *Arabian Sands* iii. 54 In Syria I..had visited the summer camp of the Rualla, a city of black tents.

ruana (ru,ā·nă). [Amer. Sp.] A type of Colombian and Peruvian cape or poncho.

1942 F. CARPENTER *Our S. Amer. Neighbors* iii. 63 Men and women dressed in elegant clothes..meet Indians wearing native straw hats, and bright home-woven ponchos which are here called 'ruanas'. **1971** *Sat. Rev.* (U.S.) 31 July 55/2 'Knee-rugs' (about 38″ × 54″) which double for long skirts, ruanas, jumpers. **1974** *Times* 22 Aug. 9/3 Macindo..[is] an ethnic shop specializing in the handwork of the Northern Andes... For the winter there will be heavier wool ponchos, ruanas and serapes. **1977** *Western Living* (Vancouver) Apr. 28/3 The women folk [in Peru], dressed in broad full length skirts, rainbow coloured *ruanas* (cape-like shawls) and a variety of masculine hats.

Ruanda, var. *RWANDA.

rub, *sb.*[1] Add: **1. d.** *Naval slang.* A loan *of.* Also const. *at.*

1914 'BARTIMEUS' *Naval Occasions* xxi. 193 'Don't you take on, Taff,' said another, pushing over his pannikin of rum. ''Ave a rub at this lot.' **1919** W. LANG *Sea Lawyer's Log* xiii. 162 'Innyone as hasn't had a letter can have a rub of mines,' says Moriarty, the big Irishman, generously. **1946** J. IRVING *Royal Navalese* 147 'The rub of a dollar' means the loan of a dollar. **1948** PARTRIDGE *Dict. Forces' Slang* 158 Give us a rub of five bob till pay day!

2. c. Also *fig.*

1931 *Times Lit. Suppl.* 31 Dec. 1048/4 If he is unfortunate in having finished his task before his problem was knocked completely out of shape by England's suspension of the gold standard, that is just the 'rub of the green'. **1962** *Guardian* 5 Nov. 2/2 If applications.. reached fantastic proportions, the Government would have to consider the matter. 'At present we treat it as a rub of the green.'

8. A sound as of rubbing.

1907 R. H. BABCOCK *Dis. Lungs* xxvi. 511 If the chest be examined a friction-rub is likely to be detected on the side corresponding to the pain... The symptoms are the result of a dry, circumscribed pleurisy. *Ibid.* xxxv. 726 The so-called pleuritic friction sound or pleuritic rub..is a succession of fine crackling sounds..produced by the separation of the two pleural surfaces or by their rubbing together when rendered sticky. **1950** *Audio Engin.* Aug. 15/3 When the cause of the noise is mechanical, as in defective or ill-designed speakers, pickups and microphones, we may hear rattles,..rub, and wheeze. **1976** *Lancet* 13 Nov. 1083/1 Bronchial breath sounds and a pleural rub were present over the right middle lobe.

9. Special Combs.: **rub resistance,** the degree to which print will withstand rubbing without becoming smudged or detached; so **rub-resistant** *a.*

1958 E. A. APPS *Printing Ink Technol.* xxvii. 431 A high standard of rub-resistance is necessary in inks used for food cartons which are jostled in transport, and for display cards which are frequently handled. *Ibid.* 432 Inks which tend to give gloss and very level films are also usually fairly rub-resistant. **1967** E. CHAMBERS *Photolitho-Offset* xvi. 240 Anti-driers retard drying on the machine and modifiers such as waxes and oils control setting, flow,..and rub-resistance.

rub, *v.*[1] Add: **I. 1. c.** (Further examples.) Also *fig.*

1922 JOYCE *Ulysses* 324 The *Times* rubbed its hands and told the whitelivered Saxons there would soon be as few Irish in Ireland as redskins in America. **1966** *Listener* 30 June 936/1 The British Government has invested half a million pounds on this display of international sport and the London hotel-keepers are rubbing their hands. **1973** *Times* 30 Apr. 5/4 Cloth manufacturers of all sorts must be rubbing their hands just now, because fashion definitely calls now for a greater volume of material per garment.

5. a. Joc. phr. *not to have two pennies to rub together*, and varr., expressing lack of money or poverty.

1929 M. DE LA ROCHE *Whiteoaks* vii. 98 George, like Finch, was always hard up. Sometimes they had not between them two coins to rub together. **1977** K. O'HARA *Ghost of T. Penry* xvii. 172 I've known Mrs Bathhurst without two pennies to rub together, and always..concerned about others.

b. *to rub shoulders with* (further examples.) Also *to rub elbows* (*with*) (chiefly *U.S.*).

1851 CARLYLE *Life J. Sterling* I. viii. 74 One right peal of concrete laughter at some convicted flesh-and-blood absurdity, one burst of noble indignation at some injustice or depravity, rubbing elbows with us on this solid Earth. **1906** U. SINCLAIR *Jungle* xxvi. 327 Young white girls from the country rubbing elbows with big buck Negroes with daggers in their boots. **1916** L. N. PARKER *Disraeli* II. 56 You would pass him in the street without the faintest idea you had rubbed elbows with one of the world's greatest powers! **1922** JOYCE *Ulysses* 497 Have we cold feet about the cosmos? No... You have that something within, the higher self. You can rub shoulders with a Jesus, a Gautama, an Ingersoll. **1946** MEZZROW & WOLFE *Really Blues* ii. 21, I got my kicks out of rubbing elbows with all those bigtime gamblers. **1956** B. HOLIDAY *Lady sings Blues* (1973)

i. 8 A whorehouse was about the only place where black and white folks could meet in any natural way. They damn well couldn't rub elbows in the churches. **1961** in WEBSTER s.v. *rub* vb., Reports on social products rub shoulders with book reviews and notes. **1976** E. MACLAREN *Nature of Belief* iii. 20 I'm rubbing shoulders with questions of religious philosophy all the time. **1979** *Yale Alumni Mag.* Apr. (Suppl.) cn20/3 As a piano played show tunes of the 1930's, Teng rubbed elbows with George Weyerhauser of the Weyerhauser Co.

d. *to rub noses* (*with*), to touch noses in greeting, in token of friendship. Also *fig.*

This custom was practised among Eskimos, Maoris (see *HONGI *sb.*), and elsewhere in the Pacific Islands. Among Eskimos it has practically died out.

1822 G. F. LYON *Jrnl.* 28 July (1824) vi. 247 When the principal [Eskimo dancer] had pretty well exhausted himself, he walked gravely up to him, and taking his head between his hands, rubbed noses with him amidst the plaudits of all present. **1832** A. EARLE *Narr. Residence N.Z.* 159 He..rubbed noses so forcibly with me that I felt his friendship for some time. **1858** R. M. BALLANTYNE *Coral Island* xx. 242 Tararo went up to Jack and rubbed noses with him... Seeing that this was their mode of salutation..we rubbed noses heartily with the whole party. **1891** [in Dict., sense 5 b]. **1945** D. LEECHMAN *Eskimo Summer* 240 Before they had much contact with white men, the Eskimos used to rub noses on meeting old friends after a protracted absence. **1964** MRS. L. B. JOHNSON *White House Diary* 15 June (1970) 169 Lynda Bird..had enjoyed Hawaii hugely, rubbing noses with Maori children. **1973** 'D. JORDAN' *Nile Green* xx. 82 He's got three daughters and an Eskimo au pair girl... It's all this rubbing noses... It gets him down.

e. *to rub one's nose in it*: see *NOSE *sb.* 9 e.

II. 11. *rub out*: **a.** (Further *fig.* examples.)

1936 E. AMBLER *Dark Frontier* vii. 111 Rovzidski rubbed out by Red Gauntlet mob... Government fail to take action against slayers. **1946** L. B. LYON *Rough Walk Home* 24 Again the random child by robot thumb Of war rubbed out. **1950** A. LOMAX *Mr. Jelly Roll* 220 The gangsters..had promised to rub him out if he didn't stop trying to hire away their star New Orleans side-men. **1957** WODEHOUSE *Over Seventy* xii. 125 The heavy goes to his asylum, and two months later is released as cured. Upon which, he dresses up as a Siberian wolf-hound and hurries off to rub out another citizen. **1961** B. FERGUSSON *Watery Maze* xiv. 333 The first task was to destroy the enemy's radar organisation, and the R.A.F. addressed themselves to the job of rubbing out as many stations as they could. **1979** E. NEWMAN *Sunday Punch* xxiv. 214, I learned what the man muttered when the fight ended and Aubrey was declared the winner. It was: 'That Philpott-Grimes. I maybe rub him out.'

13. *rub up*: **e.** Also *ellipt.* without *the wrong way.*

1882 E. W. HAMILTON *Diary* 31 Aug. (1972) I. 328 Lord Dufferin is half inclined to advise that we should concede this to them in order not to rub up the Sultan more than we can help. **1971** *Weekend World* (Johannesburg) 9 May 6/7 Judges, magistrates, prosecutors, defence lawyers treat everybody alike, but the minor officials sure know how to rub up a non-White.

f. To caress (a person) in order to excite him or her sexually. *slang.*

1656 R. FLETCHER tr. *Martial's Epigrams* II. 102 Me thinks I scarcely am wound up by thee..to the height of Venerie... Thus Phillis rub me up, thus tickle mee. **1937** PARTRIDGE *Dict. Slang* 710/2 *Rub up*,..so to caress a person that he or she becomes actively amorous. **1939** JOYCE *Finnegans Wake* (1964) I. 203 Rubbing her up and smoothing her down, he baised his lippes in smiling mood.

g. To make clean, clear, or bright (again) by rubbing.

1847 C. BRONTË *Jane Eyre* III. viii. 183 My first aim will be..to *clean down* Moor House..; my next to rub it up with bees-wax, oil, and..cloths, till it glitters again. **1859** MRS. STOWE *Minister's Wooing* xviii. 179 He rubbed up his optical instruments to see whether they were rising in right order. **1886** F. R. STOCKTON *Casting away of Mrs. Lecks & Mrs. Aleshine* III. 111 In the mornin' I'll rub up that floor till it's as bright as new. **1974** A. ROSS *Bradford Business* 76 Even the short heavy bolts had been rubbed up with a wire brush.

III. 14. d. *to rub up*: to masturbate. *slang.*

1937 PARTRIDGE *Dict. Slang* 710/2 *Rub up*, the v. corresponding to *rub off*, 2 [*sc.* a masturbation]. **1963** C. MACKENZIE *My Life & Times* II. 115 Just as I was going down the steps into our area B— asked me if I ever rubbed up... In bed that night I tried the experiment recommended by B—.

e. *to rub off*, of qualities, etc.: to have influence *on* through close or continued contact; to be transmitted to others.

1959 N. MAILER *Advts. for Myself* v. 463 He spent years hobnobbing with gentlemanly shits and half-ass operators and some of it had to rub off on him. **1965** *Listener* 11 Nov. 761/1 There is no evidence that anything of Sickert's powerful teaching rubbed off on him. **1969** 'G. NORTH' *Procrastination of Sgt. Cluff* v. 44 How long was it since he'd begun to work with the Sergeant? How much of the Sergeant had rubbed off on him? Could he think any more except as the Sergeant thought? **1971** *Times* 9 Sept. 3/2 One hopes that something of their Christian charity and principles would rub off. **1976** E. MACLAREN *Nature of Belief* ii. 16 Jews come in contact with Zoroastrians and certain ideas rub off. **1978** *Jrnl. R. Soc. Arts* CXXVI. 185/1 Morale was lower, there was apathy and this was rubbing off on new entries.

18. rub-board, (*b*) *N.Amer.*, a wash-board; **rub rail,** a rail to protect (a vehicle, etc.) against rubbing.

1964 *Amer. Folk Music Occasional* I. 28 Clifton Chenies is no doubt the best known of the so-called 'Zydeco' musicians. This music..usually features the accordion with drum or rub-board accompaniment. **1972** *Daily Colonist* (Victoria, B.C.) 13 Feb. 22/2 Building furniture and washing clothes on a rub board in a small tub. **1961** WEBSTER, Rub rail. **1969** *Jane's Freight Containers 1968–69* 550/2 Products: G-85 fifth wheel container and general purpose trailer with cushioned rub rails. **1980** *Reader's Digest* Feb. 226/2 The car skidded..25 feet up the bridge, jumped a 5½-inch-high rub rail and hurtled..into the water.

rubabah, var. *REBAB.

rubaboo, var. *RUBBABOO.

rub-a-dub, *sb.* Add: **2.** A pub, a hotel. *Austral.* and *N.Z. rhyming slang.*

c **1926** 'MIXER' *Transport Workers' Song Bk.* 81, I gazed upon the motley crowd Within this 'rub-a-dub'. **1963** H. SLESAR *Bridge of Lions* iii. 52 He could fathom why rub-a-dub meant a pub.

So **rub-a-dub-dub** (*b*) in same sense.

1941 [see *RUBBEDY]. **1945** BAKER *Austral. Lang.* xv. 270 A hotel becomes known in rhyming slang as a *rub-a-dub-dub*—by rhyme on 'pub'. **1971** *National Times* (Austral.) 13 Dec. 20 'Let's grab some Kate and Sydney and a pint of apple fritter at the rub-a-dub-dub.'..Translated: 'Let's grab some steak and kidney and a pint of bitter at the pub.'

‖ **ruba'i** (rubā·,ī). Also **ruba'iy.** Pl. **rubaiyat** (rū·bai(y)at, rū·bei(y)at). [Arabic *rubā'īyah*, f. *rubā'īy* composed of four elements.] In Persian poetry, a quatrain.

The pl. is chiefly familiar in the title of the work by Omar Khayyam (cf. *OMARIAN *a.* and *sb.*), known to English speakers in FitzGerald's version.

1859 E. FITZGERALD *Rubáiyát of Omar Khayyám* p. xii, The original Rubáiyát (as, missing an Arabic Guttural, these *Tetrastichs* are more musically called), are independent Stanzas, consisting each of four Lines of equal, though varied, Prosody; sometimes *all* rhyming, but oftener (as here attempted) the third line suspending the Cadence by which the last atones with the former Two. **1885** *Encycl. Brit.* XVIII. 656/1 Those principal forms of poetry now used in common by all Mohammedan nations—the forms of..the *rubá'í* or quatrain (our epigram, for which the Persians invented a new metre in addition to those adopted from the Arabs), [etc.]. **1934** [see *QASIDA]. **1959** *Chambers's Encycl.* X. 202/1 *Omar Khayyam*.., Persian poet and master of the *ruba'i* (quatrain)... The *Rubaiyat* have now been translated into almost all the literary languages of the world.

Rubarth's disease (rū·bāɹt). *Vet. Sci.* [Named after C. S. *Rubarth* (b. 1905), Swedish veterinary scientist, who described it in 1947 (*Acta Path. & Microbiol. Scand.* Suppl. No. 69).] An infectious disease of dogs, caused by an adenovirus, that affects chiefly the liver and is sometimes fatal; infectious canine hepatitis.

1951 *Vet. Record* 15 Dec. 833/2 We wish..to place before you certain data which we have been able to collect..regarding virus hepatitis in dogs, or Rubarth's disease, within our own kennels. **1961** C. H. D. TODD *Popular Whippet* x. 147 There are four common canine diseases, viz. two virus diseases—distemper (including hard pad variety) and Rubarths disease (a liver affection), and two bacterial diseases which affect the kidneys. **1970** A. R. JENNINGS *Animal Path.* vi. 118 Intranuclear inclusion bodies are a feature of Rubarth's disease and the inclusions have diagnostic significance.

‖ **rubashka** (ruba·ʃkă). Pl. **rubashkas, rubashki.** [Russ.] A type of blouse or tunic worn in Russia.

The pl. *rubashka* in quot. 1956 is *erron.*

1921 *Glasgow Herald* 29 Nov. 6 In North Russia during the summer the men, young and old, clean-shaven or whiskered and bearded, wear rubashkas, or blouses, of various colours, some of them even set off with touches of bright embroidery. **1924** *Blackw. Mag.* Feb. 149/1 She had opened her khaki rubashka and shown the subaltern the scar. **1956** WALLIS & BLAIR *Thunder Above* (1959) ii. 11 An orchestra, colourful in their rich silk *rubashka*, providing music that seemed to flow from the Volga. **1972** *Nat. Geographic* Sept. 401 The bearded men wore *rubashki*, the hand-embroidered blouses of old Russia.

rubato. Add: (rubā·to). (Earlier and later examples.) Also *transf.*

1883 GROVE *Dict. Mus.* III. 188/1 *Rubato*,..referring to the values of the notes, which are diminished in one place and increased in another. **1921** G. SAMPSON *English* III. 94 The natural *rubato* of civilised speech. **1925** J. A. JOHNSTONE (*title*) Rubato, or the secret of expression in pianoforte playing. **1946** J. CARY *Moonlight* ii. 9 Her old-fashioned style, indeed, with its exaggerated rubato, her swayings, murmurings, tosses of the head, might have amused or disgusted a modern audience. **1955** *Times* 9 May 3/7 His use of rubato sounded nonchalant instead of expressive in Mozart. **1977** *N.Y. Rev. Bks.* 13 Oct. 6/2 He remarks that Victorians had moved away from the hard clear notes of Wordsworth's *Lyrical Ballads*: the ballad had moved into the rubato and vibrato of the drawing room. **1979** *Early Music* July 341/1 In the Vivaldi example the *fermate* over the semiquaver rests can only be indications of *rubato*.

rubbaboo (rʊbăbū·). Also **rababoo, robiboo, rubaboo, rubeiboo,** etc. *N. Amer.* (*Obs.* exc. *Hist.*) [ult. ad. Algonquian.] A kind of soup or porridge made from pemmican.

1821 N. GARRY *Diary* 22 Aug. in *Trans. R. Soc. Canada* (1900) VI. 151 Our men are now eating Rababoo made of Pemican and Flour. **1857** P. JACOBS *Jrnl.* 72 The food that is generally prepared and eaten in these regions by voyagers is what is called 'ahrubuhboo'. I do not know what the word itself means. I spell it as I hear it pronounced. **1862** R. KENNICOTT *Jrnl.* Jan. in J. A. James *First Sci. Exploration Russ. Amer.* (1942) ii. 85 Rubbaboo is a favorite dish with the northern voyageurs, when they can get it. It consists simply of pemmican made into a kind of soup by boiling in water. Flour is added when it can be obtained, and it is generally considered more palatable with a little sugar. **1865** MILTON & CHEADLE *N.-W. Passage by Land* xv. 289 Our fare was what the half-breeds call 'rubaboo', which we made by boiling a piece of pemmican the size of one's fist in a large quantity of water thickened with a single handful of flour. **1881** E. S. FARROW *Mountain Scouting* xiii. 200 When required for use, it [*sc.* pemmican] is cut from the hard mass and either eaten cold, or is mixed with flour forming a porridge called '*robiboo*'. **1935** *Beaver* Sept. 135 One of the tastiest forms, and one more often mixed than any other for table use, was 'rubeiboo', consisting of pemmican boiled down with a mixture of potatoes, onions and other vegetables. This, when properly seasoned, was very palatable. **1969** E. W. MORSE *Fur Trade Canoe Routes* I. ii. 23 The pemmican was either sliced off and munched raw, or made with flour and water into a *potage* called 'rubbaboo'.

rubbedy, rubberdy, rubbidy (rʊ·bə(ɹ)di), altered f. *RUB-A-DUB *sb.* 2. Cf. *RUBBITY. *Austral.*

1941 BAKER *Dict. Austral. Slang* 62 *Rubberdy*.., a public house. Rhyming slang on 'rub-a-dub-dub' for 'pub'. **1957** D. NILAND *Call me when Cross turns Over* iv. 101 How about a gargle? Down to the rubberdy, come on. **1962** 'D. FORREST' *Hollow Woodheap* i. 11 'Where..is "The Eagle on the Hill"?' 'A rubbedy in South Australia.' **1969** *Melbourne Herald* 8 May 15, I was having a gargle with a cobber in a Fleet St. rubbedy. **1970** K. GILES *Murder Pluperfect* ii. 46, I met another of the Fennels down at the rubbidy. **1971** *Australian Post* 8 Apr. 40/5 There's the story of the barman in the rubbedy.

rubber, *sb.*¹ Add: **I. 4. b.** (Earlier example.)

1816 W. Y. OTTLEY *Inquiry Origin & Early Hist. Engraving* I. 81 The friction of a rubber, made of hair, or of pieces of cloth, was then applied to the paper, which was thus rubbed backwards and forwards till the impression of the engraving was transferred to the paper.

c. An article usu. consisting of a soft pad attached to a wooden handle, used for erasing chalk from a blackboard.

1880 [see *BLACKBOARD]. **1978** P. MARSH et al. *Rules of Disorder* ii. 38 They just started..chucking wooden dice at her..and blackboard rubbers.

6. Also,† a brake acting by friction on the wheels of a vehicle.

1850 R. GLISAN *Jrnl. Army Life* (1874) iv. 32 The third one [*sc.* vehicle], having no rubbers or brakes to the wheels, went so fast, down a steep hill, that the driver was thrown from his seat. **1894** T. B. SEARIGHT *Old Pike* 145 The 'rubber', called brake at this day, was not in use when the National Road was first thrown open for trade and travel. Instead,..saplings, cut at the summit of the hills, were shaped and fashioned to answer the ends of the 'rubber', and at the foot of the hills taken off and left on the roadside.

7. (Further examples.)

1898 *19th Ann. Rep. U.S. Geol. Survey* VI. B. 407 A second kind, and one much used in London for fronts, is a large, light-red brick, so soft as to be readily scratched by the knife. These are called 'rubbers'. **1977** *Listener* 20 Oct. 519/4 There is quality, too, in the kind of bricks still known as 'rubbers'—ones that have been hand-rubbed on all surfaces to achieve an immaculate join.

II. 8. a. *spec.* one who massages sportsmen or athletes (chiefly *N. Amer.*).

1895 J. L. WILLIAMS *Princeton Stories* 185 Another sub and William, the negro rubber, picked Wormsey up. **1911** *Daily Colonist* (Victoria, B.C.) 6 Apr. 9/5 A 'rubber' has been engaged by Manager Wattalet. Perhaps the use of such a term would shock the sensibilities of our ball players. The correction, therefore, is made with all haste. It is a 'masseur', who has become attached to the Victoria baseball club. **1949** *Sun* (Baltimore) 14 Oct. 27/6 The schools have hired some of the best men in the training profession. Today, they are not just rubbers; they know anatomy, physiology and chemistry. **1950** J. DEMPSEY *Championship Fighting* xxiv. 192 The 'rubber' (rub-down man) applies soothing lotions to the muscles as he kneads them with his fingers.

III. 11. a. Now also applied to any of a large range of synthetic organic polymers having properties of elasticity, etc., resembling those of natural rubber. (Further examples.)

1912 *Jrnl. Soc. Chem. Industry* 15 July 617/1 There can now be no doubt that rubber may actually be obtained synthetically by the polymerisation of isoprene and its homologues and that the synthetic product is really rubber and strictly comparable with natural rubbers. **1941** *Jrnl. R. Aeronaut. Soc.* XLV. 145 Mechanically, natural rubber is not surpassed by any synthetic rubber. However, in resistance to swelling by organic liquids.. and deterioration by sunlight or oxidising agents, synthetic rubbers have been found superior. **1961** L. MUMFORD *City in Hist.* (1966) xv. 545 The new régime was based on..new synthetic materials, like rubber, bakelite, and the plastics. **1973** *Nature* 6 Apr. 420/1 Natural rubber is still the preferred polymer for many high performance applications.

b. For *U.S.* read 'orig. *U.S.*' and add earlier and later examples; (*b*) plimsolls, esp. plimsolls worn for climbing.

(*a*) **1842** *Southern Lit. Messenger* VIII. 516/2 The younkers who would go 'a Maying', very prudently provided themselves with rubbers and tippets before encountering the rough southeaster. **1856** S. ROBINSON *Kansas* xii. 160 The snows..are fast melting, and mingling with the clayey soil. So, besides the burden of rubbers, one has to carry no little portion of the native earth. **1901** *Daily Colonist* (Victoria, B.C.) 22 Oct. 4/6 (Advt.), Special sale of rubbers today at 2.30 p.m. *c* **1921** D. H. LAWRENCE *Mr Noon* iv, in *Mod. Lover* (1934) 221 He went out to his motor-cycle and got it ready. He went indoors and put on his rubbers. **1951** E. PAUL *Springtime in Paris* iii. 53 An umbrella rack with a porcelain tray for rubbers. **1972** J. MINIFIE *Homesteader* xvii. 145 It [*sc.* gumbo] was notorious for its adhesive quality; it stuck to everything, pulled off rubbers—galoshes as people were beginning to call them [in 1914]. **1974** M. Z. LEWIN *Enemies Within* iv. 17 Snow made it look beautiful. I put on my rubbers and walked around. (*b*) **1925** *Jrnl. Fell & Rock Climbing Club* VII. 12 Arrowhead Ridge..Leader needs about 60 feet of rope. Rubbers. **1933** G. D. ABRAHAM *Mod. Mountaineering* v. 107 Rubbers are usually used, but I have also made the ascent in nailed boots, and in either footgear dry rocks are advisable. **1941** C. KIRKUS *Let's go Climbing* vi. 95 A climb of such difficulty is not done in boots, but in rubbers. These are ordinary plimsolls or gym shoes. **1950** T. LONGSTAFF *This my Voyage* xiv. 282 To the right of Kern Knotts Crack is a narrow vertical cracklet... It..is now a recognised climb. It is led, generally in rubbers, without any moral support of a rope from above. **1957** CLARK & PYATT *Mountaineering in Brit.* vii. 134 For the climb,..for which rubbers are recommended, all the party wore boots. **1968** P. CREW *Encycl. Dict. Mountaineering* 103/2 With the advent of P.A.'s and similar footwear, and their widespread use in Britain, the use of rubbers has diminished considerably.

c. Also *collect.*, the tyres of a vehicle; occas. used in colloq. phrases expressing speed or acceleration. Chiefly *U.S.*

a **1961** G. FELSEN in WEBSTER s.v., I'll road test her for you after we get new rubber on. **1976** N. THORNBURG *Cutter & Bone* iv. 91 The huddled figure..going round the car and out of sight for a few moments and then back into it almost immediately and laying down rubber again. **1977** *Hot Car* Oct. 61/1 There really is an incredibly large number of cars and vans cruisin' round on completely the wrong sort of rubber for the type of vehicle. **1980** J. BALL *Then came Violence* (1981) i. 7 Every patrol and supervisory car..would be burning rubber within seconds.

d. A piece of rubber for erasing pencil or ink marks. Also used of erasers made of other substances.

1788–9 [in Dict., sense 11 a]. **1891** *Catal. & Price List* (Waterlow & Sons Ltd.) 169 (*caption*) Artists' Rubber. Stationer's Rubber... Grey Vulcanised Rubber. **1907** A. E. ZAPF *Cycl. Drawing* I. 14 In making drawings, but little erasing should be necessary. However, in case this is necessary, a soft rubber should be used. **1928** [see *BUNGIE, BUNGY]. **1952** PRICE & BISHOP *Art School Self-Taught* II. iii. 275 For erasing errors, a harder rubber is needed. **1968** F. G. HOLLIDAY *Man. Stationery* v. 113 Erasers are often called 'rubbers', but today a surprisingly small proportion of them actually consists of rubber. **1973** M. AMIS *Rachel Papers* 139 Between my finger and thumb I take a rubber and bounce it up and down on the desk.

e. *U.S. Baseball* (*a*) The home plate; (*b*) the pitcher's plate (now the usual sense).

1891 *Chicago Herald* 5 May 6/1 Those same errors.. hustled two runs over the rubber. **1895** *Evening Star* (Washington, D.C.) 2 Oct. 9/3 Twirler Magee once more tried to 'make good', in a pitching sense, but..he was put out of the running, and Billy Dineen sent to the rubber. **1910** O. JOHNSON *Humming Bird* v. 54 In the breakaway Tyrell, the first to dust the rubber for the Chaperons, selected a hole in the circumambient and poked a buzzer over short. **1919** *Chicago Daily Tribune* 12 Apr. 19/2 An unfairly delivered ball is a ball delivered by the pitcher to the batsman with the bases unoccupied, while no foot is in contact with the rubber. **1950** A. DALEY *Times at Bat* 106 He hit the first pitch a mile... Still seething inwardly he crossed the rubber and returned to the dugout. **1975** *New Yorker* 14 Apr. 92/2 Seaver, too, restored memory—the cold, intelligent gaze; the unwasteful windup; the sudden forward, down-dropping stride off the rubber.

f. *slang.* A contraceptive sheath made of rubber; a condom. Cf. *rubber goods, shop* below.

1947 C. WILLINGHAM *End as Man* xiv. 173 Maybe next time you'll use a rubber. **1955** W. GADDIS *Recognitions* I. v. 184 What are you reading?..Malthus, for Christ sake... The next thing, you'll be peddling rubbers in the street. **1968** B. TURNER *Sex Trap* viii. 53, I need more rubbers. There's only enough for about a couple of good days left. **1978** J. IRVING *World According to Garp* iv. 71 'Oh, Garp,' Cushie said. 'Don't you have any rubbers?'

12. a. Also used of articles made of synthetic rubber. *rubber apron, band* (hence as *v. trans.*), *bed, boat, bone, boot, bullet, clothing, coat, dinghy, garment, glove, hose* (further examples), *pants, ring* (further examples), *sheet, sheeting, shoddy, shoe, sole, suit, truncheon.*

1926 *Daily Colonist* (Victoria, B.C.) 10 Jan. 7/1 (Advt.), Women's rubber aprons. Save your frocks and save your laundry bills, too. **1977** 'M. UNDERWOOD' *Fatal Trip* xviii. 102 'Caught me just in time,' the pathologist said, peeling off his rubber apron. **1895** *Montgomery Ward Catal.* Spring & Summer 117/1 Cabinet of assorted pure rubber bands for home and office use. Assortment of sizes up to one-half inch wide and 2½ inches in length. **1919** F. HURST *Humoresque* 128 'I asked you why you was like a rubber band.' 'Aw, I give up, Miss Sadie.' ''Cause you're so stretchy, see?' **1947** J. C. RICH *Materials & Methods Sculpture* v. 121 Tape or rubber bands can be used to hold sections of a mold together for casting positives. **1962** D. LESSING *Golden Notebk.* IV. 488, I found a stack of letters rubber-banded together in one corner. **1973** 'E. MCBAIN' *Let's hear It* xv. 216 He removed the rubber band from the roll, and spread the plans on the floor. **1849** N. KINGSLEY *Diary* 26 Oct. (1914) 78 Some of the fellows went in swimming this afternoon by takeing rubber beds. **1943** J. B. PRIESTLEY *Daylight on Saturday* xxix. 219 Sometimes they got into little rubber boats and so weren't found for a day or two. **1976** J. LEE *Ninth Man* 5 The billowy outlines of the rubber boat, rapidly filling with air. **1949** N. STREATFEILD *Painted Garden* v. 48 His spare collar and lead, his water bowl, his rubber bone. **1973** E. LEMARCHAND *Let or Hindrance* viii. 91 A dog basket with a rubber bone in it. **1852** S. C. DAVIS *Jrnl.* 16 Dec. in B. A. Richards *Calif. Gold Rush Merchant* (1956) 85 Arrived at San Francisco and purchased 200 books, pamphlets, magazines, &c., also some Rubber Boots, &c. **1975** *Ecology* LVI. 538/1 In very dry years the whole bog surface..may be dry enough to walk on without rubber boots. **1971** *Guardian* 14 June 1/8 The soldiers, wearing gas masks and riot helmets, fired nine rounds of rubber bullets. **1976** P. FERRIS *Detective* viii. 150 You're half hoping I've got..a patrol group coming in with gas and rubber bullets. **1980** *Jrnl. R. Soc. Arts* July 486/1 Fire hoses as favoured on the Continent or rubber bullets favoured by the Army in Ireland. **1895** *Montgomery Ward Catal.* Spring & Summer 296/3 Medium and heavy weight rubber clothing. **1967** G. FREEMAN *Undergrowth of Lit.* x. 151 Talcum powder is also useful to apply to the body before squeezing into rubber clothing. **1850** N. KINGSLEY *Diary* 21 Nov. (1914) 157, I put on my Rubber Coat and built a chimney outside the tent. *a* **1918** G. STUART *40 Yrs. on Frontier* (1925) I. 69 Rubber coats and shoes were unknown at that time. **1939** Rubber dinghy [see *DINGHY 2 c]. **1953** *News Chron.* 2 June 2/6 Nothing had been forgotten by the crowds... Even rubber dinghies had been brought to keep off the rain. **1973** E. LEMARCHAND *Let or Hindrance* v. 54 Can't we get hold of an RAF rubber dinghy, or inflatable raft? **1921** *Dict. Occup. Terms* (1927) §608 Garment maker, rubber. **1967** G. FREEMAN *Undergrowth of Lit.* x. 152 There is a wide belief among women that sweating in rubber garments makes them slim. **1895** *Montgomery Ward Catal.* Spring & Summer 297/1 Ladies' Rubber Gloves... Men's Rubber Gloves. **1914** 'E. BRAMAH' *Max Carrados* 96 Here is a rubber glove. I have cut the wire but you had better put it on. **1932** E. HEMINGWAY *Death in Afternoon* xii. 138 The doctor..picked up the pistol in his rubber gloves. **1975** *Listener* 24 July 125/3, I have peeled off my rubber gloves and put the Fairy Liquid back on the shelf. **1939** *N.Y. Sunday News* 4 June 68/3 What do you think, they're using a rubber hose on her? Piffle! **1976** H. TRACY *Death in Reserve* xii. 100 Why don't you get your bloody rubber hoses out and your hallucinogens..and your flashing lights. **1897** *Sears, Roebuck Catal.* 783/3 Rubber pants. **1936** F. M. FORD *Let.* 6 Sept. (1965) 261 She [*sc.* Pennsylvania] led the Universe in the production of rubber pants. **1895** *Montgomery Ward Catal.* Spring & Summer 537/3 Rubber rings, for Mason fruit jars. **1970** W. H. PARKER *Health & Dis. in Farm Animals* v. 50 A more recent method of 'bloodless castration' is the use of the rubber ring. **1976** H. TRACY *Death in Reserve* xix. 146 Free bucket-and-spade, beachballs, rubber rings. **1922** *Encycl. Brit.* XXXII. 300/1 In very exacting work, such as the vulcanizing of hard-rubber sheets, curing is effected by immersion of the material in hot water. **1957** *Ibid.* XIX. 610/1 The rubber sheet is firmly pressed against the prepared fabric. **1966** MAY & MOSS *New Math for Adults Only* xii. 71/2 Such geometry..is known as topology. Youngsters call it rubber-sheet geometry because the figures can be twisted and stretched and still remain the same. **1975** I. STEWART *Concepts Mod. Math.* x. 144 Topology is sometimes described as 'rubber-sheet geometry', a whimsical and somewhat misleading description. **1976** M. MILLAR *Ask for me Tomorrow* (1977) xiii. 107 She gave him a sponge bath..on a rubber sheet on the bed. **1895** *Montgomery Ward Catal.* Spring & Summer 108/2 Nursery rubber sheeting. **1965** M. THOMAS *Grannies' Remedies* 16 Another..poultice is a piece of soft thick sheet-lint..squeezed out in hot water, and laid over the part, covered with a larger piece of thin rubber-sheeting. **1907** *Sci. Amer.* 5 Oct. 240/2 Scrap rubber, or rubber 'shoddy' as it is called, is made up principally of worn-out boots and shoes. **1844** *Knickerbocker* XXIV. 287 Old rubber-shoes! old rubber-shoes! Humble theme for heavenly muse! **1931** M. ALLINGHAM *Look to Lady* xvii. 178 They heard the soft scrape of his rubber shoes on the bole of the tree. **1897** *Sears, Roebuck Catal.* 208/1 Royal rubber soles. **1901** E. W. HORNUNG *Black Mask* vi. 112 There had been no warning step.. and my suspicious eye had searched his feet for rubber soles. **1975** G. SEYMOUR *Harry's Game* iv. 66 A night patrol, their faces blackened, rubber soles on their shoes. **1948** H. INNES *Blue Ice* viii. 205 Sweating underwater in a rubber suit. **1931** M. ALLINGHAM *Look to Lady* v. 66 A small but wicked looking rubber truncheon and..[a] Colt revolver. **1959** J. BRAINE *Vodi* iv. 69 They beat him up with rubber truncheons. The marks don't show. **1973** W. FAIRCHILD *Swiss Arrangement* viii. 94 Give this one a pair of jackboots..and he'd be just like the rest... Drench you with charm first and, if that failed, slug you with a rubber truncheon.

c. *rubber addict, latex, plantation, shop, substitute.*

1967 G. FREEMAN *Undergrowth of Lit.* x. 152 To the rubber addict 'slimwear' is a key word. **1907** *Chem. Abstr.* I. 1326 (*heading*) On the action of iodine and bromine on the resins in rubber latex. **1972** *Materials & Technol.* V. iv. 80 Rubber latex is not naturally very stable: the rubber particles coagulate spontaneously in course of time. **1911** *Encycl. Brit.* XXIII. 798/2 The experience of planters in general is in favour of the complete removal of weeds from a rubber plantation. **1948** M. LASKI *Tory Heaven* i. 6 James had..been sent to try his luck on an uncle's rubber-plantation in Malaya. **1923** in M. Box *Trial of Marie Stopes* (1967) 166 A. As a matter of fact..these things have been used by the hundreds... Q. Bought at rubber shops, rubber goods' shops? *Ibid.* 254 Contraceptives are for sale at what have been called, I think, rubber shops? **1936** 'G. ORWELL' *Keep Aspidistra Flying* i. 13 He'd slink into one of the rubber-shops and buy *High Jinks in a Parisian Convent.* **1940** GRAVES & HODGE *Long Week-End* vii. 105 Contraception['s]..association with the pornographic literature of rubber-shops. **1911** Rubber substitute [see *artificial rubber* s.v. *ARTIFICIAL *a.* 5]. **1935** *Chambers's Encycl.* VI. 123/2 Paints, varnishes, rubber-substitutes, oil-cloth, soft soap, &c.

13. a. *rubber-planter.*

1937 *Discovery* May 143/2 The rubber planter uses coconut shells for collecting the raw latex from his trees.

b. Also with vbl. sbs., as *rubber-collecting.*

1910 *Blackw. Mag.* May 729/1 Rubber-collecting is less laborious, but takes you into dangerous parts.

c. *rubber-boned, -booted, -caped, -coated, -cored, -cushioned, -faced, -gloved, -insulated, -jointed, -legged, -lipped, -mounted, -mouthed, -necked, -soled* (further examples), *-stoppered, -tipped, -tired* (later *fig.* examples), *-treaded.*

1958 *New Statesman* 22 Feb. 227/2 Especially fantastic is the dance of a rubber-boned neighbour (Stephen Preston). **1943** J. W. DAY *Farming Adventure* xx. 228, I dined and went down to the quay, oil-skinned and rubber-booted. **1935** W. CATHER *Lucy Gayheart* I. ix. 75 When the rubber-caped boy was gone, Lucy stood looking at the yellow envelope. **1934** WEBSTER, Rubber-coated. **1959** *Chambers's Encycl.* XII. 30/2 Some of the earliest known products of rubber, observed in Brazil, shoes and rubber-coated garments for example, possessed these faults. **1972** *Classification of Occupations* (Dept. Employment) III. 244/1 Operates machine to wind..rubber coated wire round core. **1902** *Amer. Golfer* Apr. 102/1 The advent of the rubber-cored ball has made this [*sc.* an official golf ball] a question of practical interest to all golfers. **1929** W. DEEPING *Roper's Row* xxxv. 398 Sillocks was a golf maniac, and went from Rye to Hoylake..smiting a rubber-cored ball. **1935** *Chambers's Encycl.* VI. 122/1 Innumerable other new applications, as in rubber-cored golf-balls and vulcanite fountain-pens, have caused an enormous increase in the importation of rubber. **1971** *Flying* Apr. 40/1 The free-floating rubber-cushioned engine cowling. **1911** *Encycl. Brit.* XXIII. 803/2 The required thickness of the spread sheet is very often secured by the rubber-faced surfaces of two cloths being united before curing. **1965** F. SARGESON *Mem. Peon* vii. 241 It was more as though he aimed at captivating me with his abilities as a rubber-faced comedian. **1970** *Motoring Which?* July 107/4 Bumpers lightly mounted; rubberfaced overriders front and back. **1956** P. SCOTT *Male Child* II. vii. 174 Handled by sterilized, rubber-gloved hands. **1977** B. PYM *Quartet in Autumn* x. 91 Her pink rubber-gloved hands plunged in the washing-up water. **1965** *Motor* 17 July 6/1 The wiring was rubber insulated and in poor condition. **1934** J. A. LEE *Children of Poor* (1949) 200 People go to the circus to see the rubber-jointed wonder. **1942** BERREY & VAN DEN BARK *Amer. Thes. Slang* §702/32 '*Punch-drunk*'; *dazed*,..rubber-legged. **1950** J. DEMPSEY *Championship Fighting* xxv. 200 He becomes 'rubber-legged' as he lurches about the ring. **1960** *Times* 24 Feb. 16/5 A right to the head had Luukkonen rubber legged as he stood against the ropes. **1973** M. AMIS *Rachel Papers* 180 Once, I affectionately imitated her pout; she veered away in pained bewilderment, so I changed it to an imitation of rubber-lipped Norman, claiming I had heard him on the stairs. **1947** CROWTHER & WHIDDINGTON *Science at War* iv. 166 A rubber-mounted dome was found. **1969** *Jane's Freight Containers 1968–69* 245/1 Two rubber-mounted gantry cranes are used to transfer containers. **1968** *Times* 15 Nov. 16/1 The American designers swing-a-ding-ding with such vivacity and with such rubber-mouthed, beady-eyed professionalism as to render our homemade brand soft-in-the-head amateurs by comparison. **1928** D. H. LAWRENCE *Lady Chatterley* x. 168 There was a toughness, a curious rubber-necked toughness and unlivingness about the middle and upper classes. **1932** BLUNDEN *Face of England* 114, I came to an old tree over the stream, and crossing with some disadvantage to its rubber-necked lichens, I was in an open meadow. **1957** R. CAMPBELL *Coll. Poems* II. 126 The rubbernecked, Hell-touring Thracian. **1913** E. C. BENTLEY *Trent's Last Case* v. 123 He wore rubber-soled tennis shoes. **1973** J. LEASOR *Host of Extras* iii. 49, I had not heard his rubber-soled shoes. **1927** C. B. NEBLETTE *Photography* xx. 465 The potassium pentasulphide solution thus formed is then allowed to cool, filtered and kept in a rubber-stoppered bottle tightly closed. **1913** *T. Eaton & Co. Semi-Ann. Sale Catal.* No. 36. 17/2 Rubber tipped pencils. **1926** 'C. BARRY' *Detective's Holiday* xi. 97 A sallow, unhealthy-looking man of about thirty years, who walked with the help of a stout stick and a rubber-tipped wooden stump. **1976** M. GILBERT *Night of Twelfth* i. 8 He..walked..with the aid of a rubber-tipped stick. **1901** KIPLING *Let.* May in Ld. Birkenhead *R. Kipling* (1978) xiv. 235 We were bung full of beastly spiritual pride... We went about despising things and people, unconsciously turning our ideals to mean an easy life...soft rubber-tyred. **1980** A. CROMIE *Lucky to be Alive* i. 9, I would be leaving the rubber-tired life behind. **1936** J. STEINBECK *In Dubious Battle* i. 11 He went in a dark entrance and climbed the narrow stairs rubber-treaded, the edges guarded with strips of brass.

14. Special Combs.: **rubber boa,** a short, stout, brown snake, *Charina bottæ*, belonging to the family Boidæ and found in western North America; **rubber cement,** a cement or adhesive containing rubber in a solvent; hence *rubber-cemented* adj. (also *fig.*); **rubber cheque** *slang* (orig. *U.S.*), a cheque that 'bounces'; **rubber-chicken circuit** *N. Amer. slang*, an after-dinner-speaking circuit; **rubber dam, rubberdam:** see *DAM *sb.*¹ 4 e; **rubber fetishism,** sexual fetishism which is centred

on objects made of rubber; hence **rubber fetish; rubber goods** *pl.*, articles made of rubber; freq. *spec.* contraceptive devices; also *attrib.* and *fig.*; **rubber gum,** the sap or latex of rubber trees; **rubber ice** *N.Amer.*, thin, flexible ice; **rubber johnny** *slang*, a condom (cf. sense 11 f above); **rubber kite** *slang* = *rubber cheque; **rubber-leather** *a.*, consisting of rubber and leather; **rubber-like** *a.*, resembling or suggestive of rubber; **rubber plant,** a tree, *Ficus elastica*, belonging to the family Moraceæ, and native to south-east Asia, the juvenile form of which is widely cultivated as a house plant for the sake of its large leathery leaves which are dark green above and pale yellowish-green beneath; **rubber-proofed** *a.*, coated or treated with rubber for water-proofing; **rubber snake** = *rubber boa*; **rubber solution,** a solution of rubber, *spec.* one used as an adhesive in the repair of tyres; **rubberware,** rubber goods; **rubberwear,** rubber clothing.

1907 R. L. DITMARS *Reptile Bk.* xxv. 211 The Rubber Boa... Size moderate. Form very stout. **1977** *Westworld* (Vancouver, B.C.) May–June 46/2 Another snake common around the lake is the rubber boa. **1895** *Montgomery Ward Catal.* Spring & Summer 525/2, 1 can rubber cement. **1939** R. STOUT *Some Buried Caesar* xii. 175 I'll close it forever and seal the crack with rubber cement. **1965** ZIGROSSER & GAEHDE *Guide Coll. Orig. Prints* vii. 107 Synthetic adhesives, such as rubber cement. **1961** W. SANSOM *Last Hours S. Lee* 246 Bossom was already home, rubber-cemented to his favourite comedian on the telly. **1962** 'E. MCBAIN' *Like Love* (1964) vi. 82 Grossman turned over the lucite-encased sketch and studied the typewriter key rubber-cemented to its back. **1927** Rubber cheque [see *BOUNCE *v.* 6 c]. **1936** WALLACE & CURTIS *Mouthpiece* i. 9 By now the woman has exhausted her credit in Vienna, issued a few rubber checks and passed on to Budapest or somewhere. **1955** J. POTTS *Death of Stray Cat* xv. 157 Jimmy was going to have a lot more to explain than just a handful of rubber cheques. **1973** R. BUSBY *Pattern of Violence* vi. 94 Have you got that blighter who's been trying to put me out of business with rubber cheques? **1959** *Maclean's Mag.* 23 May 1/1 Next year's rubber-chicken circuit is being sewed up by three Toronto women with a public-speaking agency called Canadian Celebrity Bureau. **1977** *Rolling Stone* 5 May 47/3 He spent the winter making speeches on the rubber-chicken circuit. **1954** B. KARPMAN *Sexual Offender* xix. 352 Another rubber fetish in a case reported by Payne was a mackintosh, and in this case also the patient preferred one that had been stolen. **1930** S. PARKER tr. *Stekel's Sexual Aberrations* I. v. 105 During the two years that he was engaged, he continued his rubber and glove fetishism unabated. **1951** HARTWICH & BURBURY tr. *Krafft-Ebing's Aberrations of Sexual Life* x. 173 Despite this strong rubber-fetishism he had a normal sexual relationship. **1971** E. CHESSER *Human Aspects Sexual Deviation* iii. 51 Although rubber fetishism features largely in pornographic literature, if it makes for married happiness it is impossible to see how any moral issue can arise. **1853** *Pathfinder Railway Guide* Sept. (Advt.), Goodyear's patent vulcanized rubber goods. **1897** *National Police Gaz.* (U.S.) 26 May 15/4 (Advt.), T. W. Harrison's rubber, cigar, and book stores. List of all kinds of rubber goods, French & American specialities, pessaries of every description. **1922** JOYCE *Ulysses* 533 Rubber goods. Neverrip. **1928** D. H. LAWRENCE *Lady Chatterley* x. 141 She wasn't all tough rubber-goods and platinum, like the modern girl. **1951** G. GREENE *End of Affair* III. vii. 143 They ought to have opaque glass in their doors like rubber-goods shops. **1973** A. BROINOWSKI *Take one Ambassador* xii. 187 The rubber goods factory next door. **1910** *Chambers's Jrnl.* Mar. 153/2 In these forests grow the trees which yield the finest quality of rubber-gum. **1896** *Dialect Notes* I. 423 *Rubber ice*, thin ice that bends when skated upon. **1916** *Ladies' Home Jrnl.* Apr. 101/2 'Soft as Cheese!' Doctor Rolfe concluded. 'Rubber ice and air holes.' **1962** W. O. MITCHELL *Kite* xiv. 171 Jimmy Sangster had gone through rubber ice, to be found far downstream in a back-water. **1980** *Private Eye* 29 Feb. 13/1 Even the rubber johnny merchants gave him the thumbs down. **1961** *John o' London's* 30 Nov. 610/3 A worthless cheque is a *rubber kite*. **1923** *Daily Mail* 28 May 3 A new process for the manufacture of rubber-leather compounds. **1922** *Encycl. Brit.* XXXII. 299/1 Isoprene undergoes polymerization on exposure to light with the production of a rubber-like mass. *a* **1930** D. H. LAWRENCE *Last Poems* (1932) 156 The vibration of the motor-car has bruised their insensitive bottoms Into rubber-like deadness. **1959** *Times* 27 Apr. (Rubber Industry Suppl.) p. ii/4 Some of the earliest research work on a synthetic product with rubberlike properties was carried out in this country. **1888** Rubber plant [see RUBBER *sb.*¹ 12 b]. **1908** 'O. HENRY' *Gentle Grafter* 138 The little wine-stained table..between the rakish rubber plant and the framed palazzio della something. **1959** C. MACINNES *Absolute Beginners* 81 The rubber-plants in the espressos had been dusted. **1971** *New Scientist* 9 Sept. 554/1 The prime atmospheric essential that they should be places fit for rubber plants to live in. **1934** J. A. SINCLAIR *Airships* iii. 65 The envelopes were composed of rubber-proofed fabric, two fabrics being used with rubber interposed between them, and also on the inner or gas surface. **1960** *Textile Terms & Definitions* (Textile Inst.) (ed. 4) 123 Rubber-proofed sheeting. **1897** J. VAN DENBURGH *Reptiles Pacific Coast & Gt. Basin* 156 The Rubber Snake..is not rare in the moister portions of California. **1954** R. C. STEBBINS *Amphibians & Reptiles Western N. Amer.* 352/2 Rubber Snake... Usually found in moist locations, often near, or within, coniferous woods. **1894** ALBERMARLE & HILLIER *Cycling* (rev. ed.) 471 The hole [is] discovered..and a small patch of rubber stuck over it with rubber solution. **1911** *Encycl. Brit.* XXIII. 802/1 The best solvents for rubber are carbon bisulphide, benzol and mineral naphtha, carbon tetrachloride and chloroform. These liquids, either alone or mixed, are employed in making the rubber solutions used for technical purposes. **1967** E. SHORT *Embroidery & Fabric Collage* ii. 45 It is also a useful precaution to put a spot of rubber solution onto the endings of threads on the back of the work. **1972** *Materials & Technol.* V. xiv. 516 Sometimes rubber cements are made for impregnating or proofing of fabrics, or rubber solutions are prepared for dipping or adhesive purposes. **1950** *N.Z. Jrnl. Agric.* Oct. 300/1 During the war..rubberware of any kind was very scarce. **1977** *Hot Car* Oct. 71/2 A full range of butch and beefy rubberware is also stocked both Stateside and UK. **1967** G. FREEMAN *Undergrowth of Lit.* x. 150 There is also a comprehensive catalogue of the latest range of rubberwear called 'Black Panther'. **1972** *Guardian* 2 Dec. 10/1 Allen Jones's search for potent imagery has led him into a vicarious world..[of] rubberwear and lingerie catalogues.

rubber, *sb.*² Add: **2.** (Earlier example.) Also at bridge.

1744 S. FIELDING *Adventures D. Simple* I. II. i. 140 The best Expedient to be found out is, to play a Rubbers at one place, and then drive their Horses to death, to get to the other time enough not to disappoint their Friends. **1886** *Biritch* 3 After each rubber there is a fresh cut for partners. **1908** *Laws of Auction Bridge* §11 At the end of the rubber the total scores..are added up. **1930** [see *AUCTION *sb.* 2 b]. **1960** J. BETJEMAN *Summoned by Bells* vi. 56 Depositing their wraps and settling down To a nice rubber. **1965** *Listener* 20 May 758/3 Even a 33 per cent. chance of game is worth taking if the reward is a 700 rubber.

3. (Examples in *Cricket.*)

1792 in H. T. Waghorn *Dawn of Cricket* (1906) 118 The first game of the rubber. **1895** J. N. PENTELOW *Eng. v. Austral.* 76 Shaw's team thus winning the rubber in fine style. **1912** J. B. HOBBS *Recovering 'Ashes'* 120 England thus decisively winning the match and the rubber by an innings and 225 runs. **1955** *Times* 10 June 4/1 It was not a sparkling partnership, but it was a sober and extremely serviceable start to the new rubber. **1975** *Cricketer* May 8/1 His side won four to one in the recent rubber in Australia.

5. *attrib.* and *Comb.*, as (sense 2) *rubber player, saver, -value*; **rubber bridge,** a type of bridge in which the hands are not replayed and in which settlement is made after each rubber; **rubber game,** a game played to determine the winner of a series; **rubber match,** a match to determine the winner of a series; also *fig.*

1936 R. LEDERER *Mod. Contract & Duplicate* 10 At Rubber Bridge you are faced by a variety of partners. **1951** E. CULBERTSON *Bidding & Play in Duplicate Contract Bridge* xv. 194 The strategy of bidding and play in total-point duplicate is almost exactly the same as the strategy of bidding and play in rubber bridge. **1977** *Times* 29 Aug. 6/4 The real experts preferred rubber bridge at which they could win hard cash. **1908** R. FOSTER *Auction Bridge* 32 It is very important not to let players make a declaration that will put them out, especially on the rubber game. **1946** J. CARY *Moonlight* ix. 68 'Of course it was always a rubber match,' Robin was saying to Amanda. 'From the first night. It didn't seem to matter at first. The really *interesting* thing was how it went bad on us.' **1976** *National Observer* (U.S.) 20 Nov. 5/1 In the rubber match this year, he seems to have sneaked by the same Sam Young with those 201 votes. **1977** *New Yorker* 25 July 58/1 In their rubber match the following year it was Miss Sutton, 6–1, 6–4. **1974** *Times* 16 Feb. 15/2 Without..a code, even the strongest rubber player..will fail to find the perfect answers. **1928** A. E. M. FOSTER *Auction Bridge for All* 201 Z's bid was a good and cheap rubber saver. **1912** F. IRWIN *Fine Pts. Auction Bridge* 166 The rubber-value is 250 points above the line.

rubber, *v.* Add: **1. a.** (Earlier and later examples.) Also const. *around, for.*

1896 ADE *Artie* xi. 100 About a dozen ringers followed us in and stood around rubberin. **1899** W. J. KOUNTZ *Billy Baxter's Lett.* 4 Up there you are likely any minute to come face to face with an Apache or some leftover Aztec rubbering around among the trees. **1910** 'O. HENRY' *Whirligigs* viii. 108 Every few minutes he would pick up his stick rifle and tiptoe to the mouth of the cave to rubber for the scouts of the hated paleface. **1916** H. L. WILSON *Somewhere in Red Gap* ii. 58 So I rubbered awhile,..and then I forgot 'em, looking at some other persons that come in. **1929** WODEHOUSE *Gentleman of Leisure* xiii. 107 Shall I rubber around and find out where is dey kept, boss? **1930** *Living Age* 1 Apr. 183 Bill Coyote..was loping around the trail and rubbering for eats. **1950** *Chicago Tribune* 24 Jan. III. 1, I just saw Moon Mullins out in the alley rubbering up here. **1974** P. DE VRIES *Glory of Hummingbird* (1975) ii. 13 The oncoming cleric who could be seen from the curtained window at which we all rubbered to be even now approaching.

b. To listen or listen *in* (on a party telephone line). *N. Amer. colloq.*

1920 S. LEWIS *Main St.* 189 Say, did you hear me putting one over on these goats that are always rubbering in on party-wires? I hope they heard me! **1948** *Southern Folklore Q.* Sept. 191 She's always rubberin' on a party line. **1963** G. H. THOMSON *Crocus Country* xxxviii. 237 No one thought it much of a crime to 'rubber', as it helped to pass the time for isolated people.

rubberdy, var. *RUBBEDY.

rubber heel, *sb.* (*phr.*) [RUBBER *sb.*¹ III + HEEL *sb.*¹]

1. A shoe heel made of rubber.

1916 *Daily Colonist* (Victoria, B.C.) 23 July 10/6 The best rubber heel costs only 10¢ more. **1921** *Dict. Occup. Terms* (1927) §602 *Rubber heel maker*,..a moulder.. engaged in moulding rubber soles and heels. **1922** *Encycl. Brit.* XXXII. 301/2 To a large degree the rubber heel has also displaced leather in medium-grade footwear.

2. One who investigates the conduct of members of his own organization; *spec.* an internal police investigator. *slang.*

1942 BERREY & VAN DEN BARK *Amer. Thes. Slang* §458/16 '*Spotter.*' (One who spys upon employees.).. rubber-heel. **1970** G. F. NEWMAN *Sir, You Bastard* 12 It was the detail that led the Rubber Heels to Shepherds Market. **1975** *Listener* 6 Feb. 163/3 'Have those five-day wonders and rubber heels never copped a drop or fitted someone up?' (Have those graduates from the Police College investigating a complaint against an officer never accepted a bribe or planted evidence?)

3. *attrib.* and *Comb.*, as (sense 2) *rubber-heel boy, inquiry, mob.*

1962 PARKER & ALLINSON *Courage of his Convictions* iv. 152 Stamper thought he'd be clever and he phoned the rubber-heel mob at Scotland Yard. **1971** *Daily Tel.* 18 Sept. 1/7 They led to investigations into the conduct of police officers by senior detectives. To all British police forces this type of investigation is known as a 'rubber heel' inquiry. **1976** P. FERRIS *Detective* vii. 110 They were the rubber-heel boys, the policemen who investigated policemen.

Hence **ru·bber-heel** *v. intr.* and *trans.*, to investigate (a colleague), to keep (an associate) under surveillance, to spy *on*; **ru·bber-heeler** = sense 2 above; **rubber-heeling** *vbl. sb.*

1959 M. PUGH *Chancer* 91 'So,' he said, 'you fancy yourself as a rubber-heeler?' The phrase usually applied to a policeman, sent to check on another policeman, and to get his facts from the underworld. The 'rubber-heeler' was disliked by criminals as much as he was disliked by the police. **1968** 'B. MATHER' *Springers* xiv. 157 But Sonia? Was she here only for her own safety—or was she rubber-heeling on me to make certain there were no slip-ups? *Ibid.* xv. 168, I was under the closest rubber-heeling and I certainly couldn't afford to interest myself in something that was no longer in my parish. **1973** —— *Snowline* iv. 49 To a brothel? Not with me rubber-heeling on him, he wouldn't... Anyhow, as a copper he'd want it for free. **1976** R. BUSBY *New Face in Hell* viii. 114 He..had..been rubber-heeled for flagrantly fabricating evidence, demoted to sergeant and sent back into uniform. **1977** F. WEBB *Go for Out* v. 71 The Metropolitan Police owned four such vehicles for use by their crime squads, rubber-heelers, or the Special Branch.

ru·bberiness. [f. *RUBBERY *a.* + -NESS.] Rubber-like quality.

1952 E. HEMINGWAY *Old Man & Sea* 125 The old man swung the club down on him..and hit only the heavy solid rubberiness. **1959** *Times* 27 Apr. (Rubber Industry Suppl.) p.vi/6 Embrittlement caused by high temperature oxidation or by loss of rubberiness in the cold must therefore be avoided. **1972** P. W. ALLEN *Natural Rubber & Synthetics* i. 1 All [rubbers] are high polymers, but of a special type possessing distinctive chemical structural characteristics which enable them to be transformed by one means or another into materials having the property of 'rubberiness'.

rubberize (rʊ·bəraiz), *v.* [f. RUBBER *sb.*¹ + -IZE.] *trans.* To treat, coat, or impregnate with rubber. Hence **ru·bberized** *ppl. a.*; **ru·bberizing** *vbl. sb.*

1912 *Chem Abstr.* VI. 1554 Hides and skins..are rubberized by treatment with hot rubber soln. consisting of Para rubber, [etc.]. **1918** *Sphere* 2 Feb. 109/2 The rubberised cotton envelope has a capacity of 77,000 cubic ft. **1925** *Sunday at Home* Mar. 346/1 The balloon..is of rubberised fabric. **1936** *Lancet* 10 Oct. 865/2 The airman wore a two-piece suit of rubberised fabric. **1951** *Oxf. Jun. Encycl.* VII. 466/1 Putting on the rubber, or 'rubberizing', takes place in several stages. **1953** J. Y. COUSTEAU *Silent World* 8 To protect myself from cold I spent days tailoring and vulcanizing rubberized garments. **1963** A. J. HALL *Textile Sci.* v. 235 Water repellency obtained by means of silicones retains its permeability to air.. —this advantage is not possessed by textile materials which have been rubberised. **1967** *Jane's Surface Skimmer Syst. 1967–68* 2/2 The skirts can be manufactured simply by cutting up and bonding strips of standard rubberised cloth. **1972** *Materials & Technol.* V. xiv. 454 (*caption*) Fabric being rubberised on calender for use as conveyer belt carcase. **1977** *Field* 13 Jan. 66/4 What is the best way to clean a white rubberised riding mackintosh?

rubberneck, *sb.* and *v.* For *U.S. slang* read '*colloq.* (orig. *U.S.*)' and substitute for def.: **A.** *v.* **a.** *intr.* To crane the neck in curiosity, to gape; also, to look around, to sight-see. (Earlier and later examples.) **b.** *trans.* To stare at. Hence **ru·bbernecking** *ppl. a.* and *vbl. sb.*

1896 ADE *Artie* iii. 23, I stood around there on one foot kind o' rubber-neckin to find an openin. **1927** H. V. MORTON *In Search of England* ix. 173 Here's a great sight going on that hundreds of rubber-necking tourists would pay anything to see. **1932** D. L. SAYERS *Have his Carcase* iv. 59 She..could not waste time rubber-necking round Wilvercombe with Lord Peter. **1939** *Daily Mail* 12 Apr. 8/4 Thousands of people..have 'rubber-necked' this monstrosity [*sc.* Ming, the giant panda] until their

eyes ached. **1939** *Times Lit. Suppl.* 20 May 293/3 Mr. Graves resembles some of the professional guides who showed him round when he went 'rubber-necking'. **1946** *Sun* (Baltimore) 5 Nov. 10/7 The long, vaulted centra hall..was crowded with chairs for invited guests with probably five times as many more people standing behind them. Londoners love to rubberneck on tiptoe. **1958** *Observer* 27 Apr. 6/7 Mr. Gunther has the born tourist's eye, and he can put down what he sees. He carries his rubber-necking from the pavements and the cafés to every corner into which he is allowed to penetrate. **1969** *Daily Tel.* (Colour Suppl.) 21 Nov. 73/2 Hortensio was rubber-necking like an American tourist, admiring the scenery, sniffing the breeze. **1973** J. MANN *Only Security* vi. 61 'You're not itching to get your hands on the site?'.. 'Not a bit, lovely just to rubberneck for a change.' **1977** *Time* 16 May 54/1 Wisconsin motorists may never see a purple cow, but they are rubbernecking at an enormous piebald blue one emblazoned on Farmer Hilbert Schneider's 75-year-old barn at Johnson Creek.

B. *sb.* Someone who stares; an inquisitive person; a sight-seer, a tourist. (Earlier and later examples.)

1899 *Amer. Jrnl. Sociol.* May 726 Oh, no! in the language of the shop, she was only a 'rubber-neck'. **1909** G. B. MCCUTCHEON *Truxton King* iii. 41 They are the nobility —the swells. They don't hang around the streets like tourists and rubbernecks. **1918** 'I. HAY' *Last Million* xii. 188 Attended by a respectfully interested cohort of disciples, or rubbernecks. **1937** *Daily Herald* 6 Feb. 6 One of its valuable features will be to deprive the rubbernecks, who gloat over the domestic troubles of their neighbours in the local police court, of their entertainment. **1941** J. SMILEY *Hash House Lingo* 46 *Rubber neck*, tourist. **1974** P. MCCUTCHAN *Call for Simon Shard* xiii. 119 Can you clear the place up, Inspector? Move the rubbernecks on, back to bed? **1975** C. WESTON *Susannah Screaming* (1976) xxiv. 123 Without apology, Krug shoved through the rubbernecks.

b. *attrib.* and *Comb.*, as *rubberneck party, ride, tour, tourist*; **rubberneck auto, bus, car, wagon,** a vehicle for taking people on a sight-seeing tour.

1906 'O. HENRY' *Four Million* (1916) xix. 192 The Rubberneck Auto was about ready to start. The merry top-riders had been assigned to their seats by the gentlemanly conductor. **1949** *Chicago Daily News* 13 Aug. 5/6 That's the relatively harmless impression of Skid Row seen from the rubber-neck busses. **1951** E. PAUL *Springtime in Paris* x. 175 Large rubberneck buses from travel agencies drive through, packed with sightseers from various States of the Union. **1915** *Dialect Notes* IV. 245 *Rubber-neck car, n. phr.*, sight-seeing vehicle. 'We saw several *rubber-neck cars* in Yellowstone Park.' **1916** GALSWORTHY *Sheaf* 276 There exists in America a vehicle called the 'rubber-neck' car. **1925** H. L. FOSTER *Trop. Tramp with Tourists* 326 The tourists go riding through town in rubberneck parties. **1927** *New Republic* 12 Oct. 210/2 'The Manhatters' is founded upon the idea of a rubberneck ride through this island. **1915** *Chicago Herald* 8 Nov. 4/2 The black and tans from the southern states..have been taken on a rubberneck tour. **1949** *Nat. Geogr. Mag.* Dec. 783/2 Twice daily a horse-drawn stage leaves the Plaza on a 'rubberneck' tour. **1926** *Glasgow Herald* 27 July 10 As somebody has to get it in the neck, it may as well be..the rubberneck tourist. **1942** BERREY & VAN DEN BARK *Amer. Thes. Slang* §424/2 *Rubberneck tourist*, a sight-seeing tourist. **1908** G. H. LORIMER *Jack Spurlock* xi. 321 The Major inquired loudly of Horton, the Governor's secretary, whether he was 'runnin' a blank rubber-neck waggon'. **1932** *New Yorker* 11 June 38/2 The one who stepped from the rubberneck wagon happened to be the first [Japanese lady] they had ever seen. **1943** M. FLAVIN *Journey in Dark* 174 On the rubberneck wagons the fellow with the megaphone would point it out and say: 'Residence of Stanley Adams, financier and banker.'

Hence **ru·bbernecker** = *RUBBERNECK *sb.*

1934 in WEBSTER. **1942** BERREY & VAN DEN BARK *Amer. Thes. Slang* §765/8 *Rubbernecker*, a sight-seer. **1958** *N.Y. Times* 19 Apr. 16/5 He..completed a swing that lofted the ball over the barrier. He grinned apologetically at the rubber-neckers. **1969** S. HYLAND *Top Bloody Secret* i. 61 The usual crowd of rubberneckers on the far pavement. **1974** *Spartanburg* (S. Carolina) *Herald* 22 Apr. A4/2 American rubber-neckers in Moscow or Leningrad or elsewhere, with their free and easy manners, will leave as much an impression on the Russians they meet as they will take away with them.

rubberoid (rʊ·bərɔid). Also **Rubberoid.** [f. RUBBER *sb.*¹ III + -OID.] A substitute for rubber. Also *attrib.*

Formerly a proprietary name in the U.S. Quot. 1968 seems to denote a different substance (cf. *RUBEROID, with which there may have been confusion).

1884 *Official Gaz.* (U.S. Patent Office) 8 Jan. 112/1 Composition as a substitute for hard rubber—The James D. Frary & Son Co., Bridgeport, Conn. Application filed November 26, 1883. 'The word Rubberoid.' **1910** [mentioned s.v. RUBBER *sb.*¹ 11]. **1951** R. BRADBURY *Illustrated Man* (1952) 160 The city awaited the soft tread of their rubberoid boots. **1968** S. E. ROBERTS *Of Us & Oxen* x. 137 We also lost a piece of rubberoid. I don't know how such a heavy thing could have gotten away.

rubber stamp, *sb.* (*phr.*) and *v.* [RUBBER *sb.*¹ III.] **A.** *sb.* (*phr.*) **1.** (See quot. 1888.) Also, the imprint of such a stamp.

1881 *Instructions Clerks Census Eng. & Wales* (1885) 158/2 Rubber Stamp Maker. **1888** [see RUBBER *sb.*¹ 12 a]. **1954** KOESTLER *Invisible Writing* IV. xxxix. 420 Queuing up each time..to obtain the rubber-stamp which granted a further stay of a day or a week. **1957** E. H. GOMBRICH *Story of Art* xiv. 203 The result looked like any rubber stamp we use today, and the principle of printing it on paper was practically the same. **1975** J. VAN DE WETERING *Outsider in Amsterdam* (1976) ii. 24 They looked at the imprint of the rubber stamp and the signature.

2. *fig.* Used of a person or institution whose power is formal but not real; a person who or body which endorses uncritically.

1919 W. R. THAYER *Theodore Roosevelt* xxi. 334 He may have heard the exhortation 'Be your own President; don't be anybody's man or rubber stamp.' **1943** *Ann. Reg. 1942* 235 This new body..was not to have any of the traditional attributes of a Spanish Cortes. It was to be..an assemblage of Government nominees and notables, a rubber stamp. **1956** A. WILSON *Anglo-Saxon Att.* II. i. 196 The danger of the oldest of all representative bodies becoming a mere rubber stamp. **1965** *Listener* 3 June 823/2 The regional councils were attacked as being mere rubber stamps for the regional boards. **1976** *Survey* Winter 66 The governmental assemblies..are most certainly not rubber stamps for the decisions of their respective executives.

b. *attrib.* or as *adj.*

1931 *Government of Oxford* 5 Others believe that.. interest would be stimulated if Congregation could be relieved of its 'rubber stamp' duties. **1940** *Manch. Guardian Weekly* 29 Mar. 252 Lord Samuel transfixed the 1931–5 Parliament in a phrase: he called it 'The rubber-stamp Parliament'. **1946** W. S. CHURCHILL *Victory* 197 Equal opportunity for all, under free institutions and equal laws—there is the banner for which we will do battle against all rubber-stamp bureaucracies or dictatorships. **1953** *Manch. Guardian Weekly* 8 Oct. 7 The President does not want a 'rubber-stamp' Congress. **1977** *Time* 14 Mar. 23/1 This month's session of the People's Assembly, Burma's rubber-stamp parliament.

B. *v.* (With hyphen.) *trans.* **a.** To mark with the imprint of a rubber stamp; to print with a rubber stamp.

1922 *Hotel World* 13 May 2 No hotel would rubber-stamp its stationery. **1965** M. SPARK *Mandelbaum Gate* iv. 113 Abdul went to start rubber-stamping the soles of smuggled sandals. **1973** *Radio Times 50th Anniv. Souvenir* 66/1 I'd written a fan letter to Bing Crosby... I received a photo of him back—with his autograph rubber-stamped across it.

b. *fig.* To endorse or approve uncritically; to pass routinely or automatically.

1934 WEBSTER, *Rubber-stamp, v.t.* a. To sign with a rubber stamp. b. Hence figuratively, to approve, endorse, or dispose of (as a document or policy) as a matter of routine, usually without the exercise of one's judgment. **1935** *Ann. Reg. 1934* 304 Moreover the Democrats in Congress were completely at one with the Republicans.. in their dislike at appearing merely to 'rubber-stamp' measures drafted by the President. **1959** *News Chron.* 8 July 1/1 We do not believe that the purpose of a conference of our type is to rubber-stamp every declaration. **1978** S. BRILL *Teamsters* i. 15 The current trustees ..had rubber-stamped loans to mob fronts.

Hence **rubber-sta·mping** *ppl. a.* and *vbl. sb.*; **rubber-sta·mpish** *a.* [-ISH¹], quasi-automatic, almost purely formal.

1932 L. C. DOUGLAS *Forgive us Our Trespasses* (1937) ix. 183 An hour and a half was spent..in a rubber-stampish approval of the 'tentative suggestions' sent from the faculty 'as a basis of discussion'. **1952** 'M. COST' *Hour Awaits* 24 The first letter of 1919—its address almost obliterated by rubber-stampings. **1958** *Sunday Times* 20 Apr. 16/4 He is, in a sense, a dictator, for..the Constituent Assembly is virtually a rubber-stamping body. **1969** *Daily Tel.* 26 Aug. 14 A fresh round of repressive legislation, including instant loss of employment for all who question the official line in deed or word, was all ready for rubber-stamping. **1979** *China Now* Mar./Apr. 3/3 A central working conference..took decisions of far-reaching importance. These decisions were..not just a rubber-stamping of proposals.

rubbery (rʊ·bəri), *a.* [f. RUBBER *sb.*¹ III + -Y¹.] Resembling or suggestive of rubber. Also *fig.* and *Comb.*

1907 GALSWORTHY *Country House* II. v. 147 He left his hand against the animal's warm, soft, rubbery mouth. **1928** *Collier's* 10 Nov. 20/2 He..wrapped a thick hairy arm about Dan's neck in a chancery hold and squeezed his Face into the thick rubbery flesh of his side. **1935** W. CATHER *Lucy Gayheart* I. viii. 67 Even his white skin looked harder, somewhat rubbery. **1950** J. D. MACDONALD *Brass Cupcake* (1974) x. 95 He paid off with a very rubbery check. **1959** *Washington Post* 19 Jan. A12/2 Mr. Mikoyan's responses to questions on the television program, 'Meet the Press', last evening were.. as evasive and rubbery a performance as one could imagine. **1962** *Times* 9 Apr. 4/1 His legs went rubbery as Pender smashed him with lefts and rights to the head. **1973** *Country Life* 1 Nov. 1322/1 Palms and rubbery-leaved banana plants. **1977** *People Weekly* 10 Oct. 44/2 There..are the two rubbery faces mugging through those unforgettable sketches that kept America home Saturday nights from 1950 until 1954.

rubbidy, var. *RUBBEDY.

rubbing, *vbl. sb.* Add: **5. a.** *rubbing alcohol, table.*

1955 T. STERLING *Evil of Day* iii. 37 Celia checked her vanity case to see that she had enough rubbing alcohol. **1971** *Sci. Amer.* Aug. 114/3 Spray the wax joints one at a time with rubbing alcohol. **1939** 'E. QUEEN' in *Blue Bk.* Oct. 21/2 Koyle slipped from the rubbing-table, and Barney Hawks began shooing men out of the shower-room. **1976** *N.Y. Rev.* 24 June 8/3 Lyndon Johnson liked to talk to people while..lying in bed, on his rubbing table, skinny-dipping.

b. *rubbing plate, strake* (further examples).

1969 *Jane's Freight Containers 1968–69* 550/3 The front end incorporates a rubbing plate and retractable king pin. **1928** G. CAMPBELL *My Mystery Ships* iii. 36 The hinges were outboard, and had to be covered with rubber and made to look like a rubbing strake for going alongside a jetty. **1975** *Daily Colonist* (Victoria, B.C.) 24 May 7/3 Her hull mostly white down to the rubbing strake then black to the waterline.

rubbish, *sb.* (and *a.*) Add: **2. a.** (Later examples used of people.)

1885 R. HOLLAND *Gloss. County of Chester* (1886) 293 They're nowt bu' rubbitch. **1976** W. TREVOR *Children of Dynmouth* v. 114 Stringer, the headmaster, was rubbish; the P.E. man went after the girls.

3. a. Delete † *Obs.* and add: Revived in mod. colloq. use.

1979 M. BOYCE *I was There!* 83/2 A side that can be easily beaten..a rubbish side, Bedworth or Nuneaton.

b. *rubbish dump, heap* (earlier and later examples), *-pile, -tip.*

1976 W. TREVOR *Children of Dynmouth* iii. 72 It's hardly irrelevant that the country for which men were prepared to give their lives has become a rubbish dump. **1878** *Jrnl. Speculative Philos.* XII. 12 In the *failures* to 'adjust'—in the rubbish-heap, according to Spenser—lies, for them, the real key to the truth. **1932** KIPLING *Limits & Renewals* 299 He very rarely went down into what had now become a rubbish-heap. **1959** C. FREMLIN *Uncle Paul* iv. 33 The five of spades would have to go on the rubbish heap after all. **1884** 'MARK TWAIN' *Huck. Finn* xxxvii. 375 The rubbage-pile in the backyard. **1889** —— *Connecticut Yankee* xix. 235 Just a rubbish-pile of battered corpses. **1922** JOYCE *Ulysses* 422 On a step a gnome totting among a rubbishtip crouches to shoulder a sack of rags and bones. **1971** *Country Life* 24 June 1597/3 Wayside flowers..are still plentiful enough..especially in waste places like rubbish tips.

c. *rubbish-collector*; also *rubbish-dumping* ppl. adj. Also instrumental, as *rubbish-filled* adj.

1965 F. SARGESON *Mem. Peon* iii. 47 It was said that one [absent-minded scholar] had arrived at the college with his household rubbish after leaving his umbrella outside his gate to be collected by the rubbish-collector. **1937** BLUNDEN *Elegy* 84 By mysterious law each place Where Nature looks most gentle and glad Attracts the rubbish-dumping race. **1954** W. FAULKNER *Fable* 385 The corporal's body..went over backward..onto the edge of the rubbish-filled trench behind it.

d. rubbish shop, store, a junk shop.

1869 C. SCHREIBER *Jrnl.* 17 June (1911) I. 17 We found a small teapot, Venetian,..in a rubbish shop in the Spaderia. **1872** *Ibid.* 14 Apr. 156 He took us to a rubbish store.., from which we got nothing but a 'Davenport' plate.

rubbish, *v.* orig. and chiefly *Austral.* and *N.Z.* [f. the sb.] **1.** *trans.* To disparage, criticize severely. Hence **ru·bbished** *ppl. a.*; **ru·bbishing** *vbl. sb.*

1953 T. A. G. HUNGERFORD *Riverslake* ii. 20 If Verity was going to tramp you for burning the tucker..he would have rubbished you long before this. **1965** *Telegraph* (Brisbane) 26 Feb. 13, I knocked him down and I hope he dies. He rubbished me to a mate of mine. **1968** *Comment* (N.Z.) June 33/2 The paper..was recently rubbished in the Catholic *Tablet.* **1972** *Guardian* 16 Oct. 8/1 This live show had a live and participating audience; so Hockney got briefly rubbished the moment his film ended. **1975** *Observer* 12 Jan. 17/1 His plight, and that of the cricketers, have both been latched on to as a chance, not to be missed, of rubbishing the Poms. **1977** *Bulletin* (Sydney) 22 Jan. 14/3 To that extent the much rubbished figures of the CES are consistent with the quite independently determined estimates of the Bureau of Statistics. **1979** *Spectator* 14 July 28/2 A conventional rubbishing of the Left and applause of the Right.

2. *Surfing.* (Chiefly in *pass.*) To tip (a surfer) off a wave.

1962 *Austral. Women's Weekly* 24 Oct. (Suppl.) 3/3 *Rubbished*, to be thrown off wave and dumped on shore. **1963** *Sun-Herald* (Sydney) 22 Sept. 84/5 The fate the board rider dreads is the 'wipe out'. This is when he is 'rubbished' or tipped violently off a wave.

rubbishry (rʊ·biʃri). *rare.* [f. RUBBISH *sb.* + -RY.] Rubbish; a collection of rubbish.

1894 KIPLING in *Scribner's Mag.* Dec. 670 Fillin' my bunk wi' rubbishry the Chief put overside.

rubbishy, *a.* Add: **2. a.** (Example of use of a person.)

1946 [see *LEAVABLE *a.*].

b. *Comb.*, as *rubbishy-looking* adj.

1874 'MARK TWAIN' *Lett. to Publishers* (1967) 81 You notice that the Gilded Age is a rather rubbishy looking book.

rubbity (rʊ·bĭti), shortened f. next. Cf. *RUBBEDY, etc. *Austral.*

1941 BAKER *Dict. Austral. Slang* 62 *Rubbity*.., a public house. **1957** 'N. CULOTTA' *They're a Weird Mob* (1958) vii. 104 'Where's Jimmy an' Pat?' 'Down the rubbity.' **1963** *Australasian Post* 8 Aug. 47/3 The proprietor of the local rubbity was a woman. **1968** D. O'GRADY *Bottle of Sandwiches* 54 Roebourne boasted one pub..—the rubbity—we soon became aware of what was doing around the little joint. **1973** A. BUZO *Rooted* 63 'Been down to the rubbity lately?' 'No, I haven't hit the hops for a couple of weeks.'

RUBBITY-DUB

rubbity-dub (rʊ·bĭtidʊ·b), altered f. *RUB-A-DUB-DUB (b). *Austral.*

1957 'N. Culotta' *They're a Weird Mob* (1958) vii. 104 'What is a rubbity?' Joe said scornfully, 'Rubbity-dub.' **1971** *National Times* (Austral.) 13 Dec. 20/2 'Let's grab a do-or-die, have a couple of inky stinks at the rubbity dub...' Translated: 'Let's grab a pie, have a couple of drinks at the pub.'

rubble, *v.* Restrict 'Now *dial.*' to senses in Dict. and add: **1. b.** *trans.* To reduce to rubble. Also *fig.* Chiefly in *pass.* and as **ru·bbled** *ppl. a.*

1926 F. M. Ford *Man could stand Up* I. ii. 37 Things had become more rubbled—mixed up with alarums. **1945** *Daily Progress* (Charlottesville, Va.) 2 Mar. 1/8 Cologne, rubbled anew after dawn by a thousand British heavy bombers. **1953** *Encounter* Nov. 52/1 Palaces like Priam's, scarcely now to be identified among the rubbled trenches that were Ilium. **1978** *Islands* (N.Z.) Aug. 67 O Brave New World..without cities and the bombs to rubble them.

rubby (rʊ·bi). *Canad.* [f. *rubbing* (*alcohol*) s.v. *RUBBING *vbl. sb.* 5a: see -Y[6].] **1.** A habitual drinker of rubbing alcohol (see sense 2 below).

1950 A. Palmer *Montreal Confidential* 102 The police department has probably given up keeping score of rubbies they have fished out of the river. **1965** *Vancouver Sun* 18 Oct. 35/6 Most of the dinner guests were men off the street, rubbies, derelicts, the jobless, alcoholics, the lost ones, residents of Vancouver's Skid road. **1978** W. S. Avis in *Occasional Papers Dept. English R. Military Coll. Canada* No.2. 45 Both skid roads remained to become run-down, unsavoury slums.., the hangouts of drifters, rubbies, and other unfortunates.

2. Rubbing alcohol, sometimes mixed with wine, etc., used as an intoxicant.

1961 *Maclean's Mag.* 29 July 36/1 A gallon of wine and two bottles of rubby and you can throw a party in the jungles that'll last all night. **1974** D. Richards *Coming of Winter* i. 29 And there in the shacks the old men hard on rubby, telling stories of the war.

Also **ru:bby-du·b** [cf. also *DUB *sb.*[6]] = sense 1 above.

1950 A. Palmer *Montreal Confidential* 101 If the bum looks a bit plastered don't stop... Chances are he's a 'rubby-dub' and his mind is no doubt clouded with smoke. **1957** *Maclean's Mag.* 25 May 68/2 'We've got everything here from ex-cons to rubby-dubs,' says..one of Elliot's six provincial policemen. **1972** *Daily Colonist* (Victoria, B.C.) 7 Mar. 31/8 Mr. Minister, don't talk nonsense—don't suggest the rubby-dub has to gather up enough money for his own treatment.

rubby-dubby (rʊ·bidʊ:bi). *Angling.* [? f. Rub *v.*[1], Dub *v.*[1] 5: see -Y[6].] Minced fish such as pilchards, mackerel, etc., placed in a net-bag and used as a lure for shark and other large fish. Also *attrib.*

1957 R. Arnold *Compl. Sea Angler* xi. 176 As the rubby-dubby moves through the water, the oil from the broken-up bait spreads out from behind the boat, leaving an ever-widening channel down which the hungry sharks ..will cruise searching their prey. **1959** *Angling Times* 27 Feb. 6/3, I, drifting with a rubby-dubby trail, soon had a shark. **1960** *Sunday Express* 24 July 13/1 Two net bags stuffed with old pilchards and mackerel (the skipper calls it 'rubby-dubby'). **1970** *Daily Tel.* 2 May 9/3 Large fish can be attracted, like shark, with the 'rubby dubby' method. **1971** *Angling Times* 10 June 24 Ivan got over the rubby-dubby bags, and started a drift.

ru·b-down. [f. vbl. phr. *to rub down*: Rub *v.*[1] 8.] An act of rubbing down in any sense.

1885 *Boy's Own Paper* 21 Mar. 305/1 When the stick has dried in shape, trim it to taste with a sharp knife, and give it a good rub down with sand-paper. **1896** S. Hale *Let.* 4 June (1919) 299 We reached here reeking, just in time for a rubdown. **1903** [see Rub *sb.*[1] 1 c]. **1917** M. T. Hainsselin *Grand Fleet Days* xv. 104 To think that I should get a rub-down like this from the Admiral. **1936** J. Curtis *Gilt Kid* xiii. 133 Just imagine getting a rub-down at the copper-house and the bogies dragging a lump of coal out of his sky. **1936** 'P. Quentin' *Puzzle for Fools* viii. 63 We took rubdowns and other uncomfortably beneficial treatments. **1963** X. Field *Under Lock & Key* xi. 143 They and their cells are searched every fortnight or so, at irregular intervals and at an unexpected moment. The 'rub downs' usually lead to their precious belongings being removed. **1965** Mrs. L. B. Johnson *White House Diary* 5 Oct. (1970) 325 Lyndon on the table getting a rubdown and holding them in conversation. **1977** 'E. McBain' *Long Time no See* viii. 123 A hawker for one of the rubdown emporiums handed her a leaflet.

Rube, var. *REUB.

rubeanic acid (rū·biænik æ·sid). *Chem.* [tr. G. *rubeanwasserstoffsäure*, f. L. *rube-us* red + G. *-an* (as in *cyanwasserstoffsäure* hydrocyanic acid) + *wasserstoff* hydrogen + *säure* acid.] Dithio-oxamide, $[CS(NH_2)]_2$, an orange-red crystalline solid formed by reaction of cyanogen and hydrogen sulphide, and employed in analysis as a reagent to detect copper.

[**1884** *Jrnl. Chem. Soc.* XLVI. 1109 (*heading*) The so-called rubeanhydric acid (cyanogen bisulphydrate).] **1891** *Ibid* LX. ii. 1008 The following experiments show that the red compound ('rubeanic acid, rubeanwasserstoff') obtained by the combination of cyanogen and hydrogen sulphide behaves in many reactions as if it were dithi-oxamide, $NH_2{\cdot}CS{\cdot}CS{\cdot}NH_2$. **1928** *Q. Jrnl. Indian Chem. Soc.* III. 118 Rubeanic acid may be regarded as a tautomeric compound consisting of an equilibrium mixture of sym.-dithio-oxamide and sym.-di-imido-dithio-oxalic acid. **1967** *New Scientist* 2 Feb. 272/3 A test plate subjected to 500 hours accelerated weathering while protected with a polyurethane resin containing rubeanic acid shows no sign of tarnishing. **1981** *Sci. Amer.* Feb. 127/1 For a more sensitive test [for copper in silver coins] Epstein suggests using a saturated solution of rubeanic acid (dithiooxamide) in alcohol and a 20 percent solution of malonic acid.

Rube Goldberg (rūb gōu·ldbəɹg). *U.S.* The name of the American humorous artist Reuben ('*Rube*') Lucius *Goldberg* (1883–1970), used *attrib.* of any unnecessarily complicated, impracticable or ingenious device of the kind illustrated by this artist. Hence **Rube Goldbe·rgian** *a.*

1956 Rice & Steinmetz *Amish Year* 69 The whole Rube Goldberg device is hitched to a wire which runs through ringbolts attached to short poles stuck in the ground, all the way to the house. **1961** Webster, Rube Goldbergian. **1962** *Time* 22 June 38 It [sc. Bertrand Russell's *History of the World in Epitome*] consists of a page with seven words, a drawing of the Garden of Eden,..a drawing of a Rube Goldbergian battle scene, and a few final words. **1963** *Johns Hopkins Mag.* Jan. 20 Insofar as Congress' Rube Goldberg machinery is contrived to help it wait for the propitious moment, it is much to be valued. **1977** *Time* 26 Sept. 33/1 It contains a Rube Goldbergian arrangement of pulleys, ropes and rollers. **1978** *Nature* 9 Nov. 122/3 Orchids are Rube Goldberg machines; a perfect engineer would certainly have come up with something better.

rubeiboo, var. *RUBBABOO.

rubella. Add: (Further examples.) Also *attrib.*

1962 A. Sorsby in A. Pirie *Lens Metabolism Rel. Cataract* 298 Congenital cataract..can be caused by such frankly environmental disturbances as maternal rubella. **1970** *Nature* 11 Apr. 172/1 Growth retardation occurs in rabbits congenitally infected with rubella virus. **1971** *Where* Sept. 271/1 Blindness..in an increasing proportion of cases..is linked with additional handicaps such as deafness, cerebral palsy or mental retardation (for example, 'rubella' babies often have more than one handicap).

rubelliform (rube·lifǭɹm), *a.* *Med.* [f. Rubell(a + -i- + -form.] Resembling the characteristic rash of rubella.

1959 *Amer. Jrnl. Trop. Med & Hygiene* VIII. 104/1 The rash occurs..as blotchy, maculopapular, rubelliform or occasionally petechial lesions. **1969** *Amer. Jrnl. Epidemiol.* LXXXIX. 665/2 A rubelliform rash..mild upper respiratory symptoms and absence of Köplik's spots were the clinical diagnosis criteria. **1976** *Lancet* 6 Nov. 990/1 Three children had a rash, rubelliform in 2 cases and localised and purpuric in 1.

Rubenesque: see *RUBENSESQUE *a.*

Rubens (rū·bĕnz). The name of the Flemish painter Sir Peter Paul *Rubens* (1577–1640), used *attrib.* in **Rubens brown**, a brown earth-colour; **Rubens hat** (see quot. 1960); **Rubens madder**, madder brown.

1860 Rubens brown [see *CASSEL]. **1885** A. Edwardes *Girton Girl* I. xiii. 250 A distant lovely head..its waves of amber hair set off against the soft velvet of a Rubens hat. **1886** H. C. Standage *Artists' Man. Pigments* vi. 67 Rubens brown is a native earth of an ochreous character. *Ibid.* 69 Rubens madder, otherwise known as Orange Russet, [etc.]. **1934** H. Hiler *Notes on Technique of Painting* ii. 125 *Madder,..Rubens madder,..*etc. These names are now applied both to products from the genuine madder root, and also to those made from its synthetic colouring principles alizarin and purpurin. **1960** C. W. Cunnington et al. *Dict. Eng. Costume* 185/1 *Rubens hat*.., a hat with a high crown and brim turned up on one side. **1969** R. Mayer *Dict. Art Terms & Techniques* 341/1 Rubens madder is now made from synthetic alizarin. *Ibid.* 414/1 Rubens brown is a variety of Van Dyke brown.

Rubensesque (rūbənze·sk), *a.* [f. prec. + -esque.] Characteristic or suggestive of the paintings of Rubens; esp. of a woman's figure: full and rounded. Also **Rubene·sque,** *a.*

1913 *Maclean's Mag.* July 106/2 There are, no doubt, eccentric artists who prefer a Rubensesque figure, but these are the exceptions, and for most private work and school work a spare figure is far more valuable. **1925** W. Deeping *Sorrell & Son* xx. 185 He had a view of her broad back, and her robust curves... A Rubenesque figure, sumptuous and solid. **1927** *Observer* 17 July 15/4 The models of his choice are of rather Rubenesque fullness. **1952** G. Raverat *Period Piece* v. 87 She had auburn hair ..a charming Rubenesque complexion, and a deep rich voice. **1957** W. Camp *Prospects of Love* ii. xiv. 89 'Was she about as big as me?'.. 'Yes, I think she was. Slightly more Rubensesque hips, if anything. But I should think her waist was the same.' **1971** R. Hill *Advancement of Learning* i. 13 The nude was Rubensesque. **1976** L. Deighton *Twinkle, Twinkle Little Spy* viii. 78 The artless gesture of the *ingénue*, inappropriate for this Rubenesque wife and mother.

Rubensian (rube·nziăn), *a.* [f. *RUBENS + -IAN.] Of, pertaining to, or characteristic of Rubens or his work.

1890 *Athenæum* 18 Jan. 90 The composition is distinguished by the true Rubensian 'swing' and emphatic movement. **1940** *Burlington Mag.* June 193/2 This family, with all its Rubensian attributes, as plainly inherits something from each of Rubens's three masters. **1964** *Punch* 1 Apr. 490/2 A voluptuous Rubensian still-life. **1976** *Jrnl. R. Soc. Arts* CXXIV. 624/2 The *composition* of Constable's picture is perhaps the least Rubensian thing about it. **1979** *Amer. N. & Q.* Oct. 29/1 Rubensian themes that appear in the earlier part of Van Dyck's career.

rubeola. 2. Delete 'Now *rare* or *Obs.*' and add further examples.

1883 J. N. Hyde *Pract. Treatm. Dis. Skin* ix. i. 389 The distinction between rubeola and rötheln will be given later. **1909** C. B. Ker *Infectious Dis.* ii. 21 It would be simpler if every one referred to measles as 'morbilli' and to German measles as 'rubella', and if the term rubeola were allowed to drop. *Ibid.*, Unfortunately the term 'rubeola' is..freely used to designate measles. **1947** K. Wiener *Skin Manifestations of Internal Disorders* iv. 90 The latin term rubeola is used for this disease [sc. German measles] in the German literature, while in the English-American terminology, rubeola designates true measles. **1969** A. B. Christie *Infectious Dis.* xii. 346 The term rubeola still lingers on as a synonym of measles, though this usage was condemned as long ago as 1909 by Ker. **1973** *Sci. Amer.* Sept. 105/1 The same principle now allows very effective immunization against poliomyelitis, tetanus, diphtheria and both kinds of measles (rubella and rubeola).

Ruberoid (rʊ-, rū·bəroid). Also **ruberoid.** A proprietary name applied esp. to a roofing material composed of felt impregnated with bitumen. See also *RUBBEROID.

1901 *Official Gaz.* (U.S. Patent Office) 28 May 1848/1 Certain named substances of the nature of rubber. The Standard Paint Co., New York, New York. Filed Nov. 22, 1900. Ruberoid. **1902** *Trade Marks Jrnl.* 14 May 599 *Ruberoid*..Roofing pasteboard or paper and roofing felt. The Standard Paint Company Zweigfabrik Hamburg. **1910** *Ibid.* 8 June 894 *Ruberoid*... Paint and varnish included in Class 1..., and sheathing material included in Class 1 for heat insulating purposes. The Ruberoid Company Limited London. **1911** R. F. Scott *Jrnl.* 10 Jan. in *Last Exped.* (1913) I. iv. 111 On the outside [of the roof] is a matchboarding, then a layer of 2-ply 'ruberoid'. **1916** *Cornh. Mag.* Apr. 504 Myself and a chum had just returned..laden with 3″ by 4″ timbers and ruberoid which we found. **1921** H. G. Ponting *Gt. White South* 123 The roof..was covered with a thicker layer of ruberoid, and was lined with a single thickness of boards. **1925** *Glasgow Herald* 3 Aug. 5 The hut..was timber-built and roofed with rubberoid [*sic*]. **1934** *Trade Marks Jrnl.* 21 Nov. 1504/2 *Ruberoid*... Nails; and sectional sheets of ordinary metal for use in building. The Ruberoid Company Ltd. **1958** *House & Garden* Mar. 66/2 Roofs can be of shingles, clay tiles or, as shown here, Ruberoid felt. **1975** *Cricketer* May 47/1 (Advt.), Ruberoid Cricket Pitch is the year round match or practice wicket which can be used out of doors then lifted and re-laid for internal use.

Rubicon, *sb.* Add: **3.** Also applied to a variety of piquet. Also *absol.* (see quots.).

1882 'Cavendish' (*title*) The laws of Rubicon piquet, adopted by the Portland Club. **1897** R. F. Foster *Compl. Hoyle* 438 Rubicon piquet, for two players. The chief difference between this game and the usual form, Piquet au cent, is in the manner of declaring... Rubicons. If either or both players fail to reach 100 points in the six deals, the one having the most is the winner, and adds to his own score all the points made by the loser, with 100 in addition for game. **1950** *Hoyle's Games Modernized* (ed. 20) 76 It is only necessary to discuss the Rubicon Game, the game of 100 or 101 points being in disuse. *Ibid.*, There is another condition, namely, the establishment of 100 as a 'Rubicon'. **1973** J. Scarne *Encycl. Games* 604 Rubicon (*piquet*), failure of the loser of a game to reach 100 points. **1975** *Way to Play* 105/1 The procedure then depends on whether these totals exceed the 'rubicon' of 100 points.

rubicundly, *adv.* (Later example.)

1980 *Daily Tel.* 6 Oct. 3/1 'We can't go on living in the 19th century,' says the rubicundly amiable secretary of St Stephen's [Club].

rubidium. Add: [Coined in Ger. by Bunsen in *Ann. d. Chem.* (1861) CXIX. 107.] **1.** Atomic number 37; symbol Rb. (Earlier and later examples.)

1861 H. E. Roscoe in *Proc. R. Inst.* III. 326 A few days ago the speaker received a letter from Bunsen, which contains the following most interesting information:— 'The substance which I sent you as impure tartrate of Cæsium contains a *second* new alkaline metal... I propose to call the new metal "Rubidium".' **1861** *Chem. News* 27 July 44/2 Both rubidium and cæsium, the two alkali metals recently discovered by means of spectrum analysis, have a great chemical similarity to potassium. **1912** J. W. Mellor *Mod. Inorg Chem.* xix. 359 Metallic rubidium is prepared by heating an intimate mixture of the carbonate with finely divided carbon. **1946** *Nature* 2 Mar. 269/1 Minerals richest in rubidium are the lithia micas (lepidolites) which quite frequently contain as much as 2–3 per cent Rb_2O. **1950** N. V. Sidgwick *Chem. Elements* I. 65 Rubidium and caesium catch fire at once on exposure to air. **1974** *Encyl. Brit. Micropædia* VIII. 705/3 Rubidium, because of its electropositiveness, is second only to cesium as a proposed working fluid in plasma propulsion for deep-space probes.

2. a. *attrib.* and *Comb.*

1862 *Phil. Mag.* XXIV. 46 (*heading*) On the preparation of the rubidium compounds. **1911** *Encycl. Brit.* XXIII. 809/1 The rubidium salts are generally colourless, mostly soluble in water and isomorphous with the corresponding potassium salts. **1950** F. E. Zeuner *Dating Past* (ed. 2) x. 334 Other minerals like hydrothermal microclines, pollucite, and rubidium-rich varieties of muscovite, may in due course become important. **1950** *Thorpe's Dict. Appl. Chem.* (ed. 4) X. 637/2 Rubidium Sulphate, Rb_2SO_4, forms rhombic crystals. **1962** F. I. Ordway et al. *Basic Astronautics* iv. 127 A rubidium-vapor magnetometer to measure magnetic fields in space. **1971** I. G. Gass et al. *Understanding Earth* ii. 44/2 A small amount [of strontium] is usually also incorporated into calcium-poor, rubidium-bearing potassium minerals. **1977** *Broadcast* 13 June 6/3 The MSF time signals are derived from what is loosely described as an atomic pendulum but is more correctly known as a rubidium vapour oscillation.

b. Special Comb.: **rubidium-strontium**, used *attrib.* to denote a method of isotopic dating, or results obtained from it, based upon measurement of the relative amounts in rock of rubidium 87 and its beta decay product, strontium 87.

[**1946** *Nature* 2 Mar. 269/1 By means of this standard, Rb/Sr ratios of five samples of lepidolite and one of pollucite were determined spectrochemically.., the resultant ages being as follows.] **1950** F. E. Zeuner *Dating Past* (ed. 2) x. 334 Minerals suitable for the rubidium/strontium method must be rich in Rb and free from non-radiogenic Sr. **1961** *Times* 25 Apr. 2/6 The Department ..is at present using both the potassium-argon and rubidium-strontium methods. **1977** A. Hallam *Planet Earth* 184/2 Rubidium-strontium and uranium-lead measurements conclusively show that all these rocks were formed between about 3700 and 3800 million years ago.

Rubik (rū·bik). [The name of E. *Rubik*, Hungarian teacher, who patented the puzzle in Hungary in 1975.] *Rubik('s) cube*: a puzzle consisting of a cube seemingly formed by 27 smaller cubes, uniform in size but of various colours, each layer of nine or eight smaller cubes being capable of rotation in its own plane; the task is to restore each face of the cube to a single colour after the uniformity has been destroyed by rotation of the various layers.

1980 D. E. Taylor (*title*) Rubik's cube. **1980** D. Singmaster *Notes Rubik's 'Magic Cube'* (ed. 5) p. i, This edition has been retitled since the Magic Cube is now being sold as Rubik's Cube. *Ibid.* 37 Ideal [sc. the Ideal Toy Corp.] has renamed the cube as 'Rubik's Cube' on the grounds that 'magic' tends to be associated with magic. **1981** *Sci. Amer.* Mar. 14/1 Bűvös Kocka—the Magic Cube, also known as Rubik's Cube—has simultaneously taken the puzzle world, the mathematics world and the computing world by storm. **1981** *Bookseller* 4 July 45/1 Rubik's cube is the latest game/puzzle aimed at driving both parents and children to madness. **1981** *Daily Tel.* 9 July 14/1 Those who in recent months have been driven potty by the clicking of the intellectual's worry beads, the multi-coloured and multi-faceted Rubik Cube, will be glad to know that help has arrived.

‖ **rubinetto** (rūbine·tō). *rare.* [It.] A tap, a faucet (in quot. *fig.*).

a **1930** D. H. Lawrence *Last Poems* (1932) 157 The half-hidden private parts just a little brass tap, rubinetto, turned on for different purposes.

rubinglimmer (rū·bin,glimər). *Min.* [a. G. *rubinglimmer*, f. *rubin* ruby + *glimmer* mica, Glimmer *sb.*[2]] = *lepidocrocite* s.v. Lepido- in Dict. and Suppl.

1836 T. Thomson *Outl. Min., Geol.* I. 439 Rubinglimmer. **1837** J. D. Dana *Syst. Min.* vi. 380 Brown iron ore... The crystallized variety has been called Onegite, rubinglimmer, pyrosiderite, and Göthite. **1879** *Encycl. Brit.* X. 229/1 Hæmatite (peroxide of iron) occurs crystallized in veins through crystalline rocks,.. and sometimes in minute scales (rubin-glimmer) disseminated through the minerals of many crystalline rocks. **1919** [see *lepidocrocite* s.v. *LEPIDO-]. **1944** C. Palache et al. *Dana's Syst. Min.* (ed. 7) 683 Goethite... Found at numerous localities in Nassau, Westphalia, and the Rhine Provinces, Germany, especially at Siegen (does not include the rubinglimmer = lepidocrocite, found at Siegen), [etc.].

rubio (ru·bio). [Sp., lit. 'fair, blond(e), golden'.] Limonite mined in northern Spain.

1892 *Trans. Fed. Inst. Mining Engin.* III. 611 A feature of the rubio deposits is the occurrence of numerous clay 'backs' in the ore. **1901** E. H. Davies *Davies's Treat. Metalliferous Minerals & Mining* (ed. 6) xxviii. 2808 'Rubio' ore continues to be the principal class of ore produced [in Bilbao]. **1923** R. H. Rastall *Geol. of Metalliferous Deposits* xv. 340 One of the most important groups of mines is that of Somorrostro in Viscaya, west of Triano. In this region four types of ore are recognized: (1) Vena... (2) Campanil... (3) Rubio, limonite, often siliceous and aluminous. (4) Carbonato. **1935** *Economist* 2 Feb. 292/2 Business in foreign ore is quiet. Whilst best rubio is nominally 17s. per ton c.i.f. Middlesborough, the tendency is upward.

ru·b-off. *slang.* [f. vbl. phr. *to rub off*: Rub *v.*[1]] An act of masturbation or manual stimulation to orgasm.

1937 Partridge *Dict. Slang* 710/2 *Rub-off*.., a masturbation. **1969** T. Parker *Twisting Lane* 202 She charges three quid for sex, two quid for a rub-off.

ru·b-out. *U.S. slang.* Also **rubout.** [f. vbl. phr. *to rub out* s.v. Rub *v.*[1] 11 a.] A murder, an assassination, esp. of one gangster by another. Also *attrib.*

1927 D. Hammett in *Black Mask* May 11/2 The hombre she blamed for Paddy's rub-out. **1934** *Sun* (Baltimore) 21 Aug. 7/1 Another..witness is Mrs... Fontaine, alleged 'gang-girl' associate of 'Big George' Phillips wounded in one 'rub out' attempt. **1953** *Ibid.* 12 June 34/3 Police marked down the Messina murder as one of Baltimore's extremely rare gangland rubouts. **1959** *Washington Post* 15 Aug. A3/1 Two hoodlums were gunned to death on Chicago's West Side today and police said at least one of the executions was probably a crime syndicate 'rubout'. **1977** *Time* 10 Jan. 22/1 In what was clearly a political rub-out, the couple—who were discovered by Mrs. Tayyeb's sister, sometime Skyjacker Leila Khaled—had been killed by at least a dozen shots.

rubral (rū·brăl), *a.* *Anat.* [f. L. *ruber, rubr-* red + -al.] Of or pertaining to the red nucleus of the brain.

1954 [see *RUBRO-]. **1972** M. L. Barr *Human Nervous System* vii. 110/2 A few rubral efferents enter the cerebellum through the superior peduncle to end in cerebellar nuclei.

rubredoxin (rubrĕdǫ·ksin). *Biochem.* [f. L. *rub-er* red + *REDOX + -IN[1]: cf. *FERREDOXIN.] Any of a class of natural proteins having an iron atom co-ordinated to the sulphur atoms of four cysteine residues, and concerned in intracellular electron-transfer processes.

1965 Lovenberg & Sobel in *Federation Proc.* XXIV. 233/2 This protein, which we tentatively named rubredoxin has been isolated in pure form. **1970** *Nature* 4 July 16/1 An interesting set of metalloproteins, which occur in plants and bacteria, are the non-haem iron proteins, such as the ferredoxins and rubredoxins. **1977** *Jrnl. Amer. Chem. Soc.* XCIX. 3505/1 (*heading*) Theoretical studies of the oxidized and reduced states of a model for the active site of rubredoxin.

rubric, *sb.* and *a.* **I. 2. b.** Delete *rare* and add: (Further examples.) Also, an injunction, a general rule.

1891 *N.Y. Times* 28 Sept. 4/5 It is the duty of independents—the duty of all voters—..to..'weigh the merits and demerits of each candidate and each party.' ..No better rubric of conduct could be laid down. **1934** J. W. Powell in *Webster* s.v. rubric, The groups of opinion inculcated by instruction are again found to fall into five 'rubrics'—animism, cosmogony, mythology, metaphysic, and science. **1962** W. Nowottny *Lang. Poets Use* viii. 202 There are no critical rubrics or Queensberry rules about this game. **1965** G. McInnes *Road to Gundagai* i. 13 'Don't go out too far!' A censorious well-worn rubric and I barely heard it. **1970** I. L. Horowitz *Masses in Lat. Amer.* i. 3 To understand the processes that go under the rubric of social development it is necessary to study masses as well as elites.

rubrician. (Earlier example.)

1843 H. Mozley *Let.* 23 Aug. in D. Mozley *Newman Family Lett.* (1962) 135 The notion of Puseyites or Rubricians..slipping unawares into popery is too absurd.

rubricism. (Later example.)

1978 C. Howell in C. Jones et al. *Study of Liturgy* ii. iii. x. 241 Trent ushered in four centuries of rigidity and fixation; it was an era of rubricism.

rubricize (rū·brisəiz), *v.* *rare.* [f. Rubric *sb.* + -ize.] **a.** *trans.* To provide with a rubric or rubrics. **b.** *intr.* To make general categorizations. Hence **ru·bricizing** *ppl. a.*

1920 R. Harris *Testimonies* ii. vii. 65 There is no reason to alter the opinion that the *Testimony* material should be rubricized. **1951** [see *CONSTELLATE *v.* 2 b].

rubro- (rū·bro), comb. form of L. *ruber* red, forming adjs. in *Anat.* with the sense 'relating to the red nucleus of the brain and (another part)', 'passing from the red nucleus to (another part)', as *rubrobulbar, -frontal, -oculomotor, -parietal, -reticular, -spinal.*

1902 H. Morris *Treat. Human Anat.* (ed. 3) v. 769 The rubro-spinal tract is formed by a number of fibres which are scattered in the anterior part of the lateral pyramid, in the posterior part of the lateral ground bundle, and in the posterior part of Lowenthal's tract. **1937** J. H. Globus *Pract. Neuroanat.* 150 Others..descend to the brain stem and spinal cord as the rubroreticular and rubrospinal tracts. **1954** T. L. Peel *Neuroanat. Basis Clin. Neurol.* xvii. 386/2 The rubro-oculomotor fibers to the third, fourth, and sixth cranial-nerve nuclei..form the most mediodorsal part of the capsule of the red nucleus. *Ibid.* 387/1 Fibers to other cranial-nerve motor nuclei, a rubrobulbar tract, are probably included in the rubral outflow. *Ibid.*, Rubrofrontal and rubroparietal fibers are described as leaving the dorsolateral surface of the nucleus and passing..to frontal and parietal lobes. **1972** M. L. Barr *Human Nervous System* vii. 110/2 Neuronal activity in the red nucleus..influences lower motor neurons through the rubrospinal tract and through rubroreticular and reticulospinal connections. **1974** D. & M. Webster *Compar. Vertebr. Morphol.* xii. 290 The rubrospinal tract leaves the red nucleus, and the reticulospinal tract leaves the reticular formation.

ru·b-up. [f. vbl. phr. *to rub up*: Rub *v.*[1] 13 in Dict. and Suppl.] The act of rubbing up in any sense.

1928 G. Campbell *My Mystery Ships* xiii. 245 We.. went out to the Sound for a good 'rub up' in our drill and to get everything tested. **1943** 'Taffrail' *White Ensigns* 26 He would take voluntary classes of men who wanted a rub-up in gunnery or seamanship before passing for higher rating. **1953** *Chambers's Jrnl.* June 355/1 Back then to the purgatory of waiting—with no text-books for a final rub-up permitted. **1967** Karch & Buber *Offset Processes* vi. 227 Plates are repaired:.. 4. By 'rub up'—to bring back or strengthen spots or areas that may become weak from an unknown cause.

ruby, *sb.* and *a.* Add: **I. 5. d.** *ellipt.* Ruby port (see sense *11).

1938 G. Greene *Brighton Rock* i. iii. 37 'Give me a glass of Ruby,' the sombre man said. **1959** W. James *Word-bk. Wine* 148 Ruby is a young, deep-red wine, or a tawny which has been refreshed with a younger wine.

II. 8. a. *ruby laser.*

1961 *Ann. Reg.* 1960 396 One drawback of the ruby laser was that it produced light only in bursts. **1974** *Encycl. Brit. Micropædia* VIII. 707/1 The chromium atoms responsible for the ruby's colour are also responsible for the emission of red light when ruby is excited by radiation, as in the red light produced by a ruby laser. **1977** *Jrnl. R. Soc. Arts* CXXV. 765/1 The first ruby laser of Maiman in 1960.

b. Also *ruby-ripe, -sweet.*

1918 W. de la Mare *Sam's Three Wishes* in *Twelve Poets* 27 Ruby-ripe to see, The pixy-pears burn on yon hawthorn tree. **1920** E. Sitwell *Wooden Pegasus* 21 As isles of the cherry Or ruby-sweet berry.

10. a. *ruby-eyed, -tasselled.*

1919 R. C. Punnett *Mendelism* (ed. 5) ix. 95 In canaries, again, there are ruby-eyed cinnamon forms corresponding to the various green and yellow varieties. **1950** D. Gascoyne *Vagrant* 38 Fatalist, Ruby-eyed. **1920** Blunden *Waggoner* 55 And ruby-tasselled shepherd's rose.

11. ruby anniversary, a fortieth anniversary; **ruby-back**, used *attrib.* to designate fine Chinese porcelain enamelled on the reverse in pink or crimson; so **ruby-backed** adj.; **ruby-dazzler** *Austral.* and *N.Z. slang*, something exceptionally fine (cf. *BOBBY-DAZZLER); **ruby port**, port of a deep red colour, *spec.* that matured in wood for only a few years and fined before bottling; **Ruby Queen** *Forces' slang* (see quot. 1925); **ruby wedding**, a fortieth (occas. forty-fifth) wedding anniversary.

1962 *Guardian* 17 Nov. 5/2 The celebration of the BBC's ruby anniversary. **1915** R. L. Hobson *Chinese Pott. & Porc.* II. xii. 213 A ruby-back saucer dish delicately painted. **1935** *Burlington Mag.* Jan. 25/2 At that period the Chinese kilns produced the ruby-back egg-shell porcelain. **1960** H. Hayward *Antique Coll.* 115/1 A new delicate painting style began to oust that of the *famille verte*..about 1720, and was applied especially to plates, bowls and cups and saucers of 'egg-shell' thin porcelain. The 'ruby-back' variety is coloured deep rose-pink on the reverse. **1980** *Catal. Fine Chinese Ceramics* (Sotheby, Hong Kong) 180 Compare the ruby-back cups painted with fruit in the interior sold in these rooms 29th November, 1977. **1900** F. Litchfield *Pott. & Porc.* vii. 113 The most highly-prized egg-shell, which is termed 'ruby backed' china. **1970** G. C. Williamson *Bk. of Famille Rose* vii. 104 Ruby-backed pieces can be marked off as belonging to a particular group, but the division must be confined exclusively to the ruby back. **1941** Baker *N.Z. Slang* vi. 51 Expressions..in constant use by our youngsters:..bobbydazzler, rubydazzler, dag, swinjer, [etc.]. **1941** —— *Dict. Austral. Slang* 62 *Rube*,..something esp. fine. *Rubydazzler*, as for 'rube'. **1977** W. S. Ramson in *Quadrant* (Sydney) May 67/1 The *Australian Pocket Oxford*..is a real *beaut*, a *ryebuck* dictionary, a *ringer*, a *ripper*, a gem amongst dictionaries if not a *rubydazzler*. **1921** A. L. Simon *Wine & Wine Trade* v. 59 It also happens sometimes that a vintage Port..will be kept in wood for a more or less extended number of years before it is bottled. The result will be a wine with less colour and strength than the early bottled vintage Port, but with more body and colour than tawny Port. This wine is often described as 'Ruby' Port. **1938** G. Greene *Brighton Rock* i. iii. 47 Life was sunlight on brass bed-posts, ruby port. **1967** A. Lichine *Encycl. Wines* 411/2 A bottle of Vintage Port..will suffer if it is open to the air very long, though a Ruby Port or Tawny will survive better. **1925** Fraser & Gibbons *Soldier & Sailor Words* 246 *Ruby Queen*,..an occasional nickname for any young nurse or Sister of fresh complexion. **1934** Blunden *Choice or Chance* 31 With Ruby Queens We once crowned feeds of pork and beans. **1911** Webster s.v. ruby *a.*, *Ruby wedding*, the forty-fifth wedding anniversary. **1963** B. Smith *Etiquette* vii. 117 According to a former convention..the interim anniversaries are..fortieth year, Ruby Wedding. **1977** *Times* 15 Apr. 12/5 This year marks his ruby wedding as well as his retirement.

ruby-red, *a.* **b.** quasi-*sb.* (Earlier example.)

1885 [see *ISOCHROMATIC *a.* 2].

ruby-throated, *a.* Add: Also occas. used of people.

1957 O. Nash *You can't get there from Here* 68 Our ruby-throated playgirls and madcap millionaires.

rucas, ruccus, *varr.* *RUCKUS.

ruched, *ppl. a.* (In Dict. s.v. Ruche *sb.*) Add: (Earlier and later examples.) Also *fig.*

1847 E. Gray *Let.* 5 May in W. James *Order of Release* (1947) iii. 31 Cloaks of pale glacé silk with ruished frills round them. **1848** —— *Let.* 10 May in *Ibid.* v. 107 A stone silk dress with two broad flounces Ruched and a Brussells lace cape. **1900** *Ann. Rep. Board of Regents Smithsonian Inst. 1898* ii. 703 In a broad way we may distinguish as leading types [of hemipenis] the following: The smooth; the plicate, or flounced; the calyculate, or ruched; and the disk-bearing. **1923** E. Sitwell *Bucolic Comedies* 61 Ruched as their country waterfalls, The cherried maids walk beneath the dark walls. **1932** *Woman's Weekly* 19 Mar. 467/1 The..skirt is cut in wide scallops and trimmed with ruching. Little ruched pieces..give the frock quaintness and charm. **1978** *Detroit Free Press* 5 Mar. D7/5 (Advt.), This terrific coat with its softly ruched back.

ruching. (Later examples.) Also *fig.*

1914 C. Mackenzie *Sinister St.* II. iv. iv. 914 The beginning of the street ran between two high brown walls crowned with a ruching of broken glass. **1932** [see prec.]. **1971** *Homes & Gardens* Sept. 61/2 And not only have they pounced on all the lace, all those bibs and tuckers and ruchings and veils and shawls and scarves and yard after yard of flouncing.

ruck, *sb.*[1] Add: **3. b.** *in a ruck*, in *Racing*: in one group.

1840 *Spirit of Times* 10 Oct. 380 When Randal blew his bugle, away they all flew in a ruck.

d. (*a*) *Rugby Union.* (Also *loose ruck*.) = *loose scrummage* s.v. *LOOSE *a.* 9. (*b*) *Australian Rules football.* (See quot. 1969.)

1906 Gallaher & Stead *Compl. Rugby Footballer* ix. 134 What we call a loose ruck..represents the disordered state of things occurring, for example, when..a back has slipped and stopped the play when trying to block a forward rush. One man is down, and all his other colleagues in the back division are induced to come up to his assistance. **1955** *Times* 8 Aug. 2/2 With the South African pack controlling the line outs..and driving hard from the rucks the home side looked to have a firm grip on the game. **1956** V. Jenkins *Lions Rampant* xii. 180 On the muddy ground they made rush after rush, and piled into the loose rucks as if their lives depended on it. **1967** *Australian* 17 Apr. 12 Terry Waters..was moved into the ruck in the third quarter. **1969** Eagleson & McKie *Terminol. Austral. Nat. Football* iii. 10 *Ruck.* 1. The three members of a team who do not occupy fixed positions but are free to follow the play wherever it goes around the field; the ruck consists of a rover and two followers. *Ibid.* 11 *Ruck.* 2. A member of a team selected to play in the ruck, other than the rover. **1973** [see *MAUL *sb.*[1] 4]. **1979** *Times* 12 Dec. 9/1 Oxford..won most of the rucks or mauls that mattered.

4. a. (Later example.)

1922 E. Wallace *Flying Fifty-Five* xii. 70 Its jockey had given up all attempt at winning and was content to finish with the ruck.

5. (Earlier example.)

1885 'Mark Twain' *Let.* 11 Sept. (1917) II. xxv. 460 Flowers and general ruck sent to him by Tom, Dick, and Harry from everywhere.

6. *attrib.*, as (sense *3 d (a)) *ruck ball*; (sense *3 d (b)) *ruck man, -rover.*

1976 *Scotsman* 24 Dec. 16/4 The ruck ball was promptly knocked on by a centre. **1963** *Sunday Mail* (Brisbane) 24 Feb. 24/5 Dr. Eric Laithwaite,..senior lecturer in electrical engineering at Manchester University,..looks more like an uncomplicated ruckman than an ingenious scientist. **1969** Eagleson & McKie *Terminol. Austral. Nat. Football* iii. 11 *Ruckman*, a member of the ruck whose function it is to take marks, and to knock the ball to the rover when the ball is thrown up or bounced by the umpire. **1977** *Age* (Melbourne) 18 Jan. 26/4 (*caption*) Fumbling is embarrassing anytime for a famous VFL ruckman and Jones covers his face in despair. **1967** *Australian* 24 Apr. 12 The brilliant Polly Farmer is now being used as a ruck-rover. **1969** Eagleson & McKie *Terminol. Austral. Nat. Football* iii. 12 *Ruck-rover*, a mobile member of the ruck who is usually intermediate in size between a follower and rover.

ruck (rʊk), *sb.*[5] *colloq.* [Perh. f. *RUCK *v.*[6] or shortening of Ruction or *RUCKUS.] A quarrel, a row.

1958 F. Norman *Bang to Rights* i. 15, I didn't feel like having a ruck about this. **1960** *Times* 15 Mar. 7/7 All that gun business is—silly isn't it. I mean they were only having a ruck to start with. **1963** T. & P. Morris *Pentonville* vi. 144 The prisoner said that he had 'had a bit of a ruck with the instructor over this'. **1964** *Listener* 31 Dec. 1055/2 Squaddies and Teds, personal rucks forgotten, are fleeing from a common enemy—the law. **1976** 'P. B. Yuill' *Hazell & Menacing Jester* vi. 66 'I heard him and her having a ruck about Nicholas, that's all.' 'What kind of a row?'

ruck (rʊk), *v.*[5] *slang.* [Of unknown origin.] *intr.* **a.** To inform *on* a criminal. **b.** To give information about a crime or a criminal. **c.** *gen.* To abandon, to repudiate a person. With *on*.

1884 *Daily News* 20 Sept. 2/2, I told the prisoner that I was not going to ruck on an old pal. **1889** *Session Paper Cent. Criminal Court, 1729–1913* CX. 871 He said 'Has Cleasby *rucked*? If he has, I will b-y well kill him when I come out'—ruck means telling. **1898** J. D. Brayshaw *Slum Silhouettes* 207 Yer won't tell Mo that I told yer—will yer? If he knew as I'd rucked on him, he'd kill me. **1903** A. M. Binstead *Pitcher in Paradise* iv. 96 Your worthy parent..isn't going to ruck on you in the golden autumn of his life, just because you were denied the keen commercial instincts that led him to make a pile! **1906** E. Pugh *Spoilers* viii. 92 'I don't care,' said Deuce, defiantly... 'I ain't goin' to ruck on Dad.'

ruck, *v.*[6], var. Rux *v.*

1936 G. Ingram *Muffled Man* i. 13 'Oh, all right,' sulked Sonny. 'You ain't going to "ruck" me, are you?' **1959** C. MacInnes *Absolute Beginners* i. 109, I saw I mustn't keep on rucking him, because, after all, this was a party. **1966** P. Willmott *Adolescent Boys* vi. 112 The governor of my place is horrible... He rucks you if you take more than ten minutes for a quarter of an hour's job.

rucked, *ppl. a.*[1] Add: Also with *up*: rumpled; caught up.

1944 J. D. Carr *Till Death do us Part* vi. 64 A light-haired young man..lying on a rucked-up sofa. **1964** D. Francis *Nerve* i. 7 Mr Brewer pulled down his unconscious wife's rucked-up skirt. **1980** C. Fremlin *With no Crying* iv. 20 She'd..straightened her rucked-up skirt.

rucked, *ppl. a.*[2] [cf. *RUCK *sb.*[1] 3 d (a).] Passed from a loose scrummage.

1976 *Wymondham & Attleborough Express* 3 Dec. 26/3 The youthful, fit students started in an attractive manner while Diss resisted with strong tackling and counter attacking from rucked possession.

rucking, *vbl. sb.*[2] The action of Ruck *v.*[2]

a **1915** in W. H. Chantrey *Theatre Accounts* (1915) 67 Druggets or crumb cloths where used must be secured so as to be in no way liable to rucking.

rucking, *vbl. sb.*[3] [cf. *RUCK *sb.*[1] 3 d (a).] Loose scrummaging.

1958 [see *loose scrummaging* s.v. *LOOSE *a.* 9]. **1963** *Times* 31 May 3/4 He told New Zealanders: 'I think we can learn much from your game—particularly your forwards' rucking and driving over the ball, which we are trying to practise.' **1966** *Sunday Times* 2 Oct. 20/6 Their captain, Matthews, set an example with his rucking and gained them some valuable balls.

rucking, *vbl. sb.*[4] *slang.* [f. *RUCK *v.*[6] + -ING[1].] A reprimand; a scolding, telling-off.

1958 F. Norman *Bang to Rights* i. 41 I'll have to give her a right rucking about that. **1974** T. Barling *Shooter Man* iii. 23 Ask him. It'll only cost you a few coppers and a rucking for calling him back. **1976** E. Dunphy *Only a Game?* v. 146 Perhaps all the rucking he was taking was getting through to him, and he started doing a little bit more.

rucksack. Delete ‖ and add pronunc. (rʊ·ksæk, ru·ksæk). Also **rucsac, rucsack,** 9 **rücksack.** (Earlier and later examples.)

1866 *Nature & Art* I. 192/2 We therefore confidently recommend a perusal of it to all those about to grasp the 'Alpen-stock', and shoulder the 'Rücksack'. **1882** W. A. Baillie-Grohman *Camps in Rockies* 411 '*Rücksack*', or *Stalker's Bag* is..for all sporting purposes a most useful article. **1932** *Pontings Catal.* Whitsun, Rucksacks made of a..rubber proofed twill material. **1955** *Times* 31 Aug. 6/5 She wore shorts and rode a man's bicycle, on the back of which was strapped a heavy rucksack and a spare wheel. **1969** W. H. Little in C. Cullingford *Man. Caving Techniques* iii. 33 A rucsac of a suitable kind and size will often be necessary to carry the caver's needs to the cave entrance. **1976** *Liverpool Echo* 7 Dec. 5/1 An electric drill, a sanding machine and two rucksacks worth a total of £110. **1978** *Vole* No. 7 29/1, I did manage to corner a walker in Dorset, whose rucsac nearly broke my arm when I tried to lift it.

ru·cksacked, *a.* [f. prec. + -ed[1].] Provided with or carrying a rucksack.

1909 H. G. Wells *Ann Veronica* xvi. 322 To walk beside him, dressed akin to him, rucksacked and companionable, was bliss in itself. **1973** A. Price *October Men* i. 8 Holidaying couples and rucksacked students.

ru·cksackful. [f. Rucksack + -ful.] As much as a rucksack will contain.

1971 C. Bonington *Annapurna South Face* xi. 126 Ian was therefore carrying up the entire load of fixed rope left by Nick and Martin, a rucksackful weighing around forty pounds.

ruckus (rʊ·kəs). orig. and chiefly *U.S.* Also **rucas, ruccus, rucus, rukus.** [cf. Ruction and Rumpus *sb.*] An uproar, a disturbance; a row, a quarrel; fuss, commotion. Also *attrib.*

The earliest examples, spelt with a single *c* or *k*, may possibly represent the variant usually spelt *rookus*.

1890 *Dialect Notes* I. 66 *Rucus* (rūkes): for *rumpus*. [Kentucky.] **1902** *Ibid.* II. 244 Rukus. **1909** 'O. Henry' *Roads of Destiny* xiii. 210 There shall be rucuses in Salvador..and the monkeys had better climb the tallest cocoanut trees. **1923** C. E. Mulford *Black Buttes* ii. 20 Them two bummers [sc. restless cattle] was raisin' more of a ruckus than usual to-night. **1934** *Sun* (Baltimore) 17 May 10/1 The ruccus in the City Hall over discharge of a municipal employé by the Mayor. **1948** F. Blake *Johnny Christmas* ii. 69 With this Kiowa-'Rapaho ruckus and these picture-book soldiers that just showed up, we don't want anything more on our hands. **1963** *Economist* 12 Oct. 147/2 The ruckus kicked up by the outraged wives and mothers of America. **1972** *Time* 10 July 38/1 But then ruckus raising is Fischer's speciality. **1977** *Times Lit. Suppl.* 1 July 792/1 World Team Tennis.. now actively encourages..'audience participation', a polite phrase that covers barracking, beer-cans, and the kind of ruckus that England normally only sees after a Cup Final. **1979** *Dædalus* Spring 162 Like the *graeculi* of the Roman Empire, we Europeans are still capable of raising a little cultural ruckus.

ruction. (Further examples.)

1878 A. Hume *Remarks Irish Dial.* 111 When a.. ruction has been 'riz'. **1900** F. P. Dunne *Mr. Dooley's Philos.* 24 That's life in America. 'Tis a gloryous big fight, a rough an' tumble fight, a Donnybrook fair three thousan' miles wide an' a ruction in ivry block. **1905** [see *PAVVY]. **1913** [see *RIPPIT]. **1921** E. O'Neill *Diff'rent* i, in *Emperor Jones* 218 That brown gal took an awful shine to Caleb and when she saw the ship was gittin' ready to sail she raised ructions,..howlin' and screamin' and beatin' her chest with her fists. **1943** *Sun* (Baltimore) 17 Nov. 14/1 As a result of this little ruction, Baltimore is freed..from the grip of a political coalition which boded no good for the city. **1964** D. Varaday *Gara-Yaka* xii. 103 The ructions of a clash between rival tribes.

rud, *sb.*[1] **3.** (Later *Comb.* example.)

1876 G. M. Hopkins *Poems* (1967) 177 The blood-gush blade-gash Flame-rash rudred..and dingle-a-danglèd Dandy-hung dainty head.

rudaceous (rudē·ʃəs), *a.* *Geol.* [f. L. *rūd-us* rubble + -aceous.] Of a rock: composed of larger grains than is an arenaceous rock.

1904 A. W. Grabau in *Amer. Geologist* XXXIII. 242 In the further subdivision of the clastic rocks, texture or size of grain takes precedence over chemical composition. .. We commonly recognize these sizes of grain, 1st that larger than what is commonly considered the normal sandgrain, 2d, the sand-grain, and 3d, the rock flour or impalpable powder. The first texture is most appropriately called rudaceous. **1920** —— *Gen. Geol.* xviii. 570 Rocks of all textures may be argillaceous, those of rubbly (rudaceous) texture and those of arenaceous texture generally carrying the clay as an admixture or as part of the cement. **1949** F. J. Pettijohn *Sedimentary Rocks* vii. 196 The rudaceous subtypes..are marked by characteristic compositional and textural features. **1977** A. Hallam *Planet Earth* 168 Most sedimentary rocks, classified as either detrital or chemical-organic, are also classified according to their grain-size as rudaceous rocks, arenaceous rocks or argillaceous rocks.

rudbeckia (rʊd-, rudbe·kiă). [mod.L. (Linnæus *Systema Naturæ* (1735)), f. the name of Olaf *Rudbeck* (1660–1740), Swedish botanist + -ia[1].] A perennial herb of the genus so called, belonging to the family Compositæ, native to North America, and bearing yellow or orange flowers with a prominent conical disc of dark florets in the centre of each one.

1759 P. Miller *Gardeners Dict.* (ed. 7) s.v. Rudbeckia with oval, Spear-shaped, undivided Leaves, placed alternate. **1789** W. Aiton *Hortus Kewensis* III. 250 Broad jagged-leav'd Rudbeckia. Nat[ive] of Virginia and Canada. **1821** *Bot. Reg.* VIII. tab. 525 (*heading*) Eight-rayed Rudbeckia. **1870** W. Robinson *Wild Garden* ii. 86 Newman's Rudbeckia... A very showy vigorous plant. **1908** G. Jekyll *Colour in Flower Garden* ix. 79 The fine double Rudbeckia called Golden Glow is treated in the same way. **1931** *Daily Mirror* 27 Aug. 7/3 Among the best of autumn-flowering perennials for the mixed border, or the wild garden, are the rudbeckias. **1962** *Amateur Gardening* 17 Feb. 5 A flower which makes for a bold display in the garden and as a cut flower is the rudbeckia. **1974** C. Milne *Enchanted Places* vi. 49 The penstemons, the bergamots,..the rudbeckias, the dahlias ..were still looking as lovely as ever.

rudd[1]**.** For *Leuciscus* read *Scardinius*. (Earlier and later examples.)

1526 in 'Antiquary' *Forme of Cury* (1780) 177 Fresh Sammon... Great Ruds... Baken Turbuts. **1925** J. T. Jenkins *Fishes Brit. Isles* 297 The Rudd is usually deeper in the body than the Roach. **1966** *Studia Neophilologica* XXXVIII. 130 The English name of the rudd obviously refers to the red colour of the lower fins of the fish.

Rudd[2] (rʊd). [Perh. f. the name of Margaret Caroline *Rudd* (d. 1779), a notorious courtesan, for whom the table may have been invented.] Used *attrib.* and in the possessive in *Rudd('s) table*, an elaborately appointed lady's toilet table of the late eighteenth century.

1788 in R. Fastnedge *Shearer Furnit. Designs from Cabinet-Makers' London Bk. of Prices* (1962) 14 A three foot 4 inch, rudd table, all solid, with astragal, or 2 beads, and hollow round the edge of the top, the 2 outside drawers with no quadrant boxes, a glass hung to each drawer, supported by quadrants.., plain Marlbro' feet, and an astragal round the bottom of the frame. **1793** *Cabinet-Makers' London Bk. of Prices* (ed. 2) 161 *A Rudd, or Lady's Dressing Table*..Three feet four inches long, two feet wide, three drawers in front, a glass frame hing'd to each end drawer, and supported by quadrants, a moulding on the edge of the top, plain Marlbro' legs, and an astragal round the bottom of the frame. **1892** F. Litchfield *Illustr. Hist. Furnit.* vii. 186 The names given to some of these designs [in Hepplewhite's *Guide*] appear curious; for instance: 'Rudd's table or reflecting dressing table,' so called from the first one having been invented for a popular character of that time. **1902** W. H. Hackett *Decorative Furnit. 16th, 17th & 18th Cent.* x. 124 About the year 1788, Shearer also published.. 'Household Furniture'... The nineteen plates consist of designs for secretaries, bookcases, bureaux..and what was known as a Rudd or lady's dressing table. **1970** [see *lobby chest* s.v. *LOBBY *sb.* 4].

rudder, *sb.* **2. a.** Delete note in small type, and add: **d.** An analogous flat movable structure used for controlling the motion of an aircraft; now usu. a vertical flap, hinged at its leading edge, forming part of the tailplane of an aeroplane.

The 'boat' in quot. 1804 is the gondola of a balloon.

1804 G. Cayley in J. L. Pritchard *Sir G. Cayley* (1961) 220 Fixed upon a universal joint a Rudder of considerable length opposing both an horizontal and vertical surface.. intersecting each other in right angles to the air. A handle to direct this Rudder must communicate with the Boat. **1843** *Mechanics' Mag.* XXXVIII. 278 The broad horizontal rudder, or tail, H, capable of being turned on its hinge to any angle, at pleasure, gives the power of ascent and descent when the propellers are used, and forms also the chief means of stability in the path of the flight. The small vertical rudder I, is for the purpose of lateral steerage. **1879** *Encycl. Brit.* IX. 321/1 M. Pénaud succeeded in overcoming the difficulty in question by the invention of what he designates his automatic rudder. This consists of a small elastic aero-plane placed aft or behind the principal aero-plane which is also elastic. **1910** R. Ferris *How it Flies* vi. 116 The rudder for steering to left or right is mounted at the extreme rear end of the body. **1966** D. Stinton *Anat. Aeroplane* viii. 163 Aerobatic aeroplanes usually have a large portion of the fin surface lying ahead of the tailplane, or a large portion of the fin and rudder lying behind its trailing edge. **1969** K. Munson *Pioneer Aircraft 1903–14* 9 The operator lies prone on the lower surface, his hips resting in the cradle, and his hands grasping the roller, D_1, which actuates the front rudder, D.

e. Use or turning of the rudder, the extent to which the rudder is turned.

1918 W. G. McMinnies *Pract. Flying* 218 If you are turning to the right and notice wind striking your left cheek, you are side-slipping outwards, so give a little more bank or take off some rudder. **1936** W. H. McCormick *Mod. Bk. Aeroplanes* x. 86 In order to turn an aeroplane to the right, right rudder is put on by moving the right-hand end of the rudder bar gently forward by means of the right foot. **1958** 'N. Shute' *Rainbow & Rose* ii. 49 She needed quite a bit of rudder. **1978** J. S. Evans *Pilot's Manual* iv. 151 Let it be assumed that our aeroplane is in cruising flight, with sufficient rudder applied to prevent propwash-induced yaw.

5. b. The tail of an otter.

[**1903** H. Johnston *Brit. Mammals* vii. 138 The otter.. swims and dives with great facility, and propels itself with all four limbs, using the tail as a great rudder.] **1907** *Yesterday's Shopping* (1969) 694 Horses' hoofs, deer slots, fox, hare, or otter pads cured and mounted in various styles; also fox brushes and otter rudders. **1941** H. Cory *Mammals Brit. Isles* 34 When swimming submerged the animal uses the forefeet for paddling and the hind feet, assisted by the rudder, for steering. **1965** P. Wayre *Wind in Reeds* xi. 151 Canadian otters are larger and bulkier [than British ones] with broader and deeper heads and thicker rudders. **1976** *Scotsman* 24 Dec. (Weekend Suppl.) 2/2 He [sc. an otter] was coiled..with his jaws clenched, and his rudder curled round the webs of his hindfeet.

6. a. *rudder pedal, post*; **rudder-bar**, a bar operated by the pilot's feet which controls the position of an aircraft's rudder.

1912 *Q. Rev.* July 240 Machines like Dunne's, which have no separate rudder-bar allowing of foot-correction. **1918** W. G. McMinnies *Pract. Flying* 234 *Rudder post*, the upright member to which the rudder is hinged. **1919** Pippard & Pritchard *Aeroplane Struct.* v. 36 In some aeroplanes..the rudder bar is replaced by pedals and directional control is obtained by pressing the appropriate pedal. **1935** C. G. Burge *Compl. Bk. Aviation* 537 *Rudder post*, the main vertical member of a rudder to which the rudder hinges are attached. **1959** Rudder-bar [see *'N, 'N']. **1966** D. Stinton *Anat. Aeroplane* viii. 139 If hinge-moments are too high to be handled efficiently, artificial forces may be transmitted through the stick and rudder-pedals by an artificial feel-system. **1976** B. Jackson *Flameout* iv. 63 The flight data recorder..tape-recorded..the pilot's movements of the control yokes and rudder pedals, [etc.].

b. *rudder-fan, flutter, power.*

1915 S. H. Carden in M. Gilbert *Winston S. Churchill* (1972) III. Compan. i. 625 Large rudder power makes her sufficiently handy although starboard engines out of action. **1928** *Daily Tel.* 13 Mar. 11/3 The machine dived into the sea from a height of about 100 ft. after developing what appeared to be rudder flutter. **1930** J. S. Huxley *Bird-Watching* vi. 102 They became birds through the evolution of feathers out of scales... The other peculiarities of modern birds, such as..the transformation of their originally long and awkward tail, like a kite's, into an efficient rudder-fan..came later.

rudder, *v.* (In Dict. s.v. Rudder *sb.*) Delete *rare* and add: *trans.* and *intr.* Also, to use the rudder. Also *fig.* (Further examples.)

1940 'N. Shute' *Landfall* 210 He glanced quickly at the cruiser to check the direction, ruddering slightly to maintain his course. **1942** *Tee Emm* (Air Ministry) II. 65 Do a quick barrel half roll..pulling the stick back.. when you are on your side, and then rudder into a steep dive. **1952** M. Tripp *Faith is Windsock* ii. 35 Hamish pulled back on the throttles, strangling life from the engines; Bergen ruddered and braked hard. **1954** W. Faulkner *Fable* 107 He..ruddered the tracer right onto it. **1960** S. Plath *Colossus* 66 A pigeon rudders down. **1973** M. Amis *Rachel Papers* 71 That afternoon,.. ruddered by perceptive questions, encouraging smiles and apt generalizations from myself, Rachel Noyes told the story of her life.

ruddering (rʊ·dəriŋ), *a.* [f. Rudder *sb.* + -ing[2].] That acts as a rudder; that guides or steers. Cf. *RUDDER *sb.* 5 b.

1960 T. Hughes *Lupercal* 46 With webbed feet and long ruddering tail. **1960** R. W. Marks *Dymaxion World of B. Fuller* 29/2 As with the pulled (rather than pushed) wheelbarrow, the ruddering tail wheel was lifted over, rather than shoved into the traveled terrain.

rudderless, *a.* **b.** (Later examples.)

1887 W. B. Yeats *Let.* 11 Mar. (1954) 32 Please excuse this somewhat rudderless scrawl. **1977** *Oxf. Mission Q. Paper* Jan.–Mar. 15 Young folk, often rudderless in their religious thinking and experience of life.

ruddervator (rʊ·dəɹvēitəɹ). *Aeronaut.* [f. Rudder *sb.* + Ele)vator.] A control surface designed to act as both rudder and elevator.

1962 *Flight Internat.* LXXXI. 172/1 The ruddervators are controlled from a control column horizontally mounted under the right side of the couch. **1966** D. Stinton *Anat. Aeroplane* 244 Flaps, ailerons, and 'ruddervators' were designed to incorporate the minimum number of ribs.

ruddle, *v.* Add: Also *absol.*

1960 S. Plath *Colossus* 52 Imagine their deep hunger, deep as the dark For the blood-heat that would ruddle or reclaim.

ruddy, *a.* (*sb.*) and *adv.* Add: **A.** *adj.* **3.** *ruddy shelduck, turnstone.*

1852 Ruddy shelduck [see Sheld-duck]. **1954** J. Delacour *Waterfowl of World* I. 250 The Ruddy Shelduck is a strong and successful species which..occupies a very large range. **1909** J. & M. Macoun *Catal. Canad. Birds* (ed. 2) 212 Ruddy Turnstone... This species is a common migrant in Newfoundland. **1938** P. A. Taverner *Birds of Canada* 185 The turnstone is represented in America by the Ruddy Turnstone.., rather smaller than the European form. **1972** S. Burnford *One Woman's Arctic* iv. 92 Only about fifty yards away, were the.. nests of two pairs of ruddy turnstones.

3*. Orig. a euphemistic substitution for Bloody *a.* 10; now freq. used as an intensive in its own right: damnable, blasted, confounded. *colloq.*

1916 'Taffrail' *Pincher Martin* ii. 28 Go on, Ginger!.. Slosh 'im one on the ruddy boko! **1924** Galsworthy *White Monkey* ii. i. 121 Only why didn't Mr. Elderson say: 'You ruddy liar!'? **1945** [see *EUSTON ROAD]. **1968** [see *GIVE *v.* 16 c]. **1969** I. Kemp *Brit. G.I. in Vietnam* ii. 31 'Oh no!' I thought. 'The ruddy thing won't have time to open before I hit.' **1977** *Radio Times* 12–18 Nov. 69/4, I carted my ruddy topee all over India and finally abandoned it under a bed in Fort William in Calcutta.

5. a. *ruddy-bodied, -clustered, -faced* (examples), *-finned, -muzzled.*

1916 D. H. Lawrence *Twilight in Italy* 89 The many ruddy-clustered oranges beside the path remind me of the lights of a village. **1916** Blunden *Harbingers* 34 Ruddy-finned roach and bronze carp swam. **1922** Joyce *Ulysses* 291 The figure seated on a large boulder at the foot of a round tower was that of a broadshouldered..ruddyfaced sinewyarmed hero. **1923** D. H. Lawrence *Birds, Beasts & Flowers* 62 Cyclamens, ruddy-muzzled cyclamens. **1960** S. Plath *Colossus* 79 Bronze dead dominate the floor, Resistive, ruddy-bodied.

b. *ruddy-dark, -golden, -orange.*

1927 D. H. Lawrence *Mornings in Mexico* 35 Calico pantaloons round his ruddy-dark waist. *Ibid.* 126 The men are naked to the waist, and ruddy-golden. *a* **1930** —— *Last Poems* (1932) 266 Green moonlight And ruddy-orange limbs stirring the limbo Of the unknown air.

B. *adv.* Used, usu. preceding an adj., as an intensifier (orig. a euphemistic substitute for Bloody *adv.* 2: cf. sense A. 3* above): confoundedly, damnably, damned.

1914 C. Beresford *Mem.* I. xiii. 119 All I've got to say, is to say you've got a ruddy good billet. **1922** Joyce *Ulysses* 420 Lay you two to one Jenatzy licks him ruddy well hollow. **1933** M. Lincoln *Oh! Definitely* iii. 23 I'd have ruddy well..locked the door. **1959** M. Gilbert *Blood & Judgement* xiii. 139 Culver Street's been pulled down and a ruddy great block of flats put up. **1979** *Oxford Times* 28 Dec. 9/2 Most of the groups I heard there and elsewhere played too ruddy loud.

ruddy, *v.* **2.** (Later example.)

1938 W. de la Mare *Memory* 49 See, how the sun Ruddies through his filmy grey, Turns to light the dreaming one.

rude, *a.* and *adv.* Add: **A.** *adj.* **I. 5. e.** Fig. phr. *rude awakening*, a severe disillusionment or arousal from complacency.

1895 G. Allen *Woman who Did* vi. 71 Alan was often quite alarmed in his soul when he thought of the rude awakening that no doubt awaited her. **1912** T. Dreiser *Financier* v. 47 Life had given him no severe shocks nor rude awakenings. **1971** *Daily Tel.* 12 June 18/1 Anyone visiting this year's Grosvenor House Antique Fair.. with the object of buying antiques on the cheap is in for a rude awakening. **1975** Shea & Wilson *Golden Apple* iv. 248 Then comes the rude awakening: food riots, industrial stagnation, a reign of lawless looting and plunder.

II. 15. rude boy, one of a class of unemployed black youths inhabiting the poorer areas of Jamaica and typically seen as indolent and apt to commit petty crimes.

1967 *Caribbean Q.* Sept. 39 Rude bwoy is that person, native, who is totally disenchanted with the ruling system; who generally is descended from the 'African' elements in the lower class... Rude bwoys are largely centred in those urban areas that suffer from chronic depression. **1975** *Globe & Mail* (Toronto) 11 June 3/1 The rude boys, rudies or just plain rudes are the street corner toughs, hustlers, petty thieves and dealers in ganja (marijuana). **1976** D. Hebdige in Hall & Jefferson *Resistance through Rituals* 152 The exotica of Rastafarianism provided distractive screens behind which the rude boy culture could pursue its own devious devices unhindered and unseen. **1977** Logan & Woffinden *New Musical Express Bk. of Rock* 414 The rude boys (outlaws) of Jamaica's shanty towns began to move into the studios, celebrating their own chosen lifestyle, which resulted in a spate of rudeboy records.

B. *adv.* **a.** (Later examples.)

1795 J. Woodforde *Diary* 8 Sept. (1929) IV. 226 Jane behaved quite rude this Evening. **1885** G. M. Hopkins *Poems* (1967) 99 But ah, but O thou terrible, why wouldst thou rude on me Thy wring-world right foot rock?

b. *rude-masoned, -ripened, -rounded* adjs.

a **1889** G. M. Hopkins *Poems* (1967) 185 Who built these walls made known The music of his mind, Yet here he had but shewn His ruder-rounded rind. **1928** Blunden *Japanese Garland* 20 Over the rude-ripened vale. **1930** —— *Poems* 128 There is a sluice through whose rude-masoned stones And fissured planks our timid river falls.

rude (rūd), *sb.* *colloq.* [f. prec.] **a.** An impolite or unsophisticated person. **b.** = *rude boy* s.v. *RUDE *a.* 15.

1961 J. Dawson *Ha-Ha* iv. 74 No Brains' Trust will work so long as you've always got to have a gaggle of rudes and silly old sages to balance the bright young men. **1975** [see *rude boy* s.v. *RUDE *a.* 15].

ruderal, *a.* Add: (Further examples.) Also as *sb.*, a ruderal plant.

1905 F. E. Clements *Research Methods Ecol.* iv. 263 The pioneers in impoverished or exhausted fields are uniformly ruderal plants. **1929** J. W. Bews *World's Grasses* vi. 226 Most of the species of *Digitaria* tend to become ruderals. **1963** *New Scientist* 20 June 677/2 The more permanent colonisers [on Lake Kariba, East Africa] appear to fall into two groups, those of semi-aquatic habitats,..and ruderals of open ground. **1970** *Watsonia* VIII. 175 Weeds or ruderal plants. **1979** *Nature* 20–27 Dec. 780/2 An ecological classification of plants into competitors, stress-tolerators and ruderals.

rudery (rū·dəri). [f. Rude *a.* and *adv.* + -ery.] Rudeness; rude or impolite speech or behaviour; a rude remark, comment, practical joke, etc.

1932 R. Ackland *Strange Orchestra* ii. 71 Just been having a bit of rudery, dear. **1933** Dylan Thomas *Let.* 25 Dec. (1966) 77, I have..been averse to including such obvious rudery in my letters to you. **1940** 'N. Blake' *Malice in Wonderland* ii. xvi. 235 What's this? Some new rudery taken place? **1960** S. W. C. Pack *Admiral Lord Anson* i. 3 Although a superficial courtesy existed, coarseness and rudery were common. **1968** *Blackw. Mag.* CCCIII. 454/1 The lane twisted and turned, and small Sikh boys kept pace with us, their long hair tied in top-knots, chanting songs in Punjabi which I trust were not ruderies. **1979** J. Smyth *Milestones* xviii. 230 Then suddenly he would uncoil, spring to the despatch box and reply pungently to some rudery which had appeared to float over his head.

Rudie (rū·di). *Jamaica.* Also **rudie.** [f. Rude *a.* and *adv.* + -ie.] = *rude boy* s.v. *RUDE *a.* 15.

1967 *Caribbean Q.* Sept. 41 The number of rudie tunes on the air-waves reflects the increased status accorded Rudies by this other Afro-Jamaican society. **1974** *Howard Jrnl.* XIV. 48 This theme is a strong one in West Indian culture and is reinforced by the image of the rudie in Reggae—the super cool hooligan who always come[s] out on top. **1976** *Daily Mirror* 2 Apr. 21/2 Unemployment was, and is, constantly high, with the Rudies being the main sufferers. Their problems became a theme of ska. **1977** *Westindian World* 3–9 June 13/4 'Steppin Razor'..is a little reminiscent of the rudie tunes of the sixties.

rudiment, *sb.* Add: **1. d.** Pl. (With capital initial.) The name of the lowest class in certain Roman Catholic schools and colleges, freq. divided into the 'third', 'second', and 'first' class (of) Rudiments. Cf. *FIGURE *sb.* 22 b.

1716 [see *FIGURE *sb.* 22 b]. *a* **1799** in C. Butler *Acct. Life Alban Butler* (1799) 6 The year after Mr. Alban Butler's arrival at Douay, I was placed in the same school, under the same master, he being in the first class of rudiments, as it is there called, and I in the lowest. **1846** in *Stonyhurst Mag.* (1933) Dec. 415/2 July 25th Sun. Themes judged Rhet... 29 th. Themes judged Rudiments. **1885** J. Gillow *Lit. & Biogr. Hist. Eng. Catholics* II. 553 At the period of his liberation Robert Gradwell was in second-class Rudiments. **1893** B. Ward *Hist. St. Edmund's Coll.* iv. 58 The two classes of 'Figures' were changed very shortly after this into three classes of 'Rudiments', and this term has survived at St. Edmund's to the present day. **1912** B. Ward *Eve Catholic Emanc.* III. xxxiv. 2 He was a boy in the 'Second of Rudiments' [*Note*] Equivalent to the Second or Third form at an English school. **1936** M. Trappes-Lomax *Bishop Challoner* i. 5 The 'classes', or forms, were named Figures or Rudiments, Grammar, Syntax, Poetry, and Rhetoric, names which originally were related to the work of the class... This nomenclature is still retained in some of the English Catholic schools. **1972** *Publ. Catholic Rec. Soc.* LXIII. 142 The vast majority of students began their course..in one or other of the Rudiments classes.

rudist (rū·dist). Also **Rudista.** [a. mod.L. family name *Rudista* (J. E. Gray in *Synopsis Contents Brit. Mus.* (ed. 21, 1823) 62), f. L. *rudis* unformed + -*t*- + -A 4.] A fossil pelecypod bivalve mollusc belonging to the superfamily Rudistacea, which included cone-shaped reef-forming animals. Also *attrib.* Also **Ru-, rudi·stid** [-ID³], in the same sense.
1889 NICHOLSON & LYDEKKER *Man. Palæont.* (ed. 3) I. xxxvi. 734 The *Rudistæ* are..entirely extinct. 1890 *Mem. Boston Soc. Nat. Hist.* IV. 322/1 The Rudistæ are conical or cup-shaped Pelecypods with a superficially marked radial symmetry. 1935 TWENHOFEL & SHROCK *Invertebr. Paleontol.* ix. 403 The various Rudistids are especially noteworthy. 1959 *New Scientist* 1 Jan. 16/1 A good collection of fossil reef-forming corals, rudistids, and related atoll fauna. 1969 BENNISON & WRIGHT *Geol. Hist. Brit. Isles* xiv. 332 The coral-like Rudists are important, forming reef environments in Tethys. 1978 *Nature* 26 Oct. 779/2 Future editions could be improved by the inclusion of a few famous foreign examples such as the Burgess Shale fauna and a rudist reef association.

rudite (rū·dəit). *Geol.* Also **rudyte.** [f. L. *rūd-us* broken stone, rubble + -ITE¹.] Any consolidated breccia or conglomerate consisting of particles larger than sand grains; = PSEPHITE.
1904 A. W. GRABAU in *Amer. Geologist* XXXIII. 242 The consolidated rock whether conglomerate or breccia may be called a rudyte. 1920 —— *Gen. Geol.* xviii. 569 Three textural types of rock may be recognized: (1) the rubble-rock or rubble-stone, or rudyte, which when the fragments are rounded is a conglomerate and when angular a breccia; (2) the sand-rock or sandstone or arenyte; and (3) the mud-rock or mud-stone or lutyte. 1935 *Bull. Nat. Res. Council* (U.S.) XCVIII. 239 Rudite. This is a general term..for fragmental sedimentary rocks coarser than sand grains. 1959 W. W. MOORHOUSE *Study of Rocks in Thin Section* xviii. 334 The clastic sediments are classified according to size as rudytes (rudaceous), which are conglomerates, arenytes (arenaceous)..and lutytes (lutaceous). 1966 B. SIMPSON *Rocks & Minerals* xvii. 174 The quartz-rich rudites are quartz conglomerates and quartz breccias. 1971 I. G. GASS et al. *Understanding Earth* i. 31/1 The rudites are represented by such rock types as conglomerate.

rue, *sb.*¹ For 'Now *dial.* or *arch.*' read 'Chiefly *dial.* or *arch.*' and add: **1. a.** (Later examples.)
1959 *Listener* 31 Dec. 1174/2 The nature and the mixture of the ingredients in the poetry—nostalgia, bathos, irony, rue, and religious fervour. 1976 *Publishers Weekly* 8 Mar. 64/2 A mix of rue and wit that is vintage [Mort] Sahl.

ruelle. Add: **3.** In France, a small street; a lane, alley.
1908 T. E. LAWRENCE *Let.* 9 Aug. (1938) 59 Streets—mostly stairs, irregular and broken, running under archways and tunnels... Cover these ruelles with grass, heap them with refuse. 1911 O. ONIONS *Widdershins* vii. 242 He took us back along a plantain-groved street, and suddenly turned up an alley... It was a dilapidated, deserted *ruelle*.

Rueping process (rü·piŋ). Also Rüping, (*erron.*) Ruping. [Named after Max *Rüping* (fl. 1902), German timber engineer, its inventor.] An economical method of preserving wood by applying creosote to it after subjecting it to high air pressure so that the cells are permeated but not filled.
1904 S. M. ROWE *Handbk. Timber Preservation* (rev. ed.) 129 In the new Ruping process the seasoned wood is for some time (from about a half hour to an hour) exposed to a pressure of 5 atmospheres in the boiler..so that all the cells must be filled with air. 1917 A. J. WALLIS-TAYLER *Preservation of Wood* vii. 200 The Rueping Process has been patented in Great Britain... It was primarily devised with the object of reducing the cost of creosoting by preventing the heavy loss occasioned by dripping when the wood is treated by the ordinary process. 1930 H. FERGUSSON in H. Boulton *Century of Wood-Preserving* 66 The British Post Office for some little time has had all its poles done by the Rueping process. 1968 [see *empty-cell* s.v. *EMPTY *a.* and *sb.* C.].

Rufai (rufā·,i). Also † Rifa'ee, † Rufaee, Rufa'i. Pl. as sing. or -s. [Turk. *Rufai*, ad. Arab. *rifāʿī*, f. the name of Aḥmad al-*Rifāʿī* (d. 1183), the founder of this order.] A howling dervish (see quot. 1877 and DERVISH), one of an order of Muslim friars pledged to poverty and self-mortification.
1832 G. A. HERKLOTS tr. *Jaffur Shurreef's Qanoon-e-Islam* xxviii. 291 *Rufaee* or *Goorz-mar*, they originate from Syed Ahmud Kubeer, whose *fuqeers* strike the point of the *goorz* against their breasts, or into their eyes, level blows at their backs with the sword, thrust a spit through their sides, or into their eyes. 1836 E. W. LANE *Acct. Manners & Customs Mod. Egyptians* I. x. 310 The Rifa'ee durwee'shes are celebrated for the performance of many wonderful feats. 1868 J. P. BROWN *Dervishes* ii. 51 *Ahmed Sa'eed Rufâ'ee* was the founder of the Order of the *Rufâ'ees*, generally known among European travellers as the 'Howling Dervishes', from their peculiar mode of worship. 1877 *Encycl. Brit.* VII. 114/1 This leads to the Devr, or rotation, in which the Rufai, or Howling Dervishes, stand in a circle, shoulder to shoulder, each on his right foot, and swaying the body and the left leg backwards and forwards or from side to side. 1885 T. P. HUGHES *Dict. Islam* 120/1 Some..wear tall caps called *kulāhs*, made also of felt; and others, such as the Rufā'īs, use short caps called Tāqīyah, to which is added a coarse cloth. 1900 'ODYSSEUS' *Turkey in Europe* v. 192 There are several orders..the most important being the Bektashis, the Mevlevis, and the Rufais (the two latter commonly known to Europeans as dancing and howling dervishes). 1928 W. B. SEABROOK *Adventures in Arabia* xiii. 251 The objective of our pilgrimage was a monastery of the Rufai, or Howling Dervishes—a sect fundamentally different from the Whirling Melewi—in the mountains between Hama and Aleppo. 1965 *Encycl. Islam* (rev. ed.) II. 164/2 The Sa'dis, Rifā'īs and Aḥmadīs have particular feats, peculiar to each *tarīḳa*, of eating glowing embers and live serpents or scorpions and glass, of passing needles through their bodies and spikes into their eyes.

ruff, *sb.*³ Add: **3. b.** *Comb.* **ruff and discard** *Bridge*, an opportunity (usu. for declarer) to win a trick by ruffing in one hand while discarding a loser from the other.
1939 N. DE V. HART *Bridge Players' Bedside Bk.* xxxix. 120 If West leads a Heart, then declarer gets a ruff-and-discard which enables him to ruff the third round of either minor suit. 1972 R. MARKUS *Common-Sense Bridge* iii. 102 A ruff and discard disposed of a losing club. 1977 *Bridge Mag.* July 34/2 Aunt Agatha now played the king of clubs, giving West the option of conceding a ruff and discard or leading away from his king of spades.

ruff, *sb.*⁵ **1.** (Later examples.)
1927 *Melody Maker* Aug. 804/3 All the various beats used in military drumming. The stroke-and-drag paradiddle and the four-stroke ruff, for instance, are not essential. 1957 A. A. SHIVAS *Art of Tympanist & Drummer* i. 30 *Embellishments.* These are very important and much used in side drumming... The three important ones are the flam, the drag and the ruff.

Ruff (rʊf), *sb.*¹⁰ [f. the name of W. *Ruff*, founder editor of *Guide to the Turf or Pocket racing Companion* (1842–53), a twice-yearly compendium of horse-racing information, subsequently published as *Ruff's Guide to the Turf* and since 1869 incorporating various other racing periodicals.] A colloq. abbreviation of *Ruff's Guide to the Turf*.
1854 *Sporting Rev.* Jan. 63 The racing world have in Ruff and the Book Calendar two very competent expounders of the 'forms' in which thorough-breds have 'gone' this season. 1902 in Farmer & Henley *Slang* VI. I. 70/1. 1918 G. FRANKAU *One of Them* xxi. 159 Weep for a Shrunken Ruff, a tipless tout.

ruff, *v.*² Add: **2. a.** (Earlier example of present spelling.)
1813 *Hoyle's Games of Whist & Quadrille* 50 *Ruff*, and *over-ruff*, to trump a suit led, second or third hand.
b. Const. *out*. To defeat (a card, etc.) by ruffing, so as to establish master cards in the suit led; also, with suit as object.
1927 M. C. WORK *Contract Bridge* 141 *Ruffing out*, trumping the low cards of a suit before playing its high cards. 1939 N. DE V. HART *Bridge Players' Bedside Bk.* 22, I laid down dummy's Ace and King of Hearts, and then tried to ruff out the suit. 1960 T. REESE *Play Bridge with Reese* 23 If the spades are breaking as well, I can play Ace and another spade, ruff out the King of clubs and enter dummy to make two long clubs. 1967 P. ANDERTON *Play Bridge* xi. 88 South covers with his A. .. and West trumps. This procedure is known as ruffing out a suit and East now holds the master cards in the Spade suit. 1972 *Times* 20 May 13/3 She ruffed out the clubs ruffing the third round with the ♡10.

ruffer¹. For *rare*—⁰ and (See quot.) in Dict. read: **a.** = *ruffing trick* s.v. *RUFFING *vbl. sb.*¹ **b.** A card that ruffs or trumps another (see also quot. 1611 in Dict.).
1936 E. CULBERTSON *Contract Bridge Compl.* xxxix. 435 A trick made by ruffing a losing card with an otherwise worthless card of the trump suit is called a ruffer. 1974 *Country Life* 28 Feb. 453/3 This not only deprives South of a diamond ruff, but the trump trick comes back if South uses dummy's Ten as a ruffer. 1975 *Ibid.* 30 Jan. 289/3 The right way is to use his own trumps as ruffers.

ruffianing, *vbl. sb.* **2.** (Earlier example.)
1829 P. EGAN *Boxiana* 2nd Ser. II. 446 Ruffianing was all *the go* at Moulsey Hurst so slippery, Till Science took the cause in hand.

ruffianly, *a.* **1.** (Later *Comb.* example.)
1882 W. D. HAY *Brighter Britain!* I. xi. 301 A line of nine ruffianly-looking scarecrows, under review by.. head-master of the ceremonies.

ruffing, *vbl. sb.*¹ Add: Also *attrib.*, as *ruffing element, trick, value*.
1930 E. CULBERTSON *Contract Bridge Blue Bk.* iv. 60 The..characteristic of the Supporting Hand lies in the use of ruffing tricks. 1936 'LUCIAN' *Straight Bid* xii. 126 In actual play the result is materially affected by.. ruffing values, etc. 1950 *Bridge from 'Times'* ii. 63 This is a pretty example of the use of the long trump hand to make ruffing tricks. 1952 I. MACLEOD *Bridge* xii. 143 The introduction of the ruffing element, and the problems of trump management make the play in a suit contract far more difficult and intricate. 1971 *Country Life* 22 Apr. 973/1 The ruffing element which obtains in a suit contract. 1977 *Bridge Mag.* Sept. 153/2 His ruffing values being distinctly unimpressive he quickly converted to seven no-trumps.

ruffing, *ppl. a.*² [f. RUFF *v.*¹] That forms or rises in ruffs.
1865 R. D. BLACKMORE in *Macm. Mag.* XII. 33/1 The blackcocks..swell their ruffing breasts, and crow for their rivals to spar with them.

Ruffini (rufni·i). *Anat.* The name of Angelo *Ruffini* (1864–1929), Italian anatomist, used *attrib.* and with *of* to designate certain dermal sensory organs.
1900 HUBER & DEWITT in *Jrnl. Compar. Neurol.* X. 175 The author [*sc.* Ruffini] thus distinguishes this spindle, which may bear his name, from the neuro-tendinous end-organs of Golgi... The Ruffini organ is composed of connective and elastic tissue. 1928 *Amer. Jrnl. Psychol.* XL. 357 No Golgi-Mazzoni, Krause or Ruffini end-organs were discovered in the hairy parts of the human skin. 1974 D. & M. WEBSTER *Compar. Vertebr. Morphol.* x. 200 On the other hand, the end bulbs of Ruffini—flattened, bulblike endings within a fine, connective tissue network—respond to temperature increases. 1981 A. BRODAL *Neurol Anat.* (ed. 3) ii. 51 Another type of slowly adapting receptor..has a resting discharge that increases when a mechanical stimulus is applied to the skin. These receptors..are found in the dermis and appear to be Ruffini endings.

ruffle, *sb.*¹ **II. 7. b.** Substitute for def.: *pl.* Handcuffs. *slang* (? *Obs.*). For **1796** *Grose's Dict. Vulgar T.* (ed. 3) read **1785** GROSE *Dict. Vulgar T.* and add examples.
1839 W. H. AINSWORTH *Jack Sheppard* I. II. ix. 305 'I'll accommodate you with a pair of ruffles.' And he proceeded to handcuff his captive. 1840 H. COCKTON *Life Valentine Vox* xiv. 109 'Sam! here, where are the ruffles?' and the fellow addressed instantly produced a pair of handcuffs. 1912 A. H. LEWIS *Apaches of N.Y.* viii. 179 Outside they found Cohen..with the ruffles on the Ghost.
8. *attrib.* and *Comb.*, as *ruffle collar, lace*; *ruffle cuffed, headed* adjs.; **ruffle shirt** *N. Amer.*, (*a*) a shirt decorated with ruffles; (*b*) *transf.*, an aristocrat, a person of means; hence **ruffle-shirted** *a.*, **ruffle-shirter.**
1968 J. IRONSIDE *Fashion Alphabet* 52 *Ruffle collar*, a collar cut on the bias or circular so that it falls in a fluted ruffle round the neck. 1880 'MARK TWAIN' *Tramp Abroad* 399, I did not get back the same drawers I sent down... I got a pair on a new plan. They were merely a pair of white ruffle-cuffed absurdities. 1927 W. DEEPING *Kitty* xxix. 370 The ruffle-headed pianist bumping up and down on his chair. 1682 J. PINNEY *Let.* 4 Sept. (1939) 15 The remnant you sent downe shall speadily be cut & sent & a ruffel lace of 2 nailes broad. 1830 *Amer. Sentinel* (Philadelphia) 27 Aug. 2/2 Where a dinner is to be got up, a few mechanics are procured to take the first rank, and the ruffle shirts fall into the rear. 1831 *American* (Harrodsburg, Kentucky) 22 July 3/1 General Jackson and his friends are lessening the burthens of the people by.. placing the Tax, on Wines, Rum,..and fine cloth such as the *Rufle* [*sic*] *shirt* gentry wear. 1838 B. DRAKE *Tales & Sk.* 64 The colonists presented, indeed, a curiously grotesque appearance, loitering about the station in ruffle shirts and coon-skin caps. 1840 J. P. KENNEDY *Quodlibet* xii. 158 If he does get on with his business, and makes a little fortune, we can call him a..Ruffle Shirt. 1848 *Knickerbocker* XVIII. 520 It was asserted..that he wore a ruffle-shirt and overshoes. 1835 A. B. LONGSTREET *Georgia Scenes* 85 The *ruffle-shirted* little darlings of the present day. *a* 1864 [see sense 7 a in Dict.]. 1876 *Harper's Weekly* 26 Aug. 691/1 They belonged to the class which the ward politicians of to-day sneer at as ruffle-shirted and silk-stockinged. 1842 *Knickerbocker* XIX. 305 Many a taunt, hitherto repressed, was thrown at the ruffle-shirters, as the town boys called them.

ruffled, *a.* Add: **a.** *ruffled shirt*, a shirt decorated with ruffles; also *transf.*, = *ruffle shirt* (*b*) s.v. *RUFFLE *sb.*¹ 8.
1754 *Calendar Virginia State Papers* (1875) I. 249, 2 fine Ruffled shirts and 2 plain shirts for themselves..sent by the Governor to them. 1768–74 [in Dict.]. 1860 O. W. HOLMES *Prof. at Breakfast-Table* i. 19 Joe Warren, the first bloody ruffled-shirt of the Revolution, was as good as born here. 1905 A. H. RICE *Sandy* 271 A few feet farther away hung a portrait of her grandfather, brave in a high stock and ruffled shirt. 1974 J. AIKEN *Midnight is Place* iv. 120 He wore black buckled shoes and a ruffled shirt.
b. (Further examples.)
1850 *Rep. Comm. Patents: Agric. 1849* (U.S.) 289 The ruffled oat is very much cultivated, and highly esteemed. 1941 J. STEINBECK *Sea of Cortez* xvii. 170 There were many of the ruffled clams with hard, thick, wavy shells.

ruffler¹. (Later examples.)
1908 *Sears, Roebuck Catal.* 41/1 The set of attachments..consists of one ruffler, one shirring plate, one tucker. 1964 *McCall's Sewing* x. 144/2 The ruffler attachment is also capable of gathering large sections of fabric.

Rufflette (rʊ·flĕt, rʊfle·t). [f. RUFFLE *sb.*¹ + -ETTE.] A proprietary name for a kind of tape that can be sewn to the top edge of a curtain, having slits at regular intervals by which curtain-hooks may be attached, and cords threaded through that enable the tape and curtain to be gathered or 'ruffled'.

1931 *Trade Marks Jrnl.* 18 Nov. 1533 The Rufflette. Use claimed from 28th September, 1922... Cotton curtain heading tapes included in Class 25. 1947 *Radio Times* 18 Apr. 22/2 (Advt.), There's more than material in curtains..there's the art of making them drape beautifully—how simple that is with 'Rufflette' curtain tape with hooks and rings. 1964 *McCall's Sewing* xvi. 280 Two types of Rufflette are available giving gathered or pleated effects. In both cases first form hem..then baste Rufflette tape to this. 1973 *Guardian* 28 Feb. 11/3 Curtains, with Rufflette tape topping, are in two sizes.

ruffly, *a.* Add: Also **ruffley.** (Later examples.)
1909 [see *HATTY *a.*]. 1980 M. G. EBERHART *Casa Madrone* ii. 30 A ruffly blouse.

ruff-scuff, alteration of *rough-scuff* s.v. ROUGH *a.* 21. (In quot. 1902, = poor fodder.)
1902 H. F. DAY *Pine Tree Ballads* 4 Drat the man who feeds out ruff-scuff, wood and wire from the swale. 1936 *N.Y. American* 13 Aug. 17/1 There is a snug harbor.. chivvied out of a city's ruff-scuff for dreamers.

rufous, *a.* Add: **1. a.** (Later examples.)
1922 JOYCE *Ulysses* 182 Glittereyed, his rufous skull close to his greencapped desk-lamp sought the face. 1977 *Time* 3 Jan. 21/1 The flat kindly face of Mr. Earl's photographs has made no print on any of them except the rufous Billy, a wily jester.
c. *rufous bee*, a solitary bee, *Andrena fulva*; *rufous rat-kangaroo*, a small marsupial, *Æpyprymnus rufescens*, found in south-eastern parts of Australia.
1926 LE SOUEF & BURRELL *Wild Animals Australasia* 234 Rufous Rat-kangaroo... General colour above coarsely grizzled rufescent grey. 1928 *Observer* 17 June 24/3 That queen of burrowers,..the rufous bee, Andrena. 1972 *Sunday Mail Mag.* (Brisbane) 3 Sept. 4 The rufous rat-kangaroo..occurs in many parts of central and south Queensland, particularly in open forest.
2. *rufous-brown* (later examples).
1953 D. A. BANNERMAN *Birds Brit. Isles* I. 300 The black crown feathers are then mostly obscured by buff or rufous-brown tips to the feathers. 1968 B. HINES *Kestrel for Knave* 26 On a shelf behind the bars stood a kestrel hawk: Rufous brown. Flecked breast, dark bars across her back and wings.
4. b. (Later examples.)
1907 R. SOUTH *Moths Brit. Isles* 1st Ser. 299 The Small Rufous..varies from pale ochreous white, through reddish shades, to a greyish brown. 1958 W. J. STOKOE *Caterpillars Brit. Moths* (new ed.) I. 281 The Small Rufous..occurring in fens and marshes.

rufus, *a.* Add: **B.** *sb.* (Also with capital initial.) **1.** *U.S. slang.* A countryman. **2.** *colloq.* A nickname for a red-haired person.
1955 D. W. MAURER in *Publ. Amer. Dial. Soc.* xxiv. 106 A farmer or yokel is called by old-timers a *rufus* (obsolescent) or a *hoosier*. 1959 I. & P. OPIE *Lore & Lang. Schoolch.* ix. 170 Red heads attract a barrage of nick-names:..reddy, red kipper, red mop, red thatch, red paint brush, Rufus, and Rusty.

rug, *sb.*² Add: **3. a.** (Earlier and later examples.) Now freq. of shapes other than square or oblong. *to cut a* (or *the*) *rug*: to dance (esp. to jazz music); *to pull the rug out from under someone*: see *PULL *v.* 19 k; *to sweep* (or *kick*) (something) *under the rug*: to conceal (something difficult, embarrassing, or unpleasant) in the hope that it will go unnoticed or be forgotten (cf. *CARPET *sb.* 2 e).
1808 JANE AUSTEN *Let.* 1 Oct. (1952) 212 She does not doubt your making out the Star pattern very well, as you have the Breakfast-room-rug to look at. 1930 *Heal's Catal.: Furnit.* 9 Knotted Axminster Circular Rug, designed by the late Noel Simmons, 5 ft. in diameter. 1942 *Harper's Bazaar* July 21/3 Let's dance, wanta cut a rug. 1943 *N.Y. Times* 9 May II. 5/4 Why, brother, all the cats cut a mean rug to that music. 1945 T. WILLIAMS *Glass Menagerie* vii. 107 'How about cutting the rug a little, Miss Wingfield?' 'Oh, I—' 'Or is your program filled up? Let me have a look at it. Why, every dance is taken!' 1961 D. M. DISNEY *Mrs Meeker's Money* vii. 77 The whole thing..was so far outside the normal routine.. that it practically demanded being swept under the rug. 1973 *Times* 1 Dec. 20/8 Those who used to look for reds under the beds now fear bugs under the rugs. 1976 *National Observer* (U.S.) 25 Sept. 12/4, I can see numerous problems that were rather swept under the rug in the article. 1978 M. PUZO *Fools Die* xvi. 174 Unless the government covers the whole thing up, you know, kicks it under the rug.
b. *U.S. slang.* A wig.
1940 J. O'HARA *Pal Joey* 190, I even wear a little rug up front. 1967 W. & M. MORRIS *Dict. Word & Phr. Origins* II. 75 Advertisements for men's wigs invariably refer to them as *hair pieces*, but in the trade a wig may be a *doily*, a *divot* or a *rug*. *Ibid.*, And a *rug*? Well, that's the works—a wig to conceal over-all baldness. 1978 *Telegraph* (Brisbane) 18 Feb. 8/2 'Now, in fact, I do wear a hairpiece in the film I'm making.'..The film for which he has donned a 'rug' as they are called, is Meteor.
6. b. *rug-chest, -fringe, -hook, -hooking, -making* (later example), *-peddler, -wool, -work* (earlier example), *-yarn*; **rug brick**, a rough-surfaced brick; **rug-cutter** *U.S. slang*, an enthusiastic or expert dancer; also *transf.*; so **rug-cutting**, dancing; **rug-ranking** *Canad.*, the system of determining the salary of a secretary in federal employment by the status of the person for whom she works; so **rug-rank** *v.*; **rug-rat** *U.S. slang*, a child.
[1952 *Bricklaying* (Structural Clay Products Inst.) I. ii. 28 Types and Colors of Brick... The Matt faces and Rug faces can almost be placed in the same family. The degree of smoothness or roughness is almost unlimited.] 1961 WEBSTER, Rug brick. 1968 *Globe & Mail* (Toronto) 3 Feb. 43/1 (Advt.), 3 bedroom rug brick home. 1972 *Times* 19 Sept. 9/5 (Advt.), Partners desk; rug chests. 1976 *Leicester Trader* 24 Nov. 19/2 (Advt.), Rug chests... £25. 1938 *N.Y. Amsterdam News* 2 Apr. 17/1 The thousands of..rugcutters..that are being hatched daily.. are a peril. 1941 W. C. HANDY *Father of Blues* (1957) i. 6 Country gals and their..suitors got as much enjoyment.. as jitterbugs or rug-cutters get nowadays from a swing band. 1942 Z. N. HURSTON in *Amer. Mercury* July 96/1 *Rug-cutter*, originally a person frequenting house-rent parties, cutting up the rugs of the host with his feet; a person too cheap or poor to patronize regular dance halls; now means a good dancer. 1959 N. MAILER *Advts. for Myself* (1961) 107 He seemed full of strength and merriment. He would clap two geishas to him, and call across..to another soldier. 'Hey, Brown,' he would shout, 'ain't this a rug-cutter?' 1940 *Amer. Speech* XV. 205/1 *Rug-cutting*, violent, eccentric dancing. 1942 *Chatelaine* Apr. 54/2 The rug-cutting addicts discovered an older form of swing. 1947 S. LEWIS *Kingsblood Royal* 21 As I say: we don't know where Belfreda goes or what she does—rug-cutting or witchcraft or maybe she belongs to some coloured left-wing political gang. 1981 'S. CAUDWELL' *Thus was Adonis Murdered* ix. 114 He raised again the matter of the rug-cutting expedition... The only places..where there might be dancing..looked to me formidably expensive. 1897 *Sears, Roebuck Catal.* 298/3 Wool Rug Fringe, with gimp heading 3 inches deep. 1922 JOYCE *Ulysses* 691 The upturned rugfringe. 1966 *Olney Amsden & Sons Ltd. Price List* 35 Latchet Rug Hooks... 16/- Dozen. 1967 *House & Garden* June 19/1 *Hooked rugs*... You need open-mesh canvas and a rug hooking tool which is a metal hook fitted into a wooden handle. It has a hinged metal shank at the hook end which will open and close automatically to prevent the hook getting caught in the canvas. 1974 *Aiken* (S. Carolina) *Standard* 22 Apr. 2-A/6 Rug Hooking, 9:30 a.m.–2:30 p.m., Recreation Center, North Augusta. 1976 N. ROBERTS *Face of France* iii. 39 Shops with displays of equipment for tapestry work and rug-making. 1916 J. LONDON *Let.* 12 Oct. (1966) 473 'Uncle Charley'..then proceeded to shake you down in proper money-lender,..rug-peddler fashion. 1977 *Kingston* (Ontario) *Whig-Standard* 9 Feb. 7/2 Her abilities caught the eye of some of the rising lights of her time, and the story has it that she 'rug-ranked' her way up out of the secretarial classifications. 1973 *Globe & Mail* (Toronto) 21 July 1/1 Rug-ranking..refers to the Government's method of establishing the pay level of secretaries in the same way the quality of the manager's rug is determined. It goes with the title on the door. *Ibid.*, While Mr. Drury says 'there isn't much possibility' of replacing the rug-ranking system with a point-rating system, he expresses sympathy for the problems faced by the secretaries. 1968–70 *Current Slang* (Univ. S. Dakota) III–IV. 104 *Rug rat*, a small child. 1976 *Daily Tel.* (Colour Suppl.) 16 July 10/1 He is without children; he has rug-rats instead. 1926 S. T. WARNER *Lolly Willowes* II. 114 She bought an extensive parcel..of variously coloured rug-wools. 1967 E. SHORT *Embroidery & Fabric Collage* iii. 83 A soft Persian rug wool is the easiest to sew with. 1823 E. WEETON *Let.* 21 Nov. (1969) II. 243 She is fond of rug work, and has done a great deal in a superior and beautiful manner. 1895 *Montgomery Ward Catal.* 124/3 Colored Rug Yarn... ingrain carpet yarn, assorted colors, 4 skeins to pound. 1932 D. C. MINTER *Mod. Needlecraft* 224/2 The thinner qualities of rug yarn—such as Persian, Straight, Shetland and various Thrums yarns.

rug, *sb.*³ (Further examples.)
1622 DRAYTON *Poly-Olbion* II. xxvi. 124 Thicke Vapours, that like Rugs still hang the troubled ayre. 1969 G. M. BROWN *Orkney Tapestry* 29 The old Orkneymen had a range of words for every kind and intensity of rain—a driv, a rug, a murr..a hellyiefer.

rug, *v.*¹ Add: **1. a.** (Later example.)
1930 KIPLING *Limits & Renewals* (1932) 234 Old dry bites—when they get good hold and rugg you. That showed he must have dealt with the Beasts.
2. a. Also, to struggle.
1832 *Chambers's Edin. Jrnl.* I. 225/1 The gilly,..who comes into the town,..rugging and riving for a place in some writer's office. 1951 R. RENDALL *Orkney Variants* 24 Their lowan e'en are taakan tent O'chiels like Mansie o' the Bu Whose days upon the land are spent Ruggan wi' Taurus and the Pleugh.

rug, *v.*³ [f. RUG *sb.*²] *trans.* To cover with a rug. Freq. with *up*.
1818 M. EDGEWORTH *Let.* 15 Oct. (1971) 126 He hopes to have the rooms carpetted and rugged by tuesday. 1936 F. CLUNE *Roaming round Darling* iii. 26 Many sharp turns and wattle-trees. The latter, droopy after being out all night, should be rugged up this weather. 1961 C. H. D. TODD *Popular Whippet* iii. 39 Kennels of greyhounds usually have the inmates 'rugged up' at night. 1968 E. R. BUCKLER *Ox Bells & Fireflies* vi. 93 You rugged the oxen and took the double-bitted ax from its leather fastener. 1975 D. FRANCIS *High Stakes* i. 18 He was a great horse..he would soon be rugged up nice and quiet in a stable.

ruga. Add: (Further examples.)
1913 *Cunningham's Text-bk. Anat.* (ed. 4) 1298 When this muscular layer is contracted, the scrotum becomes smaller..and the skin is thrown into folds or wrinkles called rugæ. 1962 *Gray's Anat.* (ed. 33) 1514 The loose texture of the mucous layer allows the mucous coat to be thrown into folds or rugae when the bladder is empty. 1963 J. OSBORNE *Dental Mech.* (ed. 5) ii. 40 The rugae may be accentuated with blue inlay wax. 1969 *Gloss. Terms Dentistry* (*B.S.I.*) 107 *Rugae*, the irregular ridges of the mucous membrane covering the anterior part of the hard palate.
Hence **ru·gal** *a.*
1936 KANTNER & WEST *Phonetics* (ed. 3) I. iii. 48 Attention should be drawn to the alveolar or rugal ridge which is the raised line of flesh found at the point where the teeth emerge from the gums.

Rugbeian. Add: Also **Rugboean.** (Earlier and later examples.)
1714 T. CAVE *Let.* 16 June in M. M. Verney *Verney Lett.* (1930) I. xiii. 249 We also favoured this day our two young Rugbeians and found Em well. 1971 *Times* 18 Dec. 13/7 He had, in the opinion of many people with whom I have spoken (including one eminent old Rugboean), fully captured Hughes's message.

Rugby. Add: Also with small initial. **a.** (Further examples.) Also *Comb.* **Rugby League**, an association of Rugby football clubs formed in 1922 (previously the 'Northern Union'), having rules differing from those of the Rugby Union; Rugby played according to these rules; also *attrib.*; **rugby tackle**, a tackle in which the arms are used to bring an opposing player down, as in Rugby football; **Rugby Union**, an association of Rugby football clubs formed in 1871; Rugby played according to its rules; also *attrib.*
See esp. quot. 1975.
1871 *Field* 22 July 82/2 (*heading*) *The Rugby Union.* A special general meeting of the Rugby Football Union will be held at the Arundel Hotel, Arundel Street, Strand, on Monday evening. 1874 G. H. WEST *Rugby Union Football Ann.* 2 The following clubs have now enrolled themselves under the Rugby Union. *Ibid.* 11 Since the formation of the Rugby Union, the rules of football have been somewhat modified and altered... A few hints, therefore, on the general style of play in the Rugby Union game, and also an explanation of those rules which most affect it may not be out of place. 1885 [see *ASSOCIATION 11]. 1906 GALLAHER & STEAD *Compl. Rugby Footballer* xix. 265 There were fifty thousand people present—by far the biggest attendance of spectators we had ever seen at a Rugby match. 1923 *Whitaker's Almanack* 474/2 *Northern Union in 1921–22* This body is now called the Rugby League. 1926–7 *Army & Navy Stores Catal.* 721/2 Rugby jerseys (to order only). 1929 R. HARDING *Rugby* 142 It was said that a certain Rugby League Club was prepared to give me £750 to become a professional Rugby player. *Ibid.* 144 The system of professionalism which obtains in the Rugby League is the best. 1931 R. CAMPBELL *Georgiad* I. 18 Jack Squire..who ..makes a funeral of a Rugby Match. 1959 *Times* 21 Sept. 3/6 In this community of Rugby-minded souls every man..could readily appreciate the virtues of an attacking as against a defensive policy. 1959 F. GUEST *Indian Cavalryman* viii. 97 The subaltern, who was an athletic young man, immediately dived at the men in a Rugby tackle. 1960 T. MCLEAN *Kings of Rugby* viii. 57 His Rugby-playing days were over. 1961 *Times* 7 Apr. 20/7 In the matter of crowd behaviour, in fact, the Rugby-watching public can in no way afford to be smug. 1963 *Listener* 14 Feb. 300/1 David Storey's *This Sporting Life* is a *bildungsroman* of an unusual sort. Machin, its central character, is a miner turned Rugby League player. 1969 [see *LEAGUE *sb.*² 1 c]. 1969 *Listener* 20 Mar. 384/3 The 'sheer disorder' of the broken field of your Rugby Union game. 1975 *Oxf. Illustr. Dict.*, *Rugby football*, one of the two main types of football (the other being Association football), played with 15 players a side (in Rugby Union) or 13 (in Rugby League), with an elliptical football punted, dropped, or passed from hand to hand, the object being to touch down behind the opponents' line and score a try, and to kick the ball over the crossbar of the H-shaped goal. 1976 J. MCCLURE *Rogue Eagle* ii. 30 Hulk's Airtex shirt and ridiculous rugby shorts. 1976 *Western Mail* (Cardiff) 27 Nov., Mid-Wales Howells Cup first round rugby tie with Mid-Glamorgan..has been postponed. 1976 *Alyn & Deeside Observer* 10 Dec. 3/7 In the second half..Chester played some excellent rugby. 1978 *Rugby World* Apr. 45 (Advt.), The *original* long playing records of Rugby Songs in the *Jock Strapp Series*. *Ibid.* 51 (Advt.), Top quality rugby jerseys.
b. **Rugby fives**, the leading form of fives: see FIVES².
1897 *Encycl. Sport* I. 398/2 In the Eton court..the presence of the pepper-box, the hole, and the step, make the game in practice as different from Rugby Fives as in theory it is similar. 1935 *Encycl. Sports* 270/1 A Rugby fives court, called after the school of that name, is a covered four-walled building. The walls are all plain, except that on the front wall there is a ledge or board, above which the ball must be struck to be in play. 1958 *Times* 16 Dec. 4/6 Old Oundelians..beat Rugby Fives Association..by 121 points to 70. 1975 *Oxf. Compan. Sports & Games* 870/1 Rugby fives is mainly a school, old boys', and university game.

ruge, *sb.*¹ (Later example.)
1791 A. GRAHAM in *Publ. Hudson's Bay Rec. Soc.* (1969) XXVII. IV. 117 [The tusk of the unicorn fish] is quite straight, and has a double spiral ruge on its surface.

rugged, *a.*¹ Add: **4. b.** Also *colloq.* in weakened sense: tough, difficult.
1942 *Yank* 7 Oct. 7 'Rugged' the Destroyers call the Tank Hunting Course. 1943 *Newsweek* 27 Sept. 23/1 The war here is still pretty rugged, as the boys say. 1946 *News Chron.* 30 Aug. 3/8 The first night was a bit rugged, in a way (George said). There being no bed, Mrs Cain made up the bedclothes on the concrete floor. 1953 *Manch. Guardian* 31 July 4/6 They thought it not too strenuous: it had not been easy but had been 'rugged'. 1973 J. PATTINSON *Search Warrant* v. 81 If things get really rugged I just put the bite on my old man.
8. (Later examples.)
1931 *Amer. Speech* VI. 185 Frequent adjectives of encomium in book reviews are..romantic, rugged, ruthless. 1966 *Listener* 10 Mar. 363/2 Nicholas Maw's.. string quartet, a closely-knit, rugged work, product of a rich and fertile imagination. 1976 *Gramophone* Dec. 1057/2 Jochum's reading has a rugged truth combined with poetic sensibility.
9. a. *spec.* in phr. *rugged individualism*. So *rugged individualist, individuality*.
1928 H. HOOVER *New Day* 154 We were challenged with a peace-time choice between the American system of rugged individualism and a European philosophy or diametrically opposed doctrines—doctrines of paternalism and state socialism. 1937 [see *DELIQUESCENT *a.* 3]. 1937 *Education* Nov. 186/1 Each of them..is a rugged individualist doing everything to satisfy his own personal desire. 1946 G. B. SHAW *Geneva* III. 76 Your pose is that of the rugged individualist, the isolationist. 1962 [see *SAME B. 2 d]. 1973 *Guardian* 18 June 4/4 The apolitical frame of mind which is summed up in the cliche 'rugged individualism'. 1974 *Encycl. Brit. Macropædia* V. 403/1 Cypriots are a people of rugged individuality.
b. orig. *U.S.* Of a manufactured object: strongly constructed, capable of withstanding rough usage. Also *transf.*
1921 *Wireless World* 29 Oct. 477/2 The whole design has been made robust, or, as our American friends would say, 'rugged'. 1960 *Practical Wireless* XXXVI. 302/1 The mains transformer should be a rugged component capable of secure attachment to the chassis to survive the hazards of transport. 1975 J. WYLLIE *Butterfly Flood* xxvii. 128 Everybody uses Land Rovers. Good cars, too. Rugged. 1978 *Detroit Free Press* 16 Apr. (Gardening Guide) 6 (Advt.), This 19.9 hp grounds maintenance tractor is specially designed for big jobs where you need rugged, dependable power.

ruggedize (rʊ·gĕdəiz), *v.* orig. *U.S.* [f. RUGGED *a.*¹ + -IZE.] *trans.* To make rugged; to produce in a version designed to withstand rough usage. So **ru·ggedized** *ppl. a.*; **ruggediza·tion.**
1954 *Electronic Engin.* XXVI. 255 The recent introduction of 'ruggedized' valves. 1958 C. C. ADAMS *Space Flight* 198 Automatic-control mechanisms must be added to the ship so that maneuvers can be programmed into them while the men are lying all but helpless on their contour chairs. And these equipments themselves must be 'ruggedized' to withstand the very accelerations they are compensating for. 1959 *Wall St. Jrnl.* (Eastern ed.) 18 May 13/3 There are no two ways about it. Ruggedization is costly. 1962 *Daily Tel.* 21 May 10/8 A leaflet..from a New York electrical equipment firm..describes a 'ruggedised weather-proofed instrument'. 1969 *IEEE Trans. Nuclear Sci.* XVI. 314/1 An electronic subsystem packaged in a compact and ruggedized fashion suitable for operation in outer space. 1971 *Daily Tel.* (Colour Suppl.) 10 Dec. 31/4 The tubes were used extensively in the Korean War. Since then they have been miniaturised and protected—or 'ruggedised'—against the rough treatment they can expect on active service, and have spawned many instruments. 1977 *Sci. Amer.* Oct. 140/1 (Advt.), When the frequency, sinusoidal amplitude and sweep rate of the vibration levels demand the ultimate in ruggedization, consider Questar.

ruggedness. Add: **3.** Of manufactured objects: robustness, durability. Cf. *RUGGED *a.*¹ 9 b.
1936 *Physics* VII. 75/1 The ruggedness of the simplified extrusion plastometer has been established by over two years of practically uninterrupted service. 1971 *Physics Bull.* Nov. 644/1 (Advt.), Designed to provide the versatility required for laboratory use with the ruggedness of production line equipment.

rugger². Add: (Later *attrib.* examples.) **rugger-tackle** *v. trans.* = TACKLE *v.* 5 (*a*).
1914 'BARTIMEUS' *Naval Occasions* ix. 64 They earned their 'Rugger' colours together as scrum and stand-off halves. 1927 *Granta* 14 Oct. 9/1 He has had to be content with four years and a captaincy in the Magdalene Rugger side. 1929 *Mercury Story Bk.* 162 'Rubbish' I heard an eupeptic Rugger man protesting. 1930 R. CAMPBELL *Poems* 10 Nor at his Rugger-Match is Squire more gay. 1937 F. SMYTHE *Camp Six* v. 65 The football was also very popular and Nursang did his best to form the Sherpas into a rugger scrum. 1951 M. MCLUHAN *Mech. Bride* (1967) 137/1 The same contrast may be seen between the rugger crowd and the soccer crowd. 1955 *Times* 3 Aug. 7/4 Gradually strength and energy returned, and in two days the Sherpas produced the expedition rugger ball. 1967 D. PINNER *Ritual* xii. 119 He rugger-tackled the policeman. 1973 D. LEES *Rape of Quiet Town* vii. 111 It was all in the spirit of a rugger club frolic.

ruggerite (rʊ·gərəit). *rare.* [f. RUGGER² + -ITE¹.] One who plays Rugby football.
1951 R. CAMPBELL *Light on Dark Horse* iv. 75 In the end his school of 'scruffy soccerites' defeated Langley's ruggerites at their own rugger.

ruggery (rʊ·gəri), *a. rare.* [f. RUGGER² + -Y¹.] Being or resembling a typical Rugby player.
1961 A. COMFORT *Come out to Play* ii. 136 Gaudeamus was a large, ruggery man.

rugging, *sb.* (Later examples.)
1939 *Country Life* 11 Feb. p. xxxiii/2 (Advt.), Cheaper quality in rugging, 27/6, 25/- & 20/-. 1963 E. H. EDWARDS *Saddlery* xx. 149 The most simple of all is the Yorkshire boot, which consists of an oblong of stout rugging with a tape sewn along the centre.

Rugian (rū·dʒiăn), *sb.* and *a.* [f. L. *Rugii* pl., Rugians + -IAN.] **A.** *sb.* A member of an ancient Germanic tribe; the East Germanic language of this tribe. **B.** *adj.* Of or pertaining to the Rugians.
1611 CORYAT *Crudities* 106 Odoacer the Rugian that vsurped the Kingdome of Italy..by expelling Augustulus the last Emperour of Rome. 1845 *Encycl. Metrop.* XI. 268/1 Those united bands the Heruli, the Alani, the Scyrri, and the Rugians..constituted the military force of Italy. 1884 *Encycl. Brit.* XVII. 727/1 This Rugian war was probably an indirect cause of the fall of Odoacer. 1934 PRIEBSCH & COLLINSON *German Lang.* 25 A few other Germanic languages, e.g. Skirian, Rugian, [etc.]. 1935 H. A. L. FISHER *Hist. Europe* I. x. 120 There was a bigness of scale about Theodoric which redeemed many of the grosser vices... After three years' hard fighting he eliminated from Italy the Rugian army of Odovacar, and thereafter gave thirty-six years of golden peace to that much harassed land.

rugosa (rugōᵘ·ză). [L., fem. of *rūgōsus* (see RUGOSE *a.*) and the specific epithet of *Rosa rugosa* (C. P. Thunberg *Flora Japonica* (1784) 213).] A hardy shrub rose belonging to the species *Rosa rugosa*, or one of its varieties or hybrids, distinguished by dark green, wrinkled leaves and large, globular, orange-red hips. Also *attrib.*
1892 W. PAUL *Contrib. Hort. Lit.* 189 The Rugosa Rose. Where large showy Roses are valued these flowers will not fail to please. 1899 T. W. SANDERS *Cultivated Roses* 33 Madame Georges Bruant (hybrid rugosa). 1906 *Roses* 121 Rugosas, owing to their very bushy growth, should be in separate beds. 1920 *19th Cent.* July 174 Roses innumerable—teas, chinas, rugosas, briars, and their hybrids—are asserting their decorative uses. 1943 T. C. MANSFIELD *Roses* ii. 18 The rugosa hybrids make exceptionally fine hedges. 1962 R. PAGE *Educ. of Gardener* vii. 209 The rugosa rose 'Blanc Double de Coubert' goes into every garden I make. 1977 *Vole* No. 1. 34/3 The rugosas flower for weeks and have lovely hips.

rugose, *a.* Add: **a.** *rugose mosaic*, a mosaic disease of potatoes characterized by a marked wrinkling of the leaves and increased chlorosis and dwarfing compared with other mosaics.
1923 SCHULTZ & FOLSOM in *Jrnl. Agric. Res.* XXV. 52 The writers believe that most of Murphy's crinkle..is identical with the type here designated as 'rugose mosaic' with some leaf-rolling mosaic symptoms. 1967 A. E. COX *Potato* iv. 93 The two most serious of the virus diseases, leaf roll and rugose mosaic, are transmitted from infected to healthy plants by aphids.
c. [a. mod.L. order name *Rugosa* (Milne-Edwards & Haime *Monogr. Brit. Fossil Corals* (1850) I. p. lxiv).] Of a fossil coral: belonging to the extinct order Rugosa (or Tetracoralla), which includes horn-shaped corals with ridged surfaces.
1872 H. A. NICHOLSON *Man. Palæont.* I. viii. 99 It has been shown that some..abnormal Rugose corals were provided with a lid. 1935 *Geol. Mag.* LXXII. 482 The rugose corals..are a group of Palaeozoic corals. 1972 *Sci. Amer.* June 61/2 The conical or cylindrical stone tube that sheltered the second type has conspicuous external growth wrinkles on its surface; these corals are called rugose.
d. *fig.*
1942 A. L. ROWSE *Cornish Childhood* vi. 133 Old Sidney was a rugose personality... I am bound to say that, rough-edged as he was with everybody, he was always very kind to me.

rugosity. Add: **2.** (Later *fig.* example.)
1969 T. E. B. HOWARTH *Culture, Anarchy & Public Schools* iii. 54 It may..be doubted if the new sixth-former..will take kindly to the traditionally rigorous system of the old type of sixth form. He will expect his teachers to spare him the sterner rugosities.

Ruhmkorff (rū·mkǭɹf). *Physics. Obs.* except *Hist.* The name of Heinrich Daniel *Ruhmkorff* (1803–77), German-born inventor, used *attrib.*, *absol.* and in the possessive to designate a powerful type of induction coil first made by him.
[1852 W. R. GROVE in *Phil. Mag.* IV. 500, I procured one of these apparatus from M. Ruhmkorff: the size of the coil portion of the apparatus is 6·5 inches long, 4 inches diameter.] 1855 —— in *Ibid.* IX. 1 If a small Leyden phial have its coatings connected respectively with the extremities of the secondary wire of a Ruhmkorff coil (the primary being, as usual, connected with the condenser of M. Fizean, and two wires being attached to the terminals and brought within striking distance), the noise and brilliancy of the discharges are greatly increased. 1878 *Encycl. Brit.* VIII. 103/1 The type of these is the induction coil or inductorium, sometimes called Ruhmcorff's coil, after the great Parisian instrument-maker who first brought the instrument to perfection. 1908 *Rep. Brit. Assoc. Adv. Sci. 1907* 621 The aerials connected through the secondary of a peculiarly made Ruhmkorff coil constitute one oscillating system of a low frequency. *Ibid.* 622 When the swing is worked up they burst through the spark gap, short-circuiting out the Ruhmkorff. 1971 *Sci. Amer.* May 82/2 The final phase of the coil's development was reached during the 1850's. Heinrich Daniel Ruhmkorff, a German instrument maker living in Paris, turned his attention in 1851 to the construction of better and more powerful coils... The quality of his products quickly brought him fame, and such coils soon became known as Ruhmkorff coils.

ruin, *sb.* Add: **I. 4.** (Further *fig.* examples.)
1946 B. MARSHALL *George Brown's Schooldays* ii. 7 Hullo, here's Abinger. He looks a fearful ruin, doesn't he? *Ibid.* xxiv. 102 A ruin's a chap who's a swot and a punk and who's rotten at games.
III. 11. a. *ruinward* adv.
1936 A. E. HOUSMAN *More Poems* 61 And on through night to morning The world runs ruinward.
b. *ruin-mound* = TELL *sb.*²
1911 *Encycl. Brit.* XIV 741/2 There are in Irak hundreds of ruin mounds, some of them of considerable size, covering ancient Babylonian cities. 1939 P. CARLETON *Buried Empires* i. 23 Attracted by the numerous *tells*, or ruin-mounds, in his district, he set workmen to dig.

ruin, *v.* Add: **I. 3. d.** (Later examples.)
1929 E. O'NEILL *Dynamo* I. ii. 28 Pa and Ma warned me linesmen were no good... they just ruined you and went their way. 1955 *Radio Times* 22 Apr. 4/3 The sentimental blatherings of Mrs. Arbuthnot who was 'ruined' by Lord Illingworth twenty years before. 1962 E. BENTLEY tr. *Brecht's Mother Courage* vi. 51 She's not so pretty anyone would want to ruin her.

ruinate, *ppl. a.* Add: **1. c.** *Jamaica.* Of land: exhausted, abandoned. Hence as *sb.*, land which has reverted to the wild, scrubland, 'bush'.
1835 B. M. SENIOR *Jamaica* 54 Lands termed 'ruinate', which means such as have been used till worn out, and then allowed to grow up in bushes and weeds. *Ibid.* 55 In three or four years after the ruinate is cleared up, the pimento begins to bear. 1847 P. H. GOSSE *Birds of Jamaica* 11 We see it [*sc.* the Red-tailed Buzzard] all the year round, sailing deliberately in wide circles over the pastures and ruinates. 1894 R. T. BANBURY *Jamaica Superstitions* 30 We would advise parents never to allow their little ones to wander about near woods, or ruinates by themselves.

ruinate, *v.* **1. b.** (Later *arch.* examples.)
1922 E. R. EDDISON *Worm Ouroboros* xviii. 253 If I do not..remedy for you our fortunes which this bloody fool hath laboured to ruinate, spit in my face. 1935 G. BARKER *Poems* 55 Time, though slowly, ruinates Love, with which it arbitrates.

ruinous, *a.* Add: **3.** (Later example.) Also *transf.*, excessively expensive.
1897 A. BEARDSLEY *Let.* 27–28 Feb. (1970) 260 If the hotel turns out too ruinous, and our expenditure in these matters could not be controlled, then we will decide for lodgings.

‖ **rukh** (ruk). Also **ruk.** [a. Hindi *rūkh*, f. Prakrit *rukkha*- tree.] In India: a forest; a forest reserve.
1893 KIPLING *Many Inventions* 191 He made no pretence at keeping a garden, for the *rukh* swept up to his door. 1928 *Blackw. Mag.* Apr. 443/1 It was a week or two before General Devi Deen came again for our ride through the *ruk*.

rukus, var. *RUCKUS.

rule, *sb.* Add: **I. 3. b.** *rules of the game* transf., conventions in political or social relations or the like.
1910 S. E. WHITE *Rules of Game* xli. 644 Things change; and a man is foolish to act as though they didn't. He's just got to keep playing along according to the rules of the game. And they keep changing too. 1936 M. MITCHELL *Gone with Wind* xxxi. 521 The rules of the game had been changed and..honest labor could no longer earn its just reward. 1964 ROUSSEAS & FARGANIS in I. L. Horowitz *New Sociol.* 287 Operating within the rules-of-the-game of institutionalized conflict. 1974 *Daily Tel.* 15 Oct. 18/2 As leader of the Opposition Mr Heath, under the rules of the game, has the right of reply tonight to Mr Wilson's Ministerial broadcast last night.
4. b. *rules of evidence*, the legal rules that apply to the giving of evidence.
1756 *J. Gilbert's Law of Evidence* (rev. ed.) 8 The Rule of Evidence commands no farther than to produce the best that the Nature of the Thing is capable of. 1801 T. PEAKE *Law of Evidence* p. v, The chapter on Parol Testimony, also is in a great measure new; for the rules of evidence in this respect have been so much altered, and so much light has been thrown on them by modern decisions, that, comparatively, little is to be collected from ancient books. 1892 S. L. PHIPSON *Law of Evidence* p. v, I have..adhered to one uniform method of arrangement throughout—that of stating: (1) The rules of evidence..[etc.]. 1908 J. H. WIGMORE in *Sel. Ess. Anglo-Amer. Legal Hist.* II. xl. 691 (*heading*) A general survey of the history of the rules of evidence. 1942 E. M. MORGAN in *Model Code of Evidence* (Amer. Law Inst.) 5 The rules of evidence have been developed in myriads of cases. *Ibid.* 34 This has led to the invention of the hypothetical question, which, as Mr. Wigmore says, 'is one of the truly scientific features of the rules of Evidence'. 1956 E. C. CONRAD *Mod. Trial Evidence* I. i. 15 No exceptions to the general applicability of the rules of evidence as a broad proposition has been noted. 1973 *N.Y. Law Jrnl.* 4 Sept. 3/4 This committee was not bound by the rules of evidence. It was not constrained to follow courtroom procedures.
c. *rule of law*: (*a*) a valid legal proposition; (*b*) a doctrine, deriving from theories of natural law, that in order to control the exercise of arbitrary power, the latter must be subordinated to impartial and well-defined principles of law; (*c*) spec. in English law, the concept that the day-to-day exercise of executive power must conform to general principles as administered by the ordinary courts.
(*a*) *a* 1634 E. COKE *Third Part Institutes Lawes Eng.* (1644) vii. 53 In case of life the rule of law ought to be certain. 1756 *J. Gilbert's Law of Evidence* (rev. ed.) 16 The Rule of Law that requires the greatest Evidence that the Nature of the Thing is capable of. 1768 BLACKSTONE *Comm.* III. xxiii. 383 If a whole county is interested in the question to be tried, the trial by the rule of law must be in some adjoining county. 1969 *Columbia Law Rev.* LXIX. 1168 It is clear that those rules of precedent which are binding as 'rules of practice' are also rules of law.
(*b*) 1883 J. E. C. WELLDON tr. *Aristotle's Politics* iii. §16. 154 The rule of law then..is preferable to the rule of an individual citizen. 1929 LD. HEWART *New Despotism* ii. 23 What is meant here by the 'Rule of Law' is the supremacy or the predominance of law, as distinguished from mere arbitrariness, or from some alternative mode, which is not law, of determining or disposing of the rights of individuals. 1936 F. G. WILSON *Elem. Mod. Politics* viii. 207 It is of historic importance that the rule of law in the medieval and early modern sense was the rule of superearthly law..—eternal law, divine law, natural law, and human law. 1953 T. D. WELDON *Vocab. Politics* iii. 69 Strictly speaking there is nothing difficult or impressive about 'the Rule of Law'. It is merely a convenient way of referring to the fact that associations have rules and unless those rules are pretty generally kept and enforced the association breaks down and the activity which it was designed to promote becomes impracticable. 1959 E. C. S. WADE in A. V. Dicey *Law of Constitution* (ed. 10) p. xcvii, In another sense the rule of law means the recognition of certain fundamental obligations as binding upon States in their dealings with one another... The United Nations.. claims to give effect to the rule of law. *Ibid.* p. cvii, The International Commission of Jurists considers that the basic idea uniting lawyers in many different legal systems is a conception of the rule of law. 1971 *Engineering* Apr. 54/1 Industry-wide negotiations, ending in a kind of rule-of-law. 1974 J. LAPALOMBARA *Politics within Nations* iii. 106 But the difference between the *Rechtsstaat* and constitutionalism is that the rule of law in the former is based on a concession from the ruler. 1977 *Rolling Stone* 24 Mar. 31/2 Those senators..knew of the need to continue redeeming the rule of law that Edward Levi had begun.
(*c*) 1885 A. V. DICEY *Law of Constitution* v. 172 When we say that the supremacy or the rule of law is a characteristic of the English constitution, we generally include under one expression at least three distinct though kindred conceptions. We mean, in the first place, that no man is punishable or can be made to suffer in body or goods except for a distinct breach of law established in the ordinary legal manner before the ordinary courts of the land. 1923 W. S. HOLDSWORTH *Hist. Eng. Law* (rev. ed.) iv. 405 The precocious development of our common law has..given..the opportunity for the development of those two fundamental characteristics of our English constitution—the system of self-government and the rule of law. 1933 W. I. JENNINGS *Law & Constitution* 256 The 'rule of law' in this sense means that public authorities ought not to have large powers. 1959 *Polit. Stud.* VII. 114 He [*sc.* Dicey] would not have admitted for one moment that a Rule of Law followed from the mere fact that the conduct of government had a legal basis. 1971 S. A. DE SMITH *Constitutional & Admin. Law* ii. 40 Nor would it be justifiable to examine the general concept of the rule of law at length... The concept is usually intended to imply (i) that the powers exercised by politicians and officials must have a legitimate foundation and (ii) that the law should conform to certain minimum standards of justice. 1975 LD. HAILSHAM *Door wherein I Went* xxxvi. 253 The rule of law, an increasingly sophisticated idea..is essentially a province for an official with a foot in both camps, a sworn judge as well as a sworn Privy Councillor, with an independent duty towards the judiciary and the legal profession.
5. *to bend* or *stretch the rules*: to interpret the rules leniently, to overlook or allow an infringement of the rules; hence *rule-bender*; *rule-book* (later examples); also *fig.*
1910 W. M. RAINE *Bucky O'Connor* 13 The situation was one not covered in the company's rule book. 1945 F. H. HUBBARD *Railroad Avenue* ii. 10 Casey was never the type known as a 'rule-book engineer'. 1954 W. FAULKNER *Fable* 113 Germans fight wars by the rule-books. 1959 M. GILBERT *Blood & Judgement* xiv. 151 Some stuffy old Chief Superintendent, who's lived with one finger in the Rule Book. 1968 *Punch* 20 Mar. 417/3 Despite some fierce and not altogether rulebook tackling by their opponents, the Students were..taking the game right into the Police half. 1973 M. WOODHOUSE *Blue Bone* iv. 36 The Communists..were bound to loathe the guts of the big old families since that was what the rule book said. 1973 *Times* 2 Nov. 5/7 Trying to get other members of the European Community to 'bend the rules' so that exports can be resumed. 1977 'O. JACKS' *Autumn Heroes* ii. 37 He bent over backwards to be straight in all his dealings... He wouldn't stretch the rules. 1978 S. BRILL *Teamsters* vii. 272 They were rule-benders (or perhaps sometimes lawbreakers) in a rule-benders and lawbreakers' world.
b. (Also with lower-case initials.) **Rules Committee**, a committee of a house of a U.S. federal or state legislature responsible for expediting the passage of bills.

1918 H. W. DODDS *Procedure in State Legislatures* (Annals Amer. Acad. Polit. & Social Sci. Suppl. No. 1) iv. 60 Enjoying as much parliamentary power as the English cabinet, the rules committee [of the New York Assembly] nevertheless escapes any measure of responsibility before the people. 1976 *National Observer* (U.S.) 12 June 5/2 Lobbying..was so intense the Rules Committee wouldn't release the legislation.
c. Followed by a number or letter: a particular regulation imposed by an institution (see quots.).
1929 *Bookman* (U.S.) July 527/2 *Rule G*, in all railroad rule books, prohibiting the use of intoxicants. 1932 *Santa Fé Mag.* Jan. 34/2 Getting drunk is *Rule b*, failing to protect your train or to flag it is 99, attending an investigation is *going on the carpet*. 1974 *Guidelines to Volunteer Services* (N.Y. State Dept. Correctional Services) 43 *Rule 5*, when a parolee must abstain from alcohol. 1976 A. MILLER *Inside Outside* 6, I would also call on those men under Rule 43 (the segregation rule), and chat with them for a while. 1977 *Times* 11 Apr. 7/7 Over 60 prisoners are in segregation 'for the maintenance of good order or discipline' under rule 43 on any one day.
II. 7. a. Also, *rule-of-brain* (nonce-wd., after RULE OF THUMB in Dict. and Suppl.)
1948 L. MACNEICE *Holes in Sky* 25 Tom and Tessy..of themselves significant, To rule-of-brain recalcitrant.
c. (Later examples.) In modern Linguistics, usu. applied to any one of a system of rules that can be formulated in such a way that together they describe all the features of a language. Freq. *Comb.*
1953 [see *morpheme-sequence* s.v. *MORPHEME c]. 1957 N. CHOMSKY *Syntactic Struct.* x. 107 A grammar has a sequence of rules from which phrase structure can be reconstructed and a sequence of morphophonemic rules that convert strings of morphemes into strings of phonemes. Connecting these sequences, there is a sequence of transformational rules. 1965 *Language* XLI. 548 Language is rule-governed behavior, and learning a language involves internalizing the rules. 1968 J. LYONS *Introd. Theoret. Linguistics* i. 48 Learning the language 'naturally' as children, they [*sc.* the speakers of a language] come to speak it according to certain systematic principles, or 'rules', 'immanent' in the utterances they hear about them. It is the task of synchronic linguistic description to formulate these systematic 'rules' as they operate in the language at a particular time. 1968 *Language* XLIV. 735 It follows from premise 1 that from proto-language *L there will be *n* rule sequences into each of *n* daughter languages. 1971 P. KIPARSKY in W. O. Dingwall *Survey Linguistic Sci.* 612 The concept of *rule opacity*..has an important role to play elsewhere in linguistic theory. 1972 *Language* XLVIII. 83 There is every reason to believe that they will allow one to get rid of the unprincipled blocking device of extrinsic rule-ordering. 1974 G. M. GREEN *Semantics & Syntactic Regularity* vi. 194 The notions of redundancy rule, structural description feature, and deep-structure constraint were necessitated by the concepts of rule government. 1976 [see *phrase-structure* s.v. *PHRASE *sb.* 7]. 1978 *Language* LIV. 41 These features trigger rules that apply only to forms bearing the corresponding rule feature... We can call this device the 'rule-feature' theory. 1979 *Trans. Philol. Soc.* 18 The rule-environment is *arbitrary*—why not a rule deleting *no* only when it is *followed* by N we might ask?
9. *work to rule*: see *WORK *v.* and *sb.*
V. 17. c. Also, (*b*) of police: to interrogate (a suspected criminal); (*c*) of a doctor of medicine: to examine (a patient).
1948 *Free-Lance Writer & Photographer* Apr. 54/2 When a P.C. stops a suspect in the street and interrogates him, he 'runs the rule over him'. 1953 *Times* 21 Oct. 1/5 Good afternoon, doctor, I don't suppose it's anything really, just a bit off-colour..thought you'd better run the—er—rule over me.
23. c. In sense 3, as *rule formulation, system*; *rule-bound* (hence *-boundedness*), *-giving, -governed* adjs.
1905 W. JAMES *Mem. & Stud.* (1911) v. 89 You ask for a free man, and these utopias give you an 'interchangeable part', with a fixed number, in a rule-bound organism. 1950 *Mind* LIX. 391 Why not say 'rule-giving' method? 1968 *Listener* 29 Aug. 266/2 Societies have defined and structured rule systems of reward and punishment. 1977 J. D. DOUGLAS in Douglas & Johnson *Existential Sociol.* i. 39 As Becker realized, there is almost always conflict over such presentations of rule-boundedness. 1977 A. GIDDENS *Stud. in Social & Polit. Theory* iii. 144 Universal pragmatics..attempts to reconstruct the rule systems which allow actors to communicate in any type of context. 1978 C. HOOKWAY in Hookway & Pettit *Action & Interpretation* 27 If indeterminacy obtains it is likely to infect the translation of the object language into a meta-language involved in the rule-formulation. 1978 *Listener* 30 Mar. 396/2 To try to explain how the speaker's intentions, his rule-governed intentional behaviour, relates language to the world. 1979 *Dædalus* Summer 9 Rule-bound, conventional, and traditional ethics continue to hold their own. *Ibid.* 20 It remains a normal, rule-governed collective activity.
d. **rule-box**, a rectangle formed by ruled or printed lines.
1928 *Publishers' Weekly* 30 June 2605 Above the stamp ..must be *printed* the words..enclosed in a rule box.
e. *pl.* used *Comb.*, chiefly in senses 3 b and 5, as *rulesmaker, rulespeople*.
1963 *Punch* 20 Mar. 416/1 Does anyone ever know the order of the draw? Yes, the rulespeople. 1974 *Sunday* (Charleston, S. Carolina) 21 Apr. 3-A/3 Rulesmakers said politicians were put in a class by themselves, 'because they get to practice all year in their daily work'. 1978 *Detroit Free Press* 2 Apr. 6E/3 Whatever the coach can do for them, Bolinger-Boden-Markovich and all the other offensive linemen around have received very big help recently from the rulesmakers in their game.

rule, *v.* Add: **I. 1. b.** (Later example.)
c 1921 D. H. LAWRENCE *Mr. Noon* viii, in *Mod. Lover* (1934) 266 It's just like him—but there you are. Those that won't be ruled can't be schooled.
5. a. *Rule, Britannia*: the usual name for a patriotic song sometimes sung on public occasions in Britain. Also as *attrib. phr.* Hence **Rule-Brita·nnia,ism** chauvinism (*nonce-wd.*).
The name is taken from the refrain of the song (see quot. 1740).
[1740 THOMSON & MALLETT *Alfred* II. v. 42 *Ode*... Rule, Britannia, rule the waves; Britons never will be slaves.] 1806 [see BUFF *sb.*² 6]. *a* 1888 [used in def. of BRITON *sb.* 1 c]. 1898 *Academy* 8 Oct. 25/1 A preference for accuracy above Rule-Britanniaism. 1899 KIPLING *Absent-Minded Beggar* 1 When you've shouted 'Rule Britannia', When you've sung 'God save the Queen'. 1918 *Daily Mirror* 12 Nov. 2/1 When the strains of 'Rule, Britannia!' rang out his Majesty raised his naval cap. 1936 G. B. SHAW *Simpleton* I. 24 Let the whole earth be England; and let Englishmen rule it. (Singing) Rule Britannia: Britannia rules the wa— He blows his brains out. 1941 'G. ORWELL' *Lion & Unicorn* 19 In England all the boasting and flag-wagging, the 'Rule Britannia' stuff, is done by small minorities. 1968 *Listener* 18 July 86/2 Judges are good at making grand Rule Britannia statements, like Judge Salmon's in 1958... 'Everyone.. is entitled to walk the streets in peace,..and free from fear.'
d. Slang phr. —— *rule(s), O.K.*, used orig. in wall graffiti to affirm the superiority of a gang, football team, etc. Freq. in *transf.* use.
1975 S. JACOBSON in *New Society* 27 Mar. 780 (*title*) Chelsea rule—okay. 1976, etc. [see *O.K. *a.* a]. 1981 *Times* 31 June 1/1 It is a case of the tobacco industry rules, OK.
6. a. (Later example.)
1964 *Financial Times* 3 Mar. 2/3 Prices yesterday ruled fully firm for all descriptions of merino fleece and skirtings.
b. (Later examples.) Also *transf.*
1909 *Chambers's Jrnl.* June 409/1 During the past year the longest period when the wind velocity ruled below five miles per hour was only seven days. 1979 *Morning News* (Karachi) 24 May 7/1 In the jutes section Indus and Pak Jute ruled firm. Sugar shares were irregular.
II. 8. a. Also with object and complement.
1928 H. G. WELLS *Mr. Blettsworthy* iii. 154 He it was had first ruled me insane and immune from Reproof.
d. (Earlier U.S. and later examples.) Also, more generally: to eliminate as a possibility; to make impossible; to decide against.
1869 'MARK TWAIN' *Innoc. Abr.* li. 539 Though they have been ruled out of our modern Bible, it is claimed that they were accepted gospel twelve or fifteen centuries ago. 1883 —— *Life on Mississippi* xlv. 413 One of these [mules] had to be ruled out, because he was so fast that he turned the thing into a one-mule contest. 1903 J. ST. L. STRACHEY in 'Vigilans sed Æquus' *German Ambitions* p. vii, To rule out the writings of the men on whom 'Vigilans sed Æquus' has based his Letters because they are obscure..is to misunderstand the evolution of public affairs in Germany. 1925 N. E. ODELL in E. F. Norton *Fight for Everest: 1924* 335 The disadvantage of the North Col is the fact that the camp here must be pitched on snow, though under all but the worst conditions this need not rule it out. 1928 *Daily Tel.* 12 June 13/4 The possibility of a battle between the rival Southern commanders cannot be ruled out. 1966 C. MACKENZIE *Paper Lives* vi. 85 The Right Honourable Henry Upjohn thought for a moment about trying that joke at the next political meeting in his constituency but ruled it out at once. 1971 I. BUTYKAI tr. *Lukovich's Electric Foil Fencing* II. 84 Certain parts should be ruled out as being compulsory so that the combined movement should present an acceptable, applicable and expedient picture. 1976 *National Observer* (U.S.) 13 Nov. 5/2 Nothing in the group's by-laws rules out inmates, officials said, so the invitations stand. The prisoners said they were 'very pleasantly surprised' to find they could join Phi Kappa Phi.
e. With *in*, used in opposition to sense 8 d above.
1904 G. B. SHAW *Let.* 6 Dec. (1972) II. 471 It is just this personality that rules her out, whereas if we had a scrap of originality it would rule her in. 1973 *Observer* 17 June 1/1, I haven't ruled it out and I haven't ruled it in.
III. 10. b. *Comm.* With *off*: to close (the books) for the day. Also *absol.*
1977 *Times* 17 Sept. 20/4 Books were eventually ruled off within a band of 5½–6½ per cent. 1978 *Times* 15 Aug. 18/8 Closing balances were being found at sharply lower levels, so that books were eventually ruled off within a band of 4½ per cent to 6½ per cent. *Ibid.* 17 Aug. 21/3 Houses ruled off anywhere between 2 per cent and 4 per cent.
11. (Later *fig.* example.)
1924 R. CAMPBELL *Flaming Terrapin* i. 18 As he rose up, the moon with slanted ray Ruled for those rapid hoofs a shining way.

rule-joint. Add: Also *attrib.*
1966 A. W. LEWIS *Gloss. Woodworking Terms* 47 *Rule-joint hinge*, brass hinge with one long and one short leaf and with the countersinking for the screw heads on the

opposite side to the knuckle. *Ibid.* 95 Stay, metal fitting which limits the movement of a door or lid, e.g. a shad stay, or a rule-joint stay.

rulelessness. (In Dict. s.v. RULELESS *a.*) (Later example.)
1969 P. ANDERSON in Cockburn & Blackburn *Student Power* 222 Durkheim's account..produced the concept of *anomie*—the unceasing reproduction of subjective rulelessness by a society that is defined by its ensemble of objective rules.

rule-maker. Add: Hence **ru·le-making** *sb.*
1946 *Amer. Bar Assoc. Jrnl.* XII. 599 Study..of rule-making in action in those states where the courts still retain much of their rule-making power. **1946** *Nature* 21 Dec. 894/1 In Africa the rule-making power of native authorities can have a significant influence on future developments in this field. **1964** GOULD & KOLB *Dict. Social Sci.* 585/1 In sociology and related fields, *legislation* is sometimes applied to the function and products of rule-making by any agency, governmental or private, the rules of which are normally accepted by the persons to whom they apply. **1979** *Railway Age* 31 Dec. 14/2 The ICC Bureau of Operations agreed with the proposal to institute a rulemaking, but was less eager than the OPA to see IPD eliminated.

rule of thumb. Add: **1.** (Later examples.) Also, a particular stated rule that is based on practice or experience.
1906 [see *drill sb.*¹ 7 b]. **1965** C. D. ERY *Siege of Alcázar* (1966) viii. 135 In doubtful cases a rule of thumb applied: if the prisoner employed one servant in his household or two workers in his business, then he was a Fascist. **1967** G. F. FIENNES *I tried to run a Railway* ii. 14 George Jackson who lived by rule of thumb faster and as accurately as any Graybeer. **1976** *Inorg. Chem.* XV. 1032/2 The *s.t.* spectra..show no apparent deviation from the rule of thumb that vibrational spectra of mixed-valence compounds are approximate superpositions of the single-valence spectra. **1976** *Southern Even. Echo* (Southampton) 11 Nov. 3/7 The rule of thumb over the tenancy of a council home should be 'follow the children'. **1977** *National Observer* (U.S.) 1 Jan. 7/2 The usual rule of thumb in the real-estate business is that a family can afford a house 2 to 2½ times its income. **1980** *Jrnl. R. Soc. Arts* Feb. 166/2 By day that same boy's master, and overlooker, and fellow-workmen, are all teaching him..that rule of thumb is the only safe guide.

2. a. (Later examples.) Also in predicative use.
1935 E. WAUGH *Edmund Campion* ii. 55 Old-fashioned priests..came to him when they found their simple, rule-of-thumb dialectics insufficient to cope with their trained opponents. **1947** R. M. FORSTER in *Harper's Mag.* July 15/2 Virginia Wood..believed in reading a book twice. The first time she abandoned herself to the author unreservedly. The second time she treated him with severity and allowed him to get away with nothing he could not justify. After these two readings she felt qualified to discuss the book. Here is good rule of thumb advice. **1962** W. NOWOTNY *Lang. Poets Use* iii. 53 This attitude..however rule-of-thumb it may be, is reasonable enough. **1977** *N.Y. Rev. Bks.* 24 Nov. 16/1 Ridiculing the 'rule of thumb' methods used in the house-hold.

b. (Later example.)
1947 [see *NOMOGRAPHER].
Hence **rule-of-thu·mbite**, a person who works by rule of thumb (*nonce-wd.*).
1916 H. G. WELLS *Mr. Britling* I. i. 16 Ruskin and Morris..were as reactionary and anti-scientific as the dukes and the bishops. Machine haters. Science haters. Rule of Thumbites to the bone.

ruler, *sb.*¹ Add: **6. Comb.** (sense 3 b) **ruler-straight** adj.; (sense 1) **ruler-cult** *Antiq.*, worship offered to a hereditary ruler; also *transf.*
1928 A. D. NOCK in *Jrnl. Hellenic Stud.* XLVIII. [title] Notes on ruler-cult. **1955** M. P. NILSSON *Cults, Myths, Oracles, & Politics in Anc. Greece* iii. 108 The ruler cult was from the age of the Epigoni the state religion of the Hellenistic monarchies in the East. **1958** *Times* 15 Nov. 9/7 Revolutionary and critical times may produce the phenomenon we call ruler-cult. **1960** *Lebende Sprachen* V. 35/3 Ruler-straight seam, schnurgerade Naht. **1963** A. LUBBOCK *Austral. Roundabout* 14 The horizon meets the sky in a ruler-straight line.

ru·lered, *ppl. a. poet.* [f. RULER *sb.*¹ (sense 3 b) + -ED¹.] = RULED *ppl. a.* 4.
1952 L. MACNEICE *Ten Burnt Offerings* 50 Steam is a dry word; the best word is water... Best in the East..on the rulered page of a Moghul garden—Cool marrow of marble spines.

rulereen. For *rare*⁻⁰ read *rare* and add example.
1937 G. FRANKAU *More of Us* xii. 125 Was this The ruleress of waves, R.N., all-British, Who stooped to plant the Cytherean Kiss?

ru·lering, *vbl. sb. rare.* [f. RULER *v.* + -ING¹.] The action of RULER *v.*; a beating with a ruler.
1849 DICKENS *Dav. Copp.* (1850) viii. 77 Tear-blotted copy-books, canings, rulerings.

rulership. Add: **1. a.** (Later example.)
1975 F. HEER *Charlemagne* xv. 225 The Emperor dreamed that a man came to him with a present from God, a sword symbolising rulership.
c. Rulers collectively.

1964 *Listener* 11 June 945/2 The emergence of a professional Civil Service alongside an increasingly amateur political rulership.

Rules (rūlz). [pl. of RULE *sb.*] **1.** *Racing.* (*a*) Jockey Club Rules (the Rules of Racing). (*b*) National Hunt Rules.
1848 A. E. T. WATSON *Turf* vi. 128 Unauthorised meetings—that is to say, meetings not under Rules. **1965** *Austral. Encycl.* IV. 134/2 It is known as the Australian Game, National Football, or Australian Rules. **1976** *Horse & Hound* 3 Dec. 10/4 He is one of twin brothers who have both ridden several winners under Rules.
2. = *Australian rules* s.v. *AUSTRALIAN *a.* b.
1946 D. STIVENS *Courtship of Uncle Henry* 18 In those days..they played Rules in long pants that reached below the knee. **1965** *Austral. Encycl.* IV. 134/2 It is known as the Australian Game, National Football, or Australian Rules. In some parts it becomes merely 'Rules'. **1967** *Canberra Times* 17 June 27 Victoria expects to win over W.A. in Rules. **1976** *Sydney Morning Herald* 27 May 22 Rules penalty upsets Saints.

ruling, *vbl. sb.* Add: **3. a. ruling engine,** a machine for engraving equally spaced parallel straight lines on a surface.
1901 *Physical Rev.* XII. 9 It is generally assumed in treating the grating that the lines of the ruling are of equal width and are separated by equal spaces. This cannot be the case in view of..the almost inconceivable rigidity of the ruling engine which would be necessary. **1969** D. RICHARDSON in R. Kingslake *Appl. Optics & Optical Engin.* V. ii. 28 Most ruling engines use screws as the basic reference standard. **1971** *Sci. Amer.* May 126/1 A narrow strip of phosphorescent paint was applied down the middle of this detector, which could be moved along the support with the screw of a ruling engine.
b. (Later examples.) *spec.* in *Palæography*, the lines ruled by the scribe on a page or throughout a manuscript.
1944 P. HODGSON *Cloud of Unknowing* (E.E.T.S.) p. x, Single columns of text, usually 34 lines to a page, vertical and horizontal rulings, [etc.]. **1958** *Scriptorium* XII. 51 [title] The ruling of the Exeter Book. **1963** H. R. KER *Owl & Nightingale* (E.E.T.S.) p. xii, The pencil ruling is often indistinct. **1976** *Codicologica* I. 78 Other aspects of the medieval book: the nature of parchment, ink, pricking, and ruling.

ruling, *ppl. a.* Add: **1. a.** (Further examples.)
a 1832 BENTHAM *Wks.* (1843) X. 571/1 He hates the ruling few; but he does not love the subject many. **1868** R. COBDEN Let. in W. L. Burn *Age of Equipoise* (1964) ii. 69 There has been a great reaction..among that which I call the ruling class, against..humanitarianism. **1943** J. B. PRIESTLEY *Daylight on Saturday* ii. 39 Cheviot..was a good engineer... He belonged to tomorrow's ruling class. **1958** D. KELLY (*title*) The ruling few. **1963** T. H. PEAR *Eng. Social Differences* i. 15 An undiscriminating ear mistakes for 'ruling-class speech' a synthetic approximation to it. **1964** L. DAVIDSON *Rose of Tibet* xv. 280 The ex-enemy was now treating quite amiably with 'ruling circles' exiled in Chumbi. **1964** T. B. BOTTOMORE *Elites & Society* i. 6 Mosca's 'political class' is nothing but the intellectual section of the ruling group. *Ibid.* iv. 71 A second group which has attracted attention as a potential ruling élite is that constituted by the managers of industry. **1972** R. CRAWFORD' *Whip Hand* II. i. 58 Neville, for all his blue blood, was closer to the breadline than he was... How could a ruling class produce such examples of weakness? **1979** *Dædalus* Summer 1 They are the very models of ruling-class incompetence.
b. (Later examples.)
1945 J. T. COX *Practice Church of Scotland* 104 All are elders..ministers being teaching or preaching as well as ruling elders, and the others 'ruling elders' only. In the Congregational Churches 'ruling elder' are restricted to such as are members of a Kirk Session, exclusive of the minister of ministers. **1976** *Marlboro Herald-Advocate* (Bennettsville, S. Carolina) 18 Apr. 8/3 A congregational meeting has been called..for the purpose of electing one Ruling Elder to serve on the session.

rulley. (Earlier *Comb.* and later examples.)
1806 W. SHOUT in *N. & Q.* (1963) Apr. 136/1, 6 rulley load of stone—7.. 6. **1977** *Times* 14 Oct. 16/6 East Yorkshire's College of Agriculture has converted a strong cob, Folly, and four-wheel rulley to assist their shepherd on his rounds. **1978** J. CUMMINGS *Railway Motor Buses* I. 40 The third Stirling, fitted with a rulley body.. found a home with the Mechanical Engineers Department at Hull. **1979** *Bull. Yorks. Dial. Soc.* Summer 9 When the tide rose again they sailed away to Hull where they [sc. the goods] were transferred on rullies to the station and entrained to Hornsea.

ruly, *a.*¹ **1.** (Later *poet.* example.)
1952 DYLAN THOMAS *Coll. Poems* 165 And truly he Flows to the strand of flowers like the dew's ruly sea.

rum, *sb.*¹ **1. b.** (Earlier example.)
1779 J. WOODFORDE *Diary* 12 Aug. (1924) I. 258 Rum is a good spirit.
c. For *U.S.* read *N. Amer.* and add earlier and later examples.
1800 *Upper Canada Gaz.* (York, Ontario) 5 Apr. 3/2 Many have labored to calumniate Rum, and render it unpopular, by dwelling on some of its supposed bad effects. **1851** *Voice of Fugitive* (Windsor, Ontario) 5 Nov. 2/5 Rum and Negro hate [are] the two great public evils of our time. **1918** W. A. MACKAY *By Trench & Trail* 15 No one will rejoice more than Oscar Dhu to see the demon rum utterly destroyed in Canada ere many moons. **1933** E. O'NEILL *Ah, Wilderness!* II. 74 'Never marry a woman who drinks! Lips that touch liquor shall never touch yours!'..Too bad! So fine a woman once—

and now such a slave to rum! **1957** *Prairie Overcomer* (Three Hills, Alberta) Dec. 444/2 Of these two foes we cannot say whether *Rome* or *rum* is the greater adversary of the pure Gospel.

2. a. rum-bottle (later examples), *distillery, ration, still.*
1847 THACKERAY *Van. Fair* (1848) xxxix. 359 The three tumblers and the empty rum-bottle. **1967** A. LICHINE *Encycl. Wines & Spirits* 464/2, I found..gave a customer the sugar-syrup bottle, the rum bottle, a tumbler, and left him to himself. **1774** N. CRESSWELL *Jrnl.* 4 Sept. (1925) 34 In the evening went with Mr. Perkins to see Mr. Kid's Plantation. The Sugar works and Rum distilleries are very extensive. **1816** *Mass. Hist. Soc. Coll.* and Ser. IV. 124 A rum distillery was established in 1738. **1968** *Spirits* ('Know the Drink' Ser.) 32/2 Sugar factories and rum distilleries are much larger and more efficient than they were. **1923** KIPLING *Irish Guards in Gt. War* I. ii. 89 Their bivouacs..where a hot meal..and a rum-ration awaited them. **1977** *Amer. N. & Q.* XV. 135/2 Temperance campaigns, reduced rum rations, more leave, recreational programs, recruitment of higher caliber personnel, all helped to reduce alcoholism. **1914** F. C. GLASS *With Bible in Brazil* iv. 45, I recalled the big rum-still in the back-yard. **1968** *Spirits* ('Know the Drink' Ser.) 33/1 Rum stills are often fitted with 'rectifiers' which allow the least volatile elements to return to the pot.
b. rum-maker, -seller (earlier and later examples).
1902 J. MASEFIELD *Odtaa* i. 4 The northward provinces became sparsely inhabited by..sugar-growers, rum-makers, and copper-miners. **1781** J. GREENWOOD in *Maryland Hist. Mag.* (1910) V. 125 We..took in..seven passengers, who were visitors or rum-sellers to Gen. Washington's army. **1900** *Congress. Rec.* 25 Jan. 1100/2 A rumseller is as bad as a pickpocket. **1923** H. ROBERTSON *Grass Roots* iv. 76 There seems to be no limit to the number of offences these licensed rumsellers can commit.
d. adjs. *rum-bathed*; also *rum-brave.*
1972 G. B. SHAW *Let.* 13 May (1965) I. 264 His rum bathed hair. **1924** E. HEMINGWAY *Transatlantic Rev.* Apr. 119/2, I know you haven't got any guts unless you've got rum... I want you rum-brave. I don't want you useless.

3. rum baba: see *BABA²; **rum baron**, a magnate in illegal liquor traffic; **rum butter**, a hard sauce made from rum and butter; **rum chaser** *U.S.*, during the Prohibition era, a coast-guard speedboat for pursuing rum-runners; **rum-cherry** *U.S.*, the wild black cherry, *Padus serotina*, or the tree bearing this fruit; **rum cocktail**, a cocktail in which rum is the principal ingredient; **rum fleet** *U.S.*, during the Prohibition era, a 'fleet' of ships engaged in rum-running; **rum-hitting** [cf. *HIT *v.* 22* b] *vbl. sb.*, excessive drinking of rum; **rum-hound** *slang*, (*a*) = *rumpot*; (*b*) a prohibition agent (*U.S.*); **rum-jar** *slang*, a type of German trench-mortar shell; **rum jelly**, a concentrated form of rum; **rum-joint** *U.S.*, formerly, a place where illicit liquor was sold; **rum-mill** *U.S.*, a tavern or liquor shop; **rumpot** *N. Amer. slang*, a habitual heavy drinker; **Rum Rebellion** *Austral. Hist.*, the rebellion against Governor William Bligh by officers of the New South Wales Corps (noted for trafficking in rum) in 1809; **Rum Row** *U.S.* (see quot. 1927); **rum-runner**, (*a*) one who smuggles or lands illicit liquor; (*b*) = *rum ship*; **rum-running** [RUN *v.* 43 c] *vbl. sb.*, smuggling or landing prohibited liquor; also as *ppl. adj.*; **rum ship**, a ship engaged in rum-running; **rum shop** *U.S.* and *Caribbean*, a shop or tavern selling rum and other liquor; a saloon.

1923 *Westm. Gaz.* 4 Apr. 8/5 Reminiscences are involuntarily called up by bearing on barons. **1975** H. WHITE *Raincoast Chron.* (1976) 13/1 A few 'rum barons' could be apprehended in the United States. **1880** A. B. MARSHALL *Cookery Bk.* ii. 58 Rum Butter... Prepare as in foregoing recipe, using Liquid Sunshine rum instead of brandy. **1939** [see *brandy-butter* s.v. *BRANDY *sb.* 2]. **1967** 'J. MUNRO' *Money that Money can't Buy* i. 8 Shops that sold Lakeland jet, woollens and rum butter. **1973** *Country Life* 26 Oct. 1041/1 Butter was..taken to Wigton market..with a special delicacy called rum-butter. **1924** *Rudder* Jan. 40 Congress will be asked to appropriate many millions of dollars for a fleet of rum chasers. **1931** D. RUNYON in *Hearst's Internat.* May 64/2 She is riding in a big foreign automobile the size of a rum-chaser. **1829** A. H. LINCOLN *Familiar Lect. Bot.* 301 Wild-cherry, rum-cherry, cabinet cherry... In dense forests, it grows to a very great height. **1843** *Knickerbocker* XXI. 365 They had been feeding him upon that inebriating article of food, rum-cherries. **1906** N. L. BRITTON *N. Amer. Trees* 506 This well-known tree, also called the Black Cabinet, or Rum cherry, is abundant in mixed forests and cleared clearings, from Nova Scotia..southward to Florida. **1949** COLLINGWOOD & BRUSH *Knowing your Trees* 256 They have a pleasant, slightly bitter taste and are sometimes used in a beverage called 'cherry bounce' hence the name 'Rum Cherry'. **1861** *Harper's Mag.* Jan. 150/2 Measures of the most vital importance are first introduced in rum-cocktails, then steeped in whisky, after which they are engrossed in gin for a third reading. **1936** A. THIRKELL *August Folly* vi. 181 Richard..had made and drunk two rum cocktails. **1976** J. VAN DE WETERING *Tumbleweed* x. 93 The iced rum cocktail went down well. **1923** *Westm. Gaz.* 4 Apr. 8/5 Off-shore

is the rum fleet. **1975** J. GORES *Hammett* xvi. 113 Dom brings in most of the real Canadian from the rum fleet these days. **1910** J. MASEFIELD *Ballads & Poems* 34 There's..Stabbing, of course, and rum-hitting, Dirt, and drink, and stink, and crime. **1918** L. E. RUGGLES *Navy Explained* 120 Rum Hound, a boozer, or a man who likes his oil. **1920** *Ade Hand-Made Fables* 5 Just as the western Sun was ducking behind the Hills, the amateur Rum-Hounds piled out. **1951** E. PAUL *Springtime in Paris* vi. 198 What he resented was the insinuation that he was a chronic rumhound. **1918** P. MCGREGOR *Let.* 29 June in M. Moynihan *Greater Love* (1980) xi. A 'Rum Jar', the largest Hun shell known on our front, can knock in yards of trenches. **1923** KIPLING *Irish Guards in Gt. War* I. 252 Rum-jar by rum-jar, borne joyously through the dark streets. **1924** *Listener* 17 Sept. 431/1 The Germans also stepped up their mortar fire on our front-line trenches—the fearful 'rum jars'. **1976** J. VAN DE WETERING *Tumbleweed* x. 93 The rum comes from Jamaica, packed in drums, rum jelly. We mix it with water in a little factory. **1928** *Sunday Express* 24 June 8/4 One of that men started a row with one of these birds... They fought in a rum-joint and everyone joined in. **1865** *Yankee Humour & Uncle Sam's Fun* 87 Every rum-mill, groggery and tippling-shop..is a trap set by the devil to catch those who are guilty of not having over the cents. **1867** [see *TREADFALL* 2 c]. **1883** SANBORN & LELAND *Dict. Slang* I. 238/1 *Charter the bar, charter the grocery,* to (American), to buy all the liquor in a groggery or 'rum-mill' and give it away freely to all comers. **1939** R. CHANDLER *Big Sleep* xxvii. 241 I'm not! All he sees..is this rumpot bum. **1942** *Sun* (Baltimore) 17 Feb. 18/3 He admires good food and good whisky. 'All cooks,' he said, 'I mean, all good cooks, is rumpots.' **1977** T. N. RADDALL *Hangman's Beach* iv. xxii. 345, I had him moved in there as soon as that rumpot vouched for him. **1855** W. HOWITT *Land, Labour & Gold* ii. 118 From the date of this 'rum rebellion', and the forcible deposition of poor Bligh..the system of political grants went on swimmingly. **1958** H. V. EVATT [*title*] Rum Rebellion. **1966** G. W. TURNER *Eng. Lang. in Austral. & N.Z.* i. 8 The opposition of the Corps to authority culminated in a rebellion (the 'Rum Rebellion') against Governor William Bligh. **1924** *Lit. Digest* 26 May 54/2 Small consignments are carried from there down to the 'Rum Row' of ships anchored beyond the three-mile limit of the Long Island and New Jersey shores. **1927** W. E. COLLINSON *Contemp. Eng.* 81 We all know..about..Rum Row (where the liquor ships gather outside the prohibited area). **1949** JRNY & SLOCUM *Tax Dodgers* i. 16 There were three Rum Rows, one on each coast and a smaller one working in the Gulf of Mexico. **1920** *N.Y. Times* 19 Sept. 6/1 The Detroit rum runners have had a good deal of notoriety. **1923** H. L. FOSTER *Trop. Tramp with Tourists* 3 To that a rum-runner! **1941** R. SCHULBERG *What makes Sammy Run?* xii. 264 He got up to leave. **1973** *Washington Post* 13 Jan. C7/1 Waiting for a rum runner's boat. [?]... **1923** *Westm. Gaz.* 4 Apr. 8/5 The rum runners..one solitary light moving slowly along the horizon... It was a rum runner. **1969** Smithsonian Aug. 5/2 The match became known as 'the rum-runners' paradise' because so many Coast Guard patrol boats had to be diverted to control the spectator fleet. **1924** *Lit. Digest* 31 May 58/1 Dozens of small rum-running boats are laid up at least a temporary setback. **1926** *Scribner's Mag.* Aug. 166/2 Lately—on the third of mine, in the rum-running game now—will get you over the line into Canada. **1930** *Aberdeen Press & Jrnl.* 1 Feb. 7/5 Grey Ghost, one of the fastest rum-running craft on Lake Erie, has been sighted locked in the ice. **1959** N. MAILER *Advts. for Myself* (1961) 65 With the things he'd done, the Marines in Nicaragua, ..rumrunning in New Orleans, somehow he'd beaten..to death. **1924** *Sunday Express* 26 Sept. 7/1 The rum ship lay outside Yesterday's limit. **1932** D. MARTIN in... **1924** Rum-ship [see *HIJACKING *vbl. sb.* and *ppl. a.*]. **1931** F. L. ALLEN *Only Yesterday* x. 245 Rum-ships rolling in the sea outside the twelve-mile limit. **1838** W. STEPHENS *Jrnl.* 30 Dec. in Colonial Rec. Georgia (1906) IV. 242 Those pernicious Rum-Shops were become as common among the People, in Proportion, as Gin-Shops formerly at London. **1873** 'MARK TWAIN' *Gilded Age* xl. 426 Eighty per cent of any average American town could enter a rum shop in a tall saloon. **1953** M. SADLER *Windswept & Other Stories* (1969) 17 Then I bear in the rumshop. **1974** *Sunday Advocate-News* (Barbados) 24 Feb. 17/1 Well populated with rum shops and nightclubs and many stores having a horse racing betting shop, the area has become the noisiest in the country.

rum, *sb.*⁴ Also **rhum.** *U.S.* [Origin uncertain.] A form of rummy (*RUMMY *sb.*²).
1910 [see *RUMMY *sb.*²]. **1912** *Official Rules of Card Games* (U.S. Playing Card Co.) 15 Rum. (This is a combination of Conquian and Whiskey Poker.) Objects of the Game.—To get rid of the cards dealt to the player by laying them out in triplets or fours, or in sequence and suit of three or more. **1913** *Chicago Record-Herald* 1 Apr. 10/1 I never found one of them The kale I lost at rhum. **1921** M. C. WORK *Auction for Two or Three* 79 The modern game of Rum resembles Conquian in many respects and was at first called 'Coon Can'. **1974** *Encycl. Brit. Macropædia* XVI. 25/1 Among the most widely played Rummy games is 500 Rum,..and its variants including Michigan Rum.

rum, *a.*² Add to def.: Also, bad, spurious. (Further examples.)
1783 *Session Paper Cent. Criminal Court* Oct. 951 By God, that's a rum one. **c 1803** G. COLMAN in R. B. Booth *Eng. Plays of 19th Cent.* (1973) III. 70 Dang me, but he's a rum customer! **1847** G. W. M. REYNOLDS *Mysteries of London* III. xlv. 71/2 Work the bulls and coutersrum. **1870** E. J. KEWAN *Palace & Hovel* xxii. 485 'Ah! that's a rum customer,' said the policeman; 'she's hy to heavy-think.' **1895** *Wales* July 323/2 What's rum is that he is one of the best 'uns in chapel. **1930** G. B. SHAW *Apple Cart* I. 3 *Pamphilius.* He must have been a rum old bird. *Sempronius.* Not rum enough to be noticed. There are millions like him. **1942** *Gen* 1 Sept. 13/1 Anything that is good in the navy is 'scran' and if it's bad it's 'rum'. **1955** J. THOMAS *No Banners* vii. 61 'This is a rum go,' Alfred said. **1971** H. A. SMITH *View from Chivo* xix. 192 A rum cove if ever I met up with one. **1977** J. I. M. STEWART

Madonna of Astrolabe xi. 153 Some Scottish names are distinctly rum. Yours is.
Comb. (Later example.)
1955 *Times* 16 Aug. 10/5 That's a rum looking swallow.

‖**rumaki** (rumā·ki). [Perh. altered f. Jap. *harumaki* spring roll.] An appetizer, of Oriental origin, consisting chiefly of chicken livers, water chestnuts, and bacon, marinated and broiled.
1965 R. CARRIER *Cookbk.* iii. 96 To make 24 rumakis (Japanese hot canapés), you will need 4 pound chicken livers, 24 half-slices of bacon, 8 water chestnuts and 24 cocktail sticks. **1972** *Village Voice* (N.Y.) 1 June 74/1 (Advt.), Chicken Liver Rumaki. **1978** *Chicago* June 221/1 Delicious Cantonese-style appetizers include superlative king crab egg rolls and the best rumaki we've ever tasted.

Ruman (rū·măn), var. ROUMAN *sb.* and *a.*
1878 *Chambers's Encycl.* VI. 512/2 Every Ruman who possesses a small yearly income is eligible for a seat in parliament. **1977** *Encycl. Brit.* XIX. 637/1 Tradition, embodied in a local chronicle of the 16th century entitled 'History of the Ruman land since the arrival of the Rumans', gives 1290 as the date of the founding of the Walachian state.

Rumanian (rumē¹·niăn), *sb.* and *a.* [Var. ROUMANIAN *sb.* and *a.*. Until recently the dominant spelling in the twentieth century, but now *Romanian* is the officially preferred form.] **A.** *sb.* **1.** = *ROMANIAN *sb.* 1, ROUMANIAN *sb.* 1.
1878 *Chambers's Encycl.* X. 709/1 The choice of the Rumanians fell upon Prince Charles of Hohenzollern-Sigmaringen, who was proclaimed Prince of R. on April 20, 1866. **1929** *Encycl. Brit.* XXXII. 231/1 As regards nationality [in Rumania], there are Rumanians, 5,489,036; foreigners, 171,065; and nondescript (principally Jews), 272,241. **1935** HUXLEY & HADDON *We Europeans* vii. 213 The largest nation in the Balkan peninsula is that of the Jugoslavs,.. who are largely separated from their northern brethren by the Magyars and Rumanians. **1971** HOWE & TAYLOR *Dict. World Hist.* 1331/1 Rumanians expected this to lead to union and autonomy.
2. = *ROMANIAN *sb.* 2, ROUMANIAN *sb.* 2.
1889 in Cent. Dict. **1902** *Encycl. Brit.* XXX. 396/2 The urban and most of the rural Vlachs are bilingual, speaking Greek as well as Rumanian. **1948** A. L. KROEBER *Anthropol.* (rev. ed.) vi. 244 These Greek idioms and structure features recur in Latin-derived Rumanian. **1968** J. LYONS *Introd. Theoret. Linguistics* v. 204 The so-called 'postpositive' articles of Swedish, Rumanian, ..Macedonian, etc. **1974** [see *ROMANIAN *sb.* 2]. **1975** *Language* LI. 411 In Rumanian, the polite pronoun..is, used to govern the plural.
B. *adj.* = *ROMANIAN *a.*, ROUMANIAN *a.*
1878 *Chambers's Encycl.* VII. 313/1 In 1877, the entire Rumanian military force numbered 114,668 men. **1902** *Encycl. Brit.* XXXII. 313/2 The Rumanian Church is autocephalous, but holds the same dogmas as the Orthodox Greek Church. **1927** PEAKE & FLEURE *Priests & Kings* xi. 177 The painted pottery people of the Rumanian plain spread along the foot-hills on the outside of the Carpathian arc. **1938** *Oxf. Compan. Mus.* 819/2 The Rumanian development of a national music..served to some extent as an incentive to Rumanian musicians. **1950** TREVELYAN & CAMPBELL *Encycl. World Politics* 372/1 On 7 October 1940 German troops occupied Rumania to secure Rumanian oil- and agricultural supplies for Germany. **1967** M. B. PICKEN *Fashion Dict.* 270/1 *Rumanian stitch*, series of parallel stitches intersected at center by shorter stitches. **1968** J. LYONS *Introd. Theoret. Linguistics* ix. 315 The criterion of 'interceptability' (or 'insertability') which distinguishes the English article as more 'word-like' than the Rumanian or Macedonian article. **1973** *Times* 2 Apr. 6/5 (Advt.), A fine quality Rumanian rug in superb Balkan colours.
Hence **Ruma·nianism**, Romanian identity, Rumanian nationalism.
1938 *Times* 1 Jan. 11/1 The new Government believed ..in the spiritual renaissance of Rumanianism through the Christian Church.

Rumanianize (rumē¹·niănaiz), *v.* [f. *RUMANIAN *a.* + -IZE.] *trans.* To make Rumanian in character. Hence **Ruma:niani·zation.**
1922 O. JESPERSEN *Language* xi. 205 A Saxon village which had been almost completely Rumanianized. **1938** *Times* 1 Jan. 11/1 The laws for the protection of Rumanian labour, and of the Rumanianization of undertakings which..employ foreigners instead of Rumanians.

rumänite (rumē¹·nait, rū·mănait). *Min.* Also **r(o)umanite.** [ad. G. *rumänit* (O. Helm 1891, in *Schr. d. Naturforsch. Ges. in Danzig* VII. iv. 186), f. *Rumänia* Romania: see -ITE¹.] A variety of amber containing sulphur and succinic acid and found in Romania.
1892 E. S. DANA *Dana's Syst. Min.* (ed. 6) 1004 *Rumänite*... A yellow amber-like resin obtained from different points in Rumania, as in sandstone in the Buseo district, at Telage in the Bohosa district, etc. **1904** L. J. SPENCER tr. *M. Bauer's Precious Stones* II. 554 Roumanite, or Roumanian amber, is rarely yellow, but is usually brownish-yellow to brown. **1932** G. C. WILLIAMSON *Bk. Amber* 214 The ordinary phrase of Black Amber applied by dealers to Rumanite is misleading, because its colour value is a very high one, and many examples are amongst the most lovely coloured pieces of Amber that have ever been found. **1962** R. WEBSTER *Gems* I. xxiii.

442 At several places in Romania is found an amber, named appropriately roumanite, which is said to contain less succinic acid and more hydrogen sulphide.

rumba (rv·mbă), *sb.* Also **rhumba.** [Amer. Sp.] An Afro-Cuban dance; a ballroom dance imitative of this, danced on the spot with a pronounced movement of the hips. Also, the dance rhythm of the rumba; a musical composition with this rhythm. Also *transf.*
1922 J. HERGESHEIMER *Bright Shawl* 112 Her life..was incredibly, wildly, debauched. Among other things, she danced, as the mulata, the rumba, an indescribable affair. **1928** *Nation* 15 Sept. 244/1 A half-dressed couple behind a stick window dancing the rumba. **1932** *New Yorker* 9 Apr. 39/2 Cerise's tango band reminds you of every illicit Paris border you ever had. Rumbas and tangos abound. **1934** H. MICHAEL *Peace without Honour* in J. W. Marriott *Best One-Act Plays of 1933* 135 Put on a dance record... It's a rumba. Will that do? **1939** [see *BEGUINE*]. **1950** A. LOMAX *Mister Jelly Roll* (1952) ii. 79 Time seemed to flow like a dreamy rhumba. **1958** M. STEARNS *Story of Jazz* iii. 26 The rhumba, which is by far the most popular Afro-Cuban, is considerably diluted for Western ears and has become a fixture at fashionable American night clubs. **1958** E. BORNEMAN in P. Gammond *Decca Bk. Jazz* xxi. 364 Ellington..started it all in 1930 with his jazz rumba *Maori.* **1974** *Encycl. Brit. Micropædia* VIII. 716/1 Best known for the dancers' subtle side to side hip movements with the torso erect, the rumba is danced with a basic pattern of two quick side steps and a slow forward step. **1980** *Tablet* 26 Jan. 84/1 A group of liturgical dancers swathed in white, practising the Our Father dance. No doubt this particular excursion into religious rumba was absurd.
b. *slang.* A spree.
Perhaps an erroneous use.
1934 E. HEMINGWAY in *Cosmopolitan* Apr. 108/3 He'd been giving the nigger a dollar a day and the nigger had been on a rumba every night. I could see him getting sleepy already.
2. *attrib.* and *Comb.*, as *rumba band, competition, dancer, dance, orchestra, record, rhythm*; **rumba-box** (see quot. 1961).
1944 H. MCCLOY *Panic* 110 The swish of a sand-filled gourd in a rhumba band. **1967** O. LANCASTER *With an Eye to Future* v. 119 The strains of 'Peanut Vendor' played by one of the newly fashionable rumba bands. **1976** *Islander* (Victoria, B.C.) 4 Jan. 2/2 As far as we know the African Bongo Tribe made the first bongo drums for the giant use in the modern thumba band. **1961** F. G. CASSIDY *Jamaica Talk* xii. 266 As for the rhumba-box..it is a recent invention or importation which simulates the tones of the bass viol by means of four pieces of metal of different gauges attached to a box (the resonance chamber), which vibrate when they are plucked with the fingers, and are tuned to correspond to the strings. **1978** *Stars & Sun of W.I.* x. 89 The Rastafarians with their home-made drums and rumba-boxes. **1934** M. SHARP *Cluny Brown* xix. 130 She entered for a Rumba competition. **1973** *Black World* Sept. 12/1 Guillén captured the vitality of Afro-Cuban life in a series of 'sones' (songs) which deal with cane cutters, rumba dancers, and folk types. **1941** B. SCHULBERG *What makes Sammy Run?* xi. 271 He was..dancing with her out on the patio to the rhumba orchestra. **1972** J. MCCLURE *Caterpillar Cop* viii. 118 Liebel had to raise her voice..above the rhumba record to catch his attention. **1932** *Radio Times* 8 Apr. 129 There is Southern planner to the Rumba Rhythm of a tango tune. **1977** *Encycl. Brit.* VII. 213/1 The Rumba rhythm: one, two, three, pause; one, two, three, pause. **1970** W. APEL *Harvard Dict. Mus.* (ed. 2) 744/1 After 1930 rumba rhythms were incorporated in jazz.

ru·mba, *v.* Also **rhumba.** [f. the *sb.*] *intr.* To dance the rumba. Also, to move as though dancing the rumba.
1936 G. GREENE *Brighton Rock* II. i. 66 The tune the band was playing, the crowd on the floor trying to rumba. **1944** M. SHARP *Cluny Brown* xvi. 156 Belinski at once rose, grasped her girl round the waist, and then began to rumba. **1955** E. TAYLOR *Game of Hide-&-Seek* II. i. 120 'Englishwomen should never rumba,' he told her. **1961** G. DURRELL *Whispering Land* iii. 85 He [sc. a fur seal] decided that the seagull should be taught a lesson, so he humped himself up indignantly and rumbaed towards it ferociously. **1970** V. CANNING *Great Affair* vii. 111 You once rhumbaed two boys to rhumba.

rumble, *sb.* Add: **1. c.** In sound reproduction, low-frequency noise originating as mechanical vibration in a turntable. Also *attrib.* and *Comb.*
1949 FRAYNE & WOLFE *Elem. of Sound Recording* xiv. 271 An additional factor that must be considered in noise measurements is the vibration or 'rumble' of the turntable. **1968** *Times* 29 Nov. (Sound of Leisure Suppl.) p. vii/1 Background noise while a record is playing can be very disturbing and it is the elimination of this noise, appropriately called rumble, which is the main concern of the manufacturers of gramophone records and record-playing units. **1970** J. EARL *Tuners & Amplifiers* i. 19 Sometimes the high-pass filter is referred to as the rumble filter, since its basic attenuation characteristic significantly reduces the amplifier's response to low-frequency noises generated by the turntable unit and passed on via the disc and pickup. **1971** *Hi-Fi Sound* Feb. 67/2 With a rumble-free turntable, a player may cost as much as an entire low-budget sound system. **1977** *Time* 10 Oct. 43/1 (Advt.), There is rumble from the cutting machine in most record grooves.
d. A rumour. *U.S. colloq.*
1961 P. A. BRODEUR in *Webster* s.v., Picked up the rumble..and thought he'd pass it on just in case. **1966** 'E. MCBAIN' *Eighty Million Eyes* iv. 67 The neighbourhood

rumble is that he was fooling around with one of their wives. **1974** E. MCGIRR *Murderous Journey* 63 The rumble is that he works for Marcello.
2. b. A street-fight between rival gangs. Also *fig. slang* (chiefly *U.S.*).
1946 *Amer. Mercury* Apr. 480 We're going to have a rumble with the Happy Gents tonight. Gang kids call these fights rumbles. **1953** KRAMER & KARR *Teen-Age Gangs* p. v, A 'rumble'—a wild group fight—which gang youths consider the glamorous high point of their existence. *Ibid.* i. 4 A leader naturally headed his followers in a rumble with another gang. **1958** H. SALISBURY *Shook-Up Generation* (1959) iv. 64 He would do things no other boy would dare. He would sound a cop on the beat and run away laughing. In a rumble he was like a wild-cat. **1969** C. BURKE *God is Beautiful, Man* (1970) 38 By the time they got the drink there was a big rumble brewin'. The Israelites set up a big crash pad and there was another gang that didn't like it and they decided they would have a real big rumble with these Moses people. **1971** P. L. CAVE *Chopper* iv. 28 So I missed out on a rumble tonight. **1977** *Time* 31 Oct. 55/1 Singer Frank Sinatra seldom ducks a rumble with a reporter.
4. Also in a motor vehicle, = *rumble seat. U.S.*
1929 W. FAULKNER *Sanctuary; Orig. Text* (1981) vii. 82 The car drew up... The ones in the rumble said nothing. **1939** —— *Wild Palms* 90 McCord drove them up to the lake on the Saturday night before Labor Day, the hundred dollars worth of food—the tins, the beans and rice..—in the rumble. **1941** B. SCHULBERG *What makes Sammy Run?* viii. 189 I'll look in the rumble... I think I have some.
6. [Perhaps a different word.] An interruption in the course of a crime; an alarm; a tip-off. *Criminals' slang.*
1911 C. G. ROE *Horrors of White Slave Trade* iv. 80, I posed as a theatrical manager..and caught many an unwary stage struck girl... I was taking small chances of being caught and in fact did not have a 'rumble' during all the time I was there. **1913** A. STRINGER *Shadow* v. 93 'But he blew out for 'Frisco this morning,' contended the puzzled Sheiner. 'Shot through as though he had just had a rumble!' **1914** JACKSON & HELLYER *Vocab. Criminal Slang* 73 If you walk on the main stem you'll get a rumble. **1927** D. HAMMETT in *Black Mask* Feb. 17/2 The neighbors give us the rumble. **1949** A. HYND *We are Public Enemy* i. 21 The cops had gotten a rumble that..gangsters were holed up. **1949** in Wentworth & Flexner *Dict. Amer. Slang* (1960) 437/1 If there's a rumble, we do the time. **1957** *Life* 9 Dec. 57 The boys slip into town. You wouldn't think they would be noticed. But some busybody catches on and puts in a rumble.
7. *attrib.* and *Comb.*, as **rumble seat** *N. Amer.* = *DICKY, DICKEY *sb.* 9 c; **rumble strip** (see quots. 1962 and 1975).
1912 *Collier's* 23 Mar. 20/3 She's burnin' the wind out of town in a college boy's car with big May on the rumble seat behind. **1929** M. LIEF *Hangover* 301 The next morning she packed a small bag of necessaries, stowed it in the rumble seat, and drove off. **1951** T. CAPOTE *Grass Harp* iv. 95 They'd stuffed her into the rumble-seat of Big Eddie's old coupé and driven straight to the jail. **1972** *Evening Telegram* (St. John's, Newfoundland) 5 Aug. 15/1 Three old-fashioned things you seldom see on the roads these days... Running boards, rumble seats and courtesy. **1962** *Punch* 27 June 960/2 American towns have been experimenting with 'rumble strips'—coarse-textured expanses of road surface just before cross-roads. **1974** *Oxford Times* 12 July 15/3 Rumble strips put down in the road to slow traffic had no effect. **1975** *Daily Mail* 3 Jan. 11/4 *Rumble strips*: These are tiny ridges on a road surface which cause a high-pitched whine as the car tyres pass over them... They indicate a hazard ahead.

rumble, *v.*¹ Add: **2. b.** (Earlier example.)
1803 G. COLMAN *John Bull* II. ii. 19 I've rumbled on the road, all night, Frank; my bones ache.
3. d. To have a gang fight. *slang* (chiefly *U.S.*).
1959 *Listener* 29 Jan. 201/2 Do you know why a 'diddley bop' should put on a 'stenjar' to 'go down' to 'rumble' at a 'jitterbug'... That is the language of the teenage gangs of New York. **1969** S. GREENLEE *Spook who sat by Door* xiv. 121 The teenage gangs..haven't been rumbling and so they have a lot of latent hostility to get rid of. **1977** 'E. MCBAIN' *Long Time no See* viii. 116 We *was* tired, man. We been rumblin all the past month..gang-busting.
5. d. (Later examples.)
1963 *Times* 28 Jan. 9/5 Many people who even want Britain to enter the Common Market express a little joy in seeing de Gaulle rumbling your people. **1976** LD. ANNAN in *Ann. Rep. Univ. Coll. (London)* 1975–76 7 It looks..as if the Department of Education and Science has rumbled the Schools Council plan for a common system of school leaving examinations, a proposal which the universities regarded with the deepest suspicion.
e. To clean in a rumble.
1904 HARBORD & HALL *Metall. of Steel* xxxiv. 532 Small forged or stamped and malleable cast articles, which can be 'rumbled' bright in a shaking barrel, take the metal fairly well. **1957** *New Scientist* 7 Nov. 23/3 The thin parting bridges [of ball-bearings] are automatically severed and, after cooling, the balls are rumbled to remove the burr.
6. [Perhaps a different word.] *trans.* To get to the bottom of; to see through, understand, grasp; to recognize; to detect, discover, disturb. *slang.*
1886–96 in Farmer & Henley *Slang* (1903) VI. 75/2, I rumbled the tip as a matter of course. **1898** A. M. BINSTEAD *Pink 'Un & Pelican* ix. 209, I soon rumbled he was in it, when I heard Ball givin' him the 'me lord' for it. **1912** C. MACKENZIE *Carnival* (ed. 5) x. 126 I've

properly rumbled your friends. **1925** N. VENNER *Imperfect Imposter* iii. 30 He'd have rumbled me. He can't rumble me now. **1928** E. WALLACE *More Educated Evans* iii. 69, I rumbled you as soon as I took a screw through the winder. **1930** P. MACDONALD *Link* 74 If I hadn't had so many queer things happen to me in such a short time..I'd never have rumbled him. **1939** *Almanac for New Yorkers* 125 Some mugs rumbled us. **1956** 'A. GILBERT' *Riddle of Lady* viii. 115 The tobacconist..had been rumbled and compelled to give evidence. *Ibid.* xi. 178, I might have guessed you'd rumble me. **1959** *Encounter* Aug. 29/2 He evidently didn't rumble anything was at all unusual. **1966** J. BINGHAM *Double Agent* iii. 43 You've been rumbled. What's the use of a gun? **1979** E. NEWMAN *Sunday Punch* xvi. 145 'Have you any influence with him?' 'He'd rumble that. He'd think I was your agent.'

rumbled, *ppl. a.* Add: **2.** Also *fig.*
1930 R. CAMPBELL *Adamastor* 67 Nature..Admits no vegetable green... But with snarled gold and rumbled blue Must disinfect the sight.

rumbler. Add: **1. b.** A type of round bell on a harness, etc. Also *attrib.*
1953 A. JOBSON *Household & Country Crafts* xii. 130 But at Chichester the first horse had three bells, the second four and the third five. In the case of round bells they were known as rumblers. **1961** *Countryman* LVIII. III. 596 Some early ones had round 'rumbler' bells. **1971** *Country Life* 8 Apr. 839/2 Rumblers were another type of bell, circular with a ball inside to give them their characteristic sound. This type was quite common for cart horses or pack horses, but never used on a sheep as that animal did not give the bell the right kind of motion to make it rumble properly.
3. A machine for peeling potatoes.
1976 *Star* (Sheffield) 20 Nov. (Advt.), One 56 lb Triumph Superb Potato Rumbler, fully reconditioned, £300. **1977** *Grimsby Even. Tel.* 24 May 4/9 (Advt.), Potato room with an imperial rumbler and a Crypto chipper.

rumble-tumble. 4. Add: *Anglo-Indian.* (Further examples.)
1956 D. WALKER *Harry Black* xiii. 192 'What's for breakfast?' 'Keventer's sausages and rumble-tumble,' Christian said... She went off to feed the family..on sausages and scrambled eggs. **1966** J. & R. GODDEN *Two under Indian Sun* iii. 54 Rumble tumble, the Indian name for scrambled eggs. **1980** D. HART-DAVIS *Heights of Rimring* vii. 74 The rumble-tumble was a mighty omelette with strips of onion in it, accompanied by fried potatoes.

rumbling, *vbl. sb.* Add: **4.** Cleaning in a rumble. Also *attrib.*
1888 in *Lockwood's Dict. Mech. Engin.* 295. **1924** *Jrnl. Inst. Metals* XXXII. 294 Small articles, such as may be treated in rumbling barrels. **1965** G. J. WILLIAMS *Econ. Geol. N.Z.* xiv. 217/1 The relatively unaltered phenocrysts were experimentally removed by screening, cleaned by washing and rumbling, and analysed. **1976** *Western Mail* (Cardiff) 27 Nov., Further ancillary machines are avilable for drilling, tapping, rumbling and welding.

rumbly, *a.* (Earlier example.)
1874 L. TROUBRIDGE *Jrnl.* 3 Sept. in J. Hope-Nicholson *Life amongst Troubridges* (1966) 92 We..had the usual stuffy, rumbly drive.

rumbo³. Add: Also **rumbow.** (Earlier example.)
1846 *Swell's Night Guide* 130/2 *Rum bow*, rope stolen from any of the king's dock-yards.

rumbullion¹. For *Obs.* read '*Obs.* except *Hist.*' and add: Also, a glass or drink of rum.
1972 F. VAN W. MASON *Roads to Liberty* 19 Katie! A brace of rumbullions. Fast as you can brew 'em!

rumbunctious, var. *RAMBUNCTIOUS *a.*

rumbustiously (rvmbv·stiəsli), *adv.* [f. RUMBUSTIOUS *a.* + -LY².] In a rumbustious manner.
1966 *Listener* 10 Feb. 217/2 In *Horseman, Pass By!* we had a fair share of both, as well as poetry read rumbustiously by Mr O'Connor and grimly by Mr Kennelly. **1977** *Time* 5 Dec. 49/2 Lampie, the hard-drinking lighthouse keeper (played rumbustiously by Mickey Rooney), tells everyone that he has seen a dragon.

rumbustiousness (rvmbv·stiəsnės). [f. as prec. + -NESS.] Rumbustious character; boisterous behaviour.
1926 C. L. GRAVES *Hubert Parry* II. 106 In spite of his occasional exuberance and 'rumbustiousness' (a favourite word of his) he could, when the need arose, assume a wonderful dignity of demeanour. **1959** *Times Lit. Suppl.* 2 Oct. 556/4 More than one old White Russian..has felt his exiled convictions wavering a little before the rockets and rumbustiousness of the new Russia. **1977** *Gramophone* Jan. 1141/1 This come-to-the-fair rumbustiousness is well caught in the Scherzo.

rumdum, rumdumb (rv·mdvm), *a.* and *sb.* Also **rumdumm** and with hyphen. *N. Amer. slang.* [f. RUM *sb.*¹ + DUMB *a.* (*sb.*).] **A.** *adj.* **a.** Stupefied through drink; unconscious; incapacitated; stupid. **b.** Humdrum.
1891 *Brooklyn Eagle* 11 Sept. 2/4 *Rum-dumb*,..stupid with continual drinking. **1922** S. LEWIS *Babbitt* vii. 99 Don't faint with surprise if some of those rum-dumm

liars get one good swift poke from Mike. **1936** J. STEINBECK *In Dubious Battle* i. 15. Cop slugged me from behind, right in the back of the neck... I was rumdum for a long time. **1939** —— *Grapes of Wrath* xv. 215 He jus' stan's there lookin' at that dead kid. Can't get a word of 'im. Jus' rum-dumb. **1973** *Newsweek* 12 Mar. 96 Ponicsan finds surprising depth and touching delicacy in the rumdum lives he weaves together—dime-store and dinner women, odd-job truckers and coal-mine cripples. **1975** *Islander* (Victoria, B.C.) 14 Sept. 6/1 Mr. Brown had gone on a humdinger of a toot after a lucky find and wound up in bed rumdumb with no fire in his cabin.
B. *sb.* **a.** A habitual drunkard; a stupid person. **b.** Someone of ordinary ability.
1891 *Brooklyn Eagle* 11 Sept. 2/4 *Rum-dumb*,..an habitual soak. **1916** G. A. ENGLAND *Pod, Bender & Co.* 8 Why don't we lure in some rumdum of an ex-heavyweight. **1949** N. ALGREN *Man with Golden Arm* I. 16 The cell was full of a drifting flesh-coloured light and the murmuring rumdums were being let out of the cells to wash. **1960** C. HAMBLETT in J. Pudney *Pick of Today's Short Stories* XI. 137 Other drinkers..mocked at him and called him cloth-head, or rumdum, or plain moron. **1976** *Harper's Mag.* July 72/2, I beat the rum-dums but go down before quality players.

Rumelian, var. *ROUMELIAN *a.* (*sb.*).

Rumeliot, var. *ROUMELIOTE.

rumenal, var. RUMINAL *a.* in Dict. and Suppl.

rumenitis (rūmĕnəi·tis). *Vet. Sci.* [f. RUMEN + -ITIS.] Inflammation of an animal's rumen.
1905 [see *RETICULITIS]. **1963** JUBB & KENNEDY *Path. Domestic Animals* II. i. 51/1 A more common form of acute rumenitis, presumably chemical, develops after overeating on rapidly fermentable carbohydrate, usually grain. **1973** *Vet. Ann.* 1972 19 A deep haemorrhagic and necrotizing rumenitis.

rumenotomy (rūmĕnǫ·tǒmi). *Vet. Med.* [f. RUMEN + -O + -TOMY.] Incision into an animal's rumen.
1882 J. W. HILL *Bovine Med. & Surg.* xxiv. 587 Rumenotomy consists in removing the contents of the viscus through an artificial opening. **1973** [see *RUMINAL *a.* 2].

Rumford (rv·mfǭɹd). *Obs.* except *Hist.* The name of Count (von) Rumford (see RUMFORDIZE *v.*) used *attrib.* to designate kitchen articles or fireplaces designed by him or improved according to systems devised by him. Also *ellipt.* as *sb.*
1810 *Agric. Museum* (U.S.) I. 42 The dinner was principally prepared hot on the ground, by means of a portable Rumford kitchen... The utility of the portable Rumford had not probably been experienced on the field on any previous occasion in New England. **a 1817** JANE AUSTEN *Northanger Abbey* (1818) II. v. 87 The fire-place..was contracted to a Rumford. **1831** R. COX *Adventures Columbia River* II. xiv. 322 A bright brass footman..was suspended from the shining bars of a Rumford grate. **1854** THOREAU *Walden* 34 Spacious apartments, clean paint and paper, Rumford fire-place. **1937** M. LANSING *Mary Lyon through her Lett.* 230 All marveled at the Rumford oven, given by Deacon Safford. ..This Rumford oven was a sheet iron box with a compartment beneath in which the fire was built. **1951** *Dict. Americanisms* II. 1428/2 (*caption*) Rumford oven beside a fireplace.

rumgumptious, *a.* Add: Also **rumgumshus.** (Later example.)
1962 A. JOBSON *Window in Suffolk* vi. 96 A quarrelsome person was *rumgumshus*.

Rumi, var. *ROUMI.

ruminal, *a.* Add: Also **rumenal.** **2.** Of or pertaining to the rumen of an animal.
1923 [see *RETICULAR *a.* 4]. **1963** JUBB & KENNEDY *Path. Domestic Animals* II. i. 52/1 The ruminal environment is, except immediately after eating, anaerobic. **1973** HICKMAN & WALKER *Atlas Vet. Surg.* iii. 70 Rumenotomy is indicated..for the relief of rumenal impaction.

ruminant, *sb.* Add: **b.** A contemplative person. *rare.*
1940 C. S. LEWIS *Let.* 22 Mar. (1966) 179 Why should quiet ruminants like you and I have been born in such a ghastly age?

ruminate, *v.* **1. a.** (Later examples.)
1928 *Oxford Poetry* 7 About your poorly-mounted majesty Stand cow-faced women ruminating sales. **a 1961** A. HUXLEY in *Webster* (1961) s.v., Ruminating the contents of that last batch of letters she had received. **1975** *N.Y. Times* 26 Sept. 35/2 Mr. Rodgers has lived long, created much and filled an age with the sound of his music. Perhaps he has earned the right to ruminate..the art of autobiography.
ruminating *vbl. sb.* (later example).
1955 *Sci. News Let.* 16 July 36/2 Grazing animals frequently chew their cud for about nine hours out of every 24, and this ruminating is somewhat concentrated at night, Dr. Balch suggested.

rummage, *sb.* Add: **4. rummage sale** (*b*) (further examples); also *attrib.* and *fig.*
1910 M. BEERBOHM *Let.* 15 Nov. (1964) 192 It will be a different sort of exhibition..: a *retrospective* and rum-

mage-sale affair. **1912** *Strand Mag.* Aug. 155 Good wickets at rummage-sale prices. **1922** JOYCE *Ulysses* 737 Rummage sale a lot of trash. **1973** *Black World* Sept. 62 Your black skin Loose as a rummage sale coat. **1977** *Time* 24 Oct. 48/3 The *Road* shows were rummage sales of stuff out of vaudeville, burlesque—marvelously shoddy masterpieces of farce and fantasy, stitched together with clichés and ad libs.

rummage, *v.* Add: **I. 3. b.** (Later example with *up.*)
1889 J. K. JEROME *Three Men in Boat* iv. 57, I rummaged the things up into much the same state that they must have been before the world was created, and when chaos reigned.
5. b. (Earlier example with *out.*)
1803 *Lett. Miss Riversdale* II. 303 In rummaging out the contents of a *secrétaire*,..[she] had discovered a private drawer.
II. 8. b. Also const. *around.*
1883 'MARK TWAIN' *Life on Mississippi* iii. 36 He come rummaging around in the dark amongst the shingle bundles.

rummager. 2. (Further examples.)
1856 E. G. PARKER *Lesson of '76* 9 Layard, the great rummager of Nineveh. **1935** L. LUARD *Conquering Seas* iii. 44 Next time I'll let the rummagers put a stop to their nonsense.

rummish, *a.* (Earlier example.)
1760 DUCHESS OF NORTHUMBERLAND *Diary* 17 Aug. (1926) 26 Ld Dumfries very drunk, talk'd of being frisky & rummish.

rummy, *sb.*¹ Add: **2. a.** A habitual drunkard; an alcoholic. *slang* (chiefly *U.S.*).
1851 J. H. GREEN *Twelve Days in Tombs* 55 The learned counsel of the rummies opened his defence... the court adjourned, and the rummies repaired to *another bar* to congratulate each other upon the success of the morning. **1907** G. B. SHAW *Major Barbara* 170 Your Rummies of the tamest respectability pretending to a past of reckless and dazzling vice. **1939** [see *nut factory* s.v. *NUT *sb.*¹ 21]. **1962** E. LACY *Freeloaders* vi. 124 You'd go to pieces, become a rummy. **1975** *Publishers Weekly* 20 Jan. 67/1 A salty old rummy who specializes in pious letters pleading for help.
b. A stupid person; a blockhead; a sucker. *U.S. slang.*
1912 ADE *Knocking Neighbors* 108 She extracted a promise from Cousin and several other Desperate Characters that they would come out into the wilderness and give the Rummies a Touch of High Life. **1913** J. LONDON *Valley of Moon* 59 You'd better tell the rummy to beat it. **1937** *Reader's Digest* Oct. 36/2 Most rummies never discover they have been rooked.

rummy, *sb.*² orig. *U.S.* Also **rumme.** [Origin uncertain.] Any of a group of card games, similar to coon-can, the main object of which is to acquire runs or flushes of three or more cards. Also *attrib.* See also *gin rummy* s.v. *GIN *sb.*² 2 b, *OKLAHOMA.
1910 *Sun* (N.Y.) 10 Sept. II. 3/1 The leader this season seems to be a new round game that is called rum. Some persons have it rhum, rhummy and even rhumston. **1913** *Official Gaz.* (U.S. Patent Office) 13 May 536/2 Milton Bradley Company, Springfield, Mass. Filed Mar. 20, 1913. Rumme. **1915** *Chicago Herald* 30 Nov. 15/4 This gave him the idea the game was rummy and he spread the nines on the table. **1919** S. LEWIS *Free Air* xix. 202 He takes some kind of dope, and he cheats at rummy. **1928** L. NORTH *Parasites* 285 Sometimes they played écarté..or Rummy. **1934** W. SAROYAN *Daring Young Man* 213, I would sneak out of the bookie joint and run across the street to a rummy parlour and get into a game. **1964** A. WYKES *Gambling* vii. 164 Games of the rummy series appear to be the oldest. **1974** *Harrods Gift Catal.* 60/1 Packs of cards to play Bingo, Crosswords, Rummy and a variety of games.

rummy, *a.*¹ Add: **B.** as *sb.* An odd or unconventional person.
1975 G. V. HIGGINS *City on Hill* iv. 117 At least the rummies that swing it in Dorchester're *alive.* **1977** J. WAMBAUGH *Black Marble* (1978) vi. 72 She always had to look at the rummy to see if he was putting her on. *Ibid.*, It must be dope... This rummy's a doper!

rummy, *a.*² Add: (Earlier and later examples.) Also, drunken.
1834 *Jamestown* (N.Y.) *Jrnl.* 29 Jan. 1/5 The Massachusetts Masons,..like the rummy deacon, who fell from his horse—have merely 'got off to get on better'. **1843** *Amer. Pioneer* II. 372 He departed, muttering curses loud and deep, and in a voice peculiarly rummy. **1961** W. A. WHITE in *Webster* s.v., His face was blotched..his eyes were rummy, his jaw was uncertain.

rummy, *v.* [f. *RUMMY *sb.*²] *intr.* To obtain a hand that can be laid down at rummy; to say 'rummy' signifying this. Also const. *out.*
1929 *Encycl. Brit.* XIX. 658/2 Directly a player rummies that deal is finished, and all the hands are exposed and added up. **1966** L. DAVIDSON *Long Way to Shiloh* x. 144 'What about a game of cards?'..She played with immense concentration and rummied out in about five minutes.

rumorous, *a.* Add: **1.** (Further examples.)
1889 W. B. YEATS *Wanderings of Oisin* 68 Wandering of yore in forests rumorous, Beneath the flaming eyeballs of the night. **1926** GALSWORTHY *Silver Spoon* I. xii. 89 The

rumorous town still hummed; the sky was faintly coloured. **1970** I. MURDOCH *Fairly Honourable Defeat* II. xix. 371 It was dark outside, windy rainy rumorous darkness coming from far away.
Hence **rumoro·sity**, stir, public outcry. *rare.*
1906 JOYCE *Let.* 25 Sept. (1966) II. 166 Ibsen..seems to have disclaimed some of the rumorosity attaching to *A Doll's House.*

rumour, *sb.* Add: **2.** Also phr. *rumour has it.*
1912 J. N. MCILWRAITH *Diana of Quebec* xviii. 276 Rumour had it they were engaged. **1922** JOYCE *Ulysses* 190 Mr. Russell, rumour has it, is gathering together a sheaf of our younger poets' verses. **1957** D. ROBINS *Noble One* xi. 111 Rumour has it that when Brett went off on his trip, he was asked to give the pretty Juliet a lift..and they've *neither* of them come back. **1961** B. N. CARDOZO in *Webster* s.v., We make our blunders..as rumour has it that you make your own.
3. Comb. *rumour-factory, -mill, -monger* (further examples), *-mongering*; *rumour-ridden* adj.
1933 DYLAN THOMAS *Let.* 11 Nov. (1966) 63 Every country in this rumour-ridden world..is branded like Cain across the forehead. **1935** Rumour-monger [see *NON-ARYAN *sb.*]. **1953** in P. C. Berg *Dict. New Words* 138/2 A sincere and humble apology printed by the paper in response to charges of rumour-mongering. **1953** J. MASTERS *Lotus & Wind* vi. 76 He must come quickly for her sake—and for his own, to confront the rumourmongers and force them to eat their vile words. **1967** *Economist* 25 Mar. 1131/3 One decree, so far unused, imposing fierce penalties for rumour-mongering and another providing for custody without trial. **1973** *New Journalist* (Australia) July–Aug. 2/3 If there's any truth at all in what's been processed through the industry rumour mills, the.. journalists..might well be coming from the Sydney production lines. **1977** *Private Eye* 1 Apr. 18/3 The peculiar recent gyrations in the shares..may owe as much to the ambitious designs of certain mysterious figures..as to the myriad stories poured out by the well-primed City rumour factory. **1977** *China Now* July/Aug. 18/2 In rebellion against the People's Government, one of the..Grand Lamas went on a rumour-mongering rampage. **1979** *Time* 8 Jan. 28/1 Communications in Iran are unreliable, with the result that the country has become a vast rumor mill. **1979** *Railway Age* Nov. 34/1 Rumormongers have ranged from the *Des Moines Register*, the major newspaper of a major ICG state, to a federal official.
5. (Later examples.)
1889 W. B. YEATS *Let.* 3 Feb. (1954) I. 110 They always long for rest and to get away from the noise and rumour of the world. **1922** JOYCE *Ulysses* 388 His heart shook within the cage of his breast as he tasted the rumour of that storm. **1927** F. B. YOUNG *Portrait of Clare* 624 Her ears became aware of an unusual sound. At first she thought it was only the wind-swept rumour of one of the great munition trains. **a 1973** J. R. R. TOLKIEN *Silmarillion* (1977) iii. 52 But many refused the summons, preferring the starlight and the wide spaces of Middle-earth to the rumour of the Trees.

rumour, *v.* Add: **2. d.** To force through rumour *into* (an action, etc.). *rare.*
1925 F. SCOTT FITZGERALD *Great Gatsby* i. 25, I had no intention of being rumored into marriage.

rumouring, *vbl. sb.* (Later example.)
1957 T. HUGHES *Hawk in Rain* 54 You hear..through all The leafy valley a rumouring of air go.

rump, *sb.*¹ Add: † **1. d.** A type of bustle. *Obs.*
1786 E. SHERIDAN *Jrnl.* 22 Jan. (1960) iii. 79 However you may tell her as a friend gradually to reduce her Stuffing as Rumps are quite out in France and are decreasing here but can not be quite given up 'till the weather grows warmer. **1807** R. SOUTHEY *Lett. from England* II. xlix. 335 There were protruberances on the hips called bustlers, another behind which was called in plain language a rump.
2. c. (Further example.) Also, corporal punishment administered on the buttocks.
1833 J. ROMILLY *Diary* 20 Mar. (1967) 31 The original bet was 1 G[ui]n[e]a, but Sedgwick proposed a rump & dozen. **1922** JOYCE *Ulysses* 323 A rump and dozen says the citizen, was what that old ruffian sir John Beresford called it but the modern God's Englishman calls it caning on the breech.
5. a. *rump-patch, roast, -steak* (earlier example).
1902 *Nature* 14 Aug. 375/2 The author states that the gaur and the gayal have a white rump-patch. **1948** A. L. RAND *Mammals Eastern Rockies* 206 Elk... Sides of body yellowish brown, rump patch lighter. **1926** *Daily Colonist* (Victoria, B.C.) 6 July 6/1 (Advt.), Small Rump Roasts, per lb. 23c. **1976** *National Observer* (U.S.) 22 May 20/4 Right in the Rump Roast. **1747** H. GLASSE *Art of Cookery* i. 6 *To Broil* Steaks..take fine Rump Steaks about Half an Inch thick.
b. *rump caucus, government, -junta, meeting, parliament* (see also sense 3 c in Dict.), *party, state.*
1933 *Sun* (Baltimore) 19 Apr. 1/7 The incipient revolt as reflected in the rump caucus of Democratic inflationists. **1861** *Richmond* (Va.) *Examiner* 4 Dec. 3/3 It may very reasonably be doubted how far General Sherman or his officers would suffer schemes so vitally important to the Rump Government to leak out through the indiscretions of loquacious volunteers. **1937** *Nation* 6 Nov. 419/1 The Franco rump government in Spain. **1977** *Time* 15 Aug. 15/2 In 1975 the Turks declared their own Turkish Federated State of Cyprus; last week the only notice this rump government took of Makarios' passing was to announce flatly that it would not recognize his successor

as the leader of a united Cyprus. **1835** *Times* 19 June 2/2 The Westminster Rump-juntas of our own day. **1935** *Sun* (Baltimore) 10 Dec. 2/7 Mr. Berry..declared the meeting adjourned, and Dr. Haake and his confreres edged their way out to the sidewalk, where they theatened for a time to hold a 'rump meeting'. **1977** *New Yorker* 10 Oct. 50/3 On the eve of Eaton's ninetieth birthday.. he was summarily booted upstairs to chairman emeritus by his own board of directors at a rump meeting. **1838** *Ibid.* 17 Feb. 764/1 What prevents Congress from declaring itself perpetual—a rump Parliament? **1963** *Times* 23 May 13/4 This move is a direct reversal of the proposal, threatened by the Governor of the Gambia, that the remaining validly elected members of the House of Representatives constituted a 'rump parliament', capable of curing this difficulty. **1976** *New Yorker* 22 Mar. 98/2 Krishnan Kant recently made an eloquent and wide-ranging indictment of the emergency, a sort of *cri de cœur*, in what could properly be called the rump Parliament, inasmuch as so many of both its opposition and its Congress members have been jailed. **1959** *Ann. Reg. 1958* 121 Rump parties would continue to exist and split the anti-P.A.P. vote. **1938** *Sun* (Baltimore) 31 Oct. 8/3 Germany and Italy have consented to serve as arbiters in the dispute between Hungary and the rump Czechoslovakian state. **1940** *Tablet* 4 May 422/1 Herr Joseph Bühler..is at present a sort of head of the Governor-General's Government in the rump-State of Poland.
c. rumpsprung *a.*, sprung or become baggy in the seat; also *fig.*; hence **rumpspringing** vbl. sb.
1939 C. MORLEY *Kitty Foyle* xiii. 131 Pop creaking in his rumpsprung wicker chair. **1954** *Publ. Amer. Dial. Soc.* XXI. 35 *Rumpspringing: pres. part., n.*, of a skirt, the act of bagging in the seat, caused by sitting. An inner lining is sometimes used to prevent *rumpspringing*... *Rumpsprung: adj.* **1970** D. WATERFIELD *Continental Waterboy* ii. 11 'In my opinion,' Mrs. Neuberger told the reporters, 'Vancouver women are rump-sprung.' **1975** *Weekend Mag.* (Montreal) 31 May 9/1 In the hallway, his rumpsprung wife is making a blasé remark about the new labor code.

Rumpelstiltskin (rv:mpəlsti·ltskin). [ad. G. *Rumpelstilzchen.*] The name of a vindictive dwarf in German folk-tale, used allusively.
1949 G. ORWELL *Nineteen Eighty-Four* II. 181 A little Rumpelstiltskin figure, contorted with hatred. **1976** *National Observer* (U.S.) 21 Feb. 5/2 A hunch that computerized direct marketing would become the Rumpelstiltskin of American politics in the 1970s.

Rumpety, var. *RUMPTY *sb.*²

rumpless, *a.* Add: (Further example.) Hence **ru·mplessness**, the state of being without a rump.
1945 *Jrnl. Exper. Zool.* XCVIII. 65 The injection of solutions of certain chemicals into unincubated chicken eggs led to the appearance of increased numbers of rumpless embryos and chicks. *Ibid.* 67 Insulin produced a high degree of rumplessness. **1971** *N.Z. Med. Jrnl.* LXXIII. 340 (*heading*) Lumbo-sacral agenesis or rumplessness [in humans].

rumply, *a.* For *rare*⁻¹ read *rare* and add later examples.
1961 *Guardian* 21 Jan. 1/1 The rumply 35 White House staff members who had been marooned. **1967** E. B. NICKERSON *Kayaks to Arctic* xviii. 176 But the water was only slightly rumply, our boat moved easily along.

Rumpty (rv·mpti), *sb.*² Also **Rumpety.** *Air Force slang.* [f. RUMP *sb.*¹, after *BUMPETY, BUMPITY *adv.*] A Farman training aeroplane, used esp. during the war of 1914–18.
1917 A. S. G. LEE *Let.* 31 Aug. in *No Parachute* (1968) vi. 103 The Maurice Farman Rumpety I learned to fly on. **1917** in *Liberty* (1926) 28 Aug. 14/1 We are going to start on Rumptys as these Henry Farman planes are called. **1934** V. M. YEATES *Winged Victory* I. x. 83 Tom told them the first time he went up was in a Rumpty, that was to say, a Maurice Farman Shorthorn, a queer sort of bus like an assemblage of birdcages. *Ibid.* 86 After Rumpties he had gone on to Avros which really were aeroplanes, and quite different to fly. **1968** J. J. HUDSON *Hostile Skies* iii. 33 The 'Rumpty', the famous Farman primary trainer.

† **rumpty,** *sb.*³ and *a.* *Obs. Austral.* and *N.Z. slang.* [Origin uncertain.] (Something) excellent. Also **ru:mptydoo·ler.**
1941 BAKER *N.Z. Slang* vi. 51 Expressions..in constant use by our youngsters..rorter, rumpty, rumptydooler, [etc.]. **1945** *2nd N.Z. Expeditionary Force Times* 29 Jan., What a rumpty. *Ibid.* 5 Feb. 45/4 It's a rumpty. **1945** BAKER *Austral. Lang.* vi. 126 Here are some of the many synonyms for *bonzer* (it should be noted that most of them are interchangeable as nouns and adjectives).. *rumptydooler.* **1946** E. G. WEBBER *Johnny Enzed in Italy* 45 What a rumpty.

rumpus, *sb.* Add: **c.** *Comb.*, as **rumpus room** orig. *N. Amer.*, a room set aside for recreation, which does not need to be kept tidy.
1940 *Chatelaine* July 37/2 Off through a double-doored hallway can be seen the 'rumpus room', that dennish haunt of Priscilla and Rosemary. **1945** NELSON & WRIGHT *Tomorrow's House* ii. 14/2 Their daughter took over the rumpus room in the basement. **1958** J. K. GALBRAITH *Affluent Society* xiii. 151 In the more censorious social levels of American society there is already a well-developed..aversion to gadgetry... In such circles shiny rumpus rooms, imaginative barbecue pits,..and

magnificent cars no longer win acclaim. **1959** *Encounter* Sept. 50/2 Retreating to a rumpus room with ping-pong tables and do-it-yourself work-benches. **1960** *News Chron.* 30 June 6/4 How things start out on their journey to the rumpus room. **1970** J. BLACKBURN *Land of Promise* xvii. 222 Betty brought university friends home for many good sing-songs and games in the rumpus room which we fixed up in the basement. **1977** J. I. M. STEWART *Madonna of Astrolabe* i. 7 The festivity..became more and more of a romp. Indeed, not so much a romp as a rumpus. But this too was in order. The room was called the rumpus room.

rumpy. Add: **2.** A chicken without a tail. Also *attrib.*

[**1885** L. WRIGHT *Bk. Poultry* 448 It is the Rumpless or Persian Cock of Latham, and the Rumpkin of others.] **1895** *Funk's Stand. Dict.* II. 1561/3 Rumpy... 2. A variety of domestic fowl in the Isle of Man and the Hebrides. **1972** *Nat. Geographic* Sept. 438/2 We used to have a number of animals that were special to the Isle of Man... There are still some hens without tails—'rumpy hens' we call them.

ru·m-strum, *v.* [Echoic.] *intr.* To strum.

1872 HARDY *Under Greenw. Tree* II. ii. viii. 23 [He'd] want to see her young figure sitting up at that quare instrume't, and her young fingers rum-strumming upon the keys.

rumti-. Add: (Earlier and later examples in imitations of sounds.) Also, used in comb. with adjs.; *rumti-too* adj., commonplace.

(*b*) **1817** KEATS *Let.* 15 Apr. (1958) I. 129, I hope one of you will be competent to take part in a Trio..when you have said Rum-ti-ti you must not rum any more. **1898** G. B. SHAW *Perfect Wagnerite* 109 A little rum-ti-tum triplet. **1912** *World* 7 May 690/2 Mr. Cyril Maude makes a rather rumtifoolting bishop amusing and forgivable even in his most flagrant lapses from clerical circumspection. **1974** *Listener* 31 Jan. 131/1 Try translating Goethe's *Faust* with the same metre..it's apt to sound fatally rum-ti-tum. **1976** G. EWART *No Fool* I. 31 This is a convention, we know, of course, and a wistfulness in the rum-ti-tum might be detected.

(*c*) **1906** GALSWORTHY *Man of Property* I. i. 22 Did you ever see such a collection of rumty-too people? **1920** —— *In Chancery* I. i. 9 He was feeling more strongly than ever that Timothy's was hopelessly 'rum-ti-too', and the souls of his aunts dismally mid-Victorian.

rum-tum. Add: **3.** Used in imitation of a regular rhythmic sound; also *attrib.*

1898 G. B. SHAW *Perfect Wagnerite* 139 The strings play a rum-tum accompaniment. **1917** —— *London Music in 1888–89* (1937) 380 The accompaniments are a derisive rum-tum. **1922** JOYCE *Ulysses* 51 Of all the glad new year, mother, the rum tum tiddledy tum. **1958** *Times* 28 Nov. 8/4 The rum-tum bars in Malcolm Arnold's *United Nations* tone-poem. **1963** *Times* 14 May 15/1 Mr. Charles Groves..brought out the rum-tum rhythms more successfully than the great arches of melody they support.

run, *sb.*[1] Add: **I. 1. d.** (Further examples.) Also *to give* (someone) *a run for his money* (colloq.), to give (that person) satisfaction or a good return for trouble taken; to offer (him) a strong challenge.

1905 *Athenæum* 1 Apr. 397 We do not get the proper run for our money, if we may put it in sporting lingo. **1908** CHESTERTON *Man who was Thursday* xiii. 277 Since the beginning of the world all men have hunted me like a wolf... I have given them a good run for their money, and I will now. **1914** G. B. SHAW *Dark Lady of Sonnets* Pref. 108 If I had been born in 1556 instead of in 1856, I should have taken to blank verse and given Shakespear a harder run for his money than all the other Elizabethans put together. **1916** 'TAFFRAIL' *Pincher Martin* ix. 160 Most of them longed for a run for their money... The graver possibilities of war did not intrude themselves upon their minds until long afterwards. **1920** A. HUXLEY *Limbo* 83 'We'll give you a good run for your money,' said Hyman. 'I hope they'll be feeling a little uncomfortable by the time they have done with you, Greenow.' **1948** *Sun* (Baltimore) 26 Nov. 17/1 Backers of..Egretta, a stakes-winning filly, in the Endurance 'Cap did not get a run for their money. **1952** E. O'NEILL *Moon for Misbegotten* I. 14 You're a wonderful fighter. Sure, you could give Jack Dempsey himself a run for his money. **1955** *Times* 27 Aug. 6/1 Pickering said that he was going.. 'simply to satisfy the people of Bloxwich. They demand a run for their money and I will give it to them.' **1976** J. WAINWRIGHT *Bastard* i. 13 The old Beetle punches the rear wheels into the softness and with good tyres..this bus could give a snow-cat a run for its money.

e. *Cricket.* The act of running by the bowler to the bowling crease in delivering the ball; a run-up.

1836 *New Sporting Mag.* Oct. 358 The only fault is in his taking too long a run before he delivers the ball. **1891** W. G. GRACE *Cricket* ix. 240 When Smith begins his run he is behind the umpire and out of sight of the batsman... It is rather startling when he suddenly appears at the bowling crease. **1904** P. F. WARNER *How We recovered Ashes* i. 22 With a short run Relf bowls a fast medium ball. **1976** J. SNOW *Cricket Rebel* 77, I finished the match with ten wickets—the five in the second innings off a short run—for 80 runs.

f. *U.S.* A movement of settlers to new land; = RUSH *sb.*[2] 4 a in Dict. and Suppl.

1894 *Daily Ardmoreite* (Ardmore, Okla.) 30 Apr. 2/1 Buckskin Joe and his followers are camped at Marlow preparatory to making a run on the Fort Sill country tomorrow. **1901** *World's Work* June 894/1 Hitherto the settlers made a 'run' for the homesteads. **1930** *Publishers' Weekly* 8 Feb. 697 On April 22, 1889, this strip was opened up with the land rush known as the famous Oklahoma Run. **1948** *Daily Oklahoman* (Okla. City) 16 May E 3/2 The nine great land openings began in 1889 with the 'run' into the area now occupied by Oklahoma City, Guthrie, Norman, Stillwater and other cities.

2. a. (Earlier and later *Comb.* examples.)

1853 F. LILLYWHITE *Guide to Cricketers* 32 He is a splendid field anywhere, and one of the most sure run getters to be met with. **1867** *Baily's Monthly Mag.* July 250 The wickets good, and the ground in splendid order for run-getting. *Ibid.* Dec. 137 Harrow had not a great run-getting Eleven this year. **1877** *London Society* May 416/2 The run-stealer's heart would surely be broken in his first innings. *a* **1907** F. THOMPSON *Sel. Poems* (1908) p. viii, And I look through my tears on a soundless-clapping host As the run-stealers flicker to and fro. **1921** G. R. C. HARRIS *Few Short Runs* iv. 95 We..got two of their best bats caught..by George Remnant—one of the finest fields I ever saw, and in second-class matches a wonderful run-getter. **1934** BLUNDEN *Mind's Eye* 186 The pair amuse themselves and astonish us with slogging and run-stealing. **1950** *Sport* 7–11 Apr. 11/3 Our batsmen will find run-getting more easy. **1963** *Times* 17 Apr. 3/1 P. K. Thomas, a consistent run-scorer in the Colts, is expected to mature with the first XI experience. **1965** G. McINNES *Road to Gundagai* xii. 209 It was an era of run-getters. **1976** *0–10 Cricket Scene* (Austral.) 7/1 His career Test aggregate of 5187 leaves him fourth on the Australian run-gathering list, among the elite. **1977** *Sunday Times* 2 Jan. 28/3 Australia's reaction..was to score at a run-a-minute rate.

b. (Earlier examples.) Also *Comb.*

1856 *Spirit of Times* 6 Sept. 13/3 At the time of the adjournment the score stood fifteen runs in favor of the Union, and twelve runs for the Baltic. **1858** *By-Laws Knickerbocker Base-Ball Club of N.Y.* 20 The game shall consist of nine innings to each side, when, should the number of runs be equal, the play shall be continued until a majority of runs, upon an equal number of innings, shall be declared, which shall conclude the game. **1891** *Harper's Weekly* 23 May 391/4 As for Poole, he is the same 'run-getter' that he was last year. **1970** *Washington Post* 30 Sept. D1/2 But the Twins came back to tie it 11–11 in their half on run-scoring singles by Cardenas and Chuck Manuel.

c. *Croquet.* The passage of a ball under a bridge or hoop. Cf. RUN *v.* 37 d.

1863 MAYNE REID *Croquet* 34 If a ball, after running a bridge, strike an obstacle, and recoil back through the bridge, the run remains good.

4. a. Also in *Comb.*, as **run-boat** *U.S.*, a boat which collects or transports the catch made by marine fishing vessels; also *transf.*

1911 *Rudder* Aug. 49/2 The run-boats, in the local vernacular, are schooners mostly, about 60 to 70 feet on deck, and merely run back and forth between the dredging fleet and Baltimore. **1935** *Sun* (Baltimore) 6 Feb. 13/6 Most of the catch is brought to Crisfield in large run-boats, sent to the fishing grounds by fish dealers here. **1941** *Ibid.* 17 Mar. 11/3 Large dealers here go down the bay in run-boats to buy from the catchers. **1967** *Washington Star* (Sunday Mag.) 25 June 11 The Jessie Taylor out of Smith Island, Md., is typical of the 'runboats' that bring the seafood to town. **1974** *News & Observer* (Raleigh, N. Carolina) II. 13/2 He told me he'd run aground in his private 'run boat'.

c. (Earlier and later examples.) Now freq. an excursion or drive by car or bicycle. Also in phr. *run ashore* (Naut.), a brief period of shore leave; also (with hyphen) *attrib.*

1819 H. COCKBURN *Let.* 8 Oct. (1932) 18, I also took a run t'other day to Blair Adam. **1881** *Sportsman's Year-bk.* 199 Bicycle Touring Club... The members..enjoy Club runs or tours without the heavy outlay attendant on forming a small local Club. **1902** C. L. FREESTON in Harmsworth *Motors & Motor-Driving* xxi. 388 Several tours and runs also took place, the anniversary run to Southsea..being an enormous undertaking. **1912** *Motor Man.* (ed. 14) v. 163 Never start on a run without being assured that there is ample oil. **1948** PARTRIDGE *Dict. Forces' Slang* 159 *Run ashore*, a short evening's shore leave. **1959** *Motor Man.* (ed. 36) xii. 259 A party in an ordinary family car..can count its daily run, including stops, at a kilometre a minute. **1977** *Navy News* June 8/2 It was certainly a good run ashore for the ship's company, with the Principality of Monaco granting free admission to many places of interest. *Ibid.* Aug. 31/2 Also 'out of this world' are the run-ashore opportunities.

d. (Earlier example.)

1857 *Lawrence* (Kansas) *Republican* 4 June 2 The train ..made a quick and pleasant run, arriving in Jefferson City promptly in time.

e. A brisk walk or perambulation. Now usu., a dog's exercise walk.

1837 W. TAYLER in J. Burnett *Useful Toil* (1974) II. 178, I am obliged to stay within to help the sick. This is what I don't like as I like to get a run everyday when I can. **1871** 'L. CARROLL' *Through Looking-Glass* iv. 79 'O Oysters,' said the Carpenter, 'You've had a pleasant run! Shall we be trotting home again?' **1967** P. MOYES *Murder Fantastical* viii. 106 'What on earth made you go off down to the river on your own?' 'I was only giving Tinker her run... There was no need to come after me.' **1977** 'J. BELL' *Such Nice Client* viii. 83, I was giving Caesar a very short run on the lead.

f. A single trip on a toboggan, sleigh, etc., down a slope or course. Cf. sense 23 d below.

1898 *Encycl. Sport* II. 473/2 A good average run down the Cresta course takes 75 seconds. **1919** [see *LUGE *sb.*]. **1935** *Encycl. Sports* 178/2 The art of making good time on a run is acquired by long study of the ten banked turns. **1956** *Ski-ing* ('Know the Game' Ser.) 22/1 If the run is made too early, the snow will be as hard as iron. **1976** F. RAPHAEL *Glittering Prizes* 57 I'll make the first run. You grab the stop watch and bugger off down to the bottom of the hill, OK?

g. *Mil.* An offensive operation, *spec.* an attack by sea or air. See also *bomb run* s.v. *BOMB *sb.* 6, *dummy run* s.v. *DUMMY *sb.* 7 b. Also *transf.*

1916, etc. [see *dummy run* s.v. *DUMMY *sb.* 7 b]. **1941** *Flight* 13 Mar. 204/2 The bomber had successfully bombed its target on the first run; another run was then made and incendiaries started small fires. **1944** *Hutchinson's Pict. Hist. War* 12 Apr.–26 Sept. 43 (*caption*) The aircraft is seen making its second run over the target. **1948** AUDEN *Age of Anxiety* i. 18 We began our run; Death and damage darted at our will. **1963** *Listener* 4 Apr. 585/2 Our patrol car got the call as 'shots fired', with the address given. My partner and I responded. In such radio 'runs' you never know what to expect. **1975** J. GRADY *Shadow of Condor* vi. 109 The CIA agent.. tips us to another run, which we intercept. **1977** *Time* 30 May 46/3 For the climactic battle sequence, which includes dogfights in space and missile runs on the Death Star, Lucas gathered all the old war movies he could find and spliced together their aerial-combat footage.

h. A single or regular journey made by an aircraft; the distance thus travelled.

1912 KIPLING *Diversity of Creatures* (1917) 4 DeForest, whose business it is to know the out districts, told us that it..was about half an hour's run from end to end. **1944** [see *milk-run* s.v. *MILK *sb.* 10]. **1958** 'N. SHUTE' *Rainbow & Rose* i. 3, I was on the Sydney–Melbourne run. **1976** *Daily Mirror* 16 July 2/1 President Amin's jet airliner was poised last night to take off for Uganda..on a whisky run.

i. A regular round (freq. one accomplished by means of a vehicle). Also in phrases *mail-run* s.v. *MAIL *sb.*[3] 4 b, *milk-run* s.v. *MILK *sb.* 10, *paper run* s.v. *PAPER *sb.* 12.

1925 *N. & Q.* 21 Mar. 208/1 In the dairy trade phrases such as 'He has a milk-run' or, 'he has a milk-walk' or 'he has a milk-round' are common. **1946** [see *mail-run* s.v. *MAIL *sb.*[3] 4 b]. **1968** K. WEATHERLY *Roo Shooter* 38 At night when they had done the evening run on their traps they would return home. **1978** *Oxf. Diocesan Mag.* July 16/3 A Soup run was established, operating four nights a week, and we have made contact with up to forty people in derelict property in and around the town centre.

8. With advs. See main entries.

II. 9. d. A downward flow or trickle of paint when applied too thickly; the action of paint in 'running'. Cf. RUN *v.* 22 b.

1935 J. LAWRENCE *Painting A to Z* xi. 103 Don't leave the quirks swimming in paint, or it will wrinkle, or perhaps even run down into the lower mouldings, and leave a 'run'. **1951, 1958** [see *CURTAIN *sb.*[1] 1 e]. **1975** *Amer. Speech* 1969 XLIV. 24 *Run*, n., the action of paint when it is applied too heavily; it can't adhere to the wall surface and begins to stream down the wall.

III. 13. e. *Oil Industry.* (A distance drilled during) a spell of drilling with a particular bit.

1880 J. F. CARLL *Geol. Oil Regions* xxviii. 310 The engineer examines the steam and the water gauges and the fire, and then proceeds to sharpen the tool required for the next 'run'. **1946** M. C. SEAMARK in *Mod. Petroleum Technol.* (Inst. Petroleum) 94 Cores of 10–20 feet can be taken at one 'run'. **1974** R. D. GRACE in P. L. Moore et al. *Drilling Practices Man.* xiv. 354 Bit records of that time [*sc.* the late 1940s] were filled with typical runs of only five to ten feet in four to five hours at depths below 10,000 feet.

f. A length of electric wiring; a distance covered by uninterrupted cable.

1905 C. C. METCALFE *Pract. Electr. Wiring* i. 5 If the run is this length..a conductor of greater sectional area, with a negligible resistance, should be used. **1938** J. W. SIMS *Elect. Installations* vii. 128 Special care should be taken to avoid metallic obstructions inside the conduit.. and bushes should be fitted at the end of a run. **1957** A. L. OSBORNE *Elect. in Building* ii. 20 By eliminating long unbroken cable runs, voltage drop within the building is not likely to occur. **1970** J. EARL *Tuners & Amplifiers* iv. 94 Use 15-ampere cables for runs in excess of 10 ft.

g. A tear in a knitted garment or stocking; = *LADDER *sb.* 3 b. Also *attrib.* and *Comb.* Cf. *RUNNER 9 d.

1922 M. B. HOUSTON *Witch Man* xii. 146 She looked the suit over, darned a tiny run in the tights, [etc.]. **1933** *Radio Times* 14 Apr. 95, I had an awful ladder in my stocking... I scarcely ever have a run now. **1936** G. G. DENNY *Fabrics* (ed. 3) i. 104 *Run resist*, knitting process which locks stitches to reduce runs in hosiery and underwear. **1938** O. NASH *I'm Stranger here Myself* 173 She stopped to moisten her finger on account of a run in her stocking. **1938** 'E. QUEEN' *Four of Hearts* iv. 67 You've got a run in your stocking. **1938** *Knit Goods Weekly* 15 Aug. 11 These hosiery finishes bind fibre to fibre.. strengthen the fabric..make it snag-resistant, run-resistant..add miles more wear. **1939** *Business Week* 27 May 32/1 Merchants..seem not at all worried about the inroads that this really run-resistant hosiery might make in total sales volume. **1951** in M. McLuhan *Mech. Bride* (1967) 95 Ivory Flakes care helps safeguard sheerest nylons from embarrassing, eye-catching runs. **1957** M. B. PICKEN *Fashion Dict.* 279/2 *Run-proof*, a knitted construction where locked loops prevent a run. *Run resist*, a type of knitting with loops so constructed as to resist the tendency to run. **1969** *Sears Catal.* Spring/Summer 409 Run-resistant mesh-knit seamless stretch nylons. **1970** *Focus* June 15/2 The term run-resist is used on the advice of hosiery trade associations, as an 'out' under the Trade Descriptions Act, 1968. **1973** 'E. McBAIN' *Let's hear It* x. 146 She's noisy and vulgar; there are runs in her nylons. **1974** H. L. FOSTER *Ribbin'* v. 186 A female student may have a run in her stockings and will be ribbed about it.

14. d. (Earlier examples.)

1798 *Rec. Smithtown, N.Y.* (1898) 351 The grist mill house..[will] carry three run of stones with three Bolting mills. **1815** *Niles' Weekly Reg.* IX. 187/1 The whole expense in generating steam sufficient to drive two run of stones upon this principle will not exceed two hundred and twenty dollars.

f. *pl.* with *the.* An attack of diarrhoea. *colloq.*

1962 E. LACY *Freeloaders* vii. 147 I'd picked up a touch of '*la tourism*' or in basic English, the runs, from..the unwashed fruit. **1966** 'L. LANE' *ABZ of Scouse* 91 *Go like a bookie wit' ther runs*, to move very fast. **1971** B. MALAMUD *Tenants* 214 Sam Clemence, a witness from Harlem U.S.A., despite a bad case of the runs.., stands up for his friend Willie. **1976** U. HOLDEN *String Horses* ix. 112 'What can she do Lil? Please help.' 'There's nothing. Pills will only give her the runs. I don't hold with that abortion lark.'

15. a. (Further examples.) Also *spec.* a sudden movement on the part of foreign depositors to withdraw their holdings of a nation's currency by exchanging them for equivalent sums in other currencies. Freq. const. *on.*

1891 G. CLARE *Money-Market Primer* vii. 59 Country bankers,..fearing that the shock to confidence may cause a 'run' on the part of their depositors, telegraph to London for more notes. **1932** P. EINZIG *Tragedy of Pound* vii. 65 In July [1931] the failure of the Nordwolle and other commercial firms resulted in a run on several German banks. **1955** H. WILSON in *Hansard Commons* 10 June 148 More confidence has been given to the speculators, and therefore the run on sterling has temporarily stopped. **1964** S. BRITTAN *Treasury under Tories* vi. 189 Mr Thorneycroft responded to the run on the pound with his famous deflationary package. **1976** *Economist* 16 Oct. 23/3 The Bank of England reacted to the March run on sterling by using up reserves and borrowing money to try to check the exchange rate collapse.

b. (Earlier example.)

1816 JANE AUSTEN *Emma* II. vi. 97 A couple of pair of post-horses were kept, more for the convenience of the neighbourhood than from any run on the road.

19. a. *spec.* (*a*) the amount of sap drawn off when sugar maples are tapped; the amount of maple sugar produced at one time; (*b*) (*Oil Industry*) the action of transferring a quantity of oil through a pipeline, or of subjecting it to a process such as distillation; the amount of oil so treated.

(*a*) **1822** *Farmer's Diary 1823* (U.S.) sig. C 3, Sugar makers may venture to set seven or eight hundred pails to one of these pans,..in case of extra ordinary runs, which, however, do not often happen. **1890–3** E. M. TABER *Stowe Notes, Lett. & Verses* (1913) 40 The early runs are not so sweet as the later; the trees being full of frost. **1949** *Highway Traveler* Feb. 17/2 In the average season of a month..sap can be expected to run on about half of the days, while on two to five days there will be 'good runs'. **1978** N. PERRIN *First Person Rural* 84 It was no hard run—but my 104 buckets would probably yield 40 or 50 gallons [of maple sap] today.

(*b*) **1883** [in Dict.]. **1888** *Science* 12 Oct. 172/2 This past spring an oil-man..was suffocated in one of these tank-sheds while making a run of oil; viz., running the oil from the receiving-tank to the transportation or pipe-line company's tanks. **1898** *19th Ann. Rep. U.S. Geol. Survey* VI contd. 29 Usually the terms 'production' and 'pipe-line runs' are considered as synonymous, but production is always slightly in excess of runs. The expression 'pipe-line runs' means the amount of oil the pipe lines have received from the wells, and as the pipe lines do not run all the oil in the tanks at the wells, it would be remarkable if the same amount remained in the tanks at the wells at the close of each year. **1914** *Chem. Abstr.* VIII. 2247 Flushing out the vapors remaining in the still with steam so that they will not mix with the vapors from the next run. **1931** *Economist* 14 Feb. 361/1 Crude oil 'runs' to refinery stills have, therefore, been restricted to about 10 per cent. below last year's level.

c. Also, a spell of manufacturing some product; an instance or a spell of carrying out an experimental procedure, esp. one involving automatic equipment. (Further examples.)

1931 *Anatomical Rec.* XLIX. 180 In an original trial run on five albino rats..three became pseudopregnant. **1935** *Industrial & Engin. Chem.* Sept. 1074/2 In one run,.. the temperature began to rise and continued to rise after the heat input to the bomb was stopped. **1951** S. JENNETT *Making of Bks.* vii. 106 When the make-ready is completed and the machine is ready to start its run the hand-feeder takes her place at the feeding board. *Ibid.* 107 The run then commences. **1971** J. E. HARRY *Plastics Fabrication & Electrotechnol.* v. 38 Preformed materials such as sheet or tube..are sometimes used instead of moulding processes for short runs of large components. **1972** *Nature* 18 Feb. 397/1 The difference observed was found consistently in different electrophoretic runs. **1972** *National Observer* (U.S.) 27 May 11/2 At first he intended to fake the run, using plain creek water in the barrels with..foam rubber pellets floating on top to simulate the head on the fermenting mash.

d. A spell of sheep-shearing. *Austral.* and *N.Z.*

a **1910** G. MEEK in A. E. WOODHOUSE *N.Z. Farm & Station Verse* (1950) 59 The record shearing run of nineteen-nothing nine. **1933** L. G. D. ACLAND in *Press* (Christchurch, N.Z.) 18 Nov. 15/7 *Run*,..stretch of work. Shearers work for an hour before breakfast, two stretches in the morning and three in the afternoon. The stretches are divided by meal-times and *smokos*. **1956** G. BOWEN *Wool Away!* (ed. 2) 157 *Run*, the shearing time worked between official stops, smokos, or meals. *a* **1964** H. P. TRITTON in *Penguin Bk. Austral. Ballads* (1964) 227 My shearing days are over, though I never was a gun: I could always count my twenty at the end of every run.

e. *Computers.* An instance of the execution of a program or other task by a computer.

1946 *Math. Tables & Other Aids to Computation* II. 151 From a series of positive values of *x* and *y*, it could form Σx, Σy, Σx^2, Σxy, Σy^2 and check them in one run. **1952** *Rev. Electronic Digital Computers* (Amer. Inst. Electr. Engineers) 17/2 The chance that the machine would get through any particular 20-minute run was independent of its chances of getting through any other 20-minute run. **1964** F. L. WESTWATER *Electronic Computers* ix. 144 It requires two runs on the computer..to solve the problem. **1971** J. B. CARROLL et al. *Word Frequency Bk.* p. xxxvi, In the particular computer run that produced this table, the number turned out to be 609,798. **1977** *Sci. Amer.* Oct. 116/2 The program and certain aspects of the discharging procedure had to be modified to overcome the problems indicated by the first computer runs.

20. a. (Later examples of *general run.*) Also *normal run.*

1957 G. RYLE in C. A. MACE *Brit. Philos. Mid-Century* 257 This question did not begin seriously to worry the general run of philosophers until..sixty years ago. **1965** G. McINNES *Road to Gundagai* v. 87 What..set our tannings aside from the normal run..was the deliberation and the ritual. **1970** *Incorporated Linguist* IX. IV. 111 This is a book which should be of interest to the scholar and the linguistic specialist, less so to the general run of professional translators.

d. *Printing.* The total number of copies of a book, newspaper, etc., printed during a single period of press-work. Cf. *press-run* s.v. *PRESS *sb.*[1] 16 b and *print run* s.v. *PRINT *sb.* 15 a.

1909 WEBSTER s.v. *Run* n., A run of 3,000 copies on a press. **1936** *Penrose Ann.* XXXVIII. 146 Rotary printing will master any long runs. **1951** S. JENNETT *Making of Bks.* ix. 138 The film assists in the retention of moisture, and longer runs can be printed at greater speed [by collography]. **1976** *Penrose Ann.* LXIX. 132 Over half of all printing jobs involve runs under 10,000 sheets.

e. *run of the mill* (also *the mine*, etc.), the material yielded by a mill, mine, etc., as it emerges from the production process and before being sorted or inspected for quality; also *run-of-mine*, etc. Hence *fig.*, the ordinary, average, undistinguished, or mediocre type (cf. *mill run* (*d*) s.v. *MILL *sb.*[1] 10). Also in various extended and nonce-uses. Freq. as *adj.* or *attrib.* phrase.

1909 *Cent. Dict. Suppl.*, *Run of the kiln*, bricks of all kinds and qualities just as they happen to come from the kiln. *Ibid.*, *Run of the mine*, coal just as it comes from the mine, large and small sizes and all qualities together. **1930** *Daily News Record* (U.S.) 17 Mar. 19/4 Suspicion has attached to the yarn producers' definition of 'inferior'. Leading users of these yarns say they interpret the word to mean 'run of the mill'. A purchase of such yarns may include various percentages of first, second and sub-qualities of yarn, in fact all that 'run of the mill' suggests —but the grading is done by the cloth mill. **1930** *Engineering* 20 June 811/2 This machine is intended to reduce run-of-mine coal to any size between 6in. and 1½in. cube, in one operation. **1930** *Hearst's Internat.* Sept. 37/2 But level-headed as a wife and a darned sight better-looking than the run of the mill of wives. **1933** *Sun* (Baltimore) 14 Oct. 4/3 An ordinary, run-of-the-mill gravy. *Ibid.* 23 Dec. 8/7 The attitude of run-of-the-mine citizens on lynching. **1938** K. A. PORTER in *Southern Rev.* Winter 429 I've got a special job beside my usual run of the mill. **1939** EVANS & McGOWAN *Guide to Textiles* 66 Run-of-the-mill is a term which in general means that the merchandise has not been inspected... Sheets and pillowcases are frequently sold as run-of-the-mill. **1940** O. NASH *Face is Familiar* 118 And in celestial circles all the run-of-the-mill angels would rather be archangels or at least cherubim and seraphim. **1941** *Sun* (Baltimore) 30 July 10/1 The balance is composed of items which, in the main, are run-of-the-mine budget pruning. **1943** B. A. DE VOTO in *Harper's Mag.* May 645/1 But what they have to say is mostly run of the mill. **1944** *Sun* (Baltimore) 17 Feb. 17/1 The growers want 25 cents a pound for wrapper leaf..and 15 cents for 'run of the crop' as against the OPA set prices of 21 for wrapper..and 10 for ungraded. **1945** 'L. LEWIS' *Birthday Murder* (1951) i. 13 Hime..has managed out of..a run-of-the-mill story, something..unusual in its effects. **1946** *Richmond* (Va.) *News Leader* 29 Nov. 15/1 That's about 80 times the cost of a hide from a 'run-of-the-mine' cow. **1950** *Engineering* 11 Aug. 131/3 The..washery will treat 750 tons of run-of-mine coal per hour. **1951** *News Chron.* 12 Dec. 4 To the ordinary run-of-the-mill bank customer these may perhaps seem lordly examples of living on overdrafts. **1952** [see *IMPERFECT *sb.* 3]. **1952** *Amer. Speech* XXVII. 264 Fabric which is shipped to a buyer just as it comes from the loom without inspection and without elimination of weaving defects is referred to as run-of-the-loom. **1953** A. UPFIELD *Murder must Wait* x. 95 Her taste wasn't the usual run of the mill. **1960** E. DELAVENAY *Introd. Machine Transl.* vii. 106 It should free intellectual ability for more productive work than that of run-of-the-mill translations. **1967** *Gloss. Mining Terms* (*B.S.I.*) VIII. 23 *Run of mine* (R.O.M.), the product of a mine before sorting or cleaning. **1969** *Daily Tel.* 21 Apr. 17/7 No hard boundaries exist to separate jazz singers from run-of-the-mill night club performers. **1975** 'D. JORDAN' *Black Account* xii. 60 The difference between run-of-mine ores and washed fines. **1975** *Publishers Weekly* 9 June 52/1 Taylor's thoughtfully written, low-keyed text proves far superior to most run-of-the-battlefield 'popular' histories. **1977** *Amer. N. & Q.* XV. 117/1 Prices of run-of-the-mine books are often more difficult to ascertain than those of $500-plus items. **1978** R. LUDLUM *Holcroft Covenant* iii. 35 Althene was not your run-of-the-mill mother, as mothers were understood by this particular son. **1980** *Times Lit. Suppl.* 31 Oct. 1240/2 We are left with a pretty run-of-the-mill thriller.

IV. 22. (Earlier and further examples.) Also *Austral.* and *N.Z.*, a sheep station. Also *attrib.*, as *run cattle, sheep.*

The early Amer. usage in quot. 1658 is not supported by further evidence.

1658 *Rec. Brookhaven, N.Y.* (1880) 3 This land and the grass thereof for a range, or run, for to feed horses and cattle on..I have sold. **1804** *Sydney Gaz.* 12 Feb., A commodious dwelling-house [with] an extensive run for stock. **1847** A. HARRIS *Settlers & Convicts* xvi. 330 If the shepherd suffers the flock to spread, in these mountainous runs especially, they get into creeks and hollows. **1858** *Richmond-Atkinson Papers* (1960) I. vii. 421, I found that Elliot's run..was in a capital position..in a finely grassed country. **1911** C. E. W. BEAN '*Dreadnought*' *of Darling* i. 12 Huge 1,000,000-acre runs or little 20,000 acre homestead leases. **1930** L. G. D. ACLAND *Early Canterbury Runs* 1st Ser. i. 2 A number..sold or abandoned their runs, and came to New Zealand early in 1851. *Ibid.* 7 All run sheep had been Merinos in the old days. **1933** —— in *Press* (Christchurch, N.Z.) 18 Nov. 15/7 *Run* sheep, *run*-cattle, as opposed to paddock sheep or milking cows and hand-reared calves. **1936** F. CLUNE *Roaming round Darling* xviii. 178 Toorale.., headquarters of a run of 1,000,000 acres. **1950** *N.Z. Jrnl. Agric.* Sept. 215/1 With fertility declining pastures are becoming more difficult to control for sheep, and because of this, run cattle numbers are being increased. **1963** *Weekly News* (Auckland) 10 July 38/3 Run cattle..can be moved on a mob basis from one paddock to another. **1966** G. W. TURNER *Eng. Lang. Austral. & N.Z.* iii. 50 A *run* is a parcel of land as leased. A sheepfarmer might own several adjacent runs of land, in this sense, and the whole 'station' so formed might also be termed a run.

Comb. **1859** F. FULLER *Five Yrs. Residence N.Z.* ix. 162 The Runholder kept the remaining portion as the payment for his trouble and expense in looking after the sheep. **1864** *Sunday at Home* 17 Dec. 811/1 A small chapel has been built for the use of the inhabitants by a runholder near. **1911** W. H. KOEBEL *In Maoriland Bush* xxii. 284 His career as a run-holder is usually a fleeting one with an abrupt termination. **1930** L. G. D. ACLAND *Early Canterbury Runs* 1st Ser. 8 On the whole, runholding has not been much of a business in Canterbury. **1940** W. G. McCLYMONT *Explor. N.Z.* xii. 130 The country they contained had been granted to run-holders in order of application.

23. c. *pl.* A place at which wagons may be loaded or unloaded. *U.S.*

1870 *Huntington* (N.Y.) *Town Rec.* (1889) III. 585 The said land..[is] sufficient..to build two runs, so called, or three runs..to load brick at. **1923** C. R. COOPER *Under Big Top* 226 Many a man [is saved] from injury at the unloading runs. **1931** *Amer. Mercury* Nov. 354/1 *Runs*, the unloading place at the railroad.

d. A slope of snow on which tobogganing, skiing, etc., are carried out. Also *transf.* Cf. sense 4 f above.

1874 [see TOBOGGAN *v.*]. **1898** *Encycl. Sport* II. 470/2 The English and American visitors to St. Moritz and Davos introduced tobogganing as a sport, and set to work to construct both toboggans and runs. **1910** [see *BOB-SLED, -SLEIGH]. **1935** *Encycl. Sports* 178/2 The most famous toboggan run in the world, the Cresta is rebuilt every year under expert supervision. **1956** *Ski-ing* ('Know the Game' Ser.) 22/1 The *piste* is artificial and either made by a large number of skiers ski-ing down a run, or [etc.]. **1961** [see *MOGUL *sb.*[2]]. **1972** 'M. YORKE' *Silent Witness* vi. 137 She took off her skis at the bottom of the run. **1974** *Rules of Game* 253/3 The brake [on a bobsleigh] is used only in emergencies, as its serrated edge damages the surface of the run. **1976** F. RAPHAEL *Glittering Prizes* 58 As he reached the top of the run, he was conscious that someone was standing there.

V. 28. a. *on the run* (further *fig.* examples, chiefly in sense 'fleeing' or 'escaping' from justice).

1909 J. B. ATLAY *Ld. Haliburton* 184 There was a widespread impression that the assailants had got the Government 'on the run', and that a vigorous campaign would show further concessions when Parliament met. **1932** *Week-End Rev.* 9 Apr. 456/2 In 'Secret Sentence' she explores another world—the world of political assassination, of criminal investigation departments, of men on the run. **1954** X. FIELDING *Hide & Seek* xi. 139 They were fugitives from justice and had been on the run in this area for over a year. **1955** *Times* 30 June 10/7 The President strongly resisted the suggestion that, as Russia was now 'on the run' it might be possible to reduce American expenditure on foreign aid. **1957** *Times* 31 Oct. 3/1 The gangster, the No. 1 Public Enemy, on the run. **1963** T. TULLETT *Inside Interpol* i. 17 If it had not been for the men in the Rue Paul Valéry he might still be 'on the run'.

e. *to get the run*: to be dismissed from one's employment. *slang.* (chiefly *Austral.*).

1889 BARRÈRE & LELAND *Dict. Slang* I. 403/2 *Get the run, to*, (English and Australian), to be discharged. **1941** [see *running shoe* s.v. *RUNNING *vbl. sb.* 17 a]. **1959** BAKER *Drum* (1960) II. 141 *Run, get the*, to be dismissed from employment.

29. b. Also, a flow *of* speech.

a **1915** JOYCE *Giacomo Joyce* (1968) 12 She stands black-robed at the telephone. Little timid laughs, little cries, timid runs of speech suddenly broken.

e. *run-of-river* adj. phr (usu. *attrib.*), denoting (an installation employing) a water supply taken directly from a river, esp. for generating hydroelectricity, with no major attempt to store water or regulate flow.

1943 STEINBERG & SMITH *Economy Loading* v. 169 In a run-of-river plant, the flow of the river must be utilized as it comes, so that this plant would normally

supply the system base load. **1956** V. A. THIEMANN in B. G. A. Skrotzki *Electric Generation* v. 228 Hydro stations may be classified as either peaking or run-of-river. **1964** LINSLEY & FRANZINI *Water Resources Engin.* xvi. 453 Some run-of-river plants have enough storage.. to permit storing water during off-peak hours for use during peak hours of the same day. **1965** R. G. KAZMANN *Mod. Hydrol.* iv. 89 The raw water obtained from an impounded supply is generally better than that obtained from a run-of-river water source.

30. b. For *a* **1890** J. W. PALMER *New & Old* 62 (Cent.) read **1859** J. W. PALMER *New & Old* ii. 62, and add further examples. Also, *to lose* (*the*) *run of.* Now *rare.*

1872 MRS STOWE *Oldtown Fireside Stories* 29 She hed the in and out o' the Sullivan house, and kind o' kept the run o' how things went and came in it. **1893** 'THANET' *Stories of Western Town* 145 I've been in this block, Mrs Carleton and me, ever since it was built; and, some way, between us we've managed to keep the run of all the folks in it. **1893** 'MARK TWAIN' *£1,000,000 Bank-Note* 29 You couldn't afford to lose the run of business and be no end of time getting the hang of things again when you got back home. **1918** J. C. LINCOLN *Shavings* xix. 320, I kind of lost run of the time.

32. b. (Earlier example.) Also *transf.*, complete freedom of action. Also *the run of one's knife.*

1807 in *N. & Q.* (1904) 11 June 478/1 And it suits to a T, To receive as your fee, The run of your teeth And five guineas a day. **1927** *Daily Mirror* 10 Dec. 9/1 Sir Granville Ryrie..began work as a cattle driver at £30 a year and the 'run of his knife', which means his food. **1974** *Broadcast* 2 Dec. 16/3 Hugh Carleton Green..allowed—nay! encouraged a group of bright and irresponsible young men and women to have the run of their teeth in the so-called 'satire' programmes.

34. Special Comb.: **run time** *Computers*, (*a*) the time at or during which a program or other task is executed; (*b*) the length of time taken by the execution of a particular task.

1965 MITCHELL & WILLMOTT *Programming Computer in Atlas Autocode* viii. 51 Each routine or block is associated with a serial number for use in tracing faults found at run time. **1968** M. V. WILKES *Time-Sharing Computer Syst.* iv. 37 At run time, the segment number is used to enter the segment table belonging to the user whose program is running at that instant. **1972** [see *OVERLAY *sb.* 6]. **1974** ADBY & DEMPSTER *Introd. Optimization Methods* v. 178 Colville found that the number of function evaluations proved a totally unreliable guide to run time.

run, *v.* Add: **I. 1. g.** *to run counter* (*to*): see COUNTER *adv.* 1 and 3.

h. *Cricket.* To act as a runner (*RUNNER 1 f) *for* (a disabled batsman).

1855 F. LILLYWHITE *Guide to Cricketers* 7 No substitute shall in any case be allowed to stand out or run between wickets for another person without the consent of the opposite party. **1900** W. A. BETTESWORTH *Walkers of Southgate* xi. 290 Mr Rutter..was..standing behind the wicket in the usual place of a man who is running for another. **1908** W. E. W. COLLINS *Leaves from Old Country Cricketer's Diary* xiii. 223 Once only in more than forty years, when I was hit badly on the knee-cap, I had a man to run for me.

i. *colloq.* To suffer pressingly from diarrhoea. Cf. *RUN *sb.*[1] 14 f.

1966 A. E. LINDOP *I start Counting* vi. 92 'I said to her .."if you lie down on that wet grass you'll come down with the running trots"—' 'And did she?'.. 'She was run, run, run, *run*! All the time.' **1967** A. WILSON *No Laughing Matter* III. 188 Suddenly she knew by sensation the meaning of that unattractive expression 'it kept me running all night'... She was indeed kept 'running all night'.

2. b. (Further example.) Also with *in.*

1960 G. E. EVANS *Horse in Furrow* xi. 151 But the danger of undersized or ill-bred stallions 'running in' with the mares on communal..pastures had become considerably less. **1972** *Country Life* 3 Feb. 288/3 The heifers.. run with the Hereford bull to produce their first calf.

c. *to run* (*a*)*round*: to associate or consort *with* (someone, esp. of the opposite sex); to court, have an affair with; similarly with *together.* Also in general sense, to go about hurriedly with no fixed goal; to go from one place or person to another. Also *transf.* *to run* (*a*)*round in circles*: see *CIRCLE *sb.* 1 c.

1887 in *Amer. Speech* (1950) XXV. 37/1 She used to run around with Jim Reiley. **1891** J. H. PEARCE *Esther Pentreath* III. v. 194 The speaker, a big awkward slattern, had been long trying hard..to get Casy to 'run around' with her. **1920** H. CRANE *Let.* 24 Sept. (1965) 42 I've been running around talking, talking, talking and waiting for the proper persons to arrive at their offices. **1925** F. SCOTT FITZGERALD *Great Gatsby* vi. 125, I may be old-fashioned in my ideas, but women run around too much these days to suit me. **1929** D. RUNYON in *Hearst's Internat.* July 56/1 He writes about..who is running around with who, including guys and dolls. **1940** M. ALLINGHAM *Black Plumes* iii. 19 He and Phillida ran round together quite a bit. **1940** F. & R. LOCKRIDGE *Norths meet Murder* viii. 129 You could tell me about it, and why you didn't like Brent, if it wasn't because he was running around with your wife. **1952** M. LASKI *Village* xi. 161 They've been running around together for some time... She's certainly a lovely girl. **1962** [see *CRUDDY *a.* 2]. **1969** in Halpert & Story *Christmas Mumming in Newfoundland* 213 Since people know who the uncovered janney 'runs around with' (i.e., his friends), they will have a clue to the probable identity of the others.

4. a. Also *to run out on* (someone), to abandon, desert.

1920 H. C. WITWER in *Collier's* 15 May 57/1 Kin you imagine him runnin' out on me too? **1934** ADE *Let.* 8 Mar. (1973) 180, I received no invitation or notification and neither did John Golden and so we must not be accused of running out on our little pop-eyed friend. **1942** T. RATTIGAN *Flare Path* I. 25 You *were* a fool to run out on me, weren't you? **1951** M. McLUHAN *Mech. Bride* (1967) 60/2 Some of the fellows were sneering that her husband was running out on her. **1962** H. HOOD in R. Weaver *Canad. Short Stories* 2nd Ser. (1968) 210 I'm not running out on you. **1973** 'D. HALLIDAY' *Dolly & Starry Bird* xv. 221, I decided I was going back to Rome. ..Johnson, on whom I was running out, listened to me with patience.

e. In weakened sense, to leave, depart (freq. with an implication of haste). Also with *along, away. Run along* (*with you!*) imp., used esp. to children or inferiors (cf. *get along!* s.v. ALONG *adv.* 2).

1816 JANE AUSTEN *Emma* II. i. 18 'I am afraid we must be running away,' said Emma..beginning to rise... 'I, had no intention..of staying more than five minutes.' **1890** O. WILDE *Pict. Dorian Gray* i, in *Lippincott's Monthly Mag.* July 7 You are not going to run away so soon, Mr. Hallward? **1902** B. POTTER *Tale of Peter Rabbit* 15 Now run along, and don't get into mischief. **1927** A. A. MILNE *Now We are Six* 57 But every one says, 'Run along!' (Run along, run along!) All of them say 'Run along! I'm busy as can be.' **1933** M. LOWRY *Ultramarine* ii. 62 Oh well, I'll have to put up with you. Run away and play. **1934** E. BOWEN *Cat Jumps* 242 'Yes, run along with you,' she said. 'And don't be so silly again.' **1935** N. MITCHISON *We have been Warned* II. 202, I must run or the garden party will miss me. **1952** E. O'NEILL *Moon for Misbegotten* I. 56 So run along now and play with your horse, and don't bother me. **1962** I. MURDOCH *Unofficial Rose* vi. 62 Douglas Swann rose again, accepting his dismissal. 'No thank you, Ann, I must run.' **1965** G. McINNES *Road to Gundagai* xii. 211 Tell your Mother we're going to the flicks and I'll be back about eleven. Better run along now. **1975** A. BERGMAN *Hollywood & Levine* xi. 164 'Helen, we'll be running,' said Wohl... There was a final chorus of good-byes.

7. a. Also, *to run to* (also *true to, up to*) *form*: of a horse, to perform in a race consistently with its previous record. Freq. *transf.* and *fig.*

1891 G. CHETWYND *Racing Reminisc.* I. 85 The result of the Prince of Wales' Stakes was interesting, as it afforded a striking proof of the way in which horses sometimes consistently run up to their form. **1934** WEBSTER, *Run true to form, or type*, to operate as might be expected by its inherent nature. **1960** *Bedside 'Guardian'* IX. 178 This [canvassing] ran true to form until one young woman reddened and said: 'I don't think you should accost mothers on a playground.' **1970** G. F. NEWMAN *Sir, You Bastard* viii. 253 It was simply Paul running to form, never arriving anywhere on time. **1973** D. LESSING *Summer before Dark* 227 I'd be running true to form wouldn't I? A few years in the wide world and then back to the home paddock.

b. (Earlier and later examples.) Also, to stand as a candidate for office *on* a specific issue or policy.

1826 *Virginia Herald* (Fredericksburg) 22 Nov. 3/1 Mr. Pitcher is elected Lt. Governor, by a large majority over Mr. Huntington, who ran on the same ticket with Mr. Clinton. **1851** J. A. QUITMAN in J. F. H. Claiborne *Life & Corr. J. A. Quitman* (1860) II. xvi. 147 A majority of the people have declared against the course of policy.. upon which alone I had consented to run as a candidate. **1859** *Knickerbocker* Oct. 372 We have never had the misfortune to run (or 'be run', as the phrase is) for Congress. **1912** M. NICHOLSON *Hoosier Chron.* 54 I'd go into their counties and spend every cent I've got fighting 'em if they ever ran for office again. **1929** W. FAULKNER *Sartoris* II. 67, I know what I'll do: I'll run for Congress. **1950** *Manch. Guardian Weekly* 24 Aug. 15 Mr. Dewey has sworn not to run again for the Governorship. **1964** GOULD & KOLB *Dict. Social Sci.* 484/2 The Democratic candidate..ran instead on the issue, among others, that the Eighteenth Amendment should be repealed. **1967** G. F. FIENNES *I tried to run Railway* iii. 16 There's a job going as Assistant District Superintendent at Burntisland. Do you want to run for it? **1968** *Globe & Mail* (Toronto) 3 Feb. 3/8 Mr. Woodcock, who says he is running on a youth ticket, joins two other fringe candidates. **1978** *Detroit Free Press* 5 Mar. A 14/1 President Carter says he does not know whether he will run for re-election in 1980 or whether he will if he does.

e. *to run scared*: in *U.S.* political terminology, to compete for office in a manner indicating or suggesting a fear of losing, esp. to avoid over-confidence. Also in generalized and weakened senses, to be frightened, to panic. Usu. as *pres. pple.*

1960 *Newsweek* 19 Sept. 39/1 Confident as he is of winning, Nixon intends to run hard and scared until the very last minute. **1964** G. McDONALD (*title*) Running scared. **1968** W. SAFIRE *New Lang. Politics* 389/2 The phrase is directed..to the candidate who is in the position of Thomas E. Dewey in 1948, considered a 'shoo-in'... Hindsighted politicians now say Dewey should have 'run scared'—conducted a more aggressive, fighting campaign. **1969** P. F. SIMON *The Boxer* (song) 3 In the quiet rail-way sta-tion run-ning scared. **1976** *Times* 1 Mar. 7/4 President Ford..has been running scared against Reagan for many months. **1976** J. PHILIPS *Backlash* III. i. 116 He's big, but running very scared. **1978** *Time* 1 May 24/2 'People are running scared,' says..a Boston drug-abuse expert. 'A situation exists which borders on hysteria,' agrees the deputy director of the Illinois dangerous-drugs commission. **1978** *Washington Post* 12 June C2/1 Members of Congress who are already running scared (this is an election year).

8. b. (Earlier examples.)

1743 M. CATESBY *Nat. Hist. Carolina* II. p. xxxiii, Herrings in March leave the salt waters, and run up the rivers. **1806** LEWIS & CLARK *Orig. Jrnls. Lewis & Clark Exped.* (1905) IV. 95 These women informed us that the small fish began to run which we suppose to be herring from their description. **1884** G. B. GOODE *Fisheries U.S.: Nat. Hist. Aquatic Animals* 376 [Kingfish] occasionally run to a considerable distance up the rivers.

9. b. *to run off the rails*: see *RAIL *sb.*[2] 4 a, b.

10. a. Also of a torpedo: to pass through the water.

1914 F. T. JANE *Navy as Fighting Machine* xiii. 100 The 'balance chamber'. This regulates the depth at which the torpedo will run. **1942** *R.A.F. Jrnl.* 16 May 21 They dived and dropped their torpedoes, all of which ran satisfactorily.

b. Also in phr. *to run into the sand*(*s*), to peter out; to come to nothing.

1931 E. WILSON *Axel's Castle* iv. 112 We may put it down to an academic assumption that English drama ended when the blank verse of the Elizabethans ran into the sands. **1956** A. L. ROWSE *Early Churchills* xvii. 378 A naturally clever woman with genuine interests of the mind that ran into the sand.

c. Also *fig.*

1918 *Dialect Notes* V. 21 *Run foul of*, to meet. **1932** *Times Lit. Suppl.* 27 Oct. 792/4 He ran foul of the most powerful gang in New York and the police at the same time. **1971** *New Scientist* 27 May 533/1 Tamplin originally ran foul of the AEC when he was asked to demolish Ernest Sternglass's case.

11. a. (Further examples.)

1861 J. A. SYMONDS *Let.* 30 Apr. (1967) I. 287 This is just the time that I sd like to be at home... If I can, I shall run down for a day this Term. *Ibid.* 28 Oct. 314 Do you not think you cd persuade Papa to run up with me one night this week to see it? **1866** G. MEREDITH *Let.* 14 Jan. (1970) I. 356 A..letter disarranged the plan, which would have left me free of conscience to run down to you. **1866** TROLLOPE *Belton Estate* xxxii, in *Fortn. Rev.* 1 Jan. 429 I'll..just run over once or twice in the year. It would not be a nice place for you to live at long. **1871** J. BLACKWOOD *Let.* 31 Dec. in *Geo. Eliot Lett.* (1956) 230 Hamley only ran up for an afternoon and could not get to the Priory. **1885** C. M. YONGE *Two Sides of Shield* I. x. 168 He says he would run over to see me if it were not for the dragons. **1898** G. B. SHAW *Let.* 18 Oct. (1972) II. 68 Our visitors here are..Strandring and Pease, who run down occasionally, or at least intend to do so. **1902** E. NESBIT *Five Children & It* ix. 237 I'll run up to town and have some lunch at club. **1921** G. B. SHAW *Back to Methuselah* III. 136 Why not run over and join me for the afternoon? **1972** D. SUTTON in *Lett. R. Fry* I. 53 During the 1910s Fry had formed the habit of running over to Paris.

b. Also const. † *it.* Also of a company, to schedule journeys *over* a given route. Now freq. with qualifying advb., as *to run late*, (of a transport service) to be behind schedule; also *transf.* of persons.

a **1817** JANE AUSTEN *Northanger Abbey* (1818) I. vii. 84 How long do you think we have been running it from Tetbury, Miss Morland? **1869** *Bradshaw's Railway Man.* XXI. 177 The company authorized to run over the Mid-Kent, the West End of London and Crystal Palace, ..and the West London Extension. **1954** L. MACNEICE *Autumn Sequel* 161, I today, equally undefended, Not knowing if we are running fast or late, Walk through this empty train. **1956** N. MARSH *Off with his Head* (1957) vi. 113, I got called out on an urgent case and found myself running late. **1973** *Daily Tel.* 3 Feb. 14 Any attempt to discover by telephone whether an incoming flight is running late is futile. **1977** N. MARSH *Last Ditch* vi. 176 Alleyn looked at his watch. 'I'm running shamefully late,' he said.

c. To slide or travel on a sleigh or toboggan or on skis.

1887 [see COAST *v.* 13]. **1898** *Encycl. Sport* II. 472/1 Now, as each competitor is obliged to run three times,.. and as the course varies, not only day by day, but hour by hour,..a great deal of judgment is required on the part of the rider. **1935** *Encycl. Sports* 562/2 The attitudes of skiers in running vary from an almost upright one to a very low crouch. *Ibid.* 563/1 Having acquired confidence, the ability to go fairly fast, and to run straight over moderate slopes without falling, the novice can now tackle the turns. **1956** *Ski-ing* ('Know the Game' Ser.) 13/2 The fastest method of ski-ing down a slope is naturally by running straight (*schussing*).

15. d. (Earlier example.)

1731 D. EATON *Let.* 2 May (1971) 135 We shall view the saplins at Oakly Wood on Wednesday and have the sale day on Saturday, bycause the bark runns very well.

17. b. (Examples of a motor vehicle or engine.)

1912 *Motor Man.* (ed. 14) vi. 174 When a car is running badly the owner very often comes to the conclusion, [etc.]. **1952** *Chambers's Jrnl.* Apr. 208/1 With the port outer-diesel running the exciter and the other diesels cut to no load. **1939** G. B. SHAW *Geneva* III. 110 It's no use going on making motor cars that you know will never run. **1959** E. K. WENLOCK *Kitchin's Road Transport Law* (ed. 12) 112/1 The petrol tank must not be filled..while the engine is running.

c. *transf.* Of a business, household, etc.: to function or operate.

1927 E. O'NEILL *Marco Millions* II. iii. 141 Sound common sense and a home where everything runs smooth. **1939** J. B. PRIESTLEY *Let People Sing* xiv. 416 I've got this place. It's doin' well, makin' money. But I don't want it all the time, an' now it's running easily it doesn't need me all the time. **1969** J. BARZUN (*title*) The American university: how it runs, where it is going. **1974** N. FREELING *Dressing of Diamond* 84 Meals were always on time. ..The 'house' ran like silk.

d. Of a cinematographic film, recording tape, etc.: to pass between spools, to (continue to) be in motion; to be shown or played.

1931 *Discovery* Dec. 386/1 The speed at which the film was running, ninety feet per minute, made it necessary that statements should be brief. **1969** 'A. GILBERT' *Missing from her Home* vi. 84 I'd been to see a film in the afternoon, and it ran longer than I expected. **1972** *Listener* 21 Dec. 852/1 Production Assistant: 'Quiet. Going for a take. Standing by.' Director: 'Right.' Sound: 'Sound running.' Director: 'Turn over.' **1973** V. CANNING *Finger of Saturn* i. 8 The film began to run... I just watched. **1976** *Oxf. Compan. Film* 743/1 The first part, running about 3½ hours, was released as *The Wedding March.*

20. c. (Further example.) Also in *fig.* phr. *to run hot*: of persons, to become angry (cf. HOT *a.* 6 b).

1924 A. D. SEDGWICK *Little French Girl* I. iv. 31 We have our baths in the morning, and the water doesn't run very hot then. **1941** *Sun* (Baltimore) 29 July 10/7, I seen him reading across the table and called him down. He run hot and so I told him to git his money. *Ibid.*, He was so pleased with the phrase 'run hot' that he regarded himself as the gainer on balance. **1976** K. BENTON *Single Monstrous Act* v. 166 The Detective Chief Superintendent's waiting for us, and beginning to run hot, too. He's got a lot on his mind.

21. a. (Earlier U.S. examples of ice.)

c **1804** P. GASS *Jrnl.* (1807) v. 61 The ice began to run in the river. **1867** A. D. RICHARDSON *Beyond Mississippi* xi. 145 Reaching the Missouri again, I found the ice running so heavily, that it was impossible to cross.

25. a. (Further example.)

c **1865** MRS. GASKELL *Let.* 6 Oct. (1966) 777 Still the scullery tap *did not run*; & until it does that smell will go on.

d. Of a bath: to be in the process of being filled with water.

1936 J. BUCHAN *Island of Sheep* vi. 117 He's back now, for I heard his bath running. **1946** J. B. PRIESTLEY *Bright Day* x. 298 She popped her head round the door.. to tell me that a bath was running for me. **1973** 'P. REID' *Harris in Wonderland* xiv. 103 Mayer woke me at nine with a mug of tea. He was fully dressed. 'The bath's running for you,' he said. **1977** 'A. YORK' *Tallant for Trouble* xii. 184 I've a bath running.

27. c. (Further examples.) Also of a cinematographic film: (to continue) to be shown to the public.

1923 H. CRANE *Let.* 5 Oct. (1965) 149 Charlie [Chaplin] ..is here in New York at present to see that the first film he has produced in it gets over profitably... It's running now for just a week or so more at the 'Lyric' theatre. **1940** G. MARX *Let.* 5 Sept. (1967) 25 He also hates Noel Coward and even refuses to see his playlets, which are now running at El Capitan. **1976** *Oxf. Compan. Film* 646/1 Rodgers and Hammerstein's stage musical, which opened in New York in 1959 and ran for four years.

29. b. (Further examples.)

1928 R. A. KNOX *Footsteps at Lock* ix. 87 These things do run in families... In *our* family, we're always appearing when we're not wanted to. **1966** A. E. LINDOP *I start Counting* xx. 259 Runs in the family, doesn't it. Goddam bossy, both of you. You're a real little chip off the brotherly block. **1971** 'H. CALVIN' *Poison Chasers* x. 137 Curiosity..runs in the family..like wooden legs. **1973** [see *military policeman* s.v. *MILITARY *a.* 3 b].

c. Of a newspaper or magazine article: to be printed or published, to appear; to be printed without abridgement.

1928 *Amer. Speech* IV. 135 If news is 'heavy' on a 'tight day' and is permitted to 'run' in length practically as written, [etc.]. *Ibid.* The copy reader now knows whether he is to let 'copy', news articles, 'run' or must 'cut'. **1974** *Publishers Weekly* 18 Nov. 12/3 Janet Flanner's introduction to 'London Was Yesterday'.. will run in the February issue of *Travel & Leisure.*

30. b. (Later examples with *to*).

1914 G. B. SHAW *Parents & Children* p. xi, The something unpleasant may be only a look of suffering..or it may run to forcible expulsion from the room. **1939** —— *In Good King Charles's Golden Days* I. 45 Mr. Newton: your privilege with me does not run to the length of knocking my brother down. **1967** 'S. WOODS' *And shame Devil* 222 He was brilliantly attired in crimson pyjamas... Who would have thought his taste would run to the exotic?

33. b. Also const. *at*, and with a specified amount.

1924 *Sci. Amer.* Sept. 213/1 This means that the ore runs approximately four and one-half tons per gram of radium. **1940** G. MARX *Let.* 5 Sept. (1967) 24 He ran it [*sc.* a film] yesterday for the Breen office—it runs over 13,000 feet. **1960** *Bedside 'Guardian'* IX. 216 The original operetta, which runs to a bothersome two and a half hours, was compressed..into a tight, not to say breathless, hour. **1971** *Daily Tel.* 26 Oct. 1/8 Unemployment benefit is running at about £6,900,000 a week. **1973** *Country Life* 14 June 1712/2 The Historic Buildings Council grants are now running at a rate of £1½ million a year. **1978** *Nat. Geographic* Nov. 623/1 Last autumn arrests [of illegal immigrants] were running 80 a week.

II. 35. a. Also *fig.* in colloq. phr. *to run a mile*, to seek safety in flight; to evade through fear, reluctance, etc.

1949 D. SMITH *I capture Castle* v. 64 Men..run a mile from obvious fascination. **1952** 'R. GORDON' *Doctor in House* xvii. 188 The ones that run a mile if they see a nurse and talk big about staying single. **1963** A. HERON *Towards Quaker View of Sex* 67 Were a woman to whom he exposed himself to respond sexually, the average exhibitionist would run a mile. **1969** H. E. BATES *Vanished World* x. 98, I run a mile from intellectual swank words such as 'esoteric' and 'proliferate'. **1973** J. WILSON *Truth or Dare* iv. 44 Full of talk—yet if Betty gave any of them the come-on they'd run a mile. **1977** *Gay News* 7–20 Apr. 29/1 Whenever anything reasonably likely appears on the scene Cole runs a mile and wallows in neo-platonic discussions on the..differences between lust and love.

d. *esp.* to navigate (a stream, esp. a dangerous stretch of one) in a small boat. (Earlier examples.)

1805 LEWIS & CLARK *Orig. Jrnls. Lewis & Clark Exped.* (1905) III. 23 There were five shoals neither of which could be passed with loaded canoes nor even run with empty ones. **1839** J. K. TOWNSEND in R. G. Thwaites *Early Western Trav.* (1905) XXI. xv. 358 Here Mr. M'Leod and myself debarked, and the men ran the dall. **1875** 'MARK TWAIN' *Old Times on Mississippi* ii. 37 Each of our pilots ran such portions of the river as he had run when coming up-stream.

36. c. (Earlier example.) Also to score from (a stroke) by running; cf. sense *77 i (*d*).

1744 *Laws* [of Cricket] in *New & Compl. Dict. Arts & Sci.* (1755) IV. 3459/2 If in running a notch, the wicket is struck down by a throw [etc.]. **1816** W. LAMBERT *Instr. & Rules Cricket* 35 The Striker should be careful and attentive in running both his own and partner's hits. **1878** *Sussex Archaeol. Coll.* XXVIII. 80 Cricket grounds are not larger now than then, and yet the batsmen *ran* their hits.

e. *to run rings round*: see RING *sb.*[1] 14 d in Dict. and Suppl.

f. *to run interference*: in *U.S. Football*, to move in such a way as to cause interference (cf. *INTERFERENCE 1 c). Also *fig.*

1929 JONES & WESSON *Football for Fan* ii. 23 Whether he is to charge straight ahead, cross check, pull out to run interference or to protect a pass..he must always look the same to his rivals before the ball is snapped. **1932** F. OAKES *Football Line Play* xi. 135 The most difficult block the center must make occurs when both linemen on each side of him drop back to run interference. **1947** *Partisan Rev.* May–June 236 The official liberal runs interference for the Communist with a system of intellectual evasion. **1972** J. MOSEDALE *Football* iii. 39 Nagurski was described as a 'man who runs his own interference'. **1976** *National Observer* (U.S.) 1 May 5/3 He said he wasn't going to allow *his* police to run interference for employees trying to get through a union picket line.

40. d. *slang* (orig. and chiefly *N. Amer.*). To drive past (a traffic signal showing red). Cf. *JUMP *v.* 10 b.

1935 *Harper's Mag.* June 60/2 Perhaps we even 'ran' a light, relying on the waiting cars to continue to wait until we were out of their way. **1951** *Tuscaloosa* (Alabama) *News* 28 Jan. 1/7 Mitchell..had been arrested on charges of reckless driving and running a stop sign. **1953** *Birmingham* (Alabama) *News* 8 Aug. 1/8 Wilson told officers the brakes on his..truck failed, causing him to run a red light at the intersection. **1967** *Boston Traveler* 1 June 5/6 Policeman Howard Park stopped a minister for passing a red light... 'He who runs red light gets ticket.' **1972** *Even. Telegram* (St. John's, Newfoundland 27 June 14/4 Guy forgot to turn on his lights. And ran a stop sign. A policeman pulled him over. **1978** *Guardian Weekly* 15 Jan. 13/2 Cairo drivers run red lights and drive the wrong way down one-way streets.

41. b. To darn (the heel of a stocking) before wearing in order to strengthen it.

1802 D. WORDSWORTH *Jrnl.* 24 Dec. (1941) I. 186, I have been..running the heel of a stocking. **1844** S. S. ARNOLD in *Proc. Vermont Hist. Soc.* (1940) VIII. 170 Paid Mrs. Wales for knitting silk stockings and running the heels 67 cents. **1904** *Eng. Dial. Dict.* V. 186/1 It is common to run the heels of stockings with cotton before wearing them.

c. To attach (a ribbon or similar decoration) to cloth by passing it through a series of holes in the material.

1872 *Young Englishwoman* Nov. 607/2 Run a braid or ribbon through the open row. **1908** M. MORGAN *How to dress Doll* v. 48 Ribbon is run through neck and sleeves.

42. a. Also *spec.*, to chase or hunt animals (e.g. buffalo) on horseback or (occas.) with a vehicle. Chiefly *N. Amer.*

1841 G. CATLIN *Lett. on N. Amer. Indians* I. 219 On this journey we saw immense herds of buffaloes; and although we had no horses to *run* them, we successfully *approached* them on foot. **1849** F. PARKMAN *Calif. & Oregon Trail* xxiv. 386 The chief difficulty in running buffalo..is that of loading the gun or pistol at full gallop. **1900** W. F. DRANNAN *31 Yrs. on Plains* (1901) xxi. 300, I met about thirty Kiowa Indians going out to run the buffalo near there. **1949** L. NORDYKE *Cattle Empire* 264 Horses are furnished for the care of the cattle and for other useful purposes, and they must not be used to run wild horses, or buffalo, or antelope. **1963** G. F. G. STANLEY *Louis Riel* 5 There could be no room for selfish individualism when the métis ran the buffalo. **1968** K. WEATHERLY *Roo Shooter* 130 Hunter would have to run the little doe with the Land-Rover.

c. (Earlier example.)

1767 J. WEDGWOOD *Let.* 27 May (1965) 54 The Ministry were run very hard yesterday in the House of Lords. They carried their point by a Majority of three only.

e. (Earlier example.) Also, to exhaust or defeat by constant pursuit or pressure; to destroy by excessive use.

1836 W. T. PORTER in *Spirit of Times* 9 July 162/1 It's no use to run the thing into the ground. **1947** J. STEINBECK *Wayward Bus* viii. 135 Well, start feeling good, then, and don't run it into the ground. Nobody likes sick people very long. **1955** *Times* 3 Aug. 3/7 Close marking, hard tackling, and shrewd tactical kicking, until the opposition has been 'run into the ground'. **1977** *N.Y. Rev. Bks.* 31 Mar. 21/1 Crossman during his brief tenure as editor..just about ran the paper into the ground.

43. a. Also *transf.*, esp. in political use.

1880 E. W. HAMILTON *Diary* 2 Dec. (1972) I. 83 He expressed his belief that the suspension of the Habeas Corpus a month or two ago would have arrested the evil, and as things now are the only course to take is to 'run' two measures side by side the moment Parliament meets. **1898** W. S. CHURCHILL *Let.* 5 Jan. in R. S. Churchill *Winston S. Churchill* (1967) I. Compan. II. 854 We can run Tirah and Egypt in double harness.

c. (Further examples.) Chiefly *Austral.*

1862 R. HENNING *Let.* 28 Aug. (1966) 95 He is going to run some sheep on the station, and I dare say will do very well. **1901** M. FRANKLIN *My Brilliant Career* ii. 10 Mother felt dubious of her husband's ability to make a living off a thousand acres, half of which were fit to run nothing but wallabies. **1930** L. G. D. ACLAND *Early Canterbury Runs* 1st Ser. x. 241 The dry cattle were run further out than the sheep. **1966** G. W. TURNER *Eng. Lang. Austral. & N.Z.* iii. 50 The word might be used as a transitive verb. 'He runs merinos' i.e. has merino sheep as his stock.

44. a. Also *to run* (one) *off one's feet*, to occupy or overwork to the point of exhaustion (usu. *pass.*). *to run* (one) *ragged*: see *RAGGED *a.*[1] 6.

1857 C. M. YONGE *Dynevor Terrace* II. xv. 235 Charlotte was what Martha expressively called 'fairly run off her feet'. **1937** M. ALLINGHAM *Dancers in Mourning* xxii. 279 Run off his feet, poor lad. Don't know when he sleeps. **1949** N. MITFORD *Love in Cold Climate* I. x. 107 You'd never believe that woman was nearly eighty, she could run us all off our feet. **1970** W. J. BURLEY *To kill Cat* i. 7 'I expect you've got enough to do at this time of year.' 'Run off our feet, sir.'

b. Also, to drive (one) *crazy, out of one's head. U.S. dial.*

1924 L. VOLLMER *Sun-Up* I. 13 Neither one of us is got 'nough [learning] to run us crazy. **1928** J. PETERKIN *Scarlet Sister Mary* xxv. 288 It'll run you crazy if it don' kill you. **1940** J. STUART *Trees of Heaven* I. 20 Some say whiskey will run a man crazy. **1942** L. VOLLMER in *Sat. Even. Post* 22 Aug. 12/3 Fink's meanness had run his wife out of her head.

c. (Further examples.) Also with advbs.

a **1861** T. WINTHROP *John Brent* (1876) xvi. 183 But then he knows ther ain't no Utes round here to stampede his animals or run off any of his gals. **1901** W. N. HARBEN *Westerfelt* xvi. 220 He was here the night they run him off. **1911** T. DREISER *Jennie Gerhardt* I. 10 A man run us away. **1924** H. CROY *R. F. D. No. 3* xi. 189 He's got to be run out. **1946** G. FOREMAN *Last Trek of Indians* x. 195 The agent announced his intention of running out of the country any such preacher who might appear. **1949** W. GANN *Tread of Longhorns* v. 57 The city rulers felt that the marshal should have stopped the jail delivery, and for his failure to do so, he was run off the job. **1967** *Boston Sunday Globe* 23 Apr. 4/6 Something most 17-year-old girls seldom mention..rats. 'They're bad and they'd run you right out of the cellar,' she said. **1976** C. EGLETON *State Visit* xiii. 120 Some cowboy of a truck driver ran us off the road. **1977** 'E. ANTHONY' *Silver Falcon* ii. 44 Get out of Beaumont!.. There's enough of us here who loved your father to run you out.

45. a. (Later *fig.* example.)

1901 M. FRANKLIN *My Brilliant Career* xxiv. 207 The recent 'going bung' of a building society—his sole remaining prop—had run him entirely ashore.

b. Also *spec.*, to convey (someone) in a motor vehicle to a particular destination.

1909 W. J. LOCKE *Septimus* iv. 57 The chauffeur touched his cap. 'I'll run you both over to Nice,' said Clem Sypher... 'I'll run you back again.' **1924** KIPLING *Debits & Credits* (1926) 326 I'll run you out home before sun-up. I'm a haulage contractor now. **1936** L. A. G. STRONG *Last Enemy* ix. 274 'I must go over and see him.'.. 'I can't run you over to day, I'm afraid.' **1939** A. THIRKELL *Before Lunch* v. 126 'Shall I run you home?' he asked. Daphne said her bicycle was in the bottle room. **1952** 'M. INNES' *Private View* iii. 60 Better run you home first... It will save you five minutes. **1958** 'A. BRIDGE' *Portuguese Escape* viii. 128 A taxi..will take aeons. I'll run you out when it's all fixed. **1971** 'D. HALLIDAY' *Dolly & Doctor Bird* iv. 58, I ran Johnson back to my house. **1976** M. BIRMINGHAM *Heat of Sun* ix. 152 I'll run you over later... You stay and have some tea now.

49. a. (Further examples.) Also, to cause (a ball) to move rapidly in a specified direction.

1889 W. T. LINSKILL *Golf* iv. 32 In 'running' the ball with the iron..keep the hands forward in advance of the club head. **1971** *Times* 15 Feb. 9/4 He ran the ball strongly 30ft. past the hole. **1977** *Observer* 30 Jan. 24/8 Tueart made ground on the left before passing to Power, who struck a low centre and David Craig ran the ball past his own keeper.

50. a. Also *spec.* to pass (a duster, etc.) hurriedly *over* (a surface) or *under* (furniture).

1948 A. CHRISTIE *Taken at Flood* I. xvi. 95, I know service isn't up to much nowadays—but I still think they run a mop under the furniture. **1952** M. ALLINGHAM *Tiger in Smoke* xii. 190, I ought just to run a tape over the place. **1975** W. J. BURLEY *Wycliffe & Pea-Green Boat* viii. 115, I cook a meal for him occasionally and I run a vacuum over the place. **1977** J. R. L. ANDERSON *Death in City* i. 10 Emptying waste-paper baskets, running a duster over desks, and vacuum-cleaning floors.

d. *Theatr.* To move or carry (scenery) about the stage; to shift (a 'flat') along a groove. Freq. with advs., as *on, off.*

1831 J. Boaden *Life Mrs. Jordan* I. ix. 201 English play and farce, demanding a constant succession of scenes called flats, run on suddenly for the frequent changes of place. **1889** *N.Y. Tribune* 14 July 10/5 Nearly all scenes..are mounted on wheels which enable them to be easily moved upon the stage, hence the compound verbs 'run-on' and 'run-off', which are in universal use in the theatre. The word 'move' is scarcely ever heard. **1921** G. C. D. Odell *Shakes. from Betterton to Irving* I. iii. 99 A Shutter is the modern 'flat', run—in two pieces—on grooves from opposite wings and clamping together when they meet midway of the stage. *Ibid.* iv. 109 The second scene is 'Ambrosio's House', and may with equal certainty be attributed to a second 'flat' scene run in, on the second groove, behind the first... The next act begins again with 'the Street', which I am convinced was run on immediately at the end of the first act. **1959** W. C. Lounsbury *Backstage from A to Z* 102 *Running a flat*, carrying a flat.

e. *to run the rule over*: see Rule *sb.* 17 c in Dict. and Suppl.

51. a. Also, to cause (a conveyance, vehicle, etc.) to move in a particular direction, or to a specified destination.

1902 J. H. A. Macdonald in A. C. Harmsworth *Motors & Motor-Driving* xix. 373 On this run he on one occasion got down for a moment, asking his friend to steer, which the friend did by promptly running the car off the road. **1913** *Autocar Handbk.* (ed. 5) xxi. 228 The car may be run in one way and out the other. **1970** J. Porter *Rather Common Sort of Crime* iv. 42 She got back to Shangrila and ran the car into the garage.

b. (Earlier and further examples.) *spec.* to keep, use, and maintain (a road vehicle).

a **1817** Jane Austen *Northanger Abbey* (1818) I. vii. 86 What do you think of my gig, Miss Morland?.. A friend of mine..ran it a few weeks, till..it was convenient to have done with it. *a* **1877** Knight *Dict. Mech.* II. 1346/1 Attempts are being made..to run locomotives by means of liquid fuel. **1902** A. C. Harmsworth *Motors & Motor-Driving* iii. 38, I am running at present four cars of French construction. **1912** *Motor Manual* (ed. 14) vi. 174 A car owner had, for a long time, been running his car with a very defective lubricator. **1924** *Discovery* June 98/1 Simple..apparatus of this kind can be run off an ordinary lighting circuit. **1939** G. B. Shaw *Geneva* II. 45 No No: motor oil. The stuff you run your aeroplanes on. **1959** E. K. Wenlock *Kitchin's Road Transport Law* (ed. 12) 78/2 The driver of every vehicle run under an A, special A, B or C license. **1973** 'D. Jordan' *Nile Green* xxii. 87, I can't collect you. I don't run a car.

c. (Earlier and further examples.) Also in various extended uses. In *transf.* sense *esp.* to look after, manage, or control (someone, *spec.* a spy). Also *refl.* (said of a business or other organization): to function smoothly, to require little administrative interference. *to run the show*: see Show *sb.*[1] 16 in Dict. and Suppl.

1861 O. J. Victor *Hist. Southern Rebellion* I. xvii. 252/2, I suppose I will have to run the machine as I find it. **1866** *Harper's Mag.* Mar. 539/1 The real owner of a grocery.., which was 'run' nominally by another individual. **1883** *Referee* 29 Apr. 7/2 American evangelists and speculators who run salvation on much the same lines as Barnum runs his menagerie. **1888** J. Bryce *Amer. Commonwealth* II. lxii. 446 The primaries have almost always been so carefully packed, and so skilfully 'run', that a majority of trusty delegates has been secured. **1899** R. Whiteing *No. 5 John Street* i. 4 A..coral island..'run' on principles of almost primitive Christianity. **1911** G. B. Shaw *Doctor's Dilemma* p. xxvii, He may make considerable profits at the same time by running what is the most expensive kind of hotel. **1928** E. O'Neill *Strange Interlude* vii. 235 Why couldn't Mother let me run my own birthday? **1932** E. Waugh *Black Mischief* viii. 295 Can't think what you see in revolutions... I suppose you ran the whole country. **1956** H. L. Mencken *Minority Rep.* 206 Why assume so glibly that the God who presumably created the universe is still running it? **1959** *Motor Manual* (ed. 36) xii. 265 One way of running a trial of this sort is to give each competitor a list of places. **1972** *Jrnl. Social Psychol.* Dec. 180 As a result of unexpected difficulties (early summer vacation at Patna University) only 26 groups could be run. **1974** J. Mann *Sticking Place* i. 14 He was as helpful as could be, the members always said..though of course the place pretty well ran itself. **1977** J. Aiken *Last Movement* iv. 76 Our staff are highly efficient; the place runs itself almost without our interference.

transf. **1890** S. Hale *Let.* 2 May (1919) viii. 242 Cornelia is running me, and she is really just the right sort. **1904** Conrad *Nostromo* I. vi. 67 He was not running a great enterprise there... He was running a man! **1931** E. F. Benson *Mapp & Lucia* iii. 67 She wanted to run her, to sponsor her, to arrange little parties for her. **1949** *Sat. Even. Post* 23 Apr. 130/4 You're my father and all that, but I'll be damned if you run me any more. **1961** 'J. Le Carré' *Call for Dead* ix. 91 The East Germans..run their agents direct from Germany. **1967** A. Christie *Endless Night* x. 89, I felt that Ellie was dependent on Greta.. that she let Greta run her. **1972** D. Bloodworth *Any Number can Play* xx. 206 The central Chinese department..were running you for all you were worth. **1976** Scott & Koski *Walk-In* (1977) xv. 94 Major Ch'en was running this agent.

e. *U.S.* To support or provide for (a person or family).

1871 'Mark Twain' in *Galaxy* Apr. 616/1 Turnips enough to run the family for two years! **1880** —— *Tramp Abroad* 225 'Pap's so po' he cain't run me no mo', so I want to git a show somers if I kin, 'tain't no diffunce what..I don't turn my back on no kind of work.' **1909** R. A. Wason *Happy Hawkins* 280 She was in the habit of estimatin' just how little nourishment it would take to run her to the next feed.

f. orig. *U.S.* To publish or print in a newspaper or magazine; *spec.* to publish repeatedly or successively (an advertisement, article, etc., or a series of such items). Also *transf.* of broadcast items.

1884 E. W. Nye *Baled Hay* 202 The business manager.. hated to lose old Balshazzar's whole trade, for he wouldn't run any of his ads unless he would take them all according to his contract. **1912** G. M. Hyde *Newspaper Reporting* iii. 30 If..the editor decides not to print the story, he *kills* it; otherwise he *runs* it. **1916** J. London *Let.* 31 Oct. (1966) 479 Please send me as many prints or proofs.. of this letter of mine (if you run it). **1930** *Publishers' Weekly* 8 Feb. 667/2 The full page advertisement we ran in the New York Times..brought in more business than any advertisement we have ever run. **1950** *Time* 16 Jan. 65/3 With his vigorous news pages, Dana ran blistering editorials against Boss Tweed, the Credit Mobilier and the Whisky Ring. **1966** *Listener* 12 May 699/1 For Mr Allsop to say that the film would be shown next week was like a newspaper editor saying that he would postpone the headline of today's news to run it the following day. **1973** *N.Y. Law Jrnl.* 20 July 4/3 During the last gubernatorial election campaign in New York State, the incumbent ran some 3,000 television commercials on twenty-two different television stations. **1976** N. Thornburg *Cutter & Bone* iii. 62 The lady who ran the ad..evidently had liked his voice.

g. To be suffering from (a fever or high temperature).

1918 A. Woollcott *Let.* 6 July (1944) 64 Baldridge.. was running a fever which worried me a little. **1926** I. Mackay *Blencarrow* xxxii. 273, I don't like her running this temperature. **1956** A. H. Compton *Atomic Quest* ii. 82 The following morning, still running a low fever, I cleared these moves with Vice President Filbey. **1961** P. Dougherty *Mother Mary Potter* xxviii. 245 All through Lent Mother Potter was running a high temperature and suffering greatly. **1963** 'E. McBain' *Ten plus One* viii. 107, I had a little virus, I was running a small fever. **1967** C. Potok *Chosen* xi. 189, I came home from school with a fever... I was running 103·6. **1970** D. Uhnak *Ledger* i. 17 You're warm, Christie. You must be running about a hundred and one.

h. *to run a book* [Book *sb.* 10 in Dict. and Suppl.], to take bets; also *transf.*

1931 *Economist* 10 Oct. 642/1 The discount marker has been inactive, and many brokers are running narrow books, and so are needing less money than usual. **1955** *Times* 12 Aug. 5/4 Powell, who explained that he had been 'running a book' in the prison.

i. To show (a film or television recording); to set (a film camera) in action. Also with *through*.

1940 [see sense 33 b above]. **1953** E. Simon *Past Masters* III. 196 'Have you ever thought of doing anything, with that Mexican film of yours, Hamish?'.. 'If I could have it run through somewhere and have another luke at it.' **1956** H. Kurnitz *Invasion of Privacy* viii. 64 'Your film..unmistakably duplicates a heartbreaking episode experienced by my client.'.. 'We'll run the picture for your client and guarantee to substantially change any objectionable portion.' **1973** V. Canning *Finger of Saturn* i. 8 I'll run them [*sc.* films] straight through. **1974** I. Murdoch *Sacred & Profane Love Machine* 125 Harriet felt giddy and exposed as if very quietly, as in a silent film run in slow motion, the house had been hit by a bomb. **1974** *Daily Tel.* 2 May 3/4 Using a friend's projector and screen, he ran a short colour film taken at the wedding.

j. To perform (a test, analysis, experiment, or the like); to subject (something) to, or measure (a property) by means of, an experimental procedure.

1947 *Jrnl. Biol. Chem.* CLXVII. 553 Assays run in replicate of course give narrow limits of confidence, the limits decreasing with increasing replication. **1961** *Lancet* 5 Aug. 291/2 The mobility of the abnormal screen-globulin did not correspond with that of the Bence Jones protein when both were run in a starch gel containing 2-mercaptoethanol. **1964** Roberts & Caserio *Basic Princ. Org. Chem.* ii. 29 Solids are often run as finely ground suspensions. **1970** H. McLeave *Question of Negligence* (1973) xxi. 167 Could Cameron have some sort of brain lesion?.. It was imperative to run those tests. **1976** M. Machlin *Pipeline* lviii. 588 In a test he ran, oil soaked ice disintegrated and sank within five days, while adjacent ice did not melt at all. **1978** *Nature* 8 June 456/2 Curie temperatures were run for seven specimens, and they ranged from 222 to 272 °C, with an average of 248 °C.

k. *Computers.* To perform (a computation), execute (a program or other task), investigate (a problem), etc., on a computer.

1952 *Rev. Electronic Digital Computers* (Amer. Inst. Electr. Engineers) 12/3 The last problem in this field has not yet been run, but the study has shown that the entire gamut of stock control for a large supply office can be covered by the computer in approximately 3 weeks time. **1968** E. O. Joslin *Computer Selection* iv. 70 One should examine each class of programs to determine the equipment required to run that class of program through the computer. **1973** *Computers & Humanities* VII. 225 Instructors considering adopting this text can be assured that the programs accompanying the flowcharts do work since all solutions have been run. **1977** [see *program library* s.v. *program, programme *sb.* 4].

52. a. See also Face *sb.* 7 b in Dict. and Suppl.

b. (Earlier examples.)

1789 *Maryland Jrnl.* 2 Jan. 3/2 It was agreed to run the following ticket in their respective Districts. **1792** A. Hamilton *Let.* 10 Oct. in *Wks.* (1886) VIII. 286 Either Governor Clinton, or Mr. Burr,..is to be run in this quarter as Vice-President, in opposition to Mr. Adams.

1825 J. K. Paulding *John Bull in Amer.* v. 85 [They] talk of running him for the next governor.

c. (Earlier U.S. examples.)

1835 P. Hone *Diary* 16 Mar. (1889) I. 134 This is a club..where they sup, drink champagne and whiskey punch, talk as well as they know how, and run each other good-humouredly. **1860** J. G. Holland *Miss Gilbert's Career* 349 Now what's the use of running a feller?

e. *slang.* To report or hand over (someone) to the police, etc. *spec.* in *Mil.* use, to bring a charge against (someone).

1909 E. Wyrall *Spike* iii. 17 In tramp language, to be 'run' is to be handed over to the police. **1919** *Athenæum* 18 July 632/2 'Running a man' means bringing a charge against him for orderly room. **1925** Fraser & Gibbons *Soldier & Sailor Words* 247 *Run, to*, to report or charge anyone with an offence. **1930** Brophy & Partridge *Songs & Slang 1914–1918* 157 Let them spades alone or I'll run yer. **1933** 'G. Ingram' *Stir* xii. 188 Was any of you monkeying with the cocoa last night?.. If I find out who it was, I'll run 'im and make it 'ot for him. *a* **1935** T. E. Lawrence *Mint* (1955) II. iii. 107 We are supposed to have a flight-lieutenant over us. I saw one, when the Sergeant Major ran me.

f. To manipulate or falsify, esp. in phr. *to run the odds*.

1922 Joyce *Ulysses* 312, I heard So and So made a cool hundred quid over it, says Alf... He let out that Myler was on the beer to run the odds and he swatting all the time.

g. *to run one's mouth*, to talk profusely or excessively, to chatter; to complain. Cf. *to shoot (off) one's mouth* s.v. Shoot *v.* 23 g in Dict. and Suppl. *U.S.* and *Black slang*.

1940 W. Faulkner *Tomorrow* in *Sat. Even. Post* 23 Nov. 39/1 Drunk still & running his mouth. **1954** in Cassidy & Le Page *Dict. Jamaican Eng.* (1967) 388/1 Yu run yu mout fe not a ting, all de talk yu talk fe nutten. **1970** C. Major *Dict. Afro-Amer. Slang* 98 *Run (one's) mouth*, to talk excessively; to complain. **1973** *Black Panther* 24 Mar. 14/3 Maybe you call working running your mouth on these TV programs. **1977** *Time* 13 June 50/1 All there is to real estate is running your mouth a bit, knocking on doors and asking people if they want to sell their house.

h. *to run a game*: to obtain money by deceit or trickery; freq. const. *on*. *U.S. Blacks*.

1967 J. Horton in *Trans-Action* Apr. 6/2 Their reasons for disapproving of hustling were not moral. Hustling meant trouble... Others said there was not enough money on the street or that it was too difficult to 'run a game' on people. **1973** T. Kochman *Rappin' & Stylin' Out* 162 Other operators on the street who are looking for a chance to 'whup' (Chicago) or 'run' (Los Angeles, New York) 'a game' (trick someone out of some money) are known as 'slicks' or 'slicksters'. **1974** H. L. Foster *Ribbin'* ii. 30 This is..the behavior that urban blacks use to 'run a game on the man'. *Ibid.* iv. 160 He knows how to 'run a game' to get what he desires from people. **1975** *Language* LI. 246 If we look at a number of Los Angeles examples of black 'put down' terms (e.g.. *to run a game on someone*..) and K's example *to mount someone*, we can see that such expressions..are positive and kinetic.

53. f. (Earlier and later examples.)

1823 *Practical Builder* ix. 376 When the mould is ready, the process of running the cornice begins. **1893** J. P. Allen *Pract. Building Construction* xxi. 337 Cornices, and other ornamental mouldings,..are 'run' in plaster by means of 'horsed mouldings', running on a wood ground fixed on the wall, truly horizontal, the required depth of the cornice. **1966** C. Lloyd *Building Construction* 115 (*caption*) Cornice mould run in situ in coarse plaster, using a sheet zinc profile.

g. *Bridge.* To take an uninterrupted succession of tricks in (a particular suit), to take (a number of tricks) in that way, to play (one's cards in a suit) in that way.

1929 M. C. Work *Compl. Contract Bridge* iv. 21 The bidder..is insured against having his adversaries open and 'run' that particular suit. **1976** *Country Life* 29 Jan. 250/1 South drew trumps..then ran three Club tricks.

56. c. Also *spec.* without *const.*, to cause water to flow into (a bathtub); to pour out water for (a bath).

1933 R. Hichens *Paradine Case* xxvi. 277 His valet was running the bath. **1955** W. Gaddis *Recognitions* II. i. 341 When Basil Valentine got home, he ran his bath immediately. **1958** Osborne & Creighton *Epitaph for George Dillon* I. 20 I'll go and run myself a bath. **1971** *Ink* 12 June 15/1 Anna..asks if she should run her bath. **1974** 'M. Allen' *Super Tour* vi. 225 Be a good boy. You can begin by running me a hot bath.

g. To cause water to flow over (something) held *under* a tap.

1921 in *Sc. Nat. Dict.* (1968) VII. 448/2 To run one's hands under the tap. **1972** *Guardian* 18 Aug. 11/3 Boil the pasta..then drain and run under the cold tap. **1979** *Daily Tel.* 24 Aug. 14 One wine expert commented: 'This is just the same as running a bottle under the tap. It would ruin any decent wine.'

III. With prepositions, in specialized uses.

59. run across—. (Earlier and later examples.)

1880 'Mark Twain' *Tramp Abroad* xxi. 202 If I don't run across you in Italy, you hunt me up in London before you sail. **1903** 'C. E. Merriman' *Letters from Son* 151, I keep running across Job Withers. **1930** D. Runyon in *Collier's* 20 Dec. 32/3 Now in the summer of 1928 I am in Halifax.., when I run across Louie the Lug.

61. run against—. b. (Earlier and further examples in sense 'to encounter suddenly or casually'.) Also *to run up against* (fig.), to

meet with difficulty, obstruction, or opposition from (a person or thing).

1821 P. Egan *Life in London* II. v. 285 Such is the *hypocrisy* displayed on the one side, and the saucy low independence exhibited on the other, which are to be *run against* every day in 'Life in London'! **1886** [see *up against* s.v. Up *adv.*[1] 24]. **1914** 'High Jinks, Jr.' *Choice Slang* 23 Running up against a stone wall. **1924** *Truth* (Sydney) 27 Apr. 6 *Run against*, to meet. **1960** A. Munro in R. Weaver *Canad. Short Stories* (1968) 2nd Ser. 278, I had run up against the simple unprepossessing materialism which was the rock of their lives.

64. run into—. c. Also, *to run into money* (see quot. 1934). *U.S. colloq.*

1934 Webster s.v., *Run into money*, to amount to a considerable sum; to cost a considerable amount. **1973** N. Meyer *Target Practice* (1975) ii. 20 You realize this could run into money?.. I generally get a hundred dollars a day plus expenses.

h. orig. *U.S.* To encounter, meet by chance.

1902 G. H. Lorimer in *Sat. Even. Post* 22 Feb. 11/1 You're just about due now to run into a smart Aleck buyer. **1926** E. O'Neill *Great God Brown* I. ii. 35 The one time I ran into him, I thought he told me he'd destroyed all his pictures. **1934** 'A. Bridge' *Peking Picnic* vi. 60 Mrs. Leroy and the Kuniangs, walking with Derek Fitzmaurice on the City Wall, ran into Miss Ingersoll and Henri Delache. **1954** Koestler *Invisible Writing* xv. 164, I ran into Hahn as I was getting off the Number Eleven tram. **1977** A. Morice *Scared to Death* iii. 21, I expect she's run into one or two acquaintances.

66. run out of—. b. (Further examples.) Also with inanimate subject.

1929 D. Mackail *How Amusing!* 244 He had run out of tobacco the night before. **1938** Mrs. Belloc Lowndes *Diary* 29 Sept. (1971) 165, I ran out of methylated in the last war. **1966** *New Statesman* 14 Oct. 546/2 There was a popular line about the Tories running into Europe because they had run out of ideas. **1970** *Amer. Speech* 1968 XLIII. 59 Those who participated were asked to fill their cars with a small amount of gasoline and then have the cars run out of gas on one of the highways leading to the New York World's Fair. **1971** [see *let *v.*[1] 29 b]. **1973** J. Porter *It's Murder with Dover* viii. 71 Do you mind just hanging on for a second, sir? I've-er-run out of cigarettes.

c. In various colloq. phrases, as *to run out of road*: to approach the end of the roadway (usu. on failing to turn into a bend); also *transf.* and *fig.*; hence *to run out of track*, etc.; *to run out of steam*: of persons, to exhaust one's energy, ideas, etc.; also of things.

1961 Run out of steam [see *post-Christmas* s.v. *post- B. 1 a]. **1961** *Sunday Tel.* 9 July 6/3 When he [*sc.* the motorist] 'runs out of road', he gets severely 'bent'. **1965** Priestley & Wisdom *Good Driving* xii. 83 'Coming unstuck' or 'running out of road' are the light-hearted expressions used by the motor racing fraternity for an experience which can be the far-from-amusing result of attempting to take a bend with the 'wrong' camber at too great a speed. **1968** *Times* 29 Nov. 13/4 It is losing energy more rapidly than other slower pulsars, and is likely to 'run out of steam' soonest. **1969** R. V. Beste *Next Time I'll Pay* xv. 235 If Sequierez's appearance had meant anything, it was that the Cultural Attaché to Her Britannic Majesty's Embassy in Madrid had run out of rope. **1970** *Listener* 19 Nov. 710/3 The real-life situation of the superstar simply running out of road gives the work a clearly recognisable integrity of plot. **1973** D. Francis *Slay-Ride* vii. 78 When I'd run out of steam, they would begin to wonder why they listened. **1974** *Country Life* 26 Dec. 2009/2 At Zahedan in southern Iran I ran out of railway. **1975** I. S. Black *Man on Bridge* xii. 170 Munro..drove till he ran out of track. **1977** *Times* 11 June 11/3, I chose not to hire one [*sc.* a motor-bike], having run out of road on a 350 some years back. **1977** *Gramophone* Dec. 1191/1 The disc input.. tends to run out of steam at the low frequency end.

67. run over—. f. Also, to go over with a machine. Cf. sense 50 a above.

1969 *Guardian* 17 July 11/5 When you've..hoovered the landing..you could just run over Mr Neville's carpet.

g. (Earlier example.) (Cf. sense 78 i below.)

1794 J. Woodforde *Diary* 15 Feb. (1929) IV. 96 A very young Man coming back from Norwich to day with an empty Waggon, falling under it was run over by it.

h. *U.S. colloq.* To impose upon, treat with contempt; to push (someone) around.

1836 *Spirit of Times* 9 July 162/2, I would not advise any man to run over me, for I ask no man any odds further than civility. **1914** B. Tarkington *Penrod* xxv. 264 I've stood enough around here for one day, and you can't run over *me*, Georgie Bassett. **1929** W. Faulkner *Sound & Fury* 227 You may think you can run over me like you do your grandmother and everybody else.

68. run through—. f. To rehearse, repeat, go over (a procedure, role, or the like).

1975 Cowie & Mackin *Oxf. Dict. Current Idiomatic English* 269/2, I think I've grasped your main proposals, but would you mind running through them once again? **1980** K. Hagenbach *Fox Potential* xvi. 157, I ran through it for her without the grim details... When I had finished the story, Frankie put her hand on my thigh.

69. run to—. b. (*b*) (Further examples.) Now freq. const. *can, could*, etc.

1924 H. De Sélincourt *Cricket Match* iii. 76 [He] only gave it me..because he knows I couldn't run to one of my own. **1931** T. R. G. Lyell *Slang, Phrase & Idiom* 653 *B.* 'How much do they want for it?' *A.* 'Five hundred pounds.' *B.* 'I'm afraid I can't possibly run to that.' **1953** E. Simon *Past Masters* III. 156 Bits and pieces in the press, posters if we can run to them. **1970** *Alberta Hist. Rev.* Summer 1/1 Frame houses..needed to be heated and insulated with something of the modern thoroughness to be at all tolerable; and few men could run to it.

(*c*) (Earlier and later examples.)

1883 *Daily Tel.* 4 Oct. 3/2 What I *should* like is a nice pair of spectacles, and, as far as my money would run to it, everything else accordin', sir. **1900** P. White *West End* v. 40, I always had an idea that the guv'nor had some money, but I didn't imagine it would 'run' to this. **1931** T. R. G. Lyell *Slang, Phrase & Idiom* 653 I've only got ten shillings, and..it certainly won't run to twenty-four!

(*d*) To manage to provide, go so far as to have.

1880 *Punch* 25 Dec. 298/2 A red 'un [*sc.* button-hole] with maiden'air trimmings is what I consider O.K. Suits my style and complexion, yer know, so I runs to it once in a way. **1934** G. B. Shaw *On Rocks* I. 208 Weve got nothing out of this. We dont run to Spanish in the Isle. **1960** *Bedside 'Guardian'* IX. 215 Some of the others [*sc.* theatres] run to 1930-ish chrome and glass fittings. **1973** 'M. Innes' *Appleby's Answer* xv. 128 Am I right.. in remembering that Gibber [*sc.* a village] runs to a teashop?

e. Now usu. *to run to seed*. (Further example; also earlier and further *transf.* examples.)

1740 Fielding *Champion* 15 Mar., For Virtue itself by growing too exuberant, and (if I may be allowed the Metaphor) by running to Seed changes its very Nature, and becomes a most pernicious Weed of a most beautiful Flower. **1861** *N.Y. Herald* 23 Nov. 4/5 Can such a country be..permitted to run to seed? **1873** *Amer. Jrnl. Insanity* Apr. 559 We believe somewhat in moral depravity and in accepting the results of our father's sins, but carried to the extent advocated by the Doctor, it is Calvinism run to seed. **1893** *Burpee's Farm Ann.* 65/1 It comes early and stands a long while before running to seed. **1924** A. Huxley *Little Mexican* 249 He pictured a large, blonde, barmaidish personage, thirty-one and not yet married, running a bit to seed. **1953** J. Wain *Hurry on Down* iv. 66 He was plump, but not yet running to seed; aged about forty-five to fifty. **1956** G. Durrell *Drunken Forest* iii. 62 At one time she must have been a handsome woman, but now she had run slightly to seed. **1976** *National Observer* (U.S.) 27 Nov. II. 2/3 Today there is a striking resemblance among many of the 1,000 or so residents, and most of the youngsters run to towhead.

g. *U.S.* Of persons: to have a preference for or a leaning towards (something); to favour.

1873 'Mark Twain' *Gilded Age* xxxiii. 307 We had Dr. Spooner a good while, but he runs so much to emetics.. that we changed off and took Dr. Leathers. **1896** —— in *Harper's Mag.* Aug. 346/2 In my nature I have always run to pie, whilst in his nature he has always run to mystery.

71. run with—. c. orig. and chiefly *U.S.* Of persons: to associate with (a person or group), to befriend.

1909 *Dialect Notes* III. 365 *Run with*.., to associate with, go in the company of. **1914** B. Tarkington *Penrod* xiv. 126 You fellers have treated me nice—and some day you come over to my yard; I'd like to run with you fellers. You're the kind of fellers I like. **1922** E. O'Neill *Hairy Ape* vi. 64 If I can't find her I'll take it out on de gang she runs wit. **1946** Mezzrow & Wolfe *Really Blues* (1957) 378 *Run with*, associate with. **1969** G. Donaldson *Fifteen Men* xiv. 240 He ran with a crowd of kids known as 'Les Snobs'. **1975** *New Review* May 70/1 Hunter Thompson, who ran with the Hell's Angels for eighteen months to write *The Hell's Angels*.

IV. With adverbs, in specialized uses.

72. run away. b. Also *transf.* Freq. used jocularly in the negative (as, *it won't run away*) to give assurance of the permanence or fixity of something or someone.

1882 C. M. Gaskell in *Nineteenth Cent.* Sept. 460 The landowner has been credited with the..most valuable form of security;..it could not 'run away'. **1888** C. M. Yonge *Beechcroft at Rockstone* II. xxi. 191 The charms of 'the halls of Ivor'..which, after all, would not run away. **1908** A. Bennett *Old Wives' Tale* IV. iii. 515 There's no earthly reason why you should go back... The house won't run away. **1928** A. M. M. Douton *Bk. with Seven Seals* 21 Sunday will be round again in a week, and Park Chapel won't run away. **1942** A. E. W. Mason *Musk & Amber* i. 15 'What of Grest [*sc.* an estate] meanwhile?' 'Grest won't run away, Sir.' **1973** J. Porter *It's Murder with Dover* vii. 65 What's your sweat? This Tiffin-bird's not going to run away.

transf. **1920** E. O'Neill *Beyond Horizon* III. i. 152 You've spent eight years running away from yourself. **1934** —— *Days without End* I. 36 It's a rocky road.. this running away from truth in order to find it? **1944** B. Hutchison *Hollow Men* vi. 79 It's his mask. It fools nearly everybody. He's always running away from himself. **1966** *Listener* 17 Nov. 718/2 The whole of the world ran away from the pound, and if this doesn't reveal an inflationary situation, what does?

d. (Further *transf.* examples.)

1935 *Industrial & Engin. Chem.* Sept. 1074/2 'Ethylene at a temperature above 350°C. and under a pressure of 175 kg. per sq. cm...decomposes with explosive violence.' Fortunately, this was not found to be the case, but in many experiments the temperature 'ran away'. **1945** *Rev. Mod. Physics* XVII. 482 If the reaction was not to 'run away', it was essential to make use of neutrons of very low energy in the individual steps of the chain process. **1946** [see *oil operated* s.v. *oil *sb.*[1] 6]. **1953** *Times* 31 Oct. 2/7 We have no practical experience of what happens if a reactor runs away. **1959** C. Hodder-Williams *Chain Reaction* xviii. 198 When the pile 'ran away', one of the heat-exchangers burst with the increased steam pressure.

73. run down. b. (Later examples of (*b*).)

1897 A. Beardsley *Let.* 22 Nov. (1970) 396, I am abominably ill; I ran down at Paris quite alarmingly. *Ibid.* 22 Dec. 410, I had run down terribly before I came here and was quite shattered by the journey.

m. (Earlier example.)

1737 *London Mag.* Oct. 542/1 Our directing the next Payment to be made to the Bank would, I believe, possess the Generality of Mankind with an Opinion that we were resolved to abolish the Company.., which would of course run the Price of their Stock down to very near Par.

n. To reduce or bring (an activity, operation, organization, etc.) to a halt gradually or progressively.

1861 C. M. Yonge *Young Step-Mother* vii. 74 Miss Meadows began one of her tangled skeins of words.. and Mr. Kendal, knowing..that the only chance of a conclusion was to let her run herself down, held his tongue. **1976** A. Price *War Game* I. vi. 118 We're running down the Incident Room, it's true. But we're not giving up. **1977** *Times* 18 Aug. 15/8 Over the past few years, Volkswagen has been progressively running down its 'beetle' production in Germany... Only 100 cars a day are at present produced.

o. *U.S. slang.* To rehearse or perform (a piece of music); to recite (verse).

1948 *Down Beat* 1 Dec. 10 We ran down three new instrumentals and a vocal for Baubles Buxon! **1959** G. Kanin *Blow up Storm* 19, I distributed the parts and we ran it down. **1960** *Jazz Rev.* Nov. 12 When we rehearsed an arrangement that no one had seen before, we'd run it down once or twice. **1961** R. Russell *Sound* i. 10 Bernie struck off a rich chord and began running the tune down in his immaculate post-Teddy Wilson style. **1969** H. R. Brown in T. Kochman *Rappin' & Stylin' Out* (1972) 205 The teacher expected me to sit up in class and study poetry after I could run down shit like that. **1975** R. S. Gold *Jazz Talk* 228 *Run down*, ..to perform, usually in rehearsal, a piece of written music.

p. *to run it down*: to describe or explain a situation in full; to tell the whole truth about a subject. *U.S. slang.* (Freq. in Black English.)

1964 T. Cade in *Massachusetts Rev.* Summer 622, I tried to figure out the best way to run it down to this girl right quick that they didn't have to live in this town. **1967** J. Horton in T. Kochman *Rappin' & Stylin' Out* (1972) 22 Street repartee at its best is a lively way of 'running it down', or of 'jiving'. **1970** G. Jackson *Let.* 17 Mar. in *Soledad Brother* (1971) 236 Write me a letter..and run it down; school, politics, futurities. I want to know it all. **1972** B. G. Cooke in T. Kochman *Rappin' & Stylin' Out* 48 This gesture of lowering the lip is a result of the emphatic manner in which they are 'running it down'.

74. run in. d. Also without *to*. (Earlier example.) *to run in and out* (*of* one another's homes, etc.): to make frequent informal visits (to one another).

1857 [see *get *v.* Etym. note]. **1876** C. M. Yonge *Womankind* xi. 81 A little croquet, a great deal of chatter; and worse than all, much running in and out among near neighbours. **1952** M. Laski *Village* iv. 77 It's lucky it's so far away; at least they can't be running in and out of each other's homes every minute. **1958** A. White tr. *Colette's Claudine in Paris* iv. 30 Just a few yards from here, there's a delightful flat, and we'd be practically on each other's doorsteps.... We could be always running in and out... it would be nice for Claudine and for you too.

g. (Earlier and later examples.) Also in *Naval* use (see quot. 1962).

1859 G. W. Matsell *Vocabulum* 76 *Run in*, arrested. **1889** *N. & Q.* 20 July 49/1 The respectable gentleman who ..takes my part if I get 'run in' by the police. **1909** J. R. Ware *Passing Eng.* 212/1 Thus hooked he is 'run in', where..he is treated for 'D.T.'s'—the origin of most amok —when he either recovers or is passed into an asylum. **1933** J. Cary *Amer. Visitor* 35 Cottee was over the boundary, so I'm going to run him in. **1938** [see *fowl *sb.* 1 d]. **1948** Partridge *Dict. Forces' Slang* 159 I've warned you, the next time you are adrift I'll run you in to the First Lootenant. **1951** *New Yorker* 15 Dec. 94 'Am I going to have to run you in?' the policeman asked. **1962** Granville *Dict. Sailors' Slang* 98/2 *Run in*, place a man in the Commander's report or bring him before the Officer-of-the-watch.

i. (Earlier and later examples.)

1817 *Ackermann's Repository* Jan. 53/1 A row of straw-colour ribbon is run in next to the border. **1900** F. M. Ford *Let.* Oct. (1965) 13, I took hold of that young woman and ran in every bit of her charm I could think of and then smashed in all the repulsion I could think of.

k. To operate (new machinery, esp. a motor vehicle or its engine) at reduced speed or load until it has reached a normal working condition. Also *refl.*, said of the machinery. Also *fig.*

1919 W. H. Berry *New Traffic* (*Aircraft*) xv. 86 Some engineers suggested that the flight should be used to 'run in' the engines. **1925** *Morris Owner's Manual* xvi. 103 It must, of course, clearly be understood that when an engine is new and stiff much more heat is developed than is the case when it has run itself in. **1934** *Punch* 6 June 629/1 Never start a new engine on the self-starter. The battery and dynamo need running-in before you use them. **1939** N. Monsarrat *This is Schoolroom* III. xvii. 390 Anthea bought a car, we ran it in. **1953** A. White tr. *Colette's Cat* iv. 109 We're going out to Rambouillet for lunch in the forest. I've got to run the car in. **1959** *Listener* 2 Apr. 603/1 If you are running-in a new car, and conscientiously keeping down to a maximum of thirty miles an hour, [etc.]. **1961** H. Nicholson *Let.* 1 June

(1968) 395 Kennedy 'must run himself in' before he can inspire confidence. **1972** *Guardian* 18 Sept. 11/4 My car offers its apologies for curious behaviour: 'Running In' is plastered across its back. **1973** A. Behrend *Samarai Affair* iii. 31 Having thus ticked over gently during the previous ten minutes, the committee had now run itself in for the major business of the afternoon.

75. run off. c. (Further examples.)

1925 *Daily Tel.* 13 May 4/3 On purchasing that company it was decided to allow its marine business to run off. **1955** *Times* 3 May 15/1 In the first quarter of this year the surplus has tended to run off.

d. Also *to run off at the mouth*: to talk excessively; to talk nonsense. Cf. sense 52 g above. *U.S. slang.*

1909 *Dialect Notes* III. 403 *Runnin' off at the mouth*,.. loquacity; talking too much. Used of one excessively loquacious. 'He's got a bad case of runnin' off at the mouth.' **1942** Berrey & Van den Bark *Amer. Thes. Slang* §151/6 *Talk nonsense*... go *or* run off at the mouth. *Ibid.* §189/3 *Be talkative*... run (off) at the mouth. **1951** *Rochester* (N.Y.) *Democrat & Chron.* 12 Sept. 14/6 Culio... A run-of-the-mob gunsel—till he runs off at the mouth! **1962** A. Lurie *Love & Friendship* xi. 223 I'm a pig coming over here and running off at the mouth, probably boring you to hell. **1976** *National Observer* (U.S.) 28 Aug 1/1 The man they simply ran off at the mouth about here, Jimmy Carter.

k. To produce or turn out (duplicated copies, etc.).

1889 *Cent. Dict.* 5271/2 *To run off*,.. In printing, to take impressions of; print: as, this press will run off ten thousand every hour; to run off an edition. **1901** Merwin & Webster *Calumet 'K'* vi. 106 Now, we'll write to Mr. Brown—no,.. I'll do that one myself. You might run off the other and I'll sign it. **1932** A. J. Worrall *Eng. Idioms* 78 This machine will run off eighty copies per minute. **1970** H. McLeave *Question of Negligence* xxvii. 229 They had trundled in a portable X-ray machine... They ran off six plates.

l. *Austral.* and *N.Z.* (See quots.)

1933 L. G. D. Acland in *Press* (Christchurch, N.Z.) 25 Nov. 15/7 *Run off*,..(3) To separate: e.g., 'I will *run off* the strangers when we draft the mob'. **1965** J. S. Gunn *Terminol. Shearing Industry* II. 15 *Run-off*, to take a group of sheep from the flock without necessarily 'cutting out' all of this group or 'drafting' them into special lots, for example 'I'll run off some fats'.

76. run on. a. (Further examples in *Racing*.)

1971 *Rand Daily Mail* 4 Sept. 23/4 Fighting Heart was running on when he beat Desert Oil over 1 000 metres. **1977** *Field* 13 Jan. 56/1 Trainers had no way of knowing whether their fastest puppies had that ability to 'run on', an endowment with stamina so essential for Altcar honours.

h. (Further examples.) Also, to set (an advertisement) as continuous text rather than displayed matter.

1951 S. Jennett *Making of Bks.* xv. 256 Some manuscripts and some of the early printed books avoided paragraph divisions and ran all paragraphs on, indicating the commencement of each by means of a paragraph mark. **1973** F. Jefkins *Advertising made Simple* I. i. 17 By classified we mean not only that the advertisements are small and run-on but that they are grouped together under identifying headings. *Ibid.* III. xvii. 200 *Displayed classified*, or *semi-display*—advertisements in the classified section which are not merely run on, but set out and possibly illustrated.

77. run out. c. (*a*) (Later examples.)

1965 A. Nicol *Truly Married Woman* 103 'Oxygen,' Doc said. 'We ran out last week, sir, I forgot to tell you,' the nurse said. **1972** J. Wilson *Hide & Seek* vi. 106 'Haven't you got any?' asked Alice. 'No—no, I've run out,' he said.

(*c*) Of a crop variety: to lose its distinguishing characteristics in successive generations.

1890 *Bull. Cornell Agric. Exper. Station* XXI. 88 We are still confirmed in our belief that varieties of tomatoes are unstable and that they soon 'run out'. **1901** I. P. Roberts in L. H. Bailey *Cycl. Amer. Hort.* III. 1418/1 Old varieties [of potatoes] which have 'run out' often find their way into a locality where conditions are superior. **1957** Duncan & Ross *Growing Field Crops* 10 Sometimes disease and insect invasions reverse the process of plant improvement by adjustment. When this happens, varieties are said to run out.

d. (*c*) (Earlier example.)

1875 *Baily's Mag.* July 162 Mr Greenfield in particular running out almost to every ball and driving it along the ground.

†(*d*) *Cricket.* To be run out. Cf. sense 77 m below. *Obs.*

1828 *Boy's Own Bk.* 22 When a striker has run out, the notch they were running for shall not be reckoned. **1860** F. Lillywhite *Eng. Cricketers' Trip to Canada & U.S.* 20 Hayward contrived to run out. **1876** *Haygarth's Cricket Scores 1855–75* V. 368 Five ran out on the Bradford side.

h. (Further examples.) Also without complement, to win.

1928 *Daily Tel.* 26 June 11/7 He went on to lead 4–2 in the third set, and then at 5–4, had three set balls before running out. **1941** G. Heyer *Envious Casca* xii. 219 [She] miscued... 'You'll run out now.' **1959** *Times* 19 Mar. 17/5 Then Borch began to smash brilliantly and ran out for the game. **1960** E. S. & W. J. Higham *High Speed Rugby* xxi. 289 In ninety-nine cases out of a hundred, a fitter team will run out winners. **1976** *Bridgwater Mercury* 21 Dec., British Cellophane's second team ran out 4-1 winners against Bridgwater Police in the fourth division of the Somerset Squash League.

i. (*c*) (Later example.)

1977 *New Yorker* 10 Oct. 152/2 He then dropped service again, after which Connors ran out the set, 6–2.

(*d*) *Cricket.* To score from (a hit) by running. Freq. *imp.*

1856 *Househ. Words* 2 Feb. 60/2 When you holloa out... 'Run it out!' **1886** J. Pycroft *Oxf. Memories* II. xxi. 89 There were no bounds, all hits were run out. **1898** J. A. Gibbs *Cotswold Village* xi. 234 'Come six, Podder!' I shouted, amid cries of 'Keep on running!' 'Run it out!' etc., from spectators and scouts alike. **1908** W. E. W. Collins *Leaves from Old Country Cricketer's Diary* xi. 180 Run it out, sir, run it out. I hope you'll get six.

l. (*c*) (Further example.)

1911 *Chambers's Jrnl.* Sept. 596/2 When you have read it over, I will run out another copy.

m. (Earlier and further examples.) Also, of a batsman: to cause one's partner to be run out.

1750 in H. T. Waghorn *Cricket Scores 1730–73* (1899) 47 Tom Bell (run out) 0. **1823** M. R. Mitford in *Lady's Mag.* July 391/1 Joel Brent..ran out his mate, Samuel Long. **1860** *Baily's Monthly Mag.* Aug. 364 His only perceptible fault was his tendency to run himself and others out. **1900** P. F. Warner *Cricket in Many Climes* I. iv. 59 Dick Berens ran me out when I was well on my way to a hundred. **1912** —— *England v. Australia* xii. 134 Hitch and Strudwick between them ran out Tumilty. **1933** M. Lowry *Ultramarine* ii. 80 You go in first and run out the chief steward. **1974** *Times* 4 Feb. 1/1 A decision whereby Kallicharan..was given run out for 142 off the last ball of the day..was later reversed.

78. run over. i. Of a vehicle, etc.: to knock down and pass over (someone); to injure or kill by knocking down. Also *transf.* Freq. in *pass.*

The passive use is indistinguishable from that of sense 67 g.

1932 L. Golding *Magnolia St.* I. vii. 114 He is so small the driver hardly notices him till he has very nearly run him over. **1954** T. S. Eliot *Confid. Clerk* III. 109 *Eggerson:* ..Unfortunately, the father died suddenly... *Lady Elizabeth:* He was run over. By a rhinoceros In Tanganyika. **1970** [see *roar *v.* 3 e]. **1978** *Daily Tel.* 1 Nov. 19/6 Mann appeared to be more concerned about the damage to his motor-cycle than the youth he had run over.

81. run up. d. (*a*) (Later example.)

1870 [see *break *v.* 8 c].

f. (Further example.)

1970 *Field* 16 Apr. 703/1 Stanley and Michael Lunt, father and son who between them won three amateur championships and ran up in a fourth.

g. (*a*) (Further examples in *Cricket*.)

1955 *Times* 25 Aug. 3/3 The South Africans ran up 467 runs for the loss of eight wickets. **1977** *Sunday Times* 2 Jan. 28/3 They ran up 322 for four wickets despite a rain-soaked outfield.

(*d*) (Earlier example.)

1870 W. W. Fowler *Ten Yrs. in Wall St.* xxiv. 394 They stepped into the gap, and ran up the price again.

i. (*e*) To raise (a flag) to the top of a mast, etc. Also *fig.* (see quot. 1962).

1901 G. B. Shaw *Devil's Disciple* I. 28 Run up the American flag on the devil's house. **1930** E. M. Brent-Dyer *Chalet Girls in Camp* v. 76 Nearer the lake were the two flag-poles, and the flags would be run up at six o'clock the next morning. **1938** A. J. Liebling *Back where I came From* 27 He run up the American flag. **1962** S. Strand *Marketing Dict.* 639 *Run it up the flag pole*, a Madison Avenue jargon for trying out any idea. Full expression: 'Let's run it up the flag pole and see who salutes it.' **1964** Mrs L. B. Johnson *White House Diary* 21 May (1970) 143 They had already provided a flag pole and we ran up the flag while they all gave the Pledge of Allegiance. **1966** *New Statesman* 25 Mar. 409/2 The decision was made—in the admen's jargon that comes naturally to Tory strategists—to run it up the flagpole and see if anyone saluted.

(*f*) To run (an aircraft engine) quickly while it is out of gear in order to warm it up. Also *intr.*

1938 W. O. Manning *Flight Handbk.* iii. 77 These [wheel brakes] are used..for holding the aeroplane while the engine is being 'run up'. **1942** *R.A.F. Jrnl.* 3 Oct. 26 Intermittently, the sound of engines running up overcomes the wind. **1958** 'N. Shute' *Rainbow & Rose* i. 33 Billy Monkhouse had got the Auster out and was running it up outside the hangar. **1976** *Farnborough 76* (Soc. Brit. Aerospace Companies) 11/1 Inboard engines '2' and '3' are run up first, then the outboards '1' and '4' follow.

j. (*a*) (Later *transf.* examples.)

1965 *New Statesman* 19 Mar. 458/3 Were one to ask a computer to run up a composite 18th-century man, the result would be remarkably like Dr Burney. **1974** 'J. Le Carré' *Tinker, Tailor* ix. 68 In KL I had them run me up a British passport.

(*c*) Now usu. to make (a garment, etc.) by sewing quickly or simply.

1977 *Lancashire Life* Nov. 74/1 The women keep it clean, scrubbing floors, washing curtains, running up new ones on the machine [etc.]. **1979** P. Driscoll *Pangolin* iii. 38 The tailors who ran up suits in twenty-four hours.

V. 82. Collocations used attributively or as sbs. **run and fell** *Needlework* (see quot. 1968); also *attrib.*; **run-flat** *a.*, applied to a kind of tyre on which a vehicle may run after a puncture has occurred; **run-sheep(y)-run** *N. Amer.* and *Sc.*, a children's hiding game (see quot. 1909).

1882 Caufeild & Saward *Dict. Needlework* 428/1 *Run and fell*..is a method sometimes adopted in lieu of Over-sewing, and employed in making seams, either in underlinen, or in the skirts and sleeves of dresses. **1961** M. Spark *Prime of Miss Jean Brodie* iii. 69 In the worst cases they unstitched what had been done and did it again, saying 'This'll not do', or 'That's never a run and fell seam'. **1968** J. Ironside *Fashion Alphabet* 99 *Run-and-fell*, a seam similar to a flat-fell but only one row of machining shows. The two edges are seamed together, one is trimmed close to stitching and the other turned under and then laid flat against the main part of garment and machined. **1976** P. Clabburn *Needleworker's Dict.* 230 *Run and fell seam*,.. type of seam worked on fairly light fabrics, commonly used for children's garments and undergarments, but now often superseded by other varieties. **1946** R. Storrs *Dunlop in War & Peace* xi. 61 As the name *Run Flat* (*Bullet-proof*) *Tyre* suggests, a bullet may penetrate this tyre, but there is no immediate deflation. **1958** A. G. Donnithorne *Brit. Rubber Manuf.* iv. 39 New kinds of tyres were produced to meet service demands in the two World Wars, such, for example, as Dunlop's 'run-flat' tyre. **1973** *Guardian* 30 May 9/3 The Dunlop Denovo 'run-flat' tyre, designed to end the dangers of punctures at high speed, and to enable the motorist to continue his journey without changing the wheel, will be available in October. **1905** *Dialect Notes* III. 93 *Run, sheep, run*,.. a kind of hide and seek in which the participants hide together. **1909** J. H. Bancroft *Games* 6 The author found a good example of folklore-in-the-making in the game usually known as 'Run, Sheep, Run!' in which a band of hidden players seek the goal under the guidance of signals shouted by a leader. **1949** M. Mead *Male & Female* 456, I set myself to study changing patterns of run, sheepy, run or prisoner's base. **1962** W. Stegner *Wolf Willow* I. i. 15 The open field beside Down's where we used to play run-sheep-run in the evenings. **1968** *Sc. Nat. Dict.* VII. 447/1 Rin-sheep-rin. **1969** I. & P. Opie *Children's Games* iv. 173 'Run, Sheepie, Run' (Cumnock [in Scotland]). **1975** *Islander* (Victoria, B.C.) 4 May 2/1 It seems to me I played run-sheep-run there.

run, *ppl. a.* Add: **II. 9. a.** (Earlier example.)

1774 T. Pennant *Tour in Scotl.* (ed. 3) 287 Some of the walls, all of run lime, do as yet remain.

10. (Earlier example.)

1822 T. Creevey in *Creevey Papers* (1903) I. x. 236, I met..the Duke [of Wellington]... It has been a damned serious business,' he said... 'It has been a damned nice thing—the nearest run thing you ever saw in your life.'

11. (Earlier and later examples.) Also, exhausted by running; worn out.

1876 *Coursing Calendar* 147 Troapham proved herself a rare stayer, for she was fearfully run in her first course. **1917** G. Bell *Let.* 13 Jan. (1927) I. xv. 393, I wonder you have time to write me such splendid long letters! You really must not do it when you feel dreadfully run.

13. a. (Earlier example.)

1865 F. B. Palliser *Hist. Lace* xxxvi. 424 To France must be assigned the application of the Jacquard system to the net-frame, and consequently the invention of machinery lace. Shawls and large pieces in 'run lace', as it is termed, had previously been made after this manner.

14. *run-after*, sought after; popular; *run-on* (further examples); *run-out* (earlier examples); *run-over*, (*a*) of (the heels of) shoes: worn down on one side; (*b*) = *run-on*.

1878 *Trans. Illinois Dept. Agric.* XIV. 144 The long, lank hog of the old, run-out breed has given place to the improved Poland China. **1880** *Rep. Vermont Board Agric.* VI. 28 The first condition of fertility we have, even in these run-out pastures. **1882** G. M. Hopkins *Lett. to R. Bridges* (1955) 158 The question of what they call run-on lines and the rhymes or other final words belonging is difficult. **1906** 'O. Henry' in *N.Y. World Mag.* 8 July 8/1 'Smoky' was dressed in..run-over shoes, and trousers of the 'serviceable' brand. **1908** R. W. Chambers *Firing Line* vi. 75 I've heard that you are the most assiduously run-after girl at Palm Beach. **1919** *Ladies' Home Jrnl.* Mar. 89/4 '*Why* do my heels run-over?' Run-over shoe heels are more than unsightly. They are warning of incipient foot trouble. They are usually due to a weakness of the foot structure which permits the foot to 'give' outwardly or inwardly in walking. **1931** *Times Lit. Suppl.* 15 Jan. 43/2 His rules about mid-verse pause and run-over lines. **1946** B. Macdonald *Egg & I* xvi. 175 Reddish cotton stockings, run-over shoes. **1955** *Ess. & Stud.* VIII. 61 The madcap movement of the run-on lines perfectly parallels the hither-thithering of the trapped mind. **1965** R. E. Longacre in *Language* XLI. 74 Such imbedding of sentence within a phrase has special phonological characteristics (level, run-on intonation, and lack of pause before termination of the imbedding phrase). **1978** J. A. Michener *Chesapeake* 667 He wore run-over shoes, baggy trousers, torn shirt and smashed hat, items which he rarely changed. **1979** *Dictionaries* I. 15 The question arises of whether run-on entries will be defined or not.

run-about. Add: **1. a.** Also, an assistant, a dogsbody.

1957 [see *garde champêtre]. **1959** M. Summerton *Small Wilderness* i. 15 He hadn't relished my taking a job as a tea-maker and general runabout on a ritzy household magazine. **1976** J. Fraser *Who steals my Name?* xv. 186 A constable was kept on duty as a part-time secretary to the chief superintendent, a run-about, a screen protecting his privacy.

b. (Earlier example.)

1775 J. Woodforde *Diary* 13 June (1924) I. 163 His whole Face..was uncommonly ugly, not unlike one of the runabout gipsies.

4. (Further examples.)

1900 Ade *Fables in Slang* 155 He took her riding in his new Runabout every Evening. **1900** *Motor-Car World* I. 130/2 A new small car... It is known as the University Runabout. **1912** *Chambers's Jrnl.* Feb. 144/1 The 'auto-carrier'..gives everything that the small runabout motor-car can offer. **1930** *Punch* 2 Apr. 374/1 Just ahead of us was a small runabout car. **1949** *Chicago Tribune* 18 Sept. 34/2 Among some 70 old-time cars will be a one

cylinder 1904 Cadillac runabout, [etc.]. **1956** *News Chron.* 1 Nov. 8/5 It is the latest in a line of economy runabouts—the Mark E Bond Minicar. **1968** 'J. Le Carré' *Small Town in Germany* ii. 24 The pavements are obstructed by the runabout cars of British Counsellors' wives. **1980** *Times* 7 Mar. 25 Whereas the Mini is really a Town runabout, the Metro is conceived as a family car that will be more comfortable for longer runs.

5. A light aircraft.

1904 A. Santos-Dumont *My Airships* xxii. 282, I determined to build a small air-ship runabout for my pleasure and convenience only. **1922** *Westm. Gaz.* 14 Nov. 3/6 Arrangements are being made..to launch the first power-driven runabout of the air. **1932** *Flight* 13 Oct. 952 The machines would have been the forerunners of cheap aerial runabouts. **1959** *Economist* 3 Jan. 58/2 A small aircraft need not be barred from using a big airport provided that it carries radio. But many of the smallest runabouts have no radio. **1966** P. O'Donnell *Sabre-Tooth* vi. 87 The twin-engined de Haviland Dove, the six-passenger aircraft which plied as a runabout from the valley..to Kabul. **1977** 'J. Le Carré' *Hon. Schoolboy* xii. 273 'Is that a single-engined plane?'.. 'Kind of executive runabout kind of thing.'

6. orig. and chiefly *U.S.* A small motor-boat.

1932 *Daily Progress* (Charlottesville, Va.) 7 Sept. 10/3 This event will be followed by..two free-for-all outboard races and a race for runabouts not over 50 horsepower. **1946** *Sun* (Baltimore) 16 Sept. 12 (*caption*) The D, E, F and G Class service and racing runabouts leave the starting line in the second heat. **1966** T. Pynchon *Crying of Lot 49* iii. 64 'Help,' said Di Presso, looking back.. across the lake. Another runabout had appeared and was headed toward them. **1970** J. Cleary *Helga's Web* xv. 274 He had started up the motor of the runabout and cruised quietly back to the yacht. **1975** B. Garfield *Hopscotch* ii. 27 A little motor runabout zipped past the barge.

‖ **runanga** (rū·naŋa). *N.Z.* Also with capital initial. [Maori.] In Maori society, an assembly or council. (See also *whare runanga* s.v. *whare.)

1858 J. Morgan *Let.* 21 June in *Richmond–Atkinson Papers* (1960) I. 411 This arrangement was proposed by Potatau's *runanga*. **1861** A. S. Atkinson *Jrnl.* 18 Nov. in *Ibid.* 727 The King party will not..accept the new runanga system. **1862** —— *Jrnl.* 5 Feb. in *Ibid.* 744 One man, the head of the Runanga, shall have £100 a year, 7 Runanga men £70 each, [etc.]. **1905** W. Baucke *Where White Man Treads* 33 He..carves effigies of his tribe-founders and heroes on the memorial posts of his runanga house, that when the elders meet to discuss matters of state, the presence of his dear dead may preside, and guide the thoughts of the council. **1946** *Jrnl. Polynesian Soc.* June 157 *Runanga*, assembly; council; so, *whare-runanga*, meeting-house. **1967** J. Metge *Maoris of N.Z.* ii. 33 As a group, the kaumātua formed a community council (*runanga*) which advised and could influence the chief. **1975** D. Scott *Ask that Mountain* (1976) ii. 41 Discussion and debate was reserved for the *runanga*, the council of leaders and elders.

run-around. Restrict *U.S. colloq.* to sense in Dict. and add: Also as one word. **1.** (Earlier and later examples.)

1857 *Knickerbocker* XLIX. 97 There comes us a 'run-round' on the end of our pen-finger. **1913** J. London *Valley of Moon* III. iv. 352 His finger was hurting too much, he said... 'It might be a run-around,' Saxon hazarded. **1968** Leider & Rosenblum *Dict. Dermatol. Words* 364 *Run(-)around*.., is colloquial for inflammatory conditions of the soft parts about nails and conveys the idea of tendency to extend circularly.

2. *colloq.* (orig. *U.S.*). Deceit, evasion; behaviour likely to mislead or cause inconvenience, esp. in phr. *to give* (someone) *the run-around*.

1915 *Chicago Herald* 2 Dec. 13/4 Pitts is satisfied that he is the victim of the grandest run-around ever put over on a boxing promoter. **1924** H. C. Witwer in *Cosmopolitan* Jan. 84/2 If you wanted to give William a run around why not say we were Cleopatra and Salome and be done with it? **1929** J. P. McEvoy *Hollywood Girl* i. 3 All I get is the run around—that nothing today, my dear, but keep in touch with us, you never know what is liable to turn up. **1933** *Sun* (Baltimore) 24 Nov. 20/4 The fans were given the runaround in the third race here yesterday. **1934** E. S. Gardner *Case of Lucky Legs* xiii. 191 A small-town dentist.., and you think that fits you to give me a run-around in a murder case. **1938** O. Nash *I'm Stranger here Myself* 115 Humanity must continue to follow the sun around And accept the eternal run-around. **1944** 'N. Shute' *Pastoral* ii. 27 Perhaps her boy friend was giving her the run around. **1950** 'S. Ransome' *Deadly Miss Ashley* ix. 109 He had already taken more than enough of a run-around from her. **1960** W. Haggard *Closed Circuit* iv. 43 My instructions are to..give him the runaround and soften him up. **1973** J. Wainwright *Devil You Don't* 176, I don't trust you... It could be a runaround. **1979** E. Newman *Sunday Punch* xix. 164 We were seeing a deliberate run-around by Smith.

3. *Printing.* Type set in shorter measure so as to fit at the side of an illustration.

[**1934** V. Steer *Printing Design & Layout* 327/2 *Run Round Block*, type set to surround a block.] **1949** Melcher & Larrick *Printing & Promotion Handbk.* 260/2 The use of run-arounds increases the cost of type composition and is often avoided on that account. **1978** *Verbatim* May 7/1 It is extremely difficult (and expensive) to program a computer to set run-arounds.

4. *colloq.* (orig. *U.S.*). A short journey or excursion. Also *attrib.* and *fig.*

1954 *Ties* (U.S.) Dec. 14/2 Steve operates the Little Southern's passenger train and local freight on regular-style train orders, with 'meets' and 'runarounds' carefully scheduled. **1976** *Broadcast* 29 Nov. 7/3 Over 1600 [people] have said they would become 'runaround drivers', ferrying around the elderly and infirm. **1977** *Time Out* 28 Jan.–3 Feb. 15/3 LBC's runaround with Alexander Walker on the week's new films.

runaway, *sb.* (and *a.*) Add: **I. 1. c.** Also *transf.* of (part of) a railway train.

1945 G. B. Grundy *55 Yrs. at Oxf.* i. 17 The eight trucks of a luggage train which I had seen were runaways.

2. a. (Further examples.) Also (*U.S.*) of horses: an act of bolting. Also *transf.* of a railway train or wagons.

1850 L. H. Garrard *Wah-to-Yah* xxi. 291 Three of the muleteams, made handsome runaways. **1898** *Kansas City* (Missouri) *Star* 18 Dec. 1/5 Miss Agnes Peterson was hurt in a runaway caused by the driver of an express wagon. **1967** G. F. Fiennes *I tried to run Railway* iv. 32 We never had a runaway. **1971** *Leader* (Durban) 7 May 16/4 Occasional runaways by Derrick Norris and Elijah Adams looked dangerous at times, but failed to bring home any goals. **1975** *Budget* (Sugarcreek, Ohio) 20 Mar. 7/8 Jacob D. Shetler had a runaway recently. He wanted to haul wood to saw, when the horses took off.

b. = Runway 1 a. *colloq.* (chiefly *U.S.*).

1868 *Fur, Fin, & Feather* 205 After a hard pull after a deer, or a long and tiresome vigil on a 'run-away', nothing is so vexatious as a miss-fire. **1944** *Living off Land* ii. 32 Carry a few snares..set them on the runaways through the grass or bushes... Set the noose in a spot where the runaway is narrow.

c. *transf.* (*ellipt.* uses of senses 4 and 6 in Dict. and Suppl.)

1947 *Sun* (Baltimore) 8 Jan. 18/1 The [stock-market] ticker tape frequently was idle until the final hour when sufficient offerings arrived to quicken the pace. There was nothing like a run-away, however. **1967** *Boston Sunday Herald* 14 May 2-5/2 Nineteen hits, 14 walks and three big innings added up to a 26-4 runaway for Bates over Brandeis Saturday. **1976** 'G. Black' *Moon for Killers* ii. 23 That book didn't need promoting. It was a runaway.

d. *Science.* Uncontrolled departure of a system from its usual or intended equilibrium.

1955 *Sci. Amer.* Oct. 60/3 Although reactors are designed so that there is a built-in tendency to overcome runaways automatically, accidents are always possible. **1957** *Practical Wireless* XXXIII. 684/1 Damage might result due to the heating effect of the lamp raising the temperature of the junction causing a 'run away', irreparably damaging the photo-transistor. **1973** *Physics Bull.* July 411/2 The danger of runaway of a fusion reactor is minimal, for the total deuterium and tritium in the reactive zone would be only about 0·25 g per 1000 MW of output power. **1974** *Nature* 29 Mar. 399/1 When the mass accreted exceeds $\sim 10^{-4} M_{\odot}$ nuclear energy generation becomes violent enough to drive a thermal runaway typical of novae.

II. 3. a. (Further examples.) Also (*U.S.*) of slaves.

1699 *N. Carolina Colonial Rec.* (1886) I. 514 A particular law..injoyns all persons on a penalty to apprehend runaway Negroes. **1784** J. F. D. Smyth *Tour U.S.A.* II. 102 Run-away Negroes have resided in these places for twelve, twenty, or thirty years and upwards. **1804** R. Sutcliff *Trav. N. Amer.* (1811) iii. 58 Whenever he saw a Negro whom he judged to be a runaway slave, he would.. jump from his work-board. **1852** *Morning Courier* (N.Y.) 8 Oct. 2/2 She causes a reward to be offered for the recovery of a runaway slave 'dead or alive', when no reward with such an alternative was ever heard or dreamed of south of Mason and Dixon's line. **1885** 'Mark Twain' *Huck. Finn* xxxviii. 325 On the scutcheon we'll have a.. crest, a runaway nigger, *sable*, with his bundle over his shoulder. **1977** *Time* 30 May 14/2 Rhodesian officials shrugged off Kaunda's declaration as the diplomatic equivalent of a mosquito bite, but the brutal civil war in the runaway British colony continues.

b. (Earlier example of *runaway match* and later example of *runaway marriage*.)

1838 Mrs. Gaskell *Let.* 17 Aug. (1966) 27 Wm has promised..to marry you if it comes to a runaway match. **1921** Runaway marriage [see *film actress* s.v. *film *sb.* 7 c].

4. a. (Further examples.)

1957 [see *hair-raising* adj. s.v. *hair *sb.* 10]. **1958** Hayward & Harari tr. *Pasternak's Dr. Zhivago* II. ix. 268 Revolutionaries who take the law into their own hands are horrifying, not as criminals, but as machines that have got out of control, like a run-away train.

fig. **1960** *New Left Rev.* Sept.–Oct. 40/1 More important than sympathy actions for and with Negroes, the sit-ins stimulated a similar burst, a run-away brush fire of activity for all sorts of other aims. **1971** *Fremdsprachen* XV. 45 The site of an oil tanker wreck and a runaway underwater oil well. **1973** *N.Y. Law Jrnl.* 1 Aug. 4/7 The third occurrence of pacemaker failure involved the supplying of an accelerated beat by the second pacemaker. This is a situation commonly known as a 'runaway' pacemaker.

b. In general use, of economic, natural, etc., conditions: thoroughly out of control, developing at an ever-increasing rate; unrestrained, rampant.

1925 *Scribner's Mag.* July 59 All of them expressed relief that predictions of a 'runaway market' for staple products had not been fulfilled. **1936** *Sun* (Baltimore) 23 Nov. 2/4 Because such an oversupply of excess funds might provide the basis for 'runaway' credit expansion, the board is considering increasing reserve requirements. **1949** *Ibid.* 3 Mar. 2/7 Representative Eberharter.. warned against making the home relief program 'too attractive', pointedly asserting that some states already were showing a 'runaway' tendency. **1967** *Listener* 21 Dec. 807/2 Lots of aspects of the world..are..'in a run-away condition'; population growth, technological growth, the destruction of Nature, to name only three. **1974** *Ibid.* 21 Feb. 235/2 If the miners' dispute is settled on very inflationary terms..it's going to be runaway inflation in Britain. **1980** *Times* 9 Jan. 12 The private pocket, not the public purse, is the best defence against runaway inflation.

c. Of commercial sales, etc.: immeasurable, overwhelming. *spec.* in phrases *runaway best seller, success*. Also *fig.*

1953 L. Z. Hobson *Celebrity* iii. 31 If *The Good World* should develop a runaway sale in bookstores, this windfall might reach a *hundred* and fifty-two thousand. *Ibid.* xii. 180 So this was what a 'run-away best seller' meant. **1960** *Harper's Bazaar* July 18/2 The season's runaway best-seller. **1968** M. Jones *Survivor* ii. 27 *Down in Flames* had been such a runaway best-seller in 1946 that there were bound to be people who remembered it. **1971** *Sunday Times* (Johannesburg) 28 Mar. 6/5 This film was a runaway box-office success. **1976** Botham & Donnelly *Valentino* ix. 71 His plan was a runaway success.

6. (Earlier example.)

1877 *Illust. London News* 2 June 515/3 The only other noteworthy feature..was the runaway victory of Lady Lumley in the Stanley Stakes.

7. In U.S. industrial terminology, designating a plant which is transferred from one location to another in order to prevent trade-union activity or evade labour regulations. Also applied *transf.* to a ship sailing under a *flag of convenience* (see *flag *sb.*[4] 1 f).

1949 *Dict. Labor Law Terms* 71 *Runaway shop*, a plant transferred to destroy union effectiveness and to evade bargaining duties. **1952** J. A. Morris *Woolen & Worsted Manuf. Southern Piedmont* iv. 118 Even though few 'runaway plants' are involved in the southern woolen and worsted movement, the effect of unions is nevertheless felt. **1954** *West Coast Sailors* 1 Oct. 4/1 Why has Mr. Rothschild started this program to assist operations of runaway foreign flags? **1957** Clark & Gottfried *Dict. Business & Finance* 308/1 In the textile and apparel industries, especially, many such runaway shops have left the New England and Middle Atlantic states over a period of years, and relocated in the South. **1960** *Wall St. Jrnl.* 15 Mar. 14 On the high seas, according to the maritime unions, are 1,695 'runaway' flag ships which have owners of one nationality and registry of another... Registered in Panama, Liberia and Honduras..these ships are described as flying 'flags of necessity' by operators and 'flags of convenience' by seamen. *Ibid.*, The union claimed that the use of foreign technicians was the equivalent of a 'runaway shop'. **1967** *Ibid.* 12 Dec. 1 Runaway plants set up by U.S. companies just inside Mexico to utilize cheap labour will come under increasing attack from the AFL-CIO.

run-back. [f. Run *v.*] **1.** The action or fact of running backwards.

1926 *Gloss. Terms Electr. Engin.* (Brit. Engin. Stand. Assoc.) 140 *Runback preventer*, a system of connection in a tramcar controller such that, in the event of the car running backwards, the motors act as short-circuited generators and thus exert a braking action. **1929** A. T. Dover *Electric Traction* (ed. 2) viii. 183 Where the motors are cross-connected for braking..the braking positions of the controller are effective for both directions of motion of the car, and may, therefore, be used for preventing a run-back. **1973** P. Dickinson *Green Gene* ii. 41 He wanted to see whether any usable figures were available for..the mid-nineteenth century, enough at least for him to construct a crude model and attempt a run-back.

2. The additional space located at either end of a lawn tennis court.

1908 A. W. Myers *Compl. Lawn Tennis Player* 213 At Auteuil..the run-back is inadequate and the timbered roof too low. **1927** *Daily Express* 30 May 3/5 The proper run-back for a court should be 21 ft. from the base-line to the stop-netting. **1977** *Club Tennis* Mar. 15 Where economy of space and/or money are important considerations a size of 33.53m × 16.46m (110ft × 54ft) which gives a 4.88m (16ft) run back is entirely adequate for the average home player.

3. *Amer. Football.* (See quot. 1976.)

1944 *Sun* (Baltimore) 6 Oct. 14/1 The 'one-man gang's' total net gain was 275 yards, 243 yards gained in 22 times carrying the ball from scrimmage, and 37 yards gained in two kickoff runbacks. **1949** *Lafayette Alumnus* (Lafayette Coll., Easton, Pa.) 24 Oct. 1/2 Winston Williams was outkicking Delaware's Smith and the Maroon had a slight edge on runbacks. **1971** L. Koppett *Guide to Spectator Sports* ii. 59 The receiving team can attempt a 'runback' at any time, even after a fumble. **1976** *Webster's Sports Dict.* 364/1 *Runback*,..a run made to advance the ball after catching a kick or punt or after intercepting a forward pass.

runcible (rʊ·nsib'l), *a.* [Prob. a fanciful alteration of Rouncival.] A nonsense word used by Edward Lear in *runcible cat, hat*, etc., and esp. in **runcible spoon**, in later use applied to a kind of fork used for pickles, etc., curved like a spoon and having three broad prongs of which one has a sharp edge.

The illustrations provided by Lear himself for his books of verse give no warrant for this later interpretation.

1871 E. Lear *Owl & Pussy-Cat* in *Nonsense Songs*, They dinèd on mince, and slices of quince, Which they ate with a runcible spoon. **1872** —— *More Nonsense* 235 The Dolomphious Duck, who caught Spotted Frogs for her dinner with a Runcible Spoon. **1877** —— *Laugh-*

able Lyrics 24 He has gone to fish, for his Aunt Jobiska's Runcible Cat with crimson whiskers! **1888** —— *Nonsense Songs & Stories* (ed. 6) 8 His body is perfectly spherical, He weareth a runcible hat. **1895** —— *Ibid.* (new ed.) 76 What a runcible goose you are! *Ibid.* 77 We shall presently all be dead, On this ancient runcible wall. **1926** *N. & Q.* 11 Dec. 430/2 A runcible spoon is a kind of fork with three broad prongs or tines, one having a sharp edge, curved like a spoon, used with pickles, etc. Its origin is in jocose allusion to the slaughter at the Battle of Roncevaux, because it has a cutting edge. *Ibid.*, Does a 'runcible' hat mean one of the sort called a trilby? In that case a 'runcible' spoon may be one with prongs or teeth. **1949** Partridge *Name into Word* 373 'He weareth a runcible hat.' Thus Edward Lear in 'Self-Portrait', where the hat is a 'topper' with a sharp rim. Now, a *runcible spoon* (Lear, 1871) is not a spoon at all but a pickle fork, broadly and triply tined, one tine being sharp-edged and curved like a spoon... The word *runcible* has been built in the architectural style of fencible; indeed, it may constitute a blend of Roncevaux and fencible (capable of defending). **1969** R. & D. De Sola *Dict. Cooking* 195/2 *Runcible spoon*, not a spoon but a fork with three broad curved prongs, used for serving appetizers. **1979** *Washington Post* 25 Mar. N6/2 A runcible spoon..is a large, slotted spoon with three thick, modified fork prongs at the bowl's end, and a cutting edge on the side.

run-down, *ppl. a.* Add: **3.** (Further example.) Also *transf.* of appliances not run by clock-work. Cf. also quot. 1866 at sense 5 below.

1974 A. Goddard *Vienna Pursuit* iv. 141 As though a run-down gramophone had been rewound, he went on. **1977** E. Ambler *Send no more Roses* x. 231 On the bench was a trickle charger with spring-clip connectors on long leads for attaching the thing to a run-down battery.

4. (Further examples.) *spec.* of districts, etc.: decayed, shabby, seedy.

1929 T. Wolfe *Look Homeward, Angel* xxxix. 612 He was offered employment..on the teaching staff of the run-down military academy. **1938** M. Brinig *May Flavin* iv. 369 The particular district was one of cheap run-down rooming houses. **1948** H. Lawrence *Death of Doll* iv. 75 Run-down neighbourhood but respectable; rooming houses and railroad flats. **1953** K. Tennant *Joyful Condemned* xxxii. 309 A run-down little suburban house with..a broken wooden verandah. **1962** A. Lurie *Love & Friendship* iv. 71 The streets of run-down two- or four-family frame houses. **1977** *New Yorker* 27 June 24/3 The church, a structure in American-Gothic style.. dominates a run-down neighborhood.

5. (Earlier and later examples.)

1866 Geo. Eliot *Let.* 12 Feb. (1956) IV. 232 George.. was a little benefited, but only a little. He is too far 'run down' to be wound up in a very short time. **1889** *Jrnl. Mental Sci.* XXXV. 200 Her general appearance gave the idea of being 'run down' in bodily health. **1916** [see *nervy *a.* 5]. **1927** C. Connolly *Let.* c 24 Jan. in *Romantic Friendship* (1975) 219, I am sorry you are run down. **1938** E. Waugh *Scoop* I. ii. 32 Once or twice when Mrs. Salter complained of being run down, they had visited prosperous resorts on the East Coast.

ru·ndown, *sb.* Also **run-down** and as two words. [f. Run *v.*] **1.** *U.S. Baseball.* An action whereby defensive players attempt to tag out a runner caught off base between them. Also *attrib.*, as *rundown play*.

1908 *Spalding's Base Ball Guide* 69 Chance forced Tinker and then working the steal stunt for a run down was put out. **1946** *Sun* (Baltimore) 27 May 15 (*caption*) Charlie Keller, the Yankee's slugger, is shown being caught in a rundown in the ninth inning of the first game. **1971** L. Koppett *N.Y. Times Guide Spectator Sports* i. 30 Less reprehensible mixups occur on rundown plays. **1978** *Detroit Free Press* 5 Mar. c 3/1 Chilly, wet weather prevented the pirates from a scheduled workout on pickoffs, rundowns and cutoff throws.

2. *U.S. horse-racing slang.* A list of entries and betting odds. Chiefly *attrib.* (see quot. 1942).

1935 *Sun* (Baltimore) 12 Jan. 7/8 Sergt. Ignatius Benesch, who led the raiders, said the bookmaking activities were being carried out on the second floor... Sergeant Benesch and his squad took several telephones and a quantity of run-down sheets to the station as evidence. **1942** Berrey & Van den Bark *Amer. Thes. Slang* §733/2 *Rundown board*, the bookmaker's board on which is posted the sheet of entries and odds; *rundown sheet*, *slate*, a list of entries and betting odds posted by a bookmaker. **1948** *Sun* (Baltimore) 17 Dec. 12/5 He observed Yateman making off with a board containing run-down sheets, while Hoffman was attempting to flee with a money bag and race-bet slips. **1951** *Publ. Amer. Dial. Soc.* XVI. 55 *Run down*,..the current change of odds in a booking establishment, caused by fluctuations of opinion or by a flurry of betting. *Ibid.* 56 *Run down sheet*,..a printed list provided for the patrons of the book. It contains the day's entries and the morning line odds and is used by the gambler to keep track of the fluctuations in odds.

3. *slang* (orig. *U.S.*) A (usu. verbal) listing of items of information; a summary or brief account of pertinent facts; a short description.

1945 *Sun* (Baltimore) 21 Feb. 7/4 In between taking care of the general's two uniforms..Orderly Powder gave inquiring correspondents a short rundown. **1949** *Tuscaloosa* (Alabama) *News* 8 Oct. 1/8 Here's a quick run-down of major disputes which already have made idle more than a million workers. **1953** *N.Y. Times* 29 June 29/1 A run-down on market conditions for various steel products follows. **1960** *Guardian* 21 Sept. 6/2 The movie guide provides a brief run-down on each film and assesses its suitability for children. **1966** T. Pynchon *Crying of Lot 49* v. 105 John Nefastis..brought out his Machine.. 'You know how this works?' 'Stanley gave me a kind of rundown.' **1971** *Farmer & Stockbreeder* 23 Feb. 8/2 He gave a strangely fragmented rundown of British Rail policy as it affected horticulture in the West. **1977** I. Shaw *Beggarman, Thief* III. i. 184 I'm going to tell them they got to give me a complete rundown on the family.

4. A gradual and sustained reduction in the size or scope of an organization, enterprise, or activity.

1948 *Hansard Commons* 8 Mar. 931, I feel that the timing of the run-down [of the Navy] has not been very satisfactory. **1955** *Times* 5 Aug. 7/1 A few minor financial worries have also intruded. So far all have been surmounted, and the 'run down' of British forces is proceeding swiftly and smoothly. **1957** *Economist* 21 Dec. 1068/2 If informal agreements can be reached with newly independent Commonwealth countries about the pace of the run-down of their balances, so be it. **1960** *Guardian* 10 Nov. 3/3 The telegraph service again showed a deficit, but the rate of run-down in the traffic has been checked. **1973** *Listener* 14 June 785/3 By attracting people away from the city centre, they [*sc.* ringways] generate the run-down of the city centre. **1978** *Daily Mirror* 12 Jan. 2/2 They produced a report which could lead to an even swifter and more drastic rundown in the industry than has been planned.

rune[2]. Add: **1.** (Further *transf.* examples.) Also, applied to a letter or character of a non-Germanic alphabet (esp. in fictional writings) having a resemblance to the Germanic runes.

1883 I. Taylor *Alphabet* 201 An adaptation or survival of the 'Slavonic Runes', the existence of which is however entirely hypothetical. **1937** J. R. R. Tolkien *Hobbit* i. 30 Look at the map..and you will see there the runes in red. **1948** D. Diringer *Alphabet* II. v. 314 The monumental inscriptions are written in a runic character, termed Kök Turki runes. **1954** J. R. R. Tolkien *Fellowship of Ring* v. 339 They were written by many different hands, in runes, both of Moria and of Dale, and here and there in Elvish script. **1958** *Everyman's Encycl.* (ed. 4) IX. 461/1 Orkhon Inscriptions (also known as Siberian, Early Turki, Pre-Islamic Turki or Kök Turki Runes) are the earliest epigraphical monuments written in Turki. **1961** M. Savill tr. *E. Doblhoffer's Voices in Stone* ix. 289 Babinger sent a photograph..to the decipherer of the Old Turkish runes, Vilhelm Thomsen. **1968** U. K. Le Guin *Wizard of Earthsea* iv. 67 He studied the Further Runes and the Runes of Éa, which are used in the Great Spells. *a* **1973** J. R. R. Tolkien *Silmarillion* (1977) 322 *Cirth*, the Runes, first devised by Daeron of Doriath.

2. c. (Further examples.) *spec.* a cryptic or magic verse, an incantation; a lament.

1870 D. G. Rossetti *I saw Sibyl at Cumæ* in *Coll. Wks.* (1886) I. vi. 378 'I saw the Sibyl at Cumæ' (One said)... 'She hung in a cage, and read her rune.' **1900** A. Carmichael *Carmina Gadelica* I. p. xx, The wife knew many secular runes, sacred hymns, and fairy songs. **1908** Somerville & 'Ross' *Further Experiences Irish R.M.* viii. 211 She chanted..words in measured cadence... By the time this rune had been repeated three times she was in the hall. **1922** Joyce *Ulysses* 203 There he keened a wailing rune.—*Pogue mahone! Acushla machree!* **1936** W. Holtby *South Riding* I. i. 20 Curses could be lifted by spells. Midge was always trying them, inventing her own runes and incantations. **1949** *New Yorker* 22 Oct. 38/2 (*title*) Runes for an old believer. **1973** G. M. Brown *Magnus* i. 23 Tana repeats a small bridal rhyme, a rune of fertility, the meaning of which is not at all clear but she has learned it from her grandmother. **1977** P. Fitzgerald *Knox Brothers* i. 32 Eddie had begun on Kennedy's Latin Grammar; there were more inexplicable runes for Wilfred to repeat in the nursery: 'Caesar adsum jam forte—Cæsar had some jam for tea.'

3. a. *rune-carver* (further examples), *-collector*, *-cutter*, *-rister* (further *arch.* example), *-singer*, *-writer* (further example).

c **1865** E. Charlton in *Archæologia Æliana* VI. 131 The ignorance of the Rune cutter has transposed one or two of the letters. **1883** Vigfusson & Powell *Corpus Poeticum Boreale* I. 571 Bali, a Swedish Rune carver from Upland, appears to have lived about the end or middle of the eleventh century. **1898** I. M. Anderton tr. *Comparetti's Trad. Poetry of Finns* I. i. 5 The first rune collectors..only considered and published detached songs, and did not think of classifying them. **1904** *Saga-Bk. of Viking Club* III. III. 320 The..uniformity can only be explained by supposing that there were professional rune-writers, travelling over the country and inscribing stones. **1908** *Ibid.* V. II. 258 To the right of the stem is an inequality in the stone, which the rune-cutter (*rune-rister*) apparently wished to avoid. **1927** E. V Gordon tr. runic inscription in *Introd. Old Norse* 170 Biari has the temple, a wise rune-carver. **1962** C. L. Wrenn in Davis & Wrenn *Eng. & Medieval Stud. presented to J. R. R. Tolkien* 316 If the 'first fronting' had in fact not yet been completed in the dialect of the rune-cutter. **1963** S. B. F Jansson in Browne & Foote *Early Eng. & Norse Stud.* ix. 112 In all probability the rune-carver wished his rune-ribbon to start and end at the same level on the stone. **1972** *Funk's Stand. Dict. Folklore* 382/1 Lönnrot himself said, 'Because I am sure that not one of the rune-singers could surpass me in the knowledge of the runes, I used my right to put together the songs as it seemed best.'

b. *rune font*, *-stone* (further examples).

1883 Vigfusson & Powell *Corpus Poeticum Boreale* II. 589 A Swedish Rune-stone has the roasting-scene of l. 4 carved upon it. **1931** *Times Lit. Suppl.* 9 July 548/1 The discovery of the Eggjum rune-stone in 1917. **1962** P. G. Foote tr. *S. V. B. Jansson's Runes of Sweden* 163 The finest of the rune fonts is the one..carved by the Gotlander Sigraf. **1980** K. Randsborg *Viking Age in Denmark* 32 The persons mentioned on the early rune-stones were connected with royal power.

c. *rune-inscription*, *-letter*, *-lore*, *-maiden*, *-master*, *-name*, *-poem*, *-song*, *-worship*; *rune-blazoned*, *-inscribed* (further example), *-like* (further example) adjs.; **rune-ribbon**, the carved scroll on a runic stone in which the runes are engraved; **rune-row**, a runic alphabet; **rune-tree**, (*a*) = *tree-rune* s.v. Tree *sb.* 10 c; (*b*) (see quot. 1899).

1938 W. de la Mare *Memory* 76 A subtle Serpent.. Raised its rune-blazoned head. **1898** *Saga-Bk. of Viking Club* II. II. 337 Asfrid..raised a..gravehaugh over the body of her husband, whereon she set up a rune-inscribed stone in his honour. **1931** C. L. Ewen *Hist. Surnames* iii. 65 The rune-inscribed crosses. **1925** *Saga-Book of Viking Soc.* IX. II. 272 The rune-inscriptions..must be assigned to the first part of the 11th century. **1937** J. R. R. Tolkien *Hobbit* iii. 64 'What are moon-letters?'.. 'Moon-letters are rune-letters.' **1965** C. L. Wrenn in Bessinger & Creed *Medieval & Linguistic Stud.* 50 The seven rune-like symbols just mentioned. **1868** G. Stephens *Runic Mon.* I. 94 (*heading*) Rune-lore. **1959** R. W. V. Elliott *Runes* iii. 30 Other pagan rites and customs that went hand in hand with rune-lore. **1906** C. M. Doughty *Dawn in Britain* I. II. 80 Her covert image..and holy cart Shall her rune-maiden ministers, in the lake, Wash. **1965** R. Derolez in Bessinger & Creed *Medieval & Linguistic Stud.* 33 A simple formula such as..'unknown Danish runemaster > Hrabanus Maurus', is tempting but dangerous. *Ibid.* 34 One might..suppose that a 'rune master'..would resort to runes if asked to write the names in full. **1970** Foote & Wilson *Viking Achievement* ix. 312 Once the act of carving stone had developed in southern Scandinavia, Swedish sculptors and rune-masters experimented to bring it to fruition. **1879** I. Taylor *Greeks & Goths* xviii. 117 Let us compare these.. Ogham names with the corresponding rune names. **1927** E. V. Gordon *Introd. Old Norse* 161 The first letter of each rune-name gives the value of the rune. **1974** *Eng. Stud.* LV. 512 The inscriber of the Franks Casket normally represented the voiceless dorsal fricative by 'g', so that for him the rune-name would have been *eg*. **1861** D. H. Haigh *Anglo-Saxon Sagas* 16 In the Anglo-Saxon rune-poem, the following stanza occurs:- Ing wæs ærest, mid East-Denum, (etc.). **1879** Vigfusson & Powell *Icelandic Prose Reader* 457 The idea is..possibly taken from some such English poem as the Exeter Codex Rune Poem. **1962** C. L. Wrenn in Davis & Wrenn *Eng. & Medieval Stud. presented to J. R. R. Tolkien* 316 The Old English, Old Norse, and Icelandic rune-poems. **1963** S. B. F. Jansson in Browne & Foote *Early Eng. & Norse Stud.* ix. 111 All that was visible..was a short section of the rune ribbon, with some carved lines above it. **1868** G. Stephens *Runic Mon.* I. 105 All the *oldest* written Runerows are Futhorcs. **1955** J. R. R. Tolkien *Return of King* 397 It was often called *Angerthas Moria* or the Long Rune-rows of Moria. **1973** R. I. Page *Introd. Eng. Runes* xii. 190 The common English rune-row had twenty-eight [characters] or more to the Germanic twenty-four. **1868** G. Stephens *Runic Mon.* I. 105 As for there being any 'German' people whatsoever..who.. practist heathen rites and used Rune-songs and Rune-books and Rune-carvings in incantations and divinations —why the thing is ridiculous. **1892** S. A. Brooke *Hist. Early Eng. Lit.* I. 192 These phrases are from the *Rune Song*. **1927** E. V. Gordon *Introd. Old Norse* 161 The usual forms of this fuþark..are given in the Rune-Song. **1863** J. M. Mitchell tr. runic inscription in *Mesehowe* 51 Cut to our late Father these Rune Trees, (He was a) leader on the West Sea. **1879** I. Taylor *Greeks & Goths* xviii. 129 The characteristic of the five classes of the rune trees would be (1) branches; (2) forks; (3) loops; (4) crooks; (5) roots. **1899** A. H. Keane *Man, Past & Present* ix. 341 A great feature of the system were the 'rune-trees', made of pine or birch bark, inscribed with figures of gods, men, or animals, which were consulted on all important occasions. **1940** F. Scott Fitzgerald *Let.* Dec. (1964) 100 But be sweet to your mother at Xmas despite her early Chaldean rune-worship which she will undoubtedly inflict on you.

rune (rūn), *v. rare.* [f. Rune[2].] *intr.* To compose or perform poetry or songs; to lament.

1936 M. Franklin *All that Swagger* i. 7 He was wont to rune to himself as he sat alone, thrust aside by his sons as childish. **1964** Auden in *Listener* 1 Oct. 575/2 Our handful Of clients at least can rune.

Runge–Kutta (ru·ŋgə ku·ta). *Math.* The names of Carl David Tolme *Runge* (1856–1927) and Martin Wilhelm *Kutta* (1867–1944), German mathematicians, used *attrib.* to designate a method of approximating to solutions of differential equations.

1930 J. B. Scarborough *Numerical Math. Analysis* xiii. 274 In the special case where dy/dx is a function of x alone the Runge-Kutta method reduces to Simpson's rule. **1950** *High-Speed Computing Devices* (Engin. Res. Associates) vii. 128 By the Runge-Kutta method, the formulas which are applied are given below. **1975** *Nature* 9 Oct 516/2 Membrane action potentials were computed with Hodgkin-Huxley equations, modified for *Myxicola*, using a modified fourth-order Runge–Kutta algorithm, and the six-parameter model. **1980** *Daily Tel.* 16 Sept. 2 (Advt.), It has 128 program steps that fulfil practically every function a mathematician needs. From setting a program for the definite integral by the Simpson's rule.. to the Runge-Kutta method.

runic, *a.* and *sb.* Add: **A.** *adj.* **1. a.** (Further examples.)

1948 [see *rune[2] 1]. **1962** G. Clauson *Turkish & Mongolian Stud.* v. 72 The Runic alphabet is in a class by itself. **1968** W. S. Allen *Vox Graeca* i. 37 Special symbols are found only in the Old Germanic Runic and Old Celtic Ogham systems of writing. **1973** *Cassell's Encycl.*

World Lit. i. 412/1 *Orhon Inscriptions*, the most important of the oldest surviving specimens of the Turkish language. .. They are inscribed in the Turkish runic alphabet.

d. Of or pertaining to runes; concerned with runes.

1861 J. Farrer *Let.* 28 Sept. in J. M. Mitchell *Mesehowe* (1863) p. viii, I shall send one to an English Runic scholar. **1862** P. A. Munch in J. Farrer *Notice Runic Inscr. Orkneys* 29 In the later times of the Runic period. **1868** G. Stephens *Runic Mon.* I. 94 The 'Scandinavian Futhork' is..a peculiar modification and compendium of the common Runic traditions. *Ibid.* 410 This runic drama was not yet ended. **1881** *Yorkshire Archæol. & Topogr. Jrnl.* 1879–80 54 His learning was of a very varied character. 'Anglo-Saxon' and Runic lore was that by which he was best known. **1953** *Saga-bk. Viking Soc.* XIII. 281 Random runic studies in the eighteenth century. **1973** R. I. Page *Introd. Eng. Runes* i. 3 It was the scholars of the late seventeenth and early eighteenth centuries who put English runic studies on a sound basis. *Ibid.* 4 His [*sc.* George Hickes's] *Thesaurus* contained a large amount of runic material. *Ibid.*, Runic knowledge. *Ibid.* viii. 113 A living runic tradition there.

B. *sb.* **2. b.** One of the runic alphabets. Also in *pl.*, runic characters collectively (*rare*).

1863 J. M. Mitchell *Mesehowe* 32 It is probable that only one or two of the best educated in each ship could write or read the Runic. **1868** G. Stephens *Runic Mon.* I. 94, 4 of these [letters] (the H— for which the Runic prefers the Phœnician mark for CH—, the Z, the GH and the SH) are more or less wanting in the Runic. *Ibid.* 157 In Scandinavian-runics, when the W had quite died out. *Ibid.* 159 In Scandinavian-runics the old rune for Y lingers on for a time. **1886** T. Le Marchant Douse *Introd. Gothic of Ulfilas* I. 16 Of the foregoing letters, *urus* and *faihu* are runes..*bairka*, *eis*, *ôthal*, and perhaps, *quairthr*, common to runic and Greek. **1961** M. Savill tr. *E. Doblhoffer's Voices in Stone* ix. 280 The traveller discovered an impressive granite monument engraved with three inscriptions:.. the third in 'Siberian' runic. **1963** *Times* 4 June 14/6 And after thousands and thousands and thousands of years, and after Hieroglyphics, and Demotics, and Nilotics, and Cryptics, and Cufics, and Runics, and Dorics, and Ionics, and all sorts of other ricks and tricks..the fine old easy, understandable Alphabet—A, B, C, D, E, and the rest of 'em—got back into its proper shape again for all the Best Beloveds to learn when they are old enough.

runically, *adv.* [f. Runic *a.* + -al + -ly[2].] In a runic manner; with runes.

1920 *Times Lit. Suppl.* 28 Nov. 779/3 Runically inscribed objects contained in..better known public collections.

run-in. Also run in. **1.** (See Run *sb.*[1] 8.)

2. *colloq.* (chiefly *U.S.*). A quarrel, argument, or row; a clash or fight. Usu. in phr. *to have a run-in* (*with* someone).

1905 'H. McHugh' *You can search Me* 82 Sorry we had the run in but it was all my fault. **1912** C. Mathewson *Pitching in a Pinch* viii. 180 Fred Tenny has said for a long time that Mr. Klem gives him a shade the worst of it on all close ones because he had a run in with that umpire one day when they came to blows. **1920** I. Ostrander *How Many Cards?* xii. 145 The cook at the de Forests' two doors away had a run-in with that Sarah, the butler's wife, and she won't speak to any of them. **1930** D. Runyon in *Collier's* 20 Dec. 32/3, I am all tired out..from getting a slug in my chest in the run-in with Jerk Donovan's mob in Jersey. **1945** G. Marx *Let.* 16 Feb. (1967) 50 Your father and I have many run-ins these days... He can't get it into his thick skull that I have come of age. **1952** A. Baron *With Hope, Farewell* 44, I 'ad a run-in with the caterers, too. **1962** A. Lurie *Love & Friendship* ii. 31, I had a run-in with the Administration about it last spring. **1979** 'A. Hailey' *Overload* I. iii. 19, I hear you had a run-in with Nancy Molineaux.

3. The approach of an aircraft to a dropping point or landing place; = *run-up 1 d.

1943 *Combined Operations* (Min. of Information) ii. 19 Two lights—a red switched on when the pilot is beginning his run-in to the dropping zone and a green indicating that the moment to jump has arrived. **1944** *Hutchinson's Pict. Hist. of War* 27 Oct. 1943–11 Apr. 1944. 366/2 An aircraft would be mortally wounded during its attacking run in but would not crash until after that attack had been well and truly pressed home. **1958** 'Castle' & 'Hailey' *Flight into Danger* x. 137 We must have plenty [of fuel] in hand for a long run-in over the ocean, if I decide..to ditch. **1958** P. Kemp *No Colours or Crest* (1960) v. 88 On the first run-in the aircraft would loose the containers and the 'free drops'. **1971** R. Dentry *Encounter at Kharmel* ix. 158 Alden reached his bedroom window in time to see the Cherokee..lose height for its run-in on to the strip.

4. An introductory statement or event; an approach (to a subject).

1961 *Clergy Rev.* Oct. 627 Mr Derrick provides a run-in to Fr Brown's story in the form of a Prologue. **1962** A. Nisbett *Technique Sound Studio* viii. 140 Remember not to start fading up until the run-in is complete. **1966** 'A. Hall' *9th Directive* i. 13 He sensed I would try to refuse the mission... He poured some more lime to give himself a last chance of planning his run-in. 'This is a special job,' he began.

5. *Criminals' slang.* A place to which stolen goods are driven and in which they are concealed.

1959 J. Gosling *Ghost Squad* iii. 43 He tipped us off to the whole plot, gave the address of the 'run-in'—the place where the stolen goods were cached—and the names of the thieves. *Ibid.* x. 130 The normal method was to hire a van from a small lorry-owner, run the van to the warehouse, break in, load the van, take the contents to a 'run-in'—usually a shed or garage in the central London area—and return the van. **1962** D. Warner *Death of Bogey* iv. v. 163 Just waiting to hear that the lorry reached the run-in. It's late. *Ibid.* 170 Sapper Neal and a bunch of the Sparrow boys been seen cruising around this manor in a car like they was looking for something. Is the run-in round here? **1970** P. Laurie *Scotland Yard* vi. 129 It's a run-in for stolen lorries.

6. *Comb.*, as **run-in groove**, on a gramophone record, the blank groove traversing the annular area outside the grooves carrying the recording; **run-in shed** *U.S.*, an open-fronted shelter in which horses are housed.

1962 A. Nisbett *Technique Sound Studio* viii. 146 If the disc is being played from the start, or if there is no sound definite enough to be used as a cue, it will be necessary to count the revolutions from the run-in groove. **1976** *Gramophone* Sept. 510/1 It..moves the stylus to the run-in groove of the record. **1964** *Blood-Horse* 26 Dec. 1874/1 Kelly has built a pair of L-shaped run-in sheds... The run-in sheds—unlike three sided Madden sheds used on many Blue Grass horse farms—are closed on only two sides, the north and west. **1977** J. W. Evans et al. *Horse* xxiii. 723 Flies, hot weather, cold rains, and strong, cold winds seem to bother horses, and run-in sheds provide adequate protection from these conditions.

runkle, *v.* Add: Also (*rare*) in colloq. use. Also with *up*.

1929 E. Bowen *Joining Charles* 188 She..pulled down the sofa loose-cover where it had 'runkled' up. **1958** M. Allingham *Hide my Eyes* viii. 85 He looked so neglected with his green tights runkled round his ankles.

runless (rʌ·nlės), *a.* [f. Run *sb.*[1] + -less.] In *Baseball* and *Cricket*: devoid of runs; unable to score.

1921 *Daily Colonist* (Victoria, B.C.) 7 Oct. 11/4 The New York Americans made world's series history today.. leaving the New York Nationals runless. **1963** *Times* 30 May 4/7 Devereux, a left-hander of medium pace, rendered the patient Tebay runless and strokeless for lengthy periods by the simple expedient of bowling a shade short of a good length.

runnel (rʌ·n'l), *v.* [f. Runnel[1].] *trans.* To form streams or channels in (a surface); to channel or furrow. Hence **ru·nnelled** *ppl. a.* Cf. Runnelling *a.*

1933 G. Barker *Thirty Preliminary Poems* 13 And sparkling veins Escape in dark wooded places Runnelling like willow trees The lachrymose moist soil. **1947** *New Writing* XXXI. 164 Their faces were black with coal dust, runnelled with sweat. **1970** R. J. Small *Study of Landforms* ix. 309 These pediments were..usually 'runnelled', and dissections of up to 15–20 ft in depth were not uncommon. **1977** P. Scupham *Hinterland* 19 A disenfranchised demon wears His runnelled face in sour grotesque, A conduit for the tumbling skies.

runner. Add: **I. 1. d.** *N.Amer.* One who chases or hunts buffalo. Now *Hist.* Cf. Run *v.* 42a in Dict. and Suppl.

1837 W. Irving *Captain Bonneville* I. xvii. 286 'It was a beautiful sight,' says the captain, 'to see the runners, as they are called, advancing in column, at a slow trot, until within two hundred and fifty yards of the outskirts of the herd, then dashing on at full speed, until lost in the immense multitude of buffaloes.' *Ibid.* II. xi. 173 The 'runners', then, as they are called, mounted on fleet horses, and armed with bows and arrows, moved slowly and cautiously toward the buffalo. **1974** *Publishers Weekly* 11 Mar. 46/2 The early buffalo runners (hunters).

e. *Baseball.* = *base runner* s.v. *Base *sb.*[1] 20 b.

1845 [see *balk *sb.*[1] 5 b]. **1857** *Spirit of Times* 7 Feb. 372/3 Mr. Thos. Leavy..mans the first base, and rare it is, that a runner reaches the first base, if the ball is passed up quickly. **1948** *Lawton* (Oklahoma) *Constitution* 4 July 12/2 If the pitcher doesn't keep the runners close to base, the best catcher in baseball can't throw them out.

f. *Cricket.* One who runs on behalf of a disabled batsman. (Cf. *Run *v.* 1 h.)

1862 J. Pycroft *Cricket Tutor* 4 Having the luck to be lamed by a sprain, I was allowed a runner. **1908** W. E. W. Collins *Leaves from Old Country Cricketer's Diary* xiii. 219 Once, then, only in my life have I acted as runner for another batsman. **1971** *Times* 16 Feb. 7/6 His knee is stiff..and unless it improves he may need the help of a runner.

g. *N. Amer.* One who runs ahead of a dog-sledge in order to find or clear a path in snow.

1867 *Ann. Rept. Smithsonian Inst.* 1866 309 The man was a famous runner, and despite the disadvantage of small tripping snow-shoes..he would have reached the houses before them had not the line that confined the snow-shoe on his foot broken. **1921** *Beaver* (Winnipeg) June 27/2 The 'runner's' duty is to travel in front of the dogs picking out and breaking the trail. It is also his duty to clear away with an axe any trees which may have been blown across the road. **1930** L. Munday *Mounty's Wife* iv. 60 [We] were making a trip..without a runner (that is, an Indian guide who goes ahead of the dogs to break trail). **1971** T. Boulanger *Indian Remembers* 62 From Norway House..the runner was Old John Clark... In Berens River..the runner was Donald Bittern.

3. a. (Further examples.) Also used to designate one whose employment also involves the fetching and carrying of articles; an assistant; *spec.* in *U.S.* prison terminology, a prisoner entrusted with special duties; *Mil.* an orderly; a soldier who carries messages, esp. along the front line.

1830 *5th Ann. Rep. Boston Prison Discipline Soc.* 7 [At the Charlestown prison] there is, however, a class of men, consisting of ten or twelve, called *runners* and *lumpers*, whose duty consists in moving about the yard. **1912** T. Dreiser *Financier* lxvi. 679 Some of the prisoners, after long service were used as 'trusties' or 'runners', as they were locally called; but not many. **1917** G. S. Gordon *Let.* 13 Feb. (1943) 69, I was searching for the Hqs. of a Battn. and there wasn't a living soul above ground to ask except myself and my runner. **1929** R. C. Sherriff *Journey's End* III. iii. 116 Stanhope sits at the table and begins to write a short report... Stanhope calls 'Runner!' as he writes. A soldier comes from the servants' dug-out. **1933** *Sun* (Baltimore) 17 Apr. 14/4 Clifton had been assigned to the odd-jobs duties of a 'runner' on the first floor of the south wing, where both his and Kellam's cells were located. **1960** J. Grant *Come again, Nurse* xxvii. 180 The theatre runner opened the doors and said quietly: 'Mr. Spindells has arrived, sir.' **1974** *Guidelines to Volunteer Services* (N.Y. State Dept. Correctional Services) 43 *Runner*, inmate whose job is to deliver things around the prison. **1976** F. Warner *Killing Time* I. i. 6 I'd sooner be in the assembly trench waiting to go over the top than a runner. **1976** 'W. Trevor' *Children of Dynmouth* v. 100 Mrs Abigail took round Meals on Wheels with Miss Poraway as her assistant, or runner, as the title officially was.

c. (Earlier and later examples.) Also *spec.* one who provides custom for a lawyer.

1824 *Microscope* (Albany, N.Y.) 21 Feb 183/3 Our wholesale property-speculators and their gentry in livery, called *runners*. **1836** C. R. Gilman *Life on Lakes* I. 31 [At Oswego] a struggle began between the runners of the two boats. **1948** *Chelsea* (Massachusetts) *Record* 30 Nov. 8/7 Unethical lawyers, plus their hired 'runners', probation officers, jail attaches and police officers were 'selling' justice in the courthouse corridors to ignorant criminal defendants. **1951** *Life* 25 June 119/1 [They] employed 15 runners to give school children free samples of heroin and ridicule those who wouldn't try it as 'squares'. **1974** *Times* 21 Jan. 6/6 If there is a car crash, one of the first people on the scene is likely to be a 'runner', who has heard about it by tuning in to police radio frequencies. He will point out that the victim needs a lawyer immediately, and will sign him up with one on the spot.

f. A freelance antique dealer.

1969 R. Quest *Cerberus Murders* xxviii. 153 He was a free-lance—the trade term is 'runner'—picking up antiquities here and there and selling them at a profit. **1976** G. Sims *End of Web.* xii. 86 Klein is a sort of runner.. buys things in the country and sells them to West End dealers. **1978** *Observer* 16 Apr. 38/3 There are 'runners' going from shop to shop, detecting the margin of a bargain and taking the merchandise one step nearer to Bond Street.

4. b. *N.Amer.* A horse trained for hunting buffalo. Now *Hist.* (Cf. sense 1 d above.)

1858 J. Palliser *Jrnl.* 31 July (1863) 90 Breakfast finished, our 'runners' saddled and mounted, the whole party moved slowly on... Having ascended the slightly elevated ridge we then beheld our game, four or five thousand buffalo. **1971** J. McDougall *Parsons on Plains* viii. 63 From all parts of the camp riders came forth, many of them leading their runners, so as to have them as fresh as possible for the coming race.

c. *transf.* A roadworthy motor vehicle; phr. *good runner*, a motor vehicle which runs well.

1948 Partridge *Dict. Forces' Slang 1939–1945* 159 *Runner*, a vehicle that was in running order, as opposed to one that was off the road. **1972** *Fairbanks* (Alaska) *Daily News-Miner* 3 Nov. 23/7 (Advt.), '67 Chrysler Newport, stick shift, winterized, a good runner. $750. **1975** *Drive* Spring 40/2 We were asked to collect a car from a chap who had stripped it to service an identical second car... We found a car standing in the drive. So naturally we took it. But it was the runner we towed away. The wreck was in the garage. **1978** *Cornish Guardian* 27 Apr. 22 (Advt.), Peugeot 204 Saloon. Sun roof. Good runner.

5. e. Substitute for def.: Any of several carangid fishes found in tropical or temperate seas, esp. *Elagatis pinnulatus*, *Caranx crysos* (= *Hardtail a), or *C. ruber*.

1884 [see Shoemaker 2 a]. **1902** [see *hardtail a]. **1905** D. S. Jordan *Guide Study of Fishes* II. xvii. 272 Most like the true mackerel are the 'leather-jackets', or 'runners', forming the genera *Scomberoides* and *Oligoplites*. **1913** C. F. Holder *Game Fishes of World* xxii. 211 There are a number of smaller allied fishes which are game in every sense, if taken with appropriate tackle, as the Runner (*Caranx crysos*). **1952** L. L. Mowbray in J. O. La Gorce *Bk. Fishes* (rev. ed.) 210 The runner (bottom—*Caranx ruber*) ranges throughout the West Indies and Florida, and strays northward along the southern United States. *Ibid.*, The runner and other jacks are the terror of small fishes. **1966** Leim & Scott *Fishes Atlantic Coast of Canada* 247 Blue runner. *Caranx crysos* (Mitchill). *Ibid.* 248 Although there are few actual records it is believed that the blue runner occurs sparingly along the Nova Scotian coast in the late summer months. **1975** J. G. Walls *Fishes Northern Gulf of Mexico* 201 Rainbow runner. *Elagatis bipinnulata* (Quoy & Gaimard)... A popular sportfish which is not uncommon offshore.

f. *U.S.* A black snake, the racer, *Coluber constrictor*.

1795 T. Todd *Let.* Feb. in S. Williams *Nat. & Civil Hist. Vermont* (1809) I. 485 In a field in Connecticut..I approached with caution within twenty feet of a black snake, about seven feet long, having a white throat, and of the kind which the people there call runners. **1855** W. G. Simms *Forayers* xxxix. 456 Push forward, quick as a runner (black snake).

g. In full, *runner duck*. A small white or fawn duck belonging to the breed so called and distinguished by an erect posture.

1895 W. Cook *Ducks* (rev. ed.) 96 Many cross-bred Indian Runner ducks have been sold as pure. **1900** *Reliable Poultry Jrnl.* Apr. 207/2 The Runners are particularly adapted to the market poultry man's needs. **1918** E. A. Taylor *Runner Ducks* ii. 15 The Runner Duck differs from all other breeds in that it has an upright carriage, which ensures the running gait. **1921** *Daily Colonist* (Victoria, B.C.) 29 Oct. 4/3 Anyone who has used runner ducks' eggs can hardly discern in cooking the difference between these and hens' eggs. **1937** [see *Kerry[1]]. **1960** L. Bonnet *Pract. Duck-Keeping* xiii. 116 The Indian Runner was the prolific layer of the duck world.

6. Also *spec.* = *run-boat* s.v. *Run *sb.*[1] 4 a. *U.S.*

1881 E. Ingersoll *Oyster Industry* 164 Another branch of the trade conducted by vessels, generally known as runners... The runner will anchor near some tonging-ground, and an empty basket or a small flag will be hoisted to the masthead as a signal that she is ready to receive oysters. **1890** [in Dict.].

7. a. Also freq. with qualifying sb.: see *gun-runner* s.v. Gun *sb.* 15, *rum-runner* s.v. *Rum *sb.*[1] 3, etc. Also *slang* (chiefly *U.S.*), one engaged in conveying prohibited goods (as drugs, liquor) secretly.

1930 *Amer. Mercury* Dec. 457/2 *Runner*, one who transports liquor from the border to inland towns. 'He's a torpedo for a big runner.' **1963** T. Tullett *Interpol* v. 65 Members of the gang, known as 'runners', were sent to Paris, or Marseilles, to pick up the drug. **1963** *Amer. Speech* XXXVIII. 276 A student is not permitted to buy liquor; hence, should he want some, he must find someone to obtain it for him. Such an intermediary is termed a *runner*. **1971** E. E. Landy *Underground Dict.* 163 *Runner*,.. person carrying a drug between buyer and seller.

8. c. (Earlier and further examples.) Also, the leader of a (freq. Black) street gang.

1874 M. N. Forney *Catechism of Locomotive* 547 Every locomotive runner should..have an exact knowledge of the engine intrusted to him. **1893** M. Holley *Samantha at World's Fair* i. 4 His parents..[were] good respectable..people..and runners of a cheese factory. **1945** F. H. Hubbard *Railroad Avenue* ii. 9 Dispatchers regarded him as a 'fast roller', a runner who could be depended upon to get his train over the road 'on the card'. **1962** *Amer. Speech* XXXVII. 135 *Runner*,..a locomotive engineer. **1972** C. H. Fuller in W. King *Black Short Story Anthol.* 142 'He coulda' done you in,' Rosalee enjoined... 'He's the *runner* of Tenth and Montgomery—Reuben is 'Little Blood', girl!' *Ibid.* 145 'Durango,' the *runner*, stood to the rear of his troops and when they were ready nodded to 'Cornbread' the warlord, who moved to the center of the street screaming challenges. **1973** *Philadelphia Inquirer* (Today Suppl.) 14 Oct. 29/1 Often nobody even knows who the runner (gang leader) is.

II. 9. d. = *Run *sb.*[1] 13 g.

1928 L. North *Parasites* 42 His quick eye detected a neatly darned spot near the heel of one, and a laboriously-checked runner in the other. **1931** M. de la Roche *Finch's Fortune* xvii. 257 Alayne noticed a long 'runner' on the shoulder of her knitted jumper. **1942** in H. Wentworth *Amer. Dial. Dict.* (1944) 525/1 She wears my finest evening gowns, gets runners in my hose.

10. e. A tool used in decorating pottery (see quots.).

1893 E. A. Barber *Pottery & Porcelain of U.S.* i. 9 Incised ornamentation is sometimes added by the use of a small wheel, bearing an engraved device on the edge, called the 'runner', which is held in a frame. **1974** Savage & Newman *Illustr. Dict. Ceramics* 250 *Runner*, a pointed tool used to decorate the body of a piece of pottery while it is cheese-hard and revolving on the potter's wheel.

f. The rotor or wheel of a turbine.

1908 S. F. Walker *Steam Boilers, Engines & Turbines* v. 317 The moving wheels or runners consist of wrought-steel discs. **1916** R. L. Daugherty *Hydraulics* xi. 179 The part of the turbine upon which the water does its work is called the runner. **1947** G. F. Wislicenus *Fluid Mech. Turbomachinery* i. 1 The most essential part of all types of turbomachinery is a vane-carrying rotating element, the 'runner', operating inside a stream of fluid or under its influence. **1972** J. M. K. Dake *Essent. Engin. Hydraulics* vi. 172 These concepts are applied to the runner of reaction pumps.

13. b. Also *spec.* in *Mountaineering* = *running belay* s.v. *Running *ppl. a.* 23 c.

1956 C. Evans *Kanchenjunga* xii. 126 Runner—a 'running belay' made by threading the climber's rope through a ring fixed to the mountain. **1956** [see *chock *sb.*[1] 8]. **1971** D. Haston in C. Bonington *Annapurna South Face* xvii. 206 The rope ran out so I tied all my aid slings and runners together. **1973** C. Bonington *Next Horizon* xii. 179 I'm thirty feet above my last runner, time for another but the skin of ice is too thin to take ice screws.

14. a. (Earlier example.)

1747 [see *sleigh runner* s.v. Sleigh *sb.* 4].

c. orig. *U.S.* A long narrow strip of (freq. embroidered) cloth, usu. placed along or across a table as an ornament.

1889 *Harper's Bazar* 20 Apr. 284/2 (*caption*) Embroidered table runner. *Ibid.* (Suppl.) 2/4 The cream-colored canvas grenadine centre of this table runner is a yard and three quarters long and twelve inches wide. **1904** *Buffalo Commercial* 2 June 6 Two long linen runners, one each way of the table, are now used in preference to the whole cloth or doilies. **1922** *Daily Mail* 11 Dec. 14 The illustrated table runner..was decorated at each end with a simple design. **1932** *Modern Weekly* 5 Mar. 997 Such a lovely runner to add a freshening touch to your dining-table. **1974** M. Ingate *Sound of Weir* ii. 11 On the lace runner on the dressing table was a photograph.

d. A long narrow rug or strip of carpet, used esp. in a hall or on a staircase.

1901 J. K. Mumford *Oriental Rugs* vii. 95 *Makatlik*, or 'runners'.—These are what we know as 'hall' or 'stair' rugs. **1910** S. Humphries *Oriental Carpets, Runners & Rugs* 251, I have used the term 'runners', instead of 'strips' (as some call them), because the latter description might lead one to suppose that only a fragment of a carpet was intended, whereas the runner is distinct and complete in itself... Runners are, in fact, very long rugs. **1918** V. O. Freeburg *Art of Photoplay Making* 236 He tries the stairs vaguely again, and, losing his balance at the top, grasps the 'runner' which pulls loose and wraps itself around him as he rolls down. **1937** M. Allingham *Dancers in Mourning* xvi. 200 Petals lay on the imitation parquet..and on the imitation Persian runner in the hall. **1947** [see *body carpet* s.v. *Body *sb.* 30]. **1955** C. Smith *Speaking Eye* xiii. 140 The corridor was yellow with a pale green runner. **1960** *News Chron.* 12 Sept. 6/4 A good buy in Oriental carpeting is the..runner. **1972** 'H. Carmichael' *Naked to Grave* i. 8 The entrance hall had a runner of royal-blue carpet down the centre with parquet flooring on either side. **1977** *Times* 9 Sept. 16/3 In the carpet section a rare pair of Shiraz runners were bought..for £1,500.

15. a. (Earlier and later examples.)

1815 *Niles' Reg.* IX. 201/2 [We] moved the one-half of the arch off sideways, forty-six feet, on to the runners one hundred and eighty-five feet long. **1939–40** *Army & Navy Stores Catal.* 1077/2 Curtain fittings... 'Huntband' Glide... 4 Runners to the foot. **1952** Granville *Dict. Theatrical Terms* 155 *Runner*,..a curtain track. **1960** *Practical Wireless* XXXVI. 328/2, 1 nylon ball-bearing curtain runner.

b. A long horizontal beam, girder, or other member (see quots.).

1891 *Notes on Building Constr.* (new ed.) II. x. 216 The capsills or 'runners' [of a gantry]..are supported by struts. **1932** Dowsett & Bartle *Pract. Formwork & Shuttering* i. 1 Propped shuttering [for floors] consists, generally, of 'sheeting' or 'decking' laid on joists..supported by heavier timbers, called 'binders' or 'runners', resting on upright 'props'. **1960** O. Skilbeck *ABC of Film & TV* 112 *Runners*, overhead girders with their depending tackle, from which, on modern stages, lamps and even complete prefabricated sets are hung. **1968** *Gloss. Formwork Terms* (*B.S.I.*) 21 *Runner*. 1. A longitudinal member spanning across a number of support members to lace them together. 2. One of a pair of supports running parallel to the axis of the centering for a deep arch or tunnel. One runner is above the other and they are separated by folding wedges. 3. See 'ledger' [= a horizontal timber supported on posts or hangers and carrying joists].

c. (See quot. 1940.)

1891 *Notes on Building Constr.* (new ed.) II. x. 209 Sometimes in very bad soil long planks called 'runners', having sharp ends shod with iron, are substituted for the poling boards. **1928** W. Simpson *Foundations* vii. 176 When test pits are of considerable depth, say, 40 ft., it is necessary to 'double set' the timbers, in which case two, or more sets of runners are used to reach the bottom. **1940** *Chambers's Techn. Dict.* 734/2 *Runners*, a form of sheet pile much used for timbering wide excavations. It consists of short planks shaped to a chisel point at one end and usually shod with thin steel strip, so that as each runner is driven in, it wedges up against its neighbour. **1963** M. J. Tomlinson *Foundation Design & Construction* ix. 538 In water-bearing sands and silts continuous support will have to be given to the face by means of timber runners or poling boards or by steel trench sheets or sheet piling.

18*. *Typogr.* (See quot. 1956.)

1888 C. T. Jacobi *Printers' Vocab.* 117 *Runners*, figures or letters placed down the length of a page to indicate the particular number or position of any given line. **1926** W. H. Slater *What Compositor should Know* III. 16 Runners are generally used in translations, and then only when the author has written a running commentary on the peculiarities of the original. **1956** *Bookman's Conc. Dict.* 259/2 *Runners*, letters or figures placed in the margin of a page opposite lines at regular intervals for assistance in reference (especially in poetry).

runnerless (rʌ·nəɹlės), *a.* [f. Runner 12 + -less.] Of a strawberry plant, esp. an alpine one: not producing runners.

1956 *Dict. Gardening* (R. Hort. Soc.) *Suppl.* 118/2 Baron Solemacher. (Alpine.) This is runnerless and makes large plants. **1957** E. Hyams *Speaking Garden* ix. 114 The most useful of these [mutants]..was a runnerless strawberry. **1981** *Country Life* 25 June 1846/2 The new French Bordurella perpetual [strawberry], virtually runnerless, and therefore a suitable plant for flower borders.

runner-up. Add: Also runnerup; pl. runners-up, runner-ups. **1. b.** (Further examples.)

1949 *Cavalier Daily* (Univ. of Virginia) 22 Oct. 1/3 The last issue of the magazine will announce the award winner and the runner-ups. **1955** *Times* 16 May 13/4 Miss Garvey, twice runner-up was unable to play. **1976** *Burnham-on-Sea Gaz.* 20 Apr., Highbridge soccer team Bristol Bridge look like losing their chance of finishing runners-up in the second division of the Bridgwater and District Sunday Football League. **1976** *Billings* (Montana) *Gaz.* 27 June 8-F/6 Other accomplished cowboys attending the rodeo include Royce Smith..who was runnerup for the world title twice.

c. *transf.* and *fig.*

1932 *New Yorker* 11 June 24/3 Blyth Daly is the horsy young lady who serves as runner-up to the Brat in offensiveness. **1949** *Sat. Rev. Lit.* (U.S.) 11 June 38/2 It was a 'runner-up' for the *Herald Tribune* award. **1974** *Times* 9 Nov. 10/5 Shirley Gee's *Stones* was runner up in the *Radio Times* play competition.

3. *attrib.*, as *runner-up list*, *prize*.

1925 *Scribner's Mag.* Oct. 384/1 At the tournament dance..he was called out for the runner-up prize in the fourth division. **1976** *National Observer* (U.S.) 18 Dec. 6/4 Britain's Queen Elizabeth II and Prince Philip made the runner-up list, as did Alistair Cooke, Shirley MacLaine, [etc.].

running, *vbl. sb.* Add: **I. 1. a.** Also *spec.* in *Cricket*, the action of making runs; also in phr. *running between* (*the*) *wickets*.

1744 *Laws* [of Cricket] in *New & Compl. Dict. Arts & Sci.* (1755) IV. 3460/1 They [*sc.* the Umpires] are sole judges of all hindrances, crossing the players in running, and standing unfair to strike. **1833** *New Sporting Mag.* V. *Cricketers' Reg.* 12 The steady manner in which Pilch bowled rendered the running very difficult throughout. **1877** *Encycl. Brit.* VI. 580/1, 21 Or if in running the wicket be struck down by a throw..before his bat (in hand) or some part of his person be grounded over the popping crease. **1897** *Encycl. Sport* I. 225/2 The art of running between wickets, without coming to a misunderstanding with one's partner, is indispensable. *Ibid.* 226/1 Every information should be given to one's partner that may help him in running. **1906** A. E. Knight *Compl. Cricketer* ii. 87 Some of the best running between wickets ever witnessed has been shown by Australian teams, who rarely miss a possible run. **1963** A. Ross *Australia 63* vii. 151 Catching and running between the wickets apart, England had at least given as good as they had got. **1976** *Milton Keynes Express* 23 July 39/3 David Berrill and Finch with some firm strokes and quick running between the wickets took the score to 61 all out after 17.3 overs.

2. a. Also *fig.* of a person, the action of standing as a candidate or competing (*for* an office). Cf. Run *v.* 7 b. orig. *U.S.*

1830 *Wiscasett* (Maine) *Citizen* 20 Aug. 3/2 Such politicians..bet on a candidate's running for the Chief Magistracy of the Union or of a single State, precisely as they would bet on the running of a race horse. **1870** *Nation* (N.Y.) 7 July 1/1 He has never failed in getting such offices as he wanted, the record of his 'running' being about as good as that of any man in the country. **1910** *Sat. Even. Post* 24 Dec. 16/2 There was less excuse for his running on the liquor ticket. **1961** T. H. White *Making of President 1960* iv. 86 Kennedy..felt that running for the Presidency was his most important full-time business and Senate attendance took second place.

c. (*d*) (further examples).

1923 Galsworthy *Captures* 161 Dinner was certainly a disharmonic feast: little Mrs Weymouth..and the Countess subdued, Radolin artificial, our scoundrel and myself had to make the running. **1954** I. Murdoch *Under Net* v. 80 'Where did you meet Madge?' I asked. I wasn't going to let him make all the running. **1958** *Times* 13 Sept. 7/2 First Russia turns on the heat in the Middle East and then it is China's turn to make the running. **1971** *Sunday Nation* (Nairobi) 11 Apr. 18/1 When he begins courting, he is unsure of himself, half-frightened of the girl, so she has to make the running. **1977** B. Freemantle *Charlie Muffin* vi. 65 You'll have to be bloody careful. Let Kalenin make the running.

e. (Further *fig.* examples.) Similarly, *in the running*. Also const. *for* (some prize or objective).

(*a*) **1906** *Dialect Notes* III. 149 'Isn't Jack sweet on the girl I saw with that fellow?' 'No, he's out of the running.' **1918** Galsworthy *Five Tales* 304 When a man is very old and quite out of the running, he loves to feel secure from the rivalries of youth, for he would still be first in the heart of beauty. **1930** G. B. Shaw *Apple Cart* I. 34, I think I am in the running. That is why I do not feel bound to accept this ultimatum. By signing it I put myself out of the running. **1949** M. Mead *Male & Female* xvi. 341 His married daughter, who with each step that she takes towards maturity puts him more definitely out of the running.

(*b*) **1886** H. Baumann *Londinismen* 162/2. *c* **1926** 'Mixer' *Transport Workers' Song Bk.* 65 And I never strike a top-job That the other fellows get, For I'm out upon my 'lonesome', And not in the running yet. **1930** W. S. Maugham *Cakes & Ale* xi. 120 It may be that posterity will scrap all the best-sellers of our day, but it is among them that it must choose. At all events Edward Driffield is in the running. **1950** W. Cooper *Scenes from Provincial Life* IV. ii. 235 It was one of the events in which Frank was hoping to shine, since he was in the running for victor ludorum. **1975** N. Blakiston in C. Connolly *Romantic Friendship* 5, I was in the running for playing in the school eleven.

7. d. With *out*. The disappearance of the characteristics of a particular variety of a crop.

1891 *Proc. Western N.Y. Hort. Soc.* 86 By 'running out' is meant the disappearance of the characteristics of any variety... Running out, therefore, is not necessarily deterioration. **1918** Babcock & Clausen *Genetics in Relation to Agric.* xviii. 340 The so-called 'running-out' of varieties can be prevented by reasonable care to avoid mixing seed and by occasional mass selection from the field. **1949** C. C. Lindegren *Yeast Cell* xxvii. 2 The degeneration or 'running out' of hybrids showing heterosis has been one of the principal problems of hybrid vigor. **1952** tr. *Gram & Weber's Plant Dis.* 361/2 (*heading*) 'Running out' or 'decline' of strawberry crops.

e. With *off*. = *Run-off 4.

1921 *Jrnl. Pomol.* II. 160 Investigations relating to the 'Running off' or dropping of Black Currant Fruits before they are ripe. **1939** H. Wormald *Dis. Fruits & Hops* ix. 181 Running off is not caused by a parasite, but is a result of defective pollination.

II. 11. c. (Later examples.)

1832 D. J. Browne *Sylva Amer.* 232 The scraping is a coating of sap which becomes solid before it reaches the boxes, and which is taken off in the fall and added to the last runnings. **1872** *Trans. Dept. Agric. Illinois* IX. 73, I have a plantation five or six years old, parts of which still grow fine fruit, with two runnings annually.

V. 15. c. *U.S. colloq.* Teasing; scolding. Cf. Run *v.* 52 c in Dict. and Suppl.

1832 S. Smith *Life & Writings Major Jack Downing* (1833) 158, I feel a little put out with Dr Burnham for an unhansome running he gave me 'tother day. **1902** J. Corbin *Amer. at Oxford* ii. 16 The freshman breakfast is nothing in the world but a variation of the 'running' that is given newcomers in those American colleges where fraternity life is strong. **1936** *Nat. Geogr. Mag.* LXIX. 799/1 Much of this 'running', or badgering, is in the spirit of fun.

d. *running in*, the process of operating a new machine (*spec.* the engine of a motor vehicle) at reduced power in order to establish proper working.

1935 *Jrnl. R. Aeronaut. Soc.* XXXIX. 159 Tapered piston rings reduce the running-in period considerably. **1963** R. F. Webb *Motorists' Dict.* 189 The object of running-in is to enable microscopic irregularities in the working surfaces to become burnished. **1968** 'M. Finch' *Eye with Mascara* viii. 80 A little rev-up from time to time is good for running-in.

16. b. *log running*: see *Log *sb.*[1] 9.

VI. 17. a. *running clothes*, *costume*, *costs*, *drawers*, *expenses*, *ground* (earlier and later examples), *pants*, *shorts*, *suit*, *track*, *vest*; **running brand** *U.S.*, a cattle brand made with a running iron (see next); also = next; **running iron** *U.S.*, a straight branding iron used (freq. illegally) for altering cattle brands; also *fig.*; **running light**, (*a*) *Naut.* = *navigation light* s.v. *Navigation 8; (*b*) one of a set of small lights located on the front, rear, or sides of a motor vehicle that remain illuminated during the running of the vehicle (see quots.); **running order**, (*a*) a condition in which a machine, etc., will function; (also qualified by *well*, *badly*, etc.); the condition in which a road, etc. is fit to be used (cf. *working order* s.v. Working *vbl. sb.* 16 b); (*b*) *Theatre* and *Broadcasting* the sequence in which scenes or parts of a programme are presented; **running shoe**, a (freq. spiked) shoe for running (usu. *pl.*); also *fig.*, esp. in (orig. *N.Z.*) phr. *to give* (one) *his running shoes* (see quots.); **running time**, (*a*) (earlier example); (*b*) the time occupied by the running of a machine, performance, etc.; so, a schedule. Also *Running board.

1884 Sweet & Knox *On Mexican Mustang through Texas* xii. 160 The other, called a running brand, is a long piece of iron curved at the end. **1934** *Denver* (Colo.) *Post* 4 Aug. 10/3 A running brand..[is] a brand made with a straight poker called a 'running iron', and used like a pencil. *a* **1727** in *Gentl. Mag.* (1791) LXI. I. 199 No money allowed when I run any way under twenty miles. To find my own stockings and pumps, and to have my running clothes washed in the house. **1916** W. Owen *Let.* 14 Mar. (1967) 385, I shall consider my running clothes as my Birthday Present. **1907** *Sports Trades Jrnl.* X. 25 (Advt.), Running costumes. **1913** *Autocar Handbk.* (ed. 5) i. 14 Depreciation is a large item generally included amongst running costs. **1979** *Homes & Gardens* June 153/2 Keep running costs to a minimum by placing the freezer in a cool, well ventilated room and ensuring that it is full. **1903** *Sports Trades Jrnl.* III. 101 (Advt.), Running drawers, running vests. **1904** *N.Y. World Mag.* 1 May 6/6 This does not include running expenses. **1744** *Laws* [of Cricket] in *New Dict. Arts & Sci.* (1755) IV. 3459/2 When the ball is hit up, either of the strikers may hinder the catch in his running ground. **1927** E. V. Gordon *Introd. Old Norse* 196 Let the adventurous steed of the sand's heaven explore the broad running-ground of ships. **1894** *McClure's Mag.* July 101/2 The running-irons, or *guachos*,..are now considered bad form by progressive cattlemen. **1913** L. V. Kelly *Range Men* 16 But the running or round iron was most favored, as it was easily and secretly made by cutting a wagon-iron in two and using the rounded end. **1945** *Everybody's Digest* Aug. 89 Of a dying man, the puncher might say: 'Death's got the runnin' iron on him brandin' him for the Eternal Range.' **1968** R. F. Adams *Western Words* 261/1 *Running iron*, a branding iron made in the form of a straight poker or a rod curved at the end... In the 1870's a law was passed in Texas forbidding the use of this iron in branding. This was a blow aimed at the brand blotter, whose innocent single iron would tell no tales. **1881** *Naval Encycl.* 439/2 *Light*,..the term for all lamps or lanterns used on ship-board; as, *running lights*, *signal-lights*, *mast-head lights*, etc. **1948** R. de Kerchove *Internat. Maritime Dict.* 613/1 *Running lights*, a general term applied to the various lights carried from sunset to sunrise by different classes of vessels when under way. **1971** M. Tak *Truck Talk* 135 *Running lights*, a tractor-trailer's clearance lights. **1975** *Times* 21 Aug. 3/2 Volvo cars..are to be fitted with 'running lights', special side lights switched on automatically with the ignition. **1977** *Observer* 3 Apr. 37/8 The working group appointed by the Nordic Road Safety Council has proposed that the use of running lights be made mandatory in daytime all year round in Denmark, Finland, Sweden and Norway. **1978** H. Wouk *War & Remembrance* xx. 192 Unescorted, floodlights on a white hull, brilliant running lights, huge red cross painted on her side. **1850** N. Kingsley *Diary* 4 Mar. (1914) 112 Got up steam again today and tried the larboard engine and have got both in running order. **1860** A. Sherwood *Gazetteer of State of Georgia* (ed. 4) 152 Soon, say in summer of 1861, much will be in running order. **1875** *General Statutes of State of Michigan* I. (1882) 829 Every corporation owning a road in use shall..draw over the same the merchandise and cars of any other corporation..: Provided, such cars are of the proper gauge, are in good running order, and properly loaded. **1902** 'O. Henry' in *Ainslee's Mag.* Apr. 209/2 The running order of the bank was smooth and clean, and that had facilitated his work. **1939** N. Coward *Play Parade* II. p. x, *Running order*, a list of the Scenes in their correct order. **1948** [see *Runner 4 c]. **1961** G. Millerson *Technique Television Production* x. 186 A running order, prepared from this script after the technical planning meeting, will contain a breakdown of the entire studio operations throughout the show. **1977** B. Langley *Death Stalk* ii. 21 If you're editing a programme and you're not prepared for it [*sc.* an item], it can knock hell out of your running order. **1930** L. W. Olds *Track Athletics* i. 4 The running pants should have plenty of room in the hips. **1884** *Spectator* 26 July 975/2 The running-shoes must be doffed for good and all. **1913** C. Mackenzie *Sinister St.* I. I. vii. 107 Every evening there was steady practice..in spiked running-shoes on the grass-track. **1941** Baker *N.Z. Slang* vi. 53 *To give a person his running shoes*, to dismiss a person from office.., which is an extension of the phrase *to get the run*, to be dismissed or fired. **1943** J. A. W. Bennett in *Amer. Speech* XVIII. 92 Only a few current phrases can be traced to their creators. One of these is 'to give a man his running shoes', coined by a New Zealand Minister of the Crown as a vivid substitute for the English 'sack' or the American 'fire'. **1960** Wentworth & Flexner *Dict. Amer. Slang* 438/1 *Running shoes, give someone his*, to dismiss someone, as a suitor or an employee; to terminate a personal or business relationship, usu. in anger. **1963** B. Pearson *Coal Flat* vi. 111 Like Bob Semple used to say about hit-and-run drivers—give them their running shoes. **1979** *Tucson* (Arizona) *Citizen* 20 Sept. 6A/1 President Carter is losing ground by waiting to put on his official re-election running shoes. **1912** E. W. Hjertberg *Athletics in Theory & Practice* xii. 54 The running shorts should be wide and should not sit tightly anywhere, whatever movement one happens to make. **1974** *Times* 10 Apr. 13/2 A lifelessly wordless eccentric in running shorts. **1905** Graham & Clark *Pract. Track & Field Athletics* 9 Sufficient capital to purchase a running suit and a pair of spiked shoes,..and a spare hour somewhere in the course of the day are all that are required to give any one a chance to develop his latent possibilities. **1806** W. Clark in *Lewis & Clark Exped.* (1905) V. 294 Emence herds of Buffalow about..as it is now running time with those animals. **1897** Kipling *Capt. Cour.* ix. 195 Our runnin' time from San Diego to Chicago was 57.54. **1911** H. S. Harrison *Queed* xii. 143 Queed..pulled into supper only three minutes behind running-time. **1952** Granville *Dict. Theatrical Terms* 156 *Running time*, the actual time the play takes in performance, act by act, excluding intervals and final curtain calls, which are logged separately. **1962** D. R. Cox *Renewal Theory* vii. 80 Suppose that a machine is subject to stoppages and call the time necessary to restart a stopped machine a repair-time. There is thus an alternating sequence of running-times and repair-times. **1977** A. Morice *Murder in Mimicry* II. x. 176 Even in normal running time my first entrance did not come for ten minutes. Nevertheless..I got to the theatre with two hours to spare. **1883** *Harper's Mag.* Oct. 416/2 The running track, commonly used for trotting as well, has..seen some notable achievements. **1907** *St. Nicholas* June 694/1 'And a new running track', added Dick. **1903** Running vest [see *running drawers* above].

b. (Earlier and later examples with *down*.) Also with *about*, *in*.

1966 J. Derrick *Teaching Eng. to Immigrants* iv. 155 Traditional children's games..are mainly running-about games which are probably more suited to the playground..than to the classroom. **1825** *Spirit of Public Jrnls. 1823* 315 Mr. Joseph Arnold being thus foiled in his running-down scheme, placed his jarvey right across the road. **1856** Dickens in *Househ. Words* XIII. 554/2 The landsman was relating his experience..of a fearful running-down case in the Channel. **1931** *Times* 16 Jan. 14/2 He acquired a large practice, esp. in running-down cases. **1968** *Times* 29 Nov. 13/3 The running-down rate of the pulsar implies that it has a lifetime of the same order as the Crab nebula. **1930** *Engineering* 4 Apr. 439/1 In the running-in process, it has usually been necessary for the engine to be carefully watched..to prevent damage from over heating or seizing. **1957** *Railway Mag.* June 438/1 This is part of a regular two-day running-in roster from Crewe Works. **1973** 'J. Ashford' *Double Run* iii. 18 The driver ignored both the fifteen m.p.h. factory speed limit and the forty-five running-in limit.

running, *ppl. a.* Add: **I. 1. b.** Also, a constant supply of water from a tap, main, or the like.

1912 M. L. Fuller *Domestic Water Supplies for Farm* xx. 151 In very few ways, if any, may the drudgery be so readily lessened or the pleasures and comforts of rural life so increased as by the installation of running water in the houses & barns. **1936** G. Mitchell *Dead Men's Morris* i. 16 We got no runnen water endoors round this part. **1946** A. Huxley *Let.* 21 Apr. (1969) 543 There is running water, electric light and bottled gas, so that the fundamentals are all right. **1961** L. Mumford *City in History* xv. 466 A collective water system with running water available for every house. **1974** O. Manning *Rain Forest* I. iii. 52 There's no running water, the sani-cans stink.

c. *running ice*, ice which moves downstream in blocks and sheets. *N. Amer.*

1913 W. Ogilvie *Early Days on Yukon* iii. 44 In this it satisfactorily succeeded, when running ice put a stop to further mining operations. **1922** H. A. Cody *King's Arrow* xii. 101 The bark canoe seemed like a thing of life as it cut through the water... It had battled with running ice; it had been borne over innumerable portages. **1959** M. Shand *Summit & Beyond* iv. 65 The Mounted Police sent out warnings that no more small boats could be used on lakes and rivers on account of the running ice. **1968** [see *ceiling *vbl. sb.* 6 c].

3. c. (Further example.)

1975 *Times* 4 Sept. 19/3 If the water is very hot it stops people washing things under a running tap.

4. Also of the eyes or nose. *running sore*: also *fig.*, a constant nuisance or irritation; a long-lasting trouble or problem.

1807 E. Weeton *Let.* 18 Nov. (1969) I. 50 As to the running eyes and noses of which you request me to give you the exact number. **1961** C. Cockburn *View from West* ix. 117 At that time the National Union of Journalists was as a running sore to the anti-Communists of the T.U.C. **1964** J. P. Clark *Three Plays* 82 Do forgive my running nose. **1973** *Times* 21 Nov. 6 Running sore of London staff shortages may defy short-term cure.

II. 7. e. *running mate*, a horse entered in a race in order to set the pace for another horse from the same stable which is intended to win. Also in *transf.* use, a fellow candidate (*of* someone), usu. one standing for a subordinate office, and *spec.* the vice-presidential candidate in U.S. presidential elections; *gen.*, a partner, colleague, spouse, etc. orig. and chiefly *U.S.*

1868 H. W. Woodruff *Trotting Horse of Amer.* xxxvi. 284 He has been..especially great for his knack at going with a running-mate. **1883** *Illustr. Sporting & Dramatic News* 30 June 399/2 'How fast do you think she can trot with a running-mate?' the reporter asked; a 'running-mate' it is, perhaps, unnecessary to explain, being a horse harnessed to a pole, which gallops or 'runs' while the other trots. **1900** *Rev. of Reviews* Jan. 7/2 A better man could hardly be selected as Mr. McKinley's 'running mate'. **1902** B. Whitlock *13th District* 61 There were.. pictures of the candidate himself,..and pictures, too, of his 'running mate', the candidate for vice-president. **1911** *Munsey's Mag.* Mar. 865/2 His running-mate, Elizabeth Brice, in spite of the eyes she makes, also inspires liking. **1935** *N. Amer. Rev.* Mar. 275 Nast's representation of a running-mate by a tag on the coat-tail of the head of the ticket is a fair indication of popular interest and respect. **1944** *Newsweek* 31 July 25/1 The substitution of Sen. Harry S. Truman of Missouri for Vice President Henry A. Wallace as Mr. Roosevelt's running mate. **1953** *Economist* 25 July 258/1 Eisenhower and his running-mate, Senator Nixon, spoke in stern criticism. **1958** *Listener* 21 Aug. 255/1 From President Nasser's point of view, the Sudan would be a far more desirable running-mate in the United Arab Republic than Syria. **1968** R. F. Adams *Western Words* 261/1 *Running mate*, a cowboy's term for his pal or his wife. **1968** W. Safire *New Lang. Politics* 389/1 In horseracing, a single stable will often enter two horses in a race, the lesser horse used as a pacesetter and called a running mate. This second horse usually vanishes into obscurity. **1972** D. H. Laurence *Bernard Shaw: Coll. Lett. 1898–1910* 188 A similar letter had been posted..to women voters, appealing to them to vote for Shaw and his five Progressive running-mates to protect women's rights in the Borough. **1973** 'D. Halliday' *Dolly & Starry Bird* i. 2, I..shared digs with my running mate, a photographer. **1977** *Irish Times* 8 June 8/6 His running-mate, Alice Glenn, received 1,240 first preferences in Dublin North-Central in 1973.

8. f. *running dog* [tr. Chinese *zǒugǒu*, f. *zǒu* to run + *gǒu* dog]: in communist terminology, one who is subservient to counter-revolutionary interests; a lackey (see *Lackey *sb.* 1 c). Also in generalized use.

1937 E. Snow *Red Star over China* IX. iv. 325 Vanguards of young Moslems were..urging the overthrow of the 'Kuomintang running-dog'. **1961** tr. *Mao Tsetung's Sel. Works* IV. 284 Without a revolutionary party,..it is impossible to lead the working class and the broad masses of the people in defeating imperialism and its running dogs. **1968** *Guardian* 13 July 9/5 The Kremlin's fears that the Czechoslovak revolution has become a treacherous running dog of the West. **1969** R. Quest *Cerberus Murders* xxi. 119 This is how we worked—we were not a team and I was certainly not his running-dog. **1970** [see *housenigger* s.v. *House *sb.*[1] 23]. **1977** 'E. Crispin' *Glimpses of Moon* xi. 226 'Imperialist running-dogs,' said the hunt saboteuse. **1978** J. Updike *Coup* (1979) iii. 98 He admired the French, he admires the polluting Americans and their new running dogs the Chinese.

g. *running back*: in American football, a back whose function is to run carrying the ball.

1924 *Collier's* 20 Dec. 38/4 There has been no running back in football history who had these baffling, bewildering qualities to such an extent. **1962** R. Walker *Compl. Bk. Backfield Play* ii. 28 What distinguishes a back from a linesman? Many ends would make outstanding running backs. **1967** R. Graves *Guide to Mod. Football Offense* vi. 47 In discussing techniques and drills involved in backfield play, we will place the position of fullback and tailback under the classification *running backs*... Since the play of the quarterback is uniquely different from that of running backs, the fundamental requirements of his position will be discussed separately. **1971** L. Koppett *Guide to Spectator Sports* ii. 50 And the halfback and fullback are simply 'running backs'. **1979** *Arizona Daily Star* 5 Aug. (Parade Suppl.) 6/1 Susie Forton played running back and middle linebacker for the Vikings.

9. Also *running moss* = Club-moss (earlier and later examples); *running postman* = *coral-pea* s.v. *Coral *sb.*[1] 9.

1845 Running moss [see *mountain laurel]. **1898** E. E. Morris *Austral Eng.* 247/1 K[*ennedya*] *prostrata* is called the Coral Pea..or Running Postman. **1917** H. H. Richardson *Fortunes R. Mahony* I. ix. 87 The short-lived grass was picked out into patterns by the scarlet of the Running Postman. **1945** E. Step *Wayside & Woodland Ferns* (ed. 2) 129 These local names [for common club-moss] are..numerous... Others are..Lamb's-tail, Running Moss, Robin Hood's Hatband, [etc.]. **1962** Running postman [see *coral-pea* s.v. *coral *sb.*[1] 9].

11. a. Also in phr. *running rhythm*, used by G. M. Hopkins to denote common English metre.

c 1883 G. M. Hopkins in *Poems* (1967) 45 The poems in this book are written some in Running Rhythm, the common rhythm in English use, some in Sprung Rhythm, and some in a mixture of the two. Common English rhythm, called Running Rhythm above, is measured by feet of either two or three syllables. 1957 N. Frye *Anatomy of Criticism* 263 The sixteenth century was a period of experiment, mainly in verse *epos* or running rhythm, to use Hopkins's term. 1970 J. T. Shipley *Dict. World Lit. Terms* (ed. 3) 284/2 *Running rhythm*, the common Eng[lish] rhythm, measured by feet of 2 or 3 syllables... Opp[osed] by G. M. Hopkins to sprung rhythm.

III. 14. a. Also in various (chiefly *U.S. Sporting*) phrases, as *running attack, game, start* (see quots.).

1910 W. Camp *Bk. Foot-Ball* viii. 306 Probably there is greater fascination in the running game than in any other department of foot-ball... There is no play that brings the spectators to their feet to such wild enthusiasm as a good run. 1929 D. Runyon in *Hearst's Internat.* July 125/1 Dave the Dude is more corned than anybody else, because he has two or three days running start on everybody. 1961 J. S. Salak *Dict. Amer. Sports* 375 *Running play*,..a play during which there is a runner and which is not followed by a kick or forward pass from behind scrimmage line. 1971 L. Koppett *Guide to Spectator Sports* ii. 56 To stop a running attack, a defensive line must hold its ground. 1976 *Webster's Sports Dict.* 160/2 *Flying start*, auto racing, a start of a race in which the competitors are already moving as they cross the starting line or receive the starting signal (also called *running start*). 1977 *New Yorker* 10 Oct. 177/1 Princeton, which has another useful quota of real heavyweights..has as well a sturdy running attack, now that Isom is paced by Larson.

d. Also *running battle*. Also, any military engagement which constantly changes its location. Also *transf.* (in later use, perh. influenced by sense *17 a).

1916 'Taffrail' *Pincher Martin* xvii. 317 The *Mariner* and various other destroyers were present with the battle-cruisers throughout the first shock of the engagement and the running fight which ensued. 1928 G. B. Grinnell *Two Great Scouts* iv. 65 The Pawnee warriors sprang on their horses and set out in pursuit of the enemy. During a running fight, the Pawnees killed a number of Sioux. 1945 *Ann. Reg. 1944* I. 2 The raiders.. carried on a running fight with German fighter planes. 1967 in G. Marx *Groucho Lett.* 13 (*heading*) Running battle with Warner Brothers. 1974 [see *motor-cycle sb.].

e. *running repairs*, hurried, minor, or temporary repairs made to machinery, equipment, etc., while in service. Also *transf.*

1913 *Autocar Handbk.* (ed. 5) i. 16 He [*sc.* the chauffeur] should do most of the running repairs, such as tyre repairs. 1924 Kipling *Debits & Credits* (1926) 166 We'd been sent back for rest an' runnin'-repairs, back pretty near our base. 1951 N. Marsh *Opening Night* vi. 134 He..effected a number of what he called running repairs to her make-up and hair. 1957 *Encycl. Brit.* XVIII. 935/2 Most yards of any considerable size also include, or have adjacent, facilities for servicing of, and running repairs to, cars and locomotives. 1971 D. J. Smith *Discovering Railwayana* x. 59 *Running repairs*, small scale repairs carried out in a 'running' or engine shed. 1973 K. Benton *Craig & Jaguar* vii. 89 Your nose is shiny..Make some running repairs while we leave you for a moment.

f. *running jump*, a jump preceded and augmented by a run. Usu. *fig.*, esp. in phr. *to take a running jump* (*at oneself*), freq. used colloq. as an expression of hostility, contempt, or indifference to someone.

1914 E. A. Powell *Fighting in Flanders* i. 18 Thompson took a running jump. 1920 S. Lewis *Main Street* xxxv. 415 There aint a town..got a better chance to take a running jump..right up into the two-hundred-thousand class. 1933 M. Lowry *Ultramarine* ii. 76 You go and take a running jump at yourself. 1953 A. Upfield *Murder must Wait* xi. 104 Tell your Chief Commissioner to take a running jump at himself. 1954 J. B. Priestley *Magicians* vi. 120 The public can take a running jump at itself. I stopped liking people a long time since. 1959 'R. Macdonald' *Galton Case* xii. 97 Tell him to take a running jump in the Truckee River and do us all a favor. 1968 *Landfall* XXII. 22 If you think I'm subsidizing you..you can take a running jump at yourself. 1972 M. Gilbert *Body of Girl* xx. 180, I told her to take a running jump at herself. The only person who could make trouble for *me* would be old Henry Prior.

g. *running fix*, a fix obtained by determining bearings at different times and making allowance for the distance covered by the observer in the interval.

1916 S. F. Card *Navigation Notes & Examples* vii. 32 A running fix is the position obtained from two position lines by observations at different times, allowing for the run in the interval. 1942 *Tee Emm* (Air Ministry) II. 82 Here's a good tip to save yourself a running fix. 1974 K. Wilkes *Pract. Yacht Navigator* (ed. 2), ix. 119 The accuracy of the running fix depends on the correctness of the direction and distance travelled over the ground between the two bearings.

IV. 17. a. (Further examples.) Also, continually produced or maintained; constantly repeated or recurring.

1966 *This is Bill-Broking* (Allen, Harvey & Ross Ltd.) 34 *Running yield*, the interest rate on an investment, expressed in terms of a percentage on the capital invested. 1966 *Listener* 17 Nov. 732/1 A series of comic set-pieces linked into a wildly slapstick context by carefully contrived running gags. 1973 *Daily Tel.* 7 Mar 21 At the issue price of £99½ the 1980 stock will give a running yield of 9·05 p.c. and a gross redemption yield of 9·10 p.c. to 1980. 1973 *Times* 30 Oct. 4/6 As he walked the Dalai Lama gave a running audience.

e. *running commentary*, a sustained series of comments on events, actions, utterances, etc., as they occur; a continuous description of an event in progress, *spec.* a broadcast report of a game, contest, or race.

1811 C. Lamb in *Reflector* iv. 342 The writings of Fuller are usually designated by the title of quaint... But.. his way of telling a story, for its eager liveliness, and perpetual running commentary of the narrator happily blended into the narration, is perhaps unequalled. 1824 *Mirror of Literature* 17 Jan. 44/2 The Count's running commentary upon these evolutions, too, is a *chef d'oeuvre* in the art of reasoning. 1853 R. S. Surtees *Mr. Sponge's Sporting Tour* lxi. 348 His pleasure was, perhaps, damped by a running commentary he overheard through the lattice-window of the stable. 1858 *Chambers's Jrnl.* 4 Dec. 359/1 (*heading*) Bill Fustian's running commentary on the doings of the respectable classes. 1883 J. M. Barrie *Auld Licht Idylls* xii. 239 He loved to recite long screeds from Spenser, with a running commentary on the versification and the luxuriance of the diction. 1905 *Pall Mall Mag.* July 40/2 She gasped and..trembled out her tale of horrors, while..her daughter,..in the exasperating fashion of the chorus in a Greek play, kept up a running commentary, emphasising the points. 1927 *B.B.C. Hand-bk. 1928* 140/1 Running commentaries fall easily under two different headings—Sporting and purely Descriptive. 1929 *B.B.C. Year-Bk. 1930* 146 Tennis..provides excellent material for a running commentary, although the commentators find the strain of following the strokes.. with an instantaneous spoken description very great. 1931 *Discovery* Dec. 386/1 This was the first scientific film of its kind to be synchronized with the running commentary. 1946 G. N. M. Tyrrell *Personality of Man* vi. xviii. 158 Then there is Mrs. Willett's running commentary. She is always there, interjecting her own comments. 1966 B. Johnston *Armchair Cricket* 26 He has to comment rather than give a *running* commentary which is basically what happens with football or racing on television. 1969 M. Pugh *Last Place Left* xvii. 124, I drank tea with my back to the floor but Katriona gave me a running commentary until Nell rejoined me. 1977 D. Bennett *Jigsaw Man* iv. 71 I'll give you a sort of running commentary... Reorient you... The village is on the right.

f. *running set* (see Set *sb.*^{2} 14), a country dance, originating in the Appalachian Mountains, in which the dancers perform a number of figures in quick succession.

1918 C. Sharp *Country Dance Book* v. 9 The Running Set..differs materially from any other known form of the Country-dance. 1927 *Observer* 27 Nov. 14/5 He got.. from elsewhere in that district, the 'running set'—a fine dance which has been received here with outspoken enthusiasm. 1938 *Times* 10 Jan. 10/4 Two American dances, the Running Set and the Big Set, were shown. 1964 W. G. Raffé *Dict. Dance* 431/2 *Running set*, an English square dance in quick time, preserved in villages of the Appalachian Mountains, in North America. *Ibid.*, Danced by four couples (sometimes more), the *Running Set* consists of an Introduction and some fourteen figures, which follow each other without pause. 1974 *Encycl. Brit. Micropædia* IX. 501/2 Historians trace the square dance to two derivatives of English country dance: the Kentucky running set, a rhythmic, complicated figure dance derived from pre-17th-century English round dances; and the cotillon.

VI. 22. b. *running mould* (Plastering), a pattern moving on fixed guides and used to shape cornices and other mouldings.

1825 [see Run *v.* 53 f]. 1911 *Encycl. Brit.* XXI. 786/1 Plain, or unenriched, mouldings are formed with a running mould of zinc cut to the required profile. 1955 N. W. Kay et al. *Mod. Building Encycl.* 483/2 Solid cornices are..formed in the position they are to occupy, by a running mould, called a horse, which runs along guides fixed to the walls.

23. c. *running belay* (Mountaineering), a belay (see *belay *sb.*) through which the climbing rope runs freely, and which acts as a pulley if the climber falls.

1941 T. A. H. Peacocke *Mountaineering* ii. 26 Without the use of slings, running belays are unsatisfactory. 1946 Leonard & Wexler in *Sierra Club Bull.* Dec. 91 He [*sc.* G. W. Young] is fully aware of the dynamic belay, referring to it as the 'running belay', one of the 'expert belays' that only great skill and strength can hope to regulate. 1956 C. Evans *On Climbing* iii. 52 When negotiating a difficult pitch, the leader will, if possible, arrange running belays at intervals... He..drives a piton into a crack, and runs his rope through a snap-link clipped to the..piton. 1968 P. Crew *Encycl. Dict. Mountaineering* 87/1 Apart from the ease of carrying, the fact that the sling goes *through* a nut, instead of *round* a chockstone, often makes the running belay more mechanically sound. 1973 C. Bonington *Next Horizon* xx. 274 Don went up..quicky and smoothly without bothering to protect himself with running belays.

24. b. *running bowline* († *knot*), a bowline adapted to form a noose.

1726 Defoe *Four Years Voyages of Capt. George Roberts* 110, I got ready another Rope, at the end of which I made a running bowling knot, and the Noose so as to keep it open with one Hand. 1823 [see Bowline^1 2]. 1883 *Man. Seamanship* (Admiralty) 89 *Q.* What is a running bowline used for..? *A.* It is used for throwing over anything out of reach, or anything under water. 1911 *Encycl. Brit.* XV. 872/1 Running bowlines are formed by making a bowline round its own standing part... It is the most common and convenient temporary running noose. 1932 E. M. Brent-Dyer *Chalet Girls in Camp* vii. 111 The rope was swung down, and as it stopped swinging, Miss Wilson knew that Jo had it safely. 'Running bowline!' she called down. 1968 E. Franklin *Dict. Knots* 24 *Running bowline*, the knot which was universally used at sea whenever a noose was needed. Useful for commencing to tie a parcel.

25. a. (Earlier and later examples.)

1848 E. C. P. in C. H. Hartshorne *Eng. Medieval Embroidery* 128 An inner line of yellow floss silk in a running stitch. 1967 E. Short *Embroidery & Fabric Collage* iii. 86 Although traditionally carried out by hand in running or back stitch, the quilting can be done on the sewing machine.

26. b. With advbs., as *running-down* (in sense 73 a of the vb.).

1968 *Times* 29 Nov. 13/7 A running-down pulsar might be found. 1973 L. Cooper *Tea on Sunday* xxiv. 177 For what was supposed to be a running-down business it all looked remarkably active.

27. *running fit*: (see quots.).

1908 S. H. Moore *Mech. Engin. & Machine Shop Practice* vii. 184 A running fit is designed to allow the surfaces in contact to move or revolve freely over each other. *Ibid.*, Two formulas are given for running fits; one for close running fits, to be used in ordinary work.., and the other for free running fits, to be used for high-speeds, heavy pressures, rocker shafts, etc. 1953 W. H. Armstrong *Mech. Inspection* iv. 51 A running fit is one in which an allowance is made so that a shaft will be free to rotate in a bearing. 1964 S. Crawford *Basic Engin. Processes* xiv. 290 Running fits.., tolerances which allow the shaft to rotate freely in the hole.

ru·nning-board. Also **running board.** [f. Running *vbl. sb.*] † **1.** A narrow gangway on either side of a keel-boat. *U.S. Obs.*

1817 *Essex Inst. Hist. Coll.* (1866) VIII. 240 Our boat being very deep..[we] were obliged to give up after being at the expense of putting on running boards. 1834 H. Brackenridge *Recoll.* iv. 37 One night..I..lay..on the running board (a plank at the edge of the boat, on which the men walk in pushing with the pole). 1843 *Amer. Pioneer* II. 271 Keel-boats..were provided with running boards, extending from bow to stern, on each side of the boat... The crew, divided equally on each side, set their poles near the head of the boat, and bringing the end of the pole to their shoulders, with their bodies bent, walked slowly down the running board to the stern.

2. a. A foot-board extending along the side of a locomotive, railway wagon, or tram, or one extending along the roof of a railway wagon. orig. and chiefly *U.S.*

1860 Clark & Colburn *Recent Practice Locomotive Engine* 51/2 The cab, domes, 'running-board', and other matters of external finish, are very much the same on most American engines. 1874 M. N. Forney *Catechism of Locomotive* 337 The running-boards are planks..placed on each side of the boiler to enable the locomotive runner or fireman to go from the cab to the front end of the engine when it is running. 1889 *Cent. Dict.*, *Running board*,..(*a*) A narrow platform extending along the side of a locomotive. (*b*) A horizontal board along the ridge of a box freight-car or the side of an oil-car, to form a passage for the trainmen. 1903 *Electrical World & Engin.* 14 Nov. 795/2 The 'mule' has two large hooks for the towropes and has also a running board and guard hand rail. 1917 C. Mathewson *Second Base Sloan* 284 The cars that buzzed and clanged their way past Wayne were filled to the running-boards. 1930 *Amer. Speech* V. 277 *Running-board* is a bit puzzling, but the speculation that it was derived from the old summer trolleys, now almost extinct in the north, is at least permissible. 1940 *Life* 4 Mar. 50/2 The rear-end brakeman..makes an inspection tour along the 'running board', looking for loose brake beams or hot boxes. a 1966 'M. na Gopaleen' *Best of Myles* (1968) 187 Particularly if the running board of the tram was already crowded with fat women.

b. A foot-board located on either side of a motor vehicle between the front and rear mudguards.

1907 S. Krausz *Practical Automobile Dict.* 26 *Running-board*, s., marchepied, s.m. 1910 *Sears, Roebuck Motor Buggy Booklet* 28 With the addition of running boards connecting front and rear fenders, convenient for shopping or business where frequent getting in and out is necessary. 1914 E. A. Powell *Fighting in Flanders* vii. 169 A big grey car shot down the road... Clinging to the running-board was her English chauffeur. 1927 M. de la Roche *Jalna* xii. 136 Wakefield mounted the running board and held the Michaelmas daisies out to her. 1929 *Daily Express* 14 Jan. 6/3 Heath, leaning over the running-board, shouted some unintelligible words. 1932 Kipling *Limits & Renewals* 139 Phil sat down on the running-board of Mr. Haman's car. 1959 *Motor* 7 Oct. 246/2 The body sides have now been carried out towards the rear to narrow down the running boards and reduce the protruding width of the rear wings. 1965 M. Bradbury *Stepping Westward* viii. 380 They sat on the running-board of the car. 1974 *Country Life* 17 Oct. 1104/1 There is the Volkswagen Beetle... Here in the 1970s we still have a car with very rounded lines, small windows, a very cramped interior and outside running boards.

3. A device used in positioning overhead power lines which enables several conductors to be pulled simultaneously using a single pulling line.

1898 E. J. Houston *Dict. Electr. Words* (ed. 4) 911/1 *Running-board*, a device employed in the construction of a heavy overhead line, consisting in placing a number of reels of wire, usually ten or more, on a spindle, and arranging a piece of wood as a cross-arm to which ten or more wires are attached, harnessing horses to the cross-piece, and then dragging the running board away as the

wires are paid out from the reels, and passing them over their appropriate cross-arms, where they are at once secured to the insulators by line-men. 1964 E. B. Kurtz *Lineman's & Cableman's Handbk.* (ed. 4) xiv. 7 Tension stringing of bundled conductors. Usually two or three conductors are pulled simultaneously by one pulling line with the use of a unidirectional articulated running board.

running gear. orig. and chiefly *U.S.* Also **running-gear.** [f. Running *ppl. a.*; in senses 1 and 3 a the plural form is used interchangeably with the singular.] **1.** The moving parts of a mill or other large machine.

1662 *Rec. East-Hampton, N.Y.* (1887) I. 201 Mr Backer shall have seven pounds for this yeare for tendinge the mill and maintayninge the runninge geares that is coggs and rounds. 1725 *New England Courant* 18–25 Jan. 2/2 The Wind... carry'd off the Top of the Mill, with the Shaft, Vanes, and running Geer, and brake them to Pieces. 1834 in J. S. Bassett *Southern Plantation Overseer* (1925) vi. 73 The runinge geares that is hear I cant under take to pick a crop with them. 1901 Merwin & Webster *Calumet 'K'* xiv. 262 Down in the cellar putting in the running gear for the 'cross-the-house conveyors'.

2. The rope and tackle used in handling (part of) a boat; = *running rigging*.

1838 J. F. Cooper *Homeward Bound* II. iii. 55 The standing rigging are the bones and gristle; the running gear the veins in which her life circulates. 1856 E. K. Kane *Arctic Explorations* II. iii. 48 We can burn hemp cable and cast-off running-gear. 1911 J. Barten *Compl. Naut. Pocket Dict.* 165/1 Running gear. 1962 A. G. Course *Dict. Naut. Terms* 163 *Running gear*, ropes, tackles, etc., that move in the course of ship handling or cargo working.

3. a. The wheels and axles of a cart or carriage. Also *transf.* and *fig.*

1857 D. H. Strother *Virginia Illustrated* 230 A shadowy group was dimly visible, a carriage mounted on the running-gear of a wagon, and drawn by four horses. 1876 [in Dict. s.v. Running *ppl. a.* 22]. 1904 *N.Y. World* (Mag. Sect.) 1 May 6/1 The running gear is dark red and the upholstering is drab. 1916 *Dialect Notes* IV. 348 *Running gear*, the remnants or 'carcass' of a fowl served up cold. 1923 *Ibid.* V. 208 *Runnin' gears*, Those portions of a wagon other than the box or bed. 1924 F. R. Bechdolt *Tales of Old-Timers* 363 The boy was driving a span of horses hitched to the running-gear of a lumber-wagon. 1941 *Amer. Speech* XVI. 24/1 Of a skinny person. 'He's got the running-gears of a katydid.' 1948 E. N. Dick *Dixie Frontier* xix. 208 The driver..fastened a big deep box on the axle or the front wheels of a wagon running-gear. 1953 Randolph & Wilson *Down in Holler* viii. 184 When an airplane crashed and burned, my neighbor viewed the wreckage. 'It looks *like the runnin'-gears of a grasshopper*,' said he. 1972 J. S. Hall *Sayings from Old Smoky* 53 (Someone or something) 'looks like the runnin' gears of a crow's nest'.

b. The wheels, axles, and suspension of a railway locomotive, carriage, or wagon; the steering, suspension, and wheel systems of a motor vehicle.

1877 *11th Ann. Rep. Proc. Master Car-Builders Assoc.* 57 Probably every one of us who has experience in handling foreign cars can fully realize the importance of our draw bars and oil-boxes, and, in fact, of all the running gear of the car. 1889 *Nat. Car & Locomotive Builder* Mar. 35/1, I have always believed that the running gear of railway rolling stock should be constructed of such strength and with such intelligence as to give it a high factor of safety. 1900 *Motor World* 8 Nov. 100/1 The motor is located centrally of the vehicle, and is hung on the upper section of the running frame, which is spring supported from the main tubular running gear. 1905 *Motor Man.* (ed. 7) v. 85 Periodical cleaning and inspection of all the running gear is..the best possible insurance against breakdowns. 1919 Fraser & Jones *Motor Vehicles* xxvi. 270 The parts of a motor vehicle not included in developing and transmitting power are classified under the general heading of running gear. This includes such parts as frames, springs, axles, wheels, brakes, steering gear, etc. 1932 *New Yorker* 14 May 32/2 It was a club-sedan,..with red running gear. 1957 *Encycl. Brit.* XIV. 284/1 The mechanical parts include those portions of the locomotive which make it suitable as a vehicle, *i.e.*, the running gear, and the cab or superstructure. 1959 *Motor Man.* (ed. 36) v. 99 What is often referred to as the running gear comprises the steering system, the springing or suspension, the brakes and the wheels and tyres. 1969 *Northern Territory News* (Darwin) *Focus '69* 13/2 (Advt.), Semi-trailer tippers. Tough structural design, coupled with the best hoist and running gear available, assures down time reduced to periodic servicing.

runny (rʌ·ni), *a.* [f. Run *v.* + -y^1.] **a.** Tending to run or flow; having the consistency of liquid, fluid, not set; soft, melting, watery; (of eggs, etc.) soft-centred.

1817 *Niles' Reg.* XII. 165/2 This flour would prove similar to a previous baking of new flour (which was *runny*). 1904 *Eng. Dial. Dict.* V. 185/2 *Runny*,..inclined to liquefy. 1913 G. Stratton-Porter *Laddie* vii. 210 He slid in a whole plateful of bread, another of cake... Then we took some of every thing that wasn't too runny. 1935 M. Morphy *Recipes of All Nations* 73 A well-made purée should be almost 'runny'—only just sufficiently thick to be eaten with a fork. 1937 *Evening News* 23 Mar. 15/5 (Advt.), No runny butter or sour milk: no more waste. 1951 *Good Housek. Home Encycl.* 524/1 A 'runny' jelly is very difficult to manipulate. 1957 J. Frame *Owls do Cry* (1958) II. 103 Haven't I told you..to leave the egg till it's..hard, that I don't like them runny? 1972 K. Lo *Chinese Food* I. 43 Eggs are sometimes scrambled with minced meat, mushrooms, onions, lard and some good broth. The dish is called Runny Yellow Egg (*Liu Huang Ts'ai*), and is another good runny dish to eat with rice.

b. Of the nose: running, discharging mucus.

1951 J. Steinbeck *Log from 'Sea of Cortez'* p. xxxii, She was a red-eyed..woman with a runny nose. 1969 A. Lurie *Real People* (1970) 85 All he had produced so far was a slow pathetic drip and trickle, as if the fat marble cherub poised above the marble bowl had a bad runny nose.

Hence **ru·nny-no:sed** *a.*

1972 J. Wilson *Hide & Seek* ii. 32 A couple of runny-nosed children..tried to get on [the roundabout] too. 1976 M. Hartmann *Leap for Sun* i. 28 Big-arsed nannies squatting on the pavements surrounded by runny-nosed kids.

‖ **runo** (rū·no). Also **Runo**; pl. **runot, runos.** [Finnish: cf. Rune^2.] In Finland, a short poem or song on an epic or legendary subject; *spec.* one of the songs which together constitute the Kalevala, = Rune^2 2 b.

[1802 J. Acerbi *Travels* I. xxiii. 301 The species of verse is called *runic*, from the ancient Gothic word *runoot*. *Ibid.* 317 These songs, called *jauho runot* or *mill-songs*, are..sung to a slow plaintive air.] 1895 *Oracle Encycl.* II. 653/1 It was long known that there lived amongst the Finns a great number of lyrics known as *Runot* ('Runic songs'). *Ibid.*, The discovery that many of these *Runot* could be dovetailed into a true and noble epic, based on the old myths. 1898 I. M. Anderton tr. *Comparetti's Tradit. Poetry of Finns* I. i. 3 There is one word, *runo*, which characterises and distinguishes the traditional poetry of the Finns. 1944 W. Apel *Harvard Dict. Music.* 266/2 Next follow the *runos* (sung to the traditional poems of epic characters, called 'runes'), which are melodic and rhythmically vigorous. 1954 *Grove's Dict. Mus.* (ed. 5) III. 238/1 The ancient melodies of the period following are the beautiful *runo* tunes mentioned above. *Ibid.*, Sibelius..has a distinctive *runo* style of his own. 1963 B. J. Timmer tr. *De Vries's Heroic Song & Heroic Legend* vii. 143 Lönnrot was a folklorist, who set himself the task of collecting the older poetry of the Finns, in particular the epic songs or *runot*. 1974 *Encycl. Brit. Macropædia* VI. 910/2 Elias Lönnrot.., who composed this master-piece [*sc.* the *Kalevala*] by combining short popular songs (*runot*) collected by himself among the Finns, had absorbed his material so well, and identified himself so completely with the *runo* singers.

runo-. Add: **runolo·gical** *a.*, pertaining to runes or runology; **runologist** (earlier and later examples); **runology** (earlier example).

1962 *N. & Q.* Dec. 451/2 Values he could have taken from contemporary runological works. 1965 R. Derolez in Bessinger & Creed *Medieval & Linguistic Stud.* 31 No satisfactory explanation has been offered for this runological revolution. 1977 *Ann. Bibliogr. Eng. Lang. & Lit. 1974* 214 The runological placing of the Caistor-by-Norwich inscription. 1847 I. A. Blackwell in M. Mallet *Northern Antiquities* (ed. 2) i. 247 These celebrated Runologists..arrived at the same interpretation of the characters. 1971 S. E. Morison *European Discovery Amer.: Northern Voy.* iii. 76 Every leading runologist of Scandinavia and Germany who has deigned to examine the inscription has called it a clumsy forgery. 1862 J. Farrer *Notice Runic Inscr. Orkneys* p. ix, My very imperfect acquaintance with Runology.

run-off. Add: Also **runoff.** Pl. **run-offs.**
1. a. Delete *U.S.* and add: The amount of water that is carried off an area by streams and rivers after having fallen as precipitation; the water itself; also, water that runs straight off the ground without first soaking into it.

1910 *Westm. Gaz.* 19 Feb. 2/2 Iron ore, which stains all the other streams..and the run-off of the night's rain to the colour of tanyard. 1929 Weaver & Clements *Plant Ecol.* ix. 190 It [*sc.* rainfall] may be of such a torrential nature that only part of it can be absorbed and the rest is lost as run-off. 1957 G. E. Hutchinson *Treat. Limnol.* I. iv. 229 A number of independent estimates of total runoff have been made by attempting to sum the rate of discharge of the rivers of the world. 1959 *Listener* 10 Sept. 378/2 The run-off from the winter snowfields of the Australian Alps. 1969 *Physics Bull.* Oct. 410/1 Calculated by subtracting runoff from rainfall, annual evaporation ranges from about 16 in per year in the north of England to 19 in per year in the Thames valley. 1970 T. Hillerman *Blessing Way* xiii. 109 He stopped at a pool where runoff had been trapped in a pocket of rocks. 1978 J. Irving *World according to Garp* iv. 81 The runoff from the rain—washing over the Steering School, rinsing everything clean.

b. The process or fact of water, or what the water contains, running off from an area; an instance of this; (*N. Amer.*) the period when such a process occurs, esp. the spring thaw.

1935 *Discovery* Aug. 219/2 The rapid run-off causes disastrous floods at lower levels. 1944 F. Clune *Red Heart* 6 West of the Darling..the thirsty red soil soaks every particle of rain that falls, with no surplus for a run-off. 1949 W. Vogt *Road to Survival* v. 104 The rate of runoff can be reduced on even the steepest land. 1962 W. O. Mitchell *Kite* iii. 25 As soon as the sky is blue and the run-off starts—down town every Saturday morning. 1972 *Times* 26 June 12/2 Sewage and run-off of fertilizer from agricultural land is eliminating desirable fish. 1980 *Beautiful British Columbia* Summer 4 This bucolic alternative becomes impossible when the snow flies and during spring runoff, as Duffey Lake Road then becomes impassable.

c. *attrib.*, as *run-off map, rate, water.*

1892–3 Run-off map [in Dict., sense 1]. 1937 *Sun* (Baltimore) 1 July 24/3 Slowing up the run-off rate of heavy rainfall. 1939 C. W. Towne *Her Majesty Montana* 95 Our farmers are fortified by thousands of dams for the storage of run-off waters for livestock needs and to supply irrigation for feed and hay crops. 1979 A. Hailey *Overload* III. viii. 229 Hydroelectric power next year might be reduced by twenty-five percent because of the lack of runoff water.

2. a. (Earlier and later examples.) Also in other kinds of contest.

1873 *Carthusian* June 56 Hanson and Jeaffreson ran a dead heat for second place... The run-off for the second prize was won by Hanson. 1963 H. Garner in R. Weaver *Canad. Short Stories* (1968) 2nd Ser. 49 'You tryin' out for the bowling team, Eric?' he asked. 'Sure thing. You?' 'May as well. Run-offs are on Thursday night.' 1973 *Shooting Times & Country Mag.* 7 July 19/3 If the judges have difficulty in arriving at a result, they can arrange a more difficult 'run-off' for the top dogs.

b. Chiefly *U.S.* An election held to decide the issue between the two candidates who gained the largest number of votes in a previous indecisive election. Freq. *attrib.* or as *adj.*, esp. in *run-off primary* (see Primary *sb.* 6 in Dict. and Suppl.).

1924 *Lit. Digest* 6 Sept. 8/2 Texas..has a double primary. If no one has a majority in the first primary election, a later 'run-off' primary is held, in which the voters choose between the two candidates receiving the highest number of votes at the first balloting. 1933 *Sun* (Baltimore) 15 Sept. 3/4 A. H. Carmichael, of Tuscumbia, and B. L. Malone, of Decatur, will face each other in a run-off primary October 3 for the Democratic nomination to Congress from the Eighth Alabama district. 1944 *Ibid.* 31 May 7/1 (*heading*) Alabama run-offs watched for clue to group's strength. 1954 *Economist* 31 July 365/2 In Oklahoma Senator Kerr failed to gain the necessary clear majority of the votes cast, and was threatened with having to fight a 'run-off' election against a rival millionaire, Mr Roy Turner. 1959 B. & R. North tr. *M. Duverger's Polit. Parties* (ed. 2) II. i. 220 In primaries in the South where the nomination is conducted at a single ballot the Democratic party generally divides into two factions;..in the system with two successive primaries.. the second or run-off primary operating in the event of no candidate securing an absolute majority at the first primary—the factions tend to increase in number. 1965 *N.Y. Times* 7 Dec. 4/2 (*heading*) De Gaulle silent on whether he will enter runoff. 1966 Mrs. L. B. Johnson *White House Diary* 4 June (1970) 386 Today is the Democratic primary runoff and we had come home to vote. 1968 *Listener* 27 June 826/1 If General de Gaulle's victory at the polls is confirmed by the run-off vote next Sunday, he may well see this as a mandate for his projected social reforms. 1973 *Times* 31 Mar. 5/4 His nearest rival, Señor Ricardo Balbin, announced he would not contest a runoff. 1977 *Time* 21 Nov. 29/1 The gloves are expected to come off when Briscoe faces former City Councilman Jim McConn, a Houston developer, in a run-off next week.

3. a. The action or process of running off a person or thing in other senses of the vb.; a quantity run off or removed; *spec.* the material run off or produced by a mechanical process; a print run.

1843 J. H. Green *Exposure of Gambling* 96 The adversary, fearing that his hand is really the better hand, will, in preference to risking more, throw up his own hand, and forfeit what has already been bet. This is a run-off, as well as in cases where he has no money enough to meet the proposed bet. a 1948 L. G. D. Acland *Early Canterbury Runs* (1951) 393 *Run off*,..sheep counted out from a mob without being drafted; e.g., 'I cannot buy the whole line. I will take a run off of 300.' Usually *fair r.o.* 1952 *Bull. Poetry Soc. Amer.* Oct. 4 An impromptu and rather novel program was offered: a run-off of a phonograph recording of a broadcast made in 1948 of poems from the PSA *Anthology*. 1967 A. L. Lloyd *Folk Song in England* i. 27 Several of these [broadsides] were produced in massive run-offs. 1972 *Times* 30 Oct. 19/4 Insurance companies are seldom capable of an accurate assessment of the outstanding liabilities on their current portfolio of business (a 'run-off'). 1979 *Church Times* 26 Oct. 7/3 The low price for such a well-produced volume must indicate a large run-off.

b. An instance of running off a railway or road.

1855 *Chicago Western Times* 9 Aug. 1/8 The frequency of these run-offs demands the special attention of all railroad directors. 1872 W. S. Huntington *Road-Master's Assistant* 87 It is best always to keep spare [switch] rods on hand, to be used in case of a run-off. 1970 *Courier-Mail* (Brisbane) 19 Mar. 6/2 An invitation.. to the Transport Minister..to discuss whether runoff roads could be a means of preventing semi-trailer accidents in the Adelaide hills.

4. The dropping of fruit before it is ready for picking; = *running off* s.v. *running *vbl. sb.* 7 e.

1921 *Jrnl. Pomol.* II. 170 If frost caused the 'Run off' on half the bush, why did it not do so on the other? 1974 *Daily Tel.* 6 June 6/4 A grower at Cropthorne, Worcestershire, said the level of fruit 'run-off'—unformed fruit shrivelling and falling from trees—was much higher than usual through lack of moisture.

5. *N.Z.* Also *run-off paddock*. (See quot. 1933.)

1933 *Press* (Christchurch, N.Z.) 25 Nov., *Run off*, paddock used with turnips or green feed on which the sheep may camp and get what extra feed they can. 1950 *N.Z. Jrnl. Agric.* Apr. 366/3 Utilisation of the heavier pockets [of coastal land] and sand dunes is complementary, the latter being used mainly as a winter run-off. *Ibid.* 389/1 During early winter they [*sc.* ewes] are rationed turnips, fed good hay, and driven off their turnip break on to a large run-off paddock daily.

6. *slang.* An act of urination.

1961 Partridge *Dict. Slang* Suppl. 1255/2 *Run off, have a*, to urinate. 1967 H. W. Sutherland *Magnie* ix. 117 What with the cold and the beer she was bursting for a run off again... The nearest ladies she knew was at Pier Head.

ru·n-out. Also **run out, runout.** [f. Run *v.*]
1. *Founding.* † **a.** (See Run *v.* 82.) *Obs.*

b. Leakage of molten metal from a cupola or a mould.

1888 *Lockwood's Dict. Mech. Engin.* 296 *Run out*, the escape of metal from a mould during the act of pouring, due to open joints somewhere. 1901 *Shop & Foundry Practice* (Colliery Engineer Co.) IV. xxxv. 20 If there is any breaking of joints by drawing the pattern or from a straining of the cope, allowing large fins or a run out, the metal will run into the vent channels..and fill them with iron. 1928 *Proc. Inst. Brit. Foundrymen* XX. 366 In Fig. 1—an ordinary scullery copper—it is quite obvious that if there be a runout there is little chance of saving the casting. 1960 R. Lister *Decorative Cast Ironwork in Gt. Brit.* ii. 56 Possible defects in castings are manifold, and may be in the form of blow holes, unfused chaplets, wrong grain-structure, fractures, distortions, runouts, [etc.].

2. *Cricket.* (See Run *sb.*^{1} 8.)

1851 *Bell's Life* 21 Sept. 6/5 (*heading*) The 'runs out' and 'runs lost' at cricket. 1867 G. H. Selkirk *Guide to Cricket Ground* vii. 122 Never run past the wicket, unless to save a run out, when you can stop yourself. 1891, 1892 [see Run *sb.*^{1} 8]. 1930 *Morning Post* 16 July 11/6 Only once..during their many long partnerships has a run-out been recorded. 1950 W. Hammond *Cricketers' School* xv. 140 He broke the wicket from 30 yards away in one of the most startling run-outs I have ever seen. 1976 *Milton Keynes Express* 2 July 41/6 If it had not been for three run-outs they might have got nearer their mammoth target.

3. *Mountaineering.* The length of rope required to climb a single pitch; also *transf.*, a pitch climbed by means of a single length of rope.

1920 [see *belaying *vbl. sb.* 2]. 1965 A. Blackshaw *Mountaineering* vii. 195 There are exceptions such as the routes on the Idwal Slabs in North Wales where many pitches involve run-outs of eighty feet or more. 1971 D. Haston in C. Bonington *Annapurna South Face* xvii. 206 It was a long and tortuous pitch done in one run-out on one of our big ropes. *Ibid.* 214 My immediate prospect was a three hundred foot run-out to the top of the gully. 1972 —— *In High Places* ii. 28 It's slightly awkward to do full run-outs with three people on the rope.

4. An act or instance of running out, fleeing, or escaping; also *attrib.*, esp. in U.S. slang phr. *to take a run-out powder*, to withdraw; to leave, abscond; cf. *powder *sb.*^{1} 2 h.

1920 *Our Navy* Aug. 33/1 The 'Wilmington' challenged us to a boat race, but when we slapped up a sack of good Chinese taels to back our team the 'Wily Willie' took a run-out powder and called off the race. 1928 *Amer. Mercury* May 80/1 The fair charmer has taken 'a run-out with the bank roll.' 1933 D. Runyon in *Collier's* 28 Jan. 7/4 Well, The Sky says he sees no way of meeting these obligations and he is figuring the only thing he can do is to take a run-out powder. 1943 *Richmond* (Va.) *Times-Dispatch* 23 Aug. 2/1 (*heading*) Kiska forces disappointed by run-out. 1952 Pohl & Kornblath in *Galaxy Sci. Fiction* July 147/2 You crossed us up with that cowardly run-out. a 1953 E. O'Neill *Hughie* (1959) 14, I stuck it till I was eighteen before I took a run-out powder. 1968 'E. Peters' *Grass Widow's Tale* viii. 114 They came back for their money, just when she had everything planned for her run-out.

5. A mock auction. Cf. Mock *a.* 2. Usu. with def. article or *attrib.*

1934 P. Allingham *Cheapjack* vii. 72 The London Mob were working the R.O. This is short for the 'Run Out'. 1938 F. D. Sharpe *Sharpe of Flying Squad* xv. 171 'Run out' shops disposing of valueless jewellery worked in conjunction with the pickpockets. *Ibid.* xxvii. 280 Run Out Shops have given me a good deal of work. Many of the methods of the Run Out Mob and their premises on which they hold their mock auctions of worthless junk range from small and cheap set-ups on the race-courses..to elaborate dens of swindle in the West End. 1939 J. B. Priestley *Let People Sing* x. 258 He told me his old partner, Charlie, had left him to join the run-out boys from Brum—that is, the gang from Birmingham running a fake auction—and, by the way, if anybody wants to see me lose my temper just let 'em talk as if I was on the run-out game. 1943 *Police Jrnl.* Mar. 69 *Run-out mob*, a gang that conducts mock auctions. 1959 *News Chron.* 16 Nov. 5 The run-out men..are mock auctioneers who draw large crowds with their showman's patter... As well as in Petticoat Lane, they operate in many of the seaside towns during the summer.

6. *Engin.* Deviation of a wheel, drill, etc., from its proper course; the extent of this.

1946 W. H. Crouse *Automotive Mech.* xxiv. 512 Wobble or 'run-out' of the wheels can be checked by spinning the front wheels and holding a piece of chalk against the rim or side wall of the tire. 1951 C. W. Kennedy *Inspection & Gaging* ii. 28 Common terms..are squareness, parallelism, waviness,..eccentric, run-out and out-of-line. 1975 Bram & Downs *Manuf. Technol.* vii. 194 An important point of design is to reduce end thrust, to prevent bowing and flexing, producing as a consequence hole run-out and short drill-life. 1977 *Hot Car* Oct. 58/1 Subsequent to machining the wheels undergo a rigorous testing programme checking for run-out and wobble.

7. *Skiing.* = *outrun *sb.* 4.

1956 *Ski-ing* ('Know the Game' Series) 13/2 These [nursery slopes] should be of average steepness at the top, have a gentle gradient in the middle and a long flat run out so that the ski will come to a stop naturally if the skier is unable to control them. 1974 *Rules of Game* 239/1 There must be a wide, gently sloping, and unobstructed run-out at the finish [of a ski-slope].

8. On a gramophone record, (the blank groove traversing) the annular area between the label and the grooves carrying the recording. Freq. *attrib.* as *run-out groove*.

1962 A. Nisbett *Technique Sound Studio* viii. 150 If the surface noise on the run in and run out of the two records is not sufficient..a 'blank'..record can be used to lend continuity of background. 1975 *Gramophone* May 2048/3 A new TD 145 turntable..has an automatic lift and shut-off device relying on electronic sensing of the run-out groove. 1976 *Ibid.* Nov. 768/2 What puzzles me is that both these discs do in fact bear the re-make matrix numbers on both label and needle run-out.

9. Special Comb.: **run-out table** (see quot. 1948).

1948 T. Lyman *Metals Handbk.* 12/2 *Runout table*, in a rolling mill a plane area at the receiving end, for holding rolled metal. 1973 J. G. Tweeddale *Materials Technol.* II. 99 Most commonly, a hot semicontinuous extrusion press..is placed horizontally so that the extruded product can be discharged straight on to a 'run-out' table and never has to carry its own weight in tension whilst in its weakened hot state.

ru·n-over. Also **runover.** [f. Run *v.*] **1.** (See Run *sb.*^{1} 8.) (Further examples.) Also, an instance of overrunning a time limit.

1937 *Printers' Ink Monthly* May 42/1 *Runovers*, occasions when the [radio] program itself overruns its allotted time. 1947 J. Bertram *Shadow of War* vi. 203 The shifts of the P.O.W. in stowing his loot, and..the *amount* that can be carried by one man on his own person, even through a 'run-over' and a 'strip-search'. 1963 *Times* 27 Sept. 12/4 Lord Home, the Foreign Secretary, today met Mr. Dean Rusk, the American Secretary of State, for nearly three hours and had what was called 'a very full runover' of matters of mutual concern.

2. In *Printing*, etc., the action or an instance of continuing matter into a margin, or on to a subsequent line or page.

1934 Webster, *Run-over*, an extension, as of printed matter, beyond the space allotted; overmatter; also, U.S., the part of an article continued from a preceding page. 1956 N. R. Ker *Pastoral Care* 21 The runover to avoid breaking a word at the end of the last line on fos. 49v, 50v, 68r, marked by a dot at the end of the line and another before the runover itself. 1969 in Halpert & Story *Christmas Mumming in Newfoundland* 192 The placing of the speech designations has been regularized, the run-over of lines in narrow newspaper columns abandoned, [etc.]. 1976 H. MacInnes *Agent in Place* ix. 96 The typescript finished each line neatly—no runovers onto the right-hand margin.

runt, *sb.* Add: **1. b.** (Further example.)

1928 R. Campbell *Wayzgoose* ii. 58 The Sacred Carrot with the golden rind, Whose magic runt..The more one nibbled it, the larger grew.

2. d. (Further examples.) In *gen.* use, a small pig that is weakly or undernourished.

1939 V. C. Fishwick *Pigs* I. i. 19 Such piglings grow well and are a sound proposition. They are not regarded as 'runts', a term which is here used to describe a pig that is in poor condition. 1939 *Nature* 23 Sept. 552/2 Radiographic examination of a 'runt', the small starveling pig, shows evidence of arrested growth in the skeleton. 1977 P. R. English et al. *Sow* viii. 163 One can have a litter in which most piglets are thriving well but in which one or two are obviously suffering from malnutrition and are in danger of becoming nutritional 'runts'.

3. c. (Further examples.) Also used in weakened sense as a term of abuse.

1896 *Dialect Notes* I. 423 *Runt*, worthless fellow. (Cowboys.) c 1926 'Mixer' *Transport Workers' Song Bk.* 52 As a slimy runt, I'm it! 1930 *Amer. Speech* V. 119 The expression 'little runt' was merely contemptuous. 1936 *Nat. Geogr. Mag.* June 787/2 A cadet's height determines his assignment, the tall men going to the 'flanker' companies, A and M, the 'runts' to companies F and G in the center. 1956 J. Cannon *Who struck John?* 228 You're Conn McCreary, a fat runt. You're tiny. 1958 *Chicago Tribune* 9 Feb. (Comics Feature Mag.) 9 'Anyhow, who'd ever think of connecting that runt with *this* deal?' 'Maybe you're right! But just the same I'm going to keep my beady eye on young Mister Dondi.' 1969 I. & P. Opie *Children's Games* viii. 234 'Come on you miserable runts,' we say, and one of the runts runs up the steps, only to be thrown over the side of the railing.

d. (Earlier and later examples.)

1819 M. Wilmot *Let.* 8 Dec. (1935) 32 What think you of my *deeply* regretting not having brought a white Tabinet gown. 'Tis admired here beyond satin, and my old *runt* has in consequence been jinkumbobbed out of Mamselle and white satin till the poor dear old dress..is become the most admired thing I have. 1973 *Amer. Speech 1969* XLIV. 249 In official terminology, ground signals are referred to as *dwarf* signals, but railroad men call them *pots*, because they are round and silver, or *runts*, because they are small in comparison with the signals on bridges that span the tracks.

runted, *a.* Delete '*Obs.* exc. *dial.*' and add later examples.

1951 E. Paul *Springtime in Paris* iii. 50 The myriads of chimney pots and vents are like runted gargoyles which look up toward scudding clouds and mackerel skies instead of downward. 1961 R. P. Hobson *Rancher takes Wife* ii. 38 Two runted pearl-gray kittens purred and rubbed themselves against Rich's legs. 1976 T. Heald *Let Sleeping Dogs Die* v. 87 Runted little dogs hers are. No wind and dreadful temperaments.

ru·n-through. Also **runthrough.** Pl. **run-throughs, runs-through.** [f. Run *v.*] **1.** A (freq. hasty or cursory) rehearsal of a play, a radio or television programme, etc. Also *gen.* a performance or showing (of a play, film, etc.), esp. a preview.

1923 Wodehouse *Inimitable Jeeves* x. 104 It must have been about a week after this rummy little episode that George Caffyn called me up and asked me if I would care to go and see a run-through of his show. 1930 J. Dos Passos *42nd Parallel* iv. 287 She felt it in her bones that the play would be a hit..and Mr Freelby said Ike Gold.. had sat through the runthrough with the tears running down his cheeks. 1946 *Life* 2 Dec. 51 The director, handsomely played by Sam Wanamaker, talks of his theatrical troubles until the star, Mary Grey, appears. There begins a run-through of a play about Joan of Arc, with Mary in the lead. 1952 S. Kauffmann *Philanderer* (1953) xiii. 217 She had to go back for the final run-through. 1957 Duncan & Bone *Oxf. Pkt. Bk. Athletic Training* (ed. 2) v. 71 Minor adjustments being made before competition after practice runs-through. 1959 *Times* 10 July 9/1 The two gentlemen were felicitously played at this public *première* (as at the private run-through half a year ago) by Mr. Stephen Manton and Mr. Bruce Boyce. 1963 *Ann. Reg. 1962* 421 *Christopher Sly*, a chamber opera by Thomas Eastwood, heard in Britain only in a public run-through in 1960, was given its first staging at Pforzheim. 1973 *Times* 17 Oct. 11/2, I.. immediately asked about rehearsals. I was told there would be run-throughs with a full orchestra but without the chorus and other principals. 1973 E. Lemarchand *Let or Hindrance* xiv. 177 The film..has been very recently edited... Like a run-through of this last part? 1977 S. Brett *Star Trap* v. 53 The..cast assembled for a pre-tour run-through.

2. A brief survey (of facts); a summary, a concise account.

1947 *Sun* (Baltimore) 20 Mar. 1/6 The Big Four conference which, at the conclusion of the ninth session tonight, had accomplished little more than a run-through of German issues on which the Council is split. 1957 [see *fado]. 1963 *Listener* 7 Mar. 432/2, I gave a group of young Russians a quick run-through of Eng. Lit. from Hopkins and Hardy to Amis and Osborne. 1973 A. Hunter *Gently French* iv. 36 Tell me about Quarles... Give me a quick run through.

3. The fact or an instance of running trains through intermediate points without stopping for crew changing, loading, etc.

1964 *Canad. Labour* Dec. 31/3 Representatives of the non-operating railway unions met in Montreal with officials of the running trades to work out a common policy on rail abandonments, runthroughs and other technological changes. 1967 *Canad. Ann. Rev. 1966* 29 The implementation of the Freedman report on railway run-throughs. 1969 *Jane's Freight Containers 1968–69* 125/3 To improve interline service, the Frisco is co-operating with connecting rail carriers to pre-block traffic in run-through trains—avoiding intermediate terminal switching operations.

runting (rʌ·ntiŋ), *vbl. sb.* [f. Runt *sb.* + -ing^1.] **1.** The birth or development of (laboratory) animals that are small for their kind.

1959 *Jrnl. Exper. Med.* CX. 522 Runting is the result of an immunological reaction of foreign cells against a tolerant host. 1974 *Nature* 11 Oct. 548/2 In some litters all rats grew normally and there was no runting. 1978 *Ibid.* 27 July 365/2 In the colony described here there has been no evidence of the runting syndrome frequently seen in conventional nude mice.

2. (See quot.) *slang.*

1976 *Drive* July–Aug. 37/2 In the [ice-cream] trade, giving kids small portions when no parents are about is called 'runting'.

runtish, *a.* Add: **1.** Also, of human beings.

1969 N. Cohn *AWopBopaLooBop* (1970) x. 89 He grew up small, runtish, with bad hair and unhealthy skin. 1974 'R. Tate' *Birds of Bloodied Feather* ii. 30 The child.. was about thirteen, runtish and not very clean.

runty, *a.* Add: **1.** (Further examples.) Also *Comb.*

1848 in Farmer & Henley *Slang* (1903) VI. 84/2 'No indeed,' ses another little runty-lookin' feller—we've got enuff to do to take care of our own babys in these diggins. 1903 J. London *People of Abyss* xii. 143 We cannot understand the starved and runty toiler of the East End.. till we look at the strapping Life Guardsmen of the West End. 1938 M. K. Rawlings *Yearling* xiv. 156 Us Baxters is all runty and tough. 1972 *New Yorker* 22 Jan. 100/2 A young English mother..gets herself sexually awakened by a runty Italian actor... If Miranda were more intelligent and Oreste less swinish, their obsessive affair might touch us deeply. 1974 *Sci. Amer.* Nov. 61/1 The treated birds were runty and deficient in lymphocytes; all cell-mediated immune functions were suppressed.

run-up. Add: Also **run up, runup. 1. b.** (Further examples.) Also *attrib.* Chiefly in *Golf.*

1907 'I. Hay' *Pip* III. xi. 339 Anything in the shape of a run-up ball would be trapped. 1931 *Daily Express* 31 Jan. 9/5 Compston, playing a run-up shot to the first green, shouted after the ball, 'Hit the stick.' 1955 *Times* 2 May 4/1 On most of Friday he putted well, and his run-ups were often in the highest class. 1963 *Times* 14 Jan. 3/7 Agate won back the 13th, where he played a run-up to the hole.

c. A run made in preparation for jumping, throwing, etc., in *Athletics*; in *Cricket*, the bowler's approach to the bowling crease before delivery.

1897 *Encycl. Sport* I. 52/2 Pace in the run-up supplies the impetus; spring enables the jumper to lift himself into the air. 1919 F. A. M. Webster et al. *Success in Athletics* x. 83 The last stride must be a short one, so that the jumping leg may be gathered well under the body for the spring. This accounts for the run-up, which must be most assiduously practised. 1929 G. M. Butler *Mod. Athletics* ix. 120 *The run-up.* The first essential of successful long jumping is speed in the approach. *Ibid.* 122 Should the run-up be inaccurate, there will be a loss of at least a foot. 1948 K. S. Duncan *Oxf. Pkt. Bk. Athletic Training* iv. 62 Practise and standardize the run-up, cross step and throwing stance. 1959 *Times* 17 June 6/6 Horner, with his upright stance and utter immobility during the bowler's run-up. *Ibid.* 24 Aug. 4/2 She was handicapped..in the long jump by a crumbling run-up. 1966 B. Johnston *Armchair Cricket 1966* 97 If the batsmen attempt to steal a run during the bowler's run-up [etc.]. 1974 *Rules of Game* 18/3 The parallel lines may be crossed during run-up, but the competitor must be between them when the javelin is released. 1977 J. Laker *One-Day Cricket* 48 The length of a bowler's run-up is limited to 15 yards.

d. = *run-in 3.

1942 *R.A.F. Jrnl.* 27 June 8 Another Stirling and a Wellington adopted almost identically the same run-up as ourselves. 1958 'N. Shute' *Rainbow & Rose* ii. 43, I went up again and circled round... 'I'm going to do a dummy run.'.. I took a longer run-up this time. 1976 'G. Black' *Moon for Killers* vii. 99 A small, single-prop job was coming in for a landing..its turn completed, the run-up going to be towards us.

e. A period of time or series of occurrences leading up *to* some important (freq. political) event; an action which prepares the way for one on a larger scale.

1966 *Sunday Times* 20 Nov. 48/2 The Petit Palais show offers, also, invaluable evidence in its drawing section of the ways in which Picasso manoeuvred during the crucial run-up to the 'Demoiselles d'Avignon'. 1968 *Listener* 5 Dec. 761/1 The run-up to the election of Oxford's new Poetry Professor has aroused a good deal of mirthful interest. 1970 *Daily Tel.* 2 Jan. 14 The 1970s open, with the approach and run-up to the most critical General Election in a generation. 1975 M. Kenyon *Mr. Big.* v. 47 Heathrow was the run-up to the train job because there had to be capital. 1976 *Nature* 29 July 344/2 Remaining hitches in reactor technology can, it is argued, be straightened out during the 20-year run-up to a commercial FBR network. 1977 *Film & Television Technician* Jan. 1/2 During the run-up to the overtime ban..the Trade Press was uniformly critical of the employers.

3. On the *U.S. Stock Market*, a rapid increase in the price or value of a commodity. Now also in *gen.* use.

1935 *Sun* (Baltimore) 13 Apr. 17/8 Corn advanced to 1 to 1½ cents a bushel, but cotton was reactionary after Thursday's run-up. 1942 *Ibid.* 1 Oct. 21 Laclede gas preferred had a runup of 6¼ points. 1953 *Ibid.* 30 Oct. B-28/1 A fast runup in the final dealings gave the stock market one of its sharpest boosts of the year yesterday. 1958 *Wall St. Jrnl.* 3 Dec. 27/4 Referring to the recent sharp run-up and activity in Walworth Co. stock, amid talk of merger possibilities, Fred W. Belz, president, said [etc.]. 1976 *National Observer* (U.S.) 6 Mar. 4/3 The price scare and runup in the futures prices in recent weeks 'says to me that farmers deep down inside know that this crop is not lost'. 1978 *Daily Tel.* 15 Sept. 21/4 Money market analysts have been watching a rapid run-up in short-term, negotiable bank certificates of deposit and commercial paper traded by corporations. 1979 *Time* 13 Aug. 26/3 The industry most severely dented by the oil run-up is auto manufacturing.

4. The running of a motor or mechanical device until it attains normal working or speed; warming up.

1943 *Yank* 26 Feb. 6 Last summer the intensive heat raised hell with us AMs in making engine run-ups for regulation check on props, [etc.]. 1946 *Happy Landings* July 3/1 Correct use of air filters and observing precautions against dust during run up are matters for the pilot to remember. 1958 'Castle' & 'Hailey' *Flight into Danger* i. 16 In the run-up each engine in turn is opened to full throttle and each of the mags tested separately. 1959 W. S. Sharps *Dict. Cinematogr.* 126/2 *Run up*, the term given to the passage of film or a magnetic recording medium through a camera or recorder before the correct recording speed is obtained; or through a projector or other machine before the first subject image or sound is reached. 1962 A. Nisbett *Technique Sound Studio* viii. 139 The reproducer may take as much as a second or so to run up... Check this run-up time by using a recording of pure tone.

runway. For 'Chiefly *U.S.*' read 'orig. *U.S.*' and add: Also **run-way. 1. a.** (Earlier and later examples.)

1833 C. F. Hoffman *Winter in West* (1835) I. 202 The numerous deer-runways,..and innumerable tracks of rackoons, wolves, and bears, showed us that we were upon a favourite hunting ground of the Pottawattamies. 1855 *Knickerbocker* XLV. 193 The sound of the rifle has by this time brought the other hunters from their runways. 1908 A. J. Dawson *Finn* x. 168 Reynard picked up the dead rabbit and..trotted leisurely down the runway towards his own earth. 1948 A. L. Rand *Mammals Eastern Rockies* 45 Four of them [*sc.* species of shrews] are terrestrial animals, often making little runways through the moss. 1953 P. Provencher *I live in Woods* xii. 117 In winter the rabbit runway is easily distinguished in the snow. 1977 *Sci. Amer.* May 106/2 Rats in the laboratory have logged endless miles in runways, mazes and activity wheels.

b. (Earlier and later examples.)

1871 W. M. Lewis *People's Pract. Poultry Bk.* 8 The hennery should be placed in a warm, dry location..with runways ample to allow of plenty of exercise. 1913 J. London *Valley of Moon* xiv. 459 A goodly portion was devoted to white-washed henhouses and wired runways wherein hundreds of chickens were to be seen. 1949 *Sat. Even. Post* 9 Apr. 59/2 He even had a plan for one in his desk; the runways were to be painted green outside and whitewashed inside.

2. a. (Earlier and later examples.) *spec.* in *Theatr.* use (see quot. 1926); also in *Fashion*, a raised gangway on which models parade when exhibiting clothes.

1883 E. W. Howe *Country Town* iv. 20 Pushing this into my wagon with the assistance of his wife, after we had first made a run-way of boards, I hauled him to Fairview. 1901 Merwin & Webster *Calumet 'K'* xiii. 246 A runway from the hoist to the end of the building. 1912 'W. Lawton' *Boy Aviators' Flight for Fortune* ix. 102 The rolling glide down the runway was made..and at last the bow of the *Sea Eagle*'s hull struck the water. 1926 *Amer Speech* I. 437/2 *Runway*, a platform built at stage level and extending part way down the center aisle of the theater. Used in burlesque and musical comedy for the presentation of chorus numbers. 1929 *Variety* 11 Sept. 54 Muggs still going for burlesque want that close-up of flesh which the runway provides or they won't give the teasers a tumble. 1941 B. Schulberg *What makes Sammy Run?* xii. 287 The solid glass desk looked like a burlesque runway. 1961 *Sunday Times* 3 Dec. 29/2 Advertising is the next fattener of bank accounts, then fashion photography and finally live modelling on the runways at fashion shows. 1971 C. Fick *Danziger Transcript* 95 Fourteen skinny models rehearsing on a runway. 1979 *Tucson Mag.* Mar. 12/3 Producer Spots Baxter's lifeless body stretched out on the runway of his great stage.

b. A specially prepared surface on an airfield for the taking off and landing of aircraft.

1923 *Aviation* 8 Oct. 445/1 (*caption*) A wonderful landmark—Boston Airport with its T type runways. 1926 *Nat. Geogr. Mag.* Sept. 363/2 We got off the end of the runway at a terrific speed. 1930 *Flight* 7 Feb. 194/1 To make it usable for aircraft, cinders have been laid to form three runways and these are raised above the rest of the unprepared ground. 1943 *Times* 16 Dec. 3/4 In northern Burma the U.S. Air Forces destroyed runways and dumps and burned barrack areas. 1957 *Economist* 21 Sept. 922/2 The sprawling Tachikawa airfield on the northern fringes of Tokyo is regularly the scene of 'anti-base' demonstrations by Japanese who uproot boundary fences, plant flagpoles on the runways and skirmish with the police. 1977 *Whitaker's Almanack 1978* 572 There was an official death toll of 576 when two Boeing 747 jumbo jets ..collided on the runway at Santa Cruz airport.

4. (Example.)

1874 B. F. Taylor *World on Wheels* II. vii. 250 Like the dusty 'run-ways' of thy brooks, soft pulses have grown dry and dumb.

5. *attrib.* and *Comb.*, as (sense *2 b) *runway aerodrome, marker, strip*; **runway light,** each of a series of lights marking the course of a runway.

1933 *Jrnl. R. Aeronaut. Soc.* XXXVII. 3 Detroit, where a purely runway aerodrome has been developed. 1951 *Gloss. Aeronaut. Terms* (*B.S.I.*) III. 24 *Runway lights*, lights defining a runway to indicate the area of taking-off and landing. 1958 'Castle' & 'Hailey' *Flight into Danger* xi. 154 Put out your runway lights, except zero-eight. 1976 'A. Hall' *Kobra Manifesto* xv. 200 The flick-flick-flick of the runway lights, falling away. 1939 *Air Ann. Brit. Empire* 65 A new metal runway marker has been introduced in order to improve the safety of aeroplane landings. 1937 *Sun* (Baltimore) 21 Apr. 9/6 The field should be equipped with boundary and beacon lights and the area of the runway strips outside of the paved portion should be sod.

Hence **ru·nwayed** *a.*, provided with a runway; consisting of runways.

1948 *Hansard Commons* 15 Mar. 1806 Two-fifths of a runwayed airfield is covered with concrete runways. 1949 *Jrnl. R. Aeronaut. Soc.* LIII. 903/1 Large grass areas on runwayed aerodromes are a liability in upkeep, except insofar as grass drying is a revenue-earning aspect.

Runyonesque (rʌnyənε·sk), *a.* [f. the name *Runyon* (see below) + -esque.] Characteristic of or resembling Alfred Damon *Runyon* (1884–1946), U.S. journalist and author, or his writings. Also **Runyonese** (-ī·z), slang or underworld jargon characteristic or suggestive of that used in the short stories of Runyon.

1938 *Times Lit. Suppl.* 22 Jan. 59/2 A good many people appear to think that 'runyonese'..is not only vulgar but also pernicious. 1938 *N.Y. Times* 28 Feb. 19/1 For a Runyonesque panel the casting director had the marvelous good fortune to find Edward G. Robinson and Ruth Donnelly to play Mr. and Mrs. 1950 *Observer* 22 Oct. 7/4 It is hard to comment on the quality of translation without seeing the original, and Mr. Maclaren-Ross had had to cope with that appallingly difficult problem of specialised slang. Mr. Maclaren-Ross's solution is Runyonesque, probably the most suitable, but still inevitably tending to denationalise the book. 1953 P. Bonner *SPQR* xvii. 150 Runyonese..is the patois which best illustrates the swath which this character is cutting. 1955 *Time* 6 June 109/1 Ruth was singing in obscure Chicago nightclubs when she first encountered a Runyonesque character who called himself Colonel Martin Snyder. 1964 *Amer. Speech* XXXIX. 304 He gives details of murders, gang wars, robberies, and the activities and special talents of such Runyonesque members of Costa Nostra as Vincent 'Jimmy Blue Eyes' Alo, [etc.]. 1964 E. P. Hoyt *Gentleman of Broadway* xvi. 249 Librarians were surprised that Runyon continued to be popular long after his style of slang and much of the language which was called 'Runyonese' had been thoroughly absorbed in American English. 1980 *Jewish Chron.* 4 Jan. 15/5 Its characters are a little self-consciously drawn, notably a pair of Runyonesque policemen.

rupestral, *a.* Add: Also as *sb.*, a rupestral plant. (Further examples.)

1926 J. J. Walker *Nat. Hist. Oxford Distr.* 104 Many of these [alpine species] are rupestrals. 1932 G. C. Druce *Comital Flora Brit. Isles* p. xiii, A 'rupestral' may also be a wet or a dry lover. 1970 *Watsonia* VIII. 115 Unlike most species of *Hieracium* it is usually a soil plant rather than rupestral.

rupestrine *a.* (example); **rupestrean,** also **rupe·strian** *a.*, done on rock or cave walls.

1896 A. H. Keane *Ethnology* 137 The carvings on the dolmen *des marchands*, Brittany, are almost identical with those of the so-called 'rupestrian inscriptions' of Tunisia and South Algeria. 1934 'H. MacDiarmid' *Stony Limits & Other Poems* 52 Look over this beach. What ruderal and rupestrine growth is here? 1952 O. R. Gurney *Hittites* 215 The powerful rupestrian art of the later [Hittite] empire was probably stimulated by the rulers. 1967 M. Bullock tr. *Lommel's World of Early Hunters* iv. 127 The influence of rupestrian art on Negro art, particularly Negro sculpture, has not been elucidated. When we speak of African art we have grown accustomed to thinking only of Negro art, ignoring the rock-paintings or referring to them only marginally.

rupiah (rūpī·ā). [Indonesian, f. Hind. *rūpiyah*: see Rupee.] The basic monetary unit of Indonesia, equal to 100 sen.

1947 *Encycl. Brit. Bk. of Year* 307/2 The new Indonesian republic on Oct. 30 [1946] began the issue of its own currency, the rupiah, and on Dec. 1 declared its value to be 1·9 per U.S. dollar. 1952 *Ann. Reg. 1951* III. 335 The estimated deficit in the Budget was reduced to Rupiahs 987 million. 1959 'M. Derby' *Tigress* iv. 154 Traders..preferred to sell..in free markets for honest currency instead of government controlled prices in semi-worthless Indonesian *rupiahs*. 1964 *Asia Mag.* 27 Sept. 17/2 A fantastic rise in the rupiah costs of Indonesian projects. 1973 D. May *Laughter in Djakarta* ii. 33 He gave her a hundred rupiah note. 1977 *Time* 16 May 14/1 You start dropping in 10,000-rupiah ($24) notes until he says that's enough and closes the drawer.

rupt (rʌpt), *a. rare.* [f. as Rupt *v.*] Broken, craggy.

1916 Blunden *Harbingers* 66 Run, echo, up the tarn's rupt wall.

ruptured, *ppl. a.* Add: **3. ruptured duck** *U.S. Forces' slang*, (*a*) a damaged aircraft; (*b*) the discharge button given to ex-service men, with reference to its eagle motif.

1930 'W. W. Windstaff' in S. Longstreet *Canvas Falcons* (1970) xvii. 291, I began to look for a place to bring down my ruptured duck, oil spitting in my face. 1945 *Time* 29 Oct. 11/1 The design of the present discharge button is not popular (G.I.s know it as the 'ruptured duck'). 1953 R. Chandler *Long Goodbye* xxxiii. 202 He was wearing a British Army Service badge. Their version of the ruptured duck. 1955 A. Morgan *Great Man* 34 Just another guy in a sailor suit with a ruptured duck and a set of discharge papers. 1959 W. Faulkner *Mansion* xii. 333 The ex-soldier or -sailor or -marine with his ruptured duck pushing the perambulator with one hand.

rupturing, *vbl. sb.* Restrict *Bot.* to sense in Dict. and add: **2.** *attrib.* **rupturing capacity** *Electr. Engin.*, a measure of the ability of a circuit-breaker to withstand the surge produced by its operation.

1916 C. C. Garrard *Electr. Switch & Controlling Gear* ii. 54 The remaining considerations which determine the rupturing capacity of an oil circuit breaker apart from size, robustness, quickness of break and, of course, the quality of the oil used, are length and number of breaks under oil, or speaking generally the capability of the switch to bring a large quantity of oil into intimate contact with the arc so as to smother the same as effectively as possible. 1930 *Engineering* 24 Jan. 97/3 These circuit breakers were designed for a rupturing capacity of 1,000,000 kva. 1970 J. Shepherd et al. *Higher Electr. Engin.* (ed. 2) xvii. 560 It is normal practice to specify the rupturing capacity of circuit breakers in kilovolt-amperes or megavolt-amperes. This practice is well established but may be criticized as not being logical, since the breaking capacity in megavolt-amperes is obtained from the product of short-circuit current and recovery voltage.

rural, *a.* and *sb.* Add: **A.** *adj.* **7.** Special collocations, as *rural district council*, the local council of a rural district (see District *sb.* 3 b); abbrev. **R.D.C.*; *rural free delivery* (U.S.), the free delivery of mail to a rural area with limited local postal services; *rural industry*, an industry or manufacture carried out in the country; (*Women's*) *Rural Institute* (Sc.), a Women's Institute (see quot. 1958); *rural municipality* (Canad.), an administrative division of a province; *rural route* (N. Amer.), a rural mail-delivery route; *rural science*, the study of rural concerns, esp. agriculture; *rural slum*, a country dwelling in disrepair; *rural–urban* adj., designating comparison or interchange between country and town; *rural urbanization*, the investment of the country with an urban character.

1894, 1895 Rural district council [see *district-council* s.v. DISTRICT *sb.* 6]. 1929 [see *COUNCIL-HOUSE 2]. 1974 *Times* 1 Apr. 14/1 In all, 422 authorities take over the functions of the 1,385 existing authorities—counties, boroughs, urban and rural district councils. 1893 M. H. CUSHING *Story of our Post Office* 1006 A very important effect of the rural free delivery has been to increase the pay of postmasters where it has been tried. 1900 *Congress. Rec.* 16 Jan. 873/1 The rural free delivery service has come to stay. 1930 J. M. STAHL *Growing with West* viii. 109, I talked upon rural free delivery to many thousands. 1944 *N.Y. Times* 19 Oct. 23/2 Mr. Stahl first proposed the establishment of the rural free delivery service in 1879. 1892 *Pall Mall Gaz.* 23 June 2/1 There is little hope of the general establishment of rural industries as long as the Post Office treats country districts with such scant consideration. 1949 'J. TEY' *Brat Farrar* xxiv. 219 Mrs Stack,..being interested solely in rural industries, represented a Fixed Point in the flux of an agricultural show. 1958 *Listener* 6 Nov. 746/3 Are the efforts of the Rural Industries Organization being bent in the right direction? 1973 *Country Life* 28 June 1904/2 Twenty-seven small rural industries serving agriculture will be represented. 1922 *Scottish Women's Rural Institutes Handbk. 1921* 24 The Badge of the Scottish Women's Rural Institutes may be obtained from..London. 1932 'O. DOUGLAS' *Priorsford* xxvii. 242 'Have you heard how many teams are going in for the Festival?'.. 'Seven. Three of them Rural Institutes.' 'The Rurals are very good as a rule.' 1958 *Everyman's Encycl.* XII. 637/1 In Scotland there is a similar [to the Women's Institutes] but quite independent organisation, and the title 'Rural Institutes' is used. 1861 *Nor' Wester* (Red River Settlement, Canada) 15 Aug. 1/4 Such was the state of things in Canada until 1847, when the Canadian Legislature passed an act (4 & 3 Vic. cap 10) to extend the municipal system to districts (now counties) and other rural municiparities. 1904 *Univ. Toronto Stud. Hist. & Econ.* II. 140 The council of a rural municipality is made up of a reeve and not less than four nor more than six councillors, the number being fixed by by-law. 1945 G. W. BROWN *Canad. Democracy in Action* (1947) vii. 84 One of the reasons for the apparent confusion is that rural municipalities go by different names in different provinces. 1964 *Naicam* (Sask.) *Sentinel* 26 Mar. 2/3 A meeting of the council of the Rural Municipality of Pleasantdale No. 398 was held in the R.M. office Wednesday. 1898 *Ann. Rep.* (U.S. Post Office Dept.) 163 Nine rural routes were carefully laid out by special agents of the free delivery service. 1956 *Chatham* (Ont.) *Daily News* 14 June 2/6 Entrants will be accepted from Thamesville and surrounding rural routes. 1965 *Globe & Mail* (Toronto) 15 Oct. 7/1 George Harris Hees of Bay Street, Toronto, St. James Street, Montreal, and, within the last month, of Rural Route 5, Cobourg. 1914 Rural science [see *HOUSECRAFT]. 1939 *Nature* 18 Feb. 305 (*heading*) Training teachers of rural science. 1976 *Daily Times* (Lagos) 24 May 13/3 In order to promote sufficient interest in the tillage of the land subjects like nature study and rural science should be made compulsory in all primary schools. 1958 P. POLLACK *Picture Hist. Photogr.* III. 350/2 The somber, seamy existence endured by Americans living in rural slums. 1972 L. LAMB *Picture Frame* i. 13 Why..should an extremely sophisticated exponent of abstract expressionism decide to set up his easel in..this rural slum? 1975 *Times* 10 Sept. 18/3 The site enveloped a rural slum community. 1957 R. K. MERTON *Soc. Theory & Soc. Structure* (1962) xviii. 592 Differences in rural–urban distribution of the two religions ..may be seen. 1970 B. ROBERTS in I. L. Horowitz *Masses in Lat. Amer.* x. 346 The career experiences of low-income families, such as rural–urban and intra-urban migration. 1974 tr. *Wertheim's Evolution & Revolution* 198 Intensified rural–urban relationships might equally increase the revolutionary potential while..facilitating communication lines for revolutionaries. 1970 J. COTLER in I. L. Horowitz *Masses in Lat. Amer.* xii. 436 As a result of the confluence of urban ruralization and of rural urbanization, there has been a change in the patterns of social stratification of Mancha India.

B. *sb.* **3.** ellipt. for: (*a*) *(*Women's*) *Rural Institute*; (*b*) **Rural District Council* (rare).

1932 [see (*Women's*) *Rural Institute*, sense A. 7 above]. 1940 'O. DOUGLAS' *House that is our Own* viii. 82 'She takes to do wi' the Nursing and the Rural.' 'The Rural?' 'Aye, ye ken, Women's Rural Institute.' 1952 M. LASKI *Village* iv. 71 The new Housing Estate..the Walbridge Rural is going to put down at the bottom of Archery Lane. 1967 I. TAIN *Cherrycake Death* ii. 15 The Rurals were generally held to fulfil a need in the district. 1973 A. MACVICAR *Painted Doll Affair* x. 113 Jessie's out at a Women's Rural and Moira's at night school.

rurban (rǖ·ɹbăn), *a.* [f. R(URAL *a.* + URBAN *a.*] Combining the characteristics of country and town; designating an area sharing rural and urban ways of life.

1918 C. J. GALPIN *Rural Life* iii. 64 The word *rurban* is formed by blending *rural* into *urban*. 1932 *Times Lit. Suppl.* 11 Feb. 86/2 The science of towns..shows itself a science with technical terms, such as 'conurbation' and 'rurban'. 1939 O. E. BAKER in *Agric. in Mod. Life* I. ix. 165 Should..rural and urban merge in what has been called a 'rurban' civilization the cultural consequences would also be profound. 1945 H. H. BALK in *Econ. Geogr.* XXI. 108/2 A rurban area has very definite advantages over a strictly agricultural or rural area from the farmer's point of view. 1961 *New Yorker* 28 Oct. 43/3 In an article about the spread of highways and housing developments from cities to former farmland, we came upon a reference to 'the rurban explosion'. 1981 *Country Life* 2 July 7 To draw attention to his unsatisfactory rural-urban development, the Second Land Utilisation Survey has given it a distinctive name: rurban fringe.

Hence **ru·rbanism**, the properties of town and country life regarded as interacting and inseparable; **ru·rbanist**, an advocate of rurbanism; **rurbaniza·tion** the susceptibility of town life to rural influences.

1918 C. J. GALPIN *Rural Life* iii. 64 The idea of rurbanism is that..the open country is an element in the clustered town, and the town is a factor of the land, and the civilization, culture, and development of rural people are to be found in conjunction with town and small city, and not apart. *Ibid.*, The rurbanist boldly attempts to adjust anew the malrelations of the farm to the cluster. 1931 N. CARPENTER *Sociol. City Life* xiv. 453 Urbanized societies are those in which the cultural effects of 'rurbanization' are to be found. 1943 C. L. WHITE *Regional Geogr. Anglo-America* xiii. 403 Rather than have three fourths to four fifths of our population reside in great cities, O. E. Baker recommends rurbanism. 1959 *Economist* 30 May 850/1 Millions of city dwellers have swarmed into the countryside to set a new and increasingly widespread pattern of life which is sometimes called 'rurbanisation', sometimes 'urbiculture'. 1976 *National Observer* (U.S.) 10 July 6/4 He calls the young back-to-the-landers the 'vanguard of a new ruralism'. Another way to look at it is the 'rurbanization' of America, he says.

Ruritania (rūᵊritēⁱ·niă). [Name of the scene of Anthony Hope's novels *The Prisoner of Zenda* (1894) and *Rupert of Hentzau* (1898); f. L. *rūri-s*, *rūs* country + *-tania* as in *Lusitania*.] An imaginary kingdom of Central Europe: used allusively for a scene of court romance and intrigue in a modern setting, or for a petty state; more generally, any imaginary country.

[1894 'A. HOPE' *Prisoner of Zenda* i. 9 There came on a visit to the English Court a certain prince, who was afterwards known to history as Rudolf the Third of Ruritania.] 1897 G. B. SHAW in *Sat. Rev.* 30 Jan. 115/2 If Mr Alexander..had produced it and Sodom's Ende and so forth at a series of *matinées* of the 'Saturday Pop' class, financing them from the exchequer of the Kingdom of Ruritania [etc.]. 1929 —— *League of Nations* 8 Let us suppose that Ruritania is given a mandate to govern Lilliput provisionally for Lilliput's good. 1939 *Flight* 21 Sept. 249/1 One cannot expect every Ruritania to maintain an air arm equal to that of the Germans. 1956 A. WILSON *Anglo-Saxon Attitudes* 154 Inge..looked like the queen of some Northern Ruritania at the novel's happy ending. 1965 *New Statesman* 23 Apr. 638/1 Britain is being slowly pushed out of the main manufacturing export markets, and is taking refuge in the luxury trades... I suppose if we can't make our living in any other way, it will have to do. But let's not kid ourselves about a renaissance. This is national senescence, the Road to Ruritania. 1976 *Times* 9 Mar. 17/2 The right tactics..are to let the central bank of Ruritania drive the price of sterling down against itself.

Ruritanian (rūᵊritēⁱ·niăn), *a.* (*sb.*) [f. prec. + -AN.] Of, pertaining to, or characteristic of Ruritania, esp. with reference to the romantic or fanciful associations of the name; hence used with reference to any imaginary country. Also as *sb.*, an inhabitant or supporter of a Ruritania, or a person endowed with Ruritanian attributes or characteristics; more generally, an imaginary inhabitant of a country.

1894 'A. HOPE' *Prisoner of Zenda* i. 9 Prince Rudolf.. was adroitly smuggled off by the Ruritanian ambassador. *Ibid.* ii. 24 Every Ruritanian knows Duke Michael. 1896 G. B. SHAW in *Sat. Rev.* 11 Jan. 39/1 Our common sense which, if aroused, must immediately put a summary stop to the somewhat silly Ruritanian gambols of our imagination. 1902 *Daily Chron.* 21 Nov. 5/4 'The Traitors' is a spirited example of what may be called Ruritanian romance. 1920 H. G. WELLS *Outl. Hist.* xxxvii. 601 He [*sc.* Mirabeau] had..indulged in a sort of Ruritanian flirtation with the queen. 1935 *Punch* 20 Mar. 335/1 The third volume of *The Story of My Life* by Marie, Queen of Roumania..consists almost entirely of entries in the author's diaries. A book for Ruritanians and royalists. 1944 H. G. WELLS *'42 to '44* II. 52 The planners count noses and claim this or that district because there are 59 per cent. alleged Ruritanians here, or because 42 per cent. of the people there belong to the Lutheran Church. 1950 'P. WOODRUFF' *Island of Chamba* v. 71, I do enjoy it, ..the alternations between rich farce, charming comedy and Ruritanian politics. 1958 *Times* 20 May 10/5 In the Ruritanian world of mountain Lebanon almost anything can happen. 1969 G. GREENE *Travels with my Aunt* I. xii. 113 The Montreux Palace in baroque Edwardian like the home of a Ruritanian King. 1977 *Times* 5 Nov. 15/3 The modern world..invaded the 'ruritanian charm' of the travel writers.

rurp (rǖɹp). *Mountaineering.* [f. initial letters of *r*ealized *u*ltimate *r*eality *p*iton.] A type of very small piton.

1968 P. CREW *Encycl. Dict. Mountaineering* 104/2 *Rurp*,..a very small American chrome-molybdenum piton, designed for use in hair-line cracks. 1972 D. HASTON *In High Places* ix. 104 A rurp—the smallest piton, about the size of a postage stamp. 1976 *Times* 13 Nov. 12/6 A curious armoury of pegs, rurps, sky hooks, bolts, and bongs now form the armoury of the modern [rock] climber.

ruru (rū·rū). *N.Z.* [Maori.] The morepork, *Ninox novæseelandiæ*; = MOPOKE, MOREPORK 1 in Dict. and Suppl.

1859 A. S. THOMSON *Story of N.Z.* I. I. i. 25 The natives call the owl Kou-Kou or Ru-ru. 1862 A. S. ATKINSON *Jrnl.* 22 Aug. in *Richmond-Atkinson Papers* (1960) I. xiii. 782 Old Potatau had likened himself to a ruru mobbed by a flock of popokateas. 1905 W. BAUCKE *Where White Man Treads* 48, I..heard snatches of an ancient witch karakia being muttered round the corner, wherein a ruru..is the medium of destruction.

Rus (rʊs, rūs). Also 9 **Russ.** [Russ. *Rus'* (see RUSS *sb.* and *a.*), Arab. *Rūs*; cf. medieval Gr. οἱ Ῥῶς.] The name of a group of Swedish merchant warriors who established themselves around Kiev and the Dnieper in the ninth century, whose settlements gave rise to the later Russian principalities.

1845 *Encycl. Metrop.* XXIV. 225/1 The Vareghi (*conquering*) Russ, one of the enterprising Tribes of Scandinavia, whose fleets had appeared in the Bosphorus during the first half of the IXth Century. 1876 V. THOMSEN *Relations betw. Anc. Russia & Scandinavia* II. 37, I am going to..corroborate..the Scandinavian origin of the Russ. 1918 R. BEAZLEY et al. *From Varangians to Bolsheviks* I. i. 3 Both the name of *Rus* and the fact of a Russian people and Russian States are due to them [*sc.* the Scandinavians]. 1927 E. V. GORDON *Introd. Old Norse* p. xxi, From the Swedish founders of this kingdom ..Russia takes its name, for the Swedes were known in the east as Rus. The population of the kingdom of the Rus was..mainly Slavonic, and the Rus themselves gradually lost their traditions and language. 1948 G. VERNADSKY *Kievan Russia* vi. 138 In Kiev the prince's retinue..consisted of the Swedish Rus. 1965 H. M. SMYSER in Bessinger & Creed *Medieval & Linguistic Studies* 92 In 921, Ibn Faḍlān..described..a tribe of Swedish Rūs Vikings, or, more accurately, Rūs armed merchants, and..a funeral which these Rūs accorded one of their chief men. 1976 H. R. ELLIS DAVIDSON *Viking Road to Byzantium* I. iv. 56 It is in the ninth century that we first hear of the Rus, who were well known to Arab geographers, and whom the Byzantine Greeks called *Rhos*... For most western scholars, the name *Rus* is taken primarily to denote the Scandinavian settlers in Russia, particularly those established at Kiev in the ninth century. *Ibid.* 62 There is..general agreement that in the ninth century the important Rus state on the Dnieper around Kiev was formed.

rusa. Substitute for def.: Either of two deer, *Cervus equinus* or *C. unicolor*, native to southern Asia. Cf. SAMBUR. (Earlier and further examples.)

1783 W. MARSDEN *Hist. Sumatra* 94 Deer: *rooso*: *keejang*. There are variety of the deer species; of which some are very large. 1839 T. J. NEWBOLD *Straits of Malacca* I. vii. 436 Of the genus Cervus, are the Kijang.., the Rúsa or Cervus Hippelaphus. 1958 J. SLIMMING *Temiar Jungle* ii. 23 A few yards away were the tracks of a *Rusa*, a Malayan deer.

rusbank (rö·sbaŋk). *S. Afr.* Also **rus-bank, rustbank,** etc. *Pl.* **rusbanks, rusbanke, rust banken.** [Afrikaans, f. *rus(t)* rest + *bank* bench.] A wooden settle or couch, usu. with a seat of woven leather thongs or riempies. Also *attrib.*

1880 J. NIXON *Among Boers* 216 Hans was seated on the 'rustbank', smoking a short wooden pipe. 1902 W. DOWER *Early Annals of Kokstad* 25 The few very rough seats and the rust banken..were occupied by the men, the women sat on the floor. 1910 D. FAIRBRIDGE *That which hath Been* xxiii. 277 The baas and huisvrouw..came out on their stoeps and sank into the capacious chairs and rust-banks. 1935 P. SMITH *Platkops Children* 76 After that was a long white house with a big stoep an' rus'-banks at each end. 1939 S. CLOETE *Watch for Dawn* 29 How alike all these Boer houses were. Each had the same rough, home-made riempie-seated rus-banks. 1947 *Cape Times* 5 Feb. 14 The farm-house has a huge dining-room with a massive centre table and, along one wall, the district's longest rustbank. 1965 M. G. ATMORE *Cape Furniture* 77 At all times the rustbank has been a 'multiple chair' in which the form was copied from the single chair of the time. 1971 *Evening Post* (Port Elizabeth) 8 May 20 The Furniture: Round Hand-made Stinkwood and Yellowwood Table on Pedestal Leg: three Yellowwood and Stinkwood Rusbanke. 1971 *Daily Dispatch* (East London) 8 Sept. 18 Old Rusbank-type Lounge Suite. 1972 *Grocott's Mail* (Grahamstown) 22 Feb. 1 Rusbank with riempie seat.

‖ **rusé** (rüze), *a.* Also fem. **rusée**, Pl. **rusés.** [Fr.] Given to ruses, sly, cunning; deceitful, deceptive. Also as *sb.*

1761 G. COLMAN *Jealous Wife* III. 45 Your Ladyship, I hope, has no Objections to my being a little *rusé*, for I must have Her, 'pon Honour. 1847 DISRAELI *Tancred* II. IV. iii. 189 Aberdeen and Sir Peel will never give her this advice; their habits are formed. They are too old, too *rusés*. 1889 G. MEREDITH *Let.* 20 Sept. (1970) II. 980 *Rusée* that you are! 1903 A. BENNETT *Truth about Author* i. 8, I..ordered the old rusé self to exploit the self just born. 1923 G. ATHERTON *Black Oxen* viii. 33 She was certainly *rusée*. 1938 H. G. WELLS *Apropos of Dolores* ii. 31 We hunted for five of the most rusé tennis balls I have ever known... They changed colour according to their surroundings. 1940 G. ARTHUR *Concerning Winston Spencer Churchill* 143 It was a most successful, if rather rusé, coup, but when anyone spoke of it as a military measure Kitchener would always say that Winston Churchill must have a large share of the credit. 1955 A. L. ROWSE *Expansion of Elizabethan England* x. 399 As a commander, he [*sc.* Sir Francis Vere] was exceedingly *rusé*. 1968 *Listener* 13 June 779/1 The values

are unsurprising—the baby, the reliable if *rusé* Italian director, the true choice at the end. 1973 C. M. WOODHOUSE *Capodistria* v. 110 They constantly used of him [*sc.* Capodistria] the conventional epithets which seemed to fit his nationality—wily, *rusé*, supple, crafty.

‖ **ruse de guerre** (rüz də gɛ̄r). Pl. **ruses de guerre.** [Fr., lit. 'ruse of war': see RUSE *sb.*] A course of action intended to deceive an enemy in war; a stratagem. So, in extended uses, a justifiable trick.

1807 *Naval Anecdotes* vii. 185 We consider the whole of this singular passage as a kind of *ruse de guerre* to divert the public censure from Lord St. Vincent's Admiralty in regard to the supply of stores. 1814 M. EDGEWORTH *Patronage* III. xxxii. 302, I..thought even your praises of Rosamond's disposition..might only be *ruse de guerre*, or *ruse d'amour*. 1888 *Academy* 10 Nov. 300/2 Stubborn party duels, *ruses de guerre*, and all the heiving and hacking of the parliamentary fray. 1915 F. PIGGOTT *Neutral Merchant* i. 11 By a *ruse de guerre*, or stratagem of war, I understand the adoption of some means of deceiving the enemy in war, some device out of the ordinary course of fighting. 1919 G. B. SHAW *Peace Conference Hints* vii. 99 These deceptions are necessary as *ruses de guerre*. 1922 J. BUCHAN *Huntingtower* v. 84 'It was an abominable lie.'.. 'Not at all. It was a necessary and proper *ruse de guerre*.' 1938 C. S. FORESTER *Flying Colours* i. 8 That had been a legitimate *ruse de guerre* for which historical precedents..could be quoted. 1962 *Times* 24 Apr. 12/6 It [*sc.* martial music] was no longer used as a *ruse de guerre*.

rush, *sb.*[1] Add: **5. a.** *rush-bottom*.

1831 J. M. PECK *Guide for Emigrants* II. 105 In all the rush bottoms they [*sc.* cattle] fatten during the severe weather on rushes.

b. *rush-bottom* (also as *adj.*), *-house*, *-matting*, *-rope* (later example), *-seat* (also as *adj.*), *-work* (later examples).

1792 A. YOUNG *Trav. France* I. 24 Oak chairs with rush bottoms. 1809 'D. KNICKERBOCKER' *Hist. N.Y.* I. III. iii. 258 The young ladies seated themselves demurely in their rush-bottom chairs. 1866 D. G. ROSSETTI *Let.* 20 May (1965) II. 598 If there is anything besides rush-bottoms and ascetic glasses on which I should be glad to offer a fundamental remark..it is this. 1923 W. DEEPING *Secret Sanctuary* xiii. 138 He made a move to sit down, and she saw him take one of the straight-backed rush-bottoms. 1960 J. STROUD *Shorn Lamb* iii. 27 Facing the magistrates ..was a single rush-bottom chair. 1835 Rush-house [see RAUPO]. 1926–7 *Army & Navy Stores Catal.* 1103/4 Rush matting, 36 in. wide — per yard. 4/9. 1942 E. WAUGH *Put out More Flags* ii. 108 The floor was covered in coarse rush matting and in places by bright Balkan rugs. 1964 *New Statesman* 14 Feb. 271/4 (Advt.), Rush-matting made-to-measure 2s. sq. ft. Rush-seated Italian chairs 9 gns a pair. 1922 JOYCE *Ulysses* 508 Woman undoing with sweet pudor her belt of rushrope, offers her allmoist yoni to man's lingam. 1896 *Heal & Son Catal.* 156 Ebonised Rush-Seat Chair—£0 4 9 Ebonised Chair, Rush Seat £0 6 3. 1918 *Ibid.* 25 Dark Oak Rush-seat Arm chair, 35/-. 1949 R. HARVEY *Curtain Time* i. 6 And soon the audience would begin to gather, first for the rush seats in the gallery, then for the balcony and the main floor. 1971 *Country Life* 18 Feb. 366/3 The square drop-in rush seat was originally upholstered. 1934 E. BOWEN *Cat Jumps & Other Stories* 190 She had discovered that Miss Weekes morris-danced, that she did rush-work. 1959 *Observer* 15 Mar. 14/5, I have just had a rustic chair reseated... Very neat rushwork. 1977 *Vogue* Feb. 115/2 A Connemara Craft Centre with ceramics, rushwork, tweeds.

6. b. *rush-bordered* (further example), *-bottomed* (earlier and later examples), *-matted*, *-plaited*, *-seated* (further examples), *-seating*.

1912 W. DE LA MARE *Child's Day* 26 A green, rush-bordered pool. 1753 S. FIELDING *Adv. David Simple* (ed. 2) V. VII. ii. 178 His Candle falling off the Table, set fire to a Rush-bottomed Chair. 1840 C. F. HOFFMAN *Greyslaer* II. III. i. 96 The apartment..was large and rudely furnished, containing only..a small cherry-wood table and a few rush-bottomed chairs. 1867 A. D. RICHARDSON *Beyond Mississippi* xi. 131 The adjacent settlers came..in heavy ox-wagons sitting upon rush-bottomed chairs. 1902 *Chambers's Jrnl.* July 471/2 Here were..ancient rush-bottomed chairs, an old corner-cupboard with glass doors [etc.]. 1918 *Heal & Son Catal.* 2 Rush-bottomed Chair. 1976 'D. HALLIDAY' *Dolly & Nanny Bird* v. 60 A creak from the stairs..or the sounds of the rushbottomed chair I had used when feeding Benedict. 1960 C. DAY LEWIS *Buried Day* x. 226 Our stone-floored, rush-matted living-room..felt at first like an Aeolus' cave of draughts. 1939 F. THOMPSON *Lark Rise* xv. 273 His wide, rush-plaited hat. 1952 M. LASKI *Village* viii. 135 A long rush-seated oak stool. 1977 *Times* 3 Sept. 11/3 Rush-seated chairs and homely local service. 1926 *Daily Colonist* (Victoria, B.C.) 14 July 5/1 We do old-fashined rush seating also chair recaning... The Red Cross Workshop. 1979 *Jrnl. R. Soc. Arts* CXXVII. 453/2 The pieces to be shown will demonstrate the variety of the skills learned by the students..—rush-seating, marquetry, turning, [etc.].

rush, *sb.*[2] Add: **1. a.** Also, the movement of large numbers of people at a specified time or season to or from work, recreation, shops, etc.; *gen.*, haste, urgency; excessive activity.

1924 G. B. SHAW *Saint Joan* p. vii, His accuser.. might have been picked out of any first class carriage on a suburban railway during the evening or morning rush from or to the City. 1925 H. CRANE *Let.* 1 Dec. (1965) 220 Selling books in stores during the Christmas rush. 1931 H. NICOLSON *Diary* 22 Aug. (1966) 88, I have learnt that rapidity, hustle and rush are the allies of superficiality. 1932 E. BOWEN *To the North* vi. 51 She had not come down all this way..in the middle of what she and Peter considered the Whitsun rush. 1939 [see *AFTER- I. 1]. 1943 E. B. WHITE *Let.* 13 Aug. (1976) 243, I would like to discuss my publishing life with you some time... There is no rush about it, however, as I have no book ready to go. 1951 E. PAUL *Springtime in Paris* ix. 157 The easter rush of tourists..had crowded the St. Sulpice district to overflowing. 1973 [see *office worker* s.v. *OFFICE *sb.* 12].

b. (Further examples.) Also, a rushing sensation in the body; a thrill (of fear, pleasure, etc.); a drug-induced euphoria, = *FLASH *sb.*[2] 1 h. *colloq.*

1916 'BOYD CABLE' *Action Front* 113 At nine, sharp to the tick of the clock, the rush, rush, rush of a field battery's shells passed overhead. 1922 JOYCE *Ulysses* 509 He couldn't get a connection. Only, you know, sensation. A dry rush. 1971 *Frends* 21 May 11/1 When you start smoking one type of Hash, the best rushes come during the first day of smoking. (Assuming the Hash is good). 1976 *National Observer* (U.S.) 23 Oct. 6 Methadone is addictive, too, but it doesn't give the pleasurable 'rush' that heroin addicts speak of. 1979 *Washington Post* 25 Mar. N5 Never again was there anything quite like the rush we got from the simple fact of spring.

d. *pl. Cinemat.* The first prints of film resulting from a period of shooting; the preliminary showing of such film; = *DAILY *sb.* 4.

1924 G. R. CHESTER *On Set & Off* xvii. 206 Isidor Iskovitch sat very cockily exhibiting to his friend and boss ..some thousands of feet of 'rushes' on his pet picture, 'the Woman's Half'. 1927 L. FAWCETT *Films Facts & Forecasts* xiv. 130 A good many pictures are entirely remade, and sometimes sequences are ordered to be reconstructed when the 'rushes' (short lengths of film) are seen during actual production. 1934 [see *DAILY *sb.* 4]. 1940 *Manch. Guardian Weekly* 15 Nov. 355 We were shown 'stills' of Mr. Gielgud's Disraeli and..in a private theatre we saw 'rushes' of the previous day's work. 1952 [see *DAILY *sb.* 4]. 1956 B. HOLIDAY *Lady sings Blues* (1973) xv. 126 Every night after we'd finished work at six o'clock, Blondie would rush to the projection room to see the rushes. 1962 *Movie* Sept. 31/2 For Rouch, the ideal film of this title would be the four hours of 'rushes', without cuts or montage. 1969 *New Yorker* 29 Nov. 160/2 Nothing makes us more aware of staginess than actors reciting poetry outdoors—as directors of Shakespearean movies discover when they look at their first day's rushes. 1976 C. BERMANT *Coming Home* II. iv. 160 My function was to write the outline script as a rough guide for the film crew. Then, when the rushes were available, I re-wrote to fit the pictures.

e. Used *fig.* in phrs. *bum's rush*, see *BUM *sb.*[4] 1 b; *to get a rush*, of a girl or woman: to be the recipient of frequent attentions from men; *to give* (someone, *spec.* a girl or woman) *a rush*: to lavish attention on (that person) in the form of social engagements and entertainment. *colloq.*

1928 *Amer. Speech* III. 221 To say that a girl 'certainly gets a big rush' means that she has many desirable dates, and is 'cut' a lot at dances. 1934 J. O'HARA *Appointment in Samarra* (1935) i. 20 Wilhelmina Hall..was still the best dancer in the club, and was getting the best rush. 1938 E. BOWEN *Death of Heart* II. iv. 249, I don't think most girls appreciate friendship; all they want is to be given a rush. 1940 WODEHOUSE *Eggs, Beans & Crumpets* 93 He's been giving me the rush of a lifetime. 1953 H. WAUGH *Last seen Wearing* 71 She goes round with another guy..and I think he was giving her pretty much of a rush. 1956 W. H. WHYTE *Organization Man* (1957) 252 An actor..comes to town from the city for a short stay. He gives her a mild rush, and she dreams of a glamorous life with him. 1969 A. LURIE *Real People* 18 She certainly wasn't prepared for the rush she got, probably for the first time in her life... You've got to admit she's not madly attractive.

f. In *attrib.* use passing into *adj.*, denoting rapidity of movement, haste, or urgency. Also *ellipt.* as quasi-*adv.*

1879 W. WHITMAN *Daybooks & Notebooks* (1978) I. 145, I am told that Saturday is a real rush day. 1896 Rush order [in Dict., sense 9]. 1900 J. LONDON *Let.* 15 Mar. (1966) 102 This isn't sharpshooting, but repelling a rush attack of a body of men. 1901 C. MOFFETT *Careers of Danger* 381 Already the mail clerks are swarming at the pouches, like printers on a rush edition. 1901 MERWIN & WEBSTER *Calumet 'K'* vii. 126 But if you ever try to put me on a rush job, I'll quit and buy a small farm. 1904 *N.Y. Herald* 17 Sept. 1 He stated that six weeks' rush work would be required to repair the boilers to make them serviceable and the ship seaworthy. 1929 T. H. BURNHAM *Engin. Econ.* xv. 199 Rush orders are difficult to put through, even in well-organized works. 1933 D. L. SAYERS *Murder must Advertise* viii. 126 Mr. Copley..was left working overtime upon a rush series of cut-price advertisements for Jamboree Jellies. 1933 BALMER & WYLIE *When Worlds Collide* i. 24 You see, Tony, some—some things were being sent rush, by airplane. 1939 C. DAY LEWIS *Child of Misfortune* II. vi. 241 Christmas was a rush-time of services, visits to the sick, parties for the children and old people of the parish. 1946 *Ann. Reg.* 1945 40 There should be a three weeks' interval..so as to avoid a rush election. 1955 'A. GILBERT' *Is she Dead Too?* viii. 151 She was gone before he arrived on the scene. It was a rush job. 1958 [see *ONE *numeral a.* 29* b]. 1965 MRS L. B. JOHNSON *White House Diary* 2 July (1970) 292 So I got a rush appointment with Mr. Per and went over for a permanent. 1968 *Listener* 8 Aug. 176/3 It was an odd life at Oxford at the time because most people had just come out of the army and were going in for quick degrees, sort of rush degrees. *a* 1974 R. CROSSMAN *Diaries* (1975) I. 238 They are horrifyingly superficial—merely a collection of the facts available to central government with one or two rush-job social surveys and some very hasty conclusions. 1977 R. V. HUDSON in Bond & McLeod *Newslett. to Newspapers* II. 123 His expertise earned him an assignment to rush work and the highest wage among some fifty printers.

3. a. (Later examples in *N. Amer.* use.)

1970 *Globe & Mail* (Toronto) 28 Sept. 18/6 Fleming was Hamilton's leading ground gainer with four catches for 66 yards and nine rushes for 42 more. 1979 *Honolulu Advertiser* 8 Jan. C-1/1 Larry's strong rush helped keep pressure on Ram quarterback Pat Haden.

c. (Earlier and later examples.)

1860 *Yale Lit. Mag.* XXVI. 22 As a basis, a Rush tacitly assumes that it is promoting a rivalry that is proper and praiseworthy. 1916 C. A. EASTMAN *From Deep Woods to Civilization* 68 The two classes met in a first 'rush'. 1937 *Amer. Speech* XII. 156 *Cane rushes*, or encounters between freshmen carrying canes and sophomores seeking to break them were an institution at the University of Nebraska in the late '80's of the last century. They were vigorous affairs and fraternity rushing may well have had name from them.

d. *U.S.* A round of entertainment in which candidates for admission to a fraternity or a sorority participate. Also *attrib.*, as *rush party*, *week*, etc.

1899 A. H. QUINN *Pennsylvania Stories* 60 It was not long before Theta Chi gave him a bid to a rush smoker. 1918 *Dialect Notes* V. 27 The object of the many attentions of a frat-rush. 1931 *Kansas City Times* 24 Sept. 20/6 Aunt Phoebe Tilden read where so many colleges are having rush parties. 1940 *Sun* (Baltimore) 16 Sept. 2/8 (*heading*) Rush week for C. of C. 1944 *Greeley* (Colo.) *Daily Tribune* 24 Sept. 3/5 Formal rush week for all sororities on the campus will be Oct. 1 to Oct. 6. 1964 Rush week [see *pledge week* s.v. *PLEDGE *sb.* 7]. 1970 *Guardian* 23 Apr. 11/4 What is called the Rush Programme. Girls who want to be 'rushed', i.e. who want to join a sorority, put their names down for sororities (the screening process) and 'rush' to about 30 or 40 parties. 1979 *Arizona Daily Star* 19 Apr. 6/2 Jennifer Johnston.. was elected assistant rush chairman.

4. a. (Earlier and later examples.)

1848 *Morning Courier & New-York Enquirer* 11 Dec. 2/1 There is a general rush for the new found *Dorado*. 1849 *Merchant's Mag.* XX. 60 In May, the gold itself began to come into the town. And then began the rising and the rush. 1893 [see *gold-rush* s.v. GOLD[1] 10 a]. 1897 *Boston Globe* 29 Aug. 6/6 There are only about 1600 new Americans in the mines... There are plenty of supplies there for those already in but not for any big rush late in the fall. 1908 E. J. BANFIELD *Confessions of Beachcomber* II. ii. 288 A party of bushmen, fresh from the excitement and weariness of the Gilbert rush. 1911 C. E. W. BEAN *'Dreadnought' of Darling* xxxv. 306 A friend of the writer's, who was in Coolgardie from the time when the rush there first started, tells of how first the waistcoat came onto the mining field. 1935 *Chambers's Encycl.* VI. 459/2 Only in 1896 was gold found in such abundance as to create a rush [to the Klondike]. 1947 R. PEATTIE *Sierra Nevada* 60 The discovery in 1859 of a glittering silver bonanza in Washoe County, Nevada, started a frantic rush over the mountains to Virginia City. 1955 *Bull. Atomic Sci.* Mar. 88/2 Thus the uranium boom began. The rush has grown rather than quieted, but there are healthy signs of stabilization. 1966 'J. HACKSTON' *Father clears Out* 104 Following on this 'rich' find, the big Red Range rush set in. 1972 *Standard Encycl. Southern Africa* V. 227/1 By then [*sc.* 8 Sept. 1886] the rush of diggers had already set in, so that fully 3 000 people were estimated to be scattered along the Rand by this time.

b. (Further examples.)

1900 H. LAWSON *Story of Oracle* in *Stories* (1964) I. 435 My Uncle Bob was mates with him on one of those rushes along there—the Pipeclay, I think it was, or the Log Paddock. 1966 'J. HACKSTON' *Father clears Out* 52 Why is he so poor now, after finding all the rich reefs and rushes?

5. b. (Earlier and later examples.)

1841 *Daily Picayune* (New Orleans) 10 Dec. 2/2 They all travel round to the old brushing ground where they 'go it with a rush'. 1846 S. F. SMITH *Theatrical Apprenticeship & Anecdotal Recollections* 152 When you find yourself in possession of *four aces*, go it with a perfect rush. 1898 *McClure's Mag.* X. 352 The gray-backs came through with a rush. 1903 G. B. SHAW *Man & Superman* III. 138 The complete reality comes back with a rush. 1914 T. DREISER *Titan* i. 3 Chicago, when it finally dawned on him, came with a rush on the second morning. He had spent two nights in the gaudy Pullman..when the first lone outposts of the prairie metropolis began to appear. 1934 A. CHRISTIE *Parker Pyne Investigates* 53 She stared at Mr Parker Pyne with a desperate intentness. Suddenly she spoke with a rush.

c. (*all*) *in a rush* (and variants): phrs. denoting rapidity, liveliness, briskness, suddenness, or haste.

1859 HOTTEN *Dict. Slang* 84 *Doing it on the rush*, running away, or making off. 1876 'MARK TWAIN' *Tom Sawyer* xviii. 149 He is always in such a rush that he never thinks of anything. 1877 G. M. HOPKINS *Spring* in *Poems* (1967) 67 That blue is all in a rush With richness. 1890 KIPLING *Barrack-Room Ballads* (1892) 11 An 'appy day with Fuzzy on the rush Will last an 'ealthy Tommy for a year. 1901 H. JAMES *Sacred Fount* iv. 75 Last night she was on the rush. 1916 'TAFFRAIL' *Pincher Martin* I. 11 Take a pride in yourself, an' obey all orders at the rush. 1938 *Sun* (Baltimore) 8 June 8/8 Later in the season they [*sc.* soft crabs] drop off almost altogether until late July or in August, when they seem to come back all of a rush. 1962 E. B. ATWOOD *Regional Vocab. Texas* iii. 71 *To leave in a rush*. The most common single expression for hurried departure is *light a shuck*. 1971 *Cassell's Mod. Guide to Synonyms* 502 The slow, jolting pace of one who is in no rush.

9. rush dodge, the act of overcoming or disarming a person by means of a rush;

rush line (earlier and later examples); also *fig.*; **rush-release**, the action or an instance of producing and marketing a gramophone record in the shortest possible time; so *rush-release* vb. trans. Also *RUSH HOUR.

1888 'R. BOLDREWOOD' *Robbery under Arms* II. ii. 19 It's no use trying the rush dodge with them. 1887 *Century Mag.* XXXIV. 891/2 Across the field stretch the football infantry, the 'rush-line' or 'rushers'. 1906 *Life* 4 Oct. 366 We hear of a surprising prevalence among the young men..of the disposition to get into the political rush-line. 1923 R. D. PAINE *Comr. Rolling Ocean* i. 3 The unlucky young men who were left in his wake when he tore through a rush-line. 1976 *Webster's Sports Dict.* 365/1 *Rush line*, the defensive line of a football team. 1966 *Melody Maker* 16 July 4 The group's 'Pet Sounds' LP—rush-released by EMI—entered the MM's best-selling LPs chart this week at number nine. 1968 *Ibid.* 22 June 2 The Regal Zonophone label is rush-releasing the new Move E.P. 1978 *New Musical Express* 11 Feb. 4/2 Radiators From Space have just finished recording their new single..and Chiswick hope to have it ready for rush release on February 17.

rush, *v.*[2] Add: **3. b.** Also, without connotations of violence, to convey (someone or something) rapidly or urgently.

1914 S. LEWIS *Our Mr. Wrenn* 214 I'll make Tom rush us a growler of beer. 1927 U. SINCLAIR *Oil!* 264 It was our job to rush them supplies. 1935 in A. P. Herbert *What a Word!* v. 143 She was rushed to Alton Hospital, where her condition is critical. 1947 *Milwaukee Jrnl.* 29 Oct. 2 Uncle Tom doesn't scurry around to rush us a loan. 1958 [see *JERRICAN, JERRYCAN]. 1966 L. COHEN *Beautiful Losers* I. 107 Rush to me the *free* book on the Home Method of Slenderizing Heavy Legs. 1971 *Sunday Express* (Johannesburg) 28 Mar. 7/5 (Advt.), Rush me my..illustrated Guide. 1976 *Star* (Sheffield) 3 Dec. 10/7 My husband to be was rushed into hospital for a serious operation.

4. a. (Further examples.) Also, in weakened senses, to accomplish or produce rapidly; to expedite; to hurry or hustle. Also with *off*, *through*, etc.

1883 'MARK TWAIN' *Life on Mississippi* li. 452 Since there was so much time to spare that nineteen years of it could be devoted to the construction of a mere towhead, where was the use, originally, in rushing this whole globe through in six days? 1901 *Chambers's Jrnl.* Apr. 210/1 Candida rushed her news. 1918 W. OWEN *Let.* 19 Aug. (1967) 569 I rushed off a note in time for this evening's post. 1938 *Amer. Speech* XIII. 156/1 *Bootleg 'em*, to rush a special order through outside of regular channels. 1946 *R.A.F. Jrnl.* May 161 Do not through nervousness rush your replies. 1947 *Sun* (Baltimore) 8 Nov. 6/2 Representative Knutson..announced his intention of rushing the special session with a 'quickie' bill for income-tax reduction. 1949 SHURR & YOCOM *Mod. Dance* 5 Head of Developing and Printing at Willoughby's Camera Stores, Inc., who rushed prints and supplies through in record time. 1974 *Times* 1 Feb. 2/8 Nevertheless, Mr Campbell Adamson, director general, emphasized that the CBI was 'rushing' a council meeting, a somewhat unprecedented action, to consider Mr Heath's letter. 1976 M. MACHLIN *Pipeline* lxi. 517 If you ask me..he's rushing the whole thing and I think that these ULCC'S—these ultra-big tankers—are a mistake.

c. (Earlier and later examples.) Also *ellipt.*

1887 J. PAYN *Glow-Worm Tales* II. 44 That a fraud had been committed on us was certain, and a fraud of a very clumsy kind... He had 'rushed us' as, the phrase goes. 1930 BROPHY & PARTRIDGE *Songs & Slang 1914–1918* 158 *How much did they rush you?* meant 'How much did you have to pay?' 1931 T. R. G. LYELL *Slang, Phrase & Idiom* 655 *Rush a person*,..to overcharge a person; to make him pay an exorbitant price. *A.* 'How much d'you say you paid for this car?' *B.* 'Two hundred and fifty pounds.' *A.* 'My word! they rushed you, all right! It's not worth a penny more than £120.' 1973 N. W. SCHUR *British Self-Taught* 340 *Rush*,..soak. For instance: 'How much did they rush you for that sherry?' To rush is to charge, with the distinct implication that the price was too high.

d. To hurry or pressure (a person); now freq. *pass.* (passing into *ppl. a.*), of a person: to have much to do in a limited time, to be hard-pressed by shortage of time (also with the activity or the period of time as subject). Hence in colloq. phrases *to be rushed around*, *to be rushed off one's feet* (or *legs*) (cf. *run off one's feet* s.v. *RUN *v.* 44 a).

a 1890 [in Dict., sense 4 b]. 1902 W. N. HARBEN *Abner Daniel* 268 Wish I had more time at my disposal..but I really am rushed, to-day particularly. 1911 M. BEERBOHM *Let.* 2 Oct. (1964) 204, I wish you would tell Sister Loveridge..that I was so 'rushed' that I had not time to go and see anybody. 1916 E. FENWICK *Diary* 14 Feb. (1981) 111 Just rushed off my legs the whole day long. 1923 H. CRANE *Let.* 6 Feb. (1965) 118, I have been so rushed around with too much society that I have not yet got at the review for your study. 1924 *Ibid.* 5 Mar. (1965) 177 What with one's work, one's friends, books, writing, eating and sleeping, things are certainly rushed! 1937 W. H. SAUMAREZ SMITH *Let.* 20 Sept. in *Young Man's Country* (1977) ii. 91 I've been so rushed off my feet that I've not had a moment to do anything except work. 1943 D. POWELL *Time to be Born* x. 225 Amanda's too rushed right now to attend to these details. 1944 C. HIMES *Black on Black* (1973) 196 'State yo' plan, Charlie Chan—then scram!' 'Don't rush me, don't rush me.' 1947 M. MORRIS in B. James *Austral. Short Stories* (1963) 345 She..waited on the tables in Gleeson's dining-room when they were rushed. 1963 *Listener* 31 Jan. 223/2 Lentil soup requires little preparation, and, on a rushed day, takes kindly to the pressure cooker. 1965 *Listener* 4 Nov. 724/2 The elegant Harley Street consultants, the fatherly GPs, the harassed hospital house surgeons,.. those 'you can talk to' and those who are 'rushed off their feet'. 1966 'J. HACKSTON' *Father clears Out* 203 In fact, I rushed her so, that I flabbergasted her, got her rattled. 1977 *Oxford Star* 22 Dec. 1/1 Shopkeepers have been rushed off their feet rolling out the barrel at Sainsburys, the Co-op, [etc.].

e. *U.S.* Of fraternity or sorority members: to entertain (a new student) in order to assess his or her suitability for membership, or to offer him or her membership.

1896 W. C. GORE in *Inlander* Jan. 149 *Rush*,..to entertain a student in various ways, with the view of inducing him to join a fraternity. 1922 S. LEWIS *Babbitt* xxvi. 309 He was on the committee for the Freshman Hop, and..he was being 'rushed' by two fraternities. 1924 P. MARKS *Plastic Age* vii. 62 He ought to be a good man for the fraternity... We've got to rush him sure. 1946 E. B. THOMPSON *Amer. Daughter* x. 173 There were a lot of students who weren't rushed or pledged who found solace in the Y, in literary or musical clubs, but not Dora. 1970 [see *RUSH *sb.*[2] 3 d].

f. To court the affection of (a girl or woman) by means of frequent entertainment, 'dating', etc. orig. and chiefly *U.S.*

1899 F. NORRIS *McTeague* xi. 226 Marcus had 'taken up with' Salna a little after Trina had married, and had been 'rushing' her ever since. 1922 F. SCOTT FITZGERALD *Beautiful & Damned* II. i. 144 With one she had gone to New Haven..she had been flattered because 'Touch down' Michaud had 'rushed' her all evening. 1932 'B. ROSS' *Tragedy of X* 71 He had 'rushed' her, she said, for several months, and they had decided to announce their engagement. 1938 E. BOWEN *Death of Heart* II. iv. 249 That may be because you are so young that no fellow has started to rush you yet. 1955 F. A. COLLYMORE *Barbadian Dialect* 73 Who's the girl your brother's rushing now?

5. b. (Earlier and later examples.) Also, in extended uses, to attack (someone) by means of a sudden rush; to 'go for' (a person).

1863 A. S. ATKINSON *Jrnl.* 29 May in *Richmond-Atkinson Papers* (1960) II. 47 There were two sets of pits (called *rifle* pits by courtesy) the first were rushed but the Maoris ran & got all away. 1930 L. G. D. ACLAND *Early Canterbury Runs* viii. 196 The cook took up his gun and ordered him off, but the man rushed him and the cook shot him dead. 1934 WODEHOUSE *Right Ho, Jeeves* xvi. 197 Damn it, they'd rush the platform. 1937 C. HINES *Nigger* in *Black on Black* (1973) 131 He tried to shift the wire to his right hand so he could flay her with it, but she rushed him, clawing and biting. 1939 *Sun* (Baltimore) 23 Aug. 1/2 The shots were fired by Gerald Blowers.. after about fifty pickets rushed a milk truck on which he was riding. 1962 WODEHOUSE *Service with Smile* iii. 42 'Your sermon was a success, I trust?' 'Well, they didn't rush the pulpit.'

c. *to rush one's fences*: see *FENCE *sb.* 5 c.

d. (Earlier and later examples.)

1862 *Otago: Goldfields & Resources* 26 The Highlay [goldfield]..has been rushed, condemned, almost deserted, and yet survives. 1872 *Daily Tel.* 9 Feb. 3/3 The place was 'rushed'—an expressive word,..which signifies that the diggers swarmed to the spot in such crowds as to render merely foolish any resistance which an owner might be inclined to make. 1878 I. L. BIRD in *Leisure Hour* 5 Oct. 635/2 Even their [*sc.* Indians'] 'reservations' do not escape seizure practically; for if gold should 'break out' on them, they are 'rushed'. 1973 *Nation Rev.* (Melbourne) 31 Aug. (Suppl.) 1/1 It was first explored by Hume and Hovell, then opened up by cattlemen, rushed by gold seekers, and finally developed as a prosperous agricultural area.

6. e. (Further examples.) Also, to hurry, to hasten. Freq. with (*a*)*round* (hence *rush-round* attrib. phr.); *to rush round in circles*: see *CIRCLE *sb.* 1 c.

1914 'HIGH JINKS, JR.' *Choice Slang* 22 We rush off shopping. 1916 W. OWEN *Let.* 18 Mar. (1967) 386, I am obliged to rush into Romford for Running Clothes for a Run announced for the 17th. 1923 H. CRANE *Let.* 9 May (1965) 134 Of course I have been rushing around to a lot of other agencies. 1958 *Spectator* 20 June 807/2 Young Asia and young Africa delight to pull the legs of rush-round correspondents. 1965 G. MCINNES *Road to Gundagai* xii. 216, I..tore out of the room and rushed off to school. 1973 [see *NOSE-BAG 3]. 1976 F. RAPHAEL *Glittering Prizes* 23 'It's C7, Third Court, St John's. Only I've got to rush.' 'I shall be there.'

f. (Earlier and later examples of *to rush it*.)

1856 in B. H. HALL *College Words & Customs* (rev. ed.) 365 Leg it, put it, rush it, streak it, *Run* and worship God. 1976 L. SANDERS *Hamlet Warning* (1977) ix. 79 She laughed and looked up at him. 'Well, let's not rush it, Loomis.'

g. In American football, to run carrying the ball; to gain ground by running with the ball.

1949 *Lafayette Alumnus* (Lafayette College, Easton, Pa.) 24 Oct. 1/1 The Maroon had made 9 first downs rushing, three by passing and one by penalty. 1974 *Cleveland* (Ohio) *Plain Dealer* 13 Oct. C. 1/3 The Buckeyes rushed for 359 yards as quarterback Cornelius Greene and wingback Brian Baschnagel each scored twice. 1979 *Tucson* (Arizona) *Citizen* 20 Sept. 10 D/7 The Warrior running game hasn't been as effective as McKee would like, having rushed for only 133 yards in the first two games.

rushed, *ppl. a.* Add: (Further examples.) Also, made of rushes.

1918 *Heal & Son Catal.: Cottage Furnit.* 31 Jacobean Chair, in Dark Oak with rushed seat. 1957 A. CLARKE *Later Poems* (1961) 67, I think of rushed bones, Bogland, in furnaces, grown greener.

rushee (rʊʃī·). *U.S. College slang.* [f. RUSH *v.*[2] + -EE[1].] One who is 'rushed' (see *RUSH *v.*[2] 4 e); a candidate for membership of a fraternity or sorority.

1916 *Dialect Notes* IV. 279 *Rushee*,..a girl being 'rushed' for a college fraternity. 'The chapter has some good-looking rushees this year.' Widespread at Nebraska University. 1928 *Amer. Speech* III. 220 *Rushee*,..an individual who is being rushed by fraternities or sororities. 1940 *Sun* (Baltimore) 16 Sept. 2/8 It [*sc.* a telegram] requested a report on the character, family and scholarship of a rushee. 1942 *College Topics* (Univ. Virginia) 12 Oct. 1/2 (*heading*) Rushees navigate to houses by maps. 1960 *Amer. Speech* XXXV. 104 The girl rushee who does not have 'tights-omania' will be blackballed in short order.

rusher[2]. Add: **3.** (Further examples.) Also, any player who rushes (see *RUSH *v.*[2] 6 g).

1969 *Internat. Herald Tribune* 6 Nov. 13/6 Floyd Little, leading rusher in the American Football League, suffered a sprained right knee. 1974 *Cleveland* (Ohio) *Plain Dealer* 13 Oct. C. 6/2 Wellington carried 12 times for 129 yards to lead all rushers. 1979 *Tucson* (Arizona) *Citizen* 3 Oct. 1D/4 The seventh-leading rusher in the city.

rush hour. Also **rush-hour.** [f. RUSH *sb.*[2] + HOUR.] A period of the day during which the movement of people is at its height, esp. one during which large numbers of people are travelling to or from work. Also *attrib.*

1898 [in Dict. s.v. RUSH *sb.*[2] 9]. 1907 'O. HENRY' *Trimmed Lamp* 233 As solid as granite in the 'rush-hour' tide of humanity, stood the Man from Nome. 1926 *Daily Graphic* 13 May 1 (*caption*) The 'rush hour' at Earl's Court yesterday. Travelling discomforts are mitigated by much good humour and politeness. 1931 *Morn. Post* 18 Aug. 6/4 Rush-hour trains held up. 1932 D. L. SAYERS *Have his Carcase* iv. 50 The place is like the Corner House in the rush hour. 1955 *Times* 17 June 9/4 Even now, great congestion is caused by traffic entering and leaving the park, particularly in the rush hours. 1961 I. MURDOCH *Severed Head* xxvii. 221 Through the rush-hour traffic the god that protects drunken men protected me. 1973 'M. INNES' *Appleby's Answer* iii. 32 It was the first of London's evening rush-hours, and their taxi made only a tedious stop-go progress. 1977 B. PYM *Quartet in Autumn* ii. 17 A woman, slumped on a seat on the Underground platform while the rush hour crowds hurried past her.

rushing, *vbl. sb.* Add: **b.** (Later examples in Football.) Cf. *RUSH. *v.*[2] 6 g.

1966 ROTE & WINTER *Lang. Pro Football* III. 134/2 *Rushing*,..offensive yardage gained by running with ball. 1970 *Globe & Mail* (Toronto) 26 Sept. 36/5 Raimey..is leading the Eastern Football Conference in rushing. 1972 J. MOSEDALE *Football* x. 143 He led the league in rushing in 1942 and 1946.

d. *rushing bases*, a children's game, = *King Cæsar* s.v. *KING *sb.* 5.

1849, 1969 [see *KING *sb.* 5].

e. *U.S.* The process of entertaining candidates for fraternities and sororities and of selecting those who are suitable (see *RUSH *v.*[2] 4 e). Also *attrib.*

1901 *Independent* (N.Y.) 15 Feb. 392/1 The inter-fraternity contract..limited the 'spiking' or 'rushing' (terms covering all methods of competing for desirable members) to ten days. 1910 *Collier's* 23 July 16/3 My four friends..admitted that during this so-called 'rushing' for new members there was considerable rivalry among sororities. 1929 *Daily Maroon* (Chicago) 8 Oct. 2/1 No pledges are made until the fourth day of rushing week. 1931 *Kansas City Star* 10 Oct., Fraternity rushing is entirely over and the freshmen have been told their place in life. 1942 *College Topics* (Univ. Virginia) 12 Oct. 1/2 Rushing got underway and first year-men sought in the darkness for the various fraternity houses. 1946 *Life* 18 Nov. 114/2 Howard has a normally lively interest in extracurricular activities like football, swimming, college dances..fraternity and sorority rushing. 1957 *Encycl. Brit.* IX. 701/2 All of the fraternities aim to be select and to choose their members carefully from the mass of incoming students, the rushing, as the process of selection is called, being well organized and supervised by the older members. 1972 C. S. OGILVY *Tomorrow's Math.* (ed. 2) ii. 37 On college campuses where the fraternity system flourished, it was customary to allow each fraternity to choose..whom it should invite to become members. There were rules and 'codes of rushing', but even so..many undergraduates never had a chance to get into any fraternity.

rushing, *ppl. a.* Add: (Further examples.) Also *fig.*

1881 W. G. MARSHALL *Through America* (1882) 93 Each and all of these have done a 'rushing' business during the past year. 1915 *N.Y. World* 7 Aug. 1/3 All this time the soda-water stands were doing a rushing business.

rushlight. Add: **1. b.** (Earlier example.)

1827 G. GRIFFIN *Holland-Tide* 326 It was neither like sun-light, nor moonlight, nor the light of the stars, nor fire, nor rush light.

c. Also of persons.

1866 *Yale Lit. Mag.* Apr. 229 Peters told him that good scholars were looked upon here as mere rush-lights.

2. *rushlight box, holder, love.*

1934 L. B. LYON *White Hare* 34 And dowsed in dark Their little rush-light love. 1937 *Discovery* Jan. 12/1 Rushlight holders (examples of which date from the 15th century onwards but are commonly of the 18th or early

19th) are generally in iron but might be..merely split sticks. 1955 G. STEVENS *In Canadian Attic* 23 Light for domestic purposes was first supplied by the campfires; next came the rushlight holder. 1969 E. H. PINTO *Treen* 123 Rush-light boxes. These containers..usually of oak or mahogany and mostly dating from the 18th century, are rare and seldom identified correctly. 1969 *Canadian Antiques Collector* Jan. 19/2 An interesting item you may be able to pick up is a 'Rushlight-holder'.

rushy (rʊ·ʃi) *a.*[2] [f. RUSH *sb.*[2] + -Y[1].] Quick, hurried. Also as *adv.*, in a rush, hurriedly.

1908 H. G. WELLS *War in Air* iv. 153 Too soon, Bert my boy—too soon and too rushy. 1976 W. TREVOR *Children of Dynmouth* i. 34 It was all half joking, all quick and rushy, his mother laughing her shrill staccato laugh, Rose-Ann laughing also, neither of them listening to him.

‖ **rus in urbe** (rū·s in ɜ̄·ɹbe), *phr.* [a. L. *rūs in urbe* country in city.] The creating of an illusion of the countryside in a city; an urban building, garden, prospect, etc., which suggests the countryside. Also *attrib.* So **rus-in-u·rbe-ish** *a.*

1759 GRAY *Let.* 24 July (1827) II. 40, I am now settled in my new territories commanding Bedford gardens, and all the fields as far as Highgate and Hampstead..; so *rus-in-urbe-ish*, that I believe I shall stay here. 1795 tr. *C. P. Moritz's Travels* 68 In Grosvenor-square..there is a little circular wood, intended, no doubt, to give one the idea of *rus in urbe*. 1804 A. SEWARD *Life of Dr. Darwin* i. 16 To this *rus in urbe*..resorted..a knot of philosophic friends. 1841 W. M. THACKERAY in *Fraser's Mag.* XXIV. 393/2 A very handsome country place..a first-rate *rus in urbe*, as the great auctioneer called it when he hammered it down. 1873 A. J. MUNBY *Diary* 21 May in D. Hudson *Munby* (1972) 330 P. A. Taylor is going to sell this charming rus in urbe; ample widespreading old country house, with timbered lawns, and acres of garden. 1939 'N. BLAKE' *Smiler with Knife* ii. 33 This..is a remarkably non-committal room. Not so much *rus in urbe* as *suburbia in rure*. 1963 *Times* 6 Apr. 11/3 The Rumanian passion for *rus in urbe* exceeds even the English. 1968 *Times* 15 Oct. 7/1 It is a natural progression of the rus in urbe feeling of fashion this autumn. 1976 *Times* 9 Aug. 10/8 Two foxes..live in a corner of the allotments—which seems to be taking *rus in urbe* too far.

Ruski, var. RUSKY *a.* and *sb.*[2] in Dict. and Suppl.

Ruskin[4]. Add: *Ruskin work* = *Ruskin linen*. **Ruskinese** *sb.* (earlier and later examples), *a.* (earlier example); **Ruskinia·na**, memorabilia of Ruskin; **Ruski·nianly** *adv.*, in a Ruskinian manner; **Ruskinism** (earlier and later examples); **Ru·skinist** = RUSKINIAN *sb.*; **Ru·skinite** *sb.* and *a.* = RUSKINIAN *sb.* and *a.*; **Ru·skiny** *a.*, designating the style admired by Ruskin.

1863 *Macmillan's Mag.* Nov. 67/2 'Laying by', therefore, in Ruskinese, can only mean simple hoarding. 1863 G. M. HOPKINS *Let.* 10 July (1938) 55, I..hope you will approve some of the sketches in a Ruskinese point of view. 1933 *Scrutiny* II. 1/2 Almost every page is littered with clichés, floating in Ruskinese, and quotations from Horace are liberally applied. 1931 *Times Lit. Suppl.* 18 June 492/2 Every scrap of Ruskiniana is now scattered to all quarters. 1978 *Lancashire Life* July 50/4 It was the Severns' ill-luck to preside over the gradual running-down of Brantwood after 1918, a process completed after Arthur's death in 1931, in his ninetieth year, by the sale of household effects and Ruskiniana. 1974 SHERWOOD & PEVSNER *Buildings of England: Oxfordshire* 325 Capitals of the French Early Gothic foliage type but also Ruskinianly naturalistic. 1851 'AN ARCHITECT' *Something on Ruskinism* p. iv, These few pages do not pretend to exhibit a portraiture of Ruskinism, or anything like one. 1940 E. GILL *Autobiogr.* vii. 277 Anything that looks like Ruskinism in my subsequent development is chiefly due to the fact that we both accepted the same first principles. 1969 *Daily Tel.* 8 Feb. 15/4 For long there have been a few devoted Ruskinists in this country and America. 1889 *St. James's Gaz.* 31 July 9/1 (*heading*) The travels of a Ruskinite's letter. 1899 G. B. SHAW *Let.* 17 Oct. (1972) II. 113 On the whole, the wooden figures [in the Alhambra], which classical & Ruskinite tourists alike disdain, are the things best worth looking at. 1975 *Maclean's Mag.* May (B.C. Suppl.) 6 As far back as 1890, a utopian community of Ruskinites had set up shop in the Fraser Valley. 1963 *Bookseller* 20 July 248/1 It was Ruskin who introduced linen cut-work to England and it is sometimes called Ruskin work, but..the proper name [is]..Linen Cut-Work. 1977 M. GREEN *Children of Sun* (rev. ed.) iii. 135 Harold [Acton] liked the Brighton Pavilion, but not the grey Gothicism, nor the Ruskiny ruins.

Rusky, *a.* and *sb.*[2] Delete *rare* and add: *slang* or *colloq.* Also **Roosky, Ruski, Russki** (now the most usual form), **Russky.** (Earlier and later examples.) Hence **Ru·sski-land**, Russia.

1858 F. DUBERLY *Let.* July in E. E. P. Tisdall *Mrs Duberly's Campaigns* (1963) vi. 197 Was I the Englishwoman who had gone with the armies to make war against the Ruski? 1919 *Amer. Legion Weekly* 22 Aug. 22/1 Large numbers of lowly 'Rooskies' plodded through the weary days. 1919 *Our Navy* (U.S.) Nov. 15/1 The Russkis were friendly to us because we had a lot of rubles. 1920 *Amer. Legion Weekly* 12 Mar. 5 A. E. F. Siberia..en route to Russki-land. 1923 D. YORK *Company A* 50 A little Russki at one side drew our attention. 1937 G. FRANKAU *More of Us* xiii. 135 To Mussolini's braves and Hitler's huskies Left we the task of tackling Stalin's Russkies. 1948 M. LASKI *Tory Heaven* i. 14 People like the Russkis and the Yanks like dealing with gentlemen. 1957 V. NABOKOV *Pnin* 71 He and Serafima, his large, cheerful, Moscow-born wife..would throw Russki parties every now and then, with Russki hors d'oeuvres. 1959 C. MACINNES *Absolute Beginners* 55 We've got to produce our own variety, and not imitate the Americans—or the Ruskis, or anybody. 1961 *Even. Bull.* (Philadelphia) 29 Mar. 22/3 (*caption*) Keeping up with the (Russki) Joneses. 1978 I. B. SINGER *Shosha* ii. 38 A Russky with all these qualities is awaiting you there.

Rusnak, Rusniac, Rusniak, vars. *RUSSNIAK *sb.* and *a.*

Russ, var. *RUS.

Russell[1]. (Earlier and further examples of *Russell cord*.)

1873 [see *Persian cord* s.v. *PERSIAN *a.* 2]. 1896 *Woman's Life* 11 July 179/2, I cannot say that the coarse blue alpaca in various shades of navy blue and prune finds favour in my eyes... It reminds me too much of the fearsome fabric known as Russell cord, that in the far-off days of my childhood constituted my school dresses. 1940 *Chambers's Techn. Dict.* 735/1 *Russel cord* (textiles), a dress fabric of plain weave, with a cord effect; made from cotton warp and worsted or mohair weft, the warp being in tapes. 1966 *Guardian* 25 Apr. 7/2 Russell cord is the right stuff for stuff wearers.

Russell[2] (rʊ·sěl). The name of Patrick *Russell* (1727–1805), Scottish physician and naturalist, used *attrib.* and in the possessive in **Russell('s) viper** to designate a venomous snake, *Vipera russellii*, found in India, Burma, and Thailand, distinguished by a yellowish-brown skin marked with black rings or spots, and first named *Coluber russellii* in his honour by G. Shaw in 1797.

[1797 G. SHAW *Naturalist's Miscellany* VIII. pl. ccxci (*caption*) The Russelian Snake. *Ibid.*, Dr. Patrick Russel ..presented the elegant specimen here figured to the British Museum.] 1908 E. P. STEBBING *Man. Elem. Forest Zool. India* xii. 182 The Russell's Viper is one of the most deadly of all snakes. 1937 L. BROMFIELD *Rains Came* I. i. 7 With the first splattering drop of rain they would come swarming out of old roots and crannies in the wall—the cobras, the Russell's vipers, the fierce little kraits. 1940 *Lancet* 17 Aug. 195/1 It was decided to determine how far commercial preparations of Russell-viper venom would meet these requirements. 1961 *Listener* 2 Nov. 735/2 Russell's Vipers, when I have caught them, were always very sluggish. 1972 M. RICHARDSON *Fascination of Reptiles* xv. 158 Another snake which Indian snake charmers sometimes carry around in their baskets is Russell's viper.

Russell[3] (rʊ·sěl). The name of Bertrand Arthur William *Russell*, 3rd Earl Russell (1872–1970), mathematician and philosopher, used *attrib.* and in the possessive in connection with a paradox concerning the set of all sets that do not contain themselves as members: the condition for it to contain itself is that it should not contain itself.

1922 tr. *Wittgenstein's Tractatus* 57 Herewith Russell's paradox vanishes. 1937 *Jrnl. Symb. Logic* II. 31 This contradiction corresponds to Russell's paradox. 1950 W. V. QUINE *Methods of Logic* (1952) §42. 249 This difficulty is called *Russell's paradox*, for its discoverer (1901). 1963 G. T. KNEEBONE *Math. Logic* iv. 127 Russell's antinomy..this is the paradox of the class {*x*/*x*∉*x*}. 1967 *Encycl. Philos.* V. 46/1 *Russell's Paradox*,..Russell.. came upon a new paradox, that of the set of all sets that do not contain themselves as elements. A set *r*, the 'Russell set', is defined by the following condition: for every *x*, *x*∈*r* if and only if *x*∉*x*. By substitution we obtain: *r*∈*r* if and only if *r*∉*r*. 1977 BELL & MACHOVER *Course in Math. Logic* x. 462 Unfortunately..(1.2) is untenable even when *k* = 0, because it leads to the well-known Russell paradox.

Russell[4] (rʊ·sěl). The name of George *Russell* (1857–1951), English gardener, used *attrib.*, esp. in **Russell lupin**, to desginate a large perennial lupin belonging to a variety of *Lupinus polyphyllus* developed by him, introduced in 1937, and distinguished by long racemes of papilionaceous flowers in one or two of a wide range of colours.

1937 *My Garden* XI. 332 The 'Russell Lupins'..would be more correctly described as a new 'race' rather than an improvement of an old. 1957 A. BLOOM *Hardy Perennials* 229 The famous Russell strain..has superseded all others in recent years. *Ibid.* 230 Russell Lupins ..have few dislikes other than over-rich or limy soils. 1974 *Country Life* 17 Jan. 72/3 George Russell, a Yorkshire gardener, was attempting to improve the perennial lupin and found..one that had sported to a new shape of flower... Virtually all subsequent lupins have been of the 'Russell' type. 1979 *Guardian* 25 Aug. 9/6 We grow Russell lupins as a hobby.

Russell body (rʊ·sěl). *Path.* [Named after William *Russell* (1852–1940), Scottish pathologist, who described it in 1890 (*Brit. Med. Jrnl.* 13 Dec. 1356–60).] A hyaline mass of immunoglobulin produced in numbers in, and sometimes extruded by, plasma cells in excessive response to challenge by antibodies.

1913 O. C. GRUNER *Biol. of Blood-Cells* vi. 276 The intracellular Russell bodies are considered the result of myelin degeneration of the cell-substance. 1970 PASSMORE & ROBSON *Compan. Med. Stud.* II. xxv. 39/2 Sometimes the fabrication of antibody within plasma cells is so excessive that hyaline masses of protein (Russell bodies) are deposited in their cytoplasm. 1977 R. B. THOMPSON *Disorders of Blood* xxxii. 508/2 Occasionally Russell bodies are present... They are intracellular acidophil hyaline bodies which can be produced in animal cells by injections of bacteria and can be found in many organs and tissues.

Russell fence (rʊ·sěl). *Canada.* Also **Russel fence, rustle fence.** [Said to derive from the name of Mr. *Russell*, its inventor.] A fence in which the top rail lies in the crux of crossed posts and the lower rails hang suspended from it by looped wires.

1932 N. M. JAMIESON *Cattle in Stall* 75 There was the rustle fence, with boom on top and centre. 1953 *Canad. Geogr. Jrnl.* Dec. 226/2 The Russell fence has..been patented and Russell, the inventor, succeeded..in collecting royalties from people who built fences on his model. It consists of pairs of crossed posts,..but the rails are hung from the crotches and from each other in wire loops. 1962 A. FRY *Ranch on Cariboo* i. 4 It was Russel fence, an ingenious stake and rail structure held together by heavy wire and named, I'm told, after the man who invented it. 1968 *Islander* (Victoria, B.C.) 29 Dec. 16/4 The contract price for a log fence would be around $400 a mile while a Russel fence of the same length could be had for $150. 1971 W. HILLEN *Blackwater River* xiii. 122 This..falcon would have been seen sitting on nearly any Russell fence..a few years back. 1972 R. WRIGHT *Cariboo Mileposts* 14 (*caption*) A typical Cariboo scene of a Russell fence, cattle and pine trees.

Russellian (rʊse·liăn), *a.* and *sb.* [f. the name *Russell* (see *RUSSELL[3]) + -IAN.] **A.** *adj.* Designating the mathematical or philosophical ideas of Bertrand Russell; characteristic of or pertaining to Russell (in quot. 1956, *spec.* of *Russell's paradox*: see *RUSSELL[3]). **B.** *sb.* An adherent of Russell's ideas. Hence **Ru·ssellism**, the system of Russell's thought and practice.

1923 C. D. BROAD *Sci. Thought* xiii. 534 Physical objects in the Russellian sense. 1934 R. CAMPBELL *Broken Record* 145 Russellism and Waughism seem to me to be as tyrannical and brutal..as Arnold-of-Rugby-ism. 1937 *Discovery* Feb. 61/1 The Russellian 'calculus of propositions'. 1950 *Mind* LIX. 344 Neither Aristotelian nor Russellian rules give the exact logic of any expression of ordinary language. 1954 R. WELLS in *Word* X. 235 Thus Wittgenstein has played a major part in all three branches of the Russellian movement. *Ibid.* 245 Examples ..have been separately discussed by various Russellians and Wittgensteinians. 1956 G. E. M. ANSCOMBE tr. *Wittgenstein's Remarks on Found. of Math.* v. 166 The Russellian contradiction is disquieting, not because it is a contradiction, but because the whole growth culminating in it is a cancerous growth. 1972 *Listener* 27 Jan. 119/1 His interest is the more Russellian one of getting the system to work. 1977 *Language* LIII. 74 Perhaps he is taking a Russellian view of definite descriptions.

russellite (rʊ·sĕləit). *Min.* [f. the name of Arthur E. I. M. *Russell* (1878–1964), English mineralogist + -ITE[1].] A tetragonal mixed oxide of bismuth and tungsten, $Bi_2O_3.WO_3$, found as pale yellow or green fine-grained masses.

1938 HEY & BANNISTER in *Mineral. Mag.* XXV. 42 We propose for the mineral the name russellite, in honour of Mr. Arthur Russell. *Ibid.* 49 Russellite..occurs at the Castle-an-Dinas wolfram mine, St. Columb Major, Cornwall, as pale yellow fragments. 1944 *Ibid.* XXVII. 2 Two pellets of comparatively pure russellite about the size of split peas were found to show on their rounded surfaces blebs of bright gold easily visible to the naked eye. 1970 *Ibid.* XXXVII. 705 Russellite $Bi_2O_3.WO_3$ occurs in a small pegmatite near Poona, Western Australia.

Russell–Saunders (rʊ·sěl sǭ·ndəɹz). *Physics.* [The names of Henry Norris *Russell* (1877–1957), U.S. astrophysicist, and Frederick Albert *Saunders* (1875–1963), U.S. physicist; they first described the scheme in *Astrophys. Jrnl.* (1925) LXI. 38.] *Russell–Saunders coupling*, an approximation employed in a procedure for describing the possible energy states which can be adopted by a set of electrons in an atom; = *LS-coupling* (s.v. *L II. 6* b); also *Russell–Saunders scheme, state*, etc.

[1927 *Zeitschr. f. Physik* XL. 532 Im Russell-Saunderschen Schema.] 1928 *Physical Rev.* XXXI. 957 In most spectra there is a different coupling of the quantum vectors, the Russell-Saunders coupling, which is responsible for normal multiplets. 1935 CONDON & SHORTLEY *Theory of Atomic Spectra* vii. 208 The Russell-Saunders terms corresponding to the configurations s^x and p^x. 1961 WALKER & STRAW *Spectroscopy* I. i. 87 It will be assumed..that Russell-Saunders coupling applies to the individual momentum vectors associated with the electrons. 1962 COTTON & WILKINSON *Adv. Inorg. Chem.* xxvi. 574 Just as the set of five *d* orbitals is split apart by the electrostatic field of surrounding ligands to give two or more sets of lower degeneracy, so also are the various Russell-Saunders states of a d^n configuration. 1966 D. H. WHIFFEN *Spectroscopy* xi. 141 Weak transitions disobeying these selection rules..are not at all uncommon and their existence implies that the Russell–Saunders scheme is not accurately applicable. **1967** W. R. Hindmarsh *Atomic Spectra* iii. 27 The assumption of negligible spin-orbit interaction on which the Russell–Saunders coupling scheme is based.

Russenorsk (ru·sənǫɹsk). [Norw.] A pidgin of Russian and Norwegian used by fishermen.

1964 E. Palmer tr. *Martinet's Elem. General Linguistics* v. 155 *Russenorsk*, the product of contacts between Russian and Norwegian fishermen on the shores of the Arctic Ocean, which had an ephemeral existence but has been adequately described. **1974** L. Todd *Pidgins & Creoles* i. 6 Russenorsk, a pidgin now almost extinct, arose from the contact of two Indo-European languages, Russian and Norwegian, as a means of facilitating communication between Russian and Norwegian fishermen. **1974** *Encycl. Brit. Macropædia* VI. 1063/2 Pidgins..are not true languages since they are derivations from two or more parent tongues. An example is Russenorsk (from Russian and Norwegian), used by Norwegian fishermen with Russian traders.

russet, *sb.* and *a.* Add: **B.** *adj.* **1. b.** (Earlier and later examples.)

1629 J. Parkinson *Parad.* III. xix. 587 The Russet pippin is as good an apple as most of the other sorts of pippins. *Ibid.* xxi. 592 The russet Catherine is a very good middle sized peare. **1929** M. de la Roche *Whiteoaks* vii. 99 'H'm,' grunted Finch, tearing a bite from a russet apple. **1970** *Globe & Mail* (Toronto) 28 Sept. 5/1 (*caption*) Seventy-eight acres of..Russet and Tolman Sweet apple trees, all of them laden with ripe fruit.

d. *russet-brown* (later example), *-gold.*

1861 J. G. Whittier *Cobbler Keezar* in *Poet. Wks.* (1898) 84/2 Yellow and red were the apples, And the ripe pears russet-brown. **1959** E. Pound *Thrones* cii. 82 The colour..As lacquer in sunlight haliporphuros, russet-gold In the air.

e. *Comb.*, as *russet-clad, -faced, -haired, -skinned.*

1849 Thoreau *Week Concord Riv.* 19 It may be many russet-clad children lurking in those broad meadows. **1897** W. B. Yeats *Secret Rose* 80 A russet-faced boy..sat ..watching the swallows. **1936** M. H. Bradley *Five-Minute Girl* ix. 159 A small girl of eight, hazel-eyed, russet-haired. **1878** G. M. Hopkins *Lett. to R. Bridges* (1955) 48 He [*sc.* a seaman]..is..russet-of-morning-skinned With the sun, salt, and whirling wind.

Hence **ru·ssetly** *adv.*, with a russet colour; in a russet state. *poet.*

c **1864** E. Dickinson *Poems* (1955) II. 687 Though Pyramids decay And Kingdoms, like the Orchard Flit Russetly away.

russeted, *ppl. a.* (in Dict. s.v. Russet *v.*) Add: *spec.* (of fruit and vegetables) rough-skinned; (of their skin) rough. (Further examples.)

1917 [see next]. **1930** *Jrnl. Pomology & Hort. Sci.* VIII. 299 Examination of the russeted fruit revealed the presence of the fungus already mentioned. **1950** Smock & Neubert *Apples & Apple Products* iii. 29 This periderm is a cork cambium and is capable of forming cork-like cells which result in the russetted appearance of some fruits. **1978** *Sci. Amer.* June 64/3 The 'Idaho' potato..is the most prized of the North American varieties because of its large size, pulpy interior,..and russeted (textured) skin.

russeting, *vbl. sb.* (In Dict. s.v. Russet *v.*) Add: *spec.* The roughening of the skin of normally smooth-skinned fruit, esp. apples. (Further examples.)

1917 *32nd Ann. Rep. Maine Agric. Experiment Station* 86 These plots have been introduced solely..as a basis of comparison with other sprays..with respect to scab control and the production of fruit russetting and foliage injury... This treatment has increased the number of russetted apples from 5 to 10 per cent. **1928** C. E. Owens *Princ. Plant Path.* xxv. 609 Bordeaux is apt to cause leaf injury and russeting of fruit..in moist weather. **1974** *Nature* 8 Feb. 337/3 The most insidious type of damage is russeting, and fruit skins are sensitive to certain fungicides in May and June.

Russia. Add: **2. a.** *Russia braid* (earlier examples), *duck* (earlier examples), *iron, matting* (earlier example), *sheet-iron* (earlier example).

1847 *Lady's Newspaper* 11 Dec. 566 Gentleman's waistcoat. Material—blue..cloth, and green or amber Russia braid. **1873** *Young Englishwoman* Nov. 572/1 Travelling toilet of brown cashmere, braided with brown of a darker shade in Russia and Breton braid. **1761** *Essex Inst. Hist. Coll.* (1912) XLVIII. 95 Best prime Russia Duck **1822** M. Edgeworth *Let.* 6 Feb. (1971) 344 His Russia duck jacket and trowzers. **1897** *Sears, Roebuck Catal.* 127 We make it [*sc.* a stove] lined and unlined, and in smooth steel and planished iron (usually called Russia iron). **1901** *Daily Colonist* (Victoria, B.C.) 12 Oct. 8/7 (Advt.), 'Famous' Air-Tights [*sc.* stoves]. Built for light service. Will burn rough blocks of wood. Made of Russia iron. **1839** J. J. Audubon *Ornithol. Biogr.* V. 504 The nest..has uniformly been built of Russia matting. **1859** *Rep. Comm. Patents 1858* (U.S.) I. 530 The process of manufacturing sheet-iron, to possess most of the qualifications of 'polished Russia sheet-iron'.

Russian, *sb.* and *a.* Add: **A.** *sb.* **1. a.** *Great, Little, White Russian* (earlier examples). See also White *a.* 11 e in Dict. and Suppl.

1845 *Encycl. Metrop.* XXIV. 225/1 Two principal branches are distinguished, rivals of each other, and still cherishing the bitterest animosity—the Russians *great* and *little*, and the Poles. **1854** J. S. Maxwell *Czar & his People* xxiii. 125 The Great Russian is predominant among the various peoples of the empire. **1866** *Chambers's Encycl.* VIII. 380/2 The 50,500,000 Russians..are divisible into—1. Great Russians..2. Little Russians..3. White Russians.

b. Delete *rare*$^{-1}$ and add later examples.

1866 *Chambers's Encycl.* VIII. 388/2 The Russians adopt the same expedient with the Greeks, viz., of selecting the bishops from among the monks. **1963** T. Ware *Orthodox Church* viii. 165 It is not without reason that the expressions 'Soviet Church' and 'Soviet Patriarch' have now become common in the mouth of Russians.

c. (Earlier and later examples.)

1845 D. Mackenzie *Emigrant's Guide* 118 These wild *Russians*, as they are here called, will..clear at the first leap a stockyard six feet in height. **1848** H. W. Haygarth *Recoll. Bush Life Austral.* xii. 135 Though he had been among horses since he was a child, his present lot were a set of the veriest 'Russians' (Anglicè, wild things) *he* ever had anything to do with. **1945** Baker *Austral. Lang.* iii. 68 An old term worth noting, since it has been obsolete for half a century or more, is *Russians* for wild stock.

2. Also *Comb.*, as *Russian-speaking.*

1960 *Amer. Speech* XXXV. 163 The material..was gathered among the Russian-speaking population. **1976** 'M. Barak' *Secret List of Heinrich Roehm* vii. 77 You need Russian-speaking agents to infiltrate Russian circles.

3. Also *ellipt.* for *Russian cigarette.*

1892 [see *Egyptian *sb.* 5 b]. **1937** R. Chandler in *Dime Detective Mag.* Nov. 43/2 There were three long cigarettes..Russians, with hollow mouthpieces. **1963** N. Freeling *Because of Cats* x. 163 He had juju cigarettes too; like Russians, with a big mouth piece, and pretty loose.

B. *adj.* **1. a.** Also with distinguishing adjs., as *Great, Little, White Russian* (see sense A. 1 a in Dict. and White *a.* 11 e in Dict. and Suppl.).

1911 C. J. Hogarth tr. *V. O. Kluchevsky's Hist. Russia* I. xiii. 203 The Great Russian stock stands to the Little Russian in the proportion of three to one, and the Little Russian to the White Russian in a similar ratio. **1918** R. Beazley et al. *Russia, from Varangians to Bolsheviks* I. iii. 79 The principality of Moscow had become a Great Russian nation, the Prince of Moscow a Great Russian sovereign. **1942** L. B. Namier *Conflicts* 8 The conflict with Russia turned on Poland's dominion over vast stretches of land inhabited by White Russian and Little Russian peasantries. **1963** *Times Lit. Suppl.* 31 May 388/4 The growth of Great-Russian jingoism.

Comb. **1963** R. I. McDavid *Mencken's Amer. Lang.* v. v. 265 *Nudnik*..is widely used by Russian-Jewish immigrants. **1976** *Times* 15 May 14/8 Sholem Aleichem came from a middle class Russian–Jewish background.

2. a. *Russian bear* (later *fig.* examples); *Russian Blue*, a lightly built short-haired cat belonging to the breed so called, distinguished by greyish-blue fur, green eyes, and large pointed ears; *Russian long-hair(ed)* (*cat*), a stocky, long-coated cat with a relatively short tail, belonging to a breed once so called but no longer a distinct group; *Russian pony*, a small, hardy, roan pony belonging to a breed originally developed in Russia; cf. *Cossack 2 b; *Russian sable*, the heavy dark fur of the sable, *Martes zibellina*; cf. Sable *sb.*^{1} 1 a; *Russian wolfhound* = *borzoi.

1737 Pope *1st Epistle 2nd Bk. Horace Imitated* 22 No Lord's Anointed, but a Russian Bear. **1806** M. Edgeworth *Leonora* II. lxix. 121 It would really be pleasant to have a Czar at one's feet... The ancients represent Cupid riding the Numidian lion, and why should he not tame the Russian bear? **1972** C. Short *Naked Skier* xxvi. 147 All the gay, waltzing and slightly frenetic set-up [of Vienna] can be shattered at an instant by the roar of a Russian bear. **1977** W. Feaver *When We were Young* 19 The political cartoon figures—British lion, Russian bear, German eagle. **1889** H. Weir *Our Cats* 66 The Blue Cat was at first shown as the Archangel cat, then Russian blue, Spanish blue, Chartreuse blue, and..the American blue. **1933** E. Buckworth-Herne-Soame *Cats* xxxvii. 164 Russian Blues, so called because they were originally brought from Russia, are now known as 'Foreign Blues'. **1953** A. White tr. *Colette's Cat* viii. 191 I'll withdraw.. into my cold room..under the protection of..a Russian Blue cat. **1971** W. J. Burley *Guilt Edged* viii. 132 Trotsky is our cat—a Russian blue. **1976** *Loughborough Monitor* 26 Nov. 2/3 (Advt.), Ready for Christmas.. Russian Blue Kittens. **1889** H. Weir *Our Cats* 30 The Russian long-haired cat..differed from the Angora and the Persian in many respects. **1939** [see *Burmese cat* s.v. Burmese *a.* b]. **1972** Ing & Pond *Champion Cats of World* 72 The Russian Longhair..apparently had an even more woolly coat, and a shorter tail. **1898** J. D. Brayshaw *Slum Silhouettes* 151 I'm off to the Cattle Market to buy one o' those little Rooshian ponies. **1903** A. Bennett *Truth about Author* xvi. 214 Arrival of the second post on a Russian pony that cost fifty shillings. **1936** A. W. Seaby *Brit. Ponies* 144, I stopped before a strange looking old light roan pony between the shafts of a sweep's cart. It was a Russian pony the owner averred. **1871** Russian sable [see *doll-land* s.v. *Doll *sb.*^{1} 5]. **1930** M. Bachrach *Fur* xxi. 322 There probably has never been a peltry that has enjoyed such popularity..for so long a period of time as the Russian Sable. **1952** 'M. Cost' *Hour Awaits* 152 The fur was sable, and Russian sable. **1973** D. Orgill *Jasius Pursuit* x. 99 He took one of the fur coats... It was a magnificent garment—Russian sables. **1872** G. H. Lewes *Jrnl.* 28 Dec. in *Geo. Eliot Lett.* (1956) V. 352 He took us up to Lady Paget to see a superb Russian Wolfhound. **1922** R. Leighton *Compl. Bk. Dog* ix. 141 There is not a more elegant and graceful dog than the Borzoi or Russian Wolfhound... The wearer of a lovely silky coat, he is essentially a spectacular animal. **1941** B. Schulberg *What makes Sammy Run?* viii. 157 She walked.. with a haughty pride, the way one does with Russian wolfhounds. **1976** Botham & Donnelly *Valentino* viii. 59 He bought a pair of Russian wolfhounds (white).

b. *Russian olive* (U.S.), the oleaster, *Elæagnus angustifolia*, a spiny shrub with silvery leaves belonging to the family Elæagnaceæ, native to Europe and western Asia, and naturalized in parts of western North America; *Russian poplar* (Canada), a poplar native to north-east Asia, *Populus maximowiczii*, which has leathery leaves with whitish undersides; *Russian thistle* (U.S.), a tumbleweed, *Salsola kali*, a creeping prickly herb belonging to the family Chenopodiaceæ; = Saltwort 1; *Russian vine*, a fast-growing deciduous climbing plant, *Polygonum baldschuanicum*, of the family Polygonaceæ, native to southern Turkestan and bearing clusters of white or pink flowers.

1938 W. R. Van Dersal *Native Woody Plants U.S.* 119 Russian olive... A large shrub to small tree; introduced from Europe and Western Asia. **1951** T. H. Kearney et al. *Arizona Flora* 586 Russian-olive, native of the Old World, often cultivated as an ornamental in the United States. **1972** W. A. Weber *Rocky Mt. Flora* 173 The Russian Olive..is cultivated throughout the region at lower elevations. **1950** E. A. McCourt *Home is the Stranger* ii. 19 Around the farmyard ran a stunted, ill-kempt wind-break of Russian poplar. **1965** I. Reekie *Melita Trail* ii. 6 He planted the first grove of trees in the community—ash, cottonwood, Russian poplar, and Manitoba maples. **1894** *Amer. Folk-Lore* VII. 97 *Salsola Kali*, var. *Tragus*, Moguin, Russian thistle, Russian cactus. **1898** *Monthly S. Dakotan* I. 103 Only tiny triangular spots remained dry in the lee of broken corn-stalks and scattered Russian thistles. **1939** *Nat. Geogr. Mag.* Aug. 262/2 The Russian thistle..is a relative of beets and spinach in the family of the 'goosefoots'. **1971** *Country Life* 2 Sept. 583/1 In autumn the Russian Thistles..ran freely before the wind at thirty miles an hour. **1948** N. Catchpole *Flowering Shrubs & Small Trees* vii. 156 A vigorous climber, and the quickest growing one known in our gardens..is commonly called the Russian vine. **1963** *Oxf. Bk. Garden Flowers* 164/2 Russian Vine..is, in fact, a native of Bokhara, and is a rampant twining plant which will quickly smother any unsightly object in one season. **1977** K. O'Hara *Ghost of T. Penry* iv. 25 The stone arch ..was half-blocked by the ruins of a ramshackle gate overgrown with Russian vine.

d. *Russian bagatelle, braid, crash, diaper, poker.*

1850, etc. Russian bagatelle [see *cockamaroo]. **1953** R. Senhouse tr. *Colette's Gigi* 25 The Russian braid of her nightdress. **1973** *Country Life* 15 Feb. 425/1 Black wool suit, the jacket..trimmed with Russian braid. **1932** Russian crash [see *nurse cloth* s.v. *Nurse *sb.*^{1} 8]. **1827** *Hallowell* (Maine) *Gaz.* 20 June 4/4 (Advt.), Received.. Russian Diaper. **1957** Simpson & Weir *Weaver's Craft* (ed. 8) xiii. 165 *Russian diaper*.—This has a total of 26 threads to each pattern,..and it can also be used quite successfully as a border to some of the other patterns, for curtains, covers, etc. **1970** T. Lilley *Projects Section* x. 121 He plays Russian poker in the mess. *Ibid.* xvii. 232 In Russian Poker there are four players: each has thirteen cards which he arranges in three hands—two of five cards and one of three. These hands are each arranged in poker fashion... Three players play against the fourth—the banker—and the bank changes after every fourth deal.

d*. Special collocations: *Russian ballet*, a style of ballet developed at the Russian Imperial Ballet Academy and popularized in the West by Sergei Diaghilev's Ballet Russe from 1909; also a group of dancers trained in this style; *Russian Bank* (*Banker, banque*), a card game similar to solitaire but played by two persons; *Russian bath* = *Turkish bath* s.v. Turkish *a.* 2 a; also *fig.*; *Russian boot*, a leather boot that extends to the calf, usu. with a wide cuff; *Russian cigarette*, a cigarette with a hollow pasteboard filter; *Russian dancer*, one who performs a Russian folk-dance; *Russian dinner*, a style of dinner in which fruit and wine are placed at the centre of a table and courses are served from a sideboard; *Russian doll*, any of a set of hollow wooden dolls, the smallest of which fits inside the next smallest, and so up to the largest; *Russian dressing*, a savoury dressing with a mayonnaise base; *Russian Easter egg*, an artificial egg shell designed as a container for presents given at Easter; *Russian egg*, a poached egg served on a lettuce leaf with mayonnaise; *Russian* (*spring–summer*, etc.) *encephalitis*, a viral encephalitis transmitted by wood ticks; *Russian Revolution*, the overthrow of the Tsar and the eventual establishment of the Bolshevik form of government in Russia between February and October (Old Style) 1917; cf. *October Revolution* s.v. *October 3 and *Revolution *sb.* 11; *Russian roulette*, an act of bravado in which a person loads (usu.) one chamber of a revolver, spins the cylinder, holds the barrel to his head, and pulls the trigger; also *fig.*; *Russian salad*, a salad of vegetables with mayonnaise; *Russian scandal*, (*a*) a game in which a whispered message, after being passed from player to player, is contrasted in its original and final versions; (*b*) gossip inaccurately transmitted; *Russian tea*, (*a*) tea grown in the Caucasus or a drink made from this; (*b*) any tea laced with lemon or rum.

1911 *Westm. Gaz.* 3 June 2/3 She disapproved of early morning tea and auction bridge, of ski-ing and the two-step, of the Russian ballet and the Chelsea Arts Club ball. **1928** A. Christie *Mystery of Blue Train* xxxi. 261, I never saw anything in this Russian ballet... Too highbrow for me. **1937** J. Laver *Taste & Fashion* viii. 110 The overwhelming wave of Orientalism which swept over Parisian society..was due to..Paul Poiret and the Russian ballet. **1947** *Ballet Ann.* I. 68 Diaghileff, in search of inspiration, made straight for Paris, *chic* and *chi-chi*, and so Russian ballet became *Ballet Russe*. **1973** W. Tute *Resident* ii. 36 As you know the Russian ballet is the best in the world. **1915** W. Del Mar *Rules of Russian Bank* 1 The game of Russian Bank is played by two persons each with a pack of fifty-two cards. The object of the game is to dispose of the cards. **1930** A. Woollcott *Let.* 26 Apr. (1944) iv. 85 We played backgammon or Russian Bank all the way over [the Atlantic]. **1930** 'E. Queen' *French Powder Mystery* xv. 118 Not many people know how to play Russian banque. **1970** R. Lowell *Notebk.* 121 Ford, playing Russian Banker. **1804** M. Wilmot *Let.* 24 Apr. in *Russ. Jrnls.* (1934) I. 94 The true Russian Bath admits a Vapour which I cou'd not support. **1863** C. Lever *Day's Ride* in *All Year Round* 16 Feb. 455/1 It is a sort of intellectual Russian bath, in which the luxury consists in the exaggerated alternative between being scalded first and rolled in the snow afterwards. **1961** L. Mumford *City in Hist.* xiii. 386 In the seventeenth century..the bath was re-introduced as a foreign importation, a luxury..: the so-called Turkish or Russian bath. **1975** A. Hailey *Moneychangers* II. vi. 193 Mr Quartermain likes either a sauna or a Russian bath wherever he is. **1915** in C. Willett Cunnington *Eng. Women's Clothing* (1952) iv. 132 The fashionable side-lacing Russian boots with fawn cloth tops and patent leather fronts, 21/- a pair. **1926** Wodehouse *Heart of Goof* iii. 108 You bet your Russian boots I was! **1977** V. S. Pritchett *Gentle Barbarian* xiii. 212 Turgenev..wandered about in heavy Russian boots. **1905** C. Mackenzie *Diary* 30 Mar. in *My Life & Times* (1964) III. 222 A lazy young man..who used to smoke Bobbie's Russian cigarettes. **1926** C. Beaton *Diary* 24 Apr. in *Wandering Yrs.* (1961) 87 Smoking Russian cigarette after Russian cigarette. **1940** E. Hemingway *For whom Bell Tolls* ii. 20 Robert Jordan..brought out one of the flat boxes of Russian cigarettes... They were long narrow cigarettes with pasteboard cylinders for mouth pieces. **1972** J. Wainwright *Requiem for Loser* iii. 52 They sipped tea, smoked Russian cigarettes and discussed this and that. **1913** Mrs. P. Campbell *Let.* 25 Mar. in *B. Shaw & Mrs. Campbell* (1952) 102 Russian dancers were imitated—shoes kicked off—hair came down. **1931** C. Remfry-Kidd tr. *Colette's Renée Néré* I. i. 9 The Russian dancers are trying to get warm... They shout 'Yonk!' all together. **1851** *London at Table* I. 26 We have already alluded to a Russian dinner, which is the best and most economical. It is always served hot from the kitchen, and as the entrées are not exposed to the public gaze, there may be fewer of them; the joints served at the side-board by an experienced artist, are more palatable and tempting than when carved on the table. **1868** M. Jewry *Warne's Model Cookery* 56 The present fashion of Russian dinners is fast banishing the necessity for promiscuous carving. **1937** K. Blixen *Out of Afr.* ii. 135 Those Russian wooden dolls which will unscrew, and have then got another doll inside them, and another inside that, and which are sold under the name of Katinka. **1967** C. Fremlin *Prisoner's Base* xvi. 114 Each item in the dream was fitting into her interpretation like a set of Russian dolls. **1922** *Hotel World* 15 Apr. 15/1 Russian Dressing. **1938** L. Bemelmans *Life Class* II. ii. 127 The salad, covered with Russian dressing, is a mixture of endives..pineapple..cream cheese with chopped chives. **1976** *Billings* (Montana) *Gaz.* 1 July, In medium bowl, thoroughly combine ½ cup Russian dressing, ground beef, and breadcrumbs. **1949** H. C. Bainbridge *P. C. Fabergé* iv. 67 (*heading*) The Imperial Russian Easter Eggs. **1955** W. Gaddis *Recognitions* III. v. 903 Like a Russian Easter Egg, this Thing had a tiny window in one end. **1932** M. F. Daniels tr. *F. Nietlispach's Cold Dishes & Hors-D'œuvre* I. 16 *Russian Eggs.* Poach the eggs, strain and cool. Arrange on a lettuce leaf..cover with mayonnaise and sprinkle with..minced ham and a little chopped parsley. **1969** G. Lyall *Venus with Pistol* vii. 39, I yelled an order for.. Russian eggs... It comes up a salad the size of the Garden of Eden. **1943** *Science* 12 Mar. 246/1 (*heading*) Close relation between Russian spring-summer encephalitis and louping-ill viruses. **1948** Olitsky & Casals in T. M. Rivers *Viral & Rickettsial Infections of Man* viii. 192/1 Russian Far East encephalitis (Synonyms: Russian spring–summer encephalitis; Russian spring or summer encephalitis; Russian forest-spring encephalitis; Russian tick-borne encephalitis; Russian endemic encephalitis). *Ibid.*, Russian Far East encephalitis is a disease occurring in spring and early summer, mainly in the Far East provinces of the Soviet Union and less frequently in European and Siberian Russia. **1976** W. L. Drew *Viral Infections* i. 7 Group B [*sc.* arboviruses] includes the viruses of St. Louis encephalitis, yellow fever, and dengue, as well as the viruses of Russian encephalitis and hemorrhagic fever found in Europe and Russia. [**1805** C. Wilmot *Let.* 7 Dec. in *Russ. Jrnls.* (1934) II. 208 The famous 28th of June 1762, the day of the Russian Revolution.] **1907** I. Zangwill *Ghetto Comedies* 399 We are a Labour party... We have the whole Russian Revolution on our shoulders.] **1917** C. P. Scott *Let.* 25 Mar. in D. Ayerst *Guardian* (1971) xxvii. 403 Don't you feel the Russian revolution rather stirring in your bones? **1919** *Mr. Punch's Hist. Great War* 176 A 'History of the Russian Revolution' has already been published. **1922** *Encycl. Brit.* XXXII. 319/2 The history of the Russian Revolution starts with the gradual dissolution of all fundamental institutions and notions. **1945** 'G. Orwell' in *Common Wealth Rev.* Nov. 12/1 The failure of the Russian Revolution—failure, that is, in the sense that the Revolution has not fulfilled the hopes that it aroused twenty-five years ago. **1977** *Times* 26 Mar. 12/4 A history-teaching colleague suggested that it might have been better if the Russian Revolution had never taken place. **1937** G. Surdez in *Collier's* 30 Jan. 16 'Did you ever hear of Russian Roulette?'.. With the Russian army in Rumania, around 1917,..some officer would suddenly pull out his revolver,..remove a cartridge from the cylinder, spin the cylinder, snap it back in place, put it to his head and pull the trigger. **1946** *N.Y. Post* 23 Oct. 5/2 The game was 'Russian Roulette', and the odds were 5–1. **1956** 'M. Innes' *Appleby plays Chicken* I. ii. 18 'Is it done with a revolver..with one of the six chambers loaded?' 'No. That's Russian roulette.' **1960** *Guardian* 27 July 16/4 This party..had 'played Russian roulette with American strength and American progress'. **1976** *Lancet* 9 Oct. 776/2 Abusive parents are often the scarred survivors of generations of reproductive russian roulette. **1879** M. Jewry *Warne's Model Cookery* (new ed.) 456/2 *Russian Salad*... Cold boiled beetroot; cold carrots [etc.] ..smoked salmon, or white meat of chicken and tongue. Cut the vegetables into pieces all of one size, add the salmon..mix with Mayonnaise sauce. Garnish with anchovies. **1940** M. Dickens *Mariana* iii. 64 'I'll tell you something, Tich,' said Uncle Geoffrey beginning on his Russian salad. **1973** 'S. Harvester' *Corner of Playground* I. viii. 70 He had carved the cold roast chicken and served it with Russian salad. **1873** L. Troubridge *Jrnl.* 28 Aug. in J. Hope-Nicholson *Life amongst Troubridges* (1966) vi. 50 We played Russian Scandal in the train, which was very jolly. **1873** C. M. Yonge *Pillars of House* IV. xxxviii. 126 Susie has been well lectured on Russian scandal! **1893** —— *Girl's Little Bk.* 17 Do not repeat it [*sc.* gossip]. You will probably make Russian scandal of it, and the next person will add to it. **1929** H. G. Wells *King who was King* ii. 59 '*We used to play a game called Russian scandal*'... The screen shows a row of young people... The first whispers to the second, who whispers to the third, and so on. **1953** 'P. Wentworth' *Ivory Dagger* lxii. 209 There used to be a game called Russian scandal. Something was whispered from one to another, and you have no idea what it would come out like by the time even a few people had had the handling of it. **1960** G. E. Evans *Horse in Furrow* xiii. 177 Stories passed from one to another are proverbially incorrect as 'Russian scandal'. **1862** M. B. Chesnut *Diary* 25 June in C. V. Woodward *M. Chesnut's Civil War* (1981) 395 They had Russian tea, champagne, a samovar. **1884** G. Meredith *Let.* 24 Mar. (1970) II. 732 Bid him arrive by half-past five, that the thirsty troop may be refreshed by Russian tea. **1930** L. G. D. Acland *Early Canterbury Runs* v. 111 The publican offered to make him some 'Russian tea'... 'Russian tea' turned out to be ordinary tea well laced with rum. **1952** 'R. Crompton' *William & Tramp* vii. 217 She said he had lemon in his tea 'stead of milk an' I know that's called Russian tea. **1975** *Times* 1 May 15/3 There is Russian tea, a long leaf variety grown in the foothills of the Caucasian mountains. **1976** *Eastern Even. News* (Norwich) 9 Dec. 8/1 Tins of Russian, Formosa or Jasmine tea are about 68p for half a pound.

Russianism. Add: **2. b.** Soviet communism as practised by the Russians.

1933 *Catholic Times* 21 Apr. 5/1 One of the reasons for the great success of Russianism was the austerity of many of its leaders and the complete self-sacrifice of many of its rank and file.

c. A Russian custom.

1957 V. Nabokov *Pnin* iii. 71 Shy graduate students would be taught vodka-drinking rites and other stale Russianisms.

3. b. A Russian idiom.

1957 V. Nabokov *Pnin* iii. 87 Her fluent and flashy New York English, with..soft lapses into furry Russianisms. **1962** *Amer. Speech* XXXVII. 279 (*title*) Russianisms in the American press. *Ibid.*, Russianisms are divided here into three main groups: loanwords, foreignisms, and calques. **1967** *Listener* 19 Jan. 99/3 Professor Markov allows the occasional Russianism to show through in his introduction, which isn't perfectly idiomatic.

Russianist (rɒ·ʃănist). [f. Russian *sb.* + -ist.] A student of Russian language and literature.

1976 *Times Lit. Suppl.* 23 Jan. 79/4 The occupational obsessions of Russianists writing on Chekhov. **1980** *Ibid.* 19 Sept. 1018/4 Joe Andrew will be known to most British Russianists as a co-chairman of the Neo-Formalist circle.

Russianness (rɒ·ʃănˌnĕs). [f. Russian *a.* + -ness.] The quality or state of being Russian.

1937 *Sunday Times* 21 Nov. 9/2 His [*sc.* Lenin's] essential Russianness had not been weakened by culture; it had not been Westernised by foreign contacts. **1954** U. Weinreich in Saporta & Bastian *Psycholinguistics* (1961) 382/2 We may characterize the utterance by the feature of 'Russianness' or 'Englishness'. **1968** *Economist* 9 Nov. p. x/2 This book is an account of the author's life-long love affair with Russianness: not with Russia, past or present, but with a Russianness conceived in the nursery as a daydream of the trans-Siberian railway. **1973** *Observer* 4 Feb. 37/3 Wilson is worrying away about the peculiarities of the Russian language and the astonishing Russianness of Russians. **1977** V. S. Pritchett *Gentle Barbarian* ii. 27 The Russian disease..the ever shadowy figure of Russianness.

Russic, *a.* For *Obs.*$^{-1}$ read *Obs.* and add later *poet.* example.

1757 J. Dyer *Fleece* IV. 399 Culder's woofs, and those of Exe and Frome,..Thither by Russic caravans are brought.

Russification. (Later examples.)

1936 *Discovery* Feb. 50/1 The education provided by the Russian authorities was..entirely subordinated to the policy of 'Russification' uncompromisingly adopted since the suppression of the Polish rebellion in 1863. **1972** *Times* 15 May 14/4 In the non-Russian republics one can get a 'feel' for the way people reach to russification. **1976** A. Powell *Infants of Spring* x. 173 An increasing policy of Russification resulted in much unrest there [*sc.* in Finland]. **1976** *Survey* Spring 188 Ukrainians, Byelorussians, Lithuanians and Latvians..are exposed to intensive Russification. **1979** *Daily Tel.* 29 Nov. 18 The policy of the Soviet Union in the Baltic countries is to annihilate their peoples by enforced Russification and deportation.

Russify, *v.* Add: (Earlier and later examples.) Hence **Ru·ssifying** *ppl. a.*

1865 Queen Victoria *Let.* 23 May in R. Fulford *Your Dear Letter* (1971) 27 Good Alice seems quite Russified. **1924** [see *Prussification]. **1954** Koestler *Invisible Writing* x. 110 The natives were drawn into the towns, educated, Russified and Stalinised. **1960** E. R. Goodman in J. A. Fishman *Readings Sociol. of Lang.* (1968) 731 The..denigration of Stalin left the Russifying impact of Stalin's linguistic policy intact. **1973** *Listener* 5 Apr. 444/2 The Party instructions were to Russify the *Moscow News*. **1974** *Encycl. Brit. Macropædia* XVI. 47/1 His [*sc.* Ivan III's] churches, the original aspect of which has been altered by successive russifying restorations, were clearly in the Italian style.

Russki (now the most usual form), **Russky,** varr. Rusky *a.* and *sb.*^{2} in Dict. and Suppl.

Russniak, *sb.* and *a.* Add: Also **Rusnak, Rusniac, Rusniak. a.** *sb.* (Examples of the language.)

1862 [see Ruthenian *sb.* 2]. **1894** A. Lefèvre *Race & Lang.* 239 Little Russian, Rusniac, or Ruthene. **1955** R. Jakobson *Slavic Lang.* (ed. 2) 4 Ukrainian dialects are classified into Northern, Southeastern, Southwestern, and Carpathian groups; the marginal dialects (called Rusnak) of the latter group are Slovak-influenced.

Russo-. Add: **a.** *Russo-American, -Byzantine, -Chinese, -Czech, -French, -German, -Greek* (later example), *-Japanese, -Persian, -Polish* (later example), *-Slavonic, -Swedish.*

1814 tr. *G. H. von Langsdorff's Voyages & Travels* II. 99 At our arrival, we found this new settlement of the Russio-American [*sic*] Company in want of almost all the necessaries of life. **1977** *Gramophone* June 56/3 A Russian conductor..and a Russo-American concerto which I haven't heard him play before. **1889** *Cent. Dict.*, Russo-Byzantine. **1973** *Country Life* 6 Dec. 1913/1 Russo-Byzantine work in purple and gold. **1903** G. Bell *Let.* 20 May (1927) I. viii. 162 We had to go to the renowned Russo-Chinese bank to change our notes. **1949** I. Deutscher *Stalin* 418 The Russo-French and the Russo-Czech alliances were concluded. **1897** E. A. Bartlett *Battlefields of Turkey* ii. 39 Two such evenly poised camps as the Russo-French League and the German monarchies. **1928** L. Robinson *Let.* in *Lett. S. O'Casey* (1975) I. 266 The second act in the modern Russo-German manner is very fine. **1972** D. Dakin *Unification of Greece* ix. 126 Phillipon..was sent to..Athens in the hope of improving Russo-Greek relations. **1906** A. Bennett *Let.* 6 Mar. (1966) I. 70 The last trick of bringing the Russo-Japanese war into the story. **1953** A. Smith *Blind White Fish in Persia* iii. 50 Further along the coast there was a Russo-Persian Caviare industry but their fleet was not visible. **1926** Russo-Polish [see *expressionism]. **1959** G. Nandriș *Handbk. Old Ch. Slavonic* I. 20 In the later period Russo-Slavonic, Old Serbian, and Old Croatian texts show a regular epenthetic *L*. **1974** *Encycl. Brit. Macropædia* XVI. 60/2 Russo-Swedish relations were settled during the Napoleonic era.

b. Russophil(e (further examples), **Russophobe** (further examples); **Russophobia** (earlier and later examples); **Russophobism** (earlier example), **Russophobist** (earlier adj. example).

1897 E. A. Bartlett *Battlefields of Thessaly* ii. 38 There was a curious outburst of Russophile writing in a portion of the English Press. *Ibid.* iii. 46 One great factor in the game..is also ignored by our English Russophiles. **1946** R. Capell *Simiomata* III. 153 The new Foreign Minister..is a socialist and Russophil. **1967** C. Seton-Watson *Italy from Liberalism to Fascism* ix. 342 Aehrenthal..had the reputation of an extreme conservative and Russophil. **1971** *Daily Tel.* 11 May 9/4 It is true that as a result of our history we have been and still are Russophile. **1946** G. Stimpson *Thousand Things* 250 The war party, the Russophobes, who urged Prime Minister Disraeli to side with the Turks, against the Russians, became known as jingoes. **1966** R. Blake *Disraeli* xxvi. 607 The more Turcophobe Gladstone became, the more Russophobe was Disraeli. **1836** J. S. Mill in *Westm. Rev.* XXV. 276 Ministers are smitten with the epidemic disease of Russo-phobia. **1980** *Daily Tel.* 8 July 14, I fear that it might only serve to conceal the ever lurking presence of good old 19th-century British Russophobia. **1880** E. W. Hamilton *Diary* 25 Aug. (1972) I. 37 It is extraordinary what amount of Russophobism pervades the royal mind. **1882** *Ibid.* 4 June (1972) I. 284 She may have the Russophobist and anti-Russian party in power again in this country.

Russonorsk (rɒ·sonǫɹsk). [f. Russo- + Norw. *norsk* Norwegian.] = *Russenorsk.

1966 R. A. Hall *Pidgin & Creole Lang.* I. i. 12 On a basis of Russian and Norwegian, there grew up a pidgin known as Russonorsk, which was used..between Russian and Norwegian fishermen along the Arctic coast of Norway. **1977** C. F. & F. M. Voegelin *Classification & Index World's Lang.* 311 Russonorsk can be said to be a Russian-based pidgin-creole, but might just as well be said to be Norwegian-based.

rust, *sb.*^{1} Add: **9. a.** *rust-free*; *rust-bearded.*

1922 Joyce *Ulysses* 238 A sailorman, rustbearded, sips from a beaker. **1951** Whitby & Hynes *Med. Bacteriol.* (ed. 5) iii. 30 Iron in the form of rust-free filings..reduces the oxidation-reduction potential of liquid media.

b. *rustblack, -brown* (later example), *-red* (later examples).

a **1915** Joyce *Giacomo Joyce* (1968) 15, I kissed her stocking and the hem of her rustblack dusty skirt. **1977** *Lancashire Life* Nov. 56/2 By May 19 these were six [eggs] —tiny, white and rust-brown speckled. **1937** V. Woolf *Years* 297 The down was soft rust-red on its wings. *a* **1963** S. Plath *Ariel* (1965) 21 A sunken rust-red engine.

c. Special Combs.: **rust-resistant, -resisting** *adjs.*, (of a metal) made so as not to rust; (of a plant) not liable to rust disease; so **rust-resistance.**

1911 *Jrnl. Agric. Sci.* IV. 99 Have any results of a definite progressive nature in the physiology of rust-resistance been yet obtained? **1940** J. C. Hudson *Corrosion Iron & Steel* ii. 10 The use of rust-resisting steels has hitherto..been confined to definite fields of service, in which rust resistance is of primary importance. **1947** *Ann. Rev. Microbiol.* I. 78 The population shifts of physiologic races show the practical need for extensive replication in time and space in testing varieties for stem rust resistance. **1907** *Jrnl. Agric. Sci.* II. 127 In some countries a careful search has already been made for rust-resistant varieties, but, on the whole, with comparatively little success from the economic point of view. **1930** H. Goldschmidt tr. *Müller-Hauff & Stein's Automobile Steels* iv. 142 Krupp was the first to use rust-resistant steel. **1947** *Ann. Rev. Microbiol* I. 78 The case of Ceres wheat..illustrates this point. This rust-resistant variety was distributed in 1926, and by 1934 was grown on more than four million acres. **1964** *Abraham & Straus Catal.* Jan. 29 Rust-resistant, lightweight, aluminium ladder. **1891** R. Wallace *Rural Econ. Austral. & N.Z.* iv. 72 A sample of rust-resisting wheat from Queensland. **1909** *Chambers's Jrnl.* Nov. 766/1 The steel used is a special light, thin, rust-resisting, seamless metal. **1962** *Sci. Survey* XXI. 332 'Stainless steel' (more correctly 'rust-resisting steel') is now familiar in both domestic and industrial applications.

10. rust bucket *N. Amer. colloq.*, an old and rusty ship; also *Austral. colloq.*, a rusty old car; **rust disease** = sense 6 a; **rust hypha,** a hypha of a rust fungus.

1945 *Seafarers' Log* 8 June 2/2 C. M. Chaney, J. D. Riffle and R. R. Ullan were dispatched to one of the more notorious rust buckets as Quartermaster and AB's respectively. **1959** *Wall St. Jrnl.* 13 Oct. (Eastern ed.) 1/6 To try to get the jump on the weather, ore carriers.. 'will put every rust bucket that floats into the ore trade'. **1969** *Sunday Mail* (Brisbane) 9 Nov. 15/4 (*heading*) Car trade-ins fit for scrap. Dealers stuck with 'rust-buckets'. **1979** F. Forsyth *Devil's Alternative* 7 The *Garibaldi* an amiable old rust-bucket out of Brindisi. **1902** W. Watson *Thompson's Gardener's Assistant* (rev. ed.) II. 600/1 (Index), Rust disease on vines—cause and treatment. **1975** *Times* 30 May 16/5 A team of scientists at Wye College has discovered in the tobacco plant a naturally produced fungicide effective against the 'rust' diseases which are commonly destructive to important food crops, vegetables and garden flowers. **1909** W. Bateson *Mendel's Princ. Heredity* 25 Miss Marryat found that the rust-hyphae are checked before entering the stomata of the resistant plants.

rustbank, var. *Rusbank.

rusted, *ppl. a.* Add: (Later *fig.* example). Also, stuck or lodged as a result of rusting, and with *in* and *up*.

1924 J. Masefield *Sard Harker* II. 82 The catch of this beastly revolver seems to have jammed... I'm afraid it's rusted-in, or something. **1929** J. M. Ross in *Oxford Poetry* 36 The Purple Beech..brandishing aloft his burnt, dark, rusted leaves. **1967** E. Short *Embroidery & Fabric Collage* i. 18 A piece of rusted up metal..could be noted in a sketch-book. **1972** *Guardian* 15 Mar. 10/2, I don't ever feel that I get rusted-up. I don't think I could have done this play if I'd been rusted-up.

rusticated, *ppl. a.* Add: **3. b.** Of pottery: (see quot. 1936).

1936 *Proc. Prehistoric Soc.* II. 19 Rusticated pottery, *i.e.* pottery of which the surface has been roughened all over as a method of decoration. **1939** V. G. Childe *Dawn Europ. Civilization* (ed. 3) 338 Definitions of certain terms, descriptive of *ceramic decoration*, here used in a special or restricted sense... Rusticated—by roughening the surface, generally covered with a thick slip, by pinching with the fingers, brushing, etc. ('barbotine'). **1967** *Antiquaries Jrnl.* XLVII. 202 Somersham..is further known for its imposing 'pot-beaker', with rusticated finger-decoration. **1977** G. Clark *World Prehistory* (ed. 3) x. 448 Impressed and rusticated pottery was supplemented by polychrome painted Tupiguarini ware.

rustiness. 2. (Later example.)

1900 G. B. Shaw *Let.* 14 Mar. (1972) II. 136 In the old days I was always standing between Bland and the rustinesses that used to come from his Tory imperviousness to the Radical notions with which Socialism was adulterated.

rustle, *v.* **4.** For '*U.S. colloq.*' read 'orig. *U.S. colloq.*' and add: **b.** (Earlier and further examples.) Now usu. with *up*.

1844 *Spirit of Times* 14 Sept. 343/3 He nailed my thumb in his jaws, and rostled up a handful of dirt & throwed it in my eyes. **1891** *Advance* 29 Jan. 101/2 Some of the members have arranged..to go out on the hills and 'rustle up' wood. **1903** A. Adams *Log of Cowboy* xxi. 332 Honeyman being excused on agreeing to rustle the wood and water. **1919** *Punch* 29 Jan. 87/1 All George's performances in the art of rustling bivvies rank as star. **1931** 'Dean Stiff' *Milk & Honey Route* viii. 81 Kid, you go out and rustle some breakfast and meet us at the water tank. **1944** M. Laski *Love on Supertax* iii. 45 Do you think you can rustle me up something to eat? **1959** *Spectator* 11 Sept. 331/1 They put on their programme with such costumes as they could rustle up from Edinburgh's shops. **1978** G. Greene *Human Factor* v. ii. 243 We may not be able to rustle up more than an omelette, Muller. Pot luck.

c. Of animals: to forage (for). Also *absol.* and *transf.*

1881 *N.Y. Times* 18 Dec. 4/3 Cattle, in winter, 'rustle' for food by nosing through the snow to the dried grass beneath. **1913** L. V. Kelly *Range Men* 109 The [Red River] settlement took up the ranching of cattle, the turning out of herds to rustle their own living. **1916** 'B. M. Bower' *Phantom Herd* xiv. 243 He turns you out thinking he'll let you rustle for yourself awhile. **1924** R. Campbell *Flaming Terrapin* vi. 61 As shepherd winds drove forth their foamy sheep To rustle through the verdure of the deep. **1925** *Chambers's Jrnl.* Mar. 168/1 It is accustomed to rustle its living through the long severe winters of its habitat. **1955** J. C. Ewers in *Bull. U.S. Bureau Amer. Ethnol.* No. 159. 42 Unless the snow was too deep..(i.e. over ca. 2 feet) they [*sc.* horses] generally could rustle enough food..to gain a meagre subsistence. **1961** R. P. Hobson *Rancher takes Wife* vii. 111 The range horses never had to be fed hay, but rustled all year round. **1966** H. Marriott *Cariboo Cowboy* ii. 32 One year with another most of the cattle, except the calves, bulls, and thin cows, all rustled out in most of these ranges for most of, if not all, the winter.

d. To gather, round *up*; also with *in* and *out*.

1896 G. W. Dice *Life* vi. 30, I was more fortunate than ever this time, and..'rustled up' a good big herd of cattle, which we shipped to Kansas City. **1903** A. Adams *Log of Cowboy* iv. 53 Our foreman..sent Honeyman to rustle in the horses. **1924** A. J. Small *Frozen Gold* iii. 81 Why ain't you rustlin' a crowd of the boys up to corral the swabs? **1947** 'N. Blake' *Minute for Murder* v. 101 He's to rustle out all his men..and post them round the building. **1965** G. McInnes *Road to Gundagai* x. 183 Well, go and rustle up the rest of them.

e. To steal (cattle, horses, etc.) by rounding them up. Also *transf.* and *absol.*

1902 A. H. Lewis *Wolfville Nights* xv. 234, I claims that this Bowlaig b'ar is guilty of rustlin' the mails an' must.. be hanged. **1910** W. M. Raine *B. O'Connor* xix. 299 We're after them for rustling a bunch of Circle 33 cows. **1916** 'B. M. Bower' *Phantom Herd* xiv. 243 You hold a grudge against your dad, and you rustle from him mostly. **1948** *Range Riders Western* May 30/1, I ain't ever rustled a cow in my life. **1951** L. MacNeice tr. *Goethe's Faust* 163 One rustles cattle, one a wife.

rustler. Add: **2. b.** (Later examples.) Now also outside the U.S. and in extended and *transf.* uses.

1908 J. M. Sullivan *Criminal Slang* 20 *Rustler*, a horse thief. **1964** *Wall St. Jrnl* 9 Jan. 1/4 He covered the orange rustlers with a double-barreled shotgun. **1971** *Daily Tel.* 19 Apr. 2/1 East Anglian farmers were warned to beware of pig rustlers after 15 sows and five pigs.. were stolen. **1977** *Oxford Times* 16 Sept. 5/5 Police trying to round up horse rustlers in Oxfordshire have admitted the efforts often prove futile.

rustling, *vbl. sb.* Add: **3.** Stealing (esp. cattle) from farms, ranches, etc. Also *transf.* and *attrib.*

1893 *Aberdeen* (S. Dak.) *Sun* 5 Jan. 7/4 Rustling cattle is an exciting trade and very profitable, but extremely hazardous. **1907** [see *cattle-rustling* s.v. *Cattle 8 a]. **1924** C. E. Mulford *Rustlers' Valley* x. 118 There had been no signs of rustling for months. **1937** *Sun* (Baltimore) 23 Oct. 3/2 Battle front of the cotton-rustling racket..is the Fabens Island area..comprising 3,000 acres of rich cotton land. **1942** E. Paul *Narrow St.* ix. 69 These crows had learned that living in Paris, near the central markets.., was easier than rustling in the country, exposed to the farmers' shotguns. **1963** *Wall St. Jrnl.* 11 Oct. 9 Add 'orange rustling' to the list of crimes against society. **1976** *Evening Post* (Nottingham) 14 Dec. 6/2 It's the peak of the shoplifting season..a time for turkey rustling and Christmas tree thefts.

rustly, *a.* (Later examples.)

1936 M. Mitchell *Gone with Wind* xlviii. 851 She wanted a tafetta petticoat..so rustly that the Lord God would think it was made of angels' wings. **1959** *N.Z. Listener* 13 Mar. 5/4, I always wrapped my sandwiches in cloth rather than rustly paper.

ru·stproof, *a.* [See Proof *a.* 1 b.] **a.** Of metal: not susceptible to corrosion by rust; rust-resistant.

1691 [s.v. Rust *sb.* 9 a]. **1907** *T. Eaton & Co. Catal.* Spring & Summer 215/3 Three large cupboards..each one has metal rust proof bottom. **1931** *Chambers's Jrnl.* June 414/1 (*caption*) A new rust-proof coating. **1960** *Farmer & Stockbreeder* 19 Jan. 45/2 Rust-proof heavy gauge pressed steel bowl. **1972** *Sat. Rev.* (U.S.) 27 May 4/2 The early safety-razor blades rusted, so I had to dry them... But soon came rustproof steel.

b. Of a plant: resistant to infection by rust.

1931 J. S. Huxley *What dare I Think?* i. 36 We can now produce relatively rust-proof wheat.

rustproof (rɒ·stprūf), *v.* [f. the adj.] *trans.* To make rustproof.

1910 *Cycling* 2 Feb. 95 (*heading*) Rust-proofing the bicycle. **1953** *Archit. Rev.* CXIV. 393 Before despatch each link was tested for size and strength and 'rust-proofed' by immersing it, when hot, in an oil bath and heating it again on removal until the oil dried on the surface. **1977** 'E. Crispin' *Glimpses of Moon* xii. 252 Was it practicable to rustproof metallic structures, as one did cars?

Hence **ru·stproofed** *ppl. a.* Also **ru·stproofer,** one who makes something rustproof.

1925 *Morris Owner's Man.* 83 The K.L.G. is a detachable plug consisting of three parts—a rustproofed steel body and gland nut and an insulated central electrode. **1960** *Farmer & Stockbreeder* 22 Mar. 138/2 (Advt.), Storage Bin. Substantially made from rust-proofed steel. **1976** *Globe & Mail* (Toronto) 4 Dec. 1/2 Car dealers were often able to avoid liability in those situations by saying that the customer's contract was with the rustproofer alone.

ru·stproofing, *vbl. sb.* [f. prec. + -ing^1.]

1. The action or process of making something rustproof. Also *attrib.*

1918 *Aviation* 1 Aug. 40/2 The increased cost of.. processes of rust-proofing..has turned attention to the safety transparent, rust-proof and acid-proof finish. **1931** *Machinery* Oct. 111/2 News comes from England of the development of a new process known as the 'thermo-zinc' rust-proofing method. **1941** Steinbeck & Ricketts *Sea of Cortez* xiv. 135 The eventual disintegration of a stick of wood or a piece of iron..is assured, even though it may be delayed by such protection..as is afforded by painting and rustproofing. **1980** *Times* 29 Feb. 19 Rust-proofing methods are available which can delay by about five years the onset of corrosion.

2. A substance with which something is made rustproof.

1976 *Time* 20 Dec. 57/2 (Advt.), Each car receives 2 separate coats of rustproofing.

rusty, *a.*^{1} Add: **I. 4. c.** Delete 'Now *rare*' and add later examples. Also *transf.*

1924 R. Campbell *Flaming Terrapin* iii. 47 Huge carrion crows came rasping rusty jaws. **1936** J. B. Priestley *They walk in City* vii. 192 All the time his rusty voice went on and on, half jeering at himself, half mocking the audience. **1938** M. K. Rawlings *Yearling* x. 95 They [*sc.* cranes] made a great circle against the sunset, whooping their strange rusty cry that sounded only in their flight. **1967** G. F. Fiennes *I tried to run Railway* ii. 14 At 87 [he] talked in his slow, rusty voice as if he were still on the job.

II. 9. d. *rusty spot* = *Red spot 2.

1900 *Bull N.Y. Agric. Exper. Station* No. 183. 188 'Rusty Spot' is the name given to small yellowish-red points or patches scattered quite evenly throughout the mass of the cheese and having the general appearance of iron rust. *Ibid.* 189 Connell..isolated from a rusty spot cheese an organism which he called *Bacillus rudensis*. **1958** E. M. Foster et al. *Dairy Microbiol.* ii. 20 *Lactobacillus plantarum* var. *rudensis* and *Lactobacillus brevis* var. *rudensis* have been implicated as causes of rusty spot defect in Cheddar cheese.

10. b. *rusty-dusty* (later examples). Also as *sb.*, the buttocks. *Black English.*

1953 W. Burroughs *Junkie* iv. 42 A negro voice was singing, 'Get up, get up, woman, off your big fat rusty-dusty.' **1970** C. Major *Dict. Afro-Amer. Slang* 99 *Rusty dusty*, the buttocks.

c. rusty-back (fern), the scale fern, *Ceterach officinarum*; (further examples).

1908 E. Step *Wayside & Woodland Ferns* 50 The development of the sori gives a distinctly red hue to the underside and justifies the name Rustyback. **1945** A. B. Jackson *Step's Wayside & Woodland Ferns* (ed. 2) 57 The Rusty-back Fern is pretty generally distributed in England. **1960** P. Taylor *Brit. Ferns & Mosses* 120 The Rusty-back Fern is mainly confined in Europe to the Mediterranean region, extending northwards up the Atlantic coast to Great Britain. **1976** *Westmorland Gaz.* 10 Sept. 10/2 Interest was aroused..by seeing the Rusty-back and Wall-rue ferns on a wall.

11. a. *rusty-voiced*; *rusty-old.*

1912 W. Owen *Let.* 2 July (1967) 148 A taciturn, rusty-voiced man. **1917** —— *Poems* (1963) 57 Finished fields, and wire-scrags rusty-old.

rutabaga. Delete *rare* and add *U.S.* For Latin name substitute *Brassica napus* var. *napobrassica*. Add: = Swede 3. Also *attrib.* (Earlier and later examples.)

1799 J. B. Bordley *Essays & Notes on Husbandry* 30 The new turnip, called roota baga, is likely to stand our winters. **1833** W. Sewall *Jrnl.* 20 June (1930) 149/2 Finished planting potatoes. Sowed rutabaga turnips. **1916** *Yukon Territory* (Canada Dept. Interior) 213 The type shape is similar to a rutabaga, but the roots are smooth without laterals. **1951** O. Nash *Family Reunion* 107 We gobbled like pigs On rutabagas and salted figs. **1975** *New Yorker* 10 Nov. 176/2 Pertly written by pertly pretty housewives who have discovered organic gardening and how to rub two rutabagas together to feed four happy, whimsical tots—such books glut the shelves. **1976** [see *oyster plant* s.v. *Oyster 5 d].

ruthenium. Add: **b. ruthenium red,** an intensely coloured red mixed-valence complexsa it of ruthenium, $[(NH_3)_5Ru^{III}ORu^{IV}(NH_3)_4ORu^{III}(NH_3)_5]Cl_6$, obtained by air oxidation of a solution containing ammonia and ruthenium (III) chloride, and employed as a microscopic strain.

1912 *Chem. Abstr.* VI. 297 (*heading*) Differentiation of natural textiles and artificial silks by means of ruthenium red. **1950** N. V. Sidgwick *Chem. Elements* II. 1472 The colour of ruthenium red itself can be detected in solutions more dilute than one in a million. **1978** *Sci. Amer.* Jan. 86 (*caption*) In both preparations the cells were stained with ruthenium red, which is taken up by any polysaccharide glycocalyx fibers that are present.

ruther (rɒ·ðəɹ), *adv.* [Repr. a U.S. colloq. or dial. pronunc. of Rather.] = Rather *adv.*

1872 [see *Hold *v.* 42 c]. **1929** *Amer. Mercury* Sept. 47 Ruther be in cornfield workin' hard, Than be buck private in National Guard. **1938** M. K. Rawlings *Yearling* viii. 71 I'd ruther they hunted their way and leave me hunt mine. **1942** W. Faulkner *Go down, Moses* 59, I ruther never to know than to find out later I have been fooled. **1970** M. Chisholm *McAllister says No* vi. 48 I'd ruther have him in front of me than behind.

ruther, var. *Rather *sb.*

1933 Z. N. Hurston *Mules & Men* I. ii. 49 Her tongue is all de weapon a woman got... She could have had mo' sense, but she told God no, she'd ruther take it out in hips. So God give her her ruthers. She got plenty hips, plenty mouf and no brains. *Ibid.* vii. 162 You didn't figger Ah was draggin' behind you when you was bringin' dat Sears and Roebuck catalogue over to my house and beggin' me to choose my ruthers. **1949** *Richmond* (Va.) *Times-Dispatch* 26 Dec. 6/5 In the Smokies,..to take one's preference is to 'have your ruthers'.

Rutherford (rɒ·ðəɹfɔ̄ɹd). *Physics.* [The name of Ernest *Rutherford* (1871–1937), New Zealand-born English physicist.] **1.** Used *attrib.* and in the possessive to designate concepts developed by him, as **Rutherford('s) (scattering) formula** or **law,** a mathematical expression of Rutherford scattering; **Rutherford model,** a model of the atomic nucleus devised to account for Rutherford scattering; **Rutherford scattering,** elastic scattering of charged particles by the electric fields of atomic nuclei; = **Coulomb scattering*; hence **Rutherford-scatter** *v. trans.*

1931 G. Gamow *Constitution of Atomic Nuclei* iv. 85 The ratio of the observed scattering to that given by Rutherford's formula for a given angle falls to a minimum and then rises again. **1961** Powell & Crasemann *Quantum Mechanics* xii. 465 Coulomb scattering of low-energy protons, for which the classical cross section is given by the Rutherford formula. **1931** G. Gamow *Constitution of Atomic Nuclei* iv. 84 If..the potential barrier is high enough compared with the energy of the α-particle..deviations from Rutherford's law will be small. **1970** I. E. McCarthy *Nuclear Reactions* I. i. 8 The first example of nuclear information being obtained from measurements of the differential cross-section as a function of momentum transfer is the Rutherford law for elastic scattering. **1930** J. Buckingham *Matter & Radiation* iii. 59 The Rutherford model. **1968** M. S. Livingston *Particle Physics* ii. 17 The nucleus of the Rutherford model must have a diameter of less than 10^{-12} cm. **1974** G. Reece tr. *Hund's Hist. Quantum Theory* iv. 62 The formula remains true in Bohr's theory, while (4) is also valid for the Rutherford model of the atom. **1977** *Nature* 6 Jan. 35/2 A beam of ^{32}S ions is Rutherford scattered backwards from the sample and the energies of the scattered ions are measured. **1928** *Proc. R. Soc.* A CXVIII. 548 This gives the Rutherford scattering formula exactly for all velocities of the incident particles. **1935** J. Dougall tr. *Born's Atomic Physics* v. 126 It can actually be proved that Rutherford's scattering formula is strictly valid in wave mechanics also. **1977** *Nature* 6 Jan. 36/1 The Z^2 dependence of the Rutherford scattering cross section has been taken into account.

2. (Usu. written **rutherford.**) A unit of radioactivity orig. equal to one million disintegrations per second; later defined as the quantity of any particular nuclide exhibiting this degree of activity.

The curie is the more usual unit; one rutherford is approximately $2{\cdot}7 \times 10^{-5}$ curie.

1946 Condon & Curtiss in *Physical Rev.* LXIX. 673/1 Since the curie was named in honor of M. and Mme. Curie, the co-discoverers of radium, it is natural to select the name 'rutherford' for the new unit. The appropriate abbreviation is 'rd' which conflicts with the abbreviation of no other well-accepted physical unit. The micro-rutherford would become one disintegration per second. **1947** *Nucleonics* Oct. 34/2 A carefully defined new unit, the rutherford (rd), has been proposed for general use. Ambiguities as a result of choice of numerical values, failure to distinguish between beta rays per sec and distintegrations per sec, and extensions to arbitrary and undefined gamma-ray intensities, can then be avoided. **1958** S. Glasstone *Sourcebk. Atomic Energy* (ed. 2) xvii. 521 In 1948, the Committee on Standards and Units of Radioactivity of the National Research Council (United States)..favored the adoption of the proposal..that the term 'rutherford' be used to designate a quantity of radioactive material giving 10^6 disintegrations per second. **1962** H. D. Bush *Atomic & Nuclear Physics* iv. 84 The standard unit adopted is the curie... Another unit, which has not achieved universal acceptance, is the rutherford.

rutherfordine (rɒ·ðəɹfɔ̄ɹdīn). *Min.* [ad. G. *rutherfordin* (W. Marckwald 1906, in *Centralbl. f. Mineral.* 763); see *Rutherford + -ine^5.] An orthorhombic uranyl carbonate, UO_2CO_3, found as yellow fibrous masses, esp. in association with uraninite in East Africa.

1907 *Mineral Mag.* XIV. 409 *Rutherfordine*, a yellow uranyl carbonate, UO_2CO_3, resembling uranochre in appearance and resulting by the alteration of uraninite. **1955** *Science* 1 Apr. 473/1 In actual crystals of rutherfordine, faults occur in the stacking of layers; regions in which the sequence of layers corresponds to structure A are occasionally terminated by regions in which the layers follow the sequence of structure B. **1959** Hogan & Gilbert in G. J. Williams *Econ. Geol. N.Z.* (1965) xiii. 206/2 The yellow uranium mineral is rutherfordine which is consistent with the carbonate cement in the rock.

rutherfordite^1 (rɒ·ðəɹfɔ̄ɹdəit). *Min.* [f. the name *Rutherford* (see below) + -ite^1.] A name given to a poorly characterized yellow-brown form of fergusonite found in gold mines in Rutherford County, North Carolina.

1851 C. U. Shepard in *Proc. Amer. Assoc. Adv. Sci.* IV. 312 (*heading*) Rutherfordite. **1852** *Amer. Jrnl. Sci.* LXIV. 344 By its translucency, rutherfordite is readily distinguished from samarskite, which it otherwise closely resembles. **1880** *Ibid.* CXX. 57, I have detected along with the samarskite of this locality a few very small crystals..nearly identical with those found in the sands from the gold washings of Rutherford, N.C., named by me as rutherfordite, and which I now consider as belonging to the species fergusonite. **1966** Z. Lerman tr. *Vlasov's Geochem. & Mineral. Rare Elem.* II. xi. 430 Synonyms of fergusonite: rutherfordite.., bragite.., tyrite.., arrhenite..and sipylite.

ru·therfordite^2. *Min.* [f. as *rutherfordine: see -ite^1.] Used as a synonym of *rutherfordine.

1922 N. H. & A. N. Winchell *Elem. Optical Mineral.* II. v. 88 Rutherfordite (UO_2CO_3) is orthorhombic (?), finely fibrous. Soft... Color yellow, earthy. An alteration product of uraninite. Rare. **1971** *Mineral. Mag.* XXXVIII. 104 Recommendations of the Commission on minerals for which more than one name is in common use. .. Rutherfordine, not rutherfordite.

rutherfordium (rɒðəɹfɔ̄ə·ɹdiəm). *Chem.* [f. *Rutherford + -ium.] (A name proposed for) an artificially produced transuranic element, atomic number 104. Symbol Rf. Cf. *kurchatovium.

1969 *Science* 5 Dec. 1254/1 Scientists from the Lawrence Radiation Laboratory of the University of California announced results of chemical experiments on element 104 and used the occasion [*sc.* 17–19 Nov. 1969] to propose a new name. Albert Ghiorso..suggested that the element be called rutherfordium for Lord Rutherford 'the great pioneer of nuclear science'. **1970** A. Ghiorso et al. in *Physics Lett.* XXXII. B. 95/1 We have proposed that the element 104 be named rutherfordium. **1971** *Inorg. & Nuclear Chem. Lett.* VII. 1115 As not until 1969 did the American researchers at Berkeley also succeed in obtaining element 104..there is no ground to use the name rutherfordium proposed by them. **1971** *Nature* 26 Feb. 603/1 Certain questions have been raised regarding the validity of our work on the discovery of two alpha-emitting isotopes of element 104 (rutherfordium). **1975** [see *nielsbohrium].

ruthfully, *adv.* Restrict *Obs.* to sense 2 and add: **1.** (Later example.)

1936 'M. Innes' *Death at President's Lodging* xv. 246 No change, he reflected ruthfully a moment later, was to be got from Empson that way.

rutic, *a.* Add: Now † *Obs.* Substitute for def.: *rutic acid*: **a.** = Rutin in Dict. and Suppl. **b.** Capric acid.

rutilant, *a.* Delete 'Now *rare*' and add later examples. Also *fig.*

1917 A. Huxley *Let.* 30 Sept. (1969) 135 Behemoth His eyes are little rutilant stones Sunk in black basalt. **1944** S. Putnam tr. *E. da Cunha's Rebellion in Backlands* i. §4. 35 Diminutive-leafed opuntias,..bordered with rutilant flowers. **1954** *Times Lit. Suppl.* 2 July 425/4 There are certain magical elements constantly working against the proof-corrector—a being less rutilant, but not less vulnerable, than Tchaikovsky's Prince. **1956** K. Watson *Source* 46 Rutilant the trail in space Of some recurrent meteor.

Hence **ru·tilance** [-ance], rutilant quality (*rare*).

1922 Joyce *Ulysses* 691 He..ignited it in the candle-flame, applied it when ignited to the apex of the cone till the latter reached the stage of rutilance.

rutilated (rū·tilĕitĕd), *a. Min.* [f. Rutil(e + -ate^2 + -ed^1.] Of quartz: containing needles of rutile. Cf. Sagenite.

1889 [in Dict. s.v. Rutilate *v.*]. **1977** A. Hallam *Planet Earth* 136 Rutilated quartz contains orientated needles of rutile.

rutile. (Further examples.)

1951 *Chambers's Jrnl.* Sept. 568/2 Both the principal titanium minerals, ilmenite and rutile, are present in the black sands of streams and beaches in many parts of the world. **1965** *Sunday Mail Mag.* (Brisbane) 17 Jan. 11 Rutile was also needed as a flux for electric welding rods, armaments, ship building, tanks, etc. **1971** *Materials & Technol.* II. viii. 502 Still more spectacular fire is seen in synthetic rutile (titanium oxide) which can also be made by flame-fusion. **1971** *Jrnl. Oil & Colour Chemists' Assoc.* LIV. 849 Film volume measurements..showed a shrinkage of 6 per cent during six weeks of ageing at 25°C in the laboratory, for unpigmented and rutile titanium dioxide pigmented alkyd films.

rutin. Delete entry and substitute:

rutin (rū·tin). *Chem.* Also † **-ine.** [a. G. *rutin* (A. Weiss 1842, in *Pharm. Centralbl.* XIII. 903), f. L. *rūta* Rue *sb.*^{2}: see -in^1.] A yellow crystalline phenolic glycoside, $C_{27}H_{30}O_{16}$, found in several plant species (notably common rue, buckwheat, and capers) which possesses vasopressor properties and is taken to reduce blood pressure.

1857, etc. [in Dict.]. **1967** *Times Rev. Industry* Feb. 118/1 (Advt.), High blood pressure? Rutin, the natural product, has helped thousands of sufferers. **1977** *Martindale's Extra Pharmacopoeia* (ed. 27) 1697/2 Rutin was formerly used in the treatment of disease states characterised by capillary bleeding associated with increased capillary fragility but evidence of its value is inconclusive.

rutted, *ppl. a.* Add: Also *fig.*

1913 [see *routined *ppl. a.*]. **1957** G. Ryle in C. A. Mace *Brit. Philos. in Mid-Cent.* 259 Equations are not mere records of deeply rutted associations of ideas.

ru·tter^2. Also ritter. [f. Rut *v.*^{2}, Rit *v.*^{1} + -er^1.] A spade for cutting or slitting peat turf.

1877 [see Rit *v.*^{1} 1 b]. **1923** *Chambers's Jrnl.* 12 May 370/1 The rutter is a two-handed spade, the blade heart-shaped and sharp. **1975** *Times* 27 Aug. 8/5 A curved rutter is used for cutting the [peat] turf, a long-handled spade or flaughter for removing it.

ru·tter^3. *N. Amer.* [f. Rut *sb.*^{2} or Rut *v.*^{4} + -er^1.] A kind of plough used by lumberjacks for making tracks for sleighs.

1969 L. G. Sorden *Lumberjack Lingo* 100 *Rutter*, a form of plow for cutting ruts in an iced logging road for the runners of a sleigh. It was often combined with a snowplow. The roads were sprinkled with water from the water tank and frozen to make ice roads. **1972** *Islander* (Victoria, B.C.) 19 Nov. 4/3 The lumbermen had a unique system of hauling logs. In early fall, while the ground was still soft, they would build a rutter. Using the front bob of a wide logging sleigh, a small V-shaped plow was welded to the point of each sleigh runner. Then the sleigh bob was taken to the top of the proposed logging road, turned around, and twin tracks were then plowed eight inches deep and six inches wide down to the main camp.

ruttier. (Later *Hist.* examples of form *rutter.*)

1937 *Geogr. Jrnl.* XC. 386 It appears that there were existing rutters up to this point. **1962** [see *routier^1]. **1971** S. E. Morison *European Discovery Amer.: Northern Voy.* v. 138 The rutters (*routiers*), unofficial coast pilots of the period [*sc.* the sixteenth century], were written primarily for finding one's way along European shores. **1973** D. Divine *Opening of World* v. 85 An English Rutter, the northern and slightly less refined version of the *portolano*, describing a harbour entrance in 1295.

rutting, *vbl. sb.* Add: **b.** *rutting call, -wrath.*

1937 *Discovery* Oct. 314/1 The rutting call of the stag. **1893** Kipling *Seven Seas* (1896) 59 And when the first September gales have slaked their rutting-wrath, The great man-seal haul back to the sea.

ruttish, *a.* Delete *Obs.* and add later examples. Also, of or pertaining to sexual excitement.

1938 R. Graves *Coll. Poems* 158 A score of bats bewitched By the ruttish odour Swoop singing at his head. **1977** *Daily Tel.* 2 Dec. 15/5 He returns to ogling the field with his ruttish chum.

Ruwala, var. *Rualla.

rux (rɒks), *sb.*^{2} *Naut. slang.* [Origin unknown: cf. Rux *v.* in Dict. and Suppl. and *ruckus.] Disturbance, uproar.

1918 *Blackw. Mag.* CCIV. 68/1 Harker, who for fifteen months had haunted the shadows on the look-out for just such a 'rux', whose ear caught every illicit sound. **1931** Kipling *Limits & Renewals* (1932) 196 The nastiest rux I ever saw, when a boy, began with 'All hands to skylark.' *I* don't hold with it. *Ibid.* 200 I've seen worse ruxes in my time, but a quicker breeze-up—never!

rux, *v.* Add: **2.** To vex, worry. *rare.*

1887 Kipling *Plain Tales* (1888) 60 'E [was] too busy to rux 'isself about p'raids.

Ruy Lopez (rū·i lōᵘ·pez). *Chess.* [The name of *Ruy López* de Segura (fl. 1560), Spanish bishop and writer on chess, who developed this opening.] A chess opening characterized by the moves 1 P–K4, P–K4; 2 Kt–KB3, Kt–QB3; 3 B–Kt5.

1876 *Encycl. Brit.* V. 594/2 The following are given as indicative illustrations of certain of the leading openings. .. Ruy Lopez. **1894** *Yale Wit & Humor* 49/1 Our [chess] team appears to have executed a masterly flank movement in 'retaking the exchanged Pawn in the Ruy Lopez'. **1958** [see *NIMZO-INDIAN *a.*]. **1976** *Milton Keynes Express* 2 July 40/5 He opened with the Ruy Lopez and won in 20 moves.

Rwanda (rwæ·ndā, ru,æ·ndā). Also **Ruanda.** **a.** A Bantu language of East Africa. **b.** An East African people; the inhabitants of the country of Rwanda. **c.** An East African republic (founded 1961), formerly kingdom. Also *attrib.* Hence **Rwa·ndan** *sb.* and *a.*, **Rwande·se** *a.*

1902 H. H. JOHNSTON *Uganda Protectorate* II. 969 Urunyaruanda is spoken in Ruanda, or Bunya-ruanda, south of Ankole... English, *ant*... Ruanda, *entwzi*. **1924** SMITH & SHARP *Ruanda's Redemption* 18 (*heading*) Receipts for Kigezi and Ruanda work 1920–23. *Ibid.* 22 The only literature in the Ruanda language is a translation of the four Gospels. **1939** L. H. GRAY *Foundations of Lang.* 405 Homburger's classification [of Bantu languages] is as follows:..(2) *Ruanda*, north-east of Tanganyika. **1959** *Listener* 29 Oct. 740/1 The Ruanda and Urundi of that trusteeship territory, the Belgian Congo, Uganda and Tanganyika. **1969** J. C. KING *Evangelicals* vii. 60 A similar tight-knit group within Church of England Evangelicalism is the Ruanda movement. This consists of people influenced by the East African revival movement, through the Ruanda Mission. **1973** *Times* 11 Dec. (Zaire Suppl.) p. vii/9 On our return to Goma we passed a memorial to 23 wardens killed while defending the park against Zairian, Ugandan and Rwandese poachers. **1974** *Encycl. Brit. Micropædia* VIII. 737/3 In 1969, an estimated 3,600,000 Rwanda occupied an area of roughly 10,000 square miles. **1974** *Encycl. Brit. Macropædia* XVI. 109/1 The first impression given by the Rwandan landscape is that it resembles an immense green park dominated by banana plantations. *Ibid.* 109/2 Traditionally, Rwandans believe in a supreme being called Imana. **1979** *Brit. Med. Jrnl.* 15 Dec. 1560/1 We slowly came to appreciate the fabric of a Rwandan home, from the mud walls without to the complex and supportive family within.

rya (rī·ă). Also **ryiji, ryijy**. [Sw. *rya* in same sense; cf. Finnish *ryijy*.] A Scandinavian type of knotted pile rug. Also *attrib.*

1957 B. PEPIS *Guide Interior Decorating* iv. 124 The only remotely luxurious note is the small, brightly colored heavily piled 'rya' rug, an adaptation of a Finnish design. **1960** *Guardian* 20 July 4/6 Rya rugs are a very old form of Finnish folk art. *Ibid.* 4/7 Ryas..were used in everyday life up to the sixteenth century. **1960** H. HAYWARD *Antique Coll.* 245/2 *Ryijy rugs*, Finnish rugs made in the old Norse tradition of knotted pile technique, which may go back to the Danish Bronze Age. **1964** G. LYALL *Most Dangerous Game* xviii. 139 The only shot I fired hit the ryiji on the floor. **1972** *Homes & Gardens* Aug. 28/2 The choice of rugs ranges from Axminster..to the wildest, woolliest rya imaginable. **1975** 'E. LATHEN' *By Hook or by Crook* vi. 55 When rya rugs first came into fashion..she had..become a trend setter.

rybuck, var. *RYEBUCK *a.* (*adv.*) and *int.*

Rydberg (ri·dbɔ̈ɪg). *Physics.* The name of Johannes Robert *Rydberg* (1854–1919), Swedish physicist. **1.** Used *attrib.* and in the possessive to designate various concepts developed by him, as **Rydberg('s) constant,** an atomic constant, evaluated from several of the fundamental constants of physics, which appears in the formulae for the wave numbers of lines in all atomic spectra (in the case of a hypothetical atom whose nucleus has infinite mass, equal to $2\pi^2 me^4/ch^3$, where m and e are the rest mass and charge of the electron, c is the speed of light, and h is Planck's constant); see also *R III. 4; **Rydberg correction,** a correction term appearing in the formula for the energy of the single electron in the outermost shell of hydrogen-like atoms, arising because the inner shells do not screen the electron completely from the nucleus; **Rydberg('s) formula,** an empirical formula giving the wave numbers of frequencies of the lines in the spectral series of atoms and simple molecules.

1913 *Phil. Mag.* XXVI. 489 An attempt to explain the appearance of Rydberg's constant in the formula for the line-spectrum of any element. **1920,** etc. [see *R III. 4]. **1937** *Ann. Reg. 1936* II. 61 Birge..pointed out that the substitution of well-established values for e/m and h/e in the Bohr formula for the Rydberg constant gives a value for e nearly half of 1 per cent less than the others. **1955** C. G. DARWIN in W. Pauli *Niels Bohr* 7 There can be few other cases in science where a theory has been made which succeeds in yielding a particular number—here Rydberg's constant—from quantities all of which are known, without the admissibility of any adjustable constant to help in doing so. **1979** *Sci. Amer.* Mar. 74/3 Later refinements have complicated Rydberg's empirical formula for the wavelengths of spectral lines, and so the Rydberg constant is now defined as this combination of *m*, *e* and *h*. **1927** J. W. FISHER tr. *Born's Mechanics of Atom* iii. 160 Rydberg was the first to suggest this form and verified it by measurements of numerous spectra. We shall therefore denote the quantity δ as the Rydberg correction. **1936** *Discovery* Jan. 28/1 The quantisation of the Rydberg correction into multiples of a fundamental unit. **1974** G. REECE tr. *Hund's Hist. Quantum Theory* vii. 95 Schroedinger in 1921 realized that the essential point for the interpretation of the large 'Rydberg corrections' was that the s orbits dipped deep into the atom. **1913** *Phil. Mag.* XXVI. 12 The constant K entering in Rydberg's formula is the same for all substances. **1974** G. REECE tr. *Hund's Hist. Quantum Theory* vii. 97 In 1914 A. Fowler, inspired by Bohr's theory of the He⁺ lines, showed that for the doublet series of these elements Rydberg formulae held with $4R$ and that they therefore belonged to Mg^+ and Ca^+.

2. (Also written **rydberg**.) **a.** A unit of energy given by $e^2/2a_0$ (approximately $2{\cdot}425 \times 10^{-18}$ joule), where e is the electronic charge and a_0 is the radius of the first Bohr orbit for a nucleus of infinite mass. Freq. *attrib.* as *Rydberg unit.*

1935 *Jrnl. Chem. Physics* III. 563/2 The first choice would lower all [energy] values by 2·3 Rydberg units. **1944** *Physical Rev.* LVI. 336/1 The energy difference of curves III and IV is 0·44 Rydberg units more in Fig. 5 than it is in Fig. 4. **1954** *Ibid.* XCIV. 1519/2 (*caption*) Energy integrals (in Rydbergs) for diamond. **1975** *Nature* 27 Mar. 297/2 If gas accretion by this object produces all the ionising radiation required to maintain the H II region at the galactic centre, the luminosity of this radiation is $L_1 = 1{\cdot}9 \times 10^8 L$...if the energy per photon is 2 rydberg.

b. A name proposed for the unit of wave number, cm^{-1}; = *KAYSER.

1951 C. CANDLER in *Nature* 21 Apr. 649 Call 'cm.$^{-1}$' by some new name, such as 'Rydberg', however, and the difficulty disappears. Absorptions can be conveniently recorded in 'kilo-rydbergs'... The name 'Rydberg' was suggested to me many years ago by Prof. H. Dingle.

rye, *sb.*[1] Add: **3. a.** (Earlier and later examples.) Also *Canad.*

1835 J. H. INGRAHAM *South-West* II. 56 The painful effects of 'old rye' in the abstract upon the body. **1860** *Grumbler* (Toronto) 19 May 3/3 And, tho' the crowd may smile at me, I'll take some neat 'old rye'. **1873** G. W. PERRIE *Buckskin Mose* xvii. 248 But for the quantity of rye we had all of us been swallowing, the others must have seen through this impudent operation as I had done. **1913** J. LONDON *Valley of Moon* 392 Some drink rain and some champagne..; But I will try a little rye. **1930** D. RUNYON in *Collier's* 1 Feb. 12/3 Wilbur is a great hand for drinking Scotch, or rye. **1945** P. CHEYNEY *I'll say she Does!* iii. 66, I..finish off my rye an' pour myself another four fingers. **1974** E. MCGIRR *Murderous Journey* 31 He slopped along..towards the living-room bar. I took a straight rye.

b. *Comb.* in the names of drinks, as *rye-and-dry* (see *DRY *sb.* 2 c), *rye-and-ginger*, *rye-and-orange*, *rye-and-soda*, *rye-on-the-rocks*.

1909 G. ADE *Let.* 24 Mar. (1973) 45, I have just had a rye & soda. **1942** *Tee Emm* (Air Ministry) II. 127 Say? What's mine? A Rye and dry. **1956** 'N. SHUTE' *Beyond Black Stump* 5 'What's it to be?' 'Orange juice,' said the young man. Mr Johnson ordered it, with rye on the rocks for himself. **1963** R. I. MCDAVID *Mencken's Amer. Lang.* 168 Canadian topers have an array of combinations..as *rye and orange* (Canadian whiskey and orange pop). **1964** *Time* (Canada ed.) 31 Jan. 7/1 Accepting a rye and ginger, Mike Pearson then went back to writing out a personal report.

3*. *ellipt.* Rye-bread.

1941 [see *PASTRAMI]. **1969** [see *MAYO]. **1971** 'O. BLEECK' *Procane Chron.* xiv. 123 A Danish sardine sandwich..between two thick slices of German rye. **1976** H. MACINNES *Agent in Place* v. 48 A ham on rye with a gallon of coffee.

4. b. *ryebloom*; *ryehigh* adj.

1922 JOYCE *Ulysses* 261 The bag of Goulding, Collis, Ward led Bloom by ryebloom flowered tables. *Ibid.* 282 O'er ryehigh blue. Bloom stood up.

5. rye and Indian (also **Injun**) (**bread**) *U.S.*, bread made from a mixture of rye and (Indian) cornmeal; **rye brome** (**grass**), add: *Bromus secalinus*; (earlier and later examples); **rye coffee** *U.S.*, a drink resembling coffee, made from roasted rye; **rye waltz** *N. Amer.* (see quot.).

1840 *Knickerbocker* XVI. 18 There were eggs and fried ham,..rye-and-Indian bread. **1887** A. W. TOURGÉE *Button's Inn* 224 She passed around a hot plateful of toasted slices of 'rye and Indian'. **1932** L. I. WILDER *Little House in Big Woods* iv. 45 She baked salt-rising bread and rye 'n' Injun bread and Swedish crackers. **1812** W. WITHERING JR. *Withering's Brit. Plants* (ed. 5) II. 210 Smooth Rye Brome-grass...In corn-fields. **1954** C. E. HUBBARD *Grasses* 67 'Rye Brome' was no doubt introduced into the British Isles long ago with the seeds of cereals. **1769** *Boston Gaz.* 16 Oct. 1/3 And as true Daughters of Liberty, they made their Breakfast upon Rye Coffee, and their Dinner was partly made of that sort of Venison called Bear. **1877** H. RUEDE *Jrnl.* 13 June in *Sod-House Days* (1937) 99 Most people out here don't drink real coffee, because it is too expensive... So rye coffee is used a great deal—parched brown or black according to whether the users like a strong or mild drink. **1951** L. CRAIG *Singing Hills* iv. 31 Every one had coffee... When I tasted mine I thought, for a moment, that poison had been put in it; it certainly was not like anything I had ever tasted before, for never before had I drunk rye coffee. **1941** W. C. HANDY *Father of Blues* ii. 16 The waltz was popular, as was also the rye waltz, a combination of three-four and two-four tempos.

rye (rəi), *sb.*[3] *slang.* [ad. Romany *rai* gentleman; cf. Skr. *rāj* to rule.] A man, gentleman. Also *Comb.*, as **rye mort,** a lady (in quot., *attrib.*); **rye mush,** a gentleman. See also *Romany rye* s.v. *ROMANY[3] 3 b.

1851 BORROW *Lavengro* II. xxvi. 242, I had always..been a great favourite with Mrs. Petulengro, who had frequently been loud in her commendation of the young rye, as she called me. **1857** —— *Romany Rye* I. vi. 74 Gentility will carry the day, madam, even with the young rye. He will ask words of the black lass, but beg the words of the fair. **1936** J. CURTIS *Gilt Kid* 55 He did not feel choosey; why, he could be a rye mush himself for one night. *Ibid.* 232 Anyone taking a quick look at her might think she was on the up-and-up. She would give that impression too, to anyone who heard her talk and saw her act. Though.. she would have to give up that rye mort touch. **1939** —— *What Immortal Hand* xiv. 151 If she's gone and got herself tangled up with a lot of rye mushes she don't want to have nothing to do with a gaol-bird like me.

ryebuck (rəi·bɒk), *a.* (*adv.*) and *int. slang* (chiefly *Austral.*). Now *obsolescent.* Also **ribuck, rybuck, rye buck,** etc. [Origin uncertain: perh. ad. G. *reibach*, var. of *rebbach* profit, ad. Yiddish (Heb.) *revaḥ*: cf. RYBECK.]

A. *adj.* Good, excellent; genuine; *ryebuck shearer*, an expert or 'gun' shearer (see *GUN sb. 13*). Also as *adv.*

1859 G. W. MATSELL *Vocabulum* 55 My pals have got up a bene moey to send to the head bloke, and if it comes off rye buck, I shall soon vamose from the stir. **1895** *Bulletin* (Sydney) 9 Feb. 15 I'm ryebuck and the girl's okay. **1906** E. DYSON *Fact'ry 'Ands* x. 132 'It's rybuck, girls,' said Feathers. 'Yer on velvet. Ther firm's willin' t' accept responsibility fer ther actions iv it's dooly accredited cat, 'n' pays compensation.' **1916** C. J. DENNIS *Moods of Ginger Mick* 92 But the reel, ribuck Australia's 'ere among the fightin' men. **1918** R. H. KNYVETT *Over There* viii. 82 They even knew our slang, for here was 'The 'Fair Dinkum' Store', and across the way 'Ribuck Goods'. *a* **1957** in Stewart & Keesing *Old Bush Songs* (1957) 267 There's a bloke on the board and I heard him say I couldn't shear a hundred sheep a day, But some fine day I'll show him the way And prove I'm a ryebuck shearer. **1965** J. S. GUNN *Terminol. Shearing Industry* II. 16 *Ryebuck shearer*, see *gun*. *Ibid.* I. 30 *Gun*, a really fast shearer, also known as a 'ryeback [*sic*] shearer', but not the same as the fastest in the shed (see *ringer*).

B. *int.* An expression of agreement or assent.

1859 G. W. MATSELL *Vocabulum* 76 *Rybuck*, all right; straight, it will do; I am satisfied. **1898** *Bulletin* (Sydney) 17 Dec. (Red Page), *Rye-buck* (all right) is no doubt an abbreviation of 'all right, my buck'. **1911** L. STONE *Jonah* i. 11 'Oh! I don't suppose you'll be missed,' replied Chook, graciously. 'Rye buck!' cried Jonah. **1916** C. J. DENNIS *Songs of Sentimental Bloke* 21 We kin get an intro, if we've luck. 'E sez, 'Ribuck'. *Ibid.* 72 'E'en in the days when she's no longer fair She's still yer wife,' 'e sez. 'Ribuck,' sez I. **1933** *Bulletin* (Sydney) 27 Sept. 42/2 'We'll meet you at the yards.' 'Ryebuck, Boss,' said The Gov'ner civilly.

rye-grass. 1. (Earlier and later examples.)

1712 J. MORTON *Nat. Hist. Northants.* ix. 482 Rye-grass is with us accounted the best thing in the World for Wood-land in Enclosures. **1931** R. BEALE *Bk. Lawn* iv. 48 Rye grass mixtures..are recommended for heavy soils. **1961** R. M. PATTERSON *Buffalo Head* vi. 216 The rocks were all hidden and a magnificent growth of rye-grass was swaying in the wind. **1979** *Buffalo* (N.Y.) *Evening News* 18 May II. 22/2 The new turf-type perennial rye-grasses available today are real beauties.

Ryeland (rəi·lănd). The name of a district in Hereford & Worcester, where the breed was first developed, used *attrib.* and *absol.* to designate a sheep belonging to the small, hornless breed so called, which is a good producer of both wool and meat.

[**1801** J. POWELL *Let.* in *Ann. Agric.* (1808) XLV. 6 The hardiness of the Ryelanders..is proverbial, as milkers.] **1802** J. SOMERVILLE *Let.* 12 Nov. in *Facts & Observations relative to Sheep* (1803) 10 The same land, which carried forty-five breeding ewes, was immediately stocked with 150 Ryelands in their stead. *Ibid.* 12 We.. sent this Ryeland mutton to market. **1837** W. YOUATT *Sheep* vii. 258 The distinguishing breed of sheep in Herefordshire is the Ryeland, so called from a district in the southern part of the county. *Ibid.* 260 The Ryeland sheep..quickly fattens. **1861** MRS. BEETON *Bk. Househ. Managem.* xiv. 323 Eleven varieties have been reared in this country of the domesticated sheep..embracing.. the Ryeland; South-Down; the Merino. **1912** R. LYDEKKER *Sheep & its Cousins* v. 101 The modern Ryeland.. retained the diminutive proportions of the ancestral breed. *Ibid.*, Ryeland wool was formerly regarded as the finest produced in the British Islands. **1929** W. C. COFFEY *Productive Sheep Husbandry* (ed. 2) xxi. 173 The Ryeland originated in Herefordshire, early in the nineteenth century. **1971** *Farmers Weekly* 19 Mar. 83/1 Both Suffolk and Ryland [*sic*] rams have been used this season.

ryiji, ryijy, vars. *RYA.

Rylean (rəi·liăn) *a.* *Philos.* [f. the name of Gilbert *Ryle* (1900–76), English philosopher + -AN.] Of, pertaining to, or characteristic of Ryle's theories or his approach to linguistic philosophy or philosophical behaviourism.

1958 *Times Lit. Suppl.* 10 Oct. 581/1 The first part of this book gives an account, in roughly Rylean terms, of different senses of 'know', and of the relations between 'knowing that' and 'knowing how'. **1963** W. SELLARS *Sci., Perception & Reality* v. 178 What I shall call a Rylean language, a language of which the fundamental descriptive vocabulary speaks of public properties of public objects located in Space and enduring through Time. **1966** *Philos. Rev.* LXXV. 99 Farrer shrinks..from the Hobbist mortalism that would naturally go with this Rylean view of body and mind. **1971** G. J. WARNOCK in Wood & Pitcher *Ryle* 273 It was the answer which his very Rylean proforma of a solution temptingly left room for.

‖ **ryo** (ryō). Also 9 **rio, riyo.** [Jap.] A former Japanese monetary unit (see quots.).

1871 A. B. MITFORD *Tales of Old Japan* I. 70 A Japanese noble will sometimes be found girding on a sword, the blade of which unmounted is worth from six hundred to a thousand riyos, say from £200 to £300. **1876** W. E. GRIFFIS *Mikado's Empire* (1877) II. 610 In popular language, the terms *hiyaku* (hundred), *fun*, *mommé*, and even *riō* (4 mommé, 5 fun), do not represent any coin, but are used to denote values. They are expressions belonging to the period when money was computed by weight only. **1899** L. HEARN *In Ghostly Japan* vi. 103 The sum of a hundred ryō in gold. **1915** F. BRINKLEY *Hist. Jap. People* xxxi. 438 The gold *ryō* represented 2 *koku*, or 30 *yen* of modern currency, the silver *ryō* representing 3 yen. *Ibid.* xxxii. 444 Gold..was much more valuable in China than in Japan. Ten *ryō* of the yellow metal could be obtained in Japan for from twenty to thirty *kwan-mon* and sold in China for 130. **1938** D. T. SUZUKI *Zen Buddhism & its Influence on Japanese Culture* I. vii. 160 Two loads of gold were equivalent in the currency of the time to 12,000 *ryo*. **1964** *Japan* (Unesco) (rev. ed.) i. 45/2 It is said that between 1601 and 1647 about 4,800,000 *ryō* (one *ryō* contained four *me* of pure gold) of gold and 750,000 *kan* of silver were paid to foreign countries. **1972** *Mainichi Daily News* (Japan) 6 Nov. 7/4, I will kill anyone or accept a mission of the sword for five hundred ryo in gold.

‖ **ryokan** (ryōᵘ·kăn). [Jap.] A traditional Japanese inn or hostelry.

1963 *Maclean's Mag.* 9 Mar. 37 The most charming hotel I ever stayed at was a Japanese ryokan in the mountain spa of Kinugawa north of Tokyo. **1968** *Sat. Rev.* (U.S.) 23 Dec. 57/2 Stay in a 17th-century *ryokan*—inn. **1970** *Guardian* 12 Dec. 6/6 The *ryokans*, country inns, are worth the slight additional expense over Westernized hotels. **1972** *Times* 8 May (Japan Suppl.) p. viii/2 The site..contains a magnificent temple and several *ryokan*—traditional Japanese inns. **1979** *Amer. Poetry Rev.* Mar./Apr. 45/2 Several ferries, sighted from the small balcony of our private Ryokan overlooking the beach, circle Dogashima Bay from dawn to dusk.

Ryvita (rəivī·tă). [f. RYE *sb.*[1] + L. *vīta* life.] The proprietary name of a type of crispbread.

1925 *Trade Marks Jrnl.* 18 Feb. 385 Ryvita..Bread. John Edwin Garrat, 96 Southwark Street, London SE1 Manufacturer. **1926–7** [see *CRISPBREAD]. **1930** A. BENNETT *Imperial Palace* xxii. 142 Oldham softly entered with the tea-tray... 'I've brought you some hot ryvita in case you should fancy it, sir.' **1937** 'G. ORWELL' *Road to Wigan Pier* vi. 95 A millionaire may enjoy breakfasting off orange juice and Ryvita biscuits. **1953** R. FULLER *Second Curtain* v. 79 A girl..carrying a plate of Ryvita spread with paste. **1967** *Trade Marks Jrnl.* 22 Mar. 368/1 Ryvita..Bread, crispbread and biscuits (other than biscuits for animals). The Ryvita Company Limited.. London. **1974** *Times* 19 Oct. 6/6 He had inadvertently eaten the toast (possibly Ryvita).

S

S. Add: **I. 1. b.** *s-aorist* (Philol.), in certain Indo-European languages, an aorist formed from the verbal stem by adding *s* and the ending; a sigmatic aorist.

1895 CONWAY & ROUSE tr. *Brugmann's Compar. Gram. Indo-Gmc. Lang.* IV. 371 Special vowel-grades for the root-syllable, as in the *s*-aorist.., cannot be made out for the parent language. **1933** C. D. BUCK *Compar. Gram. Greek & Latin* 281 The distinctive IE aorist is the *s*-aorist formed from the root by the addition of *s* and the secondary endings. **1962** C. W. WATKINS *Indo-Europ. Orig. Celtic Verb* I. 55 The more common situation in Vedic is one where a root athematic present has an *s*-aorist associated with it.

2. c. *S-curved*, *-decorated*, *-scrolled*, *-shaped* (later examples) adjs.; *S-bend*, *-curve* (later example), *-ornament*, *-rope*, *-scroll*, *-sofa*, *-trap*, *-turn*.

1930 *Motor* 10 June 892/2 We were negotiating an S bend on the proper side of the line on a main road. **1931** D. L. SAYERS *Five Red Herrings* xi. 115 The road makes a very sharp and dangerous S-bend. **1975** R. BROWNING *Emperor Julian* x. 187 The northern section of the frontier formed a great S-bend. **1977** R. E. HARRINGTON *Quintain* xii. 109 Fronck negotiated an s-curve, and.. pulled the Ford out onto a straight stretch. **1940** *Burlington Mag.* Mar. 81/2 The wings with their S-curved shape. **1961** M. W. BARLEY *Eng. Farmhouse & Cottage* IV. i. 189 An English boat came into Boston in 1628 with 3,500 tiles aboard, and the earliest references to pantiles, the S-curved roofing tile, occur in the 1630s. **1963** G. DANIEL in Foster & Alcock *Culture & Environment* ii. 21 The S-decorated pottery which may be a degeneration of the duck motifs found on Early Iron Age pottery in Brittany and north Spain. **1934** *Burlington Mag.* Sept. 120/2 A finely-carved double-headed eagle, resting on a symmetrical S-ornament at the bottom. **1883** W. S. GRESLEY *Gloss. Coal-Mining* 234 *S-rope*, the winding rope which passes round the under side of the drum from or to the pulley; so called because it takes the form of the letter S. **1934** *Burlington Mag.* Sept. 120/2 The symmetrically inverted S-scroll. **1956** G. TAYLOR *Silver* vii. 143 The graceful and irregular S- and C-scrolls that are the chief ingredient of the style in its linear form. **1934** *Burlington Mag.* Sept. 125/2 The lambrequin 'apron' and the S-scrolled legs both, I would suggest, came to Europe from India. **1955** R. FASTNEDGE *Eng. Furnit. Styles* iii. 77 Early examples with S-scrolled legs and bun feet were frequently decorated with floral, or later, seaweed, marquetry. **1937** T. RATTIGAN *French without Tears* II. ii. 57 From sideways on it's a bit S-shaped, if you know what I mean. **1966** *Publ. Amer. Dial. Soc.* XLII. 3 *Chicane*,..an S-shaped curve of a race track. **1906** W. DE MORGAN *Joseph Vance* xxvi. 211, I found myself sitting beside Miss Spencer on a thing like an S in the back drawing-room... As I sat by Miss Spencer on the S-sofa. [**1882** S. HELLYER *Lect. Sci. & Art Sanitary Plumbing* iii. 108 About the first form of trap used for fixing under water-closets was the syphon or round-pipe trap, *i.e.*, a pipe bent and recurved in the shape of the letter *ɷ*.] **1885,** **1976** S trap [see *P trap* s.v. *P III. 1]. **1920** A. J. L. SCOTT *Sixty Squad* 56 Putting in a couple of 'S' turns, he made a good slow landing. **1973** *Times* 3 Mar. 15/2 The Labour Party has done an S-turn when the Government has merely done a U-turn.

4. a. S (*Bacteriol.*) = *SMOOTH *a.*; S., strain (of virus etc.), *spec.* in S.19, S19 = *strain 19* s.v. *STRAIN *sb.*[1]; S.A., s.a., sex appeal; S.A., S-A (*Med.*), sino-auricular or -atrial; S.A., small arms; S.A., S/A [F. *société anonyme*; also It. *società anonima*, Sp. *sociedad anónima*], in France, Italy, etc., a limited or joint-stock company; S.A. = *STURMABTEILUNG; S.A.A., small arm(s) ammunition; S.A.C., senior aircraftman; SAC (*U.S.*), Strategic Air Command; S.A.C. EUR., SACEUR, Saceur (also with pronunc. sæ·kiūɹ), Supreme Allied Commander Europe; S.A.C.W., senior aircraftwoman; S.A.E., Society of Automotive Engineers (used *spec.* to designate a scale of viscosity used for lubrication oils); S.A.E., s.a.e., stamped addressed envelope; S.A.L., South Arabian League; SAM, surface-to-air missile; S & L (*U.S.*), savings and loan (association); S and M, S–M, sadism and masochism, sado-masochism; SAR, search and rescue; S.A.S., Special Air Service; SAT (*U.S.*), scholastic aptitude test; S.B., simultaneous broadcast; S.B., Special Branch; S.B., stretcher bearer; S.B.A., sick-berth attendant; SBA (*U.S.*), Small Business Administration; S.B.A.C., Society of British Aerospace Companies (formerly Society of British Aircraft Constructors); SBM, sbm, single buoy moor(ing); SBR, styrene-butadiene rubber; S.C., s.c., self-contained; SC, structural change (in Transformational Grammar); SCAP (also with pronunc. skæp), Supreme Commander Allied Powers (in Japan); also used *transf.* of the Command Headquarters; Sc.D. [L. *Scientiæ Doctor*], Doctor of Science; S.C.F., Save the Children Fund; scf, standard cubic feet (i.e. cubic feet of gas at standard temperature and pressure); SCLC (*U.S.*), Southern Christian Leadership Conference; S.C.M., State Certified Midwife; S.C.M., Student Christian Movement; SCP, single-cell protein; S.C.R., senior common room (orig. and chiefly in the University of Oxford); SCR (*Electronics*), silicon-controlled rectifier; S.C.U.A., Suez Canal Users' Association; S.D., s.d., semi-detached (house); S.D., sequence date; S.D. = *SICHERHEITSDIENST; s.d., S.D. (*Statistics*), standard deviation; SD, structural description (in Transformational Grammar); S.D.A., Scottish Development Agency; S.D.E.C.E. [F. *Service de documentation étrangère et de contre-espionnage*], the official counter-intelligence agency in France; S.D.F., Social Democratic Federation; hence *S.D.F.er*; S.D.L.P., Social Democratic and Labour Party; S.D.O., Subdivisional Officer; S.D.P., Social Democratic Party; S.D.R., special drawing right (usu. *pl.*); S.D.S. [G. *Sozialistischer Deutscher Studentenbund*], the Federation of Socialist Students (in West Germany); SDS (*U.S.*), Students for a Democratic Society; s.e.(m.), S.E.(M.) (*Statistics*), standard error (of the mean); S.E., S/E, Stock Exchange; S.E.A.C. (also with pronunc. sī·æk), South East Asia Command; SEC (*U.S.*), Securities and Exchange Commission; SECAM [F. *séquentiel couleur à mémoire* colour sequence by memory], a colour television system developed in France and widely used; SEM, scanning electron microscope, microscopy; S.E.N., State Enrolled Nurse; S.E.T. (also with pronunc. set), selective employment tax; S.F., San Francisco; S.F., s.f., science fiction; S.F. = *SINN FEIN; S.F.A., Scottish Football Association (cf. F.A. s.v. *F III. 3); S.F.A., Sweet Fanny Adams (cf. F.A. s.v. *F III. 3, *FANNY ADAMS 2); S.F.I.O. [F. *Section française de l'Internationale ouvrière*, French section of the workers' International], the French socialist party, known since 1969 as the *Parti Socialiste*; s.h., shit-house; S.H.F., s.h.f. (*Radio*), superhigh frequency; S.H.O., Senior House Officer; s.h.p., S.H.P., shaft horsepower; SI [F. *système international (d'unités)*], International System of Units (see *SYSTÈME INTERNATIONAL and *INTERNATIONAL *a.* 1 c); S.I.D. s.i.d. (*Radio*), sudden ionospheric disturbance; SIDS, sudden infant death syndrome; S.I.N.S., ship's inertial navigation system; S.I.S., Secret Intelligence Service; S.I.W., self-inflicted wound (see also quot. 1929); S.J., Society of Jesus (cf. JESUIT *sb.*); S.L.A., Symbionese Liberation Army; SLBM, submarine-launched ballistic missile; SLCM, submarine-launched cruise missile; SLE (*Med.*), systemic lupus erythematosus; S level, Scholarship (also, Special) level (of the General Certificate of Education examination); S.L.P., Scottish Labour Party; SLR (*Photogr.*), single-lens reflex (camera); S–M: see S and M above; S.M., sergeant-major; S.M., short metre (cf. SHORT *a.*, *sb.* and *adv.* 26); S.M., s.m., stage manager; S.M.L.E., short magazine Lee-Enfield (rifle); S.M.M.T., Society of Motor Manufacturers and Traders; S.M.O.N. (see *SMON as main entry); SMPTE (*U.S.*), Society of Motion Picture and Television Engineers; SNCC (*U.S.*), Student Nonviolent Co-ordinating Committee; S.N.C.F. [F. *Société Nationale des Chemins de Fer*], the French State railway authority, also used for the railway system itself; S.N.F., s.n.f., solids, non-fat; SNG, simulated, substitute, or synthetic natural gas; S.N.O., Senior Naval Officer (cf. N.O. s.v. *N II. 1); S.N.P., Scottish National Party; SNU (sniū) *Astr.*, solar neutrino unit (see quot. 1970); S.O., standing order; S.O.B., s.o.b. (chiefly *U.S.*), son of a bitch, also silly old bastard, etc.; S.O.E., Special Operations Executive; S. of S., Secretary of State; S.O.L., s.o.l., soldier (also strictly, shit, surely: see quot. 1917) out of luck (*U.S.*); SOP, standard operating procedure (*U.S.*, orig. *Mil.*); S.P., s.p., starting price; S.P.A.B., Society for the Protection of Ancient Buildings; S.P.C.K., Society for the Promotion of Christian Knowledge; S.P.D. [G. *Sozialdemokratische Partei Deutschlands*], the Social Democratic Party in West Germany; S.P.E., Society for Pure English; S.P.G. (earlier examples); S.P.Q.R. [L. *Senatus Populusque Romanus*], the Senate and People of Rome; also in joc. adaptations, esp. = small profits, quick returns; S.P.R., Society for Psychical Research; SQ [f. stereophonic-quadraphonic], a designation (proprietary in the U.S.) of audio equipment used with reference to a system of quadraphonic recording and reproduction; S.R., Socialist Revolutionary (Party); S.R., Southern Railway; S.R., Special Reserve; sr, steradian; S–R, stimulus–response *adj.* (in Psychol.); SRBC (*Med.*), sheep red blood cell(s); S.R.M.N., State Registered Mental Nurse; S.R.N., State Registered Nurse; sRNA († S-RNA) (*Biol.*), soluble RNA; SRO (*U.S.*), single-room occupancy; S.R.O. (orig. *U.S.*), standing room only; SRS(-A) (*Med.*), slow-reacting substance (of anaphylaxis); S.S. = *SCHUTZSTAFFEL; S.S., secret service, security service; SS, social security (benefit); S.S., steamship: also s.s., ss. (examples); SSB, ssb (*Radio*), single side-band (transmission); SSBN [Submarine (symbol SS), Ballistic, Nuclear], a nuclear-powered ballistic missile submarine; S.S.N., severely subnormal; SSPE (*Path.*) subacute sclerosing panencephalitis; SSR, secondary surveillance radar; S.S.R. [Russ. *Sovétskaya Sotsialisticheskaya Respúblika*], Soviet Socialist Republic (cf. *U.S.S.R.); SSRC, Social Science Research Council; SST, supersonic transport; S.T.C., short-title catalogue, esp. *A Short-Title Catalogue of Books Printed in England, Scotland, and Ireland 1475–1640*, by A. W. Pollard and G. R. Redgrave, first published in 1926; STD (*Teleph.*), subscriber trunk dialling; STOL, stol., short take-off and landing; S.T.V., single transferable vote; SU (*Physics*), special unitary (*sc.* group): used with following numeral denoting the number of rows and of columns in the matrices that can be used to represent it, as *SU*(3): cf. *SPECIAL *a.*; SV (*Med.*), Simian virus: used, freq. with following numeral to identify the strain, as the designation of various viruses isolated from monkeys or cultures of monkey cells; s.v. = *sub verbo*, *sub voce* s.v. SUB *Latin prep.* in Dict. and Suppl.; also s.vv., *sub verbis* (followed by more than one citation); SVD, swine vesicular disease; S.W., small women('s size); S.W.A.(L.)K., SWA(L)K, sealed with a (loving) kiss; SWAT (*U.S.*), Special Weapons and Tactics; s.w.g., S.W.G., standard wire gauge; SWP, Socialist Workers' Party; S.W.R., s.w.r., standing-wave ratio. See also (as main entries) *SAGE, *SALT, *SAVAK, *SEATO, *SHAEF, *SHAPE, *SNAFU, *SOGAT, *S.O.S., *STP, *SWANU, *SWAPO.

1920 J. A. ARKWRIGHT in *Jrnl. Path. & Bacteriol.* XXIII. 359 The appearance of colonies on agar of the two forms is different. The S form makes smooth, round, domed, shiny, translucent colonies; the R form grows in colonies which have a more or less jagged outline. **1974** [see *ROUGH *a.* 1 e]. **1949** *Vet. Rec.* LXI. 318/1 Each animal was inoculated intravenously with approximately 20 times the recommended vaccine dose of S.19 in a volume of 50 ml. **1960** *Farmer & Stockbreeder* 26 Jan. 89/1 As the animals are ready to go out on the early spring grazing they are collected together on the farms

for vaccination with S19. **1978** *Amer. Jrnl. Vet. Res.* XXXIX. 884/1 Lymphocyte stimulation..was detected in 3 steers which had been vaccinated with S19 but not with lymphocytes from 5 nonvaccinated heifers. **1926** *Amer. Mercury* Dec. 465 The girl is a looker with an armful of S.A. (sex appeal). **1932** P. MACDONALD *Maze* 216 A Gallic young woman with apparently some looks and, let us say, 98 per cent. vigorous S.A. **1961** *John o' London's* 6 July 57/2 Surely one of Hollywood's finest character actresses—all this and blonde S.A. too. **1974** E. MCGIRR *Murderous Journey* 96, I saw you and the dame go into her apartment... I expected you to take longer. Losing the old s.a., Piron? [**1907** *Jrnl. Anat. & Physiol.* XLI. 175 (*caption*) *s.a.j.*, sino-auricular junction.] **1908** J. MACKENZIE *Dis. Heart* p. xix, Sino-auricular node (s.-a. node.) **1910** *Jrnl. Physiol.* XLI. 69 This observer finds no altered rhythm as the result of destruction of the S-A node by burning. **1944** C. P. ANTHONY *Textbk. Anat. & Physiol.* v. 177 It is named the sinoatrial node but is usually referred to simply as the S.A. node... It is also called the 'pacemaker' of the heart. **1974** M. C. GERALD *Pharmacol.* xxi. 386 Digitalis directly depresses the conducting tissues responsible for carrying the excitatory impulse from the S.A. node pacemaker. **1876** VOYLE & STEVENSON *Mil. Dict.* (ed. 3) p. x, *S.A.*, small-arms. **1888** *Man. Field Service—Army Signallers* (War Office) 12 Pins, linch, 3rd class (or 2nd class, steel, if for cart, ammunition, S.A.). **1924** *Regulations Equipment of Army* (*Provisional*) (War Office) II. i. 4 Eyepieces, rubber, sights, telescopic, S.A., No. 2. **1921** *London Directory* 1707/1 Geneva, Switzerland. Mondiale Express Transports S/A. **1938** E. AMBLER *Cause for Alarm* v. 83 There it was in black and white—Società Anonima Braganzetta, Torino. I had found S.A. Braga of Turin! **1977** 'J. LE CARRÉ' *Hon. Schoolboy* iv. 80 Indocharter, Vientiane SA..was an overseas Chinese company. **1931** W. LEWIS *Hitler* II. 60 The abovementioned defence-service.. received the name of *Storm-detachments—S.A.*, in memory of the 'heroic onset of the at that time mere handful'. **1934** *Ann. Reg. 1933* 168 On February 22 Goering incorporated picked S.A. men as auxiliary Police Corps. **1955** *Times* 15 Aug. 4/4 A picture of him..wearing S.A. uniform and with his right arm raised. **1968** *Listener* 19 Sept. 358/3, I do not even know the difference between the SA and the SS, so how can I make such a film. **1977** *Daily Tel.* 17 Nov. 36/7 The SS (Schutzstaffel) emerged as a powerful force after the 1934 purge which eliminated the SA (Sturmabteilung), the brown-shirted thugs who helped bring Hitler to political prominence. **1876** VOYLE & STEVENSON *Mil. Dict.* (ed. 3) p. xi, *S.A.A.*, small-arm ammunition. **1907** *Field Service Pocket Bk.* ii. 50 A total of 500 rounds S.A.A. will be maintained in the field..for every man, whether combatant or not, included in an expeditionary force. **1954** J. MASTERS *Bhowani Junction* I. xi. 95, I rummaged in the table drawers and found a list:..1,000 feet of slow-burning fuse; 12,000 rounds of SAA .303 Mark VIII Z. **1952** *R.A.F. Rev.* Jan. 11/1 Under the New Trade Structure I am now expected to pass a SAC board to qualify as a Corporal and gain the increase in pay. **1970** *Athanian* 1 Apr. 31/1 The club is an 'All Ranks' affair ranging from an S.A.C. to a Sqn Ldr. **1947** *Army & Navy Bull.* 1 Feb. 8/3 General St. Clair Streett, SAC Deputy Commander since its activation last March, received a new assignment in the War Department. **1958** *Times* 8 May 11/6 Right now, therefore, the S.A.C. crew in a S.A.C. plane is the west's number one deterrent to the Kremlin. **1974** *Publishers Weekly* 16 Dec. 22 (Advt.), H. Bruce Franklin, Melville Scholar, former SAC officer, and tugboat mate. **1951** *Army Information Digest* July 27 Supreme Allied Commander Europe (*SACEUR*). **1953** *Britannica Bk. of Year* 752/2 Saceur, Supreme Allied Commander, Europe. **1958** *Economist* 1 Feb. 393/1 Its [*sc.* Britain's] dwindling proportion of the forces at Saceur's disposal. **1959** *Times Lit. Suppl.* 13 Feb. 79/2 Proposals for placing the missile bases under the command of S.A.C.EUR...do not, as Mr. Moore says, really change the situation. **1964** *Ann. Reg. 1963* 162 Approval was given to the assignment of the British V-bomber force..to the Supreme Allied Commander Europe (SACEUR). **1979** *Observer* 25 Nov. 34/3 SACEUR, Supreme Allied Commander Europe, the military head of NATO; currently the American General Bernard W. Rogers. **1951** *R.A.F. Rev.* Sept. 34/3 The R.A.F.'s qualifying trade test leads to promotion to Senior Aircraftsman or SACW. **1977** *R.A.F. News* 30 Mar.–12 Apr. 18/1 The only WRAF rider, SACW Jennie Hye of West Drayton, put up a plucky 44·42. **1924** *Jrnl. Soc. Automotive Engineers* XV. 31/1 The fact that the present S.A.E. numbers for crankcase lubricating oil specifications were used..is of interest. **1966** *McGraw-Hill Encycl. Sci. & Technol.* XIII. 313/1 The carburizing steels..which have the greatest ability to harden..are SAE 3310 and 4320. **1967** KARCH & BUBER *Offset Processes* x. 475 The pump oiler should be filled every 25 to 50 hours of running time with ≠ 105SAE or equivalent pure mineral oil. **1974** *Encycl. Brit. Macropædia* XIV. 188/2 In the U.S., the Society of Automotive Engineers established a system of SAE numbers to indicate the viscosity at a particular temperature, 0°F (−18°C). Oils in common use have SAE numbers varying from 5 to 50. *Ibid.*, An oil designated 10W/40 has the viscosity of an SAE 10W oil at 0°F.., and of an SAE 40 oil at 210°F... Such an oil will help start an engine in winter (hence the suffix W) and will lubricate well under running conditions in summer. **1939** 'F. O'BRIEN' *At Swim-two-Birds* i. 15 To all my friends forwarding 6d. and two S.A.E.'s I will present this three-star cast-iron plunger. **1962** *Woman's Own* 15 Sept. 69/3, I will send you a leaflet on this subject on request (s.a.e., please). **1966** *Punch* 26 Jan. 116/3 That stream of SAEs for the free, illustrated brochure. **1977** *Vogue* Dec. 90/2 Send a SAE for the catalogue. **1966** *Economist* 29 Oct. 457/3 The much publicised series of talks between federal ministers, SAL leaders and a couple of dissident sheikhs..has come to nothing. **1970** H. TREVELYAN *Middle East in Revolution* 218 The original Nationalist party, the South Arabian League, known as SAL, were in decline. **1958** *Chambers's Techn. Dict.* 983/2 SAM. **1975** R. JACKSON *South Asian Crisis* v. 107 The Indian SAM missile systems were improved. **1979** P. NIESEWAND *Member of Club* xv. 122 Tanks, armoured cars, SAM missiles..are being landed at Beira. **1951** *Business Week* 22 Sept. 152/2 (*heading*) 'Thin Ice' for S & L's. **1967** *Economist* 25 Mar. 1145/2 Some Californian S & Ls suffered near runs on their accounts as savers began to doubt their solvency. **1976** *National Observer* (U.S.) 25 Dec. 7/1 The result is that institutions that finance a large volume of home purchases, as S&Ls do, may not be able to afford to be as generous on longer-term deposit rates. **1965** *Acronyms & Initialisms Dict.* (Gale Research Co.) 645 *S & M*,..sadism and masochism (generic term). **1966** *Realist* May 19/3 Remember the S–M ads: 'seeks discipline', 'seeks uniforms', 'seeks leather and rubber'. **1975** *New Yorker* 26 May 32/2 Death is by far the most controversial and hottest subject in America, ranking twenty-five shock points above transsexuality, school busing, S & M, and interracial cloning. **1977** *Time* 15 Aug. 31/1 The streets teemed with whores, transvestites and the S–M crowd dangling slave bracelets and chains. **1955** R. J. SCHWARTZ *Compl. Dict. Abbrev.* 159/1 *SAR*,..search and rescue. **1958** *Oxf. Mail* 1 Aug. 6/6 SAR..has two squadrons situated at ten stations mainly around the south and east coasts. **1977** *R.A.F. News* 5–18 Jan. 3/1 An SAR Wessex was scrambled from Manston the following morning. **1945** M. JAMES *Born of Desert* xxi. 319 As the continental offensive developed and gained weight, so the S.A.S. probed deeper and deeper into the enemy lines. **1960** B. A. YOUNG *Artists & S.A.S.* xix. 46 The S.A.S. first went into action on November 16, 1941, when their target was the enemy airfields at Gazala and Tmimi, and their object to cripple the enemy's air before General Auchinleck launched his attack a few days later. **1976** G. SEYMOUR *Glory Boys* xvi. 221 The SAS anti-hi-jack force had been lifted by Wessex helicopter from their base camp. **1961** A. ANASTASI *Psychol. Testing* (ed. 2) ix. 226 A number of tests have been specially developed for use in the admission, placement, and counseling of college students. An outstanding example is the Scholastic Aptitude Test (SAT) of the College Entrance Examination Board. **1971** E. ASHBY *Any Person, Any Study* ii. 59 Taken together with SAT scores the information is as good as any measure yet devised to predict academic performance in college examinations. **1974** A. LURIE *War between Tates* ii. 39 Until very recently, girls like her, whatever their SAT scores, didn't usually go to graduate school. **1923** J. REITH *Diary* 29 Aug. (1975) ii. 132, I read the News Bulletin at 7.00 p.m.—the first real SB. **1929** *B.B.C. Year-Bk. 1930* 310 One or two transmissions are, perhaps, being sent to the provinces via the S.B. lines... Tests are being taken of outside broadcasts or incoming S.B. **1964** L. DEIGHTON *Funeral in Berlin* xxxvii. 228 'Makes me curious about the locked room,' said the young S.B. man. **1975** O. SELA *Bengali Inheritance* iv. 34 Special Branch won't like it... But you will need some assistance from SB—the files at least. **1917** A. G. EMPEY *Over Top* 307 S.B., stretcher bearer. The motive power of a stretcher. He is generally looking the other way when a fourteen-stone Tommy gets hit. **1919** W. DEEPING *Second Youth* xix. 168 'Ere, you blitherin' S.B.'s, get a move on. **1942** PARTRIDGE *Dict. Abbrev.* 87/2 *S.B.A.*, Sick-Bay Attendant. **1964** J. HALE *Grudge Fight* iii. 44 'Got a nice new one for you,' said the sick bay attendant to Adams, meaning the needle which glittered in space for a moment before the S.B.A. rammed it into his arm and pressed the plunger. **1953** *Newsweek* 24 Aug. 62/3 The new Small Business Administration whirled into rapid action... SBA..will keep some functions formerly performed by the..Small Defense Plants Administration. **1976** *Billings* (Montana) *Gaz.* 27 June 2-D/8 They file for Small Business Administration (SBA) loans and get on the long HUD list for trailers. **1932** *Flight* 1 July 601 The Flying Display arranged by the S.B.A.C., with the co-operation of the Air Ministry, at Hendon last Monday must be counted a success. **1951** *R.A.F. Rev.* Oct. 13/2 It flew faster than any aircraft has ever flown before in an S.B.A.C. show. **1968** C. SIMS *Royal Air Force* xi. 174 An item in the S.B.A.C. show. **1973** HOBSON & POHL *Mod. Petroleum Technol.* xxix. 945 A development during recent years has been the so-called Single Buoy Mooring (SBM) which may be either a fixed tower or a large buoy to which the ship is moored bow on. **1975** *Offshore Engineer* Sept. 33 (*heading*) Pioneering new sbm system off coast of Anglesey. **1975** *Petroleum Rev.* XXIX. 324/1 Floating hose conveys the crude from the SBM into a tanker. **1956** *Rubber World* May 239/2 The use of the term 'butadiene-styrene rubber' and the coding 'SBR'. **1971** G. J. VAN DER BIE et al. in C. M. Blow *Rubber Technol. & Manuf.* iv. B. 84 Emulsion SBR, for long the only synthetic general purpose rubber, has blossomed from the few 'hot' types produced during World War II into the multiplicity of grades now available in 'cold' types. **1920** *Dalton's Weekly Advertiser* 10 Jan. 3/2 (Advt.), House or S.C. Flat wanted by married couple. **1975** *Irish Independent* 27 May 18/1 (Advt.), Newly furnished hall flat, completely s.c., own door, double bedroom. **1977** *Weekly Times* (Melbourne) 19 Jan. 58/7 (Advt.), Clean, SC, ground level flats, acc. 2 to 6. **1964** E. BACH *Introd. Transformational Gram.* iv. 61 The second part of the rule specifies the *structural change* (SC) by means of variable signs. **1966** A. KOUTSOUDAS *Writing Transformational Gram.* i. 24 There are different notational conventions for writing a T-rule; rule (8) above can also be written...SC: $x_1 - x_2 - x_3 \rightarrow x_3 - x_1 - x_2$. **1946** *Newsweek* 12 Aug. 43/3 Again SCAP (Supreme Commander for the Allied Powers, the official designation of the occupation) left the implementation up to the Japanese. **1978** C. HUMPHREYS *Both Sides of Circle* xii. 131 The formal opening took place of our own Empire Building near that of S.C.A.P. (headquarters of the American army). **1885** *Ordinances Univ. Cambr.* p. viii, *Students in Science and Letters.* Proceeding to the Degrees of Sc.D. and Litt.D. **1917** J. R. TANNER *Hist. Register Univ. Cambr.* 195 Sc.D. Robes and hood of scarlet cloth, both lined with silk shot with pink and light blue. **1979** *Oxford Univ. Gaz.* 1 Mar. 554/2 University Preachers... Sunday, 11 March, at 10.15 a.m. D. E. Broadbent, C.B.E., M.A., D.Sc. (M.A., Sc.D. Cambridge, Hon. D.Sc. Southampton). **1921** *Ann. Rep. Save the Children Fund* 4/1 During the year under review, the S.C.F. has succeeded in making the needs of the children widely known throughout the United Kingdom and the British Empire. **1967** *Punch* 1 Mar. 292/2 'A simple pleasure,' said the SCF supervisor, 'but one that flat-dwelling kids just don't have.' **1974** *Petroleum Rev.* XXVIII. 794/2 The combined recoverable reserves of the field are estimated to be in the order of 1·5 to 2·0 × 10⁹ barrels of oil, with some 3·0 × 10¹² scf of associated gas. **1959** L. D. REDDICK *Crusader without Violence* xii. 205 There was so much.. rumor about friction between the SCLC and NAACP that King scampered up to New York. **1968** L. LOKOS *House Divided* x. 375 SCLC called upon the President to de-escalate the war unilaterally. **1973** *Freedomways* XIII. 8 Recent action taken by the leaders of..SCLC..is a most welcome development. **1935** *Nursing Mirror & Midwives' Jrnl.* 2 Feb. p. xx/3 (Advt.), S.C.M. required... General training not essential. **1955** *Times* 8 July 2/5 Candidates must be S.R.N., S.C.M., and should preferably have had similar experience in a Teaching Hospital. **1924** *Fellowship of Students* (Student Christian Movement) 49 The work of the S.C.M. must be closely related to the developing work of the Churches. **1948** H. G. G. HERKLOTS in M. Warren *Triumph of God* vii. 180 Now it was men and women who had been schooled in the S.C.M., who accepted its 'interdenominational position', who were planning the Edinburgh Conference. **1976** A. LOCKLEY *Christian Communes* vi. 55 The SCM headquarters were moved out of London in 1974 to Wick Court, a Jacobean mansion near Bristol, where the central staff attempted to live communally. **1971** *Nature* 16 Apr. 430/1 The big attraction of SCP is the possibility of utilizing cheap raw materials as fermentation substrates. **1976** *Jrnl. R. Soc. Arts* CXXIV. 580/2 It might be more economical to convert part of our oil supplies into food by using it as a substrate for micro-organisms, to produce 'Single Cell Protein' (SCP) that could be used as the raw material for the textured meat substitutes that were referred to earlier. **1923** D. K. HORNE in G. Bailey *Lady Margaret Hall* v. 94 Each student..is placed under a tutor, who is almost invariably one of the resident members of the S.C.R. **1964** M. HUTT in D. Daiches *Idea of New Univ.* iii. 49 Falmer House belongs neither to the Union nor the SCR. **1965** *New Statesman* 7 May 734/1 Mr Soyinka..knows the SCRs of Ibadan and Ife. **1976** A. CROSS *Question of Max* vi. 73 If..I had one wish right now, it would be to be connected for a time with an Oxford college and dine at the high table, chat in the SCR. **1963** *New Scientist* 13 June 600/1 The SCR, at the outset, is basically an 'insulator'. It does not allow current to pass. **1975** D. G. FINK *Electronics Engineers' Handbk.* VII. 52 During forward-bias operation.., the *pnpn* structure of the SCR is electrically bistable and may exhibit either a very high impedance (OFF state) or a very low impedance (ON state). **1956** *Times* 22 Sept. 10/4 The members of the Suez Canal Users' Association (S.C.U.A.) shall be those nations which have participated in the second London Suez conference and which subscribe to the present declaration, and any other adhering nations. **1970** H. TREVELYAN *Middle East in Revolution* 100 The Menzies Mission..was followed by the second conference of the 'users', the formation of the 'users'' association and the realisation that the S.C.U.A., the skewer as we called it, had a blunt point. **1939** *London Weekly Advertiser* 7 June 6/4 £70 Exclusive.—S.D. Modern House, three beds, etc., near Dyke Road Park. **1975** *Evening Herald* (Dublin) 8 May 10/2 (Advt.), Very unusual 4 bedroomed s.d. in cul de sac. **1901** W. M. F. PETRIE *Diospolis Parva* i. 6 We now make a first division into fifty equal stages, numbered 30 to 80, termed *sequence dates* or S.D. **1939** —— *Making of Egypt* vii. 55 This period (S.D. 60–75) has no generally distinctive culture. **1944** *Jrnl. Near Eastern Stud.* III. 110 Petrie distinguished it from Gerzean by naming it the Third Predynastic of Semainean, ranging from Sequence Dates 60/63 to S.D. 76. **1950** G. E. DANIEL *Hundred Yrs. of Archaeol.* v. 176 He started his sequence with S.D. 30,.. and carried on to dynastic times at S.D. 80. [**1940** H. KOEHLER *Inside Gestapo* ii. 30 Heydrich's power is much more founded on the Security Service of the Reich Fuehrer S.S. (shortened SD. RFSS.) than on the Gestapo.] **1947** H. TREVOR-ROPER *Last Days of Hitler* i. 28 Otto Ohlendorf was head of *R.S.H.A. Amt III* (also called S.D. or *Sicherheitsdienst*). **1968** *Listener* 8 Aug. 174/3, I was wearing my old SD cap which I'd brought through the whole war. **1974** A. WILLIAMS *Gentleman Traitor* xiii. 212 The father had returned to Germany in 1938..and had served in the SD, the civil arm of the S.S. **1902** *Biometrika* I. 206, σ_1 is the S.D. of the organ. **1973** *Nature* 31 Aug. 587/2 The mean (±s.d.) IQ then was 98·7 (±7·1) with no precocity. **1978** *Jrnl. R. Soc. Med.* LXXI. 659 Twin, malformed or grossly autolysed fetuses, and any which were ≥ 2 s.d. below the mean for the gestational age group, were excluded. **1964** E. BACH *Introd. Transformational Gram.* iv. 61 The first part of the rule is a *structural description* (SD..) specifying the class of strings (in the terms of their analysis by P markers) to which the rule applies. **1975** *Studies in Eng. Lit.: Eng. Number* (Tokyo) 170 The term 'obligatory'..has generally been understood to mean that an obligatory rule must apply to a phrase-maker which meets the SD of that rule. **1975** *Glasgow Herald* 17 Dec. 1 (*heading*) S.D.A. pledge quick action. **1976** *Scotsman* 15 Dec. 5/1 Capital restructuring..will give the SDA one-third of the shares at a cost of £60,000. **1966** *Economist* 22 Jan. 301/2 His superior in SDECE admitted last week that this was true. **1966** M. WOODHOUSE *Tree Frog* viii. 58 There is Defensive Intelligence and Active Intelligence... In France, the Ministry of the Interior and S.D.E.C.E. **1972** K. BENTON *Spy in Chancery* xvii. 193 They have a.. gentleman's agreement..with both MI6 and CIA—and with the French SDECE, too, I think. **1893** G. B. SHAW *Let.* 24 Apr. (1965) I. 390 My remarks..were not levelled at the I.L.P., but at the S.D.F. **1910** CHESTERTON *George Bernard Shaw* 68 Bernard Shaw was thrown early into what may be called the cosmopolitan club of revolution. The Socialists of the S.D.F. call it 'l'Internationale'. **1957** R. HOGGART *Uses of Literacy* xi. 261 They worked for Hyndman's S.D.F. in the 'eighties, and for the I.L.P. in the 'nineties. **1980** 'FIRST' & SCOTT *Olive Schreiner* iv. 109 The Democratic Federation [of 1881] became the Social Democratic Federation (SDF) in 1883. **1936** S.D.F.er [see *MORRISITE]. **1970** *Times* 22 Aug. 3/1 'The S.D.L.P. is the only institution that can bring about reform in Northern Ireland,' he said. **1974** *Freedom* 25 May 8/1 Willing followers of the power seekers of the IRA, SDLP and all the 'loyalist' organisations. **1978** D. MURPHY *Place Apart* vi. 117 An elderly woman

S

then observed, 'What we need is a sort of cross between the Provos and the Stickies and SDLP.' **1936** W. H. S. Smith *Let.* 21 June in *Young Man's Country* (1977) i. 9 I've been in correspondence with the present S.D.O. **1947** *Civil & Milit. Gaz.* 9 Apr. 2/4 'Youngsters' are attached to S.D.O.'s to learn their jobs. **1977** W. H. S. Smith *Young Man's Country* iii. 103, I was welcomed by several clerks in the S.D.O.'s office. **1908** *Times* 18 Apr. 10/1 The delegate..declared that as a militant force the S.D.P. could desire no more suitable antagonist than the scion of the aristocracy who represented Rossendale. **1912** R. Macaulay *Views & Vagabonds* ii. 24 We stand for all the principles of the Fabian Society, the S.D.P., and..so forth. **1961** C. Tsuzuki *H. M. Hyndman & Brit. Socialism* viii. 164 In October 1907 this progress and the desire to emulate the Labour Party encouraged the executive to change the name of the Federation to the Social Democratic Party (S.D.P.). **1981** *Times* 27 Mar. 1/1 (*heading*) SDP launched with aim of 'reconciling the nation'. *Ibid.* 3 (Advt.), If you share our aims you can join the SDP by filling in the application and returning it with a subscription. **1967** *Guardian* 28 Aug. 1/7 It will be up to the managing director of the IMF..to initiate and suggest the size of any proposed allotment of special drawing rights (SDRs). **1972** *Penguin Dict. Econ.* 382 The S.D.R...is an entry in a member country's book balance with the I.M.F. **1976** 'J. Davey' *Treasury Alarm* iii. 43, I happened to hear some Treasury folk gossiping..and I find that SDRs, which baffled me, are Special Drawing Rights. **1968** *Times* 19 Apr. 10/4 The extreme left-wing Federation of German Socialist Students (S.D.S.). **1977** R. Crossman *Diaries* III. 77 In West Berlin on April 11th there was an attempt to assassinate Rudi Dutschke, left-wing leader of the militant Socialist Students' League (S.D.S.). **1961** *Mademoiselle* Aug. 335/2 'Students have a mystique about action,' says Al Haber, president of the nationwide Students for a Democratic Society... Haber has traveled to many campuses this year to establish new S.D.S. groups. **1965** *Moderator* Winter 14/2 SDS was re-formed in 1962, at which time the Port Huron statement defined its purposes. **1974** H. L. Foster *Ribbin'* iii. 92 The SDS and some third world groups sold drugs to college students to earn money and keep their cause going a few years ago. **1946** C. E. Weatherburn *Math. Statistics* vi. 110 This S.D. is usually called the standard error (S.E.) of the number of successes in a sample of size *n*. *Ibid.* 111 A deviation from the mean less than twice the S.E. is regarded as not significant. **1964** F. N. Kerlinger *Foundations Behavioral Res.* ix. 167 $SE_M = 2{\cdot}73$. **1971** *Nature* 1 Jan. 62/1 On these assumptions, the mean number (±s.e.) of skeletomotor nerve fibres in normal nerves was 191 ± 26. **1974** *Ibid.* 23 Aug. 654/1 The mean concentration of prolactin in the peripheral blood of women during the cycle is 15 ± 1 (± s.e.m.). **1927** *Financial Times* 7 May 6/5 (*heading*) S.E. Clerks' Provident Fund. **1942** Partridge *Dict. Abbrev.* 88/2 S/E, Stock Exchange. **1978** *Times* 23 Jan. 15/1 A Stock Exchange investigation..is now awaiting examination by the SE Council. **1944** *SEAC: Daily Newspaper of South East Asia Command* (Calcutta) 10 Jan. 1/1 First light is breaking over this awakening city as the birthday issue of SEAC (pronounce it See-ack) comes flying off the presses. **1945** *Daily Mirror* 15 Aug. 1/1 There is an Army warning against expecting the quick homecoming of SEAC troops. **1971** R. Russell tr. *Ahmad's Shore & Wave* xv. 159 In those days he [sc. Lord Mountbatten] was the head of S.E.A.C., and was there on leave. **1934** *Time* 16 July 46/1 The S.E.C. will take over enforcement of the Securities Act of 1933. **1955** *Times* 29 Aug. 11/3 Banks..have indicated an interest if the S.E.C. provides a means of registering them. **1966** *Economist* 1 Oct. 46/3 The SEC has steadfastly opposed the exchange's efforts to limit trading in listed stocks to members only. **1978** *Financial Times* 3 Mar. 18/2 The setting-up of a body like the SEC is a last resort. **1962** *Rep. Comm. Broadcasting 1960* 218 in *Parl. Papers 1961–2* (Cmnd. 1753) IX. 259 Our attention has been drawn to another system of colour-transmission, the 'SECAM' system. This might..prove a serious rival to the NTSC system. **1968** [see *PAL s.v. *P II]. **1978** *Gramophone* Aug. 391/1 All the major video protagonists have mounted elaborate press launches of their systems, suitably adapted to the PAL TV format commonly used in Europe instead of the SECAM standard employed elsewhere. **1968** *Proc. Symp. Scanning Electron Microscope* 3/1 The scanning electron microscope, or SEM, as I shall henceforth call it. **1974** SEM [see *macerate sb.]. **1961** *Nursing Mirror & Midwives' Jrnl.* 26 May 760/1 Male S.E.N. or Nursing Auxiliaries required for duties in acute wards & departments. **1964** G. L. Cohen *What's Wrong with Hospitals?* ix. 192 Negro domestics abounded, but the S.E.N. school was conspicuously white. **1977** *R.A.F. News* 22 June–5 July 13 (*caption*) SACW Margie Lee, an SEN, nursing a premature baby in the maternity unit at RAF Hospital Wegberg. **1966** *Financial Times* 4 May 1/1 Most Fleet Street papers seized on the payroll tax (SET) as the Budget's main news point. **1966** *Observer* 8 May 9/1 The main virtue of S.E.T. is that it is indirect. *a* **1974** R. Crossman *Diaries* (1975) I. 508 My first reaction to S.E.T. was that in terms of farming at Prescote Manor as well as in terms of building it was absolutely unbearable. *c* **1875** *Pocket Exchange Guide San Francisco* 178 S.F. Ten-Cent Parcel Delivery Co.—16 Post. **1975** B. Meggs *Matter of Paradise* ix. iii. 267 One year at the University of Wisconsin, then moved to S.F. with her boyfriend, drummer in a rock-band. **1929** *Sci. Wonder Stories* June 92/3 The S.F. Magazine. (Science-Fiction). **1948** G. Conklin *Treas. Sci. Fiction* p. ix, Many SF writers are feeling the urgent need for social controls over our physical powers. **1954** I. Asimov in *Mag. Fantasy & Sci. Fiction* Oct. 69 (*title*) The foundation of S.F. success. **1959** *Times Lit. Suppl.* 20 Mar. 166/2 Both are by O.K.-names in the s.f. world. **1968** *Punch* 10 Jan. 69/3 Let's take a step beyond in the company of Ray Bradbury..and other (mainly *sf*) bloodcurdlists. **1969** *Daily Tel.* 16 Oct. 22/7 Two regular themes in S.F. are the world dominated by a, usually evil, visionary and the world run by a Great Computer. **1973** 'D. Halliday' *Dolly & Starry Bird* viii. 108 Johnson..was looking as everybody looks when they first step into an electronic workshop. That is blasé. This is the fault of the S.F. kiddie shows on the telly. **1975** Felton & Fowler *Best, Worst & most Unusual* 44 Arguably, it is the peak of sf film in its long history. **1980** *Times Lit. Suppl.* 7 Nov. 1263/1 The reason for preferring the old established and unpretentious 'sf' need not baffle us. The use of 'sci-fi' was clearly intended to imply a jaunty insider's knowledge coupled with a modern..demonstration of slick neologism. **1922** Lady Leslie *Let.* 17 Mar. in M. Gilbert *Winston S. Churchill* (1977) IV. Compan. iii. 1809 The Union Jacks are being silently stored away—the SF flag floats on the orange Hall. **1973** *Irish Times* 2 Mar. 8/3 Sherlock J. (S.F.)... 2,488. **1882** *Football* 4 Oct. 4/1 The match would be played under the supervision of the S.F.A. **1974** *Evening News* (Edinburgh) 9 Apr. 18/8 The SFA stipulated that no game could be cancelled at such an early hour on the strength of a weather forecast. **1933** *Bulletin* (Sydney) 14 June 12/3 Ask any modern sailor who has been refused an issue of pay or rations 'What luck?' and he will be apt to reply 'Sweet Fanny Adams', or just 'S.F.A.', meaning that he received nothing. **1935** *Ann. Reg.* 1934 i. 172 Thus at Toulouse the S.F.I.O. (the French section of the Workers' International) sought..to form an entente with the Communists against Fascism. **1977** *Compar. Politics* IX. 364 The Popular Front, with its broad policies of alliance extending past the SFIO to the radical party..responded to Soviet and Comintern demands. **1949** E. Pound *Pisan Cantos* lxxvii. 52, I heard it in the s.h. a suitable place To hear that the war was over. **1948** *Prof. Papers Inst. Post Office Electr. Engineers* No. 197. 3/1 S.H.F. systems are those operating in the range 3,000 to 30,000 Mc/s, 10 to 1 cm. wavelength. **1958** *Electronic Engin.* XXX. 276/1 The radio repeater design is based on the principle of amplification at intermediate-frequency with reconversion to s.h.f. and the use of a travelling-wave amplifier to deliver the final output power. **1965** *Acronyms & Initialisms Dict.* (Gale Research Co.) 638 SHO,..Senior House Officer. **1976** *Proc. R. Soc. Med.* LXIX. 818/1 A hypothetical increase in the number of graduates to over 5000 by 1980 would require about 18 000 SHO and registrar posts in 1982. **1931** *Times Trade & Engin. Suppl.* 24 Jan. p. iv/3 Six vessels, of 10,500 s.h.p. **1972** C. Mudie *Motor Boats & Boating* 28 The most important is SHP which is Shaft Horse Power or power delivered to the propeller shaft. **1961** SI [see *international a. 1 c]. **1970** *Nature* 2 May 473/2 Editors of scientific journals have been recently accused of forcing some unwilling scientists to adopt SI units. **1971** *Inside Kenya Today* Mar. 30/1 The Engineering and Construction Industries can go ahead with the change-over to the Metric System, but in particular in its modernized version called the 'Système International d'Unités' or the SI as it is popularly known in all languages. **1973** tr. *Internat. System of Units* 1 The 11th CGPM [sc. General Conference of Weights and Measures] (1960), by its Resolution 12, adopted the name *International System of Units*, with the international abbreviation SI, for this practical system of units of measurement. **1977** *Westworld* (Vancouver, B.C.) May–June 34/1 Canada..is one of the last major countries still using the imperial system of measurements instead of the International System of Units—universally known as SI. **1955** *Sci. News Let.* 29 Jan. 72/2 Polarized radio waves may also be a solution to the problem caused by 'SID's', or sudden ionospheric disturbances, when the sun may hurl out a great tongue of flame from which hydrogen atoms bombard the earth about 20 hours later. **1968** *Radio Communication Handbk.* (ed. 4) xii. 8/2 Ionospheric storms are often preceded (by approximately two days) by what are called sudden ionospheric disturbances (s.i.d.). **1970** J. B. Beckwith in A. B. Bergman et al. *Sudden Infant Death Syndrome* 15, I personally feel the term 'Sudden Death Syndrome' should at least be amplified to include the word 'infant'... I should like, therefore, to cast my vote for the term 'Sudden Infant Death Syndrome' (SIDS). **1980** *Sci. Amer.* Apr. 52/1 The phenomenon, which is also known by its initials SIDS (pronounced as letters rather than as a word), is defined clinically as the sudden, unexpected death of an apparently healthy infant for whom a routine autopsy fails to identify the cause of death. **1958** *Listener* 13 Nov. 779/2 The 'Nautilus' used a fully integrated apparatus of this kind called the Ship's Inertial Navigation System—inevitably contracted into its initials, 'S.I.N.S.' **1979** A. Fox *Threat Warning Red* i. 2 He could check those latitude and longitude figures against the readings on the SINS dials on the bulkhead. **1939** J. Reith *Diary* 21 Apr. (1975) iv. 227 Meeting in the foreign secretary's room, where I have never been before. Present: Halifax, Chatfield, Hoare, Cadogan, Leeper, CID secretary and Admiral Sinclair, the hush-hush SIS chief. **1964** G. Lyall *Most Dangerous Game* xiii. 91 You're SIS—one of the Foreign Office boys. What the mob would call The Secret Service. **1978** R. V. Jones *Most Secret War* vii. 63 While my evenings were spent discussing cryptography, my days went in perusing the S.I.S. files. *a* **1918** W. Owen *Coll. Poems* (1963) 74 (*title*) S.I.W. **1929** *London Mag.* Dec. 629/2 'What really coopered him, was being in charge of an S.I.W. just before Armistice.' 'What is an S.I.W.?' I said. 'A hospital for self-inflicted wounds.' **1975** P. Fussell *Gt. War & Mod. Memory* viii. 294 The final two lines of 'S.I.W.' tell us how the victim of the self-inflicted wound was buried. **1822** *Catholic Miscellany* May 208 H. More *Hist. Prov. Ang. S.J.* p. 467. **1916** Joyce *Portrait of Artist* iv. 187 The Reverend Stephen Dedalus, S.J. His name in that new life leaped into characters before his eyes. **1967** *Cath. Dict. Theol.* ii. 1/2 J. A. Jungmann SJ, *Handing on the Faith*..contains much recent information. **1974** *Time* 18 Feb. 16/1 The S.L.A. emblem is a seven-headed cobra. **1978** *Maledicta* 1977 I. 123 Terrorists have come to rely on the press to deliver their graphic messages to the world—the Hanafi Muslims in Washington, the PLO at the Munich Olympics, the SLA in Oakland. **1967** *New Scientist* 9 Feb. 340/2 To achieve overwhelmingness, especially in the face of Russia's ABM effort, Mr. McNamara proposes to produce and deploy the Poseidon SLBM. **1973** *Sci. Amer.* Feb. 14/2 The primary SALT I restrictions..impose numerical ceilings on both land-based intercontinental ballistic missiles (ICBM's) and submarine-launched ballistic missiles (SLBM's). **1979** *Financial Rev.* 28 Sept. 10/3 The strategic deterrent is conceived as a triad consisting of bombers armed with nuclear weapons, submarine-launched ballistic missiles (SLBMs) and land-based intercontinental ballistic missiles (ICBMs). **1972** *Time* 19 June 17/3 The Pentagon is also asking for $20 million to get started on another submarine-based missile, the SLCM (Submarine-Launched Cruise Missile). **1975** *Bull. Atomic Sci.* XXXI. 13 *SLCM*, submarine-launched cruise missile. **1979** *Observer* 25 Nov. 34/1 Cruise Missile... Can be ground-launched (GLCMs, known as Glickums)..or launched from submarines (SLCMs). **1958** *Jrnl. Clin. Invest.* XXXVII. 876/1 The serum of patients with systemic lupus erythematosus (SLE) has been shown to contain factor(s) with a special affinity for nucleohistone (NH). **1978** *Jrnl. R. Soc. Med.* LXXI. 149 This is..the first case in which SLE has been reported in a patient with Hashimoto's thyroiditis and pernicious anaemia. **1951** *Times Educ. Suppl.* 8 June 466/1 (Advt.), Wanted for January, 1952, Senior English Mistress. The work includes G.C.E. at O., A., and S. levels. **1973** *Guardian* 28 June 13/3 Jonathan Mestel is 16 and has just finished his A and S level maths and physics exams. **1975** *Glasgow Herald* 23 Dec. 4/2 Mr Sillars said the SLP did not believe oil was a Scottish resource, but something to be used for the benefit of the entire United Kingdom. **1976** *Times* 23 Jan. 14/3 The new threat which the SLP presents..is untimely. **1964** *Colour Photogr.* Mar.–Apr. 68/3 With one manufacturer recently introducing a half-frame SLR, I wondered if this type of instrument would replace its larger brother. **1971** *Amateur Photographer* 13 Jan. 80/3 (Advt.), Nikkorex F, f2 Nikkor, w/meter, S.L.R... £69. **1977** J. Hedgecoe *Photographer's Handbk.* 14 Because of the SLR's worldwide popularity the newest and most advanced electronic and optical technology tends to be designed to suit this camera before the others. **1890** Webster 1923/2 S.M.,..sergeant major. **1953** J. Masters *Lotus & Wind* v. 70 It wouldn't affect his chances of promotion to S.M. **1973** J. Wood *North Beat* xiii. 163 He had been the first company S.M. to earn a Bar to the decoration in the division. **1764** A. Williams *Universal Psalmodist* (ed. 2) 57 [Tune] Southwell. Psalm 90th Dr W. S.M. **1832** J. Jowett *Man. Parochial Psalmody* 12 St. Bride's. S.M. **1909** W. H. Frere in *Hymns Anc. & Mod.* p. lvi/2 The whole musical balance was now altered [by the influence of T. Ravenscroft's *Psalmes* of 1621]: the D.C.M. and P.M. tunes had hitherto formed the bulk of the collection, with a few C.M. and S.M. tunes to supplement them. **1909** J. R. Ware *Passing Eng.* 212/2 S.M. (Theatrical), stage manager. **1952** Granville *Dict. Theatr. Terms* 172 When artistes have settled into their parts, the S.M. (as he is known) may hand over the book to his assistant. **1972** V. C. Clinton-Baddeley *To study Long Silence* ii. viii. 82, I was still in the theatre when the S.M. found him. **1909** *Army & Navy Gaz.* 27 Mar. 297/3 (*heading*) S.M.L.E. Rifle. **1958** J. A. Barlow *Elem. Rifle Shooting* (ed. 5) p. xii, The S.M.L.E...will shortly be replaced by a self-loading rifle. **1914** *Autocar* 2 May 817/1 The R.A.C., the A.A. and M.U., and the S.M.M.T. **1958** *Economist* 1 Nov. 435/2 If it were desired to try to stop the rush at some point there is no salient that looks capable of being held and no one body that could do it—not the Finance Houses' Association, not the Industrial Bankers, not the SMMT nor the Hire Purchase Trade Association. **1978** *Dumfries & Galloway Standard* 21 Oct. 7/6 Although organising international motor shows is but one function of the SMMT, it is best known for this role. **1950** *Jrnl. Soc. Motion Pict. & Television Engin.* Mar. 389 SMPTE Officers and Committees..are published annually in the April issue of the Journal. **1959** W. S. Sharps *Dict. Cinematogr.* 11. 129/2 In Britain, the British Kinematograph Society was created originally as the British branch of the S.M.P.T.E. **1975** G. J. King *Audio Handbk.* viii. 191 The l.f. bands adhere to the SMPTE (Society of Motion Picture and Television Engineers) requirements for IMD analysis. **1960** *Atlanta Daily World* 18 Oct. 1/3 There was no announcement of formal action on SNCC's proposed march on the polls on election day. **1961** *Commonweal* 15 Dec. 311/1 This fall the S.N.C.C. has been concentrating on recruitment on white campuses. **1971** J. Bishop *Days of M. Luther King* iv. 369 The Student Nonviolent Coordinating Committee had been in that part of Selma for eighteen months... SNCC was in Marion too. **1949** *Progress French Nat. Railways* Oct. 1948 28 (*colophon*) S.N.C.F. 1949. **1963** *Times* 23 May 13/6 French Railways (SNCF) most closely resemble B.R. in size. **1976** A. White *Long Silence* vi. 46 The S.N.C.F. did not permit the express trains to stop at Colauvin. **1944** *Jrnl. Dairy Res.* XIII. 53 Milk has been deteriorating in quality (s.n.f. content) during the last two or three years. *Ibid.*, This extra labour would detect adulterated samples which yet had over 8·5% s.n.f. **1960** *Farmer & Stockbreeder* 16 Feb. 66/3 A Milk Marketing Board survey..had shown an average of 3·7 per cent butter-fat and 8·75 s.n.f. for the breed. **1976** *Burnham-on-Sea Gaz.* 20 Apr. (Advt.), Butterfats 4%. S.N.F. 8·83%. Total 12·83%. The cows have not been prepared for sale in any way. **1972** E. N. Tiratsoo *Natural Gas* (ed. 2) i. 15 The incipient shortage of natural gas..led to the adoption of the CRG process..as the basis of what is planned to be the world's largest reforming plant producing SNG ('substitute natural gas'). *Ibid.* viii. 149 Simulated natural gas (SNG)..has been extensively used as an intermediate fuel to facilitate the changeover of industrial plants from manufactured town gas to full-scale natural gas supply. **1974** *Natural Gas* (Shell Internat. Petroleum Co.) 11 Apart from LNG, which is natural gas cooled to the liquid state, there are three main possible supplementary sources of supply, namely low-calorific-value gas from coal, synthetic natural gas (SNG) from oil or coal, and methanol derived from natural gas. **1977** K. A. D. Inglis in P. A. Stockil *Our Industry Petroleum* (Brit. Petroleum Co.) (ed. 5) 26 As supplies of natural gas run short..systems for the manufacture of substitute (or synthetic) natural gas (SNG) are being developed. **1914** A. B. Milne in *Publ. Navy Rec. Soc.* (1970) CXV. 146 Have requested S.N.O., Gibraltar, to keep special lookout for *Strassburg*. **1955** C. S. Forester *Good Shepherd* ii. 264 SNO meant senior naval officer in accordance with British usage, not one of those odd collections of letters like DSO or MBE which merely meant a decoration. **1970** A. J. Marder *From Dreadnought to Scapa Flow* V. i. ii. 17 The S.N.O. on the spot..would not allow him to have a go at the *Goeben* as soon as it was known that she was aground. **1935** *Glasgow Herald Index 1934* 231/1 Compton

Mackenzie addresses first S.N.P. mtg. in Edin., 21 Ap. **1977** M. Walker *National Front* viii. 215 It may have been, as the SNP posters said 'Scotland's Oil' but a goodly portion of it had already been used by British Governments to guarantee foreign loans. **1970** Bahcall & Ulrich in *Astrophysical Jrnl.* CLX. L58 We have expressed the counting rate in solar-neutrino units: 1 SNU ≡ 10^{-36} capture per target atom per second. **1972** *Ann. Rev. Astron. & Astrophysics* X. 28 The best present estimates suggest..that the ultimate sensitivity of the current 10^5-gallon experiment will be set by the cosmic-ray background at ~0·4 SNU. **1976** J. Kleczek *Universe* ii. 78 It is usual to specify the rate of the neutrino capture reactions..in terms of so-called solar neutrino units (SNU or 'snew') which are such that 1 SNU corresponds to 10^{-36} captures per target Cl atom per second. **1844** T. E. May *Treat. Parliament* xii. 215 A division is effected in the lords by the not-contents remaining within the bar, and the contents going below the bar. [*note*] Lords' S.O. No. 22. **1929** G. F. M. Campion *Introd. Procedure House of Commons* iv. 136 S.O. No. 11 of 1888 (the so-called 'Ten Minutes Rule'). **1976** in R. Crossman *Diaries* II. 75 It was only when the S.O. was redrafted in 1967 that M.P.s could make full use of its possibilities. **1918** H. V. O'Brien *Wine, Women & War* (1926) 286 What an S.O.B. that fellow is! **1930** E. Pound *XXX Cantos* x. 45 That monstrous swollen, swelling s.o.b. Papa Pio Secundo. **1934** C. Stead *Seven Poor Men of Sydney* iv. 120 That s.o.b. Montagu got me the job 'ere, you know. **1955** Auden *Shield of Achilles* iii. 79 And all poor s-o-b's who never Do anything properly. **1962** L. Deighton *Ipcress File* xxxii. 211 Just a simple case of a couple of well-informed S.O.B.s. **1975** 'E. Lathen' *By Hook or by Crook* xvi. 154 A stubborn SOB who doesn't give a damn. **1948** *Jrnl. R. United Service Inst.* XCIII. 212 An S.O.E. base was established alongside General Eisenhower's Headquarters in Algiers from the outset, primarily for work into France. **1968** D. Lampe *Last Ditch* xi. 113 SOE, the most special of the Special Forces, had just been set up under the cover of the Ministry of Economic Warfare. **1978** G. Greene *Human Factor* iii. i. 104 Over his head were a retired businessman who had once been connected with the rival wartime service SOE, and a retired general who had fought in the Western Desert. **1905** W. S. Churchill *Let.* 28 Oct. in R. S. Churchill *Winston S. Churchill* (1969) II. Compan. i. 402 Of course the S of S must always control the Viceroy. **1922** G. Bell *Let.* 4 Dec. (1927) II. xxii. 658, I sent you by post the yearly report to the S. of S., a very silly sort of Xmas present. **1958** L. Durrell *Mountolive* iv. 92 My dear chap..if you propose to make an issue of it with the S. of S. I can't help it. **1917** R. Lord *Captain Boyd's Battery A.E.F.* (1920) ii. 24 S.O.L.—Payroll abbreviation for Soldier, adapted to mean Soldier Out 'a Luck or Certainly Out 'a Luck, according to the way you spell it. Applicable to everything from death to being late for mess. **1921** J. Dos Passos *Three Soldiers* ii. iii. 99 'We shall have to put him down A.W.O.L. You know what that means?'..'I guess he's S.O.L.'; this from someone behind Fuselli. **1946** B. C. Bowker *Out of Uniform* iii. 48 As the phrase went, they were 'SOL' ('surely' out of luck). **1942** *Yank* 2 Sept. 14/2 Our regimental SOP in reference to any MP reports on enlisted men. **1961** B. Fergusson *Watery Maze* x. 248 The Americans had evolved what they call a Standard Operating Procedure, or S.O.P., which was thoroughly unsatisfactory. **1980** S. King *Firestarter* 335 There will be two Shop men along, partly to act as stewards and partly to keep an eye on you. SOP, you know. **1911** s.p. [see *job sb.[2] 4 e]. **1928** E. Wallace *More Educ. Evans* ix. 201 You backed that horse s.p. with every unfortunate bookmaker in England. **1974** G. F. Newman *Price* ii. 60 What's the full SP, Trevor? **1937** *Q. Rep. Soc. Protection Anc. Buildings* I. 1. 4 From that moment almost it can be said that the S.P.A.B. was born. **1943** J. Lees-Milne *Ancestral Voices* (1975) 234 At an S.P.A.B. meeting I found a small attendance considering a matter of the first importance, whether or not to protest in the press against the night bombing of historic German cities. **1978** *Jrnl. R. Soc. Arts* CXXVI. 477/1 The Canterbury Cloisters have been discussed by the SPAB recently as being a bad example of restoration. **1861** *Sat. Rev.* 23 Nov. 535/2 Strong in S.P.G. and S.P.C.K. **1892** C. M. Yonge *Old Woman's Outlook* x. 240 A charming book of my childhood, which I rather believe belonged to the S.P.C.K. **1939** J. Cary *Mr. Johnson* 10 Johnson's idea of a civilized marriage, founded on..a few novels approved by the S.P.C.K., is a compound of romantic sentiment and embroidered underclothes. **1969** *Listener* 24 Apr. 586/2 Judas looking for all the world like an SPCK Jesus, all gently waving locks and sad benevolence. **1921** *Labour Monthly* Oct. 344 We may say, without presumption, that the S.P.D. (German Social-Democratic Party) is the party which deserves above all others the title of Republican Party. **1947** *Partisan Rev.* Mar.–Apr. 143 The arduous theorists of the SPD were steeped in German pedantry. **1976** T. Allbeury *Only Good German* ix. 58 Both the SPD and the CDU had offered him party seats in the Bundestat. **1913** R. Bridges (*title*) S.P.E. [manifesto of the Society for Pure English]. **1919** —— *Let.* 22 June (1940) 147 Is there any reason for delay about the S.P.E. **1923** J. M. Murry *Pencillings* 268 Such is the appearance—neat, decorous, small, discreet—of an S.P.E. tract. **1948** *S.P.E. Tract* lxvi. 177 When I addressed myself to the honourable, melancholy task of writing the obsequy of S.P.E., I revived fond memories by turning over my file. **1839** C. Fox *Jrnl.* 22 Aug. (1883) v. 53 Mary Coleridge..read a letter from Macaulay describing the state of feeling into which one of Samuel Wilberforce's sermons had thrown him, who is now on a tour westward for the S.P.G. **1854** V. Lush *Jrnl.* 16 Feb. (1971) 153 We settled to write to the Bishop and request him in conjunction with the Revd Mr Venn and Revd Mr Hawkins, secretaries of C.M.S. and S.P.G.—to select a master for us. **1565** J. Jewel *Replie M. Hardinges Answeare* 294 Beda..expounded these foure solemne letters *S.P.Q.R.* in this wise. *Stultus populus quærit Romam. Fooliske Folke Flee to Rome.* **1621** I. Sylvester *Lacrymæ Lacrymarum* in *Du Bartas, His Diuine Weekes*, This loss (alas!) which unto All belongs!.. To all the world; except S.P.Q.R. **1881** *N. & Q.* 8 Jan. 34/2 S.P.Q.R. .. The following interpretation of these letters may amuse your readers. They form part of the decoration of the Adam ceiling of the Court Room of the Bank of England, and on a remark by a visitor that they seemed very incongruous in such a place, 'Not at all,' said one of my colleagues; 'they stand for small profits and quick returns.' **1927** E. J. P. Benn *Trade* iii. 52 Our grandfathers used to talk of small profits and quick returns, 'S.P.Q.R.', which most schoolboys of my time imagined had something to do with the early Romans, was a very favourite shop sign. **1977** L. Meynell *Hooky gets Wooden Spoon* xiv. 180 'So far he seems to have steered clear of anything big.' 'Wise man. S.P.Q.R. Small profits quick returns.' **1978** P. Finney *Crow Goddess* 112 'What does that say?' he asked, pointing at the label. 'S.P.Q.R.—the Senate and People of Rome.' **1883** *Proc. Soc. Psychical Res.* p. i, The Council of the S.P.R. have from time to time received letters from Members and Associates. **1926** A. Conan Doyle *Hist. Spiritualism* I. viii. 185 No serious attempt of any sort, up to the formation of the S.P.R., was made to understand or explain a matter which was engaging the attention of millions of minds. **1937** A. Huxley *Let.* 17 Feb. (1969) 415 Broad's presidential address to the SPR is interesting in this context. **1968** M. Collis *Somerville & Ross* xiv. 157 Arthur Balfour, who for a time was President of the S.P.R. **1971** B. Bauer et al. in *Jrnl. Audio Engin. Soc.* XIX. 639/1 We..'encode' (combine) the four channels into two by using a special 'matrix', or linear additive circuit.., which we called the SQ (for stereophonic-quadraphonic) matrix encoder. **1973** *Official Gaz.* (U.S. Patent Office) 20 Nov. TM 138/2 Columbia Broadcasting System, Inc., New York, N.Y. Filed Dec. 6, 1971... The mark represents the stylized lettering 'SQ'. For pre-recorded phonograph records, and pre-recorded tapes... First use June 10, 1971. **1976** *Listener* 23 Dec. 846/1 Discs in compatible quadraphonic/stereo pressings utilising the SQ system. **1919** *Round Table* IX. 286 The British Labour Party is described as 'worse than the right S.R.'s'. **1967** *Soviet Stud.* XVIII. 449 The SR's were able to obtain absolute majorities even in Petrograd and Moscow. **1974** T. P. Whitney tr. *Solzhenitsyn's Gulag Archipel.* I. i. ii. 30 The Left SR's had been cleverer and had gone on pretending longer that they were allies of the one and only consistent party of the proletariat. **1923** *Southern Railway Mag.* Sept. p. xvi (Advt.), For All Southern Railway Staff..Halden Estates Co., Ltd. (S.R. Proposition). **1959** *Chambers's Encycl.* XI. 490/1 Subsequently S.R. lines west of Exeter were transferred to the Western region. **1967** J. Joyce *Story Passenger Transport in Britain* vii. 186 Nationalisation came in 1948 when the 'Big Four'—the LMS, LNER, GWR and SR..became the all-embracing 'British Railways'. **1908** *Army & Navy Gaz.* 26 Dec. 1241/3 The latter officer joins the 3rd (S.R.) Batn. **1919** W. S. Churchill in M. Gilbert *Winston S. Churchill* (1977) IV. Compan. i. 466 The Regular Army and such SR & TF officers and men as must be retained. **1923** Kipling *Irish Guards in Gt. War* I. 44 Major Webber, 'S.R.' (this is the first time that the Diary makes mention of the Special Reserve), arrived the day before. **1963** *Recommendations for Letter Symbols, Signs & Abbreviations (B.S.I.)* vi. 26 Steradian... sr. **1977** J. Narlikar *Struct. Universe* vii. 223 The survey is over 3 sr..in the sky. **1935** C. L. Hull in *Psychol. Rev.* XLII. 502 A trace conditioned reaction in an S → R relationship. **1948** E. R. Hilgard *Theories of Learning* xii. 349 Woodworth gradually shifted from his S-R motto. **1967** [see neo-behaviourism s.v. *neo- 1 a]. **1977** *Dædalus* Fall 120 When Chomsky demonstrated that the simple application of S-R learning theory to chained responses is inadequate in principle as an account of grammar, he could not be ignored. **1971** *Nature* 23 Apr. 531/1 While investigating the interaction between antigen-antibody-complement complexes and lymphoid cells, we have noticed that sheep red blood cells (SRBC) adhered to a surprisingly large proportion of human peripheral blood lymphocytes forming clusters ('rosettes'). **1977** *Lancet* 19 Feb. 394/2 These eight dilutions and two control tubes..were tested for the formation of rosettes between human lymphocytes and S.R.B.C. **1946** *Nursing Times* 11 May p. ix/1 Mental Ward Sisters..required, S.R.M.N., or holding R.M.P.A. Certificate. **1965** *Nursing Mirror & Midwives' Jrnl.* 2 Apr. (Suppl.) 18/1 (Advt.), Applicants must be S.R.N., S.R.M.N. **1922** *Ibid.* 1 July 261/1 The Scottish General Nursing Council..'unanimously adopted' the resolution as to the undermentioned initials.. General Trained..[England] S.R.N...[Scotland] R.G.N. **1926** *Ibid.* 25 Dec. p. x/3 (Advt.), General, S.R.N., fever, and mental nurses Wanted for private work. **1974** R. Ingham *Yoris* ii. 5 She was wearing her stiff white nurse's uniform with..the small brooch on her left breast bearing the initials SRN. **1957** M. B. Hoagland et al. in *Biochim. & Biophys. Acta* XXIV. 216 This is apparently a low molecular weight RNA (S-RNA) with different metabolic properties from the high molecular weight RNA of the ribonucleoprotein of the microsomes. **1963** F. H. C. Crick in *Progr. in Nucleic Acid Res.* I. 196 Holley and his collaborators fractionated yeast sRNA by countercurrent distribution. **1971** D. J. Cove *Genetics* x. 146 If the sRNAs act as adaptor molecules, recognising the codon of the amino acid they carry, then it is to be expected that they will have somewhere in their sequence, three adjacent bases which are complementary to that codon, and able therefore to specifically hydrogen bond to it. In all the sRNAs whose sequence has been analysed, this has been found to be true. **1941** SRO [see *palsy sb.[2]]. **1966** *Social Work* Oct. 32/1 The clustering of unattached individuals, many of whom are economically dependent and chronically ill, in licensed SRO buildings is a recognizable pattern. **1977** *New Yorker* 27 June 85/3 Queens has only nine of New York's several hundred S.R.O. buildings (the letters stand for 'single-room occupancy', and the tenants..are often present or former drug addicts). **1890** *Texas Siftings* 15 Nov. 13/1 At the Grand Opera House Bobby Gaylor, in the Irish Arab, called out the S.R.O. sign. **1903** 'O. Henry' in *McClure's Mag.* July 333/1 After one reading of the Declaration of Independence in New York I've known the S.R.O. sign to be hung out at all the hospitals and police stations. **1970** *Islander* (Victoria, B.C.) 5 Apr. 16/2 The 1970 edition [of an annual concert], held on March 3,..was an SRO success. **1939** SRS [see *slow-reacting substance* s.v. *slow adv. 2 a]. **1955** W. E. Brocklehurst in *Jrnl. Physiol.* CXXVIII. 1 P The occurrence of a slow-reacting substance together with histamine, in the perfusate coming from the isolated lung of a guinea-pig during anaphylactic shock, has been reported previously. .. The substance has been provisionally named 'SRS-A' (slow-reacting substance in anaphylaxis) to differentiate it from the considerable number of ill-defined gut-contracting substances to which the term 'SRS' has been applied. **1958** *Pharmacol. Rev.* X. 422 When egg yolk is incubated with cobra venom a slow reacting substance (SRS-C) is formed. **1964** W. G. Smith *Allergy & Tissue Metabolism* i. 14 Evidence for the existence of chemical mediators of anaphylaxis other than histamine has existed since 1940. Recent work on one of these, the slow reacting substance of anaphylaxis (SRS-A), led to the discovery that the hypersensitive state exerts a profound influence on tissue metabolism. **1979** *Nature* 6 Sept. 14/2 SRS has now been identified as a novel cysteinyl derivative of arachidonic acid. **1932** H. Nicolson *Diary* 5 Jan. (1966) 105 The former would correspond to the S.S. or *Schutzstaffel* organisation of the Nazis. **1938** *Encycl. Brit. Bk. of Year* 125/1 They are known as the 'SS'..or Élite Hitler Protective Guard, now under the command of Heinrich Himmler. **1945** *Daily Mirror* 8 May 3/2 S.S. men went through the streets driving people out of their homes. **1958** *New Statesman* 19 Apr. 505/2 The nature of his work compelled him to form intimate relationships with members of two rival German organisations, the *Abwehr*, or Counter-Intelligence, and the SS. **1968** *Listener* 19 Sept. 359/1, I myself had only one sound-camera, and just before the opening of the Games a squad of SS men tried to take it away from me by force. **1975** W. Craig *Strasbourg Legacy* i. 4 A short, red-faced officer, resplendent in his black SS uniform, read from a memorandum. **1933** C. Mackenzie *Water on Brain* xviii. 276 Katzenschlosser, the American S.S. man. **1969** *Sun* 12 Feb. 5/5 An advance party of 60 State Department men arrived from Washington on Monday. Probably 20 of them were SS men. (They do actually call them that in the U.S.). **1979** *Maclean's Mag.* 9 Apr. 29/1 The Trudeau government was pressuring the SS for better intelligence..about suspected FLQ cells in Quebec. **1963** F. D. Fawcett *Cycl. Init. & Abbrev.* 140/1 SS,..Social Security. **1973** *Freedom* 21 July 5/2 Hardly any squatter draws SS and parasites on society. **1975** *New Society* 20 Nov. 412/3 Tez thinks he'll be a rock star.. tomorrow. Meanwhile he's having trouble getting it together and lives off the SS. **1980** L. Cody *Dupe* xxiii. 168 The father was knocked off in a pub bombing..and..the family's been on the SS ever since. **1868** *Times* 1 July 2/2 (Advt.), To sail 9th of July, the Liverpool and Australian Navigation Company's s.s. *Great Britain*. **1870** *Weekly Standard* (Buenos Aires) 12 Jan. 7/5 Departures. Per ss. Flamsteed on the 9th Jan. **1876** W. S. Lindsay *Hist. Merchant Shipping* IV. xv. 558 (*caption*) S.S. 'Victoria'. **1907** *Shipping World* 16 Jan. 111 (Advt.), S.S. 'Lusitania' is being fitted with Passenger Lifts. **1955** *Times* 5 July 11/3 Things have gone very wrong indeed when British vessels suffer the treatment received by the ss. *Anshun*. **1956** SSB [see PEP s.v. *P II]. **1976** *S9* (N.Y.) May/June 5/2 If, however, you don't want to go for ssb, then by all means buy the best radio you can find with the largest number of features. **1969** *New Acronyms & Initialisms* (Gale Research Co.) 389 SSBN, Fleet Ballistic Missile Submarine (Nuclear powered). **1973** *Washington Post* 13 Jan. A23/3 Gormley's proposals..stem from his realization that SSBN's are the only survivable, non-provocative, anti-proliferative nuclear deterrent forces, existing or planned. **1961** *Forward Trends* V. iv. 47 The Guild Art Exhibition was in two main parts. All of it, naturally, dealing with the use of art with S.S.N., E.S.N., and backward children. **1967** *Punch* 19 Apr. 557/1 Down below the plimsoll line of an IQ of fifty are the erstwhile imbeciles and idiots, now classed as SSN—severely subnormal. **1972** *Observer* 20 Aug. 7/8 A sixth of the children in the SSN (severely sub-normal) department of this school are boys like these. **1968** *Neurology* XVIII. ii. 48/2 Our data suggest that SSPE is caused by the measles virus, or a virus immunologically indistinguishable from the measles virus. **1974** SSPE [see *panencephalitis* s.v. *pan- 2]. **1962** *Aeroplane* 21 June 6/3 Secondary surveillance radar (ssr), the so-called answer to the air traffic controller's prayer, does not appear to be so near, or as technically perfect, as one is led to believe. **1977** *R.A.F. News* 22 June–5 July 9 (Advt.), Experience is required of approach and long range surveillance equipment plus SSR, HF, VHF and UHF radio equipments. **1926** *Encycl. Brit.* III. 429 The population of the Union of Socialist Soviet Republics and its principal parts was, in 1924... Ukraine S.S.R... 27,700,000. **1947** *Whitaker's Almanack* 972/1 Uzbekistan comprises the former feudal states of Bokhara and Khiva and the Kara Kalpak S.S.R. **1977** R. Perry *Dead End* vi. 77 He elected to resume his studies in the Turkmen SSR. **1967** *Economist* 11 Feb. 501/1 Under the direction of Dr Michael Young, the SSRC has begun to deliver the goods. **1975** M. Bradbury *History Man* ii. 20 He was now..a research student, with an SSRC grant. **1977** *Dædalus* Summer 62 The major funding agencies, the NSF, NIMH, SSRC, Ford Foundation, etc., should be approached to provide the basis for a series of 'summit' meetings among the leaders of the various modes of 'anthropologizing'. **1961** *Fortune* June 161/1 Now in the preliminary design stage is the supersonic transport, or SST. **1969** *Listener* 6 Mar. 313/2 Russia is talking about getting her SST into passenger use before the end of 1970. **1977** *New Yorker* 27 June 86/3 The Concorde's sponsors believe that if the plane doesn't get New York landing rights the SST enterprise will end in financial disaster. **1932** *N. & Q.* 16 July 45/2 He published seven books of airs, of which the above is the sixth, at various dates between 1604 and 1638 (S.T.C. 7460–67). **1952** J. Carter *ABC for Bk.-Collectors* 164 STC, a landmark in enumerative bibliography and one of the most frequently quoted of reference books, has recently been reprinted. **1962** Davis & Wrenn *Eng. & Medieval Stud.* 270 The edition is attributed to 1550 in *S.T.C.* under no. 4817. **1958** *New Scientist* 4 Dec. 1421/1 In planning a scheme for STD it is very desirable that the dialling procedure should be simple and easy to understand. **1963** *Engineering* 25 Jan. 176 STD is not limited to the United Kingdom—it exists on both the Continent and America. **1973** J. Wainwright *Pride of Pigs*

169 She..picked up the receiver, waited for the S.T.D. pips to stop, said 'Hello?' **1977** P. Strevens *New Orientations Teaching Eng.* viii. 105 STD code, oh-three-one. **1956** *Aeronaut. Engin. Rev.* Mar. 48 This type, or types, of airplane can..be called Short Take-Off and Landing, or STOL, airplanes. **1959** *Times Rev. Industry* July 80/1 Any s.t.o.l. aircraft can use quite short airstrips. **1974** *Sci. Amer.* Mar. 83/3 If STOL and VTOL aircraft, including helicopters, become accepted as a major means of intercity transport, traffic density will increase substantially. **1975** E. Hillary *Nothing venture, Nothing Win* xvi. 260 The answer was to build an airfield and use STOL. **1953** Batham & Thorpe *To All who are interested in Democracy* 14 The National Union of Teachers elects its officers and executive by the S.T.V. **1974** *Times* 12 Mar. 15/4 This is the single transferable vote (STV). Its effect is to ensure that every vote is of equal value, and that nearly every voter has an MP of his choice. **1955** B. Higman *Appl. Group-Theoretic & Matrix Methods* xii. 175 (*table*) Special unitary [matrix group] *SU*(*n*). **1967** G. G. Hall *Appl. Group Theory* vi. 84 *SU*(*n*). The special unitary group is the subgroup of *U*(*n*) whose matrices have a determinant of unity. **1977** *Nature* 4 Aug. 469/1 It provides a classification of a totally different nature to the canonical isospin, strangeness (SU(2) and SU(3)), beloved of group theorists which classifies objects of the same spin. **1956** R. N. Hull et al. in *Amer. Jrnl. Hygiene* LXIII. 205/1 The agents isolated will be referred to as 'Simian viruses' (S.V.) until such time as a definite association with some other host or identification can be established. *Ibid.* 214/1 A large number of production samples has not been studied for S.V. contamination but of those that have been assayed a few have been found to contain S.V.$_4$ only. **1957** *Ann. N.Y. Acad. Sci.* LXVII. 414 (*caption*) Agents received from other laboratories and temporarily classified as SV's. **1967**, etc. [see *polyoma]. **1970** *New Scientist* 29 Jan. 194/1, SV40 occurred as a contaminant in many of the earlier batches of polio vaccine.., but has had no discernible effect on Man. **1977** *Time* 18 Apr. 48/2 He hoped to insert a monkey virus, SV40, into *E. coli*. **1864** *N. & Q.* 12 Mar. 211/2 See Richardson *On the Study of Words*, and *Dict.*, s.vv. 'Lord', 'Lady'. **1962** *Ibid.* Aug. 304/2 *Lotus-eating* (O.E.D. s.v. *Lotus-eater*). **1976** *Classical Q.* XXVI. 310 For this incorrect form of the name of the elder Suetonius, cf. *OCD* s.v. Suetonius. **1981** *Times Lit. Suppl.* 2 Jan. 6/2 One receives no help if one tries to check it sv, 'music'. **1973** *Vet. Rec.* 3 Mar. 234/1 On occasion some pigs have shown no abnormal behaviour although they had widespread lesions of SVD. **1975** G. P. West *Black's Vet. Dict.* (ed. 11) 747/1 SVD has been transmitted to laboratory workers. **1980** *Times* 15 Sept. 14/4 SVD cannot be distinguished from foot-and-mouth disease (FMD) on the farm. [**1917** *Harrod's Gen. Catal.* 1385/2 S. Wm's... Wm's... O.S. Wm's.] **1926–7** *Army & Navy Stores Catal.* 674/3 White drill coat... Sizes S.W., W. and O.S. **1974** *Harrods Xmas Catal.* 7 Sophisticated wrap... SW, W, or WX. **1925** Fraser & Gibbons *Soldier & Sailor Words* 274 *SWAK*, from the initials S.W.A.K.—i.e., Sealed with a Kiss. A common superscription on the envelopes of letters to sweethearts from sailors and soldiers. **1948** Partridge *Dict. Forces' Slang* 185 *S.W.A.K.* was the commonest (Sealed With a Kiss), with the variant *S.W.A.L.K.*, the *L* in this case standing for 'Loving'. **1952** E. Waugh *Men at Arms* iii. viii. 312 The old soldiers wrote SWALK on the envelope, meaning 'sealed with a loving kiss'. **1971** R. Quest *Death of Sinner* xvi. 154 They [sc. the letters] might represent a sentence like S.W.A.K... Girls at school sometimes wrote it on the backs of envelopes. It means 'sealed with a kiss'. **1973** 'D. Halliday' *Dolly & Starry Bird* xi. 164, I posted him a long letter with SWALK on it to make him laugh. **1968** *Time* 19 July 18/3 Two hundred marksmen have been assigned to a squad named S.W.A.T. (Special Weapons and Tactics), designed to pick off snipers and to eliminate..the need for indiscriminate police gunfire. **1979** *Tucson* (Arizona) *Citizen* 28 Apr. 1A/1 Police said a SWAT team bombarded the vehicle with tear gas and gun blasts. **1911** *Motor Manual* (ed. 13) ii. 50 The 'secondary' winding, composed of many turns of fine wire (42 or 44 s.w.g.). **1953** *Electronic Engin.* XXV. 66, 1200 turns 36 S.W.G. enamelled copper. *a* **1977** *Harrison Mayer Ltd. Catal.* 50/1 The working ends are fabricated from 15swg stainless steel which will not rust. **1938** *Socialist Appeal* 22 Jan. 2/4 The S.W.P. will advocate the continuance of the class struggle. **1943** *Fourth International* Aug. 234/1 The new movement of the masses was developing outside the SWP. **1977** *Times* 8 Sept. 12/3 In true Marxist fashion, the SWP looks forward to the withering away of the state. **1961** *Amateur Radio Handbk.* (ed. 3) xiii. 358 The ratio of maximum to minimum voltage at the crest and trough of the standing wave, is called the voltage standing wave ratio (v.s.w.r., often abbreviated to s.w.r.). **1976** *S9* (N.Y.) Feb. 62/2 The column will also contain useful information about adjusting antennas, dope about SWR (standing-wave-radio) measurements, and all that good stuff.

b. S.A., South Africa(n), South America(n), South Australia(n); S.A.P., Sap, South African Party; S.E., S.W., also used to designate a London postal district.

1864 *N. & Q.* 6 Feb. 117/1 Cape Town, S.A. **1891** W. S. Churchill *Let.* 27 Sept. in R. S. Churchill *Winston S. Churchill* (1967) I. Compan. i. v. 270 Mama has got a big map of S.A. on which she follows your route. **1933** J. Cary *Amer. Visitor* iv. 39 We were in S.A. together—the yeomanry. **1967** L. Meynell *Mauve Front Door* xv. 214 A bottle of S.A. sherry. **1890** Webster, *S.A.*, South America. **1930** E. Pound *XXX Cantos* xii. 55 And the bust-up of Brazilian securities (S.A. securities). **1864** *South Austral. Advertiser* 17 Oct. 2/4 Share List... S.A. Insurance... S.A. Gas. **1944** *Living off Land* iv. 82 The S.A. family..could have been saved had they carried sufficient drinking water. **1971** *Sunday Australian* 8 Aug. 11/2 Senator Hannaford of SA suffered a heart attack. **1920** S. Black *Dorp* 9 The scornful word 'Sappers', which he knew to be a term of contempt applied by members of Hertzog's Party (the Nationalists) to all those of the Botha-Smuts element or 'SAP'. **1933** J. C. Smuts *Let.* 7 Oct. in *Sel. Smuts Papers* (1973) V. 567 It may be a case of Sap predominance, with a Nat prime minister with a small following of his own. **1935** *Ann. Reg. 1934* i. 132 Around him rallied those S.A.P. men who vowed with him that a surrender of principles was too heavy a price to pay. **1972** *Sunday Times* (Johannesburg) 3 Sept. 2 Nats, Progs, Saps all climb in to make mischief. **1977** *Jrnl. Commonwealth & Compar. Politics* XV. 7 The networks of rural and provincial notables originally fused together by the South African Party (SAP) gave it majorities in the Cape, Transvaal, and Orange Free State. **1857** *Punch* 7 Feb. 51/2 Rowland Hill has just divided London's waste of brick by ten... Pimlico is in S.W., Brompton fast, and Chelsea mild. *Ibid.*, Yonder dismal hole S.E., Southwark. **1885** *List of Subscribers, Classified* (United Telephone Co.) (ed. 6) 62 Atkinson & Co, Westminster Bridge Road, S.E. **1934** Dylan Thomas *Let.* Dec. (1966) 147, 5 Redcliffe Street, London, S.W.10... Dear Bert. **1968** *Listener* 19 Dec. 810/2 In between the bombing raids, the doodle-bugs and the V2s they'd improvised a splendid life in SE20.

d. In *Biol.* S or *S* (also s) = Svedberg unit: used after a number to denote the (often characteristic) sedimentation coefficient of a subcellular body.

1942 [see *Svedberg]. **1942** W. B. Bridgman in *Jrnl. Amer. Chem. Soc.* LXIV. 2350/2 The average values of s_{20}..varied from 60 to 70 *S* (one Svedberg unit, $S_v = 1 \times 10^{-13}$ c.g.s. units). **1944** *Jrnl. Exper. Med.* LXXIX. 310 Its sedimentation constant of approximately 800 S is higher than either of those associated with A virus. **1977** M. W. Berns *Cells* iv. 71 In the eukaryotic organism, it appears that 18s and 28s cytoplasmic ribosomal RNAs are produced from a larger 45s ribosomal RNA molecule that is cleaved in a stepwise sequence that produces several intermediate size RNA molecules. **1978** *Nature* 5 Oct. 461/1 Both reconstituted 30S and reconstituted 50S particles containing mutant 16S and (23+5S) RNA, respectively, could form 70S couples with their complementary native subunits only in the presence of higher Mg^{2+} concentration.

II. Symbolic uses. **5.** *s* or *S* (*Physics* and *Chem.*) = sharp: orig. used to designate one of the four main series of lines in atomic spectra, but now more frequently applied to electronic orbitals, states, etc., possessing zero angular momentum and total symmetry.

1890 J. R. Rydberg in *Phil. Mag.* XXIX. 335 Mg (S_2) [denotes] the (whole) second sharp series of Mg. **1922** A. D. Udden tr. *Bohr's Theory of Spectra* iii. iii. 97 He [sc. Schrödinger] assumes that the 'outer' electron in the states corresponding to the *S* terms—in contrast to those corresponding to the *P* and *D* terms—penetrates partly into the region of the orbits of the inner electrons during the course of its revolution. **1926**, etc. [see *L 6 * b]. **1930** [see *K 3 f]. **1935** Pauling & Wilson *Introd. Quantum Mech.* v. 142 Only for *S* states (with *l* = 0) is the wave function different from zero at *r* = 0. **1963** F. A. Cotton *Chem. Applic. Group Theory* viii. 193 An *s* orbital is totally symmetric in the O_h environment. **1978** P. W. Atkins *Physical Chem.* xiv. 433 Whereas the s-orbitals all have non-zero values at the nucleus, the p-orbitals vanish there.

6. [Initial letter of *secondary*.] Used, chiefly in *S wave*, to denote an earthquake wave which oscillates transversely to the direction of propagation, a shear wave; so named because secondary waves arrive at a given place later than primary waves. (See also *P III. 3.)

1908, etc. [see *P III. 3]. **1913** G. W. Walker *Mod. Seismol.* vi. 39 A pronounced movement corresponding to the arrival of the longitudinal disturbance, and..a pronounced movement when the transversal disturbance arrives, both of which have travelled by the brachistochronic path... These are..identified with the beginning of the first phase P and the second phase S of a seismogram. **1937** Wooldridge & Morgan *Physical Basis Geogr.* ii. 16 The velocities of both P and S waves increase with depth, to a depth of approximately three-tenths of the radius. **1955** *Sci. Amer.* Sept. 57/1 S waves travel at about two thirds of the speed of P waves. **1971** I. G. Gass et al. *Understanding Earth* iii. 54/1 The core is liquid..as can be shown from its inability to transmit shear waves, the *S* waves of earthquakes.

7. In *Physics*, *s* and *S* denote the quantum numbers of spin angular momentum of one electron and a group of electrons, respectively. [Introduced by F. Hund 1926, in *Zeitschr. f. Physik* XXXVI. 658.]

1926 *Bull. Nat. Res. Council* (U.S.) No. 57. 5 Electronic angular momentum in units of $h/2\pi$... s. **1932** Bacher & Goudsmit *Atomic Energy States* 6 The spin moments *s* of the individual electrons form, together, a definite resultant spin moment *S*. **1966** D. H. Whiffen *Spectroscopy* xi. 134 One must be careful not to confuse *S* meaning a state with *L* = 0, with *S* the value of the total electron spin. **1970** G. K. Woodgate *Elem. Atomic Struct.* iv. 57 There is no integration in the normalization condition.., and there is no approach to the classical limit in the sense that *s* → ∞ because *s* is confined to the value ½ only.

8. S is used to denote one of the two directions of twist (see quot. 1935); so *S-spun*, etc.

1935 *Proc. Amer. Soc. for Testing Materials* XXXV. i. 448 A yard or cord has 'S' twist if, when held in a vertical position, the spirals conform in slope to the central portion of the letter 'S', and 'Z' twist if the spirals conform in slope to the central portion of the letter 'Z'. **1950** A. V. Pringle *Theory of Flax Spinning* xiii. 45 Because the outer fibrils in flax ultimates are arranged in 'S' twist spirals it is considered that a stronger yarn or thread can be spun when the final twist is inserted 'Z'-wise. Hence yarns for weaving are always spun Z-wise, but yarns for twisting are commonly spun S-wise. **1964** H. Hodges *Artifacts* ix. 129 Thus, if the threads are S-spun the ply will normally be Z-spun.

9. [Initial letter of *slow*.] *s-process* (Astr.): a process thought to occur in giant stars by which heavy atomic nuclei are produced from other nuclei over a long time scale by a combination of neutron captures and more rapid beta decays.

1956 [see *R III. 7]. **1971** *New Scientist* 27 Apr. 248/2 The solar-system proportions of many heavy elements reflect the results expected from the s-process, but the lighter elements defy quantitative explanation. **1977** [see *R III. 7].

10. *S meter*: a meter on a radio that indicates the strength of a received signal.

1939 *A.W.A. Technical Rev.* IV. 187 It is preferable that the S-meter be available for signal strength comparisons when the receiver gain is manually controlled. **1962** *B.B.C. Handbk.* 130 The more expensive short-wave receivers include such additional features as..'comprehensive band-spread' and 'S' meter or magic eye, which facilitate accurate tuning. **1976** Perkowski & Stral *Joy of CB* ix. 96 The output is about four times the talk power (6 dB) of AM, equivalent to a difference of one 'S' meter unit, or slightly better.

11. *S-matrix* (Physics): a scattering matrix, i.e. a matrix of probability amplitudes that occurs in the expression of the initial wave functions in a scattering process in terms of all the possible final wave functions. [After G. *matrix S* (W. Heisenberg 1943, in *Zeitschr. f. Physik* CXX. 521).]

1945 C. Møller in *Matematisk-Fysiske Meddelelser* XXIII. i. 18 The discrete energy values are completely independent of the form of the *S*-matrix. **1964** W. K. Heisenberg in *Cambr. Rev.* 24 Oct. 47/1 The S-matrix elements for complicated processes will be functions of many variables. **1974** *Nature* 15 Mar. 265/1 He begins by developing the mathematical description starting from the Minkowski formalism of space-time, passing briefly through field theory, finally arriving at the S-matrix formalism which is to form the basis for the rest of the investigation.

12. *S-band*: the range of microwave frequencies between 1550 and 5200 megahertz, used for radio communication and radar.

1946 *Radar: Summary Rep. & Harp Project* (U.S. Nat. Defense Res. Comm.) 143/2 *S-band*. Refers to wavelengths of the order of 10 cm. **1952** Reintjes & Coate *Princ. Radar* (ed. 3) i. 33 Radar equipment operating..in the S, X, and K bands is called microwave radar. **1965** Filipowsky & Muehldorf *Space Communications Techniques* ii. 111 S-band high power transmitters in the 2000 to 2400 Mc range are being provided for future Earth to spacecraft deep-space communications, with power ratings from 10 KW to 100 KW. **1970** N. Armstrong et al. *First on Moon* xi. 257, I haven't heard a word from those guys, and I thought I'd be hearing them on your S-band relay. **1974** *Encycl. Brit. Macropædia* XV. 370/1 Early in the war, the British had developed an airborne S-band..radar for bombing, called the H2S.

13. *Chem.* [Abbrev. of L. *sinister* left.] *S* is used to designate (compounds having) a configuration about an asymmetric carbon atom in which the substituents, placed in order according to certain rules, form an anticlockwise sequence when viewed from a particular direction. Opp. *R* III. 6.

1956, **1971** [see *R III. 6]. **1973** *Nature* 6 Apr. 367/3 In these experiments, the R(−) isomers..were considerably more active than the alternate S(+) configuration.

14. *S* denotes the strangeness quantum number of sub-atomic particles.

1956 M. Gell-Mann in *Nuovo Cimento* IV. Suppl. 852 Since we have *S* = 0 for ordinary particles and *S* ≠ 0 for 'strange' ones we shall refer to *S* as 'strangeness'. **1965** C. M. H. Smith *Nuclear Physics* xi. 734 In the final state the total strangeness is zero as Λ^0 and K^+ have *S*-values of −1 and +1 respectively. **1973** L. J. Tassie *Physics of Elem. Particles* vi. 54 The K^- is the antiparticle of the K^+, and has *S* = −1.

15. *Particle Physics.* [Repr. *strange*; also understood as = *singlet* or *sideways*.

'The letters u and d stood for "up" and "down" (that is to say, isotopic spin projection up and down) and s stood for "strange", with "singlet" (isotopic spin singlet) as a supplementary meaning. "Sideways" was a joke that I used from time to time.'—M. Gell-Mann, let. to Ed.]

s is used to designate one of the three quarks originally postulated by Gell-Mann, viz. the 'strange' quark, which has zero isospin and charge −⅓.

1964 [see *quark sb.]. **1964** *Physics* I. 74 There is a triplet *t* of fermion fields corresponding to three spin ½ quarks: the isotopic doublet *u* and *d*, with charges ⅔ and −⅓ respectively, and the isotopic singlet *s*, with charge −⅓. **1973** *Physics Lett.* XLVII. B. 365/2 Even if there is a fourth 'charmed' quark u′ in addition to the usual u, d, and s, there are still three colors. **1975** *Physics Bull.* Apr. 177/1 There are two nonstrange quarks, u and d, a doublet under SU(2), and a strange quark s which is a singlet under SU(2). **1975** L. H. Ryder *Elem. Particles & Symmetries* xi. 192 Let us..take up the suggestion of Gell-Mann in 1964 that the three basic particles, which he called quarks, do not have the same hypercharge *Y* as *p*, *n* and Λ, but are as shown in Figure 2. (*u* and *d* stand for isospin up and down, *s* for singlet.) **1975** *Sci. Amer.* Oct. 43/1 The third quark, *s*, is needed only to construct strange particles, and indeed it provides an

explicit definition of strangeness: A strange particle is one that contains at least one *s* quark or *s̄* antiquark. **1977** *Nature* 21 July 204/1 Quark model enthusiasts have been having a field day predicting..the properties of new 'charmed' hadrons composed of c-quarks together with the old u, d and s-quarks. *Ibid.* 204/2 Each flavour of quark (u, d, s or c) comes in each of the three colours, but hadrons are always combinations of quarks with no net colour.

s'. Add: **2.** = So *adv.* Now *colloq.* (Written continuously with the succeeding word.) Cf. S'elp.

1607 Shakes. *Cor.* iv. vi. 120 You have brought A trembling vpon Rome, such as was neuer S'incapeable of helpe. **1930** M. Allingham *Mystery Mile* xxi. 200 'It anythink yer see, and 'it like 'ell—s'long as it ain't me. **1947** K. Tennant *Lost Haven* i. 20 Man brings his own booze, and gets shot out without s'much as one drink of it.

's. Add: Also written as a separate word.

1. b. = *it is*. Chiefly *poet.* or (in imitation of informal or careless speech) *colloq.*

1599 Shakes. *Much Ado* iii. iv. 9 By my troth's not so good, and I warrant your cosin will say so. **1933** H. C. Bailey *Mr. Fortune Wonders* 98 You wouldn't blame your dear boy! Your only one! 's too bad. **1951** J. Wyndham *Day of Triffids* i. 25 'S that bloody comet, b—it! Thash what done it.

c. = *that's*, esp. in phr. *'sright* (and varr.) = *that's right* (see *right a. 7 e). *colloq.*

In some uses represented erron. by *s'*.

1939 M. Harrison *What are we waiting For?* 99 'It was two years last August: wasn't it, Fred?' 'Sri,' said Fred. **1958** C. Watson *Coffin, scarcely Used* xix. 174 'I thought he had only one funeral today.' 'S'right.' **1968** S. Woods *Past Praying For* iii. 245 'Wednesday, the twenty-second of September?' ''sright.' **1969** N. Freeling *Tsing-Boum* x. 64 'Man to see you, chief. Says you're expecting him.' ''s right.' **1977** 'M. Underwood' *Fatal Trip* xxv. 153 'Did they all leave together?' ''Sright.' 'By car?' ''Sright.'

5. Now also *colloq.*, esp. in phr. *so's* = *so as* (see So *adv.* and *conj.* 29; As *adv.* 21 a). (Further examples.)

1867 J. T. Trowbridge in *Our Young Folks* Mar. 133 Soon's I've got the hang o' the thing,..I'll astonish the nation. **1942** *R.A.F. Jrnl.* 16 May 2 We heave sandbags and pull ropes and tie knots..till we get so's we don't notice the weather. **1948** D. Ballantyne *Cunninghams* i. 5 He had a system for using up the day so's time didn't drag too much. **1955** W. Moore *Bring Jubilee* iii. 28 You're a gloomy gus, Hodge. Tain't's bad's that. **1970** 'R. Llewellyn' *But we didn't get Fox* ii. 22 I'm sore's hell, but that's the situation!

6. = *does*: see Do *v.* A. 2 c. *colloq.*

1934 N. Bell *Winding Road* xxii. 611 When's Parliament reassemble, Stephen? **1938** N. Marsh *Artists in Crime* xvii. 253 What's he know about it? **1966** J. Hackston *Father clears Out* 22 That fellow was most disheartening. What's he know about gold! **1980** 'M. Underwood' *Crime upon Crime* i. 7 What's he do?.. I wondered if he belonged to one of those hush-hush outfits?

-s, *suffix*[2]. A shortened form of the hypocoristic dim. suffix -sy, added to the same classes of words, as *Babs, Toots; ducks* (see *duck sb.[1] 3 c), *moms*.

saab, sa'ab, varr. Sahib in Dict. and Suppl.

Saadian (sā·diăn), *a.* Also **Sadian, Sa'dian.** [f. Arab. *Sâadi, Sa'dī*, the name of a 16th- and 17th-cent. dynasty of sharifs in Morocco + -an.] Of or belonging to the Sa'dī dynasty.

[**1899** B. Meakin *Moorish Empire* vii. 116 In consequence of..the assertion that this family only belonged to the Beni Sâad, it was contemptuously known as the Sâadi dynasty.] **1951** W. Blunt *Black Sunrise* iv. 40 In the sixteenth century..Marrakesh had risen again under the Sadian kings. **1963** *Guardian* 27 Feb. 15/3 There are two marvellous things to see [in Marrakesh]—the Saadian tombs, and the Souks, or covered bazaars. **1971** J. M. Abun-Nasr *Hist. Maghrib* viii. 205 The Sa'dian *Sharifs* were able to organize a religio-political movement which eventually unified Morocco.

‖**saaidam** (sai·dam). *S. Afr.* Also **zaai-** and with capital initial; pl. **-damme, -dams.** [Afrikaans, f. *saai* to sow + *dam* Dam *sb.*[1]] A basin of land enclosed by artificial earthen walls, designed to receive flood-water for its irrigation. Also *attrib.*

1925 R. Deakin *Southward Ho!* vii. 79 The raising of crops with the help of *saaidams*..would transform the scene. **1937** Marais & Sim in D. J. Seymore *Handbk. for Farmers in S. Afr.* (S. Afr. Dept. Agric. & Forestry) 704 The so-called 'saaidam' system is practised. **1947** *S. & E. Afr. Year Bk. & Guide* 551 From Kotjeskolk to Sak River the branch line runs through the fertile Fish River valley along which large 'Zaaidams' have been made. Cultivation by zaaidams is carried out by diverting the flood water..into extensive areas enclosed by dams. **1953** *Cape Argus Mag.* 2 June 1/6 'Koos Nel' constructed the first large 'Saaidam' in the North-West. **1955** J. H. Wellington *S. Afr.* I. 385 Saaidam irrigation is practised along the Sak river... The sowing 'dam' is a basin bordered by low earthen walls into which the flood waters are diverted. **1972** *Stand. Encycl. S. Afr.* V. 445/1 There is considerable irrigation, including the unique system whereby flood-water is diverted into shallow basins, known as 'saaidamme'. **1975** *Ibid.* XI. 422/2 In the North-Western Cape wheat is grown on saaidams.

‖**saal** (zāl). *rare.* Also **Saal.** [Ger.; cf. Salle.] A large room or hall. (Used with reference to European countries, esp. Germany.)

1855 Geo. Eliot in *Fraser's Mag.* LI. 702/1 A more interesting place to visitors is the library, which occupies a large building not far from the Schloss. The principal *Saal*..is ornamented with some very excellent busts. **1876** —— *Dan. Der.* I. ii. xv. 291 Sir Hugo saying as they entered the large *saal*—'Did you play much at Baden, Grandcourt?' **1978** *Chicago* June 52/1 The Chicago Public Library offers tours of this splendid building—it ranks with the best of the European saals.

Saale (zā·lə). *Geol.* The name of a river in E. Germany used *attrib.* with reference to the third (penultimate) glaciation of the Pleistocene epoch in northern Europe, equivalent to the Riss glaciation in the Alps.

1937 W. B. Wright *Quaternary Ice Age* (ed. 2) x. 136 In several places two deposits of loess are separated from one another by glacial formations, indicating that loess formation preceded the advance of the Saale Ice-sheet. **1945** F. E. Zeuner *Pleistocene Period* iii. 76 The most remarkable feature of the Ukrainian succession is the moraine of the Dnjepr lobe of the Saale glaciation. **1959** Wells & Kirkaldy *Outl. Hist. Geol.* (ed. 4) xix. 368 The Gipping and Lowestoft Boulder Clays must be correlated with the Saale and Elster glaciations respectively. **1971** R. F. Flint *Glacial & Quaternary Geol.* xxiv. 628 The distribution of the extensive Saale drift is generally well defined, though in places it is not differentiated clearly from the Elster. **1979** *Nature* 18 Jan. 172/1 It lies between Saale glacial deposits and the sands of the Last (Weichselian) Glaciation (isotope stages 2, 3, 4 and 5a–5d).

Saalian (zā·liăn), *a.* (*sb.*) *Geol.* [ad. G. *saalisch* (H. Stille 1920, in *Nachr. v.d. K. Ges. d. Wissensch. z. Göttingen* (Math.-phys. Kl.) 219), f. *Saale* (see prec.): see -ian.] Designating, or pertaining to, a minor orogenic episode in Europe which is believed to have occurred in the Permian period. Also *ellipt.* as *sb.* Also **Saa·lic** *a.*

1931 Gregory & Barrett *Gen. Stratigr.* 19 The Saalian is represented in England by the Armorican which is earlier than Middle Permian. **1933** *Proc. Leeds Philos. & Lit. Soc.* (*Sci. Sect.*) II. 456 (*table*) Saalian folding. **1937** A. L. Du Toit *Our Wandering Continents* vii. 156 The relatively weak Saalian and Pfalzian Phases marked out the end of the Lower and Upper Permian respectively in the central Variscan zone. *Ibid.* xvi. 309 The Saalian was weak in Europe and the Urals, but strong in both west and east—in the Caribbean and Appalachians and along the southern margin of Asia. **1969** Bennison & Wright *Geol. Hist. Brit. Isles* x. 244 The pre-Permian unconformity of north-east England may be due to the Asturic Phase or to the (later) Saalic Phase. **1973** P. J. Burek in Tarling & Runcorn *Implications Continental Drift to Earth Sci.* II. 822 The closing of the Ural trough along with the formation of the Uralides (Saalian orogenic phase). **1974** *Encycl. Brit. Macropædia* XIV. 99/1 In Europe, orogenic and igneous activities were on a much smaller scale than they had been in the Carboniferous. Here, two minor orogenies..are generally named Pfälzian (post-Early Permian) and Saalian (Late Permian).

Saam(e, var. *Sami.

Saan, var. *San[2].

Saanen (sā·nĕn). The name of a small town in the canton of Berne, Switzerland, used *attrib.* and *absol.* to designate (a member of) a breed of white goats which was first developed in the region.

1908 'Home Counties' *Case for Goat* iii. 26 Something is now heard of the white Saanen. **1909** G. J. G. Jensen *Goat-Keeping for All* ii. 7 Saanen Goats.—Some two or three specimens of this breed were recently imported by Mr. H. E. Hughes, of Broxbourne. **1920** C. J. Davies *Goat-Keeping for Milk Production* i. 14 The white breed of Switzerland takes its name from Gessenay in Berne... In England the variety is usually known as the Saanen. **1948** A. Huxley *Ape & Essence* (1949) 125 Three-horned and robed impressively in a white Saanen soutane the great man is sitting with a couple of two-horned Familiars at a large table. **1976** *Denbighshire Free Press* 8 Dec. 17/6 (Advt.), Territone goats, Anglo Nubian British Saanen.

Saar (zāɹ). The name of a river in West Germany, a tributary of the Moselle, used *attrib.* and *absol.* to designate a white wine produced in this region.

1905 G. Meredith *Let.* 1 Mar. (1970) III. 1515 Can it be a Moselle? Or a Saar wine. **1967** A. Lichine *Encycl. Wines* 474/1 The Saars can outclass the Moselles in the best years... Saar wines..are always classed with the Moselles. *Ibid.*, *Kauzem*. Very delicious wines, heavier, fuller, rounder than most Saars.

Saarlander (zā·ɹlandəɹ). [Ger., f. *Saarland*, the name of a West German *Land*; cf. prec.] An inhabitant of Saarland. Also *attrib.* or as *adj.*

1955 *Times* 4 May 10/1 A Gaullist senator complained that the Government, in accepting the Franco-German consortium, had excluded the Saarlanders. **1975** *Times* 18 June 25/3 The iron and steel works..were returned to German control at the request of the Saarlanders. **1980** E. Leather *Duveen Let.* xv. 175 He was a Saarlander, and a dedicated Nazi. *Ibid.* 178 His ancestors had been..Saarlander plumbers.

saater, var. *saeter, setter.

sabadilla. Add: Also, a preparation of this for medicinal or agricultural use. (Further examples.)

1890 Hughes & Dake *Cycl. Drug Pathogenesy* III. 759 We have thought it better to omit the symptoms belonging to them, lest they should prove as worthless as those.. which were observed in a boy suffering from tape-worm before he took Sabadilla 30, and which (naturally) disappeared after 46 ells of the worm had been passed. **1907** *Brit. Pharmaceutical Codex* 241 Sabadilla consists of the dried ripe seeds of *Schoenocaulon officinale*,..a tall herbaceous plant growing on the low mountain slopes in Mexico, Guatemala, and Venezuela. **1946** *Richmond* (Va.) *Times-Dispatch* 4 Feb. 4/1 A powerful new insecticide has been developed from a long-known plant... The new bug killer is known as sabadilla. **1977** *Martindale's Extra Pharmacopoeia* (ed. 27) 796/2 Sabadilla was formerly used as a parasiticide, especially for pediculosis capitis, in the form of ointment..or vinegar.

sabadine (sæ·bădīn). *Biochem.* Also † **sabatine.** [ad. G. *sabadin* (E. Merck 1891, in *Arch. der Pharm.* CCXXIX. 164): see Sabadilla and -ine[5].] A veratrum alkaloid ester, $C_{29}H_{47}NO_8$, present in sabadilla seeds.

1891 *Jrnl. Chem. Soc.* LX. ii. 844 The author has isolated two alkaloids from sabadilla seeds, *Asagræa officinalis*. Sabadine is best separated as the nitrate. **1951** A. J. Hennig et al. in *Jrnl. Amer. Pharmaceut. Assoc.: Sci. Ed.* XL. 168 Evidence has been obtained of at least five additional alkaloids in the water-soluble portion, crude fraction D..of sabadilla alkaloids. One of the new alkaloids has been obtained in crystalline state and given the name 'sabatine'. **1962** *Jrnl. Med. & Pharmaceut. Chem.* V. 693 Sabatine was found to be identical with sabadine. ..On historical grounds, the names sabadine and sabine (for the ester and alkamine respectively) deserve preference. **1976** *Jrnl. Neurochem.* XXVII. 1271/2 Veratrine sulfate consisted of the alkaloids veratridine, cevadilline, sabadine and cevadine.

Sabæan, Sabean, *a.* and *sb.* Add: Also **Sabaean. A.** *adj.* Also, of or pertaining to the language of the Sabæans (see below).

1883 [in Dict.]. **1886**, **1902** [see *Minæan sb. and a.]. **1968** [see *Lihyanic sb.]. **1974** *Encycl. Brit. Macropædia* I. 620/1 The Sabaean offshoot, a graceful and elegant script consisting of 29 letters, spread into Africa, where it became the progenitor of the Ethiopic alphabet.

B. *sb.* **b.** The language of the Sabæans, a dialect of Old South Arabic.

1905 G. Bell *Let.* 23 Feb. (1927) I. 196 A mass of rocks all covered with inscriptions..one..very like the oldest script of Yemen Sabaean. **1910** J. Buchan *Prester John* xi. 183 It must have been some old sacred language—Phoenician, Sabæan, I know not what—which had survived in the rite. **1951** [see *Minæan sb. and a.].

sabal (sēi·băl). [Generic name (M. Adanson *Familles des Plantes* (1763) II. 495), perh. a. S. Amer. native name.] A fan palm of the genus so called, or a related fossil plant, belonging to the family Palmaceæ and native to tropical America. Cf. Palmetto.

1812 *Curtis's Bot. Mag.* XXXV. 1434 (*heading*) Dwarf Sabal, or Swamp Palmetto. **1902** L. H. Bailey *Cycl. Amer. Hort.* IV. 1593/1 The Sabals have proved a great success. **1975** *Islander* (Victoria, B.C.) 20 Apr. 14/2 Leaves from sabal have been found around Nanaimo.

sabalo. (Examples.)

1889 in *Cent. Dict.* [**1904** W. M. Gallichan *Fishing & Trav. Spain* xvi. 161 These traps were set for the sábalos, or shad.] **1938** V. Heilner *Salt Water Fishing* xii. 195 Look at all the big marlin Hemingway has taken. And of course sabalo. **1965** A. J. McClane *Stand. Fishing Encycl.* 924/2 Tarpon... Also known as the *sabalo* (Spanish), this species is considered by many anglers as the king of gamefishes.

sabatia, var. *sabbatia.

Sabatier (sabatye). *Photogr.* Also (*erron.* but more commonly) *Sabattier*. The name of Armand *Sabatier* (1834–1910), French physician and scientist, used *attrib.* and in the possessive to designate a process and an effect developed by him, as † **Sabatier's amphi-positive process**, the process of image-reversal giving rise to the Sabatier effect; **Sabatier effect**, partial or complete reversal of an image on film or paper, resulting from exposure to unsafe light after partial development. Cf. Solarization 1 in Dict. and Suppl., *pseudo-solarization.

1894 E. L. Wilson *Cyclopædic Photogr.* 329/1 Sabatier's Amphi-Positive Process. The peculiarity of this process consists in the pictures being the result of a superposition, or entangling of two images, one negative, the other positive. **1930** tr. *L. P. Clerc's Photogr.* xvi. 138/1 A similar phenomenon is observed when white light is momentarily admitted to the room while the normal image is still weak. (Sabatier effect.) **1939** M. Natkin *Fascinating Fakes in Photogr.* 58 Solarisation, sometimes called Sabatier effect, has been known for a very long time. **1956** [see *solarization 1 a]. **1970** C. B. Neblette *Fund. Photogr.* v. 52 If a photographic material is exposed, developed, washed but not fixed and then exposed to diffused light and again developed, a positive image or a combination of a positive and a negative image is obtained... This is known as the Sabattier effect (Sabatier, 1850). **1970** M. J. Sethna *Photography* xii. 180 Actually producing the Sabatier effect is not an easy matter. **1976** K. I. & R. E. Jacobson *Imaging Syst.* v. 105 Although the Sabattier effect has been ascribed to the screening effect of the negative image produced by the first exposure and development on the printing by the second exposure onto the underlying emulsion, desensitization by the products of development is a more likely explanation.

sabatine, obs. var. *Sabadine.

‖ **sabayon** (sabayoṅ). [Fr., ad. It. *zabaione* zabaglione.] A dessert or sauce made with egg yolks, sugar, and white wine, whipped together, thickened over a slow heat, and served hot or cold. Also *attrib.* and *Comb.*

1906 *Mrs. Beeton's Bk. Househ. Managem.* lxii. 1669 (*heading*) Sabayon..Pudding sauce, composed of cream or milk, sugar, white wine, and eggs. **1939** A. Simon *Conc. Encycl. Gastron.* I. 46/2 (*heading*) Sabayon sauce. 4 egg yolks. 1½ oz. sifted sugar. 2 gills Marsala wine. **1960** V. Nabokov *Invitation to Beheading* i. 13 Excellent sabayon! Should still like to know if it will be long now. **1973** *Sat. Rev. World* (U.S.) 18 Dec. 48/3 *Sabayon* reeling from an overdose of Grand Marnier. **1975** *Times* 31 May 7/4 There were..triumphs:..the white wine sauce for sole paillard, and the sabayon au kummel.

Sabba-day. Now *rare*. Also **Sabber-day**, etc. U.S. colloq. var. of Sabbath-day. Also *Comb.*, as **Sabba-day house**, a house used for rest in the interval between church services; = *noon-house* s.v. *noon *sb.* 6 b.

c **1772** T. Burbank in *Essex Inst. Hist. Coll.* (1920) LVI. 292 There was in the yeare 1738 a great athcak one sabbady. **1858** *Harper's Mag.* Nov. 856/2 A thousand terrible thoughts rushed into her mind;..above all, the loss of her 'Sabber-day' dinner. **1868** H. W. Beecher *Norwood* 47 Duties never conflict, you said, only Sabby-day morning last. **1876** J. E. Todd *John Todd* 40 Near by were a number of rough, stone-built 'Sabba'-day houses, where they flocked at noon, for warmth in winter (they had chimneys), and coolness in summer. **1891** [see *noon-house* s.v. *noon *sb.* 6 b]. **1935** J. C. Lincoln *Cape Cod Yesterdays* 5, I knew that, when I next dressed, it would be in the prim and stiff and spotless garments befitting what Grandmother often said her mother used to call 'Sabba' Day'.

sabbatarial (sæbătēᵊ·riăl), *a.* *rare*⁻¹. [f. L. *sabbatāri-us* (see Sabbatarian *a.* and *sb.*) + -al.] Favouring or tending to the observance of the Sabbath.

1867 Trollope *Last Chron. Barset* II. lxxiii. 294 The archdeacon had been very stoutly anti-sabbatarial when the question of stopping the Sunday post to Plumstead had been mooted in the village.

Sabbath. Add: **4.** *Sabbath dress* (later example), *-tide*; *Sabbath-dark* adj.; (objective and objective genitive) *Sabbath-breaking* sb. and adj. (further examples), *-keeping* sb. (earlier examples); **Sabbath candle**, a candle lit shortly before dusk on the eve of the Jewish Sabbath; **Sabbath goy** [*goy], a Gentile who performs for Orthodox Jews tasks forbidden to the latter on the Sabbath; = *shabbos-goy* s.v. *Shabbos b; **Sabbath lamp**, a lamp lit on the eve of the Jewish Sabbath; **Sabbath loaf**, a plaited loaf eaten on the eve of the Jewish Sabbath; **Sabbath school** (*a*) (earlier and later U.S. examples).

1815 Mills & Smith *Rep. Missionary Tour* 29 Sabbath-breaking, profanity and intemperance prevail [in New Orleans] to a fearful extent. **1918** H. G. Wells *Joan & Peter* v. 96 Secularists and socialists..planned..to..plunge the whole world into vice and rapine and Sabbath-breaking. **1958** B. Hamilton *Too Much of Water* iii. 59 Charity can do without help from the proceeds of vice and Sabbath-breaking. **1978** P. Bailey *Leisure & Class in Victorian Eng.* ii. 39 Certain magistrates..prosecuted sabbath-breaking cricket players. **1892** I. Zangwill *Children of Ghetto* I. viii. 191 She, at least, would never fail to light the Sabbath candles. **1967** *Listener* 20 July 83/3 My mother..stopped lighting the Sabbath candles..but still spoke as constantly to God. **1945** Dylan Thomas *Let.* 30 July (1966) 280, I went to the Edwinsford Arms, a sabbath-dark bar. **1977** H. Kaplan *Damascus Cover* (1978) iv. 35 Girls in white Sabbath dresses laced with colored embroidery. **1977** *Listener* 24 Mar. 382/3, I was a Sabbath *goy*; that is, for a penny or two, I lighted fires and performed other chores forbidden to orthodox Jews on Saturday, or Friday evening. **1977** *Times* 8 June 4/3 My host had been a sabbath *goy*, that is he had lighted fires for the orthodox on the sabbath. **1643** W. Walwyn *Power of Love* 32 Men are not pleased except salvation be proved to be very difficult to bee obtained, it must still depend either on our beleeving.., or repenting,..or Sabbath-keeping,..or else man is not pleased. **1832** F. Trollope *Domestic Manners of Americans* I. xviii. 284 The waiving the sabbath-keeping by the proprietor, was for his own convenience. **1850** G. Aguilar *Vale of Cedars* xxxiv. 281 It was..the Sabbath eve... The Sabbath lamps were lighted. **1892** I. Zangwill *Children of Ghetto* I. v. 139 For three things a woman dies in childbirth, for not separating the dough, for not lighting the Sabbath lamps, for not —. *Ibid.* II. xviii. 79 The Sabbath loaves shaped like boys' tip-cats—with a curious plait of crust from point to point, and thickly sprinkled with a drift of poppy-seed, and covered with a velvet cloth embroidered with Hebrew words. **1951** L. W. Leonard *Jewish Cookery* v. 26 *Challah*, in twist form or Biblical beehive coil, is the Sabbath loaf of white bread. It is customary to place two *challas* under a special napkin... The two loaves are symbolic of the 'two portions of manna' which fell for the Sabbath. **1972** H. Kemelman *Monday Rabbi took Off* xlvii. 271 The candles were already lit and the table set with the two braided Sabbath loaves. **1820** *Rec. Early Hist. Boston* (1909) XXXIX. 131 The application for liberty to use the Boylston school house on Fort hill for a Sabbath school..was granted. **1832** W. D. Williamson *Hist. State of Maine* III. 688 *Sabbath-schools*..[were established] in Philadelphia, about 1811, and have since spread over the United States. **1866** J. C. Gregg *Life in Army* 209 It was sung by the colored Sabbath School children. **1900** *Congress. Rec.* 23 Jan. 1104/2 A wonderful petition is rolled in..from Sabbath schools, sectarian churches, and societies. **1945** J. Betjeman *New Bats in Old Belfries* 48 And so my thoughts this happy Sabbathtide.

sabbatia (sæbē¹·ʃiă). Also **sabatia**. [mod.L. (M. Adanson *Familles des Plantes* (1763) II. 503 as *Sabatia*), f. the name of Constantino and Liberato Sabbati, 18th-cent. Italian botanists + -ia¹.] An annual or perennial herb of the genus so called, belonging to the family Gentianaceæ, native to eastern North America, and bearing clusters of pink or white flowers.

1814 *Curtis's Bot. Mag.* XXXIX. 1600 (*heading*) Dichotomous sabbatia. **1847** W. Darlington *Agric. Bot.* 260 Angular sabbatia. Centaury. **1902** L. H. Bailey *Cycl. Amer. Hort.* IV. 1594/2 Sabbatias require a light, sweet soil. **1949** *Natural Hist.* June 278/3 On grassy, pine-sprinkled savannas, in the company of gaudy sabbatias, insignificant burmannias, and orange habenarias. **1972** F. Perry *Flowers of World* 123/1 *Sabatia* (*Sabbatia*) are annual or biennial [*sic*] plants from North America with erect branching stems... The plants contain a bitter principle and when dried are used as a tonic.

Sabbatian, *a.* and *sb.*² Add: **A.** *adj.* (Further examples.)

1941 G. G. Scholem *Major Trends in Jewish Mysticism* viii. 284 The swift rise and the sudden collapse of the Sabbatian movement in 1665 and 1666, from Sabbatai Zevi's proclamation of his messianic mission to his renunciation of Judaism. **1974** *Times Lit. Suppl.* 20 Sept. 1024/5 The seventeenth-century mystical Messiah, Sabbatai Sevi (1626–1676), and the Sabbatian heresy named after him.

B. *sb.* (Further examples.)

1974 *Encycl. Brit. Micropædia* IX. 99/1 A sect of Muslim Sabbatians—the Dönme of Salonika—survived him [*sc.* Shabbetai Tzevi]. **1977** *N.Y. Rev. Bks.* 14 Apr. 27/1 The antinomian energies of the Sabbatians..did in fact feed the subsequent currents of enlightenment and reform in Western Europe.

Sabbatianism (further examples).

1941 G. G. Scholem *Major Trends in Jewish Mysticism* viii. 297 Sabbatianism as a movement was long identified with its more extreme, antinomian and nihilistic aspects. **1977** *N.Y. Rev. Bks.* 14 Apr. 28/1 The way was thus prepared for the mighty swell of Sabbatianism.

sabbatic, *a.* Add: **b.** = *Sabbatical *a.* 2 c. *rare*.

1905 *N.Y. Even. Post* 23 Sept. 8 Professors Hull and Durham are spending their sabbatic year in Europe. **1970** *Newslet. Amer. Dial. Soc.* Nov. 7 Budget requests may include..up to two-thirds of sabbatic, academic, or other leave pay offered by an applicant's institution.

sabbatical, *a.* Add: **2. c.** orig. *U.S.* Designating a period of leave from duty granted to university teachers at certain intervals (orig. every seven years) for the purposes of study and travel; *spec.* in *sabbatical year* (cf. sense 2 a in Dict.). Now freq. used *transf.* of rest or absence from other occupations, professions, or activities. Also *sabbatical officer*, one granted sabbatical leave (from work or study) for the performance of a certain office.

[**1880** *Ann. Rep. Pres. & Treas. of Harvard Coll. 1879–1880* 19 The Corporation adopted, on the 31st of May, 1880, new rules with regard to leave of absence for professors and assistant professors... The Corporation have decided that they will grant occasional leave of absence for one year on half-pay, provided that no professor have such leave oftener than once in seven years.] **1886** E. N. Horsford *Scheme adopted by Trustees, Wellesley Coll.* 8 To each of the heads of the above departments the Sabbatical Grant contemplates that every seventh year of her academic service from a given date, she shall be eligible to have..a year's leave of absence, to be passed in Europe, and with it her half-yearly salary. If for any reason an eligible officer declines the Sabbatical Year, the grant in her case may be offered to another equally eligible. **1892** W. James *Let.* 13 July (1920) I. 321 Only why talk of 'sabbatical' years? **1905** *N.Y. Even. Post* 23 Sept. 8 Professors Allinson, Sears and Hill are spending their sabbatical year of absence in foreign travel and study. **1926** B. Russell *On Education* III. xviii. 242 Every university teacher ought to have a Sabbatical year (one in every seven) to be spent in foreign universities or in otherwise acquiring knowledge of what is being done abroad. **1949** *Time* 18 Dec. 12/2 Kennan announced that he was leaving the State Department 'on sabbatical leave'. **1962** *Times* 12 Apr. 18/2 The break-up of his subsequent marriage impels him to escape from England and to spend a 'sabbatical year' in travel. **1972** *Nature* 4 Feb. 277/2 On sabbatical leave from the Department of Therapeutic Research, University of Pennsylvania School of Medicine. **1976** *Postmaster* (Merton Coll., Oxf.) 30 A union of university students, with clearly defined objectives, and with no sabbatical officers. **1976** *Gramophone* Apr. 1575/3 I'm doing it on May 9th with Rostropovich in the last concert before I take off a sabbatical year, promised to my wife when I reached sixty.

B. *sb.* A period of sabbatical leave; a sabbatical year (cf. sense 2 c above). Freq. in phr. *on* (*a*) *sabbatical*.

1934 in Webster. **1946** H. Howe *We Happy Few* 18 Then when Papa had his sabbatical, we went to Paris. **1958** *Manch. Guardian* 7 June 1/6 Parliament will be reconstituted after a six-month sabbatical. **1961** *Harper's Bazaar* Dec. 47/2 The ultimate in holidays is the 'sabbatical', a term which business is taking over from the academic world. **1978** L. Heren *Growing up on The Times* iii. 102 Pat found a furnished flat, which belonged to an academic on a sabbatical in the United States.

Sabber-day, var. *Sabba-day.

sabe (sā·be, sā·ve), *v.* [Re-formation after, or re-borrowing of, Sp. *sabe*: see Savey, savvy *v.* in Dict. and Suppl.] = Savey, savvy *v.* Cf. *quien sabe.

1850 *California Courier* (San Francisco) 6 Sept. 2/3 Ha! *Sabe* that? **1874** [see Savey, savvy *v.*]. **1903** A. Adams *Log of Cowboy* vi. 37 Girls, you know, sabe each other that way. **1907** S. E. White *Arizona Nights* i. 9, I sabed that they'd seen the original exhibit your Uncle Jim was making of himself.

sabe (sā·be, sā·ve), *sb.* [See prec.] = Savey, savvy *sb.*

1872 B. Harte in *Atlantic Monthly* Mar. 352/2 Did n't hev no more *sabe* than to come round yar with sickness in the house and no provision. **1892** [see Savey, savvy *sb.*]. **1913** J. London *Valley of Moon* 311 We ain't got the *sabe*, or the knack, or something or other. **1931** *Lariat* Apr. 53 You ain't got much sabe.

Sabei, var. *Sebei.

Sabellian, *a.*² and *sb.*² Add: **a.** *adj.* Also, of or pertaining to the language of the Sabellians. **b.** *sb.* Also, any of the numerous dialects of Italic spoken by the Sabellians.

1904 C. D. Buck *Gram. Oscan & Umbrian* 3 The Oscan-Umbrian group..includes also the dialects of most of the minor tribes of central Italy, which may be conveniently designated as Sabellian. *Ibid.*, Strictly speaking the Samnite tribes were Sabellian, and their language, the Oscan, a Sabellian dialect. **1939** [see *Latino-Faliscan* s.v. *Latino-]. **1939** [see *Marsian *sb.* and *a.*]. **1972** W. B. Lockwood *Panorama Indo-Europ. Lang.* 58 A few early inscriptions characterised as Sabellian show that this dialect was closely akin to Oscan.

‖ **sabha** (sābā·). [Hind. *sabhā* assembly.] In India, an assembly; a council or society (see quots.). Cf. *Lok Sabha, *Rajya Sabha.

1922 A. B. Keith in *Cambr. Hist. India* I. iv. 96 The power of the king [in Vedic India] cannot have been in normal circumstances arbitrary or probably very great. There stood beside him as the mode of expression of the will of the people the assembly, which is denoted by the terms *samiti* and *sabha* in the Saṃhitā. **1936** J. Nehru *Autobiogr.* xli. 323 All manner of allied or sympathetic or advanced organisations had been declared unlawful—kisan sabhas and peasant unions. **1950** M. Masani *Our Growing Human Family* vii. 66 Some of these ancient Indian republics were extremely democratic... They had popular assemblies of all the people called the samiti and a parliament elected by the people called the sabha. **1968** *Jrnl. Mus. Acad. Madras* XXXIX. 119 Hardly 250 to 300 kritis are sung in the concerts of today—in spite of the availability of Music Sabhas. **1974** *Encycl. Brit. Micropædia* VIII. 747/3 *Sabhā*.., an important unit of self-government in Hindu society. It is basically an association of persons who have common interests, such as members of the same endogamous groups, but may also be an intercaste group (*e.g.*, a *mazdūr sabhā*, or association of labourers). The *sabhā* differs from the *pañcāyat* (caste or subcaste council) in that its scope is much larger.

‖ **sabi** (sā·bi). [Jap.] In Zen Buddhist philosophy, a quality of simple, restrained, and mellowed beauty.

1932 B. L. Suzuki *Nōgaku* 15 The feeling of *sabi* or *shibumi* is the essence of the art of Nō. **1938** D. T. Suzuki *Zen Buddhism* I. ii. 17 *Sabi* consists in rustic unpretentiousness or archaic imperfection, apparent simplicity or effortlessness in execution, and richness in historical associations. **1948** *Introd. Classic Jap. Lit.* (Kokusai Bunka Shinkōkai) p. vi, This new spirit was restrained by the persistence of traditional ideas as..in the appreciation and cultivation of 'mellowness' (*sabi*, patina) in literature. **1965** W. Swaan *Jap. Lantern* xvi. 184 A quality most valued in architecture and art connected with the tea-ceremony is that of *sabi* or *wabi*. **1979** S. Coe in I. Webb *Compl. Guide Flower Arrangement* xvii. 227/3 The first [mood] is *sabi*, a sense of loneliness which comes from being completely detached, and seeing things as if they are happening by themselves.

Sabian, *a.* (Earlier example.)

1787 W. Jones in *Asiatick Researches* (1790) II. 8 It is generally asserted, that the old religion of the Arabs was entirely Sabian; but I can offer so little accurate information concerning the Sabian faith, or even the meaning of the word, that I dare not yet speak on the subject with confidence.

sabin (sēⁱ·bin). *Acoustics.* Also **sabine** and with capital initial. [f. the name of Wallace Clement *Sabine* (1868–1919), U.S. physicist.] A unit of sound absorption equal to the absorbing power of one square foot of perfectly absorbing surface; = *open window unit* s.v. *open *a.* (*adv.*) 22 c.

1934 *Jrnl. Acoustical Soc. Amer.* VI. 101 Total absorption—600 units (Sabine—0·50 sec.). **1936** *Gloss. Acoustical Terms* (*B.S.I.*) 22 The unit of equivalent absorption is termed an absorption unit. When the unit of area is a square foot, this unit is called a sabin. **1956** *IRE Trans. Audio* IV. 21 *A* is the absorption of the room in sabins. **1968** [see *open window unit* s.v. *open *a.* (*adv.*) 22 c]. **1969** *Daily Tel.* 5 Nov. 5/1 The Professor assembled 10 miniskirted secretaries in a physics department's reverberation chamber... The 10 girls averaged a sabine count of 2·5 each, whereas a similar test in 1964, when skirts were longer, produced an average sabine count of 4·6. Twice as much sound was therefore absorbed in 1964. **1975** G. J. King *Audio Handbk.* i. 23 A room of total surface area 111·45 m² (1200 ft²),..and 0·2 coefficient of absorbency, signifying that 20% of the sound is absorbed.., would have a total absorbency of almost 240 sabins.

Sabine, *a.* and *sb.*¹ Add: **A.** *adj.* **a.** (Earlier and further examples.)

1600 Holland tr. *Livy's Romane Hist.* i. 8 And the youth of Rome upon a token and watch-word given, fell on every side to carrie away the Sabine maidens. **1606** Jonson *Hymenaei* sig. Cᵛ, The Speare, which (in the Sabine tongue) was called Curis. **1784** Cowper *Let.* 3 July (1904) II. 219, I may..refresh my spirits by a little intercourse with the Mantuan and the Sabine bard. **1822** M. Wilmot *Jrnl.* 19 Apr. in *More Lett.* (1935) 165 We set out..to seek for Horace's Sabine Farm at the back of Mt Lucretiles. **1823** Byron *Don Juan* ix. vii. 8 You, my Lord Duke!..half a million for your Sabine farm Is rather dear!

b. Of or pertaining to the Sabine language.

1888 [see *Marrucinian *sb.* and *a.*]. **1977** *Word 1972* XXVIII. 7 They are obviously Italic (Sabine), not really Latin words.

B. *sb.*¹ **c.** The Italic language of the Sabines.

1834 W. Gell *Topogr. Rome* II. 381 Cata, in Sabine, says Varro, means pointed. **1933, 1939** [see *Marsian *sb.* and *a.*]. **1974** [see *Marrucinian *sb.* and *a.*].

d. Sabine wine. *rare*.

1863 Whyte-Melville *Gladiators* x. 153, I talked myself well-nigh hoarse, and stayed out the drinking of two flagons of sour Sabine to boot.

Sabine² (sæ·bəin). The name of Sir Edward *Sabine* (1788–1883), British explorer, soldier, and President of the Royal Society, used *absol.*, *attrib.*, or in the possessive in **Sabine('s) gull** to designate *Xema sabinii*, an Arctic gull with a forked tail, grey head, and black collar, first named *Larus sabini* in his honour by his brother Joseph Sabine in 1818 (*Trans. Linn. Soc.* XII. 522).

1852 P. C. Sutherland *Jrnl. Voy. Baffin's Bay* II. 88 Sabine and ivory gulls, and other birds,..were on their flight up the Channel. **1886** *Code Nomencl. & Check-list N. Amer. Birds* (Amer. Ornithologists' Union) 91 (*heading*) Sabine's Gull. **1958** *Evening Telegram* (St. John's, Newfoundland) 6 May 30/5 Few people have the opportunity to see Sabine's gulls as they rarely come south. **1972** S. Burnford *One Woman's Arctic* ii. 43 Once only I saw a solitary sabine.

sabine, var. *Sabin.

sabinene (sēⁱ·binīn). *Chem.* [ad. G. *sabinen* (F. W. Semmler 1900, in *Ber. d. Deut. chem. Ges.* XXXIII. 1464), f. L. (*Juniperus*) *sabin-a* (see Savin, savine) + *-en* -ene.] A colourless liquid bicyclic terpene, $C_{10}H_{16}$, found in a number of essential oils, notably oil of savin.

1900 *Jrnl. Chem. Soc.* LXXVIII. i. 454 When oil of savin is distilled, a fraction, forming 30 per cent. of the whole, boils between 162° and 170°, and consists principally of a terpene, $C_{10}H_{16}$, which the author terms sabinene. **1942** *Jrnl. Org. Chem.* VII. 399 The presence of α-terpinene in this case, is due probably to isomerization of sabinene under the conditions of the experiment. **1976** *Nature* 22 Apr. 726/2 Terpenes with an exomethylene bond (for example, sabinene, nopinene and camphene) show three strong peaks in the Raman spectrum around 920, 877 and 856 cm^{-1}.

Sabinian (săbi·niăn), *sb.* and *a.*¹ *Roman Law.* [ad. L. *Sabinianus*, f. *Sabinus* (see below).] **A.** *sb.* A follower of Massurius Sabinus, a celebrated jurist in the time of the emperor Tiberius. **B.** *adj.* Of or pertaining to Massurius Sabinus or his views.

1862 T. Mackenzie *Stud. in Roman Law* 13 Capito..was the chief of the rival sect, called after two of his followers Sabinians or Cassians. **1886** E. E. Whitfield tr. *Salkowski's Inst. & Hist. Roman Law* 47 The opposition..arises between the two schools of Law..the Proculians and Sabinians. **1903** F. P. Walton *Hist. Introd. Roman Law* xvii. 137 We frequently read that the Proculian view upon some question was so and so, and that the Sabinian view differed from it. **1907** S. L. Phipson *Law of Evidence* (ed. 4) xlvi. 558 The old controversy between the Proculians and the Sabinians, between the logical, inferential or liberal school of interpreters, and the grammatical or literal. **1953** A. Berger *Encycl. Dict. Roman Law* 687/1 Among the prominent Sabinians after Sabinus and Cassius were Lavolenus, Gaius, and Julian. **1977** A. Watson *Nature of Law* vii. 102 Even in the second century A.D. one of the two famous schools of jurists, the Sabinian, argued that barter should be included within the contract of sale.

Sabi·nian, *a.*² *rare.* [f. Sabine *a.* and *sb.* + -ian.] = Sabine *a.*

1902 Belloc *Path to Rome* 432 Rome was hidden by the low Sabinian hills.

Sabin vaccine (sēⁱ·bin). *Med.* [Named after Albert Bruce *Sabin* (b. 1906), Russian-born U.S. microbiologist who developed the vaccine in 1955.] A vaccine against poliomyelitis made from attenuated viruses of the three serological types and administered orally.

1955 *Brit. Med. Jrnl.* I. 24/1 (Index), Vaccination; human trial of Sabin (live) vaccine. **1961** *Lancet* 30 Sept. 756/2 Administration of candied live Sabin vaccine led to a sharp reduction in poliomyelitis incidence and complete prevention of a seasonal rise in incidence during summer and autumn months. **1978** *Daily Tel.* 13 June 16 The production of anti-polio vaccine in America is being imperilled because India has stopped the export of rhesus monkeys, needed for processing Salk and Sabin vaccines, in the past few months.

Sabiny, var. *Sapiny.

Sabir (sābīᵊ·ɹ, ‖ sabir). Also (in *transf.* sense) **sabir**. [Fr., a. *sabir* 'to know' in the language invented by Molière for a song in *Le bourgeois gentilhomme* (1670), prob. ad. Sp. *saber* to know.] A French-based pidgin language used in parts of North Africa; also, = *lingua franca*; also *transf.* and *attrib.*

1867 'Ouida' *Under Two Flags* III. i. 18 'You are great warriors,' he cried, in the Sabir tongue. **1939** L. H. Gray *Foundations of Lang.* 37 *Sabir*, a mixture of French, Spanish, Italian, Greek, and Arabic, which serves as a *lingua franca* for the Mediterranean ports. **1964** E. Palmer tr. *Martinet's Elem. Gen. Linguistics* v. 155 These somewhat sketchy tools of communication are often called *sabirs* after the language which long flourished in the Mediterranean ports and is also known as the *lingua franca*. **1972** R. Mayne *Europeans* iv. 58 The multinational institutions of the Common Market are gradually developing a modern administrative *sabir* compounded from French, German, Italian, and even English. **1974** *Florida FL Reporter* XIII. 17/1, I cannot agree that the 'Sabir Pidgins'..at least are special formations. Rather, they were transmitted by maritime routes—and in some frontier situations. **1978** *Language* LIV. 338 It seems clear that temporal priority must be granted to the contact system used with North Africans of the western Mediterranean ('Sabir') over that used with Black Africans (the reconnaissance language). *Ibid.*, Eastern Sabir, the pidgin used by pilgrims and merchants in the eastern Mediterranean, was, however, much more destructured than the Western Sabir recorded in the Portuguese documents.

sabji, var. Subjee in Dict. and Suppl.

sabkha (sa·bχa, sa·bka). *Geogr.* Also **sabquha, sebk(h)a**. Pl. **sabkha(s), sebakh.** [ad. Arab. *sabḵah* a saline infiltration, salt flat.] A flat, salt-encrusted depression, usu. just above the water-table, that is subject to periodic flooding and evaporation, resulting in accumulation of alternating layers of æolian clays and salts, and is found esp. in N. Africa and Arabia. Cf. *chott, *kavir, *playa 1, Shott in Dict. and Suppl.

1878 [see Shott]. **1891** [see *chott]. **1909** Groom & Balfour tr. *E. Warming's Œcol. of Plants* lviii. 233 Here also may be placed the *shotts* and *sebakh* of North Africa, depressions which contain salt water during the rainy season, but many of which are dry and covered with incrustations of salt in summer. **1911** G. Bell *Let.* 23 Feb. (1927) I. xii. 280 The ground here is what the Arabs called 'sabkha', soft, crumbly salt marsh, sandy when it is dry and ready at a moment's notice to turn into a world of glutinous paste. **1937** *Trans. Connecticut Acad. Arts & Sci.* XXXIII. 110 In the smaller sebkas, comparable in dimensions to Carson Lake, the water is less salt, and supports a more varied fauna. **1957** R. J. H. Church *W. Afr.* xiv. 231 There is, in the north, a succession of salt encrusted mud-flats (sebkhas), marshy only after rare rains, which are remnants of former lagoons. **1963** [see *kavir]. **1964** *Nature* 23 May 759/1 The coast of the Sheikdom of Abu Dhabi, Trucial States, ..is bordered for most of its length by low coastal flats which stand just above normal high-tide level. These flats are known locally as sabkha. **1972** *Sci. Amer.* Dec. 29/3 Sabkhas became an object of considerable interest to geologists soon after it was realized that certain types of ancient rock formation are practically identical with the sabkha sediments; both are characterized by the presence of nodular anhydrite and stromatolitic dolomite. **1973** *Nature* 20 July 145/1 The Triassic sabkhas were not all coastal sabkhas, or tidal flats, some may have been continental sabkhas, or playa flats. *Ibid.* 7 Sept. 29/2 The cores..show microalternations of organic matter and carbonate (sometimes replaced by chert) which closely resemble Recent supratidal deposits in the Persian Gulf (sabquha),..which are a product of a hot dry climate and hypersaline marine conditions with frequent drying out of the sediment. **1977** A. Hallam *Planet Earth* 157 Many modern desert coasts are bordered by salt marshes, usually known as sabkhas. **1980** D. Creed *Scarab* v. 39 The long flat distances of the sabkha..the time-worn cliffs, the unbroken blue of the sky.

sable, *sb.*¹ Add: **1. b.** For 'pencil' read 'brush'. (Further examples.) Cf. *kolinsky.

1899 [see *red sable* s.v. *red *a.* 19 a]. **1958** M. L. Wolf *Dict. Painting* 41 Modern painters' brushes are in two general types: bristle, or coarse hair, usually that of the pig; and fine hair, made of sable, or so-called 'camel hair'. **1973** F. Taubes *Painter's Dict.* 207 Sables are standard painting tools for all water-based mediums—watercolor, acrylic, casein, gouache, etc.—which require large, thin passages of fluid color.

2. b. Short for *sable coat*.

1975 R. Stout *Family Affair* xiv. 123, I..went to the hall with Mrs Bassett's mink or sable or sea otter and held it for her. **1977** J. Crosby *Company of Friends* xvi. 105 She eyed the sable some more... In a few more years they would be hanging that coat on the wall..like a painting.

3. (Earlier example.)

1785 *Daily Universal Reg.* 1 Jan. 4/3 About 140 tons of old Sable and Gurioff's iron saved out of the Westmoreland.

4. a. *sable coat* (later example; hence *-coated* adj.); *sable-trimmed* adj.

1928 Sable coat [see *ask *v.* 5 d]. **1978** F. Maclean *Take Nine Spies* 336 The seductive, sable-coated countess ..on the Orient Express. **1922** Joyce *Ulysses* 457 A sabletrimmed brick quilted dolman.

sable, *sb.*² and *a.* Add: **A.** *sb.* **6.** *sable-gowned, -tinted* adjs.

1848 J. G. Whittier in *National Era* 14 Dec. 198/5 The sable-gowned divine..shall prove Their trade accordant with the Law of Love. **1918** G. Frankau *One of Them* II. xvi. 121 Hat thrown aside from tresses sable-tinted.

B. *adj.* **2. a.** *his sable majesty* (also, *excellency*): applied to a dark-complexioned potentate; *spec.* the Devil.

1867 G. H. Lewes *Let.* 8 Aug. in *Geo. Eliot Lett.* (1956) IV. 384 Here he translated the Bible, and here he wrestled with Satan, flinging (like a true literary antagonist) his ink-stand at his sable majesty's head, and staining the whitewashed wall. **1875** J. D. Lang *Hist. Acct. New South Wales* (ed. 4) I. xi. 393 If Her Majesty could have commissioned the Prince of Darkness to represent her in the Colony.., I doubt not but his sable Excellency would have received a Farewell Address of respect. **1881** [see Majesty 2]. **1895** *Photos* 8 June 10 His sable majesty.. is a man of great force of character.

sablefish (sēⁱ·b'l,fiʃ). *N. Amer.* [f. Sable *a.* + Fish *sb.*¹] A grey- or black-skinned fish of the family Anoplopomatidæ, esp. *Anoplopoma fimbria*, found in the Pacific off the western coast of North America.

1936 P. S. Barnhart *Marine Fishes S. Calif.* 61 Family Anoplopomatidae. Sablefishes... *Anoplopoma fimbria* (Pallas). Sablefish... In northern waters this fish becomes very fat and is much valued for food. **1964** G. C. Carl *Some Common Marine Fishes Brit. Columbia* 48 The sablefish..is smoked and sold under the name of 'black cod'. **1972** *Islander* (Victoria, B.C.) 16 Apr. 13/3 Should black cod, or sablefish, prove to be unrewarding, she is designed to convert easily for tuna fishing. **1975** *Sci. Amer.* Oct. 88/3 The flatnose codling, the sablefish and the arctic sleeper shark are common inhabitants of the bottom off the coast of southern California and Lower California.

sabliere¹. (Canad. examples.)

1800 Henry & Thompson *Jrnl.* 23 Oct. in E. Coues *New Light Early Hist. Greater Northwest* (1897) I. iii. 123 Oak logs..contg. 100 ft. for the sablieries. **1809** D. Thompson *Jrnl.* 18 Nov. (1950) 64 Men put up the partition Wall of my Room & finished the Walls the Sablier Beams Roof Beams &c which ended the Day.

Sabme, var. *Sami.

sabot. Add: **2. c.** Any device fitted inside the muzzle of a gun to hold or support the projectile to be fired (as when they are of different calibres).

1950 Scott & Richardson *Fin Stabilized Projectile Devel. for 3 inch/70 Gun* (NAVORD Rep. 1537: AD 857–242) 3 Removal of the sabot by spin can be disregarded as the projectile acquires little, if any, spin in the smooth bore tube... The type of sabot developed by the Germans during World War II, and operating primarily by muzzle blast with the assistance of the air stream, is the simplest in design. **1954** K. W. Gatland *Devel. Guided Missile* (ed. 2) ii. 47 Models launched from guns in the new supersonic free-flight wind-tunnel are protected in the gun barrel by plastic 'sabots' which keep the models correctly aligned and act as pistons. **1957** E. Burgess *Guided Weapons* iv. 100 The models..which are being tested are launched through a smooth-bore gun by means of a discarding sabot. **1963** *Dict. U.S. Mil. Terms* (U.S. Dept. Defense) 188 *Sabot*, lightweight carrier in which a subcaliber projectile is centered to permit firing the projectile in the larger caliber weapon. The carrier fills the bore of the weapon from which the projectile is fired; it is normally discarded a short distance from the muzzle. **1975** I. V. Hogg *German Artillery of World War Two* 267 An enormous range of sabot shells was developed in Germany with the intention of either increasing the range of field guns or reducing the time of flight of anti-aircraft shells.

5. (See quot. 1966.)

1962 *Internat. Art Treasures Exhib., Victoria & Albert Mus.* 20/2 A Louis XV parquetry table à écrire..raised on cabriole legs with gilt bronze sabots. **1966** M. M. Pegler *Dict. Interior Design* (1967) 383 Sabots... Decorative metal coverings for the feet of wood furniture ..appeared in the 18th century, and were made of bronze doré, bronze, brass, etc. **1980** *Country Life* 3 July 11/2 A ravishing writing desk... The cabriole legs are framed by moulded ormolu borders reaching down to pierced sabots.

6. In baccarat and chemin de fer, a shoe: see *shoe *sb.* 5.

[**1963** C. Graves *None but the Rich* 3 Baccarat, in fact, is chemin-de-fer played with a fixed bank, chemin-de-fer taking its name from the fact that the 'shoe' (in French, *sabot*), as the deal box is known, moves like a toy railway train round the table each time the dealer loses.] **1964** A. Wykes *Gambling* vii. 177 (*caption*) The *sabot* from which the 'chemmy' cards are dealt. **1966** P. O'Donnell *Sabre-Tooth* vi. 93 The sabot containing the six packs of cards, recently shuffled and stacked by the croupier. **1977** X. Fielding *Money Spinner* 162 Finally they are placed in the 'shoe' or *sabot*, from which the banker deals them one by one.

sabotage (sæ·bŏtāʒ, ‖ sabotāʒ), *sb.* [Fr., f. *saboter* to make a noise with sabots, to perform or execute badly, e.g. to 'murder' (a piece of music), to destroy wilfully (tools, machinery, etc.), f. *sabot*: see Sabot and -age.] The malicious damaging or destruction of an employer's property by workmen during a strike or the like; hence *gen.* any disabling damage deliberately inflicted, esp. that carried out clandestinely in order to disrupt the economic or military resources of an enemy. Also *transf.*, *fig.*, and *attrib.*

1910 *Church Times* 11 Nov. 631/2 We have lately been busy in deploring the *sabotage* of the French railway strikers. **1916** *Sydney Morning Herald* 18 Oct., A shearing rouseabout,..charged..with having written a letter to Senator Lynch, threatening him and certain other Labour politicians and employers of Australia with acts of sabotage. **1918** E. S. Farrow *Dict. Mil. Terms* 528 *Sabotage*, wanton destruction of property to embarrass or injure an enemy; such as the smashing of machinery, flooding of mines, burning of wheat and grain, destroying fruit and provisions, dynamiting reservoirs and aqueducts, tying up railroads, etc. **1920** *Glasgow Herald* 26 June 7 Forces are at work in Germany for the sabotage of the Treaty. **1931** W. Martyn *Scarlett Murder* iv. 53 He was in that mood of smouldering rage which only sabotage would slake. **1948** *N.Y. Jrnl. American* (Sunday Mail ed.) 9 May 1/5 Berger and Dasch gave..'full and complete' identification of all connected with the sabotage plot. **1955** *Times* 27 Aug. 6/7 These were the most considerable sabotages of telephone lines which have yet occurred in this area. **1958** *Spectator* 20 June 791/3 The most recent attack on him has been for cultural sabotage. **1977** *South China Morning Post* (Hong Kong) 22 July 1/4 The border flare-up began on July 12, the radio said, when a four-man Libyan sabotage squad was arrested after crossing the border armed with machineguns and explosives. **1978** T. Allbeury *Lantern Network* vii. 86 Langlois had led six-man teams on fifteen sabotage missions. **1979** *Tucson* (Arizona) *Citizen* 20 Sept. 11A/5 The PLO has provided guns and sabotage devices to its IRA friends.

Hence **sa·botage** *v. trans.*, to ruin, destroy, or disable deliberately and maliciously (freq. by indirect means); **sa·botaging** *vbl. sb.*

1918 *New Appeal* 7 Dec. 1/2 Testimony..that the companies are sabotaging the government. **1920** *Glasgow Herald* 20 Aug. 7 When the miners threaten to sabotage the commerce of the country struggling to get back to pre-war prosperity. **1923** *Ibid.* 4 Jan. 4 The sabotaging of the Dual Monarchy, the revolt of the Yugoslav troops, [etc.]. **1934** C. Lambert *Music Ho!* v. 303 Technically speaking it [*sc.* the atonal school] sabotaged the moribund romantic tradition. **1941** *Sun* (Baltimore) 22 Mar. 24/1 The fireworks bill..was passed by the Senate tonight..despite eleventh-hour attempts to sabotage it or delay enactment. **1975** *Times* 14 Jan. 14/3 [Michael Foot] defended the social contract and weighed into the press for sabotaging it.

sabota lark (sābōᵘ·tă lāɹk). [f. *sabota*, native name of the bird adopted as its specific name (A. Smith *Rep. Exped. for Exploring Central Africa* (1836) 47) + Lark *sb.*¹] A buff-coloured lark, *Mirafra sabota* or *M. nævius*, of the family Alaudidæ, found in southern Africa.

1884 R. B. Sharpe *Layard's Birds S. Afr.* (rev. ed.) 526 Sabota Lark... Sir Andrew Smith procured this lark on the arid plains north of Latakoo. **1896** H. A. Bryden *Tales S. Afr.* v. 109 There, too, are the thick-billed lark, the Sabota lark, with its clear ringing call, and a few other—but not many—small birds. **1948** C. D. Priest *Eggs of Birds breeding in S. Afr.* 67 Sabota Lark... In grass on the ground, at times the cup-shaped nest not being well concealed. **1964** P. A. Clancey *Birds Natal & Zululand* 293 The sabota lark commences to breed in October.

saboteur (sæbŏtɜ·ɹ, ‖ sabotœr). Also fem. **saboteuse**. [Fr.] One who commits sabotage.

1921 tr. *W. Rathenau's New Society* 125 The saboteurs of labour. **1931** *Observer* 11 Jan. 20/2 Two managers of a dairy were dubbed saboteurs and sentenced to..imprisonment for letting two hundred tons of butter spoil. **1947** J. Mulgan *Report on Experience* xii. 148 The men who were killed had been *saboteurs*, or gleaners of information for the Allies, patriots in their own quiet and dangerous way. **1973** E. Pace *Any War will Do* iii. 199 They're too chicken to send in their paratroopers, and they're not smart enough to send in saboteurs. **1977** Saboteuse [see *pig *sb.*¹ 6 b]. **1977** *South China Morning Post* (Hong Kong) 22 July 1/3, 12 members of the 9th Libyan Armoured Division were taken prisoner along with 30 saboteurs. **1980** *Daily Tel.* 20 Mar. 1/5 Two fox-hunt saboteurs who had appeared as prosecution witnesses against hunt followers.

sabquha, var. *sabkha.

Sabra (sæ·bră). [ad. mod.Heb. *ṣābrāh* prickly pear.] **1.** (Also with small initial.) A Jew born in Palestine (see *Palestinian *a.* and *sb.*) or, after 1948, in Israel (see *Israel 3). Also *attrib.*

1945 *Zionist Rev.* 16 Nov. 8/2 Of course I was born here. No, I am not a new immigrant, not even an old one; I am a *sabra*. **1946** Koestler *Thieves in Night* i. ii. 9 They were Sabras—nicknamed after the thorny, rather tasteless fruit of the cactus, grown on arid earth, tough, hard-living, scant. *Ibid.* ii. ii. 95 These Sabra-boys regard a glass of wine as something like opium or hashish. **1949** —— *Promise & Fulfilment* 222 He is a *sabra*, with a pleasant open face and good physique. **1958** M. E. Spiro *Children of Kibbutz* p. x, My research problem, therefore, was restricted to the relationships between kibbutz child training and *sabra* (one born and raised in a kibbutz) personality. **1971** *Times* 28 Sept. (Israel Suppl.) p. iv/4 Only two of the 18 Ministers in Israel's Government are *sabras*, native Israel Jews... There is no *sabra* university president. **1977** *Time* 4 July 16/1 Begin's predecessor, Yitzhak Rabin, was a nonobservant Sabra who often seemed uncomfortable wearing a yarmulke and unfamiliar with the words of daily prayers.

2. (See quots.)

1970 *House & Garden* Nov. 139/1 Sabra is a new liqueur from Israel..a blend of Jaffa orange and chocolate. **1975** *Times* 11 Jan. 11/5 Sabra, the Israel liqueur made with bitter oranges and a very bitter Swiss chocolate.

sabre, *sb.* Add: **1. a.** (Further *fig.* examples.)

1922, etc. [see *sabre-rattling* vbl. sb. and ppl. adj., sense 4 a below]. **1928,** etc. [see *sabre-rattler*, sense 4 a below]. **1949** *Western Folklore* VIII. 112 To rattle the saber. To threaten military action. **1968** *Listener* 29 Aug. 259/2 The antique apparatus of Soviet diplomacy complete with rattling sabres and dutiful crowd noises from the Warsaw satellites, was mobilised in the hope of strengthening the Old Guard in the Czechoslovak party. **1976** *Times* 27 Jan. 1/3 At least a few Conservatives..doubt whether Mrs Thatcher was prudent to appear to rattle sabres and remind electors that women national leaders..have sometimes looked more warlike than men. **1978** J. A. Michener *Chesapeake* 704 Orators from many southern states came north to excite voters against the dangers of black franchise, and sabers rattled as ancient battles were recalled.

c. In Fencing, a weapon with a flattened blade and blunted cutting edge, either curved or straight, lighter than the *épée; the exercise of fencing with sabres.

[**1880** J. M. Waite *Lessons in Sabre* p. xi, I have had the honour of instructing the following Clubs in both Fencing and Sabre:—The London Fencing Club [etc.].] **1910** *Encycl. Brit.* X. 252/1 Just as the practice of the 'small' or thrusting sword gave rise to two rival schools, the French and the Italian, that of the sabre or cutting sword..became split up into two main systems, Italian and German. **1927** L. Bertrand *Cut & Thrust* vi. 75 He was..a resplendent figure..waving..a light and fragile silver-plated sabre. **1935** *Encycl. Sports* 531/1 In fencing with the sabre, the upper part of the body is the sole target, and the hits are made by cuts. **1952** *Fencing* ('Know the Game' Ser.) 5 To hit at foil, épée and sabre, is to strike the opponent with the point of the sword so that it fixes clearly and distinctly and has a character of penetration. **1954** R. Crosnier *Fencing with Sabre* i. 22 At sabre, the distance between two fencers is such that body or head cannot be hit when the opponent lunges fully. **1971** I. Butykai tr. *Lukovich's Electric Foil Fencing* ii. 166 This is also in support of what is described above about the application in sabre of feints with the body. **1978** G. Wright *Illustr. Handbk. Sporting Terms* 32 If, in foil and sabre, hits are equal, the bout continues until a deciding hit is landed. *Ibid.* 33 The ancestors of the sabre include the eastern sabre, the English broadsword, and the cavalry sabre.

4. a. *sabre-fencer, -fencing, -play, -player, -stroke*; *sabre-like* adj.; **sabre leg** (see quot. 1952); also *attrib.*; **sabre-rattler**, a reckless militarist; one who threatens violent action; **sabre-rattling** *vbl. sb.*, military aggression; threatening violent action; aggressive blustering; also as *ppl. adj.*; **sabre saw**, a portable electric saw with a narrow reciprocating blade, used for cutting curves.

1952 *Fencing* ('Know the Game Ser.) 26 Modern sabre fencers..have developed a technique when attacking, of directing the blade, changing its direction, and striking, by means of wrist actions and finger manipulation. **1954** R. Crosnier *Fencing with Sabre* i. 28 Sabre fencers who have progressed in technique and sword control, acquire the ability to change their grip, slightly, when attacking or defending. **1927** L. Bertrand *Cut & Thrust* vi. 77 The rules of sabre-fencing are eminently practical. **1954** R. Crosnier *Fencing with Sabre* 14 When reading this text-book, some may accuse me of having approached the subject of sabre fencing with the mind of a confirmed foilist. **1975** *Oxf. Compan. Sports & Games* 304/2 It is.. necessary to have..a president to control the bout and award hits according to the rules and conventions applicable to sabre fencing. **1952** J. Gloag *Short Dict. Furnit.* 404 *Sabre leg*, a hollow curved leg of rectangular section, so called because of its resemblance to the curve of a cavalry sabre. It was copied from the seats and thrones depicted on Greek and Roman vases, and was introduced towards the end of the 18th century. After 1815, it was sometimes called a Waterloo leg. In chairs of cheap quality the front edge is usually rounded. **1963** *Times* 2 Mar. 5/4 A small walnut kneehole desk made £220 (Quinney's), six sabre-leg Regency dining chairs £160. **1974** *Country Life* 5 Dec. (Suppl.) 78/2 A George III Sofa Table..with swept sabre legs. **1934** Webster, Saber-like. **1962** D. Nichols *Echinoderms* i. 20 *Machaeridia*, bilaterally symmetrical worm-like remains with a skeleton of imbricating plates. Greek: 'sabre-like'. **1880** J. M. Waite *Lessons in Sabre* p. vi, The English method of sabre play..could be considerably improved. **1927** L. Bertrand *Cut & Thrust* vi. 81 In..observance of this maxim [sc. *sciabola in mano*] lies the *alpha and omega* of all sabre-play. **1954** R. Crosnier *Fencing with Sabre* i. v. 26 The Hungarian principle..maintained that sabre play was a combination of finger-play and wrist work, conducive to light, rapid, and precise blade actions. **1880** J. M. Waite *Lessons in Sabre* p. vi, Sabre players, as a rule, have not been fencers, or at least have been fencers with trifling skill. **1928** *Daily Express* 6 Dec. 5/2 There is no reason for supposing that the child Napoleon will grow up a sabre-rattler. **1975** *Times Lit. Suppl.* 6 June 625/4 When he [sc. Churchill] came to the rescue of Montagu in the stormy Amritsar debate, he incurred the disgust of the sabre-rattlers. **1922** *Weekly Dispatch* 19 Nov. 8 A policy of adventure, sabre-rattling, and reckless expenditure. **1928** *Observer* 26 Feb. 16/4 A sabre-rattling gesture against a nation with whom we have been at peace for more than a hundred years. **1958** Hayward & Harari tr. *Pasternak's Dr. Zhivago* i. iv. 105 You have to swagger about in an officer's uniform too, you have to do your own bit of sabre-rattling. **1973** 'I. Drummond' *Jaws of Watchdog* x. 136 A sabre-rattling pink-hating American. **1977** C. McCullough *Thorn Birds* ii. 40 Look at the way that saber-rattling Churchill sent our men into something as useless as Gallipoli! **1977** *Time* 24 Oct. 8/1 Despite saber-rattling rhetoric, a steel war is far from inevitable. **1953** R. J. De Cristoforo *Power Tool Woodworking for Everyone* v. 179 Saber saws are usually confined to heavy cutting when curves are not too severe. **1976** M. Machlin *Pipeline* xxviii. 334 The group carried two battery operated saber-saws with hacksaw blades in their chucks. **1980** *Sci. Amer.* Mar. 29/2 A big saber saw, its diamond blade able to cut a four-foot slab at one pass (beyond the two-foot reach of the biggest rotary blades), requires some 25 horsepower, delivered by hydraulic flow from its engine trailer nearby. **1854** Tennyson *Charge of Light Brigade* in *Wks.* (1896) 222/2 Cossack and Russian Reel'd from the sabre-stroke Shatter'd and sunder'd.

b. **sabre-toothed,** *a.* (*fig.*), ferocious; **sabre-tooth(ed) cat** = *sabre-toothed lion, tiger*; **sabre-toothed lion**, for 'the genus Machairodus' substitute 'the subfamily Machærodontinæ'; **sabre-tooth** *sb.* (later example).

1906 E. Ingersoll *Life Animals: Mammals* 86 A divergent branch..developed amazingly throughout most of the Tertiary period,..to which Cuvier gave the name of 'saber-tooth cats'. **1933** A. S. Romer *Vertebr. Paleont.* xv. 294 In sharp contrast are the saber-toothed 'cats', Machaerodontinae... In the saber-tooths the upper canines were exceedingly long stabbing and slicing structures. **1968** *Times* 21 Dec. 2/3 There was a sabre-toothed scrummage of photographers. **1973** Sabretooth cat [see *machærodont *a.*]. **1975** J. G. Evans *Environment Early Man Brit. Isles* i. 21 Three species of mammal considered to have become extinct prior to the Hoxnian ..a sabre-toothed cat (*Homotherium latidens*) and two voles. **1977** *Times* 14 Mar. 5/3 Henry Howard, Earl of Surrey..fell in the sabre-toothed power struggle for the succession when Henry VIII was dying.

sabreur. Add: **a.** (Further example.) See also *beau sabreur.

1969 N. Freeling *Tsing Boum* ix. 59 Castries the cavalryman, swaggering sabreur.

b. A sabre-fencer.

1927 L. Bertrand *Cut & Thrust* vi. 81 How to acquire this touch is a question the would-be sabreur has to answer. **1952** *Fencing* ('Know the Game' Ser.) 26 Sabreurs attack more often by means of a 'Flèche', than by means of a lunge. **1958** *Oxf. Mag.* 13 Mar. 377/1 The Cambridge sabreurs challenged our supposed superiority in the sabre. **1978** G. Wright *Illustr. Handbk. Sporting Terms* 35 (*caption*) The flèche attack demonstrated by a sabreur.

sabrina neckline (săbrī·nă). *U.S.* Also with capital initial. [f. *Sabrina*, the title of a film (1954), in which the actress Audrey Hepburn appeared wearing a dress with such a neckline.] A neckline with ties at the shoulders.

1959 E. Head *Dress Doctor* ix. 119, I had to console myself with the dress, whose boat neckline was tied on each shoulder—widely known and copied as 'the *Sabrina* neckline'. **1967** *Boston Sunday Herald* 30 Apr. v. 4/2 The bride wore an 18th century gown of white satin sabrina neckline with a long bodice trimmed with pearls and crystals and a train fastened at the shoulders with bows. **1976** *Columbus* (Montana) *News* 3 June 2/4 The bride's gown fashioned with chantilly lace basque bodice with long fitted sleeves and sabrina neckline, had a full skirt with petal overskirt of organza.

sabugalite (săbiū·găləit). *Min.* [f. *Sabugal*, name of a town in Beira Province, Portugal + -ite¹.] A mineral of the autunite group, $HAl(UO_2)_4(PO_4)_4.16–24H_2O$, first found as yellow crystals in a number of mines in Portugal, and also prepared artificially.

1951 C. Frondel in *Amer. Mineralogist* XXXVI. 671 Sabugalite is a new member of the autunite group with the composition $HAl(UO_2)_4(PO_4)_4.16H_2O$. **1959** *Ibid.* XLIV. 420 The average index of refraction of the synthetic product indicated a value of about 1·57, which is within the range of indices given by Frondel for natural sabugalite. **1959** [see *saléeite].

sabulite (sæ·biuləit). *Mil.* [f. L. *sabul-um* sand + -ite¹.] A high explosive consisting of ammonium nitrate with some TNT and calcium silicide.

1914 *Daily Consular & Trade Rep.* 17 June 1641 Tests were recently made..of a new explosive called sabulite, which will be manufactured at Coquitlam, British Columbia. **1914** *Mining & Engin. World* XLI. 676/1 Sabulite, a recent invention by a Belgian explosive expert, had a thorough and satisfactory test..a few days ago. **1919** E. de B. Barnett *Explosives* iv. 114 Very similar explosives in which calcium silicide is used in place of aluminium are also manufactured, Sabulite being an explosive of this class. **1940** *Thorpe's Dict. Appl. Chem.* (ed. 4) IV. 464/1 Military sabulite contained ammonium nitrate 78%, trinitrotoluene 8% and calcium silicide 14%, the latter constituent increasing the heat of explosion.

sabzi, var. Subjee in Dict. and Suppl.

sac². Add: **3.** (Earlier example.)

1814 M. Birkbeck *Journey through France* 18 In three days the same postillion left our sac at the hotel unopened, not an article missing.

4. **sac-winged bat**, a South American bat belonging to the genus *Saccopteryx* or closely related genera of the family Emballonuridæ, distinguished by the pouch-like scent gland found in the wing membrane of the males.

1891 W. S. Dallas in P. M. Duncan *Cassell's Nat. Hist.* I. v. 313 The Striped Sack-winged [*sic*] Bat is rather a small species. **1939** G. M. Allen *Bats* ix. 139 Still more remarkable..are the South American sac-winged bats,.. in which a large pouchlike gland is present in the membrane that extends from the fore shoulder to the wrist. **1964** E. P. Walker et al. *Mammals of World* I. 235/1 Sac-winged bats generally roost in groups of 3 to 30 individuals.

sac³, **sacch** (sæk). *slang.* [Abbrev. Saccharine *a.* and *sb.*] A saccharine tablet.

1961 Partridge *Dict. Slang* Suppl. 1257/1 *Sac*, a saccharine tablet: coll. (domestic, and small traders'): heard in 1917, but not gen. until 1942. **1968** 'E. Trevor' *Place for Wicked* vi. 79 Sacchs. You couldn't get them down there.

Sac, var. *Sauk. **Saca,** var. *Saka. **sac à commis,** var. *sagakomi.

sac-à-lait. (Earlier and later examples.)

[**1877** C. Hallock *Sportsman's Gazetteer* 378 Goggle-eyed Perch;..sac-a-lac (New Orleans Creoles).] **1903** T. H. Bean *Fishes N.Y.* 463 Still other names of local application [for the calico bass] are barfish, bitter head, tinmouth, *sac-a-lait*, lamplighter, [etc.]. **1931** W. A. Read *Louisiana-French* 67 In Louisiana the final *t* of *sacalait* is silent. **1937** *Zeitschr. für Französische Sprache & Literatur* LXI. 82 *Sacalait*, the Louisiana name for the crappie.., commonly thought to have been suggested by the beautiful white flesh or the silvery appearance of this fish. The actual source of the name is Choctaw *sakli*, 'trout', French *sac à lait* being merely a typical example of folk etymology. **1949** *New Orleans Times-Picayune Mag.* 16 Oct. 20/3 If you run out of bait while the bream, sacalait, and other fish are practically jumping into the boat, then the lily is your friend. **1973** *Trailer Travel* Jan. 73/2 You can catch large-mouth bass, channel cats, breams, and sac-a-lait or crappie.

sacaline, var. *sachaline.

sacate, zacate (sākā·te, zā-). Also **Zacate**. [ad. Mexican Sp. *zacate* grass, hay, ad. Nahuatl *çacatl*, *zacatl* grass, reed.] Any of several grasses grown in Mexico, the southern U.S.A., and the Philippines, and used for hay or fresh forage; fodder made from such a grass. Cf. next.

1848 J. W. Abert *Rep. Exam. New Mexico 1846–'47* 29 As there were no pasture grounds near the village, I was forced to buy 'zacate' for my mules. **1891** G. Vasey *Illustr. N. Amer. Grasses* (U.S. Dept. Agric. Div. Bot.) I. ii. (*facing plate* XX), [*Sporobolus wrightii* is] a tall coarse grass, growing in dense tufts, commonly called Saccaton or Zacate. **1921** *Blackw. Mag.* Nov. 659/1 The jungle almost obliterated the track. He cut it down with his *machete*.., and, pushing his tired beast through *sacate* and bamboo, emerged..on a gravelly bank. **1977** A. V. Bogdan *Trop. Pasture & Fodder Plants* 44 *Axonopus affinis* Chase... Carpet grass;..Zacate amargo. *Ibid.* 45 *Axonopus compressus* (Swartz) Beauv. Carpet grass;.. Zacate amargo. *Ibid.* 94 *Cynodon dactylon* (L.) Pers... Bermuda grass;..Zacate Bermuda.

sacaton, zacaton (sæ·kătōᵘn, zæ-, ‖ sakatō·n). Also **sacatone**, † **saccato**, † **saccaton(e)**, and with capital initial. [ad. Mexican Sp. *zacatón*, augmentative of *zacate* (see prec.)] Any of several coarse, tough grasses grown in Mexico and the southern U.S.A. and used for hay, esp. species of *Sporobolus* and *Epicampes*; *alkali sacaton*, a tussock grass, *Sporobolus airoides*, able to survive drought and alkaline soils. Cf. prec.

1865 *Harper's Mag.* Jan. 147/1 The grass consists of three principal varieties; the sacatone, a coarse, thick, and strong variety, growing in bunches; the mesquit..and the grama. **1886** *Outing* Dec. 223/2 We came upon a caved-in well, a wide hollow with a black bottom, covered with high rank grass, the Mexican *zacaton*. **1891** G. Vasey *Illustr. N. Amer. Grasses* (U.S. Dept. Agric. Div. Bot.) I. i. (*facing plate* XXV), It [sc. *Muhlenbergia distichophylla*] is one of the grasses called saccato. *Ibid.* (*facing plate* XXVII), This [sc. *Epicampes macroura*] is another of the grasses called saccato, or saccatone. **1929** J. W. Bews *World's Grasses* v. 201 Two species of the S.W. States are important forage grasses in the arid or semi-arid regions of Nebraska, Arizona, and Texas—S[*porobolus*] *wrightii* Munro, 'Saccaton', and *S. airoides* Torr., 'Alkali Saccaton' or 'Alkali Drop-seed'. **1936** J. A. McKenna *Black Range Tales* 177 The Indians..crept from rock to rock; they crawled like snakes from one bunch of sacatone to another. **1942** Castetter & Bell *Pima & Papago Indian Agric.* 22 Along the edges and in the openings of the forests of these two drainages, sacaton grass (*Sporobolus Wrightii*) thrives. **1968** F. W. Gould *Grass Systematics* v. 265 *Sporobolus airoides* (Torr.) Torr., alkali sacaton, is a characteristic bunchgrass of alkaline areas in the western states. **1972** G. Durrell *Catch me a Colobus* ix. 194 The zacaton grass..is tall—as much as three feet high—a very pale golden-yellow in colour, and it grows in huge tussocks all over the soft, black, volcanic soil. **1977** A. V. Bogdan *Trop. Pasture & Fodder Plants* 181 *Panicum maximum* Jacq. Guinea grass;..Zacaton (Mexico).

sacbrood (sæ·kbrūd). [f. Sac *sb.*² + Brood *sb.*] A fatal viral disease of bee larvæ.

1913 G. F. White in *Circular Bureau Entomol., U.S. Dept. Agric.* No. 169. 1 Sacbrood..is no new disease. *Ibid.* 3 There is, therefore, a disorder attacking the brood of bees in which [the] brood dies, but in which there has not been demonstrated any microorganism to which the cause of the trouble could be attributed. For this disease the name of 'sacbrood' is here suggested. *Ibid.*, Many larvæ dead of this disease can be removed from the cell without rupturing their body wall. When thus removed they have the appearance of a small closed sac. This character suggested the name 'sacbrood'. **1928** R. W. Glaser in T. M. Rivers *Filterable Viruses* viii. 281 The number of colonies that die as a direct result of sacbrood is comparatively small; the loss of individual bees, however, in the aggregate is enormous. **1967** K. M. Smith *Insect Virol.* v. 95 Not much is known of the biology of sacbrood but at the moment the virus has only been recorded from the larvae of the honeybee.

saccacom(m)i(s), varr. *sagakomi.

saccade. Delete ‖ and add: Also with pronunc. (săkā·d). **b.** A brief, rapid movement of the eye from one position of rest to another, whether voluntary (as in reading) or involuntary (as when a point is fixated).

1953 *Jrnl. Optical Soc. Amer.* XLIII. 495/2 These [types of eye movement] include relatively large slow waves, saccades, and slow drifts of fixation. **1962** *Ibid.* LII. 571/2 The eye does not move continuously along a line of print in reading, but executes a regular alternation of rapid jumps, called saccades, and fixational pauses. **1967** *New Scientist* 20 Apr. 156/1 Apart from a rapid trembling which plays a part in the mechanism of perception itself, there are two main types of eye-movement: slow 'drifts' away from the target image, and rapid jerks or 'saccades' tending to recentre it. **1971** *Sci. Amer.* June 35/2 Each saccade leads to a new fixation on a different point in the visual field. Typically there are two or three saccades per second. **1974** *Nature* 22 Mar. 308/3 Some observers can learn to suppress small saccades completely, without decreasing the accuracy of fixation or the visibility of the target.

saccadic (săkæ·dik), *a.* [f. prec. + -ic.] **1.** Of the nature of or pertaining to a saccade or saccades (sense *b).

1916 R. Dodge in *Psychol. Bull.* XIII. 422 German and Scandinavian writers are commonly using the descriptive class term 'saccadic' to denote the rapid eye-movements for which we have only the arbitrary name of 'type 1'. I am not sure with whom the term originated, but it seems worth adopting. *Ibid.* 423 He independently rediscovers and thus confirms a number of eye-movement phenomena; such as the inability to see during saccadic movements. **1940** R. S. Woodworth *Psychology* (ed. 12) xiv. 478 The saccadic movement carries the eyes from one object to another, while the pursuit movement follows a moving object. **1948** *Brit. Jrnl. Psychol.* XXXVIII. 144 In a task such as reading, where we wish to observe different parts of the field successively, it can be shown by photography that the eye makes jerks, or 'saccadic movements', having a mean duration of 0·03 to 0·05 sec. **1954** *A.M.A. Arch. Ophthalm.* LII. 710 Most experimenters have found that all voluntary movements executed in the absence of a moving visual stimulus are saccadic. **1977** Dell'Osso & Troost in Brooks & Bajandas *Eye Movements* 52 Saccadic palsy with normal pursuit occurs in both congenital and acquired ocular motor apraxia.

2. *gen.* Jerky, discontinuous.

1937 Scott & Vlastos *Towards Christian Revolution* 247 Marxists are disposed to charge Christians with.. failure to appreciate the saccadic movement of history. **1951** J. S. Bruner *Beyond Information Given* (1974) vi. 92 The reader may object that our model of the information-confirming cycle seems too saccadic, too jumpy. **1980** *Times Lit. Suppl.* 18 Jan. 54/2 From these things—parties, cafes, trips, gigs—a saccadic inconsequential life is made.

Hence **sacca·dically** *adv.*

1962 *Jrnl. Optical Soc. Amer.* LII. 572/2 One can produce conditions of stimulation under which the saccadically moving eye will not be able to see as well as the fixating eye. **1964** *Jrnl. Physiol.* CLXXIV. 259 In Fig. 11B is also shown what can never be measured in practice, the net active-state tension needed to drive the eye saccadically. **1975** *Nature* 1 May 68/2 When the cage was rotated, the bird showed the classical optomotor response of the head: alternately stabilising in visual space and saccadically moving to a new position.

saccarist, var. *sacrist 2.

saccato, saccaton(e, varr. *sacaton, zacaton.

sacch: see *sac³.

saccharase (sæ·kărēⁱz). *Biochem.* [f. med.L. *sacchar-um* sugar + *-ase.] Invertase, sucrase.

1920 *Chem. Abstr.* XIV. 2344 Influence of the temperature and the acidity upon the formation of saccharase. **1930** J. B. S. Haldane *Enzymes* ix. 168 In the case of saccharase the amount of substrate transformed is proportional over wide ranges both to the enzyme concentration and the time. **1973** *Enzyme Nomenclature* (Commission on Biochem. Nomenclature) 217/2 β-Fructofuranosidase... Other Names: Sucrase, Invertase, Invertin, Saccharase, β-*h*-Fructosidase.

saccharescent (sækăre·sĕnt), *a.* *rare*. [f. as Saccharine *a.* and *sb.* + -escent.] Exuding sugar; sugary. Also *absol.* as *sb.* (In quots. *fig.*)

1930 E. Pound *XXX Cantos* xv. 65 The saccharescent, lying in glucose, the pompous in cotton wool. **1979** *Sunday Tel.* 15 July 12/7 A. C. Benson..dispensed saccharescent sweetness and cosy light through a number of best-sellers.

saccharhinoceros (sækărəinǫ·sĕrəs). *nonce-wd.* [Blend of Saccharine *a.* and *sb.* and Rhinoceros.] A lumbering person with an excessively effusive or affectedly sentimental manner. *So* **saccharhino·ceroid** *a.*

1951 R. Campbell *Light on Dark Horse* xvii. 240 The saccharine of false purity exuded from every pore of this saccarhinoceros advocate of virtue. *Ibid.* 241 The saccharhinoceros went off rumbling out some inarticulate phrases about 'Impertinence'. *Ibid.* 251 Sennacheribs.. lost his head completely, and seemed to be about to make a sort of saccharhinoceroid charge.

saccharic, *a.* Add: Also *loosely*: sweet.

1945 R. Hargreaves *Enemy at Gate* 138 Anything from porter and sour Crimean wine to..exalted, if saccharic champagne.

saccharide. Substitute for entry:

saccharide (sæ·kărəid). *Chem.* [f. med.L. *sacchar-um* sugar + -ide.] † **a.** A substance formed in the fermentation of melted sugar (see quot. 1862). *Obs.* [Introduced in Fr. by A. Gélis 1859, in *Compt. Rend.* XLVIII. 1062.]

Quot. 1857 in Dict. is wrongly dated. The word occurs first in the second edition (1862).

1860 *Q. Jrnl. Chem. Soc.* XII. 376 Researches on melted sugar, and on a new principle—saccharide: by A. Gélis. **1862** W. A. Miller *Elem. Chem., Org.* (ed. 2) 78 According to Gélis, when sugar which has thus been melted is dissolved in water it furnishes a solution which when fermented with yeast yields only half the quantity of alcohol that ordinary sugar would have produced, a peculiar body to which he gives the name of saccharide ($C_{12}H_{10}O_{10}$) remaining in solution. It exerts a slight rotatory power to the right upon a beam of polarized light.

† **b.** A compound formed by the action of an acid on a sugar. *Obs.* [Introduced in Fr. by M. Berthelot 1860, in *Ann. de Chim. et de Phys.* LX. 94.]

1862 H. Watts tr. *Gmelin's Handbk. Chem.* XV. 316 By heating dextro-glucose with (organic) acids, compounds are formed..which belong to the class of saccharides.

c. A sugar, esp. a monosaccharide; freq. used unsystematically to denote any mono- or oligosaccharide or a simple derivative of such a compound. [Introduced as G. *saccharid* by B. Tollens in *Kurzes Handb. d. Kohlenhydrate* (1888) 16.]

1895 Thomson & Bloxam *Bloxam's Chem.* (ed. 8) 705 The above considerations have given rise to a classification of the carbohydrates into (1) saccharides or monoses, ..(2) disaccharides or bioses,..(3) polysaccharides or polyoses. **1914** *Chem. Abstr.* VIII. 2716 (*heading*) Resolution of racemic saccharides by means of optically active amyl mercaptan, and some mercaptals. **1932** *Analyst* LVII. 374 By reducing the concentration of nitric acid, it is possible to obtain a mixture which reacts rapidly with saccharides and not with polysaccharides. **1955** *Stain Technol.* XXX. 286 Methods for the demonstration of saccharide, fatty acid, amino acid, vitamin and ketosteroid were chosen for testing. **1973** *Jrnl. Biochem.* (Tokyo) LXXIV. 144/2 Changes in the CD [sc. Circular Dichroism] bands at 295 and 305 nm on adding saccharides, acetamides or alcohols were monitored to study the

SACCHARILLA

interaction with lysozyme. **1974** *Amer. Jrnl. Physiol.* CCXXVI. 720/1 Hyperosmolarity with these saccharides [*viz.* glucose, mannitol, raffinose] caused sustained reduction of spontaneous frequency.

saccharilla (sækări·lă). *Disused.* [app. fancifully f. L. *sacchar-um* sugar.] A kind of muslin.

1851 *Illustr. Catal. Gt. Exhib.* III. 480/1 Saccharilla book muslin. *Ibid.*, Saccharilla mull muslin. *a* **1877** KNIGHT *Dict. Mech.* II. 1503/2 s.v. *Muslin*, Varieties are known as..lawn, saccharilla, harness. **1884** *Encycl. Brit.* XVII. 109/2 Plain, striped, and figured grenadines, and saccarillas.

saccharin. 2. Add to def.: *o*-sulphobenzoic imide, $C_7H_5NO_3S$. (Further examples.)

1893, etc. [see *DULCIN b]. **1918** G. FRANKAU *One of Them* xvii. 128 Swiftly and cloying-sweet as saccharine In Governmental tea, a week had melted. **1973** J. G. TWEEDDALE *Materials Technol.* II. vii. 164 Many pharmaceutical products (including the sweetening substance saccharine) originated from coal-gas waste. **1977** *N.Y. Times Mag.* 15 May 88 It seems to strike most people as absurd, even outrageous, that saccharin has been indicted, convicted and condemned as a carcinogen. **1977** *New Yorker* 27 June 24/2 Only recently, the press has devoted acres of space to the subject of saccharin.

attrib. **1926–7** *Army & Navy Stores Catal.* 480/1 Saccharine tablets. bot. 100 -/10. **1968** *Canad. Antiques Collector* Nov. 21/2 During World War II still another use was found for these decorative little boxes..as containers for saccharin pills. **1978** G. FOX *Amok* iii. 25 Lenore took a tiny saccharin pill.., grimaced as she dropped it into her coffee. **1978** E. MALPASS *Wind brings up Rain* iv. 43 He pushed the bag of sugar into a drawer. If she wanted a cup of tea she could use her saccharine tablets.

saccharine, *a.* and *sb.* Add: **A.** *adj.* **5.** (Further examples.)

1933 *Punch* 16 Aug. 178/1 Here is actually a Viennese film based not on copious draughts of The Blue Danube (with sugar), but on the crisper life which must presumably exist in that city, even in defiance of the saccharine mirage which appears to be the fondest of Hollywood's illusions. **1934** C. LAMBERT *Music Ho!* II. 106 Such a saccharine melody as 'None but the Weary Heart'. **1951** *Essays in Crit.* I. III. 289 The saccharine honeymoon by the seaside. **1955** W. GADDIS *Recognitions* II. ii. 370 A saccharine line drawing of a woman. **1970** K. MILLETT *Sexual Politics* II. iii. 92 It was enough for him to rely on sentiment, a vague nostalgia about the heroic middle ages, and saccharine assertions about The Home. **1976** *Amer. N. & Q.* XIV. 147/2 The parable is saccharine and simplistic. Its sentimental treatment..asks for the cheap pity of melodrama and offers too easy a solution.

saccharined (sæ·kărind), *a.* [f. SACCHARINE *a.* and *sb.* + -ED^2.] Excessively sweet and sugary in tone.

1962 *Punch* 13 June 916/1 The recipe here [in a BBC series] has a queasy mélange of saccharined goodies with disc jockies selecting tunes by pop composers [etc.]. **1973** O. SELA *Portuguese Fragment* (1974) xxi. 118 A saccharined voice announces that your flight will be delayed.

saccharinity. (In Dict. s.v. SACCHARINE *a.* and *sb.*) (Further *fig.* examples.)

1932 B. DE VOTO *Mark Twain's Amer.* viii. 191 Similar items in saccharinity..had created a brummagem reputation. **1971** A. BURGESS *MF* i. 15 Loewe suddenly smiled with horrible saccharinity. **1977** *Times Lit. Suppl.* 18 Feb. 176/2 Juxtapositions of venom and saccharinity, iciness and boredom.

saccharinize (sæ·kărinəiz), *v.* [f. SACCHARIN + -IZE.] *trans.* To sweeten by adding saccharin. Freq. *fig.*, to make agreeable; to render inoffensive. Hence **sa·ccharinized** *ppl. a.*

1971 S. MARCUS in *Atlantic Monthly* Apr. 95 His praiseworthy intention to bring great genius before large numbers of readers comes to seem suspect in the light of the corrupt and corrupting means he employs in censoring, simplifying, and saccharinizing it. **1977** *New Scientist* 27 Oct. 208 The House..would require such a notice to be displayed only at the shop or other retail outlet where 'saccharinised' products are actually bought.

saccharolytic (sæ:kăroli·tik), *a.* *Biochem.* [f. SACCHARO- + *-LYTIC.] Of or pertaining to the chemical breakdown of carbohydrates; able to effect this.

1908 *Jrnl. Med. Res.* XVIII. 86 Bacillus coli showed but a weak sugar-splitting power, the saccharolytic action ceasing after twenty-four and sometimes after eighteen hours. **1928** L. E. H. WHITBY *Med. Bacteriol.* xv. 154 Saccharolytic organisms are so-called on account of their marked power of producing acid and gas in a carbohydrate medium. **1975** R. R. GILLIES *Lect. Notes Med. Microbiol.* xiv. 81 The six antigenically distinct types show variation in their saccharolytic and proteolytic activities.

Saccharomycetes (sæ:kăromaisī·tīz, -ts), *sb. pl.* [mod.L., f. generic name SACCHAROMYCES (J. Meyen 1838, in *Archiv für Naturgeschichte* IV. II. 100) + MYCETES.] A group name for yeasts, esp. those now included in the family Saccharomycetaceæ.

1884 W. B. GROVE *Synopsis Bacteria & Yeast Fungi* ii. 57 The saccharomycetes, or Yeast Fungi, are unicellular plants, which multiply themselves by budding. **1902** *Encycl. Brit.* XXVIII. 560/2 No satisfactory proof has as yet been given that Saccharomycetes are derivable by culture from any higher form. **1906** G. MASSEE *Text-bk. Fungi* I. 54 In the Saccharomycetes, or Yeasts, three modes of spore-formation are known. **1958** J. LODDER et al. in A. H. Cook *Chem. & Biol. Yeasts* i. 13 The fission yeasts were placed in a subfamily of their own,..separated from the Saccharomycetes.

saccharose. Add: **a.** = *DISACCHARIDE. Now *Obs.* (Further example.)

1911 [see *GLUCOSE 1 b].

b. = *SUCROSE b.

1876 [in Dict.]. **1928** A. B. CALLOW *Food & Health* i. 15 Each molecule of saccharose is composed of one atom of glucose (also called dextrose or grape sugar), and one atom of fructose (also called laevulose or fruit sugar). **1962** *Nature* 22 Sept. 1201/1, I have isolated two melibiose-fermenting yeasts,..which..are capable of fermenting maltose, but which neither ferment nor assimilate saccharose. **1979** *Digestion* XIX. 213 The jejunal absorption of either an elemental solution (amino acids, glucose and glucose oligosaccharides), or of nonelemental diet (chicken meat,..glucose, saccharose, maltose and dextrin maltose, corn and wheat oils) were compared in 25 healthy subjects.

Saccopastore (sa:kopastō·re). The name of a village near Rome used *attrib.* in **Saccopastore cranium, skull**, to designate the remains of a Neanderthal type of *Homo sapiens* found there in 1929.

1934 S. SERGI in *Proc. 1st Internat. Congr. Prehist. & Protohist. Stud. 1932* 50 The dimensions of the Gibraltar skull are slightly larger than those of the Saccopastore skull. *Ibid.* 51 The Saccopastore cranium does not at present permit of a complete examination. **1973** B. J. WILLIAMS *Evolution & Human Origins* xi. 181/2 In terms of dimensions and overall form the Saccopastore skull is much like the earlier Steinheim skull.

sacculus. Add: **2. b.** *Microbiol.* A bag-shaped macromolecule present as a structural element in the cell walls of some bacteria.

1964 WEIDEL & PELZER in *Adv. Enzymol.* XXVI. 194 Sacculi, as we shall call bagshaped macromolecules of the kind discussed here, are objects located on that border where Organic Chemistry merges into Morphogenesis and Morphology... A sacculus is not merely a complex chemical compound; it is, in a truly biological sense, a morphological entity. **1972** *Nature* 25 Feb. 426/2 Penicillin has been shown to interfere with the biosynthesis of the structural element of the bacterial cell wall, the sacculus. **1973** R. G. KRUEGER et al. *Introd. Microbiol.* v. 189/2 The mucopeptide sacculus obtains much of its structural rigidity from the repeating *β*-1,4-glycosidic bonds between the polysaccharide monomers.

‖**sac de nuit** (sak də nwi). *?Obs.* [Fr.] A night-bag, a travelling bag.

1814 M. BIRKBECK *Journey through France* 18 An article of our baggage was missing... It was a *sac de nuit* containing sundries of some value. **1819** M. WILMOT *Let.* 3 Sept. (1935) 10 My invaluable Parisian pelisse..was only crammed, on second thoughts, to fill up a chink in my sac de nuit. **1845** R. FORD *Hand-bk. for Travellers Spain* I. I. 17 The company makes itself responsible for baggage..at relative allowances for *sacs de nuit*, portmanteaus, and trunks. **1860** *Once a Week* 8 Dec. 646/2 A little black *sac-de-nuit*.

sacerdoce. For *rare*$^{-1}$ read *rare* and add later example.

1926 R. FRY *Transformations* 58 Sir Claude Phillips was a great High Priest [of art history]..and..had to the full the sense of his sacerdoce.

sacerdos. For *Obs.* read *rare* and add later examples.

1930 E. POUND *XXX Cantos* xxix. 135 As who with four hands at the cross roads By King's hand or sacerdos' are given their freedom. **1949** —— *Pisan Cantos* lxxx. 92 Torn from the *sacerdos* hurled into unstillness.

‖**sacerdotium** (sæsəɹdō^u·ʃiŏm, sækəɹdō^u·tiŏm). [a. L. *sacerdōtium*: see SACERDOCY.] **a.** = SACERDOCY. **b.** The dominion of the Church in mediæval Europe.

1931 *Times Lit. Suppl.* 20 Aug. 628/2 The unhappy Episcopus..nearly lost his 'sacerdotium' in consequence. **1955** *Times* 2 July 7/3 When later the regular universities grew up the journey from Paris to Oxford, though arduous, was a routine event in the life of a scholar. This easy movement within the European *Studium*—as it called itself in grand contrast to the *Imperium* and *Sacerdotium*—was killed like many other good things by the rise of nationalism. **1956** *Scottish Jrnl. Theol.* IX. 174 Does the consecration of a bishop confer a new *character*? Or does it simply give him authority and grace to perform functions inherent in the *sacerdotium* or *ordo* of the presbyter, but not at present exercised by presbyters? Historically the *sacerdotium*, or ministry of the Word and Sacraments, was at one time regarded as particularly the characteristic of the bishop rather than of the presbyter.

sachaline (sæ·kalin, -īn). Also **sacaline**. [ad. the specific epithet of *Polygonum sachalinense* (F. Schmidt in C. J. Maximowicz *Primitiæ Floræ Amurensis* (1859) 233), f. *Sakhalin*, name of an island north of Japan.] A large perennial knotweed, *Polygonum sachalinense*, of the family Polygonaceæ, native to Japan and bearing clusters of small greenish flowers and very large oval leaves which are sometimes used as fodder.

[**1882** *Garden* 22 Apr. 280/2 (*heading*) The Sachalian knotweed.] **1901** L. H. BAILEY *Cycl. Amer. Hort.* III. 1393/2 Sacaline... Exceedingly vigorous plants, spreading rapidly from the tips of strong underground shoots. **1905** W. J. SPILLMAN *Farm Grasses U.S.* xv. 234 The two most prominent fads of this kind in recent years were sachaline, a well-nigh worthless representative of the smartweed family, and penicillaria. **1943** FERNALD & KINSEY *Edible Wild Plants* iii. 176 The leafy summits of young stems of Sachaline..cooked as a potherb..are as good as or superior to French Sorrel. **1951** L. & J. BUSH-BROWN *America's Garden Bk.* (ed. 2) xii. 535 Perennials blooming in September and October... Sacaline. **1975** *Daily Colonist* (Victoria, B.C.) 9 Oct. 6/1 A fast-growing weed threatening new growth in forests [is] sachaline, a bamboo-like weed from eastern Europe.

sachem. 2. (Earlier and later examples.)

1684 in *Documents Colonial Hist. New-York* (1853) I. 402 Wee have put ourselves under the Great Sachim Charles that lives over the Great Lake. **1836** O. W. HOLMES *Song for Centennial Celebr. Harvard Coll.* in *Poems* (1849) 194 And, when at length the College rose, The sachem cocked his eye At every tutor's meagre ribs Whose coat tails whistled by. **1861** *Charleston* (S. Carolina) *Mercury* 29 Mar. 1/2 The Sachems of the Black Republican party did not appreciate the peculiarity of the times when they enacted the Morrill Tariff. **1942** E. PAUL *Narrow St.* xx. 164 [André] Breton, the pontifical sachem, turned to Trotsky and became an enemy. **1972** *Science* 16 June 1222/2 Chairman of the study group was Detlev W. Bronk, former president of Rockefeller University, president of the academy from 1950 to 1962, and himself a grand sachem of the advisory system. **1973** *Caribbean Contact* Feb. 2/2 It's the customs and immigration sachems, though, who must get top billing as the real purveyors of theatre in this world within a world. **1977** *Time* 18 July 26/3 Most party sachems are lining up behind either Incumbent Abraham Beame or Governor Hugh Carey's choice.

‖**Sachertorte** (za·χərto:rtə). Also **Sacher Torte, sachertorte.** [Ger., named after *Sacher*, proprietor of a hotel in Vienna, or the hotel itself + *torte* cake.] A rich chocolate cake of a kind orig. made in Vienna.

1906 *Mrs. Beeton's Bk. Housek. Managem.* lii. 1543 *Sacher torte.* (German chocolate tart.) **1954** 'M. COST' *Invitation from Minerva* 133 That crowning achievement of the Viennese table: *Sachertorte*. **1961** W. BUCHAN *Helen All Alone* 58 Large..cups of coffee..with brioches and sachertorte. **1974** *Times* 5 Apr. 16/3 Ice-cream, butterscotch,.. Sachertorte, pears, Baked Alaska. **1978** M. DICKENS *Open Bk.* i. 4 A rather formal tea, with damp cucumber sandwiches and *Sachertorte*.

sachet. Add: **4.** A small sealed bag-like container, now usu. of plastic, for holding a liquid, a powder, or air.

1917 *Harrod's Gen. Catal.* 358/3 Shampoo Sachets... box 1/0. **1941** N. MARSH *Death & Dancing Footman* (1942) ii. 42 The sachets used in permanent waving. **1970** *Which?* June 169/1 Jackets using air-filled sachets should be very safe too. **1975** C. F. ROSS *Packaging of Pharmaceuticals* i. 4 *Powders*..are sometimes presented in single-dose containers. These may include sachets, filled automatically on suitable strip-packaging machines using paper, aluminium foil, plastics films or laminations of these.

‖**Sachlichkeit** (za·χliχ^{y}kəit). [Ger., = 'objectivity'.] Objectivism, realism; *spec.* in the fine arts = *NEUE SACHLICHKEIT.

1930 *Times Lit. Suppl.* 24 Apr. 341/4 The prevailing note is one of indifference, of complete detachment, of *Sachlichkeit*. **1938** C. FULLMAN tr. *P. Thoene's Mod. German Art* 9 The general characteristics to be found in contemporary German painting..belong to the spiritual categories of a post-war world: Despair and its concomitants—satire and irony; realism, matter-of-fact-ness, *Sachlichkeit*. *Ibid.* 95 Max Beckmann has called his form 'transcendental *Sachlichkeit*'. **1968** *Listener* 3 Oct. 436/1 A generation of excellent artist-designers, such as Bruno Paul, Hans Poelzig and the brothers Taut, all worked in the spirit which was already associated with the word *Sachlichkeit*, which—taken literally—means 'thingness'; it is usually translated as 'matter-of-fact', 'realistic', 'sober', 'objective', and gained currency in matters of art and design early in this century.

‖**Sachverhalt** (za·χferhalt). *Philos.* Pl. **Sachverhalte.** [Ger., = status rerum (Grimm).] Esp. with reference to the philosophy of Wittgenstein and phenomenology, a state of affairs, an objective fact.

1922 B. RUSSELL in *Wittgenstein's Tractatus* 9 Facts which are not compounded of other facts are what Mr Wittgenstein calls *Sachverhalte*. **1931** W. R. B. GIBSON tr. *Husserl's Ideas* 461 The 'substantive' quality attaches to the 'Substrat' underlying the 'Sachverhalt', as well as to the 'Sachverhalt' itself. **1932** A. H. GARDINER *Theory of Speech & Lang.* i. 26 The unit of speech is the sentence and hence the 'thing' signified by every such unit is always of a complex kind—a state of things, as we might say, or a *Sachverhalt*, if we prefer to use the convenient German equivalent. **1950** *Mind* LIX. 266 That new type of object called the *Sachverhalt* or State of Affairs (the Meinongian 'Objective'). **1972** J. N. FINDLAY *Meinong's Emotional Presentation* p. xv, We become aware of what Husserl called variously states of affairs (*Sachverhalte*) or propositions.

sack, *sb.*^{1} Add: **I. 1. f.** *Criminals' slang.* A pocket.

1699 B. E. *New Dict. Canting Crew*, *Sack*,..a Pocket. **1858** A. S. MAYHEW *Paved with Gold* III. iii. 265 I've brought a couple of bene coves, with lots of the Queen's pictures in their sacks. **1955** *Publ. Amer. Dial. Soc.* XXIV. 125 Rarely one hears the side coat pocket called a *sack*.

g. (*a*) A hammock; a bunk; (*b*) a bed; freq. as *the sack*; *to hit the sack*: see HIT *v.* II c. *slang* (chiefly *U.S.*; orig. *Naval*).

1829 *Sailors & Saints* II. iv. 92 There was no more to do, nor hand him below, and bundle him into his sack. **1883** MELTON & OLIPHANT *Cruise of U.S.S. Galena* 48 We were congratulating ourselves that the drills were over and retired to our 'dreaming sacks'. **1942** *Chevron* 17 Jan. 4/3 *Sack*, bunk. **1943**, etc. [see HIT *v.* II c]. **1947** *Reef Points 1947–48* (U.S. Naval Acad., Annapolis) 219 *Flake out*, to utilize one's sack between Reveille and Taps. **1950** 'D. DIVINE' *King of Fassarai* vi. 41 The first time I came on board you were lying in your goddam sack. **1952** in Wentworth & Flexner *Dict. Amer. Slang* (1960) 439/2 Let me stay in the sack all day. **1963** 'E. MCBAIN' *Ten Plus One* xv. 194 Helen seems to think a little more than necking took place... She seems to think you all crawled into the sack. **1968** J. UPDIKE *Couples* ii. 168 Women with that superheated skin are usually fantastic in the sack. **1977** I. SHAW *Beggarman, Thief* III. x. 342 Probably in the sack, he thought, with that fellow with the beard.

h. A bag, large or small, made of paper or the like; *paper sack*: see *PAPER *sb.* 12. *U.S.*

1904, etc. [see *PAPER *sb.* 12]. **1928** *Dialect Notes* VI. 60 A paper bag is always a *sack* or a *poke*, since *bag* means scrotum in the hill country. **1933** *Collier's* 28 Jan. 8/1 While he is at the ball game, he buys himself a sack of Harry Stevens' peanuts. **1956** B. HOLIDAY *Lady Sings Blues* (1973) viii. 77, I got so tired of scenes in crummy roadside restaurants over getting served, I used to..sit in the bus and rest—and let them bring me out something in a sack. **1974** M. G. EBERHART *Danger Money* v. 56 Greg paid for the food and took the sacks to the station wagon.

i. A base in baseball. Cf. *BAG *sb.* 1 c. *U.S.*

1914 LARDNER & HEEMAN *Mar. 6, 1914* 30 We've larruped out th' four-sack poke And scored among a salvo. **1922** E. J. LANIGAN *Baseball Cycl.* III. 47 Until 1920, a notable athlete..could skip around the circuit in the ninth..and, although unmolested, receive credit for a group of stolen sacks. **1938** H. E. WEST *Baseball ScrapBk.* 20 Before he reached the keystone sack the umpires flagged him down and sent him back to bat over again.

j. In American football, an act or occasion of tackling a quarter-back behind the scrimmage line before he can make a pass.

1972 S. DELUCA *Football Playbk.* 370 *Sack*, when the quarterback is thrown for a loss while attempting to pass. **1974** [see *quarterback sack* s.v. *QUARTERBACK *sb.* 3]. **1978** *Detroit Free Press* 2 Apr. 6E/3 Other changes have been made, this year and in recent years, to put juice into the offence, the feeling being that people come to see touchdowns and not quarterback sacks. **1980** *Washington Star* 3 Nov. D3, I would have to say the sacks were the difference in the game.

k. *sad sack*: see as main entry.

3. *to hold the sack*: to be saddled with an unwelcome responsibility. *U.S.*

1904 W. H. SMITH *Promoters* xxiv. 343 They are the ones that are always left to hold the sack. **1921** C. E. MULFORD *Bar-20 Three* xii. 140 Long an' Thompson are holding the sack. They're scapegoats for th' whole cussed gang. **1929** *Univ. Kansas Graduate Mag.* Apr., We will be holding the sack for an additional..deficit of nearly $1000. **1936** E. S. GARDNER *Case of Stuttering Bishop* xii. 191 Perhaps you didn't plan to drag me into the case and leave me holding the sack, but it sure looks as though you did. **1954** W. FAULKNER *Fable* (1955) 176 You might leave your own kinfolks holding the sack, but these are the sheriff's friends.

4. (Further examples.)

1913 J. STEPHENS *Here are Ladies* 102 Getting the 'sack' is an experience which wearies after the first time. **1935** D. GARNETT *Beany-Eye* I. 34 If I just give him the sack he won't get another job and will get into a brawl and be sent to prison again. **1937** 'G. ORWELL' *Road to Wigan Pier* I. i. 11 If they failed to secure a minimum of twenty orders a day, they got the sack. **1958** *Times Lit. Suppl.* 16 May 274/1 Always late, crumpled and scruffy, perpetually in debt, hourly expecting the sack, Greave takes refuge from the horrid realities of life in Mittyesque fantasies, pretending he is a high-powered American salesman.

III. 7. a. *sack-barrow* (later example), *-cart*, *-end* (example fig.), *-pile*, *-pocket*; (sense *1 j) *sack pack*. **b.** *sack-carrier*, *-hauling*. **c.** *sack-wise* adv.

1979 *Daily Tel.* 10 Nov. 10/6, I stuff the bags till I can hardly drag them, and then have to move them on a sack barrow. **1745** W. ELLIS *Agriculture Improv'd* in *Mod. Husbandman* VII. I. 124 So..that Sack-carriers or Cornporters convey the bought Wheat..to such Loft or Granary. **1966** 'L. LANE' *ABZ of Scouse* p. iv, Merseyside's prosperity depended mainly..upon the crate-handlers, the sack-carriers and the horse-whackers, or in the most up-to-date cliché, the 'service industries'. **1963** *Times* 14 Jan. 10/7 My duties, on the other hand, were many and varied. They included propelling a two-wheeled vehicle, known to the initiated as a sack-cart, for long distances, delivering parcels at the houses of well-to-do customers. **1969** *Listener* 8 May 640/3, I used to have to get them [*sc.* sacks of flour] onto what we call a sack-cart, a trolley, shoot them into a bin. **1937** E. MUIR *Journeys & Places* 26 Proud history has such sackends. **1965** A. BLACKSHAW *Mountaineering* III. xvii. 420 If sack hauling is unavoidable use a separate rope. **1976** *Time* 13 Sept. 68/2 On defense, the Colts' front four is largely unknown to fans but not to opposing quarterbacks. Pittsburgh's fearsome front four has the rep, but it was the Colts' 'Sack Pack' that led the league in dumping passers last season. **1897** 'MARK TWAIN' *Following Equator* xxviii. 273 He saw a white linen figure stretched in slumber upon a pile of grain-sacks... The form whirled itself from the sack-pile. **1938** F. D. SHARPE *Sharpe of Flying Squad* xiv. 154 Others [*sc.* shoplifters] have spacious sack pockets underneath their skirts large enough to contain a roll of cloth, a dress, or a small suitcase. **1923** D. H. LAWRENCE *Birds, Beasts & Flowers* 178 And all her weight, all her blood, dripping sack-wise down towards the earth's centre.

8. sack-bag (earlier example); **sack-bearer**, the larva of an American moth of the family Lacosomidæ, which makes cases from leaves; **sack chair** (see quot. 1970); **sack drill, duty** *U.S. Naval slang*, sleep; time spent in bed; **sack kraft**, a type of strong brown paper used esp. for making large paper sacks; **sack lunch** *N. Amer.*, a packed lunch; a lunch in a paper bag; **sack paper** = **sack kraft*; **sack race** (examples); **sack-racing** (example); **sack ship** *Canad. Hist.*, a large vessel used for transportation in the Newfoundland fisheries; **sack-shouldered** *a.* (*nonce*), carrying a sack on the shoulders; **sack time** *slang* (orig. *U.S. Forces'*), time spent in bed; sleep; bedtime; **sack-worthy** *a.*, deserving of the sack (sense 4).

1842 S. S. ARNOLD *Diary* 28 Oct. in *Proc. Vermont Hist. Soc.* (1940) VIII. 160 Mr. Gleason borrowed a sack bag to carry up his cocoons in. **1842** T. W. HARRIS *Treat. Insects New Eng. Injurious to Vegetation* 298 The Germans give these insects a more characteristic name, that of *sackträger*, that is sack-bearers. **1895** J. H. & A. B. COMSTOCK *Man. Study of Insects* xviii. 358 Melsheimer's Sack-bearer... The larva of this species feeds on oak. **1954** BORROR & DELONG *Introd. Study Insects* xxvi. 524 The Lacosomidae are called sack-bearers because the larvae make cases from leaves and carry the cases about. **1970** [see *POLYSTYRENE]. **1976** 'Z. STONE' *Modigliani Scandal* IV. v. 188 Dee was lying in a sack chair, naked. **1946** *Calif. Folklore Q.* Oct. 387 The Navy Man enjoys resting or sleeping. A sailor who retires *hits the sack, sacks in, sacks out, gets in some sack drill*,..*or gets some shut-eye.* **1954** WEBSTER Add., Sack duty. **1960** WENTWORTH & FLEXNER *Dict. Amer. Slang* 440/1 *Sack duty*, sleep; time spent sleeping. **1963** *Economist* 11 May 555/1 Reed's will take..the paper—sack kraft—into its own mills for conversion. **1972** *Daily Colonist* (Victoria, B.C.) 18 June 24/1 Others brought blankets and sack lunches early Saturday and sat sprawled on the grass. **1975** J. GRADY *Shadow of Condor* (1976) v. 91 There was still enough room for the sack lunch he would buy at the restaurant and his two thermos jugs, one for coffee, one for milk. **1957** V. S. SMITH *Introd. Paper & Papermaking* 125/1 (Index), Sack paper. **1968** Sack paper [see *KRAFT]. **1859** GEO. ELIOT *Adam Bede* II. xxv. 195 Here is the prize for the first sack race. **1945** G. MILLAR *Maquis* x. 207, I got up in the sleeping-bag and crossed the floor in it like a child doing the sack-race. **1967** Mrs. L. B. JOHNSON *White House Diary* 12 Sept. (1970) 568, I was wild about the sack races! **1887** *World Almanac* 103 (*heading*) Sack-racing records. **1732** E. FALKINGHAM *Let.* 4 Oct. in *Calendar State Papers Amer. & W. Indies 1732* (1939) 225 Which fish they sell to the British sack ships, for bills of exchange. **1907** J. G. MILLAIS *Newfoundland* viii. 160 In 1527, the little Devonshire fishing ships were unable to carry home their large catch, so 'sack ships' (large merchant vessels) were employed to carry the salt cod to Spain and Portugal. **1965** W. S. MACNUTT *Atlantic Provinces* 14 Its larger vessels, now known as 'sack ships', appeared on the scene at St. John's, taking no part in the catching of the cod, and serving primarily as freighters and transporters. **1922** JOYCE *Ulysses* 429 A sack-shouldered ragman bars his path. **1944** *Yank* 18 Feb. 4 The biggest difference between the Scouts and other doughboys is their sacktime conversation. **1945** *House Beautiful* Jan. 39 Sack Time means just lying on your cot doing nothing. **1949** in Wentworth & Flexner *Dict. Amer. Slang* (1960) 440/1, I didn't have any sack time. **1959** *Alfred Hitchcock's Mystery Mag.* Feb. 74/1 Last night, when I was just getting eyes for some sack time, this bear falls up to my pad, a type looking to score for free. **1974** L. DEIGHTON *Spy Story* xix. 204 I'll make sure they kick your ass from sun-up to sack-time. **1942** D. F. BRUCE *Dimsie carries On* xxi. 197, I can't just sack her for talking to a man in the road, even if he happens to be one for whom we have no great liking; there's nothing sack-worthy in that.

sack, *sb.*^{3} Add: **2. b.** *sack-holding* ppl. adj.

1858 W. BAGEHOT in *National Rev.* Oct. 474 Falstaff is a sort of sack-holding paunch.

sack, *sb.*^{4} Add: **2. a.** (Earlier examples of *sack coat.*)

1847 S. S. MAGOFFIN *Diary* 26 Aug. in *Down Santa Fé Trail* (1926) 253 The general was dressed in his famed old gray sack coat. **1869** S. BOWLES *Our New West* v. 100 My last winter's thick pantaloons and heavy sack coat.. completed my clothing.

b. sack suit, a suit with a straight, loose-fitting jacket; a lounge suit. Hence **sack-suited** *a.*

1895 *N.Y. Dramatic News* 6 July 14/4 Four button sack suit, $25. **1907** H. LAWSON in Murdoch & Drake-Brockman *Austral. Short Stories* (1951) 68 He wore a saddle-tweed sac suit two sizes too small for him. **1960** B. KEATON *Wonderful World of Slapstick* 116, I cleaned up, got into a natty sack suit, and brushed my hair. **1978** J. RAVEN *Triad Consignment* iii. 26 Those sack-suited characters in B-pictures.

3. (May belong under SACK *sb.*^{1}) A cut of dress, being short, unwaisted, and usu. narrowing at the hem; a dress in this style; also *sack dress*.

Fashionable during the second half of the 1950s.

1957 *Punch* 18 Sept. 333 After all, the *belted* sack-dress, in some form or another, is a perennial we have known all down the years, flowering chiefly in the suburbs and the provinces. **1957** *Daily Mail* 26 Sept. 4/2 The sack has swept London like a prairie fire. *Ibid.* 10 Oct. 10/3 A sack, however well cut, needs a tallish figure, and it *must* be very short and tight at the hemline. **1958** *Observer* 21 Sept. 9/3 If there's still a sack to be seen, next week it will acquire a drawstring below the bust. **1959** *Listener* 8 Jan. 56/2 The sack is out. Now, it's the Empire line. **1959** *Times* 25 July 7/4 Hence the rapid disappearance of the A line, the Z line, the *sac*, and the rest of the hideous devices for disguising the fact that women really look their best when they wear bright colours and bulge (moderately) in the proper places. **1969** *Listener* 14 Aug. 206/3 The next big fashion thing was the Sack, and after that the waist, if it was indicated at all, was round the knees or the hips or the diaphragm. **1973** *Guardian* 10 Apr. 13/3 Lagerfeld shows signs of the sack coming back. **1975** 'M. FONTEYN' *Autobiogr.* II. iv. 173 Elizabeth [Taylor] was wearing a 'sack' dress, the latest fashion.

sack, *v.*^{1} Add: **1. d.** In American football, to tackle (a quarter-back) behind the scrimmage line before he can make a pass.

1969 *Internat. Herald Tribune* 6 Nov. 13/4 If you're sacked it's second and 17. **1974** *Plain Dealer* (Cleveland, Ohio) 27 Oct. 7-C/3 Despite all the problems the Buckeye defense managed to sack Anderson three times and picked off three of his passes. **1976** *Washington Post* 4 Sept. D1/5 Kilmer..was sacked hard early in the second quarter by Bears tackle Ron Rydalch.

4. b. *pass.* with *in*, *out*, or *up*: to be in bed or asleep. Cf. sense 8 below.

1954 BERREY & VAN DEN BARK *Amer. Thes. Slang* (ed. 2) §251/11 Asleep,..sacked out. *Ibid.* §892/3 In bed,..sacked out. **1959** W. FAULKNER *Mansion* xii. 280, I was all right. I had had it. I had it made. I was sacked up. **1965** 'R. L. PIKE' *Police Blotter* iii. 56 His punk grandson took it when the old man was sacked in one night.

5. a. (Active examples.) Also *transf.* and *fig.*, esp. (*a*) to reject (a suitor), to jilt; (*b*) to expel from school.

1861 H. MAYHEW *London Labour* II. 469/1 Ah! she's a good kind creetur'; there's no pride in her whatsumever—and she never sacks her servants. **1882** R. D. BLACKMORE *Christowell* III. xi. 160 He had never known more than one girl, worth the end of a cigar—and that one had sacked him. **1907** G. B. SHAW *Major Barbara* II. 214 When trade is bad..and the employers az to sack arf their men, they generally start on me. **1914** 'I. HAY' *Lighter Side School Life* vii. 191 Tommy..arrives home one afternoon in a taxi in the middle of term, and announces..that he has been 'sacked'. **1929** *Amer. Speech* V. 20 When a hillman anounced that 'Lucy done *sacked* me' he meant that his sweetheart had refused him a date, or rejected his proposal of marriage. **1930** *Punch* 2 Apr. 376/3 If it doesn't turn out well I shall sack the lot of you. **1933** *Sun* (Baltimore) 8 May 8/2 The general contention that competition must be sacked in favor of some scheme of controlled coöperation. **1955** *Times* 21 July 13/4 The difficulties were due to the failure of nationalization and ..the remedy was to sack the Coal Board, [etc.]. **1970** G. F. NEWMAN *Sir, You Bastard* iii. 110 Scotty sacked the policemen who had arrived in the patrol car; they could add nothing. **1977** *Times Lit. Suppl.* 14 Jan. 24/1 He hated the two and a half terms he spent as a boy there before, aged twelve.., he was sacked.

8. *intr.* With advbs. **a.** *to sack in*: to turn in, to go to bed; also, to lie in. *slang* (orig. *U.S.*).

1946 [see *sack drill* s.v. *SACK *sb.*^{1} 8]. **1951** in Wentworth & Flexner *Dict. Amer. Slang* (1960) 440/1 Shut up and sack in. **1962** 'S. RANSOME' *Without Trace* x. 107 After she left I had some more drinks and sacked in. **1966** D. F. GALOUYE *Lost Perception* xvi. 168 'I let you sack in this morning,' he told Gregson, 'so you could stockpile your energy.' **1967** 'T. WELLS' *What should you know of Dying?* iii. 41 Benedict's call, at about nine o'clock, woke me up... I'd planned to sack in till about eleven. **1976** N. THORNBURG *Cutter & Bone* iii. 79 Listen, pal, before I sack in..why don't you tell us.

b. *to sack out*: to go to bed, to have a sleep, to doss down. *slang* (orig. and chiefly *U.S.*).

1946 [see *sack drill* s.v. *SACK *sb.*^{1} 8]. **1951** *Arkansas Democrat* 3 July 14/5 Well, it's time to sack out. **1961** 'E. LATHEN' *Banking on Death* viii. 66 The radio said the roads were closed, so I said the hell with it and sacked out on the couch. **1970** J. HANSEN *Fadeout* vii. 55, I was getting ready to sack out. I'd just had a shower. **1971** *Daily Tel.* 19 July 3/2 Many young travellers..are faced with the choice of curling up in a doorway or 'sacking out' in one of London's parks. **1977** *New Yorker* 9 May 46/1 One night we missed the last train. We sacked out in the waiting room in Grand Central.

c. *to sack down*: to go to bed. *slang.*

1956 F. HERBERT *Dragon in Sea* 84 Want me to bring up some sandwiches before I sack down? **1978** E. V. CUNNINGHAM *Case of Russian Diplomat* i. 11, I lost a night's sleep... How about I sack down for a few hours?

sackable (sæ·kăb'l), *a.* [f. SACK *v.* 5 a + -ABLE.] For which one may be sacked; justifying the sack. So **sackabi·lity**, liability to be sacked.

1975 *Financial Times* 13 Jan. 25/6 Mr. Carew thinks that to-day's average British executive has had sackability built into him from childhood. **1975** *Daily Tel.* 3 Oct. 6/5, I admit I may have been impetuous in writing what I did about the school, but every word is truth. I don't consider publication of the truth to be a sackable offence.

sackbut. Add: **1.** Now used again in the performances of some early music. (Later examples.)

1972 *Register of Early Music* Autumn 19 (*heading*) People who have expressed an interest in:— Cornetts, Serpents, Sackbuts and Early Brass. **1973** *Early Music* I. 48 (Advt.), Brass Instruments... Sackbuts, Renaissance and Baroque trumpets by Meinel & Lauber. **1978** *Early Music Gaz.* Jan. p. 11/3 *Cornett and Sackbut* is a new magazine for all players of early lip-reed instruments.

sackbutter, var. SACKBUTER.

1916 STANFORD & FORSYTH *Hist. Mus.* ix. 180 Four sackbutters were enough for her grandfather. *Ibid.* 188 The other three are playing on brass instruments with slides. One may call them simply trombones. These are the *Royal Sackbutters*.

sackcloth. Add: **1. d.** *sackcloth-bound* adj.

1843 J. G. WHITTIER *Lays of My Home* 14 And mate with maniac women, loose-haired and sackcloth-bound.

sackclothed, *a.* Add: Also *fig.* (Later examples.)

1922 BLUNDEN *Shepherd* 23 And rising floods gleam silver on the verge Of sackclothed skies and melancholy grounds. **1924** R. CLEMENTS *Gipsy of Horn* ix. 169 Half-bred negroes and Indians, sackclothed and uncivilised.

sacked, *ppl. a.*^{2} [f. SACK *v.*^{1} + -ED^1.] **1.** That has been put into a sack; stored in a sack.

1895 *Funk's Stand. Dict.* s.v. sack^1 *vt.*, Sacked grain. **1937** E. HEMINGWAY *To have & have Not* II. i. 78 The man went on slowly lifting the sacked packages of liquor and dropping them over the side. **1970** D. WATERFIELD *Continental Waterboy* i. 3 The trouble with lock gates built of sacked mud is that they do not ordinarily open easily.

2. That has been 'given the sack'; dismissed, discharged (from employment or office). Also *absol.*

1934 G. B. SHAW *On Rocks* 148 The exterminated, or, as we call them, the evicted and sacked, try to avoid starvation. **1981** *Daily Tel.* 10 Sept. 8/8 (*heading*) Pay out for sacked heart man.

sa·cker^2. *U.S.* [f. SACK *v.*^{1} 6 + -ER^1.] One engaged in sacking logs.

1902 S. E. WHITE *Blazed Trail* lii. 360 It was noon. The sackers looked up in surprise.

sa·cker^3. *N. Amer.* [*SACK *sb.*^{1} 1 i.] A baseman in baseball. (Usu. preceded by ordinal number indicating the base position.)

1914 LARDNER & HEEMAN *Mar. 6, 1914* 46 He once was the world's most famous first sacker. **1926** *Amer. Speech* I. 369/2 Basemen are 'sackers'. **1938** H. E. WEST *Baseball Scrap Bk.* 158 Wally Pipp became the Yankee first sacker in 1915, and Lou Gehrig succeeded him ten years later and is still going strong. **1944** *College Topics* (Univ. Virginia) 30 Mar. 3 Bob Bryon, first sacker from North Carolina State, seems assured of the first base position. **1958** [see *HOME-BREW 2]. **1974** *Anderson* (S. Carolina) *Independent* 22 Apr. 7A/1 As proof of his defensive prowess, Hargrove led the WCL first sackers in fielding with a .988 percentage.

sackie (sæ·ki). [Local name in Guyana.] Any of several small parrots found in northern South America, esp. *Pionites melanocephala*, which has black, blue, and green plumage.

[**1916** C. CHUBB *Birds Brit. Guiana* I. 340 The 'Macusis' call it [*sc.* the black-headed caique] *Sackuih*.] **1951** E. MITTELHÖLZER *Shadows move among Them* III. ii. 260 Sackies kept up a gay twittering. **1969** S. M. SADEEK *Windswept* 3, I would..imitate the kiss-ka-dees and sackies as they sang.

sacking, *vbl. sb.*^{1} **1.** (Later examples.)

1958 *Daily Sketch* 2 June 1/2 This will not mean sackings as the buses are 3,000 men short now. **1970** G. F. NEWMAN *Sir, You Bastard* vi. 183 The normal fracas following a sacking would bring too much attention to Sneed when he stepped into the vacancy.

sacking, *sb.*^{3} Add: **1.** Also *transf.* of other material used for the same purpose.

1843 'R. CARLTON' *New Purchase* I. xxi. 199 Next was a sacking of clapboards pinned down; and then a very thick straw bed.

3. *sacking bottom* (earlier examples), *needle*; *sacking-wrapped* adj.

1744 J. HEMPSTEAD *Diary* (1901) 425 [I was] fitting a new Bedstid that I Sent with Sacking Bottom. **1841** G. CATLIN *Lett. on N. Amer. Indians* I. 191 A sacking-bottom, made of the buffalo's hide. **1868** G. G. CHANNING *Early Recoll. Newport, R.I.* 254 Sometimes it [*sc.* the bedstead] was furnished with a 'sacking bottom'. **1970** A. H. WHITEFORD *N. Amer. Indian Arts* 67/1 Sacking needles are used to insert the final weft threads. **1952** M. ALLINGHAM *Tiger in Smoke* viii. 129 One small sacking-wrapped bundle.

Sacky, var *SAUK.

‖**sacra** (sæ·kră), *sb. pl.* [ad. L. *sacr-a* sacred things, rites, etc., neut. pl. of *sacer* sacred.] Things endowed with sacred significance.

1819 S. FLEMING *Sherburne's Misc. Poems* p. xiii, The poems are of a miscellaneous description: some being amatory, which the Author styles Erotica;..and others of a graver cast, to which he applies the titles of Ethica and Sacra. **1903** J. E. HARRISON *Prolegomena to Study of Greek Relig.* iv. 126 Probably on this day the magical *sacra* lay upon the altars where the women placed them. *Ibid.* 132 Special cakes..were provided for them, but whether to eat or to carry as *sacra* does not appear. **1945** *Mind* LIV. 77 The contemplation of *sacra* gives rise to motor attitudes—shouting, prancing, rolling on the earth —which are no doubt in the first instance self-expressive. **1959** *Listener* 14 May 853/2 Plato..included the *sacra* in his concept of play. **1964** V. W. TURNER *Forest of Symbols* (1967) iv. 102 In the Lesser Eleusinian Mysteries of Athens, *sacra* consisted of a bone, top, ball, tambourine, apples, mirror, fan, and woolly fleece.

sacral, *a.*^{2} Add: Also with pronunc. (sæ·crăl). (Later examples.) Also *gen.*, set apart for a religious purpose, sacred; pertaining to that which is sacred.

1912 J. E. HARRISON *Themis* p. xi, The *dromenon* in its sacral sense is, not merely a thing done, but a thing *re*-done, or *pre*-done with magical intent. **1958** R. F. C. HULL tr. *Jung's Psychol. & Relig.* in *Coll. Wks.* XI. 350 Any sacral action, in whatever form, works like a vessel for receiving the contents of the unconscious. **1974** R. HELMS *Tolkien's World* i. 24 Tolkien's profoundly suggestive insights into the sacral nature of the human imagination parallel Blake's rather than Arnold's. **1977** *Church Times* 10 June 10/2 A kind of apostolic succession of kingship, temporal and sacral intermingled to form a regal high priesthood. **1977** J. N. M. WIJNGAARDS *Did Christ rule out Women Priests?* vii. 66 The Old Testament priests had to offer frequently at specified sacral times. **1979** *N.Y. Rev. Bks.* 25 Oct. 15/1 Moon recently announced in that sacral third person he uses in public appearances, 'he will go to Germany.'

sacrality (săkræ·liti). Chiefly *Anthrop.* [f. SACRAL *a.*^{2} + -ITY.] Sacral character.

1958 W. R. TRASK tr. *Eliade's Birth & Rebirth* iii. 59 Sacrality, spirituality, and immortality are expressed in images that, in one way or another, signify the beginning of life. **1964** R. MANHEIM tr. *Eliade's Mystery & Spiritual Regeneration* in *Papers from Eranos Yearbks.* V. 26 It is not the natural phenomenon of birth that constitutes the mystery; it is the revelation of feminine sacrality, that is, of the mystical bond between life, woman, nature, the godhead. **1977** J. N. M. WIJNGAARDS *Did Christ rule out Women Priests?* vii. 64 Christ replaced a priesthood based on sacrality by a priesthood based on grace.

sacralization (sæ:krălaizē^i·ʃən). *Anthrop.* [f. next + -ATION.] The action or fact of endowing with sacred qualities. Also *transf.* Cf. *DESACRALIZATION.

1918 *Encycl. Relig. & Ethics* X. 897/1 This 'sacralization' is not proved for sacrifice generally, either savage or civilized. **1937** C. CAUDWELL *Illusion & Reality* vi. 112 To the capitalist commodity-fetishism takes the form of sacralisation of the common market-denomination of all commodities—money. **1954** B. & R. NORTH tr. *Duverger's Pol. Parties* I. ii. 122 Totalitarian parties are in the 'sacred' category... The Party is personified (with a capital letter: a typical characteristic of 'sacralization'), the all-powerful infallible, protective, transcendent Party. **1958** G. W. BROMILEY tr. *Barth's Church Dogmatics* IV. II. 667 It [*sc.* the Church] may fall victim either to alienation (secularisation) or self-glorification (sacralisation). **1976** *Times Higher Educ. Suppl.* 6 Aug. 7/4 [Stanley] Spencer's sacralization is often concerned with transforming profane, urban or suburban icons. **1980** *Times Lit. Suppl.* 8 Aug. 902/5 Unexpected, if partial, justifications of.. boredom in church, learning by rote, the sacralization of war.

sacralize (sæ·krălaiz), *v.* *Anthrop.* [f. SACRAL *a.*^{2} + -IZE, after F. *sacraliser* (see quot. 1899).] To endow with sacred significance (freq. through ritual); to set apart from ordinary life or use as sacred.

[**1899** HUBERT & MAUSS in *l'Année Sociologique* III. 215 Pour employer la terminologie que nous voudrions faire admettre: ils se sacralisent et, en même temps, désacralisent pour les autres l'espèce totémique.] **1933** E. E. EVANS-PRITCHARD in *Ess. Social Anthropol.* (1962) vii. 134 Exchange of blood in such situations [*sc.* blood-brotherhood] sacralizes and endows with sanctions a politico-economic transaction. **1957** V. W. TURNER *Schism & Continuity in an Afr. Society* x. 294 In the past doctors drove the uninitiated away..from areas in the bush which they had sacralized for ritual purposes. **1967** *Listener* 11 May 616/3 In spite of his call to sacralize secularity, there is little expectation of transcendence here. **1972** S. TUGWELL *Did you receive Spirit?* ix. 75 It is only where action and contemplation have become secularised (or sacralised, for that matter..), that any contradiction appears.

sacrament, *sb.* Add: **2. e.** *the last sacraments*, Holy Communion and Extreme Unction administered to the dying; (see also quot. 1920); *the sacrament of the sick*, in the Roman Catholic Church, Extreme Unction (now officially termed the Anointing of the Sick).

1760 in J. O. Payne *Old English Catholic Missions* (1889) 29 Jan. 7, William Hornby died at Middleham. He had the last sacraments. **1893** E. BELLASIS *Mem. Serjeant Bellasis* viii. 184 He left him..to go and tell the Curé.. that the Serjeant ought to have the last Sacraments without delay. **1920** *Encycl. Relig. & Ethics* XI. 574/1 At Cwm Yoy, in the Black Mountain, on the way to Llanthony, the people have at a funeral what they call 'the Last Sacrament'. The coffin is brought out and placed on trestles, and beer and cake are then partaken of by the guests and persons assembled.., before the funeral procession starts. **1966** 'HAN SUYIN' *Mortal Flower* i. 41 The priest..with a Chinese choir boy holding the implements of Extreme Unction,..myself and my sisters assembled in Father's hospital room, to witness..the last sacraments of the Church. **1972** S. TUGWELL *Did you receive Spirit?* xi. 98 It is painful,..and at times comic, to read the Fathers of Trent arguing about the sacrament of the sick. **1975** *N.Y. Times* 26 Oct. 1/5 A mass was held in the Prado Palace at which he [*sc.* Franco] took communion and received the sacrament of the sick, a religious ritual that used to be known as the last rites. **1981** *Church Times* 4 Sept. 9/4 He was the priest in the famous photograph giving the Last Sacraments (a term seldom used now) to the wounded and dying on what the Irish call 'Bloody Sunday'.

3. d. *sacrament of the present moment*, any and every moment regarded as an opportunity for the reception of divine grace.

1921 E. J. STRICKLAND tr. *de Caussade's Abandonment to Divine Providence* I. i. 3 What treasures of grace lie concealed in these moments filled, apparently, by the most ordinary events... O Bread of Angels! heavenly manna!... Sacrament of the present moment! **1930** J. CHAPMAN *Spiritual Lett.* (1935) 83 The whole point of the 'Sacrament of the present moment' is that it is a..sacrament; it is God's action, God's will. **1943** O. WYON *School of Prayer* iii. 38 God makes His will known to us through the things that happen every day... Once we see it, our whole life is lifted on to a higher plane. This way of living has been described as *The Sacrament of the Present Moment*. **1967** J. N. WARD *Use of Praying* iii. 36 There is the use of the 'Jesus Prayer'... There is the cultivation of the 'sacrament of the present moment'. **1979** *Tablet* 22/29 Dec. 1251/2 We miss the many-splendoured thing in the goings-on of daily life, but it is there, totally transforming it and bestowing the sacrament of the present moment on those who are willing to accept it.

6. sacrament day, a day on which Holy Communion is celebrated; **sacrament house**, delete † and add later examples; **sacrament Sabbath** = *Sacrament Sunday*; **Sacrament Sunday** (earlier example).

1687 W. SEWALL in *Mass. Hist. Soc. Coll.* (1878) 5th Ser. V. 176 May 15th was our Sacrament-day. **1765** T. LINDSEY *Let.* 1 Nov. in *N. & Q.* (1942) 1 Aug. 62/2 Being a sacrament-day, I could but barely ask the former how he did as he went out of the church. **1826** A. CONSTABLE *Let.* 10 Oct. in J. Constable *Corr.* (1962) 228 Golding din'd with me on Sunday (Sacrament day). **1876** C. SCHREIBER *Jrnl.* 16 July (1911) I. 464 Many objects of the rarest interest—a dance of death (1742).., a sacrament house. *Ibid.* 24 July 470, I saw a fine Sacrament house, the third I have met with. **1975** A. MAYCOCK *Malling Abbey* (rev. ed.) 15 The nuns enter their choir from the cloister.., passing..on the right a circular sacrament house on which the light falls from a conical shaft immediately above it. **1816** in *Sc. Nat. Dict.* (1971) VIII. 3/2 'Twas sacrament Sabbath and much had been laid in. **1957** E. E. EVANS *Irish Folk Ways* xviii. 253 The 'sacrament Sabbaths' of Presbyterian Ulster were great gatherings having something of the nature of fairs. **1768** J. WOODFORDE *Diary* 9 Oct. (1924) I. 80 David Maby.. dined with us, being Sacrament Sunday.

sacramentalism. Add: **2.** The theory that the natural world is a reflection or imitation of an ideal, supernatural, or immaterial world.

1936 C. S. LEWIS *Allegory of Love* ii. 45 The attempt.. to see the archtype in the copy, is what I mean by symbolism or sacramentalism. **1963** H. BLAMIRES *Christian Mind* II. vi. 175 A living Christian mind would elucidate for the young a finely articulated Christian sacramentalism which would make sense of, and give value to, the adolescent's cravings towards the grandeur of natural scenery, towards the potent emotionalism of music and art, and towards the opposite sex.

Sacramentarian. Add: **B.** *sb.* **2.** (Earlier example.)

1732 J. WESLEY *Let.* 18 Oct. (1931) I. 130 Some of the men of wit in Christ Church..made a pretty many reflections upon the Sacramentarians, as they were pleased to call us.

sacramented, *ppl. a.* *rare*$^{-1}$. Consecrated, made sacred, sealed by a sacrament.

1914 R. BROOKE in *New Numbers* I. 116 They'll..sell Love's trust And sacramented covenant to the dust.

sacre, *v.* Add: **2. a.** (Later *arch.* example.)

1976 N. ROBERTS *Face of France* xxv. 229 French kings, with rare exceptions have been sacred in Reims Cathedral.

‖**sacré** (sakre), *sb.* [Fr., cf. SACRÉ *v.*] (The utterance of) the word 'sacré' as a profane imprecation.

1828 LYTTON *Pelham* I. xix. 148 He uttered a short, low, laugh..; and, pushing through the atmosphere of *sacrés* and *mille tonnerres*..strode quickly to the door.

‖**sacré** (sakre), *a.* [Fr.: cf. prec.] Holy, sacred, used in various French oaths, as *sacré bleu* (sakre blö), also *sacre bleu, sacrebleu*, a euphemism for *sacré Dieu*; *sacré Dieu* (sakre d^{y}ö); *sacré nom* (sakre noṅ), *sacré tonnerre* (sakre tonɛ̄r).

1768 STERNE *Sentimental Journey* II. 134 If there is but a cap-full of wind in or about Paris, 'tis more blasphemously *sacre Dieu*'d there than in any other aperture of the whole city. **1869** 'MARK TWAIN' *Innoc. Abr.* xxvii. 294 'Is, ah—is he dead?' 'Oh, *sacre bleu*, been dead three thousan' year!'. **1901** S. J. WEYMAN *Count Hannibal* i. 1 Sacré nom, am I King, or a dog of a —. **1905** BARONESS ORCZY *Scarlet Pimpernel* i. 10 '*Sacré tonnerre*,' said the captain. **1923** W. L. LOCKE *Moordius & Co.* ii. 23 *Sacrebleu!* the world had changed since 1870. **1974** K. BENTON *Craig & Tunisian Tangle* vi. 63 But *sacre bleu!* you can't depend on that.

sacred, *a.* and *sb.* Add: **A.** *adj.* **3. b.** *Sacred Blood*, the blood of Christ; *sacred concert* (examples); *sacred music* (examples); *sacred orders* [eccl. L. *ordines sacri*], the holy or major orders.

[**1709** J. BINGHAM *Origines Ecclesiasticæ* II. III. i. 9 The Clergy of the Superior Orders are commonly called the ἱερώμενοι Holy and Sacred, as in Socrates and others.] **1726** J. AYLIFFE *Parergon* 184 The first [*sc.* sub-deacons, deacons, and priests] the Canon Law..stiles Sacred Orders. **1832** *Rep. Sel. Comm. Dramatic Lit. with Minutes of Evidence* 50 in *Parl. Papers 1831–2* VII. 1, I thought it would be a better thing to represent plays than to give a pretended sacred concert. **1836** *Penny Cycl.* VI. 374/1 The clerical orders of the Catholic church are divided into two classes, *sacred* and *minor* orders. **1862** in N. Longmate *Hungry Mills* (1978) viii. 113 Never was so much sacred music heard upon the streets of Manchester as during the last few months. **1884** ADDIS & ARNOLD *Cath. Dict.* 622/1 The orders of bishop, priest, deacon, and (but only since the thirteenth century) subdeacon are called 'sacred' or 'greater'. **1900** *Cornish Echo* 30 Mar. 4/4 Wesley Chapel Falmouth. *Sacred Concert* by Truro Wesleyan Church Choir. **1901** PROCTER & FRERE *Prayer Bk.* xvi. 650 The Sacramentary of Serapion gives forms of ordination only for the three sacred orders. **1920** WODEHOUSE *Jill the Reckless* (1922) xvi. 230 It is the custom of the dwellers in Atlantic City..to attend a species of vaudeville performances—incorrectly termed a sacred concert—on Sunday nights. **1922** CHESTERTON *Ballad of St. Barbara* p. vii, In the grey rocks the burning blossom Glowed terrible as the sacred blood. **1934** *Daily Gleaner* 5 Jan. 21/2 *Sacred Concert*..A fine programme of sacred songs, instrumental music and recitations will be presented at St. Thomas' Church, Bath, on Sunday. **1950** *Cornishman* 12 Jan. 4/2 Sacred Concert by Ludgvan Male Choir with Elise Harvey, guest soloist. **1965** *New Statesman* 19 Nov. 794/3 We are made aware of the ideas of the period—notably the Pope's own contributions to the Franciscan v. Dominican battle about the Sacred Blood. **1978** *Listener* 24 Aug. 244/4 By 1733 Bach had written the vast majority of his sacred music.

7. sacred circle, an exclusive company, an élite; **sacred egoism** = *SACRO EGOISMO; **sacred way**, a route used by religious processions, pilgrims, etc.

1939 *Country Life* 11 Feb. 156/1 The courses, however, which have been admitted into this sacred circle—Westward Ho! for the Amateur Championship, Carnoustie for the Open, and Troon and St. Anne's for both meetings, have all been of a certain ancient standing. **1928** H. W. SCHNEIDER *Making Fascist State* i. 11 Salandra's policy of 'sacred egoism', of bargaining with both sides to see who would promise Italy the most for her neutrality, is both disgraceful and useless. **1970** R. A. H. ROBINSON *Origins of Franco's Spain* iv. 186 Aguirre still pursued a policy of sacred egoism and held aloof from non-Basque causes. **1884** A. LANG in M. Hunt tr. *Grimm's Household Tales* I. p. xiv, Amber and jade and slaves were carried half across the world by the old trade-routes and sacred ways. **1910** *Encycl. Brit.* II. 883/2 The chain..of Aegaleos, through a depression in which was the line of the sacred way, where the torchlight processions from Athens used to descend to the coast. **1937** G. MITCHELL *Come away, Death* i. 31 Now we go to Eleusis along the Sacred Way, to penetrate the meaning of the Mysteries. **1971** GARSIDE & WILKINS tr. *Ceram's Gods, Graves & Scholars* (ed. 2) xxiii. 292 The Sacred Way of Babylon ran..from the outer city walls to the Gate of Ishtar.

sacred cow. [f. SACRED *a.* + COW *sb.*^{1}] **1.** The cow as an object of veneration amongst Hindus.

1891 J. L. KIPLING *Beast & Man in India* vi. 116 The Muhammedan..creed is in opposition to theirs [*sc.* the Hindus] and there are rankling memories of a thousand insults to it wrought on the sacred cow. **1972** E. J. SHARPE in Hinnells & Sharpe *Hinduism* xxxvii. 121 As every visitor to India knows, the 'sacred cow' is not as a rule particularly well looked after. **1979** E. H. GOMBRICH *Sense of Order* vi. 167 The couch in Tutankhamun's tomb ..was shaped like a sacred cow, ready to carry the pharaoh into the other life.

2. *fig.* (orig. *U.S.*). **a.** *Journalism.* (*a*) someone who must not be criticized; (*b*) copy that must not be altered or cut.

1910 *Atlantic Monthly* Mar. 308/1 In the office these corporations were jocularly referred to as 'sacred cows'. **1922** U. SINCLAIR *They call me Carpenter* xxvi. 92 It doesn't matter, because I couldn't use the story. Mr. Stebbins is one of our 'sacred cows'. **1923** O. G. VILLARD *Some Newspapers* 143 The editors gave me their word that there is..no list of men to be attacked and no 'sacred cows' (i.e., favourites to be spared) in their shop. **1936** W. E. HALL *Reporting News* 430 Sacred cow—copy which is not to be changed or cut. **1940** R. E. GARST *Headlines & Deadlines* 206 *Sacred cow*, slang for a subject or story in which the publisher or higher editors are interested and which must be printed. **1973** B. BROADFOOT *Ten Lost Years* xxxi. 351 Newspapers had a lot more sacred cows than they do now.

b. An idea, institution, etc., unreasonably held to be immune from questioning or criticism.

1936 M. MITCHELL *Gone with Wind* xiii. 240, I think of my brother, living among the sacred cows of Charleston, and most reverent towards them. **1955** *Bull. Atomic Sci.* Feb. 62/2 The need for widespread secrecy has become a sacred cow, a belief hedged by the deepest emotions and accepted without question by many Americans. **1961** L. MUMFORD *City in Hist.* vi. 182 Economic exploitation, slavery, war, specialized life-time labor... Plato's polis relied for daily meat and drink on these sacred but diseased cows. **1962** *Listener* 5 July 10/2 Business is the 'sacred cow' here. **1963** *Richmond* (Va.) *Times-Dispatch* 16 Dec. 19/1 Military bases..are sacred cows in Congress. **1967** COULTHARD & SMITH in Wills & Yearsley *Handbk. Managem. Technol.* 207 A new movement, variously described as 'management by objectives', 'improving management performance' or 'performance planning', now promises to sweep away some of the 'sacred cows' of management development. **1971** *Country Life* 4 Nov. 1223/3 By clearing away any sacred-cow reputation clinging to the master he stirs the reader to look for himself. **1978** *Maledicta 1977* I. 111 He has for many years attacked most of their sacred cows and revealed them to be dry and motheaten skins. **1978** L. HEREN *Growing up on The Times* iv. 152, I was not an Establishment man. I had often caused trouble for the paper because of my refusal to give proper obeisance to sacred cows.

Sacred Heart. 1. The heart of Jesus, regarded as an object of devotion; similarly, *Sacred Heart of Mary*. *Feast of the Sacred Heart* (R. C. Ch.), a festival observed on the Friday in the week following Corpus Christi; also *ellipt.*

1765 (*title*) The devotion to the Sacred Heart of Jesus. **1793** J. BERINGTON in *Panzani's Memoirs* p. xxxii, He might be busied in preparing a gay posey of devotion to the *sacred heart of Mary*. **1823** C. BUTLER *Contin. A. Butler's Lives Saints* 228 The devotion to the sacred Heart was sanctioned by all the prelates of the neighbouring country. **1833** M. ARUNDELL *Let.* 30 June (1894) vii. 47/2 You will pity me however confined to a sick bed during all my favourite feasts, Whit-Sunday, Corpus Christi, Sacred Heart! **1853** J. B. DALGAIRNS *Devotion to Heart of Jesus* i. 59 Symbols, which are also realities, as the Sacred Hearts of Jesus and Mary, the Five Wounds, or the Precious Blood of our Lord. **1881** G. M. HOPKINS *Sermons* (1959) 100 (*heading*) For Sunday June 26 1881 being the Sunday..nearest the Feast of the Sacred Heart (which this year is to be kept on Monday the 27th). **1924** E. LEAHY tr. *Bainvel's Devotion to Sacred Heart* II. ii. 103 To St Margaret Mary and her revelations is really due the inception of the devotion to the Sacred Heart in the form that has received the Church's sanction. **1945** J. BETJEMAN *New Bats in Old Belfries* 25 As Thy Sacred Heart displays Lush Kildare of scented meadows, Roscommon, thin in ash tree shadows,.. Kneeling all in silver haze? **1967** K.-H. & B. KRUGER tr. *Rahner's Theol. Investigations* III. xxii. 331 The Sacred Heart devotion in the doctrine and practice of the Church both past and present manifests itself under many aspects and forms. **1977** B. LUCAS tr. *De Foucauld's Lett. from Desert* iv. 69, I was determined to write..to tell you how close I shall be to you on the feast of the sacred Heart.

2. *transf.* **a.** A form of prayer used in private devotions to the Sacred Heart.

1815 J. MILNER in F. C. Husenbeth *Life J. Milner* (1862) xvi. 288 Spiritual Reading afternoon.—Grace, Rosary, Sacred Heart. **1881** G. M. HOPKINS *Sermons* (1959) 102 When we say the Sacred Heart it is of Christ himself we are thinking and not of his heart only.

b. A devotional picture of the Sacred Heart.

1931 *Jrnl. Amer. Folk-Lore* XLIV. 413 Pictures of Saints, etc., are used also [in voodoo magic in America]... Sacred Heart of Jesus. For organic diseases. **1941** A. WHITE *Let.* 22 Mar. in *Hound & Falcon* (1965) 108, I prefer the Catholic Church with all the awful Sacred Hearts and Little Flowers and the rest to the still more awful bleak 'tastefulness' of the others. **1977** M. KENYON *Rapist* iv. 40 Above the door..hung the Sacred Heart in lurid primaries.

3. Used as (part of) the name of several religious orders and of schools run by them; freq. *attrib.* and *ellipt.*

1885 H. JAMES *Little Tour in France* iii. 22 The modern buildings (of the Sacred Heart)..are in the vulgar taste which seems doomed to stamp itself on all new Catholic work. **1907** E. WHARTON *Fruit of Tree* ii. 20 A girl who was at the Sacred Heart in Paris with me. **1919** T. S. ELIOT *Sweeney among Nightingales* in *Poems*, The nightingales are singing near The Convent of the Sacred Heart. **1965** A. WHITE *Hound & Falcon* 20 Even in my day Tunbridge Wells was always said to be much less strict... [*Note*] Another Sacred Heart convent. **1973** *Guardian* 25 Apr. 15/3 Mr Joe Faye..of the Sacred Heart School, Redcar... The 12-man governing board of Sacred Heart (a Catholic maintained school.) **1974** V. CANNING *Mask of Memory* i. 19 Gave the sweets to the Sacred Heart children on the beach.

sacrificable, *a.* For *rare*$^{-1}$ in Dict. read *rare* and add: also, rightly or properly to be sacrificed.

1973 *N.Y. Law Jrnl.* 27 Feb., No citizen's individuality is sacrificable to expediency.

sacrifice, *sb.* Add: **3. c.** *sacrifice of praise* (*and thanksgiving*): a phr. drawn from biblical sources (e.g. Lev. vii. 12, Ps. l. 14, 23 (R.V., etc.), Heb. xiii. 15) used *gen.* for an offering of praise to God, and liturgically in the anaphora of many post-Reformation Eucharistic rites, tr. *sacrificium laudis* of the Latin Canon of the Mass.

1535 COVERDALE *Heb.* xiii. 15 Let vs therfore by him offre allwayes vnto God the sacrifice of prayse. **1549** *Bk. Common Prayer* f. cxxviii^v, Entierely desiryng thy fatherly goodnes, mercifully to accepte this our Sacrifice of praise and thankes geuing. **1611** BIBLE *Jer.* xvii. 26 Bringing sacrifices of praise vnto the house of the Lord. **1864** F. PIERPOINT *For the Beauty of the Earth* (hymn) in O. Shipley *Lyra Eucharistica* (ed. 2) 340 Christ, our God, to Thee we raise This our Sacrifice of Praise. **1877** E. DANIEL *Prayer-Bk.* 296 That sacrament in which we offer our sacrifice of praise and thanksgiving for the redemption of the world. **1884** ADDIS & ARNOLD *Cath. Dict.* 564/1 The Mass is a sacrifice of adoration, of praise and thanksgiving. **1980** *Alternative Service Bk. 1980* 132 Accept through him, our great high priest, this our sacrifice of thanks and praise.

5. a. Delete (*nonce-use*) and add later examples.

1915 *Truth* 30 June 1068/1 Being convinced that his 'trade sacrifice' will result in increased business. **1976** *Billings* (Montana) *Gaz.* 5 July 9-C/8 (Advt.), Yaesu FT101 base, desk mike, antenna & tripod. *Sacrifice.* **1976** *Evening Advertiser* (Swindon) 31 Dec. 17/2 (Advt.), Bargain: 1971 Ford Escort 1100..genuine sacrifice, £380 only.

b. *Baseball.* = *sacrifice hit* (see 6 in Dict.).

1880 *Inter-Ocean* (Chicago) 29 June 8/3 Force's winning run came off a wild throw by Ward, a sacrifice and single. **1904** R. H. BARBOUR *Bk. School & Coll. Sports* 173 When the batsman is looking for a sacrifice keep the ball high. **1912** C. MATHEWSON *Pitching in a Pinch* xii. 261 Snodgrass got a base on balls and journeyed to second on a sacrifice. **1968** *Washington Post* 4 July C2/8 Willie McCovey threw wildly trying to force the Atlanta pitcher at second on a sacrifice. **1976** *Billings* (Montana) *Gaz.* 6 July 3-C/1 Rookie Butch Hobson drove in three runs with a sacrifice and a single.

c. *Chess.* The action of *SACRIFICE *v.* 3 e.

1915 J. DU MONT tr. *Lasker's Chess Strategy* I. iv. 25 White..prefers to end up with a magnificent sacrifice. **1933** H. PHILLIPS *Week-End Problems Bk.* 310 The key-move..offers double sacrifice with a cross-check. **1952** E. LASKER *Chess Secrets* 122 With this Bishop sacrifice Janowski tears down his opponent's defenses. **1977** *Guardian Weekly* 25 Dec. 23/5 Black took 80 minutes to decide to accept this strong pawn sacrifice.

d. *Bridge.* = **sacrifice bid(ding)*.

1952 *Bridge Mag.* Apr. 39/2 Five clubs is only two down, a good sacrifice against five spades. **1964** FREY & TRUSCOTT *Offic. Encycl. Bridge* 480/1 One will earn a fat score with one's sacrifice only when most of the field is bidding game with his opponents' cards. **1974** [see *MAKE *sb.*^{2} 10].

6. *sacrifice bringer*, *sale*; **sacrifice bid(ding)** *Bridge*, (making) a bid higher than the contract that one expects to be able to fulfil, in order to prevent opponents from making a score greater than the penalty one is likely to suffer; **sacrifice bunt** *Baseball*, a bunt that puts the batter out whilst allowing a base runner to advance; **sacrifice fly** *Baseball*, an outfield fly that is caught so that the batter is put out but which allows a base runner to advance after the ball is caught; **sacrifice hit** (earlier example); **sacrifice meat**, meat eaten at a feast following the offering of a sacrifice to a deity.

1932 H. PHILLIPS *One Hundred Contract Bridge Hands* 115 This is a good example of a 'sacrifice bid'... South now bids Four Hearts as probably a cheaper 'sacrifice' (even if doubled) than that of the game. **1959** *Listener* 13 Aug. 262/1 The hidden value of a part score has a bearing also on sacrifice bidding at the game level. **1964** FREY & TRUSCOTT *Offic. Encycl. Bridge* 481/1 A hidden advantage of sacrifice bidding is the chance that the opponents will be pushed one higher and will go down. **1923** D. H. LAWRENCE *Birds, Beasts & Flowers* 149 And you, great bird... Can be put out of office as sacrifice bringer. **1974** *Anderson* (S. Carolina) *Independent* 24 Apr. 5B/2 Jack Brohamer..moved to second on Buddy Bell's sacrifice bunt. **1970** *Globe & Mail* (Toronto) 25 Sept. 31/1 Roger Freed's sacrifice fly..helped Baltimore stretch its lead to 4-1 in the fifth. **1979** *Arizona Daily Star* 22 July C5/5 Butch Wynegar also drove in a run in the sixth with a sacrifice fly to help Goltz raise his record to 9-6. **1881** *N.Y. Herald* 21 July 8/3 The Metropolitans scored another run on two pretty singles, a passed ball and a sacrifice hit. **1926** D. H. LAWRENCE *David* iv. 24 They be all there, waiting for the sacrifice meat. **1902** G. H. LORIMER *Lett. Self-Made Merchant* x. 129 It was the record-breaking, marked-down sacrifice sale of the year on dogs.

sacrifice, *v.* Add: **3. c.** To sell or get rid of at a sacrifice, esp. in commercial use. Also *absol.*

1850 *Punch* XVIII. 130/2 A newspaper advertisement announces that 'A Professional gentleman is instructed to sacrifice *three* young sound *Horses* at half their cost.' We wonder what deity horses could be sacrificed to? **1903** FARMER & HENLEY *Slang* VI. 91/1 *Sacrifice*..to sell regardless of cost. **1930** *San Antonio* (Texas) *Light* 31 Jan. 14/7 (Advt.), Owner Must Sacrifice Must sell at bargain. **1947** E. HODGINS *Mr. Blandings builds his Dream House* ii. 23 'Farm dwelling..original beams..will sacrifice', The *New York Times* advertisement had said.

d. *Baseball.* (*a*) *intr.* To make a hit which advances another player, the batter being put out. (*b*) *trans.* To advance (another player) in this way.

1905 *Sporting Life* 2 Sept. 4/1 Lumley's effort to sacrifice resulted in a short pop fly. **1912** C. MATHEWSON *Pitching in a Pinch* ix. 202 Brown sacrificed, sending Kling to second. **1971** L. KOPPETT *N.Y. Times Guide Spectator Sports* i. 31 When a batter walks, is hit by a pitch, sacrifices (by bunting so that other runners advance even though he is out),..he is not charged with an official time at bat. **1974** *News & Courier* (Charleston, S. Carolina) 22 Apr. 9-A/3 Denny Doyle..was sacrificed to second. **1975** *Verbatim* Sept. 5/1 But Joe Ferguson did something much more drastic than that: *He sacrificed Garvey home!*

e. *Chess.* To put or leave (a man) in a

position where it can be captured without equivalent loss by one's opponent, in order to gain a future advantage.

1915 J. du Mont tr. *Lasker's Chess Strategy* II. 224 White decides to sacrifice a Knight in order to open the files in the centre for his Rooks. 1952 E. Lasker *Chess Secrets* 54 White could sacrifice a piece for three Pawns. 1969 A. Glyn *Dragon Variation* vii. 193 They'd both sacrifice every piece on the board. By the middle-game they'd just have the two Kings left. *Ibid.* ix. 271 He tried to break the stranglehold by sacrificing first a Knight and then a Rook. 1974 Hartston & Keene *Karpov–Korchnoi 1974* 66 Korchnoi plucks up his courage and sacrifices his K-side in order to create a passed pawn of his own.

f. *Bridge. intr.* To make a sacrifice bid.

1952 Phillips & Reese *Bridge with Mr. Playbetter* xiv. 59 He must take all possible measures to prevent Hurry sacrificing in Five Clubs. 1959 *Listener* 22 Jan. 189/2 Is it possible, under the Laws, to sacrifice at the level of Eight? 1962 *Ibid.* 13 Sept. 410/3 Over Four Hearts North could raise to six. No doubt, in that event, East-West would sacrifice in Six Spades. 1964 Frey & Truscott *Offic. Encycl. Bridge* 480/2 Be alert to sacrifice against confident auctions when it appears that everyone else will be in game too.

4. To kill (an experimental animal) for scientific purposes.

1903 *Jrnl. Physiol.* XXIX. 83 The animal was sacrificed on the 315th day after the 1st lesion had been established. 1926 J. S. Huxley *Ess. Pop. Sci.* 282 When, after a couple of months, the dog was sacrificed, it was found..that the histological character of the cells had changed, cross-striations arising in them. 1944 *Jrnl. Immunology* XLIX. 316 The animals were sacrificed by a blow on the head and the small intestine was immediately removed. 1971 *Sci. Amer.* July 55/1 In rats we destroyed the mitral cells in the olfactory bulb by surgical intervention and, after a survival time of from three to five days, sacrificed the animal to conduct a microscopic examination of the fibers leading from these cells.

sacrificial, *a.* Add: **4.** Involving or being an anode of a metal that is used up when protecting another metal against electrolytic corrosion.

1861 *Mechanics' Mag.* 5 Apr. 229/1 A curious statement made in the House of Commons, a few days ago, induces us to offer a few remarks on what may be called, appropriately enough, 'the sacrificial function in metals'. .. Failing to achieve what was intended of it, the copper-protecting process of Davy ceased to be employed; but the failure of it is that which alone concerns us here, as illustrating what we would wish to convey by the words *sacrificial metal*. 1937 U. R. Evans *Metallic Corrosion Passivity & Protection* xii. 531 The protection method is sacrificial, the valuable iron boiler being protected by sacrificing the easily replaceable zinc. 1949 *Jrnl. R. Soc. Arts* XCVII. 598 A sufficient cathodic current density may be impressed upon structures of iron or steel by the use of external anodes of magnesium, aluminium or zinc, the anode undergoing 'sacrificial corrosion'. 1968 *Rep. Progr. Appl. Chem.* LIII. 69 It was better to use an all-nickel coating..rather than a copper undercoat where sacrificial corrosion resulted in complete loss of plate. 1976 *Daily Tel.* 25 Nov. 7/1 It is common practice to fit the pipes with collars of a zinc alloy, which act as 'sacrificial anodes', corroding in preference to the steel pipe. 1978 *Metals* (Shell Internat. Petroleum Co.) 7 Zinc is ideal as a sacrificial metal for the corrosion-protection of steel.

sacrificially, *adv.* [f. Sacrificial *a.* + -ly[2].] In a sacrificial manner.

1937 L. C. Douglas *Forgive us our Trespasses* i. 2 'Wish you was a-goin' along'... 'No,' Martha would reply, sacrificially, 'somebody's got to stay on th' place.' 1972 *Daily Tel.* 7 Sept. 18 Our officers and staff who serve so sacrificially year in and year out..are not men who have come to terms with squalor.

sacrilege, *sb.*[2] (Later *poet.* example.)

1802 W. S. Landor *Poetry* 7 Thrown prostrate on the earth, the Sacrilege Rais'd up his head astounded.

sacrilegious, *a.* Add: Now usu. with pronunc. (sækrili·dʒəs).

sacrist. Add: **2.** Also 8 **saccarist.** An officer in the University of Aberdeen (formerly King's and Marischal Colleges); orig. a cleric whose responsibilities included the furnishings of the church, later a senior janitor or head porter with some ceremonial duties.

1638 *King's Coll.* (Aberdeen) *Minutes* 27 Dec., In the visitatioun of the Kinges Colledge of the Universitie of Aberdeine..convenit..Mr. Alexander Ross doctor of divinitie, principall of the said Colledge,..Mr. Gilbert Ross, cantor, and Patrick Innes, sacrist. *a* 1670 [in Dict.]. 1732 in A. M. Munro *Rec. Old Aberdeen* (1899) I. 256 Robert Gordon, Saccarist in the King's College. 1792 J. Spalding *Hist. Troubles* I. 127 Mr. David Lindsay Parson of Belhelvie, was said to be moderator of this committee, to the which committee upon the 24th of March were summoned in name of the assembly and moderator, the principal of the King's College of Old Aberdeen, the four regents, the canonist, Doctor of Medicine, civilist, sacrist, and janitor, founded members thereof. 1825 *Aberdeen Censor* Dec. 210 Enrolled as a student in divinity, by paying six shillings to the sacrist of Marischall College and a moiety to the library. 1865 G. MacDonald *Alec Forbes* II. ii. 8 A long broom which the sacrist had been using to clear foot-paths. 1902 *Minutes Aberdeen Univ. Court* V. 250 The Joint Committee were of the opinion that appropriate costumes, including robes and hats, for the two sacrists, could be procured for £15. 1965 *Aberdeen Univ. Rev.* Autumn 70 No ceremony is complete without the presence of the sacrist (or sacrists) robed in purple, with a tricorne hat trimmed with gold upon his head, and bearing his silver, bell-headed, mace.

sacristaness. Add: Also *fig.*

1924 C. C. Martindale *St. Paul* xi. 141 The city of Ephesus was the sacristaness of the great Artemis and of the image that fell from heaven.

sacro-[2]. Add: *sacrospinous, -tuberous, -uterine*; **sa:cro-ilii·tis,** inflammation of the sacro-iliac joint.

1934 *Acta Rheumetologica* VI. xxiii. 7/1 All the cases of Spondylitis adolescens..so far examined, have radiographically shown indications of bilateral infection of both sacro-iliac joints, (Sacro-ileitis) usually in the form of ankylosis. 1936 *Brit. Jrnl. Radiol.* IX. 127 (*caption*) Complete ankylosis of the sacro-iliac joints—the end-result of a chronic sacro-iliitis. 1977 *Lancet* 17 Sept. 591/1 Around the classic centre, rigid spine with radiographic sacro-iliitis, there is a very large fringe of milder forms of spondylitis. 1910 H. W. Cattell *Lippincott's New Med. Dict.* 845/1 *Sacrospinous ligament* = sacrosciatic ligament, lesser. 1913 *Gray's Anat.* (ed. 18) 354 The sacrospinous ligament (small sacrosciatic ligament) is thin, and triangular in form. 1967 G. M. Wyburn et al. *Conc. Anat.* vi. 163/2 The short sacrospinous ligament lies anterior to the sacrotuberous ligament and extends from the spine of the ischium to the lower part of the sacrum and coccyx. 1910 H. W. Cattell *Lippincott's New Med. Dict.* 845/1 *Sacrotuberous ligament*, = Sacrosciatic ligament, great. 1925 *Jrnl. Amer. Med. Assoc.* 31 July 307/2 Von Meyer..described the rotary movement of the sacrum, and showed..the restraint put on this rotation by the sacrotuberous and sacrospinous ligaments. 1962 *Gray's Anat.* (ed. 33) 531 The sacrotuberous ligament..is placed at the lower and posterior part of the pelvis. 1967 Sacrotuberous [see *sacrospinous* above]. 1888 W. H. & H. T. Byford *Pract. Med. & Surg.* (ed. 4) ii. 83 Two fingers carried high up in the posterior fornix vaginae can usually feel the semi-circular folds of the sacro-uterine ligament extending outward, backward and upward. 1946 R. W. Te Linde *Operative Gynecol.* vii. 125/1 Pushing back the mucosa in the attempt to see the peritoneum exposes each sacro-uterine ligament.

‖**sacro egoismo** (sa·kro egoi·zmo). [It., 'sacred egoism': see *Sacred *a.* 7.] Nationalism without scruples in relationships with other states. Also *transf.*

[1914 A. Salandra in G. Fumagalli *Chi l'ha Detto?* (1958) 560 Anima sevro..da ogni sentimento che non sia quello della illimitata ed esclusiva devozione alla Patria nostra, del *sacro egoismo* per l'Italia.] 1944 *Zionist Rev.* 21 Apr. 6/3 Characteristic of our present attitude is the suggestion that my insistence on 'using every means for the pursuit of our own aims', sounds rather like the doctrine of *sacro egoismo*. 1947 'G. Orwell' *Eng. People* 15 Power politics, 'realism', *sacro egoismo* and the doctrine that the end justifies the means. 1981 J. Sutherland *Bestsellers* xi. 126 The adolescent's values..*sacro egoismo*, refusal to form lasting relationships, machismo.

sacsac (sæ·ksæk). [Pidgin.] A local name in Papua New Guinea for the sago palm, *Metroxylon sagu*. Also *attrib.*

1947 I. L. Idriess *Isles of Despair* xxiv. 163 Sago from the sac-sac palm. 1962 *Coast to Coast 1961–62* 55 We told them where they would find the sago-palms, the sacsac, to plait for walls and roof.

Sacta, Sacti, varr. *Sakta, *Sakti.

sad, *a.* Add: **III. 11.** *sad-faced* (later examples), *-garbed, -hearted* (later example), *-lidded, -looking, -making, -serene, -smiling, -sweet, -visaged*; **sad-ass** *N. Amer. slang*, used *attrib.* as a term of abuse; also **sad-assed** *a.*

1971 *Black World* Apr. 63 How is Philadelphia?.. Thats one sad-ass city..bout to sink into the ground. 1974 D. Sears *Lark in Clear Air* xiii. 158 A few general comments on sad-assed, puritanical sons-of-bitches individually and collectively. *c* 1893 A. W. Pinero in M. R. Booth *Eng. Plays of 19th Cent.* (1969) II. 285, I knew her when she was a sad-faced, pale baby. 1947 J. Mulgan *Report on Experience* xi. 126 In the streets were sad-faced men and women, still hungry and no longer happy. 1971 *Where?* Oct. 309/1 Indeed, one might argue that there is an urgent and essential need to produce this quality of communal participation and commitment if our society is to survive. Not that this means a sadfaced approach. 1848 J. R. Lowell *Poems* 2nd Ser. 167 He looks a sachem, in red blanket wrapt,..'mid some council of the sad-garbed whites. 1952 C. Day Lewis tr. *Virgil's Aeneid* I. 16 And spoke these words of comfort to his sad-hearted friends. 1921 D. H. Lawrence *Tortoises* 35 His black, sad-lidded eye sees but beholds not. 1961 C. McCullers *Clock without Hands* iv. 78 The red lamp with ragged fringes, two obviously broken chairs and other pieces of sad-looking furniture. 1930 E. Waugh *Vile Bodies* xii. 208 'My dear, isn't that rather sad-making for you?' 'I'm desperate about it.' 1955 J. D. Salinger *Franny* in *New Yorker* 29 Jan. 30/3 But just so tiny and meaningless and—sad-making. 1960 J. Stroud *Shorn Lamb* xxiii. 251 You were watching the end of an epoch; that's always rather sad-making. 1933 W. de la Mare *Fleeting* 17 How sad-serene the abandoned house. 1928 Blunden *Undertones of War* 4 And there, sad-smiling,.. were two or three of the convalescent squad. 1909 E. Pound *Exultations* 12 Ye know somewhat the strain, the sad-sweet wonder-pain of such singing. 1925 J. Gregory *Bab of Backwoods* iii. 33 He managed to get his one free arm about her, hugging her tight while he said good-bye; Bab would never forget that terribly sad-sweet moment. 1953 S. Kauffmann *Philanderer* xiv. 233 A lamp-lighted restaurant table, the distant sad-sweet music, all these flowed together in a comfortable alcoholic haze. 1869 'Mark Twain' *Innoc. Abr.* xxxiv. 370, I never saw such.. starving, sad-visaged, broken-hearted looking curs in my life.

saddhu, var. *Sadhu.

saddle, *sb.* Add: **I. 1. d.** Ellipt. for *saddle brown* in sense 10 below.

1976 *Billings* (Montana) *Gaz.* 26 June 14-A/3 (Advt.), The perfect handbag for carrying everything in Saddle, Tan or Rust. 1977 *West Briton* 25 Aug. 31/5 (Advt.), 1974 (Oct.) Lancia Fulvia 3 Coupe, finished in maroon with saddle interior.

II. 4. a. *spec.* in *Geol.*, a depression along the axis of an anticline, concave in longitudinal section and convex in transverse section.

1886 T. M. Reade *Origin of Mountain Ranges* xvi. 187 (*caption*) Saddle in lower Silurian rocks between Clarach Bay and Aberystwyth, formed by the junction of anticlinal and synclinal curves. 1928 E. R. Lilley *Geol. Petroleum & Natural Gas* xii. 293 Where the amount of oil and gas is sufficient only to fill the distinctly domed portions of the fold, the saddle is normally water-bearing. 1952 *Q. Jrnl. Geol. Soc.* CVIII. 16 Individual anticlines have lengths of up to 250 miles, following long straight courses and rising and falling into culminations and saddles. 1977 *Offshore Engineer* May 52/1 The structure is a tilted fault block with hydrocarbons trapped on the upthrown side of a normal fault in two culminations separated by a saddle.

c. *Mining.* = *saddle reef* in sense 10 below; *spec.* one that is anticlinal rather than synclinal.

1872 *Rep. Vermont Board Agric.* 630 The miners were quarreling about false and true veins, horses of rock and saddles of ore. 1908 J. M. Maclaren *Gold* II. 368 In working the saddles, prospecting for lower saddles is effected by sinking shafts designed to strike a 'leg' of an underlying saddle, from whence stopes are carried up to the crest of the anticline. 1937 W. H. Emmons *Gold Deposits of World* vii. 528 As many as 24 quartz saddles in an anticline have been passed through from the surface to a depth of 2,200 feet. 1950 David & Browne *Geol. Commonwealth of Austral.* II. xxvii. 196 Inverted saddles are smaller and less common than saddles, but they have given profitable yields in several mines.

5. j. An insulating device designed to be fitted around an electrical wire or conduit to help to hold it in place.

1888 D. Salomons *Managem. Accumulators & Private Electr. Light Installations* (ed. 3) II. ii. 107 Leather saddles answer well to keep the wires in place. 1930 F. C. Raphael *Electr. Wiring of Buildings* vi. 91 Multiple saddles are used if two or more lengths of conduit run together. 1969 H. A. Miller *Pract. Wiring* I. v. 34 It is good practice to secure single runs by clips and multiple runs by saddles.

k. *Dentistry.* The basal part of a denture, which replaces alveolar tissue and bears the artificial teeth.

1907 H. J. Goslee *Princ. & Pract. Crown & Bridge-work* xxiv. 443 The saddle should now be swaged of 30 to 32 gauge platinum, trimmed to the proper outline, fitted in the mouth with the caps in place, and then soldered thereto with platinum solder. 1930 I. G. Nichols *Prosthetic Dentistry* xli. 638 The various materials employed in rebasing saddles are: modeling compound, plaster, and wax. 1962 Blake & Trott *Periodontol.* iv. 39 The free end saddle, which cannot be supported by occlusal rests, always presents a problem.

l. A fire-clay bar for supporting ceramic ware during glazing.

1911 A. B. Searle tr. *Bourry's Treat. Ceramic Industries* (ed. 2) xii. 399 The pieces, especially plates, may be placed upright in rectangular saggers, kept up at the bottom by rectangular saddles and at the top by a series of thimbles. 1930 —— *Encycl. Ceramic Industries* III. 74/2 *Saddle*, a type of support used for plates, etc. in a saggar. It consists of bars of triangular cross-section. Two of these are laid parallel to each other on the bottom of the saggar, and the plates are stood on end across these. 1964 H. Hodges *Artifacts* i. 39 These setters vary considerably in shape, and their names are usually adequately descriptive—saddle, stilt, spur, thimble, pin. 1967 M. Chandler *Ceramics in Mod. World* iii. 102 Thimbles, saddles, and stilts are made of fireclay and so designed that the weight rests only on sharp points.

7. c. (Earlier and later examples.)

1854, etc. [see *saddle feather, saddle hackle* (*feather*), sense 10 below]. 1976 J. Batty *Understanding Old Eng. Game* (ed. 2) 58 *Saddle*, that part of the back of a male bird nearest to the tail which includes long feathers known as saddle hackles.

8*. a. A piece of leather stitched across the instep of a shoe, often in a contrasting colour.

1930 *Footwear Organiser* Jan. 37/1 The tie shoe is a development of the one-bar, with a bar or saddle fastened in the centre by means of a fancy lace or ribbon tie through eyelets. 1948 R. T. Wilcox *Mode in Footwear* 170 (*caption*) Saddle oxford—white elk—brown calf saddle—red rubber sole—worn since the 1920's. 1972 *N.Y. Times* 3 Nov. 8/1 (Advt.), In bone leather with a blue saddle..in big girl sizes 5 to 9 medium width.

b. Ellipt. for *saddle shoe* in sense 10 below.

1972 *N.Y. Times* 3 Nov. 8/1 (Advt.), Everyone loves our bumpy, bouncy saddle—the shoe that sparks up the classics.

8.** *Math.* A saddle point.

1952 W. Hume-Rothery et al. *Metall. Equilibrium Diagrams* xxx. 262 (*heading*) Diagrams involving intermediate compounds: saddles. *Ibid.*, The highest point in the saddle, *R*, will be a maximum in the direction of the eutectic valleys, and at this point the solidus line.. touches the liquidus. 1978 *Nature* 7 Sept. 76/1 They explain..Thom's listing of the ways in which critical points of functions (that is, maxima, minima and saddles) of *n* variables can coalesce as k parameters vary.

III. 9. Simple attrib., as *saddle-cover, -horn* (earlier and later examples), *-lining, -load, -pad, -paste, -pouch*; with the sense 'used for riding', as *saddle mare* (later example), *stock*; *saddle-peaked* adj.; in the names of affections incident to the use of the saddle, as *saddle boil, sore, -weariness*; objective, as *saddle-stitching*.

1816 U. Brown *Jrnl.* 20 Aug. in *Maryland Hist. Mag.* (1916) XI. 151 Cumberland's Back is Compleatly full of Saddle Biles & in a full fever. 1884 'Mark Twain' *Huck. Finn* ii. 8 He said they rode him all over the world..and his back was all over saddle-boils. 1977 *Times Lit. Suppl.* 8 Apr. 422/1 His [*sc.* Mark Twain's] more excruciating experiences (seasickness, saddle boils, the running war against vermin). 1676 S. Sewall *Diary* 27 Oct. (1878) I. 27 Saddle Cover [was] lost. 1895 M. A. Jackson *Mem. Stonewall Jackson* (ed. 2) xx. 403 A superb English saddle, bridle, holsters, saddle-cover. 1926 T. E. Lawrence *Seven Pillars* (1935) viii. xcii. 508 Afterwards we slept on our saddle-covers, the tanned fleece hooked last of all over the saddle-load to make a slippy and sweat-proof seat for the rider. 1856 A. Cary *Married* 184 The bridle rein was twisted around the saddle horn. 1926 T. E. Lawrence *Seven Pillars* (1935) vii. lxxxviii. 490 She [*sc.* a camel]..was docile and smooth to ride, turning left or right if the saddle-horn were tapped on the required side. 1971 D. C. Brown *Yukon Trophy Trails* i. 20, I hung the box by a string from the saddlehorn and climbed on. 1919 J. Masefield *Reynard* i. 5 Some.. Brushed at saddle-flaps or hove Saddle-linings to the stove. 1926 Saddle-load [see *saddle-cover* above]. 1975 *Islander* (Victoria, B.C.) 7 Sept. 16/2 His reports usually dealt with his saddlemare, Snippet, and her unexpected foal. 1750 J. Hempstead *Diary* 30 Mar. (1901) 546, I mended my old Sadle pad. 1971 J. McDougall *Parsons on Plains* xv. 135 Then I dressed, and putting a saddle-pad on, rode her [*sc.* a mare] all the afternoon. 1917 *Harrods Gen. Catal.* 537/4 Harrods' Saddle Paste. Per tin... o/6. 1919 J. Masefield *Reynard* i. 5 The savour Of saddle-paste and polish spirit. 1939–40 *Army & Navy Stores Catal.* 767/2 Propert's Saddle Paste..1/–. 1869 'Mark Twain' *Innoc. Abr.* xlix. 521 Saddle-peaked Hattin, traditional 'Mount of Beatitudes'. 1926 T. E. Lawrence *Seven Pillars* (1935) viii. xcvi. 532 We followed dragging my few things in their light saddle-pouch. 1946 M. C. Self *Horseman's Encycl.* 354 Riders with bad seats will often give a horse saddle sores. 1962 C. Storr *Lucy runs Away* v. 26 I've ridden two miles... I've got saddle sores. 1954 E. Jenkins *Tortoise & Hare* xii. 149 A small suitcase..was being repaired by the local iron-monger, who did saddle-stitching for the farmers. 1903 A. Adams *Log of Cowboy* 17 Then the entire saddle stock was driven in, so as to be at hand in case a hasty change of mounts was required. 1948 F. Blake *Johnny Christmas* ii. 75 He went directly on to the door and pushed it open, passing in among the oxen and saddle-stock of the fort. 1909 *Chambers's Jrnl.* June 347/1 In a general way, the word Bush recalls to the writer..the sun and dust and saddle-weariness of the great gray inland plains.

10. saddle-bill = *saddle-billed stork*; also *attrib.*; **saddle-blanket** (earlier and later examples); **saddle block** *Obstetrics*, the technique of anæsthetizing the perineal region by a low spinal injection; freq. *attrib.*; **saddlebred** *a.*, bred to have the gaits of an American Saddle Horse; also *ellipt.* as *sb.*; **saddle bronc** *U.S.*: in a rodeo, a bronco ridden with a saddle; freq. *attrib.* as *saddle bronc riding* (also *ellipt.* as *saddle bronc*); **saddle brown,** the tan colour of saddle-leather; **saddle-burned** *a.*, chafed by a saddle; **saddle club,** a riding club; **saddle-coloured** *a.*, (of complexion) saddle brown, tanned; **saddle embolus** *Med.*, an embolus straddling the fork of an artery, esp. the aorta, so as to block both branches; **saddle feather** = *saddle hackle* (*feather*) below; **saddle-gall** (later example); **saddle graft,** a graft made by saddle grafting; **saddle gun** *U.S.*, a gun carried on the saddle of a horse; **saddle hackle (feather),** any of the long feathers growing backwards from the saddle of a cock; so **saddle-hackled** *a.*; **saddle horse,** (*a*) (see sense 9 in Dict.); (*b*) (see quot.); **saddle-leaf,** substitute for def.: = Tulip-tree 1 a; (earlier and later examples); **saddle mark,** (*a*) a mark or patch on a horse's back produced by the friction of a saddle; (*b*) (see quot. 1963); **saddle-notch** (see quot. 1930); hence as *vb. trans.* and **saddle-notched** *a*; **saddle oxford,** a saddle shoe in the Oxford style; **saddle point** *Math.*, (*a*) a point at which a curved surface is locally level but at which its curvature in two directions differs in sign, i.e. for a surface defined by a function *f* of *x* and *y*, a point at which $\partial f/\partial x = \partial f/\partial y = 0$ and $(\partial^2 f/\partial x^2)(\partial^2 f/\partial y^2) < 0$; (*b*) in a zero-sum game, the joint outcome of the two parties following their unmixed optimal strategies, where these exist; **saddle quern** (earlier and further examples); **saddle reef** *Mining*, a reef or vein of ore between the strata in the curve of an anticline or syncline; **Saddle Rock** *local U.S.*, a large oyster; in full *Saddle Rock oyster*; **saddle-rug** (examples); **saddle scabbard** *N. Amer.* (see quot. 1944); **saddle-shaped** *a.* (further example of sense 'anticlinal'); **saddle shoe,** a shoe with a saddle (sense 8* a); **saddle shoulder** *Fashion*, a square-cut shoulder on a jersey, etc., that is an extension of the sleeve; also *attrib.*; **saddle-sore** *a.* (further examples); **saddle-stone,** (*b*) (example); **saddle thrombus** *Med.* = *saddle embolus* above; **saddle tramp** *N. Amer. slang*, a vagrant on horseback; **saddle vein** *Mining* = *saddle reef* above; **saddle wire,** (*b*) *Bookbinding*, a wire staple passed through the back fold of a single gathering; usu. *attrib.*; hence **saddle-wired** *a.*

1906 W. L. Sclater *Birds S. Afr.* IV. 43 (*heading*) Saddle-bill, or African Jabiru. *Ibid.* 44 The Saddle-bill is found all over tropical Africa. 1947 J. Stevenson-Hamilton *Wild Life S. Afr.* xxxiv. 289 The saddle bill or jabiru (*Ephippiorhynchus senegalensis*)... They are striking-looking birds; the forepart of the long bill crimson, the centre black, and the saddle or shield which comes just above the nostrils, bright yellow. 1973 *Times* 11 Dec. (Zaire Suppl.) p. vii/5, I was able to identify a saddle-bill stork. 1817 E. P. Fordham *Let.* 26 July in *Personal Narr. Travels* (1906) 98 My cloak and saddle-blanket, spread on the floor, form my couch. 1925 C. Jacobson *Life Story Jeff Davis* xxviii. 234 When I licked that gang at Little Rock during the last campaign, they went around on the streets with their faces looking as long as a saddle blanket. 1973 A. H. Whiteford *N. Amer. Indian Arts* 75 Twill weaving is done by the Navajo in making saddle blankets. 1946 Parmley & Adriani in *Southern Med. Jrnl.* XXXIX. 194/2 The term 'saddle block analgesia' is well chosen inasmuch as it is not only descriptive but is also free from the word spinal which sounds very, very dangerous to most obstetric patients. 1974 Passmore & Robson *Compan. Med. Stud.* III. xl. 17/2 Low spinal anaesthesia (saddle block) involves the injection of local anaesthetic into the C[erebro] S[pinal] F[luid] of the subarachnoid space. *Ibid.*, The incidence of serious complications is low and saddle block is popular in North America. 1974 Saddlebred [see *pleasure-horse* s.v. *Pleasure *sb.* 6]. 1976 *Horse & Hound* 10 Dec. 73/1 (Advt.), 2 beautiful colts to mature 15 hands 2 in by American saddlebred Goldmount Bourbon Genius. 1977 *Islander* (Victoria, B.C.) 18 Sept. 6/1 Lancer, an American saddlebred, who is nearly eight. 1949 G. Rounds *Rodeo* 49 (*heading*) Saddle Bronc Rodeo. 1956 *N.Y. Times Mag.* 23 Sept. 47/2 Rodeo people call them broncs, but never broncos if they are saddled. Otherwise, they're 'bareback horses'. Saddle broncs usually are larger. Neither are apt to be vicious. Most can be halter-led. *Ibid.*, A saddle bronc ride, which requires more skill, lasts ten [seconds]. 1973 *Houston Chron.* 14 Oct. (Suppl.) 2/3 Texas Prison Rodeo features bareback riding, saddle bronc riding, calf tussles, clowns and other events. 1976 *Columbus* (Montana) *News* 3 June 1/4 The rodeo picks up again at 1:00 p.m. Sunday. Events are saddle bronc, bull riding, steer wrestling, [etc.]. 1977 *New Yorker* 6 June 48/2 He rode dogies and then steers and saddle broncs. 1961 Webster, Saddle brown. 1975 *Cleveland* (Ohio) *Plain Dealer* 31 Mar. 24-D (Advt.), Perfectly matched Rocker-Recliner or Swivel Rocker in handsome, saddle-brown Masland Duran vinyl. 1941 J. Steinbeck *Sea of Cortez* xvi. 160, I removed the saddle to see whether he might not be saddle-burned. 1946 *Carleton Coll.* (Northfield, Mass.) *Bull.* Mar. 85 The Saddle Club, open to students proficient in horsemanship. 1962 A. Sampson *Anat. of Britain* xvi. 258 Guards officers..have their own club in Mayfair, their own polo club, cricket club, saddle club, flying club, shooting club. 1977 *Navy News* June 18/3 Services saddle clubs have been helped. 1854 'Logan' *Master's House* 260 That 'saddle-colored' nigger grinning at me..would be all the better for about 'forty-five', well laid on. 1900 Kipling *Land & Sea Tales* (1923) 39 The saddle-coloured sons of the soil looked down their noses. 1936 M. Mitchell *Gone with Wind* xxxiii. 553 A saddle-colored negro of middle age. 1937 *Ann. Surg.* CVI. 909 Incisions have been made very close to the aorta, in either one or the other iliac arteries, and successful removals of saddle emboli accomplished. 1974 J. D. Maynard in R. M. Kirk et al. *Surgery* xi. 236 Retrograde catheterisation to the bifurcation of the aorta will allow dislodgement of a saddle embolus. 1854 L. A. Meall *Moubray's Treat. Poultry* 128 The hackle and saddle feathers are straw colour. 1901 *Nature* 4 July 233/1 Manipulation of the tail-feathers..would not account for the likewise abnormal elongation of the saddle-feathers. 1946 Winter & Funk *Poultry Sci. & Pract.* (ed. 2) iii. 55 Castrated males grow longer neck, saddle, and tail feathers than do cockerels. 1946 K. Tennant *Lost Haven* (1947) xiv. 224 The smooth patch of grey rock that looked like a saddle gall. 1951 *Dict. Gardening* (R. Hort. Soc.) II. 917/1 (*caption*) Saddle graft. 1959 *New Biol.* XXX. 38 He.. made a saddle graft between the two plants then, when the graft had taken, he cut transversely across the graft-union so that the wound callus formed would be a mixture of tissues from the two plants. 1886 *Outing* (U.S.) VIII. 7/1, I..had with me the little forty-sixty Winchester saddle gun. 1949 *10 Story Western* May 21/1 They jerked the saddle guns from their scabbards. 1854 L. A. Meall *Moubray's Treat. Poultry* 85 The saddle hackle and back feathers. 1855 *Poultry Chron.* III. 44/1 Should the saddle-hackle feathers of the Silver Spangled cock be spangled, or perfectly white? 1951 W. H. Silk *Bantams & Miniature Fowl* iv. 26 Neck and saddle hackle are silvery-white as free from black striping as possible... Saddles and wing-bows are rich yellow or orange, shading to silvery-white in saddle-hackle. 1976 Saddle hackle [see sense 7 c above]. 1855 *Poultry Chron.* III. 209/2, I consider the really perfect hen-feathered cocks vastly superior in plumage to the long-feathered saddle-hackled. 1958 J. Hislop *From Start to Finish* viii. 67 The best way to [learn]..is to practise on a saddle-horse (a wooden stand, upon which saddles are cleaned). 1820 C. Mathews *Let.* 31 Aug. in A. Mathews *Mem. C. Mathews* (1839) III. vii. 149 If you have not got any in the grounds, a saddle-leaf tulip is beautiful. 1931 W. N. Clute *Common Names of Plants* 39 The tulip-tree (*Liriodendron tulipifera*) was called saddle-leaf because the young leafblades in the bud were bent back across the petiole in such a way as to retard the growth of the tip and make it appear as if cut square across. 1908 *Animal Managem.* (War Office) 32 Saddle marks are extremely common, in fact it is the exception to find an old troop horse without them. 1963 Bloodgood & Santini *Horseman's Dict.* 172 *Saddle mark*, hair left unclipped on a horse's back in the shape of a saddle; usual in clipping hunters. 1976 *Horse & Hound* 10 Dec. 67/2 (Advt.), Bay mare... Some saddle marks. 1930 J. Beames *Army without Banners* 12 He disdained the clumsy 'squaw notch', where one log sits simply in a shallow groove cut in the one below, and fitted them neatly into place with the 'saddle notch', a triangular ridge cut to fit closely into a deep V in the log above. 1974 *Islander* (Victoria, B.C.) 8 Sept. 4/2 After the logs were peeled David..saddle-notched each log and fitted them into place. 1976 *Amer. Speech* 1973 XLVIII. 166 In the South the saddle-notch was favored, while in the Mid-Atlantic and Midwestern areas V-notching was preferred. 1967 *Dict. Canadianisms* 651/2 *Saddle notch*, a saddle-notched joint. 1977 *New Yorker* 27 June 58/3 He and Lilly built a..cabin of unpeeled, saddle-notched logs. 1948 R. T. Wilcox *Mode in Footwear* 165 (*caption*) Saddle oxford of the period—black or brown and white. 1967 A. West in *Coast to Coast 1965–66* 212 Onto her feet she secured a pair of saddle oxfords that were too big by wrapping the laces around her ankles. 1922 G. N. Watson *Treat. Theory Bessel Functions* viii. 235 The points [u_0, v_0, $Rf(w_0)$] are saddle points, or passes, on the surface. 1946 H. & B. S. Jeffreys *Methods of Math. Physics* xvii. 472 Through any saddle-point it will be possible to draw at least two curves such that ϕ is constant along them. 1960 A. Rapoport *Fights, Games, & Debates* vii. 136 A saddle point is an entry in the game matrix which is the smallest in its row and the largest in its column. 1966 S. Beer *Decision & Control* xviii. 467 In the inadequate game-theoretic model, neither side can exploit information about the other, because the game has a fixed saddle point. 1973 *Listener* 21 June 826/2 We can conceive of a kind of space that has saddle-points in it, over which massive bodies slide in some directions more easily than in others. 1867 *Archaeol. Jrnl.* XXIV. 246 A 'saddle-quern', resembling that found at Ty Mawr, was sent to the museum of the Institute at the Hull meeting, 1867. 1938 *Proc. Prehist. Soc.* IV. 35 The true saddle-quern..was a two-handed implement allowing only a to-and-fro movement of the upper stone. 1978 A. & G. Ritchie *Anc. Monuments Orkney* 41 The original rubbing stones were found beside this massive saddle quern, together with a pile of crushed razor-shells. 1860 *Mining Surveyors' Rep.* (Mining Dept., Victoria) Aug. 216 The Wellington Reef..is what is termed by miners a saddle reef, or, in other words, a vein of quartz branching from the cap in two distinct underlies, viz. one to the east and the other to the west. 1906 J. Park *Text-bk. Mining Geol.* ii. 49 The gold-bearing veins at Cape Terawhiti, near Wellington, in New Zealand, are interesting examples of saddle-reefs which exhibit both an anticlinal and synclinal arrangement. 1975 E. Hillary *Nothing venture, Nothing Win* (1977) xviii. 351 Saddle reefs of quartz in the goldfields of Victoria, Australia, and on the west coast of New Zealand. 1852 *Lantern* (N.Y.) II. 158/1 Oyster House sages..acknowledge that for a consideration they will puff anything from Saddle Rock Oysters to Fancy soap. 1865 J. H. Browne *Four Years in Secessia* 279 The stewing of 'Saddle-Rocks' in a chafing dish, or the preparation of a lobster salad, was as far as I had ever advanced in the mysteries of the cuisine. 1881 E. Ingersoll *Oyster-Industry* 244 *Fancy Oysters.*—In New York, these are 'Saddle Rocks', 'Blue Points', etc. *Ibid.* 247 *Saddle Rock Oysters*, a trade name in New York for the largest and finest oysters. 1679 *Rec. Court of New Castle on Delaware* (1904) 361, 2 saddle Ruggs & 3 old Blancketts. 1931 A. U. Dilley *Oriental Rugs & Carpets* Pl. 20 (*caption*) Kerman Saddle Rug. 1898 H. S. Canfield *Maid of Frontier* 185 His horse came up to his ranch..with the gun still in the saddle scabbard. 1944 R. F. Adams *Western Words* 137/1 *Saddle scabbard*, a heavy saddle-leather case in which to carry a rifle or Winchester when riding. The gun fits in as far as the hammer, leaving the stock exposed. 1973 R. D. Symons *Where Wagon Led* vi. xviii. 280 He couldn't get at his rifle which was in the saddle scabbard. 1950 David & Browne *Geol. Commonwealth of Austral.* II. xxvii. 176 A few of the ore-bodies appear to be saddle-shaped. 1941 J. C. Furnass *How Amer. Lives* 272 You could paint an accurate oil portrait from those data alone, right down to the socks and saddle shoes. 1958 *Listener* 31 July 157/2 A young American boy and girl, dressed in teenage style of blue jeans, suede saddle shoes, and peach-coloured polo shirts. 1974 D. Ramsay *No Cause to Kill* i. 6 Saddle shoes. Brown and white... Of all things! Who wore saddle shoes nowadays? 1957 M. B. Picken *Fashion Dict.* 280/2 *Saddle shoulder sleeve*, sleeve with shoulder extended into neckline somewhat like raglan, but square-cut in 'saddle' effect. 1969 *Sears Catal.* Spring/Summer 20 Classic Cardigan... Saddle shoulders add a well tailored look. 1978 *Detroit Free Press* 2 Apr. (Detroit Suppl.) 21 (Advt.), Crew neck pullover with saddle shoulder. 1956 R. Braddon *Nancy Wake* xv. 178 It's just that damned bicycle. I'm so saddle-sore I could die. 1975 *Times* 8 Feb. 10/5 What if riding pales and saddlesore guests seek other diversions? 1932 G. M. Boumphrey *Story of Wheel* 42 The 'saddle-stone', which had a hollow face in which a smaller stone was rubbed backwards and forwards. 1933 *Ann. Surg.* XCVIII. 262 At about the point of bifurcation of the deep and superficial femoral is a constricted portion with a saddle thrombus which shows beginning organiza-

tion. 1937 *Ibid.* CVI. 908 The..patient was operated upon rather late, after the saddle thrombus developed, by the transabdominal route. 1942 Berrey & Van den Bark *Amer. Thes. Slang* § 913/10 *Saddle..tramp*, a cowboy who rides from ranch to ranch living on Western hospitality. 1962 E. Lucia *Klondike Kate* 7 Most of them [*sc.* prostitutes] led wretched lives,..attached to gamblers, card sharps..saddle tramps, gun-slingers and rogues. 1979 *Radio Times* 5–11 May 23/2 Kirk Douglas back on the range for King Vidor, in the one about the saddle tramp up against the barbed wire. 1935 Stočes & White *Structural Geol.* 293 Saddle veins are filled openings.. which were similarly formed in the arches and troughs of folded beds. 1977 A. Hallam *Planet Earth* 314/3 Saddle veins are lens-shaped, concave below and convex above. 1911 Webster, Saddle wire stitch. 1948 R. R. Karch *Graphic Arts Procedures* xii. 301 Saddle-Wire bound booklets are the simplest and cheapest in form... The cover and pages are held by two or more stitches on saddle-wire booklets, which allows them to lie flat and open. 1967 V. Strauss *Printing Industry* x. 659/1 Saddle wire stitching produces a completely flat-opening book. 1967 Karch & Buber *Offset Processes* xii. 492 Saddle-wired books lie flat when open, and may be folded upon themselves.

saddle, *v.* Add: **1. d.** *S. Afr. to saddle off* = Off-saddle, off-saddle *v.*

1835 J. W. D. Moodie *Ten Yrs. S. Afr.* I. 65 He.. asked us if we would 'saddle off' our horses.

e. *trans.* To enter (a horse that one has trained) in a race.

1928 *Daily Mail* 25 July 14/2 Scott will not saddle Lamintone for the Church House Handicap Plate (2.30) at Liverpool. 1947 *Sun* (Baltimore) 11 June 17/4 Palmer Sowers, of Washington, saddled two winners on the program and might have made it three had not Jockey J. Keenan lost a stirrup in the final drive of the second race. 1970 *Globe & Mail* (Toronto) 25 Sept. 32/3 Trainer Glen Magnusson..saddled three successive winners. 1976 *Southern Even. Echo* (Southampton) 13 Nov. 15/3 The Bishop Auckland trainer should start successfully by saddling Clever Prince to win the Threlkeld Handicap Chase.

6. c. To attach after the manner of a saddle.

1831 J. J. Audubon *Ornith. Biogr.* I. 303 The nests were fixed to a horizontal bough, but were not *saddled* upon it so deeply as those of the Wood Thrush are. 1881 *Amer. Naturalist* XV. 217 Our nest..was saddled to a horizontal limb after the fashion of our wood pewee. 1940 *Bull. U.S. Nat. Museum* No. 176. 321 It was about 30 feet from the ground, saddled on a horizontal branch of a maple over the trail.

saddleback, *sb.* and *a.* Add: **A.** *sb.* **4. b.** (Earlier and later examples.)

1770 G. Cartwright *Jrnl.* 2 Oct. (1792) I. 40 They returned with three shellbirds and a saddleback. 1932 J. Barbour *Forty-Eight Days Adrift* vi. 61 A 'saddle-back' coming towards us..was a good sign we were getting near land.

d. Substitute for def.: A New Zealand wattle-bird, *Creadion* (or *Philesturnus*) *carunculatus*; (later example).

1966 *Weekly News* (Auckland, N.Z.) 1 June 17/1 A pair of saddlebacks point like hunting dogs... They snap from one rigid position to another, head down, tail erect, wattles a blood crimson, the fragmented sunlight burnishing the chestnut saddle on their backs.

g. The brown and green larva of the moth *Sibine stimulea*, which has stinging spiny hairs and is found in southeastern North America.

1895 J. H. & A. B. Comstock *Man. Study of Insects* xviii. 225 The Saddle-back Caterpillar... Its most characteristic feature is a large green patch on the back. 1943 *Sun* (Baltimore) 9 Sept. 16/7 The Saddleback is of the family *Eucleidae*. 1954 [see *10[2]].

h. A parti-coloured black and white pig belonging to the breed so called. Also *saddle-back pig*.

1919 (*title*) The Wessex Saddleback Pig Society's herd book. 1927 *Daily Tel.* 6 Dec. 9/2 Mr. Douglas Vickers' herd of Wessex saddlebacks at Temple Dinsley, Hitchin. 1978 E. Downing *Keeping Pigs* iii. 25 The Large Black and the Saddleback are still hardy. 1981 *Times* 25 May 10/6 The Johnstone Cup..was won by the cider firm, Whiteways..with British saddleback pigs.

i. The African black-backed jackal, *Canis mesomelas*.

1947 J. Stevenson-Hamilton *Wild Life S. Afr.* ii. 23 The side-striped jackal..began to decrease... Its place had been taken by the black-backed jackal... It may be that it contracted and died from the same disease as affected the wild dog, and from which the saddle-back.. remained..immune. 1964 D. Varaday *Gara-Yaka* vii. 61 A pair of saddle-back jackals coming into view from the sands of an ant-bear hole.

j. The North American Arctic shrew, *Sorex arcticus*.

1948 A. L. Rand *Mammals Eastern Rockies* 51 The saddle-back shrew with its red-brown sides and tricolour pattern is the most beautiful of our shrews. 1966 R. L. Peterson *Mammals E. Canada* 38 *Sorex arcticus* Kerr—Arctic or Saddle-back Shrew. *Ibid.* 38/1 The young do not show the distinct dorsal band or 'saddle back'.

B. *adj.* **1.** (Examples of *saddleback chair.*)

1904 M. Beerbohm *Let.* 3 Jan. (1964) 155 There is nothing to do except to sit in the hall on a saddle-back chair. 1981 P. Vansittart *Death of Robin Hood* iii. iii. 142 The dulled crimson sofa and saddleback chairs.

2. Also *saddleback caterpillar, jackal, pig, shrew*: see sense A. 4 above.

saddle-backed, *a.* Add: **1.** Also of furniture.

1910 A. Bennett *Clayhanger* ii. xxi. 313 He would begin to establish himself in a saddle-backed, ear-flapped easy-chair.

b. Placed astride like a saddle.

1878 J. H. Beadle *Western Wilds* xxx. 487 Colorado is divided nearly down the center by the main chain of the Rocky Mountains—or, in miner's phrase, 'saddle-backed across the range'.

4. Also used of other animals with similar markings, as *saddle-backed shrew* = *Saddle-back *sb.* 4 j.

1927 H. H. T. Jackson *Taxon. Rev. Amer. Long-Tailed Shrews* 69 A specimen of the American saddle-backed shrew..was mentioned as early as 1772.

saddle-bag. Add: **1.** (Earlier and later examples.) Also a bag attached behind the saddle of a bicycle.

1773 H. Finlay *Jrnl.* (1867) 43 The rider had saddle bags quite full besides. 1780 W. Fleming in N. D. Mereness *Trav. Amer. Colonies* (1916) 651 Capt. Pawling ..had gone back for my saddlebags. 1901 Duncan & Scott *Allen & Woodson Counties, Kansas* 619 The doctor of 1858,..with saddle-bags like paniers to a pack mule, would make a strange comparison with the well-dressed and well-barbered M.D. of the present era. 1920 *Cycling* 12 Aug. p. xvii/3 (Advt.), Few pigskin saddle bags, 9 by 6 by 4, 17s 6d, post 9d. 1940 J. Betjeman *Old Lights for New Chancels* 49 Kant on the handle-bars, Marx in the saddlebag. 1965 D. Murphy *Full Tilt* p. xi, Apart from the normal accessories—saddle-bag, bell, lamp and pump —she carried only pannier-bag holders on either side of the back wheel. 1973 R. T. Way *Bicycle* 63/2 For normal day-to-day riding, a small saddlebag is all that is needed— it only has to take a cape, cap, tools, snack and camera.

2. Also *ellipt.*, a saddle-bag chair.

1919 C. Dane *Legend* 82 'There's your chair. Isn't that always your chair?'..But he did not take the saddle-bag near Anita's own seat.

sa·ddle-bag, *v.* *U.S.* Also (*rare*) **saddle-bags.** [f. the sb.] *intr.* To double round an obstruction.

1884 'Mark Twain' *Huck. Finn* xiii. 114 They lost their steering-oar..and saddle-baggsed on the wreck. 1898 *Derrick's Handbk. Petroleum* 32 A boat..laden with 1,500 bbls. of oil, 'saddle-bagged' on pier of Oil Creek bridge. Boat and contents a total loss. 1901 W. F. Fox in *6th Ann. Rep. N.Y. State Forest, Fish & Game Comm.* 254 Sometimes the long, floating mass [*sc.* a lumber raft] would swing in the wind and current so that it would 'saddle-bag' on the head of the bar below the dam. 1905 *Terms Forestry & Logging* (U.S. Dept. Agric. Bureau Forestry) 45 *Saddlebag*, as applied to a boom, to catch on an obstruction and double round it.

saddle seat. 1. The seat of a saddle.

1822 J. Fowler *Jrnl.* 28 Apr. (1898) 135 We then passed threw Some low Hills a little East of South Seven miles to the River and Crossing over found the Watter up to the Saddle Sceats. 1850 *Rep. Comm. Patents* 1849 (U.S.) I. 263 The bent tension springs CC, for supporting the saddle seat. 1963 Bloodgood & Santini *Horseman's Dict.* 170 *Waist*, saddle-seat at its narrowest point.

2. See Saddle *sb.* 10.

3. a. A seat made by the crossed hands of two persons.

1913 *Amer. Jrnl. Insanity* Jan. 575 She..suggested that a couple of gentlemen..should clasp their hands so as to form a 'saddle seat' for her, and thus she rode from the platform to the car.

b. A chair seat resembling a saddle (see quot. 1960); also, a chair with such a seat.

1934 in Webster. 1952 J. Gloag *Short Dict. Furnit.* 406 *Saddle Seat*, a solid wooden seat with two shallow depressions separated by a slight central ridge, suggesting the shape of a saddle. It is used on many types of Windsor chair. 1960 H. Hayward *Antique Coll.* 246/1 Some late 18th cent. chairs have a gently curving seat sloping down in the centre which is described colloquially as a saddle seat. A slightly different form is found on Windsor chairs where the centre of the wooden seat is shaped to resemble a saddle. 1976 *County Life* 27 May (Suppl.) 486/1 Late 18th century wheel-back armchairs with saddle seats and X-frame legs.

saddle stitch, *sb.* **a.** *Bookbinding.* A stitch of thread or a wire staple passed through the back fold of a booklet while it lies open on a saddle-shaped support.

1934 Webster, Saddle stitch. 1956 A. Williamson *Methods Bk. Design* xix. 331 The whole is fastened together by a saddle-stitch which passes through the spine fold. 1960 G. A. Glaister *Gloss. Bk.* 365/1 *Saddle stitch*, a method of stitching brochures or pamphlets by placing them open astride a saddle-shaped support and stitching through the back.

b. *Needlework.* (See quot. 1964.)

1962 W. Butler *Dolls' Dressmaking* 92 This trim shirt has saddle-stitch outlining the cuffs, collar and front opening. 1964 *McCall's Sewing* ii. 31/2 *Saddle stitch*, a decorative top-stitch made by taking longer stitches on top and shorter ones underneath. 1974 H. McCloy *Sleepwalker* viii. 157 Glove seams were mended, too, with a special stitch like the saddle stitch.

Hence **sa·ddle-stitch** *v. trans.*; so **sa·ddle-stitched** *a.*; **sa·ddle-stitching** *vbl. sb.* (see also *Saddle *sb.* 9). Also **sa·ddle-stitcher,** a device that performs saddle-stitching.

1923 H. A. Maddox *Dict. Stationery* 69 *Saddle Stitching*, the method of wire stitching pamphlets and single section books through the centre of the fold. 1933 J. Mason in W. Atkins *Art & Pract. Printing* V. viii. 193 Thread-stitching machines are widely used for pamphlets, exercise books, and note-books. They will saddle-stitch a single section. 1947 C. Talbot *Compl. Bk. Sewing* xxi. 144/2 Saddle-stitching is a very smart..finish for tailored clothes. It can be done in self-colour or definite contrasts. 1948 *Words into Type* 546 *Saddle stitched*, stitched through the back, the thread, silk, or wire showing on the back and in the middle fold. 1958 *Times* 6 Oct. 13/1 The next figure wears a coat designed in France and made in a supple taupe-coloured suède with saddle-stitching along the seams in a lighter shade. 1961 *Harper's Bazaar* Feb. 75 A short-sleeved white shirt, saddle-stitched..round the collar. 1964 *McCall's Sewing* xiii. 239/2 *Saddle stitching.* Take short stitches through the garment and facing an even distance from the edge. The stitches on the top side are longer than the stitches on the underside. 1967 V. Strauss *Printing Industry* x. 659/1 The cover of saddle-stitched books is usually, but not necessarily, of paper. 1973 *Country Life* 11 Oct. 1089 The suit..in 100% wool, with deep centre vent and impeccable saddle stitched lining. 1974 *McGraw-Hill Yearbk. Sci. & Technol.* 121/2 The most popular methods of binding soft-cover books are saddle stitching, side stitching, and perfect or patent binding. 1975 M. Banister *Bookbinding as Handcraft* xiii. 114/1 Like the Swingline Saddle Stitcher, this one also clinches its own staples. 1979 *Jrnl. R. Soc. Arts* July 485/2 The work involved.. collating and folding the sheets and the covers, and saddle-stitching the copies with stainless steel staples.

saddling, *vbl. sb.* Add: **2.** *saddling enclosure*; *saddling paddock*, also (*Austral. slang*) a nickname for a bar in the Theatre Royal, Melbourne, frequented by prostitutes in the nineteenth century; also, a similar bar elsewhere; hence, a known place of rendezvous; (earlier and later examples).

1876 *Argus* (Melbourne) 1 July 4/4 The stranger sees that the women, possibly picking up a male companion, all enter the apartment which was previously closed, and which is now guarded by swing doors. Curiosity will doubtless prompt him to enter, and he will find himself in the far-famed 'saddling paddock' of the Royal. *Ibid.* 4/5 The existence of the 'saddling paddocks' is a scandal established with a forethought for the courtesan's benefit which is an eternal disgrace to their promoters. 1909 J. R. Ware *Passing Eng.* 213/1 *Saddling-paddock* (*Australian*), place of amusement or rather place of assignation. 1958 G. Casey *Snowball* iii. 29 The ribald, popular name of the enclosure round the Government Dam was 'the saddling paddock'. 1969 *Sun* (Melbourne) 12 July (Turf Guide) 4/2 The AJC new grandstand in the saddling paddock at Randwick will be open on August 4. 1969 *Sun-Herald* (Sydney) 13 July 33/1 Mr Swales told us he could not get through to the stewards on the phone and gave us permission to return to the saddling enclosure to put our case to the stewards.

Sadean (sā·diăn, sēi-), *a.* Also **Sadeian, Sadian.** [f. the name of the Count de *Sade* (see Sadism) + -an.] Of, pertaining to, or characteristic of de Sade.

1960 *20th Cent.* Mar. 206 The Sadian dream of the unleashed subconscious. 1963 N. Gear *Divine Demon* 5 There were also many fantastic accounts of his life which obscured the facts in a mist of Sadeian mythology. 1978 R. Hayman *De Sade* xiv. 228 Baudelaire was the first poet to express modern alienation, but his negativity is Sadean.

sadful, *a.* For † *Obs. rare*[−1] read *rare* and add later example.

1884 'Mark Twain' *Huck. Finn* xvii. 158 She could write about anything..just so it was sadful.

sadhana (sā·danā). *Indian Philos.* Also **sadhan** and with capital initial. [Skr. *sādhanā* dedication to an aim; *sādhana* means to the goal, etc., f. *sādh* to succeed, attain.] (See quots.)

1898 K. L. Sarkar *Hindu System of Relig. Sci. & Art* vii. 137 Worship and prayer (Sadhana and Upashana) are in the main of two kinds. *Ibid.* 138 Some leaders of those sects in which *sakama sadhana* (selfish prayer) prevails, are more or less driven to give preference to that blank conclusion of Rationalism called *Nirvan Mukti* (merger in the Infinite One.) 1909 S. Tattavabhûshan *Philos. of Brâhmaism* i. 2 Brâhmaism..presents itself to us in three aspects,—(1) as a creed, (2) as a system of *Sâdhan* or spiritual culture, and (3) as a scheme of social reform. 1921 C. Eliot *Hinduism & Buddhism* II. xxxii. 282 *Siddhi* is produced by Sâdhana, or that method of training the physical and psychic faculties which realizes their potentialities... It is part of Sâdhana to arouse..energy and make it mount from the lower to the higher centres. 1932 N. K. Brahma *Philos. of Hindu Sādhanā* ii. 13 The term '*Sādhanā*' is a current Bengali expression... Its literal meaning is 'that by which something is performed' or more precisely 'means to an end'. In the sphere of religion, it is always used to indicate the essential preliminary discipline that leads to the attainment of the spiritual experience which is regarded as the *summum bonum*... *Sādhanā* includes all the religious practices and ceremonies that are helpful to the realisation of spiritual experience. 1941 K. G. Mashruwala *Practical Non-Violence* 48 Violence is born of a narrow conception of 'I' and 'Mine'. The search and discipline of non-violence (the *Sadhana of Ahimsa*) consists in a constant and progressive widening of that conception. 1958 V. Raghavan in W. T. de Bary et al. *Sources of Indian Tradition* xiii. 303 As Indian philosophy aims at experiencing the Truth, all the schools include disciplines (*Sādhanas*), practical means for the attainment of the spiritual goal. 1968 *Indian Music Jrnl.* V. 33 After seven years of Sâdhanâ..[he] was appointed the principal teacher of Lahore branch. 1972 P. Holroyde *Indian Music* vii. 252 The artistic search through feeling out the sādhana of the musical art is, as a result, inward and contemplative.

sadhu (sā·du). Also **saddhu** and with capital initial. [Skr. *sādhú* good, pious, holy man, saint, f. as prec.] In India, a holy man or sage.

1845 *Encycl. Metrop.* XXI. 672/2 When their [*sc.* spiritual guides'] sanctity is such, that they are believed to have the power of securing or withholding divine blessings, they are styled Sád'hú (saint). 1901 Kipling *Kim* xi. 291 He switched out his..turban-cloth and.. rolled it over and under about his loins into the intricate devices of a Saddhu's cincture. 1920 *Glasgow Herald* 25 May 9 He put on the saffron robe of the Sadhu. 1924 *Blackw. Mag.* Oct. 481/2 The temple's guest-chamber, kept for visiting *sadhus* and *sanyasis*. 1955 *Times* 10 Aug. 8/3 A procession of sadhus or 'holy men' is said to have left Pathankot on the Kashmir frontier. 1958 L. Durrell *Balthazar* x. 208 The old man was a judge in India... He dresses like a *saddhu*... You English are eccentrics. 1968 *Indian Music Jrnl.* V. 32 The philosophical discussions which his father had with the Scholars and sādhu-s who came to see him. 1978 *Times Lit. Suppl.* 3 Feb. 145/2 The Hindu Sadhus were pythons of the psychic world, whose slumbrous coils contained, and at the same time hid, the force of a battering ram.

Hence **sa·dhu,ism,** the principles or practices of a sadhu; **sa·dhuship** [-ship 3 b], in *his sadhuship*, a humorous title for a sadhu.

1903 J. C. Oman *Mystics, Ascetics & Saints of India* xii. 278 (*heading*) The future of Sadhuism. 1914 W. G. Lawrence *Let.* 2 Apr. in *Home Lett. T. E. Lawrence* (1954) 519 It's wonderful to see and hear the way the students listen to him owing to their respect for his old sadhuship. 1979 F. Olbrich *Sweet & Deadly* vii. 74 Holiness is a game to them [*sc.* hippies], a cheap thrill—instant Sadhuism.

Sadian, var. *Sadean *a.*

Sadian, Sa'dian, varr. *Saadian *a.*

sadic (sā·dik, sēi-), *a.* [f. the name of the Count de *Sade* (see Sadism) + -ic.] = *Sadean, Sadistic *adjs.*

1919 R. Fry *Let.* 5 Jan. (1972) II. 443 A pretty bad melodrama..a Sadic German Jew who starts the worship of Astarte. 1926 T. E. Lawrence *Seven Pillars* i. i. 5 What now looks wanton or sadic seemed in the field inevitable. 1928 *Daily Tel.* 6 Nov. 10/6 The oily warder.. finds a Sadic pleasure in flogging the life out of his helpless charges. 1976 *Times Lit. Suppl.* 17 Sept. 1147/4 There's a new genre coming into focus..which may illuminate thanatos, celebrate the death wish or merely exploit a sadic lechery.

Sadie Hawkins (sēi·di hǭ·kinz). *U.S.* The name of a character in the cartoon strip *Li'l Abner* by 'Al Capp' (Alfred Gerald Caplin, 1909–79), used *attrib.* to designate a day early in November on which, according to a 'tradition' in the cartoon series, women can propose marriage to men, demand dates with them, etc., or to designate events taking place on that day.

1939 *Daily Mirror* (N.Y.) 4 Nov. 21 Sadie Hawkins Day!! 1940 *Ibid.* 2 Nov. 12 Oh, happy Sadie Hawkins Day! Befo' the sun goes down Ah'll catch me a man, daid or alive!! *Ibid.* 8 Nov. 39 The first 'gun'! The Sadie Hawkins Day Race is on. 1941 *Sun* (Baltimore) 3 Nov. 15/2 Rulman also promised a free marriage to women who catch a man in the Sadie Hawkins footrace. 1942 *Ibid.* 27 Oct. 16 Your chances are four times as good of marrying up wid him on Sadie Hawkins Day as on Leap Year Day—Sadie Hawkins Day comes once *every year*. 1952 *Ibid.* 28 Feb. 32/2 The dean..was the only man to get away in the Sadie Hawkins Day race this afternoon. 1977 *Guardian Weekly* 20 Nov. 19/1 Much of Li'l Abner has been absorbed into American folklore—Sadie Hawkins Day, when sexual roles are reversed and girls chase and catch their men.

sad-iron. (Earlier and later examples.)

1761 *Newport* (Rhode Island) *Mercury* 3 Nov. 4/3 To be sold by Naphtael Hart, jun... sad Irons, Telescopes, [etc.]. 1815 *Niles' Weekly Reg.* IX. 94/2 Sad irons [were manufactured]. 1936 M. Mitchell *Gone with Wind* i. v. 84 Hands like sadirons when it comes to reins. 1964 F. O'Rourke *Mule for Marquesa* 99 Washday smell,.. don't forget to damp and starch, spit on the sadiron.

sadism. Add: Now usu. with pronunc. (sēi·diz'm). (Later examples.) Now understood as cruelty that evidences a subconscious craving and is apparently satisfied, sexually or otherwise, by the infliction of pain on another by means of aggressive or destructive behaviour or the assertion of power over that person; also *loosely*, deliberate or excessive cruelty morbidly enjoyed.

1924 J. Riviere et al. tr. *Freud's Coll. Papers* II. xxii. 261 We should not be astonished to hear that under certain conditions the sadism or destruction instinct which has been directed outwards can be introjected. 1937 H. G. Wells *Brynhild* vii. 116 He..with an expression of impish sadism..prodded his sceptre into young Bates. 1943 H. Read *Politics of Unpolitical* ii. 18 Sadism is the unconscious impulse to acquire unrestricted power over another person, and to test the fullness of this power by destroying that other person. 1952 *Times Lit. Suppl.* 11 Jan. 28 Feeling [in France] had been bruised by the war [of 1914–18], but it had not been forced (as happened after the Second World War) to find..an outlet in sadism or violence. 1965 H. Deutsch *Neuroses & Character Types* ix. 132 Such observations led Freud to assume that it is the same aggression whether it is attached to the libidinal impulses in the form of sadism or whether..it is incorporated in the superego and made the expression of its severity. 1974 I. Bieber in S. Arieti *Amer. Handbk. Psychiatry* III. xv. 318/2 In my view sadism is a maladaptive response to threat; it is a paranoid constellation in which the victim is a personified representative of a variety of irrationally perceived threats. 1979 M. Glasser in I. Rosen *Sexual Deviation* x. 281 The distinction I am making between aggression and sadism may be illustrated by some simple examples.

sadist. Add: Now usu. with pronunc. (sēi·dist). (Examples.) Also, more generally, someone who derives satisfaction from inflicting pain or asserting his or her power over others. Also as *adj.*

1919 H. Walpole *Secret City* i. x. 68 There was something almost sadist..in the old gentleman's observation of Markovitch's labours. 1919 M. K. Bradby *Psycho-Anal.* x. 133 The need for a greater outlet of emotion and energy in daily life where the sadist has capacity for ruling others by the strength..of his personality. 1920 *Glasgow Herald* 9 Sept. 6 The fatal policy which has given a handful of political cranks, aided by hired mercenaries and sadist abnormals, the right to stand for Russia. 1934 H. G. Wells *Exper. Autobiogr.* II. ix. 760 Those mucky little sadists, Stalky and Co. 1954 B. Karpman *Sexual Offender* xix. 355 Every sadist possesses certain elements of masochism. 1962 C. Allen *Textbk. Psychosexual Disorders* vii. 105 Those who retain a sentimental view of the blue-eyed innocence of the child naturally find it difficult to imagine..that the ruthless ferocity of the sadist originates at that time. 1974 J. Bancroft *Deviant Sexual Behaviour* v. 120 Some interesting studies of suppression of fantasy have been carried out. An example is a sadist treated by Marks.

sadistic, *a.* Add: Now usu. with pronunc. (sădi·stik). (Earlier and later examples.) Also, more generally, of or characteristic of a sadist; also *Comb.*, as *sadistic-anal* adj., relating to sadism that is typical of the anal stage of development; *sadistic-masochistic* adj. = *Sado-masochistic *a.*

1892 C. G. Chaddock tr. *Krafft-Ebing's Psychopathia Sexualis* iii. 170 The pain of tight lacing, experienced by himself or induced in women, is a delight to him,—sadistic-masochistic element. 1915 C. R. Payne tr. *Pfister's Psychoanal. Method.* 78 The sadistic-masochistic instinctive tendency. 1922 J. Riviere tr. *Freud's Introd. Lect. Psycho-Anal.* 289 Regression of..the Libido to the antecedent stage of the sadistic-anal organization. 1929 B. Russell *Marriage & Morals* 98 He drowns his dissatisfaction..by the sadistic pleasure to be derived from watching prize-fights or persecuting radicals. 1936 H. G. Wells *Anatomy of Frustration* xx. 255 He hated people who nursed 'wrongs'. The 'wrongs' of Ireland—of India—of women, roused an almost sadistic impatience in him. 1946 H. Pearson *Life O. Wilde* xvi. 319 Like all people who believe in punishment, he [*sc.* the governor of Reading Gaol] was vindictive and sadistic by nature, modelling himself on the God of his fathers as depicted in the Old Testament. 1952 *Times Lit. Suppl.* 15 Feb. 124/3 The vicarious sadistic lust for power of a disappointed man. 1954 B. Karpman *Sexual Offender* ix. 101 He admits drawing pictures of a sadistic nature; the torture of females, perversion of corpses. 1973 'E. McBain' *Let's hear It* xiv. 209 Teddy normally enjoyed films, except when she was submitted to the excesses of a sadistic *nouvelle vague* camera. 1977 A. Sheridan tr. *Lacan's Écrits* vii. 270 Regression is no more dependent on the need in demand than sadistic desire is explained by anal demand.

sadistically (sădi·stikăli), *adv.* [f. prec.: see -ically.] In a sadistic manner; cruelly.

1922 G. B. Shaw in S. & B. Webb *Eng. Prisons under Local Govt.* p. ix, When such people are..sadistically excited by reports of the White Slave traffic, they clamor to have sentences of two years' hard labor supplemented by a flogging. 1936 F. M. Ford *Let.* 6 Sept. (1965) 261, I believe that publishers should be as sadistically punished as possible. 1951 M. McLuhan *Mech. Bride* (1967) 16/2 The rest of the program fits this pattern by allowing Charlie to ride sadistically over a number of carefully selected victims. 1963 *Times* 12 Jan. 4/1 The east wind continued to blow sadistically at Littlestone yesterday.

saditty (sæ·diti), *a.* *U.S. Blacks.* Also **seditty.** [Orig. unknown.] Affecting an air of superiority.

1967 *Jet* 20 July 43 Eartha..is considered 'seditty' by many Negroes. 1971 C. Mitchell-Kernan in T. Kochman *Rappin' & Stylin' Out* (1972) 318 That's all I hear lately—soul food, soul food. If you say you don't eat it you get accused of being saditty. 1973 *Black World* Aug. 61/2 Them big man-eatin' dogs them saddity niggers had roun' the house.

Sadler (sæ·dləɪ). The name of John *Sadler* (fl. 1871–80), British botanist, used *attrib.* or in the possessive in **Sadler('s) oak** to designate *Quercus sadleriana*, an evergreen shrub with serrate leaves, found in parts of western North America and named in his honour by Robert Brown in 1871 (*Ann. Mag. Nat. Hist.* 4th Ser. VII. 249).

1908 N. L. Britton *N. Amer. Trees* 338 Sadler's oak.. is an interesting shrub of the high mountains of northwestern California and adjacent Oregon, with sharply serrate leaves. 1939 L. Rowntree *Flowering Shrubs Calif.* ix. 149 When the endemic..Sadler's Oak or Deer Oak, has room to do as it pleases,..it takes on the shape of an inverted pyramid. 1951 H. E. McMinn *Illustr. Man. Calif. Shrubs* 83 Deer Oak. Sadler Oak. An evergreen shrub, 2 to 8 feet high, with many slender flexible stems from the base.

sadly, *adv.* Add: **9. b.** As a sentence adverb: regrettably, unfortunately.

1973 *Times* 16 Feb. 19/4 The Headmaster of Winchester College asks: 'Is there any other ancient cathedral city in Western Europe with so much fast, heavy, long-distance traffic planned to run so near?' Sadly, the answer is 'Yes, York'. 1974 *Times Lit. Suppl.* 7 June 607/3 No one would dispute the pointed courage of Camus's early journalistic campaigning. Sadly, though, this does not make him a great artist. 1978 *Lancashire Life* July 44/3 Sadly, his collection was sold and dispersed throughout the world after his death.

sado- (sēi·do), *Psychol.*, comb. form of Sadism or Sadistic *a.* (in Dict. and Suppl.). Cf. also *Sado-masochism.

1935 [see *sado-masochism]. 1954 B. Karpman *Sexual Offender* ix. 131 Alcohol is the illegitimate satiation of appetite which she fights in obsessionalism based on sado-necrophilia. 1964 *Observer* 30 Aug. 28 Client prints paperbacks for all tastes from (I think) the nasty sado-snobbism of poor Fleming to Tolstoy. 1970 *Guardian Weekly* 11 Apr. 19 A kind of all-senses collage assembled from bits of girlie photos, tropical stills, and mock-ups of sado-erotic temple carvings. 1976 *New Yorker* 26 Apr. 121/1 Given the sado-erotic content of the film... one tends to make Grace a woman. 1980 R. Ludlum *Bourne Identity* xiii. 195 The sado-romantic myth turns into a brilliant, blood-soaked monster who brokers assassination.

sado-maso (sēidomæ·so), *a.* (*sb.*) Slang (chiefly *U.S.*) abbrev. of *Sado-masochist; *Sado-masochistic *a.*

1970 *Time* 23 Feb. 54/1 A gang of seminude galley slavettes..bend to the oar under a whip cracked by everyone's favorite sado-maso slave queen, Raquel Welch. 1973 *Listener* 22 Feb. 237/1 He passed..the spill-out of the sado-maso bar... At present, the sado-masos are in the ascendant. 1978 M. Puzo *Fools Die* xxiv. 277 No fantasies could be spun around them unless you were into sado-maso stuff.

sado-masochism (sēi·domæ·sŏkiz'm). *Psychol.* Also **sadomasochism.** [f. *Sado- + Masochism.] The co-existence of Sadism and Masochism in one individual; the need both to inflict and to suffer pain or to assert power over another and to be submissive combined as one psychic condition evidenced in sexual relationships (freq. in a fantasied manner) or socially, as an outlet for aggressive or destructive impulses. Also *transf.* and *fig.*

1935 L. Brink tr. *Stekel's Sadism & Masochism* I. p. v, I do not claim that I have solved the perplexing problem of sadomasochism. 1937 M. Hirschfeld *Sexual Anomalies* xvi. 302 Thus it is quite correct to speak of sado-masochism and, quite logically, many sadists are, simultaneously, also masochists. 1959 *Listener* 16 Apr. 683/3 The sado-masochism of the Christian ascetic tradition. 1963 A. Heron *Towards Quaker View of Sex* 67 Clinical instances of sado-masochism are not sufficiently numerous to constitute a threat to society. 1975 T. Allbeury *Special Collection* xvii. 114 I've been covering a vice-ring..in Mayfair... It specializes in sado-masochism..every thing from thumb-screws to a crucifix. 1977 *Early Music* July 415/3 The *Art of Fugue* is invariably presented in 'complete' performances which strike one rather as exercises in musical sado-masochism.

Hence **sa:do-ma·sochist,** one afflicted with the condition of sado-masochism; also *attrib.* or as adj.; **sa:do-masochi·stic** *a.*

1935 L. Brink tr. *Stekel's Sadism & Masochism* I. p. v, The literature concerning sadomasochistic disorders is extraordinarily abundant. *Ibid.* iv. 60 All sadomasochists are affect-hungry individuals. 1942 *Observer* 15 Nov. 3/6 Sado-masochistic fusions of instinct are an all-important factor in the unconscious development of character. 1951 M. McLuhan *Mech. Bride* (1967) 10/1 This sado-masochist mechanism of punch and get punched will be found everywhere. 1963 *Jrnl. Amer. Psychoanal. Assoc.* XII. 306 Abraham stressed the importance of the sadomasochistic elements in his patient. 1977 *Gay News* 24 Mar. 3/2 Study and encounter groups for sadomasochists, transvestites and Jewish homosexuals are the latest projects of Pastor Douce. 1980 *Times Lit. Suppl.* 25 Apr. 459/4 His [*sc.* Somerset Maugham's] relationship with Haxton, the only person with whom he established intimacy, was sado-masochistic.

sad sack (sæ·d sæk). *slang* (chiefly *U.S.*). [The name of a cartoon character invented by G. Baker, U.S. cartoonist.] A stupid and blundering member of the armed services; an inept, ineffectual, and unfortunate person; a social or occupational misfit. Also *transf.* and *attrib.*

[1942 *Yank* 17 June 7 (*caption*) The Sad Sack.] 1943 *Sun* (Baltimore) 28 Dec. 14/6 A forlorn look, a G.I. haircut, an oversized fatigue uniform and all the paraphernalia that goes with them branded me as a typical 'sad sack'. **1951** M. McLuhan *Mech. Bride* (1967) 68/2 Model mother saddled with a sad sack and a dope. **1953** *Word Study* May 5/1 Everyone knows of the sensitive misfit, the 'sad sack' who suffers a good deal of spiritual depression, the result of an unfortunate maladjustment to service routine. **1967** *New Yorker* 15 Apr. 148/3 Mr. Goldman's movie sweeps up a dustpanful of young Village sad sacks and patronizes them. **1971** J. Gray *Red Lights on Prairies* iii. 58 A sad-sack of a shack town on Pile of Bones Creek. **1973** *Observer* (Colour Suppl.) 15 July 21/4 On the whole the pre-1914 spinster had been something of a sad sack. **1974** T. P. Whitney tr. *Solzhenitsyn's Gulag Archipel.* I. I. v. 222 These sad-sack spies, with the milk hardly dry on their lips. **1978** *Listener* 31 Aug. 286/4 The sad sack of a hero, who speaks in the first person, is called Lewis Redfern.

‖ **sadza** (sæ·dzä). [Native name.] In southern and eastern Africa, a porridge made of ground maize.

1950 *Cape Times Week-end Mag.* 3 June 2/3 Manaas had gorged himself with *sadza* and his little stomach was distended. **1965** *Observer* 7 Nov. 2/3 Each family owns its few acres of land from which it produces its main diet of maize (made into a porridge called sadza) and pumpkin. **1975** M. Hartmann *Game for Vultures* vi. 79 Marunga pecked at the greasy stew and dry sadza. **1979** P. Niesewand *Member of Club* xiii. 88 The sadza—thick, starchy maize meal porridge—bubbled in tins.

‖ **saeta** (sa,e·tä). Also **saetta**. [Sp., lit. = arrow.] An unaccompanied Andalusian folk-song, sung during religious processions.

1923 *Chambers's Jrnl.* Mar. 213/1 Somewhere in the crowd a woman is singing a *saeta*, sad and undulating, like no other music on earth. **1939** Spender & Gili tr. *Lorca's Poems* 19 Among troubled *saetas* And stars of crystal. **1966** *New Statesman* 26 Aug. 297/1 Clusters of microtones which resemble nothing so much as the ululations of the *saeta* singers in the Easter Day procession in Seville. **1977** P. Somerville-Large *Eagles near Carcase* vi. 123 He hummed a high nasal tune which I recognized as a *saetta* I had last heard sung to a Seville Madonna during Holy Week.

saeter, setter (sēi·təɹ, se·təɹ). Also saater, sæter, saether, sater, seater, seter. [ad. ON. *sætr* mountain pasture; cf. Norw. *sæter, seter*; Sw. *säter*. In sense 2 a directly from Norw.

In sense 1, the word in some examples may represent ON. *setr* a homestead, a residence (see esp. quot. 1931). The two are common formative elements in placenames of the Northern Isles, and cannot always be distinguished (see J. Jakobsen *Etymol. Dict. Norn Lang. in Shetland* (1932), s.v. *seter*).]

1. *Shetland* and *Orkney*. A meadow associated with a dwelling; a summer pasture in the out-field.

1576 in D. Balfour *Oppressions 16th Cent. in Orkney & Zetland* (1859) 72 The said Magnus complenis upon the said Laird, that quhair he had ane steding, callit Sater, lying in Brassay, of four merk and ane half land..; nevertheless, quhen he had gottin bot ane ȝeiris crope thairoff, he put him furth of the same. *c* **1772** in A. C. O'Dell *Historical Geogr. Shetland Islands* (1939) II. xi. 239 Feued property and udal comprehend the lands of Shetland of all denominations Setter-lands excepted. **1795** *Statistical Acct. Scotl.* XIV. 321 As to our meadows, they are always called *Seaters*. Though I am little acquainted with the Norwegian language, I understand a *Seater* to be a place for maintaining milch cows; and these *Seaters* are to this moment properly adapted for it. **1822** S. Hibbert *Descr. Shetland Isles* 427 In the ancient Shetland language, the green pasturage attached to a dwelling was named a Setter or Seater. **1931** *Proc. Orkney Antiquarian Soc.* IX. 27/2 Just beyond the Wideford Burn..lies the three-farthing land, skatland, of Grymesetter. Next adjacent lies the 'quoyland' of Grymesquoy... Both names point to an original farmer *Grimr*. He apparently settled there on a 'setter' just before skat was imposed on the Orkney lands. **1939** A. C. O'Dell *Historical Geogr. Shetland Islands* II. xi. 246 The 'Setter Lands', or areas settled since Norwegian times, as revealed by a MSS Scatt Rental of 1824 have been mapped, and the distribution reveals mainly an intensification in the Norwegian settlement [over that of the Merk Lands]. **1952** H. Marwick *Orkney Farm-Names* III. 229 In Orkney..there is no evidence of sæters, and accordingly in the present work no derivation [of farm-names] from *sætr* is suggested.

2. a. In Scandinavia, a mountain pasture where cattle remain during the summer months. Also *attrib.*

1799 Malthus *Diary* 9 July (1966) 132 His cows are now gone to pasture on the mountains—to Saaters, as they seem to call it. **1841** H. Martineau *Feats on Fiord* vi. 161 The mountain pasture belonging to a farm is called its Seater. **1882** Lees & Clutterbuck *Three in Norway* 56 This sæter is in a most beautiful situation, perched on a little flat bit of ground on the mountain side. **1924** *Contemp. Rev.* Feb. 236 Part of a herd of sixty or seventy ..had wandered down from the *fjeld* into the *saether*. **1940** J. Buchan *Memory Hold-the-Door* viii. 191, I do not mean the Swiss alp or the Norwegian saeter pasture, for these are on too large a scale. **1968** G. Jones *Hist. Vikings* II. ii. 82 Increasingly the husbandman came to have his own upland grazing, his seter (*seter*, Swedish *säter*). Sometimes the seter was of a permanent nature.

b. A mountain dairy or farm on such a pasture.

1923 G. F. Barbour *Life Alex. Whyte* xxii. 451 He and Dr. Sutherland Black..drove seventy miles up the Saetersdal..picnicked for several days in a fishing 'saeter'. **1926** *Public Opinion* 25 June 585/2 Mountain farms were being turned into saeters. **1931** *Hardy's Anglers' Guide* 42 The angler taking up his quarters at a small farm or 'saeter'. **1955** M. E. B. Banks *Commando Climber* vi. 106 The local farmer and his wife in a neighbouring *saeter*.. always moved about their wooded farm on skis.

saetta, *var.* *Saeta.

‖ **saeva indignatio** (səi·va indignā·tio). [L.] 'Savage indignation', an intense feeling of contemptuous anger at human folly. (Orig. and in later allusive use with reference to the epitaph of Swift: see quot. *a* 1745.)

[*a* **1745** Swift *Wks.* (1841) I. p. lxxi/1 (*epitaph*) Hic depositum est corpus Jonathan. Swift... Ubi saeva indignatio Ulterius cor lacerare nequit.] **1853** Thackeray *Eng. Humourists of Eighteenth Cent.* i. 32 The 'sæva indignatio' of which he [sc. Swift] spoke as lacerating his heart..breaks out from him in a thousand pages of his writing, and tears and rends him. **1900** F. M. Ford *Let.* Oct. (1965) 12 You haven't enough contempt, enough of the *saeva indignatio*. **1928** W. B. Yeats in *Exile* Spring 5 Swift beating on his breast in sibylline frenzy blind Because the heart in his blood-sodden breast had dragged him down into mankind.. *Saeva Indignatio* and the labourer's hire. **1957** R. Speaight *Life H. Belloc* xxi. 529 The furniture of home itself, the laughter and the love of friends—must he leave them, too? Yes, he exclaimed, with a *saeva indignatio* worthy of his master Swift, he must. **1969** *Punch* 1 Jan. 34/1 There was Solzhenitsyn's *The First Circle*..which..fell short of greatness because it was too docile. It lacked *saeva indignatio*. **1972** *Eng. Stud.* LIII. 280 It lacks the poised humour which saved Aluko's earlier characters from becoming mere sitting ducks for his *saeva indignatio*.

Safaitic (safāi·tik), *a.* Also **Safahitic**. [f. the Arab. place-name *Safa* in Syria, SE of Damascus + -ITIC (see -ITE[1]).] Of or pertaining to an ancient Semitic language known only from inscriptions probably of the first centuries AD discovered near Safa.

1905 G. Bell *Let.* 24 Oct. (1927) I. xi. 225 I'm going to his house tomorrow to look over some Nabathean and Safaitic inscriptions and discuss what is to be found in Nejd. **1911** *Encycl. Brit.* XXIV. 626/1 To the first centuries of the Christian era belong the thousands of Arabic inscriptions, found in the wild, rocky districts south-east of Damascus, which are commonly termed Safaitic, after Safa, a locality in their neighbourhood. **1939**, etc. [see *Lihyanic *sb.*]. **1951** A. M. Honeyman in H. H. Rowley *Old Testament & Mod. Study* ix. 270 A new Pars Quinta has been projected to cover the Safaitic, Lihyanite, and Thamudic material. **1976** *Times* 3 Sept. (Qatar Suppl.) p. iv/9 From..the end of the first millennium BC,..a Safaitic inscription by a warrior to Du-Shara, high god of the Nabataeans.

safari (săfā·ri), *sb.* Also † sefari. [Swahili, journey, expedition, f. Arab. *safar* journey.]

1. a. A journey; a cross-country expedition, often lasting days or weeks, orig. in E. Africa and on foot, especially for hunting; now often with motorized vehicles, for tourism, adventure, or scientific investigation. Often in phr. *on safari*.

[**1860** *Harper's Mag.* Oct. 630/1 Safari! safari los! a journey, a journey to-day!] **1907** J. H. Patterson *Man-Eaters of Tsavo* vi. 61 [He] had left me and gone on *safari* (a caravan journey) to Uganda. *Ibid.* xi. 119 They join another caravan and begin a new *safari* to the Great Lakes. **1922** H. B. Hermon-Hodge *Up against it in Nigeria* iv. 54, I am an indifferent marksman both at range and on safari. **1928** *Daily Express* 16 Nov. 9 The royal safari—as a shooting expedition of this nature is described in Africa—is complete to the minutest detail. **1935** E. Hemingway *Green Hills of Africa* II. iii. 46 We had gone on a foot safari to hunt rhino in the forest. **1958** L. van der Post *Lost World of Kalahari* iv. 74 It was time we did another safari together. **1964** C. Willock *Enormous Zoo* ii. 23 Justin Tokwar's account of his historic porter safari to the Nile. **1970** *Drum* (E. Afr. ed.) Feb. 27/5 The time when safari in Tanzania meant roughing it will soon be over and visitors can enjoy the awe-inspiring scenery..and the relaxation of miles of unspoiled tropical beaches in comfort and luxury. **1976** *San Francisco Examiner* 30 May (Sunday Scene) The safari is organized to be an inside view of the naturalist's Africa.

b. *transf.* and *fig.*

1958 *Spectator* 22 Aug. 249/3 The London Studio are performing for a group of ten, for each other. And as long as they pursue this safari down a cul-de-sac, this is all the audience they will either attract or deserve. **1975** T. Dinesen *My Sister, Isak Dinesen* vi. 79, I cannot help seeing it like our safari sometime in the future, in which we shall remember all the shauries as shadows and smile at them. **1977** P. Hill *Liars* viii. 105 His educated hands went on safari down her stomach.

2. A hunter's or traveller's party or caravan.

1890 F. Lugard *Diary* 2 Feb. (1959) I. 92 A Safari is by no means an easy thing to manage, especially at first. **1892** *Daily News* 15 July 5/6 It would be a great thing if the next sefari (caravan) brought up a small Nordenfelt or Hotchkiss gun. **1901** *Ann. Rep. Board of Regents Smithsonian Inst. 1900* 433 We collected our safari of one hundred and thirty Manyema carriers. **1909** W. S. Rainsford *Land of Lion* vi. 141 Be always careful to look for signs of crocodiles,..and warn your sefari to be careful. **1928** *Blackw. Mag.* Oct. 549/1 It is seldom indeed that a safari passes through the bush without some news of it being 'telegraphed' ahead by the natives.

3. *attrib.* and *Comb.* **a.** *gen.*, as *safari accounts, coach, horn, lodge, path, plan, ranch, work*. **b.** Designating articles of clothing suitable for wearing on safari, or made in a similar fashionable style, as *safari boot, hat, jacket, kit, shirt, suit*. **c.** Of furniture, etc. (proprietary name): designed for use whilst on safari or otherwise travelling, as *Safari (camp) bed, chair, mattress*. **d.** Special Combs.: **safari ant**, a nomadic, carnivorous, African ant of the subfamily Dorylinæ, esp. one belonging to the genus *Anomma*; **safari camp** *Austral.*, a camp in the outback; **safari look** (see quot. 1968); **safari park**, an area of parkland where wild animals are kept in the open and through which visitors may drive in motor vehicles.

1890 F. D. Lugard *Diary* 30 Jan. (1959) I. ii. 89 Discussed a plan for working safari accounts &c. with Dick. **1926** D. Strickland *Through Belgian Congo* vi. 94 The driver or safari ant is perhaps, from an entomological standpoint, the most interesting. **1966** B. Kimenye *Kalasanda Revisited* 51 Safari ants..those large, shiny black insects whose jaws clamp into flesh. **1976** K. Thackeray *Crownbird* ix. 189 A crawling mass of siafu, large safari ants with big pincers. **1945** *Trade Marks Jrnl.* 7 Mar. 123/2 *Safari*... Beds (furniture). Hounsfield Limited, 81, Morland Road, Croydon, Surrey, manufacturers. **1976** M. Birmingham *Heat of Sun* vii. 110 [We] dragged a safari bed from the bottom of the linen cupboard and set it up in..my room. **1970** *Times* 20 May 7/1 Mr. Lea was said to have been wearing..khaki trousers, and brown safari boots. **1977** H. Innes *Big Footprints* III. iii. 292 [We] took our safari boots off and dabbled our bare feet in a pool. **1969** *Northern Territory News* (Darwin) *Focus* '69 97/1 More are being encouraged here by small plane links between Darwin and Arnhem Land safari camps, as well as other outback attractions. **1972** V. Canning *Rainbird Pattern* ii. 39 The inner room, smaller, contained a safari camp bed with the appropriate bedclothes. **1977** *Bulletin* (Sydney) 22 Jan. 65/1 One lives 48 kilometres away on the Nourlangie safari camp. **1967** H. Harrison *Technicolor Time Machine* (1968) viii. 75 Slithey was leaning back in her safari chair while her wig was being combed. **1973** 'S. Harvester' *Corner of Playground* III. vii. 224 A safari coach of rich Americans drew up. **1968** J. Ironside *Fashion Alphabet* 144 *Topee* or *Safari hat*, shaped like a pith-helmet with rounded crown and brim sloping down, slightly wider at the back and front. **1977** H. Innes *Big Footprints* II. iii. 179 Her eyes, shaded by the safari hat, were gazing towards the distant mountains. **1928** *Blackw. Mag.* Oct. 549/1 The sound of a safari horn came drifting across the bush. **1972** *Vogue* Feb. 86 Safari jacket, unlined seersucker. **1977** M. Allen *Spence in Petal Park* xii. 50 She was wearing a gold roll-neck sweater and a brown suede safari jacket. **1928** *Daily Express* 29 Nov. 1/1 The Prince, dressed in safari kit. **1964** C. Willock *Enormous Zoo* iii. 34 The safari lodge at Mweya was unexpectedly losing money. **1975** 'D. Jordan' *Black Account* xxxii. 163 There was a safari lodge with hot water and white sheets. [**1968** J. Ironside *Fashion Alphabet* 27 *Safari*. This is a White-hunter look that has been in and out of fashion since the beginning of the twentieth century. Details such as belted and vented jackets..imitation pith helmets and epaulettes..in both skirt and trouser suits.] **1969** *Sears Catal.* Spring/Summer 40 Cape, Top and Pants Set for the total safari look. **1974** *Times-Picayune* (New Orleans) 4 Oct. IV. 1 (Advt.), The safari look with shell and slacks in beige, belted jacket in brown. **1971** Safari mattress [see *garden umbrella* s.v. *Garden *sb.* 4 a]. **1969** *Times* 1 Oct. 14/2 Opened early this summer Windsor Safari Park covering an area of 140 acres on the north edge of Windsor Great Park has proved very successful. **1977** B. Pym *Quartet in Autumn* v. 45 There would be visits to a safari park and to the stately homes that offered the best attractions. **1920** *Blackw. Mag.* Feb. 205/1 A good safari path winds through a deep valley in the Livingstone Mountains. **1890** F. D. Lugard *Diary* 3 Apr. (1959) I. iv. 170 Had a long talk over Safari plans and worked out the details of my idea of food supply &c. **1975** 'D. Jordan' *Black Account* xix. 104 Angel Engelbrecht has a safari ranch in Northern Transvaal. **1968** *Vogue* 15 Apr. 77 Safari shirt in brave bright red Linoseta over navy blue supraline trousers. **1967** *Observer* 26 Feb. 29/3 (*heading*) Safari suit. The most convincing and adaptable fashion from Paris. **1979** P. Niesewand *Member of Club* xv. 115 A mild-mannered American.. wearing a light blue safari suit. **1890** F. D. Lugard *Diary* 17 Jan. (1959) I. i. 75 A fine body of men..ready to take service with the Company in fort and station building, shamba making, and safari work.

Hence as *v. intr.*, to go on safari; also *transf.*

1908 *Times Lit. Suppl.* 19 Nov. 413/1 Mr. Chapman then has safaried twice for pleasure to British East Africa. **1936** *Punch* 5 Aug. 164/1 I've safaried in Sahara, And I've wandered in Peru. **1971** L. Gutteridge *Cold War in Country Garden* I. iii. 55 The two men went up-country and safaried around for a while. **1977** W. McIlvanney *Laidlaw* xxxiv. 160 The receptionist was waiting... In the time it took Harkness to safari to her desk, she didn't look up once.

Safavid (sæ·făvid), *a.* and *sb.* Also **Safawid**. [ad. Pers. *ṣafawī* (see Sophy[1]) + patronymic suff. *-id*.] **A.** *adj.* Of or pertaining to a ruling dynasty in Persia (1501–1736). **B.** *sb.* A member of this dynasty.

1911 *Encycl. Brit.* XXI. 233/2 By the fall of the Safawid dynasty Persia lost her race of national monarchs. **1957** *Ibid.* XVII. 574/1 The Safawid dynasty did not actually come to an end until Nadir Shah usurped the throne in 1736. **1972** *Country Life* 20 Jan. 155/3 The numerous flasks of wine for which the Safavid Kings had such a weakness. **1975** *New Yorker* 13 Oct. 31/3 'This is not a catalogue,' Mr. Ross said, 'because many of the objects now on view will help you sort out the Timurids from the Safavids.'

safe, *sb.* Add: **1. b.** (Earlier example.)

1820 *Rec. Early Hist. Boston* (1909) XXXIX. 174 A fire proof safe in the Selectmens room for the security of the records.

3. A tray laid under plumbing fixtures to receive spilled water.

1862 *Illustr. Catal. Internat. Exhib., Industr. Dept., Brit. Div.* II. No. 6392 Patent Bath, sienna marbled inside, verdantique outside. Taps and safe fitted. **1896** T. E. Coleman *Sanitary House Drainage* xvi. 129 The floor of the bath-room should be laid with mosaic..the bath standing within a properly constructed safe, which may be made of slate, marble, glazed earthenware, or tiles. **1956** Gumbrill & Smith *Blake & Jenkins's Drainage & Sanitation* (ed. 11) vii. 215 The lead safe sometimes placed under the cistern must have a waste pipe which should be carried through an external wall.

4. A contraceptive sheath. *colloq.*

1897 *Science of Generation* xx. 235 The use of various mechanical contrivances, such as French Safes, Condom Sheaths, etc., is also objectionable. **1959** V. Packard *Status Seekers* (1960) xi. 155 Young Italian-American men..of high-school age regularly carry 'safes' or condoms. **1979** E. Koch *Good Night Little Spy* x. 94 Just in time he remembered his safe. He took it out of his pants pocket.

5. The operative position of a firearm's safety device; the state in which a gun cannot be fired. Cf. Safety 8.

1920 G. Burrard *Notes on Sporting Rifles* 71 One may ..fail to stop a dangerous charge through the rifle being at 'safe'. **1967** V. Canning *Python Project* ii. 31, I hope you've got that damned thing on 'safe'? **1978** F. Ross *Sleeping Dogs* 127 The safety catch was off. He clicked it to 'safe' and tossed it on the carpet.

6. *attrib.* and *Comb.* (chiefly sense 1 b): simple attrib., as *safe-door, -key, -robbery*; objective, as *safe burster, buster, -maker, -making, -opener, -robber*; **safe-blower** orig. *U.S.*, a safe-robber who uses explosive material to burst open safes; hence **safe-blowing** *vbl. sb.*; **safe-breaker** orig. *U.S.*, a robber who breaks open safes; hence **safe-breaking** *vbl. sb.*; **safe-cracker** orig. *U.S.* = *safe-breaker; hence **safe-cracking** *vbl. sb.*

1873 G. Lening *Dark Side N.Y. Life* 148 Namely, first those who burst open the safe with gunpowder,—'safe blowers'. **1951** Wodehouse *Old Reliable* iv. 51 Are you a safeblower magically gifted with the art of buttling, or a butler who has somehow picked up the knack of blowing safes? **1972** *Times* 12 May 2/8 A former safe-blower..claimed to have got away with a total of £10,000 at a cost of 20 years in different jails. **1928** H. Asbury *Gangs of N.Y.* x. §2. 217 [Marm Mandelbaum] also offered advanced courses in burglary and safe-blowing. **1970** 'D. Halliday' *Dolly & Cookie Bird* viii. 113 We amuse ourselves with safeblowing. **1870** M. H. Smith *20 Yrs. Wall St.* xxv. 320 A safe-breaker from Boston, a bank-robber from Philadelphia, a New York thief, have each their own way of doing things. **1977** J. Wainwright *Nest of Rats* I. vii. 46 The genuine peterman —the safe-breaker who takes a personal pride in pitting his wits against those of the safe-makers. **1934** Webster, Safebreaking. **1937** 'M. Innes' *Hamlet, Revenge!* II. ii. 115 The burglary and safe-breaking had been unsuccessful. **1981** 'M. Hebden' *Pel is Puzzled* vii. 61 Two years for attempted safe-breaking in Lyons. **1873** G. Lening *Dark Side N.Y. Life* 148 Then those who, not using powder, have recourse to mechanical means, these are 'safe bursters' *par excellence*. **1934** Webster, Safecracker. **1960** *Times* 27 May 18/6 His hero, Bob, is an (almost) reformed safe-cracker. **1977** D. Bagley *Enemy* xiv. 111 The chief safe-cracker was a man I'd met before. **1934** Webster, Safecracking. **1937** 'M. Innes' *Hamlet, Revenge!* II. iii. 130, I know something about this safe-cracking business. **1967** *Times* 28 Feb. (Canada Suppl.) 31 Montreal claimed that it had more..safecrackings than any other city in the world. **1977** D. Bagley *Enemy* xiv. 111, I..found the safe-cracking team at work. **1894** 'Mark Twain' in *Century Mag.* XLVIII. 22 The safe-door was not open. **1911** *Encycl. Brit.* XXIII. 997/1 Where larger quantities of valuables had to be preserved than a safe would conveniently hold, a safe-door of larger dimensions would be made and attached to a masonry or brick room. **1894** 'Mark Twain' in *Century Mag.* XLVIII. 22 His plan was, to..steal the safe-key..and then go back and rob the safe. **1911** *Encycl. Brit.* XXIII. 996/2 Well-authenticated experiments performed by safe-makers on their own and other makers' productions. **1977** Safe-maker [see *safe-breaker]. **1886** *Encycl. Brit.* XXI. 144/1 The ingenuity of inventors has..effected much in safe-making. **1970** H. Trevelyan *Middle East in Revolution* 251 The combination lock was stuck and only the girls, who had been evacuated, could open the safe. Miraculously, an amateur safe-opener appeared and did the trick. **1873** G. Lening *Dark Side N.Y. Life* 148 The safe robber..usually obtains access to houses by means of false keys. **1959** J. Cary *Captive & Free* xxvii. 124 He might have been a test pilot, a racing motorist, an explorer, a climber of Himalayan peaks; or perhaps a cat burglar, safe-robber, or hold-up man. **1886** *Frank Leslie's Pop. Monthly* XXI. 47/2 The Egerton safe-robbery..had baffled all the detectives in town.

safe, *a.* Add: **I. 4.** (Later example of *with (a) safe conscience*.)

a **1817** Jane Austen *Northanger Abbey* (1818) I. xiii. 231 Now we may all go to-morrow with a safe conscience.

II. 7. Also const. *for*.

1697 [in Dict.]. **1917** W. Wilson in *Sel. Addresses* (1918) 195 The world must be made safe for democracy. Its peace must be planted upon the tested foundations of political liberty. **1929** H. W. Nevinson *English* viii. 63 It was believed by some that the Great War was waged to make the world safe for democracy, and the result has been that democracy was destroyed in many European countries. **1932** J. Fortescue in *Eighteen-Sixties* 244 The pain of seeing the world made safe for that most unsafe and lowering of influences, vulgarity. **1932** A. P. Herbert in *Punch* 15 June 653/2 The last few years of the War were directed by the great brains up above to thinking out new ways of making the War safe for the infantry. **1963** J. F. Kennedy in *Evening Star* (Washington, D.C.) 10 June A-7/4 And if we cannot end now our differences, at least we can help make the world safe for diversity.

9. c. (Earlier example.) Cf. *the sure(r) side* s.v. Sure *a.* 1 e.

1811 Jane Austen *Sense & Sens.* III. iv. 78 Determining to be on the safe side, he made his apology in form as soon as he could say any thing.

11. a. Also *spec.* in *Cricket*.

1823 *Lady's Mag.* July 387/1 Samuel Long..is..so steady a [cricket] player! so safe! **1851** J. Pycroft *Cricket Field* x. 185 The safest pair of hands in England. **1897** K. S. Ranjitsinhji *Jubilee Bk. Cricket* ii. 18 'A safe field'..signifies that the fielder may be relied upon to stop hits that come within reasonable distance of him, and to hold practically all catches. **1975** *Oxf. Compan. Sports & Games* 648/2 A bulky left-handed batsman of safe and unspectacular method.

b. Also in proverbial phr. *better (to be) safe than sorry*.

[**1837** S. Lover *Rory O'More* II. xxi. 148 'Jist countin' them,—is there any harm in that?' said the tinker: 'it's betther be sure than sorry'.] **1933** *Radio Times* 14 Apr. 125/1 Cheap distempers very soon crack or fade. Better be safe than sorry. Ask for Hall's. **1958** [see *Lightship]. **1962** A. Nisbett *Technique Sound Studio* vii. 120 For tapes that are to be broadcast it is better to be safe than sorry. **1972** J. Wilson *Hide & Seek* vii. 128 It's not that I want to shut you in... But—well, it's better to be safe than sorry.

c. *to play safe*: see *Play *v.* 18 f.

14. safe area, during the war of 1939–45, an area not liable to be attacked or invaded; **safe deposit** (further *attrib.* examples); **Safe-hand, safe(-)hand**, applied *attrib.* and *absol.* to a variety of courier services available for confidential documents; also as *adv.*; **safe hit** *Baseball* (see quot. 1895); **safe house**, a place of refuge or rendezvous for those wanted by the authorities, engaged in spying, etc.; **safe lamp, light**, restrict † to sense in Dict. and add: (*b*) *Photogr.*, a translucent filter for use on a dark-room light, coloured according to the sensitivity of the materials used; also, a lamp that produces such a dim, coloured light; so **safe-lighting** *vbl. sb.*, **-lighted, -lit** *ppl. adjs.*; **safe period**, the part of the menstrual cycle during which conception is least likely; **safe seat** *Pol.*, a parliamentary seat which is likely to be retained at an election with a large majority; **safe-tray** = *Safe *sb.* 3.

1944 *Ourselves in Wartime* 175 The threat of invasion, and the air-blitz of 1940–1941 over London and the provinces stimulated evacuation afresh... Many thousands of children were removed to safe areas, and.. over 620,000 children were settled in reception areas. **1944** *Daily Tel.* 11 July 2 Married couple wanted. Safe area (Alva, Scotland). **1882** *Century Mag.* Mar. 769/1 They did not ask for the key of the safe-deposit box, or for other evidence. **1970** K. Roos *What did Hattie See?* x. 92 You don't give a dame a key to your safe deposit box. **1947** Ld. Mountbatten *Let.* 12 June in *India Office Rec.* 1450 GG 43 Coll. I. p. 31 Please follow it up with the letter, which should be sent by safehand of pilot. **1965** 'W. Haggard' *Hard Sell* xi. 113 The rumblings from London..had ceased with a Safehand letter from the Minister. **1975** N. Luard *Robespierre Serial* iv. 16 Delivered safe-hand by courier eight months before, the letter stated briefly that Darley had been approached by an individual who'd indicated he might be interested in political asylum. *Ibid.* 17 A second safe-hand letter had arrived. **1867** *Ball Player's Chron.* 6 June 2/3 Flagg afterward made his base by a safe hit. **1895** G. J. Manson *Sporting Dict.* 98 *Safe Hits*, this term is applied to high balls sent from the bat with just force enough to carry them over the heads of the infields, but not far enough out for the outfielders to catch. **1897** *Encycl. Sport* I. 77/2 Immediately the batsman hits a fair ball, he endeavours to get to first base... He may get there on a safe hit made by the succeeding batsman. **1963** J. Joesten *They call it Intelligence* I. iv. 44 A so-called 'safe house'..is usually a piece of extra-territorial property owned by a particular embassy. **1969** H. MacInnes *Salzburg Connection* xv. 212 So that is what it was: a safe house. They could shelter several people here..while new passports and identities were being faked. **1979** H. Kissinger *White House Years* xxi. 889 A seedy little apartment in an old brownstone that the CIA had used as a safehouse. **1968** *Gloss. Terms Offset Lithogr. Printing (B.S.I.)* 11 *Safelamp*, a lamp providing light of a spectral composition to which a photographic material is relatively or completely insensitive. **1978** *Amateur Photographer* 2 Aug. 131/1 With a 150-watt enlarger lamp, two 25-watt lamps in the safelamps and a 60-watt lamp in the illuminator there is no risk of overload. **1903** A. Payne *Pract. Orthochrom. Photogr.* 90 Red sensitive plates..may be used with a safe light. **1932** *Discovery* Sept. 292/1 These infra-red plates are..easy to manipulate in the dark room with a lamp screened by a greenish yellow safe-light filter. **1976** J. McClure *Rogue Eagle* vi. 98 If..he'd had the orange safe-light turned on..the film would have fogged instantly. **1977** J. Hedgcoe *Photographer's Handbk.* 51 These features make the paper convenient to handle in an orange safe-lighted darkroom. *Ibid.* 39 Printing papers and films intended for copying black and white originals have this sensitivity, allowing the use of bright orange safe-lighting. **1979** *Amateur Photographer* 10 Jan. 75/1 Electronic timers are far more accurate than relying on peering at your watch in a safelit darkroom. **1918** M. Stopes *Wise Parenthood* iv. 31 Some people..may find the comparative security of a 'safe period' sufficient. **1923** —— *Contraception* ii. 14 The proper form of contraceptive must be one available at *any* time by the pair: and so the 'safe period' often advocated by those who pose as moralists is not satisfactory. **1934** *Jrnl. Amer. Med. Assoc.* 10 Feb. 452/2 The woman determines her 'safe period' on the basis of her shortest cycle, and also of her longest cycle. The overlapping 'safe days' constitute her 'safe period'. **1936** C. G. Hartman *Time of Ovulation in Women* xviii. 183 There is an absolute Safe Period for the monkey female. *Ibid.* xix. 192 Announcement of failures of the Safe Period is a daily occurrence. **1956** A. Huxley *Adonis & Alphabet* 2E4 In the kind of society which has the most urgent need of birth control, the Safe Period Method is almost useless. **1971** *Petticoat* 17 July 6/3 The safe period should more correctly be known as the safer period. You are less likely to conceive then, but that is all. **1976** *Winter's Crimes 8* 180 The so-called safe period won only limited approval... She couldn't let herself be pregnant. **1891** W. Fraser *Disraeli & his Day* 491 A material element in the future of Constitutional Government is the non-existence of safe seats. **1939** W. I. Jennings *Parliament* ii. 27 The influence of a great landowner... May secure nomination by the local Conservative association and so enable the person nominated to acquire a safe seat. **1974** *Times* 13 Feb. 4/6 Redistribution can make a safe seat marginal. **1886** *Encycl. Brit.* XXI. 715/2 Under most plumbing fixtures it is usual to place a safe-tray to receive any water accidentally spilt. **1912** G. Thomson *Mod. Sanitary Engin.* xvi. 142 When built-up baths were in use, safe trays were an indispensable part of the installation.

15. *safe-buttressed, -enshrined, -going, -sequestered* (later example), *-swung* ppl. adjs.

1918 G. Frankau *One of Them* xxxii. 249 O Empire thrice and four times blessed by Fate, Safe-buttressed on ten thousand O.B.E.s! **1926** W. de la Mare *Memory* in Kipling & de la Mare *St. Andrews*, Keeps the for me, then, safe-enshrined—Cold of the north—those bleached grey streets. **1874** Trollope *Way we live Now* (1875) I. xlvii. 296 In this safe-going country young men perhaps are not their own masters till they are past thirty. **1930** R. Campbell *Poems* 17 Safe-sequestered in some rural glen. **1900** Kipling in *Century Mag.* Jan. 407 Safe-swung above the glassy death.

safe, *v.* Delete † *Obs. rare* and add: (Later examples.) **b.** *intr.* and *trans.* In *Mountaineering*, to belay. Also const. *up*. Hence **sa·fing** *vbl. sb.*

1940 *Tararua Tramper* July 6 The real uses for alpine work may be divided into three: Step-cutting, 'safing' (by which I mean anchoring or belaying), and control during descent. *a* **1945** E. R. Eddison *Mezentian Gate* (1958) xxxviii. 202, I am sick..of for ever climbing mountains safed with a dozen ropes held by a dozen safe men. **1960** M. Redgrove in *Pick of Today's Short Stories* XI. 194 His mountain-sense stabbed a quick reproach and he dragged his attention back to safing Creade up. **1969** *Word Study* Apr. 6/1 To avoid having any unused explosive going off in the faces of members of the post-recovery team, the capsule must be disarmed or *safed*. **1972** *New Scientist* 14 Dec. 645 'Safing' procedures were now being carried out by the astronauts while Launch Control itself tried to identify the cause of the cut-off. **1974** 'J. le Carré' *Tinker, Tailor* xxv. 215 We tossed them agents we could do without, we gave them good communications, safed their courier links.

safeguard, *v.* Add: **c.** To 'protect' (a native manufacture or industry) against foreign imports. Cf. *Safeguarding *vbl. sb.* b.

1926 *Encycl. Brit.* III. 445/2 (*heading*) Four classes of goods safeguarded. **1928** *Manch. Guardian Weekly* 10 Aug. 105/1 By appointing a Royal Commission to inquire into the expediency of safeguarding the iron and steel industries. **1929** *Morning Post* 5 Feb. 14/4 Safeguarding Wool.

safeguardance (sēi·fgā:ɹdəns). *rare.* [f. Safeguard *v.* + -ance.] Safeguarding, protection.

1908 Hardy *Dynasts* III. v. v. 213 To all eyes it is imperative That some mode of safeguardance be devised.

safeguarding, *vbl. sb.* Add: **b.** The protection of native manufactures and industries against foreign imports. Also *attrib.*

[**1903** *Daily Mail Year Bk.* 149/1 *Protection* is the name given to the system of safe-guarding from foreign competition, native industries by the imposition..of duties.] **1921** *Act* 11 & 12 *Geo. V* c. 47 Part 1, Safeguarding of Key Industries. **1925** *Times* 10 Feb. 12/4 The idea of introducing a general Safeguarding of Industry Bill has been abandoned. **1926** H. Bell in F. W. Hirst *Safeguarding* p. vi, Protection is no longer called 'Tariff Reform'. It is called 'Safeguarding of Industries' or 'Buy British Goods' or 'Merchandise Marks'. **1932** G. D. H. Cole *Brit. Trade & Industry* 366 The McKenna and Safeguarding duties.

safekeep, *v.* *rare.* [Back-formation f. Safe-keeping *vbl. sb.*] *trans.* To keep safe, protect.

1966 *Anchor Bible* XVI. *Psalms* i. 6 But Yahweh shall safekeep the assembly of the just, While the assembly of the wicked shall perish. **1972** *Harper's Mag.* Oct. 80 Banking on Dictys to safekeep her, I'd set out for Samos.. to learn about life from 'art'.

safener (sēi·fnəɹ, -fənəɹ). [f. Safe *a.* + -en[5] + -er[1].] A substance that reduces the harmfulness to plants of other substances, esp. one in an insecticide or fungicide.

1942 *Industr. & Engin. Chem.* Apr. 498/1 The principal use of zinc as a spray is for the control of peach bacterial spot and as a 'safener' for arsenate of lead sprays on peach. **1950** J. C. Walker *Plant Path.* xvi. 647 Glyceride oils are..good safeners for copper sprays. **1975** *Big Farm Managem.* June 61/2 George Moore considers Eradicane to be the important herbicide for the British market at present. This is the chemical which has a built-in 'safener' which protects maize from the herbicide which would otherwise kill it.

safety. Add: **1. a.** (Examples of 'safety in numbers'.)

1816 Jane Austen *Emma* II. i. 2 She determined to call upon them and seek safety in numbers. **1886** C. M. Yonge *Chantry House* II. xii. 112 They all came creeping down after her, feeling safety in numbers. **1914** T. Dreiser *Titan* xvii. 140 Perhaps he was beginning to run around with other women. There was safety in numbers—that she knew. **1941** E. Howie *Murder for Christmas* xi. 135 The old adage—there's safety in numbers—may very well apply here. **1973** 'S. Woods' *Yet she must Die* 115 'Lydia was flirtatious. But nobody took that seriously, least of all the men concerned.' 'Safety in numbers, in fact.'

g. (Earlier example.) *to play for safety*: see *Play *v.* 18 f.

1857 M. Phelan *Game of Billiards* (ed. 2) iv. 65 Playing for safety.—When you forego a possible advantage, in order to leave the balls in such a position that your opponent can make nothing out of them.

h. *safety first*: see as main entry in Suppl.

6. (Further examples.) Now usu. as *safety factor*.

1909 Webster, Safety factor. **1916** W. H. Molesworth *Spons' Electr. Pocket-Bk.* 482/1 Safety factor, aerial conductors. **1971** L. Pilborough *Inspection of Chem. Plant* ii. 16 Factors of safety for many metallic materials at temperatures up to 650°F may vary from 4 to 5 in the U.K. **1973** C. Sagan *Cosmic Connection* (1975) iii. 17 Its orbit [sc. that of Pioneer 10] was not disturbed by an errant asteroid—the safety factor was estimated as 20 to 1.

8. (Further examples.)

1936 Hemingway in *Hearst's Internat.* Sept. 168/1 He had the safety on and..he lowered the rifle to move the safety over. **1968** K. Weatherly *Roo Shooter* 11 The shooter picked up the smaller rifle and brought it to his shoulder, flipping the safety off with his thumb. **1972** *Shooting Times & Country Mag.* 27 May 13/3 Never push the safety off until the moment of shooting.

9*. a. (*a*) *N. Amer. Football*, an act of carrying the ball into one's own end zone; a score of two points awarded against a team for this; (*b*) *Polo* (see quots. 1905).

(*a*) **1881** *Proc. Intercollegiate Conventions Conf.* in P. H. Davis *Football* (1911) 469 If the game still remains a tie the side which makes four or more safeties less than their opponents shall win the game. **1910** W. Camp *Bk. of Foot-Ball* ii. 54 A 'safety' is made when a side are so sorely pressed that they carry the ball behind their *own goal line*. **1941** *Charlottesville* (Va.) *Daily Progress* 14 Jan. 11 If a legal forward pass is incomplete in the offensive team's end zone, it is to be ruled an incompleted pass instead of a safety. **1950** *Chicago Tribune* 26 Feb. 20/2 A blocked kick and safety can be credited against him. **1972** J. Mosedale *Football* iv. 48 The ball hit a goal post and was ruled a safety—the winning margin.

(*b*) **1905** T. F. Dale *Polo* xvi. 309 Whenever a player either accidentally or intentionally gives the ball an impetus with his mallet which carries the ball over the goal line he is defending, and it touches nothing except the goal-post or the ground after leaving his mallet, it shall be deemed a safety. *Ibid.* 310 A safety (an excellent word, by the way, to define what we describe as a hit behind to save their goal by the defending side) counts as −½. **1931** 'Marco' *Introd. Polo* II. iv. 72 In America, hitting the ball over one's own line is called hitting a 'safety'; since it is occasionally safer to do this when one can't clear the ball, than to leave it in position for a certain goal to be scored. **1959** *Times* 3 Aug. 2/1 Lucas managed to force in a safety and this was followed by Harper's run. **1973** H. Disston *Beginning Polo* x. 119 'Technical' fouls, such as hitting the ball behind your own goal line (a safety).

b. *ellipt.* for *safety match*.

1900 J. Vaizey *About Peggy Saville* v. 31, I..go in for safeties, which 'strike only on the box'. **1927** R. A. Knox *Three Taps* iv. 39 That match worries me... Those are ordinary safeties. This is a smaller kind. **1938** S. Beckett *Murphy* xii. 263 Whether..it was a Brymay safety that exploded the mixture, or a wax vesta.

c. *Baseball.* A safe hit.

1905 *Sporting Life* 9 Sept. 2/3 Harry whaled away at the ball and hit it on a line over short stop for a safety. **1917** C. Mathewson *Second Base Sloan* 105 Billy White led off with a safety to left. **1931** *Randolph Enterprise* (Elkins, W. Virginia) 9 July 5/3 The locals hammered out 15 hits on the first contest while the visitors collected eight safeties. **1968** *Washington Post* 4 July c2/8 Pappas ..gave up six safeties in the seven innings he pitched. **1976** *Billings* (Montana) *Gaz.* 5 July 3-c/1 Carlos Pimental and Scott Meade led Billings' 12-hit attack with three safeties apiece.

d. *ellipt.* for *safety razor.

1924 Kipling *Debits & Credits* (1926) 165 'You could with a Safety, though,' said Anthony. And, indeed..one might have shaved in it with comfort. **1925** *Punch* (Almanack No.) 2 Nov. p. iv, When you decided to use a 'safety', instead of the old solid hollow-ground razor, why did you do so? **1932** D. L. Sayers *Have his Carcase* iv. 62 A young man who had so much difficulty with his razor would be more likely to change over to a safety and use a new blade every few days.

e. *N. Amer. Football.* *ellipt.* for *safety man (*b*).

1931 K. K. Rockne *Coaching* iii. 19 The safety who always catches the punts, but never brings them back very far is more valuable than the 'flash' who brings them back quite a distance, but is inclined to fumble. **1969** *Eugene* (Oregon) *Register-Guard* 3 Dec. 1D/2 Washington State's Eric Dahl was supposed to be the top sophomore defensive back, but in retrospect, the writers should have given that nod to UCLA's 5-9 safety, Ron Carver. **1976** *Honolulu Star-Bull.* 21 Dec. H-3/1 'You can't cry over spilled milk,' said Chuck Foreman. 'We're just glad we're going to be there.' Minnesota strong safety Jeff Wright concurred.

f. *N. Amer.* A metal-ringed outlet for a stove-pipe in the roof of a tent, etc.

1962 M. E. Murie *Two in Far North* I. iv. 40 The pipes from the many stoves went out through the roofs through galvanized-iron drums called 'safeties'. **1968** C. Helmericks *Down Wild River North* I. v. 83 You could hear a breeze sigh across the tent, rattling the tin safety against the little stove pipe.

10. *safety bar, barrier, carabiner, device, equipment, harness, inkstand, line, lock* (examples), *-mechanism, rail* (later examples), *rope, seat, sling, snap, spring, strap*; also designating items of protective clothing, as *safety boot, helmet, jacket, shoe, suit*; gen., as *safety code, margin, measure, regulation, standard*; *safety-conscious, -related* adjs.; **safety belt**, a protective or restraining belt; *spec.* (*a*) a belt in an aeroplane to hold a passenger in his seat, esp. on take-off or landing; (*b*) a belt in a motor vehicle to hold the wearer in his seat in the event of a collision or emergency stop (cf. *seat belt* s.v. *Seat *sb.* 29); **safety boat**, delete † and add earlier and later examples; **safety box**, (*a*) a box with a surface on which safety matches can be ignited; (*b*) a safe-deposit box; **safety cab**, (*b*) a tractor cab designed with a view to safety in use; **safety catch**, a catch or stop attached to a mechanical contrivance as a safe-guard, esp. in hoisting apparatus or on the trigger of a gun; **safety chain**, a chain providing additional security; *spec.* (*a*) a subsidiary chain connecting railway-cars, etc., together; (*b*) a chain securing a watch or jewellery to the clothing; (*c*) (see quot. *a* 1877); (*d*) a chain on a door preventing opening beyond a certain point; **safety committee**, a committee appointed to deal with safety in a place of work, etc.; **safety curtain**: in theatres, a fire-proof curtain which can be lowered to protect the main body of the theatre from fire on or behind the stage; **safety deposit** (chiefly *attrib.*) = *safe deposit* s.v. Safe *a.* 14 in Dict and Suppl.; so as *v. trans.*, to place or store in a safe deposit; **safety engineer**, a person trained in accident prevention and the organization and implementation of (esp. industrial) safety measures; hence **safety engineering**; **safety-film**, a slow-burning film specially prepared for cinematographic work; **safety fuse**, (*b*) *Electr.*: see *Fuse *sb.*[5]; **safety glass**, toughened or laminated glass; **safety island, isle**, a traffic island constituting a safety zone; = Refuge *sb.* 3 c; **safety man**, a person responsible for safety; *spec.* (*a*) a person whose work is to guard a temporarily disused mine-shaft in readiness for the resumption of work; (*b*) *N. Amer. Football*, the defensive back who plays in the deepest position; **safety net**: chiefly in circuses, a net to prevent injury in the event of a fall from a height; also *fig.*; **safety officer**, a person responsible for safety in a factory, etc.; **safety paper**, paper specially prepared to guard against the tampering with or counterfeiting of banknotes, etc.; **safety play**, (*a*) *Billiards* (see sense 1 g in Dict. and Suppl.); (*b*) *Bridge* (see quots.); † **safety plug**, (*a*) a plug or stopper that allows the quick release of contents when their presence becomes unsafe, *spec.* one of fusible metal that melts when the contents become too hot; (*b*) an electrical fuse; **safety razor**, a razor in which the blade is prevented by a guard from cutting the skin during shaving; also *attrib.*; **safety representative**, a representative of the workforce on an industrial safety committee; **safety rod** *Nuclear Engin.*, a rod of a neutron-absorbing material which can be inserted into a reactor in an emergency to slow or stop the reaction; **safety switch** (see quot. 1940); **safety touch(down)** *N. Amer. Football* = *Safety 9 a; **safety vault**, a vault or strong room for the safe custody of valuables; **safety vent**, an outlet affording safety; *spec.* = sense *9* f; also *fig.*; **safety zone**, (*a*) an island or part of a road or square where pedestrians may wait in safety for buses, etc.; (*b*) an area round the Americas in which warlike activities were to be proscribed during the war of 1939–45; also *transf.*

1963 E. H. Edwards *Saddlery* xiv. 99 Numerous so-called 'safety bars'... These ingenious devices, which were hinged in various ways to open up and release the stirrup leather when occasion demanded, have largely disappeared. **1951** *Gloss. Aeronaut. Terms (B.S.I.)* III. 25 *Safety barrier*, a net or other contrivance by means of which an aircraft that misses the arresting gear is brought to rest. **1858** Safety belt [in Dict.]. **1911** *Aero* 8 Apr. 6 Safety belt made for monoplanists. **1948** 'N. Shute' *No Highway* iii. 56 Then she pulled out the safety belt from behind the seat and showed him how to clasp it round his body. **1955** *Sci. News Let.* 17 Sept. 181/1 He would not drive without a safety belt. **1962** *Which?* Jan. 5/1 Safety belts should really be an integral part of car safety. **1976** S. Barstow *Right True End* III. xiii. 195 Now he wants to sit with his mother in front. I lengthen the safety-belt and strap them in together. **1976** P. Cave *High Flying Birds* iii. 40 Blood pounding in my head and lungs bursting, I was only dimly aware of the safety-belt catch finally coming free to let me float gently towards the surface. **1840** *Niles' Reg.* 4 Apr. 71/2 Lake, sound and sea going steamers [are] to have an equipment of..safety boats sufficient to carry all the passengers and crew. **1976** *Yachts & Yachting* 20 Aug. 369/1 On the first day a race was abandoned..because one safety boat was unserviceable (despite the presence of mark boats and spectator boats which could have doubled as safety boats in an emergency). **1967** *Times Rev. Industry* Mar. 16/3 Their shoe arose out of their earlier interest in the safety boot market. **1977** *West Briton* 25 Aug. 5/2 Mr. Cock..wore safety boots, leather gloves and carried a torch as he climbed down into the tank. **1902** 'Mark Twain' *Double-Barrelled Detective Story* II. 131, I hold in my fingers a burnt Swedish match—the kind one rubs on a safety-box. **1926** J. Black *You can't Win* x. 133 If I get snared by the bulls they won't know I've got a safety box. **1965** *Farmer & Stockbreeder* 21 Sept. 58/2 (Advt.), Safety cab by Clydebuilt... For your positive safety Clydebuilt has enormous structural strength, N.I.A.E. tested under rigorous conditions. **1973** *Times* 17 Nov. 6/1 All new tractors sold to farmers after September 1, 1975, would have to be fitted with a safety cab in which the noise level did not exceed 90 decibels. **1972** D. Haston *In High Places* xii. 148 Sliding back down the ropes was something of a joke: you just fixed a safety carabiner and ran down the line. **1877** *Encycl. Brit.* VI. 75 Various forms of safety catch and disengaging hooks. *a* **1884** Knight *Dict. Mech.* Suppl. s.v. *Safety Catch*, Safety catches attached to the cage are held away from the guides while the weight of the cage hangs on the rope. **1908** *Chambers's Jrnl.* 26 Dec. 61/2 The door can be opened a few inches and yet be held by the safety-catch. **1928** *Daily Mail* 31 July 5/3 Thinking the safety-catch was fixed she handled the revolver carelessly and the trigger fell. **1962** *Daily Tel.* 6 July 1/8 An electronic 'safety catch' which could be released only by a coded radio signal from headquarters. **1970** H. Trevelyan *Middle East in Revolution* 35 He had therefore imposed a limit on trade with the Communists: what he called his safety catch. **1973** 'R. MacLeod' *Nest of Vultures* 8 The gun was a Mauser and the safety catch was off. **1841** C. H. Gregory *Managem. Locom. Engine* 10 The draw-bar connecting the Engine and Tender must be secure, and the safety-chains attached. **1845** *Business Advertiser & Gen. Directory Chicago* 122 Clocks, Jewelry, Gold Safety Chains, Gold Fob Chains. **1851** *Illustr. Catal. Gt. Exhib.* III. 674/1 Safety chain brooches, for effectually fastening a lady's dress. *a* **1877** Knight *Dict. Mech.* III. 2016/1 *Safety-chain*, a slack chain which attaches a truck to a car-body and limits the excursions of the former as it slues round. **1965** D. Francis *Odds Against* x. 138 A nervous grey haired elderly man opened the front door on a safety chain. **1972** M. J. Bosse *Incident at Naha* i. 23, I called through the door, 'Who's there?' and opened it only to the length of the safety chain. **1973** J. Stubbs *Dear Laura* i. 21, I wonder whether you could not look at the safety chain of my new brooch? **1976** *Billings* (Montana) *Gaz.* 30 June 9-D/1 (Advt.), New tow bar and safety chain for pinto Datsun, Toyota, Vega, etc. **1954** (*title*) Institute of Petroleum Marketing Safety Code. **1961** *Lancet* 12 Aug. 365/2 A safety code for workers exposed to ionising radiations in industry is laid down. **1971** *Guardian* 22 June 6/6 Moving pavements..could become a major form of city transport with the adoption of a new safety code to supersede the existing 2 mph speed limit. **1945** *Proc. Inst. Mech. Engin.* CLII. 149 Safety engineering is advancing in experience and practice, and is receiving an impetus by the appointment of safety officers and safety committees by many important firms. **1961** *Sunderland Echo* 14 Jan. 2/1, 120,000 miners each received a letter from the divisional chairman urging them to be more safety conscious. **1973** C. Bonington *Next Horizon* xxi. 291 It seemed bitterly ironic that the person in the team who was, perhaps, the most safety-conscious should have been caught out by this cruel act of fate. **1909** *Weekly Budget* 21 Aug. 4/6 The safety curtain at the Lyceum went on strike one evening last week. **1912** *Theatreland* 11 Oct. 4/2 It was left to his successor, Richard Brinsley Sheridan, to stand sponsor for the iron 'safety' curtain. **1974** J. Gardner *Return of Moriarty* 303 Dr. Night had the stage cleared, the safety curtain lowered. **1891** 'Mark Twain' *Lett. to Publishers* (1967) 280 Yes, the statement was what I wanted... I sent it to Whitmore to be safety-deposited. **1892** *Ibid.* 304 As fast as Halsey delivers the securities to you I want you to put them in a box in a Safety Deposit Vault, and keep the key yourself. **1936** L. C. Douglas *White Banners* v. 86 She maintained a safety deposit-box there which she occasionally visited. **1978** S. Sheldon *Bloodline* xxxix. 350 A safety-deposit box in Zurich, contents unknown. **1884** *Harper's Mag.* Dec. 118/1 If the elevator has a safety device. **1929** *Daily Express* 7 Nov. 8/4 All the latest safety devices, such as four-wheel or six-wheel brakes, and safety glass. **1971** *Reader's Digest Family Guide to Law* 580 An employee working in a place from which he could fall more than 6 ft 6 in. must be provided with a safety device, such as fencing, where this is 'reasonably practicable'. **1934** Webster s.v. *Safety adj.* 2, Safety engineers. **1945** *Proc. Inst. Mech. Engin.* CLII. 166/1 The management of a mechanical engineering works should be just as much safety engineers as those claiming to be specialists in that direction. **1974** *Encycl. Brit. Macropædia* XVI. 138/1 The safety engineer is concerned with reducing both the frequency with which accidents occur and the frequency with which they threaten. **1945** Safety engineering [see *safety committee*]. **1977** *Jrnl. R. Soc. Arts* CXXV. 668/2 The effective application of safety engineering must go hand in hand with an understanding of management techniques. **1969** Safety equipment [see *chastity belt* s.v. *Chastity 6]. **1971** *Reader's Digest Family Guide to Law* 612/1 If..the employer can prove that he..provided the necessary safety equipment, he may not have to pay damages. **1928** *Daily Mail* 25 July 5/5 All the film used is safety film. **1959** W. S. Sharps *Dict. Cinematogr.* 84/2 *Cellulose acetate base*.., a slow burning safety film base. **1981** *Daily Tel.* 10 Feb. 12/4 We have been able to keep up to schedule because we stockpiled safety film when the price was low. **1922** *Tatler* 4 Oct. p. xii (Advt.), Another striking testimony for the Triplex Safety Glass. **1935** [see *Armoured *ppl. a.* 3]. **1950** *Engineering* 10 Feb. 167/3 The cab is..fitted with safety-glass windows. **1964** L. Deighton *Funeral in Berlin* xix. 111 The safety glass shattered into milky opacity. **1920** *Flight* 9 Sept. 978/2 Attention is drawn to the necessity of ensuring that the fitting and maintenance of safety belts and harness in aircraft is secure and functions properly. **1937** C. Boff *Boys' Bk. of Flying* xvii. 185 The pupil, in the rear cockpit, held the aeroplane on its back, with the startled instructor, in the front cockpit, holding on for dear life... His safety harness should have been properly secured, but it wasn't. **1972** D. Francis *Smokescreen* i. 7, I sat in the driving seat of a..sports car... [It] would not start until the safety harness was fastened. **1961** J. H. Goodier *Dict. Painting & Decorating* 247 Modern safety helmets are often made from resin bonded fibreglass, with a head harness of polythene. **1973** *Daily Tel.* 6 June 14/1 Britain's turbanned Sikh motor-cyclists are not alone in having problems with the new law making it compulsory for riders to wear crash helmets (sorry, *safety* helmets). **1869** J. C. Patteson *Let.* 24 Nov. in C. M. Yonge *Life J. C. Patteson* (1874) II. xi. 391 Patent safety inkstands—these things are useful on board ship. **1873** C. M. Yonge *Pillars of House* II. xx. 180 Felix..his safety ink-stand planted in the sand. **1933** *Sun* (Baltimore) 4 Apr. 3/4 A few years ago safety islands were placed in the middle of some of Cambridge's principal thoroughfares to safeguard the lives of pedestrians. **1965** J. von Sternberg *Fun in Chinese Laundry* vii. 189, I met the man..on one of the safety islands for pedestrians in the middle of Piccadilly Circus. **1934** *Transit Jrnl.* Nov. 437/1 Serious accidents in which fast moving automobiles crashed into the ends of safety isles in Baltimore. **1971** *Rand Daily Mail* (Home Owner) 27 Mar. 16/4 (Advt.), Wetlook diving suits... Safety jackets. **1976** A. Price *War Game* I. 46 There was a cowman in the road ahead, bright in his orange-banded safety jacket. **1957** R. G. Collomb *Dict. Mountaineering* 134 *Safety Line*, an independent rope attached to a climber's waistline when he is making an abseil. **1973** C. Bonington *Next Horizon* xix. 262 Alastair Newman had swum across first and McLeod followed, after tying on a safety-line. *a* **1877** Knight *Dict. Mech.* III. 2017/1 *Safety-lock*. 1. (*Lock*). A lock so contrived as not to be opened by a picklock or without the proper key... 2. (*Fire-arms.*) One provided with a stop or catch to prevent accidental discharge. **1970** *Which?* July 217/1 Most had a safety lock to prevent you exposing the film by accident. **1928** *Sunday Dispatch* 23 Dec. 3/4 These officials, known as 'safety men', will eat their dinner in semi-darkness hundreds of feet below the surface of the earth. **1929** *Daily Express* 7 Nov. 2/4 'Although,' she said, 'I am not a safety man myself, I have lived twenty-five years with a safety man, so I think I may claim to know a little how things work.' **1931** K. K. Rockne *Coaching* iii. 19 In catching punts the safety man stands with both feet flat on the ground. **1962** C. Forsyte *Diving Death* xx. 161 As Left began to get into his diving things again he wished..that somebody could be left in the boat as safety man. **1972** J. Mosedale *Football* ix. 135 He scored from four yards out, running straight over the safety man. **1967** W. Soyinka *Kongi's Harvest* 40 Five minutes. That's enough of a safety margin isn't it? It had better be! **1934** Webster s.v. *Safety* adj. 2, Safety measures. **1959** *Petroleum Handbk.* (ed. 4) 379 The safety measures at installations and depots are..based on the elimination of all possible sources of ignition from all areas where dangerous concentrations of petroleum vapours are at all likely. **1972** *Classification of Occupations* (Dept. Employment) II. 78/1 *Safety officer.* Advises on industrial safety and organises and co-ordinates accident prevention and safety measures within an organisation. **1977** P. Johnson *Enemies of Society* xv. 197 The technique of all-purpose explanation is completed by another safety-mechanism. **1950** P. Tempest *Lag's Lexicon* 183 *Safety net*, the steel nets spread across the hall, from landing to landing on the first floor, to prevent accidents, attempts at homicide, suicide, etc. **1953** *Economist* 11 July 87/1 A genuine flexibility, 'worked out' in the market, is compatible with the security of floor prices, or more accurately of 'safety net' prices that would protect the farmer against serious losses. **1958** *Spectator* 22 Aug. 249/3 Acrobatic tricks no commercial management would risk without a safety net. **1965** 'W. Haggard' *Hard Sell* iv. 45 The fire chief was speaking into the walkie-talkie..and men were running with a safety net. They spread it and held it. **1971** *Guardian* 29 July 11/6 An open invitation..to let costs rise in expectation of a taxpayers' safety net. **1974** G. Mitchell *Javelin for Jonah* xi. 137 He had been with a travelling circus..but they dismissed him... He had begun to insist on having a safety-net for his act. **1978** D. A. Stanwood *Memory of Eva Ryker* xxiii. 215 It'll be the first time..without Dr. Stanford's help... A triple somersault, with no safety net. **1939** *Engineering* 18 Aug. 215/2 The Chief Inspector of Factories..was prepared to call together a committee representing makers of presses, ..safety officers,..and factory inspectors. **1976** *Guardian* 15 Apr. 1/4 The firm's safety officers wearing breathing apparatus went down the tunnel..to see that all the men had escaped. **1851** *Illustr. Catal. Gt. Exhib.* III. 540/1 White and coloured safety paper for bankers' cheques. **1967** Karch & Buber *Offset Processes* 553 Safety Paper, paper treated usually by printing a design in a light tint which protects the sheet against forgery. **1896** Safety play [see *Balk *sb.*[1] 9]. **1959** *Listener* 3 Sept. 370/1 The safety play, properly so called, is a play that risks the loss of a trick which can be spared to guard against the possible loss of a trick which cannot be spared. **1964** *Official Encycl. Bridge* 481/2 A safety play is the play of a suit in such a manner as to protect against an abnormal or bad break in that suit, thereby either eliminating or minimizing the danger of losing the contract. **1977** *Cleethorpes News* 6 May 29/4 Sid's superb safety play in this last frame..stood him in good stead. **1837** P. Nicholson *Pract. Masonry* II. viii. 145 In the case..of any choking up of the connection-pipes, the stoker has merely to lift the safety-plugs, and clear out the pipes, by introducing a rod of iron into them. **1869** *Appleby's Illustr. Handbk. Machinery & Iron Work* p. xiii/1 (Index), Safety plugs, fusible. **1882** *Engineering* 7 July 11/3 Every lamp in the electrolier has its safety plug. **1887** *Ibid.* 11 Nov. 503/2 The fusible safety plug illustrated.. has been adopted by the South Wales and Monmouthshire Boiler Insurance Company. **1890** J. W. Urquhart *Electr. Light Fitting* v. 163 The usual safety plugs are marked with the number of ampères of current they can carry without fusion. **1923** *Power Engineer* XVIII. 475/2 (*heading*) Safety plug for heavy oil engines. **1940** *Chambers's Techn. Dict.* 154/2 *Check rail*.., a third rail laid on a curve alongside the inner rail and spaced a little from it, to safeguard rolling-stock against derailment due to excessive thrust on the outer rail. Also called.. *safety rail*. **1964** *Eng. Stud.* XLV. 23 A pulpit is a raised safety-rail in the bows of a yacht or motor cruiser. *a* **1877** Knight *Dict. Mech.* III. 2018/1 Safety-razor. **1903** *Hardwareman* 11 July 53 (Advt.), Something new in safety razors. **1921** A. Huxley *Crome Yellow* xxiv. 262 The packet of safety-razor blades. **1973** 'R. MacLeod' *Nest of Vultures* 8 The fair-haired man had shaved that morning. A safety razor shave by the smooth shine of his cheeks. **1956** A. Toynbee *Historian's Approach to Relig.* xviii. 238 The need for safety-regulations would not be eliminated if atomic power were to be applied exclusively to pacific and beneficent uses. **1971** *Reader's Digest Family Guide to Law* 580/1 Employees must observe safety regulations so that they do not endanger themselves or other workers. **1976** *National Observer* (U.S.) 28 Aug. 9/1 General Motors Corp. has been ordered to..pay a $400,000 penalty to the United States for refusing to notify owners of a safety-related defect. **1977** *Jrnl. R. Soc. Arts* CXXV. 676/2 Safety representatives will have the legal right to paid time off from work for undertaking these functions and for undergoing training for them. **1950** *Chem. & Engin. News* 4 Dec. 4257 The Argonne heavy water reactor is equipped with two control rods, two safety rods, and three shim rods. The safety and control rods are each formed of a 3·5-inch tubular sandwich of 1/32-inch cadmium placed between two aluminium tubes. **1971** *New Scientist* 13 May 389/1 The safety rods of a nuclear reactor are for use in emergencies when the neutron flux within the reactor core has to be immediately reduced. **1845** *Times* 31 July 1/5 Carriage for the Continent.—A Travelling Britzska, with every possible travelling appendage, namely eight trunks and imperials, two drag shoes and staff, safety ropes, pair and four horse bars, solid flap and German shutter. **1935** *Discovery* Mar. 73/1 For the descent..it is essential..to make constant use of the safety-rope. **1975** G. Moffat *Miss Pink* xii. 168 He wouldn't have fallen backwards... The safety rope would have held him. **1966** *Observer* 17 Apr. 21/3 The most hopeful sign is the emergence of the 'safety seat',..bolted firmly to the floor, with belts built-in. **1976** *Star* (Sheffield) 26 Nov. (Advt.), K. L. Jeenay Safety Seat. £8. **1943** *Sun* (Baltimore) 31 Mar. 13/2 A worker will be required to fill in a form showing..that the employer does not furnish safety shoes to him, that he needs the shoes to protect his health and safety [etc.]. **1974** *Encycl. Brit. Macropædia* XVI. 144/1 Safety shoes have been developed for protection in a wide variety of situations. **1974** H. MacInnes *Climb to Lost World* xi. 192, I had a karabiner and safety sling running on the other rope. **1932** Safety snap [see *Karabiner]. **1862** Safety spring [see *Bradoon]. **1960** *B.S.I. News* June 8/2 Finally, Mr. McNeill considered the question of 'safety' standards. **1976** 'R. B. Dominic' *Murder out of Commission* i. 7 You're in the Atomic Energy Commission... You can bring Ben up to date on safety standards. *a* **1877** Knight *Dict. Mech.* III. 2018/2 *Safety-strap*.., an extra back band passing over the seat of a gig-saddle..; used as a safeguard on light trotting harness. **1938** R. G. Collingwood *Princ. Art* xi. 240 The child's finding itself..wheeled about in a perambulator with a safety-strap round its waist. **1963** *Amer. Speech* XXXVIII. 207 The safety straps that are snapped or tied..from the ski to the boot to prevent run-away skis. **1917** *Chambers's Jrnl.* Oct. 702/2 A recently-invented life-saving apparatus, known as the 'ever-warm safety-suit', goes far towards removing this danger. **1974** *Times* 21 Feb. 3/4 (Advt.), Safety suits. **1940** *Chambers's Techn. Dict.* 293/2 *Emergency stop*, a switch installed in a lift-car, or other similar piece of equipment, by means of which the power to the operating motor can be cut off. Also called a *safety switch*. **1944** *Engineering* 8 Sept. 192/3 Investigation..revealed that some time previously the safety switch had failed. **1904** 'R. Connor' *Prospector* 38 With a brilliant series of passes the 'Varsity quarters and halves work the ball through the McGill twenty-five line, and by following hard a high punt, force the captain to a safety touch. **1958** *Edmonton* (Alberta) *Jrnl.* 7 Aug. 7/2 Flying wing Jack Hill..booted three converts while guard Don Walsh picked up two points on a safety touch. **1970** *Globe & Mail* (Toronto) 25 Sept. 33/3 The Panthers conceded a safety touch for Humberside's other points. **1887** *Century Mag.* XXXIV. 889/2 A 'safety' touch-down counts two points against the side which makes it. **1957** *Encycl. Brit.* IX. 472/2 No penalty was attached to the safety touchdown until 1881. **1846** T. L. McKenney *Mem.* I. 26 One set [of vouchers was] for the Treasury Department, one for my office proper, and the third for a safety vault. **1902** A. D. McFaul *Ike Glidden* xvii. 129 The safe suddenly became the people's depository and safety vault. **1963** *Times Lit. Suppl.* 26 Apr. 297/2 Using his diary as a safety-vent. **1968** C. Helmericks *Down Wild River North* I. vi. 86 The little stovepipe rattled and scratched against the tin safety vent of the tent. **1915** *Policeman's Monthly* Oct. 3/2 (*caption*) Safety zone at near-side car stop, Detroit. **1921** *Daily Colonist* (Victoria, B.C.) 8 Apr. 9/4 Mr. E. S. Harris asked the board to make some provision for 'safety zones' on the busiest streets of Victoria. **1939** *Daily Tel.* 18 Dec. 7/2 The safety zone would help the Allies, by keeping German submarines out of, roughly, a third of the Atlantic. **1940** J. Betjeman *Old Lights for New Chancels* 56, I will labour for Thy Kingdom, Help our lads to win the war, Send white feathers to the cowards Join the Women's Army Corps, Then wash the Steps around Thy Throne In the Eternal Safety Zone.

safety first. A maxim or slogan inculcating the necessity of taking precautions for the avoidance of accident. Also *attrib.*; occas. applied to the safest kinds of investment.

Various safety-first campaigns (in factories, schools, etc.) were organized in Britain in the early-twentieth cent. The slogan is said to derive from the American railway industry (see *Encycl. Brit.* (1926) III. 446). It was widely used as a slogan in Conservative election posters in 1922 and (with reference to Stanley Baldwin) in 1929.

1873 *Cassell's Mag.* Nov. 71/2 A system that would go on the motto of safety first. **1914** G. M. Price *Mod. Factory* 138 Corporations which have within the last five years taken up the slogan of 'safety first' and have done great work in accident prevention. **1924** J. S. C. Bridge *Hist. France* II. 118 The so-called battles were conducted under the rules of a carefully framed code, of which 'safety first' was the unacknowledged watchword and inspiration. **1927** [see *narrow-beamed* s.v. *Narrow *a.* 7]. **1927** *Daily Mail* 5 Aug. 3/1 (*heading*) Rising 'Safety First' Stocks. **1931** *Daily Mirror* 27 Aug. 4 These and other safety first signs are being introduced all over the Dominion. **1932** *Daily Mail* 2 July 3/1 The volume of business transacted in 'safety-first' stocks was probably not so large as earlier in the week. **1936** A. Christie *Cards on Table* xiii. 130 The moment you begin..adopting as your motto 'Safety First'—you might as well be dead. **1944** *Living off Land* iii. 61 Where the ground appears to be in the least unsafe, it should *always* be timbered as an essential 'safety-first' principle. **1953** Earl Winterton *Orders of Day* xi. 153 Mr Davidson was accused by many Conservatives, at least in private, of being responsible for the defeat of the Government and the invention of the Party's election slogan, 'Safety First'. **1964** S. Duke-Elder *Parsons' Dis. Eye* (ed. 14) xxvi. 377 Every attempt should be made by the provision of comfortable goggles and by educative means, such as 'Safety First' notices. **1965** A. J. P. Taylor *Eng. Hist. 1914–1945* viii. 282 The cause had a strong appeal for many Conservatives who wanted something more exciting than Baldwin's Safety First. **1977** J. Wainwright *Nest of Rats* I. ix. 64 The Koh-i-noor's twin sister... The glitter deserving of all that sophisticated safety-first garbage.

Hence **safety-fi·rster**, a person unwilling to take risks.

1928 *Daily Express* 19 Nov. 5 Many women would dislike the uncertainty, so the safety-firster, the shy, the unadaptable, and the disliker of change should keep their permanent posts and be thankful they have them. **1929** E. Linklater *Poet's Pub* xi. 140 You're trying to persuade me to be cowardly and middle-aged, a safety-firster.

sa·fety-pin, *v.* [f. the sb.] *trans.* **a.** To pin *on* or attach with a safety-pin; also *fig.* **b.** To attach a safety-pin or safety-pins to; to put a safety-pin into.

1919 'K. Mansfield' *Let.* 30 Oct. (1928) I. 270 Her ears which are neatly buttonholed on to the sides of her head and not just safety-pinned on as most babies' are. **1960** P. A. Bennett in J. Pudney *Pick of Today's Short Stories* XI. 12 The buttons on my coat will have to be sewn on.. so she safety-pinned them on. **1971** *Country Life* 6 May 1087/1 Fortunately, I had sent the engineer over the side to safety-pin the bombs beforehand. **1975** *Listener* 6 Feb. 176/1 Delysia dressed me in her clothes, safety-pinned me all down the back.

safety-valve. Add: **2.** Also *attrib.*

1925 I. A. Richards *Princ. Lit. Crit.* xxxi. 232 If we do not..try to bring under this Safety-valve heading work with which it has no concern, it may be granted that in some cases the explanation is in place. **1956** 'J. Wyndham' *Seeds of Time* 100 There had been nothing worse than safety-valve grumbling. **1964** R. Miliband in I. L. Horowitz *New Sociol.* 868 What Mills condemned in Safety-valve welfarism was not the welfare.

Hence **sa·fety-valving** *vbl. sb.* (*nonce-use*), letting *off* or discharging as though through a safety-valve.

1965 K. Amis *James Bond Dossier* ix. 93 Violent films, TV shows and the like are useful in safety-valving off our private aggressions.

safflower. Add: **2.** The oil from the seeds is also used in cooking, making margarine, etc. (Further examples.)

1974 *Nature* 13 Dec. 519/2 No work is at present supported at international level on oil seeds such as sunflower, safflower and rapeseed, although their oils are important in the diets of many developing countries. **1980** *Holistic Health News* (Berkeley, Calif. Holistic Health Center) Sept./Oct. 8/3 Mix together: 1 cup of oil, safflower works well. 1½ teaspoons of sea salt.

3. (Further examples.)

1968 *Globe & Mail* (Toronto) 17 Feb. B7 Safflower seed

oil has especially good stability for cooking and frying oils. 1971 H. McCloy *Question of Time* I. iii. 28 Margarine made with safflower oil (butter is as bad for arteries as eggs).

saffra(a)n (sǎfrā·n). [Afrikaans, f. Du. *zaffraan* yellow.] A large evergreen forest tree, *Cassine crocea*, of the family Celastraceæ, found in coastal areas of south-eastern Africa, and bearing yellowish bark and clusters of greenish flowers followed by white plum-shaped fruit; also, the hard light brown wood of this tree. Also *attrib.*

1819 C. G. Curtis *Acct. Colony Cape of Good Hope* 72 Saffran hout... Close and hard. 1831 G. Greig *S. Afr. Almanac* 187 The other woods most in request, and found in Albany are..Red and White Pear, Saffran. 1854 L. Pappe *Silva Capensis* 11 Saffronwood; Saffraan-hout. Branches much spreading. 1950 *Cape Argus* 22 Apr. (Mag. Section) 2/3 He points to a..saffraan, as the oldest inhabitant of the Cape Town gardens. 1953 *Ibid.* 28 Feb. (Mag. Section) 3/7 Near the fountain were some high Saffraan pear trees. 1957 *Cape Times* 26 July 11/1 Holes are being dug..for about 80 shade trees. The species agreed upon..are saffraan and milk-wood. 1973 *Eastern Province Herald* (Port Elizabeth) 28 May 13 A typical wagon of the Great Trek period would have had..wheel falloes of hard pear or saffraan.

saffron, *sb.* and *a.* Add: **A.** *sb.* **6. a.** *saffron flower.*

1910 W. de la Mare *Three Mulla-Mulgars* viii. 108 A little bunch of faded saffron-flower. 1970 Simon & Howe *Dict. Gastron.* 332/1 The English town of Saffron Walden was an important producer [of saffron] and its town arms still have three saffron flowers pictured within the turreted walls.

b. (parasynthetic and with pa. pples.) *saffron-clad, -coloured* (later example), *-flavoured, -robed, -spotted* adjs.

1881 O. Wilde *Poems* 106 Beheld an awful image saffron-clad. 1931 W. Faulkner *Sanctuary* (1981) xviii. 175 A final saffron-colored light lay upon the ceiling. 1959 I. & P. Opie *Lore & Lang. Schoolch.* xii. 243 Simnel Cake, a rich saffron-flavoured fruit cake with almond icing. 1971 *Guardian* 5 July 18/5 The saffron-robed members of the [Hare Krishna] order. 1945 J. Betjeman *New Bats in Old Belfries* 26 Little fields with boulders dotted, Grey-stone shoulders saffron-spotted.

c. **saffron bun,** a bun flavoured with saffron; **saffron milk cap,** an edible orange-coloured funnel-shaped agaric, *Lactarius deliciosus*; **saffron rice,** rice flavoured with saffron; **saffron-wood** (earlier example); = *SAFFRA(A)N.

1852 C. M. Yonge *Two Guardians* i. 12 A feast..of saffron buns, Devonshire cream, and cyder. 1922 Joyce *Ulysses* 158 Saffron bun and milk and soda lunch in the educational dairy. 1977 *West Briton* 25 Aug. 3/4 Each child received a saffron bun and a bottle of pop. 1954 E. M. Wakefield *Observer's Bk. Common Fungi* 55 Saffron Milk Cap..is recognisable by the orange milk which quickly turns green on exposure to the air. 1972 *Times* 23 Sept. 14/5 The..'Saffron Milk Cap' is harmless and eagerly sought. 1926 T. E. Lawrence *Seven Pillars* (1935) III. xxxvii. 217 They took very long about the food and it was not till near noon that at last it came: a great bowl of saffron-rice, with a broken lamb littered over it. 1973 R. Parkes *Guardians* ii. 42 Dan helped himself to another portion of saffron rice, annointed it with curry and tabasco. 1854 Saffron-wood [see *SAFFRA(A)N].

Hence **sa·ffronic** *a.* (*rare*) = SAFFRONY *a.*

1949 E. Sitwell *Canticle of Rose* 245 Then the King who is part of the saffronic dust.

safranin. Add: **a.** Now *Obs.* **b.** Now more commonly **safranine** (-in). Also, any of a large class of azine dyestuffs (chiefly red) related to this, which are obtained typically by coupling of diazotized aromatic monoamines with aromatic diamines. Sometimes with following letter designating particular compounds. (Earlier and later examples.)

1872 *Jrnl. Chem. Soc.* XXIV. 271 (*heading*) Preparation of saffranine [*sic*]. *Ibid.* 828 Safranine when treated with aniline yields a purple dye. 1905 Cain & Thorpe *Synthetic Dyestuffs* xviii. 134 The first technical production of Safranine under this name was carried out under the French patents of Felix Duprey in 1865, but without success. 1911 I. W. Fay *Chem. Coal-Tar Dyes* xii. 298 Mauve, the very first dye prepared by Perkin in 1856, has been shown..to be a true safranine. 1952 K. Venkataraman *Chem. Synthetic Dyes* II. xxv. 766 The simplest Safranine (Safranine B; Phenosafranine; CI 840) is obtained by oxidizing a mixture of *p*-phenylenediamine and aniline to the indamine by means of dichromate and hydrochloric acid, and boiling the solution to convert the blue indamine into the red Safranine. 1971 R. L. M. Allen *Colour Chem.* viii. 124 Safranine T is used for dyeing tannin-mordanted cotton, bast fibres, wool, silk, polyacrylonitrile fibres, leather and paper.

safrol. Substitute for entry:

safrole (sæ·frōᵘl). *Chem.* Formerly **safrol** (-ǫl). [ad. F. *safrol* (Grimaux & Ruotte 1869, in *Compt. Rend.* LXVIII. 928) f. sas)*safr*(as SASSAFRAS: see -OL, *-OLE.] A colourless, liquid, bicyclic, aromatic ether, $C_{10}H_{10}O_2$, which occurs in a number of essential oils, esp. oil of sassafras of which it is the major constituent.

1869 *Chem. News* 16 July 35/1 The oil further contains safrol, $C_{10}H_{10}O_2$, boiling at between 231° and 233°. 1884 *Jrnl. Chem. Soc.* XLVI. 1338 Safrole is the main constituent of the essential oil of sassafras. 1922 [see *PINENE]. 1950 *Thorpe's Dict. Appl. Chem.* (ed. 4) X. 656/1 Oil of sassafras is obtained from *Sassafras officinale* Nees..and contains 78% safrole... Safrole is also found as a constituent of many essential oils especially those derived from the order Lauraceæ. 1970 *New Scientist* 30 July 232/2 There are very many substances used as food ingredients..which might, like the safrole in root beer, be found to be toxic. 1976 *Nature* 22 July 252/1 After many generations of people had enjoyed the natural flavour of sassafras, it turned out that safrole, the substance responsible for this, caused cancer in rats.

sag, *sb.*² Add: **3. a.** (Earlier examples.)

1727 in *Amer. Speech* (1940) XV. 387/1 Thence along the North Side of the Mountains to a Corner Several Saplins by a Sagg. 1850 *Rep. Comm. Patents 1849: Agric.* (U.S.) 443 Strawberries are met with..on the edges of 'sloughs' or 'saggs'.

b. *fig.*

1868 W. James *Let.* 15 May in R. B. Perry *Tht. & Char. W. James* (1935) I. 512 Such an event rather dislocates my mind from its habitual 'sag' in contemplating the world. 1938 E. Bowen *Death of Heart* III. i. 323 Behind the opaqueness of her features control permitted no sag of tiredness.

4. Also, in a business or programme of development.

1897 E. Hough *Story of Cowboy* 334 Then in time came ..the 'sag' in the cattle business. 1946 *Sun* (Baltimore) 20 Aug. 8/2 (*heading*) The sag in the housing program. a 1974 R. Crossman *Diaries* (1975) I. 555 The Chancellor had cut back local-authority spending on mortgages... In that case, I said, he must permit us to use public-sector building to make up for the sag.

5. *attrib.* and *Comb.*, as **Sagbag,** the proprietary name of an informal chair consisting of a large bag filled with polystyrene granules which accommodates itself to the form of the sitter; also **sagbag; sagbend,** the curved stretch of pipe below the point of inflexion in the S-shaped length of pipeline as it is lowered on to the sea bed from a barge (cf. *OVERBEND *sb.*); **sag pond,** a pond whose basin is the result of earth movement associated with a fault; **sag wagon** *Cycling* (see quot. 1961); also *transf.*

1974 *Observer* 13 Jan. 23/6 (*caption*) Polystyrene sagbag in various colours of canvas..from..Habitat. 1974 *Trade Marks Jrnl.* 18 Dec. 2520/2 Sagbag... Furniture; chairs and settees: seats and seating... cushions (not for medical or surgical purposes);..Habitat Designs Limited, Hithercroft Road, Wallingford, Berkshire; Merchants. 1978 *Evening Standard* 28 Apr. 18/4 June Mendoza's picture of La Rippon, shoeless in a denim jump-suit and reclining fetchingly in a purplish sag-bag. 1969 *Preprints 1st Ann. Offshore Technol. Conf.* II. 37/2 To prevent excessive bending in the sag bend a straight stinger must discharge the pipe very near to the bottom. 1975 *Petroleum Rev.* XXIX. 309/1 A pipe tensioning system has been provided which will avoid buckles in the sagbend. 1933 *Calif. Jrnl. Mines & Geol.* XXIX. 197 Numerous little water-holding depressions known as sag ponds mark the site of local subsidences. 1974 Gribbin & Plagemann *Jupiter Effect* x. 114 And yet in the San Francisco Peninsula, where memories of 1906 should prompt some caution, lines of so-called sag ponds, which geologists use as a clear indicator of the fault line, have been filled in to make building land! 1961 Partridge *Dict. Slang Suppl.* 1257/1 Sag-wagon, a van that, following a [cycle] race, picks up exhausted riders. 1963 *Times* 6 June 5/7 When, later, Selaru had gear trouble they both gave up the struggle for the solace of a sag waggon. 1977 C. McFadden *Serial* (1978) xlvii. 102/2 You wanna come along in the bus in case I need a sag wagon?

sag, *v.* Add: **2. c.** (Earlier example.)

1870 W. W. Fowler *Ten Yrs. in Wall St.* xxv. 393 The price grew firmer when two or three men were observed selling quietly large amounts, and then the price sagged to 250.

6. *intr.* and *trans.* To play truant (from). *Liverpool local.*

1959 I. & P. Opie *Lore & Lang. Schoolch.* xvii. 372 *Sagging.* This is definitely the prevailing term [for playing truant] amongst delinquents in all parts of Liverpool. 1965 *Woman* 28 Aug. 8, I re-visit childhood haunts in Liverpool, meet the next generation in the Cathedral grounds where we used to 'sag'—that is, play truant. 1966 F. Shaw et al. *Lern Yerself Scouse* 45 I'm saggin skewl, I am playing truant.

saga¹. Add: **1. b.** Also, a novel or series of novels recounting the history of a family through several generations, as *The Forsyte Saga.* Now freq. in weakened use, a long and complicated (account of a) series of more or less loosely connected events.

[1891 R. L. Stevenson *Let.* 19 May (1899) II. 231 Henry Shovel has now turned into a work called 'The Shovels of Newton French'.., which work is to begin in 1664..and end about 1832... I mean to make it good; it will be more like a saga.] 1895 Hall Caine *Bondman* (ed. 3) p. viii, I have called my story a Saga, merely because it follows the epic method. 1919 J. Galsworthy *Let.* 25 Nov. in H. V. Marrot *Life & Lett. J. Galsworthy* (1935) v. i. 485, I have just finished a sequel to *The Man of Property*, and, in accordance with the scheme I broached to you..have still one story and a third novel in further sequel to write, to make the whole of *The Forsyte Saga*. 1935 D. L. Sayers *Gaudy Night* iii. 51 She felt she would rather be tried for life over again than walk the daily treadmill of Catherine's life. It was a saga, in its way, but it was preposterous. 1942 'M. Innes' *Daffodil Affair* II. 89 Appleby and Hudspith were scarcely in a position to give it the dispassionate appraisal of literary critics; the saga had a sort of aura of alligator which made it uncomfortable hearing. 1952 *Times Lit. Suppl.* 1 Jan. 15/3 The latest, no doubt the logical, development of the 'life with mother' saga is the chronicle of pregnancy and childbirth. 1959 *Listener* 18 June 1074/1 The Burrell Collection..is still, after a long saga of misadventures, looking for a site. 1970 *Nature* 18 Apr. 197/1 By now, the daily newspapers will tell how the saga of Apollo 13 has been finished. 1977 'E. Crispin' *Glimpses of Moon* x. 190 Rousing themselves hastily from the morbid fascination induced by this saga, Thouless, Padmore and the Major all went into action. 1978 H. Wouk *War & Remembrance* xlix. 497 'Found her! Where?' 'In Marseilles. Told me about it for two hours over dinner. It's a saga.'

¶ **2.** (Earlier and further examples.)

1845 B. Thorpe in J. M. Lappenberg's *Hist. Eng.* I. 90 The poem of Beowulf.., in which the old Anglian saga is ennobled by an Anglo-Saxon of the eighth century. 1855 Geo. Eliot in *Fraser's Mag.* July 55/1 The libretto is founded on the old German saga of the Venusberg and the knightly minstrel Tannhäuser. 1881 H. Morley *Longer Works in Eng. Verse & Prose* I. i. 1/1 Most ancient of English poems is the old saga which tells how Beowulf rescued Hrothgar from the attacks of Grendel. 1898 T. Arnold *Notes on Beowulf* v. 71 Whether the Sigemund-Siegfried saga is of Scandinavian or German origin. 1903 L. F. Anderson *Anglo-Saxon Scop* 16 The great number of sagas learned by the scop of *Beowulf* is expressly mentioned... It was praiseworthy in a scop to have learned not only the more familiar sagas, but some not generally known. 1912 R. W. Chambers *Widsith* 15 How much of this is history, and how much saga, it is not easy to say. 1960 M. B. McNamee in *Jrnl. Eng. & Gmc. Philol.* LIX. 199 At least by the eleventh century, the mysterious serpent-infested mere of Anglo-Saxon saga had provided a means of making the story of Christ and Satan and Hell graphic to the Anglo-Saxon imagination.

3. *saga-age, -cycle, -hero;* **saga boy** *W. Indies,* [perh. f. a different word], a well-dressed lounger, a playboy.

1897 W. P. Ker *Epic & Romance* iii. 230 In the material conditions of Icelandic life in the 'Saga Age' there was all the stuff that was required for heroic narrative. 1956 Peterson & Fisher *Wild Amer.* xxxiii. 354 The..Eskimos used to drive the geese across the tundra..and net them..a method of wildfowling known..in Iceland, where it became a great art in the Saga Age. 1949 *Human Relations* II. 358/1 This change in behaviour is clearly demonstrated by men who have been to Aruba, Curacao, Trinidad, or U.S.A., and who have acquired some money. ..In this group the 'Saga-Boys' are to be found—flamboyantly dressed men with exaggerated manners and mannerisms and somewhat aggressive tendencies. 1959 V. S. Naipaul *Miguel St.* xi. 118 Eddoes was a real 'saga-boy'. This didn't mean that he wrote epic poetry. It meant that he was a 'sweet-man', a man of leisure, well-dressed, and keen on women. 1966 P. Sherlock *West Indies* xi. 143 Saga boys dressed in sheath-like saga pants, 'peg-top trousers' and saga coats called Bim-Bams. 1892 S. A. Brooke *Hist. Early Eng. Lit.* I. 104 The first saga-cycle includes the songs sung concerning the earlier deeds of Beowulf before he became king. 1899 W. H. Schofield tr. *S. Bugge's Home of Eddic Poems* 172 In the oldest reference to this saga-hero, in *Widsith*, 21, we read: *Hagena* [*wéold*] *Holmrygum*.

sagaciate (sǎgēⁱ·ʃiˌēⁱt, -æ·ʃu-), *v.* *U.S. dial.* Also **segashuate,** etc. [App. jocularly f. SAGACI(OUS *a.* + -ATE³.] **a.** *intr.* To thrive or prosper. (Freq. used when inquiring after one's health.)

1832 *Boston Transcript* 2 Aug. 2/3 Well, Clem, how do you sagatiate dis lubly wedder? 1842 *Literary Gaz.* 1 Jan. 6/3 How does your copperosity sagaciate this morning? 1880 J. C. Harris *Uncle Remus* ii. 24 'How duz yo' sym'tums seem ter segashuate?' sez Brer Rabbit, sezee. 1890, etc. [see *CORPOROSITY]. 1906 *Dialect Notes* III. 154 How are you sagashawatin'? 1976 K. Bonfiglioli *Something Nasty in Woodshed* xi. 131 My symptoms started to sagashuate again but Jock blocked my every move to slink back into bed.

b. *nonce-uses.*

1904 A. Morrison *Green Eye of Goona* v. 181 The police sagaciate that Pooley must ha' gone straight to London. 1909 'O. Henry' *Roads of Destiny* xxii. 366, I sagatiated in your associations once, if I am not mistaken.

sagakomi (sægākōᵘ·mi). *N. Amer.* Also 8 **segockimac,** 8– **sac à commis,** 9– **sacacom(m)i(s).** [a. Ojibwa *sakākkomin* bearberry.] = BEARBERRY *a, b*; also, the leaves of this plant used with, or as a substitute for, tobacco.

In quot. 1934 wrongly applied to madroño, *Arbutus menziesii*, another member of the Ericaceæ.

1703 L. A. Lahontan *New Voy. N. Amer.* II. 53 They are forc'd to buy up Brasil Tobaco, which they mix with a certain Leaf..call'd Sagakomi. 1778 J. Carver *Trav. Interior Parts N. Amer.* 31 A weed that grows near the great lakes..is called by the Indians Segockimac, and creeps like a vine on the ground,..bearing a leaf about the size of a silver penny, nearly round... These leaves, dried and powdered, they likewise mix with their tobacco. 1823 J. Franklin *Narr. Journey to Polar Sea* 741 Jackashey-puck..has received the name of *Sac à commis*, from the trading clerks carrying it in their smoking bags. 1836 G. Back *Narr. Arctic Land Exped.* ix. 257 We passed

many sandhills, variegated by the..plant, called..by the traders 'sac à commis'. 1837 *Trans. Lit. & Hist. Soc. Quebec* III. 91 Saccacommi [is] frequently used to smoke in lieu of tobacco, by the traders engaged in the fur countries. 1890 L. F. R. Masson *Bourgeois de la Compagnie du Nord-Ouest* II. 102 *Graine d'ours*, bear berry, also called *sac à commis*, a creeping plant which is smoked, and which the clerks put in their sacs. 1910 F. W. Hodge *Handbk. Amer. Indians.* II. 407/2 Sagakomi. The name of a certain smoking mixture, or substitute for tobacco, applied also to the bearberry bush..or other shrubs the leaves and bark of which are used for the same purpose. 1934 L. L. Haskin *Wild Flowers Pacific Coast* 263 The coast Indian name for it [sc. madroño] seems to have been *saccacomis*, upon which the French constructed a pun, calling it *sac-a-commis*.

sagamité. **a.** (Earlier and later examples.)

c 1665 P. E. Radisson *Voyages* (1885) 40 Then my father made a speech shewing many demonstrations of vallor, broak a kettle full of Cagamite with a hattchett. 1744 J. de Champigny *Present State Louisiana* 22 They were employed in..making Sagamité and baking it. 1880 G. W. Cable *Grandissimes* 26 They sat down to bear's meat, sagamite and beans. 1916 F. W. Waugh *Iroquois Foods* 91 Probably no corn or other food is referred to so frequently as hominy, or sagamité, as it was more familiarly known to the early French. 1931 W. Cather *Shadows on Rock* iv. iii. 193 Cécile did not want much breakfast... She had sagamite and milk. 1940 E. J. Pratt *Coll. Poems* (1958) ii. 256 It was the middle room that drew the natives, Day after day, to share the sagamite And raisins, and to see the marvels brought From France. 1963 *Beaver* Autumn 17/2 Their [sc. the Hurons'] sagamité, a kind of corn porridge that drew excruciatingly long faces from the early Frenchmen, was, nevertheless, remarkably nourishing.

sagashuate, etc., varr. *SAGACIATE.

sage, *sb.*¹ Add: **2. b.** *sage-and-onion stuffing* (earlier example).

1861 Mrs. Beeton *Bk. Househ. Managem.* 241 (*heading*) Sage-and-onion stuffing, for geese, ducks, and pork.

4. Delete '?' (Earlier example.)

1805 M. Lewis *Jrnl.* 12 May in *Orig. Jrnls. Lewis & Clark Exped.* (1904) II. 29 The wild hysop sage..and some other herbs also grow in the plains and hills.

4*. The colour of sage.

1881 C. C. Harrison *Woman's Handiwork* I. 20 A ground of sage or of Pompeian red velvet. 1971 *Vogue* 15 Sept. 129/1 Suit..sizes 10–16: colours: brown/white, burnt orange/white, olive/sage.

5. a. *sage-ash, -scrub.*

1923 D. H. Lawrence *Birds, Beasts & Flowers* 147 An eagle at the top of a low cedar-bush On the sage-ash desert. 1927 —— *Mornings in Mexico* 136 Across the grey desert..low, grey, sage-scrub was coming to pallid yellow.

b. **sage-brush** (earlier and later examples); also in *Comb.*, as **Sagebrush State,** popular name of Nevada (formerly also applied to Wyoming); **sage-bush** (later examples); **sage-green,** also as adj.; (earlier and later examples); **sage-grey** *a.* = *sage-green* in Suppl.; **sage-willow** (earlier example); **sage-wood,** (*b*) a small tree or shrub, *Buddleia salviifolia*, of the family Loganiaceæ, found in southern Africa and bearing leaves like those of common sage and racemes of white or purple flowers; also, the hard, heavy wood of this tree.

1850 K. Webster *Gold Seekers of '49* (1917) iii. 84 We were compelled to tie our mules to sage brush to keep them from straying away. 1861 'Mark Twain' *Lett.* (1917) I. 54 On the plains, sage-brush and grease-wood grow about twice as large as the common geranium. 1893 L. Wagner *Significance of Names* 35 Nevada is also called..The Sage-Brush State, from the wild artemesia covering the plains. 1907 S. E. White *Arizona Nights* 191 We began to toil in the ankle-deep sand of a little sage-brush flat. 1917 *Boston Even. Globe* 11 Apr. 16/4 Nevada has been known for many years as the Sagebrush State. 1934 G. E. Shankle *State Names* ii. 155 The sobriquet, the *Sagebrush State*, applied to Wyoming, refers to the fact that wild sage (*Artemisia tridentata*) grows on the desert sections of this State. 1946 D. C. Peattie *Road of Naturalist* v. 53 A few forms like sage-brush or creosote bush..repeat themselves for fifty miles on end. 1976 *Billings* (Montana) *Gaz.* 20 June 10-c/2 Sagebrush, Silver and Battle Born State are nicknames for Nevada, first explored by the Spaniards in 1776. 1976 *Sci. Amer.* Oct. 20/3 Strong lives with his wife and 11-year-old daughter in pine and sagebrush country overlooking the Rio Grande. 1977 J. F. Fixx *Compl. Bk. Running* ii. 25 We are in a canyon surrounded by mountains, trees and sagebrush. 1902 'Mark Twain' in *Harper's Mag.* Jan. 269/2 He started on a run, racing in and out among the sage-bushes. 1977 J. L. Harper *Population Biol. Plants* xx. 604 Woodland dominated by pinyon pine..is intimately associated with *Artemisia* sage-bush communities. 1810 *Repository of Arts* (Ackermann) Apr. 262/2 Light sage green, or cream-coloured kerseymere breeches. 1929 [see *DEMI-SEMI]. 1976 *Star* (Sheffield) 20 Nov. 10/2 (Advt.), Bed-settee with arms. Teak frame. Sage green expanded vinyl. 1923 D. H. Lawrence *Birds, Beasts & Flowers* 190 Day has gone to dust on the sage-grey desert. 1846 G. B. Emerson *Rep. Trees & Shrubs Mass.* 256 The sage willow is a slender, hoary plant, or a spreading tufted bush. 1854 L. Pappe *Silva Capensis* 31 Sage-wood... Wood hard, tough, heavy. 1932 Watt & Breyer-Brandwijk *Medicinal & Poisonous Plants S. Afr.* 140 Sagewood, Saliehout,..is possibly used medicinally by the Hottentots. 1973 *Stand. Encycl. S. Afr.* IX. 458/2 The sagewood..belongs to the rather heterogeneous family Loganiaceæ.

c. **sage cock** (earlier and later examples); **sage hare** (example); **sage hen,** substitute for def.: = *sage cock*; (earlier and later examples); **sage rabbit,** substitute for def.: one of several small hares of western North America, esp. *Sylvilagus nuttallii*; (earlier example).

1840 A. Wislizenus *Ausflug nach Felsen-Gebirgen 1839* ix. 49 Sage cock, cock of the plains. 1917 T. G. Pearson *Birds Amer.* II. 30/1 The Sage Cock has a sharp cackle. 1868 *Amer. Naturalist* II. 536 The Sage Hare..is more rare near Fort Benton. 1843 J. Williams *Jrnl.* 27 July in *Narr. Tour to Oregon* (1921) 14 The sage hen is found here also. 1917 [see *COCK *sb.*¹ 10]. 1962 E. Lucia *Klondike Kate* viii. 170 A great flight of sagehens darkening the sky. 1846 R. B. Sage *Scenes Rocky Mts.* p. iv, [The] sage rabbit..is nearly three times the size of the common rabbit.

d. *attrib.* or as *adj.* Resembling the colour of sage (sense 1). Cf. sense 4* above.

1785 E. Sheridan *Jrnl.* 5 July (1960) 59, I have one [plume of feathers] for mine [sc. a hat] of dark sage, pink and white feathers. 1820 M. Edgeworth *Let.* 8 June (1979) 160 My two tabbinets, sage and fawn ditto have done excellent service, new furbished. 1904 T. Eaton & Co. *Catal.* Spring & Summer 187/1 All-wool carpet..in red and sage colorings.

SAGE, Sage (sēⁱdʒ), *sb.* *Mil.* [Acronym f. the initial letters of 'semi-automatic ground environment'.] A name given to an early warning and air defence control system covering the United States and Canada. Freq. *attrib.*

1955 *N.Y. Times* 25 Sept. IV. 2/2 Some time in 1954—the exact date has not been disclosed—the National Security Council gave the Air Force the go-ahead on a project..called Sage. 1958 *Times* 23 July 9/6 Sage is basically a computer..which will evaluate all the information received from the early warning networks, guide missiles and aircraft to their targets, and even work out which is the best weapon to use in a particular situation. 1958 *Electr. Engin.* LXXVII. 793/1 At the present time, excluding other connecting weapons systems, there are three main types of data systems used in the SAGE system. Grouped according to use, these are ground-to-ground, ground-to-air and radar data systems. 1961 *Aeroplane* C. 115/1 Construction has started on the new SAGE (semi-automatic ground environment) defence system which is to be introduced into Canada during this year. 1971 E. Luttwak *Dict. Mod. War* 44/2 Guidance is by command direct from the SAGE Air Defence System supplemented by radar homing for final interception.

saggar, seggar, *sb.* The forms **saggar, sagger** are now usual. **1.** Also, more widely, a case made of refractory material or cast or wrought iron used to protect objects while in a furnace, esp. as in Dict. and also during annealing of iron castings. (Further examples.)

1888 *Lockwood's Dict. Mech. Engin.* 298 *Saggers*, cast-iron boxes used for packing the castings and sifted red hæmatite, in readiness for the annealing oven, in the process of manufacture of malleable cast iron. 1928 H. M. Boylston *Iron & Steel* v. 151 If the parts are small..they are packed with a mixture of rolling-mill scale or scale from saggers and brick-bats or sand. The packed pots, or saggers, are then heated in an annealing furnace. 1960 *Times Rev. Industry* July 22/1 A rich iron ore is packed, together with coke breeze (the reducing agent) and limestone, into clay containers called saggars (the term is taken from the pottery industry, and in fact the process is very similar to that used in making pottery). 1964 H. Hodges *Artifacts* i. 39 Some glazed wares need to be protected from the direct flame, and..this may be done by placing them in lidded boxes called saggars (saggers or seggers). 1967 M. Chandler *Ceramics in Mod. World* ii. 79 The ceramist uses saggers only when he must. 1977 R. Fournier *Illustr. Dict. Pract. Pottery* (rev. ed.) 196/2 With the coming of cleaner fuels, smaller kilns, and 'continuous' firing the use of the saggar has declined sharply and it is becoming difficult to purchase them.

2. *saggar-clay* (earlier example).

1786 J. Wedgwood *Let.* 13 Feb. (1965) 292 With regard to Sagar clays, they cannot be judged of from their external appearance.

saggar, *v.* Add: Hence **sa·ggaring** *vbl. sb.*

1901 W. P. Rix tr. *Bourry's Treat. Ceramic Industries* xiii. 718 The great trouble of burning porcelain, looked at from all points, is saggering.

saggy, *a.*² Add: Also *colloq.*

1977 'L. Egan' *Blind Search* ii. 30 An old saggy couch to sleep on.

sagina (sădʒəi·nă). [a. L. *sagīna* fatness, adopted as a generic name by Linnæus (*Systema Naturæ* (1735)).] A small annual or perennial mat-forming herb of the genus so called, belonging to the family Caryophyllaceæ, esp. *S. pilifera* or *S. procumbens*, which are sometimes used instead of grass as lawn plants; = PEARLWORT.

1962 R. Page *Educ. of Gardener* iv. 132 A smaller, sunken oval..set out in a chessboard design with squares of stone alternating with squares of sagina. 1972 *Country Life* 23 Mar. 676/2 During recent years continental exhibitors at flower shows have employed turves cut from prostrate, moss-like sagina, and lawns of this have met with limited success in Germany.

sagittal, *a.* Add: **3.** *Optics.* Pertaining to or designating the plane that contains the chief ray from an off-axis point source and those rays that are brought to a point in the further (radial) line image formed by an astigmatic system (in a plane at right angles to the sagittal plane).

1902 Mann & Millikan tr. *Drude's Theory of Optics* iii. 50 All the rays emitted by P..cross the axis at the same point P_2. The beam made up of such rays is called a sagittal beam. It has a focal point at P_2. 1910 J. P. C. Southall *Geom. Optics* vii. 333 Following the usage of most modern writers, we shall call the incident and refracted rays lying in the planes π, π', respectively, the Sagittal Rays. [*Note*] 'Sagittal' is a term borrowed from Anatomy... Some writers..prefer..the word 'equatorial' instead of sagittal. 1936 H. T. Flint *Geom. Optics* vii. 150 This line and P_1P_1' are the focal lines of the astigmatic reflected pencil,..the sagittal and tangential lines respectively. 1972 O. N. Stavroudis *Optics of Rays* xii. 226 These skew rays, called sagittal rays, also pass through opposite sides of both the circle on the exit pupil and the ellipse on the image plane... Any fan of sagittal rays from the fixed object point will converge to..the sagittal focus... The tangential focus and the sagittal focus are the astigmatic foci.

sagitally *adv.* (examples).

1950 *Jrnl. Compar. Neurol.* XCII. 142 The right postcentral sulcus..its lower limb running almost sagitally. 1977 *Lancet* 29 Oct. 930/2 The pineals were removed, bisected sagitally, homogenised, and stored at −20°C.

Sagittarian (sædʒitēᵊ·riăn), *sb.* and *a.* *Astrol.* [f. SAGITTARI(US + -AN.] **A.** *sb.* A person born under Sagittarius (22 November–21 December), the ninth sign of the Zodiac. **B.** *adj.* Of, pertaining to, or characterized by Sagittarius; born under Sagittarius.

1911 I. M. Pagan *Pioneer to Poet* ix. 126 The chief characteristic of the fully developed Sagittarian is his extraordinary power of mental activity. 1924 C. E. O. Carter *Conc. Encycl. Psychol. Astrol.* 11 Psychologically the progressiveness of the Sign shows as Hope, reaching forward into the future... The belief in immortality is typically Sagittarian. 1940 R. Gleadow *Astrol. in Everyday Life* ix. 210 Sagittarian luck depends very much upon Jupiter's condition. 1950 C. Fry *Venus Observed* i. 12 Your birthday? No, you're a Sagittarian. This is only October. 1964 L. MacNeice *Astrol.* iii. 96 In music the Sagittarian type is Beethoven. 1979 J. Leasor *Love & Land Beyond* i. 12 He enjoyed the ultimate privilege of freedom which, as a Sagittarian, he put..beyond price.

Sagittarius. Add: **3.** *Astrol.* = *SAGITTARIAN *sb.* Also without article.

1940 R. Gleadow *Astrol. in Everyday Life* ix. 209 No one, of course, can think more quickly than Gemini; but Sagittarius can guess; he is very intuitive. 1969 'V. Packer' *Don't rely on Gemini* (1970) i. 3 'Was Pope John a Gemini?' 'Oh no... He was a Sagittarius.' 1970 *Guardian* 27 Apr. 1/1 The Labour member for Bebington is a Sagittarius. 1979 S. Rifkin *McQuaid in August* (1980) vi. 35 It is clear to me you're Sagittarius. You're prompt, calm, and very reliable.

sago. Add: **3.** *sago pudding* (later example).

1973 'D. Jordan' *Nile Green* xxiii. 92 A notorious property developer..was spooning sago pudding into his face.

saguaro. Add: Also **sahuaro, sugarro.** Substitute for def.: A large branching cactus, *Carnegiea gigantea*, found in desert regions of southwestern North America. (Earlier and later examples.)

1856 *Wild West* (San Francisco) Oct. 4/6 There are in this region a few Indian rancharies, to which the *Papagos* resort to gather the fruit of the sugarro. 1864 S. Mowry *Arizona & Sonora* 161 Gradually appear..scattered saguaras. 1881 *Amer. Naturalist* XV. 982 By far the most conspicuous form is..the 'saguara' cactus. 1907 S. E. White *Arizona Nights* 220 [The] snake..looked just like a sahuaro stalk. 1916 E. C. Peixotto *Our Hispanic Southwest* 64 Tall saguaros reared their fluted columns like giant candelabra. 1933 *Sun* (Baltimore) 27 Sept. 15/6 The Southwest's giant tree cactus, or sahuaro, is susceptible to crown gall. 1955 [see *CHOLLA]. 1968 W. Garner *Deep, Deep Freeze* xxxii. 261 Drieter slipped behind a great saguaro with a girth matching his own. 1978 *Times* 21 Aug. 10/6 We could tell this was the real desert because we glimpsed our first saguaro cactuses—those monstrous plants, some more than 100 years old and standing up to 15 feet high. 1979 *Arizona Daily Star* 1 Apr. (Advt. Section) 16/7 This parcel [of land] is dotted with native sahuaro.

sah (sā), *colloq.* and (*U.S.*) *dial.* var. SIR *sb.* Cf. *SUH.

1893 H. A. Shands *Some Peculiarities of Speech in Mississippi* 54 *Sah* (sa), Negro for *sir*. 1901 W. Churchill *Crisis* I. i. 5 'But, Ephum! Say, Ephum!' 'Yes sah.'

Sahaptin (sāhæ·ptĭn), *sb.* and *a.* Also † **Sahapotin, Shahaptan.** [Southern Interior Salish *Sʔaptnx* Sahaptin, Nez Percé; of uncertain ulterior etymology.] **A.** *sb.* **a.** Formerly, a (member of an) American Indian tribe of the Snake River basin, also called the *NEZ PERCÉ; also, any of several groupings of

the Nez Percé and others believed to be linguistically related. Now applied to a number of closely related North American Indian peoples of the Columbia River basin. **b.** The language or language grouping of any of these peoples. **B.** *adj.* Of or pertaining to any of these peoples or their language.

1836 A. Gallatin in *Trans. Amer. Antiquarian Soc.* II. 264 (map) Sahapotins. 1841 *Jrnl. R. Geogr. Soc.* XI. 225 The first and more northern Indians of the interior may be denominated the Shahaptan Family, and comprehends ..the Shahaptan, or *Nez Percés*..; the Kliketat..; and the Okanagan. 1846 H. Hale *U.S. Exploring Exped.: Ethnogr. & Philol.* 198 The South-Oregon division. To this belong the *Sahaptin* family (Nez-percés and Wallawallas), [etc.]. 1918 J. E. Rees *Idaho Chronol.* 109 Their earliest home was upon the Columbia River and when they were pushed southward the Salish called them 'Shahaptans', meaning 'strangers from up the river'. 1918 *Internat. Jrnl. Amer. Linguistics* I. 176, I have..gathered voluminous data supporting previously expressed contentions concerning the genetic relationship between Lutuamian, Wailatpuan, and Sahaptin. 1921 E. Sapir *Language* 222 The presence of postpositions in Upper Chinook..is clearly due to the influence of neighboring Sahaptin languages. 1940 M. W. Smith *Puyallup-Nisqually* 22 If he spoke Sahaptin, it is also certain that he spoke Salish. 1947 B. A. De Voto *Across Wide Missouri* 11 Ethnologists use the name which the Flatheads bestowed on them, the Shahaptan, of uncertain meaning but perhaps a designation of the country they lived in. 1965 *Canad. Jrnl. Linguistics* X. 125 Jacobs also has published a considerable quantity of Sahaptin texts. 1971 *Language* XLVII. 840 The northernmost example, the Sahaptin and Nez Perce shift of n > l, is reversed in comparison to the more southern shifts.

Sahara. Add: **1. b.** (Earlier example.)

1855 Dickens *Holly-Tree Inn* in *Househ. Words* Extra Christmas No. 2/2 The bleak wild solitude..was a snowy Saharah.

2. A shade of brown or yellow. Also *attrib.*

1923 *Daily Mail* 9 Oct. 1/1 Colours: Lemon,..Fawn, Sahara, Mole. 1930 *Daily Express* 8 Sept. 11/5 The suit is stocked in shades of sand, Sahara Brown and Grey. 1970 'D. Halliday' *Dolly & Cookie Bird* ii. 12 He was..broad-shouldered, with that super kind of Swedish suède jacket in Sahara sand colour. 1974 *Times* 4 May 5/2 Bathroom suites in..honeysuckle, orchid, midnight blue, sahara, black. 1976 *Yorkshire Evening Press* 9 Dec. 20/2 (Advt.), 1971 Opel Rekord coupe in Sahara Gold.

Saharan, *a.* (In Dict. s.v. SAHARA.) Add: **B.** *sb.* **a.** One of a group of languages spoken in the eastern Saharan region. **b.** A member of a people living in the Sahara, *spec.* native to or inhabiting the former Spanish Sahara on the Atlantic coast.

1963 J. H. Greenberg *Lang. Afr.* vi. 130 To the..grouping which consists of Songhai, Saharan, Maban, Fur and Coman in addition to Chari-Nile, the name Nilo-Saharan is given. 1970 *Daily Tel.* 30 June 16 The men from the Spanish Sahara, known as 'Saharans', now in Egypt have two main objectives. 1975 *N.Y. Times* 8 Nov. 26/2 The referendum under international supervision, recommended by a United Nations commission, will be difficult to conduct fairly among mostly illiterate and nomadic Saharans.

Saharaui, var. *SAHRAWI.

Sahelian (sāhī·liăn), *a.* [f. *Sahel*, proper name of the region + -IAN.] Of, pertaining to, or designating the belt of land in West Africa south of the Sahara desert which comprises parts of Senegal, Mauritania, Mali, Niger, and Chad and is mostly savannah.

1973 *Nature* 28 Sept. 194/2 The present drought situation and the probable long term trends now seriously threaten the economic and political viability of the Sahelian states of West Africa. 1973 *Times* 30 Oct. 16/6 The exhibition shows the effects of the drought in the Sahelian region of West Africa. 1976 *New Society* 29 Apr. 220/2 Africa's poorest region, the Sahelian 'famine belt' from the Atlantic to Lake Chad, was once a place of fabulous wealth. 1980 *Spectator* 21 June 18 Next come the even more gruesome aid donors, whose antics Markham witnessed during the terrible Sahelian drought a few years ago.

sahib. Add: Also **saab, sa'ab** and with pronunc. (sāb). **1. a.** (Further examples.) Also affixed to Indian and Bangladeshi titles and names.

1886 Kipling *Departmental Ditties* (ed. 2) 7 Rajah Rustum..Heaped upon the Bukshi Sahib wealth and honours manifold. 1921 E. M. Forster *Let.* 1 Apr. in *Hill of Devi* (1953) 60 The Palace is inhabited by four chief people—me, H.H., Malarao Sahib, and Deolekr Sahib. 1971 *Shankar's Weekly* (Delhi) 4 Apr. 8/1 Here we are grappling with basic issues and our director saab is bothered about mixed metaphors and split infinitives. *Ibid.* 21/4 He then went to Lalaji's house outside which Vijay was furiously pacing up and down. 'Yes, sa'ab,' he meekly announced his arrival. 1977 'D. MacNeil' *Wolf in Fold* ii. 18 The native nodded. 'I understand, Ogilvie sahib. I believe also that the risaldar sahib will help.' 1978 F. Olbrich *Desouza pays Price* iii. 12 A gentleman would like to see you, Inspector saab.

b. *transf.* A gentleman; someone considered socially acceptable.

1919 W. Deeping *Second Youth* xxv. 212, I happen to know Colonel Horseley out there; he's a sahib, and quite big, one of the biggest things I've met. 1928 D. L. Sayers *Unpleasantness at Bellona Club* ix. 102 'Is the fellow a sahib?' 'Good God, no! Looks like an attorney's clerk or something.' 1952 A. Grimble *Pattern of Islands* 24 A sahib, naturally..right kind of breeding, right kind of school. 1977 *Listener* 28 July 123/1 Being a muff can be as arduous a vocation as being a sahib.

2. *Comb.*, as **sahib-log** [Urdu *log* people, caste], the European gentlefolk in India.

1848 J. H. Stocqueler *Oriental Interpreter* 199/2 *Sahib logue*, the common appellation given to European gentlemen in India. 1927 W. H. Todd *Tiger, Tiger!* vii. 117 The 'sahib-log' were after him. 1953 P. Scott *Alien Sky* I. iv. 42 The Sahib-log lived in whitewashed bungalows. 1978 'M. M. Kaye' *Far Pavilions* ii. 30 The troopers..asserted that all the Sahib-log in Meerut were dead.

Sahib-dom (sā·ibdəm). ? *Obs.* [f. SAHIB + -DOM.] The quality or condition of being a sahib.

1901 Kipling *Kim* ix. 215 'Oah!' said Kim, firmly resolved to cling to his Sahib-dom. 1909 M. Diver *Candles in Wind* iv. 45 A creature without either the birthright of caste, or the prestige of Sahib-dom.

sahibhood (sā·ibhud). [f. as prec. + -HOOD.] = prec.

1946 [see *NEGRONESS]. 1953 J. Trench *Docken Dead* vii. 104 He looked round for admiration..at..the evidence of sahibhood. 1977 A. Wilson *Strange Ride R. Kipling* i. 23 The need to assert his lost sahibhood.

‖**sahitya** (sā·hitya). [Skr., association, agreement; composition, literature; lyrical verse.] The lyrical verse which forms part of an Indian dance-song (see quots.).

1953 F. Bowers *Dance in India* 46 Three types of singing, determined by the nature of the dance, are performed in Bharata Natya: (1) Ordinary poetic songs with words for abhinaya portions, called *sahitya*. 1965 E. Bhavnani *Dance in India* v. 34 Then comes the rendering in gesture language and emotional acting, the explanation of a song or *Sahitya* which are devotional sentiments in lyrical verse form and are the text to be interpreted. 1968 *Jrnl. Mus. Acad. Madras* XXXIX. 8 The Raga chosen for the song aptly conveys the sentiment expressed by the Sahitya. 1971 *Shankar's Weekly* (Delhi) 18 Apr. 24/1 One of the sisters was out to prove that they could tackle the swaraprasthara to coincide terminally with the point of commencement of the sahitya.

Sahiwal (sā·hivăl, -wăl). Also **Sanhiwal.** [The name of a town in the central Punjab, Pakistan.] A cow or bull belonging to the breed so called, originally native to Pakistan but now used in tropical regions elsewhere, distinguished by small horns and a hump on the back of the neck; also, the breed itself. Also *attrib.*

1916 *Rep. Agric. Research Inst. & Coll., Pusa 1914–15* 10 Two herds are now being maintained at Pusa, one of selected Sanhiwal (Montgomery) cows and their descendants, the other of cross-bred Ayrshire-Sanhiwal cattle. 1919 *Rep. Progress Agric. in India 1917–18* v. 182 Experiments with crossing the ordinary *desi* cow of good stamp with the Hissar, Sahiwal and Kosi strains are in progress. 1941 *Empire Jrnl. Exper. Agric.* IX. 11 The Sahiwal has reached in 25 years a level of milking performance which foreign breeds would have taken more than a century to attain. 1959 R. B. Kelley *Native & Adapted Cattle* v. 71 Sahiwal cattle are also known as Montgomery cattle. *Ibid.* 75 Most Sahiwals are red. 1968 *Sunday Mail Mag.* (Brisbane) 7 July 5/1 We have 1200 head of cattle—Ayrshires; four horses and a few sahiwals. 1970 *Kenya Farmer* Feb. 13/1 The range of breeds in Kenya is now very considerable. For dairy, there are Ayrshire, Friesland, Guernsey and Jersey; for dual purpose, Brown Swiss, Red Poll and Sahiwal.

Sahli (sā·li). *Med.* The name of Hermann *Sahli* (1856–1933), Swiss physician, used *attrib.* and in the possessive with reference to a method he devised for determining the hæmoglobin content of the blood by converting a sample into acid hæmatin and adding water until the colour matches a standard.

1906 R. C. Cabot *Physical Diagnosis* (ed. 3) xxiii. 465 Sahli's instrument..must be obtained from one of the firms recommended by him. *Ibid.* 569/2 (Index), Sahli's test for hæmoglobin. 1931 Osgood & Haskins *Textbk. Lab. Diagnosis* II. ix. 354 The ordinary type of Sahli apparatus is worthless because the acid hematin used as the standard fades too rapidly. *Ibid.* 347 Diluted to the 100 mark in a Sahli tube. 1956 *Nature* 17 Mar. 524/1 No significant change was noted in the red-cell count; but there was a drop in the hæmoglobin-level from 81 to 70 per cent when tested by the Sahli method. 1974 Passmore & Robson *Compan. Med. Stud.* III. xxi. 2/1 The Hb of blood is measured colorimetrically after it has been converted to a stable form. The methods available use acid haematin, oxyhaemoglobin or cyanmethaemoglobin. The simplest is the Sahli method.

sahlinite (sā·linəit). *Min.* [f. the name of Carl A. *Sahlin* (1861–1943), manager of a Swedish ironworks + -ITE¹.] A basic monoclinic arsenate and chloride of lead found as pale yellow scales in dolomite at Långban, Sweden.

1934 G. Aminoff in *Geol. Foreningen Förhandlingar* LVI. 493 (*heading*) Note on a new mineral from Långban (Sahlinite). 1951 C. Palache et al. *Dana's Syst. Min.* (ed. 7) II. 775 Sahlinite... Monoclinic. In aggregates of small thin scales. 1968 I. Kostov *Mineral.* 467 Sahlinite has perfect cleavage on {010}.

Saho (sā·ho), *sb.* and *a.* Also † **Shiho, Shoho.** [Cushitic.] **A.** *sb.* **a.** A (member of a) Cushitic-speaking people of Eritrea. **b.** The language or dialect of this people. **B.** *adj.* Of or pertaining to this people or their language.

1790 J. Bruce *Trav.* III. v. iii. 68 The Shiho were once very numerous; but, like all these nations having communication with Masuah, have suffered much by the ravages of the small-pox. 1831 S. Gobat *Jrnl.* 22 May (1834) iv. 291, I have just passed three very disagreeable months, in the midst of the savage Shohos. 1842 Isenberg & Krapf *Jrnl.* 30 Apr. (1843) 521 The Governor promised this morning that he would send to the next Shoho village for a guide to take us to Arkeeko. a 1860 W. C. Plowden *Trav. Abyssinia* (1868) i. 23 The Shihos, a nomad race to the southward of Massowah. *Ibid.* xviii. 360 There are two roads, through the countries of two tribes of Shihos, leading to Adowah, the one through the tribe called Asowarta, the other, Tora... These two tribes form the Shiho nation, and occupy the mountainous tracts between Massowah and Christian Abyssinia. 1883 R. N. Cust *Sk. Mod. Lang. Afr.* I. ix. 128 (*heading*) Saho *or* Shiho *or* Shoho. *Ibid.* 129 The Saho bring down their herds in the rains to graze. 1885 [see *DANAKIL *sb.* and *a.*]. 1932 W. L. Graff *Lang.* xi. 404 The most important dialects [of Cushitic] are Bedja, Saho and Afar. 1960 E. Ullendorff *Ethiopians* iii. 40 The Saho tribes live in the coastal depression between Massawa in the north, the gulf of Zula in the east, and the escarpment of the Akkele Guzay in the west. 1962 G. A. Lipsky et al. *Ethiopia* iv. 47 Almost all the Saho-speaking tribes are located in Eritrea. *Ibid.* vii. 113 The Saho and Dānākil people also are Moslem. 1968 M. Abir *Ethiopia* vi. 132 Most of the Sahos and their Belau rulers left Arkiko and the surrounding area, and escaped into the mountains. 1972 *Language* XLVIII. 847 His list can be supplemented with examples from..Chinantec, Saho, Slave.

Sahrawi (sarā·wi). Also **Saharaui.** [a. Arab. *ṣaḥrāwī* (whence Sp. *saharaui*) of the desert, f. *ṣaḥrā'* desert, SAHARA.] An inhabitant of Western (formerly Spanish) Sahara, a Saharan; also *collect.*, the people itself. Also *attrib.*

1976 *Times* 27 Feb. 14/1 Self-determination for the Sahrawi people is a prerequisite for any settlement. *Ibid.*, Polisario should be recognised as the legitimate representative of the Sahrawis. 1976 *Times Lit. Suppl.* 16 Apr. 466/2 In either case the Saharauis will hardly be the beneficiaries. *Ibid.*, The inner working and tensions of Saharaui society. 1977 *Guardian Weekly* 6 Nov. 12/4 Two French technicians..taken away along with 24 Mauritanian workers by a Sahrawi guerrilla unit... The Sahrawi are not agents of subversion, but people who want back the homeland [sc. Western Sahara] they were forced to quit. 1980 J. Mercer *Canary Islanders* 264 In 1975 Spain handed over the Spanish Sahara and the Saharaui people to Morocco and Mauretania, in exchange for economic and other benefits. This led to the current war between the occupying neo-colonial powers and the Saharauis (Polisario).

saht-bai, var. *SAT-BHAI.

sahuaro, var. SAGUARO in Dict. and Suppl.

sahukar, var. SOUCAR.

1913 J. M. Keynes *Indian Currency & Finance* iv. 95 Notes, even of the value of Rs. 5, are looked upon with distrust by the village yokels and even by the village sahukars. 1930 *Economist* 12 Apr. 820/1 The majority..consists of rural sahukars, who, though shrewd, are primitive in their methods, and prefer the use of cash and notes to cheques and bills. 1936 J. Nehru *Autobiogr.* xxxix. 302 The tenant's position was even worse. He was also a *sahukar's* serf.

saice, var. SYCE.

Saigonese (səigonī·z), *collect. sb.* [f. *Saigon* (now Ho Chi Minh City), formerly the capital of South Vietnam + -ESE.] The people of Saigon. (No longer in official use.)

1967 [see *HIGH *adv.* 9 b]. 1975 *Daily Tel.* 1 May 1 Laughing guerillas..drove through the streets exchanging waves and banter with the Saigonese.

sail, *sb.*¹ Add: † **1. d.** *Aeronaut.* Applied to a flat aerodynamically structured part of an aircraft. *Obs.*

1808 G. Cayley *Aeronaut. & Misc. Note-bk.* (1933) 64, I tried a small square sail in one plane, with the weight nearly in the same, & I could not perceive that the centre & resistance differed from the centre of bulk. 1817 *Phil. Mag.* L. 35 The sketch..represents a side view of the arrangement of the moving and steering sails of a balloon on the wing plan. 1837 *Mechanics' Mag.* XXVI. 421/2 From the hinder mast C a sail may be conveniently braced to either side, so as to act as a rudder, and thus preserve a steady course. 1902 F. Walker *Aërial Navigation* viii. 118 A head sail *i* and stern sails *h*, h^1 had braces and halliards for steering... The sails *h*, h^1 acted as aëroplanes as well as for steering purposes. 1903 —— *Pract. Kites & Aëroplanes* ii. 25 The 'leeches', or free edges of the sails ..are double-stitched around a leech-rope.

3. c. Also *fig.*

1893 'Mark Twain' *Lett. to Publishers* (1967) 348 A well-organized business..an enterprise not experimental but under full sail.

8*. The conning-tower of a submarine.

1959 *Jane's Fighting Ships* 414/1 'The sail', as the conning tower is now called on nuclear submarines. 1963 *Guardian* 1 Mar. 1 The Ethan Allen looked like any other submarine though the conning tower—which they call the sail these days—was much larger than usual. 1968 *New Scientist* 26 Dec. 704/2 Photographs of the wreckage show that the Scorpion split in two at the point on the hull where the 'sail' (the new name for the conning tower) is mounted near the forward end. 1974 L. Deighton *Spy Story* xviii. 190 The great submarine threaded its way out through the Sound... The skipper came down from the sail.

9. a. *sail area;* **b.** *sail-stiffening* adj.

1898 W. F. Jackson in W. A. Morgan *'House' on Sport* I. i. 19 Traditions are still heard of boats lurking behind barges..to dart out at the last moment with something surprising in the way of sail area. 1976 *Oxf. Compan. Ships & Sea* 947/2 There was no form of handicap on size or sail area. 1945 P. Larkin *North Ship* 35 Increasingly to fear Sail-stiffening air.

10. **sailboard** orig. *U.S.*, a surf-board or light sail-boat which is propelled by wind caught in its sail; also as *v. intr.*; hence **sailboarder, sailboarding** *vbl. sb.*; **sail-boat:** for '? *rare*' read '*chiefly N. Amer.*' and add earlier and later examples; **sail curtain** = CURTAIN *sb.*¹ 2 a; **sail-duck** (earlier example); **sail-flying** = *SAILPLANING *vbl. sb.*; **sail-maker** (later example in aeroplane construction); **sail plan** (see quot. 1961); **sail wing,** the sail of a hang glider with its framework; (the structures described in quots. 1972, 1974 differ from one another).

1962 D. Klein *Beginning with Boats* iv. 95 Another boat that may tempt you because it can give you a great deal of fun at rather low cost is what is called a sailboard—that is, a sort of surfboard equipped with centerboard, rudder, and sailing rig. 1978 B. Webb tr. *Brockhaus & Stanciu's Sailboarding* 8 You can ski in any mountainous region where there is snow, just as you can sailboard on any water, whether it be an ocean or a reservoir. 1980 *Daily Tel.* 15 Sept. 2 (*caption*) A 22ft-long sailboard made for two..being demonstrated at Southampton. 1974 A. H. Drummond *Sailboarding* 10 The surfboarder catches a wave and uses its energy to surf along. The sailboarder does just about the same thing, except that he uses a sail to capture the energy of the wind... Thus, sailboarding is surfing using wind power. 1978 *Times* 5 Apr. 8/5 Beau Vallon is the island's most visited beach. .. Enthusiasts use it for sail-boarding, water-skiing, para-gliding, diving and goggling. 1979 *Yachts & Yachting* 9 Nov. 1433/3 Wandering sailboarders could be pleased with a complete cover for their board. 1798 C. Williamson *Descr. Genesee Country* iii. 19 The number of sail-boats have greatly increased on the Lake. 1831 M. Holley *Texas* (1833) 47 From Brazoria to Bolivar, I came in a sail-boat. 1911 J. C. Lincoln *Cap'n Warren's Wards* xxi. 333 He had gone to see the sail-boat man. 1956 M. Duggan *Immanuel's Land* 64 A flatbottomed sailboat on the slope shifted almost afloat, and settled again into the mud. 1977 E. Leonard *Unknown Man No. 89* xxi. 211 A painting..of..a sailboat with the mast broken off. 1941 J. Masefield *Gautama* 52 The red sail-curtain droops. 1776 T. Pennant *Tour in Scotl. & Voy. Hebrides 1772* II. 143 At present the manufactures have risen to a great pitch: for example, that sail-cloth, or sail-duck, as it is here called, is very considerable. 1931 A. Gymnich in V. W. Pagé *Henley's ABC of Gliding* 148 By sailflying we understand a flight without any kind of motor or other driving power in which the energy required for the flight without loss in altitude, is taken solely from the air currents. 1944 T. Horsley *Soaring Flight* 71 The chapter on soaring sites will have given an indication of the winds used in the simplest sail-flying. 1916 H. Barber *Aeroplane Speaks* 103 All is now ready for the sail-maker to cover the surface with fabric. 1953 J. Masefield *Conway* 298 As it happens, we have the sail-plan of her sister-ship. 1961 F. H. Burgess *Dict. Sailing* 178 *Sail plan*, a diagram to show a boat's rig and measurements. 1962 C. H. Gibbs-Smith *Sir George Cayley's Aeronautics 1796–1855* xlii. 129 It is interesting to find at the present time (1962) a powered aeroplane using flexible sail-wings: this is the American Ryan 'Flex Wing' which has plastic-coated nylon wings supported in a delta plan by only three rigid spars, which meet at the front; one is central and the other two spread out to form the sides of the triangle. 1972 *Daily Tel.* (Colour Suppl.) 13 Oct. 9 A sail wing is a device shaped rather like an extremely ambitious paper dart and is made from dural aluminium and nylon. From a point at its centre hangs an 'A' frame... The pilot hangs in space upon an arrangement structurally similar to a child's swing, complete with a narrow wooden seat. 1974 *Sci. Amer.* Dec. 141/1 The sail wing consists of a tubular spar that supports the leading edge of a fabric envelope and a set of short, rigid booms at the tip and foot of the spar between which a slender cable is stretched to form the trailing edge of the wing. 1978 P. O'Donnell *Dragon's Claw* xiv. 293 The sail-wing rested on the grass... They stood surveying the wing.

sail, *sb.*³ Delete † *Obs.* and *Arch.*, and add later example.

1924 H. J. Butler *Motor Bodywork* xviii. 276 Some of the lighter types of delivery van are made with a recessed rocker side... The body is then built up to the seat line by means of, say, an 1½" hardwood rocker side lapped on vertically, or with a slight sail, into the bottom side.

sail, *v.*¹ Add: **I. 4. a.** (Examples referring to aircraft.)

1897 [see *GLIDER 2 a]. 1910 *Daily Mail Year Bk.* 149/2 The Gross was compelled to descend, after sailing above the enemy's line.

b. (Earlier example.)

1866 'Mark Twain' *Speeches* (1923) 13 The Kanaka, without spur or whip,..sailed by us on the old plug.

5. b. (Earlier examples.) Also in weakened sense, to glide over a surface; to pass rapidly or smoothly.

1819 M. R. Mitford *Let.* 18 Mar. (1925) 161 Just as we were at our merriest, in sailed Madam J—, like a tragedy queen. 1836 W. Dunlap *Thirty Years Ago* I. ii. 22 Mrs. Epsom sailed majestically about the house. 1876 'Mark Twain' *Tom Sawyer* v. 58 Then there was a wild yelp of agony and the poodle went sailing up the aisle. 1909 R. A. Wason *Happy Hawkins* 10, I flopped onto a pony an' sailed out to a little glen. 1949 W. Awdry *Tank Engine Thomas Again* 50 He remembered the Level Crossing. There was Bertie fuming at the gates while they sailed gaily through. 1979 C. Egleton *Backfire* xii. 135 He sailed through Immigration and collected his suitcase.

c. (Earlier example.) Also, to launch *into* or attack; also *fig.*

1856 'Q. K. P. Doesticks' *Plu-ri-bus-tah* iv. 69 'Sailing in', without regard to Any of the laws of 'Fancy'. 1883 'Mark Twain' *Life on Mississippi* xxvi. 246 Old General Pillow..sailed in, too, leading his troops as lively as a boy. 1903 A. H. Lewis *Boss* iv. 52 Half an hour before six, blow your whistle an' sail in. 1934 R. Campbell *Broken Record* ii. 33, I sailed into him with a beauty on the ear. 1936 F. Clune *Roaming round Darling* xvii. 173, I sailed into Mrs O'Malley's cooked meat and damper.

‖**sailab** (sai·lāb). Also **sailaba.** [Hindi, Punjabi *sailāb(ā)* flood, torrent f. Pers. *sail* flowing + *āb* water.] A method of cultivation used in the Indus basin in Pakistan and northern India in which the land is irrigated by floodwater from the rivers.

1916 J. Douie *Panjab, N.W. Frontier Province & Kashmir* xiv. 142 'Unirrigated' embraces cultivation dependent on rain (*bārāni*) or on flooding or percolation from rivers (*sailāb*). 1960 *Indus Basin Devel. Fund Agreement* 30 in *Parl. Papers 1961* (Cmnd. 1527) XXXVII. 501 Pakistan may also withdraw such waters from each of the following Tributaries..for irrigation of that part of the following areas cultivated on *sailab*. 1962 *Times* 2 June 11/6 The annual migration follows a restricted round which includes short halts for cultivation by the *sailaba* method. Rough earthen bunds are built in the wadis to form a trap for the soil wash from the occasional flash floods and the sorghum and millet seeds are sown in these small patches of saturated soil. 1973 N. D. Gulhati *Indus Waters Treaty* iii. 43 The total area in the Indus basin, along different rivers, cultivated annually after inundation or *sailab*, was about 2·17 million acres... This cultivation was referred to as *sailab*.

sailable, *a.* Delete 'or *Obs.*' and add: **2.** (Later example.)

1976 *New Scientist* 16 Dec. 646/2 A sailable expanse of water.

sailcloth. Add: **3.** Also used for other garments, upholstery, etc. (Earlier and later examples.)

1873 *Young Englishwoman* Jan. 39/1 This hunting pouch consists of a back, front, and flap of grey sailcloth, lined with dark green American cloth. 1881 C. G. Harrison *Woman's Handiwork* I. 48 Among other washing fabrics used in art needlework are *crash*.., *twilled cotton, duck, sail-cloth*, [etc.]. 1962 R. P. Giles *Fabrics for Needlework* iv. 79 Sail-cloth. A very strong, firm, canvas-type fabric made in different weights... Not originally intended for a clothing fabric but nowadays the lighter weights are used for jeans, sportswear, and even summer dresses and skirts. 1979 *Arizona Daily Star* 5 Aug. (Parade Suppl.) 14/1 (Advt.), Comfortable, carefree sailcloth casuals that go their fun-loving way on soft and springy crepe soles.

sailer. Add: **3.** *Baseball.* (See quot. 1961.)

1937 *Sun* (Baltimore) 28 May 14/7 There were two strikes and three balls on Cochrane when Hadley threw his ill-fated 'sailer'. 1961 J. S. Salak *Dict. Amer. Sports* 379 *Sailer* (baseball), a pitched fast ball that takes off, that is, sails. 1975 *New Yorker* 17 Nov. 158/2 The throw, however, was a horrible sailer that glanced off Burleson's glove and went on into center field.

sailing, *vbl. sb.*¹ Add: **2. b.** fig. *plain sailing* (see main entry). Also with similar qualifying adjs.

1827, etc. [see PLAIN SAILING *sb.*]. 1841 Lytton *Night & Morning* II. viii. 118 'Oh! then it's all smooth sailing,' replied the other. 1927 H. Crane *Let.* 19 Dec. (1965) 313 After a good deal of fair 'sailing' since arriving here—I am now convinced that 'flying' is even better. Right now however..I am 'all fives' on the ground. 1959 *Daily Tel.* 15 Oct. 12/2 Brilliant sailing in the comparatively calm waters of the Post Office.

4. a. *sailing club, date, day* (earlier example), *match* (earlier example); **sailing-line,** (*b*) a line (LINE *sb.*² 22) of sailing vessels; **sailing master** (earlier U.S. example); **sailing orders** (earlier *fig.* example); **sailing rule,** a rule of the sea, to prevent the collision of ships, etc.

1810 E. Weeton *Let.* 5 Sept. in *Jrnl. of Governess* (1969) I. 293 A sailing club consisting of four or five young men of fortune, have conducted the annual Regattas. 1973 G. Moffat *Lady with Cool Eye* vii. 73 The inspector, meeting the traffic superintendent in the local sailing club, chanced to mention Mrs. Wolkoff's latest protest. 1906 J. London *Let.* 1 Dec. (1966) 227 All..that you wanted answered..was my sailing-date. 1839 in M. Johnson *Amer. Advertising, 1800–1900* (1960), The sailing days of the above ship have been altered. 1905 *Chambers's Jrnl.* May 366/1 Sailing-lines to the West Indies..give Bermuda a wide berth. 1779 in *New Hampsh. Hist. Soc. Coll.* (1863) VII. 194 Appointed—Curtis Sailing Master of the armed ship Hampden. 1810 E. Weeton *Let.* 15 Aug. in *Jrnl. of Governess* (1969) I. 284 You must not suppose that Mr. and Mrs. P. or myself were in the boat during the sailing match. 1796 W. Scott *Let.* 26 Sept. (1932) I. 56 Your sailing orders are—If the subject is casually introduced to treat it lightly. 1877 *Regulations for Government of Navy of U.S.* 185 *Steering and sailing rules*,.. Art.15. If two ships, one of which is a sailing-ship, and the other a steamship, are proceeding in such directions as to involve risk of collision, the steamship shall keep out of the way of the sailing-ship. 1976 *Oxf. Compan. Ships & Sea* 954/1 The actual sailing rules embody in general the Rule of the Road as it affects sailing vessels.

b. *sailing-boat* (earlier and later examples), *dinghy, -packet* (earlier example), *-ship* (earlier example), *vessel* (examples); *sailing-chariot* (earlier example).

1721 *New-England Courant* 14 Aug. 2/2 On the 4th Inst. at Night were drowned going to Thomsons Island in a small sailing-Boat, Mr. Heskew, [etc.]. 1976 *Oxf. Compan. Ships & Sea* 960/2 A sailing boat with masts stepped as above but sloop-rigged on the foremast would be termed a yawl. 1759 Johnson *Rasselas* I. vi. 35 He ..found the master busy in building a sailing chariot. 1930 A. P. Herbert *Water Gipsies* vi. 55 Sailing-dinghies, eights and single-scullers. 1975 *Oxf. Compan. Sports & Games* 1123/1 To take part, all a man needs is a yacht,.. or a sailing dinghy as small as 12 ft. (3.65 m.) long. 1842 Dickens *Let.* 17 Feb. (1974) III. 66 There is a sailing-packet from here to England tomorrow. 1871 D. G. Rossetti *Let.* July (1967) III. 959 They are coming back..by sailing-ship. 1748 B. Franklin *Exper. & Observations Electricity* (1751) I. 38 In the wake of every sailing vessel. 1976 Sailing vessel [see *sailing rule* above].

sailing, *ppl. a.*² (Earlier and later examples of *sailing course.*)

1807 T. D. W. Dearn *Bricklayer's Guide* 50 Then proceed to take the sailing course, and the wall on either side the chimney. 1946 Holgate & McDougall *Brick-laying* v. 63 An attractive method of making an all-brick coping more effective is by first laying on top of the wall a course of three-quarter bats as headers and after completing the coping, filleting this 'sailing' course with cement mortar.

sail-off. *N. Amer.* [f. SAIL *v.*¹ + OFF *adv.*, after *play-off, row-off.*] **a.** An additional sailing contest to decide between tied contestants. **b.** A series of sailing contests or races held to decide a championship.

1949 *Sun* (Baltimore) 19 July 14/6 The Miller Series at Gibson Island last week end also was sail-off of the home star fleet's championship tie between Ron Blizzard, with Snowflake and Dave Dunigan who sails Lodestar. 1955 *Ibid.* 27 June 13/1 There will be two unlucky sailors next week end when the tie is broken by a sudden death sail-off. 1970 *Times* 19 Aug. 6/6 The winner of the France-Australia sail-off will meet America for the Cup. 1972 *Even. Telegram* (St. John's, Newfoundland) 5 Aug. 15/5 Small-boat sailing..is gaining in popularity and the skippers might like to show what they can do in an annual sail-off.

sailor. Add: **4.** (Earlier and later examples.)

1890 *Demorest's Family Mag.* June 504/2 Boat-shaped, wide-brimmed sailors in white..are worn by either boys or girls for play-hats. 1891 *Delineator* Sept. 230/1 *Ladies' felt sailor hat*.—A stylish and dressy sailor is pictured here in a dark brown felt. 1922 H. Titus *Timber* xxix. 252 She pulled the straw sailor tighter over her golden hair. 1943 D. Powell *Time to be Born* x. 227 Her smart little toasted straw sailor with floating pink veil. 1979 D. Eden *Storrington Papers* vi. 68 Miss Featherstone had whipped off her modest sailor and arranged the light-as-air confection on her head.

5. a. *sailor-blue* adj.; (appositive) *sailor-boy* (earlier example), *-king.*

1930 J. Dos Passos *42nd Parallel* I. 91 She was waiting for him..looking like a Gibson girl with her neat sailor-blue dress. 1978 J. Krantz *Scruples* iii. 65 Perhaps his height came from his father, but the bright blond hair and sailor-blue eyes were pure Swedish Viking. 1835 J. E. Alexander *Sk. Portugal* x. 245, I..engaged a Portuguese sailor-boy..to accompany me to Africa. 1911 Fletcher & Kipling *School Hist. Eng.* 91 He [sc. Edward III] was merchant-king, sailor-king, soldier-king. 1965 Finer & Savage *Sel. Lett. J. Wedgwood* i. 38 [The Royal patronage] was again extended in 1830 by William IV, the 'Sailor King'. 1975 B. Meyrick *Behind Light* xii. 149 King George, the Sailor King, because he had served at sea.

b. **sailor collar** (see quot. 1968); **sailor hat** (examples); hence **sailor-hatted** *a.*; **sailor knot** = *sailor's knot*; hence **sailor-knotted** *a.*; **sailor-man,** (*a*) (earlier example); (*b*) a sailing-barge(man); **sailor pants** *U.S.*, flared trousers such as those worn by sailors; **sailor suit,** a suit similar to that of an ordinary seaman, worn mainly by small boys; hence **sailor-suited** *a.*; **sailor top,** a jerkin similar to that worn by sailors; also applied to a ladies' blouse of this design; **sailor trousers** *U.S.* = *sailor pants.

1895 *Montgomery Ward Catal.* 79/2 Guipure Open work sailor collars. 1932 'E. M. Delafield' *Thank Heaven Fasting* II. v. 223 A grey satin blouse, with a black bow in the front of the square sailor collar. 1968 J. Ironside

Fashion Alphabet 52 *Sailor*, A collar cut deep and square at the back, narrowing to a 'V' in the front. It is often trimmed with braid—as worn by sailors. **1974** *She* Jan. 52/2 Braided jacket with square-back sailor collar, £8·50. **1980** *Times* 22 Oct. 10/7 Sailor collar, shift shape and hip belt. **1873** *Young Englishwoman* Mar. 131/2 Brown velvet sailor hat of two shades. **1912** A. Bennett *Matador of Five Towns* 46 A quite little girl..with a short frock and long legs, and a sailor hat (H.M.S. *Formidable*). **1976** *Vogue* Jan. 48 White tunic..with white duck American sailor hat. **1909** E. Nesbit *Daphne in Fitzroy St.* x. 152 'It's only me, miss,' said the sailor-hatted charwoman. **1872** 'Mark Twain' *Roughing It* lxii. 447 Black silk neck-cloth tied with a sailor knot. **1939** T. S. Eliot *Old Possum's Bk. Pract. Cats* 14 The curtain-cord she likes to wind, and tie it into sailor-knots. **1923** W. J. Locke *Moordius & Co.* viii. 109 With deft fingers she gave his sailor-knotted tie a twist and a pull. **1761** G. Colman *Jealous Wife* iii. 45 The Irish Sailor-Man, for whom I prevailed on your Lordship to get the Post of a Regulating Captain. **1948** *Sea Breezes* VI. 337/2 From Colchester sails Francis & Gilder's large fleet of 'sailormen'. **1951** H. Benham *Down Tops'l* 187 *Sailorman*, the London River term for either a sailing-barge or a sailing-bargeman. **1961** G. Foulser *Seaman's Voice* i. 20 The winter of 1936–7 was a rough one, with a lot of windbound intervals for the 'sailormen'. **1931** H. Crane *Let.* 13 June (1965) 373 My usual household white sailor pants and shirt. **1976** *National Observer* (U.S.) 2 Oct. 18/1 Today Stramler is in white sailor pants and a T-shirt. **1880** *Harper's Mag.* Aug. 337/2 Excursionists in sailor suits were playing croquet. **1885** C. M. Yonge *Nuttie's Father* II. xii. 145 We can't persuade ourselves to cut his hair, and it looks so lovely on his sailor suit. **1946** G. Millar *Horned Pigeon* iv. 53 He wore a sailor suit that was much too small for him—clothing that had been provided when a Messerschmitt had shot his Blenheim down into the sea. **1976** *Times* 27 Feb. 10/5 Susanna Agnelli was born in 1922... She and her brothers and sisters were dressed in sailor suits, blue in winter, white in summer. **1960** *Times* 3 Aug. 5/2 A juvenile delinquent cousin who appears sailor-suited in the first act. **1977** *Times* 7 May 9/1 The sailor-suited members of the Vienna Boys' Choir. **1913** C. Mackenzie *Sinister St.* I. i. v. 80 He..wished that he were not compelled to wear a sailor-top that was slightly shabby. **1916** Joyce *Portrait of Artist* (1969) i. 12 He had a blue sailor top on. **1962** G. Avery *Greatest Gresham* i. 20 She had..a navy blue sailor top to her blue serge suit. **1971** *Vogue* Dec. 70 Gabardine trousers. Sailor top with big bow. **1851** M. Reid *Scalp Hunters* xx. 69 *Calzoneros* of green velveteen. These are cut after the fashion of sailor-trousers,—short-waist—tight round the hips, and wide at the bottoms.

c. sailor's blessing *Naut. slang*, a curse; also **sailors' blessing**, such rigging or tackle as eases the sailors' work; **sailor's farewell** *Naut. slang*, a parting curse; † **sailor's hat** *Obs.* = *sailor hat*; **sailors' home** (earlier example); **sailor's knot** (earlier example); **sailor's pleasure** *Naut. slang* (see quots.); † **sailor's suit** *Obs.* = *sailor suit* above.

1876 F. W. H. Symondson *Two Years abaft Mast* ii. 56 Poor 'doctor' not unfrequently comes in for a 'sailor's blessing' (a growl). **1944** J. Masefield *New Chum* 166 Being almost new had all the latest sailors' blessings; nothing above her royals, double topgallant yards, a spike jib boom and no spanker gaff. **1937** Partridge *Dict. Slang* 722/1 *Sailor's farewell*, a parting curse. **1974** *Listener* 10 Jan. 50/3 The sole baker there..found himself ruined, and in some anger he gave the village a sailor's farewell and announced that he was off. **1862** *Englishwoman's Domestic Mag.* V. 142/1 Two styles of hat..seem to be equally in favour this season—one, the sailor's hat with straight brim; the other, the turned-down or bell-shaped hat. **1885** *Outing* 7 Nov. 138/2 Their round straw hats, with flat-topped crowns, and shape usually termed by Americans 'sailor's hat'..were trimmed with a plain white ribbon around the crown. **1839** *New Orleans Commerc. Appeal* 18 Apr. 2/2 (*heading*) Public meeting to promote the establishment of a Sailors' Home. **1843** Poe *Mystery of Marie Rogêt* in *Ladies' Compan.* (N.Y.) Feb. 165/2 The 'sailor's knot' with which the bonnet-ribbon is tied. **1856** C. Nordhoff *Merchant Vessel* 132 Others take what is called, *par excellence*, 'sailor's pleasure', in overhauling their chests, bringing their best clothing on deck to air, and counting over their stock of tobacco and pipes. **1932** J. W. Harris *Days of Endeavour* 57 They must have a sailor's pleasure on Sunday to see what can be raked up. *Ibid.* 232 *Sailor's pleasure*, overhauling contents of sea-chest and bag, and airing go-ashore clothes. **1933** P. A. Eaddy *Hull Down* v. 122 Sunday at sea in a deep-water sailing-ship, especially if the weather is fine, and nearing port the sole topic of conversation, means 'sailor's pleasure'. **1869** G. Meredith *Let.* Dec. 19 (1970) I. 406 You should see Willie Godson in his sailor's suit.

sailorship (sēˡ·ləɹʃip). *rare.* [f. Sailor + -ship.] Seamanship; the skill of a good seaman.

1820 J. Severn *Let.* 20 Sept. in H. E. Rollins *Lett. John Keats* (1958) II. 343 Keats this Morning brags of my sailorship. **1856** C. Nordhoff *Merchant Vessel* 111 Fancy seizings and lashings bore witness to the sailorship of the mates and crew.

sailplane (sēˡ·lplēˡn). Also **sail-plane.** [f. Sail *sb.*¹ + Plane *sb.*³] A heavier-than-air aircraft without an engine (or having only a small engine which is not normally used except to take off); = *glider 2 a (but see quot. 1971).

1922 *Flight* XIV. 545/2 The gliding angle of a good 'sail-plane' might be in the neighbourhood of 1 in 16 or 1 in 18. **1933** *Sun* (Baltimore) 5 Aug. 15/5 During the day sport planes circled about his sailplane. *Ibid.* 23 Sept. 4/5 Federal officials..will gather..to witness a demonstration in sailplane flying. **1935** *Ibid.* 17 May 1/4 The cheapest flight from London to Paris was made this evening when Robert Kronfeld..landed..in his sailplane driven by a motor-cycle engine of five horsepower, having consumed $1.50 in fuel for the 210-mile flight. **1940** *Illustr. London News* CXCVII. 85/1 The type of sailplane (or, to use the popular, but less correct, term, 'glider') required for use in transporting troops must, of course, be very large. **1950** *Chambers's Jrnl.* 137/2 In tropical countries some birds utilise these thermals, as they are called by sailplane pilots, and indeed make no attempt to fly until the air has warmed up. **1961** *New Scientist* 18 May 362/2 The Olympia 460 is what is called a 'standard' class sailplane. That is to say, it has a span of only 15 metres instead of the 19 metres allowed in the open class for world championship flights. **1971** N. Ellison *Brit. Gliders & Sailplanes* 9 The terms 'gliders' and 'sailplanes' nowadays are somewhat synonymous. When these terms were first introduced a sailplane was defined as 'a glider having a sinking speed of less than 0·8 metres (2·625 feet) per second'. Today, nearly all motorless aircraft are sailplanes and capable of soaring flight, i.e. flying without loss of height or gaining height.

Hence **sa·ilplaner**, in the same sense; **sai·l-planing** *vbl. sb.*, the flying of sailplanes, gliding; also *transf.*

1923 *Flight* XV. 34/2 In order to encourage gliding and sailplaning in America, the National Aeronautic Association of U.S.A. have appointed a sub-committee to deal with this form of flying. **1930** *Daily Express* 8 Sept. 16 The first lesson in the art of sail-planing. **1962** *Punch* 15 Aug. 237/2 It [*sc.* the fulmar] makes distance at sea..by sailplaning and tacking. **1973** *Sci. Amer.* Dec. 134/2 The cockpit panel of a serious sailplaner shows 10 dials, plus radio, oxygen gear and cameras. **1977** *Maclean's Mag.* 2 May 58/2 For recreation they have turned to such exotic and often dangerous sports as..sailplaning.

sainfeldite (sēˡ·nfeldəit). *Min.* [a. F. *sainfeldite* (R. Pierrot 1964, in *Bull. de la Soc. franç. de Min. et de Crist.* LXXXVII. 180/1), f. the name of P. *Sainfeld* who collected the material: see -ite¹.] A hydrous arsenate of calcium, $Ca_5H_2(AsO_4)_4.4H_2O$, occurring as small rosettes of transparent monoclinic crystals.

1964 *Chem. Abstr.* LXI 14371 Three new naturally occurring minerals, which were already known as synthetic compounds, were weilite.., rauenthalite.., and sainfeldite. **1972** *Bull. de la Soc. franç. de Min. et de Crist.* XCV. 33/2 Sainfeldite is the least hydrated member of the group including vladimirite, $Ca_5H_2(AsO_4)_4.5H_2O$, and guérinite, $Ca_5H_2(AsO_4)_4.9H_2O$.

saint, *a.* and *sb.* Add: **A.** *adj.* **4. c. St. Augustine grass,** a coarse grass, *Stenotaphrum secundatum*, native to the southeastern United States and central America and named after a town in Florida; **St. Bees Sandstone,** a pebbly sandstone occurring in thick beds in northwest England, formerly regarded as Upper Permian but now as Lower Triassic; **St. Bernard('s) lily,** a perennial herb, *Anthericum liliago*, belonging to the family Liliaceæ and bearing racemes of white flowers; **St. Brigid('s) anemone,** a plant belonging to a garden race of *Anemone coronaria*, bearing single or double red or blue flowers; **St. Bruno's lily,** a rhizomatous perennial herb, *Paradisea liliastrum*, which resembles St. Bernard's lily but has larger flowers (cf. Lily 1 b); **St. Dabeoc's heath,** an Irish heath, *Dabœcia cantabrica* or one of its varieties, belonging to the family Ericaceæ and bearing white, pink, or purple flowers; **St. George's mushroom,** a creamy-white, flattened mushroom, *Tricholoma gambosum*; **St. Kilda (field, house) mouse,** a variety of the long-tailed field mouse, *Apodemus sylvaticus hirtensis*, or the house mouse, *Mus musculus muralis*; **St. Kilda wren,** a local variety of the wren, *Troglodytes troglodytes hirtensis*, with paler plumage; **St. Leger** (further examples); **St. Louis encephalitis** [*St. Louis*, city of Missouri, U.S.], a severe viral encephalitis transmitted by mosquitos; **St. Patrick's cabbage** (see Cabbage *sb.*¹ 2).

1905 W. J. Spillman *Farm Grasses U.S.* xiii. 196 St. Augustine grass occurs along the Atlantic coast from Charleston, S.C., southward. **1968** F. W. Gould *Grass Systematics* v. 203 St. Augustine grass is relatively coarse. [**1836** *Trans. Geol. Soc.* IV. 398 The red sandstone of St. Bees Head is unquestionably the exact equivalent of the upper red sandstone of that series.] **1865** E. W. Binney in *Mem. Lit. & Philos. Soc. Manchester* II. 373 Fine-grained red sandstone, laminated and ripple-marked, same as that seen at Moat,.. Maryport, and other places, which may be conveniently called St. Bees sandstone. **1946** L. D. Stamp *Britain's Struct. & Scenery* xxii. 224 The St. Bees Sandstone,..of New Red Sandstone age, forms the red cliffs of St. Bees Head. **1969** Bennison & Wright *Geol. Hist. Brit. Isles* xi. 265 In this case the base of the St. Bees Sandstone, of Bunter age, may also be diachronous. **1883** W. Robinson *Eng. Flower Garden* 26/2 The St. Bernard's Lily..grows from 1 foot to 2 feet high, producing single, sometimes branched flower-spikes. **1900** W. D. Drury *Bk. Gardening* x. 315 The St. Bruno and St. Bernard Lilies..are fast becoming popular. **1964** H. Ramsbotham tr. *Schauenberg's Bulb Bk.* iii. 106 St. Bernard's Lily..is a common plant in Alpine meadows. [**1894** *Jrnl. R. Hort. Soc.* XVII. p. liv, Award of Merit. To Anemone St. Brigid's strain..from Earl Cowper, Panshanger, Hertford (g[ardener] Mr. Fitt).] **1902** *Ibid.* XXVII. p. lxxxvi. Award of Merit. To the Alderborough strain of St. Brigid Anemones. **1939** W. Fortescue *There's Rosemary* lxxix. 408 We had the joy..of seeing his beautiful frail hands caress the petals of flaming St. Brigid Anemones and slender tulips which bordered the drive of the Domaine. **1971** *Country Life* 2 Sept. 543/3 St. Brigid's anemones. Sown in April they flower from August. **1795** *Curtis's Bot. Mag.* IX. 318 (*heading*) Savoy Anthericum, or St. Bruno's Lily. **1883** W. Robinson *Eng. Flower Garden* ii. 26/2 The large variety of the St. Bruno's Lily has much larger flowers than the type. **1964** H. Ramsbotham tr. *Schauenberg's Bulb Bk.* iii. 204 The English name of this lovely Alpine plant [*sc.* St. Bernard's Lily] is 'St. Bruno's Lily'. **1863** R. C. A. Prior *On Pop. Names Brit. Plants* 195 St. Dabeoc's Heath, from an Irish saint of that name, a species found in Ireland. **1978** P. Rowe-Dutton tr. *van de Laar's Heather Garden* 130 St. Dabeoc's Heath..low, evergreen Irish native with broad fresh green leaves, silvery beneath. **1891** M. C. Cooke *Brit. Edible Fungi* iv. 34 'St. George's mushroom'..makes its appearance about the time of St. George's Day. **1966** *Oxf. Bk. Flowerless Plants* 134/2 'St. George's Mushroom'..grows in rings or growth on the edges of woods..and in open grass. **1899** G. E. H. Barrett-Hamilton in *Proc. Zool. Soc.* 78 I have now before me..a fine adult pair..of the St. Kilda Mouse. **1913** —— *Hist. Brit. Mammals* II. 540 (*heading*) The St Kilda Field Mouse. *Ibid.* 661 (*heading*) The St Kilda House Mouse. **1960** M. Burton *Wild Animals Brit. Isles* 78 St. Kilda field mouse..with brown underparts. *Ibid.* 88 Since the human inhabitants left the island in 1930, the St. Kilda mouse has become extinct. **1976** *Islander* (Victoria, B.C.) 7 Mar. 3/3 The St. Kilda house mouse has become extinct. *Ibid.*, The St. Kilda field mouse is also larger. **1884** H. Seebohm in *Zoologist* VIII. 333 Those ornithologists who regard the climatic races of this bird as distinct species, will probably come to the conclusion that the St. Kilda Wren is one of them quite distinct. **1914** [see Wren 1 b]. **1944** J. S. Huxley *On Living in Revolution* ix. 96 The St. Kilda wren..was for some time classified as a separate species. **1976** *Islander* (Victoria, B.C.) 7 Mar. 3/3 The St. Kilda wren is unique. **1847** Thackeray *Van. Fair* (1848) xxxiv. 302 He and his father fell to talking about odds on the St. Leger. **1930** *Daily Express* 11 Sept. 9/5 The St. Leger was run in almost ideal conditions. **1977** *Times* 10 Sept. 1/1 Thirteen runners have finally stood their ground for this year's St Leger..at Doncaster this afternoon. [**1933** *Jrnl. Amer. Med. Assoc.* 9 Sept. 860/2 (*heading*) The St. Louis encephalitis epidemic.] **1934** *Ibid.* 18 Aug. 462/2 The virus of St. Louis encephalitis had an almost exclusively neurotropic activity. **1962** Gordon & Lavoipierre *Entomol.* xix. 130 As regards western equine encephalomyelitis and St. Louis encephalitis..the important vector appears to be *Culex tarsalis*. **1977** *Jrnl. Virol.* XXII. 608 The antigenic determinants of St. Louis encephalitis, Japanese encephalitis, and dengue virus envelope and nucleocapsid proteins were examined by solid-phase competition radioimmunoassay. **1851** C. A. Johns *Flowers of Field* I. 240 S[*axifraga*] *umbrosa* (London Pride or St. Patrick's Cabbage). **1976** *Church Times* 14 May 14/5 Other flowers with religious or curious folk-names are 'Yellow Archangel'..; 'St. Patrick's Cabbage' (one of the saxifrages); [etc.].

‖ **d.** Similarly found in various place- and personal names of French origin, as **St. Cloud** (sæṅklū), used *attrib.* to designate porcelain or faïence made at St. Cloud, Seine-et-Oise, in the late-seventeenth and eighteenth centuries; **St. Emilion** (sæntemilyoṅ), the name applied to various wines produced in the region of St. Emilion, Gironde, in south-west France; **St. Galmier** (galmye), an effervescent natural mineral water from St. Galmier, Loire, in central France; **St. Honoré** (onore) (see quot. 1964); usu. *attrib.*, as *gâteau St. Honoré*; **St. Paulin** (polæṅ), a kind of cheese (see quots.); **St. Porchaire** (pōrʃɛ̄r), used *attrib.* to designate a kind of earthenware made at Saint-Porchaire, Deux-Sèvres, France, in the sixteenth century; **St. Raphael (wine)** (rafaʸel), an aperitif wine from St. Raphael, Var, France.

[**1699** M. Lister *Journey to Paris* 138, I saw the *Potterie of St. Clou* with which I was marvellously well pleased.] **1721** M. W. Montagu *Let.* June (1966) II. 6 If you have not allready laid out that small Summ in St. Cloud ware I had rather have it in plain Lutestring. **1870** C. Schreiber *Jrnl.* 17 Feb. (1911) I. 71 We found an exquisite pâte tendre St. Cloud group. **1978** *Times* 4 Mar. 10/7 The Garrick Club have..Thomas King's cane with a fine St Cloud porcelain handle. **1833** C. Redding *Hist. Mod. Wines* v. 142 St. Emilion has plenty of body, and superior flavour. **1981** P. Fox *Satan's Messenger* ii. xviii. 133 I don't serve a Château Lafite to two hundred people... The St. Emilion would be perfectly adequate. **1883** *Encycl. Brit.* XVI. 436/1 Classes I. and II. of alkaline waters..are very abundant on the Continent, and some of the best-known ones enumerated below are ..French..St. Galmier, Pougues, Chateldon. **1919** M. Beerbohm *Seven Men* (1919) 114 'Apollinaris? St. Galmier? Or what?' I asked. He preferred plain water. **1907** *Yesterday's Shopping* (1969) 55/2 *Iced & Fancy Cakes*... Gâteaux St. Honore..each 1/5. **1964** L. de Launay *Caviare & After* 143 *Saint Honoré*, a rich, round pastry filled with cream and topped with crystallized fruits. **1968** V. Canning *Melting Man* v. 120 He..came back with a concoction that made me feel I would never want to eat again... 'It is a *Saint-Honoré*. He was, you know, once Bishop of Amiens and is the

patron saint of pastry-cooks. **1968** D. Hopkinson *Incense-Tree* i. 6 Her dinner parties were graced with..Gâteau St Honoré. **1956** A. Simon *Cheeses of World* 73 *Saint-Paulin* is a semi-hard cheese made from cow's milk. ..The Trappists of..Tamié..used to sell their cheese as *St. Paulin*, but it is now sold as *Fromage de Tamié*. **1958** *Catal. County Stores, Taunton* June 9 *Cheese*..St. Paulin—each 5/6. **1971** *Sunday Times* (Colour Suppl.) 28 Mar. 34/3 Saint-Paulin, resembles Port Salut in texture, taste and origins. First made in a Norman monastery, it is a rich yellow whole cow's milk cheese, at once soft and firm to the touch and very mildly ripe to taste. **1899** F. Glazier *Man. Hist. Ornament* 81 Henri-deux, or St. Porchaire ware, now more properly described as Oiron ware, originated at St. Porchaire in 1524. **1925** R. Hannover *Pott. & Porc.* III. i. 15 Specimens of the 'Henri II' [St. Porchaire] ware, which is also extremely rare, have repeatedly been offered for sale..in our own days. **1960** [see *Henri Deux]. **1975** *Times* 20 May 16/4 One of the greatest rarities in..European ceramics, a St. Porchaire ewer, is to be offered for sale... St. Porchaire wares also known as Henri Deux.. St. Porchaire ware, also known as *faïence de Henri II*..was rediscovered by the public, like Palissy ware, as a result of the 1862 'Special Exhibition of Works of Art' at the South Kensington Museum. **1899** Hardy *Let.* 25 Aug. in *One Rare Fair Woman* (1972) 83 I have taken one bottle of St Raphael wine—and it has picked me up. **1951** [see *Lillet]. **1977** *Guardian* 3 June 9/4 St. Raphael and Dubonnet are the sweetest [aperitifs]. **1980** E. Leather *Duveen Let.* xii. 138 Glasses of St Raphael and Vichy water were ordered.

B. *sb.* **2. a.** Also, a monk or anchorite, esp. in phr. (*island of saints*, Ireland.

1888 Chesterton *Ballad of White Horse* v. 102 His men were all as thin as saints. *Ibid.* 103 Though Ireland be but a land of saints, and Wales a land of thieves. **1904** C. Walsh in J. McCarthy *Irish Lit.* I. p. xvii, Her missionaries and her national spirit have been recognized during the last twenty years as they never were since the days when Ireland was the 'island of saints and scholars', the land of intellectual light and healing in Europe. **1938** W. B. Yeats *New Poems* 11 My father upon the Abbey stage, before him a raging crowd: 'This Land of Saints' and then..'Of plaster Saints'. **1964** *Welsh Hist. Rev.* II. 122 The emigrations of the 'saints' from Britain can be dated almost exclusively to the sixth century. *Ibid.* 123 We can but picture these early British 'saints' (monks) seeking solitary places at home and abroad to serve their God. **1979** *Guardian* 1 Oct. 4/8 It was the Pope's arrival at Dublin Airport..which truly set the distinctive character of this personal pilgrimage to his 'island of saints'.

3. a. Also used by the Plymouth Brethren of their own members.

1838 G. V. Wigram *Let.* in T. S. Veitch *Story of Brethren Movement* (1933) iv. 59 The question I refer to is 'How are the meetings for communion of Saints in these parts to be regulated?' **1866** H. Groves *Darbyism* ii. 25 God so ordered it, that the assemblies which had divided the Brethren were.. [illegible] **1975** *Times* 7 Nov. [illegible] ..the saints..

4. a. Also in colloq. use, an extremely good or long-suffering person.

1883 Thackeray *Esmond* III. iii. 92 'O how good she is,' Harry..'Beatrix went on to say, 'O what a saint she is!' **1958** R. Barrett *Unruly Son* xvii. 186 My mother.. always thought about me. She was a saint.

6. saint's day, (*b*) = Name-day 1.

1962 E. M. Almedingen *Frossia* iii. 149 It is my saint's day, we have guests coming. **1980** 'J. Le Carré' *Smiley's People* xxii. 274 Felicity had called her in..to have Russian company on her saint's day.

saint, *v.* Add: **2. b.** (Later example.)

a 1920 'Mark Twain' in C. B. Taylor *Margins on Thackeray's 'Swift'* (1935) 47 It would have been enough merely to have forgiven Swift in this paragraph—not sainted him.

sainted, *ppl. a.* Add: **2. b.** Used trivially as an expletive in phr. *my sainted aunt* (also *mother*)!

1869 'Mark Twain' *Innoc. Abr.* v. 52 'Twenty-five cigars, at 100 reis, $100 reis!' Oh, my sainted mother! **1918** M. Diver *Desmond's Daughter* ii. ii. 90 My sainted aunt! You did ought to have been in the anteroom just now. **1919** E. Hurst *Humoresque* 114 'Your sainted mother!' **1921** [see *aunt 5]. **1926** 'Sapper' *Final Count* v. 142 Oh! my sainted aunt! don't tell me that old goose bush was Carl Peterson. **1939** Wodehouse *Uncle Fred in Springtime* xvii. 256 'Oh, my aunt! Don't tell me she's changed her mind and wants the stuff after all?' .. 'Exactly.' 'Oh, my sainted baby aunt!' **1971** R. Roberts *Classic Slum* viii. 127 Self-consciously we incorporated weird slang into our own oath-sprinkled banter—'Vevoh!' 'My sainted aunt!'..

St. Elmo (sent e·lmo). Also † **St. Elm, St. Helmo, San Telmo, sant-elmo.** [A corruption, via *Sant'Ermo*, of the name of *St. Erasmus* (martyred 303), Italian bishop and patron saint of Mediterranean sailors; cf. It. *fuoco di Sant'Elmo*.] Used in the possessive, *absol.*, and with *of* to denote the luminous appearance of a naturally occurring corona discharge about a ship's mast or the like, usually in bad weather. Now usu. as *St. Elmo's fire*; = Corposant, Helena.

1561 Sant-elmo [see Corposant a]. **1621** J. Chamberlain *Let.* 21 July (1939) II. 390 His comming was taken for a goode presage, like the appearing of S. Elmo after a tempest. **1774** Fires of St. Helmo [see Fire sb. A. 10 b]. **1854** tr. *E. v. Von Langsdorff's Voy. & Trav.* II. iv. 102 In the winter months the air is often so charged with electricity, that for many hours together in the darkest nights a bluish green electrical light, called St. Helen's, or St. Elm's fire, is seen. **1863** *Encycl. Metrop.* IV. 135/1 The fire of St. Elmo, so frequently seen upon the masts of vessels in the Mediterranean, and from very early times connected with the names of Castor and Pollux, meets with a very simple explanation on the principle of a pointed conductor imbibing electricity. **1882** *Encycl. Brit.* XIV. 633/2 This glow is known to sailors as St. Elmo's (San Telmo's) fire, in old days Castor and Pollux. **1942** *Tee Emm* (Air Ministry) II. 56 St. Elmo's Fire..occasionally seen on extreme cases long streaks of fire appear at the propellers, wing-tips or nose. **1966** G. Durrell *My Family & Other Animals* xvii. 235, I tell you, we'll find the chimney covered with Saint Elmo's fire one night, and before we know where we are we'll be drowned in our beds by a tidal wave. **1969** M. A. Uman *Lightning* 244 Ball lightning and St. Elmo's fire are sometimes confused. St. Elmo's fire is a corona discharge from a pointed conducting object in a strong electric field. Like ball lightning, St. Elmo's fire may assume a spherical shape. Unlike ball lightning, St. Elmo's fire must remain attached to a conductor, although it may exhibit some motion along the conductor. Further, St. Elmo's fire can have a lifetime much greater than the lifetime of the usual ball lightning. **1976** *Scotsman* 10 Nov. (Weekend Suppl.) 1/1 The top of the mast was surrounded with an eerie pale green phosphorescence. This was St. Elmo's Fire—known and feared by seamen of old—caused by static electricity.

St. Kildan (sənt ki·ldən). Also † **St. Kildean, St. Kildian.** A native or inhabitant of the island of St. Kilda in the Outer Hebrides.

The island is now the property of the National Trust for Scotland. The last native inhabitants were formally evacuated on 29 Aug. 1930.

1764 K. Macaulay *Hist. St. Kilda* v. 77 The St. Kildans are too wise or too good protestants, to neglect their secular affairs on the festival days of Columba and Brendan. **1810** D. Webster *Topogr. Dict. Scotl.* 375/2 One of the St. Kildians was some years ago acquainted with the small-pox, and died. **1842** J. Wilson *Voyage round Coasts of Scotl.* II. i. 9 Another young St Kildean.. was at the ..[illegible] Terra-Kilda... The small group of St. Kildeans.. were cheered by our arrival. **1861** M. Chambers *Domestic Ann. Scotl.* III. 181 Mr. Macaulay..mentions..that not only is a St. Kildian's person disagreeably odoriferous to a stranger, but 'a stranger's company' is as offensive to them'. **1939** *Geogr. Mag.* X. 73 St. Kildans were killed and salted, as its name, the Saladero, implied. [illegible] **1965** T. Steel *Life & Death of St. Kilda* 10 The St. Kildan and the mainland Scot..in the island home little else than a republic. **1980** *Times* 29 July 4/1 It is 50 years since the last St Kildans elected to leave the island.

St. Lucian (sənt lū·ʃən), *sb.* and *a.* **A.** *sb.* A native or inhabitant of St. Lucia in the West Indies. **B.** *adj.* Of or pertaining to St. Lucia.

1844 H. H. Breen *St. Lucia* v. 169 The early refugees, being unwilling to gratify the curiosity of the St. Lucians had rallied their numerous inquiries. **1952** S. Selvon *Brighter Sun* viii. 149 Ah sorry for all them Grenadians and St. Lucians who come over here to make money. **1955** *Caribbean Q.* IV. ii. 99 St. Lucian life has had a consistency and continuity rare in the New World. **1971** *Advocate-News* (Barbados) 17 Sept. (Guyana Suppl.) p. 1/1 Barbadians, St. Lucians, St. Vincentians, Grenadians, Dominicans..all of them Guyanese by definition though not by birth..these are the people one meets in almost every hinterland settlement in Guyana. **1973** *Caribbean Contact* Feb. 4/1 In a brief conversation with the St. Lucian born elevator attendant.. Salkey records valuable information. **1978** *Daily Tel.* 11 Mar. 13/1 Because of the to-ing and fro-ing of the two languages, it takes a St. Lucian to understand another St. Lucian when he breaks out into the local Creole patois.

saintpaulia (sëntpǭ·liä). [mod.L. (H. Wendland 1893, in *Gartenflora* XLII. 321), f. the name of Baron Walter von *Saint-Paul* (1860–1910), German explorer + -ia¹.] A stemless perennial herb of the genus so called, belonging to the family Gesneriaceæ, native to East Africa, and bearing ovate hairy leaves and clusters of violet, pink, or white flowers; esp. a pot plant of the species *Saintpaulia Ionantha*, the African violet.

1895 Hofmarschal Baron St. Paul in *Curtis's Bot. Mag.* CXXI. 7408 The *Saintpaulia* was discovered by my son, who lives in East Africa. **1946** F. Kent *All about House Plants* xv. 127 If you carefully examine a Saintpaulia which has been growing in the house all winter you will see..that it has split up into several crowns. **1961** *Amateur Gardening* 14 Oct. 27/2 It would be best to divide the saintpaulia when it finishes flowering. **1974** *Times* 5 Oct. 12/2 Millions of African violets, saintpaulias, are sold every year, but vast numbers do not live for long in houses or flats.

Saint-Simonian, *a.* and *sb.* Add: (Earlier and later examples.) **Saint-Simonist** (earlier example); also *attrib.* or as *adj.*; **Saint-Simonianism** (earlier and later examples), **-Simonism** (earlier example).

1829 J. S. Mill *Let.* 7 Nov. in *Wks.* (1963) XII. 40, I object altogether to the means which the St Simonists propose for organizing the *pouvoir spirituel*. **1830** —— *Ibid.* 9 Feb. 45 His objections to the Saint-Simonian philosophy. *Ibid.* 48 France must pass through several states before it arrives at St Simonism. **1831** *Ibid.* 20–22 Oct. 76 A Christian would be positively less fit than a St Simonian (for example), to form part of a national church. **1833** *Ibid.* 25 Nov. 193 The great majority here retained of St Simonianism about as much as is good and true, dropping the rest. **1952** F. A. von Hayek *Counter-Revolution of Sci.* iv. 144 The greatest of the Saint-Simonians..and the medium through whom many of them had received the doctrines of the master, was Auguste Comte. **1953** S. Spender *Creative Element* ii. 54 At the end of *Une Saison en Enfer* Rimbaud seems, indeed, to wish to reconcile Christianity with Saint-Simonist socialism. **1974** *Times Lit. Suppl.* 25 Jan. 66/2 He suggests that Heine saw himself as a Saint Simonian prophet and that Saint Simonianism, not the 1830 revolution, drew him to Paris. **1976** A. W. Gouldner *Dialectic of Ideol.* xi. 274 The surprising continuities between the Saint-Simonian formulations and the Weberian.

St. Trinian's (sənt tri·niənz). The name of a girls' school invented by the cartoonist Ronald Searle (b. 1920) in 1941. Used *absol.* and *attrib.* to designate allusively the characteristic style of hoydenish behaviour, school uniform, etc., of the girls in the cartoons and the subsequent associated books and films.

Searle's daughters attended St. Trinnean's school in Edinburgh.

[**1941** R. Searle in *Lilliput* IX. iv. 313 (*caption*) Owing to the international situation the match with St. Trinian's has been postponed.] **1948** D. B. Wyndham Lewis in R. Searle *Hurrah for St. Trinian's* 8 Those typical English Roses, the girls of St. Trinian's, a nightmare synthesis of Roedean, Heathfield and Wycombe Abbey.) **1958** *Times* 20 May 11/4 How the girls of to-day, thinking at St. Trinian's taking their degrees at St. Jude's, will smile with affectionate tolerance at these vague and unpractical theories. **1959** *Guardian* 3 Mar. 10/4 A St Trinian's type of schoolgirl. **1964** C. Dale *Other People* xv. 88 She was big and fat and pasty... In her school uniform..she looked a complete St. Trinian's type. **1972** *Guardian* 25 Jan. 9/2 Louis Feraud.. includes a group of dresses called schoolgirl frocks. Lolita lives again, and one longs for the innocence of St Trinian's. **1977** 'D. Craig' *Bennett* iii. 35 His high-pitched St Trinian's giggle. **1981** R. Barnard *Sheer Torture* xi. 121 Aunt Kate..an overgrown product of St Trinian's.

sais, saïs, var. Syce. (Further examples.) Also [a. Arab. *sā'is*] in African and Asian use.

1887 Kipling *Plain Tales from Hills* (1888) 28 He..deserved a V.C., if only for saving the sais's life.. **1924** L. Eckenstein *Tutankh-amen* ii. 24 The saises running on either side of the chariots as only syces run in Egypt can run. **1935** J. N. H. Suzman *Pa Bri. Malaya Today* 143 The Malay saises will all be playing cards. **1936** W. H. Smith *Let.* 26 June in *Young Man's Country* (1977) ii. 1, I said good-bye to Peter and his sais yesterday morning. **1953** J. Masters *Lotus & Wind* viii. 113 I'll walk back to your bungalow with you. My sais can bring Beauty along. **1975** P. Dunbar *My Sister, Isak Dinesen* v. 36 The sais (horse-keeper) was to bring the horses up after us.

Saiva (ʃai·vä), *sb.* and *a.* [a. Skr. *śaiva* relating, belonging, or sacred to Siva; a worshipper or follower of Siva.] **A.** *sb.* A member of one of the three great divisions of modern Hinduism, exclusively devoted to the worship of the god Siva as the Supreme Being. **B.** *adj.* Of or pertaining to this division of Hinduism.

1810 E. Moor *Hindu Pantheon* 15 Saivas or worshippers of Siva. **1845** *Penny Cycl.* XXXI. 169/1 The great Saiva reformer, Sankara Acharya. **1876** *Encycl. Brit.* IV. 210/1 The Saiva, Vaishnava, and Sakta sects. **1974** *Encycl. Brit. Macropædia* VIII. 893/1 Most Saiva worship is not systematic but a complex amalgam of pan-Indian Saiva philosophy and local or folk worship.

Hence **S(h)ai·vism** = Sivaism; **S(h)ai·vite** *sb.* and *a.* = Sivaite.

1861 H. H. Milman *Jrnl.* 21 Nov. in F. M. Milman *Mem. R. Milman* (1879) iii. 48 This temple is reckoned..the holiest shrine in India among Shaivites. **1877** Monier Williams *Hinduism* viii. 97 Saivism and Vaishnavism are, as it were, the two constituents of Hinduism. **1889** G. Hoddling.. *Encycl. Brit.* XIV. 228/1 Saivite gods of devils. **1924** E. M. Forster *Passage to India* xxxviii. 349 They cantered..past a Saivite temple, which invited to lust, but under the semblance of eternity. **1956** R. Redfield *Peasant Society & Culture* 88 The important Vaishnavism and Shaivism are theistic and ethical. **1969** *Indo-Asian Culture* Oct. 70 Both Saivism and Vaishnavism were popular in Sthatta and the neighbouring regions during the late Gupta and medieval times. **1978** 'E. Peters' *Death to Landlords!* x. 153 A Saivite sadhu seated in contemplation.

saj (sadʒ). [Hindi.] The Indian laurel, *Terminalia tomentosa*, a tropical tree of the family Combretaceæ, native to India and Burma and bearing terminal spikes of yellow flowers; also, the dark hardwood produced by this tree and others of the genus. Also *attrib.*

1839 E. W. Lane tr. *Arabian Nights* II. xiii. 384 Its door was of sáj, adorned with brilliant gold. **1931** J. W.

Best *Tiger Days* xii. 173 Nobler trees take their place; the stately saj and the dark-limbed ebony. **1952** J. Masters *Deceivers* viii. 88 The man..started back..and began to run toward a thin line of saj trees bordering the road.

Saka (ʃɑ·kä), *sb.* and *a.* Also **Çaka, Saca.** [Skr. *Saka*; cf. Gr. pl. Σάκαι, L. pl. *Sacæ*.] **A.** *sb.* **a.** (A member of) an ancient Indo-Scythian people originating in central Asia. **b.** The language of this people. = *Khotanese *sb.*

[**1601** P. Holland tr. *Pliny's Nat. Hist.* vi. xvii. 123 Beyond the realme Sogdiana, inhabit the nations of the Scythians. The Persians were wont to call them in generall Sacas, of a people adjoining unto them, so named. **1795** J. Nott tr. *Catullus' Poems* I. xi. 35 Whether he treads Hyrcanian ground; Or seeks the gentle Arab's home; The Parthians, for the dart renown'd; Or midst the Sacæ's flood to roam.] **1880** H. W. Bellew *Races of Afghanistan* ii. 18 The province itself derived its name of Sakistân..from the Saka, who were probably the same people as the Saka Hamuvarga mentioned in the tablets of Darius. **1934** Ahmad & Aziz *Afghanistan* vii. 45 Driven from their home in Central Asia the Sakas migrated into Kashmir. **1961** [see Indo-Scyth s.v. *Indo-¹]. **1966** G. Lanz in Birnbaum & Puhvel *Anc. Indo-European Dial.* 223 Of the fifty-one words submitted as possible borrowings from Iranian, twenty-one are attested in Saca (Khotanese), or on various grounds appear to be for the most part of Saca origin. **1972** W. B. Lockwood *Panorama Indo-European Lang.* 237 The Persians are said by Herodotus to have called the various Scythian tribes Saka.. Rich manuscript remains of Saka came to light in Turkestan... The language of Khotan is called Khotanese Saka or simply Khotanese... Saka appears to survive in the mountains to the west. **1974** *Encycl. Brit. Macropædia* II. 173/1 Iranian tribes of nomadic Saka.. seem, before 130 BC, to have made a pact with the Parthians and to have settled in Sīstān whence they spread eastward..into India.

B. *adj.* **a.** Of or pertaining to this people or their language. **b.** In Indian chronology, designating or pertaining to an era reckoned from A.D. 78.

1868 [see *Mahayana]. **1886** *Encycl. Brit.* XXI. 854/1 The ancient Aryan inscriptions usually employ the Saka (Sâlivahana) era, dating from 79 A.D. **1923** *Cambr. Hist. India* I. xxiii. 585 It was in consequence of its long use by the Çaka princes of western India that the era has become generally known in India as the Çaka era—a name which effectually disguises its origin, and one which has in no small degree perplexed modern scholars in their endeavours to unravel the secret of its history. **1937** L. H. Gray *Foundations of Lang.* 371 Other Middle Iranian documents, especially known through recent discoveries, represent the Sogdian and Saka dialects. **1956** R. Pieris *Sinhalese Social Organization* ii. 92 The Saka era was made use of in all legal instruments. It is said to date from a king Saka. **1958** O. Caroe *Pathans* iv. 63 Greek or Macedonian soldiers were needed to guard the frontier marches against the Saka nomads. **1974** *Encycl. Brit. Macropædia* IX. 374/2 The Śaka, or Śālivāhana, era (78 CE), now used throughout India, is the most important of all. It has been used not only in many Indian inscriptions but also in ancient Sanskrit inscriptions in Indochina and Indonesia. The reformed calendar promulgated by the Indian government from 1957 is reckoned by this era. It is variously alleged to have been founded by King Kaniska or by the Hindu king Salivāhana or by the king Nahapāna.

Hence **Sa·kian** = *Saka *sb.* b.

1933 L. Bloomfield *Language* iv. 63 Other medieval Iranian languages, which have been identified as Parthian, Sogdian, and Sakian. **1939** [see *Khotanese *sb.* and *a.*].

sakabula (sakabu·lä). *S. Afr.* Also **sac(c)abola,** etc. [a. Zulu *isSakabuli* widow-bird.] The long-tailed widow-bird, *Euplectes progne*, of the family Ploceidæ, the male of which is black, with red patches on the wings and very long tail-feathers. Also *fig.*

1877 Lady Barker *Year's Housekeeping S. Afr.* ix. 179 Lynx tails hung down like lappets on each side of her face which was over-shadowed and almost hidden by the profusion of sakabula feathers. **1885** Rider Haggard *King Solomon's Mines* viii. 127 They wore upon their heads heavy black plumes of Sacaboola feathers, like those which adorned our guides. **1896** H. L. Tangye *In New S. Afr.* iv. 100 One of the most strange inhabitants of the Transvaal is a small black bird, the Sakabula. **1911** *E. London Dispatch* 20 July 3 They bartered the highly prized tail feathers of the sakabula bird. **1937** S. Cloete *Turning Wheels* 362 A saccabula, gorgeous in his black spring feathers, his wings blotched with red, flew past them followed by his wives that were grey and dull. **1951** R. Campbell *Light on Dark Horse* x. 144 The finest variety [of widow-bird], the 'sakabula', is quite a common sight. **1973** *Weekend Post* (Port Elizabeth) 28 Apr. 3 The long-tailed widow birds commonly known as sakabullas.

Sakai (sā·kai), *sb.* (and *a.*) Also 9 **Sakkye.** [Malay, lit. subject, dependent.] **a.** An aboriginal people of the Malay peninsula (loosely used of Malayan aborigines collectively); a member of this people. **b.** The language of the Sakai. Also *attrib.* or as *adj.*

1839 T. J. Newbold *Pol. & Statistical Acct. Straits of Malacca* I. vi. 421 The Semangs, Sakkye, or Orang Bukit, men of the hills. **1886** *Jrnl. Anthrop. Inst.* XV. 285 In this state of Perak there is at present besides the Sakais one other race, the Sêmang. *Ibid.* The Sakai race inhabits the left bank of the Perak River. **1906** [see *Jakun]. **1920** R. J. Wilkinson *Hist. Peninsular Malays* (ed. 2) i. 3 The fair wavy-haired aborigines known as the Sakai inhabit both sides of the Malayan main range. *Ibid.* 8 The grammar..of Sakai is extraordinarily complex and inflected. **1932** L. Golding *Magnolia St.* iii. vi. 538 The people seemed stranger to him than the pygmies of the African jungle or the Sakais of Malaya, who live up in the hills and make their clothes of the bark of the trees. **1952** P. D. R. Williams-Hunt *Introd. Malayan Aborigines* i. 1 Sakai, used generally for aborigines is a derogatory term which is disliked by most jungle dwellers. **1966** *Telegraph* (Brisbane) 18 Nov. 2/3 Malaysia's aborigines, the little brown jungle men called Sakai, have until recently had no modern technology. **1977** P. Theroux *Consul's File* 45 The local sakais—they might be called the Temiars—had deported some wild monkeys.

sakawinki. Add: Also **sak(k)iwinki(e).** (Further examples.) Also, a South American squirrel monkey of the genus *Saimiri*.

1954 G. Durrell *Three Singles to Adventure* i. 38 'What are they?' ..'Squirrel monkeys, but I don't know the local name for them.' 'Sakiwinkis, Chief.' **1958** J. Carew *Black Midas* vi. 102 Behind us, sakki-winki monkeys chattered, and toucans screamed. *Ibid.* vii. 155 Red howlers roared..and tinamous and saki-winkies joined in.

sake, *sb.* Add: **4*.** (See quot. 1879².) *nonce-use.*

1876 G. M. Hopkins *Wreck of Deutschland* xxii, in *Poems* (1967) 58 Five! the finding and sake And cipher of suffering Christ. **1879** —— *Henry Purcell* in *Ibid.* 80 Let him..with the sheer air..lay me! only I'll Have an eye to the sakes of him, quaint moonmarks, to his pelted plumage under Wings. **1879** —— *Lett. to R. Bridges* (1935) 83 Sake is a word I find it convenient to use... It is the sake for 'the sake of'..., I mean by it the being a thing has outside itself, as a voice by its echo, a face by its reflection,..a man by his name, fame or memory, and also that in the thing by virtue of which especially it has this being abroad,..as for a voice and echo clearness; for a reflected image light, brightness; for a man genius, great achievements... In this case it is, as the sonnet says, distinctive quality in genius.

saké. Add: Now usu. with pronunc. (sā·ki). (Further examples.)

1926 [see *brown rice]. **1927** E. Pound *Lustra* 189 We drink our parting in saki. **1931** G. B. Sansom *Japan* [illegible] the new brew. **1947** J. Bertram *Shadow of War* 273 The guards looted the [illegible] from their own stores. **1948** G. Mikes *East is East* 56 Drinking saké, watching dancing and listening to singing. [illegible] filling my cup ..I guessed. **1963** [illegible] And fill my [illegible] *Japan* v. 118 An immense sake bowl was filled with about 4 gallons of this wine. **1979** J. Melville *Wages of Zen* ii. 17 Otani [illegible].

Sakel: see next.

Sakellaridis (sækelæ·ridis), **Sakellarides** -idiz, **Sakel** (sæ·kĕl). [The name of Σακελλαρίδης, a Greek cotton-grower who originated the variety.] The name of a superior variety of Egyptian cotton, widely grown in the early 20th century.

1912 W. L. Balls *Cotton Plant in Egypt* vi. 105 The main varieties at present cultivated on a commercially important scale are Yannovitch and Sakellaridis in the 'fine-spinning' group; [etc.]. *Ibid.* 106 One of the cherished fables of the practical cotton grower..crops and fine ..of this belief, ..been proved ..*World's Cotton* ..is commonly ..from about ..Sakellaridis at ..famous Sakel- ..name in the ..parent of most ..now being ..*Growing Probl.* ..acquired pro- ..Sakel (1909).. ..even Sakel. ..in 1918 several (Sakellarides), ..variety in [illegible] **1953** *New Biol.* XIV. 49 The famous Sakellaridis variety.

sakhaite (sɑ·khe,ait). *Min.* Also **sahaite.** [ad. Russ. *sakhait* (I. V. Ostrovskaya et al. 1966, in *Zapiski vsesoyuznogo min. Obshchestva* XCV. 193), f. *Sakha*, name of the locality in Siberia where it was discovered: see -ite¹.] A hydrous borate and carbonate of calcium and magnesium, the crystals of which belong to the cubic system and occur as greyish white masses.

1966 *Chem. Abstr.* LXV. 3367 A new mineral, called sakhaite..in..magnesian skarns.. **1974** ..The formula of sakhaite was recalculated in an attempt to determine whether a relationship existed between sakhaite and harkerite. **1975** *Soviet Physics Doklady* XIX. 559/1 We have studied the synthetic analogs of sahaite..under hydrothermal conditions in the $CaO-MgO-B_2O_3-CO_2-H_2O$ system.

Sakmarian (sakmæ·riǎn), *a. Geol.* [ad. Russ. *Sakmarskii* (first used as a stratigraphic term by A. Karpinsky 1874, in *Zap. Imperatorskago Min. Obshchestva* IX. 269), f. *Sakmara*, name of a river in the Southern Urals: see -ian.] Name of a stage in the Lower Permian in the Soviet Union; of or pertaining to this stage and the rocks that characterize it, and the geological age during which they were deposited. Freq. *absol.*

1936 V. E. Ruzhentsev in *Problemý Sovetskoĭ Geologii* VI. 506 The upper Carboniferous is overlain in complete conformity by the Permian system which begins with Schwagerina beds. The writer defines by the name Sakmarian the whole of the deposits with Schwagerina princeps Ehr. and with the Ammonoid fauna described for the first time from the Sakmara river... The Sakmarian consists of sandy-argillaceous beds among which there are many conglomerates. **1960** *Bull. Geol. Soc. Amer.* LXXI. 1766/2 It is now the official usage of the Geological Survey of the U.S.S.R. to draw the base of the Permian below the Sakmarian and equivalent beds in the Ufa Plateau. **1963** D. W. & E. E. Humphries tr. *Termier's Erosion & Sedimentation* i. 20 Lakes and marshes are known in the northern hemisphere which were contemporaneous with the Stephanian and Sakmarian glaciers of the Southern hemisphere. **1974** *Nature* 8 Feb. 396/1 McLachlan and Anderson have recorded orthocerid nautiloids, the brachiopod *Attenuatella*, [etc.]..from the base of the succession near Kimberley. They favoured a Sakmarian age for this marine incursion.

Sakta (ʃā·ktā). Also 9 **Sacta.** [a. Skr. *śākta* relating to power or to the Sakti; a worshipper of the Sakti.] A member of one of the principal sects of modern Hinduism which worships the Sakti or divine energy, especially as identified with Durgā, the wife of Siva. Also *attrib.* Cf. *Saiva *sb.* and *a.* Hence **Sa·ktism,** the worship of the Sakti.

1810 E. Moor *Hindu Pantheon* 116 Those, of whatever sect, who worship exclusively the female power..are called Sactas. **1845** *Encycl. Metrop.* XXIV. 443/1 The Hindus are almost always either 1. Vaishnavas..; 2. Saivas..; or 3. Sáktas. **1877** Monier Williams *Hinduism* ix. 123 Tântrism, or Sâktism, is Hindûism arrived at its last and worst stage of medieval development. **19**[illegible] [see *nada¹]. **1931** G. Macmunn *Relig. India* 69 The Sakta groups have borrowed much from aboriginal practices and influence. *Ibid.* 160 The really secret cult of Saktism. **1974** *Encycl. Brit. Macropædia* VIII. 896/2 The Tantric movement is not rarely inextricably interwoven with Sâktism. *Ibid.* 897/2 Sâkta adepts are trained to direct all their energies toward the conquest of the Eternal.

Sakti (ʃa·kti). Also 9 **Sacti; Shakti** and with small initial. [a. Skr. *śakti* power, divine energy, f. *śak* to be able.] In Hindu religion, the female principle, esp. when personified as the wife of a god, as Durgā is the Sakti of Siva, etc.; supernatural energy embodied in the principle.

1810 E. Moor *Hindu Pantheon* 10 All the principal, and several of the secondary deities..have wives assigned to them, who are called Sacti. **1842** *Penny Cycl.* XXII. 67 That thou, united with thy Sakti, dost in sport create the universe from thy own substance. **1862** Mrs. J. B. Speid *Our Last Yrs. in India* vii. 174 Seresvati, the goddess of letters, &c., Lackshmi, of prosperity, and Kali or Parvati, of destruction... These three goddesses under the name of the Sactis, sometimes receive an exclusive worship. **1871** J. Garrett *Classical Dict. India* 540 The Sakti is said to have originated in God, the Supreme Being... There are many special forms of Sakti worship. **1918** J. Woodroffe *Shakti & Shâkta* 49 According to Shâkta doctrine each man and woman contains within himself and herself a vast latent magazine of Power or Shakti. **1922** Joyce *Ulysses* 499 It has been said by one: beware the left, the cult of Shakti. **1937** M. Covarrubias *Island of Bali* x. 339 Every Balinese believes that his body, like an electric battery, accumulates a magic energy called *sakti* that enables him to withstand the attacks of evil powers... This *sakti* is not evenly divided; some people are born with a capacity to store a higher charge of magic than others; they become the priests, witch-doctors, and so forth, endowed with supernatural powers. **1962** A. Huxley *Island* xiii. 21[illegible] Paintings of tropical animals, Bodhisattvas and their bosomy Shaktis. **1968** A. Warhol *A* 421 He doesn't have any bhakti it's all shakti. **1972** D. Bloodworth *Any Number can Play* xviii. 184 In thirty times thirty years will come one with the ears of the Buddha and with the *sakti*... Supernatural power. **1977** *N.Y. Times Mag.* 4 Dec. 144 Joya's famous *shakti*, or spiritual energy.

sakura (sakū·rä). [Jap.] A flowering cherry tree belonging to one of the many varieties bred from various species of *Prunus*; also, the blossom or wood of a tree of this kind.

1884 tr. *J. J. Rein's Japan* ii. iii. 471 *Yoshino*..once the residence of the anti-emperors, a famous old place with many Sakura (Prunus pseudocerasus). **1892** F. T. Piggott *Garden of Japan* 19 *P[runus] pseudocerasus*—*sakura*, with enormous pink double flowers. **1911** *Encycl. Brit.* XV. 175/2 The wood used is generally that of the cherry-tree, *sakura*, which has a grain of peculiar evenness and hardness. **1948** C. Ingram *Ornamental Cherries* 1[illegible] Will you please tell me why you are so very fond of our

Sakura—our Cherries? **1963** *Times* 22 Apr. 11/7 Famous songs such as '*Sakura, Sakura*' (Cherry blossom, Cherry blossom) elicited no gleam of sentiment. **1970** J. Kirkup *Japan behind Fan* 41 The season when the *sakura* or cherry blossom blooms.

† **sal³** (sæl). *Theatr. slang. Obs.* Abbrev. of Salary *sb.* 1.

1844 E. R. Lancaster *Manager's Daughter* (ed. 2) in *Oxberry's Budget of Plays* I. 110/1 Who does he suppose was to cut comic mugs before noblemen, without being paid double sals.? **1870** O. Logan *Before Footlights* xxxii. 433 'You're earning your sal easy,' says Clown to him with some reproach. **1885** *Househ. Words* 29 Aug. 350/1, I say that part of this money shall be shared among us as 'sals', and some of the remainder shall be used for mounting the guv'nor's panto.

sal⁴ (sæl). [f. S(ilicon + Al(uminium).] **1.** *Petrogr.* One of the two primary categories erected by Cross, Iddings, Pirsson, and Washington to classify igneous rocks and their characteristic minerals, and broadly including those rich in non-ferromagnesian aluminous and siliceous minerals such as quartz, feldspars, and feldspathoids. Hence **salic** (sæ·lik) *a.*², of or pertaining to this category of rocks. Cf. *femic *a.*

1902 W. Cross et al. in *Jrnl. Geol.* X. 573 To express concisely the two groups of standard minerals and their chemical characters in part, the words *sal* and *fem* have been adopted. The former is employed to designate group I, mnemonically recalling the *si*liceous and *al*uminous character of its minerals. **1902**, etc. [see *femic *a.*]. **1931** A. Johannsen *Descr. Petrogr. Igneous Rocks* I. viii. 86 The classes are determined by the salic-femic ratio. The five classes are: I. Persalic. Ratio sal:fem greater than 7·00 [etc.]. **1974** I. S. E. Carmichael et al. *Igneous Petrol.* ii. 48 The most generally used index of magmatic evolution is the differentiation index (DI) proposed by Thornton and Tuttle (1960); this is simply the weight percentage of the..salic components quartz.., albite.., orthoclase.., nepheline.., leucite.., and kalsilite.

† **2.** *Geol.* Also **Sal.** [a. G. *Sal* (E. Suess *Das Antlitz der Erde* (1909) III. ii. xxiv. 626), f. *S*(*i* + *Al*, chem. symbols for silicon and aluminium.] = *sial (now superseded by that term). *Obs.*

1909 [see *Nife]. **1922** *Geol. Mag.* LIX. 338 Wegener accepts the terminology of Suess, except that he follows Pfeffer in writing Sial instead of Sal. *Ibid.* 340 The boundary of the Sal should therefore be drawn at the foot of the continental slope, where the continental masses begin to rise from the ocean-floor. **1954** R. L. Parker tr. *P. Niggli's Rocks & Mineral Deposits* xi. 476 A granite-gneiss association takes the upper hand and is the reason for calling the entire outer crust the sial crust (sial or sal, containing Si and Al, besides alkalies, as the most important elements).

salable, var. Saleable *a.*

salad. Add: **3.** *salad bowl* (earlier and later examples), *cream* (further example), *-eater*, *fork*, *leaf*, *plate* (examples); **salad bar** chiefly *U.S.*, a servery from which a salad may be obtained; **salad basket,** (*a*) a wire basket in which superfluous moisture is shaken from the constituents of a salad after washing; (*b*) *slang* [tr. Fr. *panier à salade*], a police van, 'Black Maria'; **salad days:** also *attrib.* in *sing.*; **salad servers,** a large spoon and fork for serving salads.

1976 *Amer. Speech 1974* XLIX. 116 *Salad bar*, counter in many restaurants, with ingredients from which the diner can make his own salad. **1978** *Times* 23 Apr. 12/6 The..assistant manageress..led me to the salad bar with its two kinds of salad, four kinds of bread and four kinds of salad dressing. **1906** *Mrs. Beeton's Bk. Househ. Managem.* xxxv. 1092 Where a salad basket is not available, the materials should be well drained and shaken in a colander. **1962** P. Brickhill *Deadline* vi. 83 A row of large 'Black Marias', or, as I learned, '*paniers à salade*' (salad baskets) as the French call them. **1966** J. Dos Passos *Best Times* (1968) ii. 54 The French cooks were already out..whirling the salad around in wire salad-baskets to dry it. **1975** H. McCutcheon *Instrument of Vengeance* iii. 52 There will be a salad basket here soon... What you call, I think, a Black Maria. **1773** J. Wedgwood *Let.* 21 Nov. (1965) 156 Sa[lad] Bowles, and boats. **1867** Trollope *Last Chron. Barset* I. xxxii. 267 A bitter leaf will now and then make its way into your salad-bowl. **1921** *Daily Colonist* (Victoria, B.C.) 22 Oct. 7/7 (Advt.), China salad bowls—hand painted. **1980** *Berkeley Graduate* Oct. 5/2 Even in California, the salad bowl of the nation, thousands of people were hungry. **1976** D. Clark *Dread & Water* ii. 26 A woman..was shaking salad cream from a bottle. **1953** Dylan Thomas *Under Milk Wood* (1954) 60 She whispers to her salad-day deep self. **1963** *Times* 8 Mar. 15/4 This was a young concerto for a young pianist—it was, we have tried to suggest, not such a salad-day reading. **1947** Auden *Age of Anxiety* (1948) iii. 70 The parlour cars and Pullmans are packed also With scented assassins, salad-eaters Who murder on milk. **1917** *Harrods Gen. Catal.* 892/2 Glass salad forks... Prices on application. **1978** *Detroit Free Press* 5 Mar. A17/5 (Advt.), Stainless tableware..setting includes salad fork, dinner fork, [etc.]. **1927** Joyce *Pomes Penyeach*, The still garden where a child Gathers the simple salad leaves. **1881** C. C. Harrison *Woman's Handiwork* iii. 219 The little salad-plates were silver-gilt. **1976** G. McDonald *Confess, Fletch* (1977) xxxii. 150 Sylvia entered with salad plates. The salad consisted of..cold, canned peas. **1907** *Yesterday's Shopping* (1969) 148/2 Salad Servers, boxwood..set 1/1. **1978** 'M. Delving' *No Sign of Life* v. 94 Betsy is a carver... She carves the handles of salad servers and jugs for me.

‖ **salade niçoise** (salad nīswāz). [Fr., = salad of, or from, Nice in the south of France.] A variety of salad (see quots.).

1955 E. David *Bk. of Mediterranean Food* 160 There may be anchovies, gherkins, artichoke hearts, lettuce... *Salade niçoise*..is made with the same variety of ingredients. **1960** —— *French Provincial Cooking* 145 (*heading*) *Salade niçoise*..The ingredients depend upon the season and what is available. But hard-boiled eggs, anchovy fillets, black olives, and tomatoes, with garlic in the dressing, are pretty well constant elements in what should be a rough country salad, rather than a fussy chef's concoction. **1969** C. Irving *Fake!* vii. 86 A gallery, as Elmyr put it, 'like a *salade niçoise*, a little bit of everything, mostly for the American tourists'. **1975** *Times* 22 July 14/3 Eight dinner guests fed on *salade niçoise* made with fresh French beans.

‖ **saladero** (saladēᵊ·ro). [Sp.] In Spain and Latin America, a slaughter-house where meat is also prepared by drying or salting.

1870 *Weekly Standard* (Buenos Aires) 19 Jan. 8/5 The sales of saladero ox and cowhides during the last fifteen days. **1885** *Encycl. Brit.* XIX. 762/2 The principal prison in the capital of the kingdom [*sc.* Spain] was nothing more than a converted slaughter house where pigs were killed and salted, as its name, the Saladero, implied. **1902** *Encycl. Brit.* XXXI. 461/2 The increase in the herds of recent years has caused the owners of *saladero* establishments in Argentina and Uruguay to try the working of factories in Paraguay for the preparation of *tasajo* (jerked beef). **1930** C. F. Jones *S. Amer.* xviii. 403 As a lean animal served the purposes of the saladero, or salting establishment, the native cattle proved quite satisfactory. **1960** H. S. Ferns *Brit. & Argentina in 19th Cent.* xiii. 416 The old-fashioned *saladeros* supplying the domestic market with fresh meat and foreign markets with dried and salted meat were faced with severe difficulties. **1973** M. Kochan tr. *F. Braudel's Capitalism & Material Life* iii. 135 *Charque*, boned and dried meat produced in the saladeros of Argentina (once again intended for slaves and the European poor), was to all practical purposes invented at the beginning of the nineteenth century.

‖ **Salagrama** (ʃālagrā·ma). Also **Salagram** and with small initial. [a. Skr. *śālagrāma* (see Shalgram).] = Shalgram. Freq. *attrib.*

1801 H. T. Colebrooke in *Asiatick Researches* VII. 240 A Sâlagrâma stone ought to be placed near the dying man. **1833** R. Everest in *Ibid.* XVIII. ii. 111, I have several times looked for such among the Salagrams in the Hindoo temples. **1913** J. N. Farquhar *Crown of Hinduism* 267 If he recognizes Vishnu, he may possess a discus, a *salagrama* stone, a conch shell or a *tulsi* plant. **1920** —— *Outl. Relig. Lit. India* vii. 293 The more usual symbols are: Vishnu, the Śālagrāma pebble.

salak (sala·k). Also **salac.** [Malay.] A thorny palm tree belonging to the genus *Salacca*, native to tropical south-east Asia, esp. *S. edulis*, or its pear-shaped edible fruit.

1820 J. Crawfurd *Hist. Indian Archipelago* I. iv. 445 The *Salak*, affords a fruit about the size of a pullet's egg, which consists of a hard stone, enveloped by a firm white pulp, which is covered by thin husks. **1856** B. Seemann *Pop. Hist. Palms* 345 Nothing is recorded of the other species of this genus,—for instance,..the Salak of Penang. **1937** M. Covarrubias *Island of Bali* v. 105 *Salak*, a pear-shaped fruit that grows on a palm, tastes like pineapple, and is covered by the most perfect imitation snakeskin. **1952** W. March *October Island* x. 125 He made up a poem from the names of the palm trees..and..would recite it... Salak, pigafettia, orania palindan,... And the great royal palm. **1981** *Oxf. Encycl. Trees of World* 259/2 Many palms have edible fruits though rather few are widely cultivated for this product, amongst them Salac (*Salacca edulis*) of southeast Asia.

salal. Substitute for def.: An evergreen shrub, *Gaultheria shallon*, belonging to the family Ericaceæ, native to western North America, and bearing racemes of pink or white flowers followed by edible purple berries. Also *attrib.* (Earlier and further examples.)

1825 D. Douglas *Jrnl. Trav. N. Amer.* (1914) 104 *Gaultheria shallon*; called by the natives 'Salal' not 'Shallon'. **1833** W. F. Tolmie *Jrnl.* 29 Aug. (1963) 230 Have supped on Sallal & at dusk, shall turn in. **1884** C. Phillipps-Wolley *Trottings of Tenderfoot* 140 In front lay in the river-bed a grove of cottonwood, and the bush I think British Columbians call 'sal lal'. **1926** *Daily Colonist* (Victoria, B.C.) 11 July 16/3, I caught my foot in a trailing vine and ploughed head first into the salal bushes. **1946** [see *black-cap 5]. **1952** *Beaver* Sept. 7/1 Fireweed blazes in the rear and salal sprouts out of the unpainted totem poles. **1977** J. Gillis *Killers of Starfish* (1979) xii. 105 She started to lead the way through Mike's pile of salal cuttings.

salamander, *sb.* Add: **3. d.** (Earlier and later examples.)

1755 H. Glasse *Art of Cookery* (ed. 5) 331 Put it in the Oven to brown, or do it with a Salamander. **1943** F. Thompson *Candleford Green* iii. 54 The smith then heated red-hot one end of a large, flat iron utensil known as the 'salamander' and held it above the plate until the rashers were crisp and curled. **1958** *Observer* 18 May 10/5 Caramelise the sugar by passing a red hot salamander very close to the surface till the sugar melts.

e. (Earlier example.) Also (*N. Amer.*), a workman's brazier.

1873 *Chicago Tribune* 3 Feb. 1/7 It caught fire from the 'salamander' used in drying the plaster. **1944** S. Bellow *Dangling Man* 107, I warmed myself at a salamander flaming in an oil drum. **1971** R. Lewis *Fenokee Project* viii. 151 They caught a glimpse of twinkling lights.. 'Salamanders... The workers over there have set up fire pots made out of punctured oil drums.'

4. (Earlier and later examples.)

1805 M. Lewis *Jrnl.* 9 Apr. in *Orig. Jrnls. Lewis & Clark Exped.* (1904) I. 289 Their work resembles that of the salamander common to the sand hills of the States of South Carolina and Georgia. **1834** J. J. Audubon *Ornith. Biogr.* II. 264 Thousands of 'mole-hills', or the habitations of an animal called 'the salamander'..presented themselves. **1885** *S. Florida Sentinel* 8 Apr. 1/6 The gophers (Florida salamanders) proved its [*sc.* the garden's] destruction. **1943** A. G. Powell *I can go Home Again* 225 The small burrowing rodent..which others call the gopher, we called the salamander. **1964** W. H. Burt *Field Guide to Mammals* (ed. 2) 136 Southeastern pocket gopher..(Salamander).

6. salamander safe (earlier examples); now *obs.*

1840 *Merchant's Mag.* (U.S.) II. 280 The Salamander Safe. **1845** in C. Cist *Cincinnati Misc.* I. 194/2 These Salamander safes are made of stout, wrought bar and plate iron,.. lined with a chemical preparation, which is a non-conductor of heat, and is indestructible by fire. **1852** *Hunt's Merchants' Mag.* XXVI. 256 In April, 1833 I [*sc.* C. J. Gayler] patented my 'Double Fire Proof Safe'. The same year the name 'Salamander' was applied to it, for the reason that one had been subjected to a very intense heat for a long time, and fully protected its valuable contents.

Hence **salama·nderish** *a.* (*rare*⁻¹).

1921 W. de la Mare *Mem. Midget* xxxii. 225 Even my salamanderish body sometimes gasped like a fish out of water.

salamander, *v.* Add: **c.** *Cookery.* To brown by means of a salamander. Hence **sala·mandering** *vbl. sb.*

1878 *Amer. Home Cook Bk.* 65 When it is cooked, glaze the top and salamander it. **1943** F. Thompson *Candleford Green* iii. 54 Another cooking process..which perhaps may have been peculiar to smithy families was known as 'salamandering'.

salame. Delete ‖. **1.** Now usu. in form salami (sălā·mi), constr. *sing.* (Further examples.)

1937 *Time & Tide* 11 Sept. 1209/1 Everyone carried a basket with their food for the day—red wine, long rolls of bread, salami, cherries. **1956** A. Wilson *Anglo-Saxon Att.* ii. iii. 363 Tea at Slough was a curious meal. There was salami and *mortadella* and caraway bread. **1973** C. Bonington *Next Horizon* x. 146 We had a mass of high protein food; nuts, cheese, salami [etc.].

2. *attrib.* and *Comb.*, as *salami sandwich*, *sausage*; **salami tactics,** the piecemeal attack on or elimination of (esp. political) opposition (see quot. 1952).

1925 N. Coward *Fallen Angels* in *Three Plays* 258 At the last moment he said he wanted a Salami sandwich. **1977** C. McFadden *Serial* (1978) xlviii. 102/2 I'm not going back to the same old lifestyle with..me making salami sandwiches all the time. **1946** G. Millar *Horned Pigeon* iii. 50 The driver handed out a bit of Salami sausage and a small flask of wine. [**1947** *Time* (Latin Amer. ed.) 9 June 25/1 In Budapest, the citizens considered that the Smallholders' Party had been wrecked. 'Rakosi has eaten the last of the salami', was the word.] **1952** *Times* 19 May 7/3 Mr. Rákosi describes one stage in it as 'salami tactics', by which slices of the Small-holders' Party were cut away and its strength worn down, even while the Small-holders' leader was Prime Minister and Mr. Rákosi his deputy. **1964** *Spectator* 29 May 731/2 Castro's skilful use of 'salami tactics' was helped by their prevailing reluctance to be considered 'witch-hunters'. **1978** *Times* 28 Apr. 17/7 If these salami tactics are continued it will not be long before they [*sc.* Kew Gardens] are closed on every public holiday.

salarian, *a.*² Restrict † *Obs.* to sense in Dict. and add: **b.** *Salarian Way*, the name of an ancient road, the *Via Salaria*, running from Rome north-east to Reate (now Rieti) and later extended to the Adriatic.

[*a* **900**: see Way *sb.*¹ 1 c.] **1866** tr. *P. Guéranger's Life St. Cecilia* viii. 87 Two figures in the cemetery of Priscilla in the Salarian Way, have been reproduced by Agincourt. **1945** R. Hargreaves *Enemy at Gate* 38 The Salarian Gate stood upon the Salarian Way, the road by which the Romans had been accustomed to carry sea-salt up to the country of the Sabines.

salariat (sălēᵊ·riăt). [a. F. *salariat*, f. L. *salārium* (see Salary *sb.*) after *proletariat* Proletariate, -at.] The salaried class; salary-earners collectively.

1918 Reckitt & Bechhofer *Meaning of Nat. Guilds* iv. 85 Hypnotized by the round 'O' in the figure of their pay, the salariat feel that they really are important members of the industry. **1922** *Q. Rev.* Apr. 288 The 'salariat' is almost as much enslaved as the proletariat. **1926** *Glasgow Herald* 2 Feb. 8/2 Departmental economies, involving..savings on the salariat. **1937** *Daily Herald* 8 Feb. 13/2 Sir Walter Citrine..stressed the fact that technological progress had enormously increased the importance of the 'salariat'. **1965** *Sunday Times* 17 Jan. 4/1 Mr. Iain Macleod..said..'The age of the Salariat is here, and the age of the wage earner is passing.' **1971**

Oxford Times 26 Nov. 7/2 Most of its members came from the British salariat—he [sc. Clive Jenkins] preferred this to the usual term white collar workers. 1978 Listener 26 Jan. 106/1 The professions and the salariat.

salary, *sb.* Add: **3.** *salary bracket, -earner, man, officer, scale; salary-fixing* vbl. sb.
1969 L. Hellman *Unfinished Woman* vi. 62 We were in what was called 'the same salary bracket'. 1926 *Socialist Rev.* Oct. 47 A minority of salary-earners receive also unearned incomes of varying sizes. 1961 *Guardian* 25 Oct. 1/7 The machinery for salary-fixing in the universities is complicated. 1719 in A. McF. Davis *Tracts Currency Mass. Bay* (1902) 193 Salary Men, Ministers, School-Masters, [etc.]..are pincht and hurt more than any. 1962 *Spectator* 29 June 846/2 Expensive cameras are being crowded out as the ultimate dream of what the Japanese call 'salarymen'. They are being replaced by a little bubble of an automobile. 1816 *Deb. Congress U.S.* 4 Dec. (1854) 240 The only difference between a salary officer and a per diem, is simply in the mode of payment, and not in the amount. 1940 R. S. Lambert *Ariel & all his Quality* xi. 302 Grade and salary scales were defined, and every employee informed where he stood.

‖ **salaud** (salo). [Fr., f. *sale* dirty.] A French term of abuse: filthy beast, 'swine', 'bastard'.
1962 D. Lessing *Golden Notebk.* III. 374 Jules said he would only pay me three hundred dollars for it. Salaud! 1967 C. L. Markmann tr. *F. Fanon's Black Skin White Masks* 11 When in the words of a gang of *salauds* it is no longer possible to find the sense of non-sense. 1971 E. Paul *Reluctant Cloak & Dagger Man* xv. 177 'Salaud,' Jean whispered... 'For that you are going to die.' 1977 Fontana & Van de Water in Douglas & Johnson *Existential Sociol.* iii. 109 In *Nausea* Sartre referred to the others as *salauds* ('swine'). They live a smug existence, feel no anguish, and easily find meaning and justification in their lives.

salbutamol (sælbiū·tămǫl). *Pharm.* [f. Sal(icyl + But(yl + Am(ine + -ol.] A white crystalline sympathomimetic agent which is used esp. as a bronchodilator in the treatment of asthma and is given as tablets of the sulphate or as an aerosol; 1-(4-hydroxy-3-hydroxymethylphenyl)-2-t-butylamino-ethanol, $C_{13}H_{21}NO_3$.
1969 *Brit. Jrnl. Dis. Chest* LXIII. 173 Salbutamol is a new, metabolically stable adrenergic stimulant apparently more specific than either isoprenaline or orciprenaline for adrenergic $β_2$-receptors. 1977 *Lancet* 23 Apr. 908/2 Aerosolised salbutamol 1500 μg had no significant effect on lung function. *Ibid.* 13 Aug. 354/2 In 208 women in premature labour oral salbutamol 8 mg 6-hourly prolonged pregnancy for more than 2 days in 90% of patients. 1980 *Brit. Med. Jrnl.* 29 Mar. Advt. between pp. x and xi, A metered-dose aerosol delivering 100 mcg salbutamol BP per actuation.

salchow (sæ·lkǫv, sæ·lkǫ). *Skating.* Also **Salchow.** [f. the name of Ulrich *Salchow* (1877–1949), Swedish figure skater, who invented it.] In full, *salchow jump.* A jump in which the skater takes off from the inside back edge of one skate and lands, after a complete rotation, on the outside back edge of the other.
1921 B. Meyer *Skating* 113 *Salchow jump*, outside forward three with jump from the back inside edge to the outside back edge of the other foot. 1930, 1959 [see *axel]. 1968 *Daily Tel.* 6 Dec. 15/6 An inspired Haig Oundjian took the men's title. He never put a foot wrong and his brilliant triple salchow was the jumping highlight of this meeting. 1976 *Times* 19 Jan. 9/7 Beginning with a perfect triple salchow and a soaring double axel, Miss Pötzsch went through her programme with rare charm.. marred only by a two-footed landing of a double salchow. 1980 *Times* 25 Jan. 9/3 The British champion did play for safety by eliminating the two triple jumps he is not totally sure of, the lutz and toe salchow.

Saldanha (sældā·nă). The name of a bay in western Cape Province, South Africa, used *attrib.* in **Saldanha man, skull,** to designate a fossil hominid belonging to an archaic form of *Homo sapiens* or the fragments of it found at Hopefield by Singer and Jolly in 1953.
1953 M. R. Drennan in *S. Afr. Jrnl. Sci.* L. 8 (*caption*) Side view of the skull-cap of Saldanha Man. *Ibid.* 8/2 The Saldanha skull is thus somewhat shorter..than..the Rhodesian skull. 1954 *Amer. Jrnl. Physical Anthropol.* XII. 349 Fluorine tests also revealed that *Mesochoerus* and *Paleoloxodon* lived contemporaneously with Saldanha Man. *Ibid.* 352 The Saldanha skull..at present consists of a fairly complete 'cap' or vault. 1959 J. D. Clark *Prehist. S. Afr.* iv. 83 Saldanha Man may..be considered to be representative of the kind of 'proto-Australoid' individual who was responsible for the final expression of the Earlier Stone Age cultures in southern Africa at the end of the Middle and beginning of the Upper Pleistocene. 1973 B. J. Williams *Evolution & Human Origins* xi. 184/2 The later find of the Saldanha skull provided another specimen almost identical to that of Rhodesian Man.

Saldanier (sældānīə·ɹ). *S. Afr. Hist.* Also **Saldanhar** (-ā·ɹ). [Afrikaans, f. the name of *Saldanha* Bay in Cape Province (cf. prec.).] A member of a Hottentot group that, in the seventeenth century, inhabited the region of Saldanha Bay; an African cattle-dealer.
[1607 W. Keeling in R. Raven-Hart *Before Van Riebeck* (1967) 36 Saldanians alias Cafares.] 1838 D. Moodie tr. *J. van Riebeck's Jrnl.* in *Record* I. 16 In the evening some of the Saldania Ottentoos came to the Fort... These two Saldaniers were much bolder and livelier men than the Strandlopers who daily live with us, but still having the same language and clothing. *Ibid.* 22 The Saldaniers..lay in thousands about Salt River with their cattle in countless numbers. 1900 A. H. Keane *Boer States* p. xviii, *Saldaniers*, originally the Hottentots of the grassy Saldanha Bay district, who had always plenty of cattle to sell to the Dutch East India Company's people; later, any native livestock dealers. 1972 *Stand. Encycl. S. Afr.* V. 606/2 As soon as they arrived in South Africa the Portuguese..followed by the Dutch colonists in 1652, came into contact with a yellowish brown pastoral people at the Cape. The colonists at first called them Kaapmans and Saldanhars, but later on the name 'Hottentot'..became firmly established.

sale, *sb.*[2] Add: **1. c.** (Earlier examples.)
1866 *Chambers's Jrnl.* 30 June 402/2 (Advt.), Enormous and incredible sale.., for ten days only!!! 1875 L. Troubridge *Life amongst Troubridges* (1966) 124 We.. found a vague little shop where a sale was going on and everything was too ridiculously cheap. We bought some-little silk scarves for a penny three farthings each. 1880 [see *clearance sale* s.v. *clearance 10].
d. *Bookselling.* The ordinary trade rate.
1900 *What will it Cost?* 48 [Trade phrases] *Sale*, 30% discount off published price.
2. f. Now usu. *sale or return*, and freq. *attrib.* (Later examples.)
1952 E. Coxhead *Play Toward* iv. 100 The tickets.. were distributed on a sale-or-return basis to every child in the school. 1954 L. Durrell *Let.* in *Spirit of Place* (1969) 122 The local bookseller..has been pestering me to help him re-arrange his shop... Is there any sale or return system? 1973 *Times* 17 Apr. 23/2 A clause forcing direct-sales firms to offer their goods on a 'sale or return' basis. 1978 S. Hodges *Gollancz* vii. 154 Reg Dignum, the London traveller, persuaded Victor to let him sell it [*sc. Guilty Men*] 'on sale or return', a practice which the firm normally never agreed to.
g. *sale of work*, a sale of articles that have been made by members of an association, congregation, or the like, held on behalf of some charitable, religious, or political object. Also, a commercial sale of handiwork.
1859 in F. K. Prochaska *Women & Philanthr. 19th-c. Eng.* (1980) 258 (*title*) Second annual report of the association for the sale of work by ladies of limited means. 1873 *Young Englishwoman* May 258/1 Can the Editor inform M.A.B. of any repository where needle or network by distressed gentlewomen is removed and sold for their benefit? (New Society for Sale of Work, North Audley Street W). 1890 *New Road Chapel Monthly Visitor* Feb. 18b/1 Sale of work and mothers' meetings. 1905 *Grand Mag.* June 810 Ladies..are informed that..a shop or gallery for the sale of work is shortly to be opened. 1917 F. Klickmann *Between Larch-Woods & Weir* ii. 21 The vermilion satin cushion embroidered with yellow eschscholtzias, that had lain in a trunk in the attic since the last Sale of Work but two. 1969 Joyce *Finnegans Wake* (1964) 446 'Tis post purification we will, sales of work and social service, completing our Abelite union by the adoption of fosterlings. 1973 A. Behrend *Samarai Affair* iv. 54 A ride round the farm, a coffee morning or a Conservative sale of work. 1976 M. Hinxman *End of Good Woman* vii. 99 The success of the last sale of work.
h. *sale and lease-back*: see *lease-back.
3. a. *sale catalogue, -goer, shop* (earlier and later examples); **sale day,** (*a*) the day on which a sale is held; (*b*) *Austral.* and *N.Z.*, a market-day; **sale-leaseback** = *sale and lease-back* (*lease-back); **sale-yard** *Austral.* and *N.Z.*, an enclosure in which livestock is sold.
1792 J. Lackington *Mem. First 45 Yrs.* xxxi. 329, I soon after this proposed printing a sale catalogue. 1852 *Fraser's Mag.* June 723/2 When he [*sc.* a wholesale bookseller] subscribes a book, or issues a sale catalogue. 1910 *Quaritch's Catal.* No. 286 (*title*) Sale-Catalogue of the library of David Garrick. 1840 *Spirit of Times* 25 Apr. 90/2 Sale days. 1898 *Bulletin* (Sydney) 26 Mar. 31/1 Tuesday was sale-day. Monday afternoon was devoted to the yarding of cattle and the yarding and drafting of innumerable sheep. 1937 *Burlington Mag.* Nov. p. xix/1 Let us hope..that this game..will end on the sale-day. 1948 N. Scanlan *Rusty Road* i. 12 Thursday was Sale Day..market day, they would call it in England, but there was no market in these small New Zealand towns. 1927 *Daily Express* 4 July 3/3 Sale-goers are advised to remember the date. 1973 *N.Y. Law Jrnl.* 1 Aug. 5/3 Private placement of mortgages, joint ventures, sale-leasebacks on income properties and land, [etc.]. 1978 *Detroit Free Press* 5 Mar. B1/2 The sale-leaseback arrangement, which enables the farmer to raise money for new equipment despite low farm prices. 1757 *Connoisseur* (ed. 2) III. 151, I am sure we have cast-off cloaths sufficient to furnish a sale-shop. 1922 Joyce *Ulysses* 285 In Lionel Marks's antique saleshop window..candlestick melodeon oozing maggoty blowbags. 1957 *Beaver* Autumn 38/1 The 'Saleshop' classification marked a modest type of urban transition, from fur trade to general store operation. 1976 *Derbyshire Times* (Peak ed.) 3 Sept. 18/5 (Advt.), Self service grocery stores with modern detached house... Spacious living accommodation..plus saleshop 31ft.×19ft. 6 ins. fully fitted for the trade. 1901 M. Franklin *My Brilliant Career* iv. 18 He was a familiar figure at the Goulburn sale yards every Wednesday. 1934 [see *back-ing ppl. a.]. 1975 *N.Z. Jrnl. Agric.* Sept. 61/1 Normally stud stock are sold either from yards on the farm itself, or by auction at recognized centrally situated saleyards.
b. *sales appeal, area, campaign, chart, correspondent, curve, figures, force, girl, -goer, graph, -lady* (earlier and later examples), *-manager, -master* (later example), *message, outlet, -people, -person, presentation, promoter, promoting, promotion, volume*; **sales clerk** *N. Amer.*, a shop asistant; **sales drive,** an energetic effort to sell goods extensively; hence **sales-drive** *v. trans.*; **sales engineer,** a salesman with technical knowledge of his goods and their market; hence **sales engineering**; **sales pitch** [*pitch *sb.*[2] 5 b] = *sales talk*; hence **sales pitchery**; **sales rep**: colloq. abbrev. of next; **sales representative,** one who represents a commercial firm to prospective customers and solicits orders; a traveller (cf. Representative *sb.* 4 a in Dict. and Suppl.); **sales resistance,** the ability or disposition to resist buying something offered for sale; also *fig.*; hence **sales-resistant** *a.*; **sales room** (examples); **sales slip,** a slip of paper recording the price of an article and other details of its sale; **sales talk,** persuasive rhetoric designed to promote the sale of goods or (*transf.*) the acceptance of an idea; **sales tax,** a tax levied on the retail sales of commodities.
1931 C. Bedell *Seven Keys to Retail Profits* iii. 36 Instead of using a $20 bill to give him *two* profit opportunities, a double sales appeal, many a retailer spends the entire twenty for a quantity of one item. 1936 *Jrnl. R. Aeronaut. Soc.* XL. 289 In the case of commercial aircraft, at any rate, by the gain in 'sales appeal' resulting from the general air of cleanness. 1966 *B.B.C. Handbk.* 39 Another sales area which has great potential—the distribution of programmes for non-theatric use in schools, universities, training colleges. 1969 D. C. Hague *Managerial Econ.* III. xiii. 288 We talk of price wars, sales campaigns, marketing strategies. 1959 'F. Newton' *Jazz Scene* iv. 72 Rhythm and blues have not only swamped ordinary pop music in America and Britain, at least in terms of the sales-chart, [etc.]. 1934 Webster, Sales clerk. 1968 *Globe & Mail* (Toronto) 17 Feb. 39 A 19-year-old Toronto sales clerk. 1979 *Honolulu Advertiser* 8 Jan. D-3/9 Sales Clerk..full time and part time. Apply at B.S. Co. Ward Warehouse. 1951 in M. McLuhan *Mech. Bride* (1967) 41/3 A book that ought to be read by all advertising writers, sales correspondents, editors and business-paper writers. 1961 *Evening Standard* 14 July 20/3 Sales Correspondent in an expanding Mail Order Organization. 1946 Sales curve [see *roof *sb.* 2 a]. 1969 'J. Morris' *Fever Grass* ix. 81 You know I'm worth it. Just watch your sales curves. 1951 M. McLuhan *Mech. Bride* (1967) 144/2 Every success drive and sales drive is committed to erasing this [*sc.* resistance] in all its varieties. 1962 *Punch* 21 Nov. 754/1 To..sales-drive their dish-washing machines. 1942 *Sun* (Baltimore) 16 July 2/6 Three self-styled 'sales engineers' stood to garner commission on millions of dollars of Government war work. 1969 *Sales Engineer* Mar. 29/1 (Advt.), A Sales Engineer is any person who is directly or indirectly selling technical products to industry. Sales Engineering is not a trade, it is a profession, and the readers of *Sales Engineer* are professionals. 1966 G. N. Leech *Eng. in Advertising* x. 99 The only criterion of success known to the advertising profession—sales figures. 1934 Webster, *Sales force*, the sales clerks or sales agents of an establishment. 1974 *Times* 9 Mar. 24/6 (Advt.), Opportunity for a girl..to join Sales Force in the exciting new ski development of Anzère. 1980 M. Babson *Queue here for Murder* ii. 21 Soon the Bonnard's sales force would start clocking in, and..after that the customers. 1887 *Courier-Jrnl.* (Louisville, Kentucky) 2 Feb. 4/7 In order to cripple his old partner, he offered superior inducements to the sales girls to go with him. 1978 M. Kenyon *Deep Pocket* xi. 136 The squeak of a salesgirl flattened against a wall. 1925 *Glasgow Herald* 6 Jan. 7/2 The large number of men among the sales-goers. 1967 R. Jeffries *Deadly Marriage* i. 8 I've returned with firm orders for three parlour-sheds... That'll put the old graph up... The sales graph. 1856 *Daily Alta California* (San Francisco) 29 Oct. 4/3 (Advt.), *Wanted*—By a young lady, a situation as *saleslady* in a dry goods, trimming, or millinery store. 1928 *Sunday Dispatch* 5 Aug. 5/6 An amatory porter and a sales-lady sitting on some dirty steps on the Underground. 1976 *Billings* (Montana) *Gaz.* 20 June 8-D/5 (Advt.), Mobile Lot-Imperial Park. All city utili. Call Real Estate saleslady Geri Erickson, 252-0264. 1913 *Writer's Mag.* Nov. 184/2 *The Accountant*, Detroit, Mich., is in the market for interesting business stories—material of interest to business managers, advertising and sales managers, [etc.]. 1933 H. Nicolson *Diary* 5 Jan. (1966) 131 We are then met by..the sales-manager of Doubleday Doran. 1979 R. Perry *Bishop's Pawn* i. 13 My cover as sales manager for a multinational electronics firm. 1922 Joyce *Ulysses* 392 A worthy salesmaster that drove his trade for live stock and meadow auctions hard by Mr Gavin Low's yard in Prussia street. 1966 G. N. Leech *Eng. in Advertising* iii. 30 The kernel of the sales message..has to be in some way special and different for each product. 1957 C. Smith *Case of Torches* iv. 46 We must..keep the Belgian company as healthy as possible otherwise we stand to lose their valuable sales outlets. 1977 *Times* 5 Nov. 12/7 For.. the purchasers of holidays, there will be..a wider choice of sales outlets. 1876 *Scribner's Monthly* Feb. 599/2, I walked through the crowds of purchasers and salespeople. 1976 *Evening Standard* 14 June 24/8 (Advt.), 2 salespeople required to manage small gift shop. 1978 *Tucson Mag.* Dec. 33/1 If you find pleasure in being the only customer in the midst of a convention of used car salespeople, you'll love buying stereo equipment in Tucson. 1920 *Harper's Mag.* June 86 We have long been familiar with *salesman* and *saleswoman*—even, alas! with *saleslady*; and the latest member of the family to whom we have been introduced to, *salesperson*, a name intended to apply to employee of either sex. 1928 *Publishers' Weekly* 10 Nov. 1962/2 We shall be glad to send a complimentary copy of the novel on request to any retail salesperson to read. 1955 *Sun*

(Baltimore) (B ed.) 12 Sept. 10/7 The 'pencil box' she bought for her grandson and which the salesperson called a 'companion', wasn't a box at all. 1976 *Evening Standard* 14 June 24/5 (Advt.), Salespersons required for expanding Northern based home improvement company. 1980 *Times* 28 Feb. 12/6 One of those cheap department stores where you may browse for several weeks without even locating a salesperson at all. 1962 *Listener* 18 Jan. 133/2 Ditchburn went through a rigorously mastery sales pitch. 1976 *National Observer* (U.S.) 19 June 1/5 In fact, as things turned out, it was an extraordinary sales pitch for Reagan himself. 1980 *Jrnl. R. Soc. Arts* Feb. 145/1 Managers are impatient and practical people, who having accepted our sales pitch, will want to learn how to use what they have bought. 1968 *Punch* 7 Aug. 206/1 ... a piece of crypto-sexual sales-pitchery. 1947 *Fortune* Nov. 175/1 (Advt.), They bring real 'theater' to a sales presentation. 1981 W. H. Hallahan *Trade* iv. 111 Myers is doing the sales presentation for the Essen Arms Company. 1935 *Punch* 2 Sept. 264/1 The great advantage of being a Sales Promoter is that the working hours are short. *Ibid.*, Sales Promoting is one of those lovely jobs in which it is impossible to judge by results. 1916 (*title*) Sales promotion by mail: how to sell and how to advertise. 1964 A. Wykes *Gambling* iii. 70 A young Italian business-man...went to Tokyo on a sales-promotion visit for his firm. 1979 *Jrnl. R. Soc. Arts* CXXVII. 346/1 ... 1969 *Observer* (Colour Suppl.) 23 Mar. 23/3 Sales reps in their company cars are my number-one headache. 1979 *Business Traveller* Nov.–Dec. 46/1 ... 1949 *Daily Tel.* 21 Nov. 2/1 Experienced Sales Representatives. 1981 'E. Ferrars' *Experiment with Death* iv. 83 He's a sales representative for a firm of confectioners. 1928 *New Yorker* 1 July ... 'Beggar on Horseback' presents no sales resistance problem... 1932 P. Fleming *Brazilian Adventure* I. x. 88 ... their sales resistance. 1972 M. Babson *Murder on Show* xvi. 185 Heaven help you the day some woman gets her hooks into you—you've no sales resistance at all. 1979 E. H. Gombrich *Sense of Order* i. 25 ... 1957 *Times Lit. Suppl.* 22 Mar. 174/3 ... makes a reader sales-resistant. 1840 *Knickerbocker* XVI. 226 ... from his sales-room ... 1929 W. Faulkner *Sanctuary* ... The block was filled by a row of automobile sales-rooms. 1980 *Times* 20 July 18/5 ... 1962 *Lebende Sprachen* VII. 35/3 Sales slip, ... 1964 G. Jackson *Let.* 21 Feb. in *Soledad Brother* (1971) ... sales slip ... 1976 *New Yorker* 23 Feb. 35/1 Do you have a sales slip? 1926 *Amer. Speech* II. 92/2 Sales talk ... 1933 *Punch* 1 Feb. 122/2 'No sales-talk!' ... 1968 Mrs. L. B. Johnson *White House Diary* 1 Aug. (1970) 697 ... a sales talk about coming to Chicago. 1974 N. Marsh *Black as he's Painted* ii. 32 Motivated by sales talk and embarrassment, he bought...a cat bed-basket. 1932 *Daily Colonist* (Victoria, B.C.) 27 Oct. ... sales tax ... 1942 *Economist* 31 Aug. ... the sales tax has been raised ... 1978 *N.Y. Times* 30 Mar. ... 1959 *Listener* 26 Mar. ... to increase the sales-volume of a new shade of lipstick.

sale, *v.* For *rare*—[1] read *rare* and add later *trans.* example.
1922 Joyce *Ulysses* 555 Lovely ladies saling gloves.
2. *intr.* To hold a sale; to shop at the sales. Hence **sa·ler,** a person who frequents sales; **sa·leing** vbl. sb. All now *rare* or *Obs.*
1901 *Sketch* 3 July 443/1 To go 'saleing' in Bond Street. 1902 ... All London is 'saleing' at the present moment. 1928 *Daily Express* 19 June 3/2 Men were 'sale-ing' at lunch time. 1928 *Morning Post* 25 June 8 Many experienced salers will tell you that it is an excellent plan to go to the sales with an open mind. 1928 *Daily Express* 31 Dec. 5/3 'Saleing' has become a specialised art. 1929 *Ibid.* 8 Jan. 3/4 The great furniture houses are 'saling'.

saleability. Add: Also irregularly **salability.** (Further examples of the various spellings.)
1940 E. Gill *Autobiogr.* vii. 195 ... salability of what he buys. 1949 ... What has not...wanted to say anything...until some verdict has been reached..as to its salesability. 1972 *Nature* 28 Jan. 232/2 One is forced to the conclusion that the title was chosen with an eye to salability. 1975 *Language* LI. 447 It was a common practice..for grammatical works by well-known scholars to be more or less extensively emended and refashioned by later publishers, in order to enhance their usefulness and saleability. 1976 *Publishers' Weekly* 1 Nov. 70/1 Undoubtedly it enhances gift book salability. 1979 *SLR Camera* Jan. 14/2 Subjects of universal and eternal—as far as saleability was concerned—content.

saleable, *a.* Add: **1. a.** Also *absol.* or as quasi-*sb.*
1945 *Sun* (Baltimore) 4 Aug. 8/7 Five hundred salables were offered and 2,500 went directly to packers. 1946 *Ibid.* 15 Jan. 10-O/2 Salables amounted to 2,500 head, compared with 5,000 head marketed a week ago.

‖ **sale Boche, sale boche** (sal bɔʃ). [Fr., f. *sale* dirty + *Boche.] A French term of abuse for a German.
... I get called sale boche if I open my mouth. 1934 D. L. Sayers *Nine Tailors* ... called him *sale Boche*—but Jean knocked him down. 1938 L. Bemelmans *Life Class* I. iv. ... 'Sale Boche!' I called him a French pig. 1979 D. Robinson *Eldorado Network* vi. 53 ... 'Sales boches!' he spat.

saléeite (sæ·lēɪ,aɪt). *Min.* orig. **saléite**; also without accent. [ad. F. *saléite* (Thoreau & Vaes 1932, in *Bull. de la Soc. géol. de Belgique* XLII. 96), f. the name of Achille *Salée* (d. 1932), Belgian palaeontologist: see -ite[1].] A hydrated phosphate of magnesium and uranium, $Mg(UO_2)_2(PO_4)_2·10H_2O$, which occurs as yellow crystals in association with torbernite as an oxidation product of uranium minerals.
1934 *Chem. Abstr.* XXVIII. 5372 Saléite is the Mg analog of autunite. 1940 *Mineral. Mag.* XXV. 643 ... 1951 *Amer. Mineralogist* XXXVI. ... Under the microscope, ... 1959 ... The yellow coating of secondary uranium mineral..forms small plates and is an unidentified member of the saléeite-autunite-torbernite group.

Salem (sēɪ·lĕm). [Name of a place in Gen. xiv. 18 (Heb. *Shālēm*), understood to be another name for Jerusalem and to mean 'peace' (Heb. *shālōm*).] Occasionally (chiefly in the nineteenth century) adopted by Methodists, Baptists, Independents, etc., as the name of a particular chapel or meeting-house. Hence used as synonym for 'nonconformist chapel'. Cf. Bethel 2, Ebenezer 2, Zion.
1857 Geo. Eliot in *Blackw. Mag.* July ... The Independent chapel, known as Salem, stood red and conspicuous in a broad street. 1880 Trollope *Duke's Children* ... 1963 W. H. Boore *Valley & Shadow* ... 1970 *Guardian* 1 Aug. 9/8 The grey chapels called Salem and Zion.

Salempore. Add: (Later examples of spelling *salampore.*)
1851 R. Mitford *Through Zulu Country* xv. 189 On shelves against the walls are arranged blankets, Salampore cloth, [etc.]. 1928 E. Sitwell *Five Poems* 15 Gaze d' Ispahan and bolts of salampore.

sale price. [Sale *sb.*[2]] **a.** Retail price. **b.** A price fetched at auction. **c.** A price reduced for a sale (sense 1 c).
1765 [see Sale *sb.*[2] 3]. 1866 *London Society* Mar. 258/2 ... 1897 (*title*) The sale prices of 1896. ...
Hence **sa·le-price** *v. trans.*, to set at a price for sale.
1959 *Time* 2 Nov. 6/2 In Seattle, the suburban Grinnell & McLean furniture store ballyhooed 'Mother-in-law-Mattresses', sale-priced at $9.95. 1978 *New York* 3 Apr. 54 ... Sofa into a queen-sized sleeper convertible! Of course, we've also sale-priced the pieces separately.

saleratus. (Earlier examples.)
1837 S. Graham *Bread-Making* 46 Pearlash or saleratus is also used by them in considerable quantities. 1846 *Knickerbocker* XXVII. 310 The white sal-æratus cake and the 'water bewitched' are quickly devoured. 1853 Mrs. A. L. Webster *Improved Housewife* 130 Salaeratus Biscuit.

Salesian (sălī·ʃăn, sălī·ʒăn), *a.* and *sb.* [ad. *Salésien*, f. the name of St. François de *Sales* + -ien -ian.] **A.** *adj.* Of or pertaining to St. Francis of Sales (1567–1622), Roman Catholic mystic, or to communities founded by him or living according to his rule, as the nuns of the order of the Visitation founded in 1610 under his direction, and societies founded by St. John Bosco for the rescue of poor and neglected children. **B.** *sb.* A follower of St. Francis of Sales or a member of a Salesian order; a Brother or Sister of one of the orders founded by St. John Bosco.
1836 *Account Conversion of L.T.H. to Holy Catholic Church* 18 Till I had examined the Rules and visited a Convent of Salesian Nuns. 1884 *Month* Jan. 46 His [*sc.* Bosco's] institution henceforth went by the name of 'The Oratory of St. Francis of Sales', and his co-labourers were called 'Salesians'. 1884 H. B. Mackey *St. Francis de Sales's Treat. Love of God* p. xxxiii, S. Francis also had his special characteristics, which, therefore, are not French but Salesian. 1890 Lady Martin tr. *Villefranche's Life Dom Bosco* xx. 232 His first missionary expedition.. included ten priests and coadjutor Salesian Brothers. 1912 *Catholic Encycl.* XIII. 399/1 The Salesians established themselves at Battersea in London. 1928 J. Brodrick *Life & Work Cardinal Bellarmine* I. ix. 182 There is an added quality in his work, a characteristic Salesian grace which is the best part of it. 1930 *Tablet* 4 Oct. 425/1 This learned and zealous Salesian is already known as one of the ablest men in the Church to-day. 1964 F. D. Parker *Central Amer. Republics* vi. 179 Five Franciscans and one Salesian. 1974 *Oxf. Dict. Chr. Ch.* (ed. 2) 190/2 *St. John Bosco* (1815–88)... In 1859 he founded the 'Pious Society of St. Francis de Sales', commonly known as the 'Salesians'.

salesite (sēɪ·lzəit). *Min.* [f. the name of Reno H. *Sales* (1876–1969), U.S. geologist + -ite[1].] A very rare basic iodate of copper, $CuIO_3(OH)$, found as bluish-green orthorhombic prisms at Chuquicamata, Chile, and also prepared synthetically.
1939 Palache & Jarrell in *Amer. Mineralogist* XXIV. 388 Salesite is an iodate of copper first found by the junior author in 1936 on the west side of Bench E-4 at the south end of the open pit at Chuquicamata, Chile. 1962 *Acta Crystallographica* XV. 1106/1 Rotation and Weissenberg photographs of the artificial $CuIO_3(OH)$..indicate that it is identical with salesite in cell dimensions and the distribution of intensities.

salesman. Add: **c.** In various *transf.* senses.
1912 J. Sandilands *Western Canad. Dict.*, *Salesman*, a commercial traveller, a drummer, a store counter-man, or a man who canvasses real estate. 1930 *Amer. Speech* VI. 134 *Salesman*, confidence man. 1937 *Daily Herald* 15 Jan. 12/8 Increased employment means that people are replacing old-fashioned furniture for [*sic*] new, and the attractive designs made possible by veneering are good salesmen. 1942 Berrey & Van den Bark *Amer. Thes. Slang* § 507/3 *Pimp*; *procurer*..crack salesman,.. salesman. 1944 *Amer. N. & Q.* IV. 10/2 A transportation company in New Jersey now refers to its drivers and motormen as 'salesmen'. 1945 Mencken *Amer. Lang. Suppl.* I. 588 American milk-wagon drivers are called *milk-salesmen* and bakers' deliverymen *bread-salesmen*. 1968–70 *Current Slang* (Univ. S. Dakota) III–IV. 105 *Salesman*, n. Pimp; one who sells anything.

salesmanship. Add: (Later examples.) Also *fig.*
1930 *Economist* 20 Dec. 1164/2 The Prince of Wales, who is shortly leaving for Buenos Aires, where he will open the British Empire Trade Exhibition, delivered an outspoken address on salesmanship. 1936 *Discovery* Apr. 129/2 Consult any of the books on psychology and salesmanship so numerous on the market. 1937 Wodehouse *Ld. Emsworth & Others* v. 172 When a woman is to all intents and purposes waiting for a demon lover, it requires super-salesmanship to induce her to accept on the this-is-just-as-good principle an Ernest Plinlimmon. 1960 *Farmer & Stockbreeder* 12 Jan. 102/2 Salesmanship in food is liable to be overrated because nothing is more distasteful than..food, when you have had enough.

saleswoman. Add: (Later examples.)
1916 D. H. Lawrence *Amores* 97 She puts me away like a saleswoman whose mart is Endangered by the pilferer on his quest. 1932 L. Golding *Magnolia St.* III. vi. 548 She's one of the leading saleswomen in our firm.
Hence **sa·leswomanship,** the position of a saleswoman; the character of being a (good) saleswoman.
1973 M. Amis *Rachel Papers* 28 Gloria held the assistant pet-food saleswomanship in, handily, a Shepherds Bush emporium. 1977 *Church Times* 29 Apr. 11/2, I felt that a little saleswomanship might be a good thing.

salgram, var. Shalgram. **salic,** *a.*[2]: see *sal[4] 1.

salic (sēɪ·lik), *a.*[3] *Soil Sci.* [f. L. *sal* salt + -ic.] Applied to a soil horizon which is at least 15 cm. thick and is enriched with salts more soluble in water than gypsum (see quot. 1971).
1960 *Soil Classification* (U.S. Dept. Agric.) v. 60/1 A salic horizon is a horizon 6 inches or more thick with secondary enrichment of salts more soluble in cold water than gypsum. 1970 E. M. Bridges *World Soils* iii. 24/2 These soils develop a surface encrustation of salt... Such soils possess salic horizons. 1971 *Gloss. Soil Sci. Terms* (Soil Sci. Soc. Amer.) 26/2 A salic horizon is 15 cm or more in thickness, contains at least 2% salt, and the product of the thickness in centimeters and per cent salt by weight is 60% cm or more.

salicaceous, *a.* Add: Also *transf.* (*joc.*), made of willow.
1963 *Times* 13 June 13/3 This makes one wonder whether any of the aforetime Latinists ever called a bat a salicaceous implement.

salicetum (sælisī·tǝm). Also **salictum.** Pl. **saleceta, -cetums.** [f. L. *salix, salic-* willow + *-etum.] A plantation of willows, esp. a collection of different species and varieties of willow.
1776 A. Hunter *Evelyn's Silva* xx. 252 In order to raise a Salictum, or a plantation of Willows for timber, the ground must be dug or plowed; and the cuttings for this purpose should be of the last year's shoot. 1838 J. C. Loudon *Arboretum* I. i. 129 Woburn Abbey, where a salicetum, or salictum..was planted in 1825. *Ibid.* III. 1477 A Salictum is the only scene in which a complete collection of willows can be displayed to advantage. 1853 C. G. B. Daubeny *Oxford Bot. Garden* (ed. 2) 16 Salicetum. On the opposite side of the garden outside of the walls, and bordering upon the river, there existed a few years ago a tolerably good collection of Willows. 1875 *Encycl. Brit.* III. 422/1 Mr. William Scaling..cultivates a salictum of about 100 acres. 1926 *Nat. Hist. Oxford District* 73 Many interesting Willows which were formerly grown in the Salicetum in the Botanic Gardens. 1952 G. Taylor *Victorian Flower Garden* x. 153 The Willow garden, or salicetum, as it was called, slightly antedates the pinetum. 1972 S. C. Warren-Wren *Willows* ii. 32 A salictum should be a sizeable area of ground set aside for the express purpose of growing a fully representative group of willows. *Ibid.* 33 It may be considered wise to have two salicta. 1978 A. Huxley *Illustr. Hist. Gardening* ix. 307 The numerous species of *Salix* were gathered into salicetums or sally gardens.

salicyl. Add: **sa:licyla·ldehyde,** *o*-hydroxybenzaldehyde, $C_7H_6O_2$, a colourless volatile liquid having an odour of bitter almonds, which is found in oil from meadowsweet and related species, and is used esp. in perfumery.
1869 [in Dict., s.v. Salicyl *attrib.*]. 1896 W. T. Brannt *Animal & Vegetable Fats* (ed. 2) II. xix. 274 Salicylaldehyde..exists in the oils of meadow sweet and of other species of *spiraea*. 1973 *Nature* 4 May 37/2 Several compounds with specific odours, such as 2,4-pentanedione, morin, salicylaldehyde and acetic acid, are well known chelating agents for heavy metals.

salicylate. Substitute for def.: A salt or ester, or the anion (o-$C_6H_4(OH)COO^-$), of salicylic acid. (Further examples.)
1964 W. G. Smith *Allergy & Tissue Metabolism* iii. 40 Salicylate inhibits both the 'in vitro' activation of permeability globulin and its action on capillaries. 1972 *Sci. Amer.* Jan. 92/3 Physicians have found that most patients coming to the emergency room of a hospital with massive bleeding of the upper gastrointestinal tract have taken salicylates within the preceding 24 hours. 1977 *Addictive Dis.* III. 284 Researchers have called attention to the hazard to pregnant women of the easy accessibility of over-the-counter drugs such as salicylates.

salie (sa·lǝ). Also **saliehout, zalie.** [Afrikaans, a. Du. *salie* sage.] = *sage-wood* (*b*) s.v. *sage *sb.*[1] 5 b.
1819 C. G. Curtis *Acct. Colony Cape of Good Hope* 72 Saly hout..Hard and heavy. 1908 F. C. Slater *Sunburnt South* 11 Wild-willows and feathery-flowered *zalie* trees grew in..profusion. 1932 [see *sage-wood* s.v. *sage *sb.*[1] 5 b]. 1952 *Cape Times* 2 Aug. 9/3 Among indigenous trees, three are milkwood, salie and Kafir plum. 1973 *Stand. Encycl. S. Afr.* IX. 459/1 Saliehout... Tall, much-branched shrub..with large, simple, opposite, grey-green leaves.

salience. Add: **2. b.** *Social Psychol.* The quality or fact of being more prominent in a person's awareness or in his memory of past experience.
1938 H. D. Spoerl tr. *Stern's Gen. Psychol. from Personalistic Standpoint* iv. 74 The different proportions of salience and embedding give the process and content of every experience its special character. 1938 G. W. Allport *Personality* xx. 553 At other times..consciousness is embedded..more deeply; there is less clearness, less salience. Salience represents an act of pointing, a directedness of the person toward something that at the moment has special significance for him. 1953 C. I. Hovland et al. *Communication & Persuasion* v. 155 We shall refer to the degree to which..a specific group is present and prominent in a person's 'awareness' as the *salience* of that group. 1958 W. C. Schutz *FIRO* vii. 147 If the reaction to the anxiety is withdrawal from interchange in that area, the area acquires a negative salience in that the actor tries to avoid it. 1965 T. M. Newcomb et al. *Social Psychol.* iii. 58 The difference between the centrality of an object to an individual and the closely related matter of its *salience*. 1972 *Jrnl. Social Psychol.* Aug. 256 Relatively low Salience problems..produced shifts predominantly towards greater risk.

saliency. Add: **2. b.** *Social Psychol.* = *salience 2 b.
1965 T. M. Newcomb et al. *Social Psychol.* ii. 37 The notion of saliency has an interesting counterpart in the information storage of modern 'thinking machines' or large computers.

salient, *a.* Add: **A.** *adj.* **5. b.** Also *Psychol.*, standing out or prominent in consciousness.
1938 H. D. Spoerl tr. *Stern's Gen. Psychol. from Personalistic Standpoint* iv. 74 Dissonance is constant by being augmented or diminished. All experience consequently tends to become either *salient* against or *embedded* with the totality. 1938 G. W. Allport *Personality* xx. 553 The most important of all facts about consciousness is that it is graded; sometimes it stands out, as it were, against the diffuse background of personal life. It is *salient*... The more salient an experience, the greater its objective meaning. 1953 C. I. Hovland et al. *Communication & Persuasion* v. 161 A communication will produce more immediate change when the opposing group norms are at a low level of salience than when they are highly salient. 1965 T. M. Newcomb et al. *Social Psychol.* ii. 37 We shall use the term 'salient' to describe stored information that has been prompted to the forefront of the individual's conscious thought.
6. *Electr.* *salient pole*, a type of field pole used in electrical machinery in which the energizing coil is wound on a pole-piece projecting inside the yoke of a stator assembly or outside the cone of a rotor assembly.
1886 S. P. Thompson *Dynamo-Electric Machinery* (ed. 2) vii. 121 This pattern differs from that of the better known 'A' Gramme in using salient poles instead of having the 'consequent poles' at the middle points of the electro-magnets. 1920 *Whittaker's Electr. Engineer's Pocket-Bk.* (ed. 4) 169 The turbo-alternator is now the standard a.c. generator, and is almost invariably built with a cylindrical (or non-salient pole) rotor, the salient pole construction being confined to slow-speed alternators and water turbine-driven alternators. 1962 [see *alternator]. 1970 J. Shepherd et al. *Higher Electr. Engin.* (ed. 2) x. 331 An alternative arrangement to having uniform slotting on both sides of the air-gap is to have salient poles around which are wound concentrated coils to provide the field winding. The salient poles may be on either the stator or the rotor.
B. *sb.* **2. a.** A narrow projection or spur of land extending from a larger feature; a spur-like area of land, esp. one held by a line of offence or defence, as in trench-warfare; *spec.* (freq. with *the* and capital initial) that at Ypres in western Belgium, the scene of severe fighting in the war of 1914–18.
1864 W. G. Mitchell in *War of Rebellion* (U.S. War Dept.) (1891) 1st Ser. XXXVI. I. 359 Conducted General Wright to a point near the Salient we had captured. 1903 A. F. Mockler-Ferryman *Milit. Sketching & Reconnaissance* ix. 88 Select..a line of level to be assumed as a crest-line, so situated that when drawn-in it will show the shape of all the principal salients and re-entrants. 1914 *War Illustr.* 5 Dec. 366/1 The British salient at Ypres fascinated the Kaiser. 1915 *N.Y. Tribune* 8 May 9/2 The salient at Ypres always has been dangerous. Formerly it made a semi-circular loop, with Ypres a little above the centre. After this successful movement of the Germans it took the shape of the eye in a hook and eye. 1927 R. H. Mottram *Spanish Farm Trilogy* 238 Poperinghe was the railhead for that essentially English battle-field, the Ypres Salient. 1944 *Daily Progress* (Charlottesville, Va.) 2 Oct. 9/4 The British drove five miles north of the village of Oss at the north-western corner of their salient to the Maas. 1972 K. Bonfiglioli *Don't point that Thing at Me* xix. 172, I had almost succeeded in becoming..'Mad Jack' Mortdecai, V.D. and Scar, the ice-cool toast of the Ypres Salient. 1974 *News & Courier* (Charleston, S. Carolina) 28 Apr. A-1/6 An officer identified in a national radio interview as Yoav, commander of the southern salient.
b. *fig.*
1936 [see *industrial psychologist* s.v. *industrial *a.* e]. 1969 *Daily Tel.* 31 Oct. 18 With this week's pamphlet on 'The Police and the Citizen'..he will be taking the council into a very hot salient indeed.

salientian (sēɪli,e·ntʃiăn, -e·ntiăn), *a.* (and *sb.*) *Zool.* [f. mod.L. name of order *Salientia* (J. N. Laurenti *Synopsis Reptilium* (1768) 24), f. L. *salient-em* (see Salient *a.* and *sb.*) + -ia[1]: see -an.] = *anuran *a.* Also as *sb.*
1948 *Evolution* II. 29/2 The general trend of modified salientian ontogeny is towards withdrawal of development from water. 1951 [see *calcar[2] 2]. 1956 *Nature* 18 Feb. 342/2 How, then, can the salientian trends so clearly shown in the hind limbs and girdle..be correlated with the absence of a functional sacrum? 1973 Estes & Reig in J. L. Vial *Evolutionary Biol. Anurans* i. 43 This hypothesis does not afford an answer..to the question of placement of salientian origins.

saligram, var. Shalgram.

salina. Delete ‖ and add: Also, a low, marshy area of land near the coast (orig. *Jamaican*). (Further examples.)
1756 P. Browne *Civil & Nat. Hist. Jamaica* 356 The Samphire of Jamaica... This plant is common in all the Salinas on the south side of Jamaica: it abounds with alkalious salts, but the manufacture of this commodity has not been yet attempted in that island. 1774 E. Long *Hist. Jamaica* I. II. iv. 474 In making roads to traverse the salinas, or level grounds adjacent to the sea, and in swampy places, a *stratum* should first be laid. 1811 W. J. Titford *Sk. Hortus Bot. Amer.* 33 Herbaceous marsh samphire..grows in great plenty in Jamaica, on the Salinas and Marshes near the sea coast. 1889 *Nat. Geogr. Mag.* I. iv. 334 Beyond the narrow gateway in the hills, less than three miles of level swampy *salinas* reach to the surf of the Pacific.

Salina[2] (sălǝi·nă). *Geol.* The name of a town (now a part of Syracuse) in New York State, used *attrib.* and *absol.* to designate a group of sub-stages of the upper Silurian in New York State and adjacent areas, characterized by thick shale formations that contain beds of rock-salt; of or pertaining to this group or the time when it was deposited.
1863 J. D. Dana *Man. Geol.* III. ii. 246 With the opening of the Salina period there was a change by which shales or marls and marly sandstones, with some impure limestones, were formed over a portion of New York. 1905 H. Ries *Econ. Geol. U.S.* vi. 129 The vast beds of rock salt which occur in the Salina (Monroe) are exploited along the Detroit and St. Clair rivers. 1906 Chamberlin & Salisbury *Geol.* II. 388 Gypsum..is present in the Salina series. 1949 C. O. Dunbar *Hist. Geol.* ix. 193 In central New York the salt-bearing shales of the Salina group..succeed the Niagaran limestone. 1960 R. L. Bates *Geol. Industr. Rocks & Minerals* vi. 216 An aggregate salt thickness of at least 1800 feet has been penetrated in oil test wells in the central part of the Michigan Basin, where the top of the Salina is some 8000 feet below the surface. 1974 *Encycl. Brit. Micropædia* VIII. 810/3 The Salina Group consists of two shale formations that attain a thickness of more than 300 metres (1,000 feet).
Hence **Sali·nan** *a.*
1909 *Jrnl. Geol.* XVII. 245 (*heading*) The Middle Siluric or Salinan. 1924 C. Schuchert *Textbk. Geol.* (ed. 2) II. xxi. 264 Cayugan or Upper Silurian..Salinan.

saline, *a.* and *sb.* Add: **A.** *adj.* **1. b.** (Earlier and further examples.)
1789 in J. M. Brown *Polit. Beginnings Kentucky* (1889) 255 Kentucky in general appears to be a limestone soil.. abounding in..saline springs, which by simple evaporation plentifully supply the country with salt. 1840 in *Trans. Michigan State Agric. Soc.* (1855) VI. 289 Several saline springs and deer-licks were examined in the valley and vicinity of Maskego river.
5. b. *saline solution* = *physiological saline* s.v. *physiological *a.* 2 b.
1833 J. Forbes et al. *Cycl. Pract. Med.* II. 213/2 In extreme cases, or when the practitioner is not called in till the very last stage of fever, Dr. Stevens thinks life may be occasionally saved by injecting a saline solution into the veins. We have lately adopted this saline treatment in some cases of typhous fever. 1890 F. Taylor *Man. Pract. Med.* 105 The intravenous injection of saline solutions has appeared to do good in some cases of profound collapse. 1932 L. N. Katz in *Practitioners Libr. Med. & Surg.* I. xxv. 1170 Isotonic saline solution injected subcutaneously or intravenously is valuable. 1971 A. C. Guyton *Basic Human Physiol.* xx. 223/2 The arterial pressure remained normal until the animals were required to drink 0·9 per cent saline solution.
B. *sb.* **3. b.** = *physiological saline* s.v. *physiological *a.* 2 b. Also *attrib.*
1926 S. Wright *Appl. Physiol.* vi. 245 If saline is injected intravenously into a normal animal, a condition of hydræmic plethora results. 1951 [see *hypertonic *a.* 2]. 1952 E. F. Davies *Illyrian Venture* ix. 160 Saline injections followed, bottles hung above me, needles feeding into my arm. 1956 A. C. Guyton *Textbk. Med. Physiol.* xxvi. 304/2 If the sodium chloride solution is isotonic with the body fluids (that is, the injected saline has exactly the same crystalloidal osmotic activity as do both the extracellular and intracellular fluids), it does not increase or decrease the crystalloidal osmotic pressure of the extracellular fluid. 1971 *Nature* 11 June 344/2 Cholera can be treated by killing the bacteria with antibiotics such as tetracycline and replacing the body fluid lost through diarrhoea with saline.

salinely (sēɪ·lǝinli), *adv. rare.* [f. Saline *a.* + -ly[2].] In a saline or salty manner.
1929 W. Faulkner *Sartoris* III. ix. 268 He still felt nausea, and he drank long of the tepid water from the tap. Immediately it welled salinely within him.

salinification (săli:nifikēɪ·ʃǝn). [f. Saline *a.* + -ification.] The action or process of becoming, or causing to become, saline.
1911 Webster, *Salinification*, process of making salt. *Rare.* a 1961 in *Webster*, s.v., The..salinification of many agricultural soils. 1979 B. L. C. Johnson *Pakistan* v. 78/2 It may be necessary for Sind to restrict its irrigated area in order to ensure that whatever water is applied is given in sufficiently copious quantities to avoid salinification.

salinity. Add: **2.** Special Comb.: **salinity crisis** *Geol.* and *Geogr.*, a period of increased evaporation and salinity in the Mediterranean at the end of the Miocene epoch which resulted in the local disappearance of marine life.
1967 C. Ruggieri in Adams & Ager *Aspects Tethyan Biogeogr.* 286 The Gibraltar straits (probably the true asylum for the Indo-Pacific relicts during the salinity crisis of the Upper Miocene). 1977 A. Hallam *Planet Earth* 231/1 This evaporative phase..is known as the 'salinity crisis' because of the extreme effect which it had upon the marine fauna and flora present in the late Miocene Mediterranean.

salinization (sæ:linǝizēɪ·ʃǝn). [f. Saline *a.* and *sb.* + -ization.] The accumulation of salts in the soil.
1928 A. A. J. de 'Sigmond in *Proc. & Papers 1st Internat. Congr. Soil Sci.* I. 334 Circumstances under which only the salinization took place, giving no chance for alkalization. 1951 W. P. Kelley *Alkali Soils* iv. 77 Salinization is the initial step in the formation of an alkali soil. 1973 *Nature* 12 Jan. 105/2 With the available evidence strongly indicating salinity to be the primary cause of woodland mortality, it is necessary to offer some reason for the rapid and widespread salinization of Amboseli basin.

salinometer. Delete the clause beginning '*esp.*' and add further examples.
1963 G. L. Pickard *Descriptive Physical Oceanogr.* vi. 86 One of the great advantages of the electrical salinometer is that it uses a null-balance method. 1977 M. G. Gross *Oceanogr.* (ed. 2) v. 127 Salinometers are commonly used for salinity determinations on oceanographic ships and at shore-based laboratories.
Hence **salino·metry,** the use of a salinometer; measurement of the salinity of water.
1907 in Webster. 1964 *Oceanogr. & Marine Biol.* II. 104 Magazine loading of samples and automatic print-out of results have been achieved for radioactive samples, and similar developments in salinometry would not be impossible.

Salisbury steak (sǫ·lzbǝri stēɪk). *U.S.* Also with small initial. [f. the name of J. H. *Salisbury* (1823–1905), American physician specializing in the chemistry of foods + Steak.] A variety of hamburger steak initially promoted by Salisbury.
1897 A. K. Eccles *Man. what to Eat* 9 To cook the Salisbury steak..place the Cakes on the broiler, turning frequently until done. 1914 D. C. C. L. Roper *Scientific Feeding* ii. 61 *Salisbury steak.* Secure some fresh, thick, sliced round steak. Scrape or grind in a meat-cutting machine, and mould into flat, round cakes. Have an iron spider very hot and oiled... Lay the meatcake in, and turn from side to side till cooked sufficiently. 1945 Mencken *Amer. Lang.* Suppl. I. 429 During World War I an effort was made by super-patriots to drive all German loans from the American vocabulary. *Sauerkraut* became *liberty cabbage*, *hamburger steak* became *Salisbury steak*. 1953 R. Chandler *Long Goodbye* xix. 117 After a while I ..ate one of Rudy's 'world-famous' salisbury steaks, which is hamburger on a slab of burnt wood, ringed with browned-over mashed potato, supported by fried onion rings and one of those mixed-up salads which men will eat with complete docility in restaurants. 1966 L. J. Braun *Cat who could read Backwards* vii. 80 'I'll eat with you' said Quilleran... Odd ordered Salisbury steak. 1970 T. Coe *Wax Apple* xxiii. 162 Two plates of that kind of outsize hamburger usually called Salisbury steak and frequently served in places where large numbers of people are being fed without a choice of menu.

Salish (sēɪ·liʃ). Also † **Salisk, Selish.** [Southern Interior Salish *sé?liš* Flat-heads, Northern Okanagan *sįylx* Salish: of uncertain ulterior etym.] **1. a.** Formerly, an American Indian tribe of N.W. Montana, also called the Flat-heads (see Flat-head 1); now used to designate a group of American peoples, including the Flat-heads, inhabiting the N.W. United States and S.W. Canada.
The group is freq. subdivided geographically into *Coast(al)* and *Interior Salish.*
1831 W. A. Ferris *Life in Rocky Mts.* (1940) v. 88 They [*sc.* Flat-head Indians] call themselves in their beautiful tongue, 'Salish', and speak a language remarkable for its sweetness and simplicity. 1881 *Encycl. Brit.* XII. 826/2 Selish or Flat Heads. 1910 F. W. Hodge *Handbk. Amer. Indians* II. 415/2 Salish... Formerly a large and powerful division of the Salishan family, to which they gave their name, inhabiting much of w. Montana and centering around Flathead lake and valley. 1933 W. Schmidt *High Gods in N. Amer.* vii. 111 Of the three Amerindian groups whose religions include a High God, the Selish are the most recent. 1978 *Amer. Poetry Rev.* Sept./Oct. 15/3 The organization of the animal kingdom by a lunar divinity occupies a predominant place..in the myths of the Salish of North America.
b. The name of a group of languages spoken by the Salish. (In quot. 1848, the language of the Flat-heads.)
1848 R. G. Latham in *Jrnl. Ethnol. Soc. London* I. 158 *The Salish.*—This is an anonymous vocabulary from Duponceau's collection... It is evidently closely akin to the Okanagan. 1923 A. L. Kroeber *Anthropol.* v. 120 Chinook and Coast Salish, indeed, are in contiguity, and one may therefore have taken up the trait in imitation of the other. 1929 [see *Mosan *sb.*]. 1940 M. W. Smith *Puyallup-Nisqually* 20 Although the language of the Puyallup-Nisqually is classified as Salish, the people themselves used no special language names. 1977 C. F. & F. M. Voegelin *Classification & Index of Worlds' Lang.* 302 The argument for leaving Salish unaffiliated with respect to phylum classification is given by Voegelin and Voegelin (1967).
2. *attrib.* or as *adj.*
1849 in *Ex. Doc. 31st U.S. Congress 1 Sess. Senate* (1850) No. 52. 170 The *Salish* or *Flat Head* Indians occupy from Bitter Root river, a fork of the Columbia, all the country that is drained by that stream down to what is called the Hell Gate. 1902 *Encycl. Brit.* XXV. 373/1 The Shoshone, Shahaptin, and Salish tribes are of middle stature. 1933 L. Bloomfield *Language* 470 Quilleute, Kwakiutl, and Tsimshian..distinguish between visibility and invisibility in demonstrative pronouns; the latter peculiarity appears also in the neighboring Chinook and Salish dialects. 1965 *Canad. Jrnl. Linguistics* Spring 159 Several other Coast Salish languages distinguish by sex of referent in the older generation. 1977 *Islander* (Victoria, B.C.) 2 Oct. 14/1 She has in her studio..a Salish loom.

Salishan (sēɪ·lifăn), *sb.* and *a.* [f. prec. + -an.] **A.** *sb.* = *Salish 1 b. **B.** *adj.* Of or pertaining to the Salish people or language group.
a 1886 J. W. Powell in *7th Ann. Rep. U.S. Bureau Amer. Ethnol.* (1891) 104 Eastern Vancouver Island to about midway of its length was also held by Salishan tribes. 1897 [see *Kwakiutl]. 1902 G. W. James *Indian Basketry* (ed. 2) v. 51 They are of the Salishan stock. 1937 H. H. Turney-High *Flathead Indians of Montana* i. 11 That much of the North Pacific Coast is inhabited by Salishan people is a well-known ethnological fact. That a portion of western Montana is also inhabited by a Salishan group of tribes is also well-known. 1940 H. Vogt *Salishan Stud.* 1 During my stay with the Kalispel Indians..I had the opportunity of taking down some material on Spokan and Colville, Salishan languages related to Kalispel. 1959 E. Tunis *Indians* viii. 112/2 Not far from the Kutenai, in southern Canada, lived the Salishan tribes, speaking a language that has no traceable connection with any other Indian speech. 1965 *Canad. Jrnl. Linguistics* Spring 88 In Salishan, Kutenai, Quileute and Nootka a *k-* or *q-* seems to be added in the numeral 'three'. 1973 *Amer. Speech* 1969 XLIV. 232 Chinook jargon is a pidginized Chinukan-derived contact vernacular with Nootkan, Salishan, English, French, and Algonkian lexical elements. 1977 *Language* LIII. 502/2 In the Salishan languages, as elsewhere on the Northwest Coast, much of the burden which in other languages falls on the syntax is shouldered by the principles of word structure.

‖ **salita** (salī·ta). Pl. **salite, salitas.** [It.: see Sally *sb.*[1]] In Italy, an upward slope or incline, a stretch of rising ground.
1910 H. G. Wells *Hist. Mr. Polly* i. 35 Other country-sides have their pleasant aspects... Italy gives salitas and wayside chapels, and chestnuts and olive orchards. 1937 E. Pound *Fifth Decad Cantos* xliii. 16 The kalypygous Sienese females Get that way from the *salite* That is from continual plugging up hill. 1949 —— *Pisan Cantos* lxxx. 89 And Italy one eucalyptus pip From the salita that goes up from Rapallo. 1967 P. E. H. Durston *Mortissimo* xii. 100 A short, bulky Italian..strolled casually toward the steep *salita* leading up to the Pincio.

saliva. Add: **1.** Also *fig.*
a 1957 R. Campbell tr. *F. García Lorca's Romance de la Guardia Civil* in *Coll. Poems* (1960) III. 63 The Virgin cures the children With the saliva of the stars.
2. (See quot.)
1969 P. Hollister *Encycl. Glass Paperweights* 303 *Saliva*, unwanted string or conglomerate of small bubbles that may be the result of insufficient expulsion of internal air or cooling of the gather during assembly.
3. *attrib.*, as *saliva gland*; **saliva ejector,** † **extractor** *Dentistry*, a device incorporating a suction pump, for removing saliva from the mouth during a dental operation; **saliva test,** any scientific test performed on a sample of saliva.
a 1884 Knight *Dict. Mech.* Suppl. 778/1 *Saliva ejector*, an instrument for carrying off the accumulating saliva in dental operations. 1897 E. C. Kirk *Textbk. Operative Dentistry* vii. 157 An excessive flow of saliva is uncomfortable to the patient, by its accumulation it impedes the operation... During the preparation of accessible cavities ..the accumulation may be carried off by the use of a saliva ejector. 1931 N. Bennett *Dental Surg.* (ed. 2) xix. 774 The saliva ejector when first introduced met with considerable opposition, but its use is now almost universal, and if simple aseptic precautions are taken no objection can possibly be found to its adoption. 1963 C. R. Cowell et al. *Inlays, Crowns, & Bridges* v. 53 This permits a saliva ejector to be used, which retracts the tongue and keeps the teeth dry. 1877 J. Taft *Operative Dentistry* (ed. 3) vi. 176 Various pumps were devised for removing the saliva... All..have been superseded by the introduction and use of the rubber dam and saliva extractor... The profession is indebted..for the latter to Dr. J. E. Fisk. 1915 W. Owen *Let.* 13 Mar. (1967) 327, I noticed that my saliva-glands were a trifle addled. 1939 *New Yorker* 8 July 77/1 You might like to know something about just what the New York Racing Commission does to prevent tampering with horses. Every winner gets a saliva test. 1973 J. Thomson *Death Cap* i. 9 The saliva test will prove that the person who licked the stamp belongs to the 'O' blood group. 1975 *Times* 21 May 3/4 Detectives have taken their first saliva test in the hunt for the Cambridge rapist.
Hence **sali·va** *v. intr.*, to salivate; **sali·vaed** *a.*, flecked or covered with saliva.
1939 G. Greene *Confidential Agent* I. ii. 84 He felt her hand rest on his knee: she wasn't romantic, she had said: this was an automatic reaction, he supposed, to the deep seats and the dim lights and the torch songs; as when Pavlov's dogs saliva'd. 1975 J. Goulet *Oh's Profit* xxx. 187 Oh's salivaed middle finger had stumbled across a pocket of ants directly in front of the television.

salivarian (sælivē^ə·riăn), *a.* *Biol.* [f. mod.L. *Salivaria*, name of a section of the genus *Trypanosoma* (C. A. Hoare 1964, in *Jrnl. Protozool.* XI. 203/1), fem. of L. *sālivā·rius* (see Salivary *a.*): see -an.] Used to designate those species of *Trypanosoma* which occur in the bloodstream of the secondary host, and are transmitted from its mouth when it bites a vertebrate. Cf. *stercorarian *a.*
1969 *Jrnl. Protozool.* XVI. 466 (*heading*) A new organelle of bloodstream salivarian trypanosomes. 1971 P. C. C. Garnham *Progr. Parasitol.* iii. 27 The earlier evolution of these mammalian trypanosomes... At first.. eliminated in the faeces on to the inhospitable ground..; next, in the 'stercorarian' trypanosomes..passed in the faeces and..transferred to the mucous membrane of the new host; finally and best, in the 'salivarian' trypanosomes..the organisms..reach..salivary glands and when the insect next bites, the infection inevitably enters the new host. 1977 Soltys & Woo in J. P. Kreier *Parasitic Protozoa* I. vi. 241 *Trypanosoma (N.) congolense* is a small salivarian trypanosome and varies in length between 8 and 24 μm. 1977 J. M. Mansfield in *Ibid.* viii. 310 Unlike other stercorarian trypanosomes, *T. rangeli* infects the hemolymph and salivary glands as well as the alimentary canal of its intermediate host. 1980 *Nature* 24 Jan. 383/2 Infection rates of salivarian trypanosomes (subgenera *Nannomonas*, *Duttonella* and *Trypanozoon*) in the tsetse fly *Glossina*.

salivarium (sælivē^ə·riǝm). Pl. **salivaria.** [f. Saliva + -arium; cf. med.L. *salivarium* a linen cloth used to catch discharged spittle (DuCange).] A spittoon, esp. one genteelly disguised with a lid, ornamental casing, etc.
1883 *Graphic* 25 Aug. 194 (Advt.), Inlaid Walnut Salivarium, 6s. 9d. 1939 Joyce *Finnegans Wake* (1964) ii. 286 With his primal handstoe in his sole salivarium. 1960 [see *non-U *a.* and *sb.*].

salivate, *v.* Add: **2. c.** *fig.* To display one's relish *at* some prospect or anticipated event, to 'lick one's lips'.
1970 *Guardian* 11 May 10/3 On May 1, the American military were delighted to be unleashed into the Cambodian sanctuaries. One officer said of his colleagues: 'They've been salivating at the prospect of this for months.' 1977 *Times* 1 Nov. 14/6 The double LP set has sold two million copies in America. Pye, who are marketing the records in Britain, are salivating at the sales prospects.

salix (sēɪ·liks, sæ·liks). [a. L. *salix* willow.] = Willow *sb.* 1.
1775 T. Blaikie *Diary Scotch Gardener* (1931) 61, I found..several sorts of *salixes*. 1965 P. Wayre *Wind in Reeds* viii. 96 A hazel hen feeding quietly among the salix and lichens.

Salk vaccine (sǫlk). *Med.* [Named after Jonas Edward *Salk* (b. 1914), U.S. virologist, who developed the vaccine in 1954.] The first vaccine developed against poliomyelitis, made from viruses of the three immunological types inactivated with formalin.
[1954 *Brit. Med. Jrnl.* 6 Mar. 593/2 This is..the first large-scale trial of Dr. Salk's vaccine.] 1954 *Jrnl. Amer. Med. Assoc.* 10 July 1021/1 There is no chance that injections of Salk vaccine will cause human Rh-negative subjects to produce Rh antibodies. 1958 *Oxford Mail* 22 Aug. 1/9 Chicago has been completely free from polio this year for the first time, states the Health Department, which credited this to the extensive use of Salk vaccine. 1964 [see *killed *ppl. a.* 1 c]. 1976 M. Grossman in W. L. Drew *Viral Infections* ix. 246 The Salk vaccine is no longer manufactured nor used in the United States.

salle. Add: **1. a.** (Earlier and later examples.)
1765 H. Walpole *Let.* 5 Dec. (1904) VI. 375 You may go into the *petit cabinet*, and then into the great *salle*, and the gallery. 1913 H. James *Small Boy & Others* xxv. 359, I..enjoyed the commemorative show of Delaroche given ..in one of the rather bleak *salles* of the École des Beaux-Arts.
b. = *salle de jeu* (see sense 2 below).
1886 C. M. Yonge *Chantry House* II. xv. 144 Martyn was doing his best for him..while Lady Peacock was at the *salle*. 1966 G. Greene *Comedians* I. iii. 89, I watched him leave the *salle*. He had over three hundred dollars to change now. 1970 'J. Morris' *Candywine Devel.* xxiii. 247 He stood at the big roulette table in the main *salle*.
c. = *salle d'armes* (see sense 2 below).
1961 F. C. Avis *Sportsman's Gloss.* 197/1 *Salle*, the fencing hall or studio, often open to the public. 1973 *Where* Mar. 73/3 Among the luxuries enjoyed by school C..a fencing salle. 1975 *Oxf. Compan. Sports & Games* 306/2 A few schools, such as that of the famous *Angelos* and the *London Fencing Club*, founded in 1848, kept the sport alive in a few London *salles*, some public schools, and the universities.
2. salle d'armes (sal darm), a fencing-room, school or club; **salle d'attente** (earlier and later examples); **salle d'audience** (sal dodyãns), a court-room; **salle d'eau** (sal do), a wash-room, shower-room; **salle de jeu** (sal dǝ ʒö), a gambling house or room; **salle des pas perdus** (sal de pā pęrdü), a waiting-hall (at a law-court, station, etc.), lobby; **salle privée** (sal prīve), a private gambling room in a casino.
1885 E. Castle *Schools & Masters of Fence* x. 159 How different a 'salle d'armes' in Paris or London in those days from the old Italian schools of Queen Bess and Henri III. 1902 G. B. Shaw *Let.* 4 Mar. (1972) II. 269 There should be a salle d'armes where stage combats & wrestlings could be practised. 1952 *Fencing* ('Know the Game' Ser.) 19/1 In a friendly encounter in the Salle d'Armes (Fencing Room) or Club, the sporting tradition of acknowledging a hit has been jealously preserved. 1863 *Miss Jemima's Swiss Jrnl.* 26 June (1963) i. 9 Passengers..are locked in the *salle d'attente* until the arrival of the train. 1909 E. Nesbit *Daphne in Fitzroy St.* iv. 44 The rout of dark-skinned, browbent, hurrying, preoccupied French folk.. in the *salle d'attente* at the station. a 1666 Evelyn *Diary* an. 1644 (1955) II. 98 Within are severall Chambers, Courts, Treasures &c above that most rich and glorious Sale de l'Audiens. 1957 L. Durrell *Spirit of Place* (1969) 138 Of course no lavatories and *salle d'eau* a rarity. Even in this lovely villa we wash from a bucket. 1964 *Punch* 14 Oct. 573/3 Town-dwellers in France who have a *salle d'eaux* [*sic*]..of their own. 1968 D. Torr *Treason Line* 130 They were in the *salle des jeux* [*sic*], the hushed sanc-

tuary of the temple of chance. **1901** V. Bethell *Monte Carlo Anecdotes* 4 In the year 1858 a grand banquet was held to inaugurate the opening of his *Salle-de-Jeux*. **1839** *Indispensable Eng. Vade Mecum Paris* 135 The most remarkable hall is that named *la salle des Pas-Perdus*, being 222 feet long, by 84 wide. **1885** H. James *Little Tour in France* xvii. 120 The curious *salle des pas perdus*, or central hall, out of which the different tribunals open.. is a feature of every French court-house. **1977** *Listener* 10 Feb. 183/1 The image is growing on me of Limbo as a large railway terminus.. where the dead hang about in a *salle des pas perdus*. **1930** E. Waugh *Labels* ii. 35 The cinema producer's version of the *salles privées*, with jewelled courtesans and ribboned grand-dukes, is a thing of the past. **1976** H. MacInnes *Agent in Place* xiv. 153 There were two wings.. the left one consisted of the *Salle Privée*.

3. With varying pronunc. (sāl, sǫl). Also † **saul.** The finishing department of a paper-mill, in which sheets of paper are examined, sorted and packed.

1819 *Rees's Cycl.* XXVI. s.v. Paper, The paper, being sufficiently dried for the last time, is carried to the building where it is examined, finished, and pressed: this is called the *Saul*. **1854** C. Tomlinson *Cycl. Useful Arts* II. 364/1 The paper.. is taken down, carried to a building called the Saul.. where it is examined, finished, and pressed. **1888** Cross & Bevan *Paper-Making* 175 The sheets of paper are now ready to be examined before being finally sent away from the mill. This is done in the 'Finishing-house', or 'Salle' as it is sometimes called. **1946** H. Whetton *Pract. Printing & Binding* xxviii. 345/1 If the paper is being sold in sheets it goes to the 'salle' or finishing department, where each sheet is examined top and bottom.. and sorted. **1976** *Oxford Times* (City ed.) 12 Mar. 1/7 Sogat members at Wolvercote Mill could not recommend acceptance of the management's latest proposals for the 70 men in the salle.

sallee, var. *Sally *sb.*⁴

sallow, *sb.* Add: **4. b. sallow (wattle),** one of several Australian acacias that resemble willows in habit or foliage.

1884 A. Nilson *Timber Trees New South Wales* 21 *A[cacia] dealbata*.—Silver Wattle; Sallow. **1965** *Austral. Encycl.* VII. 539/2 *A[cacia] longifolia*, *A. mucronata* and several related species with long flower-spikes are known as sallow wattles in Victoria.

sallow, *a.* Add: **c.** Also *sallow-hued, -thinking*.

1910 W. de la Mare *Three Mulla-Mulgars* 81 There came spindling along an old sallow-hued Earth-mulgar. **1606** Marston *Parasitaster* iii. sig. E1, A blacke hayred, pall-fac'de, sallowe thinking Mistresse.

† **Sally** (sæ·li), *sb.*³ *Obs.* Corruption of Sal enixum. Also **Sally Nixon.**

1879 G. Lunge *Sulphuric Acid* II. ii. 19 Sulphate known as 'nitre-cake', 'salonix' (= sal enixum), or 'sally'. **1882** W. Crookes *Dyeing & Tissue-Printing* 81 The crystallized sulphate of soda, known.. in many dye-houses as Sally Nixon.

sally (sæ·li), *sb.*⁴ *Austral.* Also **sallee.** [Variant of Sallow *sb.*] One of several eucalypts or acacias that resemble willows in habit or appearance; (see quot. 1965).

1884 A. Nilson *Timber Trees New South Wales* 22 *A[cacia] falcata*.—Hickory; Sally;.. Willow. **1889** J. H. Maiden *Useful Native Plants Austral.* 149 *Acacia falcata*,.. 'Hickory'. 'Lignum-Vitae'. 'Sally'. *Ibid.* 250 *Eucalyptus stellulata*,.. 'Sally' or 'Black Gum'. *Ibid.* 335 *Acacia falcata*... Called variously 'Hickory',.. and 'Sally' or 'Sallee'. **1932** R. H. Anderson *Trees New South Wales* 58 Snow Gum or White Sally. *Ibid.*, Black Sally.. Also known as Sally or Muzzlewood. **1941** Baker *Dict. Austral. Slang.* 62 Sally: an acacia. **1949** J. Wright *Woman to Man* 17 In the olive darkness of the sally-trees Silently moved the air. **1957** *Forest Trees Austral.* (Austral. Forestry & Timber Bureau) 96/2 Swamp gum or broad leaved sally.. occurs in cold and damp situations. *Ibid.* 144/1 White sallee is usually only 30–60 feet in height. **1965** *Austral. Encycl.* VII. 539/2 Sallee, or sally, a corruption of the English 'sallow' which is applicable to certain willow species.. and commonly used for Australian eucalypts and wattles that are supposed to resemble them in habit or foliage. Black sallee and white sallee are the names standardized in the timber trade for the cold-loving *Eucalyptus stellulata* and *E. pauciflora* respectively. *Acacia floribunda* and *A. prominens* are among the eastern wattles which have been called sally.

Sally (sæ·li), *sb.*⁵ *colloq.* [Alteration of Salvation (Army).] **1. a.** The Salvation Army. Also with *the*, and *attrib.*

1915 *N.Y. World Mag.* 9 May 14/3 *Sally*, nickname for Salvation Army. **1931** 'D. Stiff' *Milk & Honey Route* v. 52 The Salvation Army, more intimately known in Hobohemia as 'The Sally'. **1977** *Gay News* 7–20 Apr. 7/3 (*heading*) Sally soldier... A Salvation Army social worker who indecently assaulted young boys was sent to jail for three years at the Old Bailey recently.

b. A member of the Salvation Army; usu. *pl.*, the Salvation Army.

1936 I. L. Idriess *Cattle King* xx. 189 The surest place to find Sid Kidman, when in town on a Saturday night, was among the crowd around the 'Sallies'. **1942** Berrey & Van den Bark *Amer. Thes. Slang* §327/4 *Sally*, a Salvation Army girl. **1957** D. Niland *Call me when Cross turns Over* ii. 31 The woman that runs it, she used to be some sort of a high-up with the Sallies down in Sydney. **1966** A. La Bern *Goodbye Piccadilly* iv. 43 The constable recommended the Salvation Army hostel...

Why not? Better men than Dick Blamey have slept with the 'Sallies'. **1977** C. McCullough *Thorn Birds* iii. 65 It's a hotel for the workingman run by the Sallies.

2. A Salvation Army hostel.

1931 'D. Stiff' *Milk & Honey Route* 213 *Sallies*, Salvation Army hotels and industrial workshops. **1966** *New Statesman* 1 Apr. 479/2 Julie Felix sang against the Salvation Army—and we were.. miles away from the sad Sally where the meth-drinkers are deloused. **1977** *Church Times* 18 Nov. 9/1 He knew that the only other places to find a bed—the 'Sally', the Cyrenian shelter, even the fairly distant 'Spike'—would not have him that night.

3. *Comb.*, as **Sally Ann(e)** [colloq. alteration of *Army*], the Salvation Army; a Salvation Army hostel; **Sally Army,** the Salvation Army.

1927 *Amer. Speech* II. 387/1 *Sally Ann* is the sobriquet for Salvation Army. **1961** W. A. Hagelund *Flying Chase Flag* iii. 48 Now you go see the Major at the Johnson Street Sally Anne about some meal tickets and beds. **1976** *New Society* 5 Aug. 290/3 The Salvation Army?.. You'd never get me sleeping there... Everyone knows you pick all sorts of things up from the Sally Ann. **1961** E. Williams *George* xxiii. 386 Your dear Brother Tom has celebrated his thirteenth birthday with buying a uniform for the Sally Army. *Ibid.* xxvi. 441 Tom sat uneasily polishing his Sally-Army trumpet. **1978** *Guardian* 9 Aug. 7/5 At Christmas, the Sally Army gave her a slap-up lunch.

sally, *v.*² Add: **4. a.** Also **saully.** To move, sway, or run from side to side (see quot. 1887 and cf. Sally *sb.*¹ 7); to progress by making a rocking movement from side to side. *dial.* and *Naut.*

1825 J. T. Brockett *Gloss. North Country Words* 181 *Sally*, to move or run from side to side; as is customary with the persons on board of a ship after she is launched. **1887** D. Donaldson Jamieson, Suppl. 210 To *Sally, Saully*,.. to move or run from side to side, as children do in certain games, and as workmen do on board a ship after it is launched; to rock or swing from side to side, like a small boat at anchor; also, to rise and fall, like a ship on a rough sea. **1972** *Daily Tel.* 15 July 3/8 He told the court that he was 'sallying' down Lowther Street when a policeman stopped him. Asked by the Judge what 'sallying' meant, he said: 'I was just sitting on the saddle pushing the bike along with my foot on the kerb.'

b. *trans.* To rock (a stationary or slow-moving ship) by running from side to side in order to assist its progress. *Naut.*

1919 E. Shackleton *South* i. 33 The engines running full speed astern produced no effect until all hands joined in 'sallying' ship.

Sally Lunn. Add: **1.** (Earlier example.)

1780 P. Thicknesse *Valetudinarian's Bath Guide* (ed. 2) iii. 12, I had the misfortune to lose a beloved brother in the prime of life, who dropt down dead as he was playing on the fiddle at Sir Robert Throgmorton's, after drinking a large quantity of Bath Waters, and eating a hearty breakfast of spungy hot rolls, or *Sally Luns*.

2. Applied loosely to several varieties of yeast and soda bread, esp. in the southern United States. Also *attrib.*

1901 *Picayune's Creole Cook Bk.* (ed. 2) 407/2 Sally Lunn is nothing more than the old breakfast dish known to the Creoles for generations as 'Pain à la Vieille Tante Zoé'. **1933** F. M. Farmer *Boston Cooking-School Cook Bk.* (rev. ed.) 53 Sally Lunn Tea Cakes... Make like Raised muffins. **1976** M. G. Eberhart *Family Fortune* vii. 70 Alice was tucking food away.. Alice asked for more Sally Lunn.

Sally Nixon: see *Sally *sb.*³

sallyport. 3. (Earlier example.)

1814 Jane Austen *Mansf. Park* III. vii. 161 The three boys.. determined to see their brother.. to the salley-port.

Salmanazar (sælmănē·ĭ-zāɹ). Also **Salmanasar.** [ad. *Salmanasar*, the form in the Vulgate of the name of *Shalmaneser*, King of Assyria (II Kings xvii, xviii).] A large size of wine-bottle. Cf. *Balthazar, Jeroboam, *Rehoboam 2.

1935 A. L. Simon *Dict. Wine* 225 *Salmanazar*, the fancy name given to a fancy bottle large enough to hold a dozen reputed quarts, or 9·60 litres, equal to 338·025 fluid ounces. **1959** *Gloss. Terms Packaging* (B.S.I.) 28 *Salmanazar*, a wine bottle—capacity 12 reputed quarts. **1962** [see *Balthazar]. **1978** *Daily Tel.* 13 June 16/2 Edward Heath is to be given a salmanazar of champagne... A salmanazar contains the equivalent of 14 ordinary bottles.

salmine (sæ·lmīn). *Biochem.* Also **-in** (-in). [ad. G. *salmin* (A. Kossel 1896, in *Zeitschr. f. physiol. Chem.* XXII. 180), f. L. *salm-o* salmon: see -ine⁵.] A protein, one of the protamines, isolated from the sperm of the salmon and related species.

1896 *Jrnl. Chem. Soc.* LXX. i. 582 The sulphate [of the protamine] from salmon sperm has the formula $C_{16}H_{31}N_9O_3,H_2SO_4$. That from sturgeon sperm has rather different solubilities in sodium chloride solutions, and the names *salmine* and *sturine* are suggested by [*sic*] the two protamines. **1949** *Proc. Soc. Exper. Biol. & Med.* LXX. 494/1 Salmin, an acid hydrolysis, was shown to yield arginine, proline, serine, valine, alanine, and isoleucine. **1963** F. Haurowitz *Chem. & Function of Proteins* (ed. 2) ii. 18 Heterogeneities have also been discovered.. in the protamines clupein and salmine of fish sperm.

salmon, *sb.*¹ and *a.* Add: **A.** *sb.* **3.** (Earlier example.)

1873 [see *paper *sb.* 8 b].

4. a. *salmon boat, fishery* (earlier example), *mousse, paste* (Paste *sb.* 1 d), *river* (later examples); *salmon fly* (later examples), *gaff*.

1894 *Rudder* Mar. 77 White Class—Salmon boat, Canthelpit, Captain Jacobsen. **1905** J. London *Tales of Fish Patrol* 23 The salmon boat got out its oars. **1732** *Calendar State Papers: Colonial Ser.* (Publ. Rec. Office) (1939) XXXIX. 226 The salmon fishery is still carried on in the several rivers and to advantage. **1856** 'Stonehenge' *Man. Brit. Rural Sports* v. ii. 246/1 Salmon-flies are made on the same principle as the trout-flies. **1927** M. Asquith *Lay Sermons* v. 106 The Durham Ranger and Black Dog are salmon-flies. [**1907** *Yesterday's Shopping* (1969) 669/3 Gaffs.. Salmon and Pike.] **1922** Joyce *Ulysses* 558 Follow the footpeople with knotty sticks, salmongaffs. **1936** Lucas & Hume *Au Petit Cordon Bleu* 43 Put a little of the aspic into a pan... Run a thin coating of this jelly on the top of the salmon *mousse*. **1972** K. Stewart *Times Cookery Bk.* vi. 84 Salmon mousse... Allow several hours for mousse to chill. [**1917** *Harrods Gen. Catal.* p. lxvii/6 Salmon.. and Shrimp Paste.] **1939** T. S. Eliot *Old Possum's Bk. Pract. Cats* 45 You might now and then supply.. Some potted grouse, or salmon paste. **1771** G. Cartwright *Jrnl.* 29 May (1792) I. 127 At the head of this place we found a very fine salmon river. **1886** *Critic* 16 Oct. 183 A map and an annotated list of salmon-rivers locate them chiefly north of the St. Lawrence. **1968** R. M. Patterson *Finlay's River* 88 The Yukon, a salmon river with a name nobody had ever heard of and which was not to be found on any map.

b. *salmon-fisher* (earlier and later examples).

a **1670** [see Coble¹ 1]. **1771** T. Pennant *Tour in Scotl.* 1769 148 Near is a cave, where the Salmon-fishers lie during the season. **1925** F. Scott Fitzgerald *Great Gatsby* vi. 118 He had been beating his way along the south shore of Lake Superior as a clam-digger and a salmon-fisher.

4. c. salmon bass *S. Afr.* = *kabeljou; **salmon berry,** substitute for def.: *N. Amer.*, one of several species of *Rubus*, esp. the white-flowered *R. chamæmorus* and *R. parviflorus* or the pink-flowered western raspberry, *R. spectabilis*; also *attrib.*; (earlier and later examples); **salmon coble** (further examples); **salmon-colour** (earlier example); **salmon disease,** (*a*) a fatal epidemic skin disease of salmon; (*b*) = *salmon poisoning* below; **salmon gum** (see quot. 1883); **salmon-pink,** an orange-pink shade (cf. sense B in Dict.); **salmon poisoning,** a fatal disease of dogs on the Pacific Coast of North America which affects lymphoid tissue and the central nervous system and is caused by rickettsias present in flukes infesting ingested salmon; **salmon pool** (further examples).

1929 *Hardy's Anglers' Guide* (ed. 51) 48 The Kabeljaauw, known in Natal as Salmon Bass.. runs to as much as 150 lbs. **1957** S. Schoeman *Strike!* iii. 70 The very big ones are variously referred to as.. salmon, salmon bass and often even Cape salmon if caught from East London to Durban. **1844** A. Sylvester *Jrnl.* in *Oregon Hist. Q.* (1933) XXXIV. 359 A salmon berry.. being put into the mouth of a fish [*sc.* a salmon], destroys the charm. *a* **1861** Salmon-berry [see *hiaqua]. **1901** J. Grinnell *Gold Hunting in Alaska* 16 The other day we picked three quarts of salmon berries. **1971** *Islander* (Victoria, B.C.) 14 Mar. 16/2 Directing their steps toward the beach.. they hurried.. through the salmonberry thicket. **1977** J. Gillis *Killers of Starfish* (1979) xxii. 217 A precipitous tangle of salmonberry and alder. **1881** W. Gregor *Notes Folk-lore N.-E. Scotl.* 146 In going past a salmon cobble in the harbour, a fisherman would not have allowed his boat to touch it. **1973** W. Elmer *Terminol. Fishing* iii. 78 The salmon coble.. differs in structure from the rest of the cobles. **1813** J. Constable *Let.* 30 June (1964) II. 109 The paper will be a sort of salmon color and the sofa & chairs crimson. **1880** *Proc. R. Soc. Edin.* X. 242, I am led to believe that the so-called salmon disease does not depend upon a pre-diseased condition of the fish. **1950** *Amer. Jrnl. Path.* XXVI. 617 (*heading*) The pathology and etiology of salmon disease in the dog and fox. **1964** G. W. Stamm *Dog Owner's Vet. Guide* 79 Salmon disease has been successfully treated with certain sulfa drugs and with penicillin. **1971** D. Mills *Salmon & Trout* iii. 91 The salmon is subject to a number of diseases... The diseases include furunculosis, Dee disease, kidney disease, salmon disease,.. and columnaris. [**1883** F. von Mueller *Eucalyptographia* ix, *Eucalyptus salmonophloia*... A tree, when aged, attaining to fully 100 feet in height, known vernacularly as the 'Salmon-colored Gumtree', in allusion to the smooth grey and somewhat purplish bark of an oily lustre.] **1934** *Bulletin* (Sydney) 24 Oct. 20/3 A Digger mate and myself saw three salmon gums, trees of the Westralian wheatbelt and eastern goldfields. **1969** Chippendale & Johnston *Eucalypts* 72/1 The salmon gum has been regarded as an indicator of good, loamy soil on which much of the West Australian wheatbelt is developed. **1884** Salmon-pink [see *crevette]. **1979** *Country Life* 24 May 1618/3 The salmon-pink of dawn. **1925** *Jrnl. Amer. Vet. Med. Assoc.* LXVI. 638 A microscopic cyst has been found in the muscle of 'sore-back' salmon. When these fish were fed to dogs typical symptoms of so-called salmon poisoning were produced. **1974** T. McGinnis *Well Dog Bk.* 74 This fluke is host to an organism (a rickettsia) which causes a severe disease called salmon poisoning. **1874** W. Lennox *My Recoll.* II. 72 My guide then informing me that within three miles there were several salmon pools, I lost no time in proceeding there. **1892** Salmon pool [*Easy B. 4 b].

Salmonella (sælmŏne·lă). *Bacteriol.* Also **salmonella.** Pl. **-ellæ, -ellas,** (*erron.*) **-ella.** [mod.L. (coined in Fr. by J. Lignières 1900, in *Bull. de la Soc. centrale de Méd. Vét.* XVIII. 389), f. the name of Daniel Elmer *Salmon* (1850–1914), U.S. pathologist + L. *-ella* (see -ell².)] A member of the genus of pathogenic, Gram-negative, rod-shaped bacteria so called, which includes some causing food poisoning, typhoid, and paratyphoid in man and various diseases in domestic animals.

1913 H. J. Hutchens tr. *Besson's Pract. Bacteriol.* 442 Lignières proposed to designate all those organisms which had the morphological and cultural attributes of the bacillus of hog-cholera.. by the name Salmonella after Salmon. **1929** *Lancet* 12 Jan. 96/2 So long as there appeared to be but two types in this group of the Salmonellas.. there was little need to find a name in common for them. **1933** J. H. Dible *Rec. Adv. Bacteriol.* (ed. 2) iv. 79 White.. adduces evidence of the presence of common antigenic complexes, relating these to the enteric and food-poisoning salmonellas. **1944** L. R. Thompson *Introd. Microorganisms* xvi. 269 Salmonella have been ingested with meats, fish,.. dairy products,.. and drinking water. **1951** *Chambers's Jrnl.* Oct. 588/1 Dr. Williams Smith and Professor J. C. Cruickshank have been enquiring into the danger cats and dogs may convey by acting as reservoirs of the salmonellæ, noxious bacteria which cause food-poisoning in man. **1959** *Times* 29 Sept. 12/8 Twenty-seven of 64 samples of poultry manure yielded salmonella. **1980** *Brit. Med. Jrnl.* 29 Mar. 928/1 Zoonotic pathogens, such as salmonellas,.. may be present in any type of slurry.

2. = *Salmonellosis.

1962 *Telegraph* (Brisbane) 27 Aug. 32/2 He believed there was a risk in the sale of kangaroo meat. It.. carried Q fever and salmonella. **1977** *Shooting Times & Country Mag.* 13–19 Jan. 27/3 There is still the possibility of such diseases as forms of Salmonella (now a notifiable disease) being spread over a wide area.

3. *attrib.*

1916 A. I. Kendall *Bacteriol.* xv. 344 These organisms are variously known as the hog cholera, Salmonella, Gärtner, enteritidis, intermediate, paratyphoid or paratyphoid group. **1925** J. W. Bigger *Handbk. Bacteriol.* xxvii. 259 There are two closely related types of bacilli which are responsible for isolated cases or epidemics of food poisoning... Both these are exceedingly closely related to B[acillus] paratyphosus B, and together with it form what is called the 'Salmonella' group. **1963** *Lancet* 19 Jan. 161/1 Most outbreaks of salmonella infection result from the contamination of a single article of food which is then eaten by a number of individuals. **1979** *Daily Tel.* 9 Aug. 7/3 A woman.. died in hospital after an outbreak of salmonella food poisoning.

salmonellosis (sæ:lmŏnelō·sis). *Path.* Also **Salmonellosis.** [f. Salmonell(a + -osis (J. Lignières 1901, in *Recueil de Méd. Vét.* VIII. 416), f. prec.: see -osis.] Infection with or a disease caused by salmonellæ.

1913 in Dorland *Med. Dict.* (ed. 7). **1931** *Nomencl. Diseases* (Min. of Health) (ed. 6) 104 *Bacterium* (*Salmonella*) *enteritidis*.. *suipestifer*. Causes of epidemic food-poisoning in man (Salmonellosis) and occasionally of paratyphoid fever. [illegible] classical case of the salmonellosis.. is that of a generalized infection, namely that of typhoid fever. This human clinical picture has its close parallel in animal salmonellosis. **1965** *N.Z. News* 13 Apr. 3/1 Several thousand sheep have died since Christmas in salmonellosis outbreaks among stock in the Rotorua and Waikato districts. **1970** W. H. Parker *Health & Dis. in Farm Animals* xiii. 218 The breaks of Salmonellosis in stock are a matter of concern for public health.

salmonid. Delete *rare* and add later examples. Also *attrib.* and *fossil*.

1895 B. Dean *Fishes, Living & Fossil* viii. 186 Eggs of Salmonids are deposited loosely in 'nests' on a clean, gravelly bottom. **1931** J. R. Norman *Hist. Fishes* xiii. 268 When the climate was considerably colder the range of migratory Salmonids extended much further south. **1964** *Oceanogr. & Marine Biol.* II. 378 Further changes have been reported to occur after fertilization of the ova in another salmonid, the rainbow trout. **1970** *New Scientist* 19 Feb. 353/1 Infectious dermal necrosis.. attacks salmon and other salmonid fish. **1975** *Nature* 14 Aug. 528/1 Fish farming in the UK has grown slowly in the freshwater and salmonid fisheries.. **1978** *Ibid.* 2 Mar. 77/1 Evidence of large-scale gene loss following tetraploidy has been reported in both salmonid and catostomid fishes.

salmonize, *v.* Add: Also, to (attempt to) introduce salmon into (a river, etc.). So **sa:lmoniza·tion;** also **sa·lmonizing** *vbl. sb.*

1870 *Sat. Rev.* 30 Apr. 576/1 Why should I be 'chaffed' about the salmonization of the Thames? The Thames once produced plenty of salmon; why should not the Thames produce salmon again? **1901** *Chambers's Jrnl.* Sept. 585/2 It does not seem quite fair that a few gentlemen should be able to prevent the salmonising of such a large extent of water as is here indicated.

salmon-trout. 2. (Earlier and later Amer. examples.)

1705 *Boston News-Let.* 15–22 Oct. 2/1 Our men were refresh'd with variety of Fish, especially Salmon Trouts, some whereof 2 foot long. **1806** W. Clark *Jrnl.* 13 Mar. in *Orig. Jrnls. Lewis & Clark Exped.* (1905) IV. 166 The Salmon Trout are seldom more than two feet in length. **1848** E. Bryant *What I saw in California* xi. 158 He had taken with his hook about a dozen salmon-trout. **1939** *Nat. Geogr. Mag.* Feb. 212/2 Both of these species [*sc.* Dolly Varden and blackspotted trout] are known in some localities as 'salmon trout'.

salmony (sæ·məni), *a.* [f. Salmon *sb.*¹ and *a.* + -y¹.] Somewhat salmon-coloured.

1935 E. Farjeon *Nursery in Nineties* 237 She has one evening dress.. a salmony-pink brocade with bunches of lemon-coloured flowers. **1948** V. S. Pritchett in E. Bowen et al. *Why do I Write?* 14 Lichfield.. a nice, dull little place in glazed salmony Midland brick. **1974** R. Crossman *Diaries* (1977) III. 805 Anne looked fresh and exquisite, too, in her lovely salmony red suit.

salon. Add: Now also with pronunc. (sæ·lɒn, sæ·lən). **1. a.** (Earlier example.)

1699 M. Lister *Journey to Paris* 196 The Castle is.. most commodious. The Great Salon and the Gallery are extremely well Painted.

3. b. || **salon des refusés** (*de* rəfüze). [Fr., exhibition of rejected work], an exhibition ordered by Napoleon III in 1863 to display pictures rejected by the official Salon; also *fig.*

1896 J. C. Beckwith in J. C. Van Dyke *Mod. French Masters* iii. 220 His [*sc.* Manet's] works became known.. at the exhibition of pictures refused at the Salon, which were for several years gathered together and shown in a building generously provided by the government, and called the Salon des Refusés. **1932** Kondov & Lathom *Introd. Fr. Painting* xiii. 194 *Déjeuner sur l'herbe* [by Manet].. rejected by the Salon.. was exhibited in the *Salon des Refusés*, and frowned on by Louis Napoleon. **1981** *Listener* 1 Jan. 4/3 Roy Jenkins plus Shirley Williams.. plus the non-reselected MPs in the salon des refusés.

4. An establishment in which the trade of a beauty specialist or hairdresser is conducted.

1913 *Vogue* 1 June 106/1 (Advt.), Firming the skin is the new process used exclusively by Elizabeth Arden... **1927** *Harrods Gen. Catal.* 397 Enshrined in an atmosphere of refinement and artistic comfort, Harrods Hairdressing Salons are a favourite resort with ladies. **1932** *New Yorker* 9 Apr. 68/3 See the telephone book for nearest Salon. **1956** Ashley & Stevenson *Hair Design & Colour* i. 12 The salon itself, as a background to modern hair-styling, must inevitably play a highly important part in creating the right atmosphere. **1973** A. MacVicar *Painted Doll Affair* vii. 82 My wife swears it's much better than the expensive 'salon' she used to go to in Glasgow.

5. a. *attrib.*, as (sense 2) *salon philosopher, science, volume, -writer*; (sense 3) *salon furniture, norm, -piece, -vocabulary*; (sense 4) *salon facial, service, treatment*.

1974 *Times* 17 Aug. 9/2 All the products Marisa uses in the salon facials can be bought. **1973** R. Hayes *Hungarian Game* ii. 18 An incredibly tasteless collection of Regency and Salon furniture in the drawing room. **1942** Wyndham Lewis *Lett.* (1963) 334 The artist is labelled 'decadent' who departs from the Salon norm.. by the Hitlerite pundit of 'sanity'. **1947** A. Einstein *Music in Romantic Era* xvi. 304 'The salon philosopher'. **1974** *Impressionism* (R.A. Catal.) 8 Daubigny.. moved to narrow the distinction between outdoor study and Salon-piece. [illegible] **1974** *Harrods Christmas Catal.* p. 13 Make a Gift of Beauty with a Gift Token.. to the value of whichever Salon Service or Treatment you require. [illegible] The work has higher claims than that of being merely a salon volume designed for presentation. **1944** L. MacNeice *Christopher Columbus* 73 Radio's contemporary triangle.. insists on a function of words which salon-writers are perhaps too apt to forget; this function is communication.

b. *attrib.* (passing into *adj.*) and *Comb.* with (occas. derogatory) reference to light music played as in a fashionable salon.

1914 *Étude* Oct. 708/1 Nearly all compositions for the piano by modern composers are Salon Music. **1935** *Vanity Fair* (N.Y.) Nov. 58/1 He's not making a salon man of me. [illegible] It is salon music. [illegible] **1947** A. Einstein *Music in Romantic Era* xvii. 331 He was a somewhat eccentric salon-composer. **1948** Mencken *Amer. Lang. Suppl.* II. 706 A performer who.. undertakes conventional music is a *commercial, salon-man, long-underwear* or *long-hair*. **1949** Koestler *Promise & Fulfilment* iii. i. 301 Middle-class families in the Tel Aviv cafés applauded.. the Russian marches played by salon orchestras in the Viennese style. **1950** Blesh & Janis *They all played Ragtime* iv. 77 It is unsyncopated and in a light salon vein. **1955** *Times* 12 July 5/5 A Prelude and Fugue for string orchestra by Moszkowski.. combined learned and *salon* styles with surprising success. **1979** *Guardian* 5 May 14/5 Jones.. wrote the music for such skittish romances as Girl from Utah.. and San Toy. Excerpts from these propped up many a salon orchestra's repertoire.

|| **salone** (salō·ne). [It.: see Saloon.] **1. a.** = Salon 1 a. **b.** = Salon 1 b. (Only with reference to Italy.)

1902 H. James *Wings of Dove* X. xxxiii. 515 She received me.. in that glorious great *salone*. **1912** Beerbohm *Christmas Garland* 134 A wintry Venetian sunshine poured in through the vast windows of his *salone*. **1960** E. Bowen *Time in Rome* v. 140 The cavalier.. charged through the *salone* and out again on to the balcony. **1969** 'I. Drummond' *Man with Tiny Head* v. 74 He had despised her at their first meeting, in the huge Uccello salone.

|| **salonfähig** (zalo·nfε:iç), *a. rare.* [Ger.] Fit for (polite) society; socially respectable.

1905 W. James in *McClure's Mag.* May 39/2 Neither in dress nor in manner did he ever grow quite 'gentlemanly' or *Salonfähig* in the conventional and obliterated sense of the terms. **1980** *Encounter* May 41/1 The Austrian initiative, whatever its underlying motivations, serves to lend an air of cultural respectability to a blood-stained struggle and to render Arafat *salonfähig* in Western Europe.

|| **salonnière** (sălonyε·r). [Fr., f. Salon.] A woman who holds a salon; a society hostess.

a **1922** T. S. Eliot *Waste Land Drafts* (1971) 27 Fresca's arrived (the Muses Nine declare) To be a sort of can-can salonnière. **1925** A. Huxley *Those Barren Leaves* ii. i. 86 Mrs. Aldwinkle the salonnière, the hostess, the planner of literary parties and agapes of lions—is she not classical? a household word? a familiar quotation? **1964** *New Statesman* 10 Apr. 572/2 For a time she [*sc.* Beatrice Elvery] worked in the Co-operative Stained Glass Studio run by Sarah Purser, a patriotic salonnière. **1976** S. J. Darroch *Ottoline* iv. 60 Being a hostess gave many women a chance to use talents and abilities that would otherwise have had no outlet. They turned their houses into cultural or intellectual oases, or merely centres of gossip. And a few did so in such a style that they qualified for the exalted title of salonnière.

saloon. Add: **4. a.** (Earlier and further examples.) Also, the passenger cabin of an aeroplane. Also quasi-*advb.* in *to go* (etc.) *saloon*.

c **1800** in M. Johnson *Amer. Advertising, 1800–1900* (1960), Fare to Bristol—In main Saloon, and cuddy state rooms, Thirty-Five Guineas; in fore and lower saloons, Thirty Guineas. **1882** W. D. Hay *Brighter Britain!* I. i. 57 If you can only afford to travel 'saloon'—the extra comfort on a long voyage is well worth the extra price. **1884** *Whitaker's Almanack* (Advt. section) 20 These large, highest classed and full-powered Steamships.. are fitted up in the latest and most approved fashion to ensure the comfort of Passengers, having the Saloon on Upper Deck. **1842** [see Steerage 5]. **1931** *Daily Mail Year Bk.* 27/1 Eight passengers, in armchair seats in a draught-proof saloon. **1930** *Daily Express* 6 Oct. 2/3 [illegible] swept around in a wide circle, visible only by her red and green navigating lights and the glow of the illuminated saloons.

b. (Earlier single example.)

1842 *Illustr. London News* 18 June 89/1 Previous to the departure from Paddington, the Royal Saloon, the fittings of which are upon a most elegant.. scale, were tastefully improved by bouquets. **1850** C. Schreiber *Jrnl.* 17 June (1952) 14 We had a saloon carriage. **1859** *First Impressions New World* 214 There were four of these [state rooms], besides a general saloon in the middle; but the whole was greatly inferior to the elegance of Mr. Tyson's car on the Baltimore and Ohio Railway. **1886** *Encycl. Brit.* XX. 247/1 Saloon carriages are occasionally used, so called because two or more of the ordinary compartments are thrown into one.

c. A type of motor car with a closed body for four or more passengers. Cf. *sedan 1 c.

1908 *Motor Manual* (ed. 11) iii. 92 Other forms of bodies fitted to more expensive cars include the brougham, landaulet, saloon, double phaeton, [etc.]. **1913** R. N. Seymour *Three Wheels* i. x. 157 He.. secured the services of a Buick saloon. **1933** Auden & Isherwood *Dog beneath Skin* 18 Brought in charabanc and saloon along arterial roads. **1955** *Times* 6 June 7/7 A chauffeur-driven saloon draws up with a single passenger in the back—a prim little boy reading a school book. **1971** *Daily Tel.* 13 Apr. 2/4 A new saloon with front-wheel drive and transversely-mounted engine.. is announced today by Fiat. **1976** Royham & Donnelly *Valentino* vii. 48 The street, where a gleaming new four-door Ford saloon was parked.

5. a. (Earlier example.)

1851 [see *ice-cream *attrib.*].

b. = *saloon theatre* below.

1864 G. A. Sala *Robson* 14 The place was a 'saloon'—that is to say, drinking and smoking went on during the performance, but the pieces put upon the stage were all of a high class. **1902** *Encycl. Brit.* XXXI. 465/2 The principal 'saloons' were the [Eagle] in the Whitechapel Road, the 'Bower' in the Lower Marsh, Lambeth, [etc.]. **1974** *Encycl. Brit. Macropædia* VIII. 330/1 'Saloon' became the name for any place of popular entertainment; 'variety' was coming at the time to describe a program offering a varied bill that featured a mixture of musical and comic entertainment.

6. (Earlier and later examples.) Also, in British use, a refreshment bar in a theatre; a separate bar in a public house (as opposed to *public bar*), = *saloon bar* below.

1841 *Southern Lit. Messenger* VII. 764/1 After going into the saloon (grog-shop) to 'freshen the nip'—.. they led me into the upper tier of boxes. **1841** Dickens *Let.* 28 Dec. (1969) II. 454 This note is about the saloon... The refreshments are preposterously dear... There ought to be a boxkeeper to ring a bell or give some other notice of the commencement of the overture to the afterpiece. **1854** *Harper's Mag.* Apr. 586/2 As I re-entered the bar-room labeled 'saloon', of mine inn. **1902** 'N. Gubbins' *Dead Certainties* 106 Exactly thirty days from the day upon which I first entered the accursed swing-doors of the Bull and Beehive, late one night, a stranger entered the 'saloon'. **1946** *Amer. Speech* XXI. 277 The English saloon-keeper was the keeper of the 'saloon', or as it would

now be termed 'refreshment bar', in a London theatre. **1949** *Columbus* (Ohio) *Sunday Dispatch* 16 Oct. c1/3 He returned to Westerville in 1887 and opened a saloon at a new location on State St. **1969** House & Storey *Lett. Charles Dickens* II. 454 Macready had gone to great lengths to civilize the saloon. **1976** *National Observer* (U.S.) 28 Aug. 13/1 People have been tossed out of saloons in downtown Utica, N.Y., with more style.

7. b. saloon bar, a separate bar in a public house offering more comfort, services, etc. than the public bar; **saloon car,** (*b*) = 4 c above; **saloon-keeper** (earlier and later examples); also in British use, the keeper of a refreshment bar in a theatre; **saloon man** *U.S.*, one who frequents drinking saloons; **saloon theatre:** see Theatre *sb.* 2 in Dict. and Suppl.

1902 G. Hill in G. Sims *Living London* II. 292/3 The distinction between the 'private' bar and the 'saloon' bar is subtle... The saloon bar is the ante-chamber of the billiard room. **1932** L. Golding *Magnolia St.* i. iii. 45 The Public Bar, nothing like so grand as the Saloon Bar, nothing like so cosy as the Private Bar. **1977** 'J. Gash' *Judas Pair* ii. 17 The saloon bar was crowded. **1915** *Motor Manual* (ed. 18) xi. 135 Landaulets, cabriolets, and saloon cars. **1931** D. L. Sayers *Five Red Herrings* xv. 167, I observed Mr. Gowan's saloon car standing before the door. **1974** *Country Life* 17 Oct. 1112/1 In post-war years we have had some very exciting saloon-car racing. **1849** *Theatrical Mirror* 10 Sept. 21 The 'Mirror' is supplied at most moderate prices to the Saloon Keepers, in order to enable them to sell it to advantage. **1873** 'Mark Twain' *Gilded Age* lix. 530 Leave the true source of our political power.. in the hands of saloon-keepers. **1944** B. A. Botkin *Treas. Amer. Folklore* i. 131 He followed the construction of a new line on the Southern Pacific Railroad as camp saloon-keeper. **1977** *Times* 9 July 9/1 His grandfather was a poor tenant farmer.. who became a saloon keeper. **1870** J. W. McClung *Minnesota* 213 Spring Valley, with 400 population.. and no saloon. 'Saloon men cannot live in Spring Valley.' **1915** J. London *Star Rover* ii. 9 You can weave the political pull of San Francisco saloon-men and ward heelers into a position of graft.

saloonist. a. (Earlier and later examples.)

1870 *Territorial Enterprise* (Virginia, Nevada) 3 Mar. 3/2 (*heading*) New saloonists. **1946** *Chicago Daily News* 8 Nov. 18/2 Saloonists voted out of business in the Woodlawn local option election talk of going to court to upset the vote.

|| **salopette** (salopɛt). [Fr.] A pair of overalls or dungarees of a kind worn orig. in France by workmen and later introduced for general wear, esp. as a skiing garment. Also in *pl.*

1972 *Guardian* 31 Oct. 11/2 Basically, the salopette is a Frenchman's overalls with a high waist, a bib front and adjustable shoulder straps, modified for skiing only by a snow cuff fitting snugly round your ankles. **1973** *Country Life* 21 June 1847/2 A summer salopette suit in glazed cotton. **1974** N. Freeling *Dressing of Diamond* 130 The old man.. in his sleeveless vest and bib-and-brace *salopette*. **1977** *Guardian* 16 Mar. 9/1 Peter Blacklay is a doctor and he made a pair of bright red and blue nylon salopettes in a weekend. **1978** *Daily Tel.* 2 Feb. 15/4 Salopettes are far warmer and more comfortable.. than stretch ski pants. **1980** *Woman's Jrnl.* Jan. 35/1 He wears a navy blue salopette.

Salopian, *a.* and *sb.* Add: **A.** *adj.* **b.** Designating a variety of porcelain made at the former Caughley manufactory (closed 1814) near Broseley, Shropshire in the late eighteenth and early nineteenth centuries. Also *transf.* Cf. *Caughley.

1850 J. Marryat *Coll. Hist. Pott. & Porc.* ix. 182 The Salopian ware is very similar to the Derby in pattern and colouring. **1857** —— *Hist. Pott. & Porc.* xii. 297 The early Salopian porcelain was originally made.. at Caughley. **1910** J. F. Blacker *ABC of Collecting Old Eng. China* (ed. 3) iii. 49 Salopian blue is somewhat similar in tone to that of Worcester. **1933** W. B. Honey *Eng. Pott. & Porc.* ii. xii. 190 The Caughley porcelain was known as 'Salopian'. *Ibid.* 191 The Salopian blue-painting.. was mostly cribbed from Worcester and Chantilly. **1957** *Encycl. Brit.* XVIII. 354/1 Some Liverpool factories and that at Caughley (the 'Salopian' factory).. may be regarded as offshoots of Worcester.

c. *Geol.* Of, pertaining to, or designating an alternative division of the Silurian comprising the Wenlockian and (lower) Ludlovian. Freq. *absol.*

1879 C. Lapworth in *Ann. & Mag. Nat. Hist.* III. Table facing p. 455 Silurian System: Middle Division (Salopian). **1880** —— in *Ibid.* V. 48 The second natural division of the Silurian system is undoubtedly Murchison's Great Mudstone series, which includes the so-called Wenlock and Lower Ludlow groups... In Shropshire this great mudstone or *Salopian* formation is by far the most important physical group in the Silurian. **1883** J. E. Marr *Classification of Cambrian & Silurian Rocks* 42 No higher Salopian beds, and no Downtonian beds occur in North Wales, partly perhaps owing to subsequent denudation. **1929** O. T. Jones in Evans & Stubblefield *Handbk. Geol. Gt. Brit.* iii. iv. 92 In certain areas Lapworth's term Salopian is sometimes useful, since the line of separation between Wenlock and Ludlow, represented by that between Lower and Upper Salopian, is not easy to determine. **1940** *Q. Jrnl. Geol. Soc.* XCV. 335 (*heading*) The geology of the Colwyn Bay district: a study of submarine slumping during the Salopian period. *Ibid.* 374 The strike of these structures.. indicates that the floor of the Salopian area in the Colwyn Bay region sloped from north to south. **1971** *Jrnl. Geol. Soc.* CXXVII. 104 Lapworth's.. term *Salopian* was eventually extended upwards by O. T. Jones.. to include the whole of the Ludlow as well as the Wenlock. It is thus still employed from time to time... It finds no place in our classification.

B. *sb.* **b.** A pupil of Shrewsbury School.

1866 *Blackw. Mag.* Apr. 432/2 He has left us specimens of Latin verses of which even modern Salopians might be proud. **1898** *Public School Mag.* Dec. 487/2 'Swilling', a substitute for baths, is still an institution, of which every Salopian is proud. **1932** Pendlebury & West *Shrewsbury School* xi. 82 Nearly 2,000 Salopians saw active service... A sum of money was set aside to assist in the education of the sons of Old Salopians who had been killed. **1964** P. Cowburn *Salopian Anthol.* p. vi, This book.. is to enable Salopians.. to know what has been written about this particular school.

|| **salotto** (salǫ·tto). Also (erron.) **salotta;** pl. **salotti.** [It. dim. f. Sala¹.] In Italy, a drawing-room, reception room; a lounge.

1918 G. Frankau *One of Them* xii. 87 Can we repay those.. midnight cocktails of your flowered *salotti*. **1924** D. H. Lawrence in M. Magnus *Mem. Foreign Legion* 51 So we went into the salotta. 'Oh, what a beautiful room,' he cried. **1930** E. Pound *XXX Cantos* xxvii. 126 In the salotto of that drummer's hotel. **1932** *Times Lit. Suppl.* 21 Jan. 38/3 Her prototype.. can always be found in the cheerless *salotto* of a certain kind of Florentine pension.

Salpausselkä (sa·lpɑuselkä). *Physical Geogr.* Also **-selka.** [Finnish.] Each of two long, wide end moraines in southern Finland that are regarded as marking the last readvance of the ice sheet at the end of the Pleistocene.

1923 *Bull. de la Commission Géologique de Finlande* No. 60. 8 The southern slope toward the Gulf of Finland and Lake Laatokka (Ladoga).. consists of two faces of different grade: (1) The larger inner part, the Lake District of Finland, extending to the belt of the recessional moraines of Salpausselkä ('the damming ridge') and (2) the narrower bow-shaped Coast zone between the Salpausselkä and the coasts of Lake Laatokka and the sea. **1937** Wooldridge & Morgan *Physical Basis Geogr.* xxii. 390 The great sand and gravel ridge which extends from east to west through Finland and is known as the Salpausselka. **1957** J. K. Charlesworth *Quaternary Era* II. xxxi. 675 The Salpausselkä readvance,.. of 12,000 years ago, has been linked with a temporary increase of snowfall. *Ibid.* xlii. 1172 The First or Outer Salpausselkä has a maximum altitude of 70–80 m, an average width of 2·5 km. **1968** R. W. Fairbridge *Encycl. Geomorphol.* 919/1 The upper boundary of the Pleistocene.. must lie somewhere between 10,000 and 10,500 years B.P., and may be defined as.. the time equivalents of sediments overlying the terminal moraines of the Valders (in North America) or the Salpausselkä (in Europe).

salpicon. Delete ? *Obs.* and add earlier and later examples. Also used as a garnish for vol-au-vents and the like.

1723 J. Nott *Cook's & Confectioner's Dict.* sig. Gg7, Make a Hole in your Piece of roast Meat.. and pour the Salpicon into the Hole. **1877** E. S. Dallas *Kettner's Bk. of Table* 81 *Bouchées*—Morsels—These are small Vol-au-vents.. filled with a salpicon of chicken, game or fish. **1906** *Mrs. Beeton's Bk. Househ. Managem.* lxi. 1651 Salpicon.—This name is applied to the various mixtures used in filling timbales, bombs, patty-cases. **1936** Lucas & Hume *Au Petit Cordon Bleu* 44 Mix the sauce into the salmon, mushrooms, and onions and pour this *salpicon* on to the bottom of the dish. **1965** E. David *French Provincial Cooking* (ed. 2) 99 *Salpicon*. May be one of a score of mixtures comprising flavouring vegetables, herbs, ham, veal, fish or meat but always cut into very small dice and bound into a thick white or brown sauce. Used as a stuffing, or as a garnish for little tartlets or *vols-au-vent*. **1977** *Time* 25 Apr. 17/2 He makes.. a saddle of boned lamb stuffed with a mousse of chicken, accompanied by a salpicon of kidneys, brains and sweetbreads.

salpiglossis. For 'mod.L.' read 'mod.L. (H. Ruiz & J. Pavon *Floræ Peruvianæ et Chilensis Prodromus* (1794) 94)' and substitute for def.: An annual or perennial herb of the genus so called, native to Chile and bearing funnel-shaped flowers of various colours. (Earlier and later examples.)

1827 W. J. Hooker *Exotic Flora* III. 229 (*heading*) Straw-coloured Salpiglossis. **1915** H. H. Thomas *Bk. Hardy Flowers* 392 The great improvement in the size and colour of the flowers.. renders the Salpiglossis very desirable for beds and groups in the border. **1931** A. N. Scott tr. *Carossa's Boyhood & Youth* v. 72 A whole host of the most beautiful salpiglosses. **1962** R. Page *Educ. of Gardener* xii. 328, I remember a fantastic planting of yard-high salpiglossis in warm reds. **1979** *Daily Tel.* 26 May 30/2 The handsome salpiglossis.. will do very well from a May sowing out of doors.

salpingo-. Add: **salpinge·ctomy** [*-ectomy], excision of a Fallopian tube; **salpi·ngogram,** an image of the Fallopian tubes obtained with X-rays or ultrasound; **sa:lpingogra·phic** *a.*, of or pertaining to salpingography; **salpingo·graphy** [ad. G. *salpingographie* (F. Schoker 1925, in *Zentralbl. f. Gynaekol.* XLIX. 290)], the process or technique of obtaining salpingograms; **salpingo·lysis** [mod.L. (coined in Fr. by P. E. Goullioud 1914, in *Lyon Médicale* CXXII. 689): see *-lysis], the removal of adhesions that constrain the Fallopian tubes in abnormal positions with respect to the ovaries and hence prevent conception; **salpingo-oöphorectomy** (examples).

1888 W. H. & H. T. Byford *Pract. Med. & Surg.* (ed. 4) 816/2 (Index), Salpingectomy. **1897** *Amer. Jrnl. Med. Sci.* CXIV. 497 (*heading*) The stump after salpingectomy. **1978** G. Vidal *Kalki* i. 5 In perfect health and with maximum publicity at the Marie Stopes Clinic in Daly City, I underwent a bilateral partial salpingectomy, better known as 'Band-Aid Surgery'. **1927** *Surg., Gynecol. & Obstetr.* XLV. 140/2 By means of roentgenological study after the injection of iodized oil, an accurate uterogram and salpingogram, visualizing the entire internal female generative tract, may be obtained in cases in which the Fallopian tubes are not occluded. **1964** Browne & McClure Browne *Postgrad. Obstetr. & Gynaecol.* (ed. 3) xii. 158 If the salpingogram seems normal and tubercle is suspected, an endometrial biopsy should be done. **1927** *Surg., Gynecol. & Obstetr.* XLV. 132/2 Rosenblatt.. reported his salpingographic observations on three women who had submitted to the Alexander-Adams operation for sterility. **1935** *Ibid.* LX. 228/1 Salpingography is.. of value in determining the presence or absence of tubes in patients who have had a previous operation on the adnexa but are uncertain of its nature. **1976** G. Berci *Endoscopy* xvii. 236/1 The use of combined laparoscopy and intra-operative salpingography employing television fluoroscopy with aimed spot films allows the most complete evaluation of uterine and tubal anatomy conducted during a single procedure. **1937** *Amer. Jrnl. Obstetr. & Gynecol.* XXXIII. 39 When the occlusion is at the fimbriated end [of the fallopian tubes], simple release of adhesions may suffice to restore the patency of the tube (salpingolysis). **1980** *Fertility & Sterility* XXXIV. 223/1 In Table 1 are listed the pregnancy rates after bilateral salpingolysis, bilateral salpingostomy, [etc.]. **1890** *Johns Hopkins Hosp. Bull.* I. 57/1 Eight cases of salpingo-oöphorectomy for fibroids are symptomatically relieved of their pressure symptoms. **1977** *Proc. R. Soc. Med.* LXX. 189/2 Fig 1 shows the ureter deliberately exposed in relation to a clamp placed across the infundibulopelvic ligament during the course of a hysterectomy and left salpingo-oophorectomy.

salsa (sa·lsă). [Sp.; cf. Sauce *sb.*] || **1.** *Cookery.* A variety of sauce served with meat. Also *Comb.*

1846 R. Ford *Gatherings from Spain* xi. 132 What sort of a stew is it? Let me smell and taste the *salsa*. **1935** J. Steinbeck *Tortilla Flat* xvi. 289 Her two sons.. carried a wash-tub of salsa pura between them. **1973** *Listener* 18 Jan. 98/2 Serve the meat... Horse-radish, *salsa verde*, spicy tomato sauce.. all go very well. **1978** *Tucson Mag.* Dec. 84/3 Steak and salsa rate high.

2. [Amer. Sp.] A kind of dance music of Latin American origin which incorporates elements of jazz and rock music; a dance performed to this music.

1975 *New Yorker* 29 Sept. 41/3 The group I play with consists of some Latin kids who are not afraid to break out of pure salsa. **1975** *N.Y. Times* 11 Nov. 61/5 Take the sound and percussion rhythm of an eight-piece Latin band; add a mixture of mambo, cha cha and merengue dance steps;.. simmer gently for about 10 to 15 minutes on a crowded dance floor; add a pinch and a squeeze and you have it—salsa. **1976** *Monitor* (McAllen, Texas) 21 Oct. 5B/3 The Caribbean has given us gentle calypso and Trinidad's brash steel bands,.. the spicy latin 'salsa' of Puerto Rico and the whimsical chants and big beat of Jamaican reggae. **1978** *Detroit Free Press* 5 Mar. D16/5 San Juan's major hotels have lounge bands that specialize in salsa (Latin rhythm with a jazz beat). **1981** *Weekly Guardian* 12 July 17/1 Salsa music drifts out of the bar as a group of grease-spattered youths tinker with the engine of a new Toyota.

salt, *sb.*¹ Add: **2. d.** (Further examples.) Now freq. *with a pinch of salt*.

1948 F. R. Cowell *Cicero & Roman Republic* xvi. 243 A more critical spirit slowly developed, so that Cicero and his friends took more than the proverbial pinch of salt before swallowing everything written by these earlier authors. **1949** V. Grove *Language Bar* ii. 29 Even if we accept such a statement with a pinch of salt, it is an undisputable fact that its writer did look upon Latin as a guiding mistress. **1965** M. Shadbolt *Among Cinders* xxvi. 258, I take what he says with a half-pound of salt, after his review of that play. **1981** J. S. Bratton *Impact of Victorian Children's Fiction* ii. 41 We must take William Jones's enthusiasm about the eag erness of [tract] readers with a pinch of salt.

i. *to rub salt in one's wounds*: to behave or speak to someone so as to aggravate a hurt already inflicted.

1944 [see *curl *v.*¹ 1 c]. **1967** Wodehouse *Company for Henry* x. 182 He could see that Henry was deeply stirred, and he had no wish to rub salt in his wounds. **1973** *Guardian* 16 Feb. 13/8 Mr Nixon's treatment for war wounds is rubbing salt in them.

3. a. (Further examples.) Now also applied to a person or persons of great worthiness, reliability, honesty, etc.

1916 G. B. Shaw *Androcles & Lion* p. xv, They may not be the salt of the earth, these Philistines; but they are the substance of civilization. **1931** T. R. G. Lyell *Slang, Phrase & Idiom* 659 If he's a friend of yours, you're a lucky man, for if ever a fellow was one of the salt of the earth, he is. He's the best man I've ever met, in every way. **1948** E. S. Gardner *D.A. takes Chance* x. 103 Eve was a mighty fine girl, and her mother is the salt of the earth. **1951** E. M. Forster *Two Cheers for Democracy* i. 56 If you don't like people, kill them, banish them, segregate them, and then strut up and down proclaiming that you are the salt of the earth. **1953** Wodehouse

Performing Flea 78 You dine with the President on Monday, and he slaps you on the back and tells you you are the salt of the earth, and on Tuesday morning you get a letter from him saying you are fired. **1976** N. Thornburg *Cutter & Bone* vi. 148 And such *good* friends they were too. Real salt of the earth.

5. b. *salt of lemon*, potassium hydrogen oxalate, used to remove ink-stains and iron-mould from linen (earlier and later examples); *Carlsbad* (or *Karlsbad*), *Vichy salts*, salts prepared from the mineral springs in these places, or imitations of them; *Everitt's salt* (see quot. 1939); † *Preston salts*, a variety of smelling-salts.

1810 *New Family Receipt-bk.* 349 Essential Salt of Lemons. **1858** P. L. Simmonds *Dict. Trade Products*, *Preston-salts*,.. smelling-salts.. containing carbonate of ammonia in small pieces, with a drachm of the following mixture added, viz. oils of bergamot, cloves, and lavender, and the strongest solution of ammonia. **1866** *Chambers's Encycl.* VIII. 453/2 The celebrated Preston smelling-salts are scented with oils of cloves and pimento. **1868** *Ibid.* X. 75/2 Ink-stains.. require to be taken out with.. the essential salts of lemon. **1890** Billings *Med. Dict.* I. 482/1 *Everitt's salt*, a compound of cyanide of iron and potassium, formed when potassium ferrocyanide is decomposed by sulphuric acid. **1895** *Army & Navy Co-op Soc. Price List* 15 Sept. 696/1 Carlsbad Salts. *Ibid.* 710/2 Vichy Salts, Effervescing. **1901** *To-Day* 1 Aug. 38/1 'Eisiklene Hat Wash', which I find far superior to oxalic acid, salts of lemon, or any of the usual articles used for the purpose. **1908** *Chem. Abstr.* II. 3126 Artificial crystallized 'Karlsbad salts' as sold on the market is really impure Na_2SO_4. **1939** *Thorpe's Dict. Appl. Chem.* (ed. 4) III. 471/2 Ferrous potassium ferrocyanide, $K_2Fe[Fe(CN)_6]$, (Everitt's salt) is produced by heating saturated potassium ferrocyanide solution for 40 hours at 90°C with an equal volume of 20% sulphuric acid. **1960** *Chem. Abstr.* LIV. 8120/2 Hexametaphosphate.. combined with 34% Vichy salts.. gives a detergent which restores the original whiteness of superpolyamide textiles. **1977** *Martindale's Extra Pharmacopoeia* (ed. 27) 1459/1 Artificial Carlsbad Salt... A crystallised preparation of sodium sulphate 55, potassium sulphate 1, sodium chloride 10, and sodium carbonate 35. *Ibid.*, Artificial Vichy Salt. Anhydrous sodium sulphate 40, anhydrous sodium phosphate 20, potassium bicarbonate 35, sodium chloride 75, sodium bicarbonate 830.

c. (*a*) (Earlier example.)

1741 Richardson *Pamela* II. 247 Mrs. Jewkes held her Salts to my Nose, and I did not faint.

(*b*) Also, *like a dose of salts*: see *dose *sb.* 2 c.

6. Also, † *ethereal salt*, an ester.

1876 *Encycl. Brit.* V. 553/2 The thio-acids also form ethereal salts. **1905** Goodchild & Tweney *Technol. & Sci. Dict.* 633/2 Salts like ethyl acetate, derived from an organic acid and an alcohol, or from an alcohol and an inorganic acid, are called ethereal salts or esters.

12. a. *salt-spoon* (earlier example), *-spoonful* (earlier example), *-warehouse*. **b.** (objective, etc.) *salt-burner*; *salt-blue, -bright, -caked, -eaten, -free, -licked, -strewn, -tanged, -wavy, -white* (further examples), *-worn* adjs.

1922 Joyce *Ulysses* 51 Brown eyes saltblue. **1930** E. Pound *XXX Cantos* xvii. 79 And in her hands seawrack Salt-bright with the foam. **1910** G. T. Zoëga *Conc. Dict. Old Icelandic* 346/2 [*Salt*]-*karl*,.. salt-burner. **1975** C. Fell tr. *Egil's Saga* iv. 5 Those who worked in the forests and the salt-burners and all those who hunt.. had to pay his taxes. **1903** J. Masefield *Ballads* 19 Dirty British coaster with a salt-caked smoke stack. **1916** Joyce *Portrait of Artist* (1969) iv. 170 Picking a pointed salteaten stick out of the jetsam among the rocks, he clambered down the slope of the breakwater. **1909** *Practitioner* Dec. 867 When nephritis occurs, the child is given milk for some days, and then a salt-free diet, or at least one poor in salt. **1977** J. Cheever *Falconer* 49 A salt-free diet.. no salt added. **1962** A. Sampson *Anat. of Britain* xvi. 264 In the past the air force has been led by aviators, as the navy has been led by saltlicked admirals. **1820** M. Edgeworth *Let.* 4 June (1979) 144 Salt spoons never to be seen. **1837** Dickens *Pickw.* xlviii. 518 Tom Smart beat him in the drinking by about half a salt-spoon-full. **1892** W. B. Yeats *Countess Kathleen* i. 24 My curse upon the salt-strewn road of monks. **1933** W. de la Mare *Fleeting* 119 This wide salt-tanged vast of air. **1883** 'Mark Twain' *Life on Mississippi* xli. 423 The old brick salt-warehouses clustered at the upper end of the city. **1912** E. Pound *Ripostes* 27 That I on high streams The salt-wavy tumult traverse alone. **1922** Joyce *Ulysses* 50 A corpse rising saltwhite from the undertow, bobbing landward, a pace a pace a porpoise. **1961** A. Sillitoe *Key to Door* xxvii. 426 Water foamed into salt-white patches below the stern. **1921** W. de la Mare *Veil* 78 And the ocean water stirs In salt-worn casemate and porch.

c. salt bath, a bath of a molten salt or salts, as used in annealing; **salt bridge** *Chem.*, (*a*) a tube containing an electrolyte (freq. in the form of a gel) which provides electrical contact between two solutions; (*b*) a structure linking parts of a large molecule by means of a polar bond; *spec.* one formed between an acidic and a basic group; **salt-burn** = **salt-sore*; **salt bush** (further examples); also found in arid regions elsewhere; **salt-cake,** (*a*) (earlier example); **salt cedar,** a tamarisk, *Tamarix gallica*, growing as a shrub or small tree in warm parts of the United States; **salt dome,** a dome-shaped geological structure formed around and over a salt plug, often the source of oil or other minerals; also, a salt plug; **salt glaze:** also *transf.*, ceramic objects to which salt glaze has been applied; hence as *v. trans.*; **salt lake** (see main entry below); **salt-like** *a.*, *spec.* in *Chem.*, ionic; applied esp. to those hydrides which contain the anion H⁻; **salt mine,** also *joc.* (esp. in *pl.*) with allusion to the practice of sentencing offenders to labour in a salt mine; *spec.* one's work or place of employment; **salt plug,** an approximately cylindrical mass of salt, typically a mile in diameter and several miles deep, which has been forced upwards by subterranean pressure, distorting the overlying strata and forming a salt dome; **salt-raker** (earlier example); **salt-shaker** *U.S.* = *salt-sprinkler*; **salt sore,** a sore caused by exposure to salt water; **salt-spreader,** a vehicle that spreads salt on roads in order to melt snow and ice; hence **salt-spreading** *vbl. sb.* and *ppl. a.*; **salt tablet,** a tablet of salt that is swallowed, usu. to replace salt lost in perspiration.

1913 *Lockwood's Dict. Mech. Engin.* (ed. 4), *Salt bath furnace*, a type of hardening furnace in which the temperature is regulated by the employment of fused salts. **1925** *Jrnl. Iron & Steel Inst.* CXI. 536 The purification of fused salt baths composed of equal parts of sodium and potassium chlorides by the addition of boric acid and charcoal is also dealt with. **1980** *Railway Gaz. Internat.* Jan. 59/2 Molten salt bath nitriding and induction hardening caused bore distortion. **1915** *Jrnl. Amer. Chem. Soc.* XXXVII. 2781 Bjerrum's method of extrapolation.. is to add, to the voltage obtained by using 3·5 *M* KCl as a salt bridge, the difference between this voltage and that obtained by using 1·75 *M* KCl as the salt bridge. **1929** H. T. S. Britton *Hydrogen Ions* viii. 109 These two solutions are connected through the 'salt bridge', a narrow inverted U-tube, containing saturated KCl solution. **1965** *Jrnl. Molecular Biol.* XIII. 656 This arrangement would allow the α-amino group of one β-chain to form a salt-bridge with the α-carboxyl group of its symmetrically related partner, resulting in the formation of two salt bridges on either side of the dyod axis. **1978** P. W. Atkins *Physical Chem.* xii. 347 Another way of eliminating the junction potential is to connect the two half-cells with a salt bridge formed by dissolving potassium chloride in a water-soluble jelly. **1978** *Nature* 23 Nov. 362/1 Protein subunits in the two layers of the disk of tobacco mosaic virus have very similar conformations. Much of the bonding between subunits is polar, including salt-bridge systems. **1917** D. H. Lawrence *Look! We have come Through!* 37 Nevertheless, once, the frogs, the globe-flowers of Bavaria, the glow-worms Gave me sweet lymph against the salt-burns. **1901** M. Franklin *My Brilliant Career* xxii. 185, I listened with interest to stories of weeks and weeks spent.. crossing widths of saltbush country. **1909** Coulter & Nelson *New Man. Bot. Rocky Mts.* (ed. 2) 165 Atriplex L. Saltbush. Orache. **1911** C. E. W. Bean *'Dreadnought' of Darling* xv. 144 The grass might die off and the salt bush wither up. **1936** I. L. Idriess *Cattle King* ii. 10 He had never seen saltbush before. He felt strangely attracted by this little grey bush; its sombre colouring typical of the area. **1940** E. C. Jaeger *Desert Wild Flowers* 53 It [*sc.* the hoary saltbush] is one of the most widely distributed of American saltbushes. **1944** *Living off Land* ii. 42 Lucerne leaves, nettles, saltbush and milk thistles can all be used as substitutes for spinach. **1973** *Stand. Encycl. S. Afr.* IX. 480/1 Several species of *Atriplex*.. are known as saltbush. *c* **1702** C. Fiennes *Journeys* (1947) I. 49 The thinner part [of the salt] runns through on Moulds they set to catch it which they call Salt Cakes. **1881** *Harper's Mag.* Apr. 731/1 Salt cedars and stunted live-oaks.. were the only trees growing from the thin soil. **1973** *Tucson* (Arizona) *Daily Citizen* 22 Aug. 58/3 We wound up tramping.. through the mud and salt cedars. **1908** *Science* 28 Feb. 348/1 The expansive force of the salt from the crystallizing source will be very circumscribed and the salt domes local in character. **1928** E. R. Lilley *Geol. Petroleum & Nat. Gas* xvi. 376 The salt dome.. is known in areas where it does not appear to be associated with oil. **1945** M. F. Glaessner *Princ. Micropalaeont.* ix. 232 Lower Tertiary, Cretaceous, and Upper Jurassic microfossils (foraminifera and ostracodes) have been described from the salt-dome area.. between the northern shore of the Caspian sea and the southern foothills of the Ural Mountains. **1964** W. C. Putnam *Geol.* vi. 134/1 Many of the Gulf Coast salt domes are crowned with an irregular covering of limestone, anhydrite, gypsum, and occasionally suphur, termed the cap rock. **1976** *Billings* (Montana) *Gaz.* 5 July 4-A/2 It not only will transmit needed crude oil to the Midwest, it also will make usable the vast salt domes of the Williston basin for strategic storage of crude. **1967** M. Chandler *Ceramics in Mod. World* ii. 52 Porous drainpipes are still often salt-glazed, a process that is unique among glazing processes. **1968** J. Arnold *Shell Bk. Country Crafts* 236 The studio potters produce various kinds of terracotta.. and saltglaze. **1977** *Ashmolean Mus. Rep. Visitors 1975–76* 23 A selection of white salt-glaze from the Church bequest. **1928** *Chem. Abstr.* XXII. 3343 (*heading*) Salt-like hydrides. **1952** D. T. Hurd *Introd. Chem. Hydrides* iii. 23 The salt-like hydrides are very susceptible to hydrolysis in aqueous solution. **1965** Phillips & Williams *Inorg. Chem.* I. xvii. 619 The non-interstitial carbides are, in some senses, intermediate in character between the metallic interstitial carbides and the reactive salt-like carbides. **1963** *Times* 13 May 3/1 Rhodes is back in favour after a year or two in the saltmines for throwing. **1966** L. Deighton *Billion-Dollar Brain* xvii. 186 We finished our milk. 'Back to the salt mines,' said Harvey. **1975** B. Garfield *Hopscotch* xxvii. 281 I'd better get back to the salt mines. I've got a lot of unfinished jobs. **1977** *Listener* 10 Nov. 616/2 Harding was summoned by Sir John Reith and.. sent off to the salt-mines of Manchester. **1918, 1944** Salt plug [see *plug *sb.* 2 l (ii)]. **1967** M. T. Halbouty *Salt Domes* vi. 87/2 Oil and formation waters migrated from sediments surrounding the salt plug and were trapped in porous sections of the cap rock. **1837** A. Mallory *Let.* 20 Apr. in J. J. Audubon *Ornith. Biogr.* (1839) V. 257 Several of the fishermen, and salt-rakers,.. frequent the keys to the windward of this place. **1895** Salt shaker [see *pepper shaker* s.v. *pepper *sb.* 5]. **1931** W. Cather *Shadows on Rock* ii. i. 50 His ragged jacket was as much too tight as the trousers were too loose, and this gave him the figure of a salt-shaker. **1977** B. Rouéché *Fago* (1978) i. iv. 72, I.. picked up the kitchen salt shaker and rubbed it clean. **1908** N. Duncan *Every Man for Himself* v. 140 [*Armenian loq.*] An' thee salt-sores from thee feeshin' is on thee han's. **1979** F. Forsyth *Devil's Alternative* 7 Those parts submerged in sea water soft and white between the salt-sores. **1951** *Sun* (Baltimore) 21 Dec. B32/6 The Board of Estimates is expected to approve today the purchase of 25 latest-type salt spreaders. **1962** *B.S.I. News* Feb. 8/1 One London council whose salt-spreading was hindered because supplies had become 'rock-hard'. **1962** *Times* 27 Nov. 13/2 For the motorways, a fleet of snow ploughs and heavy salt-spreading vehicles is at constant readiness, day and night. The salt-spreaders can cover the whole of the M.1 at 40 to 50 m.p.h., within an hour. **1944** *Living off Land* v. 102 The cure is a pinch of salt, or one of the salt tablets now provided for the purpose, on the back of the tongue before each drink. **1976** A. Price *War Game* ii. iv. 230 A heavy leather buff-coat.. trapped the sweat and delayed the dehydration... So even though the salt tablets.. were necessary, the discomfort was endurable.

SALT (sǫlt). Also **S.A.L.T., Salt.** [Acronym f. the initials of *S*trategic *A*rms *L*imitation *T*alks.] Negotiations, involving esp. the U.S.A. and the Soviet Union, aimed at the limitation or reduction of nuclear armaments. Freq. *attrib.*

The last element, which is freq. redundant in *attrib.* uses, is also understood as *Treaty*.

1968 Mrs. L. B. Johnson *White House Diary* 1 July (1970) 693 When and where the talks would start, we do not know. They are being referred to as Strategic Arms Limitation Talks (SALT). **1969** *New Scientist* 14 Aug. 314/2 The progress of SALT is likely also to be slow. **1972** *Guardian* 6 June 4/4 The Secretary for Defence.. told Congress today that the United States could not afford to relax its defence effort in spite of the SALT agreement. **1973** E. Osers tr. *Waldheim's Austrian Example* xv. 196 The first ceilings set by the Salt Talks may prove to be an important landmark in limiting the arms race. **1975** *Daily Tel.* 23 Sept. 14/3 Whether the SALT discussions were a success or not is a matter of embittered controversy. **1976** *Survey* Summer-Autumn 24 The need for a further agreement in SALT remains paramount, given the threat to human survival posed by the nuclear arms race. **1979** *Sci. Amer.* Feb. 30/1 As the Senate prepares to debate the ratification of the new treaty emerging from the second round of strategic-arms-limitation talks (S.A.L.T. II) between the two superpowers.

salt, *a.*¹ Add: **1. a.** *salt finger*, one of a number of alternating columns of rising and descending water produced when a layer of water is overlain by a denser, more salty layer; so *salt fingering*, the occurrence of salt fingers; *salt spray*, used *attrib.* to denote a test in which an article is subjected to a spray of salt water, and the associated apparatus.

1918 *Proc. Amer. Soc. Testing Materials* XVIII. 237 (*heading*) Method of making the salt-spray corrosion test. **1945** *Electroplated Coatings of Nickel & Chromium on Steel & Brass* (*B.S.I.*) 18 Salt spray cabinet. **1962** *B.S.I. News* Feb. 18/2 A frequently-used test for determining resistance to corrosion is the salt spray test. **1967** *Deep-Sea Res.* XIV. 599 The opposite situation of a stable temperature gradient made unstable with a little salt leads to the formation of 'salt fingers'. *Ibid.* 606 Because of salt fingering, salt will escape across the bottom of this layer faster than heat. **1970** *Materials & Technol.* III. ix. 704 Exposure to a continuous mist of salt water, the so-called salt-spray test,.. does not truly simulate atmospheric exposure. **1977** *Sci. Amer.* Oct. 147/1 The warm salty water of the Mediterranean sets up the conditions for salt fingering as it flows through the Straits of Gibraltar and over the fresher, cooler waters of the Atlantic. **1978** J. A. Knauss *Introd. Physical Oceanogr.* ix. 187 It would appear that.. at least some of the microstructure in the ocean is caused by salt fingers.

c. Also *salt flat*, a flat expanse of land covered with a layer of salt; *salt meadow* (chiefly *N. Amer.*), a meadow liable to be flooded by salt water (further examples).

1656 *New Haven* (Connecticut) *Town Rec.* (1917) I. 288 It was don.. by the cattell hurrying downe in to ye salt meddows. **1789** J. Morse *Amer. Geogr.* 287 There are large bodies of salt meadow along the Delaware. **1873** J. L. Crawford in D. Eagan *6th Ann. Rep. Commissioner of Lands, Florida* (1874) 97 Hundreds of salt-works were erected upon the 'salt-flats' along the sea-shore within the limits of Wakulla. **1881** *Harper's Mag.* Jan. 254/2 The sluggish river winds through tracts of salt-meadow. **1931** *Amer. Speech* VII. 5 Sometimes the hunter found that he could make his best 'killings' at the 'salt licks' or 'salt flats' frequented by the buffalos. **1952** E. F. Davies *Illyrian Venture* i. 20 Why was the plain white? Was it snow? No, it looked more like salt flats. **1966** T. H. Raddall *Hangman's Beach* iii. xix. 286 A fringe of farms and salt meadows along the shore. **1972** *Guinness Bk. Records* (ed. 19) 128/2 The highest speed attained by any wheeled land vehicle is 631·368 m.p.h... on the Bonneville Salt Flats, Utah, on 23 Oct. 1970... The highest speed attained by a wheel-driven car is 429·311 m.p.h... on the salt flats at Lake Eyre, South Australia, on 17 July 1964.

2. a. (Further examples.) *salt rising*: see RISING *vbl. sb.* 15 in Dict. and Suppl.; *salt side* (U.S.), salt pork (cf. SIDE *sb.*[1] 3).
1892 O. WISTER *Jrnl.* 25 Nov. in *Out West* (1958) 143 We fried some bread..and I cooked some salt side. **1961** *Amer. Speech* XXXVI. 266 The term *salt side* is probably a similar blend of Northern *salt pork* and Midland *side meat*, terms for bacon.
in phr. used attrib. or Comb. **1747** H. GLASSTONE *Art of Cookery* ix. 114 A Salt-Fish Pye. Get a Side of Salt-Fish, lay it in Water all Night [etc.]. **1966** M. WOODHOUSE *Tree Frog* x. 76 We fought our way through thick salt-beef sandwiches.

b. *salt horse* (earlier example); also *transf.*, a naval officer with general duties.
1836 MARRYAT *Mr. Midshipman Easy* III. i. 11 Why you stay in Midshipman berth—eat hard biscuit, salt pig, salt horse? **1914** F. T. JANE *Navy as Fighting Machine* viii. 69 A non-specialist officer (known colloquially as 'salt horse') serves as a watch-keeper. **1917** 'TAFFRAIL' *Sub.* v. 115 Next came Lieutenant Hinckson, the senior 'salt horse', two and a half striped Lieutenant. **1946** J. IRVING *Royal Navalese* 149 *Salt horse, A*, an officer who has not specialised in gunnery, torpedo, etc. and does not intend to. **1957** D. MACINTYRE *Jutland* ii. 33 Here was a simple 'salt-horse', indeed, and such were not often selected, in time of peace, for the higher ranks of the Service. **1960** J. BISSET *Commodore* 17 Officers in big ships called destroyer-officers 'salt horses'—meaning non-specialists, a term of disdain.

3. b. Also *salt grass* (U.S.), one of a number of grasses growing in salt meadows or dry plains, esp. *Distichlis spicata* and several species of *Spartina* (earlier and later examples); *salt hay* (U.S.) hay made from salt grass (earlier and later examples).
1648 in *Mass. Hist. Soc. Coll.* (1852) 4th Ser. I. 204 Salt hay and fresh there thousands are of acres I do deeme. **1704** *Early Rec. Providence, Rhode Island* (1894) V. 224 The which sd Cove is a place of Salt Grass called Thatch. **1732** J. HEMPSTEAD *Diary* 23 Sept. (1901) 252, I went to Mamacock & fetcht a L[oa]d of Salt hay alias Rushes. *a* **1816** B. HAWKINS *Sk. Creek Country* (1848) 43 Such is the attachment of horses to this moss, or as the traders call it, salt grass. **1843** *Knickerbocker* XXII. 34 Range your eye along the summits of the salt hay-stacks. **1910** J. HART *Vigilante Girl* xxv. 350 The little stream..ran from the spring through bunches of salt grass. **1952** L. & J. BUSH-BROWN *America's Garden Bk.* (ed. 2) xii. 446 Salt hay is one of the most satisfactory materials mentioned [for winter mulching]. **1972** R. G. KAZMANN *Mod. Hydrol.* (ed. 2) v. 175 Salt grass will survive when the water table is as much as 12 ft below the land surface.

salt, *v.*[1] Add: **1. b.** (Earlier and later examples.)
1849 N. P. WILLIS *Rural Lett.* viii. 355 'Calm as the shadow of a rock across the foam of a cataract', would be a neat thing to 'salt down' for Calhoun or Van Buren. **1931** *Kansas City* (Missouri) *Star* 19 Sept. 12/5 It is a well known fact that all gamblers salt away their ill-gotten gains and die inordinately rich. **1952** *New Statesman* 17 May 578/2 Many palms itched for the millions that the Nationalists had salted away. **1959** *Times* 22 Apr. 8/4 Undisclosed profits were 'salted away' in banks in Eire and Rhodesia. **1966** *Economist* 9 Apr. 172/3 Members of previous governments, some of them now restricted to their homes, have salted away enormous sums of hard currency in foreign banks during their period of office. **1974** *Socialist Worker* 26 Oct. 3/1 The press, the experts and the pontificators see nothing wrong or hypocritical in the fact that the Banks can salt away these millions and make still more in this time of crisis.

d. Also in active use: *trans.*, to render (an animal) immune by inoculation; *intr.* of an animal: to become immune by suffering a disease.
1898 *Cape of Good Hope Agric. Jrnl.* 9 Jan. 6 The expression *to salt a beast* means to render the animal immune to the disease, to immunize him. **1906** *Rep. Brit. Assoc. Adv. Sci. 1905* 545 Dr. Edington..reports that..by inoculating mules with Heart-water blood he has been able to salt them against Horse-sickness. **1912** *S. Afr. Agric. Jrnl.* July 54 All farmers agree that cattle which recover [from Lamziekte] do not *salt* from the disease, in other words, there is no immunity.

2. c. Also, to sprinkle (a roadway) with salt in order to melt snow or ice (later example).
1977 *Oxford Jrnl.* 2 Dec. 12/4 Roads will only be salted when it is absolutely certain a cold snap is on the way.

3. Delete *Obs.* and add later examples.
1747 H. GLASSTONE *Art of Cookery* i. 3 Never salt your roast Meat before you lay it to the Fire, for that draws out all the Gravy. **1882** MME. BOUCHARD *How to live on Nothing* 17 All roasts should be peppered as well as salted, very little flour dredged over, and they should be served with a thick gravy. **1931** E. WEIR *When Madame Cooks* v. 55 After cleaning the fish..Salt and pepper the inside of each half and then grill them like a steak. **1965** *New Statesman* 5 Nov. 692/3 He..took up his knife and fork. He carefully salted his egg.

5. b. *U.S. colloq.* To reprimand or dress *down*.
1904 *Springfield* (Mass.) *Weekly Republ.* 9 Sept. 6 Senator Depew salts down William Allen White, who has stated that the senator tried to bully the president. **1913** J. LONDON *Valley of Moon* viii. 61 You're too fresh to keep... You need saltin' down.

6. c. Also more generally in *Chem.*, to reduce the solubility of, or precipitate (an organic substance) by adding an electrolyte to the solution; similarly *to salt in*, to increase the solubility of (an organic compound) by adding an electrolyte to the solvent.
1928 [see *CORTIN]. **1933** *Chem. Rev.* XIII. 91 There are numbers of cases in which the addition of certain salts increases the solubility of particular non-electrolytes causing them to be 'salted in'. **1939** *Thorpe's Dict. Appl. Chem.* (ed. 4) III. 286/2 The power of these electrolytes in 'salting out' organic compounds from their solutions. **1966** MAHLER & CORDES *Biol. Chem.* ii. 58 From the data in this figure, it is clear that at low ionic strengths the protein is salted in and at high ionic strengths the protein is salted out.

d. For 'cattle' read 'livestock' and add earlier and later examples. *N. Amer.*
1783 'J. H. ST. JOHN DE CRÈVECŒUR' *Sk. 18th-Cent. Amer.* (1925) 111 We..salt our cattle regularly once a week... From the horses to the sheep everyone must have a handful given them. **1819** E. DANA *Geogr. Sk. Western Country* 234 It is rare in this country that cattle are either fed, salted, or sheltered. **1852** [see *lick-log* s.v. *LICK *v.* 8]. **1931** *Amer. Speech* VI. 359 The absence of a salt sage diet on the summer range necessitates 'salting mutton'... Every second or third day one or two fifty-pound sacks of salt for every fifteen hundred sheep will be emptied into 'salt troughs' on the 'bed grounds'. **1968** R. M. PATTERSON *Finlay's River* 240 The packer..decided to leave those two [horses] here on the meadows to fill up and recuperate. He would salt them here.

8. (Further example.)
1977 *New Yorker* 29 Aug. 54/3 That made it easy for me to salt their accounts, and that's what I did. I began putting checks from company accounts into their personal accounts, and from there into oblivion via dummy companies.

9. (Earlier and later examples.) Also *transf.* and *fig.*
1852 in *Pioneer* (San Francisco) (1855) Mar. 146 The quicksilver which was procured at the Ranch, for the testing of the quartz, the victims declared was 'salted'; and they accused the *Rancheros* of conniving at the fraud. **1863** W. H. GOODE *Outposts of Zion* III. 415 The grounds have been 'salted'—gold dust scattered to deceive. **1880** *Harper's Mag.* Dec. 88/1 The deacon had stuck in a bit of Scriptur so's to salt it like. **1924** G. B. STERN *Tents of Israel* vii. 114 The Nong-Khan mine had been cleverly 'salted'... Only spinel sapphires, of practically no value, were to be found in it. **1951** *Times* 13 Dec. 4/6 (*heading*) Gold samples 'salted'. **1966** W. S. RAMSON *Austral. Eng.* vii. 148 One interesting and now probably obsolete expression is *to salt a claim*, meaning 'to sprinkle salt over the dirt', the salt having the appearance of gold-dust and giving the impression that the miner concerned has 'struck it rich'. **1968** A. S. ROMER *Procession of Life* xviii. 296 The gravel pit it would seem, was 'salted' by someone (? Dawson) with specimens to be later excavated as seeming authentic fossils. **1977** J. B. HILTON *Dead-Nettle* ii. 20, I shall want to see some evidence that there really is a seam. No salting it, no faking..your first job is to collect your showing.

salta (sæ·ltă). [f. L. *saltāre* to leap, perh. imitating HALMA.] A game played on a checkerboard of 100 squares by two persons with fifteen pieces each, with the object of occupying the opponent's side of the board.
1901 *Daily Express* 23 Mar. 8/7 Salta is played on a board of 100 squares, each player having fifteen pieces. *Ibid.*, Like in the first international salta tournament.., a chess master has again held his own against the draughts and salta experts who competed. **1904** E. B. TWEEDIE *Behind Footlights* viii. 153 She [*sc.* Sarah Bernhardt]..plays Salta with her son. This game is a kind of draughts. **1969** R. C. BELL *Board & Table Games* II. iii. 59 Salta was invented about 1900, and is played on the black squares of a continental draughtsboard of 10 × 10 squares.

sa·lt-and-pe·pper, *a.* Applied to things and materials (esp. hair) which are of two or more colours, one being light. Cf. PEPPER-AND-SALT in Dict. and Suppl. Also applied *transf.* to places, schemes, etc., in which black and white persons are mixed. orig. and chiefly *U.S.*
1915 *Sat. Even. Post* 2 Jan. 8/3 Hattie Krakow ran her hand over her smooth salt-and-pepper hair. **1959** *Wall St. Jrnl.* 12 Aug. 19/2 Houston is considering the 'salt and pepper' plan which has been widely suggested but not yet used. It calls for initial integration in schools where there is least objection from parents and expansion into other areas later. **1966** J. S. COX *Illustr. Dict. Hairdressing* 130/2 *Salt and pepper hair*, a head of hair in which the hairs are of at least two different colours, one of which is white. **1971** *New Yorker* 21 Aug. 3 (Advt.), Braid-bound suit of salt and pepper tweed. **1972** *Ibid.* 23 Dec. 38/3 Detroit is a salt-and-pepper situation. A great mix of black and white. **1973** M. AMIS *Rachel Papers* 52 In common with every American over eight and under twenty-five, he looked like a middle-aged American sports-writer: freckled pinhead, cropped salt-and-pepper hair. **1978** R. LUDLUM *Holcroft Covenant* vi. 79 Thick eyebrows, the coiled, matted hair an odd mixture of black and white. Salt-and-pepper eyebrows.

saltarello. Add: Pl. **saltarelli, -ellos. 1.** (Further examples.)
1876 STAINER & BARRETT *Dict. Mus. Terms* 383/2 Saltarelli are frequently found as movements or separate pieces in harpsichord and pianoforte music. **1928** E. CANZIANI *Through Apennines & Lands of Abruzzi* iv. 55 At Mascione, when the *saltarella* is danced, if lovers quarrel, the man or the woman kneels and asks, '*Cosa hai fatto?*' **1968** *Listener* 22 Aug. 249/3 The orchestra takes over from the voices and provides what might be regarded as a cue for dancing—as in the *saltarello* episode (in 'Sloth'). **1976** *Early Music* Oct. 457/1 Two of them are included on the Telefunken record already mentioned: the third of the manuscript's four saltarellos and 'Chominciamento di gioia'. **1980** *Ibid.* July 406/3 Apart from some isolated examples which have appeared in various anthologies, including some of the saltarelli and the popular *Lamento di tristano*, transcriptions have been restricted to scholarly editions.

Saltash (sǫltæ·ʃ). The name of a fishing-port in Cornwall used *attrib.* in **Saltash luck** (occas. **catch**), a thankless or fruitless task that involves getting wet through. *Naut. slang.*
1914 'BARTIMEUS' *Naval Occasions* xxiii. 225 One of the securing chains wants tautening... 'Saltash Luck' for some one! **1946** J. IRVING *Royal Navalese* 149 *Saltash luck*, a wet and thankless task such as securing up a bower anchor's slips in a seaway with the forecastle streaming with spray. **1962** GRANVILLE *Dict. Sailors' Slang* 99/2 *Saltash catch or Saltash luck*, 'a wet arse and no fish'. This West Country phrase has long been in use at sea, both in the Royal and Merchant Navies... It is believed to have originated in the lucklessness of fishermen at Saltash near Plymouth who sit on the bridge and catch nothing but the tide.

saltate, *v.* Restrict *rare* to sense in Dict. and add: **2.** *Physical Geogr.* To move by saltation (sense *1 d); also *trans.* (causatively). Chiefly as **saltating** *ppl. a.* (further examples).
1941 R. A. BAGNOLD *Physics Blown Sand & Desert Dunes* viii. 104 The energy supplied to the saltating grains by the wind. **1961** N. D. OPDYKE in A. E. M. Nairn *Descript. Palaeoclimatol.* iii. 47 Millet seed sand grains..show very high sphericity and roundness values due to their mode of transport which tends to round off the individual grains while they are being saltated. **1969** *Nature* 23 Aug. 792/2 Larger particles may be moved, not by the wind itself, but by momentum exchange with saltating grains. **1976** R. C. SELLEY *Introd. Sedimentol.* vi. 172 In a situation such as a river channel,..gravel will be rolling along the bottom, sand will sedately saltate, and silt and clay will be carried in suspension.

saltation. Add: **1. d.** *Physical Geogr.* A mode of transport of hard particles over an uneven surface in a fluid stream (as a wind or river), in which they progress in leaps, and on falling to the surface either bounce up for another leap or impart their momentum to other particles which on rising are accelerated forward by the stream. Cf. *SALTATE *v.* 2.
1908 W. J. MCGEE in *Bull. Geol. Soc. Amer.* XIX. 199 Transportation may be regarded as the general movement of earth matter seaward by streams; it comprises carriage of material (*a*) in solution, (*b*) in suspension, and (*c*) in what may be denoted saltation. **1941** R. A. BAGNOLD *Physics of Blown Sand & Desert Dunes* ii. 20, I shall use the name 'saltation' for the motion of sand in air, but without prejudice to the question of whether or not the mechanism which causes the grain to jump from the surface is the same in the two fluids. In air it is certainly the impact of a grain with the surface; but this is rarely so in water. **1962** READ & WATSON *Introd. Geol.* I. iv. 206 The mechanisms of transport in the sea are similar to those already described in connection with rivers, namely, suspension, rolling and saltation. **1977** A. HALLAM *Planet Earth* 50 The sand grains suspended in the air are the smaller ones, movement of larger particles being along the ground by saltation—by a series of jumps.

3. *Biol.* **a.** A mutation, esp. one with marked effects on several characters.
The 'saltations' studied by de Vries (see quot. 1906) are now known to have been translocations, which in *Œnothera* with its unusual system of chromosomes lead to large phenotypic changes.
1870 [see sense 1 c in Dict.]. **1906** *Pop. Sci. Monthly* June 485 The name 'saltation', or in recent years 'mutation', has been applied to extreme fluctuation, the immediate cause of which is unknown. *Ibid.*, Experiments of Dr. Hugo de Vries on the saltations of the descendants of an American form of evening primrose. **1919** *Jrnl. Exper. Zool.* XXVIII. 381 In our opinion, the attempted distinctions between 'saltations', 'mutations', and 'variations of slight degree' have led rather to confusion of thought than to clearer thinking. To us these are all a single class, 'mutations'. **1930** R. A. FISHER *Genetical Theory Nat. Selection* vii. 163 Unless such a resemblance formerly existed a gradual mimetic evolution is precluded, and we should be forced to admit that the mimetic females arose as sports or saltations totally unlike their mothers. *Ibid.* 164 A single saltation from a male of the same species. **1963** E. MAYR *Animal Species & Evolution* xv. 435 The sudden origin of new species, new higher categories, or quite generally of new types by some sort of saltation has been termed macrogenesis.

b. Change of phenotype occurring within a fungal colony.
1922 F. L. STEVENS in *Bull. Illinois Nat. Hist. Surv.* XIV. v. 157 The existing differences in definition and usage of the term mutation, as also our very limited knowledge of cytological conditions in the genus Helminthosporium and our ignorance as to whether it has sexual stages, have led me to select the term saltation for the variations here discussed. **1926** *Ann. Bot.* XL. 223 Changes of a more lasting nature may be conceived as arising gradually as a response or adaptation to certain growth conditions, or by sudden jumps. The latter type of phenomenon, which is known to occur in a considerable number of fungal genera,..is usually described as a 'mutation', or more conservatively as a saltation. **1940** J. RAMSBOTTOM in J. S. Huxley *New Systematics* 414 The morphological range is often so great that a single saltation will give what would be considered as a new species. **1978** *Nature* 29 June 755/1 The common and poorly understood phenomenon of frequent

somatic variation in certain supposedly haploid fungi (saltation) may perhaps be due to the loss of extra chromosomes that had been acquired previously.
Hence **salta·tional** *a.*, of, pertaining to, or occurring by means of saltation.
1963 E. MAYR *Animal Species & Evolution* xv. 435 The reorganization of the gene pool, required for successful speciation, is (except in the case of polyploidy) never saltational. *Ibid.* 437 Some saltational postulates are based on the assumption of essentially invariant evolutionary rates. **1978** *Sci. Amer.* Sept. 41/1 Even T. H. Huxley..could not accept the gradual origin of higher types and new species; he proposed a saltational origin instead.

saltationist (sæltēɪ·ʃənist), *a.* and *sb. Biol.* [f. SALTATION + -IST.] **A.** *adj.* Of or pertaining to saltationism. **B.** *sb.* One who supports or advocates saltationism.
1954 R. A. FISHER in J. S. Huxley et al. *Evolution as Process* 93 Darwin's criticism of the saltationist theory of M. Mivart. **1978** *Sci. Amer.* Sept. 44/1 They were essentialists and saltationists, and they looked on mutation as the probable driving force in evolution. **1980** *Nature* 4 Dec. 430/1 T. H. Huxley himself was unable fully to accept Darwin's gradualism, and preferred the saltationist camp.
So **salta·tionism**, the theory that new species arise suddenly as a result of major mutations.
1975 KELLY & MCGRATH *Biology* vii. 213/2 DeVries..insisted that a new species could arise by the introduction of a single mutation in an organism. His theory, called saltationism.., has been disproved, with one exception. **1978** *Sci. Amer.* Sept. 41/1 Saltationism was also popular with such biologists as Hugo De Vries, one of the rediscoverers of Gregor Mendel's laws of inheritance. **1979** M. RUSE *Darwinian Revolution* ix. 249 There were scientific reasons why many favored saltationism.

saltative (sæ·ltătiv), *a. rare.* [f. SALTATE *v.* + -IVE.] = SALTATORY *a.* 2 a.
1829 [implied in SALTATIVENESS]. **1911** *Law Rep., King's Bench* I. 654 These Scotch sheep are of a peculiarly wandering and saltative disposition.

saltatory, *a.* and *sb.*[1] Add: **A.** *adj.* **2. a.** (Example in *Physical Geogr.*)
1908 *Bull. Geol. Soc. Amer.* XIX. 199 The coarser particles due to corrosion..and to washing move forward at ever varying rates in saltatory fashion, the variable or leaping movements arising largely in combinations of friction with inertia.

d. *Physiol.* Used to designate the mode of transmission in a myelinated nerve in which the nerve impulse 'jumps' from node to node.
1934 *Amer. Jrnl. Physiol.* CX. 308 The pictures could be accounted for if progression were saltatory and by a process such as Lillie (1925) has described as occurring in the iron wire model... Here, due to reactivation by eddy currents flowing around the segments, activity progresses in jumps from node to node and consequently is more rapid than in the simple model. **1949** *Jrnl. Physiol.* CVIII. 339 The finding..that a large decrease in node spacing can occur without a drop in conduction velocity is shown not to conflict with the theory of saltatory conduction. **1977** *Proc. Nat. Acad. Sci.* LXXIV. 211/1 Myelinated nerve conducts by transmission of electrical excitation from node to node through local electrical circuits. This 'saltatory' mode of conduction results from a discontinuity in the excitability properties of the axon: excitable regions (nodes) alternate with nonexcitable passive core conductors (myelinated internodes).

e. *Biol.* Of the movement of small particles within cells: proceeding in directed jerks.
1964 L. I. REBHUN in Allen & Kamiya *Primitive Motile Syst. in Cell Biol.* 503 Particles may at one time undergo Brownian movement and suddenly undergo a process converting this to sudden, discontinuous motion, i.e., saltatory motion. **1970** *Nature* 7 Feb. 559/1 It may well be..that microtubules in brain function in the saltatory transport of material and vesicles from their site of formation in the cell body to their site of utilization at the synaptic endings. *Ibid.* 5 Sept. 1006/2 Translocation has been pictured as a saltatory interaction between enzyme-containing vesicles and fibrous proteins, chiefly microtubules.

3. Biol. *saltatory replication*, a hypothetical evolutionary event in which very many identical copies of a short section of DNA are added to a genome.
1968 R. J. BRITTEN in *Carnegie Inst. Year Bk. 1966–7* 72/2 *Saltatory replications*, the hypothetical events by which families of hundreds of thousands of similar nucleotide sequences are produced in the DNA of an organism. ..Families are produced in a time short compared to the time required for their loss by divergence (a few hundred million years). **1968** —— & KOHNE in *Ibid.* 84/1 Events in which very many copies [of a DNA segment] are made in a short time interval (saltatory replication). Evidence is now available which clearly indicates saltatory replication. *Ibid.* 88/1 A saltatory replication producing 100,000 copies of the right sort of gene is a candidate for a genetic event with immense potentiality. **1970** *Nature* 12 Dec. 1043/2 Such gene expansion has been designated saltatory replication and is illustrated in Fig. 1 *C*.

salt-box. Add: **1. c.** *U.S.* Used *attrib.* or *absol.* to designate a kind of frame-house which resembles a salt-box in shape, having two storeys at the front and one at the back.
1876 J. S. INGRAM *Centennial Exposition* 717 One of the chief oddities of the Exhibition—the Hunter's Cabin. It was built of logs in the 'salt-box' style and entirely open in front. **1900** J. DE F. SHELTON *Salt-Box House* i. 17 Colloquially, it was called a 'salt-box house', its lines repeating those of the wooden salt-box that hung in the kitchen chimney. The ridge-pole was set far to the front, from which a short roof sloped..down to the outer line of the ceiling of the ground floor. **1934** *Sun* (Baltimore) 14 Aug. 10/6 The first of the salt boxes were almost always made by adding the lean-to to the two-room house. **1944** *Sat. Rev.* (U.S.) 2 Sept. 30/1 (Advt.), New England saltbox in scenic New York setting. **1952** F. ALLEN *Big Change* II. viii. 126 New England salt-box-type houses with attached garages. **1967** V. SILTER *Biltmore Call* 57 Some were remodelled farm houses..and some were old saltboxes..and some were just plain old country houses. **1976** *New Yorker* 22 Mar. 125/1 Cunningly combining painted backcloths, a two-story saltbox frame, and picturesque detail in the way of furniture and properties, Ming Cho Lee's decor for the six different settings was at once varied, realistic, and romantic.

salt-cellar. Add: **c.** *colloq.* Each of the pronounced hollows at the base of a thin neck. (Usu. with reference to young women.)
1870 O. LOGAN *Before Footlights* 26, I was a child of the most uninteresting age..a tall scraggy girl, with red elbows, and salt cellars at my collar-bones, which were always exposed, for fashion at that time made girls of this age uncover neck and arms. **1880** F. BELTON *Random Recoll. Old Actor* vi. 87 The bones of her elbows were painfully prominent, with enormous salt-cellar hollows in her neck. **1913** 'O. ONIONS' *Story of Louie* I. i. 25 The copper-haired girl with the long thin neck and the 'salt-cellars' showing through her white flannel blouse. **1913** *Queen* 17 May 35 (Advt.), 'Saltcellars' and thinness of the neck and shoulders. **1964** P. WHITE *Burnt Ones* 162 She was so thin, but he loved her even for her salt-cellars.

salt chuck. *N. Amer. colloq.* [Chinook jargon, f. SALT *a.*[1] + *CHUCK *sb.*[6]] In western Canada and north-western U.S.: the sea, the ocean.
1868 F. WHYMPER *Trav. Alaska* iv. 45 An Indian, paddling in his 'frail kanim' on the great 'salt chuck' or sea, was swallowed—canoe and all—by a great fish. **1874** C. HORETZKY *Canada on Pacific* 132 A thick heavy mist hung over the valley, completely hiding the Cascade range which we had now to enter and pass through before reaching the 'salt-chuck'. **1909** E. I. DENNY *Blazing the Way* I. vii. 120 The fish, of many excellent kinds, from the 'salt chuck', brought fresh and flapping to our doors, in native baskets by Indian fishermen. **1938** G. CASH *I like Brit. Columbia* 61 Unless you are camped near a log dump—which means where a logging company is dumping logs into the salt chuck—you have quite a time gathering enough. **1958** R. G. LARGE *Skeena* (ed. 3) x. 65 Sailing the salt-chuck easily, o'er an oft familiar route. **1964** L. LINTON *Of Days & Driftwood* iv. 24 Even the gulls, screeching over the gray saltchuck..were giving their last accolade to summer. **1975** *Islander* (Victoria, B.C.) 27 July 14/2 In 1905, most people lived close to the saltchuck and along Rainey Creek.
Hence **sa·lt-chucker**, a sea-water angler.
1958 in R. E. Watters *Brit. Columbia* 216 It is the spirit that counts, and that spirit extends to trying to make life happier for thousands of scattered salt-chuckers. **1963** *Sun* (Vancouver) 20 July 15/1 Now, however, with an average of almost two fish per short outing of a few hours each trip, I'm wearing the saltchucker's smug smirk.

salted, *ppl. a.* Add: **2. a.** Now used esp. of prepared foods, as *salted almond, peanut*, etc. (Further examples.)
1892 *Encycl. Pract. Cookery* I. 15/1 Salted and 'Devilled' Almonds. **1921** A. HUXLEY *Crome Yellow* xix. 202 Georgiana ate only an olive, two or three salted almonds, and half a peach. **1935** M. MORPHY *Recipes of All Nations* 775 Salted Green Peas, first cooked in cinders and then salted like almonds, are among Persian delicacies. **1954** 'R. CROMPTON' *William & Moon Rocket* iv. 85 Salted nuts..potato crisps..celery. **1970** E. DAVID *Spices, Salt & Aromatics in Eng. Kitchen* 231 Salted almonds, whatever the promises held out by the words vacuum-sealed or oven-fresh on tins and jars are not to be bought. **1972** A. MACVICAR *Golden Venus Affair* v. 49, I ordered a Pym's No. 1... We munched salted peanuts.

4. (Earlier and later examples.)
1864 T. BAINES *Explorations in S.W. Afr.* xv. 418 He asked carefully 'whether the horse was salted' (i.e. acclimatised by having recovered from the horse sickness). **1977** BUXTON & FRASER *Animal Microbiol.* II. xlviii. 634/1 Horses and mules that have recovered from a natural attack of horse sickness are generally more resistant to disease than other equines and are known as 'salted', as are animals that have survived for a number of years in badly infected areas without ever showing obvious signs of the disease.

5. (Earlier and later examples.)
1862 *California Mag.* Jan. 355/1, I lost my $2,000 by buying a 'salted' claim. **1949** *This Week Mag.* 15 Oct. 27/4 They are occasionally called upon by unscrupulous characters whose main object is to sell them a 'salted' mine.

saltery. Add: **3.** *N. Amer.* A factory where fish is prepared for storage by salting. Now chiefly *Hist.*
1903 *Sci. Amer. Suppl.* 21 Mar. 22751/3 During 1900 there was but one saltery operated solely as such in this district. It is situated on the Nushagak and had an output of 7,186 barrels of red-fish and 536 barrels of king salmon for the season. **1960** M. SHARCOTT *Place of Many Winds* viii. 132 A few bricks and a couple of rotted and barnacled pilings tell of a long-forgotten cannery or saltery. **1972** L. HANCOCK *There's a Seal in my Sleeping Bag* ix. 218 Alert Bay was once a small saltery to preserve salmon prior to shipment to Victoria.

‖ **saltimbocca** (sæltimbǫ·kă). [It., f. *saltare* to leap + *in* IN *prep.* + *bocca* mouth.] A dish consisting of rolled pieces of veal and ham cooked with herbs. Also in *Comb.*, as *saltimbocca (alla) Romana*.
1937 M. MORPHY *Good Food from Italy* 89 (*heading*) Veal and Ham à la Romana [Saltimbocca alla Romana]. **1959** *Good Food Guide* 224 Escalope Cordon Bleu, 'rather like a Roman Saltimbocca only deep fried in batter'. **1960** *Harper's Bazaar* Oct. 154/2 *Saltimbocca* combines paper-thin slices of veal with *prosciutto* and a sage leaf. **1969** G. GREENE *Travels with my Aunt* I. xiii. 126 He put a lot of *saltimbocca Romana* into his mouth. **1977** C. MCCULLOUGH *Thorn Birds* xvii. 447 I'll have pâté, some scampi and a huge plate of saltimbocca. **1978** *Chicago* June 237/1 [There are] half a dozen veal dishes (Saltimbocca alla Romana—with prosciutto, butter, herbs, and marsala—is a specialty), [etc.].

saltine (sǭ·ltīn). orig. and chiefly *U.S.* [f. SALT *sb.*[1] + -INE[4].] A salted cracker or thin crisp biscuit.
1907 *Grocery World* 4 Nov. 40/2 Crackers and cakes... Orange Cookies... Quaker City Mixed... Salted Strips... Saltines [etc.]. **1914** H. C. SHERMAN *Food Products* viii. 287 Crackers,..Pretzels... Saltines... Soda crackers (etc.). **1933** E. O'NEILL *Ah, Wilderness!* II. 63 Mrs. Miller. (as Norah comes back with a dish of saltines—begins ladling soup into the stack of plates before her). *c* **1938** *Fortnum & Mason Price List* 19/1 Southern American Biscuits..Saltines..2/3. **1958** E. S. WARNER *Silk-Cotton Tree* xvii. 177 The Head was passing around a box of soggy-looking saltines. **1969** 'E. LATHEN' *Murder to Go* xiv. 134 'Would anybody..like some crackers?'.. He delayed his own departure until the appearance of a dish of saltines. **1975** *New Yorker* 14 Apr. 104/3, I sought him out in his office at Hi Corbett Field (where he was lunching on two Cokes and some saltines crumbled into a cup of soup). **1980** R. L. DUNCAN *Brimstone* v. 89 Have my lunch brought in. Milk and saltines.

saltiness. (Later example in *fig.* sense.)
1934 A. WOOLLCOTT *While Rome Burns* 26 Hansoms have the advantage of semi-privacy, and what their drivers lack in chic they make up in saltiness.

salting, *vbl. sb.* Add: **2. a.** (Earlier and later examples in sense 9 of the vb.)
1856 *Santa Barbara* (Calif.) *Gaz.* 21 Feb. 2/5 The best yield I have seen is eighteen cents to the pan, and this was without any 'salting'. **1869** 'MARK TWAIN' *Lett.* (1917) I. 164 When it was discovered that those lumps were melted half dollars and hardly melted at that, a painful case of 'salting' was apparent. **1949** *Sun* (Baltimore) 31 Oct. 3/4 Farrell and others pointed out that 'salting'..along nine miles of river shore would be pointless and profitless. **1951** *Times* 13 Dec. 4/6, ——, works manager, of Malvern, Johannesburg, was found Guilty at the Rand criminal sessions to-day on two counts of *falsitas* in the 'salting' (fraudulent enrichment) of the basal and leader reef third deflection core samples of the Erdeel 5 mine. **1972** *Courier-Mail* (Brisbane) 20 June 5/3 (*heading*) Cutler denies nickel salting.

b. Chem. *salting in, out* (cf. *SALT *v.*[1] 6 c).
1857 [in Dict., sense 2 a]. **1905** *Jrnl. Physiol.* XXXII. 329 The only method which, according to our present knowledge, leaves proteids absolutely unaltered is that of 'salting out'. **1926** R. WRIGHT in *Jrnl. Chem. Soc.* 1203 The mutual lowering of solubility which takes place when an electrolyte and an organic substance are dissolved together in water..is the basis of the process of 'salting out' when an organic compound is driven out of aqueous solution by the addition of a salt. What may be termed 'salting in' is the reverse phenomenon, that is, a mutual increase in solubility of electrolyte and organic compound when added to the same solvent, which in this case is not pure water but aqueous alcohol. **1939** *Thorpe's Dict. Appl. Chem.* (ed. 4) III. 286/2 The 'salting out' effect of electrolytes on hydrophilic colloids is due to their dehydrating action as well as to their power of neutralising the charge. **1957** G. E. HUTCHINSON *Treat. Limnol.* I. ii. 183 Salting out of charged silt particles by water of compensation currents. **1970** A. L. LEHNINGER *Biochem.* vii. 133 Salts containing divalent ions..are far more effective at salting-in than salts such as NaCl, NH_4Cl, and KCl.

4. (sense 1) *salting-pan* (earlier example), *-shed*.
1816 JANE AUSTEN *Emma* II. iii. 43 My mother was so afraid that we had not any salting-pan large enough. **1889** W. B. YEATS *Wanderings of Oisin* 82 Times from the saltin' shed..I scarce could drag my feet. **1961** N. FROUD et al. tr. *Montagné's Larousse Gastronomique* 493/2 The fish is transported from the boat to the salting sheds.

saltire. Add: Also, a cross having this shape.
1970 H. BRAUN *Parish Churches* viii. 104 The 'saltire' or diagonal cross formed of two struts crossing, was nearly always formed of two serpentine timbers. **1974** *Northern Times* (Golspie, Sutherland) 2 Aug. 3/4 The gift was a saltire—a St. Andrew's Cross in blue and white with the arms of the cross outlined in gold thread.

salt lake. [SALT *a.*[1]] A saline lake, usu. one with no outlet to the sea so that salts brought in by rivers accumulate in it; *esp.* one which is not particularly alkaline (cf. *bitter lake* s.v. *BITTER *a.* 1 c).
1763 J. BELL *Trav. from St. Petersburg* I. 289 We set up our tents near a lake of brackish water, called Solonoy-Osera, or the salt lake. *Ibid.* 326 The 22d, we quitted the salt lake. **1836** *Penny Cycl.* VI. 343/2 A great number of smaller and larger salt lakes. **1882** [see *BITTER *a.* 1 c]. **1923** J. S. HUXLEY *Ess. Biologist* i. 34 A salt-lake shrimp could tolerate an even higher concentrate of brine. **1970** [see *EXOGENETIC *a.* 2 b].

saltlessness. (Later example in *fig.* sense.)
1867 QUEEN VICTORIA *Let.* 13 Feb. in R. Fulford *Your Dear Letter* (1971) 121 There is great bitterness in the constant depression..and total saltlessness of my life.

salt-lick. Add: (Further examples.) Also *fig.* Now chiefly *N. Amer.*
1859 J. PALLISER *Jrnl.* 16 Feb. (1863) 129 A splendid ram..had been caught by setting a snare in a path leading to a 'salt-lick'. **1922** *Beaver* May 7/2 They [*sc.* bighorn sheep] being in the habit of seeking the salt-licks early in the morning and again late in the evening. **1948** C. DAY LEWIS *Poems 1943–47* 75 The sea rolled up like a blind, oh pitiless light Revealing, shrivelling all! Lacklustre weeds My hours, my truth a salt-lick. **1965** R. MCKIE *Company of Animals* vii. 113 Jim went at first light to check which animals were visiting a small salt-lick in the jungle. **1976** N. THORNBURG *Cutter & Bone* viii. 202 Immediately she was weeping in his arms, her face a lovely saltlick to his mouth.

salt-marsh. Add: **b.** (Further examples.)
1862 *Harper's Mag.* Nov. 737/2 'Salt-marsh fly'—is a nuisance found everywhere..near salt marshes. **1932** *Sun* (Baltimore) 23 Aug. 4/7 The salt marsh mosquito causes intense discomfort. **1972** SWAN & PAPP *Common Insects N. Amer.* xxii. 592 Salt-marsh Mosquito: *Aedes sollicitans*. *Ibid.*, The California Salt-marsh Mosquito..breeds in salt marshes and tide pools along the Pacific coast.
c. *attrib.* in general use.
1937 *Discovery* Apr. 98/2 The occupation was brought to an end with the onset of salt-marsh conditions. **1960** J. J. ROWLANDS *Spindrift* 91 The salt-marsh hayfields are favorite stopping-places for geese and ducks on their northward flight. **1975** J. G. EVANS *Environment Early Man Brit. Isles* vii. 180 Later stages in the saltmarsh succession form good sheep pasture.

‖ **salto** (sa·ltǫ). [It., leap; cf. SALTUS.] **1. salto mortale** (mǫrtā·le) [It., = fatal jump, somersault], a daring or flying leap (as of a trapeze artist, etc.); also *fig.*, a step that involves risk; an unjustified inference, a 'leap of faith'.
1896 W. CALDWELL *Schopenhauer's Syst.* vii. 361 He really solved it [*sc.* the question of altruism] only by a *salto mortale*. *a* **1910** W. JAMES *Ess. Radical Empiricism* (1912) ii. 67 The transcendentalist..holds knowing to consist in a *salto mortale* across an 'epistemological chasm'. **1937** J. M. MURRY *Necessity of Pacifism* vii. 115 England will take this glorious *salto mortale* into a more human future. **1952** R. MANNING-SANDERS *Eng. Circus* IV. xvii. 237 Let us look..at the act of the two Codonas,..after Alfredo, in 1922, had mastered that wonderful feat, the *salto mortale*, or triple somersault. **1968** M. GUYBON tr. *Solzhenitsyn's First Circle* (1971) xxviii. 164 He was escorted up another flight of steps—where, as in a circus during the *salto mortale*, there were nets to catch him if he jumped off. **1977** *Language* LIII. 44 While the enclitic nature of the copula is beyond question in itself, inferring 'aphaeresis' as a phenomenon consequential upon it has involved an epistemological 'salto mortale' which has not been very successful.
2. *Gymnastics.* A somersault.
1972 B. TAYLOR et al. *Olympic Gymnastics* iii. 35 With more advanced movements (such as a double back salto or double twisting back layout), a spotting belt is sometimes used. **1974** *Rules of Game* 36 *Compulsory exercises*... Arms backward, two or three running steps into forward piked salto, land on one leg. **1980** *Sunday Times* 20 July 28/2, I will show you a new dismount off the beam... It is a double Salto off one leg with half a turn.

Saltoun (sǫ·ltən). [Proper name: see quot. 1886.] A variety of artificial trout fly (see quots.).
1886 F. M. HALFORD *Floating Flies* v. 90 Saltoun. *Wings.* Palest starling. *Body.* Black silk, ribbed with silver wire. *Hackle and Whisk.* Pale ginger cock. *Hook.* 00 or 000. A very useful summer fly, invented by and named after the late Lord Saltoun [prob. Alexander Fraser, 17th Lord Saltoun, d. Feb. 1886], a prominent member of the old Stockbridge Club. **1892** M. O. MARBURY *Favorite Flies* (ed. 2) 379 (*heading*) Frederic M. Halford's floating flies for dry-fly fishing... No. 223 Saltoun. **1926** *Chambers's Jrnl.* 13 Feb. 164/1 There's a two-pounder at the stream-mouth that has risen twice to the saltoun. **1931** *Hardy's Anglers' Guide* (ed. 53) 66 Lake and Sea Trout Flies... No. 44. Saltoun. **1961** A. C. WILLIAMS *Dict. Trout Flies* (ed. 3) 302 *Saltoun*, an old pattern and a one-time favourite on the chalk streams. After falling into disuse, it has been revived in recent years as a lake fly.

salt rheum. 2. For *U.S.* read *N. Amer.* and add: **a.** (Earlier and later examples.)
1809 E. KENDALL *Trav. Northern Parts U.S.* I. 325 In the neighbourhood, the greater number of patients that it attracts appear to be such as labour under scrofulous diseases. That, of which I heard the name in everyone's mouth, is the *salt rheum*. **1877** R. J. BURDETTE *Rise & Fall of Mustache* 291 'Centennial Cordial and American Indian Aboriginal Invigorator'..has absolutely no equal for the cure of..salt rheum. **1901** *Daily Colonist* (Victoria, B.C.) 26 Oct. 8/1 This preparation seems to have magnetic powers in stopping the dreadful itching, burning sensations of salt rheum and eczema.

salt river. *U.S.* [SALT *a.*[1]] †**1.** A river which is tidal a considerable distance from its mouth. *Obs.*
1659 *Early Rec. Providence, Rhode Island* (1892) I. 97 A percell of land..lieth upon the salt River at the furthermost side of the towne boundes. **1704** *Ibid.* (1894) V. 224 Sd Cove..lieth adjoyneing to the North side of the salt River called Pautuckett. **1791** W. BARTRAM *Trav. N. & S. Carolina* I. iv. 29 Numerous small rivers and their branches: these they call salt rivers, because the tides flow near to their sources.
2. a. The name of a river (perh. one in Kentucky) used as *attrib. phr.* to designate the inhabitants of the American backwoods region, esp. with reference to their uncultivated manner of speech. Also applied to the speech, etc., of these people. Now only *Hist.*
1828 *Western Intelligencer* (Hamilton, Ohio) 26 Dec. 1/4 A 'Salt River Roarer'. One of these two fisted backwoodsmen, 'half horse, half alligator, and a little touched with the snapping turtle'. **1835** *Knickerbocker* V. 403 They [*sc.* speeches in Congress] are chiefly made up of extracts from the common school collection of lessons for reading and speaking, sprinkled with scraps of dog-Latin, and a sort of patois, called Salt-river roaring. **1835** T. FLINT in *Athenæum* July 511/2 There is, in fact, a well-known rivalry between the collectors of the Downing dialect of New England, and the Crocket or Salt River dialect of the South and West. **1947** J. CONROY *Midland Humor* p. x, The ring-tailed roarers and Salt River screamers of the half-horse and half-alligator breed, both male and female, were ordinarily combinations of physical might and mother wit which enabled them to outsmart invaders from other regions.
b. *fig.* In slang phr. *to row* (someone) *up Salt River* and varr., to defeat (a political opponent); to overcome, send to oblivion. Also with intransitive vb., to be defeated or overcome, to go to oblivion; to get drunk. Freq. in allusive and proverbial uses. Now *rare*.
The simplest of the numerous explanations offered for this usage is that which connects it with sense 2 a; see H. Sperber and J. N. Tidwell in *Amer. Speech* (1951) XXVI. 241–7.
1828, etc. [see ROW *v.*[1] 8 a in Dict. and Suppl.]. **1830** *Cincinnati Chron.* 2 Jan. 1/2 He replied he didn't 'smoak me', and unless I cut cable in short order, he'd roar me up salt river. **1832** *Washington* (Ohio) *Herald* 17 Nov. 3/4 The Jackson boys of Ohio have been enabled to give them another ride 'up Salt River'. **1838** *Bentley's Misc.* IV. 588, I can drink till the world gets too old to move. While another man rows up Salt River, I'm only putting the fire out in the forest. *Ibid.*, Rowing up Salt River is a slang term for getting intoxicated. **1852** *Chicago Democrat* 11 Nov., One Thomas Holt, lately a clerk in The Chicago Post Office, when last seen,..was on his way up 'Salt River' with Gen. Scott. **1880** in J. C. Andrews *Pittsburgh Post Gaz.* (1936) xvi. 218 For Salt River—The River Boat Democracy left its Wharf Tuesday, Nov. 2, 1880 bound up Salt River in search of the late lamented Samuel J. Tilden. **1941** L. D. BALDWIN *Keelboat Age on Western Waters* 97 It'd shore be harder'n rowin' up Salt River to find a cleverer parcel o' fellers 'n them keelers.

salt spring. [SALT *a.*[1]] A flow of salt water or brine out of the earth; a brine-spring, brine-well. Also as *attrib. phr.*
1601 HOLLAND tr. *Pliny's Nat. Hist.* II. XXXI. vii. 416 In some parts of Spaine there be salt springs. *a* **1647** [see SALT *a.*[1] 1]. **1683** J. PETTUS *Fleta Minor* I. 321 Of Salt-Petre, Vitriol, Allum and Salt Springs. **1748** J. HILL *Hist. Fossils* 382 The Sea-water and Salt-springs sustain it [*sc.* alimentary salt]..in a liquid form. **1782** T. PENNANT *Journey Chester to London* 27 The Britons, who had, in several places, plenty of salt-springs. **1834** *Phil. Mag.* IV. 31 The comparative strength of the salt springs of that country at different depths. **1839** G. ROBERTS *Dict. Geol.*, *Salt Springs*, which contain a large quantity of common salt, obtained from them by mere evaporation. **1852** J. REYNOLDS *Hist. Illinois* 86 They discovered in the present county of Galatin, salt springs. **1853** *Trans. Mich. Agric. Soc.* IV. 9 The twenty-two sections of salt spring lands now unappropriated.

saltus. Add: Now used in unnaturalized Latin form (further examples). Also in *Comb.* Cf. *salto mortale* s.v. *SALTO 1.
1894 A. C. FRASER *Locke's Essay Annotated* II. IV. xii. 348 The inductive *saltus*, which transcends this datum. **1913** E. W. HOBSON *Squaring the Circle* ii. 18 There is no jumping to the limit as the supposed end of an essentially endless process, to be reached by some inscrutable *saltus*. **1923** G. B. SHAW in *Nation & Athenæum* 10 Feb. 714/2 He [*sc.* Wright] was hampered not only by the mistakes of Pasteur, but by a remarkable *saltus empiricus* made by a famous bacteriological acrobat..named Metchnikoff. **1934** A. C. EWING *Idealism* viii. 407 One can..pass from one to the other without a *saltus in aliud genus*. **1951** J. HOLLOWAY *Lang. & Intelligence* iii. 55 There must be a *saltus naturae*, an innate idea of symbolization must come to fruition.

salt water, *sb.* and *a.* Add: **B.** *attrib.* as *adj.* **a.** Also *salt-water taffy* (TAFFY[1], var. form of TOFFEE), a type of confectionery made chiefly from corn syrup and sugar, freq. sold at North-eastern (chiefly New Jersey) seaside resorts. *U.S.*
1894 *Official Gaz.* (U.S. Patent Office) 17 July 410/1 (*caption*) The representation of a four masted schooner with the words '*The Original Atlantic City Salt Water Taffy*'. **1910** H. T. PECK *New Baedeker* II. vi. 309 And there are also itinerant venders of every sort of edible..from 'salt-water taffy'..down to peanuts and 'hot dogs'. **1933** *Nat. Geogr. Mag.* May 520/2 Next to the visitor, Atlantic City's biggest 'Industry' is the making and shipping of 'salt-water taffy'. Legend says that in the early eighties a man had a candy stand on the beach. One day an unusually high tide splashed over a batch of old-fashioned, pulled taffy on a slab. Being an enterprising person, he told his customers that he had something new—'salt-water taffy'. **1954** W. RICHMOND *Choice Confections* xxi. 385 This formula produces a salt water taffy or kiss of very fine quality... The formula can be used for regular kiss-shaped pieces or long sticks of salt water taffy. **1960** J. J. ROWLANDS *Spindrift* 65 Through the grimy windows of the salt-water taffy counter you see the cold steel arms of the taffy puller motionless and empty-handed. **1979** *United States 1980/81* (Penguin Travel Guides) 48 Vermont cheese and maple syrup, salt-water taffy along the New Jersey shore..are all specialties of their respective regions.
c. *U.S.* and *W. Indies.* Used to designate a recent, usu. black, immigrant (see quots.).
1774 E. LONG *Hist. Jamaica* II. III. iii. 410 The Creole Blacks differ much from the Africans, not only in manners, but in beauty of shape, feature, and complexion. They hold the Africans in the utmost contempt, stiling them, 'salt-water Negroes', and 'Guiney birds'; but value themselves on their own pedigree. **1818** H. B. FEARON *Sk. Amer.* 93 If I had my will there should never be a salt-water man employed in the States. *a* **1820** B. H. LATROBE *Jrnl.* (1905) iii. 63 The ferryman..is one of several who are children of a man and woman, negroes, brought from Africa—called here salt-water negroes. **1855** F. DOUGLASS *My Bondage* 323 The salt water slave who hung in the guards of a steamer..has, by the publicity given to the circumstance, set a spy on the guards of every steamer departing from southern ports. **1961** F. G. CASSIDY *Jamaica Talk* viii. 156 A sort of half-way condition between the creole Negro and the salt-water Negro was the *salt-water Creole*—one born during the voyage to Jamaica. **1966** *Publ. Amer. Dial. Soc.* 1964 XLII. 39 Irish informants use *turkey* and *saltwater turkey* to designate a recent immigrant.

salty, *a.*[1] Add: **A.** *adj.* **3.** (Later example.)
1978 J. A. MICHENER *Chesapeake* 359 When Captain Turlock learned that his mate had studied with the rector, there was salty discussion of that churchman's habits.
4. *U.S. Naut. slang.* Of a sailor: tough; hard-bitten; aggressive. Cf. SALT *sb.*[1] 11.
1920 H. R. CHAMBERS *U.S. Submarine Chasers in Mediterranean* ii. 12 We were all very 'salty' and 'rolled' fore and aft along the deck instead of walking. **1926** ANDERSON & STALLINGS *Three Amer. Plays* III. 73, I lived with a Spanish girl at Cavite back in '99... In those days I was salty as hell, a sea-going buckaroo. **1926** J. W. CROSLEY *Bk. Navy Songs* II. 24 A salty bunch of Ensigns we, from the great Atlantic Fleet, And we're here to learn the reason why a valve must have a seat. **1939** *Sat. Even. Post* 23 Dec. 6/1 He was a salty old regular, with one of those wedge-shaped figures and an ugly underslung face of the texture and color of seamed leather. **1941** M. GOODRICH *Delilah* iii. 210 The consensus was that Delilah's men now, for some reason, thought they were 'salty' and were looking for trouble.
5. *U.S. slang.* Angry, irritated; hostile. *to jump salty*: to undergo a sudden change of mood or outlook; to become annoyed or angry (with someone).
1938 *Amer. Speech* XIII. 314/1 *Jump salty*, implies an unexpected change in a person's attitude or knowledge. The person may become suddenly angry, or an unhipped person may become hipped. **1938** *N.Y. Amsterdam News* 26 Feb. 17/2 Let's sound a high C on the postoffice man whose Girl Friday is 'jumpin' salty' 'cause he won't Reno the wife who thinks but isn't sure. **1944** C. CALLOWAY *Hepsters Dict.*, *Salty*, angry, ill-tempered. **1952** C. BROSSARD *Who walk in Darkness* xi. 67 Why do you have to get so salty when people want to have fun? **1958** *Partisan Rev.* XXV. 292 That man jumped salty on me. **1967** J. A. WILLIAMS *Man who cried I Am* xvi. 187 Oops! The dozens, is it? I made you salty eh? **1975** P. G. WINSLOW *Death of Angel* vi. 137 He was furious when I said I didn't have any [money] and got very salty.
B. *sb.* Also **saltie.** A sea-going ship (as opposed to LAKER[1] 4). *N. Amer.*
1959 *Ottawa Citizen* 29 Apr. 53/1 Sixty or more ocean ships—called 'salties' by lake seamen—and inland ships were expected to be in transit today. **1961** *Times* 24 Apr. 16/6 Hundreds of miles eastward again the 'salties' are converging from all over the world, soon to thread the canals and locks linking our vast ocean-like lakes, and bringing a nostalgic Atlantic tang into the very heart of the Dominion. **1966** *Kingston* (Ontario) *Whig-Standard* 5 Jan. 19/7 The only saltie to visit Kingston that year, the 17,170 ton Malmanger of Norway, sailed with her holds only half full of grain. **1971** *Cleveland* (Ohio) *Plain Dealer* 14 Dec. C7 (*heading*) British salty will be last in Cleveland this season.
Hence **sa·ltily** *adv.*
1926 R. MACAULAY *Crewe Train* II. ix 172 Arnold's old flannel trousers were rolled above his knees; his white, slim, long legs glistened saltily beside Denham's firm, brown ones. **1945** C. MANN in B. James *Austral. Short Stories* (1963) 77 After a time he did not so much hear and saltily smell those myriad fish. **1955** *Times* 7 July 5/1 Parents should teach a straightforward, 'saltily realistic' approach to sexual questions. **1958** *Times* 24 Dec. 3/6 The drawings pay marked attention to the arts. Constable..is accompanied by..the young Brangwyn, saltily caught by Phil May.

‖ **salud** (salu·ð), *int.* [Sp., = (good) health: see SALUTE *sb.*[1]] A toast before drinking: 'cheers!', 'good health!'
1938 E. HEMINGWAY *Fifth Column* I. ii. 7 Salud, Comrade Stamp Collector. **1940** G. GREENE *Power & Glory* II. ii. 139 'I will have a little brandy.' 'Salud!' **1961** J. WELCOME *Beware of Midnight* xi. 140 'Salud,' she said, lifting her glass. **1973** G. GREENE *Honorary Consul* III. iii. 148 'That is a very large whisky.'.. 'Large? Why, it is only half as big as mine. Salud!'

saluki (sălū·ki). Also **selugi, sleughi; slogie, slokee, sloug(h)i, slughi.** [ad. Arab. *selūḳi*, f. *Saluḳ* the name of a town in the Yemen.]

A large, lightly built hound belonging to the breed so called, with feathered tail and feet and large pendant ears; formerly called the Persian greyhound. Also *attrib.*
1809 J. G. JACKSON *Acct. Empire of Marocco* v. 31 They often hunt the gazel with the (slogie) African greyhound. **1844** J. H. D. HAY *Western Barbary* xiii. 89/2 The beaters kept good and steady line, and woe to the wild ones that showed themselves to..the swift-footed slokees on the plain. **1891** 'OUIDA' in *N. Amer. Rev.* Sept. 316 The Siberian and the Persian greyhounds are one and the same breed; called *sleughi* in Persia and Arabia. **1913** *Dress & Vanity Fair* (N.Y.) Oct. 110/2 Among them is a Saluki or gazelle hound. **1924** *Blackw. Mag.* Jan. 24/2 Among them..a few Selugis, Persian greyhounds of as ancient and pure a strain as our own. **1924** G. BELL *Let.* 13 Feb. (1927) II. xxiii. 684 When I came in at 4 from the office I found Marie sitting in the garden looking like a female St. Jerome, with a needle for a book, a slughi dog for a lion and a tame red-legged partridge standing solemnly beside her instead of a quail. **1926** *Public Opinion* 30 July 102/2 A tall great sloughi came out of the house, beating his tail against the posts of the verandah. **1928** *Evening News* 5 May 9 He was requested by the Bey to bring him back a really fine English slougi. **1931** C. S. JARVIS *Yesterday & To-day in Sinai* xi. 212 A Saluki hunt on camel-back. **1938** J. W. DAY *Dog in Sport* i. 18 The fleet gazelle-hound of the desert, ancestor of the graceful, tassel-eared Saluki of to-day. **1945** C. L. B. HUBBARD *Observer's Bk. Dogs* 134 Being a member of the Greyhound family, the Saluki is extremely old and of the purest descent. It traces back to about 5,000 years B.C., when it was little different to the modern dog of the Arabs of to-day. **1953** A. SMITH *Blind White Fish in Persia* ii. 41, 100 miles along it there was a tea house, just a hut with a man outside smoking a hookah and a boy with two Salukis. **1973** *Country Life* 8 Feb. (Suppl.) 33/3 (Advt.), Afghans and Salukis, also Ibizans and Pharoahs (with whom we will be at Crufts). **1978** *Times* 7 Jan. 12/2 Bahrain's pure-bred saluki hound, which recently came close to extinction as a pedigree strain, is making a comeback.

‖ **salumeria** (salumeri·a). [It., grocer's or pork-butcher's shop, f. *salume* salted meat f. *sale* salt (L. *sal* salt).] A delicatessen.
1926 R. HALL *Adam's Breed* I. v. 42 There was Fabio's salumeria..his wares—the sausages, the paste, the rich yellow oil, the..Chianti. **1967** P. JONES *Fifth Defector* ii. 10, I was just coming out of the Salumeria in Via Canzotti... Where we get that delicious smoked cheese?

saluresis (sæliurī·sis). *Med.* [f. as next + DI)URESIS.] The renal excretion of a greater quantity of salts than is usual. Cf. next.
1959 *Lancet* 25 Apr. 866/1 The urinary output of salt is raised... The saluresis soon abates. **1975** *Aviation, Space & Environmental Med.* XLVI. 1358/1 Exogenous mineral-corticoid prevents the diuresis, saluresis, and kaluresis.

saluretic (sæliure·tik), *a.* and *sb. Med.* [f. L. *sal* salt + DI)URETIC *a.* and *sb.*] **A.** *adj.* Promoting the renal excretion of salts. **B.** *sb.* A saluretic drug. Cf. prec.
1959 *Lancet* 25 Apr. 866/1 Recently a more potent saluretic agent, hydrochlorothiazide, has become available. .. The saluretic and antidiuretic actions were interdependent. **1964** L. MARTIN *Clin. Endocrinol.* (ed. 4) i. 53 Restriction of protein and salt in the diet has been advised, and hydrochlorthiazide may be used for its saluretic effect. **1975** *Jrnl. Pediatrics* LXXXVI. 831/2 In newborn infants the saluretics are the diuretics of choice. *Ibid.* 832/1 Mannitol has..been used successfully, either alone or with one of the saluretic agents.

‖ **salus populi suprema lex (esto).** Latin phr. (occurring in Cicero *De Leg.* III. iii. 8): the safety of the people must be the supreme law. Also *ellipt.* as *salus populi*. Similarly *salus rei publicae*.
1612 BACON *Essaies* xxvii, Judges ought above all, to remember the Conclusion of the Romaine twelve Tables; *Salus populi suprema lex*, and to know that Lawes, except they bee in order to that end, are but things captious, and Oracles not well inspired. **1617** J. CHAMBERLAIN *Let.* 10 May (1939) II. 74 Necessitie hath no law, and yf *salus populi be suprema lex*, in this case *salus regis* was included too. **1788** GIBBON *Let.* 29 Nov. in *Wks.* (1796) I. 193 In so new a case the *salus populi* must be the first law. **1794** [see *political science* s.v. *POLITICAL *a.* 6]. **1836** J. F. DAVIS *Chinese* I. vi. 251 These are contained in their sacred books, whose principle is literally, *salus populi suprema lex*. **1845** H. BROOM *Legal Maxims* I. 1 *Salus Populi Suprema Lex*... Hence there are many cases in which individuals sustain an injury for which the law gives no action. **1910** CHESTERTON *G. B. Shaw* 89 The real and ancient emotion of the *salus populi*, almost extinct in our oligarchical chaos. **1963** *Times* 8 Mar. 13/5 In matters concerning the safety of the state, the definition of which can be safely left to our Courts of Law, *salus rei publicae* must surely still be *suprema lex*? **1978** *Times* 10 May 17/5, I do not wish to pursue the arguments for and against birching..except to express doubts as to whether the vague and potentially dangerous maxim 'salus populi suprema lex'..is really the best possible guideline to be recommended to a court of law for the determination of a legal issue.

‖ **salut** (salü), *int.* [Fr., lit. 'health'; cf. *SALUD *int.*, SALUTE *sb.*[1]] A toast: 'cheers!', 'good health!'
1933 E. HEMINGWAY *Homage to Switzerland* in *Scribner's Mag.* Apr. 206/2 'Prosit,' said Johnson... The other two porters said 'Salut'. **1938** L. DURRELL *Spirit of Place* (1969) 53 All the Best. I hope to look you two up soon..Salut—Larry. **1966** L. COHEN *Beautiful Losers* I. 4 Salut F., old and loud friend! **1976** 'TREVANIAN' *Main* (1977) xiii. 253 She lifts her glass. 'Salut?' 'Salut.'

‖ **salut** (salü), *sb.* Also with capital initial. [Fr., ellipt. for *salut du Saint Sacrement*, salutation (or benediction) of the Blessed Sacrament.] In French Roman Catholic churches: an evening service of Benediction (at which the Host is exposed and the hymn 'O Salutaris Hostia' is sung).
1694 J. DRUMMOND *Let.* 30 Mar. (1845) 17 One may be either offering the Holy Sacrifice in conjunction with the priests of God, or singing the praise of the Almighty God at the Saluts, or hearing exhortations. **1815** E. WYNNE *Diary* 15 Sept. (1952) xxxi. 536 We went into church while the *Salut* was going on. **1843** C. BRONTË *Let.* 2 Sept. in W. Gérin *C. Brontë* (1967) xiv. 241, I found myself opposite to Ste Gadule, and the bell..began to toll for evening 'salut'. **1853** —— *Villette* xii. 43/2 At sunset or the hour of *salut*, when the externes were gone home. **1901** *Month* Sept. 268 The word *salut*, which is still in French-speaking countries the name most commonly employed to designate the service of Benediction, preserves the memory of an institution which most probably must be regarded as the primitive stock, upon which the Exposition of the Sacred Host and the blessing imparted with It are only an excrescence. **1967** W. GÉRIN *C. Brontë* xiii. 188 The great bell..ringing to Matins, Vespers, and to 'Salut' throughout the catholic year.

salutation. Add: **1. c.** *esp.* in the Church of England: 'The Lord be with you'. (Later examples.)
1929 E. C. THOMAS *Lay Folk's Hist. Liturgy* II. v. 182 In 1552 the Salutation and Kyrie [in Morning Prayer] were postponed to the Creed. **1978** D. M. HOPE in C. Jones et al. *Study of Liturgy* II. III. ix. 231 The people said 'Amen' at its [*sc.* the Epistle's] conclusion and the Gospel continued..after the salutation by the celebrant.

salutatorian. (Later examples.)
1943 *Lafayette Alumnus* (Lafayette Coll., Easton, Pa.) Nov. 1/1 Lloyd Felmly..was salutatorian of his class. **1977** *Transatlantic Rev.* LX. 130 Elwood was valedictorian; I was salutatorian.

salutatory, *a.* and *sb.* Add: **A.** *adj.* **b.** Also in high schools.
1940 W. L. FINK *Evaluation Commencement Pract. Amer. Public Secondary Schools* ii. 25 Time would not permit all of the members of the class to speak. Accordingly, certain pupils chosen on the basis of scholarship alone were given the honor of delivering the salutatory and valedictory addresses. **1947** E. A. KAUMP *High School Commencement Bk.* (rev. ed.) 93 (*heading*) The Salutatory. *Ibid.*, The Salutatory address is another honor speech..given by the student who makes the second highest average during the high school years.
B. *sb.* **2. a.** (Earlier example.)
1869 'MARK TWAIN' in *Buffalo* (N.Y.) *Morning Express* 21 Aug. 2/3 Your new editor feels called upon to write a 'salutatory' at once.
b. (Earlier and later examples.) Also in high schools.
1779 *Pennsylvania Packet* 7 Oct. 1/1 John Woodword [gave] the salutatory in Latin. **1864** *Harper's Mag.* Sept. 501/1 Still another is the burlesque philosophical oration and the half Latin, half Saxon Salutatory. **1905** *N.Y. Even. Post* 12 June 12 The annual class day exercises of the University of Pennsylvania were held to-day. H. B. Taylor delivered the salutatory. **1932** *School Life* May 165/1 On 85 occasions fond..parents had listened to the same old story: Salutatory, oratory, valedictory. **1947** [see A. b above].

salute, *sb.*[1] Add: **3. d.** With defining term prefixed, denoting the attitude adopted by the saluter, or his affiliation, as **raised-arm salute**, a salute made with the arm outstretched at an angle of about 45° from the vertical; **clenched fist salute**, a raised-arm salute with fist clenched (chiefly in communist use); **Hitler** or **Nazi salute**, a raised-arm salute with hand outstretched.
1935 [see *HITLER]. **1937** V. BARTLETT *This is my Life* x. 165 Hundreds of arms went out in the Hitler salute, hundreds of voices yelled the *Horst Wessel Lied*. **1943** D. GASCOYNE *Poems 1937–1942* 5 The centurions..Greet one another with raised-arm salutes. **1959** *Chambers's Encycl.* XII. 173/2 Special forms of salute, the clenched fist salute of the Communists, the 'Roman salute' of the Fascists and the Hitler salute, have been a feature of modern political life. **1969, 1974** [see *NAZI *adj.*]. **1976** *Times* 13 Nov. 4 (*caption*) Clenched fist communist salutes from a group of Madrid car workers who had earlier voted to join the strike. **1977** *Times* 27 Jan. 6/8 Riot police looked on impassively here [*sc.* Madrid] today as a massive crowd of mourners..gave the clenched fist Marxist salute at the funeral of five lawyers gunned down on Monday night by right-wing terrorists.

saluting, *vbl. sb.* Add: **b.** *saluting-base.*
1961 *John o' London's* 19 Oct. 447/2 All the Queen's horses..gallop right past the saluting-base. **1976** C. EGLETON *State Visit* ix. 88 The Queen had climbed on to the saluting base... The Queen left the dais to inspect the Guard of Honour.

Salvadorean (sælvădǭ·riăn), *a.* and *sb.* Also **Salvadoran, -ian.** [f. El *Salvador* (see below) + -AN.] **A.** *adj.* Of or pertaining to El Salvador, a republic in Central America. **B.** *sb.* A native or inhabitant of El Salvador.
1886 *Encycl. Brit.* XXI. 268/2 The tree from which it [*sc.* Balsam of Peru] is obtained grows naturally nowhere else in the world except in a limited part of the Salvadorian seaboard known as the Balsam coast. **1887** *U.S. Consular Rep.* XXIII. No. 82. 292 However great the advantage given to the Salvadorean debtor by English commercial codes,..American is worth 3 per cent. more than English gold. **1895** *Handbk. Salvador* (*Bull. Bureau Amer. Republics No. 58*) 79 Salvadorians are such either by birth or naturalization. **1909** 'O. HENRY' in *McClure's Mag.* July 330/1 For a Salvadorian he was not such a calamitous little man. **1941** C. M. WILSON *Central Amer.* (1942) iii. 52 The distinctly Salvadorean Feast of the Holy Savior. **1947** M. LOWRY *Let.* Nov. (1967) 159 Mlle. Zaza, wife of Salvadorean new passenger. **1969** *Guardian* 19 July 8/2 The quarter of a million Salvadoreans in Honduras. **1979** *Daily Tel.* 20 Apr. 4/4 Another hostage, a Salvadorean who was the Israeli honorary consul. **1980** *Ibid.* 11 July 19/7 A Mexican and a Salvadoran have been charged..with illegally smuggling a group of aliens..into the United States. **1981** *Times* 24 Feb. 13/4 The victory of the Salvadorians is certain.

salvage, *sb.* Add: **2. c.** In wartime, esp. the war of 1939–45: the saving and collection of waste material, esp. paper, for recycling; also *transf.*, those who organized and carried out this collection.
1918 *Times* 2 Mar. 3/5 A National Salvage Council has been set up with the approval of the War Cabinet to deal with the problems of civil salvage and the recovery of waste products generally. **1942** *Oxford Mag.* 29 Jan. 147/1 Next week sees the end of the great drive for salvage of waste paper. **1943** *Punch* 20 Jan. 51/3 Careless of salvage we tore wildly at the wrapper and turned eagerly to the last page. **1944** M. LASKI *Love on Supertax* i. 13 A large pile of empty bottles bore witness to the family's constant failure to remember which day the salvage called. **1946** R. LEHMANN *Gipsy's Baby* 118 Found last week in turning out old papers for salvage. **1961** E. S. TURNER *Phoney War* xx. 291 Some notable gestures were made that summer [in 1943] by persons whose idea of sacrifice was not fulfilled by lending money to the State at interest or putting out old love-letters for salvage.
3. c. Waste material, esp. paper, suitable for recycling. (Cf. sense 2 c above.)
1939 *Times* 11 Nov. 8/4 The salvage department will collect and organize the use of salvage. **1942** *Times Lit. Suppl.* 9 May 229/2 Recently,..a perfect copy of Von Gerning's 'Tour Along the Rhine', with colour plates by Ackermann, was sent to his firm as salvage, together with other fine volumes. **1943** G. WINN in S. Briggs *Keep smiling Through* (1975) 187 Queen Mary.., whenever she sees salvage lying around unclaimed—bones, bottles, scrap iron—Her Majesty stops the car, has it picked up, and taken home in triumph to the village dump. **1945** 'R. CROMPTON' *William & Brains Trust* xi. 204 We'll say we're collectin' salvage if anyone comes. **1951** *Good Housek. Home Encycl.* 237/2 The local Councils in many districts still undertake the collection of salvage. **1959** *Chambers's Encycl.* XII. 176/1 Industrial salvage arises in some form at nearly all factories. Apart from waste paper and canteen scraps, there are textile and chemical wastes, used oils, metal scrap, sawdust [etc.]. **1978** CADOGAN & CRAIG *Women & Children First* x. 213 Older children could help the war effort; they..collected salvage, joined fire-watching rotas.
4. *salvage brigade, campaign, collector, -drive, -dump, man, operation, sack*; *salvage-minded* adj.; **salvage archæology, excavation** = *rescue archæology, excavation* s.v. *RESCUE *sb.* 3 c.
1967 G. H. GROSSO *Cave Life on Palouse* in *Encycl. Sci. Suppl.* (Grolier) 30 'Salvage archaeology' became a way of life for anthropologists in Washington after Grand Coulee Dam created Roosevelt Lake more than 20 years ago. **1977** *Jrnl. R. Soc. Arts* CXXV. 199/1 If the Canada Council spent 90 per cent of its funds on salvage archaeology in California, for example, defenders would be very hard to find. **1890** W. BOOTH *In Darkest England* II. ii. 115, I propose to establish in every large town..'A Household Salvage Brigade',..entrusted with the task of collecting the waste of the houses in their circuit. **1942** P. JEPHCOTT *Girls growing Up* iii. 47 Ordinary time-table lessons are supplemented by..salvage campaigns. **1941** 'R. CROMPTON' *William does his Bit* vii. 165 (*heading*) William—the Salvage Collector. **1975** S. BRIGGS *Keep smiling Through* 187 The salvage collector assured her that the letters would not be read, but suggested that she could tear them into small pieces. **1942** *Ann. Reg. 1941* 335 Such special occasions as a War Weapons Week or a Salvage Drive. **1952** R. A. KNOX *Hidden Stream* p. vii, My store of back-numbers is full to bursting again, and calls for a fresh salvage-drive. **1943** *Punch* 14 Apr. 321/1 Since picking out of the salvage-dump a book entitled Half-Hours with the Stars, my father, a municipal dustman, has become keen on astronomy. **1972** *Even. Telegram* (St. John's, Newfoundland) 29 June 14/3 Provision for the preservation or salvage-excavation of archaeological and historical sites. *a* **1945** in S. Briggs *Keep smiling Through* (1975) 187 The war is driving Hitler back But here's one way to win it: Just give your salvage men the sack And see there's plenty in it. **1942** *R.A.F. Jrnl.* 13 June 23 (*caption*), I want all you hut orderlies to *get* salvage-minded and *stay* salvage-minded. **1919** 'SAKI' *Fate* in *Toys of Peace* 200 The billiard table..was not the best place to have chosen for the scene of salvage operations. **1975** *Globe & Mail* (Toronto) 4 June 2/2 Even filet mignon and Spencer steaks from Mr. Dumais' meat salvage operation went into hamburger. **1942** *Times Lit. Suppl.* 9 May 229/2 Before dropping books into the salvage sack, owners have been urged..to consult the nearest public or university librarian or literary friend.

salvage, *v.* Add: **2.** *U.S.* and *Austral.* To take (esp. *euphem.* by misappropriation) and make use of (unemployed or unattended property).

1918 *Stars & Stripes* 8 Feb. 2 *Salvage*, to rescue unused property and make use of it. 1919 S. Prentice *Padre* xv. 266 When he came out five minutes later it was gone; someone had 'salvaged' it again. 1919 K. D. Morse *Let.* 1 Jan. (1920) vi. 206 The boys were setting off pyrotechnics of all sorts 'salvaged' from the dump. 1920 Riggs & Platt *Hist. Battery F* 15 We manœuvered around and got a loaf of bread and anything else we could 'salvage' before the M.P.'s were put guarding it. 1928 J. B. Wharton *Squad* i. 40 If you two'll collect up all the canteens, we'll go off an' see what we can salvage. 1941 Baker *Dict. Austral. Slang* 63 *Salvage, to*: to steal, purloin.

3. To save and collect (waste material, esp. paper) for recycling.

1943 *Ann. Reg. 1942* 313 The great national campaign to salvage paper for re-pulping resulted in..the destruction of many..irreplaceable volumes.

Hence **sa·lvaged** *ppl. a.*, **sa·lvaging** *vbl. sb.*

c 1920 J. F. McGrath *War Diary* 171 Salvaged rabbits, chicken, beer, and wine to add to the rations. 1951 *Manch. Guardian* 20 Apr. 6/7 His salvaging rather from the morgue of the Rules Committee of the Marshall Plan. 1969 R. Emerson *Judging Delinquents* x. 275 Hard-core, discredited delinquents most in need of salvaging.

salvageable (sæ·lvėdʒăb'l), *a.* [f. Salvage *v.* + -able.] Capable of being salvaged. Also *fig.*

1976 *New Yorker* 24 May 115/1 Some urban experts suggest that the blacks who were salvageable were swept up and out into the suburbs—into decent jobs, and all that. 1977 *Custom Car* Nov. 20/1 Result: one completely junked digger with only the engine salvageable. 1981 *Times* 2 Mar. 10/4 The party is still salvageable..but it will not be saved if its best members leave it.

salvar, var. *shalwar.

Salvarsan (sæ·lvăɹsæn). *Pharm.* Also **salvarsan.** [a. G. *salvarsan*, f. L. *salv-āre* to save + G. *ars-enik* Arsenic *sb.*[1] + *-an* -an.] A former proprietary name for *arsphenamine. Now chiefly *Hist.*

1910 *Official Gaz.* (U.S. Patent Office) 23 Aug. 987/2 Farbwerke vorm. Meister Lucius & Brüning, Höchst-on-the-Main, Germany. Filed Feb. 23, 1909. *Salvarsan.* 1910 *N.Y. Med. Jrnl.* 3 Dec. 1137/1 Ehrlich followed up the arsenic compounds, and after many, many trials, changes, and improvements, placed, carefully and well prepared, dioxydiamidoarsenobenzol or 'salvarsan', to give it its trade name, before the profession. 1911 *Brit. Med. Jrnl.* 14 Jan. 100/1 The remedy has rapidly undergone improvements and successively become '606 ideal', '606 hyperideal', and now 'Salvarsan'. 1913 A. B. Reeve *Poisoned Pen* iii. 91 In these tubes I have the now famous salvarsan. 1913 *Times* 9 Aug. 3/1 The spirillum of relapsing fever, cultures of which after treatment with salvarsan were no longer capable of infecting mice. 1927 *Glasgow Herald* 29 Dec. 4/1 Setting right a joint by manipulation is different from killing a spirochaete with salvarsan. 1937 M. Covarrubias *Island of Bali* (1972) x. 352 But the reluctance of the Balinese to undertake foreign treatment, the forbidding cost of Salvarsan, and the natural promiscuity do not help the situation. 1956 B. Holiday *Lady sings Blues* (1973) ii. 24 At the hospital they were giving everybody shots of salvarsan for syphilis—only it was called 'bad blood' then. 1973 *Sci. Amer.* Sept. 106/3 In 1907 Paul Ehrlich of Germany, after long travail, succeeded in synthesizing Salvarsan, an arsenic compound that would kill *Treponema pallidum*, the microorganism that causes syphilis.

salvation. Add: **1. e.** Phr. *to work out* (*one's own*) *salvation*; freq. *fig.*, to be independent or self-reliant in striving towards one's goal.

1535 [see sense 1 a in Dict.]. 1678 S. Butler *Hudibras* III. i. 86 With Crosses, Relicks, Crucifixes, Beads, Pictures, Rosaries and Pixes: The Tools of working out Salvation, By meer Mechanick Operation. 1818 Keats *Let.* 9 Oct. (1931) I. 243 The Genius of Poetry must work out its own salvation in a man: It cannot be matured by law and precept, but by sensation and watchfulness in itself. 1881 T. R. Davids tr. *Buddhist Suttas* I. vi. 114 Decay is inherent in all component things! Work out your salvation with diligence! 1891 [see sense 1 d in Dict.]. 1911 L. T. Hobhouse *Liberalism* iv. 80 Let every people be free to work out its own salvation. 1948 A. J. Toynbee *Civilization on Trial* x. 210 It is for other Muslims to work out their salvation for themselves as may seem good to them. 1957 A. Thwaite *Ess. Contemp. Eng. Poetry* ix. 142 His [*sc.* Robert Graves'] self-imposed isolation from English literary life has left him free to work out his own poetic salvation and to take an idiosyncratic view of what everyone else is writing. 1981 *Daily Tel.* 12 Jan. 14/3 There are those..who resist the radicals' attempts to force their 'rights' upon them, and prefer to work out their own salvation.

f. (With initial capital.) *ellipt.* for (a member of) the Salvation Army.

1889 *Longman's Mag.* Feb. 407 My father says he is shamed to be called an Inglishman when he sees how the Salvation is knocked about and prossecuted. He says people will hold a drunken man up, but will knock a Salvation down.

4. *salvation banner*; *salvation-contemning* adj.; **salvation history** = *Heilsgeschichte; **Salvation Jane** = *Paterson's curse* s.v. *Paterson; **Salvation lassie** = *lass 1 d.

1931 R. Campbell *Georgiad* II. 33 One or two whose love is not unfurled Like a salvation banner to the world. 1919 Kipling *Years Between* 63 Drunk with enormous, salvation-contemning Love for a tinker. 1959 *Times Lit. Suppl.* 20 Mar. (Relig. Bks. Section) p. vi/4 Such people are also rejecting the Incarnation, Crucifixion and Resurrection as events in 'salvation-history'. 1962 [see *kerygma]. 1977 E. Quinn tr. *Küng & Lapide's Brother or Lord* 20 We ought to ask..how far the Jew..can help us to reach a more authentic understanding of Jesus: an understanding which brings home to us afresh the continuity of salvation history. 1911 *Jrnl. Dept. Agric. S. Austral.* XV. 305 (*heading*) Salvation Jane. 1912 *Ibid.* XV. 679, I went to considerable trouble..in cutting down every plant of 'Salvation Jane' on a portion of my farm. 1937 Lady Rockley *Some Canadian Wild Flowers* 5 There are tracts in New South Wales and South Australia covered with a Viper's Bugloss..but there it is a noxious weed reprobated by the names of 'Paterson's Curse' and 'Salvation Jane'. 1970 P. W. Michael in R. M. Moore *Austral. Grasslands* xxiii. 356/2 *Phalaris tuberosa* has been shown to give excellent control of Salvation Jane or Paterson's curse in south-eastern South Australia. 1895 A. James *Diary* 7 Apr. (1965) 188 Lifting up her voice in prayer as she knelt among the Salvation lassies. 1972 F. M. Hubbard *Whisper in Glen* iv. 34 These slight, intense men had to have their pound of flesh, whether it was a blowsy trollop..or a Salvation lassie.

Salvation Army. Add: **2.** *attrib.*

1881 W. Corbridge *Salvation Mine* (recto front cover), Salvation Army Stores, 101 Queen Victoria Street. 1910 'Saki' *Lost Sanjak* in *Reginald in Russia* 16 The corpse was that of a Salvation Army captain. 1920 H. Begbie *Life W. Booth* II. ii. 17 Blasphemous handbills, supposed to be circulated by Salvation Army Officers. 1921 M. L. Carpenter *Angel Adjutant* v. 40 The wives of Salvation Army Bandsmen make their sacrifices. 1921 *Lady Curate's Promise* xv. 133 He saw..two officers..go into a large Salvation Army Hostel for women. 1928 H. Crane *Let.* 27 Mar. (1965) 320, I finally had to finish the night in a ward of the Salvation Army Hotel. 1966 [see *Rowton]. 1966 [see *Sally *sb.*[3] 1 b]. 1972 F. M. Hubbard *Whisper in Glen* iv. 34 A Salvation Army lass with the body of a Rubens Venus. 1978 *Lancashire Life* Sept. 90/1 In her latest picture..she shows a Salvation Army band playing hymns in a northern town some time between 1920 and 1930.

salvationist. Add: **2.** One who rescues from peril; a saviour.

1971 *Daily Tel.* 4 May 14 Our great wartime leader, and I think salvationist, Sir Winston Churchill.

3. *attrib.* or as *adj.*

1934 Webster, *Salvationist*, n. & adj. 1943 J. S. Huxley *Evolutionary Ethics* vii. 56 What I may call salvationist ethics, aimed at achieving salvation in a supernatural other life. 1959 A. F. Wright *Buddhism in Chinese Hist.* iv. 81 Salvationist Buddhism..is neither an anomaly nor a temporary aberration of an otherwise 'rational' people.

salva·tionize, *v.* [-ize.] *trans.* To convert, save, preach salvation to.

1927 *Scribner's Mag.* Feb. 118/2 Molly never could get over bein' salvationized. Some persons, when they get salvationized, get more joyous and happy. *a* 1930 D. H. Lawrence *Phoenix II* (1968) 439 Power..isn't bossing, or bullying, hiring a manservant or Salvationizing your social inferior.

Salvatorian (sælvătō³·riăn), *sb.* [f. L. *salvātor* (It. *salvatore*), saviour + -ian.] A member of a Roman Catholic congregation, the Society of the Divine Saviour, founded in Rome in the late nineteenth century. Also *attrib.* or as *adj.*

1903 F. M. Steele *Monasteries & Relig. Houses Gt. Brit. & Ireland* 188 (*heading*) The Fathers of the Society of the Divine Saviour, or the Salvatorians. 1909 *Catholic Encycl.* V. 53/2 The Salvatorians have establishments in Italy, Sicily, Austria, Poland. 1931 *Tablet* 22 Aug. 252/2 Father Melchior Geses, a German Salvatorian of the mission of Shaowu in the Vicariate of Foochow. 1962 L. Smith *Salvatorians* 7 The chief ways in which the Salvatorians play their part in the work of the Church for the salvation of souls, is by contributing their particular spirit to the ministry of external work, in preaching and teaching, in assisting the parochial clergy, in the Mass and administration of the sacraments, as well as in the specialized tasks of educational and youth work. 1979 *Tucson* (Arizona) *Citizen* 20 Sept. 5A/1 In 1970, the Salvatorian Fathers filed for bankruptcy, the first instance of legal insolvency in the history of the Church in this country... The Salvatorians, LaSalettes and Paulines—essentially unregulated by virtue of their religious status—were able to float bonds in violation of normal Securities and Exchange Commission requirements that money raised be spent on the advertised purpose.

salvatory, *a.* (Later examples.)

1921 *Challenge* 18 Feb. 241/2 Salvatory and reconstructive work. 1922 J. Y. Simpson *Man & Attainment of Immortality* xiv. 334 The fact of Christ remains, solitary and salvatory. 1958 J. Lodwick *Bid Soldiers Shoot* III. vii. 222 The murderer has but a single advantage: the patient police, who, in extended order, beat the bush in search of him, or of the macabre *trouvaille* of his hastily buried victim, do not..believe that they will encounter either, personally. Therefore, when they do..a salvatory hiatus follows.

‖ **salva veritate** (sæ·lvă ve:rītā·te), *advb. phr.* [L.] Saving the truth, without infringement of truth.

1930 P. P. Wiener tr. *Nicod's Found. Geom.* III. v. 141 It will be found in all cases necessary and sufficient that these two relations be *equivalent*, that is to say,..that one can be replaced by the other and the field of one by the field of the other, *salva veritate*, in every proposition containing nothing besides logical or mathematical expressions. 1957 [see *inter-substitutability* s.v. *inter- 2 a]. 1963 J. Lyons *Structural Semantics* iv. 56 Attempts to handle synonymy in terms of substitutability throughout the language *salva veritate* are generally regarded as unsatisfactory.

salve, *sb.*[1] Add: Now freq. with pronunc. (sælv). **c.** (Later examples.)

1896 *Leeds Mercury Weekly Suppl.* 21 Nov., Put plenty o' sauve on th' back; that'll get owght abt on him 'at iver tha wants to. 1908 J. M. Sullivan *Criminal Slang* 21 *Salve*, getting on the right side of the arresting officer. 1926 Maines & Grant *Wise-Crack Dict.* 14/1 *Spread the salve*, soft, conciliatory talk.

salve, *v.*[1] Add: Now freq. with pronunc. (sælv).

salvia. For (Tournefort 1700) substitute (J. P. de Tournefort *Institutiones Rei Herbariæ* (1700) I. 180), and add: Also *attrib.* and *fig.* (Later examples.)

1923 D. H. Lawrence *Birds, Beasts & Flowers* 66 If there were salvia-savage Bolshevists To burn the world back to manure-good ash, Wouldn't I stick the salvia in my coat! 1941 E. F. O'Donnell *Great Big Doorstep* ii. 28 There were humming-birds twirling around a wild salvia bush in the grove. 1963 W. Blunt *Of Flowers & a Village* 222 Mrs. Stringer insisted upon my going round to see her salvias.

salvific, *a.* Delete † *Obs. rare* and add later examples.

1946 R. A. Knox *Epistles & Gospels* 223 A salvific law, something like to Israel only, might have seemed to contravene them [*sc.* the promises of God]; not a purely domestic Law like that of Sinai. 1958 *Times Lit. Suppl.* 17 Oct. 599/3 Stephen would have seen in the Crucifixion nothing 'salvific', but only the latest in a series of crimes committed against the prophets of the pure religion of Moses and Abraham. 1967 E. R. Fairweather in Clark & Davey *Anglican/R.C. Dialogue* (1974) iv. 49 Anglican theology has revealed no sympathy with..any other doctrine which would seem to infringe upon the salvific role of Christ's human will. 1979 J. Hick in M. Goulder *Incarnation & Myth* vi. 190 It is no longer acceptable..to assume the salvific uniqueness of one's own religion.

salvo, *sb.*[2] Add: **2. b.** (Later examples in wider applications.)

1924 'W. Frank' (*title*) Salvos, an informal book about books and plays. 1955 *Times* 26 May 4/3 It [*sc.* a broadsheet] was intended to be the 'hush-hush' weapon, which by triumphant revelation at the last moment and, by its powerful propaganda salvo, would bring to submission any wavering voters still about. 1971 *Daily Tel.* (Colour Suppl.) 21 May 18/3 They can stay on deck,..watch salvoes of gannets plummet in white streaks to the sea. 1977 W. M. Spackman *Armful of Warm Girl* 43 He bought her the Hindu nose-jewel..and gently slipped it on (which with little salvoes of apologetic kisses she had at once slipped off, and never worn again).

d. Of bombs dropped from aircraft.

1942 *R.A.F. Jrnl.* 27 June 36 A..change of course saved the ship from a direct hit from the salvo dropped by the leading aircraft. 1949 *Sun* (Baltimore) 17 Oct. 1/5 These loads are dropped either in 'chain' (a trail of bombs, blasting out a path between two and three miles long) or in 'salvo', where the scores of 500-pounders tumble out of the bays together in an 'area' bombing operation.

Salvo (sæ·lvo), *sb.*[3] *Austral. colloq.* [f. Salv(ationist + -o[2].] A member of the Salvation Army; *pl.*, the Salvation Army.

1896 *Bulletin* (Sydney) 31 Oct. 27 (*title*) The Salvo's Error. 1908 C. H. S. Matthews *Parson in Austral. Bush* xxvii. 256 Well, I was rared a Carthlick, but I haven't followed it up much. To tell ye the truth, I class 'em all alike—priests, parsons, 'salvos', and all the lot of 'em. 1942 J. Sweeney in Murdoch & Drake-Brockman *Austral. Short Stories* (1951) 382 We come to the Salvo hut where there is a big joker sitting on a form drinking coffee and eating biscuits. 1952 J. Cleary *Sundowners* iii. 144 I've only met one other Rupert... That was when I was in the Salvos. 1962 A. Upfield *Will of Tribe* ix. 87 Can't help bringing out old clichés. The Salvo padre at Derby was down on them but we learned them quick. 1968 *Telegraph* (Brisbane) 29 June 16/6 Hundreds of former Diggers have similar stories of the morale-boosting work done by the 'Salvos'. 1978 R. McKie *Bitter Bread* 77 When workers everywhere got their notices and the slump showed every sign of lasting, the Salvos decided to open a doss house.

salvo, *v.* Add: Also *transf.* and with sense 'to drop a salvo of (bombs)'.

1895 H. G. Hutchinson *Peter Steele, Cricketer* vii. 155 He had just made his century, and been salvoed with applause. 1943 *Yank* 17 Dec. 5 The pilot feathered the props and kept on; the Fort limped in over the target and salvoed its bombs.

‖ **salvoconducto** (salvokondu·kto). [Sp.: see Safe-conduct *sb.*] A pass, safe-conduct.

1955 W. Gaddis *Recognitions* II. vi. 545 What happened! What happened to Huss? John Huss, enticed by a salvoconducto up to Constance, where three bishops sat on his case, and he was burned. 1957 P. Kemp *Mine were of Trouble* i. 13 Just beyond the Spanish barrier we halted; Vicuña went into the control hut to report and to collect my *salvoconducto*.

salvy, *a.* (Earlier U.S. example.)

1861 *Trans. Ill. Agric. Soc.* IV. 103 Care should be taken not to work it too much, as it will hurt the grain of the butter and make it salvy.

salwar, var. *shalwar.

Salyrgan (sæ·ləɹgăn). *Pharm.* Also **salyrgan.** A proprietary name (orig. used in Germany) for *mersalyl.

1924 *Official Gaz.* (U.S. Patent Office) 10 June 270/1 H. A. Metz Laboratories, Inc., New York,..*Salyrgan*.. Preparations for the treatment of spirochetal and other infectious diseases and as a diuretic. Claims use since about Apr. 11, 1924. 1924 *Trade Marks Jrnl.* 15 Oct. 2285 *Salyrgan*... Farbwerke vorm. Meister Lucius & Brüning... Hoechst am Main, Germany; Manufacturers.—16th August 1924. 1928 *Canad. Med. Assoc. Jrnl.* XVIII. 45/1 Salyrgan was first introduced as an antiluetic mercurial agent. 1956 *Internat. Rev. Cytol.* V. 223 Experiments..indicate that salyrgan strongly inhibits the hydrolytic splitting of the high-energy phosphate bonds of ATP. 1976 *Amer. Jrnl. Clin. Path.* LXV. 686/1 High concentrations of salyrgan..shortened the lag period.

sam, *sb.*[1] Add: **2.** (Later examples.) Also without contract.: on oath, a promise.

1939 J. Masefield *Live & Kicking Ned* 115 On that I swop my solemn sam. 1960 M. Allingham *Black Plumes* xii. 138 Upon my Sam I think you're both mad. 1966 'J. Hackston' *Father clears Out* 71 He'd see that things were righted, upon his Sam he would.

Sam, *sb.*[2] Abbrev. of Sambo (sense 2) in Dict. and Suppl.

1867 W. H. Dixon *New Amer.* II. ii. 13 Sam—all negroes there are Sam—may be a Methodist. 1877 L. Hearn *Genius Loci* in *Cincinnati Commercial* 11 Aug. 6/4 In Nag-a-back Sam. And I don't care a d—n..For I sooner be a nigger than a poor white man. 1928 *Amer. Speech* XIII. 151/1 *Sam*, a negro who demeans himself to secure favor from white people. 1964 L. Nkosi *Rhythm of Violence* i. Black Sams! Why don't they do somethin' so we can handle this once and for all! 1973 K. Johnson in T. Kochman *Rappin' & Stylin' Out* 148 Sam, a common name of black males, it is used to refer to any black male. In addition, the story character, Sambo, was black; perhaps the label derives from 'Little Black Sambo'.

sam, *v.*[1] **2.** (Further examples.)

1824 W. Carr *Horæ Momenta Cravenæ* 12 If shoe nobbud cud git a bit a naturable rist, shoe wod sam up strang. 1942 J. B. Priestley *Eden End* i. 10 Ma sam up in, b'care, samming up these old clothes for the doctor.

sam, *v.*[2] Add: Also **samm.** Hence **sammed** *ppl. a.*, **sa·mming** *vbl. sb.*

1885 A. Watt *Art of Leather Manuf.* xii. 151 The butts are next piled in a heap to sam, or, as it is also termed, for several days, by which the leather becomes tempered, or, in other words, attains a uniform moist and softened condition. 1909 H. G. Bennett *Manuf. of Leather* xx. 256 Samming is an exceedingly important operation by which leather is brought into a uniformly half-dry condition, this state being quite necessary for many of the finishing operations... The 'sammed' condition may be obtained in three ways—by drying out completely and then wetting back by dipping through water (often tepid) and leaving 'in pile' for some hours; by drying the wet goods in samming to the required consistency and further, wetting back any parts that have dried drier than the bulk and leaving in pile for a time to become uniform; and by machine samming, in which case the superfluous moisture is removed by the pressure of machine rollers.

‖ **samadh** (sămā·d). [See next.] The tomb of a holy man or yogi who is assumed to have achieved samadhi rather than to have died. Cf. *Samadhi 1 b.

1888 *Asiatick Researches* XVI. 39 A temple, sacred to the deity whom they worship, or the *Samādh*, or shrine of the founder of the sect, or some eminent teacher. 1888 Kipling *Departmental Ditties* (1890) 16 Pray for the Raja's wish in his honour, A mark for his resting-place. 1934 Monier Williams *Brāhmanism & Hinduism* 179 A native of Oudh, whose *samādh* or tomb is at Kotwa. 1969 A. Swinson *Six Minutes to Sunset* i. 16 A dilapidated domed *samadh* (or tomb).

‖ **samadhi** (sămā·di). *Indian Philos.* [Skr. *samādhi* a placing together, f. *sam* together + *ā* prefix + *dhā* to place (see Do *v.*).] **1. a.** The state of union with creation into which a perfected yogi or holy man is said to pass at his apparent death. **b.** The voluntary burial of such a person before death in anticipation of this state; the site of the burial of a holy man (cf. prec.).

1795 *Asiatick Researches* IV. 218 Dhritara'shtra, in the state of *samādhi*, quitted his terrestrial form to proceed to the mansion of the blest, leaving no remains behind. 1861 Monier Williams *Brāhmanism & Hinduism* 261 When such a man dies in India, his body is not burnt but buried, because in fact he is not supposed to die at all. He is believed to lie in a kind of trance, called Samadhi. 1935 *Glasgow Herald* 14 Sept. 7 The Sadhu did not commit suicide, but performed the religious rite of Samadhi. 1968 *Jrnl. Mus. Acad. Madras* XXXIX. 103 Something more beautiful and commodious must be built at his Samadhi to commemorate his work. 1969 *Times of India* 17 Aug. 34 The Janata party had betrayed the oath taken at Gandhi's samadhi that they would fulfil his dreams of a regenerated, socialist India by forming connections with the RSS, he said.

2. The highest state of meditation, in which the distinctions between subject and object disappear and unity with creation is attained; the last stage of yoga.

1827 *Trans. R. Asiatic Soc.* I. 25 The collection of *Yóga-sútras*..is distributed into four chapters or quarters..: the first on contemplation (*samádhi*)..[etc.]. 1850 [see *dhyana]. 1913 [see *raja yoga]. 1939 A. Huxley *After Many a Summer* II. i. 189 Baby..is now walking about in a state of perpetual *samadhi*. 1958 J. Sykes *Quakers* I. i. 31 A moment almost of group samadhi, of the displacement of all by God's Being, and Becoming. 1960 J. Hewitt *Yoga* i. 7 By a programme of bodily and mental self-discipline we who move on lower levels of consciousness can achieve Samadhi (union with divine consciousness). 1965 *New Statesman* 16 Apr. 616/1 Ramakrishna's ability to pass into a trance-like state of 'higher spiritual awareness' known as samadhi. 1971 *Shankar's Weekly* (Delhi) 11 Apr. 22/2 From time to time, the Swamiji would come out of his samadhi to frown at the opening and closing of the compartment door. 1977 L. A. Govinda *Creative Meditation* III. vi. 135 Though samadhi may be the culmination in the meditational experience, we cannot remain in that state..but have to return to the world.

Samain, var. *Samhain.

Samaj (sămā·dʒ). Also **Somaj.** [a. Hindi and Bengali *samāj* society, f. Skr. *samāja* a meeting with, f. *sam* together + *aj* to drive.] An assembly or congregation in India; a church or religious body, as in *Brahmo Samaj* (see *Brahmoism).

1867 W. H. Davies *Unorthodox London* 2nd Ser. (ed. 2) 185 The present representative of the Brahmo Samaj in London, was to preach at Mr. Conway's chapel. 1876 *Encycl. Brit.* IV. 201/1 He gave a printing-press to the Samaj. *Ibid.* 201/2 They encourage the establishment of branch Samājes in different parts of the country. 1884 [see Progressive *a.* 4]. 1913 J. N. Farquhar *Crown of Hinduism* 76 The truths of religion which they find there are the doctrines taught by the Samāj. 1968 [see *Ārya Samaj]. 1978 W. de Bary et al. *Sources of Indian Tradition* xxii. 629 The Ārya Samāj ('Society of the Aryas', or noble men') which he [*sc.* Dayānanda] established at Bombay in 1875 has since reflected the militant character of its founder, and from its stronghold in the Punjab has contributed to the rise of Hindu nationalism.

Saman[1] (sā·măn). [a. Skr. *sāman* chant.] A sacred text or verse forming the third of the four kinds of Vedas; the name of the Veda thus formed. Also *attrib.* So **Samaveda** (sā·mă,vē·dă), the name of the third Veda.

1798 *Asiatick Researches* V. 364 Prayer..is beginning a lecture of the Samaveda. 1843 *Penny Cycl.* XXVI. 171/1 There are the Rich, Yajush, Sāman, and Atharvan... *Ibid.*, The Sāmaveda contains songs of lyrical character to be recited with melancholy. 1886 *Encycl. Brit.* XXI. 277/1 The strictly technical collection of Sāman-texts, the *Sāmaveda-samhitā*, or collection of texts (*ṛich*) used for making up the sāman-hymns, and the *Gānas*, or tune-books. *Ibid.* 278/1 The *Vamśa-brāhmaṇa*, a mere list of the Sāmaveda teachers. 1913 J. G. Frazer *Golden Bough* (ed. 3) I. 92 A particular hymn of the ancient Indian collection known as the Samaveda. 1913 J. N. Farquhar *Crown of Hinduism* 77 The Sāman, Yajus, and Atharvan exhibit the same polytheism. 1954 *Grove's Dict. Mus.* (ed. 5) IV. 456/1 The Samans (sacrificial chants) may, though rarely, be heard nowadays. 1968 *Jrnl. Mus. Acad. Madras* XXXIX. 105 Sambhutana is in itself Brahman and is greater than Sama-veda.

saman[2] (samā·n). Also **samaan, samang.** [a. Amer. Sp. *saman*, f. Carib *zamang*.] = *guango, Zamang. Also *attrib.*

1888, etc. [see *monkey-pod (tree)* s.v. *monkey *sb.* 17 b]. 1951 J. C. Penny *Sonnet in Bottle* II. iii. 46 The huge trees of the saman trees were islands of peace in a pale, unpuzzling moonlight of night air. 1958 G. Lamming *Of Age & Innocence* III. vi. 117 The tall black saman tree at the curve of the hill-top. 1960 *Times* 10 Mar. 18/5 The saman trees..were full of holders of tree tickets. 1965 [see *guango]. 1968 E. Lovelace *Schoolmaster* viii. 120 Christiana was sitting on the trunk of a fallen saman tree in the shade near the cocoa house. 1974 *Times* 1 Feb. 12/4 Queen's Park Oval is unlike its namesake at Kennington. Where the gasholders should be there are some fine samaan trees, with a range of low, wooded hills right behind them.

Samang, var. *Semang.

samango (samæ·ŋgo). [Native name.] In full, *samango guenon* or *monkey*. An African monkey, *Cercopithecus mitis*, which has blue-grey fur with black markings.

1888 *Proc. Zool. Soc.* 564 The most notable additions during the month were:—..The Small-clawed Otter..the Samango Monkey. 1894 H. O. Forbes *Handbk. Primates* II. 72 (*heading*) The Samango Guenon. 1922 J. Stevenson-Hamilton *Animal Life in Afr.* xvi. 260 The Samango monkey..extends from the eastern part of Cape Colony, through Portuguese East Africa, and the whole of Rhodesia. 1933 S. Zuckerman *Social Life Monkeys & Apes* xi. 185 The Samango monkey is encountered in the forests of eastern South Africa. Very near where the Samango monkeys were seen, I came across a party of seven Vervet monkeys. 1967 J. R. & P. H. Napier *Handbk. Living Primates* 104 (*caption*) Samango monkey..grooming.

samara. Add: Also with pronunc. (sămā·ră). (Further examples.)

1960 [see *achene]. 1976 *New Yorker* 22 Jan. 66/2 The best-known samara is the maple key, which is much larger than the fruit of the birch.

Samaritan, *sb.* and *a.* Add: **A.** *sb.* **a.** (Later examples.)

1957 *Oxf. Dict. Chr. Ch.* 1211/1 *The Samaritan Pentateuch*, a slightly divergent form of the Pentateuch in Hebrew, current since pre-Christian times among the Samaritans. It is the only part of the OT accepted by the Samaritans. 1965 M. Spark *Mandelbaum Gate* iv. 91 Those Israelites, Samaritans, those boys. 1977 *Sci. Amer.* Jan. 100/1 Although the kingdom of Samaria vanished long ago, the Samaritans still survive today as perhaps the smallest ethnic minority in the world.

b. Freq. in full, *good Samaritan*; also *transf.*, a kind and helpful person; hence (nonce-wds.) *good Samaritanism*, *good Samaritanship*.

1640 N. Rogers (*title*) The good Samaritan, or an exposition on that parable, Luke x. vv. xxx–xxxviii. 1840 J. Ruskin *Let.* 1 July in *Lett. to College Friend* (1894) 11 You have sacrificed half a Good Samaritanship to indulge your friends with letters of brown paper. 1861 [in Dict.]. 1898 'A. Hope' *Rupert of Hentzau* iii. 42 Good Samaritans but not men of war, they returned to where I lay senseless on the ground. 1919 M. Beer *Hist. Brit. Socialism* I. i. v. 74 The new order would act as a good Samaritan and pour oil and wine into the wounds of the victims. 1923 *Virginia Law Rev.* Apr. 423 It is unreasonable that the priest and the Levite should go free while the good Samaritan should be forced to undergo the ordeal and expense of a trial. 1925 A. Huxley *Those Barren Leaves* I. vi. 156 On the faces of all my good Samaritans I noticed an expression of child-like eagerness. 1930 H. Kenwood *God is Shame* 14 A co-opted partner in every kind of Good Samaritanism. 1950 T. S. Eliot *Cocktail Party* I. ii. 49 Don't you realise how lucky you are To have two Good Samaritans? 1953 E. O'Neill *Touch of Poet* (1957) i. 9 Sure, the good Samaritan was a crool haythen beside you. 1963 *Reader's Digest* May 89/1 The best known and most effective curb of the malpractice-suit racket in California is so-called Good Samaritan law. 1977 *Times of Zambia* 7 Sept. 1/6 Shawa started as a Good Samaritan, trying to separate a fight in which Mr Sichinga was involved.

c. *the Samaritans*, an organization founded in London in 1953 that offers counselling by telephone to those in distress or contemplating suicide; hence as *sing.*, a member of this organization. Also (in *sing.*) *attrib.*

1953 *Church Times* 27 Nov. 874/5 The Rev. Chad Varah..intends to open a 'Good Samaritan' centre—on the telephone. 1960 *Times* 17 Nov. 6/6 The vast majority of those who came to the Samaritans were not mentally ill... The Samaritans had helped people of all types, from a duke to a dustman. 1967 *Guardian* 8 June 3/5 The most common reasons for people calling the Telephone Samaritans for help in the Greater Manchester area..were depression, anxiety, and mental illness. 1969 *Listener* 10 Apr. 508/1 The girl..works for Oxfam and wants to be a Samaritan. 1973 J. Seabrook *Loneliness* 115 I'd break down, I'd be all trembling. I used to ring the Samaritans. I don't know what I'd have done without the man who was my counsellor. 1977 *Hongkong Standard* 14 Apr. 13/3 (Advt.), Discouraged/Depressed? Dial the Samaritans, day or night. 1978 M. Dickens *Open Bk.* xxi. 186 The Samaritans is a world-wide fellowship of men and women of all ages, creeds and races, dedicated to befriending people who are desperate enough to want to kill themselves. *Ibid.*, I went to the Samaritan centre in London to talk about the isolated and lonely people who I knew must be among their callers.

B. *adj.* **c.** Also *Comb.*, as (sense b of the sb.) *samaritan-like* adj.

1973 E.-J. Bahr *Nice Neighbourhood* vi. 64 We callously discussed how Samaritan-like we'd been, having John over for dinner.

Samaritanism. 1. (Later example.)

1973 *Sci. Amer.* Jan. 80/1 Of all the multifarious forms of Jewish religious expression that arose, only two have survived and flourish today... A marginal survivor is Samaritanism, maintained by a tiny group of Samaritans..who continue to worship on their holy Mount Gerizim.

samarium. Substitute for entry:

samarium (samē³·riəm). *Chem.* [mod.L., coined in Fr. (P.-É. Lecoq de Boisbaudran 1879, in *Compt. Rend.* LXXXIX. 214): see *samarskite and -ium.] A hard grey metallic element of the lanthanide series, found in small quantities in monazite sand, samarskite, and other rare earth minerals. Symbol Sm; atomic number 62.

1879 *Jrnl. Chem. Soc.* XXXVI. 890 A new metal to which the author gives the name samarium. 1907 *Athenæum* 31 Aug. 244/3 A sulphide of calcium containing a trace of the rare element samarium. 1923 I. R. Evans *Metals & Metallic Compounds* IV. i. 13 If an attempt is to be made to obtain pure compounds of samarium, europium, or gadolinium, the double magnesium salts are most satisfactory. 1955 *Sci. News Let.* 12 Mar. 164/2 Rare earth metals, such as samarium and europium, have long remained a mystery, simply because there was not enough of them available to find out what they could be used for. 1969 *New Scientist* 28 Aug. 430/2 Storage densities of 100 000 bits of information per square inch may be achieved with an orthoferrite containing samarium and terbium. 1974 *Encycl. Brit. Micropædia* VIII. 829/2 In addition to its more stable trivalent state, samarium..has a +2 oxidation state... Trivalent samarium..forms a series of yellow salts and solutions. 1977 *Gramophone* Oct. 590/1 (Advt.), It has an unusually tiny, samarium cobalt (rare earth) magnet of remarkably high power. 1980 *Sunday Times* 13 Apr. 16/2 It is a ring of about 250 tiny magnets made of samarium cobalt.

sama·ria, substitute for def. in Dict.: the oxide Sm_2O_3, a cream-coloured solid; (earlier and later examples).

1885 *Jrnl. Chem. Soc.* XLVIII. II. 1025 The spectrum of a phosphorescent mixture of samaria 90 parts, and yttria 10 parts, in high vacua, shows none of the lines of yttrium, but is almost a facsimile of the spectrum of pure samarium. 1974 *Encycl. Brit. Macropædia* XV. 516/1 Mosander's didymia was resolved into several oxides—samaria (samarium; 1879), praseodymia..., neodymia.., and europia.

Samarra[2] (sămā·ră). The name of a city in northern Iraq, used in phr. *an appointment in Samarra* to indicate the inevitability of death. Also *transf.*

In Maugham's play (see quot. 1933), the servant to a merchant meets Death in the market-place at Baghdad, and flees to Samarra to escape his clutches. When questioned by the merchant, Death explains his surprise at seeing the servant, replying as in quot. 1933.

1933 W. S. Maugham *Sheppey* III. 112, I was astonished to see him in Bagdad, for I had an appointment with him to-night in Samarra. 1934 J. O'Hara (*title*) Appointment in Samarra. 1971 A. Price *Alamut Ambush* xi. 133 He had ridden out innocently..and had set up his own appointment in Samarra. 1973 *Times* 28 June 16/8 All Mr Heath's justified complacency as he watches the Labour Party destroying itself will avail him little if, come the next General Election, his own rendezvous with destiny turns out to be an appointment in Samarra.

samarskite. Substitute for entry:

samarskite (sæ·măɹskəit). *Min.* [ad. G. *samarskit* (H. Rose 1847, in *Ann. d. Physik* LXXI. 166), f. the name of Col. M. von *Samarski*, 19th-cent. Russian mining official: see -ite[1].] A complex niobate and tantalate of yttrium, uranium, and iron, with small quantities of other metals including lanthanides, which is found as velvet-black or dark brown monoclinic prisms in granite pegmatites.

1849 J. Nicol *Man. Mineral.* 285 Samarskite... Rhombic; isomorphous with columbite..mostly imbedded in flat, somewhat polygonal grains. 1947 *Proc. Indian Acad. Sci.* A. XXV. 405 The samples of samarskite were secured from the Kodanda Rama mine in Nellore District. 1973 T. Moeller in J. C. Bailar et al. *Comprehensive Inorg. Chem.* IV. xliv. 49 Minerals of lesser importance include..samarskite, an yttrium earth–iron–calcium–uranyl niobate-tantalate, found in North Carolina, the Ural Mountains, and Madagascar.

Samaveda: see *Saman[1].

samba (sæ·mbă), *sb.*[1] [Pg., of Afr. origin.] A Brazilian dance of African origin; a ballroom dance imitative of this; also, a piece of music such as accompanies this dance. Also *attrib.*

1885 W. Moberly *Rocks & Rivers Brit. Columbia* 17 It was here I first saw the graceful South American dance—the Zemba Queca (I am not certain how it is spelt). 1911 B. Miall tr. *P. Denis's Brazil* xiii. 324 It is during these festivals that the negro dances are performed; the *Coco* and the *Samba*. 1929 H. Miles tr. *P. Morand's Black Magic* p. i, 1919—Darius Milhaud arrives from Brazil. He..plays me those Negro *Sambas* which are shortly to serve for the music of his *Bœuf sur le Toit*. 1939 *Britannica Bk. of Year* (U.S.) 201/1 The samba is the national dance of Brazil, much as the fox trot is in the United States... In 1938, the samba was just beginning to make its way in the New York night clubs. 1942 D. Pierson *Negroes in Brazil* ix. 248 It was of these *Bahianas* that Carmen Miranda sang when she recently captivated Broadway with the staccato notes of the rollicking *samba* by Dorival Caymmi, 'Que é a Bahiana tem?' *Ibid.* 249 The *samba*, or *samba batida*, a regional form of the old *batuque*, although it has now been taken over by the upper classes and in a modified form become not only one of the most characteristic musical forms but also one of the favorite dances of Brazil, is still enjoyed in its primitive simplicity by the Bahian lower classes. 1949 M. Dickens *Flowers on Grass* ix. 240 Mervyn and Wanda were doing a Samba. 1950 J. Vedey *Band Leaders* 132 He [*sc.* Edmundo Ros] states that many of the numbers published as Sambas are not really Sambas at all... In 1940..the Rumba was the only known dance of its kind, played in either slow or fast tempo. The Samba, which Ros himself actually introduced, followed. *Ibid.* 144 Who are these thousands of people packing the Palais de Danse and jiving wildly to the sambas and Bop arrangements of the popular bands? 1954 J. Steinbeck *Sweet Thursday* xxviii. 199 The crazy trumpet put a samba beat to the 'Wedding March'. 1965 W. Soyinka *Road* 21, I have not seen any other tout who would stand on the lorry's roof and play the samba at sixty miles an hour. 1974 *Down Beat* 18 July 26/3 Zoot, Jaki and Al all get their say in a medley of sambas. 1977 *Gramophone* Aug. 353/3 Barlow specializes in the more traditional sequined ballroom fare of quicksteps, waltzes, foxtrots, a tango, cha cha chas, a samba, slow rumbas, a pasodoble and jive. 1979 P. Fox *Mantis* iv. 62 Great driving-music: hard rock Samba, plenty of guitar.

samba (sæ·mbă), *sb.*[2], var. Sambo (sense 1).

1958 J. Carew *Wild Coast* ix. 126 It please me eyes to see you growing up into a proper samba man. 1959 J. Morris *Adversary* i. 9 A couple of upcountry *corregidors*..had three girls between them... One of them looked like a *samba*. 1974 *Black World* Aug. 55 The Dirty Tricks store window featured a grotesque, black-purple mask of a 'samba' sister with a bone through its nose.

samba (sæ·mbă), *v.* [f. *samba *sb.*[1]] *intr.* To dance the samba. Also *fig.*

1950 in Webster Add. 1959 'J. Drummond' *Black Unicorn* xviii. 128 He put his arm round my waist, and started trying to make me samba. 1972 *Time* 22 May 9/3 Brazil had sambaed away with the talks. 1975 *Times* 14 Apr. 12/4, 37 couples..were required to cha-cha, samba, rumba and paso doble. 1979 C. Wood *James Bond & Moonraker* xi. 111 How do you kill five hours in Rio if you don't samba?

samba, var. Sambur in Dict. and Suppl.

‖ **sambal** (sæ·mbăl, ‖ sambal). Also **sambaal, sambel.** [Malay.] A highly seasoned condiment, of Malayan and Indonesian origin, consisting of raw vegetables or fruit prepared with spices and vinegar and used as a relish; found also in other (esp. *S. Afr.*) cookery. Cf. *pol sambol.

1815 A. Plumptre tr. *Lichtenstein's Trav. S. Afr.* II. iv. xxxiii. 84 *Sambal* is a mixture of gherkins cut small, onions, anchovies, Cayenne pepper, and vinegar. 1817 S. Raffles *Hist. Java* I. 98 The most common seasoning..is the lombok; triturated with salt, it is called *sámbel*. 1839 T. J. Newbold *Brit. Settlements in Straits of Malacca* II. xii. 178 The ordinary food of Malays..is rice, and in times of scarcity, sago seasoned with a little salt fish, Blachang, the caviar of the East, made with acid fruits, &c., into a variety of condiments termed Sambals. 1871 *Cape Monthly Mag.* June 334 They make a sort of chutnee out of quinces, which they call 'sambal'. 1933 L. Ainsworth *Confessions Planter in Malaya* 145 The usual small side-dishes containing what are known as 'sambals', which consist of such things as fried ground nuts, shredded cucumber, burnt grated coconut, Bombay duck and red and green chillies. 1942 S. Cloete *Hill of Doves* xxiii. 328 He thought of food once more—bobotee; breede, made of mutton ribs; cucumber sambal. 1950 *Cape Times* 7 June 16/1 Her letter has reference to *melktert*, brandied peaches, *kreef frikkadels*, quince *sambaal*—all things which are becoming a lost art to us. 1953 du Plessis & Lückhoff *Malay Quarter* i. 15 The spicy stew is enhanced by means of various *sambals* or condiments. 1971 L. Charteris *Saint & People Importers* iii. 24 Order me some samosas, lamb curry, pilau rice, dhal, and all the sambals you can crowd on the table. 1978 *Courier-Mail* (Brisbane) 26 Oct. 25/3 We ordered..a made-on-the-premises sambal mixing pineapple, capsicum and mild chillies in soy sauce, sugar and vinegar.

‖ **sambaquí** (sambakī·). [Tupi.] A form of shell heap, found on the S. Brazilian coast, resulting mainly from the action of the wind and the sea, in which remains of prehistoric and historic cultures have been found; also *attrib.* (See quot. 1946.)

1944 S. Putnam tr. *E. da Cunha's Rebellion in Backlands* ii. 50 The pre-Columbian of the 'Sambaquis'. 1946 A. Serrano in J. H. Steward *Handbk. S. Amer. Indians* I. iii. 401 The word 'sambaquí' is of *Tupi-Guarani* origin and means 'hill of shells'. *Ibid.* 403 The sambaquis are littoral cordons or concentrations of shells, broken and reshaped by natural forces. *Ibid.* 404 Artifacts in the most ancient sambaquis, which are farthest from the sea, correspond to the primitive culture of Lagoa Santa. *Ibid.*, The prevailing idea..has been that of a cultural unity—a single sambaqui culture—that is distinctive and characteristic of these deposits. It is no longer possible to maintain this. 1953 *Jrnl. R. Anthrop. Inst.* LXXXIII. 60 On the coast, the classic *sambaqui* culture, called the 'Southern Phase', may succeed the chipped-axe phase. 1977 G. Clark *World Prehist.* (ed. 3) x. 447 There is evidence from the shell mounds or sambaqui sites that intensive exploitation of coastal resources had begun.

‖ **sambar** (sa·mbar). Also **sambhar.** [Tamil.] In South Indian cookery, a highly seasoned lentil gravy. Also *attrib.*

1957 S. Rangarao *Good Food from India* vii. 68 Sambar powders go well into meat curries. 1967 M. Waldo *Internat. Encycl. Cooking* II. 541/2 *Sambar*, (Indian), a highly seasoned vegetable and lentil dish. 1972 *Indian Express* 28 Dec. 10/1 South Indian dishes—idli, dosa and sambhar—have become popular. 1973 *Times* 19 June (Bombay Suppl.) p. xv/3 The food of South India, especially the idli (spongy rice cakes) with sambar (liquid lentil juice) should not be missed. 1976 *Sunday Standard* (India) 11 Jan. 10/4, I distribute bread to these children everyday. I've also brought sambar to go with it. 1977 *Sunday Times* (Colour Suppl.) 27 Nov. 35/4 The food is as authentic as it is in Madras—..*sambar* (the thick lentil gravy cooked with vegetables and tamarind juice).

Sambo. Add: **2.** (Earlier and later examples.) Now used only as a term of abuse. Also *attrib.*, esp. with reference to the appearance or subservient mentality held to be typical of the black American slave.

1704 *Boston News-Let.* 2 Oct. 2/2 There is a Negro man taken up supposed to be Runaway from his Master,..calls himself Sambo. 1735 J. Atkins *Voy. to Guinea, Brazil & W. Indies* 170 If you look strange and are niggardly of your Drams, you frighten him; *Sambo* is gone, he never cares to treat with dry lips. 1781 I. Jackson *Divorce* II. 34 So then, Sambo, you want to be in the fashionable world, I see?.. Timothy, show the black Gentleman downstairs. 1818 'A. Burton' *Adventures J. Newcome* IV. 222 His Steward was a scoundrel Sambo, And in his own conceit a d—d beau; A true Barbadian being born, He others held in utter scorn. 1922 Joyce *Ulysses* 322 *Black Beast Burned in Omaha, Ga.* A lot of Deadwood Dicks in slouch hats and they firing at a sambo strung up on a tree with his tongue out and a bonfire under him. *Ibid.* 436 Tom and Sam Bohee, coloured coons in white duck suits, scarlet socks, upstarched Sambo chokers and large scarlet asters in their buttonholes leap out. 1927 G. B. Shaw *Doctors' Delusions* (1932) 137 When a vivisector says, in effect, 'I have a dread secret to wrest from Nature: so you must license me to sacrifice a guinea pig', the Sambo in us assents. 1957 [see *boot *sb.*[3] 1 e]. 1959 S. Elkins *Slavery* 227 What, then, of the 'reality' of Sambo? Did the Sambo role really become part of the slave's 'true' personality? 1962 L. Deighton *Ipcress File* xix. 123 'I'd just better be right about you pale-face,' he said. 'You'd just better had, Sambo.' 1969 *N.Y. Rev. Bks.* 13 Mar. 3/1 The reasons for the development of the 'Sambo' response of the Negro slave to his environment which help to explain the paucity of slave revolts in America. 1969 *Guardian* 1 Apr. 7/4 The brothers wore Afro costume... Their loose jigging and gestures, open-mouthed, Sambo style, reverberated with their long solid jammed numbers. 1973 *Times Lit. Suppl.* 2 Mar. 230/2 The 'Sambo' stereotype of the loyal, lazy, affectionate and child-like slave. 1977 *Times* 10 June 8/5 A white Zambian..had called him a black sambo during the struggle for independence.

sambok, var. Sjambok.

sambook. Add: Also **sambuq.** (The usual spelling is now *sambuk*.) Also *attrib.* (Later examples.)

1906 H. W. Smyth *Mast & Sail* 307 The Red Sea *sambuk* is generally from 18 to 20 tons only. 1938 F. Stark *Jrnl.* 1 Mar. in *Winter in Arabia* (1940) 201 A king..sent for one thousand virgins from Somaliland across the sea; they were all shipped in a sambuq. 1942 [see *mashwa]. 1963 *Times* 1 Feb. 14/6 They want a sambuq to meet them at Bir Sukaiya at noon next Wednesday—it's three hours' sail by sambuq from there to Perim. 1974 *Nat. Geographic* Sept. 333 (*caption*) *Sambuk*, once the most common of Arab dhows, has ferried generations of Moslem pilgrims from Africa towards Mecca. 1975 *Financial Times* 31 Oct. 5/4 The monsoon was pushing the seas too high, even for the sturdy *sambuq* fishing boats.

Sam Browne (sæm braun). [The name of Sir *Sam*uel James *Browne* (1824–1901), British general, who invented it.] In full, *Sam Browne belt*: a belt with a supporting strap that passes over the right shoulder, worn by commissioned officers of the British Army and also by members of various police forces, etc. Also *transf.*, a commissioned officer.

1915 *Punch* 6 Oct. 288/3 Military Wedding Equipment. Sam Browne belt, single brace and frog, best bridal leather. 1916 E. C. Middleton *Aircraft* xvi. 114 Should he be posted to the Army wing he will probably present himself to an astonished and apoplectic adjutant wearing two cross straps to his 'Sam Browne'. 1919 *Amer. Legion Weekly* 5 Sept. 27 It wasn't the privates or the acting corporals or the full-fledged Sam Brownes who had a monopoly on this particular quality. 1933 J. Cary *Amer. Visitor* xvii. 278 Stoker in a Sam Browne and Gore with a huge Webley strapped to his waist were interrogating Sam and Henry. 1942 E. Waugh *Put out More Flags* i. 55 Taking in every detail of his uniform, the riding boots, Sam Browne belt, the enamelled stars of rank. *Ibid.*, I heard they had stopped wearing cross straps on the Sam Browne. 1972 J. Wambaugh *Blue Knight* (1973) i. 16, I loosened my Sam Browne for the joy of eating. 1977 'A. York' *Tallant for Trouble* i. 13 He wore the khaki shorts and bush jacket, and the Sam Browne belt, of a police officer.

‖ **sambuca**[2] (sambu·kă). Also **sambucca,** and with capital initial. [It., ad. L. *sambūc-us* elder tree: see Sambucene.] An Italian liqueur resembling anisette.

1971 P. Purser *Holy Father's Navy* xv. 121 They stared at me, coffee cups and brandies and *sambuccas* half-raised to lips. 1975 *Times* 11 Jan. 11/5 Sambuca, from Italy..contains liquorice. 1977 *New Yorker* 20 June 26/1 It got so cold that..when they tried to pour anisette or Sambuca into their coffee in the early mornings to warm up, they sometimes found it frozen in the bottle.

sambunigrin (sæmbiunəi·grin). *Chem.* [ad. F. *sambunigrine* (Bourquelot & Danjou 1905, in *Compt. Rend.* CXLI. 598), f. mod.L. *Sambucus nigra*, taxonomic name of the common elder (f. L. *sambūcus* elder + *niger* black): see -in[1].] A colourless crystalline glycoside of the nitrile of *d*-mandelic acid, found in the leaves of the elder and having the formula $C_6H_5·CH(CN)·O·C_6H_{11}O_5$.

1905 *Jrnl. Chem. Soc.* LXXXVIII. i. 912 The leaves of *Sambucus nigra* contain only traces of emulsin; it is therefore possible to extract from the air-dried or the fresh leaves the glucoside sambunigrin $C_{14}H_{17}O_6N$, which crystallises from ethyl acetate in long, colourless needles. 1965 Ansell & Gigg in S. Coffey *Rodd's Chem. Carbon Compounds* (ed. 2) Ic. ix. 104 It [*sc.* hydrogen cyanide] is widely distributed in plants in the form of..glycosides of the cyanohydrins of various aldehydes and ketones (amygdalin, prunasin, sambunigrin,..etc.).

sambuq, var. Sambook in Dict. and Suppl.

sambur. Add: Also **samba.** Substitute for def.: Either of two large deer, *Cervus unicolor* or *C. equinus*, native to southern Asia. (Further examples.)

1874 H. H. Cole *Catal. Objects Indian Art S. Kensington Mus.* ii. 107 Strips of sappan wood.., ebony and samber horn dyed green. 1913 L. Woolf *Village in Jungle* ii. 25 He showed them the sambur lying during the day in the other great caves. 1950 J. H. Williams *Elephant Bill* xi. 166, I once jumped into a creek, ten yards from a tiger, that was..eating a freshly killed samba deer. 1964 Ld. Medway in Wang Gungwu *Malaysia* I. iii. 57 This large group includes several animals widespread in South-east Asia such as the..Sambar and Barking Deer. 1969 J. Leasor *Week of Love* v. 92 A dapper man..with sambur skin shoes.

Samburu (sæmbu·ru), *sb.* and *a.* [Native name.] **A.** *sb.* **a.** A pastoral people of mixed Hamitic stock inhabiting northern Kenya; a member of this people. **b.** The Nilotic language of this people. **B.** *adj.* Of or pertaining to this people or their language.

1896 W. A. Chanler *Through Jungle & Desert* vii. 281 They said they had originally belonged to the Berkenedji or Samburu tribe. *Ibid.* 306 On our way to the new zeriba we were approached by a band of 100 Samburu. 1927 W. M. Ross *Kenya from Within* xxiv. 436 Heavy pressure was brought to bear on the local Government to grant land in actual use by the Samburu tribe to one or more of the European sheep-ranchers. 1947 [see *Masai]. 1959 A. Moorehead *No Room in Ark* iv. 100 The Samburu..are a tall fine slender people with something of the ancient Egyptians about them. 1964 J. Hillaby *Journey to Jade Sea* 81 Lelean said something uncomplimentary in Samburu. 1965 P. Spencer *Samburu* p. xxii, Pardopa clan was typical of less than a half of the Samburu clans. 1976 D. Topolski *Muzungu* xiv. 222 He answered, just as chattily in Samburu. *Ibid.* 223 An uncircumcised man is still considered to be a boy amongst the Samburu.

same, *a.* (*pron.*, *adv.*). Add: **I. 6.** *same difference*, the same thing, no difference. *colloq.*

1945 E. Wilson *I am gazing into my 8-ball* xx. 106 'That fluff from my office.' 'Fluff?' laughed Miss Lawrence. 'Fluff, doll, same difference.' 1951 J. Cornish *Provincials* II. i. 130 'I found you.' '*I* found *you*.' 'Same difference.' 1976 A. Hill *Summer's End* viii. 115 'In the first place,' he said, 'these'm *boats*, not barges.' 'Same difference,' Noggie insisted.

II. 7. a. Colloq. phr. (*the*) *same but* (or *only*) *different*: almost the same; subtly different.

1942 Berrey & Van den Bark *Amer. Thes. Slang* §16/9 *Same but different*, nearly the same. 1977 *Lancashire Life* Nov. 60/1 More seasoned observers may remark that the scene's the same, only different.

8. (Later examples.)

1914 G. B. Shaw *Pygmalion* (1916) III. 157 *Pickering.* We have taken her to classical concerts and to music..halls; and its all the same to her: she plays everything..she hears right off when she comes home. 1962 L. Deighton *Ipcress File* 8 If it's all the same to you, Minister, I'd prefer you to make a note of the questions, and ask me afterwards.

III. 10. *same-aged, -named, -natured, -sexed, -sidedness, -sized* (later example); also *same-day, -sex, -size* attrib.; **same-level** *Social Science*, analogous; that uses an established principle in one field of research for the explanation or analysis of phenomena in another field.

1949 M. Mead *Male & Female* xiv. 285 He will frown upon the same-aged youth who has a reputation for active premarital sex relations. 1967 *Punch* 22 Feb. 258/3 And such cleaning and pressing, a same-day service like you never seen. 1934 J. T. Wisdom in *Aristotelian Soc. Suppl. Vol.* XIII. 66 When the psychologist says 'I am in awe of you' means 'I fear and admire you' he is giving a more ostensive but still same-level translation of the first sentence. 1936 *Mind* XLV. 442 All material analysis is 'same-level analysis'. 1958 M. Argyle *Relig. Behaviour* xii. 141 A third kind of theory explains an empirical result by showing that it is an example of a law in another field of research—this will be called a 'same-level' theory. *Ibid.* 143 This is clearly a 'same-level' explanation, postulating that religion is learnt by the same processes of socialization as are other attitudes and beliefs. 1954 S. Duke-Elder *Parsons' Dis. Eye* (ed. 12) xxviii. 473 In vertical palsies the paresis is due to failure of the 'same-named' rectus muscle (in the left superior area, the left superior rectus) or the most 'crossed-named' oblique muscle (right inferior oblique). 1696 J. Sergeant *Method to Sci.* I. ii. 20 The same Causes upon the same-natur'd Subjects, must work the same Effects. 1949 M. Mead *Male & Female* xiv. 284 Their sex, lightly anchored to the model of the same-sex parent. 1917 Same-sexed [see *fraternal *a.* c]. 1977 *Lancet* 24 Sept. 657/2 There is sometimes same-sidedness in familial breast cancer. 1967 Karch & Buber *Offset Processes* v. 151 Set the camera for same-size reproduction. 1951 W. de la Mare *Winged Chariot* 32 On every nut there swelled the same-sized husk.

B. *absol.* and as *pron.* **2. d.** (*the*) *same again*: another drink of the same kind as the last; *same here*: the same (thing) applies to me; my case is similar; I agree; (*the*) *same to you*; I say the same thing to you (as you have just said); freq. used as a retort.

1896 W. C. Gore in *Inlander* Jan. 150 *Same here!* I agree. 1907 A. P. McKishnie *Gaff Linkum* xi. 59 'I've enj'yed th' ride in th'moonlight jest as much as I enj'yed th' singin' school.' 'Same here,' said Mr. Goosecall. 1911 G. B. Shaw *Blanco Posnet* 404 *Blanco.* Dearly beloved brethren—*A Boy.* Same to you, Blanco. 1913 Kipling *Diversity of Creatures* (1917) 288 'Do you know I've broken this man's neck?' 'Same here,' I says. 1925 *New Yorker* 17 Oct. 12/2 The same to you. 1929 D. H. Lawrence in *Star Rev.* Nov. 624 It is as if the young girl said to the young man today: I rather like you, you know. You are so thrillingly repulsive to me.— And as if the young man replied: Same here! 1938 L. MacNeice *Earth Compels* 22 What will you have now? The same again? 1949 G. B. Shaw *Buoyant Billions* II. 21 A chain shopkeeper, not a country squire. *She.* Same here: my father is a famous lucky financier. 1959, etc. [see *knob *sb.* 1 e]. 1962 *Sunday Times* 19 Aug. 18 'I'm a rugged individualist: I think for myself.' 'Me too.' 'Same here.' 1972 H. Kemelman *Monday Rabbi took Off* xxi. 136 'To tell the truth, I think it was the *rebbitsin* that wrote it and he signed it.' 'Same here.' 1975 D. O'Sullivan in D. Marcus *Best Irish Short Stories* (1977) II. 90 A thump on the counter brought the barmaid... 'Same again, ladies?'

4. a. (Later examples.)

1901 M. Franklin *My Brilliant Career* viii. 56 A big red-bearded man..had received a letter from Mrs. Bossier instructing him to take care of me. He informed me also that he was glad to do what he termed 'that same'. 1926 in H. W. Fowler *Mod. Eng. Usage* 512/1 Sir,—Having in mind the approaching General Election, it appears to me that the result of same is likely to be as much a farce as the last. 1966 G. W. Turner *Eng. Lang. in Austral. & N.Z.* vi. 135 A different influence of written language is seen in the use of *same* as a pronoun equivalent to *it*, as in 'put the tailboard up and secure same with a length of wire' from New Zealand (Wally Crump, 1964), a facetious borrowing of lawyer's English which is quite common. 1973 *N.Y. Law Jrnl.* 24 July 4/4 The following sentence in a brief is typical of its misuse as a noun: 'Waldbaum purchased the soda..then stacked it on the shelves in order to sell the same.'

5. b. pl. *Linguistics.* Features or utterances that are identical.

[1926 Bloomfield in *Language* II. 155 Within certain communities successive utterances are alike or partly alike... That which is alike will be called *same*. That which is not same is *different*. This enables us to use these words without reference to non-linguistic shades of sound and meaning. 1948 B. Bloch in *Ibid.* XXIV. 10 Successive phonotations composed wholly of the same articulations are *the same*. Other aspects are different.] 1961 R. B. Long *Sentence & its Parts* xvii. 378 The evidence of history warrants our regarding the italicized words in the following pairs as 'sames' in spite of the differences in meanings. We rode in the day *coach*. We went with the football *coach* [etc.]. 1962 [see *form *sb.* 5 c]. 1964 Crystal & Quirk *Syst. Prosodic & Paralinguistic Features in Eng.* iv. 49 We should..only subsequently look for the correlations between postulated 'sames' of tension and formal items in the linguistic and situational context which will enable us to make statements of meaning. 1977 *Trans. Philol. Soc. 1975* 9 Certain configurations in languages typically result from the principled ('lawful') divergence over time of original sames.

C. *adv.* and in adverbial phrases. **1. a.** (Examples with omission of *the*.) Also, in weakened sense: just as, as.

1857 'S. Sondnokkur' *Ryde fro Ratchda to Manchistur* (ed. 2) iv. 9 Aw kuddunt elp wundurin..wether it wur to put iz grund coffi in, saym uz wi dun o whoam. *Ibid.* vi. 14 Thir wur o rattlin saym uz uv o lot a peawur looms. 1884 'Mark Twain' *Huck. Finn* ii. 10 Strange niggers would..look him all over, same as if he was a wonder. 1930 W. Faulkner *As I lay Dying* 4 She ought to taken those cakes when she same as gave you her word. 1933 M. Lowry *Ultramarine* i. 16 He knows bloody well same as myself it doesn't pay to shout and be unkind to youngsters. 1957 L. P. Hartley *Hireling* viii. 65 But I shouldn't be able to serve them personally, same as I do now. 1975 *Listener* 6 Feb. 174/1 There was no work... They were all bad years, because, same as I say, there was nothing.

c. *same like*: just like, the same as, in the same manner as. *dial.*, *illiterate*, or *joc.*

1898 W. P. Ridge *Mord Em'ly* x. 142 Beef Pudding same like Mother makes! 1922 E. O'Neill *Anna Christie* II. 134 Two my bro'der dey gat lost on fishing boat same like your bro'ders vas drowned. 1928 J. Peterkin *Scarlet Sister Mary* iv. 47 E weddin-dress fits em same like a green shuck fits a young ear o corn. *Ibid.* xix. 207 I'll lay down on de ground an' holler same like a dog. *Ibid.* xxi. 227 'How you do today?' 'Fine. Same like a lamb a-jumpin.' 1959 A. Christie *Cat among Pigeons* ix. 107 'See no evil, hear no evil, think no evil. Same like the monkeys,' observed Sergeant Percy Bond. 1968 'L. Egan' *Serious Investigation* vi. 78 But same like the gent in Holy Writ, Beware the anger of a patient man. 1973 G. Mitchell *Murder of Busy Lizzie* xv. 185 Ain't going to be no share-out. Same like the boy with the apple-core, if you happen to know that story. 1980 I. Murdoch *Nuns & Soldiers* vii. 382, I have rich friends, same like you.

samey (sē¹·mi), *adv.* and *a.* *dial.* and *slang.* [f. Same *a.* (*pron.*, *adv.*) + -y[6].] **A.** *adv.* *all the samey* = Same *adv.* 2.

1897 Kipling *Captains Courageous* x. 221 All the samey, something's got to be done about it.

B. *adj.* Identical, characterized by sameness; lacking in variety, monotonous. Hence **sa·meyness.**

1929 E. Raymond *Family that Was* iii. 49 The days that followed, becoming 'samey'.., sank out of memory's sight. 1959 *Sunday Times* 19 July 10/6 Many of his pictures of expensive men and women on expensive horses seem samey. 1962 *Listener* 11 Jan. 90/3 All that 'samey' food and the lack of service. 1969 M. Tripp *Malice & Maternal Instinct* iv. 25 Arthur never varied his approach or technique. Arthur was samey. 1977 *Oxf. Times* 9 Dec. 17/3 Their thick sound tends towards sameyness, but the songs have enough character to retain one's interest. 1978 *Illustr. London News* Nov. 142/2 She moves beautifully and does all that a dancer could do to differentiate steps that are samey.

Samfrau (sæ·mfrau). *Geol.* [See quot. 1937.] The name of a geosyncline postulated to have extended across *Gondwanaland.

1937 A. L. Du Toit *Our Wandering Continents* iv. 62 A major geosyncline..traversing Bolivia, north and central Argentina, Cape, Weddell Sea, passing east of King Edward VII Land and through Edsel Ford land, crossing Tasmania and the eastern part of Australia to New Guinea... This feature, which seems to have played so vital a rôle during the evolution of Gondwana,..can conveniently be called the 'Samfrau' Geosyncline—a contraction of the words 'South America—South Africa—Australia'. 1959 *New Biol.* XXIX. 14 A further example of this type is the Samfrau geosyncline, of which Du Toit has suggested the remnants now occur in South America, South Africa, and Australia. 1971 M. H. P. Bott *Interior of Earth* vii. 202 Examples [of good fits of tectonic features on the assumption of continental drift] include..the fitting together of the Samfrau orogenic belt of Gondwanaland.

samfu (sa·mfū). Also **samfoo.** [Cantonese *sāam-fu*.] A suit consisting of jacket and trousers worn by Chinese women, particularly in Malaysia and Hong Kong; also worn by men. Freq. *attrib.*, as *samfu jacket, trousers.*

1955 D. Moore *We live in Singapore* 41 Her thin, pendulous breasts hung down inside her *samfoo* jacket like malignant deformities. 1963 J. Kirkup *Tropic Temper* ii. 23 Chinese girls in samfu, a kind of flowered pyjama suit with short flaring jacket. 1966 D. Forbes *Heart of Malaya* iv. 47 A Chinese woman dressed in a white peasant smock and blue *samfu* trousers. 1967 A. Cordell *Bright Cantonese* iv. 51 The people..came..to the bath, pulling their samfoo jackets over their heads. *Ibid.* vi. 69 He brought white samfoo trousers and a jacket heavily stained with crimson flowers. 1969 J. Bennett *Dragon* ii. 19 She was wearing a *samfu*, the blue or black suit which looks like the Viet Cong pyjama uniform. 1975 O. Sela *Bengali Inheritance* ii. 20 The witness was a frail, elderly Chinese... He wore a tattered black *samfoo*.

Samgha, var. *sangha.

‖ **Samhain** (saun; sau·in, sa·win). Also **Samain, Samhainn.** [a. Ir. *samhain* (Sc. Gaelic *samhuinn*), OIr. *samain*.] The first day of November, celebrated by the ancient Celts as a festival marking the beginning of winter and of the new year according to their calendar; All Saints' Day or Hallowmass. Also *attrib.* Cf. Beltane.

The OIr. form *samain* is used only with reference to the ancient Celts. 'Samhain Eve' (quot. 1904) and 'the night of Samhain' (quot. 1910) are different renderings of Ir. *oidhche Shamhna* 'Hallowe'en'.

1888 J. Rhys *Lect. Orig. & Growth Relig. as illustr. by Celtic Heathendom* v. 518 The Samhain feast..was, like the Greek Apaturia, partly devoted to business..otherwise the feast, which occupied, not only Samain or the first of November, but also the three days before and the three days after it, was given up to the usual games. 1904 W. B. Yeats *Stories of Red Hanrahan* 1 The barn where some of the men were sitting on Samhain Eve. 1910 J. M. Synge *Deirdre of Sorrows* I. 5 And it raining since the night of Samhain. 1917 J. M. Clark *Vocab. Anglo-Irish* vii. 27 Irish folk-lore has kept alive words of such classic associations as..Samhain and shanahus..which mean..'All-Hallowtide' (Nov. 1) and 'a friendly chat' respectively. 1949 J. A. MacCulloch *Celtic & Scandinavian Religions* I. viii. 58 Samhain, which means 'summer end', naturally pointed to the fact that the powers of blight, typified by winter, were beginning their reign. But it may have been partly a harvest festival. 1957 W. R. Kermack *Scottish Highlands* 153 At Samhain (Hallowe'en, 31st October, the beginning of Winter) the Lewismen made libation to the sea-god Shony, who could send them plenty of seaweed to manure their fields. 1958 T. G. E. Powell *Celts* iii. 117 At *Samain*, sacrifices were certainly offered although no material descriptions have survived. 1968 *New Larousse Encycl. Mythol.* 236/1 The [Celtic] year began on what is now the first of November with the feast of *Samain*... The ordinary people felt less sanguine about the possibility that on the eve of *Samain* the people of the *side* left their domain and wandered in the world of man. 1970 *Q. Rev. Guernsey Soc.* XXVI. 60 These four were the feast of Beltainn the great Sungod in May; mid-summer, mid-August..and Samhainn or Hallowmass (November 1).

Sam Hill (sæm hil). *N. Amer. slang.* Also **sam hill, samhill.** [Orig. unknown.] A euphemism for hell; used especially in expressions of impatience or irritation preceded by *in* or *the* with an interrogative word.

1839 *Havana* (N.Y.) *Republican* 21 Aug. 1/4 What in sam hill is that feller ballin' about? 1868 J. T. Trowbridge *Three Scouts* vi. 26 When you might a'married!—why in Sam Hill didn't ye, then? 1894 'Mark Twain' in *St. Nicholas* Jan. 257/2 Hateful people..giving me Sam Hill because I shirked. 1909 *N.Y. Even. Post* 10 Apr. (Sat. Suppl.) 3/5 How in Sam Hill can she do it? She's just as hot when she gets to bilin' p'int as she'll ever be. 1918 M. E. Freeman *Edgewater People* 314 What in Sam Hill made you treat him so durned mean fur? 1927 W. James *Cow Country* 77 What the Sam Hill do you think we are out here, servants? 1948 *Salt Lake Tribune* 18 Dec. 10/7 He wondered who the Sam Hill the 'senator' was. 1962 H. Green *Time to pass Over* xii. 147 Why in the samhill didn't you step in and stop them, Mike. 1973 B. Broadfoot *Ten Lost Years* xxiii. 262 He probably never could have figured what the Sam Hill was going on.

Samhita (sa·mhitā). Also **sanhita.** [Skr. *saṃhitā* union, connection, f. *sam* together + *dhā* to place.] A text treated according to sandhi; a version of the vedas which is the continuous text formed from the pada or separate words by the appropriate phonetic sound changes. Also *attrib.*

1805 H. T. Colebrooke in *Asiatick Researches* VIII. 476 Tradition..reckons sixteen Sanhitas of the Rigveda. 1843 *Penny Cycl.* XXVI. 171/1 The Rigveda is the first in order and its Sanhitā contains mantras..to the elemental deities. 1887, 1917 [see *pada b]. 1920 J. N. Farquhar *Outl. Relig. Lit. India* i. 26 The Veda as handed down in the various schools..soon showed considerable differences.

It has come down to us in four distinct forms called *Saṁhitās.* **1953** in K. W. Morgan *Relig. of Hindus* vii. 265 The Upaniṣhads are the philosophic and mystical elaboration of the truths first revealed to the Seers and recorded in the Saṁhitās. **1974** *Encycl. Brit. Micropædia* X. 375/3 The foremost collection, or Saṁhitā, of such hymns..is the Ṛgveda.

Sami (sā·mi, sām). Also †**Salme-Same; Saam(e, Sabme,** etc. [Lappish *Sami* (in earlier orthography, *Sabme, Samek*) of uncertain ultimate etym.; cf. also Sw. and Norw. *Same.*] The native name of the Lapps; occas. *sing.*, a Lapp.

This word is preferred to *Lapp* by scholars.

1797 *Encycl. Brit.* IX. 572/1 The Laplanders call themselves *Salme-Same*, and *Salmen-Almatjeh.* **1842** *Penny Cycl.* XXIII. 390/1 They [*sc.* the Laplanders] call themselves Sami. **1864** *Chambers's Encycl.* VI. 38/1 The Lapps, who call themselves the *Sami* or *Sahmelads*, are a physically ill-developed, diminutive race. **1935** S. J. Beckett *Wayfarer in Norway* xx. 145 The Lapps call themselves Sami or Sahmelads, whilst they call the Finns Suomi (which, like the name Finn, means 'Fen', or marsh-dwellers). **1957** R. Paine *Coast Lapp Society* I. 1. i. 3 The people whom we know as Lapps have their own name for themselves—*sabme*, plural *sāmek*. In academic circles inside Scandinavia, the Lappish term is now replacing any other... Outside Scandinavia, however,..*sabme* is not widely understood. **1964** S. Dunn et al. tr. E. D. Prokof'yeva in Levin & Potapov *Peoples of Siberia* 547 Some scholars have compared the name 'Samoyed' with the Lappish (Saam) words 'same-yedne' ('land of the Saams'). This is based on the fact that the territory settled by the Nentsy.. was in earlier times inhabited by the Lapps (Saams). **1968** [see *pot *sb.*[1] 13 f]. **1977** *Daily Colonist* (Victoria, B.C.) 19 June 22/3 The..reindeer herder is one of the Saame, better known as Lapps. **1980** *Times* 8 Feb. 14/7 The 2,000 or so Lapps, or more accurately Sami, who live in this area.

Samian, *a.* and *sb.* Add: **A.** *adj.* (Earlier and later examples of pottery.) Also *ellipt.* and with small initial. **Samian ware** (earlier example).

1779 T. Pownall in *Archæologia* V. 287 The one [*sc.* vessel] is a red sort, the Ionian, or particularly the Samian, which is most commonly found. *c* **1841** W. T. P. Shortt *Sylva Antiqua Iscana* 110 The great quantity of fragments of Roman Red Ware, especially of that beautiful description, known to the ancients by the generic term of Samian, is not by any means the least interesting of the curiosities dug up in the city of Exeter, of late years. **1844** *Gentl. Mag.* July 35/2 Whether that singularly beautiful red glazed earthenware.. of which such vast quantities have been since exhumed in every part of England and France where their respective records have assigned a Roman station, be really the identical Samian pottery of Pliny, is, I think, a question yet to be decided. **1848** *Jrnl. Brit. Archaeol. Assoc.* Apr. 2 The Samian ware is found throughout this country almost wherever Roman remains are met with. **1958** Stanfield & Simpson *Central Gaulish Potters* 1. 52 Black slip samian has a pinkish-red or buff-coloured core. **1967** *Antiquaries Jrnl.* XLVII. 192 It seems most likely that the later samian has come from pits dug into the rampart. **1981** P. Salway *Roman Britain* 202 Mr B R Hartley's study of the dies for the makers' stamps on samian pottery reveals that the amount of samian bearing the same stamps which comes from the two Walls is negligible.

samiel. Add: Now with pronunc.: (sa·miĕl). (Later examples.)

1832 J. Bell *Syst. Geogr.* IV. 231 The most dreadful of all winds is the famous semoum or samiel,..which prevails in the desert bounded by Bassora, Bagdad, Aleppo, and Mekka, and the effects of which are suffocation and immediate putrefaction of the body. **1962** [see *leveche].

samisen. Add: Also 7 **shamshin,** 9 **samsi, samishen,** 9– **shamisen; shamisan.** (Earlier and further examples.)

1616 R. Cocks *Diary* 9 Oct. (1883) I. 188 The *tuerto* that plaid on the *shamshin.* **1822** F. Shoberl tr. *Titsingh's Illustr. Japan* 94 Several young females came to bear them company, playing on the *samsi*, and dancing. **1840** *Chinese Repository* Dec. 630 The *samishen* is a three-stringed guitar, and is usually played with a plectrum. **1871** A. B. Mitford *Tales of Old Japan* I. 243 The *shamisen*, a sort of banjo. **1880** I. Bird *Unbeaten Tracks in Japan* I. 134 Yuki plays the *samisen*, which may be regarded as the national female instrument. **1936** K. Sunaga *Japanese Music* i. 19 The instrument employed as the accompaniment for the songs of geisha girls..was the *samisen*... It might be described as a three-stringed, rectangular banjo. **1955** E. Pound *Classic Anthol.* II. 115 And words soft as the shamisan Distinguish the thick-faced man. **1964** I. Fleming *You only live Twice* i. 18 [Bond was] far from being..bewitched by the inscrutable discords issuing from the catskin-covered box of the three-stringed *samisen.* **1970** J. W. Hall *Japan* x. 227 The music of the *shamisen.* **1972** *Times* 18 Sept. 1/4 Guests knelt on tatami and used chopsticks to eat while geisha girls played the samisen.

samite. Add: Also *fig.*, *attrib.*, and *Comb.*

1938 R. Graves *Coll. Poems* 63 Into their many-shielded, samite-curtained, Jewel-bright hall where twelve Kings sit at chess. **1971** 'A. Burgess' *MF* xvii. 191 There were fireworks out tonight, thudding and searing the samite air.

samiti (sæ·mĭti). Also **samity.** [Hind. *samiti* meeting, committee.] In India and Bangladesh, an assembly or committee.

1930 M. L. Darling *Rusticus Loquitur* v. 124 The 250 Mahila Samitis or Women's Institutes founded in Bengal by Mrs G S Dutt. **1950** M. Masani *Our Growing Human Family* 41 The village assembly, known in Europe as the Folk-moot and in India as the Samiti. **1962** *Times* 26 Jan. (Survey of India) p. vi/6 Presidents of the various block *samitis* in a district form the *zila parishad.* **1975** *Bangladesh Times* 19 July 8/8 All the members of the Board of Directors of Bangladesh National Bidi Sramik Samabaya Samity.. have also applied for membership of the national party. **1976** D. Hiro *Inside India Today* 50 What then emerged was a three-tiered system whereby the old district boards.. were replaced by Zilla parishads (i.e. district councils) with responsibility for co-ordinating development plans to be channelled through panchayat samitis (i.e. council committees) consisting of a number of popularly elected panchayats encompassing one or more villages—all interlinked through indirect elections. This system, popularly known as the panchayat raj, was first introduced..in 1959.

‖ **samizdat** (sæ·mizdæt, səmizda·t). Also with capital initial. [Russ., abbrev. of *samoizdátel'stvo* self-publishing house, f. *samo-* self + *izdátel'stvo* publishing house.] The clandestine or illegal copying and distribution of literature (orig. and chiefly in the U.S.S.R.); an 'underground press'; a text or texts produced by this. Also *transf.* and *attrib.* or as *adj.* Phr. *in samizdat*, in this form of publication.

1967 *Times* 6 Nov. (Russia Suppl.) p. xxii/4 A vast and newly educated [Soviet] population..do not pass around the precious *samizdat* (unpublished) manuscripts. **1968** tr. I. A. Yakhimovich *Let.* in *Probl. of Communism* July–Aug. 48/1 One must not speculate with the honor of the state, even if a certain leader wants to end *samizdat.* **1968** *Time* 27 Sept. 22/2 Those lines [of Solzhenitsyn] have not been published in the Soviet Union. But they are nonetheless read and passed from hand to hand in *samizdat*, the readers' answer to Soviet censorship. **1970** *New Statesman* 20 Feb. 241/1 The underground distribution of manuscripts and their publication abroad means that the *samizdat* writers have—at least in the eyes of the authorities—opted out of the Soviet scheme of things. **1971** *Guardian* 15 July 13/8 Nicolae Ceausescu's latest puritanical damper on 'Bourgeois Influences' in Rumania coincides with the first case of underground 'Samizdat' literature to come from there. **1973** R. Rosenblum *Mushroom Cave* 55 I've told you how effective the samizdat network has been in circumventing the repression of criticism. **1977** M. Walker *National Front* vii. 182 The NF *samizdats* which did so much to pollute the atmosphere of NF life during the year [*sc.* 1975]. **1977** *Time* 28 Nov. 30/2 An exhibit of clandestine *samizdat* in the Correr Museum. **1978** *Manch. Guardian Weekly* 27 Aug. 7 Jiri Hrusa's novel 'The Questionnaire', which was printed by the Prague Samizdat. **1980** *Times Lit. Suppl.* 3 Oct. 1094/4 The strongest works to have come out since 1962—Solzhenitsyn's *The First Circle* [etc.].. —have appeared, and could only appear, in *samizdat.*

Hence **samizda·tchik** [Russ. *-chik*, agent suffix], one who takes part in the writing, copying, and distribution of *samizdat* material (pl. *samizdatchiki*).

1972 *N.Y. Times Mag.* 10 Sept. 92 To fill their reserves.. the *samizdatchiki* seek ties with other cities... They arrive with copies of the originals, which have been given abroad. **1979** *N.Y. Times Bk. Rev.* 20 May 3/2 He assiduously collects information for the samizdat journals..writes pseudonymous articles for samizdat and spends weeks on end retyping the Chronicle and other materials from Moscow in multiple copies. He is the quintessential *samizdatchik.*

Samkhya, var. *Sankhya.

‖ **samlor** (sæ·mlǭɹ). Also **samlo.** [Thai.] Chiefly in Thailand, a three-wheeled vehicle, freq. motorized, used as a taxi.

1955 *Times* 6 May 11/6 Let him be insidiously towed in a motor *samlor* round this town of Bangkok. **1960** R. Kirkbride *Innocent Abroad* xi. 82 We crossed a humped bridge.. in the city, dodging about amongst cycle-rickshaws and samlors. **1963** 'Han Suyin' *Four Faces* 54 Peter would photograph her in front of the temples of Angkor, stepping down from a *samlo*, the three-wheeled vehicle, man-propelled, in use in Cambodia. **1974** *Time* 7 Jan. 50/2 The Assembly even includes a *samlor* driver, who intends to park his three-wheel smoke-belching minitaxi at the National Assembly building.

Sammy (sæ·mi), *sb.* Also **Sammie.** [Familiar dim. of the name *Samuel*: see -y[6].] † **1.** *slang.* A ninny, simpleton. Also in *Comb. Obs.*

1837 E. Howard *Old Commodore* II. iii. 54 You have been sammy-foozled by a rascally swindler. **1838** R. B. Peake *Quarter to Nine* I. ii. 10 What a Sammy, give me a shilling more than I axed him! **1897** F. T. Jane *Lordship Passen and We* xv. 165 Simple Sammy, as we called Mr. Pote, the new pastor.

2. *slang.* In British use: an American soldier in the war of 1914–18, so called from the name *Uncle Sam* (see Uncle *sb.* 2 c). Now *rare* or *Obs.*

1917 *Punch* 13 June 384/2 As a term of distinction and endearment [for the American 'Tommies'] Mr. Punch suggests 'Sammies'—after their uncle. **1917** *Nation* (N.Y.) 16 Aug. 164/1 The 'Sammies' whom the headlines are featuring. **1918** *Stars & Stripes* 29 Mar. 4/1 A Sammie may be defined as an American soldier as he appears in an English newspaper or a French cinema. It is a name he did not invent, does not like, never uses and will not recognize. **1921** *Glasgow Herald* 8 July 7/2 While a French soldier costs on average 13 francs 37 per day,..a 'Tommy' costs 31 francs 69, and a 'Sammy' 59 francs 30.

sammy, *v.* Add: (Earlier and later examples.) Also, to dampen (leather that has been allowed to dry out) slightly.

1885 C. T. Davis *Manuf. Leather* xxix. 502 The eleventh step, which consists in 'sammying' the hides, is then carried into effect. **1922** A. Rogers *Pract. Tanning* xv. 449 After the leather has been dried out, in order to set the fiber it must again be dampened back or sammied before carrying out the finishing process. **1974** P. W. Blandford *Country Craft Tools* xv. 199 The currier used a 'sleaker' to force out dirt, then the hide was 'sammied' by rolling either between a pair of rollers or under a heavy brass roller.

Samnite, *sb.* Add: **b.** A type of gladiator. **c.** The language of the Samnites.

1600 Holland tr. *Livy's Romane Hist.* ix. 344 The Campaines upon a pride, and inveterate hatred that they bare against the Samnites, used to arme their swordplayers and fensers at the sharpe (which was a solemne sight and pastime they had at their great feasts) with this same attire, and termed them in mockerie, by the name of Samnites. **1859** B. W. Dwight *Mod. Philol.* I. 187 The Umbro-Samnite Dialects: Umbrian; Samnite or Oscan; Volscian; Marsian. **1882** [see *Marsian *sb.* and *a.*]. **1957** *Encycl. Brit.* X. 383/2 The Samnites fought [in gladiatorial contests] with the national weapons—a large oblong shield, a vizor, a plumed helmet and a short sword. **1971** M. Grant *Cities of Vesuvius* iii. 74/1 Down to the first century BC, 'gladiator' and 'Samnite' were synonymous terms; and then the latter became the name of a particular type of gladiator.

Samoan, *sb.* **a.** (Earlier examples.)

1856 J. C. Patteson *Let.* in C. M. Yonge *Life J. C. Patteson* (1874) I. vii. 262 Another crew arrived with a Samoan teacher... I rode out pick-a-back on the Samoan, Leonard following on a half-naked Anaitean. **1871** C. M. Yonge *Pioneers & Founders* ix. 250 These Samoans, though they deified many animals, had no temples, idols, priests, nor sacrifices.

samoleon, var. *simoleon.

samosa (sămō[u]·să). Also **samoosa, samusa.** [Hind.] A triangular pastry fried in ghee or oil, containing spiced vegetables or meat.

1955 R. P. Jhabvala *To whom she Will* ix. 67 Another plate was filled with..samusas. **1960** —— *Householder* i. 54 'They have made vegetable samusas with our tea,' Romesh told his father. **1971** *Weekend* (Colombo) 12 Sept. 6/3 (Advt.), Rizwana for cool cool Faluda and fresh hot Samosa. **1974** N. Gordimer *Conservationist* 131 What'd you find to buy?—..Samoosas.—..He puts the neat, crisp, greasy triangle whole into his mouth. **1978** [see *pakora].

Samoyed, *sb.* and *a.* Add: Also **Samoyede.** **A.** *sb.* **1.** (Earlier and further examples.)

1589 A. Jenkinson in Morgan & Coote *Early Voy. Russia & Persia* (1886) I. 36 The Tartars and Gentiles, called Samoydes. **1841** [see *Ostyak]. **1911** *Encycl. Brit.* XXIV. 118/1 The language now spoken by the Samoyedes belongs to the Finno-Ugrian group, and is allied to Finnish but has a more copious system of suffixes. **1944** [see *Nenets]. **1972** *Language* XLVIII. 206 The Samoyeds make up only one small group of scattered tribes among the many non-Russian peoples who have inhabited Siberia.

2. Also with small initial. A white or buff dog belonging to the breed so called, once used as working dogs in the Arctic, and distinguished by a thick, shaggy coat, stocky build, pricked ears, and a tail curled over the back. Also *attrib.*

1889 *Pall Mall Gaz.* 30 Apr. 6/2 A beautiful brown silky-haired sharp-eared Samoyed dog. **1905** [see *laika]. **1914** N. Newnham-Davis *Gourmet's Guide to London* liii. 340 [He] brought me in..to look at a delightful little Samoyede puppy. **1922** R. Leighton *Compl. Bk. Dog* vii. 98 The white Samoyed is one of the most beautiful of all dogs. **1934** [see *malamute]. **1954** M. K. Wilson tr. *Lorenz's Man meets Dog* ix. 90 Pointed muzzles, obliquely set Mongolian eyes and pricked ears pointing sharply upwards..that fascinating expression which distinguishes Greenland sledge-dogs, Samoyeds and Huskies. **1977** G. Marton *Alarum* 61 The well-fed passengers..probably expected to be carried across immense ice fields by rough Samoyed dogs.

B. *adj.* (Earlier and later examples of use as a quasi-*sb.*)

1822 tr. *Malte-Brun's Universal Geogr.* I. xxiii. 571 The Tunguse is a dialect of the Mantchou; the Samoyede differs from it. **1956** J. Whatmough *Language* 28 In the north, Samoyede, a member of the same family as the Finnish dialects.

sampan[1]. Add: **3.** *Comb.*, as **sampan-wallah** [Wallah], a boatman in charge of a sampan.

1932 *Times Lit. Suppl.* 29 Sept. 693/3 In time he became a sampan-wallah. **1934** 'G. Orwell' *Burmese Days* vii. 123 The successful sampan-wallah turned and discharged at his rival a mouthful of spittle.

sampan[2] (sæ·mpăn). [Khoi-khoin *samban.*] = Tampan in Dict. and Suppl.

1898 W. C. Scully *Between Sun & Sand* i. 8 The ground beneath is full of the dreaded 'sampans', which bury themselves in the flesh and cause serious injury. **1920** *Glasgow Herald* 2 Sept. 4 The minor plagues of scorpions and 'sampans'.

samphire. Add: **3.** *samphire-bush* (later example), *-gatherer*, *-greens.*

1928 V. Woolf *Orlando* iv. 148 Closer and closer they drew, till the samphire gatherers, hanging half-way down

the cliff, were plain to the naked eye. **1941** I. L. Idriess *Great Boomerang* ii. 10 Out among the samphire bushes lay huddled the little cloud of sheep. **1970** S. Trueman *Intimate Hist. New Brunswick* iii. 57 Come home with thoughts of periwinkles, dulse, maple cream, samphire greens. **1971** *Country Life* 28 Oct. 1132/3 The reference to samphire, which for most of us means *King Lear*, the blinded Gloster and the wretched samphire-gatherer on Dover cliff.

sampi (sæ·mpəi). Also **sanpi.** [Late Gr. σαμπῖ, prob. f. ὡς ἂν πῖ like pi.] The modern name for an ancient Greek numeral (ϡ) = 900, which has been hypothetically identified with one of several sibilants in early Greek alphabets.

1833 *Penny Cycl.* I. 385/2 The letter *tsadi* has no representative in the Greek alphabet, unless, indeed, it bear any relation to the Greek figure called *sanpi*, which, however, was never used, as far as it is known, for an alphabetic character. **1875** *Encycl. Brit.* I. 609/2 Herodotus..speaks of the 'same letter which the Dorians call σάν, the Ionians σίγμα'; and though *san* was no letter of the Ionic alphabet, the compound *sampi* (= σαν+πι) denoted 900. **1912** E. M. Thompson *Introd. Greek & Lat. Palaeogr.* vii. 91 A symbol derived from the old letter *san*,..which, from its partial resemblance to *pi*, was called *sampi* (= *san*+*pi*), for 900. **1968** W. S. Allen *Vox Graeca* i. 58 This stage [*sc.* affricate stage of [ts]] is probably represented by some early Asiatic Ionic inscriptions which show in such cases a special letter **T**.., which may be derived from the Semitic '*tsade*' (and perhaps survives in the numeral symbol ϡ = 900, now known by the late Byzantine name of σαμπῖ < ὡς ἂν πῖ).

sample, *sb.* Add: **2. c.** A specimen taken for scientific testing or analysis.

1878, 1882 [in Dict., sense 2 a]. **1895** J. C. Guernsey *Urinalysis including Blanks* 11 If a sample of urine cannot be analyzed immediately upon its receipt, add ten to fifteen grains of salicylic acid. **1938** Lundell & Hoffman *Outl. Methods Chem. Analysis* iii. 21 In a chemical analysis, the first consideration is the use of a sample that truly represents the material under test. **1950** Race & Sanger *Blood Groups in Man* i. 3 The reactions when different cell samples are tested against parallel titrations of the same antisera. **1973** J. G. Dick *Analytic Chem.* ii. 32 Samples of impure acidic substances were analyzed by a neutralization method.

d. *Statistics.* A portion drawn from a population, the study of which is intended to lead to statistical estimates of the attributes of the whole population.

1903, 1922 [see *population[2] 2 d]. **1944** H. G. Wells *'42 to '44* 42 He would get answers to his questions from Samples of his Consumers. **1951** [see *mid-range 1]. **1961** *Listener* 9 Nov. 780/2 There is the social survey of, say, the Young and Wilmott kind, with its planned interviews of samples. **1979** *Church Times* 9 Mar. 2/2 A nationally representative sample of 956 people was interviewed for the survey.

6. a. (Further examples.)

1895 W. Schlich *Man. Forestry* III. 1. iv. 66 Having ascertained the volume of the sample plot, that of the whole wood can be calculated. **1970** G. A. & A. G. Theodorson *Mod. Dict. Sociol.* 361 The extent to which generalizations based on sample data may be considered applicable to the total population from which the sample was drawn depends on the method used to select the cases included in the sample and the size of the sample. **1978** C. H. Stoddard *Essent. Forestry Pract.* (ed. 3) vi. 119 Timber estimators also measure and tally the trees in the strips or sample plots. **1978** R. V. Jones *Most Secret War* xix. 156 Maita had stayed in Salisbury because she was nervous about London, but she wanted to come for a sample weekend, and when would I recommend?

a*. General *attrib.*, as *sample investigation, method, study, survey* (hence *sample-survey* vb. trans.).

1930 *Economist* 1 Nov. 801/2 Even so, his impressions are inevitably based on 'sample' investigations and must be read with these limitations in mind. **1944** H. G. Wells *'42 to '44* 43 The sample method of dealing with human affairs is exemplified by various uses to which we can put a jury. **1965** J. Meuvret in Glass & Eversley *Population in Hist.* xxi. 516 Other sample-studies in parish registers..have revealed analogous results. **1966** *Economist* 12 Nov. 682/3 The development of economic knowledge sample-surveyed in these articles has been matched by a growing use of economists in business and government. **1975** *Listener* 6 Feb. 187/3 The study of human behaviour by..sample surveys.

a.** Used *attrib.* to denote various statistical attributes of a sample, as *sample average, mean, range*, etc.

1939 A. E. Treloar *Elements Statistical Reasoning* x. 137 The standard error of the sample mean. **1941** *Ann. Math. Statistics* XII. 91 (*heading*) Determination of sample sizes for setting tolerance limits. **1947** O. L. Davies *Statistical Methods Res. & Production* ix. 217 One [chart] on which the sample averages $\bar{x}$ are recorded and the other on which sample ranges *w* are recorded. **1971** Hickman & Hilton *Probability & Statistical Analysis* ix. 153 The sample variance..is said to be a point estimator of the population variance in the same sense that the sample mean..is a point estimator of the population mean.

b. sample book, a book containing samples of fabrics for prospective buyers; **sample bottle,** a bottle in which samples of fluid from the body may be collected; **sample case,** a case containing samples carried by a travelling salesman; **sample room,** (*a*) and (*b*) (earlier examples).

1938 *Burlington Mag.* Apr. 200/2 The distinguished firm of weavers, whose sample-books of 100 years and more ago are still in existence. **1976** P. Clabburn *Needleworker's Dict.* 232/1 Old sample books still in existence are of the greatest value in telling later generations what was in fashion at a particular date. **1977** *Belfast Tel.* 22 Feb. 8/6 Doctors' hands.., little sample bottles, having your arm draped in black as your blood pressure is taken. **1875** *North Alabamian* (Tuscumbia, Alabama) 30 Sept. 3/3 We were not glad to see him, as he had left his sample case at home. **1935** [see *keister 1 a]. **1971** D. E. Westlake *I gave at the Office* 123 A salesman rapping his knuckles on his sample case in a waiting room. **1865** G. A. Sala *My Diary in Amer.* II. 46 Sometimes the bar is at the side, screened off, and genteelly disguised under the name of 'sample room'. You enter ostensibly to purchase cherries, and immediately 'put yourself outside' a 'tot' of Bourbon. **1869** W. H. Brewer *Rocky Mountain Lett.* (1930) 10 'Saloons', 'barrooms', 'sample-rooms', 'liquor stores', 'lager beer', etc., furnish most of the signs on the places of business. **1887** *Grip* (Toronto) 21 May 10/2 One of the drug travellers insisted..that the clerk..had, in the north sample room, first nicknamed Albendis 'Chippy'. **1892** *Hist. Rev. York County* (Pa.) 62/1 To the side is the reading and sample-rooms for the commercial traveler.

sample, *v.* Add: **5. a.** (Examples of use with inanimate subj.)

1974 Harvey & Bohlman *Stereo F.M. Radio Handbk.* v. 119 The composite audio signal voltage is sampled at the midpoints by a train of short pulses and the sampled voltage level is held between pulses. **1978** *Nature* 13 July 135/2 A synchronous gating circuit..samples the V_2 signal at a selectable phase ('phase lock'), converting it to a proportional d.c. voltage.

6. To provide with samples.

1935 A. P. Herbert *What a Word!* iii. 83 From a firm of 'Publishers and Educational Contractors for Handicraft Materials':..'We shall welcome the opportunity of *sampling* you with anything you would like to see.' 'We are *sampling* Norway with the new articles.' **1946** K. T. Keller in *Chrysler Corp. Ann. Rep.* XXI, The limited production to date of our new models has been inadequate to properly sample our dealers.

7. *Comb.*: **sample-and-hold** *adj. phr. Electronics*, applied to a circuit or technique in which a varying voltage is sampled periodically and the sampled voltage is retained in the interval until the next sampling.

1966 M. Schwartz et al. *Communication Syst. & Techniques* vi. 244 One difficulty in PAM systems used for time-division switching is that the short samples do not deliver very much average signal power to the individual receiving channels. The difficulty can be remedied by the use of a sample-and-hold circuit. **1974** Harvey & Bohlman *Stereo F.M. Radio Handbk.* v. 119 Using this sample-and-hold technique, good channel separation, low distortion and low subcarrier breakthrough may be realized. **1979** C.-T. Chen *One-Dimensional Digital Signal Processing* 435 The conversions between analog and digital signals are performed by sample-and-hold (S/H) circuits, analog-to-digital (A/D) converters, and digital-to-analog (D/A) converters.

sampled, *ppl. a.* Add: **2. sampled data,** data supplied at regular intervals, rather than continuously; freq. *attrib.*, designating a system whose behaviour is modified by such data.

1951 *Trans. Amer. Inst. Electr. Engin.* LXX. 1779/1 The design of a sampled-data servo system is as direct as the design of a conventional system. *Ibid.* 1779/3 A control system makes use of sampled data when it is impossible to supply continuous data to all its parts. **1955** J. G. Truxal *Automatic Feedback Control System Synthesis* ix. 500 Servo-mechanisms which operate on sampled data: *i.e.*, systems for which the input (or the activating signal) is represented by samples at regular intervals of time, with the information ordinarily carried in the amplitudes of the samples. **1968** *Brit. Med. Bull.* XXIV. 252/1 The breathing mechanisms must include a form of 'memory' and a so-called 'sampled data' system.

sampler, *sb.*[1] **5.** Delete † *Obs. rare* and add later examples.

1972 T. Kochman *Rappin' & Stylin' Out* p. xv, Minimally necessary would be a comparative sampler of the diverse preaching styles that exist in the black community. **1975** *Booksellers Weekly* 15 Sept. 55/3 The authors also include a sampler of foreign menus: continental, Italian, Greek, Mexican, oriental, Indonesian. **1976** *National Observer* (U.S.) 17 Jan. 14/1 (Advt.), Try this Vermont Sampler... We'll send you, on approval, our Vermont cob-smoked ham..and 1 lb. of our delicious cob-smoked bacon.

6. That which contains a sample or representative selection; *spec.* a gramophone record of examples of a performer, type of music, etc.

1969 *Nature* 10 May 599/1 This paperback is a sampler of letters, periodicals, and reports in the United States relating to the publication late in 1859 of Charles Darwin's 'Origin'. **1970** *Melody Maker* 20 June 27/4 We are promised jazz releases from A. & M. Records commencing shortly with a jazz sampler. **1977** *Linlithgowshire Jrnl. & Gaz.* 15 Apr. 6/6 And one of the best ways to start is with Atlantic Record's sampler containing numbers from the albums 'War Babies', 'Abandoned Luncheonette' and 'Whole Oates'.

7. *attrib.* and *Comb.*, as (sense 3 b) *sampler rhyme*; (sense *6) *sampler album* (*album[1] 6), *collection, record.*

1977 *Zigzag* Mar. 28/3 He's also doing a sampler album, with sleeve notes too, I think. **1973** A. Dundes *Mother Wit* p. xii, There are already sampler collections of raw folklore data available. **1975** *Gramophone* May 2024/1 Gustav Leonhardt Portrait. Sampler Record. **1951** W. de la Mare *Winged Chariot* 51 My cross-stitch sampler-rhyme.

sampler, *sb.*[2] Add: **1.** Also, one employed in any other form of sampling. (Further examples.)

c **1950** G. van Delden *I have Plan* i. 15 He..came to the mine.., so they put him on the staff as a sampler. **1971** J. B. Carroll et al. *Word Frequency Bk.* p. xviii, The responsibility for judging whether or not a textual segment was an essentially English sentence was left to the sampler.

2. A device for obtaining samples for scientific study.

1902 *Bull. U.S. Fish Comm.* 1901 XXI. 58 (*caption*) Soil sampler, after Delbecque. **1927** *Bull. Nat. Res. Council* (U.S.) No. 61. 237 A modification of the Davis peat-sampler. *Ibid.* 238 In deeper water the other sampler had to be used. **1946** *Geogr. Jrnl.* CVII. 164 The core-sampler gave relatively short cores or none at all... The piston-sampler constructed by Dr. Kullenberg—a modification of the original vacuum core-sampler—secured practically undisturbed cores down to a maximum depth of over 3600 metres. **1959** *Jrnl. Sci. Instruments* XXXIV. 3 (*heading*) Impaction sampler for size grading air-borne bacteria-carrying particles. **1974** *Nature* 25 Oct. 678/2 The air samplers do not register this excess presumably because the size of the spray particles is beyond the upper limit of collection of the sampling duct.

sampling, *vbl. sb.* Add: **2.** (Further examples in *Statistics.*)

1924 J. Stamp *Stud. Current Probl. Finance & Govt.* 12 A second development of statistics, along the lines of the theory of probability, is in the important principle of 'sampling'... Under certain conditions 20 per cent., or even 5 per cent., samples may yield satisfactory and reliable results. **1935** *Brit. Birds* XXVIII. 332 Sampling is an attractive labour-saving device, but it presents a number of obstacles and pitfalls, and needs to be used with considerable caution. **1973** *Jrnl. Genetic Psychol.* CXXII. 249 This wide sampling yields a measure of a generalized expectancy of reinforcement.

3. *attrib.*, as *sampling method, rate, survey*; **sampling distribution,** the theoretical frequency distribution of a statistic, as calculated from a sample, over all samples of the same size and kind; **sampling error,** error due to the use of a sample which does not perfectly characterize the population from which it is drawn.

1928 *Proc. R. Soc.* A. CXXI. 654 (*heading*) The general sampling distribution of the multiple correlation coefficient. **1967** R. C. Campbell *Statistics for Biologists* ii. 32 A sample statistic..has a sampling distribution. This last term is slightly misleading, because all the distributions we have considered arise from sampling; the name is however usually reserved for the distribution from sample to sample of a statistic calculated from each sample. **1914** *Psychol. Rev.* XXI. 109 The correlation (compensated for sampling errors) between any two columns. **1955** *Times* 8 July 7/3 It is anticipated that..the estimated total population of the Sudan will have a sampling error of substantially less than 1 per cent. **1974** *Times* 11 Feb. 15/3 Polls are subject to considerable sampling error. **1943** *Ann. Math. Statistics* XIV. 289 The accuracy of a sampling method may be measured by the variance of the estimate of the quantity which is of interest. **1975** *Listener* 6 Feb. 187/3 We accept sampling methods in our everyday life. **1947** *Bell Syst. Techn. Jrnl.* XXVI. 396 Any input wave can be represented by a series of regularly occurring instantaneous samples, provided that the sampling rate is at least twice the highest frequency in the input wave. **1978** *Gramophone* Apr. 1789/2 The Sound Stream recorder, which sells for $70,000, is a full 6-bit binary conversion and recording system, with a sampling rate of 48K, frequency response to 17kHz,..and a tape speed of 30ips. **1960** *Amer. Speech* XXXV. 176 The picture is probably as true as a sampling survey can give. **1972** H. Kurath *Stud. Area Linguistics* 76 A sampling survey carried out on a modest scale can reveal important aspects of the dialectal structure.

‖ **sampot** (sa̋po). [Fr., ad. Cambodian *sampuet.*] A kind of Cambodian sarong.

1931 *N. & Q.* 22 Aug. 127/1 The women of Cambodia make *sampots*. These are the long and wide sashes of silk of many colours which they bind around their waists. **1957** *Encycl. Brit.* IV. 641/2 Both sexes wear the *sampot* (a copious sort of loincloth) which the men supplement with a short jacket, the women with a long scarf draped around the figure or a long clinging robe. **1963** 'Han Suyin' *Four Faces* 100 The woman in a Cambodian *sampot* and blouse.

samprasarana (sæmprăsā·rănă). *Philol.* [Skr. *samprasāraṇa*, lit. 'a stretching out, extending', f. *sam-* together + *pra-* forth + *-sāraṇa* extension.] In Sanskrit, the interchange between the vowels i, u, ṛi, lṛi and their corresponding semi-vowels y, v, r, l; hence, a similar process in other Indo-European languages.

1861 T. Goldstücker *Pāṇini: his Place in Sanskrit Lit.* 169 It is probable, therefore, that Pāṇini did not invent these terms, but referred to them as of current use. On the other hand, he distinctly defines..*upadhā, lopa, samprasāraṇa*, and *abhyāsa.* **1888** J. Wright tr. *K. Brugmann's Elements Compar. Gram. Indo-Gmc. Lang.* I. 473 Vowel absorption often happens in languages with predominantly expiratory accentuation... If the absorption happens in such a manner that the syllable retains its value as a syllable, which is only possible, if another sound is able to undertake the part as bearer of the syllabic accent, we call the process samprasāraṇa (after the Indian grammarians). **1916** A. A. Macdonell *Vedic Gram. for Students* 5 The Samprasāraṇa series. Here the accented high grade syllables *ya, va, ra*..interchange with the unaccented low grade vowels *i, u, ṛ.* **1933** L. Bloomfield *Language* xxi. 384 When a relatively sonorous phoneme is non-syllabic, it often

acquires syllabic function; this change is known by the Sanskrit name of *samprasarana.* **1968** *Language* XLIV. 278 The forms *x[u]*arta-* and **u̯arta-* are correct: they are doubtless full grade replacements for the zero grade in samprasarana roots.

‖ **Samsam** (sæ·msæm). Now chiefly *Hist.* [Malay.] A person of mixed Malayo-Thai origin from the west coast of the Malay peninsula (see quot. 1961).

1836 J. Low *Diss. Soil & Agric. Penang* viii. 293, I believe there are some converts also amongst the Samsams, or mixed descendants of Siamese and Malays. **1839** T. J. Newbold *Pol. & Statistical Acct. Straits of Malacca* I. 420 The Samsams are a race of Malays who have adopted the religion and language of the Siamese. **1883** *Encycl. Brit.* XV. 322/2 A mixed Malayo-Siamese people, commonly known as Samsams, form the bulk of the population in the lower parts of Ligor and Sengora, and in the north of Kedah. **1961** L. D. Stamp *Gloss. Geogr. Terms* 403/1 *Samsam*.., a person of mixed Siamese–Malay origin, especially characteristic of the State of Kedah under Siamese suzerainty from 1821 to 1909.

‖ **samsara** (samsā·ra). *Indian Philos.* Also **sangsara.** [Skr. *saṃsāra*, a wandering through, f. *sam* prefix expressing completeness + *sṛ* to run, glide, move.] The endless cycle of death and rebirth to which life in the material world is bound; also *attrib.* Hence **samsa·ric** *a.*

1886 *Encycl. Brit.* XXI. 289/1 The notion of saṃsâra has become an axiom, a universally conceded principle of Indian philosophy. **1913** J. N. Farquhar *Crown of Hinduism* v. 213 All souls, whether living as gods, demons, men, animals, or plants, are afloat on the stream of transmigration (*saṁsāra*). **1928** W. Y. Evans-Wentz *Tibet's Great Yogi* p. xvi, The golden fish..symbolizes sentient beings immersed in the Ocean of Sangsaric (or Worldly) Existence. **1930** S. N. Dasgupta *Yoga Philos.* 67 The metaphysics of the samsāra cycle in connection with sorrow, origination, disease, rebirth. **1935** W. Y. Evans-Wentz *Tibetan Yoga* 16 The Sangsāra, or external universe, is a psycho-physical compound of mind; matter, as we see it, being crystallized mental energy. **1963** 'Mayananda' *Tarot for Today* xi. 140 It [*sc.* Stellar Power] can be distributed and assimilated by the Earth and Solar System, generally, thus producing all the minutiæ of Samsaric detail. **1966** R. F. C. Hull tr. *Jung's Ulysses* in *Coll. Wks.* XV. 127 Ulysses..is for Joyce.. the higher self who returns to his divine home after blind entanglement in *samsara.* **1977** L. A. Govinda *Creative Meditation* I. x. 43 The basic qualities of human individuality binding us to our worldly existence (*samsāra*) are at the same time the means of liberation and enlightenment.

samsi, var. Samisen in Dict. and Suppl.

‖ **samskara** (sanskā·ra). *Indian Philos.* Also 9 **sanscara, sanskara.** [Skr. *saṃskāra* a making perfect, preparation, f. *sam* together + *kṛ* to make, perform.] **1.** A purificatory ceremony or rite marking a stage or an event in life; one of twelve rites enjoined on the first three classes of the Brahman caste.

1807 *Asiatick Researches* IX. 288 The *Jainas*..admit the same division into four tribes, and perform like religious ceremonies, termed *sanscaras*, from the birth of a male to his marriage. **1832** *Ibid.* XVII. 309 Some of the original rites are still preserved..in such of the Sanskaras, or purificatory ceremonies as are observed at the periods of birth, tonsure, investiture, marriage, [etc.]. **1891** Monier Williams *Brāhmanism & Hindūism* (ed. 4) 353 Twelve purificatory rites, called Sanskāras were prescribed in the ancient collections of domestic rules..for the purification of the three higher castes. **1913** J. N. Farquhar *Crown of Hinduism* ii. 104 Debendranath Tagore..rebelled against the polytheistic and idolatrous character of the sacraments (*saṁskāras*) of the Hindu family. **1962** R. Zaehner *Hinduism* vii. 201 Saṁskaras or sacraments play an important part throughout the life of a Hindu. **1977** B. Saraswati *Brahmanic Ritual Trad.* p. xii, Of all the social institutions, the institution of the *samskaras* serves as the corner-stone of the total cultural complex of the brahmanic society.

2. A mental impression, instinct, or memory.

1827 *Trans. R. Asiatic Soc.* I. 562 Thence comes passion (sanscára), comprising desire, aversion, delusion, &c. **1875** Monier Williams *Indian Wisdom* iii. 79 Saṅskāra, implying —*a.* impetus as the cause of activity; *b.* elasticity; *c.* the faculty of memory. **1896** 'Swāmi Vivekānanda' *Yoga Philos.* 233 *Samskāra*, impressions in the mind-stuff that produce habits. **1930** N. Dutt *Aspects Mahayana Buddhism* iii. 94 They have been blinded by avidyā (ignorance of the Truth), from which have followed the saṃskaras (impressions). **1952** H. Zimmer *Philos. of India* III. ii. 324 The noun saṁskāra, signifying 'impression, influence, operation, form, and mold', is one of the basic terms of Indian philosophy. **1977** J. Hewitt *Yoga & Meditation* v. 42 The other is that which consists only of *Samskaras*, being brought on by the practice of the cause of complete suspension.

Sam Slick (sæm slik). *U.S.* The name of a peddling clock-seller, hero of a series of stories by T. C. Haliburton (1796–1865), Nova Scotian judge and political propagandist, used *transf.* of a type of smooth-spoken and sharp-practising New Englander, and hence *gen.* of any resourceful trickster or 'spiv'. Also *attrib.*

1897 R. G. Haliburton in *Haliburton: a Centenary Chaplet* 26 Sixty years ago the Southern States were familiar with the sight of Sam Slicks. **1916** M. Aiken *Canada in Flanders* I. 118 A 'hyphenated' voice..cried out peevishly next evening: 'Say, Sam Slick, no dirty tricks tonight.' **1944** B. A. Botkin *Treas. Amer. Folklore* III. 358 For Yankee trickiness or slickness the name Sam Slick has become proverbial. **1962** *Amer. Speech* XXXVII. 84 Other items of the standard vocabulary of this 'Sam Slick' American were suggested rather than directly quoted.

Samsoe (sæ·mso). Also **Samso, Samsø.** In full, *Samsoe cheese*: a firm, buttery cheese from the Danish island of Samsoe.

1953 G. P. Sanders *Cheese Varieties & Descriptions* 124 Swiss Cheese is made in many other countries besides Switzerland... Danish Swiss is called Samso. **1955** *Times* 10 May 12/4 Samsoe cheese..takes its name from the island of Samsoe. **1968** *Vogue* 15 Apr. 42/2 You need 8 oz. Danish Blue cheese, 4 oz. Samsoe cheese. **1968** L. Deighton *Continental Dossier* 8 Local dishes are rare—specialities are found country-wide, like..'Samsø'—the Cheddar of Denmark. **1976** M. Patten *Barbecue & Outdoor Eating* 27/2 Grate 12 oz (350 g) Danish Samsoe cheese and slice 2–3 tomatoes.

Samson. Add: Also, except in senses 1 and 6, with small initial. **1.** *Samson-like* adj. and adv. (later examples); *Samson-passion.*

1796 Southey *Joan of Arc* ix. 359 By experience rous'd shall man at length Dash down his Moloch-gods, Samson-like And burst his fetters. *a* **1821** Byron *Don Juan* (1956) III. lvii. variant line 8 And make him Samsonlike—more fierce with blindness. **1929** Blunden *Near & Far* 49 Joy's masque and fashion of Time's Samson-passion Deceived no lark that springs from weed and clod.

3. For ? *Obs.* in Dict. read 'Now *Hist.*' and add later examples.

1965 E. Tunis *Colonial Craftsmen* iv. 95 The wheelwright ..pulled the joint hard together with a large threaded clamp called a samson. **1968** J. Arnold *Shell Bk. Country Crafts* 163 There was a samson, for drawing felloes together when the strakes were being nailed on.

5. *Logging* (see quots.). Hence **sa·mson** *v.*

1905 *Terms Forestry & Logging* (U.S. Dept. Agric. Bureau Forestry) 45 *Sampson*, an appliance for loosening or starting logs by horsepower. It usually consists of a strong, heavy timber and a chain terminating in a heavy swamp hook. *Ibid.*, *Sampson a tree, to*, to direct the fall of a tree by means of a lever and pole. **1913** [see *kilhig]. **1971** F. C. Ford-Robertson *Terminol. of Forest Sci., Technol. Pract. & Products* 148/2 *Killig*.., *Pushpole*.. = *Sampson* (USA). A stout pole, sometimes notched into the tree stem at one end and braced against the base of a peavey handle at the other, used to push a small tree manually in the desired direction.

6. Samson fox [in allusion to Judges xv. 4], a fox belonging to a variety of the North American red fox, *Vulpes fulva*, in which the fur lacks guard hairs and so has a scorched appearance. Also *absol.*

1910 E. T. Seton *Life-Hist. Northern Animals* II. xxxii. 709 Another freak is the 'scorched' or 'Samson Fox'. **1921** N. M. W. J. McKenzie *Men of Hudson's Bay Company* xvii. 160 Foxes that were burned like these were [what] we called 'Samsons', and were useless. **1933** E. Merrick *True North* 305 He said he weren't goin' to..go clear into Canada to trap a few weasels and samson foxes. **1948** A. L. Rand *Mammals Eastern Rockies* 105 The Samson fox is a freak, in which the guard hairs are lacking.

Samsonite[1] (sæ·msənəit). Also **samsonite.** [f. Samson + -ite[1].] **1.** A variety of dynamite having an inert base of borax and salt.

1909 *Jrnl. Soc. Chem. Industry* 31 Aug. 915/2 The Secretary of State has made an Order adding the following explosives to the schedule..Nobel Ammonia Powder.. Samsonite..Titanite No. 1. **1915** A. Marshall *Explosives* 213 A charge of undoubtedly hard frozen Samsonite exploded whilst being rammed home with a wooden rammer. **1921** *Glasgow Herald* 28 Feb. 11 Illegal possession..of 2980 gelignite cartridges, 10 samsonite cartridges. **1936** E. Hart *Shotfirer's Man.* viii. 95 He then charged the shothole with 8 ozs. of Samsonite No. 3 and fired it.

2. A proprietary term in the U.S. for a make of suitcases, briefcases, and other items of luggage, etc. Chiefly *attrib.*

1939 *Official Gaz.* (U.S. Patent Office) 21 Feb. 538/2 Shwayder Bros., Inc., Denver, Colo. Filed Oct. 17, 1938. *Samsonite Streamlite.* For trunks, suitcases, and traveling bags. Claims use since Apr. 18, 1938. **1963** *Times* 24 Apr. 16/4 After this generous present, it is a bit of a come-down for the eight regional winners to receive 'Samsonite bridge tables and chairs'. **1969** J. Gardner *Compl. State of Death* iv. 40 The file was dropped into a slim brown Samsonite brief-case. **1971** D. MacKenzie *Sleep is for Rich* vi. 196 The samsonite case would carry a hundredweight without collapsing. **1977** *Time* 28 Feb. 47/1 Just now, Linda is cooling out in Los Angeles after months of bashing about in planes and buses like a piece of lost Samsonite. **1977** C. McFadden *Serial* (1978) xliv. 94/2 She dragged the Samsonite over the threshold.

samsonite[2] (sæ·msənəit). *Min.* [ad. G. *samsonit* (Werner & Fraatz 1910, in *Centbl. f. Mineral.* 331), f. the name *Samson* (see quot. 1910) + *-it* -ite[1].] A sulphide of silver, antimony, and manganese which occurs as black prisms with a metallic lustre.

1910 *Mineral. Mag.* XV. 430 Samsonite... Found with pyrargyrite and pyrolusite in the Samson mine, St. Andreasberg, Harz. **1968** I. Kostov *Mineral.* 173 Samsonite ($Ag_4MnSb_2S_6$) is a mineral of peculiar composition, containing up to 5·96% Mn and crystallizing in the monoclinic system. **1969** *Acta Crystallographica* B. XXV. 1004/2 The presence of two atomic polyhedra—the squat pyramids, SbS_3, and slightly deformed octahedra, MnS_6,—is the most characteristic feature of the crystal structure of samsonite.

Samson's post. 2. b. (Earlier and later examples.)

1865 *Harper's Mag.* Apr. 573/2 The walking-beam is a heavy horizontal piece of timber, supported in the centre by a Samson-post. **1960** C. Gatlin *Petrol. Engin.* iv. 45/1 The walking beam is supported by the sampson post, and imparts the reciprocating motion to the drilling line.

Samuelite (sæ·miuĕləit). [f. the name of Sir Herbert Louis *Samuel*, first Viscount Samuel (1871–1963), Liberal politician + -ite[1].] A supporter of Sir Herbert Samuel; used *spec.* to designate a member of the official Liberal Party, which was led by Samuel, subsequent to the secession in 1931 of the Liberal National Party under Sir John Simon (see *Simonite). Freq. *attrib.*

1931 *Times* 13 Oct. 14/4 The Liberal Party was split into at least three well-marked divisions. There were the Simonites, who had thrown in their lot boldly with the national cause; the Samuelites, about whom he [*sc.* Churchill] was unable to give any correct information; and the Lloyd Georgeites. **1931** A. Sinclair *Let.* 3 Nov. in J. Bowle *Visct. Samuel* (1957) xvii. 286 If you will forgive me saying so..we don't want to be called..Samuelite Liberals as opposed to Simonite Liberals. **1936** *Ann. Reg. 1935* 87 Against it [*sc.* the Government] were..the Liberals without prefix, led by Sir Herbert Samuel, and commonly known as 'Samuelites'. **1952** Visct. Simon *Retrospect* x. 180 The Samuelite Liberals supported the vote of censure. **1976** C. Cook *Short Hist. Liberal Party 1900–1976* x. 118 In March 1932, the National Government's decision to introduce the Import Duties Bill provoked a rebellion by the Samuelite Liberals.

Samuel-Smilesian: see *Smilesian.

samurai. Delete ‖ and '(Unchanged in the plural)' and add: Pl. **samurai,** occas. **samurais.**

1. a. (Earlier and later examples.)

1727 J. Scheuchzer tr. *Kaempfer's Hist. Japan* II. i. 396 'Tis from thence they are call'd *Samurai*, which signifies persons who wear two swords. **1795** tr. *C. P. Thunberg's Trav. Europe, Afr., & Asia* (ed. 2) III. 123 The people in office at this place, who wore two sabres, were called *Samrai.* **1841** *Chinese Repository* X. 17 Class 4 is that of the *samorai*, or military, and consists of the vassals of the nobility. **1896** L. Hearn *Kokoro* x. 172 The fear of the dead was held not less contemptible in a samurai than the fear of man. **1898,** etc. [see *bushido]. **1904,** etc. [see *heimin]. **1972** *Mainichi Daily News* (Japan) 6 Nov. 7/7 The Samurai were distinguished in dress most easily by the swords they wore.

b. *transf.* and *fig.*

1905 H. G. Wells *Mod. Utopia* ix. 259 These people constitute an order, the *samurai*, the 'voluntary nobility', which is essential in the scheme of the Utopian State. **1918** G. Frankau *One of Them* xx. 151 Stern mitred prelates; Law-lords; back-woods Samurais Who flung to consequence a scornful 'Damn your eyes'. **1934** H. G. Wells *Exper. Autobiogr.* II. ix. 735, I have told already..how I tried to make the Fabian Society into an order of the Samurai. **1977** *Time* 24 Jan. 17/2 Yukio Mishima, the right-wing literary samurai who committed spectacular hara-kiri in 1970.

2. *attrib.* and *Comb.*, as *samurai code, ethic, order, spirit, sword, warrior*; *samurai-minded* adj.

1971 *Times Lit. Suppl.* 20 Aug. 984/1 The samurai code embraced more than the practice of Zen and the ethics of Japanese forms of Confucianism. **1970** *Newsweek* 7 Dec. 32/2 The Japanese militarists of the 1930s twisted the ancient samurai ethic into the ideology of Fascism. **1938** *Times* 17 Feb. 16/1 In private conversation business men will unequivocally express disapproval of the course of events, a minority of *samurai*-minded ultra-patriots being the only exceptions. **1906** G. B. Shaw *Let.* 24 Mar. (1972) II. 614 A proposal for a set of observances of the Samurai order. **1923** Samurai spirit [see *bushido]. **1961** I. Murdoch *Severed Head* xix. 155 Was it..when I saw her cut the napkins in two with the Samurai sword? **1977** *National Observer* (U.S.) 15 Jan. 5 Hayakawa..went on to even more fame as the 'samurai warrior' president of San Francisco State University.

samusa, var. *samosa.

samyama (samya·ma). *Indian Philos.* Also **sanyama.** [Skr. *saṃyama* restraint, control of the senses, f. *sam* together + *yam* sustain, hold up (*yāma* rein, bridle, self-control).] The name given to the three final stages of meditation in yoga, which lead on to *samadhi*, or the state of union.

1828 *Trans. R. Asiatic Soc.* III. 164 That which removes sin is Brahmanhood. It consists of..*Neyama*, and *Ripavas*, and *Dan*,..and *Sanyama.* **1884** R. C. Bose *Hindu Philos.* 160 Three internal subservients, attention, contemplation, and meditation, collectively called by the name of 'subjugation' (sanyana). **1899** Max Müller *Six Syst. Indian Philos.* vii. 459 It is difficult to find a word for Samyama, firm grasp being no more than an approximate rendering. It is this Samyama, however, which leads on to the Siddhis, or perfections. **1959** E. Wood *Yoga* xiii. 237 *Sanyama* is then a definite tool of mind, which can be used for gaining knowledge of various kinds. **1978** D. N. Bradshaw tr. *Oki's Meditation Yoga* iii. 84 People usually assume that there are many preparations to be made before entering samyama, but the method of Yoga is simple and clear.

san[1] (sæn). [Gr. σάν.] The name (first recorded by writers of the sixth century B.C.) for a sibilant (M) found in early Doric scripts (later displaced by sigma), which has been compared with various Semitic sibilants and *sampi.

1584 B. Rich tr. *Herodotus' Famous Hystory* I. f. 44[v], All the wordes in theyr language which consist of 4 or more sillables do commonly end in one letter: which letter the Dores cal San the Iones Sigma. **1709** I. Littlebury tr. *Herodotus' Hist.* I. 1. 89 All Names representing the Person or Dignity of a Man, terminate in that letter which the Dorians call San, and the Ionians Sigma. **1860** *Chambers's Encycl.* I. 169/2 In accommodating itself [*sc.* the Phoenician alphabet] to the necessities of the Greek tongue..the name *Sigma* was transferred to *San.* **1912** [see *sampi]. **1915** J. Sandys tr. *Pindar's Odes* 559 In olden days, the lay of the dithyramb was wont to wind its straggling length along, and the sibilant *san* was discarded. **1933** [see *koppa]. **1961** L. H. Jeffery *Local Scripts Arch. Greece* I. ii. 33 By the second half of the fifth century, the sign of *san* was no longer in use, except in conservative Crete, and as an emblem on the coins of Sikyon.

San[2] (sān). Also **Saan.** [Bushman, app. of Khoikhoi (Hottentot) origin: cf. Nama *sā-* to inhabit.] **a.** The name used for themselves by the Bushmen of southern Africa (see Bushman 1); also *attrib.* **b.** The principal language of the Bushmen.

1876 *Encycl. Brit.* IV. 575/1 Bushmen..so named by the British and Dutch colonists of the Cape, but calling themselves *Saab* or *Saan*, are an aboriginal race of South Africa. **1878** K. Johnston *Africa* xxiv. 440 The Bushmen or Saan are the nomads of the Kalahari. **1881** [see *Nama *a.*]. **1907** *Rep. Brit. Assoc. Adv. Sci. 1906* 689 They are called Baroa by the Basuto, Abatwa by the Kafirs, San by themselves. **1930** [see *Khoikhoi]. **1944** M. Oldevig *Sunny Land* v. 50, I had the rare good fortune to come upon a Saan Bushman, one of the few who still inhabit parts of the Namib desert. **1967** D. S. Parlett *Short Dict. Lang.* 73 The Khoin or 'Click' languages..comprising to the south Bushman (San), to the north Hottentot (Nama) [etc.]. **1974** J. Flint *Cecil Rhodes* i. 9 South Africa was the home of the San (the so-called Bushmen). **1977** C. F. & F. M. Voegelin *Classification & Index World's Lang.* 201 South African Khoisan. Central... 36. San = Saan.

san[3] (san). [Jap.: a contraction of the more formal *sama.*] A Japanese honorific title, equivalent to Mr., Mrs., etc., suffixed to personal or family names as a mark of politeness; also *colloq.* or in imitation of the Japanese form, suffixed to other names or titles (cf. *mama-san).

When suffixed to a female personal name, and in more polite endearment, *san* is often coupled with the prefix *O-* (see quot. 1922).

1878 C. Dresser in *Jrnl. Soc. Arts* XXVI. 175/1 Mr. Sakata, or, as they would say Sakata San, who was appointed ..as one of my escort through Japan. **1891** A. M. Bacon *Japanese Girls & Women* xi. 304 He is a person to be treated with respect,—to be bowed to profoundly, addressed by the title San, and spoken to in the politest of languages. **1922** Joyce *Ulysses* 321 The fashionable international world attended *en masse* this afternoon at the wedding... Miss Grace Poplar, Miss O Mimosa San. **1952** T. J. Mulvey *These are your Sons* vii. 146 'You go away, O'Reilly-san?' the little girl asked. **1964** I. Fleming *You only live Twice* i. 16 'Bondo-san,' said Tiger Tanaka, Head of the Japanese Secret Service, 'I will now challenge you to this ridiculous game.' **1968** *Guardian* 23 Feb. 11/4 Corpsman Kenneth Corner..told her [*sc.* a Vietnamese girl]: 'It's going to be all right baby-san, it's going to be all right.' **1972** J. Ball *Five Pieces of Jade* xiv. 188 It would make me the greatest pleasure, Nakamura san.

San[4] (sæn). Also **san.** Colloq. abbreviation of Sanatorium (esp. in sense *3).

1906 R. Brooke *Let.* 1 Apr. (1968) 47, I started this disease..rather badly, and as the San. was full, we were put into a room in the house. **1914** 'I. Hay' *Lighter Side School Life* iii. 71 Broken neck, inflammation of the lungs, ringworm, and leprosy, old son... You are going to the San. **1936** M. Kennedy *Together & Apart* III. 174, I was in quarantine for mumps, so I stayed in our school San. all the holidays. **1945** [see *know *v.* 1 b]. **1976** 'D. Fletcher' *Don't whistle 'Macbeth'* 45, I sounded like some old-fashioned matron, soothing the felled captain of the First Eleven in the san.

sanakatowzer (sæ:nākătau·zəɹ). *Naut. slang. rare.* Also with capital initial. [Of uncertain origin: cf. Towser *sb.* and *bandowzer* in *D.A.E.*] An extremely forceful blow; something particularly large or powerful, such as a heavy wave.

1903 Kipling in *Collier's Weekly* 15 Aug. 9/3 Mr. Ducane catches 'im a sanakatowzer of a smite over the 'ead with the flat of 'is sword. **1920** *Blackw. Mag.* Apr. 501/2 She shipped one Sanakatowzer that nearly swamped her.

sanaphant (sæ·năfænt). *Electronics.* [f. *sana(tron + *phant(astron.] (See quots.)

1949 B. Chance et al. *Waveforms* v. 200 Although somewhat more complex than the screen-coupled phantastron, the sanatron and sanaphant can generate waveforms..as short as 1 μsec. **1955** *Electronic Engin.* XXVII. 397/2 This undesirable loading may be avoided by the use of the sanaphant circuit, in which the gating waveform is obtained by amplification of the voltage developed across a small resistor inserted in the cathode circuit of the charging valve. **1960** Cooke & Marcus *Electronics & Nucleonics Dict.* 413/2 *Sanaphant*, a linear time-delay circuit similar to the sanatron, differing chiefly in the connections between the two pentodes.

Sanatogen (sănæ·tŏdʒĕn). A proprietary name for a tonic wine.

[**1898** *Official Gaz.* (U.S. Patent Off.) 14 June 1657/2 Dietetic albuminous preparations. Bauer, Cie Berlin... Sanatogen... Used since October 25, 1897.] **1924** G. B. Stern *Tents of Israel* xii. 173 She dispensed a share in her jellies and Sanatogen and grapes. **1936** H. Nicolson *Let.* 28 Apr. (1966) 259 Tell Gwen I do not need Sanatogen at present. **1939** *Trade Marks Jrnl.* 1 Mar. 290/1 *Sanatogen Tonic Wine...* Genatosan Limited, 43, Regent Street, Loughborough, Leicestershire; manufacturers.

sanatorium. Add: Also with pl. **sanatoriums.** **1.** (Earlier and later examples.) Also *fig.*

1839 *London Med. Gaz.* XXV. 406/2 Dr. Southwood Smith, Dr. Arnott, and some other gentlemen, have it in contemplation to establish, under the name of 'Sanatorium', an institution..where patients are provided with board and medical treatment on the payment of a certain sum per week. *Ibid.* 407/2 We anticipate..that the parties who have set about the Sanatorium will abandon the scheme before it has been brought into actual operation. **1934** Dylan Thomas *Let.* 15 Apr. (1966) 104, I don't want to see my books; a library is a sanatorium of sick minds. **1973** *Sci. Amer.* Sept. 130/2 The care of the mentally ill in distant upland sanatoriums.

3. A room or building in a boarding school for the accommodation of the sick. Cf. *San[4].

1860 *Eton Gloss.* 30 Sanatorium. The Hospital—a modern improvement—where a boy seized with any infectious and dangerous illness is at once sent. **1901** *Eton Boy's Lett.* 98 They dont take measels to the Sanatorium as they arent dangerous. **1914** 'I. Hay' *Lighter Side School Life* iii. 71 When dragged from the scrummage he was in a half-fainting condition. He revived as he was being carried to the Sanatorium. **1981** E. North *Dames* iv. 75 Should Sister move the general's daughter to the sanatorium?.. Polio was about at Eton..where many girls had brothers.

sanatron (sæ·nătrǫn). *Electronics.* [Perh. irreg. f. Sanitary *a.* (cf. quot. 1951): see *-tron.] A circuit which generates a saw-tooth output waveform on receipt of a short trigger pulse, used in time-bases and similar applications.

1946 *Jrnl. Inst. Electr. Engineers* XCIII. IIIa. 1191/1 The circuit, known as the Sanatron, is illustrated in Fig. 5. **1951** O. S. Puckle *Time Bases* (ed. 2) ix. 181 The Royal Air Force used many slang terms. The name 'Sanatron' has been derived from the term 'sanitary', meaning satisfactory. **1955** *Electronic Engin.* XXVII. 397/2 Examination of the sanatron circuit also shows that the gating waveform is obtained by partial differentiation..and amplification of the sweep waveform. **1966** *McGraw-Hill Encycl. Sci. & Technol.* XIII. 645/1 The basic sanatron delay circuit, of which there are a number of variations, combines in two pentode tubes the function of a gate waveform generator, clamp and linear saw-tooth generator.

sanbornite (sæ·nbǫɹnəit). *Min.* [f. the name of Frank *Sanborn* (d. 1945), U.S. mineralogist + -ite[1].] A triclinic silicate of barium, $BaSi_2O_5$, which occurs as white or colourless plates at a locality in California, and has been artificially prepared.

1932 A. F. Rogers in *Amer. Mineralogist* XVII. 161, I am indebted to Mr. Frank Sanborn of the Division of Mines, Department of Natural Resources, State of California..for the specimen which contains the new mineral, sanbornite, herein described. **1950** *Jrnl. Amer. Ceramic Soc.* XXXIII. 43/1 Point *M*, the quintuple point for the fields of sanbornite, tridymite, and mullite..is therefore a eutectic. **1968** I. Kostov *Mineral.* 381 The structures of apophyllite, sanbornite, and gillespite are similar and correspond to a layered type with basic silicon sheets of condensed wollastonite chains.

Sancerre (sañsę̈r). The name of a town in the Cher department of central France, used *attrib.* and *absol.* to designate a light white (occas. red) wine produced in its neighbourhood.

1787 A. Young *Jrnl.* 1 June in *Trav. France* (1792) I. 13 We are now in Berri... We drank there excellent Sancerre wine, of a deep colour, rich flavour, and good body. **1946** A. L. Simon *Conc. Encycl. Gastron.* VIII. 148/2 *Sancerre*, a noted white wine..sold..under the name of *Château de Sancerre.* **1962** P. Brickhill *Deadline* iv. 61, I ordered a bottle of Sancerre and it came well chilled. **1977** C. McCarry *Secret Lovers* vii. 86 Cathy would take a half-bottle of Sancerre.

Sancho[2] (sæ·nʃo). The forename of *Sancho* Panza, the squire of Don Quixote (see Don *sb.*[1] 1 c, Quixote *sb.*) used allusively of one who is a companion or foil to a quixotic person.

1870 D. G. Rossetti *Let.* 15 Mar. (1965) II. 817 He [*sc.* Stillman] is a complete Don Quixote in every way, and with such a Sancho as myself to back him, we ought not to lack for adventures. **1934** R. Campbell *Broken Record* 10 Humanity can be divided roughly into two classes, the Quixotes and the Sanchos.

‖ **sancocho** (sanko·tʃo). [Amer. Sp., a. Sp. *sancocho* half-cooked meal, f. *sancochar* to parboil.] In South America and the Caribbean: a rich soup containing meat, plantain, yucca, etc. (see quot. 1969).

1939 C. Brown *S. Amer. CookBk.* 78 Sancocho, a truly native dish with its tropical ingredients, takes the place in the Dominican Republic of the pucheros encochidos in cooler Latin-American countries. **1954** M. Waldo *Compl. Round-the-World Cookbk.* 361 The wonderful soup-stew of Latin countries, *sancocho*, is undoubtedly the [Dominican Republic] people's choice for a national dish. **1969** R. & D. De Sola *Dict. Cooking* 199/2 *Sancocho*,..Latin-American souplike stew containing fish, fowl, meat, seafood, vegetables, and spices. **1977** *Time* 22 Aug. 23/1 Following a meal of *sancocho* (Panama's national soup) and hot chili sauce, Torrijos offered the following comments.

‖ **sancta simplicitas** (sæ·ŋktă simpli·sitæs, sa·ŋkta simpli·kitās), *Latin phrase.* [L. 'holy simplicity'.] An expression of astonishment at another's naïvety. Also used *substantivally.*

These are said to have been the dying words of John Huss (1373–1415), Bohemian religious reformer and martyr, provoked by the sight of a simple peasant adding wood to the fire about his stake.

1847 F. A. Kemble *Let.* Dec. in *Rec. Later Life* (1882) III. 278 Miss L.—ingenuously replied, 'Oh dear! that she'd never thought of that..' *Sancta Simplicitas!* **1889** G. B. Shaw in *Star* 13 July 4/4 She..thinks it would be too much to ask the public to listen to two sonatas. *Sancta simplicitas!* too much! **1894** M. Beerbohm in *Yellow Bk.* Apr. 65 The day of sancta simplicitas is quite ended. **1936** *Times Lit. Suppl.* 31 Oct. 870/1 Setting and character perfectly fused..the *sancta simplicitas* of the Reverend Micah Balwhidder. **1963** L. Meynell *Virgin Luck* iv. 84 'Me? I've never had a bet in my life. I don't even know how to.' '*O Sancta simplicitas.* I wish I didn't.' **1980** —— *Hooky & Prancing Horse* xi. 187 'How on earth did you get in?' '*Sancta simplicitas*..which means you are still wet behind the ears.'

sanctification. Add: **4.** *slang.* Blackmail, esp. the extortion of political favours from a diplomat. Cf. *sanctify *v.* 9.

1975 *Observer* (Colour Suppl.) 23 Nov. 25/3 *Sanctification*, blackmail for the purposes of extracting political favours from a victim, not money. **1977** J. Gardner *Werewolf Trace* x. 87 He told himself to be careful. They were not above trying a bit of sanctification.

sanctify, *v.* Add: **9.** *slang.* To blackmail (a person), esp. for the purposes of extracting political favours. Cf. *sanctification 4.

1977 J. Gardner *Werewolf Trace* vii. 71 Can't you sanctify him, or give him a dose of measles? Isn't that how you people talk about blackmail and murder? *Ibid.* xiv. 127 They've sanctified Maubert... It's what they call it. They've made him holy, separated him. Blackmailed him.

sanction, *sb.* Add: **2. d.** *Pol.* Esp. in *pl.*, economic or military action taken by a state or alliance of states against another as a coercive measure, usu. to enforce a violated law or treaty.

1919 G. B. Shaw *Peace Conference Hints* vi. 84 Such widely advocated and little thought-out 'sanctions' as the outlawry and economic boycott of a recalcitrant nation. **1935** *Punch* 25 Dec. 728 'And you,' we replied in great excitement, 'are the very man to give it to him. Come, now, put on your beard, fly over to Italy, and—sanctions or no sanctions—put into his stocking your One Hundred and Eighty-Ninth Volume.' **1937** A. Huxley *Ends & Means* ix. 109 Military sanctions *are* war. Economic sanctions, if applied with vigour, must inevitably lead to war-like reactions on the part of the nation to which they are applied, and these war-like reactions can only be countered by military sanctions. **1943** H. A. Wallace in *N.Y. Times* 26 July 10/6 He witnessed the collapse of sanctions under the League of Nations. **1948** P. D. Whitting in M. Beloff *Hist.* 356/1 Abyssinia was annexed by Italy in May, 1936. Sanctions were dropped two months later. **1965** *New Statesman* 9 Apr. 562/2 Given sufficient pressures to ensure the cooperation of British firms and banks operating in Rhodesia..sanctions could work if they were maintained for an extended period. **1981** *Guardian* 20 July 12/2 If Israel is to be stopped from riding roughshod over Western interests in the Middle East.., American sanctions may have to be a lot more convincing.

9. a. *attrib.* and *Comb.*, as (sense *2 d) *sanction-breaker, -buster, -busting; sanction-induced* adj.

1968 *Guardian* 25 Apr. 1/8 British citizens would be able to come to Britain from Rhodesia 'unless they are known sanction breakers or supporters of the illegal regime'. **1973** *Times* 8 June 27/1 (*heading*) Dutch move to stop the sanction busters. **1973** *Guardian* 16 Apr. 1/6 The Smith regime in Rhodesia has carried out its most spectacular coup in sanction-busting..with the triumphant announcement that three Boeing-707 jet airliners have been delivered. **1974** A. Williams *Gentleman Traitor* xii. 186 He studied the.. South African and Rhodesian economies, and how these interlocked with the complex methods of Sanction-Busting. **1970** D. Goldrich et al. in I. L. Horowitz *Masses in Lat. Amer.* v. 192 We can project the possibility..of sanction-induced parochialism on the part of formerly more highly politicized actors.

b. *attrib.* and *Comb.* in *pl.* (sense *2 d), as *sanctions-breaker, -breaking, -buster, -busting; sanctions-busting* adj.

1973 R. Lewis *Blood Money* viii. 110 *Scathe* would not be publishing an exposé on the German businessman, sanctions-breaker or not. **1935** *Times* 7 Nov. 14/6 It may be taken for granted..that the German conception of neutrality does not permit of what might be described as 'sanctions-breaking'. **1976** P. Driscoll *Barboza Credentials* I. ii. 29 Countries whose laissez-faire attitudes had encouraged sanctions-breaking. *Ibid.* III. i. 92, I had one immediate

SANCTION

concern: the British consulate. Sanctions-buster or not, I was in desperate need of their help. **1970** *Observer* 1 Mar. 4/4 It is disappointed that so little is being done to..promote the campaign against the sanctions-busting ships. **1975** M. Hartmann *Game for Vultures* ii. 22 He had started seriously in the sanctions busting game.

sanction, *v.* Add: **4.** To impose sanctions upon (a person), to penalize.
A use of doubtful acceptability at present.—Ed.
1956 *Universe* 27 July 1/1 (*heading*) Let Church sanction road killers. **1978** *Daily Mail* 29 Nov. 9/1 Sir Geoffrey Howe..referred to Ford's being 'sanctioned'... Nobody..made a protest about this violence being done to the English language (or about normal meanings being stood on their head).

sanctionable, *a.* Delete *rare* and add examples.
1927 A. Kocourek *Jural Relations* 441 *Sanctionable acts*, unlawful acts which are visited by a sanction. **1944** *Scrutiny* XII. 155 The only sanctionable activities unconnected with religion are parlour games. **1976** *Interdisciplinary Sci. Rev.* I. 182/1 It was our visit to the Flower Children..that suggested to me the need for an alternative to the polar position—the need for a totally new and socially sanctionable drug.

sanctioneer (sæŋkʃənɪəˑɹ). [f. SANCTION *sb.* + -EER.] = *SANCTIONIST 1.
1937 G. Frankau *More of Us* v. 53 Ask not of him—my noble sanctioneers Whose peaceful intents of such warlike mood are. **1965** *Observer* 21 Nov. 3/2 The 'sanctioneers', as they are coming to be called, are highly satisfied with Mr. Heath. **1967** *Economist* 7 Jan. 19/1 South West Africa would offer the sanctioneers a far more permanent bridge-head, the chance of applying sanctions, in effect, against apartheid itself.

sanctioning, *ppl. a.* Add: **1. b.** That imposes or maintains sanctions. Cf. *SANCTION *sb.* 2 d. *rare.*
1976 *Individualist* Dec. 66/2 South Africa will surely fall, and another great satellite state will have been created in a powerful strategic position. Have the 'sanctioning' countries considered this?

sanctionism (sæˑŋkʃənɪz'm). *rare.* [f. SANCTION *sb.* + -ISM.] The theory of economic or military sanctions; advocacy of such sanctions.
1938 *Nation* (N.Y.) 29 Jan. 115/2 The struggle against the 'highly civilised hordes of sanctionism'.

sanctionist (sæˑŋkʃənɪst), *sb.* (and *a.*). [f. SANCTION *sb.* + -IST.] **1.** One who advocates or supports the employment of sanctions. Cf. *SANCTION *sb.* 2 d.
1935 *Observer* 6 Oct. 18/3 The 'News Chronicle', a sanguinary sanctionist, had a displayed article last week called 'Christmas is coming'. **1937** A. Huxley *Ends & Means* ix. 111 Sanctionists reply by asserting that the mere display of great military force by League members will be enough to deter would-be aggressors.
2. *attrib.* passing into *adj.*
1935 *Observer* 6 Oct. 18/3 British policy and the sanctionist mania were originally based on the delusion that Signor Mussolini was bluffing. **1936** *Empire Rev.* LXIII. 145 Sanctionist policy. **1937** A. Huxley *Ends & Means* ix. 112 According to sanctionist theory, the League is to take military action in order to bring about a just settlement of disputes.

Sanctoral. Delete † *Obs.* and add: Also with small initial. (Later examples.)
1955 A. A. King *Liturgies of Relig. Orders* iii. 195 The mediaeval sanctoral was similar to that in many of the calendars of the time. **1975** *Church Times* 7 Mar. 8/4 Priests of the Society of Retreat-Conductors gave him a desk and something described as a coffee-table calendar of the Church's year and sanctoral.

sanctuary, *sb.*[1] Add: **II. 5. d.** An area of land within which (wild) animals or plants are protected and encouraged to breed or grow.
1879 A. P. Vivian *Wanderings in Western Land* xiii. 299 The suggestion..of setting apart certain districts as 'sanctuaries', within which the buffalo should never be molested, is one well worthy of consideration. **1887** [see *bird sanctuary* s.v. *BIRD *sb.* 9]. **1897** *Cornh. Mag.* Jan. 37 The national forests will become, as the New Forest is now in some measure, sanctuaries for all the animals *feræ naturæ* of England. **1909** *Bull. N.Y. Zool. Soc.* June 511/2 Around the coast there is gradually being extended a chain of insular bird sanctuaries that means much to the avifauna of North America. **1943** J. S. Huxley *TVA* 54 Game management areas and game refuges or sanctuaries have been set up. **1975** M. Russell *Murder by Mile* iii. 26 The glen's by way of being something of a bird and animal sanctuary. **1978** *Country Life* 16 Nov. 1632/1 Rare and vulnerable plants and animals will be protected by setting aside 'sanctuaries'.

sand, *sb.*[2] Add: **1. h.** *Soil Sci.* Applied *spec.* to particles whose sizes fall within a specified range, and to soils having a specified proportion of such particles (see quots.). Hence *sand-size* sb. (adj.).
1873 E. W. Hilgard in *Amer. Jrnl. Sci. & Arts* CVI. 337 (*table*) Coarse Sand, 80–90 (1/180) mm... Finest Sand 20–22 (1/180) mm. **1900** R. Warington *Lect. Physical Properties Soil* i. 8 Coarse sand 0·5–1·00 mm... Fine sand 0·1–0·25 mm. **1925** P. Emerson *Soil Characteristics* i. 6 The different soil particles are designated according to size as follows... Very coarse sand 2·0 to 1·0 millimeters... Very fine sand 0·1 to 0·05 millimeter. *Ibid.* 7 The United States Bureau of Soils recognizes the following classes [of soil]:..*Sand*: more than 25 per cent very coarse, coarse and medium sand, less than 50 per cent fine sand, more than 20 per cent silt and clay. **1952** L. M. Thompson *Soils & Soil Fertility* ii. 8 Based on size of soil particles there are three fractions, sand, silt, and clay. **1957** Sand-size [see *SEDIMENTOLOGICAL *a.*]. **1964** K. W. Butzer *Environment & Archeol.* x. 158 The modified Wentworth grade scale..is most widely used in North America. It has the following logarithmic subdivisions:..sand 0·064–2 mm., silt 0·004–0·064 mm... The non-logarithmic, modified Atterberg scale widely used in Europe has slightly different nomenclature... coarse sand 0·2–2·0 mm... fine sand 0·02–0·06 mm., silt 0·002–0·02 mm. **1971** *Gloss. Soil Sci. Terms* (Soil Sci. Soc. Amer.) 14/2 *Sand*, a soil particle between 0·05 and 2·0 mm in diameter. *Ibid.* 18/1 *Sand*, soil material that contains 85% or more of sand; percentage of silt, plus 1·5 times the percentage of clay, shall not exceed 15. **1972** J. G. Cruickshank *Soil Geogr.* ii. 55 The products of physical weathering are usually large on the particle size scale; that is, they are stone, gravel, or sand size and less commonly as small as silt size.

i. A fashionable shade resembling the colour of sand.
1923 *Daily Mail* 13 Feb. 13/2 (Advt.), Artificial silk hose ..in black, white, beaver, nude, cinnamon, sand, suede. **1930** *Daily Express* 6 Oct. 5/6 (Advt.), Imitation nutria fur sets... In dark grey, fawn, beaver, sand, and nutria. **1971** *Guardian* 28 Sept. 11/2 (*caption*) Quilted raincoat... In sand, orchid, or damson. **1979** *Country Life* 24 May (Suppl.) 55 (Advt.), The new Renault 5..comes in black, silver, blue or sand.

2. b. (Later examples.)
1873 Trollope *Phineas Redux* I. vi. 53, I complain of no injustice. Our castle was built upon the sand. **1905** G. L. Dickinson *Mod. Symposium* 77, I have been watching..one building after another laboriously raised by each speaker in turn, only to collapse ignominiously at the first touch administered by his successor. And why? For the ancient reason, that the structures were built upon the sand. **1920** Galsworthy *In Chancery* ii. iii. 151 She put out her hand to him. 'I feel you're a rock.' 'Built on sand,' answered Jolyon. **1963** *Times* 9 Jan. 4/2 On slower courts the story with Hughes would be different, but here, where even the best stroke is not an outright winner until it has died, his game is indeed built on sand.

d. *to bury* (or *hide*) *one's head in the sand* (and allusive varr.): to ignore unpleasant realities.
In some quots. with direct reference to the legendary belief that an ostrich buries its head in sand when threatened.
1844 [see *OSTRICH[1] 2 a]. **1899** W. H. D. Rouse in North tr. *Plutarch's Lives* VI. 345 Like the ostrich that hides his head in the sand. **1916** W. Wilson in *N.Y. Times* 2 Feb. 1/1 America cannot be an ostrich with its head in the sand. **1929** L. MacNeice in *Oxford Poetry* 24 Asking..Whether it would not be better To hide one's head in the warm sand of sleep. **1937** F. P. Crozier *Men I Killed* vii. 137 Our new system of rearmament is at least..encouraging our Colonel Blimps to hide their heads, stupidly like the ostrich, in the sand! **1946** E. O'Neill *Iceman Cometh* iii. 201 He thrusts his head down on his arms like an ostrich hiding its head in the sand. **1976** *Star* (Sheffield) 29 Oct. 10/4 The people of England should not bury their heads in the sand and say it can't happen here.

7. a. Delete † *Obs.* and add later examples.
1918 L. E. Ruggles *Navy Explained* 20 Bread is called 'punk'; sugar, 'sand'. **1935** A. J. Pollock *Underworld Speaks* 86/1 *Pass the sand*, pass the sugar. **1945** *California Folklore Q.* 19 Oct. 46 Joe with cow and sand. **1971** M. Tak *Truck Talk* 100 *Load of sand*, a cargo of sugar.

b. (Earlier and later examples.)
1867 G. W. Harris *Sut Lovingood* 102, I tell yu he hes lots ove san' in his gizzard; he is the best pluck I ever seed. **1872** *Newton Kansan* 5 Dec. 3/3 We hope to see Mr. Pettibone with sufficient 'sand in his craw' for this new position [*sc.* police judge]. **1875** B. Harte *Tales of Argonauts* 71 Blank me if I didn't think he was losing his sand, till he walked to position. **1924** Galsworthy *Forest* iv. ii. 120 By Jove, Mr. Farrell, there's sand in you. Tell me, isn't he ever ashamed of himself? **1933** J. Buchan *Prince of Captivity* iii. i. 264 A plain face with nothing showy about it, but all the horse-sense and sand in the world. **1954** 'W. Henry' *Death of Legend* 4 You losing your sand, Buck?

e. *to raise sand* (U.S.): to create a disturbance; to make a fuss.
1892 *Dialect Notes* I. 231 'To raise sand' is slang [in Kentucky] for to get furiously angry, the same as 'to raise Cain'. **1893** H. A. Shands *Some Peculiarities of Speech in Mississippi* 74 *Raise sand*,..to create a disturbance, to raise a row. **1948** *Sun* (Baltimore) 1 Dec. 17/4 Boudreau raised sand but the decision stuck. **1970** C. Major *Dict. Afro-Amer. Slang* 96 *Raise sand*,..to make an outcry; to brawl; to fight.

9. a. *sand-barge, -beach* (earlier examples), *-canyon, -cart, -flat* (earlier and later examples), *-heap* (later examples), *-island, -knoll, -land* (later examples), *-line, -mound, -pile, -reef, -sack, -sea, -spit, -stretch, -vein.*
1840 R. H. Dana *Two Yrs. before Mast* 225 We were as deep as a sand-barge. **1887** S. Samuels *From Forecastle to Cabin* 197 My ship was loaded as deep as a sand barge. **1709** J. Lawson *New Voy. Carolina* 151 The Sand-Birds..frequent our Sand-Beaches. **1728** J. Comer *Diary* 7 Apr. (1893) 50 A schooner..was cast on shore on a sand beach at Westport. **1806** *Deb. Congress U.S.* (1852) 9th Congress 2 Sess. App. 1117 They passed a number of sand-beaches, and some rapids. **1939** Auden & Isherwood *Journey to War* 120 Sand-canyons, guarded by fantastic sandy spires and pinnacles. **1788** Cowper *Let.* 1 Feb. in R. Southey *Life & Wks. W. Cowper* (1836) VI. 117 Thinking myself an ass, and my translation a sand-cart. **1825** J. Constable *Let.* 1 Aug. (1966) IV. 97 A scene on Hampstead Heath, with broken foreground and sand carts. **1834** *Chambers's Edin. Jrnl.* III. 233/3 It was like subjecting a pampered palfrey all of a sudden, to the sorrows of the sand-cart. **1923** *Glasgow Herald* 30 Jan. 9 There is generally a so-called sandcart, a sort of squat fly with an awning for two. **1773** in E. W. McMullen *Eng. Topogr. Terms in Florida* (1953) 190 From this point runs a sand flat 1¼ mile from the shore of Anastasia Island. **1794** *Trans. Soc. Promotion Agric., Arts, & Manuf.* (U.S.) I. 143 He..kept him in a very poor pasture adjoining a creek where creek-thatch grew on sand-flats. **1826** J. J. Audubon *Ornithol. Biogr.* II. 41 The dead fish that frequently are found about the sand-flats of rivers. **1922** Joyce *Ulysses* 41 Unwholesome sandflats waited to suck his treading soles. **1854** C. M. Yonge *Heartsease* II. iii. xv. 327, I hope she will take her down to the sand-heap, where the children have been luxuriating all morning. **1974** *Times* 5 Oct. 12/2 That sand-heap played a large part in his method of teaching. **1840** Poe *Jrnl. of Julius Rodman* in *Compl. Wks.* (1902) IV. 43 Sand-island. **1975** *Offshore Engineer* Dec. 16/3 A sand island could engulf a conventional steel or concrete platform. **1916** Joyce *Portrait of Artist* (1969) iv. 172 A ring of tufted sandknolls. **1963** *Times* 10 June 7/1 This is 73 percent above the average of 16 other sandland farms carrying cattle and sheep as well as growing corn. **1972** *Plant Dis. Reporter* LVI. 695 This pathogen spread rapidly into all the tomato sand-land areas of Florida. **1891** W. B. Yeats *John Sherman & Dhoya* ii. 185 By the..edge of the lake..there suddenly stood before him a slight figure, at the edge of the narrow sand-line, dark against the glowing water. **1872** 'Mark Twain' *Roughing It* v. 51 He..climbs the nearest sand-mound, and gazes into the distance. **1921** *Daily Colonist* (Victoria, B.C.) 8 Apr. 4/2 Organized playgrounds were a valuable asset to any city—a playground in which there were sandpiles and wading pools for the little ones. **1976** *National Observer* (U.S.) 30 Oct. 16/5 She recalls playing 'kick the can' and burying each other in sand piles. **1883** 'Mark Twain' *Life on Mississippi* xxiv. 267 You can tell a sand-reef—that's all easy. **1973** *Publ. Amer. Dial. Soc.* lx. 8 The mainland is..cut off from the Atlantic by the long lines of sand reefs called the Outer Banks. **1889** W. B. Yeats *Wanderings of Oisin* iii. 49 But prone on the pathway, prone struggling, They lay 'neath the sand-sack at length. **1936** M. H. Mason *Paradise of Fools* xix. 218 When we finally get stuck in the middle of the Sand Sea..you'll have to carry everything. **1976** L. Deighton *Twinkle, twinkle, Little Spy* ii. 13 This road skirted the edges of the Sahara's largest sand-seas. **1854** V. Lush *Jrnl.* 5 Feb. (1971) 151 The boat beat about all the afternoon and towards evening ran fast upon the sandspit off the mouth of the Mungamungaroa Creek. **1910** S. P. Hyatt *Diary of Soldier of Fortune* xv. 161 The town..stands on a little sandspit which juts out from a mangrove-circled bay. **1934** *Discovery* May 130/1 One result of the storm was that a sand spit was built out across a bay. **1974** *Nat. Geographic* Dec. 785/1 Its reef supported two islets, one a mere sandspit and the other some 350 yards long. **1930** E. Pound *XXX Cantos* ii. 9 Glare azure of water, cold-welter, close cover. Quiet sun-tawny sand-stretch. **1922** Blunden *Shepherd* 28 Where the sand-vein still bubbles its clear spring.

b. *sand-castor, -shaker, -strewer; sand-loving, -teasing* adjs.
1897 'H. S. Merriman' *In Kedar's Tents* xxv. 281 Vincente was writing at the table... He smiled as he shook the small sand-castor over the paper. **1924** [see *BATTERSEA]. **1940** R. Graves *Sergeant Lamb of Ninth* 206 The chest was filled with pens, ink, paper, sand-castors. **1915** R. Lankester *Diversions of Naturalist* 17 The rare sand-loving plants of the dunes. **1967** *Oceanogr. & Marine Biol.* V. 505 Sand-loving species such as the tectibranch gastropod *Philine aperta*. **1958** *Washington Post* 26 June A1/8 They [*sc.* microphones] would be located where the old and now empty 'sand shakers', once used as blotters, are placed on each desk. **1972** *Country Life* 3 Feb. 272/3 It [*sc.* a 1652 inkstand] opens to reveal..on the right a sand-shaker. **1975** *New Yorker* 26 May 105/3 (Advt.), Sterling Silver Salt and Pepper Reproductions of the original sand shakers used by George Washington at Mt. Vernon. **1922** Joyce *Ulysses* 428 Through rising fog a dragon sandstrewer, travelling at caution, slews heavily down upon him, its huge red headlight winking. **1865** G. M. Hopkins *Poems* (1948) 33 Eye greeting doves bright-counter to the rook, Fresh brooks to salt sand-teasing waters shoaly.

c. *sand-blanched, -built* (earlier and later examples), *-buried* (later example), *-cleaned, -faced, -laden, -obliterated, -rubbed, -silted, -smothered, -stained.*
1932 W. Faulkner *Light in August* v. 105 A smooth, sandblanched floor. **1788** T. Dwight *Triumph of Infidelity* 6 As sand-built domes dissolve before the stream,..The structure fled. **1916** Joyce *Portrait of Artist* (1969) iv. 160 The music passed..over the fantastic fabrics of his mind, dissolving them painlessly and noiselessly as a sudden wave dissolves the sandbuilt turrets of children. **1960** Auden *Homage to Clio* 58 A sand-buried site. **1891** W. B. Yeats *John Sherman & Dhoya* 17 Our sand-cleaned door-steps. **1931** *Times Lit. Suppl.* 3 Sept. 668/2 Hand-made and sand-faced [tiles]. **1976** *Liverpool Echo* 7 Dec. 11/2 They were hand-made, sand-faced Flemish bricks, mellowed by time and totally irreplaceable. **1902** D. G. Hogarth *Nearer East* 72 The chief ranges run north and south, weathered to fantastic outlines by the sand-laden winds and keen frosts of winter nights. **1955** P. Larkin *Less Deceived* 41 Those few forbidding signs Of the continuous coarse Sand-laden wind, time. **1938** D. Gascoyne *Hölderlin's Madness* 47 The sand-obliterated face. **1922** V. Woolf *Jacob's Room* i. 13 Wind-swept, sand-rubbed, a more unpolluted piece of bone existed nowhere. **1945** C. Mann in Murdoch & Drake-Brockman *Austral. Short Stories* (1951) 259 It broke through the sand-silted block. **1924** Lawrence & Skinner *Boy in Bush* 11 Clogged,..sand-smothered, that's what we are. **1916** A. Huxley *Burning Wheel* 50 Who marked the land-weeds and the sand-stained foam.

d. *sand-rimmed, -roofed, -wharfed.*
1857 J. G. Whittier *Poetical Wks.* II. 231 Mine the sand-rimmed pickerel pond. **1845** Longfellow *Belfry of Bruges* 50 Whole villages of sand-roofed tents. **1930** Blunden *Poems* 318 So unexpected and so beautiful That they live on in the sand-wharfed pool.

e. adverbial, as *sand-blond, -sized, -toned* adjs.; locative, as *sand-groping* vbl. sb.;

sand-bogged, -burrowing, -dwelling, -marooned, -mounded adjs.
1953 C. Day Lewis *Italian Visit* ii. 32 The hills are sand-blond. **1959** A. Upfield *Bony & Black Virgin* xi. 88 Lots of drift sand now. We'd find it rougher in the ute. Be sand-bogged a lot. **1963** R. P. Dales *Annelids* i. 29 Such protonephridia..are found in phyllodocids and in the sand-burrowing nephthyids. **1911** F. O. Bowen *Plant-Life on Land* 128 Certain sand-dwelling plants. **1963** R. P. Dales *Annelids* ii. 43 In lugworms, in the fusiform sand-dwelling opheliids. **1924** Lawrence & Skinner *Boy in Bush* 21 They walked off the timber platform into the sand, and Jack had his first experience of 'sand-groping'. **1946** W. de la Mare *Traveller* 19 Meagre his saddlebag as camel's hump When, sand-marooned, she staggers to her doom. **1921** —— *Veil* 24 Rent hull, and broken mast, She sprawls sand-mounded. **1965** G. J. Williams *Econ. Geol. N.Z.* xx. 365/2 In them [*sc.* sandstones] the clay mineral occurs as large sand-sized aggregates. **1977** A. Hallam *Planet Earth* 24/2 Somewhat larger particles, sand-sized grains, offer sufficient air resistance to be briefly heated to incandescence by friction before being entirely destroyed in the upper atmosphere. **1916** *Chambers's Jrnl.* Sept. 635/2 In the midst of the mass of sand-toned uniforms.

10. a. sand-ball (earlier example); **sand-bar** (earlier and later examples); also, a sandbank in the course of a river or close to a beach; **sand-bar willow,** a North American shrub or small tree, *Salix longifolia*; **sand blow,** the removal or deposition of large quantities of sand by the wind; a place where this has occurred; **sand-body** *Geol.*, a permeable underground mass of sand or sandstone (which may contain oil); **sand boil** *U.S.*, an eruption of water through the surface of the ground; **sand-castle,** a structure of sand resembling the form of a castle, of the kind made by a child on the beach; also *fig.*; **sand cay** [CAY], a small sandy island, usu. elongated parallel to the shore, freq. found on a coral reef and there composed of fine coral debris; = **sand key*; **sand-clock** = SAND-GLASS 1; **sand-club,** (*a*) (example); (*b*) orig. *U.S.*, = *sand-iron* (b); **sand core,** a compact mass of sand that is dipped into molten glass and withdrawn, so as to serve as a core in the making of a hollow vessel; freq. *attrib.*; **sand-crack,** (*a*) substitute for def.: a fissure in a horse's hoof; (without *a* and *pl.*) a condition so characterized; (later examples); **sand crater** (earlier example); **sand culture** *Bot.*, a hydroponic method of plant cultivation in which the plants are rooted in beds of purified sand supplied with nutrient solutions, used esp. to determine their mineral requirements; a culture of this kind; usu. *attrib.*; **sand-devil,** in Africa, a small whirlwind; **sand drown,** chlorosis of plants caused by magnesium deficiency in the soil; **sand filter,** a filter used in water purification consisting of layers of sand arranged with coarseness of texture increasing downwards; **sand garden,** in Japanese landscape gardening, an open space covered with sand, the surface of which is raked into a pattern; so **sand gardening,** the practice of this style of landscape design; **sand glacier** *Geomorphol.* (see quot. 1972); **sand grain** *Printing* (see quot. 1906); also *attrib.*; **sand-groper** (earlier and later examples); **sand-grown** *a.*, designating a native of Blackpool; **sand-happy** *a.* (see *-HAPPY); **sand-hog** *U.S.*, a man who works underground, as in a caisson or in foundation-work; also *fig.*; **sand-hole,** (*c*) a hole in sand; **sand-iron,** (*b*) (earlier example); **sand key** *U.S.* [KEY *sb.*[3]] = **sand cay*; **sand-lime,** used *attrib.* to denote a type of brick made by baking sand with a proportion of slaked lime under pressure; **sand-painting,** the technique used esp. by the Navajo Indians of painting with coloured sands; an instance of this; **sand-picture** (examples); also more *gen.*, a design made in sand; **sand pie,** wet sand formed by a child into the shape of a pie; **sand-plain** (earlier example); **sand-pump** (earlier example); **sand ripple,** one of a series of small parallel ridges or undulations in the surface of sand; **sand shadow,** an accumulation of sand to the lee of an obstruction; **sand-shoe** (later examples); **sand-slinger** *Founding* (see quot. 1948); **sand-smoke,** a whirlwind or sandstorm; **sand-stock (brick)** (later examples); **sand-storm,** now usu. written as one word (later examples); also *fig.*; **sand-table,** (*a*) a sand-covered surface on which letters or designs can be drawn and erased or models placed and removed; (*b*) = *SAND-TRAP 1; **sand-tray,** (*a*) = **sand-table* (a); (*b*) = *SAND-BOX 2 e; **sand-wash** *U.S.*, a sloping surface of sand spread out by an intermittent stream; **sand wave,** a wave-like formation in sand; *spec.* in *Physical Geogr.*, an undulation similar to a megaripple but on a larger scale; **sand-wedge** = *sand-iron* (b).
1846 *Jewish Manual, or Pract. Information Jewish & Mod. Cookery* iv. 212 Sand-balls are excellent for removing hardness of the hands. **1766** J. Bartram *Jrnl.* 29 Jan. in W. Stork *Descr. E. Florida* 55 Towards the opposite shore there is a sand-bar. **1782** T. Jefferson *Notes State of Virginia* ii. 9 The Missisipi, below the mouth of the Missouri, is always muddy, and abounding with sand bars, which frequently change their places. **1796** A. Ellicott *Jrnl.* (1803) 14 The fog was so thick that we could neither discover sand-bars nor logs. **1897** *Outing* (U.S.) XXX. 50/2 This one sheet of water formed a small harbor to the lee of a sand-bar. **1935** M. M. Atwater *Crime in Corn-Weather* i. 2 The little river—at this season no more than a network of shallow runnels between thirsty sand bars. **1968** W. Warwick *Surfriding in N.Z.* 10/3 At a beach break..the takeoff area is always changing due to drifting sand-bars. **1884** C. S. Sargent *Rep. Forests N. Amer.* 168 Sand-Bar Willow... Very common throughout the Mississippi River basin. **1975** M. C. Davis *Near Woods* v. 64 A natural hedge of sandbar willows accompanied us for twenty yards or so into the lake. **1922** *Chambers's Jrnl.* XII. 428/2 The drifting sand held sway... Towns and villages were devastated by it... Sand-blow alone did not complete the desolation. For months great areas were covered with water. **1934** *Antiquity* VIII. 182 Vast sand-blows begun by cattle breaking down the dunes. **1980** *National Trust* Spring 15/1 They were isolated from the sea by the extraordinary thirteenth- and fourteenth-century sand-blows. **1910** R. H. Johnson in *Oil Investors' Jrnl.* 20 Feb. 70/3 The necessity of conceiving the shape of the sand body as something different from the shape of the actual oil-containing reservoir is of great importance. *Ibid.*, I have found this of considerable value in predicting the shape of a 'sand-body'. **1911** —— in *Econ. Geol.* VI. 809 In order to emphasize the importance of shape I have suggested that the term sand-body be adopted, from the analogy of the word ore-body, to describe the reservoir, i.e., continuous mass of sand or sandstones sufficiently porous to be capable of containing oil and gas in commercial quantities. **1927** *Petroleum Devel. & Technol.* 1926 (Amer. Inst. Mining Engin. Petroleum Div.) 202 He is also enabled to determine such vital subsurface conditions as (1) porosity, (2) density, (3) saturation, and (4) thickness of sandbodies. **1937** *Daily Progress* (Charlottesville, Va.) 2 Feb. 1/8 Dread 'sand boils' bursting up in the heart of..Cairo [Illinois] forewarned of deeply undermined barriers guarding the..city today... The eruptions sprang from the terrific pressure of the flooded Ohio River waters slowly eating their way beneath the..levels. **1939** W. Faulkner *Wild Palms* 24 Even those who..had probably never before seen more water than a horse pond..could (and did) talk glibly of sandboils. **1954** *Encounter* Oct. 9/1 The owners of the..plantations along the Big River confederated..to hold the sandboils and the cracks. **1976** C. S. Brown *Gloss. Faulkner's South* 167 A sandboil must be neutralized promptly. This is done by building a wall of sandbags around it so that a column of water will be built up above it to equalize the pressure. **1854** C. M. Yonge *Castle Builders* v. 63 The children are..dabbling after sea-weed and shells, and building sand castles. **1925** H. G. Wells *Christina Alberta's Father* i. iv. 95 They had..camped on the beach while Mr. Preemby and Christina Alberta had made sand-castles. **1975** C. A. Haddad *Moroccan* i. 5 We tried to build a sandcastle romance out of our few short months in the [desert] sand. **1980** D. Newsome *On Edge of Paradise* vii. 228 Playing like children on the beach..making sand-castles. **1934** T. Wood *Cobbers* xvii. 219 You do not see it [*sc.* the Barrier Reef]... You see instead islands... Islands which are sand-cays covered with birds. **1937** *Geogr. Jrnl.* LXXXIX. 138 Sand-cays may occur on almost any reef, but they are most typical of the inner reefs of the outer barrier. **1968** R. W. Fairbridge *Encycl. Geomorphol.* 972/2 During hurricanes, sand cays are liable to be swept clear of vegetation and may disappear completely in a single storm. **1865** *Student & Schoolmate* June 177 One evening, fifty years ago, the noiseless 'sand-clock' in Squire Allen's bar-room was fast running down. **1964** *Listener* 24 Dec. 1011/3 The watch makers of Nuremberg were still turning out sand clocks on the egg-timer principle. **1873** *Winfield* (Kansas) *Courier* 11 Sept. 1/7 A weapon of a peculiarly dangerous and for a time mysterious nature..is a sand club, formed by filling an eel skin with sand. **1912** *Punch* 15 May 380/2 Incidentally I am pleased to know that Americans call a niblick a sand-club. **1977** P. Alliss *Play Golf with P. Alliss* 57 If you play on a heavy course with hard muddy bunkers then you will need a sand club with a sharpish leading edge. **1894** W. M. F. Petrie *Tell El Amarna* iv. 27 A tapering rod of metal was taken..; on the end of this was formed a core of fine sand... The rod and core were dipped in the melted glass... When the whole was finished, the metal rod in cooling would contract loose from the glass; it could then be withdrawn, the sand core rubbed out, and the vase would be finished. **1933** *Antiquity* VII. 421 In the technique of glass-manufacture..the process of pressing into a mould as distinct from modelling on a sand-core came into vogue. **1934** *Greece & Rome* May 140 Vessels of glass made by the sand-core technique, a process well known in Egypt during the eighteenth dynasty. **1962** D. Harden *Phoenicians* xi. 154 From the seventh to the third century sand-core fabrics made up the bulk of existing glass vessels. **1903** Somerville & 'Ross' *All on Irish Shore* 82 The glow from the fire illumined the smith's sardonic grin of remembrance. 'She had a sandcrack in the near fore that time, and there's the sign of it yet.' **1934** A. Russell *Tramp-Royal in Wild Austral.* xix. 120 This in a country where the hooves of horses develop sandcrack. **1976** *Horse & Hound* 3 Dec. 53 (Advt.), Daily use after sand-crack, seedy-toe, brittle or contracted feet, encourages the natural growth of healthy horn. **1856** Thoreau *Jrnl.* 9 Apr. (1949) VIII. 268, I..sit on the edge of that sand-crater near the spring by the railroad. **1916** *Soil Sci.* II. 208 The sand culture solutions giving low yields of tops are characterized by a wide range in the Mg/Ca ratio. **1936** *Phytopathology* XXVI. 279 Soil cultures were similarly prepared and kept with the sand cultures under the same conditions. **1940** [see *gravel culture* s.v. *GRAVEL *sb.* 9]. **1978** *Fluoride* XI. 76 In *Helianthus annus* seedlings grown in sand culture for five weeks the concentration of fluoride in the root and shoot was generally proportional to the concentration in the substrate. **1901** *Lancet* 16 Mar. 771/1 A number of small whirlwinds, called 'sand-devils', which would pass slowly along sucking up quantities of sand and any light articles such as pieces of paper. **1977** H. Innes *Big Footprints* iii. ii. 282 There was nothing visible..except here and there the dancing whirl of a sand devil. **1922** *Science* 22 Sept. 341/2 The popular name of this chlorosis is 'Sand Drown', a term referring to the fact that the disease is likely to occur in aggravated form in the more sandy portions of the field after heavy rainfall. **1968** B. C. Akehurst *Tobacco* v. 96 Magnesium deficiency (called sand drown) is shown by a characteristic chlorosis that starts with the tips of the bottom leaves, spreads across them and moves up the plant in a similar manner. **1894** Rafter & Baker *Sewage Disposal in U.S.* xiv. 267 Sand filters have considerable capacity for storing the nitrogenous matter at one period and later on converting it into nitrates. **1977** F. M. Middleton in H. I. Shuval *Water Renovation & Reuse* i. 13 Sand filters have been used for many years. **1936** T. Tamura *Art of Landscape Garden in Japan* 225 (*caption*) A sand garden carefully raked to print lines and waves. **1965** 'S. Harvester' *Assassins Road* iii. 32 The lighted windows showed patches as desolate as a Japanese sand-garden. **1975** R. L. Duncan *Dragons at Gate* (1976) iii. 89 Calder only half heard what she was saying,..fixing his attention on the sand garden. **1960** *Spectator* 16 Feb. 261/1 It's an uneasy, foreign respect—the sort one feels for minor, inscrutable Japanese arts such as Noh or sand-gardening. **1875** *Encycl. Brit.* III. 599/1 Among the less ordinary geological phenomena [of the Bermudas] may be mentioned the 'sand glacier' at Elbow Bay. **1897** *Geogr. Jrnl.* IX. 286 Wind blowing outwards from a deep sand tract forms a horizontal plateau terminated by a talus as steep as the sand can rest. Under these conditions the encroachment of sand recalls the manner of advance of a glacier, and to this formation I restrict the term 'sand glacier'. **1919** *Proc. R. Soc. Victoria* XXXI. 416 The typical forms of sand accumulation known as 'sand glaciers', which have been described in various parts of the world are due to sand being blown up the sides of hills or mountains, thence finding a passage through any passes or saddles, and spreading out on the opposite sides to form wide fan-shaped plains. **1972** *Gloss. Geol.* (Amer. Geol. Inst.) 627/2 *Sand glacier.* (*a*) An accumulation of sand that is blown up the side of a hill or mountain and through a pass or saddle, and then spread out on the opposite side to form a wide fan-shaped plain. (*b*) A horizontal plateau of sand terminated by a steep talus slope. **1906** Goodchild & Tweney *Technol. & Sci. Dict.* 203/1 Sand Grain... A ground is laid as for etching; a sheet of sandpaper is then laid face downwards on the plate, which is passed through the printer's press with sufficient pressure for the grains of sand to pierce the ground. **1960** H. Hayward *Antique Coll.* 248/1 A sand-grain aquatint is obtained from a plate which has been pulled through the press with a piece of sand paper to roughen its surface. **1896** H. Lawson *Let.* 3 Sept. (1970) 62 W[estern] A[ustralia] is a fraud... The old Sand-gropers are the best to work for or have dealings with. **1934** [see *BANANALAND]. **1946** K. S. Prichard *Roaring Nineties* 214 'I'm a sand-groper,' she snapped... 'Don't know anything about London or Paris.' **1974** *Sunday Tel.* (Austral.) 30 June, Mining millionaire Lang Hancock has a sizeable number of sandgropers prepared to support his view that Western Australia should be detached from the rest of the nation. **1969** *Listener* 6 Mar. 300/1 Natives of Blackpool are called sand-grown men. **1972** *New Society* 16 Nov. 394/2 The 'sand-grown-'uns' (the Blackpool-born). **1943** *Fortune* Dec. 268 A British Tommy on the North African desert..may have gone..'sand happy'. **1944** J. Gunther *D Day* 129 Many are what the officers call 'sand-happy'; this is a phrase almost equivalent to punch-drunk, except that it does not mean lack of fighting instinct. **1961** *Times* 14 Sept. 15/2 Captain Scott, weathered, expatriate, sand-happy. **1903** *Century Mag.* Nov. 43/1 The tunnel workers, or 'Sand Hogs', enter the lower chambers of the shield. **1904** *N.Y. Even. Post* 11 Jan. 3 The men who are employed as 'sandhogs' or excavators in the caisson for the new Manhattan Bridge. **1940** R. Chandler *Farewell, My Lovely* xiii. 98 He just got through working as a sandhog on the San Jack tunnel. **1965** *National Observer* (U.S.) 13 Dec. 12/1 Those who view Mr. Sweeney and his Appalachian Commission associates as 'sandhogs' are the other poverty operations. **1977** N. Hynd *Sandler Inquiry* xvii. 130 George McAdam was a 'sandhog'. *Ibid.* 131 The sandhogs were the British agents in oil intelligence. **1897** *Encycl. Sport* I. 457/1 Golf may be played..where the..whins, sand-holes and banks, supply the conditions which are essential to the proper pursuit of the game. **1910** W. de la Mare *Three Mulla-Mulgars* xx. 267 Home he goes to his leaf-thatched huddle or sand-hole. **1935** W. Empson *Poems* 22 By jackal sandhole to your air flung wide. **1862** Sand-iron [see NIBLICK]. **1775** B. Romans *Conc. Nat. Hist. E. & W. Florida* App. p. xli, We found ourselves surrounded by three very small low sand keys (full of prickly pears). **1829** in *Amer. State Papers: Naval Affairs* (U.S. Congress) (1861) IV. 968 An effort is now making to form a naval establishment on the insulated cluster of sand keys called the Dry Tortugas. **1837** J. W. Williams *Territory of Florida* 23 Anclote Sound is sheltered on the west, by Anclote, Jacs and Sand Keys. **1880** G. W. Cable *Grandissimes* v. 34 A beautiful land of low, evergreen hills..[looked] out across the pine-covered sand-keys of Mississippi Sound. **1930** J. F. Dobie *Coronado's Children* xviii. 308 They landed the Laffites on a barren sand key with just enough provisions to keep them alive a few days. **1937** *Geogr. Jrnl.* LXXXIX. 143 The reefs which bear a sand-key, and on which there is no sub-aerial accumulation of coral-shingle, have a least depth of water of 3 feet. **1910** *Encycl. Brit.* IV. 521/1 The so-called sand-lime bricks are now made on a very extensive scale in many countries. **1933** *Archit. Rev.* LXXIV. 225/2 (*caption*) The whole of the internal walls are faced with cream sand-lime bricks. **1966** W. G. Nash *Brickwork* I. i. 30 There are four classes of sand-lime bricks. **1902** W. Hough in *Rep. U.S. Nat. Museum 1900* 467 The ceremonial sand painting of the Hopi and Navajo, where the most beautiful effects are secured by allowing sand in slender streams of different colors to fall from the hand guiding it over the surface

to form designs. **1908** *Encycl. Relig. & Ethics* I. 826/2 The sand-paintings..may be regarded as actual pictorial prayers. **1963** G. S. Maxwell *Navajo Rugs* (1973) iii. 47 Sandpainting rugs are woven copies of actual sandpaintings. **1978** T. Hillerman *Listening Woman* i. 3 Tell me more about how these sand paintings got messed up. **1957** J. Kirkup *Only Child* xiv. 188 There was a man who made wonderful sculptures in the damp sand... Once,..he made a low-relief sand-picture of the Shields Town Hall. **1970** G. Savage *Dict. Antiques* 369/2 Apart from the work of Zobel, sand-pictures are rarely signed, and must be identified from their characteristics. **1975** *Times* 6 Dec. 11/5 A collection of sand pictures, mostly made in the Isle of Wight. **1835** C. F. Hoffman *Winter in West* I. 148 A bevy of rosy little girls..were making 'sand pies' on the bank of the river. **1980** M. Drabble *Middle Ground* 181 Girls in a concrete playground, making sand pies. **1818** A. Eaton *Man. Bot.* (ed. 2) 291 On the sand plains, at the foot of Pine-rock, in New-Haven, a [juniper] root..often sends off shoots. **1865** *Harper's Mag.* Apr. 573/2 A sand-pump is a metal case from five to ten feet in length, constructed with a valve at the bottom. **1879** T. D. Forsyth in E. D. Morgan tr. *Prejevalsky's From Kulja to Lob-Nor* 27 The upheaval of the Gobi..causes an entirely independent direction of profile..to that of the sand-ripples which cover it. **1897** *Geogr. Jrnl.* IX. 279 The uniformity of the wind-ripple pattern is at all times remarkable. In water-formed sand-ripples no such uniformity has been recorded. **1941** R. A. Bagnold *Physics of Blown Sand & Desert Dunes* xi. 144 A sand ripple is merely a crumpling or heaping up of the surface, brought about by wind action, and cannot be regarded as a true wave in a strict dynamical sense. **1960** B. W. Sparks *Geomorphol.* xi. 248 The formation of sand ripples is closely connected with the process of saltation. **1941** R. A. Bagnold *Physics of Blown Sand & Desert Dunes* xiii. 188 Deposits caused directly by fixed obstructions in the path of the sand-driving wind... These sand shadows and sand drifts are dependent for their continued existence on the presence of the obstacle. **1971** I. G. Gass et al. *Understanding Earth* xiii. 184/2 Left behind protecting shells or pebbles are elongate mounds of sand ('sand-shadows') which give the beach a distinctive appearance. **1916** J. B. Cooper *Coo-oo-ee* xvi. 235 In the circumscribed space of the vessel, the men, clad in their blue dungarees, wearing white sand-shoes, prepared themselves for their future battles. **1931** V. Woolf *Waves* 16 Those are Louis' neat sand-shoes firmly printing the gravel. **1948** J. Betjeman *Sel. Poems* 79 Don't empty children's sand-shoes in the hall. **1965** S. T. Ollivier *Petticoat Farm* vii. 96 Rather than walk the dusty road in their freshly cleaned sparkling white sandshoes the girls took a short cut across the paddocks. **1979** *Guardian* 23 May 31/4 The sand shoe and school sandal look which was justifiably popular last summer. **1928** *Jrnl. Iron & Steel Inst.* CXVII. 805 Stripping machines are mounted on turntables, which bring the flasks within range of a sand-slinger, and then delivers them to the mould conveyor. **1948** J. E. Garside in H. W. Baker *Mod. Workshop Technol.* I. iii. 65 For the ramming of sand moulds, a machine known as the 'sand slinger' is often used. It ejects a stream of sand vertically downwards at a high speed, so that the sand is rammed by impact with the pattern. **1930** T. S. Eliot tr. *St.-J. Perse's Anabasis* 49 These sandsmokes that rise over dead river courses. **1956** *Archit. Rev.* CXIX. 257/2 Leicestershire sand-stock bricks are used in the panel on the west elevation. **1973** *Parade* (Austral.) Oct. 28/3 'Sandstock' (handmade) bricks were made from clay in the valley. **1928** H. Crane *Let.* 27 Apr. (1965) 325 Efforts for a foothold in this sand-storm [*sc.* Hollywood] are still avid. **1966** 'J. Hackston' *Father clears Out* 139 We missed the old..weather... Missed our blinding sandstorms even. **1978** A. & G. Ritchie *Anc. Monuments Orkney* 43 The people who were forced to abandon their homes in the final sandstorm had been using essentially the same sort of pottery vessels as their ancestors who founded the settlement. **1812** N. J. Hollingsworth *Address Madras Syst. Educ.* p. ix, To the finger and sand-table may succeed the pencil and slate. **1911** *Encycl. Brit.* XX. 728/2 To get rid of them [*sc.* impurities] the esparto pulp when washed and bleached is run from the potcher into storage chests, from which it is pumped over a long, narrow serpentine settling table or 'sand-table'. **1928** *Daily Tel.* 7 Aug. 4/4 A thorough groundwork of tactical knowledge has been formed by sand-table and week-end schemes during the winter. **1955** F. G. Patton *Good Morning, Miss Dove* 13 One group..modelled clay caribou for the sand table. **1963** R. R. A. Higham *Handbk. Papermaking* ii. 67 With rifflers and sand tables the stock is passed at approximately 0·5% consistency along narrow channels. **1969** E. H. Pinto *Treen* 423 The sand table is a very ancient device and may be referred to by Isaiah 'Now go write it before them in a table'. **1971** J. Wainwright *Last Buccaneer* ii. 243 'What..is a sand-table?'..'It's usually a tray, filled with sand. The army uses them. It's possible to mould the sand into the contours of geographic locations for demonstrating military tactics.' [**1817** A. Bell *Instructions for Conducting Schools through Agency of Scholars Themselves* ii. i. 68 For writing on sand, smooth and level (trays or) boards, ten inches wide, with ledges on every side of an inch deep..are prepared.] **1893** *N. & Q.* 25 Mar. 233/1 Economy being a great feature in the plan, the sand trays..were adopted. A full account of the system was published by the S.P.C.K. in 1840. **1968** *Guardian* 23 Aug. 7/6 A livid deputation approached me, waving the kitten's sand-tray. **1972** *Country Life* 6 Jan. 31/2, I was also interested in the 19th-century sand tray or abacus in the north aisle. This was used for teaching children to write with a wooden stick on the sand. **1901** *Science* 4 Jan. 38/1 From this point the party worked down the sandwash of Rio San Ignacio (or Rio Altar) to the coast of the Gulf of California, where the Tepoka Indians lived until recently. **1937** *Discovery* Jan. 24/1 The sand-washes surrounding the wells in the Gobi. **1948** *Sierra Club, S. Calif. Chapter, Schedule* No. 129. 69 The campsite will be in the sand wash at the mouth of the Fan Hill Canyon. **1819** Sand-wave [in Dict., sense 9 a]. **1899** *Geogr. Jrnl.* XIII. 624 The sand-waves which corrugate the beds of streams and rivers. **1902** 'Mark Twain' in *Harper's Mag.* Jan. 269/2 He started on a run, racing in and out among the sage-bushes a matter of three hundred yards, and disappeared over a sand-wave. **1917** *Bull. Geol. Soc. Amer.* XXVIII. 915 Cross-bedding..probably represents in many instances one phase of a phenomenon called sand waves, which are nothing more than current-made ripple-mark[s] of mammoth proportions... The crests are often 15 to 35 feet apart and rise from 2 to 3 feet above the troughs. **1939** W. H. Twenhofel *Princ. Sedimentation* vi. 190 The sand waves or antidunes move up-current as the individual sands move down-current. **1978** *Nature* 14 Sept. 101/2 Sandwaves are the largest scale of bedform.., with average heights and wavelengths markedly larger than those of megaripples. **1937** H. Longhurst *Golf* i. xxii. 196 No chapter on bunker play would be complete without a description of..the..sand wedge. **1952** *Chambers's Jrnl.* May 298/1, I couldn't use a sand-wedge in a bunker because I hadn't the strength to swing it. **1971** 'D. Halliday' *Dolly & Doctor Bird* xv. 215 Wallace Brady..landed in the long, pale trap in front of the green and stayed there doing explosive shots with a sand-wedge.

b. sand bird (earlier and later examples); **sand boa,** a snake of the genus *Eryx*, found in north and east Africa and south and east Asia; **sand crab,** also *fig.*; (earlier and further examples); **sand dab,** (*a*) substitute for def.: either of two eastern North American flatfishes, the American plaice, *Hippoglossoides platessoides*, or the windowpane, *Scophthalmus aquosus*; (earlier and later examples); **sand dollar,** substitute for def.: a flattened, irregular sea urchin belonging to the order Clypeastroida; (further examples); **sand fiddler** *U.S.*, a small burrowing fiddler crab of the genus *Uca*; **sand-fish,** (*c*) *S. Afr.* = *MOGGEL; (*d*) *S. Afr.*, the beaked salmon, *Gonorhynchus gonorhynchus*; **sand goanna,** an Australian monitor lizard, *Varanus gouldii*; **sand goby,** the common goby, *Pomatoschistus minutus*; **sand-lance** = *sand-launce*; **sand lizard,** (*a*) (later examples); (*b*) *U.S.*, a fringe-toed lizard of the genus *Uma* or the striped race-runner, *Cnemidophorus sexlineatus*; **sand-mason,** substitute for def.: a burrowing polychæte tube-worm belonging to the genus *Lanice*; also *attrib.*; (later examples); **sand monitor,** (*b*) = **sand goanna*; **sand perch,** substitute for def.: a small bass, *Roccus americanus*, found in marine and fresh water in eastern North America; (earlier and later examples); **sand-runner** (examples); **sand-shark,** (*a*) one belonging to the family Carchariidæ, esp. *Carcharias taurus*; (later examples); (*b*) = *guitar-fish* s.v. *GUITAR *sb.* b; (later examples).
1709 J. Lawson *New Voy. Carolina* 151 The Sand-Birds are about the Bigness of a Lark, and frequent our Sand-Beaches. **1917** T. G. Pearson *Birds Amer.* I. 234 White-rumped Sandpiper..Sand-bird. **1910** R. L. Ditmars *Reptiles of World* iv. 233 The Sand Boas, *Eryx*, are degenerate burrowing species,..with a flat body, very stumpy tail, a small head,..and tiny eyes. **1970** *E. Afr. Standard* 23 Jan. 6/4 These [snakes] include..a sand boa and two boa constrictors. **1844** J. E. DeKay *Zool. N.Y.* vi. 6 This [*sc.* *Platycarcinus irroratus*] and the succeeding species [*sc.* *P. sayi*] are both designated by our fishermen as the Spotted Crab and Sand Crab. **1883** Sweet & Knox *On Mexican Mustang through Texas* 24 The calling of each other names, such as 'sand-crabs' and 'mud-turtles', is one of the harmless ways in which they ventilate their spleen. **1946** K. Tennant *Lost Haven* (1947) xii. 199 The little cream sand-crabs swift as impatient foam. **1952** W. J. Dakin *Austral. Seashores* xv. 190 The sand bubbler-crab... This little crab may be found..resting at the bottom of a vertical chimney-like burrow. **1955** V. Palmer *Let Birds Fly* 108 No, you ol' sandcrab, you don't know Charlie. **1839** D. H. Storer *Rep. Ichthyol. Mass.* 143 *Platessa dentata*..known by the fishermen as the 'Sand-dab' in the Boston market. **1903** T. H. Bean *Fishes N.Y.* 726 Sand Dab..is also known as the rusty dab. **1924** J. A. La Gorce et al. *Bk. Fishes* 15/1 The Sand Dab, lying on the sand, has harmonizing blotches imprinted all over the upper part of its body. **1954** J. Steinbeck *Sweet Thursday* xxiv. 155 Joe Elegant ordered sand dabs for supper. **1884** *Bull. U.S. Nat. Museum* No. 27. 123 The so-called 'sand dollar'..inhabits the east coast. **1923** *N. & Q.* 18 Aug. 133/1 The stone pies appear to be the fossilized remains of certain echinoderms kindred to the North American sand-dollar. **1962** [see *KEYHOLE *sb.* 4]. **1969** R. Lowell *Notebk. 1967–68* 70 His face an azure sand-dollar on the pail of a child. **1976** *National Observer* (U.S.) 6 Nov. 17-A/4, I stare down at the water-stained sand, hoping to find a sand dollar. **1852** C. H. Wiley *Life in South* 30/1 Sand-fiddler,..the local name for a small animal of the shell-fish kind. **1973** *Publ. Amer. Dial. Soc.* lx. 1 The long beaches are left to the sun and the surf, the sand fiddlers, the gulls and the pelicans. **1925** *Ann. S. Afr. Mus.* XXI. 125 Beaked Salmon or Sand Fish... Greyish brown above, silvery below. **1946** L. G. Green *So Few are Free* x. 135 The sandfish..migrates at spawning time. **1947** [see *MOGGEL]. **1949** Vesey-Fitzgerald & Lamonte *Game Fish of World* v. 375 The sandfish, a species of *Labeo* characterised by the inferior position of the mouth, is another common inhabitant of this river system [*sc.* the Olifants river]. **1953** J. L. B. Smith *Sea-fishes S. Afr.* 87 Sandfish or Beaked Salmon (Austral.). **1968** K. Weatherly *Roo Shooter* 119 A sand goanna..has no respect for snakes at all; he would give most of them a very rough time of it. **1911** F. Ward *Marvels of Fish Life* ii. 13 The sand goby..merely scoops out a hollow. **1935** D. B. Wilson *Life of Shore & Shallow Sea* viii. 88 Sand gobies..could not possibly see the bait. **1971** *Nature* 21 May 150/2 Other workers have found that the scarcity of the sand goby in inshore waters is matched by an increase offshore. **1905** D. S. Jordan *Guide to Study of Fishes* II. xxix. 521 The small family of sand-lances..comprises small, slender, silvery fishes, of both Arctic and tropical seas. **1975** *New Yorker* 12 May 80/3 The sand lances had both the length and the diameter of standard pencils. **1910** R. L. Ditmars *Reptiles of World* iii. 173 The Sand Lizard or Striped Race-Runner..is the only species of its genus ranging into the southeastern portion of the United States. **1915** E. G. Boulenger *Reptiles & Batrachians* i. iv. 81 The Sand Lizard..is a very local creature with us, confined to sandy heaths. **1928** *Bunker's Mag.* Jan. 73 The little sand lizards so common in West Texas possess the same ability to snap off their tails when they get into a tight corner. **1954** R. C. Stebbins *Amphibians & Reptiles Western N. Amer.* 224/1 Buried sand lizards can sometimes be frightened from the sand. **1979** *Jrnl. R. Soc. Arts* CXXVII. 405/2 The heathland..is the habitat of reptiles such as the smooth snake and sand lizard. **1935** E. G. Boulanger *Nat. Hist. Seas* v. 77 Another common worm is the Sand Mason.., the tubes of which few can have overlooked. **1977** *Radio Times* 12–18 Nov. 19/1 Now he has photographed the denizens of mudflats: sea urchins, sand-mason worms, and the dog-whelk. **1975** H. G. Cogger *Reptiles & Amphibians Austral.* 236/1 Gould's Goanna or Sand Monitor... A widespread species subject to considerable geographic variation in colour, pattern and size. **1878** C. Hallock *Sportsman's Gazetteer* 378 Sand Perch, or Bachelor Perch;..Apparently a cross between the yellow belly and silver perch. **1946** *Richmond* (Va.) *Times-Dispatch* 4 Aug. iv. 4-D/2 There is always the likelihood of catching..sand perch and blue-nosed perch. **1965** A. J. McClane *Stand. Fishing Encycl.* 737/1 The sand perch..is one of the small sea basses distributed from North Carolina to Texas. **1894** A. Newton *Dict. Birds* iii. 813 Sand-runner, like the foregoing [*sc.* sand-plover], but perhaps sometimes used more for Sandpiper. **1913** H. K. Swann *Dict. Eng. & Folk-Names Brit. Birds* 205 Sand Runner: The Dunlin. Also the Ringed Plover and the Sanderling on the Humber. **1979** *Bull. Yorks. Dial. Soc.* Summer 7 We would find eggs on the sand at the sea side of the Point laid by a bird we called a sand runner. **1938** A. H. Verrill *Strange Fish* ix. 92 Certain species of sharks..may be considered harmless to man. Such are the sand-sharks and dogfish. **1949** W. W. Small in Vesey-Fitzgerald & Lamonte *Game Fish of World* v. 381 A sandshark (really a shovelnose skate)..can give an angler hell. **1961** E. S. Herald *Living Fishes of World* 17/2 Sand sharks—Family Carchariidæ. **1968** D. O'Grady *Bottle of Sandwiches* 51 He said it was only a sand-shark, or shovel-nose.

c. sand cherry, substitute for def.: a shrub or small tree, *Prunus pumila*, of central North America, or a related species, *P. besseyi*, of the western states; also, the fruit of these plants; (earlier and later examples); **sand flower** = SANDWORT; **sand grass,** (*b*) *N.Z.* = *PINGAO; **sand lily,** (*a*) *U.S.*, a stemless rhizomatous herb, *Leucocrinum montanum*, belonging to the family Liliaceæ and bearing clusters of fragrant white flowers; (*b*) a bulbous plant, *Pancratium maritimum*, belonging to the family Amaryllidaceæ, native to the Mediterranean region, and bearing fragrant white flowers; = *sea-daffodil* s.v. SEA *sb.* 23 f; **sand myrtle,** substitute for def.: a small evergreen shrub, *Leiophyllum buxifolium*, of the family Ericaceæ, native to eastern North America and bearing pink or white flowers; (earlier and later examples); **sand-oat** = *sand-reed* in Dict. and below; **sand pear,** substitute for def.: an oriental species of pear, *Pyrus pyrifolia*; (earlier and later examples); **sand-reed,** substitute for def.: the marram grass, *Ammophila arenaria*; cf. MARRAM 1; (later examples); **sand spurry** (earlier and later examples); **sand verbena** *N. Amer.*, a trailing herb of the genus *Abronia*, belonging to the family Nyctaginaceæ, found in western North America, and bearing clusters of fragrant red, yellow, or white flowers; **sand-weed** = SANDWORT.
1778 J. Carver *Trav. N. Amer.* 30 Near the borders of the Lake [Michigan] grow a great number of sand cherries. **1800** A. Henry *Jrnl.* 17 Aug. in E. Coues *New Light Early Hist. Greater Northwest* (1897) I. ii. 40 We found an abundance of sand-cherries, which were of an excellent flavor. **1970** J. H. Gray *Boy from Winnipeg* 55 When we tired of that [*sc.* swimming] we would go picking sand-cherries. **1916** W. de la Mare *Songs of Childhood* (rev. ed.) 80 Alliolyle where the sand-flower blows Taught three old apes to sing. **1937** Dylan Thomas in *Life & Letters* Spring 70 He stumbled on over sand and sandflowers like a blind boy in the sun. **1905** Sand grass [see *PINGAO]. **1959** A. H. McLintock *Descr. Atlas N.Z.* 31 Planting of sand grass, lupins, and, in places, pines..is needed to protect farm land. **1909** Webster, *Sand lily*, a white-flowered scapose liliaceous plant..of the western United States. **1929** *Encycl. Brit.* XIX. 939/1 Sand Lily..native to plains and mountain valleys from South Dacota and Nebraska west to California. **1951** T. H. Kearney et al. *Arizona Flora* 177 The star-lily or sand-lily..is to be looked for in northern Arizona. **1956** G. Durrell *My Family & Other Animals* xvi. 215 The smooth curve of the dune..was the only place on the island [*sc.* Corfu] where these sand lilies grew, strange, misshapen bulbs buried in the sand, that once a year sent up thick green leaves and white flowers above the surface. **1973** Hitchcock & Cronquist *Flora Pacific Northwest* 691 Fl[ower]s white, rather showy, borne in clusters... Sand lily, star lily. **1814** F. Pursh *Flora Americana* I. 301 *Ammyrsine buxifolia*..known by the name of Sand-myrtle among the inhabitants of New Jersey. **1882** *Harper's Mag.* June 71 Of the smaller shrubs now in bloom we find the sand-myrtle, with its terminal umbel-like clusters of small pinkish flowers. **1943** R. Peattie *Great Smokies & Blue Ridge* 266 Tangled growths of rhododendrons..with some amounts of mountain laurel, blueberry, smilax, and occasionally sand myrtle. **1881** *Encycl. Brit.* XII. 60/1 The dunes show a tendency, except where the Dutch prevent it by planting wood or sand-oats, to wear away on the side towards the sea. [**1629** J. Parkinson *Parad.* iii. xxi. 593 The Sand peare is a reasonable good peare, but Small.] **1880** [see *KIEFFER]. **1951** *Dict. Gardening* (R. Hort. Soc.) IV. 1722/2 Sand Pear... Edible var[ietie]s are grown in China and Japan. **1849** W. H. Harvey *Sea-Side Bk.* i. 12 The sand-reed..naturally grows on the sandy shores of Europe. **1879** *Scribner's Monthly* Sept. 651/1 After laboriously cleaning their fish, they laid them among the sand-reeds. **1910** *Encycl. Brit.* XIII. 590/2 The most common plant here is the stiff sand-reed. **1975** M. C. Davis *Near Woods* i. 3 On a wave-lashed slope, this sand reed measures land's end. **1866** Sand spurry [see SPURREY 2]. **1960** *Oxf. Bk. Wild Flowers* 112/2 The Cliff Sand Spurrey (*S. rupicola*), found on rocky coasts in the south and west, has glandular hairy stems... Sand Spurrey (*S. rubra*), common in sandy and gravelly places, is a rather hairy plant. **1898** A. M. Davidson *Calif. Plants* 174 The wild four o'clock and the sand verbena are classed in this group [of beautiful weeds]. **1929** *Encycl. Brit.* XIX. 940/1 The white sand-verbena,..with very numerous fragrant flowers, occurs from Iowa to Idaho. **1946** D. C. Peattie *Road of Naturalist* i. 16 Pervading the sunny waste with fragrance, rose sprawling sand-verbenas. **1849** D. G. Rossetti *Let.* 18 Oct. (1965) I. 78 Curse the big mounds of sand-weed!

sand, *v.* Add: **3. a.** (Later examples with *up*.)
1918 Galsworthy *Five Tales* ix. 61 They would..sand up his only well in the desert. **1956** Peterson & Fisher *Wild Amer.* xxxiv. 369 Novashtoshnah, which means 'the new growth' (newly sanded up from island to peninsula), is the northeast point of St. Paul.
5. a. Also in phr. *to sand and canvas*, to clean thoroughly. Also *fig.* orig. *Naut. slang.*
1912 J. Masefield in *Eng. Rev.* Oct. 345 Unless you're clean we'll sand-and-canvas you. **1914** *Dialect Notes* IV. 151 *Sand and canvas*,..to clean. **1933** P. A. Eaddy *Hull Down* 187 The Mate was anxious to get on with the 'sand and canvasing' of the bright work.
b. = SANDPAPER *v.*
1928 E. W. Hobbs *Mod. Furnit. Veneering* vii. 84 The wood finish..is sprayed on, allowed about three hours to dry, and sanded lightly with No. 400 waterproof paper and water. **1939** Pattou & Vaughn *Furnit.* ii. vi. 197 Sand all first coaters with the grain and do not lap the sanding more than necessary. **1958** *Listener* 11 Sept. 399/1 After sanding the piece of furniture, you will be using oil paint to give a hard, durable surface. **1976** F. E. Sherlock *Enjoying Home Carpentry & Woodwork* xi. 116 When the project has been glued and cleaned-up.., it must be sanded.
6. *intr.* To become clogged or bunged *up* with sand.
1926 *Summary of Operations, California Oil Fields* (Calif. State Mining Bureau) Oct. 9 The well..stopped of its own accord, probably sanding up.

sandal, *sb.*[1] Add: **4.** *sandal-footed* adj.; *sandal-mark, -shoe*; **sandal-foot,** used *attrib.* and *absol.* to designate a kind of stocking with a non-reinforced heel, suitable for wearing with sandals.
1959 *Vogue* June 71 Coming in..are the sandal-foot stockings... Aristoc have fully-fashioned sandal-foots. **1970** *Focus* June 10/2 Sandalfoot is used to indicate a vision or non-reinforced heel. **1978** *Detroit Free Press* 2 Apr. 2B (Advt.), Sheer, sandalfoot pantyhose with bone or self-colour panty knit right in. **1927** D. H. Lawrence *Mornings in Mexico* 83 A white, sandal-footed man following with the silent Indian haste. **1949** R. Campbell tr. St. John of the Cross in *Coll. Poems* I. 167 Tracking your sandal-mark The maidens search the roadway for your sign. **1603** Sandal-shoe [see *sandal shoon* in Dict.]. **1882** W. D. Hay *Brighter Britain!* II. 127 Sandal-shoes upon their feet.

sandalwood. Add: Also **sandal-wood.** **3.** A perfume derived from sandalwood oil.
1865 E. Rimmel *Bk. Perfumes* viii. 143 Indra..appears very partial to scent, for he is always represented with his breast tinged with sandal-wood. **1973** G. Butler *Coffin for Pandora* viii. 161 Her heavy scent of heliotrope and sandal-wood.
4. A fashion shade resembling the colour of sandalwood, a light yellowish brown. Also as *adj.*
1926 *Daily Express* 1 Sept. 10 (Advt.), Shades of mulberry, sandalwood, purple, [etc.]. **1927** *Ibid.* 26 Feb. 5 This attractive model is designed in sandalwood face-cloth. **1937** [see *MIST *sb.*[1] 1 e]. **1976** *Country Life* 26 Feb. 502/3 Stockings..in..a browny colour called Sandalwood.
5. *Comb.*, as **Sandalwood English** = *BEACH-LA-MAR; **sandalwood oil**; a strongly aromatic oil obtained by distillation of sandalwood (*Santalum*), used in perfumes and cosmetics and formerly as a genito-urinary antiseptic.
1922 Jespersen *Language* 216 The so-called *Beach-la-mar* (or Beche-le-mar, or Beche de mer English); it is also sometimes called Sandalwood English. **1936** S. Robertson *Devel. Mod. Eng.* iv. 89 Beach-la-Mar or Sandalwood-English, spoken..all over the Western Pacific. **1950** J. C. Furnas *Anat. Paradise* v. 355 *Beach-la-mar* (sometimes Sandalwood English) is the specific name of this pidgin. **1971** I. F. Hancock in D. Hymes *Pidginization & Creolization of Lang.* vii. 523 Melanesian Pidgin English, also known as Neo-Melanesian, Sandalwood English, Bêche-de-mer, Beach-la-mar, etc. **1851** *Illustr. Catal. Gt. Exhib.* iv. 878/2 Sandal-wood oil..from Mangalore and Canara. **1901** W. H. White *Text-bk. Pharmacol. & Therapeutics* 586 Sandal-wood oil is very similar in its action to the oils of copaiba and cubebs. **1952** Kirk & Othmer *Encycl. Chem. Technol.* IX. 589 The Australian sandalwood oils and the West Indian sandalwood oils are distilled from different species. **1965** F. Sargeson *Memoirs of Peon* v. 109 He was so knowledgeable about the virtues of copaiba and sandalwood oil,..not to mention a chemist who would be of great assistance to me.

sandar, pl. of *SANDR, SANDUR.

Sandawe (sandā·we). Also **Sandawi.** [Native name.] The name of a tribe in central Tanzania having racial, cultural, and linguistic affinities with the Hottentots; a member of this tribe; their language; also *attrib.* or as *adj.*
1924 *Jrnl. Afr. Soc.* XXIV. 26 Farther to the south-east, another 'click' language is met with, the Sandawi. **1925** *Ibid.* XXIV. 219 A section of Wanyaturu some 5,000 strong, who for generations have lived with them, accepting the rule of the Sandawi headmen. *Ibid.* 226 The Sandawi is a bowman. *Ibid.* 334 For a description of the Sandawi language I must refer to some notes by Father Lemble... Its most interesting feature is its clicks. **1947** *Jrnl. R. Anthrop. Inst.* LXXVII. 61/1 The Sandawe are a tribe, some 21,000 strong, inhabiting part of the Kondoa Irangi District..in the Central Province of Tanganyika. **1958** J. P. Moffett *Handbk. Tanganyika* v. 158 The members of the Sandawi tribe occupy the south-western part of the Kondoa District. **1963** in Oliver & Mathew *Hist. E. Afr.* iii. 62 It should be added that the pastoralism of the Hottentots *seems* to be older and more deep-seated than that of the Sandawe. **1974** *Encycl. Brit. Macropædia* XVII. 1029/2 During the Stone Age, bands of hunter-gatherers of the Bushmen type inhabited parts of the country [*sc.* Tanzania]. The Sandawe are vestiges of this early group.

sand-bag, sandbag, *sb.* Add: Now usu. as one word. **2. e.** (Earlier example.)
1808 E. Weeton *Let.* 8 Nov. in *Jrnl. of Governess* (1969) I. 123 Scarce a window or a door was permitted to be opened. My room window was fastened down, and stuffed with sand-bags.

sandbag, *v.* Add: **1. b.** *intr.* To attend to sandbags.
1928 *Sat. Even. Post* 4 Feb. 100/2 One of the chauffeurs had just finished fueling the plane. 'You fly her,' said Andy. 'I'll sandbag.'
2. (Earlier and later examples.) Also *fig.*, to bully or coerce; to criticize or lambaste.
1887 *Courier-Jrnl.* (Louisville, Kentucky) 2 Feb. 6/2 The next day Claytor turned up at Central Station with a fairy story that he had been sand-bagged on his way home. **1901** *Congress. Rec.* 23 Jan. 1345/1 [This district] is lying in wait, as it were, from one year's end to the other, awaiting an opportunity to sandbag the public. **1903** 'O. Henry' in *Ainslee's Mag.* Feb. 59/2 About what figure had you and the kalsominer agreed to sandbag the state for? **1919** *Daily News* 12 Mar. 8/1 While the [German] revolution was being side-tracked in Parliament it was being sand-bagged in the proletariat. **1973** *Globe & Mail* (Toronto) 4 May 6/1 Each will attempt to sandbag the Liberals into adopting its policies. **1974** *Listener* 27 June 818/1 Mr Heath and Mr Wilson sandbagging each other at televised press conferences.
3. *Poker.* To refrain from raising at the first opportunity in the hope of raising by a greater amount later.
1940 O. Jacoby *On Poker* v. 36 The time to sandbag is when you have three of a kind or better. **1950** G. S. Coffin *Poker Game Compl.* vi. 71 Jacks back sometimes offers a fine chance to sandbag. **1977** D. Anthony *Stud Game* i. 7 He fondled his stack of blue chips. He was sandbagging me. I gave him the same dose of silence. **1978** *Sci. Amer.* July 112/3 By under-representing a strong hand (sand-bagging) and thus keeping his opponents from folding a player may increase the pot he expects to win.

sa·ndbagged, *ppl. a.* [f. SANDBAG *v.* + -ED[1].] Having or equipped with sandbags. Also *fig.*
1916 *Blackw. Mag.* May 615/2 He had betaken himself..to a blockhouse which guarded a section of the..railway... The outlook from this sand-bagged sanctuary was extensive and curious. **1930** Wodehouse *Very Good, Jeeves!* viii. 223 The Snettisham..was standing there with a sand-bagged look watching her nominee pass right out of the betting. **1940** *Economist* 20 July 91/2 Claims under personal accident ..were higher, for the black-out and sand-bagged pavements produced many minor injuries. **1952** Dylan Thomas *Coll. Poems* 43 Man-in-seed, in seed-at-zero, From the star-flanked fields of space, Thunders on the foreign town With a sand-bagged garrison. **1959** I. Jefferies *Thirteen Days* ii. 26 The Yehudi convoys used to form up..with iron-clad buses and sand-bagged lorries. **1977** *Time* 21 Feb. 14/1 The demilitarized zone in Nicosia..separates sandbagged Turkish- and Greek-Cypriot gun emplacements.

sandbagger. Add: **1.** (Earlier and later examples.) Also *fig.*
1882 G. W. Peck *Peck's Sunshine* 203 Suppose all the men that have been robbed in the past year by cowardly sand-baggers, could have 'put up their hands'. **1893** *Chicago Tribune* 26 Apr. 6/4 One of the Chicago papers recently complained that Illinois had no first-class highwaymen. It must have overlooked the legislative 'sand-baggers'. **1929** C. E. Merriam *Chicago* 343 A matter to be carefully watched here [in subcommittees of the city council] is room for blackmail, even in the case of worthy measures unless the sandbaggers are offset by those of an opposite persuasion. **1981** P. McCutchan *Shard calls Tune* xiii. 148 Senglea..had had its quota of sandbaggers once, evil men who lurked upon roofs and swung heavy sandbags to strike sailors on the head so that their pockets could be rifled.
2. *Poker.* One who sandbags. Cf. *SANDBAG *v.* 3.
1940 O. Jacoby *On Poker* v. 36 In this event the sandbagger intends to raise. **1950** G. S. Coffin *Poker Game Compl.* v. 56 We have bet aces up so many times in last position when all checked after the draw, and butted into triplets and sandbaggers.

sand-blast. Add: **2.** Also, a blast of sand-laden liquid. (Further example.)
1913 V. B. Lewes *Oil Fuel* iii. 69 A big gusher would, by sand-blast action, cut through the chilled steel shields in a few days.
sand-blasted *a.*, delete *nonce-wd.* and add further examples; **sand-blaster** (further examples); also, = SAND-BLAST 1.
1920 *Public Health Rep.* (U.S. Public Health Service) XXXV. 518 (*heading*) The efficiency of certain devices used for the protection of sand blasters against the dust hazard. **1937** U. R. Evans *Metallic Corrosion Passivity & Protection* xiii. 545 A roughened (sand-blasted) surface appears necessary for good adhesion. **1974** *Nature* 5 Apr. 502/1 Abraiding freshly made implements in a tumbling mill, or sandblaster. **1975** M. Bradbury *Hist. Man* iii. 44 Another [student] brought a sand-blaster and cleaned off the walls of the basement. **1975** *New Yorker* 19 May 11/3 (Advt.), Polished and sandblasted stainless-steel sculptures. **1976** 'Trevanian' *Main* (1977) iv. 76 The sandblasters have cleaned..a façade that used to bear the comfortable patina of soot... For months now, they have been sand-blasting the building.

sand-blast, *v.* [f. SAND *sb.*[2] + BLAST *v.*; cf. SAND-BLAST.] To subject to a blast of sand or the like, esp. so as to clean or polish. Also *fig.*
1888 *Texas Siftings* 6 Oct. 6/3 'Sleigh-bells! Well, I'll be sand blasted!' said the business man. 'What do you mean by trying to sell sleigh-bells in this section of the country? Don't you know it never snows here?' **1924** *Jrnl. Inst. Metals* XXXII. 294 The present-day practice is to sand-blast almost every article which is to be metal sprayed. **1939** A. K. Lobeck *Geomorphol.* xi. 376 (*caption*) The sand grains have been sand-blasted away but the more durable binding silica..has resisted the attack. **1972** *Timber Trades Jrnl.* 3 June 47/1 The metal surfaces to be coated are first sand-blasted to remove grease and impurities. **1979** H. McCloy *Smoking Mirror* 176 The old buildings..had been sand-blasted to preserve their fabric.
Hence **sa·nd-blasting** *vbl. sb.*
1904 Goodchild & Tweney *Technol. & Sci. Dict.* 257/1 Sand blasting is another method of producing an etched effect upon glass. **1935** H. R. Simonds *Finishing Metal Products* xii. 105 The term 'sand blasting' is commonly used to describe the application of an abrasive material under pressure to surfaces to be cleaned or otherwise treated. Even when steel grit is used as the abrasive, the term 'sand blasting' is frequently retained. **1977** *New Yorker* 24 Oct. 42/1 Dissolution, leaching, sandblasting, cracking and melting of fireproof doors.

sand-blind, *a.* (Examples in *poet.* use.)
1864 G. M. Hopkins *Poems* (1967) 15 Are you sand-blind? Slabs of water many a mile Blaze for him all this while. **1938** W. de la Mare *Memory* 46 Hope..Led sand-blind Despair To a clear babbling wellspring And laved his eyes there.

sand-box. Add: **2. d.** A small low-sided sand-pit (cf. *SAND-PIT 3). Chiefly *U.S.*
1937 [see *HUH *int.*]. **1968** *Globe & Mail* (Toronto) 17 Feb. 39 Divided they stand not unlike urchins defying each other in the sandbox. **1969** [see *MICKEY MOUSE 1]. **1976** *National Observer* (U.S.) 8 May 20/4 All the men in her life, from sandbox playmate to lip-smacking savage.
e. A box kept indoors and filled with sand or other material for a cat to defecate in.
1967 L. J. Braun *Cat who ate Danish Modern* viii. 73 Qwilleran showed Koko the new location of his sandbox and gave him his old toy mouse. **1971** J. McClure *Steam Pig* viii. 98 The *Daily Post*..an evening rag not worth putting in the cat's sand-box. **1974** M. G. Eberhart *Danger Money* (1975) iv. 40 I've fixed up a sandbox for the cat.

sand-boy. Add: **1.** (Earlier and further examples of the proverbial phr.) Now commonly *as happy as a sandboy.*
1821 P. Egan *Life in London* ii. v. 289 Logic..appeared to be as happy as a sand-boy, who had unexpectedly met with good luck in disposing of his hampers full of the above-household commodity. **1928** *Daily Express* 17 Mar. 3/1 The King was in his element here... He was happy as a sandboy. **1958** *Daily Sketch* 2 June 11/3 Brimming with health, polished like a Derby cup, happy as a sandboy. **1973** *Perthshire Advertiser* 17 Feb. 18/3 It isn't hot, but they're as happy as sandboys.

sand-bur (sæ·nd,bɜ̄ɹ). *U.S.* Also **sand-burr.** [f. SAND *sb.*[2] + BUR *sb.*] The small prickly fruit of any of several plants, esp. a bur-grass of the genus *Cenchrus* or an annual herb, *Franseria acanthicarpa*; also, any of several plants bearing such a fruit. Also *attrib.*
1830 W. A. Ferris *Life in Rocky Mts.* (1940) vi. 28 These grass-knots, are called 'Sand-burrs'. **1834** A. Pike *Prose Sk. & Poems* 48 To add to our comforts, the ground here was covered with sand-burs. **1867** E. Eggleston in *Little Corporal* Sept. 37/1 A bad name..sticks to you like a sand-burr. **1896** [see SAND *sb.*[2] 10c]. **1904** *Topeka* (Kansas) *Daily Capital* 11 June 4 A sandbur will grapple on to a man's coat tail and stay there all day just to get a chance to fall into his bed at night. **1948** F. Blake *Johnny Christmas* i. 39 Weatherby stalked back to his bed, knocked the sandburs from his socks, and..pulled his blankets over his ears. **1957** L. Eiseley *Immense Journey* 69 There passed before my eyes the million airy troopers of the milkweed pod and the clutching hooks of the sandburs. **1971** *Country Life* 4 Nov. 1193/1 These [impurities] include the long spiky seeds of

shepherd's needle. .and the viciously armed burs of the sand-bur grasses. .from South America.

sa·nd-cast, *v. Founding.* [f. Sand *sb.*[2] + Cast *v.*] *trans.* To make (a casting) by pouring molten metal into a sand mould.

1949 C. J. Smithells *Metals Ref. Bk.* 596 Nearly all alloys can be sand cast, including relatively hot short materials. **1952** Wood & von Ludwig *Investment Castings for Engineers* xvii. 371 The preference for location of bosses in parts which are to be sand cast is on internal surfaces rather than external ones.

So **sa·nd-cast** *ppl. a.*; **sa·nd casting** *vbl. sb.*, (*a*) an object cast in a sand mould; (*b*) the process of casting in a sand mould.

1934 *Jrnl. Inst. Metals* LIV. 103 A sand-cast ingot. **1939** *Light Metals* II. 361 (*heading*) The production of aluminium-alloy sand castings. **1949** C. J. Smithells *Metals Ref. Bk.* 596 Sand casting offers the widest scope of all the casting processes. *Ibid.*, In weight, sand castings range from less than an ounce up to more than 100 tons. **1960, 1964** [see *gravity die-casting* s.v. *gravity 8 b]. **1967** A. H. Cottrell *Introd. Metall.* xiii. 184 In sand casting a wooden pattern of the required shape, slightly enlarged to allow for shrinkage of the casting, is firmly packed in sand in a moulding box... A green sand casting is made in sand bonded with clay. **1981** *Pop. Hot Rodding* Feb. 22/1 The only engine components that should be changed are the pistons, and only if they are of sand-cast manufacture.

sande (sa·nde). Also **sandee**, and with capital initial. [W. Afr.] The name of a cult for women based on secret rites of initiation, etc., widespread amongst tribes in Sierra Leone and Liberia. See also *Poro.

1803 T. Winterbottom *Acct. Native Africans Sierra Leone* II. 235 In the river Sherbró,. .which is inhabited by Bulloms, there is a society of girls called Sandee girls, who, besides being initiated into various mysteries, are instructed to dance in public. **1930** *Harvard Afr. Exped. 1926–27* I. v. 83 Most of the tribes in Liberia still practice certain ceremonies connected with the initiation of both boys and girls in the bush schools... The. .girls' school is known as *sande*. **1954** R. Lewis *Sierra Leone* i. 9 Quite a number of the mauve- and green-uniformed but nubile-looking schoolgirls who walk in crocodile through the streets will be rushed through Sande in the holidays. **1968** Harris & Sawyer *Springs of Mende Belief* i. 2 *Sande*. .plays a more restricted role in the community. Its main concern is to cultivate in adult women the qualities of wifehood and motherhood. *Ibid.* vi. 104 This accusation. .includes. .any attempts to watch the masked *sande* dancers remove their head-piece.

Sandemanian, *sb.* and *a.* Add: Also erron. **Sandimanian, Sandymanian.** (Earlier and further examples.)

1766 B. Stevens *Let.* 21 Jan. in E. Stiles *Extracts from Itineraries & Other Misc.* (1916) 566 In the late times of oclocracy some small damages have been done to the Sandimanian meeting house. **1773** *Massachusetts Gaz.* 8 Apr. 3/1 The Fire likewise communicated to the Sandemanian Meeting House. **1926** A. Huxley *Let.* 10 Aug. (1969) 272 Faraday. .was at once a Sandemanian and a Fellow of the Royal Society. **1935** B. Russell *Relig. & Sci.* vii. 171 Faraday was a Sandymanian, but the errors of that sect did not seem, even to him, to be demonstrable by scientific arguments.

sander, *sb.* Add: **3.** A sand-papering machine.

1895 in *Funk's Stand. Dict.* **1930** *Engineering* 23 May 688/1 The combination consists of a saw-table,. .a disc sander. **1975** M. Bradbury *Hist. Man* iii. 43 Students. . with a rented sander exposed, and then waxed yellow with a rented waxer, the good old wood of the floors.

sandesh (sa·ndeʃ). [a. Bengali *sandesh* a sweetmeat.] An Indian sweetmeat resembling cheese fudge.

1944 in D. K. Gupta *Best Stories Mod. Bengal* I. 72 She. . pushed a plate full of Sandesh towards me. **1953** R. Godden *Kingfishers catch Fire* xi. 122 *Sandesh*, which is like toffee, and *jilibis* that are rings of clear sugar. **1966** J. & R. Godden *Two under Indian Sun* iv. 103 Indian sweets, jillipis or sāndesh. **1973** *Times* 19 June (Bombay Suppl.) p. xv/3 West Bengal's rosogollas and sandesh.

sand-fly. 1. a. Substitute for def.: A small blood-sucking fly belonging to the family Simuliidæ or Psychodidæ or a biting midge of the family Ceratopogonidæ. (Later examples.)

1907 Fountain & Ward *Rambles Austral. Naturalist* xi. 121 The sand flies. .irritated us greatly. **1932** [see *owl fly* s.v. *owl *sb.* 7 b]. **1947** I. L. Idriess *Isles of Despair* xxxii. 214 Barbara was fortunate that it was beautiful weather;. .no hellish mosquitoes or burning sandflies to torture her naked hide. **1962** Gordon & Lavoipierre *Entomol. for Students of Med.* xxi. 141 Members of the family Simuliidæ are widely known as 'buffalo-flies' and 'black flies', but in some parts of the world, as in Australia, they are . .designated by other names such as 'sandflies' or 'midges'. **1972** Swan & Papp *Common Insects N. Amer.* xxii. 595 The biting midges. .include some very annoying pests, variously called no-see-ums, punkies, sand flies, moose flies, and gnats.

3. sand-fly fever, an acute viral fever transmitted by flies of the genus *Phlebotomus*.

1911 Dorland *Med. Dict.* (ed. 6) 629/2 *Phlebotomus papatasii* [printed *papatassii*]. .is thought to convey by its bite an infection known as sandfly fever. **1936** *Indian Jrnl. Med. Res.* XXIII. 870 The lesion in dengue fever. . differs to some extent from that in sandfly fever. **1962** Gordon & Lavoipierre *Entomol. for Students of Med.* v. 26 There remain certain virus infections, such as sandfly fever and dengue, which cause disease in man, but for which no animal reservoir has, as yet, been demonstrated.

sandfracing, sand fracing (sæ·ndfrækiŋ), *vbl. sb. Oil Industry.* Also **-fraccing, -fracking.** [f. Sand *sb.*[2] + Frac(tur)ing *vbl. sb.*] A method of stimulating production from an oil field by forcing fluid containing sand grains into the reservoir rock. So **sa·ndfrac** *sb.*, a name for this process; **sand frac(k** *v. trans.* and *absol.*, to apply this treatment to (an oil field).

1953 *Petrol. Engineer* XXV. B108/2 This well had two sandfrac treatments, using. .round grained sand. **1957** *Times* 11 Dec. 16/4 It is believed that sandfraccing and drilling in proven areas will maintain the present rate of production. **1960** *Oil & Gas Reporter* XII. 1034 After the well was completed, Phillips contracted with *D* to sand frack. *Ibid.* 1035 Phillips arranged. .to sand frack the producing formation to increase productivity. **1961** *Ibid.* XIV. 111 On December 9, 1959, Gregg et al. requested the Commission to enter field rules including one which would authorize any operator to 'sand frac' by using a maximum of 20,000 gallons of fluid and 40,000 pounds of sand per well. **1961** *Texas Law Rev.* XXXIX. 359 It is suggested that compulsory pooling is a workable and effective method of solving the small tract problem and, more specifically, the issues arising from an alleged trespass caused by sandfracing.

sandhi (sa·ndi). *Philol.* Also †**sundhi,** and with capital initial. [a. Skr. *saṃdhi* junction, combination, f. *sam* together + *dhā* to place (see Do *v.*).] The term applied orig. by Sanskrit grammarians to assimilative changes occurring in Sanskrit in the final and initial sounds of words in a sentence (*external sandhi*), and in the final sounds of stems in word-formation (*internal sandhi*); extended by modern philologists to analogous phenomena in other languages.

1806 W. Carey *Gram. Sungskrit Lang.* I. iii. 15 (*heading*) Of Sundhi, or the permutation of letters occasioned by the junction of syllables. **1841** H. H. Wilson *Introd. Gram. Sanskrit Lang.* ii. 8 The changes to which letters are subject for the sake of euphony are numerous, and carefully defined, forming that part of Sanskrit grammar which is termed. .Sandhi, 'a holding together', 'a function'. **1888** H. Sweet *Hist. Eng. Sounds* 15 An equally primitive stage is preserved in the Sanskrit sandhi, only here it is generally the end of a word that is modified. **1888** J. Wright tr. *Brugmann's Elem. Compar. Gram. Indo-Europ. Lang.* I. 501 Owing to the scantiness of the Umbrian–Samnitic materials, handed down to us, it remains doubtful at what period certain processes of sandhi took place in Latin. **1901** A. A. Macdonell *Sanskrit Gram.* p. ii, The rules of Sandhi are based chiefly on the avoidance of hiatus and on assimilation. **1933** *Eng. Stud.* XV. 41 O[ld] E[nglish] *æt þam ende*, which became *atten ende*, and then, through sandhi, *at an ende*. **1939** [see *morphophonemics *sb. pl.*]. **1952** [see *nuclear *a.* (and *sb.*) 1 c]. **1976** *Language* LII. 212 There is ample evidence that much of Sanskrit external sandhi is the result of sweeping generalizations.

2. *attrib.* and *Comb.*

1888 J. Wright tr. *Brugmann's Elem. Compar. Gram. Indo-Europ. Lang.* I. 488 The fettered language of the Vedas already furnishes the proof that the sandhi-system of the classical Sanskrit is not a thing of natural growth. **1933** L. Bloomfield *Language* xxiii. 418 There resulted sandhi-alternants of words like *water*:. .['wɔtə]. .['wɔtər iz]. . . This. .resulted in the sandhi-form *the idea-r is*... English sandhi-alternation is limited largely to cases like the above. **1935** *Amer. Speech* X. 86/1 The sandhi customs are different in English from what they are in Dutch. *Ibid.*, A preliminary statement of the sandhi rules of the dialect or language studied. **1945** *Mod. Lang. Notes* Dec. 539 The sandhi-affricate formed in the sequence *right here*. **1961** R. E. Keller *German Dialects* 264 This sandhi-lenition is not indicated in the orthography of the *Lux*[*emburger*] *W*[*ör*]*t*[*er*]*b*[*uch*]. **1969** *Eng. Stud.* Suppl. p. lxxxi, At a conservative estimate twenty-two of the above twenty-eight instances support a spoken *sandhi*-distribution of voiced and voiceless final consonants. **1975** *Language* LI. 551 Words may have more than one form, due to sandhi processes such as French liaison and elision.

sand-hill. Add: **b. sand-hill crane,** substitute for def.: a North American crane, *Grus canadensis*; also *absol.*; (earlier and later examples).

1805 W. Clark *Jrnl.* 31 Oct. in *Orig. Jrnls. Lewis & Clark Exped.* (1905) III. 176 Jo killed a Sand hill Crane. **1834** J. K. Townsend *Narr. Journey Rocky Mts.* (1839) i. 12 We observed great numbers of the brown, or sandhill crane,. . flying over us. **1907** W. O. Lillibridge *Where Trail Divides* 115 He can. .stalk a sandhill crane where there isn't cover to hide your hat. **1938** C. H. Matschat *Suwanee River* 186 He seen the sandhills a-dancin' their matin' dance. **1949** *Natural Hist.* Oct. 378/1 Once heard. the far-reaching call of the sand-hill crane is a sound that can never be forgotten. **1960** R. T. Peterson *Field Guide Birds of Texas* 79 Sandhill Crane. .A long-legged, long-necked gray bird with a bald red forehead. **1977** *New Yorker* 9 May 113/1 He had later seen a pair of sandhill cranes.

d. *Canad. pl.* A region of southeastern Alberta; in the mythology of Plains Indians, the abode of departed spirits.

1949 J. G. MacGregor *Blankets & Beads* 113 Nothing marks the spot where some mighty chief or minor brave sleeps, while his spirit travels the trails of the Great Sand Hills. **1957** *Camsell Arrow* (Edmonton, Alberta) Christmas 77/1 The sun dance site is in the heart of the 50-mile-square Blood reserve about 40 miles south of Lethbridge. There are situated the sacred sand hills and the happy hunting grounds for departed spirits. **1959** N. Sluman *Blackfoot Crossing* 13 Little Tree would have to go unadorned to the Sand Hills, for her daughter could not part with the red glass beads. **1963** R. D. Symons *Many Trails* xiii. 138 He [*sc.* a missionary] had been saying that it would not be long now before he [*sc.* an Indian] would be called to the Sand-hills. **1975** *Alberta Hist.* Spring 16/2 Indians tell that a blizzard came up and blue and yellow lightning coloured the sky when Wolf Collar's ghost departed for the Sand Hills, the home of the dead.

sand-hiller (earlier and later examples).

1848 *Congress. Globe* 30th Congress 1st Sess. App. 137/1 The thing is whispered even among the sandhillers of South Carolina. **1850** E. P. Burke *Reminisc. Georgia* 205 These people are known at the South by such names as crackers, clay-eaters, and sand-hillers. **1872** [see *piney-woods cracker* s.v. *piney wood b]. **1944** [see *piney-woods tacky* s.v. *piney wood b]. **1958** H. Babcock *I don't want to shoot an Elephant* 155 Barefooted and shirtless, the sandhiller was sprawled listlessly on the porch when I arrived.

sandhya (sa·ndyā). [a. Skr. *saṃdhyā* a holding together, junction: cf. *sandhi.] **a.** Twilight. **b.** The period which precedes a yuga or age of the world. **c.** Morning or evening prayers.

1868 *Chambers's Encycl.* X. 327/1 A long mundane period of years, which is preceded by a period called *Sandhya*, 'twilight'. **1876** Monier Williams *Indian Wisdom* (ed. 3) 248 The two Sandhyas of sunrise and sunset. **1891** —— *Brāhmanism & Hindūism* 401 The first act of the Morning Sandhyā Service. .is sipping water. **1913** J. N. Farquhar *Crown of Hinduism* 164 The daily devotions (sandhya) are restricted to the three castes. **1971** *Leader* (Durban) 7 May 9/4 They were all saying the 'sandhya', the evening prayers. In perfect Hindi! **1974** *Encycl. Brit. Micropædia* VIII. 850/1 *Sandhyā*. ., Hindu religious acts performed by the twice-born (the three higher castes) at the three divisions of the day (morning, noon, and night).

Sandinista (sa:ndini·stă), *sb.* (*a.*) [Sp., f. the name of Augusto César *Sandino* (1893–1934), Nicaraguan nationalist leader + *-ista* -ist.] A supporter of Sandino; a member of the revolutionary Nicaraguan guerrilla organization founded by him or of a similar organization founded in his name in 1963. Also *attrib.* or as *adj.*

1928 *Nation* (N.Y.) 29 Feb. 232/2 Everybody from here on was a Sandinista; the trail was full of Sandinistas. **1931** *Foreign Affairs* (N.Y.) IX. 499 A purely objective view of the facts hardly warrants calling the Sandinistas 'bandits'. **1954** *Southwestern Social Sci. Q.* (U.S.) Sept. 140 The Guardia's inability to destroy the *Sandinista* movement did not permit the early withdrawal of the marines. **1967** N. Macaulay *Sandino Affair* iii. 55 On November 2 [1926] he led this force in an attack on the two-hundred-man government garrison at Jícaro, near San Albino. The Sandinistas killed some of the defenders. **1974** *N.Y. Times* 29 Dec. 1/4 The guerrillas, members of the so-called Sandinista Front, were said to have demanded the release of some 40 political prisoners. **1977** *Time* 31 Oct. 16/1 Others, like the Sandinista guerrillas of Nicaragua or the Islamic Marxists of Iran, have specific targets. **1980** *Ann. Reg. 1979* 66 A new Government, formed by the Sandinistas, was quickly recognized by the US.

sand lot. Add: Also with hyphen and as one word. **1.** (Earlier and later examples.) Also used *absol.* in the political sense; in literal use, a plot of empty or undeveloped land, esp. in a town or suburb.

1878 *N.Y. Tribune* 14 Aug. 4/3 Mr. Kearney, the 'sandlot orator' of California. .came East with the prestige of a victorious leader. **1880** *San Francisco News Let.* 3 July 10/1 The Sand-lot barely escaped a Kilkenny fight on Sunday last. **1885** *Mag. Amer. Hist.* Feb. 201/2 One Dennis Kearny [*sic*]. .made his headquarters in what were known as the 'Sand lots', near San Francisco. **1898** G. F. Atherton *Californians* 37 She drew Helena into a sand lot opposite. **1913** J. London *Valley of Moon* 78 I've known [about his bad thumb] since he first got it as a kid fightin' in the sandlot at Watts Tract. **1930** J. Dos Passos *42nd Parallel* 192 The scorched sandlots and pinebarrens laid out into streets. **1949** *Sun* (Baltimore) 8 Jan. 9/5 The survey revealed that of the 18 direct fatalities [in football games] occurring in 1948, 6 were in 'sandlots'. **1978** *Verbatim* May 13/1 The sandlot in Washington Square has a sign 'Sandlot reserved for children and their guardians'.

2. Used *attrib.* (now usu. as one word) with reference to sports and games of the kind played by amateurs in a sand lot.

1890 *Breeder & Sportsman* (San Francisco) (Base Ball Suppl.) 7 June 3/3 Why. 'skates' and 'wafters' are kept in the team simply because at one time they were alleged good players by some sand lot critic. **1921** *Daily Colonist* (Victoria, B.C.) 10 Apr. 10/1 The National Baseball Federation, the governing body of the sandlot baseball, today went on record as opposing the 'Black Sox'. **1932** *Sun* (Baltimore) 6 Sept. 14/4 The Bugle nine will be composed of the pick of sand lot teams. **1942** *Short Guide Gt. Brit.* (U.S. War Dept.) 12 'Village cricket' which corresponds to sandlot baseball. **1954** *Encounter* Oct. 8/2 Interpret what this goddamn cotton market is going to do tomorrow, and we can both quit chasing this blank blank sandlot ball team. **1964** R. Murphy *Pond* iv. 64 He played sandlot football in the afternoons. **1979** *Amer. Poetry Rev.* Mar./Apr. 24/1 Just once the kid with bad eyes hit a home run in an obscure sandlot game.

sand lotter, also with hyphen and as one word; (*b*) one who plays in a sandlot team.

1889 *Breeder & Sportsman* (San Francisco) 14 Dec. 485/1 The local sports kicked like old sand-lotters; they visited the judges stand. .and told the poolseller that their money would never be paid. **1979** *N.Y. Times* 6 Aug. c2/1 These White Sox. .are stocked with college players, sandlotters and, as Lonborg says, 'guys who love the game but, well, have some deficiencies'.

‖ **sandolo** (sæ·ndŏlo). Also **sandalo** (pl. -i), *erron.* **sandola.** [It.] A flat-bottomed rowing-boat of the kind used in the waterways of Venice.

1928 D. H. Lawrence *Lady Chatterley* xvii. 314 He was a sandola man, a sandola being a big boat that brings in fruit and produce from the islands. **1940** V. Woolf *Roger Fry* iv. 100 Row out across the lagoons in a sandolo. **1962** N. Mitford *Water Beetle* 123 Young men. .ferry tourists from the steamer to the village in sandolos. **1966** J. G. Links *Venice for Pleasure* 223/2 Take a gondola or the smaller *sandolo* from Burano. **1969** B. Malamud *Pictures of Fidelman* vi. 154 Sandali sailed under bridges, heaped high with eggplants. **1974** *Encycl. Brit. Macropædia* XIX. 72/2 There are almost as many different kinds of watercraft in Venice as there are surface vehicles in a mainland city, from the dainty little *sandolo*, rowed standing with crossed oars, to giant ocean liners.

Sandow (sæ·ndo). The name of Eugen *Sandow* (1867–1925), Russo-German exponent of physical culture, used as the type of a strong man; also applied *attrib.* and in the possessive to exercises, an exercise machine, and societies endorsed by him. Also *fig.*

1898 *Physical Culture* I. 112/2 If 'Cantab' had proposed the theory that Sandow's system would have produced as good a ten-stone oar as Oxford training made of Mr. Kent, he would have stated a definitely arguable proposition; for if Sandow's system is to be applied to rowing at all, this is one of the results we shall immediately ask it to produce. **1905** W. B. Yeats *Let.* July (1954) III. 454, I have got into my routine here... To this I have added Sandow exercises twice daily. **1911** L. Stone *Jonah* I. ix. 100 He threw down the hammer with the air of a Sandow. **1914** C. Mackenzie *Sinister St.* II. III. ii. 531 They talked instead of Sandow exercises and mountain-climbing. **1932** A. Huxley in *Lett. D. H. Lawrence* p. xiii, How bitterly he [*sc.* Lawrence] loathed the Wilhelm-Meisterish view of love as an education, as a means to culture, a Sandow-exerciser for the soul. **1947** N. Coward *Peace in our Time* I. iv. 51 Nora: You're thinner than you were when—when you went away. Stevie: I'm Sandow to what I was when I left the prison camp. **1947** C. Gray *Contingencies* i. 21 A complete fallacy. . that it is possible for aesthetic sensibility to be imparted . .by any such methods of spiritual jerks or intellectual Sandow exercisers. **1952** D. Davie *Purity of Diction in Eng. Verse* 175 He [*sc.* G. M. Hopkins] has no respect for the language, but gives it Sandow-exercises. **1962** *Listener* 2 Aug. 166/2 The founding of hundreds of Sandow physical-culture clubs throughout England and Wales. **1965** F. Sargeson *Mem. Peon.* v. 116 Anyone who engaged in Sandow exercises.

Hence **Sa·ndowism,** the principles of physical culture advocated by Sandow.

a **1930** D. H. Lawrence *Phoenix* (1936) 656 Physical training and Sandowism altogether is a ridiculous and puerile business.

sand-paper, sandpaper, *sb.* Add: Now usu. as one word. **1. b.** *fig.* Chiefly *attrib.* or as *adj.*, rough, abrasive, aggressive.

1953 'N. Blake' *Dreadful Hollow* II. xi. 145 The voice which had that sandpaper timbre of the overdriven. **1976** *Time* 20 Dec. 22/1 With his sandpaper style and naked drive for power, Burton had quite a few enemies.

Hence **sa·ndpapery** *a.*, resembling sandpaper, rough.

1957 V. Nabokov *Pnin* 22 The sandpapery side of his head. **1970** E. McGirr *Death pays Wages* iii. 53 He. . massaged a sand-papery jaw. **1975** T. Allbeury *Special Collection* xi. 77 His hand was dry and sandpapery.

sand-pit. Add: Also as one word. **3.** A space in a garden or park enclosed by low walls and filled with sand in which children may play.

1898 G. B. Shaw *Candida* I. 80 A park. .containing. .a sandpit. .imported from the seaside for the delight of children. **1908** G. Jekyll *Children & Gardens* xi. 90 You will find out endless ways of playing with the sand-pit. **1937** T. Adams *Playparks* 49 Sandpits are not desirable in crowded playgrounds of large towns. **1959** *Oxf. Mail* 11 Mar. 6/4 Nurseries featuring merry-go-rounds, sand-pits, and attendants dressed as clowns are supplied free of charge. **1960** F. G. Lennhoff *Exceptional Children* vii. 138 He found some release through child-like games. .including sandpits and mud. **1976** 'D. Halliday' *Dolly & Nanny Bird* ix. 118 Grover. .was given half an Italian Easter egg and was sick in the sandpit.

‖ **sandr, sandur** (sæ·ndŭr, sæ·ndŭı). *Physical Geogr.* Pl. **sandar**; also **sandr, sandrs, sandurs.** [a. older Icel. *sandr* (pl. *sandar*) Sand *sb.*[2] In mod. Icel. the sing. is spelt *sandur*.] A broad, flat or gently sloping, sheet of glacial outwash.

1893 *Proc. Boston Soc. Nat. Hist.* XXVI. 172 This would lend much support to the theory that the sand and gravel plains of the Cape, of Long Island, and of the outlying islands, are to be regarded as confluent fan-deltas built up by streams issuing from the ice sheet, at times when its edge lay along the northern margin of the plains. They are therefore homologous with the sandr of Greenland,. .and the gravel fans of Alaska. **1899** *Geogr. Jrnl.* XIII. 299 Here they [*sc.* glacial streams] cover vast areas with gravelly or sandy deposits, the equivalent of the 'sandr' of Iceland. **1937** W. B. Wright *Quaternary Ice Age* x. 138 The outer moraine is distinguished from the inner by its great development of *sandr* or outwash sand plains. The inner moraine here lacks the *sandr* altogether. **1946** F. E. Zeuner *Dating Past* v. 113 Analysis of the zones of moraines and sandrs of the Scandinavian Glaciation confirmed that there were at least three major glaciations. **1969** J. L. Davies *Landforms Cold Climates* xi. 180 The Icelandic sandurs have an overall gradient of about 1 : 200 to 1 : 250. **1976** H. M. French *Periglacial Environment* viii. 177 Periglacial sandar are particularly well developed in the broad valleys which drain towards the Beaufort Sea in the western Arctic.

sa·nd-trap. [f. Sand *sb.*[2] + Trap *sb.*[1]] **1.** A device for separating sand and other impurities from a stream of water or pulp passing through it, esp. in the manufacture of paper.

a **1877** Knight *Dict. Mech.* III. 2027/2 *Sand-trap*, a device for separating sand, etc., from water flowing through a pipe. **1885** *Encycl. Brit.* XVIII. 221/2 From them the pulp is pumped into the supply-box, which communicates with the sand-traps by means of a regulating cock. **1927** T. Woodhouse *Artificial Silk* 22 These sand traps are long, shallow, wooden troughs, the bottoms of which are covered by suitable rough-haired felt and baffle plates. **1963** R. R. A. Higham *Handbk. Papermaking* ii. 68 Sand traps. .are the small recesses in the trough of a beater or breaker which are covered with a perforated metal plate.

2. *Golf.* A bunker. Also *fig.*

1922 Wodehouse *Clicking of Cuthbert* iv. 99 As for the deep sand-trap in front of the seventh green, he spent so much of his time in it that there was some informal talk. .of charging him a small weekly rent. **1927** *Daily Express* 29 Oct. 3/3 We cannot eliminate from the game that part of it which is played within the confines of the sandtraps. **1971** 'D. Halliday' *Dolly & Doctor Bird* xv. 212 He doesn't like soft sculpted sand-traps, which he likes to call bunkers. **1980** *Amer. Speech* LV. 127 Sixteen pages of notes and documentation attest to the authors' concern for accuracy and good scholarship, even though the inevitable sandtraps occur where nonspecialists attempt specialized courses.

sandveld (sæ·ndvelt, ‖ sæ·ntfelt). Also **sand veld, sand-veld, zandveld.** [a. Afrikaans, f. *sand* Sand *sb.*[2] + *veld* Veldt, veld.] In southern Africa, (the name of) a region of light sandy soil.

1824 W. J. Burchell *Trav. S. Afr.* II. ix. 242 The plains on the other side [of the Langberg], are called by the name of *Zandveld*. **1873** *Cape of Good Hope Blue Bk.* App. p. JJ3, Several hundred trees have been planted along the main road to Cape Town. .on 'Zandveld'. **1919** *S. Afr. Geogr. Jrnl.* III. 73 The Free State farmer in the early spring-time treks to the 'sand veld' with his stock. **1937** Marais & Sim in D. J. Seymore *Handbk. Farmers in S. Afr.* (Dept. of Agric. & Forestry, S. Afr.) 704 In the sandveld the production of a grain crop is not easy—the soil is a light sand, not particularly fertile. **1939** 'D. Rame' *Wine of Good Hope* I. xiii. 158 They left the last of the wheat and came to a queer sand-veld. **1944** M. Oldevig *Sunny Land* vi. 51 The Kalahari, the waterless sandveld of Bechuanaland and South West Africa. **1953** D. Lessing *Five* iii. 121 This. .was farming country,. .a pocket of good, dark, rich soil in the wastes of the light sandveld. **1959** [see *Africander, Afrikander 3]. **1964** *Listener* 6 Aug. 192/1 He plants the first rose on the burnt sandveld.

Sandwich, *sb.*[1] Add: **2. Sandwich tern,** a black, grey, and white tern, *Sterna sandvicensis*, found in Europe and Africa.

1785, 1888 [see Tern *sb.*[1]]. **1914** *Chambers's Jrnl.* May 308/1 These birds are the ring-ouzel. .and the sandwich tern. **1934** *Discovery* Oct. 293/1 Many species like the. . Sandwich tern, and puffin are growing scarcer. **1968** [see *machine *sb.* 3]. **1971** *Country Life* 18 Feb. 356/2 Two birds, at least, which breed nowhere else in Spain, have important colonies here: the black-headed gull and the Sandwich tern.

sandwich, *sb.*[2] Add: **1. a.** Now made of almost any filling or spread, occas. with only one slice of bread, as in *open* or *open-faced sandwich* (see *open *a.* (*adv.*) 22 c), or with biscuits, sliced buns, or cake. The specifying word now freq. denotes form as well as contents: *club* (see *club *sb.* 20), *Dagwood, Denver, hero* (see *hero *sb.* 5), *peanut butter* (see *peanut b), *poor boy* (see *poor *a.* (*sb.*) 8), *submarine sandwich* (see *submarine *sb.*). (Earlier *fig.* example.)

1925 S. Lewis *Arrowsmith* xxiv. 280 You might bring me a Denver sandwich from the Sunset Trail Lunch. **1932** G. Greene *Stamboul Train* I. i. 11 Get me a sandwich... I'm so empty I can hear my stomach. **1954** *Good Housek. Cookery Bk.* (rev. ed.) 443/1 Buns and cakes, provided they are not too sweet may be cut in thin slices and used for sweet sandwiches. **1977** *Rolling Stone* 16 June 12/1 Our past albums were like Dagwood sandwiches because you had to listen to them 30 or 40 times on very sophisticated equipment to hear everything we'd dub in. **1978** G. Mitchell *Mingled with Venom* iv. 37 Take Diana the smaller of my two sponge sandwiches.

fig. **1790** T. Wilkinson *Mem.* III. 154, I will, by way of a sandwich, halt for a few minutes refreshment, and present the reader [etc.].

b. A form of training involving alternate periods of practical and theoretical instruction. Freq. *attrib.* or as *adj.* (cf. sense 3 below).

1913 Fleming & Bailey *Engineering as Profession* ii. 113 A sandwich arrangement comprising short alternating periods of technical and practical training until the full course in each is completed. **1955** *Times* 14 July 2/6 This professional training scheme is organized over four or five years on a 'sandwich' basis. **1961** *Technology* Aug. 197 For its support of the sandwich principle *Technology* has often been taken to task by teachers in the colleges of advanced technology. **1965** *Listener* 2 Dec. 887/2 An undergraduate comes up in October, spends two terms in the college. In April he begins his first six-month spell in industry... This is the Brunel sandwich evolved over the past eight years. **1972** *Guardian* 20 June 18/6 The polytechnics. .have a special interest in sandwich degree courses . ., either the thick sandwich, with one year out of a total of four spent in an industrial job, or the thin sandwich, during which the student spends alternating periods. .in college and in industry. **1980** *Jrnl. R. Soc. Arts* Feb. 157/1 We have a lot to learn and the sandwich graduates may possibly help to bridge that gap.

c. A laminated board or panel consisting of a layer of light-weight material situated between and bonded to two thin sheets of a strong material, used in light constructions, esp. in aircraft.

1944 *Use of Wood for Aircraft in U.K.* (U.S. Forest Products Lab. Publ. No. 1540) 21 The fuselage. .is composed of a plywood and balsa wood sandwich about 1/2 inch thick. **1946** *Rep. & Mem. Aeronaut. Res. Comm.* No. 1987. 2 The various kinds of sandwich considered are those in which the faces are of steel or duralumin, and the fillings of onazote, balsa wood or plywood. **1954** D. M. Desoutter *All about Aircraft* viii. 134/2 Corrugated metal. .makes a good filling for sandwiches. In this kind of sandwich two thin sheets are held apart by the corrugated metal between, and they are attached to it by welding or any other means. **1976–7** *Sea Spray* (N.Z.) Dec./Jan. 86/3 The Adelaide boat, built of foam/glass sandwich in a highly sophisticated layup technique was designed as a good all-rounder.

d. Used *attrib.* in *Chem.* to denote (complexes having) a structure in which a metal atom is bonded between two parallel cyclic ligands in different planes, as in ferrocene.

1952 *Jrnl. Amer. Chem. Soc.* LXXIV. 4971/2 A projection on the (*x*, *y*) plane concomitant with the 'sandwich' structures proposed by Wilkinson [et al.]. .immediately appeared. **1966** Phillips & Williams *Inorg. Chem.* II. xxvii. 336 It does not appear that the benzene ring can act as a bridging sandwich-ligand, but two metal atoms can be 'sandwiched' between two benzenes. **1973** *Nature* 2 Nov. 3/1 Wilkinson used the name 'sandwich compounds' for the metallocenes. **1974** *Ibid.* 11 Jan. 85/1 In the so-called 'sandwich' molecules $(C_6H_6)_2Cr$ and $(C_5H_5)_2Fe$, the metal atoms are symmetrically placed between the rings.

3. (in sense 1 a) *sandwich bar, bell, bread, counter, grill, loaf* (also ellipt.), *lunch, paper, shop, spread, supper* (later example); (in sense *1 b) *sandwich course, student, system, training*; **sandwich-board man** = sense 2; also *fig.*; **sandwich boy,** (*a*) = sense 2; (*b*) a student on a sandwich course; **sandwich cake** = *layer cake* s.v. *layer *sb.* 5; **sandwich case** (later examples); **sandwich construction,** the structure or method of fabrication of sandwich panels; **sandwich flag,** a miniature flag that identifies the filling of a sandwich; **sandwich panel,** a panel constructed as a sandwich.

1955 H. Smith *Making Money in Catering Business* vi. 51 (*heading*) Analysis of operating costs of a small provincial snack and sandwich bar. **1971** E. Paul *Reluctant Cloak & Dagger Man* x. 122, I found a sandwich bar, settled on a stool and ordered sandwiches and beer. **1977** *Lancashire Life* Mar. 101/1, I once met a chap who worked behind the counter of a sandwich bar at a railway station. **1922** Joyce *Ulysses* 284 Under the sandwichbell lay on a bier of bread one last, one lonely, last sardine. **1890** W. Booth *In Darkest England* p. xv, The expense of providing boards for 'sandwich' boardmen. **1936** W. B. Yeats *Lett. on Poetry* (1940) 124 When I excluded Wilfred Owen. .I did not know I was excluding a revered sandwich-board Man of the revolution. **1961** K. Reisz *Technique Film Editing* (ed. 9) ii. 199 Sandwich-board man carrying airline advertisement placard. **1835** *Bell's Life in London* 11 Oct. 1/1 The Sandwich boy took the hats and bonnets at the street-door. **1958** *Daily Mail* 19 Sept. 11/3 The first of the 'sandwich boys'. .have won diplomas in technology. **1971** B. Malamud *Tenants* 103 A loaf of sandwich bread on the table. **1911** C. E. W. Bean *'Dreadnought' of Darling* xxv. 221 The thin layer of jam or chocolate in a sandwich cake. **1929** J. B. Priestley *Good Companions* I. ii. 54 Mrs. Chillingford said this with immense gusto, then went slap into a piece of sandwich cake. **1968** 'P. Hobson' *Titty's Dead* vi. 70 At her elbow stood a pot of strong Indian tea and half a sandwich cake. **1908** Sandwich-case [see *patch pocket* s.v. *patch *sb.*[1] 8]. **1948** F. Thompson *Still glides Stream* ii. 40 'That fool of a groom'. .had carried off with him his mistress's sandwich case. **1944** *Use of Wood for Aircraft in U.K.* (U.S. Forest Products Lab. Publ. No. 1540) 3 The sandwich construction so effectively used in the Mosquito fuselage consists of birch plywood faces and a balsa core, affording a relatively thick section of high strength and rigidity, and good sound and thermal insulating qualities. **1946** *Rep. & Mem. Aeronaut. Res. Committee* No. 1987. 2 Considerable interest has recently been shown in the possibilities of the so-called 'sandwich' construction in the design of stressed-skin wings and fuselages. **1963** H. R. Clauser et al. *Encycl. Engin. Materials & Processes* 587/1 The largest single reason for the use of sandwich construction and its rapid growth to one of the standard structural approaches during the past 10 years is its high strength or stiffness-to-weight ratio. **1913** S. Story *Spirit of Paris* I Cafés. .have been elbowed away by vulgar bars and automatic sandwich counters. **1960** R. E. Wolf tr. *T. Henrot's Belgium* 189 Department stores with low-price sandwich counters. **1978** 'A. Stuart' *Vicious Circles* 3 The sandwich counter of the Bar Roma. .Russian salad, prosciutto,

baby pizzas. **1955** *Times* 15 July 9/7 This can be arranged in the 'sandwich' course, which alternates periods of study in college with periods of training in industry. **1957** *Technology* Apr. 44/4 Up to fifteen of these students will be Vickers undergraduate apprentices using the 'thick' sandwich course. **1966** *New Scientist* 13 Oct. 8/2 Most of the 2000 first-degree students are on sandwich courses which generally last 4½ years. **1972** *Accountant* 5 Oct. 436/1 The first paper was on sandwich courses. **1980** *Jrnl. R. Soc. Arts* Feb. 155/1 The apathy of senior management to design was felt in our BA (Hons) 4-year Sandwich course. **1907** *Yesterday's Shopping* (1969) 352 B/3 (*heading*) Sandwich flags. *Ibid.*, Stamped in Gold and Colours, with different names such as—Anchovy. .Tongue. .Foie Gras. **1950** *Vogue* Aug. 100/3 Intellectuals. .spend very little on. .sandwich flags. **1962** F. T. Day *Introd. to Paper* viii. 87 Sandwich flags are designed to distinguish party dishes. **1955** Sandwich grill [see *master *sb.*[1] 30]. **1937** D. L. Sayers *Busman's Honeymoon* iv. 83 How many loaves would you be wanting? . . A cottage and a sandwich. And a small brown? **1943** C. Milburn *Diary* 30 Jan. (1979) 166, I got two sandwich loaves. **1978** F. Weldon *Praxis* xxii. 194 On the estate bread was a sandwich loaf and the cheese cheddar or processed. **1932** D. L. Sayers *Have his Carcase* i. 9 She carried . .little. .beyond a pocket edition of *Tristram Shandy*, a vest-pocket camera, a small first-aid outfit and a sandwich lunch. **1959** *Economist* 3 Jan. 26/2 About 15 per cent of men eat a sandwich lunch. **1973** K. Benton *Craig & Jaguar* vi. 67 There will be a sandwich lunch for us on the way. **1946** *Rep. & Mem. Aeronaut. Res. Comm.* No. 1987. 2 A sandwich panel is one in which a thick sheet of a relatively weak 'filling' is interposed between two thin sheets of a more orthodox structural material, such as steel, duralumin or plywood. **1953** *Archit. Rev.* CXIV. 132/3 Walls are sandwich panels made up of two asbestos cement sheets with a cellular core. **1963** H. R. Clauser et al. *Encycl. Engin. Materials & Processes* 586/2 When a sandwich panel is loaded as a beam, the honeycomb and the bond resist the shear loads while the facings resist the moments due to bending forces, and hence carry the beam bending as tensile and compressive loads. **1923** T. S. Eliot *Waste Land* iii. 14 The river bears no empty bottles, sandwich papers. **1924** [see *carton[2] b]. **1970** 'D. Halliday' *Dolly & Cookie Bird* vii. 108 The empty packets of cigarettes, the greasy sandwich paper. **1948** Mencken *Amer. Lang.* Suppl. II. 580 Eat shop, sandwich shop. **1967** A. Bailey in L. Deighton *London Dossier* 55 Sandwich shops abound to feed the mid-day lunch-hungries. **1978** *Detroit Free Press* 2 Apr. 11A/2 Business at the sandwich shops and stores on the edge of the campus was brisk. *c* **1938** *Fortnum & Mason Price List* 37/2 Sandwich Spread—per glass 10½d. **1950** A. Wilson *Such Darling Dodos* 134 Dainty bridge rolls filled with sandwich spread. **1972** R. P. Jhabvala *New Dominion* I. 69 Your ketchup—and this is something new—sandwich spread—I thought you'd like to try it for your tea. **1963** *Times* 24 May (London Underground Suppl.) p. xv/4 The quaintly described dip. tech. sandwich student has his place. **1975** *Times* 1 Sept. 10/8 The difficulty of finding places for sandwich students. **1954** J. Betjeman *Few Late Chrysanthemums* 26 Settles down to sandwich supper and the television screen. **1919** *Proc. Inst. Automobile Engineers* XII. 450 This training should be taken along with their apprenticeship... The sandwich system has been in existence in Glasgow for over 70 years. **1940** *Nature* 21 Dec. 812/2 Some large firms testify highly to the value of the product of such a 'sandwich' system. **1956** *Ibid.* 3 Mar. 412/1 All its departments have increased their facilities to students,. .and in particular the 'sandwich system' has been established. **1971** *New Scientist* 1 Apr. 36/2 Industry is not yet prepared to cooperate sufficiently with educational establishments to make the sandwich system work as it should. **1957** *Technology* Mar. 10/3 Student apprentices, on completing their first two years sandwich training, are also transferred to the main works. **1978** *Jrnl. R. Soc. Arts* CXXVI. 347/1 The Tech's part-time study, which was a form of sandwich training.

Sandwich (sæ·ndwitʃ), *sb.*[3] The name of a town on Cape Cod, Massachusetts, U.S.A., applied to a factory and to glass produced there from 1825 to 1888.

1881 C. C. Harrison *Woman's Handiwork* III. 227 American finger-bowls. .are made at the sandwich factory in Massachusetts. **1922** *Antiques* Feb. 57/2 Until recently no one had taken the trouble to look into the sources of sandwich glass. **1935** J. C. Lincoln *Cape Cod Yesterdays* 164 The buttery shelves of every house in our town were filled with Sandwich glass at that period. **1947** R. P. Coffin *Yankee Coast* 276 Long shelves across the north windows, every inch of them covered with sandwich glass drinking the pure north light. **1964** J. Cleary *Flight of Chariots* vi. 251 She bought four-poster beds, Windsor chairs, Sandwich glass, hooked rugs.

sandwich, *v.* Add: **2.** (Later *fig.* examples.) Also, to place (different elements) alternately.

1900 *Times* 7 July 10/1 To offend the ear still further these calls of screeching boys are sandwiched by 'Any seat, Sir, but the first four rows'. **1924** H. De Sélincourt *Cricket Match* iv. 104 He liked to sandwich the weak and the strong, the swift and the slow. **1937** 'G. Orwell' *Road to Wigan Pier* ii. 29 The miner does that journey to and fro, and sandwiched in between. .are seven and a half hours of savage work. **1942** *R.A.F. Jrnl.* 27 June 30 Since the D.H. was sandwiched between them. .he was almost pulverised. **1957** *Technology* Mar. 16/2 Mechanical engineering students at Hendon Technical College sandwich eight weeks of study with eight weeks of factory work. **1977** *Time* 30 May 40/3 He sits in a rocker sandwiched between speakers blaring the hard rock music of the Grateful Dead.

intr. for *refl.* **1931** *Times Lit. Suppl.* 18 June 484/2 Tobogganing and their other misdemeanours agreeably sandwich with the humours of the always optimistic Waterall.

sandwiching *vbl. sb.* (earlier example).

1877 E. S. Dallas *Kettner's Bk. of Table* 334 In puff paste the butter and the paste are separate and there is no mixing or kneading—only what may be called fine sandwiching.

sandy, *a.* Add: **5. a.** *sandy-haired* (earlier example).

a **1817** Jane Austen *Persuasion* (1818) IV. iii. 49 Colonel Wallis's companion. .certainly was not sandy-haired.

b. sandy hill crane = *sand-hill crane* s.v. Sand-hill b in Dict. and Suppl.; **sandy pear** = *sand pear* s.v. Sand *sb.*[2] 10 c in Dict. and Suppl.

1819 D. Thomas *Trav. Western Country* 210 A bird inhabits this country called the sandy hill Crane. **1825** Sandy hill crane [see *ley var. Lye *v.*[2]]. **1884** tr. *A. de Candolle's Orig. Cultivated Plants* 233 Sandy Pear, Chinese Pear.

sa·nd-yacht. Also **sand yacht, sandyacht.** [f. Sand *sb.*[2] + Yacht *sb.*] A sail-driven craft mounted on a three- or four-wheeled chassis, used for sailing on sand.

1912 *Car* 6 Nov. 458/2 Some of the sand-yachts have three wheels; others have four wheels. **1924** F. M. Ford *Joseph Conrad* IV. 227 Knocke was just within the Belgian border. You could run in a sand-yacht in front of the dunes, right to Sluys. **1960** L. Lamplugh *Sixpenny Runner* ii. 21 By sailing he meant sand-yacht sailing. .along the four-mile stretch of sands. **1967** *Daily Tel.* 21 Feb. 15/1 Sand yachts in the first race across the Sahara. .are 70 miles from Tindouf. **1970** R. Marr in *Sports Stories for Boys* 309 The *Sandsprite* was a sand yacht, a sleek, streamlined vessel of gleaming aluminium and scarlet enamel, mounted on three motor-cycle wheels. **1980** *West Lancs. Even. Gaz.* 23 June 5 Action from the sandyacht racing at Fylde International Sand Yacht Club.

Hence **sand-yachter,** one who uses a sand-yacht; **sand-yachting** *vbl. sb.* and *ppl. a.*; **sand-yachtsman** = *sand-yachter* above.

1922 *Car* 6 Nov. 458/2 (*heading*) Sand-yachting on the Belgian coast. **1937** *Illustr. London News* 21 Aug. 315 Sailing the desert at 45 m.p.h.: the thrilling sport of sand-yachting at an R.A.F. station in Egypt. **1960** L. Lamplugh *Sixpenny Runner* vii. 76 Keith had spent most of the week with the sand-yachters. .on Brenstowe beach. *Ibid.* xv. 157 His sand-yachting pals. **1967** *Sunday Times* 28 May 19 Forty sand yachtsmen from six countries had a disappointing day at St. Annes, Lancashire, yesterday when races for the international. .sand yacht championships were unable to be started. **1970** *Daily Tel.* 23 May 9 Sensible sand-yachters wear a crash helmet. **1973** *Country Life* 25 Jan. 218/1 Ideal wind speeds for sand yachting are between 12 and 20 m.p.h. **1980** *West Lancs. Even. Gaz.* 23 June, The following Fylde sandyachters were chosen for the British team.

sanely, *adv.* (Earlier example.)

1803 M. Edgeworth *Let.* 19 Mar. (1979) 101, I am not famous for judging sanely of strangers.

san fairy ann (sæn fē·ᵊri æn). *slang.* Also **san ferry ann,** etc. [Jocular form repr. F. *ça ne fait rien* 'it does not matter', said to have originated in army use in the war of 1914–18.] An expression of indifference to, or resigned acceptance of, a state of affairs. Also *ellipt.* as **Fairy Ann.**

1919 W. H. Downing *Digger Dial.* 43 *San ferry ann*,. .it doesn't matter. **1921** *Amer. Legion Weekly* 8 Apr. 14 Son fairy Ann. **1922** B. A. Colonna *Hist. Company B, 311th Infantry* 78 If he did not sign he did not get paid and often when he does sign he don't get paid. So 'sanferriens'. **1924** *Radio Times* 19 Dec. 589/1 My mottoe's still san fairy Han. **1927** H. Kimber *San Fairy Ann* ix. 312 'There is a magic charter,' he whispered. 'It runs, "San Fairy Ann".' **1930** Kipling *Thy Servant Dog* III. 88 We said we were wonderful brave dog... He said 'Fairy Ann! Fairy Ann!' **1941** W. P. Crozier *Diary* Mar. in D. Ayerst *Guardian* (1971) xxxiv. 544 Gradualness, san faery ann. **1956** F. B. Vickers in *Coast to Coast 1955–6* 72 'Ya. Good night.' 'San ferry ann, Joe.' 'Which means black you, Jack, I'm all right,' Tom shouted. **1965** L. Brain *It's Free Country* xx. 181 'I wish you'd thought of my ulcer before you—' he began, and then broke off. 'Oh, san fairy anne!' **1973** *Times* 22 June 20/1 (Advt.), San fairy Ann... It doesn't matter to us whether it is fixed wing or helicopter because we sell the best of both.

‖ **Sanfan** (sa·nfan). Also **San-fan, San Fan** and with small initial. [Chinese *sānfǎn*, f. *sān* three + *fǎn* anti-, against.] Used *attrib.* to designate an official campaign conducted in China in 1951–2 against corruption, waste, and bureaucratism in State affairs. Cf. *Wufan.

1956 *Contemp. China 1955* I. 63 The *san-fan* movement directed against the 'three evils' of corruption, waste and bureaucracy in state institutions and enterprises. **1966** F. Schurmann *Ideol. & Organization in Communist China* v. 318 The regime resorted to terror to enforce controls. This took the form of the Three-Anti (*Sanfan*) and Five-Anti (*Wufan*) movements. The *Sanfan* campaign which started in the winter of 1951 was directed against corruption, waste, and bureaucratism. **1971** H. Trevelyan *Worlds Apart* viii. 98 The early campaigns were followed by the 'San-fan' and 'Wu-fan' movements, the so-called Three Antis and Five Antis, directed against corrupt Government officials and businessmen, but doubtless also against the politically unreliable. **1974** tr. *Wertheim's Evolution & Revolution* 333 In the early fifties, the *San Fan* (three-anti) campaign was directed at all kinds of malpractices in the newly built state apparatus.

Sanfedist (sæ·nfĕdist), *sb.* (and *a.*) [ad. It. *sanfediste* (also used), f. *santa fede* holy faith, used in the title of the society known as the Bande della Santa Fede + -ist.] A member of an Italian political and military organization of the late 18th and early 19th centuries loyal to the Papacy and hostile to republicanism. Also *attrib.* or as *adj.*

1842 F. W. Faber *Foreign Churches* II. 268 It has been asserted. .that among the higher orders of society in the Papal States another party has been formed, which includes within itself a few of the princes of the Church and affords some disquiet to the Austrian embassy at Rome. The members of this society call themselves Sanfedists. It is indeed but a revival of old Guelphic principles... The Sanfedists are said to have arisen in 1780 as an anti-Austrian party at Turin, Gregory VII, and Sixtus V, being the great objects of their admiration. **1881** *Encycl. Brit.* XIII. 486/1 In a short while, the Carbonari societies, with Sanfedisti and many other revolutionary associations, had extended their organization through the length and breadth of the peninsula. **1920** J. P. Trevelyan *Short Hist. Ital. People* xxvii. 475 During the reign of. .Leo XII. .Romagna was ravaged by the blood feuds of Carbonari and Sanfedisti. **1960** E. E. Y. Hales *Revolution & Papacy* vii. 123 The most interesting. .Sanfedist leader was Cardinal Ruffo. *Ibid.* xvi. 264 During Leo's. .reign—he died on February 10, 1829—the discontent in the Papal States. .grew more serious. The policy of using the Sanfedists. .was now encouraged from Rome, with a view to suppressing. .the secret societies. **1965** C. Hibbert *Garibaldi & his Enemies* I. iv. 48 At Ancona twenty-eight *Sanfedisti*, murderous anti-liberals who acted in the name of the faith, were assassinated.

Sanforized (sæ·nfŏrəizd), *a.* Also **sanforized.** [f. the name of *Sanford* L. Cluett (1874–1968), U.S. inventor of the process + -ize + -ed[1].] **a.** A proprietary name for cotton and other fabrics which have been preshrunk by a special process. Also *transf.*

1930 *Official Gaz.* (U.S. Patent Office) 30 Sept. 737/1 Cluett, Peabody & Co., Inc., Troy, N.Y... *Sanforized* for piece goods of cotton, linen, woolen, silk, rayon, and combinations thereof. **1938** *Times* 21 Feb. 11/1 A water-repellant finish is demonstrated at the stand of the Bleachers' Association, where Sanforized cotton and linen cloths are shown. **1939** *Trade Marks Jrnl.* 24 May 706/2 *Sanforized*. . . Piece goods of textile material but not including cotton piece goods. Cluett, Peabody & Co., Inc.,. .Troy, New York. .; manufacturers. **1944** T. D. Clark *Pills, Petticoats & Plows* 221 A frugal backwoods customer adequately sized up the complications of the 'sanforized' era in men's clothing when he sauntered into the Harbour Pitts store. **1952** M. Steen *Phoenix Rising* I. 21 If this were the States. . you'd be in a tuxedo and a sanforized shirt. **1963** *New Yorker* 15 June 78 Made in the Orient of fine Sanforized cotton. **1970** *Which?* Nov. 340/1 Some jeans carried a Sanforised label. **1975** G. Howell *In Vogue* 104/2 In shops people were asking for uncrushable fabrics like zingale, and for cottons, linens and spun rayons which were Sanforized—preshrunk. **1978** *Church Times* 25 Aug. 8 Soap-operas featuring sanforised nurses or grubby Lancastrians—every-day stories of boring folk.

b. *fig.*

1968 *Observer* (Colour Suppl.) 11 Feb. 24 (Advt.), The Sanforized Big Car. £567. The Imp is the big car shrunk small before it reaches you. **1970** S. J. Perelman *Baby, it's Cold Inside* 130 The next time I go shopping for naïve art, I'll make bloody well sure it's Sanforized.

Hence **Sa·nforizing,** this process.

1948 *Time* 11 Oct. 91/3 The company was also lucky in its Vice President Sanford Cluett, the original families' only remaining executive. Cluett was an experiment-minded man. His tinkering had turned up Sanforizing. **1963** A. J. Hall *Textile Sci.* v. 241 Recently the inventors of the Sanforizing machine have been able to use a 2-inch thick rubber belt in a modified machine and so enable fabrics to be shrunk up to 20% in length.

San Franciscan (sæn frænsi·skăn), *sb.* and *a.* [f. *San Francisc*(*o* (see below) + -an.] **A.** *sb.* A native or inhabitant of San Francisco in California, U.S.A. **B.** *adj.* Of or pertaining to San Francisco.

1875 *Scribner's Monthly* July 277/2 San Franciscans are remorseless critics. **1886** F. C. Baylor *On Both Sides* iv. 227 The glasses rattled as if in a San Franciscan earthquake. **1899** Kipling *From Sea to Sea* II. xxv. 8 It may be this sense of possible disaster. .that makes San Franciscan society go with. .a whirl. **1949** *Los Angeles Times* 6 June 2/5 San Franciscans wear overcoats and furs even in the summer. **1960** Partridge *Charm of Words* 55 *Beatniks*, as the San Franciscan press christened members of the Beat community. **1973** S. Cohen *Diane Game* (1974) iv. 43 I'm afraid I've become a San Franciscan. **1977** W. Marshall *Thin Air* ii. 16 His soft San Franciscan accent.

sang (sæŋ), *sb.*[2] *U.S.* colloq. abbrev. of Ginseng.

1843 'R. Carlton' *New Purchase* I. xxvii. 256 The storekeeper was obliged to book the nine and a quarter cents, to be paid in 'sang'. **1886** *Harper's Mag.* June 58/2 Formerly, digging 'sang', as they call ginseng, was a general occupation. **1897** W. E. Barton *Sim Galloway's Daughter-in-Law* 20 The sang was short this year. **1948** E. N. Dick *Dixie Frontier* 32 He spent some time digging ginseng, or 'sang' as they called it. **1978** *Nat. Parks & Conservation Mag.* Feb. 18/1 Hunters of 'sang', as ginseng is known in Virginia and West Virginia, can tell. .exciting stories about finding the 'big root' or 'patch'.

attrib. **1859** Bartlett *Dict. Amer.* (ed. 2) 379 *Sang-hoe*, the implement used in gathering ginseng. **1878** C. B. Coale *Life & Adventures Wilburn Waters* xxi. 124 These hill-sides are a godsend to 'sang-diggers'. **1899** M. G. Kains *Ginseng* 31 The average 'sang' digger has very little conscience. **1927** K. Eubank *Horse & Buggy Days* 53 The trail of death which lasted for twenty years started over the ownership of a 'sang-digger' hog. **1949** J. Nelson *Backwoods Teacher* xxii. 233 Thar I was in them deep woods huntin' sang roots. **1975** C. Bogue in E. Wigginton *Foxfire 3* 247 A man could go 'sang' hunting and return with a fortune.

Hence as *v. intr.*, to gather ginseng; **sa·nging** *vbl. sb.*

1848 Bartlett *Dict. Amer.* 282 *Sang*,. .is or was also used in Virginia as a verb; *to go a sanging*, is to be engaged in gathering ginseng. **1859** *Ibid.* (ed. 2) 379 In Alleghany Co., Maryland, is Sang Run near which is a well-known 'sanging ground'. **1877** *Field & Forest* III. 40 Why, I have sanged all over it [*sc.* the mountain]. **1892** J. L. Allen *Blue-Grass Region of Kentucky* 249 In the wildest parts of the country. . entire families may still be seen 'out sangin'. **1975** C. Bogue in E. Wigginton *Foxfire 3* 247 With some domestic sale, as well as a continuing foreign market, 'sanging' became a business.

‖ **sang** (sɒŋ), *sb.*[3] Also **srang**; pl. -, (anglicized) **-s.** [Tibetan *s*(*r*)*ang* ounce.] A former Tibetan unit of currency, consisting of 10 *sho*; a coin or note of this value.

1902 S. C. Das *Journey to Lhasa & Central Tibet* vii. 182 The Government revenue for each *kang* is, on an average, fifty *srang* (125 rupees), or about one hundred and fifty *khal* of grain. **1947** *Whitaker's Almanack* 886/2 The present currency [of Tibet] is reckoned in *sangs*... The 1939 value was about 8 sangs = 1 rupee. **1962** R. A. G. Carson *Coins* 545 Since 1935 on various srang values in silver. .has been the lion with a background of mountains. **1962** L. Davidson *Rose of Tibet* ix. 174 The current yuan went 330 to the Tibetan sang: the sang six and a half to the rupee. **1970** R. D. Taring *Daughter of Tibet* xviii. 242 Thubtenla lent me six hundred *sangs* (about £6). **1974** D. Norbu *Red Star over Tibet* ii. 36 His profits and premium from Chang Thang amounted to 600 *sang*.

sang, var. *sheng[1].

Sanga (sæ·ngă). [Amharic.] A bull or cow belonging to the East African breed so called, distinguished by large, lyre-shaped horns. Also *attrib.*

1814 H. Salt *Voy. Abyssinia* 258, I was gratified by the sight of the Galla oxen, or Sanga. **1862** *Chambers's Encycl.* IV. 583/1 Galla Ox, or Sanga, a remarkable species or variety of ox inhabiting Abyssinia. **1912** R. Lydekker *Ox & Its Kindred* vii. 160 These Galla or Sanga cattle are generally white and have small or no humps, their muzzles being black... In stature these oxen are very large. **1959** J. D. Clark *Prehist. S. Afr.* xi. 283 The indigenous long-horned African breed—the Sanga cattle. **1970** W. J. A. Payne *Cattle Production in Tropics* I. ii. 46 East Africa is the most likely centre of origin of the earliest Sanga cattle.

sangar: now the usu. form of Sungar in Dict. and Suppl.

sangaree, *v.* (Earlier example.)

1835 J. H. Ingraham *South-West* I. 115 [Devotees of domino are] clustered around the tables, with a tonic, often renewed and properly sangareed, at their elbows.

sang-de-bœuf. Add: (Earlier and later examples.) Also *transf.*, a ceramic glaze of this colour; porcelain bearing such a glaze.

1881 C. C. Harrison *Woman's Handiwork* II. 104 A number of antique Chinese vases in high glaze, sang de bœuf, céladon, gray, rose, mandarin, yellow. **1900** F. Litchfield *Pott. & Porc.* iv. 45 The pottery made in self-colour, such as *sang de boeuf*. **1957** Mankowitz & Haggar *Conc. Encycl. Eng. Pott. & Porc.* 88/2 The monochrome copper-red glazes of the Chinese were successfully imitated in the closing decades of the nineteenth century by Bernard Moore and the Burtons. These glazes were designated *Sang de Boeuf* and 'Flambé'. **1960** O. Manning *Great Fortune* II. 121 She had seen an Italian tea-set of fine *sang-de-bœuf* china. **1965** D. Torr *Diplomatic Cover* vi. 102 A scholarly, balding man. .was examining a plain deep red vase. Janine was saying,. .'It's a genuine Lang Yao sang de boeuf, seventeenth century.' **1972** *Trans. Oriental Ceramics Soc.* XXXVIII. 47 All these transmutations can be seen. .in the *sang de boeuf* bowl No 231 where the almost colourless rim shows faintly green. **1974** Savage & Newman *Illustr. Dict. Ceramics* 254 *Sang-de-bœuf*,. .a brilliant red glaze which exhibits patches resembling the coagulation of ox-blood... The colouring agent was copper oxide fired in a reducing atmosphere, and it was developed in China during the Ch'ing dynasty.

sanger, var. Sungar in Dict. and Suppl.

Sängerfest (se·ŋəɹfest). *U.S.* Also **Saenger-fest,** *erron.* **Sangfest.** [a. Ger. *sängerfest*, f. *sänger* singer + *fest* *fest.] A choral festival.

1865 *Harper's Weekly* 5 Aug. 490/2 Arrangements were made for the Saengerfest, which will be celebrated at Philadelphia in 1867. **1903** *Forest & Stream* 24 Jan. 78 It is thought that the event will attract several hundred shooters from all over the United States, as the Saengerfest will be in progress here the week of the shoot. **1950** 'D. Divine' *King of Fassarai* xxvii. 241 They've got a *Sangfest* on down at the *falu*... Teresa's leading the choir. **1966** *Amer. Speech* XLI. 13 In 1853 the first *Sängerfest* took place in New Braunfels [Texas], at which singing societies from neighboring towns participated. Several *Sängerfests* still take place annually.

‖ **sangha** (sa·ŋa). *Buddhism.* Also 9 **Thanga**; **samgha** and with capital initial. [Hind. *saṅgha*, Skr. *saṃgha*, f. *sam* together + *han* to come in contact.] The community or order of monks. Also *transf.*

1858 P. Bigandet *Life or Legend of Gaudama* 234 The Budhist Religious constitute the Thanga, or assembly of the Perfect. They are the strict followers of Budha. **1876** *Encycl. Brit.* IV. 429/2 The Sangha, or Society, as Buddha's order of mendicants was called. **1921** C. Eliot *Hinduism & Buddhism* III. xxxvi. 71 The Sangha has always shown a laudable reserve in interfering directly with politics... In 1886, when the British annexed Burma, the Head of the Sangha forbade monks to take part in the political strife. **1951** E. Conze *Buddhism* ii. 53 The core of the Buddhist movement consisted of monks... The entire 'brotherhood' of monks and hermits is called the *Samgha*. The Samgha naturally always formed only a small minority of the Buddhist community. **1968** T. Wolfe *Electric Kool-Aid Acid Test* xxi. 292 Boise in that moment is in the tiny knot of Perfect Pranksters, the inner circle, ascending into the the *sangha* for good! **1978** C. Humphreys *Both Sides of Circle* xii. 135 The adoption of Twelve Principles became my major Buddhist activity in Japan, and their later presentation to the Sanghas and leading Buddhists of Thailand, Burma and Ceylon was. .my major interest in those countries.

sanglier. (Further examples.)

1805 J. Sibley *Let.* 10 Apr. in *Deb. Congress U.S.* (1852) 9th Congress 2 Sess. App. 1104 There were innumerable quantities of. .deer, foxes, sangliers, or wild hogs [etc.]. **1842** W. Tolmie *Jrnl.* 7 May in *Physician & Fur Trader* (1963) 356 The Sanglier or wild boar is still found there. **1896** C. M. Yonge *Release* II. xiii. 198 It is like having to do with a set of tame pigs. .turned loose among the wild sangliers.

Sango (sæ·ŋgo). [Native name.] An African language of the Adamawa-Eastern group of the Niger—Congo family, *spec.* that pidginized version of Sango spoken as a lingua franca in the Central African Republic and elsewhere in central Africa.

1948 M. Guthrie *Classification Bantu Lang.* 74 Full classified list of the Bantu languages... Zone B... Group 10... B. 14 Cira, i- (Saŋgo). **1955**, etc. [see *Ngbandi]. **1967** W. J. Samarin *Gram. of Sango* 17 Sango is a creolized language because it stands in somewhat the same relationship to vernacular Sango as Haitian Creole to French. **1971** B. Mafeni in J. Spencer *Eng. Lang. W. Afr.* 112 Pidgin languages have been known to develop in circumstances where no master–servant relationship existed between the groups in contact. Pidgin Sango, spoken in the Central African Republic, is a good example. **1977** C. F. & F. M. Voegelin *Classification & Index World's Lang.* 128 *Sango*,. .a lingua franca. .with many French and Bantu words.

Sangoan (sæŋgōᵘ·ăn), *a. Archæol.* [f. the place-name *Sango* Bay in Uganda + -an.] Of, pertaining to, or designating a palæolithic culture in central Africa, roughly contemporary with the Mousterian culture in Europe, and the people and tools associated with it. Also *absol.*, this culture.

[**1924** *Man* XXIV. 169 The large tools of the Sângo types. .are now known to occur practically wherever extensive beds of quartzite crop out.] **1931** L. S. B. Leakey *Stone Age Cultures of Kenya Colony* x. 232 Kamasian pluvial. Sub-divisions unknown but may include Mr Wayland's Kafuan and Sangoan. **1952** *Geol. Survey Uganda Memoir* VI. II. 64 The typical Sangoan pick is an elongated steep-sided, double-ended implement with small flat dorsal and large ventral faces... The most finely finished product is somewhat canoe-like in shape. **1959** J. D. Clark *Prehist. S. Afr.* ii. 40 The cultures of the First Intermediate period. . can be subdivided into two contemporary cultures—the Fauresmith and the Sangoan. **1969** *Geol. Survey Uganda Memoir* X. 87 The occurrence of tranchets and pressure flaked lances, together with mint-fresh, steeply flaked duck-head and other small hand axes. .invites correlation with the latest Sangoan or upper Lupemban. **1977** G. Clark *World Prehist.* (ed. 3) i. 34 The leading artefacts of the Sangoan continued to be bifacial, including core-axes, picks and narrow lanceolate forms.

Sangrado. (Earlier *fig.* example.)

1812 Byron *Works* (1898) II. 429 After feeling the pulse and shaking the head over the patient, prescribing the usual course of warm water and bleeding—the warm water of your mawkish police, and the lancets of your military—these convulsions must terminate in death, the sure consummation. .of all political Sangrados.

‖ **sangre azul** (sa·ŋgre aþū·l). [Sp.] The 'blue blood' of the old and aristocratic Spanish families (see note s.v. Blood *sb.* 8).

1834 [see Blood *sb.* 8]. **1846** F. Ford *Gatherings from Spain* xix. 259 *Sangre azul* is the ichor of demigods which flows in the arteries of the grandees. **1876** *Gentl. Mag.* Nov. 601 The sephardim. .once contained the *sangre azul* of the nation. **1975** H. McCloy *Minotaur Country* vii. 74 Carlos. . was the embodiment of *sangré azul*.

Sangria (sæŋgrī·ă). Also **sangria.** [a. Sp. *sangría* (see Sangaree).] A cold drink of Spanish origin composed of red wine variously diluted and sweetened.

1961 'J. Welcome' *Beware of Midnight* x. 119 Hugo ordered a dry Martini for himself and a jug of *Sangria* for the others. .'It's a sort of Spanish Pimms.' **1966** *House & Garden* Dec. 79/3 Visitors to Spain soon become familiar with sangria—the national iced wine cup. The simplest form consists of slices of fruit. .soaked in a rough Spanish red wine and with a little water. .and ice added. **1972** D. Lees *Zodiac* 107, I ordered a pitcher of sangria to go with the couscous. **1978** *Times* 23 Apr. 12/8 There was time for a glass of the house Sangria (a sweet wine tasting of Cherryade).

sangsara, var. *samsara.

sanguinaria. For 'mod. L.' in etym. read 'mod. L. (J. J. Dillenius in Linnæus *Systema Naturæ* (1735))' and add earlier example.

1808 [see *Jersey tea* s.v. *Jersey[1]].

sanguinary, *a.* (and *sb.*) ¶ **4.** (Examples.)

1890 Kipling in *Macm. Mag.* LXI. 155/1 This is sanguinary. This is unusual sanguinary. Sort o' mad country. **1891** —— *Lett. of Marque* xv. 110 'Eres this sanguinary down mail a stickin' in the eye of the Khundwa down! **1910** G. B. Shaw *Lett. to Granville Barker* (1956) 168 The inhabitants raise up their voices and call one another sanguinary liars. **1942** *Tee Emm* (Air Ministry) II. 131 Lovely crate, but lousy on the approach with that sanguinary great nose sticking up in front of you.

sanguine, *a.* and *sb.* Add: **A.** *adj.* **5.** *sanguine-flowered* adj.

1922 Joyce *Ulysses* 44 His fustian shirt, sanguine-flowered, trembles its Spanish tassels at his secrets.

sanguinity. 2. (Later example.)

1979 *Time* 8 Jan. 72/2 Nevertheless, Lasch, a history professor. ., legitimately finds cracks of doom in our sanguinity.

sanhita, var. *samhita. **Sanhiwal,** var. *Sahiwal.

‖ **san hsien** (san ʃyen). *Mus.* Also 9 **san heen, hien**; **san-hsien.** [Chinese *sānxián*, f. *sān* three + *xián* string of musical instrument.] A Chinese three-stringed plucked instrument with a long neck and oval-shaped body. Cf. Samisen.

1839 *Chinese Repository* May 43 The *san heēn*. Three-stringed guitar... The *san heēn* is played as an accompaniment to the *pepa*, as its sounds are low and dull. **1848** S. W. Williams *Middle Kingdom* II. xvi. 169 The *san hien*, or three stringed guitar, resembles a rebeck in its contour. **1874** *Jrnl. North-China Branch R. Asiatic Soc.* VIII. 115 The *San-hsien* is usually played as an accompaniment to the *P'i-p'a*, its sound being low, and dull, and deficient in character. **1917** *Encycl. Sinica* 388/2 Hsien tzŭ or San hsien is a three-stringed instrument with a small oval body covered above and below with snake-skin, and a neck about thirty inches long. There are no frets. It is played with a plectrum of jade. **1933** N. Waln *House of Exile* 204 An orchestra. .played serpent-bellied *san hsien*. **1954** *Folk Arts of New China* 34 Before he came to Peking, the blind minstrel Han Chi-hsiang, as he frankly admitted, was somewhat complacent about his technique on the *san hsien* (a Chinese three-stringed musical instrument). **1975** C. P. Mackerras *Chinese Theatre in Mod. Times* 22 The other principal plucked instruments are the *yüeh-ch'in* and the *san-hsien*, both of which function as secondary accompanying instruments in many dramatic styles... The name *san-hsien* means 'three strings'... It produces a characteristic twanging sound.

Sanibin (sæ·nibin). Also **Sani-bin** and with small initial. [f. Sani(tary *a.* + Bin *sb.*] The proprietary name of a receptacle for refuse.

1921 *Trade Marks Jrnl.* 16 Mar. 550 Sanibin. 411,006. Bins and the like Receptacles... Robert Bailey and Son Limited,. .Stockport, Cheshire; Manufacturers of Surgical Dressings. **1963** *Spectator* 1 Mar. 273/2 Cosi-jade sanibins. **1966** A. E. Lindop *I start Counting* xviii. 219 Under the sink where he kept the Sani-bin. **1975** *Listener* 25 Dec. 879/1 A blaze in one of the downstairs sani-bins.

sanidinite. (Examples.)

1887 *Mineral. Mag.* VII. 227 The blocks of sanidinite and laacher-trachyte occur of all sizes up to masses measuring two feet in diameter... The sanidinite consists principally of sanidine, or of sanidine and nosean. **1916** J. A. Thomson in David & Priestley *Brit. Antarctic Exped. 1907–9: Geol.* II. 139 Trachytes appear to have the power of converting inclusions of such rocks as older trachytes, gneisses and granulites into sanidinites. **1962** *N.Z. Jrnl. Geol. & Geophysics* V. 395 The finest examples of sanidinites. .were obtained from a zone of yellowish-green glass, in some places as much as half an inch in thickness, formed at the contact between porcellanite and overlying basalt.

sanification (sænifikēⁱ·ʃən). *rare.* [f. Sanify *v.*: see -fication.] The action or process of making healthy.

1895 W. James *Let.* 16 June (1920) II. 21 Just about to get a little health into me, a little simplification and solidification and purification and sanification.

Sanio (sæ·nio). *Bot.* The name of Gustav *Sanio* (1832–91), German botanist. † **a.** [First designated, as *Sanio'sche balken*, by C. Müller 1890, in *Ber. d. Deutsch. Bot. Ges.* VIII. 23.] Used in the possessive and with *of* to designate a thickening of the primary wall or medial lamella separating or inclosing pits in wood, esp. of the radial walls of tracheides in gymnosperms. *Obs.*

1891 *Jrnl. R. Microsc. Soc.* 488 'Sanio's Bands' in the Coniferæ.—By this term (*Sanioische Balken*) Herr C. Müller proposes to designate the beams or thickenings commonly found in the xylem-elements, chiefly in the tracheids of Coniferæ. **1915** *Jrnl. Linn. Soc.: Bot.* XLI. 462 All the Indian species of *Pinus* show Sanio's rims. **1916** *Ann. Bot.* XXX. 425 Sanio's bars are small rods crossing the tracheides, cambium, and phloem elements in many conifers. **1920** *Bot. Gaz.* LXX. 431 Considerable importance was attached for some years to the presence or absence of 'bars' or 'rims' of Sanio. **1935** C. J. Chamberlain *Gymnosperms* xi. 245 A cytological study of the origin and development of the bordered pit and the bars and rims of Sanio would be interesting.

b. *Sanio's law*: any of a set of empirical results that describe the growth of tracheides in conifers.

1903 Mudge & Maslen *Class-bk. Bot.* ii. 51 It follows from Sanio's law that the elements of the phloem and xylem, unless subjected to subsequent disturbances, should be arranged in radial rows. **1915** Bailey & Shepard in *Bot. Gaz.* LX. 66 (*heading*) Sanio's laws for the variation in size of coniferous tracheids. *Ibid.* 70 It is evident from Table V that Sanio's second law is applicable to *Picea rubens* as well as to *Pinus sylvestris*. **1961** *Forestry* XXXIV. 125 The foundation of the study of tracheid and fibre-length variation was laid by Sanio (1872) who presented his results of studies on Scots pine (*Pinus sylvestris*) in a set of five conclusions which, for some time now, have been regarded as 'Sanio's Laws'. **1975** S. Carlquist *Ecol. Strategies of Xylem Evol.* 5 The variations noted in Sanio's laws have been confirmed for a number of conifers and dicotyledons.

‖ **sanitar** (sænitā·ɪ). [Russ.] In Russia, a hospital attendant; *spec.* a medical orderly in the army.

1916 H. Walpole *Dark Forest* I. i. 33 Then I came to Petrograd and through the English Embassy found a place in one of the hospitals, where I worked as a sanitar for three months. **1927** —— in *Daily Express* 21 Dec. 8/6 He was my servant during part of 1915, when I was a sanitar in the Russian army. **1933** —— *Vanessa* IV. 674 The sanitars began to dig a grave. **1974** F. Farmborough *Nurse at Russian Front* II. 30 The 1st *Letuchka*, (Flying Column)..was staffed with four surgical sisters,..two doctors,..about 30 *sanitars* (ambulance orderlies) and an officer.

sanitarian, *sb.* and *a.* Add: **A.** *sb.* **b.** *U.S.* A public health officer.

1946 *Richmond* (Va.) *News* 20 Mar. 4/2 A field trip to observe high-temperature, short-time pasteurization was one of the high-lights today of the closing session of the Virginia Association of Milk Sanitarians. **1974** *Index-Jrnl.* (Greenwood, S. Carolina) 23 Apr. 12/1 J. D. Kirby, chief sanitarian with the health department, said the vaccine is 'the safest, most reliable yet perfected and is specified by the Public Health Services'. **1976** *National Observer* (U.S.) 23 Oct. 10/2 The boss is Dale Reeves, senior public-health sanitarian and head of the consumer-protection program.

sanitary, *a.* Add: **1. a.** Also *sanitary reform, reformer.*

1850 C. Kingsley *Alton Locke* I. i. 4 A sanitary reformer would not be long in guessing the cause of my unhealthiness. **1857** —— *Two Years Ago* II. iv. 68 Sanitary reform is thrust out of sight, simply because its necessity is too humiliating to the pride of all. **1884** *Times* 4 Oct. 4/5 Dr Jaeger's sanitary woollen system has been adopted by some of our most eminent sanitary reformers. **1966** N. Longmate *King Cholera* xi. 112 The sanitary reformer triumphed over the cleric: the churches were shut for one whole Sunday. **1974** H. R. F. Keating *Underside* iv. 41 There's an immense amount to be done in sanitary reform.

b. (Further examples.)

1877 T. L. Nichols *Herald of Health Almanack* 21 (Advt.), O for the muse of Dryden, or of Pope To hymn thy praises, *Sanitary Soap*! **1934** A. Huxley *Beyond Mexique Bay* 2 The last word in cocktail bars and peach-pink sanitary fittings. **1940** *Chambers's Techn. Dict.* 740/1 *Sanitary ware*, ..glazed earthenware used for some sanitary fittings. **1977** *Times* 30 July 10/4 Plain colours..in bath and sanitary ware.

2. Delete small-type note and add further example.

1861 Trollope *Orley F.* (1862) I. xxxi. 229 The judge, though he rode everyday on sanitary considerations, had not a sportsman's celerity in leaving and recovering his saddle.

3. Special collocations: **sanitary belt,** a belt to which a sanitary towel is attached; **Sanitary Commission** *U.S. Hist.*, one of various commissions established to supervise matters of health and sanitation, *spec.* that set up by the U.S. government in 1861 to care for soldiers and their dependants during the Civil War; **sanitary engineer,** one whose profession is the design, construction, or maintenance of sanitary appliances or sewerage; a plumber; hence **sanitary engineering**; **sanitary inspector,** an officer appointed to inspect sanitary conditions, a public health inspector; **sanitary napkin** (*U.S.*), **pad, towel,** a pad worn by women to absorb menstrual flow.

1908 *Sears, Roebuck Catal.* 998/2 The EZ Sanitary Belt.. fits the body so smoothly that it is not felt when either worn over or under garments. **1969** B. Malamud *Pictures of Fidelman* ii. 55 He trotted to get her anything she had run out of—drawing pencil, sanitary belt, safety pins. **1861** *N.Y. Times* 25 June 4/4 A week ago we noticed the formation in Washington of the Sanitary Commission for the volunteers, and its approval by the Army Medical Bureau and by the Government. **1898** *Kansas City* (Missouri) *Star* 19 Dec. 2/5 The sanitary commission's work can all be done by a state veterinarian. **1949** J. B. Herrick *Mem. 80 Yrs.* I A clearer war memory is that of the fair of the Sanitary Commission, held in Chicago in the summer of 1865. **1873** B. Latham *Sanitary Engin.* Pref., The whole range of works in which the Sanitary Engineer is engaged. **1901** *Daily Colonist* (Victoria, B.C.) 5 Oct. 6/2 The septic tank system of sewerage..is highly recommended by some of the most eminent sanitary engineers in Europe and America. **1974** 'M. Yorke' *Mortal Remains* v. iv. 156 Her grandfather had been a sanitary engineer, making lavatory basins. **1868** B. Latham (*title*) Inaugural address..before the Society [of Engineers]..upon..the results of sanitary engineering. **1957** *Encycl. Brit.* XVIII. 740/2 A barrister by profession, he [*sc.* Edwin Chadwick] mastered the elements of sanitary engineering, then a little-understood science. **1863** *Times* 24 June 7/5 We shall leave it to the report of the sanitary inspector... He was astonished..to find the rooms so nice and clean. **1897** *Act* 60 & 61 *Vict.* c. 38 The expression 'sanitary inspector' means a sanitary inspector appointed by the local authority. **1907** *Nature* 21 Feb. 400/1 Among the recommendations are..the provision of an expert staff of inspectors under the medical officer, whose title shall be altered from that of 'Inspector of Nuisances' to 'Sanitary Inspector'. **1943** *Our Towns* (Women's Group on Public Welfare) iii. 88 A Sanitary Inspector gave evidence that the majority of houses have outside W.C.'s only. **1956** *Act* 4 & 5 *Eliz. II* c. 66 § 1 Sanitary inspectors appointed under the Local Government Act, 1933, or the London Government Act, 1939, shall henceforth be designated public health inspectors. **1977** *Lancashire Life* Dec. 92/3 Not only did Christopher become Burnley's first fire brigade chief and first sanitary inspector. **1917** W. J. Robinson *Sex Knowledge for Women & Girls* vi. 45 Menstrual blood..is discharged from the uterus..to the outside, where it is caught on cotton, sanitary napkins or some other pad. **1975** D. Ramsay *Descent into Dark* 74 She had experienced no traumas over the change from sanitary napkins to tampons. **1926** *Daily Colonist* (Victoria, B.C.) 2 Jan. 7/7 (Advt.), The hazards of the old-time sanitary pad have been supplanted with a protection both absolute and exquisite. **1974** *Times* 27 Apr. 1/6 One [line]..will always have to be sold cheaply.. denture powder, sanitary pads and tampons, nappies. **1881** *Trans. Obstetr. Soc.* XXII. 188 Dr. Galabin showed the new ladies' sanitary towels manufactured by Messrs. Southall, Barclay, & Co., of Birmingham. They were extremely light and soft, and contained a pad of absorbent cotton wool. **1896** *Eng. Illustr. Mag.* Aug. 8/2 (Advt.), Ladies will find the use of Southalls' Sanitary Towels in assorted sizes to be a great convenience and a great saving. **1917** *Lancet* 28 July 145/2 (*heading*) The destruction of sanitary towels and surgical dressings. *a* **1935** T. E. Lawrence *Lett.* (1938) 503 You'd think they'd have had some other place for their sanitary towels. **1977** B. Freemantle *Charlie Muffin* iii. 33 On the wall..there was still a white outline where the sanitary-towel dispenser had been.

sanitation. Add: **1.** Also *spec.* (the provision of) toilet facilities.

1901 V. Bethell *Monte Carlo Anecdotes* p. xii (Advt.), *Hotel Metropole*... Sumptuous private suites, excellent cuisine, perfect sanitation. **1934** M. V. Hughes *London Child* x. 117 Sanitation was not known at Reskadinnick, neither earth nor water nor any such thing.

2. In *Comb.*, designating a person or vehicle employed in the removal and disposal of domestic refuse, as *sanitation man, truck, van* (*U.S.*).

1939 *N.Y. Times* 16 Sept. 19/2 Sweepers and drivers..will be known as 'sanitation men'. **1975** *New Yorker* 8 Sept. 111/1 The city paid the sanitationmen $713,500 at time-and-a-half rates to clean up the debris that had accumulated because the strikers had refused to collect it. **1958** *N.Y. Times* 18 Nov. 26/6 'Some kind of job action' by the crews of sanitation trucks can be expected. **1974** *Anderson* (S. Carolina) *Independent* 23 Apr. 3B/4 Among the critically hurt was a crewman on a sanitation truck that happened to be picking up refuse at the..building. **1973** *Times* 27 Aug. 5/7 My wife explained that this caper was weightier than that by defining dust cart as Garbage truck, or, in Current American, sanitation van.

sanitize, *v.* Delete *rare* and add: **1.** (Further examples.)

1950 C. A. Lawrence *Surface-Active Quaternary Ammonium Germicides* vi. 90 It can be said that dishes are sanitized by adequate cleaning. **1968** *National Observer* (U.S.) 22 Apr., His demand: an injunction directing the companies to sanitize their smoke or close down. **1971** *Listener* 11 Nov. 659 Air-conditioning sanitises the air. **1978** *Detroit Free Press* 5 Mar. 20/2 (Advt.), This formula permanently and completely removes urine and sanitizes your carpet.

2. *transf.* and *fig.*, esp. (*U.S. slang*) to render more acceptable, clean up, as by the removal of undesirable, improper, or confidential material.

1934 *N.Y. Times* 7 July 2/4 New words are being manufactured at NRA's code factory... Leon Henderson, economic adviser, has just turned out two which may some day find their way to dictionaries. 'Sanitize' is one. Mr. Henderson says it means putting 'sanity and sanitation in business'. **1966** *Amer. Speech* XLI. 300 After it [*sc.* a document] has been *sanitized*, or redrafted to remove the secret information. **1974** *News & Courier* (Charleston, S. Carolina) 28 Apr. A-10/4 Aides claim the transcripts are accurate, but they have been 'sanitized' to delete Nixon's profanity and character slurs spoken in confidence. **1977** *Rolling Stone* 13 Jan. 38/1 The Congressional Joint Committee on Atomic Energy reacted to the criticism by trying to sanitize the report. **1978** *Listener* 30 Mar. 394/3 The language of the Pentagon is designed (if I may use an Americanism) to sanitise disagreeable realities and disreputable motives. **1978** *Guardian Weekly* 27 Aug. 15/4 NBC also has said that the rape scene was essential to the film's artistic integrity, although the network sanitized the scene after the furor created by the initial showing. **1980** *Times Lit. Suppl.* 7 Nov. 1258/4 A writer has no duty to sanitize his imagination; if defilement is what Potter most wants to write about, write about it he should. But..we do not have to like the taste of greasy hamburger.

Hence **sa·nitized** *ppl. a.*

1950 C. A. Lawrence *Surface-Active Quaternary Ammonium Germicides* vi. 93 In those tests in which the number of bacteria did not exceed 100 per swab, the glasses were considered adequately sanitized regardless of the chemical agent used. **1970** *Nature* 17 Oct. 203/2 The commission found no consensus..that explicit sexual materials (the commission's sanitized word for 'pornography') should be banned. **1973** *Ibid.* 5 Oct. 231/1 The 'sanitised' transcripts of the proceedings have contained virtually no discussion of the possible implications of the programme. **1973** *Philadelphia Inquirer* (Today Suppl.) 7 Oct. 44/2 The towels are in place, the bed's made, even the 'sanitized' slip is still over the toilet. **1977** N. Freeling *Gadget* IV. 173 A hotel with..sanitized lavatory seats. **1979** G. Robertson *Obscenity* 274 The press published the original text, juxtaposed with the sanitised version prepared by the Corporation, after the director and dramatist had publicly protested against 'this new and Orwellian form of political censorship'.

sanitizer (sæ·nitəizəɹ). [f. Sanitize *v.* + -er¹.] A substance which sanitizes: a disinfectant, or a preservative of food.

1950 *Jrnl. Milk & Food Technol.* XIII. 63/2 The most recent trend has been toward the use of the quaternary ammonium detergent sanitizer. **1968** Shaffer & Stuart in Lawrence & Block *Disinfection, Sterilization & Preservation* x. 160/1 Acceptances of chemicals as sanitizers have been based in the past on tests conducted under conditions of actual use. **1979** *Nature* 19 Apr. p. xvii/1 The Brentchem range of 14 specifically developed detergents, sanitisers and ancillary products for use in laboratories has been recently improved.

San Joaquin (sæn wǫkī·n). [The name of a river in southern California.] *San Joaquin Valley fever*: = *coccidioidomycosis.

1958 *New Biol.* XXVII. 65 [Coccidioidomycosis] was originally known as the San Joaquin Valley fever from the part of California with which fatalities from the disease were associated. **1974** M. C. Gerald *Pharmacol.* xxvi. 450 (*table*) *Coccidioides*—San Joaquin Valley fever.

Sanka (sæ·ŋkā). Chiefly *U.S.* [Repr. abbrev. form of F. *sans caffeine* without caffeine.] The proprietary name of a make of decaffeinated coffee.

1923 *Official Gaz.* (U.S. Patent Office) 14 Aug. 251/1 Société Anonyme Fabriques de Produits de Chimie Organique de Laire, Issy, France... *Sanka*... Teas and coffees; tea and coffee extracts, both dry and liquid, and tea and coffee substitutes. Claims use since Mar. 19, 1910. **1933** *Ibid.* 18 Oct. 642/1 Sanka Coffee Corporation, New York... *Sanka.* For coffees. **1952** *Trade Marks Jrnl.* 17 Dec. 1176/1 *Sanka*... Coffee. General Foods Corporation.., New York. **1964** H. Campbell *Why did They name It...?* 10 Dr. Roselius named the new product *Sanka*—a contraction of the French phrase *sans caffeine*. **1975** *New Yorker* 26 May 28/1 For breakfast, he ordered orange juice, a boiled egg (five minutes), toast and Sanka. **1978** G. Vidal *Kalki* viii. 192 Giles served Jason Scotch. Prager asked for Sanka. 'I have an ulcer,' he said.

‖ **Sankaracharya** (ʃaŋkarātʃā·riā). Also **sankaracharya.** [Skr.] The name and title of Sankara Acharya, a famous teacher of Vedānta philosophy (prob. of the eighth century A.D.), used as the title of one of various Indian religious teachers and leaders.

1947 K. M. Pannikar *Survey of Indian Hist.* xii. 133 Soon, however, especially after the disappearance of Buddhism, the Mutts became centres of luxury like the great abbeys and the Sankaracharyas who presided over them assumed pontifical dignities. **1960** Koestler *Lotus & Robot* i. 54 The nearest to an authoritative position..was attributed to the five Sankaracharyas, leaders of an important Traditionalist sect. **1969** *Cultural News from India* Nov. 43 The annual *Agama Silpa Vidwat Sadas* inaugurated a few years ago by the Sankaracharya of Kanchi Kamakoti Peetam bids fair to be an event of increasing importance and to establish a wholesome tradition. **1977** *Times* 20 Jan. 6/8 India's mass spectacle of faith during the Kumbh Mela Fair started before dawn... Many sages—sankaracharayas, nahatmas [*sic*] and guru—rode on elephants or in chariots.

Sankhya (sā·ŋkiă). Forms: 8–9 **Sanchya**; 9– **Sankhya, Samkhya.** [Skr. *sāṃkhya*, lit. = relating to number, prob. referring to the 'enumeration' of the twenty-five principles of the philosophy.] One of the six orthodox systems of Hindu philosophy, based on a dualism of matter and soul.

1788 G. Caul in *Asiatick Researches* I. 344 Both these works contain a studied and accurate enumeration of natural bodies and their principles; whence this philosophy is named Sánc'hya. **1808** H. T. Colebrooke *Amarasimha's Cósha; or, Dict. Sanskrit Lang.* III. iv. § xxvi, The third quality, according to the Sán'chya system of philosophy: darkness or illusion, contrasted to truth and passion. **1838** *Penny Cycl.* XII. 234/1 The *Sânkhya* system of philosophy..maintains that true knowledge can alone secure perfect deliverance from evil. **1877** Monier Williams *Hinduism* 193 The Sānkhya philosophy, founded by a sage named Kapila, though probably prior in date, is generally studied next to the Nyāya, and is more categorically dualistic. **1934** J. Baillie *And Life Everlasting* v. 117 In opposition to such monism, arose the dualistic Sankhya system. According to this teaching human souls are not all united in one impersonal *atman*, but exist separately, each in its own right. **1957** *Encycl. Brit.* XII. 251/1 The *Sāmkhya* is a reaction against the idealistic monism of the Upanishads. It believes in a real matter and an infinite plurality of individual souls which are not emanations of a single world-soul. **1959** *Listener* 17 Sept. 431/2 This [religion] again has resemblances to Brāhmanism and to Sānkhya. **1977** J. Hewitt *Yoga & Meditation* iii. 19 Yoga borrows extensively from the Samkhya (or Sankhya) system.

San Luiseño: see *Luiseño.

sanmartinite (sænmā·ɹtinəit). *Min.* [f. *San Martín*, name of a town in San Luis Province, Argentina + -ite¹.] Monoclinic zinc tungstate, $ZnWO_4$, usu. also containing some iron, calcium, and manganese, found as dark brown microcrystalline aggregates in quartz, in association with scheelite.

1948 Angelelli & Gordon in *Notulae Naturae* (Acad. Nat. Sci. Philadelphia) 9 Apr. 1 (*heading*) Sanmartinite, a new zinc tungstate from Argentina. *Ibid.* 2 While usually compact, the sanmartinite may be quite porous, and in color varying from dark brown to dark gray depending upon the amount of admixed scheelite. **1968** I. Kostov *Mineral.* 483 Sanmartinite is a zincian member $(Zn,Fe)WO_4$, containing about 18% ZnO.

sannyasi. Now the most usual form of Sunnyasee, sunnyasi in Dict. and Suppl.

Sanocrysin (sẽⁱnokrəi·sin). *Pharm.* Also **sano-, -chrysin.** [a. Da. *sanocrysin*, irreg. f. L. *sān-us* healthy, Sane + -o + Gr. χρυσ-ός gold + -in -in¹.] A colourless crystalline complex salt of gold, sodium (dithiosulphato)-aurate(I), $Na_3[Au(S_2O_3)_2].2H_2O$, formerly used in the treatment of tuberculosis.

A proprietary term in the U.S.

1924 H. Møllgaard *Chemotherapy of Tuberculosis* ii. 24 The compound built by this synthesis is registered under the name: Sanocrysin. **1924** *Brit. Med. Jrnl.* 8 Nov. 870/2 Arrangements have been made for producing it in bulk in Denmark under the name 'sanocrysin'. **1925** *Official Gaz.* (U.S. Patent Office) 10 Mar. 255/1 *Sanocrysin*... Medicine for phthisis. Claims use since Mar. 16, 1924. **1926** D. Masters *How to conquer Consumption* 103 Møllgaard's treatment is generally referred to..as the 'gold treatment', because it consists of a metallic compound of gold, which the discoverer has named 'sanocrysin'. **1929** *Daily Express* 7 Jan. 4 In the treatment of consumption..by sanochrysin. **1943** *Thorpe's Dict. Appl. Chem.* (ed. 4) VI. 117/2 Sodium aurothiosulphate..has been of considerable interest since 1924 when it was introduced..under the name 'Sanocrysin' for the treatment of tuberculosis. **1965** *Biochem. Pharmacol.* XIV. 1174 Gold sodium thiosulphate ('Sanochrysin'), Platinum Chemicals Ltd., Asbury Park, N.J., U.S.A.

‖ **sanpaku** (sanpa·ku). [Jap., lit. 'three white', f. *san* three + *haku* white.] Visibility of the white of the eye below the iris, as well as on either side. Also *attrib.* or as *adj.*

1963 *N.Y. Herald Tribune* 18 Aug. 23/1 George Ohsawa, the Japanese philosopher and prophet of the Unique Principle walked through the streets of New York yesterday... There were many beautiful girls... But so many *sanpaku.* **1964** T. Wolfe in *N.Y. Herald Sunday Mag.* 12 Jan. 15/2 Abdul Karim Kassem, President Ngo Dinh Diem and President Kennedy; all *sanpaku* and, now, all shot to death, all destroyed by the fate of the *sanpaku.* **1965** W. Dufty tr. *Sakurazawa Nyoiti's Macrobiotics* (1972) 60 Any sign of sanpaku meant that a man's entire system—physical, physiological and spiritual—was out of balance. **1970** W. Burroughs Jr. *Speed* 162 Had I had a rose, I'd have held it in my teeth all morning with sanpaku eyeballs.

San Pellegrino (sæn pelegrī·no). The name of a village in Lombardy, used *attrib.* and *absol.* to designate a mineral water obtained from springs there; a bottle or glassful of this water.

1924 W. Stormont *Summer in Italy* (ed. 2) 174 The following is an abridged classification of the principal waters:.. Affections of the Stomach and Intestines: Agnano,..San Pellegrino, Telese. **1953** G. Cooper *Your Holiday in Italy* (ed. 2) II. 49 Mineral waters *aranciata* (made from oranges—ask for 'San Pellegrino'). **1964** H. Rose *Your Guide to N. Italy* II. 94 San Pellegrino mineral water is mentioned in historical literature as early as the twelfth century. **1965** 'W. Haggard' *Hard Sell* xi. 116 To drink?... A bottle of Recoardo. No Recoardo? Then San Pellegrino. **1971** M. McCarthy *Birds of Amer.* 283 He had had noodles, a salad, and a small San Pellegrino. **1981** R. Thomas *Mordida Man* xxviii. 242 In front of Abedsaid was a small bottle of San Pellegrino mineral water.

sanpi, var. *sampi.

sans, *sb.* Add: Also **Sans** (esp. as the proper name of particular type-faces). (Examples.)

1927 A. J. Watkins *Advertisement Lay-Out & Copy-Writing* 116/2 Serifs,..not present on block letters or sans type. **1932** H. A. Maddox *Printing* (ed. 2) iii. 40 (*caption*) Sans-serif (Gill Sans in light, medium, and bold). **1959** O. Mills *Stairway to Murder* ii. 16 Brash cardboard notices in harsh Sans letters. **1966** Berry & Poole *Annals Printing* 208/1 Modern sans include Futura (1927), Cable (1927), [etc.]. **1969** J. Wainwright *Big Tickle* 124 The double-column headline was in Sans Heavy Italic. **1978** *Antiques & Art Monitor* 28 Oct. 19/2 The result was a series of type-faces, 'Perpetua' and 'Sans', which remain some of the noblest and least fussy in the world.

sans, *prep.* Add: **1.** Freq. *joc.* (Further examples.)

1901 G. B. Shaw *Admirable Bashville* III. 324 And my blows unpaid, Sans stakes, sans victory, sans everything I had hoped to win. **1922** L. Strachey *Let.* 6 Feb. in *Let.: V. Woolf & L. Strachey* (1956) 97, I am sans eyes, sans teeth, sans prick, sans..but after that there can be no more sanses. **1929** *Oxford Poetry* 45 Here, foundling and cheat, my Lord Parvenu suns His armorial lie, sans remorse and sans pother. **1942** *Tee Emm* (Air Ministry) II. 78 *Sans* rank, *sans* aircraft, *sans* everything to show off about, he'll be just a foolish little figure. **1970** *Nature* 28 Feb. 781/2 The specimen, though sans its right hind paddle, is a fine example of *Ichthyosaurus platyodon* (Conybeare). **1975** *Publishers Weekly* 18 Aug. 58/2, 75 relevant illustrations sans razzle-dazzle. **1977** *Rolling Stone* 30 June 68/1 It offers anxiety enough for the Rhodesians themselves, ..to face the prospect of starting life afresh in some harsher, colder country, sans servants, sans swimming pool, sans sunshine, sans supremacy. **1979** A. Hailey *Overload* III. i. 196 The result was a high-quality recording, sans commercials, which the adults and other families watched later at their leisure.

2. a. sans blague, you don't say! I don't believe it!; **sans dire,** without saying anything, without mentioning (something specified); **sans doute,** doubtless, no doubt; **sans-pareil,** restrict † to subst. use in Dict. and add: (*b*) (as two words) unique, unequalled; **sans phrase,** (further examples); also, without exceptions or qualifications; **sans recours** (Law) [cf. Recourse *sb.*¹ 4 b], 'without recourse (to me)', an endorsement on a bill of exchange absolving the endorser or any other party from liability as such party; **sans-souciant** *a. rare*, carefree, unworried.

1922 Joyce *Ulysses* 398 A drenching of that violence, he tells me, *sans blague*, has sent more than one luckless fellow in good earnest posthaste to another world. **1967** *New Yorker* 25 Feb. 39/1 *Aristide*..I happen to be only forty-four. *Auto-coiffeur. Sans blague?* I would have put you at twice that age. **1881** Trollope *Ayala's Angel* II. xxviii. 68 I ain't. You might as well let that accident pass, sans dire. **1890** E. Dowson *Let.* 17 June (1967) 154 *Sans doute* you know your way by this time. **1918** 'K. Mansfield' *Let.* 11 Jan. (1977) 90 A pimp getting *in* [a train] to hold a seat for some super-pimp gave me such a blow in the chest that it is blue today. I thought: 'This is Marseilles, *sans doute.*' **1962** *John o' London's* 20 Sept. 287/1 This conducted tour..must be quite *sans pareil* for..candour. **1885** *Encycl. Brit.* XVIII. 793/2 This study gives us the science of empirical psychology, or, as it is now termed, psychology *sans phrase.* **1919** D. Russell *Let.* Nov. in *Tamarisk Tree* (1975) v. 78, I got him to accept sans phrase for the moment. **1956** *Ann. Reg. 1955* 8 If they chose to expel him, he would become Independent *sans phrase.* **1961** J. Wilson *Reason & Morals* iii. 161 Thus we might say (almost *sans phrase*, as some early Christian authorities did) 'sex is bad'. **1976** *Times Lit. Suppl.* 26 Mar. 337/1 A. J. P. Taylor is the representative historian of our century... He is a historian *sans phrase*, not a man using history as the vehicle for other gifts. **1874** J. B. Byles *Treat. Law Bills of Exchange* (ed. 11) v. 38 A safe and proper mode in which an agent may indorse, so as to avoid personal responsibility, is by adding the words, *sans recours* or *without recourse to me.* **1974** D. W. Fiddes *Business Terms, Phr. & Abbrev.* (ed. 14) 184 Sans recours is a phrase used in the endorsement of bills and notes. When an endorser wishes to free himself of responsibility, he adds the words *Sans Recours*, or *Without recourse to me.* **1826** W. Scott *Jrnl.* 10 Mar. (1972) 109, I have in my odd sans souciant character a good handful of meal from the grist of the Jolly Miller who—once Dwelld on the river dee.

‖ **sansa** (sæ·nsa). Also **sanse, zanza, zanze.** [Marungu (Bantu), ad. Arab. *ṣanj*, Pers. *sinj* cymbals.] An African musical instrument consisting of a wooden box having at the top tongues of bamboo or iron which the performer vibrates with his thumb and forefingers. Cf. *Kaffir piano, *marimba.

1864 C. Engel *Music Most Anc. Nations* 14 Nos. 4, 5, and 6 show the notes of three *zanzes.* **1874** —— *Descr. Catal. Musical Instruments S. Kensington Museum* (ed. 2) 297 The *zanze*, or sansa, is to be found principally among the Negro tribes of upper and lower Guinea. **1876** Stainer & Barrett *Dict. Mus. Terms, Zanze*... Known also by the names of mambira, ambira, marimba, ibeka, vissandschi, in different parts of Africa. **1909** *Cent. Dict.* Suppl., Sanse. **1929** *N. & Q. Anthropol.* (ed. 5) II. 299 The 'musical-box' is an elaborated mechanical analogue of the *sansa.* **1970** *Guardian* 24 Apr. 9/1 They improvise together on the xylophone, sansas (thumb pianos), guitar, or piano. **1975** S. Marcuse *Mus. Instruments* (rev. ed.) 455/1 *Sansa*, linguaphone consisting of tuned metal or split-cane tongues fitted to a wooden board or resonator, so that one end of the lamellae can vibrate freely. *Sansa* is the name of the linguaphone among the Marungu people of the Congo; by extension it has come to be used in a generic sense for all similar instr[ument]s.

sanscara, var. *samskara.

sansculotte. 1. (Earlier and later *gen.* examples.)

1794 Ld. Torrington *Diary* 5 May (1938) IV. 10 Nor do I hope to live to see the Sans Culottes of this land laying all distinction waste. **1927** G. B. Shaw *Gt. Composers* (1978) I. 18 Mozart was still to him the master of masters.. but he was a court flunkey in breeches while Beethoven was a Sansculotte. **1940** [see *deflate *v.* 2 b]. **1955** *Times* 18 July 6/1 So it is the crowd of *sans autos*—the modern *sans-culottes*?—who are left to swarm over the streets, empty of all but the buses (each with its little crest of Tricolor flags for the occasion) and taxis, to celebrate with a certain fervour their annual rites. **1969** *N.Y. Rev. Bks.* 30 Jan. 8/1 The term plebs is convenient for the *sans-culottes* and similar movements made up mainly of small shopkeepers, artisans, journeymen; proletariat for factory workers.

sansculottism. (Earlier example.)

1794 J. B. S. Morritt *Let.* 22 May (1914) ii. 32 His enemies charge him with sansculottism.

sansei (sæ·nsēⁱ). [Jap., f. *san* three, third + *sei* generation.] An American born of nisei parents (see *nisei); a third-generation Japanese American.

1945 in Webster Add. **1950** *Amer. Speech* XXV. 242 Further distinctions lie in other colloquialisms such as *sansei* ('third generation'), the few descendants of *nisei.* **1971** *Newsweek* 19 Apr. 108/1 Aoki says bluntly that nisei and sansei (second and third generation) are too educated and don't work hard enough. **1975** *Time* (Canad. ed.) 20 Oct. 39/1 Carl Takamura, a young sansei (third generation) state legislator.

sanserif. Add: Also **sans serif.** Cf. Ceriph. (Further examples.)

1961 *Guardian* 17 June 14/6, I saw a..truck bearing in white sans serif capitals the name Fay Improvement Company. **1970** *Brit. Printer* July 77/2 Bold sans serifs have remained popular on posters up to the present day. **1976** *Visible Language* X. 88 Sans serif, to him, was the best, if not the only kind of type suitable for the modern world.

sansevieria (sænsivī³·riă). Also **sanseveria, sanseviera.** [mod.L. (C. P. Thunberg *Prodromus Plantarum Capensium* (1794) I. 65), f. the title of Raimond de Sansgrio, Prince of Sanseviero (1710–71) + -ia¹.] A herbaceous perennial of the genus so called, belonging to the family Liliaceæ, native to tropical Africa or south-eastern Asia, and bearing racemes of white or greenish flowers and rosettes of stiff, erect, variegated leaves yielding a strong white fibre; also called bowstring hemp. Also *attrib.*

1804 *Curtis's Bot. Mag.* XIX. 739 (*heading*) Chinese Sanseviera. **1851** *Illustr. Catal. Gt. Exhib.* IV. 882/2 Liliaceous plants, such as..the Sanseviera, the pine-apple, and even the plantain. **1899** F. V. Kirby *Sport E. Central Afr.* xxiv. 268 Quantities of the sansevieria plants..yield a valuable fibre. **1955** *Sci. News Let.* 2 Apr. 213/2 Kenaf is seen as a substitute for jute, sanseveria for manila hemp. **1959** J. D. Clark *Prehist. S. Afr.* ix. 248 The knife or scraper used by the Hukwe Bushmen in Northern Rhodesia for shredding *sansevieria* leaves to obtain fibre for rope and string is also made of wood. **1961** [see *mother-in-law 4]. **1976** *Hortus Third* (L. H. Bailey Hortorium) 1002/2 Sansevierias are commonly grown as durable porch and house plants.

Sansi (sā·nsi). Also **Sansiya, sansya.** [Origin uncertain (see quot. 1896).] A low-status caste group of the Punjab, India; a member of this group. Also *attrib.*

1882 E. J. Gunthorpe *Notes on Criminal Tribes* xiii. 78 Kunjurs are..a branch of the great family of Sansya robbers, who claim their descent from Sainsmull. **1883** D. C. J. Ibbetson *Outl. Panjáb Ethnogr.* vi. 311 The thieving Sánsis are said to admit any caste to their fraternity except the Dhedhs and Mhangs; and the man so admitted becomes..a Sánsi. **1896** W. Crooke *Tribes & Castes North-Western Provinces & Oudh* IV. 277 Sânsiya. A vagrant thieving tribe... Of their name no satisfactory account has been given. Some derive it from the Sanskrit *swāsa*, 'breathing', or *srasta*, 'separated', others with *svagânika*, 'one who has to do with dogs', or *svapâka*, 'dog-cooking', a person of a degraded and outcaste tribe, who, by the older law, was required to live outside towns, to eat his food in broken vessels, to wear the clothes of the dead, and to be excluded from all intercourse with other people... The Sânsiya is no doubt the near kinsman of the other degraded wandering races who occupy the same part of the country, such as the Kanjar. **1901** Kipling *Kim* iv. 86 They meet a troop of long-haired, strong-scented Sansis, with baskets of lizards and other unclean food... The Sansi is deep pollution. **1931** E. A. H. Blunt *Caste System N. India* ix. 149 The Beriya, Bhantu, Habura, Karwal, and Sansiya..may be regarded as offshoots of a single nomadic race. **1972** S. R. Sharma in F. Singh *Hist. Punjab* III. xvi. 366 Prostitution had come to be associated with certain castes—Kanjai, Bangali, Sansi and Pema.

Sanskara, var. *samskara.

Sanskritic, *a.* (Earlier example.)

1848 Max Müller in *Rep. Brit. Assoc. Adv. Sci. 1847* XVII. 336 There is still another more Sanscritic termination in *e*, for the locative of words ending in a consonant or the vowel *a*.

Sanskritist. (Earlier example.)

1853 W. D. Whitney in *Jrnl. Amer. Oriental Soc.* III. 292 This little work..gave perhaps the most powerful impulse to that movement which has since carried all Sanskritists irresistibly to the study of the Vedas.

Sanskritize, *v.* Add: **b.** *trans.* To adapt to the beliefs or practices of a high Hindu caste. **Sanskritization** (examples corresponding to sense *b of the vb.).

1952 M. N. Srinivas *Relig. & Society among Coorgs of S. India* ii. 30 A low caste was able..to rise to a higher position in the hierarchy by adopting vegetarianism and teetotalism, and by Sanskritizing its ritual and pantheon. In short, it took over, as far as possible, the customs, rites, and beliefs of the Brahmins... This process has been called 'Sanskritization' in this book, in preference to 'Brahminization', as certain Vēdic rites are confined to Brahmins and the two other 'twice-born' castes. *Ibid.* 38 Some Coorg families are more Sanskritized than their neighbours. **1964** *Diogenes* XLV. 99 (*heading*) Sanskritization and cultural mobility. **1974** tr. *Wertheim's Evolution & Revolution* iii. 240 Even if a sub-group within a caste..achieved a certain measure of prosperity and aiming at a higher status through 'sanskritization', has constituted itself as a separate sub-caste, [etc.].

Sansya, var. *Sansi.

Santa Ana (sæ·ntă æ·nă). *U.S.* Also **Santa Anna, Santana** (sæntæ·nă). [Sp., = Saint Anne.] A hot, dry, föhn-type wind of desert origin, freq. strong and dust-laden, which blows on the coastal plain of southern California after being channelled and heated adiabatically during its descent of the Santa Ana Mountains. Also *Santa Ana wind.*

The suggestion made in some dicts. that this is named after Antonio Lopez de *Santa Anna* (?1795–1876), Mexican revolutionary leader and president, seems without foundation.

1887 *Ann. Meteorol. Rev. Calif. 1886* (Calif. State Agric. Soc.) 128 Another health-giving, but extremely disagreeable wind, is the 'Santa Ana', or 'norther'. *Ibid.*, The 'Santa Ana' wind receives its name, because it frequently issues from the Santa Ana Pass. **1889** G. Davidson *Pacific Coast: Coast Pilot of Calif.* (U.S. Coast & Geodetic Survey) (ed. 4) 40 In November, 1888, when the Santa Ana wind had passed its greatest strength, a reverse current of wind was drawing along the shore from Anaheim Landing towards Newport Bay. **1915** *Nature & Sci. Pacific Coast* (Amer. Assoc. Adv. Sci.: Pacific Coast Comm.) 22 Known locally as Santa Anas, these wind storms constitute the most disagreeable feature of the weather in the great valley of the south. **1931** A. A. Miller *Climatol.* viii. 150 The Santa Annas of southern California and the Northers of the Sacramento Valley are hot, dry winds. **1941** B. Schulberg *What makes Sammy Run?* iv. 56 The music came at us like a Santa Ana wind. **1970** Koenig & Dixon *Children are Watching* xxi. 181 Out in the sunset the Santana scoured the beach and hurled fine sand with enough force that hands cupped protectively over eyes. **1972** C. Weston *Poor, Poor Ophelia* xi. 58 Santa Ana wind, he thought. No fog tomorrow, desert heat instead. **1973** R. Hayes *Hungarian Game* xlii. 248 A Santa Ana, that dry desert wind, blew away the smog.

Santa Claus. Add: Also *dial.* and *colloq.* **Santy. a.** (Earlier examples.) Also, a person wearing a red cloak or suit and a white beard, to simulate the supposed Santa Claus to children, esp. in shops or on shopping streets. Also *transf.*, *fig.*, *attrib.*, and *ellipt.* as *Santa.*

1773 *N.Y. Gaz.* 26 Dec. 3/1 Last Monday the Anniversary of St. Nicholas, otherwise called St. A Claus, was celebrated at Protestant-Hall. **1808** *Salmagundi* 25 Jan. 407 The noted St. Nicholas, vulgarly called Santaclaus—of all the saints in the kalendar the most venerated by true hollanders, and their unsophisticated descendants. **1886** P. Stapleton *Major's Christmas* 201 Papas and mammas..planned the Santa Claus performance which was to come when the inquisitive eyes were closed in slumber. **1821** *Weekly Visitor* IV. 262/1 For time immemorial the Dutch had a tradition, that there existed a being of no *earthly birth*, who was called *Santa Claus.* **1909** *Chicago Daily News* 10 Aug. 8/3 Uncle Sam is by no means an impartial Santa Claus. **1913** *Sat. Even. Post.* 6 Dec. 50/1 If you want to act the part of Santa this Christmas. **1925** T. Dreiser *Amer. Trag.* (1926) I. II. xxix. 356, I know something Santy has brought my Dad that he'll like. **1932** J. Beames *Gateway* vi. 108 You're just as kiddish as what you was when you'd be up at three in the mornin' to see what Santy had brung you. **1934** *Amer. Mercury* May 5/2 The Santa Claus theory of relief may be appropriate to a genuine emergency like an earthquake or a big fire. **1943** K. Tennant *Ride on Stranger* iii. 24 Come on down, Ma. Come and see what Santa's brought you. **1956** H. Gold *Man who was not with It* (1965) xxxii. 310 It was practically Christmas, too, with all the Santy Clauses peddling in the streets. **1957** [see *good-time *a.*]. **1973** 'D. Halliday' *Dolly & Starry Bird* i. 2 The Zodiac Trust is the Santa Claus of worldwide astronomy. A private foundation richly funded.., it makes grants to struggling centres. **1975** *Times* 10 Dec. 4/4 Being a man was a genuine occupational qualification for a Santa Claus. **1976** M. Machlin *Pipeline* ix. 103 A huge, heavy-set man,..with a bushy unkempt Santa Claus beard, walked unsteadily toward their table. **1976** *Scotsman* 24 Dec. (Weekend Suppl.) 1/1 Stop rakin', Rikki. Santy says ye've had enough. **1976** *Scottish Daily Express* 27 Dec. 2/8 She was one of nine women charged with prostitution in Dallas, Texas, for propositioning Vice Squad officers disguised as Santas. **1977** *Times* 24 Dec. 16/5 Santa must have been updated over the years. Presumably girls hang out their tights now, instead of a solitary stocking.

b. (*collect. sing.*) Christmas presents; Christmas delicacies. *U.S. dial.*

1929 W. Faulkner *Sound & Fury* 107 Buy yourself some Santy Claus. **1939** *These are our Lives* (Federal Writers' Project, U.S.) 22 One Christmas we ask him for fifty dollars for some clothes and a little Santy Claus for the chil'en.

Santa Gertrudis (sæ·ntă gəɹtrū·dis). The name of the Santa Gertrudis division of the King Ranch, Kingsville, Texas, used to designate a breed of large red-coated beef cattle suitable for hot climates, developed there between 1910 and 1940 by crossing Brahmans and Shorthorns; an animal of this breed. Also *attrib.*

c **1946** R. J. Kleberg *Santa Gertrudis Breed of Beef Cattle* 8 The bull on the ranch known as 'Monkey'..marks the real beginning of the improved breed of Santa Gertrudis cattle. *Ibid.* 11 Santa Gertrudis calves at eight months of age will average over 500 pounds. **1949** *Jrnl. Heredity* XL. 115/1 One of the most noteworthy contributions to livestock breeding has been the creation and development of the Santa Gertrudis breed of beef cattle. **1955** *Times* 16 July 11/5 The Zebu and Santa Gertrudis breeds are suitable in country where grasses lack nutriment for a great part of the year. **1960** *Times* 1 Oct. 7/7 A remarkable artificial insemination programme [in Queensland]..aims to transform a herd of 30,000 Shorthorn cattle into one of 70,000 high grade Santa Gertrudis. **1962** *Listener* 6 Dec. 956/1 Mr Joyce had introduced Santa Gertrudis cattle from Texas. **1978** J. B. Friend *Cattle of World* 148/2 The Santa Gertrudis today carry approximately ⅝ Shorthorn blood and ⅜ Brahman blood.

Santal² (sæ·ntal). Also **Santhal, Sonthal.** [Native name.] A Kolarian people of north-eastern India; a member of this people. Also, the language of this people (see next). Also *attrib.*

1852 J. Phillips *Introd. Sántál Lang.* i. 1 *The Sántál*, having been hitherto an unwritten language, has..no characters of its own. *Ibid.* 3 Pronouns in Sántál, are.. completely interwoven with the declension of nouns. **1866** [see *Munda *sb.* and *a.*]. **1873** L. O. Skrefsud *Gram. Santhal Lang.* p. iii, Santhali is the language spoken by a people called by foreigners the Santhals or Santals, inhabiting the western frontier of Lower Bengal. *Ibid.* i. 8 *Semi-consonants.* k', ch', t', p'. The sounds, which these letters represent, are peculiar to the Santal language... The 'Schnalz-laute' (click-sounds) mentioned by Dr. Lepsius, as existing in some of the African languages, appear to have some similarity to these Santal jerks. **1891** Monier Williams *Brahmanism & Hinduism* (ed. 4) xxii. 578 We came to what appeared to be a good typical example of a Santal village-community. **1891** Kipling *City of Dreadful Night* 85 We have any amount of Sonthals besides Mahomedans and Hindus of every possible caste. **1936** *Times Lit. Suppl.* 3 Oct. 788/4 The pictures of Sonthal life are evidently based on personal and intimate knowledge. **1941** J. H. Hutton in L. O'Malley *Mod. India & West* xii. 422 The ignorance and honesty of the Santal enabled the first adventurous traders from the plains to make rapid fortunes out of the hill-men. **1969** *Illustr. Weekly of India* 27 July 29/1 Santal women do not wear much jewellery. **1971** *Ibid.* 25 Apr. 42/2 (*caption*) Resettled Santhals in the Malkanagiri zone of Dandakaranya Project. The Santhals are also found in the border districts of West Bengal. **1974** W. G. Archer *Hill of Flutes* i. 19 Flanked by rows of tall palmyra palms, Santal villages have an air of genial comfort. *Ibid.* 24 Within this neat and ordered setting most Santals lead calm and happy lives.

Santali (sæntā·li), *sb.* and *a.* Also 9 **Santalee**; **Santhali, Sonthali.** [f. prec. + adj. suff. -*i*.] **A.** *sb.* The Munda language of the Santals. **B.** *adj.* Of or pertaining to the Santals or their language.

1873 [see *Santal²]. *c* **1875–9** E. L. Brandreth *On Non-Aryan Lang. India* 7 The Santali verb..has twenty-three tenses. **1891** Kipling *City of Dreadful Night* 96 Sonthali..is more elaborate than Greek. **1927** *Other Lands* July 138/2 At one house I was presented with a burnt arm to examine, and having no Santali with which to explain that I was not a doctor, the only thing to do was to look at it carefully and express, by nods and smiles, great satisfaction at the state of its progress. **1961** Webster, Santhali. **1969** [see *madal]. **1974** *Times* 18 Apr. 16/6 Elmhirst took his first group of students into the Moslem, Hindu and Santali villages of the district early in 1922. **1974** W. G. Archer *Hill of Flutes* 343 The Santali Language. According to Grierson's Linguistic Survey, Santali is an Austro-Asiatic Language.

santalol (sæ·ntălǫl). *Chem.* [f. Santal + -ol.] Either of two isomeric terpenoid alcohols, $C_{15}H_{24}O$ (known respectively as α- and β-*santalol*), which are fragrant liquids found in sandalwood oil.

1895 *Pharm. Jrnl.* I. 118/1 Chapoteaut, after a very careful study of the oil, announced some time ago that it consisted almost entirely of two bodies, $C_{15}H_{26}O$ [*sic*], an alcohol termed santalol, and $C_{15}H_{24}O$, probably the corresponding aldehyde. **1935** *Jrnl. Chem. Soc.* 312 Tautomerism is observed also in the two santalols themselves, since they yield on ozonolysis both formaldehyde and acetylcarbinol. **1966** *McGraw-Hill Encycl. Sci. & Technol.* XIII. 497/2 Santalol occurs in sandalwood oil and is used in perfumery. **1976** *Nature* 5 Aug. 487/2 The essential oils D-bornyl acetate, α- and β-santalol and several plant sesquiterpene hydrocarbons have been shown to induce sexual excitement in male American cockroaches.

Santa Lucia (sæ·ntă lutʃī·ă). The name of a range of mountains in southwestern California, used *attrib.* in **Santa Lucia fir** to designate the bristlecone fir, *Abies bracteata,* which is native to the region.

1905 *Occasional Papers Calif. Acad. Sci.* IX. 7 Santa Lucia Fir is found in only a few cañons of the Santa Lucia Mountains in Monterey County. **1948** *Sierra Club Bull.* (San Francisco) Mar. 137 Among these were the Santa Lucia fir..and hosts of others. **1965** *Listener* 20 May 742/3 You mentioned a Santa Lucia fir just now, and you were saying that in its original habitat it is sadly depleted. **1977** *Daily Colonist* (Victoria, B.C.) 7 Aug. 1/1 The Marble Cone fire has destroyed two-thirds of the Ventana Wilderness Area, home of the Santa Lucia fir which grows nowhere else.

‖ **santé** (sañte), *int.* Also in anglicized form **santy** (sæ·nti). [Fr., lit. 'health'.] An exclamation used as a salutation before drinking. Cf. Health *sb.* 6.

1903 Kipling *Traffics & Discoveries* (1904) 43 Here's santy to us all! **1952** P. Frankau *Wreath for Enemy* I. 18 She cried, '*Santé, santé*' raising her glass to each of us. **1966** J. Fowles *Magus* iv. 21 She had poured herself a whisky. '*Santé*.' **1975** D. Gray *Ride on Tiger* iv. 30 He raised his own glass of champagne and said, '*Santé!*' **1980** P. Harcourt *Tomorrow's Treason* II. ii. 141, I watched him..pour us each a generous tot. '*Santé!*' He grinned.

‖ **santeria** (santerī·ă). Also **Santeria.** [Sp., lit. 'holiness, sanctity'.] An Afro-Cuban religious cult with many Yoruba elements.

1950 *Southwestern Jrnl. Anthropol.* VI. 64 The worship of African deities, as it is practised in Cuba today, is known as *santeria.* The deities and the men and women who work with them are known by the Spanish words *santos, santeros,* and *santeras,* or by the Yoruba words *orisha, babalorisha,* and *iyalorisha.* **1953** [see *santero 2]. **1956** *Publ. Amer. Dial. Soc.* XXVI. 34 In Cuba the practitioners of a religion known as Santeria use a variety of African language called Lucumí in their religious services. **1972** W. R. Bascom *Shango in New World* 20 A second major center of santeria in the United States, perhaps even more important than Miami, is New York City, where a Shango temple has been established. Apparently this has attracted not only Cuban refugees but also Puerto Ricans and New York Negroes.

‖ **santero** (santē·ro). Fem. **santera.** [Sp.]

1. In Mexico and Spanish-speaking areas of south-western U.S.: a maker of religious images.

1931 R. L. Barker *Caballeros* xiii. 333 As the colonists became more securely rooted on the prados and mesas of Nuevo México certain men developed greater proficiency as carvers and painters. They were called santeros, the saint-makers, who signed and dated their work on the backs of the santos de retablos. **1944** *Horizon* Jan. 23 At about the same period [*sc.* 18th and 19th centuries] the *santero* of New Mexico produced for religious needs starkly primitive paintings and sculpture (*retablos* and *bultos*). **1951** *Western Folklore* Apr. 153 The following material about *santeros* was collected in the San Luis Valley, where there still remains a definite *santero* tradition... Southern Colorado is..the one remaining place where something may still be learned of the *santero* and his art.

2. A priest (or priestess) of a religious cult, esp. *santeria.*

1950 [see *santeria]. **1953** *Language* XXIX. 157 Lucumí is a term here used to refer to a language spoken in Cuba by practitioners of the religion known as *Santeria.* In Cuba both the *santeros* and the language are often called Lucumí. **1972** W. R. Bascom *Shango in New World* 20 Refugee santeros and santeras have..spread to many other parts of the New World. **1977** *N.Y. Rev. Bks.* 4 Aug. 27/3 He did many odd jobs for his neighbors,..served as a *santero*, or priest, in two religious cults.

santo. Add: **2.** A wooden representation of a saint or other religious symbol from Mexico or south-western U.S.

1834 A. Pike *Prose Sk. & Poems* 146 The santos and other images had been brought from Mexico. **1948** F. Blake *Johnny Christmas* I. 19 The lines of prayer benches, the altar with its tapestry and candlesticks, the religious santos set in niches. **1976** *National Observer* (U.S.) 1 May 18/1 *Folk sculpture USA*... Santos from the Southwest, trade signs, voodoo cult objects.

Santobrite (sæ·ntǒbrəit). [f. the name of the Mon*santo* Chemical Company.] A proprietary name for preparations of sodium pentachlorophenate (s.v. *penta-), used as a fungicide, wood preservative, insecticide, etc.

1936 *Official Gaz.* (U.S. Patent Office) 2 June 14/2 Monsanto Chemical Company, St. Louis, Mo... *Santobrite.* For preservatives for wood, lumber, and other cellulosic materials. Claims use since Apr. 2, 1936. **1938** [see *pentachlorophenate* s.v. *penta-]. **1944** *Trade Marks Jrnl.* 29 Nov. 567/1 *Santobrite*... Insecticides, germicides, algicides, [etc.]. **1959** [see *pentachlorophenate* s.v. *penta-]. **1971** N. E. Hickin *Wood Preservation* 89 The technical materials are also known as 'Santobrite' (the Monsanto Chemical Co.) and 'Dowicide G' (the Dow Chemical Co.).

Santo Domingan (sæ·nto domi·ŋgăn), *a.* [f. *Santo Domingo* (see below) + -an.] Of or pertaining to Santo Domingo, former name of the Dominican Republic, and also the name of a district, and of the capital city of the Dominican Republic.

1934 in Webster. **1947** J. C. Rich *Materials & Methods Sculpture* x. 291 The West Indian *Swietenia mahogani* is marketed as Cuban, Santo Domingan, or Spanish mahogany. ..Santo Domingan mahogany is one of the hardest and heaviest varieties. **1976** B. Lecomber *Dead Weight* x. 118 A little Santo Domingan whore.

santon. Add: **3.** Chiefly in Provence: a figurine adorning a representation of the manger in which Christ was laid.

1926 E. I. Robson *Wayfarer in Provence* xviii. 226 The little home-made crèches, the simple figures known as the Santons,..the pastoral ceremonies..at the famous midnight mass of Les Baux, are all witnesses to the way in which the Nativity story went home to the hearts of the Provençals. **1952** *Spectator* 10 Oct. 478/1 The pretty *santons*, the traditional clay figures of the Holy Family and the ancient trades of Provence. **1963** G. K. Wilkinson *Guinea-Pigs* xii. 189 Christmas will soon be on us and I hope that you will arrange the Holy Crèche and the Santons in the church. **1976** N. Roberts *Face of France* iii. 40 These days *santons*, the little pottery crib figures introducing characters from daily life..into the Nativity scene, are known far beyond their native Provence.

Santonian (sæntōᵘ·niăn), *a. Geol.* [ad. F. *Santonien* (H. Coquand 1857, in *Bull. de la Soc. géol. de France* XIV. 749) f. *Santon*, native or characteristic of Saintes, a town in Charente-Maritime Dept. (f. L. *Santoni* or *Santones*, ancient name of a people of Aquitania), + *-ien* -ian.] Name of a stage in the Upper Cretaceous in France and adjacent areas, corresponding to the middle Senonian and to part of the Upper Chalk in Britain; of or pertaining to this stage and to the strata which characterize it, or the geological age during which it was deposited. Freq. *absol.*

1869 H. Coquand in *Q. Jrnl. Geol. Soc.* XXV. 239, I have divided the Cretaceous formation of the south-west [of France] in the following manner..Santonian stage [of Saintes]. **1885** A. Geikie *Text-bk. Geol.* (ed. 2) 833 This stage [*sc.* the Senonian]..consists mainly of white chalk separable into the two divisions of: 1st, Micraster (Santonian) sub-stage composed of chalk beds [etc.]. **1924** C. Schuchert *Text-bk. Geol.* (ed. 2) II. xxxviii. 537 Upper Cretaceous [of Europe]... Senonian... Santonian. **1971** *Nature* 18 June 439/2 Misfit motion in the African plate closed this rift in Santonian time. **1974** *Encycl. Brit. Micropædia* VIII. 886/2 In northern Europe the Santonian is represented by the Granulaten Chalk, the equivalent of portions of the Upper Chalk in Great Britain.

Santorin (sæntǒrī·n, sæ·ntǒrin). Also **santorin, Santori·ni.** [ad. Gr. Σαντορίνη Santorini, former name for Thira, ad. It. *Sant' Irene* St. Irene, Italian name for the island.] In full *Santorin earth.* A natural volcanic ash, similar to pozzolana, found on the island of Thira in the Cyclades.

1868 Watts *Dict. Chem.* V. 191 *Santorin*, an argillaceous mineral, occurring on the island of Santorin, which yields an excellent cement. **1876** *Min. Proc. Inst. Civil Engineers* XLV. 291 For hydraulic works on the Mediterranean coast Santorin earth recommends itself through facility of excavation and lading and economical transport at sea. **1951** Ladoo & Myers *Non-Metallic Minerals* (ed. 2) 407 Santorini or Santorin earth is a variety of pumice mined on Santorin Island (Greece), used in making pozzuolana cement. **1971** *Materials & Technol.* II. ii. 106 Cements ground together with siliceous materials other than pozzolan—such as santorin earth (Greece), trass..and diatomaceous earth (USA)—are also named pozzolan cement.

Santos (sæ·ntǫs). The name of a port in Brazil, used *ellipt.* and *attrib.* of coffee exported from there.

[**1885** C. F. van D. Laërne *Brazil & Java: Rep. Coffee-Culture* x. 267 The second or Santos zone lies..between 21 and 24 degrees S.L. but a little further landwards, as it does not begin till about 150 kilometers from Santos. *Ibid.* 289 The coffee-shrub in the Santos zone is much larger than the coffee-shrub in the Rio zone, besides being almost twice as productive.] **1888** C. G. W. Lock *Coffee* vi. 112 As to impurities:—San Domingo is usually very dirty; Ceylon, East India, Rio, Santos, Martinique, and Java, generally well prepared and clean. **1899** [see *Jamaica b]. **1956** A. E. Haarer *Mod. Coffee Production* xvii. 398 Though an ever-increasing number are beginning to pick ripe coffee and prepare it by the wet method,.. most planters still follow the old dry method. Of the latter there are two kinds, those who process only ripe cherry and take more care in its preparation,..and those who strip the crop from the trees at the stage when most of it is ripe... The first of these two methods produces a softer and milder product such as Santos coffee. **1975** *Times* 6 Mar. 7/7 Blue Sumatra, 82 p per pound..very strong aroma. Santos, 72 p, the best Brazilian, rich flavour.

‖ **san ts'ai** (san,tsəi). Also **san-ts'ai.** [Chinese *sāncǎi*, f. *sān* three + *cǎi* colour.] Chinese pottery, esp. of the Tang dynasty, decorated in three colours; decoration in three enamel colours applied to pottery and porcelain. Also *attrib.*

1901 C. Monkhouse *Hist. & Descr. Chinese Porc.* I. caption facing p. 35 Squirrel and grapes: *San-ts'ai* water pourer. **1906** S. W. Bushell *Chinese Art* II. II. viii. 37 For a typical example of the *san ts'ai*, or 'three-coloured', decoration *sur biscuit*, see the pictures..of a fish-shaped water pourer, which is painted with the brownish-purple, green, and yellow enamels of this *genre.* **1915** R. L. Hobson *Chinese Pott. & Porc.* II. x. 151 The Dresden collection is peculiarly rich in this kind of *san ts'ai.* **1959** H. Chevalier tr. *Grousset's Chinese Art & Culture* vi. 292 The main categories of Ming ceramics are..: 1. *San-ts'ai* or 'three-colours', namely green.., yellow.., and aubergine-violet. **1972** *Trans. Oriental Ceramics Soc.* XXXVIII. 29 The combination of green, blue and yellow-brown glaze which constitutes the famous T'ang polychromes ('three-colour ware', *san-ts'ai*) appear to belong..to the first half of the 8th century. **1978** P. Van Greenaway *Man called Scavener* xi. 159 A priceless san ts'ai bowl.

santy: see *santé. **Santy (Claus:)** see *Santa Claus. **Sanusi, Sanusiya(h:** see *Senussi; **sanyama**: see *samyama.

sanyas(s)i, sanyas(s)in, varr. Sunnyasee, sunnyasi in Dict. and Suppl.

s-aorist: see *S I. 1 b.

Saorstát Éireann (sī·rstāt ē·rʸən, sē·r-). [Ir., = The Free State of Ireland.] = *Irish Free State* s.v. *Irish *adj.* 2 c. Also *ellipt.* as **Saorstát.**

1922 *Bille um Bun-reacht Shaorstáit Eireann* (*Constitution of Saorstát Eireann Bill*) 2 Article 1. The Irish Free State/Saorstát Eireann is a co-equal member of the community of nations. **1923** *Glasgow Herald* 18 June 9/8 One was Article 12 where the Northern Government was entitled to pass a resolution against entrance into the Saorstat. **1924** W. B. Yeats *Senate Speeches* (1961) 69 Your Committee is gravely impressed by the responsibility now laid upon the Saorstat towards the Irish people. **1938** S. Beckett *Murphy* 197 Turf is compulsory in the Saorstat, but one need not bring a private supply to Newcastle.

sap, *sb.*¹ Add: **1. b.** (Later examples.)

1865 G. M. Hopkins *Poems* (1967) 169 My sap is sealed, My root is dry. **1942** T. S. Eliot *Little Gidding* i. 7 Between melting and freezing The soul's sap quivers. **1961** B. J. Chute *Moon & Thorn* iv. 37 An old man..gave her a more than reflective look as she passed, the sap still plainly rising in his branches.

d. Cytology. *cell sap* [tr. G. *zellsaft*] (see quot. 1875); *nuclear sap*, the fluid within the nuclear membrane.

1875 Bennett & Dyer tr. *J. Sachs's Text-bk. Bot.* I. i. 62 The term Cell-sap may be understood in a wider or in a narrower sense. In the former it would express the collective mass of all fluids by which the cell-wall, the protoplasm-body, and all other organised structures of the cell are saturated, and would also embrace the fluids contained in the vacuoli of the protoplasm; in a narrower sense the latter only is ordinarily designated as cell-sap. **1884** *Jrnl. Bot.* XXII. 124 The rich, violet-coloured cell-sap in the flower of *Justicia speciosa*..crystallizes very easily into minute slender prisms. **1887** *Jrnl. R. Microsc. Soc.* 979 Linin and paralinin, the substance respectively of the nuclear threads..and of the intermediate matrix or 'nuclear sap'. **1955** *Internat. Rev. Cytol.* IV. 293 Another suggestion for the origin of nucleolar material is that it is formed from nuclear sap. **1971** Villee & Dethier *Biol. Princ. & Processes* vi. 152 The activation of amino acids for protein synthesis, the process of glycolysis and many other reactions occur in the soluble cell sap. *Ibid.* xvi. 499 The plant cell, inside its cellulose wall, has one or more large vacuoles filled with cell sap. **1975** *Nature* 4 Sept. 21/1 Similar preparations were..made from rat liver chromatin but after previous removal of 'nuclear sap' which contains soluble nuclear proteins. **1978** B. S. Beckett *Illustr. Biol.* xxxi. 62/1 As root hairs take up water their cell sap is diluted and soon becomes a weaker solution than the sap of cells deeper inside the root.

4. b. *U.S. slang.* A club; a short staff. So *saps* (see quot. 1899).

1899 'J. Flynt' *Tramping with Tramps* 396 *Saps*, a clubbing with weapons made from saplings. **1915** *N.Y. World Mag.* 9 May 14/3 *Sap* or *sapstick*, a crutch, cane or club. **1926** J. Black *You can't Win* vii. 83 The town marshal would then appear with a posse armed with 'saps', which is short for saplings, young trees. **1932** J. Dos Passos *1919* 436 He could hear the crack of saps on men's skulls. **1940** R. Chandler *Farewell, my Lovely* xxvi. 116 He had the sap out this time, a nice little tool about five inches long, covered with woven brown leather. **1955** W. Foster-Harris *Look of Old West* vii. 218 Its [*sc.* a quirt's] handle, or butt, would probably be loaded with an iron spike or with buckshot, thus giving you a handy sap when you needed one. **1974** D. Sears *Lark in Clear Air* iv. 49 His main staff of office was a lead sap that must have weighed two pounds.

7. a. *sap-flow, -pressure*; *sap-clear, -filled, -rife* adjs.; **sap-rot** (further examples); **sap-stain,** discoloration of sap-wood, esp. a bluish discoloration by fungi; so **sap-stained** *a.*, **sap-staining** *sb.* and *a.*; **sap-sucker** (earlier and further examples); **sap-whistle**: delete † and add later example.

1953 E. Sitwell *Gardeners & Astronomers* 31 The gardener plays upon his sap-clear flute. **1915** D. H. Lawrence *Rainbow* xiii. 383 Her own world of warm sun and growing, sap-filled life was turned into nothing. **1935** C. Day Lewis *Time to Dance* 64 We remember them as the glowing fruit remembers Sap-flow and sunshine. **1976** *Sci. Amer.* May 104/3 Hales measured the springtime sap pressure by placing open mercury manometers on a cut vine. **1942** W. Faulkner *Go down, Moses* 326 Wet and saprife spring in their ordered immortal sequence. **1918** J. W. Harshberger *Mycol. & Plant Path.* xxxv. 545 Sap-rot (*Polystictus versicolor* (L.), Fr.).—*Polystictus versicolor* is one of the most cosmopolitan species of fungi known... It grows on the sapwood of every species of deciduous tree known. It is the most serious of all the wood-rotting fungi, destroying probably 75 per cent. of the timber used for railroad ties. *Ibid.* 558 Sap-rot (*Daedalea quercina* (L.) Pers).—One of the most important enemies of structural oak, produces a soft, mushy decay of the wood. **1971** *Country Life* 4 Nov. 1224/2 The chestnut for the frames is cleft..soon after cutting to prevent sap rot. **1910** *Bot. Gaz.* L. 142 The examination of microscopic sections of this sap-stained lumber reveals the fact that the colored substance, produced by the chemical reaction, is most conspicuously developed in the wood rays and wood parenchyma cells. *Ibid.*, Favorable conditions for sap-staining are found during warm weather. *Ibid.* 147 Sap stain is in general produced in two ways, by the attacks of fungi and by chemical discoloration. **1921** *Phytopathology* XI. 214 As a sap-staining organism *Lasiosphaeria pezizula* has been previously reported by Humphrey. **1953** F. T. Brooks *Plant Dis.* (ed. 2) xii. 199 Several species of Ceratostomella and allied genera, together with many Fungi Imperfecti, cause sap-stain or blueing of the sap-wood of soft and hard timber felled for lumber, and of pulp-wood... Affected wood is reduced in marketability as the stain is unsightly in timber used for certain purposes. **1976** B. M. Bakshi *Forest Path.* III. 280 Sap staining fungi..do not cause any wood decay. *Ibid.* 281 The fungi causing soft rot, like those causing sap stain, belong to the Ascomycetes and Fungi Imperfecti. **1805** Lewis & Clark *Jrnl.* 8 Apr. in *Orig. Jrnls. Lewis & Clark Exped.* (1905) VI. 187 [I saw] the small woodpecker or sapsucker as they are sometimes called. **1834** J. J. Audubon *Ornith. Biogr.* II. 81 The Downy Woodpecker..is best known in all parts of the United States by the name of Sap-sucker. **1941** *Sun* (Baltimore) 25 Jan. 6/1 The cardinals have been flashing to and fro, and the flickers and sapsuckers and the tiny snowbirds. **1962** T. A. Imhof *Alabama Birds* 329 These far-ranging woodland birds are called Peckerwoods and Sapsuckers in the South. **1971** *Islander* (Victoria, B.C.) 13 June 13/2 A sapsucker tapped out an accompaniment on his favorite tree. **1979** *Bull. Yorks. Dial. Soc.* Summer 7 Here's a sap whistle, lads er aw alike, Here's en aad knife, en a nut off a bike [in a boy's pocket].

b. *N. Amer.* With *spec.* reference to the sap of the sugar maple, as *sap beer, -boiling, bucket, -cider, -gatherer, -house, -kettle, pail, pan, season, sled, syrup, trough, tub, weather, works, yield*; **sap-boiler** (in Dict., sense 7 a); **sap-bush,** a grove of sugar-maples; **sap neckyoke** = *sap yoke; **sap orchard** = *sap bush; **sap porridge** (see quots.); **sap run,** an increased flow of sap in a sugar-maple tree; **sap spout,** a spout through which sap is drawn from a sugar-maple tree; **sap sugar** = *maple sugar* s.v. Maple 3; **sap tree,** the sugar maple, *Acer saccharum*; **sap weather,** the kind of weather that encourages the flow of sap in a sugar-maple tree; **sap yoke,** a yoke used for carrying sap pails.

1950 H. & S. Nearing *Maple Sugar Bk.* ix. 202 The other maple product is sap beer. **1876** Sap-boiling [see sense 7 a in Dict.]. **1845** S. Judd *Margaret* I. iii. 12 [Here were] frows, sap-buckets, a leach-tub. **1969** E. H. Pinto *Treen* 94 A maple sap bucket of coopered pine,..is shown... The wire loop, for suspending it on a nail below the sap incision in the tree, can be seen in the photograph. **1980** *Blair & Ketchum's Country Jrnl.* (Brattleboro, Vermont) Oct. 102/1 I've used mine [*sc.* a wooden packboard] to carry 200 sap buckets up the washed-out road to the sugarhouse and to carry finished gallons of syrup back down. *a* **1882** T. Weed *Autobiogr.* (1883) I. ii. 12, I now look with great pleasure upon the days and nights passed in the sap-bush. **1845** J. F. Cooper *Chainbearer* II. v. 60, I don't think anything of bringing you..a little water,.. nor should I had we any beer or sap-cider. **1874** *Rep. Vermont Board Agric.* II. 719 The 'sap-gatherer' or 'draw-tub', as it is called, is a hogshead containing from one hundred to one hundred and fifty gallons. **1917** D. Canfield *Understood Betsy* vii. 110 The sap-house, where Cousin Ann and Uncle Henry were making syrup. **1939** I. B. Wolcott *Yankee Cook Bk.* 338 Any one who.. returns to the sap house. **1904** M. E. Waller *Wood-Carver of 'Lympus* ii. 51 [I] drew trees and sheep and loggers' camps on the flat stones beneath the crotch set for the sap-kettles. **1968** E. R. Buckler *Ox Bells & Fireflies* iv. 77 You thought..about the sap kettle in the cool green shadow, waiting to be emptied at noon. **1905** W. M. Webb in A. E. Cowles *Past & Present City of Lansing & Ingham County, Michigan* 441 One neighbor whittled out brooms... Another gauged the sap neckyokes and another made ox yokes. **1861** *Boston Herald* 12 Apr. 2/6 Owners of sap orchards can afford to work day and night. **1947** K. M. Wells *Owl Pen Reader* (1969) I. 44 Jim..followed him, hanging sap pails to the already dripping spouts. **1874** *Rep. Vermont Board Agric.* II. 729 Russia iron is the best material for home made sap pans as the niter can be removed from it more easily. **1842** *Amer. Pioneer* I. 346 'Sap porridge',..when made of sweet corn meal, and the fresh sacarine juice of the maple, afforded both a nourishing and a savory dish. **1948** E. N. Dick *Dixie Frontier* 290 Corn-meal mush was a regular supper dish. In the spring it was made with maple sap and was known as sap porridge. **1876** J. Burroughs *Winter Sunshine* 119 A 'sap-run' seldom lasts more than two or three days. **1950** H. & S. Nearing *Maple Sugar Bk.* ix. 202 Maple vinegar..is made of sap run at the end of the season. *Ibid.* iii. 48 Much of the boiling was done far from home, and the sugar makers camped out in the deep woods until the sap season was over. *Ibid.* v. 98 The loaded sap sled..moves down rather easily. **1878** *Rep. Vermont Board Agric.* V. 105 We now have the Eureka sap spout, the tin bucket, [etc.]. **1949** *Highway Traveler* Feb. 16/2 A sap spout, or 'spile' as your boss may call it, is driven into the opening with a few taps of a hammer. **1800** C. D. R. D'Eres *Mem.* 63 The squaws in particular, would make me many and valuable [presents]..consisting of sap sugar. **1895** S. O. Jewett *Life of Nancy* 105 [She] handed us sap sugar on one of her best plates. **1951** T. Capote *Grass Harp* i. 11, I could hear the tantalizing tremor of their voices flowing like sapsyrup through the old wood. **1843** *Knickerbocker* XXII. 161 One felled the proper trees, taking care to leave the sap-trees, the sugar-maple, untouched. **1804** T. G. Fessenden *Orig. Poems* (1806) 41 Your love I well repaid By..a sap-trough neatly made. **1840** [see sense 7 a in Dict.]. **1897** R. E. Robinson *Uncle Lisha's Outing* x. 84 These 'ere boots... They're stiffer'n sap troughs. **1872** *Rep. Vermont Board Agric.* I. 215 When I was a boy I purchased one hundred sap tubs, and commenced sugaring on my own hook. **1950** H. & S. NEARING *Maple Sugar Bk.* vi. 137 The 20-degree-night and the 45-degree-day, sunny-days and cold-night formula for sap weather is very far from telling the whole story. **1832** J. J. STRANG *Diary* 19 Feb. in M. M. Quaife *Kingdom of St. James* (1930) 202, I expect to dismiss my school soon and leave the place..for the people want their boys to work in the sap works. **1849** *Knickerbocker* XXXIII. 279 '*The Sugar Bush*' has vividly recalled to memory..the pale blue smoke curling up from the 'sap-works'. **1950** H. & S. NEARING *Maple Sugar Bk.* iv. 82 There is some evidence that length of trunk plays a part in sap yield. **1878** *Rep. Vermont Board Agric.* V. 105 The sap was lugged with sap yoke and pails on their shoulders.

sap, *sb.*[4] Add: **2.** Study, book-work. *Eton College slang.*

a **1862** Q. HOGG *Let.* in E. M. Hogg *Quintin Hogg* (1904) ii. 32 The night before last I..worked the whole night... I hope I shall take well after all my sap. **1901** *Quiet Evening* in *Eton Echoes* 13 Soon a drowsiness steals o'er you, and all thought of 'sap' is banished.

sap, *sb.*[5] (Later examples.)

1930 *Sat. Even. Post* 26 July 145/1 In some ways Angelo's a sap, but I never thought he'd get himself in a spot like that. **1940** WODEHOUSE *Quick Service* xix. 240 You were a sap to come away. **1945** 'N. SHUTE' *Most Secret* vii. 154 But when you come to think of it, I'd have been a sap. **1959** I. & P. OPIE *Lore & Lang. Schoolch.* x. 181 The word 'sap'..the children define as meaning a sissy or a softy ('soft in that he does not do anything wrong'), and suggest other moist alternatives, as 'milksop', 'soppy date', a 'wet', or a 'drip'. **1968** *Globe & Mail* (Toronto) 3 Feb. 35/1 Bobby Mull.., is a sap if he accepts less than $ 100,000 from the tight-fisted..management. **1973** 'H. HOWARD' *Highway to Murder* vi. 72 My brother was a prize sap... Guess he knows better now.

sap, *v.*[1] Add: **2. d.** To erode by glacial sapping (*SAPPING *vbl. sb.*[1] 2 b).

1910 *Geogr. Jrnl.* XXXV. 269 Lack of glacial scratches or polish in uplands sapped by this process should not be allowed to weigh too heavily in reconstructing the glacial history of the district. **1940** *Geogr. Rev.* XXX. 81 Whether these glaciers, when at their maximum thickness, were able to sap vigorously the very bottom of the head walls.. is a little doubtful.

sap (sæp), *v.*[4] *U.S. slang.* [f. SAP *sb.*[1]] *trans.* To hit or club (someone) with a sap (see *SAP *sb.*[1] 4 b). Also with *up* and *intr.* in *to sap up on* (someone).

1926 J. BLACK *You can't Win* vii. 83 The posse fell upon the convention and 'sapped up' on those therein assembled and ran them..out of town. **1926** *Clues* Nov. 162/1 *Sapped*, beaten up. **1931** 'D. STIFF' *Milk & Honey Route* 213 *To get sapped* means to be clubbed by the bulls. **1935** A. J. POLLOCK *Underworld Speaks* 101/1 *Sapped*, struck with a club or billy by a police officer. **1940** R. CHANDLER *Farewell, my Lovely* xxxviii. 178 He slumped sideways and clawed at a corner of the desk, then rolled on his back. It was nice to see someone else get sapped for a change. **1971** *Black World* Apr. 65 My eye was swole... I remember how you sapped me up somethin awful.

Sapei, var. *SEBEI.

sapele (săpī·li). The name of a port on the Benin River, southern Nigeria, used to designate the reddish-brown hardwood timber of *Entandophragma cylindricum*, a large West African forest tree belonging to the family Meliaceæ. Also *attrib.*

1904 *Timber Trades Jrnl.* 2 Apr. 740/2 A fair amount of the African wood [*sc.* mahogany]..was sold. There were submitted 346 lots of Lagos..69 logs of Sapeli, Benin, padouk, birch and Gaboon logs. **1914** E. W. FOSTER *Notes Nigerian Trees & Plants* 20 The wood..has been exported to Europe under the name of 'Unscented Mahogany' presumably to distinguish it from the 'Sapele Scented Wood'. **1928** *Sunday Express* 29 July 15/4 We take the most handsome pieces of burr walnut, of rosewood and sapele mahogany... We place these on the surface of our furniture in such a way that they provide all the beauty and decoration that is needed. **1936** *Nature* 9 May 790/1 The following woods amongst others have been used: laurel wood,..Sapele. **1954** *Archit. Rev.* CXV. 189/3 Display shelves are of sapele wood supported by light steel rods, cellulosed black and orange. **1958** [see *MANSONIA]. **1960** *News Chron.* 21 Apr. 6/2, 11 steps, with treads of sapele (an African hardwood), lead to a landing above the hall. **1964** R. W. J. KEAY et al. *Nigerian Trees* 265 This [*sc. Entandophragma cylindricum*] is one of Nigeria's largest and finest trees, producing the well known Sapele Wood. **1972** 'K. ROYCE' *Miniatures Frame* ix. 117 A board room..with steel tubular chairs running the length of a sapele mahogany table. **1981** *Times* 24 Apr. 12/2 Great sapele logs are stacked along the banks of the Oubangy river.

sap-head. (Earlier U.S. example.)

1798 T. G. FESSENDEN in *Farmer's Weekly Museum* 2 Jan. 4/1 The poet nimbly trips it back—Over the Union courses rapid, And squibs each Jacobinick saphead.

‖ **saphir d'eau** (safir do). Also **sapphir(e) d'eau.** [Fr., lit. 'sapphire of water'.] A translucent blue variety of cordierite occurring in Sri Lanka; = *water-sapphire* s.v. WATER *sb.* 29.

In Fr. the term was orig. used by mineralogists to refer to blue quartz.

1820 R. JAMESON *Syst. Mineral.* (ed. 3) I. 174 The sapphire d'eau of collectors. **1897** L. FLETCHER *Introd. Study Mineral.* 106 Cordierite is a silicate of magnesium, iron and aluminium; its transparent variety is the *Saphir d'eau* of jewellery. **1925** KRAUS & HOLDEN *Gems & Gem Materials* II. 165 Ceylon is the most important locality, and the gems from that country have sometimes been called *saphir d'eau*, or 'water sapphire'. **1936** H. P. WHITLOCK *Story of Gems* 175 Iolite. This silicate of magnesium, aluminium and iron is better known as *water-sapphire* (saphir d'eau).

sapient, *a.* and *sb.* Add: **A.** *adj.* **3.** *Anthrop.* Of, pertaining to, or characteristic of modern man, *Homo sapiens.*

1971 *Nature* 28 May 213/1 At sites in East Africa can be seen evidence of the various stages of human evolution—the older levels have the remains of the australopithecines and the younger levels have, in succession, early hominines and, finally, fully sapient types. **1976** *Ibid.* 5 Aug. 487/1 It [*sc.* the Ndutu cranium] differs from Swanscombe and Steinheim in its occipital curvature and in that the mastoid of Steinheim is sapient in form.

sapiential, *a.* Add: **2.** Also applied to similar writings in Old English.

1970 *N. & Q.* Dec. 445/1 Old English sapiential poetry has received a good deal of scholarly attention.

sapin. Delete † *Obs.* and add later examples.

1793 E. WYNNE *Diary* 14 Sept. (1935) I. xii. 216 Not such Sapin and Pine woods as in the Country of St. Gall but Oak's. **1813** A. HENRY *Jrnl.* 14 Dec. in E. Coues *New Light Early Hist. Greater Northwest* (1897) II. xxiv. 772 The place is deeply shaded with spruce, pine, sapin, etc. **1927** *Brit. Weekly* 1 Sept. 470/3 Across the valley..are the mountain slopes, with the valiant *sapins* sending their spear points, in massed formation, to the highest level at which a tree can grow.

Sapiny (sæ·pini). Also **Sabiny, Sapin, Saviny.** [Native name.] = *SEBEI.

1909 A. C. HOLLIS *Nandi* I. 2 It seems probable that the tribes allied to the Nandi who live on or near Mount Elgon (the Lako, Kony, Mbai, Sabaut, Sapin, Pŏk, and Kâpkara) are only a section of the migrants. **1964** Sabiny [see *KIPSIGIS]. **1977** C. F. & F. M. VOEGELIN *Classification & Index World's Lang.* 323 *Nandi*... Sapiny = Sabei = Savei = Sebei = Saviny = Kamecak (572,000; Uganda).

Sapir–Whorf hypothesis (sāpiᵊ·ɪ hwɔ̄ᵊɪf). [f. the names of Edward *Sapir* (1884–1939) and Benjamin Lee *Whorf* (1897–1941), American linguists.] A hypothesis, first advanced by Sapir in 1929 and subsequently developed by Whorf, that the structure of a language partly determines a native speaker's categorization of experience. Cf. *WHORFIAN *a.*

1954 H. HOIJER *Language in Culture* I. 93 The central idea of the Sapir–Whorf hypothesis is that language functions, not simply as a device for reporting experience, but also..as a way of defining experience for its speakers. **1954** —— in *Mem. Amer. Anthropol. Assoc.* LXXIX. 95 Differences..which reflect a people's habitual and favorite modes of reporting, analyzing, and categorizing experience, form the essential data of the Sapir–Whorf hypothesis. **1956** J. B. CARROLL in B. L. Whorf *Lang., Thought, & Reality* 27 Whorf's principle of linguistic relativity, or, more strictly, the Sapir–Whorf hypothesis (since Sapir most certainly shared in the development of the idea) has..attracted a great deal of attention. **1976** *Word 1971* XXVII. 242 This is 180 degrees different from what has been known about the Sapir–Whorf hypothesis, which advocates that it is language that has the power to dictate man's world view in a tyrannical way.

sapogenin. Add: In mod. use, a generic term for any of the steroid aglycones of the saponins. (Further examples.) [Coined in G. by P. A. Bolley 1854, in *Ann. d. Chem. u. Pharm.* XC. 216.]

1916 *Jrnl. Biol. Chem.* XXVIII. 443 Hydrolysis [of a new saponin] yielded a sapogenin. **1955** *Sci. Amer.* Jan. 57/1 The steroid part of a saponin is called a sapogenin. **1977** LEWIS & ELVIN-LEWIS *Med. Bot.* ii. 19/2 Some of the plants having useful steroidal sapogenins include *Dioscorea* spp. (yams, Dioscoraceae), *Agave* spp., and *Smilax* spp. (Liliaceae).

saponaria (sæpǒnēᵊ·riă). [med.L. *sāpōnāria* (see SAPONARY *a.* and *sb.*), adopted as a generic name by Linnæus (*Systema Naturæ*, 1735).] = SOAPWORT I. Cf. SAPONARY *sb.* I, SAPONER.

1865 M. EYRE *Lady's Walks S. of France* vii. 87, I gathered..eye-bright, saponaria, and ling. **1900** J. M. ABBOTT in W. D. Drury *Bk. Gardening* vii. 237 Saponarias.. are dwarf hardy annuals..of the Pink family. **1951** *Dict. Gardening* (R. Hort. Soc.) IV. 1865/2 Saponarias mostly grow readily in well-drained soils.

saponarin (sæpǒnēᵊ·rin). *Chem.* Formerly also **-ine.** [a. G. *saponarin* (G. Barger 1902, in *Ber. d. Deut. Chem. Ges.* XXXV. 1296), f. med.L. *sāpōnār-ia* (see below and SAPONARY *a.* and *sb.*) + *-in* -IN[1].] A white or pale yellow crystalline flavonoid diglycoside, $C_{27}H_{30}O_{15}$, first found in soapwort, *Saponaria officinalis.* Hence **saponare·tin** [*-ETIN], a monoglycoside derived from this by hydrolysis.

1902 *Jrn. Chem. Soc.* LXXXII. I. 387 Saponarin dissolves in about 1000 parts of hot water and crystallises on cooling in minute, birefringent needles. **1905** G. BARGER in *Rep. Brit. Assoc. Adv. Sci. 1904* 531 Unless the solution be dilute, a second product of hydrolysis separates as a thick yellow oil, which has not yet been obtained crystalline. The name saponaretin is suggested for it. **1923** *Nature* 25 Aug. 304/2 The formation of a glucoside (saponarine) in the mitochondria. **1950** *Thorpe's Dict. Appl. Chem.* (ed. 4) X. 687/1 Saponarin..dried in air is a white powder, but after drying *in vacuo* becomes pale yellow. **1967** *Chem. Abstr.* LXVI. 10485/2 Saponaretin..was obtained by chromatog[raphy] of flavonoids on a polyamide column. **1969** *Acta Chemica Scandinavica* XXIII. 2910/2 At the time of Molisch's investigation the constitution of saponarin was not known. It is now identified as isovitexin-7-glucoside (apigenin-6-C-7-O-diglucoside).

saponification. Add: **b.** *saponification equivalent* (further examples); now usu. defined in terms of the amount of alkali required to saponify a particular quantity of oil, *spec.* the number of milligrammes of potassium hydroxide required by one gramme of oil; also *saponification number, value.*

1895 PEARMAIN & MOOR *Aids to Analysis of Food & Drugs* 89 The saponification value of an oil may be stated in terms of alkali absorbed per cent., or the number of grammes of the oil which would be saponified by one litre of normal solution of alkali, which is usually known as the 'saponification equivalent'. **1896** *Analyst* XXI. 192 The ether value thus obtained, added to the acid value, gave the saponification number. **1944** *Industrial & Engin. Chem.* (*Analytical Ed.*) XVI. 53 The indicator..is therefore recommended for use in the determination of acid numbers and saponification equivalents. **1946** F. SCHNEIDER *Qualitative Organic Microanalysis* vi. 163 For the identification of the acid portion of the ester use the titrated solution from the determination of the saponification equivalent. **1964** C. J. BONER *Gear & Transmission Lubricants* iii. 68 The saponification number divided by two will give a close approximation of the percentage of fat in a compounded oil. **1975** *Materials & Technol.* VIII. i. 17 The saponification value..expressed as the number of milligrams of potassium hydroxide required to saponify one gram of fat.

saponin, -ine. Add to def.: In mod. use, any of a large class of steroid glycosides obtained from plants, which are usu. toxic (esp. to fish), causing hæmolysis, and are characterized by the property of foaming in aqueous solution. (Further examples.)

The spelling *saponin* is now usual.

1891 *Jrnl. Chem. Soc.* LX. II. 1531 The author [*sc.* R. Kobert] considers that there are a series of saponins of the general formula $C_nH_{2n-8}O_{10}$, several of which are known. **1916** *Jrnl. Biol. Chem.* XXVIII. 443 A new saponin, $C_{24}H_{40}O_{14}$, was isolated from the rootstock of *Yucca filamentosa.* **1953** C. W. & E. SHOPPEE in E. H. Rodd *Chem. Carbon Compounds* II. B. xix. 1035 Saponins..are haemolytic when injected into the bloodstream of animals and therefore highly toxic intravenously but comparatively harmless when ingested. **1977** LEWIS & ELVIN-LEWIS *Med. Bot.* ii. 19/2 The recent importance of plant steroidal compounds, especially the steroid saponins, is their suitability as cortisone and hormone precursors.

sapotoxin (sæ·potǫksin). *Chem.* Also **-ine.** [a. G. *sapotoxin* (R. Kobert 1887, in *Arch. f. exper. Path. u. Pharm.* XXIII. 241), f. med.L. *sāpōnāria* (see below and SAPONARY *a.* and *sb.*): see TOXIN.] A saponin found in the bark of the Chilean soap-bark tree, *Quillaja saponaria*; also, any markedly toxic saponin.

1891 *Jrnl. Chem. Soc.* LX. 1532 The sapotoxin of *Agrostemma githago* (corn cockle)..is absorbed both by the subcutaneous tissues and by the intestinal canal, and thus acts as a dangerous poison. **1892** *Ibid.* LXII. 350 The sapotoxin of *Agrostemma* has the same composition as those of *radix saponariæ albæ* and of quillaja bark, but differs from them in its physiological properties. **1924** C. T. KINGZETT *Chem. Encycl.* 486 Saponins from quillaya bark..are stated to consist of one-third quillajic acid and two-thirds of a body named sapotoxin. **1927** *Glasgow Herald* 11 June 4 The saponine and sapotoxine in effluents from beet-sugar factories. **1941** *Martindale's Extra Pharmacopœia* (ed. 22) I. 910 Quillaia... Contains quillaic acid..and sapotoxin, $C_{17}H_{26}O_{10}$..closely allied to saponin.

sapped, *ppl. a.* Add: Also, eroded or broken off by glacial sapping.

1972 J. G. MCCALL in C. Embleton *Glaciers & Glacial Erosion* xi. 220 The general lack of any further frost action on the 'sapped' blocks of rocks which formed the moraine and screes in the area.

sapphir(e) d'eau, var. *SAPHIR D'EAU.

sapphire. Add: **1. f.** A sapphire used as a stylus for gramophone records.

1943 *Electronic Engin.* XVI. 121/2 The portion of the wire between the ribbon and the sapphire provides sufficient vertical compliance to minimise mechanical noise. **1957** [see *playing-life* s.v. *PLAYING *vbl. sb.* 2]. **1964** P. J. GUY *Disc Recording & Reproduction* vii. 99 In the author's experience some sapphires have a very much shorter life.

2. b. A sapphire mink (see sense *3 c).

1951 *Genetics* XXXVI. 575 Several color phases result from the combination of two or more of these mutant genes, the sapphire..and the 'red-eyed' pastel..being of most commercial importance at the present time.

3. a. *sapphire needle, point, stylus* (all = sense *1 f); *sapphire-shot* adj.; **sapphire quartz,** a rare indigo-blue variety of quartz; = SIDERITE[1] 5.

1940 *Chambers's Techn. Dict.* 740/1 Sapphire needle. **1943** *Gramophone* Dec. 107/2 In the last two years of manufacture of radiograms in this country practically all of them came thru with sapphire needles as standard equipment. **1899** *T. Eaton & Co. Catal.* Spring & Summer 191 Graphophone Supplies... Recorder, with sapphire point, $5.00. Reproducer, with sapphire point, $5.00. **1972** *Country Life* 28 Dec. 1777 We..listened avidly to Mr Alfred Heather's rendering of *I'll sing thee songs of Araby* on our sapphire-point Pathé gramophone. **1868** J. D. DANA *Syst. Min.* (ed. 5) 193 Siderite, or Sapphire-quartz. Of indigo or Berlin-blue color. **1904** L. J. SPENCER tr. *M. Bauer's Precious Stones* 488 Sapphire-quartz (azure-quartz or siderite) is a blue, crystalline quartz... It is used to a very small extent and is correspondingly low in price. **1971** *Country Life* 3 June 1382/3 Quartz, however, may be variously tinted:..sapphire quartz—blue; [etc.]. **1883** G. M. HOPKINS *Poems* (1967) 95 Yet such a sapphire-shot, Charged, steepèd sky will not Stain light. **1947** *Gramophone* Oct. 74/1 This..has the..replaceable sapphire stylus, for which the makers claim 1,000 playings before replacement is necessary. **1974** *Encycl. Brit. Macropædia* XVII. 54/1 When, between 1933 and 1935, attempts were made to use sapphire styli with electrical pickups weighing 50 to 150 grams.., record wear was found to be excessive.

c. sapphire mink, a variety of mink with blue fur (see also sense *2 b); also, the fur of this animal.

1960 *Guardian* 26 Aug. 6/4 A generously wide sapphire mink stole. **1974** *Genetika* XII. II. 109 The interactions between genes are demonstrated to result in a significant rearrangement of cells and the development of a new pigmentation type in sapphire minks. **1976** *S. Wales Echo* 25 Nov. 22/7 (Advt.), Full-length Sapphire Mink, £850.

Sapphist. Add: (Examples.) So **sapphi·stically** *adv.*, in the manner of a Sapphist.

1913 R. BROOKE *Let.* 13 Dec. (1968) 547 A woman.. who loved Lulu sapphistically. **1923** V. WOOLF *Diary* 19 Feb. (1978) II. 235 She is a pronounced Sapphist, & may..have an eye on me. **1925** [see *PÆDERAST]. **1975** 'M. ORR' *Rich Girl, Poor Girl* (1977) xviii. 247 Winifred knew herself to be..a Sapphist on the prowl for a desirable *jeune fille.*

sappiness. 2. (Example.)

1943 *New Yorker* 20 Feb. 22/1 She was..convinced.. that a floppy feminine hat was a symbol of celluloid sappiness.

sapping, *vbl. sb.*[1] Add: **2.** *Physical Geogr.* **a.** Undercutting by water, esp. backward erosion by a waterfall of softer layers of rock at its base; headward erosion of hillsides by springs.

1863 [in Dict.]. **1902** W. M. DAVIS in *Bull. Mus. Compar. Zoöl.* XXXVIII. 328 Whatever flood plains may have been produced during the excavation of the present basin floor, the streams have now so well taken advantage of their opportunity for lateral corrosion or 'sapping' that terraces at high and intermediate levels are everywhere obliterated. **1932** W. H. EMMONS et al. *Geol.* vi. 133 As the swirling water back of the falls loosens the soft, shaley formation it removes it piecemeal and undermines the capping limestone, until finally it remains as an inadequately supported overhanging ledge from which large masses of rock plunge into the pool at the bottom of the falls. This process of undercutting is termed sapping. **1936** *Proc. Geologists' Assoc.* XLVII. 40 A coombe formed in jointed chalk by the sapping back of springs. **1957** *Ibid.* LXVIII. 31 There remains the curious series of right-angled bends in the Ravensburgh Valley system, which has been attributed to sapping along major joints. **1970** R. J. SMALL *Study of Landforms* ii. 53 On rocks such as chalk and limestone the actual sources of streams are extended into escarpments and steep slopes by the process known as 'spring sapping'. This involves underground chemical erosion, surface stream erosion, and slumping of moistened debris around the springhead.

b. Undermining by glacial erosion; (*loosely*) plucking; *spec.* erosion of rock slopes by frost action under the margins of a glacier.

1899 W. D. JOHNSON in *Science* 20 Jan. 106/1 An unrecognized process was set forth, that of sapping, whose action is horizontal and backward... The tendency of the sapping process is to produce benches and cliffs. **1938** *Geol. Mag.* LXXV. 261 As the wall at the head of the cirque retreats under the action of sapping and plucking, immediately downstream the ice abrades and smoothes. **1954** *Jrnl. Glaciol.* II. 421 In accounting for these features [*sc.* roches moutonnées] the assumption ordinarily made is that rock has been removed by plucking or sapping from the downstream side, leaving that face steep and irregular. **1968** R. W. FAIRBRIDGE *Encycl. Geomorphol.* 741/1 The walls have been kept steep and caused to retreat by the collapse of unsupported rock faces as they have been undercut by the process of glacial 'sapping'... The explanation of sapping appears to be found in rending and disintegration of rock by the freeze-and-thaw process. **1972** J. G. MCCALL in C. Embleton *Glaciers & Glacial Erosion* xi. 217 The term sapping, as used here, implies frost-riving on the rock slopes under the margins of a glacier. It is produced by the freezing of any water which flows in under the 'cold' glacier and, in the case of cirques, it results in a horizontal retreat of the headwall.

sapping, *vbl. sb.*[2] (Earlier and later examples.)

1821 *Salt-Bearer* No. 26. 303 When at Eton, boxing, rowing, cricket, and even *sapping*, had by turns the honour of possessing a stall in his hobby stable. **1922** S. LESLIE *Oppidan* iv. 48 That..was why *sapping* was unnecessary.

sappy, *a.* Add: **2. b.** (Later example.)

1948 F. R. LEAVIS *Great Tradition* ii. 111 *The Portrait of a Lady* belongs to the sappiest phase of James's art, when the hypertrophy of technique hadn't yet set in.

7. Also as *sb.* Hence **sa·ppyhead,** a foolish person. Cf. SAP-HEAD.

1922 JOYCE *Ulysses* 114 Martin could wind a sappyhead like that round his little finger without his seeing it. **1930** D. H. LAWRENCE *Nettles* 23 You know that they've got to think that they're happy... Oh so happy, you sappy.

sapric (sæ·prik), *a. Soil Sci.* [f. Gr. σαπρ-ός rotten, putrid + -IC.] Of a soil or soil horizon: characterized by the presence of highly decomposed organic material.

1965 FARNHAM & FINNEY in *Adv. Agronomy* XVII. 138 In the classification of organic soils presented here, only three types of horizons are considered diagnostic... These are the fibric, mesic, and sapric horizons, listed in order of increasing decomposition. **1972** J. G. CRUICKSHANK *Soil Geogr.* vi. 187 At the other extreme, sapric.. soils or horizons are well decomposed and contain a high proportion by weight of mineral material.

† **saprine.** *Chem. Obs.* Also **-in.** [ad. G. *saprin* (L. Brieger *Untersuchungen über Ptomaine* (1885) II. 46), f. as prec.: see -INE[5].] A ptomaine of doubtful identity isolated from putrefying flesh.

1887 [see CADAVERINE]. **1894** *Watts's Dict. Chem.* IV. 346/2 Saprine is isolated by means of its platinochloride. **1910** *Practitioner* June 830 Ptomaines obtained from putrid meat and other albuminous bodies:..Saprin $C_5H_{14}N_2$.

‖ **sapristi** (sapristi), *int.* [Fr., corruption of *sacristi* in same sense.] An exclamation of astonishment, exasperation, etc.; a mild oath.

1839 THACKERAY *Cox's Diary* in *Comic Almanack 1840* 33 Shouting out, 'Aha!' and '*Sapprrrristie!*' **1867** 'OUIDA' *Under Two Flags* II. i. 3 Sapristi! And what did he say? **1932** A. CHRISTIE *Peril at End House* xxii. 244 And the card—*my* card! *Ah! Sapristi*—she has a nerve! **1957** O. NASH *You can't get there from Here* 97 So when I sight my island home I'll salvage but a single tome, Which is—what should it be, *sapristi*, But any book by Agatha Christie? **1966** A. CHRISTIE *Third Girl* ii. 10 Ah *Sapristi!* That must be a woman—undoubtedly a woman. **1972** A. MACVICAR *Golden Venus Affair* i. 9 Sapristi, what a condition he's in!

saprobe (sæ·proᵘb). *Biol.* [f. Gr. σαπρό-ς putrid + β-ίος life; cf. G. *saprobie* (Kolkwitz & Marsson 1902, in *Mittheilungen aus der K. Prüfungsanstalt f. Wasserversorgung und Abwässerbeseitigung* I. 46).] Any organism that derives its nourishment from decaying organic matter.

1932 G. W. MARTIN in *Bot. Gaz.* XCIII. 427 The word *saprophyte* and its derivatives, implying that a fungus is a plant, can be replaced by saprobe (σαπρός + βίος), which is without such implication. **1952** C. J. ALEXOPOULOS *Introd. Mycol.* i. 30 Fungi obtain their food either as parasites or saprobes. Some are obligately parasitic or saprobic. **1971** G. C. AINSWORTH *Ainsworth & Bisby's Dict. Fungi* (ed. 6) 518 Saprobe is the preferred usage for fungi.

saprobial (sæprōᵘ·biăl), *a. Ecol.* [f. *SAPROB(IC *a.* + -IAL.] Serving as a measure of saprobity.

1965 *Hydrobiologia* XXV. 523 The secondarily introduced methods of statistical calculation and expression of saprobial indices may only pretend an exact mathematical basis. **1970** J. SCHWOERBEL *Methods Hydrobiol.* vii. 155 The saprobial valency is better characterized than the allotment to a single zone of the saprobic system.

saprobic (sæprōᵘ·bik), *a. Ecol.* [f. G. *saprob-ie* *SAPROBE + -IC.] **a.** Characterized by the prevalence of decaying organic material; *spec.* = *polysaprobic* adj. s.v. *POLY-1; *saprobic system,* a system by which a body of polluted water is divided into zones characterized by the presence of certain organisms that are treated as indicators of the degree of pollution.

1913 *Bull. Illinois State Lab. Nat. Hist.* IX. x. 498 We will distinguish..three stages of impurity, by use of the following terms applicable both to the waters themselves and to the characteristic organisms, given here in the order of a diminishing impurity, namely: (1) septic or saprobic [etc.]. **1925** [see *polysaprobic* adj.]. **1967** A. F. BARTSCH in Olson & Burgess *Pollution & Marine Ecol.* VI. 294 Various North American biologists have expressed doubt as to the applicability of the saprobic system to coastal and estuarine environments. **1971** R. J. BENOIT in L. L. Ciaccio *Water & Water Pollution Handbk.* I. iv. 255 The general relationship between the saprobic zones and the zones shown by oxygen sag curves has been illustrated. **1975** D. F. WESTLAKE in B. A. Whitton *River Ecol.* iv. 126 In general macrophytes are not good indicators for use in the saprobic system of classifying rivers.

b. Pertaining to or characteristic of a saprobe; deriving nourishment from decaying organic matter.

1932 *Bot. Gaz.* XCIII. 429 The nutrition of the Phycomycetes is saprobic or parasitic. **1960** H. B. N. HYNES *Biol. Polluted Waters* xiii. 161 Complex organic..molecules encourage the growth of saprophytic plants and saprobic animals. **1976** *Nature* 27 May 336/2 The free-living saprobic form found in soil is mycelial.

Hence **sapro·bical** *a.* = *SAPROBIAL *a.*; **saprobi·city** = *SAPROBITY.

1961 *Arch. Hydrobiol.* LVII. 405 Tables of the saprobical valency of species studied by the authors as well as examples of the saprobiological evaluation are appended. **1971** *Ann. Rev. Microbiol.* XXV. 565 The rate of multiplication of the test organism, i.e., optimum biomass produced, is considered to be an integrated biological measure of the content of biologically active organic nitrogen in the water and thus a measure of the saprobicity. **1975** G. A. COLE *Textbk. Limnol.* iv. 61/2 Saprobicity, the total of all the processes that are antithetical to primary production, is another classification of heterotrophy.

saprobiology (sæprobəi,ǭ·lodʒi). [f. as next + BIOLOGY.] The study of saprobic environments.

1958 *Ecology* XXXIX. 547/2 It can be stated in terms used in saprobiology that the polysaproby was changed into alpha-mesosaproby. **1965** *Hydrobiologia* XXV. 524 The applicability in practice is..the main criterion of an applied scientific branch, as saprobiology is.

Hence **sa:probiolo·gical** *a.*; **saprobio·logist.**

1960 *Biol. Abstr.* XXXV. 3801/2 The saprobiological analysis is based on the determination of approximately 8,500 organisms. **1965** *Hydrobiologia* XXV. 526 Such problems as the influence of different amino acids..on the stimulation of the growth of *Sphaerotilus* and *Leptomitus* cannot be solved by the saprobiologists alone. **1971** *Ann. Rev. Microbiol.* XXV. 574 Caspers & Karbe..have proposed a saprobiological classification of waters.

saprobiotic (sæprobəi,ǫ·tik), *a. Biol.* [f. Gr. σαπρός putrid + βιωτικ-ός pertaining to life.] = *SAPROBIC *a.*

1940 *Chambers's Techn. Dict.* 740/2 *Saprobiotic*, feeding on dead or decaying animals or plants. **1950** P. D. F. MURRAY *Biology* xlvii. 52 A saprobiotic organism (the term covers 'saprozoic' and 'saprophytic') is one which lives on the dead bodies of other organisms or their inanimate products. **1960** E. N. WILLMER *Cytol. & Evolution* ix. 151 Many flagellates..depend for their energy supplies on the extraction of materials dissolved in the fluid in which they are swimming, i.e. they lead a saprobiotic existence.

Hence **saprobio·tically** *adv.*

1957 G. E. HUTCHINSON *Treat. Limnol.* I. ix. 623 Living plankton metabolizing its reserves or living saprobiotically.

saprobity (sæprōᵘ·biti). *Ecol.* [f. *SAPROB(IC *a.* + -ITY.] The degree to which the decomposition of organic material is occurring in an aquatic environment.

1956 *Archiv für Hydrobiol.* LI. 389 Utilization of the running water Macroorganism as Indicators of the water saprobity degrees (sensu Kolkwitz-Marsson). **1965** *Hydrobiologia* XXV. 523 The saprobity system is applicable only to organic pollution undergoing bacterial decomposition and it is useless for the assessment of the effects of poisons or other pollutional matters. **1973** BAYLY & WILLIAMS *Inland Waters & their Ecol.* xii. 254 The 'saprobity indices' of several European workers..are indirect measures of pollution based upon a combination of chemical, bacteriological, and biological features.

saprolite (sæ·proləit). *Geol.* [f. Gr. σαπρός putrid + -LITE.] Soft, clay-rich, thoroughly decomposed rock formed *in situ* by chemical weathering of igneous and metamorphic rocks.

1895 G. F. BECKER in *16th Ann. Rep. U.S. Geol. Survey* III. 289, I propose the term *saprolite. Ibid.* 290 The deposits referred to..are gold-bearing saprolites. **1935** *Jrnl. Geol.* XLIII. 745 In the Appalachian Piedmont of the southern states, weathering has reduced the granitic rocks to an extensive mantle of incoherent clay, or saprolite, that in places extends to a depth of over 100 feet. **1948** *Prof. Papers U.S. Geol. Survey* No. 213. 125/2 The upper parts of the lodes were in the saprolite zone and were worked as residual placer deposits, but at the base of the saprolite these deposits graded into solid lodes. **1977** A. HALLAM *Planet Earth* 48/1 Chemical weathering can produce a rotted rock-form known as a saprolite, which is the product of chemical changes which have taken place *in situ.*

Hence **saproli·tic** *a.*; **sa:prolitiza·tion,** the process of formation of saprolite.

1904 L. J. SPENCER tr. *Bauer's Precious Stones* 361 Several crystals of rhodolite were found..embedded in a decomposed saprolitic rock. **1970** D. CARROLL *Rock Weathering* iii. 20 Extensive areas in arid Western Australia are underlain by saprolitic rocks on which lateritic profiles have developed (probably in the Pliocene). *Ibid.*, Saprolitization also occurs in rocks that are covered by later deposits that protect them from erosion. An example is saprolitization under a cover of river gravel or sand through which water percolates.

sapropel (sæ·propel). *Geol.* [a. G. *sapropel* (H. Potonié 1904, in *Sitzungsber. Ges. naturforsch. Freunde Berlin* 13 Dec. 243), f. as next.] An unconsolidated nitrogen-rich slime or sludge, formed of incompletely decomposed

aquatic micro-organisms, esp. algæ, found in anaerobic environments on the bottoms of lakes and seas.

1907 *Rep. Brit. Assoc. Adv. Sci. 1906* 748 The sapropel is formed from the excrements and bodies of completely aquatic animals and plants which have lived in stagnant water, and therefore, because the water is stagnant, do not decay completely. *Ibid.*, Cannel coal..is a fossil sapropel. **1929** H. B. MILNER *Sedimentary Petrogr.* (ed. 2) 335 The coal-substance has been regarded by Potonié as 'sapropel', a solidified jelly-like carbonaceous slime. **1970** *Nature* 17 Oct. 200/1 Cores from the Mediterranean's three deep basins also yielded dolomites,..diatomites and organic sapropels. **1978** *Ibid.* 16 Nov. 259/2 The preservation of non-siliceous algae is uncommon, although found in the sapropel deposits of the USSR.

sapropelic (sæprope·lik), *a. Geol.* and *Zool.* [ad. G. *sapropelisch* (R. Lauterborn 1901, in *Zool. Anzeiger* XXIV. 50), f. Gr. σαπρός putrid + πηλός mud, earth, clay: see -IC.] Found in, characterized by, or derived from sapropel.

1901 *Jrnl. R. Microsc. Soc.* 144 'Sapropelic' Fauna... Dr. R. Lauterborn uses the term '*sapropelische*' to denote the organisms found in the muddy debris covering the bottom of stagnant fresh-water pools. **1918** [see *HUMIC *a.*]. **1963** D. W. & E. E. HUMPHRIES tr. *Termier's Erosion & Sedimentation* xi. 239 The presence of a sapropelic bottom inhibits aerobic life over the whole of the lower part of the basin. **1966** [see *HUMIC *a.*]. **1971** *Nature* 31 Dec. 508/1 They [*sc.* labyrinthodonts] are never found in humic coals, but usually in sapropelic coals, laid down in the deep anaerobic mud which formed in stagnant conditions.

saprophile, *sb.* (Example of *fig.* use.)

1934 S. BECKETT *More Pricks than Kicks* 67 A little saprophile of an anonymous politico-ploughboy setting him off.

sapskull. (Later example.)

1974 J. AIKEN *Midnight is Place* i. 11 Idiot! Sapskull! How dared you write?

Saraband[2] (sæ·răbænd). Also **Sarabend, Serabend,** etc. [ad. *Saravand*, name of a district in western Iran.] A kind of Persian rug characterized by a pattern of leaf or pear forms. Also *attrib.*

1901 J. K. MUMFORD *Oriental Rugs* vi. 68 The 'pear' [*sc.* a motif] seems to have..original association with Persia... In the Sarabands..it covers the whole field. *Ibid.* xi. 197 The Saraband rugs are made in the district of Sarawan. **1913** W. A. HAWLEY *Oriental Rugs* ix. 130 In Mir-Sarabends one of two threads encircled by a knot is doubled under the other at back. In Royal Sarabends each is equally prominent. **1931** [see *palm-leaf pattern* s.v. *PALM-LEAF c]. **1943** *Burlington Mag.* May 130/2 A Saraband carpet, the property of Mrs. Gilbert Russell. **1962** C. W. JACOBSEN *Oriental Rugs* 278 Choicest antique Sarabends will have as many as 350 knots to the square inch. **1975** 'E. LATHEN' *By Hook or by Crook* viii. 78 An old and valued customer was..closing a deal for an old and valued Saraband. **1975** *Oxf. Compan. Decorative Arts* 612/1 Serabend rugs with all-over cone designs.

Saracen. Add: **4.** Saracen's corn, also = *Saracen corn* or buckwheat, *Fagopyrum esculentum.*

1600 R. SURFLET tr. *Stevens & Liebault's Maison Rustique* I. xi. 53 Let her cause to be ground amongst her corne beanes, pease, fetches or sarrasins corne in some small quantitie. **1804** M. WILMOT *Russ. Jrnls.* (1934) I. 123 So many different sorts of Corn..the Sarazens' Corn so white, the flax with its blue flowers, the peas so green.

Sarah (sēᵊ·ră). [f. *s*earch *a*nd *r*escue *a*nd *h*oming.] Name given to a portable radio transmitter used by wrecked airmen to signal their position to rescue ships or aircraft. Also *attrib.*

1955 *Times* 31 Aug. 8/3 'Sarah', the device demonstrated yesterday, weighs only about 3lb. and can be carried in a Mae West. It contains a beacon battery of 24 hours operating capacity which can send signals to a Shackleton aircraft 75 miles away. A wrecked airman can start 'Sarah' working quite easily and can speak on it to those who are searching for him. **1956** *Times* 18 July 10/7 Tryout for 'Sarah'..'Sarah', the R.A.F.'s new air-sea rescue system, had its first real test yesterday. **1962** S. CARPENTER in *Into Orbit* 59 One of these beacons, a British invention called 'Sarah'—for 'Search and Rescue and Homing'—put out the signal that helped tell the search planes exactly where I was. *Ibid.* 60 There was one Sarah beacon aboard the capsule.

saralasin (săræ·lăsin). *Pharm.* [Contraction of 1-*sar*-8-*ala*- angiotens*in*, f. SAR(COSINE + ALA(NINE + *angioten*)*sin* (f. ANGIO- + *HYPER)TENSIN).] A synthetic octapeptide which blocks the pressor action of hypertensin II, thereby reducing high blood pressures.

1974 *Lancet* 28 Dec. 1535/1 Angiotensin-II blockade by the competitive antagonist sar[1]-ala[8]-angiotensin II ('Saralasin')... Blockade of angiotensin II with saralasin has been advocated. **1977** *Ibid.* 24/31 Dec. 1317/1 Two types of renin-angiotensin inhibitor have been tested in renal hypertension—competitive antagonists of angiotensin II, such as saralasin, and inhibitors of the converting enzyme.

Saramaccan (særămæ·kăn), *sb.* and *a.* Also **Saramak(k)an.** [f. the name of the river *Saramacca* in Surinam.] **A.** *sb.* **a.** A native or inhabitant of the upper reaches of the river Saramacca. *rare.* **b.** A creole language of this region; = *Jew Tongo* s.v. *JEW *sb.* 3 c. **B.** *adj.* Of or pertaining to the people or language of this region.

1959 J. VOORHOEVE in *Word* XV. 436 (*title*) An orthography for Saramaccan. *Ibid.* 437 The phoneme analysis was based on a Saramaccan story and a series of test-words. *Ibid.*, The Rev. Schmidt is a Saramaccan by birth. **1961** *Compar. Stud. Society & Hist.* III. 278 Saramakkan is spoken only by the Bush Negroes on the upper reaches of the Surinam or 'Saramakka' river. **1968** [see *Jew Tongo* s.v. *JEW *sb.* 3 c]. **1970** *Language* XLVI. 408 (*title*) A Saramaccan narrative pattern. **1976** *Amer. Speech 1974* XLIX. 141 Saramakkan and other creoles of the Americas with a higher proportion of African linguistic content are recognized tone languages.

Saran (sărǽ·n). orig. *U.S.* Also **saran.** A proprietary name for PVC, esp. as a film. Also *Saran Wrap* (hence *Saran-wrapped* adj.).

1940 *Official Gaz.* (U.S. Patent Office) 26 Nov. 809/2 The Dow Chemical Company, Midland, Mich. Filed Sept. 27, 1940. *Saran* for thermoplastic synthetic resins comprising polymers and co-polymers derived from vinylidene chloride. Claims use since Aug. 21, 1940. **1942** [see *POLYVINYLIDENE]. **1948** *Textile Colorist* Feb. 46/3 Plastic Sales Division—of the Dow Chemical Company has formally released its trademark rights to the name 'saran' permitting it to become the descriptive name of the product. **1958** *Trade Marks Jrnl.* 26 Nov. 1214/2 *Saran*.. Wrapping (packaging) materials included in Class 16 in the form of films. The Dow Chemical Company..Manufacturers. **1966** N. SIMON *Odd Couple* II. i. 64 After the.. leftovers have been Saran-Wrapped—what do we do? **1968** T. WOLFE *Electric Kool-Aid Acid Test* xxvii. 391 Dresses made out of..supermarket Saran Wrap. **1969** W. R. R. PARK *Plastics Film Technol.* vi. 161 The three plies, 'Saran' 18 (outer), PVC 88 (center), and 'Saran' 22 (inner) each serve a specific function. **1974** D. E. WESTLAKE *Help* ii. 15, I was stretching Saran Wrap over the toilets. **1979** *Maclean's Mag.* 21 May 9/1 Clark began by granting interviews in a separate compartment aboard his plane but by the fifth week of the campaign, Tory campaign manager Lowell Murray sensed that the party was taking the heat for a Saran-Wrapped strategy. **1980** *Yachts & Yachting* 29 Feb. 656/2 Blown saran to produce an easily worked filler.

‖ **sarangi** (sā·raŋgi). Also 9 **sarungee.** [Skr.] An Indian musical instrument resembling a violin. Cf. *SARINDA.

1851 *Illustr. Catal. Gt. Exhib.* IV. 913/2 Sarungee and bow, or Hindoostanee fiddle. **1886** GONDAL *Jrnl. of Visit to Eng. in 1883* 155 Those niceties of sweet sounds which a *sitar* or a *sárangi* can alone give. **1891** C. R. DAY *Mus. & Mus. Instruments S. India* vi. 93 The use of Sârangi in Southern India..is rapidly being discontinued. **1921** [see *ESRAJ]. **1929** *Radio Times* 4 Jan. 38/3 The *sarangi* has a sweeter, slightly deeper tone than the violin. **1969** [see *DILRUBA]. **1980** *Early Music* July 351/2 This, as can be heard when listening to recordings of the sarangi, does not chop a note from the string but leaves a small *bruyard* to escape.

Saratoga. Delete 'prob.' in etym. and add: **1.** (Earlier examples.)

1858 *N.Y. Tribune* 26 July 3/1 The Saratoga Trunk is an article that has been a theme of story for some time. **1874** B. F. TAYLOR *World on Wheels* I. ix. 72 It is not a carpet-bag, nor a valise nor a Saratoga.

2. In Combinations: **Saratoga chips, (fried) potatoes** *U.S.*, thinly-sliced fried potato served cold, potato crisps; **Saratoga water** *U.S.*, a mineral water obtained from the springs at Saratoga.

1880 F. M. A. ROE *Army Lett.* (1909) 262 The Saratoga chips were delicate and crisp. **1947** *Reader's Digest* Feb. 95/2 She compromised on..a broiled lobster drenched with butter, Saratoga chips, and a fancy ice cream. **1973** *Daily Colonist* (Victoria, B.C.) 4 Feb. 24/2 An Indian—an American Indian—was a chef in Saratoga, N.Y., in 1853 when he had an order for French fries—sliced thin. He sliced them too thin and they came out crunchy, and for the next 40 or 50 years, they were known as Saratoga chips. **1876** M. N. HENDERSON *Cooking* 194 Nothing deteriorates more by getting cold or keeping than fried potatoes (with the exception of Saratoga fried potatoes, which are served cold). **1877** *Golden Hours* Apr. 187/2 (*heading*) Saratoga Potatoes. **1911** *Oysterman & Fisherman* Mar. 25/2 Serve with oysters... French-fried or Saratoga potatoes. **1829** *Amer. Advertiser* (Philadelphia) 29 July 3/6 (Advt.), Fresh Saratoga or Congress Spring Water. **1893** *Harper's Mag.* Jan. 323/1 In front of me was the sign: 'Saratoga water. All you wish for five cents.' **1969** R. & D. DE SOLA *Dict. Cooking* 200/1 *Saratoga water*, any of several mineral waters bottled at their source in Saratoga Springs, New York.

sarbut (sā·ɪbǒt). *local slang.* Also **sarbot, sarbutt.** [App. a proper name.] In Birmingham: a police informer. Also as *v. intr.* = INFORM *v.* 7 b.

[**1896** *Birmingham Daily Mail* 17 Apr. 2/5 'Old Sarbot' says that..the Corporation have no legal powers to superannuate them from the rates.] **1897** *Ibid.* 5 Aug. 3/1, I knew him as one of those men who were engaged by the police for the purpose of putting up robberies and then giving information about them... They are called touts, 'sarbuts' or something else. **1928** F. C. TAYLOR *Language of Lags* in *Word-Lore* Oct. 122 Should one of the fraternity turn informer, he is for ever afterwards known as a *nark*, a *sarbot*, a *copper. Ibid.* 124 May be she'll sarbot to the D. who clobbered the kids. **1969** R. BUSBY *Robbery Blue* iii. 24 Your sarbut's story wasn't good enough... We were fooled. **1976** —— *New Face in Hell* viii. 110 The hand-picked city crime squad..recruited their 'sarbuts', the city slang for informants. **1978** *Daily Mail* 25 Jan. 12/2 In Birmingham an informer..is a 'sarbutt'.

sarc (sāɪk). *rare.* Abbrev. of SARCASM. Cf. *SARKY *a.*

1926 E. WALLACE *Square Emerald* xv. 236 She always knew when her young lady was indulging in what Lucretia described as 'sarc'.

Sarcee (sā·ɪsi), *a.* and *sb.* Also † **Sursee, Sussee, Sarsee, Sarsi.** [ad. Blackfoot *saaxsiiwa*; 18th-c. forms ad. Cree *sasiw*, pl. *sasiwak.*] **A.** *adj.* Of or pertaining to the Sarcee or their language (see below). **B.** *sb.* **a.** An Athapaskan people of Alberta in Canada; a member of this people. **b.** Their language.

[**1772** M. COCKING *Jrnl.* 1 Dec. in *Trans. R. Soc. Canada* (1909) II. II. 111 There are 4 Tribes, or Nations, more, which are all Equestrians Indians, Viz... Pegonow or Muddy-water Indians & Sassewuck or Woody Country Indians.] **1790** E. UMFREVILLE *Present State of Hudson's Bay* i. 78 Those Indians from whom the Peltries are obtained are known to us by the following names, *viz.* The Ne-heth-aw-a Indians. The Assinne-poetuc Indians. The Fall Indians. The Sussee Indians [etc.]. **1801** A. MACKENZIE *Voy. from Montreal* p. lxxi, The Sarsees, who are but few in number, appear from their language, to come.. from the North-Westward, and are of the same people as the Rocky-Mountain Indians. **1820** D. W. HARMON *Jrnl. Voy. & Trav. Interior N. Amer.* 313, I have been acquainted with fifteen different tribes of Indians, which are the.. Black feet Indians, Blood Indians, Sursees [etc.]. **1904** *Jrnl. Amer. Folklore* July–Sept. 180 (*heading*) Traditions of the Sarcee Indians. *Ibid.*, The Sarcee Indians of Alberta, N.W.T., Canada, claim to have belonged at one time to the Beaver Indians. **1915** *Univ. Calif. Publ. Amer. Archaeol. & Ethnol.* XI. III. 190 The Sarsi are an Athapascan-speaking group of Indians who have been closely associated with the Northern Blackfoot of Alberta. **1919** *Anthropol. Papers Amer. Mus. Nat. Hist.* XVI. IV. 273 The text itself was dictated by Eagle-ribs..a younger son of the head chief of the Natsilt'inna, one of the four Sarsi bands. *Ibid.*, A.. running account of the sun dance was recorded as a text in Sarsi. **1921** E. SAPIR *Language* 213 The buffalo culture of the Plains (Sarcee). **1933** L. BLOOMFIELD *Language* iv. 72 The Athabascan family covers all but the coastal fringe of northwestern Canada (Chipewyan, Beaver, Dogrib, Sarsi, etc.). **1936** D. MCCOWAN *Animals Canad. Rockies* ix. 81 Amongst the Stony and Sarcee Indians there was formerly a vague superstition imposing a sort of taboo on the cougar. **1965** *Language* XLI. 171 Harry Hoijer and Janet Joël, 'Sarsi nouns'. **1965** [see *ATHAPASCAN, -PASKAN *sb.* 2]. **1973** A. H. WHITEFORD *N. Amer. Indian Arts* 92 Checks, diamonds, and terraced triangles were old patterns among the..Sarcee. **1977** T. A. SEBEOK *Native Lang. Americas* II. 316 The term Mountain..has been applied also to groups speaking Beaver, Chipewyan, Kaska, Sarsi, Slave, Tsetsaut and Yellowknife.

sarcenchyme. Add: Hence **sarcenchy·matous** *a.*

1888, 1900 [see *CHONDRENCHYMA].

sarco-. **sarcomere, sarcosome**: delete defs. and see as main entries.

sarcococca (sāɪkokǫ·ksă). [mod.L. (J. Lindley 1826, in *Bot. Reg.* XII. 1012), f. SARCO- + Gr. κόκκος seed.] A small evergreen shrub of the genus so called, belonging to the family Buxaceæ, native to India, China, and Malaysia, and bearing clusters of white, often fragrant, flowers followed by black or red berries.

1914 W. J. BEAN *Trees & Shrubs Hardy in Brit. Isles* II. 500 The hardy Sarcococcas, all Chinese, are neat and pleasing shrubs,..the flowers white, fragrant. **1972** *Country Life* 16 Mar. 624/2 Equally valuable for winter flowering are the evergreen sarcococcas.

sarcocyst (sā·ɪkosist). *Microbiology* and *Vet. Sci.* [f. SARCO- + CYST.] **a.** A cyst in muscle tissue containing spores or sporoblasts of sarcosporidia. **b.** An individual of the genus *Sarcocystis* of sarcosporidia.

1892 G. FLEMING tr. *Neumann's Treat. Parasites & Parasitic Dis.* VI. i. 662 The Sarcocysts of Miescher are very frequent—at least, in certain countries and at certain periods. **1932** GAIGER & DAVIES *Vet. Path. & Bacteriol.* xxii. 333 The sarcocyst in the muscles is the only stage of the parasite which is known. *Ibid.* 334 Sarcocysts may die *in situ.* **1938** SOUTHWELL & KIRSHNER *Guide Vet. Parasitol. & Entomol.* (ed. 2) ii. 19 Sarcocysts are included in the Sporozoa, though it is doubtful if they really belong to this class. **1970** [see *SARCOSPORIDIUM].

Hence **sarcocy·stic** *a.*

1927 [see *SARCOSPORIDIAL *a.*]. **1979** *Acta Leidensia: Scholae Medicinae Tropicae* XLVII. 46 Schizogonic and sarcocystic stages.

sarcoid, *a.* and *sb.* Add: **A.** *adj.* **2.** *Path.* Pertaining to or resembling sarcoidosis.

1935 *Proc. Soc. Exper. Biol. & Med.* XXXIII. 403 A sarcoid lesion of the skin..was removed. **1962** *Lancet* 26 May 1107/2 Sarcoid tissue obtained from a skin lesion was suspended in saline solution. **1976** EDINGTON & GILLES *Path. in Tropics* (ed. 2) xi. 522 The presence [in Crohn's disease] of a sarcoid reaction in the tissues of the bowel wall, and regional lymph nodes.

B. *sb.* **2. a.** *Path.* Sarcoidosis; also, a tumour resembling a sarcoma.

1899 C. BOECK in *Jrnl. Cutaneous & Genito-Urin. Dis.* XVII. 543 (*heading*) Multiple benign sarkoid of skin. **1941** *Arch Ophthalm.* XXVI. 358 The term sarcoid was adopted by Boeck in 1899 for lesions simulating sarcomas and leukemic conditions of the skin. He believed the condition to be one only of the skin. Later he recognized his mistake in the term and changed it to multiple miliary lupoid. **1963** JUBB & KENNEDY *Path. Domestic Animals* II. x. 565/1 Sarcoids are usually multiple and occur most frequently about the base of the ear, on the neck, and on the lower limbs. **1977** *Proc. R. Soc. Med.* LXX. 484/2 Greenberg *et al.* (1964) found parotitis in 23 (6%) of 388 patients suffering from sarcoid. **1978** *Price's Textbk. Pract. Med.* (ed. 12) III. 275/2 Subcutaneous telangiectases or haemangiomata become visible in some patients. The latter may develop into small tumours or 'sarcoids' up to a centimetre across.

b. *Comb.*, as *sarcoid-like* adj.

1943 *Arch. Dermatol. & Syphilol.* XLVII. 62 Sarcoid-like lesions have..been produced by the injection of bovine tubercle bacilli of low virulence into rabbits. **1968** A. ROOK et al. *Textbk. Dermatol.* xxvi. 937/1 Beryllium causes either a local or a systemic sarcoid-like reaction.

sarcoidal (sāɪkoi·dăl), *a. Path.* [f. prec. + -AL.] = *SARCOID *a.* 2.

1961 *Jrnl. Amer. Med. Assoc.* 4 Nov. 476/1 Employing a suspension of human sarcoidal tissue as test material. **1962** *Lancet* 26 May 1108/1 If sarcoidosis were produced by many agents..then a series of sarcoidal test-tissues would presumably be necessary.

sarcoidosis (sāɪkoidōᵘ·sis). *Path.* [f. as prec. + -OSIS.] A chronic disease characterized by the widespread appearance of sarcoid granulomata derived from the reticuloendothelial system.

1936 *New England Jrnl. Med.* CCXIV. 346 (*heading*) Hutchinson–Boeck's disease (generalized 'sarcoidosis'). **1955** *Lancet* 26 Mar. 640/2 There is no general agreement on whether any form of treatment is effective for pulmonary sarcoidosis. **1975** *Guardian* 25 Feb. 6/1 He recorded a verdict of death from natural causes after hearing that Mrs Rogers died of acute adrenal insufficiency due to sarcoidosis, a chronic illness which can affect all the organs of the body.

sarcolemma. Add: Hence **sarcole·mmal** *a.*, of or pertaining to the sarcolemma.

1912 *Brain* XXXIV. 370 In some places the sarcolemmal nuclei were much increased in numbers. **1974** *Nature* 1 Mar. 69/2 Our data favour the hypothesis that sarcolemmal sensitivity to ACh is regulated by a mechanism located in or near the muscle membrane.

sarcoma. Add: Also with pl. **sarcomas.**

1. b. Now applied to almost any malignant tumour not derived from epithelial tissue. (Further examples.)

1971 *Nature* 21 May 147/2 In human populations only about 10 per cent of cancers are sarcomas; the other 90 per cent are carcinomas—cancers of tissues of epithelial, not mesodermal, origin. **1975** *Sci. Amer.* Nov. 64/2 The cancers are divided into three broad groups. The carcinomas arise in the epithelia... The much rarer sarcomas arise in supporting structures such as fibrous tissues and blood vessels. The leukemias and lymphomas arise in the blood-forming cells.

sarcomere (sā·ɪkomiᵊɪ). *Anat.* [f. SARCO- + *-MERE.] A unit of a myofibril in striated muscle, consisting of a dark band and the nearer half of each adjacent pale band.

1891 E. A. SCHAFER in *Proc. R. Soc.* XLIX. 281 The segment of a sarcostyle comprised between two transverse membranes may be termed 'muscle-segment' or 'sarcomere'. **1897** *Jrnl. Anat. & Physiol.* XXXI. 336 [Schäfer] regards the sarcostyle as divided at regular intervals by Krause's transverse membranes into 'muscle segments' or 'sarcomeres', which are only new names for Krause's 'muscle caskets'. **1930** W. BLOOM *Maximow's Text-bk. Histol.* viii. 205 In the invertebrates (arthropods)..sarcomeres as long as 17μ have been found. The length of the sarcomere and the physiologic peculiarities of the muscle have not..been correlated. **1970** T. S. & C. R. LEESON *Histol.* (ed. 2) ix. 162/2 It is customary to consider the muscle fibril as composed of structural units. Each unit extends between adjacent Z lines and is termed a sarcomere. **1980** CRAWFORD & JAMES in R. Owen et al. *Sci. Foundations Orthopaedics & Traumatol.* x. 68/1 Changes in sarcomere length occur by movement of the thick myosin filaments along the thin actin filaments, their own length remaining unchanged.

sarcomic (saɪkōᵘ·mik), *a. nonce-wd.* [f. SARCOMA + -IC.] = CANCEROUS *a. fig.*

1958 J. STEINBECK *Once there was War* p. xx, We are poisoned in our souls by fear, faceless, stupid sarcomic terror.

sarcophagize, *v.* Add: Also *intr.* for *pass.* (*nonce-use*).

1953 E. SITWELL *Gardeners & Astronomers* 9 The hue of honey sarcophagising or of sard.

sarcoplasmic (sāɪkoplæ·zmik), *a. Anat.* [f. *sarcoplasm* s.v. SARCO- + -IC.] Of, pertaining to, or containing sarcoplasm; *sarcoplasmic reticulum,* the characteristic endoplasmic reticulum of striated muscle.

1891 *Internat. Monatsschrift für Anat. u. Physiol.* VIII. 229 The optical effect produced by the enlarging sarcoplasmic accumulations will involve more and more of the segment. **1902** *Encycl. Brit.* XXXI. 733/1 The muscle-cells of the ventricles are thicker, less sarcoplasmic, and more clearly striated than the auricular muscle. **1948** *Jrnl. Neurol., Neurosurg. & Psychol.* XI. 78/1 The dense areas showed an accumulation of pale oval sarcoplasmic nuclei. **1953** BENNETT & PORTER in *Amer. Jrnl. Anat.* XCIII. 69 The disposition of the larger masses of this sarcoplasmic reticulum is interpreted as evidencing an arrangement entirely analogous to the cross-fiber reticulum of Thin (1876), Melland (1885)..and others. **1970** T. S. & C. R. LEESON *Histol.* (ed. 2) ix. 164/2 Sarcoplasmic reticulum corresponds to the endoplasmic reticulum of other cell types, but its membranes are not associated with ribosomes. The sarcoplasmic reticulum comprises an extensive, continuous system of membrane-limited sarcotubules enclosing each myofibril as in a net. **1970** *Nature* 31 Oct. 417/2 The group of muscle fibres in the centre of the section show sarcoplasmic basophilia.

sarcopterygian (sā:ɪkǫptəri·dʒiăn). [f. mod. L. name of subclass *Sarcopterygii* (A. S. Romer 1955, in *Nature* 16 July 126/2) + -AN: see SARCO- and PTERYGO-.] A fossil or living fish belonging to the subclass Sarcopterygii, distinguished by fleshy fins.

1966 A. S. ROMER *Vertebr. Paleontol.* (ed. 3) v. 71/2 Many later sarcopterygians have simple bony scales. **1974** D. & M. WEBSTER *Compar. Vertebr. Morphol.* v. 96 Among the sarcopterygians there are striking differences between the pectoral girdles of the living lungfish, Dipnoi, and the living coelacanth, *Latimeria.*

sarcoptid (sāɪkǫ·ptid). [f. mod.L. family name *Sarcoptidæ*, f. generic name *Sarcoptes* (P. A. Latreille *Hist. Nat. Crustacées & Insectes* (1802) III. 67): see SARCOPTES and -ID[3].] An itch or mange mite of the family Sarcoptidæ. Also *attrib.* or as *adj.* Also **sa·rcopt** in the same sense.

1870 A. S. PACKARD *Guide Study of Insects* 666 Various Sarcoptids occur on birds. **1892** G. FLEMING tr. *Neumann's Treat. Parasites Domestic Animals* I. v. 121 The sarcopt of scabies..has the body slightly oval. **1932** L. VAN ES *Princ. Animal Hygiene* xliii. 736 Sarcopts are the most common cause of mange in dogs. **1962** GORDON & LAVOIPIERRE *Entomol. for Students of Med.* xliv. 264 (*caption*) The life-cycle of a sarcoptid mite.

sarcosome (sā·ɪkosōᵘm). *Biol.* [ad. G. *sarcosom* (G. Retzius 1890, in *Biol. Untersuchungen* Neue Folge I. 76): see SARCO- and *-SOME[4].] A large mitochondrion found in striated muscle.

1899 [in Dict. s.v. SARCO-]. **1912** *Amer. Jrnl. Anat.* XIV. 5 The 'exoplasmic granules' (J granules and Q granules) and the 'endoplasmic granules' of Holmgren correspond to the 'Sarcosomes' of Retzius which in turn correspond to Kölliker's true interstitial granules. It is possible that Retzius and Holmgren may have occasionally confused fat droplets with sarcosomes. **1919** *Anat. Rec.* XVI. 217 The wing muscle of the mantis furnishes an exceptionally favorable material for the investigation of the interfibrillar sarcoplasmic granules, or 'sarcosomes', characteristic of insect wing muscle. **1956** *Physiol. Rev.* XXXVI. 3 It is proposed in this article..to use the term sarcosome in its original sense as a general term to describe the lipoprotein granules which lie between the myofibrils and which can be seen with the light microscope. **1970** *Sci. Amer.* Feb. 101 The specialized mitochondria of the myocardial cell are unusually large; they are called sarcosomes.

sarcosporidiosis (sā:ɪkosporidiōᵘ·sis). *Vet. Sci.* [f. next + -OSIS.] Infection with, or a disease caused by, sarcosporidia.

1893 *Bull. Bureau Animal Industry, U.S. Dept. Agric.* No. 3. 80 Barrows' description renders it almost certain that this was a case of sarcosporidiosis. **1953** R. P. HALL *Protozool.* vi. 326 Sarcosporidiosis of man is apparently rare, although cases are reported occasionally. **1978** AYERS & JONES in K. Benirschke et al. *Path. Laboratory Animals* I. i. 14/1 Sarcosporidiosis is probably the most common parasitic disease of the heart seen in laboratory animals.

sarcosporidium (sā:ɪkospori·diǒm). *Microbiology* and *Vet. Sci.* Also **Sarco-.** Pl. **-sporidia.** [mod.L., ad. F. *sarcosporidie* (G. Balbiani, in *Jrnl. de Micrographie* (1882) VI. 262, (1883) VII. 87): see SARCO- and SPORIDIUM.] A spore-forming protozoan of the genus *Sarcocystis* that is a common parasite in the muscle tissue of many vertebrates, esp. domestic and laboratory mammals. Usu. *pl.*

1891 *Jrnl. Compar. Med. & Vet. Arch.* XII. 693 The small cysts found in the muscular fibres of various animals and known as Sarcosporidia. *Ibid.*, The negative results obtained..from feeding meat infected with sarcosporidia to various animals. **1927** *Indian Jrnl. Med. Res.* XV. 142 It is suggested that this parasite is an undescribed species of sarco-sporidium infecting the human host. **1930** *Jrnl. Parasitol.* XVI. 111 From the economic standpoint Sarcosporidia are of chief interest to the veterinarian rather than the physician. **1970** JUBB & KENNEDY *Path. Domestic Animals* (ed. 2) I. ii. 121/2 Of those parasites with an affinity for muscle, the ubiquitous Sarcosporidia are the most common. The sarcocysts may be found in the Purkinje cells as well as in the myocardial fibres and normally appear to be of little detriment.

Hence **sa:rcospori·dial** *a.*, **-spori·dian** *a.* and *sb.*

1903 E. A MINCHIN in E. R. Lankester *Treat. Zool.* I. II. 301 The dangerous effects of the Sarcosporidian parasites. **1913** *Proc. Cambr. Philos. Soc.* XVII. 221 (*heading*) *Sarcocystis colii*, n. sp., a Sarcosporidian occurring in the red-faced African mouse bird. **1924** HEGNER & TALIAFERRO *Human Protozool.* xi. 372 (*caption*) Sarcosporidian spores. **1927** *Indian Jrnl. Med. Res.* XV. 142 Sarcocystic (Sarcosporidial) infection is common in cattle. **1949** C. A. HOARE *Handbk. Med. Protozool.* xiv. 271 Sarcosporidial infection has been reported in the muscles of the heart, larynx, tongue and the extremities. **1957** SMITH & JONES *Vet. Path.* xx. 704 Myocarditis... Sarcosporidial.. cysts are common in the heart muscle.

Sard, *a.* and *sb.*[2] Add: Also ‖ **Sarde. B.** *sb.* **1.** (Further examples.)

1845 *Encycl. Metrop.* XXIV. 318/2 The Sards are greatly attached to the pleasures of the table. **1889** C. EDWARDES *Sardinia & Sardes* vi. 147 The foreman was a Sarde of an advanced type. **1932** [see *SARDINIAN *sb.* 2]. **1968** *Listener* 29 Feb. 267/1 No Sard will betray another... There's the unwritten law of *omerta*, of silence.

2. = *SARDINIAN *sb.* 2.

1885 [see *LOGUDORO]. **1889** C. EDWARDES *Sardinia & Sardes* iii. 59 Modern Sarde is what Sardinia's conquerors made it—a language much more nearly kin to Latin than Italian. **1957** *Whitaker's Almanack 1958* 899/1 Sard, the dialect of Sardinia, is accorded by some authorities the status of a distinct Romance language. **1975** *Times Lit. Suppl.* 25 Apr. 452/4 Gramsci was a humane and intelligent man, but in no sense an 'authority' on anything except Mussolini's prisons and Sard.

‖ **sardana** (saɪdā·nă). [Sp.] A popular Catalan dance performed to pipes and drum.

1922 *Glasgow Herald* 28 Apr. 8 The music played by amateur orchestras, even in small villages, to the Sardanas or national Catalonian dances is delightful. **1934** C. LAMBERT *Music Ho!* III. 172 The Catalan sardanas..have added to their primitive basis sophisticated and foreign elements. **1953** *Observer* 13 Sept. 9/2 To describe the new local dance, the Sardana from Spanish Catalonia, Mr. O'Brian has essayed a poetic prose. **1965** *Listener* 10 June 877/3 Spain retains small keyless shawms..and also has modern keyed forms in the sardana bands of Catalonia. **1976** D. MUNROW *Instruments Middle Ages & Renaissance* 40/3 The Catalonian *coblas* of north-eastern Spain which play for the *sardana*, the 'national' dance of the region.

sardine[2]. Add: **1. b.** (Earlier example.)

1870 L. M. ALCOTT *Old-Fashioned Girl* xiii. 266 We've got sardines, crackers, and cheese.

d. In colloq. phr. *to be packed (in) like sardines*: to be crowded or confined tightly together, as sardines in a tin.

1911 W. OWEN *Let.* 12 Sept. (1967) 80 The entrance hall.. where for half an hour the boys stand waiting packed like sardines. **1922** *Dialect Notes* V. 172 We were packed in there like sardines in a box. **1974** *Daily Mirror* 11 Nov. 4/3 Lodgers at a lorry drivers' digs hit by a horror blaze were 'packed in like sardines', it was claimed yesterday.

e. *pl.* (const. as *sing.*) A party game of hide-and-seek, in which each seeker joins the hider upon discovery until one seeker remains. Also *sardines-in-the* (also *a*)-*box* (U.S.).

1924 in Mendel & Meynell *Weekend Bk.* 241 *Sardines* is gaudier still. Only one player hides, all the others seek; the first to find him hides with him, the next..squashes in alongside,..till everybody's hiding in the same spot but one Seeker. **1925** F. SCOTT FITZGERALD *Great Gatsby* v. 82 'Hide-and-go-seek' or 'sardines-in-the-box' with all the house thrown open to the game. **1935** N. MARSH *Enter Murderer* xx. 242 Give us all the light in the house. I refuse to play sardines with Mr. Hickson. **1959** J. BYROM *Take only as Directed* xiii. 147, I remembered the big linen-chest... I had once hidden there playing Sardines. **1960** N. HALE *New England Grilhood* 113 We used to play hide-and-go-seek, and a game called sardines-in-a-box. **1962** B. COBB *Murder: Men Only* iv. 37 That game—'Sardines,' isn't it?—in which men hide with girls in cupboards. **1974** N. FREELING *Dressing in Diamond* 116 Tomorrow is a holiday... So we weekend... And play sardines. **1980** G. M. FRASER *Mr. American* xiii. 259 The festivities were strictly of the nursery variety..musical chairs, 'sardines', and hide-and-seek.

2. *sardine boat, fishing, fleet; sardine can, sandwich, tin; sardine-packed* adj.; **sardine box** (earlier example).

1927 L. RICHARDSON *Brittany & Loire* 128 The early type of sardine boat had no overhang—a long, straight keel, straight stem. **1976** F. GREENLAND *Misericordia Drop* II. xiii. 161 A converted sardine-boat. **1855** *Harvard Mag.* I. 266 O ghosts of innumerous sardine-boxes, and emptied cracker-kegs. **1977** *Modern Railways* Dec. 484/1 The first run was with an eight-car formation of this stock forming the 18.00 down Clacton packed to sardine-can condition. **1979** P. DRISCOLL *Pangolin* ii. 22 The tram..was more crowded than usual..a clanking sardine can. **1775** J. SCHAW *Jrnl. Lady of Quality* (1921) iv. 220 Above a hundred boats engaged in the sardine fishing. **1939** H. M. MINER *St. Denis* ii. 23 There is still some commercial eel- and sardine-fishing, but this has declined. **1942** 'A. BRIDGE' *Frontier Passage* iv. 65 The many-coloured dancing shapes

SARDINE

of the sardine-fleet. **1917** Wyndham Lewis *Let.* Sept. (1963) 92, I am now absolutely sardine-packed with the quintessence of the prosperous slums of a Protestant country. **1954** B. Malamud in *Partisan Rev.* Nov.–Dec. 587 Leo fixed tea and a sardine sandwich. **1978** F. Weldon *Praxis* vii. 42 She had lit the fire and made sardine sandwiches. **1890** W. Booth *In Darkest Eng.* II. ii. 121 Most of the toys which are sold in France on New Year's Day are almost entirely made of sardine tins. **1933** M. Allingham *Sweet Danger* xv. 187 'Leave that smelly little sardine tin [*sc.* a motor car] alone.'... 'The exhaust smells a little, but that's nothing.' **1973** 'A. Hall' *Tango Briefing* x. 124 A rip-string and I pulled it, opening the polyester like a sardine-tin.

sardine (sɑ̄ɹdī·n), *v. colloq.* (orig. *U.S.*). [f. Sardine[1].] *trans.* To pack closely, as sardines in a tin; to crowd, cram, press tightly.

1895 W. C. Gore in *Inlander* Dec. 114 *Sardine*,..to pack closely, side by side. 'We sardined ourselves in front of the Law Building and howled.' **1896** *Advance* 24 Dec. 916/2 There are 350 people outside.., and in some way we are going to sardine them in. **1940** H. Walpole *Roman Fountain* vii. 124 We were pressed back and sardined together. **1953** Dylan Thomas *Under Milk Wood* (1954) 69 Mrs Probert..is the one love of his sea-life that was sardined with women. **1968** *N.Y. Times* 22 Apr. 36 Hundreds of thousands of people..will be sardined into the famous amusement park. **1977** *New Yorker* 11 July 79/1 Once sardined in place, they are subject to terrifying hazards in case of fire.

Sardinian, *a.* and *sb.* Add: **A.** *adj.* **c.** Designating a Romance language (or group of dialects) spoken by the Sardinians.

1835 G. C. Lewis *Origin & Formation Romance Lang.* i. 48 Niebuhr..says that 'specimens of the Sardinian language from the civilized districts exhibit peculiarities which are more than varieties of dialect'. **1960** W. D. Elcock *Romance Lang.* v. 474 In..perceiving that the 'outlandish' character of Sardinian speech lay in its approximation to Latin the poet-philologist [Dante] had almost divined the truth concerning the origin of the Romance languages. **1974** *Encycl. Brit. Macropædia* XV. 1029/1 It is sometimes said that these last two [*sc.* Sassarian and Gallurian] are not Sardinian dialects but rather Corsican.

d. *Sardinian warbler*, a black, brown, and white warbler, *Sylvia melanocephala*, found in the Mediterranean region.

1909 C. Whymper *Egyptian Birds* 209 Sardinian Warbler ..Rare. **1954** D. A. Bannerman *Birds Brit. Isles* III. 150 In Sicily..the Sardinian warbler is the commonest warbler. **1971** *Country Life* 1 July 27/3 Our most familiar sounds by day were..the chatter of Sardinian warblers among the mastic bushes.

B. *sb.* **2.** A Romance language (or group of dialects) spoken in Sardinia.

1813 *Q. Rev.* Oct. 259 Bolognese.., Sicilian.., Sardinian. **1841** *Penny Cycl.* XX. 427/1 A book was published at Cagliari, in both Sardinian and Italian, called 'Moriografia Sarda'. **1894** W. M. Lindsay *Latin Lang.* ii. 34 Short ŭ and ŏ of Latin are distinguished not only in Sardinian.., but also in Roumanian and in the Latin element of the Albanian language. **1932** G. F.-H. Berkeley *Italy in Making* I. iv. 52 French was permitted for the Savoyards and Valdostani, Genoese for the Ligurians, and Sardinian for the Sards. **1965** W. S. Allen *Vox Latina* i. 25 At what period such a [phonetic] change took place it is impossible to say, but Sardinian suggests that it was very late. **1974** *Encycl. Brit. Macropædia* XV. 1026/1 On linguistic grounds Sardinian (not the language of an independent nation since the 14th century) and Occitan (the medieval Provençal) are usually regarded as languages rather than dialects. *Ibid.* 1029/1 The first documents in Sardinian are legal contracts dating from about 1080.

sardonic, *a.* Add: **c.** *Comb.*, as *sardonic-looking* adj.

1921 D. H. Lawrence *Tortoises* 29 She is..A little sardonic-looking, as if domesticity had driven her to it.

Hence **sardo·nicism**, the quality or state of being sardonic; an instance of this; a sardonic remark.

1928 *Daily Express* 6 Jan. 8/3 The old Spartan régime has gone, but there is a relentlessness about the public school system that engenders secret terrors at every turn. It may be the fear of ridicule, or the sardonicisms of a satiric master, or one of a dozen things. **1930** W. de la Mare *On Edge* 197 A corrosive sardonicism had come into her voice. **1940** W. Faulkner *Hamlet* II. i. 100 He would speculate now and then with cold sardonicism. **1964** *Listener* 29 Oct. 667/2 Because familiarity with the role has made Sean Connery feel able to play Bond more relaxedly, an agreeable sardonicism has been added to the earlier deliberately overdone Superman masculinity.

Sardoodledom (saɹdū·d'ldəm). [f. blend of the name of Victorien *Sardou* (1831–1908), French dramatist + Doodle *sb.* + -dom.] A fanciful word used to describe well-wrought, but trivial or morally objectionable, plays considered collectively; the characteristic milieu in which such work is admired.

1895 G. B. Shaw in *Sat. Rev.* 1 June 725/2 (*heading*) Sardoodledom. **1897** —— in *Ibid.* 17 Apr. 410/2 It is rather a nice point whether Miss Ellen Terry should be forgiven for sailing the Lyceum ship into the shallows of Sardoodledom for the sake of Madame Sans-Gêne. **1931** *Times Lit. Suppl.* 1 Jan. 2/1 The 'cup-and-saucer' comedy of Robertson and what Mr. Shaw christened 'Sardoodledom' ..opened a new phase of theatrical history. **1959** *Listener* 30 July 186/3 Sardoodledom is not forgotten. **1960** *Times* 15 Jan. 16/1 We do not want to try to rebut Shaw's criticism of 'Sardoodledom'.

saree. Restrict ‖ to sense b and add: The form *sari* is now common, esp. in writings outside India. (Further examples.)

1908 [see *choli]. **1930** *Aberdeen Press & Jrnl.* 6 May 6, I was struck yesterday by the international character of London. Indians in gay saris, a group of Japanese..all passed me. **1960** [see *choli]. **1969** *Hindu* 3 Aug. 12/4 A petition on behalf of over 4,000 handloom weavers of Salem and other parts of Tamil Nadu was presented to the Rajya Sabha on Tuesday urging the immediate implementation of the order reserving manufacture of coloured sarees for the handloom sector. **1969** *Femina* (Bombay) 26 Dec. 8/4 The piece-de-resistance of the show, a smart zip-up sari, found great favour with the foreigners in the audience. **1971** [see *Manipuri *a.* and *sb.*]. **1971** R. Russell tr. *A. Ahmad's Shore & Wave* i. 14 Love for the dark-skinned maidservants in their grubby sarees. **1976** *Leicester Trader* 24 Nov. 22/1 (Advt.), Also we clean..dresses, bedspreads, sarees, etc.

2. *attrib.* and *Comb.*

1936 J. Flanner in *New Yorker* 22 Aug. 65/1 The Hindu hockey team, and its handsome, sari-clad womenfolk. **1955** R. P. Jhabvala *To whom she Will* xiv. 167 Radha.. would walk into a sari-shop, because she was so fond of looking at nice silks. **1968** *Guardian* 19 Sept. 9/1 Embroidered sari silks for evening dresses. **1978** 'M. M. Kaye' *Far Pavilions* xxvi. 392 There was little chance of seeing her..save as a sari-shrouded figure on the occasion of her marriage.

Hence **sa·reed** *a.*, wearing a saree.

1958 E. A. Robertson *Justice of Heart* xv. 212 The sari'd head wagged slower. **1975** O. Sela *Bengali Inheritance* vii. 53 Oval-eyed, smiling sareed girl with skin like brown silk.

Sargasso. Add: **a.** Also *fig.*, esp. in sense 'a confused or stagnant mass'.

1934 Dylan Thomas *18 Poems* 11 The dry Sargasso of the tomb Gives up its dead to such a working sea. **1968** A. C. Clarke *2001: Space Odyssey* xlii. 206 It had swept him across the Galaxy, and dumped him..in this celestial Sargasso. **1976** *Globe & Mail* (Toronto) 11 Feb. 18/5, I started the week with a careful schedule and ended in bed with flu, lost in a Sargasso of phone-ins, pop, news, avant-garde operas and the reminiscences of David Niven. **1977** *Mystery Writers' Choice* 62, I waited..adrift in a sargasso of conflicting feelings.

b. *Sargasso Sea*: also *fig.*

1961 P. Solomon in B. E. Flaherty *Psychophysiol. Aspects Space Flight* iii. 275 Deprived of sensory input, the mind is cut adrift and regresses inexorably into that Sargasso Sea of the primary process, where time disappears, ..where' vivid, multicolored hallucinations swirl and befuddle the senses. **1966** C. H. Hapgood *Maps of Anc. Sea Kings* ii. 25 We found ourselves in a veritable Sargasso Sea of uncertainties. **1966** J. Rhys (*title*) Wide Sargasso Sea. **1979** P. O'Connor *Into Strong City* ..., I was experiencing a severe London pea-soup fog... I swam through an impenetrable ochre sargasso sea.

sargassum (sɑɹgæ·səm). [mod.L. (G. E. Rumpf *Herbarium Amboinense* (1755) VI. tab. 76): see Sargasso.] A large floating seaweed of the genus so called, found in masses in warm or temperate seas. Cf. Gulf-weed, Sargasso.

1905, etc. [see sense b below]. **1951** G. M. Smith *Man. Phycol.* vii. 19 *Sargassum*..forms immense floating masses in the Sargasso Sea, between the West Indies and the coast of North Africa. **1969** J. M. Kingsbury *Seaweeds of Cape Cod* 118 The berry-like bladders..serve to identify *Sargassum* or 'gulfweed' as it is sometimes called.

b. sargassum angler, fish, a small toadfish, *Pterophryne histrio*, which lives in clusters of sargassum; sargassum weed = sense a above.

1961 Sargassum weed [see *sargassum weed* below]. **1905** D. S. Jordan *Guide to Study of Fishes* II. xxxi. 549 (*caption*) Sargassum-fish. **1928** W. Beebe *Beneath Tropic Seas* iii. 26 An unexpected performance was suddenly staged in the jar of sargassum fish. **1962** K. F. Lagler et al. *Ichthyol.* iv. 119 The sargassum fish (*Pterophryne*) and the alga-resembling sea dragons are the most frequently cited examples of the extension of the skin into flaps. **1928** W. Beebe *Beneath Tropic Seas* iii. 24 On several days great masses of sargassum weed drifted into the bay. **1962** E. S. Herald *Living Fishes of World* 189/1 The sargassum angler can change its shades in colour pattern to match its background, but seems to be limited to those shades found in the sargassum weed.

sarge² (sɑ̄ɹdʒ). orig. *U.S.* Also **Sarg(e)**, **serg.** Colloq. abbrev. of Sergeant *sb.* (Freq. used as a term of address.) **a.** *Mil.* = Sergeant *sb.* 9 a. Also *Comb.*

1867 W. L. Goss *Soldier's Story* 98 You look hungry too, Sarg. *Ibid.* 128 Sarge, the Colonel has got his wind up, and you'll be sent into the stockade. **1913** *Sat. Even. Post* 5 Feb. 6 'Sergeant Tanner?' asked the bartender incredulously. 'The sarge,' replied Kennedy with some satisfaction. **1919** W. H. Downing *Digger Dial.* 43 *Sarge* (n.), Sergeant. **1929** F. A. Pottle *Stretchers* (1930) 238 But sarge, I've been out since five without a bite. **1940** Partridge *S. P. E. Tract* LV. 191 The Regular Army's pre-1914 slang.. consisted mainly of words from Hindustani and Arabic..and abbreviations (..*sarge* 'sergeant'). **1958** M. K. Joseph *I'll soldier no More* ix. 166 Hey, sarge, there's another bugger out in the middle of the field. **1973** *Jewish Chron.* 19 Jan. 14/2 We are never allowed to forget the grim and earnest purpose behind the farcical square-bashing and sarge-baiting.

b. = Sergeant *sb.* 10.

1926 *Scribner's Mag.* Aug. 193/2 'Quiet, serg,' volunteered the desk man. 'Too quiet,' corrected the sergeant. **1934** J. M. Cain *Postman always rings Twice* x. 115 Just a few minutes, sarge. **1938** G. Greene *Brighton Rock* v. vi. 254 'You aren't pulling my leg, are you?' the sergeant said. 'Not this time, sarge.' **1959** M. Gilbert *Blood & Judgement* i. 15 Garn, Sarge, this isn't Guy Fawkes, it's Father Christmas. **1977** 'J. Bell' *Such a Nice Client* xi. 96 'I want you over here,' said Sergeant Thomas.. 'Right, sarge,' answered the constable.

Sargonid (sɑ̄ɹgǫnid), *sb.* and *a.* Also **-ide.** [f. the Akkadian royal name *Sargon* + -id, after Seleucid, etc.: cf. -id³.] **A.** *sb.* A member of the Assyrian dynasty founded by Sargon II (ruled 722–705 B.C.), which remained in power until the fall of Assyria in 607 B.C.

1887 A. A. Ragozin *Assyria* ix. 295 (*heading*) The Sargonides. **1913** H. R. Hall *Anc. Hist. Near East* x. 516 Sennacherib was the first Sargonid who no longer went forth to war himself.

B. *adj.* **1.** Of, pertaining to, or designating this dynasty.

1913 H. R. Hall *Anc. Hist. Near East* x. 517 (*heading*) Sargonide dynasty. **1925** *Camb. Anc. Hist.* III. ii. 43 The days of the Sargonid dynasty. *Ibid.* 45 A fanciful genealogy of the Sargonid house.

Sarik, var. *Saryk.

Sarin (sɑ̄·rin). Also **sarin.** [Ger., of unknown origin.] The name of an odourless organophosphorus nerve gas.

1951 *Acta Physiol. Scandinavica* Suppl. No. 90. 106 (*table*) Isopropoxy-methyl-phosphoryl-fluoride (sarin). **1967** *New Scientist* 26 Jan. 196/3 At Newport, Indiana, there is a plant producing sarin (GB) to fill into rockets, land mines and artillery shells. **1968** *Observer* 15 June 9/1 By the end of the war, three of these gases—compounds of phosphorus—were known in Germany: tabun, sarin and soman. **1978** A. Melville-Ross *Blindfold* ii. 18 That stuff 'Sarin'. Nerve gas containing fluorine and phosphorus... Absorption through the skin..means paralysis and death.

‖ **sarinda** (səri·ndă). Also **sarindah.** [Hind., dial. var. of *sārangī* *sarangi.] An Indian stringed musical instrument played with a bow (see quots.).

1851 *Illustr. Catal. Gt. Exhib.* iv. 913/2 Musical Instruments..Sarindah, or fiddle, from Moorshedabad. **1872** *Catal. Special Exhib. Anc. Musical Instruments* (S. Kensington Mus.) 35 Sarinda. A kind of Violin. **1921** H. A. Popley *Music of India* vii. 109 The Sārindā is another variety of the sārangī, peculiar to Bengal. **1944** W. Apel *Harvard Dict. Music* 601/1 In India the fiddles called *sarinda* have truly fantastic shapes such as only the Indian fancy could have produced. **1964** S. Marcuse *Musical Instruments* 456/2 *Sārindā*, folk sārangī of India, with thin wooden body of irregular shape, and skin belly covering only the lower part of the body, short neck, 3 gut or hair strings that are bowed. **1966** [see *dilruba]. **1977** G. Wells in *Early Music* Apr. 250/2 A sarinda from Northern India.

sark, *v.* **2.** (Later example.) Also, to cover with sarking felt or boards.

1961 *Guardian* 21 Feb. 5/7 (Advt.), Other kinds of Sisalkraft will insulate buildings, cure concrete, sark roofs. **1977** *Belfast Tel.* 19 Jan. 24/2 (Advt.), Roofspace partly floored, sarked and felted.

Sarkese (sɑ̄ɹkī·z), *sb.* (and *a.*) [f. the place-name *Sark* (see below) + -ese.] **A.** *sb.* **a.** *collect.* Also **Sarkees.** The inhabitants of the Channel Island of Sark. **b.** The language of Sark, a variety of Norman French. **B.** *adj.* Of or pertaining to Sark.

1840 Q. W. Aves *Sark Guide* 78 Most of the Sarkese now manufacture their own lobster pots... The Sarkese certainly love money even to a fault. **1882** D. H. *Channel Islands* 73 The kind-hearted Sarkese thought the sentence too severe... The good Sarkese women kept her company. **1928** L. E. Hall H. *J. V. Cachemaille's Island of Sark* 105 This..harbour..has a history of its own, interesting to the Sarkese. **1950** L. H. [illegible] Sarkese. **1957** *Sunday Times* 10 Feb. 3/8 The debates in the Parliament of Sark are conducted in Sarkese which is apparently a slightly modernized version of Ancient Norman. **1958** J. W. Day *Lady Houston* ix. 127 No other race of people but the Sarkese need apply. **1978** *Times Lit. Suppl.* 26 May 572/5 Some of the Sarkese did sympathise with her fraternization with the Germans.

sarking, *vbl. sb.* Restrict *Sc.* and *north.* to sense 2 and add: **1.** (Further examples.) Also, the action of Sark *v.* 2 (in Dict. and Suppl.).

1926–7 *Army & Navy Stores Catal.* 308/2 Ruberoid Sarking Felt. Rolls contain 24 sq. yds. **1957** *Archit. Rev.* CXXI. 354/4 In addition to these they will show their extensive range of general purpose papers for sarking and damp-proofing. **1958** *N.Z. Timber Jrnl.* July 73/1 *Sarking boards*, close boarding to carry roof tiles, shingles, or slates. Thin boards used as a lining. *Ibid.*, *Sarking felt*, a bituminous underlining placed beneath slates or tiles.

sarky (sɑ̄·ɹki), *a. colloq.* [f. abbrev. of Sarcastic *a.* + -y¹: cf. *sarc.] Sarcastic. (Widely used amongst schoolchildren.) Also in *Comb.* Hence **sa·rkily** *adv.*; **sa·rkiness.**

1912 D. H. Lawrence *Let.* 1 Feb. (1962) I. 97 Why are you so sarky? **1924** H. de Sélincourt *Cricket Match* iii. 46 He says it sarky-like and sneering. **1930** *Diary of Public School Girl* 76 Made some currant buns. Bob very sarky about them. **1949** E. Taylor *Wreath of Roses* vii. 107 She's funny with Ernie, very sarky sometimes the way she answers him back. **1958** C. Watson *Coffin, scarcely Used* iii. 25 The bland and (he had heard) 'sarky' inspector. **1965** *New Statesman* 30 July 163/1 John's saturnine profile, George's sarkiness, Paul's ageing chorister naughtiness and Ringo's deadpan outsider appeal are well brought out by David Watkin's restless camera. **1967** M. Waddell *Otley Pursued* xv. 139 'Eating it would have been bad for your digestion, I suppose,' she said sarkily. **1977** 'J. Bell' *Such Nice Client* xvi. 161 You needn't be sarky, I've never refused you.

Sar-Major (sɑ̄ɹmē¹·dʒəɹ). Also **Sarmajor**, etc. and with small initial. Mil. colloq. abbrev. of Sergeant-major 2. (Freq. used as a term of address.) Cf. *Sarge²; *Sarn't.

1919 W. H. Downing *Digger Dial.* 43 *Sarmajor*, Sergeant-major. **1958** P. Scott *Mark of Warrior* i. 26 Thank you, Sar-Major. I congratulate you on your staff. **1969** D. Clark *Nobody's Perfect* iii. 109 There's some..think because a man's been a sar' major he'll want to turn the place into a training depot. **1974** P. McCutchan *Call for Simon Shard* i. 6 What's up, Sar-Major?

‖ **sarmale** (saɹmā·le), *sb. pl.* Also **sarmalas** and in sing. **sarmala.** [Romanian.] A Romanian dish of forcemeat and other ingredients wrapped in leaves, esp. cabbage or vine leaves.

1945 A. L. Simon *Conc. Encycl. Gastron.* VII. 107/2 *Sarmalas.* Rub a little garlic on some raw beef and mince the beef with a little ham, a scrap of onion, parsley and other seasonings. Dip some spinach or young vine leaves in hot water and roll up the mince in them... Braise very slowly. **1958** W. Bickel tr. *Hering's Dict. Classical & Mod. Cookery* 509 *Sarmale*, Saurkraut Rolls: ground beef and pork mixed with boiled rice, seasoned with garlic, salt, pepper and finely chopped onions, wrapped in leaves of cabbage pickled in whole heads and rolled together. **1969** *Listener* 2 Jan. 31/1 Local dishes include sarmale (meat and rice in pickled cabbage leaves). **1970** 'M. Underwood' *Silent Liars* ii. xii. 131 First we have mamaliguta... Then Sarmale which are meat balls in cabbage leaves.

Sarmatian, *sb.* Add: **b.** The language of the Sarmatians, known only from Greek inscriptions in the southern U.S.S.R., and now regarded as a member of the Iranian group.

1922 O. Jespersen tr. R. Rask in *Language* ii. 39, I divide our family of languages in this way: the Indian.. Iranic..Thracian..Sarmatian (Lettic..and Slavonic),.. Gothic..and Keltic. **1939** L. H. Gray *Foundations of Lang.* 320 Old Sakian..and Old Sarmatian are preserved only in a few proper names and glosses. **1972** W. B. Lockwood *Panorama of Indo-Europ. Lang.* 235 The exiguous records of the Median language are of the same character as those of Scythian and Sarmatian.

sarmientite (sɑ̄ɹmie·ntəit). *Min.* [f. the name of Domingo Faustino *Sarmiento* (1811–1888), Argentinian educator and statesman + -ite¹.] A monoclinic hydrated basic arsenate and sulphate of ferric iron, $Fe_4(AsO_4)(SO_4)(OH).5H_2O$, found as pale yellow-orange microcrystalline nodules.

1941 Angelelli & Gordon in *Notulae Naturae* (Acad. Nat. Sci. Philadelphia) 16 Sept. 1 (*heading*) Sarmientite, a new mineral from Argentina. **1941** *Science* 26 Sept. (Suppl.) 9/1 Called sarmientite.., the new mineral is found in fair-sized nodules of great purity, of a pale yellow-orange color, in iron sulfate deposits of the Santa Elena mine. **1968** *Amer. Mineralogist* LIII. 2081 When heated at 300° C for one hour sarmientite yields a buff colored product, amorphous to X-rays.

sarnie (sɑ̄·ɹni). *slang.* Also **sarney.** [Prob. f. *sarn-*, repr. colloq. or (north.) dial. pronunc. of initial element of Sandwich *sb.*² + -y⁶, -ie.] = Sandwich *sb.*² 1. Freq. in *pl.*

1961 Partridge *Dict. Slang* Suppl. 1259/1 *Sarnies*, sandwiches. **1966** F. Shaw et al. *Lern Yerself Scouse* 39 *Sarneys, abnabs*, sandwiches. **1973** *Observer* 22 Apr. 27/5 Most people clamour for tea and sarnies within an hour, but I'm funny where dope's concerned. **1980** *Times* 11 Sept. 8/1 Questions like the protein content of bacon butties..and the vitamin rating of corned beef sarnies.

Sarn't (sɑ̄ɹnt). Also **Sarnt, sar'nt,** etc. and with small initial. Mil. colloq. abbrev. of Sergeant *sb.* 9 a. (Freq. used as a term of address.) Also **Sarn't-major** = Sergeant-major 2. Cf. *Sar-Major.

1930 Brophy & Partridge *Songs & Slang* 159 *Sarnt*, a smart and soldierly pronunciation of sergeant. Only used before the N.C.O.'s surname, e.g. 'Sarnt Smith', but 'Here's the Sergeant'. Also in *Sarnt-Major*, but here it could be used without the surname. **1945** *Gen* 30 June 50/1 An erb would turn up from no-where, come up to the sarnt and 'Flight Sarnt So-and-So's compliments, Sarnt, and can A. C. Actor be released from polo practice'. **1946** [see *pee *v.*² 2 b]. **1959** I. Jefferies *Thirteen Days* i. 21 'Ah, Sar'nt,' he said, nodding to my salute. **1972** G. Bell *Villains Galore* vii. 85 'Get out of there. Sarn't major...' They got out and were searched. The sergeant-major was thorough. **1972** F. Durbridge *Bat out of Hell* v. 157 Let me put you in the picture, Sar'nt. **1978** R. Mark *Office of Constable* iii. 39 A bugler whose lip split whilst blowing the single-note half-hour call provoked the falsetto scream, 'Sarn't major... Take his name for idle blowing of the 'orn.'

‖ **sarod** (sarǒᵘ·d). Also **saroda, sarode,** etc. [Hindi.] An Indian stringed musical instrument of Persian origin, variously bowed or plucked. Also *attrib.* and *Comb.*

1865 *Proc. R. Irish Acad.* IX. I. 115 *Sarrooda*, may be called the tenor or second fiddle... The *sarrooda* is..powerful but..difficult of execution; and it combines the effect of a guitar..and the violin. **1898** B. A. Pingle *Indian Music* (ed. 2) ii. 58 On the..Saroda..the limit of a Ghasita is not fixed. **1921** H. A. Popley *Music of India* vii. 109 The *Sāroda* or *Sarrawat* is a sārangi played with the plectrum instead of the bow. **1957** *New Oxf. Hist. Music* I. iv. 224 The *sarod*, played with a plectrum held between the fingers, has no frets. **1961** *Observer* 26 Nov. 28/1 Two sarod recitals. Srimati Sharan Rani. Foremost woman sarod player, acc. by Chatur Lal (tabla). **1961** *Guardian* 8 Dec. 10/6 Sharan Rani, the famous Indian sarode player, was.. in London today. **1975** R. P. Jhabvala *Heat & Dust* 96 They had a tape playing of sarod music.

‖ **saron** (sā·rǫn). [Javanese.] An Indonesian musical instrument, normally having seven bronze bars which are struck with a stick.

1817 T. S. Raffles *Hist. Java* I. viii. 470 The *sáron*.., the *démong*.., and *selántam*.., are *staccátos* of metallic bars, and a sort of bells placed on a frame. They contain a regular diatonic scale, and nearly two octaves. **1940** C. Sachs *Hist. Mus. Instruments* (1942) xii. 239 The Javanese *saron* cannot have been constructed much earlier than 900 A.D. The modern saron has a wooden resonance box which frequently is carved in the shape of a crouching dragon...Sarons are constructed in four main sizes an octave apart. **1961** [see *metallophone* s.v. *metallo-]. **1964** S. Marcuse *Mus. Instruments* 457/1 *Saron*, metallophone of Bali and Java, first depicted at Borobudur (*ca.* 800). In modern sarons the bars are set above a wooden trough resonator.

sarong. Restrict ‖ to sense 2. **1. a.** Substitute for def.: The Malay national garment, resembling a skirt, which consists of a long strip of (often striped or brightly-coloured) cloth worn round the waist and sometimes the chest by both sexes. (Its use is not restricted to Malaysia.) (Further examples.)

1911 *Encycl. Brit.* XVII. 483/1 The silk is imported raw and is re-exported in the form of Malay clothing (*sarongs*) of patterns and quality which are widely celebrated. **1923** D. H. Lawrence *Kangaroo* i. 6 Somers had..opened the bags, so she fished out an Indian sarong of purplish shot colour, to try how it would look across the table. **1953** G. M. Durrell *Overloaded Ark* ix. 166 Here he removed his sarong and proceeded to bathe. **1965** R. McKie *Company of Animals* iv. 77 They..pulled up their checked sarongs to spit between their crossed legs. **1971** *Sun* (Colombo) 20 Sept. 5/2 When it comes to crossing a stream groin deep the man in sarong has an advantage over the one in long trousers.

b. *attrib.* and *Comb.*

1913 L. Woolf *Village in Jungle* vii. 193 In the roof between the thatch he found the two sarong cloths. **1944** *Film Star Parade*, Dorothy Lamour... Paramount tested her for the leading part in 'Jungle Princess' and it was thus that she first became the 'sarong girl'. **1972** M. Sheppard *Taman Indera* 40 Thin cut out panels with another sarong design were fitted into the end wall at the same level. **1979** W. H. Canaway *Solid Gold Buddha* xiii. 95 Both men wore sarong-like lower garments.

Hence **saro·nged** *a.*, wearing or attired in a sarong.

1934 R. V. C. Bodley *Jap. Omelette* iii. 19 The good-natured smile of the saronged Malays and their cousins the Sudanese and Madoerese concealed no sinister thoughts. **1962** *Punch* 18 July 106/1 We barely have time to settle down from the last lot of celluloid raptures enacted beneath our palms when some other super-colossal unit arrives to shoot further bouts of saronged amour.

Saronic (sarǫ·nik), *a.* [ad.L. *Saronicus*, Gr. Σαρωνικός.] Of, pertaining to or designating the *Saronic Gulf*, a part of the Aegean Sea between Attica and the Peloponnese. Also † **Saro·nian** *a.*

1601 Holland tr. *Pliny's Nat. Hist.* I. iv. iv. 73 The one side thereof is called the Corinthian gulfe, the other, the Saronian. **1845** *Encycl. Metrop.* XIX. 725/1 To the Myrtoan Sea belonged the deep Saronic Gulf, (Σαρωνικός). **1890** J. G. Frazer *Golden Bough* I. i. 6 Hippolytus..had been killed by his horses on the sea-shore of the Saronic Gulf. **1956** A. Toynbee *Historian's Approach to Relig.* I. 44 The pinnacle of Acrocorinthus a stone's throw away, just across the Saronic gulf. **1977** *Times* 11 June 11/3 Spetses..one of the Saronic islands.

Sarouk (sarū·k). Also **Saruk.** The name of a village near Arak in Iran, used *attrib.* or *absol.* to designate various types of rug made there.

1900 J. K. Mumford *Oriental Rugs* xi. 204 Persian magnates..never demur at the loose colours which are the only drawback to the Saruks. **1913** W. A. Hawley *Oriental Rugs* ix. 126 Probably not one in a score..of the Sarouks now offered for sale in this country was woven there. **1920** [see *Kashan]. **1931** A. U. Dilley *Oriental Rugs & Carpets* iv. 121 The best Saruks are now all woven in Sultanabad. **1962** C. W. Jacobsen *Oriental Rugs* iii. 32 In the City of Arak..and in the surrounding villages, a good many thousand Sarouks have been woven each year, especially for the American market. *Ibid.* II. 281 The early Sarouks were very fine and short pile rugs... No large Sarouks were made until about the turn of the century. **1975** 'E. Lathen' *By Hook or by Crook* x. 97 A Sarouk, gleaming on the wall like a Rembrandt. **1977** *Times* 10 Sept. 14/6 A Saruk carpet of about 1930 fetched £2,100 in Sotheby's sale..yesterday.

‖ **sarpanch** (sa·ɹpantʃ). [Hindi–Urdu, = head arbitrator, foreman of a jury or council.] In India: the head of a panchayat or village council.

1963 F. G. Bailey *Politics & Social Change* I. i. 55 In order to run the panchayat one of the..members is selected as head and another as his assistant. The head is called the 'Sarpanch' and his assistant is the 'Naib Sarpanch'. **1971** *Hindustan Times* (New Delhi) 7 Apr. 12/4 They went around with Girdhari Lal, the village sarpanch. **1976** D. Hiro *Inside India Today* 50 Forty to fifty panchayats are banded together to form a panchayat samiti (covering a population of 30,000 to 100,000). Its membership consists of the sarpanches of the constituent panchayats and ten co-opted members.

sarrusophone. (Further examples.)

1894 G. B. Shaw in *World* 7 Mar. 23/1, I want a craftsman to take the matter up, with the object, not of inventing some new instrument like the saxophone or sarrusophone which nobody wants, but of giving us back the old instruments. **1926** Whiteman & McBride *Jazz* ix. 196 The sarrusophone, which is made in seven or more sizes, is named with the wood winds although it is metal. For this reason, it is sometimes mistakenly called a metal oboe. **1975** *Gramophone* Oct. 611/3 Since the nine encyclopaedic columns..nowhere spell out the forces required by Schmitt's score, I feel justified in giving them here:.. two bassoons, sarrusophone, two E flat clarinets [etc.].

sarsartie, Sarsi, varr. *sosatie, *Sarcee.

Sart (sɑ̄ɹt), *sb.*² and *a.* [Turki.] **A.** *sb.* **a.** A member of a settled people of mixed Turkoman and Iranian descent, living as town-dwellers and traders in Turkestan and parts of Afghanistan. Cf. *Tajik. **b.** The Eastern Turkic dialect of Uzbek spoken by the Sarts. **B.** *adj.* Of or pertaining to this people or their language.

The name is widely used, with varying degrees of exactness, to designate the sedentary people of this region. They are thus contrasted with the Kurds and others, by whom the term is considered derogatory. Other commentators view the Sarts as an ethnologically distinct people.

1871 R. B. Shaw *Visits to High Tartary* ii. 26 All the Khokandees whom I met with in Eastern Toorkistân agreed in affirming that Sart is merely a word used by the Kirghiz to denote all who do not lead a nomad existence like themselves, whether they be Tajiks or Oosbeks. **1879** *Encycl. Brit.* IX 85/2 Tajiks..in the chief towns and central districts, who are known as Sarts, show a large infusion of Uzbeg and other Turki blood. **1898** Bealby & Hearn tr. *Hedin's Through Asia* v. 61 Guided by some Sart boys, I threaded my way through a labyrinth of narrow lanes. **1900** 'Odysseus' *Turkey in Europe* iii. 101 Sart, though now commonly used as a name for the Jagatai Turkish spoken in those provinces [sc. Fergana, Turkestan, etc.] is, strictly speaking, not a linguistic designation, but denotes a dweller in cities and a merchant, as distinguished from a countryman and agriculturist, called Tajik. **1920** *Glasgow Herald* 31 Aug. 8 It is among the highly intelligent Sarts and Tadjiks, speaking Persian and Arabic fluently and many of them conversant with Hindustani, that the Bolshevists find their cleverest agents. **1946** F. M. Bailey *Mission to Tashkent* iii. 36 Sart writers sometimes refer to themselves as Turks but this word is..misleading. **1953** O. Caroe *Soviet Empire* iii. 34 It was to sedentary dwellers of this kind, whether bilingual or speaking only Tajik, that the true Turks formerly applied the pejorative appellation of 'Sart'. **1954** Pei & Gaynor *Dict. Linguistics* 190 *Sart*, an Asiatic language; member of the Central Turkic group of the Altaic sub-family of the Ural-Altaic family of languages. **1964** R. A. Pierce in N. J. Couriss tr. *Pahlen's Mission to Turkestan* 10 Pahlen..regarded the Sarts as a distinct ethnic group with their own language. Originally..the word was applied to the sedentary, and.. urban, population of Turkestan without any reference to race or language. A Sart might..be of Tadzhik.., Uzbek.., or of mixed Iranian and Turkic stock; and..speak.. Tadzhik or Uzbek... There is no such thing as a Sart language. During the Soviet régime the word acquired a derogatory significance and is now no longer used.

sartorially (sɑ̄ɹtō·riăli), *adv.* [f. Sartorial *a.* + -ly².] With regard to clothes.

1905 W. J. Locke *Morals of Marcus Ordeyne* xii. 146 When she puts her foot upon my sartorially immaculate knee. **1916** —— *Wonderful Year* xvii. 245 Like a woman clothes-starved for years..Martin ran sartorially mad. **1928** *Daily Express* 16 Apr. 3/4 Sartorially magnificent in all-over woolly tights. **1970** *Daily Tel.* 30 Dec. 9/2 Sartorially speaking, men are at last catching up with the women. **1972** *Ibid.* 14 Mar. 13/7 Sartorially, the beginning of the Chinese 'thaw' is to be seen in the gaily-coloured clothes worn by girls and women. **1974** 'M. Underwood' *Pinch of Snuff* ii. 12 The club's most sartorially elegant member.

Sartrean, Sartrian (sā·ɹtriăn), *a.* Also **Sartreian.** [f. the name of *Sartre* (see below) + -an, -ian.] Of, pertaining to, or characteristic of the French writer and philosopher

Jean-Paul Sartre (1905–80), his writings, or his existentialist philosophy. Hence as *sb.*, an admirer of the ideas of Sartre.

1948 [see *commitment 6 c]. **1949** E. L. Mascall *Existence & Analogy* vi. 126 Given their atheist dogma, the Sartrians are quite right in asserting that existence is absurd, that the world does not make sense. **1951** N. Annan *Leslie Stephen* viii. 247 In the 'forties Sartrian Existentialism..was a bizarre attempt to justify the duties to society which are inescapably binding upon individuals, who, whether they like it or not, are forced to commit themselves. **1958** *Spectator* 25 July 141/2 That unique Sartrean blend of intellectual and moral disintegration. **1961** *Encounter* June 42/2, I cannot claim to be either a Sartrian or a Thomist. **1962** *Listener* 24 May 920/2 From the Sartrian standpoint, Rousseau went about this enterprise in the wrong way. **1970** J. D. Caute *Fanon* iii. 35 Fanon's prose reverberates with Sartreian concepts, phrases, dialectical juxtapositions, paradoxes and essentialist abstractions. **1975** J. Symons *Three Pipe Problem* xix. 218 Were they both dead, joined in a permanent squabble in some Sartrean hinterland? **1977** *New Yorker* 16 May 147/1 Soyinka discusses material..with ample cross-references to Greek drama, Nietzschean aesthetics, Jungian philosophy and Sartrean opinionizing.

Saruk, var. *Sarouk.

Sarum. (Further examples.)

1929 S. Leslie *Anglo-Catholic* x. 117 His ritual was simple: two lights, Sarum Use and Sarum colours with simple vestments of linen. **1954** O. Chadwick *Founding of Cuddesdon* v. 133 This was just the time when the old-fashioned cassock was giving way to the Sarum cassock, which was intruding from about 1887 onwards and conquering by 1897. **1957** *Oxf. Dict. Chr. Ch.* 1209/1 In the years preceding the Reformation the output of Sarum books was enormous. The much increased knowledge which has followed their discovery..has led to the revival of Sarum customs and ornaments in many English cathedral and parish churches. **1966** J. Betjeman *High & Low* 55 And there we'll sing the Sarum rite Tae English Hymnal airs. **1972** C. Stephenson *Merrily on High* iv. 64 SS. Philip & James, very Sarum, and St. Margaret's close by very western, were the centre of great turmoil at one stage when the vicar of Phil & Jim was making it western and the vicar of St. Margaret's was busy taking off the six candles from the altar and substituting two while introducing Sarum practices. **1974** *Encycl. Brit. Micropædia* VIII. 908/3 The Sarum chants resemble Gregorian ones in the use of free rhythm, modes.., psalm tones.., musical form, and the addition of tropes.

‖ **sarvodaya** (saɹvo·daya). [Skr., f. *sárva* 'all' + *udayá* 'uplift, prosperity'.] The welfare of all; the name given to the new social order advocated by the Indian leader M. K. Gandhi (1869–1948) and his followers. Also *attrib.*

[**1908** M. K. Gandhi (*title*) Sarvodaya.] **1919** *Bombay Chron.* 8 Apr. 713 The committee has selected the following prohibited books for dissemination:..*Sarvodaya or Universal Dawn* by M. K. Gandhi. **1941** K. G. Mashruwala *Practical Non-Violence* (1946) 45 This is the civilization of Sarvodaya (the wellbeing of all). **1948** *Harijan* 4 Apr. 54/2 *Samaj*, which corresponds more to brotherhood than to association. .. The Sarvodaya Samaj has been established to strive.. towards a society based on Truth and Non-violence, in which there will be no distinction of caste or creed. **1954** B. Kumarappa *Sarvodaya* (1958) p. iii, Sarvodaya, as the welfare of all, represents the ideal social order according to Gandhiji. Its basis is all-embracing love. **1962** B. Smith *Portrait of India* vi. 45 Vinoba..became at Gandhi's death, the leader of the *Sarvodaya* movement of selfless service. *Ibid.* 46 Gandhi's *sarvodaya* embodied the idea of regeneration in the individual and in society. **1965** E. Linton *World in Grain of Sand* ix. 163, I thought of the Gandhians in their present dilemma as expressed later at the Sarvodaya Conference at Vedchi. **1971** *Peace News* 10 Sept. 8/2 The concept of the freedom march was born in the minds of Indian Sarvodaya workers. **1974** *Times* 7 Dec. 5/4 The village is conceived of in Gandhian or *sarvodaya* terms as a miniature, self-governing republic. **1978** *Times Lit. Suppl.* 3 Feb. 121/2 India's own home-spun village socialist movement known as *Sarvodaya* acquired its name, significantly, from the word used by Gandhi to translate the title of Ruskin's book, *Unto This Last.*

sarwan, var. Surwan.

Saryk (sari·k). Also **Sarik.** [Native name.] **a.** One of several Turkic tribes inhabiting the Turkmen Soviet Socialist Republic; a member of this tribe. **b.** *attrib.* Used to designate a carpet or rug made by this tribe, similar in design to a Bokhara carpet. Also *absol.*

1885 E. W. Hamilton *Diary* 9 Apr. (1972) II. 831 The news reached me that the Russian Colonel—Alikhanoff—attacked the Afghans at Penjdeh on the 30th,..and that Alikhanoff had actually instigated the Sariks (a Turcoman tribe in that neighbourhood) to attack the English party... The Sariks fortunately declined the offer. **1889** G. Curzon *Russia in Central Asia* v. 96 Transcaspia..includes.. the minor oases inhabited by the Sarik and Salor Turkomans. **1899** Skrine & Ross *Heart of Asia* II. iv. 268 The Merv oasis was inhabited by the Sâriks,..who were engaged in a struggle with the Khivans. **1922** H. Clark *Bokhara, Turkoman & Afghan Rugs* p. xiv, In the course of years I acquired..specimens of Salor, Saryk..and Afghan Turkoman rugs... A Saryk Turkoman rug of the early 18th century, nearly square in shape and of wonderful colouring. **1957** C. W. Hostler *Turkism & Soviets* ii. 69 The Turkmens are divided into seven main tribes (the Chauders,..Sariks, Salors and Ersaris). **1960** H. Hayward *Antique Coll.* 248/2 *Saryk rugs*, Turkestan rugs, generally not so fine in texture as Bokhara rugs.., but bearing a similar design. **1962** C. W. Jacobsen *Oriental Rugs* II. 286 Even today nine out of ten dealers have never heard of the name 'Saryk'. **1964** *Sunday Times* (Colour Suppl.) 19 Jan. 25 Sarik. Carpet from a Turcoman tribe almost destroyed by the Tekkes in the last century. **1974** *Encycl. Brit. Macropædia* XVIII. 800/1 The number of people in other tribes (the Salyr, Saryk, Groklen, and Choudor) fluctuated between 20,000 and 40,000.

sas, var. *sass *sb.* **sasaitie,** var. *sosatie.

Sasak (sā·sak), *sb.* and *a.* Also 9 **Sassak.** [Native name.] **A.** *sb.* One of the Malay inhabitants of the island of Lombok. Also, the language of the Sasaks. **B.** *adj.* Of or pertaining to the Sasaks.

1817 T. S. Raffles *Hist. Java* II. p. cxcviii, Comparative vocabulary of the Bugis, Makasar, Mandhar, Búton, Sásak, Bíma, Sembáwa, Tembóra and Endé Languages. **1869** A. R. Wallace *Malay Archipel.* I. xi. 256 Beyond Mataram..is Karangassam, the ancient residence of the native or Sassak Rajahs before the conquest..by the Balinese. *Ibid.* 270 The aborigines of Lombock are termed Sassaks. They are a Malay race... They are Mahometans. **1897** E. J. Taylor tr. *W. Cool's With Dutch in East* III. 121 The language of the Sassaks is totally different from that spoken by the Balinese; although the Sassaks have borrowed many words from their neighbours, still they are unable to understand each other's language. **1937** M. Covarrubias *Island of Bali* ii. 30 In 1885 there was a rebellion of Sasaks, the vassals of the Balinese in Lombok... The Sasak chiefs complained to the Dutch, asking to be freed from the tyranny of the Balinese princes. **1954** E. D. Laborde tr. *C. Robequain's Malaya, Indonesia, Borneo & Philippines* II. xii. 24 Balinese Hinduism is isolated between Java and Lombok, for Islam has..almost wholly won over Lombok... Sasak converts to Islam form the great majority..of the population. **1961** P. Kemp *Alms for Oblivion* xi. 163 Almost nine-tenths of this population are Sasaks, a simple agricultural people of Malay stock. *Ibid.* 166 We enlisted the Eurasian as an extra interpreter, for..he spoke Dutch, Malay, Balinese and Sasak. **1965** *Language* XLI. 294 Madurese, like some of its neighboring languages (Sundanese, Javanese, Balinese, and Sasak), has socially determined choices of words. **1974** *Encycl. Brit. Micropædia* VI. 308/3 The population of Lombok is composed largely of Sasaks of Malay origin.

Sasanian, var. Sassanian *a.* and *sb.* in Dict. and Suppl.

Sasanid, var. Sassanid *sb.* and *a.* in Dict. and Suppl.

sasanqua (sāsæ·nkwă, -kă). Also **sasank(w)a.** [Jap. *sasankwa* mountain tea-flower.] An evergreen shrub, *Camellia sasanqua*, belonging to the family Theaceæ, native to Japan, and bearing fragrant white or pink flowers and seeds yielding an edible oil also used in the production of silk and soap.

1866 Lindley & Moore *Treas. Bot.* I. 207/2 *C. Sasanqua* (Sasanqua is the Japanese name of the plant) is found in many parts of China and Japan. **1878** *Trans. Asiatic Soc. Japan* VI. 216 A kind of evergreen with poplar-like leaves..is called *Sasanka* by the Japanese. **1884** tr. *J. J. Rein's Japan* 441 In November and December Sasankwa and Cha..blossom. **1962** J. L. Threlkeld *Camellia Bk.* i. 3 In the gardens of Japan the sasanqua predominates.

sasatie, var. *sosatie.

sash, *sb.*² Add: **3. sash cramp** (see quot. 1964); **sash-door** (earlier examples); **sash-pane**, each of the panes of glass in a sash-window.

1964 J. S. Scott *Dict. Building* 275 *Sash cramps*, cramps between 2 and 5 ft long used for clamping sashes during gluing. **1969** E. H. Pinto *Treen* 381/2 Wooden cramps. The general run of both G cramps and sash cramps are too familiar to need any special description. **1726** D. Eaton *Let.* 25 Sept. (1971) 60, I think the sash door at Little Deen ought to be oak, and these planks we have will do very well. **1739–40** Richardson *Pamela* (1740) I. 95 In this green Room was a Closet, with a Sash-door and a Curtain before it. *c* **1806** D. Wordsworth *Jrnl.* (1941) I. 311 The dwelling-house was distinguished from the outer buildings..by a chimney and one small window with sash-panes.

sashay (sæʃē¹·), *sb. N. Amer.* [f. next.] **1.** A venture, a sally; an excursion, trip, or expedition.

1900 G. Ade *More Fables* 184 Lutie never got out of her Dream until she made a bold Sashay with a Concert Company. **1935** H. L. Davis *Honey in Horn* 15 If you yank him out for any all-night sashay on these roads, you ought to be ashamed of yourself. **1941** *Sat. Even. Post* 16 Aug. 68/3 On my first sashay into the flying field. **1952** E. B. White *Let.* 6 Apr. (1976) 355 Spring is making little sashays about coming to town, but it has been a fairly unconvincing demonstration so far. **1961** R. M. Patterson *Buffalo Head* v. 183 He could make a long sashay north along the foot of the range. **1968** —— *Finlay's River* 147 Swannell, Copley and Alexander and the two dogs..set out for a three-day sashay up the strong creek that flowed into the Ingenika.

2. A step in square dancing (see quot. *c* 1940). Also *transf.* and *attrib.*

c **1940** *Square Dance* (Writers' Program, Illinois) 40 The Sashay is a series of short quick steps directly to the side, either to the right or to the left... The gent holds the lady's left hand in his right, and her right hand in his left. **1941** R. J. McNair *Western Square Dances* xix. 78 The Sashay step is a quick side step. **1956** R. Holden *Contra Dance Bk.* iv. 45 Sashay, the American chassez. **1971** *Flying* (N.Y.) Apr. 49/1 It's got a marvelous sort of sashay movement because it's such a big airplane that when you roll you can feel those booms kind of rolling around behind you. **1974** 'J. Marks' *Mick Jagger* 32 The juvenile Jagger... Trying a few quick sashays and eating a banana.

sashay (sæʃē¹·), *v. colloq.* (chiefly *U.S.*). Also **sasshay.** [Now the dominant form of Sashy *v.*] **1.** *intr.* **a.** To perform a chassé, esp. in square dancing; freq. *transf.*, to perform a movement similar to the chassé. **b.** To glide, walk, or travel, usu. in a casual manner. **c.** To move diagonally or sideways; to travel an irregular path; to wander or saunter. **d.** To move or walk ostentatiously, conspicuously, or provocatively; to strut or parade. Freq. with *adv.*

1836 *Franklin Repository* (Chambersburg, Pa.) 4 Oct. 1/3 If you don't sashay across, button your lip, and go home quietly, you and I will have to promenade all around, and swing corners into the watch house. **1865** 'Mark Twain' in *Californian* 18 Mar. 8/1 For all they're so handy about keeping her sasshaying around from shanty to shanty ..none of 'em's ever got a good word for her. **1878** F. H. Hart *Sazerac Lying Club* 83 S'pose, gentlemen, that we sashay up to the bar. **1888, 1891** [see Sashy, sas(s)hay, *v.*]. **1905** *Dialect Notes* III. 64 They sashayed back and forth to beat the band. **1913** C. E. Mulford *Coming of Cassidy* v. 80 Logan..is about thirty miles east. You must 'a sashayed some to get only this far in four days. **1917** H. Garland *Son of Middle Border* xv. 163 At dancing parties they balanced or 'sashayed' in *Honest John* or *Money Musk.* **1935** Z. N. Hurston *Mules & Men* I. v. 113 John was callin' for de new set: 'Choose yo' partners.'..'Sashay all.' **1942** E. Paul *Narrow St.* ix. 74 He staggered eastward toward the Panthéon and I sashayed westward to the rue Lafayette. **1944** C. Himes *Black on Black* (1973) 201, I picked up my sack and sorta sashayed off. **1949** K. M. Wells *Owl Pen Reader* (1969) III. 246 Fireflies danced by Moonstone Creek. They dipped and cavorted, they sashayed like a million wee stars gone mad. **1951** E. Paul *Springtime in Paris* v. 114 Instead of continuing toward the rue de la Huchette, Christophe hopped and sashayed to the left. **1960** F. Raphael *Limits of Love* I. iii. 41 A large Negro..was sashaying through the crowd towards them. **1968** J. Updike *Couples* iv. 311 Sashaying from the shower nude, her pussy of a ferny freshness. **1973** S. Alsop *Stay of Execution* (1974) II. 201 Stewart brought a pretty..girl friend home. As she sashayed through the living room, Andrew remarked, 'I like the way she wiggles her things.' **1978** J. A. Michener *Chesapeake* 270 He hoped that Nelly Turlock would not sashay in, demanding dividends for her family. *Ibid.* 545, I see her sashayin' past in a dress I know she stole from Miss Susan.

2. *trans.* To cause (someone or something) to sashay; to walk or parade (a person); to carry or convey (an object); to manœuvre (a vehicle).

1928 L. H. Nason *Sergeant Eadie* 130 What the hell good a rifle does to me sashayin' these jugheads up an' down the road, I don't know. **1944** J. S. Pennell *Hist. Rome Hanks* 189 Take them guns thar—tuck 'em from the Yanks at the fust battle of Manassas, an' been a-farin' 'em eveh since an' sashayin' 'em all oveh hell an Vuhginny. **1963** T. Pynchon *V.* i. 22 Rachel would gee and haw this MG around Route 17's bloodthirsty curves and cutbacks, sashaying its arrogant butt past hay wagons. **1977** J. Gardner *Werewolf Trace* i. 17 'James Bond rules. Okay?' chuckled Bud, sashaying the car neatly between a pair of taxis.

Hence **sasha·ying** *vbl. sb.* and *ppl. a.*

1935 R. Stout *League of Frightened Men* xx. 272 It did mean his sashaying out of the house twice in two days, which was an all-time record. **1976** *New Yorker* 8 Mar. 109/1 He'd start with a fusillade of rim shots, sink into a sashaying figure that strode back and forth between his tomtoms.

sashed, *ppl. a.*¹ Add: *sashed window* (earlier examples).

1816 Jane Austen *Emma* II. vi. 98 He stopt for several minutes at the two superior sashed windows which were open. *Ibid.* III. xiv. 260 A brick house, sashed windows below, and casements above.

sashed (sæʃt), *ppl. a.*² [f. Sash *v.*¹ or *sb.*¹ + -ed¹,².] Dressed or adorned with a sash.

1869 'Mark Twain' *Innoc. Abr.* vii. 69 Turbaned, sashed and trowsered Moorish merchants. **1894** [see Sash *v.*¹]. **1970** *Daily Tel.* 27 Apr. 14 A good sashed white midi coat for £25.

‖ **sashimi** (sæ·ʃimi). [Jap., f. *sashi* pierce + *mi* flesh.] A Japanese dish consisting of thin slices of raw fish served with grated radish or ginger and soy sauce. Also *attrib.* or as *adj.*

1880 I. L. Bird *Unbeaten Tracks in Japan* I. 239 The preparation of raw fish cut into oblong strips called *sashimi*. **1920** *Japan Advertiser* 22 Aug. 5 Sashimi or arai..is raw tai, tunny or kare, served with horseradish. **1933** P. Peto *Recipes Rare from Everywhere* 29 *Sashimi.* The fish is skinned, cleaned and cut into fillets about ¼ inch thick; it is arranged on a dish and garnished with fresh thinly sliced vegetables, and is eaten with Shoyu blended with Japanese shredded horseradish. **1936** K. Tezuka *Jap. Food* 14 *Sashimi* (raw sea-bream, flounder, tunny, etc. cut into thin slices). **1959** R. Kirkbride *Tamiko* vii. 54 They had hors d'oeuvres of raw wild vegetables, sashimi, thin slices of raw tuna [etc.]. **1967** *Guardian* 8 Dec. 8/2 Finding fish fresh enough to serve sashimi (raw) is very difficult. **1969** R. Howe *Far Eastern Cookery* 189 *Sashimi*..is a truly Japanese speciality... I took myself to a small *sashimi* bar..and ordered *sake.* **1973** J. Gores *Final Notice* x. 60 Waiting for the Japanese waitress to arrive with the sukiyaki and sashimi. **1978** *Maclean's Mag.* 13 Nov. 47/1 Each spring the tiny fishing village 20 miles south of Halifax prepares to satisfy the yearnings of 100 million Japanese for sashimi.

sashy, sas(s)hay, *v.*: see *sashay *v.*

sasquatch (sæ·skwǫtʃ). *Canad.* Also **Sasquatch.** [Salish.] A name for a huge, hairy, man-like monster supposedly inhabiting the north-west of the U.S. and Canada. Also *collect.* and *attrib.*

1929 *Maclean's Mag.* 1 Apr. 61/1 The strange people, of whom there are but few now—rarely seen and seldom met— ..are known by the name of Sasquatch, or, 'the hairy mountain men'. **1950** C. P. Lyons *Milestones on Mighty Fraser* 28 Indian lore has it that a mysterious race of giants, known as the Sasquatch, live in the high mountains around Harrison Lake. **1958** *Encycl. Canadiana* IX. 233/1 Known originally to the Indians..as Saskehavas (wild men), they are called by the..whites Sasquatch (hairy men). **1966** *Globe Mag.* (Toronto) 11 June 3/3 Most villagers relate every Sasquatch sighting to the amount of alcohol they insist must have been consumed immediately prior to the monster's appearance. **1971** W. Hillen *Blackwater River* xi. 108 Stories of strange lights and huge, wild, hairy men, or 'sasquatch', circulate periodically, usually toward spring. **1972** L. Hancock *There's a Seal in my Sleeping Bag* vi. 123 We scanned the steep forested slopes of the pass for Sasquatch. **1974** *New Yorker* 25 Feb. 92/2 The Northwest's legendary Sasquatch, a huge, humanoid seven-or-so-foot creature akin to the Abominable Snowman of Tibet. **1976** *Toronto Star* 31 Jan. 7/1 A nine-man team, using computerized information and electronic detection gear, will go sasquatch hunting in British Columbia in April or May. **1977** *New Yorker* 20 June 72/2 It was lumpy, pitted, pocked, rough, ugly—an apparent filling from the tooth of a Sasquatch. **1979** T. Gifford *Hollywood Gothic* (1980) v. 53, I feel like a sasquatch has been using me for a soccer ball.

Sasquehanno, etc., obs. varr. *Susquehannock.

sass (sæs), *sb. U.S. colloq.* Also **sas.** [var. Sauce *sb.*] **1. a.** = Sauce *sb.* 4 a.

1775 J. Stevens *Jrnl.* 5 May in *Essex Inst. Hist. Coll.* (1912) XLVIII. 43 Steven Barker come down & brought us som sas. **1836** B. Tucker *Partisan Leader* II. xxxv. 124 The fellow talked to me about living at home on codfish, and potatoes, and cider, and pies, and *all sorts of sass.* **1860** *Knickerbocker* July 102 White turnip, yellow turnip, or any sort of sass, long sass, or short sass. **1945** M. Lyon *Fresh from Hills* iv. 46 A family could get along without garden sass.

b. = Sauce *sb.* 4 b.

1913 H. Kephart *Our Southern Highlanders* xiii. 293 Your hostess, proffering apple sauce, will ask, 'Do you love sass?'

2. = Sauce *sb.* 6 b.

1835 [see *chunk *v.*¹ 1]. **1853** G. C. Hill *Dovecote* 88 I've a precious good mind to duck you for your sass! **1876** 'Mark Twain' *Tom Sawyer* i. 23 If you give me much more of your sass I'll take and bounce a rock off'n your head. **1880** J. C. Harris *Uncle Remus* iv. 31 Brer Rabbit wuz bleedzed fer ter fling back some er his sass. **1897** 'O. Thanet' *Missionary Sheriff* 21, I shall take more advantage of it if you give me any sass. **1935** J. T. Farrell *Judgement Day* ii. 25 It must have been something of the old Studs Lonigan left in him that led to his not taking sass, risking a fight. **1967** P. Welles *Babyhip* iii. 46 Is this what we get? Sass? No gratitude. **1977** *Time* 3 Jan. 21/1 If she's mostly given over now to laughter, pride and sass, she has earned her fun.

sass (sæs), *v. U.S. colloq.* [var. Sauce *v.*] **1.** *trans.* = Sauce *v.* 4 d.

1856 'J. Phoenix' *Phoenixiana* xvi. 125 While the squire..sasses all respectable persons With his talk of pills he's invented. **1867** 'Mark Twain' *Celebr. Jumping Frog* 166 You ought never to 'sass' old people—unless they 'sass' you first. **1887** [see Sauce *v.* 4 d]. **1896** *N.Y. Dramatic News* 18 July 2/3 When he was requested to desist he 'sassed' the officer. **1920** S. Lewis *Main Street* ix. 118 There had to be one man in town independent enough to sass the banker! **1929** W. Faulkner *Sound & Fury* 67 Dont you sass me, nigger boy. **1956** W. H. Whyte *Organization Man* (1957) vii. xxvi. 358 If little Johnny sasses Mrs. Erdlick just once more. **1966** K. L. Morgan in A. Dundes *Mother Wit* (1973) 602/1 She wanted to know if it was all right to 'sass' the woman the way she did since she was 'trash'. **1978** J. A. Michener *Chesapeake* 536 'But, Missy, I done clean it.' 'Don't you sass me!' she screamed.

2. *intr.* To speak impertinently; *to sass back*, to reply impertinently, to 'answer back'.

1880 J. C. Harris *Uncle Remus* iv. 29 You been runnin' roun' here sassin' after me a mighty long time. **1884** 'Mark Twain' *Huck. Finn* (1885) xxvii. 237 The king sassed back, as much as was safe for him. **1891** O. W. Holmes *Over Teacups* 154, I suppose Me-Number-Two will 'sass back'. **1976** *National Observer* (U.S.) 24 July 10/3 No teacher chooses to teach students who are sleeping or throwing chalk or sassing.

Hence **sa·ssing** *vbl. sb.*

1962 W. H. Gass in Foley & Burnett *Best Amer. Short Stories* 112 Listen to me, Jorge, I've had enough to your sassing. **1967** P. Welles *Babyhip* iii. 48 'Don't you get hysterics,' Mrs Green warned, 'or I'll throw cold water on you. I'll have none of this sassing.' **1977** *New Yorker* 6 June 48/3 Far worse than any welts he ever got when his father or Abel beat him in the shed for misbehaving or sassing.

sassafras. Add: **1. b.** The wood or timber of this tree.

1728 *Rec. Early Hist. Boston* (1883) VIII. 222 No Popler, ..Sassifax, Black ash, Basswood, or Ceder Shall be Corded up. **1900** *19th Ann. Rep. U.S. Bureau Amer. Ethnol.* *1897–98* I. 422 Sassafras is tabued as fuel among the Cherokee..perhaps for the practical reason that it is apt to pop out of the fire when heated. **1921** C. C. Deam *Trees of Indiana* 165 Floors were made of sassafras to keep out the rats and mice.

2. a. (Later examples.)

1863 *Rio Abajo Weekly Press* (Albuquerque, New Mexico) 14 Apr. 2/3 Sassafras.—Those who use this drink will find [etc.]. **1871** E. Eggleston *Hoosier Schoolmaster* 88 He drank his glass of water, having declined even her sassafras. **1912** M. Nicholson *Hoosier Chron.* 44 Sassafras in the spring, and a few doses of quinine in the fall,.. were all the medicine that any good Hoosier needed.

3. *sassafras-bush*; **sassafras soap** (earlier example); **sassafras tea** (later U.S. examples).

1848 G. C. Furber *Twelve Months Volunteer* 54 The field, or the larger part of it, growing up with tall weeds and sassafras bushes. **1944** T. D. Clark *Pills, Petticoats & Plows* 261 The graveyard is scraped bare of crab grass,.. Johnson grass and sassafras bushes to give them a 'cared-for' appearance. **1860** J. G. Holland *Miss Gilbert's Career* 108 Arthur took his accustomed seat at the head of the table, with Leonora at his right hand,.. [in an] atmosphere of sassafras-soap. **1817** T. Dean in *Indiana Hist. Soc. Publ.* (1918) VI. 324 We took some bread and sassafras tea. **1960** I. Wallach *Absence of Cello* 41 Perry sipped a cup of sassafras tea.

Sassanian, *a.* and *sb.* Add: **a.** *adj.* (Further examples.) Also, of, pertaining to, or characteristic of the period of this dynasty.

1928 C. Dawson *Age of Gods* iv. 84 In historical times we find the influence of the Sassanian culture of Persia following very much the same paths. **1929** E. C. Thomas *Lay Folks' Hist. Liturgy* I. v. 22 Persia..was conquered by the Mohammedans in 651, thus ending the native Sassanian dynasty, which was founded in 223 B.C. **1940** *Burlington Mag.* July 31/1 An important feature in Sassanian and Muslim architecture. **1958** A. Toynbee *East to West* 167 On bas-reliefs of the Sassanian age one is shocked to find the goddess Anahita holding her ground beside Ormuzd. **1971** R. Russell tr. *Ahmad's Shore & Wave* viii. 93 Strange houses. One like a dreadnought..one in Japanese style, one displaying Sassanian arches. **1977** *Ashmolean Mus. Rep. Visitors 1975–76* 17 (*title*) Parthian and Sassanian metalwork in the Bomford collection. **1980** J. Lees-Milne *Harold Nicolson* xiv. 303 The Sassanian city of Istakh, destroyed during the Arab conquest in the seventh century A.D.

b. *sb.* (Further examples.)

1931 A. W. Seaby *Art in Life of Mankind* 80 The Sasanians were as bitter enemies of the Christian eastern or Byzantine empire as they had been of pagan Rome. **1976** *Nature* 10 June 472/1 The Venus–Jupiter conjunction of 650 was one recorded by the Chinese; this conjunction was later regarded by Masha'allah as signifying the fall of the Sasanians and the rise of the Arabs.

Sassanid, *sb.* and *a.* Add: Also **Sasanid.**

a. *sb.* (Later examples.) **b.** *attrib.* and *adj.* (Earlier and later examples.) Also = *Sassanian *a.*

1867 C. M. Yonge *Pupils of St. John* xvii. 270 The Sassanid princes had taken up all the traditions of their supposed ancestry. **1904** F. C. Burkitt *Early Eastern Christianity* 25 The rise and decay of Christianity in the Sasanid Empire. **1910** [see *aquamanile]. **1929** E. C. Thomas *Lay Folks' Hist. Liturgy* I. v. 22 In A.D. 202 Abjar IX., Prince of Edessa, adopted Christianity, but this Church [in Persia] was barely tolerated by the Parthians and often persecuted by the Sassanids. **1958** A. Toynbee *East to West* 165 The capital of the Sasanid Persian Empire's Arab wardens of the Arabian marches. **1977** *Field* 31 Mar. 525/1 For the early Persian dynasties such as the Archamedians and Sassanids, the rug may have been used as a gift from one ruler to another.

sassatje, var. *sosatie.

Sassella (sæse·lă). Also **sassella.** [It.] The name of a red wine from the Valtellina district, Lombardy, in Italy.

1935 A. L. Simon *Dict. Wine* 227 *Sassella*, one of the best red wines of the Valtellina (Lombardy). **1967** A. Lichine *Encycl. Wines & Spirits* 330/1 In this district [*sc.* Valtellina], centred around the town of Sondrio, the three almost identical red wines are Sassella, Grumello, and Inferno, all from Nebbiolo grapes grown on the more manageable Alpine slopes. **1970** *House & Garden* May 140/2 Of all the Lombardy wines..I have most enjoyed..sassella— a red wine. **1975** 'S. Marlowe' *Cawthorn Jrnls.* xix. 166 A bottle of the dark, velvety Sassella stood on the table.

sasshay: see also *sashay *v.*

sassy (sæ·si), *a. colloq.* (orig. and chiefly *U.S.*). [Var. Saucy *a.*¹] Impudent, saucy, 'cheeky'; outspoken, provocative; conceited, pretentious; self-assured, spirited, bold; vigorous, lively; stylish, 'chic'. Also quasi-*adv.*

[**1815** D. Humphreys *Yankey in Eng.* I. 22 'Ah, you sly boots. Don't be saucy.' 'Saisy!'.] **1833** S. Smith *Life & Writings J. Downing* 128 If I should give out now.., them are sassy chaps in Portland would laugh at me. **1862** C. F. Browne *A. Ward his Book* 200 A hansum yung gal, with..a sassy little black hat tipt over her forrerd, sot in the seat with me. **1870** 'Mark Twain' *Lett. to Publishers* (1967) 38 And then I talked sassy to him for a page or two. **1880** J. C. Harris *Uncle Remus* ii. 24 Brer Rabbit pacin' down de road..dez ez sassy ez a jay-bird. **1908** J. H. Shinn *Pioneers & Makers Arkansas* xxxii. 258, I have seen sassy people, but of all sassy people in the world Arkansassy people are the worst. **1917** T. H. Comstock *Man thou Gavest* 12, I kept the sassy little hen. **1936** Wodehouse *Laughing Gas* vii. 77 Have you ever had to look after a sassy, swollen-headed, wisecracking child star? **1945** L. Saxon et al. *Gumbo Ya-Ya* xii. 233 My ma was hard-headed and sassy, and she'd talk right back to anybody, Massa or nobody. **1958** J. Kerouac *On Road* I. x. 75 You can fill your filthy belly and get fat and sassy right before my eyes. **1961** *New Statesman* 11 Aug. 193/2 The film..is big and sassy, full of generous visual effects. **1969** N. Cohn *A Wop Bopa Loo Bop* (1970) xviii. 169 They looked at the things the Who did and analysed them and thought up sassy names for them. If the Who smashed up their instruments..was that violence? Certainly not: it was auto-destruction. **1974** K. Millett *Flying* (1975) III. 368 Celia's voice sweet brave sassy, 'Of course I'll be okay.' **1977** *Time* 7 Feb. 57/1 He has..a stand-up lush of an ex-wife..whose sassy words rain mockery on all. **1977** *Spare Rib* July 50/1, I learned a lot of things from the Beatles about sassiness. I always thought they were sassy, that was my label for them. **1979** *Arizona Daily Star* 5 Aug. I. 10/4 She plays a leading character, Persona Non Grata, a hip, wise, slightly sassy new friend of Alic. **1980** W. Safire in *N.Y. Times Mag.* 21 Sept., The Oxford American Dictionary, a sassy and helpful addition to any library. *Ibid.* 9 Nov., He initials 'MFC', a sadly sassy signal that means 'measure for coffin'.

Hence **sa·ssily** *adv.*; **sa·ssiness.**

1976 C. Weston *Rouse Demon* (1977) xii. 56 'Go ahead, ask me something,' he urged sassily. **1976** 'Trevanian' *Main* vi. 128 Now that his first panic is over, something of his haughty sassiness returns.

sastruga (sæstrū·gă). Chiefly as pl. **sastrugi** (-i). [a. G. *sastruga*, f. dial. Russ. *zastrúga* small ridge, furrow, f. *zastrugát'* to plane, smooth, f. *strug* plane.] One of a series of irregular ridges formed on a snow surface by wind erosion and deposition, aligned parallel to the direction of the prevailing wind.

1840 E. Sabine tr. *von Wrangell's Narr. Exped. Polar Sea* vii. 146 We were guided by the wave-like stripes of snow (sastrugi) which are formed, either on the plains on land or on the level ice of the sea, by any wind of long continuance. *Ibid.* 147 It often happens that the *true* permanent sastruga has been obliterated by another produced by temporary winds. **1878** E. L. Moss *Shores of Polar Sea* vi. 42 The sloping shore hills are barred with 'sastrugi'—wind-made ridges of snow—but the abrupt scooped-out rifts between them are smothered over with fleecy powder in gentle undulations. **1911** R. F. Scott *Jrnl.* in *Last Exped.* (1913) I. 517 The hard surface gave place to regular sastrugi. **1937** *Geogr. Jrnl.* LXXXIX. 195 He remarks on sastrugi 6 inches high, then for three days very few. **1960** *Times* 17 Feb. 10/3 Much of the traverse was conducted over a difficult surface rippled with *sastrugi*. **1975** E. Hillary *Nothing venture, Nothing Win* xiii. 243 The surface, which had appeared so smooth from above, was..liberally peppered with large sastrugi—some of them up to three feet in height. **1979** R. Fiennes *Hell on Ice* v. 70 The sastruga ripples we encountered along the summit ridges.

satai, var. *satay.

Satan. Add: **3.** Also *Comb.*, as *Satan-mad*; **Satan monkey**, substitute for def.: the black saki, *Chiropotes satanas*, which is found in dense forest in parts of South America and has thick reddish-black fur; (examples).

1918 W. de la Mare *Motley* 51 Not simple happy mad like me,..But that foul Satan-mad. **1906** E. Ingersoll *Life of Animals: Mammals* 44 (*caption*) Black Saki, Cuxio, or Satan Monkey. **1941** J. S. Huxley *Uniqueness of Man* ix. 205 Others, like..the Satan monkey with his fine beard, are curiously reminiscent of ourselves.

Satanist. 1. Delete 'Now *rare*' and add later examples.

1926 C. Connolly *Let.* 16 May in *Romantic Friendship* (1975) 126, I think he's a satanist. **1974** *Encycl. Brit. Macropædia* XIX. 899/2 Such modern satanists as Aleister Crowley and Gerald Gardner. **1976** *Eastern Even. News* (Norwich) 29 Nov., The sisters..are on the trail of a group of Satanists, believed to have caused a young man's death.

4. A writer of the 'Satanic school'.

1921 *Glasgow Herald* 9 Apr. 6/3 Thus he [*sc.* Baudelaire] is a Satanist in the Miltonic sense of a rebel against stifling power.

satay (sæ·tē¹). Also **satai, saté.** [Mal. *satai, sate*, Indonesian *sate.*] An Indonesian and Malaysian dish, consisting of small pieces of meat grilled on a skewer and usually served with a spiced sauce.

1934 *Willis's Singapore Guide* 149 'Satai' I am given to understand was introduced into this Country by the Chinese, the word being spelt 'Satae', meaning three pieces of meat. **1937** M. Covarrubias *Island of Bali* v. 108 The *saté* can be made of pork or chicken, but turtle remains the favourite of the Balinese of Den Pasar. **1955** P. Anderson *Snake Wine* II. vi. 163 The Malays crouch over their portable stoves, fanning the embers below sticks of spicy broiled goat known as *satay*. **1967** L. Deighton *London Dossier* 56 You can eat Malay Satay in the Singapore restaurant in Allen Street, W8. **1971** *Carry Singapore in your Pocket* (Singapore Tourist Promotion Board) (ed. 3) 30 One of the most famous Malay dishes is satay which is tenderised and spiced mutton, chicken or beef barbecued over charcoal and dipped in a chilli-hot peanut sauce. They are served skewered. **1971** *National Geographic* Jan. 16/2 Saté consists of bits of meat skewered on bamboo slivers, grilled over charcoal, and served with a spicy peanut sauce. **1976** *Outdoor Living* (N.Z.) I. II. 64/1 The sate is Asia's answer to the shishkebab. The sate is usually all meat, beautifully spiced and traditionally served on small wooden skewers. **1980** *Times* 5 July 11/2 A menu that ranges from Indonesian satay..to Persian khoresh faisinjan.

‖ **sat-bhai** (sātbāi). Also **saht-bai, sathbhai.** [Hind. *sātbhāi*.] An Indian jungle babbler, *Turdoides striatus*, a large brown bird with a long tail and slightly curved bill; = Seven sisters 4.

1863 T. C. Jerdon *Birds of India* II. 65 It [*sc.* the large grey babbler] leaves the jungles and wilds, and becomes the familiar and unscared..*Sat bhai*. **1886** Kipling *Departmental Ditties* (ed. 2) 62 The blue jay screams and flutters where the cheery *sat-bhai* dwell. **1928** H. Whistler *Pop. Handbk. Indian Birds* 32 The vernacular name [of the jungle babbler] is 'Sathbhai', the Seven Brethren. **1953** S. Ali *Birds of Travancore & Cochin* 28 A frowzled, untidy-looking earthy brown bird..invariably in flocks of half a dozen or so, whence its popular Hindustani name Sātbhai (= seven brothers). **1978** 'M. M. Kaye' *Far Pavilions* II. 27 The normal noises of an Indian morning:..the harsh cry of a peacock..and the chatter and chirrup of tree-rats, *saht-bai* and weaver-birds.

satchel, *sb.* Add: **2. satchel charge** (see quot. 1973).

1961 in Webster. **1969** *New Yorker* 20 Sept. 145/1 Setting off satchel charges and other explosives at police stations. **1973** J. Quick *Dict. Weapons* 385/1 *Satchel charge*, a number of blocks of explosive taped to a board fitted with a rope or wire loop for carrying and attaching. **1977** *Time* 20 June 6/3 The troops used satchel charges to widen the gap made by the armored car, causing thunderous explosions that awoke sleeping villagers.

sate, *v.* Add: **1. e.** *intr.* (for *refl.*). To become sated. *rare.*

1869 Browning *Ring & Bk.* IV. XI. 179 Let me turn wolf, be whole, and sate, for once.

sate (sēit), *sb. Blacksmithing.* [Var. Set *sb.*[1]] A heavy chisel or punch used for cutting metal. Cf. Set *sb.*[1] 33 in Dict. and Suppl.

1906 T. Moore *Handbk. Pract. Smithing & Forging* II. 15 The cold sate..is a very simple tool in itself, and easy to make. *Ibid.* 17 The hot sate..is made in much the same way as the cold sate. **1942** W. H. Atherton *Workshop Pract.* (ed. 2) V. 198 Making two small holes..by slitting with the hot sate and opening out slightly..will widen the hole sufficiently to take a drift of the size required. **1962** [see *set *sb.*[1] 33].

saté, var. *satay.

sateen. Add: **2.** *Comb.*, as *sateen-backed* adj.

1939–40 *Army & Navy Stores Catal.* 629/2 Down quilts.. Figured Rayon Marocain, Sateen backed. **1960** *Farmer & Stockbreeder* 15 Mar. (Suppl.) 4/1 This wool-lined, sateen-backed quilted pad, with elastic waist belt, fits snugly.

sateless, *a.* Add: Also const. *in*.

1935 L. Luard *Conquering Seas* 6 The heedless voice of the land sateless in greed.

satellite, *sb.* Add: **2. c.** A man-made object placed (or designed to be placed) in orbit around an astronomical body (usu. the earth).

[**1880** W. H. G. Kingston tr. *Verne's Begum's Fortune* xiii. 180 A projectile, animated with an initial speed twenty times superior to the actual speed, being ten thousand yards to the second, can never fail! This movement, combined with terrestrial attraction, destines it to revolve perpetually round our globe... Two hundred thousand dollars is not too much to have paid for the pleasure of having endowed the planetary world with a new star, and the earth with a second satellite.] **1936** *Discovery* Sept. 299/2 The scheme for building a metal outpost satellite and propelling it in a fixed orbit 600 miles above the earth's surface. **1945** A. C. Clarke in *Wireless World* Oct. 305/2 This 'orbital' velocity is 8 km per sec. (5 miles per sec), and a rocket which attained it would become an artificial satellite, circling the world for ever with no expenditure of power. **1955** *Times* 30 July 6/1 The satellite is expected to be about the size of a basketball, and will be shot into the upper atmosphere by a rocket, where it will circle the earth at an altitude of between 200 and 300 miles at a speed of about 18,000 miles an hour. **1956** *Spaceflight* I. 6/2 After the Earth satellite stage, the next target will almost certainly be the Moon. **1957** *Ibid.* 49/1 Each satellite will be launched into its orbit by being ejected from the third stage of a multiple stage rocket. **1957** *Times* 7 Oct. 8/1 The Russian satellite soaring over the United States seven times a day has made an enormous impression on American minds. **1961**, etc. [see *communication(s) satellite* s.v. *communication 12]. **1964** *Ann. Reg. 1963* 185 Among other notable American achievements in space during the year was the launching of a communications satellite. **1972** *Computers & Humanities* VII. 49 An experiment..was conducted during the fall of 1971 at Stanford, where users were able to communicate with a computer by using NASA's ATS-1 experimental satellite. **1977** *Times* 16 Dec. 16/1 Killer satellites are small spacecraft. They carry an explosive charge which destroys itself and any nearby satellite on detonation.

6. a. A country or state politically or economically dependent upon and subservient to another.

[**1776** T. Paine *Wks.* (1796), II. 24 In no instance hath nature made the satellite larger than its primary planet; and as England and America..reverse the common order of nature, it is evident that they belong to different systems: England to Europe, America to itself.] **1800, 1827** [in Dict., sense 2b]. **1930** *Economist* 8 Nov. 844/2 Do they portend a military alliance against France between a Fascist Italy and a Fascist Germany, with a bevy of East European satellites—Bulgaria, Albania, Hungary, Austria—to balance Poland and the Little Entente? **1936** *Pacific Affairs* Sept. 404 Outer Mongolia may well be called a satellite of the Soviet Union. **1941** *Ann. Reg. 1940* 204 This [*sc.* the Tripartite Pact of the Axis Powers] made Hungary a mere satellite of Germany. **1948** *Sun* (Baltimore) 9 Jan. 1/2 Several of the Soviet Union's satellites. **1974** M. B. Brown *Econ. of Imperialism* xii. 286 Cuba is not a satellite of the USSR in the same sense that other Latin American States are satellites of the USA. **1977** *Time* 21 Feb. 8/1 In Czechoslovakia, East Germany, Poland and even some of the less volatile satellites, the Russians and their local rulers are being forced to put out brushfires of discontent.

b. A community or town that is economically or otherwise dependent on a nearby larger town or city.

1912 G. R. Taylor in *Survey* (N.Y.) 5 Oct. 14/2 In some sections of the South scarcely a city of any size lacks one or more satellites thrumming with spindle and shuttle. **1935** *Archit. Rev.* LXXVII. 188 (*caption*) 19th Century. Came the railways and with them the first general exodus, suburbs and satellites springing up round the railway stations. **1947** [see *overspill *sb.* a]. **1958** *Manch. Guardian* 30 June 6/2 And if Manchester itself is some way from Tatton, Manchester's proposed satellite at Lymm is much nearer. **1977** *R.A.F. News* 27 Apr.–10 May 8/2 No. 50(B) Squadron was then based at Skellingthorpe, west of Lincoln (a satellite of Swinderby).

7. *Spectroscopy.* A spurious or subordinate spectral line; *spec.* one caused by an irregularity in the positions of lines in a diffraction grating. Also *satellite line*.

1904 *Astrophysical Jrnl.* XIX. 118 The appearance and disappearance, according to circumstances, of the satellite lines still remains a most curious fact. **1924** *Phil. Mag.* XLVIII. 501 On moving the eyepiece back, the line broadened and a faint black 'satellite' split off from it, moving slowly across the grating. **1945** R. A. Sawyer *Exper. Spectrosc.* vii. 175 It often happens that satellites or diffuse edges will be observed for strong lines at the best obtainable focus. **1969** [see *Rowland ghost*]. **1971** *Physics Bull.* July 388/3 The centre line is due to Rayleigh scattering and the satellites arise from transverse (T) and longitudinal (L) phonons.

8. *Anat.* Chiefly as *satellite cell*. Each of the cells that go to make up the membrane surrounding the nerve cell bodies in many ganglia, analogous to the Schwann cells that surround their axons; also, formerly, a Schwann cell.

[**1908** G. Marinesco in *Compt. Rend. Hebdom. des Séances et Mém. de la Soc. de Biol.* LXV. 99 De toutes ces recherches, il résulte qu'il existe à l'état normal un équilibre entre la nutrition des cellules satellites et celle des cellules des ganglions sensitifs.] **1928** W. Penfield in E. V. Cowdry *Special Cytol.* II. xxx. 1055 Specific stains showed the perivascular and perineuronal oligoglia satellites to be definitely increased. **1954** M. Singer in R. O. Greep *Histology* xi. 216 Each cell body of spinal, cranial, and autonomic ganglia is completely encapsulated by a thin membrane composed of so-called satellite cells which contains small, scattered, and flattened nuclei. **1958** *Exper. Cell Res. Suppl.* V. 33 The structural characteristic which is present in all fibers so far studied..is the Schwann or satellite cell which..appears everywhere to enclose the axon. **1960** G. Causey *Cell of Schwann* v. 69 The regeneration of nerve fibres and their satellite cells in the tail of the tadpole. **1971** W. M. Copenhaver et al. *Bailey's Textbk. Histol.* (ed. 16) x. 259/1 When these companion cells are in association with a nerve cell body.., they are called satellite cells; when they provide ensheathment for axons, they are called neurilemma cells, or cells of Schwann.

9. *Cytology.* A short section of a chromosome demarcated from the rest by a constriction (if terminal) or by two constrictions (if intercalary). [The sense is due to S. G. Navashin, who used Russ. *spútnik* satellite (*Izvestiya Imper. Akad. Nauk* (1912) VI. 378).]

1926 C. D. Darlington in *Jrnl. Genetics* XVI. 246 Chromosome 'G' is seen to be approaching the pole with the satellite foremost; this means that the satellite is endowed with special responsiveness to the attraction of the pole. **1960** *Lancet* 14 May 1063/2 In some chromosomes the additional criterion of the presence of a satellite is available (table 1), but in view of the apparent morphological variation of satellites, they and their connecting strands are excluded in computing the indices. **1975** A. & D. Löve *Plant Chromosomes* I. i. 26 A secondary constriction may demarcate a short part of the chromosome, either intercalary or, most frequently, terminally. Such a terminal piece is called a satellite.

10. *Bacteriol.* A bacterial colony growing in culture near a second colony which is the source of a diffusible substance which promotes the growth of the first but is not produced by it; it consequently shows accelerated growth, or resists a substance which would otherwise poison it. Usu. *attrib.*

1938 in Dorland & Miller *Med. Dict.* (ed. 18) 1243/1. **1940** M. Frobisher *Fund. Bacteriol.* (ed. 2) xxv. 355 (*caption*) 'Satellite' formation by *Hemophilus influenzae* on 'chocolate-agar' plate. **1943** *Jrnl. Bacteriol.* XLV. 522/1 The development of satellites depended upon the concentration of sulfonamide, the susceptibility of the satellite strain, the temperature of incubation, and the size of the inoculum of both satellite and inhibitor. **1975** *Jrnl. Clin. Microbiol.* I. 90/2 The satellite growth of *Haemophilus* species around a colony of *Staphylococcus* can be attributed not only to NAD but also to catalase, which is produced by staphylococci.

11. *Molecular Biol.* A portion of the DNA of a genome distinguished from the rest of the genome by its distinctive base composition and density. Freq. *attrib.*

1961 S. Kit in *Jrnl. Molecular Biol.* III. 711 The mean buoyant densities of the principal and the satellite mouse DNA bands were 1·701 and 1·690 g cm⁻³, respectively. **1962** *Ibid.* IV. 439 Calf thymus satellite was found at the same position in each of three different DNA preparations isolated from thymus tissue obtained from different animals. **1970** *New Scientist* 27 Aug. 406/1 Discovered originally in the mouse, where it constitutes some 10 per cent of the total DNA in each cell of the animal, satellite DNA can be distinguished from the rest by its different density, and by the fact that it apparently consists of repeating base sequences—i.e., multiple copies of a given sequence repeated again and again. **1977** Rees & Jones *Chromosome Genetics* ii. 22 Exceptional DNA segments may have an unusually high or low G + C content. When plotted, these fractions appear as heavy or light satellites respectively at the tails of the 'main-band' DNA. Heavy satellites are found in the guinea pig and in human DNA. Light satellites ..are less common.

12. Used *attrib.* to designate a computer or computer terminal distant from, but connected to and serving, a main computer.

1966 C. J. Sippl *Computer Dict. & Handbk.* 278 As a satellite system the real-time system relieves the larger system of time consuming input and output functions as well as performing preprocessing and postprocessing functions. **1970** O. Dopping *Computers & Data Processing* vi. 95 Input data in cards or paper tape are converted to magnetic tape by the satellite computer. **1971** E. F. Schoeters in B. de Ferranti *Living with Computer* viii. 68 The way in which their huge networks of small satellite computers, or calculating terminals, connected to big machines in London behave..will show just how much more work has to be done.

13. *attrib.* and *Comb.*, as (sense *2 c) *satellite camera, communication(s), killer, launcher, navigation, observatory, programme, -tracking*; *satellite-borne* adj.; *satellite-to-home* adj. phr.; (sense *6) *satellite city, community, country, government, nation, state, town, township*; **satellite airfield,** an airfield auxiliary to and serving, if necessary, as a substitute for a larger airfield; **satellite photo(graph),** a photograph taken from an artificial satellite; so **satellite photography**; **satellite picture,** a satellite photograph; **satellite station,** (*a*) an artificial satellite; *spec.* (see quot. 1950); (*b*) a secondary radio station which receives and retransmits programmes, so as to improve local reception; **satellite telescope,** a telescope in orbit beyond the range of atmospheric distortion; **satellite television,** television in which the signal is transmitted via an artificial satellite.

1941 F. H. Joseph *Lett. Home from Brit. at War* (1942) 38 Clear skies over West Raynham's satellite airfield, Massingham. **1951** O. Berthond tr. *P. Clostermann's Big Show* I. 20 We spent the last three weeks of our training at Montford Bridge, a small satellite airfield lost in the hills. **1968** *Wall St. Jrnl.* 25 Sept. 36/1 Flight delays at World Chamberlain and the satellite airfields are almost non-existent. **1962** W. B. Thompson *Introd. Plasma Physics* i. 4 Recently, rocket- and satellite-borne counters have detected belts of energetic radiation, electrons and ions, high above the earth's atmosphere. **1974** *Sci. Amer.* June 132/2 Within less than a decade the bulk of transoceanic telephony (and all transoceanic television) has become satellite-borne. **1963** Satellite camera [see *satellite picture* below]. **1966** P. O'Donnell *Sabre-Tooth* xiv. 185 The end of the journey..was on neutral ground, in an area where spy-plane or satellite cameras would never seek. **1912** G. R. Taylor in *Survey* (N.Y.) XXIX. 5 Oct. 23/1 Like camp sutlers, the traffickers in demoralization are quick to follow the trail of satellite cities. **1960** *Washington Post* 20 Dec. A14 They urge that the growth of this region from some 4 million to 9 million persons in the remainder of this century be organized in a pattern of some 50 new satellite cities, each of 75,000 to 150,000 population. A dozen of them would fill the corridor between Baltimore and Washington. **1977** *New Yorker* 13 June 94/2 The new Taichung port..is to include a separate satellite city. **1959** J. H. Straubel et al. *Space Weapons* 243 (Index), Satellite communication. **1960** *Signal* XIV. 32/1 A means of communication is needed that will immediately provide several hundred channels linking key cities throughout the world. This requirement will be filled by a satellite communication system. **1961** *Times Rev. Industry* Feb. 26/3 Last autumn a team of British experts visited the United States to discuss with their opposite numbers the feasibility

of establishing a satellite communications system. **1964** *Economist* 1 Aug. 481/2 Complex legal controversies arising from satellite communications systems. **1946** *Nature* 13 July 39/2 The Manchester request for compulsory powers to buy land for the creation of satellite communities. **1970** R. Stavenhagen in I. L. Horowitz *Masses in Lat. Amer.* vii. 254 Not only in the city but also in the 'satellite communities' is commerce usually in Ladino hands. **1956** *Times* 7 Feb. 8/5 Dropping leaflets over the satellite countries..was begun by Radio Free Europe in April, 1954. **1969** A. G. Frank *Latin Amer.* (1970) i. 4 Relations between the satellite underdeveloped and the now developed metropolitan countries. **1976** B. Freemantle *November Man* iv. 43 The Americans actually believe we [*sc.* the Russians] are going to withdraw all our troops from the satellite countries. **1949** Koestler *Promise & Fulfilment* I. xii. 133 Experts of the Foreign Office..tried to set up a puppet Jewish Agency as a kind of satellite Government. **1977** *Guardian Weekly* 2 Oct. 15/2 A new weapon that could destroy Soviet satellites in space... Vought is expected to have a battle version of the satellite killer ready to test in space in about two years. **1977** *Time* 17 Oct. 32/1 The U.S. will now emphasize efforts to design an American satellite killer to defend against the Soviet version. **1959** *Daily Tel.* 2 July 5/5 This satellite launcher is about 110 ft long and 15 in in diameter at the base. **1961** *New Scientist* 19 Jan. 133/1 Several of these countries will discuss the specific proposal for the development of a satellite-launcher based on *Blue Streak*. **1916** C. M. Meredith tr. *F. Naumann's Central Europe* vi. 180 What is meant by a satellite nation..? We might also say a planet State. Such States have their own life. **1956** E. E. Cummings *Let.* 26 Nov. (1969) 253 Urging (via night & day broadcasts) the socalled satellite nations to revolt from colossal Russia. **1967** *Oceanogr. & Marine Biol.* V. 145 In February 1965 *Atlantis* II returned to the area to carry out a hydrographic and coring survey of this area using a satellite navigation system and ship-board computer for the location of this small area. **1975** *Offshore Progress—Technol. & Costs* (Shell Briefing Service) 7 With satellite navigation, however, the rig can fix its own position by computer, processing signals received from orbiting satellites. **1953** J. N. Leonard *Flight into Space* 159 They suspect that the human intellect is approaching a boundary of mystery which its present tools cannot penetrate. Some of them feel that the satellite observatory may be the necessary tool. **1976** H. Kemelman *Wednesday the Rabbi got Wet* xii. 61 The noon broadcast had been almost entirely devoted to..Hurricane Betsy. There were..satellite photos of the eastern coast. **1963** Van Dijk & Rutherford in Wexler & Caskey *Rocket & Satellite Meteorol.* 305 Satellite photographs were obtained of a cut-off low over southeast Australia. **1977** A. Hallam *Planet Earth* 43 (*caption*) A satellite photograph of the Andes. **1971** P. O'Donnell *Impossible Virgin* v. 107 I'll have it checked by our own Map Section... There's *something* there which is detectable by satellite photography. **1963** Van Dijk & Rutherford in Wexler & Caskey *Rocket & Satellite Meteorol.* 305 Facility in interpretation of meteorological satellite pictures can best be achieved by exercises in which clouds of known type and distribution are charted and compared with pictures of the same cloud taken by satellite camera. **1977** L. P. White *Aerial Photogr. & Remote Sensing for Soil Survey* vii. 73 Early examination of coverage of this kind did, however, serve to indicate the possibility of using automatic satellite pictures for purposes other than meteorology and oceanology. **1959** *Daily Tel.* 13 May 1 Britain has decided to take the essential steps to enable scientists here to participate in a satellite programme. **1916** C. M. Meredith tr. *F. Naumann's Central Europe* vi. 181 Round about the satellite States there still exists a certain mass of unorganised national material. **1943** *Ann. Reg. 1942* 176 Their [*sc.* Pan-Germans'] plan was that Germany..should carve out in the Danube basin several satellite states. **1950** *Sun* (Baltimore) 17 July 11/2 Fortifications toughening the ragged western borders of central Europe's satellite states. **1976** *Survey* Summer–Autumn 41 Here was the authentic voice of the unconscious Western desire to believe that the satellite states of the Soviet Union were free. **1945** *Wireless World* Oct. 306 (*caption*) Three satellite stations would ensure complete [radio] coverage of the globe. **1950** W. Proell *Handbk. Space Flight* 174 *Satellite station*, synonym for space station... *Space station*, a habitable vehicle placed in a satellite orbit around a planetary body, for use in refueling of space ships, communications relaying, or military use. **1954** E. Pangborn *Mirror for Observers* (1955) I. i. 21, I understand men will have their first satellite station in a very short time, four or five years. **1959** *Times Lit. Suppl.* 30 Oct. 631/4 The cost of building a moon rocket at a satellite station, including the fuel of the rockets carrying the materials, he estimates as £40m. **1959** *Proc. Inst. Electr. Engineers* V. 416/1 A number of low-power satellite stations are therefore planned... They will be designed to..pick up signals from an existing B.B.C. station and retransmit them on a different channel for local reception. *Ibid.*, The B.B.C.'s plan for extending and improving the coverage of the television service and of..sound services on v.h.f. by building low-power satellite stations in various parts of the country. **1962** *Rep. Comm. Broadcasting 1960* 197 in *Parl. Papers 1961–2* (Cmnd. 1753) IX. 259 It is possible to provide low-powered relay stations..to extend coverage still further... These satellite stations..have been planned as a stage by stage project. **1951** J. P. Marbarger *Space Med.* 26 If we turn such a satellite telescope to the outer reaches of the universe, the planets and the stars, we shall find observation conditions which no terrestrial observatory could equal. **1960** *Aeroplane* XCIX. 358/1 It turns out that this is a design study into a stabilised platform for a small satellite telescope. **1966** *B.B.C. Handbk.* 53 The BBC's first satellite television transmissions were shown in 1962. **1971** L. Koppett *N.Y. Times Guide Spectator Sports* xii. 194 Satellite television. **1967** *Economist* 1 July 32/2 What about lasers? What about direct satellite-to-home broadcasting?..Perhaps the only way in which the federal government could expect to keep abreast of the developments in communications technologies would be to set up a Department of Communication. **1973** *Computers & Humanities* VII. 226 The uses of such wonders as switched data networks, computer terminals, mobile radio transceivers, and satellite-to-home-receiver television transmission.

1925 C. B. Purdom (*title*) The building of satellite towns. **1929** *Times* 17 July 17/6 Since neither complete decentralization nor the proposal to 'departmentalize' the government of Greater Paris is found to give general satisfaction, the system of 'satellite towns' has been suggested as a way out. **1933** *Archit. Rev.* LXXIV. 166/2 The proposed formation of a ring of satellite towns around the immediate radius of London. **1946** F. J. Osborn *Green Belt Cities* I. 182 *Satellite Town*. This term was first used in Great Britain in 1919 as an alternative description of Welwyn Garden City... Some planning writers have thoughtlessly renewed the old confusion by using the term Satellite Town to describe an Industrial Garden Suburb. It is better reserved for a Garden City or country town, at a moderate distance from a large city, but physically separated from that city by a Country Belt. **1955** *Sci. Amer.* Jan. 40/3 As population continues to move from cities out to ever more distant suburbs and satellite towns [etc.]. **1971** *Rand Daily Mail* 27 Mar. 3/7 A giant new satellite township near Pretoria.. will provide housing..for about 200 000 White people. **1958** A. Budrys in Aldiss & Harrison *Decade 1950s* (1976) 68 I'm assigned to the satellite-tracking station. **1969** *Listener* 20 Feb. 233/2 Satellite tracking is not as easy as it appears.

14. *attrib.* passing into *adj.* That is a satellite to something else; subsidiary, subordinate; associated; ancillary.

1892 B. Potter *Jrnl.* 8 Aug. (1966) 245 We..found the thirteen or fourteen vans drawn up in the town square, and covered with a tarpaulin, with several satellite peep shows. **1923** N. Shaw *Forecasting Weather* v. 115 Two detached secondary or satellite depressions. **1931** *Economist* 17 Oct. 699/1 The Indian currency and..the various 'satellite' currencies of the Crown Colonies and Possessions. **1939** *Oxoniensia* IV. 13 Post-holes 1.., 3 and 6 were also provided with from two to four satellite sockets and slots for supports. **1949** *Caribbean Q.* I. III. 43 A central model farm..would carry on intensive dairy farming... The satellite farms would be run by skilled farmers. **1957** *Observer* 8 Sept. 7/3 When fashion makes a decisive move innumerable satellite trades are affected. **1965** B. Sweet-Escott *Baker St. Irregular* iii. 77 This was to be their home for the next four years and became in due course surrounded by a series of satellite premises. **1967** *Boston Sunday Globe* 23 Apr. (Mag.) 33/1 Satellite clinics for children and pregnant mothers..run jointly by several Harvard affiliated hospitals and the City of Boston. **1969** *Wall St. Jrnl.* 1 Dec. 9/1 Pan Am..is trying to sell passengers on use of the 'satellite' terminal facilities around the New York metropolitan area. **1972** *Accountant* 26 Oct. 518/2 Satellite reports, or supplementary reports, would be prepared for the particular interests of particular users. **1976** *NBR Marketplace* (Wellington, N.Z.) III. 37/2 The satellite seminar was joined by dozens of doctors and nurses. **1976** *Offshore Engineer* July 20/3 A cluster of 10 wells with four satellite wells for water and gas injection.

satellite (sæ·tėləit), *v.* [f. prec.] **1.** *intr.* To orbit like a satellite.

1959 *IRE Trans. Military Electronics* III. 62/2 Mission periods of the order of one year (including a brief period.. of satelliting about the target planet).

2. *trans.* To transmit by way of a communications satellite.

1974 *Listener* 14 Dec. 826 The telephone woke me. It was Peter Lynch, our contact in Tel Aviv (from where our film was being satellited). **1976** A. Davis *Television* iv. 50 During the war in Cyprus in 1974, film shot by British cameramen was flown to Tel Aviv where it was processed, then satellited to Rome, where it was fed into the Eurovision network. **1978** *Broadcast* 23 Oct. 5/1 BBC TV News reporter Bob Friend..satellited the pictures to London from Tai Pei.

satellited (sæ·tėləitėd), *a.* [f. Satellite + -ed[2].] **1.** (In Dict. s.v. Satellite.)

2. *Cytology.* Having a satellite or satellites (*satellite *sb.* 9).

1934 L. W. Sharp *Introd. Cytol.* (ed. 3) xviii. 319 (*caption*) Synaptic configurations..after deletion of portion of shorter arm of satellited chromosome. **1938** *Jrnl. R. Microsc. Soc.* LVIII. 103 Every primary diploid plant or animal generally has one pair of satellited chromosomes which..produce by fusion a single nucleolus. **1971** *Nature* 21 May 195/1 In the Australasian superfamily Dasyuroidea, all species have seven pairs of chromosomes, the autosomes including..two small pairs of chromosomes of which one has a satellited short arm.

satellitism (sæ·tėləitiz'm). [f. Satellit(e + -ism.] **1.** *Bacteriol.* The occurrence of satellites (*satellite *sb.* 10); the promotion of bacterial growth by the proximity of a colony of different bacteria.

1951 Whitby & Hynes *Med. Bacteriol.* (ed. 5) xvi. 282 Staphylococci secrete enough of the factor to stimulate growth of *H[æmophilus] influenzæ*; colonies of the latter on a blood plate are always larger when they lie near a staphylococcal colony (satellitism). **1975** *Jrnl. Clin. Microbiol.* I. 89 (*heading*) New satellitism test for isolation and identification of *Haemophilus influenzae* and *Haemophilus parainfluenzae* in sputum.

2. *Pol.* The fact or condition of being a satellite (state); the role of a satellite.

1955 O. Lattimore *Nationalism & Revolution in Mongolia* 41 (*heading*) Anatomy of satellitism. **1962** —— *Nomads & Commissars* viii. 155 Stalin..and Mongolia's loyalty to the Soviet alliance bring up the question of satellitism. **1964** *Economist* 10 Oct. 101/1 Only by helping to create a state much closer in strength to the superpowers can we escape satellitism or neutralism. **1969** D. Widgery in Cockburn & Blackburn *Student Power* 137 Wilson's satellitism to Washington forces an attack on Government spending.

satellitosis (sætlėəitō[u]·sis). *Path.* [f. Satellit(e + -osis.] A proliferation of neuroglial cells around nerve cells in the brain.

1928 W. Penfield in E. V. Cowdry *Special Cytol.* II. xxx. 1055 From this satellitosis a perivascular and even a perineuronal leucocytic infiltration must be carefully distinguished. **1969** Brown & Bertke *Textbk. Cytol.* xxiii. 549 Oligodendrocytes..may increase in number during aging to produce a satellitosis, which is often seen in aging organs. **1979** *Jrnl. Compar. Path.* LXXXIX. 490 Some of the ganglion cells showed satellitosis.

satellize (sæ·tėləiz), *v.* [f. Satell(ite + -ize.] **1.** *intr.* To cluster *about*. *rare.*

1916 E. V. Lucas *Variety Lane* 60 A little band of important men hurried up, satellizing about a quiet, gentle-looking but distinguished man.

2. *trans.* To make into a political or economic satellite.

1951 *Melbourne Herald* 16 Apr., Dr. W. E. Stanner.. introduced..the verb *to satellise*. Dr. Stanner used it when referring to other countries which, in given conditions, Russia might *satellise*. **1965** *Observer* 19 Sept. 2/3 Pakistan.. will not become a satellite of India; but..she will not be satellised by China either.

Hence **satelliza·tion**, the action of making into a satellite; the condition or process of being satellized; **sa·tellized** *ppl. a.*

1958 *Times* 13 May 8/5 Mr Rountree, Assistant Secretary of State..told the committee:..'Satellization of the Middle East now seems less a danger than it did a few months ago.' **1962** *Economist* 12 May 551/2 A small communist country which..wants to escape satellisation by China. **1968** 'Han Suyin' *Birdless Summer* I. iii. 56 Japan's terms were the permanent satellization of China. **1969** A. G. Frank *Latin Amer.* (1970) i. 7 The satellized national, regional, and local metropoles in Latin America find that their economic development is at best a limited or underdeveloped development. **1976** *Globe & Mail* (Toronto) 18 June 7/3 In spite of the somewhat unsure character of its national identity and its excessive satellization by the American economic and cultural empire, Canada-without-Quebec has enough 'difference' left, [etc.].

satelloid (sæ·tėloid). [f. Satell(ite + -oid.] A craft designed to follow approximately a free-fall orbit, but to expend power to overcome air resistance or to change its course.

1955 *Times* 3 Aug. 8/5 He has constructed a special earth satellite, the 'Satelloid' which goes to an altitude of 100 miles and from there is moved forward by an engine. **1956** *Jrnl. Brit. Interplanetary Soc.* XV. 166 Dr. Krafft A. Ehricke, of the guided missile group of Convair, has suggested that a weakly-powered satellite might be placed in a lower orbit than that required for an unpowered satellite, and used the word 'satelloid' to describe such a vehicle when discussing it at the I.A.F. meeting in Copenhagen. He said that the satelloid might be placed in an orbit at 80 miles altitude. **1960** *Aeroplane* XCVIII. 496/1 The authors examine various statements which have been made by U.S. military leaders on the merits of arching ballistic missiles, jump-down bombs, variable orbit satelloids, and boost-glide devices.

satem (sā·təm). *Philol.* Also **satəm.** [f. Avestan *satəm* hundred, from its pronunciation with (s), as opposed to *centum: first used by P. von Bradke 1890 in *Über Methode und Ergebnisse der arischen Alterthumswissenschaft* I. iv. 63.] A name given by philologists to one, chiefly eastern, group of Indo-European languages, distinguished by their use of sibilants where the corresponding sounds in cognate words in the western group (cf. *centum) are velar stops.

1901, etc. [see *centum]. **1933** L. Bloomfield *Language* xviii. 316 Many scholars suppose that the earliest traceable division of the Primitive Indo-European unity was into a western group of so-called '*centum*-languages' and an eastern group of '*satem*-languages'. **1952** O. R. Gurney *Hittites* vi. 119 The main characteristics of the Indo-Iranian (or so-called 'Satem') languages (change of original *k* to *s*, *qu* to *k*, and *e* and *o* to *a*). **1973** *Word 1970* XXVI. 3 The time when the back velar stops moved forward in satem languages.

sater, var. *saeter, setter.

sathbhai, var. *sat-bhai.

'satiable, colloq. reduced form of Insatiable *a.* Used in phr. *'satiable curiosity* in allusion to Kipling's *Just So Stories* (see quot. 1900).

1900 Kipling *Just So Stories* (1902) 63 There was one Elephant..—an Elephant's Child—who was full of 'satiable curtiosity, and that means he asked ever so many questions. **1963** L. Egan *Run to Evil* i. 6 Talk about the Elephant's Child... Nobody could really dislike the Brandon boy, even with his 'satiable curiosity. **1974** K. Benton *Craig & Tunisian Tangle* v. 52 She's like the Elephant's Child, full of 'satiable curiosity.

satiated, *ppl. a.* Add: (Later examples, esp. in sense of *satiation b and *satiety 1 e.)

1935 Adams & Zener tr. *Lewin's Dynamic Theory of Personality* viii. 255 Both agreeable and disagreeable tasks are comparatively more rapidly satiated than neutral ones. **1969** J. D. Davis et al. in *Jrnl. Compar. & Physiol. Psychol.* LXVII. 407 Intake of milk by fasted rats was reduced 50% below normal after their blood had been

transfused with that of satiated rats. **1975** Schneider & Tarshis *Physiol. Psychol.* xvi. 283 These studies have shown that the size of the cells in the ventromedial hypothalamus are larger and thus presumably more active in satiated animals than in deprived animals.

satiation. Add: **b.** *Psychol.* The point at which satisfaction of a need or familiarity with a stimulus reduces or ends an organism's responsiveness or motivation. Also *attrib.*

1935 Adams & Zener tr. *Lewin's Dynamic Theory of Personality* viii. 254 The progressive process of satiation is evidenced by such typical criteria as variation, dissolution of the whole.., inattention, forgetting. **1944** Köhler & Wallach in *Proc. Amer. Philos. Soc.* LXXXVIII. 276/1 We propose to call only the alterations of *T*-objects 'figural after-effects' and to refer to the affection of the medium as 'satiation'. **1954** Woodworth & Schlosberg *Exper. Psychol.* (rev. ed.) xiv. 426/1 Satiation..is not offered as an explanation of the illusion itself, but as a cause of its reduction and final destruction. **1967** J. R. Millenson *Princ. Behavioral Analysis* (1969) xv. 367 There are drive operations that reduce or eliminate reinforcing value... The most universal of these is *satiation*—repeatedly presenting the reinforcer until it loses its power to reinforce. **1975** Schneider & Tarshis *Physiol. Psychol.* xvi. 276 Studies have confirmed the notion that the ventromedial hypothalamus comes into play during satiation to inhibit eating. *Ibid.* 283 The transfused rats no longer seemed to be hungry... Davis took this to mean that the blood does carry an off, or satiation, signal. **1978** F. Leukel *Essent. Physiol. Psychol.* xii. 197/2 Satiation stimuli are more readily aroused.

satiety. Add: **1. e.** *Psychol.* Satisfaction of a need (esp. hunger) as it is registered physiologically; also *attrib.* and *Comb.*, as *satiety hormone, mechanism, process*; **satiety centre**, an area of the brain concerned with the regulation of food intake.

1951 *Amer. Jrnl. Physiol.* CLXIV. 186 The physiological release of enterogastrone is apparently not involved in the production of satiety. **1962** *Science* CXXXV. 374/2 The so-called 'feeding center' of the lateral hypothalamus and the 'satiety center' of the medial hypothalamus are well known. **1969** J. D. Davis et al. in *Jrnl. Compar. & Physiol. Psychol.* LXVII. 407/1 Is food intake..regulated by a 'satiety hormone' which terminates feeding when it reaches a threshold level? **1971** K. H. Pribram *Lang. of Brain* x. 192 Somehow the lesion had impaired the patient's *feelings* of hunger and satiety and this impairment was accompanied by excessive eating! *Ibid.* 195 The term 'motivation' can be restricted to the operations of appetitive 'go' processes..and the term 'emotion' to the operations of affective 'stop' or satiety processes of equilibrium. **1974** J. Olds in W. R. Adey et al. *Brain Mechanisms* vii. 379 In one of these areas, known as the 'satiety center', destruction of tissues caused animals to overeat and become obese. **1975** F. P. Valle *Motivation* xii. 227 There are several hypotheses regarding the variables that govern the activity of the 'satiety center' in the ventromedial nuclei. **1977** N. R. Carlson *Physiol. of Behav.* xii. 324 The fact that we stop eating before a significant amount of food is digested makes it necessary to postulate a satiety mechanism. *Ibid.* 325 Satiety has many sources, from several kinds of detectors. **1978** F. Leukel *Essent. Physiol. Psychol.* xii. 203/1 The first center is the ventromedial nucleus of the hypothalamus. This nucleus appeared to function as a satiation, or satiety, centre.

satin, *sb.* (and *a.*) Add: **I. 1. c.** A woman's satin dress.

1787 'T. Wignell' *Contrast* I. 2 She is to be married in a delicate white sattin. **1866** Mrs. Gaskell *Wives & Daughters* I. xxvi. 287, I remember the time when Mrs Kirkpatrick wore old black silks..and now she is in a satin. **1932** [see *low-cut* s.v. *low *adv.*]. **1958** J. Cannan *And be a Villain* iv. 100 A high-waisted pomegranate satin with gold lace sleeves.

4. (Earlier and later examples.)

1845 J. R. Planché *Golden Fleece* I. 13 An ardent spirit, known By several names..Some 'Cupid's eye water' the liquor call, 'White Satin' some. **1934** T. S. Eliot *Rock* ii. 66, I brought you along a drop o' satin. Four glasses and all.

5*. A domestic rabbit belonging to the breed so called, developed in America during the early 1930s by Walter A. Huey and distinguished by smooth fur with a satin-like sheen. Also *attrib.*

1934 W. L. Cotta in *Fur Animals* Aug. 3/1, I take great pleasure in describing, for the first time publicly, the most amazing rabbit of all time, the Satin Havana. **1935** *Small Stock Mag.* Aug. 7/2 Anything in the nature of a boom will do the satin more harm than good. **1946** *Amer. Rabbit Jrnl.* XVI. 44/2 In 1936 the American Satin Rabbit Breeders Association was organized. *Ibid.* 45/1 With the exception of the Satin Havanas, none of the Satin breeds have an Approved Working Standard. **1947** *Fur & Feather* 9 May 191/3 The Satin..a beautiful animal..comes in various colours, white, an orange, blue, black... Its fur feels like satin. It is a breed about nine years old and was started from a freak litter of Havanas. **1957** J. C. Sandford *Domestic Rabbit* i. 2 A second mutation of a coat character is the Satin. *Ibid.* 3 The Satin coat has also been combined with a number of colours. **1979** G. R. Scott *Rabbit Keeping* i. 26 The Satin rabbit is another mutation. *Ibid.*, The early Satins were ivory in colour.

II. 6. b. (Later examples.)

1913 C. Mackenzie *Sinister St.* I. i. vii. 103 Boys emerged from the tuckshop, sucking gelatines and satin pralines and chocolate creams. **1930** E. Pound *XXX Cantos* vii. 27 Square even shoulders and the satin skin, Gone cheeks of the dancing woman. **1975** P. Moyes *Black Widower* v. 56 A single big tear ran down her black satin cheek. **1977** *Hot Car* Oct. 59/2 The finish will be a nice satin which is a sod to keep clean.

c. Delete † *Obs.* and add later example.

1912 W. de la Mare *Listeners & Other Poems* 8 Her satin bosom heaving slow With sighs that softly ebb and flow.

7. a. *satin-like* adj. (earlier and later examples); **b.** *satin-clad, -frilled, -purfled, -sandalled, -shimmering* adjs.

1881 'Mark Twain' *Prince & Pauper* xxxii. 349 Satin-clad officials are flitting and glinting everywhere. **1949** Blunden *After Bombing* 25 Enchanting poppies satin-frilled. **1699** M. Lister *Journey to Paris* 59 A very smooth Sattin-like Skin. **1919** E. Pound *Quia Pauper Amavi* 16 There is a satin-like bow on the harp. **1862** G. M. Hopkins *Poems* (1967) 10 And trample and tread The satin-purfled smooth to foam. **1917** Blunden *Poems* (1930) 44 Satin-sandalled Chloes glimmering. **1952** R. Campbell tr. *Baudelaire's Poems* 89 On satin-shimmering, downy avalanches.

8. a. satin-finish, also any effect resembling satin in texture or surface produced on materials in various ways; **satin leather, satin oil,** leather finished so as to resemble satin; **satin-paper** (earlier examples); **satin weave** (see quot. 1897); **satin wire** (see quot. 1925).

1865 Mrs. Stowe *House & Home Papers* 157 For satin finish,..American papers equal any in the world. **1929** *Encycl. Brit.* XXIX. 7/2 Frequently the surface [of glass] had been dulled by acid so as to produce a 'satin' finish. **1959** *Gloss. Packaging Terms* (*B.S.I.*) 32 *Satin finish*, a decorative matt finish mechanically or chemically applied to aluminium and tinplate sheets. **1969** *New Yorker* 27 Sept. 92/3 (Advt.), It's Norway Pewter with the gleaming, never-tarnish satin finish. **1972** *Homes & Gardens* Mar. 106/2 They [*sc.* paints] are obtainable in gloss, semi-gloss, eggshell and satin finishes. **1974** *Harrods Christmas Catal.* 8 Housecoat in washable satin-finish flocked nylon. **1802** *Monthly Mag.* XIV. 203/2 White and chamois leather.. are evidently in danger of being beat out of the market by the English satin-leather. **1903** L. A. Flemming *Pract. Tanning* xiv. 264 Wax calf and satin leather are finished upon the flesh or inner side. **1971** T. C. Collocott *Dict. Sci. & Technol.* 1033/1 *Satin leather*.., leather with a perfectly smooth finish and without grain marks. **1895** *Montgomery Ward Catal.* Spring & Summer 517/3 Men's Satin Oil Congress Gaiters. **1897** C. T. Davis *Manuf. Leather* (ed. 2) xxviii. 424 This blacking is for satin oil, glove grain, plow grain, oil grain and dongola. **1834** M. Edgeworth *Tour in Connemara* (1950) 55 Mr. Jones wrote me as elegant a note as ever you saw on satin paper. **1840** Thackeray in *Fraser's Mag.* XXI. 684/1 I'll keep everything: the red wax, because it's like your lips; the black wax, because it's like your hair; and the satin paper, because it's like your skin! **1897** Stephenson & Suddards *Text Bk. Ornamental Design Woven Fabrics* 104 What is known in textile manufacturing as a satin weave, which is a construction of cloth where the weft comes to the surface in greater proportion than the warp, or *vice versa*, in a certain definite order. **1964** *McCall's Sewing* iv. 52/2 Satin weaves produce smooth, lustrous fabrics. **1969** *Sears Catal.* Spring/Summer 20 Blazer stripes in a satin weave on sand beige. **1899** in A. Adburgham *Shops & Shopping* (1964) xxii. 261 Satin wire. **1925** G. E. Martin *Make your Own Hats* (rev. ed.) i. 4 *Satin wire*, the thickest wire used in millinery, covered with a padding of cotton and then wrapped with silk; sometimes used for head line and edge wires. **1966** Satin wire [see *millinery 3].

b. satin bell = *mariposa lily; **satin-flower,** (*e*) a small herb of the genus *Sisyrinchium*, esp. *S. douglasii*, which is native to western North America and has grass-like leaves and small blue or purple flowers; **satin walnut** (earlier and later examples); **satin wave,** a white moth, *Sterrha subsericeata*; **satin-wood,** add: esp. *Fagara flava*; also, the similar yellowish wood of any of several African or Australian trees, esp. *Daphnandra micrantha* or *Zanthoxylum brachyacanthum*; also, any of the trees producing this timber; the colour of this timber; (earlier and later examples).

1898 A. M. Davidson *Calif. Plants* 123 Mariposas are.. sometimes called globe tulips,..the satin-bell or fairy's lantern. **1925** W. L. Jepson *Man. Flowering Plants Calif.* 237 White Globe Lily..Also called Snow-drops, Indian Bells, and Satin Bells. **1882** G. P. Lathrop *Echo of Passion* iv. 76 Marigolds and satin-flowers..were growing in the midst of rank weeds. **1971** *Daily Colonist* (Victoria, B.C.) 18 Apr. 22/2 Numerous clumps of satin flower blend their purple hued petals with the rosy shooting star. **1897** G. B. Sudworth *Nomencl. Arborescent Flora U.S.* 205 Sweet Gum..Satin Walnut. **1949** Collingwood & Brush *Knowing your Trees* 247/1 Sweetgum..is frequently marketed as satin walnut. **1908** R. South *Moths Brit. Isles* 2nd Ser. 117 The Satin Wave... The wings of this species are glossy whitish. **1958** W. J. Stokoe *Caterpillars Brit. Moths* II. 29 The Satin Wave..is widely distributed throughout England and Wales. **1792** G. Imlay *Topogr. Descr. W. Territory N. Amer.* 214 Satin-wood tree. Not classed. **1799** *Times* 1 June 4/1 (Advt.), Cabinet articles.. in mahogany, satin, and other woods. *Ibid.*, Valuable, and seasoned stock of Mahogany and satin wood in lots, planks, boards, and veneers. **1884** A. Nilson *Timber Trees New South Wales* 50 D[*aphnandra*] *micrantha*.—Satin-wood; Light yellow-wood... Timber fragrant, quite yellow when fresh. *Ibid.* 125 Z[*anthoxylum*] *brachyacanthum*. Satinwood; Thorny Yellowwood. **1902** G. S. Boulger *Wood* v. 97 About 1750, Satinwood..became fashionable for coach-panels. **1907** *Yesterday's Shopping* (1969) 145/1 Stains.. as used by the Working Ladies Guild, colours:—Rosewood, Satinwood, Oak, [etc.]. **1908** [see *obeche]. **1920** [see *afrormosia]. **1926–7** [see *maple 2 b]. **1936** R. H. Anderson *Trees New South Wales* 127 Socket Wood.. is also known as Light Yellow-wood, Satin Wood, and occasionally as Sassafras. **1958** *N.Z. Timber Jrnl.* July 73/2 There is a great variety of satinwoods. **1962** S. Wynter *Hills of Hebron* v. 73 The indent where the pulse beat was smooth, like satinwood.

c. satin beauté, a soft finely woven material with a dull crêpe back and brilliant satin finish; **satin de chine,** a silk fabric with a silk finish; **satin de laine** (earlier example); **satin de Lyon(s)** (see quots.).

1922 *Daily Mail* 18 Dec. 8 Her gown, in the Early Italian style, will be of cream satin beauté. **1928** *Times* 9 May 10/6 A draped gown of lavender satin beauté, embroidered with silver. **1880** L. Higgin *Handbk. Embroidery* ii. 14 '*Satin de Chine*', and other silk-faced materials of the same class. **1895** *Army & Navy Co-op. Soc. Price List* 15 Sept. 1095/1 Satin de Lyon..Satin de Chine, for dress linings. **1969** R. T. Wilcox *Dict. Costume* 303/2 *Satin de chine*,..was known in medieval Europe... Because of its exquisite texture, it became a court favorite. **1851** Satin de laine [see Mousseline 1 b]. **1881** C. C. Harrison *Woman's Handiwork* II. 115 Satin de Lyons, of a fine close quality, may be used with water-colors. **1915** L. Harmuth *Dict. Textiles* 137/2 *Satin de Lyon*, silk satin made with a twilled back, and finely striped face, used for lining. **1969** R. T. Wilcox *Dict. Costume* 303/2 *Satin de lyon*, satin with a ribbed back. Used for masculine evening wear trim such as top hat, waistcoat, lapel or trouser stripes.

satined, *ppl. a.* Add: **a.** (Later example.) Also, having a satin-finish. **b.** Clothed in satin.

1817 Jane Austen *Venta* in *Minor Wks.* (1954) vi. 457 The Lords & the Ladies were sattin'd & ermin'd. **1897** *Sears, Roebuck Catal.* 415 Solid Sterling Silver [bracelet], chased satined links.

satinize, *v.* Add: Hence **sa·tinized** *ppl. a.*

1972 *Guardian* 18 July 11/1 Satinised cotton trousers. **1975** *Harper's & Queen* June 96 Shocking pink shawl in satinised cotton.

satire (sæ·taiəɹ), *v.* [f. the sb.] *trans.* = Satirize *v.* 2 a.

1905 S. Joyce in *Lett. J. Joyce* (1966) II. 104 He doesn't think the critics will approve, or the people satired. **1961** in *Amer. Speech* XXXVI. 138 Hawthorne in his story 'Earth Holocaust' satires Emerson's idea of books.

satirism. Delete † *Obs.* and add later example.

1950 *Scrutiny* XVII. II. 145 What strikes one in reading, however, is not so much the variety of these satirisms..but simply their ubiquity.

satisfaction. Add: **5. e.** *Psychol.* The satisfying of a need or desire as it affects or motivates behaviour.

1911 E. L. Thorndike *Animal Intelligence* vi. 244 The Law of Effect is that: Of several responses made to the same situation, those which are accompanied or closely followed by satisfaction to the animal will..be more firmly connected with the situation, so that when it recurs, they will be more likely to recur. **1922** R. S. Woodworth *Psychol.* xix. 488 Dancing also gives a chance for muscular activity which is obviously one source of satisfaction in the more active games. **1951** J. M. Fraser *Psychol.* xiv. 161 Another group of satisfactions can be drawn, not perhaps from the work itself, but from the surroundings in which it is carried out. **1966** Katz & Kahn *Social Psychol. of Organizations* xii. 363 If there is one confirmed finding in all the studies of worker morale and satisfaction, it is the correlation between the variety and challenge of the job and the gratifications which accrue to workers. **1976** R. H. Moos *Human Context* viii. 265 Once workers feel competent with the transition, they often report long term gains in satisfaction or morale.

III. 7. satisfaction note *Insurance*, an acknowledgement of satisfaction with repairs made to a car signed by one claiming repair costs from an insurance company; **satisfaction theory** *Theol.* = *doctrine of satisfaction* (see sense 3 in Dict.).

1971 *Reader's Digest Family Guide to Law* 533/1 When repairs [to a car] are finished, the policy-holder is usually asked to sign a satisfaction note... Before signing, inspect the vehicle carefully and, if possible, take it for a test drive. **1973** *Times* 15 Dec. 19/7 Normally you will have to sign a 'satisfaction note' for the repairers before you can regain possession of your car after it has been repaired. **1932** Satisfaction theory [see *Anselmic *a.*]. **1969** *Dict. Christian Theol.* 23/1 Anselm..in his work *Cur Deus Homo?*,..interpreted the doctrine [of atonement] in terms of the 'satisfaction' or 'juridical' theory.

satisfiable, *a.* Restrict *rare* to sense a and add further examples of sense b.

1942 W. S. Churchill *End of Beginning* (1943) 228 All these conditions were satisfiable around 23rd October. **1944** *Annals Math. Stud.* XIII. 91 Formulas which are valid (or satisfiable) in every domain of individuals. **1952** R. L. Wilder *Introd. Foundations Math.* ii. 26 An axiom system Σ is satisfiable if there exists an interpretation of Σ. **1978** *Sci. Amer.* Jan. 106/3 When the machine is given a yes instance of a problem in *NP*, its operation is described by a satisfiable sentence, whereas the operation of a machine given a no instance is described by a sentence that cannot be satisfied.

Hence **sa:tisfiabi·lity.**

1944 *Annals Math. Stud.* XIII. 90 We may study the decision problem from the point of view either of validity or of satisfiability, instead of that of provability. **1952** R. L. Wilder *Introd. Foundations Math.* ii. 26 Where a system [of axioms] is consistent, we are usually unable to tell the fact

from 1.1. But..we have a very simple test showing 'satisfiability' in the sense of 1.3. **1977** *Word 1972* XXVIII. 285 Something which, with Hockett, we may term the 'productivity' of the description, one of the fundamental conditions of its acceptability or 'satisfiability'.

satisfice, *v.* Restrict *Obs.* exc. *north.* to sense in Dict. and add: **2.** *intr.* To decide on and pursue a course of action that will satisfy the minimum requirements necessary to achieve a particular goal. Hence **sa·tisficer**; **sa·tisficing** *ppl. a.* and *vbl. sb.*

1956 H. Simon in *Psychol. Rev.* LXIII. 129/2 Evidently, organisms adapt well enough to 'satisfice'; they do not, in general, 'optimize'. *Ibid.* 136/1 A 'satisficing' path, a path that will permit satisfaction at some specified level of all its needs. **1957** —— *Models of Man* iv. 205 The key..appeared to lie in substituting the goal of satisficing, of finding a good enough move, for the goal of minimaxing, of finding the best move. **1958** March & Simon *Organizations* vi. 141 To optimize requires processes several orders of magnitude more complex than those required to satisfice. **1963** G. P. E. Clarkson in A. R. Oxenfeldt *Models of Markets* II. 340 Two important innovations..have occurred... The first of these is the modified concept of rational behavior known as 'satisficing'... Important changes in the theory of the firm have been brought about by the introduction of the satisficing concept of behavior. **1967** H. Simon in N. Rescher *Logic of Decision & Action* i. 19 It is easy to see how GPS can be made into a satisficer. **1973** *N.Y. Times* 11 Feb. III. 1/2 Big business executives don't really try to maximize profits but '*satisfice*'—that is, they try to make enough profit to keep stockholders and boards of directors happy without bringing the wrath of government regulators, consumer groups or business competitors down on them. **1977** P. N. Khandwalla *Design of Organizations* xi. 404 To the seat-of-the-pants 'satisficer', scientific analysis may be acceptable in dealing with relatively trivial problems. **1977** Janis & Mann *Decision Making* ii. 32 A much more serious flaw of this complex form of satisficing lies in its failure to ensure that the alternatives retained are..superior to those eliminated.

‖ **satori** (sātō·ri). *Zen Buddhism.* [Jap., = spiritual awakening.] A sudden indescribable and uncommunicable inner experience of enlightenment. Also *transf.* Hence **sato·ric** *a.*, pertaining to or inducing satori.

1727 J. G. Scheuchzer tr. *Kæmpfer's Hist. Japan* I. III. vi. 242 This profound Enthusiasm is by them call'd *Safen*, and the divine truths revealed to such persons *Satori*. **1921** A. Waley *Nō Plays of Japan* 58 The only escape from this 'Wheel of Life and Death' lies in *satori*, 'Enlightenment', the realization that material phenomena are thoughts, not facts. **1921** D. T. Suzuki in *Eastern Buddhist* May 33 The power to see into the nature of one's own being lies also hidden here [in the subconscious]. Zen awakens it. The awakening is known as *Satori*, or the opening of a third eye. **1933** —— *Ess. Zen Buddhism* 2nd Ser. I. 21 When you have satori you are able to reveal a palatial mansion made of precious stones on a single blade of grass; but when you have no satori, a palatial mansion itself is concealed behind a simple blade of grass. **1949** C. Humphreys *Zen Buddhism* ii. 33 *Satori*, the immediate experience of truth as distinct from understanding about it. **1957** *New Yorker* 31 Aug. 35/1 It takes at least ten years of meditation and *koans* to attain even one flash of *satori*. **1968** T. Wolfe *Electric Kool-Aid Acid Test* viii. 102 It was as if Cassady..was in a state of satori, as totally into this very moment, Now, as a being can get. **1970** W. Burroughs Jr. *Speed* vii. 154 His music was improvised to fit short declarations and imaginary rhyming words, prayers and questions to the audience that he fit together in a satoric sound that brought me back to and into the running water. **1974** *Sci. Amer.* Oct. 140/2 Thought, like the kingdom of life, grows and evolves slowly; the book focuses on 'the inner work of synthesis' and not on the claimed *satori* of one chance page.

satrangi, satranji, satringee, satrunjee, varr. Sitringee in Dict. and Suppl.

satsang (satsa·ŋ). *Indian Philos.* Also **Satsang.** [ad. Skr. *satsaṅga* association with good men, f. *sat* good man + *saṅga* association.] A spiritual discourse, a sacred gathering.

1929 J. N. Farquhar *Mod. Relig. Movements in India* iii. 171 As in Theosophy, you may be a Rādhā Soāmi and yet remain a Hindu, a Muḥammadan or a Christian... Yet it is definitely stated that the religion is for all, and that outside the Satsang there is no salvation. **1971** *Shankar's Weekly* (Delhi) 4 Apr. 5/4 Local communists had been watching with uneasiness at this priest's continual attendance at kirtans and satsangs and were even of the mind that he should be denounced as a CIA spy. **1972** *Times* 23 Oct. 12/1 There [in Delhi] they will sit at the boy's feet, listen to his spiritual discourses (*satsangs*) or just enjoy his physical presence. **1977** *New Society* 30 June 672/3 The 'satsangs' (spiritual discourses) given at the recent Wembley festival.

Satsuma. Delete ‖ and add: Now (esp. in sense 2) freq. with pronunc. (sætsū·mă). Also **Satzuma. 1.** Also *absol.*

1880 T. W. Cutler *Grammar Jap. Ornament* 16 Modern Satsuma is largely decorated at Tokio and elsewhere. **1909** M. Diver *Candles in Wind* ix. 86 Roses..filling every available bowl, even the sacred Satzuma. **1974** Savage & Newman *Illustr. Dict. Ceramics* 255 True Satsuma is comparatively rare outside Japan.

2. (Freq. with small initial.) A small tangerine belonging to a variety of *Citrus reticulata* so called; also, the variety itself. Also *attrib.* as *Satsuma orange*.

1882 E. S. Hart in *Proc. 18th Session Amer. Pomological Soc. 1881* 67/1 One [variety of tangerine] from Japan called Satsuma, bore a temperature of 16°. **1905** *Flora & Sylva* III. 66/1 Satsuma, an early fruiting Mandarine. **1909** *Circular Bureau Plant Industry U.S. Dept. Agric.* XLVI (*title*) The limitation of the Satsuma orange to trifoliate-orange stock. **1922** [see *mikan]. **1926** H. H. Hume *Cultivation of Citrus Fruits* xxix. 477 Satsuma oranges are susceptible to the disease. **1943** Webber & Batchelor *Citrus Industry* I. v. 551 The Satsuma was first introduced into the United States in 1876 by Dr. George R. Hall... It is characteristic of Satsuma fruits that although they mature and fill with juice..the rind frequently remains green or shows only slightly colored. **1967** [see *clementine]. **1980** 'M. Yorke' *Scent of Fear* vii. 64 She bought..some tangerines—or satsumas, as they were called nowadays.

sattrangee, var. Sitringee in Dict. and Suppl.

sat-upon (sæ·t,ǝpǫn), *ppl. a.* *colloq.* [See Sit *v.* 26 d.] Downtrodden, humiliated, 'squashed'.

1892 *Times* 30 July 8/2 In his concession, where he is an apologetic and much sat-upon importation, the foreign resident does no harm. **1893** *Chambers's Jrnl.* 25 Feb. 128 With that sat-upon sort of man..you never know where he may break out.

saturability. (In Dict. s.v. Saturable *a.*) (Example.)

1979 *Nature* 19 Apr. 747/2 If the direct linkage EGF-receptor complex is a complex of EGF and its physiological receptor, the saturability for reversible EGF binding and direct complex formation should be similar.

saturable, *a.* Add: (Later examples.)

1966 *Electronics* 31 Oct. 44 The transmittance of the saturable absorber increases with the light flux. **1979** *Nature* 19 Apr. 748/2 A stereospecific, saturable, high affinity binding site for ³H-diazepam has recently been characterised in membrane fractions isolated from the brains of mammals.

b. Of magnetic systems: capable of retaining a saturating magnetic field (see Saturate *v.* 4 b). *Saturable reactor*: an iron-cored coil whose impedance to alternating current can be varied by varying the direct current in an auxiliary winding so as to change the degree of magnetization of the core.

1944 W. D. Cockrell *Industr. Electronic Control* x. 84 Another device by means of which we make deliberate use of the saturating effect for loads both large and small is the saturable reactor. **1956** [see *ferro-resonance* s.v. *ferro- 1 c]. **1962** F. I. Ordway et al. *Basic Astronautics* v. 187 The flux gate magnetometer..consists essentially of a flux gate of two, identical, high-permeability, saturable coils, oppositely wound with identical coils. **1966** *McGraw-Hill Encycl. Sci. & Technol.* XII. 37/1 Saturable-core reactors are used to control large alternating currents where rheostats are impractical. Theater light dimmers often employ saturable reactors. **1975** D. G. Fink *Electronics Engineers' Handbk.* XIII. 82 The response time of saturable reactors is the combination of the time constants of the control circuit and of the gate-winding circuit.

saturate (sæ·tiurět), *sb. Chem.* [f. the vb.] A saturated fat or fatty acid.

1959 R. H. Potts in E. S. Pattison *Industr. Fatty Acids* ii. 13 In selecting a raw material, one always considers that more saturates can be made if required, but the unsaturated requirements must be purchased with the raw material. **1977** *Nature* 3 Nov. 2/2 Pursuit of the lipid hypothesis does not mean just swapping polyunsaturates for saturates.

saturate, *v.* Add: **2. c.** *Mil.* To overwhelm (enemy defences) by aerial attack, esp. by intensive bombing.

1942 *Times* 1 June 4/3 The plan for saturating the defences of Cologne was an undoubted success. **1943** *Times* 12 Mar. 8/4 Air Marshal Sir Arthur Harris and his commanders and staffs have displayed extraordinary fertility in tactical ideas. The monster raids saturating the enemy's active and passive systems of defence is one example. **1944** *Ann. Reg. 1943* I. 74 The ultimate possibility of saturating the enemy's defences both on the ground and in the air. **1956** A. H. Compton *Atomic Quest* 228 The areas attacked were saturated with bombs.

d. To supply (a market) to the point of over-satisfaction of demand for a product.

1958 *Engineering* 4 Apr. 435/1 The Swiss vehicle market, if not saturated, seems to be reaching a certain stabilisation of demand. **1976** 'G. Black' *Moon for Killers* i. 18 The market was saturated, and Robert bought two thousand of them at a throw-away price. **1978** *Times Lit. Suppl.* 27 Jan. 84/5 *Man watching* will saturate the market and maintain its well-deserved primacy.

3. a. Also, to cause to become saturated (see *saturated *ppl. a.* 3 b).

1866 *Notices Proc. R. Inst. Gt. Brit.* IV. 419 This new molecule—we call it hypochlorous acid—we open again: again two attraction units are liberated and saturated by a second atom of bivalent oxygen. **1926** H. G. Rule tr. *J. Schmidt's Text-bk. Org. Chem.* 26 Thiele assumes that all such unsaturated compounds possess a double bond, but that the two affinities do not completely saturate one another, leaving a certain residual affinity or partial valency in excess on each carbon atom. **1972** *Materials & Technol.* V. x. 282 Such isomers may arise from the addition of hydrogen at a double bond which is normally not saturated by natural processes. **1977** *Lancet* 20 Aug. 401/1 Many dietitians tell patients not to re-use the oil more than once because reheating is thought to saturate the double bonds.

b. *Physiol.* To cause (tissues of the body) to retain the greatest amount of inert gas possible at the given pressure during a saturation dive.

1965 *Jrnl. Appl. Physiol.* XX 1269/2 The decompression schedule after and while breathing helium takes longer than with nitrogen because the helium saturates a greater proportion of the body tissues. **1971** J. Salzano et al. in C. J. Lambertsen *Underwater Physiol.* 347 (*heading*) Arterial blood gases, heart rate, and gas exchange during rest and exercise in men saturated at a simulated seawater depth of 1000 feet. **1974** *Encycl. Brit. Macropædia* X. 926/1 For any given depth,..there is a saturation point, at which body tissues are saturated with inert gas; after that, no matter how long a worker stays under pressure his decompression time does not increase.

4. b. (Further examples in *Magnetism*.)

1891 S. P. Thompson *Electromagnet* iv. 151 The iron is.. more saturated round the edge than at the middle... If the edge is already far saturated you cannot by applying higher magnetizing power increase its magnetization much. **1928** [see sense 6 below]. **1962** R. D. Pettit in G. A. T. Burdett *Automatic Control Handbk.* v. 18 The core flux is initially saturated negatively.

5. *Electronics. trans.* To cause or maintain a state of saturation in (a device or a current); *pass.*, to be in a state of saturation. Cf. *saturation 3 d, f.

1919 J. A. Crowther *Ions, Electrons, & Ionizing Radiations* ii. 17 The effects when the current is not saturated are in general very complex. **1956** J. C. Logue in L. P. Hunter *Handbk. Semiconductor Electronics* xv. 30 It is possible to have a high degree of saturation or just barely to saturate the transistor. **1962** Simpson & Richards *Physical Princ. Junction Transistors* xvi. 388 The base current chosen must be sufficient to saturate the transistor. **1969** J. J. Sparkes *Transistor Switching* i. 18 This equation applies only when the transistor is saturated. **1976** Millman & Halkias *Electronic Fund. & Appl.* iv. 81 A knowledge of h_{FE} tells us the minimum base current..which will be needed to saturate the transistor.

II. 6. *intr.* To reach or exhibit a condition of saturation, in any sense; to reach a state in which no further change or increase is possible.

1928 *Observer* 17 June 26/3 The essential thing is the current that can be carried without any danger of saturating the core... If the core saturates there will at once be a falling-off in the quality of reception. **1947** F. G. Spreadbury *Electronics* iv. 184 The thermionic current does not truly saturate, but continues to increase slowly. **1953** *Physical Rev.* XCI. 632/2 As the rf level is increased, the peak amplitudes labeled M_1 and M_2 in Fig. 2 saturate quite readily. **1957** R. D. Middlebrook *Introd. Junction Transistor Theory* ii. 28 The hole flow very soon reaches a limit as the potential is increased. This occurs when all the available holes are being drawn out of the *p*-region. The electron current saturates in a similar way. **1962** R. D. Pettit in G. A. T. Burdett *Automatic Control Handbk.* v. 21 When the flux in element A saturates, that in element B is unsaturated. **1969** J. J. Sparkes *Transistor Switching* iii. 74 The circuit can be designed so that the output transistor saturates. **1975** *Nature* 6 Nov. 85/1 Figure 1 shows that the steady-state amplitude saturates at a relatively low stimulus level. **1976** [see *saturation 3 a]. **1977** *Nature* 21 Apr. 709/1 This ratio tends to saturate for crystallite sizes less than 40 Å.

saturated, *ppl. a.* Add: **2. b.** *transf.* Filled to capacity; *spec.* in *Econ.*, of a market in which demand is completely satisfied.

1962 S. Strand *Marketing Dict.* 653 *Saturated market*, the ultimate point of absorption of a product or service within a territory. Now limited to parts replacement. **1965** *Monthly Economic Let.* (First National City Bank, N.Y.) Apr., Household durables with 'highly saturated markets'—those which the vast majority of families already have, such as refrigerators or black-and-white television sets—achieved sales gains through growing replacement demand.

3. a. (Further examples.) In mod. use, applied to solutions containing as much solute as is possible in equilibrium conditions (in contrast to those that are supersaturated).

1939 *Thorpe's Dict. Appl. Chem.* (ed. 4) III. 452/2 Removal of solvent from a saturated solution at this point results in the solution becoming supersaturated. **1978** P. W. Atkins *Physical Chem.* viii. 220 If a lump of solid is left in contact with a solvent it will dissolve until the solvent has become saturated. The saturated solution corresponds to the case in which the chemical potential of the pure solid is equal to the chemical potential of the solute in the saturated solution.

b. Orig., applied to compounds which contained the greatest possible proportion of some element, and to (the chemical 'affinities' of) atoms, radicals, etc., which had entered into chemical combination to the maximum extent. Now applied to organic compounds, molecules, groups, etc., which have structures containing the greatest possible numbers of hydrogen atoms, and hence have no multiple bonds between carbon atoms; occas. applied also to carbon atoms in such structures.

1866 *Notices Proc. R. Inst. Gt. Brit.* IV. 429 We have thus been led..to a distinction of a novel kind, that of finished and unfinished molecules; or, to use the more frequently employed expression, that of saturated and non-saturated compounds. **1876** *Phil. Mag.* II. 167 The group OH is related in one case to a carbon atom a large number of whose affinities are already 'saturated' (to use a common term). **1888** [in Dict., sense 3]. **1935** A. K. Anderson *Essent. Physiol. Chem.* iv. 67 Chemically, fats differ from oils in that fats contain saturated fatty acids whereas oils contain rather large quantities of unsaturated fatty acids. **1949** *Thorpe's Dict. Appl. Chem.* (ed. 4) IX. 6/2 In the higher land-animals the most abundant component acids are always the monoethenoid oleic and the saturated palmitic acid. **1961** [see *POLYUNSATURATED *a.*]. **1968** MURTHY & NATHAN *Org. Chem. made Simple* vii. 122 A saturated carbon atom may be represented by a model showing only the tetrahedrally directed linkages. **1971** *Jrnl. Gen. Psychol.* LXXXV. 155 Increasing the amount of saturated fat..resulted in a similar increase in the excitatory process. **1976** *Sci. Amer.* Mar. 35/2 Such multiple-ring, or polycyclic, compounds are said to be saturated if all the bonds of the carbon atoms, beyond the minimum needed for carbon–carbon bonding, are linked to hydrogen atoms.

c. *Min.* and *Petrol.* Of a mineral (see quot. 1913). Of a rock: containing neither free quartz (or some other specified oxide) nor any undersaturated minerals.

1913 S. J. SHAND in *Geol. Mag.* Decade V. X. 508 Of the various minerals which enter into the composition of igneous rocks, about one-half are capable of forming in presence of free silica... These may..be termed saturated minerals. *Ibid.* 510 A rock which contains only saturated minerals may be termed a saturated rock. **1947** [see *OVERSATURATED *ppl. a.*]. **1951** TURNER & VERHOOGEN *Igneous & Metamorphic Petrol.* iii. 54 Saturated minerals are those which are compatible with excess silica under magmatic conditions, and are therefore commonly associated with quartz. **1968** B. BAYLY *Introd. Petrol.* vi. 53 All saturated rocks fall within the shaded area in Fig. 6·1. The commonest such rocks are made of feldspar with pyroxene or amphibole.

4. b. *Electronics.* Characterized by or exhibiting saturation (senses *3 d, f); of or pertaining to a device in such a state.

1896 *Phil. Mag.* XLII. 394 For a given intensity of radiation the current through the gas does not exceed a certain maximum value whatever the electromotive force may be, the current gets, as it were, 'saturated'. **1899** *Ibid.* XLVII. 160 The gas tends to become more readily saturated with diminution of pressure. **1933** *Proc. IRE* XXI. 1667 The practical limitation of this 'saw-tooth' generator lies in the fact that there is no such thing as a completely saturated thermionic tube. **1956** J. C. LOGUE in L. P. Hunter *Handbk. of Semiconductor Electronics* xv. 11 It is necessary to impose an upper limit on r_c in the saturated region. This is to ensure that the voltage drop between the emitter and collector terminals is small when the transistor is in a saturated state. **1967** *Electronics* 6 Mar. 122/2 They permit a whole spectrum of products with the highest speed possible with saturated logic. **1977** TAUB & SCHILLING *Digital Integrated Electronics* i. 18 When a base current I_B is supplied, the transistor is able to furnish a current $I_C = h_{FE}I_B$. If the current I_C is actually less than $h_{FE}I_B$, the transistor is said to be in saturation. However, such is the case because of the constraint imposed by the circuit and not by the transistor. Hence, strictly, we should speak of a saturated circuit and not a saturated transistor.

c. *saturated diving* = *saturation diving* s.v. *SATURATION 5.

1968 *New Scientist* 17 Oct. 125/2 The important element in saturated diving is that after six days or six months of exposure to a given depth or pressure, the diver requires a single, fixed decompression period. **1971** *Petroleum Rev.* July 248/1 Saturated diving requires a considerable increase in equipment sophistication and diver training.

saturation. Add: **3. a.** (Further examples in *Chem.* Cf. *UNSATURATION.) More widely in *Physics*, a condition or phenomenon in which a quantity (usu. the value of some property) no longer increases in response to an increase in the magnitude of some external influence, or ceases to alter in the usual way; *spec.* in *Spectroscopy* (see quot. 1976). See also senses 3 c, d, f following.

1866 *Notices Proc. R. Inst. Gt. Brit.* IV. 422 The saturation of these two units [of attraction] by the trivalent nitrogen atom. **1902** J. B. COHEN *Theoret. Org. Chem.* xvii. 240 The saturation of one unsaturated carbon atom necessitates that of the other. **1964** N. G. CLARK *Mod. Org. Chem.* vi. 89 By partial saturation of this triple bond with hydrogen, an olefin is produced. **1968** A. A. BAKER *Unsaturation in Org. Chem.* vi. 71 His formulas..illustrate the progressive saturation of a diatomic carbon molecule to acetylene, to ethylene, and to completely saturated ethane. **1948** *Physical Rev.* LXXIII. 683/1 As H_1 is increased, the thermal contact between spin system and lattice eventually proves unable to cope with the energy absorbed by the spin system, the spin temperature rises, and the relative absorption..diminishes. It is the onset of this saturation effect which has been used to measure the spin-lattice relaxation time. **1953** *Ibid.* XCI. 206/2 From the saturation of the absorption and a measurement of the rf field, a spin-lattice relaxation time of approximately one millisecond is calculated. **1959** G. TROUP *Masers* iii. 37 If the energy density of radiation falling on an assembly of molecules having an excess upper state population is increased, there comes a time when the energy of induced emission is no longer linearly dependent on the incident radiation energy density. This phenomenon is known as saturation. **1961** G. R. CHOPPIN *Exper. Nuclear Chem.* v. 62 The higher the atomic number of the scattering material, the larger is f_b [backscattering factor]. Also, f_b increases with thickness up to a saturation thickness beyond which it is a constant. **1972** MCFARLANE & WHITE *Techniques of High Resolution N.M.R. Spectroscopy* v. 55 The gross observable effects of saturation are a general broadening of the spectrum with associated loss of peak height and resolution. **1976** D. SHAW *Fourier Transform N.M.R. Spectroscopy* ii. 20 Saturation is the equalisation of the population in the ground and the excited state which occurs because relaxation from the excited state is slow and with a strong exciting field a dynamic equilibrium can be set up. In this equilibrium the number of nuclei in the upper and lower states become [*sic*] equal, and the signal saturates, or disappears.

c. *Magnetism.* The condition of being as strongly magnetized as possible, or so strongly magnetized that an increase in magnetizing force produces no appreciable increase in magnetization.

1837, 1864 [in Dict., sense 3]. **1920** *Whittaker's Electr. Engineer's Pocket-bk.* (ed. 4) 144 In addition to the limitations imposed by saturation, the parts of the magnetic circuit where the flux is continually changing in value are further restricted by losses due to eddy-currents and hysteresis. **1962** *Newnes Conc. Encycl. Electr. Engin.* 463/2 The coercivity is the magnetizing force necessary to remove the magnetism completely from a specimen which has been magnetized to saturation. **1974** *Encycl. Brit. Macropædia* XI. 333/2 It was suggested in 1907 that a ferromagnetic material is composed of a large number of small volumes called domains, each of which is magnetized to saturation.

d. *Electronics.* The condition in which increase in the potential difference between two electrodes in a gas-filled or evacuated vessel leads to no increase in the current flowing between them, owing to the limitations of the gas as a current-carrier or the electrode as an electron-emitter.

1896 *Phil. Mag.* XLII. 394 It is evident that this saturation must occur if the current destroys the conducting power of the gas. **1899** *Ibid.* XLVII. 158 The great difficulty in producing complete saturation, i.e. to reach a stage when all the ions produced reach the electrodes, may be due to one or more of three causes. **1947** R. LEE *Electronic Transformers & Circuits* v. 116 In plate-modulated class C amplifiers, sufficient excitation must be applied so that grid saturation still obtains at 100 per cent modulation; otherwise output would not be proportional to plate voltage. **1962** D. F. SHAW *Introd. Electronics* x. 192 Other cathode materials, such as metallic oxides, do not exhibit full saturation.

e. *Psychol.* A term used in mental testing based on the theory of two-factor analysis put forward by C. S. Spearman (1863–1945) for the degree to which the general factor (*g*) saturates the specific factor or ability in question; also *attrib.*

1904 C. S. SPEARMAN in *Amer. Jrnl. Psychol.* XV. 276 Intellective saturation, or extent to which the considered faculty is functionally identical with General Intelligence. *Ibid.* 277 Mathematics, for example, has a saturation of 74 and Common Sense has one of about 96. **1927** *Psychol. Bull.* XXIV. 392 Slocombe..applies the intellective saturation formula of Spearman to nine group tests. **1940** C. L. BURT *Factors of Mind* xii. 299 Let us suppose that both the variances for the different factors and the saturation coefficients for the different tests are everywhere equal. **1951** R. H. THOULESS *Gen. & Social Psychol.* (ed. 3) xxiii. 367 This degree of dependence on the general factor was called by Spearman the saturation with *g* of the ability in question; the term more commonly used at the present time is the general factor loading.

f. *Electronics.* The state of operation of a transistor in which the collector current becomes independent of the base voltage, arising when the base-collector junction becomes forward-biased.

1956 J. C. LOGUE in L. P. Hunter *Handbk. of Semiconductor Electronics* xv. 48 In pulse-type computer systems, the length of time that the transistor is driven into saturation is controllable. **1962** SIMPSON & RICHARDS *Physical Princ. Junction Transistors* xvi. 389 'External' control causes the 'on' position to be largely independent of transistor parameters..and makes heavy base overdrive possible without danger of saturation. **1975** D. G. FINK *Electronics Engineers' Handbk.* xvi. 13 When the collector current I_C reaches its maximum possible value.., saturation occurs and the collector junction becomes forward-biased.

g. The retention by the blood of the greatest amount of inert gas possible under the given pressure, as during a saturation dive (see sense 5 below); also *transf.*, a saturation dive.

1971 J. K. SUMMITT et al. in C. J. Lambertsen *Underwater Physiol.* 519 A study of five trained men during compression to a simulated depth of 1000 FSW, during subsequent saturation at this pressure for 77 hr and 30 min, and during decompression. **1974** *Encycl. Brit. Macropædia* X. 926/1 Reasonably safe and efficient decompression from saturation at depths up to 600 feet..can be accomplished by a decompression at the rate of 15 minutes per foot.., or about 100 feet..per day. **1975** *BP Shield Internat.* May 5/1 In excess of 14,000 diver man hours were spent in saturation without a single decompression problem or lost time accident. **1975** *Offshore Engineer* Dec. 7/2 A 17-day saturation involving six divers at depths of up to 260m carried out by Strongwork Diving (International) has given a British company a new record.

4. (Further examples.) Cf. *HUE *sb.*[1] 3 c.

1966 [see *HUE *sb.*[1] 3 c]. **1967** E. SHORT *Embroidery & Fabric Collage* i. 12 The hues are all used at their full strength or saturation, i.e. they are not diluted in any way by black or white. **1970** *Nature* 19 Sept. 1183/1 Discrimination tests revealed that sorters identified stamps most easily if seven colours were used each at two distinct levels of saturation, for example, dark blue and light blue, dark green and light green. **1978** *Sci. Amer.* Mar. 87/3 The color pictures were generated by first determining the spectral irradiance of Mars in each of the regions and then computing the hue, brightness and saturation of color for the range of wavelengths to which the human eye is sensitive.

b. *transf.* The name of a control on a colour television set used to adjust the quality of colours in the picture.

1964 M. S. KIVER *Color Television Fundamentals* (ed. 2) v. 144 There is also a color saturation control to adjust the vividness or depth of color. **1967** *Punch* 12 Apr. 532/3 It is good to be able to report that the colour sets shown at the Ideal Home Exhibition in London..had one colour control only. For reasons impossible to conjecture it is labelled 'Saturation'. Twisting this knob does not release a jet of water, however; it simply changes the picture from black-and-white to any strength of colour desired. **1968** *Guardian* 5 July 8/5 There is a secondary colour knob marked either 'saturation' or 'colour' which enables you to control the shade you receive. **1974** A. G. PRIESTLY *Receiving PAL Colour Television* v. 103 A good saturation control is not easy to design. The control itself is usually situated on the front of the receiver for use by the viewer.

5. *saturation charge, recording, time, weapon*; **saturation current,** the greatest current that can be carried by a gas or electronic device (cf. senses 3 d, f above); **saturation dive,** a dive made with the diver's blood-stream saturated with an inert gas, usu. helium or nitrogen, at the pressure of the surrounding water, so that the time required for decompression afterwards is independent of how long the dive lasts; so **saturation diver, diving** *vbl. sb.*; **saturation point,** the state or condition at which saturation begins; the limit of acceptance; freq. *fig.*; **saturation (vapour) pressure** *Physics* (see quot. 1969).

1969 Saturation charge [see *saturation time* below]. **1896** *Phil. Mag.* XLII. 403 The saturation current depends only on the number of conducting particles produced by the rays. **1929** *B.B.C. Year-bk. 1930* 450/2 As the anode voltage applied to a three-electrode valve is increased, the anode current also increases up to a point, when a further increase in anode voltage does not increase the anode current. This maximum value of the current is called the 'saturation current'. **1954** L. M. KRUGMAN *Fund. Transistors* iii. 43 The saturation current is composed of two components. The first is formed by thermally generated carriers which diffuse into the junction region. The second component is an ohmic characteristic which is caused by surface leakage across the space charge region. **1976** MILLMAN & HALKIAS *Electronic Fund. & Appl.* iv. 79 In addition to the variability of reverse saturation current with temperature, there is also a wide variability of reverse current among samples of a given transistor type. **1966** *Sci. Amer.* Mar. 27/1 It is clear..that the 'partial pressure' of oxygen should be kept between about 150 and 400 millimeters of mercury during the at-depth phase of a long saturation dive. **1974** *Daily Tel.* 22 Feb. 7/6 The record saturation dive in the North Sea was 621ft. The diver took a day to get down, and after surfacing spent 3½ days in a decompression chamber. **1970** *Sci. Jrnl.* Feb. 15 The *Argyronète*, a self-propelled submersible combining a house in which saturation divers can live (under sea bottom pressure) and a conventional submarine with a crew at normal atmospheric pressure. **1974** *Encycl. Brit. Macropædia* X. 926/1 In practice, saturation divers are compressed slowly to working pressure, generally in the deck chamber, and are then transferred as needed to and from the work site sealed in the diving bell. **1975** *BP Shield Internat.* May 5/4 It is..not uncommon for saturation divers to spend from two to three weeks under saturation conditions. **1966** *Sci. Amer.* Mar. 27/1 Although this 'saturation diving' is efficient, it imposes an extra technical burden, because the schedules for the ultimate decompression must be calculated and controlled with particular care. **1970** *New Scientist* 26 Mar. 617/2 The experiment successfully demonstrated the feasibility in scientific research of saturation diving—a technique which relies on the fact that once the body tissues become saturated with gases breathed under pressure, the time to remove them during decompression remains the same no matter how much longer the person stays at that pressure. **1974** *Daily Tel.* 22 Feb. 7/6 Bone necrosis is a growing fear connected with saturation diving. **1976** *Offshore Platforms & Pipelining* 122/3 Any diving inspection beyond the 140-ft. depth, which requires any appreciable time on the bottom, is carried out by using saturation diving. **1858** Saturation point [in Dict.]. **1902** *Encycl. Brit.* XXXIII. 631/2 It is a fair inference that similar behaviour would be observed up to the saturation-point if surface condensation could be avoided. **1927** *Sunday Times* 13 Feb. 2 Those controlling the industry realize that the world production of motor-cars has by no means reached saturation point. **1932** WODEHOUSE *Louder & Funnier* 71, I rather fancy that sinister jewel-trackers have about reached saturation-point. **1977** *Times* 30 Apr. 3/3 The popularity of the forest may have reached saturation point. **1884** Saturation pressure [in Dict.]. **1902** *Encycl. Brit.* XXXIII. 631/1 The values of the saturation-pressure have been very accurately determined for the majority of stable substances. **1975** D. G. FINK *Electronics Engineers' Handbk.* xxiii. 56 There are two methods of recording called saturation and nonsaturation. With saturation recording, material under the head is fully saturated throughout the material thickness. **1969** J. J. SPARKES *Transistor Switching* i. 23 The turn-off time is divided into two parts. First the saturation time,..during which the saturation charge..is used up and the collector current does not significantly change. Second, the fall time. **1955** *Sci. Amer.* Mar. 74/2 For every temperature there is a 'saturation vapor pressure' at which the rates of escape and of deposit at a step balance. Under these conditions the crystal does not grow. It can grow only when the vapor is supersaturated. **1969** *Gloss. Terms Vacuum Technol.* (*B.S.I.*) I. 9 *Saturation vapour pressure*, the pressure exerted by a vapour when in equilibrium with its solid or liquid phase. **1955** *Bull. Atomic Sci.* Jan. 14/3 The construction of 'saturation weapons' became possible when it was discovered that under certain circumstances a tiny amount of matter transforms into a tremendous amount of energy.

b. Designating an activity intended to achieve the complete saturation of its object; orig. *Mil.*, referring to intensive bombing operations, esp. in *saturation bombing*; hence *saturation bomb* vb.; more widely, applied to

an intensive operation in the fields of marketing, advertising, security, and the like.

1942 *Sun* (Baltimore) 15 Oct. 13/5 The fact that only nine bombers were lost..was taken to mean that the 'saturation technique' was used to crowd so many planes over the area in a short raid that the strong defenses..were swamped. **1943** *Time* 7 June 29/3 According to U.S. testimony, the precision bombing of the American forces is more effective, ton for ton, than the saturation bombing of the R.A.F. *Ibid.* 30 Aug. 33/2 The greatest air force the world has known: a combination of the daylight precision bombing planes of the U.S. Eighth Air Force and the heavy night-time saturation raiders of the R.A.F. *Ibid.* 6 Sept. 36/3 Of the 73 raids Berlin had experienced, this was the worst, the first of the kind of saturation raids that had wrecked Hamburg. **1944** *Times* 28 Mar. 4/5 A great weight of high-explosives and incendiaries pounded Essen in a 'saturation' attack which lasted just under half an hour. **1957** CLARK & GOTTFRIED *Dict. Business & Finance* 314/2 *Saturation* selling involves making a product available in every outlet in an area, and using every possible means of sales promotion. **1958** *Listener* 5 June 950/3 Mrs. Ancsa's father had been killed in a saturation air-raid. **1962** *Economist* 2 June 920/3 The large number of copies of each film necessary for this so-called 'saturation release'. **1966** *Times* 12 July 11/3 It is simply not true to say that America has engaged in saturation bombing. **1971** *Wall St. Jrnl.* 13 Aug. 14/1 He came to realize such journalism is possible through 'saturation reporting'. **1975** R. H. RIMMER *Premar Experiments* (1976) i. 19 The days when we believed we could change things—like the draft, or the saturation bombing of Vietnam, or the Pentagon running the universities. **1977** *Time* 28 Nov. 29/2 As the deadline arrived.., West Germany's national airline responded with a policy of saturation security for its 411 daily scheduled flights worldwide. **1979** P. NIESEWAND *Member of Club* xxi. 165 The Cubans were saturation bombing the camp. **1981** I. A. GORDON in *N.Z. Listener* 18–24 Apr., The big idea behind the saturation-bombing of consumers with a trade-name is to persuade you that X (and not Y or Z) is the brand to remember.

Saturday. Add: **3.** *Saturday-afternooner*; **Saturday penny,** a penny or small sum of money given to a child on Saturday as pocket-money.

1906 Saturday-afternooner [see *early-closer* s.v. *EARLY *a.* 7]. **1972** *Homes & Gardens* Apr. 60, I am old enough to remember the small child's pocket money called the 'Saturday penny'. **1979** *Church Times* 27 Apr. (Mayflower Suppl.) p. iii/2 When I was in trouble with my Mum and Dad and they wouldn't give me my 'Saturday penny'..I had at least twelve other homes where there were relations where I could go and 'con' them for a penny.

Saturdaying (sæ·təɹdē[ⁱ]iŋ, -di[ⁱ]iŋ), *vbl. sb.* [f. SATURDAY + -ING[1], after Russ. *subbótnik*.] An English rendering of *SUBBOTNIK. So **Sa·turdayite.**

1920 *Manch. Guardian* 5 Feb. 9/7 In Moscow it has been found worth while to set up a special bureau for 'Saturdayings'. **1920** *Contemp. Rev.* Oct. 504 For members of the Bolshevik party, 'Saturdaying' had become compulsory. **1932** C. HOGARTH tr. *Kollontai's Free Love* 233 She will persuade you..that it is necessary..to deny oneself everything that gives joy, to live only for the 'Saturdayites'.

Saturday night. [SATURDAY 3.] **1.** Used *attrib.* of activities taking place on or as on a Saturday night, esp. some form of revelry.

[**1847** H. MELVILLE *Omoo* xii. 49 The evening of the last day of the week was always celebrated by what is styled on board of English vessels, 'The Saturday-night bottles'. Two of these were sent down into the forecastle, just after dark.] **1896** 'M. RUTHERFORD' *Clara Hopgood* xii. 121 Saturday-night drunkenness and looseness in the relations between the young men and young women. [**1938** G. GREENE *Brighton Rock* III. 124 'Saturday,' he thought, 'today's Saturday,' remembering the room at home, the frightening weekly exercise of his parents which he watched from his single bed. *Ibid.* VII. 320 The Boy was shaken again with his nocturnal Saturday disgust. He couldn't blame his father now... You couldn't even blame the girl.] **1942** BERREY & VAN DEN BARK *Amer. Thes. Slang* § 509/17 *Saturday-night habit, week-end habit*, indulgence in small amounts of narcotics at irregular intervals. **1951** *Evening Sun* (Baltimore) 27 Mar. 4/1 The graduate 'hype' was a 'student' or 'hoosier fiend' who 'dabbled' with drugs occasionally. He had what is known as 'chippy habit', a 'Saturday night habit', or an 'ice cream habit'. **1963** R. I. McDAVID *Mencken's Amer. Lang.* xi. 742 Most cats consider it necessary to probe the mystic depths with the assistance of wine, a joint of pot.., peyote buttons and large infusions of invigorating jazz music—..in any event indulged in with friends as part of the Saturday night kicks. **1964** *New Statesman* 17 Apr. 606/2 Is the Saturday-night blind..any less characteristic of the modern urbanised proletariat than of the traditional rural peasantry? **1976** *N.Y. Times Mag.* 10 Oct. 111/2 [In the southern States of the U.S.] there were all those cross burnings, lynching bees and Sairday Nite Socials.

2. *spec. attrib.* uses: *Saturday night palsy* or *paralysis*, temporary local paralysis of the arm, esp. wrist drop, after it has rested on a hard edge for a long time, as during sleep following a bout of drinking (*colloq.*); *Saturday night pistol* (*U.S. colloq.*) = *Saturday night special*; *Saturday night soldier*, a member of a volunteer army, as opp. a regular soldier; *Saturday night special* (*U.S. colloq.*), a cheap, low-calibre pistol or revolver such as might be used by a petty criminal.

1927 I. S. WECHSLER *Textbk. Clin. Neurol.* III. 249 The frequent occurrence of wrist drop in alcoholics who fall asleep and lean heavily on the arm has given rise to the common designation of 'Saturday night palsy'. **1942** *Sun* (Baltimore) 23 Apr. 22/2 A similar ailment is called 'shelter paralysis'—formerly known as 'Saturday night paralysis' because its victims were generally payday tipplers. **1951** E. PAUL *Springtime in Paris* xii. 216 Berthe was suffering from what is known in the United States as Saturday-night paralysis,..when drunken men go to sleep in gutters, with one arm across a sharp kerbstone. **1974** PASSMORE & ROBSON *Compan. Med. Stud.* III. xxxiv. 35/1 Wrist drop thus produced is known as a 'Saturday night palsy'. **1929** M. A. GILL *Underworld Slang*, *Saturday night pistol*, 25 automatic. **1917** A. G. EMPEY *Over Top* 311 '*Terrier*', Tommy's nickname for a Territorial or 'Saturday-night soldier'. **1974** *Maclean's Mag.* Oct. 30/1 My husband was a Saturday Night soldier, the militia, and he couldn't wait for the war and when it started, zoom, he was called up and then he was happy. **1968** *N.Y. Times* 17 Aug. 1/1 Title IV of that law bans the importation of the cheap, small-caliber 'Saturday night specials' that are a favorite of holdup men. **1976** *Pioneer* (Big Timber, Montana) 30 June 4/2 A ban on 'Saturday Night Special' handguns. **1977** C. MCFADDEN *Serial* xlvi. 98/1 I'm not packing a Saturday-night special, *really*.

Hence **Saturday nighter,** a person who attends an entertainment on a Saturday night; **Saturday-night** *v. intr.*, to spend a Saturday night in enjoyment or revelling.

1962 D. LESSING *Golden Notebk.* IV. 462 The fellows were out Saturday-nighting true-hearted, the wild-hearted Saturday-night gang of true friends. **1966** *Listener* 24 Mar. 422/2 The Korean script announced that *Dr No* was showing inside. So he was..and half the population of Korea was inside, too..all of us lapping up James Bond like Surbiton Saturday nighters.

Saturn. 2. (Further examples.)

1964 R. H. BAKER *Astronomy* (ed. 8) viii. 225 Saturn is encircled by three concentric rings... There is no gap between the bright ring and the crape ring. **1974** *Encycl. Brit. Macropædia* XVI. 274/2 Saturn has ten satellites... Janus, the most elusive and closest to the planet, was found by A. Dollfus in 1966.

Saturnalia, *sb. pl.* Add: **2.** (Earlier example.)

1775 *Answer to Pamphlet, entitled Taxation no Tyranny* 61 Thus you would establish a Saturnalia of cruelty, and expose these devoted men to the brutality of their own slaves.

Saturnian, *a.* Add: **1. a.** (Earlier example.)

1612 J. SELDEN in Drayton *Poly-olb.* sig. A4, This later age..hath, in our greatest Latine Critiques..so receiued that Saturnian Language, that, to Students in Philology, it is now grown familiar.

3. a. Delete † and add later example in the sense 'due to the baleful influence of Saturn'.

1922 W. B. YEATS *Seven Poems & Fragment* 6 Stretch out your limbs and sleep a long Saturnian sleep.

c. *Physics.* Of or pertaining to a model of the nuclear atom in which electrons are assumed to orbit in rings around a central nucleus, thus resembling the appearance of Saturn. Now *hist.*

1904 H. NAGAOKA in *Phil. Mag.* VII. 445 The system differs from the Saturnian system considered by Maxwell in having repelling particles instead of attracting satellites. *Ibid.* 455 There are various problems which will possibly be capable of being attacked on the hypothesis of a Saturnian system, such as chemical affinity and valency. **1911** *Phil. Mag.* XXI. 688 Nagaoka has mathematically considered the properties of a 'Saturnian' atom which he supposed to consist of a central attracting mass surrounded by rings of rotating electrons. **1967** D. TER HAAR *Old Quantum Theory* iii. 31 Nagaoka (1904) had considered earlier the properties of a 'Saturnian' atom. **1974** G. REECE tr. *Hund's Hist. Quantum Theory* iv. 56 Nuclear types of atom included..the 'Saturnian system' of H. Nagaoka (1904).

saturnian (sătɜ̄·rniăn), *sb.*[3] [f. mod.L. generic name *Saturnia* + -AN.] = *SATURNIID *sb.*

1842 T. W. HARRIS *Treat. Insects Injurious to Vegetation* 276 These insects..belong to a family called Saturnians.

saturniid (sătɜ̄·rni[ⁱ]id), *a.* and *sb. Ent.* Also **Saturniid.** [f. mod.L. family name *Saturniidæ*, f. generic name *Saturnia* (F. von P. Schrank *Fauna Boica* (1802) II. 149): see SATURNIAN *a.* and *sb.*[2]] **A.** *adj.* Of, pertaining to, or belonging to the family Saturniidæ, which includes large, mainly tropical moths with a few species of temperate regions. **B.** *sb.* A moth of the family Saturniidæ.

1892 W. L. DISTANT *Naturalist in Transvaal* 122 The fine Saturniid moth *Urota sinope*. **1928** G. H. CARPENTER *Biol. Insects* xii. 378 The large Chinese Saturniid silk-moth..is represented in Japan and Java by readily distinguishable forms. **1952** *Bull. Amer. Mus. Nat. Hist.* XCVIII. 355/1 The saturniid moths appear to be most closely related to the small South American families Oxyteridae and Cercophanidae. *Ibid.* 365/2 The eyes of saturniids are large in relation to the head. **1964** [see *LASIOCAMPID *sb.* and *a.*]. **1979** *Smithsonian* X. 68 (*caption*) Saturniid moth..is one of the many colorful insects harbored in Costa Rica's forests.

‖ **satya** (sa·tya). [Skr.] In Indian philosophy: truth, truthfulness.

1943 C. S. LEWIS *Abolition of Man* i. 10 Righteousness, correctness, order, the *Rta*, is constantly identified with *satya* or truth, correspondence to reality. **1956** E. WOOD *Yoga Dict.* 139/2 *Satya.* (Truthfulness). The second of the abstinences. **1974** *Encycl. Brit. Macropædia* VIII. 900/1 The ancient ideals of *ahimsā*, chastity, observances, and *satya* (Truth, which he [*sc.* Gandhi] identified with God) were the main principles of his undogmatic doctrine and social and political practice. **1975** DASTUR & AIYAR in H. M. Patel et al. *Say not the Struggle Nought Availeth* 192 In Gandhi's view, truth (*satya*) and non-violence (*ahimsa*) were inter-related.

‖ **satyagraha** (satyā·graha). [a. Skr. *satyāgraha* insistence on truth, f. *satya* truth + *āgraha* pertinacity.] The Indian form of passive resistance, as formulated by M. K. Gandhi. Also *transf.* and *attrib.*

1920 M. K. GANDHI *Non-Co-operation* (1921) 46 But all the painful experience that I then gained did not in any way shake my belief in Satyagraha or in the possibility of that matchless force being utilised in India. **1928** V. G. DESAI tr. *Gandhi's Satyagraha in S. Afr.* xii. 173 A small prize was therefore announced in *Indian Opinion* to be awarded to the reader who invented the best designation for our struggle... Sr. Maganlal Gandhi..suggested the word 'Sadagraha', meaning 'firmness in a good cause'. I..corrected it to 'Satyagraha'. Truth (Satya) implies love and firmness (Agraha) engenders and therefore serves as a synonym for force. I thus began to call the Indian movement 'Satyagraha', that is to say, the Force which is born of truth and love or non-violence, and gave up the use of the phrase 'passive resistance'. **1929** *Daily Express* 10 Jan. 1/5 The ex-soldiers expressed their intention of performing satyagraha until their leader was released. **1930** *Aberdeen Press & Jrnl.* 7 Apr. 7/1 The 'Untouchables', dissatisfied with his campaign, have threatened counter satyagraha. **1955** *Times* 5 July 10/5 But there were signs of official uneasiness; Sikh dignitaries were arrested before offering satyagraha, Press censorship was imposed in Amritsar, [etc.]. **1958** *Economist* 26 July 280 Delay in producing this Bill was one of the grievances advanced by the Tamil minority when they launched their satyagraha campaign in the spring. **1963** *Times* 3 May 12/2 Negro and white youngsters, organized by bodies such as the student non-violent coordinating committee and the south Christian leaders' conference, will be offering their own kind of satyagraha throughout the deep south. **1969** *Pioneer* (Lucknow) 13 Aug. 7/7 A batch of 12 girls were taken into custody..when they staged satyagraha in support of a separate Telengana. **1976** *Times* 23 Jan. 9/2 The opposition..is pursuing its campaign of *satyagraha* (non-violent demonstration) in traditional Indian style. **1980** *Times Lit. Suppl.* 21 Nov. 1339/1 For him [*sc.* Michael Scott], the world desperately needs a moral force which will work for justice. He finds this in *satyagraha*, the soul-force, more positive than non-violence or passive resistance, an active weapon of good in its perennial fight against evil.

So **satya·grahi** (with pl. **satyagrahi** or **satyagrahis**), **satya·grahist,** an exponent or practitioner of *satyagraha*.

1928 V. G. DESAI tr. *Gandhi's Satyagraha in S. Afr.* xx. 233 Rama Sundara was the first Satyagrahi prisoner. *Ibid.* xlv. 468 Only thus could the Satyagrahis..bring their struggle to a triumphant end. **1930** *Aberdeen Press & Jrnl.* 6 Mar. 8/6 He [*sc.* Gandhi] will be accompanied by a band of satyagrahists or home rule volunteers, who will march on foot. **1934** H. MILLER *Tropic of Cancer* 102 The little band of Satyagrahists imitated the devotion of their master. **1968** H. J. N. HORSBURGH *Non-Violence & Aggression* i. 22 Belligerents do, and satyagrahi do not, claim finality for their interpretation of a just settlement. **1976** *New Yorker* 24 May 43/1 According to Gandhi, a satyagrahi, or votary of satyagraha, is governed by the belief that the soul can be saved from evil in the world, and so helped along in its search for Brahma, by truth and truth alone.

satyr. Add: **6.** *satyr-brood, forest, -spring, -talk*; *satyr-charming, -hairy, -haunted, -like* (earlier example), *-shrewd* adjs.

1924 E. SITWELL *Sleeping Beauty* i. 11 Smiling dim as satyr-broods. **1883** J. G. WHITTIER *Bay of Seven Islands* 31 Calm as the hour, methinks I feel A sense of worship o'er me steal; Not that of satyr-charming Pan. **1933** E. SITWELL *Five Variations* 2 Mowhair for satyr forests. **1953** —— *Gardeners & Astronomers* 29 Like the first budding of the small red satyr-hairy leaves upon the fruit-boughs. **1924** —— *Sleeping Beauty* xv. 53 From satyr-haunted caverns drip These lovely airs on brow and lip. **1835** POE in *Southern Lit. Messenger* I. 637/2 Satyr-like figure of Mentoni himself. **1928** BLUNDEN *Retreat* 38 And almost catch the horned and rude Woodgod at gaze ere satyr-shrewd He dodges by. **1922** E. SITWELL *Façade* 7 Like red Furred buds of satyr-springs long dead. **1944** L. MACNEICE *Springboard* 49 Not smut but satyr-talk, not clever but wise.

satyress. (Later examples.)

1952 [see *NIXIE[1]]. **1978** *Daily Tel.* 12 Apr. 14/4 A Tiepolo drawing, 'A Centaur with a Satyress', was bought for £4,000.

saty·rically, *adv. rare.* [f. SATYR.] In the manner of a satyr.

1887 SWINBURNE *Let.* 14 Aug. (1962) V. 209, I have written a poem..called 'Pan and Thalassius'... Pan is figured in all his different shapes or phases..lord of the mystery of earth and immanent godhead of—or in—the terrene All: only not of the human soul, the stars, Urania, and the sea—on whose general behalf the intruder in his domain has the last word—while recognizing the folly and falsehood of the cry that 'Pan is dead'..over which premature cry the old wood-god chuckles satyrically.

Satyrid (săti·rid), *sb.* and *a.* [a. mod.L. family name *Satyridæ*, f. generic name *Satyrus* (P. A. Latreille & J. B. Godart *Encyclopédie Méthodique* (*Insectes*) (1819) IX. 11): see SATYR and -ID[3].] **A.** *sb.* A small,

usually brown, butterfly belonging to the subfamily Satyrinæ of the family Nymphalidæ. **B.** *adj.* Of or pertaining to a butterfly of this kind.

1901 D. SHARP in *Cambr. Nat. Hist.* VI. vi. 348 The species of the genus *Pierella* connect these transparent Satyrids with the more ordinary forms. **1912** H. ROWLAND-BROWN *Butterflies & Moths* viii. 79 The Continental Satyrids..and our 'Grayling',..well-nigh invisible on the tree trunks where they love to perch. **1936** *Discovery* Dec. 370/2 A shining silvery insect..quite different from the general run of Satyrid butterflies. **1963** V. NABOKOV *Gift* ii. 126 His father accompanied him up a trail through the pinewoods in order to show him, with a smile of condescension for this European trifle, the Satyrid recently described by Kuznetzov, which was flitting from stone to stone. **1975** *Zool. Jrnl.* CLXXVII. 333 *Zethera hestioides* Felder, a Philippine Satyrid..shows unimodal mimicry.

satyrish (sæ·tiriʃ), *a.* [f. SATYR + -ISH[1].] Characteristic of a satyr (sense 1); erotic, sensual.

1932 W. FAULKNER *Sartoris* III. vi. 233 Simon chuckled again, unctuously, a satyrish chuckle rich with complacent innuendo. **1937** J. C. POWYS *Maiden Castle* v. 198 His satyrish pleasure in the exposed curves of her limbs.

satyromaniac (sæ:tiromē[ⁱ]·niæk), *a.* and *sb.* [f. SATYR + -O + -MANIAC.] **A.** *adj.* Of a man: exhibiting excessive sexual desire. **B.** *sb.* A man who exhibits excessive sexual desires; a sex maniac.

1889 *Cent. Dict.*, *Satyromaniac*, *a.* and *n.* I. *a.* Affected with satyromania. II. *n.* A person affected with satyromania. **1892** G. B. SHAW *Let.* 12 Aug. (1965) I. 360, I hear from Oxford that de Mattos is ravishing every maiden in the country... York Powell writes to me privately urging the importance of dissociating ourselves from the satyromaniac W. S. de M. **1909** —— *Let.* 22 June (1972) II. 847, I have read the play... It is made impossible by your nymphomania. There are two men in it.., one a satyromaniac, the other a mere imaginary male figment to focus the nymphomania of all the women. **1944** D. L. SAYERS *Let.* 18 Oct. in J. Brabazon *D. L. Sayers* (1981) x. 112 All Satyromaniacs, sadists, connoisseurs in rape.

Satzuma: see *SATSUMA.

sauce, *sb.* Add: **1. a.** Also with qualifying *adj.*, as *black, brown, hard, white sauce*.

Occas., in the names of sauces taken unchanged from French into English, found with the qualifying word following; in such cases the Fr. pronunc. (sos) may be heard. *sauce Robert* (sos robɛr), now the usual form of *Robert sauce* (in Dict.) See also *ALLEMANDE *sb.* 3, *BÉARNAISE, *MORNAY, SOUBISE 2.

1573 'C. HOLLYBANDE' *French Schoole-maister* 114 Cut some of these loynes of the hare, drest with a blacke sauce. **1723** J. NOTT *Cook's & Confectioner's Dict.* sig. Bb6, *To dress Pikes* à la Sainte Robert [*sic*]..make your Sauce Robert in the following manner. *Ibid.* sig. D[v], *Artichokes with white Sauce*... Make a Sauce for them with the Yolks of Eggs, a Drop or two of Vinegar, and a little Gravy. **1806** J. SIMPSON *Compl. Syst. Cookery* 293 Pigs feet au gratin, ears shredded, and sauce robert. **1845** E. ACTON *Mod. Cookery* iv. 116 *Bechamel.* This is a fine French white sauce, now very much served at good English tables. *Ibid.* 130 *Sauce Robert*... Large onions,..butter,..flour... Gravy... Mustard. **1884** White sauce [in Dict.]. **1909** *Cent. Dict.* Suppl., *Hard sauce*, a creamy sauce of butter and sugar, usually flavoured with vanilla or the like. **1911** WEBSTER, Brown sauce = Espagnole sauce. **1928** S. LEWIS *Man who knew Coolidge* I. 103 A..Plum Pudding..with both hard and soft sauce. **1935** 'R. HULL' *Keep it Quiet* xxix. 279 A brown substance..called generally 'Sauce Robert', which disfigures cutlets and suchlike. **1939** A. L. SIMON *Conc. Encycl. Gastron.* I. 29/2 In U.S.A., a Hard Sauce is made with one measure of fresh butter to two of castor sugar... A squeeze of lemon is then added... It is usual, in some States,..to add some Brandy or Rum... In England, a similar sauce is called Brandy Butter or Rum Butter. **1960** *Good Housek. Cookery Bk.* (rev. ed.) 196/1 The foundation of all brown and white sauces in which flour is the thickening agent is the roux, formed by cooking the butter and flour together. For white sauces the butter should be melted, the flour added and the two stirred and cooked together until well incorporated. The liquid should then be added by degrees. **1974** E. MCGIRR *Murderous Journey* 90 His man had a certain way with Sauce Robert which gave it an added piquancy. **1981** M. C. SMITH *Gorky Park* III. 304 She'd brought cartons of spaghetti with meat, clam and white sauces.

d. *U.S. slang.* (See quot.)

1919 E. V. RICKENBACKER *Fighting Flying Circus* p. xi, *Sauce*, petrol or gasoline.

e. *slang* (orig. *U.S.*). Alcoholic liquor; occas., a narcotic drug.

1940 J. O'HARA *Pal Joey* 114 It made him sad and he almost began hitting the sauce. **1953** W. BURROUGHS *Junkie* (1972) xiii. 134 The first thing you have to do is cut down on the sauce and build up your health. You look terrible. **1960** WODEHOUSE *Jeeves in Offing* xvii. 176 Her first husband,..was..a constant pain in the neck to her till one night he most fortunately walked into the River Thames while under the influence of the sauce and didn't come up for days. **1970** M. BRAITHWAITE *Never sleep Three in Bed* vi. 66 Which means any occasion when any group of the brothers and sisters..have got into the sauce. **1975** N. FREELING *What are Bugles blowing For?* xii. 74 Castang found a narcotics squad cop... Patricia was known, but not well. 'She got off the sauce for nearly a year.' **1976** W. TREVOR *Children of Dynmouth* v. 114 'You often get loonies in joints like that,' he remarked on the street. 'They drink the sauce and it softens their brains for them.' **1978** H. C. RAE *Sullivan* I. ii. 25 You're not in debt, on the sauce, going gay... I can't blackmail you.

2. *fig.* Also in Fr. phr. *sauce piquante*.

1821 HAZLITT *Table-T.* Ser. I, *Character of Cobbett* 121 How fine were the graphical descriptions he sent us from America:..what a fine *sauce piquante* of contempt they were seasoned with! **1934** C. LAMBERT *Music Ho!* III. 206 They are only thorns protecting a fleshy cactus—a sauce piquante poured over a nice juicy steak.

7. *sauce-bottle, -bowl, -tureen* (earlier and later examples); *sauce-stained* adj.

1925 HODKIN & COUSEN *Textbk. Glass Technol.* v. 49 Glasses..of the type usually used for ordinary white flint glass, for medical, paste, and sauce bottles, and for those used in machines with automatic feeding devices. **1973** *Country Life* 1 Nov. 1313/1 The autumn gathering [of mushrooms] went to make ketchup, put up in old sauce bottles. **1765** J. WEDGWOOD *Let.* 2 Mar. (1965) 29, I have sent the Green & Gold Sauce bowles and stands..in a box. **1922** JOYCE *Ulysses* 44 His breath hangs over our saucestained plates, the green fairy's fang thrusting between his lips. **1772** J. WEDGWOOD *Let.* 17 Feb. (1965) 119, I thank you..for the hint respecting the sauce Terrine. **1776** [see TUREEN β]. **1971** *Country Life* 1 Apr. 765/2 At the dining table, the classical urn was, of course, readily applied..to the now popular sauce tureen.

sauce, *v.* **1. a.** Delete *arch.* and add later examples.

1973 *Jewish Chron.* 2 Feb. 19/1 If..I choose to sauce them, then I find the ordinary four-to-a-fish fillets quite suitable. **1975** *Times* 4 Oct. 12/4 A sole dish..said to be sauced with cream, wine and egg... The pale yellow sauce tasted sour.

4. d. (Earlier and later examples.) Also *transf.*

1862 H. ADAMS *Let.* 10 Jan. in N. Longmate *Hungry Mills* (1978) iv. 61, I found myself this morning sarsed through a whole column of *The Times* and am laughed at by all England. **1892** B. POTTER *Jrnl.* 6 Oct. (1966) 274 He puts on wrong postage..and will sauce anybody who is unprovided with small change; he wants reporting. **1962** D. LESSING *Golden Notebk.* II. 274 He sauced her with his eyes; sitting up broad, solid, pink-cheeked; very sure of himself.

sauce-boat. Add: **2.** *Archæol.* A vessel of the Early Helladic and Early Cycladic cultures resembling a sauce-boat and prob. used for drinking or pouring liquids.

1967 R. HIGGINS *Minoan & Mycenean Art* ii. 55 Sauceboats like the popular Mainland variety..were decorated with an all-over wash. *Ibid.* iii. 67 Favourite shapes are now the so-called 'sauceboats', a very common type whose function is unknown. *Ibid.* iii. 70 Only one form of gold or silver plate has been recorded from mainland Greece for this period. That form, known in two surviving examples, is a translation into gold of the common pottery 'sauceboat' shape. **1977** G. CLARK *World Prehist.* (ed. 3) iv. 157 Another unusual ceramic vessel common to the three areas is the sauce-boat which also occurs in Early Helladic Greece in gold.

saucebox. Add: Also *attrib.*

1825 in C. E. Pearce *Life & Times Madame Vestris* (1923) 116 We thought that the stamping sort of sauce-box air with which she marched away to the tune of the 'Dashing White Sergeant' was too much in keeping with her notorious male-attire exhibitions. **1969** V. C. CLINTON-BADDELEY *Only a Matter of Time* 89 He hadn't used 'camp' for several weeks—not since his sauce-box notice of *Idomeneo*.

saucepan. Add: **2.** *saucepan brush*; **saucepan lid,** rhyming slang for (*a*) a 'quid', a one-pound note; (*b*) a 'kid', a child.

1926–7 *Army & Navy Stores Catal.* 181/1 Steel saucepan brush. Each—/6. **1944** C. MILBURN *Diary* 30 Dec. (1979) 260 A few oddments at the ironmonger's..a dish mop, a baking tin, a colander, a saucepan brush. **1952** *Observer* 12 Oct. 5/3 Won't dish-mops, saucepan-brushes and swabs, rubbers, and all kitchen cloths one day be of nylon? **1951** P. HOSKINS *No Hiding Place!* xvii. 191/2 *Saucepan lid*, £1 note. **1960** J. FRANKLYN *Dict. Rhyming Slang* 119/1 *Saucepan lid*,..kid.

saucepanful (sǭ·spǎnful). [f. SAUCEPAN + -FUL.] The contents of a saucepan; the amount a saucepan will hold.

1868 DICKENS *Holiday Romance* II, in *All Year Round* 8 Feb. 206/2 The other Princes and Princesses were squeezed into a..corner to look at the Princess Alicia turning out the saucepan-full of broth, for fear..they should get..scalded. **1976** *Horse & Hound* 3 Dec. 34/4 Two or three saucepansful may be needed simultaneously and it is a good idea to put a spoon in the glass when dispensing. **1980** J. O'FAOLAIN *No Country for Young Men* iv. 71 Judith boiled a saucepanful [of water].

saucer. Add: **2. c.** = *flying saucer* s.v. *FLYING *vbl. sb.* 3. Also *attrib.*

[**1878** *Denison* (Texas) *Daily News* 25 Jan., in C. & J. Lorenzen *UFOs* (1969) i. 10 When directly over him it [*sc.* a flying object] was about the size of a large saucer and was evidently at a great height.] **1947** *Daily Progress* (Charlottesville, Va.) 5 July 1/4 Describing what they saw as flat, translucent plates 12 to 15 inches in diameter, several Port Huron, Mich., residents reported seeing the 'saucers'. **1958** *Times Lit. Suppl.* 16 May 274/3 The author declares he was arrested while camping out in a fertile saucer district and narrowly escaped a mental examination court. It all affords a good occasion to re-tell certain saucer stories the author inquired into. **1966** *McGraw-Hill Encycl. Sci. & Technol.* V. 363/2 Light reflections from material objects account for most reports of saucers. **1978** D. A. J. SEARGENT *UFO's* vi. 122 People frequently shy away from traditional religions and look for salvation from the 'saucers'.

3. (Earlier example.)

c **1702** C. FIENNES *Journeys* (1947) III. v. 177, I went to this Newcastle in Staffordshire to see the makeing the fine tea-potts cups and saucers of the fine red earth.

4. So (*slang*), an eye. Also *attrib.*

1864 M. LEMON *Jest Bk.* 185, I always know when he has been in his cups by the state of his saucers. **1958** *Spectator* 22 Aug. 246/2 Nor were they wasting any saucer stares on National Savings or 'Taking up a Career in the Midland Bank'.

7. b. *saucer-barrow, -brooch, -cloud, -hat*; **saucer bath,** a wide shallow bath usu. kept in a bedroom and used for sponging oneself down; **saucer-buried** *a.*, formerly in the southern U.S., applied to a Black person whose burial was paid for by donations placed in a saucer laid on or near the corpse; hence **saucer-burial**; **saucerman,** a being imagined or believed to be the pilot or passenger of a flying saucer. **c.** *saucer-blue* adj.

1941 *Proc. Prehist. Soc.* VII. 88 The saucer-barrow may be defined as a low mound, generally one or two feet high..enclosed in a ditch and outer bank. **1899** SOMERVILLE & 'ROSS' *Some Experiences Irish R.M.* ii. 38 A conspicuous object outside the door was a saucer bath full of something that looked like flour. **1927** G. MURRAY *Classical Trad. in Poetry* 4 He used a sponge and a tooth-brush and a saucer bath. **1951** C. V. WEDGWOOD *Last of Radicals* i. 19 The patriarchal old gentleman, who rose every morning to a cold saucer-bath. **1925** W. DE LA MARE *Miss Jemima* 12 She was staring about her..with her saucer blue eyes. **1912** *Archaeologia* LXIII. 167 The find included no less than four saucer brooches, one decorated with the star with incurved sides..and three with spirals. **1965** A. H. SMITH in Bessinger & Creed *Medieval & Linguistic Stud.* 61 The presence of artifacts in the Avon valley cemeteries, like applied, disc, and saucerbrooches..has been interpreted as a mixed culture. **1963** P. POLLACK *Photography* xxvii. 351/1 A picture of a saucer burial taken in Alabama. **1925** DU B. HEYWARD *Porgy* I. 25 It had even become a grievous reproach to have a member of the family a 'saucer-buried nigger'. **1911** H. S. WALPOLE *Mr. Perrin & Mr. Traill* iv. 72 Faint blue skies, dim and shining like clear glass with a hard yellow sun stuck like a tethered balloon between saucer-clouds. **1940** M. SADLEIR *Fanny by Gaslight* I. 30 My own tartan frock..and tiny saucer hat. **1965** J. POTTS *Only Good Secretary* iv. 68 Her head, topped with its black saucer hat. **1967** *Time* 4 Aug. 40/2 Barney and Betty Hill..whose 'abduction' by saucermen during an auto trip was described in the fast-selling book [etc.] **1971** *New Scientist* 30 Sept. 722/1 Visiting saucermen from Mars might well report back to base that all our Gods must be hard of hearing.

sau·cer, *v.* [f. the sb.] **1.** *intr.* To be saucerlike; to take the shape of a saucer, be as shallow as a saucer.

1925 W. DE LA MARE *Broomsticks* 112 The immense starry sky that saucered in the wide darkness of the Moor. **1977** *Times* 19 Nov. 1/6 They prophesy that the rate is now 'saucering' and that after a brief dip into single figures it will rise again.

2. *trans.* To make saucerlike; to shape (something) like a saucer.

1934 WEBSTER, *Saucer*, *v.t.* & *i.* To make or be saucerlike. **1977** *Whitaker's Almanack 1978* 1058/1 The site has been successfully 'saucered' to disguise the bulk and reduce the overall height.

3. *trans.* To pour (a liquid) into a saucer, esp. from a cup.

1938 *Atlantic Monthly* Oct. 552/2 'Want a sasser o' sorghum?... The visitor would 'sasser' some sorghum. **1944** A. CLARKE *Coll. Plays* (1963) 246 Mind you don't utter A word..Until you have eaten six slices of bread With plenty of butter—and saucered your tay! **1951** H. GILES *Harbin's Ridge* x. 100 Granny saucered her coffee and blew on it. **1978** *New Yorker* 9 Jan. 41/3 Have you ever heard the old Texas expression 'saucered and blowed'?.. If a cowboy's coffee is too hot, he puts some in a saucer and blows on it. A cowboy will say to a friend, 'Take mine, it's already saucered and blowed.' Jim needs to get the energy bill saucered and blowed.

saucer eye. Add: (Later examples without supernatural associations.)

1970 'D. HALLIDAY' *Dolly & Cookie Bird* vi. 78 She still had the huge saucer eyes I remembered, with false eyelashes and then spikes drawn in under the lashes. **1976** G. MOFFAT *Short Time to Live* v. 48 'This is the astonishing thing—' she turned to Miss Pink with saucer eyes.

saucer-eyed, *a.* Add: (Further examples.) Also *transf.*, of an expression, emotion, etc.

In quot. 1968 the sense is 'susceptible to seeing flying saucers'.

1843 *Ainsworth's Mag.* IV. 5 The frightful, open-mouthed, saucer-eyed expression of wonder. **1934** A. WOOLLCOTT *While Rome Burns* 57 He rushed at me in saucer-eyed excitement. **1968** *Listener* 27 June 823/1 As if people haven't tended in such matters to see the expected thing in the expected form, as if they were unlikely to go saucer-eyed to their vigils. **1978** J. IRVING *World According to Garp* xvii. 361 Garp looked for the strange saucer-eyed girl. **1979** N. FREELING *Widow* xvii. 108 I've been shot at... Don't look so saucer-eyed..don't let's dramatize.

saucerful. Add: Also, the contents of a saucer, and *fig.* (Earlier and further examples.)

1852 MRS. GASKELL *Cranford* (1853) viii. 156 She..mixed a saucer-full for him, and put it down for him to lap. **1917** D. CANFIELD *Understood Betsy* vi. 123 She herself ate three saucerfuls. **1927** [see *grape-nuts* s.v. *GRAPE *sb.*[1] 9]. **1944** M. LASKI *Love on Supertax* viii. 77 A saucerful of margarine. **1973** K. GILES *File on Death* ii. 48 Here, sit at the table, and I'll give you a saucerful.

saucerian (sǭsī^ə·riăn), *a* and *sb.* [f. SAUCER + -IAN.] **A.** *adj.* Of or pertaining to a flying saucer (see *FLYING *vbl. sb.* 3). **B.** *sb.* **a.** A believer in the existence of flying saucers. **b.** An entity which travels by flying saucer.

1950 *Jrnl. Brit. Interplanetary Soc.* IX. 300 This passage..surely deserves immortality. We..happen to believe it is true—but perhaps not in the saucerian sense that the author intended. **1965** *New Society* 9 Sept. 14/4 The definition of the situation in occult terms began in 1950 with the publication of..the first saucerian book complete with little green men..looked upon by saucerians as the beginning of the tradition. **1973** C. SAGAN *Cosmic Connection* (1975) vi. 43 Likewise, the category of contact story, now quite fashionable in some UFO enthusiast circles, of sexual contact between human and saucerian..must be relegated to the realm of improbable fantasy. Such crossings are about as reasonable as the mating of a man and a petunia.

saucerization (sǭ:sərəizē[ⁱ]·ʃən). *Surg.* [f. SAUCER + -IZATION.] The surgical excision of bone or flesh so as to leave a shallow saucer-shaped cavity. So **sau·cerize** *v. trans.*

1928 P. LEWIN *Orthopedic Surg. for Nurses* xv. 252 In chronic osteomyelitis the entire infected area must be removed, the so-called 'saucerization'. **1940** *Lancet* 13 July 32/1 No method of treating such a case can compare with closed plaster after saucerisation, which may include laying open all heavily infected areas. **1946** *Jrnl. Bone & Joint Surg.* XXVIII. 19 The wounds are saucerized and then packed. **1964** W. A. LARMON in L. Davis *Christopher's Textbk. Surg.* (ed. 8) xxv. 1105/1 This operation is termed saucerization because the bone cavity is made as shallow and broad as possible by the surgeon.

‖ **saucier** (sosye). [Fr.] A sauce cook.

1961 *Evening Standard* 14 Sept. 29/3 (Advt.), Saucier required for first class London Club. Interesting position with good wages. **1976** *National Observer* (U.S.) 10 Apr. 5/1 She..gained 10 pounds that could be attributed almost entirely to the magic of the finest sauciers in town. **1980** J. CARTWRIGHT *Horse of Darius* xvii. 269 Chef Leon..took three under-chefs, a *saucier*, a vegetable cook, and his desert chef.

saucisson. Add: **1.** (Later examples.) Freq. with Fr. qualifying words designating spec. types of sausage. Also *attrib.*

Following Fr. usage the term usu. describes a sausage which does not need to be cooked, as opp. to a *saucisse*.

1958 W. BICKEL tr. *Hering's Dict. Cookery* 31 *Marinated sausage*: saucissons à l'huile: sliced cervelat, marinated in French dressing with chopped onions. *Viennese sausage*: saucissons de Vienne: heat up in boiling water, serve with grated horseradish, mustard, saurkraut or goulash sauce. **1962** *Harper's Bazaar* Aug. 69/1 A few slices of *saucisson d'ail* and *saucisson sec*. **1965** *House & Garden* Jan. 60 *Saucisson à l'ail* (*garlic sausage*), a type that includes many of the large sausages such as saucisson de Lyon and cotechino. *Saucisson d'Arles* (*Arles sausage*)..is dried and delicately seasoned. *Saucisson de Lyon*, a sausage from Lyon made of pork, with fat and lean mixed. **1972** *Guardian* 11 Mar. 15/3 [The] Brasserie du Nord..is noted..for its *saucisson* and *rognon* dishes. **1975** *Women's Jrnl.* Sept. 73/3 We started off with a plate of beetroot and tomato and saucisson. **1980** J. DITTON *Copley's Hunch* I. ii. 60 A little *saucisson sec*, bread and apples.

sauconite (sǭ·kǫnəit). *Min.* [f. *Saucon*, name of a valley near Bethlehem, Pennsylvania + -ITE[1].] A clay mineral of the montmorillonite group containing a high proportion of zinc.

1875 F. A. GENTH *Prelim. Rep. Mineral. Pennsylvania* v. 120 The first [peculiar clay] occurs at the Ueberoth Zinc Mine near Friedensville, Lehigh county, where it has been discovered by Prof. W. Th. Roepper, who named it 'Sauconite'. **1946** *Amer. Mineralogist* XXXI. 414 Sauconite proves to be a member of the montmorillonite group of minerals in which three bivalent zinc ions proxy two trivalent alumina ions in octahedral positions in the lattice structure. **1968** I. KOSTOV *Mineral.* 373 Additional representatives of the sub-groups are the following corresponding varieties:..Zn-saponite (sauconite) with about 35 % ZnO.

saucy, *a.*[1] Add: **2. b.** Now freq. in coy use: 'daring', smutty, suggestive.

1962 *Times* 13 Apr. 18/4 The comedy is all reduced to relentlessly 'saucy' sniggering farce. **1975** *Radio Times* 3 Apr. 17 George Formby..died 15 years ago. His songs, especially the saucy ones, have passed into legend. **1977** *News of World* 17 Apr. 9/7 He [*sc.* Mozart] won a reputation as a bed-hopping gambler and earned a fortune... Experts unearthed the saucy truth when they studied the great man's personal accounts.

3. (Later example.)

1904 A. BENNETT *Great Man* vi. 50 He ate a little of the lean, leaving a wasteful margin of lean round the fat..; then he unobtrusively laid down his knife and fork. 'Come, Henry,' said Aunt Annie, 'don't leave a saucy plate.'

‖ **saudade** (sa[ⁱ]uda·di). [Pg.] Longing, melancholy, nostalgia, as a supposed characteristic of the Portuguese or Brazilian temperament.

1912 A. F. G. BELL *In Portugal* i. 7 The famous *saudade* of the Portuguese is a vague and constant desire for something that does not and probably cannot exist, for something other than the present, a turning towards the past or towards the future; not an active discontent or poignant sadness but an indolent dreaming wistfulness. **1936** R. GALLOP *Portugal* xi. 262 In a word *saudade* is yearning: yearning for something so indefinite as to be indefinable: an unrestrained indulgence in yearning. **1957** R. CAMPBELL *Portugal* p. ix, It [*sc.* Portugal] is an intensely poetic country, and it is the country of *saudade*, that mysterious melancholy which sighs at the back of every joy. **1976** *Gramophone* Aug. 320/1 The vigour and the *saudade*, the two Brazilian qualities with which the Preludes are imbued, are here replaced by a *gaucho* nostalgia.

Saudi (sau·di, sǭ·di, sa[ⁱ]ū·di) *sb.* and *a.* [ad. Arab. *sa'ūdī*, f. the name *Sa'ūd* (see below) + *-I.] **A.** *sb.* **a.** A member of the Arabian Sa'ūd dynasty, the rulers of Nejd since the eighteenth century and of the kingdom of Saudi Arabia since 1932. **b.** = *SAUDI ARABIAN *sb.* **B.** *adj.* **a.** Of or pertaining to the Sa'ūd dynasty. **b.** = *SAUDI ARABIAN *a.*

1933 K. WILLIAMS *Ibn Sa'ud* i. 18 Muhammad took Riyadh... The Sa'udis could neither forget nor forgive their humiliation. *Ibid.* 23 Was the star of the Sa'udi scion..not to appear? *Ibid.* ii. 28 Arabia knew that a Sa'udi was..again master of Riyadh. **1949** *Britannica Bk. of Year* 51/1 A fraternal declaration which it was hoped would be the beginning of friendlier relations between the Saudi and Hashimi dynasties. **1957** *Encycl. Brit.* II. 169/1 The name Arabia is quite often used, incorrectly, when reference is intended to the Saudi kingdom only. **1959** W. THESIGER *Arabian Sands* xii. 227 At that time I was dressed as a Saudi. *Ibid.* xiii. 245 They were dressed in Saudi fashion, in long white shirts, gold-embroidered cloaks, and white head-cloths. **1962** *Listener* 5 Apr. 587/1 The Saudis could not bear the thought of an expansion of Hashimite power into Syria. **1974** *Encycl. Brit. Macropædia* XVI. 275/2 The 10,000 Saudi Arabs employed by Aramco..are exposed to modern industrial skills, technology, [etc.]. **1976** *Daily Record* (Glasgow) 22 Nov. 15/2 While I was there, Lebanese and Saudis in exquisite white silk suits, sipped pink champagne and peeled off £20 tips to English waitresses. **1976** *Star* (Sheffield) 3 Dec. 14/7 Sheffield workers got a pat on the back..for helping to save a Saudi customer waiting.

Hence **Sau·dian, Sau·dite** *adjs.*

1949 [see *HASHIMITE *a.* and *sb.*]. **1950** W. THEIMER *Encycl. World Politics* 378/1 Some experts believe that the Saudian oilfields contain one-half of the world's oil reserves.

Saudi Arabian (sau·di ărē[ⁱ]·biăn, sǭ·di ărē[ⁱ]·biăn), *a.* and *sb.* Also **Sa'udi Arabian,** and with hyphen. [f. *Saudi Arabia* (see below): cf. *SAUDI *sb.* and *a.*, ARABIAN *a.* and *sb.*] **A.** *adj.* Of or pertaining to Saudi Arabia, a kingdom founded in 1932 by Abdul Aziz ibn Sa'ūd (1882–1953), comprising the greater part of the Arabian peninsula. **B.** *sb.* A native or inhabitant of Saudi Arabia.

1934 *Times* 23 Nov. 13/7 Sheikh Hafiz Wahba, the Saudi Arabian Minister to the Court of St. James's, is about to return to Mecca. **1947** K. PHILBY *Let.* in F. Maclean *Take Nine Spies* (1978) vii. 251 Ignorance and arrogance make a bad combination, and the Saudi Arabians have both. **1951** *Britannica Bk. of Year* 48/1 Syria undertook to supply Syrian goods for Saudi-Arabian consumption. *Ibid.*, The Saudi Arabian province of Hasa on the Persian Gulf. **1959** *Chambers's Encycl.* I. 504/2 The financial strength of the Sa'udi Arabian government has been greatly increased by the granting of foreign concessions for oil and minerals. **1959** G. A. LIPSKY *Saudi Arabia* i. 2 Loyalty to the family..and loyalty to the tribe are the strongest bonds felt by most Saudi Arabians. **1976** *Alyn & Deeside Observer* 10 Dec. 9/6 A group of Saudi Arabians in Chester, celebrated one of their own religious feasts last week.

sauerbraten (sau^ə·ɹbrā:t'n, ‖ zau·ərbrā:tən). *U.S.* [Ger., f. *sauer* sour + *braten* roast meat.] A dish of German origin consisting of oven- or pot-roasted beef that has been marinaded in vinegar with peppercorns, onions, garlic, and bay-leaves before being cooked.

1889 B. K. KRAMER *'Aunt Babette's' Cook Bk.* 62 (*heading*) Sauerbraten. **1923** *Ladies' Home Jrnl.* Mar. 133/1 A demure little Mennonite maid..will invite you cordially to 'sit up' to a table arrayed with the wealth of cup cheese and pot cheese and sugar cakes and *sauerbraten* and noodles and all the rest of the savory dainties..on the menu of a Pennsylvania Dutch family. **1931** *Better Homes & Gardens* Mar. 44/3 The Sauerbraten mit Kartoffelklossen—that pot roast with the wonderful sauce. **1938** L. BEMELMANS *Life Class* I. i. 29 They lived on a diet of sauerbraten and cabbage. **1964** S. BELLOW *Herzog* 80 We eat in twenty minutes. Good chow. Sauerbraten. **1966** N. FREELING *King of Rainy Country* 84 There was a very good delicate sauerbraten, with almonds and raisins in it, not too vinegary. **1978** *Detroit Free Press* 16 Apr. (Detroit Suppl.) 28/2 Or the sauerbraten, the German version of roast beef, is always good, marinated with a vinegar mixture which later is used to make the rich brown gravy.

sauerkraut, sourcrout. Add: For the full history of the anglicized form, see SOUR CROUT, SOUR-CROUT. The usual form is now *sauerkraut*. (Further examples.)

1863 P. S. DAVIS *Young Parson* 48 [You] eat the best of roast beef, while I have to put up with sauerkraut and spec. **1892** S. BARING-GOULD *Strange Survivals* vi. 130 Such an umbrella..was sufferable as spread over an old woman vending *sauerkraut*. **1945** C. A. PRICE *German Settlers in S. Austral.* ii. 14 Various Silesian recipes such as..Sauerkraut (pickled cabbage), etc., made their appearance and even penetrated to the English colonists. **1973** *Times* 29 Dec. 10/2 The stereotypes of the German as a man born with a monkey wrench in his hand and eating vast quantities of sausage and sauerkraut..should really be discarded. After all, the French eat more sauerkraut.

2. *U.S. slang.* (Often with capital initial.) A German. Cf. *KRAUT 2.

1841 H. J. MERCIER *Life in Man of War* 232 Yes, old sour-crout, and you'd be *dirty too* if you were..in that infernal shot-locker as long as these fellows have been. **1858** J. A. STONE *Put's Golden Songster* 41 Sauer-Kraut was looking for a Justice of the Peace. **1909** *Sat. Even. Post* 3 July 7 I'll expurgate you, you old Dutch Sauerkraut!

3. *attrib.* and *Comb.*, as *sauerkraut barrel, cutter*; **sauerkraut-eater** *slang* = sense 2 above.

1888 *Century Mag.* Mar. 807/1 The representative Americans of the present day..[are] the Micks and the Pats, the Hanses and the Wilhelms, redolent still of the dudeen and sauerkraut barrel. **1969** *Canadian Antiques Collector* Aug. 21/1 Another..bargain is a primitive sauerkraut cutter, which she..purchased for twenty-five cents. **1918** R. BRAMBLETT *Let.* 9 June in K. Cowing *Dear Folks at Home* (1919) 217 We will scatter those 'sauerkrauter eaters' [*sic*] before the summer is over.

‖ **sauf** (sōf), *prep.* [Fr.: cf. SAVE *quasi-prep.* and *conj.*] Used for: except for, apart from.

c **1844** H. TAYLOR *Let.* in F. A. Hayek *John Stuart Mill & Harriet Taylor* (1951) 115 Your liability to take an over large *measure* of people—sauf having to draw in afterwards. **1847** J. S. MILL *Let.* 9 Mar. in *Wks.* (1963) XIII. 708, I have had a book to write..which I have now..completed, sauf the revising. **1864** G. MEREDITH *Let.* 1 June (1970) I. 259 Her Papa can't bear to lose her, though he always lets his daughters have their way in this matter, *sauf* the guarantee of moral character.

‖ **saugrenu** (sogrənü), *a.* Also fem. pl. **saugrenues.** [Fr.] Absurd, preposterous, ridiculous.

1876 W. JAMES *Let.* 5 July in R. B. Perry *Tht. & Char. W. James* (1935) I. 371 The *saugrenu*, comic Shakespeare scenes. **1889** E. DOWSON *Let.* 18 Oct. (1967) 110, I have many adventures to tell you of—some assez saugrenues. **1908** 'ODYSSEUS' *Turkey in Europe* (ed. 2) xiii. 440 The Great Powers of Europe are very like ordinary prejudiced individuals, not to say the lower animals. What they shy at one day, what they denounce with diminishing invective as impossible, revolutionary *saugrenu*, or crude, they accept a few years later as a matter of course. **1933** *N. & Q.* CLXV. 378/2 If we take 'chevalier' as addressed by Christ to the poet, do we not get merely a retort—frigid, and, by reason of the historical association a little *saugrenu*?

Sauk (sǭk). Also **Sac**; 8 **Sacky, Sax.** [ad. Canad. F. *Saki*, f. Ojibwa *osākī*; cf. Sauk *asākiwa* person of the outlet.] An Algonquian Indian people inhabiting parts of the central United States, formerly in Wisconsin, Illinois, and Iowa, now in Oklahoma and Kansas; a member of this people. Also, the language of this people, a dialect of Fox. Also *attrib.* or as *adj.*

1722 D. COXE *Descr. Carolana* 48 The Nations who dwell on this River, are Outogamis,..Sacky, and the Poutouatamis. **1762** [see *MENOMINEE]. **1789** *Deb. Congress U.S.* 25 May (1834) 41 The treaties..with the sachems and warriors of the Wyandot, Delaware,..and Sac nations,..appear to have been negotiated [etc.]. **1810** Z. M. PIKE *Acct. Expeditions Sources Mississippi* App. I. 20 The Sauks and Reynards are planting corn. **1810** in *Deb. Congress U.S.* (1853) 12th Congress 1 Sess., App. 1858 A considerable number of Sacs went..to see the British superintendent. **1835** [see *MENOMINEE]. **1836** J. HALL *Statistics of West* 53 On this prairie is a small village of the Sauk and Fox Indians. **1877** L. H. MORGAN *Anc. Society* II. vi. 169 The Shawnees had a practice, common also to the Miamis and Sauks and Foxes, of naming children into the gens of the father or of the mother or any other gens. **1881** *Encycl. Brit.* XII. 832/1 The *Sacs* and *Foxes*, now one tribe, located in Indian Territory, were originally separate, living near Green Bay, Wisconsin... A few still remain in Iowa, Nebraska, and Kansas. **1933** L. BLOOMFIELD *Language* iv. 72 The Algonquian family..includes the languages of..the Great Lakes region (..Menomini, Sauk, Fox, Kickapoo, [etc.]). **1946** G. FOREMAN *Last Trek of Indians* 187 Treaties were thus made with the following tribes: Delawares, Kansas, Sauk and Foxes of the Mississippi, Sauk and Foxes of the Missouri, [etc.]. **1972** J. MOSEDALE *Football* i. 3 The famous Sac and Fox warrior, Chief Blackhawk. **1974** *Encycl. Brit. Micropædia* VIII. 921/1 In the 1970s there were about 1,000 Sauk. **1978** *Handbk. N. Amer. Indians* XV. 654 Organized as the Sac and Fox tribe of Indians of Oklahoma under the Oklahoma Indian Welfare Act of 1936, the Sauk had an elected chief and business committee.

sauna (sǭ·nă, ‖ sau·na), *sb.* [Finn.] A bath-house or bathroom in which the Finnish steam bath is taken; the steam bath itself, taken in very hot steam produced by throwing water on to heated stones. Also *attrib.* and *Comb.*, as *sauna bath, heat, stove, suite*; **sauna-like** *a.*, oppressively hot and steamy.

1881 P. B. DU CHAILLU *Land of Midnight Sun* II. xvii. 206 One of the most characteristic institutions of the country is the *Sauna* (bath-house), called *Badstuga* in Swedish. **1897** E. B. TWEEDIE *Through Finland in Carts* iii. 42 Every house in the country, however humble that house may be, boasts its *bastu*, or bath-house, called in Finnish *Sauna*. **1936** *Discovery* Apr. 110/1 A speciality of Finland which everyone who visits the country ought to try is the *Sauna*—the special steam-baths which Finnish people from time immemorial have been in the habit of taking. **1939** *Daily Tel.* 18 Dec. 1/5 The Finnish soldiers..continue to take their celebrated 'sauna' steam baths wherever they are stationed. **1957** A. BUCHWALD *I chose Caviar* 31 But in Finland a sauna is not just a bath—it is a way of life. A sauna is to a Finn what a pub is to a Britisher, what a café is to a Frenchman, what a television set is to an American.

1959 *Times* 2 Dec. 5/4 A move to make British business men sauna bath conscious begins next week with the opening of a Finnish-style sauna in the City of London... City Wall sauna, as it is called, has the requisite little wooden rooms with a stove containing a pile of heated stones on to which water is sprinkled to produce the sauna heat, and showers and rest cubicles. Bunches of leafy birch twigs will also be available, price 2s. 6d., for bathers to whisk up their circulation; the sauna itself costs 15s. **1971** *Country Life* 26 Aug. 512/3 It stands in six acres, with frontage to the river, and has a..sauna bath, stables, [etc.]. **1975** *N.Y. Times* 13 Apr. x. 1/2 The sauna, otherwise known as the Finnish bath, is a wood-lined room with benches built up toward the ceiling. A special sauna stove (today, usually electric) heats small rocks piled atop it and the rocks in turn radiate a dry heat. **1976** *Times* 22 July 4/4 The sauna-like conditions of the Oxford court during the [last] five weeks. **1978** *Morecambe Guardian* 14 Mar. 15/3 Preparing for the play involved the cast in a trip to the sauna suite at Lancaster Baths.

sau·na, *v.* [f. the sb.] *intr.* To take a bath in a sauna, to visit a sauna.
1967 J. Eastwood *Little Dragon from Peking* iv. 28 Do you sauna? **1972** *Vogue* June 118/1 You could retire here to..shower, sauna, ring your friends.

saunter, *v.* Add: **2. b.** Also, to travel by vehicle in a slow and leisurely manner.
1932 R. Fry *Let.* 6 June (1972) II. 671 We sauntered through North Italy and saw a lot of lovely things.
3. Delete † and add: Also *trans.* (Further examples.)
1926 D. H. Lawrence *Plumed Serpent* ix. 153 Sauntering the day away. **1970** *Daily Tel.* 14 Nov. 9/3 If you're not fond of boats you soon will be..sauntering the sunny quay, watching the gulls.

saurian, *a.* and *sb.* Add: **A.** *adj.* **3.** Also *fig.*
1864 W. Bagehot in *National Rev.* XVIII. 525 Much of *Tristram Shandy* is a sort of antediluvian fun, in which uncouth Saurian jokes play idly in an unintelligible world. **1929** G. Mitchell *Mystery of Butcher's Shop* i. 8 Mrs. Beatrice Lestrange Bradley..smiled the saurian smile of the sand lizard and basked in the sun. **1940** —— *Brazen Tongue* vi. 67 Lady Selina had never approved..the deep affection of her daughter Sally for this oddly saurian aunt. **1970** [see *morning-glory 1].
B. *sb.* **2.** Also *fig.*
1923 H. G. Wells *Men like Gods* i. i. 13 A car with the voice of a prehistoric saurian warned him. **1953** A. Huxley *Let.* 25 Sept. (1969) 684 The mesozoic reptiles of the Ford Foundation are being as mesozoic as ever... Hutchins has recently flown to New York and has promised to do what he can with the saurians. **1974** V. Nabokov *Look at Harlequins* (1975) ii. i. 130 He was one of the very few larger saurians in the *émigré* marshes.

saurischian, *a.* and *sb.* Insert in etym. after *Saurischia*: (H. G. Seeley 1887, in *Proc. R. Soc.* XLIII. 170). (Later examples.)
1933 A. S. Romer *Vertebr. Paleontol.* ix. 181 We shall here use a conservative classification which divides the saurischians into two suborders. *Ibid.* 184 (*caption*) The manus in saurischian dinosaurs. **1970** *Nature* 11 Apr. 109/1 It was carnivorous and is thus classified in the Theropoda within the saurischian dinosaurs. **1973** J. Updike *Museums & Women* 197 The two saurischians entered his party with the languid confidence of the specially cherished. **1977** *Radio Times* 17 Dec. 45 One of the liveliest of these disputes concerns the two great groups of dinosaurs, the saurischians (with hip-bones like those of lizards) and the ornithischians.

sauropod, *a.* and *sb.* Insert in etym. after *Sauropoda* (O. C. Marsh 1884, in *Nature* 20 Nov. 68/2). **b.** *sb.* For 'order' substitute 'suborder'. (Later examples.)
1933 A. S. Romer *Vertebr. Paleontol.* ix. 189 A first step in sauropod development is perhaps illustrated by *Anchisaurus*. *Ibid.* 190 The sauropods were massively built. **1971** *Nature* 15 Jan. 153/1 The great herbivorous dinosaurs known as sauropods are inevitably the most impressive exhibit in any natural history museum. **1976** *Ibid.* 8 Apr. 559/2 The laminar bone of sauropod dinosaurs is indistinguishable from that of some of the larger artiodactyls and it is presumed that this indicated a similar metabolism.

sausage, *sb.* Add: **2. c.** = *sausage-balloon.
1858 *Househ. Words* 30 Jan. 168/1 Down came the grand royal blue sausage. **1874** *Belgravia* Aug. 170 This sausage was incased in the ordinary net-work and dependent shrouds, encircled by the ordinary hoop, and sustaining the ordinary car—a big circular basket capable of containing four persons comfortably. **1916** J. Buchan *Battle of Somme* 20 Captive balloons, the so-called 'sausages', glittered in the sunlight. **1916** J. R. McConnell in *World's Work* Nov. 53/2 Norman Prince became obsessed with the idea of bringing down a German 'sausage', as the observation balloons are called. **1928** C. F. S. Gamble *Story N. Sea Air Station* xx. 356 While the first pilot brings the boat down to 1,000 feet and flies over the air station to have a careful look at the 'sausage' to confirm the wind direction. **1929** Hall & Niles *One Man's War* 164 A balloon job is either a success or a failure the very first time you try, as the crew on the ground haul in their 'sausage' at the first note of warning from the observers. **1940** [see *obbo].
d. *slang.* A German. Also *attrib.* ? *Obs.*
1890 Barrère & Leland *Dict. Slang* II. 203/2 *Sausage game* (billiards), a German game. **1909** *Sat. Even. Post* 3 July 30 The durned old beer-swillin' sausage! **1919** *Athenæum* 8 Aug. 727/2 The German was known by several names, as 'Jerry',..'Sausage', [etc.]. **1923** J. Manchon *Le Slang* 255 *Sausage*..sobriquet de l'Allemand. **1929** E. A. Dolph *Sound Off!* 186 In the World War..our soldiers not only sang about the 'Huns', 'Krauts', and 'sausages', but they even took a fling at the..French.
e. *slang.* A German trench-mortar bomb, so called because of its shape. ? *Obs.*
1915 [see *Bath Oliver* s.v. *Bath *sb.*[1] 2 a]. **1918** H. W. McBride *Emma Gees* 164 At first we called them 'sausages', then 'rum-jars'..then they became 'flying pigs'. **1926** F. M. Ford *Man could stand Up* ii. v. 184 What the Germans called *Minenwerfer* might project what our people called sausages.
f. *colloq.* A person, esp. in phr. *silly old sausage* and varr.
[**1900** *Dialect Notes* II. 57 *Sausage*, 1. A person easily imposed upon. 2. An easy-going, inoffensive person.] **1934** W. Gibson *Fuel* 72 His mother's stopped Waving, to wipe her eyes, the silly old sausage! **1955** 'A. Gilbert' *Is she Dead Too?* ii. 38 Dr Grieve..was a silly old sausage. **1972** K. Bonfiglioli *Don't point that Thing at Me* v. 54 Very good customer of mine..Very nice old sausage. **1977** *Harper's & Queen* Nov. 308/4 He's only had five letters, the dear old sausage.
g. colloq. phr. *not a sausage* (and varr.), nothing at all.
1938 M. Allingham *Fashion in Shrouds* xix. 349 I've been..to Ben's and I dropped in at Conchy Lewis's. Not a sossidge [*sic*] anywhere. **1943** P. Brennan et al. *Spitfires over Malta* 29 Nothing happened, & we came back very brassed off, not having seen a sausage. **1955** J. Bingham *Paton Street Case* viii. 139 Don't go and quarrel with the old geezer, or he'll cut you off without a sausage. Hang on, and you'll get the lot. **1963** V. Nabokov *Gift* iii. 179 Time flies, he gets older, she blossoms out—and not a sausage. Just walks by and scorches you with a look of contempt. **1970** P. Laurie *Scotland Yard* iii. 69 We do this for three nights and don't get a sausage—we stop lots of people but they're all relatively straight. **1978** J. Wainwright *Ripple of Murders* 134 'Anything?' 'Not a sausage, Dick.' **1981** *Times* 29 June 12/6 Mr Healey said the press did not print Labour's actual policies. 'Not a sausage.'
h. A length of padded fabric that can be placed at the foot of a door to stop draughts.
1961 Partridge *Dict. Slang Suppl.* 1259/1 *Sausage*,..a draught-excluder placed at foot of a door. **1962** *Times* 10 Feb. 11/3 Red twill coated, sand filled sausages along window ledges. **1977** *Times* 30 Apr. 20/1 Keeping the maximum heat indoors by..using sandfilled sausages against gaps under doors.
4. a. *sausage-shop* (earlier examples); **b.** *sausage-eating* adj.; **c.** *sausage-finger*; *sausage-pink*, *-shaped* (later examples) adjs.
1913 'Saki' *When William Came* xii. 206 A highly civilized race like ours..is not going to be held under for long by a lot of damned sausage-eating Germans. **1922** Joyce *Ulysses* 324 And as for the Prooshians and the Hanoverians,..haven't we had enough of those sausageeating bastards? **1910** *Practitioner* Jan. 33 The fingers..as large at their tips as at their base—the so-called sausage fingers. **1922** Joyce *Ulysses* 59 The ferreteyed porkbutcher folded the sausages he had snipped off with blotchy fingers, sausagepink. **1926** J. S. Huxley *Ess. Pop. Sci.* 251 It will become simpler..and finally be converted into a sausage-shaped semi-opaque mass of tissue. **1956** *Nature* 18 Feb. 320/2 Dr. Dessens mentioned a small sausage-shaped (presumably organic) type of particle. **1767** Sterne *Tr. Shandy* IX. v. 12 A Jew who kept a sausage shop in the same street. **1873** Sausage-shop [see *corner *sb.*[1] 2 b].
d. **sausage balloon,** (*a*) an elongated aeronautical balloon; † (*b*) *slang*, a kite balloon used for observation (*obs.*); **sausage board,** a surf-board rounded at both ends; **sausage-burger** [*burger], a hamburger made with sausage meat; **sausage curl** (later examples); also, *esp.* a horizontal curl (see quots.); **sausage dog** *colloq.*, a dachshund; † **sausage-eater** *slang*, a German (*obs.*); **sausage machine,** a machine for manufacturing sausages; also *fig.*, esp. with reference to an institution that is held to 'process' its members so that their views, outlook, etc., are routinely identical; also *attrib.*; **sausage-meat** (earlier examples); also *attrib.*; **sausage roll** (earlier examples); **sausage toad** *colloq.* (see quot. 1937); **sausage-tree,** an evergreen tree, *Kigelia pinnata*, belonging to the family Bignoniaceæ, native to tropical Africa, and bearing red, bell-shaped flowers followed by pendulous, hard-shelled fruits shaped like large sausages.
1874 *Belgravia* Aug. 170, I am not, at this length of time, quite certain as to whether the body of the 'sausage' balloon was provided with two valves—one at each end of the cylinder—or whether there was but a solitary trap for the emission of gas at the convexity of the summit. **1916** F. M. Ford *Let.* 28 July (1965) 67 The air is full of sausage balloons, swallows, larks & occasional aeroplanes. **1917** 'Sapper' *No Man's Land* 97 A row of sausage balloons like a barber's rash adorned the sky. **1930** Blunden *De Bello Germanico* 79 Daylight relieving still prevailed, despite the hovering sausage-balloons. **1965** J. Pollard *Surfrider* ii. 18 Or it might be a 'sausage board'—straight for most of its length and rounded at both ends. **1970** *Studies in English* (Univ. of Cape Town) I. 28 Older designs [of surfboard] include the *sausage board*; rounded at both ends. **1942** *Better Homes & Gardens* Aug. 41/3 (Advt.), Sausageburgers. Add 1 tsp. Heinz Horseradish (soaked 10 minutes in 1 tbs. water) to 1 lb. bulk pork sausage. Shape into four cakes. Pan-broil, turning often. **1979** *Good Housekeeping* Nov. 367/2 *Sausage burgers.* 450g..pork sausagemeat. 125g..fresh white breadcrumbs [etc.]. **1966** J. S. Cox *Illustr. Dict. Hairdressing* 131 *Sausage curl*, a wide, croquignole-wound curl. Not to be confused with a spirally-wound drop or hanging curl. **1968** J. Ironside *Fashion Alphabet* 198 *Sausage curls*, similar to ringlets but laid horizontally. **1974** *Country Life* 28 Mar. 712/3 Pearls, ringlets and sausage curls. **1938** J. W. Day *Dog in Sport* v. 77 From Royal circles the snaky 'sausage dog' permeated downward through the aristocracy to the ranks of the common or show-bench exhibitors. **1958** L. Durrell *Mountolive* xv. 298 The door..opened and a dispirited-looking sausage-dog waddled into the room. **1972** *Country Life* 21 Dec. 1727/3 They poke fun at my toy German sausage dog. **1918** *Sat. Even. Post* 22 June 70 The sausage eaters decided to drop a few samples on our escadrille. *c* **1840** C. Webb *Vagrant* i. i. 14 *Coco.* [*Furiously.*] Why you infernal old Tomahawk!—you Patent Mangler!—you Sausage Machine to young men! **1850** *New England Farmer* II. 379 Sausage or Mincing Machine. This is a small, compact machine, remarkably strong and durable. **1860** [in Dict., sense 4 b]. **1889** Kipling in *Pioneer Mail* 20 Nov. 647/3 They will be sorry that they began tampering with the great sausage-machine of civilization. **1934** R. Mackenzie *Maitlands* ii. 64 When I became a schoolmaster I was full of hope... But I soon saw I was just part of a sausage-machine. **1960** *Encounter* Jan. 40/2 Producing a stock of plays and playwrights to feed the relentless sausage-machines of the drama departments. **1976** *Howard Jrnl.* XV. i. 55 Rise in the incidence and severity of juvenile delinquency may increase pressures towards an even more 'sausage-machine' and delinquency-orientated approach.., with no better results. **1723** J. Nott *Cook's & Confectioner's Dict.* sig. Hh4v, Lay in..some Sausage-meat fry'd. **1741** E. Smith *Compl. Housewife* (ed. 10) 66 Slice a penny white loaf..and work it in well with your Sausage-meat. **1845** E. Acton *Mod. Cookery* xi. 301 (*heading*) Sausage-meat cake; or, pain de porc frais. **1861** Mrs. Beeton *Bk. Househ. Managem.* x. 249 (*heading*) Sausage-meat stuffing, for Turkey. **1852** *1st Rep. Commissioners Exhib. 1851* App. xxix. 150 Sausage Rolls [consumed] 28,046. **1875** V. Lush *Jrnl.* 30 Jan. (1975) 157 Mrs O'Keefe and Mrs Spencer sent a large quantity of peaches and Mamma sent sausage rolls for the teachers. **1937** Partridge *Dict. Slang* 728/1 *Sausage toad*, sausage toad-in-the-hole: eating-houses' coll[oquialism]: late C. 19–20. **1958** B. Pym *Glass of Blessings* xiv. 159 Would you even have sausage toad if I ordered it? **1915** L. H. Bailey *Stand. Cycl. Hort.* III. 1738/1 The 'fetish-tree' and 'sausage-tree', is offered in S[outhern] Calif[ornia], and specimens may be expected in botanical collections in the W. Indies. **1944** *Sun* (Baltimore) 6 Dec. 8–D/3 An 'Admirer Visiting in Florida' sends me a colored picture postal-card view of a sausage tree... There they hang, the sausage-like seed pods, amid a background of wonderful green foliage. **1956** E. E. Evans-Pritchard *Nuer Relig.* xii. 298 The man who has committed incest..cuts in two the fruit of a sausage-tree. **1962** *Times* 9 Oct. (Uganda Suppl.) p. viii/4 The incredible sausage-tree with its dangling woody fruits. **1977** D. Beaty *Excellency* xii. 133 The sausage trees with heavy fruits shaped like giant loofahs.

sausage, *v.* *rare.* [f. prec.] *trans.* To subject (a person or thing) to treatment reminiscent of the manufacture or shape of a sausage.
1922 Joyce *Ulysses* 500 He is sausaged into several overcoats. **1949** Dylan Thomas *Let.* 13 Oct. (1966) 329 So that I won't..have at once to set into motion again the..little machines that sausage out crumbs and coppers for me. **1951** N. Mitford *Blessing* ii. ii. 168 'Sometimes they only sausage them.' 'They what?' 'Tie them up like sausages, brr round and round.' **1965** *Sunday Times* (Colour Suppl.) 11 July 9/2 Once or twice we had a bit of an indiscretion, might sausage a motor into an island, or over a muddy pasture.

Saussurean (sosʊ̈·riăn), *a.* Also **Saussurian.** [f. the name *Saussure* (see below) + -an.] Of, pertaining to, or characteristic of the Swiss scholar Ferdinand de *Saussure* (1857–1913) or his linguistic theories. Hence as *sb.*, an adherent of these theories; also **Saussu·reanism.**
1937 J. Orr tr. *I. Iordan's Introd. Romance Linguistics* iii. 194 His [*sc.* Gilliéron's] linguistic is descriptive, or, in the Saussurian terminology, 'synchronic'. **1939** P. Christophersen *Articles* 16 A word has thus two aspects: verbal image and meaning (in the well-known Saussurean terms: *signifiant* and *signifié*). **1943** *Language* XIX. 55 In its essence it is Saussurean, but differs in certain respects also from the practices of the major European groups that follow the teachings of Saussure. **1952** *Word* VIII. 264 Eleven papers are devoted to general problems and methodology, among them..the Saussurean opposition between synchrony and diachrony. **1954** *Word* X. 391 Orthodox Saussureanism of the Geneva school. **1968** J. Lyons *Introd. Theoret. Linguistics* ix. 429 This fact is expressed in Saussurean terms by saying that each language imposes a specific *form* on the *a priori* undifferentiated *substance* of the content-plane. **1971** [see *parole *sb.* 2*]. **1975** Lass & Anderson *Old Eng. Phonol.* iv. i. 117 Without adhering to the Saussurean dichotomy one can still realize that diachronic evidence does not crucially determine choices for synchronic ordering. **1977** *Language* LIII. 391 The history of Saussureanism—perhaps one of the most challenging topics that a historian of linguistics could undertake. *Ibid.* 398 As a Saussurean, K views with suspicion any suggestion 'that linguistics might not yet have reached the status of an autonomous science'.

saussurite. Add: Also **sau·ssuritized** *ppl. a.*, converted into saussurite, or having component minerals converted into saussurite.
1907 J. S. Flett in W. A. E. Ussher *Geol. Plymouth & Liskeard* 101 There are..saussuritized residues of felspar. **1954** *Mineral. Mag.* XXX. 525 The high density..of the rock..distinguishes it from saussuritized gabbros. **1974** *Nature* 25 Jan. 195/2 The plagioclase of the gabbros is often saussuritised.

‖ **saut** (so). [Fr., = 'leap'.] **1. Saut Basque** (also in pl. and with small initials), a dance of the French Basque provinces (see quots.).
1895 L. Grove *Dancing* x. 313 The *Mutchico*, or *Saut Basque*, of the French Basque provinces, is held during winter nights in large kitchens or on threshing-floors. **1930** R. Gallop *Bk. of Basques* iv. 56 A hundred years ago it was

not unusual to see the village priest lead the *Saut Basque* on a Sunday evening. *Ibid.* vi. 104 The *jantziak* or *sauts basques*..are neither wholly ritualistic nor yet purely recreational. **1948** 'La Meri' *Spanish Dancing* iv. 39 The *Sauts Basques* is also danced by men and women and is now recreational, although its origin is ritualistic... The Sauts is better known in the French provinces of the Basque country than in the Spanish. **1964** W. G. Raffé *Dict. Dance* 445/1 *Saut Basque*, a dance of French Basque provinces, especially Basse Navarre, where it has two forms—a recreational dance in a large kitchen or on a threshing-floor; and a more ceremonial form, out of doors.
2. *Ballet.* A leap in dancing; chiefly used in the names of special steps, as *saut de Basque*, *saut de l'ange* (see quots. 1957).
1948 A. Chujoy tr. *Vaganova's Basic Princ. Classical Ballet* vii. 91 Saut de basque... Both legs in this pas should be fully turned out. **1952** Kersley & Sinclair *Dict. Ballet Terms* 84 *Saut*, a jump in which the dancer springs off both feet and lands in the same position. **1957** G. B. L. Wilson *Penguin Dict. Ballet* 240 *Saut de Basque*, lit. a Basque jump. Turning step performed in the air with one leg straight and the other in a retiré position. *Ibid.* 241 *Saut de l'ange*, lit. angel's jump. Similar to a temps de poisson..but the body is held obliquely to the ground in the direction of travel. **1972** H. J. Summers *Guide to Ballet* 155 *Saut de l'ange*, an angel's jump, or a forward leap with the body obliquely to the ground and arms *en couronne* and legs slightly bent. **1976** *New Yorker* 24 May 146/1 From this one performance, I seem to recall Baryshnikov landing from a double saut de basque in a split on the floor.

sauté, *sb.* (Later examples of *sauté pan.*)
1846 A. Soyer *Gastron. Regenerator* 341 Melt two ounces of butter in a sauté-pan. **1960** E. David *French Provincial Cooking* 68 Few English kitchens seem to possess sauté pans.

sauté, *v.* (Later examples.)
1907 [see *jardinière 2]. **1953** Rombauer & Becker *Joy of Cooking* 73/1 Dice bread and sauté it in butter. **1968** *Globe Mag.* (Toronto) 13 Jan. 16/3 Halve frankfurters lengthwise. Melt butter in heavy skillet, add onion and saute over low heat until just tender but not brown.

Sauterne. Add: The correct Fr. form, now usual in the U.K., is *Sauternes*. (Later examples.)
1895 *Army & Navy Co-op. Soc. Price List* 165 Sauternes... 18/0. **1908** [see *Sauvignon a]. **1959** W. James *Word-bk. Wine* 168 The sauternes-types made in Australia, South Africa, and California are most frequently a well-sulphured mixture of white wine and fortified grape juice. **1967** A. Lichine *Encycl. Wines* 486/2 The wines to be met with in many other places in the world which call themselves sauternes—or even sauterne, as if to justify bad practice by bad spelling—are not what they claim to be.

‖ **sautoir** (sotwār). [Fr.: cf. Saltire.] A long necklace consisting of a fine gold chain usu. set with jewels.
1936 *N.Y. Times Mag.* 6 Dec. 10/1 It was the era of sunbursts, sautoirs, immense brooches and flaming diamonds. **1957** M. B. Picken *Fashion Dict.* 284/1 *Sautoir*.., long jeweled chain of gold. **1960** *Times* 29 Mar. 22/7 A pearl and diamond sautoir. **1969** R. T. Wilcox *Dict. Costume* (1970) 304/1 *Sautoir*, a long, fine gold or silver chain upon which women carried a watch, or a small gold or silver chain purse, or, perhaps, a medallion. **1980** *Times* 18 Oct. 7/4 Sautoirs, or long neckchains..were very popular at the beginning of the century.

sauve-qui-peut. Add: (Further examples.) Also as a phrase in the original Fr. sense. Hence as *vb.*, to stampede or scatter in flight.
1875 *Encycl. Brit.* III. 321/2 *Sauve qui peut* was the universal cry; and..in less than six weeks above seventy banking establishments were swept off. **1939** tr. *E. N. Marais's My Friends the Baboons* iii. 35 All the baboons do in such a case is *sauve qui peut* with an alarm-call that makes the mountains echo. **1964** *Reading Teacher* Dec. 211/1 Working-class whites, themselves anthropologically unsophisticated, join the *sauve qui peut* in search of a suburban haven. **1973** *Times* 26 Nov. 15/4 It is difficult to understand the Government's present policy, or indeed that of any of the oil users. *Sauve qui peut* will serve no one well in the long run. **1980** *Guardian* 11 Nov. 10/8 It is in those hallowed halls of the UN..that I feel most keenly the theatre of anarchy; of sauve-qui-peut.

Sauveterrian (sōuvte·riăn), *a.* *Archæol.* [ad. F. *Sauveterrien* (E. Octobon 1930, in *Actes XV Congr. Internat. d'Anthrop. & Archéol.* (1931) 332), f. *Sauveterre* (see below) + -ian.] Of, pertaining to, or designating the mesolithic culture of which remains were first discovered at Sauveterre-la-Lémance, in Lot-et-Garonne, France. Also *absol.*
1940 C. F. C. Hawkes *Prehist. Foundations Europe* iii. 50 We are..in the region of Azilian tradition, but the industry might almost equally well be a 'Middle Tardenoisian', with its crescentic and angular microliths and its moderate but not excessive development of micro-burin technique, and the French now propose to call such industries in general Sauveterrian, reserving the name Tardenoisian in a strict sense for the later stage. *Ibid.* 66 In Britain,..microlithic technique was never carried beyond the Middle Tardenoisian or Sauveterrian stage. **1952** *Proc. Prehist. Soc.* XVIII. 109 Since then, the discovery of mesolithic tools which preceded the Tardenoisian on the Causses west of the Massif Central has added 'Sauveterrian' to the terminology. **1963** E. S. Wood *Collins Field Guide Archæol.* iv. 50 The character of the mesolithic of all but the south-east and east of England shows affinities with the French mesolithic culture called *Sauveterrian*. *Ibid.* 103 The mesolithic people of the Sauveterrian were particularly fond of rock-shelters. **1975** *Nature* 3 July 33/2 This radiocarbon evidence from the Pennines indicates conclusively that simple 'broad blade' microlithic industries identical to those of Thatcham and Star Carr precede 'narrow blade' (sometimes termed 'Sauveterrian') industries with small scalene triangles and rod-like microliths. **1975** J. G. Evans *Environment Early Man Brit. Isles* v. 92 On the high moors of north-east Yorkshire another group of industries, also Mesolithic.., occurs. These..contain..a profusion of microliths. The name Sauveterrian—after the type site in France of Sauveterre-la-Lémance—has been applied to them.

‖ **Sauvignon** (sovinʸoñ). [Fr.] **a.** A white grape of France; the white wine made from this grape. Also *attrib.* or as *adj.*
1846 C. Cocks *Bordeaux* ii. 142 The following are the names of the most esteemed white wines. 1. The *Sauvignon*, of a yellowish or greyish brown-spotted wood. **1875** H. Vizetelly *Wines of World* 18 The fine white wines of the Gironde are produced from the Sauvignon and Semillon grapes, the former of which yields a limpid, perfumed, delicate-flavoured, amber-coloured, heady wine. **1888** *Encycl. Brit.* XXIV. 604/2 The principal vines used in the Médoc are..for white wines, the Semillon, the Sauvignon, and the Muscatelle. **1908** E. & A. Vizetelly *Wines of France* 82 Sauternes and the better Graves..are chiefly the produce of the Sauvignon and Semillon grapes. **1935** Schoonmaker & Marvel *Compl. Wine Bk.* v. 142 Corvo, a..dry white wine, made..out of two native Sicilian grapes blended with the famous Sauvignon grape of Sauternes. **1959** W. James *Word-bk. Wine* 168 *Sauvignon*, one of the three white grape varieties used in Sauternes and grown to a limited extent in South Africa, Australia, and California for high-quality white wines. **1969** V. Rose *Loire* 30 Pouilly-Fumé..is made from the Sauvignon grape. **1973** *Times* 30 June 11/4 Wines made solely from the Sauvignon Blanc grape..are now to be found in many of the French wine regions. *Ibid.* 11/5 Many merchants list Sauvignons. **1976** *Times* 6 Mar. 13/5 A classic Sauvignon..comes from Haut Poitou..with the steely flavour of this great great..wine of astonishing quality.
b. Short for Cabernet-Sauvignon, a black grape of France; also the red wine made from this grape.
1846 C. Cocks *Bordeaux* ii. 140 The *Carmenère*, or grosse *Vidure*, called also *grand Carmenet*, *Carbonet*, or *Sauvignon*, has..grapes..of a bright colour. **1895** *Army & Navy Co-op. Soc. Price List* 166 'Imperial', from finest Sauvignon grapes. **1907** *Yesterday's Shopping* (1969) 97/3 Sauvignon..(Burgoyne) 16/0. **1917** *Harrods Gen. Catal.* 1289/2 Sauvignon, full bodied..1/11. **1952** A. Lichine *Wines of France* 46 The wine..comes from a vineyard planted two-thirds in Cabernet Franc and Sauvignon... No white wine is made.

sav, abbrev. of Saveloy.
1936 J. Curtis *Gilt Kid* 75 Cup o' tea, sav and a slice. **1969** C. Drummond *Odds on Death* vi. 130 Some home-made savs—not the shop kind.

Savage (sæ·vėdʒ), *sb.*[2] The name of Arthur *Savage*, inventor, of Brooklyn, N.Y., used *attrib.* and *absol.* as a brand name (proprietary in the U.S.) of a repeating rifle produced by him in 1894, and of other firearms produced by the Savage Arms Company.
1892 *Ann. Rep. Chief of Ordnance to Secy. of War* (U.S.) App. IX. 224 *Savage*... This arm was brought before the board by Mr. Arthur Savage, of Brooklyn, N.Y. **1902** *Encycl. Brit.* XXXII. 657/2 The Savage magazine rifle, model 1899, is a 'hammerless', lever-action repeating arm. **1903** *Kynoch Jrnl.* Feb.–Mar. 62/1, I had my ·301 Savage. [**1914** *Official Gaz.* (U.S. Patent Office) 18 Aug. 989/2 Savage Arms Company, Frankfort, N.Y. Filed Apr. 25, 1913... Particular description of goods.—Rifles, Pistols, and Cartridges. Claims use since Jan. 1. 1906.] **1964** H. L. Peterson *Encycl. Firearms* 30/2 This system was later used by the Mexican Obregon pistol and in a slightly modified form by Savage pistols. **1968** *Globe & Mail* (Toronto) 17 Feb. 1/1 Police seized a battered, old ·303-calibre Savage hunting rifle.

savage, *v.* Add: **4.** Also *transf.* and *fig.*
1923 *Public Opinion* 2 Sept. 103/2 Human lust and hatred have first savaged them to death. **1926** *Bulletin* 9 June 13 He is much too severe on the form of novels—the Cogglesby comedy in 'Evan' is savaged, for example. **1929** Chesterton *Poet & Lunatics* 107, I can no more see him savaging somebody like poor young Saunders than I can see him kicking a crippled child. **1962** I. Murdoch *Unofficial Rose* xxxiv. 319 Once he stroked it [*sc.* a picture] absently, as he had done when it was his, and was savaged by an attendant. **1963** [see *cut *ppl. a.* 6]. **1977** *Time* 26 Dec. 36/1 Minnelli is only the latest in a long line of actresses savaged by Simon.

† **savagerous** (sæ·vėdʒerəs, sǣvæ·dʒərəs), *a.* *U.S. dial. Obs.* Also **sawagerous, servagerous, sevagerous.** [f. Savage *a.* + Dang)erous *a.*] Fierce, wild, violent, dangerous. Also as quasi-*adv.*
1832 F. Trollope *Dom. Manners Amer.* I. xiii. 186 The visitor took it [*sc.* a dagger] up, and examining it with much emotion, exclaimed, 'What! do you really jab this into yourself sevagarous?' **1837** R. M. Bird *Nick of Woods* I. iv. 71 The strongest men in Kentucky, and the most sevagarous at a tussle. **1845** W. T. Porter *Big Bear of Arkansas* 121 They war mighty savagerous arter likher. **1850** *Wilmington* (N. Carolina) *Commercial* 7 Mar. 1/6 Of all the untiring, unaccountable, and unspeakable 'Savagerous' rumpuses ever kicked up Cape Horn takes the banner. **1859** 'Dow, Jr.' *New Patent Sermons* 263 A very sawagerous creature called the Youknowcan. **1866** C. H. Smith *Bill Arp* 54 It [*sc.* Habeas Corpus] is, perhaps, *when suspended*, the most savagerous beast that ever got after tories and traitors. **1927** *Amer. Speech* II. 363/2 *Servagerous* (adj.), very active. 'That is a servagerous coon dog.'

SAVAK (sæ·væk, sa·vak). [Acronym f. the initial letters of Persian *Sāzmān-i-Attalāt Va Amniyat-i-Keshvar* National Security and Intelligence Organization.] The secret intelligence organization of Iran, established in 1957 and disbanded in 1979.
1967 *Time* 6 Oct. 47/2 All candidates must be approved by SAVAK, his powerful security police. **1975** *New Yorker* 8 Dec. 128/2 They speak matter-of-factly of a conspiracy that includes the C.I.A. and SAVAK (the Iranian secret police) and the Sheriff of Cole County. **1976** *Maclean's Mag.* 17 May 41/3 This is a police state ruled over by one of the most ruthless secret police forces in the world, the dread SAVAK. **1977** *Time* 28 Nov. 36/2 Documented charges by both Amnesty International and the International Red Cross that Iran's secret police organization, SAVAK, had systematically persecuted dissidents. **1979** M. McCarthy *Cannibals & Missionaries* i. 18 SAVAK, the Shah's secret police, would scarcely have the same sociable attitude as the New York State wardens and guards. **1981** *Times* 13 Aug. 2/2 A former member of Savak, the Shah of Iran's secret police, killed himself..after being told he was to be deported back home.

savannah. Add: The hitherto obs. form **savana** now has some currency. **1.** (Further examples.) In mod. use, an open plain of long grass, freq. with scattered drought-resistant trees, such as is characteristic of certain tropical and subtropical regions having distinct wet and dry seasons; grassland or vegetation of this kind.
1836 N. Isaacs *Trav. & Adventures E. Afr.* I. vi. 88 This we did for the purpose of calling at some hamlets and savannas, in our course, to obtain cattle and curiosities. **1903** W. R. Fisher tr. *Schimper's Plant-Geogr.* 261 Tropical grassland, wherever it has not been modified by human agency, occurs chiefly as savannah, more rarely as steppe. **1920** M. E. Hardy *Geogr. of Plants* iii. 142 The treeless savana is called 'campo vero';..if the savanas are strewn with clumps of low trees, they are 'serrados'. **1926** D. H. Campbell *Outl. Plant Geogr.* viii. 292 The outstanding feature of this savanna was a noble fan-palm..which formed groves of considerable extent. **1955** *Times* 28 May 7/6 The Rupununi river flows almost due north through Southern British Guiana. On either side are the wide open savannahs, broken only here and there by small clumps of stunted sandpaper bushes and groups of anthills. **1957** P. Dansereau *Biogeogr.* ii. 73 The somewhat drier types [of climate]..show a very uneven distribution of rainfall and generally support woodland or savana. **1958** L. van der Post *Lost World of Kalahari* vii. 123, I met a man..walking out of the bush into a long savannah of buffalo grass. **1968** R. W. Fairbridge *Encycl. Geomorphol.* 979/2 Savanna passes in drier regions to steppe or desert, and in wetter areas into savanna woodland. **1969** S. M. Sadeek *Windswept & Other Stories* 29 The cart rolled on..into the savannah of sagebush and beezie-beezie reeds and razor-grass. **1974** H. F. Garner *Origin of Landscapes* v. 267/2 The small forest areas on savannahs differ botanically in no great measure from more continuous rain forest elsewhere. **1976** West & Augelli *Middle Amer.* (ed. 2) ii. 47/1 One of the most puzzling features of the natural vegetation in the tropical rainy areas of Middle America is the presence of large expanses of grassland, called 'savannas', in areas that receive as much as 80 to 100 inches of rain annually, with no dry period or a quite short one... The largest of the humid savannas is found along the Caribbean margin of Nicaragua and northeastern Honduras.
b. *spec.* In the West Indies and Guyana, a particular tract of such land within definable limits; a meadow, a paddock.
1934 J. Rhys *Voyage in Dark* i. i. 4 When the black women sell fishcakes on the savannah they carry them in trays on their heads. **1952** S. Selvon *Brighter Sun* i. 13 Opposite the school was a large savannah on which cattle and donkeys grazed. **1960** *Tamarack Rev.* XIV. 48 Mittelholzer..took a walk every evening about five or six around the Port-of-Spain savannah. **1964** S. M. Sadeek *Windswept & Other Stories* (1969) 19 You don't have to go galavanting the settlements and savannahs like some coot.
2*. *U.S.* A tract of low-lying damp or marshy ground.
1671 in *S. Carolina Hist. Soc. Coll.* (1897) V. 333 You will finde..great Creeks, mar[s]hes, or Savanoes on the other side. **1737** J. Wesley *Jrnl.* 2 Dec. (1910) I. 401 There is a little [soil] of a better kind, especially in the savannahs,..so they call the low, watery meadows, which are usually intermixed with pine-lands. **1895** *Dialect Notes* I. 380 *Savannah*, stretch of bog or moorland. **1905** *Bull. Bureau of Forestry* (U.S. Dept. Agric.) No. 64. 7 Loblolly is the first pine to take possession of the savannas, or marshy pairies. **1938** J. R. Carpenter *Ecol. Gloss.* 236 *Savannah*, a tract of damp level land with a growth of grass or reeds (S[outhern] U.S.).
3. c. Special Combs.: **savannah forest, woodland,** grassland similar to savannah but with a denser growth of trees, though not enough to provide continuous cover; **savannah grass,** a stoloniferous carpet grass, *Axonopus compressus*, native to tropical and subtropical America.
1903 W. R. Fisher tr. *Schimper's Plant-Geogr.* 260 The Savannah-forest..is more or less leafless during the dry season, rarely evergreen, is xerophilous in character, usually, often much, less than twenty meters high, park-like, very poor in underwood, lianes, and epiphytes, rich in terrestrial herbs, especially in grasses. **1958** G. Lienhardt in Middleton & Tait *Tribes without Rulers* 99 Boundaries between different political communities are often not apparent to the eye in such savannah-forest areas. **1756** P. Browne

Civil & Nat. Hist. Jamaica 137 The small Savannah Grass with echinated valves..grows in the Savanna about Kingston. **1859** G. W. Perry *Turpentine Farming* 9 Every kind of turf should be turned over, such as..wire grass, savanna grass, and broom-sage grass. **1954** *Farmer's Guide* (Jamaica Agric. Soc.) 232 Savannah Grass—Carpet Grass (*Axonopus compressus*)... In the West Indies it is an important pasture grass. **1970** A. T. Semple *Grassland Improvement* viii. 177 In Malaya, fertilizer trials with savanna grass..showed a marked response. **1976** P. D. Driscoll *Barboza Credentials* v. iv. 233 A parade-ground, now overgrown with savanna grass. **1903** W. R. Fisher tr. *Schimper's Plant-Geogr.* 836/2 (Index), Savannah woodland. **1960** N. Polunin *Introd. Plant Geogr.* xiv. 442 Savanna-woodland..is found very widely in tropical and subtropical regions including much of Cuba and elsewhere in the Caribbean, Brazil and northern Argentina, East and central Africa.., and occupying much of India and China as well as of northern and eastern Australia. **1968** Savannah woodland [see sense *1].

‖ **savarin** (savaræ̃n). [f. the name of Anthelme Brillat-*Savarin* (1755–1826), French gastronome.] A light, ring-shaped cake made with yeast, soaked in syrup flavoured with liqueur, and served with fruit and cream. Also *attrib.*
1877 E. S. Dallas *Kettner's Bk. of Table* 402 Little has been said about the Bath bun, the Banbury cake, the Scotch shortbread, the Brioche, the Baba, the Savarin, the Gauffre. **1894** G. du Maurier *Trilby* I. ii. 127 The cakes were of three kinds—Babas, Madeleines, and Savarins... The Savarin..is shaped like a ring, very light, and flavoured with rum. **1928** J. Rhys *Postures* xviii. 180 A savarin, an éclair, two meringues—the ones you like, and I've ordered tea. **1943** A. L. Simon *Conc. Encycl. Gastron.* IV. 115/1 Savarin paste. **1958** [see *baba]. **1963** R. Carrier *Great Dishes of World* xv. 252 The *savarin* cake mixture..is the basis of rum baba as well as many other famous sweets. *Ibid.* 253 Butter a deep cake tin or savarin mould and half-fill it with dough. **1969** *Daily Tel.* 12 Nov. 15/6 A hinged cake-tin with two interchangeable bases, one for deep sponges, the other—fluted, with a funnel—for savarins.

savate. Add: Also *Comb.*, as *savate kick.*
1969 J. Fredman *Fourth Agency* xi. 104 He came at me in a crouching horizontal leap and dealt me a great big savate kick. **1975** P. Audemars *Nightmare in Rust* xi. 157 He..launched a tremendous *savate* kick at the base of the old man's spine.

save, *sb.*[2] Add: **2.** (Later examples.) Now. usu. such an action performed by the goal-keeper.
1942 *Sun* (Baltimore) 26 Jan. 4/1 Gil Schuerholz..made astounding saves all afternoon. **1954** *Encounter* Feb. 58/2 The highlights of a [football] game, a spectacular save, a balanced evasive run..become evocative images. **1960** B. Liddell *My Soccer Story* x. 68 One save of Bert's..was of the truly miraculous type... The ball..sped like a bullet towards the left-hand corner..but with a marvellous leap..Bert turned it over the bar. **1977** *News of World* 17 Apr. 23/4 Arsenal lost the match the precise second that Liverpool's England goalkeeper Ray Clemence made a world-class save from Frank Stapleton.
3. *Bridge.* = *sacrifice *sb.* 5 d. Freq. in phr. *cheap save.*
1927 *Observer* 31 July 14/5 Now consider the position if Z had doubled 'Six Hearts' instead of going on with Spades..which would have saved the game and rubber. A cheap save and well worth while! **1928** A. E. M. Foster *Auction Bridge* iv. 200 (*heading*) A good save on majority bidding. **1974** *Country Life* 3 Oct. 975/3 A hand from a recent session...Trying for a cheap save.

save, *v.* Add: **I. 1. b.** Also used colloq. in *fig.* phr. *to save* (someone's) *life*, to give timely assistance, esp. a stimulating drink.
1914 [see *gesundheit]. **1938** E. Waugh *Scoop* i. ii. 14 God bless you, Julia. You've saved my life. **1950** 'J. Tey' *To Love & be Wise* xii. 153 Saved my life, you have! I missed the bus. **1955** M. Allingham *Beckoning Lady* iv. 62 Tea, darling? Bless you, you're saving my life. **1977** D. Bagley *Enemy* xxviii. 218 'A sherry,' she said. 'A sherry, to save my life.'
f. Hyperbolically in trivial use, as *to save* (one's) *life* (or occas. *soul*): usu. following statement in negative, denoting lack of ability or intention to do something.
1848 Trollope *Kellys & O'Kellys* III. v. 106, I shan't remain long. If it was to save my life and theirs, I can't get up small talk for the rector and his curate. **1873** C. M. Yonge *Pillars of House* III. xxvii. 88 'Does she go to their church?' 'Oh no, she wouldn't to save her life—she thinks it quite shocking.' **1893** Yonge & Coleridge *Strolling Players* iii. 21, I couldn't act to save my life. **1916** A. Bennett *These Twain* iii. xix. 436 'What will you have to eat?' said Maggie. 'Nothing. I couldn't eat to save my life.' **1920** E. O'Neill *Beyond Horizon* iii. i. 128, I couldn't get to sleep to save my soul. **1941** J. Cary *Herself Surprised* xxxiv. 82 It took even Bill six months to get her into a motor, when motors came in, and she wouldn't telephone now to save her life. **1973** E. Berckman *Victorian Album* 192 She must have..dressed in record time, but to save my life I couldn't tell you how she looked or what she had on.
8. f. Hence *save-face* adj. = *face-saving* ppl. adj. s.v. *face *sb.* 27. Also *absol.* as *sb.*
1917 *Chambers's Jrnl.* Jan. 13/2 The civilian native staff had bolted at the first sign of trouble, 'going to report to the authorities' being their 'save face' for it! **1935** *Times* 7 Oct. 9/4 The closing phase of the War—namely, a save-face, patched up peace. **1966** R. Standish *Widow Hack* i. 8 A save-face formula to enable Janet to plead *force majeure*.
9. e. *to save the situation*, to avert disaster.
1907 W. Raleigh *Shakespeare* v. 135 If Cordelia had been perfectly tender and tactful, there would have been no play. The situation would have been saved. **1908** A. Bennett *Old Wives' Tale* iv. ii. 467 Those dogs saved the situation, because they needed constant attention. **1922** J. Williamson *Short Hist. Brit. Expansionism* v. iii. 514 Starvation more than once threatened annihilation, but on each occasion the timely arrival of food-ships saved the situation.
12. a. For † *Obs.* read *Obs.* exc. *Hist.* and add later examples.
1946 A. Huxley *Let.* 3 Sept. (1969) 547 My primary preoccupation is the achievement of some kind of over-all understanding of the world, directly and, at one remove, through the building up of some hypothesis that accounts for the facts and 'saves the appearances'. **1957** O. Barfield (*title*) Saving the appearances. **1981** *Country Life* 26 Feb. 528/3 His single professional aim is to perceive order in the physical world, not merely to save the appearances but to discover an ordered reality.
II. 17. b. (Further examples.) Now used esp. with reference to or in exhortations concerning the purchase of savings certificates, etc., instead of consumer goods.
1916 *War Savings* Oct. 12/2 A large number of circulars headed 'Save for England' have been distributed by the school children. *Ibid.* 13/1 Men and women are saving in Gloucestershire who never saved before because they have been taught that their 6d. per week..will help to end the War. *Ibid.* 16/1 Men are encouraged to save and help their country by joining the Association. **1942** J. A. Schumpeter *Capitalism, Socialism & Democracy* xviii. 210 Nor am I going to ask the reader to rely on the individual comrades' propensity to save. **1948** G. Crowther *Outl. Money* (ed. 2) v. 169 By every imaginable device of publicity people are exhorted to save. **1961** E. S. Turner *Phoney War* xx. 292 This was merely an ingenious way of getting people to save. **1969** *Whitaker's Almanac 1970* 353/1 The Chancellor..went on to introduce a contractual savings scheme—for which he said he was glad to appropriate (from the Conservatives) the title 'Save As You Earn'. **1978** *Times* 15 Mar. 21/8 The publication of Keynes' *General Theory* by its emphasis on the propensity to save (rather than the propensity to import) as the major cause of the insufficiency of demand, diverted attention from Harrod's approach.
18. d. *to save one's breath* or *wind*, to refrain from wasting one's argument or energy on a lost cause. (Perh. an ellipt. use of the proverbial phr. *to keep* (*save*, etc.) *one's breath to cool one's porridge*: see Porridge *sb.* 4 in Dict. and Suppl.)
1926 F. W. Crofts *Inspector French & Cheyne Mystery* xi. 146 If your story's going to be more lies about St John Price and the Hull succession you may save your breath. **1941** Mencken in *New Yorker* 24 May 22/1 He might very well have saved his wind, for Bill soon had him. **1952** E. Caldwell *Lamp for Nightfall* iv. 36 Now stop making me mad, talking about a new dress that you haven't any need of. Save your breath for something dearer. *Ibid.* x. 101 You'd better be saving your wind for road work, and for doing chores.

saved, *ppl. a.* Add: **1. b.** *saved by the bell* (Boxing) (see quot. 1971); hence *fig.* in general use, saved (as from an unpleasant occurrence) by timely interruption.
[**1932** *Ring* Nov. 3 Floored in the first session by a terrific right to the jaw, the bell saving the Jersey boy at the count of seven. **1954** F. C. Avis *Boxing Dict.* 98 *Saved by bell*, a boxer saved from being counted out because the end of the round is signalled.] **1959** A. Sillitoe *Loneliness of Long-Distance Runner* 31 'Ain't it next door to a pub, then?' I wanted to know. He answered me sharp: 'No, it bloody well aint.'..'Then I don't know it,' I told him, saved by the bell. **1963** *Times* 18 May 8/5 If, in future, the bell interrupts a count, the count will continue until the boxer is counted out—unless he gets up in the meantime... The expression 'saved by the bell' will, therefore, become an anachronism. **1971** L. Koppett *N.Y. Times Guide Spectator Sports* v. 116 If a man is knocked down in the closing seconds of a round, so that the bell rings ending the round before the count of 10 has been reached, he can be 'saved by the bell'. **1976** G. Sims *End of Web* i. 13 Had he been saved by the bell... Was there still a chance of some lovers' games?

Savei, var. *Sebei.

saver. Add: **5.** (Further examples.) Also *transf.*
1917 A. B. Paterson *Three Elephant Power* 17 'I had a quid on,' he says. 'And..I had a saver on the second, too.' **1950** N. Cardus *Second Innings* 163, I..suggest a saver each way on Gunga Din. **1958** G. Casey *Snowball* xvii. 168 A lot of people who had bet on Benny—and made sure of a saver on the Negro—put on a few shillings more at the ringside. **1974** Rather & Gates *Palace Guard* ii. v. 51 Nixon..decided to slap a deuce or two on a couple of long shots, as a 'saver'—just in case.

savey, savvy, *sb.* (and *a.*). Add: The form **savvy** is now usual. **A.** *sb.* (Further examples.)
1923 R. Cummins *Sky-High Corral* 31, I don't just get the savvy of this. **1936** W. R. Titterton *Chesterton* ii. iii. 138 Which idea..Armstrong actively disliked because, having more savvy than I had, he saw it meant death to his doctrine. **1951** K. Crichton *Marx Brothers* x. 134 He had bounce, stage savvy, and the optimism of a Rotarian. **1964** E. B. White *Let.* 1 Feb. (1976) 515, I felt deeply envious of their skills, their savvy, their self reliance, and their general deportment. **1974** *Publishers Weekly* 11 Feb. 58/1 Full of baseball savvy, the book is also at times very funny. **1978** J. Carroll *Mortal Friends* v. i. 496 Kennedy's reputation was for more savvy than that. He knew his history, didn't he, and its humbling lesson?
B. *adj.* Of persons, etc.: having practical sense, quick-witted; knowledgeable, wily, experienced. Also wise *to* (something).
1905 K. Inglewood *Patmos* i. ix. 124 'How very savvy of you to think of that,' he said. **1946** *Calif. Folklore Q.* Oct. 377 From the safe landing of an airplane which has followed the homing radio beam (beacon), a person who is thinking clearly, performing an act correctly, that is, who is *savvy*, is *in the groove* or *on the beam*. **1964** H. Waugh *Missing Man* xiii. 65 The kid might give himself away and Lambert's savvy enough to pick it up. **1974** *Publishers Weekly* 27 May 65/1 Norman Two Bull is a modern and savvy 15-year-old Sioux. **1975** Byfield & Tedeschi *Solemn High Murder* iv. 77 She's older and been around and savvy to a lot of things the rest of them aren't. **1978** *Guardian Weekly* 5 Feb. 15/5 They are savvy hands, and if they do not speak persuasively they do speak precisely. **1980** *Economist* 16 Aug. 51/3 A savvy tenant putting a deposit on his house gains a 12-month option to buy at the price ruling when he made the deposit.

savey, savvy, *v.* Add: Also **savee;** the form **savvy** is now usual. Also, to understand, comprehend. Freq. used in the interrogative (= 'do you understand?') following an explanation to a foreigner or to one considered slow-witted. Also *absol.*
1850 L. H. Garrard *Wah-to-Yah* 105 You've got so much 'fofarraw' stuck 'bout you, this child didn't savvy at fust! **1897** A. H. Lewis *Wolfville* 45 You've got to quit; savey? **1908** E. J. Banfield *Confessions of Beachcomber* ii. iii. 315 'You savee?' The 'savee' touched Harry's dignity. 'What for you say savee? You take me for a blurry Chinaman?' **1914** S. Lewis *Our Mr. Wrenn* iv. 59 Gotta do what I say, savvy? **1917** [see *chair *sb.*[1] 1 d]. **1920** D. H. Lawrence *Touch & Go* ii. 49 *Gerald.* Yes, I want to be told. *Anabel.* That's rather mean of you. You should savvy, and let it go without saying. *Gerald.* Yes, but I don't savvy. **1933** M. Lowry *Ultramarine* iii. 128 Let's have two starboard lights. Savee starboard lights? **1949** *True* Jan. 61/3 When there are ladies present, we say it in Mexican. The hounds savvy either. **1955** *Times* 27 July 10/6 The secretary was a literate man who 'savvied book'. **1964** E. Palmer tr. *Martinet's Elem. Gen. Linguistics* v. 155 Everywhere we find the word *savvy* 'know', which..is automatically used by a monoglot English speaker who tries to make himself understood by a foreigner.

Savile Row (sæ·vil rōᵘ). The name of a street in London celebrated for fashionable and expensive tailoring establishments, used *attrib.* to designate such tailors, their styles, or wares, esp. men's suits.
1896 G. B. Shaw in *Sat. Rev.* 13 June 597/2 A suit turned out by a Savile Row tailor. **1934** *Cornh. Mag.* Sept. 366 Sinai, like a Savile Row tailor,..does not display its goods in the shop window for all to see. **1946** A. Christie *Hollow* xxiv. 206 She took in the Savile Row cut of Edward's clothes. **1948** 'J. Tey' *Franchise Affair* xii. 126 'I'll come along and help.' 'Not in that Savile Row suit, you won't.' **1955** T. H. Pear *Eng. Social Differences* vii. 172 The tendency to conform with fashion, but not its extremes, marks the Savile Row tailor. **1972** M. Farhi *Pleasure of your Death* ii. 43 Van Loon..looked like a foetus—stillborn in a Savile Row suit.

saving, *vbl. sb.* Add: **4.** Now usu. with *pl.* **savings: savings account,** a deposit account; **savings and loan** *U.S.*, used *attrib.* to designate a co-operative association which operates in the manner of a building society, though now offering additional services, as loans for purchases other than houses, and the issue of cheques to account-holders; also *absol.*; **savings book,** a book in which an official record is kept of sums deposited and withdrawn by the holder and of interest accrued; **savings-box** = *saving-box*; (**war**) **savings certificate,** introduced February 1916, renamed 1920, (**national**) **savings certificate,** a certificate declaring that the holder has invested a small sum in government funds, encashable at any time with accrued interest, and usually maturing after five or ten years. Cf. Savings bank.
1911 *Daily Colonist* (Victoria, B.C.) 29 Apr. 8/6 Encourage your boy to save by opening a savings account for him. **1978** S. Sheldon *Bloodline* xxxix. 350 A London savings account with a balance of twenty-five thousand pounds. **1978** *Washington Post* 16 Feb. F5/4 (Advt.), You earn a high 6% on regular savings accounts. **1980** *Travel & the TSB* (Trustee Savings Bank Central Board), You'll need to show your Savings Account passbook and some identification. [**1877** *Acts & Resolves Gen. Court Mass.* ccxxiv. 613 The words 'coöperative saving fund and loan association' shall form a part of the name. **1882** *Chicago Business Directory* 86/1 Union Savings, Loan & Building Assn.] **1884** *Lakeside* (Chicago) *Ann. Directory of Business 1884–5* 1489/1 Sharpshooters' Building, Savings and Loan Association. **1887** *Laws of State of N.Y.* dlvi. 720 All associations formed under the provisions hereof shall be known as co-operative savings and loan associations; and the name of every association, so formed, shall contain as a part thereof the words Co-operative Savings and Loan Association. **1921** *Proc. 34th Ann. Convention N.Y. State League of Savings & Loan Assoc.* 7 They believe in the State organization of savings and loan associations. **1962** J. H. Ewalt *Business Reborn* i. 3 The typical predepression association [was] known more often as a building and loan than as a savings and loan. **1975** *New Yorker* 5 May 98/3 The Gibraltar Savings Association, of Houston, which is a subsidiary of the Imperial Corporation of America, a holding company that owns a number of savings-and-loan companies in four of the Western states. **1936** N. Streatfeild *Ballet Shoes* xv. 232, I was quite ashamed of your savings book... I care..that you have a nice lot saved for when you are grown up. **1977** 'J. Fraser' *Hearts Ease in*

Death ix. 103 The proprietor..was standing beside the Post Office counter when Aveyard came in, pushing a savings book..through the grille. *c* **1863** T. Taylor *Ticket-of-Leave Man* ii. 32 I've put away a shilling every week out of my savings... It's all here. (Goes to table, and..puts a savings-box into his hand.) **1922** Joyce *Ulysses* 30 Three, Mr Deasy said, turning his little savingsbox about in his hand. **1978** *Church Times* 27 Jan. 14/1 (*heading*) Savings box for Lent. **1916** *Times* 19 Feb. 5/1 The new War Savings Certificates, which can be bought from today for 15s. 6d. each at any money-order office. **1919** *Saving* 3 Dec. 140/2 Leyton school children have bought Savings Certificates to the value of over £48,000. **1920** *Act* 10 & 11 *Geo. V* c. 12 (*title*) An Act..to extend to National Savings Certificates the enactments relating to War Savings Certificates. **1927** W. Deeping *Doomsday* xix. 209 Seventy-five pounds in Savings Certificates. **1932, 1941** [see *National Savings Certificate* s.v. *national *a.* 5]. **1961** E. S. Turner *Phoney War* x. 131 If I buy three Savings Certificates at 15s. each the State will have to pay me interest and eventually repay my capital. **1978** F. Maclean *Take Nine Spies* vii. 235 Into his Foreign Office black leather briefcase..he crammed nearly £300 in notes and a bundle of Savings Certificates.

saving, *ppl. a.* Add: **3.** Now freq. in phr. *saving grace.*
1910 W. G. Collingwood *Dutch Agnes* 168 She plied me with questions until I was very nearly tormented into confession. But I had the saving grace, I trust, to remember John Bell's adage of *Vir sapit qui pauca loquitur*. **1932** J. B. Priestley *Self-Selected Ess.* 282 Here, in its plain lack of ideas, is the saving grace of this dull company. **1960** C. Day Lewis *Buried Day* viii. 170 Tchehov..has indeed said, but with all the saving grace of his felicitous compassion, that we are not put on the earth to be happy. **1978** *Lancashire Life* Oct. 36/1 In all the shouting, the bitter recriminations, there was the saving grace of native good humour.

Saviny, var. *Sapiny.

saviour. Add: **4. b.** Special combinations: **saviour's blanket, flannel,** in Sussex and Kent, a local name for several plants with greyish downy leaves, esp. lamb's ears, *Stachys lanata*, or mullein, *Verbascum thapsus*.
1882 H. Friend *Gloss. Devonshire Plant Names* 10 In Sussex the small plant (*Stachys lanata*) with a similar leaf is called 'Saviour's Blanket'. **1927** V. Woolf *Jrnl.* 4 July (1980) III. 144 They [*sc.* slightly furred cheeks] are like saviours flannel, of which she picked me a great bunch, in texture.

‖ **savoir** (savwaɪ). [Fr., = to know.] Knowledge. Used *ellipt.* for Savoir faire or Savoir vivre.
1823 Lady Blessington *Jrnl.* 12 Aug. in E. Clay *Lady Blessington in Naples* 64 Glad as I was to profit by the *savoir* of Sir William Gell.., yet I could have wished to ramble alone. **1911** A. Bennett *Card* x. 236 He had latterly acquired a considerable amount of social *savoir*. **1952** W. Stevens in *Nation* 6 Dec. 519 It is as if We had come to an end of the imagination, Inanimate in an inert savoir.

savoir faire. (Later examples.)
1897 E. A. Bartlett *Battlefields of Thessaly* iv. 74 He was a fine powerful man, with plenty of courage and *savoir faire*. **1924** *Granta* 25 Apr. 361/2 He had, it seems, spent previously some months at Deauville and Paris..and there acquired that polished French and developed that *savoir-faire*, both so typical of him. **1965** A. Renoir in Bessinger & Creed *Medieval & Linguistic Stud.* 159 If he is indeed the speaker's husband and Wulf her lover, we could hardly expect him to stretch the limits of *savoir faire* to the point of offering the latter a cup of mead at the family table. **1974** J. Betjeman *Nip in Air* 28 A luncheon and a drink or two, a little *savoir faire*——.

Savonarola (sævənărōᵘ·lă). [The name of the Dominican monk, Girolamo *Savonarola* (1452–98), famed for his fierce opposition to ecclesiastical, moral, and political licence and corruption.] **1.** Used allusively to designate someone considered puritanical in attitude, esp. in regard to the arts.
1916 G. B. Shaw *Androcles & Lion* p. xv, They save society from ruin by criminals and conquerors as well as by Savonarolas and Knipperdollings. **1963** W. K. Rose in *Lett. Wyndham Lewis* (1963) iii. 122 This might change abruptly to an angry, mind-scourging Savonarola. **1980** R. Ludlum *Bourne Identity* xvi. 252 He was a Savonarola, but without religious principles, only his own odd morality.
2. In full, *Savonarola chair*. A kind of folding chair typical of the Italian Renaissance (see quots.).
1918 G. L. Hunter *Italian Furnit. & Interiors* (1920) I. p. iv, Of Italian chairs there are more types than were until recently known to exist:..folding 'X' chairs of the type sometimes called 'Savonarola', wonderful 'Dante' chairs. **1927** Eberlein & Ramsdell *Pract. Bk. Ital., Span. & Port. Furnit.* 71 The really correct name for both the so-called 'sedia Dantesca' and the so-called 'sedia Savonarola' is *sedia del campo* or *field chair*... Being readily portable when folded up, they were carried on campaigns... These chairs were likewise commonly used by the Florentines for resting, dining, writing and reading... The finest of the 'Savonarola' chairs were made of walnut. **1969** *Observer* 11 May (Colour Suppl.) 17 Savonarola's folding chair..gave its name to this characteristic piece of Renaissance furniture—a savonarola. **1972** *Country Life* 23 Mar. 723/1 Knowledgeable inspection of the oak will reveal these shams which have been sold under such names as Dante and Savonarola chairs.
Hence **Savonaro·lan** *a.*
1960 K. Clark *Looking at Pictures* 184 Savonarolan puritanism has made Botticelli renounce the physical beauty which he still thought appropriate to the blessed spirits in his Dante drawings. **1976** *Jrnl. R. Soc. Arts* Mar. 167/2 Michelangelo himself, of course, had a strong strain of Savonarolan puritanism in his background.

Savonius (săvōᵘ·niəs). The name of Sigurd J. *Savonius* (*fl.* 1930), Finnish engineer, used *attrib.* to designate a device developed by him consisting of two opposed semicylindrical blades (see quot. 1948), used in various forms to measure the speed of air and water currents and as a windmill rotor in the generation of electricity from wind power.
1925 *Mech. Engin.* Nov. 912 In slow winds the Savenius [*sic*] wing rotor might have to be provided with an auxiliary motor... In strong winds, however, its own turning moment would be sufficient. **1948** P. C. Putnam *Power from Wind* vi. 101 The Savonius rotor was a vertical cylinder sliced in half from top to bottom, the two halves being pulled apart by about 20 per cent of the diameter... In principle it resembled a cup anemometer... The Savonius design possessed fairly high efficiency. **1955** E. W. Golding *Generation of Electricity by Wind Power* xii. 197 The Savonius rotor has the two halves of the bent sheet displaced so that the wind can pass between them... A number of Savonius type machines were built some years ago but only in small sizes. **1963** G. L. Pickard *Descriptive Physical Oceanogr.* vi. 79 An alternative..is the Savonius rotor which is not sensitive to vertical motion. **1974** *Undercurrents* Mar.–Apr. 3/3 A double Savonius rotor windmill directly driving a screw pump.

Savonnerie (savɒnəri). [Fr., lit. 'soap factory', f. *savon* soap.] The name of a factory established in a former soap works in Paris in the 17th century, used *attrib.* and *absol.* to designate hand-knotted pile carpets made there. Also used of similar products from elsewhere in France.
1876 *Encycl. Brit.* V. 129/2 The most celebrated and artistic textures of this class are the Aubusson, Savonnerie, and Beauvais carpets of France. **1899** R. Glazier *Man. Hist. Ornament* 118 About 1590, some carpets called Savonnerie were made in the Louvre, the technique being somewhat similar to the Persian carpets. **1912** *Loan Exhib. Tapestries, Carpets & Silk Fabrics from Mobilier Nat. Paris* (Victoria & Albert Mus.) 4 The chief characteristic of the Savonnerie carpets was the application of the technical methods used in..Oriental specimens to designs prepared in the contemporary style of French decorative art. **1922** Kendrick & Tattersall *Hand-Woven Carpets* I. i. vii. 73 The factory of the Savonnerie, which has provided a generic name for all French hand-knotted carpets, was founded in 1626. **1933** *Burlington Mag.* Dec. p. xxiv/2 Savonnerie panels in silk, brocades etc. **1949** N. Mitford *Love in Cold Climate* ii. iv. 225 My Savonnerie, my Sèvres, my sanguines, all my treasures gone and I confess I am very low about it. **1966** M. Jarry *Carpets of Manuf. de la Savonnerie* 18 Le Château de Fontainebleau..is fitted up with Savonnerie carpets. **1972** K. Bonfiglioli *Don't point that Thing at Me* i. 2 That..is a valuable Savonnerie rug. **1977** *Times* 11 Oct. 17/6 The sale will contain..16 antique Oriental carpets and one Savonnerie.

savoursome (sēɪ·vəɹsəm), *a.* Also 6 **savorsome.** [f. Savour *sb.* + -some[1].] Full of savour (in various senses).
1595 Chapman *Ovid's Banquet of Sence* xxxii, Come soueraigne Odors, come..Wax hotter ayre, make them more sauorsome. **1922** *19th Cent.* Sept. 513 Hot savoursome shellfish the inn people gave us. **1958** *Times* 9 Oct. 7/1 Mr. Derek Francis is a savoursome Casca.

Savoy. Add: **2.** (Earlier example.)
1723 J. Nott *Cook's & Confectioner's Dict.* sig. Fv, *Savoy Biskets*... Eggs..Rose-water..Sugar..beaten as thick..as Cream... The finest flour... Bake.
3. The name of the *Savoy* Theatre in London, used *attrib.* to designate the Gilbert and Sullivan operas originally presented there by the D'Oyly Carte company.
1889 G. B. Shaw in *Star* 13 Dec. 2/4 A new Savoy opera is an event of no greater artistic significance than..a new oratorio by Gounod. **1893** —— in *World* 11 Oct. 23/2 The announcement of a new Savoy opera always throws the middle-aged playgoer into the attitude of expecting a surprise. **1902** *Encycl. Brit.* XXXIII. 56/2 The Savoy operas did not aim at intellectual or emotional grandeur, but at providing innocent and wholesome pleasure. **1907** W. S. Gilbert in *Daily Mail Year Bk.* 90/1 Savoy opera..was snuffed out by the deplorable death of my distinguished collaborator, Sir Arthur Sullivan. **1930** *Times* 22 Mar. 13/4 Savoy Opera is a tree deeply rooted in our national fantasy. **1961** *Sunday Times* 30 Apr. 12/3 Today only the Savoy operettas and the Bab Ballads remain alive to judge him [*sc.* W. S. Gilbert] by.

Savoyard, *sb.* and *a.* Add: **A.** *sb.* **3.** A member of the D'Oyly Carte company which originally played at the Savoy theatre in productions of the Gilbert and Sullivan operas. **b.** A devotee of the Savoy operas. Cf. *Savoy 3.
1890 W. S. Gilbert (*title*) Songs of a Savoyard. **1893** G. B. Shaw in *World* 11 Oct. 24/2, I enjoyed it [*sc. Utopia Limited*] and..the majority of Savoyards will share my appreciation of it. **1908** R. Barrington *Rec. 35 Years' Exper. Eng. Stage* xxi. 265 To have been an 'old Savoyard', that is to say, one of the original company, seems to confer not only a great measure of dignity but..a greater natural activity in old age. **1922** *Glasgow Herald* 26 July 9/1 The death is announced of Mr George Thorne, the famous Savoyard..well known to the older generation of Glasgow admirers of Gilbert and Sullivan. He..appeared regularly with the [D'Oyly Carte] company at the Royalty Theatre, Glasgow. **1930** *Times* 24 Mar. 15/5 As an old Savoyard and senior vice-president of the Gilbert and Sullivan Society, I..heard with dismay Mr. Henry Lytton's tentative announcement of his possible retirement. **1961** *Times* 12 Dec. 5/3 In *Trial by Jury*..we remarked with horror..an anachronism as horrible to the designers of Printing House Square as is the fear of a rock 'n roll *Mikado* to good Savoyards. **1977** *Times* 14 July 12/7 While..the words and music of Gilbert and Sullivan are the main attraction, Savoyards have a powerful respect for the spirit of the original productions. **1978** *Lancashire Life* Feb. 62/1 My music teacher was the mother of Martyn Green the world-famous Savoyard, who was a Boltonian.
B. *adj.* (Earlier and later examples.)
1741 M. W. Montagu *Let.* 15 Nov. (1966) II. 259 This Town [*sc.* Chambéry]..is wholly inhabited by the poor Savoyard Nobillity. **1905** Ld. Coleridge *Story of Devonshire House* xvi. 239 You will find the girl in the garden with a coarse Savoyard straw hat. **1975** P. Topping in Setton & Hazard *Hist. Crusades* III. v. 154 The Savoyard prince secretly intrigued with Theodore despite his agreements with Venice.

savvy: see also *savey, savvy *sb.* (and *a.*), *v.*

saw, *sb.*[1] Add: **1. c.** A flexible saw used as a musical instrument, played with a bow.
1931 *Daily Mail* 6 Oct. 16/3 Saw solos. **1938** *Oxf. Compan. Mus.* 872/1 *Singing saw*. This is an ordinary hand saw which is held between the player's knees and played on by a violin bow; its blade is meanwhile bent, under a lesser or greater tension, by the player's left hand, so producing the different pitches. **1961** *Times* 18 Jan. 15/5 An instrument believed to be wholly new in the orchestra pit, the musical saw. **1977** *Times* 14 Dec. 14/8 The Anal Zephyr Trio *does* exist..(apart from the pianist) it includes a saw and bottles.
5. c. *saw-backed* adj.
1903 Kipling *Five Nations* 176 The same old saw-backed fever-chart. **1924** R. Campbell *Flaming Terrapin* v. 77 The angel cowboys..Vaulting on the saw-backed ridges Where they tear the sky to strips. **1961** C. H. D. Todd *Popular Whippet* 33 One is often asked about a 'saw-backed' dog and what can be done about it.
5. d. saw-doctor, (*b*) a craftsman who maintains saws in an efficient condition; **saw-grass,** (*b*) (earlier example); **saw-log** (earlier and later examples); **saw palmetto,** also, a small cluster palm, *Acoelorrhaphe wrightii*, of southern Florida and central America; (earlier and later examples); **saw-scale** = *saw-scaled viper*; **saw-scaled viper,** a small venomous rough-scaled snake, *Echis carinatus*, of the family Viperidæ, found in Africa and southern Asia; **saw-shark,** substitute for def.: a small shark of the family Pristiophoridæ, found in southern seas from Africa to Australia and distinguished by a saw-like flattened snout; (later examples); **saw-timber,** timber suitable for sawing into boards or planks; **saw-whet,** for *U.S.* read *N. Amer.* and substitute for def.: a small dark brown owl, *Ægolius acadica*, found in eastern North America; (earlier and later examples).
1936 A. M. Rust *Whangarei Early Reminisc.* 163 Timber was being got..along its..foreshore. Hundreds of bushmen..were employed besides stackers, saw doctors, benchmen and mill-hands in the different sawmills. **1949** J. L. Carvel *One Hundred Years in Timber* ix. 140 No sawmill can function long without efficient tool-rooms, and at the City Saw Mills the saw-shop and grinding-shop supply these essentials. These are supervised by the saw-doctor. **1977** *Belfast Tel.* 22 Feb. 22 (Advt.), C. D. Monninger Ltd. require Saw Doctor to take charge of the day-to-day running of their new Belfast Service Centre. **1822** W. H. Simmons *Notices E. Florida* ii. 24 They were obliged to defend their horses' feet with wrappings of cow-hide, in order to prevent their being injured by the sharp saw grass. **1799** D. W. Smyth *Short Topogr. Descr. Upper Canada* 32 The saw logs are conveyed to this mill in a very remarkable manner. **1916** *Daily Colonist* (Victoria, B.C.) 1 July 6/6 The timber returns for the month of May..show that the total scale of sawlogs for the Province amounted to 94,771,871 ft. [etc.]. **1971** *Timber Trades Jrnl.* 14 Aug. 38/1 It is estimated that quantities from British forests should increase significantly in the next decade and with the improving quality of sawlogs home producers can look forward to obtaining an increasing share of consumption of sawnwood. **1797** B. Hawkins *Let.* 18 Feb. in *Georgia Hist. Soc. Coll.* (1916) IX. 85 The whole country was a pine barren, with wiregrass and saw palmetto. **1894** B. Torrey *Florida Sketch-Bk.* 3 The ground [was] covered thickly with saw palmetto. **1938** M. K. Rawlings *Yearling* xxv. 317 The bears were..eating the berries of the saw palmetto. **1942** S. Kennedy *Palmetto Country* 4 Shrub-like saw palmetto underlies the pine flat-woods. **1964** J. Hillaby *Journey to Jade Sea* 121 Saw-scales sound like kettles of boiling water. **1935** N. L. Corkill in *Sudan Notes & Records* XVIII. 245 The Carpet or Saw-scaled Viper is usually considered to be a form restricted to a sandy habitat. **1966** C. Sweeney *Scurrying Bush* xii. 168 A very violent saw-scaled viper crawled out into the open, hissing and rustling its scales against each other. **1906** D. G. Stead *Fishes Austral.* xii. 236 The Little Saw-Shark..is a small species, having a somewhat flattened body, and attaining a length of about 4 feet. **1931** J. R. Norman *Hist. Fishes* iii. 35 In..one of the Saw Sharks..there may be as many as six or seven [gill-clefts]. **1961** E. S. Herald *Living Fishes of World* 49/1 The four known species of saw sharks have small pectoral fins with the gill openings just ahead of these fins. **1932** *Sun* (Baltimore) 17 Sept. 4/6

SAW

The cutting is always done selectively, large trees being taken for saw timber for new buildings and repairs, and weed trees and defective trees for fuel. **1979** *Sci. Amer.* Feb. 65/3 In the Western national forests, which constitute.. 50 percent of the nation's entire supply of standing saw-timber. **1834** J. J. Audubon *Ornith. Biogr.* II. 567 The Little Owl is known in Massachusetts by the name of the 'Saw-whet', the sound of its love-notes bearing a great resemblance to the noise produced by filing the teeth of a large saw. **1894** *Outing* XXIII. 406/1 The little 'saw whet' under his tiny glass globe. **1949** *Amer. Forests* Oct. 23/1 The saw-whet owl has a peculiar voice. **1959** W. R. Bird *These are Maritimes* vi. 183 Now I rather like the little fellows [*sc.* owls], especially the saw-whets. **1977** *New Yorker* 5 Sept. 24/1 Saw-whet owls and long-eared owls roost in evergreens in winter.

saw, *v.*¹ Add: **2. c.** (Later example of a cellist.)

1977 J. Crosby *Company of Friends* v. 36 Czernowski sawed away at Mozart.

d. *trans.* Phr. *to saw wood*, to attend to one's own affairs; to continue working steadily. *U.S. colloq.*

1894 *Congress. Rec.* 24 Jan. 1347/2 Is it possible that the framers of the bill hold a grudge against the voters who 'sawed wood' last November? **1909** 'O. Henry' *Options* 75 During all these wintry apostrophes, Barbara, cold at heart, sawed wood—the only appropriate thing she could think of to do. **1913** F. H. Burnett *T. Tembarom* xxix. 359 Say nothing and saw wood... It means 'shut your mouth and keep on working'. **1933** J. Buchan *Prince of Captivity* III. i. 264 He sees the next job and sits down to it—stays still and saws wood, as Lincoln said.

e. Phr. *to saw a chunk* (*length, piece*) *off*, to copulate. *slang.*

1961 Partridge *Dict. Slang* Suppl. 1259/2 *Saw off a chunk* or *a piece*, to coït: Canadian: since ca. 1920. **1977** J. Wainwright *Do Nothin'* v. 86 The act is..known, in polite circles, as 'copulation'. Known, in less polite circles, as..'sawing a length off'.

2*. *transf.* With reference to the sound made by sawing; *to saw gourds*, etc., to snore loudly. *slang* (orig. *U.S.*).

1870 F. H. Ludlow *Heart of Continent* ii. 91 In five minutes..we were all 'sawing gourds' together in the land of Nod. *a* **1897** 'R. Sanders' *Sk. Country Life* (1898) xxx. 188 When the day's work is done..he can draw his bobtail night shirt about him..knowin that while he sleeps and dreams and saws gourds his worldly possessions are growin. **1939** J. Worby *Spiv's Progress* ii. 12 I've been in the town and got the grub while you've been sawing them off. **1946** *Penguin New Writing* XXVIII. 184 The deaf-mute was asleep and sawing them off horribly. **1961** Partridge *Dict. Slang* Suppl. 1259/2 *Saw them off*, to snore; to sleep soundly... Ex the noise made with a saw clumsily handled. **1980** A. Fox *Kingfisher Scream* iii. 49 Rosemary would be asleep too now, with Don sawing wood beside her.

sawagerous, var. *savagerous *a.*

‖ **sawah** (sā·wă). Also 8 **sawoor; sawa.** [Malay.] In Malaysia and Indonesia: an irrigated rice-field.

1783, 1839 [see *ladang]. **1937** M. Covarrubias *Island of Bali* iv. 71 The most striking element of the Balinese landscape is the ever present ricefield, the *sawa*, a patch of land filled with water held by dikes cut out of the red earth. **1961** P. Kemp *Alms for Oblivion* vii. 105 The slopes are terraced with superb skill in tier after tier of *sawas*, or small paddy-fields, that produce each year two crops of the finest rice in South-East Asia. **1978** *Times* 25 Mar. 13/1 All around us were the..rice-fields. We had watched the ploughing of the water-logged *sawahs*.

sawbill. a. (Earlier and later examples.)

1763 tr. *A. S. Le Page du Pratz's Hist. Louisiana* II. II. ii. 235 We are disturbed in the night, by the hideous noise of the numberless water-fowls,..such as cranes, flamingo's, wild geese, herons, saw-bills, ducks, &c. **1833** W. F. Tolmie *Jrnl.* 1 Sept. (1963) 232 Saw the Sawbill Duck once or twice riding down on a log. **1835** *Ibid.* 12 June 311 Shot a sawbill with rifle at the upper end of lake. **1894** A. Newton *Dict. Birds* III. 814 Sawbill, a name commonly given to the Goosander and Merganser. **1973** *Nature West Coast* (Vancouver Nat. Hist. Soc.) 167 The 'toothed' bill, a necessity for holding slippery fish, has earned this bird [*sc.* the red-breasted merganser] the name 'sawbill'.

sawbuck (sǭ·bɒk). *U.S.* [ad. Du. *zaagbok* trestle, saw-horse: cf. Saw *sb.*¹ and Buck *sb.*⁷]

1. a. = Buck *sb.*⁷

1862 *Rep. Comm. Patents: Agric. 1861* (U.S.) 141 The sheep is then laid upon his back in a kind of saw-buck. **1869** [see Saw *sb.*¹ 5 d]. **1877** R. J. Burdette *Rise & Fall of Mustache* 308 You might as well tell a joke to a sawbuck as to his wife. **1906** W. Churchill *Coniston* 390 He was standing with his foot upon the sawbuck and the saw across his knee. **1920** S. Lewis *Main Street* 83 In back yards their sawbucks stood in depressions scattered with.. flakes of sawdust. **1948** *Sat. Even. Post* 10 July 83/2 I'd roped everything around the ranch—calves, hounds, horses, fence posts, sawbucks.

b. In full, *sawbuck* (*pack*)*saddle*. A pack-saddle shaped like a sawbuck.

1881 E. W. Nye *Bill Nye & Boomerang* 67 This summer, however, I will get me a little blue jackass and put a sawbuck on his back. **1907** S. E. White *Arizona Nights* ii. 12 We skirmished around and found..a sawbuck saddle with kyacks. **1913** *Outing* Jan. 425/1 The most practical equipment for pack animals is the ordinary crosstree or sawbuck pack saddle for all-round use. **1933** F. H. Cheley *Camping Out* 461 While the Government has adopted the aparejo as its pack saddle, the cross-tree or sawbuck is the best one for ordinary use. **1938** M. Thompson *High Trails of Glacier National Park* 138 If you are going on a camping trip you utter an instinctive protest as your packer cinches up the 'sawbuck' packsaddle and loads his animal.

2. *slang.* **a.** Ten dollars; a ten-dollar note. Also *double sawbuck* (*a*) s.v. *double *a.* A. 6.

In allusion to the x-shaped (Roman x = 10) ends of the sawyer's buck: cf. also *buck *sb.*⁸ dollar.

1850, etc. [see *double sawbuck* (*a*) s.v. *double *a.* A. 6]. **1852** *Oregon Statesman* 13 Nov. 1/1 Dod rabbit it, there goes another 'saw-buck', on the plag'uey jack. **1870** J. H. B. Nowland *Early Reminisc. Indianapolis* 315 In former years he was ever ready to..risk what he called a 'sawbuck' (a ten dollar note), on his success. **1933** [see *bite *sb.* 1 i]. **1973** J. Wambaugh *Blue Knight* ii. 41, I gave him a ten, which was just like folding up a sawbuck and sticking it in his arm. He'd be in the same shape twelve hours from now.

b. A ten-year prison sentence. Also *double sawbuck* (*b*) s.v. *double *a.* A. 6.

1925 *Flynn's* 25 Mar. 511/1 *Sawbuck*,..a ten-year sentence. **1929** *Sat. Even. Post* 13 Apr. 50/4 A prisoner with ten years brings in a saw-buck. **1938** D. Castle *Do Your Own Time* iii. 28 'I'm doing two saw-bucks.' 'Oh, yeah? Whatever that is.' 'Two ten spots. Twenty years.' **1945, 1950** [see *double sawbuck* (*b*) s.v. *double *a.* A. 6].

‖ **Sawbwa** (sǭ·bwa). Also **Chobwa, Tsaubwa,** etc. [Burmese.] The hereditary ruler of a Shan state in Eastern Burma.

1800 M. Symes *Acct. Embassy to Kingdom of Ava* xvi. 375 We were told that there were fifty-six Chobwas dependent on the Birman state; if it be true, their territories must be very inconsiderable. **1829** J. Crawfurd *Jrnl. Embassy to Court of Ava* xv. 395 The only class of public officers which can be called hereditary under the Burmese Government, are the Thaubwas, or Saubwas, the tributary princes of the subjugated countries. **1858** H. Yule *Narr. Mission sent to Court of Ava* xiii. 303 The Tsaubwas..retain all the forms and appurtenances of royalty. **1875** H. A. Browne *Jrnl.* 18 Jan. in *Reminisc. Court of Mandalay* (1907) 67 Kut-Loon.. is in the jurisdiction of the Maing-maw Tsawbwa, or Chinese Shan chieftain. **1911** *Encycl. Brit.* XXIII. 892/1 Politically, where not under the direct control of Chinese magistrates, the tribes are organized under their own chiefs, who are recognized by the Chinese government and endowed with official rank and title. In Burmese such native chiefs are termed *Sawbwa*. **1929** F. T. Jesse *Lacquer Lady* III. 276 The Shan Sawbwas were in open revolt. **1962** *Listener* 25 Oct. 646/2 Some of the Shans, led by some of the Sawbwas, agitated for separation from Burma. **1973** *Dict. World Hist.* 1378/2 The Shans have retained their racial identity and a high degree of separatism, with numbers of small states each until recently having its own ruling chief. The chiefs were known as Sawbwas, Myosas, or Ngwegunhmus, according to rank.

sawder, *sb.* Add: Also without *soft*.

1854 D. G. Rossetti *Let.* 11 May (1965) I. 193 MacCrac.. offers £50 for the water-colour, with all manner of soap and sawder into the bargain. **1880** 'C. E. Craddock' in *Atlantic Monthly* Jan. 103/2 That ain't the right sort o' sawder fur a candidate.

sawdust, *sb.* Add: **3. a. sawdust game** *U.S. slang*, a type of confidence trick.

1872 G. P. Burnham *Mem. U.S. Secret Service* 404 A new device for skilful robbery of the uninitiated has been introduced..known as the 'Sawdust' or 'Circular' Game. **1939** *Times Lit. Suppl.* 9 Sept. 530/2 We hear all about..the 'sawdust game', (selling bad notes).

b. With reference to the use of sawdust for strewing the floor of a place of public entertainment (as a circus, etc.) or (U.S.) the arena used by a travelling evangelist.

1864 P. Paterson *Glimpses of Real Life* xii. 120 As good as the general run of sawdust plays. **1883** Sawdust artist [in Dict., sense 3]. **1883** Sawdust ring [see Ring *sb.*¹ 12 a]. **1913** *Collier's* 26 July 7/3 And down the aisle, 'hitting the sawdust trail', they come in ones and twos and dozens, until 476 have stood before that multitude to shake the evangelist's hand and signify their intention of starting another life. **1915** T. S. Eliot in *Catholic Anthol.* 2 One-night cheap hotels And sawdust restaurants with oyster-shells. **1946** S. H. Holbrook *Lost Men of Amer. Hist.* 312 Many of these suddenly patriotic pleaders..like repentant sinners at a revival, hurried down the sawdust path. **1964** A. Wykes *Gambling* vii. 170 The terms 'carpet joint' and 'sawdust joint' meant broadly the degree of luxury or squalor to be expected in American gambling saloons. **1977** *Time* 11 July 41/2 Sawdust Evangelist Rex Humbard, likes to exhort: 'You'd better straighten out and fly right with God.' **1978** M. Puzo *Fools Die* xiv. 152, I spent the day going through all the casinos in town on the Strip and the sawdust joints in the center of town.

sawed, *ppl. a.* Add: **1. b.** *sawed-off*, short, undersized. Freq. of persons, etc.: below average height. Also *ellipt.* Cf. *sawn *ppl. a.* 2. *U.S. colloq.*

1887 C. B. George *40 Yrs. on Rail* 22, I remember..the little sawed-off cars jolting along the uneven track. **1901** S. E. White *Westerners* 220 Most marvellous was a clean-limbed, deep-chested, slender running horse, accompanied by a sawed-off English groom. **1902** G. H. Lorimer *Lett. from Self-Made Merchant* 160, I didn't understand football, but understood that little sawed-off. **1919** *Dialect Notes* V. 65 A tall girl never looks well dancing with a sawed-off. **1930** J. Dos Passos *42nd Parallel* 100 Two soldiers on guard, toughlooking sawedoff men. **1947** *Richmond* (Va.) *Times-Dispatch* 9 Nov. B7/7 Grover Jones, the sawed-off fullback who played one war year for Penn. **1973** *Black World* Jan. 63/1, I..heard her say to that little sawed-off runt she calls a man.., 'I think she tried to communicate with me.'

c. *sawed-off* (U.S.): used to designate a (shot)gun of which the barrel has been specially shortened to make it easier to handle and give a wider field of fire. Also *ellipt.* Cf. *sawn *ppl. a.* 2 a.

1898 *Scribner's Mag.* Jan. 86/2 There was another roar from the messenger's sawed-off shotgun. **1912** W. M. Raine *Brand Blotters* 80 The 'shotgun messenger' was indolently rolling a cigarette, his sawed-off gun between his knees. **1930** *Sat. Even. Post* 26 July 145/1 The other laughed harshly. 'Did they knock him off?' he grunted. 'Nothin' but a sawed-off full in the chest an' half a dozen shots from an automatic as a chaser!' **1935** 'L. Ford' *Burn Forever* 255 He'd have used a revolver, a sawed-off shotgun or a submachine gun. **1962** A. Lurie *Love & Friendship* xiv. 280, I suppose a sawed-off shot-gun would be more his speed. **1977** *Time* 16 May 18/1 He pulled out a sawed-off submachine gun.

sawfish. Add: Now applied esp. to the saw-sharks of the family Pristiophoridæ (cf. *saw-shark* s.v. Saw *sb.*¹ 5 d in Dict. and Suppl.). (Later examples.)

1880 A. C. L. G. Gunther *Introd. Study of Fishes* 335 These Sharks [*sc.* Pristiophoridæ] resemble so much the common Saw-fishes as to be easily confounded with them. **1934** *Sun* (Baltimore) 6 July 1/7 A saw fish measuring ten feet nine inches..was landed. **1978** O. White *Silent Reach* xii. 124 Sometimes you catch sawfish and barramundi there.

sawmill. Add: (Earlier Amer. *attrib.* examples.)

1654 *Suffolk County* (Mass.) *Deeds* (1883) II. 26, I Edward Colcott..doe hereby giue..vnto Thomas Rucke..one third pte of a saw mill worke. **1716** *Duxbury* (Mass.) *Rec.* (1893) 113 We began at the waste gate belonging to the saw mill.. and run from said gate Easterly as the old saw mill dam stood. **1818** T. G. Fessenden *Ladies' Monitor* 35 His elbows, hoofs and paws That rip and rend and rive like saw mill-saws.

sawmiller (earlier example).

1845 Thoreau *Jrnl.* 5 July (1949) I. 361, I lodged at the house of a saw-miller last summer.

sawn, *ppl. a.* Add: **2.** *sawn-off* (now more usu. than *sawed-off* exc. in *N.Amer.*). **a.** Of a (shot)gun: = *sawed *ppl. a.* 1 c.

1915 A. Conan Doyle *Valley of Fear* vi. 113 In the latter was a sawn-off shot-gun, so he came with the deliberate purpose of crime. **1937** N. Marsh *Vintage Murder* ii. 14 A salute of two sawn-off shotguns. **1959** *Encounter* July 59/1 A fifteen-year-old highschool boy who had taken a sawn-off shotgun into the classroom and blown off the head of a classmate. **1978** R. Westall *Devil on Road* xx. 186 A sawn-off shotgun ..sprays lead like a hose.

b. = *sawed *ppl. a.* 1 b. *colloq.*

1936 R. Campbell *Mithraic Emblems* 162 Yet could I trudge in sawn-off trousers, And redden up like logs at Yule. **1944** *Coast to Coast 1943* 56 He was a sawn-off little bloke, and they reckoned there couldn't have been much grass about when he was born. **1954** 'J. Christopher' *22nd Cent.* 104 It's a little sawn-off town up in Scotland. **1960** J. Mortimer *Call me Liar* 431 Found him, have you?..That sawn-off, bald, damp-eyed old hundred per cent British duodenal with..no convictions known.

sawn (sǭn), *sb.* Austral. slang abbrev. of Sawney *sb.* 2.

1953 K. Tennant *Joyful Condemned* xvii. 145 I'm always getting into trouble through sawns. **1961** Partridge *Dict. Slang* Suppl. 1259/2 *Sawn*, a softy, a 'dope': low Australian.

sawoor, var. *sawah.

saw-pit. Add: **b.** *N.Amer.* A wooden framework serving the function of a saw-pit.

1876 H. W. Ravenell in *Yale Rev.* (1936) XXV. 763 The saw-pit was a rude structure about seven feet high, made of strong posts set in the ground wide enough apart to hold one or two pieces of heavy pine timber, and the sawyers, one above and one beneath, sawed out one hundred feet per day. **1961** J. W. Anderson *Fur Trader's Story* x. 87 Next they would erect, from smaller trees in the vicinity, what we used to call a saw-pit, which was not really a pit at all but a frame set entirely above the ground.

saw tooth. Add: **1. c.** *Electronics.* A waveform showing a slow linear rise and rapid linear fall, or the reverse; a voltage or current varying in this way. Usu. *attrib.*, as *saw-tooth generator, waveform.*

1933 *Proc. IRE* XXI. 1666 The variation of intensity of both horizontal and vertical deflecting fields plotted against time is of a 'saw-tooth' shape. **1935** M. G. Scroggie *Television* iv. 37 The sudden charge of the condenser, followed by a slower discharge, yields a saw-tooth wave-form. **1940** [see *keystone *v.*]. **1942** *Electronic Engin.* XIV. 666/2 The time-base circuit is..adapted for single sweep operation and the saw-tooth generated can be expanded symmetrically with respect to the centre of the tube. **1947** R. Lee *Electronic Transformers & Circuits* ix. 251 Probably the most common application of sawtooth amplifier transformers is to provide a linear sweep to horizontal plates of a cathode ray oscilloscope. **1969** J. J. Sparkes *Transistor Switching* iii. 63 All sawtooth-wave generators integrate a constant voltage or current with respect to time.

2. saw-tooth roof, a roof with a serrated profile incorporating windows in the steeper sides, which face in the direction of the equator.

1900 *Engineering* 9 Feb. 173/2 The factory consists of four bays, each covered with a saw-tooth roof running east and west, and glazed on the north side only. **1942** Asher & Heal *Send no Money* 58 It had plenty of windows, a sawtooth roof to provide light, and it was airy and spacious. **1966** L. Cohen *Beautiful Losers* I. 10 The sun was just coming up over the sawtooth roof of the factory next door.

sawyer. 3. (Earlier example.)

1786 E. Beatty *Diary* 6 Sept. in *Mag. Amer. Hist.* (1877) I. 312/2 Arrived at Guyandot this evening and lay all night off its mouth in rapid water—obliged to make fast to a sawyer.

sax, *sb.*¹ **1.** (Later examples.)

1968 *Medium Ævum* XXXVII. 130 The Hailfingen, Württemberg, sax. **1972** G. Jones *Kings, Beasts, & Heroes* I. i. 20 Wiglaf pierces the dragon's unarmoured underbelly, and..draws his sax and severs him at the middle.

sax (sæks), *sb.*² Colloq. abbrev. of Saxophone. **1.** = Saxophone 1 in Dict. and Suppl. For *alto sax, tenor sax*, etc., see under first element in Suppl.

1923 *N.Y. Times* 7 Oct. IX. 2 *Sax*, a saxophone. **1926** *Picture-Play Mag.* July 3/2 (Advt.), How I used to envy Laura playing beautifully mellow notes on her sax. **1931** *Amer. Mercury* Dec. 426/1 'Mom,' he said, 'is my old sax still around here?' **1943** J. B. Priestley *Daylight on Saturday* xi. 68 The works dance band, that *Elmdown Six* in which Jack Brimber played the tenor sax. **1955** L. Feather *Encycl. Jazz* 64 A bass sax might be used as a rhythm instrument. **1976** N. Roberts *Face of France* ix. 102 The moan and scream and shudder of sax and trumpet and drums in the band.

2. = Saxophonist.

1926 *Melody Maker* Mar. 4 Then, for a certainty, you have heard some bad saxes! **1943** J. B. Priestley *Daylight on Saturday* iv. 21 In the canteen tomorrow..the Elmdown Six will perform. And I'm one of them. Jack Brimber—tenor sax. **1975** J. McClure *Snake* vi. 86 I'm the tickler. Pianist... Drums and sax were here, but they've gone..to get pissed.

3. *attrib.* and *Comb.*, as *sax-man, -player*; **sax section,** the wind section of a dance or jazz band.

1955 L. Feather *Encycl. Jazz* 118 Christy, June, *singer*... Married Kenton's tenor saxman, Bob Cooper. **1972** *Jazz & Blues* Sept. 10/1 Clarence Ford, sax-man with Fats Domino, is typical in this respect. **1926** Whiteman & McBride *Jazz* iii. 67 'Well,' said the biggest sax player, 'we didn't know what you would want us to do.' **1980** M. Booth *Bad Track* ix. 158 The sax-player could see a tired desperation..in her eyes. **1932** *Melody Maker* Jan. 11/2 Eddie Pratt, Sid Cole, and Stanley Quiddington, saxophones, the latter being previously together as Jay Whidden's sax section. **1977** J. Wainwright *Do Nothin'* viii. 124 The sax section—Ric..fills it out, with the tenor..he doubles clarinet (like most sax men).

Hence **sa·xist,** a saxophonist.

1939 *Melody Maker* 13 May 3 (*heading*) Dutch saxist collects band. **1952** B. Ulanov *Hist. Jazz in Amer.* (1958) xv. 178 Harry Carney joined the band..first as an alto saxist. **1969** *Guardian* 23 Aug. 6/4 No tenor saxist of the fifties would take the stand without a one-note stutter in his vocabulary. **1975** *Gramophone* Aug. 375/3 'The Foremost!' is devoted to three tenors and a baritone saxist from the bop era.

Sax, var. *Sauk.

saxe. Add: Also **Saxe.** (Further examples.) Also *ellipt.* for *saxe blue*. **saxe blue** (further example).

1917 in G. Howell *In Vogue* (1975) 24/1 Afternoon Gown... In grey, saxe, navy, nigger, rose, and black. **1922** Joyce *Ulysses* 441 In smart Saxe tailormade, white velours hat and spiderveil. **1939–40** *Army & Navy Stores Catal.* 607/3 Slippers...Red, Green, Saxe, Navy. **1959** G. D. Painter *Proust* I. vii. 87 Comparing the what-not on which she kept her Saxe figurines to an altar. **1974** *Harrods Christmas Catal.* 20 Cashmere cardigan... Natural, vicuna-colour, saxe, Harvard blue. **1980** *Radio Times* 29 Nov.–5 Dec. 34/1 Pyjamas... *Grey* trimmed *Wine, Saxe Blue* trimmed Navy.

sax-horn, saxhorn. (Earlier and later examples.)

1844 *Illustr. London News* 14 Dec. 384/2 The Sax Horn.. unites the powers of the French horn and those of the cornet-à-piston. **1939** [see *althorn]. **1977** *Gramophone* May 1680/2 The piece works very well because the sax-horn family is able to offer a suitably mellifluous sound.

Saxin (sæ·ksin). Also **saxin.** A proprietary name for artificial sweeteners and other products (see quots. 1897, 1964).

1897 *Trade Marks Jrnl.* 28 July 696 *Saxin*... Chemical substances prepared for use in medicine and pharmacy, but not including those prepared for use in the cure of corns and warts and not including any goods of a like kind to any of these excluded goods. Henry Solomon Wellcome,.. London, E.C.; manufacturing chemist. **1918** 'K. Mansfield' *Let.* 17 Feb. (1928) I. 130 I've just made myself a glass of boiling tea, very weak, with saxin. **1964** *Trade Marks Jrnl.* 19 Aug. 1366/2 *Saxin*... Flavourings and essences, none being essential oils; and sweetening materials included in Class 30.

saxist: see *sax *sb.*²

saxitoxin (sæksitǫ·ksin). *Biochem.* [f. mod.L. *Saxi-domus*, name of a genus of clams (f. L. *sax-um* rock + -i- + *domus* home) + Toxin.] A toxic alkaloid ($C_{10}H_{17}O_4N_7\cdot 2HCl$) synthesized by dinoflagellates of the genus *Gonyaulax* (which cause 'red tides') and accumulated by molluscs which feed on these, which thereby become toxic to man.

1962 *Jrnl. Amer. Chem. Soc.* LXXXIV. 2266/1 Saxitoxin, the paralytic poison isolated from toxic Alaska butter clams (*Saxidomus giganteus*),..is among the most toxic known substances. **1968** *New Scientist* 27 June 706/2 PSP [paralytic shellfish poison] is now known as saxitoxin and there is an extensive literature which dates back at least to the year 1778. *Ibid.*, It would appear that saxitoxin is a perhydropurine derivative into which are incorporated two guanidino moieties. **1977** *Sci. Amer.* Dec. 30/3 A tiny dose of saxitoxin—the 'red tide' toxin—can kill a man who weighs 500 million times as much as it does.

Saxo- (sæ·kso), combining form of Saxon *sb.* and *a.* [L. *Saxo-*], prefixed to ethnic adjs. in the sense 'Saxon and —', as in *Saxo-Danish, -Norman* adjs.

1798 tr. J. C. Adelung in A. F. M. Willich *Elem. Crit. Philos.* II. p. cxxiii, In the Saxon and Saxo-Danish periods, the national taste..was still much too rude to exhibit this corruption. **1932** C. J. W. Messent *City Churches of Norwich* 25 The base is supposed to be Saxon, or Saxo-Norman, that is built by Saxon labour under Norman direction. **1980** *Rescue News* No. 23. 2/3 A rim of Saxo-Norman pottery.

Saxon, *sb.* and *a.* Add: **A.** *sb.* **1. b.** (Further example.) Also, an Englishman as distinct from a 'Latin'.

1908 M. Beerbohm *Let.* 23 Dec. (1964) 180 The Latins are born actors, while the Saxons have to train themselves up to the scratch. **1977** *Times Lit. Suppl.* 1 Apr. 394/3 In 1962 Ewart Milne returned to Ireland after more than twenty years in the land of the Saxon.

B. *adj.* **3. b.** *Saxon blue, green* (later examples).

1968 E. Brill *Old Cotswold* v. 85 It is sometimes mixed with indigo, or in the old days with woad, to give what dyers call Saxon Green. **1976** *Southern Even. Echo* (Southampton) 12 Nov. (Advt. Suppl.) 14/3, 1973 Vauxhall Viva. Saxon blue... £1095.

Saxonism. Add: **1. b.** The doctrine or practice of employing English words of purely Anglo-Saxon derivation in preference to words of foreign origin.

1926 Fowler *Mod. Eng. Usage* 514/2 Saxonism is a name for the attempt to raise the proportion borne by the originally & etymologically English words in our speech to those that come from alien sources. **1952** W. D. Jacobs *William Barnes Linguist* ii. 45 If Latinism had its failings, Saxonism manifested great excellences.

Saxonist. Add: **b.** An advocate of the use of English words of purely Anglo-Saxon origin. Cf. *saxonism 1 b.

1926 Fowler *Mod. Eng. Usage* 228/1 While the plain Englishman is content that *events* should *happen*, the Saxonist..requires that there should be *happenings*, & the anti-Saxonist..that things should *eventuate*. **1934** J. J. Hogan *Outl. Eng. Philol.* II. vii. 68 The Saxonists failed with *wheelman* 'cyclist'.

Saxony, *sb.* Add: **1. a.** (Earlier example.)

1842 *Punch* III. 74/2 House-painters, and others, will obstinately refuse to do their daily work in superfine Saxony.

b. *spec.* This wool used in making carpets. Also, a synthetic material resembling Saxony used similarly.

1910 S. Humphries *Oriental Carpets* iv. 300 *Saxony Pile Carpets*.—Made in precisely the same way as the Brussels variety... The Saxony Brussels and Saxony Velvet yarns. **1924** R. Beaumont *Carpets & Rugs* viii. 298 Examples in Saxony or the longer variety of 'velvet' carpet... In.. Saxony velvets the design and colour element may be as clearly delineated as in Wiltons. **1933** *Heal & Son Catal.* Carpet..; seamless 'Saxony', various colours. **1976** *Daily Colonist* (Victoria, B.C.) 3 Oct. 3/3 (Advt.), A well-constructed, full-bodied saxony nylon that is versatile and long-wearing.

saxophone, *sb.* Add: **1.** Also preceded by qualifying adj. (or quasi-adj.), as *soprano, alto* (see *alto *a.* b), *tenor* (see *tenor *a.* 1), *baritone*, and *bass saxophone*, in descending order of pitch. (The instrument is widely used in modern dance and jazz bands.) Cf. *sax *sb.*²

1927 *Melody Maker* Aug. 767/1 Then Mr. Billy Childs proved his excessive lung power by the force he put into blowing the soprano saxophone. **1934** S. R. Nelson *All about Jazz* ii. 57 The other saxophones in common use in the band are the tenor, baritone, soprano and bass. **1954** *Grove's Dict. Mus.* (ed. 5) VII. 434/2 Occasionally in the years just before 1917 the tenor saxophone did supplement the trombone and the soprano the clarinet, particularly in the rather larger and more highly organized bands on the Mississippi river boats. **1969** *Punch* 12 Feb. 245/3 A vaguely modal thrash headed by Lynn Dobson on flute and soprano saxophone. **1977** *Listener* 17 Feb. 215/3 Modern tenor-saxophone playing.

2. One who plays the saxophone; a saxophonist.

1929 H. Miles tr. *P. Morand's Black Magic* I. 48 The saxophone was a handsome tall fellow. **1938** D. Baker *Young Man with Horn* III. i. 141 Rick met them... In the order of their presentation they were drums, saxophone, and trombone. **1949** N. Marsh *Swing, Brother, Swing* vi. 119 The first saxophone muttered something about hitting the high spots.

3. *attrib.* and *Comb.*

1927 *Melody Maker* May 489/2 Can you imagine anything worse than a saxophone section playing a nice legato movement and the banjo plonking away for all he is worth,.. and killing the good work of the saxes. **1954** *Grove's Dict. Mus.* (ed. 5) VII. 434/1 From time to time saxophone quartets..have appeared on the concert platform. **1973** *Advocate-News* (Barbados) 24 Feb. 3/6 (Advt.), Attention all musicians..Just arrived:..Trombone Stands..Saxophone Stands. **1976** A. White *Long Silence* i. 10 We'd.. have a jolly time dancing to saxophone music until the small hours.

Hence **saxopho·nic** *a.*, of or pertaining to a saxophone; **saxophonist** (further examples).

1926 Whiteman & McBride *Jazz* ii. 34 Sleep for nights became a saxophonic mockery. *Ibid.* iii. 81 He noticed that a saxophonist was absent one night. **1958** *Times* 3 Dec. 14/6 The bassoon of Mr. Karl Kolbinger, which achieves an old fashioned diapason tone or a modern saxophonic reediness as he requires. **1970** *Daily Tel.* 19 May 16/4 Johnny Hodges, Duke Ellington's masterful alto saxophonist. **1976** *New Yorker* 15 Nov. 6/1 Alto saxophonist James Vass..will front his own quartet.

saxophone (sæ·ksŏfōᵘn), *v.* [f. the sb.] *intr.* To play on the saxophone. Also *fig.*, to produce a loud and raucous noise. Hence **sa·xophoning** *vbl. sb.*

1927 *Sunday Express* 28 Aug. 5/2 America's noise was gramophoned everywhere, and bawled and saxophoned. **1928** *Daily Express* 17 Mar. 9/7 Ten pairs of Communist lungs gave vent to a chorus of as syncopated invectives as one could wish. The parties of the Centre and Right saxophoned back. **1952** B. Ulanov *Hist. Jazz in Amer.* (1958) xxi. 288 Other boppers' trumpeting or saxophoning just doesn't fit.

say, *sb.*¹ Add: **3. say-cast** = *cow-tail 2.

1940 *Chambers's Techn. Dict.* 742/2 *Say-cast*, the coarse part of a fleece, at the tail end. **1945** [see *cow-tail 2].

say, *sb.*⁴ **2.** For † *Obs.* read 'now chiefly *Sc.*' and add later examples.

1704 S. Knight *Jrnl.* 4 Oct. (1972) 14 So I remembred the old say, and suposed I knew Sarah's case. **1880** W. T. Dennison *Orcadian Sketch-Bk.* 7 A' to' hid's an' auld say an' a true say. **1923** R. L. Cassie *Heid or Hert* xii. 52 A' the says o' her deid midder wud come back tull her.

3. b. For 'U.S.' read 'orig. U.S.' and add earlier and later examples.

1838 *Jamestown* (N.Y.) *Jrnl.* 11 July 1/5 One thing I am determined on, and that is, that the folks who succeed best in hauling the Two Pollies in the stream shall have *the say* in rigging on her up for the voyage. **1944** M. Paneth *Branch Street* 99, I had the 'say' now.

say, *v.*¹ Add: **B. 1. a.** Also *fig.*, of things: to suggest, to indicate. Phrs. *I won't* (or *wouldn't*) *say no to* (something, usu. a food or drink): I would like; *to say the word*: see Word *sb.* 7; *who says —?*, with an item of food as object: who would like —?

1898 J. D. Brayshaw *Slum Silhouettes* 158 'Who says pudden? Mister What's It—a little piece?' **1910** H. G. Wells *Hist. Mr. Polly* vi. 193 Sit down, everyone... Who says steak-and-kidney pie? **1939** A. Thirkell *Before Lunch* iv. 85, I wouldn't say no to toast and honey. **1958** V. H. Collins *Second Bk. Eng. Idioms* 194, *I won't say no*, I won't refuse..often only a genteel way of saying 'Thank you'. **1970** P. Laurie *Scotland Yard* iii. 68 To me drugs say beatniks, layabouts..kids going to ruin. **1972** A. Ross *London Assignment* 33 His shirt said custom-made silk even at that distance.

c. With an inanimate item as subject: to communicate or represent; *esp.* of a clock, calendar, etc., to show (a certain time or date); of a notice, to state (a certain message).

1930 W. Faulkner *As I lay Dying* 237 The clock said twenty past twelve. **1944** M. Laski *Love on Supertax* xi. 103 On the door..Clarissa found a notice saying, 'Welfare Officer. Knock and enter.' **1951** W. Faulkner *Requiem for Nun* II. i. 112 A clock on the wall says two minutes past two. **1973** W. J. Burley *Death in Salubrious Place* v. 105 The perpetual calendar said Wednesday August 25th. **1975** S. Johnson *Urbane Guerilla* I. 23 A sign said, 'Statue of Liberty—ticket office other side of building.' **1975** *Language for Life* (Dept. Educ. & Sci.) vi. 88 To teach a child that 'kuh-a-tuh' says 'cat' is to teach him something that is simply incorrect.

2. a. (Later *fig.* examples.) *spec.*, with a sum of money as subject, used as a formula to bet or wager *that* (something is the case).

1954 W. Tucker *Wild Talent* xii. 184 A dollar says you won't come back. **1962** D. Lessing *Golden Notebk.* II. 230 The set of his shoulders said that he was listening, so she went on. **1974** L. Deighton *Spy Story* xviii. 194 'A quid,' I said. 'You're on,' said Ferdy... 'And I've got a pound that says you're wrong,' said Schlegel. That's how I lost two quid. **1975** J. Gores *Hammett* iii. 28 I've got twenty at four-to-seven that says the semifinal is a draw. **1976** *Listener* 8 Apr. 427/3 This same man has since been in contact, and wants to go on another job with us..—which, to me, says that he is happy that what could be done was done under the circumstances at the time.

b. Also *transf.* and *fig.*, to convey, communicate; to mean; to indicate.

1881 H. James *Portrait of Lady* I. xviii. 222 I'm afraid there are moments in life when even Beethoven has nothing to say to us. **1893** E. Saltus *Madam Sapphira* 57 What would a Scotch and soda say to you? **1932** J. Buchan *Sir W. Scott* xii. 333 Venice, Tirol, Munich, Heidelberg said nothing to him. **1932** R. Campbell *Pomegranates*, They change and tremble As the lips they most resemble When one red kiss is all they say. **1951** M. McLuhan *Mech. Bride* (1967) 80/2 By juxtaposition and contrast he is able to 'say' a great deal. **1955** M. Laski *Apologies* 14 No, not actually like it, but—..it just doesn't say anything to me. **1966** *Listener* 10 Nov. 694/1 A Californian who knew the difference between summer and fall, no matter what the skies and the thermometer say. **1977** H. Fast *Immigrants* v. 302, I raised a hundred and sixty thousand dollars of San Francisco money that says so. **1977** *Jrnl. R. Soc. Arts* CXXV. 602/1 Titian, in the nature of what he can and does 'say' is at least as close to Cézanne or Francis Bacon..as he is to Sannazaro or Aretino.

Proverbial phrase. (Further examples.)

1736 Gray *Let.* Dec. (1900) I. 4 Though I say it, that should not say it, there positively is not one that has a greater esteem for you. **1817** Keats *Let.* 4 Sept. (1958) I. 150 This here Beast though I say it as shouldn't..can sing. **1818** *Blackw. Mag.* II. 214/2 My adversary might find it, however, (though I say it that shouldn't say it) in the vulgar phrase, rather a *tough job*. **1842** Dickens *Let.* 1 May (1974) III. 229, I do believe, though I say it as shouldn't, that they [*sc.* Dickens's children] are good 'uns. **1863** H. E. P. Spofford *Amber Gods* 148 Though I say it that shouldn't say it. **1889** E. Dowson *Let.* 5 Mar. (1967) 45, I recognize in it, thou' I say it as shouldn't what Pater calls 'a delicate tact of omission'. **1892** C. M. Yonge *Cross Roads* I. 13 Ours is reckoned one of the best choirs.. though I say it as should not say it.

c. (Later examples.)

1840, etc. [see *it *pron.* 3 f]. **1894** 'R. Andom' *We Three & Troddles* xv. 130 Giants are always wicked people. It says so in the children's books. **1900** B. Pain *Eliza* 54 'You told me it was port!' 'So it is.' 'It says tonic port on the label.' **1977** S. Brett *Star Trap* xii. 134 'Christopher Milton is thirty-eight, at least.' 'But it says in the programme—' 'Charles, Charles, you've been in the business too long to be so naïve.'

l. *when all is said and done* (and slight varr.): after all, in the long run, nevertheless, on balance.

c **1560** T. Ingelend *Disobedient Child* sig. A iii, Whan all is saide and all is done, Concernynge all thynges both more and lesse. **1583** B. Melbancke *Philotimus* sig. S iij, It must be as yᵉ woman will, when all is said & done. *a* **1785** J. H. Stevenson *Wks.* (1795) I. 137 And yet, when all is said and done, This something's nothing but a Pun. **1886** [see Rumourer]. **1928** M. Wilkinson *Edict of Nantes* (C.T.S.) 29 When all is said Bâville was responsible for a good deal of cruelty. **1930** 'Sapper' *Finger of Fate* 162 But when all is said and done, a prospective son-in-law is as important as any letter. **1937** 'G. Orwell' *Road to Wigan Pier* iv. 73 When all is said and done, the most important thing is that people shall live in decent houses and not in pigsties. **1952** M. Laski *Village* v. 98 After all, Friday's pay-day when all's said and done. **1981** R. Barnard *Mother's Boys* iv. 49, I know. Still, when all's said and done—.

m. *what do you say to —?*: what is your response to —?; *fig.*, how would you like —?, how would — suit you?

1592 [in Dict., sense 2 b]. **1833** J. Constable *Let.* 11 Jan. (1966) IV. 391 What do you say to all or any of Mr. White's 'says'—his dogmatical manner has force. **1851** Mrs. Stowe *Uncle Tom's Cabin* (1852) II. xxiii. 77 What do you say to a game of backgammon? **1929** *Melody Maker* Jan. 20/2 What do you say to a beaker of 'the bov'? **1948** M. Laski *Tory Heaven* vi. 84 I'm getting a bit peckish. ..What do you say to us going out and looking for a bite? **1980** M. Gilbert *Death of Favourite Girl* ii. 23 What do you say we go outside and get a breath of fresh air?

n. *that is saying* (*little, much*, etc.) (and varr.): that is to concede (little, much, etc.); used to qualify or intensify a previous statement; *it says much for* (and varr.): it is much to the credit of; *to say that* (or *one thing*) *for*: to concede (the previous or following statement) as one point in favour of.

1806 C. Wilmot *Let.* 23 Mar. in *Russ. Jrnls.* (1934) II. 223 Her Lenity makes their Lot better perhaps than that of others, but that's saying very little for the System. **1840** C. Brontë *Let.* 5 Apr. in C. Shorter *C. Brontë & her Circle* (1896) xvi. 440, I cannot perceive that she is feebler now than she was a month ago, though that is not saying much. **1853** Lytton *My Novel* III. IX. ix. 48 No, I will say one thing for English statesmen, no man amongst them ever yet was the richer for place. *Ibid.* x. xx. 202 They beat the New Yorkers in manners. I'll say that for them. **1876** J. Blackwood *Let.* 18 May in *Geo. Eliot Lett.* (1956) VI. 253 She remarked that..if people were no wiser in their speculations about more serious subjects..it did not say much for human wisdom. **1917** E. Fenwick *Diary* 13 Nov. in *Elsie Fenwick in Flanders* (1981) 183 The worst and hardest day I've had for weeks and that's saying a good deal. **1942** E. Paul *Narrow St.* vii. 50 He had with him a battery of the stuffiest lawyers in the Paris bar, and that is saying a lot. **1946** E. O'Neill *Iceman Cometh* II. 138 Sure. Harry's the greatest kidder in this dump and that's saying something! **1956** B. Holiday *Lady sings Blues* (1973) xix. 154 Fishman had been around before the concert was a sellout, you could say that for him. **1965** *New Statesman* 30 Apr. 670/1 It says a good deal for Mr Evre that he..is the one Mr Powell himself seems to have favoured most. **1969** K. Giles *Death cracks Bottle* vi. 64 The most impecunious peer in Ireland, which is saying something. **1975** *New Yorker* 1 Dec. 47/3 Houtek was a Railroad Baron and acted the part, but he liked to make others feel important too, I will say that for him.

o. *you('ve) said it*: you are absolutely right; you have got the point completely; I agree with you entirely.

1919 C. H. Darling *Jargon Bk.* 50 You said it, you said the right thing and I agree with you. **1925** E. Hemingway *Undefeated* in *This Quarter* I. II. 208 'If you stand in with Retana..you're a made man.'.. 'You said it,' the other waiter..said. 'You said it then.' **1929** E. Linklater *Poet's Pub* ii. 34 'Peace is too exciting..' said Joan. 'You've said it, Miss Benbow.' **1947** 'N. Blake' *Minute for Murder* i. 9 'What do they find?' 'Chay-oh [i.e. chaos],' replied Nigel... 'You said it.' **1970** N. Streatfeild *Thursday's Child* vii. 52 'It is a big place, there must be a lot of servants needed.'.. 'You've said it.'

p. *to say it with* (something): to express one's feelings, make one's point, etc., by the use of (that thing); *esp.* and orig. in phr. *say it with flowers*, advertising slogan of the Society of American Florists, freq. in general and *fig.* use.

1918 *Florists' Review* 3 Jan. 12/2 The slogan will be 'Say It With Flowers', and every florist who deals with the public should make that phrase a conspicuous feature of his advertising from the day the first S.A.F. page appears. **1921** I. Berlin (*song-title*) Say it with music. **1925** *New Yorker* 21 Feb. 8 (*heading*) Say it with scandal. **1928** C. Sandburg *Good Morning, America* 17 Behold the proverbs of a people, a nation... Say it with flowers. Let one hand wash the other. The customer is always right. **1932** Wodehouse *Hot Water* vi. 114 Here's this Gedge bird shoutin' about the plumbing of this Chatty-o and not saying it with flowers, neither. **1960** G. Mikes *How to be Inimitable* 33, I used to say it with flowers... More gallant, no doubt... But with *cognac* it is so much quicker. **1974** G. Mitchell *Javelin for Jonah* xiv. 175 'Why did you knife your science master?' 'We disagreed... So I say it with knives.'

q. *you can say that again*, phr. expressing whole-hearted agreement with a previous speaker's statement. *colloq.* (orig. *U.S.*).

1942 *Richmond* (Va.) *Times-Dispatch* 29 Dec. 11/5 Arthur Murray keeps in step with his hobby, Broadway idiom... If you agree [to something said] you nod and add, 'You can say that again, brother.' **1950** *Sun* (Baltimore) 1 May 12/2 The Senator wrote..that he did not 'believe that savings caused by decreases in essential services constitute constructive economy.' Senator Lehman can say that again. **1960** *Observer* 20 Mar. 10/4 *Mary*:..Andy, it's serious! *Andy*: You can say that again! **1973** *Nature* 12 Oct. 339/2 'I feel that here is an area that has not been thought out completely', he writes; he can say that again. **1974** 'E. Lathen' *Sweet & Low* xi. 102 'Everybody here is waiting for Dreyer..to put some support into this market.'.. 'You can say that again!' The fervent statement came from a total stranger. **1981** R. Barnard *Mother's Boys* vii. 70 'These teenagers are all alike, aren't they?' 'You can say that again,' snarled Lill.

3. a. *you don't say so!* (earlier and later examples); similarly *you don't say!* (orig. *U.S.*), occas. also used sarcastically; *as they say*: phr. used to mark a preceding or following expression as being proverbial or hackneyed; *if you say so*: phr. denoting acceptance of a statement or an order, usu. with grudging or placatory overtones.

1779 F. Burney *Diary* Feb. (1842) I. 183 No, you don't say so? **1842** S. Kettell *Quozziana* 14 'We shall have an explosion before long, that will shake the State of Massachusetts to its uttermost foundations.' 'You don't say so!' exclaimed I, in unfeigned alarm. **1899** R. Whiteing *No. 5 John St.* xiv. 128 You don't say so; why, I'm going to a meeting at his mother's house. **1912** Mulford & Clay *Buck Peters, Ranchman* iv. 84 'An' I could never see how he done it.' 'You—don't—say,' was Buck's thoughtful comment. **1930** A. P. Herbert *Water Gipsies* xxii. 321 Ernest, as they say, 'saw red'. **1932** L. Golding *Magnolia Street* I. x. 171 'Father, indeed!.. As much 'is father as I'm Queen Alexandra!' 'You don't say!' murmured Mr. Briggs. **1955** L. P. Hartley *Perfect Woman* xiii. 121 She lets me go, and then catches me again. It's a game, as they say. **1956** H. Kurnitz *Invasion of Privacy* iii. 30 'Okay. We've got a deal.' 'If you say so, George. Anything you say.' **1959** E. H. Clements *High Tension* iii. 40 'Didn't you have a lodger, though, some time last year?' The factor..was obviously..troubled at having told a lie. 'If you say so, Kilmorrin.' **1962** N. Marsh *Hand in Glove* ii. 67 'The Scorpion's not here, George.' 'You don't say.' Mr. Copper bitterly rejoined. **1976** J. Bingham *God's Defector* vii. 101 'You can..watch who goes in, can't you?' 'If you say so.' 'I do say so.' **1977** J. Thomson *Case Closed* iii. 43 Water under the bridge, as they say. **1979** R. Jeffries *Murder begets Murder* xiii. 83 'Heard the latest, Bert?.. That young filly was murdered.' 'You don't say, sir!'

b. *says who?*: 'who says so?', used to challenge a previous speaker's remark. Occas. with retort ¶ *says me*; cf. ¶ *says you* below. *slang* (chiefly *U.S.*). Also parenthetic phr. *shall we say* (in quot. 1973, *attrib.* with ironic force).

1914 Kipling *Let.* 15 Sept. in Ld. Birkenhead *Rudyard Kipling* (1978) xviii. 279 Much water, or shall we say much blood, has flowed under the bridges since they were written. **1931** M. Gilman *Sob Sister* x. 143 We can park a car there and spoon—says who! **1932** 'Spindrift' *Yankee Slang* 32 *Says who?*, challenge to a remark—what right have you to 'say so'? **1938** C. B. Kelland *Dreamland* vii. 86 'Miss Higg, you are guilty of reprehensible waste.' 'Says Who?' 'Says me.' **1968** *Listener* 30 May 699/1, I think the play may, shall we say, amplify light which does already exist but doesn't seem to have been noticed. **1971** *Black World* June 81/2 'I just asked.' 'Had no business asking.' 'Says who?' 'Me, stupid!' **1973** E.-J. Bahr *Nice Neighbourhood* x. 104 Joe Walsh, Jack's shall-we-say housemate. **1977** J. Crosby *Company of Friends* viii. 116 It's not one [*sc.* a news story] of ours..I read it with—shall we say, total astonishment. **1977** J. Porter *Who the Heck is Sylvia?* xvi. 151 'One should never break promises to children.' 'Sez who?'

¶ *says you*: also used in the present tense to convey doubt about, or contempt for, the remark of a previous speaker (freq. in form *sez you*). *slang* (orig. *U.S.*).

1927 Dunning & Abbott *Broadway* II. 108 Steve's a fine fellow and he's just out for some innocent fun—Says you— Says I—. **1931** *Amer. Speech* VI. 205 *Says you*, you say no, but I don't believe you. 'Says me' is the answer. **1931,** etc. [see *sez]. **1932** J. Brophy *English Prose* v. 61 Oh yeah! Says you!—an expression of scornful disbelief. **1951** Wodehouse *Old Reliable* iv. 53 Says you, if I may use a homely phrase indicating doubt and uncertainty. **1981** M. C. Smith *Gorky Park* III. iii. 328 'He's a murderer.' 'Says you.'

5. a. Delete † *Obs.* and add: In modern use: (*a*) const. *for*; (*b*) without const., the personal object being understood from the context.

1906 *Dialect Notes* III. 154 The doctor said for me not to eat a pickle. **1929** E. Hemingway *Farewell to Arms* xii. 87, I woke Georgetti, the other boy who was drunk, and offered him some water. He said to pour it on his shoulder and went back to sleep. **1934** D. L. Sayers *Nine Tailors* 72 'Why is that kept locked, Mr. Godfrey?'.. 'So Rector said to fix a lock the way they couldn't get the trap-door open.' **1946** *Publ. Amer. Dial. Soc.* VI. 26 She said for us to be there by eight o'clock. **1955** W. Denlinger *Complete Boston* I. 158 Without asking the price, the woman said to buy the dog. **1959** *Times* 20 June 7/7 Father said for Chris to take one of the lanterns. **1965** *New Statesman* 30 Apr. 687/1 On no other terms than as a parody could the book [*sc.* N. Mailer's *American Dream*] carry conviction. Its first sentence pals up with Jack Kennedy; its last paragraph includes a message from the grave from Marilyn Monroe ('Marilyn says to say hello').

10. (Further examples.) **d.** Immediately following a word or phrase to show that it represents a supposition, an instance, an approximation, or the like.

1927 *New Republic* 12 Oct. 208/1, I daresay the drummer sees no difference between Gary and, say, Newark. **1937** 'G. Orwell' *Road to Wigan Pier* vi. 100 If he were, say, an Indian or Japanese coolie, who can live on rice and onions, he wouldn't get fifteen shillings a week—he would be lucky if he got fifteen shillings a month. **1938** W. Stevens *Connoisseur of Chaos* in *Parts of World* (1942) 49 An upper, particular bough in, say, Marchand. **1940** W. Faulkner *Hamlet* I. ii. 40 In Ratliff it was that hearty celibacy as of a lay brother in a twelfth-century monastery—a gardener, a pruner of vines, say. **1944** S. Bellow *Dangling Man* 85 Little since then has worked upon me with such force as, say, the sight of a driver trying to raise his fallen horse. **1951** W. Faulkner *Requiem for Nun* iii. 231 To boil for an instant to the surface like a chip or a twig—a match-stick or a bubble, say, too weightless to give resistance for destruction to function against. **1966** *Listener* 15 Sept. 388/3 A production volume of say, 20,000 units a year. **1977** L. Meynell *Hooky gets Wooden Spoon* iii. 40 Come in about six, say. **1977** *Proc. Classical Assoc.* LXXIV. 14 In very special circumstances, you might be pressured into parenthood; say, you came from a particularly respected royal line which your subjects felt should continue.

11. a. (Earlier and later examples of *so to say*.)

1823 M. R. Mitford in *Lady's Mag.* Sept. 501/2 My flowers..withered and faded and pined away; they almost, so to say, panted for drought. **1966** *Listener* 10 Feb. 210/1 The part of the picture so to say nearest you, the foreground, the front plane, is painted to represent a doorway..which frames the main subject of the picture beyond.

c. *to say nothing of* . . .: used to refer in passing to subjects that might be used to strengthen the speaker's case; cf. *not to mention* (*so-and-so*) (Mention *v.* 1 a).

1934 Webster, *Say nothing of*, not to take into consideration (something too important to be neglected). **1962** *Home Managem.* (Homecraft Ser.) 27 Much damage is caused to dressing-table and bed-side table tops by spilled cosmetics and perfumes, to say nothing of marks..caused by that early-morning cup of tea. **1966** *Listener* 28 July 126/1 In an industry that has experienced Northcliffe, Hearst, and Beaverbrook, to say nothing of Bartholomew and Cudlipp, this seems unlikely. **1976** J. Crosby *Nightfall* xxxii. 191 Elf was her revolutionary sister-in-arms... To say nothing of her lover.

12. b. *I say* (later examples); also (theatr.) *I say, I say, I say*, formula used to introduce a joke; also as *attrib. phr. say*: for 'U.S.' read 'N. Amer.'; (earlier and later examples.)

1830 F. Trollope *Notebk.* in *Domestic Manners Americans* (1949) 427 Say! **1852** *Lantern* (N.Y.) I. 122/1 Say—d'you run with our machine? **1913** J. London *Let.* 20 Nov. (1966) 410 The galley stove kept going..and hot coffee—say! **1931** *Punch* 24 June 692 (*caption*) Patient (being shown into very modern consulting-room): 'I say, I didn't come to be operated on.' **1932** W. Faulkner *Light in August* viii. 172 Well, say. Can you tie that. **1967** *Listener* 3 Aug. 154/3 The sort of performers who, every summer up and down the coasts of England, bounce cheerfully on to a number of creaking stages, shouting 'Hello, hello, hello!' or 'I say, I say, I say!' **1968** M. Richler in R. Weaver *Canad. Short Stories* 2nd Ser. 191 The middle-aged couple alighted from the car. 'Say,' Mr Cooper said, 'you've got quite a baby here.' **1968** in Partridge *Dict. Catch Phrases* (1977) 104/2 A character, mid-stage, is interrupted by a 'comic' rushing up to him yelling 'I say, I say, I say'. First character shushes him off with 'Kindly leave the stage'; intruder persists with some fatuous question. **1969** *Listener* 6 Mar. 314/1 Making idiotic jokes—'I say, I say' jokes. **1976** *Times* 3 Feb. 14/3, I say, I've been to the ballet. **1976** P. Dickinson *King & Joker* viii. 114 They..grinned inanely with heads bent..and legs in the pose of a comedy routine duo. 'I say I say I say,' said Louise, 'your public face isn't as good as mine, darling.'

d. *I'll say*: used to denote enthusiastic assent (either *absol.* or with object or dependent clause). Also *I'll say so*.

1924 *Dialect Notes* V. 276 *Say*: I'd —, I'll — (both approv.). *Ibid.* 277 *So*:..I'll say — (agreement). **1926** *S.P.E. Tract* XXIV. 123 *I'll say it is*, it's my opinion, certainly. **1926** Maines & Grant *Wise-Crack Dict.* 10/1 *I'll say so*, emphatic agreement. **1943** N. Marsh *Colour Scheme* vi. 99 'Does he want to keep him quiet?'.. 'I'll say! Too right he wants to keep him quiet.' **1945** P. Cheyney (*title*) I'll say she does. **1954** E. McLeod tr. *Colette's Vagabond* I. iv. 35 'Hullo, Stephen! Good house?' 'I'll say!' **1960** N. Hilliard *Maori Girl* 93 'Do you miss home much?' 'I'll say. Not so much now, though.' **1972** G. Durrell *Catch me a Colobus* v. 95 Would we, by any chance, be interested in a pair of leopards? 'I'll say we would! Why? Do you know where there are some?' **1974** S. Woods *Done to Death* 218 'You've taken what might have been a knock down blow with a good deal of courage.' 'I'll say she has,' said Hugh. **1979** 'J. le Carré' *Smiley's People* (1980) iv. 53 'He was a declining asset, as all ex-agents are.'.. 'I'll say,' said Strickland *sotto voce*.

‖ **saya** (sa·ya). [Sp.] In Spain and Spanish-speaking countries, a dress or outer garment worn by women (see quots.).

1841 G. Borrow *Zincali* I. II. v. 305 This female Gypsy fashion..is more properly the fashion of Andalusia, the principal characteristic of which is the saya, which is exceedingly short, with many rows of flounces. **1845** R. Ford *Hand-bk. Spain* I. ii. 196 This male *sagum* is the type of the modern *saya*, Arabic *sayah*, a long outer garment, which is always black, and is put over the indoor dress on going out. **1846** —— *Gatherings from Spain* xxiii. 323 The transparent, form designing *saya* of the lady, heightens the charms of a faultless symmetry which it fain would conceal. **1857** C. M. Yonge *Dynevor Terrace* II. ix. 132 A full dark purple satin skirt..was plaited low on the hips, and girded loosely with a brightly striped scarf. The head and upper part of the person were shrouded in a close hood of elastic black silk webbing, fastened behind at the waist, and held over the face by the hand... 'Ah, you found me out,' cried Rosita... 'I have the like *saya y manto* ready for you. Come, we will be on the Alameda [in Lima].'

sayable, *a.* Add: **B.** *sb.* That which can be said; a statement which it is possible to make.

1937 *Essays & Studies* XXII. 136 The meanable crystallized and fixed in the sayable. **1957** G. Ryle in M. Black *Importance of Lang.* (1962) 169 It is the foreign relations, not the domestic constitutions of sayables that engender logical troubles and demand logical arbitration. **1969** J. S. Cunningham *Powers that Be* 3 Infant sayables Along the seamy permeable Undersides of words.

Saybolt (sē¹·bǫlt). The name of George M. *Saybolt* (d. 1924), U.S. chemist, used *attrib.* to designate an apparatus he invented for measuring the kinematic viscosity of liquids, esp. oils, by measuring the time taken by a fixed quantity of liquid to pass through a standard capillary tube under specified conditions; so *Saybolt viscosity*, the viscosity so measured, usu. expressed as *Saybolt seconds*.

1886 B. Redwood in *Jrnl. Soc. Chem. Industry* 29 Mar. 124/2 The viscometer designed by Mr. G. M. Saybolt, inspector to the Standard Oil Company of New York, is before you... The Saybolt viscosimeter is not,..as at present constructed, suitable for use at very high temperatures. **1925** A. B. Thompson *Oil-Field Exploration & Development* I. xi. 520 The viscosity of oils is generally measured in one or other of three types of instruments, the Engler, Redwood and Saybolt viscosimeters, or viscometers. **1955** Kirk & Othmer *Encycl. Chem. Technol.* XIV. 763 Viscosity is normally reported simply as Saybolt Universal or Saybolt Furol seconds, but it may be converted to centistokes by means of appropriate tables. **1968** Snell & Hilton *Encycl. Industr. Chem. Analysis* VI. 295 The Saybolt Furol viscosity of a bituminous material is the time, in sec, that it takes 60 ml of sample to flow through the calibrated Furol orifice of the Saybolt viscometer tube, measured under carefully controlled conditions.

sayee (sē¹,ī). *rare.* [f. Say *v.*¹ + -ee¹.] A person to whom something is said.

a **1902** S. Butler *Ess. Life* (1904) 183 It takes two people to say a thing—a sayee as well as a sayer... The belief on A.'s part that he had a *bonâ fide* sayee in B., saves his speech *quâ* him, but it has been barren and left no fertile issue.

‖ **sayonara** (sayōnara, saiŏnā·rā). [Jap.] Good-bye. As *sb.*, a farewell, a leave-taking; also *attrib.* As *v. trans.*, to say 'sayonara' to.

1875 *Colburn's United Service Mag.* Oct. 185 'Sionara!' (good bye), is your answer. **1880** *Golden Days for Boys & Girls* 3 Apr. 71/4 After this speech they all cried: 'Sayonara (farewell), Momotaro!' **1892** Kipling *Lett. of Travel* (1920) 51 A traveller who has been 'ohayoed' into half-a-dozen shops and 'sayonaraed' out of half-a-dozen more. **1908** Lady R. Churchill *Reminiscences* (1973) xiii. 252 Many *sayonaras* were exchanged. **1910** *Pacific Monthly* XXIII. 259/2 He is a bad man. You go away! *Sayonara*! **1952** T. J. Mulvey *These are your Sons* vii. 146 The Sisters had arranged the children in the stiff and formal formation for the '*sayonara*'. **1965** *This is Japan 1966* 106 The Honourable Sex Shop then rescued me from an embarrassing and even disastrous *sayonara* at Kobe. **1972** *Mainichi Daily News* (Japan) 6 Nov. 7/4 The International Camera Club of Japan will hold a special Sayonara party for outgoing Chairman John Thorpe, Tues., Nov. 8. **1977** J. Wambaugh *Black Marble* (1978) iii. 25 If I ever knew for sure what I suspect about you, Philo, it'd be sayonara, baby.

Say's law (sē¹z lǭ). *Econ.* [f. the name of Jean Baptiste *Say* (see below).] The theory propounded by the French economist Jean Baptiste Say (1767–1832) that supply creates its own demand.

[**1817** D. Ricardo *Princ. Polit. Econ.* xix. 401 Is the following quite consistent with M. Say's principle?] **1934** *Encycl. Social Sciences* XII. 351/1 At any given time in one market area there could be only a single price relationship of each good to any other good there offered in exchange, so long as competition held sway. This tendency involved the fixation not only of an identity of offering price among different sellers of the given good to the same (prospective) buyer but also of an identity of the offering price to different (prospective) buyers of the given good from the same seller. This is the substance of what has been denominated Say's law. **1936** J. M. Keynes *Gen. Theory Employment* iii. 26 Say's law, that the aggregate demand price of output as a whole is equal to its aggregate supply price for all volumes of output. **1969** *Daily Tel.* 21 Apr. 14/2 If Say's Law had been true there could have been no unemployment. **1972** T. Sowell *Say's Law* i. 3 The idea that supply creates its own demand—Say's Law—appears on the surface to be one of the simplest propositions in economics.

say-so, *sb.* Delete 'Now *dial.* and *U.S.*' and add further examples. Also in extended senses: (*a*) an affirmation or assertion; (*b*) authority, authorization; (*c*) the right of consultation, a 'voice' (in some decision). Freq. in phr. *on the say-so of* (a person): according to, on the authority of (that person).

1824 *Niles' Reg.* 10 Apr. 84/2 The whole number of republican members in 1824 (on the *say-so* of Messrs. Gales and Seaton) is 216. **1902** W. N. Harben *Abner Daniel* 5, I think I've got a right..to have a say-so in this kind of a trade. **1902** S. E. White *Blazed Trail* 195 In questions of policy mine is the say-so every trip. **1924** 'W. Fabian' *Sailors' Wives* xvi. 186 'Give 'em to me.' 'Not without Bob's sayso.' **1937** *N. & Q.* CLXXII. 305/1 The labour of scholars is nothing to them; they prefer the say-so of some casual maker of legend long since disproved. **1947** J. Mulgan *Report on Experience* v. 53, I expect major-generals and upwards have a good deal of say-so. **1956** D. Meadows *Eliz. Quintet* iii. 204 Perhaps he truly believed the rest of the story—on the say-so of his assistant. **1967** *Boston Sunday Herald* 30 Apr. (Bedding Suppl.) 9/1 Another place to give your husband more say-so is in the selection of a bed. **1978** S. Brill *Teamsters* ii. 69 Giacalone and Provenzano had set up the meeting and the subsequent murder on the say-so of higher-ups in organized crime.

‖ **saz** (saz). [Turk., ad. Pers. *sāz* musical instrument.] A stringed instrument similar to the tamboura, found in Turkey, North Africa, and the Near East.

1870 C. Engel *Descr. Catal. Musical Instr. in S. Kensington Museum* 26 *Saz*, a small kind of *tamboura*... The *saz* is chiefly used by the *Sho'ara*, i.e. 'Poets', who are itinerant musicians and bands of the Mussulmans. *Double saz*, inlaid with various woods and mother-of-pearl. **1918** A. A. Stanley *Catal. of Stearns Collection of Musical Instr.* IV. 150 *Saz*, tanbur type... Algeria... Slender neck with two flat heads. Wire strings. *Ibid.* 155 *Saz*... Egypt. Pear-shaped body of some soft wood. **1957** T. Slessor *First Overland* iv. 48 Umtaz played his *saz* and sang. It was an instrument like an Elizabethan mandolin, and gave a strumming, jangling accompaniment to the folk-songs. **1969** J. Rathbone *With my Knives I know I'm Good* xiii. 103 We.. sang Turkish and Russian songs to a saz which he played well. **1977** *Early Music* July 437/1 The Early Music Consort of Melbourne (medieval harp, saz, lute, rauschpfeife, [etc.]). **1980** M. Bar-Zohar *Deadly Document* ix. 158 The music of the strolling accordion and *saz* players.

‖ **saza** (sa·za). [ad. Luganda *'ssaza.*] In Uganda, an administrative area; a county.

1950 *Times* 13 Feb. 3/2 Exempted from payment are those persons who, in the opinion of the Resident, are entirely free from blame—that is to say, non-natives and the inhabitants of eight sazas which throughout the disturbances were trouble-free. **1955** *Times* 25 Aug. 6/5 Mr. Kintu is a saza chief. **1958** E. Winter in Middleton & Tait *Tribes without Rulers* 158 Toro as a whole, is divided into seven large administrative areas called sazas which may be translated as counties. **1964** C. Willock *Enormous Zoo* iv. 58 The *saza* chief—an important man in the local [Ugandan] hierarchy—turned out a posse.

Sazarac (sæ·zəræk). orig. and chiefly *U.S.* Also **Sazerac.** [Origin unknown.] A cocktail consisting of whisky, pernod or absinthe, bitters, and syrup, served usu. with a slice of lemon. Also *attrib.*, as *Sazarac cocktail*.

1941 *Louisiana: Guide to State* (Writers' Program) 230 The most celebrated of New Orleans cocktails—the Sazerac—is a mixture of whisky, bitters, and sugar, served in a glass mixed with absinthe. **1946** C. H. Baker *Gentleman's Companion* II. 122 The best drinks produced in New Orleans stick to the ancient simple formula—and please, please, never try to vary it; for if you do you'll not be drinking a true Sazarac. **1958** E. Dundy *Dud Avocado* I. i. 18 So many marvellous new drinks..sazaracs and slings and heaven knows what else. **1961** F. Crane *Reluctant Sleuth* viii. 66 Regan had liked Sazarac cocktails. **1963** M. Malim *Pagoda Tree* xxii. 145 Then came the sazaracs. I remember having a word with Canthrop B beside the bar quite early on. He took charge of the bar, to superintend the mixing of this Fine Old Southern cocktail. 'I'm doubling up on the absinthe' he said gleefully. **1978** G. Vidal *Kalki* iv. 93 We had each polished off a pair of Sazerac cocktails, a local killer [in New Orleans].

‖ **S-bahn** (e·s bān). [Ger., abbrev. of (*stadt*) *schnellbahn* (urban) fast railway.] In some German cities. a fast (sub)urban railway line or system.

1962 I. Fleming *Living Daylights* in *Octopussy* (1966) 79 Feeling more encouraged, he took the S-Bahn back into the city. **1974** *Encycl. Brit. Macropædia* II. 851/2 East Berlin runs the S-Bahn..elevated railway system started in 1871 as a connecting system to a rail net in and out of the city. **1976** P. R. White *Planning for Public Transport* iv. 74 The creation of links across city centres..to enable S-bahn trains to offer better accessibility within the central area. **1980** G. Seymour *Contract* v. 63 We should take the U-Bahn to Alexander Platz, then the S-Bahn.

S-band: see *S 12. **S-bend:** see *S 2 c.

scab, *sb.* Add: **4. b.** (Earlier and further examples.) Also, in extended uses: a person who refuses to join a strike or who takes over the work of a striker; a blackleg; a strike-breaker.

1777 *Bonner & Middleton's Bristol Jrnl.* 5 July, To the Public. Whereas the Master Cordwainers have gloried, that there has been a Demur amongst the Men's and Women's Men;—we have the Pleasure to inform them, that Matters are amicably settled... The Conflict would not been [*sic*] so sharp had not there been so many dirty Scabs; no Doubt but timely Notice will be taken of them. **1792** in A. Aspinall *Early Eng. Trade Unions* (1949) 84 What is a scab? He is to his *trade* what a traitor is to his *country*... He first sells the journeymen, and is himself afterwards sold in his turn by the masters, till at last he is despised by both and deserted by all. **1806** *Trial of Boot & Shoemakers* (Federal Soc. Journeymen Cordwainers, U.S.) 74, I concluded at that time I would turn a *scab*, unknown to them, and I would continue my work and not let them know of it. **1889** C. H. Salmons *Burlington Strike* 259 The man who takes the place of another when that other engages in a struggle with a corporation, is a 'scab'. **1903** W. T. Mills *Struggle for Existence* xxxv. 493 The 'scab' is no longer the unorganized and hungry worker, waiting at the factory gate. **1926** [see *bitter-ender* s.v. *bitter- *a.* and *adv.*]. **1938** *Sun* (Baltimore) 8 Sept. 3/1 He had instructed pickets not to call non-strikers 'finks' or 'scabs' or other epithets. **1974** *Socialist Worker* 26 Oct. 13/2, 180 women walked out. But 70 stayed in... The scabs soon found out what it was like to be hated.

attrib. and *Comb.* **1850** *Morning Chron.* 11 Feb. 5/6 Having thus given the characteristics and conditions of the 'legal', or honourable trade, I next turn my inquiry to the state of the labouring men, women, and children employed by the slop-masters, who are distinguished from the 'wages' (or legal) shops by the terms '*illegal*', '*scab*', or '*slaughtershop*' *keepers*. **1926** *Socialist Rev.* June 10 The Labour Press.. cannot descend to 'scab' printing. **1940** M. Lowry *Let.* 7 May (1967) 31 Two years as a scab lavatory attendant in Saskatchewan. **1958** *Spectator* 15 Aug. 225/2 British writers, forced to become scab-labour, are undermining it, completely against their wishes. **1977** C. McCullough *Thorn Birds* iii. 60, I suppose some scab contractor undercut me.

5. scab weed *N.Z.*, a low-growing plant of the genus *Raoulia*, adapted to poor conditions.

1927 L. Cockayne in R. Speight et al. *Nat. Hist. Canterbury* 143 *Raoulia lutescens*... Scabweed. **1933** *Discovery* Sept. 292/1 The bare land patchily covered by flat 'scab weed' looks horribly diseased. **1955** J. K. Baxter *Fire & Anvil* iii. 78 It survives many droughts..like the scabweed in Central Otago.

scab, *v.* Add: **3. a.** (Earlier and later examples.) Also with *it* as quasi-obj. and *trans.* in phr. *to scab a job*: to perform, or employ another to perform, the job of a striking worker.

1806 *Trial of Boot & Shoemakers* (Federal Soc. Journeymen Cordwainers, U.S.) 75 Their business was to watch the *Jers* [*sc.* journeymen] that they did not *scab* it. **1889** C. H. Salmons *Burlington Strike* 357 The men..declared that they had never scabbed a day in their lives. **1895** *Rep. on Chicago Strike June–July, 1894* (U.S. Strike Commission) 308 If there is a strike ordered I will be damned if I am going to scab. **1898** *Scribner's Mag.* Oct. 445/2, I won't scab any man's job. **1932** E. Wilson *Devil take Hindmost* xxi. 223 Several speakers protest..that the companies only want to get them out so that they can scab the job. **1969** *Times* 30 Oct. 10/7 Frantic calls to friends.. summoned..a driver who was prepared to scab as a special favour. **1969** *Daily Tel.* 16 May 27/7 Peaceful pickets outside all entrances will discourage all students from scabbing on the strike.

b. *trans.* To treat or label (a person or a firm employing scab labour) as a scab; to ostracize (a person who is a scab). *rare.*

1806 *Trial of Boot & Shoemakers* (Federal Soc. Journeymen Cordwainers, U.S.) 73 They told me if I did not come to the body, I was liable to be *scabb'd*. *Ibid.* 77 In a little time after this his shop was scabbed. **1888** *Montreal Daily Herald* 21 Feb. 1/5 Engineers and others who refused to hoist or handle coal during the late effort to 'scab' the collieries. **1922** F. B. Young *Pilgrim's Rest* vi. 409 [The rioting strikers] went away, saying they'd come back again and scab us to-night.

scabbard, *sb.*¹ Add: **1. a.** Also, a sheath in which a rifle, submachine gun, or similar firearm is kept.

1923 *Dialect Notes* V. 220 *Scabbard*, holster, any leather sheath for a weapon. **1941** E. Hemingway *For Whom Bell Tolls* xxi. 264 From the scabbard on the right of his saddle projected the stock and the long oblong clip of a short automatic rifle. **1979** *Navajo Times* (Window Rock, Arizona) 24 May 19/2 (Advt.), Truck seat cover. Rifle scabbard & map pouch!

scabbing, *vbl. sb.* Add: **3. a.** The action of Scab *v.* 3 in Dict. and Suppl.; refusal to strike on the part of a worker or employment of scab labour by a firm. Also *fig.*

1944 *Sun* (Baltimore) 28 Nov. 9/2 The worker who strikes while the war is on is guilty of scabbing. **1956** *Ibid.* 2 Feb. 18/1 Under a PSC order..the company would not be in a position of 'scabbing'. 1973 Telegraph (Brisbane) 28 July 5/2 In trades union circles the deadliest of sins is 'scabbing' while your union is on strike.

scabby, *a.* Add: **1. a.** *scabby mouth* (Austral. and N.Z.), a viral disease of sheep characterized by ulceration around the mouth.

1938 J. R. Greig et al. Hutyra's Special Path. & Therapeutics (ed. 4) 579 (heading) Lip and leg ulceration, scabby mouth. 1950 N.Z. Jrnl. Agric. Aug. 100/2 On farms where scabby mouth occurs each new crop of lambs should be vaccinated, and this is most conveniently done at marking. 1966 V. G. Cole Dis. Sheep 217 Scabby mouth can be transmitted to the hands of persons handling affected sheep.

scabies. Now always with pronunc. (skē[1]·bīz).

scabland (skæ·blænd). *U.S. Physical Geogr.* [f. Scab *sb.* + Land *sb.*] Flat, elevated land consisting of igneous rock with a patchy covering of poor, thin soil and little vegetation, and deeply scarred by channels of glacial or fluvioglacial origin; *spec.* that forming part of the Columbia Plateau, Washington State, U.S.A. Freq. *pl.*

1923 J. H. Bretz in Bull. Geol. Soc. Amer. XXXIV. 577 The 'Scablands' are lowlands among the groups of 'Palouse Hills', plane in a general way, but diversified by a multiplicity of irregular and commonly anastomosing channels and rock-basins eroded in basalt, and containing meadows, swamps, and lakes... The local name refers to the absence of soil over much of these tracts, the basalt outcropping in ledges and over considerable level areas. 1923 — in Jrnl. Geol. XXXI. 617 The terms 'scabland' and 'scabrock' are used in the Pacific Northwest to describe areas where denudation has removed or prevented the accumulation of a mantle of soil, and the underlying rock is exposed or covered largely with its own coarse, angular débris. Ibid. 620 The channeled scablands are the erosive record of large, high-gradient, glacier-born streams. 1943 Science 10 Sept. 229/1 The rock basins of the scablands are found in the wider channels particularly, and rock basins are an almost universal feature of glaciated regions. It is the channels with their included low mesas which are the unique feature of the scablands. 1956 C. Relander Drummers & Dreamers 235 The River People were virtually forgotten in their deep desolation of sagebrush, basaltic cliffs, raw umber hills and scabland. 1966 N.Z. Jrnl. Geol. & Geophysics IX. 130 (heading) Antarctic scablands. 1976 C. L. Matsch N. Amer. & Great Ice Age vi. 74 At peak stage Lake Missoula had a surface area of about 7,500 km^2 and contained an estimated 2,000 km^3 of water. All this water is thought to have discharged westward in a matter of a few days... This great flood moved boulders with diameters greater than 10 m and scoured a system of coulees across the Columbia Plateau. This great tract of flood-eroded topography is called the channeled scablands.

scabrous, *a.* Add: Now freq. with pronunc. (skæ·brəs). **1. c.** Encrusted, begrimed. Chiefly *U.S.*

1939 Listener 19 Jan. 157/1 A once bewitching villa, now scabrous, awaits the knacker. a 1961 J. Reynolds in Webster s.v., [The] shell of the house is scabrous with lichen and mildew. 1962 P. H. Johnson Error of Judgement xxxiii. 240 In this early glow, the tattered and scabrous paintwork on the porticos looked like a covering of dead leaves, ivy, or virginia creeper, brittle at the end of autumn. 1967 T. Keneally Bring Larks & Heroes ii. 16 In its [sc. a hut's] bay of scabrous timber, it was altogether a poor comment on Halloran's vehemence. 1969 N.Y. Rev. Books 2 Jan. 14/1 Trudging over countless guts of cement that ran like slag in Gehenna, I stuffed my scabrous shoes with newspapers.

4. (Later examples.) Now freq. used in various extended senses: nastily abusive, disgusting, repulsive.

Cf. quot. 1862 under sense 1 b.

1951 M. Kennedy Lucy Carmichael II. i. 79 One shouldn't believe a word Emil says. I ventured to ask them.. about Terrific Charles, because Emil is always particularly scabrous about him. 1969 N.Y. Rev. Books 16 Jan. 32/4 Without going into scabrous detail, might he not have given us just a teeny hint as to *why* 'the experience convinced me the union was indeed for decentralization'? 1973 Times 24 May 19/1 [Scandals] create hysteria because they appeal to a scabrous and irrational element in the human mind. 1979 London Rev. Bks. 25 Oct. 10/1 His propaganda pieces grow more outrageously scabrous.

scabrously, *adv.* (Later example.)

1977 N.Y. Rev. Books 14 Apr. 8/2 The first of the book's three sections, in which a non-existent and uninhabited Ibansk is carefully and at times scabrously described.

scacchic (skæ·kik), *a. rare.* [f. It. *scacchi* chess + -ic.] Of or pertaining to chess.

1860 in W. Fiske Chess Tales (1912) 159 Stern old fellows were these scacchic sages! They considered the laws of chess as inviolable as those of the Medes and Persians. Ibid. 163 Since first the scacchic art was brought from the land of India. 1959 Information Bull. Libr. Congress 27 Apr. 238 The Chess Club has elected the following officers to guide its scacchic destinies during 1959.

scad[7] (skæd). *colloq.* (orig. *U.S.*). Also **skad.** [Origin unknown.] **a.** A dollar. Usu. *pl.* in sense 'money'.

1858 Hutching's Mag. Aug. 85/2 Why he seed Bill and lifted him two scads. 1884 E. W. Nye Baled Hay 59 We have mercenary motives... We desire the scads. 1902 W. Harben Abner Daniel ix. 70 Ef he kin possibly raise the scads to pay the tax. 1909 Amer. Mag. Nov. 1 This land of our dads..is a dinger at nailing the scads. 1933 J. V. Allen Cowboy Lore IV. 154 He would deal for you both day and night Or as long as he had a scad. 1959 E. Pound Thrones xcvii. 22 Canute opposing Byzantium, 20 scads to the dinar, 100 scads to the mark (of accountancy).

b. Chiefly *pl.* A large amount; 'heaps'.

1869 Overland Monthly III. 131 A Texan never has a great quantity of anything, but he has 'scads' of it..or 'Scadoodles'. 1904 W. H. Smith Promoters ii. 52 What did England do when she found she could raise scads of opium in India, but had no market for it? 1923 M. S. Watts Luther Nichols II. iv. 214 The old girl surely did have it—scads of it. 1931 E. Linklater Juan in Amer. II. xvi. 176 And the pay? Skads of dough. Oodles and oodles of money. 1950 O. Nash Family Reunion 89 There's a scad o' things that to make a house a home it takes. 1956 'N. Shute' Beyond Black Stump x. 297 It's water... Skads and skads of it, under Lucinda Station. Clear, cool water. 1977 D. Bagley Enemy xv. 121 He's installed a scad of microprocessors in that control board. 1980 Telegraph (Brisbane) 9 Apr. 53/6 They supply pay envelopes in scads to clients.

scaf. Restrict † *Obs.* to sense 2 and add: Also **scaffie, scaffy, scaph, skaffie. 1.** (Later examples.) Also *attrib.* in *scaffy boat.* Now *Hist.*

1781 Aberdeen Jrnl. 29 Oct., A large boat or scaff was put ashore two miles to the eastward of this place. 1877 E. W. H. Holdsworth Sea Fisheries 168 The Buckie boats, known as 'Scaffs' or 'Scaffy boats', are of an entirely different build from the other Scotch craft. 1906 H. W. Smyth Mast & Sail v. 100 From Portsoy westwards along the Banff and Moray coasts, and round the eastern seaboard of Ross-shire, until within the last forty years, the 'Skaffie' or 'Buckie Skaffie', as it was often known, was universally used in the herring fishing. Ibid. 436 Skaffie, or scaith, a type of Scotch lugger with raked stem and stern posts, used principally on the coastline between Frasersburgh and Dornoch, and apparently of Norse origin. 1914, 1927 [see *Fifie, fifie]. 1959 Banffshire Jrnl. 6 Jan. 4 Open sailing boats, lug-sail rigged, called herring luggers or more familiarly 'Scaffies'.

scaffold, *sb.* Add: **B. 1. a.** (Later *fig.* example.)

1889 G. M. Hopkins Poems (1967) 107 But man—we, scaffold of score brittle bones.

7. Also, a framework upon which tobacco is dried.

1784 J. Smyth Tour U.S.A. II. 134 When the tobacco plants are cut and brought to the scaffolds. 1886 C. G. W. Lock Tobacco 75 Some prefer hanging the tobacco on scaffolds in the field until it is ready to be put in the barn and cured by the fire. 1888 Encycl. Brit. XXIII. 424/2 Red shipping qualities [of tobacco] are prepared by leaving the cut stems either in the field or hung on scaffolds in the barns for a few days to wilt and wither in the air.

scaffy (ska·fi). *Sc. colloq.* Also **scavvy.** [dim. of Scavenger *sb.*] A street sweeper; a dustman. Also *attrib.*

1853 W. Blair Chron. Aberbrothock 19 Hecklers, an' wabsters, an' baxters, an' scaffies, an' wives, an' bairns, dowgs an' cats. 1876 J. Smith Archie & Bess 25 Scaffies and leeries crackin' like pea-guns. 1892 W. M. Adamson Betty Blether's Corr. 74 Tin cans intendit for the scaffy cairt. 1918 Kelso Chronicle 1 Nov. 2 She often is too late for the Scaffy Bucket. 1931 J. Hall Holy Man iii. 37 Geordie, the road scavvy, was wearily trundling his little hand-cart up the steep slope of the village street. 1933 J. Gray Lowrie 41 Dere's da scaffy fur takkin awa ony coarn o' bruck an' ess. 1967 Buchan Observer 7 Feb. 2 Not up in the morning early enough to catch the 'scaffy cairt'. 1978 Scotsman 30 June 10/7 The scaffies are now under the Environmental Health Department.

scaffy: see *scaf.

scag (skæg). *U.S. slang.* Also **skag.** [Origin unknown.] **1.** A cigarette; a cigarette stub.

1915 Dialect Notes IV. 235 Scag, cigarette stub. 1928 Amer. Speech III. 454 Skag, a cigarette; to smoke. 1936 Nat. Geogr. Mag. LXIX. 778/2 A cigarette is a 'skag' (and cadets may not smoke in public).

2. Heroin.

1967 'G. Bagby' Corpse Candle (1968) ix. 121 Acid, grass, skag? 1973 E. Bullins Theme is Blackness 152 Most of the guys that we usta swing with are gone, man. In jail, on wine or scag. Ibid. 157 This scag they been sellin' me lately makes me hear funny. 1976 R. Condon Whisper of Axe I. iv. 18 Addicts, prostitutes, skag merchants..the amoral and the lost. 1977 N. Adam Triplehip Cracksman xiii. 138 I'm no junkie myself, never touched the scag, never even used the White Dragon Pearl.

scala (skē[1]·lă). *Anat.* [L., = 'ladder'.] Each of two passages (the *scala tympani* below and the *scala vestibuli* above) into which the spiral tube of the cochlea is divided by a bony spiral lamina and which communicate at the apex of the spiral; also, the *scala media* or central duct of the cochlea, situated between these two passages and shut off from them by two membranes.

1712 Bibliotheca Anat., Med., Chir. II. 214/2 We have discover'd two Channels into which the Cochlea is divided by the Septum, called Scalæ, or Ladders; one of wnich..is called the Scala Tympani: But the other..is called the Scala Vestibuli. 1803 C. Bell Anat. Human Body III. 430 Or it [sc. the vibrating motion] must pass from the scala vestibuli into the scala tympani. 1872 Gray's Anat. (ed. 6) 596 The space between the membrane of Reissner and membrana basilaris is generally described as the Scala media. Canalis membranacea, or Canalis cochleæ, and this is the nomenclature which will be used here. 1902 D. J. Cunningham Text-bk. Anat. 717 The two scalæ communicate with each other through the opening of the helicotrema at the apex of the cochlea. 1945 [see *Reissner]. 1974 Encycl. Brit. Macropædia V. 1123/2 The interior of the cochlea is divided longitudinally into three spiral ramps or scalae: the scala vestibuli,..the scala tympani,..and the scala media.

scalable, *a.* Restrict *rare* to sense in Dict. and add: **II. 2.** Capable of being measured or graded according to a scale.

1936 Psychol. Monogr. XLVII. i. 15 A few [traits] seem common enough to be regarded as comparable from one individual to another. These might be called common or scalable traits. 1944 Amer. Sociol. Rev. IX. 147/1 The example..of desire to go to school is a fictitious version of data that have actually proved scalable for the Army. 1968 W. A. Scott in Lindzey & Aronson Handbk. Social Psychol. (ed. 2) II. xi. 222 A measure of the degree to which the set of items is scalable, that is, represents a unidimensional attribute. 1977 R. H. Brown in Douglas & Johnson Existential Sociol. ii. 81 The questionnaire was particularly attractive as a measuring device because it standardizes responses, making them easily scalable and retrievable, in principle by anyone.

3. Capable of being changed in scale. *rare.*

1977 Jrnl. R. Soc. Arts CXXV. 770/1 Such lasers are scaleable since large volumes could be pumped uniformly.

Hence **sca·lability,** the property of being scalable.

1944 Amer. Sociol. Rev. IX. 141/2 It may well be that the formal analysis for scalability may help clarify uncertain areas of content. 1959 Psychol. Rev. LXVI. 51/2 Ordinarily, when scalability is found, it is assumed that a unidimensional continuum exists. 1960 Brown & Gilman in J. A. Fishman Readings Sociol. of Lang. (1968) 270 We tested all 28 items for scalability and found that a subset of them made a fairly good scale. 1978 Sci. Amer. Nov. 44/2 It took demonstrations of the scalability of the technology and tests of improved beam focusing..to catalyze an effort that led to support by the AEC.

scalar, *a.* and *sb.* Add: Now usu. with pronunc. (skē[1]·lăɹ). **A.** *adj.* **2.** (Earlier and further examples).

1846 W. R. Hamilton in Phil. Mag. XXIX. 26 The algebraically real part may receive, according to the question in which it occurs, all values contained on the one scale of progression of numbers from negative to positive infinity; we shall call it therefore the scalar part, or simply the scalar of the quaternion, and shall form its symbol by prefixing, to the symbol of the quaternion, the characteristic Scal., or simply S. 1911 Encycl. Brit. XXVII. 962/2 The mass of a body, the pressure of a gas, the charge of an electrified conductor, are instances of scalar magnitudes. 1932 R. Gans Vector Analysis ii. 58 Let W..be a scalar..property of a field, and let it be regarded as a function of the position and of the time. 1964 N. N. Hancock Matrix Analysis Electr. Machinery ii. 18 A 'scalar' matrix is a diagonal matrix in which all the elements on the principal diagonal are equal.

3. Of or pertaining to a musical scale (Scale *sb.*[3] 4).

1928 G. Cooke Theory of Music ii. 18 One cannot.. over-emphasise the importance of these groups of notes in the theoretic study of scalar development. Ibid. vi. 77 The variety inherent in modulation and scalar variety. 1946 R. Blesh Shining Trumpets ii. 25 The basic material is recast in its scalar compass and its tonal intervals. 1959 M. T. Williams Art of Jazz (1960) xi. 106 Sliding tones peculiar to the scalar and harmonic structure. 1966 New Statesman 11 Feb. 204/1 The integration of triadic and scalar elements within a serial or non-tonal field.

4. Of or pertaining to a graduated scale (Scale *sb.*[3] 9).

1959 G. D. Mitchell Sociology 130 Very often there is an identity of functional and scalar status. 1974 G. Leech Semantics ii. 21 A selection from indefinitely many possible scales, which in any case would only provide for associative meaning in so far as it is explicable in scalar terms.

B. *sb.* **1.** (Earlier and later examples.) More widely, a quantity having magnitude but no direction, and representable by a single real number.

1846 [see sense 2* of the adj.]. 1903 [see *non-directional a. (sb.)]. 1932 R. Gans Vector Analysis i. 2 We shall denote scalars by ordinary type and vectors by heavy type. 1965 Patterson & Rutherford Elem. Abstr. Algebra v. 145 By a scalar we shall mean an entity determined by a single real number and by a vector we shall mean an entity determined by both a positive real number, measuring magnitude, and a direction in space.

2. *attrib.* (some of the following may be regarded as collocations of the adj.): **scalar field,** a map from a space to the real line (see quot. 1932); **scalar function,** a function whose value is a scalar; **scalar multiplication,** multiplication of a vector by a scalar to give another vector; **scalar product** = *inner product* s.v. *inner *a.* (*sb.*[2]) 1 k; **scalar triple product,** a scalar function of three three-vectors $((a_1, a_2, a_3), (b_1, b_2, b_3), (c_1, c_2, c_3))$ which can be calculated as $(a_1 b_2 c_3 + b_1 c_2 a_3 + c_1 a_2 b_3 - a_1 c_2 b_3 - b_1 a_2 c_3 - c_1 b_2 a_3)$, being the volume of the parallelepiped which has the three vectors as three coincident edges.

1932 R. Gans Vector Analysis i. 1 The field is called a scalar field or a vector field according as the quantity

associated with the field is a scalar or a vector. 1959 M. R. Spiegel Vector Analysis i. 3 The temperature at any point within or on the earth's surface at a certain time defines a scalar field. 1974 G. Reece tr. Hund's Hist. Quantum Theory xv. 207 It was therefore a major advance when Pauli and Victor Weisskopf developed the quantum theory of a scalar field. 1956 A. A. Townsend Struct. Turbulent Shear Flow iii. 35 The double-correlation function depends only on a single scalar function. 1972 A. G. Howson Handbk. Terms Algebra & Anal. xxvi. 126 Functions such as *f* are often referred to as vector-valued functions and are denoted by symbols printed in bold type so as to distinguish them from real-valued or scalar functions. 1901 Gibbs & Wilson Vector Analysis i. 13 The laws which govern addition, subtraction, and scalar multiplication of vectors are identical with those governing these operations in ordinary scalar algebra. 1968 A. P. Armit Advanced Level Vectors ii. 25 (heading) Scalar multiplication of a vector..in terms of cartesian components. 1878 Scalar product [see Vector 2]. 1932 R. Gans Vector Analysis i. 17 By the scalar product of two vectors **A** and **B** we mean a scalar of magnitude equal to the product of the absolute values and the cosine of the angle between the vectors. 1941, 1968 Scalar product [see *inner a. (sb.[2]) 1 k]. 1901 Gibbs & Wilson Vector Analysis ii. 68 The second triple product is the scalar product of two vectors, of which one is itself a vector product, as $\mathbf{A}\cdot(\mathbf{B}\times\mathbf{C})$ or $(\mathbf{A}\times\mathbf{B})\cdot\mathbf{C}$. This sort of product has a scalar value and consequently is often called the scalar triple product. 1959 M. R. Spiegel Vector Analysis ii. 17 The product $\mathbf{A}\cdot(\mathbf{B}\times\mathbf{C})$ is sometimes called the scalar triple product or box product and may be denoted by [**ABC**]. 1964 E. Œ. Wolstenholme Elem. Vectors ii. 38 If **a, b, c** are three vectors, any pair of them may be multiplied vectorially to form a new vector **d**, the third of the original vectors may then be multiplied by **d**, either scalarly to form what is known as a scalar triple product, or vectorially to form..the vector triple product.

scalarly (skē[1]·lăɹli), *adv. Math.* [f. Scalar *sb.* + -ly[2].] In such a way as to yield a scalar.

1964 [see *scalar triple product* s.v. *scalar sb. 2].

scald, *sb.*[5] (Later examples.)

1909 G. B. Shaw Press Cuttings 37 G'lang, you young scald: if I had you here I'd teach you manners. 1919 — O'Flaherty V.C. 179 What do you mean, you lying young scald, by telling me you were going to fight agen the English?

scalded, *ppl. a.*[1] Add: **a.** Also in proverbial phr. *like a scalded cat.* Hence *scalded-cat raid* (see quot. 1945).

1934 [see *mark sb.[1] 12 e]. 1943 Times 6 Nov. 2/1 The Luftwaffe..were now using a new type of twin-engine fighter-bomber... From the French coast to Westminster Bridge was a distance of 85 miles. In and out they flashed across it at their fastest speeds—like scalded cats. 1945 L. E. O. Charlton Britain at War IV. 80 When a small force of enemy raiders crossed the coast and penetrated towards London on November 8th, 1943,..three of them were destroyed. One of the aircraft brought down was..one of the new twin-engine fighter-bombers described by Sir Archibald Sinclair, Secretary of State for Air, a few days previously when he spoke of 'scalded cat' raids on London and the South Coast. 1977 Hot Car Oct. 75/3 We have driven a converted V6 and it certainly went like a scalded cat. 1977 C. McCullough Thorn Birds ii. 45 Meggie hopped out like a scalded cat and dressed herself without even asking for help. 1980 Herald (Melbourne) 9 Apr. 7/4 How does it go, old boy? Like a scalded cat.

c. Of land: so poor as to support little if any vegetation. *Austral.*

1936 K. C. McKeown Insect Wonders of Australia xx. 163 The eggs are deposited, as a general rule, upon the 'scalded plains' of the interior, but the insects will avail themselves of almost any area of hard bare ground. 1948 N. C. W. Beadle Vegetation & Pastures of Western New South Wales vi. 58 Scalded surfaces are in general devoid of vegetation, even in the best seasons. 1977 Weekly Times (Melbourne) 19 Jan. 17/2 Deep gullies and scalded country are evidence of the worst abuses of valuable farming country.

scald-headed, *a.* (Earlier (*fig.*) example.)

1802 C. Wilmot Let. 19 Oct. in Irish Peer on Continent (1920) 102 Grim scaldheaded Mountains.

scalding, *vbl. sb.* **1. d.** (Earlier U.S. example.)

1865 Trans. Illinois Agric. Soc. V. 208 Here is no swaying of trees to the east, no scalding of the west side of the trunks in the sun.

scale, *sb.*[1] Add: **1.** For † *Obs.* read *Obs.* except *S. Afr.* and add later examples.

1946 P. Abrahams Mine Boy iii. 26 Joseph nodded, slapped Xuma heartily on the back and offered him a scale of beer... He smiled and took the scale. Xuma put the scale to his lips, then passed it to Daddy. 1953 P. Lanham Blanket Boy's Moon v. iii. 274 Drink a scale of fine home-brewed kaffir beer with us. 1969 Post (Golden City, S. Afr.) 6 Apr. 14 Gave her R1 and told her to buy a scale of KB from Mathebula. 1970 Drum Oct. 8, I found myself firmly grasping a plastic scale.

4. *to turn the scale* (earlier example).

1777 P. Thicknesse Year's Journey I. iii. 18 As he is a good seaman, and has a clean, convenient, nay an elegant vessel, I would rather turn the scale in his favour.

5. (Later example.)

1935 [see *Algol[1]].

6. *attrib.* and *Comb.*, as *scale maker* (earlier example), *man*; **scale house** *U.S.*, a place in which large scales, as for weighing animals, are kept.

1754 South Carolina Gaz. 5 Feb. 3/1 A Scale-House Beam, Scales and Weights, compleat. 1870 Trans. Illinois Agric. Soc. VII. 442 In this division of the stock yards there are three scale houses. 1885 Rep. Indian Affairs (U.S.) 80 To the southeast..is our large cattle corral..with scales and scale-house. 1655 in Suffolk County (Mass.) Deeds (1885) III. 209, I John Saers of Casco bay scale maker..Haue bargained & Sold..one Island. 1783 in L. Chalkley Chron. Scotch-Irish Settlement Virginia (1912) I. 232 It is certified that the scale man is Peter Hane. 1930 Amer. Speech VI. 13 [Sugar beets] first go to the washer man, then to the hopper which rests upon the weighing apparatus, operated by the scale man.

scale, *sb.*[2] Add: **7. c.** *U.S. slang.* A coin; money.

1872 Schele de Vere Americanisms 296 Among the less generally known terms [for money] are..*wherewith, shadscales*, or *scales* 'for short'. 1874 B. F. Taylor World on Wheels 28 Promise him a 'scale'—scale, skilling, shilling. 1889 J. S. Farmer Americanisms 472/2 Scales, a common term for money; an abbreviation of Shadscales. 1929 Amer. Speech V. 152 The waitress received much scale at the hotel.

12. scale-fish, (*a*) (later examples); (*b*) (earlier example); **scale-reading,** the interpretation of the pattern of scales on a fish as an indicator of its age, history, etc.; an examination of scales for this purpose; so **scale-reader.**

1856 J. Reynolds Peter Gott xix. 254 Four hundred quintals of fish, heavily salted, such as are in demand for the use of the negroes on the plantations. These fish are called scale fish; they consist of hake and haddock. 1936 Discovery Jan. 16/1 The food of the natives consisted of various animals..but never scale fish, which seem to have been the object of a curious taboo. 1967 Nat. Fisherman Nov. 11-c The term 'scalefish' is used in the Bahamas for fish proper as opposed to shellfish and crustaceans. 1930 G. H. Nall Life Sea Trout iii. 28 It is the business of the scale reader to decipher how it [sc. the scale] reflects the growth, and to explain how this provides a clue to the life history of the individual fish. 1968 B. Vesey-Fitzgerald World of Fishes ii. 30 An expert scale-reader can tell the age of a fish accurately. 1912 Salmon & Trout Mag. No. 4. p. i (Advt.), The latest and most authoritative publication on the new science of scale reading. 1938 B. Curtis Life Story Fish iii. 29 Using scale-readings, he can construct the life-history of a species with far fewer specimens than he could in any other way. 1971 D. Mills Salmon & Trout xii. 281 The data from such scale readings can then be incorporated into the construction of growth curves.

scale, *sb.*[3] Add: **5. c.** *Psychol.* A graded series in terms of which the measurements of such phenomena as sensations, attitudes, or mental attributes are expressed; sometimes preceded by the name of the person to whom a particular scale is attributed (as *Binet scale*; cf. *Guttman scale), or some other qualifying word.

1898 G. F. Stout Man. Psychol. I. ii. § 5. 31 Thus, if we have a scale of increasing gradations of intensity, we may take as our point of departure any given intensity in the scale. We can then arrange other intensities in relation to this, proceeding by intervals which we judge to be equal. 1917 Pintner & Paterson (title) Scale of performance tests. Ibid. i. 11 The Stanford Revision adheres more closely to the original Binet Scale. 1929 Thurstone & Chave Measurement of Attitude ii. 22 A list of 130 statements was prepared, expressive of attitudes covering as far as possible all gradations from one end of the scale to the other. Ibid. iv. 59 The scale-values represented by the 45 statements. 1960 Jrnl. Pol. XXII. 647 Scale analysis is now common enough in political science to justify omission of the details. 1966 T. M. Newcomb et al. Social Psychol. xiv. 429 The scale was a revision of the original Bogardus scale. Ibid. 498 The Likert scale may seem..a natural way of drawing attitude measurements and combining them. Ibid. 523 'Neutral' items in Thurstone scales are a source of considerable nonvalidity. 1972 Jrnl. Social Psychol. LXXXVI. 105 The scale dimension of like-dislike was used as the source of names. 1977 K. G. Shaver Princ. Social Psychol. v. 193 We have grouped the respondents in terms of a nominal scale: a scale of measurement by which the observations can be classified, but not ordered. Ibid. 194 Regardless of the distance between scores, when the data can be rank ordered (usually from the most favorable to the least favorable) they constitute what is known as an ordinal scale of measurement. Ibid. 196 When the numbers we assign to identify observations *do* tell us something about the distances between observations (while also providing us with a logical order), those numbers are said to constitute an interval scale of measurement. Ibid. 198 If an interval scale is constructed with an absolute zero point, rather than with an arbitrary one, that scale becomes..a ratio scale.

6. c. *scale of* (*two*, etc.): a scale of arithmetical notation having as radix the number given, used *attrib.* and *absol.* to designate a form of scaler (see *scaler[3] 4) in which an output pulse is produced when a number of input pulses equal to the specified radix has been received.

1871 [in Dict., sense 6 b]. 1932 C. E. Wynn-Williams in Proc. R. Soc. A. CXXXVI. 318 As the recording.. values of the 'dial' units are, respectively, 2^0 or 1, 2^1 or 2, and 2^2 or 4, and since the meter indicates the total number of groups of 2^3 or 8, the counting is carried out according to a 'scale of two', the three thyratron dials recording 'units', 'twos' and 'fours' and the meter 'eights', instead of units, tens, hundreds, and thousands. 1933 Ibid. CXXXIX. 621 The impulses are then applied to a 'scale of two' thyratron counting circuit. 1948 Nucleonics Nov. 49/1 Scale-of-*N* circuits are important tools for counting radiations in nuclear physics, as well as for various other applications. 1950 Progr. Nuclear Physics I. 109 A scale of five can be made by the use of a form of ring circuit with five valves with their cathodes connected together. 1963 B. Fozard Instrumentation Nuclear Reactors viii. 75 A cascade arrangement of six scales-of-two gives an over-all scale factor of 2^6 or 64.

7. a. (Earlier example.)

1780 Acts & Resolves Massachusetts (1886) V. 1413 The following scale shall be the rule..for settling the rate of depreciation on all contracts.

b. *spec.* A graduated table of wage or salary rates; *transf.*, a wage or salary in accordance with such a table.

1921, etc. [see *Burnham]. 1930 [see *beggar v. 3]. 1957 [see *label sb.[1] 7 c]. 1968 New Yorker 18 May 45/2 Pookie's Pub..is not the highest-paying club in town. I make about scale, or about a hundred and fifty a week. 1977 Times Educ. Suppl. 21 Oct. 2/5 There seems to be a case for possible demotion from scales.

12. b. *Photogr.* The range of exposures (defined as the product of the light intensity and the time) over which a photographic material will give an acceptable variation in density. Also *transf.*

[1891 Jrnl. Soc. Chem. Industry 28 Feb. 104/1 By variations in the time of development it is possible to produce secondary negatives in which the scale of tones is either contracted or extended. 1920 L. A. Jones in Jrnl. Franklin Inst. CLXXXIX. 480 If this scale of negative densities is too great for printing on the papers which are available, we can reduce the scale by lowering the contrast of the negative.] Ibid. 482 The total scale of the paper may be defined as the range of light intensities, expressed either in log exposure or exposure units, which can be reproduced by the paper as perceptibly different densities. 1942 C. E. K. Mees Theory Photographic Process xix. 736 If all different gradations on the negative are to be rendered as different gradations in the print, the scale of the paper must be at least as great as the difference between the maximum and minimum densities of the negative. 1967 Electronics 6 Mar. 127/1 The persistence of the scope that was necessary for a raster scan took five seconds from top to bottom, and did not have enough grey scale for good pictures. 1970 G. L. Wakefield Practical Sensitometry viii. 83 A medium speed film is likely to have an exposure scale of at least 1,000 to 1 and it can be even bigger. On a log basis this is a range of 3·0 and higher. As a rule, the faster the material the larger the exposure scale.

c. *economy* (*economies, economics*) *of scale*, the relative gain in output or saving of costs derived from an increase in the size of plant or of a firm.

1944 A. Cairncross Introd. Econ. vi. 61 The economies of large-scale production—called for short 'economies of scale'—may be either 'internal' or 'external'. Ibid. xv. 195 Economies of scale, and economies of scale alone, make costs fall as output increases. 1953 Stonier & Hague Textbk. Econ. Theory x. 221 Over relatively low levels of output it is likely that increasing returns to outlay will occur, because with larger output there are economies of scale to be reaped. 1966 A. Battersby Math. in Management ix. 220 A picture of the familiar 'economy of scale' which results from spreading the fixed costs over a large number of items. 1972 Observer 20 Aug. 9/7 The economics of scale, that much-abused phrase, used to justify any increase in size.

13. b. Also with ellipsis of *adj.* Also with *sb.*, as *on a world scale.*

1904 H. James Golden Bowl I. i. ii. 26 Maggie's too wonderful—her preparations are on a scale! 1968 Times 15 Oct. 16/7 Possible arrangements on a world scale are affected by the telescopes available.

16. *attrib.* and *Comb.*, as (sense 9) *scale-bar, -reading* (earlier and further examples); (sense 11) *scale model*; (sense 7) *scale fee*; **scale effect,** an effect occurring when the scale of something is changed, as a result of contributory factors not all varying in proportion; *spec.* (see quot. 1940); **scale factor,** a numerical factor by which each of a set of quantities is multiplied; **scale height,** the vertical distance over which an atmospheric parameter or other quantity decreases by a factor *e* (= 2·718 . . .).

1974 Nature 18 Oct. 647 (caption) Fully developed vegetative colonies (1 month old) on liquid surface (scale bar, 0·5 cm). 1917 Rep. & Mem. Advisory Comm. Aeronaut. (1921) No. 374 (heading) Report of the scale effect subcommittee. 1930 Engineering 20 June 802/2 The skin friction of the plate gives a slightly erroneous velocity distribution under the model car... Still, the errors due to these imperfections are hardly likely to be so much greater than other unavoidable uncertainties, arising from scale effect and the varying conditions of full-scale operations. 1940 Chambers's Techn. Dict. 743/1 Scale effect, the effect of a change in Reynolds number upon the measured results in the performance of aerodynamic bodies. 1978 H. C. H. Armstead Geothermal Energy xv. 244 With conventional thermal power plants the capital cost per kilowatt installed is sensitive to what is generally known as the 'scale effect'; that is to say, a very large plant will tend to cost less per kilowatt than a small plant of similar type. 1979 Daily Tel. 15 Aug. 12/5 If you were trying to apply this concept to a Jaguar, you would need about a 20-litre engine—it only works because of the scale effect.. on a very small car. 1948 Electronics Apr. 127/1 The corresponding initial voltages must be computed and the integrators set accordingly, using the correct scale factor. 1963 [see sense *6 c]. 1968 P. A. P. Moran Introd. Probability Theory v. 244 Thus S_n has the same distribution as the X_i but increased by the scale factor $n^{\frac{1}{2}}$. 1975

Sci. Amer. Nov. 120/2 The price paid for conformality is a distortion of the scale factor that increases with distance from the centre of the map. 1970 Which? Mar. 72/2 They saved the solicitor's scale fee on the price of the house they were buying or selling. The higher the price of the house, the higher the fee. 1937 S. Chapman in Rep. Progr. Physics III. 44 *H* may then be interpreted as a unit of height-measurement relative to which, at the given level, the rate of upward decrease of log *p* is unity... The term 'height of the homogeneous atmosphere' is clearly not appropriate when *H* varies with height, and the name (local) 'scale-height' may be suggested. 1976 Sci. Amer. Mar. 53/1 In the sun or in the earth's atmosphere the size of the dominant energy-carrying cells is on the order of one scale height. 1978 Nature 26 Oct. 726/1 Suppose that 10^{38} ergs^{-1} of X rays are emitted by the pulsar... Assume the scale height of the photons is 3×10^8 cm. 1934 Planning I. xxii. 6 This is not, therefore, a scheme but a scale model for one, intended to show precisely what is involved. 1952 'T. Hinde' Mr. Nicholas v. 87 He had..small features ..as neat as a scale model. 1868 Scale-reading [see *electrometry]. 1962 L. S. Sasieni Optical Dispensing v. 110 A slight turn..will have the effect of moving both scale-readings in the same direction.

scale, *v.*[2] Add: **2. c.** *Austral.* and *N.Z. slang.* To defraud or cheat (someone), to steal (something). In phr. *to scale a train* or *tram*, to ride without paying on public transport; also *intr.*

1916 A. Wright Under Cloud 32 'How'd that happen,' asks Bill Odzon. 'Didn't think anyone could scale you.' 1941 Baker N.Z. Slang vii. 62 When we are taken down financially we are *scaled.* 1945 — Austral. Lang. v. 103 A *steel jockey* is a tramp who scales a train or rides without paying. Ibid. 106 One can *get scaled*, in the sense of being done down, when overcharged for goods. 1953 'Caddie' Sydney Barmaid xiv. 132 Better..than for them to be getting about the streets with snotty noses, and scaling trams. 1953 D. Cusack Southern Steel 3 Bumping in on the back of the old steam trams, too often scaling on the footboards because he hadn't the money to pay the penny fare.

scale, *v.*[3] Add: **II. 4. b.** (Later examples.) Also *loosely*, to reduce.

1933 Sun (Baltimore) 5 Apr. 8/2 The indebted farmer gets his mortgage debt scaled down, but with that scaling down the payment of interest again becomes the vogue. 1934 [see next sense]. 1937 Physical Rev. LI. 1027/1 (heading) Vacuum tube circuits for scaling down counting rates. 1952 M. Laski Village v. 95 Hospitality had been empirically scaled down to a universally possible level. 1979 Daily Tel. 19 May 2/1 The original pay claim for a 30 per cent. rise has been scaled down to 16 per cent.

c. With *up*: to increase in amount or size according to a fixed scale or standard; to increase from a small scale to a larger scale. Also *absol.*

1891 Daily News 17 Jan. 2/5 The scaling up instead of scaling down the London, Chatham, and Dover stock. 1934 W. Nelson Seaplane Design vi. 64 Scaling the size of existing floats and hulls up and down can be done to arrive at the dimensions of a new design. 1972 Aquaculture I. 182 During the summer of 1971, the project was scaled up in size and moved out-of-doors. 1973 Times 28 Nov. 19/5 If the pilot plant can be scaled up at this figure it offers great hopes for the development of these abundant fuel reserves. 1975 Nature 17 Jan. 149/3 There will probably be no need to scale up since the existing plant can cope with 50 tons every 24 hours. 1977 Undercurrents June–July 7/1 It remains doubtful whether the process..can work safely and effectively when 'scaled-up' to commercial size. 1979 Sci. Amer. Jan. 45/1 Several organizations are currently scaling up from laboratory-size cells to units of demonstration size.

d. To measure or represent (a quantity) in exact proportion to its absolute size or according to an arbitrary defined scale.

1885 W. Penman Land Surveying ix. 127 An area to the scale of 1 chain = 1 inch was scaled and found to give 12 ac. 1 ro. 01 pls. 1898 F. E. Dixon T. Baker's Rudimentary Treat. Land & Engin. Surveying (ed. 17) xii. 182 It sometimes happens that a distance is scaled on a plan using..a wrong scale. 1923 Rep. Internat. Air Congr., London, 1923 63 Not only is it difficult to scale the printed forms with accuracy, but there is no assurance that the silhouette corresponds closely with the model tested in the wind channel. 1940 Amer. Jrnl. Psychol. LIII. 336 (caption) The curve shows how pitch, scaled in subjective units.. varies with frequency. 1951 S. S. Stevens Handbk. Experim. Psychol. i. 23/1 These operations are limited ordinarily by the peculiarities of the thing being scaled. 1951 H. P. Bechtoldt in Ibid. xxxiii. 1240/2 Multiple-category qualitative variables representing intensive dimensions are 'scaled' in various ways, and numerical scores are determined. 1966 T. Newcomb et al. Social Psychol. (ed. 2) 506 In a most interesting approach to problems of scaling attitudes..Guttman..began to examine items apparently ordered on the basis of 'difficulty'. 1971 J. B. Carroll et al. Word Freq. Bk. p. xxvii, The base line of Graph 1 is scaled, not in terms of ϕ, but in terms of a further transformation of ϕ to the Standard Frequency Index. 1976 B. S. Phillips Social Res. (rev. ed.) ix. 211 Select or construct those items that you wish to scale.

e. To alter (a quantity or property) by changing the units in which it is measured; to change the size of (a system or device) while keeping its parts in constant proportion.

1954 Computers & Automation Dec. 20/2 Scale, computation. To change the scale (that is, the units) in which a variable is expressed so as to bring it within the capacity of the machine or program at hand. 1966 R. C. Carter Introd. Electr. Circuit Analysis vii. 239 Once the desired design performance has been achieved in the low-frequency prototype laboratory model, all factors involving frequency and impedance may be scaled to the desired operating range. 1974 Physics Bull. Mar. 98/3 The symmetry transformation consists of scaling the physical dimensions *d* of the system according to $d \to \lambda d$. If the equilateral triangle of figure 1 is scaled then although the size is changed, the geometric shape and all the dimensionless properties of the triangle such as the angles remain unchanged. 1978 Sci. Amer. Dec. 128/2 The radio waves, completely unattenuated by the intervening dust, can be scaled several orders of magnitude in frequency to predict the true intensity of the optical radiation.

f. *intr.* Of a quantity or property: to vary according to a defined rule or principle.

1974 Physics Bull. Mar. 98/3 The invariance of all dimensionless properties can be used to determine whether the figure scales or not. 1978 Nature 20 Apr. 737/3 Surprisingly the limiting torque, even at optimised pressures, scales only at [*recte* as] $T^{\frac{1}{2}}$.

7. *trans.* Of a scaler (see *scaler[3] 4): to count (electrical pulses). Also *absol.*

1938 Rev. Sci. Instruments IX. 221/1 The circuit either scaled correctly or no counts were registered. 1947 Ibid. XXIV. 322/1 Although not developed as a high-frequency instrument, the model will scale a regular pulse input up to frequencies of the order of 100 kc/s.

scaled, *ppl. a.*[4] Add: **2. a.** That has been measured by a scale or varied in a determined proportion.

1885 W. Penman Land Surveying ix. 127 The scaled area is less than the actual one, indicating shrinkage of the paper. 1938 Rev. Sci. Instruments IX. 221/1 The fluctuational analysis of the scaled counts occurring at an average rate of ten per minute indicated a scaling factor of 22 to 1. 1960 Rogers & Connolly Analog Computation in Engin. Design x. 261 The following scaled voltage equivalents are suitable for representing the variables of the problem; $2\ddot{x}$, $20\dot{x}$, $200x$, $10h$, $10t$. 1976 Attewell & Farmer Princ. Engin. Geol. vii. 525 (caption) Scaled distance attenuation relationships for blasting in a rock.

b. *scaled-up*: that has been increased proportionately in amount or size in all its parts. Similarly *scaled-down.*

1944 P. Wilkinson Aircraft Engines of World 1944 38 The 12-cylinder 1,300 h.p. Jumo 211-J is a scaled-up version of the Jumo 210. 1947 A. E. Slater in A. C. Douglas Gliding & Advanced Soaring i. 12 The earliest recorded attempts to fly, as well as the legendary ones, usually began with an attempt to reproduce a scaled-up bird's wing, often to the extent of putting feathers on it, in the belief that such a structure was inherently able to keep itself up, once it got well aloft. 1953 Trans. Soc. Instrument Technol. V. 126/2 The pilot unit is not usually constructed unless the manufacturer is satisfied that he will eventually proceed to the erection of a full-scale plant. It is a scaled-down version of such a plant. 1963 Bird & Hutton-Stott Veteran Motor Car 49 B.S.A. cars..were virtually scaled-down Daimlers. 1973 Lebende Sprachen XVIII. 7/2 Go-ahead for a scaled-up version of the engine giving a 30 percent increase in original thrust. 1977 Time 19 Sept. 43/1 Anyone with a driver's license and a few dollars can safely savor some of the adrenaline-pumping, gut-clutching fever of Grand Prix racing on a minitrack, in a scaled-down Formula I speedster.

scalene, *a.* and *sb.* Now usu. with pronunc. (skē[1]·līn). **A.** *adj.* **2.** (Later examples.)

1934 [see *scalenotomy]. 1962 Gray's Anat. (ed. 33) 600 The scalene muscles, in particular the Scalenus medius, are important accessory muscles of inspiration.

B. *sb.* **2.** (Example.)

1978 [see *scalenotomy].

scalenotomy (skē[1]lĭnǫ·tŏmi). *Surg.* [f. Scalen(us + -o + -tomy.] Division or section of a scalene muscle.

1934 Romanis & Mitchiner Sci. & Practice Surg. (ed. 5) II. xi. 449 To produce further collapse of the apex, the operation of phrenic avulsion may be supplemented by a section of the scalene muscles. This is easily done through the same incision (scalenotomy). 1978 J. E. Bateman Shoulder & Neck (ed. 2) xv. 633/1 In some instances..the involvement of the paraspinal muscles, and the scalenes in particular, is prominent. Considerable relief..may be obtained by simple scalenotomy.

scaler[2]**.** Add: **3.** *Austral.* and *N.Z. slang.* (See quots.)

1924 Truth (Sydney) 27 Apr. 6 Scaler, a fraud. c 1926 'Mixer' Transport Workers' Song Bk. 5 (title) The Scaler... He waits until his dues are due, The bloke who does a scale. 1932 C. Wills Rhymes of Sydney (1933) 13 See the shoppers, toppers, tabs, Scalers by the score, Hopping off, Dropping off, Darting into shore. 1945 Baker Austral. Lang. v. 106 A *scaler* is a person who rides in a vehicle without paying, or one who decamps with money with which he has been entrusted.

scaler[3]**.** Add: **4.** An electronic pulse-counter, suitable for high count-rates, in which a display or recording device is actuated after a fixed number of pulses has been received and added electronically.

1945 H. D. Smyth Atomic Energy for Military Purposes 1939–45 140 The scaler was set at zero. 1953 Sci. Amer. Mar. 105/3 As a series of pulses flows into the scaler a voltage builds up step by step... When the cut off point is reached, the tube begins to conduct and the condenser discharges, sending a single pulse from the tube's output. 1964 Analytical Chem. XXXVI. 2221/1 Development of the pipping scaler was stimulated by..experimental work in which it was necessary to determine time *vs.* concentration curves having a duration of a second or less. 1977 N. Freeling Gadget II. 87 That's a PM—sorry, photomultiplier tube... Sends signals here, to the amplifier, through here, that's the discriminator, to here, the scaler. 1980 J. W. Hill Intermediate Physics xxiii. 220 These are connected to about 400V obtained usually from a scaler, a piece of electronic apparatus which can count very rapidly using either 'dekatrons' or a digital display.

scale-up (skē[1]·lɒp), *sb.* (*a.*) [f. vbl. phr. *to scale up* (*scale v.[3] 4 c).] The action or result of increasing the scale of something. Also as *adj.*

1945 H. D. Smyth Gen. Acct. Devel. Atomic Energy Mil. Purposes viii. 82 Should the several steps in the separations process have to be developed partly by the empirical approach, there would be less risk in the scale-up of a precipitation process. 1953 Industr. & Engin. Chem. May 990/1 All the dimensions of the larger are *x* times those of the smaller. We call *x* the 'scale-up factor'. 1965 Amer. Scientist LIII. 280 These tanks are modeled.. with a scaleup factor of about two. 1967 Jane's Surface Skimmer Systems 1967–68 49/1 In general layout, the craft represents a 'scale-up' of the configuration tested with the Raduga. 1979 Sci. Amer. July 79/3 The design of the 1899 kite was the basis for a scale-up version that was small enough to be flown as a kite but large enough to support a man.

scalewise (skē[1]·lwəiz), *a.* and *adv.* [f. Scale *sb.*[3] 4 + -wise.] **A.** *adj.* = *scalar *a.* 3. **B.** *adv.* In the manner of a scale; in respect of a scale.

1931 G. Jacob Orchestral Technique ix. 81 Eighteenth-century trumpet parts were written very high because of the impossibility of obtaining scale-wise passages on the natural instrument except amongst the very high harmonics. 1959 Listener 8 Jan. 80/2 The opening cantabile theme, descending scalewise, continuously flows into more ornate melismata. 1977 Early Music Oct. 535/2 The interaction of *tirades* (scale-wise flourishes)..always requires some arbitrary adjustment by the player.

scaley, var. *scaly *sb.*

scalic (skē[1]·lik), *a. Mus.* [f. Scal(e *sb.*[3] 4 + -ic.] = *scalar *a.* 3.

1933 Times Lit. Suppl. 2 Mar. 139/1 'The Rebel Stranger' ..shows in its seven versions how a tune may develop by blending scalic with figural features of melody. 1960 Times 5 Mar. 9/6 A scalic tune, that can sound quite unpromising. 1971 Daily Tel. 5 Apr. 10/3 Kempff will sometimes pedal scalic passages and is apt to be crowded out by violin accompaniments. 1979 Early Music Oct. 545/1 Babell's elaborations are always scalic, and on the first page of 51 the same passage, repeated sequentially, is embellished three times in an identical manner.

scaling, *vbl. sb.*[3] Add: **2.** (Further examples.) Also, measurement or grading of attributes; variation of size or scale; the action of a scaler.

1807 G. Gregory Dict. Arts & Sci. II. 757/3 The plan being laid down, the content of the field may be found by scaling. 1929 F. N. Freeman in C. Murchison Found. Experim. Psychol. xviii. 721 The purpose of such scaling may be merely to secure items which are equally spaced in difficulty or it may also be to weigh the pupil's performance in terms of the difficulty of the items which he passes. 1938 Rev. Sci. Instruments IX. 221/1 No variations in the scaling factor were found for pulses varying in amplitude by a factor of six. 1949 Nucleonics Feb. 67/2 Several scale-of-2 circuits in tandem provide net scaling factors of 4–8–16–32–64, etc. Other designs, utilizing 'ring scalers' or modified scale-of-16 scalers yield decimal scaling ratios (10–100–1000). 1968 Fox & Mayers Computing Methods for Scientists & Engineers v. 89 It is desirable that all rows and columns of [matrix] **A**, and also of **b**, should be of reasonable size... This can always be arranged..by appropriate scaling of the rows and columns. 1975 IEEE Trans. Nuclear Sci. XXII. 1580/1 Simple start-stop scaling, count = 8, frequency 53 MHz. 1979 Sci. Amer. July 120/3 One such manifestation of movement into a marginal niche is the scaling down of body size.

scaling (skē[1]·liŋ), *ppl. a.*[2] [f. Scale *v.*[3] + -ing[2].] That scales, in the senses of Scale *v.*[3] II in Dict. and Suppl.

1937 Rev. Sci. Instruments VIII. 414/1 The ultimate efficiency..is determined by the resolving power of the first stage of the scaling circuit. 1937 Physical Rev. LI. 1027/1 The ultimate efficiency that can be reached is fixed by the resolving power of the scaling down circuit. 1950 Atomics Sept. 255/1 The simplest scaling circuit is the 'scale of two'. 1961 G. R. Choppin Exper. Nuclear Chem. iii. 43 The scaling system selects every *n*th pulse to pass on to the mechanical registers.

scallion. a. For Now *dial.* read *U.S.* **c.** Add: = *spring onion* s.v. Spring *sb.*[1] 7 b. **d.** *U.S.* = Leek 1. (Later U.S. examples.)

1902 L. H. Bailey Cycl. Amer. Hort. IV. 1622/2 Scallion, a name for the Shallot; also used for onions that do not make good bulbs but remain with thick necks. 1943 H. M. Fox Gardening for Good Eating ix. 135 When the [onion] seedlings are thinned, they can be eaten as scallions. 1963 [see *dandelion greens* s.v. *dandelion 3]. 1965 P. De Vries Let me count Ways xii. 159 Several stalks of crisp celery and a scallion or two left over from my lunch. 1969 Yearbk. Agric. (U.S. Dept. Agric.) 189/2 Green onions, shallots, and leeks are sometimes called 'scallions'. 1978 Chicago June 217/1 Other delicacies might be fresh crab legs..and transparent noodles studded with shrimp, scallions, and black mushrooms.

scallom (skæ·ləm), *sb. Basket-making.* [Of obscure origin: see Scallum *v.* in Dict. and Suppl.] A stake or rod, of which a thin or spliced end is wrapped round another stake

to form a base or frame of a basket; the method of weaving baskets thus.

1912 T. Okey Art of Basket-Making vii. 75 Bottoms and covers may also be made on hoops and scalloms. Ibid. Gloss. 154 Scallom, a method of forming the rigid inner frame of a bottom or cover, or of staking up a basket. 1929 A. G. Knock Fine Willow Basketry 18 Scallom, a stake, or the equivalent of a bottom-stick or lid-stick which has been affixed by looping its thinned end round a hoop or the outside stick of a bottom. 1959 Gloss. Terms Packaging (B.S.I.) 16 Scallom, the spliced end of a stake which is wrapped round the bottom outside stick and woven into the next two scalloms. 1959 D. Wright Baskets & Basketry iv. 115, 5 scalloms of No. 12 cane run from end to end. Ibid. vi. 136 Scallom, method of fixing stakes to a ring of willow or cane.

scallom, *v.*: see Scallum *v.* in Dict. and Suppl.

scallop, *sb.* Add: **3.** (sense 1) *scallop bed, boat, dredge* (earlier example), *-fishery, net*; *scallop-edged* adj.

1977 N.Z. Herald 8 Jan. 1–2/3 Scallop beds around the Coromandel Harbour could be wiped out within five years if the onslaught of spiked dredges used mainly by holiday-makers continued. 1977 New Yorker 15 Aug. 46/1 The Sniktaw III, a forty-foot *scallop boat*, is moving rapidly south-ward down the channel. 1884 U.S. National Mus. Bull. No. 27. 268 Implements [used in shellfish fishery include]..Scallop-dredge. 1856 W. Whitman Leaves of Grass (ed. 2) xi. 214, I saw..the scallop-edged waves in the twilight. 1967 R. S. Churchill Winston S. Churchill II. viii. 274 A splendid scallop-edged silver tray presented by all his colleagues in the Government. 1886 Amer. Naturalist XX. 1001 It is only between Cape Cod and New Jersey that any commercial scallop-fishery exists. 1881 E. Ingersoll Oyster-Industry 247 Scallop Net, the small dredge used in catching scallops.

scalloped, *ppl. a.* Add: **2.** (Further examples.)

1877 Home Keeper's Guide 45 Scalloped tomatoes. 1884 New Kentucky Home Cook Bk. 157 Scalloped potatoes. 1925 W. G. R. Francillon Good Cookery (ed. 3) 435 Scalloped Oysters... Butter a scallop shell. Arrange layers of crumbs, oysters, and butter alternately... Bake. 1936 Farmhouse Fare 10 Scalloped Meat with Macaroni... Arrange..cooked macaroni, meat, and gravy in alternate layers...cover the top with browned breadcrumbs...bake. 1960 Woman 23 Apr. 51/1 Scalloped Ham... Layer potatoes, onions and ham in the casserole... Pour in milk... Bake in moderate oven. 1975 B. Wood Killing Gift (1976) v. iv. 243 Scalloped potatoes..next to the roast.

scallopini (skælopī·ni). Also in It. form ‖ **scaloppine.** [ad. It. *scaloppine*, pl. of *scaloppina*, dim. of *scaloppa* *escalope.] A dish consisting of very thin slices of meat (esp. veal) sautéed or fried.

1950 E. Hemingway Across River & into Trees 103 The scaloppine with Marsala. 1957 F. & R. Lockridge Tangled Cord (1959) vi. 73 The waiter..heated and served scallopini. 1975 Times 6 Sept. 9/1 In Cardiff..it is easier to find well-cooked scaloppine and cannelloni than anything in the native tradition. 1977 J. Wambaugh Black Marble (1978) iv. 46 Sal Moroni got arrested for throwing a Sicilian cook out the window of an Italian restaurant for overcooking his scallopini.

scallum, *v.* Add: The more usual form is now **scallom.** (Later examples.) Hence **sca·llomed** *ppl. a.*, **sca·lloming** *vbl. sb.*

1912 T. Okey Art of Basket-Making vii. 80 When the side stakes are scallomed on. 1929 A. G. Knock Fine Willow Basketry 61 To the curved part of the hoop, eight stakes are now scallomed. Ibid. 62 In scalloming a long tongue is formed at the butt-end of the stake. 1959 D. Wright Baskets & Basketry ii. 45 The sticks are scallomed, that is: thinned down to a long, flat point and taken round the frame... Scalloming is easier to work with willow than with cane because the rods kink and stay rigid when dry. Ibid. iv. 114 This basket and its lid are made on a scallomed base.

scallywag, scallawag. Add: **1.** Also *attrib.*

1885 G. B. Shaw Let. 4 Sept. (1965) I. 138 Any socialist of the plentiful 'scallawag' type. 1926 Glasgow Herald 10 Sept. 11 Go back to your scallywag union. 1957 Listener 17 Oct. 608/2 Voyez, that rather scallywag wanderer who was dismissed by Wedgwood.

2. (Earlier and further examples.) Also *attrib.*

1862 Charleston (S. Carolina) Mercury 9 Aug. 1/3 This invaluable class is composed..of ten parts of unadulterated Andy Johnson Union men, ten of good lord and good devil-ites, five of spuss and seventy-five of scallowags. 1867 Nation (N.Y.) 12 Dec. 470/1 The Macon News has to print in full the names of thirteen persons..described (as having 'voted the Scalawag ticket'). 1888 J. Bryce Amer. Commonwealth II. II. xliv. 164 A group of such 'scallawag' members.. increase their legislative income.

Hence **scallywa·ggery,** (*a*) roguery, (*b*) political opportunism; **sca·llywagism** = *scallywaggery* (*a*); **sca·llywagging** *vbl. sb., ppl. a.*

1897 Daily News 9 Dec. 7/1 The stages of accumulating merit for the fighting man, as Lord Charles gives them, appear to be first robbing orchards, next hatred of a life at the desk, and finally scallywagism. Ibid., Robbing orchards and general scallywaggery is not within them [sc. disqualifications for military service]. 1911 H. S. Harrison Queed iv. 45 The morning *Post* was an old paper... It had crucified carpet-baggism and scalawaggery upon a cross of burning adjective. 1915 W. J. Locke Jaffery xii. 158 He was fed up with scalliwagging all over the place. He wanted a season in town! 1962 Punch 9 May 735/1 Wilkes is worth writing about..for all his scallywaggery. 1977 Times 18 Aug. 6/1 Mr Frank Johnson.. has unimpeachable credentials as a civil rights defender. Governor George Wallace of Alabama once denounced him as 'a scallywaggin', integratin', carpet-baggin' liar'.

scalogram (skē[1]·logræm). *Psychol.* [f. Scal(e *sb.*[3] 6 + -o + -gram (perh. by analogy with *cardiogram* s.v. *cardio-).] A diagram showing the numerical values assigned to responses and persons in an attitude test, designed esp. to analyse whether the questions relate to the same factor and the results are scalable. Also *attrib.* and *Comb.*: **scalogram analysis,** the analysis of results revealed by a scalogram; **scalogram board,** a board with movable slats on which the results are recorded.

1944 L. Guttman in Amer. Sociol. Rev. IX. 139/2 The results of the analysis are presented and easily assimilated in the form of a 'scalogram', which at a glance gives the configuration of the qualitative data. Ibid. 144/1 The scalogram boards used in practical procedures are simply devices for shifting rows and columns to find a scale pattern if it exists. 1950 S. A. Stouffer Measurement & Prediction i. 9 The approach which was developed in the Research Branch under..Louis Guttman has been named scalogram analysis. Ibid., The scalogram hypothesis is that the items have an order such that, ideally, persons who answer a given question favorably all have a higher rank on the scale than persons who answer the same question unfavorably. 1970 E. J. Wilkins Introd. Sociol. v. 81 The items in this type of scalogram are accorded points, both positive and negative, and they are also of a cumulative nature. 1973 Times Lit. Suppl. 2 Nov. 1332/2 The second [kind of historical explanation] sublimates personality, thrives upon roll-call votes and produces a scalogram to measure regularity or deviance as its finest achievement. 1978 T. H. Poister Public Program Anal. x. 362 The development of the coefficient of reproducibility (also referred to as scalogram analysis) is useful for assessing the..internal consistency of a Guttman scale.

scaloppine, var. *scallopini.

scalp, *sb.*[1] Add: **2. c.** *U.S.* The skin from the head of an animal preserved as proof of its death (usu. in order to obtain a bounty).

1703 Narragansett Hist. Reg. (1884–5) III. 162 All persons who shall kill any Sheep or Lambs..shall be obliged to carry in the Skalp with Ears of the same. 1847 J. S. Robb Streaks of Squatter Life 80 He can git a bonus for wolf-scalps. 1890 Stock Grower & Farmer 22 Feb. 3/1 The bounty law must be fixed up so that scalps will be paid for. 1901 Duncan & Scott Hist. Allen & Woodson Counties, Kansas 15 [The county board] offered a bounty of twenty-five cents for wolf scalps.

b. (Earlier and later examples.)

1759 W. Mason Let. 25 Jan. in Corr. of Thomas Gray (1935) II. 612 Criticks like Indians are proud of the number of scalps they make in a Manuscript. 1928 T. E. Lawrence Lett. (1938) 571 It riles me unbearably to lose my scalp to a lot of fellows round whom I can make rings. 1977 R.A.F. News 11–24 May 19/1 Convincing wins..for the RAF under-21 hockey team... The Navy provided the first scalp.

6. b. *scalp-massage.*

1930 A. Bennett Imperial Palace xxiii. 143 An electric scalp-massage. 1977 J. Aiken Last Movement vii. 125, I always gave her scalp massage in the evenings..her shaved hair was taking its time about growing back.

c. *scalp-dance* (earlier example), *-hunter* (earlier and later examples); **scalp-money** (earlier example); **scalp ticket** orig. *U.S.*, a ticket sold by a scalper (see Scalper[2] 2 a); **scalp yell,** a shout celebrating the taking of a scalp.

1791 J. Long Voyages 35 The dances among the Indians are many and various,..[including] the scalp dance. 1835 R. M. Bird Hawks of Hawk-Hollow I. 79 He acquired a singular reputation as a bold and successful scalp-hunter. 1937 T. Rattigan French without Tears II. i. 37, I can't quite see what my novel has got to do with the machinations of a scalp-hunter. 1975 Observer (Colour Suppl.) 23 Nov. 25/3 Once the scalp-hunters get the word that such-and-such a diplomat wants to defect or to become an out-and-out agent they enjoy priority over the sanctifiers and all the other categories of black operations people. 1704 in G. Sheldon Hist. Deerfield, Mass. (1895) I. 299 That the sum of Sixty Pounds be allowed and Paid to the Petitioners..as Scalp money. 1880 G. A. Sala Amer. Revisited 201/1 There are 'round trip' tickets which are something more than return, tickets; and finally, there are 'scalp' tickets, which you can deal in and discount. 1941 Baker Dict. Austral. Slang 68 Scalp ticket, the return half of a train ticket. 1792 H. H. Brackenridge Mod. Chivalry I. v. ii. 113 A warrior..separates it [sc. a scalp] from the head, giving, in the mean time, what is called the scalp yell. 1913 J. London Valley of Moon 465 He drew his finny prize to the bank..with the scalp-yell of a Comanche. 1947 National Geogr. Mag. July 108/1 The hundreds of scientists being marshaled there are pioneers more potent than any who fought when war drums rolled along the Mohawk, scalp yells quivered on the valley air, and the frontier was aflame.

scalp, *v.*[2] Add: **1. a.** (Earlier and later *absol.* and *fig.* examples.)

1759 W. Mason Let. 25 Jan. in Corr. of Thomas Gray (1935) II. 612 If you don't let them [sc. critics] scalp they'll do you no service. 1939 'A. Bridge' Four-Part Setting ii. 16 Henry is plain sailing, of course—he's quite simply scalped... He's always being scalped. It's his own fault—he will chase women so. 1973 D. Kyle Raft of Swords (1974) x. 99 Calder took a taxi to Heathrow airport. Inevitably he would be scalped on the cab fare.

2. a. (Earlier example.)

1825 J. Lorain Pract. Husb. 335 The Yankee farmer first chops the fallen timber, then scalps off the grubs level with the ground.

scalp, *v.*[1] Add: **2. c.** *Metallurgy.* To remove the surface layer of (metal); to remove (the surface) *from* metal.

1922 Brass World & Plater's Guide XVIII. 96 After the slabs are cooled they are sent to the overhauling machines where a thin layer of metal is scalped from the surface. 1922, 1949 [implied in *scalping vbl. sb.[3] 2]. 1958 A. D. Merriman Dict. Metallurgy 305/1 Other methods used to scalp the ingot are by chipping, milling, planing or by means of the oxyacetylene torch.

4. (Earlier and later examples.) Also *absol.*

1886 Harper's Mag. July 213/2 [The scalper buys] any quantity of grain that may be offered, sells it at an advance of 1/8 cent per bushel, thus scalps the market. 1897 Boston Globe 29 Aug. 39/5 The broker himself would be selling the stock at 104 in New York, thereby 'scalping' one-fourth and making a handsome profit at no risk. 1902 G. H. Lorimer Lett. Merchant 201, I saw what looked like a safe chance to scalp the market for a couple of cents a bushel. 1948 Sun (Baltimore) 26 Nov. 18/2 The Stadium attendants told me they are the same men.. who scalp at other games,..selling 60-cent tickets for $1. 1977 Time 19 Dec. 66/1 The generous benefactor to down-and-out friends wore the same loud waistcoats as the pinchpenny negotiator who scalped outmatched publishers.

scalped, *ppl. a.*[1] Add: **c.** *Metallurgy.* Having had the surface layer removed.

1958 Times Rev. Industry June 53/1 Scalped wire bars.. seen in the place of the familiar round extrusion billet. 1965 Gloss. Terms Copper, Zinc & Alloys (B.S.I.) 15 Scalped stock*(for other than tube), stock intended for further fabrication from which the surface has been removed by machining to improve the quality of the final product.

scalper[2]**.** Add: **1.** (Earlier example.)

1760 S. Niles in Mass. Hist. Soc. Coll. (1837) 3rd Ser. VI. 174 This reminds me of an account we had of a notable old scalper among [the Indians].

2. a. (Earlier example.)

1875 Chicago Tribune 8 Dec. 12/3 The new town grew up to be..the great commercial centre of rail-road 'scalpers'.

b. (Earlier example.)

1886 Harper's Mag. July 213/2 The 'Pit' is the scalper's delight.

c. *slang* (orig. *U.S.*). A speculator who obtains tickets for a popular entertainment and sells them for more than their face price.

1869 Harper's Mag. Sept. 623/2 Where theatres are all the run, And bloody scalpers come to trade. 1935 Time 26 Aug. 27/1 Comedian Joe Brown..is locked out of his dressing room by mistake on his opening night and is compelled to pay $20 to a ticket scalper to get into the theatre in time for his entrance cue. 1948 Sun (Baltimore) 26 Nov. 18/2 Since these 60-cent tickets are sold, or supposed to be sold, only to students, the question is how the scalpers obtained them. 1969 Truth (Melbourne) 12 July 24/7 I'm sure scalpers wouldn't buy lottery tickets. 1977 Rolling Stone 16 June 12/2 The Palladium shows sold out in a few hours and scalpers have been getting up to $75 per ticket. 1978 G. Vidal Kalki vi. 137 One-third of the tickets for the rally..are now in the hands of scalpers who are selling the most desirable seats..for as high as one thousand dollars a-piece!

d. *U.S.* (See quots.)

1874 J. G. McCoy Hist. Sketches Cattle Trade 292 So soon as an incoming train is announced nearing the stock yards, the hurrying tramps of solicitors, called 'Scalpers', may be heard hustling toward the unloading platform. If there is a shipper on the train whose stock is not consigned, they..[present] the business cards of the commission firms which have the Scalpers employed. 1930 Amer. Speech X. 271/2 Scalper, one who buys feeder cattle and resells to farmers and feeders at a profit. A speculator.

3. (Earlier and later examples.)

1837 R. M. Bird Nick of Woods II. xviii. 245 Captain Ralph Stacpole did..meet another Injun-savage in the woods..with gun, axe, and scalper. 1947 B. De Voto Across Wide Missouri 32 The Company is sending..100 dozen 'common scalpers' and 55 dozen more expensive knives for murder with style.

4. (Later example.)

1950 Engineering 13 Jan. 30/1 [In flour milling] the endosperm released..and sifted out in the scalpers consists of particles of various sizes.

scalpette (skælpe·t). [f. Scalp *sb.*[1] + -ette.] (See quots. 1887, 1960.)

1881 in J. Lichtenfeld Princ. Mod. Hairdressing 35 (Advt.), J. Lichtenfeld's Illustrated Catalogue contains Illustrations and Description of..Invisible Scalpette Fringes, Bébé Scalpette, [etc.]. 1887 E. Creer Board-Work vii. 97 What is a scalpette?.. I..consider it to signify an artificial covering for concealing a deficiency of hair, or to cover a bald place upon the female head—but not a wig. 1924 Chambers's Jrnl. Oct. 669/1 Skilled hair-workers who will make up wigs, scalpettes. 1960 C. W. Cunnington et al. Dict. English Costume 190/1 Scalpette,..a false front of invisible net to which luxuriant tresses are attached. 1961 E. S. Turner Phoney War vii. 54 My private drill consists of taking off my scalpette, putting it into the sponge bag and then slipping the latter with its closed outlet downwards into the hip pocket.

scalping, *vbl. sb.*[3] **2.** (Further examples.)

1922 Jrnl. Inst. Metals XXVIII. 881 Rolling data for brass and bronze, scalping, annealing, and pickling.

1949 J. E. GARSIDE *Process & Physical Metall.* viii. 123 It is becoming general practice in the case of non-ferrous alloys to subject slabs and billets to a surface machining operation known as 'scalping' prior to cold-rolling. 1960 *New Scientist* 19 May 1269/2 The machine..is used for scalping. By scalping, the quarryman means separating the dirt from the mine output before the stones are passed into the crusher. 1967 *Gloss. Highway Engin. Terms* (*B.S.I.*) 25 *Scalpings*, hard material extracted as being unsuitable for crushing and screening. 1975 *Bristol Evening Post* 19 Feb. 1/4 The firm plan to protect Portway and services underneath from impact, possibly with a blanket of steel plating topped by scalping and sand.

scaly, *a.* Add: **8. scaly-bark (hickory),** substitute for def.: the shagbark hickory, *Carya ovata*, or its edible nuts; cf. HICKORY 1; (earlier and later examples); **scaly-tail** = *scale-tail* s.v. SCALE *sb.*[1] 12; so **scaly-tailed** *a.*

1775 J. ADAIR *Hist. Amer. Indians* 360 Filberts..are as sweet and thin-shelled, as the scaly bark hiccory-nuts. 1785 T. JEFFERSON *Notes on Virginia* vi. 63 Scaly bark hiccory. 1814 F. PURSH *Flora Amer.* II. 637 This useful tree is known by the name of..Scaly-bark Hickory, on account of its bark, which is torn in loose fragments. 1852 MAYNE REID *Desert Home* 198 The tree is known among backwoodsmen as the 'scaly bark'. 1906 'O. HENRY' *Rolling Stones* (1912) 8, I saw..a little flaxen-haired man with a face like a scaly-bark hickory-nut. 1921 *Brit. Mus. Return* 97 in *Parl. Papers* XXVII. 651 A West African Scaly-tail (*Anomalurus erythronotus*), and an Ituri Scaly-tail (*Anomalurus pusillus*). 1964 L. S. CRANDALL *Management of Wild Mammals in Captivity* 229 The life-histories of the scaly-tails are not well known. 1962 M. BURTON *Syst. Dict. Mammals of World* 121 Scaly-tailed Flying Squirrels..not related to true squirrels. 1964 E. P. WALKER et al. *Mammals of World* II. 750/2 Scaly-tailed squirrels den in hollow trees. 1975 P. W. HANNEY *Rodents* ii. 30 There are no flying squirrels in Africa, but in the west of the continent their niche is filled by..the Anomaluridae or scaly-tailed squirrels.

scaly (skē̆ⁱ·li), *sb. S. Afr.* Also **scaley.** [f. the adj.] A large yellow-fish, *Barbus natalensis*, of the family Cyprinidæ, found in certain rivers in Natal.

1947 K. H. BARNARD *Pict. Guide S. Afr. Fishes* 56 The well-known Scaley..of Natal is a near relative of the Yellow-fish. 1971 *Rand Daily Mail* (Johannesburg) 27 Mar. 23/3 An interesting observation last week was the presence of shoals of scalies in the Bushmans river. 1975 *Stand. Encycl. S. Afr.* XI. 563/1 The Natal scaly..reaches 5 kg and is restricted to the Pongola system and the rivers of Natal.

scam (skæm), *sb. slang* (orig. and chiefly *U.S.*). [Origin obscure.] **1. a.** A trick, a ruse; a swindle, a racket (*sb.*[3] 3). Also *attrib.*

1963 *Time* 28 June 48/2 He..worked..as a carny huckster... 'It was a full scam.' 1971 *Harper's Mag.* Feb. 89 A gambling house is a sitting duck to every con man or outlaw who comes through; he is invariably convinced that he has a scam that you have never seen before. 1972 *Sunday Mail* (Brisbane) 2 July 19/6 It was necessary to the success of the latest 'scam' that it be worked in places where $25 chips were constantly in play. 1975 J. F. BURKE *Death Trick* (1976) iv. 64 Hustling of any kind he could live with in his hotel, dope-dealing, selling ass, almost any scam, even burglary. 1976 M. MACHLIN *Pipeline* v. 58 Gamblers, pimps, whores, conmen, and scam artists of every persuasion were drawn to the scene like sharks. 1978 M. PUZO *Fools Die* xii. 131 The bribe-taking scam had been going on for nearly two years without any kind of hitch.

b. *spec.* A fraudulent bankruptcy (see quot. 1966). Also *attrib.*

1966 *Wall Street Jrnl.* 9 Sept. 1/1 (*heading*) 'Fat Man' Scolnick & 'scams'... They're known as 'scam' operators, promoters who set up ostensibly legitimate businesses, order large amounts of merchandise on credit, sell it fast and strictly for cash—and then go 'bankrupt', leaving their creditors unpaid. 1968 J. M. ULLMAN *Lady on Fire* (1969) xiv. 181 'The main plan's to go bankrupt... The suppliers will be stuck with unpaid bills for millions. There's a name for that—' 'Scam game,' Forbes said. 1974 *N.Y. Times* 8 July 26/1 Organized crime is stealing millions of dollars from the public through planned fraudulent bankruptcies, called 'scams' by the underworld.

2. A story; a rumour; information.

1964 *Guardian* 8 July 7/6 'People want the 1930s all over again: a thousand naked chorus girls dancing in a pink smog under crystal chandeliers on a revolving staircase on an Alp.'.. 'Didn't someone tell us once that Hollywood went bust with that scam?' 1966 *Amer. Speech* XLI. 281 *Lowdown, scam, the word*, information. 1972 W. MCGIVERN *Caprifoil* (1973) viii. 137 There's been a security break... He's scheduled a press conference... The scam is he's going to break what we know on Spencer. 1972 J. WAMBAUGH *Blue Knight* (1973) i. 28, I paid them [*sc.* informers] from my pocket, and when I made the bust on the scam they gave me, I made it look like I lucked on to the arrest. 1976 *New Musical Express* 17 Apr. 10/2 No, still no scam on Donny and Marie.

scam (skæm), *v. slang* (orig. and chiefly *U.S.*). [Origin obscure: cf. prec.] *intr.* and *trans.* To perpetrate a fraud; to cheat, trick, or swindle. Hence **sca·mming** *vbl. sb.* (in sense 1 b of *SCAM *sb.*).

1963 *Time* 28 June 48/2 My boss was scammin' from the public, and I was scammin' from him. 1966 *Wall Street Jrnl.* 9 Sept. 1/1 'Scam' originally was a carnival term meaning 'to fleece the public'. 1974 *Whig-Standard* (Kingston, Ont.) 9 Apr. 4/1 Scamming..is a form of criminal bankruptcy in which a front man buys out a legitimate firm and then uses the credit rating of the firm to buy large quantities of merchandise. *Ibid.* 4/3 Scamming, he said, ranks second only to bookmaking in financial importance to criminals. 1977 *New Yorker* 30 May 96/2 Local citizens..try to avoid being scammed by the familiar tergiversations of city politicians.

scamphood. For *nonce-wd.* in Dict. read *rare* and add earlier example.

1845 J. S. LE FANU *Cock & Anchor* I. xvii. 257 He was ripe for the domestic virtues, and ought to renounce scamphood.

scampi (skæ·mpi), *sb. pl.* [a. It. *scampi.*] **1.** Also in *sing.* **scampo.** = *Dublin Bay prawn* s.v. *DUBLIN.

1928 RUSSELL & YONGE *Seas* xiv. 316 It is extremely plentiful in the Adriatic and is sold in the Italian ports under the name of 'Scampo'. 1953 P. BONNER *SPQR* viii. 70 Those little *scampi* are not enough for hungry fishermen. 1966 *Punch* 28 Sept. 483/3 The mysterious scampo which we see in this country, an animal which appears to have no head, is in fact the tail of the Dublin Bay Prawn. 1972 *Daily Tel.* 5 May 2/6 Scampi..make deep burrows in the mud..and thus easily escape the trawling nets of fishermen.

2. a. (A dish of) these prawns eaten as a delicacy, usu. coated with breadcrumbs and fried in oil, or boiled and served with (garlic) sauce.

1930 E. WAUGH *Labels* vi. 158, I ate *scampi* at Cavaletto and felt no ill effects. 1951 N. BALCHIN *Way through Wood* xii. 176 You look like a man who's been gazing on the scampi when they're brown. 1958 A. WILSON *Middle Age of Mrs Eliot* II. 268 It's sure to be scampi or snails or something I couldn't eat. 1962 D. LESSING *Golden Notebook* II. 244 This theme takes us through scampi and the main course. 1978 *Times* 11 Apr. 16/4 Bartolomeo Calderoni...introduced scampi to Britain. *Ibid.* 16/5 As for the scampi, he imported them from Venice's Grand Hotel..when he was head chef at Quaglino's in the 1930s.

b. *attrib.* and *Comb.*

1959 *Good Food Guide* 46 Scampi Provençales and duck Grand Marnier can be arranged to order. 1960 *House & Garden* July 60/3 Such memorable dishes as scampi risotto. 1966 D. SKIRROW *It won't get You Anywhere* vi. 32 What about a little schmaltzy restaurant down the King's Road? Or..maybe further up the river in the scampi belt? 1977 *Chicago Tribune* 2 Oct. VI. 19/1 The scampi marsala and the baked clams also were worthy antipasti choices as was the cannelloni bechamel. 1980 *West Lancs. Evening Gaz.* 7 Jan. 11 (Advt.), Vacancies for full-time/part-time scampi processors.

scan, *sb.* Add: **1. a.** (Further examples.)

1970 O. DOPPING *Computers & Data Processing* xvii. 277 If the computer were to continue the forward scanning, four scans would be needed. 1973 W. MCCARTHY *Detail* ii. 90 The air marshals scanned their bodies with their eyes. Ben passed through. I guess this scan works, he thought.

b. The action or practice of scanning with a beam, aerial, or detector. Cf. *SCAN *v.* 6 f.

1937 *Discovery* Nov. 330/1 This..scheme is modified by leaving out alternate lines during alternate scans, a technique which improves the definition and reduces flicker. 1955 *Sci. Amer.* June 41/1 When Hey published his discovery after the war, radio astronomers began an intensive radio scan of the Sun. 1958 *Times* 2 May 7/2 One of these provided the long-range warning, while the others made a coordinated scan of various sections of the target area as the structure rotated. 1966 M. WOODHOUSE *Tree Frog* xxi. 155 Say that echo's your drone up there... Then you get your vertical scan radar for altitude. 1972 *Sci. Amer.* Jan. 57/1 The rate of scan that produces a micrograph is often much lower than the scanning rate in television.

2. A single line or sweep produced by or in a scanning action (cf. *SCANNING *vbl. sb.* 2 b); also, an entire raster.

1934 J. H. REYNER *Television* ix. 103 The separation between the centres of the lenses was equal to the width of the picture scan. 1945 *Electronic Engin.* XVII. 689 The scan to fly-back ratio is constant for all time base velocities. 1952 *Jrnl. Lab. Clin. Med.* XXXIX. 153 The counter is moved alternately back and forth, with an ¼ inch vertical displacement for each sweep or scan over the area occupied by the thyroid gland. 1966 *Electronics* 17 Oct. 114 The large spike at the beginning of each scan is a turnaround transient. 1967 [see *FIELD *sb.* 16 d]. 1975 D. G. FINK *Electronic Engineers' Handbk.* xx. 7 The half line, left over at the end of the field scan, displaces the next field downwards by a full line, and interlacing is achieved.

3. An image, diagram, etc., obtained by scanning; *spec.* in *Med.* = *SCINTISCAN.

1953 *Nucleonics* Nov. 45/1 Fig. 13 presents the coincidence scan and unbalance scan of a patient who showed a regrowth of tumor beneath an area of previous resection. 1956 *Jrnl. Neurosurg.* XIII. 347 (*heading*) This scan is in the posterior-anterior orientation of the head. 1969 M. CRICHTON *Andromeda Strain* I. 22 We'll want a flyby over that town... And a complete scan. 1971 *Guardian* 6 Feb. 1/7 There were, as the first scan of the [lunar] landscape showed, a few very large boulders. 1976 *Woman's Day* (U.S.) Nov. 164/2, I might ask for bone, liver and brain scans to make sure there had not been any metastasis to other parts of my body. 1978 *Nature* 14 Dec. 733/2 (*caption*) Absorbance scans of..SDS-polyacrylamide gels.

4. Special Comb.: **scan-column index,** a tabular representation of coded information concerning or contained in a set of documents, for use in information retrieval.

1962 J. O'CONNOR in *Amer. Documentation* XIII. 205/1 Place the document number in the left-hand column. Then, for each indexing term assigned to that document, look up the column and character abbreviation for the term, and in that column enter that character. I call an index of this form a Scan Column index. 1965 M. E. STEVENS *Automatic Indexing* vi. 118 Tabledex, the Scan-column Index, and similar tools provide to some extent a display of prior associations between index terms. 1971 A. GILCHRIST *Thesaurus in Retrieval* 140 The scan-column index. This is another book-form coordinate information retrieval system..in which all the item numbers are listed numerically in the first column, the other columns containing descriptors, allotted to those items. A separate table indicates which column would be searched for a particular descriptor. To facilitate searching, descriptors have been reduced to symbols.

scan, *v.* Add: **6. b.** To search (literature, a text, a list, etc.) quickly or systematically for particular information or features.

1926 *Rec. Geol. Surv. India* LIX. 202 On scanning this table it will be observed that the pyrope molecule is present in quantity..only in one garnet. 1950 *Amer. Documentation* I. 81 The rapid selector employs an optical-electronic system for scanning a reel of motion picture film on which are entered both abstracts and corresponding index entries. 1966 *Computers & Humanities* I. 12 Some [articles] are so superficial that the reader for whom the volume is designed would do better to scan the most recent ACLS list of computerized research projects in the humanities. 1967 C. BERNERS-LEE in Wills & Yearsley *Handbk. Management Technol.* 7 Some computer manufacturers supply..suites of statistical programs for scanning files to accumulate the required statistics and then to analyse them in one of a number of ways. 1967 *Times Rev. Industry* July 89/2 Without guide lines as to where the company wants to go, scanning environmental information becomes directionless. 1970 O. DOPPING *Computers & Data Processing* xvii. 277 The computer first scans the table from the beginning to the end comparing the first record with the second, the second with the third, etc. 1972 *Computers & Humanities* VII. 19 Dilligan examines the extent to which linguistic orientation toward prosody serves as the basis for computer programs to scan large bodies of English verse. 1973 *Nature* 31 Aug. p. xiii/1 (Advt.), He or she will be required to scan incoming literature, undertake literature searches.

c. To cause (an area, object, or image) to be systematically traversed by a beam or detector; to convert (an image) into a linear sequence of signals in this way for purposes of transmission or processing.

1928 *Television* Nov. 9/1 One feature which is wrongly quoted by critics relates to how a scene is scanned. 1933 *Proc. Wireless Section Inst. Electr. Engineers* VIII. 219/2 Nipkow in 1884 proposed..to transmit the picture point by point, or to scan the picture. 1953 AMOS & BIRKINSHAW *Television Engin.* I. iv. 52 The electron beam is made to scan the target in a series of nearly horizontal lines. 1954 *Nucleonics* Jan. 60/1 By placing tracing paper and carbon paper between the stylus and the drawing table, the distribution of radioactivity in an area being scanned is recorded. 1962 A. NISBETT *Technique Sound Studio* iv. 81 A replay head scans a slightly greater length of tape than would be suggested by the size of the gap. 1967 *Nursing Times* 18 Aug. 1093/1 Not so well known is the use of radioisotopes in radiography, to enable various organs of the body to be 'scanned' to investigate function, or the presence of tumours. 1968 *Brit. Med. Bull.* XXIV. 191/2 Radiographs..and so on..can also be digitally structured for insertion in a computer... They are scanned line by line (television-wise) by a 'flying-spot scanner', and passed through an analogue-to-digital converter which..encodes the contrast level of each point, as a row of holes. 1969 *Times* 15 Mar. 7/8 The photographs are scanned point by point by a photoelectric device. 1975 D. G. FINK *Electronics Engineers' Handbk.* xx. 79 Hard-copy facsimile systems generate a signal by systematically scanning the subject copy and producing a current corresponding to its light-intensity variations.

d. *intr.* To carry out scanning. Const. various preps.

1934 J. H. REYNER *Television* viii. 95 By causing the spot on the cathode ray screen to scan over a suitable area the image of the spot traverses the whole of the film. 1948 'N. SHUTE' *No Highway* v. 147 What interested me most, however, as in every technical paper that one scans through quickly, was the paragraph headed 'Conclusions'. 1953 A. T. STARR *Radio & Radar Technique* i. 46 For the purpose of homing on a ship or aircraft, it is sufficient to scan through a relatively small angle in azimuth, say ±30°. 1961 G. MILLERSON *Technique Television Production* ii. 20 A gun in the picture-tube..produces such a stream of electrons, and this is made to scan over the powdered screen in a regular series of sweeps. 1965 'J. LE CARRÉ' *Looking-Glass War* xviii. 204 He may start with the wrong crystal... It's safest for base to scan with so many crystals. 1975 *Physics Bull.* July 327/1 As the beam repeatedly scans across the faceplate, charge is accumulated. 1979 *Sci. Amer.* Mar. 82/1 Given a source of light that is monochromatic but tunable, an absorption spectrum can be measured by passing the light through a sample of the gas and scanning continuously through the frequencies surrounding a line in the spectrum.

e. *trans.* To traverse or light upon (a constituent element) as part of the scanning of the larger whole.

1937 A. M. TURING in *Proc. London Math. Soc.* XLII. 233 The machine moves so that it scans the square immediately to the right of the one it was scanning previously. 1937 *Discovery* Nov. 329/1 When the dots are being scanned, the transmitted signal depends on the relative brightness of the dots in turn. 1961 G. MILLERSON *Technique Television Production* ii. 19 As each element is scanned and gives up its information, it becomes 'wiped clean'.

f. To cause (a beam, etc.) systematically to traverse an area; to cause (an aerial) to rotate or oscillate to this end.

1960 E. V. TRUEFITT in R. F. Hansford *Radio Aids to Civil Aviation* v. 328 The nodding heightfinder is so called because the aerial performs a nodding motion which scans the radar beam in elevation. 1972 *Sci. Amer.* Nov. 40/3 The beam is positioned and focused by scanning the beam over the sample surface and detecting the change in the emission of secondary and reflected electrons as the beam passes over surface detail. 1973 MEYER & MAYER *Radar Target Detection* i. 15/1 If the antenna is scanned sufficiently slowly, more than one pulse may be transmitted and received while the antenna beam sweeps across a given reflecting point. 1976 *Physics Bull.* Oct. 437/1 The proton beam was scanned right across the annulus and hole. 1977 *Sci. Amer.* Sept. 123/3 A much smaller area..is exposed, and the exposure is repeated by either stepping or scanning the image over the wafer.

scanned *ppl. a.* (later examples).

1937 *Proc. London Math. Soc.* XLII. 231 We may call this square the 'scanned square'. 1953 AMOS & BIRKINSHAW *Television Engin.* I. iv. 52 To avoid keystone effect and obtain a true rectangular scanned area, the line saw-tooth current is modulated by the field saw-tooth current so that the angular sweep in the horizontal plane is decreased as the beam moves up the mosaic. 1975 D. G. FINK *Electronics Engineers' Handbk.* xx. 8 Color television standards use the back porch to position the color burst, an eight-cycle burst of color subcarrier..that synchronizes the color-subcarrier oscillator at the end of each scanned line.

Scand (skænd). Colloq. abbrev. of SCANDINAVIAN *sb.*

1930 J. MASEFIELD *Wanderer of Liverpool* 14 The others, all Scands, from North Europe, not knowing a word Of English. 1965 *Sun* 28 Sept. 4/5 She is still reckoned a world beauty. How does she do it, this middle-aged Scand? 1973 H. MILLER *Open City* ix. 87 She had spent four days among the stodgy, unsubtle Scands on his behalf.

scandal, *sb.* Add: **1. c.** *scandal of particularity* [tr. Ger. (see quots. 1930, 1936)], the difficulty of seeing the particular man, Jesus, as the universal Saviour. Cf. *PARTICULARITY 1.

1930 tr. G. Kittel in Bell & Deissmann *Mysterium Christi* ii. 31 The scandal of particularity..is the problem of history. Can a particular historical happening be peculiar? Can it be significant *sub specie aeternitatis*? And above all, can this particular occurrence be either peculiar or significant? 1936 C. H. DODD *Apostolic Preaching & its Development* iv. 219 'Like a strange people left on earth After a judgment day.' This view of the historical status of the events comprised in the coming of Christ introduces us at once to what Professor Gerhard Kittel, in *Mysterium Christi*, calls '*das Ärgernis der Einmaligkeit*', 'the scandal of particularity'. 1961 *Listener* 9 Mar. 435/2 We do no service to religion by reducing either term of the problem, the total mystery of the Godhead or the scandal of particularity. 1979 C. F. D. MOULE in M. D. Goulder *Incarnation & Myth* iv. 86 The 'scandal of particularity' is by no means a denial but rather a confirmation of the ubiquity and continuity of God's activity.

7. scandal sheet, a newspaper that is notorious for publishing scandalous or sensational stories.

1904 ADE *True Bills* 110 The Scandal Sheets never show up my Family History. 1939 R. CHANDLER *Big Sleep* xi. 82 The deal has to be closed to-night or they give the stuff to some scandal sheet. 1974 M. HOUSE et al. *Lett. Charles Dickens* III. 363/2 The *Age* and *Satirist*, though infamous indeed, were mere weekly scandal-sheets of no influence or political import. 1981 C. R. LAJEUNESSE *Dead Man Running* xi. 33 Nobody pays attention to that scandal sheet, let alone reads it.

Scandihoovian (skæ:ndihū·viăn), *sb.* (and *a.*) Also **-huvian.** *slang* (chiefly *N. Amer.*). Arbitrary jocular alteration of SCANDINAVIAN *sb.* Also as *adj.*

1929 F. BOWEN *Sea Slang* 117 *Scandihoovian*, any Scandinavian; used as an alternative to *Scandiwegan* or *Scowegian*, but generally in mild contempt. 1966 *Publ. Amer. Dial. Soc.* 1964 XLII. 39 Applied to all Scandinavians..were *scoop*, *Scandihuvian*, and *Scandie*. 1968 *Amer. Speech* XLIII. 303 There's plenty of color in his [*sc.* the logger's] language too... His 'snuff' is *Scandihoovian dynamite*. 1973 B. BROADFOOT *Ten Lost Years* xi. 120 Salt cod! Ugh! Even a Scandihoovian couldn't take that.

Scandiknavery (skæ:ndinē̆ⁱ·vəri). *nonce-wd.* [Fanciful blend of SCANDINAVIAN *a.* or *sb.* and KNAVERY.] Deceit or trickery by Scandinavians.

1939 JOYCE *Finnegans Wake* 47 We'll have a free trade Gaels' band and mass meeting For to sod the brave son of Scandiknavery. 1971 S. E. MORISON *European Discovery Amer.: Northern Voy.* iii. 72 Michael A. Musmanno's *Columbus was First* (1966) is an amusing, emotional assault on what he calls 'Scandiknavery'.

Scandinavian, *a.* Add: **2.** Applied to a style of furnishing, etc., in a Scandinavian manner, esp. as characterized by simplicity of design and the use of pine-wood.

1959 R. CONDON *Manchurian Candidate* ii. 19 All of the furniture was made of blond wood in mutated, modern Scandinavian design. 1964 L. DEIGHTON *Funeral in Berlin* vi. 42 There was Scandinavian-style East German furniture in the room. 1968 S. B. HOUGH *Sweet Sister Seduced* xxviii. 163 He looked around the room, at the Scandinavian chairs, at the window curtains, and the Hi-fi in the corner. 1972 C. FREMLIN *Appointment with Yesterday* xi. 83 Visions of colourful teenage rooms in the Sunday colour-supplements, with Scandinavian wood window-seats, and bright cushions. 1979 M. EDEN *Document of Last Nazi* xxix. 171 A neat, cold-looking room, with..Scandinavian furniture.

B. *sb.* **2.** The various languages of the Scandinavian peoples considered as a unit; *spec.* North Germanic, a subdivision of the Germanic group of Indo-European languages spoken principally in Scandinavia.

1766 J. CLELAND *Way to Things by Words* 63 A sense which it also specifically has in the old Scandinavian. 1822 tr. *Malte-Brun's Universal Geogr.* I. 568 The Mœso-Gothic,..the Icelandic and modern Scandinavian, in its two principal dialects the Swedish and the Danish, constitute the Gothic branch. 1888 J. WRIGHT tr. *Brugmann's Elem. Compar. Gram. Indo-Gmc. Lang.* I. 10 Norse (or Scandinavian)..down to the Viking period (800–1000 A.D.) was practically a single language. 1933 L. BLOOMFIELD *Language* iv. 59 While the language of the Lombards seems to have been of the West Germanic type, the others, including Gothic, were closer to Scandinavian. 1954 PEI & GAYNOR *Dict. Linguistics* 148 *North Germanic*, a branch of the Germanic group of the Indo-European family of languages; it comprises Icelandic, Swedish, Danish, Norwegian, Faroese and Gotlandic (or Gutnian). Also called *Scandinavian*. 1966 W. P. LEHMANN in Birnbaum & Puhvel *Anc. Indo-Europ. Dial.* 18 The occurrence of a third singular form without *-t* in the three coastal dialects of West Germanic and in Scandinavian gives evidence of interrelations between these dialects subsequent to the earliest dialect division of Proto-Germanic. 1978 W. WHITE in W. Whitman *Daybooks & Notebooks* I. 69 Rasmus B. Anderson..Professor of Scandinavian at the University of Wisconsin.

Scandinavianize (skæ:ndinē̆ⁱ·viănəiz), *v.* [f. SCANDINAVIAN *a.* + -IZE.] *trans.* To render (place-names, etc.) Scandinavian in form or character. So **Sca:ndina·vianized** *ppl. a.*; hence **Sca:ndina:vianiza·tion.**

1924 MAWER & STENTON *Introd. to Survey of Eng. Place-Names* iv. 60 English names often appear in a Scandinavianised form. 1933 *Times Lit. Suppl.* 20 Apr. 271/3 The distribution of place-names..is conclusive proof that all but the south-western Corner of Northamptonshire was Scandinavianized to a far greater extent than has been realized. 1937 *Harvard Stud. & Notes in Philol. & Lit.* XX. 155 *Cyninges-clif*, High Coniscliffe... The modern form shows adaptation of the first el[ement] to ON *konungr* for OE *cyning*; for further examples [of] a similar Scandinavianization see..'Connington'. 1956 I. S. MAXWELL in D. L. Linton *Sheffield* 131 Those settlements sited somewhat farther from the rivers whose names also contain Scandinavian elements or have been Scandinavianized. 1959 C. L. WRENN *Word & Symbol* (1967) 24 It would seem..that the Irish word [*cros*] came into Old English rather through Scandinavianised Irish settlers than direct. 1962 H. R. LOYN *Anglo-Saxon England* (1963) i. 60 Only occasionally, as in the Wreak valley..is there overwhelming Scandinavianization of the place-name structure. 1970 *Jrnl. Eng. Place-Name Soc.* II. 12 Rudston YE 98 (originally OE *rōd-stān) appears in the Bruce fief with the second element scandinavianised to *-stein*, *Rodestein* 332v beside DB *Rodestan*. *Ibid.*, The second element has probably been scandinavianised to *-heim*. 1981 *N. & Q.* Apr. 177/1 In the Danelaw the Grimston(e) names have been taken to be partial scandinavianizations of earlier English names.

scandium. Substitute for def.: A silvery white metallic element, the 'eka-boron' of Mendeleev, which is found in small quantities in association with rare-earth metals (among which it is often classified) and in some tin and tungsten ores, and forms colourless salts in which it is trivalent. Symbol Sc; atomic number 21. (Add further examples.) Hence **sca·ndia,** the white oxide, Sc_2O_3. [*scandium* first formed in Sw. (L. F. Nilson 1879, in *Öfversigt af K. Vetenskaps-Akad. Förh.* XXXVI. III. 47).]

1880 *Jrnl. Chem. Soc.* XXXVIII. 7 Scandium forms but one oxide, scandia, Sc_2O_3. 1922 *Nature* 17 June 799/1 The extraction and purification of scandium from thorveitite of Madagascar. This mineral, which contains 42 per cent of scandium oxide, is fused with soda and the silica removed by washing. 1955 *Sci. News Let.* 11 June 374/3 Studies made by Britain's Hydraulics Research Station and the Atomic Energy Research Establishment have shown that very finely ground glass containing radioactive scandium oxide moves with the mud when mixed with it in the River Thames. 1974 *Encycl. Brit. Micropædia* VIII. 944/3 Nilson discovered..its oxide, scandia, in the rare-earth minerals gadolinite and euxenite. *Ibid.*, Scandium is now produced on a small scale mostly as a by-product of uranium extraction from the mineral davidite (about 0·02 percent scandium oxide). Very few uses..have been developed.

Scanian (skē̆ⁱ·niăn), *a.* (*sb.*) [f. med.L. *Scania*, ad. ON *Skáni* or *Skáney*, the province of Skane in south Sweden + -AN.] **1.** Of or pertaining to the province of Skåne.

1895, etc. [see below]. 1932 *Times Lit. Suppl.* 28 Apr. 306/3 The unrounded *o*, which sometimes becomes the 'sorry caterwaul' of Scanian Swedish. 1963 J. SAHLGREN in Brown & Foote *Early Eng. & Norse Studies* 176 The formation of Scanian place-names.

2. a. Designating the first glaciation of the Pleistocene in northern Europe, roughly corresponding to the Günz glacial in the Alps. Also as *sb.* Now *rare*.

1895 J. GEIKIE in *Jrnl. Geol.* III. 246 *Scanian*. The earliest glacial deposits of northern Europe occur in Skåne—the old division of southern Sweden—hence the provisional name I suggest. *Ibid.* 263 Not a trace of the Scanian bowlder-clay has been recognized in Britain 1903 A. GEIKIE *Text-bk. Geol.* (ed. 4) II. 1313 Scanian or 1st Glacial Epoch, represented only in the south of Sweden (Scania), which was overridden by a large Baltic glacier. To this period may belong..the oldest terminal moraines and fluvio-glacial gravels of the Arctic lands. 1910 *Encycl. Brit.* XII. 59/1 Although it is admitted that no strict correlation of the European and North American stages is possible, it has been suggested that..the Kansan may represent the Saxonian;..the Jerseyan, the Scanian; [etc.]. 1957 J. K. CHARLESWORTH *Quaternary Era* II. xxxvi. 921 J. Geikie, a constant advocate of multiplicity, postulated six glaciations, named Scanian, Saxonian, Polandian, [etc.]. 1972 R. G. WEST *Pleistocene Geol. & Biol.* (ed. 2) xi. 219 (*table*) Scanian.

b. Designating a stade in the retreat of the ice-sheet at the end of the last Pleistocene glaciation in northern Europe, (corresponding to the end of the Würm glacial in the Alps) and the resulting stadial moraines. Also as *sb.*

1937 WOOLDRIDGE & MORGAN *Physical Basis Geogr.* xxiii. 413 The retreat [of the Scandinavian ice-sheet] was punctuated by pauses, marked by well-developed stadial moraines. We have thus the Pomeranian Moraine, the Scanian Moraine (13,700 B.C.) and the Salpausselka of Finland. 1963 R. A. DALY *Changing World of Ice Age* ii. 54 During the third substage, which will be referred to as the Scanian, the front retreated to the position of one of the strong moraines in central Finland, where it bears the name 'First Salpausselkä Moraine'.

scannable, *a.* (Later examples.)

1936 W. DE LA MARE *Wind blows Over* 37 How narrow a circle of its waters was actually scannable from where she stood. 1975 *Nature* 28 Aug. 703/1 The charges can be liberated optically, in which case the device acts as a scannable photo-detector array, already on the market in the form of hand-held TV cameras.

scanner. Add: **1.** (Further example.)

1967 *Times Rev. Industry* July 89/2 A really sensitive scanner can pick up a bit of information which to most people would be..irrelevant but which to him assumes significance.

3. a. Any device for scanning or systematically examining all parts of something.

1927 *Public Opinion* 18 Feb. 152/3 Place the 'telegraph card' on an endless band passing at a fixed rate under the 'scanner', while at the other end a reproduction soon tumbles into a basket. 1952 *Progress* Spring 34 The Time-Springdale electronic scanner was used to make, from an Ektachrome transparency, the negatives needed for the colour reproduction. 1958 *Times Rev. Industry* July 25/3 A beam-scanner with an output window of pure aluminium foil is fitted to the accelerator. 1970 *New Scientist* 27 Aug. 420/3 Trials with an airborne infrared scanner over the leaking oil well..demonstrated that oil shows up clearly on the scanner. 1977 *Listener* (N.Z.) 15 Jan. 10/1 Scanners are an important innovation because of their superior diagnostic capabilities over alternative techniques, and the enhanced degree of patient safety and comfort they offer. 1977 *Time* 27 June 25/3 Using elaborate 'scanners' to monitor police radio channels, reporters were often at the spot of a reported sighting before the guards and dogs.

b. *Television.* Any of several devices that permit the sequential transmission of an image or its subsequent reconstruction in a receiver.

1929 SHELDON & GRISEWOOD *Television* xiv. 162 Another decided advantage of the drum-scanner is its compactness. 1958 *Observer* 12 Oct. 1/3 There is..a television-type scanner which can transmit pictures of the perpetually unseen far side of the moon. 1975 D. G. FINK *Electronics Engineers' Handbk.* xx. 31 Flying-spot scanners are used in broadcast stations in Europe for both color and monochrome film reproduction.

c. A transmitting and receiving radar aerial, usu. one that rotates or oscillates in order to scan a large area.

1946 *Electronic Engin.* XVIII. 360/2 The speed of rotation of the scanner when manually controlled can be varied from 0 to 4 r.p.m. in either direction. 1965 *New Scientist* 15 July 130/1 The radar scanner continuously sweeps the sky, sending out its impulses and receiving back the reflected impulses. 1970 H. A. TAYLOR *Airspeed Aircraft since 1931* 172 The raised fuselage also permitted the aircraft to be loaded with the retractable air-to-surface-vessel (ASV) scanner, below and aft of the control cabin.

d. *fig.*

1959 *Listener* 8 Jan. 83/2 The selectivity of your mind—your mental scanner—will quickly reveal what there is to be thought about. 1964 *Ann. Reg.* 1963 403 Work by an Oxford psychologist..suggested that the mechanism..involved the use of a 'scanner' in the brain which sifted through a large number of possible names.

4. Special Comb.: **scanner fo(u)nt,** a type-writing font that can be read by an optical character-recognition device.

1968 *Amer. Documentation* Jan. 74/2 A secretary types each record for scanner input, using a standard typewriter fitted with an ASA scanner font golfball. *Ibid.*, The optical character reader..reads standard ASA scanner font: 26 alphabetic characters (all upper case), 10 digits, and 25 punctuation and special characters. 1969 *Computers & Humanities* III. 132 Each item to be entered into the file is coded; typed in ASA scanner font; read by an optical scanner..onto magnetic tape; and, finally, entered into an INFOL file.

scanning, *vbl. sb.* Add: **2. b.** The action of systematically traversing with a beam or detector, esp. in *Television*.

1927 *Bell System Technical Jrnl.* VI. 552 We have thus available in television the same artifice..that is, of scanning, or running over the elements of the image in sequence. 1933 *Discovery* May 156/2 As much as 120-line scanning was used, thus permitting very fine detail indeed. 1936 *Electr. Commun.* XV. 187/1 The most recent demonstration of television in Italy was at Milan in April, 1936. The equipment employed electronic scanning for transmission. 1971 *Amat. Photographer* 13 Jan. 65/2 The system [for reading videotape] was later superseded by transfer scanning, using four magnetic heads on a 2 in diameter drum rotating at 14,400 rpm almost at right angles across 2 in tape, pulled past at 15 ips. 1975 D. G. FINK *Electronics Engineers' Handbk.* xx. 7 Interlaced scanning is achieved by making the horizontal (line-scanning) rate an odd multiple of one-half the vertical (field-scanning) rate.

c. The rapid or systematic searching of textual material for particular information or features.

1937 *Discovery* Sept. 256/2 A random scanning of the list reveals many names familiar to the British Association. 1954 *Amer. Documentation* V. 18/2 Speeds of operation are such as to permit scanning and correlating of generic and specific aspects of indexes in a reasonable time. 1967 *English Studies* XLVIII. 60 (*heading*) An archive of older Scottish texts for scanning by computer. 1970 O. DOPPING *Computers & Data Processing* xvii. 277 The misplaced record has only been moved one step, and if the computer were to continue the forward scanning, four scans would be needed. 1975 *Language for Life* (Dept. Educ. & Sci.) viii. 115 The intermediate skills, so essential in word attack in the early stages, are at work in skimming, scanning, and the extraction of meaning in the more complex reading tasks of the later stages.

4. *auditory scanning*: the emission of short pulses of sound and detection of echoes from nearby objects, thought to be used by dolphins for the location and ranging of submerged objects.

1960 W. N. KELLOGG in *Psychol. Record* X. 26 Since the noises which make up the echoes are emitted by the dolphin itself, the activity as a whole amounts to a kind of scanning by sound. We suggest the term *auditory scanning*, therefore, as a good name for both the acoustic and the general behavior comprising this elaborate pattern of activity. 1963 *Language* XXXIX. 464 The dolphin's auditory scanning is shown to consist of the emission of a continuous series of sound signals for echolocation plus binaural localization.

5. *attrib.* and *Comb.*, as *scanning movement, speed*; **scanning coil,** any of four coils arranged in pairs around the neck of a cathode-ray tube, the magnetic field of which is varied so as to cause the electron beam to trace out a raster pattern on the screen of the tube; **scanning disc,** a rotating disc having a spiral of holes near the edge, used in mechanical systems of television to provide a sequential scan of a scene by optical means for transmission and to permit reconstruction of the scene at the receiver; **scanning electron microscope,** a form of electron microscope in which an electron beam is scanned in a raster pattern across the specimen; an electrical signal is obtained by collecting and amplifying secondary electrons emitted by the specimen and is applied to a cathode-ray tube scanned in synchronism with the electron beam; hence *scanning electron micrograph, microscopy*; **scanning field** = *RASTER *sb.*[2] a; **scanning line** = *LINE *sb.*[2] 7 i; **scanning raster** = *RASTER *sb.*[2] a; **scanning spot,** the spot where an incident beam (usu. of electrons or light) strikes the surface it is scanning.

1938 J. H. REYNER *Testing Television Sets* iv. 41 The function of the transformer is to step-down the voltage applied to the scanning coil which operates with a correspondingly larger current. 1978 *Broadcast* 27 Nov. 15/2 Camera heads are still stuck with bulky camera tubes, scanning coils, splitter blocks and such paraphernalia. 1927 *Wireless World* 20 Apr. 685/1 This film was then repeated for an observer by means of a receiving equipment involving the use of a suitable neon tube and a scanning disc. 1975 D. G. FINK *Electronics Engineers' Handbk.* xx. 82 An exciter lamp illuminates the subject copy via a curved mirror, and an objective lens images the reflected light to an aperture plate in front of a scanning disk. The scanning disk is opaque except for a transparent spiral, which curves outward from the center of the disk. 1962 *Nature* 6 Oct. 82/1 Figs. 1 and 2 are scanning electron micrographs. 1979 *Sci. Amer.* Sept. 30/3 Wonderful drawings..complemented by scanning electron micrographs. 1953 *Proc. Inst. Electr. Engineers* C. II. 246/2 The main advantage of the scanning electron microscope for transparent specimens is that the resolution is not affected by energy losses of the electrons in the specimen, which in the conventional electron microscope give rise to chromatic aberration. 1972 *Sci. Amer.* Jan. 55/2 The scanning electron microscope is capable of a range of magnifications that overlaps the range of the light microscope or hand magnifying glass at the low end and the range of the transmission electron microscope at the high end. 1966 D. G. BRANDON *Mod. Techniques Metallogr.* 51 Image formation by..scanning electron microscopy. 1975 J. I. GOLDSTEIN et al. in Goldstein & Yakowitz *Practical Scanning Electron Microsc.* i. 3 The purpose of this brief historical introduction is to point out the pioneers of scanning electron microscopy and in the process trace the evolution of the instrument. 1935 *Television Today* I. 247/1 Such a scanning field is known as a 'raster'. 1975 D. G. FINK *Electronics Engineers' Handbk.* xx. 5 The lines of the second scanning field fall between the lines of the first field. 1929 Scanning line [see *LINE *sb.*[2] 7 i]. 1933 *Discovery* Oct. 318/1 The new German standard picture, consisting of 180 scanning lines, is officially considered sufficient for the opening of regular transmissions. 1960 in *Rep. Comm. Broadcasting* 334 in *Parl. Papers* 1961–2 (Cmnd. 1753) IX. 259 There was..a significant difference in the visibility of the scanning lines—the 625-line pictures being..noticeably better than the 405-line pictures. 1958 *Observer* 12 Oct. 15/4 Every few minutes, the radio telescope makes small scanning movements, up and down and from side to side. This helps to fix the direction of the radio signals from the rocket to within half a degree. 1935 *Television Today* I. 247/2 The production of a scanning raster on the cathode-ray tube of a television receiver by electrical means involves the application of two voltages of saw-tooth wave form to the two pairs of deflecting plates of a cathode-ray tube. 1975 D. G. FINK *Electronics Engineers' Handbk.* xx. 33 The starting point for generating color pictures is the optical and electronic superposition of the red-, green-, and blue-tube scanning rasters. 1929 SHELDON & GRISEWOOD *Television* xii. 126 The scanning speed may be greatly increased by use of a series of oppositely rotating lens-discs. 1934 *Sun* (Baltimore) 20 Aug. 2/3 A tiny metal mirror, mounted on a slender rod and vibrated at scanning speeds, was presented..as..the solution of one of television's major problems. 1929 SHELDON & GRISEWOOD *Television* xiii. 139 Since the scanning spot has finite dimensions, its response to an abrupt change in the surface being viewed will be less sharply defined than the original. 1975 D. G. FINK *Electronics Engineers' Handbk.* xx. 5 The electron beams that create the scanning spots are approximately circular, but their intensity is not uniform.

scant, *v.* **II. 7.** For 'Now *rare*' read 'Now chiefly *U.S.*' and add later examples.

1969 *New Yorker* 6 Sept. 111/1 Several thousand..men were on duty in the streets that day, while, presumably, Securitate was not scanting its duties elsewhere. 1977 *N.Y. Rev. Books* 14 Apr. 5 (Advt.), No thinker or movement is dismissed as too radical, no issue is scanted as too controversial.

scanting, *ppl. a.* Add: **b.** Decreasing, diminishing. *rare.*

1916 KIPLING *Tales of 'The Trade'* 107 It was necessary to go down at once and waste whole minutes of the precious scanting light.

scantling, *sb.* Add: **2. b.** Also with reference to aircraft.

1933 [see *FRAME *sb.* 11 i]. 1978 *Jrnl. R. Soc. Arts* CXXVI. 681/2 Figure 3 indicates the comparative scantlings for compression structures having the same load carrying capacity.

7. a. (Further examples.)

1958 *Chambers's Techn. Dict.* 743/2 *Scantling*..., a piece of timber of thickness from 2 to 4 in. and of width from 2 to 4½ in. 1965 'LAUCHMONEN' *Old Thom's Harvest* ii. 15 The young girl walked round the pickets that had fallen off the rotten scantling runners of the wooden part of the fence. 1972 *Gloss. Terms Timber* (*B.S.I.*) 21 *Scantling*. 1. *Softwood*. A piece of square-sawn timber 50 mm to under 100 mm thick and 50 mm to under 125 mm wide. 2. *Hardwood*. Timber converted to an agreed specification such as waggon oak scantlings. Otherwise any squared-edged piece of dimensions not conforming to other standard terms.

b. (Earlier examples.)

1703 tr. *H. van Oosten's Dutch Gardener* IV. xii. 225 You must keep your Scantling or Boares whereon your Pots stand very neat. 1743 *Colonial Rec. Georgia* (1906) VI. 68 The Reverend Mr. Bolzius [petitioned] this Board to allow him a Quantity of Boards, Planks, and Scantling. 1785 T. JEFFERSON *Notes on Virginia* xv. 279 The private buildings are very rarely constructed of stone or brick; much the greatest portion being of scantling and boards.

scanty, *a.* Add: **B.** *sb.* Now only *pl.* Underwear, esp. short knickers or panties for women. *colloq.* (orig. *U.S.*).

1928 J. P. MCEVOY *Show Girl* (*title-page*), The hottest little wench that ever shook a scanty at a tired business man. 1929 M. LIEF *Hangover* 269 There's no law in New Jersey forcing a husband to look at his wife's scanties, is there? 1934 T. SMITH *Bishop's Jaegers* 5 Whereas men..still struggle along with the old-fashioned..name of drawers..woman have far outstripped them. Theirs must be known now by such frivolous..appellations as panties, scanties..step-ins..and other similar..terms. 1944 E. CARR *House of All Sorts* 101 A puff of wind from the open door caught and ballooned the scanties. 1951 M. DICKENS *My Turn to make Tea* iv. 73 No don't go, dear. You've seen me in my scanties, anyway. 1959 'O. MILLS' *Stairway to Murder* vii. 75 'Now you've got some midnight-blue scanties.' He held up Charles's underpants apologetically. 1964 J. HALE *Grudge Fight* I. i. 22 Bennet, who always looks after number one, is wearing Scapa scanties next to the skin. Long underpants and a long-sleeved vest made of thick, oily wool. 1977 *Time* 24 Jan. 46/1 Maddie's blue scanties emerge from the M.P.s' briefcases at inauspicious moments and whip through the air like naval pennants.

scapa, var. *SCARPER *v.*

scape, *sb.*[1] Add: **7. scape-pipe** *U.S.* = *escape pipe* s.v. ESCAPE *sb.*[1] 8.

1838 E. FLAGG *Far West* I. 51 The stern roar of the scape-pipe, gave evidence of the fearful power summoned up to overcome the flood. 1949 E. HUNGERFORD *Wells Fargo* 22 This craft, in her neat coat of immaculate white, and her yellow stacks, 'scape pipes and upper works, and her gayly striped paddle-houses, was a pretty sight.

scape, *sb.*[3] Add: (Further examples as second element of combs.) See also *CITYSCAPE, *LUNARSCAPE, *MOONSCAPE, *ROOFSCAPE, etc.

1908 'O. HENRY' *Gentle Grafter* 6 The third day of the rain it slacked up awhile in the afternoon, so me and Andy walked out to the edge of the town to view the mudscape. 1930 *Sat. Rev. Lit.* 27 Dec. 486/3 One may..strive to..obtain a meager impression—starved, wry glimpses—of the private mindscape beyond. 1972 G. S. FRASER in Cox & Dyson *20th-Cent. Mind* II. xi. 382 Stephen's associations [in Joyce's *Ulysses*] are not really loose, he composes elaborate moodscapes in sub-Paterian prose. 1973 *Ar Internat.* Mar. 49/2 Raffael's minutely dabbed garnish color..has more in common with the jungle-scapes and frottages of Max Ernst. 1975 *Times Lit. Suppl.* 28 Nov. 1409/3 The 'largest oil painting in the world' (a sea-cum-strandscape by the rightly overlooked Jacob Mesdag). 1977 'J. MCVEAN' *Bloodspoor* xvii. 208 The two figures..were as much part of the desert winterscape now as the thorn barbs or the wheeling constellations in the sky above.

scape (skē̆ⁱp), *sb.*[4] [Origin unknown: perh. f. SCAPE *sb.*[3] (see *INSCAPE *sb.*).] In the terminology of G. M. Hopkins: a reflection or impression of the individual quality of a thing or action. Hence **scaped, sca·pish** adjs.; **sca·ping.**

1868 G. M. HOPKINS *Jrnls. & Papers* (1959) 170 The types of the two thieves..were in the wholeness and general scape of the anatomy original and interesting. (The prominence of the peculiar square-scaped drapery etc. in Holbein and his contemporaries is remarkable.) 1869 *Ibid.* 194 It is just the things which produce dead impressions, which the mind..has made nothing of and brought into no scaping, that force themselves up in this way afterwards. 1874 *Ibid.* 245, I saw also a good engraving of his *Vintage Festival*, which impressed the thought one would also gather from Rembrandt..of a master of scaping rather than of inscape. For the vigorous rhetorical but realistic and unaffected scaping holds everything but no arch-inscape is thought of. *Ibid.* 247 W. L. Wylie—*Goodwin Sands*—Fiery truthful rainbow-end; green slimy races of piers; all clean, atmospheric, truthful, and scapish. 1883 —— *Sermons & Devotional Writings* (1959) II. ii. 136 Our action leaves in our minds scapes or species, the extreme 'intention' or instressing of which would be painful. *Ibid.*, The soul then can be instressed *in* the species or scape of any bodily action..and so *towards* the species or scape of any object, as of sight, sound, taste, smell. 1948 W. A. M. PETERS *Gerard Manley Hopkins* i. 2 The suffix 'scape' in 'landscape'..posits the presence of a unifying principle which enables us to consider part of the countryside..as a unit..but so that this part is perceived to carry the typical properties of the actually individed whole...'Scape' comes to stand for that being which is an exact copy or reflection of the individual whole on which it is dependent for its existence.

scape, *int.* Add: Also used for the brambling's call.

1962 *Times* 6 Nov. 14/4 The bramblings' harsh and nasal call-note, usually written 'scape'.

scapegoat (skē̆ⁱ·pgōᵘt), *v.* [f. the sb. or back-formation from *SCAPEGOATING.] *trans.* To make a scapegoat of (someone); to subject to scapegoating. Hence **sca·pegoated** *ppl. a.*; **sca·pegoater.**

1943 *Jrnl. Abnormal & Social Psychol. Clin. Suppl.* XXXVIII. 143 Persons who had been inclined to scapegoat him originally. *Ibid.* 151 The immediate and desired objective of the scapegoaters was to relieve their feelings of frustration, of fear [etc.]. 1972 *Guardian* 27 Dec. 12/3 We either scapegoat the individual..or we scapegoat society. 1974 S. G. SHOHAM *Society & Absurd* IV. iv. 162 The child becomes a receptacle for the ressentiment of the scapegoater. 1976 *Child's Guardian* Winter 13/3 Oliver's problems illustrate one of the great difficulties in trying to help a scapegoated child. Often the parent/child relationships are so complicated that they seem to need each other in order to continue hurting each other. 1977 R. L. DUNCAN *Temple Dogs* (1978) I. ii. 55 A company is really too large to scapegoat.

scapegoating (skē̆ⁱ·pgōᵘtiŋ). [f. SCAPEGOAT + -ING[1].] The action or practice of making a scapegoat of someone; *spec.* in *Psychol.*, aggressively punitive behaviour directed for whatever reason against other (weaker) persons or groups.

1943 VELTFORT & LEE in *Jrnl. Abnormal & Social Psychol. Clin. Suppl.* XXXVIII. 138 Scapegoating is a phenomenon wherein some of the aggressive energies of a person or group are focused upon another individual, group, or object. 1950 T. ADORNO et al. *Authoritarian Personality* xi. 409 Lack of insight into one's own shortcomings and the projection of one's own weaknesses and faults onto others..probably represents the essential aspect of..scapegoating. 1962 *Listener* 7 June 1002/2 Speaking of scapegoating in new housing blocks, she tells of patients who have actually been robbed finding it difficult to obtain a hearing because they were suspected of paranoid delusions. 1977 C. HUSBAND in H. Giles *Lang., Ethnicity & Intergroup Relations* ix. 234 The intervention of Powell at a singularly propitious moment..propelled the already vigorous scapegoating process into an unmanageable level.

scapegrace, *sb.* Add: **2.** *N. Amer.* The red-throated loon or diver, *Gavia stellata*.

1835 J. J. AUDUBON *Ornith. Biogr.* III. 24 In the neighbourhood of Boston, and along the Bay of Fundy, they are best known by the names of 'Scape-grace' and 'Cape-racer'. 1917 T. G. PEARSON *Birds Amer.* I. 15 Red-throated Loon..Cape Racer; Scape-grace. 1957 W. L. MCATEE *Folk-Names Canad. Birds* 2 Red-throated Loon..scape-grace (Rationalization of Cape Race [where it is often seen]).

scapeless (skē̆ⁱ·plės), *a.*[3] [f. *SCAPE *sb.*[4] + -LESS.] In the terminology of G. M. Hopkins: lacking scape, without distinctive and individual quality.

1874 G. M. HOPKINS *Jrnls. & Papers* (1959) 245 Scapeless aimless background of tapestry, a cannon, and so on. *Ibid.* 248 The feet not inscaped but with a scapeless look they sometimes no doubt have..and veined too, which further breaks their scaping.

scaph, var. *SCAF.

scaphopod (skæ·fopǫd). *Zool.* [ad. mod.L. class name *Scaphopoda* (H. G. Bronn *Klassen & Ordnungen des Thier-reichs* (1862) III. II. 524), f. SCAPHO- + Gr. πούς, ποδ- foot.] (See quot. 1935); = *tusk-shell* s.v. TUSK *sb.*[1] 3. Also *attrib.*

1913 B. B. WOODWARD *Life of Mollusca* iii. 47 A true Scaphopod (*Dentalium*) and representatives of the more primitive Ammonoidea..likewise came into existence in the Devonian epoch. 1935 TWENHOFEL & SHROCK *Invertebr. Paleontol.* ix. 360 Scaphopods are small, marine, bilaterally symmetrical mollusks with an external, curved and tapering tubular shell open at each end. *Ibid.* 362 The scaphopod shell is composed of aragonite. 1975 *Nature* 11 Dec. 555/3 Molluscs, primarily bivalves but also scaphopods and various gastropods,..progress through soft substrates.

scapigerous, *a.* (Earlier example.)

1859 D. BUNCE *Travels with Dr. Leichhardt* 29 *Xanthoria*, or grass-tree, three species of which enlivened the landscape with their scapigerous white blossoms.

scapolitization (skæ:pǫlitəizē̆ⁱ·ʃən). *Geol.* [f. SCAPOLITE + -IZE + -ATION.] The alteration of alumino-silicate minerals of igneous rocks into, or their replacement by, minerals of the scapolite group. Also **scapo·litize** *v. trans.*; **scapo·litized** *ppl. a.*

1909 A. HARKER *Nat. Hist. Igneous Rocks* xii. 302 (*caption*) Apatite vein with scapolitized borders. *Ibid.* 383/1 (Index), Scapolitization. 1924 *Mineral Abstr.* II. 227 Chemical analyses by Pisani on fresh and scapolitized material from Pouzac show that the alteration is accompanied only by addition of sodium chloride. 1932 A. HARKER *Metamorphism* xvi. 255 This widespread scapolitization..is doubtless related to the mechanical conditions proper to regional metamorphism, which facilitate the permeation of the rocks by volatile bodies. 1936 *Nature* 29 Feb. 366/2 Gabbroid rocks were intruded..and were followed by late-stage solutions which scapolitised the sediments, turned the gabbro into epidiorite and chloritised the Basement schists. 1954 *Jrnl. Geol. Soc. Australia* I. 6 The current-bedded sediments are scapolitized calcareous rocks. *a* 1965 A. W. G. WHITTLE in G. J. Williams *Econ. Geol. N.Z.* (1965) xiv. 227/1 A large amount of titanite was introduced as anhedra during scapolitization, commonly aggregated to form elongated veins within the rock.

scapulimancy. Also: Also **scapulomancy.** (Further examples.)

1911 J. HASTINGS *Encycl. Relig. & Ethics* IV. 817/1 Scapulomancy is mentioned by Jāḥiẓ together with palmistry and another mode of augury..., viz. divination by the gnawing of mice. 1937 R. H. LOWIE *Hist. Ethnol. Theory* xi. 184 Speck and Cooper have traced scapulimancy from northern Europe through Asia to Eastern North America. 1961 G. CLARK *World Prehistory* viii. 199 It will be recalled that the practice of scapulomancy..can be traced back to the 'Neolithic' Lung-shan culture. 1973 T. R. TREGEAR *Chinese* i. 19 The bones used in their scapulimancy were incised by markings which are the earliest form of Chinese writing.

scar, *sb.*[2] Add: **1. a.** (Later examples of *transf.* use.)

1884 'MARK TWAIN' *Huck. Finn* (1885) ii. 24 We..pulled down the river..to the big scar on the hillside, and went ashore. 1929 W. FAULKNER *Sartoris* IV. 305 He sat his horse in the faint scar of the road. 1946 *R.A.F. Jrnl.* May 172 Their repair work had been so rapid that we could find few scars in the main part of the city.

3. scar tissue, the fibrous connective tissue of which scars are formed; also *fig.*

1875 T. HOLMES *Treat. Surg.* xxi. 386 When the scar-tissue remains permanent, although the scar is ugly and of lower organisation than the natural parts, yet it causes no important inconvenience. 1932 F. BEEKMAN *Office Surg.* xii. 291 Keloids appear most frequently in individuals of races who have a predisposition for the formation of excessive scar tissue. 1957 A. HUXLEY *Let.* 12 Jan. (1969) 815, I have just embarked on a new treatment aimed at getting rid of some of the scar tissue on my corneas. 1975 *New Yorker* 1 Dec. 55/2 'It leaves scar tissue,' one former campaign manager said. 'There's no way it can't have a deep impact on the candidate's psyche and physical condition.' 1978 J. IRVING *World according to Garp* iii. 53 The brave face was naked, the eyes clear and challenging, the scar tissue everywhere.

scarab. Add: **1.** Now also *scarab beetle*.

1921 C. A. EALAND *Insect Life* vi. 179 We do not possess in this country [*sc.* Great Britain] any of the Scarab Beetles, sacred to the ancient Egyptians. 1958 'W. HENRY' *Seven Men at Mimbres Springs* vii. 73 Young Sanchez [packed] a cottonwood slingshot and six or eight smooth stones suitable for anything up to bullfrogs or scarab beetles.

scarbroite. Now regarded as a carbonate, not a silicate. (Further example.)

1960 *Mineral. Mag.* XXXII. 353 Scarbroite, a fine-grained but compact deposit obtained from fissures in the sandstone on the north Yorkshire coast, is shown by chemical analysis to have an idealized formula $Al_2(CO_3)_3 \cdot 12Al(OH)_3$.

scarce, *adv.* **2. d.** (Further examples.)

1915 G. FRANKAU *Tid'apa* ii. 14 A scarce-breathed, flickering soul-wave, discoded but conscience-deep. 1921 W. DE LA MARE *Veil* 35 There came, scarce-heard, Claws, fluttering feathers, Of deluded bird. 1922 BLUNDEN *Shepherd* (ed. 2) 53 In the scarce-glimmering boles. 1935 C. DAY LEWIS *Time to Dance* 33 Like a bird scarce-fledged they flew, whose flying-hours are few. 1951 W. DE LA MARE *Winged Chariot* 57, I match that child with this scarce-changed old man.

scarcity. Add: **2. b. scarcity value** (later examples); *scarcity price* (example).

1883 Scarcity value [see *monopoly value* s.v. *MONOPOLY 7]. 1920 *Times* 5 June 15/3 Profits made on selling commodities at scarcity prices since then would escape the levy. 1936 J. M. KEYNES *Gen. Theory Employment* xxiv. 377 Our aim of depriving capital of its scarcity-value. 1972 'G. BLACK' *Bitter Tea* (1973) x. 162 She was a blonde. They have a great time in the Orient, scarcity value.

7. *scarcity root* (example).

1787 J. WOODFORDE *Diary* 29 June (1926) II. 330 Sr. Willm Jernegan sent me by Mr. Custance a Treatise on the Plant called Scarcity Root.

scare, *sb.*[2] Add: **4.** *scare-headline, -story*; **scare-buying** *U.S.* = *panic buying* s.v. *PANIC *sb.*[2] 3 b; **scare-head** (earlier and later examples); hence as *v. trans.*, to furnish with a scare-head; to display as a scare-headline; **sca·re-headed** *ppl. a.*; **scare tactic,** a stratagem or ruse which seeks to manipulate public reaction by the exploitation of fear; usu. *pl.*

1944 *Sun* (Baltimore) 23 Nov. 15/3 The sharply restricted supply of cotton goods..comes at a time when so-called 'scare' buying of such commodities is in boom proportions. 1959 *Wall St. Jrnl.* 4 May 1/1 Steel customers have been buying heavily for weeks, in anticipation of a strike... This 'scare buying'..boosted steel-making to a scheduled 94·4% of the industry's rated capacity last week. 1887 *Courier-Jrnl.* (Louisville, Kentucky) 15 Feb. 6/4 The 'scare' head which follows..is an evidence that the country paper tries hard to keep pace with the times and its metropolitan contemporary. 1902 F. NORRIS *Responsibilities of Novelist* (1903) 300 The name of the leading lady or leading man is 'scare-headed' [on theatre bills]. 1911 H. S. HARRISON *Queed* xviii. 219 The..penny evening paper..scare-headed a jaundiced account of the affair. 1926 *Scribner's Mag.* Sept. 251/1 If he is at all impressionable, a glance at the scare-heads will utterly ruin what otherwise might have been a successful day. 1951 E. PAUL *Springtime in Paris* v. 95 Metal workers were uneasy, having been stamped with scareheads against the Marshall Plan. 1971 *Sci. Amer.* May 10/2 Professor Reuterdahl's recent article in the *Dearborn Independent* is given its real place by the scare-head of the cover, which asks, in 3/4-inch letters, 'Is Einstein a Plagiarist?'. 1889 W. D. HOWELLS *Hazard of New Fortunes* II. 281 He read..the deeply scare-headed story of Conrad's death. 1892 J. KIRKLAND *Story of Chicago* I. xxxii. 381 The newspapers blazed with what are technically called 'scare headlines'. 1912 KIPLING *Uses of Reading* in *Book of Words* (1928) 87 The other made bad worse by shouting what was no better than a newspaper scare head-line. 1960 *Guardian* 11 Apr. 1/1 Rumours circulated..that the Government might introduce martial law, but this appeared to have been purely a scare story. 1977 P. JOHNSON *Enemies of Society* vii. 94 The technique of the lobby is to put out a scare-story, and then move on quickly to a fresh one when scientific investigation proves the first one unfounded. 1979 *Time* 8 Jan. 40/1 The scare stories are based on phony evidence or plain prejudice. 1967 *Punch* 8 Nov. 719/1 This alleged address from Zinoviev, the President of the Comintern..left an *Angst* about Tory scare-tactics from which Transport House has never recovered. 1973 *Black Panther* 17 Mar. 8/1 (*caption*) Boxes of poisoned lettuce have had to be destroyed. This is no 'scare tactic', it is for real. 1976 *Survey* Summer–Autumn 191 The slickers in the Pentagon are using their annual scare tactics in support of bigger budgets.

scare, *v.* **1. d.** For *U.S.* read 'orig. and chiefly *U.S.*' and add earlier and later examples. Also (*fig.*), to procure, obtain, 'rustle up'. *colloq.*

1846 *Spirit of Times* 25 Apr. 97/1 He is also to send us the rattles of the biggest snake ever scared up in 'Old Norf Caline'. 1852 H. C. WATSON *Nights in Block-House* 169 Ad was equal to two or three common men in scarin' up and shootin' red-skins. 1890 *Stock Grower & Farmer* 1 Feb. 4/2 A country the like of which can not be 'scared up' in many thousands of miles travel. 1913 J. LONDON *Night-Born* 262 Los Angeles must be on the dink when this is the best you can scare up. 1922 GALSWORTHY *Loyalties* II. ii. 71 Let's cut it and get out to Nairobi. I can scare up the money for that. 1940 *New Yorker* 13 Jan. 31/2 A young woman who had somehow contrived to scare up a permit to leave the country. 1952 J. JONES *From Here to Eternity* xii. 145 Maybe I can scare you up some [work]. 1961 *Listener* 2 Nov. 738/1 Professor Ford has always managed to scare up a few distinguished contributors. 1976 H. NIELSEN *Brink of Murder* ii. 21 Why don't you relax..and then we'll scare up some dinner.

2. (Later U.S. example.)

1869 'MARK TWAIN' *Innoc. Abr.* 440 This creature has scared at everything he has seen to-day.

3. (Further examples.) Freq. in negative, esp. with *easily* or *easy*.

1941 *Sun* (Baltimore) 4 Jan. 6/2 Whatever else they do or fail to do, the Irish don't scare easily. 1951 'M. SPILLANE' *One Lonely Night* iv. 61 They're the kind of people who scare easily. 1967 O. RUHEN in *Coast to Coast 1965–6* 192 The horse won't scare, but take it easy. 1972 *Village Voice* (N.Y.) 1 June 5/2 'We don't scare easy,' his cousin said as I went out the door.

scarecrow, *sb.* Add: **2. c.** *Mil. slang.* Used in the war of 1939–45, to designate weapons or manœuvres which had a purely deterrent effect (see quots.).

1943 T. DUDLEY-GORDON *Coastal Command at War* iv. 41 This was the squadron..which flew the Scarecrow Control...No one knows how many times a U-Boat captain was forced to keep submerged because a Tiger Moth, which *might* be dangerous, was doing a scarecrow on him. 1952 M. TRIPP *Faith is Windsock* ii. 41 Two daylight attacks on Solingen... Gigantic blobs of oily smoke hung in the sky... It was their first experience of the German terror weapon, the scarecrow. 1966 L. MIALL *Richard Dimbleby, Broadcaster* 39 A great gush of flame and smoke showed the bursting of a 'scarecrow', the oddity designed by the Germans to simulate a heavy bomber being shot down, and so to put any of our less experienced pilots off their stroke.

3. Delete † (*obs.*) from sense 'a lean, gaunt figure' and add later examples.

1932 E. MUSPRATT *Wild Oats* v. 96 He was a great gaunt scarecrow, bent and crippled by disease. 1959 I. & P. OPIE *Lore & Lang. Schoolch.* ix. 169 Thin people inspire almost as many names and jokes as fat people, but..the names..are merely descriptive, as:..scarecrow, scraggy, skin and bones.

scaredly (skē̆ᵊ·ɹdli), *adv.* [f. SCARED *ppl. a.* + -LY[2].] In a scared manner.

1901 G. B. SHAW *Devil's Disciple* I, in *Three Plays for Puritans* 18 *To Essie*. Essie: did you say amen? *Essie* (scaredly). No. 1978 G. VAUGHAN *Belgrade Drop* xi. 74 Savka said suddenly, scaredly: 'I hope to God you're sure!'

scaredy-cat (skē̆ᵊ·ɹdi,kæt). *slang.* [f. SCARED *ppl. a.* + -Y[6] + CAT *sb.*[1]] A timorous person, a coward; = *fraidy cat* s.v. *'FRAID *a.* Also as *adj.*, scared. Also *ellipt.*, as *scaredy*.

1933 D. PARKER *After Such Pleasures* (1934) 86 It's so nice to meet a man who isn't scaredy-cat about catching my beri-beri. 1948 D. BALLANTYNE *Cunninghams* 173 Sydney called them scaredy-cats because they wouldn't run like he had. 1959 I. & P. OPIE *Lore & Lang. Schoolch.* x. 185 The boy..who will not take part in a prank..is a 'scaredy', a 'scare-baby'. 1965 'LAUCHMONEN' *Old Thom's Harvest* xii. 149 You can play hard-to-get but don't look so scaredy-cat. 1980 H. R. F. KEATING *Murder of Maharajah* iii. 57 You know your mother, always was a scaredy-cat.

scaremonger. Add: Hence as *v. intr.*, to spread alarming reports; **sca·remongering** *ppl. a.*

1966 *New Statesman* 14 Jan. 38/3 The new Home Secretary, Roy Jenkins, commented on TV that the *Express* story was 'premature and slightly scaremongering'. 1976 'D. HALLIDAY' *Dolly & Nanny Bird* ii. 28 You scaremongered all the way through. You created panic.

scarer. Add: *spec.* (usu. as *bird-scarer*) a person or thing (other than a traditional scarecrow) for frightening birds away from crops.

1930 H. H. THOMAS *Pop. Gardening Ann.* 24 A good cheap scarer on the market is obtainable in the shape of a black cat's head. 1953 R. GODDEN *Kingfishers catch Fire* xiii. 157 The bird-scarers had come to watch over the cherry crop. 1961 *Times* 7 Jan. 8/6, I could not make out whether the contents were a bird-scarer or a child's rattle. 1971 *Country Life* 16 Sept. 682/1 We were much troubled by an explosive bird-scarer in a field of barley adjoining our house.

scarf, *sb.*[1] Add: **3. e.** (Earlier example.)

1823 C. MATHEWS *Let.* 17 Feb. in A. Mathews *Mem. C. Mathews* (1839) III. 368 And also two scarfs, I think they are called.

7. *scarf-end, -tie.*

1868 G. M. HOPKINS *Note-bks. & Papers* (1937) 115 Fine afternoon with snow-white flying scarf-ends in the clouds. 1922 JOYCE *Ulysses* 440 In an oatmeal sporting suit..., tony buff shirt, shepherd's plaid Saint Andrew's cross scarftie. 1976 *Billings* (Montana) *Gaz.* 16 June 9-c/5 (Advt.), This dress is a breeze—buttons up one side to the flutter of a scarf tie.

scarf, *sb.*[2] Add: **3. scarf-joint** (further examples).

1919 S. F. WALKER *Electr. Mining Machinery* xx. 154 A scarf joint is..good if it is well made and very carefully bound. 1948 F. WIGHTMAN *Wind is Free* ii. 33 It had to be hoisted with a block & tackle on sheerpoles to bring it up to where its scarph joint fitted into the one cut on the forward end of the keel timber.

scarf, *sb.*[5] Add: **2.** *Forestry.* A V-shaped incision cut in a trunk during felling, to govern the direction in which the tree is to fall; also, the sloping surface left by such an incision.

1863 *8th Ann. Rep. Maine Board Agric.* 36 The bark of the stock opposite the scarf with a thin sliver of wood is cut down. 1887 J. D. BILLINGS *Hardtack & Coffee* (1888) 180 When an army first went into camp trees were cut with the scarf two or three feet above the ground. 1903 R. J. CLOW *Pillar of Salt* iii. 55 It meant a bit of work to cut down a tree seven feet in diameter... Stello cut in the inside scarf and I put in the back chip. 1926 K. S. PRICHARD *Working Bullocks* xxxii. 296 Half-dozen men..stood on their rough-barked logs... The scarf showed ruddy as a wound in the logs. 1962 J. N. WINBURNE *Dict. Agric. &*

Allied Terminology 673/1 Scarf,..the beveled cut on a log or stump which results from undercutting a tree in felling.

scarf (skɑːf), *sb.*[4] U.S. slang var. SCOFF *sb.*[2] **1932** *Evening Sun* (Baltimore) 9 Dec. 31/5 Scarf, food. **1944** D. BURLEY *Orig. Handbk. Harlem Jive* 81 'Pick up on the scoff, cherub.'.. The 'scoff' or 'scarf' in the above simple statement is dinner food, meals. **1961** RIGNEY & SMITH *Real Bohemia* p. xvi, Scarf, food; eat, believed to have come from a French chef, Scarfannelli. **1973** L. SNELLING *Heresy* II. iv. 89 How's for a bit of scarf, my tummy's anguished.

scarf, *v.*[2] **1. a.** (Later example of var. form.) **1976** *Yankee* Apr. 109/1 He forced me to scarph the keel timbers in watertight sections.

scarf, *v.*[3] Restrict *Whaling* to sense in Dict. and add: **2.** *N.Z. Forestry.* To cut a scarf in (timber). Also *back-, belly-scarf* (see quot. 1928). Cf. *SCARF *sb.*[5] **1899** J. BELL *Shadow of Bush* xiv. 83 The smaller trees ..had been 'scarfed', or cut partly through in readiness, and skilfully, so that each, when struck, might again in its turn strike and bring down another. **1904** 'G. B. LANCASTER' *Sons o' Men* 164 He..scarfed the timber for the saw. **1928** P. T. KENWAY *Pioneering in Poverty Bay* v. 38 He will 'belly-scarf' and 'back-scarf' the lot, that is to say he will cut about a third through on both the lower and higher sides.

scarf (skɑːf), *v.*[4] U.S. slang var. SCOFF *v.*[2] 1. Also *absol.* and const. *up* and *down*. **1960** R. G. REISNER *Jazz Titans* 164 Scarf, eat. **1968** C. ARMSTRONG *Balloon Man* viii. 98 They don't want to faint from hunger, so..they scarf up what they call a bite before they go. **1974** *Black World* June 77/1 King Dust would sit there, 'scarfing', as he called it, in silence. **1975** *High Times* Dec. 80/3, I can pick jimsonweed and chop it up and scarf it down as well as the next guy. **1976** R. CONDON *Whisper of Axe* II. xviii. 265 Let's..scarf up some of that osso bucco.

scarfed, *ppl. a.*[1] Add: (Further examples.) Cf. SCARVED *ppl. a.* in Dict. and Suppl. **1920** BLUNDEN *Waggoner* 53 The lispering aspens and the scarfed brook grasses With wakened melancholy writhe the air. **1967** *Boston Sunday Herald* 30 Apr. v. 5/1 (Advt.), Shaped and scarfed for cool summer perfection in rayon and silk.

scarfed, *ppl. a.*[2] Add: Also **scarphed.** (Later example.) **1975** *Anglo-Saxon England* IV. 187 D. M. Wilson has noted that long ships with scarphed keels were built in Scandinavia in the thirteenth century.

scarify, *v.*[1] Add: **scarifying** *vbl. sb.* (earlier example in sense 3 a of vb.); **sca·rifyingly** *adv.*[1] **1865** W. WHITE *Eastern England* xvi. 222 The engine.. is ready for ploughing, scarifying or drain-cutting. In working the scarifier, the large iron pully is anchored on the edge of the ditch. **1921** D. CANFIELD *Brimming Cup* II. xi. 182 How scarifyingly he would laugh at me.

scarify (skɛ̄ᵊ·rifəi), *v.*[2] *slang* (orig. *dial.*). Also scarrify. [Irreg. f. SCARE *v.* + -IFY, perh. after TERRIFY *v.*] *trans.* To scare, frighten; to terrify. **1794** A. THOMAS *Newfoundland Jrnl.* (1968) 107 If a Clergyman was to make his appearance in his Canonical Robes at one of the Outharbours I have little doubt but the Weomen and Children would be scarified out of part of their senses. **1897** G. FORD *Larramys* xxxii. 231 Vine rider! Scarify mos' volks to death, 'er wüd, I reckon. **1901** 'A. FORBES' *Odd Fish* 149 It will be more likely to scarrify 'er if I tells 'er nigh twelve o'clock. **1961** 'F. O'BRIEN' *Hard Life* vii. 53 You want to scarify the divils in the town of Kinnegad? **1966** *New Statesman* 14 Jan. 51/3 There are almost forgotten casualties like *disinterested* and *jejune* and *scarify* (which was once used to mean *to wound*, not as a smart synonym for *to scare*).

Hence **sca·rified** *ppl. a.*[2], **sca·rifying** *ppl. a.*; **sca·rifyingly** *adv.*[2] **1895** MEREDITH *Amazing Marriage* II. xxix. 330 Here I'm like a cannon for defending the house, needs be, and all inside flies off scarified. **1916** M. WEBB *Golden Arrow* iv. 26 Fixing a scarifying gaze on the truant. **1924** N. COWARD *Rat Trap* I. 12, I suppose it's silly nerves, but to be on the brink of a great happiness is a scarifying feeling. **1963** *Times* 13 June 16/7 A comic role (scarifyingly overdone). **1973** *Times* 17 Jan. 17/5, I would support the comparison with a historical rationale which may be deeply disturbing, even scarifying, but it is certainly not motivated by sensationalism or propaganda.

scarily, *adv.* Add: **1.** (Earlier example.) **1845** W. G. SIMMS *Wigwam & Cabin* 1st Ser. 107 My heart as cold as ice, and jumping up and down as scarily as a rabbit's.

2. Frighteningly, unnervingly. **1967** *Economist* 19 Aug. 664/2 Shell is running a scarily fine operation with a smaller proportion than most companies of its tanker needs provided by its own fleet and with relatively short charters. **1978** L. BLOCK *Burglar in Closet* xi. 94 He'd come scarily close to the truth.

scarlet, *sb.* and *a.* Add: **B.** *adj.* **3. a.** **scarlet-chested (grass) parrakeet, parrot,** a small blue and green parrot with a red breast, *Neophema splendida*, found in parts of southern Australia. **1901** A. J. CAMPBELL *Nests & Eggs Austral. Birds* II. 654 (*heading*) Scarlet-chested Grass Parrakeet. **1931** N. W. CAYLEY *What Bird is That?* 152 Scarlet-chested Parrot... Rarely recorded, then only as isolated pairs. **1938** —— *Austral. Parrots* 283, I had the pleasure of seeing the Scarlet-chested Parrakeet living happily and breeding freely. **1977** *Weekly Times* (Melbourne) 19 Jan. 23/3 Mrs Jones (West Hobart) would be interested to hear from any reader who would exchange orange-breasted wax-bills for scarlet-chested parrots.

4. a. scarlet letter chiefly *U.S.*, a representation of the letter *A* in scarlet cloth which persons convicted of adultery were condemned to wear, as described in the novel by Hawthorne (see quot. 1850); also in *fig.* and allusive use (cf. BRAND *sb.* 4 b); **Scarlet Pimpernel** (see also sense 4 c in Dict.), the name assumed by the hero of a series of novels by Baroness Orczy (1865–1947), a dashing but elusive Englishman who rescued potential victims of the French Reign of Terror, used allusively; also *attrib.*; cf. *PIMPERNEL 3*; † **scarlet runner** *Obs. Mil. slang*, a soldier, with reference to his scarlet jacket; also *pl.*, a scarlet military uniform; **scarlet woman,** now used to mean: a notoriously immoral woman; a prostitute. **1850** N. HAWTHORNE (*title*) The scarlet letter. **1872** *Cincinnati* (Ohio) *Times & Chron.* 28 May 2/1 A grand mass meeting in Gotham the other night consecrated Apollo Hall by unfurling therein the scarlet letter—we mean banner—of Woodhull and Free Love. **1882** *Internat. Rev.* Mar. 301 Polygamy is the scarlet letter upon the brow of this young commonwealth which proclaims her deep shame and forbids her entrance into the sisterhood of States. **1944** W. J. CARRINGTON *Safe Convoy* 112 However, a few minutes later when the unwanted visitor arrived, she directed her venomous tongue against the daughter whom she branded from head to foot with verbal scarlet letters. **1965** M. DRABBLE *Millstone* 20, I walked around with a scarlet letter embroidered upon my bosom..but the A stood for Abstinence, not for Adultery. **1977** D. ANTHONY *Stud Game* xxi. 132 You hard-shelled Baptist prig. You can't see past the scarlet letter, can you? **1958** E. H. CLEMENTS *Uncommon Cold* vii. 178 As for cloak-and-dagger work on the moor, what price your family of Scarlet Pimpernels? **1958** *Observer* 25 May 15/5 George Baker..appears as a Scarlet Pimpernel type. **1961** *Guardian* 24 May 11/3 A wartime Scarlet Pimpernel organisation which rescued thousands of East European Jews from the Nazis. **1977** M. DRABBLE *Ice Age* III. 287 The image of Anthony as Scarlet Pimpernel, flying out to rescue stepdaughter in distress. **c 1864** BROUGH & HALLIDAY *Area Belle* 7 Who are you calling bluebottle?—you scarlet runner! **1920** G. FRANKAU *Peter Jackson* VII. 78 'A few of our old militia uniforms.' 'Not the old scarlet-runners?' .. 'The identical, sir, with the old white facings.' **1853** T. PARKER *Theism, Atheism, & Popular Theology* 131 Atheism turns the soul out of doors, and the flesh has no better time of it; no, has a worse time, with its scarlet woman 'tinging the pavement with proud wine too good for the tables of pontiffs'. **1924** in H. Havelock Ellis *Stud. Psychol. Sex* (ed. 3) II. 124, I sought out a scarlet woman in the streets of —— and went home with her. **1977** M. KENYON *Rapist* x. 115 'Is this me?' She was holding..a turtle-neck jersey dress... 'Or would you hazard it's..old-fashioned for the scarlet woman of the bogs?'

b. scarlet ibis (later examples); **scarlet rose-finch,** *Carpodacus erythrinus* (cf. *rose-finch* s.v. *ROSE *sb.* 24 b). **1835** J. J. AUDUBON *Ornith. Biogr.* V. 62, I have found the Scarlet Ibis less numerous than even the Glossy Ibis. **1971** *Country Life* 22 July 220/1 The vivid colouring of the scarlet ibis is as expressive of the South American tropics as the bright colours of macaws and toucans. **1884** H. SEEBOHM *Hist. Brit. Birds* II. 47 The Scarlet Rose-finch is not particularly interesting at its breeding-grounds. **1976** J. T. R. & E. M. SHARROCK *Rare Birds in Britain & Ireland* 284 Scarlet Rosefinch..breeds from Germany and southern Sweden eastwards to Kamchatka.

c. scarlet maple (earlier and later examples); **scarlet runner,** substitute for def.: a red- or white-flowered climbing bean, *Phaseolus coccineus*, or its edible pods; (earlier and later examples). **1768** P. MILLER *Gardeners Dict.* (ed. 8) s.v. *Acer*, I have observed, upon cutting off branches from the scarlet Maple in February, a great quantity of a very sweet juice hath flowed out. **1813** H. MUHLENBERG *Catal. Plant. Amer. Sept.* 95 Scarlet, white, red, or soft maple. **1916** E. T. SETON *Woodcraft Man. for Girls* 292 Red, Scarlet, Water, or Swamp Maple... Noted for its flaming crimson foliage in fall, as well as its red leaf-stalks, flowers, and fruit earlier. **1786** J. ABERCROMBIE *Gardener's Daily Assistant* p. vii, A list of kitchen-garden plants... Kidney Bean (Dwarf)..Scarlet Runner. **1806** B. MCMAHON *Amer. Gard. Cal.* 580 Bean, The Dwarf Kidney... Running kinds..Scarlet Runners. **1908** *Garden* 25 Apr. 205/2 Possibly there is not a vegetable grown that is a more general favourite among amateurs than the climbing bean known as the Scarlet Runner. **1969** *Oxf. Bk. Food Plants* 36/1 The Scarlet Runner is by far the most popular green bean in Britain.

scarlet fever. Add: † **c.** *joc.* A passion for soldiers, with reference to their scarlet uniforms. *Obs.* **1861** B. HEMYNG in H. Mayhew *London Labour* (1862) Extra vol. 235 Nurse-maids..are always ready to succumb to the 'scarlet fever'. A red coat is all powerful with this class. **1890** BARRÈRE & LELAND *Dict. Slang.* II. 206/1 Ladies who run after military society are said to have *scarlet fever*.

scarper (skā·ɹpəɹ), *v. slang.* Also **scapa, scarpa.** [Prob. ad. It. *scappare* to ESCAPE, get away; reinforced during or after the war of 1914–18 by *scapa* from Cockney rhyming slang *Scapa Flow*, to go.] **a.** *intr.* To depart hastily, run away; to escape, make one's get-away. **1846** *Swell's Night Guide* 43 He must *hook it* before 'daylight does appear', and then scarper by the back door. **1861** H. MAYHEW *London Labour* III. 48/1 When I was scarpering with my culling in the monkey. **1931** M. ALLINGHAM *Look to Lady* xxiv. 253 Round up this lot now and scarpa yourselves. **1933** G. INGRAM *'Stir'* iii. 45 'I'll be punching you up the belly if you don't scarper,' threatened Smith. **1954** M. PROCTER *Hell is City* II. i. 44 Wi' my record I thought I'd better scarper before I got dragged into trouble. **1970** *Private Eye* 27 Feb. 16 Take this lolly and scarpa for lawd's sake. **1972** A. DRAPER *Death Penalty* v. 37 Ben..shouted, 'Scapa. Everyone scapa.' **1974** *Sunday Post* (Glasgow) 21 July 16/4 His panic became unbearable. He jumped out of bed and scarpered! **1977** 'E. CRISPIN' *Glimpses of Moon* vii. 111 He's downstairs now with the others—and they're keeping a sharp eye on him; he won't have a chance to scarper again.

b. *trans.* To depart or escape from (a place); usu. in phr. *to scarper the letty*, to leave one's lodgings without paying the rent (cf. *LETTY). **1937** PARTRIDGE *Dict. Slang* 731/2 Scarper,..to decamp from... *Scarper the letty*, to leave one's lodgings without paying. **1957** [see *LETTY].

Hence as *sb.* in phr. *to do a scarper*, to run away, 'do a bunk'. **1958** F. NORMAN *Bang to Rights* 63 We had all planned to do a scarper.

‖ **scarpetti** (skaɹpe·ti), *sb. pl.* [It., pl. of *scarpetto*, a small shoe.] Rope-soled shoes worn for rock-climbing, esp. in the North Italian Alps. Cf. *KLETTERSCHUH. **1897** O. G. JONES *Rock-Climbing in Engl. Lake District* p. xxiv, The Cumberland crags are too smooth to make *scarpetti* (*Kletterschuhe*) worth trying. **1907** G. D. ABRAHAM *Compl. Mountaineer* xxviii. 463 If the conditions become wet or icy, rubber soles are a snare and a delusion; scarpetti should be carried as a reserve. **1923** —— *First Steps to Climbing* vi. 71 They [*sc.* rubber-soled shoes] were preferable to the scarpetti, or rope-soled boots, which are the standard Dolomite wear. **1931** *Times Lit. Suppl.* 19 Feb. 129/3 On it [*sc.* Mont Blanc], climbing guideless, they practise every modern refinement, the use of crampons, scarpetti and the rappel. **1941** C. KIRKUS *Let's go Climbing!* vi. 96 Scarpetti—rope soled shoes used in the Dolomites—are coming into favour in this country. **1956** C. EVANS *On Climbing* ii. 32 In the Alps there have been other kinds of friction footwear... The rag-soled scarpetti of the Dolomites.

scarved, *ppl. a.* Add: (Further examples.) Cf. SCARFED *ppl. a.*[1] in Dict. and Suppl. **1958** M. STEWART *Nine Coaches Waiting* v. 54 Philippe and I went out..coated and scarved against the breeze. **1972** F. WARNER *Lying Figures* III. 21 Scarved, laughing children, scuffing the leaves! **1976** *New Yorker* 8 Mar. 41/2 The lottery sellers were gloved and scarved.

scary, *a.*[1] Add: **1.** (Further examples.) **1938** J. STEINBECK *Long Valley* 14 It would be a lonely life for a woman, ma'am, and a scarey life too, with animals creeping under the wagon all night. **1955** E. COXHEAD *Figure in Mist* iv. 123 We're over the scary part now. **1960** *Guardian* 27 Oct. 9/5 This is real scarey. **1961** G. GREENE *Burnt-Out Case* VI. iii. 237 Goodness, I'm glad to be here. It was really scary driving all the way alone. **1975** *New Yorker* 21 Apr. 105/1 It's something scary you don't have to believe in. **1975** *University* (Princeton Univ.) Winter 9/1 The latter showed that energy consumption, per person, could easily double by 2000. That's pretty scary, with energy resources almost sure to decline. **1977** *Times Lit. Suppl.* 11 Feb. 145/4 (Advt.), Set against a background of sinister ritual, vengeance and murder, a novel to be read at one scarey sitting. **1978** *Nature* 20 July 199/1 The procession was headed by figures dressed in radioactive protection gear, clearly intended to be scary. **1981** *Listener* 1 Jan. 23/3 The threat..is pretty scary.

2. (Earlier and later examples.) orig. and chiefly *N. Amer.* **1800** M. L. WEEMS *Let.* 29 Dec. in *Works & Ways* (1929) II. 160, I have always been very scary about our monies. **1903** J. VAIZEY *Pixie O'Shaughnessy* ix. 107 She was too frightened to own up last night—you know what a scarey little thing she is. **1907** J. M. SYNGE *Let.* 25 Apr. (1971) 128, I have been getting a little bit scary about your extravagance. **1951** L. CRAIG *Singing Hills* vi. 45 He'd a been right smart proudified of your not being scary. **1970** N. STREATFEILD *Thursday's Child* xxxii. 218 He was as scary of being seen as a wild deer.

scat (skæt), *sb.*[5] *U.S. slang.* [Origin obscure.] Whisky. **1914** JACKSON & HELLYER *Vocab. Criminal Slang* 73 *Scat, noun*, general circulation. Whiskey. Derived by suggestion from 'skey' (skee), the termination of 'whiskey'. **1949** PARTRIDGE *Dict. Underworld* 597/2 *Scat*, whiskey... Perhaps proleptic: it causes intelligence to *scat*, to scatter, to vanish. **1955** *Publ. Amer. Dial. Soc.* xxiv. 161 Peter men don't punch muff guff as a rule, but sometimes the scat will loosen them up for some good yarns.

scat (skæt), *sb.*[6] (and *a.*) *Jazz.* [Prob. imitative: see quot. 1929.] **a.** A style of improvised singing in which meaningless but expressive syllables, usu. representing the sound of a

musical instrument, are used instead of words. Freq. *attrib.* passing into *adj.* (see also b below). **1929** *Melody Maker* Apr. 369/1 This particular type of vocalism is known as 'Scat' singing. This name undoubtedly owes its origin to the almost inevitable way of starting any line with 'Scat-da-doo'. A very fine example of this 'Scat Singing' is in 'Candy Lips' by Louis Armstrong's Washboard beaters.., the label rightly describing it as 'Scat' chorus by Clarence Williams. **1933** D. RUNYON in *Collier's* 28 Jan. 41/1 She has to play against a scat band. **1937** *Amer. Speech* XII. 182/2 *Scat*, a style of singing in which the vocalist scorns the lyrics, substituting meaningless but expressive syllables of his own improvisation. **1946** R. BLESH *Shining Trumpets* x. 229 The pattern..was derived.. from attempts of white singers of popular tunes to imitate the rhythmic Negro scat song. **1963** *Times* 27 Dec. 4/7 The exhilarating and often quasi-instrumental vocal duetting, sometimes in scat or in falsetto, behind the melodic line. **1977** *Rolling Stone* 24 Mar., Jarreau is a sophisticated cabaret artist whose vocal mimicry and jazz-man scat account for much of his onstage success.

b. *Comb.*, as **scat-singing** *sb.*, singing in this style; also as adj.; hence **scat-singer** and (as a back-formation) **scat-sing** *v. trans.* and *intr.* **1929** [see above]. **1934** A. BOWLLY *Modern Style Singing* xxiv. 118 Current records should be the best guide of how..to 'scat' sing. **1936** *Amer. Mercury* May p. x/2 *Scat singer*, a hi-de-ho shouter. **1949** L. FEATHER *Inside Be-Bop* v. 39 The swing era produced such notable 'scat' singers as Leo Watson. **1952** B. ULANOV *Hist. Jazz in Amer.* (1958) xx. 252 She pressed the full impact of her scat-singing personality into record grooves. **1957** *Amer. Speech* XXXII. 275 Many bop phrases seem to derive from the nonsense syllables of scat-singing, which, in turn, is simply the voice imitating the sound of an instrument, the first known instance of which, so the story goes, occurred when Louis Armstrong dropped his lyric sheet in the middle of a 1926 recording date and was forced to improvise the words. **1962** 'K. ORVIS' *Damned & Destroyed* iv. 29 A hot-jazz man.. with..a misplaced confidence in his ability as a scat-singer. **1968** P. OLIVER *Screening Blues* vi. 205 With a fierce line in 'scat' singing which had the 'dirty tone' of a muted trumpet, Mary Dixon sang with no apparent restraint. **1974** *New Yorker* 29 Apr. 73 Scat-singing Ella Fitzgerald doesn't just seen an audience. **1976** *National Observer* (U.S.) 20 Nov. 24/3 He didn't invent scat singing any more than Louis Armstrong did, but it's a technique he perfected. **1978** *Fanfare* (Toronto) 10 May 11/2 She sings Twisted, scat sings it, tosses her head back, shakes her lion's mane, pushes her voice into a falsetto that does no little damage to the eardrums. **1978** *Maledicta* 1977 I. 222 *Fang Dang* would *scat-sing* the melody (i.e. using nonsensical words or 'vocalese' to 'sing' the parts of the instrumentation).

scat (skæt), *sb.*[7] [ad. Gr. σκατ-, σκῶρ dung.] **1.** Dung; (*pl.*) droppings. **1950** in WEBSTER Add. **1959** E. COLLIER *Three against Wilderness* xx. 207 Whenever I travelled the game trails, my eyes were alert for any coyote scat (manure) deposited on them. **1966** C. SWEENEY *Scurrying Bush* iv. 48 The speculation when finding a spoor or scat. **1977** *Devon Wetlands* (Devon County Council) xix. 74 The two signs of Otters most likely to be found are their footprints and their droppings (usually known as scats or spraints)... Recognising spraints requires some practice particularly to avoid confusing them with Mink scats. **1977** *New Yorker* 27 June 70/3 We avoid a mound of bear scat.

2. *slang.* Heroin. Cf. *SHIT *sb.*[1] 1. **1970** *Lebende Sprachen* XV. 103/2 *Scat*, heroin. **1972** D. E. WESTLAKE *Cops & Robbers* (1973) ii. 39 You're dealing in machismo, man, just like I'm dealing in scat.

scat (skæt), *v.*[4] *Jazz* (chiefly *U.S.*). [f. *SCAT *sb.*[6]] **a.** *intr.* To perform scat-singing; to sing or improvise with meaningless syllables. **1935** *Metronome* Apr. 54/3 Cab scats through this pair in his best Harlem manner. **1941** *Daily News* (Chicago) 11 June 24/1 Johnny..didn't know the words to the second verse. Instead he sang 'sho-ho-ho', and discovered he liked it that way. Since his audience liked it too, he..has been 'scatting' ever since. **1975** *New Yorker* 26 May 6/1 He and Buddy Rich..launch a series of fusillading four-bar breaks, in which..Torme scats in the Ella Fitzgerald mode.

b. *trans.* To sing or improvise (a song) by replacing the words by meaningless syllables. **1946** MEZZROW & WOLFE *Really Blues* (1957) 104 Louis Armstrong riffed and scatted them. **1958** *Gramophone* Dec. 331/2 Only a couple of songs are scatted. **1973** *Black World* Aug. 58/1 Could scat all Prez's solos note for note in the right key.

Hence **sca·tting** *vbl. sb.* **1946** MEZZROW & WOLFE *Really Blues* (1957) viii. 119 The first time Old Gatemouth ever put his scatting on wax. **1952** B. ULANOV *Hist. Jazz in Amer.* (1958) xx. 252 In 1946 she coined the whole new scatting vocabulary. **1973** S. HENDERSON *Understanding New Black Poetry* 57 The most interesting technical feature of the poem, however, is the singing and scatting of two songs connected with Coltrane.

scat, *int.* and *v.* Delete 'jocularly' and add earlier and later examples. Also in phr. *quicker than scat.* **1838** 'T. TITTERWELL' *Yankee Notions* 52 Drive her away! 'scat her away! *Ibid.* 56 Stop, there! whisht! scat! **1860** J. S. JONES *Green Mountain Boy* I. iii. 13 I'll have the square discharge him quicker than s'cat. **1896** J. F. B. LILLARD *Poker Stories* ix. 210 We chucked him two watches and 380 dollars in cash quicker'n scat. **1917** D. CANFIELD *Understood Betsy* x. 229 Ann and I hitched up quicker'n scat. **1931** M. ALLINGHAM *Look to Lady* xiv. 145 Shoo! Shoo! Scat! We've got a policeman coming. **1950** 'D. DIVINE' *King of Fassarai* xviii. 147 Get the hell out of it! ..I told you kids to scat. **1977** H. GREENE *FSO-1* xvii. 152 Set the breakfast table out here in the drawing room. And then, scat!

scatback (skæ·tbæk). *U.S. Football.* [BACK *sb.*[1] 21: see SCAT *int.* in Dict. and Suppl.] A fast-running backfield player. **1946** *Sun* (Baltimore) 16 Dec. 15/1 They made it 14-0.. with an intercepted pass by Dante Magnani, the scatback from St. Mary's of California. **1948** *Sport Life* Nov. 70/1 The Bears have signed scatback J. R. Boone. **1976** *National Observer* 13 Nov. 1/1 He is slim and muscular-looking—a scatback set to run for daylight.

Scatchard (skæ·tʃɑːd). *Biochem.* [The name of George *Scatchard* (1892–1973), U.S. physical chemist, who published a form of such analysis in 1949 (*Ann. N.Y. Acad. Sci.* LI. 660–72).] *Scatchard plot*, a graph of the concentration of a solute absorbed by a protein, membrane, cell, or the like against its concentration in the surrounding medium; *Scatchard analysis*, the use of such graphs to deduce the number and nature of the binding sites on the protein, etc. **1958** EDSALL & WYMAN *Biophys. Chem.* I. xi. 617 (*heading*) The Scatchard plot of $\bar{v}$/(A) against $\bar{v}$. **1970** *Arch. Biochem. & Biophysics* CXLI. 623/2 (*caption*) Scatchard plot for binding of cytochrome *c* to normal rat liver mitochondria. **1975** *Nature* 13 Nov. 154/2 (*caption*) Scatchard plot of insulin-membrane interaction in two representative preparations from 10 control..and 10 diabetic.. animals. *Ibid.* 27 Nov. 339/2 Scatchard analysis of the binding data revealed a class of receptors for 5-α-dihydrotestosterone of uniform affinity. **1978** *Ibid.* 12 Oct. 553/2 (*caption*) Scatchard analysis of interaction between diazepane and 'endogenous inhibitor' of Na^+-independent ^{3}H-GABA binding.

scatological (skæːtǫlǫ·dʒikăl), *a.* [f. SCATOLOGY: see -ICAL.] Of or pertaining to scatology (senses 1 and 3); characterized by a preoccupation with obscenity. **1924** F. M. FORD *Some do Not* II. ii. 237 The late Mr. Duchemin was a scatological—afterwards a homicidal—lunatic. **1959** J. KIRKUP tr. *S. de Beauvoir's Mem. Dutiful Daughter* I. 82 There was one phrase grown-ups were always using: 'It's not proper!'.. At first I had taken it to have a scatological connotation. **1960** G. MAXWELL *Ring of Bright Water* ix. 123 As a useful by-product of his [*sc.* the otter's] impish sense of humour, the cattle tended to keep farther from the house, thus..reducing the number of scatological hazards to be skirted at the door. **1969** *Daily Tel.* 13 Mar. 23/2 The scatological streak in the man, his disgusted-disgusting jokes about excreting nymphs. **1979** *London Rev. Bks.* 25 Oct. 8/2 One might almost assume from a few of these scatological diatribes that he thought there was something intrinsically disgusting about physical love.

scatology. **3.** Delete *rare* and add later examples. **1936** R. QUINTANA *Mind & Art of Jonathan Swift* VI. ii. 360 From scatology one turns with relief to the capital verses entitled *Helter Skelter*. **1959** N. O. BROWN *Life against Death* xiii. 179 The most scandalous pieces of Swiftian scatology are..*The Lady's Dressing Room, Strephon and Chloe, Cassinus and Peter*. **1975** *Publishers Weekly* 22 Sept. 134/1 Funny, albeit unintentional scatology in the [zodiacal] Sign abbreviations.

scatter, *sb.* Add: **1. b.** *transf.* in Linguistics. **1934** J. R. FIRTH *Papers in Linguistics 1934–51* (1957) ii. 4 All the common phonetic contexts of each phoneme should be stated, and the contextual spread or 'scatter' of the phonemes compared. This knowledge of the contextual scatter of a phoneme will be found of the greatest importance for the statement of our future sound laws. **1935** —— in *Trans. Philol. Soc.* 45 The frequency of reference to sex had necessarily extended what I term the formal scatter of the word, and we now have *sexed, sexless, sexy, sexiness*, even *sexology*. **1963** J. LYONS *Structural Semantics* vii. 178 One point that seemed to be of relevance in the inquiry was the defective formal 'scatter' of the lexeme εἰδέναι.

2. Delete 'small' and *rare* and add later examples. Also *spec.* in *Archæol.* **1943** V. SACKVILLE-WEST *Eagle & Dove* iii. 17 A sick woman with a scatter of high-spirited children to control would welcome any method of keeping them quiet. **1954** J. B. MITCHELL *Historical Geogr.* iii. 73 A scatter of Scandinavian settlers in a district primarily English. **1959** *Listener* 12 Mar. 449/2 The ascendancy of the U.S.A., along with that of the U.S.S.R., has relegated the scatter of European nations to subsidiary status. **1974** C. TAYLOR *Fieldwork in Medieval Archaeol.* ii. 27 Much of it [*sc.* the information] will probably be vague, such as notes of pottery scatters, low banks, water-filled ditches and possible old quarries. **1977** *Christian* IV. 109 The human race is not a scatter of individuals.

3. *Statistics.* The degree to which repeated measurements or observations of a quantity differ; that which is measured by the variance. **1921** R. S. WOODWORTH *Psychol.* (1922) xii. 273 Usually there is some 'scatter' in the child's successes. **1923** *Proc. R. Soc.* A. CII. 357 The question arises as to how much of the 'scatter' of the Gaussian curve is due to error of observation, inexperience in making the readings, accidental variations, etc., and how much is due to a real difference in the physiological equipment of the observer. **1934** *Brit. Jrnl. Psychol.* XXIV. 344 The I.Q.'s of the boys showed a wider scatter than those of the girls. **1963** B. FOZARD *Instrumentation Nucl. Reactors* vii. 70 A commonly used measure of the dispersion or scatter of a number of observed values about the central values is the standard deviation. **1968** *Brit. Med. Bull.* XXIV. 246/2 Most observations are subject to considerable scatter, especially where mammalian systems are used, and statistical procedures of varying complexity are called for.

4. a. The scattering of light or other radiation. **1942** *Tee Emm* (Air Ministry) II. 145 Preventing the 'light scatter' which comes from scratched Perspex or slightly dirty windscreens. **1962** H. C. WESTON *Sight, Light & Work* (ed. 2) vi. 206 It is better that the increased illumination required by older eyes should be provided by 'warmer' illuminants so that 'hazing' due to scatter within the eyes is minimised.

b. *spec.* with reference to radio waves, freq. denoting the use of scattering within the atmosphere to extend the range of radio communication. Freq. *attrib.* **1950** *Proc. IRE* XXXVIII. 412/2 For two-directional antennas of beam width Θ facing one another, the greatest angle of scatter that need be considered is Θ. **1956** *Ann. Reg. 1955* 155 It was announced that a revolutionary new system of communications, known as 'scatter', which was not susceptible to jamming,..would be introduced. **1958** *Times* 30 Apr. 6/6 A range of tropospheric scatter transmitting and receiving equipment. **1966** *McGraw-Hill Encycl. Sci. & Technol.* XIII. 439/1 Radio relay systems are usually more suitable than the scatter systems for overland use where intermediate radio relay stations can be constructed. **1977** *Lancs. Life* Nov. 83/2 On the radar screen shown here, Manchester is permanently blacked-out because of 'scatter' from buildings and nearby high ground, to avoid masking approaching precipitation.

5. *Comb.* **scatter diagram, plot** *Statistics*, a diagram having two variates plotted along its two axes and in which points are placed to show the values of these variates for each of a number of subjects, so that the form of the association between the variates can be seen. **1925** F. C. MILLS *Statistical Methods* x. 366 The equation to a straight line, fitted by the method of least squares to the points on the scatter diagram, will express mathematically the average relationship between these two variables. **1937** YULE & KENDALL *Theory of Statistics* (ed. 11) xiv. 275 The scatter diagram in two dimensions may be generalised to three dimensions, and may also be used as a mental construct for higher dimensions, though no actual model can of course be made. **1960** [see *KARYOGRAM b]. **1971** *Nature* 9 Apr. 390/2 Scatter diagrams were drawn to show mean concentrations of albumin, γ-globulin, fibrinogen and cholesterol against age. **1971** *Jrnl. General Psychol.* LXXXV. 266 Inspection of the scatter plot..indicates that any index of relationship would be misleading. **1973** *Jrnl. Genetic Psychol.* CXXII. 45 Guilford's triangular scatterplot conceptualization of intelligence-creativity relationship seemed most congruent with the present..data.

scatter, *v.* Add: **4. c.** *Baseball.* Of a pitcher: to yield hits only at intervals and so restrict scoring. **1892** *Chicago Herald* 25 May 6/1 Young kept the hits well scattered. **1954** *Post-Herald* (Birmingham, Alabama) 7 June 7/2 Winning pitcher was Dave Benedict, who relieved in the first inning and scattered four hits the rest of the way. **1976** *Billings* (Montana) *Gaz.* 27 June 2-F/4 Joaquin Andujar scattered 10 hits Saturday to pace the Houston Astros to a 3-0 victory over the Cincinnati Reds.

5. e. More widely, to deflect, diffuse, or reflect (radiation, particles, or the like) in a more or less random fashion. Also *absol.* (Further examples.) **1878** LD. RAYLEIGH *Theory of Sound* II. xv. 139 If the primary sound be a compound musical note, the various component tones are scattered in unlike proportions. **1891** HURTER & DRIFFIELD in W. B. Ferguson *Photogr. Res. F. Hurter & V. C. Driffield* (1920) 146 Captain Abney has discovered that negatives 'scatter' so much light that our instrument cannot possibly measure all the light which a negative transmits. **1911** *Phil. Mag.* XXI. 675 In these calculations, it is assumed that the α particles scattered through a large angle suffer only one large deflexion. **1938** R. W. LAWSON tr. *Hevesy & Paneth's Man. Radioactivity* (ed. 2) vii. 75 Hydrogen nuclei alone behave differently, for they scatter [neutrons] very much more strongly than would be expected from the magnitude of the cross-section of hydrogen nuclei. **1955** HUETER & BOLT *Sonics* vi. 232 This limits the sound pressure that can be transmitted beyond the point where cavitation first occurs since the bubbles present will scatter and dissipate a part of the sound energy. **1955** C. G. DARWIN in W. Pauli *Niels Bohr* 6 There were a few α-particles scattered through such broad angles, even right backwards, that no conceivable compound effect could possibly explain them. **1959** *Listener* 18 June 1057/1 The distortions introduced by scattering a signal from such a surface might not be too serious. **1971** *Sci. Amer.* June 61/2 Since the neutron and the proton respond to the electromagnetic force, they scatter electrons aimed at them.

f. *intr. Physics.* Of radiation, particles, etc.: to undergo scattering. **1971** *Nature* 16 July 167/2 The double reflexion mechanism gives way to multiple reflexions, that is, a ray is trapped in surface cavities before scattering out, randomizing the polarization. **1975** *Ibid.* 25 Sept. 275/1 The majority of the energy is carried by phonons which inelastically scatter at the interface. **1980** *Sci. Amer.* July 57/1 Inside it an entering gamma-ray photon typically scatters off several electrons in succession.

7. a. scatter bomb, a bomb that scatters its material over a wide area; also *fig.*; **scatter bombing** *vbl. sb.*, bombing carried out haphazardly over an area; **scattershot** orig. and chiefly *N. Amer.*, the shot contained in a scatter-charge; also used *fig.* (chiefly *attrib.*)

to designate something of a random, haphazard, or indiscriminate character (cf. *SCATTER-GUN 2); **scatter-site** *a.* *U.S.* = *scattered-site* s.v. *SCATTERED *ppl. a.* 2 b. **1961** WEBSTER, Scatter bomb. **1973** J. QUICK *Dict. Weapons & Mil. Terms* 386/1 Fragmentation bombs or fragmentation clusters, as well as certain incendiary bombs equipped with bursters, are scatter bombs. **1977** *Rolling Stone* 16 June 43/2 Okay, Scorsese is a violent scatter-bomb. **1977** *Daily Tel.* 3 Aug. 5/5 West Germany's new scatter bomb..comprises more than 1,000 mini-bombs which can be fired in different patterns by rockets triggered from the cockpit. **1940** *Aeroplane* 13 Sept. 314/1 The scatter-bombing..must at times have sorely tried pilots who had seen the effects of it. **1961** WEBSTER, Scattershot. **1965** *Economist* 19 June 1393/2 The President's..scatter-shot efforts to reduce the government's spending. **1967** *Boston Globe* 5 Apr. 51/2 Jack Nicklaus is more concerned over his scattershot driver than the threat of mumps. **1972** *Publishers' Weekly* 10 July 42/2 Shirley Green brings more scattershot curiosity than serious learning to her 'history'. **1974** *State* (Columbia, S. Carolina) 27 Feb. 18-A/6 When demagogic politicians ride the land firing scatter shot, nobody..is safe unless he shares their prejudices. **1978** R. STEVENS *Law & Politics* 505 The future of the judicial role in England..may lie far more with subtle use of judicial restraint than with scattershot judicial activism. **1972** *N.Y. Times* 3 Nov. 16/2 The scatter-site housing dispute in Forest Hills. **1977** *New Yorker* 27 June 85/2 Jimmy Carter's mention of his belief in ethnic purity..in response to a *News* reporter's question about scatter-site housing.

b. *attrib.*, passing into *adj.* Designating one of a number (intended to be) scattered decoratively here and there, as *scatter cushion, pin, rug*, etc. orig. *U.S.* **1933** 'E. QUEEN' *Siamese Twin Mystery* I. ii. 30 A living-room..dotted with armchairs and small scatter-rugs. **1946** *Negro Digest* Aug. 51/1 Its large living room has a vaulted ceiling and arched beams, and the floor is covered with deer skins and scatter rugs. **1957** J. D. SALINGER in *New Yorker* 4 May 123/1 Three domestic Oriental scatter rugs, extremely worn, were on the floor. **1960** *Woman* 5 Mar. 19/1 Scatter cushions have become a favourite furnishing accessory. **1960** I. WALLACH *Absence of Cello* 13 She stopped first at the jewelry counter where she sneered at some scatter pins. **1966** T. PYNCHON *Crying of Lot 49* ii. 36 Bracelets then, scatterpins, earrings, a pendant. **1974** J. IRVING *158-Pound Marriage* 104 The bed..had pitched the mattress and us across the scatter rug. **1976** L. DEIGHTON *Twinkle, twinkle, Little Spy* xviii. 185 There [were]..scatter-cushions on the floor. **1980** P. HARCOURT *Tomorrow's Treason* I. i. 31 The floor was..wood with a couple of bright scatter mats.

scatteration. Delete *rare* and add later examples. Also, the fact or condition of being scattered. **1880** 'MARK TWAIN' *Tramp Abroad* xix. 183 A raft.. hit the pier in the center and went all to smash and scatteration like a box of matches struck by lightning. **1892** KIPLING *Lett. of Travel* (1920) 40 A household spreads itself over plots, maybe, a quarter of a mile apart. A revenue map of a village shows that this scatteration is apparently designed. **1910** *Blackw. Mag.* July 24/1 Here there is a scatteration, but the tufters..are stopped and laid on to the line of a single stag. **1930** R. FRASER *Rose Anstey* xlix. 328 At night she stared from her bed at a great scatteration of stars. **1936** *Burlington Mag.* June 261/1 The gilt frame of the mirror is not a strong enough colour to correct the consequent scatteration. **1965** *Economist* 6 Feb. 509 A growing desire to end what Mr Walter Lippmann calls 'globalism and scatteration' in foreign policy in favour of a concentration on America's 'primary vital interests'. **1965** D. OWEN *Eng. Philanthropy* IV. xx. 559 'Scatteration' philanthropy—spending too small amounts on too many agencies or individuals.

scatter bomb, bombing: see *SCATTER *v.* 7 a.

scattered, *ppl. a.* Add: **2. b.** *scattered-site* (U.S.), used *attrib.* to designate public housing (esp. for low-income families) distributed throughout a city rather than concentrated in a few areas. Also *absol.* in *pl.* (unhyphenated) as *sb.* **1956** *Jrnl. Housing* May 163/1 (*heading*) Scattered site projects. *Ibid.* 'We have inaugurated a new policy that not only permits but encourages the use of small scattered sites,' PHA Commissioner Charles E. Slusser told delegates. **1958** *Ibid.* Jan. 11 (*caption*) Diagrammatical sketch of the City of Cedartown, showing the seven sites on which the Cedartown Housing Authority constructed..the first scattered-site project in the nation. **1959** *Ibid.* Nov. 359/2 In many areas the row house has been adapted for scattered-site use. **1966** *Daily Progress* (Charlottesville, Va.) 12 Aug. 10/1 'Scattered-site' housing is an alternative to high-rise, low-income apartments.

4. c. More widely, of electromagnetic radiation generally and sub-atomic particles: subjected to scattering. (Further examples.) **1878** LD. RAYLEIGH *Theory of Sound* II. xv. 139 If a number of small bodies lie in the path of waves of sound,.. the exaltation of the higher harmonics in the scattered waves involves a proportional deficiency of them in the direct wave after passing the obstacles. **1906** *Phil. Mag.* XII. 144 From measurements of the width of the band due to the scattered α rays, it is easy to show that some of the α rays in passing through the mica have been deflected from their course through an angle of about 2°. **1926** R. W. LAWSON tr. *Hevesy & Paneth's Man. Radioactivity* iii. 39 The whole of the scattered radiation also enters the electroscope. **1966** *McGraw-Hill Encycl. Sci. & Technol.* XIII. 438/2 Although radio waves at these frequencies are not reflected by the ionosphere, it has been found that if large amounts of power are radiated, scattered energy will be received over relatively long distances beyond the horizon. **1970** I. E. MCCARTHY *Nuclear Reactions* i. 5 In addition to elastically scattered α particles, two groups of protons.. were observed. **1978** P. W. ATKINS *Physical Chem.* xvii. 546 If the molecules are excited by the light during the collision they withdraw some energy from the photons, and so the scattered light emerges with a lower frequency than the incident light.

scatterer. Add: **1.** Also, something which scatters; a device for broadcasting seed. **1868** *Rep. Iowa State Agric. Soc. 1867* 227 The seed is scattered by a vibrating scatterer.

2. *Physics.* Anything which scatters radiation, particles, or the like. **1930** A. B. WOOD *Textbk. Sound* III. 282 The amplitude of the secondary waves varies directly as the volume of the 'scatterer'. **1931** [see *IMPURITY 3 b]. **1936** *Nature* 1 Feb. 185/2 The product of the 140 sec. half-period is enhanced in the case of all scatterers [of neutrons] investigated except carbon and aluminium. **1959** *Listener* 18 June 1057/1 If it [*sc.* the moon] were behaving as a uniform scatterer of radio waves. **1973** *Nature* 7 Sept. 38/2 The beam [of protons]..is made to strike a scatterer of lead or copper. **1977** R. KATZ *Ziggurat* (1978) vi. 59 Beryllium..was an excellent reflector, or 'neutron scatterer', because it had an atomic structure more dense than any other element. **1977** *Sci. Amer.* Oct. 90/2 In synthetic-aperture radar the interference pattern from each scatterer on the terrain is recorded..as a narrow broken line parallel to the edge of the data film.

scattergram (skæ·təɹgræm). *Statistics.* A contraction of *scatter diagram* s.v. *SCATTER *sb.* 5. **1938** A. E. WAUGH *Elem. Statistical Method* ix. 235 This is the method of plotting the data on a scatter diagram, or scattergram, in order that one may see the relationship. **1966** *Jrnl. Neurophysiol.* XXIX. 812 To see how latencies and phases change with frequency, scattergrams can be made comparing each interval with the latency or phase of the spikes that occur during the interval. **1973** B. J. WILLIAMS *Evolution & Human Origins* xii. 215/2 In Figure 5.1 we saw a scattergram of a spurious correlation between R_0 and heart disease. **1980** *Amer. Speech* LV. 226 The reader is told to calculate..r's for inter-group differences, scattergrams, and correlation coefficients.

scatter-gun. For *U.S. colloq.* read 'orig. and chiefly *N. Amer.*' and add: **1.** (Earlier and later examples.) **1836** H. R. HOWARD *Hist. V. A. Stewart* 140, I have a choice scatter-gun. **1910** *Blackw. Mag.* Feb. 285/1 All round Muttra shooting with a scatter-gun is varied and good. **1923** J. H. COOK *Fifty Years on Old Frontier* I. 4 Pigeon shooting was good..for anyone who owned or could borrow a 'scatter-gun'. **1932** 'D. YATES' *Safe Custody* ix. 198 We've thirteen men, and between us we've got six pistols and three scatter-guns. **1968** *Punch* 1 May 624/1 He hands his trunk to this Puerto Rican who's carrying..a sawn-off scatter-gun. **1973** R. D. SYMONS *Where Wagon Led* I. iv. 51 Once in a while one of us would pack a scatter gun and get a brace or two of prairie chicken.

2. *fig.* (*attrib.* in quots.). Cf. *scattershot* s.v. *SCATTER *v.* 7. **1952** J. STEINBECK *East of Eden* II. xiv. 150 A scattergun method for dealing with unpleasant facts. **1963** *Daily Progress* (Charlottesville, Va.) 6 Feb., Republicans agreed.. to a 'scattergun plan'. Each representative wrote the name of his choice..on a secret ballot. There were no nominations and no debate. **1974** *Publishers Weekly* 4 Feb. 70/1 Farson sets it all down with a scattergun assertiveness that inevitably turns up contradictions. **1980** *Times Lit. Suppl.* 15 Aug. 913/1 Stuart Holland's scatter-gun polemic embodies many of the misconceptions which now threaten to dominate the Labour Party's attitude to Europe.

Hence as *v. intr.* (*fig.* in quot.) and **sca·tter-gunner.** **1968** R. M. NIXON in W. Safire *Before the Fall* (1975) I. vi. 72 If we scatter-gun too much we are not going to have an impact. **1969** *Daily Colonist* (Victoria, B.C.) 8 Nov. 20/1 Two traps will be in operation and all scattergunners are invited to compete. **1980** *Outdoor Life* (U.S.) (Northeast ed.) Oct. 104/2 Scatter-gunners bag approximately 50 million of them each hunting season.

scattering, *vbl. sb.* Add: **1.** (Further examples in *Physics.* Cf. *SCATTER *v.* 5 e.) **1911** *Proc. Manchester Lit. & Philos. Soc.* LV. p. xviii (*heading*) The scattering of the α and β rays and the structure of the atom. **1942** J. D. STRANATHAN *'Particles' Mod. Physics* xi. 405 On the theory of multiple scattering an entirely negligible number of particles should be scattered at large angles. **1950** *Nature* 30 Dec. 1103/2 It is..probable that turbulent scattering..plays an important part in determining the signals received from high-power metre-wave transmitters at distances greater than about 100 miles. **1955** HUETER & BOLT *Sonics* iii. 85 Scattering at the grain boundaries is one important cause for the absorption of ultrasonic waves in metals. **1974** G. REECE tr. *Hund's Hist. Quantum Theory* vi. 56 In 1903 J. J. Thomson worked out from the intensity of scattering of X-rays that the number of electrons must be roughly equal to the atomic weight. **1975** D. G. FINK *Electronics Engineers' Handbk.* XVIII. 91 At frequencies in the 30- to 100-MHz region, regular but weak propagation by ionospheric scattering is obtained.

3. Special Comb.: **scattering angle** *Physics*, the angle through which a scattered particle or beam is deflected. **1913** *Phil. Mag.* XXVI. 711 It is a different matter.. when the scattering angle is only about 1/10 of a degree, as in the present experiment. **1950** *Nature* 30 Dec. 1102/2 A great simplification of the results is effected by restricting the discussion to small angles (beam-widths and scattering angles). **1970** I. E. MCCARTHY *Nuclear Reactions* I. i. 9 We see that for a given scattering angle particles of higher energy come closer to the nucleus.

scattering, *ppl. a.* **2. a.** Substitute for def.: *Physics.* That causes scattering (of light, radiation, particles or the like). (Further examples.) **1911** *Phil. Mag.* XXI. 675 It is essential that the thickness of the scattering material should be so small that the chance of a second encounter involving another large deflexion is very small. **1938** R. W. LAWSON tr. *Hevesy & Paneth's Man. Radioactivity* (ed. 2) ix. 95 A strong pencil of α-particles of definite velocity is allowed to strike a thin sheet of the scattering substance. **1958** *Times* 30 Apr. 6/6 The forward scatter technique..in this case uses the troposphere as the scattering medium. **1970** D. W. TENQUIST et al. *University Optics* II. ii. 84 If the incident light photon of energy *hν* impinges upon a molecule of the scattering medium and the energy state of this molecule changes from E_1 to E_2, the energy of the Raman scattered photon is given by $h\nu - (E_2 - E_1)$.

b. *scattering layer* (Oceanogr.), any of a number of layers in the sea which give rise to strong acoustic echoes owing to the presence of a high concentration of living organisms. **1942** *Reverberation Stud. at 24 Kc* (Univ. Calif. Div. War Res. Rep. U7) 48 (*heading*) Deep scattering layers. *Ibid.* 49 Observations indicate that deep scattering layers, in a given area, may appear and disappear and yet persist for periods as long as a month or perhaps even longer. **1948** *Nat. Geogr. Mag.* Sept. 277/2 So incredibly numerous are such sea creatures that this layer of ocean life actually returns an echo of the sound sent down by the Fathometer. The echo from this so-called 'scattering layer' is sometimes so strong that it causes navigators to think they are sailing over a shoal. **1972** J. WILLIAMS *Oceanogr.* 53 At night this deep scattering layer..is centered near the sea surface... In the morning it moves down into the depths again. **1977** CLAY & MEDWIN *Acoust. Oceanogr.* vii. 237 A great deal of data, particularly the frequency dependence of scattering layers, are obtained by using explosive sources.

scatterometer (skætərǫ·mitəɹ). [f. SCATTER *sb.* + -OMETER.] A radar designed to provide information about the roughness or the profile of a surface from the way it scatters the incident microwaves. **1966** *Electronics* 14 Nov. 44 A spacecraft, using a Scatterometer, has to fly over a pock-marked area of the lunar surface only once to get a detailed profile of the terrain. **1978** *Nature* 22 June 586/2 Seasat-A is primarily a 'proof of concept' mission designed primarily to discover how effectively the microwave equipment which it carries—a scanning radiometer, a radar scatterometer, a synthetic aperture radar, and a radar altimeter—can provide useful scientific information for oceanographers, meteorologists and commercial sea-users.

scattershot, -site: see *SCATTER *v.* 7.

scattery, *a.* Add: **b.** Scatter-brained. *rare.* **1924** J. GALSWORTHY *White Monkey* I. v. 33 The scattery enthusiasm of the sucking publisher. **1928** —— *Swan Song* III. i. 226 He himself knew how to wait, but did this modern young man, so feather-pated and scattery?

scattiness (skæ·tinės). [f. *SCATTY *a.*[2] + -NESS.] The quality or condition of being scatty or scatter-brained. **1959** S. GIBBONS *Pink Front Door* iv. 52 For all her scattiness, Daisy was a good child. **1959** A. BUCHAN *Spare Chancellor* v. 118 Catherine Gladstone with the characteristic scattiness of the Glynnes and the Lytteltons always referred to Lowe's followers as 'The Dolomites'. **1976** E. BERCKMAN *Be All & End All* v. 62 A peculiar combination of scattiness accompanied by a desperate concentration. **1977** 'M. INNES' *Honeybath's Haven* viii. 82 The spectacle of [Hamlet's] random and intermittent scattiness is confusing.

scatty (skæ·ti), *a.*[1] *U.S. Underworld slang.* [Of unknown origin; cf. SCOTTY *a.*] Bad-tempered. **1909** W. H. DAVIES *Beggars* xxvi. 205 Nearly all men that live in common lodging-houses..are..more or less short-tempered, or as they say—'scatty'. **1927** *Dial. Notes* V. 461 *Scatty*, ill-natured.

scatty (skæ·ti), *a.*[2] *colloq.* [Prob. f. SCATT(ERBRAINED *a.* + -Y[1].] Of a person: scatterbrained; driven distracted, mad; of a story, etc.: illogical and absurd. **1911** J. W. HORSLEY *I Remember* xi. 254 Cockney slang was far more familiar to me than most..'scatty' for mad. **1934** *Punch* 17 Jan. 74/1 Simpson, who lives opposite, says it [*sc.* the house] was preying on his mind... 'Another week and I should have been scatty,' he explained. **1951** M. KENNEDY *Lucy Carmichael* II. i. 83 She is amusing in a breathless, scatty sort of way. **1956** J. DICKSON CARR *Patrick Butler for Defence* iii. 26 If we tell this scatty story about an impossible murder, they won't believe one word we say. **1972** 'J. BELL' *Death of Poison-Tongue* i. 11 Do you mean you know who the person is who spreads wicked lies about the neighbours? Don't the ones attacked do anything if they know the person? It sounds utterly scatty. **1977** *News of World* 17 Apr. 16/7 My scatty friend..had later gone into the butchers and loudly asked for a shepherd to make a pie with. **1980** J. MCCLURE *Blood of Englishman* i. 11 The scatty receptionist had looked at him with twinkling eyes.

scaurie. Add: Also **scory, scurrie.** (Earlier and further examples.) **a 1795** G. LOW *Fauna Orcadensis* (1813) 122 The Brown and White Gull..Orc. Scory..is the scarcest of the Gull-kind in Orkney. **1822** [see *ICELAND]. **1918** T. MANSON *Humours of Peat Commission* I. 125, I mind haein a tame

scorie..whin I wis a boy. **1960** *People's Jrnl.* (Dundee) 12 Mar. 9/2 Ah'm fair deav't wi' the awfa scraichin' o' the scurries.

scavenge, *v.* Add: **2. b.** To extract and collect (anything that can be used or eaten) from discarded material. **1922** JOYCE *Ulysses* 158 Saw her in the viceregal party when Stubbs the park ranger got me in with Whelan of the *Express*. Scavenging what the quality left. High tea. Mayonnaise I poured on the plums thinking it was custard. **1971** J. S. WEINER *Man's Natural History* v. 199 Big-game hunters, living by means of scavenging dead mammoths. **1977** *Times Lit. Suppl.* 7 Jan. 9/5 In Kingston [Jamaica].. 'scuffling' (dealing in whatever can be scavenged).

3. (Later examples.) Also, to borrow; to thieve; to search through rubbish *for* (left-overs or unwanted objects). In quot. 1960 simply 'to search thoroughly'. **1938** *Sun* (Baltimore) 17 June 3/4 A woman relief investigator said aged men whom she recognized as relief clients were 'scavenging' at a market for discarded sprigs of celery. **1941** BAKER *Dict. Austral. Slang* 63 *Scavenge, to*, to borrow: to act the petty thief. **1960** P. S. BEAGLE *Fine & Private Place* xiv. 252 She..scavenged frantically in her purse, trying to hold back a sneeze until she found a handkerchief. **1978** S. TENNENBAUM *Rachel, the Rabbi's Wife* (1979) x. 340 Rachel worked hard to refurbish her studio... She scavenged for pieces of furniture, and found an old armchair, a small table, and a low couch without a cushion.

4. a. *trans.* To remove (the combustion products) from the cylinders of an internal-combustion engine. Also *absol.*, and with the engine or cylinder as object. Cf. *SCAVENGING *vbl. sb.* 2. **1894** [implied in *SCAVENGING *vbl. sb.* 2 a]. **1903** *Amer. Inventor* 15 Aug. 78/3 The engine under description scavenges thoroughly and completely upon the return stroke of the pistons. **1954** E. J. KATES *Diesel & High-Compression Gas Engines* ii. 23 Just as before, this helps to get the exhaust gases out, or scavenges them. **1961** K. ČASLAVSKÝ tr. *Mackerle's Air-Cooled Motor Engines* xviii. 376 With a mixing ratio $\lambda > 1$ the cylinder charge is increased by the amount of air contained in the exhaust gas not scavenged from the cylinder. **1966** *McGraw-Hill Encycl. Sci. & Technol.* VII. 208/2 Most medium and large two-cycle diesel engines are usually equipped with blowers to scavenge the cylinders after the working strokes and to supply the air required for the subsequent cycles. **1975** M. J. NUNNEY *Automotive Engine* x. 246 The development of the two-stroke cycle of operation is generally attributed to Dugald Clerk who, in 1878, adopted this principle for a successful design of engine that was scavenged by a separate pumping cylinder.

b. *Chem.* To combine with or remove (free radicals, electrons, or other species). **1955** *Jrnl. Amer. Chem. Soc.* LXXVII. 3245/2 This would mean that all of the radicals which are being scavenged are swept from the solution by the mercaptan. **1966** W. A. PRYOR *Free Radicals* xxi. 324 An added free radical species will inhibit the process..if it scavenges S· but does not react with S to convert it to S·. **1974** *Sci. Amer.* Dec. 71/1 Z is eventually restored to neutrality by scavenging four electrons from two water molecules. **1978** *Nature* 1 June 374/1 Newly formed amorphous iron hydroxides seem to scavenge phosphate and silicate from solution.

Hence as *sb.*, = *SCAVENGING *vbl. sb.* 2 a. Freq. *attrib.* **1912** A. P. CHALKLEY *Diesel Engines* vi. 156 On the up stroke the scavenge ports..are closed before the exhaust ports. **1925** *Glasgow Herald* 1 Apr. 11/2 This new type of engine, with its straight through scavenge and absence of air and exhaust valves. **1930** *Engineering* 21 Nov. 645/3 The scavenge pumps for the Junkers engine are mounted on the locomotive frame. **1949** T. D. WALSHAW *Diesel Engine Design* xviii. 338 Typical figures for an engine supercharged to give 50 per cent. increase in available B.H.P. are: 30 per cent. through scavenge (i.e. a volume of air equal to 30 per cent. of the cylinder volume is swept through the exhaust valve), and the amount of overlap would be about 135°. **1955** *Know your Tractor* (Shell) i. 11 The air for combustion assists removal of the exhaust gases; it is therefore known as 'scavenge' air, and its admission to the engine as 'scavenging'. **1957** [see *LUBE *sb.* and *v.*]. **1975** A. J. WHARTON *Diesel Engines: Questions & Answers* 19 Even in slow running engines, this allows only a very short period of time for scavenge to be completed.

scavenger. Add: **3*.** *Chem.* A substance or species which scavenges (sense *4 b) free radicals or other species. **1955** *Jrnl. Amer. Chem. Soc.* LXXVII. 3244 The demonstration that various radical scavengers such as butyl mercaptan, α, α- diphenyl-β-picrylhydrazyl.., iodine and oxygen do not capture the decomposition products quantitatively. **1961** G. R. CHOPPIN *Exper. Nuclear Chem.* xii. 196 The presence of even small amounts of impurity, especially if it has a high affinity for radicals (a scavenger), causes decomposition. **1970** *Financial Times* 13 Apr. 20/4 Manganese is probably the most important 'minor' metal used in the steel industry, being used as a de-oxidiser and scavenger to combine with sulphur. **1974** C. C. PATTON in P. L. Moore et al. *Drilling Practices Manual* xv. 397 Scavengers can also be added to a drilling fluid to remove small amounts of hydrogen sulfide. **1978** *Nature* 23 Nov. 347/2 Hydroxyl is the most reactive trace species in the troposphere and is therefore the dominant scavenger of many anthropogenic substances.

4. scavenger hunt orig. *U.S.*, a game in which people try to collect certain miscellaneous objects from the neighbourhood. **1940** *Sun* (Baltimore) 1 Nov. 7/3 Eight young persons on a scavenger hunt sponsored by a Westport High School sorority went to the Sixty-third street police station willingly to obtain signatures of policemen. **1963** 'E. MCBAIN' *Ten Plus One* xi. 136 We're two college kids on a scavenger hunt... We're supposed to bring back a hibernating bear. **1977** *Times* 24 Dec. 10/2 The outdoors scavenger hunt is a good exercise after overeating. **1980** *Jewish Chron.* 18 July 25/5 Sunday, July 27. Car Rally/Scavenger Hunt.

scavenging, *vbl. sb.* Add: **2. a.** Removal of combustion products from the cylinders of internal-combustion engines. Also as *ppl. a.* **1894** *Work* 17 Feb. 73/3 Questions such as late ignitions, scavenging, varying explosive charges [etc.]. **1896** B. DONKIN *Text-bk. Gas, Oil, & Air Engines* (ed. 2) I. xix. 269 The increase in economy obtained with the new (1894) 'scavenging' Crossley-Atkinson engine. **1915** *Illustr. London News* 13 Mar. 340 Scavenging-pump for expelling used gases at the end of each stroke. **1924** *Times Trade & Engin. Suppl.* 29 Nov. 250/3 As is usual with large Sulzer engines the scavenging air is supplied from electrically driven turbo blowers installed in the engine-room. **1954** E. J. KATES *Diesel & High-Compression Gas Engines* ii. 24 Instead of rotary blowers, many two-cycle diesels employ what is called crankcase-scavenging. *Ibid.* 25 The outside atmospheric pressure then pushed open the scavenging valve and permitted a fresh supply of air to enter the crankcase. **1962** J. M. DOHERTY *Diesel Locomotive Practice* ii. 25 To obtain efficient scavenging in two-stroke engines, the air is always admitted to the cylinder under pressure. **1975** M. J. NUNNEY *Automotive Engine* x. 248 The mean effective pressures developed by a two-stroke engine..depend upon its scavenging efficiency.

b. *Chem.* The action of *SCAVENGE *v.* 4 b. **1955** *Jrnl. Amer. Chem. Soc.* LXXVII. 3245/2 There must be two reactions which produce the dinitrile, one which is subject to scavenging by the mercaptan and one which is not. **1978** *Nature* 1 June 374/1 Lal *et al.* have shown that from 12 to 15% of dissolved silicate can be removed from seawater by scavenging during precipitation of finely dispersed ferric hydroxide.

scavvy, var. *SCAFFY.

scawtite (skǭ·təit). *Min.* [f. the name *Scawt* (see quot. 1930) + -ITE[1].] A hydrated carbonate and silicate of calcium which occurs as minute, colourless, monoclinic crystals. **1929** *Nature* 7 Dec. 896/1 C. E. Tilley: On scawtite... This new monoclinic mineral, with composition $6CaO.4SiO_2.3CO_2$, occurs in the contact zone between the chalk and the dolerite. **1930** C. E. TILLEY in *Mineral. Mag.* XXII. 224 It is proposed to designate this new mineral scawtite, from the original locality, Scawt Hill, Co. Antrim. **1957** *Amer. Mineralogist* XLII. 387 A zone about ten inches thick between the larnite zone and the limestone consists essentially of scawtite. The scawtite rock is greyish-white, dense and flinty. **1973** *Acta Crystallographica* B. XXIX. 73 Scawtite contains 2 units of $Ca_7(Si_6O_{18})(CO_3).2H_2O$ in a monoclinic cell.

sceau, erron. form of *SEAU.

‖ **Sceaux** (sōᵘ). The name of a town near Paris used *attrib.* and *absol.* to denote tin-enamelled faience made there in the latter part of the eighteenth century, often painted with floral and figure subjects and modelled in the form of figures. **1884** GASNAULT & GARNIER *French Pottery* 182 (Index), Sceaux mark on faïence. **1903** M. L. SOLON *Old French Faïence* 125 An anchor, in allusion to the dignity of the Duke de Penthièvre, High Admiral of France, or the stencilled name: 'Sceaux', are the marks of the productions. **1948** A. LANE *French Faïence* x. 40 In the 1770's the factory apparently looked for inspiration to Sceaux faïence and Sèvres porcelain. *Ibid.* 42 The Sceaux faïence-painting was of a very high quality. **1960** R. G. HAGGAR *Conc. Encycl. Continental Pott. & Porc.* 421/1 Enamelling..gives a distinctive quality to Sceaux faïence. **1971** L. A. BOGER *Dict. World Pott. & Porc.* Pl. 283 (*caption*) Cruet frame and two cruets, enameled faïence. French, Sceaux, c. 1760. **1974** *Country Life* 24 Jan. 129/3 (*caption*) Sceaux faïence duck tureen.

scena. Add: **1.** In quot. 1825 for SCENE 13 read SCENE 5 d.

scenario, *sb.* Delete ‖ and add: Now usu. with pronunc. (sĭnā·rio, sĭnɛ̄ᵊ·rio). **1. a.** (Earlier and later examples.) Also, in extended use, a sketch or outline of the plot of a ballet, novel, opera, story, etc. Also *transf.* and *fig.* **1878** G. H. LEWES *Jrnl.* 28–29 Apr. in *Geo. Eliot Lett.* (1956) VII. 13 Schemed a scenario from *Daniel Deronda*. **1903** A. M. BINSTEAD *Pitcher in Paradise* ii. 51 The small card of *data* which forms the 'scenario' from which these stories are being constructed. **1911** O. ONIONS *Widdershins* 279, I myself have drafted a rough scenario of the form it appeared to me the 'Life' might with advantage be cast in. **1923** WODEHOUSE *Inimitable Jeeves* xviii. 230 'Jeeves!' 'Sir?' 'I'm in the soup.' 'Indeed, sir?' I sketched out the scenario for him. 'What would you advise?' **1924** —— *Bill the Conqueror* ix. 159 A young man in a vivid check suit came out, a small young man with close-set eyes and the scenario of a moustache. **1929** C. K. S. MONCRIEFF tr. *Proust's Captive* iii. 493 No doubt the scenario [of a series of events in the narrator's life] was not merely different but almost opposite. **1947** A. EINSTEIN *Music in Romantic Era* xvi. 284 He outlined a scenario, and..obtained a completed libretto from Somma. **1953** WODEHOUSE *Performing Flea* 69 Today I reached page 254 and have a very detailed scenario of the rest. **1955** W. DEAN in H. Van Thal *Fanfare for Ernest Newman* 59 Trianon..an inveterate compiler of librettos and ballet scenarios for all three Paris opera houses. **1977** *Dædalus* Summer 73 Thus the dramatic scenario—frequently the enactment of a sacred narrative—now becomes a performative mode sui generis.

b. *Cinemat.* A film script with all the details of scenes, appearances of characters, stage-directions, etc., necessary for shooting the film. **1911** [see *picture-play* s.v. *PICTURE *sb.* 6 a]. **1919** F. HURST *Humoresque* 184 So many times it comes up in the scenarios and the picture-plots..how money don't always bring happiness. **1922** WODEHOUSE *Girl on Boat* ix. 144 Fate, thought Sam, had constructed a cheap, mushy..five-reel film scenario. **1926, 1930** [see *CONTINUITY 6]. **1934** *Punch* 18 Apr. 426/2 The film is still full of real characters, not the pasteboard subsidiaries we meet so often in modern American scenarios of love and murder. **1937** A. HUXLEY *Let.* 15 Dec. (1969) 429 Unless in the interval I get any news about a scenario I wrote while out in Hollywood. **1941** *Spectator* 10 Oct. 355/1 Miss Bette Davis..has proved her genius for breathing life into scenarios which have been synthesized from the more extravagant..of ancient theatrical situations. **1950** T. S. ELIOT *Cocktail Party* I. i. 14 They did a film But they used a different scenario. **1969** M. STEINBECK *On Stage* 165 Strictly speaking a scenario is a film script. It is not used very much. Usually one hears.. film script.

2. A sketch, outline, or description of an imagined situation or sequence of events; esp. (*a*) a synopsis of the development of a hypothetical future world war, and hence an outline of any possible sequence of future events; (*b*) an outline of an intended course of action; (*c*) a scientific model or description intended to account for observable facts. Hence, in weakened senses (not easily distinguishable from sense 1 a *transf.* and *fig.*): a circumstance, situation, scene, sequence of events, etc. The over-use of this word in various loose senses has attracted frequent hostile comment.—Ed. **1962** H. KAHN *Thinking about Unthinkable* v. 143 A scenario results from an attempt to describe..some hypothetical sequence of events... Scenarios may explore and emphasize an element of a larger problem such as..the process of 'escalation' of a small war. *Ibid.*, The scenario is an aid to the imagination. *Ibid.* 146 The scenario begins by assuming a crisis; everybody is on edge. A Soviet missile is accidentally fired. **1965** 'R. L. PIKE' *Police Blotter* xi. 185 If you hadn't tried to build up a big scenario with that self-defense crap, if you had just kept your big mouth shut, it might have held us up. **1966** 'W. COOPER' *Memoirs of New Man* I. viii. 103, I admired the beauty and simplicity of his plan—or 'scenario', as the case might be. **1968** *Guardian* 21 Feb. 7/1 Germany then plans to produce a so-called 'scenario' for arriving at an arrangement with Britain. **1968** *Listener* 20 June 791/1 The Hudson Institute..is an organisation largely devoted to preparing what it likes to call 'scenarios of the future'. **1969** M. CRICHTON *Andromeda Strain* viii. 87 The President would face four circumstances (scenarios) in which he might have to issue the Cautery order. **1971** *Observer* 27 June 1/3 Several of the computer 'scenarios' include a catastrophic and sudden collapse of population. **1974** *Nature* 15 Feb. 445/2 As a possible scenario we assume the previously reported pulses to be the chance superposition of more frequent, randomly occurring subpulses. **1975** *Sci. Amer.* Jan. 29/2 Some meteoriticists boldly construct multistage scenarios of condensation, agglomeration, accretion, heating, metamorphism and differentiation to explain the accumulated facts. **1975** *N.Y. Times* 29 Mar. 11/1 There is a certain narrative element in this whacky art, but it would be a brave man who tried to extract a single coherent scenario from any single picture or construction. **1976** *Daily Tel.* 13 Feb. 1/8 Speculation..about the likely scenario when the Cuban-armoured units reach the point..when they will encounter South African forces. **1976** *Sci. Amer.* Oct. 79A/2 Many of the models we have mentioned here are better characterized by the term scenario... There is so little detailed information that the proposals should not be dignified by the term model. Nevertheless, a good scenario can sometimes lead to a good model. **1977** C. MCCARRY *Secret Lovers* ix. 112 I'll give you the scenario... You're free to modify it..in the light of conditions in the field. **1977** *Time* 18 Apr. 46/2 By escaping from the lab and multiplying, their scenario goes, it could find its way into human intestines. **1978** J. IRVING *World According to Garp* viii. 157 The good-byes that Garp imagined conducting with Alice were violent scenarios. **1980** *Jrnl. R. Soc. Arts* July 474/2 The best scenario..that we can envisage is one in which all those who want to do formal work will have an opportunity of doing two or three days a week.

3. *attrib.* and *Comb.* in sense 1, as *scenario department, editor, picture, production, sketch, writer, writing.* **1921** B. TARKINGTON *Let.* 2 July in *On Plays* (1959) 65 About your scenario dep't [*sic*] friend's suggestion. **1959** W. S. SHARPS *Dict. Cinematogr.* 127/1 *Scenario Editor*... The title usually applied to the person in charge of the story department of a film producing company. **1929** W. S. CHURCHILL *World Crisis* V. vii. 122 Mr. Baker detracts from the vindication of his hero by the absurd *scenario* picture which he has chosen to paint. **1945** *Amer. Cinematographer* Mar. 122 (*heading*) Trials of making a scenario production. **1921** B. TARKINGTON *Let.* 30 Mar. in *On Plays* (1959) 50, I am sending you the scenario sketch for a picture of 'Beaucaire'. **1914** R. GRAU *Theatre of Science* 224 We have seen the last of the amateur scenario writer. **1939** C. ISHERWOOD *Goodbye to Berlin* 101, I thought you took an interest in the cinema? He's miles the best young scenario writer. **1976** BOTHAM & DONNELLY *Valentino* xi. 82 The woman considered to be the best scenario writer of the day, June Mathis. **1928** H. CRANE *Let.* 27 Mar. (1965) 321 Maybe scenario writing eventually.

Hence **scena·rio** *v. trans.*, to make a scenario of (a story, book, or idea); to sketch *out*; also **scena·rioi:ze, sce·narize.**

1918 *Dial. Notes* V. 13 'The *scenarioizing* of a drama.' Moving Picture advertisement. **1922** A. Bennett *Let.* 17 Jan. (1966) I. 300, I had attempted to scenario-ize the story and had failed to do it properly. **1922** *Moving Picture Stories* 14 July 26/3 'Clarence'..is already scenarioized and requires only the producer's final approval. **1927** *Sunday Express* 21 Aug. 4 The films were scenarioised, directed, cut, edited, distributed, and exploited by him. **1946** *Amer. Speech* XXI. 304/2, 1946 Press-sheet of RKO Radio Pictures, Inc. Geoffrey Homes is now at the RKO studios scenarizing his best seller, 'Build My Gallows High'. **1953** Wodehouse *Performing Flea* 23 So far I have scenarioed it out to about the 40,000 word mark. **1974** *Daily Tel.* (Colour Suppl.) 6 Dec. 42/1 Tonight's entertainment is a fashion show. 'I hear they've got it all scenarioed out,' says Bernie.

scenarist (sīnā·rist). *Cinemat.* [f. Scenar(io + -ist.] A scenario writer.
1920 *N.Y. Times* 24 May 20/4 'Old Lady 31', taken by June Mathis as scenarist and John E. Ince as director from Rachel Crother's play. **1925** *New Yorker* 28 Nov. 26/2 You never can tell just what happened to the tale when it fell into the hands of the gifted scenarists. **1932** A. Buchanan *Films* v. 92 Every shot..is recorded by the scenarist in his script. **1941** B. Schulberg *What makes Sammy Run?* vii. 153 Sammy Glick, prominent scenarist and playwright. **1958** *Times Lit. Suppl.* 15 Aug. p. xxviii/1 The opening passages of *Bleak House* are a supreme evocation of atmosphere which no contemporary scenarist could equal. **1966** *New Statesman* 30 Sept. 488/2 Miss McCarthy could yet prove to be a born scenarist. Film largely robs her work of its overbearing, destructive omniscience. **1977** *Times* 10 June 15/4 Wenders explains that he and his scenarist, the playwright Peter Handke, both happened to read the book at the same time. **1977** *Time* 12 Sept. 60/1 Can an English playwright turned Hollywood scenarist find, in his late 40s, happiness and the right woman?

scene. Add: **1.** (Earlier *transf.* and later examples.)
1612 W. Strachey *Trav. Virginia* (1953) I. vi. 78 By their howses, they have sometymes A Scæne or high Stage raised like a Scaffold..covered with Matts, which..is a Shelter and serves for such a covered place, where men vsed in old tyme to sitt and talke. **1924** A. Huxley *Let.* 29 Apr. (1969) 229 A Palladian theatre with fixed scene and various other delights.
7. b. Also (with hyphens) as *attrib. phr.*
1955 H. Roth *Sleeper* ix. 69 His behind-the-scenes directors must have trusted him. **1959** *Manch. Guardian* 23 July 6/6 One version of the behind-the-scenes nterview. **1959** 'S. Ransome' *I'll die for You* xii. 134 He was quietly directing his official resources into a wide, behind-the-scenes investigation.
8. a. (Further examples.)
1926 *Melody Maker* Sept. 61 Since 'Nelly Kelly's Cabaret' came on the scene, it's put fresh kick into dancing. **1936** W. H. Saumarez Smith *Let.* 26 June in *Young Man's Country* (1977) ii. 11, I rode out after breakfast to the scene of action. **1946** Rosenthal & Zachery *Jazzways* 16 By 1907, Bolden had disappeared from the scene, confined to an insane asylum. **1963** D. Ogilvy *Confessions Advt. Man* (1964) ii. 26 By the time I came on the scene, the big advertisers had grown more cautious. **1968** *Jazz Monthly* Apr. 8/1 People like Buddy Collette, Red Callender,..were the big time musicians on the scene then. **1979** 'E. Ferrars' *Witness before Fact* xv. 150, I don't know what things were like for you before he arrived on the scene, perhaps not so good.
c. *the scene of the crime*, the place where a crime has been committed. Also *attrib.*, as *scene(s)-of-crime*, used *esp.* to designate (a member of) a civilian branch of the police force concerned with the collection of forensic evidence.
1923 A. Christie *Murder on Links* iv. 51 Now, Monsieur Poirot, you would without doubt like to visit the scene of the crime. **1931** D. L. Sayers *Five Red Herrings* xvi. 175 He didn't take the body with him... Now he's got to get back to the scene of the crime. **1943** G. Greene *Ministry of Fear* i. v. 67 A few elderly men in the C.I.D...might..visit the scene of the 'crime'. **1954** F. Cherrill *Cherrill of the Yard* iii. 38 Scenes of Crime prints. *Ibid.* 39 By this arrangement it became much easier to carry out a search with a single Scenes of Crime mark which had been classified in accordance with the single fingerprint system. **1961** *Observer* 21 May 5/3 The War Office have placed an order.. for thirty-eight 'Scene of Crime Kits', to issue to their security-men. **1971** R. Lewis *Error of Judgment* i. 38 The scene-of-crime unit upstairs have discovered nothing, but one of the constables..came up with a glove. **1972** *Police Rev.* 8 Dec. 1601/2 Certain duties such as those of scenes-of-crime officers..were not performed by the R.U.C. **1977** P. Hill *Liars* iii. 33 The Scene of Crime man went over.. the cottage for fingerprints.
d. Some portion of human activity (as delimited by a preceding *adj.* of place, time, etc.); the realm or sphere (of an activity or interest indicated by a preceding *attrib. sb.*).
1931 *Times Lit. Suppl.* 15 Oct. 786/3 Mr. Masters..is no optimistic observer of the contemporary American scene. **1938** D. Baker *Young Man with Horn* i. v. 52 If Rick had grown up in the present scene he'd probably have had his head perpetually inside a walnut radio cabinet listening to this one or that one playing a tea dance. **1943** H. Read *Politics of Unpolitical* vii. 98 There have been times when he was bored with the social scene, and 'doodled' while he stared hopelessly into the future. **1949** *Ebony* Nov. 24 (*heading*) The jazz scene. **1959** L. Lipton *Holy Barbarians* i. i. 40 Something was happening on the poetry scene in Venice West. **1970** *Daily Tel.* 9 Jan. 3/6 Wilson was not mixed up with the drug scene. **1974** *Howard Jrnl.* XIV. 108 (Advt.), One of the most forceful and controversial writers on the magisterial scene. **1977** *Listener* 17 Feb. 214/2 Without that little building at Swiss Cottage..London's theatre scene would be much duller.

e. *slang* (orig. *U.S. jazz* and *beatniks'*). A place where people of common interests meet or where a particular activity is carried on. Hence, more loosely, an activity or pursuit (esp. a fashionable or superior one); a situation, event, or experience; a way of life. Freq. in phrases, as *a bad scene*, an unpleasant experience; *to make the scene*, to participate in an event or activity; to arrive (somewhere); (*to go*) *on the scene*, (to become) involved in some activity, esp. drug-taking; (*not*) *one's scene* (and varr.), (not) what one enjoys or finds interesting.
There is some overlap with sense d above.
1951 E. Paul *Springtime in Paris* vi. 125 'Nobody comes on this scene wearin' any green,' said another taller Negro. **1957** *N.Y. Times Mag.* 18 Aug. 26/3 *Scene*, any place where musicians play or gather; by extension, any place where people meet or any event they attend. Thus, 'Let's make the country scene this week-end.' **1958** G. Lea *Somewhere there's Music* xxi. 179 Something on the scene you don't dig. *Ibid.*, It was a bad scene. It scared me, man. **1958** *Look* 19 Aug. 65/2 The regulars who 'make the scene'. 'The scene', geographically, is a narrow area running about four blocks along Grant Avenue in San Francisco's North Beach district. *Ibid.* 67/1 Like many on 'the scene', she is attracted to Zen Buddhism. **1964** *New Society* 20 Feb. 8/2 What happens to the young drug taker? It can be described ..by actually going through the experience, going 'on the scene'. **1966** *New Statesman* 1 July 26/2 Her final surrender to Clive Francis seems unlikely: his jeans are too baggy for her scene. **1966** *Melody Maker* 15 Oct. 6/6, I decided I wanted to play jazz more than any other scene. **1967** *Punch* 18 Oct. 574/3 They come here to work because it's exciting and new and because it's the scene. **1968** M. Richler *Cocksure* xiii. 74 Like we're having a scene on Saturday night. At Timothy's pad. **1969** *Oz* Apr. 32/1 We've all got different scenes. The whole thing is to get to know each other's trips... Are you on an acid scene? **1970** [see *make *v.*[1] 65 b]. **1975** D. Lodge *Changing Places* ii. 84 Washing up was more his scene than body language. **1977** I. Shaw *Beggarman, Thief* i. viii. 101 He could take a look at the scene and blow if he didn't like it.
11. *to make a scene* (later examples). Also, *to create a scene, to have a scene.*
1957 *Sunday Mail* (Glasgow) 10 Feb. 11 *Kick up a storm*—to cause trouble, or create a scene. **1958** [see *brace *v.*[1] 5 c]. **1959** T. S. Eliot *Elder Statesman* ii. 61 I've made him understand That the doctors want you to be free from worry. He won't make a scene. **1970** G. F. Newman *Sir, You Bastard* viii. 255 Two of them rose and followed him out. 'Don't let's have a scene,' one of the Rubber Heels said... The other ran his hands over Sneed's jacket.
13. (senses 5 and 6) *scene-change*; **scene-dock** (earlier and later examples); **scene-painted** *a.*, painted with scenes; **scene-painter** (further examples); also *transf.*; **scene-painting** (earlier example in *fig.* sense); **scene-plot** (examples); **scene-room** (earlier examples); **scene-setting**, *vbl. sb.* and *ppl. a.*, setting a scene; usu. *transf.* and *fig.*; so **scene-setter**; **scene-steal** *v.*, to appropriate more than one's fair share of attention by one's performance in a scene; so *scene-stealer* (also *transf.*), *scene-stealing* ppl. adj.
1952 W. Granville *Dict. Theatrical Terms* 158 *Scene-change*, the striking of one scene and the erection of another. **1962** A. Nisbett *Technique Sound Studio* ix. 153 When a situation calls for a scene change the simplest form that this can take is a slow fade to silence over about ten seconds, a pause of three or four seconds, and an equally slow fade in. **1871** E. L. Blanchard *Diary* 9 Mar. in Scott & Howard *Life E. L. Blanchard* (1891) II. 395 Then to Standard [Theatre]..go behind the scenes and see the wonderful scene dock. **1916** [see *back stage, backstage *sb.* and *adv.*]. **1977** *Times* 1 Nov. 14/6 The [Wexford] Opera House..[has] no scene-dock, no workshops, no adequate dressing-rooms. **1918** W. Owen *Lett.* (1967) 558 Scene-painted boulders, and all the arts and deceitful devices of Victoria. **1824** Scene-painter [see *master *sb.*[1] 24 d]. **1853** [see *churrigueresque *a.*]. **1821** H. C. Robinson *Diary* 2 Dec. (1967) 71, I have finished *Waverley*... Its merit lies in portrait and scene painting. **1847** W. C. Macready *Diary* 20 Oct. (1912) II. 375 Made one scene plot of 'Van Artevelde', and sent it with note to Stanfield. **1933** P. Godfrey *Back-Stage* i. 19 The stage-manager, with every detail of the scene-plot in his head, stands directing the whole. **1737** *Daily Advertiser* 4 Feb., And the Scene-Rooms, Green and Dressing Rooms, to be on the outside of the last mention'd Measure. **1826** J. O'Keeffe *Recoll.* II. 39 The author is often brought into the scene-room to give his opinion on the progress of their work. **1859** E. Fitzball *35 Yrs. Dram. Author's Life* II. 124 The celebrated Mr Grieve, and his two sons, Thomas and William, the most perfect scene painters in the world..in their scene-room, genius always found a welcome footing. **1974** *Times* 16 Apr. 16/3 Miss Tanburn will kick-off one of the panel discussions with a half-hour scene-setter. **1978** *Language* LIV. 353 Only three functions—*subject*, *topic*, and *relator*—are assigned by the rules of H's sample and DDG of English, although a fourth (*scene-setter*) is mentioned in the text. **1963** *Times Lit. Suppl.* 17 May 358/3 A biographer with a sense of character, an eye to scene-setting. **1968** P. Foot *Politics of Harold Wilson* 11 The two most important scene-setting subjects are food and the weather. **1972** M. Gilbert *Body of Girl* xx. 183 With a little care and scene-setting it could be made to look very convincing. **1977** D. Williams *Treasure by Degrees* iii. 34 The Prince's finery and the size of his entourage—natural scene-setting for an important Arab. **1976** *Woman's Weekly* 6 Nov. 6/2 Trish Van Devere, who not only plays Beauty in the movie (and guess who scene-steals as the Beast!) but in real life also happens to be Mrs George C. Scott. **1978** *Radio Times* 18–24 Mar. 16/3 Director Jules Dassin's wife Melina Mercouri turns every trick to scene-steal from Morley and Ustinov. **1955** T. Sterling *Evil of Day* xviii. 193 These lousy actors are all scene-stealers. **1960** *Vogue Pattern Book* No. 4. 51 The addition of demure puffed sleeves makes it the scene-stealer of more sedate occasions. **1977** M. Hinxman *One-Way Cemetery* xiii. 94 It's not the leading role, but it's a scene-stealer. **1963** *Times* 29 Jan. 11/1 Mr. Craig's is a good, workmanly Cavaradossi, a little stiff and never scene-stealing. **1980** *Times Lit. Suppl.* 21 Mar. 323/2 A small but scene-stealing knockabout part.

‖ **scène à faire** (sɛn a fɛ̄r). *Theatr.* Pl. **scènes à faire.** [Fr., lit. 'scene for action'.] The most important scene in a play or opera, made inevitable by the action which leads up to it. Also *transf.*
1893 *Manch. Guardian* 24 Oct. 8/3 The subject of the 'Dame aux Camellias' trying to begin life over again, and to live as if the past had never been, has often been essayed, and the *scène à faire* of her confrontation..with the inexorable reality of things has been often and sometimes admirably composed. **1921** P. Lubbock *Craft of Fiction* vii. 102 Thackeray's skill betrays him... His climax, his *scène à faire*, has been insufficiently prepared for. **1922** W. S. Maugham *On a Chinese Screen* xlviii. 188 He was asking for the *pièce bien faite*, the *scène à faire*, the curtain, the unexpected, the dramatic. **1948** F. R. Leavis *Great Tradition* ii. 112 The brilliant art with which James, choosing his *scènes à faire*, works in terms of dramatic presentation. **1965** *New Statesman* 10 Dec. 943/2 A big added *scène à faire* in the council chamber gave Verdi a chance for the creation of an ensemble that looks forward very clearly to the third act of *Otello*. **1969** *Listener* 13 Feb. 220/2 Robert Hoffman acts badly, and the *scène à faire* in a wobbling rowing boat..is a triumph of embarrassment. **1980** *Times* 14 Mar. 13/8 They do, eventually, get a *scène à faire* (which, so often, proves to be a *scène à ne faire*) in which she tries to treat him as Louis XV.

scenery. Add: **2. a.** Also, that used in film and television.
1959 W. S. Sharps *Dict. Cinematogr.* 127/1 *Scenery*, the various parts and accessories used on the set to represent the actual scene of an action. **1960** O. Skilbeck *ABC of Film & TV* 11 Scenery which may be viewed in close-up must be more convincing than that of the theatre. **1961** G. Millerson *Technique Television Production* 142 By special electronic equipment, we can place one camera's performers and/or scenery in the background picture provided by another picture source.
b. So phr. *part of the scenery.*
1970 N. Marsh *When in Rome* viii. 221 He..must often hang about the premises... Part of the scenery as it were. **1971** 'H. Calvin' *Poison Chasers* v. 65 Dai had mentioned her to me as an interesting part of the local scenery. **1977** *Times* 4 July 12/4 Inexperienced Mabel..was allowed to sing the old things in harmony. I was, at first, only part of the scenery.

scene-shifter. (Later examples.)
1908 [see *legit., legit]. **1957** L. Durrell *Justine* ii. 102 Quick as a scene-shifter the station packs away advertisement after advertisement. **1978** *Lancashire Life* Apr. 35/3 A saviour came in the shape of little Alfie Gee—part-time electrician and scene-shifter... 'Shuffle off the stage sideways,' he whispered, 'and don't drop it, man.'

scenic, *a.* Add: **1. d.** (Earlier examples.)
1824 R. Humphreys *Mem. J. Decastro* 16 It is that [part] of the scenic department from whence the borders of chambers or clouds drop, to complete each different scene. **1827** J. Boaden *Mem. Mrs. Siddons* II. xix. 292 A benefit proportioned to the pains that have been taken in the scenic department of our stages. **1854** C. A. Mowatt *Autobiography of Actress* 48 Costumes and rehearsals and scenic effects.
e. *scenic artist*, a painter or designer of scenery for the stage. orig. *U.S.*
1840 *Spirit of Times* 21 Nov. 456/3 C. L. Smith..is the scenic artist of the Theatre. **1877** W. R. Alger *Life Edwin Forrest* II. 581 John Wiser, a scenic artist, arranged and painted it. **1919** G. B. Shaw *Great Catherine* 114 It was quite easy for Patiomkin to humbug Catherine as to the condition of Russia by conducting her through sham cities run up for the occasion by scenic artists. **1930** Selden & Sellman *Stage Scenery & Lighting* ii. 31 Before the scenic artist can start to make scenery it is necessary that he learn thoroughly the form of scenery. **1971** Burris-Meyer & Cole *Scenery for Theatre* (rev. ed.) ii. 22 The designer is a member of the scenic artists' union.
3. a. (Later examples.) Also, of a window or the like: designed to afford a landscape view. Now chiefly *N. Amer.*
1937 *Discovery* Oct. 306/2 Small-holders in scenic areas. **1967** *Boston Sunday Herald* 26 Mar. vi. 3/1 (Advt.), See the scenic glories of our great continent. **1970** *Globe & Mail* (Toronto) 26 Sept. 30/1 (Advt.), The Canadian, one of the world's great trains... Soft music. Air conditioning... Scenic Domes..all the way. **1971** *New Yorker* 9 Oct. 170/3 (Advt.), Golf on scenic course. **1978** *N.Y. Times* 30 Mar. B17/2 (Advt.), Floor to ceiling scenic windows.
b. Applied to a road that has been planned and landscaped so as to provide fine views. orig. and chiefly *N. Amer.*
1914 H. MacNair (*title*) Scenic motorway; a motor tour de luxe. **1916** *Road Maps & Tour Bk. Western N. Carolina* (N. Carolina Good Roads Assoc.) 149 The Asheville-Murphy Scenic Highway through Swain County will afford scenery unsurpassed by any section of the country. **1934** *Popular Mechanics* Aug. 238/1 The modern de luxe highway cruiser..may take you and your baggage safely and inexpensively anywhere along historic and scenic highways. **1935** *Nature Mag.* Mar. 101 Let us hope that there will be a policy of scenic road construction. **1943**

J. S. Huxley *TVA* ix. 60 The Norris Freeway..is a scenic highway on which access is limited to a very few points, and where no building is allowed within several hundred yards on either side. **1959** W. B. Snow *Highway & Landscape* 111 For specialized types of highway, scenic parkways particularly, the national standards may not always be entirely appropriate. **1967** *Boston Sunday Herald* 26 Mar. ii. 9/1 A 7·4-mile Appleton Ridge Scenic Drive became a women's project in 1966. The women got the town to bulldoze the rough spots on a dirt road, cut bushes to open up magnificent views, and provide a stretch for 'slow drivers who really want to enjoy the scenery'. **1979** D. Clark *Heberden's Seat* i. 7 Masters had suggested that they should find a scenic route and take their time.
4. a. (Earlier example.)
1845 *Punch* VIII. 247/1 To criticise a Picture by Stanfield.—Begin by unqualified praise; then commence detracting,..on the score of..'scenic effect of the figures'; and conclude by a wish he had never been a scene-painter.
b. With reference to wallpaper: creating a continuous scene or landscape on the walls of a room.
1924 N. McClelland *Historic Wall-Papers* xii. 279 (*heading*) Some famous scenic papers and their owners. **1929** C. C. Oman *Victoria & Albert Museum: Catal. Wall-Papers* 63 The earliest scenic wall-papers..were produced by hand-painting. **1951** L. & W. Katzenbach *Pract. Bk. Amer. Wallpaper* vi. 61 This scenic wallpaper pictures a tropic Haiti. **1976** *National Observer* (U.S.) 25 Sept. 9/1 Now about all that remains is the neon-red carpet and 'scenic' wallpaper that once surrounded a bathtub.

scenic (sī·nik), *sb.* [f. the adj.] **1.** = Scene 6 *fig. rare*[−1].
1891 G. Meredith *One of Our Conquerors* III. vi. 125 She passed into music, as she always did under motion of carriages and trains, whether in happiness or sadness: and the day being one that had a sky, the scenic of music swung her up to soar.
2. A scenic film or photograph; a film or photograph the subject of which is natural scenery.
1918 *N.Y. Times* 25 Nov. 11/3 Robert C. Bruce has a scenic at the Rivoli entitled 'A Wee Bit Odd', which is entertaining pictorially in spite of labored wit in the subtitles. **1922** *Ibid.* 2 July vi. 3/3 The short comedies, scenics, travel films and other so-called non-dramatic productions are so much better than the photoplays when they are at all good. **1971** *Amateur Photographer* 3 Mar. 23/1 A cine columnist's thoughts thankfully turn from the interiors he had intended to shoot..to the spring scenic he has for years been intending to make... I like scenics and am not put off by objections that they are old-fashioned... So are trees and meadows. **1979** *SLR Camera* June 56/2 Scenics, particularly townscapes, at night are best shot while there is still some tone in the sky.
3. Short for 'scenic wallpaper' (see *scenic *a.* 4 b).
1951 L. & W. Katzenbach *Pract. Bk. Amer. Wallpaper* vi. 65 While the composition of this scenic is traditional, it is executed in a technique that is distinctly modern. **1966** M. M. Pegler *Dict. Interior Design* (1967) 393 *Scenic*, a wallpaper mural usually made up of three or four panels that create a continuous scene, vista, or design. **1972** E. A. Entwhistle *French Scenic Wallpapers 1800–1860* v. 35 *Les Monuments de Paris*..was different from most of the other scenics. **1976** B. Greysmith *Wallpaper* 92 The most striking examples of the new French manner were the 'scenics', the term used to describe *trompe l'oeil* landscapes on a grand scale, not repeating but creating a complete scene around the walls of a room.
4. Short for *scenic railway.
1968 D. Braithwaite *Fairground Architecture* viii. 125 In structural form there was little difference between the 'Scenic' and the earlier switchback.
5. A scenic pattern or design.
1977 *Chicago Tribune* 2 Oct. v. 9 (Advt.), Make slipcovers, draperies of 100 % cotton prints in florals, scenics, geometrics.

scenic railway (sī·nik rē[i]·lwe[i]). [f. Scenic *a.* + Railway *sb.*] A switchback or miniature railway running through artificial representations of beautiful or spectacular scenery, as an attraction at fairs, etc.
1894 *Official Guide Calif. Midwinter Exposition* 130 (*heading*) Scenic railway. *Ibid.*, It must not be supposed that gravity alone is the motive power in the Scenic Railway. **1908** [see Scenic *a.* 3]. **1917** *Jrnl. Exper. Psychol.* II. 158 The popularity of such amusements as 'scenic railways', the sole attraction of which lies in the fright on the steep inclines, suggests that fear may be pleasant—at least retrospectively. **1923** H. C. Witwer in *Cosmopolitan* Aug. 46/1 You're a woman, a good looker with more curves than a scenic railway. **1926** T. E. Lawrence *Seven Pillars* (1935) iii. xxxii. 184 About their crests ran narrow veins of granite-coloured stone, generally in pairs, following the contour of the skyline like the rusted metals of an abandoned scenic railway. **1930** E. Waugh *Labels* 200 There were switch-backs and scenic railways on which empty cars swooped and swerved through breath-taking descents. **1968** [see *galloper 1 b]. **1973** 'M. Innes' *Appleby's Answer* iii. 33 Shaftesbury Avenue..was like going through a 'scenic' railway in a Brobdingnagian fun-fair. **1980** *Times* 4 Oct. 12/8 One of the sights worth seeing was the drunks on the new 60 mph 'Super-looping' scenic railway.

Scenicruiser (sī·nikrū:zəɹ). *U.S.* Also with small initial. [f. Sceni(c *a.* + Cruiser.] The proprietary name of a line of luxury coaches equipped for long-distance travel, esp. for touring areas of scenic beauty.
1954 *Business Week* 17 July 33/2 Greyhound is wagering heavily that the Scenicruiser will rejuvenate its business. **1955** *American Mag.* Jan. 92/2 We were riding in a new Greyhound 'Scenicruiser'. **1959** *Official Gaz.* (U.S. Patent Office) 13 Jan. TM77/2 The Greyhound Corporation, Chicago..Scenicruiser. **1965** M. Bradbury *Stepping Westward* viii. 403 They ate fast..and returned to the scenicruiser. **1974** *Encycl. Brit. Macropædia* XVIII. 721/2 The Scenicruiser introduced in the United States in 1954 for trans-continental use..utilizes air suspension and has six wheels. **1976** *Yellowstone Explorer* July 8/2 (*heading*) Scenicruiser mini voyages. *Ibid.*, Your captain will keep you posted as to the areas you are viewing.

scent, *sb.* Add: **5. scent-bottle,** (*a*) (further example); (*b*) a bottle designed to contain scent.
1856 C. M. Yonge *Daisy Chain* i. xxv. 262 She flew for the scent-bottle, while her father bent over Margaret. **1895** *Army & Navy Co-op. Soc. Price List* 714/2 Scent Bottles... Fancy... A large assortment in Stock. **1917** *Harrods Gen. Catal.* 219 Sterling silver and cut glass scent bottle. **1930** T. S. Eliot tr. *St.-J. Perse's Anabasis* 27 And a man strode forth at the threshold of the desert—profession of his father: dealer in scent-bottles. **1975** J. O'Faolain *Women in Wall* iii. 55 Translucent scent-bottles of glass and alabaster.

scenter. (Later example.)
1977 *Hiawatha* (Victoria, B.C.) 13 Nov. 3/1 And now, in November, the rain is..a scenter of soil, a painter of stones.

scentless, *a.* Add: **3.** Also of a day on which there is no scent for the hounds to follow.
1885 *Field* 4 Apr. 428/1 That dry, scentless cycle of days. **1921** *Ampleforth Jrnl.* Jan. 137 On October 16th we hunted the Kyloe country, after a scentless day at Tom Smith's Cross on the previous Wednesday. **1976** *Horse & Hound* 3 Dec. 30/3 They had an exciting, if rather scentless, morning, catching a brace of foxes, the last one within two miles of the kennel gates!
4. scentless mayweed, a perennial herb, *Tripleurospermum maritimum* (formerly *Matricaria inodora*), belonging to the family Compositæ and bearing white, yellow-centred flowers and finely divided leaves.
1800 J. E. Smith *Eng. Bot.* x. 676 (*heading*) Corn feverfew. Scentless may-weed. **1875** A. Pratt *Flowering Plants & Ferns Gt. Brit.* III. 325 Scentless Mayweed..puzzles the young botanist by belying its name, and having an odour which, though not aromatic, is powerful and unpleasant. **1914** A. R. Horwood *Story of Plant Life Brit. Isles* I. 140 The Scentless Mayweed..has a leafier flower and a generally darker green foliage. **1932** M. Grieve *Mod. Herbal* II. 524/1 The Scentless Mayweed owes its generic name to its reputed medicinal properties. **1975** E. J. Gibbons *Flora of Lincolnshire* 232 Scentless Mayweed... Native. Weed of cultivation.

scenty (se·nti), *a. rare.* [f. Scent *sb.* + -y[1].] Smelling of scent; scented.
1937 G. Frankau *More of Us* i. iii. Yet, ere he handed scenty lace-edged Rax back, Long to his seat in knightly honour rooted The hero cleaved. **1963** D. Ballantyne *And the Glory* 148 There was the warm and scenty smell of her body.

sceptic, skeptic, *a.* and *sb.* Now usually spelt *sceptic* in the U.K. and British Commonwealth and *skeptic* in the U.S. Similarly all the derivatives, *scepticism*/*skepticism*, etc.

scepticism, skepticism. **1.** (Earlier example.)
1652 N. Culverwel *Light of Nature* 150 He [*sc.* Pyrrho] perswades men to encline to his Scepticisme.

† **Sceptism.** Add. Also 6 **sceptisme.** (Earlier and later examples.)
1652 N. Culverwel *Light of Nature* 153 Des-Cartes the French philosopher..will be fain to stop and stay in Sceptisme... He that will not cast Anchor upon these, condemnes himself to perpetual Scepticisme. **1737** A. Baxter *Enq. Hum. Soul* (ed. 2) II. 21 That kind of Scepticism called Egoism.

sch. In words derived from Yiddish in which initial (ʃ) precedes a consonant, there is much variation in written English between *sch-* and *sh-*: however (following the German usage) *sch-* seems to be the prevailing spelling, except before *l*, where German would use simple *s-*: here *sh-* is the usual form, as it is before vowels. For the two main types see *schlemiel, *schmo, *schnook, etc., and *shtick. Such words are extremely common in the U.S. but are rarely encountered in Great Britain.

schaalstein, schalstein (ʃā·lʃtəin). *Geol.* Also **schaalstein**, and partly translated as **schaalstone** (ʃā·lstōᵘn). [Ger., f. *schale* (formerly *schaale*) skin, shell + *stein* stone. See also Shale *sb.*[2] and Scale *sb.*[1]] † **a.** = Wollastonite. *Obs.*
1804 R. Jameson *Syst. Min.* I. 519 (*heading*) Schaalstein.—Werner. *Ibid.* 520 It is named schaalstone, which in German intimates that it is composed of lamellar distinct concretions. I have not been able to find any English word synonymous to the German, so that I am under the necessity of adopting it. **1819** W. Phillips *Elem. Introd. Mineral.* (ed. 2) 300/1 (Index), Schaalstein. **1836** T. Thomson *Outl. Min., Geol.* I. 129 Bisilicate of Lime. Table spar, schaalstein, grammite, wollastonite of Hauy.
b. Any of several basic or calcareous tuffaceous rocks, usu. laminated in structure, affected by low-grade metamorphism; a slaty or sheared greenstone.
1876 P. H. Lawrence tr. *von Cotta's Rocks classified & Described* ii. iv. 212 Some part at least of what has been called schalstein belongs to the tufa formation. **1882** A. Geikie *Text-bk. Geol.* 165 Schalstein.—Under this name German petrographers have placed a variety of rocks which consist of a green, grey, red, or mottled diabase-tuff impregnated with carbonate of lime and mixed with calcareous and argillaceous mud. **1897** — *Ancient Volcanoes of Gt. Brit.* II. xxix. 36 Some layers of this tuff assume a finely foliated appearance by the development of pale leek-green folia, which show slickensided surfaces parallel with the bedding. The rock then presents one of the usual appearances of schalstein. **1909** W. A. E. Ussher et al. *Geol. Country around Modbury* [illegible] the Middle Devonian Slates are associated with 'schalsteins'. **1974** E. Lehmann in G. C. Amstutz *Spilites & Spilitic Rocks* 23 In the Lahn syncline..of Western Germany, the central rock complex, of upper Middle Devonian and lower Upper Devonian age, consists of so-called 'schalstein'. Any consistency, however, suggested by that old miner's term, is not reflected in the petrographic character of the rock.

schaapsteker, var. *skaapsteker.

Schabzieger (ʃā·pˌtsīgəɹ). Also **Chapsager, -ziger, Schabzeiger, Schabziger,** etc. [Ger.: see Sapsago.] A kind of hard green cooking cheese made in Switzerland from curds, and flavoured with melilot. In full, **Schabzieger Käse.**
1837 *Penny Cycl.* VII. 15/2 The green Swiss cheese, commonly called Schabzieger, which is made in the canton of Glarus. **1846** *Chapsiger* [see Sapsago]. **1866** Lindley & Moore *Treas. Bot.* s.v. *Melilotus*, Schabzieger or Chapsiger. **1879** *Encycl. Brit.* X. 636/1 Glarus... The Schabzieger, Schotter Käse, Kräuterkäse, or 'green cheese,' made of skim milk, whether of goats or cows, mixed with butter-milk and coloured with powdered *steinklee* (*Melilotus coerulea*), is still largely manufactured. **1887** R. Bentley *Man. Bot.* (ed. 5) 534 They [*sc.* flowers and seeds of *Melilotus caerulea*] are used to give flavour to the 'Schabzieger'. **1892** *Mrs. Beeton's Bk. Househ. Managem.* 865 Schabzieger..a cheese exceedingly strong both in smell and taste. **1950** *Chambers's Encycl.* VI. 371/1 Glarus... The green cheese called 'Schabzieger' is wholly made here. [illegible] **1969** R. & D. De Sola *Dict. Cooking* 202/1 *Schabzieger Käse* (German—scraped whey cheese), hard greenish cheese used in cooking and somewhat like sapsago, its American counterpart.

schadchen, var. *shadchan.

‖ **Schadenfreude** (ʃā·dənfrɔidə). Also with small initial. [Ger., f. *schaden* harm + *freude* joy.] Malicious enjoyment of the misfortunes of others.
[**1852** R. C. Trench *Study of Words* (ed. 3) ii. 29 What a fearful thing is it that any language should have a word expressive of the pleasure which men feel at the calamities of others; for the existence of the word bears testimony to the existence of the thing. And yet in more than one such a word is found... In the Greek ἐπιχαιρεκακία, in the German, 'Schadenfreude'. **1867** Carlyle *Shooting Niagara: & After?* iii. 22 Have not I a kind of secret satisfaction, of the malicious or even of the judiciary kind (*Schadenfreude*, 'mischief-joy', the Germans call it, but really it is justice-joy withal) [illegible].] **1895** C. Lowe *German Emperor William II* ix. 236 But the *Schadenfreude*, or malicious joy, of the French was premature. [illegible] Part of Wood i. 8 Arthur Rackham..certainly had a vein of *schadenfreude* (what is now misleadingly described as sadism) and took an intense delight in scraggy fingers. **1977** 'E. Crispin' *Glimpses of Moon* iv. 62 Solidarity or no solidarity, Widger was not wholly without *Schadenfreude* at seeing his informative colleague discomfited for once. **1978** 'A. Stuart' *Vicious Circles* 15 For a Russian..there is a curious fascination, mixed with *Schadenfreude*, about..titles and honours lists.

schafarzikite (ʃăfă·ɹzikəit). *Min.* [ad. G. *schafarzikit* (J. A. Krenner 1921, in *Zeitschr. f. Kristallogr.* LVI. 198), f. the name of Ferenc *Schafarzik* (1854–1927), Hungarian mineralogist: see -ite[1].] A tetragonal antimonite of iron, first found as red to red-brown prismatic crystals with a metallic lustre in a stibnite mine in Slovakia.
1922 *Mineral. Mag.* XIX. 348 Schafarzikite... Red, tetragonal crystals found with kermesite (and resembling this in appearance) at Pernek, Hungary, contain iron and phosphorus. **1955** M. H. Hey *Index Min. Species* (ed. 2) 271 Schafarzikite... Isostructural with artificial tetragonal $FeSb_2O_4$, but the one analysis approaches $Fe_5Sb_4O_{11}$. **1975** *Tschermaks Mineral. und Petrogr. Mitteilungen* XXII. 236 The crystal structure of schafarzikite, $FeSb_2O_4$,..has been refined... The Fe atoms are surrounded octahedrally by six oxygen atoms, the Sb atoms form with three oxygen atoms a flat trigonal pyramid.

schairerite (ʃē[ə]·rərəit). *Min.* [f. the name of John F. *Schairer* (1904–70), U.S. geochemist + -ite[1].] A sulphate and fluoride of sodium, Na_3FSO_4, usu. also containing chlorine, first found as colourless rhombohedral crystals in the salt crust of Searles Lake, San Bernardino Co., California.
1931 W. F. Foshag in *Amer. Mineralogist* XVI. 134 For this new species, a sulfate and fluoride of soda, the name schairerite is proposed in honor of Dr. J. F. Schairer of the Geophysical Laboratory of the Carnegie Institution, who studied the quaternary system Na_2SO_4–NaF–NaCl–H_2O, in which this compound plays a prominent part. **1963** *Doklady Earth Sci.* CXXXIX. 839/1 We succeeded in discovering schairerite in totally different conditions—nepheline syenite pegmatites of Alluayv Mountain in the Lovozero massif (Kola Peninsula). **1971** *Amer. Mineralogist* LVI. 177 All but the smallest and most perfect schairerite crystals are affected by voids or imperfections.

‖ **schalet(e** (ʃa·lĕt, ʃale·t). *Jewish Cookery.* Also **schaleth.** [app. a Ger. variant of Yiddish *tsholnt*.] **a.** A kind of baked fruit pudding. **b.** A Sabbath dish of meat, potatoes, and vegetables, prepared on a Friday and baked slowly overnight.
1943 A. Simon *Conc. Encycl. Gastronomy* IV. 27/1 *Schaleth* is an old favourite among Jewish cookery recipes.. a pudding made of..apples..raisins and sultanas..spices.. and baked under a cover of a hard wheat paste. **1949** *Housewife* May 2/2 The traditional Saturday lunch-dish is the Schalet, a stew of meat and beans, prepared on the Friday and left to cook at the back of the stove. **1956** L. Blanch *Around World in Eighty Dishes* 99 Schalète (from Israel)... An apple and raisin cake. **1966** J. Gardner *Amber Nine* iii. 40 'Best Schalete I've ever tasted.'..'Kosher, of course.' **1970** L. M. Feinsilver *Taste of Yiddish* ii. 172 Heine, in discussing the merits of *schalet*, or *tsholnt*, expressed regret that 'the Christian Church, which borrowed so much that was good from ancient Judaism, should have failed to adopt *schalet* as its own'.

‖ **Schallanalyse** (ʃa·lanalǖ:zə). *Philology.* [Ger., lit. 'sound analysis'.] (See quot. 1931.)
1930 [see *motoric *a.* a]. **1931** *Year's Work Mod. Lang. Studies* I. iii. 126 Sievers's 'Schallanalyse'..is a method of restoring the accentuation of a given textual record by registering and analyzing the reaction of a trained observer, who responds instinctively and directly to the psychological compulsion exerted by the text on any one who reads it aloud. **1939** L. H. Gray *Found. Lang.* ii. 44 One must exclude from linguistics proper all consideration of rhythm (including the so-called *Schallanalyse*). **1947** C. L. Wrenn *Poetry of Cædmon* 3 But it is made very clear at the outset of this [*sc.* Sievers's] exposition of *Schallanalyse* that only those who possess certain qualities in their motor nerves can participate in such experiments or judge of their effects. **1953** K. Sisam *Studies in Hist. of Old Eng. Lit.* vi. 103 Sievers, using arguments from 'Schallanalyse' which I cannot follow, concludes that the translator of *Genesis B* produced the whole work *Genesis A* and *B* by a process of compilation and revision.

schallerite (ʃæ·lərəit). *Min.* [f. the name of Waldemar T. *Schaller* (1882–1967), U.S. mineralogist + -ite[1].] A reddish-brown basic silicate, arsenate, and chloride of iron and manganese crystallizing in the rhombohedral system.
1925 R. B. Gage et al. in *Amer. Mineralogist* X. 9 The name schallerite is proposed for the mineral after Dr. Waldemar T. Schaller, of the United States Geological Survey, Washington, D.C. *Ibid.*, Schallerite occurs in seams or on cleavage faces in the massive zinc ore. **1970** *Soviet Physics: Crystallogr.* XV. 40/1 This has made it possible to carry out a systematic derivation of the possible polytype modifications of pyrosmalite and to determine the structures of schallerite and friedelite.

schalstein, var. *schaalstein.

schanse. Add: Also **schans, schanz, skans.** (Further examples.)
1885 J. Nixon *Complete Story Transvaal* xi. 200 They found the Boers intrenched in a series of *schanses* (stoneworks, breast high), along the northern ridge of the valley. **1894** B. Mitford *Renshaw Fanning's Quest* xxii. 177 Lucky, I took the precaution of building a *schanz*,—eh? **1929** D. Reitz *Commando* ix. 75 We were sustaining heavy casualties from the English *schans* immediately in front of us. **1969** J. Selby *Boer War* 15 Boers digging defense works and building stone schanzes. **1974** in J. Branford *Dict. S. Afr. English* (1978) 218/1 The British thought the Boers would be hiding behind these skanses or heaps of stone.

schappe (ʃæp, ‖ ʃa·pə). [a. G. *schappe* silk waste.] A fabric or yarn made from waste silk (orig. by removal of the gum by fermentation). Hence **schappe** *v. trans.*, to ferment (waste silk) in order to remove gum; **scha·pping** *vbl. sb.*
1885 *Harper's Mag.* July 246 Now they [*sc.* waste cocoons] are spun into yarn,..and made into *schappe* or 'spun silk' fabrics, not as lustrous as reeled silk goods, but stronger and cheaper. **1909** Webster, Schapping, *n.* **1921** Beaumont & Hill *Dress, Blouse, & Costume Cloths* 94 The 'Schappe' or 'steeping practice' consists in placing the supply of waste silk in jacketed pans. **1957** *Textile Terms & Defs.* (Textile Inst.) (ed. 3) 124 *Schapping*, a continental method of de-gumming, applied to silk waste, that removes part of the gum by a fermentation process. Up to 10 per cent of gum may remain on the fibre. **1969** A. J. Hall *Stand. Handbk. Textiles* (ed. 7) iii. 129 Spun or schappe yarns which are short fibred and free or partly free from silk-gum are made from silk waste.

schapska (ʃæ·pskă). Also **chapska.** [Fr. *chapska, schapska*, ad. Pol. *czapka* cap.] A flat-topped cavalry helmet.
1894 *Daily News* 27 Mar. 5/4 Helmets, shakos, chapskas, and other head coverings were to be had in great profusion. **1909** Webster, Schapska. **1918** E. S. Farrow *Dict. Mil. Terms* 536 *Schapska*, a military helmet or shako, first worn by the Polish lancers. **1930** *Times Educ. Suppl.* 19 July (Home & Classroom Section) p. iv/2 Spahis, in their baggy trousers, their burnouses,..their chapskas, their boléros, and all the other items of equipment,..followed. **1936** C. S. Forester *General* ii. 16 Put on again the glories of blue and gold, schapska and plume, lance pennons and embroidered saddlecloths. **1951** J. Masters *Nightrunners of Bengal* xxiv. 331 A horseman galloped up... He wore..a black schapska with a gold bag tied to its side.

schatchen, var. *shadchan.

‖ **Schatz** (ʃats). Also **schatz.** [Ger., lit. 'treasure'.] In Germany: a term of endearment for a woman; a (German) girl-friend or female companion. Also dim. **Schätzi, Schatzi(e.**
1907 M. A. von Arnim *Fräulein Schmidt & Mr. Anstruther* xlii. 174 The trumpeter and his *Schatz* sat quietly in the kitchen. **1956** *Amer. Speech* XXXI. 142 A sizable body of German words and idioms has entered the lingo of Army troops in Germany and Austria... *Schatzi*, or sometimes *shots* (= German *Schatz*), is a sweetheart. **1966** E. West *Night is Time for Listening* vi. 190 He sat at a table in Pommler's..the place bulging with GI's, *Schatzis*, and miscellany. **1970** L. Sanders *Anderson Tapes* xxi. 30 Oh, *Schatzie*, I stopped wanting many years ago. Now I just accept. **1971** D. MacKenzie *Sleep is for Rich* vi. 198 We've been through all this before, *schatz*. **1972** L. P. Bachmann *Ultimate Act* xxiii. 209 'Are you all right, *Schätzi*?' she asked... 'What's *Schätzi*?' 'You. It's an old-fashioned translation of 'chéri' into German.' **1976** P. Henissart *Winter Quarry* viii. 83 *Schatz*, I know my business.

Schaumann (ʃau·măn). *Med.* [Name of J. *Schaumann* of Stockholm, who described them (*Acta Med. Scand.* (1941) CVI. 239, etc.).] *Schaumann('s) body*: a rounded, laminated body containing iron and often calcium, numbers of which are common inside giant cells in sarcoidosis tissue.
1955 P. A. Herbut *Pathology* xii. 342 Schaumann bodies are deeply bluish staining, concentric, lamellated, multiple contoured, iron- and calcium-positive concretions of variable sizes. **1976** Robbins & Angell *Basic Path.* (ed. 2) xii. 421/1 The distinctive..morphologic feature of sarcoidosis..is the noncaseating granuloma... In 80 to 90 per cent of these granulomas, laminated concretions of calcium and proteins, known as Schaumann's bodies, can occasionally be found within giant cells.

schchi, var. Stchi in Dict. and Suppl.

sched (ʃed, sked), colloq. abbrev. of *schedule *sb.* 4 b.
1958 R. Stow *To Islands* iv. 88, I took your telegram..to send to the doctor, but it was too late for the sched... I felt the wireless and it was cold. **1963** L. Diack *Labrador Nurse* iii. xviii. 91 There was the radio-telephone..and there was always a daily 'sched' at twelve noon.

schedule, *sb.* Add: **2.** (Earlier and later examples with reference to the British income tax.)
1803 *Income Tax Act* 43 *Geo. III* c. 122 s. 1 in *Statutes United Kingdom* (1804) I. 1012 During the Term herein mentioned, there shall be raised, levied, collected, and paid, throughout Great Britain, the several Duties and Contributions in the Schedules contained in this Act, marked (A) (B) (C) (D) and (E). **1902** Kipling *Traffics & Discoveries* (1904) 29 'You'll only be an additional expense to me as a taxpayer. Think of Schedule D,' he says, 'and take parole.' **1966** B. E. V. Sabine *Hist. Income Tax* ii. 35 The tax was for the first time divided up into the well-known five schedules. Schedule A charged tax on the amount of land and buildings; Schedule B covered farming profits; Schedule C taxed fundholders in respect of annuities payable out of any public revenue... Schedule D was divided into the six cases which are still familiar today and brought into charge various forms of profit and interest..and Schedule E embraced the charge on income from offices and employments of profit and annuities and pensions. **1970** *Money Which?* Mar. 4/2 Schedule A. Income from rents and other receipts from property which is unfurnished (formerly Case VIII of Schedule D, but reclassified as Schedule A as from 6 April 1970).
4. a. For 'chiefly *U.S.*' read 'orig. *U.S.*' and add earlier and further examples. Also in extended sense, a programme or plan of events, operations, etc. Freq. in phrs. *according to, before, behind, on*, etc., *schedule* (*time*).
In the sense 'a printed time-table of arrivals and departures of trains, buses, aeroplanes, etc.', the use remains chiefly *N. Amer.*
1863 O. W. Norton *Army Lett.* (1903) 282 That is all that ever caused the name to be printed on anything but time-tables and schedules of a one-horse railroad. **1866** C. H. Smith *Bill Arp* 21 We tried our durndest to comply with your schedule. **1881** A. A. Hayes *New Colorado* vii. 94 As he [*sc.* the engineer] rounded the curves in about half of schedule time. **1884** Schedule time [in Dict.]. **1901** O. Wister in *Lippincott's Monthly Mag.* Aug. 193 As a delayed train makes the last few miles high above schedule speed. **1904** *Newark Evening News* 13 June 6 It is on the schedule for the new Equal Taxation Commission to organize to-morrow. **1906** 'O. Henry' *Rolling Stones* (1912) 22 Tuesday, the day set for the revolution, came around according to schedule. **1909** *Springfield* (Mass.) *Weekly Republican* 19 Aug. 10 The train was running exactly on schedule when the party left it. **1911** C. E. Persons *Labor Laws & their Enforcement* 109 Most important of these enforced concessions was the temporary reduction to a ten-hour [factory] schedule at Fall River. **1927** *Daily Tel.* 1 Mar. 6/4 The material must be finished on time, routed on schedule, and delivered at exactly the psychological moment. **1961** *Lancet* 29 July 230/1 In investigations into new treatment schedules, close co-operation between the clinician and the laboratory is essential. **1968** *Globe & Mail* (Toronto) 3 Feb. 35/9 The standing in the Metro Junior B Hockey League is beginning to appear like a jig-saw puzzle as the schedule enters its final week. **1975** M. Russell *Murder by Mile* ix. 93 How far behind schedule are you now? **1977** I. Shaw *Beggarman, Thief* ii. i. 119 He looked up the schedule of the planes flying out of Brussels to New York. **1980** *Nature* 24 Apr. 654/1 Preparations for the launch, begun on 2 April, are going ahead on schedule following the arrival of the L02 launcher at Kourou.
b. An agreed period of time during which a radio transmission may be made; time allocated to listening for transmissions.
1958 'N. Shute' *Rainbow & Rose* i. 9, I should say they've closed down for the night. They'll be speaking on the morning schedule, at seven o' clock. **1974** D. Kyle *Raft of Swords* xiii. 143 We'd better watch this six o'clock schedule like hawks.

schedule, *v.* Add: **1.** (Later examples.) Hence, in extended uses: to place (something) on a programme of future events; to arrange for (a person or thing) to do something or *for* an event.
1898 T. N. Page *Red Rock* 478 The trial would come off as already scheduled. **1904** *N.Y. Even. Post* 30 Sept. 1 The archbishop is scheduled to speak this afternoon at the Academy of Music. **1922** Joyce *Ulysses* 313 It was a historic and a hefty battle when Myler and Percy were scheduled to don the gloves for the purse of fifty sovereigns. **1931** H. F. Pringle *Theodore Roosevelt* i. xiv. 190 The advance took place as scheduled. **1958** 'N. Shute' *Rainbow & Rose* i. 7 There was a Dakota freighter scheduled to leave for Hobart..at one o'clock. **1968** *Globe & Mail* (Toronto) 17 Feb. B 7 (Advt.), *Wanted* Masonry Superintendent, to take complete charge of a young masonry construction firm. Must be able to..schedule jobs, assist in bidding etc. **1976** *Columbus* (Montana) *News* 3 June 2/3 Kristy McFarland.. is scheduled for back surgery June 4. **1976** *Sunday Times* (Lagos) 1 Aug. 22/2 Two top Nigerian lawn tennis players.. are scheduled for the Zambian Open Championships. **1979** *Tucson* (Arizona) *Citizen* 20 Sept. 34/1 A spokesman for St. Paul [Hospital] said an autopsy was scheduled.
3. To include (a building, etc.) on a list of buildings that are to be preserved and protected for architectural or historic reasons.
1921 *Report of Ancient Monuments Advisory Committee* 5 The Commissioners of Works are bound to prepare and publish a list of all monuments the preservation of which is reported by any of the three Ancient Monuments Boards to be of national importance, and to inform the owners of their intention to include them and of the penalties hereinafter mentioned. This is called scheduling a monument. **1960** *Twentieth Century* Nov. 480 A decaying polygon.. scheduled as being of architectural or historical interest. **1971** P. Gresswell *Environment* 23 Ancient Monuments are 'scheduled' by the Department of the Environment.

scheduled, *ppl. a.* Add: **a.** (Later examples.)
1911 G. B. Shaw *Doctor's Dilemma* p. xx, Treatment varies widely from doctor to doctor, one practitioner prescribing six or seven scheduled poisons for so familiar a disease as enteric fever. **1921** *Report of Ancient Monuments Advisory Committee* 6 A scheduled monument.. whether public or private may not be in any way damaged. **1931** H. F. Pringle *Theodore Roosevelt* iii. v. 368 He insisted..upon making a scheduled speech. **1952** 'J. Tey' *Singing Sands* ix. 130 Most of us fly scheduled routes, but some fly tramps. **1970** *Guardian* 31 Dec. 18/3 BEA will be able to offer seats on scheduled flights to package tour holidaymakers..at knock-down prices. **1976** *Southern Even. Echo* (Southampton) 16 Nov. 3/7 A scheduled meeting with Mr. Ford will not take place until the following week.
b. In specific collocations: **Scheduled Caste** (or **class**), in India, a category of persons in the lowest castes, or Untouchables; **scheduled territory**, between 1947 and 1972, any of a group of countries, mostly within the British Commonwealth, with currencies linked to

sterling; the sterling area; after 1972 (see quot. 1977; the Republic of Ireland has now ceased to be part of the sterling area); **Scheduled Tribe**, in India, a group of aborigines who do not observe the taboos of caste.
1935 *Government of India Act* 25 & 26 *Geo. V* c. 42 1st. sched. § 26 The 'scheduled castes' means such castes, races or tribes or parts of or groups within castes, races or tribes, being castes, races, tribes, parts or groups which appear to His Majesty in Council to correspond to the classes of persons formerly known as 'the depressed classes'. **1943** B. R. Ambedkar *Mr. Gandhi & Emancipation of Untouchables* iii. 15 Under the Government of India Act of 1935 the Untouchables are designated as 'Scheduled Castes'. **1975** Y. B. Damle in H. M. Patel et al. *Say not the Struggle Nought Availeth* 143 Students belonging to the scheduled and backward classes tend to be less than ten per cent of the total students in most of the states. **1947** *Act* 10 & 11 *Geo. VI* c. 4 § 1 (3) In this Act..the expression 'the scheduled territories' means the territories specified in the First Schedule to this Act, so, however, that the Treasury may at any time by order amend the said Schedule, either by the addition or exclusion of territories. **1964** *Financial Times* 12 Mar. 19/5 The Scheduled Territories are the British Commonwealth (except Canada), the Irish Republic, British Trust Territories, British Protectorates and Protected States, Burma, Iceland, the Hashemite Kingdom of Jordan, Kuwait, Lybia, South Africa and South West Africa, Western Samoa. **1972** *Statutory Instruments* II. i. 2926 The Scheduled Territories now consist only of the United Kingdom, the Channel Islands, the Isle of Man, and the Republic of Ireland. **1977** *Guide to United Kingdom Exchange Control* (Bank of England) 6 The Scheduled Territories at present comprise the United Kingdom, including the Channel Islands and the Isle of Man, the Republic of Ireland and Gibraltar. **1957** G. S. Ghurye *Mahadev Kolis* i. 4 Many sections of Kolis describe themselves as Mahadev Kolis in order to be able to claim the special benefits of the Scheduled Tribes. **1972** *Times of India* 28 Nov. 13/2 (Advt.), This post is unreserved, however, preference will be given to Scheduled Caste|Scheduled Tribe candidates.

scheduler (ʃe·diuləɹ, *U.S.* ske·diuləɹ). [f. Schedule *v.* + -er[1].] **1.** One who draws up a schedule or arranges activities in accordance with one.
1952 *Antioch Rev.* Dec. 426 After school there are music lessons, skating lessons, riding lessons, with mother as chauffeur and scheduler. **1957** *Electronic Engin.* XXIX. 179/1 The two schedulers interpret the required data-processing operations in terms of machine functions. **1978** *Sci. Amer.* Mar. 124/2 The priority list *L* is an ordering of the tasks according to the preferences of the scheduler. **1979** H. Kissinger *White House Years* xxii. 923 The schedulers had arranged for a visit to the Vatican in the afternoon. **1980** *Daily Tel.* 18 June 8/7 So far the Radio 3 schedulers have mostly been able to replace these concerts with records of the works that would have been played.
2. a. A machine, esp. a computer, that can arrange a number of planned activities into the order in which they should take place.
1962 *Times* 26 Oct. (Spencer Steelworks Suppl.) p. xiv/3 The finishing end scheduler is the coordinating and planning authority for the whole finishing process. *Ibid.*, The finishing end scheduler prepares and revises production schedules.
b. *Computers.* Any of several control programs that arrange jobs or the computer's operations into an appropriate sequence; also, a part of the hardware designed to perform a similar function.
1966 C. J. Sippl *Computer Dict. & Handbk.* 279/2 The scheduler is called at regular intervals to decide which program in memory is to be run... A program may be terminated temporarily by user intervention to the scheduler, or it may suspend its own operation. **1968** *Communications Assoc. Computing Machinery* XI. 349/1 The part of the system responsible for handling block and wakeup instructions will be called the scheduler. *Ibid.* 357/2 All the external interrupt lines are directed into the scheduler, which.. loops constantly, examining them. **1973** C. W. Gear *Introd. Computer Sci.* iv. 168 When a job is terminated, the scheduler uses the space for another job and puts the output on a work list for the output processor.

scheduling, *vbl. sb.* Add: The action of entering in or drawing up a schedule; *esp.* the preparation of a timetable for the completion of the various stages of a complex project; the co-ordination of many related actions or tasks into a single time-sequence. (Further examples.)
1957 *Proc. Conf. Operations Res., Computers, & Management Decisions* (Case Inst. Technol.) 63/1 A minimum of constraints were used to define the models of the cut-and-fill operation and of construction scheduling. **1959** *Naval Research Logistics Q.* (U.S.) VI. 131 There is an indication of the possibility of constructing special algorithms to exploit the structure of certain of the 'classical' scheduling problems. **1964** A. Battersby *Network Analysis* ix. 138 A fairly simple arrow diagram..usually contains about 150 to 200 activities and forms the basis for the overall scheduling. **1967** *Times Rev. Industry* Mar. 15/1 Scheduling times on much plant and equipment are long and one can do little about rephasing such spending in the short term. **1976** P. R. White *Planning for Public Transport* viii. 170 A graphic timetable of this type can be used for scheduling purposes. **1977** *Rep. Comm. Future of Broadcasting* iv. 35 To our mind an executive Broadcasting Commission..would be bound to be drawn into the details of scheduling..in making decisions about individual programmes. **1978** *Sci. Amer.* Mar. 129/3 In critical-path scheduling the tasks are assigned to processors according to the length of the various precedence chains they head in the diagram of precedence constraints. **1980** *Times* 29 Nov. 5/8 The scheduling of an interim stop in Athens after the airliner took off.

Scheele's green (ʃē[i]·ləz grīn). *Chem.* [f. the name of Karl Wilhelm *Scheele* (1742–1786), German-born Swedish chemist, who first prepared it.] A hydrated form of copper arsenite, $Cu_3(AsO_3)_2.xH_2O$, formerly used as a pigment in calico printing and wallpaper manufacture.
1819 W. T. Brande *Man. Chem.* v. 274 Mixed with a solution of sulphate of copper, a precipitate of a fine apple-green colour falls, called from its discoverer, Scheele's green, and useful as a pigment. **1935** *Discovery* Sept. 261/2 The dark green background of a so-called Holbein portrait was recently found, on..analysis.., to consist of Scheele's green, a copper arsenate discovered in 1778. **1967** *Jrnl. Colour Group* 92/2 Scheele's green and emerald green, both copper arsenates, lost their popularity [with artists] because of their poisonous effects. **1973** J. D. Smith in J. C. Bailar et al. *Comprehensive Inorg. Chem.* II. xxi. 609 Yellow silver arsenite..and copper arsenite (Scheele's green) may be precipitated from neutral solutions.

schefflera (ʃe·flərā). [mod.L. (J. R. & G. Forster *Characteres Generum Plantarum* (1776) 45), f. the name of J. C. *Scheffler* of Danzig + -a 2.] An evergreen shrub or small tree of the genus so called, belonging to the family Araliaceæ, native to many tropical or subtropical regions, and bearing large compound leaves and clusters of small white, greenish, or red flowers, followed by small berries.
1954 F. Kingdon-Ward *Berried Treasure* xviii. 164 The slim rigid spikes of Schefflera, like tall black candles, give a Gothic dignity to this little palm-like tree. **1976** *National Observer* (U.S.) 25 Sept. 9/2 We couldn't buy a schefflera that big for $50. **1978** *Homes & Gardens* Apr. 32/1 Some philodendrons and schefflera..do well in most places, given reasonable light.

Scheherazade (ʃĕhe:rāzā·d, ʃĕhī[ə]:r-, -zā·də). The name of the female narrator of the *Arabian Nights*, used allusively as the type of a (usu. young and attractive female) teller of long or numerous stories.
1851 Dickens *Let.* 25 Oct. in W. Gérin *E. Gaskell* (1976) xii. 123 My dear Scheherazade,—For I am sure your powers of narrative..must be good for at least a thousand nights and one. **1872** O. W. Holmes *Poet at Breakfast-Table* iii. 87, I had noticed that the Young Girl—the story-writer, our Scheherazade, as I called her—looked as if she had been crying or lying awake half the night. **1896** G. B. Shaw *Let.* 5 July (1965) I. 634 Or are you only a flattering storytelling Scheherazade? **1946** L. P. Hartley *Sixth Heaven* i. 4 'I'm afraid it will be a long story,' he said...'Waste no time in self-depreciation, Scheherazade, but..take up your tale.' **1973** G. Butler *Coffin for Pandora* viii. 192 You're a teller of tales, young lady... Quite a Scheherazade. **1978** M. Puzo *Fools Die* xxix. 343 During that happy time, a blond Scheherazade, she told me the story of her life. **1981** A. Fraser *Splash of Red* i. 11 That's *another* story I shall tell you... I shall be Scheherazade.

Scheiner (ʃəi·nəɹ). *Photogr.* The name of Julius *Scheiner* (1858–1913), German astrophysicist, used, usu. *attrib.*, with reference to a way of measuring and expressing the speed of photographic emulsions that he devised, as *Scheiner degree, scale, sensitometer, speed, system*; **Scheiner number**, a number depending on the logarithm of the least exposure that will give a visible image on development.
1900 *Astrophysical Jrnl.* XI. 91 In a simple experiment with the aid of Scheiner's sensitometer equal degrees of blackening were produced by continuous exposures of 96, 72, 48, 24, 12 secs. *Ibid.* 98 Remarks on the Scheiner sensitometer. **1911** A. Watkins *Photography* iii. 47 In the case of Wynne, Scheiner, and Warnerke numbers..there is not the same direct proportion between the numbers. *Ibid.* 332 (Index), Scheiner speeds. **1918** J. R. Roebuck *Science & Practice of Photogr.* 225 'n' is the Scheiner degree and 'A' has the value in this case of about 4·4. **1936** *Discovery* June 192/2 The rating [of the exposure meter] agreeing more or less with the Scheiner system. **1938** S. G. B. Stubbs et al. *Modern Encycl. Photogr.* II. 1105/1 The so-called 'Scheiner speeds' quoted by Continental manufacturers are in reality not Scheiner speeds at all, as all 'Scheiner' measurements are now carried out by the Eder–Hecht method. **1942** C. B. Neblette *Photography* (ed. 4) xiii. 419 The ratio of the exposures between consecutive steps on the Scheiner sensitometer is as 1:1·27. **1962** *Ibid.* (ed. 6) xx. 267/1 The Scheiner speed number was obtained originally by exposing the negative material in a sensitometer with a sector wheel having exposure steps numbered from 1 to 20, with a log exposure difference of 0·15. *Ibid.* 267/2 In 1931, the German photographic industry replaced the then meaningless Scheiner numbers by a new German standard (DIN) speed. **1963** Jerrard & McNeill *Dict. Sci. Units* 106 The Scheiner scale was devised in 1898..and was first used commercially by the Secco Film Company of Boston, Mass. in 1899. **1973** *Focal Dict. Photogr. Technol.* 544 Scheiner speed was expressed in degrees, every increase of 3° corresponding to a doubling of the working speed.

‖ **Scheitholt** (ʃəi·tˌhǫlt). [Ger., f. *scheit* log + dialectal *-holt* wood.] A former stringed instrument of central Europe, a precursor of the zither.
1961 A. Baines *Musical Instr. through Ages* 210 The simplest forms [of zither], as Alpine *Scheitholt*, *épinette des Vosges*, and the Dutch *hummel*..have a long narrow hollowed-out sound-box, placed on the knees or on a table. **1976** D. Munrow *Instr. Middle Ages & Renaissance* 33/1 Various names have been used for the string drum including *Scheitholt* and *tambourin*.

Schellingian (ʃeli·ŋiăn), *a.* [f. *Schelling* (see below) + -ian.] Of or pertaining to the German philosopher, F. W. J. von Schelling (1775–1854), or to his doctrines. Hence as *sb.*, a follower of Schelling. Also **Schellingism** (ʃe·liŋiz'm) [ad. G. *Schellingismus*], the system of philosophy taught by Schelling; **Sche·llingist**, a disciple of Schelling.
1865 tr. *Strauss's New Life of Jesus* I. 190 Similar instances may be brought forward from the history of the Schellingian philosophy. **1865** J. H. Stirling *Secret of Hegel* I. i. v. 275 Once in Jena, we have to see him a declared Schellingian. **1865** W. Pater *Appreciations* (1889) 75 Schellingism, the 'Philosophy of Nature', is indeed a constant tradition in the history of thought. **1874** Morris & Porter tr. *Ueberweg's Hist. Philos.* II. 114 Kantism, the renewed Spinozism (Schellingism), and Herbartism lay conjoined and undeveloped in the doctrine of Leibnitz. **1894** C. S. Peirce *Let.* 28 Jan. in R. B. Perry *Tht. & Char. of W. James* (1935) II. 416 If you were to call my philosophy Schellingism transformed in the light of modern physics, I should not take it hard. **1895** C. Garnett tr. *Turgenev's On the Eve* iv. 30 My father was a learned man, a Schellingist. **1967** *Encycl. Philos.* VII. 260/2 The most important of the Russian Schellingians were Professor D. M. Vellanski..and Prince V. F. Odoyevski. *Ibid.* 261/2 In his early Schellingian period he [*sc.* Belinski] stressed aesthetic activity.

schelm, var. Skellum in Dict. and Suppl.

schema. Add: Also pl. **schemas.** **1. a.** (Earlier and later examples.)
1796 F. A. Nitsch *View of Kant's Princ.* 103 The Schema of a Category is no picture of anything. **1961** B. M. Milmed *Kant & Current Philos. Issues* iv. 81 For both Kant and Lewis..the image is empirical, a reproduction of past experience, while the schema, through which the image becomes part of a criterion of empirical meaning, is a priori in its role as a definition of the experience to be interpreted by it. **1963** A. Pap *Introd. Philos. of Sci.* vi. 102 The schemas correspond to the following principles of logic: the principle of the hypothetical syllogism..; a statement implied by a true statement is true [etc.]. **1966** E. S. Casey tr. *Dufrenne's Notion of A Priori* viii. 156 Now, if the schema is the *a priori* in its original state, is it not the *a priori* in its corporeal state as well?
b. *Neurol.* and *Psychol.* An automatic, unconscious coding or organization of incoming physiological or psychological stimuli, giving rise to a particular response or effect.
1920 H. Head *Stud. in Neurol.* II. iv. v. 605 For this combined standard, against which all subsequent changes of posture are measured before they enter consciousness, we propose the word 'schema'. **1926** M. Gabain tr. *Piaget's Lang. & Thought of Child* v. 236 This schema may be thought to apply only to 'whys', but it is obvious that other types of question..are more or less incorporated in it. **1932** —— tr. *Piaget's Moral Judgment of Child* ii. 20 The child is undoubtedly trying..to understand the nature of the marbles and to adapt its motor schemas to this novel reality. **1950** W. R. Brain in D. Richter *Perspectives in Neuropsychiatry* 138 The schema would then develop by becoming a resonator to a pattern received from any part of the corresponding sensory cortex and 'learned' by repetition, and would thus be the basis both of simple recognition and of abstraction. *Ibid.* 139 The schema is a neurophysiological disposition..which plays an essential part in perception and action, speech and thought... It may prove to be the bridge between body and mind. **1964** *Listener* 25 June 1029/1 Again, Koestler uses the idea of the 'schema' to discuss memory, but he does not mention that Bartlett..wrote a whole book..precisely to develop that very idea. **1971** J. Z. Young *Introd. Study Man* xxi. 277 Many, however, are very useful, especially the concept of a 'schema'. In Piagetian language this is described as a 'cognitive structure which has reference to a class of similar action sequences'. **1978** Hochberg & Brooks in J. W. Senders et al. *Eye Movements & Higher Psychol. Functions* v. iv. 295 If visual momentum is the impetus to obtain sensory information, and to formulate and test a schema, it should be reflected by the frequency with which glances are made.
2. a. (Later examples.) Also in extended use.
1895 J. Sully *Stud. of Childhood* x. 353 Number is here as little attended to as in the radial arrangements. It is worth noting that this *schema* seems to be widely diffused among children of different nationalities. **1943** H. Read *Educ. through Art* v. 121 All previous writers on the subject have attempted to trace the evolution of the schema, from the first chance recognition of a resemblance in the child's scribblings..to an outline or two-dimensional schema. **1960** E. H. Gombrich *Art & Illusion* v. 168 We shall never know what Rubens' children 'really looked like', but this need not mean we are forever barred from examining the influence which acquired patterns or schemata have on the organization of our perception. **1971** E. Kramer *Art as Therapy* vi. 127 A five- or six-year-old child who is in the process of discovering various schemata that unmistakably denote for him men, women,..or animals is..enormously increasing his power of expression. **1981** *Times Lit. Suppl.* 10 July 783/2 He painted what the schemata of Rembrandt and J. R. Cozens enabled him to see.
b. In *gen.* use, a hypothetical outline or plan; a theoretical construction; a draft, design.

1939 E. Muir *Present Age* i. 30 When he [*sc.* H. G. Wells] tried to reinstate society again his society was a schema, not an actual society such as Fielding described. 1947 *Partisan Rev.* XIV. 231 In the countries where capitalism really triumphed, it has yielded with far better grace..than the Marxist schema predicted. 1978 N. Marsh *Grave Mistake* iv. 123 The gardens today bear little resemblance in concept to this exquisite *schema*.

3. *Eccl.* A draft canon or decree submitted to either of the Vatican Councils for discussion.

1870 T. Mozley *Let.* 24 Mar. (1891) II. 273 The Council has been sitting on three successive days... Today makes the fourth given to the amended *Schema* on matters of faith. 1930 E. C. Butler *Vatican Council* I. x. 199 Two months elapsed during which the deputation worked at the remodelling of the schema. 1963 *Ann. Reg. 1962* 370 The first schema presented for discussion, Liturgy, seemed relatively innocuous.

schematic, *a.* Add: **B.** *sb.* A schematic representation; a diagram.

1929 R. T. A. Dennison *Private Automatic Branch Exchanges* vii. 187 In Fig. 5, a general P.A.B.X. schematic is given. 1949 *Electronic Engin.* XXI. 366/1 The amplifier, the schematic of which is shown here, consists of a two-stage unit in which the pulses are indicated by flashes on a neon light. 1961 *New Scientist* 16 Mar. 684/2 (*caption*) A schematic of a rotary copying machine working on the distillation principle. 1971 H. A. Whitaker in W. O. Dingwall *Survey Linguistic Sci.* 154 Further differentiation..may be seen in the schematic of the pyramid and extra-pyramid motor systems in Figure 3. 1978 R. Ludlum *Holcroft Covenant* v. 68 It was not out of the ordinary for long-range projects in faraway places to employ consulting architects, men whose names would not appear on schematics or blueprints but whose skills would be used.

schematism. Add: **4.** (Earlier example.) Also in *Psychol.* (cf. *SCHEMA 1 b).

1796 F. A. Nitsch *View of Kant's Princ.* 103 Our conceptions of figures and pictures originate in the schematism of the pure intellect. 1951 Gattegno & Hodgson tr. *Piaget's Play, Dreams & Imitation* viii. 220 Two peculiarities..remind us once again of the sensori-motor schematism of stage VI, but this time on the new plane of concepts in the process of formation. 1974 *Nature* 8 Mar. 177/1 This kind of knowledge (or understanding) is conceptualised as being based on a hierarchy of operative schemes (Piaget calls it 'schematism').

schematization (skī:mătəizē¹·ʃən). [f. SCHEMATIZE *v.* + -ATION.] **1.** The act or process of reducing to a scheme or formula; formulation in a regular order, organization according to a conventional pattern or preconceived system.

1904 W. James *Let.* 31 July in R. B. Perry *Tht. & Char. of W. James* (1935) II. 151 To me the whole Munsterbergian Circus seems a case of the pure love of schematization running mad. 1937 *Proc. Prehistoric Soc.* III. 36 A natural tendency towards schematisation accounts for the rendering of the lowest zone. 1953 *Trans. Philol. Soc. 1952* 16 Over-schematization..has also left its mark on the study of Old English. 1962 W. Nowottny *Lang. Poets Use* vi. 123 'Formal relationships' have already appeared..in connection with verse structure, not with schematization at the level of diction. 1979 *London Rev. Books* 25 Oct. 21/4 It is a curiously French combination: on the one hand, a Cartesian ..sense of schematisation, [etc.].

2. A hypothetical organization of schemata; an analytical or tabular representation of data.

1940 *Mind* XLIX. 320 The schematization of our primitive space to the more precise form is evidently correlated in some way with the conformation, structure, and distribution of our sense-organs. 1956 *Scottish Jrnl. Theol.* IX. 399 The sharply hostile and antithetic presentation of the debates [of Jesus with the Jewish authorities], where a dualistic schematisation points to a presupposition other than the relativism of merely human squabbles. 1973 A. J. Pomerans tr. *Piaget & Inhelder's Memory & Intelligence* xix. 341 It seems clear that the 'raw' memory plays no more than a limited role in these responses while schematizations are of considerable importance. 1979 *Amer. Pol. Sci. Rev.* Mar. 168/1 It rests on images, predicative assimilations, schematizations and imaginative illustrations of various sorts.

schematize, *v.* **2.** (Earlier and later examples.)

a 1866 J. Grote in *Jrnl. Philology* (1872) IV. 56 The *phonism* of one language differs from that of another in, 1st. The different radical phones used in it. 2nd. The different distribution of these among the noems, and 3rd. The different laws and ways in which the phones are schematized. 1913 A. S. Pringle-Pattison *Idea of God* (1917) 293 When we do try to schematize the fact [*sc.* that there can be no barrier between the finite consciousness and the Being in which its existence is rooted] for ourselves, we either eliminate the characteristics of selfhood..or..lose hold of the creative unity. 1954 *Circulation* X. 14/2 These reactions which seem to be enzymatic transformations may be schematized as follows.

schematized, *ppl. a.* (Later example.)

1946 R. G. Collingwood *Idea of Hist.* 109 The whole world of events in time is thus a schematized representation of the world of logical or conceptual relations.

scheme, *sb.*¹ Add: **5. d.** Also, an outing or excursion (*obs.*).

1789 Jane Austen in *Loiterer* 12 Sept. 6 That glorious achievement, A Scheme to Town. 1813 —— *Pride & Prejudice* III. ix. 166, I did not once put my foot out of doors... Not one party, or scheme, or any thing.

7. b. *scheme of colour*: now chiefly = *colour scheme* (in both senses) s.v. *COLOUR *sb.*¹ 18. Freq. *ellipt.*

1897 *Private Life of Queen* ii. 15 The general scheme of colour is crimson and cream and gold. This scheme of paint prevails throughout the suite. 1925 R. W. G. Hingston in E. F. Norton *Fight for Everest: 1924* 265 Certain of the little birds are decidedly conspicuous, and in some cases we see the obvious reason why they do not require a protective scheme. *Ibid.* 267 Its [*sc.* a locust's] scheme of colour was grey and black with delicate transverse bands across its thighs. 1969 J. Cheever *Bullet Park* ii. 31 Nubbly stretchy reps look completely out of place in my decorating scheme.

schemozzle, schepsel, varr. *SHEMOZZLE, *SKEPSEL.

Scherbius (ʃɜ·ɹbiŏs). *Electr.* The name of Arthur *Scherbius* (*fl.* 1906), German engineer, used *attrib.* with reference to a method which he devised for regulating and changing the speed of large a.c. induction motors, in which the voltage applied to the rotor is altered according to the load by means of a separate commutator motor and flywheel assembly wired in series with it.

1910 *Electrician* 8 July 513/2 In the Scherbius motor the compensating winding is connected to the brushes in opposition to the armature. 1928 *Engineering* 24 Aug. 247/2 The speed regulation of the motors is controlled by a Scherbius set. 1962 *Newnes Conc. Encycl. Electr. Engin.* 515/1 A considerable number of variable-speed induction motors with Scherbius control has been built in the past for outputs up to several thousand horse-power, particularly for steel-mill drives. 1973 J. M. D. Murphy *Thyristor Control of A.C. Motors* x. 160 In the Scherbius system..a rotary converter rectifies the slip power, and the rectified output drives a d.c. motor which is mechanically coupled to a squirrel-cage induction generator.

Schering (ʃē³·riŋ). *Electr.* The name of Harald Ernst Malmsten *Schering* (1880–1959), German engineer, used *attrib.* and in the possessive with reference to an alternating-current bridge circuit which he devised for measuring the capacitance and power factor of insulating materials.

1926 *World Power* V. 238/2 Of the many bridges so far developed, the Schering bridge is the most suitable for high voltage work. 1928 *Engineering* 13 Jan. 50/1 The Schering bridge is operated by a fixed-frequency valve oscillator, with a frequency of 800 periods per second. 1958 J. Shepherd et al. *Higher Electrical Engin.* iii. 64 The Schering bridge was developed to measure the loss resistance of dielectrics, line insulators, cables and high voltage capacitors under high voltage conditions (up to 100kV). 1975 D. G. Fink *Electronics Engineers' Handbk.* xvii. 29 Schering's bridge is widely used for measuring capacitance and dissipation factors.

scherm. Add: [Also Afrikaans *skerm*.] (Earlier and further examples.) Also, a temporary dwelling used by nomads.

1835 A. Smith *Diary* 4 Nov. (1940) II. 272 Have neither cattle nor chiefs, cut all the hair off, use red clay, have no fixed residence, make skerms under a bush. 1864 T. Baines *Explor. in S.-W. Afr.* 131 Two or three scherms for night-shooting had been thrown up. 1894 E. Glanville *Fair Colonist* xxiv. 186 The gentlemen, you know, will take a tent, and our dining-room will be a skerm. 'Good gracious! what is that?' 'A large canvas drawn over the waggon-top and stretched out to some trees, with canvas sides and an open front.' 1936 C. Birkby *Thirstland Treks* x. 118, I saw white men living in *skerms*—huts made of matting laid over frames of thorn-bush boughs. 1943 D. Reitz *No Outspan* 70 Lion roared about our skerm. 1960 *Africa* 4 Oct. 343 It always amuses me to speak of residence when I visualize the nomadic !Kung..building their nest-like grass shelters (scherms) for a stay of a few weeks. 1963 R. Lewcock *Early 19th Cent. Architecture in S. Afr.* viii. 137 In the small frontier farmhouses cooking was done in the open air, behind a simple screen shelter, or 'skerm'.

Schermuly (ʃɜ·ɹmuli). The name of William *Schermuly* (1857–1929), English inventor, used *attrib.* and *absol.* as proprietary names of apparatus comprising a line-carrying rocket fired from a pistol, used in life-saving at sea.

1922 *Life-Boat* Feb. 243/1 The Line-Throwing Gun... The two appliances to which chief attention was given were the Coston gun, an American invention..and the Schermuly Portable Rocket Apparatus. 1933 R. B. Chenevix-Trench *Jrnl.* 28 Sept. in *Mariner's Mirror* (1979) LXV. 274 One line was fired by a Schermuly pistol. 1947 *Trade Marks Jrnl.* 30 July 443/2 *Schermuly*... Pyrotechnic articles, cartridges, and apparatus for firing rockets. The Schermuly Pistol Rocket Apparatus Limited,..Newdigate, Surrey; Manufacturers. 1960 E. L. Delmar-Morgan *Cruising Yacht Equipment & Navigation* ix. 111 The Schermuly..are of sealed metal case construction and embody a sealed-in mechanical ignition device. 1973 B. Callison *Web of Salvage* x. 134 With the Schermuly pistol angled upward and to windward. 1979 P. Ferris *Talk to Me about England* II. 103 Jarre picked his way aft... Hansen [was] tearing the wrapper from a Schermuly rocket.

schertelite (ʃɜ·ɹtĕləit). *Min.* [f. the name of Arnulf *Schertel* (1841–1902), Bavarian chemist: see -ITE¹.] A hydrated acid phosphate of ammonium and magnesium, $(NH_4)_2Mg_2H_4(PO_4)_4.8H_2O$, found as small water-soluble orthorhombic crystals in deposits of bat guano in caves near Ballarat, Victoria.

1902 R. W. E. MacIvor in *Chem. News* 9 May 217/1 To prevent all future confusion, I have now decided to call this interesting mineral Schertalite [*sic*]. 1963 *Amer. Mineralogist* XLVIII. 639 Schertelite dissolves rapidly and incongruently in water with the formation of struvite. Exposure of schertelite to the atmosphere for several months results in alteration of the surface of the crystals, apparently to an intimate mixture of struvite and monoammonium phosphate.

scherzetto, scherzino: see next entry.

scherzo. Add: Now usu. with pronunc. (skē³·rtso). (Earlier and later examples.) Also *Comb.*, as *scherzo-like* adj.

1852 Geo. Eliot *Let.* 13 Nov. (1954) II. 67, I went to one of Jullien's concerts..and endured the Polkas for the sake of Zampa and Mendelssohn's Scherzo. 1931 *Times* 19 Feb. 10/1 The contrasts of solemnity and recklessness in the slow movement and scherzo were made particularly vivid. 1962 *Times* 20 June 15/2 The scherzo-like middle movement. 1976 *Scotsman* 20 Nov. 9/5 It was an exquisitely controlled and subtle performance,..delightfully airy in the scherzo-like variation.

fig. 1911 O. Onions *Widdershins* 265 'Scherzos in Silver and Grey!' he chuckled. 1955 *Sci. News Let.* 25 June 411/3 The mockingbird [will]..ring in bits from the repertories of other birds, with catcalls and rusty-hinge squeaks by way of scherzo interludes. 1964 *Listener* 12 Mar. 447/1 The play is a ruthless little fantasia (well named a *scherzo*).

Hence in dim. forms **scherzetto** (-e·to), **scherzino** (-ī·no), a short passage or piece of music with the character of a scherzo.

1884 F. Niecks *Conc. Dict. Mus. Terms* 273 *Scherzino*, a short or light scherzo. 1907 T. S. Wotton *Dict. Foreign Mus. Terms* 169 *Scherzettino*, *scherzetto*.., a little scherzo. 1954 *Grove's Dict. Mus.* (ed. 5) VII. 480/2 *Scherzino* or *Scherzetto*... The words are occasionally used for a short or very slight piece in the character of a scherzo. 1961 *Times* 3 Jan. 3/4 Fricker's octet..was well chosen..for the sociability of its scherzetto. 1963 *Times* 28 Jan. 5/2 Nothing could have been..more playful than the little scherzino movement in the *faschingsschwank aus Wien*. 1978 *Gramophone* June 95/2 There is quite a well articulated performance of the Mab *scherzetto* by Rémy Corazza.

scheulie, var. *SCHOOLIE.

schiacciato, var. *STIACCIATO.

Schick (ʃik). *Med.* The name of Bela *Schick* (1877–1967), Hungarian-born U.S. pædiatrician, used *attrib.* and *absol.* to designate a test he devised consisting in the intradermal injection of diphtheria toxin: the absence of an erythematous reaction indicates previously acquired immunity to diphtheria. [Described by Schick in *Münchener med. Wochenschr.* (1908) LV. 504–6.]

1916 *Jrnl. Immunol.* I. 203 This principle is applied today in the so-called Shick [*sic*] test of immunity to diphtheria. 1927 R. Muir et al. *Man. Bacteriol.* (ed. 8) xvii. 478 If a positive Schick is present in addition, the reaction due to the unheated toxin will be more marked. 1955 *Sci. News Let.* 9 Apr. 229/3 The Schick test, familiar to many school children, tells whether or not 'shots' to protect against diphtheria have been effective. 1971 D. Lambert in C. Bonington *Annapurna South Face* 290 A test known as the Schick Test may have to be done beforehand.

Hence **Schick-positive** (**-negative**) *adjs.*, showing (failing to show) an erythematous reaction in the Schick test.

1927 R. Muir et al. *Man. Bacteriol.* (ed. 8) xvii. 476 If the suspected carrier is Schick-positive, i.e. non-immune, the organism is likely to be non-virulent. 1932 *Ibid.* (ed. 9) xvii. 507 The proportion of Schick negative reactions increases with age much as in Europe; also the blood of Schick negative reactors contains diphtheria antitoxin. 1944 L. E. H. Whitby *Med. Bacteriol.* (ed. 4) xiii. 237 Hospital nurses should always be actively immunized if they are Schick-positive. 1951 Whitby & Hynes *Ibid.* (ed. 5) xiii. 239 Infants born of Schick-negative mothers are themselves immune to diphtheria for the few months during which maternal antibodies persist in the circulation.

Schiff (ʃif). *Chem.* The name of Hugo *Schiff* (1834–1915), German chemist, used *attrib.* and in the possessive to designate things he devised or investigated, as **Schiff('s) base**, any organic compound having the structure R¹R²C=NR³; **Schiff('s) reaction**, the action of aldehydes of restoring the magenta colour to Schiff's reagent; **Schiff('s) reagent**, an acid solution of fuchsin (magenta, rosaniline) decolorized by sulphur dioxide or potassium metabisulphite; **Schiff('s) test**, the Schiff reaction employed as a test for aldehydes.

1892 *Jrnl. Chem. Soc.* LXII. II. 1189 Schiff's bases, derived from aromatic aldehydes, have a similar constitution to benzylidenaniline, PhN:CHPh, which serves as the type of these compounds. 1915 P. E. Spielmann tr. *V.*

von Richter's Org. Chem. 383 Hydrocyanic acid attaches itself similarly to the oximes..and to the Schiff bases. 1951 I. L. Finar *Org. Chem.* I. xiii. 257 Primary amines combine with aromatic aldehydes to form Schiff bases. 1971 *Nomencl. Org. Chem.* (I.U.P.A.C.) (ed. 2) 258 Compounds R¹R²C=NR³ have the class name 'azomethines'. When the nitrogen atom is substituted, this class of compound has the generic name 'Schiff's bases'. 1975 *Nature* 30 Oct. 823/2 In visual pigments, retinal is bound by way of a Schiff base linkage to the protein. 1894 Perkin & Kipping *Org. Chem.* I. viii. 122 Aldehyde may be detected..by the 'magenta' or 'rosaniline test' (Schiff's reaction). 1897 *Chem. News* 9 July 23/2 (*heading*) Schiff reaction applied to acid fuchsine. 1951 I. L. Finar *Org. Chem.* I. viii. 123 Ketones do not give Schiff's reaction. 1897 *Jrnl. Chem. Soc.* LXXII. II. 468 The author [*sc.* B. von Bittó] has also studied the behaviour of Schiff's reagent (a 0·025 per cent. solution of magenta decolorised by passing sulphurous anhydride through it)..with a number of aldehydes and ketones. 1929 Evers & Elsdon *Analysis of Drugs & Chemicals* 279 The proportion of formaldehyde may be determined by the use of Schiff's reagent. 1964 M. Hynes *Med. Bacteriol.* (ed. 8) xxvii. 412 Schiff reagent. 1902 J. B. Cohen *Theoret. Org. Chem.* ix. 128 A further reaction for aldehydes is known as Schiff's test. 1949 English & Cassidy *Princ. Org. Chem.* xi. 202 Advantage is taken of the greater reactivity of aldehydes toward bisulfite in the Schiff test for aldehydes. 1972 Norman & Waddington *Mod. Org. Chem.* xii. 174 The addition of an aldehyde to this colourless solution restores the pink colour of the dye (Schiff's test).

Schilder's disease (ʃi·ldəɹz). *Path.* [The name of Paul Ferdinand *Schilder* (1886–1940), U.S. neurologist and psychiatrist, who described the disease in 1912 (*Zeitschr. f. die gesammte Neurol. u. Psychiatrie* X. 1–60).] A disease characterized by degeneration of the neurones of the brain, esp. in the occipito-temporal lobes, leading to blindness, deafness, and death.

1940 Hinsie & Shatzky *Psychiatric Dict.* 475/1 Schilder's disease or encephalitis periaxialis diffusa, is a slowly progressive degenerative disease of the brain occurring mainly in children and young people. 1961 R. D. Baker *Essent. Path.* xxii. 599 Diffuse cerebral sclerosis, one form of which has been called Schilder's disease, is a widespread demyelination of the cerebral hemispheres alone. 1966 Wright & Symmers *Systemic Path.* II. xxxiv. 1284/2 A feature of Schilder's disease is that the subarcuate fibres are spared, as in some cases of multiple sclerosis.

schilling¹. Add: Now, an Austrian unit of currency, equivalent to 100 groschen; a coin or note of (multiples of) this value.

1924 *Times* 23 June 11/1 The Austrian new Schilling..which is being issued to the public over the counters of the Austrian National Bank since Monday last [*sc.* 16 June]. 1932 *Daily Tel.* 8 Oct. 2/3 New bonds will not be issued for a smaller amount than 50 schillings. 1948 G. Crowther *Outl. Money* (ed. 2) ix. 312 In the countries that suffered the worst inflation, entirely new currencies were introduced (the reichsmark in Germany, the schilling in Austria, the pengö in Hungary, in place of marks and crowns). 1978 J. Irving *World according to Garp* v. 94 'It costs five hundred schillings,' the whore said.

Schilling² (ʃi·liŋ). *Med.* The name of Victor *Schilling* (1883–1960), German hæmatologist, used *attrib.* and in the possessive to designate a method of classifying and counting white blood cells, and the results so obtained; (proposed by Schilling in *Deutsch. med. Wochenschr.* (1911) XXXVII. 1159).

1922 *Jrnl. Amer. Med. Assoc.* 11 Mar. 769/2 (*heading*) The Schilling differential blood count. 1924 *Ibid.* 20 Dec. 2055/1 (*heading*) Schilling's hemogram. 1927 A. Piney *Rec. Adv. in Hæmatol.* 276/1 (*Index*), Schilling index. 1935 Whitby & Britton *Disorders of Blood* iv. 77 In Schilling's method all the data of an ordinary total and differential leucocyte count, as well as a simplified nuclear count, are correlated and considered in the form of a 'hæmogram'. 1972 F. Nour-Eldin *Haematol.* iv. 20/2 In practice, this method is more useful than the Schilling haemogram which is based on dividing the granulocytes into four groups.

Schilling³ (ʃi·liŋ). *Med.* [The name of Robert Frederick *Schilling* (b. 1919), U.S. physician, who described the test in 1953 (*Jrnl. Lab. & Clin. Med.* XLII. 946–7).] *Schilling test*, a test, used esp. for pernicious anæmia, in which a small oral dose of radioactively labelled vitamin B_{12} is followed by a much larger unlabelled dose administered intramuscularly: subsequent excretion of the label in the urine is reduced if there is malabsorption by the gut.

1955 *Gastroenterol.* XXIX. 654 The radioactive material..which appears in the urine under the conditions of the Schilling test has the same distribution coefficient between ammonium sulfate saturated urine and *n*-butanol as pure vitamin B_{12}-Co^{60}. 1976 *Lancet* 13 Nov. 1087/2 The Schilling test was repeatedly normal.

‖ **Schimpfwort** (ʃi·mpfvō³rt). Pl. Schimpfwörter (-vörtəɹ). [Ger., f. *schimpf* insult + *wort* WORD.] An insulting epithet, a term of abuse.

1949 R. K. Merton *Social Theory & Social Structure* v. 153 The community at large, however, evidently emphasizes the imperfections of bureaucracy, as is suggested by the fact that the 'horrid hybrid', bureaucrat, has become an epithet, a *Schimpfwort*. 1974 *Amer. Speech 1971* XLVI. 84 Reinhold A. Aman..lists under *Schimpfwörter*: emotive language and verbal aggression, including cuss words, swear words, terms of abuse, insults, [etc.]. 1978 *Verbatim* Winter 3/2 For the Nazis, the word provided a wonderfully Protean term of abuse, a *Schimpfwort* of unparalleled virtuosity.

‖ **schinken** (ʃi·ŋkən). [Ger.; cf. SCHINKEL.] German ham. Also in *Comb.*, as **schi·nkenwurst** (-vūrst), ham sausage.

1848 Thackeray *Van. Fair* lxii. 563 The little boy..consumed schinken, and braten, and kartoffeln, and cranberry jam. 1957 S. Strong *Good Food from Vienna* 89 (*heading*) Baked ham (Gebackener Schinken). 1962 [see *PICON]. 1967 M. Waldo *Internat. Encycl. Cooking* II. 549/1 *Schinkenwurst*.., ham sausage. 1978 *Sunday Times* (Colour Suppl.) 21 May 78/2 Schinkenwurst, fleischwurst and herb leberwurst will provide a wealth of tastes. 1979 P. Friedman *Termination Order* (1980) viii. 129 He got a plate of *schinken* and salad.

Schiötz (ʃyöts). *Ophthalm.* Also **Schiøtz.** The name of Hjalmar *Schiøtz* (1850–1927), Norwegian physician, used *attrib.* and in the possessive to designate a type of tonometer he devised for measuring the tension of the sclera, and to denote readings made with such a tonometer.

1913 Török & Grout *Surg. of Eye* vi. 173 More accurate information can be had by the use of a Schiötz tonometer. 1917 A. Duane *Fuchs's Text-bk. Ophthalm.* (ed. 5) II. i. 83 In Schiötz's tonometer a collar..bears at its lower end a concave plate..which is fitted to the curvature of the cornea. 1918 [see *HYPOTONIC *a.* 1 b]. 1964 [see *LACHRYMATE *v.*]. 1964 S. Duke-Elder *Parsons' Dis. of Eye* (ed. 14) x. 113 The type of tonometer should always be cited and the reading expressed in this form—20 mm. Hg (Schiötz).

schipperke. Substitute for def.: A small black dog belonging to the breed so called, distinguished by pointed, erect ears, a large ruff of longer fur on neck and chest, and usually a docked tail. (Later examples.)

1912 'Saki' *Unbearable Bassington* xiv. 263 A small black dog, something like a schipperke,..ran from behind my chair. 1950 A. C. Smith *About our Dogs* xxii. 327 The Schipperke Club standard states that the head is of the foxy type. 1976 A. Powell *Infants of Spring* iii. 56 The 'odd' lady..used to breed schipperkes, small black dogs from the Netherlands, with sharp ears and curly tails.

Schirmer (ʃɜ·ɹməɹ). *Ophthalm.* [The name of Otto *Schirmer* (1864–1917), German ophthalmologist, who proposed the test in 1903 (*Archiv f. Ophthalm.* LVI. 197).] *Schirmer('s) test*: a test in which the end of a strip of filter paper is placed on the surface of the eye over the lachrymal duct: the rate at which it is moistened indicates the rate of lachrymal secretion.

1935 *Trans. Amer. Ophthalm. Soc.* XXXIII. 428 Schirmer's test read 3 mm. O.S., 5 mm. O.D., in five minutes. 1941 *Amer. Jrnl. Ophthalm.* XXIV. 21/1 The Schirmer test shows zero to 6–8 mm. in 5 minutes, whereas the lower limit of the normal is 15 mm. according to Schirmer. 1977 *Lancet* 12 Nov. 1027/2 Schirmer's test, which is often reported in the assessment of eye complaints in patients who are taking beta-adrenergic-receptor blocking drugs, is misleading and inaccurate.

schism. The pronunc. (skiz'm), though widely regarded as incorrect, is now freq. used for this word and its derivatives both in the U.K. and in North America.

schismogenesis (sizmodʒe·nĕsis). *Anthrop.* [f. SCHISM *sb.* + -O + -GENESIS, after *biogenesis*, *parthenogenesis*, etc.] A term proposed for the origin of differentiation between groups or cultures caused by the reciprocal exaggeration of behaviour patterns and responses that may result in the destruction of social balance. Hence **schismoge·nic** *a.*

1935 G. Bateson in *Man* XXXV. 181 A position is set up in which the behaviour X, Y, Z, is the standard reply to X, Y, Z. This position contains elements which may lead to progressive differentiation or *schismogenesis*. 1936 R. Firth *We, the Tikopia* p. vii, Attempts are made to analyse cultures in terms of Schismogenesis. 1940 *Brit. Jrnl. Psychol.* Oct. 133 The growth or divergence between the two kinds of game [*sc.* rugby football]..is an excellent example of what Bateson has called 'schismogenesis'—the development of cultural traits in opposition and divergence. 1949 G. Bateson in M. Fortes *Soc. Structure* 47 In schismogenic theory it was tacitly assumed that the individuals would maximize intangible..variables such as prestige, self-esteem, or even submissiveness. 1969 B. McLaughlin *Stud. in Soc. Movements* 477 Norman Miller, 'Formal Organization and Schismogenesis', unpublished paper.

schistic, *a.*² For † *Obs.* read *rare* and add later example.

1931 *Discovery* Nov. 355/2 Around the melting snow in schistic soil *Androsace hedraeantha* was flowering in hundreds.

schistosity. (Earlier and later examples.)

1885 *Nature* 8 Oct. 558/2 The Arnaboll (Hebridean gneiss) can be traced..from spots where it retains its original strike and petrological characters, to others where it acquires the normal strike and mineralogical features of the ordinary Sutherland schists. The old planes of schistosity become obliterated, and new ones are developed. 1908 *Mineral. Mag.* XV. 145 To the imperfect substitution of dissolved kaolin by mica and secondary quartz and resultant gravitation under the pressure of superincumbent masses may, in like manner, be attributed schistosity in Zinnwald greisen. 1919 *Amer. Jrnl. Sci.* XLVII. 203 Geinitz called attention to the delicate ruffling of the surface of schistosity of many slates, which may represent an analogous phenomenon of friction. 1951 *Jrnl. Geol.* LIX. 68 The ice..forms along joints, planes of schistosity, or any other available fractures. 1977 A. Hallam *Planet Earth* 175/3 Deformation and the development of schistosity may result in the complete obliteration of primary planar features such as sedimentary bedding.

schistosome (ski·stosōᵘm). *Zool.* [ad. mod.L. *Schistosoma* (D. F. Weinland *Tapeworms in Man* (1858) 87), f. Gr. σχιστό-ς divided + σῶμα body; cf. *-SOME⁴.] Any member of the trematode genus *Schistosoma* (formerly *Bilharzia*), of which the cercariæ are parasitic on fresh-water snails and the adults of certain species are parasitic in man, inhabiting the blood vessels; a blood fluke.

1905 *Brit. Med. Jrnl.* 7 Jan. 11 (*caption*) Two male schistosomes. *Ibid.* 12/1 A distinctive feature of this schistosome is the absence of ciliated warts on the integument. *Ibid.* 12/2 The habitat of the new schistosome is mainly arterial. 1931 Blacklock & Southwell *Guide to Human Parasitol.* xvii. 144 Unlike all other digenetic trematodes, no rediae are produced at any time of the life history of the schistosomes, the asexual multiplication taking place in the sporocyst stage. 1955 *Nature* 19 Nov. 981/2 An important snail host in one country may be refractory or only slightly susceptible to the same species of schistosome from another region. 1968 *Sci. Jrnl.* Oct. 7/1 An infected human excretes anything from a few to a million schistosome eggs a day. 1970 G. R. Taylor *Doomsday Bk.* 92 The high schistosome infection rate in Lower Egypt.

Hence **schistoso·mal** *a.*; also **schistoso·micide** *Pharm.* [-CIDE], a substance which kills schistosomes; **schi:stosomici·dal** *a.*

1931 Blacklock & Southwell *Guide to Human Parasitol.* iii. 13 Protozoal, e.g. amoebic, and helminthic, e.g. schistosomal, dysenteries, are not the only forms which may occur in the tropics. 1954 *Trans. R. Soc. Tropical Med. & Hygiene* XLVIII. 446 (*heading*) A new series of schistosomicides. *Ibid.*, No other known schistosomicide shows a corresponding degree of activity. *Ibid.*, During the last seven years a search for drugs with schistosomicidal activity has been carried out in these laboratories. 1965 *Ann. Trop. Med. & Parasitol.* LIX. 304 (*heading*) Bacteriological and immunological findings in the presence of schistosomal infection. 1978 *Nature* 22 June 628/1 Despite the efforts ever since in the search for more effective schistosomicides, few drugs can be considered as antischistosomal agents of proven value. *Ibid.* 628/2 Although further studies are needed, oxamniquine seems to be an effective schistosomicidal oral drug suitable for use in endemic areas.

schistosomiasis (ski:stosomai·ăsis). *Path.* [f. mod.L. *Schistosoma* (see prec.) + *-IASIS.] Disease caused by infection with parasites of the genus *Schistosoma*, characterized by chronic symptoms esp. of the digestive and urinary systems, and sometimes by fever.

1906 *Philippine Jrnl. Sci.* I. 89 A second Chinese case of schistosomiasis has been recorded. 1963 O. Breland *Animal Life & Lore* ii. 110 The parasite causing snail fever (or schistosomiasis) must pass through stages in freshwater snails before it can attack man. 1977 J. Didion *Bk. Common Prayer* I. iv. 25 Isabel's children suffered gastrointestinal bleeding from..schistosomiasis. 1979 *Jrnl. R. Soc. Arts* Dec. 51/2 Zinc losses from the body are abnormally high due to a high incidence of hookworm infestation and schistosomiasis causing bleeding and excessive losses of sweat.

Hence **schistoso·mial** *a.*, pertaining to or characteristic of schistosomiasis.

1934 R. Girges *Schistosomiasis* v. ii. 194 It [*sc.* bladder irritability] may be mistaken for enlargement of that gland and remain undiagnosed as schistosomial for a considerable period.

schistosomulum (skistosǫ·miŭlŏm). *Zool.* Pl. **-somula.** Also anglicized as **-somule.** [mod.L., f. as prec. + L. *-ulum*, neut. of *-ulus*, diminutive ending.] A parasite of the genus *Schistosoma* which has entered its adult host but is not yet mature. Cf. prec.

Also erroneously used as *schistosomula* sing., *schistosomulæ* pl.

1924 Faust & Meleney *Studies on Schistosomiasis* I. 4 To test his hypothesis he attempted to recover schistosomula from the peripheral vein. 1934 R. Girges *Schistosomiasis* II. ii. 56 Entering the lymphatics or blood-vessels, schistosomula proceed..to the liver of the host. 1961 *Exper. Parasitol.* XI. 209/2 The gland cells appeared to collapse after exhaustion of their contents and no replacement of secretion was encountered even in schistosomules which had been in skin for as long as 11 days. 1961 *Jrnl. Parasitol.* XLVII. 891/2 This interval..allowed the schistosomulae to reach the hepatic portal system. 1975 *Nature* 6 Mar. 17/3 The stage in the life cycle most susceptible to the immune response is the schistosomula, the young form that penetrates the skin and migrates to the blood vessels

in which the adult develops. 1975 *Jrnl. R. Soc. Arts* May 363/2 The cercaria..then penetrate the skin, lose their tails and change to yet another stage called a schistosomule, which is much like an adult worm but smaller. 1978 *Parasitology* LXXVII. 282 The number and location of schistosomulum deaths will have a considerable influence on the pattern of migration out of the skin.

Hence **schistoso·mular** *a.*

1975 *Nature* 28 Aug. 727/1 Sera from five patients infected with *S*[*chistosoma*] *mansoni* were used as sources of anti-schistosomular antibody in different experiments.

schiz (skits), *sb.* and *a.* *slang* (chiefly *N. Amer.*). Also **schitz.** [Abbrev. of *SCHIZOID *a.* and *sb.* or *SCHIZOPHRENIC *a.* and *sb.*]

A. *sb.* A schizophrenic person; *spec.* one who experiences a drug-induced hallucination.

1955 [see *NEEDLE *sb.* 3 b]. 1967 A. Lurie *Imaginary Friends* xii. 174 How can you tell what a schiz like her is going to do? 1973 T. Pynchon *Gravity's Rainbow* 131 There's a long-time schiz..who believes that *he* is World War II.

B. *adj.* Schizophrenic.

1960 A. Huxley *Let.* 28 July (1969) 894, I imagine you cd find out a great deal—given your special knowledge of schiz symptoms. 1964 C. Hodder-Williams *Main Experiment* II. xvi. 207 Typically schitz, you know. They never can resist symbolic compulsions. 1969 —— *98·4* xii. 155, I took their damn drugs and..I nearly went schitz.

Hence **schi·tzy, schi·z(z)y** *a.*, schizophrenic; *spec.* exhibiting or suffering from the effects of hallucinogenic drugs.

1968 'R. Macdonald' *Instant Enemy* xxv. 157 What's the matter with her? Schitzy? 1972 D. Anthony *Blood on Harvest Moon* xxvii. 237, I feel schitzy. The grieving widow's mask—well, sometimes it's no mask. 1975 *New Yorker* 20 Jan. 31 This friend of mine—a bit of a schizzy dude, to be sure—has been telling me that if we go on muddying up the ecosphere, [etc.]. 1975 Shea & Wilson *Golden Apple* 142 The awkwardness of their first efforts would be published in all the psychiatric journals as proof of the regressive and schizzy nature of their unsocial and unnatural impulse toward walking. 1977 *Time* 3 Jan. 56/2 So does Director Pierson, as he captures the schizzy, druggy, enclosed, exploding tension of rock superstardom. 1979 D. Anthony *Long Hard Cure* xvi. 130 If you stay here long enough, you can get too dependent on the Retreat... You become a little schitzy yourself.

schizanthus (skitsæ·nþŏs). [mod.L. (H. Ruiz & J. Pavon *Floræ Peruvianæ et Chilensis Prodromus* (1794) 6), f. SCHIZO- + Gr. ἄνθος flower.] = *poor man's orchid* s.v. *POOR MAN 4 a.

1823 *Curtis's Bot. Mag.* L. 2404 (*heading*) Wing-leaved Schizanthus. 1908 *Garden* 30 May 263/3 Much has been done in the way of improving the Schizanthus. 1931 *Daily Tel.* 21 May 17/1 The pansy-flowered schizanthus. 1959 [see *poor man's orchid* s.v. *POOR MAN 4 a]. 1963 *Times* 1 May 14/4 The County Borough of East Ham occupies the middle of the hall with a group of schizanthus varieties and sweetly scented stocks. 1979 *Daily Tel.* 19 May 15/5 It is schizanthus, a thick bushy plant with leaves like parsley and a riotous show of flowers, pink, crimson, white and purple.

schizo (ski·tso), *sb.* and *a.* Slang abbrev. of *SCHIZOPHRENIC *a.* and *sb.* **A.** *sb.*

1945 N. Balchin *Mine Own Executioner* vi. 92 He just sank back into that queer not-really-there mood that schizos have. 1955 'E. C. R. Lorac' *Ask Policeman* vii. 87 Let's assume a split personality..a schizo, as they say nowadays. 1961 J. I. M. Stewart *Man who won Pools* III. 203 He might have been a schizo..for all the tie-up there seemed to be between the Phil of this rational conversation and the Phil who wanted Jean Canaway. 1972 'L. Egan' *Paper Chase* xii. 194 He had a long history of violence and was diagnosed as a schizo.

B. *adj.*

1957 M. Gair *Sapphires on Wednesday* x. 120, I think he must have a split mind—be schizo or something. 1958 A. Huxley *Let.* 16 Dec. (1969) 858 I'm glad to hear that your schizo research goes forward satisfactorily. 1960 *Harper's Bazaar* Aug. 50/2 He had gone stereo in April... Cheaper than going schizo. 1977 J. Aiken *Last Movement* v. 89 Gertrude was a kind of belle Otèro-cum-Phaedra..and Hamlet perfectly epicene and schizo.

schizo-. Add: **schizochro·al** *a.* *Palæont.* [Gr. χρώς skin], applied to certain trilobite eyes in which the cornea is divided to form several discrete lenses; **schi·zocœly** (-sīli) *Zool.*, schizocœlic mode of formation (of a cœlom).

1889 J. M. Clarke in *Jrnl. Morphol.* II. 254 The character of the visual area in the trilobites is twofold; (*a*) it may be covered by a smooth, continuous epithelial film or cornea, through which the lenses of the ommatidia are visible by translucence, and (*b*) the cornea may be transsected by the protrusion of the sclera and limited to the surfaces of the ommatidia.. The first group may be designated by the term Holochroal; the second group by the term Schizochroal. *Ibid.* 266 The schizochroal eyes of the Trilobites are aggregated and not properly compound eyes. 1976 *Nature* 13 May 130/1 Trilobites of the suborder Phacopina had schizochroal eyes, in which comparatively few large separate lenses are distributed over the eye surface. 1962 D. Nichols *Echinoderms* i. 14 A coelom..can arise as a split in the mesoderm (schizocoely) or as an outgrowth of the gut cavity or enteron (enterocoely). 1978 *Nature* 4 May 23/2 In this context, the mode of formation of the coelom (enterocoely, schizocoely, gonocoely) is of secondary importance.

2. *Psychol.* With pronunc. (skitso-, skidzo-). Used to repr. *SCHIZOPHRENIA, as in **schizota·xia** [Gr. τάξις order, arrangement], a genetically determined defect in the functioning of the nervous system which has been suggested as predisposing to schizophrenia; hence **schizota·xic** *a.* and *sb.*; **schi·zothyme** *sb.* and *a.* [Gr. θυμός mind, temper], (characteristic of) a person who is introverted and imaginative, and so regarded as tending to schizophrenia rather than to manic-depressive illness; hence **schizothy·mic** *a.*; also **schizothy·mia**, schizothymic constitution or temperament; **schi·zotype**, a personality type in which schizophrenia is potentially or actually present; hence **schizoty·pal, -ty·pic** *adjs.*; **schi·zotypy.**

1962 P. Meehl in *Amer. Psychologist* XVII. 830/1 This neural integrative defect, which I shall christen schizotaxia, is all that can properly be spoken of as inherited. *Ibid.*, The imposition of a social learning history upon schizotaxic individuals. *Ibid.* 831/1 All schizotaxics become, on all..existing social learning regimes, schizotypic in personality organization. 1966 I. B. Weiner *Psychodiagnosis in Schizophrenia* i. 7 Persons with schizotaxia acquire a personality organization called schizotypy that is characterized by four core behavior traits... These schizotypic traits are universally learned by all schizotaxic persons... Whereas most schizotypes remain compensated, those who are confronted with certain causal environmental influences ..are likely to decompensate into clinical schizophrenia. 1974 S. Arieti *Interpretation of Schizophrenia* (ed. 2) xlv. 697 A minority of schizotaxics..are 'potentiated into clinical schizophrenia'. *Ibid.*, Schizotaxia is a necessary but not sufficient condition in the etiology of schizophrenia. 1925, 1932 Schizothyme [see *cyclothyme* adj. and sb. s.v. *CYCLO-]. 1936 A. Huxley *Eyeless in Gaza* viii. 87 'What a lot of ribs you've got!' she said at last. 'Schizothyme physique,' he answered. 1952 H. Read *Philos. Mod. Art* iv. 84 If in the end we describe..Michelangelo as a typical 'schizothyme', the common reader is not much the wiser. 1964 I. M. Smith *Spatial Ability* vii. 229 He found the creative significantly more schizothyme, self-sufficient, withdrawn, sophisticated, desurgent and radical. 1972 *Encycl. Psychol.* III. 180/1 The schizothyme is characterized by..'a conscious contrast between the ego and the outside world', 'a touchy or indifferent withdrawal from the mass of his fellow men', the predominance of 'dreams, ideas or principles'. 1940 H. G. Wells *Babes in Darkling Wood* IV. ii. 335 Schizothymia, the psychoanalysts would have called this sort of dreaming. 1964 I. M. Smith *Spatial Ability* ix. 287 The hyperactivity.., nervousness and anxiety seem..more closely related to introversion or schizothymia than to extraversion. 1925 W. J. H. Sprott tr. *Kretschmer's Physique & Char.* xiii. 223 The group of wits..ironists and satirists whose nature is indicated by the names, Heine, Voltaire,..Nietzsche. This group belongs quite decidedly to the schizothymic side. 1951 *Mind* LX. 287 The ethical question is not whether one should be cyclothymic like Goering or schizothymic like Himmler in one's destructiveness; rather it is whether one should be destructive at all, and, if so, towards what. 1961 *Lancet* 23 Sept. 712/1 Hereditary factors were more important for excitability, the cyclothymic–schizothymic scale, and super-ego strength. 1953 S. Rado in *Amer. Jrnl. Psychiatry* CX. 409/2 In this sense the patient suffering from an open schizophrenic psychosis is a schizophrenic phenotype, engendered by a schizophrenic genotype in its interaction with the environment... For psychodynamic purposes I shall abbreviate the term schizophrenic phenotype to schizotype. *Ibid.* 410/1 The ensemble of psychodynamic traits peculiar to the schizotypes may be called schizotypal organization. 1962, 1966 Schizotypic, -typy [see *schizotaxia* above]. 1962 *Amer. Psychologist* XVII. 830/2 The most important research need here is development of high-validity indicators for compensated schizotypy. 1965 G. E. Daniels et al. *New Perspectives in Psychoanal.* 109 Variants of the schizophrenic disorders like—schizoid personality, schizotypal,..and pseudo-neurotic schizophrenia. 1974 S. Arieti *Interpretation of Schizophrenia* (ed. 2) xlv. 697 All schizotaxics become schizotypic in personality organization, but most of them do not decompensate and never develop a psychosis. 1978 P. O'Brien *Disordered Mind* iv. 75 Such syndromes are now officially classified as Schizotypal Personality Disorders.

schizo-affe·ctive, *a.* (*sb.*) *Psychol.* Also without hyphen. [f. SCHIZO- + AFFECTIVE *a.*] Exhibiting symptoms of both schizophrenia and manic-depressive psychosis. Also as *sb.*, a schizo-affective person.

1933 I. L. Polozker in *Amer. Jrnl. Psychiatry* XC. 123 We have been in the habit of labeling these cases as psychopathic personalities with schizoid make-up... I think the name of schizoaffective is more appropriate. 1933 J. Kasanin in *Ibid.* 126 My cases are not necessarily schizo-affective psychoses but schizoaffective personalities. 1965 J. Pollitt *Depression & its Treatment* iii. 38 Another criterion of schizo-affective disorder is the appearance of typical features of schizophrenia in circumscribed episodes. 1974 *Nature* 18 Jan. 160/2 All of the schizo-affectives had a history of at least five episodes of mania, hypomania or depression. 1976 Smythies & Corbett *Psychiatry* x. 185 Cases of schizoaffective psychosis where there is much admixture of depressive symptoms may require a tricyclic anti-depressant in addition. 1979 *Daily Tel.* 19 May 3/6 Dowdeswell had had a schizo-affective psychosis which had been cured but could recur.

schizoid (ski·tsoid, skidz-), *a.* and *sb.* *Psychol.* [a. G. *schizoid* (E. Kretschmer *Körperbau und Charakter* (1921) ix. 96): see *SCHIZO- 2 and -OID.] **A.** *adj.* Resembling or tending towards schizophrenia, but with milder or less developed symptoms, e.g. an absence of delusions.

1925 W. J. H. Sprott tr. *Kretschmer's Physique & Char.* xii. 208 One may for convenience call the transitional stages between illness and health, and the pathological abortive forms, 'schizoid' and 'cycloid'. 1931 *Times Lit. Suppl.* 17 Sept. 692/4 Professor Kretschmer manages to convey the impression that all philosophers and tragedians are schizophrenic, or at least 'schizoid'. 1938 *Oxford Times* 8 Apr. 23/5 He said Phillips was of what would be called 'schizoid type' but he could not agree that in the case of a split mind the subject could not distinguish between right and wrong. 1949 Koestler *Insight & Outlook* xxiv. 343 The frequent occurrence of infantile and schizoid features in the psychic make-up of poets. 1957 A. Huxley *Let.* 18 Nov. (1969) 830 Dr. Abram Hoffer..has treated several hundred patients under his care with 3 to 4 grammes of niacin—with striking success in many cases of schizoid neurosis. 1960 R. D. Laing *Divided Self* ix. 149 It is..not always possible to make sharp distinctions between sanity and insanity, between the sane schizoid individual and the psychotic. 1964 M. Argyle *Psychol. & Social Probl.* iii. 35 Withdrawn, schizoid people..produce a tense, uneasy atmosphere. 1976 Smythies & Corbett *Psychiatry* vi. 95 The 'schizoid' individual has usually been a lone wolf since childhood. 1977 A. Sheridan tr. *Lacan's Écrits* i. 5 The schizoid and spasmodic symptoms of hysteria.

b. *transf.* and *fig.*, freq. = *SCHIZOPHRENIC *a.* b.

1955 G. S. Fraser in J. Wain *Interpretations* 233 It is a kind of poem which could only have been written in the age that invented the phrase 'dissociation of sensibility' and that thought of the schizoid state as the typical occupational risk of intellectuals. 1959 *Times* 20 Feb. 14/5 It was a schizoid programme... On the one hand Daniel Jones's new fifth symphony..; on the other, a bizarre coupling of two..sets of variations on Paganini's celebrated A minor caprice. 1959 N. Mailer *Advts. for Myself* (1961) 173 'It's all schizoid,' Sam said. 'Modern life is schizoid.' 1960 *Spectator* 6 May 652 (*heading*) The schizoid state [*sc.* South Africa]. 1964 J. Jackson et al. *Rayden's Practice & Law of Divorce* (ed. 9) iii. 155 Such schizoid situations reflect little credit on the law. 1974 'R. Tate' *Birds of Bloodied Feather* ix. 176 He was schizoid..partly clever, partly stupid... I think he wanted to be found out. 1977 *Ripped & Torn* vi. 10/1 The best track..is 'Energy', a piece of schizoid trash. 1977 *Proc. R. Soc. Med.* LXX. 398/1 The principle..serves to emphasize that, if the tumour is to be cured, the surgeon must approach the problem in an almost schizoid frame of mind.

B. *sb.* A schizoid person; also *loosely*.

1925 W. J. H. Sprott tr. *Kretschmer's Physique & Char.* x. 149 We sometimes find schizoids, who look just as if they had already been through a schizophrenic psychosis before they were born. 1938 S. Beckett *Murphy* ix. 168 An emaciated schizoid..his left hand rhetorically extended.., his right, quivering and rigid, pointing upward. *a* 1941 F. Scott Fitzgerald *Tender is Night* (rev. ed., 1953) I. ix. 48 She's a schizoid—a permanent eccentric. *a* 1957 J. Cary *Captive & Free* (1959) xii. 57 Preedy has been taken to pieces by experts. They say, 'The typical schizoid—a little Hitler. You find him everywhere—the village boy who goes from Mass to do murder is the basic type.' 1970 *Science* 16 Jan. 251/1 Though unsatisfactory, the only means of identifying many—perhaps most—schizoids remains genealogical, and a clinical understanding of the schizoid can best be gained by reading descriptions of abnormal relatives of schizophrenics. 1975 D. Lodge *Changing Places* i. 6 Flown by pilots long gone over the hill, alcoholics and schizoids.

Hence **schizoi·dal** *a.*

1938 S. Beckett *Murphy* iv. 49 'That long hank of Apollonian asthenia,' groaned Neary, 'that schizoidal spasmophile.' 1973 F. Johnson *Alienation* ii. 63 The discovery of latent schizoidal themes can be found routinely in individuals whose social functioning would in no way suggest..the existence of such 'splitting'.

schizoidia (skitsoi·diă, skidz-). *Psychol.* [ad. G. *schizoidie* (E. Bleuler 1922, in *Zeitschr. f. die gesammte Neurol. u. Psychiatrie* LXXVIII. 373), f. *schizoid* *SCHIZOID *a.* and *sb.*: see -IA¹.] The schizoid state, esp. when regarded as caused by the same genetic disorder as schizophrenia.

1940 in Hinsie & Shatzky *Psychiatric Dict.* 475/1. 1970 *Science* 16 Jan. 253/1 By including schizoid disease (schizoidia), this hypothesis extends that of Slater... The view that schizoidia and schizophrenia are a single disease genetically is supported by their clinical similarity. 1973 McClearn & DeFries *Introd. Behavioral Genetics* xi. 275 Twin data are utilized to support the hypothesis that schizoidia and schizophrenia are manifestations of the same underlying genetic disease. 1976 Ehrman & Parsons *Genetics of Behav.* xi. 276 Schizoidia may be defined as a pre- or potentially schizophrenic mental state.

schizont (skəi·zǫnt). *Zool.* [a. G. *schizont* (F. Schaudinn 1900, in *Zool. Jahrb., Abt. f. Anat. u. Ontogenie* XIII. 213), f. Gr. σχίζ-ειν to split (cf. SCHIZO-) + ὀντ-, ὤν, pres. pple. of εἶναι to be, exist.] In Protozoa, a cell that divides asexually to form daughter cells; *esp.* in Sporozoa, a multinucleate cell that divides asexually to form merozoites.

1900 *Jrnl. R. Microsc. Soc.* June 336 In the author's [*sc.* Schaudinn's] nomenclature this process of asexual multiplication is known as schizogony, the mother cells are schizonts, and the daughter cells merozoites... They may grow rapidly.., and become converted into schizonts. 1912, 1957 [see *GAMONT]. 1974 *Nature* 22 Nov. 268/1 Three different sorts of vaccine are at present being investigated: irradiated sporozoites from the mosquito, extracts from schizonts (developing stages in the blood)

and emulsified merozoites (the stages which pass between blood cells).

schizonticide (skəiz-, skizǫ·ntisəid). *Pharm.* Also **schizonto-.** [f. prec. + *-icide*, as in *parricide, tyrannicide*, etc., or + -O + -CIDE.] A substance that kills schizonts.

1943 *Jrnl. Infectious Dis.* LXXIII. 11/2 In textbooks..quinine and atabrine are considered to be 'schizonticides', whereas plasmochin is called a 'gametocide'. 1944 W. N. Bispham *Malaria* x. 133 Quinine is not as satisfactory as atabrine as a schizonticide in the treatment of *P*[*lasmodium*] *falciparum*. 1963 E. Pampana *Textbk. Malaria Eradication* viii. 211 Drugs that act on the asexual forms, as schizontocides. 1970 W. Peters *Chemotherapy & Drug Resistance in Malaria* v. 139 The main outcome of this investigation was to pinpoint mepacrine as a safe and potent schizontocide, superior to quinine against all human malaria species. 1977 *Martindale's Extra Pharmacopoeia* (ed. 27) 343/2 The 8-aminoquinolines..have a marked effect on gametocytes..but are not effective blood schizonticides.

Hence **schizontici·dal, -oci·dal** *adjs.*

1963 P. F. Russell et al. *Pract. Malariol.* (ed. 2) xix. 503 Recent trials with relatively non-toxic schizonticidal drugs..have had some more promising results. 1963 *Terminol. Malaria* (World Health Organization) iv. 66 Schizontocides or blood schizontocides ('schizontocidal drugs'..) act on asexual erythrocytic stages of the parasite. 1979 *Amer. Jrnl. Trop. Med. & Hygiene* XXVIII. 937/1 By the early 1970's over 200,000 compounds had been screened for blood schizonticidal activity.

schizophrene (ski·tsofrīn, ski·dz-). *Psychol.* [ad. G. *schizophren*, f. *schizophrenie* *SCHIZOPHRENIA.] A schizophrenic, or a person with a predisposition towards schizophrenia. Also *attrib.* and *loosely*.

1925 W. J. H. Sprott tr. *Kretschmer's Physique & Char.* x. 147 In the schizophrene group, still less than in the circular, can we separate the healthy from the diseased. 1936 *Scrutiny* V. 248 Comparing the psychology of the schizophrene with that of the modern artist. 1945 *Times* 26 Apr. 6/5 No one would claim Beethoven as an example of the highly integrated personality but he is no schizophrene, as Dr. Carner calls Mahler outright. 1968 *Listener* 29 Aug. 260/2, I have heard a saloon-bar customer, a postman by profession, address the barmaid as: 'You dreamy schizophrene.' 1977 'D. Cory' *Bennett* ii. 76 Schizophrenes are often held to be people of exceptional charm.

schizophrenese (skitsofrĕnī·z, skidz-). *Psychol.* [f. next + -ESE.] Disordered speech as manifested by a schizophrenic.

1964 *Internat. Psychiatry Clinics* I. 829 The speech deficiencies are not consistent from one schizophrenic child to another. One is thus not justified in referring to 'schizophrenese'—that is, a specific and positive schizophrenic speech pattern. 1976 Smythies & Corbett *Psychiatry* v. 67 It is helpful to do an amytal or pentothal interview. The schizophrenic patient often loosens up and talks 'schizophrenese' (the typical speech disorder of schizophrenia).

schizophrenia (skitsofrī·niă, skidz-). *Psychol.* [ad. G. *schizophrenie* (E. Bleuler 1910, in *Psychiatrisch-Neurol. Wochenschr.* XII. 171), f. Gr. φρήν mind: see SCHIZO- and -IA¹.] A mental disorder occurring in various forms, all characterized by a breakdown in the relation between thoughts, feelings, and actions, usu. with a withdrawal from social activity and the occurrence of delusions and hallucinations.

Used in the U.S. with a broader meaning than in Britain (cf. quots. 1979, 1980).

The pronunc. (skits-), i.e. with short (i) and with (ts), is prob. influenced by the Ger. pronunc. (sxⁱits-).

1912 *Lancet* 21 Dec. 1730/1 This little volume is a translation of a series of articles by Professor Bleuler which appeared..during 1910 and 1911, in which he advances a theory of the negativism so frequently met with in dementia praecox or schizophrenia. 1925 J. Riviere tr. *Freud's Unconscious* in *Collected Papers* IV. 129 In schizophrenia a great deal is consciously expressed which in the transference neuroses can be demonstrated to exist in the Ucs only by means of psycho-analysis. 1944 E. A. Strecker *Fund. Psychiatry* vii. 123 In spite of the fact that schizophrenia and manic-depressive are divergent and alien to each other in psychopathology, there are clinical situations in which the differential diagnosis is difficult. 1945 W. Sadler *Mod. Psychiatry* xxxix. 464 Among the trends of schizophrenia is the persistent tendency to shun reality. 1958 M. Argyle *Relig. Behaviour* ix. 107 By the psychoses we mean the clinical conditions of schizophrenia, mania and depression, paranoia, epilepsy, together with certain organic states. 1964 *Internat. Psychiatry Clinics* I. 743 While many of these cases can be grouped into the classic forms of schizophrenia, such as simple, catatonic, hebephrenic, or paranoid, others cannot be so classified and will, therefore, be diagnosed as mixed or undifferentiated types of schizophrenia. 1979 *Internat. Rehabilit. Med.* I. 79/1 It was found that hospital psychiatrists in New York included under 'schizophrenia' part of what British hospital psychiatrists diagnosed as mania, psychotic depression, and personality disorder. 1980 J. Ashton *Everyday Psychiatry* v. 33 His [*sc.* Bleuler's] use of the word in a wide sense has influenced the practice of Swiss and American psychiatry to the present day, so that 'American schizophrenia' ranges from apparently minor personality disorders with a range of emotional reactions, through to the major deterioration of personality that is recognized as schizophrenia by British psychiatrists.

b. *transf.* and *fig.*

1933 T. S. Eliot *Use of Poetry & Use of Criticism* v. 99 For a poet to be also a philosopher he would have to be virtually two men; I cannot think of any example of this thorough schizophrenia, nor can I see anything to be gained by it. 1945 'G. Orwell' in *Polemic* I. 40 Some nationalists are not far from schizophrenia, living quite happily amid dreams of power and conquest which have no connection with the physical world. 1949 *Here & Now* (N.Z.) Oct. 32/2 There are few alien hills in Mr Witheford's poems and he is not preoccupied with cultural schizophrenia. 1958 *Listener* 9 Oct. 557/1 They admire big dams and high buildings..and the *contrast* these afford to the familiar buildings... It may occur to you that the character of our environment is likely to be split in two by this schizophrenia. 1969 *Daily Tel.* 6 Oct. 9/5 Bristol's Little Theatre illustrates the same provincial schizophrenia, with the farce 'One for the Pot' next in the bill after Ibsen's 'Master Builder'.

schizophrenic (skitsofre·nik, skidz-), *a.* and *sb.* [f. prec. + -IC.] **A.** *adj.* **a.** *Psychol.* Characteristic of or having schizophrenia.

1912 [see *AUTISM]. 1927 Henderson & Gillespie *Text-bk. Psychiatry* ix. 218 It is now generally recognised that although a schizophrenic type of disturbance is always most serious, there are certain cases which can, and do, readjust themselves. 1931 [see *SCHIZOID *a.* a]. 1945 *Times* 28 Sept. 7/5 He was schizophrenic long before the thing became fashionable, half of him being entirely rational, the other half living in a world in which it was taken for granted that pigs have wings. 1973 I. L. Child *Humanistic Psychol.* ix. 137 Laing and Esterson..argue that schizophrenic behavior appears in these patients as a somewhat sensible response to an extremely difficult situation. 1974 Passmore & Robson *Compan. Med. Stud.* III. xxxv. 55/2 Other examples are the 'schizophrenic smile', which appears without obvious external cause and is presumed to be a response to an internal hallucinatory stimulus, and the 'schizophrenic handshake', the patient's hand when grasped remaining limp. 1981 *Brit. Med. Jrnl.* 24 Jan. 313/3 While drugs have certainly facilitated the extramural care of schizophrenic patients the minimisation of prolonged inpatient treatment has, to a large extent, been due to social measures and to changes in attitude within the psychiatric services.

b. *transf.* and *fig.*, freq. with the implication of mutually contradictory or inconsistent elements.

1955 *Sci. Amer.* Oct. 113/1 The behavior of the puzzled Board reflected its schizophrenic task. The members performed as part jury, part judge, and then as part administrative agency, engaged in a part rule-making, part quasi-judicial proceeding. 1960 *Times* 13 June 14/1 It was a schizophrenic day when nearly every player seemed to live two lives. 1962 A. Lurie *Love & Friendship* viii. 155 You're not living two different lives that don't match... For me it's absolutely impossible. It's schizophrenic. *a* 1974 R. Crossman *Diaries* (1977) III. 71 We are all deeply schizophrenic on this Bill, hate the interference, hate the break with the trade unions, yet we can see that without it there must be a higher level of unemployment than we can tolerate. 1978 M. Shanks *What's Wrong with Mod. World?* iii. 45 In their reaction to inflationary pressures government have been..schizophrenic. On the one hand they have sought..to fight them. On the other hand they have felt obliged to compensate the victims. 1980 *Daily Tel.* 24 July 11/5 The work is schizophrenic in its switches of style from genuine opera-drama to operetta and then to the typical vehicle for a soprano anxious and able to sing Ophelia's mad scene.

B. *sb.* A person with schizophrenia.

1926 W. McDougall *Outl. Abnormal Psychol.* xxiii. 384 The delusions and hallucinations of the schizophrenic so commonly concern his body. 1953 W. Burroughs *Junkie* (1972) x. 111 One young schizophrenic had both hands fastened in front with a bandage so he could not bother the other patients. 1956 A. Huxley *Heaven & Hell* 84 Many schizophrenics pass most of their time..in a shadowy world of phantoms and unrealities. 1958 M. Argyle *Relig. Behaviour* ix. 109 Schizophrenics..are more chaotic and harbour a number of unrelated fantasies and identifications simultaneously. 1979 N. Scheper-Hughes *Saints, Scholars & Schizophrenics* iii. 69/1 Interviews with Irish schizophrenics support the hypothesis that the later age of onset of the disease in rural Ireland is related to the postponed adulthood..of the Irish bachelor.

Hence **schizophre·nically** *adv.*, in a manner suggestive or characteristic of schizophrenia.

1963 *Times* 23 Apr. 16/1 Ionesco's hero—perpetually exhausted, always eating, schizophrenically incapable of action. 1975 *Gramophone* Nov. 790/1 Then there's what one might call a strange psychological world in which almost schizophrenically Sibelius uses brightness and lightness juxtaposed with the darkest and most ferocious gestures. 1979 *Times* 27 Dec. 11/5 Schizophrenically Janus-like, we offer at least two different faces towards a policeman.

schizophreniform (skitsofre·nifǭɹm, skidz-), *a.* *Psychol.* [f. as prec. + -FORM.] Resembling schizophrenia.

1937 G. Langfeldt *Prognosis in Schizophrenia* i. 17 The author..is of opinion that it will be advantageous to separate *atypical* conditions, and give them their own description, such as 'Schizofreni-form' (or 'Schizophrenic reaction types'). 1951 *Practitioner* Aug. 135 Schizophreniform conditions.—Usually acute, dramatic and often running a benign course, even without much active treatment. 1976 Smythies & Corbett *Psychiatry* iv. 37 Even normal people can experience a short-lived schizophreniform psychosis as the result of unbearable stress—as in battle exhaustion—or severe sleep deprivation. 1978 P. O'Brien *Disordered Mind* ii. 35 If the disorganized state of mind lasts..less than 6 months, but more than one week, it is called a schizophreniform disorder.

schizophrenogenic (ski:tsofrīnodʒe·nik, ski:dz-), *a.* *Psychol.* [f. as prec. + -O + *-GENIC.] Tending to give rise to schizophrenia.

1949 F. Fromm-Reichmann in *Psychiatry* XI. 265/2 The schizophrenic is painfully distrustful and resentful of other people, due to the severe early warp and rejection he encountered in important people of his infancy and childhood..mainly in a schizophrenogenic mother. 1956 *Behavioral Sci.* I. 263/1 Whenever the system is organized for hospital purposes and it is announced to the patient that the actions are for *his* benefit, then the schizophrenogenic situation is being perpetuated. 1975 Hirsch & Leff *Abnormalities in Parents of Schizophrenics* vi. 95 This study provides somewhat stronger evidence against the concept of the cold, aloof, hostile schizophrenogenic mother. 1979 B. Ineichen *Mental Illness* ii. 41 The question of whether schizophrenics are downwardly mobile socially, or whether lower-class culture is schizophrenogenic, remains an open one.

schizostylis (skizostəi·lis). Pl. **-stylis.** [mod. L. (Backhouse & Harvey 1864, in *Curtis's Bot. Mag.* XC. 5422), f. SCHIZO- + L. *stilus* (*stylus*) (see STYLE *sb.* 8), in allusion to the split styles of the plant.] A rhizomatous herb of the genus so called, belonging to the family Iridaceæ, native to South Africa, and bearing linear leaves and spathes of red or pink flowers. Cf. *Kaffir lily* s.v. *KAFFIR 4.

1864 *Curtis's Bot. Mag.* XC. 5422 Crimson Schizostylis... This lovely Iridaceous plant..inhabits eastern rivers of South Africa. 1961 *Amateur Gardening* 16 Sept. 7/4 Less romantic in association but flowering later and valuable for October and November colour, are the Caffre lilies, or schizostylis. 1979 *Daily Tel.* 27 Oct. 31/2 Earlier flowering herbaceous plants, which the schizostylis have outshone.. with the onset of autumn.

schizzy, schizy: see *SCHIZ *sb.* and *a.*

‖ **schlag** (ʃlāg, ʃlāk). Abbrev. of *SCHLAGOBERS or *SCHLAGSAHNE.

1969 A. Arent *Laying on of Hands* xii. 138 She debouched at Demel's for pastry and *Schlag*. 1977 *Time* 24 Oct. 44/3 A slavic sour cream lay over the proceedings in place of Viennese *schlag*.

schlag, var. *SCHLOCK.

‖ **schlagobers** (ʃlā·gōbers). [Ger. dial., f. *schlagen* to beat + *obers* cream.] Whipped cream; coffee with whipped cream. Also *fig.*

1938 J. Flanner in *New Yorker* 10 Sept. 55/1 You can now get a seat and *Schlagobers* on Tomaselli's terrace. 1967 *Listener* 27 July 123/3 This *Schlagobers* made one thirsty for pop. 1969 *Harper's Bazaar* Oct. 12/1, I consumed huge quantities of their *Torte* heaped high with *Schlagobers*.

‖ **schlagsahne** (ʃlā·gzānə). [Ger., f. *schlagen* to beat + *sahne* cream.] Whipped cream.

1907 M. A. von Arnim *Fräulein Schmidt & Mr. Anstruther* 9 We are poor... If we were not..we should have different sorts of puddings..with *Schlagsahne* on their tops. 1936 D. Barnes *Nightwood* i. 12 The inevitable are produced by heavy rounds of burgundy, schlagsahne, and beer. 1972 J. Eastwood *Henry in Silver Frame* xviii. 155 The delicious fraulein..become[s] a devotee of *schlag-sahne*.

‖ **schlamperei** (ʃla·mpĕrəi). [Ger.] Indolent slovenliness, muddleheadedness; esp. designating a supposed south German and Austrian characteristic.

1961 *Economist* 30 Dec. 1282/2 A horrific picture of Austrian *Schlamperei* in a provincial court. 1966 F. Spiegl in F. Shaw et al. *Lern Yerself Scouse* 12 Here the Scouser's sense of humour outweighs his *Schlamperei*, for he will never use a single word when he can think of some more or less long-winded picturesque phrase in its place. 1974 *Times Lit. Suppl.* 28 June 687/3 Surely it was not all *sachertorte* and *schlamperei* among the villas of Gringing?

schlemazl, var. *SCHLIMAZEL.

schlemiel (ʃləmī·l). *colloq.* Also **schlemihl, shlemiel.** [Yiddish, possibly ad. Heb. *Shelumiel*, name of a person in the Bible (Num. i. 6) said by the Talmud to have met with an unhappy end; perh. influenced by the name of the eponymous hero of A. von Chamisso's *Peter Schlemihls wundersame Geschichte* (1814).] An awkward, clumsy person, a blunderer; a 'born loser'; a 'dope' or 'drip'. Also *attrib.*

1892 I. Zangwill *Childr. Ghetto* I. I. i. 30 The withered old grandmother..cursed her angrily for a *Schlemihl*. 1898 A. M. Binstead *Pink 'Un & Pelican* xi. 247 He also was what the Yids call a shlemiel; no matter what he turned his hand to, nothing ever came of it. 1932 *N.Y. Times* 10 Nov. 23/7 If they expect to beat me by having their names writ in, they're schlemiels—saps, if you get me. 1941 B. Schulberg *What makes Sammy Run?* iv. 68 Don't talk like a schlemiel, you schlemiel. Sounds like you're letting them push you around. 1959 *Times Lit. Suppl.* 6 Nov. p. xxxv/2 Bellow's..free-swinging translation of 'Gimpel, the Fool'—probably the best *schlemiel* story in the literature. 1963 T. Pynchon *V.* i. 37 Only something that, being a schlemihl, he'd known for years: inanimate objects and he could not live in peace. 1969 L. Michaels *Going Places* 21 A hundred fifty-five pounds of stomping shlemiel. 1972 *Listener* 14 Sept. 339/3 A schlemiel is a man who falls on his back and breaks his nose. Or you can say 'When a schlemiel leaves the room, you feel as if someone came in.' 1973 *New Society* 11 Oct. 95/1 The choice of making a fool of himself or being made a fool of by others, being a schmuck or a schlemihl. 1978 I. B. Singer *Shosha* iii. 50 You should have taken the whole five hundred. To him that's a trifle. He'll think you're a shlemiel.

schlemozzle, var. *SHEMOZZLE.

schlenter (ʃle·ntəɹ), *sb.* and *a.* Also **schlanter, shlanter, shlenter, shlinter, sl-**. [Poss. ad. Afrikaans or Du. *slenter* knavery, trick. The history of this word is obscure; the Austral. and N.Z. forms are possibly borrowed from S. Afr. English, but by what route is not clear.]

A. *sb.* **1.** *Austral.* and *N.Z. colloq.* A trick.

1864 C. R. Thatcher *Invercargill Minstrel* 15 'Twas a 'shlinter' for the tenant one morning departed Without paying his rent. 1919 W. H. Downing *Digger Dial.* 45 *Slanter* (or *schlanter*), a trick. 'To run a schlanter'—to make no genuine effort to win a game. 1925 A. Wright *Boy from Bullarah* 133 'A shlanter' he bellowed. 1934 *Bulletin* (Sydney) 20 June 47/1 'You worked a schlenter on me, laddie,' he said, grinning at me in the wings, 'but you're forgiven.' 1945 *N.Z. Geographer* I. 1. 24 Most [shearing] sheds have somebody articulate to voice their worries if any slinters are feared. 1959 G. Slatter *Gun in Hand* xii. 166 Wilkinson..worked a slinter at the end. Ref shoulda penalised him. 1965 F. Hardy *Yarns of Billy Borker* 70 (*title*) The greatest slanter in the history of the racing game.

2. *S. Afr.* Something counterfeit; *spec.* a counterfeit diamond.

1892 J. R. Couper *Mixed Humanity* 263 A new branch of industry had started in Kimberley, the manufacture of 'schlenter' stones, a name given to diamonds made of glass. 1898 *Cape Argus* (Weekly ed.) 16 Mar. 35 (Pettman), A small sack containing bars of gold or schlenter. 1937 H. Klein *Stage-Coach Dust* x. 112 Schlenters were also useful to the individual digger to drop into the pans of their rotating washing machine, to test the honesty of their native boys; and they were also useful to a more unscrupulous class to 'salt' diamondless claims. 1946 L. G. Green *So Few are Free* ix. 127 That is the trade in 'schlenters', bits of glass shaped roughly from bottle stoppers to resemble diamonds. They have none of the peculiar soapy feel of the genuine diamond, but they pass muster sometimes in a hurried deal at night. 1969 J. M. White *Land God made in Anger* 131 *Schlenters*, or *slenters*, are false diamonds. The best Schlenters in South West are made from the marbles in the necks of the lemonade or mineral-water bottles that can be found in dozens at the old German diggings.

B. *adj.* Dishonest, crooked; pretended, counterfeit, fake. *Austral., N.Z.* and *S. Afr. colloq.*

1889 Williams & Reeves *Colonial Couplets* 51 Broke! Broke! Broke! At the will of the C.J.C. For the slenter race with the favourite dead Will never come back to me. 1891 A. de Brémont *Gent. Digger* viii. 99 'Of course,' whispers the seller who had pushed his way to the side of the buyer, 'this sale was only *shlenter*.' 1900 J. Scott *Tales Colonial Turf* 35 [These race-course rogues] can draw deductions so beautifully, piecing together imaginary 'schlenter goes', and 'put-up jobs' with the cleverness of a whole courtful..of lawyers. 1916 C. J. Dennis *Songs Sentimental Bloke* 55 The slanter game I'd played wiv my Doreen—.. I seen wot made me feel fair rotten mean. 1924 L. Cohen *Reminiscences of Johannesburg & London* viii. 166 Confidence men found customers in plenty for schlenter gold bricks and amalgam. 1932 *Zionist Record* 25 Our courts employ schlenter as a word requiring no further definition, in the sense of fake when applied to mineral products. 1974 *Sunday Times* (Johannesburg) 24 Nov. 4 What makes the event more gratifying still is the fact that they sold schlenter uranium.

schlep (ʃlep), *v. colloq.* Also **schlepp, shlep.** [Yiddish *shlepn*, ad. G. *schleppen* to drag.] **a.** *trans.* To haul, carry, drag. Also *transf.* and *fig.*

1922 Joyce *Ulysses* 48 She trudges, schlepps, trains, drags..her load. 1931 L. Steffens *Autobiogr.* I. xix. 137 By this means the tuglike *Schlepper* schlepped a string of cargo boats up the Neckar to Heilbron. 1966 *New Statesman* 19 Aug. 261/3, I have a dread of being a martyr. Let them *schlep* Sonny Liston instead. 1973 *Jewish Chron.* 19 Jan. 11/1 The first thing you remember to do when shown the studio floor is to schlepp the book—out of the briefcase.. and at a right moment casually hold up the volume. 1973 *Publishers Weekly* 26 Feb. 125/1 The one thing you would not want to schlep along on a backpacking trip is this book, which runs to over 340 pages. 1975 *New Yorker* 11 Aug. 32/1 When her husband, Sidney, was alive he sustained a rupture, and Mrs. Singer says she had to schlepp him in and out of bed several times a day. 1975 R. H. Rimmer *Premar Experiments* (1976) i. 68 Merle schleps cocktails at the Persian Room in the Sheraton between six and midnight. 1977 G. Marton *Alarum* 189 The CIA schlepped you from Moscow to Washington.

b. *intr.* To toil, to 'slave'; to go or travel with effort, to traipse. Also with quasi-*obj.*

1963 'R. L. Pike' *Mute Witness* x. 172, I waste a whole evening *schlepping* around with him. 1964 W. Markfield *To Early Grave* iii. 54 My destiny, my fate..to *shlepp* for her. 1964 S. Bellow *Herzog* 136 Why should I *schlepp* out my guts? 1972 D. E. Westlake *Cops & Robbers* (1973) 137 We don't both have to hang around. Why don't you shlep on back to the station. 1978 J. Pascall *Illustr. Hist. Rock Music* 15 As he schlepped his weary way from date to date.

Hence **schle·pping** *vbl. sb.*

1977 *New Society* 3 Mar. 454/3 The endless flat-footed schlepping you have to do at Gatwick or Chicago O'Hare.

schlep (ʃlep), *sb.*[1] *U.S. colloq.* Also **schlepp, shlep.** Abbrev. of *SCHLEPPER.

1939 *News Letter & Wasp* 23 June 13 The name of the radio character known to thousands, Schlepperman, is evidently a personification of 'schlep', which means a poor slob. 1963 T. Pynchon *V.* iv. 104 'Quiet, shlep,' said the doctor, scrubbing. 1977 *New Yorker* 19 Sept. 80/3 My teacher can just zero in on one phrase, and it's immediately obvious that what I've done is so immature it makes me feel like an absolute schlepp.

schlep (ʃlep), *sb.*[2] *colloq.* (chiefly *U.S.*). Also **schlepp.** [Yiddish, prob. f. *SCHLEP *v.*] A troublesome business, a piece of hard work.

1964 *Economist* 1 Aug. 449/3 It was a schlep to find out. 1973 L. Snelling *Heresy* ii. ii. 68 Who thought up this *schlepp* with the sign, anyway? 1976 *National Observer* (U.S.) 4 Dec. 19-B/3 Anybody who has ever tried to make even a small amount of a classic brown sauce from scratch would probably agree with Liederman's assessment that 'it's the ultimate schlep'.

schlepper (ʃle·pəɹ). *colloq.* (chiefly *U.S.*). Also **shlepper.** [Yiddish, f. *SCHLEP *v.*: see -ER[1].] A person of little worth, a fool, a 'jerk'; a pauper, a beggar, a scrounger; an untidy person; (see also quot. 1934).

1934 *Amer. Speech* IX. 284/1 A customer who shops from store to store continually trying on shoes but not buying is known as a *shlepper*. 1949 S. J. Perelman *Westward Ha!* i. 13 In vain I protested that my dependents would be reduced to beggary; the editor's face remained flinty. 'About time those *schleppers* went to work,' he grunted. 1950 G. Marx *Let.* 20 Mar. (1967) 72 The paupers, or schlepper crowd, still hang on to their portable radios, but unfortunately they're not the ones who buy Chryslers. 1954 *Ibid.* 4 Aug. 59 Women always seem so much more joyous than men when another schlepper gets hooked. 1968 L. Rosten *Joys of Yiddish* 346 Hike up your slip; straighten your seams; you look like a shlepper. 1973 *Jewish Chron.* 19 Jan. 10/4 A 'star', you should pardon the expression, is never short of schleppers. And schleppers are like the tides of the ocean. If you make a hit film, they come in and almost drown you. If you make a flop, they recede into the distance. 1977 *Rolling Stone* 24 Mar., I've got a message for the Penelopes of this world. It's high time they say to their Ulysseses, 'Okay Schlepper, you've been around the world, your turn to keep the home fires burning, I'm splitting on my own trip for a while.'

schlich (ʃlɪχ). *Metallurgy.* [Ger.: see Slike *sb.*] = Slick *sb.*[2]

1677 E. Browne *Travels Germany* 135 They have also.. *slich*, or pounded and washed ore. 1757 tr. *Keysler's Travels* IV. 65 This method of burning of the *Schlich* saves considerable charges. 1839 Ure *Dict. Arts* 814 Water..is made to flow with greater or less velocity and abundance over the schlich or pasty mud spread on a table of various inclination. 1855 J. R. Leifchild *Cornwall* 207 Should the product (called *schlich*) seem tolerably rich, the operative turns the table round its axis. 1920 A. H. Fay *Gloss. Mining & Mineral Industry* 595/1 Schlich, finely pulverized ore; mud.

schlicht (ʃlɪχt), *a. Math.* [a. G. *schlicht* simple, plain.] (See quot. 1944.)

1944 J. E. Littlewood *Lect. Theory of Functions* i. 120 A function is called 'schlicht' in *D* if $f'(z) \neq 0$ in *D*, and $f(z_1) \neq f(z_2)$ for distinct points z_1, z_2 of *D*. Or: if $f(z) - \alpha = 0$ has never more than one solution (counting multiplicities) for *z* of *D*. 1968 E. T. Copson *Metric Spaces* vii. 86 An important instance of a bijection is the schlicht function of complex variable theory.

Schlieffen (ʃli·fən). The name of Alfred, Graf von *Schlieffen* (1833–1913), German general, used *attrib.* of a plan for the invasion and defeat of France that was formulated by him before 1905 and applied, with modifications, in 1914.

1919 A. P. F. von Tirpitz *My Memoirs* II. xvii. 289 The Schlieffen plan of attacking France through Belgium was intended to stave off from Germany the first vital danger. 1926 *Encycl. Brit.* Suppl. III. 479/1 How armies are to be handled in the Schlieffen spirit the war on the Eastern front showed. 1931 W. S. Churchill *World Crisis* VI. vi. 89 He drew up the celebrated 'Schlieffen Plan' in which the whole strength of Germany was to be directed from the outset with the utmost rapidity upon France by means of a wheeling movement through Belgium. 1965 A. J. P. Taylor *Eng. Hist. 1914–45* xiv. 484 He planned to attack on the extreme right, according to the Schlieffen model. 1977 *Listener* 4 Aug. 140/2 In the Schlieffen plan, the railways..took troops to Belgium and northern France.

schliere (ʃli·rə). *rare* in *sing.* Also **Schliere.** Pl. **-n**, and erron. **schliere, schlierin.** [Ger., f. regional *schliere* (fem.) striæ, streaks, corresp. to *schlier* (masc.) marl, f. early new HG. *schlier* (masc. and neut.), f. MHG. *slier* mud, related to MHG. *slier, sliere* ulcer, f. OHG. *sclierrun* (dat. pl.).] **1. a.** *Petrol.* An irregular streak or mass in igneous rock differing transitionally from its surroundings in texture or composition, and usu. elongated by flow.

[1885 A. Geikie *Text-bk. Geol.* (ed. 2) 94 Streaked, arranged in streaky inconsistent lines (Germ. Schlieren), either parallel or convergent, and often undulating.] 1888 J. J. H. Teall *Brit. Petrogr.* ii. 40 The differential motion of the lava will tend to drag out any parts of exceptional composition into the form of streaks or elongated lenticles (schliere). 1898 *Jrnl. Geol.* VI. 794 In the granites of Essex county are found in abundance streaks (Schlieren) and rounded rock masses of darker color and of finer grain than the surrounding rock. 1937 *Mem. Geol. Soc. Amer.* No. 5. 25 Sheet-like bodies, in which certain minerals appear in abnormal proportions, are called flow layers, or schlieren. 1966 *McGraw-Hill Encycl. Sci. & Technol.* VII. 15/2 Schlieren may represent early segregation drawn out by magma flow. Some may be xenoliths more or less digested and reworked by magma. Others may represent residual magmatic liquors of different composition injected into already crystallized portions. Schlieren formed in solid rocks are more properly metamorphic or metasomatic features.

b. A zone or stratum in a transparent medium whose density differs sufficiently from that of the surrounding medium for it to be detectable by refraction anomalies, usu. in consequence of pressure or temperature differences or composition inhomogeneities.

1895 C. S. Palmer tr. *Nernst's Theoret. Chem.* I. v. 121 If one adds, by means of a capillary pipette, a drop of a strong solution of potassium ferro-cyanide to a moderately strong solution of copper sulphate, one can see with the naked eye that a *schliere* (i.e. thin layer) of concentrated solution of copper sulphate flows downwards. 1946 F. Schneider *Qualitative Organic Microanalysis* ii. 22 The appearance of schlieren indicates the presence of impurities. 1949 *Proc. R. Soc.* A. CXCVII. 485 However imperfect an electron lens may be from the point of view of theoretical optics, it can contain neither dust nor 'schlieren', as the electromagnetic field smoothes itself out automatically. 1965 G. J. Williams *Econ. Geol. N.Z.* x. 149/1 The chromite occurs as..sporadic small narrow schlierin paralleling the enstatite crystal lamination... 1 to 3 mm schlierin in dunite show a local concentration of chromite. 1967 *Oceanogr. & Marine Biol.* V. 107 These variations may appear as 'Schlieren', that is, as thin bands of water at the surface or at any depth.

2. *attrib.* uses of pl. *schlieren*, with reference to an experimental method for the observation and recording of schlieren in transparent media, in which the specimen is illuminated with a collimated beam of light, and the diffraction pattern resulting from localized refraction of light rays by the schlieren is photographed or displayed on a screen, as *schlieren apparatus* [ad. G. *schlieren-apparat* (A. Töpler *Beobachtungen nach einer neuen optischen Methode* (1864) 16)], *illumination, method, photograph, photography, picture, system, technique.*

1895 C. S. Palmer tr. *Nernst's Theoret. Chem.* I. v. 121 Tammann observed the osmotic stream produced by the changes of concentration by means of a so-called *schlieren* apparatus. [*Translator's note*] This term, for which I find no concise English equivalent, is in common use in Germany to denote a delicate apparatus of Töpler used to detect small differences in the refractive power of the different layers ('schlieren') of heterogeneous media. 1933 *Jrnl. Scientific Instr.* X. 381 (*caption*) General arrangement of 'schlieren' apparatus set up for photographic or screen observation. 1971 *Sci. Amer.* May 118/1 In its simplest form the schlieren apparatus consists of a light source, two lenses, a pair of knife-edges and a sheet of photosensitive film. 1966 D. G. Brandon *Mod. Techniques Metallogr.* 18 Some increase in sensitivity can be obtained if the stop is displaced from the image of the condenser aperture, so that the direct beam is merely reduced to the same intensity as the diffracted beam and not completely eliminated. This system is known as schlieren illumination and can be used to give accurate information on surface tilt. [1899 *Phil. Mag.* XLVIII. 218 (*heading*) Photography of sound-waves by the 'Schlieren-Methode'.] 1933 *Jrnl. Scientific Instr.* X. 378 The 'Schlieren' method is an old but little-known method of rendering visible either colourless fluids, which have a different refractive index from their surrounding medium, or variations of refractive index or thickness of transparent solids. 1940 *Nature* 29 June 1021/1 The differentiation of the native proteins in the [egg] white was attempted on the basis of ionic mobilities by the method of Tiselius. The migration of the boundaries was followed optically by the 'schlieren' method using a sodium vapour lamp. 1962 *New Scientist* 6 Dec. 576/1 The effects that make optical schlieren methods feasible become negligible in gases at very low pressures. 1953 *Proc. Inst. Mech. Engineers 1951–2: Automobile Div.* 97/2 Observations by Miller of schlieren photographs taken during a very violent knocking combustion show a normal progression of the flame about three-quarters of the way across the combustion chamber. 1970 *New Scientist* 18 June 581/1 Ultrasonic frequencies up to 40kHz radiated at the base of a roaring jet of burning gas have a marked effect on the flame, altering its appearance and cutting down the noise produced. This is clearly illustrated in the two sets of spark schlieren photographs of town gas diffusion flames... An acoustic frequency of 38kHz is responsible for the remarkable alteration seen in the schlieren pictures. 1931 *Trans. Inst. Mining Engin.* LXXX. 18 Experiments carried out with Schlieren photography at Buxton are being supplemented at the U.S. Bureau of Mines Explosives Station at Pittsburgh. 1937 *Jrnl. R. Aeronaut. Soc.* XLI. 621 By using 'Schlieren' photography the air flow, fuel injection and flame formation were recorded simultaneously. 1979 *Nature* 29 Mar. 384/2 Drs Clark and B. J. Mullan used Schlieren photography to look at the air flow in

and around cabinets running with and without an operator. 1957 Liepmann & Roshko *Elem. Gasdynamics* vi. 161 Schlieren pictures are seldom used for a quantitative evaluation of density. They are, however, indispensable for obtaining qualitative understanding of flows. 1966 *McGraw-Hill Encycl. Sci. & Technol.* XII. 68/1 Numerical values of density can be obtained only from schlieren pictures of airflow about two-dimensional models or about simple axisymmetric models. 1970 Schlieren picture [see *schlieren photograph* above]. 1949 O. G. Sutton *Science of Flight* 203 Two other methods, the schlieren system and the interference method, are also in common use... The schlieren system uses either lenses alone or in combination with a concave mirror. 1956 *Nature* 10 Mar. 485/1 A schlieren system has been combined with a rotating mirror camera so that the shock propagation can be recorded in the regions outside the arc as well as in the arc channel. 1966 *McGraw-Hill Encycl. Sci. & Technol.* XII. 67/2 The schlieren system is used particularly in supersonic wind tunnels because it clearly shows the density gradients created by the shock and expansion waves of the airflow around the wind tunnel model. 1962 *New Scientist* 6 Dec. 576/1 The Schlieren technique is an optical method of studying changes of density, and hence of refractive index, in transparent media.

Hence **schli·eric** *a.*

1921 *Geol. Mag.* LVIII. 550 *Mixed rocks*, rocks which must be regarded as mixtures of carbonate magma of sövite type, and of the silicate rocks already enumerated are developed as schlieric intrusions, or as dykes cutting other members of the Fen group.

schlimazel (ʃlimǫ·z'l). *colloq.* (chiefly *U.S.*). Also **schlimazzel, schlimazl, shl-**, etc. [Yiddish, f. MHG. *slim* crooked + Heb. *mazzāl* luck.] A consistently unlucky, accident-prone person, a 'born loser'. Hence as *v. trans.*, to make a schlimazel of (a person) (*nonce-use*).

1948 N. Ausubel *Treasury of Jewish Folklore* iii. i. 344 Sholom Aleichem drew endless amusement out of the misadventures of his irrepressible, daydreaming *schlimazls*. 1960 *Encounter* May 84/1 In the *schlimazl* of Jewish tradition, I found the ancestors of Bellow's 'Angie March'. If the *schlimazl* went into the hat business, babies would be born without heads. 1962 J. Ish-Kishor *Tales from Wise Men of Israel* 199 She shrugged. What could one make of such a *shlimmazzel*? 1963 T. Pynchon *V.* i. 24 It seemed sometimes that he put himself deliberately in the way of hostile objects, as if he were looking to get schlimazzeled out of existence. 1968 L. Rosten *Joys of Yiddish* 347 A *shlimazl* wryly sighed: 'From *mazel* to *shlimazl* is but a tiny step; but from *shlimazl* to *mazel*—oy, is that far!' 1972 J. Wambaugh *Blue Knight* (1973) i. 15 Just bring me a cold drink, you old schlimazel. 1980 *Times* 12 June 16/8 When a waiter spills soup on a customer, the waiter is a *shlemiel* and the customer is a *shlemazl*.

schlock (ʃlǫk). *colloq.* (chiefly *N. Amer.*). Also **schlag, shlock.** [Yiddish, app. f. *shlogn* to strike.] Cheap, shoddy, or defective goods; inferior material, junk, 'trash' (freq. applied to the arts or entertainment). Also *attrib.* or as *adj.*, and *Comb.* in **schlo·ckmeister, -master** [G. *meister* master], a purveyor of cheap merchandise, 'special offers', and the like.

1915 *N.Y. Tribune* 25 July 12/1 Damaged articles..are sold..to the..'schlock' store proprietors. 1916 *Ibid.* 10 Jan. 14/1 There is nothing 'schlock' about the Goodell method. 1939 *Amer. Speech* XIV. 80/2 Schlag describes a skirt which has scant length, tightness where it should be full, is off size, has many loose threads, defective buttons, and off size button-holes. 1963 T. Pynchon *V.* iv. 110 She loves my rhinoplasty But the others are schlock. 1965 J. M. Ullman *Good Night, Irene* i. 20 Public relations, an elastic term that encompasses everything from crude schlockmeisters operating out of phone booths to high-powered representatives of billion-dollar corporations. 1966 L. Deighton *Billion-Dollar Brain* xv. 142 The schlock-shops were afire with sale signs and smiling suckers. 1970 *Toronto Daily Star* 24 Sept. 30/1 The most successful.. have substituted sociological satire for sentimental schlock. 1972 *Publishers Weekly* 21 Aug. 71/3 Shlock fiction with all the necessary ingredients, the result is mindlessly entertaining, if rather tasteless. 1976 *New Mus. Express* 31 July 37/1 Presley was already showing..an inclination to go in for schlock rather than rock. 1978 M. Puzo *Fools Die* xii. 131, I knew it [*sc.* magazine writing] was schlock, but still I loved it. *Ibid.* xxxiii. 388 He had signed a long-term contract with Tri-Culture and become the ace schlockmaster for Jeff Wagon. 1981 *Listener* 26 Feb. 294/1 Atkinson is more fun away from showbiz schlock.

Hence **schlo·cky** *a.*, characterized by schlock; that is schlock; shoddy, trashy.

1968 *N.Y. Times* 25 July 26 Playing the 'special guest star' in a series of schlocky European films. 1970 *Wall St. Jrnl.* (Eastern ed.) 10 Sept. 1/6 The schlocky corner gas station import dealer is gone. 1975 *Publishers Weekly* 1 Dec. 67/2 Just what the marketplace doesn't need, one more schlocky Gothic series. 1981 *Spectator* 24 Jan. 7/2 The concentration on Sinatra arises out of the suspicion that the Reagan entourage of friends and hangers-on is loaded with shabby, shady, schlocky, smarmy, shyster millionaires.

schloss. (Earlier and later examples.)

1820 D. Wordsworth *Jrnl.* 29 July (1941) II. 74 The rest of the company had proceeded..to the Schloss.. to be spectators of the moonlight festivities of the ruined castle. 1855 Geo. Eliot in *Fraser's Mag.* LI. 700/1 We saw the Schloss, and discovered the labyrinthine beauties of the park. 1974 *Times Lit. Suppl.* 23 Aug. 894/1 The incessant cold, the huge, half-ruinous, unheated schloss. 1980 *Early Music* Jan. 136 (Advt.), Our inclusive prices provide the course, room and board at Breiteneich, use of the facilities at the schloss.

schlub (ʃlɒb). *U.S. slang.* Also **shlub.** [Yiddish, perh. ad. Pol. *żłób* blockhead.] A worthless person, a 'jerk', an oaf.

1964 'E. McBain' *Ax* viii. 149 'Kaplowitz,' I say, 'are you a janitor or a schlub?' I'm a janitor... And such a dirty basement I can't stand.' 1969 D. E. Westlake *Up your Banners* (1970) v. 39 When a man..doesn't know the facts and nobody will tell him..and people keep throwing apples and unkind remarks at him, he has no choice but to look like a *shlub*. 1970 R. H. Greenan *Nightmare in Colour* (1971) xxxii. 114 He backed out—can you imagine? Hired a couple of college shlubs. 1978 *N.Y. Times Book Rev.* 2 Apr. 22 After bearing two children of the real-estate shlub, Earl Jr.

schm- (also **shm-**). *colloq.* (chiefly *U.S.*). An element, derived from the numerous Yiddish words that begin with this sequence of sounds, fused with or replacing the initial letter(s) of a word, so as to form a nonsense-word which is added to the original word in order to convey disparagement, dismissal, or derision.

1929 I. Goller *Five Bks. of Mr. Moses* v. ii. 215 'I know he made Davy go to the Palace to-day with the idea of hastening on the crisis in his illness.' .. 'Crisis-shmisis!' mocked Barnett disparagingly. 1935 A. Kober *Thunder over Bronx* 28 Now alluva sudden is fency-shmency with forks. *Ibid.* 48 So who you rushing to see, Miss Hurry Shmurry? 1952 *Jrnl. Eng. & Gmc. Philol.* LI. 226 The morphological pattern of the jocular repetition of a word or word-ending prefixed by the cluster ʃm- seems to have become quite generalized... I have heard.. moon-schmoon, etc. 1953 I. Asimov *Second Foundation* xviii. 183 'Time; schmime,' said Pappa irritably. 1963 T. Pynchon *V.* xii. 354 'It's murdering your own child, is what it is.' 'Child, schmild. A complex protein molecule, is all.' 1966 *N.Y. Herald Tribune* 20 Mar. (Sunday Mag.) 33/3 Trotsky-shmotsky, Lesbian- or adultery-wise—any way you slice *The Group* on screen it's the same old baloney about The Girls. 1967 *New Yorker* 28 Oct. 105/2 Two early Christians chanced to meet in Heaven... 'Saul of Tarsus, yet!' cried one. 'What are you doing here?' 'Tarsus-Schmarsus,' replied the other, 'I'm Paul already.' 1969 *Listener* 24 Apr. 569/1, I was surprised to find René Cutforth retelling the old story of the psychiatrist and the fond mother without specifying that she's a *Jewish* mother. ('I have to tell you, madam, that your son is suffering from an Oedipus complex.' 'Oedipus, Schmoedipus! What does it matter so long as he loves his mother?') 1971 D. Heffron *Nice Fire & Some Moonpennies* xv. 140 Gods, schmods! You can have them. 1978 F. Ross *Sleeping Dogs* 110 'Listen, honey—' 'Listen schmisten! I tell you I won't be here.'

schmagagi, var. *SCHMEGEGGY.

schmaltz (ʃmǫlts, ʃmalts), *sb.* Also **schmalz, shmaltz**, etc. [a. G. and Yiddish *schmalz* fat, dripping.] **1.** Melted chicken fat; *schmaltz herring*, a form of pickled herring.

1935 L. Zara *Blessed is Man* ii. ii. 232 Two or three other kegs of *schmalz* herring and such for pickling. 1951 L. W. Leonard *Jewish Cookery* vi. 42 (*heading*) Rendering chicken or goose fat (schmaltz). 1959 *20th Cent.* June 583 Shops all choked with..schmaltz herring. 1960 A. Wesker *I'm talking about Jerusalem* i. 13 All right, so it's shmultz herring and plum pudding. 1968 M. Richler *Cocksure* viii. 46, I don't want this apartment stinking of schmaltz herring. 1974 *New Yorker* 3 June 80/2 If a diner thinks the mashed potatoes might be improved by a bit of schmalz—liquid chicken fat—he pours some out of a dispenser. 1976 *Ibid.* 16 Feb. 58/1 The newest supermarket in Washington Avenue specializes in Cuban food instead of schmalz herring and stuffed kishke.

2. *colloq.* Sentimentality, emotionalism; excessively sentimental music, writing, etc. Also *attrib.*

1935 *Vanity Fair* (N.Y.) Nov. 71/2 *Schmaltz* (cf. the German *schmalz*, meaning grease) is a derogatory term used to describe straight jazz. 1938 *Manch. Guardian Weekly* 2 Sept. 188/3 Sometimes they play 'schmaltz' or 'salon' (ordinary jazz). 1944 [see *HAM sb.[1] B. 1]. 1950 *Here & Now* (N.Z.) Nov. 27/2 Howard Wyatt has an impressive technique for his age, and his greatest lack is what has come to be known as schmalz. A certain amount..of this quality is necessary, and..I'd recommend a serious study of the 'white' jazz exponents. 1956 [see *DISNEYESQUE a.]. 1957 J. D. Salinger *Zooey* in *New Yorker* 4 May 37/1 Will you be content with that standard box-office schmalz? 1960 *Guardian* 7 July 6/3 A purveyor of 'Schmalz' in long-winded repetitive symphonies. 1967 *Spectator* 24 Nov. 634/2 Some Presidents could turn..these frustrations to good account by retailing their moral and physical struggles... Lyndon Johnson spares us none of this schmaltz. 1977 *Spare Rib* June 46/4 She..is saying with appalling schmaltz that 'Josh's warm, funny smile was where I lived now'. 1978 *Observer* 19 Nov. 31/1 'What we call honest sentiment,' he says in equally honest puzzlement, 'you call schmaltz.'

schmaltz (ʃmǫlts, ʃmalts), *v. colloq.* Also **schmalz, shmaltz.** [f. prec.] *trans.* To impart a sentimental atmosphere to; to play (music) in a 'corny' or sentimental manner. Also with *up*.

1936 *Amer. Mercury* May p. x, *Schmalz it*, play it long-haired. 1966 D. Skirrow *It won't get you Anywhere* xxxi. 143 She was like the white light of early morning, before the hot sun schmalzes up the scene. 1968 L. Rosten *Joys of Yiddish* 351 To *shmaltz* ('to *shmaltz* it up'): to add 'corn', pathos, mawkishness. 1969 A. Laski *Dominant Fifth* ii. 41 He..tried to lighten his touch; no use giving this—visitor—the notion that they schmaltzed it up.

schmaltzy (ʃmǫ·ltsi, ʃma·ltsi), *a. colloq.* Also **schmalzy.** [f. as prec. + -Y[1].] Sentimentalized, over-emotional; 'corny'. Hence **schma·l(t)ziness.**

1935 [see *GROOVE v. 5]. 1949 L. Feather *Inside Be-Bop* iii. 22 Edgar Hayes, a pianist whose schmaltzy record of *Stardust* had made him a Harlem juke box favorite. 1952 B. Malamud *Natural* 170 A heavy-set German with a schmaltzy accent. 1959 *Guardian* 27 Oct. 7/6 I'm working on something real schmaltzy for one of your women's magazines. 1962 *John o' London's* 5 July 19/1 A few weeks ago she [*sc.* the All-American Mum] turned up in *All Fall Down*, embodied with searing schmalziness by Angela Lansbury. 1974 T. P. Whitney tr. *Solzhenitsyn's Gulag Archipelago* I. i. v. 218 Yuri painted for nothing schmaltzy pictures such as *Nero's Feast* and the *Chorus of Elves* and the like for the German officers on the commandant's staff. 1978 P. Griffiths *Conc. Hist. Mod. Music* vii. 101 In the opera *Lulu* the effect of Berg's half-tonal serialism is an over-ripe schmalzy quality. 1980 [see *SCHMUTZIG a.].

schmatte (ʃma·tə). *U.S. colloq.* Also **shmatte, schmottah**, etc. [a. Yiddish *schmatte*, ad. Pol. *szmata* rag.] A rag, a ragged garment; any garment. Also *fig.*

1970 L. M. Feinsilver *Taste of Yiddish* ii. 121 A 1969 sale catalog of the Ktav Publishing Company, New York book dealers, listed Philip Roth's licentious novel *Portnoy's Complaint* with the comment: 'A shmatte.' 1972 H. Kemelman *Monday Rabbi took Off* xxii. 144, I mean when they wear those checkered *shmattes* around their heads, then they're Arabs. Right? 1973 J. Marks *Mick Jagger* 128, I ran away from home in San Bernardino when I was fifteen... All I took was this *schmottah* I wore Halloween. 1977 *New Yorker* 24 Oct. 39/2 A young woman, pale, in a Victorian schmotta.

schmeck (ʃmek). *slang.* Also **smeck.** Pl. **schme·cken.** [a. Yiddish *schmeck*, sniff.] A drug; *spec.* heroin.

1932 *Evening Sun* (Baltimore) 9 Dec. 31/5 *Smeck*, dope. 1941 M. U. Schappes *Lett. from Tombs* 104 'Shmeck'—dope, a drug. 1966 *Sunday Times* (Colour Suppl.) 13 Feb. 35/4 *Schmeck*, heroin. 1967 M. Calpan *In Deadly Vein* ix. 196 'He was always wild... Anything for kicks... In the end it was schmeck.' 'Heroin?' 'Yes. Hooked.' 1970 L. Sanders *Anderson Tapes* xxxi. 86 She's hustling right now—schmeck, tail, abortion—the whole lot. *Ibid.* xcii. 218, I have some drugs. Some schmeck. Do you want a shot? 1971 *Oz* No. 36. 40/1 Shoot enough schmeck into them and they won't even *think* of burning and looting.

Hence **schme·cker**, a drug-addict, esp. one who takes heroin.

1953 W. Burroughs *Junkie* (1972) viii. 77 He went on talking about some old acquaintances who got their start in junk and later turned respectable. 'Now they say, "Don't have anything to do with Sol. He's a *shmecker*".' 1955 *Publ. Amer. Dial. Soc.* xxiv. 193 If they are all schmeckers, or narcotic addicts, they have a prearranged time to fix or take a bang. 1966 C. Himes *Heat's On* xix. 145 The skin-poppers and the schmeckers (those who used the needle and those who sniffed the powder).

schmeer (ʃmi·ɹ). *N. Amer. colloq.* Also **shmear, shmeer.** [ad. Yiddish *schmirn* to smear, grease, flatter.] **1.** Bribery, corruption, flattery.

1961 A. Berkman *Singers' Gloss. of Show Bus. Jargon* 78 *Shmear*,..payola; graft. 1962 E. Lacy *Freeloaders* ii. 42 Our lad didn't want the *shmear* to start with, so he ain't greedy. 1978 *Amer. Film* Apr. 57/2 He knew the *shmeer* was on when the producer invited him to lunch.. and began the meal by ordering caviar and champagne.

2. *the whole schmeer*, everything, everything possible or available, every aspect of the situation.

1969 E. Stewart *Heads* 48 Why couldn't you burrow around and ferret out the whole shmear yourself? 1970 L. Sanders *Anderson Tapes* v. 23, I want a complete list... Any thing and everything... The whole shmear. 1971 K. Wheeler *Epitaph for Mister Wynn* (1972) xxix. 374, I picked you because you know the whole schmeer. 1972 H. Kemelman *Monday Rabbi took Off* xxii. 146 Some special kind of prayer maybe where you could ask for the success of our enterprise...especially the financing, but I was thinking of the whole *shmeer*. 1978 *Maledicta* 1977 I. 282 Eventually, the whole schmeer was declared a Mexican stand-off.

schmegeggy (ʃmĕge·gi). *U.S. slang.* Also **schmagagi, shmegegge**, etc. [Origin obscure; see quots. 1968, 1970 for sense 1.] **1.** A contemptible person, an idiot.

1964 S. Bellow *Herzog* 29 He better get it this afternoon, that ludicrous schmegeggy! 1968 L. Rosten *Joys of Yiddish* 353 *Shmegegge*,..Ameridish slang. Origin: unknown; probably, a dazzling onomatopoetic child of the Lower East Side. 1. An unadmirable, petty person. 2. A maladroit, untalented type. 3. A sycophant, a *shlepper*, a whiner, a drip. 1970 L. M. Feinsilver *Taste of Yiddish* 121 *Shmegegge* (or *shmegeggi*), a galoot, a bird-brain, a stupid character... The disdain involved prompts me to suggest that the term may be a combination of two other words for 'fool': the vulgar *shmok*..and *yeke* or its German antecedent *Gecke*. 1971 *Observer* 23 May 36/3 He says he's a schlemiel which is..better than being a schmagagy... Schlemiels..drop things and..they drop on schmagogys.

2. Rubbish, nonsense.

1968 L. Rosten *Joys of Yiddish* 353 *Shmegegge*,..a lot of 'hot air', 'baloney', a *cockamamy* story. 'Don't give me that *shmegegge*!' 1970 L. M. Feinsilver *Taste of Yiddish* 121 *Schmegegge* (or *shmegeggi*),..as picked up in American theatrical circles, this is sometimes used in the sense of 'malarkey' or 'bushwa'. 1973 Boyd & Parkes *Dark Number* ii. 23 There was a bunch of students... They had the lot, the full schmagagi: girls got up like camp grannies.. boys in kaftans.

Schmeisser (ʃmai·səɹ). The name of Louis and Hugo *Schmeisser*, German small-arms designers, used *attrib.* or *absol.* to designate various German types of submachine gun, in use from 1918 onwards.

1950 G. Wilson *Brave Company* 7, I saw that he [*sc.* a German] carried a Schmeisser. 1963 D. Bagley *Golden Keel* ix. 268, I opened the locker under my berth and took out the Schmeisser machine pistol and all the magazines. 1976 *Valiant* 8 May 3 (*caption*) In the United States, gun shops sell every kind of weapon, from shotguns to schmeissers. 1981 E. Ward *Baltic Emerald* xiv. 117 Modern Schmeissers..the most reliable..of all the medium-range killing devices.

schmelz (ʃmelts). Also erron. **schmel(t)ze.** [a. G. *schmelz* enamel.] Any one of several varieties of decorative glass; *spec.* a variety coloured red with a metallic salt, used to flash white glass. Also *attrib.*

[1854 C. Tomlinson *Cycl. Useful Arts* I. 784/1 Smetz [*sic*: ? *read* Smelz] glass is formed by fusing lengths of coloured glass into each other, so that the section shall resemble carnelian and the agates.] 1859 R. Hunt *Guide Museum Pract. Geol.* (ed. 2) 111 The Bohemian ruby is thus prepared:—a preparation called schmelze is made; it is composed of silica 500, minium 800, nitre 100, calcined potash 100. 1866 *Christie, Manson & Woods Sale Catal.* 9 Feb. 1867 65 A vase, on foot, of tortoiseshell Schmeltze... A green basket, mounted with or-molu; and a Schmeltze ditto... A ball of variegated Schmeltze. 1879 *Encycl. Brit.* X. 652/1 That peculiar kind of glass usually called schmelz, an imperfect imitation of calcedony, was also made at Venice in the 15th century. 1882 *Hamilton Palace Collection Catal.* No. 846 A Fluted Tumbler, of red and white schmelz. 1907 E. Dillon *Glass* xii. 207 There are a few exceptionally fine early examples of this schmelz at South Kensington. 1961 E. M. Elville *Collector's Dict. Glass* 183/1 Variegated or marbled opaque glass, commonly known by the German word *schmeltz*.

Also ‖ **Schmelzglas.**

1935 W. A. Thorpe *English Glass* v. 148 Measey and Greene ordered and sold the following lines, but the list applies generally to the members of the Company, and excepting the items of opaque-white and *calcedonic* (marbled glass or *Schmelzglas*) it may be taken as a production list. 1960 H. Hayward *Antique Coll.* 10/2 '*Agate*' *glass*, a glass of several colours which have been allowed to mingle before the vessel is formed, in imitation of agate. This type of glass was popular during the Renaissance.., particularly in Venice and Germany, and is sometimes known as *Schmelzglas*. 1975 *Oxf. Compan. Decorative Arts* 398 Not only did they [the Venetians] reproduce the Roman 'mosaic' and *millefiori* glass and the material made of blended opaque colours in imitation of natural stones (*calcedonio*, sometimes miscalled *Schmelzglas*), but they seem even to have copied..typical Roman shapes.

schmendrik (ʃme·ndrik). *U.S. slang.* Also **schmendrick, shmendrik.** [The name of a character in an operetta by Abraham Goldfaden (1840–1908).] A contemptible, foolish or immature person; an upstart, a 'sucker'.

1944 M. Samuel *Harvest in Desert* xii. 115 The colonists called the workers *Shmendriks*, tatterdemalions, n'er-do-wells. 1951 A. Hirschfeld *Show Business is No Business* 47 A schmendrick with a noodle for a brain. 1970 S. Ellin *Bind* xxx. 151 This boy is no *shmendrik*... Believe me, he knows from the real thing. 1976 L. Deighton *Twinkle, twinkle, Little Spy* x. 93 Maybe if I'd been at college with Andrei Bekuv, I could even feel sorry for *that* schmendrik.

‖ **schmerz** (ʃmerts). Also **Schmerz.** [a. G. *schmerz* pain.] Grief, sorrow, regret, pain.

1911 'P. Harding' *Corner of Harley Street* xxx. 260 So white-coat gives him a swiftly helping hand, and within five minutes is removing a decayed semitic molar that has been giving its owner *schmerz* indescribable. 1925 R. Frost *Let.* 20 June (1964) 174 What lies at the bottom of your *Schmerz* is your own dereliction. 1977 *Times* 13 July 11/8 There is much more to Schiele than terrified mirror-gazing and sexual *Schmerz*. He was a splendid..portraitist.

Schmidt[1] (ʃmit). *Org. Chem.* [The name of Karl Friedrich *Schmidt* (b. 1887), German chemist, who first employed a reaction of this kind in 1923 (*Zeitschr. f. angew. Chem.* XXXVI. 511).] *Schmidt('s) reaction*: a widely-used synthetic method in which a carbonyl compound is treated with hydrazoic acid in the presence of mineral acid, the product(s) depending on the kind of carbonyl compound used (e.g. an aldehyde gives a mixture of a nitrile and a formyl derivative of an amide, a ketone gives an amide, and a fatty acid gives an amine).

[1936 *Proc. R. Soc.* A. CLIV. 54 The introduction of the Schmidt method leaves the Hofmann method of great historical interest but deprives it of importance for costly and delicate synthetic work.] 1937 *Jrnl. Amer. Chem. Soc.* LIX. 2658/1 Although the Schmidt reaction has been used in a few instances it has never been extensively studied. 1963 I. L. Finar *Org. Chem.* (ed. 4) I. ix. 178 Schmidt's reaction with acids is a modification of the Curtius reaction. 1967 L. F. & M. Fieser *Reagents for Org. Synthesis* I. 447 Application of the Schmidt reaction to cyclohexanone effects ring enlargement to ε-caprolactam. 1976 Streitwieser & Heathcock *Introd. Org. Chem.* xxviii. 825 The Schmidt reaction may also be used for the synthesis of simple amino acids if it is applied to an alkylated malonic acid.

Schmidt[2] (ʃmit). *Astr.* The name of Bernhard Voldemar *Schmidt* (1879–1935), Estonian-born German optician, used *attrib.* with reference to an optical system invented by him, as **Schmidt camera**, an astronomical telescope, used exclusively for wide-field photography at the primary focus, in which a Schmidt correcting lens is placed at the centre of curvature of a spherical primary mirror, the combination having no spherical aberration and little chromatic aberration; **Schmidt correcting lens, corrector, (correcting) plate**, an aspheric lens of complex figure used in the Schmidt camera and other catadioptric systems that utilize the same principle; **Schmidt telescope** = *Schmidt camera*. Also *ellipt.*, = *Schmidt camera*.

1939 Skilling & Richardson *Astron.* iii. 82 The great advantages of the Schmidt telescope are that it can photograph a very large area in the sky, giving sharp focus clear to the edge of the picture; and that it is very fast. *Ibid.*, These telescopes are sometimes called Schmidt cameras, for they cannot be used visually. 1946 *Nature* 17 Aug. 222/1 A Schmidt plate presents a very different problem, since the highest optical homogeneity is required for this, and a low-expansion glass has never yet been produced in the requisite optical quality. 1961 Miczaika & Sinton *Tools of Astronomer* iii. 99 (*caption*) Possible shapes of Schmidt correcting lenses. *Ibid.* 100 Such solid Schmidts, as they are called, may be made with *f*-numbers as small as *f*/0·6. 1964 *Listener* 21 May 831/1 The Armagh Schmidt is employed mainly on variable star research. 1966 *McGraw-Hill Encycl. Sci. & Technol.* XIII. 452/1 A similar correction of the principal defects of the paraboloidal reflector can be obtained by replacing the thin Schmidt correcting plate with a weakly diverging meniscus lens. 1973 *Sci. Amer.* Aug. 111/1 Schmidt correctors and other aspheric surfaces are sold by the tens of thousands in quality zoom lenses for film and television. 1978 Pasachoff & Kutner *University Astron.* iv. 90 The 1·2-meter Schmidt has been used to map the entire sky that is visible from Palomar. *Ibid.*, The new 1-meter Schmidt camera at the European Southern Observatory and the British 1·2-meter Schmidt camera at Siding Spring, Australia, are now being used in a joint project to extend the survey to incorporate the one-quarter of the sky that cannot be seen from Palomar. 1978 *Sci. Amer.* Dec. 90/3 The lens of the scallop eye appears to perform much the same function as the corrector plate of the Schmidt telescope.

Schmidt number (ʃmit). *Physics.* [Named after Ernst Heinrich Wilhelm *Schmidt* (b. 1892), German engineer.] A dimensionless number, analogous to the Prandtl number, used in the study of convective mass transfer and evaluated as the ratio of kinematic viscosity to mass diffusivity.

1955 D. B. Spalding *Some Fundamentals of Combustion* vi. 239 Other experimental procedures can easily be conceived with various advantages or disadvantages, the guiding principle being that the Peclet numbers based on flow velocity and on flame speed must be kept constant, and that if possible the Mach number, Prandtl number and Schmidt number..should all have equal values in the model and the original. 1957 Jakob & Hawkins *Elements of Heat Transfer* (ed. 3) xvi. 293 The heat transfer equation may be altered to represent the mass transfer equation by replacing the Nusselt number by the corresponding Nusselt number for mass transfer, and replacing the Prandtl number by the Schmidt number. 1975 Croome-Gale & Roberts *Airconditioning & Ventilation of Buildings* iii. 92 It can be shown that the well-known result expressing *Nu* [*sc.* the Nusselt number] as a function of the Reynolds and Prandtl numbers, i.e. $Nu = f(Re, Pr)$, has an analogous form in mass transfer $Sh = f(Re, Sc)$ where *Sc* is the Schmidt number.

‖ **schmierkäse** (ʃmi·r,kēzə). Also **Schmierkäse.** [G.: see Smear-case.] = Smear-case. Also *fig.*

1905 W. Wittigschlager *Minna* 104 She carried some schmierkase (cream cheese), and butter that smelt so oily..I had to turn my head away. 1931 F. Hurst *Back Street* i. 13 Sturdy, unstylish women with enormous busts, who ate and drank with relish, but knew, to the penny, for how much less they could spread their groaning home-table with these luxuries of *Schmierkäse*. 1949 *Sat. Even. Post* 23 Apr. 80/3, I took large helpings of ham and potatoes, *schmierkase*, and green salad with tomatoes. 1955 H. Kurnitz *Let.* Dec. in G. Marx *Groucho Lett.* (1967) 249, I am.. whipping up a schmierkase about light love and dark doings. 1969 R. & D. De Sola *Dict. Cooking* 202/2 *Schmierkäse*, (German—soft cheese). To Germans, any soft cheese; to Pennsylvania-Germans, cottage cheese.

Schmitt (ʃmit). *Electronics.* The name of Otto Herbert *Schmitt* (b. 1913), American biophysicist and electronics engineer, used *attrib.* and *absol.* to designate a bistable circuit devised by him, in which the output increases to a steady maximum when the input rises above a certain threshold, and decreases almost to zero when the input voltage falls below another threshold (usu. lower than the first).

1946 *Jrnl. Inst. Electr. Engineers* XCIII. IIIA. 306/1 Positive feedback can be used to reduce the effective grid base, or even to make it negative; an example is shown. [*Note*] This is the 'Schmitt circuit'. 1953 von Tersch & Swago *Recurrent Electr. Transients* viii. 272 If both plate-to-grid coupling and cathode coupling are utilized another trigger circuit is obtained. This circuit is called the Schmitt trigger circuit. 1962 Simpson & Richards *Physical Princ. Junction Transistors* xvi. 418 The Schmitt trigger is neither bistable nor monostable in the ordinary sense. Its behaviour is similar to that of a non-regenerative switch but it has the advantages that it switches regeneratively at very high speed and can be designed with an accurate adjustable trigger threshold. 1967 *Electronic Engin.* XXXIX. 752/1 A theory was required to account for the existence of a minimum ionization current below which the Schmitt fails to trigger, and a maximum above which the Schmitt fails to reset. 1975 D. G. Fink *Electronics Engineers' Handbk.* xvi. 45 Schmitt bistables, also called Schmitt triggers, are suitable for detecting the moment when an analog signal crosses a given dc level. They are widely used in oscilloscopes to provide time-base synchronization pulses... In some Schmitt trigger circuits it is possible to modify the switching level by electrically changing the operating points of the transistors.

schmo (ʃmōu). *U.S. slang.* Also **shmo(e).** [f. *SCHMUCK.] An idiot, a fool.

1948 *Life* 15 Mar. 23/2 Schlump is a friendlier, more sympathetic term than 'schmo', which has completely replaced 'jerk'. A schmo, of course, is a person who stands watching a machine make doughnuts, and 1) cannot understand the process, 2) cannot get up will power to leave. 1955 'H. Robbins' *Stone for Danny Fisher* i. vii. 56 Let some other shmoe wet-nurse a bunch of kids. 1957 S. J. Perelman *Road to Miltown* 125 A couple of shmos like you and me, we can't even get up our rent, whereas them dukes and earls..are rolling in dough. 1970 D. Francis *Rat Race* ii. 27 'Who,' he said crossly, 'is going to give that schmo a thousand quid for breaking his ankle?' 1979 'H. Howard' *Sealed Envelope* xi. 159, I was feeling like a shmoe... Paul Ingram had outsmarted me.

schmock, var. *SCHMUCK.

schmoll (ʃmǫl). *slang.* [app. ad. Yiddish *shmol* narrow.] An idiot, a fool.

1967 J. Wainwright *Worms must Wait* xl. 101 Let's say..he was killed by some *schmoll* who wanted to rob him. 1973 —— *Pride of Pigs* 103 These hot-shot scientists... They're schmolls—every last one of 'em..but they get away with it.

schmooze (ʃmūz, ʃmūs), *v. U.S. colloq.* Also **schmoos(e), schmuss, shmooz**, etc. [ad. Yiddish *shmuesn* to talk, converse, chat, f. as next.] *intr.* To chat, gossip, engage in a long and intimate conversation. Hence **schmoo·zer**; **schmoo·zing** *vbl. sb.*

1897 *N.Y. Times Weekly Mag.* 14 Nov. 4/1 He loves dearly to stop and chat (*Schmoos*, he calls it). 1921 J. Anthony *Gang* 28 When Mrs. Sinbaum comes, we *schmoos*. 1928 *Amer. Speech* III. 364 The presence of a Jewish contingent of 'producers' and managers is responsible for such New York expressions as 'mazuma' (money), 'schmuss' (talk). 1939 *New Yorker* 4 Feb. 30/1 'Schmooze' (pronounced 'shmooss') is related to the Yiddish verb 'schmooze', which means 'to talk'. But schmoozing in the garment district is more than just a lot of idle chatter. Schmoozing is a careful tradition, dear to the hearts of everyone in New York's most thickly populated business section. *Ibid.* 30/2 Everybody in the district eats fast, the better and more to schmooze. 1939 *Reader's Digest* May 106/1, The schmoozers gulp down lunch in 15 minutes and then arrange themselves according to caste and craft. 1966 H. Kemelman *Saturday Rabbi went Hungry* x. 60 On Friday nights or Saturdays, don't we stand around after the services and *schmoos* a while? 1973 *New Yorker* 3 Feb. 56/2 We would schmoose all afternoon, with her talking in that funny, high Pennsylvania Dutch voice: 'Dat's gute,' or 'Dat's humbug.' 1977 *Time* 25 Apr. 47/2 Neil Diamond's beach house, Linda Ronstadt's $325,000 clapboard and the sprawling nine-bedroom house Guitarist Robbie Robertson took over from Carole King are all within schmoozing distance. 1977 *New Yorker* 27 June 29/1 Had she worked here part time, returning today out of sentimentality to schmooze with the boss? 1980 W. Safire in *N.Y. Times Mag.* 18 May, A 'stoop', from the Dutch word for 'step', is a description of the porch and front steps on which Brooklynites sit and schmooze.

schmooze (ʃmūz, ʃmūs), *sb. U.S. colloq.* Also **shmoos.** [ad. Yiddish *shmues* chat, gossip ad. Heb. *shĕmūʻah* rumour.] Chat; gossip; a long and intimate conversation.

1939 *Reader's Digest* May 106/2 Because of schmooze, the garment district is the most hypersensitive city of 200,000 in the world. 1956 B. Holiday *Lady sings Blues* (1973) xix. 156 [Lena Horne] insisted on taking me out with her and bought me lunch, and we had a wonderful schmooze about the old days in Hollywood. 1970 S. Ellin *Man from Nowhere* (1971) xxxix. 194 Ready to order now..or do I wait until the end of the *shmoos*? 1977 *Zigzag* Aug. 24/1 The general demeanor and schmooze level of the crowd indicates it's a predominantly invitational radio/press group, more disposed to be open-minded.

schmottah, schmozzle, varr. *SCHMATTE, *SHEMOZZLE.

schmuck (ʃmɒk). *slang.* Also **schmock** (ʃmǫk), **shmock, shmuck.** [Yiddish; originally a taboo-word meaning 'penis'.] A contemptible or objectionable person, an idiot. Hence **schmu·cky** *a.*, objectionable, obnoxious.

1892 I. Zangwill *Childr. of Ghetto* II. i. xvi. 45 Becky's private refusal to entertain the addresses of such a *Shmuck*. 1945 G. Marx *Let.* 16 Feb. in *Groucho Lett.* (1967) 51 He doesn't know I can write, in fact, he thinks I'm a complete schmuck. 1958 F. Norman *Bang to Rights* iii. 132 But as I'm no shmock I deside to play along with her. 1963 *Globe & Mail* (Toronto) 10 July 9/1 Only the pay-TV concept..can break the hold of the 'Madison Ave. schmucks, the Gestapo of the television industry'. 1967 D. Skirrow *I was following this Girl* xxxix. 241, I know that one schmucky swallow doesn't have to spoil the barrel. 1971 B. Malamud *Tenants* 50 Art is the glory and only a shmuck thinks otherwise. 1972 J. Caine *Hamlet, My Boy* xi. 161 As soon as the *shmock* of a witness offered to testify on television. 1975 *Harpers & Queen* May 128/3 Schmucky agents and flacks and show-biz parasites. 1978 M. Puzo *Fools Die* xxi. 239 Cully felt some anger that this guy was treating Merlyn like such a schmuck. 1981 *Times* 2 July 15/2 Mary Gordon is extremely funny about the beautiful Robert..and about the Woody Allen-like schmuck in the apartment below whom she sleeps with.

schmutter (ʃmɒ·təɹ). *colloq.* Also **shmuter, shmutter.** [ad. Yiddish *schmatte*, rag; cf. *SCHMATTE.] Clothing; also *fig.*, rubbish. Also *attrib.*, esp. in *schmutter trade, business*, etc.

1959 C. MacInnes *Absolute Beginners* 138 Coming down the steps, wearing some very fancy schmutter: mauve, button-two tuxedo, laced shirt, varnished pumps with bows, and, on his arm, a nameless dame. 1962 F. Norman *Guntz* i. 9 There ain't all that many birds who are loaded in the shmuter trade. 1965 *New Statesman* 30 July 152/2 Jews..may prefer their son to go into the family schmutter business. 1967 G. Sims *Last Best Friend* xiii. 114 They said it was like Buck House but it was a right load of old schmutter! You see, everyone's an antique dealer today. 1972 *Bookseller* 27 May 2358/1 Several dresses (at trade terms) were bought for Mrs. Wolfe..from small shmutter merchants. 1980 *Times* 22 July 10/6 You can always dump a load of old schmutter destined for the California leisure set onto the unsuspecting women of Nottingham.

schmutz (ʃmuts). *slang.* Also **shmutz.** [Yiddish or Ger.] Dirt, filth, rubbish. Also *fig.* So **schmu·tzig, -ik** *a.*, filthy.

1967 P. Welles *Babyhip* xxiv. 161 She was the one at your party wearing the *schmutzik* suit. 1968 M. Richler *Cocksure* xix. 116 'Of my son's ability there is no question.' '—and, em, the contents of your son's novel. You see—' '*Shmuts*,' Daniels shouted at Katansky. 'Pardon?' 'Filth. Today nothing sells like filth.' 1971 O. Norton *Corpse-Bird Cries* vi. 113 It means dropping this driver in the *schmutz* insurance-wise. I was trying to avoid that. 1972 *Last Whole Earth Catalog* 178/1 It delights them to watch us rummaging around in the schmutz. 1980 *Times Lit. Suppl.* 5 Sept. 968/3 His [*sc.* Kurt Weill's] jazz is *schmutzig*, not *schmalzy*.

schnapper. For (ʃnæ·pəɹ) substitute (snæ·pəɹ). Delete 'The original form is retained in some English books, but is obsolete in Australia and New Zealand.' For Latin name substitute *Chrysophrys guttulatus* or *C. auratus*. **a.** (Later examples.)

1908 E. J. Banfield *Confessions of Beachcomber* ii. i. 243 When maybe they have caught schnapper..they drift among the turtle. 1917 *Chambers's Jrnl.* Apr. 237/2 The schnapper, a sea-bream, is a splendid fish. 1947 K. Tennant *Lost Haven* vi. 89 The deck was littered with the pink-bronze bodies of schnapper. 1971 *Sunday Australian* 8 Aug. 5/6 Sir Henry's best catch this time was a 10 lb schnapper.

b. (Later examples.)

1944 *Living off Land* vii. 133 Cotton schnapper line. 1947 K. Tennant *Lost Haven* i. 14 The schnapper boat moored beside..the grey skeleton of the half-built ship.

schnauzer (ʃnau·tsəɹ). [G.] A black or pepper-and-salt wire-haired terrier belonging to the breed so called, which includes large, standard, and miniature dogs distinguished by a stocky, robust build, docked tail, blunt, bearded muzzle, and ears that droop forwards; formerly called the wire-haired pinscher.

1923 *Dog World* Aug. 14/2 A new breed has come to America—the Schnauzer. 1930 *Observer* 9 Feb. 13/2 The German Schnauzers have sterling qualities, though they may not be particularly showy in their close wiry coats. 1957 *New Yorker* 5 Oct. 34/1 For rainy weather, this miniature schnauzer is wearing our ready-made gabardine raincoat. 1968 [see *PINSCHER]. 1970 *Manch. Guardian Weekly* 14 Mar. 4/4 She bought a schnauzer, a placid beast with large brown eyes and a yelp like a police car siren. 1977 *Time Out* 28 Jan.–3 Feb. 64/2 (Advt.), Has own home, Central London and owns beautiful little schnauzer.

schneider[1] (ʃnai·dəɹ). *Skat.* Also **Schneider.** [G., lit. 'tailor'.] (See quot. 1949.) Also as *adj.* and *v. trans.*

1886 E. E. Lemcke *Skat* 7 With 60 points only he loses; with 30 points he is '*Schneider*' or '*geschnitten*' (cut); with no count at all he is '*Schwarz*' (black = whitewashed). Consequently the two hands in opposition to the 'player', scoring jointly 60 points, win the game from player; scoring 30 are out of *Schneider*, but are *Schwarz* with no count. 1909 R. F. Foster *Foster's Compl. Hoyle* 420 If he can get 91 points, he wins a double game, which is called *schneider*. *Ibid.* 437 It may be played out to see if he can make schneider or schwarz. 1935 *Encycl. Sports, Games & Pastimes* 553/1 If he scores 91 points he makes his opponents schneider. 1947 *New Compl. Hoyle* 385 If the player wins 91 or more points in play, he is said to *schneider* the opponents and the value of his game is increased. 1949 A. A. Ostrow *Compl. Card Player* 645 Official laws of American skat... The player to be out of schneider must have at least 31 points, the opponents 90. 1975 *Way to Play* 109/1 If the bidder has named suits or grand, he may before the opening lead, declare: a) schneider, ie he aims to win at least 91 trick points; or b) schwarz, ie he aims to take every trick. 1976 *National Skat & Sheepshead Q.* Mar. 18 A grand scores 80 points and possibly 100 if the hand is schneidered.

Schneider[2] (ʃnai·dəɹ). The name of Jacques *Schneider* (1879–1928), French flying enthusiast, used *attrib.* in *Schneider trophy, cup*: the Jacques Schneider Maritime Cup, presented in 1913 by Schneider to the winner of an international competition for seaplanes comprising an air race and seaworthiness trials, and contested annually (with certain exceptions) until won outright by Great Britain in 1931.

[1912 *Flight* 14 Dec. 1182/1 M. Schneider has offered for international competition a trophy of the value of £1,000, to go to the club which the winning pilot represents.] 1913 *Ibid.* 5 Apr. 395/1 The French team for the forthcoming international contest for the Schneider Cup at Monaco. 1927 A. Huxley *Let.* 8 Oct. (1969) 291 Time rushes past as though it were trying to win the Schneider Cup. 1929 *Radio Times* 8 Nov. 395/3 We were testing all the arrangements for the Schneider Trophy relay, making sure that the loudspeaker system at various points round the coast could pick up our broadcast. 1933 *Ann. Reg. 1932* i. 24 After paying a tribute to all who had been concerned in the winning of the Schneider trophy, [the minister] remarked that though in size the Royal Air Force took only fifth place..there was no other better equipped. 1977 *Times* 23 Sept. 12/4, I hope the Schneider Trophy will be given an extra loving dust down at its home in the Science Museum.

schnitzel (ʃni·tzəl). [G.] A veal cutlet, esp. in *Wiener* (vi·nəɹ) *schnitzel*, one coated with egg and breadcrumbs, fried and often garnished with lemon, capers, anchovies, etc., in the Viennese style.

1854 *Pioneer* (San Francisco) Nov. 318 Eggs, coffee, toast, and now and then, a chop or a 'snitzel' is the order given for thousands of people. 1862 *Temple Bar* Nov. 63 After a dinner such as Vienna only can furnish; *e.g.* a delicate soup..a *Wiener Schnitzel* (a savoury cutlet greatly to be recommended)..we take rail at half-past two p.m. 1904 *Adventures of Elizabeth in Rügen* 262 Her eyes were discreetly fixed on a *Wiener Schnitzel* that she was eating with a singular mincingness. 1911 A. Filippini *International Cook Bk.* 144 Arrange a thin slice of lemon, with a twisted anchovy in oil placed over each slice of lemon, on top of each schnitzel. 1936 E. Ambler *Dark Frontier* vii. 114 They do know how *wiener schnitzel* should be cooked. 1956 [see *ESCALOPE]. 1960 *News Chron.* 23 Feb. 3/2 Swop the porridge and oatcakes for schnitzel and strudel. 1978 *Chicago* June 210/2 The entrées perk one up immediately, though. The schnitzel à la Holstein came with a perfect fried egg and a golden puff of crust.

schnockered (ʃnǫ·kəɹd), *ppl. a. U.S. colloq.* Humorous var. of *SNOCKERED *ppl. a.*

1955 *Amer. Speech* XXX. 303 *Schnockered*; *way up*,.. drunk. 1976 *Verbatim* Feb. 15/1 George really got schnockered at Judy's party. 1977 B. Garfield *Recoil* iii. 45 Bradleigh took the empty glass. 'That's probably enough. You don't want to get schnockered.'

schnook (ʃnuk). *U.S. colloq.* Also **schnuck, shnook.** [app. Yiddish: perh. repr. Yiddish *shnuk* snout, or f. G. *schnucke* a small sheep.] A dupe, a sucker; a simpleton, a 'dope'; a pitiful wretch.

[1943 S. J. Perelman *Let.* 7 Apr. in G. Marx *Groucho Lett.* (1967) 190 It's the story of a small schnükel of a barber who accidentally brings a statue of Venus to life.] 1948 H. L. Mencken *Amer. Lang.* Suppl. II. 757 *Schnuck*.., a customer easily persuaded, a sucker. 1955 N. Mailer *Deer Park* xii. 136 I'd be making a stinking seven hundred and fifty a week now like all those poor exploited schnooks. 1959 R. Chandler in Gardiner & Walker *R. Chandler Speaking* (1962) 262 Why does he want to see me so badly that he has to send a couple of shnooks after me? 1964 S. Bellow *Herzog* 29 This shnook of a chiropodist—what a hellcat *he* married. 1975 A. Bergman *Hollywood & Le Vine* xiii. 187 It was all pretty fascinating for a Sunnyside schnook like me. 1980 W. Safire in *N.Y. Times Mag.* 2 Aug. 8 To be self-conscious about the possibility of error..is to be a nerd, a schnook and a wimp.

Schnorchel, schnorkel, Schnorkel, varr. *SNORKEL.

schnorrer. Delete ‖ and add: Also **shnorrer.** Now in extended *U.S.* use, a beggar, layabout, scrounger, good-for-nothing.

1934 E. Pound *Eleven New Cantos* xxxv. 24 The tale of the perfect schnorrer. 1959 [see *LAYABOUT]. 1962 J. D. Salinger *Franny & Zooey* 136, I had lunch with him one day a couple of weeks ago. A real schnorrer, but sort of likable. 1977 *New Yorker* 24 Oct. 38/3 Investigate your own pants, you schnorrer. 1981 J. Barnett *Firing Squad* xiv. 190 A right pair of miserable schnorrers I've got here.

So **s(c)hnorr** *v. trans.* and *intr.*, to obtain by begging; to beg, sponge (*off*).

1892 I. Zangwill *Childr. Ghetto* III. ii. vii. 125 Your father..stood in the Lane with lemons, and *schnorred* half-crowns of my father. *Ibid.* xii. 221 But isn't it *schnorring* to be dependent on strangers? 1894 —— *King of Schnorrers* iii. 67 Even if you can prove you can *schnorr* enough to keep a wife, I do not bind myself to consent. 1964 W. Markfield *To Early Grave* iv. 76 Box after box. *Shnorred*, with cunning and craft, from tough-minded Cousin Schmeilick. 1968 *Encounter* Sept. 30/1, I can go out and within an hour *shnorr* the entire amount I owe you. 1975 *Publishers Weekly* 19 May 90/1, I hope to shnorr off a couple of Scottish landowners I've met here.

schnozz (ʃnǫz). *U.S. slang.* Also **schnoz.** [app. Yiddish: cf. G. *schnauze* snout, and see next.] **a.** The nose, nostril.

1942 A. Kober in *New Yorker* 13 June 19/1, I see she's not occupied excep' she's powderin' her schnoz. 1940 L. Shelly *Hepcats Jive Talk Dict.* 17/1 *Schnozz*,..the nose. 1967 P. Welles *Babyhip* xx. 131 Mr Cox stuffed a rusty paper clip up his schnozz and broke his nose. 1973 R. Hayes *Hungarian Game* iii. 28 'You remember what our boy looks like?' 'Gray hair, widow's peak, big schnozz, red ski parka and no luggage.'

b. In fig. phr. (*right*) *on the schnozz*, precisely, exactly, on the dot (of time).

1949 W. R. Burnett *Asphalt Jungle* xx. 130 Headlights flashed into the parking-lot, and then went out. 'This is us, I think,' said Louis, 'and right on the schnoz.' 1967 'E. Queen' *Face to Face* xxx. 140 Twenty minutes to twelve on the schnozz.

schnozzle (ʃnǫ·z'l). *U.S. slang.* [pseudo-Yiddish: cf. Yiddish *shnabl* beak, and see prec.] The nose. Similarly (*joc.*) **s(c)hnozzo·la** [cf. *-OLA].

Esp. applied as a nickname to the U.S. entertainer James Francis ('Jimmy') Durante (1893–1980).

1930 *Variety* 26 Feb. 24/5 It's the medium for the screen debut of Jimmy Durante, he of the large schozzola [*sic*]. *Ibid.* 56/2 'Roadhouse Night'...brings Jimmy Durante to the screen. Admirers of his peculiar madness usually fear for its reception by a general public..but the Schnozzle's first screen appearance removes any doubts that might have been entertained. 1937 J. Durante in *Amer. Mag.* May 61/1 A youngster like me whose schnozzle could be seen two blocks away. *Ibid.* 61/2 When we admit our schnozzles..we begin to laugh. 1937 in Wentworth & Flexner *Dict. Amer. Slang* (1960) 448/1 A broken nose epidemic hit Dennison. In early contests 5 players broke their schnozzolas. 1959 J. Ludwig in *Tamarack Rev.* Summer 24 What a way to louse up this new magenta outfit—streaming eyes, a shiny shnozzola! 1977 *Listener* 9 June 746/2 Hebrew amens are breathed through Yiddish schnozzles. 1981 *Times Lit. Suppl.* 15 May 535/2 He is sunk in the multiplied particularities of his vivid surroundings. Not least, in the pongs to which his great schnozzle.. is peculiarly susceptible.

‖ **Schnurkeramik** (ʃnū·rkerā:mik). *Archæol.* Also with small initial. [G., f. *schnur* string, cord + *keramik* ceramics, pottery.] = *corded ware* s.v. *CORDED *ppl. a.* 3 b.

1902 J. Abercromby in *Jrnl. Anthrop. Inst.* XXXII. 391 In Germany there is a class of ceramic which goes by the name of *Schnurkeramik*, from its being almost exclusively ornamented by cord-impressions. 1928, 1950 [see *CORDED *ppl. a.* 3 b]. 1954 S. Piggott *Neolithic Cultures* xi. 344 Comparable ornament appears on pots in a *schnurkeramik* context from Switzerland.

Schoenbergian (ʃönbɛ̄·ɹgiăn), *a.* and *sb.* Also **Schönbergian.** [-IAN.] **A.** *adj.* Of, pertaining to, or characteristic of the Austrian composer Arnold *Schoenberg* (1874–1951) or his music. **B.** *sb.* An admirer or adherent of Schoenberg; an exponent of Schoenberg's music. Hence **Schoe·nbergism** *rare*, the advocacy or practice of Schoenbergian techniques of musical composition.

1922 C. Gray in *Mus. & Lett.* III. 79 Side by side with a daring experiment like Op. 6, No. 1, *Traumleben*, with its characteristic late-Schönbergian voice part. 1931 [see *DIATONICISM]. 1934 C. Lambert *Music Ho!* v. 330 His earlier works..show signs of a Schönbergian ruthlessness. 1947 *Penguin Music Mag.* Dec. 21 An extraordinary combination of the traditional Italian lyrical *cantilena* writing with the Schönbergian technique. 1951 Schoenbergism [see *DODECAPHONIC a.]. 1959 *Times* 13 Feb. 13/4 Other names in these programmes are those of Egon Wellesz, a lapsed Schönbergian..and Karlheinz Stockhausen. 1976 *Gramophone* Aug. 324/1 He had been taking a crash course in early Schoenbergian expressionism. 1978 P. Griffiths *Conc. Hist. Mod. Music* iv. 46 Apart from that the work is not at all Schoenbergian.

Schoenflies (ʃö·nflis). *Cryst.* The name of Arthur *Schoenflies* (1853–1928), German mathematician, who listed the 230 space groups in 1891 (*Krystallsysteme und Krystallstruktur*), used *attrib.* with reference to the system of nomenclature which he devised for them.

1934 W. P. Davey *Study of Crystal Struct.* viii. 222 Of the four sets of symbols listed in the table it is recommended that only the Schoenflies and the Wyckoff be used... The Schoenflies symbols have the great advantage of world-wide use. **1962** TERPSTRA & CODD *Crystallometry* iv. 132 The symbol I thus corresponds to the Schoenflies C1. **1970** A. J. WILSON *Elem. X-Ray Crystallogr.* 225 For space groups the Schoenflies symbols are quite inconvenient. *Ibid.*, There is also a Schoenflies notation for the Bravais lattices, which is used even less than the symmetry notation.

schoepite (ʃəʊpaɪt). *Min.* [f. the name of Alfred *Schoep* (1881–1966), Belgian mineralogist + -ITE¹.] A hydrated form of uranium trioxide found as yellow to brown tabular or prismatic orthorhombic crystals as an alteration product of uranium ore; three phases (schoepite I, II, and III) are known, differing slightly in composition, colour, and morphology.

1923 T. L. WALKER in *Amer. Mineralogist* VIII. 69 As this mineral appears to be quite distinct in optical and crystallographic properties from all other uranium minerals, the writer proposes to name it schoepite in honor of Professor Alfred Schoep, of the University of Ghent, who has contributed so much to our knowledge of uranium minerals from the Congo. **1960** *Ibid.* XLV. 1026 Crystals of schoepite that are apparently single yield multiple diffraction patterns... These correspond to the presence of two out of three possible distinct orthorhombic phases in parallel intergrowth in the crystal. The three phases are designated schoepite I, II and III. **1965** *Ibid.* L. 196 Crystals of schoepite commonly occur with an amber-brown core completely or partially surrounded by a derivative golden-yellow rim which retains the morphology of the original crystal... The brown part consists chiefly of schoepite I, and the yellow part mostly of schoepite II or schoepite III.

schoff, var. SCOFF *sb.*² in Dict. and Suppl.

schol (skɒl). Colloq. abbreviation of SCHOLARSHIP (sense 2).

1899 *Captain* Nov. 115/2 Wardour had licked Eccles and forfeited the 'schol'. **1958** B. HAMILTON *Too Much of Water* xi. 247, I won a schol to the House. **1969** J. SYMONS *Belting Inheritance* ii. 35 The old thing's *delighted* about the schol, Uncle Miles said.

‖ **schola cantorum** (skəʊlə kæntɔːrʊm). [med.L. = school of singers.] **a.** The choir-school attached to a cathedral or monastery (orig. the Papal Choir at Rome, established by Gregory the Great (*c* 540–604)). **b.** Used as the title of various groups of singers.

1782 C. BURNEY *Gen. Hist. Mus.* II. i. 16 Fleury, in his *Hist. Eccl.*..gives a circumstantial account of the Schola Cantorum, instituted by St. Gregory. **1887** E. L. TAUNTON *Hist. & Growth Church Music* iv. 39 The elder members of the Schola Cantorum, as it was called, had the title of Subdeacons. **1902** E. DICKINSON *Music in Hist. of Western Church* v. 181 The Schola Cantorum of Paris..is exerting a strong influence upon church music. **1929** E. C. THOMAS *Lay Folks' Hist. Liturgy* II. iv. 223 The members of the schola cantorum to which the lectors belonged had no other function than that of singing. **1942** G. GRAHAM *Music of Spain* xi. 168 Morera..has written..choral arrangements of Catalan folk songs, some of which have been performed by the Schola Cantorum of New York. **1964** P. F. ANSON *Bishops at Large* x. 475 The ladies of the *schola cantorum* looked fetching in their red gowns and caps.

scholar. Add: **5. a.** Further appositive examples, as *scholar-official*, *-performer*, *-poet*, *-publisher*.

1978 *Nagel's Encycl.-Guide: China* 323 All the prestige and importance of the scholar-officials came from their knowledge of characters. **1978** C. HOGWOOD in J. M. Thomson *Future of Early Music in Britain* 16 Much of the scholarly evidence is so easily assimilated by the performer that you have to invent a halfway category of the scholar-performer or the research-performer. **1948** J. BAILEY *Let.* 23 July (1935) 289 Do you know Hölderlin, the scholar-poet? **1979** R. P. GRAVES (*title*) A. E. Housman: the scholar-poet. **1963** *Times Lit. Suppl.* 26 Apr. 312/2 Dr. Mardersteig's position in typographical history as at once an artist-printer and a scholar-publisher.

scholarment (skɒləmənt). *nonce-wd.* [-MENT.] Scholardom; scholars collectively.

1922 JOYCE *Ulysses* 416 Toil on, labour like a very bandog and let scholarment and all Malthusiasts go hang.

scholarship. Add: **2. b.** *spec.* (though loosely) The 'eleven-plus' examination or the entrance to a grammar school made possible by reaching a satisfactory standard.

1959 I. & P. OPIE *Lore & Lang. Schoolch.* xi. 227 On the day I went to sit the scholarship I took the little owl and wrapped it up in a handkerchief in my pocket for luck. **1959** I. & P. OPIE *Ibid.* xvi. 356 Today the sharpest feeling is between the grammar schools and the secondary moderns, that is, between those who have gained a scholarship and those who have not in the eleven-plus examination. **1966** J. PARTRIDGE *Middle School* iv. 59 In Middle School the eleven plus is still viewed as 'the scholarship'.

3. *attrib.* and *Comb.*, as (sense 2) *scholarship boy*, *-candidate*, *child*, *class*, *exam*, *kid*, *paper*, *system*; **scholarship level** = *S-level* s.v. *S I. 4 a.

1959 T. S. ELIOT *Elder Statesman* I. 31 A scholarship boy from an unknown grammar school. **1980** R. F. FOSTER in Lyons & Hawkins *Ireland under Union* 254 'Scholarship boys' in politics. **1965** N. COGHILL in J. Gibb *Light on C. S. Lewis* 65 What it was learned to know in 1950 will be expected of scholarship-candidates in 2000. **1964** D. HOLBROOK *English for the Rejected* 4 The attempt to turn every child into a 'scholarship child' fit for academic education. **1966** J. PARTRIDGE *Middle School* v. 79 In the Junior School a 'scholarship' class soon emerges. **1959** in I. & P. Opie *Lore & Lang. Schoolch.* xi. 227, I took it to the scholarship exam for the grammar school and I passed. **1977** M. WALKER *National Front* 8 My education as a scholarship kid who went to grammar schools and won a scholarship to Oxford. **1963** Scholarship level [see *ORDINARY *a.* 5 e]. **1834** J. ROMILLY *Diary* 26 Apr. (1967) 51 Viewed all the Scholarship papers all day. **1960** CARR-SAUNDERS & JONES *Social Structure Eng. & Wales* xi. 119 We must also attempt to analyse the working of the free-place and scholarship system.

scholasticized (skəlæstɪsaɪzd), *ppl. a.* [f. as SCHOLASTICIZING *ppl. a.*: see -ED¹.] Imbued with or influenced by scholasticism.

1923 C. SINGER in *Edin. Rev.* Jan. 101 Dioscorides, the drug-monger, appealed to scholasticized minds for centuries. **1937** W. R. INGE *Proclamation* 12 This philosophy, already scholasticized by Proclus, became..a coherent body of doctrine.

scholzite (ʃɒltsaɪt). *Min.* [ad. G. *scholzit* (H. Strunz 1948: see H. Strunz *Mineralogische Tabellen* (ed. 2, 1949) 164, and in *Fortschritte der Mineral.* (1950) XXVII. 31), f. the name of Adolf *Scholz*, 20th-cent. German mineral collector and industrialist: see -ITE¹.] A hydrated basic phosphate of calcium and zinc, $CaZn_2(PO_4)_2.2H_2O$, occurring as a secondary mineral in colourless to greyish monoclinic crystals.

1950 *Chem. Abstr.* XLIV. 9306 (*heading*) Scholzite, a new mineral. **1974** *Mineral. Mag.* XXXIX. 686 Scholzite is the most common and conspicuous phosphate mineral in the mineralized zones at Reaphook Hill in the Flinders Ranges, South Australia. It occurs in voids as sprays of radiating white to colourless prismatic needles up to 3 cm long.

Schönbergian, var. *SCHOENBERGIAN *a.* and *sb.*

Schönlein (ʃɜːnlaɪn). *Path.* The name of Johann Lucas *Schönlein* (1793–1864), German physician, used in the possessive and occas. attrib. to designate a form of purpura associated with arthritis (described by him in 1837 (*Path. und Therapie* II. 48–49)); also used in combination with the name of *Henoch* (see *HENOCH).

Schönlein's disease is now regarded as one form of Henoch-Schönlein purpura.

1892 W. OSLER *Princ. & Pract. Med.* iii. 318 The diagnosis of Schönlein's disease offers no difficulty... Schönlein's peliosis is thought by most writers to be of rheumatic origin. **1937** *Arch. Dermatol. & Syphilol.* XXXV. 847 (*caption*) Photomicrograph of a section from a lesion of purpura affecting the leg only of a patient with Schönlein-Henoch's purpura. **1943** ORMSBY & MONTGOMERY *Dis. Skin* (ed. 6) ii. 501 Peck describes the histological changes of four cases of Schönlein-Henoch's purpura as resembling those produced by venom. **1946** [see *HENOCH]. **1974** *Encycl. Brit. Micropædia* VIII. 972/1 He [sc. Schönlein] was the first to describe the minute hemorrhages of the skin occurring in cases of anaphylactoid (allergic) purpura (Schönlein-Henoch purpura) and purpura rheumatica (Schönlein's disease; 1837), characterized by the appearance on the skin of small purple spots, by swelling pain, and tenderness of joints, and frequently by swelling of the hands, feet, or eyelids.

school, *sb.*¹ Add: **I. 1. c.** *to go to school with*: for † read *rare* and add later examples.

1959 *Listener* 3 Dec. 1005/1 Even those who cannot accept it entirely must assuredly go to school with him.

3. b. High school. (Examples of use in England.) Cf. *-SCHOOLER.

1875 A. MCDOWALL *Let.* 9 Oct. in V. E. Stack *Oxford High School* (1963) I. 1 At a special Council meeting..it was resolved..to open the school to be called The Oxford High School for Girls. ... but the High School had been for a long time almost unique among schools represented on the Headmasters' Conference. **1974** M. SPACKMAN *Hist. Oxford Central Girls' School & Cheney Girls' Grammar School* iii. 43 In most years, two or more girls..entered for and gained one of these places at either the High School or Milham Ford.

4. b. *the school of hard knocks*, the experience of a life of hardship, considered as a means of instruction. *U.S. slang.*

1912 ADE *Knocking Neighbors* 24 They had been brought up in the School of Hard Knocks. **1931** *Kansas City Star* 23 Oct. 36/5 Fraternity brothers in the school of hard knocks. **1953** *Sun* (Baltimore) 5 Sept. 10/6 He has been through the school of hard knocks, and battled his way up with his fists to the top of fistiana. **1980** G. V. HIGGINS *Kennedy for Defense* xx. 178, I learned my business in the school of hard knocks.

5. a. (Later examples.) Also, in descriptions of works of art, in phr. *school of* (an artist), used to designate an anonymous work produced in the school of a particular artist.

1891 R. FRY *Let.* 17 May (1972) I. 145, I find..the Venetian School of painting far more instructive than the Florentine. **1903** *Ibid.* 16 Mar. 207 This, which was called 'School of Lorenzo', is Piero all over. **1958** *Spectator* 15 Aug. 219/2 An American school-of-Chayevsky drama about a jailbird's wife. **1976** D. FRANCIS *In Frame* ix. 135 Although they were original oil paintings, they were basically second rate. The sort sold as 'school of' because the artists hadn't bothered to sign them. **1981** M. SPARK *Loitering with Intent* ii. 56 She said, 'Is that a real Degas you have in your room?' 'School of,' I said.

b. (Earlier example.)

1749 SMOLLETT tr. *Le Sage's Gil Blas* IV. i. 6 Mr. Doctor..as I am a grand nephew to a physician of the old school give me leave to revolt with you against chymical medicines.

c. In extended use of sense 5 a, in phr. *school of thought* (also † *opinion*). Also used (freq. *absol.*) with 'school' considered impersonally, a particular type of doctrine or practice as followed by such a body of persons.

1864 [see sense 5 a in Dict.]. **1873** *Illustr. London News* 26 July 70/2 It will not be necessary to utter a single word that need occasion offence to either of those 'schools of thought' into which The Church of England is divided. **1892** *New Review* May 571 He is a 'gentleman and scholar', ..'trained in a liberal school of thought'. **1909** A. BERGET *Conquest of Air* II. v. 230 We are confronted by two schools of aviating apparatus: the American school..which demands everything of the aviator, and the French school.. which requires..the minimum from the pilot. **1927** *Public Opinion* 28 Feb. 179/1 There is in philosophy a school of thought christened by Professor William James with the name Pragmatism. **1940** *Manch. Guardian Weekly* 5 Apr. 270 With two schools of thought existing in France on the subject of Russia, Molotoff's speech,..has produced two different sets of reactions. **1977** R. WILLIAMS *Marxism & Lit.* II. iv. 97 The theory became at once a cultural programme and a critical school. **1979** L. KALLEN *Introducing C.B. Greenfield* xii. 148 There's a school of thought that considers benevolent paternalism a little sick.

6. a. Delete † *Obs.* and add later examples.

b. (Earlier example.)

1842 *Impositions practised by Vagrants* 12 These lurkers generally go in schools, (companies) and will obtain from One to Two Pounds daily. **1882** *Sydney Slang Dict.* 7/2 *School*, company of gamblers, mob of sharpers, and those who prey on the public. **1911** L. STONE *Jonah* II. vi. 213 He could think of nothing but the two-up school, which had swallowed all his spare money before he was married. **1946** A. MARSHALL *These are my People* 83 If I got into a school with some of the mugs round here they'd be penniless in two hours. **1952** H. INNES *Campbell's Kingdom* II. 230 Four of the boys had started a poker school. **1976** J. R. L. ANDERSON *Death in Desert* v. 87 Sometimes a few of the chaps would get a card school going after supper.

c. A group of persons drinking together in a bar or public house, and taking turns to buy the drinks.

1890 BARRÈRE & LELAND *Dict. Slang* II. 206/2 *School*,..any small gathering of people generally bent on pleasure, as a school of drinkers in a public house or canteen. Much used by soldiers. **1911** R. MACAIRE *Disease & Remedy* 11 A 'school' got more from those that did not drink. **1951** *Landfall* V. 22 [She] goes across to join another school by the wall. **1963** W. H. PEARSON *Coal Flat* i. 21 He came up to the school Rogers was drinking with. **1971** D. LEES *Rainbow Conspiracy* v. 72, I..ordered a pint of bitter for myself. I didn't want to get into a school and I needed to think.

II. 7. a. Now in revived use within U.S. and some British universities (esp. those of recent foundation), a department, faculty, or course of study in a college or university. (Perh. influenced by senses 9 and 10 in Dict.)

In the U.S. *school* is often used to designate either a department devoted to one subject or a grouping of several subject departments. It is also the standard designation for an institution providing postgraduate instruction in a particular subject (as *law school*, *medical school*, etc.). In recently founded universities in the U.K. *school* has been used to designate a department teaching a range of subjects traditionally taught separately.

1727 *Statutes Wm. & Mary Coll.* in Hofstadter & Smith *Amer. Higher Educ.* (1961) I. I. x. 43 Let there be four schools assigned within the college precincts. *Ibid.* 44 In the philosophy school we appoint two masters or professors. **1772** J. WITHERSPOON *Address Inhabitants Jamaica* in *Ibid.* II. x. 144 Two at least of the Professors of the justly celebrated Medical School lately founded in Philadelphia. **1835** J. MARTIN *Descr. of Virginia* 82 The different branches of science and literature..taught [at the University of Virginia] are styled *schools*. **1871** L. H. BAGG *Four Years at Yale* 32 Connected with the college are four professional 'schools' or 'departments', of which..the oldest is the Theological. **1894** *Rep. Commissioners Gresham Univ. London* p. xix, in *Parl. Papers 1893–4* (C. 7259) XXXI. 807 We propose that each of the teaching institutions which complies with the necessary conditions shall be admitted, either as a whole or in certain departments, as a School of the University [of London], that is as a School at which University courses of instruction are to be pursued. **1910** *Encycl. Brit.* XIII. 39/1 The medical school [of Harvard University]..dates from 1782, the law school from 1817, the divinity school..from 1819, and the dental school.. from 1867. **1949** *Cavalier Daily* (Univ. of Va.) 22 Oct. 1/3 Williams graduated from the University in 1949 and is now in his second year of medical school at Johns Hopkins University in Baltimore. **1964** A. BRIGGS in D. Daiches *Idea of New University* iv. 62 The Schools [of the University of Sussex] were envisaged not as super-departments, to

which 'subjects' were attached, but as centres of linked studies, some of which would be shared with other Schools. **1971** E. ASHBY *Any Person, any Study* ii. 71 Some British universities, in their enthusiasm to 'redraw the map of knowledge', have abolished departments and put in their place 'schools of study' (e.g. European studies, African studies, which include the history, economics, politics, language and literature, and geography of these regions). **1972** J. BEN-DAVID *Amer. Higher Educ.* vi. 87 Intellectually the graduate school had become the decisive influence in higher education by the beginning of this century. **1976** *Bull. Yale Univ.* 30 Dec. 163 The courses of study in Yale University are offered in twelve schools, as follows. Yale College (1701), which is the undergraduate school, the Graduate School (1847), School of Medicine (1810), Divinity School, [etc.].

f. *U.S.* A college or university. Also in phrases *to go* († *put*) *to school*, to attend (send to) college or university.

1767 P. V. FITHIAN *Jrnl. & Lett.* (1900) 1 A letter to my Father, begging him to put me to School. **1904** *Delineator* Oct. 657 College pillows..of crimson, with 'Harvard', n white letters; of orange, with 'Princeton' in black, and similarly with the names and colours of other schools. **1957** A. BUCHWALD *Brave Coward* 54, I am more American than you are. I even went to school on the GI Bill of Rights, in Rome. I got an honorable discharge from the Army. **1962** —— *How much is that in Dollars?* p. ix, When friends.. assured me the streets of Paris were paved with mattresses, I decided to finish up my last year of schooling there... But while we were going to school..Congress passed the monumental Marshall Plan. **1967** *Boston Sunday Herald* 26 Mar. II. 5/6 (*caption*) Oxford crewman J. K. Mullard waves jubilantly after victory over traditional rival Cambridge... Oxford won by three lengths in 113th meeting between the schools. **1977** I. SHAW *Beggarman, Thief* I. vi. 76 The proms at which he played the trumpet in the band, to help pay his way through school. **1978** *Sci. Amer.* July 15/2 The latter experience convinced him that his interest lay in research; he therefore went back to school, acquiring his Ph.D. from Stanford University in 1965.

VI. 16. a. *school-age* (earlier and later examples), *assembly*, *atlas*, *bag*, *beret*, *blazer*, *blouse*, *bus* (also *-busing*), *cap* (hence *-capped* adj.), *curriculum*, *-desk*, *dinner*, *education*, *fee* (later examples), *holiday*, *librarian*, *-life*, (earlier example), *lunch*, *mag*, *magazine*, *meal*, *nurse*, *party*, *play*, *-poem*, *prefect*, *prize* (earlier example), *reader*, *register*, *rule*, *satchel*, *scarf*, *secretary*, *slang*, *song*, *subject*, *system*, *-teacher* (later examples) (hence *-teacherish* adj.), *tie*, *treat*, *trunk*, *uniform*, *wear*, *year* (later examples).

1741 S. RICHARDSON *Familiar Lett.* cxxx. 168 Nor is the Consequence of this Defect confin'd to the School-age, as I may call it. **1939** M. S. RICE *Working-Class Wives* iii. 66 The children of school age come under the School medical services. **1972** G. SERENY *Case M. Bell* II. i. 80 Fernwood Reception Centre..is used for children of school age who come there primarily because of sudden family emergencies. **1974** *Sat. Rev. World* (U.S.) 2 Nov. 24/3, 95 per-cent of school-age Eskimos are in school. **1932** MRS. J. MURRAY (*title*) Incidental music for use at school assembly. Arranged by I. R. Davies. **1977** J. AIKEN *Last Movement* ix. 166 Opera and large gatherings ran each other close for first place among her dislikes. How did she stand school assemblies? **1815** J. A. CUMMINGS (*title*) A school atlas, accompanying ancient and modern geography. **1885** C. M. YONGE *Two Sides of Shield* I. vi. 100 The elder boys' old school atlases. **1979** H. MCLEAVE *Borderline Case* xvi. 155 On a school atlas Dr. Li charted the progress of the disease. **1895** *Montgomery Ward Catal.* Spring & Summer 118/3 Water-proof School Bags; made of enameled cloth, with flap and leather shoulder strap. **1913** P. GEDDES *Masque of Anc. Learning* 3 Boy enters, swinging his school-bag. **1977** C. FREMLIN *Spider-Orchid* xiii. 90 She'd dumped her school bag on the floor. **1967** M. DRABBLE *Jerusalem the Golden* iii. 51 The girls in her class..regarded her as relatively plain.. with no notion of how to twist a school beret or hitch a school skirt. **1975** 'J. BELL' *Victim* xiii. 140 A small girl stepped out. She held a school beret in one hand. **1913** J. VAIZEY *College Girl* v. 62 The boys wore flannel trousers with school blazers and caps. **1978** A. PRICE *'44 Vintage* vi. 74 His school blazer..had been too small for him. **1932** D. C. MINTER *Mod. Needlecraft* 253/1 *School or Gym Blouse*... Long sleeves for school type, short for gym blouse. **1979** K. CONLON *Move in Game* I. i. 14 Mrs Brennan wrote: 'butter, eggs, shoe polish, school blouse, [etc.].' **1908** *Suburban Life* July 48/1 (*caption*) The school bus. **1939** G. HOUSEHOLD *Rogue Male* 110, I saw the school-bus and an occasional car. **1976** P. R. WHITE *Planning for Public Transport* i. 25 A local authority..may also be an important customer in its own right, by allocating subsidies, and contracts for school bus services. **1974** *Times* 25 Oct. 10/5 The sensitive issue of school busing in Boston. **1908** *Magnet* I. 1, His hair was thick and curly, and there was a school-cap stuck on the back of his head. **1930** AUDEN *Poems* 18 The rest as jury, wearing school caps. **1975** *Listener* 4 Dec. 747/3 Tell us that story about going to St John's Wood. Well, I had a letter, and went up wearing a school cap. **1933** M. LOWRY *Ultramarine* ii. 97 Mothers with warm-smelling furs are fussing with their school-capped sons. **1973** M. RUSSELL *Double Hit* ii. 15 A school-capped boy bicycled out of the main gates. **1913** C. MACKENZIE *Sinister St.* I. II. i. 156 He taught Geography and English History and English Literature, so far as the school curriculum allowed him. **1981** *Listener* 1 Jan. 22/3 His comments on the school curriculum contain a germ of truth. **1842** DICKENS *Amer. Notes* I. iii. 75 A little enclosure, made of school-desks and forms. **1953** A. CLARKE *Moment next to Nothing* I. ii. 29 I'll clear the table, have our school-desk ready. **1835** DICKENS *Sk. by Boz* (1836) 1st Ser. I. 26 They were three long graces in drapery, with the addition—like a school-dinner—of another long grace afterwards. **1963** *New Society* 22 Aug. 5/1 That inevitable horror, the school dinner. **1973** J. BURROWS *Like Evening Gone* x. 115 She used to help with school dinners..serving and washing up. **1731** J. CREICHTON *Mem.* 10 Having lost the Benefit of a thorough School-Education..the Reader cannot reasonably expect to be much pleased with my Style. **1848** MILL *Pol. Econ.* I. II. xiv. 463 The earnings of.. any labour which requires school education, are at a monopoly rate. **1873** C. M. YONGE *Pillars of House* I. iii. 64 The school-fee was a mere trifle, but Mr. Ryder would willingly have boarded and lodged the boy. **1958** 'CASTLE' & 'HAILEY' *Flight into Danger* 17 Pay off the bills—the new water tank, school fees, instalments on the Chev. **1973** J. LEASOR *Host of Extras* ii. 32 He'd be selling the car to pay his son's school fees. **1939** School holiday [see *nursery tea* s.v. *NURSERY 8 a]. **1981** J. ROBIN *Elmdon* xi. 224 A private house, occupied in the school holidays by the wife of a business-man in Baghdad. **1920** B. M. PEACOCK (*title*) A school and club librarians' handbook. **1978** J. IRVING *World according to Garp* ii. 28 It was a habit among the school librarians, upon recognizing that they didn't have a book which someone sought, to say, 'Perhaps the infirmary has it.' **1721** M. CAVE *Let.* 27 Nov. in M. M. Verney *Verney Lett. of Eighteenth Cent.* (1930) II. xxiii. 71 The apprehension of Tommy's weak Constitution I find very grevious, inferring that he is unable to undergo a School Life. **1949** M. MEAD *Male & Female* xvi. 330 As school-lunches develop, the home with school-age children is deserted all day long. **1980** *Times* 10 Sept. 8/1 Changes [are] taking place in the school meals service... Many parents are weighing up the relative merits of school lunches versus packed ones. **1960** L. DURRELL *Let.* in *Spirit of Place* (1969) 153 The boys of King's School..asked for an article for the school mag. **1976** *Listener* 8 Apr. 452/4 It is very difficult to write about school, and the tone of the school mag is not wholly avoided. **1856** C. M. YONGE *Daisy Chain* II. v. 383, I got leave to send a ballad..to that school magazine... It was actually inserted. **1939** C. ISHERWOOD *Goodbye to Berlin* 311 The newspapers are becoming more and more like copies of a school magazine. **1963** A. HERON *Towards Quaker View of Sex* iii. 23 The occasional poems which seek entry in the columns of the school magazine. **1948** F. LE G. CLARK (*title*) The social history of the School Meal service. **1973** *Times* 5 Oct. 4/5 A [parliamentary] resolution called for..free school meals. **1912** *Q. Rev.* July 57 Enormous improvements.. have been effected in the environment of the nation since that time. A full account of these is here impossible but it may be said that they include..the appointment of..district and school nurses. **1976** J. PHILIPS *Backlash* (1977) III. i. 122 If..the school nurse..came, I was to tell her to give me some first aid. **1803** T. LAWRENCE *Let.* 28 Jan. in D. E. William *Life & Corr. Sir T. Lawrence* (1831) I. 231 We all sat down like a Rugby school party, but rather more vociferous. **1968** J. SANGSTER *Touchfeather* ii. 11, I went [to Berlin] with a school party... We stayed for five days..visiting the museums. **1976** H. TRACY *Death in Reserve* xviii. 136 He was told that there were parties of Boy Scouts..and a school party due next week. **1933** E. K. CHAMBERS *Eng. Folk-Play* 187 It is not to be supposed that, after the Reformation and the growth of the professional travelling companies, local plays..ceased to be performed... Some are school-plays produced by the local Holophernes. Some are May games. **1972** *Guardian* 17 Aug. 10/6 It is impossible not to wish him well..like you would a child on sports day or in a school play. **1973** J. R. L. ANDERSON *Death on Rocks* iii. 62 She's gone with the kids to a school play. **1922** JOYCE *Ulysses* 167 That last pagan king of Ireland Cormac in the schoolpoem choked himself. **1933** R. TUVE *Seasons & Months* iii. 75 *Cuculus*, above all *Philomela* (familiar as the subject of various 'school-poems'). **1949** E. COXHEAD *Wind in West* vii. 178 He who had been the naughty child was now the school prefect. **1975** P. D. JAMES *Black Tower* ii. 35 The old insistent arguments spoken in that confident school prefect's voice. **1853** C. BRONTË *Villette* I. iii. 54 Graham, it chanced, was at that time greatly preoccupied about some school-prize, for which he was competing. **1835** H. A. HANSARD (*title*) Souter's second school reader. **1940** J. BUCHAN *Memory Hold-the-Door* 194 *Prester John*.. has since become a school-reader in many languages. **1981** E. HAY *Sambo Sahib* viii. 110 Helen [Bannerman].. was delighted..that some of her books had been selected as school readers. **1973** J. BURROWS *Like Evening Gone* vi. 73 There were forty-three names on the school register. **1943** G. GREENE *Ministry of Fear* III. ii. 192 Excited like a boy breaking a school rule. **1978** F. WELDON *Praxis* vii. 36 School rules forbade conversation between girls of different age groups. **1907** *Yesterday's Shopping* (1969) 325/2 School Satchels. Waterproof Brown Canvas. **1972** J. FLEMING *Alas Poor Father* i. 7 Picking up the school satchel, he hooked it over his arm. **1907** E. NESBIT *Enchanted Castle* iii. 90 The crimson school-scarf that had supported his white flannels. **1971** A. PRICE *Alamut Ambush* x. 119 Carelessly hung coats and school scarves on the row of wooden pegs. **1958** J. TOWNSEND *Young Devils* vi. 52, I was shown into his room by a cheerful middle-aged school secretary. **1977** J. AIKEN *Last Movement* i. 24 Mother..had asked if Gina would be interested in the job of school secretary. **1900** FARMER *Public School Word-Bk.* p. v, It would, however, seem almost necessary to emphasise that this Word-Book is not, *per se*, a dictionary of school slang. **1934** PRIEBSCH & COLLINSON *German Lang.* II. v. 263 A word or two may be added on German school-slang (*Pennälersprache*). **1975** D. DURRANT *With my Little Eye* xviii. 181 Patty's coarse, cruel, school slang bitchery. **1934** M. V. HUGHES *London Child of Seventies* vi. 68 Another treat to me was the school song ('Homo plantat'). **1974** *Listener* 17 Jan. 84/1 E. E. Bowen and John Farmer started the collection of Harrow School Songs, in the 1870s. **1922** H. E. PALMER *Everyday Sentences in Spoken Eng.* p. v, English is no longer either an abhorred school-subject nor a fascinating literary hobby. **1977** *Grimsby Even. Tel.* 5 May 5/2 She said..that it could easily be possible to make road safety a school subject on its own merit. **1814** JANE AUSTEN *Mansfield Park* III. iii. 63 Common neglect of the qualification..in the ordinary school-system for boys. **1869** C. L. BRACE *New West* 79 The general school system of California..is more centralized. **1911** C. E. PERSONS et al. *Labor Laws & their Enforcement* 218 We should know how many children..the school system could no longer control, as well as those it still retains. **1976** *National Observer* (U.S.) 1 May B4/4 In America, school systems have expanded in the atmosphere of an anticipated tug-of-war with parents. **1932** G. GREENE *Stamboul Train* I. i. 6, I see his passport. Richard John. Schoolteacher. **1950** C. S. BELSHAW *Island Admin. in S.W. Pacific* xii. 122 In one or two cases Councils unofficially..levied their own funds, which were to be put to such purposes as paying school-teachers [etc.]. *a* **1930** D. H. LAWRENCE *Phoenix* (1936) 361 The heroine being one of the old-fashioned school-teacherish sort. **1978** H. WOUK *War & Remembrance* xiii. 126 She shook a schoolteacherish finger at him. **1932** School tie [see *OLD SCHOOL TIE]. **1937** G. BARKER *Poems* 24 O long lost upward in the dream descending, The flying pig and the school-tie anaconda! **1977** *Times* 23 Nov. 12/8 Western society..puts children..in school uniform and school ties to make them conform. **1888** MRS. H. WARD *Robert Elsmere* II. III. xxv. 274 Counting up the engagements of the next few weeks—the school-treat, two club field-days, a sermon. **1934** D. L. SAYERS *Nine Tailors* 121 An importunate child at a school treat. **1915** KIPLING *Diversity of Creatures* (1917) 429 We'll get his old school trunk to-morrow and pack his civilian clothes. **1978** *Times* 3 Aug. 9/2 Cash's name-tapes are very widely available... But get a move on if..the first school trunk is looming. **1933** A. WHITE *Frost in May* ii. 47 She trotted sedately behind the lay-sister, wearing her school uniform. **1976** W. TREVOR *Children of Dynmouth* ii. 37 They were still in their school uniforms—Stephen's grey with touches of maroon, Kate's brown and green. **1939–40** *Army & Navy Stores Catal.* p. lv/1 School wear. **1976** *Evening Post* (Nottingham) 14 Dec. 15/8 (Advt.), Derby's leading boyswear and schoolwear specialists. **1961** *Guardian Jrnl.* (Nottingham) 14 Nov. 4 The intention here is to reduce these dates in the school year from three to two. **1965** *New Society* 16 Sept. 4/2 The new school year has begun and millions of parents have heaved a sigh of relief. **1975** *Language for Life* (Dept. Educ. & Sci.) xx. 293 An additional benefit was the informal contact between teacher and parent, a valuable foundation for the coming school year.

b. *school bully*, *friend*, *-kid*, *-urchin*.

1907 'MARK TWAIN' in *N. Amer. Rev.* Jan. 7, I had had a quarrel with a big boy who was the school-bully. **1956** 'C. BLACKSTOCK' *Dewey Death* vii. 154 Mark..had something of the school bully in him... It seemed he derived a cruel satisfaction from the young man's palpable fear. **1853** C. BRONTË *Villette* I. iii. 43 Graham is busy with his school-friends. **1973** 'M. INNES' *Appleby's Answer* x. 97 He recalled Judith's school friend as soon as he set eyes on her. **1938** School-kid [see *FEIS 2]. **1976** J. WAINWRIGHT *Who goes Next?* 164 He went to live with his boy-friend—little more than a schoolkid. **1922** JOYCE *Ulysses* 201 Antiquity mentions that Stagyrite schoolurchin and bald heathen sage.

c. *school-bell* (earlier and later examples), *building* (later example), *gate*, *hall*, *library*.

1702 S. SEWALL *Diary* 9 Aug. (1879) II. 61 Set out from Salem as the School-Bell rung. **1779** J. WEDGWOOD *Let.* 23 Nov. (1965) 247 Rise at 7 in winter, when I shall ring the school bell. **1976** C. DEXTER *Last seen Wearing* xvii. 136 The school bell rang at 4.00 p.m., and the last lesson of the day was over. **1975** *Language for Life* (Dept. Educ. & Sci.) xix. 280 The class is likely to be held in a school building. **1847** C. M. YONGE *Scenes & Characters* xxiii. 280 William walked to the school gate with them. **1973** J. MANN *Only Security* iii. 23 When she left the school gates behind her, she was finished with the problems for the day. **1933** A. THIRKELL *High Rising* viii. 161 Amy took Laura over to the school hall. **1980** J. THOMSON *Alibi in Time* viii. 104 The school hall opened off the entrance foyer. **1854** *Rep. Trans. Pennsylvania State Agric. Soc.* 276 Another great reform would be the introducing of a school library into every district school. **1860** J. A. SYMONDS *Let.* 18 Aug. (1967) I. 260, I shall..get to Shrewsbury at about half past one. That will allow me time to see the MS in the School Library. **1941** M. TREADGOLD *We couldn't leave Dinah* i. 20 A discerning headmaster..had directed..her vivid imagination to the excellent school library. **1971** J. B. CARROLL et al. *Word Freq. Bk.* p. vi, Some of these publications are normally found in the classroom, others in the school library.

e. Produced by the pupils or assistants of a master of a school of art (see sense 5 a), as *school painting*, *-piece*, *-work*.

1903 R. FRY *Let.* 16 Mar. (1972) I. 207, I have found..a tondo..which I can't help still fancying a schoolpiece. Anyhow this, which was called 'School of Lorenzo', is Piero all over. **1905** MRS. H. WARD *Marriage of William Ashe* I. ii. 31 It was an old low-ceiled room, panelled in white and gold, showing here and there an Italian picture—Saint, or Holy Family, agreeable school-work. **1937** *Burlington Mag.* Feb. 77/1 The accidental meeting of northern and southern art-forms, as it were, in a school-piece. **1979** R. COX *Auction* iii. 58, I would certainly not say priceless. As School paintings go, yes, it's valuable.

17. *school-based* adj.

1975 *Language for Life* (Dept. Educ. & Sci.) xxiii. 339 Up to a third of the time is normally given to school-based teaching practice. **1977** *Times Educ. Suppl.* 21 Oct. 7/1 Many people..believe school-based assessments are important in giving a comprehensive picture of the candidates' achievements.

18. *school desegregation*, *governor*, *management*, *manager* (earlier example), *-teaching* (later examples).

1961 J. W. PELTASON (*title*) Fifty-eight lonely men: Southern federal judges and school desegregation. **1976** *Billings* (Montana) *Gaz.* 20 June 11-E/6 President Ford heard pro and con views on busing as a remedy for school desegregation from school superintendents and principals Saturday. **1976** L. HENDERSON *Major Enquiry* xvii. 116, I was attending a meeting of the Branton Education Committee, I am one of the school governors. **1883** J. LANDON (*title*) School management. **1975** *Language for Life* (Dept. Educ. & Sci.) xii. 193 It would not be easy to argue for another post to be added to the senior level of the school management structure. **1862** *Edin. Rev.* Apr. 415 That fortunate individual has dined at the house of a school manager. **1854** C. M. YONGE *Castle Builders* v.

65 The example of their sister..made them think school-teaching the most dignified and delightful of tasks. **1950** *Sport* 7–11 Apr. 14/1 Defensive stability has been added by the signing of schoolteaching goalkeeper, Alec Grant. **1981** E. NORTH *Dames* vi. 101 You did the right thing getting out of schoolteaching when you did.

19. school attendance, attendance at a school, used *attrib.* of persons or things involved in the enforcement of compulsory school attendance; **school(s) broadcast**, a radio or television broadcast for the instruction of children in school; also **school broadcaster, broadcasting**; **school-butter**, (*c*) (examples): **School Cert.**, abbrev. of next; **School Certificate**, in one of several (public) examination systems, a certificate of proficiency in subjects learned at school; **school colours**, the distinctive colours of a school, esp. as conferred as a sign of sporting achievement (see *COLOUR *sb.*¹ 6 c); **school committee**, (*a*) *U.S.* = *SCHOOL BOARD 2; (*b*) *N.Z.*, a group of the parents of primary school-children elected to assist the headmaster of that school; **school crossing**, a supervised road-crossing for school-children near the entrance to a school; **school district** *N. Amer.*, a unit for the local administration of schools; † **school-feast**, a tea-party or picnic for village school-children; **school inspector**, an officer appointed to inspect and report on the condition of schools and the teaching therein; hence **school-inspectorship**; **school journal** *N.Z.*, a booklet prepared by the Department of Education and issued to all primary schools at regular intervals; **school land** *N. Amer.*, land set apart for the financial support of schools (cf. *school section*); **school-leaver**, one who is about to leave or has just left school (cf. *LEAVER); **school leaving age** = *leaving age* s.v. *LEAVING *vbl. sb.*; **school method**, the teaching system to be followed by a teacher in training; the practice or theory of school-teaching; **school milk**, milk provided at reduced cost or free of charge to children in school; **school phobia** *Psychol.*, excessive anxiety about or fear of attending school; so **school-phobic** *a.* and *ellipt.* as *sb.*; **school report** = *REPORT *sb.* 2 e; **school section** (earlier and further examples); also *Canad.*; **school ship** (earlier U.S. example); **schools programme** = *school broadcast*; **schools television**, a television broadcast for schools; **school story**, a story treating of life in a school; **school-years** = *school-time* (*b*).

1876 *Act* 39 & 40 *Vict.* c. 79 § 7 The provisions of this Act..shall be enforced—(1.) In a school district within the jurisdiction of a school board, by that board; and (2.) In every other school district by a committee (in this Act referred to as a school attendance committee). **1911** G. B. SHAW *Getting Married* Pref. 185 If you pay less than £40 a year rent, you will sometimes feel tempted to say to the..school attendance officer, and the sanitary inspector: 'Is this child mine or yours?' **1944** *Act* 7 & 8 *Geo. VI* c. 31. 252 The authority shall serve upon the parent an order in the prescribed form (hereinafter referred to as a 'school attendance order') requiring him to cause the child to become a registered pupil at a school named in the order. **1971** *Reader's Digest Family Guide to Law* 219/2 The education authorities sent the school attendance officer to the home and..he found that the child had received no tuition that day. **1931** *4th Ann. Rep. B.B.C.* in *Parl. Papers 1930–31* (Cmd. 3863) X. 291 It is believed that 5,260 schools followed the school broadcasts in the year. **1949** *Radio Times* 15 July 6/1, I asked the Chief Wireless Operator if it would be possible for me to listen to the Schools Broadcast. **1962** A. NISBETT *Technique Sound Studio* 246 The BBC permits the recording of schools broadcasts. **1974** School broadcaster [see *school broadcasting* below]. **1927** *B.B.C. Handbk. 1928* 138/1 Many thousands have their school set with its loud-speaker, and School Broadcasting has become a subject for educational research. **1928** *1st Ann. Rep. B.B.C.* 6 in *Parl. Papers* (Cmd. 3123) VII. 121 The Kent Education Committee undertook an enquiry into the efficacy of schools broadcasting. **1974** *Time* 8 Apr. 13/1 This autumn radio will celebrate 50 years of school broadcasting... School broadcasters use much more sophisticated material to complement the work of the teacher. **1935** A. B. LONGSTREET *Georgia Scenes* 84, I fell down.., running after that fellow that cried 'school-butter'. **1912** *Dialect Notes* III. 588 When he yelled *school butter* at us, we yanked him off the wagon and blacked his eyes. **1937** *Discovery* Jan. p. ii/1 (Advt.), *School Cert.* and *Army*. Quick easy way if *Latin* taken for former only. **1967** H. W. SUTHERLAND *Magnie* vii. 92 She could have taken the one year course herself, but she thought you'd need a school cert. at least. **1977** D. MAY in P. Collenette *Winter's Tales 23* 90 We were doing a run-through of School Cert. **1888** KIPLING *Wee Willie Winkie* 75 They were an educated regiment, the percentage of school-certificates in their ranks were high, and most of the men could do more than read and write. **1911** *Rep. Consultative Comm. Exam. Secondary Schools* 106 in *Parl. Papers* (Cd. 6004) XVI. 159 We would suggest that the examination should be called the examination for the Secondary School Certificate. **1931** 'G. TREVOR' *Murder at School* ii. 36 He was in my junior form... I expect he'd have taken his School Certificate. **1948** *Min. of Educ. Circular* No. 168. 23 Apr. 3/1 In 1951 the Minister proposes that the existing School and Higher School Certificate examinations should be discontinued and that in their place there should be introduced an examination for the 'General Certificate of Education'. **1966** G. W. TURNER *Eng. Lang. in Australia & N.Z.* viii. 173 He [sc. a New Zealander] is likely to sit School Certificate (approximately equivalent to English GCE Ordinary Level). **1978** A. PRICE *'44 Vintage* iv. 46 The acquisition of School Certificate German had been the limit of his ambition. **1913** C. MACKENZIE *Sinister St.* I. II. xiv. 382 He respected the quest of School Colours. **1924** A. HUXLEY *Little Mexican* 3 Holding to my head..a speckled straw [hat], gaudy with the school colours. **1972** L. P. DAVIES *What did I do Tomorrow?* v. 63 A scarf in the school colours of narrow emerald and gold stripes on black. **1787** in C. O. Parmenter *Hist. Pelham, Mass.* (1898) 226 Voted Not to Devid the School Quarter where Dea. John Crawford is School Committee Man. **1877** *Statutes N.Z.* XXI. § 58. 122 For every school district constituted under this Act there shall be a School Committee consisting of seven householders within the school district, to be elected as hereinafter provided. **1945** *Suburban List* (Essex Junction, Vermont) 8 Feb. 10/3 The school committee could not keep the buses running. **1947** 'A. P. GASKELL' *Big Game* 87 To crown it all the damned School Committee had to pick on this Saturday for their school picnic. **1951** *Sunday Pictorial* 21 Jan. 4/3 The warning signs brandished by Bristol's school-crossing wardens are so large that wardens find it hard to keep both feet on the ground in a strong wind. **1979** *Hampstead & Highgate Express* 22 June 10/3 Most of the school crossings in the borough were without lollipop men and women. **1809** E. A. KENDALL *Travels through Northern Parts U.S.* I. 128 There are thirteen school districts [in Berlin, Conn.]. **1876** [see *school attendance* above]. **1903** A. B. HART *Actual Govt.* 542 The smallest unit of school administration is the school district, which in many States has its own board, raises its own taxes, and appoints its own teachers. **1978** *N.Y. Times* 29 Mar. A13/1 In Duarte,..the school district and a medical center are the two major employers. **1849** C. BRONTË *Shirley* II. vi. 137 (*heading*) The school-feast. **1879** M. E. BRADDON *Vixen* I. xvii. 325 The school-feast was fixed..for the Wednesday in Whitsun week. **1822** *Missionary Reg.* Dec. 501/2 (*heading*) School inspectors and village readers. **1873** C. M. YONGE *Pillars of House* IV. xli. 192 He's a school inspector! Don't you have inspections here? Not under Government? O thrice happy people! **1924** M. KENNEDY *Constant Nymph* xiv. 190 He knows too much about everything..being a school inspector. **1979** D. COOK *Winter Doves* I. 26 The School Inspectors came round..and they saw the state of the house, so they sent someone from the Council. **1911** H. S. WALPOLE *Mr. Perrin & Mr. Traill* iii. 47 He saw himself at Eton or Harrow, or a school-inspectorship. **1907** *Append. Jrnls. House Reps. N.Z.* E.1E. 6, I might mention..the *School Journal*, because it will give an opportunity of explaining the place it should occupy in the school system. **1935** J. GUTHRIE *Little Country* v. 102 The word [sc. Australasia] was expurgated from school journals. **1648** *Suffolk Co.* (Mass.) *Deeds* (1880) I. 91 Humphrey Johnson of Roxbury granted unto William Chenie of Roxbury twenty Acres of land in Roxbury bounded with..the school lands & Richard Peacocks northwest. **1775** *Let.* 28 Feb. in *Coll. New Hampshire Hist. Soc.* (1889) IX. 89, I might..lay out for the Clearing the School Lands to the amount of £500 Sterlg. **1885** *Rep. Indian Affairs* (U.S.) 147 Others claim they have purchased their lands from the State of Nevada under the school-land grant. **1952** D. F. PUTNAM *Canad. Regions* 372/2 Another factor in the land pattern was the reservation of certain parcels as school lands. **1925** *Contemp. Rev.* May 634 The problem of the unemployed 'school-leaver' complicates in many ways the problem of the boy at work. **1955** *Times* 14 July 2/6 The Student Training Scheme is designed to enable public and grammar school-leavers to qualify professionally, having particularly in view careers in design, development, production, or commercial engineering. **1980** *Listener* 19 June 803/1 Wakefield was a miner's son and his parents did not expect him to be a late school-leaver. **1920** *Circular* (Board of Educ.) No. 1180. 12 Oct. 3 The Board are prepared to consider proposals for making a byelaw under the subsection raising the school leaving age to 15. **1946** *Ann. Reg. 1945* I. iii. 75 On September 28 the Minister of Education announced that the school-leaving age would be raised to 15 on April 1, 1947, and that no attempt would be made to postpone the change beyond that date. **1955** *Times* 9 July 2/6 Many people thought that fewer pupils were now staying after school-leaving age, but that was not so. **1972** *Times* 15 Jan. 2/5 Mrs Thatcher, Secretary of State for Education and Science, said in London last night that she had signed the order-in-council to raise school-leaving age to 16 in September. **1877** F. J. GLADMAN (*title*) School method. **1917** BERESFORD & RICHMOND *W. E. Ford* ix. 194 A description of a typical staff-meeting discussion of school method. **1927** J. ADAMS *Errors in School* 35 School-method books. **1934** *Milk-in-Schools Scheme* (Milk Marketing Board) 6 Communications should be addressed to the 'School Milk' Dept., Thames House, Millbank. **1964** L. LEE *Firstborn* 11 I'd ask her to accept her faults..and not blame them..on..school-milk, or the British Railways. **1977** *Times* 10 May 4/3 Ministers will consider claiming an EEC grant worth 3p a pint on school milk. **1941** A. M. JOHNSON in *Amer. Jrnl. Orthopsychiatry* Oct. 702 The syndrome, often referred to as 'school phobia', is recognizable by the intense terror associated with being at school. **1959** *Times* 24 Nov. 13/3 Wherein is school-phobia different from the traditional reluctance which was met by old-fashioned compulsion? **1980** *Daily Tel.* 19 Nov. 15/5 By that time, the more timid boy had been brought to the verge of school-phobia by it all. **1977** *Daily Colonist* (Victoria, B.C.) 22 Oct. 28/7 A psychiatrist..said mothers of school-phobic children are over-protective. **1981** *Lancashire Life* Jan. 25/2 We now have a new word for it... The 'school-phobic' says Lancashire Education Authority, is clearly intimidated by being required to attend school. **1874** C. M. YONGE *Lady Hester* ix. 205 Feeling very happy over the best school report of our boy we had ever had. **1958** J. CANNAN *And be a Villain* iv. 109 As his school reports revealed..he was useless at games. **1975** T. ALLBEURY *Palomino Blonde* xi. 71 There were a few school reports showing that Kristina was doing average well. **1835** *Indiana Mag. Hist.* XXII. 438 This was an action brought by the Trustees of a school Section for money due on two years rent. **1881** *Edmonton Bull.* 5 Nov. 3/2 As the surveys in Manitoba are made it is found that sections which should be available for school sections are already occupied..by the Syndicate for station grounds and other purposes. **1841** *Southern Lit. Messenger* VII. 7/2 The means of creating officers [for the navy]..are to be derived from the school-ship. **1971** C. STORR *Thursday* viii. 92 'Heaps of people do say it [sc. 'bloody']. Even on television.' 'But not on schools programmes.' **1973** *Listener* 31 May 707/1 Of the 30 channels in the system, three are to be made available..for schools television. **1974** Schools television [see *PIP *v.*³ 1 c]. **1895** C. M. YONGE *Long Vacation* vii. 66 He had heard enough school stories to be wary of boasting of his title. **1914** 'I. HAY' *Lighter Side School Life* vi. 151 Whereas school stories were formerly written to be read by schoolboys, they are now written to be read..by grown-up persons. **1971** 'S. SMITH' *Grave Affair* iv. 52 'Some boys from the Fifth. I don't know their names,' he lied in the best tradition of school stories. **1922** JOYCE *Ulysses* 669 Moral apothegms (e.g. *My Favourite Hero* or *Procrastination is the Thief of Time*) composed during schoolyears.

school, *v.*¹ Add: **1. b.** *intr.* To attend school. *rare.*

1934 in WEBSTER. **1972** *Straits Times* 23 Nov. 15/4 'It's incredible,' says the amiable 32-year-old Globe Silk Store proprietor who has schooled in England.

8. *intr.* To gamble in a 'school' (cf. SCHOOL *sb.*¹ 6 a). *slang.*

1935 A. J. CRONIN *Stars look Down* I. ii. 17 Some colliers.. that made up the gambling school in ordinary times—squatted upon their hunkers against the wall. They were not schoolin' now, they had no coppers for schoolin'.

school board. Add: **2.** In countries other than Great Britain, a board charged with the provision and maintenance of schools.

1836 *Chambers's Edin. Jrnl.* 27 Aug. 244/2 Over these provinces [in Prussia] is placed a consistory, or council, divided also into three sections, one of which is termed the school-board. **1838** F. B. HAWKINS *Germany* xii. 201 Every circle and parish has also its school-board. **1857** *Harper's Mag.* Sept. 571/2 Can you inform me where the president of your school board resides? **1911** *Daily Colonist* (Victoria, B.C.) 11 Apr. 7/3 The school board which recently submitted its estimates for the year to the city council will have to revise them. **1972** *Even. Telegram* (St. John's Newfoundland) 24 June 6/1 The government..tried to save a few dollars on the busing of children to school. There were immediate cries of pain from the Opposition and from some school boards. **1976** *National Observer* (U.S.) 2 Oct. 15/11 It was my job to cover the..school-board meetings of several communities.

school-book. 1. a. (Earlier U.S. example.)

1745 B. FRANKLIN *Let.* 11 Dec. in *Writings* (1905) II. 296 At present I only send for a few school books, and books of navigation.

schoolboy. Add: **1. b.** In phr. *every schoolboy knows*, referring to a matter of factual information, supposed to be elementary and generally known.

1654 [in Dict. sense 1 a]. **1721** SWIFT *Poems* (1958) I. 281 How haughtily he lifts his Nose, To tell what ev'ry School Boy knows. **1840** MACAULAY in *Edin. Rev.* Jan. 295 Every schoolboy knows who imprisoned Montezuma, and who strangled Atahualpa. **1966** *Listener* 8 Sept. 365/3 Tallis's motet *Spem in alium nunquam habui* was for years more often written about than heard. A *tour-de-force* in forty voice-parts: so much every schoolboy knew. **1977** *Times* 15 Oct. 2/8 Every schoolboy knows that the No. 3 bus from Piccadilly Circus comes to Valley Fields, Wodehouse's familiar pseudonym for Dulwich.

3. *Comb.*, as *schoolboy code* (*of honour*), *English*, *French*, *honour*, *humour*. (Not clearly separable from sense 2 in Dict.)

1874 C. M. YONGE *Lady Hester* vii. 169 The boy endured all the rage and scorn that a threat so contrary to all schoolboy codes of honour and friendship might deserve. **1977** P. G. WINSLOW *Witch Hill Murder* II. xviii. 248 His blue gaze fell on Capricorn expecting him to understand and accept his schoolboy code. **1835** J. ROMILLY *Diary* 6 July (1967) 82 P. George of Camb. delighted me by his returning thanks, because it was good simple schoolboy English. **1955** E. BLISHEN *Roaring Boys* III. 158 He delighted in mimicking my schoolboy French. **1977** J. CLEARY *High Road to China* ii. 41 A six months' affair with a girl in Auxi had improved his schoolboy French. **1876** C. M. YONGE *Womankind* xviii. 138 To keep up a standard of real honour, above schoolboy honour, is most needful. **1970** P. Y. CARTER *Mr Campion's Falcon* xxii. 166 He has a sense of schoolboy honour, a perfectly straightforward code. **1962** G. K. HUNTER *John Lyly* iv. 237 The inane schoolboy humour of his part, and the tradition of Ralph Roister Doister to which it belongs, both point to a juvenile rather than an adult actor. **1977** R. PERRY *Dead End* viii. 99 Just forget the schoolboy humour and do as you're told.

schoolboyish, *a.* Add: Hence **schoo·lboyishly** *adv.*, in the manner of a schoolboy; **schoo·lboyishness**, the conduct or manner of a schoolboy.

1898 G. B. SHAW *Let.* ? 2 May (1972) II. 39 Irving & I are too eminent to indulge in such schoolboyishness in public. **1901** W. J. LOCKE *Usurper* xviii. 247 He..was so schoolboyishly happy the next morning on starting for his holiday. **1972** J. POTTER *Going West* 159 His step was jaunty and his manner schoolboyishly affable. **1976** *Daily Tel.* 9 Sept. 13/1 As an example of the schoolboyishness, I would cite the fact that Ransome and a friend..had coded signals which they displayed on their respective houses..to give notice when they were going fishing.

school doctor. 3. (Examples.)

1906 R. BROOKE *Let.* 1 Apr. (1968) 47 Dukes, the school doctor, was paying us his hurried visit. **1933** A. WHITE *Frost in May* xiii. 333 The school doctor..told them..they were now perfectly well. **1963** M. KENDON *Ladies College, Goudhurst* 19 Dr. Mapleton, the school doctor,..would drive up in his dog-cart, with his wooden stethoscope inside his silk hat. **1976** J. PHILIPS *Backlash* (1977) III. ii. 148 He had a sinecure for me. School doctor.

-schooler (skuːlə(r)). *U.S.* [f. SCHOOL *sb.*¹ + -ER¹.] As the second element in Combinations, designating a pupil at a specified type of school, or stage of school-life, as *grade*, *high schooler*; *pre-schooler*.

1954 *Recreation* Apr. 241/1 Remember that pre-schoolers are people. **1960** *Parents' Mag.* July 92/2 Brevity..produces good results with preschoolers. **1971** *Sci. Amer.* Dec. 114/3 Martin Gardner's well-known learning and mathematical depth are worn lightly in this friendly, comical book for grade schoolers. **1972** *Newsweek* 25 Sept. 106/2 (*caption*) Harvard tutor..and high schoolers. **1973** *Black World* June 44/1 Ronald and Wayne are high schoolers. **1977** *Rolling Stone* 30 June 60/2 High schoolers will dye their hair gray, and buy iron-on wrinkles, and yearn for the day when their bodies begin to sag.

schoolgirl. Add: **1. a.** (Earlier examples.)

1777 E. DRAPER *Let.* 10 July in *N. & Q.* (1944) CLXXXVII. 12/2 The pertness, of the consequential School Girl, has given place to softer Manners. **1778** F. BURNEY *Evelina* I. xi. 36, I did not choose to tell him it [sc. my fear] was owing to my never before dancing but with a school-girl.

2. *Comb.*, as *schoolgirl complexion*, *crush*, *English*, *French*, *passion*. (Not clearly separable from sense 1 b in Dict.)

1922 *Woman's Home Compan.* Oct. 35 (Advt.), Better than jewels —— that schoolgirl complexion... Choose Palmolive, because its action is soothing. **1924** WODEHOUSE *Ukridge* x. 241 A man like myself, who finds at least eight hours of sleep essential if that schoolgirl complexion is to be preserved. **1938** G. GREENE *Brighton Rock* v. ii. 196 The long parade of posters..Guinness is Good for You, Try a Worthington, Keep that Schoolgirl Complexion. **1973** A. MACVICAR *Painted Doll Affair* iii. 39 She was an enthusiast for soap and water, as her schoolgirl complexion showed. **1952** Schoolgirl crush [see *CRUSH *sb.* 2 d]. **1963** P. MOYES *Murder à la Mode* iv. 69 Olwen had a sort of schoolgirl crush on her. **1978** C. STORR *Winter's End* xiii. 143 Bran.. wondered what she'd..meant when she'd told Rosey that she loved Philip. A schoolgirl crush? A romantic fantasy? **1939** C. ISHERWOOD *Goodbye to Berlin* 32 Frl. Hippi..speaks schoolgirl English with a slight American accent. **1967** W. G. CORP tr. *L. Oriol's Short Circuit* ii. 11 She spoke a schoolgirl English which nobody could understand. **1909** W. J. LOCKE *Septimus* iv. 49 'Will a hundred francs be of any use to you?' she asked, in her schoolgirl French. **1977** J. CLEARY *High Road to China* iv. 119 Mustafa Kemal said in French,..'I apologize for not speaking English.' Eve said..'I speak only schoolgirl French.' **1936** M. MITCHELL *Gone with Wind* xvii. 312 You still cherish a romantic schoolgirl passion for him.

schoolie (skuːli). Also **scheulie**. [f. SCHOOL *sb.*¹ + -IE.] **a.** *north. dial.* and *Austral.* A schoolteacher.

1901 in *Eng. Dial. Dict.* (1904) V. 250/2 That only three children out of a hundred have been absent on an average throughout twelve months will strike the ordinary scheulie ..with astonishment. **1907** N. SPIELVOGEL *Cocky Farmer* 33 The prettiest of all the girls was the schoolie, and didn't she lead the lads a dance. **1951** E. COXHEAD *One Green Bottle* vii. 179 'What a lot I've got to teach you!' said Christopher smiling. 'Schoolie.' 'Well, of course I'm a schoolie. What else could you expect?' **1960** S. H. COURTIER *Gently dust Corpse* xiii. 189 She was away at college, being taught to be a schoolie all that time. **1980** *Globe & Laurel* July/Aug. 198/2 At Lydd and Hythe, we enjoyed such epics as the formation of the Nelson brick (Doctor, Dentist, Padre and Schoolie).

b. *slang.* In the Navy, a classroom instructor.

1946 J. IRVING *Royal Navalese* 151 *Schoolie*, a naval schoolmaster. **1964** J. HALE *Grudge Fight* vii. 97 The E.R.A. instructors, the P.T. instructors, the gunnery instructors, the schoolies began to..brace themselves for another day of ramming drill and P.T. and lathe work and chipping and filing and maths., mechanics, machine drawing, naval history..into the minds and bodies of eight divisions of apprentices. **1977** *Navy News* Dec. 12/2 The official announcement says that in meeting the needs of the Navy during the past 20 years the role of instructor officers —the schoolies—has changed significantly.

school-keeping. (Earlier examples.)

1651 *Early Rec. Dedham, Mass.* (1892) III. 191 The time of covenant in ye schoole keeping being expired. **1829** A. SHERWOOD *Gazetteer Georgia* (ed. 2) 193 Why is it that school-keeping is so disreputable an employment in our State? **1854** C. M. YONGE *Castle Builders* v. 70 All we have ever had to do with school-keeping, was in that short visit to my sister.

school-ma'am. For *U.S.* read 'orig. *U.S.*' and add: Also **-ma'm**. The form **-marm** is now usual. **1.** (Earlier and further examples.) Now freq. implying the conventionally prim and correct behaviour of a school-mistress. Also *fig.*

1831 *Ladies' Mag.* (Boston) IV. 557 [It] obliged me to stay the longest in the houses where..there was the most work to do, and the least time to make the school Ma'am comfortable. **1840** *Spirit of Times* 8 Aug. 276/2 Them mirrors.., why what you got agin 'em? Cost me twenty-five dollars for the set—they be busters! open like a School-marm, by Jerusalem! **1841** *Picayune* (New Orleans) 23 Feb. 2/1 What will the 'school marm' say when she reads the following extract of a letter? **1888** C. M. YONGE *Our New Mistress* xii. 107 He said he supposed he should be a startling visitor for the school marm. **1924** A. E. HOUSMAN *Let.* 10 Mar. (1971) 218 A French school-ma'm wrote to me wanting to translate *A Shropshire Lad*. **1929** D. H. LAWRENCE *Phoenix II* (1968) 579 Now the funny thing is that nobody, not even the most conscientious father, ever questions the absolute rightness of these school-marms. **1951** M. MCLUHAN *Mech. Bride* (1967) 69/2 Rigid with the social cocksureness of the schoolmarm. **1974** *Times* 14 Mar. 16/1 Mrs Margaret Thatcher..had the bearing of a school ma'am, an inability to suffer fools. **1977** M. EDELMAN *Political Lang.* v. 90 Schoolmarms of both sexes behave like teachers in the living room and when reacting to novels or to public affairs.

attrib. **1965** *New Statesman* 7 May 719/1 The schoolma'am tone that husbands are quick to notice. **1972** WODEHOUSE *Pearls, Girls, & Monty Bodkin* xii. 181 Less than the dust beneath his chariot wheels, if he remembered the quotation correctly from his school-marm days. **1978** M. PUZO *Fools Die* xxix. 333 Going through a bedroom, I saw a couple head to toe and I heard a woman's very schoolmarm voice say, 'Get *up* here.'

2. *N. Amer. slang.* (See quots.)

1939 H. O'HAGAN *Tay John* 217 It was a pine. Long ago its trunk had been broken off by a slide or by the wind. Two stout branches had grown up instead, lightly tufted, to form a crotch. It was what the men there call a 'school-marm tree'. **1958** *Scope Weekly* 22 Oct. 7/1 The same situation may occur in felling a 'schoolma'am' which is essentially a forked tree, having two main trunks. **1965** M. MCINTYRE *Place of Quiet Waters* iv. 82 The 'schoolmarm' turned out to be a tree that had branched out into two separate trunks.

Hence as *v. trans.*, to treat (someone) in the manner of a school-marm, to instruct or guide patronizingly; **schoo·l-marming** *vbl. sb.*, the occupation of being a school-marm; **schoo·l-marmish**, **schoo·l-marmy** *adjs.*, like or suggestive of a school-marm; **schoo·l-marmishly** *adv.*, in the manner of a school-marm.

1887 H. FREDERIC *Seth's Brother's Wife* 24 She was held to be too serious and 'school-ma'am-ish' for pleasant company. **1914** KIPLING *Egypt of Magicians* iv, in *Cosmopolitan* Sept. 458/1 Our trouble in America is we're being school-marmed to death. **1920** 'O. DOUGLAS' *Penny Plain* xii. 124 Heaps of girls would think school-marming very dull, but Elspeth makes it into a sort of daily entertainment. **1921** R. MACAULAY *Dangerous Ages* vii. 132 The W.E.A. was a practical body... Dowdy, schoolmarmish, extension-lecturish, it might be. **1941** *Scrutiny* X. 115 The priggishness of the book [sc. *Mansfield Park*] is of a special kind, not just the occasional schoolmarmy effects of *Sense and Sensibility* which there are only the result of artistic inexperience. **1943** W. S. CHURCHILL *Second World War* (1951) IV. 824 Considering..that it was the Americans.. who led the world astray, it is pretty good cheek of them now coming to school-marm us into proper behaviour. **1945** R. HARGRAVES *Enemy at Gate* 234 The Radicals' itch to continue 'school-marming' the native populations of the former Boer territories. **1959** K. VONNEGUT *Sirens of Titan* (1967) x. 174 'This way, please, We haven't got all day, you know,' said Rumfoord school-marmishly. **1967** *Economist* 15 Apr. p. xvii/1 Typical of all her encounters was her inability to find the disapproving school-marmy guide, Miss Tsu, anything but likeable. **1977** good schoolmarm. **1979** *Guardian* 23 Oct. 8/1 The remarks tend to sound school-marmy and pontifical.

schoolmaster, *sb.*¹ Add: **1. f.** An experienced horse used to train horses or riders at a riding-school.

1937 in PARTRIDGE *Dict. Slang*. **1938** H. WYNMALEN *Equitation* ix. 40 Moving away from other horses must be taught him. To this end we shall ride him beside another horse, a schoolmaster. **1976** *Horse & Hound* 10 Dec. 68/4 (Advt.), This pony is one of the finest schoolmasters jumping in 12 hands 2 in. classes.

3. b. Special Comb.: **schoolmaster studentship**, in Oxford colleges, a studentship tenable by schoolmasters; hence **schoolmaster student**.

1957 *Oxf. Univ. Gaz.* 20 June 1142/1 Balliol College Elections. *To Schoolmaster Studentships.* For Michaelmas Term. **1978** *Times Educ. Suppl.* 3 Feb. 68/3 (Advt.), Merton and St. Peter's Colleges Schoolmaster Studentships 1978/79... The two colleges above intend to elect four *schoolmaster students* between them.

Hence **schoo·lmastery** *a.* = SCHOOLMASTERISH *a.*

1928 *Observer* (Apr. 7 In the earlier days the Staff College did not justify the expectations founded on it. It was unreal, academic, and 'schoolmastery'. **1942** J. LEES-MILNE *Ancestral Voices* (1975) 17 What an unattractive, schoolmastery fellow.

schoolmaster, *v.* Add: **1.** (Earlier example.)

1850 J. OXENFORD tr. *Eckermann's Conversations of Goethe* I. 377 [Schlegel] is permitted, upon such high authority, to fall foul of this mighty ancient [sc. Euripides], and to schoolmaster him as much as he can.

2. (Later examples.)

1966 *Listener* 5 May 659/2 Nicholas Urfe is schoolmastering on a Greek island, seeking escape from..an oppressive love affair. **1977** *Times* 15 Nov. 14/6 Mr Rogers ..schoolmastered for a time.

schoolmistress. Add: Hence **schoolmistressy** *a.*, characteristic of or resembling a school-mistress.

1915 D. H. LAWRENCE *Rainbow* x. 251 Miss Grey..had a certain silvery, school-mistressy beauty of character. *a* **1974** R. CROSSMAN *Diaries* (1975) I. 339 In that grating voice, she gave her clear, schoolmistressy, common-sense view of the White Paper.

schoolroom. Add: **1.** (Earlier and further examples.) Also, in *fig.* phr. *in the schoolroom*: of a young lady, not yet 'out' (cf. OUT *adv.* 26 b(a).

1773 P. V. FITHIAN *Jrnl. & Lett.* (1900) 61, I have to myself in the Evening..my Liberty, either to continue in the School room, in my own Room, or to sit over at the great House. **1812** E. WEETON *Let.* July (1969) II. 58, I breakfast with Mr. & Mrs. Armitage, and then return again to the children till 9, we go into the school-room till 12. **1857** C. M. YONGE *Dynevor Terrace* I. viii. 126 'I suppose her daughters are not come out yet?' 'Her own are in the school-room; but there is a step-daughter who is much admired.' **1952** M. LASKI *Village* ii. 35 The younger daughter in the schoolroom, the elder about to blossom forth. **1977** C. FREMLIN *Spider-Orchid* vii. 51 She's not 'out' yet, she's only in the schoolroom still.

3. *attrib.*

1814 JANE AUSTEN *Mansfield Park* I. xviii. 353 Very good school-room chairs, not made for a theatre, I dare say. **1847** A. BRONTË *Agnes Grey* xv. 242 There was the bell—the odious bell for the school-room dinner. **1857** C. M. YONGE *Dynevor Terrace* I. xiv. 227 The school-room maid..was busy unpacking in a corner of the room. *Ibid.* xv. 238 She repaired to the school-room tea. **1875** L. TROUBRIDGE *Jrnl.* 25 Dec. in J. Hope-Nicholson *Life amongst Troubridges* (1966) 133, I..found everyone congregated round the school-room table. **1923** W. J. LOCKE *Moordius & Co.* ii. 17 The family can always come up if it likes for schoolroom tea. **1948** F. THOMPSON *Still glides Stream* iv. 91 They should have heard the music, the schoolroom piano and two violins. **1959** I. & P. OPIE *Lore & Lang. Schoolch.* xvi. 343 Children whose schoolroom attitude to history is anti-pathetic. **1969** D. HOLMAN-HUNT *My Grandfather* xviii. 232 She would get rid of the schoolroom maid and order some new dresses. **1972** W. LABOV *Lang. in Inner City* i. 30 The subjects are asked to change certain sentences to correct schoolroom English. **1975** R. PLAYER *Let's talk of Graves* iv. 118 We had schoolroom tea in deathly silence.

Hence **schoo·lroomy** *a.*, characteristic of or resembling a schoolroom.

1895 W. S. CHURCHILL *Let.* 3 Aug. in R. S. Churchill *Winston S. Churchill* (1967) I. Compan. I. viii. 581 A very lovely—but stupid and school roomy girl—to whom I talked a good deal. **1975** D. DANIELL *Interpreter's House* ii. 20 A bit schoolroomy, possibly, and a little too much the work of a *belle-lettrist*.

schoon (skuːn), *v. rare.* [See etym. note at SCHOONER *sb.*¹ The modern examples represent a fanciful back-formation from this sb.] **a.** *intr.* To sail or skim over the water, esp. in the manner of a schooner. **b.** *trans.* To run or glance (one's mind) *over* (something). Hence **schoo·ning** *vbl. sb.*

With quot. 1836 cf. etym. note in Dict. s.v. SCHOONER *sb.*¹

1836 J. R. NEWHALL *Essex Memorial* 100 Capt. R[obinson] had constructed a vessel which he masted and rigged in the manner that schooners at this day are, and on her going off the stocks into the water, a bystander cried out, 'O how she scoons!' Robinson instantly replied, 'A schooner let her be.' **1937** O. NASH in *New Yorker* 13 Feb. 20/2 Where the schooner schoons, I schoon. **1959** I. JEFFERIES *13 Days* ix. 115, I spent my time schooning my mind over the calculations.

schooner, *sb.*¹ Add: **1. b.** *schooner on the rocks* (see quots.). *Naut. slang.*

1916 'TAFFRAIL' *Carry On!* 28 A 'schooner on the rocks' does not refer to a nautical disaster, but to meat and potatoes baked in a peculiar way. **1922** *Mariner's Mirror* VIII. 222/1 *Schooner on the Rocks*. This dish consists of a joint baked in a sea of batter. **1927** P. RILEY *Memories* ii. 11 Dinner..varied from salt beef,..'Schooner on the Rocks', i.e., joint of meat roasted on potatoes, or 'toad in the hole'.

3. *schooner-rigged* (later *fig.* examples); *schooner-man*, *-sail* [tr. G. *schonersegel* fore-sail]; **schooner barge**, (*a*) *U.S.*, a short-masted vessel designed to be towed; (*b*) a flat-bottomed vessel rigged as a topsail schooner; **schooner yawl**, a variety of two-masted schooner.

1819 *Western Rev.* I. 361 The River is navigated by steam boats, barges, keel boats, schooner barges. **1867** *Mitchell's Maritime Reg.* 1620 On Monday the fine schooner-barge Edith was launched. **1900** *Bath* (Maine) *Daily Times* 22 May 5/3 The new three-masted schooner barge *Flora* for the Commercial Towboat Co. of Boston was launched yesterday. **1945** *Amer. Neptune* V. 139 In the East Coast schooner-barge fleet, only a few have been built with five masts. **1951** F. G. C. CARR *Sailing Barges* 126 As far as the hulls of these big barquentine and schooner barges were concerned, they were like very large boomies. **1914** W. D. STEELE *Storm* 270 Then he scrutinized the rank of schooner-men flanking me. **1972** F. E. BOWKER *Blue Water Coaster* 30 We thought that he had picked up an old schoonerman, but it wasn't long before we discovered that he was an unemployed shoemaker. **1924** R. CLEMENTS *Gipsy of Horn* iii. 47 All hands were working schooner-rigged, going at it with their blood up. **1935** *Amer. Speech* X. 79/1 *Schooner rigged*, unequipped with proper clothes or other necessities. **1946** R. E. HIGGINBOTHAM *Wine for My Brothers* vi. 126 The Dane travelled schooner-rigged, and philosophically heaved his mail overboard. **1930** D. MARTIN *Boy Scout with Sea Devil* 48 We hoisted up the Fores'l and the Schooners'l. **1952** G. COWAN *Log of Pelican* vi. 41 We cleaned out lockers,..bent the schooner sail and got the fore-canvas up in stops, and found a place for everything. **1889** *Forest & Stream* 4 Apr. 227/3 Adding a jigger mast..cuts off the nasty big boom and large mainsail..making the yacht a schooner yawl. 1970 *Amer. Neptune* XXX. 196 Not counting the schooner-yawl *White Cap*, which was discussed among the schooners, yawls made up 8.7 percent of American sailing yachts in 1902.

schooner, *sb.*[2] Add: **1. b*.** *Austral.* and *N.Z.* A large beer-glass of locally variable capacity (see quots. 1966 and 1973); the (measure of) beer contained in such a glass.

1934 *Bulletin* (Sydney) 21 Feb. 10/1 In Brisbane, a standard pint served in a long glass is a 'schooner'. 1947 D. M. Davin *Gorse grows Pale* 126 Sitting in the pub with a schooner under his nose. 1966 G. W. Turner *Eng. Lang. in Austral. & N.Z.* viii. 163 A *schooner* in New South Wales is a fifteen-ounce glass, in Adelaide a nine-ounce glass. 1969 *Advertiser* (Adelaide) 12 May 5/4 Just because someone wants to spend an arvo sinking a few schooners in his own way. 1973 *Courier-Mail* (Brisbane) 1 Dec. 17/7 The traveller finished up at the Federal with 128 schooners (the local term for an eight ounce glass). 1977 *Bulletin* (Sydney) 22 Jan. 27/2 He calculated he would consume eight schooners (15oz glasses) of beer, plus some spirits, over three hours. 1981 *Advertiser* (Adelaide) 2 July 6/6 Mr Connelly and Mr McKenzie said the second stage of the change was planned for October when the 285 ml (10 oz) glass would be introduced to replace the 255 ml (9 oz) schooner.

2. A tall, waisted sherry glass; the measure contained by this.

1967 J. Potter *Foul Play* xvii. 204 What about joining me in a schooner of sherry? 1973 *Times* 20 Oct. 14/3 The abominably proportioned waisted Elgin glass, sometimes used for sherry, or its vulgar outsize version, the schooner. 1975 [see *noshery]. 1977 *Habitat 1977/78 Catal.* 121 *Elgin schooner.* For large sherries. 3½ oz.

Schopenhauer (ʃǫᵘ·pənhauəɹ, ʃǫ·p-). The name of the German philosopher, Arthur *Schopenhauer* (1788–1860), used allusively, esp. for the pessimism and concept of will for which his philosophy is noted. Hence **Scho·penhaueresque** *a.*, resembling, of the same type as, the ideas of Schopenhauer; **Scho·penhauerian** (also as *sb.*), **Scho·penhauerish** *adjs.*, characterized by the doctrines or ideas of Schopenhauer; **Scho·penhauerism,** the pessimistic and atheistic philosophy of Schopenhauer, according to which the world is governed by a blind cosmic will entailing suffering from which man finds release only through knowledge, contemplation, and compassion; **Scho·penhauerist, Scho·penhauerite,** a follower of Schopenhauer or his doctrines.

1882 W. S. Lilly in *19th Cent.* May 713 Schopenhauerism.. is little more than Buddhism vulgarized. 1882 *Mind* VII. 561 Thought, with Hegel, stands for something objective and unconscious (like the Schopenhaurian Will). 1891 G. B. Shaw *Let.* 25 Oct. (1965) I. 317 This does not make me a Schopenhaurist, or Ibsen one. 1898 —— *Perfect Wagnerite* 101 Wagner's determination to prove he had been a Schopenhaurite all along. 1906 *Academy* 10 Mar. 233/1 It is a shallow philosophy that issues in Schopenhauerism. 1906 *Daily Chron.* 26 Sept. 4/4 You would say at one glance that he is a pessimistic ass, a Schopenhauer of donkeys. 1908 *Edin. Rev.* Apr. 423 He is a Schopenhauerian. 1928 C. E. M. Joad *Diogenes* 96 We shall all be living Schopenhauerian lives. 1959 K. F. Leidecker tr. *Nietzsche's Lett.* (1960) 50 To infuse into my presentation of the science this new blood, to transfer to my hearers that Schopenhauerian seriousness which is impressed on the forehead of this grand man,—this is my desire, my audacious hope. 1965 *New Statesman* 18 June 971/1 The shuddering Schopenhaueresque preoccupation with personal annihilation. 1968 *Guardian* 30 July 4/1 The Schopenhauerish misanthropy of the essays. 1976 *Amer. N. & Q.* XV. 57/1 A rather gloomy Schopenhauerian melancholy and despair in *The White Peacock* and *The Trespasser.*

schottische, *sb.* **a.** (Earlier examples.)

1849 *Theatrical Programme* 9 July 48 The aim of whose existence appears to be that of rattling through the polka or schottische with the velocity of a spinning jenny. 1855 J. E. Cooke *Ellie* 151 This abominable German usage we have imported—the polka and the schottish too.

schottische, *v.* (Earlier example.)

1865 O. W. Norton *Army Lett.* 27 Aug. (1903) 277, I could only schottische a little.

Schottky (ʃǫ·tki). [The name of Walter *Schottky* (1886–1976), German physicist.] **1.** Used *attrib.* in *Electr.* and *Electronics*: **Schottky barrier,** an electrostatic depletion layer formed at the interface of a metal and a semiconductor in contact, causing the junction to act as an electrical rectifier; freq. *attrib.*; **Schottky diagram** = *Schottky plot*; **Schottky diode,** a solid-state diode having a metal-semiconductor junction, used in fast switching and voltage-clamping applications; **Schottky effect,** the increase in thermionic emission of a solid surface resulting from the lowering of its work function by the presence of an external electric field; *esp.* the increase in anode current in a thermionic valve beyond that predicted by the Richardson equation because of the electric field produced by the anode at the surface of the cathode; **Schottky line,** the straight line on the Schottky plot predicted by the Schottky theory; **Schottky plot,** a diagram used to illustrate the Schottky effect, obtained by plotting the logarithm of the current density against the square root of the applied electric field at constant emitter temperature; **Schottky slope,** the gradient of the Schottky line; **Schottky theory,** the theoretical basis of the Schottky effect.

1949 *Proc. Inst. Electr. Engineers* XCVI. i. 258, F_s is the field of the Sc[h]ottky barrier at the contact with the metal. 1957 H. K. Henisch *Rectifying Semi-Conductor Contacts* vii. 195 It is desirable to examine to what extent tunnel penetration near the top of a Schottky barrier (where the barrier is thin) determines the effective barrier height. 1964 *Bell System Technical Jrnl.* XLIII. 215 GaAs Schottky barrier varactor diodes constructed on epitaxial films may be designed to yield a high cutoff frequency. 1975 Schottky barrier [see *Schottky diode* below]. 1967 *Brit. Jrnl. Appl. Physics* XVIII. 629 The saturation values were obtained from Schottky diagrams in which the logarithm of the current is plotted against the square root of the anode voltage. 1968 *Proc. Inst. Electr. & Electronics Engineers* LVI. 232/2 (*heading*) Integrated Schottky-diode clamp for transistor storage time control. 1969 *Electronics* 21 July 76/2 The Schottky diode storage time is effectively zero, in contrast to typical values of 6 nsec for the gold-doped junction diode and 30 nsec for the junction diode without gold doping. 1975 D. G. Fink *Electronics Engineers' Handbk.* viii. 39 In Schottky barrier diodes current flow is by majority carriers, rather than by minority carrier diffusion. Thus switching speeds of Schottky diodes are not limited by storage-time delays. 1925 J. B. Johnson in *Physical Rev.* XXVI. 71 When current is limited by space charge the Schottky effect decreases because of the interaction of the electrons. 1949 *Rev. Mod. Physics* XXI. 226/2 Theory and the periodic Schottky effect both indicate at most a small reflection effect for clean tungsten. 1975 D. G. Fink *Electronics Engineers' Handbk.* i. 34 The increase in current as the anode voltage is increased beyond the value at which the normal saturation value of emitted current occurs (the Schottky effect) results from reduction by the applied field of the work function. 1930 *Rev. Mod. Physics* II. 151 A quantitative estimate of the field strength that is needed to cause the emission to depart appreciably from the Schottky line..may be made by considering the..simplified case. 1949 *Ibid.* XXI. 200/1 The well-established low voltage deviation from the Schottky line for thermionic emission from polycrystal surfaces of clean metals indicates a variation in thermionic properties of the various surfaces of crystals. *Ibid.* 204/1 Generally, the experimental Schottky plots deviate from the theory..in the low voltage region. 1967 *Brit. Jrnl. Appl. Physics* XVIII. 629 The zero-field emission is obtained by prolonging the straight asymptote of the Schottky plots to the $\sqrt{V} = 0$ line. 1939 *Physical Rev.* LVI. 664/2 The slope of the reference line..was the Schottky slope. 1963 J. J. Brophy *Electr. Processes in Materials* xi. 317 At the larger fields, the data points fall on a straight line having the Schottky slope.., and extrapolation of this line to zero field gives the value of *J*. 1930 *Rev. Mod. Physics* II. 155 Reynolds..found a variation with voltage in excellent agreement with the Schottky theory for field strengths exceeding about 10,000 volts/cm. 1958 Condon & Odishaw *Handbk. of Physics* viii. vi. 77/2 The small periodic deviations of the emission current from that predicted by the Schottky theory are..a field effect but come about through the interference effect of electron waves as the shape of the barrier is changed by means of the applied field.

2. Used *attrib.* and in the possessive with reference to the **Schottky defect,** a vacancy in a crystal lattice in which the missing atom is not an interstitial one and the number of anion and cation vacancies is such as to preserve electrical neutrality; also, the smallest possible group of such vacancies that preserves neutrality.

1938 *Trans. Faraday Soc.* XXXIV. 861 In the case of Schottky-disorder an equivalent number of anions and cations have been removed from normal lattice positions leaving holes of both 'signs'. 1940 Mott & Gurney *Electronic Processes in Ionic Crystals* ii. 26 There are two ways in which..vacant lattice points and interstitial atoms or ions can arise; we shall call them 'Frenkel defects' and 'Schottky defects'. *Ibid.* 30 We now consider a crystal containing *n* Schottky holes. 1958 K. M. Hornsby tr. P. *Glafkidès's Photographic Chem.* iii. 25 Schottky's defects consist of shallow vacant sites of Br^- and Ag^+ ions, in equal quantity. 1958 Condon & Odishaw *Handbk. of Physics* viii. iii. 48/2 In a crystal containing a divalent cation impurity,..there will be a temperature below which the number of Schottky cation vacancies will be smaller than the number of additional free cation vacancies. 1966 *McGraw-Hill Encycl. Sci. & Technol.* III. 583/2 In $CaCl_2$, the Schottky defect is one positive-ion vacancy and two negative-ion vacancies. 1966 C. R. Tottle *Sci. Engin. Materials* iv. 90 The Schottky defect..is not limited to the migration of an atom to the surface, but refers to the production of a vacancy whenever a migrating atom moves to some position that does not create disturbance in the remaining lattice, i.e. the surface of a void or other sink of disordered atoms. 1967 F. C. Brown *Physics of Solids* x. 303 By comparing the lattice parameter..with the observed mass and volume of the sample, it is possible to distinguish between Frenkel and Schottky disorder. 1972 B. Henderson *Defects in Crystalline Solids* i. 11 The change in thermal entropy favours the formation of Schottky vacancies.

schradan (ʃrā·dǎn). Also **Schradan.** [f. the name of Gerhard *Schrader* (b. 1903), German chemist, who first prepared it + -an.] A viscous liquid organophosphorus compound, bis(bisdimethylamino)phosphonous anhydride, $((CH_3)_2N)_2PO.O.PO(N(CH_3)_2)_2$, used as a systemic insecticide in the form of an aqueous solution.

1951 *Jrnl. Sci. Food & Agric.* II. 310 The systemic insecticide Schradan is slowly broken down in the plant by enzymic reaction. 1953 *New Biol.* XIV. 108 Schradan..renders plants highly toxic to sucking insects but has no appreciable effect on bees. 1964 A. H. Burgess *Hops* i. 17 Systemic insecticides, such as schradan, which was first introduced for commercial use on hops in 1949, have revolutionized the control of pests on hops. 1977 M. B. Green et al. *Chemicals for Crop Protection & Pest Control* vii. 54 The only other compound with..any commercial utility was schradan, but this..is not now used because of its high mammalian toxicity.

Schrader (ʃrēɪ·dəɹ). The name of George H. F. *Schrader* (fl. 1895), of New York, used as a proprietary term to designate air valves of a type introduced by him and used esp. on tyres.

1895 *Official Gaz.* (U.S. Patent Office) 30 Apr. 750/2 Pneumatic and tire valves. George H. F. Schrader, New York, N.Y... Essential feature.—The words 'Schrader Universal'. Used since February 12, 1895. 1920 R. T. Nicholson *Bk. of Ford Van* xvi. 100 (*caption*) Section of the Schrader tyre valve. 1921 —— *Bk. of Ford* (ed. 6) xi. 252 Most tyres fitted to the Ford have Schrader valves, which want understanding. 1940 E. Molloy *Landing Legs, Wheels, & Brakes* i. 8 (*caption*) The cap of the special Schrader valve is screwed on in reverse to release the air pressure. 1950 C. A. H. Pollitt *Air Systems for Aircraft* iv. 38 The valve has the usual Schrader-type screwed fixing sleeve. 1967 *Trade Marks Jrnl.* 10 May 588/1 Schrader. 1977 *Lancet* 23 July 175/1 Thinking, reasonably, that this was a colour-coding marker indicating nitrous oxide, she reconnected this hose to the spigot of the blue-painted nitrous-oxide Schrader valve.

Schrage (ʃrā·gə). *Electr.* The name of H. K. *Schrage* (fl. 1914), Swedish engineer, used *attrib.* and in the possessive to designate a type of three-phase a.c. motor invented by him, in which a commutator motor is combined with an induction motor to provide speed variability at high torque.

1919 R. M. Walmsley *Electr. in Service of Man* (rev. ed.) II. ii. v. 780 (*caption*) Connections of Schrage's three-phase shunt commutator motor. *Ibid.*, The brush mounting..of a Schrage motor. 1927 V. Karapetoff *Exper. Electr. Engin.* (ed. 3) II. 439 The Schrage motor belongs to the class of adjustable-speed compensated polyphase induction motors. 1945 E. Molloy '*Electr. Engineer*' *Ref. Bk.* viii. 51 The Schrage system of speed control requires that the primary winding (which is connected to the supply system) shall be on the rotor and the secondary winding on the stator. 1962 G. A. T. Burdett *Automatic Control Handbk.* i. 12 A Schrage motor operates on the moving brush or rotor fed principle.

Schrammel (ʃræ·měl). Also with small initial. The name of Johann (1850–97) and Josef (1852–94) *Schrammel*, Austrian musicians, used *attrib.* in **Schrammel quartet** [G. *Schrammelquartett*, also used], a Viennese light-music ensemble comprising two violins, guitar, and accordion (orig. clarinet) popularized by the Schrammels. Also || **Schrammelmusik** [G. *Schrammelmusik*], music played by or arranged for a Schrammel quartet or orchestra; so **Schrammel band, orchestra.**

1924 E. Wellesz in A. Eaglefield-Hull *Dict. Mod. Music & Musicians* 445/2 Nearly all the comp[ositions] of popular Viennese comp[osers], especially operetta-comp[osers] are arr[anged] for this combination which is called *Schrammel quartet.* 1938 *Oxf. Compan. Music* 848/2 *Schrammel quartet*, this is a popular Viennese type of instrumental quartet for light music... Sometimes the combination expands somewhat, into a 'Schrammel Orchestra'. 1963 E.-L. Wuorio *Woman with Portuguese Basket* xix. 167 We can go to a 'nobelheurige' which is a big place with schrammel band and singers. 1967 R. M. Stern *Kessler Legacy* ii. 24 In one corner..a stringed orchestra played *Schrammel-musik.* 1969 A. Arent *Laying on of Hands* xii. 138 The new wine and *Schrammelmusik* were available in Grinzing. 1974 F. Gore-Booth *With Great Truth & Respect* 72 We sought the '*Heuriger*', a traditional garden restaurant in Grinzing, where people drink new wine to the sound of the Schrammel-Quartett, a small ensemble with violin and harmonica.

|| **Schrecklichkeit** (ʃre·kliχkəit). [Ger., = 'frightfulness'.] = *frightfulness. Also *transf.* and *fig.*

1917 G. B. Shaw in *New Republic* 6 Jan. 274/1 As to the deliberate *Schrecklichkeit* of the Germans in Belgium..no man should judge unless he knows the military history of all invasions, and of that very British institution, the punitive expedition. 1944 —— *Everybody's Political What's What* xxxv. 307 The British frightfulness of 1943 has left the German Schrecklichkeit of 1915 far behind. 1969 R. Lowell *Notebk. 1967–68* 66 Mohammed..smashed the celibates... Changed their non-activist Buddhistic rote to his clans' strict laws of *schrecklichkeit* and honor. 1972 K. Bonfiglioli *Don't point that Thing at Me* ii. 10, I embarked on the quotidian *schrecklichkeit* of getting up. 1976 *Listener* 6 May 588/3 The *Schrecklichkeit* in which the relations between parents and children are so often conducted in Britain.

schreibersite. Substitute for def.: † **a.** A chromium sulphide, Cr_3S_4, supposed to have been found in a meteorite. *Obs.*

1846 C. U. Shepard in *Amer. Jrnl. Sci.* LII. 383 Schreibersite, (Shepard).—Named in honor of the late Carl von Schreibers, Director of the Imperial Cabinet at Vienna, and a well known author on meteorites.

b. A strongly magnetic phosphide of iron and nickel, $(Fe,Ni)_3P$, usu. with small amounts of cobalt, that is present in iron meteorites and forms lustrous white tetragonal crystals that tarnish to yellow or brown. Cf. Rhabdite 3. [ad. G. *schreibersit* (A. Patera 1847, in *Österreichische Blätter f. Lit. und Kunst* 23 July 694/2).]

1849 *Amer. Jrnl. Sci.* LVIII. 440 Something..similar is found in the meteoric iron of Arva. Patera was enabled to collect a sufficient quantity of it to make three analyses... As Berzelius had given no name to this substance, Haidinger, in concurrence with M. Patera, proposed for it the specific name of Schreibersite. At a subsequent session.. Haidinger..says he has since learned that the American mineralogist and chemist, Prof. Shepard,..had given..*this same name* to a mineral, also of meteoric origin, which occurs in small brown striated prisms in the meteoric stone of Bishopsville, S.C., which fell in March, 1843... Haidinger..would be pleased to continue the name of Schreibersite to the Arva Species, and would propose for Shepard's new species, the name of Shepardite. 1868 [in Dict.]. 1968 *Jrnl. Geophysical Res.* LXXIII. 6963/2 The most notable feature of schreibersite is its compositional variability. 1975 *Sci. Amer.* Feb. 35/1 At 1,294 degrees gaseous molecular phosphorus reacted with the surface of the metal grains, thus forming the mineral schreibersite. 1977 A. Hallam *Planet Earth* 27/3 Other highly significant features which must be explained are the presence within some tektites of minute grains of the meteoritic minerals kamacite, troilite and schreibersite, [etc.].

|| **schreierpfeife** (ʃraɪ·ərpfaifə). Also Schreierpfeife. Pl. schreierpfeifen. [G., lit. 'screamer pipe'.] A musical instrument of the variety collectively known as *schryari.

1939 [see *schryari]. 1957 A. Baines *Woodwind Instr. & their Hist.* x. 258 Several German inventories..mention a consort of *Schreierpfeifen* or *Schryari* ('crying' or 'screaming' pipes..). These do not seem to occur in any known account of a musical performance, but Praetorius says that they could be used either alone or with other instruments. 1976 D. Munrow *Instr. Middle Ages & Renaissance* 51/1 The origin of the name *Schreierpfeife* (Italian *schryari*) is unclear: perhaps it had something to do with the instrument's 'screaming' tone quality.

schreik, var. *skrik.

Schreiner (ʃraɪ·nəɹ). *Textiles.* Also **schreiner.** The name of Ludwig *Schreiner* (fl. 1900), German textile manufacturer, used *attrib.* with reference to a method of finishing mercerized fabrics by passing them through a calender, one of whose rollers has engraved upon it many fine, evenly-spaced, parallel lines which are impressed on to the fabric imparting lustre to the material.

1904 *Dyer & Calico Printer* XXIV. 9 The Schreiner finish is daily growing in importance and will..become of vaster importance than is at present dreamt of. 1946 A. J. Hall *Stand. Handbk. Textiles* iv. 241 Whilst quite a high lustre can be secured in this way there is another kind of calender which can achieve even more. It is the Schreiner calender. 1960 *Times Rev. Industry* Nov. 46/2 Among the ways of finishing a standard grey rayon material of this type are durable schreiner finishes, permanent glazed finishes, and embossed finishes. 1962 J. T. Marsh *Self-Smoothing Fabrics* xiii. 201 For the schreiner effect, pressures of 20 tons or so are commonly utilised and a temperature of 170 to 180 °C; many finishers calender the impregnated and dried goods before putting them through the schreiner machine. 1963 A. J. Hall *Textile Sci.* v. 257 A light Schreiner calendering can very effectively and desirably make more opaque the gossamer sheer and transparent knitted nylon nightwear fabrics... The passage of this fabric through a Schreiner calender..just flattens the thermoplastic nylon threads so as to fill out the interstices in the fabric yet without..impairing the fabric lustre.

Hence **Schrei·ner** *v. trans.*, to finish (fabric) by this method; also *absol.*; **Schrei·nered** *ppl. a.*; **Schrei·nering, Schrei·nerizing** *vbl. sbs.*

1905 Bean & McCleary *Chem. Finishing* 375 When the 'schreinering' process is combined with 'mercerising', it gives the nearest approach to silk ever obtained. *Ibid.* 376 Plain woven Calico goods may also be 'schreinered' with advantage. 1906 *Dyer & Calico Printer* XXVI. 17 During the last few years Schreinerising has made rapid strides, and is now a very general method of finishing dyed cotton piece goods. 1929 E. Midgley *Finishing of Woven Fabrics* xii. 160 The type of lustre or reflection obtained from a broken surface, such as in the case of..schreinered sateens, is shown in Fig. 51. 1945 M. D. Potter *Fiber to Fabric* v. 96 Schreinerizing is an inexpensive method for imparting luster to low-priced cottons. 1946 A. J. Hall *Stand. Handbk. Textiles* iv. 242 The method of schreinering fabric is comparatively simple. 1962 J. T. Marsh *Self-Smoothing Fabrics* xiii. 198 The general method is to impregnate the fabric, dry to definite requirements of moisture content, and then glaze, emboss, or schreiner, before completing the final condensation of the resin. *Ibid.* 202 The final handle of the goods is affected by the amount of moisture in the fabric at the moment of schreinering.

|| **Schriftsprache** (ʃri·ft,ʃprāχə). *Philol.* [G., = literary or standard language.] The conventional and standardized written variety of a given language (or occas. a dialect).

1931 K. Malone in *Mod. Lang. Notes* XLVI. 8 Caxton's importance for the English language lies chiefly in the part which he played in the standardization of our *schriftsprache.* 1934 C. L. Wrenn in *Trans. Philol. Soc. 1933* 85 There was a common and universally used West-Saxon *Schriftsprache* in the late tenth and early eleventh centuries, as well known in York as in Canterbury. 1935 *Ess. & Stud. in Eng.* (Univ. Michigan Publ. Lang. & Lit. XIII) 281 The rise of a literary language, divorcing to some degree spelling and pronunciation, arouses our curiosity in regard to the extent of adoption of these voiceless inflections within the *Schriftsprache.* 1959 A. Campbell *Old Eng. Gram.* 11 The vernacular ninth-century charters show a steady tendency towards the development of a local *Schriftsprache*, with increasing avoidance of Anglian spellings, and care to express local sound-changes.

schrijk, schrik, varr. *skrik.

schröckingerite (ʃrǫ·kiŋərəit). *Min.* Also **schroeckingerite.** [ad. G. *schröckingerit* (A. Schrauf 1873, in *Mineral. Mitt.* 137), f. the name of Baron J. von *Schröckinger*, 19th-cent. Austrian mineralogist: see -ite[1].] A hydrated carbonate, sulphate, and fluoride of uranyl, calcium, and sodium found as greenish-yellow scales, usu. as an alteration product of uraninite.

1875 G. J. Brush *Dana's Syst. Min.* (ed. 5) App. ii. 50 Schröckeringite [*sic*]... Occurs at Joachimsthal on uraninite, in small, six-sided tabular crystals. 1921 *Bull. U.S. Geol. Survey* No. 679. 131 A number of other specimens labeled 'schroeckingerite' were examined, but they proved to be uranothallite, or some other uranium mineral. 1967 *Mineral. Abstr.* XVIII. 247/1 The uranium is located only in a small mineralization zone and is primarily in pitchblende form. Secondary uranium minerals are: schröckingerite, Ca-nováčekite, and meta-zeunerite.

Schröder (ʃrǫ·dəɹ). Also **Schroeder.** The name of H. G. F. *Schröder* (1810–85), German mathematician and physicist, used *attrib.* and in the possessive to designate an optical illusion described by him (see *Ann. der Physik und Chem.* (1858) CV. 307), in the form of a line drawing of a staircase drawn without convergence of receding parallel lines, so that one appears successively to look down at the top and up at the underside of the staircase as the perspective reverses.

1898 E. C. Sanford *Course in Exper. Psychol.* II. vii. 256 'Schröder's Stair Figure'..generally appears first as the upper flight of steps. 1901, etc. [see *Necker[1]]. 1925 J. P. C. Southall tr. J. von Kries in tr. *Helmholtz's Treat. Physiol. Optics* III. 597 A similar reversal of the impressions of distance occurs in looking at Schroeder's 'staircase' diagram.., especially if it is turned round. 1957 *Acta Psychologica* XIII. 86 With the Schröder stairs, now, a new means was found to ask the subject without using the words 'Up' or 'Down'. The S was simply asked: 'From which end would you approach the stairs in order to mount them, from Right or from Left?' 1974 *Sci. Amer.* July 101/1 The Schröder stairs, another 19th-century reversible-perspective illusion,..is the theme of Escher's 1953 lithograph *Relativity.*

Schrödinger (ʃrǫ·diŋəɹ). *Physics.* Also **Schroedinger.** The name of Erwin *Schrödinger* (1887–1961), Austrian-born physicist, used *attrib.* and in the possessive to designate concepts developed by him, as **Schrödinger('s) (wave) equation,** a differential equation whose solution is the Schrödinger function; this equation became the basis of the quantum-mechanical description of matter; **Schrödinger (wave, ψ-) function,** a complex function ψ of space and time such that the square of its absolute value is a measure of the local spatial probability density for a particle in the state (or with the probability amplitude) ψ, i.e., $|\psi|^2$ represents the average particle density at a given location in space and time.

1927 *Proc. R. Soc.* A. CXIV. 251 The Hamiltonian function will now provide a Schrödinger wave equation. 1935 Pauling & Wilson *Introd. Quantum Mechanics* iii. 53 The function $\Psi(x, t)$ is called the Schrödinger wave function including the time. 1938 R. W. Lawson tr. *Hevesy & Paneth's Man. Radioactivity* (ed. 2) viii. 90 In wave mechanics a vibration law is formulated for the atom which is quite similar to the law of the mechanical vibrations of strings. But here we are not dealing with the motion of material particles, but with that of an abstract quantity called the Schrödinger ψ-function, which is only mathematically intelligible. 1955 W. Heisenberg in W. Pauli *Niels Bohr* 23 His point of attack is..the fact that the wave function representing the system changes discontinuously when the observer takes cognizance of a result of measurement. Janossy asserts that this reduction cannot be deduced from Schrödinger's equation. 1961 Powell & Crasemann *Quantum Mechanics* ii. 59 Many of the properties of ψ which are of physical interest are brought out more clearly by the Schrödinger equation than by the direct representation of ψ in terms of its harmonic components. 1968 M. S. Livingston *Particle Physics* iii. 51 The Schrödinger wave equation is based on the well-known differential equation for a traveling wave in a homogeneous medium. 1968 G. Ludwig *Wave Mech.* i. iii. 42 Hence..the Schrödinger functions φ(x) are nothing other than the representation of the Hilbert space corresponding to the position operator. 1974 Gill & Willis *Pericyclic Reactions* i. 17 Now the energy can be calculated by substitution into the Schrödinger equation appropriate to the system.

schronch, var. *scronch.

† **schrötterite.** *Min. Obs.* [ad. G. *schrötterit* (E. F. Glocker, *Grundr. d. Mineralogie* (1839) 536), f. the name of Anton *Schrötter* (1802–75), German chemist and mineralogist: see -ite[1].] A name formerly applied to greenish opaline specimens of allophane.

1844 J. D. Dana *Syst. Mineral.* (ed. 2) 531 Schrötterite. Resembles allophane, and has been called opal allophane. 1858 *Amer. Jrnl. Sci.* LXXVI. 79 The fragment of rock upon which the Schrötterite occurs is a dark-colored bituminous slate. 1934 *Prof. Papers U.S. Geol. Survey* No. 185. 146/2 The analyses reported in this paper and earlier studies by others have discredited all the known reported occurrences of schrötterite, including the type material.. Schrötterite should be entirely discarded as a mineral name.

schrund. Substitute for pronunc.: (ʃrunt). Delete || and add: *spec.* = *bergschrund; *schrund line, schrundline*: (see quot. 1904[2]). (Earlier and later examples.)

1870 A. G. Girdlestone *High Alps without Guides* ii. 36 We could see into the great blue 'schrunds', in which innumerable and gigantic icicles depended from the roof. 1904 G. K. Gilbert in *Jrnl. Geol.* XII. 573 Among the numerous crevasses or schrunds of several diverse systems sharply lining the snow surface.., one master opening, the *Bergschrund* of the Swiss mountaineers, paralleled the amphitheater wall... My instant surmise..was that the curving great schrund penetrated to the foot of the wall. *Ibid.* 582 Usually in viewing a cirque it is possible to trace about its wall a somewhat definite line separating a cliff or steeper slope above from a gentler, usually scalable, slope below. This line I conceive to mark the base of the bergschrund [*sic*] at a late stage in the excavation of the cirque basin. I have called it in my notes 'the schrund line'. 1938 *Geol. Mag.* LXXV. 262 This scarp..passes into the great vertical cliffs below Y Lliwedd so that the schrundline was present only on the side wall. 1939 *Geogr. Jrnl.* XCIV. 4[illegible] Above the crevasse the surface of the glacier was covered by scree; and the material embedded in the ice, as seen in the bergschrund, is probably derived in some way from the surface material... A snow-bridge occurred 25 feet below the surface in the open part of the schrund. 1968 R. W. Fairbridge *Encycl. Geomorphol.* 741/2 Below the schrund line, where there is one, the cirque floor begins as a slope that bears marks of glacial abrasion. 1971 C. Bonington *Annapurna South Face* xii. 152 They were both worried by the huge overhang of snow that reared over the schrund like a breaking wave.

|| **schryari** (ʃriā·ri). Usu. *collect.* [Of uncertain origin (see note below).] A variety of wooden double-reed wind instrument used in the sixteenth and seventeenth centuries and recently revived for the playing of early music.

Our knowledge of the *schryari* comes principally from descriptions and drawings furnished by Praetorius, who regards the term as a plural form related to *schreierpfeife. Later authorities have suggested an Italian or Oriental origin, but satisfactory evidence is lacking.

[1618–20 M. Praetorius *Syntagma Musicum* II. xviii. 42 Schryari (Auff deutsch Schreyerpfeiffen) sennd starck vnnd frisch am Laut/können vor sich alleine/vnd auch mit andern Instrumenten gebraucht werden [etc.].] 1939 A. Carse *Musical Wind Instr.* xi. 129 Still more shadowy are the *schryari* or *schreierpfeifen* of which Praetorius gave a brief description... Not a single specimen is known. 1940 C. Sachs *Hist. Mus. Instr.* xv. 322 *Schryari* were loud, shrill, double-reed instruments with a tapering bore and a reed-concealing cap; they had seven fingerholes in front and two in back for the two thumbs. 1964 S. Marcuse *Mus. Instr.* 463/1 Schryari were made in consorts of soprano, alto/tenor, and bass; they had conical tubes with double reed protected by a reed cap, 7 front fingerholes, and 2 rear thumbholes. 1976 D. Munrow *Instr. Middle Ages & Renaissance* 52/2 The most surprising feature however is that the exterior of the schryari is conical.

schtick, var. *shtik.

schtschi, var. Stchi in Dict. and Suppl.

Schubertiad (ʃubə·ɪtiæd). Also || **Schubertiade** (ʃubertiā·də), pl. -n. [ad. G. *Schubertiade*: see next and -ad.] A concert party or recital devoted solely to the performance of music and songs by Schubert.

1869 A. D. Coleridge tr. *K. von Hellborn's Life of Schubert* I. ix. 223 The 'Schubertiaden',—social unions of Schubert's friends, where..Schubert's own compositions formed the staple of the entertainment. 1905 E. Duncan *Schubert* 93 Many of Schubert's friends and acquaintances.. banded together in a kind of social union, which became known by the name of *Schubertiaden.* Games were played, dancing and speech-making were cultivated, while the heart of the whole entertainment was discovered in the performance of Schubert's latest songs, and others of his compositions. 1945 A. Hutchings *Schubert* iv. 39 Netty Hönig, in whose home so many happy Schubertiads had been held. 1963 *Times* 12 Mar. 14/4 Last night he and Mr. Gerald Moore gave us a Schubertiad in the Festival Hall. 1967 M. J. E. Brown *Schubert Songs* v. 40 His [*sc.* Schubert's] time was occupied in Vienna by..the increasing demands of the very popular *Schubertiaden*—evenings devoted to the performance of his songs. 1977 *Times* 1 Feb. 9/2 Yesterday was Schubert's birthday..suitable occasion for a Schubertiad.

Schubertian (ʃubə̄·ɪtiăn), *a.* and *sb.* Also **Schubertean.** [f. the name of Franz Peter *Schubert* (1797–1828), Austrian composer + -ian.] **A.** *a.* Of, pertaining to, or characteristic of Schubert or his music. **B.** *sb.* An admirer or adherent of Schubert; a (skilled) exponent of Schubert's music.

1866 E. Wilberforce *Franz Schubert* iv. 83 The present biographer discovered it in a pile of Schubertian MSS. 1911 J. A. Fuller-Maitland *Brahms* iv. 90 The fourth ballade, after its Schubertian waverings between minor and major, strikes the same mysterious note. 1927 *Observer* 2 Oct. 14/4 The work was well chosen to follow the Schubert Quintet, for in its trio there is the germ from which sprang what is now recognised as a truly Schubertian progression. 1928 B. Marshall tr. *K. Kobald's F. Schubert* 276 In the streets of the city the faithful Schubertians followed their adored genius to the grave. 1945 A. Hutchings *Schubert* iii. 28 Anselm had two brothers, Josef and Heinrich, who were ardent Schubertians. 1959 *Times* 9 Nov. 6/4 The long, smooth lines of the allegretto were clouded by no more than a Schubertian wistfulness. 1971 *Daily Tel.* 25 Jan. 6/2 He [*sc.* Wilhelm Kempff] plays a lovely short A-major sonata..but he cannot, I conclude, be counted among the true-blue Schubertians.

schuchardtite (ʃū·χāɹtait). *Min.* [ad. G. *schuchardtit* (A. Schrauf 1882, in *Zeitschr. für Krist.* VI. 386), f. the name of Theodor *Schuchardt*, 19th-cent. mineral dealer: see -ite[1].] An ill-defined green hydrated silicate of nickel, resembling chlorite.

1885 *Jrnl. Chem. Soc.* XLVIII. i. 32 Schuchardtite... This mineral is of an apple-green colour, soft, and is disintegrated by water. 1966 *Amer. Mineralogist* LI. 292 The analytical data for schuchardtite are inconclusive; they do not fall near to the composition of the serpentine group where they would be expected to be situated but suggest admixture with pimelite. 1975 *Mineral Mag.* XL. 152 We hesitate to label the Jacupiranga minerals as schuchardtites mainly because the type material was insufficiently defined.

|| **Schuhplattler** (ʃū·platləɹ). Also **schuh-** and erron. **Schuplaettler, -platter,** etc. [G., f. *schuh* shoe + south G. dial. *plattler* (f. *platteln* to slap).] A lively Bavarian and Austrian folk-dance, characterized by the slapping of the thighs and heels. Also **Schu·hplattltanz** (*irreg.*); **Schu·hplatteln** [G. *schuhplatteln*, to perform this dance].

1874 K. Stieler in Schmid & Stieler *Bavarian Highlands* 108 The idea of the 'Schuhplattltanz' is taken from hunting-life—from the movements of the moor-cock and wood-grouse. 1895 L. Grove *Dancing* ix. 317 An old German dance..called the 'Schuhplatteln'. 1905 W. D. McCrackan *Tyrol* x. 82 The dancer extemporized as he threw down his money for the musicians. This pay gave him the privilege of the floor for his *Ländler* (waltz), or his *Schuhplattler.* 1920 D. H. Lawrence *Women in Love* xxix. 456 They were dancing all together, dancing the Schuhplatteln. 1958 M. West *Second Victory* i. 5 The orchestras played Strauss waltzes and the peasant troupes came in to dance the *Schuhplattler* and play the zither for local colour. 1960 *Guardian* 12 Apr. 8/7 Besides the yodelling..there are the frolicsome Schuplaettler dances. 1962 *Times* 10 Nov. 11/7 The dancing of the *schuhplattler* forms a cheerful accompaniment to a stein of beer. 1976 *Michigan Holiday News* (West Michigan Tourist Assoc.) May 11/1 The Schuhplatter Dancers, the Bavarian contribution to the art of dance, occupy a unique place in the central Michigan area.

Schüller–Christian (ʃüləɹ kri·stiăn). *Path.* The names of Artur *Schüller* (1874–1958), Austrian neurologist, and Henry Asbury *Christian* (1876–1951), U.S. physician, who each described the condition (in *Fortschritte a.d. Geb. d. Roentgenstrahlen* (1916) XXIII. 12 and *Contrib. Med. & Biol. Res.* (1919) I. 390 respectively), used *attrib.* to designate a pathological condition, often associated with diabetes insipidus, in which masses of lipid-laden histiocytes develop, usu. in the bones. Also in comb. with the name of Alfred *Hand* (1868–1949), U.S. pædiatrician.

[1925 C. Q. Thompson et al. in *Arch. Internal Med.* XXXVI. 650 As Christian's excellent paper has formed and will continue to form the basis for studies of this baffling syndrome, it is suggested for the sake of simplicity the name Christian's syndrome be used.] 1935 *Brit. Jrnl. Surg.* XXII. 810 It is that which is known to the medical profession as Schüller-Christian's syndrome. *Ibid.* 811 An excess of cholesterol in certain body tissues is the primary factor in the production of the clinical syndrome of lipoid granulomatosis (Schüller-Christian's disease). 1953 Schüller-Christian disease [see *histiocytosis]. 1960 Hand-Schüller-Christian disease [see *lipid storage* s.v. *lipid 2]. 1974 R. M. Kirk et al. *Surgery* vii. 151 Schuller-Christian disease. Histiocytes contain cholesterol.

schultenite (ʃu·ltěnəit). *Min.* [f. the name of August Benjamin Friherre at *Schultén* (1856–1912), Finnish chemist and mineralogist: see -ite[1].] A native lead hydrogen arsenate, $PbH(AsO_4)$, found as colourless, transparent monoclinic crystals.

1926 L. J. Spencer in *Mineral. Mag.* XXI. 149 His 'monétite arséniée de plomb' is the subject of the present note, and for it, as a mineral, the name schultenite is suggested. 1968 I. Kostov *Mineralogy* 467 Schultenite is found in gypsum-like crystals with distinct {010} cleavage.

Schultz–Charlton (ʃults tʃā·ɹltǫn). *Med.* The names of Werner *Schultz* (1878–1948) and Willy *Charlton* (b. 1889), German physicians, together used *attrib.* to denote the test made by intradermal injection of antibody to scarlet fever toxin, or of serum containing this; and to denote the phenomenon, characteristically diagnostic of scarlet fever, whereby such an injection causes local extinction of a rash.

1922 *Jrnl. Amer. Med. Assoc.* 12 Aug. 594/1 Rojo's experience has confirmed the specific and reliable nature of the Schultz–Charlton phenomenon in the differential diagnosis of scarlet fever. 1925 *Jrnl. Clin. Invest.* I. 275 (*heading*) The Schultz–Charlton phenomenon. *Ibid.* 293 Serum from convalescent scarlet fever patients produced the Schultz–Charlton rash extinction phenomenon in twenty-four of twenty-seven cases tested. 1974 *Encycl. Brit. Micropædia* VIII. 948/1 When an injection of a small amount of immune serum..is made into the skin while the rash is at its height, a blanched area results at the point of injection within 18 hours if the rash in question is caused by scarlet fever toxin. This reaction is known as the Schultz–Charlton test.

Schumannesque (ʃūmane·sk), *a.* [f. the name of Robert Alexander *Schumann* (1810–1856), German composer + -esque.] Resembling the compositions or technique of Schumann. So **Schumannism,** a musical element in the style of Schumann; **Schumannite,** an admirer or interpreter of Schumann.

1901 G. B. Shaw in *Anglo-Saxon Rev.* Mar. 229, I doubt whether even Puccini really studies Schumann, in spite of his harmonic Schumannisms. 1947 A. Einstein *Music in Romantic Era* xiv. 191 Another follower of Schumann was Robert Franz (1815–1892), a pure specialist in song—a 'Schumannite'. *Ibid.* 196 It is significant that his first, Schumannesque songs were composed to German texts. 1958 *Listener* 18 Sept. 441/2 The richly coloured, evocative scores of Lalo are few. The 'Symphonie espagnole' is among them but not his Schumannesque symphony. 1961 *Times* 20 Feb. 6/1 His account of the first movement of Beethoven's Waldstein Sonata seemed wayward and at times almost Schumannesque in its fluctuations of tempo. 1977 *Gramophone* Sept. 503/3 Several passages, in particular the long, Schumannesque interlude before the eighth (of twelve) song, that add something to our knowledge of Strauss's abilities.

Schu mine (ʃū məin). [App. an Eng. shortening of G. *Schützenmine* *S-mine, but see quot. 1945[2].] A type of German anti-personnel mine used in the war of 1939–45.

1945 *Finito! Po Valley Campaign* (15th Army Group) 41 The 10th Mountain Division pushed forward..across a valley studded with Teller mines, Schu mines and the glass-topped Topf mines that fooled the mine-detectors. 1945 [see *anti-personnel* s.v. *anti-[1] B. 4 (iii)]. 1961 W. Vaughan-Thomas *Anzio* ix. 209 The Germans..discovered the path..and sowed it with *Schu* mines—those vicious, light-weight, anti-personnel mines, cased in plastic.

Schumpeterian (ʃumpētī·riăn), *a.* *Econ.* [f. the name of the Moravian-born economist, Joseph Alois *Schumpeter* (1883–1950) + -ian.] Applied to the economic doctrines put forward by Schumpeter, esp. those dealing with the rôle of the entrepreneur, interest, and business cycles in the capitalist system. Hence as *sb.*, an advocate of these doctrines.

1950 Clemence & Doody *Schumpeterian System* vii. 57 The innovation of mild prosperity is the innovation of the Schumpeterian System. 1970 C. Furtado in I. L. Horowitz *Masses in Lat. Amer.* ii. 32 These urban elements were indeed the Schumpeterians of the development, 'the forward marchers' of Latin America. 1975 *New Society* 2 Oct. 28/1 The negative, Schumpeterian defence of democracy stresses the political skills of the leaders rather than the rank and file..but sees that being chosen by an appropriate constituency is the great source of legitimacy in the modern world. 1976 *Times Lit. Suppl.* 20 Feb. 206/4 Japanese businessmen have not been typical exponents of the Weberian 'spirit of capitalism' or even Schumpeterian individualists.

schungite, var. *shungite.

Schuplaettler, ¶ var. *Schuhplattler.

|| **Schupo** (ʃū·po). Also **schupo.** [G., colloq. abbrev. of *Schutzpolizei* and *Schutzpolizist* security police(man).] In Germany, a policeman; also *collect.*, the police force.

1923 *Glasgow Herald* 20 Mar. 9/5 In the events which are taking place in the Ruhr daily there is hardly one in which the schupos or green police are not concerned. 1934 *New Republic* 18 July 249/2 Similar to the *Gestapo* was the *Schupo*, the former municipal police forces of Germany. 1966 *Economist* 5 Nov. 568/1 Many of these former 'schupos' hold fairly prominent positions in the Austrian police and gendarmerie today. 1979 G. Markstein *Traitor for Cause* 14 As usual, two green-uniformed Schupos were at their post, the token presence..of West German authority.

|| **schuss** (ʃus), *sb.* *Skiing.* [G., lit. 'a shot'.] A straight, downhill run; the slope on which such a run is executed. Also *transf.* and *attrib.*

1937 O. Lang *Downhill Skiing* ii. 19 The straight running, or 'schuss', position should be very elastic. 1947 F. A. Smythe *Again Switzerland* 37 Down to the broad-backed ridge beneath..the last hundred feet in a glorious 'schuss'. 1958 L. Whishaw *As Far as you'll take Me* iii. 31 Jack [*sc.* a lorry driver] started us on another downhill schuss. 1961 *Times* 10 Jan. 14/7 There is then a final *schuss*, which will provide sufficient speed to make the bumps before the finish. 1966 M. Catto *Bird on Wing* iv. 62 Skis close—the whang of the wind eddying in a blast up the *Schuss.* 1977 C. Wood *James Bond* ii. 21 He..dropped to the schuss position as soon as he began to pick up speed.

|| **schuss,** *v.* *Skiing.* [f. prec.] **1.** *trans.* To ski down (a slope, etc.) or cover (a certain distance) by means of a schuss.

1937 O. Lang *Downhill Skiing* v. 27 In practice you will find that it is impossible to take everything 'straight' or to 'schuss' it as we express it. 1947 *Sun* (Baltimore) 28 Mar. 13/3 In April, 1939, in the annual 'inferno' races from the summit, Toni 'schussed' 3.8 miles with a vertical drop of 4,300 feet in 6¼ minutes. 1972 T. McHugh *Time of Buffalo* xii. 147 Within minutes we were schussing cornices. 1976 *National Observer* (U.S.) 20 Nov. 15/1 Schussing the Italian Alps is the newcomer on many skiers' dream list.

2. *intr.* To effect a schuss; to ski downhill. Also with *down.*

1963 I. Fleming *On H.M. Secret Service* xvii. 188 Bond schussed easily downwards..resting his limbs. 1969 R. Petrie *Despatch of Dove* xii. 181 Would she..assume Zoe had *schussed* down on her own? 1973 *Times* 28 Sept. 36/5 Forty miles of pistes where you can schuss, trek, slalom, langlauf. 1979 N. Slater *Falcon* ix. 160 When he *schussed* down the shallower *pistes*, his skis were closely parallel.

Hence **schu·ssing** *vbl. sb.*, the action of the vb.

1961 R. Skepper *Tackle Ski-ing this Way* vi. 74 Schussing is probably where the pleasure skier and the racer come into closest contact. Everybody enjoys trying a schuss. 1969 R. Petrie *Despatch of Dove* xii. 176 The Grand Finale will be mass *schussing.* 1977 *Time* 21 Feb. 52/2 He gave up downhill schussing, lest an accident keep him away from the boardroom, but enjoys cross-country skiing.

|| **schussboomer** (ʃu·sbūməɹ). *U.S.* [f. as prec. + Boom *v.*[1] + -er[1].] A fast downhill skier. Also *attrib.* Hence **schu·ssbooming,** fast downhill skiing.

1959 *Washington Post* 11 Dec. c6/4 Expert schussboomers appear to be coming down the slopes in effortless motion. 1961 *Ski* Nov. 33 (*heading*) Can schussbooming be stopped? *Ibid.* 34/2 The various solutions proposed for the 'schussboomer problem' in the past have tended to place most of the burdens on the ski area operators. 1967 *N.Y. Times* 20 Jan. 33/6 Cervenia continues to be the schussboomer's paradise.

|| **Schutzbund** (ʃu·tsbunt). [G., lit. 'defence alliance'.] In full *Republikanischer Schutzbund*, an Austrian Social Democratic paramilitary organization, dissolved in 1933. Also *attrib.* Hence **Schu·tzbü:ndler,** a member of the Schutzbund.

1927 *Daily Tel.* 19 July 11/5 In an encounter between the Schutzbund and Communists and criminals six persons were killed. 1955 Koestler *Trail of Dinosaur* 44 The few thousand foreign workers—mainly Austrian *Schutzbund* people and German Communists—who were admitted into Soviet Russia. 1973 E. Osers tr. *K. Waldheim's Austrian Example* ii. 27 From them [*sc.* workers' militias] grew the *Republikanischer Schutzbund*, the army of the Social-Democratic Party. 1974 T. P. Whitney tr. *Solzhenitsyn's Gulag Archipelago* I. i. ii. 59 There were Schutzbündlers who had lost the class battles in Vienna and had come to the Fatherland of the world proletariat for refuge.

Schutzstaffel (ʃu·tsʃta:fəl). Also pl. **Schutzstaffeln,** and with lower-case initial. [G., lit. 'defence squadron'.] The internal security force of the Nazis in Germany, more usually known by its initials S.S. (see *S 4 a). Also *transf.* and *attrib.*

1930 *Times* 18 Sept. 11/1 If the Storm detachments form the potential army of an imaginary Hitler State, the defence squads (Schutzstaffeln) are its police. 1932 [see S.S. s.v. *S 4 a]. 1946 E. Linklater *Private Angelo* iv. 40 The Count was a prisoner of the Schutzstaffel. 1968 *Punch* 26 June 939/3 Certainly Richelieu was both cruel and sinister... But then, nobody runs a country for 18 years—single-handed except for cronies, spies and *schützstaffel* [*sic*]—without being ruthless. 1974 F. Nolan *Oshawa Project* xv. 94 The black *Schutzstaffel* uniforms.

schvartze, schvartzer (ʃvā·ɹtsə, -əɹ). *slang.* Also **schw-, shv-, -ze(r), -tza,** etc. [Yiddish, f. *shvarts* black: cf. Swart *a.* (*sb.*).

Strictly, Yiddish *shvartser* represents the masculine, and *shvartse* (occas. written *schvartza*, etc.) the feminine form, but these distinctions have become blurred. The pl. *schvartzes* could, in English, correspond to a sing. form *schvartz*, but there is no evidence for the latter.]

A Negro, a Black; *spec.* (with the ending -*a* or -*e*) a black maid (in the U.S.).

Somewhat derogatory.

1961 A. Smith *East-Enders* vi. 92 All der young generation flocks rahnd de Schwarzers. 1963 *Spectator* 19 July 79 Where Perec Rachman gained his first experience of putting in the schwarzes. 1967 P. Roth in *Esquire* Apr. 191/4 She [*sc.* my mother] sews, she knits, she darns—she irons better than the *shvartze.* 1967 P. Welles *Babyhip* xvi. 114 Imagine those *schvartzas.* Pretty soon they'll be living next door. *Ibid.* 115 'She's just the *schwartza*,' Mrs Green said, 'I don't call her a nigger, it isn't nice.' 1967 *Times* 11 Nov. 17/3 Marrying a girl he hardly knew, an honest-to-god peasant, and a Haitian one at that, a schvartsa yet. 1969 L. Greenbaum *Out of Shape* xii. 78 I'm not robbing Wheeler of his civil rights or picking on him because he's a *schvartze.* 1971 B. Malamud *Tenants* 213 An irritable old man.., says to his sweating son: 'You should be ashamed to dance like a shvartzer, without any clothes on.' 1975 R. H. Rimmer *Premar Experiments* i. 140, I couldn't forget the whole mad business of bringing a black girl home to my unsuspecting parents. If I arrived with one who stank of sweat, it would only prove what they already suspected about *schvartzers*, that they not only look different, but they smell different too. 1976 R. B. Parker *Promised Land* xvii. 102 'Were you aiming at an exclusive Jewish clientele?' 'Huh? Jewish? Why Jewish? Anybody was welcome. I mean we wouldn't be thrilled if the Shvartzes moved in, but we didn't care about religion.' 1979 *Guardian* 17 Mar. 13/1 'The schwartzes can't spell,' he lapsed into Yiddish, 'but..the schwartzes have good taste.'

schwa (ʃwā). Also **shwa.** [G.: see Sheva.] The central vowel sound (ə), typically occurring in weakly stressed syllables, as in the final syllable of 'sofa' and the first syllable of 'along'; = Sheva 2. Occas., the symbol of an inverted 'e' used to represent this sound. Also *attrib.* and *Comb.*

1895 [see Sheva 2]. 1933 Bloomfield *Language* 519 Linguists sometimes speak of this phoneme by the name *shwa*, a term taken from Hebrew grammar. 1934 Priebsch & Collinson *German Lang.* i. iii. 51 These overshort vowels are called 'Schwa-vowels' (from the Hebrew *ševa*). 1954 W. F. Leopold in Saporta & Bastian *Psycholinguistics* (1961) 354/1 Central [ə] (schwa) was learned in unstressed syllables during the second half of the second year, because its neutral character made it suitable in such a position. 1956 D. Jones *Outl. Eng. Phonetics* (ed. 8) viii. 30 The sound known as the 'neutral vowel' or 'schwa'. 1963 *English Jrnl.* May 393/1 The inverted *e* or schwa for the neutral vowel used in weakly stressed syllables. 1964 D. Ward in D. Abercrombie et al. *Daniel Jones* 393 The plosives and j being registered with a following shwa vowel for present purposes. 1973 A. H. Sommerstein *Sound Pattern Anc. Gr.* iii. 87 Though Greek has had a stress accent for about 1,600 years, unstressed vowels have firmly resisted reduction to schwa. 1975 *Language* LI. 265 The syncope of a penultimate unaccented vowel and the deletion of final shwa lead to a system in which stress invariably falls on the last syllable. 1978 *Canad. Jrnl. Ling.* 1977 XXII. 226 In the treatment of German phonology..shwas (in, e.g., *Zunge, geöffnet*, etc.) are phonemicized as /e/ with no explanation. 1979 T. Burrow (*title*) The problem of shwa in Sanskrit. 1980 *Amer. Speech* 1976 LI. 272 The schwa some speakers have in the third syllable of *medicine* is produced by a low-level phonetic rule that reduces unstressed short vowels.

Schwabacher (ʃvā·bāχəɹ). [G., f. *Schwabach*, name of a town in central Bavaria.] A German black-letter type-face, a simplified, lighter version of bastarda, used in the late fifteenth and sixteenth centuries. Also *attrib.*

[1910 *Encycl. Brit.* VII. 723/2 For these scanty writings the German so-called 'Schwabach' characters were used.] 1922 D. B. Updike *Printing Types* i. iv. 62 The smaller type of the Indulgences, which is a rounder black-letter, has certain peculiarities later found in 'schwabacher' fonts. 1926 [see *fraktur]. 1934 A. F. Johnson *Type Designs* i. 31 Schwabacher has the usual Bastarda characteristics, the closed, one-storeyed a, and pointed descenders to s and f; the tail of the g is open. 1969 [see *fraktur]. 1972 P. Gaskell *New Introd. Bibliogr.* 18 The Schwabacher group..tended towards the rotundas.

schwaerm, ¶ var. *schwarm *sb.*

Schwann (ʃvan). *Anat.* The name of Theodor *Schwann* (1810–82), German physiologist, who described the neurilema in 1839 (*Mikroskop. Untersuchungen ü. d. Uebereinstimmung in d. Strukt. u. d. Wachsthum d. Thiere u. Pflanzen*), used *attrib.*, in the possessive, and with *of*: **a.** as *sheath of Schwann, Schwann's sheath*: = *neurilema, neurilemma c.*

1874 A. E. J. Barker tr. *Frey's Histol. & Histochem. of Man* 680/1 (Index), Schwann's sheath of nerve fibre. 1882 T. E. Satterthwaite *Man. Histol.* (ed. 2) ix. 110 A delicate membrane or envelope, the sheath of Schwann or primitive sheath. *Ibid.*, The neurilemma, which by some is spoken of as synonymous with Schwann's sheath (Frey). 1892, etc. [see *neurilema, neurilemma c]. 1898 *Jrnl. Compar. Neurol.* VII. 183 The terminal branches of the axis cylinder are not invested with a continuation of the sheath of Schwann. 1925 Elwyn & Strong *Bailey's Text-bk. Histol.* (ed. 7) viii. 197 (*caption*) Ring-like thickening of Schwann's sheath at node of Ranvier. 1972 Matzke & Foltz *Synopsis of Neuroanat.* (ed. 2) ii. 11 The axon may have one or two coverings: myelin and sheath of Schwann.

b. to designate the cells which enwrap the axons of peripheral nerve fibres and form the myelin sheath (when it is present); formerly, the parts of these cells containing the nucleus and cytoplasm.

1904 *Amer. Jrnl. Anat.* III. 261 [He] referred to them as 'nerve corpuscles' and 'half-moon cells', and they have since been called 'Schwann's corpuscles' from their relation to the sheath of Schwann and their supposed identity with the nucleus of that sheath in the adult peripheral nerve. 1906 *Ibid.* V. 121 If one examines a developing nerve, one sees that there are numerous spindle shaped cells (cells of Schwann, sheath cells) throughout its course. 1931 W. Bloom *Maximow's Text-bk. Histol.* xii. 254 The nuclei of the Schwann cells usually are flattened and oval. 1960 G. Causey *Cell of Schwann* i. 5 The differentiation between Schwann membrane and Schwann cell that has been made by many histologists since 1839 is not a distinction that existed in Schwann's original description. *Ibid.* iv. 45 The term Schwann cell will be now used for any cell that enfolds a nerve fibre within its cytoplasm, whether..in the somatic or visceral parts of the peripheral nervous system. 1966, etc. [see *neurilema, neurilemma c]. 1967 Matzke & Foltz *Synopsis of Neuroanat.* ii. 11 The Schwann cell contains a scanty amount of cytoplasm but a prominent nucleus. 1976 *Path. Ann.* XI. 355 Studies of regenerating nerve provided morphologic data suggesting that Schwann cells are capable of manufacturing fibrillar collagen.

Schwannoma (ʃvanǫᵘ·mă). *Path.* Also **schwannoma.** [f. prec. + *-oma.] A tumour derived from a Schwann cell.

1932 P. Masson in *Amer. Jrnl. Path.* VIII. 367 (*heading*) Experimental and spontaneous schwannomas. 1948 [see *neurilemoma, neurilemmoma]. 1961 R. D. Baker *Essent. Path.* xxii. 604 Neurilemmomas (neurinomas, Schwannomas) are tumors which arise from the nerves and are composed of tissue like that of the sheath of Schwann. 1976 J. H. Lizuka in G. Berci *Endoscopy* lix. 756/1 Schwannomas arising from the peripheral nerve sheath..can be stereoencephaloscopically diagnosed.

|| **schwarm** (ʃvarm), *sb.* Also erron. **schwaerm, schwärm.** [G.: see Swarm *sb.*] An enthusiasm, a 'craze'; *spec.*, an erotic attachment, as of one woman or adolescent girl for another; a 'crush'.

1926 F. M. Ford *Man could stand Up* i. iii. 58 Your *schwaerm* for my father's memory and all. 1931 E. F. Benson *Mapp & Lucia* vi. 151 Irene..had developed a violent *schwärm* for Lucia. *a* 1956 F. Lawrence *Mem. & Corr.* (1961) 121 The adoration, the *Schwarm*, for somebody or other exclusively. 1968 N. Marsh *Clutch of Constables* iv. 87 The wretched woman..had developed a *schwarm* for Troy herself.

|| **schwärm** (ʃvērm), *v.* [ad. G. *schwärmen*: see next.] *intr.* To feel or display enthusiasm or passion. Also **schwä·rmer,** (fem.) **schwä·rmerin,** an enthusiast, a zealot.

1884 G. Meredith *Let.* 17 Sept. (1970) II. 745 The enclosed, from a Wagner Schwärmer, may induce you to visit Munich. 1913 R. Brooke *Let.* 17 Dec. (1968) 553 They dance so well, the Fijians... I *schwärm* for it; and no white in or near Fiji cares twopence for anything except money-making. 1927 M. Sadleir *Trollope* 211 Kate Field..developed first into a *Schwärmerin* for all the arts, then into a blue-stocking, and finally into a champion of woman's rights. 1946 J. Cary *Moonlight* xxxiv. 276 'Oh, I saw it, a great bare barrack, and not even allowed to speak to other girls in the passages, or, walk arm-in-arm, and nowhere to go by yourself...' Amanda, mildly surprised by this explosion of anger, said, 'It's true we weren't allowed to schwärm—.'

|| **schwärmerei, Schwärmerei** (ʃvē·rmərəi). [G., f. *schwärmen* to swarm, to display enthusiasm, to rave: see Swarm *v.*[1]] Religious zeal, fanaticism, extravagant enthusiasm for a cause or a person; an erotic attachment, esp. of one woman or adolescent girl for another; a 'crush'.

1845 *Edin. Rev.* LXXXII. 453 His [*sc.* Lessing's] mind is both clear and strong, free from *schwärmerei*, (a word untranslatable, because the thing itself is un-English,) free from cant and affectation of all kinds. 1857 G. H. Lewes *Biogr. Hist. Philos.* (ed. 2) ii. 531 Kant's..energetic contempt for Swedenborgianism and all other *Schwärmerei* is unequivocally expressed. 1863 Crown Princess of Prussia *Let.* 21 Mar. in R. Fulford *Dearest Mama* (1968) 183, I did wrong in allowing my feelings vent in writing to you about England; I thought afterwards it would bore you and my 'schwärmerei' would make you impatient. 1880 G. Gissing *Workers in Dawn* I. xii. 261 He has no belief whatever in the heroic woman, laughing to scorn women's rights, and speaking almost as disrespectfully of that *schwärmerei* of which you are yourself such an exalted instance. 1886 *Athenæum* 3 Apr. 451/3 A few hours' *schwärmerei* over what Joan [of Arc] must have felt under certain circumstances. 1927 F. B. Young *Portrait of Clare* i. vi. 63 The expression of liberty and exultant youth that her mother's mid-Victorian fantasias and Miss Boldmere's *Schwärmerei* denied her. 1930 E. Scott *Forgotten Image* xiii. 98 Her idiotic, schoolgirlish schwärmerei attachment. 1937 *Times Lit. Suppl.* 8 May 358/2 It would be easy to dismiss her account of how she 'read Schopenhauer and was blissfully happy' as the Schwärmerei of a pretentious blue-stocking. 1958 *Observer* 26 Jan. 16/6 Philhellenism, when it is not a mere student Schwärmerei, may sometimes develop along much the same lines as a certain type of love affair. 1971 *Times Lit. Suppl.* 23 Apr. 470/1 There were two Watts sisters, the husband of one of them and a young son for whom Swinburne had at one time a *schwärmerei.* 1976 W. Gérin *Elizabeth Gaskell* x. 92 Mr. Gaskell shielded himself as best he could from the Miss Winkworths' *schwärmerei.*

|| **schwärmerisch** (ʃvē·rmərıʃ), *a.* [Ger.: see prec.] Extravagantly enthusiastic; infatuated.

1894 Mrs. H. Ward *Marcella* III. iv. iii. 295 Betty.. wrote..wild, '*schwärmerisch*' letters. 1927 D. L. Sayers *Unnatural Death* ii. xvi. 179 It is natural for a schoolgirl to be *schwärmerisch*—in a young woman of twenty-two it is thoroughly undesirable. 1933 J. Buchan *Prince of Captivity* iv. i. 321 The type of American whose mind had two compartments, realistic business and *schwärmerisch* dreams. 1980 *Country Life* 10 Dec. 2278/1 Many had adored her [*sc.* George Eliot], and not only the *schwärmerisch* women she so forcefully attracted.

schwartz (ʃvāɹts). *Skat.* Also **Schwartz, schwarz.** [a. G. *schwarz* black.] (See quot. 1880.) Also as *adj.* and *v. trans.*

1880 W. B. Dick *Amer. Hoyle* (ed. 13) 103 If he should propose to make no less than a hundred and twenty points, he would call his bid a *Schwartz.* *Ibid.*, With a very strong hand, a player may bid to *Schwartz* his opponents, that is, prevent them from making a single point. 1886 E. E. Lemcke *Skat* 7 With no count at all, he is '*Schwarz*' (black = whitewashed). 1908 A. D. Granger *Skat & How to play It* i. 35 All the ten tricks must be taken to score Schwarz. *Ibid.* 36 Some authorities consider that in Schwarz if a trick which contains no points is lost by the player it ought not to count against him. 1949 A. A. Ostrow *Compl. Card Player* 628 If declarer wins every trick, opponents are said to be 'schwarz'. *Ibid.* 645 To make schwarz, he must take every trick. 1975 *Way to Play* 109/1 If the bidder has named suits or grand, he may before the opening lead, declare:..schwarz, ie he aims to take every trick. 1976 *National Skat & Sheepshead Q.* Mar. 18 The correct answer is to declare a club solo schwartz announced which scores 96 points.

schwartza, schwartze: see *schvartze, schvartzer.

Schwarz (ʃvāɹts). *Math.* Also (*erron.*) **Schwartz.** The name of Hermann Amadeus *Schwarz* (1843–1921), German mathematician, used *attrib.* and in the possessive to designate the various forms of the theorem which states that the square of the sum of a set of products of two quantities cannot exceed the sum of the squares of the first terms multiplied by the sum of the squares of the second terms.

1955 M. Loève *Probability Theory* ix. 156 Hölder's inequality with $r = s = 2$, is called the Schwarz inequality: $E^2|XY| \leqslant E|X|^2 . E|Y|^2$. 1962 W. B. Thompson *Introd. Plasma Physics* viii. 222 This makes use of the Schwartz inequality $\int f^2 dx \int g^2 dx \geqslant [\int fg\, dx]^2$. 1964 McCord & Moroney *Introd. Probability Theory* ix. 155 Let X_1 and X_2 be any two jointly distributed random variables which have finite, positive variances... If *a* and *b* are any real constants deduce that $[E(X_1-a)(X_2-b)]^2 \leqslant [E(X_1-a)^2][E(X_2-b)^2]$, which is a form of Schwarz's inequality. 1965 Patterson & Rutherford *Elem. Abstr. Algebra* v. 176 In a unitary space $\|a\|\,\|b\| \geqslant |a\cdot b|$... The inequality is known as Schwartz's inequality. 1975 Karlin & Taylor *First Course Stochastic Processes* (ed. 2) ix. 452, $E[|YZ|] \leqslant \sqrt{(E[Y^2]E[Z^2])} = \|Y\|\,\|Z\|$. This is known as Schwartz' inequality.

schwarze, schwarzer: see *schvartze, schvartzer.

|| **schwarzlot** (ʃvā·rtslǫt). Also with capital initial. [G., lit. 'black lead'.] A type of decoration used on Dutch and German glass of the seventeenth century, and subsequently on German and Austrian pottery and porcelain, consisting wholly or chiefly of black enamel.

1925 B. Rackham tr. *E. Hannover's Pottery & Porcelain* I. iv. iii. 357 At Nuremberg and Augsburg,..there lived enamel-painters who had white wares delivered to them from the factories to be decorated by them at home.., designs in purple or pictures in the so-called *Schwarzlot.* *Ibid.*, The glass and *Schwarzlot* painter *Johann Schaper*.. lived at Nuremberg between 1640 and 1670. 1952 J. F. Hayward *Viennese Porcelain of Du Paquier Period* ix. 107 The combination of hunting subjects and schwarzlot decoration must have been one of the more successful ventures of the [Vienna] factory. *Ibid.* xi. 123 The big schwarzlot-decorated services date from the 1730s. 1954 G. Savage *Porcelain through Ages* v. 172 Daniel Preussler and his son Ignaz, were Bohemian decorators whose work was principally executed in black enamel—*Schwarzlot*—which was a characteristically Bohemian practice. 1971 *Daily Tel.* 6 Apr. 10/4 An early Meissen Hausmaler coffee pot and cover. ..It was decorated in schwarzlot by J. Auffenwerth at Augsburg. 1975 *Oxf. Compan. Decorative Arts* 335/2 Johann Anton Carli of Andernach (d. 1682), painted in schwarzlot a fine goblet bearing a view of Andernach and a beautifully rendered hunting scene.

Schwarzschild (ʃvā·ɹtsʃilt, ʃwǭ·ɹtstʃaild). The name of Karl *Schwarzschild* (1873–1916), German astronomer, used *attrib.* and in the possessive to designate various concepts developed by him or arising from his work. **1.** *Photogr.* Used with reference to a quantitative law of reciprocity failure in emulsions.

1920 *Jrnl. Optical Soc. Amer.* IV. 272 If Schwarzschild's law is correct, the reciprocity law does not hold for any value of the intensity, the error being the same in percentage amount for all intensities. 1942 C. E. K. Mees *Theory of Photographic Process* vi. 236 Schwarzschild (1899) confirmed Abney's results that the reciprocity law is not valid and concluded that a constant effect is produced so long as the condition It^p = constant is satisfied, in which *p* is constant and equal to about 0.8. This relation, with *p* constant, came to be generally known as Schwarzschild's law and is frequently referred to by this name today. 1960 G. E. Lockie tr. *K. S. Lyalikov's Chem. of Photographic Mechanisms* i. i. 58 The Schwarzschild equation is not valid for the range of exposures used in this investigation.

2. *Physics.* Denoting concepts arising out of the exact solution of Einstein's field equations described by Schwarzschild soon after the publication of the general theory of relativity (*Sitzungsber. der k. preuss. Akad. der Wissensch.* (1916) 189, 424), as *Schwarzschild coordinate, field, geometry, horizon, solution, space-time, surface*; **Schwarzschild black hole,** a static, non-rotating, and uncharged black hole, i.e., an object postulated to result from the complete gravitational collapse of an electrically neutral and non-rotating body, and which has a physical singularity at the centre of its Schwarzschild sphere to which the infalling matter inevitably proceeds and at which the curvature of space-time is infinite; **Schwarzschild line element,** (*a*) a scalar representation of the Schwarzschild metric, being an expression for the separation of two adjacent points in the space-time of Schwarzschild geometry; (*b*) *loosely* = *Schwarzschild metric* (*a*); **Schwarzschild metric,** (*a*) a mathematical description of the geometry of space-time exterior to a non-rotating body, usu. expressed as a tensor in differential geometry; (*b*) *loosely* = *Schwarzschild line element* (*a*); **Schwarzschild radius,** the radius of the Schwarzschild sphere; **Schwarzschild singularity,** a singularity in coordinates, but not a physical singularity in space-time, occurring at the Schwarzschild radius; **Schwarzschild sphere,** the effective boundary or horizon of a Schwarzschild black hole, which infalling matter reaches in an infinite time as seen by an external observer but a finite time in the reference frame of the matter, and at which the escape velocity is infinite, so that the escape of matter or radiation from the inside is impossible except by a postulated quantum-mechanical process.

1927 G. D. BIRKHOFF *Relativity & Mod. Physics* (ed. 2) xv. 255 The most general solution can be obtained from the Schwarzschild solution by a proper choice of coördinates. **1934** R. C. TOLMAN *Relativity, Thermodynamics & Cosmology* 208 There are..three consequences which can be obtained from the Schwarzschild line element which can be used to distinguish between the relativistic and Newtonian theories of gravitation. **1939** *Ann. Math.* XL. 924 In the case of a Schwarzschild field a particle is bound to follow a path with a radius greater than $(2+\sqrt{3})$ times the radius of the Schwarzschild singularity. **1957** *Physical Rev.* CVIII. 1067/2 This transformation is not acceptable..because it assumes Euclidean rather than Schwarzschild geometry for the displacement. *Ibid.*, We get the difference between the Schwarzschild metrics for the two reduced masses. *Ibid.* 1068/2 The effective potential starts from 0 at the Schwarzschild radius, rises to a maximum and then falls off again to zero at very large *r*. **1965** B. K. HARRISON et al. *Gravitation Theory & Gravitational Collapse* 157 Introduce Schwarzschild coordinates ds^2 = [etc.]..as well as the baryon number coordinate. **1966** R. AKERIB tr. *M. A. Tonnelat's Einstein's Unified Field Theory* v. 82 This is the Schwarzschild solution of the field equations. It defines completely the gravitational field in the neighborhood of attractive masses and permits the determination of the trajectories of particles moving in it. **1968** ROBERTSON & NOONAN *Relativity & Cosmology* ix. 236 The Schwarzschild line element has at the radius 2μ a singularity known as the Schwarzschild singularity. **1968** SEARS & BREHME *Introd. Theory of Relativity* xi. 200 If the Schwarzschild singularity did exist, the Schwarzschild radius would be the radius of a spherical surface which separates the universe into two parts which are isolated from one another by the fact that local time does not elapse at the bounding surface. **1968** *Commun. Math. Physics* VIII. 245 In the special case where the gravitational coupling of the electromagnetic energy density is neglected..all solutions are computed explicitly, thus extending an earlier result of Ginzburg for a magnetic dipole in Schwarzschild's space-time. **1969** *Nature* 16 Aug. 690/1 Nothing can ever pass outwards through the Schwarzschild sphere of radius $r = 2GM/c^2$. **1970** *Ibid.* 4 Apr. 64/2 The metric used to describe the geometry of space-time in the vicinity of the collapsed object in this and other papers..has been the spherically symmetric Schwarzschild metric, which is valid only if the collapsed object has zero angular momentum. **1971** *Jrnl. Math. Physics* XII. 1846/1 We consider the problem of a point charge slowly lowered into a Schwarzschild black hole as a simple example where the final outcome can be investigated. **1973** C. W. MISNER et al. *Gravitation* xxiii. 597 The above discussion identifies the Schwarzschild coordinates.. by their intrinsic geometric properties. Not only are *r* and *t* radial and time variables, respectively (in that $\partial/\partial r$ and $\partial/\partial t$ are spacelike and timelike, respectively..), but they have particular properties..that distinguish them from other possible coordinate choices... No claim is made that they are the only coordinates that might reasonably be called *r* and *t*. **1973** *Physics Bull.* Nov. 654/3 An observer falling with the surface of the collapsing star has his light cones squashed as he reaches the Schwarzschild surface of radius $2GM/c^2$; he finds it ever more difficult to signal to distant observers as collapse proceeds. He will appear to them to fall ever more slowly as he approaches the critical surface, and never actually reach it. **1974** *Nature* 5 July 37/2 There is no creation of massless particles in the exterior region of a Schwarzschild black hole, which is the static end state reached as a result of spherically symmetric gravitational collapse. *Ibid.* 17/1 In essence the significance of the Schwarzschild surface at $r = R_s$ must have been known to Eddington, certainly by the early 1930s. *Ibid.*, Once inside the Schwarzschild sphere, one cannot communicate with the world outside; and moreover, one would inexorably be propelled towards the centre: not all the King's horses nor all the King's men can prevent it from happening. **1974** *Encycl. Brit. Macropædia* XV. 587/1 The most conspicuous feature of the Schwarzschild field is that if the total mass is thought of as concentrated at the very centre, then at a finite distance from that centre, the Schwarzschild radius, the geometry of space-time changes drastically from that to which we are accustomed. **1977** *Sci. Amer.* Jan. 34/3 For a star of about 10 solar masses the Schwarzschild radius is about 30 kilometres.

Schweik (ʃwaik). A character in *The Good Soldier Schweik* by Jaroslav Hašek (1883–1923), Czech writer, pictured as an unlucky and simple-minded but resourceful little man oppressed by higher authorities; a person of this type. Hence **Schweik** *v. intr.*, to behave in the deferential, crafty manner of Schweik; **Schwei·kism,** behaviour characteristic of a Schweik; **Schwei·kist** *a.*, typical of a Schweik.

1952 M. McCARTHY *Groves of Academe* vi. 128 They're the expression of a certain reactionary Schweikism which we've seen also in faculty meetings. **1965** *New Statesman* 7 May 708/1 The Berlin Battle Groups..paraded in Bebel Square to be given their medals—a bored and happy collection of shuffling, grumbling Schweiks. **1968** *Economist* 14 Sept. 27 Censorship is now operating... In a somewhat Schweikist manner, the Czech papers succeeded in getting around this by reporting, poker-faced and without comment, what the 'socialist' press is saying, trusting to their own readers' ability to read between the lines. **1973** *Libertarian Education* XI. 17/2 They will make some kind of bitter psychological adjustment and Schweik their way to retirement.

‖ **schwein(e)hund** (ʃvai·n(ə)hunt). Also **schwine-, -hundt** and with capital initial. [G. *schweinehund*, f. *schwein* pig + *hund* dog: cf. *pig-dog* s.v. *PIG *sb.*¹ 13.] A German term of abuse: filthy dog, 'swine', 'bastard'.

1941 [see *KNOB *sb.* 1 e]. **1959** B. MATHER *Achilles Affair* I. ix 105, I heard a curt command in German—'Spread out, you *schwinehund*—spread out.' **1975** I. MELCHIOR *Sleeper Agent* III. vii. 154 Himmler had turned traitor!.. That back-stabbing *Schweinehund*! **1978** T. L. SMITH *Money War* I. 67 He farted long and loud. Mundt giggled, 'Schweinhundt!'

‖ **Schweinerei** (ʃvai·nərai). Also with small initial. [G., lit. 'piggishness'.] Obnoxious behaviour, a repulsive incident or object, a scandal.

1906 G. B. SHAW *Let.* 7 May (1972) II. 620 Reinhardt took not the smallest notice of either your letter or Barker's. Barker naturally regards this as a Schweinerei of the first order. **1938** L. BEMELMANS *Life Class* I. vi. 98 Whenever a student brought in a reproduction of one of her paintings, Thaddeus would..shout: 'Take it out, out with this *Schweinerei*!' **1965** *Economist* 25 Dec. 1433/3 Some Japanese producers were discovered selling out of line at the recent Canton Fair. But this *schweinerei* was swiftly stamped on. **1975** *Times* 7 Oct. 12/8 India..will outline the petty *Schweinerei* of Mrs. Gandhi.

schweinfurt green (ʃvai·nfūᵊɹt). *Chem.* [f. *Schweinfurt*, name of a city in Germany.] A toxic green pigment that is a mixed acetate and arsenite of copper, $Cu_3(AsO_3)_2.(CH_3COO)_2Cu$, and was formerly used as an insecticide. Also called *Paris green, Vienna green.*

1852 [see *Vienna green* s.v. VIENNA a]. **1874** W. CROOKES *Dyeing & Calico-Printing* II. ii. 156 Schweinfürt green or aceto-arsenite of copper. **1879** H. CARR *Our Domestic Poisons* 7 Experiments were made on a paper coloured with Schweinfurt-green, an aceto-arsenate of copper. **1930** I. J. KLIGLER *Epidemiology & Control of Malaria* vii. 147 Paris green or Schweinfurt green (copper aceto-arsenite) used so much as an insecticide is an efficient larvacide.

‖ **Schweizerdeutsch** (ʃvai·tsərdoitʃ), **Schwyzertütsch** (ʃvi·tsərtütʃ). Also **Schwei(t)zer-Deutsch, Schwyzerdütsch,** etc. [G. *Schweizerdeutsch*, Swiss G. *Schwyzertütsch*.] **a.** Swiss German dialect. **b.** A Swiss German; also *collect.*

1934 PRIEBSCH & COLLINSON *German Lang.* vii. 326 The Alemannic group includes Swabian..; High Alemannic (with 'Schwyzerdütsch' and the dialects of the southern parts of the Black Forest and Vorarlberg)..; Low Alemannic. **1953** U. WEINREICH *Languages in Contact* ii. 14 Schwyzertütsch, an Alemannic dialect,..spoken in the village of Thusis. **1961** [see *LUXEMBURGISCH]. **1963** I. FLEMING *On H.M. Secret Service* xxiv. 257 Swiss Air Control, in thick Schwyzerdütch, asked them politely to identify themselves. **1963** *Guardian* 22 Aug. 6/6 The cast of the Zurich opera doing 'Porgy and Bess' in black faces *and* Schweizerdeutsch. **1963** *Punch* 18 Sept. 390/2 Your single word of Schwyzertütsch will trigger a multitude of pleased smiles. **1964** 'P. QUENTIN' *Family Skeletons* II. 61 The pleasure steamer on Lac Leman..the drunken Schweizer-Deutsch trying to get fresh with her. **1969** *Beaver* Summer 6/2 Three of the sailors come in and talk the harsh Schweitzer-Deutsch of Basle. **1974** F. NOLAN *Oshawa Project* xxxi. 186 You speak Schweizer-Deutsch very well. **1978** LD. HAILSHAM *Dilemma of Democracy* xxv. 163 Schwyzer Dütsch, Suisses Romands, and Ticinese can form one Switzerland.

schwinehund, var. *SCHWEIN(E)HUND.

science. Add: **2. c.** fig. *to blind with science* (slang): to confuse by the use of polysyllabic words or involved explanations (see also quot. 1937).

1937 PARTRIDGE *Dict. Slang* 64/2 *Blinded with science*, a catch-phrase applied by brawn defeated by brains: Australian and New Zealand: C. 20. **1948** —— *Dict. Forces' Slang 1939–45* 18 *Blind with science*, to explain away an offence, a mistake, by talking at great length and very technically, thus dazzling one's interlocutor into non-pursuance of the matter. (Mostly Army.) **1973** *Daily Tel.* 17 Oct. 14/6 We are also more familiar..with the tendency for people to be blinded by science and to succumb to 'expert' medical opinion, however quackish. **1977** *Time Out* 17–23 June 11/3 It's very easy to coast and blind them with science.

4. b. Also with preceding sb., as *life science*, and combined with a prefix, as *bio-, geo-, neuroscience*. (See under the first element.)

c. In phrases: *science of art, of expression, of mind, of religion(s)*, denoting esp. the application of scientific methods in fields of study previously considered open only to theories based on subjective, historical, or undemonstrable abstract criteria.

1828 J. S. MILL in *Westm. Rev.* IX. 140 The impugners of the school logic, as they term it, may be divided into two classes. The first class consists of men not untinctured with philosophy, including even some writers of considerable eminence in the science of mind. **1869** W. JAMES *Let.* 21 Jan. in R. B. Perry *Tht. & Char. W. James* (1935) I. 291 Some weeks ago I read the three last articles on 'Science of Religions' by Émile Burnouf in the *Revue des deux mondes*. **1886** T. PATERSON *Mental Sci.* 4 This confusion of opinion has led many to deny the possibility of any science of mind, beyond the physical or material facts of life. **1902** W. JAMES *Var. Relig. Exper.* xviii. 433 Of late, impartial classifications and comparisons have become possible... We have the beginnings of a 'Science of Religions', so-called. **1909** D. AINSLIE tr. *Croce's Aesthetic (subtitle)*, As science of expression and general linguistic. **1933** *Burlington Mag.* May 248/2 The great problem as to whether the science of art really is a science in the sense that the word is used in relation to natural science remains, however, unsolved. **1937** H. READ *Art & Soc.* vii. 233 Though based on the science of art and a deduction from the whole range of relevant material, the facts in question are relative to the aesthetic sensibility. **1944** J. S. HUXLEY *On Living in Revol.* iv. 45 The science of mind developed later than biological science. **1973** N. SMART (*title*) The science of religion and the sociology of knowledge. **1976** F. McDONAGH tr. *Pannenberg's Theol. & Philos. of Sci.* iv. 256 Theology then comes under the general heading of a science of religion.

5. b. (Further examples.) This is now the dominant sense in ordinary use.

1913 C. MACKENZIE *Sinister St.* I. II. vii. 253 Science is all the go nowadays... And Science is what we want. Science and Religion. **1946** R. J. C. ATKINSON *Field Archaeol.* 12 One more problem..remains to be mentioned, the problem of co-operation between archaeologists and workers in other sciences. **1955** *Bull. Atomic Sci.* Apr. 141/1 Science has become a major source of the power of civilized man. **1976** *Norwich Mercury* 17 Dec. 3/8 Second year prizes—English,..mathematics,..science,..history,..geography,.. music. **1978** *Nature* 10 Aug. 522/1 Funds for lunar sample analysis have remained roughly constant over the past few years and the programme has received praise for the high quality of the science conducted.

d. Personified.

1742 [see sense 5 a in Dict.]. **1862** G. H. LEWES *Let.* 30 Aug. in *George Eliot Lett.* (1955) IV. 52 If the passions and impertinences of public speakers, and newspaper writers on *both* sides of the Atlantic are madly widening the wounds which each ought to strive to heal, it is some comfort to reflect that Science keeps aloof from such misplaced and unjustifiable criticisms. **1894** A. LANG *Cock Lane & Common-Sense* 328 It is in this way that Science makes herself disliked. **1975** J. PLAMENATZ *K. Marx's Philos. Man* viii. 218 Science recognizes that its hypothesis and theories are provisional and has criteria for deciding whether or not they should be discarded for bette r ones.

e. (Usually with capital initial.) *U.S.* = *CHRISTIAN SCIENCE.

1902 'MARK TWAIN' in *N. Amer. Rev.* Dec. 768 Does the Science kill a patient here and there? **1915** E. B. HOLT *Freudian Wish* 21 The 'Science' healer was immediately consulted. **1916** H. CRANE *Let.* 26 Jan. (1965) 3 Carry the *science* as far as you can. **1919** —— *Let.* 2 Apr. (1965) 15 Concerning me and my attitude toward Science. **1946** *Christian Sci. Jrnl.* Dec. 616 We called on a practitioner to learn what this Science was. **1980** A. WILSON *Setting World on Fire* II. i. 51 Servants..live in a world of doctors and illnesses and death...Of course I wasn't in Science then. I believed all their nonsense.

7. *attrib.* and *Comb.*, as (sense 5 b) *science-based* adj.; **science park** orig. *U.S.*, an area of land devoted to scientific reasearch or to industrial enterprises connected with the physical sciences.

1962 *Economist* 14 Apr. 187/1 An industry can be science-based, said Lord Hailsham, and yet do little or no actual research. **1965** A. FARRER in J. Gibb *Light on C. S. Lewis* 28 Scientific formulae may be empirically verified, but no science-based picture of the sum of things is better than a symbol. **1970** *Daily Tel.* 27 Apr. 3/8 Trinity College, Cambridge, is proposing to create a 'science park' on the north-east outskirts of the city. **1973** *Nature* 22 June 430/2 In the United States, there are over 80 science parks,..but 27 of them are wholly limited to science-based industry. **1981** *Daily Tel.* 31 July 8/3 A 116-acre science park to attract high technology-based firms, and provide hundreds of jobs, is to be established in Peterborough. Lynch Wood Science Park will also include conference and sports centres and a hotel.

sci·ence fi·ction. [f. SCIENCE + FICTION.] Imaginative fiction based on postulated scientific discoveries or spectacular environmental changes, freq. set in the future or on other planets and involving space or time travel. Also *attrib.*

Quot. 1851 shows an isolated use. The expression did not come into general use until the end of the 1920s.

1851 W. WILSON *Little Earnest Bk. upon Great Old Subject* x. 137 (*heading*) Science-Fiction. *Ibid.*, We hope it will not be long before we may have other works of Science-Fiction, as we believe such works likely to fulfil a good purpose, and create an interest, where, unhappily, science alone might fail. *Ibid.* 139 Campbell says that 'Fiction in Poetry is not the reverse of truth, but her soft and enchanting resemblance.' Now this applies especially to Science-Fiction, in which the revealed truths of Science may be given, interwoven with a pleasing story which may itself be poetical and *true*—thus circulating a knowledge of the Poetry of Science, clothed in a garb of the Poetry of Life. **1929** *Science Wonder Stories* June 89 The editor of this publication [*sc.* H. Gernsback] addressed a number of letters to science fiction lovers. The editor promised to pay $50.00 for the best letter each month on the subject of 'What Science Fiction Means to Me.' **1933** *Astounding Stories* Dec. 142/1 The..science-fiction fan does not care for stories of the supernatural... Intelligent people, as a rule, will read science fiction. **1949**, etc. [see *FANTASY, PHANTASY *sb.* 4 f]. **1954** A. HUXLEY in *Encounter* Feb. 5/1 These make up a tale which no self-respecting reader, even of Science Fiction, should be asked to swallow. **1958** *Listener* 20 Feb. 334/3 *The Naked Sun* is a happy wedding of the two great pseudo-literary forms of the century—science fiction and the 'tec. **1964** C. S. LEWIS *Discarded Image* vii. 142 The theory of the Four Zones taught that the equatorial region was too hot for life. The other hemisphere of the Earth was to us wholly inaccessible. You could write science-fiction about it, but not geography. **1972** *Sci. Amer.* Sept. 38/2, I for one would rather command a computer through a keyboard than talk to it, even if that science-fiction dream were possible. **1977** *N.Y. Rev. Bks.* 13 Oct. 13/4 The mind produces meaning like a plant branching out in a science-fiction movie. **1979** *Guardian* 18 Aug. 10/1 Science Fiction fans..have chosen April 1926 as their sacred date and nominated Hugo Gernsback for the title of Father of Science Fiction. **1980** N. BABSON *Dangerous to Know* viii. 56 The long open-plan Newsroom..always gave me a science-fiction feeling of being the last man alive.

Hence **science-fictional, science-fictive** *adjs.*, pertaining to or characteristic of science fiction; **science-fictionalized** *a.*, made into science fiction; **science-fictioneer, science-fictionist,** a writer or connoisseur of science fiction; **science-fictioner,** a film script upon a science-fictional theme.

1939 *Astounding Science-Fiction* Oct. 155 The Jekyll-science-fictionist stands for experimental truth, for logic, for *proof*. **1950** *Jrnl. Brit. Interplanetary Soc.* IX. 197 In a 'science-fictionalized' review of Arthur Clarke's book 'Interplanetary Flight', the *Daily Mirror* gave the astonishing news that British atomic scientists are now waiting for the go-ahead..to build an atomic engine which could be used as a rocket propulsion unit. **1953** C. RYAN *Conquest of Moon* i. 3 The ships the explorers will use for the long journey through space will bear little resemblance to those depicted by the science-fictionists. **1954** J. W. CAMPBELL in *Astounding Science Fiction* Aug. 5 That science-fictional device 'the planet-wrecker' bomb. **1955** *Britannica Bk. of Year* 490/1 *Science Fictioner* was coined to describe a film play of the science fiction type. **1959** C. FADIMAN in A. C. Clarke *Across Sea of Stars* p. ix, Some science fictioneers are plain old-style typewriter hacks. **1960** K. AMIS *New Maps of Hell* v. 129 That science-fictional uneasiness appears, attaching itself to an existent or incipient neurosis about overcrowded streets and buildings as well as to the rational fear of global overpopulation. **1976** I. MURDOCH *Henry & Cato* II. 222 The glossy hexagonal glasses which looked here like the appurtenances of some science fictional spaceman. **1977** *Times Lit. Suppl.* 14 Jan. 26/1 Put one science-fictioneer on a desert island and he will start a magazine. **1980** *Ibid.* 7 Nov. 1265/4 Yet Silverberg's Majipoor, where humanity..finds its future home, is about as science-fictional as Tolkien's Middle Earth.

scientaster (saiĕntæ·stəɹ). *rare.* [f. SCIENT(IST + -ASTER after POETASTER.] A petty or inferior scientist.

1899 M. FOSTER *Claude Bernard* ix. 232 We may recognise a salient difference between..the false scientaster and the true inquirer. **1969** *Sci. Jrnl.* Sept. 91/1 Zahlen made the neat distinction that in Lebanon most advisory scientists were 'not scientists at all but just had some scientific education'. Michael Foster coined the word 'scientaster' for such people; it deserves reintroduction.

scientific, *a.* and *sb.* Add: **A.** *adj.* **4. b.** Also, more loosely: systematic, methodical. (Further examples.)

1863 GEO. ELIOT *Let.* 18 July (1956) IV. 94 He [*sc.* Renan] has always seemed to me remarkable as a French mind that is at once 'scientific' (in the German sense) and eminently tender and reverent towards the forms in which the religious sentiment has incarnated itself. **1976** *National Observer* (U.S.) 13 Mar. 1/6 The Observer tally on abortion, however, is consistent with the findings of other recent, more 'scientific' polls.

c. (Earlier and later examples.)

1792 in G. B. Buckley *Fresh Light on 18th Cent. Cricket* (1935) 231 Brighton v. Lord Winchilsea, Hon. Mr. Bligh, Mr. Smith & Mr. Hale with 7 approved scientific men from the County of Hants. *a* **1817** JANE AUSTEN *Persuasion* (1818) IV. viii. 155 She had feelings for the tender, spirits for the gay, attention for the scientific, and patience for the wearisome; and had never liked a concert better. **1833** J. NYREN *Young Cricketer's Tutor* 29 In this accomplishment lies the distinction between the scientific player and the random batsman. **1851** H. MELVILLE *Moby Dick* II. xxv. 181 This accomplished swordsman..once more makes a scientific dash at the mass. **1891** W. G. GRACE *Cricket* xi. 300 From that year [*sc.* 1859] until 1876 he [*sc.* R. Daft] was the most scientific batsman amongst the professionals.

5. Of, pertaining to, or inspired by Christian Science. *U.S.*

1875 M. B. EDDY *Science & Health* viii. 428 The spirituality that abstracts all attention from the body, never manipulates and is the only positive position of scientific healing. *Ibid.* 429 To be able to discern the cause of sickness after the scientific mode of our Master, depends on your spirituality. **1919** H. CRANE *Let.* 7 Mar. (1965) 13, I feel quite certain that Mrs. Brooks is afflicted with consumption against which she is doubtless putting up a strenuous Scientific fight.

6. Special collocations: **scientific farming,** farming conducted according to theories based on science rather than on tradition; also **scientific farmer; scientific fiction** now *rare* = *SCIENCE FICTION; **scientific humanism,** a theory that humanism should be based on scientific empiricism (see quot. 1909); a doctrine that man should direct the future and the welfare of the human race by using the scientific methods he applies to other species and to the material environment; so **scientific humanist; scientific management** orig. *U.S.*, management of a business, industry, etc., according to principles of efficiency derived from experiments in methods of work, production, payment, etc., and esp. from time-and-motion studies; **scientific method,** a method of procedure that has characterized natural science since the 17th century, consisting in systematic observation, measurement, and experiment, and the formulation, testing, and modification of hypotheses; **scientific notation,** a system of representing numbers as a product of a number between 1 and 10 (or 0·1 and 1) and a power of 10; **scientific revolution,** a rapid and far-reaching development in science; *spec.* the developments occurring in the twentieth century that have involved the introduction of automation, atomic energy, electronics, etc.

1850 C. KINGSLEY *Alton Locke* II. iv. 43 He had one scientific farmer after another, staying in his house as a friend. **1789** A. YOUNG *Jrnl.* 19 June in *Trav. France* (1792) I. 115, I wish my brethren to stick to their *scientific* farming, and leave the practical to those that understand it. **1886** C. M. YONGE *Chantry House* I. xvii. 159 [He] worked off his superfluous energy in scientific farming. **1902** A. BENNETT *Anna of Five Towns* ix. 205 A great landowner is exhibiting the beauties of scientific farming for the behoof of his villagers. **1876** W. H. L. BARNES in W. H. Rhodes *Caxton's Book* 7 The great master of scientific action, Jules Verne. **1937** *Discovery* Oct. 318 'The Man in the Moone', the fantasy of Bishop Godwin.., is an early excursion into the realms of scientific fiction. **1909** W. JAMES *Meaning of Truth* iii. 59 'Energetics', measuring the bare face of sensible phenomena so as to describe in a single formula all their changes of 'level', is the last word of this scientific humanism. **1931** J. S. HUXLEY *What dare I Think?* iv. 148 The only way in which the conflict between science and human nature can be ended is by combining science and the other fruits of the human spirit in a new alliance, a new attitude, to which we may give the name of Scientific Humanism. **1941** —— *Uniqueness of Man* xiii. 274 Scientific humanism..insists that the same scientific procedure can be applied to human life as has been applied to lifeless matter and to animals and plants—scientific survey, study, and analysis, followed by increasing practical control. **1963** V. BROME *Problem of Progress* vii. 144 If the modern scientific humanist would have no truck with the religious tinge in Huxley's creed he equally rejects any divine inspiration in Buddhism, Christianity, [etc.]. **1903** F. W. TAYLOR in *Trans. Amer. Soc. Mech. Engineers* XXIV. 1366 The choice must be made between some of the types of management in common use.. and the more modern and scientific management based on an accurate knowledge of how long it should take to do the work. **1910** L. D. BRANDEIS in *N.Y. Times* 22 Nov. 8/2 As an alternative to the practice of combining to raise rates and hence to increase prices, we offer cooperation to reduce costs... This can be done through the introduction of scientific management. **1911** F. W. TAYLOR in *Amer. Mag.* Mar. 571/2 The best management is a true science, resting upon clearly defined laws, rules, and principles, and..these fundamental principles of Scientific Management are applicable to all kinds of human activities. **1949** GILBRETH & CAREY *Cheaper by Dozen* i. 1 Dad always practised what he preached and it was just impossible to tell where his scientific management company ended and his family life began. **1972** *Scientific Management in American Industry* (Taylor Soc.) i. 2 The body of interlocking procedures which resulted from these investigations came to be known as the 'Taylor System', and to the doctrine and principles later derived from them was given the name 'Scientific Management'. **1854** T. H. HUXLEY *Educational Value of Nat. Hist. Sciences* 13 The man of business must as much avail himself of the scientific method..as the veriest bookworm. **1871** J. A. FROUDE *Short Studies on Great Subjects* (ser. 2) 485 Neither history, nor any other knowledge, could be obtained except by scientific methods. **1889** 'L. CARROLL' *Sylvie & Bruno* xviii. 255 That, I believe, is the true Scientific Method. **1908** W. McDOUGALL *Introd. Soc. Psychol.* i. 4 When..the modern principles of scientific method began to be generally accepted. **1927** J. S. HUXLEY *Relig. without Revelation* iii. 83 There was a great outcry when scientific method was applied, in the form of the so-called 'Higher Criticism'. **1955** *Bull. Atomic Sci.* Oct. 295/1 Scientists possess a technique which they call the scientific method of thought, and they are impelled by circumstances to use it with the force of a new inspiration. **1959** L. W. H. HULL *Hist. & Philos. Sci.* vii. 194 The subtle blend of observation, hypothesis, mathematics and planned experiment in the Scientific Method is a more effective procedure than that of Bacon. **1961** WEBSTER, Scientific notation. **1963** W. H. WARE *Digital Computer Technol. & Design* I. ii. 22 The power of the base appearing in an expression which is in scientific notation in effect indicates the position of the point. **1973** C. W. GEAR *Introd. Computer Sci.* ii. 61 The number ·00000001 is represented as $\cdot 1 \times 10^{-7}$... We call this floating-point or scientific notation for numbers. **1975** *Physics Bull.* Mar. 135/3 The most important [feature of the calculators]..is the provision of exponential or 'scientific' notation. **1803** S. MILLER *Brief Retrospect of Eighteenth Cent.* I. II. 416 The frequency and rapidity of scientific revolutions may be accounted for in various ways. **1946** *Amer. Jrnl. Sociol.* Jan. 267/1 The use of atomic energy appears to be a beginning of the 'scientific revolution'. **1959** C. P. SNOW in *Encounter* July 22/2, I believe the industrial society of electronics, atomic energy, automation, is in cardinal respects different in kind from any that has gone before... It is this transformation that, in my view, is entitled to the name of 'scientific revolution'. **1977** G. CLARK *World Prehistory* (ed. 3) ii. 41 A Neolithic Revolution comparable in importance with the Industrial and Scientific Revolutions.

C. quasi-*adv.*, as *scientific-minded* adj.

1946 J. CARY *Moonlight* xxiii. 179 Our admirals are uneducated men who despise science, and the Germans are really scientific-minded men. **1976** I. LEVIN *Boys from Brazil* iii. 78 He's hardly a scientific-minded man.

scientifically, *adv.* Add: **2.** Systematically, methodically, thoroughly.

1922 WODEHOUSE *Jill Reckless* i. 28 Freddie poked the fire scientifically, and assisted it with coal. **1965** *Listener* 30 Dec. 1077/2 Three friends scientifically mete out to a bully the same bullying he administered to a small boy.

3. *Comb.*, as *scientifically-minded* adj.

1927 B. RUSSELL *Analysis of Matter* xiv. 130 Levers and pulleys, falling bodies, collisions of billiard balls, etc., are all familiar in everyday life, and it is a pleasure to the scientifically minded youth to find them amenable to mathematical treatment. **1931** H. N. SHENTON et al. *Internat. Communication* i. 63 Associations of scientifically-minded persons can continually bring to the problem of verbal communication the rapidly growing possibilities of social engineering. **1959** I. & P. OPIE *Lore & Lang. Schoolch.* ix. 174 Wisecracks which..scientifically minded boys indulge in when, at about twelve years old, they begin to take up their subject in earnest. **1977** P. FITZGERALD *Knox Brothers* iv. 107 It was neither weakness nor compromise to try to reach this kind of unity with millions of the half-persuaded and the scientifically-minded.

scienti·ficism. *rare.* [f. SCIENTIFIC *a.* + -ISM.] Analysis or explanation which only admits what is considered to be scientifically demonstrable.

c **1875** W. JAMES in R. B. Perry *Tht. & Char. of W. James* (1935) I. 523 In a rough way materialism or 'scientificism' gratifies no. (1) [*sc.* an explanation of things by their cause]. **1884** —— *Will to Believe* (1897) 165 Subjectivism has three great branches,—we may call them scientificism, sentimentalism, and sensualism, respectively.

scientificity (saiˌĕntifi·sĭti). [f. SCIENTIFIC *a.* + -ITY.] The quality of being scientific; scientific character.

1970 B. BREWSTER tr. *Althusser & Balibar's Reading Capital* I. 49 The form of order required at a given moment in the history of knowledge by the existing type of scientificity, or, if you prefer, by the norms of theoretical validity recognized by science..as scientific. **1973** *Screen* Spring/Summer 209 Science..in its efforts to set itself off from *opinion* (the act of break which is scientificity itself).. is led to criticise most often the opinions which it meets most often. **1976** T. EAGLETON *Crit. & Ideology* i. 32 A mistaking of scientificity for positivism..links him..with the Romantic 'anti-scientism' of Lukács and the Frankfurt school.

scientifiction (saiˌĕntĭfi·kʃən). [Blend of SCIENTIFIC *a.* and *sb.* + FICTION.] Science fiction. Hence **scientifi·ctional** *a.*

1916 H. GERNSBACK in *Electr. Experimenter* Jan. 474/1, I am supposed to report Münchhausen's doings; am supposed to be writing fiction, *scientifiction*, to be correct. **1929** *Amazing Stories Q.* Fall 575, I wish to compliment you on your choice of 'scientifictional' stories. **1930** *N. & Q.* 10 May 339/1 This class of literature is having a tremendous vogue in America just now. Quite a number of popular magazines are devoted to what they have dubbed 'Scientifiction'. **1940** *Illustr. London News* CXCVII. 32/3 'Dr. Cyclops', at the Carlton, applies Technicolor to what is called 'scientifiction'. **1940** 'G. ORWELL' in *Horizon* I. 191 H. G. Wells..is the father of 'Scientifiction'. **1943** C. S. LEWIS *Perelandra* vi. 91 He was a man obsessed with the idea which is at this moment circulating all over our planet in obscure works of 'scientifiction', in little Interplanetary Societies and Rocketry Clubs. **1955** —— *Surprised by Joy* ii. 41 That the ordinary interest in scientifiction is an affair for psychoanalysts is borne out by the fact that all who like it, like it thus ravenously. **1970** *New Scientist* 5 Feb. 264/2 In this scientifictional milieu I can still sit on the lakeshore and rationally speculate that sounds heard across the water are messages from the past.

scientism. Add: **2.** A term applied (freq. in a derogatory manner) to a belief in the omnipotence of scientific knowledge and techniques; also to the view that the methods of study appropriate to physical science can replace those used in other fields such as philosophy and, esp., human behaviour and the social sciences.

1921 G. B. SHAW *Back to Methuselah* p. lxxviii, The iconography and hagiology of Scientism are as copious as they are mostly squalid. **1937** J. LAVER *French Painting in Nineteenth Cent.* i. 73 It really appeared to many educated people that at last all the secrets of the universe would be discovered and all the problems of human life solved. This superstition..we may call 'Scientism'. **1938** G. REAVEY tr. *Berdyaev's Solitude & Society* i. 12 Science has not only progressively reduced the competence of philosophy, but it has also attempted to suppress it altogether and to replace it by its own claim to universality. This process is generally known as 'scientism'. **1942** F. A. VON HAYEK in *Economica* IX. 269 We shall wherever we are concerned, not with the general spirit of disinterested inquiry but with that slavish imitation of the method and language of science, speak of 'scientism' or the 'scientistic' prejudice. **1953** A. H. HOBBS *Social Problems & Scientism* ii. 17 Scientism, as a belief that science can furnish answers to all human problems, makes science a substitute for philosophy, religion, manners, and morals... It is a pattern of beliefs..a creed that shapes thinking and affects behavior. **1956** E. H. HUTTEN *Lang. Mod. Physics* vi. 273 This belief in the omnipotence of science is..making a mockery of science: for this scientism represents the same, superstitious, attitude which, in previous times, ascribed such power to a supernatural agency. **1957** W. H. WHYTE *Organization Man* iii. 23 *Scientism*,..the promise that with the same techniques that have worked in the physical sciences we can eventually create an exact science of man. **1969** *Encounter* Jan. 23/2 There is an aberration of science..which has come to be known as 'scientism'... It stands for the belief that science knows or will soon know all the answers. **1972** K. R. POPPER *Objective Knowl.* iv. 185 The term 'scientism' meant originally 'the slavish imitation of the method and language of (natural) science', especially by social scientists. *Ibid.* 186 But I would go even further and accuse at least some professional historians of 'scientism'. **1977** A. SHERIDAN tr. *J. Lacan's Écrits* iii. 76 The early development of psycho-analysis..expresses..nothing less than the re-creation of human meaning in an arid period of scientism. **1980** *Times Lit. Suppl.* 26 Sept. 1072/2 Naturalism, in David Thomas's usage, is equivalent to what many know as scientism: the doctrine that there is no reason to think that the study of human agents, and the study of the social systems to which human agents give rise, cannot be pursued according to a methodology drawn from natural science.

scientist. Add: (Earlier example.)

1834 *Q. Rev.* LI. 59 Science..loses all traces of unity. A curious illustration of this result may be observed in the want of any name by which we can designate the students of the knowledge of the material world collectively. We are informed that this difficulty was felt very oppressively by the members of the British Association for the Advancement of Science, at their meetings..in the last three summers... *Philosophers* was felt to be too wide and too lofty a term,..; *savans* was rather assuming,..; some ingenious gentleman proposed that, by analogy with *artist*, they might form *scientist*, and added that there could be no scruple in making free with this termination when we have such words as *sciolist*, *economist*, and *atheist*—but this was not generally palatable.

2. (Usu. with capital initial.) A Christian Scientist.

1875 M. B. EDDY *Science & Health* viii. 428 The Scientist sees more clearly the cause of disease in mind, than the anatomist can in body; the latter examines the body to learn how matter is committing suicide, and the former reads the mind to find what beliefs are destroying the body. **1902** 'MARK TWAIN' in *N. Amer. Rev.* CLXXV. 763 Where can you purchase it, at any outlay of any sort, in any Church or out of it, except the Scientist's? **1903** —— in *Ibid.* CLXXVI. 509 The Scientist hastened to Concord and told Mrs. Eddy what a disastrous mistake had been made. **1938** M. MUGGERIDGE *In Valley of this Restless Mind* ii. 8 'There's a Congregational Chapel..and a Church of England third on the right.'..'Do many people go to them?' 'Not many, I think... We're Scientists.' **1980** *Country Life* 17 July 243/1 There is the dowager, American..a Scientist (of the Christian kind).

3. Appositively in *Comb.*, as *scientist-administrator, -astronaut, -dietician, -philosopher.*

1964 M. GOWING *Britain & Atomic Energy 1939–1945* iii. 106 The two most influential American scientist-administrators..were positively anxious to have a joint Anglo-American project. **1965** M. STOKE *Man in Space* (rev. ed.) 15 A second large group of astronauts..are a different breed. These newcomers are scientists... Some of these scientist-astronauts will go along on trips to the moon with the pilot-astronauts. **1971** *New Scientist* 18 Mar. 596/1 Dr Philip Chapman, the scientist-astronaut who served as mission scientist for Apollo 14. **1961** *Ann. Reg. 1960* 14 A 'scientist-dietician' and fanatical vegetarian, she believed that if we could discover the right diet we should live for ever. **1943** BLUNDEN *Return to Husbandry* 32 Whitehead, A. N... The most eloquent of modern scientist-philosophers. **1977** *Dædalus* Fall p. v, The contributors were humanists, natural scientists, and social scientists who had met to present their papers in homage to the work of two distinguished colleagues, the scientist-philosophers P. W. Bridgman and Philipp Frank.

scientistic, *a.* Restrict *rare* to sense in Dict. and add: **2.** Of or pertaining to scientism (sense *2).

1942 [see *SCIENTISM 2]. **1943** [see *HISTORICISM 3]. **1952** K. R. POPPER *Open Society & its Enemies* (ed. 2) I. 286 A typical and influential scientistic argument in favour of historicism is, in brief, this: 'We can predict eclipses; why should we not be able to predict revolutions?' **1969** *Nature* 2 Aug. 541/1 We must apply scientific method to social studies without being besotted by a scientistic philosophy and pragmatism. **1972** *Observer* 30 Apr. 36/6 They regard his [*sc.* Russell's] kind of piecemeal, logically technical, scientistic philosophy as a covert ideological support for technological civilisation. **1980** *Times Lit. Suppl.* 26 Sept. 1072/5 My remarks..should be put down to my own lack of sympathy with the scientistic vision which Thomas upholds.

scientize (sai·ĕntaiz), *v.* [f. *scient-* (see SCIENTIST) + -IZE.] *trans.* To make scientific; to give (something) a scientific character, basis, or rationale; to organize on scientific principles.

1921 M. CORELLI *Secret Power* ix. 104, I was just crazy to help all the scientists..and started 'scientising' myself. **1957** W. H. WHYTE *Organization Man* iii. 29 If ethics is to be scientized, some specific people will have to do it. **1966** H. DAALDER in R. A. Dahl *Political Oppositions* vi. 209 Marxism..'scientized' hope, and thus made life more bearable in what was otherwise a relatively weak social and political position. **1976** *Survey* Spring 75 The hitherto forbidden fields of cybernetics and sociology were called upon to improve and scientize the management of Soviet society.

Hence **sci·entized** *ppl. a.*; also **scientiza·tion,** the action or result of scientizing.

1971 J. J. SHAPIRO tr. *Habermas's Toward Rational Society* v. 62 The scientization of politics is not yet a reality, but it is a real tendency. *Ibid.* 66 The decisionistic model.. approximates the actual procedures of scientized politics. **1976** *Amer. Speech 1973* XLVIII. 288 As occupations become more mechanized and scientized, folk terms are often displaced by standard ones.

Scientologist (saiˌĕntǫ·lŏdʒist). Also with small initial. [f. *SCIENTOLOGY + -IST.] An adherent or practitioner of Scientology; a member of the 'Church of Scientology'. Also *attrib.* and *appositively.*

1952 L. R. HUBBARD *Scientology: 8–80* vi. 24 The E-Meter is available from The Hubbard Association of Scientologists. **1954** *Notes on Lectures given by L. Ron Hubbard* iv. 22 A scientologist is expected to be able to resolve problems in a great many specialized fields, of which auditing is the first field he addresses. **1956** J. F. HORNER *Summary of Scientology* i. 9 Scientologists work toward a world in which men cheerfully and willingly work together as fully free individuals able to co-operate toward the increased understanding and improvement of themselves, the race and the universe. **1965** L. R. HUBBARD *Scientology Abridged Dict.* 30 *Scientologist*, one who knows he has found the way to a better life through Scientology and who, through Scientology books, tapes, training and processing, is actively attaining it. **1968** *Time* 23 Aug. 40/3 By watching the fluctuations of a needle, Scientologist 'auditors' can supposedly discern when a student has become 'clear' and has attained 'total awareness and freedom'. **1971** *Times* 20 Nov. 3/2 After Mr Vosper had left the institution he was declared to be in a condition of enemy and fair game for scientologists. **1977** *Daily Colonist* (Victoria, B.C.) 7 Aug. 9/2 Scientologists.. can hardly be termed medical researchers. **1978** G. VIDAL *Kalki* i. 18 They were perfect 'clears', to use Scientologist jargon.

Scientology (saiˌĕntǫ·lŏdʒi). Also **s-.** [f. *scient-* (in L. *scientia* knowledge) + -OLOGY.] A system of beliefs based on the study of knowledge and claiming to develop the highest potentialities of its members, founded in 1951 by L. Ron Hubbard (b. 1911).

Scientology is registered in the U.S. as a proprietary term.

[**1937** A. NORDENHOLZ *Scientologie* 7 Die Scientologie oder Eidologie, als eine Wissenschaft vom Wissen selbst, stellt sich ihrer Anlage nach in einen Gegensatz zu den Wissenschaften von den Dingen, die ins Wissen eingehen.] **1951** L. R. HUBBARD (*title*) Handbook for Preclears: Scientology. **1952** —— *Scientology: 8–80* 8 Scientology means knowing about knowing, or science of knowledge. **1960** *Daily Tel.* 29 Nov. 13/2 Meanwhile, I toured the town trying to discover the meaning of 'scientology' and 'creative learning', the system under which the children were instructed to imagine they were dead. **1965** L. R. HUBBARD *Scientology Abridged Dict.* 30 *Scientology*, an applied religious philosophy dealing with the study of knowledge, which, through the application of its technology can bring about desirable changes in the conditions of life. **1969** *Wall St. Jrnl.* 30 July, The Court of Claims ruled that the Founding Church of Scientology failed to show its net income didn't benefit private individuals. **1970** *Official Gaz.* (U.S. Patent Office) 1 Sept. TM 52/2 [Reg. no.] 898,018. L. Ron Hubbard, Washington, D.C...*Scientology*. For Bulletins, Books and Newsletters (Int. Cl. 16). First use Nov. 21, 1951. **1971** J. G. FOSTER *Enquiry into Pract. & Effects Scientology* iv. 42 in *Parl. Papers 1971–2* XXXVI. 917 Scientology departs from the mechanistic psychology of Dianetics by introducing a new causative agent... More usually..and especially in recent works..it is called the 'spirit' or 'thetan'... Among the goals of Scientology processing are to increase the beingness of the thetan and thus increase the creative potential of the individual personality and its analytical mind. **1973** *Daily Tel.* 31 Oct. 14/4 There was Lord Soper grudgingly admitting that Scientology was entitled to be called a religion even if it was the worst one he had come across. **1975** D. LODGE *Changing Places* v. 176 A young man distributing, without conviction, leaflets about courses in Scientology. **1980** *Daily Tel.* 14 July 3/1 A Scientology spokesman said the removal of the ban would be 'in keeping with Mrs Thatcher's beliefs in individual freedom and human rights'.

sci-fi (sai fai). Also **scifi, sci fi.** Colloq. abbrev. of *SCIENCE FICTION.

1955 *Britannica Bk. of Year* 490/1 The popularity of science fiction was reflected in the contracted form Scifi. **1957** *MD Medical Newsmag.* June 62/1 Modern sci-fi writers follow an honorable tradition. **1961** B. WELLS *Day Earth caught Fire* viii. 123 'I'm not up on my sci-fi,' hesitantly. 'So we're orbiting to the sun.' **1974** *Observer* 27 Oct. 1/7 The SF fan world abounds in language..that can baffle the novice... Most important of all, you must not say 'sci fi'—it's always SF. **1978** *N.Y. Times* 30 Mar. c 22/3 A 10-part series based on what Mr. Kotlowitz called 'speculative fiction', stories that go beyond sci-fi and deal with 'ethical and moral demands' made in new worlds to come. **1980** *Verbatim* Autumn 10/1 'Sci fi' is a term used to describe bad Hollywood science fiction movies, trashy science fiction novels, and bad science fiction written by mundane writers. **1981** 'D. JORDAN' *Double Red* xiv. 61 There was a sci-fi film we didn't watch.

sci-fic (sai fik). *rare.* = prec.

1963 *Guardian* 4 Jan. 5/3 (*heading*) Psychic sci-fic. **1979** *Now!* 14 Sept. 6/4 Arthur C. Clarke is first of five sci-fic writers to talk about their work.

Scillonian, *sb.* (Earlier and later examples.)

1794 A. THOMAS *Newfoundland Jrnl.* (1968) ii. 19 The Scillonians (as they call themselves) have very little to fear as to a viset from a foreign power. **1976** *Sunday Post* (Glasgow) 26 Dec., The council are determined to protect Scillonians against the tourist element!

scimitar. Add: **3.** **scimitar-babbler,** a northern Indian or Australian bird belonging to the genus *Pomatorhinus* or *Pomatostomus*, and distinguished by a long curved bill.

1863 T. C. JERDON *Birds of India* II. 31 (*heading*) The Southern Scimitar-babbler. **1928** H. WHISTLER *Pop. Handbk. Indian Birds* 38 This Scimitar-Babbler is a gregarious species going about in small parties. **1964** R. PERRY *World of Tiger* iv. 58 Such small pests as scimitar-babblers, whose clear ringing cries are audible half a mile away in the hills.

scindapsus (sindæ·psǒs). [mod.L. (H. W. Schott *Meletemata Botanica* (1832) 21), f. Gr. σκινδαψός a plant resembling ivy.] A tropical climbing plant of the genus so called, belonging to the family Araceæ and native to Malaysia, esp. *Scindapsus pictus*, which has large variegated leaves and is often cultivated as a house plant. Cf. POTHOS, the former name of the genus *Scindapsus*.

1946 M. FREE *All about House Plants* xiii. 103 Scindapsus, Hoya, and others..attach themselves to supports by aerial roots. **1959** *Listener* 17 Dec 1094/3 Variegated scindapsus must have plenty of light. **1980** A. HUXLEY *Huxley's House of Plants* 102/1 Philodendrons, and scindapsus (devil's ivy), will do better growing around a moss cylinder.

scintigram (si·ntigræm). *Med.* Also **scinto-.** [f. SCINTI(LLATION + -GRAM.] An image or other record of part of the body obtained by measuring radiation from an introduced radioactive tracer by means of scintillation or an analogous detection method.

1952 F. K. BAUER et al. in *Jrnl. Lab. Clin. Med.* XXXIX. 153 It is suggested that this type of picture or visualization of a radioactive area be called a 'scintigram' and that specifically in this case they be called 'thyroid scintigrams'. **1963** LEADER & STELL in M. F. Campbell *Urol.* (ed. 2) I. vii. 238 The renal scintogram does not distinguish cyst from tumour, and lesions smaller than 3 cm. in diameter cannot be picked up. **1971** EMMETT & WITTEN *Clinical Urogr.* (ed. 3) III. xx. 2041/1 To localize radioactivity graphically by moving external counting probes or by scintillation camera-type detectors, producing a cartographic image known as a scintigram or scintiscan. *Ibid.* 2064/2 The hippurate scintigram reveals the presence of abnormal kidneys. **1974** R. M. KIRK et al. *Surgery* iv. 66 A scanner or gamma camera may be used to map out the distribution of radioactivity in a body, producing a diagram or 'scintigram'. **1980** *Brit. Med. Jrnl.* 29 Mar. 883/2 Bone metastases may be detected early as areas of increased isotope uptake ('hot spots') on scintigrams.

Hence **sci·ntigraph,** (*a*) a device for producing scintigrams; (*b*) a scintigram; **scintigra·phic** *a.*, of, done by, or pertaining to scintigraphy; **scintigra·phically** *adv.*, by means of scintigraphy; **scinti·graphy,** the production and use of scintigrams.

1958 *Strahlentherapy* CV. 257 Scintigraphy and the use of collimators provide the means of portraying true size and form of the thyroid. **1960** *Radiology* LXXIV. 913/1 (*caption*) X-ray exposure is made during the scintigraphic procedure. *Ibid.* 914/1 Metastases in the right hepatic lobe can be scintigraphically demonstrated with only slightly less definition than in the left. **1961** *Med. Radiol.* VI. x. 76 Scintigraph. *A. Stefanovich.* Summary. A short description is given of a design of the apparatus used to investigate the localisation of radioactive isotopes in the human body. **1975** *Sci. Amer.* July 42/3 A scintigraph is an image constructed by a computer from the signals of many scintillation detectors arranged to detect the annihilation gamma rays in coincidence. For example, a patient can inhale nitrogen containing the positron-emitting isotope nitrogen 13. The scintigraphs clearly show how the gas enters the windpipe, passes through the bronchi and finally reaches the alveoli..in the lung. **1976** *New England Jrnl. Med.* CCXCV. 1/1 We have previously reported on scintigraphic visualization of myocardial infarction in man with use of thallium-201. **1977** *Lancet* 8 Jan. 92/2 (*caption*) Scintigraphs of left thigh showing localisation of Tc-99 diphosphonate in soft tissue. *Ibid.* 7 May 1012/2 Amyloidosis should be suspected when scintigraphy with Tc-99m diphosphonate shows a positive activity of soft tissues. **1978** *Jrnl. R. Soc. Med.* LXXI. 39 Lesions in the pubis and ischium may be very difficult to detect scintigraphically.

scintillantly (si·ntilăntli), *adv.* [f. SCINTILLANT *a.* + -LY².] In a scintillating manner.

1900 H. HARLAND *Cardinal's Snuff-Box* xix. 191 The.. buildings stood out..the white marble, palely, scintillantly amethystine. **1928** A. L. FLEMING *Dwellers in Arctic Night* 151 Flashes of light from the Aurora Borealis move scintillantly in the sky.

scintillate, *v.* Add: **1. c.** *intr. Nucl. Physics.* Of a phosphor: to fluoresce momentarily when struck by a charged particle or high-energy photon.

1958 O. R. FRISCH *Nuclear Handbk.* xiv. 20 The recent discovery that some gases scintillate will undoubtedly have many future applications. **1966** *McGraw-Hill Encycl. Sci. & Technol.* XII. 76/1 The liquid organic solvent scintillates satisfactorily. **1971** *Sci. Amer.* June 61/2 The box was.. provided with a zinc sulfide screen that would scintillate when it was struck by an alpha particle.

scintillating, *ppl. a.* Add: *scintillating scotoma* (Path.), hallucinatory flickering patterns and gaps in the visual field as seen in migraine.

1883 W. B. HADDEN tr. *J. M. Charcot's Lect. Localisation of Cerebral & Spinal Dis.* xi. 122 A particular form of megrim..characterised especially by the co-existence of scintillating scotoma. **1918** J. H. PARSONS *Diseases of Eye* (ed. 3) xix. 384 Scintillating scotomata of various kinds occur in migraine... A positive scotoma appears in the field of vision; while obscuring sight it has a peculiar shimmering character. **1950** BERENS & SIEGEL *Encycl. of Eye* 37 Visual hallucinations, such as scintillating scotoma.

scintillatingly (si·ntilēⁱtiŋli), *adv.* [f. SCINTILLATING *ppl. a.* + -LY².] In a scintillating manner; sparklingly.

1927 *Sunday Express* 6 Feb. 4 A scintillatingly funny burlesque.

scintillation. Add: **1. e.** *Nucl. Physics.* A small flash of visible or ultraviolet light emitted by fluorescence in a phosphor when it is struck by a charged particle or high-energy photon.

1903 W. CROOKES in *Science* 26 June 1002/1 Bringing the radium nearer the screen the scintillations become more numerous and brighter. **1915** *Arch. Radiol. & Electro-therapy* XX. 183 The phosphorescence observed by the naked eye is..found to consist of individual instantaneous flashes or 'scintillations', each produced by the impact of a single α particle. **1963** B. FOZARD *Instrumentation Nuclear Reactors* vi. 68 The scintillations must pass from phosphor to photocathode with minimum absorption at intervening surfaces. **1971** *Sci. Amer.* June 61/2 The screen could be moved to intercept particles scattered at any angle, and the scintillations were counted one at a time with the aid of a low-power microscope.

3. *attrib.* and *Comb.* in *Nucl. Physics*, as *scintillation fluid, method, screen*; **scintillation counter,** a particle counter consisting of a scintillation detector and an electronic counting circuit; hence **scintillation counting** *vbl. sb.*; **scintillation detector,** a detector for charged particles and gamma rays in which scintillations produced in a phosphor are detected and amplified by a photomultiplier, giving an electrical output signal; **scintillation spectrometer,** a form of scintillation counter with which the incident energy of the particle or gamma ray may be determined.

1948 *Physical Rev.* LXXIII. 1406/1 We have prepared some transparent crystalline slabs of both materials, and their behavior as scintillation counters has been compared. **1968** *New Scientist* 15 Aug. 338/2 The receiver consists of four scintillation counters, each shielded from the others and each covering a quadrant of the azimuth circle. **1975** K. H. GOULDING in Williams & Wilson *Biologist's Guide to Princ. & Techniques Pract. Biochem.* vi. 180 The fact that the pulse is directly related to the energy of the original radioactive event is a considerable advantage of scintillation counters over Geiger counters. **1949** *Nucleonics* Oct. 30/2 During the past year the technique of scintillation counting has been considerably advanced and the applications to nuclear research have become widespread. **1979** *Nature* 29 Mar. 410/1 Individual wood samples were finely chipped in preparation for chemical pretreatment and subsequent conversion to benzene for liquid scintillation counting. **1955** A. E. S. GREEN *Nuclear Physics* v. 133 In recent scintillation detectors a photomultiplier is used to change the light pulse into a large burst of electrons. **1977** *Dædalus* Fall 42 The detectors used in the balloon flights were mainly scintillation detectors, which are particularly useful for the detection of photons with energies greater than about 15 kev. **1979** Scintillation fluid [see *scintillation spectrometer* below]. **1909** *Proc. R. Soc.* A. LXXXII. 496 For the observation of the reflected particles the scintillation method was used in all experiments. **1929** *Ibid.* CXXIII. 375 An intense beam of α-particles of definite speed falls on a thin sheet of matter and the number of α-particles scattered through an angle of about 135° is counted by the scintillation method. **1953** GAYNOR & ZEPPELIN tr. *Heisenberg's Nuclear Physics* vii. 141 We shall begin with the instruments of detection and study. The oldest method is the scintillation method. **1938** R. W. LAWSON tr. *Hevesy & Paneth's Man. Radioactivity* (ed. 2) ii. 31 If we allow the α-particles from a point source to pass in a narrow pencil through thin metal foils,..and then to fall on a scintillation screen.., we find that a fraction of the α-rays is deflected through a small angle from their original direction. **1955** W. HEISENBERG in W. Pauli *Niels Bohr* 24 Schrödinger cannot hereby remove the element of discontinuity from the world, which is found everywhere in atomic physics (very obviously, for instance, on the scintillation screen). **1949** JORDAN & BELL in *Nucleonics* Oct. 38/1 The..fact that the amount of light emitted in each flash is very nearly proportional to the energy of the particle opens up the possibility of using the instrument for measurement of gamma- and beta-ray energies. We have developed such an instrument and call it a scintillation spectrometer. **1952** *Ann. Rev. Nucl. Sci.* I. 226 A γ-ray scintillation spectrometer, in conjunction with a magnetic lens spectrometer, has proven very valuable in determining decay schemes and beta–gamma angular correlations. **1979** *Nature* 25 Jan. 313/1 (*caption*) Radioactivity was assayed in a scintillation spectrometer after addition of 5 ml ACS scintillation fluid.

scintillator. Add: **2.** *Nucl. Physics.* **a.** A material that fluoresces when struck by a charged particle or high-energy photon.

1950 *Physical Rev.* LXXVIII. 81/2 Terphenyl crystals ..appear to be among the most durable of presently known organic scintillators. **1955** *Nucleonics* Feb. 10/1 (*heading*) Gaseous scintillators. **1963** B. FOZARD *Instrumentation Nuclear Reactors* vi. 63 Even in crystalline or liquid scintillators of high transparency it is usual to surround the phosphor with a reflecting surface. **1971** *Nature* 20 Aug. 574/2 Cosmic ray muons..were selected by a counter telescope consisting of one or two 2·5 cm diameter disks of plastic scintillator and a 58 cm diameter tank of liquid scintillator placed approximately 90 cm below. **1975** DAVIS & SIMPKINS in Williams & Wilson *Biologist's Guide to Princ. & Techniques Pract. Biochem.* iv. 112 The supporting medium may be cut into small sections which are then immersed in a suitable scintillation solution.

b. = *scintillation detector* s.v. *SCINTILLATION 3.

1952 *Ann. Rev. Nuclear Sci.* I. 188 The increase in scintillator signal when the plate is grounded is a measure of the beam which travels more than once around the orbit before being lost. **1958** *Times* 28 July 8/5 The radiation counters comprise two geiger counters and two scintillators, the geiger counters recording radiation within the satellite and the scintillators measuring exterior intensities. **1964** J. A. RANSOM *Range Guide to Mines & Min.* ii. 26 Anomaly maps were made originally by..flying airplanes over regions of suspected radioactivity with scintillators on long cables below the planes registering areas of abnormal gamma-ray count. **1977** *Kuwait Times* 1 Nov. 9/5 (Advt.), On display: Diagnostic and therapeutic X-ray equipments. Surgical steel instruments. Scintillators.

scintillogram (sinti·logræm). *Med.* [f. SCINTILL(ATION + -O + -GRAM.] A scintigram.

1958 *Proc. 2nd U.N. Internat. Conf. Peaceful Uses of Atomic Energy* XXVI. 245/1 Indefiniteness can be eliminated by making scintillograms in three mutually perpendicular planes. **1966** G. M. BERLYNE *Course in Renal Dis.* xvii. 348 The scintillogram using ^{203}Hg labelled mercurial diuretics is a useful way of diagnosing small infarcted areas of the kidney.

Hence **scinti·llograph,** a scintigraph; **scintillogra·phic** *a.*, **scintillo·graphy.**

1958 *Proc. 2nd U.N. Internat. Conf. Peaceful Uses of Atomic Energy* XXVI. 243/1 Scintillography is principally used to examine objects of which the shapes and location are approximately known. *Ibid.* 245/2 A systematic search for thyroid tissue precedes all scintillographic examination. *Ibid.* 248/1 The scintillograph reveals only lesions with a diameter greater than 2 or 3 centimetres, unless they are confluent. **1965** *Biol. Abstr.* XLVI. 4558/1 (*heading*) Scintillographic study of the spleen. **1975** *Nature* 2 Oct. 426/1 The possibility that the phenomenon was due to physical migration of isotope was investigated by serial gamma camera scintillography of two specially prepared T tubes.

scintillometer. Add: **2.** A device containing a scintillator for detecting and measuring low intensities of ionizing radiation.

1955 *Times* 17 Aug. 7/6 An R.A.A.F. spokesman said today that Beaver aircraft with scintillometer and other radio gear would fly from Mawson early next year to search for a radioactive minerals map of the Antarctic coastline. **1956** *Proc. Internat. Conf. Peaceful Uses of Atomic Energy* VI. 660/1 An efficient scintillometer is about 50 times more sensitive to terrestrial gamma rays than a Geiger-Müller counter. **1977** A. HALLAM *Planet Earth* 111 The most useful geophysical surveys for mineral deposits are magnetic.., radiometric (the main prospecting tool for uranium deposits, using Geiger counters, scintillometers and gamma-ray spectrometers), electromagnetic, electrical, gravitometric, and seismic.

Hence **scintillo·metry,** study by means of the scintillometer.

1960 *Los Alamos Sci. Lab. Rep.* LAMS 2445. 337 (*heading*) Clinical applications of whole body scintillometry. **1974** *Nature* 15 Mar. p. xvi/2 (Advt.), The department has excellent modern facilities for work on tissue and organ culture, including..scintillometry.

scintilloscope (sinti·loskōᵘp). Also **scintillascope, scintilliscope.** [f. L. *scintill-a* spark + -O + -SCOPE.] An instrument in which alpha rays are detected by the flashes of light which are emitted when they strike a fluorescent screen.

1904 *Nature* 29 Sept. 535 The little instrument, which is called the 'Scintilloscope', consists of a simple magnifier of adjustable focus, as in the spinthariscope, but instead of the fixed screen and particle of radio-active substance a small double plate of glass is used. **1916** *Yukon Territory* (Canada Dept. Interior) 173 The scintilloscope is a much more convenient instrument. **1949** *New Gould Med. Dict.* 912/2 *Scintillascope*, an instrument for observing minute flashes of light upon a fluorescent screen struck by alpha particles, emitted from a small source of radioactive material. **1951** I. L. IDRIESS *Fortunes in Minerals* (ed. 2) xxxix. 250 These tiny flashes are known as scintillations, and can be seen through a magnifying glass. This is best done by a simple little instrument called a scintilliscope or spinthariscope. It is merely a small tube, probably of brass. In one end is fitted a glass prism. The outer side of the glass is coated with zinc sulphide powder... At the other end of the tube is fixed an eyepiece, which is a small magnifying glass.

scintiscan (si·ntiskæn). *Med.* [Back-formation from next.] An autoradiograph obtained with a scintiscanner.

1960 *Radiology* LXXV. 821/1 (*caption*) Bilateral renal scintiscan superimposed on abdominal radiograph. **1971** [see *SCINTIGRAM]. **1977** *Lancet* 6 Aug. 261/2 Liver scintiscans with ^{99m}Tc sulphur colloid were performed in 52 patients.

scintiscanner (si·ntiskænəɹ). *Med.* [f. SCINTI(LLATION + SCANNER.] A radiosensitive device which scans the body or part of it and creates an image of the distribution of radioactivity therein.

1953 *Radiology* LXI. 88/1 The point-by-point technique of plotting the frontal area occupied by the thyroid gland.. has been simplified by the introduction of the 'scintiscanner' for obtaining an actual size scintigram of the gland. **1956** *Jrnl. Neurosurg.* XIII. 345 (*caption*) Scintiscanner showing patient beneath the focusing collimator. **1968** *New Scientist* 12 Dec. 617/1 The computer is connected 'on-line' to a scintiscanner instrument, which measures radiation intensities as it is moved over a patient who has ingested a weakly radioactive isotope.

So **sci·ntiscanning** *vbl. sb.*, the production and use of scintiscans.

1954 *Amer. Jrnl. Roentgenol.* LXXII. 881/2 One tube stand suspends a shielded Geiger tube for uptake studies, the other tube stand supports the scintiscanning device. **1967** *Nursing Times* 18 Aug. 1095/2 The diagnostic uses of scintiscanning are extending rapidly with technical advances in radiobiology. **1980** *Nature* 17 Apr. 619/1 When bonded to the γ-emitter technetium-99..it is used clinically for scintiscanning of functioning renal cortex.

scio- (saiˌo-), comb. form of Gr. σκιά shadow, as in **scio·philous** *a. Bot.* [-PHILOUS], thriving best in shade; **sci·ophyte** *Bot.* [-PHYTE], a plant that thrives best in shade; hence **sciophy·tic** *a.* Also SCIOMANCY etc. in Dict.

1905 F. E. CLEMENTS *Res. Methods Ecol.* iii. 140 The slight development of hairs in sciophilous plants is an advantage. **1932** FULLER & CONARD tr. *Braun-Blanquet's Plant Sociol.* v. 107 In general the lower layers of stratified communities..are designated as sciophilous. **1905** F. E. CLEMENTS *Res. Methods Ecol.* iii. 144 (*heading*) Heliophytes and sciophytes. **1947** R. F. DAUBENMIRE *Plants & Environment* v. 234 Sciophytes may be at a disadvantage in full sunlight if they cannot manufacture chlorophyll at a rapid rate. **1974** *Nature* 23 Aug. 623/1 For many years it has been conventional to classify plants into sun-demanding (heliophytes) and shade-requiring (sciophytes). **1976** *Ibid.* 22 July 281/1 We provide evidence for the existence of 'sun' (heliophytic) and 'shade' (sciophytic) differences in the net photosynthesis-radiant flux intensity response of individuals of one coral species..on the same lagoon.

scioness (sai·ŏnes). *joc. rare.* [f. SCION 2 + -ESS¹.] A female heir or descendant.

1928 'BRENT OF BIN BIN' *Up Country* xi. 36 Harriet Mayborn, scioness of the English aristocracy, was dumbfounded, but kept her head. **1969** T. SOUTHERN *Magic Christian* (ed. 2) xv. 120 A venerable scioness of Roman society.

Sciote (ʃi·ot), *a.* and *sb.* [f. It. *Scio* SCIO² + -OTE] **a.** *adj.* = CHIAN *a.* **b.** *sb.* A native or inhabitant of Chios.

1718 M. WORTLEY MONTAGU *Let.* 31 July (1965) I. 419 The ruins of this great City is now inhabited by poor Greek peasants who wear the Sciote habit. **1837** H. MARTINEAU *Jrnl.* 8 Sept. in *Autobiogr.* (1877) III. 190 Eastlake..must be a metaphysician to have painted his Sciote picture. **1866** *Chambers's Encycl.* VIII. 549/1 A number of the Sciotes having, in 1822, joined the Samians,..the inhabitants.. were indiscriminately massacred... Subsequently..many of the Sciote families returned.

scissile, *a.* Add: (Later examples.) Also in *Chem.*, capable of being broken. Cf. *SCISSION 3 a.

1967 *Listener* 10 Aug. 187/2 The first [play] in the series.. was...as scissile by commercials as anything on ITV. **1978** *Nature* 5 Jan. 94/3 The scissile bond in a peptide substrate.

scission. Add: **3. a.** *Chem.* Breakage of a bond, esp. in a long chain polymer such that two smaller chains result.

1923 *Jrnl. Chem. Soc.* CXXIII. I. 85 The scission of the ring..with elimination of nitrogen seems possible. **1944** *Jrnl. Appl. Physics* XV. 389/2 For the case of Hevea and GR-S, the number of places along the chain that are subject to scission is not very large. **1952** TURNER & HARRIS *Org. Chem.* xxvii. 524 Another apparently reasonable alternative is that hydrazobenzene undergoes scission at the N—N link. **1975** *Nature* 1 May 31/2 mRNA was treated with 0.M NaOH for 4 min at 0°C, which gave on average about one scission per molecule.

b. *Nuclear Physics.* The event of separation of the parts of a nucleus undergoing fission, as opposed to the process as a whole.
1956 *Physical Rev.* CII. 440/1 The corresponding deformed shape roughly approximates the egg-shaped fragment resulting from scission of a dumbbell-shaped parent nucleus. **1958** *Rev. Mod. Physics* XXX. 555/1 The separation of charge centers at the moment of scission. **1964** L. Wilets *Theories Nuclear Fission* ii. 21 Of greater relevance to the fission process is the energy release to the point of scission, the time at which the fragment masses are unalterably determined. **1975** *Physics Bull.* July 307/1 Dr Specht..concentrated on a description of those fission phenomena which seem to be decided by conditions in the nucleus at the time of scission rather than at the saddle point.

scissor, *v.* Add: **2. b.** *fig.* To excise.
1890 G. B. Shaw in *Star* 19 Apr. 2/6 The first act was vigorously scissored. **1968** *Listener* 10 Oct. 462/1, I was invited by the BBC to cut a single sentence from a broadcast talk I had recorded previously... I refused to do this, and accordingly the talk was hastily scissored out of the Third Programme. **1977** *Time* 24 Oct. 23/3 The Finance Committee scissored the entire wellhead tax scheme out of the bill.
3. a. To cause (one's legs) to move like scissors. **b.** To fix (a person) in the scissors hold or with a grip resembling it (cf. *scissors *sb. pl.* 2 a).
1961 *Rogue* May 14 Feathertop watched the smooth scissoring of her slim, trim legs as she walked to the bags. **1968** A. Keith *Compl. Guide to Championship Wrestling* v. 76 Scissoring the bottom leg frees the hips. *Ibid.* 77 A scissors his right leg through underneath the left. **1973** *Funk & Wagnalls New Encycl.* XXII. 366 The legs are then scissored while the upper arm pushes toward the feet and the lower arm returns to the chest. **1974** J. Irving *158-Pound Marriage* i. 12 When he rode you with a cross-body ride—your near leg scissored, your far arm hooked—Severin said Jones cut off your circulation somewhere near your spine. **1975** R. H. Rimmer *Premar Experiments* (1976) ii. 162 With her legs scissored around me, I found it impossible not to pat her smooth black behind.
4. *intr. Rugby Football.* To execute a scissors movement. Cf. *scissors *sb. pl.* 2 d.
1970 *Financial Times* 23 Mar. 3/8 Robertson and Turner scissored impeccably for Turner to score a try that Brown converted. **1975** *Sunday Times* 23 Feb. 28/2 Smaje scissored with Aitchison to get the Lancashire try, converted by Gullick.

scissors, *sb. pl.* Add: **1. c.** (Later example.)
1976 *National Observer* (U.S.) 29 May 11/1 Each without the other is only half a scissors.
e. (Earlier and later and *attrib.* examples.) Also *transf.*
1809 *Monthly Pantheon* Apr. 266 He was to..take the scissors and paste brush in hand. **1925** T. E. Lawrence *Let.* 21 Apr. (1938) 475, I haven't much desire to undertake so difficult a scissors and paste job. **1936** [see *collage]. **1946** R. G. Collingwood *Idea of Hist.* 257 History constructed by excerpting and combining the testimonies of different authorities I call scissors-and-paste history. **1951** [see *cradle *sb.* 2]. **1977** A. Giddens *Stud. in Social & Polit. Theory* ii. 97 'Scissors and paste' ethnology of the sort which..assembled together examples from numerous different societies without regard to the social context in which they were embedded.
2. a. Now usu. a grip with the legs or ankles crossed like a pair of scissors. Also *body scissors.*
1909 Webster, *Scissors*, a hold in which one contestant clasps the other's head or body with his legs. **1921** *Daily Colonist* (Victoria, B.C.) 13 Oct. 10/7 He [*sc.* a wrestler] took the..second [fall]..with a body scissors. **1940** R. Chandler *Farewell, my Lovely* 173 The Indian threw me sideways and got a body scissors on me as I fell. **1961** J. S. Salak *Dict. Amer. Sports* 381 *Scissors hold* (wrestling), a hold which is secured by locking the legs at the ankles around a part of the opponent's body, pressure being applied. It formerly meant a grip with the wrists crossed like a pair of scissors.
b. *High jumping.* (See quots. 1961, 1976.) Also *attrib.* as *scissors jump.*
1897 *Encycl. Sport* I. 50/2 The methods of jumping are various, but two main types predominate—viz., the straight jump, and the side-way or scissors jump. **1959** *Times* 1 Oct. 3/3 Her legs..flashed over the bar in the old-fashioned scissors style. **1961** F. C. Avis *Sportsman's Gloss.* 61/2 *Scissors jump*, in high jumping, that method in which the body is in a virtually upright sitting position, and the legs move in an up-and-down motion. **1964** M. Watman *Encycl. Athletics* 79/2 There are four basic styles of high jumping: scissors, eastern cut-off, western roll and straddle... The ordinary scissors, which is taught to most schoolchildren, is the least effective of the four styles. **1976** *Webster's Sports Dict.* 371/1 *Scissors*,..a method of high jumping in which the jumper leads with the leg nearest the bar, crosses the bar in a sitting position, and then brings the trailing leg up over the bar as the lead leg is brought down on the other side.
c. *Swimming.* A movement in which the legs, held rigid, are parted slowly and brought together forcefully. Usu. *attrib.*
1904 R. Thomas *Swimming* 418/1 The simile of the scissors clip is accurate..for the breast stroke (and would be also for the English sidestroke). **1973** *Funk & Wagnalls New Encycl.* XXII. 367/1 Sidestroke. This stroke, employed on either the right or left side with a scissors kick,..is of particular value for lifesaving technique, but is not used in competition. **1974** 'G. Black' *Golden Cockatrice* xii. 200, I saw their legs..the girl doing a scissors kick that seemed to weaken as I swam towards them.

d. *Rugby Football.* (See quot. *c* 1915.) Also *attrib.* and *transf.*
c **1915** R. A. Lloyd in E. B. Poulton *Life R. Poulton* (1919) 218 The 'Scissors' trick was this: when I had the ball, and Ronald was running beside me just as if he was going to take an ordinary pass, he would suddenly change his direction and come racing straight across at me and practically take the ball out of my hands, and breaking clean through would run right across to the opposite wing. **1927** Wakefield & Marshall *Rugger* 229 The two [*sc.* a centre and wing three-quarter] may also combine when the centre still has the ball, when..they exploit the scissors movement. *Ibid.* 230 This scissors, and dummy scissors, attack may be tried also by two centres or by a centre and stand-off half. **1960** V. Jenkins *Lions down Under* 106 One forty yards' run of his, after he and Malcolm Thomas had worked a perfect dummy scissors, was a gem. **1976** *Wymondham & Attleborough Express* 3 Dec., They worked one of their excellent set pieces including two dummies, a well taken scissors and a Gary Owen.
e. *fig.* A progressive divergence between two kinds of price or income, so called from the appearance of a graph of the two indices plotted against each other; *orig.* and *spec.* used *attrib.* of a crisis in the Soviet Union in 1923 (see quots. 1926, 1965).
1924 M. Farbman *After Lenin* vii. 125 The economic crisis of the autumn and winter of 1923–24 is known as the crisis of the scissors. **1926** *Encycl. Brit.* III. 425/1 The first of these was a crisis which from an image used by Trotsky came to be known as the crisis of the 'scissors'. The two blades of the scissors represented the prices of agricultural products and the price of manufactured goods. **1965** A. Nove in B. Pearce tr. *Preobrazhensky's New Economics* p. xi, In 1923 the Soviet economy faced the so-called 'scissors crisis': the terms of trade between town and country had become so unfavourable to the latter that the peasants were reluctant to sell their produce. **1974** J. White tr. *Poulantzas's Fascism & Dictatorship* iv. ii. 193 The index of labour income shows that the scissors between the income of skilled and semi-skilled workers widened considerably. **1979** *China Now* Mar./Apr. 25/1 The closing of the price scissors (the gap between the price paid for agricultural foods and the prices paid by the peasants for manufactured goods) has not gone far enough.
f. In phr. *scissors and stones, scissors cut paper, scissors game*, a game for two players using three postures of the right hand (see quot. 1934).
Or the left hand if one is left-handed.
1934 P. Fleming *One's Company* ii. ii. 198 From a room downstairs came that sound which so often accompanies meals in China—the staccato, competitive ejaculations of a party playing the 'scissors' game. In this you and your opponent shoot out your right hands at each other simultaneously, the fingers being arranged in one of three postures. A clenched fist means 'stone'; two fingers extended mean 'scissors'; all five fingers extended mean 'paper'. Scissors cut paper but are broken on stone, and paper wins against stone because stone can be wrapped up in paper. **1952** J. B. Pick *Phoenix Dict. Games* 291 Scissors and stones. **1964** I. Fleming *You only live Twice* i. 18 It was the old game of Scissors cut Paper, Paper wraps Stone. Stone blunts Scissors, that is played by children all over the world. **1976** *Times* 2 Dec. 16/5 The Chinese hand-game Scissors Cut Paper.
5. *scissor-leg*; *scissor-cut, -legged* adj.; **scissor(s)-bill,** (*b*) *slang* in various senses; *esp.* a foolish, incompetent, garrulous, or objectionable person; **scissor-cut** [tr. G. *scherenschnitt*], a silhouette that has been cut freehand with scissors; also as vb.; **scissor(s)-grinder** (earlier examples); **scissor-hold** = Scissors 2 a; **scissor(s)-lift,** a surface that is raised or lowered by the closing or opening of crossed supports pivoted like the two halves of a pair of scissors; **scissor-man,** a man who wields scissors, *spec.* a censor, a surgeon, or a tailor.
1871 *Atlantic Monthly* Nov. 566/2 Pootiest band of hogs in Tulare County! There's littler of the real sissor-bill nor Mexican racer stock than any band I have ever seen in the State. **1913** *Industrial Worker* (Spokane, Washington) 1 May 5/3 Scissorbill is a localized slang term. Here it refers to the 'home-guard' worker, who is filled with bourgeoise [*sic*] ideas and ethics. It ordinarily describes a worker who has some source of income other than his wages. **1926** J. Black *You can't Win* x. 129 When a bums' 'convention' is to be held, the jungle is first cleared of all outsiders such as 'gay cats',..'jungle buzzards', and 'scissors bills'. **1931** B. Starke *Touch & Go* xvi. 259 Dick praised me for not saying anything. 'You're not a scissor-bill.' A scissor-bill was a woman who gossiped and nagged and was bad generally. **1931** 'D. Stiff' *Milk & Honey Route* xiv. 163 The line in waiting [for prostitutes] is usually monopolised by the village scissorbills. *a* **1944** J. Conroy in B. A. Botkin *Treas. Amer. Folklore* (1944) iv. 548 Some sign painters couldn't dot the letter 'i' without a pounce to go by. It was enough to make a dog laugh to see some poor scissorsbills wrastling around with a pounce. **1961** R. P. Hobson *Rancher takes Wife* iii. 56 The hell you did, you big scissorbill, you stepped on my bum leg and my hand both. **1931** *Times Lit. Suppl.* 25 June (Suppl.) p. iii/2 His many brilliant students have perhaps done more service to the book-jacket than to the page by some of their shadow, scissor-cut and engraved letter forms. *Ibid.* p. iv/2 Professor D. P. Sterenberg portrays objects of daily life in flat lithographs that resemble scissor-cuts. **1931** V. Woolf *Waves* 126, I see.. Neville, scissor-cutting, exact. **1960** *Times* 11 Feb. 3/4 The graceful and elegant animated scissor-cuts of Lotte Reiniger. **1976** *Times* 26 Nov. 4/7 An octagonal scissor-cut emerald is set in a gold and enamel ring. **1841** N. Hawthorne *Amer. Notebks.* (1932) 88 The squirrel..frequently uttered a sharp, quick, angry noise, like that of a scissors-grinder's wheel. **1855** 'Q. K. P. Doesticks' *Doesticks* 155 The loving accents of the scissor-grinder's wheel. **1974** D. Sears *Lark in Clear Air* iii. 43 She pulled my head down close, jimmied her knees around my leg so that she had a scissor hold. **1923** W. de la Mare *Riddle & Other Stories* 178 Before him stood a kind of gaping wallet, of cracked American cloth, held yawningly open by its scissor-legs. **1947** —— *Collected Stories for Children* 26 His lank scissor-legs. **1967** Flakoll & Alegria tr. *M. A. Asturias' Cyclone* ii. 22 The scissor-legged cot. **1961** *Aeroplane & Astronautics* CI. 568/2 The mobile scissor-lift intermediate-base loading platform developed by Canadair is seen here being demonstrated with a Seaboard World Airways' CL-44. **1970** *Times* (Aviation Suppl.) 4 Sept. p. xiv/8 Ready-prepared meals for the galleys on board the aircraft are..loaded by mobile scissors-lift vehicles direct to the galley hatches. **1980** *BSI News* June 5/3 Most types of scissor lift present trapping hazards to persons employed on or about them. **1848** tr. *Hoffmann's English Struwwelpeter* (ed. 4) 16 The door flew open, in he ran, The great, long, red-legged scissar-man. Oh! children, see! the tailor's come. **1932** Auden in *Rev. Eng. Stud.* (1978) Aug. 301 The hump-backed surgeons And the scissor-man. *a* **1953** Dylan Thomas *Quite Early One Morning* (1954) 22 Struwelpeter—oh! the baby-burning flames and the clacking scissorman! **1968** *Listener* 5 Sept. 292/1 Arts censorship in Britain lives on mainly through the sheer personality of its few remaining scissormen.

sclaff (sklæf), *sb. Golf.* [See Sclaff *v.*] A stroke in which the club scrapes the ground before hitting the ball. Hence **scla·ffy** *a.*
1893 H. Hutchinson *Golfing* 82 'Tops', and 'sclaffs', and misses. **1896** W. Park *Game of Golf* 91 The sight of bare earth..gives the impression that contact between it and the club-head, which might happen with a sclaffy shot, would inevitably result in damage to the club. **1903** W. J. Travis *Pract. Golf* 20 If..the head is allowed to move, the chances are that a sclaff or a top will result. **1948** Dante & Diegel *Nine Bad Shots of Golf* x. 104 There is one other swing that will produce a sclaff. **1973** A. MacVicar *Painted Doll Affair* vii. 84 My drives would be hooks and slices, my irons sclaffy travesties.

sclareol (sklɛ̄ə·riǫl). *Chem.* [ad. F. *sclaréol* (Volmar & Jermstad 1928, in *Compt. Rend.* CLXXXVI. 519), f. *sauge sclarée* Clary *sb.*² (mod.L. *Salvia sclarea*): see -ol.] A colourless crystalline diterpenoid alcohol, $C_{20}H_{36}O_2$, found in the leaves of clary; also, one of the two constituent epimers of this.
1928 *Chem. Abstr.* XXII. 1828 The essence [of clary].. contains 42·2% of crystallizable sclareol. **1959** *Chem. & Industry* 1379/1 Since sclareol has been dehydrated to a mixture of manool and manoyl oxide, all three compounds must have the same absolute configuration at C(13). **1975** *Nature* 22 May 328/2 An epimeric mixture of the diterpenes sclareol and 13-epi-sclareol (sclareol)..has been shown to constitute 10% of the surface exudate on leaves of *Nicotiana glutinosa*. *Ibid.*, In replicate experiments, applications of.. sclareol consistently gave good control of rust on French bean, broad bean and wheat, reducing infection to less than 10% of control.

scleractinian (skliə·ræktiˑniăn), *sb.* and *a.* [f. mod.L. order name *Scleractinia* (coined as *Scleractineae* by G. C. Bourne, in E. R. Lankester *Treat. Zool.* (1900) II. vi. 55), f. Gr. σκληρ-ός hard + ἀκτῖν-, ἀκτίς ray + -ia¹, -an: cf. Actinia, *actinian.] **A.** *sb.* A coral of the order Scleractinia or Madreporaria, which is characterized by compact calcareous skeletons and includes all living true corals. **B.** *adj.* Of or pertaining to such a coral or the group as a whole.
1900 G. C. Bourne in E. R. Lankester *Treat. Zool.* II. vi. 61 The anatomy of any Scleractinian resembles, in essential points, that of an Actinia. *Ibid.* 80 In *Heliopora* the skeleton is not spicular but lamellar, resembling in structure that of the Scleractinian corals. **1943** *Spec. Papers Geol. Soc. Amer.* No. 44. 1 This revision is the result of a study of the genotype species..of nearly every described scleractinian genus... The distribution of fossil and recent scleractinian faunas is broadly analyzed. *Ibid.* 90 The forerunners of most of the groups of the scleractinians are found in the Middle and Upper Triassic rocks. **1952** R. C. Moore et al. *Invertebrate Fossils* iv. 143/2 The scleractinians differ from rugose corals chiefly in the mode of addition of new septa. **1973** *Nature* 27 July 201/1 The thirty-six scleractinian species of coral found on the actively growing, fringing reefs along the western coast of Barbados are complemented by varied populations of sponges, anemones,..and cucumbers.

sclereid. Add: Also **scler(e)ide.** [a. G. *sclereïd* (A. Tschirch 1885, in *Jahrb. f. Wissensch. Bot.* XVI. 308), irreg. f. Gr. σκληρ-ός hard: see -id³.] = *stone cell* s.v. Stone *sb.* 20. (Examples.)
1896 *Ann. Bot.* X. 11 The walls of the sclereids..acquire during the ripening an increasingly dark brown colour. **1914** M. Drummond tr. *Haberlandt's Physiol. Plant Anat.* iv. 160 The coarse villi or shaggy hairs of the Melastomaceae often contain whole bundles of sclereides, which penetrate below into the mesophyll. **1919** F. O. Bower *Bot. of Living Plant* ix. 145 Such stone-cells or sclereids, give a hard gritty texture to the parts where they occur, as in the bark or pith of various woody plants. **1934** *Jrnl. Arnold Arboretum* XV. 247 Clusters of sclerides or stone cells..are of not infrequent occurrence in the pith of the redwood. **1969** E. G. Cutter *Plant Anat.* I. vi. 57 Sclerenchyma may be sub-divided into sclereids and fibres.

sclerema. Add: Also *sclerema neonatorum* [gen. pl. of mod.L. *neonātus* (cf. *neonate)]. (Further examples.)
1889 J. E. Graham in J. M. Keating *Cycl. Diseases Children* II. i. 90 Sclerema neonatorum..is distinguished by a peculiar, œdematous, corpse-like hardening of the skin. **1962** *Lancet* 27 Jan. 226/1 Pneumonia, hæmorrhagic disease of the new-born, or sclerema neonatorum may be diagnosed. **1974** Passmore & Robson *Compan. Med. Stud.* III. xlv. 31/1 The baby's rectal temperature..falls... He becomes less active, less hungry and less vocal. The skin reddens and grows cold and the subcutaneous tissue slowly becomes hard... This state is known as sclerema.

sclero-. Add: **1.** **scle:roblaste·ma** *Anat.* [Blastema], the embryonic tissue which gives rise to bone; **sclerœde·ma** (also scleredema) *Path.* (see quot. 1976); **scleropro·tein** *Biochem.*, any insoluble structural protein; **sclerothe·rapy,** the treatment of varicosities by the injection of a substance which induces clotting.
1934 Webster, Scleroblastema. **1968** Passmore & Robson *Compan. Med. Stud.* I. xviii. 11/2 Each vertebra is laid down as a densely cellular mesenchymal precursor, the scleroblastema; this transforms into a cartilage model which is subsequently replaced by bone by the process of endochondral ossification. **1932** *Jrnl. Amer. Med. Assoc.* 3 Sept. 822/1 Scleredema adultorum is characterized by progressive induration and swelling of the deeper portions of the skin and subcutaneous tissues. *Ibid.* 825/1 Scleredema always involves spontaneously without subsequent atrophy of the affected tissues. **1946** *New England Jrnl. Med.* 15 Aug. 209/1 The appearance of a patient with scleredema is so striking as to suggest the diagnosis immediately. **1976** *Proc. R. Soc. Med.* LXIX. 844/2 Sclerœdema is characterized by sudden onset of diffuse symmetrical hardening of the skin of the face, neck and upper arms. **1907** Halliburton & Hopkins in *Jrnl. Physiol.* XXXV. p. xix, Sclero-proteins. This new word takes the place of the word albuminoid... It includes such substances as gelatin and keratin; the prefix indicates the skeletal origin and often insoluble nature of its members. **1958** *Immunology* I. 49 Some of the scleroproteins comprising the scale plate are antigenic, but are only slowly digested by homologous recipients. **1970** R. M. Black *Elements Palaeont.* ii. 8 The matrix of bone consists mainly of collagen (a fibrous scleroprotein) hardened by mineral salts. **1977** A. Hallam *Planet Earth* 241 More usually, however, only the most resistant and stable organic materials can survive long after death, as with the lignified tissues of fossil land plants and the scleroprotein skeleton of the extinct fossil graptolites. **1944** *Amer. Jrnl. Surg.* LXVI. 363 The advantages of sclerotherapy are low morbidity, almost no mortality and no necessity for hospitalization... The disadvantage is the very high incidence of recurrence. **1977** *Lancet* 25 June 1343/1 F. Bezzouni of Russia has an approach similar to that of most surgeons in Britain—injection sclerotherapy for small, below-knee varices and high ligation and stripping for gross main-stem incompetence.

scleroblast. **2.** Substitute for def.: A spicule-forming cell in sponges.

scleromyxœdema (skliə:romiksīdi·mă). *Path.* Also **-myxedema.** [ad. G. *skleromyxödem* (H. H. Gottron 1954, in *Arch. f. Dermatol. u. Syphilis* CXCIX. 71): see Sclero- and *myxœdema* s.v. Myxo-.] A disease characterized by the extensive proliferation of fibroblasts and deposition of mucopolysaccharides in the skin, causing distortions of the features and lichenous eruptions.
1964 *Arch. Dermatol.* LXXXIX. 446/1 Lichen myxedematosus, also known as..scleromyxedema, is a well-known clinical entity. **1968** R. J. Cairns in A. Rook et al. *Textbk. Dermatol.* II. lv. 1621/2 A variant of lichen myxoedematosus is scleromyxoedema..—the Arndt–Gottron syndrome—in which diffuse thickening of the skin underlies the papules... The features may be distorted by the exaggeration of the facial ridges and flexion of the fingers may be limited. **1977** *Lancet* 4 June 1208/2 Ten years previously this patient has been treated for scleromyxœdema..with melphalan.

sclerophyll (skliə·rofil), *sb.* and *a.* *Bot.* [ad. G. *sklerophyll* (A. F. W. Schimper *Pflanzen-geographie* (1898) v. 538: see next word.] **A.** *sb.* A sclerophyllous plant.
1911 J. M. Coulter et al. *Textbk. Bot.* II. iii. 710 In the inclement period they [*sc.* the leaves of deciduous trees] are as well protected as are the cacti and better protected than are the sclerophylls. **1923** *Jrnl. Ecol.* XI. 287 Northwards.. the forest becomes more important, but is soon supplanted by conifer forest..into which many of the sclerophylls, both shrubs and trees, pass as subordinate members. **1939** *Nature* 11 Mar. 412/2 The 'rhenosterbush'..represents the most arid kind of sclerophyll. **1975** *Sci. Amer.* Jan. 133/1 The text itself considers one by one the 10 vegetational zones, from the evergreen tropical forests to the regions of winter-rain sclerophylls.
B. *adj.* = *sclerophyllous *a.*
1926 Tansley & Chipp *Aims & Methods in Study of Vegetation* vi. 99 Where the rain mainly falls in the late autumn and winter, but is sufficient in quantity, and there is a hot dry summer, the vegetation is of the evergreen sclerophyll type—trees or shrubs with rather small leathery leaves. *Ibid.* 100 Sclerophyll regions always abut on desert regions. **1970** Gay & Calaby in Krishna & Weesner *Biol. Termites* II. ix. 413 Symmetrical dome mounds..are found only occasionally in the sclerophyll woodland and mallee. **1978** *Nature* 9 Mar. 160/1 A general and sustained increase in sclerophyll vegetation at the expense of drier rain forest types, probably a result of aboriginal man's activities, can explain some..of the evolutionary changes.
So **sclero·phylly,** the fact of being sclerophyllous.
1903 W. R. Fisher tr. *Schimper's Plant-Geogr.* i. 9 With increasing physiological dryness, the leaves become smaller in surface but proportionally thicker, more leathery (sclerophylly). **1909** E. Warming *Œcol. Plants* xlvi. 194 Sclerophylly is frequent and is due to thickness of the epidermal wall, as in Andromeda polifolia. **1973** J. Wieser tr. *Walter's Vegetation of Earth* iv. iii. 121 The ecological significance of sclerophylly is..to be seen in the ability of sclerophyllous species to conduct active gaseous exchange.. in the presence of an adequate water supply, but to cut it down radically by shutting the stomata when water is scarce.

sclerophyllous (skliə·rǫ·filəs), *a.* *Bot.* [f. Gr. σκληρό-ς hard + φύλλ-ον leaf + -ous.] Pertaining to or designating woody evergreen plants having leaves that are hard and tough, and usu. small and thick, so reducing the rate of loss of water; characterized by such plants.
1903 W. R. Fisher tr. *Schimper's Plant-Geogr.* v. 507 The mild temperate districts with winter rain and prolonged summer drought are the home of evergreen xerophilous woody plants, which, owing to the stiffness of their thick, leathery leaves, may be termed sclerophyllous woody plants. *Ibid.* 516 The best known districts inhabited by sclerophyllous woods are the coasts of the Mediterranean Sea. **1926** Tansley & Chipp *Aims & Methods & Study of Vegetation* ii. 26 The dominance of [s]clerophyllous trees or shrubs (rather small hard evergreen leaves) indicates a moderate rainfall mainly concentrated in the winter half of the year, which is mild, and a hot dry summer. **1961** *Times* 12 Apr. 3/2 (Advt.), The Department is noted for its research in..plant ecology (with emphasis on sclerophyllous and arid plants). **1973** J. Wieser tr. *Walter's Vegetation of Earth* iv. iii. 127 The roots of sclerophyllous species reach far down into the ground because the upper soil layers are usually completely dried out in summer.

sclerosant (skliə·rōᵘ-zănt, sklĕr-), *sb.* and *a.* *Med.* [f. Scleros(is + -ant¹.] **A.** *sb.* A sclerosant agent. **B.** *adj.* Producing sclerosis or hardening of tissue.
1956 *New Gould Med. Dict.* (ed. 2) 1076/1 *Sclerosant*, a chemical irritant producing an inflammatory reaction and subsequent fibrosis. **1962** *Lancet* 8 Dec. 1221/2 McEvedy seems to have had good results after the injection of a *sclerosant* solution such as ethanolamine. **1977** *Brit. Med. Jrnl.* 12 Feb. 434/2 When ligamentous pain and tenderness seem to dominate, 'sclerosant' mixtures of phenol, dextrose, and glycerine may be injected into the tender area. **1979** *Ibid.* 22 Sept. 704/2 We used two sclerosants—namely, phenol..and tetradecyl sulphate.

Scleroscope (skliə·rǫskōᵘp). Also **scleroscope.** [f. Sclero- + -scope.] An instrument for measuring the hardness of a material, this being indicated by the height of rebound of a small diamond-tipped hammer dropped from a standard height on to the material. Also *attrib.* Hence **sclerosco·pic** *a.*
Scleroscope is a proprietary term in the U.S.
1907 A. F. Shore in *Amer. Machinist* 14 Nov. 748/1 The instrument was named scleroscope, from the Greek words *sclerotus*, meaning hardness, and *scope*, because it is direct reading. **1908** [see *Shore *sb.*³]. **1915** [see *Brinell]. **1921** *Glasgow Herald* 23 Sept. 9 The use of the scleroscope on light specimens of metals. **1936** P. F. Foster *Mech. Testing of Metals & Alloys* viii. 143 A dynamic hardness test is provided by the Shore scleroscope in which a small pointed tup weighing about 0·0052 lb. is allowed to fall freely from a height of 10 in. on to the test piece. **1950** *Engineering* 31 Mar. 371/2 Scleroscopic hardness values are approximately 85 and 65. **1961** *Official Gaz.* (U.S. Patent Office) 29 Aug. tm 149/1 The Shore Instrument & Mfg. Company, Inc.,.. N.Y...*Scleroscope*... For instruments used to test hardness of metals and other substances. **1977** R. B. Ross *Handbk. Metal Treatments & Testing* 166 The Scleroscope test is of limited use and accuracy but, because of its extreme portability, has certain advantages over other more conventional tests.

sclerosis. Add: **3.** *fig.* Rigidity, excessive resistance to change.
1954 B. & R. North tr. *M. Duverger's Pol. Parties* i. ii. 89 Such a drying-up of new recruits is the symptom of a serious sclerosis. **1958** *Times* 11 Aug. 2/5 All the world knows that he was faced with the problem of revitalizing a good tradition that was beginning to suffer from sclerosis. **1966** S. H. Beaver tr. *J. Beaujeu-Garnier's Geogr. Population* x. 228 Research work..has shown the parallelism that exists between the sclerosis of social structures and the high proportion of marriages between first cousins. **1977** *N.Y. Rev. Bks.* 14 July 35/2 Popovic saw his harassment as a symptom of the ideological sclerosis which is increasing with Tito's age.

sclerotic, *a.*¹ Add: **5.** *fig.* Unmoving, unchanging, rigid.
1965 *Listener* 20 May 737/2 Why is it, I asked myself, that so many theatre people in Russia call the Moscow Arts old-fashioned and sclerotic? **1968** *Daily Tel.* (Colour Suppl.) 13 Dec. 61/1 The sidewalks of New York seem to get harder year by year, the skyscrapers more inhumanly overbearing, the traffic more sclerotic. **1979** *Washington Star* 8 May a15/7 She [*sc.* Margaret Thatcher] has to deal with tacky little things like the secondary boycotts, repealing sclerotic tax laws, [etc.].

sclerotin (skliə·rǫtin). *Biol.* [f. Sclero- + -*tin* after Chitin, Keratin, etc.] Any of a class of structural proteins which form the exocuticles of insects and harden and darken by a natural tanning process in which protein chains become cross-linked by quinone groups.
1940 M. G. M. Pryor in *Proc. R. Soc.* B. CXXVIII. 391, I therefore propose the name 'sclerotin' as a general term to describe proteins such as that of the cockroach ootheca, which owe their stability to a process of natural tanning. **1957** Richards & Davies *Imms's Textbk. Ent.* (ed. 9) i. 11 Polyphenols..are the precursors of the quinones which link the arthropodin molecules to form sclerotin. **1969** R. F. Chapman *Insects* xxii. 434 Part of the protein may later be tanned.., stabilised by cross-linkages between the molecules, to form a hard, inflexible and usually darkened structure. Such tanned arthropodin is called sclerotin, and this produces the hardness of the sclerites. **1976** *Sci. Amer.* Apr. 134/3 Such quinone-linked proteins, called sclerotins, are usually coloured brown or black, accounting for the hard beetle look we all know.

sclerotinia (skliə·roti·niă). [mod.L. (L. Fuckel 1870, in *Jahrb. d. Nassauischen Vereins f. Naturk.* XXIII–XXIV. 330), f. Sclerotium + *-inium*, arbitrary suffix.] The name of a genus of parasitic fungi, used *attrib.* and *absol.* to designate plant diseases caused by them.
1926 *Misc. Publ. Min. Agric. & Fish.* LII. 26 Sclerotinia disease..reported from Lancashire [potatoes]. **1950** *N.Z. Jrnl. Agric.* July 79/2 Causing rapid decay, sclerotinia rot (*Sclerotinia* spp.) attacks many plants, including petunias, zinnias, stocks, and wallflowers. Dead plants are usually black, but are covered by a white growth. **1976** E. Scarrow *N.Z. Veg. Gardening Guide* 64 Fungal diseases include early and late blight sclerotinia (or white mould disease) and various other stem and leaf blights.

sclerotized (skliə·rotəiz'd), *ppl. a.* *Zool.* [f. Sclerot(ic *a.*¹ and *sb.* + -iz(e + -ed¹.] Hardened by conversion into sclerotin.
1928 Ferris & Chamberlin in *Entomol. News* XXXIX. 215 We might very logically extend the use of the root which appears in the word 'sclerite'. This word alone is hardly sufficient to meet all our needs and we could speak of 'sclerotic areas' or of 'sclerotized areas'. **1955** P. A. Buxton *Nat. Hist. Tsetse Flies* iii. 53 The upper wall of the labium is also sclerotized and forms the labial gutter. **1975** *Nature* 8 May 142/2 A sclerotised edge of the left wing [of the tree cricket] serves as the scraper.
So **scle:rotiza·tion**; **scle·rotize** *v. trans.*
1957 Richards & Davies *Imms's Textbk. Entomol.* (ed. 9) i. 9 In most insects..the greater part of the cuticle undergoes a process of sclerotization whereby it becomes hardened and darkened to form more or less tough, rigid sclerites separated from each other by membraneous zones of unchanged soft cuticle. **1963** R. P. Dales *Annelids* ii. 41 The cuticle of the earthworm gizzard is like that of the epidermis and is simply collagenous, but in polychaetes the stomodeal cuticle may be sclerotized to form teeth or jaws. **1974** *Nature* 30 Aug. 799/2 An early effect of ecdysoids at metamorphosis in some caterpillars is conversion of tryptophane into red ommochrome pigments; in fly larvae, the conversion of tyrosine into quinones to sclerotise the puparium. *Ibid.* 20 Dec. 710/2 The sclerotisation and tanning of insect cuticles is generally thought to result from a crosslinking of the cuticular proteins by quinonoid derivatives of tyrosine.

sclerotome. Add: Hence **scleroto·mal,** **scleroto·mic** *adjs.*, of or pertaining to a sclerotome.
1890 Webster, Sclerotomic. **1925** J. S. Kingsley *Vertebrate Skeleton* 21 There is one great difference between Elasmobranches and higher Vertebrates; in the former cells from the sclerotomic elements..break through the elastica externa.., invade the notochordal sheath, and may chondrify there. **1971** A. J. Waterman *Chordate Struct. & Function* vi. 211 The sclerotomic cells form the perichordal tube. In addition, a mass of sclerotomic tissue migrates to the myoseptum to form the neural and haemal arch anlagen. *Ibid.* (*caption*) Precartilage stage; further compaction of sclerotomal tissue with establishment of basic vertebral shape... Sclerotome stage; sclerotomal cells migrate from somite and form sclerotome. **1974** D. & M. Webster *Compar. Vertebr. Morphol.* v. 87 The posterior half of one sclerotomal segment and the anterior half of the segment just behind it join together, so that each presumptive vertebra forms on a level overlapping two somites.

scobberlotcher. Delete † *Obs.* and add later examples. Also **scobolotcher.**
1933 C. Day Lewis *Dick Willoughby* 10 Thou bed-worm, thou scobberlotcher! **1956** *Bournemouth Daily Echo* 21 Apr. 10/4 A scobolotcher, said Mr. Moore, was an undergraduate walking around a quadrangle hands in pockets and deep in thought.

scobby. Add: Also **scobbie, skobby.** (Earlier and later examples.)
1800 D. Wordsworth *Jrnl.* 17 May (1941) I. 39 The Skobby sate quietly in its nest. **1976** *Jrnl. Lakeland Dial. Soc.* No. 38. 39 She wes gitten weel on afoor she kent et a scobbie wes oot but a scobbie.

scoff, *sb.*² Add: Also **schoff, skoff.** (Earlier and later examples.) Also in colloq. use outside South Africa. Also *attrib.*
1846 *Swell's Night Guide* 51 It vas hout-and-hout good scoff, and no flies. **1855** J. W. Colenso *Ten Weeks in Natal* 54 The *meat* and other *scoff* (food), which the Kafirs are so fond of. **1863** J. S. Dobie *Jrnl.* 6 Jan. (1945) 60 The best one was consigned to the Kafirs for 'scoff'. **1902** 'Coldstreamer' *Ballads of Boer War* vii. 66 They gives 'im 'schoff' an' treats 'im kind, Instead o' striking 'im be'ind. **1926** *Variety* 29 Dec. 5/3 Slang, in addition to providing me

with seven flops weekly and three scoffs daily, has saved me from night school. **1928** *Daily Express* 14 May 10/6 While you've had me locked up, I've eaten your scoff! **1934** *Detective Fiction Weekly* 21 Apr. 109/2 Where the criminal eats he says he scoffs, and if he goes to a restaurant it is called a beanery, chow joint or scoff joint. **1955** J. Cope *Fair House* v. 62 He treated them familiarly, shared his skoff-tin with them. **1960** [see *bevvy]. **1969** in Halpert & Story *Christmas Mumming in Newfoundland* 84 One of the men might suggest to those in his group that everyone come over to his house for a 'scoff'. **1976** *Australasian Express* 11 June 25/3 A particularly memorable scoff was had on Colitzani beach. **1977** J. Wainwright *Do Nothin'* xi. 182 A dance, all of her own, with guests and scoff and booze of her own choice. **1981** *Guardian* 24 Aug. 8 Ah! Scoff ahoy! I spy Florida Cocktail and Gammon Steak Hawaii!

scoff, *v.*¹ Add: **2. c.** To utter in a scoffing manner (with the spoken words as obj.).
1862 Mrs. H. Wood *Channings* I. v. 70 'A senior do it!' scoffed Roland Yorke. **1894** E. Fawcett *New Nero* v. 66 'I no more ruined her,' scoffed Egerton, 'than the Sultan of Turkey did!' **1898** Skeel & Brearley *King Washington* xxv. 155 'Not I,' scoffed Anne, shaking her taffeta flounces. **1921** [see *poof *int.* A]. **1976** J. Wainwright *Bastard* i. 23 'In *this* weather?' I scoff. **1977** P. G. Winslow *Witch Hill Murder* ii. xvii. 219 'Oh, come on, Supe,' Jed scoffed. 'You're really not trying to pin that murder on me, are you?'

scoff, *v.*² Add: **1. a.** (Earlier and later examples.) Also with *up, down.* Also *fig.*
1846 *Swell's Night Guide* 48 He scoffed weed; that is, chewed tobacco. *Ibid.* 50 You must grub with the grunters, and scoff cabbage without salt. **1942** *Tee Emm* (Air Ministry) II. 148 Fluorescence which he said tasted horrible, and of which he scoffed quite a fair amount while he was waiting to be lifted out of the water. **1956** I. Murdoch *Flight from Enchanter* 125, I wonder what happens to it [*sc.* a magazine]. Fay must scoff it up in her room. **1967** E. Gilzean *Murder on Sundays* ii. 33 Come on, Janet. They'll have scoffed all the beer and cider if we don't hurry. **1972** R. K. Smith *Ransom* v. 231 Scoff it up, chillun..and you'll grow up big and strong. **1973** J. Porter *It's Murder with Dover* v. 45 In the dining room the reporters ..were..noisily scoffing down everything that was put in front of them. **1979** *Daily Tel.* 13 Oct. 12/3 Should the farmer wish to supplement the rations of his cattle or sheep in bad weather then he must stand guard over the feeding trough to stop the horses scoffing the lot.
b. (Earlier and later examples.)
1798 A. Barnard *Jrnl.* 24 May in A. W. C. Lindsay *Lives of Lindsays* (1849) III. 464 [The Boer] concludes of course that the passengers want to *scoff* (to eat). **1850** H. Melville *White-Jacket* xv. 73 Bear a hand, and '*scoff*' (eat) away... Some of you fellows keep *scoffing* as if I had nothing to do but..look on. **1855** G. H. Mason *Life with Zulus of Natal* xvi. 193 A Caffre..entered our service... It soon became manifest that our new servant was a madman... He would commence a war-song, or call for us to get up and 'scoff' (eat) *with him*. **1926** *Clues* Nov. 158/2 Let's scoff. Get the duffer. **1931** 'D. Stiff' *Milk & Honey Route* 213 *Scoffing*, to eat. *To scoff regularly* means to miss no meals. **1944** D. Burley *Orig. Handbk. Harlem Jive* 70 Really knock yourself out as you scoff. **1965** R. Erskine *Passion Flowers in Business* xiii. 164 Can we please go and scoff? **1973** C. Himes *Black on Black* 133 Go on, baby, you can be back in an hour with 'nuff bread so we can scoff.

scoffer² (skǫ·fəɹ). *colloq.* [f. Scoff *v.*² + -er¹.] One who eats greedily.
1935 *Amer. Speech* X. 20/1 *Scoffer*, a glutton; one who has no control over his appetite. (Obs.) **1959** I. & P. Opie *Lore & Lang. Schoolch.* ix. 154 A pair of Hertfordshire 14-year-olds,..when asked about their classmates and invited to set down their epithets for them, promptly came to life and wrote: 'Phumph, lumber bontts [bonce],..hog, scoffer, flippin kid [etc.]'. **1976** *Times* 21 Aug. 12/6 The scoffers of doughnuts, bananas and raw eggs.

scoffing (skǫ·fiŋ), *sb. U.S. Tramps' slang.* [f. Scoff *v.*²] Usu. *pl.* Food, something to eat.
1907 J. London in *Cosmopolitan* May 17/1 A hard town for 'scoffings', was what the hoboes called it [*sc.* Reno] at that time. **1914** *Sat. Even. Post* 4 Apr. 11/3 Got to throw your feet if yuh want scoffin's. *Ibid.*, You'll have to batter for handouts this mornin'. I'll get my own scoffin's.

scofflaw (skǫ·flǭ). Chiefly *U.S.* [f. Scoff *v.*¹ + Law *sb.*¹] One who treats the law with contempt, esp. a person who avoids various kinds of not easily enforceable laws. Also *attrib.*
1924 *Boston Herald* 16 Jan. 1/2 Delcevare King of Quincy last night announced that 'scofflaw' is the winning word in the contest for the $200 he offered for a word, to characterize the 'lawless drinker' of illegally made or illegally obtained liquor. 'Scofflaw' was chosen from more than 25,000 words, submitted from all the states and from several foreign countries. The word was sent by two contestants, so the prize will be equally divided between Henry Irving Dale..and Miss Kate L. Butler. **1936** Mencken *Amer. Lang.* (ed. 4) 174 The announcement that *scofflaw*.. had won was made on Jan. 15, 1924. The word came into immediate currency, and survived until the collapse of Prohibition. **1956** *N.Y. Times* 17 Jan. 27/3 The maximum fine of $50 a ticket was imposed yesterday upon a woman scofflaw who had accumulated fifty-one parking tickets. **1961** *Observer* 1 Oct. 30/6 An unenforceable law which automatically transformed every wet citizen into a scofflaw. **1965** P. De Vries *Let Me count Ways* iv. 57 A scofflaw in such a jam could easily insist that the facts in his possession during his lifetime justified his unbelief. **1971** J. Gray *Red Lights* iii. 78 Moose Jaw became the happy hunting ground for Regina gamblers, philandering husbands..and unclassified scofflaws. **1973** D. E. Westlake *Cops & Robbers* 25 He had New York plates. Good. If I gave him a ticket he couldn't be a scofflaw, fade away into some other state and thumb his nose at me. **1977** *Sat. Rev.* (U.S.) 3 Sept. 6/1 The illegal phone-dialing devices called 'blue boxes' are about to be put out of business... By..pressing its rewired dial-tone buttons, a scofflaw could bypass phone company billing systems.

scoldingly, *adv.* (Later examples.)
1912 J. Stephens *Crock of Gold* v. xiv. 208 As they approached the door the sound of a female voice came to them scoldingly. **1933** E. O'Neill *Ah, Wilderness!* iv. iii. 152 She goes on scoldingly.

scoleces, pl. of Scolex.

scolices, pl. (*erron.*) of Scolex.

Scoline (skōᵘ·līn). *Pharm.* Also **scoline.** [f. *s(uccinyl)c(h)oline.] A proprietary name for succinylcholine.
1952 *Trade Marks Jrnl.* 30 Jan. 89/2 *Scoline*... Pharmaceutical preparations and substances. Allen & Hanburys Ltd, Three Colts Lane, Bethnal Green, London, E.2; wholesale Chemists and Druggists. **1952** *Lancet* 21 June 1226/2 Succinylcholine chloride ('Scoline') was given intravenously. **1965** J. Pollitt *Depression & its Treatment* iv. 50 A modern relaxant such as succeryl choline (Scoline) is given intravenously. **1965** *Daily Express* 14 Oct. 5/7 The two-inch long ampoules of scoline and of distilled water..were kept together in a plastic bowl in a refrigerator. **1976** Smythies & Corbett *Psychiatry* xvii. 291 American psychologists treated some criminals in jail by asking them to imagine themselves reliving the circumstances of their criminal activities and then giving an injection of scoline.

scoliosis. Add: Pl. **scolioses** (-ōᵘ·sīz). (Further examples.) **scoliotic** *a.* (example).
1939 H. H. Jordan *Orthopedic Appliances* iii. 90 A group of scolioses which are suitable for treatment by forces which we can introduce by means of an active correcting brace. **1958** *Jrnl. Bone & Joint Surg.* XLa. 553 Correction and fusion of the scoliotic spine. **1976** *Lancet* 4 Dec. 1234/1 In the causation and progression of scoliosis spinal muscles may have a major role. *Ibid.*, On electronmicroscopy there are signs of dystrophy and atrophy in the spinal muscles of scoliotic patients.

scolopale (skǫ·lǫpēᵘl). *Ent.* Formerly also **-pala** (pl. **-palæ**). [ad. G. *scolopal* adj. (in *scolopale körperchen*) (V. Gräber 1882, in *Arch. f. mikrosk. Anat.* XX. 516), f. Gr. σκολοπ-, σκόλοψ spike + -al -al.] The rod-like structure inside the sheath of a scolopidium; also, the sheath itself.
1912 J. H. Comstock *Spider Bk.* iii. 169 The failure of other observers to discover scolopalæ or auditory pins..has made this conclusion doubtful. **1917** *Ann. Entomol. Soc. Amer.* X. 69 Its distal portion penetrates the center of the enveloping cell into the proximal end of the cap cell, where the nerve enlarges to form the peg-shaped body or scolopale. **1925** [see *scolopophore]. **1932** *Parasitology* XXIV. 457 These cells are connected with the main antennal nerve, whereas the other end gives off a long scolopala, or the sense rod. **1948** *Sci. News* VII. 19 One finds in all insects structures which consist of parallel elements—elongated spindle-shaped sensory cells whose axes are prolonged into nerve fibres on the central side while on the other side they are in contact with the so-called 'scolopales' (pointed stakes). **1964** [see *scolopophore]. **1969** R. F. Chapman *Insects* xxx. 630 The most fully studied contact chemoreceptors are the trichoid sensilla on the legs and mouth-parts of *Phormia*. They are from 30 to 300 μ long. From the tip the scolopale is invaginated. **1978** H. V. Daly et al. *Introd. Insect Biol. & Diversity* vi. 107/2 In the usual arrangement [of sensilla], the dendrite of a single bipolar neuron is attached to a movable part of the body, often by a minute cuticular sheath called a scolopale.

scolopendrine, *a.* Add: Also *fig.*
1963 V. Nabokov *Gift* iv. 257 He left a scolopendrine trace in literature as the translator of foreign poets.

scolophore, var. Scolopophore in Dict. and Suppl.

scolopidium (skǫlǫpi·diəm). *Ent.* Pl. **-idia.** [mod.L., coined in Ger. (F. Eggers 1923, in *Zool. Anzeiger* LVII. 239), f. Gr. σκολοπ-, σκόλοψ spike, after Ommatidium.] An elongated sensory end-organ in insects consisting chiefly of the nucleus and dendrite of a sensory nerve cell and a tubular sheath enclosing the dendrite; *spec.* each of those that compose a chordotonal organ.
1939 V. B. Wigglesworth *Princ. Insect Physiol.* vii. 135 Chordotonal sensilla or scolopidia—These sensilla are generally believed to be derived from campaniform sensilla through their component parts becoming elongated and deeply sunk within the body. **1957** *New Biol.* XXIII. 38 Applied to the inside of the membrane [of a locust's tympanal organ] are the ends of a number of special sense cells, or scolopidia, which transmit the movements of the membrane as impulses along the auditory nerve. **1971** [see *scolopoid *a.*]. **1978** H. V. Daly et al. *Introd. Insect Biol. & Diversity* vi. 107/2 Chordotonal organs are completely internal and formed by units or scolopidia consisting of three cells: a bipolar neuron, a scolopale cell, and an attachment cell.

scolopoid (skǫ·lŏpoid), *a.* *Ent.* [f. Gr. σκολοπ-, σκόλοψ spike + -oid.] = *scolopophorous *a.*
1963 V. G. Dethier *Physiol. Insect Senses* ii. 19 (*caption*) Different forms of scolopoid sensilla. A. Terminal peg from a grasshopper sensillum. [Etc.] **1971** E. O. Wilson *Insect Societies* (1972) xi. 202/2 Typical chordotonal sensilla—or scolopoid sensilla, or scolopidia as they are often alternatively labelled.

scolopophore. Add: Also **scolophore.** Also, the sensory end-organ of which this sheath is part, comprising in addition the enclosed rod and neurone. (Further examples.) [ad. G. *scolopophor* (V. Graber 1881, in *Zool. Anzeiger* IV. 452).]
1917 *Ann. Entomol. Soc. Amer.* X. 66 Schwabe (1906).. first showed that the nerve end-organ or scolopophore is composed of three cells with definite boundaries. **1925** A. D. Imms *Gen. Textbk. Entomol.* i. 82 An auditory or chordotonal sensilla consists of a nerve end-organ or scolophore, enclosing a hollow peg-like structure or scolopale. **1933** *Jrnl. Cellular & Compar. Physiol.* IV. 80 The scolophores are either directly connected with the tympanum or lie on a secondary membrane so situated as to move with it. They contain a chitinous rod or scolopale, one end of which projects from the cell body into one or more supporting cells which abut against the body wall. **1964** R. M. & J. W. Fox *Introd. Compar. Entomol.* vi. 193 The unit of the chordotonal organ is the scolopophore..., composed of an apical cap cell attached to the body wall and an envelope cell; in the central part of the cap cell is a terminal ligament which forms a functional extension of the sensory rod (scolops or scolopale) in the envelope cell. **1967** C. P. Hickman *Biol. Invertebrates* xxv. 459/1 Many Orthoptera..have sound receptors, which are spindle-shaped bundles of chordotonal sensilla attached to the integument and called scolophores.

scolopophorous (skǫ:lǫpǫ·fōrəs), *a.* *Ent.* [f. prec. + -ous, after G. *scolopofer* (V. Graber 1881, in *Zool. Anzeiger* IV. 450).] Of a sensory end-organ: having the elongated tubular form of a scolopidium.
1935 R. E. Snodgrass *Princ. Insect Morphol.* xviii. 527 Scolopophorous sense organs are widely distributed in insects, but until recently they have not been reported in other arthropods. **1973** W. S. Romoser *Sci. of Entomol.* v. 122 A type of sensillum that is rather dramatically different from those already described is the scolopophorous or chordotonal organ.

scolops (skǫ·lǫps). *Ent.* [a. Gr. σκόλοψ spike, adopted in this sense in Ger. by F. Eggers 1923, in *Zool. Anzeiger* LVII. 239.] The rod-like structure inside the sheath of a scolopidium.
1935 R. E. Snodgrass *Princ. Insect Morphol.* xviii. 526 The distinguishing feature of sensilla scolopophora is the presence of a well-differentiated, peg-shaped 'sense rod', or scolops, at the apex of each cell. **1964** [see *scolopophore].

scolytid, *sb.* and *a.* Insert in etym. after *Scolytus* (E. L. Geoffroy *Hist. Insectes de Paris* (1762) I. 309), f. Gr. σκολι-ός bent, curved. Substitute for def.: A small cylindrical bark- or wood-boring beetle of the family Scolytidæ; of or pertaining to a beetle of this kind or the family as a whole. (Earlier and later examples.)
1890 *Proc. Entomol. Soc. Washington* II. 77 (*title*) Notes on the breeding habits of some Scolytids. **1909** *Bull. U.S. Bureau Entomol.* LXXIII. 1 (*title*) Practical information on the Scolytid beetles of North American forests. **1925** A. D. Imms *Gen. Textbk. Entomol.* iii. 510 Scolytid larvæ are apodous. **1959** E. F. Linssen *Beetles Brit. Isles* II. 256 The best-known scolytid is probably..the Large Elm Bark Beetle. **1972** *Oxf. Univ. Gaz.* CII. No. 3496 (Suppl.* 2) 22 Investigations on Scolytid beetles of timber. **1976** *Nature* 24 June 696/2 Most scolytids..occupy temporary habitats.

scone, *sb.* Add: **1. drop-, dropped scone** (examples).
1899 E. Heddle *Marget at Manse* 100 She..would bake drop-scones, and carry in my tea with her own hands. **1899** R. Wallace *Country Schoolmaster* 20 Potato scones, soda scones, 'droppet' scones, treacle scones. **1942** C. Spry *Come into Garden, Cook* xv. 213 Most people have a good recipe for dropped scones... Drop the batter from a spoon on the hot girdle and turn once. **1956** E. Grierson *Second Man* ii. 44 Some tea-cake and drop scones and jam. **1977** *Age* (Melbourne) 18 Jan. 13/4 The cheese souffle looked more like a cheese drop-scone.
3. a. (Always with pronunc. skǫn) *to do one's scone*, to lose one's head, temper. Hence *scone-doer, -doing. N.Z. slang.*
1942 *2nd N.Z.E.F. Times* 20 Apr. 6 Scone-doer. A person subject to sudden fits of excitement and irritation. *Ibid.* 19 Oct. 5 'Don't do your plurry scone, Dig!'..'Who's sconing?' **1944** F. I. Cooze *Kiwis in Pacific* i. 8 The camp at Pahantanui was much as all military camps. Tedious training, fatigues, and 'scone-doing' from 6 a.m. to 4 p.m. **1952** *Here & Now* (N.Z.) II. iv. 20 Everyone question Rangi. Everyone do the Scone. **1957** M. K. Joseph *I'll Soldier no More* (1958) ix. 167 Gillies finds him a bit of a nagger, but likes him for being efficient and not doing his scone.
b. The head. *Austral. slang.*
c **1945** in S. J. Baker *Austral. Lang.* (1966) viii. 172 *Scone*, head. **1957** D. Niland *Call Me when Cross turns*

Over v. 138, I can just see you running a house. I'd give you a week before you went off your scone. **1968** D. O'Grady *Bottle of Sandwiches* 58 He reckoned we weren't right in the scone to be travelling so far on a Sunday just to chase a ball around a paddock.
4. *Comb.*, as **scone-hot** *a.* *Austral. slang*, in phr. *to go* (someone) *scone-hot*, to reprimand (someone) severely, to lose one's temper at (someone); see also quot. 1941.
1938 X. Herbert *Capricornia* 530 Halfcaste Shillingsworth goes Copra Co scone-hot! **1941** Baker *Dict. Austral. Slang* 63 *Scone-hot*, an intensive to describe great vigour of attack, scolding or speed, e.g., 'Go for someone scone-hot', to reprimand severely. (2) Exorbitant, unreasonable. (3) Expert, proficient, e.g., 'He's scone-hot at shearing'. **1944** *Coast to Coast 1943* 116, I don't want Reg going me scone hot because his wife's not capable of looking after herself. **1967** K. Tennant *Tell Morning This* (1968) xvii. 139 When my big brother Jim come home from work, he went Dad scone hot. **1974** D. Ireland *Burn* 136 When he finds out he'll go me scone-hot.

scone (skǫn), *v.* *Austral.* and *N.Z. slang.* [f. dial. *scon, scun*: see *Eng. Dial. Dict.*] *trans.* To hit.
1948 *Coast to Coast 1947* 187 The bottle broke. Damn! he hadn't meant to scone the bottle first go-off. **1958** I. Cross *God Boy* iv. 30 Joe was worried in case he had really sconed the girl.

scoop, *sb.*¹ Add: **7. scoop bonnet,** a woman's bonnet shaped like a scoop; also *scoop-shovel bonnet.*
1846 E. W. Farnham *Life in Prairie Land* 102 Sometimes her scoop bonnet covered half my field of vision. **1901** W. Churchill *Crisis* i. iv. 40 Her face was in that most seductive of frames, a scoop bonnet of dark green velvet. **1905** J. C. Lincoln *Partners of Tide* 30 [Portraits] of ladies in flowered scoop bonnets..gazed down upon him with rigid disapproval. **1884** 'Mark Twain' *Huck. Finn* xvii. 154 One was a woman in a slim black dress..and a large scoop-shovel bonnet. **1941** L. I. Wilder *Little Town on Prairie* xix. 222 She wore a sweeping black gown and a scoop bonnet.

scoop, *sb.*² Add: **1. a.** Also, a quantity scooped up.
1832 J. P. Kennedy *Swallow Barn* I. iii. 34 Nine scoops of water in the hollow of the hand, from a sycamore spring.., will break an ague.
b. *Mus.* = Portamento.
1911 W. K. Smith *Training Village Choirs* 8 The scoop is made on the commencing note of a tune or phrase. It consists in attacking it by way of a chromatic slide from the 'fourth' below. **1967** A. L. Lloyd *Folk Song in Eng.* i. 64 The sundry ways the folk singer has of passing from note to note by means of scoops, slides, hovers and such. **1975** *Gramophone* Aug. 316/1 He pulls the *Rosamunde* 'Entr'acte' about horribly and there are all sorts of period scoops that modern listeners will find intolerable. **1977** *Early Music* July 343/2 Special techniques and playing styles that can be developed are..'scoops' on a note, e.g. D–C♯–D completely slurred and glided simply by relaxing the breath pressure and increasing it again, keeping the fingering for the upper note held all the time.
2. b. *Film* and *Television.* (See quots.)
1940 *Chambers's Techn. Dict.* 747/2 *Scoop*, one or more suspended broadsides, which are special types of incandescent flood-lights for use in motion-picture studios. **1974** *Some Technical Terms & Slang* (Granada Television), *Scoop*, 500 watt lamp suspended from studio ceiling.
3. a. (Earlier example.)
1871 F. C. Burnand *More Happy Thoughts* xxxiv. 248 'Both the nautical Cockalorums have been going on the scoop, and are slightly moppy.' By which we understand him to mean, that the two naval officers have had as much as is bad for them.
b. For *U.S.* read orig. *U.S.* and add earlier and later examples. Also *transf.*
1874 *Macomb* (Illinois) *Eagle* 23 Nov. 1/2 Owing to a slight misunderstanding, the *Sentinel* found itself without a copy of the decision, and for a time a terrible scoop seemed imminent. **1906** *Daily Chron.* 29 Jan. 4/6 The feat escaped the notice of the representatives of all other newspapers; so the item is what in Fleet-street language is styled an 'exclusive' or 'scoop'. **1913** E. Pound *Let.* Mar. (1971) 16 It's our second scoop, for I only found the man [*sc.* Robert Frost] by accident and I think I've about the only copy of the book that has left the shop. **1917** Chesterton *Short Hist. Eng.* xiv. 181 One of these scares and scoops (not to add the less technical name of lies) was the Popish Plot. **1920** *Times* 25 Oct. 15/3 The edition would have been on the streets..leaving the heartbroken editor to bewail the death of his great 'scoop'. **1930** 'Sapper' *Finger of Fate* 127 The first thing to do was to get on the 'phone to his editor, because he had intended returning to London that night. He knew there would be no difficulty—especially if he gave a hint over the wire that he was on a scoop. **1940, 1969** [see *beat *sb.*¹ 15 c]. **1973** D. Barnes *See the Woman* 75 We've got some scoop that our outstanding suspect is holed up in the Rocket Motel.
4. *attrib.* and *Comb.*, as **scoop neck,** a rounded, low-cut neck on a garment; also (with hyphen) *attrib.*; so *scoop-necked* adj.; **scoop neckline** = *scoop neck.*
1953 *New Yorker* 20 June 64/2 At Rosette Pennington, 20 East 56th Street, there's a cool little sleeveless black cotton dress with a scoop neck and a full, flounced skirt. **1956** *Ibid.* 28 Jan. 68/3 There are short beach coats and scoop-neck dresses, all of the same material. **1972** *Vogue* Jan. 22 Bell sleeves, scoop neck, pintucks on jacket. **1978** *Detroit Free Press* 16 Apr. d 14 (Advt.), The tab-closing jacket with a skirt and scoop-neck shell in pink or beige. **1955** *New Yorker* 17 Sept. 96/1 De Pinna has acquired another sleeveless, scoop-necked Trigère dress. **1977** *Daily Tel.* 4 Apr. 15/4 The scoop-necked three-quarter mohair-knit coat over scooped-necked long-torso fine-knit sweater. **1959** *Times* 21 Sept. 12/4 A gown of deep cream satin with a fitting bodice, a scoop neckline, and a full skirt.

scoop, *v.*¹ Add: **5. a.** (Earlier and later examples.) Also in various extended uses; *esp.* to defeat, destroy, get the better of. Phr. *to scoop the kitty* (or *pool*), in *Gambling*, to win all the money that is staked; also *transf.*, to gain everything, to be completely successful.
In some uses difficult to distinguish from sense *6.
1850 W. Colton *Three Years in Calif.* xxxiii. 440 A faith that could scoop up whole tribes of savages,..impressing them with the conviction that submission to the padres was obedience to God. **1866** *Harper's Mag.* Oct. 680/1 Tell him he'll have to send this other fellow some more beans, for I've got him scooped [at draw-poker]. **1867** A. D. Richardson *Beyond Mississippi* xi. 134 'Scooped' was an importation from Wall Street. 'I am badly scooped' meant [in Kansas]: 'I am used up' or 'defeated'. **1872** 'Mark Twain' *Roughing It* xlvii. 333 'It ain't no use. They've scooped him.' 'Scooped him?' 'Yes—death has.' **1903** A. Bennett *Leonora* vii. 194 Milly had shown a straight flush and scooped the kitty. **1903** Kipling *Stellenbosh* in *Five Nations* 195 The Boojers scooped the crowd, To the last survivin' bandolier an' boot. **1916** J. Buchan *Greenmantle* xxii. 297 We have won any way; and if Peter has had a slice of luck, we've scooped the pool. **1929** H. Maclaren *Private Opinions of Brit. Blue-Jacket* 100, I haves everey intentuons to make a short spich, as scoops in the old man also a fairly wheard tipe of two passingers. *Ibid.* 101 This phrase from 'scoops in the old man' is plainly to be interpreted, by any one familiar with the ways of the fo'c'sle, as 'we have persuaded the captain to come, also two weird passengers'. **1937** H. C. Bailey *Clunk's Claimant* xlvi. 315 A million to one some side-line of a next of kin would bob up and pinch their share. Josh wouldn't scoop the pool. **1939** Wodehouse *Uncle Fred in Springtime* v. 70 There was a bit of unpleasantness at the Ball, and they scooped me in. **1944** *Jrnl. R. Aeronaut. Soc.* XLVIII. 363 De Havilland engines—the Gipsy, Gipsy Six and Gipsy Twelve..—together with the Cirrus, have almost completely 'scooped the pool' for light aircraft. **1953** K. Tennant *Joyful Condemned* viii. 69 There wasn't a girl..to touch her. She could walk right in and scoop the pool. **1959** *Encounter* Aug. 37/2 The rest of them were struggling..with razors and stakes... And soon I got scooped into the thing. **1972** Wodehouse *Pearls, Girls, & Monty Bodkin* ii. 27 You will give your consent to my scooping in the girl I love. **1973** 'P. Malloch' *Kickback* xxv. 164 You've scooped the pool. If you watch your step, you shouldn't have any more worries. **1976** *Evening Post* (Nottingham) 15 Dec. 23/9 Radford Swimming Club ended the 1976 season in fine style by scooping all but one award at the Notts. ASA Medley Team Swimming League. **1978** G. Mitchell *Mingled with Venom* iii. 23 Unless we all take care, that black boy the other lot adopted is going to scoop the pool.
b. (Further examples.) Also *transf.* Also with the news as object, and occas. the person from whom information is derived.
1937 *Printers' Ink Monthly* May 42/1 *Scoop*, to gainfully outwit a rival network or station in the broadcasting of a special event or public interest program. **1938** E. Waugh *Scoop* i. v. 88 He told..how Wenlock Jakes, highest paid journalist of the United States, scooped the world with an eye-witness story of the sinking of the *Lusitania* four hours before she was hit. **1939** R. Campbell *Flowering Rifle* vi. 143 Then if some British pressman should be handy—From a safe distance, priming him with brandy, To scoop their story in his red receivers. **1948** G. V. Galwey *Lift & Drop* v. 88 The *Voice* must scoop you when you retire from Scotland Yard. **1968** J. M. Ziman *Public Knowledge* v. 98 Many scientists are so obsessed with the fear of being 'scooped'..that they issue a long succession of scrappy communications instead of waiting until the work is complete. **1974** *Times* 17 Apr. 16/7 The Israeli press..is sometimes scooped by the foreign press... The scooping..often results from ministerial indiscretions overseas. **1978** G. McDonald *Fletch's Fortune* vi. 49 Do you realize what it would be worth to a person's career to scoop the murder?.. A handful of Pulitzer Prizes.
6. To propel or to take by or as by a scooping movement. Also with *up.*
1867 *Australasian* 19 Jan. 76/3 Davis scooped a slow to Dan Wilkie, who..held it. **1882** *Sat. Rev.* 2 Sept. 313/1 The last comer scooped his first ball round to leg for 2. **1886** F. R. Stockton *Casting away of Mrs. Lecks & Mrs. Aleshine* 50 I'll never leave this place if I have to scoop myself out to sea with an oar. **1910** *Blackw. Mag.* Feb. 269/1 A very dark-coloured little man, with his arms and legs cut off, short at the knees and elbows,..scooping himself along on his stumps. **1916** 'Boyd Cable' *Action Front* 257 And he moved as if to scoop the German's head under his arm again. **1960** *Daily Tel.* 6 Dec. 1/1 Helicopters flew to the rescue of villagers trapped by floods at Hampton Bishop, near Hereford, last night. Forty men, women and children were 'scooped' from their cottages as swirling water crept towards their upstairs rooms. **1961** 'E. Lathen' *Banking on Death* xiii. 104 'I have a dinner date.' 'So do I,' replied Nicolls, hastily scooping up the letter. **1963** G. H. Thomson *Crocus Country* xix. 127 Mother never allowed anyone to 'scoop' the ball, that is, push it ahead with the mallet. **1966** *Listener* 17 Mar. 384/2 When she moves off, either she scoops the infant up to help him cling to her or else he springs to catch hold of her. **1973** N. Graham *Murder in Dark Room* ix. 60 The phone rang and I scooped up the receiver and said, 'Solo Malcolm here.' **1978** H. Wouk *War & Remembrance* ii. 20 We'd be scooped up as we stepped off the gang-plank.
7. *Mus. intr.* To perform a scoop (*scoop *sb.*² 1 b).
1927 H. J. Wood *Gentle Art of Singing* 35 They are very apt to make a slow slur, to connect the notes by scooping and dragging the voice. **1958** A. Jacobs *New Dict. Music* 333 *Scoop*, in singing, to glide up to a note disagreeably from below instead of attacking it cleanly. **1975** *Gramophone* Nov. 846/3 In the thirties Busch was frequently praised because he scooped so rarely; as opposed to Lener who did it all the time. **1977** *Ibid.* Jan. 1154/2 Both violin and cello scoop heavily from note to note.

scooped, *ppl. a.* Add: **1. b.** *scooped neck, neckline* = *scoop neck(line)* s.v. *scoop *sb.*²
1956 *New Yorker* 14 Jan. 53 The scooped neckline blouse has a soft bow and short sleeves. **1959** *Harrods News* Summer 5 A matalessé cocktail frock with scooped neck. **1969** *Sears Catal.* Spring/Summer 3 Long, full skirt gathered at waistline, scooped neckline on bodice.

scooping, *vbl. sb.* (Examples corresponding to *scoop *v.*¹ 7.)
1960 C. Day Lewis *Buried Day* ii. 38 She sang her favourite arias..with much *portamento*, or luscious scooping. **1978** *Amer. Speech* 1975 L. 301 *Scooping*, sliding into a tone, hitting a note on the flat side and sliding up to the proper pitch, an undesirable practice on the part of one voice in a quartet.

scoopy (skū·pi), *a.* *Fashion slang.* [f. Scoop *sb.*² or *v.*¹ + -y¹.] Of the neck of a garment: rounded and low-cut. Cf. *scoop *sb.*² 4, *scooped *ppl. a.* 1 b.
1970 *Daily Tel.* 1 June 13 This summer's dresses are heaven-sent for this event. The voiles are in full swing, the necks are scoopy. **1974** *Country Life* 17 Jan. 107/1 A range of knitwear..that incorporates low, scoopy necklines. **1976** *Ibid.* 22 Feb. 377/1 Evening dresses have scoopy or drawstring necklines.

scoot, *sb.*² Add: **2.** A bout of drunkenness, a drunken spree; chiefly in phr. *on the scoot. Austral.* and *N.Z. colloq.*
1924 *Truth* (Sydney) 27 Apr. 6 *Scoot*, to clear out; also continued bout of drunkenness. **1936** I. L. Idriess *Cattle King* xiv. 131 'He's a man who likes his meat raw, he chews his glass when he empties it.' 'I'm sorry to hear Eureka is on the scoot.' 'He's not. They don't go on the scoot out there. They drink dynamite and bust.' **1959** G. Slatter *Gun in Hand* iii. 42, I suppose you left the wife up there [on the farm] and you're down on the scoot. **1962** S. Gore *Down Golden Mile* vi. 120 Make mine a glass this time, seein' I have to go on the scoot with you booze artists to-night. **1975** X. Herbert *Poor Fellow My Country* 1019 We could've.. gone on a proper scoot.

scoot (skūt), *sb.*³ *slang.* [Abbrev. of Scooter *sb.*] A motor-cycle or motor-car (see also quot. 1943).
1943 *Amer. Speech* XVIII. 169/1 *Scoot*, shuttletrain. **1968–70** *Current Slang* (Univ. S. Dakota) III–IV. 105 *Scoot*, *n.* A motorcycle, the type often used by the Hell's Angels Motorcycle Club. **1977** *Custom Car* Nov. 64/1 For this season he's gone over to a radical Volvo-engined scoot.

scoot, *v.*¹ Add: **3. b.** *trans.* To move or convey suddenly or swiftly. *slang* or *colloq.*
1905 *Automobile Topics* 27 May 462 Basle's engine had all the power necessary to scoot him up the hill on his fourth speed. **1947** J. Steinbeck *Wayward Bus* ii. 22 Juan put his little platform behind the bus and he lay on it on his back and scooted himself under with his feet. **1968** *Globe & Mail* (Toronto) 5 Feb. 22/1 Larush scooted his charge to the front from the outset to reach the quarter pole in a swift 0:30 2–5. **1975** N. Freeling *What are Bugles blowing For?* vi. 35 She scooted her wheelchair across the room.

scoot (skūt), *v.*³ Colloq. abbrev. *scooter *v.*
1951 N. Mitford *Blessing* i. vii. 72 The happy crowd of scooting, skating children in the Tuileries gardens. **1962** A. Huxley *Island* ix. 140 'Scooters are going to become a major political issue.' Vijaya laughed. 'To scoot or not to scoot, that is the question.'..'Wherever I've been..they've opted wholeheartedly for scooting.'

scooter. Add: **3.** (Earlier and further examples.) In full *scooter plough.*
1820 in *Henderson's North Carolina Almanack* (1823) 25 The ridges are opened with a small plough called a scooter, something like a shovel plough. **1842** in J. A. Turner *Cotton Planter's Manual* (1857) 55 The next operation to be performed..is to plough out the middles well, the wide way, with a good shovel-plough, having first run around the young plant with a scooter-plough. **1938** M. K. Rawlings *Yearling* xxx. 385 He brought old Caesar and the scooter plow and turned in to the field, laid off and bedded up ready for the corn, to open the furrows for the planting. **1944** T. D. Clark *Pills, Petticoats & Plows* 281 By colloquial designations the various strange shapes were known to the trade as sweeps, shovels, scooters, twisters,..scrapers and subsoilers.
4. a. A boat, propelled by sails, capable of being used both on ice and in water. *N. Amer.* **b.** A fast motor-boat, used in the war of 1914–18. **c.** A motorized pleasure boat resembling a motor-scooter. In full, *sea scooter, water scooter.*
1903 *N.Y. Times* 13 Dec., The 'scooter'..is built with a bottom and a deck which are duplicates of each other. **1909** *Cent. Dict.* Suppl. s.v. *Ice-scooter.* These scooters may be run alternately through water and over ice. **1919** *Times* 21 Feb. 11/2 The war has produced 'P' boats and 'Q' boats and 'U' boats, but the wildest of all wild things is the 'scooter', professionally known as the coastal motor-boat. **1927** G. Bradford *Gloss. Sea Terms* 151/1 *Scooter*, an amphibious craft, shallow and beamy, equipped with runners beneath and rigged with a jib and mainsail. It is used as an ice boat, particularly on Great South Bay, Long Island. It is

capable of crossing patches of open water. **1929** F. C. Bowen *Sea Slang* 118 *Scooter*, a coastal motor boat in the war. **1948** J. Steinbeck *Russian Jrnl.* vi. 116 There were boat races on the river, little water-scooters with outboard motors. **1958** *Times* 21 Jan. 8/5 (*heading*) Man on sea scooter believed drowned. *Ibid.*, Mr. John Penn.., believed to have been drowned..while testing a water scooter at West Mersea, Essex. **1966** *Kingston* (Ont.) *Whig-Standard* 13 Jan. 2/6 A provincial police diver today located an ice-scooter owned by a local insurance agent who vanished here last night. **1976** *Vacation Fun in Dearborn* (Dearborn, Mich., Times-Herald) Summer, Pedal boats and water scooters on Lakes Three and Six.

5. a. A child's toy consisting of a footboard mounted between two tandem wheels with a long handle attached to the front wheel, operated by resting one foot on the footboard while pushing with the other and steering by the handle.

1919 *Times* 21 Feb. 11/2 The 'scooter' we knew before the war was a new terror to the pavement. **1921** *Spectator* 2 July 8/1 Must you not use first one foot then another on your scooter, lest you get 'scooter leg'. **1939** Joyce *Finnegans Wake* 191 A youth those reporters so pettitily wanted as gamefellow that they asked his mother for ittle earps brupper to let him tome to Tindertarten, pease, and bing his scooter 'long. **1943** D. Powell *Time to be Born* iv. 79 The first twenty years of their existence which had been wasted in marbles, dolls, hoop-rolling, and scooter-racing. **1961** *Toys & Fancy Goods* Aug. 22 Two pavement scooters.. with red frame, yellow wheels and white grips.

b. = *motor-scooter* s.v. *Motor *sb.* 6.

1917 *Autocar* 20 Jan. 60/1 For some months past it has been known in this country that the 'scooter' in America has developed into something rather beyond the child's plaything so popular in the British Isles. Until quite recently, however, the American motor-driven 'scooter' has not been seen in London. **1919** *Model Engineer & Electrician* 27 Feb. 142/1 A scooter of this type can cover 100 miles on a gallon of petrol. **1944** R. Chandler *Lady in Lake* v. 30 An anxious-looking bird thumped past on a power-scooter. **1957** *Times* 19 Nov. 11/3 The rising popularity of new types of machines—the scooter and the moped. **1971** *Daily Tel.* 16 Dec. 1/3 Safety helmets are to be made compulsory for riders and passengers of solo motor cycles, scooters and mopeds.

c. In various extended and slang uses: see quots.

1917 *Little Folks* Sept. p. vi. (Advt.), The free-wheel auto-scooter propelled by pedal. **1919** I. F. Marcosson *S.O.S.: America's Miracle in France* vi. 154 The vastness of these Depots is such that an inspection on foot..is out of the question. They are so criss-crossed with rails that you must use a 'Scooter', which is a motor-driven hand-car fitted for standard-gauge tracks... Every important official has his own 'Scooter' and you can see them scooting over the place at all hours of the day and night. **1930** A. Armstrong *Taxi!* v. 48 There were still a large unmber of two-cylinder Renaults (called 'scooters') plying for hire. **1935** N. Ersine *Underworld & Prison Slang* 64 *Scooter*, a rum-running car. **1948** 'J. Evans' *Halo for Satan* vi. 78 'We'll use your scooter, Mac... Where's she parked?'..I wondered how they knew I had a car. **1953** *Sun* (Baltimore) (B ed.) 15 June 4/2 A new $1,000 'flying scooter' powered by a 12-horse-power engine for the 'people's car' (*Volkswagen*), has been making test flights from a forest clearing south of Hamburg. *Ibid.*, The scooter has five or eight vanes, radiating from a circular passenger cabin which stands on three wheels. **1961** Partridge *Dict. Slang* Suppl. 1261/1 *Scooter*.., a single-deck bus; a driver-only bus: busmen's: since ca. 1945. **1963** *Daily Progress* (Charlottesville, Va.) 28 June 1/6 An aerospace firm has come up with something it calls a space scooter, a one-man platform with handlebars designed for crater hopping and crag climbing on the moon. **1971** A. Diment *Think Inc.* iv. 70 A scooter truck, that strange bastard little vehicle with a bulbous cab married to a small, pick-up body. **1972** *N.Y. Times* 3 Nov. 14/4 Fleets of flag-bedecked scooter-buses.

6. *attrib.* and *Comb.* (sense *5 b), as *scooter-man, rider, -traffic.* (See also senses *3 and *5 c.)

1960 *Guardian* 12 Dec. 2/4 The irresponsible way in which many scootermen (and ladies) wind their way through traffic. **1959** *Times* 16 May 7/6 As a scooter-rider of some seniority I ventured to write to you some months ago about the apparent disinterest of the authorities in the parking of scooters. **1976** *Daily Mail* (Hull) 16 Dec. 1/4 Scooter-rider J. B...was admitted to Hull Royal Infirmary..after his vehicle was involved in an accident with a car. **1960** *Daily Tel.* 14 June 1/1 Accidents involving motor-scooters showed the greatest percentage increase, 61 per cent, but the scooter traffic was estimated to have gone up by 64 per cent. during the year.

scoo·ter, *v.* [f. the sb.] *intr.* To travel by scooter (senses *4 and *5). Hence **scoo·tering** *vbl. sb.*

1911 Webster, *Scooter*.., a strongly built sailboat... Hence: *scootering, n.* **1957** C. Brooke-Rose *Languages of Love* 15 He climbed on to his Lambretta and scoctered off towards Oxford Street. **1957** *New Yorker* 26 Oct. 35/3 Scootering is the most economical and practical form of transportation available in New York today. **1960** *Housewife* May 46/1 When scootering, slacks are just about permissible. **1961** *Times* 9 Mar. 21/3 Scootering in Great Britain has also become a hobby.

scooterist. [f. Scooter + -ist]. One who drives, or travels on, a scooter (sense *4 or *5).

1919 *Model Engineer & Electrician* 27 Feb. 142/1 The 'scooterist' is Mr. Franklin Gunther, First Secretary U.S. Diplomatic Service. **1956** *New Yorker* 8 Dec. 44/3 One of the owners..was a young man wearing a white crash helmet and lying supine on the pavement under his blue Vespa... As each new scooterist arrived, he would extend an arm upward to shake hands. **1959** *Times* 16 May 7/6 Does a scooterist pay his sixpence or a shilling and take up the space that will hold a mammoth Cadillac? **1976** R. Hill *Another Death in Venice* I. ii. 31 A flotilla of motor-scooters went by... There was something so utterly careless about the scooterists that he felt a pang of envy. **1981** *Times* 27 July 26/2 They withstood extreme provocation by large numbers of these scooterists.

scop. Add: Now usu. with pronunc. (ʃɒp). (Further examples.)

1774 R. Henry *Hist. Gt. Brit.* II. 437 Whether this similarity was owing to the Welsh bards having imitated the Saxon scops and Danish scalds,..it is not easy to determine. **1826** J. J. Conybeare in W. D. Conybeare *Illustrations of Anglo-Saxon Poetry* 245 The following lines [from The Exile's Complaint] may therefore be considered as an unique specimen of an original attempt of this kind [*sc.* elegiac] by an Anglo-Saxon Scop. **1839** T. Wright *Ess. Lit. & Learning under Anglo-Saxons* 1 The heroic song in which the *scóp* or poet told the venerable traditions of the foreworld to the chieftains assembled on the 'mead-bench'. **1860** G. Stephens *King Waldere's Lay* 27 The less remarkable Hero names more or less connected with this Legend which may be found in Beowulf, the Scóp's Song, the Traveler's Lay, the Codex Dipl. and elsewhere. **1893** *Trans. Philol. Soc. 1891–4* 379 To compose with such a Prosody would imply at once the greatest crudity and the greatest subtlety in the ancient 'scop'. **1898** T. Arnold *Notes on Beowulf* ii. 16 Hroðgar..gives rich gifts to Beowulf, and his *scóp*, or poet, recites the lay of Hnæf and Hengest, and their great fight in Friesland. **1903** L. F. Anderson *Anglo-Saxon Scop* 5 The poem itself is an aggregation of several interesting specimens of the scop's art. **1928** W. W. Lawrence *Beowulf & Epic Trad.* 281 What, in a Christian era, were the court-poets, the *scops*, to do, except to fall in with the new ways? **1948** K. Malone in *English Studies* XXIX. 164 The scops kept the old ideals strong by singing the heroes of the past. **1968** E. B. Irving *Reading of Beowulf* iv. 169 The story of Finn, which Hrothgar's scop tells at the Great Banquet.

‖ **scopa²** (skō·pa). *rare.* [It.] An Italian card-game.

1965 'W. Haggard' *Hard Sell* iii. 26 There were cafés and men inside them. They were playing *scopa*. **1977** *Time* 3 Jan. 40/3 Premier Giulio Andreotti's Christmas gift to his staff last week was a single playing card—the seven of diamonds, which in the Italian game of *scopa* is worth double and thus is considered the luckiest card in the deck.

scope, *sb.²* Add: **8. b.** The ability of a horse to extend its stride or jump.

The semantic resemblance to Scope *v.*¹ must be coincidental in view of the chronologies of the words.

1970 A. Fielder *Vibart & Friends* xiv. 115 Britain has got to produce horses of scope over big courses..if we want to bring home more Olympic medals. **1971** Broome & Murphy *Jump-Off* x. 79 Sunsherpa..had a big jump in him ..but, unfortunately, he was nothing like his half-brother as far as style and scope were concerned. **1975** B. Froud *Better Show Jumping* viii. 63 The average horse with reasonable scope can clear a low fence of say three feet high from two feet or twelve feet away from the base. **1980** *Times* 11 July 11/1 The final Liverpool fence of sloping poles at 6ft required more scope than most of the contenders possessed.

scope (skōᵘp), *sb.³* *colloq.* Also **'scope. a.** A shortened form of many words terminating in -scope, as *cystoscope, horoscope, microscope, periscope, telescope*, etc.

1603 [see Scope *sb.*¹ 12]. **1872** [see -scope]. **1914** *Dialect Notes* IV. 131 *Scope*, from *microscope*. Student slang. 'Have you a slide in your scope?' **1933** Partridge *Slang To-day & Yesterday* III. iii. 190 *Scope*, the cystoscope, an instrument used for examining the bladder. **1937** V. Woolf *Let.* 17 Aug. (1980) VI. 159 Now I must..have out the scope and see if I can pry into your bedroom. **1968** C. Helmericks *Down Wild River North* I. ii. 32, I selected a good four-power scope and a carrying sling. **1976** J. F. Panish in G. Berci *Endoscopy* xxi. 296/2 We lubricate the scope with mineral oil. If examination is going to include the right side of the colon, we begin with the longer colonoscope. **1978** W. F. Buckley *Stained Glass* xxi. 206 They can peer into the bowels of the scope all they want to.

b. Also *scope sight.* A telescopic sight for a gun.

1934 in Webster. **1966** 'A. Hall' *9th Directive* x. 96 The dealer had sent it [*sc.* a rifle]..with the scope-sight already mounted. **1968** K. Weatherly *Roo Shooter* 93 All he had to do was put on the 'scope off the ·303. **1976** *Shooting Times & Country Mag.* 16–22 Dec. 7 (Advt.), The BSA Scorpion is a super accurate air pistol even without its 'scope sight. **1978** R. Ludlum *Holcroft Covenant* xxxi. 366 Automatic repeating rifle and scope are sewn into the mattress of the bed nearest the window.

c. An oscilloscope or visual display unit; *spec.* a radar screen.

1945 *Army & Navy Jrnl.* 18 Aug. 1534/1 In using the PPI, the operator knows that the plane is the center of the circular scope and that the map which forms shows by the intensity of its light the terrain below and buildings or other targets. **1948** M. H. Nicolson *Voyages to Moon* 3 Two and one-half seconds later a returning pulse was clearly detected on a radar scope. **1958** P. Bryant *Two Hours to Doom* 106 Goldsmith peered closely at his scope. **1960** *Practical Wireless* XXXVI. 401/1 (Advt.), Compact portable 'scope ideal for servicing and general work. **1964** *Ann. N.Y. Acad. Sci.* CXV. 659 There are two distinct display channels which may be connected either to the display scope..or else to remote standard oscilloscopes. **1965** *Wireless World* July 359/1 The oscilloscope (or 'scope' as it is now commonly called) is an instrument that lets you see what is going on inside an electrical circuit. **1968** *Amer. Documentation* Jan. 72/1 Editing will be done on-line with a display scope and keyboard. Information from the central file will be retrieved, displayed on the scope, edited, and then stored in a new file which will go directly to a printer for publication. **1970** *Times Lit. Suppl.* 23 July 821/3 With his text stored on magnetic tape, the linguist can..have printed out on paper or displayed on a visual display unit (a 'scope'), the parts of the text that he wants to inspect. **1971** R. Sale *Man who raised Hell* I. i. 16 A big fat green carnation popped up on the scope where the blip had been. **1974** *Sci. Amer.* Sept. 18/2 (Advt.), Its flexible controls allow the cardiologist to keep a waveform on the scope for as long as 40 seconds.

-scope. Add: Also added to L. stems, as in *fluoroscope, oscilloscope*, and to Eng. words, as in *radarscope, sniperscope*.

scopey, var. *scopy *a.*

Scophony (skǫ·fōni). *Television.* [Perh. f. Gr. σκο-πεῖν to look at, examine + *-phony*, after *telephony*, etc.] A proprietary name for a television system employing an optical and mechanical method of picture scanning Freq. *attrib.*

1932 *Trade Marks Jrnl.* 20 Apr. 488/2 *Scophony*... Philosophical instruments, scientific instruments and apparatus for useful purposes; instruments and apparatus for teaching. Scophony Limited,..London W.1.; Manufacturers. **1934** J. H. Reyner *Television* xi. 135 One of the most ingenious alternative methods proposed is the Scophony system devised by G. W. Walton. **1935** *Television Today* I. 197/1 In the Scophony system the image to be transmitted is reflected on to a special stepped prism or reflector which so displaces the image laterally that the picture is spread out into a continuous line. The line is then scanned by a vibrating light spot. **1940** D. G. Fink *Princ. Television Engin.* x. 511 The heart of the Scophony system is the so-called 'supersonic light valve'. **1957** —— *Television Engin. Handbk.* iii. 49 The Scophony system employs a liquid cell containing a piezoelectric quartz crystal as the light modulator.

scopine (skǫ·pin). *Chem.* [ad. G. *scopin* (J. F. Eijkman 1892, in *Ber. der Deut. Chem. Ges.* XXV. 3078), f. L. *Scop-olia* (see Scopol-) + *-in* -ine⁵.] A colourless crystalline alkaloid, $C_8H_{13}NO_2$, formed by hydrolysis of scopolamine (a tropyl ester of scopine) and yielding scopoline on further hydrolysis.

1923 *Chem. Abstr.* XVII. 3189 According to the investigations of Gadamer & Hammer..and of Hess and Wahl.. the basic component (I) of scopolamine (II) (which, following a suggestion of Eykman, is designated scopine) is yet unknown, the known scopoline (III) being formed from it by rearrangement of the α-oxide into a γ-oxide ring. **1957** K. W. Bentley *Alkaloids* I. i. 20 Scopine, which is optically inactive and cannot be resolved, is readily converted into oscine by acids or alkalis. **1960** A. R. Pinder in E. H. Rodd *Chem. Carbon Compounds* IVc. xxiii. 1857 Scopolamine..is the (−)- or (±)-tropyl ester of scopine.

scopol-. Add: † **scopo·leine** [ad. G. *scopoleïn* (A. Langgaard 1876–80, in *Mitth. der Deutsch. Ges. für Natur- und Völkerkunde Ostasiens* II. (Beilage III) 267)], a crystalline alkaloid said to have the formula $C_{17}H_{21}NO_4$; **scopoletin,** add: [ad. F. *scopolétine* (J. F. Eijkman 1884, in *Rec. des Trav. chim. des Pays-Bas* III. 171)], 7-hydroxy-6-methoxycoumarin, $C_{10}H_8O_4$; (earlier and further examples); **sco·polin,** a glycoside of scopoletin; **sco·poline,** an alkaloid, $C_8H_{13}NO_2$, obtained from scopolamine on hydrolysis; also called *oscine*.

1885 *Jrnl. Chem. Soc.* XLVIII. I. 404 The author has isolated three principles from the root: skopoletin, $C_{12}H_{10}O_5$, ..skopoleïne, a crystalline alkaloid;..skopolin, $C_{24}H_{30}O_{15}$ + $2H_2O$, the glucoside of skopoletin. **1911** *Chem. Abstr.* V. 2155 Halogen alkylates and alkyl nitrates of the alkaloids of the tropeine and scopoleine series are obtained by forming addition products..of the bases of the tropeine and scopoleine series with the sulfurous acid dialkyl esters. **1885** Skopoletin [see *scopoleine* above]. **1931** *Jrnl. Chem. Soc.* 1244 On cooling, the dark bluish solution deposited scopoletin in pale yellow needles. **1959** N. Campbell in E. H. Rodd *Chem. Carbon Compounds* IVb. viii. 881 Scopoletin.. occurs in the free state and as the glucoside scopolin, $C_{22}H_{22}O_{14}$..in *Solanaceae* and *Scopolia* species. **1963** T. Robinson *Org. Constituents of Higher Plants* iv. 51 Scopoletin is the most common coumarin of higher plants. **1885** Scopolin [see *scopoleine* above]. **1933** *Chem. Abstr.* XXVII. 2685 Methylated with CH_2N_2, cichorün yields a Me ether identical with scopolin. **1959** Scopolin [see *scopoletin* above]. **1892** *Jrnl. Chem. Soc.* LXII. II. 1255 The identity of scopoline is somewhat uncertain; its boiling point agrees with that of oxytropine. **1919** *Ibid.* CXV. 476 Oscine (or scopoline)..is capable of resolution into its constituents *d*- and *l*-oscine. **1960** A. R. Pinder in E. H. Rodd *Chem. Carbon Compounds* IVc. xxiii. 1856 Scopoline, $C_8H_{13}O_2N$, contains an N-methyl group and is a secondary alcohol.

scopolamine (skǭpǫ·lămin). *Chem.* and *Pharm.* Formerly also **-in.** [ad. G. *skopolamin* (E. Schmidt 1891, in *Apotheker Zeitung* VI. 522): see Scopol- and Amine.] A syrupy liquid alkaloid, $C_{17}H_{21}NO_4$, having powerful narcotic and sedative properties, which is found in plants of the family *Solanaceæ*,

notably the thorn-apple, *Datura stramonium*; hyoscine.

1892 *Jrnl. Chem. Soc.* LXII. II. 1255 Inasmuch as the name hyoscine has been misapplied to tropine, it has become somewhat ambiguous, and the author [*sc.* E. Schmidt] proposes to call the hyoscine from hyoscyamus, scopolamine, a name which is in accord with the recent isolation of this alkaloid from *Scopolia atropoides*. **1899** [in Dict. s.v. Scopol-]. **1925** F. J. Reynolds *Marvels of 1924* 44 Dr. R. E. House has experimented with inmates of prisons, who were reduced by scopolamin to a state in which only their memories functioned. **1940** R. Chandler *Farewell, My Lovely* xxxiii. 156 There's a drug called scopolamine, truth serum, that sometimes makes people talk without their knowing it. **1945** A. Huxley *Let.* 13 Oct. (1969) 535 The adumbrations of future possibilities are to be seen in the practices of contemporary dictatorships—..use of scopolamine and other drugs to extract confessions and make people more susceptible to propaganda, [etc.]. **1976** Smythies & Corbett *Psychiatry* vii. 140 Many proprietary sleeping pills.. contain small doses of scopolamine. **1977** Lewis & Elvin-Lewis *Med. Bot.* ix. 223/2 *Datura fastuosa* and *D. metel* are abundant sources of scopolamine. **1981** T. Barling *Bikini Red North* v. 114 Sedating her with enough scopolamine to keep her comatose.

scopophilia (skǫpofi·liă). *Psychol.* Also **scoptophilia** (but see quot. 1968), **skoptophilia.** [A formative element f. Gr. -σκοπία observation (cf. *-scopy) + *-philia.] Sexual stimulation or satisfaction derived principally from looking; voyeurism. Hence **scopo-, scoptophi·liac** *a.* and *sb.*, **scopo-, scoptophi·lic** *a.* and *sb.*; also **sco·ptophile** *a.*; **scopto·philist.**

1924 J. Riviere tr. *Freud's Psychogenic Visual Disturbance* in *Coll. Papers* II. 111 The obscure psychical processes implicit in the repression of scoptophilia and in the outbreak of psychogenic visual disturbance. **1928** H. H. Ellis *Stud. in Psychol. of Sex* VII. vi. 362 The failure to react to sex attractions..is a well-defined sexual perversion, with relationships to other perversions, especially scoptophilia. **1930** W. Empson *Seven Types of Ambiguity* ii. 69 Shakespeare's partly scoptophile desire to see him settled in love. **1931** *Times Lit. Suppl.* 21 May 402/3 An eventual future in which..nobody will wear anything at all;..and the scoptophilists will have their day. **1931** J. C. Flügel in W. Rose *Outl. Mod. Knowl.* ix. 374 Those [*sc.* component instincts] connected with the activities of vision..the active or 'scoptophilic', and the passive or 'exhibitionistic'. **1937** M. Hirschfeld *Sexual Anomalies & Perversions* xxviii. 621 One of the principal criteria of pathological scopophilia is the *dominant* character of the urge. **1940** C. Allen *Sexual Perversions* iv. 75 The scoptophiliac pervert may occur in two varieties. **1957** J. Strachey tr. *Freud's Instincts & Viciss.* in *Compl. Wks.* XIV. 129 The instincts whose respective aim is to look at and to display oneself (scopophilia and exhibitionism, in the language of the perversions). *Ibid.* 130 For the beginning of its activity the scopophilic instinct is auto-erotic. **1960** R. Heppenstall *Four Absentees* xxiii. 198 Gill was a bit of a scopophiliac, a 'voyeur'. **1960** *Spectator* 8 Apr. 506/3 Perhaps I am going ga-ga through skoptophilia. **1960** *Times Lit. Suppl.* 24 June 394/2 There appears to be a..scopophiliac streak in both. **1968** C. Rycroft *Crit. Dict. Psychoanal.* 148 Scopophilia.., the spelling 'scop*t*ophilia' dates from a mistake made by Freud's first translators. **1971** *Psychol. Abstr.* XLV. June 1131/2 A sexually inhibited male scopophilic.

scoptic, *a.* and *sb.* Restrict † *Obs.* to B in Dict. and add: **A.** (Later example.)

1972 P. M. Fraser *Ptolemaic Alexandria* I. x. 571 The last of the trio, Hedylus, seems to have specialized particularly in scoptic epigrams on gluttons.

scoptophile, -philia, -philiac, -philic, -philist: see *scopophilia.

scopy (skōᵘ·pi), *a.* Also **scopey.** [f. Scope *sb.*² + -y¹.] Having or displaying plenty of scope (*scope *sb.*² 8 b).

1976 *Horse & Hound* 21 May 44/4 (Advt.), This very attractive pony has a big scopey jump. **1976** *Sunday Times* 30 May 31/7 Her big, scopy Mr Vee..has shown himself a bit strong for a woman rider. **1977** *Horse & Hound* 14 Jan. 37/2 (Advt.), Bay mare..proving to be fast and exceptionally bold with a big scopy jump.

-scopy, a formative element f. Gr. -σκοπία observation (f. σκοπεῖν to examine, look at: see -y³), used to form sbs. denoting: (*a*) (formerly) divination by inspection of something (*ooscopy, ornithoscopy*); (*b*) scientific examination by means of some instrument (*stethoscopy, telescopy*); (*c*) medical examination of some part of the body (*gastroscopy, peritoneoscopy*).

scorable (skōᵊ·răb'l), *a.* [f. Score *v.* + -able.] Capable of being scored; from which a score may be made.

1964 J. Jaffe in Rioch & Weinstein *Disorders Communication* xxvii. 389 All the indices used are completely scorable by the computer. **1977** *Linlithgowshire Jrnl. & Gaz.* 15 Apr. 16/5 Each time the threat of a scorable free kick came about, they were watching for the curve shot.

scorch, *sb.*¹ Add: **1. b.** A scorched appearance of foliage, symptomatic of various plant diseases.

1906 *Misc. Publ. Board Agric. & Fisheries Dis. Fruit* 13 Cherry leaf-scorch. A disease which every now and then proves destructive to the cherry crop. The leaves are attacked by a minute fungus, which causes them to turn brown and die, often quite early in the season. **1926** *Misc. Publ. Min. Agric.* LII. 63 Leaf scorch (physiological) [of apple trees]... In Lancashire the trouble appeared largely to be due to lack of potash, which is a contributory factor in many cases. **1933** *Discovery* Nov. 350/1 Scorch, due to the fungus *Kabatiella caulivora*, a disease which has come into prominence in recent years, causes considerable destruction in pure stands of red clover. **1974** *Nature* 8 Feb. 338/1 An experimental pirimiphos-methyl formulation produced localised scorch on citrus fruit.

4. *scorch-mark.*

1952 'M. Cost' *Hour Awaits* 112 She would..wash this scorch-mark off her thumb. **1974** M. Birmingham *You can help Me* ii. 38 The whole landmark came down in spectacular flames. There is still rubble and scorch marks. **1978** R. Barnard *Unruly Son* viii. 83 If she has her eyes on someone, they show the scorch-marks pretty fast.

scorch, *v.*¹ Add: **1. a.** Also (esp. in phr. *to scorch the earth*), to subject (an area) to a scorched earth policy (see *scorched *ppl. a.*¹ 1 b). Also *transf.*

1941 H. G. Wells *You can't be too Careful* v. ii. 245 The Russians, falling back slowly upon their main line of defence, 'scorching the earth' before this last convulsive thrust of the Nazi. **1943** *Ann. Reg. 1942* I. 193 Enormous quantities of petrol, which could not be made available until the Russian oil wells, also 'scorched', produced again. **1944** *Return to Attack* (Army Board, N.Z.) 9/2 There were neither women nor children, neither villages nor farms to be destroyed. Long ago nature had scorched the earth. **1945** *Yorkshire Post* 19 Apr. 1/1 The Germans are scorching towns in the way of the great armoured thrusts now threatening Hamburg.

b. (Later examples.)

1882 'Mark Twain' *Prince & Pauper* 225 An' I tell him this, he will scorch thee finely for it. **1884** —— *Huck. Finn* xvi. 135 It hadn't ever come home to me before, what this thing was that I was doing. But now it did; and it staid with me, and scorched me more and more. **1934** Ade *Let.* 22 June (1973) 183 To me he continues to be a revelation and a marvel although he would scorch anyone who tried to put either of those labels on him. **1965** M. Spark *Mandelbaum Gate* v. 122 Gardnor's hushed confidence continued to scorch Freddy's ear-drums.

3. Also in extended use, and with *away, up.*

1906 Somerville & 'Ross' *Irish Yesterdays* 150 The priest who was to have performed the Funeral Office scorched up on his bicycle, scarlet-faced, and half an hour late. **1957** A. C. Clarke *Deep Range* iv. 48 By keeping the torp tail-heavy and nose-up he was able to scorch along on the surface like a speed-boat. **1972** *Shooting Times & Country Mag.* 27 Mar. 22/2 The favourite scorched away to win by four lengths.

scorched, *ppl. a.*¹ Add: **1. b.** *scorched earth*: used *attrib.* of a policy of destroying all means of sustenance or supply in a country that might be of use to an invading enemy, or of orders, operations, etc., designed to effect this policy; also *transf.* and *fig.*, and *absol.*

Apparently a translation of Chinese *jiāotŭ* (*zhèngcè*) scorched earth (policy).

1937 C. McDonald in *Times* 6 Dec. 12/2 The populace.. are still disturbed, in spite of official denials, by wild rumours of a 'scorched earth policy' of burning the city before the Japanese enter. **1938** *Times* 21 Feb. 15/6 (*heading*) Scorched earth. **1941** *Hutchinson's Pictorial Hist. of War* 1 Oct.–23 Dec. 115 The Soviet have left nothing but scorched earth and derelict, burned-out buildings to the invaders. **1941** E. Snow *Scorched Earth* II. iv. 60 The 'scorched-earth' policy was credited to General Pai Tsung-hsi, the ablest strategist on Chiang's staff. **1945** *Daily Herald* 20 Apr. 1/3 A special 'scorched earth' order issued by Hitler. **1959** *Listener* 12 Nov. 818/2 Remember the scorched earth, too. How can Russia forget the menace of Germany? **1960** *Twentieth Century* July 63 The so-called concentration camps..were part of Kitchener's 'scorched earth' strategy. **1963** Wodehouse *Stiff Upper Lip, Jeeves* xvii. 135 The kitchen maid..always adopts the scorched earth policy when preparing a meal. **1976** H. Wilson *Governance of Britain* iii. 44, I was announcing in Parliament the discovery of documents envisaging an IRA offensive and virtual 'scorched earth' policy.

2. b. *Comb.*, as *scorched-looking* adj.

1970 T. Hughes *Crow* 11 Who owns this bristly scorched-looking face?

scorcher. Add: **3. a.** (Earlier examples.)

1842 R. W. Griswold *Let.* 7 Sept. (1898) 120 The review in The Examiner..is a 'scorcher'. **1869** 'Mark Twain' *Innocents Abroad* 453 Every time they read me a scorcher of a lecture I mean to talk back in print.

b. (Earlier and later examples.) *spec.* a very attractive girl or woman, a 'smasher'.

1881 *Punch* 29 Oct. 204/2 She was 'a Scorcher', was Lady O. **1898** [see *hot *a.* 6 c]. **1935** Wodehouse *Luck of Bodkins* ix. 88 When I'd had a look at the young lady next door and seen what a scorcher she is.

c. In *Sport*, an extremely fast shot or hit.

1900 *Dialect Notes* II. 57 *Scorcher*,..a swiftly batted ball. **1943** *Amer. Speech* XVIII. 104 A batter who hits a line drive (also called a *liner* or a *scorcher*..) is said to *line it out*. **1977** *Belfast Tel.* 14 Feb. 21/8 He..diverted a scorcher from Pat Spence later in the game.

d. Something licentious or risqué (esp. a book or play).

1942 Berrey & Van den Bark *Amer. Thes. Slang* §281/6 *Risqué joke or story*,..scorcher. *Ibid.* §590/4 *Scorcher, sexer*, a risqué play. **1974** P. Cave *Dirtiest Picture Postcard* ii. 12 Then he produced a scorcher which managed to get itself banned by the country's leading booksellers and nominated for public prosecution at the Old Bailey. **1978** *Morecambe Guardian* 14 Mar. 15/3 (*heading*) It's a scorcher at the Duke's. A scorching new show opened at Lancaster's Duke's Playhouse on Friday.

scorching, *ppl. a.*¹ Add: **1. c.** Applied *transf.* to a period of excessive sunshine and heat.

1940 'Gun Buster' *Return via Dunkirk* II. xix. 220 It is a pale, steady dawn, breaking with a slight haze that presages another scorching day. **1962** A. Wesker *Chips with Everything* I. i. 12 This hut..is going to be your home for the next eight scorching weeks.

2. *colloq.* Astounding, sensational; licentious, risqué; in *Sport*, of a shot or hit: exceedingly fast, 'blistering'.

1890 *St. Nicholas* Sept. 945/1 The first senior to the bat made first-base on a scorching grounder past third. **1896** A. Beardsley *Let.* c 20 Sept. (1970) 167 Your joke is charming and I shall do you some scorching drawings for No. 8 [of *The Savoy*]. **1897** *Referee* 24 Oct. 3/1 A said-to-be 'scorching' play entitled 'At the Foot of the Altar'. **1963** A. Ross *Australia 63* iii. 88 The two scorching catches by Cowdrey and Jarman. **1976** *Ilkeston Advertiser* 10 Dec. 18/2 Garbett scored with a scorching left foot drive. **1978** [see *scorcher 3 d].

3. quasi-*adv.*, in *scorching cold, hot.*

1873 E. Hooper *Nurseries & School Rooms* 197 The sand so scorching hot that one could not bear one's hand upon it. **1876** E. W. Heap *Diary* 8 Sept. in *Publ. Amer. Dial. Soc.* (1969) LII. 54 Another scortching cold morning. **1883** *Century Mag.* July 428/1 The sun was scorching hot and the shade chilly.

score, *sb.* Add: **I. 1. c.** *local.* 'A vertical indentation in a hill; a gangway down a cliff; a cutting through a ridge of hills' (E.D.D.); *spec.* in East Anglia, a narrow, steep path or street leading to the sea. Used esp. in place-names. Cf. ON. *skor* in sense 'a rift in a rock or precipice'.

The place-names *Syrithescore* and *Scourton* are recorded from the 13th century and c 1550 respectively (A. H. Smith, *Place-Names of East Riding of Yorkshire* (1937) 328; E. Ekwall, *Place-Names of Lancashire* (1922) 164).

1790 E. Gillingwater *Hist. Acct. Lowestoft* viii. 356 There are several of these passages in Lowestoft called scores, leading from the High-Street to the sea side, such as the Swan Score, Salter's Score, Rant's Score, &c. **1807** J. Grierson *Delineations of St. Andrews* iii. 104 That space of ground which is now converted into a public walk, and known by the name of the *Scores*. **1835** J. D. Carrick *Laird of Logan* II. 271 The hail place was in a perfect fizz.. frae the head of the Causeyside till the Score. **1858** *Hist. & Topogr. Handbk. Lowestoft* i. 3 On the land side are many narrow streets or lanes branching off into the country; whilst seaward there are, at short intervals, steep and narrow passages down the cliff, formed into steps, and leading to the *Denes*. These passages are known locally and technically as *Scores*. **1929** H. Meredith *E. Anglia* iii. 95 The Scores are Lowestoft's counterpart of Yarmouth's more famous Rows. **1958** *E. Anglian Mag.* Feb. 193/1 East Anglian cities and towns have each and all their picturesque narrow ways... The scores of Lowestoft have a unique character added to their picturesqueness in that they are steep as well as narrow. **1961** *Scottish Studies* V. 14 The Score is the downfall of the west edge of Edinburgh Castle.

6. c. *spec.* (A piece of) music composed for a film; the musical part of the sound-track of a film; formerly, the background music and effects of a silent film.

1927 *Kinematograph Year Book* 32 Scores to films can be recorded by the world's greatest orchestras, under the baton of conductors impossible to obtain for motion picture houses. **1935** R. Spottiswoode *Grammar of Film* v. 191 The score composed by Edmund Meisel for *Ten Days*. **1957** Manvell & Huntley *Film Music* i. 23 Among the more celebrated film scores..are those by Edmund Meisel for Eisenstein's films *The Battleship Potemkin* and *October*. **1965** *Movie* Summer 40/2 Jerry Bresler had re-edited the film..adding a score that was far from Peckinpah's choosing. **1976** R. Sanders in D. Villiers *Next Year in Jerusalem* 212 Irving Berlin..wrote the best over-all score of his career, *Annie Get Your Gun*.

II. 10. c. In *colloq.* phrases, as *to go over the score* (chiefly *Sc.*): to act (esp. drink) immoderately; *to have a few* (see *few *a.* 2 f) *over the score*: to drink more at one time than one should.

1768 A. Ross *Fortunate Shepherdess* II. 100 She thinks ye hae ga'en o'er the score. **1851** W. Anderson *Rhymes, Reveries, & Reminiscences* 50 Lest some o' the nickums should gang owre the score. **1915** J. L. Waugh *Betty Grier* 157 He gangs fairly ower the score baith wi' drinkin' himsel' an' treatin' ithers. **1951** N. M. Gunn *Well at World's End* xviii. 145 'You know how, when you have had a few over the score and you may not trust your legs, your brain remains as clear—' 'I know,' said Peter.

11. b. Also *to pay off, rub out*, etc., *old scores.*

1787 'P. Pindar' *Ode upon Ode* (ed. 5) 25 A pretty Way of rubbing out old Scores! c **1863** T. Taylor *Ticket-of-Leave Man* III. 56 There's the satisfaction of doing one's duty.. but there's something better than that...Paying off old scores. **1913** E. Phillpotts *Widecombe Fair* xxx. 236 This evening..promised good opportunity to pay off old scores. **1918** L. Strachey *Eminent Victorians* 67 The old scores, they found, were not to be paid off, but to be wiped out.

14. c. *transf. Psychol.* A numerical record of the marks allotted to individuals in the measurement of abilities, capacity to learn, or in the assessment of personality.

1910 E. L. Thorndike in *Amer. Jrnl. Psychol.* XXI. 485 (*caption*) Scores reduced to single variables by allowance for examples wrong. **1929** F. N. Freeman in C. Murchison *Found. Exper. Psychol.* xviii. 722 These two measures..do not give the same learning curve, or the same curve when the scores are plotted by ages or grades. **1951** T. Hunt in J. S. Gray *Psychol. in Use* (rev. ed.) x. 421 This test underwent extensive validation by study of the relationship between the test scores of students and their subsequent performance in the medical schools. **1977** P. J. Dunham *Exper. Psychol.* ix. 240 A score of 10 representing a very anxious individual.

d. *fig.* The essential point or crux of a matter; the state of affairs, the (present) situation; how matters stand; the full facts (*about, on*, etc. someone or something); freq. in phrases, as *to know the score*; *to ask, realize*, etc., *what the score is*; *what's the score*; etc. *colloq.*

1938 D. Nowinson in *Better English* Oct. 8/1 Dope..a guy who doesn't know the score. **1939** *Time* 16 Oct. 101/2 But when Holger begins to long for home and daughter, Anita, realizing what the score is, runs off to Paris to study. **1948** G. H. Johnston *Death takes Small Bites* i. 16 Why don't you speak to some people who can really tell you the score? **1950** E. Hemingway *Across River* xxxiii. 250 It leaves a core of certain un-killed characters who know what the score is. **1953** W. Burroughs *Junkie* xii. 121, I asked Ike what the score was on pushing in Mexico city. He said it was impossible. **1958** P. Kemp *No Colours or Crest* iv. 68 You were fully justified in breaking off the action when you did, in view of the score at the time. **1958** P. Scott *Mark of Warrior* II. 176 'What's the score about Havildar Baksh?' 'He's a prisoner.' **1959** N. Mailer *Advts. for Myself* iii. 234, I was out of fashion and that was the score; that was all the score. **1962** J. D. Salinger *Franny & Zooey* 167 You've been around schools long enough to know the score. **1971** N. Stacey *Who Cares?* xvii. 284 At least he had the courage to tell me the score as far as I was concerned, so that I did not waste time yearning and hoping. **1977** A. Scholefield *Venom* III. 86 You didn't ask the Boss what the score was, he told you.

15. c. The money or goods obtained by means of a successful crime. *Criminals' slang.*

1914 in Jackson & Hellyer *Vocab. Criminal Slang* 74. **1930** D. Runyon in *Collier's* 13 Sept. 7/4 We have a business proposition for Big Butch. It means a nice score for him. **1936** [see *have *v.* 14 i]. **1956** H. Kurnitz *Invasion of Privacy* xi. 76 He's just a few months out of the jug and he hasn't turned a trick or made a score anywhere. **1977** *New Yorker* 22 Aug. 38/1 A million dollars from a computer crime is considered a respectable but not an extraordinary score.

d. The action or process of obtaining a supply of narcotic drugs; a supplier of narcotic drugs. Cf. *score *v.* 16 d. *slang* (orig. *U.S.*).

1951 [see *hit *sb.* 1 b]. **1953** W. Burroughs *Junkie* x. 97 'It's hard to find a score now,' I said. 'Most of them have gone away.' **1976** Deakin & Willis *Johnny go Home* ii. 47 The whole day passes..going from fix to score, to ripping off enough money to support the habit.

e. A prostitute's client (cf. *score *v.* 16 f); also in homosexual use. *slang.*

1961 J. Rechy in *Evergreen Rev.* July–Aug. 15, I could spot the scores easily—the men who paid other men sexmoney. **1969** *Jeremy* I. III. 23/1 The boy will then deliberately reveal and manipulate his erect penis, thereby exciting the score. **1972** G. Baxt *Burning Sappho* ix. 153, I..got my hot tail out of there. I heard the score yelling. **1976** 'Trevanian' *Main* iv. 66 She won't be able to make a score until dark, if then.

III. 20*. *Criminals' slang.* **a.** Twenty dollars; a twenty-dollar bill. *U.S.* **b.** Twenty pounds sterling (esp. in banknotes).

1929 G. L. Hostetter *It's a Racket!* 237 *Score*, twenty dollar bill, or units thereof—hundred, two hundred. **1933** G. Ingram *Stir* xiv. 231, 'I got about £10 out of the first, then £2 and then another "score".' 'That makes £32.' **1941** *Coast to Coast 1941* 225 They only owe me a couple of quid since Christmas now. I was holdin' a score but I dropped most of it. **1958** F. Norman *Bang to Rights* III. 152 When they turned me over I had about a score on me. **1979** K. Bonfiglioli *After you with Pistol* vii. 39 You'll have to give me a score to buy an old throwaway shooter.

IV. 21. (sense 15 d) *score dough* (*dough *sb.* 2 b), *money*; **score-bid** *Contract Bridge*, a bid by a player whose side has a part-score, sufficient to give his side game; **score-board,** (*b*) also *gen.*, a master board displaying the score of any contest; also *fig.* and *attrib.*; **score-book** (earlier and later examples); **score-box** *Cricket*, a room or hut in which the official scorers work and (usu.) the telegraph is operated; **score-card,** (*a*) (earlier and later examples); also in extended uses, esp. a card issued to a competitor before a contest, on which his score (or that of his opponent) is to be recorded, or one held by a referee or judge for the same purpose; (*d*) *U.S. slang.*, a menu; **score draw,** a non-goalless draw (Draw *sb.* 5) counting for three points on a football-pool coupon; **scoreline,** (a line, or part of one, in a newspaper, etc., giving) the intermediate or final score in a sports contest between two persons or teams; **score-reading,** the action or process of reading a musical score; hence *score-reader*; **score-sheet** (earlier and later examples); also *transf.* and *fig.*, esp. in phrases, as *to add one's name to the score-sheet*, to score a goal (in Association Football and the like).

1928 M. C. Work *Contract Bridge* (ed. 2) IV. 76 If my side has a contract score of 60, I must put a construction on my partner's minor two bid different from the construction put upon such a bid at no score... 'Score-bids' are exceptions to the general rules. **1936** F. D. Roosevelt in *N.Y. Herald Tribune* 2 Oct. 10/2 From where I stand it looks as if the game was pretty well in the bag... It's just plain scoreboard arithmetic... Now, when the present management of your team took charge in 1933 the national scoreboard looked pretty bad. **1963** J. Joesten *They call it Intelligence* I. v. 51 What kind of record has the CIA?.. The scoreboard: 'Soviet satellites—Excellent'... 'Missiles —Good.' **1977** *Rolling Stone* 13 Jan. 43/3 There were shouts of delight as Texas lit up in red on one of the network's scoreboards, but it was still a close race. **1977** J. Laker *One-Day Cricket* 72 A few narrow escapes kept the score-board officials busy. **1851** J. Pycroft *Cricket Field* iv. 69 'Seventy-two runs,' said Fennex, and the score-book attests his accuracy, 'was Beldham's first and only innings.' **1921** P. F. Warner *My Cricketing Life* vi. 111 Sixteen centuries stand to his credit in the Middlesex score-book. **1977** J. Laker *One-Day Cricket* 113 Gone are the days of the old green bound Club scorebook. **1890** in W. A. Bettesworth *Walkers of Southgate* (1900) xvi. 335 Pressmen were expected to..keep running to the score-box to ask for any information they required. **1934** W. J. Lewis *Lang. Cricket* 226 Underneath (the score-box was) a room for the printers. **1877** C. Box *Eng. Game of Cricket* xxvi. 459 *Score card*, a printed card, with the names of the players and the results of each person's innings. **1918** E. S. Farrow *Dict. Mil. Terms* 538 *Score cards*, pasteboard cards issued to competitors at competitions, giving the number of the target of each competitor firing,..and containing a blank space for the record of the shots fired. **1930** J. Dos Passos *42nd Parallel* 160 He handed her the menu. 'Here's the scorecard.' **1958** *People* 4 May 19/7 How much is a quarter of a point worth on a fight referee's score-card? **1976** *Cumberland & Westmorland Herald* 4 Dec. 13/6 The other [*sc.* dart-players'] score cards were not in at the time of writing. **1978** *Cornish Guardian* 27 Apr. 23/5 (Advt.), When you call in at our showrooms and test drive the Austin Morris range, we'll provide a detailed scorecard. First test our cars then try to match them against the competition. **1942** Berrey & Van den Bark *Amer. Thes. Slang* §510/1 *Connection* or *score dough*, the price of a 'bindle' of narcotics. **1970** *Sporting Life* 2 Nov. 12 Percentage is based on three points for a correct score draw and two for a correct no-score match against the total number of points possible. **1977** *Daily Mirror* 15 Mar. 27/1 Plan 6..guarantees a line of at least seven score draws if any eight of your selections result as score draws. **1969** B. James *England v Scotland* iii. 64 The score line was a far from accurate guide to the run of play. **1971** *Rand Daily Mail* 27 Mar. 23/6 Had Arcadia grabbed their chances the scoreline could have been reversed. **1977** *Sunday Times* 9 Jan. 30/6 It was only when he..scored three times, that the scoreline became slightly more respectable. **1953** W. Burroughs *Junkie* vi. 61 Nick had just arrived at my apartment with some score money when I was called to the hall phone by the buzzer. **1946** *Penguin Music Mag.* Dec. 75 Music does not exist until it is *performed*, whatever our armchair score-readers may say to the contrary. **1961** J. A. MacGillivray in A. Baines *Musical Instruments through Ages* 247 Music is written for the player, not for the score-reader. **1909** *Cent. Dict.* Suppl. II. 1183/2 Score-reading. **1931** G. Jacob *Orchestral Technique* i. 4 To facilitate score-reading we give below the English, Italian, French, and German names for the instruments. **1977** *Listener* 23 June 822/2 Score-reading involves two quite different activities. First, you must learn to read music... The second element..is the ability to hear in imagination, in the mind's ear. **1859** in W. A. Bettesworth *Walkers of Southgate* (1900) v. 54 (*plate*) 'Bell's Life in London' Score Sheets, &c. &c., may be had at the Tent. **1944** W. W. Elton et al. *Guide Naval Aviation* iv. 73 Dive bombers caused much of the Jap grief at Midway, where the score sheet revealed four Jap carriers sunk and other craft damaged and sunk. **1976** *Cumberland & Westmorland Herald* 4 Dec. 12/6 Ullswater managed to keep the score sheet blank up to half-time. **1976** *Norwich Mercury* 10 Dec. 8/3 They..made sure of the points when Stew Reynolds added his name to the scoresheet.

score, *v.* **I. 5.** For *U.S.* read *N. Amer.* and add earlier and later examples. Now esp. used in newspaper headlines.

1812 J. K. Paulding *John Bull & Brother Jonathan* xiv. 102 She..fell upon Beau Napperty, and scored him at such a rate, that if poor Beau had heard her, he would have been mad enough I warrant you. **1912** J. Sandilands *Western Canad. Dict. & Phrase-Bk.* s.v. *Scored*, An Opposition newspaper came out with the heading 'Government Legislation Scored'. **1930** *Publishers' Weekly* 8 Mar. 1331/2 (*heading*) Smoot's secret session scored. **1967** *N.Y. Times* (Internat. ed.) 11–12 Feb. 3/3 (*heading*) Professor scores Reagan.

II. 9. c. To write the score for (a film). Cf. *score *sb.* 6 c.

1934 Webster, *Score*,..to add music to a picture that already has sound effects. **1967** H. Harrison *Technicolor Time Machine* (1968) xv. 156 'Is it true you scored a couple of films?' 'It is true I did the music for a ragged piece of class-X crap.' **1969** *Daily Progress* (Charlottesville, Va.) 15 May 1-c/7 Poet-singer-composer Rod McKuen has scored three movies.

III. 13. b. *Biol.* and *Med.* To examine (experimentally treated cells, bacterial colonies, or the like), making a record of the number showing some character.

1964 *Virol.* XXIII. 118/1 Subconfluent monolayers were infected with 0·5 ml of virus and transferred the following day at an inoculation density of 100 and 500 cells per plate. Transformed colonies were scored 14 days later. **1971** *Nature* 20 Aug. 559/1 After 2–3 weeks the plates were fixed, stained and the colonies scored.

15. d. *trans. Baseball.* To cause (a team-mate) to score.

1912 C. Mathewson *Pitching in Pinch* v. 109 Schlei made a base hit..and scored both men. **1976** *Billings* (Montana) *Gaz.* 17 June 2-B/4 Mark Belanger singled to score May before DeCinces' fourth home run of the season.

e. *Psychol.* To obtain (results in a test designed to measure abilities, capacities, or personality traits); to record results in (a test). Also *intr.* Cf. *score *sb.* 14 c.

1922 *Jrnl. Experim. Psychol.* V. 101 College students scoring 88 to 195 in the first trial. **1952** A. G. Wesman in N. E. Gronlund *Readings in Measurement* (1968) xx. 201 It is important to know the extent of agreement between the persons who score them [*sc.* tests]. *Ibid.*, Such a correlation coefficient yields important information—it tells us how objectively the test can be scored. **1968** P. McKellar *Experience & Behaviour* xi. 277 Engineers tend to score highly on the economic (applied science) value trait. **1977** P. J. Dunham *Experim. Psychol.* ix. 240 We will not deal with the details of how the TAQ is scored. *Ibid.*, The calm collected type of person who would score around 1 on the TAQ measure.

16. a. *to score a point* (or *points*) *off* (a person) = *to score off* (see sense 16 b in Dict.).

1956 R. Braddon *Nancy Wake* xiii. 140 Fournier was ecstatic with pride and pleasure—and with delight at having scored a point off Gaspard! **1957** *Practical Wireless* XXXIII. 558/1 The episode I heard, 'Rumour is a Lying Jade', proved very amusing, with both stars scoring points off each other with satisfactory frequency.

c. *trans.* and *intr.* To make a (freq. dishonest) gain; spec. *Criminals' slang*, to commit a theft or robbery; to steal, filch, or purloin (something), esp. from an open counter or display. orig. *U.S.*

1914 Jackson & Hellyer *Vocab. Criminal Slang* 74 *Score*,..to successfully negotiate; to 'make a touch'... 'We scored seven times in the same joint by ringing up,' i.e., disguising. **1926** J. Black *You can't Win* xiv. 191 [The thief] throws a few dollars on the bar just to..let them guess where he 'scored' and how much he got. **1930** [see *play *sb.* 10 g]. **1942** Berrey & Van den Bark *Amer. Thes. Slang* §490/8 *Steal*,..salvage, score, shark. **1972** *Last Whole Earth Catalog* 49/3 She was already plotting in her mind to stash part of their supper in her bag so they'd have something to eat the next day. She'd already scored a can of beer and a handful of cashew nuts. **1976** D. Topolski *Muzungu* vi. 99, I spotted a sugar factory, drove in, and scored a couple of kilos. **1977** D. Mackenzie *Raven & Kamikaze* xii. 146 'Where did you get it [*sc.* a newspaper]?' ..'Nicked it... It was too early to score any milk.'

d. *intr.* and *trans.* To buy or otherwise obtain a narcotic drug; by extension, to take a narcotic drug. *slang* (orig. *U.S.*). Cf. *score *sb.* 15 d.

1935 A. J. Pollock *Underworld Speaks* 101/2 *Scored*, made a purchase of dope. **1953** W. Burroughs *Junkie* 9 Junk wins by default. I tried it as a matter of curiosity. I drifted along taking shots when I could score. **1959** *Alfred Hitchcock's Mystery Mag.* Feb. 68/1 To get you out of my pad I'll let you score for a low, low forty. **1969** *Guardian* 3 Dec. 9/1 She had needed the money to score H up in the West End. **1972** J. Brown *Chancer* ii. 30 The weekend ravers and joy-poppers..who maybe score half a pill of H for kicks. **1972** *Daily Tel.* 25 Feb. 17/3 Mick the Pimp asked me if I wanted to 'score' and gave me a tablet from a matchbox and I gave him £1. **1977** *It* June 18/1 (*caption*) I can score better shit in Hendon on an off night!

e. *intr.* Of a racehorse: to win a race.

1941 *Sun* (Baltimore) 13 June 21/1 Their choice scored by a head from Epindel. **1977** *Evening Gaz.* (Middlesbrough) 11 Jan. 13/4 Shifting Gold had gone on to score again in the Tote Northern Chase at Haydock Park.

f. *intr.* and *trans.* Of a man: to achieve intercourse (*with* a woman); to have (casual) intercourse with (a woman); also occas. of a prostitute: to obtain (a client). *slang* (chiefly *U.S.*).

1960 R. G. Reisner *Jazz Titans* 164 *Score, to*, to attain success, to get what you want. Example: I scored with that chick. **1961** J. Rechy in *Evergreen Rev.* July–Aug. 19 You wanna score?.. See that old cat over there... He wants us both to come over to his house. **1970** G. Greer *Female Eunuch* 249 The boys used to go to the local dance halls and stand around..until the..sexual urge prompted them to *score a chick*. **1973** W. H. Canaway *Harry doing Good* I. 36 They might begin to ball later on... He would like to score with the Cheryl chick. **1976** D. Craig *Faith, Hope & Death* ix. 42 They talk about 'taking' a woman... Or, 'Did you score last night?'—like some great goal, scheming and forcing. **1976** 'Trevanian' *Main* (1977) ii. 39 He feels particularly sorry for the whores..who can only score drunks.

scored, *ppl. a.* Add: **2.** *Mus. rare.* Of a film, or part of a film: provided with a score (see *score *sb.* 6 c); of a piece of music: written down in a score.

1957 Manvell & Huntley *Technique Film Music* i. 21 Cueing was unusually elaborate and some of the more closely-scored scenes began to approach the techniques later developed in sound film recording. **1962** *Observer* 27 May 27/1 By the late 1920s Henderson was learning how to blend scored passages into a succession of solos.

scoreless, *a.* Add: **2.** Also, of a game, a period of play, etc.: from which no score results; involving no score. Phr. *to hold* (a person or team) *scoreless*: to prevent (a stronger opponent) from scoring.

1961 *Sun* (Baltimore) 18 Mar. 18/4 Hartman..allowed no hits in his scoreless three-inning appearance. **1972** J. Mosedale *Football* iii. 41 The team that had held them scoreless was weak against counterplays. **1974** *Sumter* (S. Carolina) *Daily Item* 23 Apr. CA/1 Bill Paschall.. stretched his scoreless string to 23 innings. **1977** *Arab Times* 13 Dec. 9/1 N.C.C. started the first inning against Foster Wheeler with two runs but then were held scoreless throughout the remainder of the game. **1978** *Rugby World* Apr. 25/1 France..have never held Wales to a scoreless draw in the championship.

scorer. Add: **3. a.** (Earlier example.)

1732 *Applebee's Orig. Weekly Jrnl.* 16 Sept. 3/3 There wanted six Minutes of the Time by the Scorer's Watch.

b. One who allots marks or records the scores obtained in the measurement of ability, capacity, or personality traits.

1922 *Jrnl. Exper. Psychol.* V. 107 Provision must be made to free the results from the personal equations of the scorers. **1952** A. G. Wesman in N. E. Gronlund *Readings in Measurement* (1968) xx. 201 The scorer is required to make a judgment as to the correctness or quality of the response. *Ibid.*, Two scorers should agree perfectly.. in assigning scores.

4. b. A winner, esp. of a horse-race. Cf. *score *v.* 16 e.

1974 *Marlboro Herald-Advocate* (Bennettsville, S. Carolina) 18 Apr. 5/7 After dinner six tables of bridge were played. Scorers for the ladies, Mrs. Richard Fletcher..and Mrs. Robert Lockey. **1976** *Scottish Daily Express* 27 Dec. 10/3 I'm Alright Jack..a smooth Devon and Exeter scorer last month, can put up a repeat performance in the opener at Newton Abbot.

5. One who scores (i.e. composes a score for) a film. Cf. *score *v.* 9 c. *rare.*

1969 *Daily Progress* (Charlottesville, Va.) 15 May 1-c/6 The talkies brought a demand for 'scorers', composer-orchestrators of background music to enhance scenic mood.

scoring, *vbl. sb.* Add: **1.** (Further examples.)

1801 T. Busby *Dict. Mus.*, *Scoring*, the art of forming a score by collecting and properly arranging under each other the several detached parts of any composition. **1851** C. Box *Cricketer's Man.* (ed. 5) 51 Printed forms..for scoring are not procurable. *Ibid.* 53 The annexed score will serve to illustrate the principles of scoring. **1876** Geo. Eliot *Dan. Der.* I. II. xi. 201 Shall we go now and hear what the scoring says? **1904** *Daily Chron.* 21 Nov. 7/2, I viewed the match from the pavilion behind the goal where all the scoring was done. **1922** *Jrnl. Experim. Psychol.* V. 107 The scoring requires judgment. **1940** *Chambers's Techn. Dict.* 747/2 *Scoring*,..the preparation of the sound-script, in which are described all music and noises to be introduced into a motion-picture. **1952** A. G. Wesman in N. E. Gronlund *Readings in Measurement* (1968) xx. 201 Many objective tests..are not very reliable, yet the scoring is by definition objective. **1967** *Daily Mirror* 14 Aug. 9/1 This buying of heroin ('scoring' in junkie language) is being watched closely by London Drug Squad detectives. **1978** *Time* 3 July 45/1 Warren stays with a picture through editing, mixing and scoring.

3. scoring block *Card-playing*, a pad of printed score-sheets; **scoring board** (earlier and later examples); **scoring-book** (earlier example); **scoring-booth, -box** *Cricket* = *score-box* s.v. *score *sb.* 21; **scoring-card** = *score-card* (*a*) s.v. Score *sb.* 21 in Dict. and Suppl.; **scoring-paper,** (*a*) = *score-paper* s.v. Score *sb.* 21; (*b*) *Mus.* printed paper on which a musical score may be entered; **scoring-sheet** (earlier example).

1907 *Yesterday's Shopping* (1969) 375/2 *Table*..with drawer divided to take cards and bridge scoring blocks, whist cards and markers, piquet scoring blocks and cards. **1933** G. D. H. & M. Cole *End of Ancient Mariner* ii. 18 On the table stood a decanter, flanked with packs of cards and scoring-blocks. **1882** *Bell's Life* 20 May 5/1 An excellent novelty was exhibited on the ground during the Oxford Match, and that was a patent scoring board. **1904** P. F. Warner *How We recovered Ashes* iv. 60 The Melbourne Cricket Ground, with its..huge scoring-board. **1851** F. Lillywhite *Guide to Cricketers* 71 'Lillywhite's Registered Scoring Book' will be found extremely useful to Clubs, for the purpose of keeping the exact averages of all the members. **1867** F. Gale *Public School Matches* 29 And now let us go to the scoring booth and get a 'true and correct score to the end of the first innings'. **1960** *Cricketer Spring Annual* 58/2, 1851 A scoring booth was on view in the Great Exhibition. **1877** C. Box *Eng. Game Cricket* xxvi. 459 *Scoring box*, a small enclosure, so situated as to command a full view of the play. **1908** W. E. W. Collins *Leaves from Old Country Cricketer's Diary* xii. 201 When..I passed the scoring-box, *en route* for the pavilion, I found the small telegraph boy in the act of returning my score as sixty-nine. **1891** W. G. Grace *Cricket* iii. 94 We kept in constant touch..posting the scoring-card at the end of every day's play. **1907** *Yesterday's Shopping* (1969) 386/1 Whist scoring cards. **1912** A. A. Lilley *Twenty-four Years of Cricket* 124 No scoring-card could be invented to equal it in its completeness. **1851** C. Box *Cricketer's Man.* (ed. 5) 53 Some clubs make it a rule to mark the number of 'overs' that each bowler gives, at the foot of the scoring papers. **1851** F. Lillywhite *Guide to Cricketers* 79 Scoring Papers, per dozen... 2/6. **1908** G. B. Shaw *Let.* 27 May (1972) II. 788 You will only waste a good deal of scoring paper which you might employ far better by trying to deal, as Strauss does,.. with the modern world in a crisp and powerful style. **1859** in W. A. Bettesworth *Walkers of Southgate* (1900) v. 54 (*caption*) Scoring Books and Sheets.

scorp (skǭɹp). *Mil. slang.* Abbrev. of Scorpion (sense 7); an inhabitant of Gibraltar. Also *Rock-scorp.*

1912 *Jock Scott, Midshipman: His Log* iv. 32 By 'scorps' he meant rock scorpions, the name bestowed on the inhabitants of Gibraltar. **1957** W. Tute *Rock* I. 16 Millingham.. married a Rock Scorp. *Ibid.* 19 Perks and privileges for the ruling classes. Fifteen in a room for the poor-quality 'Scorps' whose Rock it was. **1973** *Publishers Weekly* 17 Sept. 59/3 Covering the rock's social strata, from the native 'scorps' to the British Governor.

Scorpio. Delete ‖, for *Astr.* read '*Astr.* and *Astrol.*' and add: **2. a.** *attrib.* or as *adj.*, born under or ruled by the sign of Scorpio (24 Oct.–21 Nov.).

1894 E. Kirk *Influence of Zodiac upon Human Life* xviii. 156 When Scorpio people live on the higher plane, they are very superior individuals. **1901** M. Mayo *Our Fate & Zodiac* 104 The..astute Scorpio man is..clever in..taking advantage of the upward revolutions of the wheel. **1930** W. Wilson *Astrology* iii. 60 Scorpio people are often found devoting themselves to art. **1964** L. MacNeice *Astrol.* iii. 95 Some modern Scorpio types excel at skin diving. **1970** 'D. Halliday' *Dolly & Cookie Bird* iii. 23 He was Scorpio; I asked him.

b. A person born under the sign of Scorpio.

1968 T. Wolfe *Electric Kool-Aid Acid Test* i. 5 Black Maria, a Scorpio herself, rummages through the Zodiac. **1972** *Guardian* 15 Jan. 9/4 He and Rommel and Montgomery all shared the same birth date, November 17, and all were Scorpios. **1976** M. Millar *Ask for Me Tomorrow* xv. 122, I thought Scorpios were supposed to be creative.

Hence **Sco·rpian** = *Scorpio 2 b.

1951 M. E. Hone *Mod. Text Bk. Astrol.* iv. 68 Just as this rulership brought energy and initiatory force to the Arien, it brings it to the Scorpian. **1980** R. Rendell *Lake of Darkness* i. 7 Scorpians are magicians, astrologers, alchemists, surgeons, bondsmen and undertakers.

scorpion. Add: **1. c.** (Later example.)

1924 R. Campbell *Flaming Terrapin* iii. 42 But life, a scorpion of tenacious hold, Fastened upon their spirits.

8. a. *scorpion-kiss.*

1961 R. Graves *More Poems* 33 Lady Morphia—Her scorpion kiss and dark gyrating dreams.

b. *scorpion whip* (earlier example).

1824 Lady Blessington *Jrnl.* May in E. Clay *Lady Blessington at Naples* (1979) 104 Byron..was..lashed into satire by the scorpion whips of envy.

c. scorpion orchid, an orchid belonging to the genus *Arachnis*, esp. *A. flos-aeris*, native to Malaysia; = *scorpion-plant* (*a*).

1897 H. N. Ridley in *Jrnl. Straits Branch Roy. Asiatic Soc.* xxx. 68 *Bunga Kasturi. Renantha moschifera* Lindl. (*Orchideae*). 'Musk-flower.' The scorpion orchid. **1937** M. Covarrubias *Island of Bali* x. 336 Their garden is filled with golden flowers that grow side by side with the pandanus, the scorpion orchids, the..pineapples. **1961** A. D. Hawkes *Orchids* 108 The multicolored flowers are generally produced in considerable numbers, and their strange form has given them the common name of 'Scorpion Orchid'. **1971** *Ceylon Observer Mag.* 19 Sept. 2/6 (Advt.), Epidendrums & Scorpion Orchids, several colours.

scorpionic, *a.* Add: **2.** *Astrol.* (With capital initial.) Of, pertaining to, or characterized by the sign of Scorpio.

1924 C. E. O. Carter *Conc. Encycl. Psych. Astrol.* 145 Scorpionic afflictions often cause nasal obstructions, especially when the mutable element is prominent in the horoscope. **1972** *Mainichi Daily News* (Japan) 6 Nov. 12/1 Today's natives are not truly 'Scorpionic' in nature as each reflects unique qualities.

scortatory, *a.* (Later examples.)

1922 Joyce *Ulysses* 199 Twenty years he dallied there between conjugal love and its chaste delights and scortatory love and its foul pleasures. **1942** Dylan Thomas *Let.* 30 Aug. (1966) 259, I hope..that the Monico made up..for the absence of one ventripotent scortatory Krut.

scorzalite (skǭ·ɹzălait). *Min.* [f. the name of E. P. *Scorza* (b. 1899), Brazilian mineralogist + -ite¹.] A basic phosphate of aluminium, ferrous iron, and magnesium, $(Fe^{2+},Mg)Al_2(PO_4)_2(OH)_2$, that forms an isomorphous series with lazulite and occurs in masses of brittle, blue, monoclinic crystals.

1947 Pecora & Fahey in *Bull. Geol. Soc. Amer.* LVIII. 1217 The new minerals are named in honour of Dr. Evarists Scorza and Dr. Antonio José Alves de Souza... Scorzalite is a massive, blue hydrous iron magnesium aluminum phosphate. **1949** *Amer. Mineralogist* XXXIV. 83 The Corrego Frio pegmatite, Minas Gerais, Brazil,..has yielded three new phosphate minerals since its discovery in 1942. Two of these new minerals, scorzalite and souzalite, are described in this paper. **1975** *Fortschritte der Mineral.* LII. Suppl. 288 Scorzalite from the Angarf-Sud pegmatite [in Morocco] is always observed intimately associated with muscovite.

Scot, *sb.*¹ Add: **4.** (Further example.) Also as *adj.*

1823 'J. Bee' *Dict.* 155 *Scot*, a butcher's designation of a fractious man, the small Scots oxen coming to their doom with little resignation to fate; indeed, all animals try harder to retain life than man. **1916** W. Owen *Let.* 9 Dec. (1967) 417 Major Melville, a snotty, acid, scot, impatient, irritated wretch.

Scotch, *a.* and *sb.*³ Add:

In recent years the word *Scotch* has been falling into disuse in England as well as in Scotland, out of deference to the Scotsman's supposed dislike of it; except for certain fixed collocations, *Scottish* (less frequently *Scots*) is now the usual adjective, and to designate the inhabitants of Scotland the pl. sb. *Scots* is preferred (see Gowers/Fowler *Mod. Eng. Usage* (1965)).

A. *adj.* **1. a.** (Later examples.)

1943 *Sun* (Baltimore) 25 June 12/7 My father came from Invernesshire and certainly never restricted the use of Scotch to the whiskey. It is only in recent years that certain Anglo-American friends have made me feel guilty of committing a particularly bourgeois *faux pas* by using the word. We always looked on Scottish as rather affected, overly poetic. **1976** *Times* 11 May 15/3 Professor Trevor Roper.. tries to irritate and provoke by using the word 'Scotch' knowing well that many decent Scots..have come to regard this as a demeaning adjective.

d. (Later examples.)

1906 E. Dyson *Fact'ry 'Ands* xvi. 209 Well, he's touched me three times in a week, and I'm as Scotch as most people. **1912** J. N. McIlwraith *Diana of Quebec* iii. 49 'You would be the first to throw me a penny..?' 'A sixpence, truly, if he be not too Scotch,' said Nelson. **1932** *Amer. Speech* VII. 403 Mr. —, if you weren't so Scotch we could have a good time here.

e. (Earlier example.) Also, with similar connotation, *Scotch sister.*

1861 J. S. Mill *Repr. Govt.* xviii. 335 The most important offices would be thrown to Scotch cousins and adventurers. **1866** Mrs. Gaskell *Wives & Daughters* I. xix. 219 She called her a sister, but whether it was a Scotch sister, or a sister *à la mode de Bretagne*, would have puzzled most people.

2. a. (Further examples); *Scotch ale* (later examples), *ballad, carpet* (earlier and later examples), *reel, salmon, snuff, whisky* (earlier and later examples).

1733 *S. Carolina Gaz.* 7 Apr. 4/2 (Advt.), To be sold..cut Tobacco, Scotch Snuff, and Pigtail. **1774** Scotch carpet [see Wilton]. **1792** F. Burney *Jrnl.* May (1972) I. 153 Miss Cooper..gave me a relation of her having been..at Mrs. Broadhead's masquerade as *Jenny*, from the Scotch ballad. **1835** Dickens *Sk. Boz* (1837) 2nd Ser. 39 Mr. Thomas Potter ordered the waiter to bring two goes of his best Scotch whisky, with warm water and sugar. **1891** S. M. Welch *Home Hist.* 183 It was not uncommon to see a couple of portly old gentlemen meeting on the street offering their boxes of Maccaboy, Rappee or Scotch, each to the other. *Ibid.* 376 [In] the 'Scotch Reel',..each lad must needs have two lassies. **1893** T. Hardy *Let.* 6 Oct. in Hardy & Pinion *One Rare Fair Woman* (1972) 28 Lady J[eune].. played and sang at least a dozen Scotch ballads to me. **1953** *Word for Word* (Whitbread & Co.) 10/2 *Scotch ale*, a draught or bottled ale of the Burton type, brewed in Scotland. **1960** *Connoisseur's Handbk. Antique Collecting* 251/1 *Scotch carpets*, double-cloth or ply weavings for the floor, also known as Kidderminster or Ingrain. **1965** V. Canning *Whip Hand* xii. 143 We had..Scotch salmon with a cucumber salad. **1969** *Daily Tel.* 30 Apr. 29/7 New definitions of whisky..are contained in the Finance Bill, published last night... 'Scotch whisky' is to mean whisky which has been distilled in Scotland. **1978** J. Mann *Sting of Death* iii. 25 Alex would probably be ordering..gulls' eggs and Scotch salmon. **1980** R. Lewis *Certain Blindness* iii. 80 Parton put a pint of Scotch Ale in front of him.

c. (Further examples.)

1726 D. Eaton *Let.* 31 Dec. (1971) 87, I think the Scotch cattell were dearer than if they had been bought in our country [i.e. district]. I mean cattel of their size might have been bought in our markets for less mony. **1980** A. Bell *Sydney Smith* 117 Scotch sheep provided material for Sydney's only contribution to agricultural literature.

e. *Typogr.* The epithet of a variety of modern typefaces deriving from one sent from an Edinburgh foundry in 1837 to the printing firm of S. N. Dickinson in Boston, Massachusetts. So *Scotch-cut* adj.

1847 S. N. Dickinson *Hand-bk. Specimen Printing Type* Pref., Our Scotch Faces were selected from the very extensive Foundry of Alexander Wilson and Sons of Edinburgh and also from an eminent letter cutter of that city. *Ibid.*, The symmetry of the Scotch cut figures. a **1863** *Specimens of Borders, Ornaments, Rules, Dashes, etc. from Dickinson Type Foundry* (Boston, U.S.) Back cover, A superior collection of the Scotch-cut letter, so highly appreciated by the trade. **1900** T. L. De Vinne *Practice of Typogr.* vi. 212 As first made the Scotch-face was a small, neat, round letter, with long ascenders, and not noticeably condensed. **1922** D. B. Updike *Printing Types* II. xx. 193 As produced by Wilson it is a very handsome and serviceable letter, and in it we have another English type-family—the Scotch modern face. **1951** S. Jennett *Making of Bks.* xiv. 248 Bodoni is an excellent letter, but capable of great degeneration, and in Scotch Roman we see the degeneration commencing. **1966** H. Williamson *Methods Bk. Design* (ed. 2) viii. 99 The original Scotch faces were a vaguely defined class, and generalization about their letter-forms is impossible.

4. a. Scotch Baronial *a.* = *Scottish Baronial* s.v. *Scottish *a.* 5; **Scotch Blackface,** a sheep belonging to the breed so called, developed in mountain and moorland regions of Scotland and northern England, and distinguished by black legs and muzzle and long wool; **Scotch boiler** [so called from its having been introduced in Scottish shipyards], a fire-tube boiler in which combustion takes place inside the shell; **Scotch** († **barley**) **broth,** a soup containing meat, vegetables and pearl barley; **Scotch egg,** a hard-boiled egg enclosed in sausage-meat; **Scotch mist,** (*b*) something insubstantial, unreal, freq. used sarcastically in a retort or rhetorical question to imply that someone has imagined or failed to comprehend something; (*c*) a drink of whisky served with a twist of lemon; **Scotch pancake** = *drop-scone* s.v. SCONE 1; **Scotch peg,** rhyming slang for 'leg'; **Scotch prize** (earlier example); **Scotch terrier,** for '(see TERRIER)' read: a small stocky terrier of the breed so called, usually black or brindle, with thick, shaggy fur, erect, pointed ears and tail, and a square, bearded muzzle; formerly, a terrier belonging to one of several other Scottish races, now treated as separate breeds (cf. TERRIER[2]); **Scotch woodcock**: see WOODCOCK *sb.* 3 d; **Scotch yoke,** a mechanism by which a steady circular motion can be transformed into a linear simple harmonic motion, consisting of a crank bearing a peg which, as the crank revolves, slides in a straight slot constrained to move to and fro along a straight line in a plane at right angles to the plane of the slot.

1880 J. J. STEVENSON *House Archit.* I. xiv. 360 The Scotch 'Baronial' architecture, as it is called, resembles that of the Renaissance châteaux of France. **1931** E. SACKVILLE-WEST *Simpson* II. vii. 144 Salathiel held up a glass globe, inside which was a miniature Scotch-baronial castle in china. [**1888** J. & C. SCOTT *Blackfaced Sheep* i. 1 The origin of the Scotch blackfaced sheep is shrouded in mystery.] **1945** J. F. H. THOMAS *Sheep* ii. 30 It deserves the title Scotch Blackface because in that country it is of paramount importance. **1903** H. DE B. PARSONS *Steam-Boilers* v. 97 (*caption*) Scotch boiler, single-ended, with common combustion-chamber. **1966** *McGraw-Hill Encycl. Sci. & Technol.* VIII. 119/2 Later boilers are of the express or water-tube type, burning fuel oil, though Scotch boilers are still used in some cases and may burn either coal or fuel oil. **1747** H. GLASSE *Art of Cookery* vi. 65 (*heading*) To make Scotch barley broth. **1834** T. HOOD *Tylney Hall* I. xv. 175 We shall have an ounce of mutton swimming in a tureen of barley-water—I've heard of their Scotch broths. **1969** R. & D. DE SOLA *Dict. Cooking* 203/1 *Scotch broth*, rich soup made of beef or mutton and vegetables, thickened with barley. **1809** M. E. RUNDELL *New Syst. Domestic Cookery* (new ed.) VIII. 207 *Scotch eggs*. Boil hard five pullet's eggs, and without removing the white, cover completely with a fine relishing forcemeat. *c* **1965** A. CHRISTIE *Autobiogr.* (1977) XI. iv. 525 He fed us entirely on..'Scotch eggs'; excessively indigestible. **1972** P. D. JAMES *Unsuitable Job* i. 19 Pushing a half pint of shandy and a Scotch egg across the counter. **1977** D. WILLIAMS *Treasure by Degrees* xv. 145 A lonely Scotch egg..was the only visible justification for the plastic proclamation 'Snacks at the Bar'. **1943** HUNT & PRINGLE *Service Slang* 57 *Scotch mist*, sarcastic comment on your eyesight, inferring that you are seeing things. **1962** *New Statesman* 18 May 708/2 'Are yer married?' 'Course she is. What do yer think that is? Scotch mist?' Rube points to my wedding-ring. **1965** O. A. MENDELSOHN *Dict. Drink* 303 *Scotch mist*, cocktail of Scotch whisky and lemon peel. **1974** *Pacifist* Feb. 13/1 What are all these price-rises we are suffering now? Scotch mist? **1977** W. H. MANVILLE *Good-bye* iii. 27 You can start by ordering me a Teacher's Scotch Mist on the rocks. **1930** BENNION & STEWART *Cake Manufacture* xiv. 122 Soda scones, Scotch pan cakes, and milk scones,..can be baked on the hot plate. **1977** D. WILLIAMS *Treasure by Degrees* viii. 71 Tea..could be quite a different matter: He recalled some truly outstanding Scotch pancakes. **1857** 'DUCANGE ANGLICUS' *Vulgar Tongue* 17 *Scotch peg*, leg. **1917** W. MUIR *Observat. Orderly* xiv. 225 If he had occasion to allude to his leg he would probably have called it 'Scotch peg'. **1818** 'A. BURTON' *Johnny Newcome* III. 170 'Tis but a Scotch prize he has stolen! **1810** *Sporting Mag.* XXXVI. 61/1 Portraits of a poodle and Scotch terrier. **1847** H. D. RICHARDSON *Dogs* vii. 70 There are two varieties of the common Scotch Terrier. **1863** [see TERRIER[1] 1]. **1880, 1889** [see *Aberdeen terrier* s.v. *ABERDEEN 2]. **1927** E. C. ASH *Dogs* II. ii. 422 Two kinds of terriers are described—the rough-haired Scotch and the smooth-haired English. **1927** HAM & CRANE *Mechanics of Machinery* ii. 27 Figure 36 shows an application of the Scotch yoke as it has occasionally been used on small engines and steam pumps. **1959** KARPLUS & SOROKA *Analog Methods* (ed. 2) viii. 242 Mechanical Harmonic Synthesizers... The sine and cosine components are almost invariably generated by Scotch yoke mechanisms. **1966** *McGraw-Hill Encycl. Sci. & Technol.* I. 377/2 The modification of the Scotch yoke shown in Fig. 25 can be used to generate a tangent function over a limited range of the argument.

b. Scotch elm, substitute for def.: the wych-elm, *Ulmus glabra*; (earlier and later examples); **Scotch rose** (later examples).

1799 W. NICOL *Pract. Planter* i. 34 The Scotch Elm may with propriety be reared for this purpose [*sc.* ship-building] on thin gravelly soils. **1838** J. C. LOUDON *Arboretum et Fruticetum Britannicum* III. 1399 The Scotch elm has not so upright a trunk as the English elm. **1969** T. H. EVERETT *Living Trees of World* xiv. 131/1 The Scotch elm or wych elm..forms a rather open, broad-headed specimen. **1892** C. M. YONGE *Old Woman's Outlook in Hampshire Village* 161 The little thorny Scotch roses..are creeping over the cottages. **1972** *Country Life* 23 Mar. 695/1 Possibly the toughest of the wild roses is..the Scotch or Burnet Rose which flourishes in the sand dunes of Great Britain and Germany.

B. *sb.* **1. a.** (Earlier and later examples.)

1743 M. W. MONTAGU *Let.* 16 Sept. (1966) II. 310 Several Scotch pass here often. **1979** *Jrnl. R. Soc. Arts* Jan. 107/2 Since then I have had the greatest admiration for the education offered to, or seized by, the Scotch—Miss Murray's word and her grandfather's.

3. a. (Earlier and later examples.) Also = *Scotch ale* (in sense A. 2 a), *face* (*A. 2 e).

1778 S. FOOTE *Cozeners* III. ii. 76, I have a box of Scotch in my pocket. **1906** *Linotype Bull.* Oct.–Dec. 6/2 (*heading*) 8-Point Scotch. **1945** O. SIMON *Introd. Typogr.* iii. 12 The roman lower-case letters of Scotch and Baskerville..are wide and generous. **1962** S. CHAPLIN *Watchers & Watched* x. 199 The beer was as bitter as bile... 'Is there anythin' wrong?' 'Your Scotch doesn't taste too well.' **1964** S. JENNETT *Making of Books* (ed. 3) xiv. 251 Though Scotch is a portent, it is not itself as poor as its descendants became. **1966** H. WILLIAMSON *Methods Bk. Design* (ed. 2) viii. 99 The first Scotch to be cut for machine composition was produced by the Merganthaler Linotype Co. in 1902. **1976** *Observer* (Colour Suppl.) 1 Feb. 9/3 All Scotches are blended spirits: a mix of malt and grain whiskies. The ratio can vary between 2 to 3 and 3 to 2, and a bottle of Scotch can be a mixture of 30, 40 or even more whiskies. **1977** *Listener* 3 Mar. 275/1 You could order 'a pint of Scotch'—Scotch Ale, because it is the cheapest beer, is still the majority drink on Tyneside.

b. *ellipt.* for *Scotch peg* (see sense A. 4 a above).

1859 HOTTEN *Dict. Slang* 87 *Scotches*, the legs. **1962** R. COOK *Crust on its Uppers* ii. 30 Down to wearing my head in its proper place and not between my scotches like a sporran.

scotch, *v.*[1] Add: **2. c.** To refute conclusively or stamp out (a rumour, report, etc.); to frustrate (a plan or hope); to quash, destroy, bring to nothing. (Perh. influenced by SCOTCH *v.*[2] 1 b.)

1926 in H. W. Fowler *Dict. Mod. Eng. Usage* 518/2 The contradiction of a rumour affecting any particular company ..is seldom entirely scotched by directorial statements. *Ibid.*, We hope the proposal for a Government news service for the Colonies is finally scotched by the debate. **1947** H. S. GLADWIN *Men out of Asia* xxvi. 270 The question has been raised as to whether they [*sc.* helmets] were not copies from late European models, but this idea was scotched by Captain Cook who found them in fashion in Hawaii. **1955** *Times* 27 July 13/1 He did so with sufficient force..to scotch once and for all any lingering doubts or rumours that the pound is to be devalued by stealth. **1966** *Listener* 2 June 792/2 The closing words of his book firmly scotch any hope we may have of evading the central question. **1976** *Australian* 30 June 1/7 The Prime Minister..is to meet the Russian Ambassador..next month to scotch reports of a serious rift in Soviet–Australian relations.

Scotch bonnet. 1. (Earlier example.)

1759 L. WOOD *Jrnl.* 27 June in *Essex Inst. Hist. Coll.* (1882) XIX. 70 It Came out in order this Day that no officer in yᵉ Rigement Should wear a Scotch bonet.

Scotch cart. Chiefly *S. Afr.* Also **scotch cart.** [Prob. f. SCOTCH *a.* + CART *sb.*; S. Afr. uses may represent derivation from Afrikaans *skotskar*, ad. G. *schuttkarren*, but this is unproved.] A light and strongly built two-wheeled cart, used chiefly for transporting rough materials such as gravel, manure, etc.

1845 *Cape of Good Hope Almanac* (Advt.), Best Scotch Carts and wheelbarrows made to order. **1850** *Mary Wedlake's Priced List Farming Implements* 33 (*heading*) Scotch Carts. These very light carriages, so superior in point of draught to the old heavy dung carts used in most counties of England, are daily substituted for those ponderous machines. **1895** R. CHURCHILL *Men, Mines & Animals in S. Afr.* xiii. 210, I accordingly borrowed a Scotch cart (a light two-wheeled covered waggon) and a span of eight oxen, with which went also two 'salted' horses. **1938** D. FORBES *My Life in S. Afr.* vi. 87 He also fixed up a scotch cart to carry the alluvial ground to the stream. **1949** *Cape Argus Mag.* 14 May 2/6 At first a few skins were sent in from farms, then..they began to arrive in sugar pockets,.. and finally by the Scotch-cart load. **1970** G. E. EVANS *Where Beards wag All* i. 31 A Scotch cart was popular round here... They'd carry about a ton. *Ibid.* 34 The extra spokes and felloes on a Scotch cart were probably needed on its home ground where a strong wheel was essential on a rougher terrain than is to be found in East Anglia. **1974** *Standard Encycl. S. Afr.* X. 571/1 The Scotch-cart, with or without springs, was always popular with farmers because it could be tilted backwards to enable its load to be discharged.

Scotchgard (skǫ·tʃgāɹd). A proprietary term in the U.S. for a series of organofluorine chemicals employed as waterproof grease- and stain-resistant finishes for textiles, suede, leather, etc.

1956 *Official Gaz.* (U.S. Patent Office) 1 May TM 7/1 Minnesota Mining and Manufacturing Company, St. Paul, Minn... *Scotchgard.* For chemical composition for application to various surfaces to repel grease and oil therefrom. First use Aug. 26, 1955. **1959** *Times* 12 Jan. 11/5 Scotch-Gard: Finish..for use on wool, cotton, or synthetic fibres to improve oil, grease, and water stain-resistance. Durable for dry-cleaning. Shortly to be marketed. **1962** *N.Y. Times Mag.* 9 Sept. 106 (Advt.), Lightweight suede 'Gliders' [*sc.* shoes] Scotchgard treated to resist stains. **1969** A. J. HALL *Stand. Handbk. Textiles* (ed. 7) v. 334 One such fluorochemical is Scotchgard F.C. 208 and when this, together with..water-repellent Velan, is dried into cotton fabric and followed by a curing..at about 130°C. for 5 min., the fabric acquires a combined water and oil repellency which withstands repeated washing with soap.

Scotch-Irish, *a.* Add: **a.** (Earlier and later examples.) **b.** = *SCOTS-IRISH *a.* b. Scotch-Irishman (example).

1744 W. MARSHE *Jrnl.* 21 June in *Coll. Mass. Hist. Soc.* (1801) 1st Ser. VII. 177 The inhabitants [of Lancaster, Pa.] are chiefly High-Dutch, Scotch-Irish, some few English families, and unbelieving Israelites. **1789** J. MORSE *Amer. Geogr.* 313 [The Irish of Pennsylvania] have sometimes been called Scotch-Irish, to denote their double descent. **1903** J. FOX *Little Shepherd of Kingdom Come* x. 117 Broadcast, through the people, was the upright sturdiness of the Scotch-Irishman, without his narrowness and bigotry. **1916** J. WEBSTER *Dear Enemy* 187 That Scotch-Irish ancestry of mine. **1948** H. MACLENNAN *Precipice* (1949) i. 5 The Scotch and the Scotch-Irish who had flooded into Ontario. **1963** W. K. ROSE *Lett. Wyndham Lewis* I. 1 An English girl of Scotch-Irish descent. **1980** G. M. FRASER *Mr. American* xvii. 312 Reason is the last thing you can look for in a Scotch-Irish Protestant. *Ibid.* xix. 361 The Scotch-Irish who saw their freedom threatened.

Scotchlite (skǫ·tʃləit). The proprietary name of a light-reflecting material containing a layer of minute glass lenses.

1941 *Official Gaz.* (U.S. Patent Office) 6 May 29/2 Minnesota Mining & Manufacturing Company, St. Paul, Minn... *Scotchlite* for light reflecting material in sheet form. Claims use since Dec. 1, 1939. **1947** *Daily Progress* (Charlottesville, Va.) 28 Oct. 4/1 A Minnesota motorist..conceived the idea of covering bicycle frames with Scotchlite, a material used on reflectorized highway signs. **1957** *Times Survey Brit. Aviation* Sept. 8/5 Scotchlite is a tough, plastic reflective film. **1964** *Times* 7 Feb. p. iv/1 (Advt.), Scotchlite reflective sheeting..consists essentially of a white or coloured reflecting surface covered with minute, optically-perfect glass lenses. On top of the lot goes a clear plastic protective coating. **1970** *Trade Marks Jrnl.* 17 June 961/1 *Scotchlite*... Sign faces made of or incorporating plastics embedded with light reflective substances. Minnesota Mining and Manufacturing Company.., United States of America; manufacturers. **1972** *Police Rev.* 10 Nov. 1444/1 'Scotchlite' markings on Police vehicles.

Scotchman. Add: **1. a.** (Later example.)

SCOTSMAN is now the preferred form on both sides of the border (see small-type note s.v. *SCOTCH *a.* and *sb.*[2]).

1977 K. M. E. MURRAY *Caught in Web of Words* xi. 209 For a Scotchman James was certainly extraordinarily lacking in hard-headedness.

b. (Earlier example.) Cf. SCOTSMAN in Dict. and Suppl.

1873 J. BLACKWOOD *Let.* 6 Jan. in *Geo. Eliot Lett.* (1956) V. 365 'The Flying Scotchman', the stoker's name for the train that goes between London and Edinburgh in little more than 9 hours! **1913** D. H. LAWRENCE *Sons & Lovers* vii. 165 You should see the Flying Scotchman come through.

c. (Earlier examples.)

1719 T. MARCHANT *Jrnl.* 10 June in *Sussex Arch. Coll.* (1873) XXV. 184 In all 15s. 9d., to John Gracie, a Scotchman, for M. Balcombe. **1793** C. SMITH *Old Manor House* I. vi. 138, I had not enough money..to buy my new cotton gown, when Alexander Macgill the Scotchman called here.

Scotch tape. Also **scotch tape.** The proprietary name of a make of adhesive tape; also applied *loosely* to any adhesive tape. Hence **Scotch-tape** *v. trans.*, to affix or join with adhesive tape; **Scotch-taped** *ppl. a.*, affixed or made fast with adhesive tape.

[**1945** *Official Gaz.* (U.S. Patent Office) 16 Oct. 373/1 Minnesota Mining & Manufacturing Company, St. Paul, Minn. ..Scotch for pressure-sensitive adhesive tape. Claims use since January 1928.] **1947** R. LEE *Electronic Transformers & Circuits* ii. 28 Electrical grade scotch tape is widely used for anchoring leads. **1949** R. CHANDLER *Little Sister* ix. 54 To the lining of the toupee a piece of..paper was fastened by Scotch tape. **1955** 'J. WYNDHAM' in 'E. Crispin' *Best SF* 81 There was a wire, scotch-taped to the upper side of the bag. **1957** V. NABOKOV *Pnin* iv. 99 Carrying his purchase, wrapped in brown paper and Scotch-taped, he entered a bookstore. **1961** WODEHOUSE *Ice in Bedroom* xxvi. 222 Her lips..shall be sealed, if necessary with Scotch tape. **1961** J. H. GRIFFIN *Black like Me* 86 The whites frequently..Scotch-tape these notices to the wall. **1969** K. AMIS *Green Man* i. 26 Coloured photographs.. Scotch-taped to the walls. **1976** *Observer* 24 Oct. 28 It [*sc.* Sellotape] still has 75 per cent of the market over here, though in America Scotch Tape has become the generic term. **1977** *Time* 31 Jan. 24/1 Carter wrote at least three more drafts, sometimes spreading the paragraphs out like pieces of a jigsaw puzzle and Scotch-taping them into a new arrangement.

Scotchy, *sb.* (Later example.)

1949 E. COXHEAD *Wind in West* ii. 59 Are you really such a prim little Scotchy that you don't see the difference between one chap and another?

Scotland Yard (skǫ·tlǝnd yāɹd). The name of the head-quarters of the Metropolitan Police, situated from 1829 to 1890 in Great Scotland Yard, a short street off Whitehall in London; from then until 1967 in New Scotland Yard, on the Thames Embankment; and from 1967 in New Scotland Yard, Broadway, Westminster: used allusively to designate the detective department of the Metropolitan Police force. Also *attrib.*

1864 M. E. BRADDON *Henry Dunbar* II. xiv. 260, I have called again upon the Scotland-Yard people, and I gave them a minute description of the scene. *Ibid.* III. iv. 49 Not that anybody would try to thwart me,..if they knew that I was detective officer Henry Carter, of Scotland Yard. **1864** [see BLUEBOTTLE 2]. **1881** *Punch* 9 July 6/2 As in all great crimes, they [*sc.* criminals] are too much for Scotland Yard and the Seldom-at-Home Secretary. **1907** [see REWARD *sb.*[1] 4 d]. **1909** [see *PHONE *v.* a]. **1926** E. WALLACE *Ringer* 1 Nobody knows, but Scotland Yard and—Henry Arthur Milton. **1939** T. S. ELIOT *Old Possum's Bk. Pract. Cats* 33 He's the bafflement of Scotland Yard, the Flying Squad's despair. **1962** J. MCCABE *Mr. Laurel & Mr. Hardy* i. 36 A backer for the troupe, a Scotland Yard detective. **1974** N. FREELING *Dressing of Diamond* 87 Just forget all the detective stories of bumbling Scotland Yard Inspectors.

Scot Nat (skǫt næt), *sb.* and *a.* Abbrev. of *SCOTTISH NATIONALIST *a.* and *sb.* Cf. *SCOTS NAT.

1970 [see *NAT[2]]. *a* **1974** R. CROSSMAN *Diaries* (1976) II. 550 He'd told me that the Scot. Nat. woman might win and the Tories would certainly lose their deposit. **1974** *Sunday Post* (Glasgow) 28 Apr. 5/5 Even by Willie Ross's standards, last weekend's attack on the Scot. Nats. was vitriolic. **1975** *Times* 8 Sept. 12/7 Both Tory and Labour politicians in Scotland..will be tempted to enter into an auction with ScotNats over devolution.

Scoto-[1]. Add: **Scotopho·bia**[2], a morbid dread or dislike of the Scots or things Scottish; hence **Sco·tophobe.**

1974 *Listener* 25 Apr. 520/3 There is undoubtedly a strong streak of Scotophobia in the English character. **1976** *Times Lit. Suppl.* 2 July 823/3 This 'never ending clan of Macs and Donalds upon Donalds', as one Scotophobe put it in the 1760s. **1980** B. LENMAN *Jacobite Risings in Britain* 289 A glance at the huge correspondence which Sir Everard organized so meticulously for Cumberland is very revealing about the origins of his royal master's sustained Scotophobia.

scoto-[2] (skōᵘ·to), comb. form repr. Gr. σκότος darkness, as in **sco·tophase** *Biol.*, an artificially imposed period of darkness; an artificial night; **sco·tophobia**[1] *Psychol.*, fear or dislike of the dark; hence **scotopho·bic** *a.* See also SCOTOSCOPE.

1971 *Nature* 6 Aug. 401/2 Bovines restrained inside environmentally controlled stalls (24±3°C; 70±7% relative humidity; 12 h photophase: 12 h scotophase per 24 h photocycle). **1975** *Ibid.* 25 Dec. 711/2 We considered whether the insects perceive the actual duration of photophases (or scotophases). **1938** *Brit. Jrnl. Psychol.* Apr. 372 Photo-phobia (in some animals) sufficiently strong to make them go always to D; the opposite trait, 'a-photo-phobia' or 'scoto-phobia' being present in an equal proportion of the animals. **1971** *New Scientist* 3 June 559/3 The scotophobic effect seems to be very specific for this structure.

scotoma. Add: Pl. also **scotomas.** Also *fig.* (Further examples.)

1943 *Horizon* Oct. 257 As with all neurotics, the confessions of Kierkegaard only contain a grain of the truth; the analytic scotoma constantly intervenes. **1957** F. B. WALSH *Clin. Neuro-Ophthalm.* (ed. 2) ix. 606/3 Retinal lesions are not rare in cases of Leber's optic atrophy. There may be complaint of chromatopsia and ring scotomas.

scotometer (skotǫ·mitəɹ). *Ophthalm.* [f. SCOTO(MA + -METER.] An instrument for diagnosing and measuring scotomata.

1890 G. FERDINANDS in *Brit. Med. Jrnl.* 27 Sept. 741/1 Those who frequently meet with cases of toxic amblyopia.. must have found the small coloured squares used in detecting scotomata both an inadequate and inconvenient test... To obviate these disadvantages I have had made for me a little instrument..which I propose to call a scotometer. **1932** *Optician* LXXXIII. 397/1, I did not take the fields for white, relying on the evidence of the scotometer. **1961** S. VAN WIEN tr. *Huber's Eye Symptoms in Brain Tumors* i. 75 For a quick survey to determine the presence of a scotoma for color, the so-called scotometer is suitable.

Hence **scoto·metry,** the use of a scotometer; **scotome·tric** *a.*

1921 *Proc. R. Soc. Med.* XIV. (Ophthalm. Sect.) 49 The purpose for which scotometry is of such importance is the determining of the increase of the blind spot in cases of suspected glaucoma. **1944** *Amer. Jrnl. Ophthalm.* XXVI. 349 (*heading*) The form and character of rod scotometry. **1955** *Jrnl. Neurol., Neurosurg. & Psychiatry* XVIII. 224/2 Scotometric studies suggested small infarctions in each infracalcarine striate cortex.

scotomization (skōᵘtǒmaizēⁱ·ʃǝn). *Psychol.* [a. F. *scotomisation* (Pichon & Laforgue in R. Laforgue *Le Rêve et la Psychoanalyse* (1926) vii. 184), f. Gr. σκοτ-οῦν to darken, make dim-sighted: see -IZATION.] (See quot. 1927.) So **sco·tomize** *v. trans.*; **sco·tomized** *ppl. a.*

1927 R. LAFORGUE in *Internat. Jrnl. Psycho-Anal.* VIII. 473 In an earlier work I have defined scotomization (or the forming of mental 'blind spots') as a process of psychic depreciation, by means of which the individual attempts to deny everything which conflicts with his ego. *Ibid.* 477 If he scotomizes them [*sc.* such stimuli as do not belong to the field of consciousness],..they seek for gratification in narcissistic compensations, and force him to a compensatory activity whose scotomized sources are hidden from him. **1954** *Brit. Jrnl. Psychol.* XLV. 233 This criticism [of horrific test pictures] does not imply any scotomization of the capacity of the child..to conceive of the horrible. **1969** P. A. ROBINSON *Freudian Left* 145 The functionalists concluded from the mere fact that a culture existed that it.. functioned harmoniously. They would thus 'scotomize' all of those psychoanalytic facts which drew attention to the terrible price we pay for civilization. **1977** A. SHERIDAN tr. *J. Lacan's Écrits* ii. 22 Freud seems suddenly to fail to recognize the existence of everything that the ego neglects, scotomizes, misconstrues in the sensations that make it react to reality.

scotophase: see *SCOTO-[2].

scotophil (skōᵘ·tofil), *a.* *Biol.* Also **skoto-, -phile.** [ad. G. *skotophil* (E. Bünning 1944, in *Flora* CXXXVIII. 95): see *SCOTO-[2] and -PHIL, -PHILE.] Applied to that phase of the circadian cycle of a plant or animal during which light inhibits, or does not influence, reproductive activity; opp. *photophil, -phile* s.v. *PHOTO- 1.

1952, etc. [see *photophil* adj. s.v. *PHOTO- 1]. **1959** F. W. WENT in R. B. Withrow *Photoperiodism & Related Phenomena in Plants & Animals* VII. 554 The leaf angle..is small in the scotophil and large in the photophil phase. **1960** *Cold Spring Harbor Symp. Quantitative Biol.* XXV. 257 In this part of the rhythm the endodiurnal system is in the scotophile state according to Bünning... The second scotophile phase is hardly weaker than the first one. **1971** *New Scientist* 29 July 254/2 The circadian rhythm [of house finches] was imagined to comprise two half cycles.., one of which was reckoned to be dark-requiring (scotophil). **1972** *Nature* 21 Apr. 407/1 The state..during which light has a promotive effect on flowering..and the state during which light is innocuous..could be respectively the photophil and skotophil phases of Bünning.

Hence **scotophi·lic** *a.*, scotophile; **scoto·phily** the state of an organism in a scotophile phase.

1960 *Cold Spring Harbor Symp. Quantitative Biol.* XXV. 265/1 According to this view long and short day effects depend on whether the photoperiod of light break coincides with the 'scotophilic' phase during the second half-cycle. **1960** Scotophily [see *photophily* s.v. *PHOTO- 1]. **1976** *Sci. Amer.* Feb. 115/2 He proposed that the measurement of the length of the day or the night was accomplished by an endogenous, or built-in, daily rhythm that consisted of two half-cycles, one photophilic ('light-loving') and the other scotophilic ('dark-loving').

scotophobia: see *SCOTO-[2]; **Scotophobia**: see *SCOTO-[1].

scotophobin (skotofōᵘ·bin). *Biochem.* [f. *SCOTO-[2] + φόβ-ος fear + -IN[1]; cf. *scotophobia*[1] s.v. *SCOTO-[2].] An oligopeptide isolated from the brains of rats which have been trained to avoid darkness, and which is claimed to induce dark avoidance in untrained rats and possibly also in animals of other species.

1970 G. UNGAR et al. in *Proc. Western Pharmacol. Soc.* XIII. 150 We propose to give the name 'scotophobin' to the new substance. **1971** *New Scientist* 3 June 559/3 The peptide, called scotophobin, contains only 15 amino acids. **1975** *Behavioral Biol.* XV. 470 Acceptance of these assumptions forces us to reject the hypothesis that 'fear of the dark' was encoded in scotophobin. **1976** F. WARNER *Killing Time* II. ix. 61 We could inject Scotophobin and induce a fear of the dark.

scotopic (skotǫ·pik), *a.* *Physiol.* [f. *SCOTO-[2] + *-OPIA + -IC.] Of, pertaining to, or designating vision in dim light, believed to involve chiefly the rods of the retina. So **scoto·pia,** the condition of scotopic eyes.

1915 J. H. PARSONS *Introd. Study Colour Vision* ii. 17 If the eye has been kept completely free from light for a considerable period it is said to be dark-adapted. I shall speak of vision under these circumstances as scotopia..and the dark-adapted eye as a scotopic eye. **1924** J. P. C. SOUTHALL tr. W. Nagel in *H. von Helmholtz's Treat. Physiol. Optics* II. 345 The so-called *Dämmerungssehen* (or twilight vision, scotopia), when the eye is dark-adapted and the light stimulus is weak. **1937** *Nature* 6 Mar. 409 It is generally accepted that visual purple plays an essential part in the process of scotopic vision. **1946** *Ibid.* 31 Aug. 303/2 S. Hecht also arrives at the conclusion that the absorption of the visual purple is less than 20 per cent, by comparing the scotopic luminosity curve with the absorption curve of visual purple. **1972** H. J. EYSENCK et al. *Encycl. Psychol.* III. 182/1 *Scotopia*, twilight vision... With twilight vision a person is insensitive to color and his peripheral vision is better for fine detail than his central vision, since there are no rods in the fovea. **1973** 'A. HALL' *Tango Briefing* xiv. 169 My eyes were adapting to scotopic vision, the torchlight growing brighter.

scotoscope. For † *Obs.* read *rare* and add later example.

1964 *Applied Optics* III. 671 The scotoscope can be arranged to give a color presentation; however, when this is done, it is at the expense of a fairly high percentage of the photons incident from the scene.

Scots-Irish, *a.* **a.** = SCOTCH-IRISH *a.* Also as *sb.* **b.** Of mixed Scots and Irish descent.

1972 *Listener* 21 Dec. 854/2 The hostility of the Catholic Irish and the Protestant Scots-Irish. *Ibid.* 854/3 The Protestant Scots-Irish community. **1973** *Guardian* 27 June 13/3 The USA has taken to calling itself 'Scots-Irish' rather than 'British'. **1973** 'D. SHANNON' *No Holiday for Crime* (1974) iv. 62 Once in a while my Scots-Irish wife uses a little ESP. **1980** —— *Felony File* i. 35 The feudal household his Scots-Irish girl had wished on him.

Scotsman. Add: **b.** (Also *Flying Scotsman*.) Now the more usual form of (*Flying*) *Scotchman*: see SCOTCHMAN 1 b in Dict. and Suppl.

One of British Rail's express trains from London to Edinburgh still bears this name, though the particular steam locomotive so named was withdrawn from service on 14 Jan. 1963.

1879 [in Dict.]. **1932** P. BLOOMFIELD *Imaginary Worlds* 34 Pretending..that our O gauge railway round the nursery floor is really the line taken by the 'Flying Scotsman'. **1936** J. MASEFIELD *Let. from Pontus* 57 On the railway beside us the Scotsman went by. **1952** A. ANDERSON *Flying Scotsman* 3 Our train is the Flying Scotsman which has left King's Cross at 10 a.m. daily for over 90 years. **1962** C. H. ELLIS *Flying Scotsman* i. 11 The Special Scotch Express sooner or later became the 'Flying Scotch Express', which in turn was transformed into 'Flying Scotchman' and later still, probably when English people began to read Robert Louis Stevenson, into 'Flying Scotsman'. That eventually became the official title. **1974** *Times* 7 Dec. 3/2 The Flying Scotsman..the majestic old LNER puffer... The Flying Scotsman..is at present stabled steamless at Carnforth.

Scots Nat (skǫts næt). *colloq.* [f. SCOTS *a.* + *NAT[2]; cf. *SCOT NAT *sb.* and *a.*] A member of the Scottish National Party (see *SCOTTISH *a.* 5). Hence **Scots Na·ttery** (*nonce-wd.*) Scottish Nationalism.

1974 *Undercurrents* July–Aug. 3/2 The British did take the precaution..of incorporating Rockall into the county of Inverness—which may or may not be a smart move depending on which way the Scots Nats jump. **1977** *Times* 23 June 16/1 The delectable MP for East Dumbartonshire, Margaret Bain..a good Scots Nat and true. **1978** *Times* 21 Jan. 14/4 Successive waves of Scots Nattery going back to J. M. Bannerman.

Scott (skǫt). *Electr. Engin.* [The name of Charles F. *Scott* (1864–1944), U.S. electrical engineer, who devised the connection in 1894 (*Electrician* 6 Apr. 640).] *Scott connection*: a way of connecting two single-phase transformers to convert a three-phase voltage to a two-phase one (or to two single-phase ones), or vice versa: on the three-phase side the midpoint of the main transformer is connected to one terminal of the second transformer; the remaining three terminals form the terminals for the three-phase supply; the two-phase supply is taken from the two pairs of terminals on the other side. So *Scott-connected* adj.

[**1911** BOHLE & ROBERTSON *Transformers* xi. 321 (*heading*) Scott's two-phase to three-phase connection.] **1926** J. L. BEAVER *Elem. Alternating Currents* ix. 200 The so-called 'Scott' connection is an arrangement of *two* single-phase transformers, whereby three-phase power may be obtained from a two-phase circuit or vice versa. **1935** MONSETH & ROBINSON *Relay Systems* x. 304 (*caption*) Scott-connected transformer differential protective scheme. **1947** R. LEE *Electronic Transformers & Circuits* viii. 214 When autotransformers are used on three-phase supply lines, they may be connected the same as two-winding transformers in star, delta, open-delta, or Scott connections. **1966** BROSAN & HAYDEN *Adv. Electr. Power & Machines* vi. 243 The Leblanc system was invented about five years after the Scott connexion when the latter had obtained a firm foothold in Britain and its use was therefore confined to the Continent. *Ibid.*, The general case of determining the regulation of a Scott-connected group is somewhat involved.

Scottie (skǫ·ti). Shortened f. *Scotch terrier* s.v. SCOTCH *a.* 4 in Dict. and Suppl.

1907 F. T. BARTON *Terriers* xviii. 131 If a Scottie has not a sound jacket to keep out the mountain dew and rain, he ought not to take a leading place at any show. **1939** *Country Life* 11 Feb. p. xxi/2 (Advt.), For Sale.—Exceptionally strong Scottie puppies. **1945** A. CHRISTIE *Sparkling Cyanide* I. iv. 40 She came out one morning with a small black Scottie dog. **1957** R. MASON *World of Suzie Wong* II. ii. 124 Her dog..was a Scottie with an absurd, sad, long face. **1971** 'L. EGAN' *Malicious Mischief* (1972) i. 4 Scotties are nice dogs. **1973** M. AMIS *Rachel Papers* 159 Glancing downwards, my rig, in its pink muff, looked unnatural, absurd, like an overdressed Scottie dog.

Scottish, *a.* and *sb.* Add: **A.** *adj.* **5.** *Scottish-American* adj.; **Scottish Baronial** *a.*, designating a style of architecture typical of the semi-fortified houses of the medieval Scottish nobility, and revived in the nineteenth century (cf. *Scotch Baronial* s.v. *SCOTCH *a.* 4 a); **Scottish Blackface** = *Scotch Blackface* s.v. *SCOTCH *a.* 4 a; **Scottish Chaucerians,** the distinguishing epithet applied to a number of fifteenth- and sixteenth-century Scottish poets influenced by and imitating the work of Geoffrey Chaucer; **Scottish National Party,** a political party formed in 1934 by an amalgamation of the National Party of Scotland and the Scottish Party, which seeks autonomous government for Scotland (cf. *SCOTTISH NATIONALIST *a.* and *sb.*); **Scottish terrier** = *Scotch terrier* s.v. *SCOTCH *a.* 4 a.

1905 W. JAMES in *McClure's Mag.* May 3/1, I wish to pay my tribute to the memory of a Scottish-American friend of mine who died five years ago. **1978** N. GOSLING *Paris 1900–14* 49 Debussy had picked the Scottish-American star Mary Garden. **1938** L. MACNEICE *I crossed Minch* vii. 98 Oban has many hotels in the Scottish Baronial style. **1956** L. E. JONES *Edwardian Youth* iv. 87 The newly-built Scottish Baronial building which contains the Main Gateway [of Balliol College, Oxford]. **1979** R. LAIDLAW *Lion is Rampant* vi. 49 The house..was a massive structure, built..in the Scottish Baronial style. **1937** A. FRASER *Sheep Farming* ii. 22 Only a few years ago I was concerned with the export of

Scottish Blackface sheep to Palestine. **1974** *Times* 23 Feb. 14/2 Several farmers may turn out their Swaledale or Scottish Blackface, Herdwick or Lonk sheep onto one moor. **1902** G. GREGORY SMITH *Specimens of Middle Scots* p. xlv, To say this of the 'Scottish Chaucerians' is almost a platitude. **1927** E. P. HAMMOND *Eng. Verse between Chaucer & Surrey* p. xi, Rhythm in Chaucer and the English Chaucerians—Verse-Forms—The Scottish Chaucerians—Vocabulary. **1935** A. BAUGH *Hist. Eng. Lang.* vi. 192 The fifteenth century is sometimes known as the Imitative Period since so much of the poetry..was written in emulation of Chaucer... In the north the Scottish Chaucerians, particularly Henryson, Dunbar, Gavin Douglas, and Lindsay, produced significant work. **1937** *Oxf. Compan. Eng. Lit.* (ed. 2) 154/1 *Chaucerians, Scottish*, name given to a group of 15th cent. Scottish writers (of whom Dunbar and Henryson are the chief) who imitated Chaucer in some of their work. **1966** *Amer. N. & Q.* May 139/1 Robbins-Cutler now include ..the so-called Scottish Chaucerians. **1934** *Times* 26 Feb. 16/5 Resolutions were unanimously passed at a conference of the National Party of Scotland in Glasgow approving of a union with the Scottish Party, subject to that Party's agreement, the united parties to be called the Scottish National Party. **1973** *Scotsman* 12 Jan. 9/2 Mr Douglas Drysdale, a former vice-chairman of the Scottish National Party, has been appointed..for liaison among other nationalist bodies in Scotland. **1980** BUTLER & SLOMAN *Brit. Polit. Facts 1900–79* (ed. 5) ii. 162 Scottish National Party. The party was formed in 1928 as the National Party of Scotland. In 1933 it merged with a body called the Scottish Party (founded 1930) and the name was then changed to the Scottish National Party. **1837** T. BELL *Hist. Brit. Quadrupeds* 230 The other [terrier] is called the Scottish or Wire-haired Terrier. **1894** R. B. LEE *Hist. & Descr. Mod. Dogs: Terriers* xi. 251 It was about the year 1874 that a newspaper controversy brought the Scottish terrier prominently before the public. **1956** D. CASPERSZ *Popular Scottish Terrier* i. 17 The Scottish Terrier descends directly from a race of small terriers of great antiquity.

Sco·ttish Na·tionalist, *a.* and *sb.* **A.** *adj.* Of or pertaining to the Scottish National Party (see *SCOTTISH *a.* 5) or its programme. **B.** *sb.* A member of this party.

1936 'H. MACDIARMID' in *Lucky Poet* (1943) iii. 145 The long confused and inefficient Scottish Nationalist groping. **1953** E. SIMON *Past Masters* IV. 246 This Scottish Nationalist you've got there. **1968** *Daily Progress* (Charlottesville, Va.) 11 July C14/1 Winifred Ewing was a noted narker in Glasgow before she became the only Scottish Nationalist in the House of Commons. **1976** LD. HOME *Way Wind Blows* xv. 206 A Scottish Nationalist Party had grown up which had begun to advocate separation from England.

So **Sco·ttish Na·tionalism,** the political programme or ideals of the Scottish National Party.

1935 N. MITCHISON *We have been Warned* I. 73 'Don't you think..there's something in Scottish Nationalism?' 'Not while it's run by ladies and gentlemen.' **1953** E. SIMON *Past Masters* IV. 246 Scottish Nationalism, as I see it, is an absurdity.

Scottishness. (Later examples.)

1933 *Times Lit. Suppl.* 3 Aug. 523/1 Henryson was sage and serious. It was part of his Scottishness. **1956** N. PEVSNER *Englishness of Eng. Art* (rev. ed.) 125 It is easy to recognize Scottishness in the Scottish castle of the seventeenth century. **1976** *Listener* 19 Feb. 209/3 The BBC in Scotland should be asked..to abandon the excessive concern with Scottishness.

Scottishry (skǫ·tiʃri). [f. SCOTTISH *a.* + -RY; cf. IRISHRY, WELSHRY, etc.] Scottish character or nationality; a Scottish trait, Scottishness.

1958 C. WATSON *Coffin, scarcely Used* iv. 44 'Ye hear tha' frae the wee booy!' he chortled... 'For heaven's sake, drop that phoney Scottishry, Rupert.' **1973** *Daily Tel.* 24 Nov. 11/1 The whole business of Highland Scottishry is so technical that an Englishman may find himself rebuked on a point of Gaelic or a question of optics. **1979** *Ibid.* 17 Aug. 14/3 Fake Scottishry of the kind which reduces the noble name of the last Lord of the Isles, Donald Dubh, to the almost indistinguishable name of the cartoon duck.

Scotty, *a.* Add: **b.** (With small initial.) (Earlier and later examples.)

1896 E. TURNER *Little Larrikin* xvii. 191 I'm *blessed* if I know what I've done this time... Don't be scotty, Marcia. **1936** M. FRANKLIN *All that Swagger* xxxv. 334 Uncle William is as scotty as a French hen with her feathers the wrong way.

scour, *sb.*[2] Add: **2.** Also, the abrading or transporting action of a current of any other material. (Further examples.)

1904 *Jrnl. Geol.* XII. 575 With these destructional effects assigned to glacial agency, a novel possibility is at once suggested as to the part played in their persistent development by glacial scour, or coarse abrasion. **1933** SCHUCHERT & DUNBAR *Textbk. Geol.* (ed. 3) xix. 425 The ice and the scour of the last glaciers removed all weak materials. **1954** *Jrnl. Geol. Soc. Australia* I. 77 The wind scour is able to drive the sand into heaps which migrate slowly down wind. **1975** *Offshore* Sept. 49-17/1 Scour is probably the greatest menace to offshore structures and pipelines.

5. By 'cattle' understand 'livestock'. Also *pl.* (chiefly *U.S.*). (Further examples.)

1848 *Rep. Comm. Patents 1847* (U.S.) 507 They say the disease called the 'scours' is the principal one to which sheep are liable. **1950** [see *BEESTINGS 1]. **1970** W. H. PARKER *Health & Dis. in Farm Animals* xvii. 226 In sheep the disease causes the same wasting but without the scour. **1973** M. CROWELL *Greener Pastures* 16 Rameses II [*sc.* a sheep] has lately been having scours, or loose-bowel trouble. **1975** *N.Z. Jrnl. Agric.* Sept. 3/1 The hens..also appear to have a green scour. **1981** 'E. PETERS' *St. Peter's Fair* i. 18 They're having trouble..with scour among the calves.

7*. *Austral.* and *N.Z.* A building in which wool is scoured.

1925 L. G. D. ACLAND *Early Canterbury Runs* (1930) 1st Ser. vi. 123 The Creek Station..was leased to T. P. Bartrum from 1879 onwards, and he established a wool scour there. **1934** T. WOOD *Cobbers* xvi. 195 The scour was a long open shed on stilts, with sheep-pens leading into it and out of it.

scour, *v.*[1] **2. b.** (Earlier example.)

1882 'MARK TWAIN' *Prince & Pauper* xxv. 300 Scour and scan me to thy content.

scour, *v.*[2] Add: **1. g.** *U.S.* Of a plough, to pass through the soil easily, without earth adhering to the mould-board; freq. in negative contexts. Also *fig.*, to succeed.

1871 *Northern Vindicator* (Estherville, Iowa) 6 May 3/1 The contemptible wretch who stole the collar to the saw at the steam mill a few weeks ago, has come to the conclusion that his meanness did not 'scour', as he anticipated, and hence he placed the collar under a board pile in town where it was discovered on Monday last. **1881** J. PERIAM *Amer. Encycl. Agric.* 742/2 In the average soil there [*sc.* in eastern U.S.] the cast-iron plow would scour perfectly. **1887** W. H. LAMON in *Washington Critic* 3 Sept. 3/1 He [*sc.* Lincoln] said to me on stand, immediately after the [Gettysburg] speech: 'Lamon, that speech won't scour. It is a flat failure, and the people are disappointed.' **1948** *Sat. Even. Post* 7 Feb. 109/1 Then his old moldboard plow wouldn't scour, and after we'd sharpened it he broke the beam.

7. d. By 'cattle' understand 'livestock'.

scourer[2]. **4.** (Earlier example.)

1859 *Rep. Comm. Patents 1858* (U.S.) I. 378 Scourer G, blast spouts EF, and fan C, [are] combined and arranged relatively with each other.

scourge, *sb.* **2.** (Examples of phr. *the Scourge of God.*)

1611 CORYAT *Crudities* 162 Came that *Flagellum Dei* that scourge of God into Italy, Attila, King of the Hunnes. **1781** GIBBON *Decl. & F.* III. xxxiv. 368 If Attila equalled the hostile ravages of Tamerlane, either the Tartar or the Hun might deserve the epithet of the *Scourge of God*. **1925** D. H. LAWRENCE *Reflections on Death of Porcupine* 157 Attila, the Scourge of God, who helped to scourge the Roman world out of existence, was great with power.

Scourian (skuˑriăn, skauˑriăn), *a.* *Geol.* [f. *Scourie*, name of a crofting village on the west coast of Sutherland + -AN.] Of, pertaining to, or designating the earlier metamorphism undergone by the Lewisian rocks of the Pre-Cambrian in NW Scotland, the rocks formed by this metamorphism, and the structures to which they belong. Also *absol.*, these rocks and structures.

1950 SUTTON & WATSON in *Q. Jrnl. Geol. Soc.* CVI. 243 In the following pages the Lewisian is..regarded as belonging to two metamorphic complexes, the first, or Scourian, being older and the second, or Laxfordian, younger than the dolerite dykes. These names are taken from localities in Sutherland where the relations of the two complexes are particularly clearly displayed. **1959** *Nature* 5 Dec. 1793/1 Three samples came from the Outer Hebrides, three from the Laxfordian, and two from the Scourian. **1961** *Q. Jrnl. Geol. Soc.* CXVII. 241 The Scourian metamorphic rocks in the vicinity of Scourie are primarily hornblende- or pyroxene-bearing gneisses containing very little mica or potassium-feldspar. *Ibid.* 242 The Scourian metamorphism occurred at least 2460 m.y. ago. **1965** A. HOLMES *Princ. Physical Geol.* (ed. 2) xiii. 370 (*caption*) The peaks are of Torridonian Sandstone, resting unconformably on Lewisian Gneiss (Scourian division). **1969** BENNISON & WRIGHT *Geol. Hist. Brit. Isles* iii. 42 The trend of the Scourian is not known since no really large structures have been found associated with the granulites. **1979** *Nature* 22 Feb. 643/1 The ~2,400 Myr age for the emplacement of the Scourie dyke swarm indicates that the Scourian tectonic episode had ceased by the beginning of the Proterozoic.

scouring, *vbl. sb.*[2] Add: **4. b.** By 'cattle' understand 'livestock'.

9. *scouring cloth, machine*; **scouring powder,** an abrasive powder used for cleaning kitchenware, etc.

1907 *Yesterday's Shopping* (1969) 120/3 Scouring Cloth—yard 0/5½ Scouring or Paint Cloths—each 0/4. **1976** W. TREVOR *Children of Dynmouth* iii. 58 He reached for a scouring cloth on a line that stretched above the sink. **1882** *Encycl. Brit.* XIV. 387/2 Another machine now largely used by curriers is the scouring machine. **1969** A. J. HALL *Stand. Handbk. Textiles* (ed. 7) iii. 110 The construction and operation of scouring machines..are such as to disturb the wool as little as possible. **1949** D. SMITH *I capture Castle* v. 61 She..scrubbed her hands until she got it [*sc.* dye] all off. She used our last grains of scouring powder. **1975** N. FREELING *What are Bugles blowing For?* vi. 37 She tacked off to write 'scouring powder' on her shopping list.

scouring, *ppl. a.*[2] Add: **4.** Of livestock: suffering from diarrhœa. Cf. SCOUR *sb.*[2] 5 in Dict. and Suppl.

1752 T. BOULT *Vet. Recipe Bk.* in *Henry Bristow Ltd. Catal.* (1974) No. 206. 31 To cure a Scowering Cow. **1973** *Country Life* 8 Feb. 360/1 A scouring cow is a very highly stressed animal.

5. *U.S.* Of a plough: see *SCOUR *v.*[2] 1 g.

1856 *Rep. Comm. Patents: Agric. 1855* (U.S. Dept. Agric.) 170 It was ploughed as near it as possible with a double-shovel scouring plough. **1943** C. CROW *Great Amer. Customer* 71 He bought old sawmill blades with which he made self scouring plows which cut through the soil as clean as a razor.

Scouse. Add: **2.** Transferred uses. *slang.* **a.** A native or inhabitant of Liverpool.

1945 *Southern Daily Echo* 27 Dec. 4/3 He was stopped by his Lordship and asked to explain the meaning of three words—'oppo', 'Geordie' and 'scouse'. His interpretations were: 'oppo' slang for opposite number, friend or colleague, 'Geordie'—a native of Newcastle-upon-Tyne; 'scouse'—a native of Liverpool where they eat 'scouse' (stew). **1960** O. MANNING *Great Fortune* II. 148 'I'm a scouse,' he said. 'From the dregs of the Liverpool soup.' **1969** R. AIRTH *Snatch!* ix. 91 'Where's home?' 'Liverpool.' 'A scouse. Fancy that.' **1980** *Times* 20 June 11/6 A roly-poly, amiable Liverpudlian, with the Scouse's seemingly god-given gift of being able to send up an overblown..occasion.

b. The dialect of English spoken in Liverpool. Also, the manner of pronunciation or accent typical to the 'scouse'.

1963 *Guardian* 3 June 10/5 This rock group suddenly made Liverpool fashionable in the entertainment world. After their first two records it became necessary for people in the business in London to learn a few words of Scouse. **1966** 'L. LANE' *ABZ of Scouse* Introd., As a dialect—for it is a dialect and not just a regional accent—Scouse has many curious features. **1979** *Times* 20 Nov. 4 A touch of Scouse in the pronunciation will be entirely acceptable.

c. *attrib.* or as *adj.*

1960 *Spectator* 14 Oct. 565 A horrifyingly plausible spiv, even down to that awful 'scouse' accent. **1965** G. MELLY *Owning Up* vi. 67 Albert Kinder, a scouse promoter who intended to tie up jazz in the North. **1969** I. & P. OPIE *Children's Games* x. 276 'Film Stars' is the most popular guessing game in Britain... Other names: 'Initials', 'Pop Stars', 'TV Stars', and, in Liverpool, 'Filmy', a typical scouse apocope. **1973** *Guardian* 1 Aug. 1/8 Scouse House was the tongue-in-cheek name given to the Merseyside Development Office. **1976** *Observer* 8 Aug. 11 (Advt.), Ar Alf sez darrevry Scouse Big'ead's brood special fer d'Pool, like.

Scouser (skau·səɹ). *slang.* [f. prec. + -ER[1].] = *SCOUSE 2 a.

1959 *Times* 8 Dec. 13/6 Their [*sc.* Liverpool workers'] catarrhal speech would identify them as 'Scousers' wherever English is recognized. **1966** 'L. LANE' *ABZ of Scouse* 93 *Scouser*, an inhabitant of Merseyside, not necessarily a Liverpudlian. This book assumes a narrower definition, namely, a Merseysider who speaks Scouse. **1966** P. MOLONEY *Plea for Mersey* 22 A scouser lass known as R. Mury. **1973** *Guardian* 5 Mar. 8/2 Here was Ted Whitehead, born in Scotland Road and therefore a scouser of the scousers, with a real Liverpool play. **1976** *Liverpool Echo* 22 Nov. 6/6 It's pretty well established that where there's a ship there you'll find a Scouser.

scout, *sb.*[4] Add: **1.** Also, an instance of this; a scouting or reconnoitring expedition. Also *scout-round*.

1906 'MARK TWAIN' in *Harper's Mag.* Aug. 328/2 We are back at Fort Paxton once more, after a forty-day scout. **1975** P. DICKINSON *Lively Dead* xxii. 137 They'd enough reason to send a bloke to do a preliminary scout round. **1978** F. BRANSTON *Sergeant Ritchie's Conscience* i. 12 'Organized a scout-round for the weapon?' 'Only in the immediate area.' **1980** A. PRICE *Hour of Donkey* iv. 55 Wimpy's scout through the wood must..be..completed... Bastable contented himself with cautious peering round each blind bend.

2. c. (*a*) *Boy Scout* (also with small initials): now, a member of the Scout Association, or one of its associate, or parallel, organizations whose ideals of good citizenship and a healthy active life are promoted at regular meetings of scout groups in Great Britain, the British Commonwealth, the United States, and elsewhere throughout the world. Various specialized divisions of the movement are distinguished, as *cub scout, sea scout* s.v. *SEA *sb.* 23 a, *venture scout* s.v. *VENTURE *sb.*, etc. (Later examples.)

The term 'Boy' has now been officially omitted from the title of the organization in the U.K., U.S., and elsewhere. *Girl Scout* has been the official name for the U.S. equivalent to the Girl Guide since 1912: see *GUIDE *sb.* 2 d.

1910 'SCOUTMASTER' *Boy Scout* i. 1 General Sir Robert S. Baden-Powell, the founder of the Movement, recognised this when he first propounded his scheme for boy scouts... The Movement was started less than three years ago. **1924** 'A. D. SEDGWICK' *Little French Girl* I. v. 39 Alix heard of a Women's Institute, of Boy Scouts and Girl Guides. **1959** *Listener* 27 Aug. 304/1 There's something of the boy scout about M. Debré. **1978** *Broadcast* 27 Mar. 20/3 We wish that ..people would stop calling us 'Boy' Scouts. Ten years ago we adopted the title, 'Scouts', in a process of updating our appeal to young people.

(*b*) *fig.* and *transf.*, freq. with reference to the honesty, preparedness, or supposed inexperience of a (Boy) Scout (see quots.).

1918 I. S. COBB *Glory of Coming* p. xiii, The Poilus called our soldiers 'Boy Scouts' and spoke of our effort as 'The Second Children's Crusade'. **1929** F. A. POTTLE *Stretchers* 60 The noncoms (who for weeks had been calling us 'Boy Scouts') hung a blue ribbon on the bulletin board. **1945** L. SHELLY *Jive Talk Dict.* 22 *Boy scout*, an immature male. **1962** *Amer. Speech* XXXVII. 267 *Boy scout*,..a traffic policeman who spends much of his time in helping motorists with flat tires, stalled cars, empty gas tanks, and so on.

1965 A. NICOL *Truly Married Woman* 84 Bandele had said, threateningly, that he had better not mention his name, but he could go and be a Boy Scout on his own. **1969** *Sunday Mirror* (Sydney) 13 Apr. 9/4 The accused officer has always been known as a 'boy scout'—a policeman who goes strictly by the book. **1977** J. I. M. STEWART *Madonna of Astrolabe* iii. 57 It was Mark's idea. Mark's absolutely the Boy Scout.

(*c*) *attrib.* and *Comb.*

1909 *Daily Chron.* 9 July 8/2 The youth of the three kingdoms of the boy scout movement. **1914** W. OWEN *Let.* 1 June (1967) 257 Certain of my Boy-Scout acquaintance. **1936** W. R. F. COLLIS *Silver Fleece* v. xvi. 266 Above all he hated..the 'boy scout' mentality, the modern tendency to march about in mobs, wearing coloured shirts, shouting, and beating up anybody who doesn't shout with you. **1950** 'E. CRISPIN' *Frequent Hearses* iv. 198 'What about the knife?' 'An oversized boy-scout affair..ground razor-sharp.' **1967** J. PORTER *Chinks in Curtain* v. 53 All right, chum, and what would you have done? Given the Boy Scout salute? **1978** S. BRILL *Teamsters* ii. 43 His father was a teetotaler, a boy-scout husband who came home every night.

(*d*) Hence *Boy Scoutery, Scoutism*, the activity or attitude of a Boy Scout; also *fig.*; *Boy Scoutish* adj., characteristic of a Boy Scout. (Freq. with pejorative connotations.)

1937 WYNDHAM LEWIS *Blasting & Bombardiering* v. ii. 254 He never got us under canvas it is true—we were not the most promising material for Ezra's boyscoutery. **1938** 'G. ORWELL' *Homage to Catalonia* iv. 49 We and they used to make daylight patrols there. It was not bad fun in a Boy Scoutish way. **1942** WYNDHAM LEWIS *Let.* (1963) 325 A curiously beastly case of boyscoutism, of arrested development or cretinism. **1962** L. DEIGHTON *Ipcress File* xiv. 84 Calling me 'boy-scoutish' which he knew would hit me where it hurt. **1963** J. VAIZEY *Educ. in Class Society* 6 These and other [travel and cultural] schemes..would.. tend to counterbalance people's enthusiasm for the Boy Scoutism of the Duke of Edinburgh's award. **1967** E. GRIERSON *Crime of One's Own* xi. 96 You must be mad... You've had that look about you ever since you started this boy-scoutery.

d. A bee searching for a new site for a swarm to settle or a new source of food.

1835 *Penny Cycl.* IV. 153/1 It is said that bees send out scouts before leaving the hive, to search for a convenient situation for their new abode. **1909** S. L. BENSUSAN *Children's Story of Bee* vii. 108 The scouts..might have been seen following their aërial roads to where the swarm was stationed. *Ibid.* 110 The last of the old queen's scouts had come bringing news of a hive—clean, sweet-scented and empty—in a garden across the valley. **1954** D. ILSE tr. *von Frisch's Dancing Bees* v. 28 While the main swarm hangs from a branch in quiet idleness, its 'scouts' are busily at work, searching in all directions to find a suitable abode.

e. One sent out by an organization (as a sports club, recording company, etc.) to look for suitably talented persons with a view to their employment by that organization; a talent scout.

1905 *Sporting Life* 2 Sept. 25/4 Padden..is the official scout of the St. Louis Club. **1926** WHITEMAN & MCBRIDE *Jazz* iii. 65 Vaudeville scouts approached us. Our pictures were in the papers. **1948** *Sporting Mirror* 19 Nov. 2/2 As chief scout for Derby County he will make sure that no young Midlander with real talent fails to get a chance to develop his soccer. **1952** A. LOMAX *Mister Jelly Roll* 291 'Fritz Pollard' introduced me to..Williams who was then a scout for the Paramount Recording Company. **1968** *Blues Unlimited* Sept. 15 They had their scouts out looking for anyone who could make records. **1976** E. DUNPHY *Only a Game?* iii. 92 He was being watched by Manchester City. Their scout left before the end.

f. An official of the A.A. or R.A.C. employed to assist motorists on the road. (No longer in use.)

1909 *Q. Rev.* Jan. 143 The scouts have, beyond doubt, done a great deal to check reckless driving. **1929** E. LINKLATER *Poet's Pub* xviii. 200 They passed a scout of the Automobile Association. 'You should have returned that A.A. man's salute.'

g. *slang.* A fellow, chap, person. Freq. in approbatory use, as *good scout*, etc., and as an affectionate term of address.

1912 M. NICHOLSON *Hoosier Chron.* 129 Dad's a good old scout and he's pretty sure to do it. **1921** WODEHOUSE *Indiscretions of Archie* vii. 61 You'll never be lonely with Peter around. He's a great scout. Always merry and bright. **1922** J. A. DUNN *Man Trap* xii. 168 You didn't tell me your name, old scout. **1933** E. O'NEILL *Ah, Wilderness!* (1934) III. i. 96 Nat Miller's a good scout. **1950** A. WILSON *Such Darling Dodos* 198 She had only two roles with men—tomboy and good scout. **1953** 'N. BLAKE' *Dreadful Hollow* 112 'It's time I sent him a report.' 'Pop? No need, old scout.' **1965** 'J. LE CARRÉ' *Looking-Glass War* iv. 38 I've got nothing against old Adrian. He's a good scout.

3. Delete † and add: Now only *U.S.* (Later examples.)

1798 I. ALLEN *Hist. Vermont* 92 He sent a scout of about 300, mostly Indians, to hunt at the mouth of Otter Creek. **1867** J. N. EDWARDS *Shelby & his Men* xxiii. 412 At Current river a scout of fifty were encountered. **1940** W. FAULKNER *Hamlet* II. i. 127 A scout of two or three would lurk about the Varner fence.

4. a. (Further examples.) Also *spec.*: in oil-drilling operations, one employed by a company to keep watch on the activities of other companies; in *Sport*, one employed to observe and report on the performance of rival teams or clubs. Also, † a policeman.

1789 [see *HORNY *sb.* 2]. **1821** P. EGAN *Life in London* II. iii. 231 Turning the corner of Old Bedlam, A scout laid me flat upon my face. **1838** DICKENS *O. Twist* III. xlvi. 213 They'll have him yet, for the scouts are out, and by to-morrow night there'll be a cry all through the country. **1883** [in Dict.]. **1883** *Derrick's Handbk. Petroleum* (1898) I. 357 Scouts have squatted on the Reed and Brenneman lease..and are keeping a vigilant watch on the well; efforts to dislodge the scouts have proved unavailing. **1904** *Dialect Notes* II. 388 *Scout*,..a man sent to obtain information regarding a mystery. **1949** *Athletic Jrnl.* Oct. 20/1 The scout should familiarize himself long before the season starts with the types of defense that have been used by opponents in the past. **1973** C. CALLOW *Power from Sea* i. 14 The..oil industry employs men to keep tabs on the competition and has given them the euphemistic term of 'scouts'. **1976** M. MACHLIN *Pipeline* ii. 32 There had to be a scout on that plane.

5. b. An airship or aeroplane used for reconnoitring; a lightly-armed fighter aeroplane. Also *attrib.*

1909 A. BERGET *Conquest of Air* II. vii. 260 Airships or aeroplanes? As 'combatants' or 'scouts'? **1914** *Daily Express* 31 Dec. 3/4 We have 'scouts' which can beat anything the enemy can bring against us. **1916** H. BARBER *Aeroplane Speaks* Pl. xi, The little Gnome-engined scout biplanes. **1928** E. BLUNDEN *Undertones of War* viii. 82 On account of the aforementioned ceremonial parade, with the gleaming bayonets and accoutrements not unnoticed by German flying Scouts, the town was shelled by heavy guns on the day that we departed. **1942** *R.A.F. Jrnl.* 2 May 1 Among the planes..is an approximately equal number of bombers, pursuit planes, and scouts. **1978** H. WOUK *War & Remembrance* xxx. 304 The air raid proved to be only some old-type scout bombers buzzing a battleship of the screen and then running away from the Zeroes into the light clouds.

6. a. (Earlier example in *Baseball* and later examples in *Cricket*.)

1856 *Spirit of Times* 27 Dec. 276/3 One of these swiftly-delivered balls, when stopped by a skillful batsman, is sure to give the outmost scout employment. **1898** J. A. GIBBS *Cotswold Village* xi. 241 So also one may say..after the famous Gloucestershire hitter [*sc.* Grace] has made things merry for spectators and scouts alike. **1924** N. CARDUS *Days in Sun* 49 Supposing the fieldsmen were set..with still a number of them idle on the off-side, with great gaps between the leg-side scouts.

8. *scout boat* (earlier examples), *-craft* (later examples), *hut, knife*; **scout bee** (sense *2 d); **scout car,** (*a*) *U.S.*, a police patrol car; (*b*) *Mil.*, a fast armoured vehicle used for reconnaissance and liaison; **Scout Law,** a code of conduct enjoined upon (Boy) Scouts; **Scout's honour,** the honour on which a (Boy) Scout promises to obey the Scout Law; freq. *transf.*, as an expression of one's good faith.

1924 A. M. STURGES *Pract. Beekeeping* 306/1 Scout-bees. **1935** J. C. KENLY *Cities of Wax* xvi. 165 A scout-bee..had just brought in news to her hive that she had discovered a honey gold-mine. **1963** T. A. SEBEOK in J. A. Fishman *Readings Sociol. of Lang.* (1968) 23 [M. Lindauer] traces how the scout bees announce the location of suitable nesting places by means of the dance in the cluster. **1975** *Country Life* 20 Feb. 448/2 This swarm..was resting while scout bees looked for a suitable site. **1717** in *Statutes at Large S. Carolina* (1838) III. 24 For the scout boat on Port Royal Island, a Captain and six private men. **1862** F. MOORE *Rebellion Record* V. II. 182 The scout-boats of Com. Montgomery notified him of the presence of the Federals. **1933** *Sun* (Baltimore) 5 May 11/2 Scout car No. 7..answered fifty-five calls in the Pimlico section of the Northern district. *a* **1944** K. DOUGLAS *Alamein to Zem Zem* (1946) 40 A Daimler scout car, flying a red cross, was moving and halting,..collecting wounded. **1960** R. M. OGORKIEWICZ *Armour* xxxi. 434 A far more advanced Daimler scout car was also designed before the war, and the first built in December 1939. *Ibid.* 435 Originally the Daimler scout cars were intended for liaison within the tank regiments of the armoured divisions,..but after the 1940 campaign their use was extended, as was that of armoured cars. **1977** H. INNES *Big Footprints* I. ii. 47 More craters. A burned-out scout car, some lorries gaping holes, then we were clear of the battlefield. **1910** *Chambers's Jrnl.* Feb. 114/2 They are taught scout-craft, which includes the art of stalking wild creatures. **1937** *Sun* (Baltimore) 2 July 6/1 Thousands of Boy Scouts gathered in a giant arena tonight to watch a pageant of scoutcraft and history. **1977** N. ADAM *Triplehip Cracksman* v. 54 Using my entirely non-existent knowledge of scoutcraft, I snuck up it in the closing shadows. **1974** *Times* 10 Jan. 18/7 The Scout huts in New Zealand. **1976** L. HENDERSON *Major Enquiry* ix. 54 He got bored with working in the scout hut. **1937** E. GARNETT *Family from One End Street* v. 99 Hadn't he wanted a scout knife as long as he could remember. **1977** J. PORTER *Who the Heck is Sylvia?* xvi. 150 The kid..clipped his scout knife back on his scout belt. **1908** R. S. S. BADEN-POWELL *Scouting for Boys* 49 The Scout Law. **1922** *Encycl. Brit.* XXX. 487/2 The Scout Promise, to carry out, on his honour, as far as in him lies, the Scout Law, is the binding disciplinary force. **1931** E. WAUGH *Remote People* 134 A Somali boy presented himself for examination in scout law. **1972** P. BLACK *Biggest Aspidistra in World* I. vi. 51 An establishment striving..to do its best according to the Scout Law, continued to pervade the spirit of Children's Hour. **1908** R. S. S. BADEN-POWELL *Scouting for Boys* 49 A Scout's Honour is to be trusted. **1956** 'E. MCBAIN' *Cop Hater* (1958) xx. 168 'Provided it's not for publication.' 'Scout's honour,' Savage said. **1959** [see *BROWNIE[1] 2]. **1974** A. ROSS *Bradford Business* 175 'I'll try my best, Charlie,' I said, 'scout's honour.'

scout, *sb.*[6] (Later examples of sense 'a female servant'.)

1935 D. L. SAYERS *Gaudy Night* v. 91 'The scouts are all women of excellent character, so far as I know,' said the Bursar. **1972** *Oxford Times* 26 May 1/3 Miss Bootes, who has been a scout at St. Hilda's College for 25 years, was presented with the teapot on Wednesday.

scout, *v.*[1] **2.** For '? Now *dial.*' read *Obs.* and add earlier and later examples.

1786 *County Mag.* Nov. 171/3 One that can throw well, likewise scout, He for a long stroke must stand out. **1928** *Observer* 1 July 29/4 An arrangement by which Tate is required to rest from his bowling by scouting at deep square leg.

scouter. Restrict † *Obs.* to sense in Dict. and add: **1.** (Later example.)

1867 J. N. EDWARDS *Shelby & his Men* xiv. 232 One of the truest scouters who ever fired a pistol.

2. An adult member of the (Boy) Scout movement. Cf. *GUIDER s.v. *GUIDE *sb.* 2 d.

1930 H. W. BENSON *Summer Camp* 27 [The Patrol Leader] must be taught..to decentralize... A Scouter or Rover will probably be required to show him how to set about this. **1948** *Lawton* (Okla.) *Constitution* 4 July 3/5 A large number of cubbers and scouters were present with their families. **1976** *Burnham-on-Sea Gaz.* 20 Apr., A strong committee should be established, to be responsible for fund-raising activities, leaving the various scouters to carry on their roles as cub-scout leaders.

scouting, *vbl. sb.*[1] Add: **1. a.** *attrib.* (Earlier and later U.S. examples of *scouting-party*.)

1756 R. ROGERS *Jrnl.* 28 Apr. (1765) 17 They discovered a scouting party of three or four hundred Indians. **1895** A. C. HAMLIN in M. A. Jackson *Mem. Stonewall Jackson* (ed. 2) 548 A Federal scouting party could have come up the Hazel Grove road and seized him as a prisoner of war. **1941** B. SCHULBERG *What makes Sammy Run?* xi. 199 The other members of the Wall Street scouting party were punctual.

b. (Later examples.) Also, the Scout movement itself.

1966 *Listener* 20 Oct. 570/1 'Scouts go mod', the headline said—as if the cut of the new mushroom-coloured, tapering long trousers was the most important feature of the new pattern of Scouting. **1976** *Eastern Even. News* (Norwich) 9 Dec. 6/4 I'm sure there must be lots of men with souvenirs from their Scouting days. **1977** *Times* 27 June 16/8 Olave, Lady Baden-Powell..exemplified the potential of Scouting and Guiding for world peace.

c. The activity of a scout (*SCOUT *sb.*[4] 2 e or 4 a).

1908 *Baseball Mag.* Nov. 1/1 There was the reconnoitering, scouting, feints, retreats, invasions, and then preparations made by all the ball-and-bat generals for the final big battle. **1961** J. S. SALAK *Dict. Amer. Sports* 382 *Scouting*, viewing an opposing team in action before playing them. **1968** *Blues Unlimited* Nov. 5 Joe Bihari..asked him to do some scouting for Modern Reds, as well as playing on further second dates.

2. For '? *Obs.* exc. *dial.*' read *Obs.* and add earlier and later examples.

1815 *Suffolk Chron.* 2 Sept., The Needham players are remarkable for excellent scouting. **1885** F. GALE *Life of Hon. Robert Grimston* vii. 77 Lords' forty years ago was practically a country ground... There were no nets, but ground boys did the scouting. **1908** W. E. W. COLLINS *Leaves from Old County Cricketer's Diary* xi. 191 Quite a young soldier..did most of the scouting in the far country.

scoutmaster, scout-master. Add: **1. a.** Now usu. in latter sense. (Later examples.)

1910 *Chambers's Jrnl.* Feb. 117/2 Officers, called scout-masters, must be over eighteen. **1928** R. KNOX *Footsteps at Lock* v. 42 The scout-master, a man of some age and education. **1942** E. WAUGH *Put out More Flags* i. 14 His binoculars which she remembered vaguely having lent to the scout-master. **1977** S. BRETT *Star Trap* xi. 126 He's a scout-master and tends to be off camping..most weekends.

Hence **scou·tmastering,** the occupation of a scoutmaster; **scou·tmasterish, scou·tmasterly** *adjs.*, resembling or characteristic of a scoutmaster.

1937 'G. ORWELL' *Road to Wigan Pier* x. 192 Shouted out of existence with a few scoutmasterish bellows of good-will. **1954** E. HYAMS *Stories & Cream* 9 The stout, genial Scoutmasterly fellow. **1957** L. DURRELL *Justine* II. 124 I've done quite a bit of scoutmastering. **1964** K. G. GRUBB *Layman looks at Church* v. 152 Any form of voluntary activity from local government to scout-mastering is in the same difficulty, namely that the pressures and claims of business make voluntary work impossible. **1979** K. BONFIGLIOLI *After You with Pistol* v. 21 He patted me on the shoulder in a scoutmasterly way.

scow[2]. Add: **1. b.** *U.S.* A small flat-bottomed racing yacht.

1929 B. HECKSTALL-SMITH *'Britannia' & her Contemporaries* viii. 82 The result of these changes was that ten years after the Britannia was built the type of racing yacht had developed into a scow with a fin keel. **1932** *Sun* (Baltimore) 13 Aug. 18/7 Pitted against the High Tide was the Inland Laker type sailing scow Elusive. **1970** *Globe & Mail* (Toronto) 25 Sept. 30/1 The Australian tub, Gretel, hit the American scow, Intrepid. **1976** *Oxf. Compan. Ships & Sea* 762/2 *Scow*,..used in the U.S.A. today to describe a small flat-bottomed racing yacht fitted with bilge boards or retractable bilge keels.

c. *transf.* Applied to one of several containers or vehicles used for transporting loads (see quots.).

1942 *Amer. Speech* XVII. 104/2 *Scow*, truck with extraordinary capacity for a big load. **1961** F. H. BURGESS *Dict. Sailing* 180 *Scow*,..A shallow tray for hoisting small packages of cargo. 1971 M. Tak Truck Talk 136 Scow,..a low-sided trailer used for hauling pipe, steel, stone, gravel, scrap and similar cargo. 1973 Amer. Speech 1969 XLIV. 208 Scow, low-sided truck or rig used for hauling pipe or steel. 1977 New Yorker 18 July 23/2 There is even, in a projected television series, a pilot of a spaceship (an interplanetary garbage scow) who is called Adam Quark.

3. scow schooner, sloop *U.S.* (see quot. 1885).

1885 17th Ann. List. Merchant Vessels U.S. p. xxx, Scows are built with flat bottoms and square bilges, but some of them have the ordinary schooner bow. They are fitted with one, two, and three masts, and are called scow-sloop or scow schooner, according to the rig they carry. 1913 J. London Valley of Moon 269 At the foot of Castro street.. the scow schooners, laden with sand and gravel, lay hauled to the shore in a long row. 1951 H. I. Chapelle Amer. Small Sailing Craft 334 A few scow schooners were built with round bilges, but they were comparatively rare. 1885 Scow sloop [see scow schooner above]. 1941 H. I. Chapelle Boatbuilding 28 The New Jersey oyster garvey, the Maine scow sloop, and the San Francisco scow schooner represent examples of the practical use of such hull forms. 1953 Sunday Sun Mag. (Baltimore) 18 Oct. 24/4 An oddity in the sloop rig was the scow sloop, once common at the head of the Chesapeake near Havre de Grace... The last in service..was abandoned about 1940.

scow, *v.*[1] **b.** (Examples.)

1751 J. MacSparran Diary 1 Oct. in Letter Book (1899) 58 He and a Boy..were Scowing wood. 1929 W. Heyliger Builder of Dam 39 From this point I will scow the supplies over to the job.

scow (skau), *v.*[3] *north. dial.* [Origin uncertain; prob. related to Scowbanker.] *intr.* To loiter, idle; to shirk work, play truant. Hence **sco·wing** *vbl. sb.*

1901 F. E. Taylor Folk-Speech of S. Lancs. s.v. Scow, to idle about. 1905 in Eng. Dial. Dict. V. 265/2 Now then, you're always scowing. 1959 I. & P. Opie Lore & Lang. Schoolch. xvii. 372 Sagging..is definitely the prevailing term [for playing truant] amongst delinquents in all parts of Liverpool. A student..adds 'scowing' as a Liverpudlian expression. 1966 F. Shaw et al. Lern Yerself Scouse 58, I wuz scowing, I was having an unofficial spell of leisure time.

scowbanker. Add: Also 8 **scou-, -bancker.** Also, † one who engages in unfair business practices, a dishonest or unscrupulous trader. (Earlier examples.)

1750 G. Beekman Let. 4 Dec. in P. H. White Beekman Mercantile Papers 1746–1799 (1956) I. 136 You may thank a Sett of People Called Scowbanckers..that Seed has Run so high this two years past. Our town is full of them and there is Scarce a Vessell Comes along the wharffe but there is Imediately a half a Dozen of Them aboard bidding against Each other. 1764 —— Let. 30 Nov. in Ibid. 478 Our Vandue houses are Crowded with Linens for Sale belonging to the Scoubankers who are Offering of it from house to house for Less then Prime Cost which hurts the merchant much.

scowly (skau·li), *a.* [f. Scowl *sb.*[1] or *v.* + -y[1].] Given to scowling; sullen, morose.

1951 H. Giles Harbin's Ridge 64 He did take to drinking mighty heavy, and he got to acting sully and scowly. 1970 Daily Tel. (Colour Suppl.) 28 Aug. 21 Her scowly freckled face lights up when she laughs.

scrabble, *sb.* Add: **2.** *U.S.* A scramble; a confused struggle, a 'free-for-all'.

1794 Gazette of U.S. 21 Feb. 3/2 The Frenchman..in a scrabble swore he would have another hem to his ruffle, and in the very scrabble lost his shirt. 1849 T. T. Johnson Sights in Gold Region 66 We often got caught by the waves, and had a grand scrabble to reach dry land. 1911 R. D. Saunders Colonel Todhunter of Missouri 43 Whoever wins will win after the toughest scrabble you and me ever saw in Missouri politics.

3. The action or sound of scrabbling (Scrabble *v.* 2 a).

1894 T. B. Aldrich Two Bites at Cherry 145 The next sound I heard was the scrabble of the animal's four paws as he landed on the gravelled pathway. 1946 D. C. Peattie Road of Naturalist i. 12, I could not hear her breathing, but I heard another sound... Someone else was trying, with a faint scrabble, to find his way out.

Scrabble (skræ·b'l), *sb.*[2] Also **scrabble. a.** The proprietary name of a game in which players use tiles displaying individual letters to form words on a special board.

1950 Official Gaz. (U.S. Patent Office) 10 Jan. 334/2 The Production and Marketing Corporation, Newtown, Conn... Scrabble. For Game including Board and Playing Pieces. Claims use since Dec. 1, 1948. 1953 New Yorker 30 May 17/2 We present for your edification the history of Scrabble, the biggest thing in games since Monopoly and maybe the biggest thing ever. Ibid. 18/1 It was as if everyone alive were suddenly clamoring to play Scrabble. 1954 Trade Marks Jrnl. 21 July 736/1 Scrabble... 'Board games.' Production and Marketing Corporation (a Corporation organised and existing under the laws of the State of Connecticut, United States of America; Merchants). 1957 T. Girtin in Pick of Punch 150/1 My suspicions were first aroused while I was losing to my wife at 'Scrabble'. 1959 C. Spry Favourite Flowers iii. 25 For relaxation I sometimes play the spelling game of Scrabble and in consequence am wearing to ribbons the unwieldy volumes of the Shorter Oxford English Dictionary. 1962 A. Sampson Anat. of Britain xxvii. 450 He likes going home early,..and plays bridge or scrabble in the evenings. 1971 C. Bonington Annapurna South Face ix. 107 After the meal we played liar dice or Scrabble. 1978 J. Matson Dear Osborne xxii. 151 Scrabble, Shove Ha'penny and Draughts indicate the levels of skills and activities.

b. *attrib.* and *Comb.*

1954 Newsweek 26 Apr. 57 To help Scrabble fans, crossword-puzzle addicts, and other persons troubled for a word ending in 'x', 'y' or 'z', a 'reverse' dictionary has been compiled at the University of Massachusetts. 1956 N. Streatfeild Judith ii. 117 Cynthia sprawled over the Scrabble board. 1960 Sunday Express 11 Sept. 6/6, I leave it to Oscar, the Scrabble-playing cat, to dredge up obscure words. 1967 Sci. Amer. Sept. 268/1 The Double-Crostic and games of the Scrabble type can be thought of as combinatorial play in which 26 elements (letters) are arranged into sets (words). 1977 B. Garfield Recoil x. 103 Anna made a word on the Scrabble board and watched him enter the score.

scrabbling, *vbl. sb.* (Later examples.)

1958 Washington Post 31 Oct. A 3/1 The kind of digging, scrabbling and clawing that accomplished the rescue of the 12 who were brought out alive early today. 1974 D. Sears Lark in Clear Air v. 60 I'd never known anything other than hard times. Nor did rough scrabbling impose other than normal conditions on Brulé Township.

scrag (skræg), *sb.*[3] *slang. rare.* [f. Scrag *v.*: see sense 1 b, quot. 1897.] In Rugby football, a rough tackle.

1903 Wodehouse Tales of St. Austin's 105 There's all the difference between a decent tackle and a bally scrag like the one that doubled Tony up.

scrag, *v.* Add: **1. c.** To treat (someone) roughly, to manhandle.

1835 Sessions Paper of Central Criminal Court May 87 He did not take him by the collar and shake him—he did not collar him at all till after the blow was struck, nor push him at all—I did not hear Emerson say, 'You b——, I'll scrag you.' 1901 Daily Colonist (Victoria, B.C.) 31 Oct. 4/3 'What makes the crowd get up and yell?' inquired the fairy maid. 'They've scragged a man, they've scragged a man,' the woolly rooter said. 1938 [see *knocking-shop]. 1947 N. Balchin Lord, I was Afraid 52 Before he could say another word they scragged him. 1959 I. & P. Opie Lore & Lang. Schoolch. x. 198 The term 'scragging' is recurrent everywhere, and seems in fact to be different from giving someone a 'beating up' or 'bashing'. One boy makes the distinction: 'To scrag is a more gentle way of having a kind of hurtful revenge. You pull his hair and take his tie off and that sort of thing.' 1969 —— Children's Games vii. 219 The first one to get off, gets scragged by the other lads. 1977 H. Fast Immigrants iii. 193 Now they've scragged me, ruined me, destroyed me.

d. To kill, murder. *U.S.*

1930 D. Runyon in Collier's 20 Dec. 13/4 John the Boss is a very fine character, and it is a terrible blow to many citizens when he is scragged. 1938 —— Furthermore iii. 51, I see by the papers where three Brooklyn citizens are scragged. 1950 Reader's Digest Nov. 57 If they aim at me they will overshoot or undershoot and scrag some scared civilian.

2. (Not *slang.*) To subject (a spring or suspension system) to scragging (see below). Also with *out*: to shorten the normal length of a spring by (a specified amount) by means of scragging.

Orig. in a different sense (see quot. 1909).

1909 Webster, Scrag, Mech., to bend, as spring steel to test it. 1923 T. H. Sanders Laminated Springs xi. 89 That spring would be subjected to probably another 3½ ins. or even 4 ins. test to 'scrag out' the unwanted ½ in. Ibid. xxxvii. 396 (caption) The finished spring being scragged. 1958 A. D. Merriman Dict. Metall. 308/1 The spring is wound somewhat longer than the required length and then scragged by compressing it to closure several times. 1969 Maxi Workshop Man. (Brit. Leyland Motor Corp.) x. A7 After fitting a new displacer unit to the front or rear suspension, the system should be scragged by raising the fluid pressure, to above its normal pressure, for a short period. 1972 Pract. Motorist Oct. 87/2 If the displacer isn't scragged, it takes up a 'set' with the car's weight upon it—it becomes permanently compressed and the car assumes a list.

scragging *vbl. sb.* (later examples); also *spec.* the process of extending a new spring beyond the desired normal length, and then compressing it, in order to improve its strength and set; an analogous process applied to a hydraulic suspension system in a motor vehicle.

1923 T. H. Sanders Laminated Springs xi. 90 American practice invariably indulges in scragging machines of the 'bull-dozer' type. Ibid. xxxvii. 395 An illustration of 'scragging' as carried out in this country is shown by Fig. 201, which shows a 12-plate..spring undergoing its test. 1936 Horner & Sprague Dict. Terms Mech. Engin. (ed. 6) 486 Scragging, the process of testing carriage and locomotive springs by impulsive loading. 1949 A. Hynd We are Public Enemies 79 The Ash Brothers had committed the scragging. 1959 'M. Innes' Hare sitting Up ii. iii. 61 We absolutely soaked them with our water jugs, and they gave us a wonderful scragging afterwards. 1969 Maxi Workshop Man. (Brit. Leyland Motor Corp.) x. A7 (heading) Displacer unit 'scragging'. 1977 R. B. Ross Handbk. Metal Treatments & Testing 338 Scragging... The process is that the spring, when initially formed, is made longer than the design requirements. By applying the necessary compression load, the length of the spring is reduced and at the same time compressive stresses are applied to the surfaces of the spring.

scraggly, *a.* Add: (Earlier and later examples.) Chiefly *U.S.*

1869 Mrs. Stowe Oldtown Folks xlii. 534 That's all we scraggly old people are good for. 1946 C. McCullers Member of Wedding ii. 57 A street preacher..was preaching on a corner to a group of warehouse coloured boys and scraggly children. 1959 T. Griffith Waist-High Culture (1960) 172 We are like an animal that casts off its skins too quickly; no wonder we look scraggly. 1976 National Observer (U.S.) 13 Mar. 12/3 At least I'm not spread-eagled any more in my skivvies and scraggly black socks.

scram (skræm), *sb.*[1] Var. Scran *sb.* 2

1831 S. Lover Legends & Stories Ireland 96 Bad scram to you, you thick-headed vagabone. Ibid. Gloss. Bad scram, bad food. 1881 J. Sargisson Joe Scoap 148 He cot a model eh what he thowt t'shap on't sud be, oot of a lump eh baykin-scram. a 1935 [see *aid sb. 1 b]. 1936 F. Clune Roaming round Darling xxiv. 246 After unloading flour, spuds, tea, sugar—every kind of scram, we lobbed inside the house.

scram (skræm), *sb.*[2] *Nucl. Physics.* [f. *scram *v.*[3]] The rapid shutting down of a nuclear reactor, usu. in an emergency. Freq. *attrib.*

Both this word and *scram v.[3] are possibly derived from *scram v.[2]

1953 Nucleonics June 40/2 Momentary-contact types [of push button] used to operate..scram circuits. 1955 Ibid. Sept. 53/2 Scram is initiated if preset power level is exceeded by 20%. 1959 New Scientist 26 Mar. 696/3 The [Nautilus submarine] Mark I had a constant plague of 'scrams' from such slight causes as vibration from a crew member's walking through the reactor compartment. 1968 F. Kertesz Lang. Nuclear Sci. (Oak Ridge Nat. Lab. TM 2367) 21 During the experiment that culminated on December 2, 1942 in the accomplishment of the first controlled nuclear chain reaction, a safety rod was held by a rope running through the pile and weighted on the opposite end. The young physicist in charge was told to watch the indicator; if it exceeded a certain value he was to cut the rope and scram. Since then the term scram is used to designate the emergency shutdown of a reactor. Today the urgency is lost and the word scram indicates simply a fast-shutdown operation. 1973 D. R. Inglis Nuclear Energy iv. 117 Emergency shutdown or scram equipment must be very sure to function properly.

scram (skræm), *v.*[2] *slang* (orig. *U.S.*). [Prob. abbrev. of Scramble *v.*; but cf. G. *schramm* imp. sing. of *schrammen* to go, depart, run away.] *intr.* To depart quickly. Freq. *imp.*

1928 W. Winchell in N.Y. Evening Graphic 4 Oct. 23 His [sc. Jack Conway's] popular slang creations include.. 'scram', meaning 'git out!' 1933 Punch 11 Jan. 29/3 Son, beat ut, d'ya get me?—Gwawn—S-C-R-R-A-M! 1937 D. L. Sayers Busman's Honeymoon iv. 84 Well, I must scram. 1940 N. Mitford Pigeon Pie iv. 75 She gave a sort of shriek ..and scrammed. 1947 D. M. Davin Gorse blooms Pale 172 You tell her to scram. 1952 J. Cannan Body in Beck vii. 146 'Perhaps you would be good enough to withdraw.'.. Sebastian said, 'He means scram.' 1962 Wodehouse Service with Smile vii. 97 'Go away, boy!' he boomed. 'You mean 'Scram!', don't you, chum?' said George, who liked to get these things right. 1973 A. Hunter Gently French xi. 96 Kindly hook it... I just want you to scram.

scram (skræm), *v.*[3] *Nucl. Physics.* [Etym. unkn., but see note s.v. *scram *sb.*[2]] **a.** *trans.* To shut down (a nuclear reactor), usu. in an emergency.

1950 Amer. Speech XXV. 27 The point of neutron intensity at which the reactor is 'scrammed'—shut down, automatically or otherwise. 1953 Nucleonics Jan. 40/2 The operator is provided with a control console from which he can change the position of rods, switch into automatic control, and scram the reactor. 1959 New Scientist 26 Mar. 695/3 A highly sensitive system of eighty different control circuits was designed to anticipate any dangerous instability of the reactor and within a fraction of a second initiate an emergency shutdown, or in the jargon of the nuclear engineer, 'scram' the reactor. 1973 D. R. Inglis Nuclear Energy iv. 95 The current can also be used, when it gets too strong, to trigger the emergency control rods and 'scram' or shut down the reactor. 1975 Nature 16 Oct. 526/1 At 1251, the operator decided to shut the reactor down by inserting the control rods into the core, thereby cutting off the chain reaction (in operator's parlance, he manually 'scrammed' the reactor).

b. *intr.* Of a nuclear reactor: to shut down, usu. in an emergency.

1957 Nucleonics Feb. 56/3 After a reactor scrams, the question immediately arises: What circuit caused the scram and what happened during shutdown? 1979 New Scientist 19 Apr. 174/1 At 2350 lb/sq. in, the reactor automatically 'scrammed' and seconds later the pressure began to drop.

Hence **scra·mming** *vbl. sb.*

1958 Nucleonics May 64 The entire basis for scramming ..may well need to be re-examined for future power reactors.

scramasax. (Later examples.)

1917 W. M. F. Petrie Tools & Weapons v. 27 Examples from Mainz.., termed scramasax, are likewise equal-curved. 1923 C. Fox Archaeol. Cambridge Region vi. 301 To the period 950–1066 probably belongs a fine scramasax from Barrington..with damascened blade. 1936 Antiquity X. 374 Typological studies of Saxon scramasaxes and spear-heads. 1962 H. R. Ellis Davidson Sword in Anglo-Saxon Eng. 41 Mention may be made of the short dagger or dirk (scramasax or handseax). Ibid. 43 A scramasax from the Thames bears the twenty-eight characters of the runic alphabet. 1977 J. I. M. Stewart Madonna of Astrolabe

xvii. 243 Although he possessed a mass of material of great archaeological interest and considerable value, the scramasax was his only major treasure.

scramble, *sb.* Add: **2. b.** A motor-cycle race across rough and hilly ground.

1926 in H. Golding Wonder Bk. of Motors 177 Such races as the 'T.T.' and the various other Trials and 'Scrambles' organized by the larger clubs afford manufacturers an opportunity of submitting their machines..to..severe tests. 1935 Encycl. Sports 539/2 Scramble, form of motor-cycle trial in which the competitors..traverse a course marked out over moorland or heath... Among a certain section of motor cyclists scramble events are very popular. 1959 New Statesman 14 Nov. 658/1 About 50,000 people now turn out every week to watch the dozen or so scrambles organised throughout the country. 1969 Daily Tel. 25 Oct. 8/8 A scramble can best be described as a motorised form of a cross-country race over a short, rough course sometimes reserved for the purpose and often adjacent to a road-racing circuit. 1977 'E. Crispin' Glimpses of Moon xi. 213 The motor-cycle scramble had arrived.

c. *Mil. slang.* A rapid or operational take-off by a group of aircraft. Cf. sense *4 of the vb.

1940 G. Barclay Diary 2 Sept. in Fighter Pilot (1976) 44, I came on the state after this scrap and we had three scrambles. [1954 I. Jones Tiger Squadron xxii. 228 To Sailor's dismay, his scramble order had not been received quickly enough.] 1957 G. Wallace R.A.F. Biggin Hill xi. 139 Three scrambles a day were common, often more. 1963 Times 11 June 7/1 The royal visitors watched a 'scramble' of four R.A.F. Vulcan bombers of the quick-reaction alert force. 1969 P. Richey Fighter Pilot p. xii, He wanted to publish it if I would finish it. I did so..in the evenings after a day spent instructing on fighters in Gloucestershire and, later, between 'scrambles' and fighter sweeps. 1976 Derbyshire Times (Peak ed.) 3 Sept. 24/4 Featured in the spectacular flying display will be a scramble (operational take-off) by a pair of Vulcan bombers.

d. *U.S. Football.* An impromptu movement by a quarterback to evade tacklers. Cf. sense *1 d of the vb.

1971 Tarkenton & Yates Broken Patterns 52 Naturally the scramble plays were the most spectacular. 1972 J. Mosedale Football vi. 89 A man who played as though he invented the scramble.

3. *Cookery.* A dish composed of hastily-mixed ingredients; an informal meal of such dishes. Cf. Scrambling *ppl. a.* 1.

1893 Yonge & Coleridge Strolling Players xxii. 187 Selva's Irish hospitality could allow no one to depart in the rain, and her Irish happy-go-luckiness saw nothing to be ashamed of in a scramble. 1898 J. D. Brayshaw Slum Silhouettes 42 Some of the ladies of the district, deeming the short cut to a poor man's soul was through his vitals, invited them to a free 'muffin scramble'. 1938 E. Waugh Scoop i. i. 18 The recipe for a dish named 'Waffle Scramble'. 1958 Woman's Own 17 Sept. 15/1 Halve the rolls... Fill with the tuna scramble.

4. *Comb.*, as **scramble net** *Naut.*, a heavy net down which persons may climb from a ship in an emergency; also in *gen.* use, the webbing of a child's climbing frame.

1944 Amer. Speech XIX. 106 The scramble net (a new term in this war; it is an oversize cargo net hung over the side in times of imminent peril as an aid to getting down to water level without breaking your back). 1948 A. Baron From City from Plough 134 Sailors pulled at cords and the wet scramble nets thumped over the sides. 1953 Physical Educ. in Primary School (Min. of Educ.) II. vi. 32 (caption) Infants using a scramble net which is supported on a fixed tubular metal frame. 1976 Outdoor Living (N.Z.) I. ii. 49 (caption) Other attachments, such as a scramble net could be put on the frame as children outgrow the present equipment. 1979 'G. Black' Night Run from Java xiii. 162 The patrol boat crew [was] going down the scramble net into the lifeboat.

scramble, *v.* Add: **1. c.** Also (freq. with advbs.), to cause (something) to move in the direction or manner indicated without proper control; to deal with hastily or ill-advisedly.

1869 Punch 3 July 270/2 One 'Lord Hamilton' who had been scrambling away his money, at a low public in Shadwell. 1911 Q. Rev. July 218 Amendments hastily scrambled through committee in a House of Commons. 1931 Times 28 Feb. 4 The putter scrambled the ball in from a foot away. 1976 Oadby & Wigston (Leics.) Advertiser 26 Nov. 15/1 His shot hit the bar, and eventually Snee and his defence scrambled the ball free.

d. *intr.* *U.S. Football.* (See quot. 1976.)

1964 Birmingham (Alabama) News 26 Oct. 20/1 Campbell had to scramble to get off passes to Jimmy Martin, Scotty Long or Ronnie Baynes. 1968 N.Y. Times 13 Aug. 31 It was Tarkenton who engineered this monumental upheaval, mainly because he bewildered the Packers with his scrambling. 1976 Webster's Sports Dict. 373/2 Scramble, ..of a football quarterback, to run around behind the line of scrimmage dodging would-be tacklers after initial pass protection has broken down before passing or running.

3. b. *fig.* To jumble or muddle (something).

1927 G. Ade et al. Let. 4 Mar. (1973) 118 When you are in the native quarter [of Algiers] you can well imagine you are in the Old Testament which has been scrambled, stood on edge and saturated with all the disagreeable odors in the world. 1950 Times Lit. Suppl. 27 Jan. 55/4 The characters have been 'scrambled' so that none shall be recognizable. 1962 Listener 5 Apr. 603/1 The different arts are being run together and the borders of art and nature are being scrambled. 1977 Time 10 Jan. 43/1 Their landing permits had been deliberately scrambled by the Cuban government in league with the Nazis, who wanted the ship to sail from port to port searching for asylum.

c. To make (a telephone or radio signal of a voice) unintelligible by means of a scrambler (see *scrambler 3 a); to render (a television transmission) usable only by a subscriber equipped with a suitable unscrambling device. Also *transf.* and *absol.*

1927 Gen. Electr. Rev. XXX. 84/2 A Hammond multiplex system may be used with seven intermediate carrier waves which are scrambled and sent out by a single transmitter and then unscrambled at the receiving station so that each controls one of the seven light beams. 1929 Times 9 Nov. 12/2 An improved means for ensuring the secrecy of radio-telephonic conversations by 'scrambling' the words of the message—that is, by changing high frequencies to low frequencies and vice versa, so that the conversation is completely unintelligible until the 'inverted' conversation has been retranslated. 1949 F. Maclean Eastern Approaches iii. xii. 445 When we resumed our conversation [on the telephone], the Prime Minister was off on a new tack. 'Shall we scramble?' he said gaily. I replied that I thought I was scrambled. 1955 Times 4 June 6/6 All three work on the principle of a device attached to the subscriber's television set which 'scrambles' the programmes to be televised until a fee is paid to unscramble them. 1959 E. H. Clements High Tension v. 74, I was beginning to wonder..whether we ought to scramble the telephone. 1978 G. Greene Human Factor ii. iv. 89 There was the usual confusion: one of them pressing the right button too soon and then going back to normal transmission just when the other scrambled.

4. *Mil. slang.* **a.** *intr.* Of an aircraft (as a fighter plane, etc.) or crew: to effect a rapid take-off; to become airborne quickly. orig. *R.A.F.*

1940 G. Barclay Diary 2 Sept. in Fighter Pilot (1976) 44 The squadron scrambled and intercepted some Do215s and Me110s. 1941 [see *angel v. 2]. 1942 Brennan & Hesselyn Spitfires over Malta 15 The signal to scramble came at about eleven o'clock... We rushed to our aircraft and in less than two minutes were off the ground. 1944 Daily Tel. 15 May 5/3 Hardly were they past the carrier than two Corsairs 'scrambled' off the deck to 'intercept an enemy plane'. 1952 Times 22 Feb. 6/4 A red Very light was fired as a warning to the pilots to 'scramble', and exactly 80 seconds later the first fighter was in the air. 1962 R. W. Clark Rise of Boffins ii. 53 Another great time-saver was the use of a code for passing instructions to the fighters, and such R.A.F. terms as 'scramble' (for take-off)..were invented during these experiments [on radar interception, 1936]. 1977 R.A.F. News 5–18 Jan. 1/1 A Wessex SAR helicopter of 22 Squadron's E flight was scrambling from Manston.

b. *trans.* To cause (an aircraft) to become airborne quickly.

1940 G. Barclay Diary 3 Sept. in Fighter Pilot (1976) 46 The squadron was off the ground which was the main thing, but they were scrambled too late to intercept. 1953 Sun (Baltimore) 18 Sept. 2/2 Col. Richard T. Hernlund..showed reporters he could 'scramble' the interceptors and get them into the air almost instantly. 1971 Daily Tel. 20 July 8/7 The final decision to scramble fighters or launch nuclear missiles is..made by..highly trained officers. 1975 Radio Times 14 Aug. 38/1 A call to the coastguard and the nearest rescue helicopter can be scrambled within seconds. 1978 R. V. Jones Most Secret War xli. 383 It was clear that their technique was to 'scramble' their fighters on a raid warning and instruct them to orbit one of a number of visual and radio beacons. 1981 Times 9 June 6/8 The Iraqis.. gave no indication whether Iraqi jet fighters had been scrambled in an attempt to shoot down the Israeli planes.

5. *Comb.*, as **scramble button**, a button which activates a scrambler (sense *3 a) when pressed.

1962 L. Deighton Ipcress File xxx. 194 The wall phone rang... I saw Jay push the 'scramble' button.

scrambled, *ppl. a.* Add: **a.** (Later examples.)

1930 Daily Express 6 Sept. 4/6 A series of mixed or 'scrambled' wave lengths, the key to which would only be in the possession of individual subscribers to the television service. 1949 Jrnl. R. Aeronaut. Soc. LIII. 29/1 Subjected to the rapid change of temperature when simulating 'scrambled take-offs'. 1951 Good Housek. Home Encycl. 170/2 Pack..loosely rolled or 'scrambled' sheets of newspaper..over the surface. 1962 L. Deighton Ipcress File xi. 70, I will want to use a phone in private—a scrambled line if possible. 1975 J. I. M. Stewart Gaudy xii. 218, I pictured Mogridge..in some high-powered but unobtrusive car—one equipped, no doubt, with telephonic devices enabling him to hold scrambled conversations with various quarters of the globe as he went along.

c. *scrambled egg(s)* fig., the gold braid or insignia worn on an officer's dress uniform (esp. the cap); hence by metonymy, an officer. *slang* (chiefly *Forces'*).

1943 C. H. Ward-Jackson It's Piece of Cake 52 Scrambled egg,..an officer of the rank of Group Captain or above. Ibid. 53 Scrambled eggs, the gold braid or oak leaves on the peak of the dress service cap of an officer of the rank of Group Captain or above. 1946 J. Batten Dirty Little Collier iv. 32 Most collier men hate their 'scrambled egg' caps, and won't wear them at all. 1958 M. Dickens Man Overboard iv. 60, I don't care about the scrambled egg, but it may be a bit tough at first, not being an officer. 1961 C. Brooke-Rose Middlemen i. 7 A blue-grey uniform with three rings round his wrist. Then four. Then a big broad one and scrambled eggs. 1968 J. Lock Lady Policeman ix. 84 The car drivers.. don't know which one to obey..being intimidated by all that scrambled egg on their caps. 1973 R. Dougall In & Out of Box xiii. 157 An older pilot with some 'scrambled-egg' on his cap was produced. 1978 Detroit Free Press 5 Mar. (Parade Suppl.) 14A/4 (Advt.), Capture the soaring spirit of American history—proud cap has smart military styling with golden 'scrambled eggs' & braid trim hand-sewn on the visor.

scrambler. Add: **I. 1.** (Later sporting examples.)

1954 M. Connolly Power Tennis 66 Many times I have seen a scrambler unnerve a much better player merely by returning her best placements. 1958 Oxford Mail 27 Aug. 4/9 Most of Britain's leading motor-cycle scramblers will be at Brill on October 5. 1961 Times 29 Apr. 3/2 Can it be that his consummate skill as a scrambler is beginning to weigh on him? 1967 Time 17 Mar. 55 He is known in the trade as a 'scrambler', who would just as soon run as throw, who can turn a potential 10-yd. loss into a 50-yd. gain. 1972 J. Mosedale Football vi. 86 Quarterbacks usually identified as scramblers got that reputation because they had to run after inept teammates failed to block for them.

II. That which scrambles. **2.** A plant, often a climbing one, depending on the support of others.

1902 L. H. Bailey Cycl. Amer. Hort. IV. 1935/2 There are many useful climbers among the scramblers. 1953 Brit. Commonwealth Forest Terminol. (Empire Forestry Assoc.) I. 116 Scrambler. A plant which, owing to lack of rigidity in its stem, and absence of special climbing organs, uses other vegetation as its support. 1974 Country Life 21 Mar. 642/3 This large-flowered scrambler [sc. a Cretan Aristolochia] is probably not very hardy.

3. a. An electronic device used, esp. in telephony and radio, to make speech signals unintelligible, usu. by dividing the signal into distinct frequency ranges which are separately inverted and displaced in frequency.

1950 G. Hackforth-Jones Worst Enemy i. 24 This line, which linked me directly with the Rear-Admiral, was fitted with a device known as a 'scrambler' which was completely secure against listening in and it was therefore possible to speak freely and at length at all times. 1968 New Scientist 19 Dec. 657/3 A simple scrambler that will turn high speech frequencies into low speech frequencies and vice versa can be bought for about £100. 1973 'I. Drummond' Jaws of Watchdog ii. 31 The radio..had a scrambler, so that if the message was picked up accidentally it sounded like static. a 1974 R. Crossman Diaries (1975) I. 41 Having made a big fuss about national security to George Wigg I have decided to be extremely careful in everything I do personally so I've had scramblers and big safes installed in London as well as here at Prescote. 1981 A. Melville-Ross Tightrope vi. 36 You can get me..the Minister's Private Secretary..on the scrambler.

fig. 1958 Listener 13 Nov. 791/3 Your work of art..has to suffer a further change as it goes through the scrambler of your reader's prejudices.

b. *attrib.*, as *scrambler line, phone, system, telephone.*

1958 L. Durrell Balthazar ii. 29 His work was invisibly dictated by a scrambler telephone. 1958 Listener 25 Sept. 462/2 It is a variant of the scrambler system which has been long in use for confidential telephonic communication. 1965 Times 16 Feb. 12/2 The 'scrambler' telephone can now be bought for £50 by individual companies. 1966 J. Bingham Double Agent xi. 162 He lifted the scrambler telephone. ..'A scrambled phone is safe, but not entirely safe.' 1974 G. Markstein Cooler lxviii. 234 He..made a call on the scrambler line to London. 1975 'M. Sinclair' Long Time Sleeping ix. 115 Pringle..switched on the scrambler phone and started writing. 1977 C. Forbes Avalanche Express ii. xi. 118 Scholten took the call on his scrambler phone.

scrambling, *vbl. sb.* Add: **a.** (Later examples.)

1930 Engineering 14 Nov. 626/1 The apparatus used for this scrambling, as it is called, is installed at the Central Telegraph Office. 1942 V. E. R. Blunt Use of Air Power viii. 72 Wireless telegraphy and radio telephony..by 'scrambling' can now be made secret. 1955 'N. Shute' Requiem for Wren 166 We were in the process of scrambling when the Jerries came over. 1959 New Statesman 14 Nov. 658/1 The simplest definition of scrambling is: the racing of motor bikes over rough ground. 1978 Guardian Weekly 24 Sept. 22/5 Scrambling, as distinct from fell walking and rock climbing, is a Cinderella of a sport.

b. *scrambling club*; **scrambling net** *Mil.* = *scramble net* s.v. *scramble *sb.* 4; also *transf.*

1961 Guardian 17 Mar. 3/3 Motor-cycle scrambling clubs. 1974 G. Moffat Corpse Road iv. 64 She belonged to a scrambling club, which means walking... It doesn't mean rock climbing. 1959 New Scientist 30 July 125/1 Home grown seeds are extracted from cones, sometimes collected by means of a 'scrambling net'..thrown over a tall tree. 1964 C. Willock Enormous Zoo v. 75 The long-forgotten sensation of climbing down the scrambling net of a troop transport into a landing craft. 1973 A. Ross Dunfermline Affair 36 The scrambling net which the Hermione put over her side.

scramblingly, *adv.* (Later examples.)

1923 D. H. Lawrence Ladybird 242 For some time.. Alexander gingerly and scramblingly led the way. The slope of ice was steeper, and rounded, so that it was difficult to stand up. 1949 D. L. Sayers tr. Dante's Inferno xxiv. 221, I..came Scramblingly up and sat down.

scrambly (skræ·mbli), *a.* [f. Scramble *sb.* + -y[1].] **1.** Of a person, limb, etc.: that scrambles, clambers, or claws. Also applied to an informal meal. Cf. Scrambling *ppl. a.* 1.

In quot. 1900 applied to a jumble of people at a meal.

1900 R. Proctor Diary 26 Mar. in V. Scholderer Fifty Ess. 15th- & 16th-Cent. Bibliogr. (1966) 34 A company of about 16... A most amusing scrambly supper in a room calculated to hold four at most. 1943 C. McCullers in Harper's Bazaar Aug. 140/3 The hunchback reached in the box with his scrambly little fingers. 1977 Time 5 Dec. 49/1 Little scrambly front legs and big thumping back legs.

2. Characterized by scrambling or clambering over rough terrain; that necessitates such action.

1900 G. Bell Let. 28 Feb. (1927) I. v. 65 We had a very scrambly walk back. 1932 A. Christie Peril at End House ii. 34 There's a scrambly cliff path down to the sea.

scramjet (skræ·mdʒet). *Aeronaut.* [f. the initial letters of *s*upersonic *com*bustion + *ramjet.] A ramjet in which combustion takes place in a stream of gas moving at supersonic speed.

1966 New Scientist 19 May 429 (caption) Supersonic combustion ramjets ('scramjets') theoretically could extend flight speeds to at least Mach 14. 1972 D. G. Shepherd Aerospace Propulsion iv. 110 There is considerable interest in supersonic combustion as this is the key to the scramjet.

scran, *sb.* Add: **2. a.** Also *spec.* in *Naut. slang*, food, rations.

1916 'Taffrail' Pincher Martin i. 8 Them two's on watch now, but they'll be down at eight bells clamourin' for their scran like a lot o' wolves. a 1935 T. E. Lawrence Mint (1955) 135 'Scran up!' he called in his sailor's belling tone against my ear. 1974 Sentinel (Ottawa) X. ii. 6/3 He's the chief cook on board, responsible for the preparation and serving of food—or 'scran', according to the hands—to 280 hungry mouths about three times a day.

scrannel, *a.* (Further examples.)

1927 E. F. Benson Lucia in London ii. 60 It was strange.. to hear..the foe of all modern music..producing these scrannel staccato tinklings that had so often made her wince. 1934 [see *mimsey a.]. 1951 Auden Nones (1952) 54 His scrannel music-making. 1976 New Yorker 1 Mar. 89/1 But the music Berlioz heard in St. Peter's was scrannel stuff, and it was years before he himself received the commissions to compose.

scrap, *sb.*[1] Add: **2. a.** (*b*) *scrap of paper*: applied contemptuously to a document containing a treaty or pledge which one does not intend to honour.

The phrase is said to have been used by the German Chancellor, Bethmann-Hollweg (1856–1921), in connection with German violation of Belgian neutrality in August 1914 (cf. G. ein Fetzen Papier). Some later examples allude to this.

1840 Chambers's Edin. Jrnl. 11 Apr. 94/1 He no more dreamt of..honouring his scraps of paper..than of paying the national debt. 1914 E. Goschen Let. 8 Aug. in Coll. Diplomatic Documents rel. Outbreak Europ. War (1915) 111 The Chancellor said that..just for a word—'neutrality',—just for a scrap of paper Great Britain was going to make war on a kindred nation. 1918 Daily Mail Year Bk. 1919 62/1 Those familiar with the 'scrap of paper' theory need hardly be told that the pledges given by the German Emperor.. were not observed. 1932 K. Campbell Sarah, Duchess of Marlborough 83 James made it plainer every day..that, compared with his Church, the constitution of England and his own coronation oaths were mere scraps of paper. 1954 W. K. Hancock Country & Calling iv. 111 The British Empire, not so many years back, had professed itself to be at war with the doctrine that a treaty was only 'a scrap of paper'. 1974 M. Gilbert Flash Point vi. 50 The First World War was fought over a small thing. A scrap of paper. 1980 Times 3 July 17/2 The Treaty of Union..wasn't a sacrosanct document, but in empirically English fact, just a 'scrap of paper'.

d. A small person. *colloq.*

1898 H. James Two Magics 60 'Perhaps she likes it!' 'Likes such things—a scrap of an infant!' 1928 E. P. Oppenheim Chron. Melhampton v. 146, I wasn't here for long, and I was a scrap of a fellow those days. 1939 N. Streatfeild Luke 109, I didn't know the poor little scrap could look so radiant. 1958 Woman's Jrnl. Mar. 77/2 'The woman?'.. 'They picked her up late last night. Poor little scrap.'

6. *scrap dealer, dealing, merchant, -metal*; **scrap basket**, a waste-paper basket; **scrap-ground** = *scrapyard* below; **scrap man**, one whose business is the collection and sale of scrap-metal and its salvageable accessories; **scrap paper**, paper that may be repulped or used again; rough paper for casual jotting; **scrap screen** chiefly *Hist.*, a screen or divider (as in a nursery) decorated with scraps (sense 2 c); **scrapyard**, the site of a scrap-heap; *spec.* a place where disused motor vehicles, etc., are scrapped.

1872 C. M. Yonge P's & Q's ix. 94 If she put it in the scrap basket, Persis herself might look in and see the writing. 1912 E. Pound Let. Dec. (1971) 13, I won't quarrel with you over what you see fit to put in the scrap basket. 1976 Loughborough Monitor 26 Nov., He had worked as a scrap dealer while claiming supplementary benefit. 1977 Belfast Tel. 22 Feb. 4/1 (caption) For scrap dealing..two heads better than one. 1927 Observer 21 Aug. 19/2 Six years is about the maximum age of the cars taken for scrapping in America. Some reach the scrap-ground much earlier. Ibid., The scrap man's interest in these vehicles is purely that of a replacement part merchant. 1977 Custom Car Nov. 5/1 When it comes to fridge pumps, beware. The scrapman is out to con you. 1978 R. V. Jones Most Secret War i. 5 It certainly was tough, the future of my contemporaries encompassing everything from barrow boy to millionaire scrapmerchant and trade union peer. 1941 Proc. Prehist. Soc. VII. 130 The Bronze Age pieces must be ..scrap-metal. 1962 A. Battersby Guide to Stock Control 3 The petty cash box resembles the scrap-metal example in reverse. 1885 Encycl. Brit. XVIII. 228/2 The materials for the commoner classes of work are old waste and scrap paper, repulped... For very delicate relief ornaments, a pulp of scrap paper is prepared. 1960 M. Spark Bachelors xi. 195 Marlene walked solemnly downstairs and demanded some scrap paper from the hall porter. 1969 C. Irving Fake! (1970) iii. 39 He..made some preliminary sketches for several hours on scrap paper. 1873 Young Englishwoman Jan. 51/3 Lizzie would be glad if the Editor could give her any information as to making a scrap screen. 1899 M. Beerbohm More 173 They will make the scrap-screen their background. 1962 N. Marsh Hand in Glove v. 148 The room was masked from its entrance by an old-fashioned scrap screen. 1964 S. Nowell-Smith Edwardian England iv. 201 The dark, cosy Victorian nursery.. brightened by the varnished scrap-screen. 1963 Times 11 Jan. 10/3 (heading) Tow breaks on way to scrapyard. 1978 T. Allbeury Lantern Network iv. 59 A scrapyard with big double gates.

scrap, *sb.*[2] Add: **2. a.** (Earlier and later examples.) Also *gen.*, a contest.

1846 Swell's Night Guide 75 By way of varying the slang, the mock combat turns into a right good scrap. 1916 [see *bandbox b]. 1959 [see *go sb. 4 b]. 1973 Times 10 Dec. 9/5 In a final that provided a keen scrap rather than a match of high quality, they beat the Etonians. 1977 J. Cleary High Road to China v. 158 My chaps..[are] itching for a scrap, y'know.

b. A contest of words; a row, quarrel, squabble; a heated discussion.

1890 Barrère & Leland Dict. Slang II. 210/1 Having a scrap up is having a quarrel, a row. 1900 Dialect Notes II. 57 Scrap,..a quarrel of words, sometimes good-natured. 1903 N.Y. Tribune 6 Sept. 2/2 In directing the proceedings ..Mr. Hill was careful to sidetrack anything containing the germ of a 'scrap'. 1928 J. Galsworthy Swan Song ii. xi. 199 It was his impression that they'd been having a scrap.

scrap, *v.*[2] Add: **b.** Also *fig.*

1936 L. C. Douglas White Banners xvi. 335, I have given him until June first to scrap it out with himself.

c. *intr.* To quarrel, squabble; to engage in heated argument or angry dispute.

1895 W. C. Gore in Inlander Nov. 65 Scrap,..to quarrel. 1900 Dialect Notes II. 57 Scrap,..to quarrel, sometimes good-naturedly. 1923 Daily Mail 28 June 5 Are you going on scrapping over this garden fence for the rest of your lives? 1941 B. Schulberg What makes Sammy Run? vii. 120 The play..was one of those things about two red-blooded guys who are always scrapping and loving each other.

scrap-book. Add: **b.** *transf.* A loosely-constructed documentary review programme, normally covering a particular year or period, presented on radio by the B.B.C.

1933 Radio Times 8 Dec. 719/2 Scrapbook for 1913. Ibid. 740/2 This is no history book—just a scrapbook of cherished fragments. 1939 H. Nicolson Diary 3 Dec. (1967) 47 After the news there is a B.B.C. scrap-book for 1910. 1947 L. MacNeice Dark Tower 69 Features..have..a great diversity of form. Some of them are as loosely constructed as scrap-books. 1959 Listener 23 July 150/2 The programmes ..were almost identical exercises in the well-proven genre of biographical scrap-book. 1972 P. Black Biggest Aspidistra in World i. v. 43 The famous Scrapbooks, in popularity and appreciation the most successful dramatised radio series ever produced in Britain..began in 1933... The title Scrapbook first appeared in Radio Times in 1932.

c. *attrib.*

1897 H. James Spoils of Poynton i. 5 Trumpery ornament and scrapbook art. 1934 C. Lambert Music Ho! i. 75 Diaghileff..was able to invest with a revolutionary glamour the scrapbook mentality which in his later years he exploited with so marked a success.

scrap-book *v.* (earlier example); hence **scra·p-booking** *vbl. sb.*

1879 'Mark Twain' Let. 12 Nov. (1917) I. 369 Put the enclosed scraps in the drawer and I will scrap-book them. c 1898 —— Autobiogr. (1924) I. 139 He usually postponed the scrap-booking until Sunday.

scrape, *sb.*[1] Add: **I. 2. d.** (Earlier example.)

c 1807 Jane Austen Watsons in Minor Wks. (1954) 327 No sound of a Ball but the first Scrape of one violin.

h. A dilatation of the cervix and curettage of the womb; *spec.* an induced abortion. *slang.*

1968 J. Hudson Case of Need iii. i. 172 The word got around..that she got a bad scrape. 1972 Rochdale's Alternative Paper No. 6. 10/2 The most common method [of abortion] in Britain for pregnancies of less than three months is D. & C. (better known to most women as a scrape). 1980 M. Drabble Middle Ground 62 She was having a D and C, a routine scrape.

4. c. *Ornith.* A shallow pit in the ground excavated by a bird, usu. during a courtship display; also, the action of making such a pit.

1926 Ibis II. 7 All the scrapes noted were within about fifty or sixty yards from a nest. 1940 H. F. Witherby et al. Handbk. Brit. Birds IV. 385 In more advanced scrapes female with back to male will peck in bottom of scrape. 1942 E. A. Armstrong Bird Display ii. 27 An unmated female red-necked phalarope makes scrapes in the herbage, and from the first day of finding a mate this 'ceremony' often follows coition... Before laying, the female visits the various scrapes and lays an egg in one of them. 1961 [see scrape ceremony s.v. *scrape v. 10 b]. 1967 B. Campbell J. Hanzák's Pictorial Encycl. of Birds 254/2 The nest is a shallow scrape lined with small stones or shells.

6*. On a woodwind instrument, the part of the cane that is scraped to a narrow edge in the production of a reed. Also, the style of this scraping.

1954 Grove's Dict. Mus. (ed. 5) VI. 161/2 According to the quality of the cane from which it is made, and the character of its 'scrape', a reel [of an oboe] may be responsive or unyielding. 1961 Sprenkle & Ledet Art of Oboe Playing 95/2 The French scrape has a rather long tip. 1980 Early Music July 363/2 There are 3 basic scrapes..of which no. 1 is the most common.

scrape, *v.* Add: **2. f.** To clean or empty *out* by scraping.

1894 'R. Andom' We Three & Troddles xxi. 198 When you have done with that jam you might let me have the pot to scrape out. 1919 E. Shackleton South xii. 237 As the cook and his 'mate' had the privilege of scraping out the saucepans, there was some anxiety to secure the job. 1925 Morris Owner's Man. 14 Thoroughly scrape out and oil cams until they work quite freely.

g. To draw (hair) tightly *back* from the forehead. Cf. *scraped *ppl. a.* 2 b.

1926 'O. Douglas' Proper Place xxxi. 280, I couldn't have believed she had such pretty, soft hair for she wore it scraped back.

5. d. In *fig.* phr. *to scrape (the bottom of) the barrel* (or *bucket*): to collect with difficulty something inferior; to obtain something (as by necessity) from a poor source. *slang* (orig. *U.S.*).

1942 Time 12 Jan. 57/1 The medical profession..[is] closer to scraping the bottom of the bucket..than any other occupation, trade or profession. 1955 N.Y. Times Bk. Rev. 30 Oct. 1/1 It was built in the spring of 1864, when the Confederacy was scraping the bottom of its barrel of men and resources. 1957 Essays in Criticism VII. 342, I was scraping the barrel for evidence. 1961 B. Fergusson Watery Maze viii. 203 The insistence of the Americans that the Casablanca assault should be much the strongest.. meant scraping the bottom of the barrel to find extra ships, craft and crews. 1970 Times 5 Nov. 14/4 Professor Barlow explains how necessary it is to 'scrape the barrel' for even the most minor scraps of information. 1981 Times 18 Aug. 1/3 We shall have to pay something. I hope to be able to scrape the barrel and come up with something.

9. a. Also with other advs. or advb. phrases in similar senses, as *to scrape by, home, in*, etc.

1927 Observer 18 Sept. 17/3 Mr. Blythe, who at the last election scraped home in Monaghan. 1951 Sport 6–12 Apr. 17/2 The Airmen just scraped through with a 3–2 win. 1958 Times 16 Dec. 4/4 (heading) Chigwell scrape home. 1958 [see *million 2 f]. 1966 Listener 1 Dec. 829/1 Osbert Lancaster..depends on words rather than drawing, understandably since his drawing..only just scrapes by. 1973 M. Woodhouse Blue Bone iv. 35 The family didn't get on with the Germans and they only just barely scraped by with the Communists. 1978 'M. M. Kaye' Far Pavilions vii. 114 As long as he can shoot and ride, I suppose he'll scrape past.

c. To acquire or obtain (something) with difficulty. *colloq.*

1963 Guardian 9 Aug. 7/1 He read English at Oxford. 'But I only scraped a third.' 1967 Listener 2 Mar. 283/1 The Congress Party, which has ruled the country since independence in 1947, has just managed to scrape a majority in the central parliament. 1980 Early Music Apr. 234/1 Nor does castrato appear in the index (though Farinelli scrapes a mention).

10. b. scrape ceremony *Ornith.*, a display by a bird, involving the excavation of a shallow pit in the ground and the pressing of the bird's breast into this, freq. performed during courtship; hence **scrape-ceremonial** *a.*

1926 Huxley & Montague in Ibis II. 10 Nervous picking at grass..may be seen during incubation, scrape-ceremonies and before coition. 1949 British Birds XLII. 8 Both sexes indulge in the 'scrape' ceremony and in many of the actions characteristic of nest-building. 1961 D. A. Bannerman Birds Brit. Isles IX. 15 The main type of sexual behaviour preceding coition is the tail-display... At this stage the scrape-ceremony is very common, a typically male performance by which he lures the female to one scrape after another. Ibid. X. 247 Dotterel and red-necked phalarope..share very similar scrape-ceremonial and egg-laying behaviour.

scraped, *ppl. a.* Add: **2. a.** Also *scraped together.*

1965 Listener 3 June 828/1 In June 1942 the hastily scraped together force called 23rd Indian Division, with which I was serving, was isolated from the rest of the world.

b. Designating women's hair that has been drawn back tightly from the forehead. Also *scraped-back.* Cf. *scrape *v.* 2 g.

1970 'D. Halliday' Dolly & Cookie Bird viii. 118 Her deep-set eyes smiling gratefully under the grey, scraped-back hair. 1977 Times 16 Sept. 7/8 Her scraped hair, self-effacing manner, and busy hands. 1978 I. Murdoch Sea 156 Her scraped-back hair revealed her bulky rounded brow.

scraper. Add: **II. 4. f.** More fully *cabinet scraper.* A thin rectangular piece of metal whose sharpened long edge is pushed over the surface of wood to smooth it.

1909 Wells & Hooper Mod. Cabinet Work v. 72 A carefully sharpened scraper frequently permits of about twelve resharpenings in all. 1924 H. G. Phillips Cabinetmaking i. 14 A very fine shaving is taken off with the scraper, which leaves the surface ready to be glasspapered. 1970 Canadian Antiques Collector Jan. 27/2 The only satisfactory method is the use of a cabinet scraper. 1977 Reader's Digest Bk. Do-it-Yourself Skills & Techniques ii. 77/1 Cabinet scrapers give a satin-smooth finish to hardwood. Ibid., If a scraper becomes hot and produces dust instead of shavings during use, it needs resharpening.

g. *Mus.* A simple percussion instrument.

1953 J. G. Moore in Dict. Jamaican Eng. (1967) 396/1 The scraper, a corrugated stick across which is rubbed a plain stick [in pocomania and revivalist services]. 1956 M. Stearns Story of Jazz (1957) v. 53 The typically African instruments, such as drums, gourd rattles, and scrapers. 1958 E. Borneman in P. Gammond Decca Bk. Jazz xxi. 275 One of the many..indigenous African string instruments, hand drums, scrapers, shakers and gong-gong. 1961 A. Baines Musical Instruments through Ages i. 27 Scrapers have survived into modern times, for instance in the folk music of Venezuela. 1976 D. Munrow Instr. Middle Ages & Renaissance 32/2 Most of the instrumental types described are of very ancient origin indeed, drums, rattles, and scrapers being the commonest instruments of primitive man.

7. a. In mod. use *spec.* an earth-mover, either self-propelled or towed, that works on the principle of a scoop. (Earlier and later examples.)

1815 T. B. Hazard Nailer Tom's Diary (1930) 442/2 Delivered C. R. Potter p[ai]r Scrapers and a Chain to hich horseis with. 1823 New England Farmer II. 9 The most expeditious, effectual, and economic mode of making a drain would undoubtedly be to use oxen, and a scraper or ox-shovel, as it is sometimes called. 1930 Engineering 7 Mar. 306/1 There are three of these scrapers in each warehouse, their function being to drag down the potash from the several heaps into the central longitudinal opening in the floor. 1939 C. W. Towne Her Majesty Montana 120 Abolishing the back-breaking labor of mucking, power driven scrapers and mechanical mucking machines are now usual. 1958 Engineering 14 Feb. 219 For outside work the three most important types of machine—all rubber tyred—are the self-propelled scrapers, mechanical shovels and the cranes. 1974 Encycl. Brit. Micropædia VIII. 996/2 The scraper is the dominant tool in highway construction.

c. (Earlier example.)

1881 E. Ingersoll Oyster-Industry 247 Scraper, a small dredge. Chiefly spoken of with reference to scallops.

7*. = *pig *sb.*[1] 8*.

1897 B. J. Crew Practical Treat. Petroleum xiv. 449 Under ordinary circumstances the scraper passes rapidly through the lines, cutting off all the sedimentary matter that has adhered to the pipes. 1959 Petroleum Handbk. (Shell Internat. Petroleum Co. Ltd.) (ed. 4) 330 A radioactive cartridge may be fitted to the scraper as a means of locating it, should the scraper become stuck in the line. 1976 M. Machlin Pipeline xlviii. 497 The oil itself, at seven miles an hour, took four and a half days to make the trip from Prudhoe to Valdez, pushing the scraper before it to separate it from the test water in the pipes.

III. 8. (Earlier example.)

1792 S. Burdy Life of Rev. Philip Skelton 84 The militia men..took to their scrapers to save themselves.

9. (Earlier example.)

1818 'A. Burton' Johnny Newcome i. 24 And John in Uniform arrayed: Behold him! with his dirk and scraper, And new Coatee, as stiff as paper.

IV. 10. scraper board *Art* (see quot. 1972); also, a piece of this material; **scraper ring**, a piston ring whose function is to scrape oil off the cylinder wall.

1895 E. J. Wall tr. Fritz's Photo-Lithography i. 25 A smooth white scraper board made by Angerer and Göschl of Vienna, which has a very even film of chalk, and which takes the lines clean and vigorously, is especially suitable for pen and ink work. 1925 Art & Publicity (Studio: Special Autumn No.) (Advt., verso front cover), Scraper boards with plain, embossed, or tinted surface for drawing for reproduction of line or half-tone. 1945 J. C. Tarr Printing To-day viii. 95 Line engravings are also made from scraper-board originals. This board is covered with a thick layer of china clay and size upon which indian ink can be drawn or brushed, and scratched away with a knife. 1972 P. Croy Graphic Design & Reproduction Techniques (ed. 2) 162 Scraper board technique is an interesting method of producing white-and-black drawings. Scraper board consists of a base card coated with white and black chalk layers. The surface layer is scratched..exposing the underneath layer. Scraper board can be bought with a white surface and a black underlayer. 1918 W. E. Dommett Dict. Aircraft 35 The lowest ring, when placed at the bottom of the skirt, is known as a 'Scraper Ring'. 1928 C. F. S. Gamble Story North Sea Air Station xiii. 216 In the early days considerable difficulty was experienced with these engines owing to the rapid deterioration of the scraper rings fitted on the piston heads. 1980 Haynes & Legg Citroën CX Owners Workshop Man. i. 29/1 The oil control ring is fitted to the bottom groove, the scraper ring to the middle groove, and the compression ring to the top groove.

scrap-heap. (Further examples.)

1937 V. Bartlett This is my Life xv. 258, I don't know who bought that car in the end or how soon it reached the scrap heap. 1956 Railway Mag. Nov. 749/2 No. 6 was rescued from an inglorious demise on the scrap heap. 1967 G. F. Fiennes I tried to run Railway iv. 40 Engines long past their time for the scrap-heap were being kept going. 1972 Times 16 May 2/2 He felt he had been thrown onto the scrap heap after 20 years' service. 1977 Times 7 Oct. 15/6 A socialist determination to drive fewer people onto the scrap heap. 1980 A. Clarke Last Voyage ix. 100 They didn't send me to the scrap-heap straight away. Gave me a little job.

scrapiana. For † *Obs.* read *rare* and add later example.

1958 E. Blunden War Poets 1914–1918 27 That denomination 'Eye-Witness', used in the first stage of the war to cover official scrapiana from the observation posts.

scrapie (skrē[i]·pi). Also † **scrapy.** [f. Scrape *v.* + -ie, -y[6].] A subacute, invariably fatal, disease of sheep and goats, characterized by degeneration of the central nervous system, leading to uncoordinated gait and itching.

1910 Vet. Jrnl. LXVI. 711 Shepherds and farmers..class more than one disease with totally different symptoms under the head of Scrapy. 1913 Jrnl. Compar. Path. & Therapeut. XXVI. 317 The term 'scrape' or 'scrapie' is a popular one which has been applied to a disease of sheep on the borders of England and Scotland. 1914 J. P. McGowan Investigation into Dis. of Sheep called 'Scrapie' i. 11 The disease has existed in Britain since before the middle of the eighteenth century under such names as 'scrapie', 'scratchie', 'rubbers', 'rickets', 'goggles', 'shakings', 'shrew-croft', and 'cuddie-trot'. 1925 Glasgow Herald 22 Apr. 7 Scrapie has mainly been confined to a certain area of the country. 1952 I. E. Newsom Sheep Dis. iv. 120 Scrapie is a chronic neurosis of sheep and goats characterized by intense pruritus, progressive incoordination, weakness, paralysis and death. 1975 Times 25 Aug. 8/5 Scrapie, an incurable disease of the nervous system, has afflicted various breeds of sheep for many years. 1979 Nature 11 Jan. 127/1 Scrapie, a subacute neuromuscular disease of middle-aged sheep, has been attributed variously to an infection or to heredity. The clinical disorder follows progressive symmetrical decay of neurones in localised brain sites,..associated with a terminal axon dystrophy. 1981 Brit. Vet. Jrnl. CXXXVII. 108/2 Although scrapie is caused by a virus-like agent, the development of clinical disease depends on genetic factors.

scraping, *vbl. sb.* Add: Also in *fig.* phr. *scrapings of the barrel* (see *scrape *v.* 5 d).

1959 Listener 22 Jan. 154/1, I think Bill Slim was a very great man, because he made do with practically the scrapings of the barrel.

scrap iron. Add: **2.** *fig.* An alcoholic drink of poor quality. *U.S. slang.*

1942 Z. N. Hurston in Amer. Mercury July 85 Maybe a shot of scrap-iron or a reefer. 1958 Washington Post 1 Nov. 1/1 A trio of investigators warned the drinking public yesterday to beware of a new bootleg concoction, 'scrap iron', noted more for its voltage than vintage. 1970 C. Major Dict. Afro-Amer. Slang 101 Scrap iron, bad liquor.

scraplet. (Further examples.)

1891 A. James Diary 7 May (1964) 203 We have a good scraplet of garden. 1972 Time 17 Apr. 63/1 He tests every anthology to see if some scraplet of Chandler's small output will turn up.

scrappage (skræ·pėdʒ). [f. Scrap *sb.*[1] or *v.*[3] + -age.] = Scrapping *vbl. sb.*[2]

1949 Sun (Baltimore) 24 Mar. 8/3 Scrappage of passenger cars in 1948 was approximately half of the normal scrappage rate. 1950 Engineering 6 Jan. 26/3 Internal stresses may lead to serious scrappage on account of cracking. 1960 Economist 22 Oct. 359/1 Apart from natural growth in the population, sales are wholly for replacement and are related to the rate of scrappage and obsolescence. 1972 Guardian 29 Mar. 14/2 He says that the scrappage rate for US cars is roughly stable at 40 per cent of the new registration increase. 1976 Nature 17 June 540/2 The causes are recognised engineering factors such as shaft seals, maintenance operations, permeation and eventual scrappage.

scrapper[2]. Add: **1. b.** *transf.* in *gen.* use.

1976 National Observer (U.S.) 20 Nov. 13/4 On auction day, Harry Miner..said with his country twang, 'Vivien was a scrapper.' 1977 'D. Ramsay' You can't call it Murder ii. 124 'She was a real scrapper, Judy was.' O'Shea went on to give instances of her capacity for scrapping. 1979 Dædalus Summer 157 This is the Dewey..who had a long and honorable record as a scrapper for the rights of citizens of the democracy.

2. *N. Amer.* A fish that is hard to land once caught.

1959 Moosehead Gaz. (Dexter, Maine) Feb. 18/3 Tiny dry flies in drab patterns..brought the silvered scrappers with a rush. 1968 Globe & Mail (Toronto) 3 Feb. 40/1 The large-mouth is a right obliging and powerful scrapper. 1974 Sunday (Charleston, S. Carolina) 28 Apr. (Cartoon Suppl.) 8 (Advt.), This palm-size powerhouse has all the guts in the world to wear down any scrapper you're liable to hook.

scrappet. Delete *rare* and add: Also **scrappit.** (Further examples.)

1901 L. Malet Hist. Sir Richard Calmady v. v. 415 There's selfishness now, if you like—to appropriate a virtue en bloc, not leaving a rag, not the veriest scrappit of it for anybody else! 1941 E. R. Eddison Fish Dinner in Memison xiii. 209 'A scrappet of ham: just half of that littlest slab,' she said. 1971 C. Bonington Annapurna South Face xv. 194 They warned me that they had very little food—just the odd scrappets that Dougal had taken up that morning.

scrapping, *vbl. sb.*[1] Add: Also *transf.* in *gen.* use.

1937 Times Lit. Suppl. 2 Jan. 4/3 One is apt to consider that the campaign ended, except for some hardly necessary 'scrapping', with the capture of Baghdad. 1977 [see *scrapper[2] 1 b].

scrapple, *sb.*[2] (Earlier and later examples.)

1855 Rural New Yorker 10 Feb. 47/3, I observe a call for a recipe for making 'Scrapple', and some other homely dishes. 1910 'O. Henry' Whirligigs x. 130, I never cared especially for feuds, believing them to be even more over-rated products of our country than grapefruit, scrapple, or honeymoons. 1942 H. W. Van Loon Van Loon's Lives 632 Dante has become a taste that has to be acquired like a love for figs or scrapple. 1943 [see *ponhaus]. 1975 R. Stout Family Affair (1976) ii. 20 A plate of slices of home-made scrapple.

scrappy, *a.*[1] (Further examples, in sporting use.)

1930 Daily Express 6 Oct. 16/6 Both sides resumed their previous rough tactics, and the play became scrappy in consequence. 1976 Milton Keynes Express 28 May 54/3 Inter City were unlucky at the start of the second half and then play became scrappy and several unnecessary fouls were committed.

scrappy (skræ·pi), *a.*[2] orig. *U.S.* [f. Scrap *sb.*[2] or *v.*[2] + -y[1].] Inclined to scrap or fight; aggressive, pugnacious, quarrelsome.

1895 W. C. Gore in Inlander Nov. 65 Scrappy,..quarrelsome. 1935 F. Scott Fitzgerald Let. 11 May (1964) 264 It was fine seeing you but I was in a scrappy mood about Tom Wolfe. 1941 B. Schulberg What makes Sammy Run? v. 83 All the instinct for self-preservation of a scrappy kitten. 1977 Rolling Stone 7 Apr. 52/1 She was only five feet tall, but she was scrappy—her sister Rebecca Julian remembers her once beating up a bully to protect their little brother. 1977 Time 27 June 33/1 'This puts transatlantic air travel in the pocket of the workingman,' proclaimed Freddie Laker, the scrappy founder of Britain's Laker Airways.

scratch, *sb.*[1] Add: **I. 4. b.** Money, esp. paper money. *slang* (orig. *U.S.*).

1914 Jackson & Hellyer Vocab. Criminal Slang 74 Scratch,..paper currency... 'He's got a bundle of scratch'. 1916 Lit. Digest 19 Aug. 424/2 Money is given a score of names; the most used is 'kale', 'scratch', or 'dough'. 1931 Writer's Digest Oct. 29 Don't mess with that iron money; get the scratch. 1939 New Yorker 1 Apr. 19/2 She..also had plenty of scratch, being the bank president's daughter. 1941 J. Smiley Hash House Lingo 47 Scratch, coins. 1955 Pohl & Kornbluth Space Merchants xiv. 123 Here's some scratch, and shop when you get a chance. 1957 N. Marsh Off with his Head xii. 269, I wouldn't have done it only I wanted the scratch like hell. 1967 I. Hamilton Man with Brown Paper Face vii. 102 Alfred Mays..had enough scratch to run two homes. 1972 D. Anthony Blood on Harvest Moon xiv. 118 She runs some kind of talent agency. Probably a tax write-off... She doesn't need the scratch. 1978 G. McDonald Fletch's Fortune xv. 106 As soon as Fletch got the story from each girl..he found himself..getting up the scratch to bus her home. 1980 Private Eye 6 June 7/1 This state-funded legal nonsense—which is..putting even more scratch into the bulging wallets of the lawyers.

5. a. (*Fig.* examples of *up to (the) scratch*, up to the required standard.) Now usu. without *the.*

1843 Dickens Let. 17 June (1974) III. 513 Pray, as a Member of the Committee, come up to the Scratch. 1848 A. Brontë Tenant of Wildfell Hall II. xi. 202 Your uncle and aunt have long been wanting us to go there, you know; but somehow, there's such a repulsion between the good lady and me, that I never could bring myself up to the scratch. 1861 C. M. Yonge Young Step-Mother xiii. 156 The Vicar..was meditating a fresh one [sc. attempt], if..he could bring his churchwarden up to the scratch. 1911 G. B. Shaw Getting Married 226 It's about the wedding... We cant get our man up to the scratch. Cecil has locked himself in his room and wont see or speak to any one. 1934 'G. Orwell' Burmese Days ix. 142 If they won't come up to scratch you can always get hold of the ringleaders and give them a good bambooing on the Q.T. 1953 D. Garnett Golden Echo vi. 133 All Edward's friends were mobilised and came loyally up to scratch. 1960 C. Day Lewis Buried Day ii. viii. 164 Those of us who had pretensions to brain Maurice brought up to scratch by loosing on them the full force of his personality. 1978 Taxi 16 Feb. 11/4 (Advt.), Cold weather can cause you a lot of problems if your cab's electrics aren't up to scratch.

b. Also *fig.*; esp. in phr. *from scratch*, from a position of no advantage, knowledge, influence, etc., from nothing.

1922 Joyce Ulysses 454 A poor foreign immigrant who started scratch as a stowaway and is now trying to turn an honest penny. 1936 Economist 2 May 233/1 Nazi Germany, starting her rapid re-armament 'from scratch' in 1933, was fortunate enough to have a surplus capacity in all sections of her heavy industries. 1939 'G. Orwell' Coming up for Air ii. v. 103 We'd no fishing tackle of any kind, not even a pin or a bit of string. We had to start from scratch. 1953 S. Kauffmann Philanderer v. 77 He took one look at her and thought: If I want that, I must begin all over again right from scratch. 1957 L. F. R. Williams State of Israel 53 Another branch of communications which has been built up from scratch to a degree of efficiency. 1962 Guardian 21 Mar. 2/5 The whole Treasury block could..have been rebuilt from scratch for the money. 1978 Peace News 25 Aug. 7/2 The daily routine was a crash programme of tuition provided by civilians, mainly Russian or other Slavic emigrés, in Russian from scratch to A-level standard, which was achieved in 10 months. 1979 Fortune 29 Jan. 77 NASA is not exactly starting from scratch out there in space; it is building on promising experiments done on prior space flights.

c. *Sporting.* A horse or other animal withdrawn from the list of entries for a race or other competition.

1938 Mr. Dec. 128/2 Scratch,..a horse withdrawn from a race. 1947 Sun (Baltimore) 20 Dec. 12/1 The overnight favorite..was a late scratch. 1960 Wentworth & Flexner Dict. Amer. Slang 450/2 Scratch,..a horse that has been withdrawn from a race after midnight of the night before the race. 1979 Arizona Daily Star 8 Apr. c2/3 We kept hoping there would be a scratch in the fast heat... I must have asked a hundred times if there were any scratches.

6. b. A rough hiss heard from the loudspeaker (or horn) when a record is played and caused by the friction of the stylus in the groove.

1908 Talking Machine News I. 9/1 Scratch seems to be filtered out of the reproduction. 1926 Punch 2 June p. iii (Advt.), Columbia new process records. The only records without scratch. 1942 [see ground noise s.v. *ground sb. 18]. 1949 G. A. Briggs Sound Reproduction xix. 117 Cutting out

SCRATCH

a slice of scratch also removes a slice of music or whatever is being reproduced. 1961 E. N. Bradley *Records & Gramophone Equipment* ii. 43 Possessors of old 78 r.p.m. records who play these on new lightweight equipment may find a quite distressing amount of scratch and surface noise.

II. 8. a. Delete *rare* and add further example.

1932 H. C. Wyld *Universal Dict. Eng. Lang.* 1068/3 Dogs enjoy a good scratch.

IV. 10*. In Billiards and related games: (*a*) a lucky stroke, a fluke ? *obs.*; (*b*) a shot that incurs a penalty. Cf. *scratch *v.* 11*.

a. 1850 M. Phelan *Billiards without Master* 12 It is amusing to observe the effect produced on some players by what is technically called a 'scratch', or fortuitous stroke. 1859 G. W. Matsell *Vocabulum* 122/1 When a player wins a stroke or count by accident, without deserving it, he is said to have made a scratch. 1869 'Mark Twain' *Innocents Abroad* xii. 116 We had played billiards..on an ancient table that made the balls..perform feats in the way of..almost impossible 'scratches'. 1907 *N. Amer. Rev.* Nov. 333, I saw nothing there in the way of science and art that was more wonderful than shots [in Billiards] which I had seen Texas Tom make..all calculated shots, and not a fluke or a scratch among them.

b. 1913 J. T. Stoddard *Science of Billiards* vii. 153 One ball is forfeited for a failure to hit any ball, or for pocketing the cue ball ('scratch'). 1974 *Rules of Game* 85/3 Scratches are also incurred during safety play on a ball frozen to a cushion, and when a player's cue ball jumps off the table. 1975 *Way to Play* 195/3 At his third scratch in succession, a player loses one point for the third scratch, plus 15 points for the three successive scratches.

V. 11. a. *scratch boat, machine, runner.*

1896 *Rudder* July 220/2 A table of time allowances has been figured out..using the 130-foot boat as 'scratch boat' (to use a foot-racing term). 1950 *Sun* (Baltimore) 18 Mar. 12/1 The Ticonderoga..was named the scratch boat. This means she is favored to finish first in the long race. 1955 *Times* 22 Aug. 4/5 The race..resolved itself in the closing stages into an exciting tussle between Mr. Clifford, in the scratch machine, and Mr. Peter Vanneck, in the longest handicap aircraft. 1976 *Star* (Sheffield) 30 Nov., Bert Oliver (Kelty), British professional 110 metres champion, is the scratch runner in the Skol Sprint 110 metres handicap on January 1 and January 3.

b. scratch-coat (later example); hence *scratch-coating* vbl. sb.; **scratch dial**, a set of marks found on the walls of old churches, usually considered to be an ancient form of sundial; **scratch filter** *Electr.*, a filter designed to reduce the audibility of scratches and hiss in sound reproduction; **scratch hardness**, the hardness of a metal or mineral as estimated by measuring the width of a scratch made on the material by a diamond point under a specified load; **scratch hit** *Baseball* (see quot. 1976); **scratch hole**, a hole or trench scratched out of the ground; **scratch paper** *N. Amer.*, scribbling paper; **scratch-plough** *v. trans.*, to plough very shallowly; **scratch sheet** *U.S. Sporting*, a printed list of the entries in the day's races and their odds; also *transf.*; **scratch stock** (see quot. 1966); **scratch-work**, (*a*) (later example); (*b*) scratched lines on an engraving plate.

1953 Van den Branden & Knowles *Plastering Skill & Practice* i. 5 Of the three coats, the first coat, or scratch coat, is a thin coat... The purpose of the scratch coat..is to ..provide a good base for the plaster coats that will follow. 1911 *Encycl. Brit.* XXI. 785/2 For the first coat a layer of well-haired coarse stuff..is put on with the laying trowel. This is termed 'pricking up' in London, and in America 'scratch coating'. 1914 *Proc. Somerset Archæol. Soc.* LIX. ii. 26 The name Scratch Dial has been given to this ancient form of sundial. 1938 *Times Lit. Suppl.* 15 Jan. 39/1 The woodwork follows—roofs, benches, pulpits; and then scratch-dials and aumbries, mural paintings and windows bring up the rear. 1960 J. Betjeman *Summoned by Bells* v. 49 Was that the reason why the pale grey slides Of tympana, scratch dials and Norfolk screens So pleased me at his lectures? 1967 C. A. Vearncombe *Hist. of Church of St. Lawrence the Martyr* (Lydeard St. Lawrence, Somerset) 19 This doorway has a scratch dial on the east, 3 ft. 9 in. above the plinth. 1929 K. Henney *Princ. Radio* xvi. 289 Similar filters are used in phonograph reproduction to eliminate the needle noise. They are called scratch filters and may tune somewhere between 3000 and 5000 cycles. 1935 Nilson & Hornung *Pract. Radio Commun.* viii. 349 It will be necessary to attenuate the high frequencies... This may be accomplished by introducing a series-resonant circuit similar to that used for scratch filters. 1977 *Rolling Stone* 5 May 80/2 Watch for this spec when you consider the usefulness of rumble and scratch filters on your next amp or receiver. 1928 *Jrnl. Iron & Steel Inst.* CXVII. 893 Annealing these cold-rolled single crystals at 250°C..gave no measurable rise of scratch hardness. 1962 R. Webster *Gems* II. 488 For the gemmologist, scratch hardness, based on the standard minerals in Mohs's list, is the only practical basis for experiment. 1972 *Jrnl. Physics* D. V. 1293 Scratch hardness measurements reflect a greater degree of anisotropy in the properties of crystals than the corresponding indentation techniques. 1917 C. Mathewson *Second Base Sloan* 166 Four men faced Chase in the third, the first getting a scratch hit. 1935 *Encycl. Sports* 63/1 Scratch-hit, a weak hit into the infield. 1943 *Amer. Speech* XVIII. 103 A 'single' in the third inning becomes a 'scratch hit' in the fourth. 1976 *Webster's Sports Dict.* 374/1 Scratch hit, a batted ball that enables a batter to reach base safely but that is neither an error nor a clean base hit. 1923 Kipling *Irish Guards in Great War* I. 6 The German trenches, which were rather in the nature of scratch-holes. 1969 G. Coppard *With Machine Gun to Cambrai* xxi. 87 We lived a mean and impoverished sort of existence in lousy scratch holes. 1899 B. Tarkington *Gentl. Indiana* xiii. 223 Sheets of blank scratch-paper lay before them, and they relaxed not their knit brows. 1934 J. O'Hara *Appointment in Samarra* (1935) vii. 214 He wanted to work.., and he tried to the extent of getting out some scratch-paper and pencils. 1971 *Islander* (Victoria, B.C.) 12 Sept. 9/5 When I take down a recipe from someone it is usually on a piece of scratch paper. 1979 *Farmington* (New Mexico) *Daily Times* 27 May 6c/4 (Advt.), Newsprint roll end paper 20c. a pound.., or else we will cut scratch paper, any length & width desired for 35c a pound. 1926 D. H. Lawrence *Plumed Serpent* x. 170 The land was being scratch-ploughed by a pair of oxen and a lump of pointed wood. 1939 *Sun* (Baltimore) 30 Nov. 24/6, I noticed that one of the girls was looking at a scratch sheet. 1956 T. Betts *Across Board* 170 William Armstrong..published the first scratch sheet that ever appeared on the newsstands of New York. The year was 1917. 1964 L. Hairston in J. H. Clarke *Harlem* 288, I..took the resumé scratch-sheet.. background..workin' experience..and such particulars. 1973 *N.Y. Law Jrnl.* 8 May 4/4 The person who reads the *New York Times* or *Daily News* or even the scratch sheets. 1934 P. A. Wells *Design in Woodwork* iii. 28 Lines or bandings round a panel,..can be pleasant..if not overdone. These are grooved in with a scratch stock, a simple tool made in the workshop. 1966 A. W. Lewis *Gloss. Woodworking Terms* 85 *Scratch stock*, tool for making small beads, mouldings, or grooves for inlays by scraping along the grain of the wood. 1976 F. E. Sherlock *Enjoying Home Carpentry & Woodwork* vii. 76 The scratch stock is very useful for the fine woodworker who wishes to inlay veneer lines. 1910 W. de la Mare *Three Mulla-Mulgars* i. 6 She is shown veiled on the rude pots of Assasimmon and in Mulgar scratch-work. 1977 *Times Lit. Suppl.* 14 Jan. 40/2 Reworked by Blake with scratchwork and blank ink... In this intermediate proof Blake has made these alterations through scratchwork and pen lines, in preparation for reworking the plate.

scratch, *a.* Add: **2.** (Earlier and further examples.) Of a game or match: impromptu, played by scratch teams. Freq. also of a meal.

1851 J. Pycroft *Cricket Field* x. 189 That is the time that some..batsman, whose eminence is little seen amidst the loose hitting of a scratch match, comes..to the wicket and makes a stand. 1851 *London at Table* i. 23 The butler.. giving directions to what the four-in-hand club used to call 'a scratch team' of servants. 1874 E. J. M. Collins *Frances* III. x. 234 Frances and Cecilia, coming down, found a hasty luncheon, and everybody busy at it... When this scratch luncheon was over, everybody went out. 1903 [see *cart *v.* 1 d]. 1923 J. Manchon *Le Slang* 258 *A scratch breakfast*, un déjeuner improvisé. 1942 *R.A.F. Jrnl.* 18 Apr. 10, I then commanded a scratch squadron of rather ropey machines. 1944 *Return to Attack* (Army Board, N.Z.) 15/2 Some of the men were keeping warm..by playing scratch football. 1952 [see *organize *v.* 2 d]. 1953 S. Kauffmann *Philanderer* vi. 103 They sat down to a scratch meal at about nine. 1973 'J. Patrick' *Glasgow Gang Observed* xviii. 146 Facilities for 'scratch' games of football. 1981 J. R. L. Anderson *Death in High Latitude* v. 85 If you don't mind a scratch meal I'd be delighted if you could stay to supper.

scratch, *v.* Add: **1. a.** (Later examples: cf. sense *3 f.)

1968 B. Evans *Dict. Quotations* 602/1 Scratch a Russian, and you will wound a Tartar. 1977 'D. Cory' *Bennett* ii. 64 Scratch a Spaniard, Hunter thought, and he oozes an offended formality instead of blood.

2. a. *to scratch one's head*: also *fig.*

1961 J. Baker *Cottage by Springs* xvii. 100 The supervisors, who occasionally arrived in shining saloon cars, scratched their heads over the problem. 1963 *Observer* 24 Nov. 21/4 We wait for the report, read it, and then scratch our heads. 1969 *Listener* 13 Nov. 667/1 We thought we'd take this year off and scratch our heads and see whether we can do something better. 1973 *Times* 26 July 33/1 The advent of the school holidays means that millions of children will soon be embarking on long car journeys to the seaside or the Continent and just as certainly many parents will be scratching their heads over the car sickness.

c. *you scratch my back and I'll scratch yours* and varr. Cf. Claw *v.* 5 b.

1704 E. Ward *All Men Mad* 18 Scratch me, says one, and I'll scratch thee. 1858 'A. Ward' Let. 27 Jan. in *Maine: Guide 'Down East'* (1937) iii. 363 You scratch my back and i will scratch your back. 1885 [in Dict., sense 2 a]. 1928 *Manch. Guardian Weekly* 10 Aug. 104/1 He goes on to spoil the effect by accusing Liberals of hypocrisy and being false to the principle of justice embodied in the phrase 'Scratch me and I'll scratch you'. 1937 'G. Orwell' *Road to Wigan Pier* ii. 33 In order that..the Nancy poets may scratch one another's backs, coal has got to be forthcoming. 1954 M. Ewer *Heart Untouched* viii. 132 It's the advertisers getting extra publicity. It's everybody scratching everybody else's back. 1961 J. Heller *Catch-22* (1962) iv. 33 A little grease is what makes this world go round. One hand washes the other. Know what I mean? You scratch my back, I'll scratch yours. 1978 P. Hill *Enthusiast* v. 61 Local farmer, got 'is own slaughter 'ouse. 'Ee scratches my back, I scratch 'is, know what I mean?

3. a. Also *fig.* Esp. in phr. *to scratch the surface (of)*: to make only slight progress in understanding, taking effective action (on), etc.; not to penetrate very far (into). See also sense *3 f.

1915 *New Republic* 13 Feb. 41/1 With all his earnest intention Amherst merely scratches the surface of the immense field of American social endeavor. 1932 Wodehouse *Louder & Funnier* 269 But this has merely scratched the surface. 1936 L. Hellman *Days to Come* III. 90 You haven't seen anything. They didn't scratch the surface here. 1969 *Listener* 13 Nov. 654/3 When it [sc. contraception] is attempted on a mass scale, as in India, it barely scratches the surface of the problem. 1971 D. Potter *Brit. Eliz. Stamps* xiii. 147 This simplified account can only scratch the surface of a story which is as intriguing and rewarding as the collecting of stamps themselves. 1977 *Time* 9 May 48/2 The industry..has been on a hot sales streak since 1973, when energy consciousness-raising really began. And the market has barely been scratched.

f. *scratch a — and find a —* and varr.: suggesting the true or fundamental character of any particular group, nation, etc.

In some cases *scratch* is interpreted as meaning 'to wound superficially': see quots. under sense *1 a.

1823 J. Gallatin *Diary* 2 Jan. (1914) 229 Very true the saying is, 'Scratch the Russian and find the Tartar.' c 1863 J. R. Green *Let.* in *N. & Q.* (1965) Sept. 348 They say, if you scratch a Russian you always find the Tartar beneath. 1888 Mrs. Oliphant *Second Son* I. xiv. 242 I don't put any faith in Russians... 'Scratch a Russian and you'll come to the Tartar.' 1892 I. Zangwill *Children of Ghetto* III. ii. vi. 81 Scratch the Christian and you find the pagan—spoiled. 1924 G. B. Shaw *St. Joan* iv. 52 Scratch an Englishman, and find a Protestant. 1926 D. Parker *Enough Rope* 60 Scratch a lover, and find a foe. 1966 *Listener* 10 Feb. 217/3 Scratch a Muse and as often as not you find nothing you can fathom, not even a woman. 1973 *Freedom* 2 June 3/4 'Scratch a liberal and you find a fascist,' says Westall. What bloody nonsense is this? Did he invent the saying himself to fit his present convenience? And what do you get when you scratch a Tory, a Fabian, a Social Democrat, a Marxist-Leninist? 1977 C. McCullough *Thorn Birds* xvii. 445 Scratch Justine's surface and you find a rebel.

5. a. (Later *intr.* examples.) Also *transf.*; *to scratch for oneself* (orig. *U.S.*), to fend for oneself; *to scratch (around) for* (something), to struggle for, to labour to achieve or find, to experience difficulty in acquiring, etc.

1850 H. C. Watson *Camp-Fires of Revolution* 30 Then each one had to scratch for himself. 1856 A. Cary *Married* 304 Shaking off the other child, [she] told him to scratch for hisself a time, while she began to prepare the supper. 1930 V. Palmer *Passage* i. viii. 65 He and Bob had to scratch for a living the best way they could. *Ibid.* 160 We'll have to scratch for another year or two to pay off the new boat. 1947 K. Tennant *Lost Haven* ii. 42 How much better did it look when you went off with that..moll, and left me and the kids to scratch for ourselves? 1953 T. A. G. Hungerford *Riverslake* 202 If his mob gets in next election they'll whip up a nice old depression, just like they did the last time, and we'll all be scratching for jobs again. The only difference is that there'll be a million or so of these bludgers scratching with us. 1960 Wentworth & Flexner *Dict. Amer. Slang* 451/1 *Scratch for (something)*..*scratch around for (something)*, to look for an object, to try to obtain something, esp. money or a much-wanted object; *fig.*, to scratch in the same way a chicken does in searching for food. 1961 Webster s.v., Turned out at an early age to scratch for themselves. 1962 A. Marshall *This is Grass* 202 Not that I read much. I've been too busy scratching for a crust. 1970 *Globe & Mail* (Toronto) 25 Sept. 3/5 They moved to hake and really did a job on that. Now, they're scratching for what's left of the hake. 1976 *Laurel* (Montana) *Outlook* 9 June 16/3 This leaves the city scratching for a means to financing garbage disposal. 1976 J. Snow *Cricket Rebel* 104 Walters scratched around for 42 during which he was given a life by another blunder by Rowan. 1979 'A. Hailey' *Overload* iii. x. 239, I scratched around for more details... Here are the exact dates of the convention and a preliminary program.

c. *intr.* To depart in haste, to make off with all speed. Freq. const. *for*. *U.S. colloq.*

1847 J. S. Robb *Streaks of Squatter Life* 109 I'm cussed if I hadn't to turn round, too, and scratch for the snag agin! 1875 'Mark Twain' in *Atlantic Monthly* Mar. 285/2 The moment it splits at the top..I know I've got to scratch to starboard in a hurry. 1887 *Outing* (U.S.) May 120/1 As I fired the gun and the horses scratched away from the mark. 1904 S. E. White *Blazed Trail Stories* i. 5 This little town will scratch fer th' tall timber..when the boys goes in to take her apart.

d. With *up*: to produce with difficulty, to scrape up.

1922 H. Crane *Let.* 24 Dec. (1965) 110, I am growing bald trying to scratch up new ideas in housekeeping and personal hygiene. 1930 'Sapper' *Finger of Fate* 188 It was six o'clock before the police arrived, and by that time we had scratched up a bit of breakfast and were feeling better.

7. c. (Earlier and later examples.)

1841 *Politician's Register for 1841* 3 Messrs. Ritner and Shulze, the Harrison Senatorial Electors, were scratched by a number of voters, and ran behind their colleagues. 1847 *Knickerbocker* Apr. 382 (Th.), He never scratched the regular ticket. 1880 *Scribner's Monthly* Oct. 909/1 They sometimes take the liberty of scratching a name, but they prefer, when the nominations are not too bad, to vote the regular ticket. 1904 *N.Y. Tribune* 8 Nov. 6 Vote the straight Republican ticket, without scratching. 1949 *Western Polit. Q.* Mar. 107 Thousands of voters scratched their ballots.

10. b. To forge (banknotes or other papers). *U.S. slang.*

1859 G. W. Matsell *Vocabulum* 77/2 *Scratch*, to write; to forge. 1926 *Flynn's Mag.* 6 Nov. 518/2 Well, scratch th' note an' we'll blow. 1935 N. Ersine *Underworld & Prison Slang* 65 *Scratch*, v. to forge checks or other papers.

11*. *U.S.* In billiards and related games: (*a*) *intr.* to make a stroke that incurs a penalty; *spec.* to hit the cue ball into a pocket; (*b*) *trans.* to hit (the cue ball) badly, incurring a penalty; *spec.* to hit (the cue ball) into a pocket. Cf. *scratch *sb.*¹ 10*.

1909 in Webster s.v. scratch *v.i.* 1959 N. Mailer *Advts. for Myself* (1961) 75 He shot poorly, hit the wrong ball and scratched. 1964 Sullivan & Crane *Young Sportsman's Guide to Pocket Billiards* ix. 91 *Scratch*, a playing stroke in which the player forfeits his playing turn. Most often caused by 'scratching' a ball unintentionally into a pocket. 1974 *Rules of Game* 85/3 A player may scratch the cue ball into a pocket at the break shot or during continuous play. 1977 *New Yorker* 4 July 24/2 This kid asked me, 'Do you ever scratch?'.. I said, 'I ain't never scratched in my life.'.. Just then, I took this shot and the cue ball went right in the pocket. He said, 'Well, you've scratched now.'

scratch-back. **1.** (Earlier example.)

1842 *Ainsworth's Mag.* I. 20 What is it but..a fan, lazy-tongs, parasol, or scratchback?

scratch blue (skrætʃ bluː). [f. Scratch *v.* + Blue *sb.*] Used *attrib.* and *absol.* to designate a decoration of incisions filled with blue pigment found on eighteenth-century stoneware or stoneware so decorated. Cf. **scratched blue*.

1924 Rackham & Read *Eng. Pottery* vi. 88 In the Liverpool museum there is a mug of 'scratch blue'. 1957 Mankowitz & Haggar *Conc. Encycl. Eng. Pott. & Porc.* 28/1 A salt-glazed mug with 'scratch blue' decoration is inscribed with..the date 1742. 1960 H. Hayward *Antique Coll.* 251/1 'Scratch blue' ware, a class of white salt-glazed stoneware decorated with..ornaments and inscriptions incised upon the wares in the unfired 'green state'. 1969 G. Wills *Eng. Pott. & Porc.* 88 The incised wording was emphasised by dusting it, before firing and glazing, with powdered cobalt-blue; a type of decoration..known as 'scratch blue'. 1971 L. A. Boger *Dict. World Pott. & Porc.* 308/2 It is thought that this scratch blue ware may also have been made at Liverpool and in other parts of England.

scratchboard (skræ·tʃ,bōᵊɹd). *Art.* Also scratch board. [f. Scratch *v.* + Board *sb.*] Cardboard specially treated and coated so that the surface can be scratched away to create drawings, etc.; a board of this type. Cf. *scraper board* s.v. *scraper 10.

1930 C. E. Wallace *Commercial Art* iv. 119 Scratchboard Drawing.—Scratchboard is a cardboard with an enamel surface of special finish. Drawings are made on it with ink and are afterwards scratched with a knife to obtain lights and special effects. 1942 *School Arts* (U.S.) Mar. 229/1 Why not try some scratchboard in your commercial art classes?.. Scratchboard is a lightweight cardboard, coated on one side with a clay-like composition. 1948 H. Missingham *Student's Guide in Commercial Art* ii. 57 Scratch boards are available in a great variety of surfaces and tones. The surface is coated with clay preparation and is drawn or scraped away where desired. 1964 Turnbull & Baird *Graphics of Communication* xii. 223 Using a bristol board coated with chalk, an artist can produce a scratchboard drawing by covering an area with ink and then scratching the ink and chalk from the surface of the board. 1976 *National Observer* (U.S.) 24 Jan. 20/1 Scratchboard—clay-coated cardboard on which you draw and scratch away.

scra·tch-build, *v.* Also scratchbuild, scratch build. [f. Scratch *sb.*¹ + Build *v.*] *trans.* To build (a model) from scratch, using no specially prepared components. So **scra·tch-building** *vbl. sb.*; **scra·tch-built** *ppl. a.*

1961 C. J. Freezer *Railway Modelling* x. 95 (*heading*) Scratch building. 1967 *Railway Modeller 'Shows You How' Booklet* No. 19 (title-page) Our cover photograph shows one of the finest scratchbuilt model railways in the world, Peter Denny's Buckingham Branch. *Ibid.*, Scratchbuilding is the craft of modelmaking from raw materials and basic components. 1974 *Times* 17 Apr. 14/2 Parren said he hoped to attract young members by encouraging the scratch building—that is, building to scale from scratch—of modern locomotives. 1975 G. Scarborough *Tank & AFV Modelling* (Airfix Magazine Guide No. 5) ii. 7/1 Conversion and scratch-built projects of most of the popular subjects. *Ibid.* 19/1 This leaves us with the basic hull to scratch-build.

scratched, *ppl. a.* Add: **1.** *scratched blue* = *scratch blue.

1883 L. M. Solon *Art of Old Eng. Potter* vi. 79 Some.. processes deserve special attention. The 'scratched blue' for instance enjoyed a successful run. 1940 *Chambers's Techn. Dict.* 748/2 *Scratched blue*.., incised ornament on unbaked clay, sprinkled with cobalt glass, then fused in the kiln.

scratcher. Add: **1. c.** (Earlier example.)

1880 *Scribner's Monthly* Feb. 621/2 Mr. Evarts will be obliged to look among the 'scratchers',..for the indorsement of..Civil Service Reform.

e. For *U.S.* read orig. *U.S.* and add earlier and later examples.

1859 G. W. Matsell *Vocabulum* 77/2 *Scratcher*, a forger; a copyist. 1927 *Writer's Monthly* Nov. 387/2 Forgers, and floaters of bad money, checks or commercial paper, are 'scratchers', 'scratchmen' and 'shovers'. 1941 V. Davis *Phenomena in Crime* iv. 48 The actual forger, known by such names as 'the scratcher', 'the scribe', 'the penman', may consider himself extremely fortunate if his period of office exceeds two years.

2. d. A device put down an oil or gas well to clear the bore or create turbulence mechanically.

1877 J. F. Carll *Oil Well Records & Levels* iv. 90 A 'scratcher' is a round brush..made of steel wire. When it is to be used the tubing is drawn from the well, a few barrels of benzine are poured in and the scratcher is..run down to the oil rock, where it is worked up and down..to scratch or scrub the walls of the well. 1974 D. K. Smith in P. L. Moore et al. *Drilling Practices Manual* xvi. 426 Scratchers or wall cleaners. 1. Rotating... 2. Reciprocating. *Ibid.* 427 Install scratchers spaced according to location of permeable zones.

scratching, *vbl. sb.* **b.** scratching post (earlier and later examples).

1890 J. Macdonald *Light in Africa* iii. 37 Cattle got among their tent ropes, and they erected 'scratching posts' to keep them away from their camp. 1968 *New York City* (Michelin) 73 Abercrombie and Fitch... One can find almost anything here, from a ten-cent fish hook to..a scratching post for your cat.

scratchless, *a.* Add: (Earlier example.) Also, without scratch (*scratch *sb.*¹ 6 b).

1829 P. Egan *Boxiana* 2nd Ser. II. 750 But Jack came off quite scratchless. 1978 *Gramophone* June 31/2 After the war Sterling pursued improvements on a big scale and one recalls in particular what were called the Columbia 'scratchless' records.

scratch pad. Also scratch-pad, scratchpad. [f. Scratch *sb.*¹] **1.** A scribbling block. Also *attrib.* and *fig. colloq.* (orig. and chiefly *U.S.*).

1895 *Montgomery Ward Catal.* Spring & Summer 111/2 Desk or Scratch Pads, made from white laid, smooth finish paper for pen or pencil. 1906 *Dial. Notes* III. 155 Can you let me have a nickel to get a scratch-pad? 1931 W. G. McAdoo *Crowded Years* xv. 220 He called it a 'scratch-pad draft'..that..was not to be taken as final. 1939 R. Chandler in *Dime Detect. Fict.* Aug. 48/1 The paper was from a scratch pad. It would have been very nice if it had had a message on it. 1960 *News Chron.* 4 Feb. 6/4 Desk and telephone and scratch-pad and paper-clips. 1966 *Listener* 9 June 838/3 A boy in bellbottoms lamented (or was it bragged?) he had a 'scratchpad memory'. 1970 'W. Haggard' *Hardliners* iii. 26 She struck out a line from her scratch-pad and the editor watched her. 1978 W. F. Buckley *Stained Glass* xix. 190 Blackford sat on the couch and made motions requesting a scratch pad.

2. *Computers.* A small, very fast memory for the temporary storage of data or for indirect addressing of the main memory; usu. *attrib.*

1965 *Proc. Conf. Amer. Federation Information Processing Societies* XXVII. i. 667/2 General-purpose commercial processors with scratchpad memories did not appear on the market until 1959. *Ibid.* 676/2 The magnetic core scratch-pad has an access limitation of one register per cycle. 1970 O. Dopping *Computers & Data Processing* x. 135 The small, fast memory, which is well suited for data and instructions that are going to be used very often, is sometimes called a scratch pad memory. 1977 *Design Engin.* July 77/2 It incorporates a 2k×8-bit mask-programmable read-only memory, a 64×8-bit scratchpad random-access memory, four 8-bit input/output ports and a binary timer. 1979 *Personal Computer World* Nov. 83/2 The user program runs in a stack whose pointer is extracted from TGTSTK (FOFA in the scratchpad RAM).

scratchy, *a.* Add: **4. b.** Of sound: rough, grating. Of a sound-recording: characterized by scratch (*scratch *sb.*¹ 6 b).

1889 *Cent. Dict.* s.v., A scratchy noise. 1961 Webster s.v., Scratchy tune came from the phonograph. 1961 L. Hughes *Ask your Mama* 3 In the quarter of the negroes Where the doors are doors of paper Dust of dingy atoms Blows a scratchy sound. 1976 W. Trevor *Children of Dynmouth* ii. 43 He told Stephen to sit on it, in a voice that wasn't as scratchy as usually it was. 1977 *Film & Television Technician* Jan. 5/2 He precariously wound up the oldest gramophone this side of the Urals so that we might hear a scratchy 78 recording of some Russian choir singing the Creed. 1978 *Oxford Times* (City ed.) 17 Feb. 19 The mono recordings are primitive and scratchy.

c. *fig.* Ill-tempered, peevish, catty.

1925 E. H. Young *William* xxxii. 277 I'm no heroine. I'm a nasty, scratchy, impatient little beast. 1936 L. C. Douglas *White Banners* viii. 180 But if she was going to write him any more scratchy letters..it might turn out badly. 1949 N. Marsh *Swing, Brother, Swing* ix. 214 We're both scratchy. I told her I thought the unfortunate Rivera was ghastly and she thinks I'm shaking my curls at Mr. Alleyn. 1958 *Spectator* 27 June 835/2 The small, scratchy, pert, unhappy son. 1971 P. Purser *Holy Father's Navy* i. ii. 11 Irby..was a bit scratchy. She said, 'Why is it we can never go anywhere on time?' 1977 'J. le Carré' *Honourable Schoolboy* i. 31 Nor did the dwarf's own..version meet with much success, which made him very scratchy.

5. Also *transf.*

1933 J. B. Priestley *Wonder Hero* ii. 36 He had taken her away from her bed-sitting room and scratchy living, and had made her queen it in a fine service flat in Knightsbridge.

Hence **scra·tchily** *adv.*

1927 *Daily Express* 27 May 13/2 Wethered,..reaching the turn rather scratchily in forty-two, stood one down. 1975 R. L. Duncan *Dragons at Gate* (1976) i. 27 Chamber music drifting scratchily from an old Victrola.

scrattle, *v.* Delete † and add: **2.** (Earlier example.)

1817 H. L. Piozzi *Let.* 4 Jan. in *Autobiogr. Mrs. Piozzi* (1861) II. 187 Coal carts scrattling up the hill often used to make me think—'Hinc ex audiri gemitus, et sæva sonare Verbera; tum stridor ferri, tractæque catenæ.'

Hence **scra·ttling** *ppl. a.* (example *fig.*).

1913 D. H. Lawrence *Sons and Lovers* iv. 75 In this flamin', scrattlin' place.

scrawled, *ppl. a.* Add: Also *transf.* and *fig*

1895 Mrs. H. Ward *Bessie Costrell* ii. 27 A wide plain travelled into the sunset, its level spaces cut by the scrawled elms and hedgerows of the nearer landscape. 1910 J. Buchan *Prester John* xxi. 351 There..was the body of Henriques, lying scrawled on the sand. 1939 Dylan Thomas *Map of Love* 4 This weak house to marrow-columned heaven, Is corner-cast, breath's rag, scrawled weed,..crow stalk, puffed, cut, and blown.

scrawliness. (Earlier example.)

1867 W. James *Let.* 17 Sept. (1920) I. 103 Excuse the scrawliness of this too hurriedly written letter.

scrawny, *a.* Delete *U.S.* and add later examples.

1946 D. C. Peattie *Road of Naturalist* iv. 43 The one where my scrawny, bronchitic body would find itself was stone-cold, a dirty grey, so I would mentally adjourn to the other. 1977 J. F. Fixx *Compl. Bk. Running* vi. 80 Frank Shorter, who won an Olympic gold medal in the 1972 marathon, is 5 feet 10½ inches and weighs a scrawny 134. 1977 *New Yorker* 27 June 59/2 This beautiful beast that I, a scrawny little thing, am destroying.

screaky. Delete *rare* and add: Also **skreaky, skreeky.** (Earlier and later examples.)

1884 'Mark Twain' *Huck. Finn* xxvii. 272 A melodeum —a sick one..pretty skreeky and colicky. 1909 *Dialect Notes* III. 404 'My shoes are skreaky.' 'I hate skreaky doors.' a 1961 in Webster, Bats..making their screaky sounds.

scream, *sb.* Add: **c.** A cause of laughter; a very amusing person or situation. *colloq.*

1888 *Boston Herald* 24 Jan. 5/1 It [sc. 'Bewitched'] is something more than a sketch, and a good deal less than a comedy, and its designation on the bill, 'farcical scream', perhaps conveys a good idea of its character. 1906 H. Green *At Actors' Boarding House* 209, I thought I'd die laughing at his making love..and me with a husband doing his bit back in Auburn. It was a perfect scream, wasn't it, Kate? 1915 *Home Chat* 9 Oct. 45/2 'He's married.'.. 'It's a fact. His bailiff told our chauffeur... Isn't it a scream?' 1919 [see *perfect *a.* B. 5 f]. 1929 R. C. Sherriff *Journey's End* iii. ii. 71 Oh, skipper, you are a scream—and no mistake! 1946 E. O'Neill *Iceman Cometh* (1947) i. 68 Listen, it was a scream. 1974 *Guardian* 30 Jan. 11/1 Like the bearded lady, Lorna has curiosity value... 'Yes, isn't it a scream?' she says. 1977 J. Fleming *Every Inch a Lady* iii. iv. 131 They're good... That Tommy Raffles is a scream!

d. The giving of information or evidence, *spec.* against one's accomplices in crime. *slang.*

1925 E. Wallace *Melody of Death* vii. 113 'Look here, George,..is it a scream?' 'A scream?' Mr. Wallis was puzzled innocence itself. 'Will you turn King's evidence?'

e. An urgent message. *slang.*

1920 'Seamark' *Down River* i. 22 'Smuggling?' queried the surgeon. 'That's the line, sir. Had a scream from Headquarters about it only this morning.'

f. *Jazz.* The sound produced when a high note is played loudly on a wind instrument. Also *attrib.*

1933 *Metronome* Jan. 34 A scream is produced somewhat the same way as the rip, only in the rip the note is cut off shortly, but in the scream it is held. 1955 B. Ulanov *Hist. Jazz in Amer.* (1958) xvi. 275 The sustained scream notes. 1962 Carter & Kunstadt *Jazz: Hist. N.Y. Scene* xvi. 200 Such selections showcased the other in 'screams' (chords) was the feature of 'Tiger Rag'.

scream, *v.* Add: **1. c.** *spec.* to travel swiftly with a noise like a scream; also hyperbolically and *transf.*, of a person.

1943 Hunt & Pringle *Service Slang* 57 *Screaming downhill*, making a power dive in a fighter aircraft. 1954 *Amer. Speech* XXIX. 101 'It screams'!, i.e., it really moves. 1957 J. Moore *Science & Fiction* 60 Airmen screaming towards the ground in a steep dive. 1975 E. Hillary *Nothing Venture, Nothing Win* x. 135 We had discussed the..likelihood of no one being able to..pull the boat onto the bay and finally decided that Tenzing and I should scream up in a booster party to make sure that the Sherpas got there. 1976 *Wymondham & Attleborough Express* 10 Dec. 24/4 Alan Green..unleashed a full-blooded volley from just inside the area that screamed into the top of the net. 1976 A. White *Long Silence* vii. 53 One of the lighters screamed down to take a closer look. 1981 *Economist* 24 Jan. 97/2 A thrill of a second is about as long as a star falling into a black hole would be expected to 'scream'.

e. To turn informer; to give evidence against one's accomplices in crime.

1935 E. Wallace *Melody of Death* viii. 114 'I don't want to hear any more about your conscience,' said the officer wearily. 'Do you scream or don't you?' 'I don't scream,' said Mr. Wallis emphatically. 1967 J. Morgan *Involved* 114 He never got paid..and my information is he's ready to scream.

2. b. *fig.* To cry out for.

1906 E. Dyson *Fact'ry 'Ands* viii. 95 Ther job's simply screamin' fer a statesman lv your sort. 1930 'Sapper' *Finger of Fate* 147 With every nerve in his body screaming for the stuff [sc. whisky]. 1978 R. Ludlum *Holcroft Covenant* xi. 133 She did not walk; she glided—an extraordinary body screaming for observation as a prelude to invasion and satisfaction.

c. *fig.* To communicate (something) strongly.

1957 [see *beddable *a.*].

d. *to scream on* (someone), to insult in 'playing the dozens' (see *play *v.* 16 e). *U.S. Blacks'.*

1970 H. E. Roberts *Third Ear* 12/1 *Screaming on*, telling someone off;..embarrassing someone publicly. 1974 H. L. Foster *Ribbin', Jivin', & Playin' Dozens* v. 198 Sometimes, 'loud mouthing' or 'loud talking', 'sounding', 'screaming on someone' or even 'bogarding' are synonyms for woofing.

3. (Further examples with direct speech as object.)

1866 C. M. Yonge *Dove in Eagle's Nest* I. ix. 189 'Peace, thou fool!' screamed the old lady. 1976 B. Freemantle *November Man* x. 131 'The servants can't hear, Jocelyn,' she screamed.

screamer. Add: **4.** *slang.* **b.** (Earlier example, in sense 'a person who tells exaggerated stories'; also, earlier example of sense 'a "screaming" farce'.)

1831 *Louisville Public Advertiser* 17 Oct. 2/3 The principal character in this production, is to use his own elegant language, a screamer. 1849 *Theatrical Programme* 4 June 16/1 At the Adelphi crowds muster nightly to see..Wright and Bedford in a 'screamer'.

c. An exclamation mark.

1895 in *Funk's Stand. Dict.* 1908 *Bohemian* XIV. 643/2 Few of us have forgotten..the origin of 'yellow' as descriptive of that class of journalism addicted to 'screamers'. 1927 *Amer. Speech* II. 239 Exclamation points may be 'screamers', 'astonishers', or 'shouts'. 1933 D. L. Sayers *Murder must Advertise* viii. 132 'Waste Nerve-Power!' Capital N, capital P, and screamer. Got that? 1960 *Guardian* 17 Sept. 12/6 Some newspapermen call the ! a screamer. 1972 *Ibid.* 18 Aug. 11/1, I once worked for an editor who cut out all the screamers—that's what you and I call exclamation marks.

d. A very powerful shot in a game.

1896 W. Park *Game of Golf* 269 *Screamer*, a very long stroke, so called from the whistling noise made by the ball. 1926 Wodehouse *Heart of Goof* i. 13 He stepped off the sixteenth, after hitting a screamer down the centre of the fairway. 1959 *Sunday Times* 9 Aug. 28 (Advt.), When you hit a screamer..you'll be glad you played Slazenger 279. 1963 V. Canning *Limbo Line* xvii. 227 Amadeo hit a screamer, dead straight and slightly left of the middle of the fairway. 1977 *Tennis World* Sept. 17/1 A 'heavy serve', one might think, means a fast serve. Wrong again. The term to denote velocity is 'big' or occasionally 'cannonball' or 'screamer'.

e. An informer, a tale-teller; a complainer.

1903 Farmer & Henley *Slang* VI. 126 *Screamer*,..a thief who, robbed by another thief, applies to the police. 1961 *John o' London's* 30 Nov. 610/3 An informer..is now more often referred to..as a singer or a screamer. 1968 *Telegraph* (Brisbane) 19 June 58/7 The man I was talking to said, 'Hang on a minute, I've got a screamer coming in.' When he came back I asked him what a screamer was. It turned out to be someone who complains about defects in a car he has bought.

f. In full *screamer headline*. A large headline.

1926 [see *rap *v.*¹ 1 d]. 1945 L. Shelly *Jive Talk Dict.* 17/1 *Screamers*, newspaper headlines. 1975 *New Yorker* 4 Aug. 20/1 'Gifts flow profusely at 124 West 42nd St. and 625 Madison Ave' was the screamer on a flyer handed to us on Forty-second. 1979 J. Drummond *I saw him Die* ii. 24 The murder was on the front page. Screamer headline.

g. *Jazz.* A passage featuring loud high notes played on a wind instrument; such a note.

1940 *Swing* Nov. 28 It's another riff tune..plus (or minus) a screamer featuring the leader's horn. 1948 *Down Beat* 1 Dec. 13 *Minor* is a screamer but not without change of pace. 1958 K. Goodwin in P. Gammond *Decca Bk. of Jazz* xiii. 149 His ability to produce stratospheric screamers with apparent ease was utilized to add bite and drive to the brass sections.

h. A bomb that makes a screaming sound as it drops.

1942 'R. Crompton' *William carries On* ii. 42 Her reactions to 'screamers'. 1943 Hunt & Pringle *Service Slang* 57 *Screamer*, a whistling bomb, i.e. a bomb with a device attached to cause a screaming sound as it descends.

i. *the screamers* = *the screaming habdabs* s.v. *habdabs.

1948 Partridge *Dict. Forces' Slang* 163 *Screamers, the*, an evident dislike of operational flying. 1952 M. Tripp *Faith is Windsock* i. 17 'Cut it out, you two,' said Bergen, 'you give me the screamers.'

j. *two-pot screamer*, one who easily shows the effects of alcohol. *Austral.*

1959 D. Hewett *Bobbin Up* 21 Look at Lou. She's a two-pot screamer, always 'as been. 1972 J. de Hooc *Skid Row Dossier* 95 It says experienced and sober, ya bloody two-pot screamer.

screaming, *ppl. a.* Add: **1. b.** *screaming eagle* (U.S. slang) = *ruptured duck (b).

1946 *Newsweek* 18 Mar. 34/1 'Ruptured duck': GI for the discharge button which ex-service men wear in their lapels, also, 'homecoming pigeon' and 'screaming eagle'. 1948 A. M. Taylor *Lang. World War II* 172 *Ruptured Duck*... Also nicknamed Screaming Eagle.

2. b. (Further examples.) Also, blatant, obvious.

1922 J. Hergesheimer *Bright Shawl* (1923) 205 The shawl ..was malevolent, screaming in color. 1944 'G. Orwell' *Coll. Essays, Journalism & Lett.* (1968) III. 168 The 'screaming' advertisement started some time in the nineteen-twenties. 1963 *Australasian Post* 14 Mar. 51/1 I'd be a screaming nong if I didn't recognise you as a creep. 1965 *Listener* 9 Dec. 941/1 None of..the anarchy of competing posters and screaming shop signs. 1968 *Globe & Mail Magazine* (Toronto) 13 Jan. 6/3 The commonly held stereotype of a homosexual is called, in gay jargon, a screaming queen. 1972 B. Rodgers *Queens' Vernacular* 177 *Screaming*, flagrantly homosexual. 1976 *National Observer* (U.S.) 25 Sept. 13/1 Such segregation was..a screaming mockery of the Bill of Rights. 1977 *New Yorker* 15 Aug. 22/1 The News and the Post ran screaming headlines. 1981 *Daily Tel.* 20 Feb. 17/1 Spring colours are bright pink and screaming green with khaki chino skirts for women and khaki chino trousers for men.

d. *screaming habdabs*, etc.: see *habdabs. *screaming meemies*, etc.: see *meemies *sb. pl.*

3. Comb., as *screaming-scared* adj.

a 1963 C. S. Lewis *Poems* (1964) 106 My body awakes in bed Screaming-scared.

screech (skriːtʃ), *sb.*² *slang.* [ult. ad. Sc. dial. *screigh* whisky.] **a.** Whisky. **b.** Any strong alcoholic liquor, freq. one of inferior quality. **c.** *Newfoundland.* A specific rum, or a specific mixture of rums.

1902 Farmer & Henley *Slang* VI. 126/1 *Screech*, subs. (common), whiskey. 1944 T. H. Wisdom *Triumph over Tunisia* viii. 68 The famous and kindly Peres Blanc from the Monastery at near-by Thibar had supplied them with drink from the monks' own cellars, and the popular drink was one that had been aptly christened 'Screech' by '111'. 1945 W. H. Pugsley *Saints, Sinners & Ordinary Seamen* 231 [The rating] gets hold of some bootleg scotch—'high life', they call it on the West Coast, and 'screech' in Newfie—and then he's away to..Cells or Detention. 1957 B. Hutchison *Canada: Tomorrow's Giant* 24 He is a little addicted to the Island's national drink called 'Screech'. 1958 *Maclean's Mag.* 27 Sept. 63/3 Screech is a mixture of rums now sold by the liquor board under a new label that displays..the legend 'Newfoundland's Famous Screech'. 1959 *Manch. Guardian* 7 July 7/4 There has been some concern at the violence during fights ashore between servicemen following the drinking of a local concoction known in the service [the Navy] as 'Screech'. It is made of a local wine, 'Imbeet', mixed with Coca Cola. 1963 [see *Newfie]. 1964 C. Rougvie *Medal from Pamplona* vi. 80 'Beer and screech.' 'Screech?'.. 'It's a term embracing all cheap Canadian wines.' 1973 *Daily Colonist* (Victoria, B.C.) 29 Aug. 2/2 The taste of beer must rank somewhere between buttermilk and Newfie Screech. 1979 *Globe & Mail* (Toronto) 17 Oct. 6/1 But allow us to explain. Screech, the hairy-chested Newfoundland libation, is based on Jamaican rum which arrives on the tight little island in 40-gallon, fire-charred oak barrels.

screech, *v.* Add: **1.** (Later example.)

1919 [see *grab *v.* 3 d].

3. (Later example.)

1972 'M. Sinclair' *Norslag* iv. 35 The ancient lift operative..had been less than polite as he had screeched the gates open for him.

screeching, *vbl. sb.* and *ppl. a.* (Later example.)

1976 *National Observer* (U.S.) 4 Dec. 15/2 A lot of this stuff is going to come to a screeching halt quickly, and we're not going to do the screeching.

screech-owl. Add: **1. b.** A small North American owl of the genus *Otus*, esp. *O. asio.*

1671 J. Ogilby *America* 147 The Birds both common and peculiar [to New England] are thus recited:..The long-liv'd Raven, th' ominous Screech-Owl. 1812 A. Wilson *Amer. Ornithol.* V. 83 Red Owl..is..well known by its common name, the Little Screech Owl. 1884 *Cent. Mag.* Nov. 121 The screech-owl would shake and shiver in the depths of the wood. 1949 *Amer. Forests* Oct. 23/2 The weird call of the more or less familiar screech owl is probably the best known of all the owls. 1960 [see *jumby-bird]. 1975 *Islander* (Victoria, B.C.) 2 Nov. 3/1 The screech owl is the one I hear and see most often.

screed, *sb.* Add: **3. a.** More generally in *Building*, a level strip of material formed or placed on any surface (e.g. a floor or a road) as a guide for the accurate finishing of it. Also, a levelled layer of material forming part of a floor or other horizontal surface. (Further examples.)

1937 S. C. B. Stubbs *Building Encycl.* IV. 1241/2 *Screeding*. A cement and sand floating on a floor, laid in preparation for a subsequent paving on or a wall for wall tiling, is known as a screeding. Often it is called simply a screed, because it is brought up level by the use of screeds. *Ibid.*, In the case of floors..it is usual to use the batten itself as a screed without forming the floated strip alongside, and hence we find the battens often referred to as the screeds. 1952 D. Nield *Building Constr. Illustr.* iii. 35 Cross reinforcement is laid over the beams and the whole covered with a layer of fine concrete (screed). 1956 Davies & Petty *Building Elements* ix. 270 A screed of cement and sand is laid later to provide a smooth and level surface for whatever floor finish is to be used. 1961 *Times* 3 July (Archit. Suppl.) p. viii/3 Heating by electric wires embedded in the concrete screed is becoming quite common. 1974 W. E. Kelsey *Building Construction* v. 87 Although the term screed is applied to the whole of the final surface, it is also used to describe the narrow strips of wet cement used as a guide to the thickness of the top layer. 1978 *Cornish Guardian* 27 Apr. 33/6 (Advt.), Uneven floors made smooth with latex screed.

b. *screed board, rail, strip.*

1949 A. G. Geeson *Gen. Building Repairs* (ed. 11) I. vi. 335 By moving the screed board with a slight tamping motion, the surface will be slightly roughened. 1968 *Gloss. Formwork Terms* (B.S.I.) 25 A tamper may be constructed so that it also acts as a screed board. *Ibid.* 22 *Screed rail*, a guide fixed at the perimeter of a concrete pour to act as a datum and support for a screed board. 1977 *Club Tennis* Mar. 13/3 The actual laying of the surface is done by screed strips and straight edge and the court is rolled and trowelled to a perfect level.

screed, *v.* Add: **4.** *Building.* **a.** To level (a surface) by means of a screed; to apply (material) as a screed to a floor surface. (Cf. Screed *sb.* 3 a in Dict. and Suppl.)

[1825: see Screeding *vbl. sb.*] 1898 F. W. Macey *Specifications in Detail* 68 State if screeded in lime and hair mortar instead. 1944 E. Lucas in R. Greenhalgh *Pract. Builder* ii. 104/2 The second coat must be screeded to bring the surface level. 1949 A. G. Geeson *Gen. Building Repairs* (ed. 11) I. vi. 333 The surface is finished by screeding it with a straightedge. 1970 *Daily Tel.* (Colour Suppl.) 5 June 35 Concrete can also be screeded to floors in old houses. *Ibid* 39/2 How to screed a concrete floor.

b. With advbs.: *screed in*, to finish off a surface around (an object, as a frame) by means of a screed; *screed off*, to take *off* (excess material) from a surface by means of a screed.

1898 F. W. Macey *Specifications in Detail* 63 Hair mortar in brickwork is only used for screeding in door and window frames. 1949 K. S. Woods *Rural Crafts of England* iv. xi. 180 The plaster was laid on very evenly, and then 'screeded off' with a long straight-edge known as a 'screeding-rule'. 1950 *N.Z. Jrnl. Agric.* Jan. 58/3 The surplus mix [of concrete] being screeded off with a straight-edge.

screeder (skriː·dəɹ). [f. *screed *v.* + -er¹.] A person employed to lay floor screeds.

1976 *Derbyshire Times* (Peak ed.) 3 Sept. 11/7 (Advt.), Floor screeders and quarry tilers.—One or two reliable layers required. 1977 *Evening Post* (Nottingham) 24 Jan. 11/2 (Advt.), Floor Screeders required (male–female) for long term contract in Nottingham.

screeding, *vbl. sb.* Add to def.: The action of *screed *v.* 4. Also *concr.*, the material of a screed. (Further examples.)

1898 F. W. Macey *Specifications in Detail* 68 The mortar to be used in the screeding in of frames. 1936 *Archit. Rev.* LXXIX. 8 The floors are reinforced concrete slabs with two layers of cork separated by a layer of hard screeding. 1937 [see *screed *sb.* 3 a]. 1956 Davies & Petty *Building Elements* ix. 272 Screeding is carried out by the use of fine mix, generally one part of cement to three or four of sand.

screef (skrif), *sb.* Also scrief, 9 skreef. [dial. var. of Scurf *sb.*¹, Scruff *sb.*¹] **1.** *Sc.* and *Forestry.* A layer of vegetation on the surface of the ground.

1817 J. Christie *Instructions for Hunting* 39 Without a bit of screef aboon, But bare and naked craigs o' stane. 1866 W. Gregor *Dialect of Banffshire* 163 There's a fine skreef o' girs on that shift. 1925 R. L. Cassie *Gangrel Muse* 33 An' swack little feeties trip licht o' the screef, I' the reelin' an' furlin' o' fairies. 1934 *Forestry* VIII. 21 Where there is a skin of peat..it is preferable to plant in a ploughed furrow, or in a screef with loosened soil. 1952 *Buchan Observer* 7 Oct., You may weel spier fa's to cast yer divots, gin sae be that ye ken o' a bittie o' gweed screef.

2. Special Comb.: **screef-mark**, an area from which surface vegetation has been cleared.

1950 R. Jenkins *So Gaily sings Lark* xxii. 203 With the spade the vegetation was scraped away, and in the black 'scrief-mark' as it was called an L notch was made, into which the tree was carefully inserted, the earth being firmed again by careful pressure of the heel.

screef (skrif), *v.* *Sc.* and *Forestry.* Also scrief. [f. prec. Cf. Scurf *v.*, Scruff *v.*¹] *trans.* To clear (surface vegetation) from the ground.

1926 *Trans. Buchan Field Club* XIII. 82 Small places had the turf skinned off or 'scriefed'. 1930 *Forestry* IV. 20 Screef the surface lightly with the flat end [of a mattock], and then loosen up the patch with the pick end. 1948 *Country Life* 8 Oct. 739/3 To screef is to clear the ground of surface vegetation (such as heather) with a mattock or comparable tool.

So **scree·fing** *vbl. sb.*

1930 *Forestry* IV. 85 Short of removing the surface layers of the fire traces by 'screefing' or turning it in with the plough ..it is usual to get rid of the surface vegetation by controlled burning. 1953 H. L. Edlin *Forester's Handbk.* ix. 151 It is seldom necessary to remove weed growth completely, though screefing, or the paring away of surface vegetation with a mattock, may be practised where trees are very slow to get away..or where manure is applied. 1962 *Finnish Paper & Timber* XIII. 162 When the weight and efficiency of the tractor was increased and the screefing equipment made more manoeuvrable over rough going suitable machinery for the work had been found.

screel (skril), *sb.* *dial.*, *Sc.*, and *Barbados.* [f. next.] The cry of a bird, a child, etc.; the piercing note of a whistle.

1876 C. C. Robinson *Gloss. Mid-Yorks.* 117/2 *Screel* (skri·h'l), v. n. and sb., to cry, in a shrieking manner; gen. 1922 *Glasgow Herald* 8 Aug. 6 With nothing to mar your peace so serene Save the screel of the curlew or sunset's red sheen. 1953 G. Lamming *In Castle of my Skin* v. 91 At the same hour every morning the whistle screel shot up like an alarm through the rumbling of cart wheels.

screel (skril), *v.* *dial.*, *Sc.*, *N. Ir.*, and *Barbados.* Also **skreel.** [Prob. imitative or ad. ME. *skrille* scream: see Skirl *v.*¹] *intr.* To screech, to scream, to utter a high-pitched or a discordant cry. Occas. used *transf.* of inanimate things. So **scree·ling** *vbl. sb.*

1875 W. D. Parish *Dict. Sussex Dial.* 106 *Skreel*, e[ast], to scream. 1889 M. Peacock *Taales fra Linkisheere* 103 What wi' yammerin' bairns, an' what wi' screälin' wimmin. 1934 *Punch* 14 Nov. 543 Though kelpies walk both bold and free And icy winds skreel off the sea. 1953 C. Day Lewis *Italian Visit* vi. 61 Again again again, the frogs are screeling Down by the lilypond. 1953 G. Lamming *In Castle of my Skin* vi. 113 Mother..lashed me thoroughly. Bob heard the screeling, and came down to our house listening. 1965 *Dundee Courier* 10 Apr. 8 Herring gulls skreeled.

screen, *sb.*¹ Add: **1. d.** (Examples of the sense 'surface for the reception of images'.) *spec.* (*a*) a usu. large white surface for receiving the image from a film projector; (*b*) a small fluorescent screen, esp. one in a television set (so *little screen*). Also *transf.* (usually with definite article), moving pictures collectively; the cinema; the film world.

1810 *New Family Receipt-bk.* 257 To make Transparent Screens for the Exhibition of the Phantasmagoria. 1846 *Penny Cycl.* Suppl. II. 254/2 Magic lantern is a species of lucernal microscope, its object being to obtain an enlarged representation of figures, on a screen in a darkened room. 1881 [see Zoetrope]. 1902 [see sense 6 in Dict.]. 1910 *Moving Picture World* 19 Feb. 249/1 People..like to see on the screen what they read about. 1915 *N.Y. Times* 15 Nov. 11/1 Unlike the legitimate stage, the screen does not have to wait for a dramatist to become inspired before it may present the topic of the hour. 1920 Mrs. P. Campbell *Let.* 20 Dec. in *B. Shaw & Mrs. Campbell* (1952) 215, I am much too aged for Eliza on the Screen! 1926 *Nature* 3 July 19/2 Every possessor of a 'televisor' will be in a position to see on his screen the performers in operas and plays as well as hearing them. 1928 E. Wallace *Double* i. 11 'What is her name?' 'Mary Dane... Mary Dane—sounds like something off the screen, doesn't it?' 1932 *Ann. Reg. 1931* 48 Death robbed the screen of Lya de Putti, best remembered for her performance in 'Variety', and Tyrone Power, veteran character actor. 1943 K. Tennant *Ride on Stranger* iv. 37 He's marvellous!.. Six feet tall and fair wavy hair. He ought to be on the screen. 1946 *B.B.C. Year Bk.* 20 A badly-produced programme may make you feel that the screen is small and cramped, but if the programme is good enough you will look at the screen not as a picture within a frame but as a view seen through a window. 1949 *Radio Times* 15 July 13/4 Music from the movies. Melodies from the screen in swingtime and symphony. 1956 R. M. Lester *Towards Hereafter* v. 63 Personages very well known in the world of industry, politics, stage, screen and radio. 1961 I. Murdoch *Severed Head* xxi. 179 Violence, except on the screen, is always pathetic, ludicrous and beastly. 1975 *Listener* 2 Jan. 23/1 The cumulative effect of watching the little screen for hours on end. 1976 A. Davis *Television: First Forty Years* 121 In the beginning, television had little or no time to devote to religion, for there were only two hours of transmissions on weekdays, and on Sundays the screen stayed blank. 1977 *Private Eye* 1 Apr. 10/3 It's all here at-a-glance—everything you want to see and know about the glamorous, dynamic world of the little screen.

f. *Photogr.* More fully *focusing screen.* A flat piece of glass on which the image formed by a camera lens is focused prior to making the exposure.

1858 [see Focusing *vbl. sb.* 2]. 1879 [see sense 6 in Dict.]. 1902 A. Watkins *Photogr.* 19 With the lens full open you will probably notice the image on the screen is not quite so sharply defined at the extreme corners as it is in the centre. 1962 A. Günther *Microphotogr. in Libr.* 23 Focusing is rather critical, and a precision camera with focusing screen should therefore be used. 1977 J. Hedgecoe *Photographer's Handbk.* 14 Some photographers find focusing on a screen more difficult than focusing with an image-coinciding rangefinder.

g. Any thin extended surface set up to intercept shot in gunnery trials.

1879 *Manual Artill. Exerc.* i. 17 The shot passes through two screens placed a certain distance apart. *Ibid.*, The velocity of the shot at the various screens [is] calculated from a comparison of the screen and time records. 1880 *Encycl. Brit.* XI. 300/1 The shot, after leaving the gun, cuts the wire of the first screen, and subsequently the wire of the second screen.

h. *Cricket.* An erection of canvas or wood placed behind the bowler, outside the boundary of the playing area, to provide a white background and a shield from moving objects behind the bowler's arm. Cf. *sight-screen* s.v. *sight *sb.*¹ 17.

1894 N. Gale *Cricket Songs* 31 O Bowler... He sends you clean beyond the screen. 1908 W. E. W. Collins *Leaves Old Country Cricketer's Diary* ix. 153 To be sure the light—this was his account—was all wrong. Anyhow we moved the screen three times to accommodate him, and even so he was not altogether happy. 1977 J. Laker *One-Day Cricket* 107, I eventually emerged from behind the screens.

i. *N. Amer.* and *Austral.* A frame covered with a fine netting of wire or the like, used in a window or a doorway to exclude insects.

1895 *Montgomery Ward & Co. Catal.* Spring & Summer 389/1 Adjustable Window Screen, so constructed as to form a perfect joint with the parting strip, so that it is not necessary to remove the screen in order to close the window. 1956 W. R. Bird *Off-Trail in Nova Scotia* ii. 51 As Saturday was a warm day everyone along the road was busy, putting up screens. 1971 *Sunday Australian* 8 Aug. 8A/6 (Advt.), Insect screens and screen doors that are custom-made. 1977 *Detroit Free Press* 11 Dec. 21-A 1, I opened the window, but I couldn't get the screen off and it was getting very hot.

2. e. *Geol.* A roughly tabular body of older rock separating two intrusions.

1910 W. B. Wright in *Summ. Progr. Geol. Surv. 1909* 33 About a quarter of a mile further inland, in the midst of the granophyre, a vertical screen of lava occurs... This screen varies much in thickness, being as little as 10 feet in one place, but reaching 300 feet in others... The granophyre on the outside of this screen is a distinct intrusion from that inside. 1942 M. P. Billings *Structural Geol.* xv. 284 If the central block subsides several times.., a number of concentric ring-dikes will form. A remnant of the older country rock left between two ring-dikes is called a screen.

3. e. *Meteorol.* A shelter that surrounds meteorological instruments and protects them from direct sunlight and precipitation, usu. painted white and louvred to provide indirect ventilation.

1881 W. Marriott *Hints to Meteorol. Observers* 10 The screen should be placed over short grass in a freely exposed situation. 1902 *Encycl. Brit.* XXX. 699/1 Various forms of open lattice work and louvre screens have been devised and used.., in all of which the wind is supposed to blow freely through the screens, while the latter cut off the greater part of the direct sunshine. 1923 F. Wild *Shackleton's Last Voy.* i. 12 One large screen, containing hair hygrograph, standard thermometer and thermograph. 1975 J. Scott *Fun with Meteorol.* 36/1 Ideally the thermometer bulbs should be about 4 ft. above ground level and the screen should have a north opening door to eliminate direct sunlight when it is opened.

f. A windscreen of a motor vehicle; also formerly, (*a*) a secondary screen to shield the occupants of the back seat in an open car; (*b*) a screen of celluloid for protecting the sides of an open car.

1904 A. B. Filson Young *Complete Motorist* vii. 176 When a cover is used it should have a removable glass screen in front. 1912 *Motor Manual* (ed. 14) iii. 99 Most cars now have adjustable and detachable glass or celluloid windscreens as a protection against the weather, dust, etc; screens can also be made of wire gauze and waterproof material. 1925 *Morris Owner's Man.* p. xi (Advt.), There are..rear screens and rear screens. 1955 *Times* 10 May 7/7 Perhaps the only fault from the driver's point of view is that his windscreen wiper is badly located and does not clean enough of the right-hand side of the screen. 1959 *Motor Manual* (ed. 36) vi. 186 Whatever the type of motor, it is usually combined with a suitable speed-reducing mechanism so that the wiper blades shall move reasonably slowly across the screen.

g. *U.S. Sports.* (See quots. 1961 and 1976.)

1939 *Sat. Even. Post* 7 Oct. 89/2 These are the components of a good passing game: the pitcher, receivers, and screen behind which the thrower can successfully operate. 1951 [see *pick *sb.*³ 10]. 1961 J. S. Salak *Dict. Amer. Sports* 383 *Screen* (basketball), term used to describe a maneuver of the offensive team in which one player, by moving in front of opponent, 'screens' or 'screens out' that opponent from his teammate. A screen generally is worked in an effort to free one of the offensive team members for a shot at the basket. *Ibid.* 384 *Screen* (handball), an assumed or retained position on the court which prevents the opponent from getting to, or playing the ball. 1967 B. Starr *Quarterbacking* 173 Screen passes are effective countermoves by the offense when the defense is applying heated pressure on the passer. 1975 *New Yorker* 7 Apr. 100/3 On offense, they zipped the ball around fast and moved with purpose, setting the picks and screens that their principal plays.. called for until someone got open for a good shot. 1976 *Webster's Sports Dict.* 374/1 *Screen*, a maneuver in various sports by which an opponent is legally cut off from the play.

6. (Further examples from *Electricity* and *Sound*.)

1915 Hawkhead & Dowsett *Handbk. Wireless Telegr.* (ed. 2) 263 Some valves are fitted with an additional screen of copper gauze covering the outside of the glass bulb... This screen protects the valve from heavy spark discharges in the neighbourhood. 1931 B. Brown *Talking Pictures* v. 127 A special sound-porous screen is used when the speakers are placed directly behind the picture. Most of the sound screens used in this country are of the woven type and of loose construction. 1950 *High Voltage Cables* (British Insulated Callender's Cables Ltd.) (ed. 2) i. 4 The screen functions as an earth conductor in close contact with the insulation. 1978 A. M. Portis *Electromagnetic Fields* xiv. 544 We wish..to discuss the scattering by various kinds of apertures in a two dimensional dielectric screen.

b. *Radio.* An arrangement of parallel wires located between a transmitting aerial and the earth's surface, serving to reduce the loss of power from the aerial to the earth.

1922 R. Keen *Direction & Position Finding by Wireless* i. 5 Experiments were being carried out in Germany..in connection with the screening of a vertical aerial... In other countries work..led to the complete wire screen or reflector. 1952 E. A. Laport *Radio Antenna Engin.* ii. 123 It is desirable to bring the ground wires to the surface a short distance from the radiator base so as to form a good ground screen above the soil near the antenna base where the electric field strengths are high. 1961 H. Jasik *Antenna Engin. Handbk.* xxiv. 16 If a screen is designed for both high and low bands in the VHF range, the over-all size should be determined by the lowest frequency.

c. *Electronics.* = *screen grid.

1927 *Amateur Wireless* XI. 269/1 The presence of the outer grid between the inner grid and the plate or anode naturally acts to some extent as a screen, and since this is connected to H.T. which is effectively at earth potential, we have a capacitative screen between the two electrodes. 1933 *Jrnl. Franklin Inst.* CCXVI. 418 Close to this first grid, and coaxial with it and the cathode, is a second spiral used to screen the control grid from the fields of the plate and called therefore the 'screen grid' or simply the 'screen'. 1942 *Electronic Engin.* XIV. 639/3 This formula..gives less accurate results for pentodes and tetrodes unless the transconductance to the screen is taken into account. 1962 D. F. Shaw *Introd. Electronics* xi. 234 The defect in the tetrode characteristics..is eliminated by the insertion of a third grid, called the suppressor grid, between the anode and the screen.

6*. *Printing.* **a.** A transparent plate, covered with two crossing sets of closely spaced parallel lines or with a uniform pattern of fine dots, behind which a photosensitive surface is exposed to obtain a half-tone image or as a step in forming the image carrier in a gravure process; also, in *Photogr.*, a patterned transparent plate or film that is combined with a negative during printing to give a textured appearance to the finished print.

1894 *Amer. Dict. Printing & Bookmaking* 465/2 Half-tone plates are made by passing the rays of light from a negative through a screen which is ruled or dotted. 1902 *Encycl. Brit.* XXIX. 411/1 This was finally accomplished by the insertion of a screen, in the camera, between the lens and the plate—the effect of which was to break up the whole surface of the negative into dots. 1940 [see *contact screen* s.v. *contact *sb.* 6]. 1946 H. Whetton *Practical Printing & Binding* xxv. 299/2 When the tissue is dry it is ready for screening. The cross-lined screen used in photogravure differs from those used in the production of half-tones. 1967 Karch & Buber *Offset Processes* v. 158 Coarser screens, such as the 65-line screen, are used in newspapers printed by letterpress. 1977 J. Hedgecoe *Photographer's Handbk.* 255 The picture, below left, was made by sandwiching the screen with a 2¼ ins sq..negative so that the pattern appeared relatively small.

b. The fine gauze or mesh (orig. of silk: cf. *silk screen *sb.*) used in screen printing.

1934 F. A. Baker *Silk Screen Practice* xvii. 114 Doubtless most screen operators have had visions of the ideal screen shop. 1938 Biegeleisen & Busenbark *Silk Screen Printing Process* v. 105 Either organdy or silk may be used as the screen for film stencils. 1957 *Screen Printer & Display Producer* July 3/2 The mesh is coated with a solution to form the screen for the photographic stencil. 1967 V. Strauss *Printing Industry* vii. 521/1 After the screen is ready, it may either be proofed or be used for running without proofing.

6.** An instance or the action of screening (see *screen *v.* 4).

1954 [see *screen test* (sense *7 a)]. 1974 M. C. Gerald *Pharmacol.* iv. 77 In a general blind screen, a range of doses of the compound a reinjected into test animals..and gross behavioral observations are made with an eye toward detecting any activity. 1975 *Language for Life* (Dept. Educ. & Sci.) xvii. 263 The majority of children scoring low on the group intelligence test have already been the subjects of consultation between head teachers and psychologists; consequently this second screen is now regarded as serving no more than a 'mopping up' purpose.

7. a. (sense 1 d) *screen image*; (sense 6*) *screen test*; **screen-cloth**, (*a*): delete † and add: metal or plastic mesh, esp. for covering a window or door screen (Webster), or for sifting material; (further examples); **screen current** *Electronics*, the current flowing in the screen grid of a valve; **screen-door**, a metallic or textile outer door of a pair, used for protection against insects or storms; also *Naut.*, for protection against explosions; **screen grid** *Electronics*, a grid placed between the control grid and the anode of a valve to reduce the capacitance between these electrodes; **screen memory, -memory** *Psychol.*, a Freudian term, orig. tr. as 'concealing memory', for a childhood memory whose apparently indifferent content screens from consciousness some (usu. previous) significant emotional event; **screen-painting**, pictorial decoration of church screens; **screen (†forward) pass** *N. Amer. Football* (see quot. 1976); **screen-perch** (later examples); **screen plate** *Photogr.*, an obsolete form of colour plate in which minute filters in primary colours are incorporated in the plate itself; freq. *attrib.*; **screen porch** *N. Amer.*, a veranda protected by a screen against insects; **screen print** *sb.*, (*a*) a picture or design produced by screen printing; (*b*) screen-printed fabric; **screen printer**, one who works at screen printing; **screen printing**, a printing process used esp. for pictorial matter in which the ink is transferred to the surface to be printed through a fine screen (sense 6* b) stretched in a frame, the non-printing parts of the screen having been rendered impervious; so **screen-print** *v. trans.*, to print (a surface or a design) in this way; **screen-printed** *ppl. a.*; **screen process**, the process of screen printing; freq. *attrib.*; **screen table** (see quot. 1794); **screen temperature** *Meteorol.*, the temperature as measured by a thermometer in a screen (sense *3 e); **screen voltage** *Electronics*, the voltage applied to the screen grid of a valve; **screen wall, -wall**, a wall that serves as a screen; so **screenwalling**; **screen-wash**, the cleaning of a windscreen automatically; also *attrib.*, of a substance added to water used in screenwashing; **screen-washer**, a device which washes a windscreen by directing a jet of water on to its exterior from below; **screen-wiper**, a windscreen wiper.

1946 *Sun* (Baltimore) 7 Oct. 2/5 Sales of bronze and copper screen cloth were allowed on an adjustable prices basis. 1974 G. S. Ormsby in P. L. Moore et al. *Drilling Practices Manual* vi. 152 The particle size a shale shaker can remove depends almost completely upon the size and the shape of the mesh openings in the screen cloth. 1936 E. D. McArthur *Electronics & Electron Tubes* v. 72 In this region, the screen-current characteristic is the exact opposite of the plate-current characteristic. 1962 D. F. Shaw *Introd. Electronics* xi. 232 The characteristics of a tetrode are more complex than those of a triode because of the additional variables of screen voltage and screen current. 1975 D. G. Fink *Electronics Engineers' Handbk.* xiv. 8 Multi-grid tubes require screen-grid modulation in conjunction with the control-grid modulation to achieve space-charge modulation and to minimize screen current. 1840 Screen door [in Dict.]. 1889 I. M. Rittenhouse *Maud* (1939) vii. 225 And after he'd gone I stood staring and staring out of the screen-door at nothing. 1914 'Bartimeus' *Naval Occasions* vi. 39 The screen-door..opened from the battery to the quarter-deck. 1933 E. O'Neill *Ah, Wilderness!* (1934) i. 16 *Mrs. Miller.* That boy! (*She rushes to the screen door and out on the porch, calling:*) Tommy! You mind what your Pa told you! 1974 *News & Courier* (Charleston, S. Carolina) 19 Apr. (Wickes Lumber Advt. Suppl.) 6 Screen doors in many styles, sizes & finishes. 1977 *Gay News* 24 Mar. 13/4 Such matters as who should fix the screen door, or do the cooking ..seem to fall into place as if by prearrangement. 1979 *Sunset* Apr. 56/1 The Boulevard Cafe, with open-air lunch counter, fly fans, and a banging screen door to an indoor counter, has been serving highway customers the same way for over 32 years. 1928 G. E. Sterling *Radio Man.* 167 The connection to the screen grid is made to the regular grid connection on a standard socket. 1930 *Manch. Guardian* 20 Sept. 15/7 A remarkable constructors' set known as the 'Osram Music Magnet Four', comprising two screen-grid stages, detector, and low frequency stages. 1942 *Electronic Engin.* XV. 10/1 When used as a triode the suppressor and screen grids are connected to anode. 1974 Harvey & Bohlman *Stereo F.M. Radio Handbk.* ii. 15 Carrier voltage is applied in push-pull via T_1 to the control grids of the two valves, whilst the modulating signal is supplied to the screen grids from T_2 which also provides push-pull drive. 1897 *Knowledge* 1 Sept. 217/2 Any mixture of indefinite light with the screen image has the effect of so much fog. 1937 *Discovery* Feb. 45/1 For production of the screen image a high intensity automatic arc is being used. 1924 J. Riviere tr. *Freud's Recollection, Repetition* in *Coll. Papers* II. xxxii. 368 In many cases I have had the impression that the familiar childhood-amnesia, which is theoretically so important to us, is entirely outweighed by the screen-memories. 1940 *Chambers's Techn. Dict.* 749/1 *Screen memory*, early childhood impressions and ideas which break through into consciousness, but are distorted and condensed into something which is unrecognisable to the individual. 1957 L. Durrell *Justine* i. 78 It is perhaps what the Freudians would call a screen-memory of incidents in her earliest youth. 1962 J. Strachey tr. *Freud's Screen Memories* in *Compl. Wks.* III. 320 A screen memory may be described as 'retrogressive' or as having 'pushed forward' according as the one chronological relation or the other holds between the screen and the thing screened off. 1967 Kanzer & Blum in B. B. Wolman *Psychoanal. Techniques* iv. 107 An examination of the painful episode as a nucleus (screen memory) with an infinity of ramifications, which involved the entire relationship of the patient with his father. 1906 G. E. Fox in *Victoria Hist. County Norfolk* II. 542 The screen paintings at Barton Turf, Edingthorpe, Harpley, Potter Heigham, Houghton-le-Dale, Lessingham, Ranworth, and Walpole-St.-Peter, have all been assigned to the fifteenth century. 1937 A. G. Little *Franciscan History & Legend in English Mediaeval Art* ii. 15 (*heading*) Screen Paintings... The painting of the lower panels of the rood and parclose screens of English churches with saints was confined almost entirely to the latter end of the 15th century. 1934 C. Walsh *Intercollegiate Football* 345/2, 1908-'Screen' Forward Pass (no longer legal). 1955 E. Pope *Football's Greatest Coaches* xxviii. 326 Zuppke originated the system of pulling back guards to protect the passer, the screen pass. 1970 *Globe & Mail* (Toronto) 26 Sept. 39/1 Ernie Carnegie gave the Redmen the lead with an 80-yard touchdown from a screen pass. 1976 *Webster's Sports Dict.* 374/2 *Screen pass* or *screen football*, a short forward pass to a back in the flat in front of whom a wall of interference has been formed by linemen who have moved over after the snap. 1979 *Arizona Daily Star* 5 Aug. c 9/1 Sproul.. found Ivery on a screen pass and Ivery weaved his way down the left sideline for the deciding touchdown. 1965 P. Wayre *Wind in Reeds* x. 142 Trained birds of prey are often tied to a screen-perch in their mews at night, this is a padded perch from the underneath of which hangs a sheet of thick canvas or hessian well-weighted at the bottom to keep it taut. 1971 *Country Life* 8 Apr. 799/3 Hawks have to be set to roost on a screen perch which prevents them from bating off and entangling themselves. [1898 *Sci. Trans. R. Dublin Soc.* VI. 134 The lined screen which can bring about this will only show its individual colours when placed under the microscope. It is then seen to consist of closely ruled adjacent lines in reddish-orange, yellowish-green, and blue tints. This screen, applied closely to the sensitive surface, analyses the image in the camera.] 1909 G. L. Johnson *Photogr. Optics & Colour Photogr.* v. 238 This screen plate is covered with a panchromatic emulsion. 1930 O. Wheeler *Photogr. Printing Processes* xvii. 218 Screen-plate processes for the production of transparencies..are undoubtedly of great merit. 1970 M. J. Sethna *Photography* 8 John Joly's 'screen-plate' method led to the 'autochrome' plates made in 1907 by the Lumière brothers. 1973 D. A. Spencer *Focal Dict. Photogr. Technol.* 548 *Screen plate process*, additive colour process in which the image is both recorded and viewed through a mosaic or reseau of microscopically small colour filter elements. The emulsion is exposed through the mosaic and reversal processed to yield an additive colour transparency. 1962 M. E. Murie *Two in Far North* i. iv. 36 Mother baked pies, many of them, and doughnuts. These were put out into the cache or the screen porch and frozen. 1970 *New Yorker* 28 Feb. 48/2, I went around to the side and up on the screen porch, lifted the window to the library, and climbed in. 1979 *Arizona Daily Star* 5 Aug. (Advt. Section) 14/9 Screen porch, sun deck, 2 car garage, 3 car carport. 1938 Biegeleisen & Busenbark *Silk Screen Printing Process* viii. 184 If a photograph or wash drawing is to be reproduced, it is first sent to a photoengraver, who makes a 'screen print' from it. 1957 *Observer* 1 Dec. 11/5 Chinese 'Sampan' screen-prints of water, reeds and flowers made full-skirted dresses. 1973 *Country Life* 21 June 1808 The characteristic screen print..is an edition of 100 published by Christie's Contemporary Arts at £65 each. 1976 *Dumfries & Galloway Standard* 25 Dec. 11/4 (Advt.), Screen prints 120 cm. wide. Modern designs. Metre usually £1·60. Now 99p. 1952 *Archit. Rev.* CXI. 194 (*caption*) Curtains and chair-cover privately screen-printed for Jane Drew. 1970 *Daily Tel.* (Colour Suppl.) 15 May 14 We made our real breakthrough when we screen-printed ceramic colours on to flat ware. 1957 *Screen Printer & Display Producer* July 4/2 The older type of screen printed transfers, which have been available to potters for some ten years.., have become well established as a medium for high quality multi-colour relief decoration. 1969 *Sears Catal.* Spring/Summer 43 Long-sleeve screen-printed Sweatshirts. 1938 Biegeleisen & Busenbark *Silk Screen Printing Process* i. 12 The screen printer may make prints which so closely resemble wood-block prints that only an expert can distinguish the difference. 1977 *Daily Times* (Lagos) 11 Jan. 22/5 (Advt.), Wanted. Experienced Screen-Printers. Apply..to Clem Advertising Productions. 1934 F. A. Baker *Silk Screen Printing* iii. 21 (*heading*) Silk and other gauzes used in screen printing. 1936 [see *hand block* s.v. *hand *sb.* 63]. 1958 *Observer* 17 Aug. 7/3 The firm of Taco use beautiful hand screen-printing of fruit and leaves on cotton ottoman, a process that takes twelve screens. (These days one has to say 'hand screening' because so much screen-printing is being done with the Swiss Buser automatic multicolour machine.) 1980 *Daily Tel.* 11 July 15/4 The exhibition confirms..that screen-printing has become a major medium for the living artist. 1930 B. Zahn *Silk Screen Methods of Reproduction* 37 The specification for a paint for screen process work. 1938 Biegeleisen & Busenbark *Silk Screen Printing Process* i. 7 Applicability of the screen process to the fine and applied arts has been recognized rather belatedly. 1967 Karch & Buber *Offset Processes* ii. 35 (*caption*) Screen process presses are often used for printing on glass bottles. 1970 *British Printer* July 69/1 Screen process had always been particularly versatile in handling a variety of surfaces and shapes which could not easily be printed by any of the three main printing processes. *Ibid.* 74/2 All screen-process inks contain inflammable solvents. 1794 T. Sheraton *Cabinet-Maker & Upholsterer's Drawing-Bk.* II. 395 *Of the Screen-Table.* This table is intended for a lady to write or work at near the fire; the screen part behind securing her face from its injuries. 1971 *Country Life* 30 Sept. (Suppl.) 29 (Advt.), A rare Sheraton period mahogany screen table, 17" wide. £245. 1913 *Rep. Brit. Assoc. Adv. Sci.* 1912 740 The explanation lies in the removal of air which has been chilled by radiation from the plant, and its replacement by air at 'screen-temperature'. 1972 *Daily Tel.* 1 Sept. 12/4 In Scotland the screen temperature fell to 32 degrees F (0 degrees C) at Tummel Bridge, Perthshire. 1954 *Cancer* VII. 1184/2 Large numbers of women could receive a screen test for carcinoma of the cervix with minimal utilization of professional personnel. 1971 *Nature* 9 July 105/1 Observations suggest that it is possible to devise an *in vitro* 'screen' test for potentially carcinogenic substances. 1936 E. D. McArthur *Electronics & Electron Tubes* v. 72 This..is fulfilled when the anode voltage becomes less positive than the screen voltage. 1945 *Electronic Engin.* XVII. 332/1 To obtain best results the control grid and screen voltages must be correctly chosen. 1962 Screen voltage [see *screen current* above]. 1900 Screen wall [in Dict.]. 1936 A. W. Clapham *Romanesque Archit.* iii. 53 This eastern screen-wall..appears only in the Apulian school. 1971 *Country Life* 30 Sept. 819/3 To guard against possible intrusion a screen wall was raised. 1976 *West Lancashire Even. Gaz.* 13 Dec. 9/5 (Advt.), Very low-priced..quality fencing, wood and concrete posts, sheds, timber, screen-walling. 1970 *Times* 5 Mar. 16 Another new feature is the 'cyclic' wipers which give not only slow and fast speeds but..eight wipes in conjunction with the screen-wash. 1976 *Scotsman* 24 Dec. 11/2 Sachets of screenwash additive are useful, however, not only in preventing washers from freezing in cold weather but in dissolving the road grime and grease that can smear or even scratch the windscreen. 1958 *Observer* 17 Aug. 15/7 The test car also had the simplest and most efficient screen washer I have seen so far. 1962 [see *accessory *sb.* 1]. 1977 *Lancashire Life* Mar. 118/4 It still has cloth seats, heated rear window and electric screenwashers. 1928 E. Wallace *Double* iii. 32 With his screen-wiper swinging madly, his mackintosh black with driving rain, Dick Staines came to Brighton. 1930 *Punch* 22 Jan. 92/3 Aggie must have something, a new mascot, a screen-wiper.., a new wing. 1970 *Railway Mag.* Oct. 558/1 With the aid of..efficient screen-wipers on the locomotive, there was no difficulty in sighting the signals. 1977 *Lancashire Life* Mar. 118/4 The screenwipers have not been re-set for right hand drive.

b. *spec.* with reference to the cinema or television screen, esp. in the *transf.* sense (see *1 d above), as *screen actor, actress, adaptation, beauty, credit, début, fan, fever, kiss, rights, set, star, story, version, world, worthiness*; *screen-filling, -struck* [after *stage-struck*], *-worthy* adjs.; **screenplay**, the script from which a motion picture film is produced; formerly, the film itself; also *attrib.*; **screen test**, a filmed test of the performing abilities of a prospective film actor, or the film shot on such an occasion; hence **screen-test** *v. trans.*; **screen time**, the time allotted to or occupied by a film or television production; **screen writer**, a writer of film scripts; hence **screen-writing** *vbl. sb.* and *ppl. a.*

1915 *Film Fun* Dec. 1 The screen actor has the best of it in holiday season. 1977 *Times* 7 Oct. 11/4 Valentino..was actually a very good screen actor. 1916 *N.Y. Times* 8 May 7 It is rumored that never again is this clever screen actress to play such a rôle. 1939 A. Huxley *Let.* 18 Nov. (1969) 448, I am working at present on the screen adaptation of *Pride and Prejudice.* 1919 H. L. Wilson *Ma Pettengill* iii. 38 It is the face of one of our famous screen beauties. 1922 Screen credit [see *credit *sb.* 13 f]. 1977 R. Ludlum *Chancellor MS.* vi. 78 They're willing to..remove your name from the screen credits..not the title, of course. 1915 *N.Y. Times* 22 Nov. 12 'The Martyrs of the Alamo', the Triangle picture in which Douglas Fairbanks made his screen début. 1923 T. Lane *What's Wrong with Movies?* vi. 100 The general run of screen fans want to do very little thinking when they go to the cinema. 1915 Screen fever [see *screen-struck* adj.]. *a* 1963 J. Lusby in B. James *Austral. Short Stories* (1963) 231 The Eccentric's reel..concluded with a screen-filling close-up of the stolid face of the armourer. 1951 M. McLuhan *Mech. Bride* (1967) 101/2 Shirley Temple gets her first screen kiss in a picture you'll never forget. 1971 *Petticoat* 17 July 28/3 Peter..was signed up by Walt Disney to give Hayley Mills..her first screen kiss. 1916 *N.Y. Times* 7 Feb. 9/3 (*heading*) Anna Held's début in a screen play. 1925 *Ladies' Home Jrnl.* Jan. 37/3 Screen play by Walter Woods. 1938 A. Huxley *Let.* 18 Nov. (1969) 437 They have followed their usual procedure and handed my treatment over to several other people to make a screen-play out of. 1945 R. Chandler *Let.* 13 Oct. in *R. Chandler Speaking* (1966) 43 He has gone so long without writing—unless you count a couple of screenplay jobs. 1977 *Times Lit. Suppl.* 24 June 750/3 A screenplay..is subsumed in the completed movie. 1980 *Times* 22 Nov. 4/8 Mr Cimino has written his own screenplay. 1920 *Q. Rev.* Apr. 185 The feverish haste exhibited at the present time to secure the screen-rights of classics in the world of letters. 1974 *She* Jan. 83/1 Bought screen rights to *Boy Shoots Gull*... Now to find writer to do Treatment. 1948 'T. Claymore' *Nest of Vipers* x. 195 Listening critically, I felt that it needed a Hollywood screen set and a background of soft music for these speeches. 1914 R. Grau *Theatre Science* xii. 211 So here we have the unique spectacle of an idolized screen star earning a prima donna's honorarium for her appearances at night only. 1977 J. Aiken *Last Movement* i. 167 Seeing a screen star for the first time in the flesh. 1914 R. Grau *Theatre Science* xii. 256 Thousands are impatiently awaiting the appearance of those publications which fictionize the screen stories. 1946 D. L. Sayers *Unpopular Opinions* 124 Wishing..that they too could live like the heroes and heroines of these witless million-dollar screen stories. 1915 *Film Fun* Dec. 1 Screen-struck. Everybody wants to get into motion pictures. It is an epidemic of screen fever. 1922 I. & H. Klumph *Screen Acting* xvi. 89 Then..she went back to her work of checking up on the scenes..and gave her screen test no more thought. 1927 *Sat. Even. Post* 17 June 14/1 Harry Rapf, one of the M-G-M executives, happened to see me dancing at the Winter Garden in New York City and asked me to make a screen test. 1952 M. McCarthy *Groves of Academe* (1953) vi. 57 The cold peroxided beauties who..were here while waiting for a screen-test. 1970 *Sunday Times* 26 Apr. 29/2 He screen-tested me... And I was offered one of the three main parts in the production. 1948 *Hansard Commons* 21 Apr. 219 Overseas producers..enjoy the same proportion of British screen time they have had in the past few years. 1973 *Listener* 20 Sept. 391/1 In a typical year, the BBC sells 11,000 hours of television screen-time abroad. 1978 *Broadcast* 29 May 10/1 The problem of screen-time for groups on the extreme political fringe. 1915 *N.Y. Times* 20 Sept. The screen version of 'Peer Gynt' begins with the reckless ride and ends with the rescue of Peer. 1933 *Radio Times* 14 Apr. 75/1 In *Cavalcade* (both stage and screen versions). 1981 *Listener* 1 Jan. 22/1 The screen version of *David Copperfield*..went way beyond the accepted running time of movies of that period. 1915 *Film Flashes* 11 Dec. 4 Are we to live only for ourselves, forgetting our brothers and sisters of the screen world? 1928 *Daily Express* 7 May 9/2 The production ..has been booked by Provincial Cinematograph Theatres, a booking generally regarded throughout the world as the hallmark of screen-worthiness. 1928 *Daily Tel.* 12 June Rachel... Her brief meteoric career needed no fanciful embellishments to make her story screenworthy. 1980 *Times Lit. Suppl.* 23 May 582/2 Though the screen-play of *The Tin Drum* was 'revised and augmented' by Günter Grass, Volker Schlöndorff's success has relatively little to do with making phrases screenworthy. 1921 *Moving Picture Stories* 12 Aug. 26/3 A Robertson–Cole picture..written by..two well-known screen writers. 1958 *Times Lit. Suppl.* 15 Aug. p. xxviii/5 How many genuine screen-writers have achieved any success in this country? 1974 *Listener* 17 Jan. 68/1 My first opportunities as a screenwriter were..in this country, but my career..has been in Hollywood. 1941 B. Schulberg *What makes Sammy Run?* vi. 124 Have some screen-writing job for you. 1973 [see *pot-boiler 2 a]. 1977 *Listener* 20 Oct. 498/1 In the 1930s, it was fairly easy to get a handle on the politics of the screenwriting community.

screen, *sb.*² (Earlier examples.)

1789 G. Parker *Life's Painter of Variegated Characters* xv. 153 *Rum screen*, a bank note. *Ibid.* 179 *Screen*, a bank note. 1795 H. Potter *New Dict. Cant & Flash Lang.* 53 *Skreen*, a bank note. 1811 *Lexicon Balatronicum* s.v. *Screen*, Queer screens; forged bank notes.

screen, *v.* **1. b.** Delete *rare* and add: Now usu. with *out*; freq. *fig.* (Further examples.)

1943 *Sun* (Baltimore) 3 Aug. 11/1 The stalks are put through a mechanical disintegrator which reduces them to a juicy puree and screens out the toughest fibers. 1946 *Cancer Res.* VI. 490/1 In resorting to histologic and cytologic studies to screen out the inactive compounds we have made the assumption that damage induced by active compounds would become evident within 48 hours after injection. 1955 *Publ. Amer. Dial. Soc.* xxiv. 51 There is an increasing tendency to screen out all argot and slang in the presence of outsiders. 1967 M. Argyle *Psychol. Interpersonal Behaviour* x. 195 The method..releases extremely powerful emotional forces, and those not able to profit from them should be screened out. 1968 *International Herald Tribune* 3 Sept. 7/3 The FBI has improved its methods of screening out inaccurate reporting. 1971 *Sci. Amer.* Oct. 42/3 The detector was located underground to screen out relatively low-energy particles produced in the atmosphere. 1975 *New Yorker* 21 Apr. 54/2 The company's instructions to its managers do seem to indicate an employment policy favoring people of conventional outlook and screening out people who might harbor tendencies towards nonconformist intellectualizing. 1979 *Bull. Amer. Acad. Arts & Sci.* Feb. 17 The committee will not screen out applications.

c. *Electr.* To protect from external electric or magnetic fields; to cover (a wire or circuit) in order to prevent it from radiating electrical interference.

1922 *Wireless World* 1 July 416/1 The problem is to screen the receiving apparatus from the effects induced directly by the oscillator. 1931 *B.B.C. Year-bk.* 1932 422 It will..be advisable to screen the coils L_1, L_2, the tuning condenser K_1, and the secondary circuit, L_3, K_2. 1950 *Engineering* 3 Feb. 140/2 On wireless-carrying vehicles electrical equipment must be screened. 1960 *Practical Wireless* XXXVI. 393/1 The lead from the input socket is screened. 1971 L. T. Agger *Introd. Electr.* viii. 119 It is sometimes necessary.. to screen a space from external electrostatic influence, as in protection against lightning of buildings containing explosives.

2. d. To surround (a nuclear reactor or other source of ionizing radiation) with a mass of material intended to absorb the radiation.

1915 Colwell & Russ *Radium, X Rays & Living Cell* vi. 160 If the radium applicator is screened by the interposition of ·5 mm. of lead..the results are different. 1931 G. E. Birkett *Radium Therapy* ii. 36 The radium in solution should be heavily screened to protect people working in adjacent rooms. 1946 *Ann. Reg.* 1945 354 The pile was not screened well enough to protect the personnel from the injurious effects of the intense radiation emitted by the unstable fission products.

e. *U.S. Sports.* To shield (a team-mate) from attack by opponents; to act as a shield against (opponents). Also *intr.* Cf. *screen *sb.*¹ 3 g.

1922 P. D. Haughton *How to watch & understand Football* 7 To stop the runner who is so thoroughly screened by interferers. 1951 *Sun* (Baltimore) 24 Dec. (B ed.) 13/2 Watch when they screen for a shooter [in Basketball]. 1961 [see *screen *sb.*¹ 3 g].

4. c. To examine systematically in order to discover suitability for admission or acceptance; *spec.* (*a*) to examine (a person) for unwanted attributes or objects, esp. political disloyalty; (*b*) to test (chemicals) for their suitability for use as drugs.

1943 *Sun* (Baltimore) 14 May 1/3 These offices 'screen' a list of prospects for the employers. 1945 *Manch. Guardian* 18 July 8/1 The recruits had come forward from the disbanded Guardia del Popolo and from most various sources and all would be screened carefully. 1949 *Cancer Res.* IX. 625/1 More than 1,000 chemical agents have been screened against Sarcoma 37 *in vivo*. 1956 W. Graham *Sleeping Partner* x. 82 When you said you were bringing an assistant to Harwell, of course we had to have her screened. 1958 *Listener* 19 June 1015/2, I am within a few yards of the Customs desk... My wife had packed all the declared trinkets in one bag, and that is all he wants to see. He screens it in fifteen seconds flat. 1962 *Sunday Times* 14 Jan. 1/7 Every flight arriving from Europe was screened by medical and immigration officials for Pakistani immigrants. 1970 *New Scientist* 11 June 538/2 Drug companies are trying to weed out drug-takers from their staffs and to screen applicants so as to avoid taking on more. 1971 *Daily Tel.* 19 Apr. 15/7 Electronic equipment at airports to 'screen' passengers for weapons and so on. 1974 M. C. Gerald *Pharmacol.* iv. 77 Of the 15,000 compounds our government screened as substitute antimalarials, only two..were found to be superior to quinine. 1979 *Daily Tel.* 21 May 12/7 Mr Corliss describes only those events which were reported in reputable scientific journals, where all material is rigorously screened, and 'mis-identifications and hoaxes are kept to a minimum'.

d. To select or separate by means of a screening process.

1943 *Sun* (Baltimore) 10 Dec. 6/7 The Attorney General said he believed it was possible to screen loyal from disloyal Japanese. 1976 *National Observer* (U.S.) 5 June 8/3 The experiment involves 20 communities, screened from an original list of 250 where some citizen efforts at decision-making already have cropped up.

e. To examine (a person, esp. as one of a large group) for disease or defects other than as a response to a request for treatment.

1944 [implied in *screening *vbl. sb.* 2 c]. 1950 *Amer. Jrnl. Public Health* XL. 275/1 A population group in one city is screened for tuberculosis. A separate program is conducted..to screen a population group for diabetes. 1970 *Observer* 12 Apr. 25/5 We could therefore soon be in a position to screen the whole population to see which recessive genes they carry. 1970 *Daily Tel.* 10 Oct. 8/4 Mass radiography is the easiest way for the man in the street to be screened.

f. To examine or search (data or an article) for any content of particular relevance or interest.

1956 A. H. Compton *Atomic Quest* 27 The committee had begun to function that soon afterwards was screening physics news for items of possible military importance. 1964 *Ann. N.Y. Acad. Sci.* CXV. 569 The system proposed.. allows the raw experimental data to be screened and digested directly by a small fast hybrid computer. 1977 *Jrnl. R. Soc. Arts* CXXV. 228/2 For the genetic resources material to be of any value to the breeder it must be evaluated or 'screened'.

5*. *Printing.* To obtain an image of (a picture, type, etc.) through a screen (*screen *sb.*¹ 6*).

1948 R. R. Karch *Graphic Arts Procedures* ix. 247 Both type matter and illustrations are screened. 1952 R. W. & E. W. Polk *Practice of Printing* (rev. ed.) xxiv. 198 In display, sometimes a heading or a block of type is screened to create a desired effect. 1972 *Physics Bull.* Sept. 532/1 Continuous tone pictures are 'screened' to allow reproduction by normal printing methods.

7. *trans.* To show (a picture) on a screen; to project on to a screen as with a magic lantern or film projector; to exhibit as a production for the cinema or television. Also *intr.*, to be (well or ill) suited for reproducing on a cinema or television screen.

1913 *Writer's Mag.* Nov. 188/2 Because you fail to sell your story, in spite of the fact that you see others of the same type screened, will not be proof that editors are prejudiced against you. 1915 *Durham County Advertiser* 18 June 8/7 'Tommy Atkins', a stirring patriotic picture..will be screened at an early date. 1919 H. L. Wilson *Ma Pettengill* ii. 67 She'll screen well, and she's one of the few that can turn on the tears when she wants to. 1962 *Rep. Comm. Broadc.* 1960 66 Programmes of national appeal screened by all or most of the companies. 1973 *Guardian* 10 Apr. 1/6 A revised version of Granada Television's controversial documentary about Mr John Poulson..will be screened on April 30. 1976 *National Observer* (U.S.) 18 Dec. 9/2 When the movie was screened, the key obscenity standard was whether a sex work was 'utterly without redeeming social value'.

screenage (skrī·nėdʒ). [f. Screen *sb.*¹ + -age.] The material used as a screen for ionizing radiation; such screens collectively; the action or the efficiency of screening.

1929 S. Cade *Radium Treatment of Cancer* i. 5 The necrosis is in inverse proportion with the screenage. 1930 *Brit. Med. Jrnl.* 8 Feb. 234 The screenage, amounting to the equivalent of just over 1 mm. of platinum, consists of monel metal, brass, aluminium, and wood. 1933 Ward & Durden-Smith *Recent Advances in Radium* vi. 67 The screenage for seeds most commonly used at the Radium Institute is 0·5 millimetre of gold. 1940 S. Cade *Malignant Dis. & Treatment by Radium* viii. 177 Screenage of eye applicators by substances of low or medium atomic weight, such as wax, rubber, and palladium, may..still further reduce the incidence of eye complications. 1956 C. W. Wilson *Radium Therapy* (ed. 2) x. 239 When lead screenage is added to a radium source filtered by 0·5 mm. platinum so as completely to surround the source, the transmission curve is virtually a straight line up to 20 cm. thickness of lead.

screened, *ppl. a.* Add: **1. a.** Also freq. with adverbs, as *screened-in, -off.*

1833 [in Dict.]. 1946 J. O'Hara in *New Yorker* 6 July 18 Eben Townsend was sitting on the screened-in porch smoking his after-dinner cigar. 1968 R. M. Patterson *Finlay's River* i. 39 We slept that night on the floor in the screened-in veranda, very comfortable and protected against the rain and the mosquitoes. 1973 J. Wainwright *Pride of Pigs* 78 They led him..to the ward, and pointed out the screened-off bed. 1978 P. Niesewand *Underground Connection* 1 The screened-off telephone switchboard. 1979 *Arizona Daily Star* 5 Aug. (Advt. Section) 18/3 Enjoy the breezes and mountain views from large screened-in Arizona room. 1981 P. Mallory *Killing Matter* iii. 33 A broad screened-in porch where Holly Devereaux sat in a rattan chair.

c. *Meteorol.* Placed in or measured in a screen (see *screen *sb.*¹ 3 e).

1894 [in Dict., sense 1 a]. 1920 *Westm. Gaz.* 2 Dec. 6/4 At Greenwich, a minimum screened temperature of 27 deg. was registered.

2. b. In the sense of *screen *v.* 4 c.

1957 *Economist* 26 Oct. 321/1 A drive for Party members among discreetly screened intellectuals, who responded 'sincerely' to the 'blooming-flowers' campaign by seeing, hearing and speaking no evil, is also under way; the first 25 recruits are proudly announced from the professorial staff of Tientsin university. 1962 L. Deighton *Ipcress File* xv. 86 The screened personnel available to us. 1967 *Times Rev. Industry* July 89/3 The procedures of some 41 companies suggest that the higher the level of responsibility the less frequently a manager gains important external information from publications compared with other sources. But this may merely mean that information from publications is received in a screened form from other folk. 1979 J. Gardner *Nostradamus Traitor* xlix. 242 The screened call came in from Washington.

4. a. *Electr.* Of a wire, circuit, or appliance; having a conducting cover or shield, intended to reduce the radiation or reception of interference.

1922 *Wireless World* 1 July 416/2 A design for a screened oscillator was prepared. 1926 S. O. Pearson *Dict. Wireless Techn. Terms*, *Screened Aerial*, an aerial beneath which is spread a network of wires to act as a counterpoise. 1927 *Amateur Wireless* XI. 269/1 Following on these lines two practical forms of screened valve have been designed. *Ibid.* 269/3 The valve can then be built into a screened circuit,.. if a slot is cut in the external screen. 1933 *Practical Wireless* 4 Feb. 962/2 The ingenious screened wiring kit manufactured by Remax Radio is the very thing for temporary or permanent screening. 1936 *Ibid.* 16 May 232/2 (*heading*) Screened leads. 1943 *Gloss. Terms Electr. Engin.* (B.S.I.) 82 *Screened cable*, as used for electricity supply. A multicore cable in which the insulation of each conductor is separately enclosed in a conducting film in order to ensure a radial electric field surrounding the conductor, the films being in electrical connection with one another and with the metallic sheath of the cable and usually earthed. 1970 J. Earl *Tuners & Amplifiers* vi. 134 Screened cable must be used on these low-level inputs to avoid excessive hum coupling.

† b. Electronics. *screened grid*: in a thermionic valve, a control grid having a screen grid around it. Usu. *attrib.* *Obs.*

1929 *Radio Times* 8 Nov. 409 (Advt.), You could not get a better 3-valve set than the Brown Screened Grid Receiver. 1930 *Manch. Guardian* 20 Sept. 15/7 Thanks to the screened-grid valve, the old monster multi-valve set is now practically a back number. 1943 C. L. Boltz *Basic Radio* x. 164 The pentode has ousted the screened-grid valve from radio circuits.

5. Projected on a screen; shown at the cinema or on television.

1917 C. N. Bennett *Kinematography* ix. 160 (*heading*) Distortion of the screened image. *Ibid.*, [It] shows the normal proportions of the screened picture. 1966 *Listener* 24 Nov. 775/3 Selectivity in exposure, perception, attention, retention, etc., means that the effect of screened violence will not be constant and will vary from person to person.

6. *Printing.* Of an image or surface: obtained with the aid of a screen (*screen *sb.*¹ 6*); bearing a pattern formed by a screen.

1946 H. Whetton *Practical Printing & Binding* xxv. 299/2 Type or pictures have now to be printed upon the screened tissue. 1952 R. W. & E. W. Polk *Practice of Printing* (rev. ed.) xxxi. 232 (*caption*) Example of surprinting on 120-line screened background of varying densities. 1973 M. Astrua *Man. Colour Reproduction* i. 106 For the preparation of screened positives or negatives, a study has been made of suitable screens.

screener. Add: **2.** In other senses of the verb (see quots.).

1913 *Dialect Notes* IV. 57 *Screener*, a person who 'screens' cranberries. 1951 *Sun* (Baltimore) 24 Dec. (B ed.) 13/2 Watch the defensive man as he hooks his arm around the screener and swings around [in Basketball]. 1967 D. Francis *Blood Sport* iv. 48 'A screener,' he said. 'How come Teller found you?' 1977 *Lancet* 16 July 131/1 In 1974 McCarthy and Widmer calculated that screening by consultants of recommended elective surgical procedures could reduce the number of operations performed, with great savings in cost. Orthopædic, urological, and gynæcological surgical procedures were the ones most usually contested by a second opinion screener.

screenful (skrī·nful). [See -ful.] As much or as many as can be displayed at one time on the screen of a cinema or of a television set or VDU, or similar device.

1966 C. Mackenzie *Paper Lives* viii. 116 Nigel was watching the last ten minutes of 'Deadwood Gulch' and a screenful of Indians. 1969 *Listener* 30 Jan. 133/2 *Sword of Swords*,..in Mandarin with English and Chinese subtitles (a screenful). 1974 *Nature* 10 May 102/1 A page is a screenful of data; there are 24 rows on each page and each row has 40 characters. 1977 *Times* 30 Nov. 16/8 The Post Office's Viewdata project..might be used to enable customers to dial up screensful of information.

screening, *vbl. sb.* Add: **1. b.** The action or practice of shielding from electric and magnetic fields, esp. by means of an enclosing cover of conducting or magnetic material. Freq. *attrib.*

1840 *Annals Electr., Magn., & Chem.* IV. 293 The screening might, in some way, be connected with an instantaneous current in the plate. *Ibid.*, A certain thickness of metal is.. required to produce the screening effect. 1891 *Electrician* 17 Apr. 722/1 Electrostatic screening is of fundamental significance throughout electric theory. 1922 *Wireless World* 1 July 416/1 Magnetic screening from a steady field can be accomplished by surrounding the apparatus by a heavy iron screen. 1929 *B.B.C. Year-bk.* 1930 339 A copper sheet shield may be fixed to the rear of the panel for screening purposes. 1951 *Good Housek. Home Encycl.* 232/2 Most modern refrigerators are provided with suitable screening arrangements to prevent interference with other electrical appliances. 1962 A. Nisbett *Technique Sound Studio* 241 Although it has no screening effect its thickness ensures that the physical separation of successive layers of the magnetic coating is sufficient to maintain printing at a low level. 1970 J. Shepherd et al. *Higher Electrical Engin.* (ed. 2) vii. 225 The object of shielding (or screening) is to prevent a magnetic field from existing at some particular point.

c. The action of an obstruction (such as a hill or a building) in attenuating or blocking broadcast signals.

1907 J. Erskine-Murray *Handbk. Wireless Telegr.* xvii. 231 That the bending of the lines and not their absence is, in part at least, the cause of this apparent screening, is rendered still more probable by the observation that though no signals could be received close to the land, in one instance, they could be obtained at a greater distance from the transmitter by moving the ship away from the land in a straight line. 1920 *Discovery* Apr. 116/1 Mountains..close to a wireless station may produce serious screening in that direction. 1926 R. W. Hutchinson *Wireless* ix. 154 An aerial passing along a passage between the gable ends of two houses..is not efficient owing to the screening effect of the houses. 1967 G. J. King *Pract. Aerial Handbk.* iv. 102 There are invariably pockets of low signal field round the area, due to heavy screening etc. (depending on frequency), where lofty outside aerials are needed to secure the full advantage of the local transmitter.

d. *Physics.* The reduction of the electric field about an atomic nucleus by the space charge of the surrounding electrons. Also in *Comb.*, as **screening constant**, the difference between the atomic number of a nucleus and its effective charge, reduced by screening.

1922 A. D. Udden tr. *Bohr's Theory of Spectra* iii. iv. 121 The effect of the presence of the electrons in inner groups upon the motion of the electrons in outer groups as a first approximation may be expected to consist in a simple screening of the nucleus. 1925 G. A. Lindsay tr. *M. Siegbahn's Spectrosc. of X-Rays* vi. 163 These doublets may be

roughly explained by the screening of the nuclear charge. *Ibid.* 166 The screening constants..must depend in general on the quantum number, and on the magnitude and form of the orbit determined by it. 1965 Phillips & Williams *Inorg. Chem.* I. ii. 54 The magnitude of the screening constant indicates the extent to which the full nuclear charge of an element is hidden from the electrons in a particular shell of the core. 1970 G. K. Woodgate *Elem. Atomic Struct.* v. 82 The direct Coulomb interaction e^2/r_{12} raises the degeneracy in *l* because the amount of screening depends on the eccentricity of the orbit of the outer electron.

e. *Physics.* = *shielding *vbl. sb.* 2 b.

1922 F. E. Simpson *Radium Therapy* vi. 45 It is best to use rather simple and uniform screening until familiarity is obtained with the effects of radium on the tissues.

f. *Basketball.* Obstruction. See *screen *sb.*¹ 3 g.

1951 *Sun* (Baltimore) 24 Dec. (B ed.) 13/2 In the old American League, the ball changed hands if the referee called screening or blocking.

2. b. (Further examples.) *spec.* in *pl.*, an inferior grade of wheat or polished rice.

1824 'A. Singleton' *Letters from South & West* 111 Their usual fare, is, a peck of corn in the ear a week, which they must break in their hand-mills; and the *grit*, or refuse, a Rice, like the western *screenings* of wheat. 1867 P. L. Simmonds *Commerc. Dict. Trade Products*, *Screenings*, a name in the United States for the inferior wheat that is removed by the screens and fans. 1901 *Yearbk. U.S. Dept. Agric.* 1900 135, 9,400 pounds of screenings, and 3,500 pounds of brewer's rice. 1906 *Chambers's Jrnl.* June 453/1 Tons of rock are..crushed for screenings and ballast. 1923 A. Bruttini *Uses of Waste Materials* ii. ii. 133 By the term screenings are understood the siftings of cereals and other grains. *Ibid.*, Screenings are..made into dog biscuits.

c. Systematic examination of a large number of subjects, esp. for the detection of unwanted attributes or objects.

1944 *Sun* (Baltimore) 8 Jan. 1/1 The preliminary screening physical examination will be discontinued, except for individuals with obvious physical defects. 1946 *Cancer Res.* VI. 490/2 Compounds found in the first screening to possess potency in damaging or killing cells of sarcoma 37..were subjected to further experimentation. 1958 *New Statesman* 6 Sept. 263/3 Some 3,000 Algerians, arrested immediately after the wave of terrorism, were brought there for screening. 1960 *Guardian* 11 Mar. 7/4 Dr. Glover..advocates a psychiatric 'screening' of the entire child population in order to detect pathological tendencies to violence. 1964 Hueper & Conway *Chem. Carcinogenesis & Cancers* v. 177 The highest priority for carcinogenic screening should be assigned to those chemicals with which large parts of the general population have contact. 1970 *Nature* 31 Oct. 416/1 Electrophoretic mass screening of blood proteins for new mutations..would be formidable. 1974 *Times* 27 Apr. 5/7 Herr Brandt..said he was generally satisfied with the screening of Federal employees... Herr Guillaume had been thoroughly screened, but not given the highest grade of security clearance. 1974 M. C. Gerald *Pharmacol.* iv. 77 There has been very exhaustive screening of soil samples from almost all parts of the world in an attempt to discover new antibiotics.

d. With *out*. Cf. *screen *v.* 1 b.

1943 *Sun* (Baltimore) 15 Mar. 5/6 The 'screening out' of school children who need medical attention.

4. The action of *screen *v.*¹ 7; *spec.*, a particular showing of a film.

1923 E. S. Van Zile *That Marvel—the Movie* 121 Mr. Harding..has suggested the screening of Wells's 'Outline of History'. *Ibid.* 198 The actual screening of the story was begun. 1928 H. Crane *Let.* 5 Feb. (1965) 316 We have met some movie actors, attended some studio screenings, etc. 1954 *Recorder* 8 Jan. 1/2 The telegram..'solemnly protested' against the screening of a Mass. 1960 *Guardian* 25 Aug. 8/5, I was present ten days prior to its [*sc.* a film's] screenings at Karlovy-Vary. 1974 'G. Black' *Golden Cockatrice* v. 81 The first screening was an old Disney short. 1977 *Times* 23 June 20/4 Many of the journalists who did attend the screenings revealed an uneasiness.

5. *Printing.* The process of exposing a photo-sensitive surface or forming an image through a screen (*screen *sb.*¹ 6*).

1946 [see *screen *sb.*¹ 6*]. 1967 Karch & Buber *Offset Processes* v. 168 A device..now permits direct screening of halftone copy that removes the screen dots and reduces the graininess from enlargements. 1973 M. Astrua *Manual of Colour Reproduction* i. 100 For printing continuous tone we have to resort to the optical device of 'screening', that is, to the conversion of the various densities of the image..into small dots.

6. *attrib.* and *Comb.*, as (sense 2 c) *screening device, experiment, method, operation, procedure, process, programme, technique, test*; (sense 4) *screening room*; **screening clinic**, a clinic at which medical screening is carried out.

1966 *Listener* 4 Aug. 151/1 Last year saw the growth of municipally run screening clinics. In Rotherham and Glasgow there were queues of citizens in the streets waiting to be X-rayed and have other simple examinations. 1977 *Spare Rib* Sept. 19/4 The aim of developing a well woman screening clinic. 1952 M. McCarthy *Groves of Academe* iv. 63 Despite a high tuition and other screening devices.. something..had worked to give the college a peculiarly plebeian and subversive tone. 1971 J. Anderson in B. de Ferranti *Living with Computer* vii. 61 New screening experiments are under way at present. 1946 *Jrnl. Amer. Med. Assoc.* 1 June 377/2 [The smear test] can be carried out on a large scale as a screening method. 1950 *Hansard Commons* 7 Nov. 769 Mr Shepherd asked the Prime Minister what inquiries he has made into the efficiency of the screening methods of M.I.5. 1974 *Times* 19 Dec. 4/3 Downing Street refused to say how long the screening operation took. 1964 *Observer* 12 July 4/6 Mammography..is thought by some to be useful as a 'screening' procedure for women without symptoms to pick up breast cancer at a very early stage. 1949 *Cavalier Daily* (Univ. of Va.) 22 Oct. 4/2 Applicants could not get their visas approved by the Hungarian government without undergoing a screening process by the American Festival Committee. 1975 *Language for Life* (Dept. Educ. & Sci.) xiv. 215 The screening process might extend across the point of transfer. 1954 *Cancer* VII. 1183/2 A screening program to detect cancer of the cervix. 1974 M. Taylor tr. *Metz's Film Lang.* vi. 156 Inside the screening room. 1978 *Detroit Free Press* 16 Apr. 23 A/1, I absolutely refused to allow 'Rabbit Test' to be shown to anyone in a screening room. 1945 *Amer. Assoc. Adv. Sci. Res. Conf. Cancer* 318/2 Utilizing a number of 'screening' techniques to evaluate the place of a variety of compounds as to their ability to impair the metabolism of malignant cells. 1942 *Nation* 27 Apr. 41 Since Jan. 1 Selective Service has given only what is called a 'screening test'. The registrant strips, walks to and away from the doctor and if he has all his limbs and his eyesight and no immediately apparent organic defect, is passed. 1951 *Jrnl. Amer. Med. Assoc.* 11 Aug. 1401/1 The detection center serving..as a laboratory facility to which the physician can refer his patients for screening tests. 1969 *Times* 15 Mar. 7/8 A swift and reliable system is necessary to meet the ideal of offering every woman an annual screening test for signs of cervical cancer.

screening, *ppl. a.* (Further examples.)

1936 *Practical Wireless* 16 May 232/2 Provided that the screening cans are connected to the earth terminal no interaction should be experienced between those components. 1942 *Electronic Engin.* XV. 284/3 Grid 3 is situated between two screening grids maintained at a positive potential of 100 volts. 1943 *Gloss. Terms Electr. Engin.* (B.S.I.) 58 *Line choking coil* (*screening reactor*), an inductor connected in series with electrical plant and serving to reduce the effects of high-frequency or steep-fronted surges by absorption or reflection. 1966 *McGraw-Hill Encycl. Sci. & Technol.* XII. 81/2 Military screening smokes.

screenless (skrī·nlės), *a.* [-less.] Having no screen; having had no screen used in its production; unprotected.

1921 *Blackw. Mag.* Sept. 348/2 In that screenless life friendship frankly condoled with weak nerves. 1976 *National Observer* (U.S.) 3 July 14/2 As the day progressed the screenless windows admitted less breeze and more heat and insects. 1980 *Times Lit. Suppl.* 12 Sept. 988/2 A scrupulously faithful facsimile of the lithograph edition, made by a new Screenless Printing Process.

screw, *sb.*¹ Add: **I. 2. b.** (*a*) (Earlier and later examples.) Also, used of other kinds of pressure, e.g. the pressure of competition. (*b*) Also, occas., used of blackmail.

1834 C. A. Davis *Lett. J. Downing* xiv. 96 And if they don't they put the screws on 'em. 1894 P. L. Ford *Honorable Peter Stirling* xli. 241 Then I can put the screws on him safely, you think? 1917 W. J. Locke *Red Planet* xxiii. 298 Gedge's nocturnal waylaying of him..was another unsuccessful attempt to tighten the screw. 1938 E. Ambler *Cause for Alarm* xi. 170 Everything..was prepared. It was only a question of waiting for Vagas to begin to turn the screw. 1977 *Navy News* Sept. 39/3 David Stracey kept the screws on, lunching with figures of three for 15 off ten overs. 1981 A. Morice *Men in her Death* viii. 93 She worked out this scheme for a phoney kidnap, to put the screws on.

3. b. (Further examples.) Now usu. with reference to persons or their mental faculties, esp. in colloq. phr. *to have a screw loose*: to be eccentric, insane, or mentally retarded. *slang.*

1873 Trollope *Eustace Diamonds* III. lxiii. 128 Something crooked about Lizzie,—a screw loose, as people say. *Ibid.* lxix. 215 Folks as would have a screw loose somewheres. 1884 *St. Louis Post-Dispatch* 4 June 7/4, I really think this wonderful woman has a screw loose in her mental organization. 1928 [see *nut *sb.*¹ 7 c]. 1959 I. & P. Opie *Lore & Lang. Schoolch.* x. 179 A person who is 'wanting in the upper storey'..has a screw loose. 1963 *Times* 1 Feb. 8/7 He asked Mr. Galbraith if when he came across a person who was 'limited intellectually' he normally referred to him as 'having a screw loose'. 1974 S. E. Morison *European Discovery of America: Southern Voyages* xxx. 725 His idea of England's opening a traffic with China independent of Spain and Portugal was sound. But there was a screw loose somewhere in Cavendish. 1977 *Lancashire Life* Nov. 63/1 An endearing little chap with a screw loose.

8. (Earlier example.)

1861 in M. Willson Disher *Cowells in Amer.* (1934) 330 Destroyed the line-of-battleship New York, on the stocks, besides scuttling the Merrimac, first-class screw, the Germantown, sloop of war.

10. b. (Earlier and later examples.)

1812 P. Egan *Boxiana* 1st Ser. I. 122 Where *flash* has been *pattered* in all that native purity of style, and richness of elocution, which would have startled a *High Toby Gloque*, and put a *Jigger Screw* [*i.e.* a prison warder] upon the alert. 1902 *Chambers's Jrnl.* June 367/1 Should there be a superfluity of 'screws' (warders) on the spot..your door is opened and the regulation bun..is handed in. 1933 [see *gold braid* s.v. *gold¹ 10]. 1948 [see *bent *ppl. a.* 5 a]. 1970 G. F. Newman *Sir, You Bastard* viii. 223 The lights never out, pervy screws watching every movement. 1977 *New Yorker* 24 Oct. 68/2 Men..call their keepers 'guards', 'officers',..'screws'.

II. 11. c. (Earlier and later examples.) Also a ball to which a spin has been imparted at its delivery. Also in *Lawn Tennis.*

1840 *Bell's Life* 2 Aug. 2/2 Morewood joined Morrier, who at length received a 'Winchester screw', which shattered his timber. 1865 J. Pycroft *Cricketana* ix. 169 Clarke could put on a decided screw..with a ball well pitched up. 1868 *J. Lillywhite's Cricketers' Compan.* 62 Southerton's 'screws' were the main cause of Kent's discomfiture. 1931 A. Powell *Afternoon Men* iii. xxiii. 203 She served underhanded screws that Pringle could not take.

11*. *coarse slang.* **a.** A prostitute; a woman considered in sexual terms; a (good, bad, etc.) sexual partner (in this use, prob. *transf.* from sense b).

1725 *New Canting Dict.*, *A screw*, a Strumpet, a common Prostitute. 1937 Partridge *Dict. Slang* 738/1 *Screw*,..a woman *qua* sexual pleasure. 1942 Berrey & Van den Bark *Amer. Thes. Slang* § 507/2 *Prostitute*,..screw. 1966 'L. Lane' *ABZ of Scouse* 93 'A bloody good screw' might refer to an attractive girl. 1969 S. Coulter *Embassy* xi. 120, I like to figure you're my regular screw, see. A whole lot more exciting. 1976 M. Machlin *Pipeline* xlix. 507 As a matter of fact, he's not such a great screw, but at least he isn't a nag, the way *you* are.

b. An act of sexual intercourse, esp. of a hasty and casual nature. Also *fig.*

1929 F. Scott Fitzgerald *Let.* 9 Sept. (1963) 307 Here's a last flicker of the old cheap pride: the *Post* now pays the old whore $4000 a screw. 1937 Partridge *Dict. Slang* 738/1 *Screw*,..an act of copulation. 1967 A. Wilson *No laughing Matter* iii. 387 He felt randy as hell but he hadn't even got the price of a quick screw. 1971 P. L. Cave *Chopper* ii. 12 Five or six Angel birds sat around over cold cups of coffee waiting for a fast ride or a quick screw. 1978 G. Greene *Human Factor* iv. ii. 209, I like a good screw as much as the next man, but it's not all that important, is it?

13. b. *the screws*: rheumatism (cf. *screwmatic *a.* and *sb.*). *slang.*

1897 G. Bartram *People of Clopton* 51 In bed roarin' mad wi' the screws. 1970 G. E. Evans *Where Beards wag All* ix. 107 Now I know all about the east wind, and I can't move my left leg without having the *screws*. 1976 'L. Black' *Healthy Way to Die* ii. 11 Any rheumatism? An occasional touch of the screws, she admitted.

16*. A look, stare, or gaze; esp. in phr. *to have a screw at*: to look at. *slang* (orig. *Austral.*). Cf. *screw *v.* 13*.

1919 W. H. Downing *Digger Dialects* 44 *Screw* (vb. or n.), look. 1928 [see *rumble *v.*¹ 6]. 1933 *Bulletin* (Sydney) 23 Aug. 12/2 (*caption*) *Election Canvasser*: 'Is your wife a Feminist?' *The Worm*: 'S-sh have a screw at me.' 1934 T. Wood *Cobbers* vi. 84 Have a screw at that bullick. *a* 1966 'M. na Gopaleen' *Best of Myles* (1968) 57 And of a Sunda the Frenchmen do be walkin' around the gardens havin' a screw at the statues. 1969 *New Society* 13 Nov. 762/3 The skinhead contribution to their parents' beliefs is this characteristic rigour. 'If we see any hippies, you know, they give you the screw, you know. I don't like it,' says Bill, using 'screw' to mean 'stare'.

III. 18. (Earlier and later examples.)

1858 D. Beveridge *Let.* in Ld. Beveridge *India called Them* (1947) ii. 26 Their delay in announcing an augmentation of screw. 1917 A. Huxley *Let.* 8 Apr. (1969) 123, I go there next week—screw, they tell me, from £200 to £250. 1939 D. L. Sayers *In Teeth of Evidence* 91 Is he in a good way of business? Good screw, I mean? Comfortable, and all that? 1959 T. S. Eliot *Elder Statesman* iii. 95 He's offered me the job With a jolly good screw, and some pickings in commissions. 1981 'M. Innes' *Lord Mullion's Secret* ii. 20 Cyprian would have to be found... 'A niche with a good screw to it.'

IV. 20. *screw-driven* (example).

1955 *Times* 6 June 6/6 Experience has shown that paddle tugs are more efficient than screw-driven tugs for work in confined basins.

22. a. **screw axis** *Cryst.*, an axis such that a combination of rotation about it and translation along it constitutes a symmetry operation, but neither does so alone; **screw-barrel** *a.*, (*c*) of a microscope, having a threaded barrel by means of which the microscope is focused; **screw-block** *Basketry*, a device for holding stakes rigid during rectangular work; **screwbound** *a.*, (*a*) fastened or held by a screw or screws (sense 10 a in quot.); (*b*) (see quot. 1966); **screw bulb**, an electric light bulb having a threaded base enabling it to be screwed into a socket; **screw-cap** (later examples); also more generally, = *screw top*; hence **screw-capped** *a.*; **screw compressor** (see quot. 1967); **screw dislocation** *Cryst.*, a form of crystal defect characterized by a unit distortion of the lattice in a particular direction such that the lattice planes perpendicular to that direction form continuous spiral sheets; **screw fly** *U.S.*, a blow-fly of the genus *Cochliomyia*, *C. hominivorax* or *C. macellaria*, which deposits eggs on animal carcasses or open wounds; **screw-hammer**, an adjustable spanner with a heavy, hammer-like head; **screw-man** *U.S. Hist.*, a worker who packed bales into cotton-ships; **screw tail**, a dog's tail which is twisted or crooked; **screw top**, a round cap or lid that can be screwed on to a bottle, jar, or the like; also *attrib.*; hence **screw-topped** *a.*; **screw worm** (*b*) read: *U.S.*, the larva of a screw fly, which has spiny hairs encircling each segment; **screw worm fly** = *screw fly*; **screw-wrench** (earlier example).

1903 H. Hilton *Math. Crystallogr.* xvi. 146 The combination of a rotation about an axis and a translation parallel to it is called a screw about that axis; and if such a combination brings a figure U to self-coincidence the axis is called a screw-axis of symmetry for U. 1937 W. L. Bragg *Atomic Struct. Minerals* i. 13 It is the possibility of screw axes and glide planes, in addition to rotation axes and reflection planes, which gives rise to the large number of space-groups. 1974 *Nature* 11 Jan. 85/2 Dark-field observation could establish the presence of centres of symmetry, glide planes and screw axes, which could lead to the establishment of the space group. 1926 *Catal. Optical & Gen. Sci. Instrum. Optical Convention* 1926 287 Ivory screw-barrel microscope: by J. Wilson, with eight powers (unsigned, *circa* 1706). 1956 *Nature* 7 Jan. 8/1 Another contemporary scientist interested in optics was the Dutch microscopist Nicolaas Hartsoeker, born on March 26, 1656. He published in 1694 an 'Essai de Dioptrique' in which he illustrated his invention, the screw-barrel microscope, generally associated with the name of Wilson, who introduced it to England. 1924 C. Crampton *Cane Work* 34 *Oblong Cane Base*... This kind of base cannot be made without using a 'screw lock', which acts as a vice for holding the sticks or stakes in an upright position. The screw block consists of two wooden blocks with thumbscrews for tightening purposes. 1959 D. Wright *Baskets & Basketry* ii. 45 *A Rectangular Base* is made in a screw-block. 1892 G. B. Shaw in *Pall Mall Gaz.* 22 Feb. 2/3 In order that they might secure the door on the outside and so retain my audience screwbound to the last syllable of the vote of thanks. 1966 A. W. Lewis *Gloss. Woodworking Terms* 85 Hinge is screwbound when the heads of the screws are not sunk correctly into their countersinking. 1960 *Practical Wireless* XXXVI. 302/2 A 500mA fuse..takes the form of a 6V, 0·5A screw bulb. 1897 *Sears, Roebuck Catal.* 35/2 Ointment pots. Flint glass. Nickle screw cap. 1936 *Lancet* 3 May 1160/2 The United Glass Bottle Manufacturers, Ltd., have produced a double-shell metal cap for bottles and pots. This cap has all the advantages of the ordinary screw-cap. 1972 *Gloss. Electrotechnical, Power Terms* (B.S.I.) iv. iii. 17 *Screw cap*, cap..in the form of a screw thread. 1898 *York Glass Company* (*Ltd.*) *Price List* 3 Pomade Bottles..Metal Screw Capped. 1964 M. Hynes *Med. Bacteriol.* (ed. 8) iv. 35 They are conveniently disposed for use in quantities of about 3 ml. in 1 oz. screw-capped bottles. 1958 *S. Afr. Mining & Engin. Jrnl.* LXIX. 243/1 The rotary screw compressor is built on the principle of an invention made by Professor A. Lysholm of Stockholm. 1967 *Gloss. Terms Materials Handling* (B.S.I.) iii. 17 *Screw or worm type compressor*, a rotary compressor having left hand and right hand worms in close engagement, which entrain the air or gas and eject it at a higher pressure. 1975 *Offshore Engineer* Dec. 57/1 Atlas Copco is to supply six ZR_4 screw compressors and ancillary equipment to be built into two modules by the fabricating engineers. [1940 J. M. Burgers in *Proc. Physical Soc.* LII. 25 Dislocation lines of this character will be said to be of the screw type.] 1948 *Rep. Conf. Strength of Solids, 1947* (Physical Soc.) 46 We may take the simpler case of a screw dislocation (Burgers' second type), lying along the *x* axis. 1966 *McGraw-Hill Encycl. Sci. & Technol.* III. 585/1 Screw dislocations have been shown to be important for crystal growth from the vapor phase. 1978 P. W. Atkins *Physical Chem.* xxviii. 931 The surface defect formed by a screw dislocation is a ledge, possibly with some kinks, where growth can occur. 1884 R. Aldridge *Life on Ranch* 191 We were a good deal troubled ..by what is called 'screw fly'. 1945 J. J. Mathews *Talking to Moon* 20 Sometimes their hides were torn, thus inviting screw flies. 1896 H. G. Wells *Wheels of Chance* iv. 24 Just then the screw-hammer slipped off the nut. 1909 *Chambers's Jrnl.* Jan. 61/1 The inner screws are then driven into the board to be lifted by a screw-hammer to secure a firm purchase. 1975 R. A. Salaman *Dict. Tools* 530/2 This smith-made example was sometimes called a 'Screw Hammer' because the upper jaw of the Wrench was made in the form of a Hammer and could be used as such. 1856 C. Nordhoff *Merchant Vessel* iv. 38 A lighter-load of cotton came down, and with it, a stevedore and several gangs of the *screw men*, whose business it is to load cotton-ships. 1950 Blesh & Janis *They all played Ragtime* ii. 39 The fellows who put bales in place were screwmen. 1894 R. B. Lee *Hist. & Descr. Mod. Dogs* (*Non-Sporting*) ix. 239 The screw tails, which are so peculiar to the [bulldog] breed, are objected to by a few authorities as indicating excessive in-breeding. 1965 Johnson & Galin *Compl. Bk. Dogs* (1968) vi. 255 If your dog has a screw or twisted tail..he may suffer from infection or sores under the skin. 1895 *Montgomery Ward Catal.* Spring & Summer 195 Large size pocket flask with collapsion cup, cover screw top. *Ibid.*, Screw top, satin engraved pocket flask. 1907 E. Nesbit *Enchanted Castle* xi. 314 A beer bottle with a screw top. 1937 G. Greene *19 Stories* (1947) 59 There's a bottle in my pocket. Have a drink... It has a screw top. 1951 *Good Houseк. Home Encycl.* 335/2 Boil the angelica..dry off..store in screw-top jars. 1963 *Times* 3 June 11/6 Make a French dressing with oil, tarragon-flavoured wine vinegar, dry mustard, salt and pepper and a pinch of castor sugar and pour into a screw-topped container. 1972 Screw-topped [see *marble-stoppered* s.v. *marble *sb.* 8 c]. 1879 *Investigation of Diseases of Swine* (U.S. Dept. Agric.) 208 Ticks, screw-worm, and the large horse or cow fly have destroyed many animals. 1936 E. Caldwell in *New Yorker* 22 Aug. 22/1 He hated weeds worse than he did boll weevils or screwworms. 1955 *Sci. Amer.* Oct. 50/3 The screwworm is a major pest of cattle in the U.S. Southeast. 1973 *Nature* 20 Apr. 494/1 The formidable task of re-eradicating the screw worm from the United States. 1908 V. L. Kellogg *Amer. Insects* 344 A flesh-fly of serious importance is the terrible screw-worm fly,..which lays its eggs on flesh..and often in the nasal passages of domestic animals and human beings. 1955 *Sci. News Let.* 29 Jan. 78/1 The screwworm fly may be eradicated from Florida and controlled in Texas, where its maggots cause millions of dollars loss to livestock men each year. 1978 *Nature* 22 June 606/2 The well publicised eradication of the screw-worm fly from Florida. 1850 *Amer. Agriculturist* Sept. 285/2 *Adjustable Screw Wrench.*—This is just about one of the most useful little farming tools ever purchased.

screw, *v.* Add: **II. 6. d.** *slang* (chiefly *N. Amer.*). To defraud (a person, esp. of money), to cheat; to deceive, to 'rook'; freq. as *pa. ppl.* in *to be* (or *get*) *screwed*.

1900 *Dialect Notes* II. 58 *Screwed*,..in phrase 'to get screwed'..deceived. 1936 J. Steinbeck *In Dubious Battle* vi. 94 'What you want to strike for?' ''Cause we're gettin' screwed..the company's store is takin' five per cent house-cut.' 1959 J. Osborne *Paul Slickey* i. v. 48 We want to screw, screw, screw the Income Tax Man. 1966 H. Kemelman *Saturday Rabbi went Hungry* xxxiii. 213 In the business dealings between Hirsch and Goralsky, it wasn't Goralsky that got screwed. It was the other way around. 1974 *Saturday Night* (Toronto) Feb. 12/3 Your chances of being screwed by a Canadian factory owner then were just as good as your chances of being screwed by an American factory owner now. 1979 *Tucson Mag.* Jan. 24/2 The Richard Nixon school of thought on public scandal, that being that it's all right to screw the people as long as you were given a large mandate in the previous election.

III. 10. b. (*b*) (Earlier example.)

1821 P. Egan *Life in London* v. 278 A well-known dashing Prig, whose *Head* was considered to have been *screwed on* the right way.

IV. 12. b. To spoil, ruin; to pervert; to upset, disturb mentally. *U.S. colloq.*

1938 'E. Queen' *Four of Hearts* iv. 54 'For gossakes!' yelled Lew, jumping up. 'That screws everything!' 1955 W. Gaddis *Recognitions* i. v. 183 She got fed up with him screwing the Sunday roast, so she shot herself. 1968 *Win* 15 Oct. 4/1 Democracy has gotten screwed, not just in Chicago but long before that. 1976 *National Observer* (U.S.) 14 Aug. 1/2 Your parents' divorce can screw you all over. It did me. I was shocked.

c. *colloq.* (orig. and chiefly *U.S.*). *to screw up*: (*a*) *intr.*, to blunder, make an error; (*b*) *trans.*, to make a mess of, spoil, ruin; to confuse, upset, disturb mentally.

This use may have originated as a euphemism for *to fuck up* (see *fuck *v.* 3) after sense 12* below.

1942 *Yank* 23 Dec. 19 You screw up on the drill field! You goof off at inspection. 1943 M. Hart *Winged Victory* i. ix. 90 My father-in-law says the OPA is screwing everything up. 1946 *Amer. Jrnl. Sociol.* Mar. 419 The common obscene expression which has the meaning in some way or another to bungle a job or to make a bad choice... There are a few acceptable substitutes such as 'screw up'. 1951 J. D. Salinger *Catcher in Rye* xix. 176 It really screws up my sex life something awful. 1955 W. Gaddis *Recognitions* i. v. 182 He's a drunk... He gets all screwed up with religion. 1967 *Melody Maker* 16 Dec. 8/6 Those people who are supposed to be propagating the Lord's word—they're screwing it all up. 1972 M. J. Bosse *Incident at Naha* 83 Did I screw up by admitting that you knew about the package? 1978 J. Irving *World according to Garp* xviii. 382 He said that women's lib had screwed up his wife so much that she divorced him. 1979 'A. Hailey' *Overload* i. i. 6 But you and your people really screwed up today! 1981 P. Niesewand *Word of Gentleman* xxvii. 188 Military men usually screw things up..and the people are bloody glad to see the back of them.

12*. *coarse slang.* **a.** *intr.* To copulate, have sexual intercourse (*with* a person). **b.** *trans.* Usu. of a man: to copulate with, have sexual intercourse with (someone).

1725 *New Canting Dict.*, *To Screw*, to copulate with a Woman. 1796 F. Grose *Class. Dict. Vulgar Tongue* (ed. 3), *To screw*, to copulate. 1927 O. W. Holmes *Let.* 1 July in *Holmes–Laski Lett.* (1953) II. v. 958 It is enough to mention his emulating a spider by screwing a woman while he killed her by biting and, put in as an extra, chewing her throat. 1937 J. T. Farrell *Can all this Grandeur Perish* 147 Him.. picking up bums in public dance halls and screwing them in hallways and taxicabs. 1945 G. Endore *Methinks the Lady* vi. 120 She thinks just because she married a sailor she can screw the whole Navy. 1952 S. Kauffmann *Philanderer* (1953) iv. 66 The first thing we do is..to run a few signed stories in the book, instead of all that anonymous 'I-got-screwed' stuff. 1958 N. Levine *Canada made Me* 16 Those who cry the most saying goodbye, screw the first. 1963 T. Pynchon *V.* i. 10 Santa's bag is filled with all your dreams come true: Nickel beers that sparkle like champagne, Barmaids who all love to screw. 1968 *Southerly* XXVIII. 38 'We have a free relationship,' Joe said. 'She's gone off to screw some old friend.' 1972 'G. Harding' *Skytrap* iii. 48 You've spent the afternoon screwing with *him*, haven't you? 1975 D. Lodge *Changing Places* i. 7 All women longed to be screwed by a god, it was the source of all religion.

c. In phr. *to screw around* (*around *adv.* 5 a): to be sexually promiscuous, to 'sleep around'; hence in weakened sense, to mess or fool about. orig. *U.S.*

1939 J. Steinbeck *Grapes of Wrath* ii. 14 Goin' all over the world drinkin' and raisin' hell and screwin' around. 1950 H. E. Goldin et al. *Dict. Amer. Underworld Lingo* 186/2 *Screw around*,..to clown and play the fool, paying scant attention to business. 'Don't you screw around when you're hustling (stealing) with me.' 1964 *New Statesman* 17 Apr. 610/3 He drinks.., screws around, lives in debt, cannot get his work published. 1972 D. S. Viscott *Making of Psychiatrist* iii. 43 Her husband is screwing around and she feels abandoned. 1974 *Times* 1 Apr. 6/8 All right—I am going to get him over because I am not going to screw around with this thing. 1978 R. Ludlum *Holcroft Covenant* iv. 51 They're honest guys. They wouldn't screw around. 1981 T. Heald *Murder at Moose Jaw* vi. 67 I've been sort of screwing around a little... I don't want to upset my husband, but a girl only has one life.

d. Used in imprecations and exclamations, as an equivalent to *fuck *v.* 2.

1949 A. Miller *Death of Salesman* i. 61 'In the business world some of them think you're crazy.'.. 'Screw the business world!' 1960 R. Dahl *Kiss, Kiss* 298 'Don't shout. There might be keepers.' 'Screw the keepers!' he cried. 1962 'E. McBain' *Like Love* vii. 102 'You sore?' 'Yes.' 'Screw you,' Kling said. 1979 'A. Hailey' *Overload* ii. v. 129 She was drowned out by a chorus of, 'screw the profiteers!' and 'power belongs to the people!'

13*. **a.** *trans.* To look at, watch (a person); *spec.*, to eye (a person) before a fight. **b.** *intr.* To look. *slang* (orig. *Austral.*).

1919 [see *screw *sb.*¹ 16*]. 1938 F. D. Sharpe *Sharpe of Flying Squad* 333 *Screw*..can also mean 'to look'. ('Screw over there', 'look over there'.) 1960 *Guardian* 29 Dec. 3/1 The accused..told them to stop 'screwing' him, which meant apparently to stop looking at him. 1964 *New Statesman* 10 Apr. 555/2 'No, no,' the Mods in the dance hall shouted 'screw..means to look you up and down.' 1978 P. Marsh et al. *Rules of Disorder* iv. 104 You get someone screwing you (staring) or just standing there all cocky like.

17*. To depart hastily, go away; to get *out*, push *off*. *slang* (orig. *U.S.*).

1896 Ade *Artie* iii. 26, 'Look here,' I says, 'you screw right away from here.' 1903 A. H. Lewis *Boss* ii. 18 'Screw out!' cried he... 'We don't want any of your talk!.. Put him out!' 1912 —— *Apaches of New York* iv. 84 As I don't want no part of it, I screws out. 1947 *Horizon* Sept. 205 Come on, let's screw out of here and find something. 1974 D. Richards *Coming of Winter* i. 23 Now if you don't screw off out of here, I'll use the phone.

V. 18. Also, more generally, to break into (a house, safe, etc.), to burgle.

1938 F. D. Sharpe *Sharpe of Flying Squad* 333 *Screw*, to break open houses and safes. 1953 H. Clevely *Public Enemy* xxvii. 219 You want to go inside for screwin' that ware'ouse. 1958 [see *bung *sb.*⁴]. 1973 'J. Patrick' *Glasgow Gang Observed* x. 88 Yir a brave wee boay that'll screw three shoaps in the wan night.

VI. 22. a. **screw-in, -on** *adjs.*, that may be attached by screwing into or on to something else; also as *sb.*

1924 G. L. Mallory *Let.* 12 Apr. in E. F. Norton *Fight for Everest: 1924* (1925) 215 Pukka wooden tables with three-ply wooden tops and screw-in legs. 1966 P. O'Donnell *Sabre-Tooth* iii. 41 Thin steel shafts made from short screw-in sections. 1976 *Alyn & Deeside Observer* 10 Dec. 3/2 A player can cost his club almost £65 a season in boots alone! Screw-ins [*i.e.* screw-in studs] cost about £17, rubbers about £14 and flats (training shoes) £7. 1928 A. L. Matthison *Stoving Finishes* 54 A screw-on cap for instance, involving the operations of stamp, screw and knurling machine, is easily withstood by a high grade coating lacquer. 1935 'G. Orwell' *Clergyman's Daughter* i. 13 The communion bell had had a screw-on clapper, which had come loose. 1967 [see O.D. s.v. *O 5 d]. 1979 *Amat. Photographer* 10 Jan. 74/1 Buy enough storage bottles, with screw-on caps, to accommodate all the liquids you use.

screwball (skrū·bǫl), *sb.* and *a.* Chiefly *U.S.* Also **screw-ball, screw ball.** [f. Screw *sb.*¹ + Ball *sb.*¹; for sense 2 cf. *oddball.] **A.** *sb.* **1.** † **a.** *Cricket.* A ball bowled with 'screw' or spin. *Obs.* **b.** *Baseball.* A ball pitched with reverse spin against the natural curve. Also *fig.* and *attrib.*

1866 'Capt. Crawley' *Cricket* 35 A 'screw' ball, which in slow bowling would describe the arc of a circle from the pitch to the wicket, becomes in fast bowling a sharp angle. *Ibid.* 36 The dotted line shows the direction of a slow screw ball *screwing* in from the leg. 1928 *N.Y. Times* 7 Oct. xi. 2/3 Haines is a large, healthy individual with..a 'screw ball' that ducks under many a well-meant swing with a hickory bludgeon. 1933 *Ibid.* 2 Aug. 20/1 Hubbell pitched his customary shrewd game for five innings then his deceptive screw ball lost its baffling influence. 1937 *Sun* (N.Y.) 1 June 24/3 'I thought Joe Robinson was mentioned [for the Supreme Court].' 'Yes, but it seems the President insists on screwball pitching.' 1949 *Sun* (Baltimore) 15 Oct. 12/5 Buxton, a 35-year-old screwball artist, was purchased from Oakland. 1960 *Time* 3 Oct. 47/2 Spahn started to perfect a screwball and a slider. 1971 L. Koppett *N.Y. Times Guide Spectator Sports* i. 15 Baseball men use the term screwball for either type of reverse curve, but the pitch is used primarily by left-handers.

2. a. An eccentric; a madman, a 'nut-case'; a fool. Freq. as a term of mild abuse. *slang.*

1933 P. Gallico in *Sat. Even. Post* 12 Aug. 56/3 McKabe was already heading for the door. He heard Billers say: 'Who is that screwball?' 1939 Wodehouse *Uncle Fred in Springtime* viii. 115 You are going to Blandings Castle now, no doubt, to inspect some well-connected screwball? 1944 H. S. Truman *Let.* 18 Aug. in M. Truman *Harry S. Truman* (1973) ix. 184 He should have been arrested as a screwball but wasn't. 1954 J. Steinbeck *Sweet Thursday* xxv. 163 He was a scientist, but whether brilliant or a screwball nobody ever knew. 1956 E. Pound tr. *Sophocles's Women of Trachis* 20 No use bothering with this screw-ball. 1956 W. H. Whyte *Organization Man* (1957) xviii. 239 To talk of the problem in terms of the lone genius or the screwball is to confuse the issue. 1978 S. Brill *Teamsters* v. 186 The word one got on Carey at Teamsters headquarters..was that he was a 'screwball', 'a weirdo', 'a strange guy', or..'a naïve kid'.

b. *spec.* Used, chiefly *attrib.* or as *adj.* (esp. as *screwball comedy*) of a kind of fast-moving, irreverent comedy film produced in the U.S. in the 1930s, of which eccentric characters were the chief feature, or of persons, etc., connected with such films.

1938 *Collier's* 26 Feb. 58/3 Wellman was named 'Screwball Bill' six or seven years ago and has, beyond any doubt, lived up to his title. 1938 *N.Y. Times* 2 Sept. 21/2 Metro-Goldwyn-Mayer..has popped up with another of those screwball comedies—this one called 'Three Loves Has Nancy'. 1939 L. Jacobs *Rise of Amer. Film* 536 Among the women Carole Lombard is the most outstanding in her

screwing, *vbl. sb.* Add: **a.** (Later examples, in sense *12* a and b of the verb.)

1952 S. Kauffmann *Philanderer* (1953) xv. 248, I have eaten up my honour with random screwing. I have defiled myself. **1958** A. Wilson *Middle Age of Mrs. Eliot* 262 Married to an old man. You've never had a proper screwing. **1971** C. Fick *Danziger Transcript* 70, I fail to see how an ancient screwing can be of current interest.

screwing, *ppl. a.* (Later examples, in sense 18 of the verb.)

1960 [see *blag *sb.*]. **1966** L. Southworth *Felon in Disguise* v. 81 A geezer called Teagueman..does a bit of driving for the screwing mobs.

screwless (skrū·lės), *a.* [f. Screw *sb.*[1] + -less.] Without a screw or screws.

1913 in Webster. **1962** L. S. Sasieni *Dispensing* i. 35 The 'Ilford' screwless mounting. **1976** *Golf International* 13–29 May 32/2 Most manufacturers change to what's termed the screwless face insert, for wooden clubs.

screwmatic (skrumæ·tik), *a.* and *sb.* *colloq.* [A humorous perversion of Rheumatic *a.* and *sb.*, after Screw *sb.*[1] (see sense *13 b).] **A.** *adj.* = Rheumatic *a.* 4 b. *rare.*

1893 P. H. Emerson *On Eng. Lagoons* 50, I had the screwmatic fever three times.

B. *sb. pl.* = Rheumatic *sb.* 1.

1895 P. H. Emerson *Birds, Beasts, & Fishes Norf. Broadland* 396 'Wiper's oil' is a reputed specific for 'screwmatics'. **1916** E. V. Lucas *Vermilion Box* 209 Wet, and rats,..and dirt and screwmaticks. **1974** P. Wright *Lang. Brit. Industry* iv. 44 Sometimes corruption has been helped by folk etymology, where the speaker supplies an imagined source for the word from what it seems to say; e.g. *screwmatics* (rheumatism).

screwsman. Add def.: A thief; a house-breaker, a burglar; also, a safe-breaker. (Further examples.)

1879 [see *dead *a.* A. 18 b]. **1928** *Daily Express* 19 Sept. 1/5 When released he came to London, and was among those whom the police call 'screws men', who break into houses to steal articles easily sold. **1936** [see *dragging *vbl. sb.*]. **1955** P. Wildeblood *Against Law* iii. 137 Suppose there's some screwsman that the law's got an eye on. **1963** 'J. Prescot' *Case for Hearing* iii. 49 What does our imaginary screwsman do? He gets his hands on the keys..to take impressions. **1976** J. O'Connor *Eleventh Commandment* iv. 63, I took up my old profession of being a screwsman again.

screw·-up. *colloq.* (orig. *U.S.*). Also **screwup.** [f. vbl. phr. *to screw up*: see *Screw *v.* 12 c.] A blunder, muddle, or mess; a state or situation of confusion or mental disturbance.

1960 Wentworth & Flexner *Dict. Amer. Slang* 453 *Screw-up*,..a chronic blunder. **1967** G. Legman *Fake Revolt* 23 The inability to *feel*, and the fear of touch, especially in sex... This is the key to the whole sexual screw-up of our time. **1971** Laver & Collins *Education of Tennis Player* v. 84 Bad courts were just one more factor in a general screw-up. **1975** *New Yorker* 19 May 24/2 Stewart Stern..tries to equate the American woman's supposed incapacity for love with the whole American screwup of recent years. **1977** C. McFadden *Serial* iv. 14/1 Everybody knew..that the rational mind was a screw-up.

screwy, *a.* Add: **5.** *slang* (orig. *U.S.*). Mad, crazy; eccentric; foolish; ridiculous.

1887 *Lantern* (New Orleans) 8 Oct. 3/3 Do, now, please to stop them notes which are so full of screwy talk. They make me tired. **1930** D. Runyon in *Collier's* 22 Mar. 54/2 'I am wondering how much you will take to hold still and let me shoot you, Jack?' 'Why,' Handsome Jack says, very much astonished, 'you must be screwy.' **1942** E. Waugh *Put out More Flags* iii. 163 They like the little pattering feet about the house—I know it sounds screwy but it's the truth. **1948** M. Allingham *More Work for Undertaker* xiii. 166 We feel our clients are screwy but not bloodstained. **1959** 'J. Bell' *Easy Prey* viii. 87, I think the whole idea is screwy... I can't see her risking her own life to save a murderess. **1966** R. H. Rimmer *Harrad Experiment* 42 Sheila was Tom's date and I had Tom's sister, Ruth, for a date. Sound screwy? **1978** J. Wainwright *Jury People* liv. 199 The prison psychiatrist ..can be told crap. He'll believe it... If anybody's screwy, it's *him*.

scribbledehobble (skri·b'ldĭhǫ:b'l). James Joyce's nonce-formation on Scribble *sb.* or *v.*[1], prob. influenced by such a word as *hobbledehoy*, the etymology of which is obscure. Hence, the name given to one of Joyce's notebooks (see quot. 1961).

1922 Joyce in T. E. Connoly *James Joyce's Scribbledehobble* (1961) 5 Scribbledehobble..I'm feeling so funny all over the same. **1939** —— *Finnegans Wake* ii. 275 That royal pair..have discusst..why lui lied to lei and hun tried to kill ham, scribbledehobbles, in whose veins runs a mixture of, are head bent and hard upon. **1961** *Times Lit. Suppl.* 20 Oct. 754/3 Of the fifty *Finnegans Wake* notebooks now in the Lockwood Memorial Library the *Scribbledehobble* book is the largest... It contains words, phrases, clichés, anecdotes, ideas, scraps of information and other memoranda. **1977** J. Garvin in D. Ó Muirithe *Eng. Lang. in Ireland* 113 *Scribbledhobble* [sic], a notebook..compiled by him [*sc.* Joyce] in 1923..contain[s] simple phrases probably culled from his wife's conversations.

scribbler[1]. Add: **2.** A scribbling-book or pad. Chiefly *N. Amer.*

1906 *Daily Colonist* (Victoria, B.C.) 11 Jan. 10/5 (Advt.), 1000 Scribblers and Exercise Books..2c each. **1913** *T. Eaton & Co. Semi-Ann. Sale Catal.* No. 36. 17/2 *School Outfit*..2 Exercise Books, 2 Scribblers, 1 Writing Pad, ruled. **1916** Joyce *Portrait of Artist* (1969) iii. 105 Stephen, leaning back and drawing idly on his scribbler, listened to the talk about him. **1939** L. M. Montgomery *Anne of Ingleside* xxxix. 309 We write each other letters in our scribblers and exchange them. **1964** L. Edel *Diary of Alice James* p. v, She kept the record of her sickroom world in two closely-written scribblers. **1969** K. Giles *Death cracks Bottle* vi. 57 [He] had taken a page from his scribbler and written on it.

scribble-scrabble, *sb.* **2.** (Earlier example.)

1760 Lady M. W. Montagu *Let.* 12 Feb. (1967) III. 232, I see you laugh..at the vanity of my supposing any thing valuable in my scribble scrabbles.

scribbling, *vbl. sb.*[1] Add: **4.** *scribbling pad.*

1938 N. Marsh *Artists in Crime* xiii. 189 Nigel had been left to write a very guarded story..on one of Troy's scribbling-pads. **1967** C. Drummond *Death at Furlong Post* i. 5 The Chairman looked doubtful; his Board appeared absorbed in their scribbling pads.

scribe, *sb.*[1] Add: **7. c.** *U.S.* A newspaper reporter.

1929 D. Runyon in *Hearst's Internat.* July 58/1 Dave grabs the scribe..and is taking him out for an airing!.. Taking a newspaper guy..out for an airing is apt to cause talk. **1930** *Amer. Speech* VI. 120 *Scribe*,..reporter, writer: Judge Quashes Police Charges Against Scribe (here, a reporter). **1962** *John o' London's* 25 Jan. 82/3 A reporter is a *scribe* or *scribbler*.

scribe, *v.* Add: **1. a.** Now done on other materials, and by means of a fine laser beam as well as pointed instruments. (Further examples.)

1967 *Electronics* 6 Mar. 218 (Advt.), Their 4-sided diamond scriber often leaves rough, chipped lines when it scribes crystalline wafers for dicing. **1975** J. B. Harley *O.S. Maps* i. 11 A second sheet of plastic material..is placed in exact registration with the first, and is then used for plotting and scribing the contours. **1977** *Engin. Materials & Design* Aug. 30/1 Blank sheets of fired alumina ceramic are accurately located on the table and, working from a datum point, the laser scribes a grid matrix system of close and regularly spaced blind holes.

scribed, *ppl. a.* **1.** (Further examples.)

1971 *Physics Bull.* July 405/2 Conventionally the individual units are separated from a block by scribing with a diamond wheel and then breaking the ceramic along the scribed line. **1977** *Engin. Materials & Design* Aug. 29/3 The new installation..will provide the electronics industry with a high speed service based on..the supply of scribed wafer from its own stock of ceramic materials.

scriber. Add: **2.** (See quot. 1968.)

1968 'B. Mather' *Springers* v. 52 A scriber is a circular piece of paper-thin copper which fits out of sight under the moving part of the [telephone] dial which has a tiny sharp point set in it. The pressure of the dialler's finger is sufficient to bring the point down on the copper... The resultant marks..can tell..what numbers have been called. **1978** J. Barnett *Head of Force* xvi. 153 We installed scribers under the dials of public phones adjacent to suspect premises.

scribing, *vbl. sb.* Add: **1.** (Further examples.)

1969 G. C. Dickinson *Maps & Air Photographs* v. 75 Scribing..consists in producing a negative-type drawing by scraping away an opaque coating applied to glass or dimensionally stable plastic sheets. **1975** J. B. Harley *O.S. Maps* i. 14 The use of scribing instead of traditional drawing was an especially important development. **1977** *Engin. Materials & Design* Aug. 29/3 The first half of 1977 also saw the introduction by Laser Cutting Ltd of what is believed to be Europe's first facility specifically designed to provide a high quality service for ceramic scribing by laser.

3. (Further examples.)

1969 *R. & E. Coordinator* Apr. 9/2 A new glass-based scribing plate for the generation of microphotography masters, precision printed circuits, and other applications requiring extremely clean, sharp lines. **1971** *Physics Bull.* Dec. 743/3 The scribing speed is 360 in min^{-1}.

Scriblerian (skribli°·riăn), *sb.* (and *a.*) [f. the name of Martinus *Scriblerus*, a character invented by members of the Scriblerus Club (see below); cf. Scribbler[1] + -ian.] A member of the Scriblerus Club formed *c* 1713 by Pope, Swift, Arbuthnot, and others, who produced the *Memoirs of Martinus Scriblerus* (publ. 1741) in order to ridicule lack of taste in learning. Also as *adj.*

1935 L. M. Beattie *John Arbuthnot* iv. 271 Swift's experience as a Scriblerian must have affected the turn given to numerous observations in *Gulliver*. *Ibid.* 276 In its humor for humor's sake it is typically Scriblerian. **1950** C. Kerby-Miller *Mem. Martinus Scriblerus* 1 The activities which may be labeled Scriblerian spanned a period of almost three decades. *Ibid.* 31 The Scriblerians began collecting material of all sorts. **1969** P. Köster in *Philol. Q.* Apr. 207 Although the subtlety of this satiric segment may have been *caviare* to all but the refined taste of Arbuthnot's fellow Scriblerians, the 'String of Epithets' could not be ignored by any but the grossest ear. **1977** —— in R. A. Wisbey *Computer in Literary & Linguistic Res.*

iv. 133 Unfortunately.., there are no examples of Scriblerian collaboration in which the shares of Arbuthnot and Swift are already known.

scrick, scrief, varr. *Skrik, *Screef.

scrieve (skrīv), *v.*[2] *Shipbuilding.* [Dialectal var. Scrive *v.* (*Eng. Dial. Dict.*); cf. Screeve *v.*[2]] = Scrive *v.* 2. Freq. in *Comb.* as **scrieve-board** = Scrive-board.

1898 T. H. Watson *Naval Archit.* vi. 54 The Scrieve Board is a platform of well-seasoned deals... Sometimes both sides of the ship are scrieved in. **1901** T. Walton *Steel Ships* vii. 180 He then transfers from the mould loft floor to the scrieve board the midship section of the vessel... He also scrieves in all decks, stringers, keelsons, and floors. **1921** *Flight* XIII. 214/1 Many engineers and aircraft builders do not realise to what fine limits naval architects have to work when scrieving out the lines. **1951** *Engineering* 15 June 730/1 Beside this striking exhibit is the scrieve board corresponding to it, showing all the details of the construction. **1957** [see *loftsman].

scrieve, var. Screeve *sb.*

scrike, scriker, scriking. (Later examples s.vv. *Skrike, *Skriker, *Skriking.)

scrim. Add: **1.** Now freq. made of muslin, sacking, or similar material. Also in *Mil.* use. (Later examples.)

1936 R. Hyde *Passport to Hell* xiii. 199 The huge spoil-dumps were camouflaged with green boughs and scrim. **1942** H. A. Maddox *Dict. Stationery* (ed. 2) 97 Scrim. The coarse textured net fabric used for attaching to the glued backs of cheap cased-in books. **1947** D. M. Davin *Gorse blooms Pale* 179 A man sat screened by scrim on the latrine. **1961** *Wall St. Jrnl.* 24 Jan. 1/4 Most of the testing.. involves use of knitted paper in such unglamorous jobs as backing, or 'scrim', for more costly synthetic fibers. **1964** *Weekly News* (Auckland) 29 Apr. 41/2 Later there will be shrubs that have been wrenched and their roots balled up in scrim. **1970** *New Society* 5 Mar. 386/1 One technique, SRM (scrim-reinforced material), uses heat to laminate a scrim of random-laid fibres to wadding; the heat reactivates the bonding glue.

2. A piece of scrim used as a window-covering; *spec.* a thin, gauze-like, curtain material. Usu. as *window scrim.*

1915 E. G. Pilling *Anzac Memory* (1933) 119 Look out of the broken, scrim-covered window across fields white with snow. **1969** Widdowson & Halpert in Halpert & Story *Christmas Mumming in Newfoundland* 149 Some would put on what they used to call muslin, you know. Muslin or scrim, window scrim.., you know, like you'd have for a curtain; you could see out through it. **1970** G. F. Newman *Sir, you Bastard* ii. 61 Outside on the balcony was a small piece of window scrim.

3. a. *Theatr.* and *Cinemat.* Gauze cloth used for screens or for filtering theatrical lighting; a screen of this material. orig. and chiefly *U.S.*

1928 A. E. Krows *Equipment for Stage Production* vii. 90 The foliage..is painted on a canvas drop..and, after being cut away, is mounted..to a curtain made of a coarse netting called scrim. **1937** F. Napier *Curtains for Stage Settings* vi. 101 The space below the aperture can be filled in with brown paper, decorated with panels outlined in paint or chalk, but it must be pasted on to butter muslin or scrim to make it more durable. **1939** N. Coward *Play Parade* II. p. x, *Scrim*, American term for a Gauze Cloth. **1950** *People* (Austral.) 15 Mar. 46/1 She was told she would be seated behind a 'scrim' (stage jargon for a gauze screen). **1957** V. J.-R. Kehoe *Techn. Film & T.V. Make-Up* i. 17 A spotlight to add highlight to the cheekbone, with a scrim over the light to soften the beam. **1975** *New Yorker* 26 May 90/2 The Prince and the Lilac Fairy get into a boat that remains motionless as a scrim painted with leaves and branches moves sideways across the stage. **1977** *Time* 27 June 45/1 The vast (5,000 sq. ft.) shuffle area is a stage, with theatrical lighting, scrims and backdrops rising as high as 85 ft.

b. *fig.* A veil or screen; something that conceals what is happening.

a **1963** S. Plath *Crossing Water* (1971) 37 The salt Scrim of a sea breeze. **1970** H. & F. Schreider *Exploring Amazon* vii. 154 The city of Manaus shimmered through the great scrim of rain like a stage setting from the pageant of its own lost glory. **1972** *Publishers' Weekly* 12 June 16/1 France and the French did more than serve as a scrim for vagabonding Americans. **1977** *Time* 6 June 10/2 The full story of Podgorny's dismissal may remain forever behind the scrim that veils the Kremlin's backstage dramas.

scrimmage, scrummage, *sb.* Add:

This is now used primarily as a sporting term. The older *i*-form is common in all senses, and has become predominant in American Football, whilst the *u*-form is preferred in Rugby Football.

3. a. (Later examples.) Also *fig.* (freq. after sense 4 in Dict. and Suppl.).

α. **1855** [see sense 3 b in Dict.]. **1900** H. A. Jones *Mrs. Dane's Defence* i. 17 Have you made a mess of your life?.. I wonder how many poor women have been sacrificed in the—scrimmage? **1930** E. M. Brent-Dyer *Chalet Girls in Camp* vi. 93 After something like a scrimmage they got the thing right. **1957** L. Durrell *Justine* iii. 185 The whole portentous scrimmage of sex. **1979** *This England* Winter 28/2 The doors of St. Paul's revolve like Marks and Spencer's onto the scrimmage round its book-stalls.

β. **1959** A. Sillitoe *Loneliness of Long-Distance Runner* 23 Mam had forgotten to buy me one in the scrummage of shopping.

4. c. *Amer. Football.* (*a*) A sequence of play which is started when two lines of opposing

2. a. intr. To seek to obtain by irregular means, as by stealth or begging; to hunt about or rummage (for something).

1909 J. R. Ware Passing Eng. 217/2 Scrunging (Country Boys'), stealing unripe apples and pears—probably from the noise made in masticating. 1915 W. H. L. Watson Adv. Despatch Rider v. 58 George and I..'scrounged' for eggs and bread. 1918 G. Goodchild Behind Barrage vi. 94 You may scrounge for rations, kit, pay, or leave. Signallers..usually scrounge for wire. Scrounging for wire is legitimized by the War Office. 1930 Brophy & Partridge Songs & Slang 1914–18 160 To scrounge about, to go seeking an opportunity of stealing. 1961 'E. McBain' 'Til Death xiii. 153 Facing the world outside the police department, scrounging for a job when I'm no longer a boy. 1973 M. & G. Gordon Informant xliv. 165 Scrounging around in her case for a freshly laundered slip, she cast curious glances at Chris.

b. trans. To appropriate; to acquire by irregular means, by stealth, or by begging; to 'pinch', to 'cadge'.

1917 A. G. Lee Let. 24 Nov. in No Parachute (1968) viii. 172 Now to scrounge the watch from its casing! 1919 [see *buckshee sb. and a.]. 1923 G. H. McKnight Eng. Words & their Background 67 British supplies were scrounged. 1939 Star 2 Dec. 4/1 The Southern Railway gave a staggering figure for the specially dimmed bulbs which had been stolen (I beg pardon, scrounged) from their carriages in the first weeks of the war. 1945 Sun (Baltimore) 14 Dec. 6/5 Food, cigarettes, chocolate, clothing, flour and canned meat which the supply team has 'scrounged' from excess military stores. 1958 Times Lit. Suppl. 10 Oct. 573/3 The crude overtures of Moondoggie and the other Huck Finn louts who scrounge a lazy summer from any foolish young woman whose parents can provide them with a meal. 1976 National Observer (U.S.) 31 Jan., Some of these [newspapers] I picked up free in the press room; others I scrounged at the lower-lobby newsstand.

Hence scrounge sb., the action of scrounging; scrounged ppl. a.; scrou·nger, one who scrounges; scrou·nging vbl. sb. and ppl. a.

1909 Webster, Scrunger. 1918 E. S. Farrow Dict. Mil. Terms, Scrounger, a slang term for a soldier with plenty of resource in getting what he wants. 1919 tr. A. L. Vischer's Barbed Wire Dis. 44 The complaints about 'scrounging', which are nothing but outbreaks of loss of moral judgment. 1927 Daily Express 17 Aug. 3 (heading) Suffolks on the scrounge. Village trek for recruits. 1941 New Statesman 29 Mar. 316/2 'Scrounged' cups, plates, cutlery and even food. 1946 Sun (Baltimore) 23 Oct. 4/3 There is a blunt reminder that 'pilfering' by a native is indistinguishable from 'scrounging' by an American soldier, and that 'chiseling' and resale of Post Exchange supplies is not an act peculiar to Filipinos. 1950 Landfall Mar. 127, I drained my fifth warm bottle-full ages ago and have been on the scrounge ever since. 1956 L. Godfrey in Pick of Today's Short Stories 94 'Besides,' added Trouncer.. 'it's a good scrounge.' 1956 A. L. Rowse Early Churchills viii. 151 The King, who sank back into the more consoling, if hardly less scrounging, arms of the Duchess of Portsmouth and Nell Gwynn. 1959 Times Lit. Suppl. 26 June 382/4 A curious collection of notes assembled under the title 'Autolycism', after Autolycus, an Athenian of scrounging habits. 1968 Science 3 May 522 He was a talented scrounger who in the early stages of the development of the cyclotron was able to find an available 85-ton magnet. 1974 Listener 7 Nov. 593/3 Reading an old, scrounged Daily Mirror. 1978 P. Marsh et al. Rules of Disorder ii. 31 You learn to scrounge. Anybody's a good scrounger around here. 1981 'J. Gash' Vatican Rip i. 7 I'm an antique dealer... I was on the scrounge and feeling very sorry for myself.

scrounge (skraundʒ), v.^{2} U.S. colloq. [Cf. Scrouge v., but perh. related to dialectal scringe, scrunge to rub with force (Eng. Dial. Dict.: see prec.).] trans. To move with a rubbing or squeezing action.

1939 J. Steinbeck Grapes of Wrath x. 123 Ma chuckled lightly and scrounged the clothes in and out of the bucket. 1954 —— Sweet Thursday xxii. 139 You keep an old lemon rind, and every time you wash your hands you scrounge your fingernails around in it.

scrub, sb.^{1} Add: I. 2. a. (Earlier and further examples.) Also, in Austral. and N.Z. usage, any tract of heavily wooded country, whether bearing small or large bushes or trees.

1805 P. G. King in Hist. Records Australia (1915) 1st Ser. V. 586 A Scrub—consists of Shrubs of low growth, Soil of a bad quality with small Iron gravelly Stones, in general Rocky Scrub and Brush may..be called the Underwood of the Forest, but it is not infrequent on the Sea Coast for Scrubs to be void of trees. 1841 N.Z. Jrnl. II. xlviii. 285 Every part is covered with vegetation, fern, scrub, copse and forest. 1860 J. McD. Stuart Jrnl. 9 Apr. (1864) 153 At four miles arrived on the top, through a very thick scrub of mulga. 1911 E. M. Clowes On Wallaby i. 5 These [prisoners] were packed off next day in boats, and let loose in the dense scrub where St. Kilda and Prahan now stand. 1947 K. Tennant Lost Haven vii. 105 She had been leading the children in botany expeditions through the scrub. 1966 'J. Hackston' Father clears Out 16 Chester tried to cannon off the road and pocket us in the scrub. 1977 Weekly Times (Melbourne) 19 Jan. 34/1 In silence the two men rode towards the river but, turning left into the scrub before the bridge, they skirted the town.

c. The Scrubs: ellipt. for Wormwood Scrubs Prison in Greater London. Also erron. Scrubbs.

The element Scrubs in the place-name is app. identical with Scrub sb.^{1} (see Conc. Oxf. Dict. Eng. Place-Names (1936) 510/1).

1923 in J. Manchon Le Slang. 1930 G. Baker Soul of Skunk ii. ii. 161 At the end of my first temporal month, I gibed at the Scrubbs... The broadest of my prison grins must have been that which I bestowed upon the Scrubbs' librarian. 1941 G. Greene When Greek meets Greek in 19 Stories (1947) 171 Before his first stay at the Scrubs he had held a number of positions. 1966 A. Prior Operators vi. 64 He had..taken his medicine, which had turned out to be three years in the Scrubs. 1976 M. Maguire Scratchproof iv. 58 'Catherine put him in the Scrubbs for twelve months,' Gibson continued. 'It was in all the papers, you must have read about it.'

II. 4. a. Now in extended sense (N. Amer.): an animal of inferior breed or pedigree; a beast of poor physique or performance.

1812 Columbia Centinel 31 Oct. 2/3 May the usefulness of our Institution be acknowledged;—its speed drive scrubs from the course. 1858 C. L. Flint Milch Cows 28 We meet with good milkers of all forms, from the round close-built Devon to the coarsest-boned scrub. 1888 Harper's Mag. Jan. 325/1 The colonel's horse—an old 'scrub' he had borrowed—'bucked'. 1901 Daily Colonist (Victoria, B.C.) 26 Oct. 8/2 Mr. Wilson, manager of the Toronto Poultry farm, says he can not get enough [chickens] of superior quality, and many others say the same. No one wants 'scrubs', the days of which are numbered. 1934 J. M. Cain Postman always rings Twice xiv. 157 All the really fine pumas come from Nicaragua. These California..things are just scrubs compared to them. 1936 M. Mitchell Gone with Wind xxx. 509 They knew thoroughbred horses from scrubs. 1972 Fredrickson & East Silence of North vii. 52 We had a dog team of sorts, two scrubs that weren't worth much but could pull a load of traps and other gear on a homemade toboggan.

5. b. slang. A disreputable woman; a prostitute, tart.

1900 Dialect Notes II. 58 Scrub,..a disreputable woman who frequents the streets. 1964 New Statesman 10 Apr. 555/2 A 'scrub' is a Rocker girl; that is, someone not fond of washing, according to the Mods, and a bit of a tart.

c. U.S. Sport. (a) A player belonging to a second or weaker team (freq. in pl.); a team composed of such players. Also fig. Cf. *scrub a. 4.

1892 College Index (Agric. & Mech. Coll. Alabama) Nov. 23 Arranged similarly, but with darker stockings, stand the inimitable 'scrubs', and although their name is rather depreciatory, they themselves are not to be scoffed at. 1903 N.Y. Even. Post 28 Oct. 9/5 The halfback tries his mettle against the scrubs. 1920 W. Camp Football without Coach 62 Do not let the absence of a scrub disturb you in the least. Many a team is better off without a second eleven. 1956 B. Holiday Lady sings Blues (1973) xxi. 173, I wouldn't have known the first team from the scrubs, but Ehrlich told me the prosecutor and the judge were the best they had. 1961 J. S. Salak Dict. Amer. Sports 385 Scrub, a player of the second, or weaker, team; one not good enough to be on the first team.

(b) (See quot. 1910.)

1892 Dialect Notes I. 214 'Scrub' in New England is that form of base ball played when there are too few players to have opposing sides. 1896 W. A. White Real Issue 66 Just before school was called Piggy Pennington was playing 'scrub'. 1910 Dialect Notes III. 447 Scrub, a game of baseball played by a half dozen or more persons (when there are not enough to 'choose up' for two nines), in which the players move up as a batter is retired. 1917 C. Mathewson Second Base Sloan 126 At the end of a week or so they were playing 'scrub' every noon hour.

III. 6. a. (sense 2) scrub bull, bush, fire, horse, jungle, -land (earlier and later examples); scrub-covered adj.; scrub-cutter Austral. and N.Z., (a) a machine for cutting scrub; (b) one who cuts scrub; hence scrub-cutting vbl. sb.; scrub-dashing vbl. sb. Austral. (see quot. 1941); scrub-rider Austral., one who rides in search of cattle that have escaped into the scrub; scrub tick, either of two small, brown, hard-bodied ticks found in Australia, Hæmaphysalis bispinosa or Ixodes holocyclus, the bush tick; scrub typhus, an acute rickettsial fever transmitted to man by mites normally parasitic upon small rodents; cf. mite(-borne) typhus s.v. *mite^1 3.

1908 A. Gunn We of Never-Never xviii. 238 Tales of scrub-bulls, maddened cow-mothers. 1946 A. Marshall in Murdoch & Drake-Brockman Austral. Short Stories (1951) 317 He was an old scrub bull,..who roamed the timbered hills beyond the Murray. 1954 [see *centre sb. 11 h]. 1977 Listener (N.Z.) 15 Jan. 34/3 Jeremy Delacy, the 'scrub bull', eccentric conservationist station-owner who has parted company with the local 'establishment'. 1897 D. McK. Wright Station Ballads 64 The scrub bushes sheltering round. 1959 Tararua (N.Z.) XIII. 45 One curious term is scrub bush. So far as I can make out it is applicable to the individual plants which go to make up tall scrub. At any rate a tall plant of teatree may be called a scrub bush. 1900 W. S. Churchill in Morning Post 19 Mar. 5/7 The proper left of this position rests on the rocky scrub-covered hill of Hlangwani. 1980 S. Wilson Dealer's War i. ii. 26 Grey scrub-covered hills. 1886 R. Stout Notes Progress N.Z. 28 The following..are manufactured in the colony—viz., ploughs, chaff-cutters..disc-harrows..scrub-cutters. a 1930 H. Stone in Austral. Short Stories (1951) 115 But them scrub-cutters... They do need constant watchin. 1937 J. West Sheep Kings ix. 87 He purchased a new block of bush land, and employed another gang of men to break it in—timber-men and scrub-cutters and post-splitters. 1965 M. Shadbolt Among Cinders xxii. 210 Probably the place had been built by scrub-cutters. 1911 'Kiwi' On Swag iii. 7 We were on a job of scrub cutting. 1968 Wanganui (N.Z.) Chron. 15 Nov. 10/5 (Advt.), Scrub-cutting contract for 120 acres. 1977 C. McCullough Thorn Birds vi. 117 The grass had lasted just long enough, eked out by scrub-cutting from the more juicy trees. 1941 Baker Dict. Austral. Slang 64 Scrub-dashing, riding through bush or scrub, esp. after strayed cattle or brumbies. 1946 F. D. Davison Dusty xv. 166 Fred had offered Tom work; odd jobs around the homestead..while Fred did the scrub-dashing. 1944 Living off Land vii. 148 Slower-burning scrub fires..call for strategy. 1965 S. T. Ollivier Petticoat Farm i. 1 Harry stood at the roadside and watched the white pumice dust..hanging in the air like smoke from a scrub fire on a fine day. 1893 D. Ferguson Bush Life in Australia & N.Z. 301 The scrub confused and handicapped [the thoroughbred] whilst Selina was a scrub horse. 1912 Blackw. Mag. Jan. 113/1, I saw my first tiger in a scrub jungle two miles from the Nepal frontier. 1934 'G. Orwell' Burmese Days iv. 69 It was scrub jungle at first, with dense stunted bushes. 1953 D. A. Bannerman Birds Brit. Isles I. 256 The buntings spread over the plains of India, chiefly affecting cultivation and scrub-jungle. 1779 W. McKendry Jrnl. 4 Oct. in Proc. Mass. Hist. Soc. (1886) 2nd Ser. III. 472 Came over skrub land this day. 1955 H. Klein Winged Courier xiv. 90 All around he lay scrubland, marsh and swamp. 1955 J. Thomas No Banners xxiii. 230 The road and the lane make a kind of elongated 'V', with the two arms separated by this scrub land. 1976 K. Royce Bustillo iv. 47 Across the scrubland the market was preparing its stalls. 1977 'J. le Carré' Hon. Schoolboy xvii. 404 The perfect tarmac road ran..over the flat scrubland. 1881 A. C. Grant Bush-Life in Queensland (1882) xv. 150 A favourite plan amongst the bold scrub-riders. 1891 Queenslander 3 Jan. 36/3 The scrub tick is a small animal with eight legs when mature, flat, brown in colour. 1936 Discovery Oct. 306/2 If the camp is in the scrub..scrub ticks have to be faced. 1965 Austral. Encycl. VIII. 499/1 In the genus Haemaphysalis are..two introduced species: a scrub tick..originally from India, and a dog tick. 1929 W. Fletcher et al. in Trans. R. Soc. Trop. Med. & Hygiene XXIII. 61 The K. form has a patchy distribution, and its virus, like the virus of the tsutsugamushi disease, has its home in circumscribed areas of untilled open country, particularly in land which after being cleared of jungle has been allowed to grow up in weeds and scrub... Because cases of the K. form have their origin in such places, we propose that this kind of tropical typhus should be called scrub-typhus. Ibid., The epidemiology of this rural or scrub-typhus, is very similar to that of the sporadic typhus-like disease of India which Megaw attributes to the bites of ticks. 1961 R. D. Baker Essent. Path. ix. 241 During World War II tsutsugamushi disease, scrub typhus, was prevalent among our troops in the Far East, and many fatalities occurred. 1978 Jrnl. R. Soc. Med. LXXI. 501 Scrub typhus is a febrile illness, endemic in much of the roughly triangular area bounded by Japan, Pakistan and Australia.

b. scrub-bird: see also noisy scrub bird s.v. *noisy a. 1 b; scrub-fowl, a grey and brown mound-building bird, Megapodius freycinet, found in coastal areas of northern Australia; scrub-hen, substitute for def.: = *scrub fowl; scrub jay U.S., a blue jay with no white markings, Aphelocoma cœrulescens cœrulescens, found only in parts of Florida; scrub-turkey, substitute for def.: a large mound-building bird, Alectura lathami, found in Australian forests and having a red head and brown body; (later examples); scrub wallaby, one of several wallabies belonging to the genus Macropus and living in woodland; scrub wren, a small Australian bird belonging to the genus Sericornis.

1908 Scrub fowl [see *dead-alive a.]. 1943 C. Barrett Austral. Animal Bk. xvii. 151 The scrub-fowl burrows into the mass to deposit each egg. 1938 M. K. Rawlings Yearling xxiv. 302 Scrub jays flew across the road. Their solid blue feathered coats..were prettier than the bluebirds'. 1947 R. T. Peterson Field Guide to Birds (ed. 2) 159 Florida, or Scrub, Jay... Look for this crestless Jay only in the stretches of 'scrub' in Florida. 1976 Southern Evening Echo (Southampton) 15 Nov. (Advt. Suppl.) 4/2 Scrub jays..feed their brothers and sisters. 1889 Scrub-turkey [see *chucky-chucky]. 1940 F. D. Davison Woman at Mill 86 The scrub turkey had her home, her nesting mounds of leaf and forest debris. 1967 Courier Mail (Brisbane) 26 June 8 Normal mounding activity by feverishly active scrub turkeys in South-Eastern Queensland begins late in June, ending the following March. 1896, 1926 Scrub wallaby [see brush-wallaby s.v. *brush sb.^{1} 4]. 1947 K. Tennant Lost Haven xii. 199 The marks of a scrub wallaby fossicking the high-tide drift. 1970 W. D. L. Ride Guide Native Mammals Austral. 46 Scrub wallaby..inhabiting woodland, forest edges, and coastal scrub. 1901 A. J. Campbell Nests & Eggs Austral. Birds I. 249 This smart Scrub Wren possesses chiefly a western distribution. 1943 C. Barrett Austral. Animal Bk. xxxii. 278 The Australian scrub-wrens..are fussy, plain-coloured little birds, which spend most of their time on or near the ground, keeping to the undergrowth. 1965 Austral. Encycl. VIII. 48/2 Most of the scrub-wrens build domed nests of soft bark and fibre in thick vegetation.

c. scrub oak, substitute for def.: (a) one of several North American dwarf oaks; (earlier and later examples); (b) Casuarina cunninghamii; scrub palmetto, a small, slow-growing palm of the genus Sabal; scrub pine, add: spec. one of several North American dwarf pines, esp. Pinus virginiana, or its wood; (earlier and later examples).

1766 J. Bartram Jrnl. 13 Jan. in Trans. Amer. Philos. Soc. (1944) XXXIII. 42/1 We came to Round-Lake,..almost surrounded with palmetto, pine, and scrub-oak. 1918 W. Cather My Antonia ii. xiv. 272 The dogwoods and scrub-oaks began to turn up the silvery underside of their leaves. 1947 V. H. Cahalane Mammals N. Amer. 365 One pair of these pockets can carry as many as twenty-

seven scrub-oak acorns. 1954 R. Murphy Pond i. 9 The second-growth pine woods, had gone back to brush, greenbrier, scrub oak. 1938 M. K. Rawlings Yearling iv. 40 Pushing through the low..scrub palmettos..was less laborious. 1968 Mrs. L. B. Johnson White House Diary 23 Nov. (1970) 740 The landscape was low and flat, clotted with scrub palmettos. 1791 P. Fidler Jrnl. 30 Oct. in Publ. Champlain Soc. (1934) XXI. 517 A high point of Rocks & scrub pine. 1818 [see *New Jersey]. 1832 [see grey pine s.v. *grey, gray a. 8]. 1949 Sat. Even. Post 9 Apr. 162/3 In front of the fireplace was a coarse-haired bearskin, scarred with burns from the snapping embers of scrub pine and cotton wood. 1976 M. & G. Gordon Ordeal (1977) xi. 71 They walked slowly through a forest of scrub pine.

scrub, sb.^{2} Add: Also scrubb. 1. a. (Further examples.) Also spec. with up: see *scrub v.^{1} 3 d; freq. attrib.

a 1902 S. Butler Way of All Flesh (1903) lxxii. 331 Ellen had given it another scrub from top to bottom. 1937 Archit. Rev. LXXXI. 52 (caption) A detail in one of the surgeon's 'scrub-up' lobbies, looking through an observatory window into an operating theatre. 1953 K. Tennant Joyful Condemned xxxix. 383 Rene eyed May's house possessively. 'First thing..this gets a good scrub.' 1964 G. L. Cohen What's Wrong with Hospitals? vii. 125 The surgeon..spared me five minutes between a thyroidectomy and his next scrub-up. 1973 Daily Tel. 27 July 3/5 The theatre suite, consisting of the operating room, the anaesthetic room, the scrub-up room and the doctors' rest room.

b. Movement of part of a tyre over the road surface while in contact with it.

1936 Proc. Inst. Automobile Engineers XXX. 733 Features in independent springing..tending to prevent 'scrub' when the suspension is functioning. 1959 Manch. Guardian 27 July 2/3 Braking and acceleration also cause scrub. 1973 Country Life 11 Oct. 1077/3 Hard cornering produces front-tyre scrub, but the back wheels stay firmly on the road.

c. slang. A cancellation or abandonment, spec. of a flying mission. Cf. *scrub v.^{1} 3* a.

1952 M. Tripp Faith is Windsock i. 14 We are marking time at the moment, three scrubs in a row. 1958 N.Y. Times Mag. 16 Mar. 10/2 The backstage crew is made up of engineers and technicians who work themselves to a frazzle during the long countdown..which may end not in a firing but in a series of 'holds' or a 'scrub'—cancellation. 1962 V. Grissom in Into Orbit 125, I was prepared for the scrub, and it was not long in coming.

2. Also spec. in Glass-painting, a brush used to scrape out lights in a coat of paint. Cf. *scrub v.^{1} 4 b.

1896 H. Holiday Stained Glass as Art i. 23 The lights are taken out..with a hoghair brush with the hairs cut short, called a scrub. 1902 E. R. Suffling Treatise on Art of Glass Painting v. 89 Hog-hair fitches are converted into what glass painters call 'scrubs'... Scrubs are made in a variety of shapes—skew, round, flat, square, pointed. 1972 R. & G. Metcalf Making Stained Glass 134 Employing some of the longer-haired scrubs to stipple the edges of the remaining matt.

5. b. (Transfer this entry to *scrub a. 4 a).

scrub, a. Add: 1. (Later U.S. examples.)

1840 J. P. Kennedy Quodlibet 158 If he..makes a little fortune, we can call him a..Scrub Aristocrat. 1881 Harper's Mag. June 88/2 Her little scrub-class in the Sunday-school. 1901 M. E. Ryan That Girl Montana ix. 125 There are always a lot of scrub whites ready to take advantage of war signals.

3. Chiefly U.S. a. Of vegetation: low-growing, stunted.

Not clearly distinguishable from Combs. s.v. Scrub sb.^{1} 6 a.

1749 [see scrub-tree s.v. Scrub sb.^{1} 6 a]. 1779 W. McKendry Jrnl. 29 Aug. in Proc. Mass. Hist. Soc. 2nd Ser. III. 465 Their breastwork was made of pine Logs coverd with green skrub bushes. 1816 U. Brown Jrnl. 9 June in Maryland Hist. Mag. (1915) X. 266 Pines of a scruby kind, Jack Oaks and other Scrub wood. 1872 Rep. Vermont Board Agric. I. 78 Men are as choice of a little scrub apple tree..as they would be were it classed among the favorite varieties of the day. 1904 G. Stratton-Porter Freckles ix. 196 There was a swarm of wild bees settled on a scrub-thorn only a few yards away. 1975 A. Bergman Hollywood & Le Vine (1976) viii. 104 A residential street that ended in sand and scrub bush.

b. Of livestock: of inferior breed or physique.

1744 W. Ellis Mod. Husbandman Apr. xviii. 118 A petty Dealer,..keeping a scrub Horse, for carrying Fish about the Country. 1839 Jrnl. Indiana Ho. Representatives 8 Jan. 232 The half-blooded calves of the improved Durhams will sell, at weaning, for $20, while those of our scrub breed will only bring 3. 1868 Rep. Iowa State Agric. Soc. 1867 130 The general idea pervades the minds of our farmers that a larger..animal of blooded stock can be produced by the same amount of feed, than can be made with the same feed fed to scrub stock. Ibid., Our stock is scrub. 1930 C. Addison in Hansard Commons 30 Oct. 269 Imported Irish stock is of an enormously higher standard than it was a few years ago. That is due to the fact that they have eliminated the 'scrub' bull. 1948 Minneapolis Morn. Tribune 28 Sept. 11/5 She couldn't resist givin' him a Home, even though she had to admit that he was a very ugly lookin' scrub cat. 1973 B. Broadfoot Ten Lost Years i. 7, I slept in the barn with about six of the sorriest looking scrub horses you have ever seen. Broom tails.

4. a. (Transferred from Scrub sb.^{2} 5 b; cf. *scrub sb.^{1} 5 c.) scrub-nine (earlier example); scrub-race, also fig.; (earlier and later examples).

1868 N.Y. Herald 11 Aug. 9/4 A club..presenting on the field the appearance of a 'scrub nine'. 1791 Address of Lad who carries Connecticut Courant (broadside), Did not our pious father S—n Run a scrub race with Mr. Chairman? 1804 Fredericktown (Maryland) Herald 10 Mar. 3/3 His antagonists seem sanguine enough for any bet, that he is either to be distanced, or will make but a scrub race for the amusement of the Gentlemen of the turf. 1894 Outing XXIV. 145/1 In a scrub race the helmsman cracks on until the lee gunwale is almost on a level with the water. 1947 C. Price Trails I Rode 190 He had put in most of his life travelling around the country with some kind of an old scrub race horse.

b. Hence in general sporting use. Of a team or player: not first-class, not of regular standing; of a game: played by scrub or scratch teams. Also fig.

1867 Ball Player's Chron. 7 Nov. 1/1 A scrub match was arranged with seven of the Star nine and two others against ten in the field. 1892 J. L. Ford Dr. Dodd's School i. 5 The school eleven..were playing a practice game of football with a scrub eleven enrolled for the occasion. 1920 W. Camp Football without Coach 63 You will have your regular center playing against a scrub center. 1947 Chicago Tribune 29 Jan. 29/2 Perhaps football could be cleaned up if it had more scrub teams. 1951 Daily Progress (Charlottesville, Va.) 19 Mar. 1/3 The Chinese have employed in the past second-rate troops in the front line. Behind them are superior troops ready to take advantage of any breakthrough made by the scrub team.

scrub, v.^{1} Add: 3. c. absol. or intr.

1870 J. P. Smith Widow Goldsmith's Daughter xxvi. 416 She began to clean the boots..while she whistled a jig and scrubbed for dear life. 1895 'G. Mortimer' Like Stars that Fall ii. 22, I must scrub and clean for you the rest of my life.

d. intr. for refl. To wash (usually with a brush) and disinfect the hands and forearms prior to performing or assisting at a surgical operation. Usu. with up.

1900 Gould & Warren Internat. Text-bk. Surg. I. xi. 283 While scrubbing, it is best to keep the hands and arms immersed in hot water, and particular attention should be given to the finger-nails. 1919 E. W. H. Groves Surg. Operations i. 10 The sister scrubs up, covers herself in sterile gown, cap, gloves, and mask. 1944 Brit. Jrnl. Surgery XXXII. 25/2 The insertion was carried out by a member of the theatre staff who had 'scrubbed up' and donned the usual cap, mask, gown, and gloves. 1966 I. Jefferies House-Surgeon x. 185, I scrubbed, with Bernard alongside me asking questions about operative technique. 1976 Lancet 25 Dec. 1402/2 His asepsis was extraordinary. When he had spent the requisite ten minutes scrubbing up, he would dip his hands routinely into three successive bowls containing fluids of different colours.

e. intr. Of a horse-rider: to rub the arms and legs urgently upon a horse's neck and flanks to urge the horse to move faster.

1958 J. Hislop From Start to Finish viii. 68 If you are riding a long-striding horse, you will find that you must scrub more slowly than on a short-striding horse. 1961 F. C. Avis Sportsman's Gloss. 236/2 Scrub, of a jockey, to move the arms and legs, particularly as the end of a race is approaching. 1977 Horse & Hound 14 Jan. 21/1 By now the field was spreadeagled and scrubbing to keep in touch with the hounds.

3*. a. trans. To cancel, scrap, call off; to eliminate, erase; to reject, dismiss. Also with out. colloq.

The current widespread use was reinforced by the popularity of the expression amongst servicemen in the war of 1939–45 (see quot. 1945).

1828 W. Scott Jrnl. 22 Mar. (1941) 212 If I were alone, I could scrub it [sc. a visit to London], but there is no doing that with Anne. 1943 H. E. Bates There's Something in Air 77 He was worked up to a very high state of tension.. when Control informed him that the whole show would be scrubbed. 1944 Yank 30 June 8 At 1400 hours there was a briefing; at 1500 the mission was scrubbed. 1945 Spectator 25 May 478/1 The author can possibly justify the inclusion of the term 'scrub', meaning 'to cancel', in a collection of R.A.F. slang. The expression is in common use in the Royal Navy and has been for many generations. It derives from the days when all signals and orders were written on a slate. When the signals were cancelled or orders executed, the words on the slate were 'scrubbed out' or, equally correctly 'washed out'. 1953 Sun (Baltimore) 4 May 2/2 What do you mean that my mission is scrubbed? It's my mission and no one scrubs my mission but me. 1958 'J. Brogan' Cummings Report xix. 202 He might have told the operator to scrub it from the record. 1962 Listener 8 Feb. 247/2 At the end of the war some people realized that the best thing to do would be to scrub it [sc. the national debt] out. 1965 J. Porter Dover Three xiii. 148 She doesn't sound the type of woman we're looking for. Scrub her! 1974 'P. B. Yuill' Bornless Keeper ii. 17, I suggest scrubbing that thing on the Kent miners, can't see any foreign sales there. 1980 News & Observer (Raleigh, N. Carolina) 28 Oct. 10/1 Metropolitan Opera House musicians voted Monday to accept a new contract with the opera company, ending—at least temporarily—a strike that forced the Met. to scrub the 1980 season.

b. intr. To manage with difficulty, to 'scrape' along. Also with on. colloq.

1831 M. Edgeworth Let. 29 Mar. (1971) 507 He..has run through two large fortunes and is now scrubbing on upon a few thousands. 1889 W. Davidson Stories N.Z. Life ii. 48 Dennis O'Brien had scrubbed along for many years, a miserable kind of existence, saving and hoarding, and living on the 'smell of an oil rag'. 1901 Merwin & Webster Calumet 'K' xi. 202 The rest of the road had to scrub along as best it could. 1905 G. Bell Let. 17 Apr. (1927) I. x. 212, I hope in a week or so I shall begin to scrub along.

c. trans. To reprimand severely; to punish Mil. slang (chiefly Naut.).

1911 [see *matelot 1]. 1916 'Taffrail' Pincher Martin v. 76, I..jolly nearly got badly scrubbed for exceeding my duty and abducting the General. 1949 J. R. Cole It was so Late 62 That was my first station after they scrubbed me.

d. intr. Const. round. To dispense with, ignore; to drop (a subject). (See also quot. 1943.) slang (orig. Mil.). Cf. sense 3* a above.

1943 Hunt & Pringle Service Slang 58 Scrub round, to wash off the slate, to agree to forget, to let bygones be bygones. 1948 Partridge Dict. Forces' Slang 163 One declines an invitation to a party with 'Thanks very much, but you'll have to scrub round me, I'm Duty Boy to-morrow'. 1962 J. Wain Strike Father Dead v. 222 'I just said I didn't want to break the contract we had at present,' I said. 'I felt it was no good trying to scrub round it.' 1964 T. White tr. P. Leullietté's St. Michael & Dragon 189, I was required to do no less than fifteen days' cells. Reason: disobedience. Luckily, the captain had a sense of humour and finally scrubbed round it.

4. a. Substitute for def.: To treat (a material, esp. a gas or vapour) so as to remove impurities, usu. by bringing it into contact with a liquid; to wash out or remove (impurities) in such a way. (Further examples.)

1931 Hoffert & Claxton Motor Benzole viii. 211 In this type of washer, the gas is scrubbed by the oil in the form of a fine spray in six or more superimposed sections or chambers, through which the gas ascends in turn. 1941 Thorpe's Dict. Appl. Chem. (ed. 4) V. 461/1 The [coal] gas is cooled by passage through condensers before scrubbing out the ammonia. 1961 G. Claxton Benzoles viii. 213 The greater proportion of the benzole is scrubbed out of the gas by means of wash oil. 1972 Sci. Amer. Oct. 26/1 Cooled and scrubbed with water to remove dust, the clean gas could be burned itself to provide the desired clean heat. 1974 Daily Tel. 22 Feb. 7/6 The diver inside the suit operates at ordinary surface pressures, and breathes oxygen which is continuously scrubbed and recirculated. 1979 Sci. Amer. Oct. 74/2 Consideration is being given to adding nitrogen-removing devices to the procedures that now 'scrub' sulfur dioxide..from stack gases.

b. Glass-painting. To scrape away (paint) or to scrape out (lights) with a scrub. Cf. *scrub sb.^{2} 2.

1897 L. F. Day Windows vi. 65 The practice in the sixteenth century was mainly, by a process of scrubbing lights out of matted or washed tints of brown, to get very considerable modelling. 1910 Encycl. Brit. XII. 106/2 The modelling was got by scrubbing away the paint with a dry hog-hair brush.

5. scrub-man, -water, -woman (earlier and later examples); (sense *3 d) scrub nurse, room.

1905 Cleveland (Ohio) Plain Dealer 24 Jan. 3 Once a soldier in the army of the great white czar, now a scrubman in one of the large department stores. 1927 Amer. Speech II. 312/2 The 'scrub nurse' is she who handles the instruments and works within the sterile field, differing from the 'dirty nurse' who may touch only contaminated or unsterilized things. 1958 F. G. Slaughter Daybreak i. vii. 54 He stepped out of the hard white cone of the operating lights in response to a scrub nurse's signal. 1972 M. Crichton Terminal Man ii. i. 57 Two scrub nurses were working in the cavernous gray-tiled space. They were setting out sterile tables and drapes. 1927 Amer. Speech II. 312/1 In the Operating Room one finds 'scrub rooms' where the surgeon and his assistants literally scrub their arms and hands with brushes and green soap. 1977 D. Bennett Jigsaw Man 13 The surgeon..strode briskly to the scrub-room... The scrub-nurse removed his mask and gloves. 1935 Z. N. Hurston Mules & Men 336 It is put in scrub water to scrub the house. 1975 New Yorker 28 July 31/2 He catches her scent of gray scrubwater as she passes. 1873 N.Y. Herald 16 Sept. 8/5 We have a specimen of this watch-dog policy in the case of a poor scrub woman. 1942 E. Paul Narrow St. i. 2 Eugénie, a pale, brown-eyed scrubwoman not yet forty. 1973 E. McGirr Bardel's Murder i. 6 A scrubwoman did what was necessary in the cleaning line. 1980 G. M. Fraser Mr American xxii. 428 There's one way of treating a suffragette who's a scrubwoman..and another of treating a peer's daughter.

scrubbable (skrʌ·băb'l), a. [f. Scrub v.^{1} + -able.] That may be scrubbed without damage or injury; capable of being cleaned by scrubbing.

1923 Daily Mail 29 May 10 Matone's 'scrubbable' matt finish is fadeless. 1960 House & Garden Oct. 150/3 Wallpaper..guaranteed scrubbable for seven years. 1976 Woman's Day (U.S.) Nov. 123/2 Vinyl wall covering is scrubbable.

scrubbed (skrʌbd), ppl. a. [f. Scrub v.^{1} + -ed^1.] Cleaned by scrubbing. Also fig.

1870 J. P. Smith Widow Goldsmith's Daughter ii. 19 Chandos and Christabel were presentable children: both.. were kept scrubbed and combed, and 'cleaned up' within an inch of their lives. 1905 E. M. Albanesi Brown Eyes of Mary xvi. 213 The sight of the old kitchen, with its scrubbed boards and red tiles. 1958 Listener 18 Dec. 1050/2 'People Today' gave us a fresh, scrubbed and shiny picture of Hawick.

scrubber^1. Add: 2. b. More widely, any apparatus or installation for scrubbing (Scrub v.^{1} 4 in Dict. and Suppl.). (Further examples.)

1948 Petroleum Handbk. (ed. 3) xvi. 231 The crude product is then neutralized in a caustic scrubber and fractionated to remove light gases. 1974 L. Deighton Spy Story xv. 154 The CO_2 scrubber that cleans the air in an atomic submarine before recirculating it. 1975 N.Y. Times 14 Apr. 49/2 The air standards..require utilities..to

install costly pollution-control systems called scrubbers to clean emissions from coal with a high sulphur content. 1977 Lancet 9 July 76/2 The gas containing benzene was.. passed through a wet scrubber, and channelled into an activated-charcoal absorber unit, where benzene was recovered and recycled.

scrubber^2. Add: 1. Austral. and N.Z. a. (Further examples.)

1869 in Occas. Papers Univ. Sydney Austral. Lang. Res. Centre (1980) No. 17. 55 'Were there any scrubbers—croppies—out here thin?'..'Four—murra [= very] wicked fellows!' 1897 D. McK. Wright in Chapman & Bennett Anthol. N.Z. Verse (1956) 42 New fences climb the warm brown spurs to guard the scrubber ewes. 1966 Sunday Mail Mag. (Brisbane) 9 Oct. 4/2 At the start of every cattle movement the 'scrubbers' (delinquent cattle) are noted, and are disciplined fast. 1978 O. White Silent Reach v. 54 Scrubber bulls have been turned into stud paddocks.

b. fig. An ill-bred or degenerate animal; an ill-favoured, despicable person.

1876 D. Kennedy Colonial Travel xviii. 249 We four adventurers..each..mounted on the shaggiest of small 'scrubbers', with a pannikin and a coil of rope dangling.. at his saddle-bow. 1941 Baker Dict. Austral. Slang 64 Scrubbers, cattle or horses that have run wild in the scrub and have deteriorated in condition. (2) Any weedy or unpleasant person. 1966 —— Austral. Lang. (ed. 2) iii. 66 Scrubber, a poor-looking, ill-bred horse. 1966 G. W. Turner Eng. Lang. Austral. & N.Z. iii. 54 Scrubbers, 'cattle that have run wild and deteriorated in condition', suggested figurative uses. It may mean urchin: I remember as a child hearing the phrase 'dirty little scrubber', and, not knowing the bush sense of the word, thinking it illogical.

c. The grey kangaroo, Macropus giganteus.

1968 K. Weatherly Roo Shooter 137 Scrubber is the name that shooters give to the grey kangaroo. 1977 J. L. Harper Population Biol. Plants xiii. 422 The grey kangaroo or scrubber..and..the red kangaroo..are regarded as pests by some sheep farmers.

3. slang. [Perh. properly related to Scrubber^1.] A prostitute, a tart (see also quot. 1965); an untidy, slatternly girl or woman.

1959 Encounter May 30 'The scrubbers': very young girls who follow jazz bands round the country. 1962 R. Cook Crust on its Uppers ii. 29 This aged scrubber, Mrs. Marengo ..she was so old, forty. 1965 G. Melly Owning Up xiv. 172 The word 'scrubber' has cropped up quite frequently in this story, and perhaps the time has come to attempt a precise definition of what it means, or rather meant, for I understand that in the beat world it has become debased and now means a prostitute. In our day this was not the case. A scrubber was a girl who slept with a jazzman but for her own satisfaction as much as his. 1968 J. Mitchell Undiscovered Country i. 134 'She's only a scrubber.' 'A what?' 'It's the new word for "short-term sexual partner".' 1970 G. Greer Female Eunuch 264 The most recent case in which contempt for menial labour has devised a new term of abuse for women is the usage of scrubber for a girl of easy virtue. 1973 B. Mather Snowline viii. 95 'She looked a scrubber. That means—' 'A mare that runs wild in the scrub country, copulating indiscriminately with stray stallions. Derivation Australian, but also applied to women of similar propensities in other parts of English-speaking world.' 1974 H. J. Parker View from Boys 213 Scrubber, used instead of 'tart' which has a non-derogatory meaning. 'A right scrubber' is a girl who's rough-looking, whore-like.

scrubbily (skrʌ·bili), adv. rare. [f. Scrubby a.^{1} + -ly^2.] In a mean or paltry manner; shabbily.

1891 G. Gissing New Grub Street I. vi. 142 'By-the-by, how has The Study been in the habit of treating you?' 'Scrubbily.' 'I'll make an opportunity of talking about your books to Fadge.'

scrubbing, vbl. sb. Add: 2. d. Surg. The action of *scrub v.^{1} 3 d. Also with up.

1898 Wharton & Curtis Pract. of Surg. viii. 156 This scrubbing should be employed for several minutes; the hands are then rinsed to remove the soap, and are soaked for two minutes in a 1 to 1000 bichloride of mercury solution. 1910 H. A. Haubold Preparatory & After Treatm. in Operative Cases vi. 127 The object of the scrubbing is not to destroy the bacteria, but to remove them. 1937 'J. Bell' Murder in Hospital vi. 109 The 'scrubbing up' process occupied about ten minutes. 1969 B. Weil Dossier IX xviii. 142 The space between the sides of the theatre and the walls had been fitted out for scrubbing up with washbasins. 1976 J. Archer Not Penny More, not Penny Less xii. 140 'Jean-Pierre, you scrub up as instructed.' ..Jean-Pierre appeared from the scrubbing-up room. 1978 Roaf & Hookinson Basic Surg. Care ix. 141 All preparation must be made beforehand to ensure that it is unnecessary to touch any unsterile surface once scrubbing is complete.

4. The action of Scrub v.^{1} 4 in Dict. and Suppl.; removal of impurities from a material (usu. a gas or vapour).

1896 B. Redwood Treat. Petroleum II. 424 The scrubbing process consists in passing the gas through a series of coke towers in which it is exposed to streams of oil and water, the volatile hydrocarbons being removed by the oil, and the ammonia by the water. 1921 W. H. Fulweiler in A. Rogers Industr. Chem. (ed. 3) xx. 499 The gas is forced to pass in contact and bubble through the scrubbing liquid by a series of partitions arranged across the flow of gas. 1959 Engineering 23 Jan. 99/3, I am insufficiently informed of the details of gas scrubbing practice to say whether the systems at present in use are satisfactory. 1976 Offshore Engineer Apr. 61/1 There is continuous scrubbing of the air (many other subs have manual bleed-in and scrubbing) with direct readouts of O^2 and CO^2 levels as well as silica gel drying agents to remove moisture. 1976 Physics Bull. Apr. 160/2 A reduction of η to about zero or even slightly negative can be achieved by ion 'scrubbing' of the vacuum chamber surfaces in a high pressure argon glow discharge.

5. = *scrub sb.^{2} 1 b.

1936 Proc. Inst. Automobile Engineers XXX. 739 Tyre wear due to lateral reaction, where location is high, offsets that which occurs in independently sprung cars due to 'scrubbing' when cornering. 1959 Manch. Guardian 27 July 2/3 Wear is the result of relative movement between tyre and road—'scrubbing' in other words. 1979 Truck & Bus Transportation Sept. 72/1 A third axle was placed between the front and rear axles to act as a turning pivot for the whole trailer. The result—no more scrubbing.

6. Comb., as scrubbing-board.

1889 H. S. Edwards in Century XXXVIII. 84/1 Her great black, muscular arms drooped towards the scrubbing-board that reclined in the tub. 1969 E. H. Pinto Treen 155 It is believed that ribbed wooden scrubbing boards.. originated in Scandinavia and the manufacture spread to other countries during the 19th century.

scrubbing, ppl. a.^{2} [f. Scrub v.^{1} + -ing^2.] That scrubs (a floor, wood, etc.).

1868 M. H. Smith Sunshine & Shadow in N.Y. 362 The crowd is composed of the millionaire and the hod-carrier..madame flashing jewels, and the scrubbing-woman who cleans paint and washes linen. 1936 M. de la Roche Whiteoak Harvest xvi. 198 Two scrubbing women were sent to prepare the house for Sarah.

scrubble (skrʌ·b'l), v. [App. var. Scrabble v.] = Scrabble v.

c 1854 Mrs. Gaskell Lett. (1966) 274 It will..cost two guineas,..and so I must scrubble up money for that. 1920 D. H. Lawrence Lost Girl vi. 105 Country..now scrubbled all over with mining villages. 1927 J. Elder Thomasina Toddy i. 11 The blanket..comes up and scrubbles on your face, all rough and horrid. 1957 H. Nicolson Let. 26 Dec. (1968) 342 They played God Save the Queen, and all the old English boys scrubbled up on their crutches.

scrubby, a.^{1} Add: 2. (Later examples.) Also, consisting of or in the form of scrub.

1901 M. Franklin My Brilliant Career iii. 14 The school was situated on a wild scrubby hill. 1936 D. McCowan Animals Canad. Rockies xvi. 144 It finally reached a haven in the scrubby thicket. 1957 M. Spark Comforters ix. 233 He saw the bits of paper come to rest, some on the scrubby ground, some among the deep marsh weeds. 1971 Sci. Amer. Sept. 130/1 It takes less energy to support a pound of biomass in a mature tropical rain forest than it does in the grassy or scrubby forest stages that precede maturity. 1979 D. Kyle Green River High vi. 82 We were edging on to scrubby land, patches of low, tangling heather.

3. (Earlier and later examples.)

1754 J. Sackville Let. 4 Sept. in 16th Rep. R. Comm Hist. Manuscripts: Rep. MSS. Mrs. Stopford-Sackville I. 40 in Parl. Papers 1904 (Cd. 1892) 1 He still continues to persecute me, and acts in regard to me in a most scrubby manner. 1913 [see *honest a. 4 d]. 1967 Southerly XXVII. 75 It would work out, as it always did, under a cover of scrubby banality. 1975 New Yorker 5 May 18/3 (Advt.), A great deal of talent has been badly used, though James Caan has some good scenes as scrubby, anxious Billy Rose.

scruff, sb.^{1} Add: 4. c. A scruffy person, an oaf, a layabout; a contemptible or inferior person, someone of no breeding. Also collect., scum, riff-raff.

1836 W. Carleton Traits & Stories of Irish Peasantry (ed. 4) II. 342 Oh, you scruff of the earth. 1896 G. F. Northall Warwickshire Word-Book 203 Scruff, a wastrel, raffish rogue. 1905 J. Wright Eng. Dial. Dict. V. 290/1 The village is well enough but for the scruff that comes in. 1958 People 4 May 12/4 'A nice class of fellow, too,' he said. 'Not one of the scruff.' 1960 D. Storey This Sporting Life i. v. 116 Every scruff in town's crept in. I don't like it. 1968 J. Lock Lady Policeman iii. 22 The 'scruff' might merely be an arty or beat type being deliberately scruffy and the very young face might belong to a grown woman. 1973 'H. Carmichael' Too Late for Tears x. 121 Nature gives some of us the wrong heredity... So you're a scruff and John Piper's a gentleman. 1977 Listener 20 Jan. 72/2 Writing books or theatre plays is the only proper pursuit for a literary lady or gent, and..people who write for the new forms are money-grubbing scruffs who have sold their souls to the ghastly mass media.

scruff, v.^{1} 1. Delete † Obs., for 'Sc. and north.' read 'orig. Sc. and north.' and add later lit. examples. Also spec., (a) in Golf, to graze (the turf) when striking the ball; (b) in Painting, to stroke (oil colour) lightly over a rough surface. Also absol. or intr. Hence scru·ffing vbl. sb.

Examples in Sc. use with various spellings (as screef, scrief, scriff, etc.) may be found in S.N.D. s.v. scruif, n., v.

1857 H. B. Farnie Golfer's Man. (1947) 74 Scruff, slightly razing the grass in striking. 1862 St. Andrews Gaz. 3 Oct. 3/3 The boy was sleeping on the cart and fell down, and..his head was a little 'scruffed' on the wheel. 1876 C. Clough Robinson Gloss. Dial. Mid-Yorks. 118/2 One will be told to get a besom and scruff the snow off the doorstone. 1896 P. A. Graham Red Scaur 347, I felt it scruff his chafts. 1920 C. Sandburg Smoke & Steel 25 Wearing leather shoes scruffed with fire. 1926 Amer. Speech I. 633/1 Scruff, to graze the grass with the club in striking. 1950 Chambers's Encycl. X. 310/2 The 'scruffing' trick: the brush laden with dry (light) paint is rapidly dragged over a darker tone. 1961 M. Levy Studio Dict. of Art Terms 100 Scruffing, an application of paint which skims the surface of a painting area, but does not take in the depressions of the panel or canvas texture. 1970 Oxf. Compan. Art 1055/1 A scumble must..be applied irregularly—dragged or scruffed—in such a way that small areas of the under colour show through.

2. [As a back-formation from Scruffy a.] to scruff oneself up: to make oneself scruffy.

1970 G. Lord Marshmallow Pie ii. 19 Scruff yourself up a bit over the weekend. You know, sweat a bit and that. 1970 Guardian 6 Apr. 9/4 The hardest thing..was to make those girls look really scruffy... You don't look like a plumber's wife, we kept saying... She didn't scruff herself up enough..for what the part demanded.

scruff, v.^{2} Add: b. To seize and hold (a calf) while it is being branded or castrated. Austral.

1881 A. C. Grant Bush-Life in Queensland I. xvi. 228 The smaller calves are scruffed. 1909 Chambers's Jrnl. Dec. 810/2 After the calves are separated from their mothers..the former are one by one 'scruffed'—that is, seized by a couple of men and held down while knife and branding-iron are applied. 1931 F. D. Davison Man-Shy (1934) v. 71 It was the red heifer's turn.. 'Get the ropes. She's too big to scruff,' he said.

c. To push roughly; manhandle. rare.

1926 J. Black You can't Win ix. 121 After they got done scruffing me around, two of them took me by each arm. 1941 Baker Dict. Austral. Slang 64 Scruff, to, to attack, manhandle a person.

scruffo (skrʌ·fo). slang. = *scruff sb.^{1} 4 c.

1959 C. MacInnes Absolute Beginners 183 One of the scruffos turned and looked at his choice companions. 1976 Tel. (Brisbane) 10 June 40/1 They are not scruffoes, layabouts, or dole cheats.

scruffy, a. Add: b. Shabby, mean, dirty; slovenly, messy, untidy. Also Comb., as scruffy-looking adj.

1871 'Mark Twain' Screamers ii. 16 When he'd got the blues, and feel kind o' scruffy, aggravated, and disgusted.. he would curl up..and go to sleep. 1925 Fraser & Gibbons Soldier & Sailor Words 253 Scruffy, dirty: slovenly: untidy in appearance. 1931 Star 8 May 6/3 Anyone who has travelled through lottery countries and seen the hundreds of scruffy ticket-shops in the cities. 1935 Punch 5 June 656/1 'Mine,' said the scruffy-looking chap who had started by borrowing a match, 'is a tragedy of jealousy.' 1940 Blunden Poems 1930–40 204 While I leisured it so, from the verge of the street Those scruffy old weeds in a flash had me beat. 1951 Auden Nones (1952) 29 A rather scruffy-looking god Descends in a machine. 1958 Times Lit. Suppl. 16 May 274/1 Always late, crumpled and scruffy, perpetually in debt, hourly expecting the sack, Greare takes refuge..in Mittyesque fantasies. 1967 A. N. Sherwin-White Racial Prejudice in Imperial Rome i. 4 It is because they..live..in a scruffy fashion, following the impulses and necessities of beasts. 1974 N. Freeling Dressing of Diamond 122 His chin was badly shaved: it gave him a dirty look, and sort of scruffy.

Hence scru·ffily adv.; scru·ffiness.

1974 Times 5 Oct. 13 That general 'scruffiness' could easily be rectified. 1977 Listener 4 Aug. 145/2 Making herself look scruffily bizarre is a time-consuming business.

scrum, sb. Add: 1. (Later examples.) Also ellipt. for scrum-half.

1921 [see *fly sb.^{1} 1 e]. 1930 R. Campbell Poems 11 See the fat nouns like porky forwards sprawl Into a scrum that never heels the ball. 1978 Church Times 23 Mar. 7/3 He pictured that duty as being like a forward's in a rugger scrum; he must put his head down and shove blindly.

2. transf. A confused, noisy throng (at a social function or the like).

1950 J. Cannan Murder Included ii. 23, I kept wondering where you were..in that awful scrum. 1959 P. Moyes Dead Men don't Ski i. 11 The handsome, fair-haired young man emerged from the scrum at the bar. 1965 P. O'Donnell Modesty Blaise iii. 32 She looked towards the manœuvring scrum at the bar. 1976 Eastern Daily Press (Norwich) 19 Nov. 1/4 Cindy, as the new Miss World likes to be called, was surrounded by the traditional scrum of over 100 press photographers. 1979 Globe & Mail (Toronto) 4 May 9/5 But he warned reporters after the second scrum yesterday, 'We're going to have to stop having these impromptu press conferences.'

3. Comb., as scrum-cap, a cap worn to protect the head in a scrum; scrum-half, the half-back who puts the ball into the scrum; also, by extension, the scrum-half's position in a team.

1917 Harrods General Catal. 449/4 Scrum Caps. All Crochet Work, 3/3 each; all netting, 2/0 each; and Padded Ear Caps, 2/6. 1933 C. Day Lewis Magnetic Mountain in Coll. Poems (1935) 118 But will it suffice To wear a scrum-cap against falling skies? 1976 Field 18 Nov. 986/2 The headgear is authentic—more sensible than the top-hats once worn, less so than the scrum-caps which some favour now. 1906 Gallagher & Stead Compl. Rugby Footballer v. 69 Wallace played in every position except that of scrum half. 1922 Somerset County Herald 11 Feb. 4/3 As scrum-half [he] manfully overcame his disadvantage in weight. 1951 Sport 30 Mar.–5 Apr. 6/3 Another surprise 'cap' is that of Ike Proctor at scrum-half. 1978 Rugby World Apr. 3/2 The most exciting confrontation that day will be the scrum-half battle between Gareth Edwards and Jérôme Gallion.

scrum (skrʌm), v. [f. the sb.] intr. To jostle, crowd.

1925 A. S. M. Hutchinson One Increasing Purpose i. xxv. 153 The trouble with me is..feeding and frivoling at night and weekends where the masters live and where we scrum at shows. 1938 P. Lawlor House of Templemore ix. 98 Young calves 'scrumming' to dip their heads in the

long troughs of milk. 1939 G. Greene Confidential Agent i. i. 3 A rugger team was returning home and they scrummed boisterously for their glasses. 1948 C. Day Lewis Otterbury Incident iv. 49 Everyone was scrumming around behind him.

scrum (skrʌm), a. School and College slang abbrev. of Scrumptious a. ? Obs.

1895 W. C. Gore in Inlander Nov. 65 Scrum, prob. from scrumptious, with which it is synonymous. 1913 J. Vaizey College Girl xviii. 250 'Good cakes?' 'Scrum!'

scrumble (skrʌ·mb'l), v.^{1} rare. [Perh. a blend of Scrape v. or Scratch v. + Crumble v.] trans. To scrape or scratch out of or from (something).

The two examples perhaps represent independent nonce-formations.

1906 W. B. Yeats King's Threshold in Poems, 1899–1905 223 I'll scrumble the ermine out of his skin! 1975 P. Lively Going Back iv. 43 We scrumble the soft innards from the loaf and hold it high above us and now it snows bread upon the snow.

scrumble (skrʌ·mb'l), v.^{2} [App. alteration of Scumble v.] trans. To produce a smeary or grainy effect on (paint). Hence scru·mbled ppl. a.

1921 Spectator 9 Apr. 454/2 The paint has been scrumbled, i.e. if you look into it, it does not present a flat surface, it shows variations something like the grain of wood. 1937 Sunday Times 17 Jan. 30/3 Dining room designed in the Tudor style with scrumbled walls and beamed ceilings. 1959 Spectator 8 May 652/2 In his later large decorations.. the light colours, the scrumbled paint and the botanist's eye meet in something very far from mythology on one hand, and neurosis on the other.

scrummy (skrʌ·mi), a. colloq. [f. Scrum(ptious a. + -y^1.] Excellent, fine, 'smashing'; enjoyable, delicious.

1915 Mrs. H. Ward Eltham House i. 14 You've got to change and rest..before dinner!.. You've got to put on a scrummy frock too! 1918 Galsworthy Five Tales 122 He's promised to take mother and me to the theatre and supper afterwards. Won't it be scrummy! 1923 'R. Crompton' William Again viii. 147 The cakes had been scrummy. 1928 —— William—the Good viii. 220 'Does it [sc. the trap-door] go out on to the roof?' called the Outlaws... 'Yes, it does. It's scrummy. Right on the edge of the roof.' 1960 News Chron. 8 July 8/5 Out spring the five white tubers of a corpse's hand. Scrummy! 1977 Harper's & Queen Nov. 52/2 Scrummy French food in cosy surroundings.

scrump, sb. Add: Also spec., a withered or stunted apple.

1887 Parish & Shaw Dict. Kentish Dial. 138 Scrump, a stunted, badly-grown apple. 1896 G. F. Northall Warwickshire Word-Bk. 203 Scrumps, sb. pl. Apples.

scrump (skrʌmp), v. dial. or slang. [f. the sb.] trans. To steal (apples), esp. from orchards. Also transf. and absol.

1866 R. Hallam Wadsley Jack iv. 17 Dick Greasy.. ax'd me if I'd mak' one to goa a scrumpim', that is, fetchin' apples off sumboddy's trees. 1931 'G. Orwell' Hop-Picking in Coll. Essays (1968) I. 71 Scrump, to, to steal. 1945 B. Naughton in C. Madge Pilot Papers I. 101 They'd come scrumping..in Woolworths, or over at the market, or from the street stalls. 1951 A. Baron Rosie Hogarth iii. 137 Scrumping apples, remember?.. God didn't put 'em on that tree for Mr. Moggeridge. 1966 M. Torrie Heavy as Lead xiv. 169 His lordship was going scrumping... You know. Kids scrump apples, Sir G. scrumped rock garden plants. 1972 K. Bonfiglioli Don't point that Thing at Me i. 3 English policemen..dare not even spank the bottoms of little boys caught scrumping apples nowadays.

Hence scrumped ppl. a.; scru·mper, one who scrumps; scru·mping vbl. sb.

1946 Scrutiny XIII. 293 There is to be, after all, no development in Fant's moral powers: only, as it were, a schoolboy scrumping of all the redeeming fruits of experience. 1957 Times 20 Aug. 5/1 Scrumping is an offence which perhaps no one in this court-room has not committed. 1969 M. Wiggin Cottage Idyll ii. 26 One scoundrel sent me a..message to the effect that he would like his cherished scrumping stick back. 1973 D. Orgill Jasius Pursuit i. 9 If the police had to jail every hippy fruit-scrumper, the prisons would soon be overcrowded. 1976 A. Hill Summer's End vii. 98 Scrumped apples always taste better than bought ones. 1981 Daily Tel. 3 Mar. 3/1 (heading) Rector used stick on girl scrumper.

scrumple, v. Delete 'Obs. exc. dial.' and add later examples. Also freq. with up.

1939 A. Ransome Secret Water xxii. 261 She jumped up, grabbed the message, scrumpled it up and poked it into the stove. 1954 M. Procter Hell is a City i. ii. 23 It [sc. money in notes] came out of my bag all scrumpled up anyhow, just as I'd stuffed it in. 1961 R. M. Dashwood Provincial Daughter 83, I discover shirt scrumpled up in polythene bag at bottom of wardrobe. 1971 G. Ewart Gavin Ewart Show i. 25 His secretary has a habit of scrumpling the top copies.

scrumpled, ppl. a. (Earlier example, and later example with up.)

1813 E. Weeton Jrnl. Dec. (1969) II. 115 She thrust something into my hand... It felt like a small parcel of scrumpled paper. 1947 A. Ransome Great Northern? xvii. 210 He pulled the scrumpled up paper out of his knapsack and spread it out again.

scrumptious, a. Add: 2. b. Now esp. of food: delicious. So scru·mptiousness, the state or condition of being scrumptious.

1881 Punch 30 July 47/2 There is a certain exquisite scrumptiousness and goloptiousness about Real Turtle. 1894 Somerville & 'Ross' Real Charlotte II. xxxii. 247 The cake was scrumptious. 1922 Joyce Ulysses 740 You will always think of the lovely teas we had together scrumptious currant scones and raspberry wafers I adore. 1930 Magnet 25 Jan. 6/2 'It's lovely butterscotch—scrumptious!' 1976 A. L. Rowse Cornishman Abroad 14 The scrumptious meal she cooked, Cornish duck and Californian avocado stuffed with shrimp, our own cream from the farm with the delicious sweet.

scrumptiously (skrʌ·mpʃəsli), adv. [-ly^2.] In a scrumptious manner, excellently, deliciously.

1844 A. S. Stephens High Life in N.Y. I. xvi. 237 The frocks answered just as well to make bonnets out on, arter she'd dashed out in 'em once or twice, and the sleeves and waist cut up scrumptiously for ruffles and furbelows. 1927 Observer 1 May 8 He shows us 'Life' and 'High Life' simultaneously, so scrumptiously and yet with O such a touch of wanton wistful weariness! 1976 New Yorker 17 May 167/1 It is all scrumptiously pretty.

scrumpy (skrʌ·mpi). dial. or colloq. [f. Scrump sb. + -y^6.] Rough cider, made from small or unselected apples. Also attrib. in scrumpy cider.

1904 in Eng. Dial. Dict. s.v., These apples are of no good but to make scrumpy of. 1932 [see *nobble v. 3]. 1962 G. Compton Too Many Murderers xii. 87 Scrumpy was only eightpence a pint. 1973 C. Bonington Next Horizon viii. 114 A brisk five-minute walk took you to the Coronation Tap, where you could down a pint of scrumpy cider and eat home-made pies. 1977 Times Lit. Suppl. 4 Mar. 242/4 Another [pub] sold evil-smelling 'scrumpy', producing its own extensive Saturday-night network of vomit.

scrunch, v. Add: 2. c. U.S. intr. for refl. To squeeze oneself into a compact shape; to huddle up, together; to cower or crouch down.

1884 'Mark Twain' Huck. Finn i. 4 Miss Watson would say,..'Dont scrunch up like that, Huckleberry—set [sic] up straight.' 1939 J. Steinbeck Grapes of Wrath ii. 11 'Scrunch down on the running board till we get around the bend,' he said. 1951 T. Capote Grass Harp ii. 63 We scrunched together to make a place for Riley. 1972 D. Delman Sudden Death ii. 48 Wally knew something bad was about to happen to him; you could tell from the way he scrunched up, as a man will, sometimes, when a punch is telegraphed. 1974 K. Millett Flying (1975) iv. 394 In the North Terminal I scrunch into a bench and wait for Nell. 1978 J. Irving World according to Garp xv. 311 He let her lean against him, though she was slightly taller..and in order to rest her head against him, she had to scrunch down.

scrunch, var. *scronch.

scrunched, ppl. a. (and pa. pple.) Add: a. Also freq. with up.

1910 C. E. Montague Hind let Loose xi. 219 Some ebb.. would leave..the foreshore only littered with..tobacco-ashes, used matches, scrunched-up envelopes. 1963 C. D. Simak They walked like Men iii. 13 Balls of scrunched-up copy paper tossed onto the floor. 1974 T. P. Whitney tr. Solzhenitsyn's Gulag Archipelago I. ii. 504 All that was left in that scrunched-up wad the engine room of the law had spewed out into the prisoner transport was a greed for life, and no understanding whatever.

b. Hunched, huddled, cowering. Usu. with up; occas. with back, down, etc. Chiefly U.S.

a 1902 S. Butler Way of All Flesh (1903) lv. 253 He looks that worried and scrunched up at times. 1905 N.Y. Sunday World 11 June (Mag. Section) 8/2 You sit there with your shoulders scrunched up, giving an imitation of Reginald Vanderbilt driving his coach. 1931 Virginia (Louisiana) Q. Rev. Jan. 106 The twins lay upon the spare bed... 'Look at dis'n all scrunched up.' 1962 W. Faulkner Reivers viii. 158 Otis was scrunched back against the wall. 1966 N. S. Hayner New Patterns in Old Mexico ix. 167 One of the men approached the padre with his back scrunched up. 1976 National Observer (U.S.) 12 June 22/3 The wolves stayed where they were, scrunched motionless against the far wall. 1977 Rolling Stone 7 Apr. 58/1 That night Carl Sagan and I sat scrunched down in a sofa in his Pasadena apartment.

scru·nching, vbl. sb. [f. Scrunch v. + -ing^1.] The action of the verb Scrunch.

1869 S. R. Hole Bk. about Roses 263 That yellow-bellied abomination, the grub which produces the saw-fly, in this month attacks the Rose... The process of 'scrunching' is disagreeable, but it must be done. 1927 A. Clarke Son of Learning i. 9 The sloppiness of custard, the sourness of green apple, With crunching, munching, scrunching.

scrunchy (skrʌ·nʃi), a. [f. Scrunch v. + -y^1.] That scrunches; that emits a crisp, crunching sound when crushed.

? 1905 Eng. Dial. Dict. Suppl. 168/2 Scrunchy, of frozen grass: emitting a crisp, crunching sound when trodden on. 1907 W. de Morgan Alice-for-Short xxvi. 274 Of course you may have scrunchy toast if you like. 1927 Daily Express 6 Dec. 11 The delicious, scrunchy crispness of 'Ovaltine' Rusks. 1937 John o' London's Weekly 29 Jan. 722/2 The minor roads are paved with scrunchy white shells. 1974 Country Life 13 June 1580/3 The vegetable..runs to flower without first making the desired, juicy, yummy, scrunchy, foliage.

scrunty (skrʌ·nti), a. orig. Sc. and north. dial. [f. Scrunt.] Stunted, shrivelled, stumpy.

1811 A. Scott Poems 59 [A bird] wha, on his native scrunty thorn, 'Mang birds o' song bude hail the morn. 1849 in C. Brontë Shirley vii. 148 Then whudder awa' thou bitter biting blast, And sough through the scrunty tree. 1868 W. Shelley Wayside Flowers 55 He'd been sae scrimpit o' his corn His scrunty banes stood brent in sight. 1947 A. McCormick Galloway 219 The scrunty aul' buddy has nae smeff. 1951 Auden Nones (1952) 54 A scrunty beggar With one glass eye and one hickory leg. 1963 S. Plath Bell Jar i. 10 A short, scrunty fellow detached himself and came into the bar with us.

scrupulant (skrū·piulănt). Eccl. [f. L. scrupul-us + -ant^1.] One who is over-scrupulous in confessing his sins; one who suffers from scrupulosity of conscience.

1938 'H. Kingsmill' English Genius 199 George Fox was not what the Catholic Church calls a scrupulant. 1961 J. B. Sheerin Sacrament of Freedom x. 133 The scrupulous person..breaks contact with reality when it comes to the matter of sin. In some cases, a scrupulant is sane about sin in general and has a blind spot only in regard to one particular type of sin. 1961 Theol. Stud. June 232 When dealing with a scrupulant..persuade him.. of the pathological element in his personality.

‖ **scrutin** (skrütæ̃n). [Fr., vote.] In Fr. combinations, referring to contrasting electoral systems: scrutin d'arrondissement (darondismañ) [lit., electoral district vote], a system of voting in France by which votes are cast for a single representative of an electoral district; scrutin de liste (list) = list vote s.v. *list sb.^{6} d.

1851 Ann. Reg. 1850 i. 230/2 According to Article 30, the election is effected in the department and by Scrutin de liste. 1911 J. H. Humphreys Proportional Representation viii. 172 List systems of proportional representation are based upon the block vote or scrutin de liste. 1921 J. Bryce Mod. Democracies I. xx. 270 Three times this method was dropped and replaced by the Scrutin d'arrondissement (the scheme of one-membered constituencies). Now the Scrutin de liste has returned once more. 1954 [see list vote s.v. *list sb.^{6} d]. 1960 Macridis & Brown De Gaulle Republic xiii. 236 The cabinet decided to adopt the scrutin d'arrondissement with ballottage.

scrutineer, sb. Add: 1. (Later examples in gen. use.) Also spec. in Motor Racing and Motor-Boat Racing, an official who inspects a car or boat in order to ensure that it complies with the regulations.

1932 S. C. H. Davis Motor Racing xxii. 270 Caracciola's Mercédès was rejected by the scrutineers. 1963 Times 4 June 14/7 The scrutineers' protest was upheld and both cars ruled ineligible. 1968 Guardian 19 Mar. 9/3 In the windy entrance lobby..there is a photograph of Mr Novotny... There is also a lady of the militia... Once past these two scrutineers you cross a yard..., and enter the works proper. 1972 [see next]. 1976 Church Times 16 July 6/5 A passport to immortality consisting of 'Cranford', 'Cousin Phillis' and 'Wives and Daughters' is convincing enough in all conscience for the most demanding of literary scrutineers.

2. (With capital initial.) A contributor to Scrutiny, a literary journal edited by F. R. Leavis (see *Leavisian sb. and a.), and others, between 1932 and 1953; a follower of Leavis. rare.

1958 J. Raymond in Times Lit. Suppl. 15 Aug. p. xxxii/1 The difficulty of writing on a general theme like this is to decide at the beginning just what we are attempting to discuss... The Situation (desperate of course) of the Contemporary Writer? The Collapse of the Essay?.. Scrutineers All—and After? 1978 J. Marcus in Ibid. 12 May 528/5 The Scrutineers have taught generations of British students that Virginia Woolf was a snob.

scrutineer (skrūtiniə·ɹ), v. [f. prec.] trans. In Motor Racing and Motor-Boat Racing: to inspect (a car or boat) in order to ensure that it complies with the regulations. Hence scrutinee·ring vbl. sb.

1930 E. Waugh Vile Bodies x. 175 Changed the whole engine over after they'd been scrutineered. Anyone else would have been disqualified. 1932 S. C. H. Davis Motor Racing xxii. 270 That..served to emphasize..the weakness of scrutineering on the day before the race. 1971 Sunday Express (Johannesburg) 28 Mar. 7/1 The scrutineering area for checking whether cars comply with regulations. 1971 E. Afr. Standard (Nairobi) 13 Apr. 1/3 The margin on the road was four points, but the gap was narrowed when the two Datsuns went to scrutineering. 1972 C. Mudie Motor Boats & Boating 150 Other boats will begin to arrive with tall tales of fantastic practice performances and all will build up to scrutineering the day before the race. Each boat has to present itself to a team of scrutineers who will check the hull and machinery. Ibid., (caption) The scrutineering team check each boat before an important race.

scrutinization (skrū:tinəizēi·ʃən). [f. Scrutinize v. + -ation.] The action of scrutinizing.

1922 Autocar 10 Nov. (Advt. Suppl.) p. lxxxv, Our unique methods of scrutinisation and the introduction of every detailed improvement. 1976 *Kybernetes* V. 36/1 If a particular scrutinization fails a logico-grammatical test, then its rejection requires the reformulation of a Postulated Hypothesis.

scrutoire. (Later examples.)
1853 J. G. M. Ramsey *Annals of Tennessee* 132 These issues of the North-Carolina Treasury..are still found in great abundance in the scrutoires and chests of the old families. 1978 W. M. Spackman *Armful of Warm Girl* 33 And he himself unpacked..a manila folder of assorted private relics, which he stowed one by one in the upstairs scrutoire.

scrutty (skrʌ·ti), *a. rare.* [Origin unknown.] Dirty, dusty, scruffy.
1914 M. Beerbohm *Let.* 27 Apr. (1964) 234 The Arnold Bennetts—very dusty and scrutty but nice—alighted from a motor-car here yesterday. 1970 T. Hughes *Crow* 68 He tried hating the sea But instantly felt like a scrutty dry rabbit dropping on the windy cliff.

scuba (skiū·bă, skū·bă). Also SCUBA. [f. the initials of *s*elf *c*ontained *u*nderwater *b*reathing *a*pparatus.] Self-contained apparatus designed to enable a swimmer to breathe while under the water. Also (rarely) *collect.* and *ellipt.* for *scuba-diving* vbl. sb. Freq. *attrib.* and *Comb.*, esp. in **scuba-dive** *v. intr.*, to swim under water using such apparatus; so **scuba-diver, scuba-diving** *vbl. sb.* and *ppl. a.*
1952 Hahn & Lambertsen *On using Self Contained Underwater Breathing Apparatus* (U.S. Nat. Acad. Sci.) 1 Within the last 3–5 years we have witnessed..a rapid increase in the numbers of self contained underwater breathing apparatus (SCUBA) in use... SCUBA are now in relatively large scale use by spearfishermen and sports swimmers. 1957 *Time* 25 Feb. 49/1 Most types of scuba are of the open-circuit design which supply air on demand, and discharge exhaled air into the water. 1962 (*title*) The new science of skin and scuba diving. 1963 G. L. Pickard *Descriptive Physical Oceanogr.* v. 57 In clear ocean water the superior penetration of blue and green light is evident..when SCUBA diving. 1963 *Today's Health* June 18/2 The scubacide victim is the person who tries to become a scuba diver in one fatal lesson (self-taught). 1966 T. Pynchon *Crying of Lot 49* ii. 31 It [*sc.* a housing development] was to be laced by canals with private landings for power boats, a floating social hall,..all for the entertainment of Scuba enthusiasts. 1973 P. O'Donnell *Silver Mistress* v. 82 Under the hull..two scuba-suited figures clung to magnetic limpets clamped to the steel plates. 1975 *New Yorker* 26 May 17 (Advt.), Swimming, scuba and long beautiful beaches. 1977 *Ibid.* 4 July 83/1 In 'The Deep'..Nick Nolte plays a scuba-diving hero called David. 1980 *Nature* 4 Sept. 12/1 Scuba dive over the lost road of Atlantis.
Hence as *vb.*; also **scu·ba,ing** *vbl. sb.* (in quot. spelt *scubering*).
1973 B. Mather *Snowline* xix. 234 Some of the boys are keen on scubering, water-skiing and fishing. 1977 *Rolling Stone* 16 June 74/5 (Advt.), Hike, swim, scuba, snorkel, sail.

scud, *sb.*¹ Add: **2. d.** Also of snow.
1969 N. W. Parsons *Upon Sagebrush Harp* xv. 85 Usually, at dawn the wind died and a knee-high scud sharp as glass would skitter sullenly along the surface of the hard-packed snow.
e. Also *Comb.*, as **scud-like** *a.*
1866 G. M. Hopkins *Jrnls. & Papers* (1959) 138 A 'dirty' looking kind of clouds, scud-like, rising.

scud, *sb.*² Add: **3.** *Tanning.* Dirt, lime, fat, and fragments of hair which must be removed from a hide. Cf. Scud *v.*³ 2.
1885 A. Watt *Art of Leather Manufacture* xxvi. 324 The 'scud' is removed by working the pelt upon the beam with the blunt knife. 1969 T. C. Thorstensen *Pract. Leather Technol.* vi. 96 The hair-destruction system may result in uneven swelling and in the formation of scud (surface dirt) on the hides. *Ibid.* 98 The strong oxidizing action of the chlorine dioxide and chlorine results in the bleaching of the hair, and there is no dark scud left on the hide.

scud, *v.*¹ Add: **6.** *Sc.* To slap, beat, strike, spank; to beat down.
1814 W. Nicholson *Tales in Verse & Miscellaneous Poems* 123 And farmers, keen to cut the crap, Lest win's should scud it. 1866 J. Smith *Merry Bridal* (ed. 2) 23 Lassie, when I get ye I'll scud ye till I'm sair. 1925 *United Free Church Mission Record* Dec. 569/2 The risen wind scudded my cheek—wet, stinging, and with the bite of the sea. 1976 *Scotsman* 24 Dec. (Weekend Suppl.) 1/1 Any more cracks and I'll scud yer hint end for ye.

scuddy, *a.*² Delete † *Obs.* and ? and add later examples.
1872 Thudichum & Dupré *Treat. on Wine* xx. 633 The wines are spoiled during fermentation, become acidified, scuddy. 1964 R. Braddon *Year Angry Rabbit* xi. 94 Jacks..at once flung himself gratefully into a chair, spilling half his cold, scuddy tea into his lap as he did so.

scuff, *sb.*¹ Add: **2. b.** A mark made by scraping or rubbing.
1954 J. Steinbeck *Sweet Thursday* v. 35 Brown calf shoes.., scuff on the right toe. 1976 B. Lecomber *Dead Weight* i. 11 A thousand scuffs and scratches in the shabby wood and leather.
4. A type of slipper or sandal without a back. Chiefly *U.S.*
1909 in *Cent. Dict. Suppl.* 1938 *Sears, Roebuck Catal.* Fall/Winter 324/3 Adorable Slip-on 'Scuffs'. No trouble at all to slip in or out of these cunning, snug 'scuffs'. 1945 *Creative Footwear* Apr. 106/2 (Advt.), Shearling scuff, leather sole. 1953 S. Ransome *Drag Dark* xiv. 140 In the snow in her stocking feet, her red scuffs lost behind her. 1968 J. Ironside *Fashion Alphabet* 135 *Mule*, loose slipper with front vamp only, no back. Also called a 'scuff'. 1974 *Spartanburg* (S. Carolina) *Herald* 18 Apr. (Kmart Advt. Suppl.) 1 Soft-stepping flowered cotton terry cloth scuffs, molded rubber sole.

scuff, *v.* Add: **5.** *intr.* for *pass.* **a.** To become marked, worn, or damaged by rubbing or scraping.
1934 Webster, s.v., Soft bindings scuff easily. 1978 *Radio Times* 18–24 Mar. 80 (Advt.), For kids who play rough, shoes that won't scuff.
b. Of a metal part: to undergo scuffing (*scuffing *vbl. sb.* 3).
1959 *Engineering* 23 Jan. 117/3 The untreated mild steel rings scuffed shortly after being put under test. 1970 H. J. Watson *Mod. Gear Production* xvi. 283 The peaks [of helical gears] were prone to scuff or pit in service largely owing to the high local loading on the restricted areas.
6. *Comb.*, as **scuff-resistant** adj., resistant to scuffing; hence **scuff resistance.**
1967 *Times Rev. Industry* May 84/3 The growing demand for higher gloss and better scuff and product resistance has led to the development of synthetic resin based types [of varnish]. 1959 *Spectator* 21 Aug. 219 (Advt.), Everything from scuff-resistant flooring and unbreakable gramophone records to transparent polyethylene wrapping. 1978 *Radio Times* 18–24 Mar. 80 (Advt.), A shoe that's an astonishing 30 to 40 times more scuff-resistant than normal leather.

scuffed, *pa. pple.* and *ppl. a.* Restrict '*Sc., Anglo-Irish,* and *U.S.*' to sense in Dict. and add: **1. b.** Of shoes, a floor, etc.: worn or marked by rubbing, scraping, or treading. Also with *up*.
1927 *Scribner's Mag.* Apr. 381/2 It wasn't a large room but everything in it, from the scuffed leather slippers to the stout..easy chairs, proclaimed a man who knew how to put himself at ease. 1973 R. Thomas *If you can't be Good* (1974) xii. 99 The beat-up desks..and the scuffed-up floor. 1975 J. Gores *Hammett* (1976) vii. 51 The hardwood floor waxed but well-scuffed, ready for dancers. 1978 *Morecambe Guardian* 14 Mar. 17/3 Generally speaking there are two categories of small boy..the studious, eyes-down-in-a-book type and the outdoor scuffed shoes clothes-in-a-mess variety.
c. *Engin.* Of a metal part: worn by scuffing.
1934 *Jrnl. R. Aeronaut. Soc.* XXXVIII. 310 Cases have come to one's notice where engines have suffered from troubles in the form of scored, or, as our friends in America term it, 'scuffed' pistons. 1941 [see *scuffing *vbl. sb.* 3].

scuffer[1] (skʌ·fəɹ). *dial.* or *slang* (chiefly *north.* and *Sc.*). Also **scufter.** [Origin obscure; perh. f. Scuff *sb.*² or *v.*] A policeman.
1860 Hotten *Dict. Slang* (ed. 2) 209 *Scufter*, a policeman. 1886 A. G. Murdoch *Scotch Readings* 1st Ser. (ed. 2) 78 The policeman..was familiar [to]..him as..'the Scufter'. 1959 I. & P. Opie *Lore & Lang. Schoolch.* xvii. 369 In Penrith children still commonly use the old northern name 'Scufty' or 'Scufter', a term which had been thought to be obsolete. 1961 Partridge *Dict. Slang* Suppl. 1263/1 *Scuffer*, a policeman: Liverpool: C. 20. Ex. dial. *scuff*, to strike.. Cf. *scufter.* 1966 P. Moloney *Plea for Mersey* 45 Scuffer! Scuffer! on the beat, With thy elephantine feet, You can't see the way to go Cos yer 'at comes down too low. 1967 J. Wainwright *Talent for Murder* 17 Are you from the slops, sonny?.. The scuffers. The jacks. Are you from the coppers, sonny? 1970 T. Lewis *Jack's Return Home* 49 What do you think I should do? Go to the scuffers? 1978 *Daily Mail* 25 Jan. 12/2 The strange language of a group who call themselves 'bogeys', 'bobbies' or 'scuffers'. They are, of course, regional variations describing policemen.

scuffer[2] (skʌ·fəɹ). *N. Amer.* [f. Scuff *sb.*¹ or *v.* + -er¹.] = *scuff *sb.*¹ 4 (see also quot. 1939).
1911 *Daily Colonist* (Victoria, B.C.) 5 Apr. 7/1 (Advt.), Correct spring styles in America's best footwear..Broadwalk Scuffers for children. 1935 *Amer. Speech* X. 9/2 *Scuffers* and *ghillies* are words of the fashion page. 1939 M. P. Picken *Lang. of Fashion* 130/3 Scuffer..child's sandal-like play shoe, light-weight and flexible with sturdy sole. Also used for sportswear by adults.

scuffing, *vbl. sb.* (In Dict. s.v. Scuff *v.*) Add: **2.** (See quot. 1928.)
1928 Holt & Cooke in *Bur. Standards Jrnl. Res.* I. 25 A considerable part of tread wear may be caused by the slipping of portions of the tread over the road surface in changing from the normal to the deflected condition and vice versa. This might be termed a scuffing action. 1955 W. H. Crouse *Automotive Chassis & Body* vi. 157 High-speed operation causes much more rapid tire wear because of the high temperature and greater amount of scuffing and rapid flexing to which the tires are subjected.
3. *Engin.* The roughening of a metal surface designed to rub against another when the lubrication is inadequate to prevent local fusion and tearing of the two surfaces.
1941 *Engineering* 11 Apr. 286/1 Should a local breakdown of the lubricating-oil film occur, the two surfaces may momentarily become fused. With the reciprocating movement of the parts, a shearing action takes place,..causing disruption of the surfaces. An example of this effect, which is known as piston-ring 'scuffing', is seen in Fig. 1, which shows the surface of a scuffed piston ring. 1950 *Engineering* 17 Mar. 310/1 Failure of a gear lubricant to prevent the teeth from making metallic contact may result in destruction of the tooth surfaces by 'scuffing', 'seizing', or 'welding': all different degrees of welding. 1961 *New Scientist* 26 Jan. 218/3 'Scuffing' of valve gear cams and the tappets that run against them occurred. 1966 G. W. Michalec *Precision Gearing* viii. 354 Scoring and scuffing are associated with welding (or seizing) and plastic deformation.

scuffle, *sb.*² **2.** (Earlier examples.)
1797 S. Deane *Newengland Farmer* (ed. 2) 95/2 *Dutch Hoe*, sometimes called a *Scuffle*; an iron instrument, with a sharp steeled edge, nearly in the shape of the letter D. 1825 J. Lorain *Nature & Reason harmonized in Practice of Husbandry* 191 The scuffle (or D hoe as it is sometimes called) will destroy weeds growing on a level surface.

scuffle, *v.*¹ Add: **2. b.** *trans.* To obtain, collect, raise (money). Also with *up* and *intr.* (const. *up on*). *slang* (chiefly *U.S.*).
1946 Mezzrow & Wolfe *Really Blues* (1957) 378 *Scuffle up*, raise, collect, get together. 1956 B. Holiday *Lady sings Blues* (1973) vii. 66, I stayed around Philly a couple of days before I could scuffle up enough to get back to New York on the bus. 1965 'Malcolm X' *Autobiogr.* (1966) xvi. 389 Trying to scuffle up on some bread. 1973 *Brit. Jrnl. Sociol.* XXIV. 203 It is surely of immense sociological significance that when a Ras-Tafarian 'scuffles' a fare (he might beg, borrow or steal it), he seeks to migrate not to Ethiopia, as we would expect, but to Britain or the United States.
3. b. *intr.* To survive with difficulty, to make a bare living by uncongenial or degrading means. *slang* (chiefly *U.S.*).
1939 W. Hobson *Amer. Jazz Music* (1940) 173 At the bottom of the economic pile are those musicians who have nothing which could accurately be called a job but are taking whatever one-night stand happens along; this is called 'scuffling'. 1956 S. Longstreet *Real Jazz Old & New* xviii. 147 *Scuffle* is to get by. 1956 M. Stearns *Story of Jazz* (1957) xvii. 212 The Basie band was scuffling. 'It was a cracker town but a happy time,' Basie recalls. 1961 Rigney & Smith *Real Bohemia* p. xvi, *Scuffle*, to live by one's wits, not by a gig. 1972 T. Kochman *Rappin' & Stylin' Out* 164 'Scuffling' in the idiom means barely making it from day to day, generally by engaging in nonprestigious..activities such as begging, collecting and returning pop bottles for the deposit, working at odd jobs for minimum wages, etc.

scuffler[1]. (Later examples.)
1961 F. G. Cassidy *Jamaica Talk* x. 215 A *scuffler* is a thief. 1965 H. Williamson *Hustler* vi. 169 He said he was a hustler, but he really wasn't nothin' but a goddamn scuffler.

scufter, var. *scuffer¹.

scug, *sb.*³ Add: Also in extended use. Also *attrib.* Hence **scu·ggish, scu·ggy** *adjs.*
1911 R. Nevill *Floreat Etona* iii. 98 Once it began to be considered 'scuggish', the fate of Eton pugilism was sealed. 1916 E. F. Benson *David Blaize* v. 101 These are all college houses, in-boarders, and rather scuggy compared to out-boarders. *Ibid.* viii. 143 You were such a scug, you see, that you didn't do those things when it was scuggish not to. 1922 S. Leslie *Oppidan* iv. 48 A *Tug* was something between a *scug* and a hireling chorister. *Ibid.* v. 57 The sad sight of a *Pop* wearing a *scug*-cap. 1928 *Observer* 15 Apr. 29/4 A band of what I can only describe as 'Scugs' in bowler hats. 1940 E. F. Benson *Final Edition* ii. 27 Mr. Luxmore..wrote to a friend in withering disdain of him and his official purple as a Monsignor, declaring that he was just the same 'sharp insignificant little scug as he had been at Eton'. 1962 J. P. Carstairs *Pardon my Gun* ii. 28 He was a bit of a scug but what the hell. 1980 D. Marlowe *Rich Boy from Chicago* xxi. 314 Many refused to talk..dismissing Lambert as a 'scug', a loathsome queer.

sculduddery. (See also *skulduggery.)

scull, *sb.*¹ Add: **1. a.** Also in *Comb.*, as **scull-hole** = *sculling-hole* s.v. Sculling *vbl. sb.* b.
1843 G. Little *Life on Ocean* 63 The crew got the bight of the rope fore and aft the boat, leading it from the stern over the scull-hole. 1973 W. Elmer *Terminol. Fishing* iv. 125 Many of the smallest boats are not fitted with rowlocks or tholes, but have a notch in the transom for sculling (*sculling-notch* or *scull-hole*).

scull, *v.* Add: **4.** *intr.* To skate without lifting the feet from the ice.
1895 in *Funk's Stand. Dict.* 1938 [implied in *sculling *vbl. sb.* a]. 1976 *Webster's Sports Dict.* 376/1 *Scull or skull*,..to propel oneself forward or backward by alternately moving the heels or the toes apart and together changing from an outer edge on the outward movement to an inner edge on the inward movement.
5. a. *to scull about,* to lie about; *spec.* of objects left on the deck of a ship instead of being put away. *colloq.*
1917 'Taffrail' *Sub* v. 136 You went round..with a large bag. In this you placed all..articles..found 'sculling' about. 1938 C. Morgan *Flashing Stream* III. 263 Don't leave it [*sc.* a key] sculling about. 1943 C. S. Forester *Ship* 12, I want those mess-traps brought back... Don't leave them sculling about on the decks.
b. *to scull around* (or *about*), to move about aimlessly; also *fig.* *colloq.*
1921 *Daily Colonist* (Victoria, B.C.) 20 Mar. 19/7 My opposite number..has one or two questions pertaining to his own branch sculling around in his mind. 1935 M. Egan

Dominant Sex i. 12 *Angela.* Where have you been these ages? *Alec.* Oh, sculling round the country on business. 1950 [see *number *sb.* 5 c]. 1961 B. Fergusson *Watery Maze* vi. 151 While these ideas were sculling around Whitehall, there arrived from Moscow..M. Molotov. 1981 'J. Ross' *Dark Blue & Dangerous* xxiii. 137 What with Wiffen and one or two others who were there sculling about, the picture is a little confusing.

sculldugggery, see *skulduggery.

sculling, *vbl. sb.* Add: **a.** (Later examples, in sense *4 of the verb.)
1938 D. Cummings *Figure Skating as Hobby* iii. 19 You can try sculling. Feet together, put your weight on the inside of both your skates..bend your knees, push down, move your feet apart, straighten slightly and bring them together again. 1973 R. S. Ogilvie *Basic Ice Skating Skills* ii. 49 This progression across the ice by moving both feet in and out is known as *sculling*. Sculling..can be done forward as well as backward. *Ibid.* 175 *Sculling*,..a method of two-footed progression forward or backward by an in-and-out movement of the feet.
b. sculling-notch, sculling score = *sculling-hole*.
1933 Bamfield & Palmer *Art of Sailing* ix. 76 Have a sculling notch cut in the stern, shod with a strip of brass. 1973 Sculling-notch [see *scull *sb.*¹ 1 a]. 1946 F. B. Cooke *Cruising Hints* (ed. 6) xxvii. 244 Every yacht's dinghy should have a sculling score in the transom. 1960 —— *Yachting with Economy* xxi. 146 When leaving the yacht to get the kedge, place the warp in the sculling score.

sculp, *sb.*¹ **2.** (Later example.)
1883 *Daily News* 18 Jan. 5/7 Perhaps no statue, except the unfortunates in Trafalgar-square, and the melancholy meeting of 'sculps' in Parliament-square, was more sharply criticised at the time of its erection.

sculp, *sb.*² Add: Also, in early or *arch.* use, a human scalp.
1743 J. Isham *Observations on Hudsons Bay* (1949) 93 They make an offering, putting a painted Stick up, some with a cross hanging a hatchet,..or Ice Chissel, or what Else they have on the top, with the sculp of their Enemies, when they go to Warr. 1758 in *Essex Inst. Hist. Coll.* (1881) XVIII. 180 They obtained fifty-two Sculps and two Prisoners. 1804 Lewis & Clark *Orig. Jrnls. Lewis & Clark Exped.* (1905) VII. 64 They took the 65 of the Mahars sculps and had them hung on Small poles. 1845 W. G. Simms *Wigwam & Cabin* 1st Ser. 53 A pretty fellow..at his time of life to be looking after sculps of women and children. 1921 J. Buchan *Path of King* xii. 242 Maybe the Indians have got his sculp.

sculp, *v.*² Add: **a.** Also *U.S.*, to scalp (a person). Now *arch.*
1758 in *Essex Inst. Hist. Coll.* (1881) XVIII. 109 Taring his Nails out by ye Roots, Sculping alive and such like torments, they wou'd shout and yell. 1759 in *Ibid.* (1882) XIX. 188 [He] retook one of ye Prisoners and killed and sculpt one of ye Indians. 1834 W. A. Carruthers *Kentuckian in New York* I. 24 But as to shootin and sculpin Injins, that's a thing there is no bones made about. 1845 W. G. Simms *Wigwam & Cabin* 1st Ser. 44 They'll be sculped, every human of them, in their beds. *Ibid.* 51 We heard of murders and sculpings on every side. 1884 Sweet & Knox *On Mexican Mustang through Texas* xviii. 246 I'm a scout from the Far West, whar..the coyote sleeps in the deserted wigwams of the skulpt Indian. 1921 J. Buchan *Path of King* xii. 243 The Shawnees cotched me and Jim... They'd ha' sculped us if it hadn't been for Jim.

sculpt, *v.* Delete '*rare* exc. in jocular use' and add: **1.** (Later examples.)
1928 *Daily Express* 16 June 4/5 He sculpts in almost every material. 1931 *Sun* (Baltimore) 13 Mar. 12/7 Somebody proposed that instead of being depicted as astride a horse (steed, charger) this general be sculpted as a figure seated in a motor car. 1966 J. Richardson *George IV* 332 The statue (known as the Copper Horse) had been sculpted by Westmacott. 1977 *Times* 5 Sept. 5/7 My uncle, my mother's brother, draws beautifully and sculpts.
2. *transf.* To shape, form, mould.
1967 *Times Rev. Industry* Apr. 86/3 Numerical control of machine tools is obtainable in a number of forms, from the simple point-to-point positioning system, through straight-line machining systems, to the complex continuous path systems which can sculpt any shape capable of being expressed numerically. 1972 *Sci. Amer.* Mar. 45/1 (Advt.), High-level signal lights, neatly sculpted into the rear deck of the car.

sculpted (skʌ·lptĕd), *ppl. a.* [f. Sculpt *v.* + -ed¹.] = Sculptured *ppl. a.* in Dict. and Suppl. Also *fig.*
1961 in Webster. 1976 *Listener* 22 Apr. 510/1 The dialogue was full of sculpted pauses, a gain for poetry but a loss for credibility. 1978 P. Porter *Cost of Seriousness* 17 Seeing grief in formal state Upon a sculpted angel group. 1978 A. & G. Ritchie *Anc. Monuments Orkney* 5 The landscape is typical of that produced by Old Red Sandstone, predominantly gentle and rounded, but rising to spectacularly sculpted cliffs along the west and north coasts.

sculptured, *ppl. a.* Add: **3.** Shaped in a manner resembling sculpture.
1966 J. S. Cox *Illustr. Dict. Hairdressing & Wigmaking* 132/2 *Sculptured curl*, a firmly and smoothly formed curl showing the comb-teeth lines. *Sculptured hair style*, a hairdress with hard, firm, definite lines in its constituent parts; not fussy, light or tapered. 1970 *Which?* Sept. 268/1 Now, however, you will also find twists, loop-pile and sculptured pile—a pattern formed by mixing loops of different heights or looped and cut pile. 1974 *Times Lit. Suppl.* 24 May 544/3 The second president of General Motors invented the scheme of elaboration and development, bright colours, sculptured lines and rising prices. 1974 *Times-Picayune* (New Orleans) 15 Aug. v. 6/1 The bride..wore a peau de soie gown styled with a sculptured yoke of re-embroidered lace and a cameo neckline.

sculpturing, *vbl. sb.* Add: **2.** *Bot.* The structural ornamentation of the surface of a pollen grain or spore.
1943 G. Erdtman *Introd. Pollen Analysis* v. 43 The outer surface of the exine may sometimes be provided with some sculpturing or ornamentation. The ornamentations of sculptured pollen grains are exceedingly varied. 1967 M. E. Hale *Biol. Lichens* ii. 37 The spores are unornamented, although exospore sculpturing is reported in *Tholurna*. 1970 *Watsonia* VIII. 4 The sculpturing of the perispore of *D*[*ryopteris*] *assimilis* differs from that of *D. dilatata* and *D. carthusiana* in density and shape and size of projections.

scum, *sb.* Add: **2. b.** More generally, any undesirable surface layer or deposit, usu. but not necessarily on a liquid. (Further examples.)
1940 *Chambers's Techn. Dict.* 750/1 *Scum*,..a surface formation of lime crystals appearing on new cement work. 1941 *Thorpe's Dict. Appl. Chem.* (ed. 4) V. 573/1 Silica scum is sometimes found on the top of tank-melted glass if the melting temperature is not very high. 1967 E. Chambers *Photolitho-Offset* xiv. 211 The albumen image..may look clear and free from ink when the grain is full of scum, which will readily ink-up on the press run with the soft machine ink.
d. *coarse slang* (chiefly *U.S.*). Semen.
1967 Wentworth & Flexner *Dict. Amer. Slang* Suppl. 703/1 *Scum* (taboo),..semen... *Scumbag* (taboo),..a condom. 1972 R. A. Wilson *Playboy's Bk. of Forbidden Words* 257 *Scum*, the semen.
4. scumspittle *nonce-wd.*, ? scummy or frothy spittle.
1922 Joyce *Ulysses* 446 The bulldog growls,..a gobbet of pig's knuckle between his molars through which rabid scumspittle dribbles.

scumbag (skʌ·mbæg). *coarse slang* (chiefly *U.S.*). Also **scum bag.** [f. Scum *sb.* + Bag *sb.*]
1. A condom.
1967 [see *scum *sb.* 2 d]. 1968–70 *Current Slang* (Univ. S. Dakota) III–IV. 106 *Scum bag*, a condom. 1974 *Time Out* 6 Dec. 21/1 Young blades carried their sheaths or condoms or..'scumbags' in their wallets.
2. A base, despicable person. Also as a term of vulgar abuse.
1971 *Courier-Mail* (Brisbane) 23 Dec. 5/2 Another called him a 'scumbag' and said he should have been killed. 1973 E. Bullins *Theme is Blackness* 80 [Ann] No, you can't think that about me! [Peter] Why can't I, scumbag? 1976 G. V. Higgins *Judgement of Deke Hunter* iv. 29, I had three scumbags that went to trial. 1977 *Zigzag* Apr. 34/2 What little scumbag would say something like that?

scumble, *v.* **3.** (Later example.)
1974 V. Nabokov *Look at Harlequins* (1975) vi. i. 227 The summer tan..would scumble, I knew, the liver spots on my temples.
scumbled *ppl. a.*, **scumbling** *vbl. sb.* (Later *fig.* examples.)
1967 *Listener* 12 Jan. 48/3 A verb 'to scumble', which means to blur and soften the outlines. A great deal of our national life seems to me to be scumbled. 1977 *Times* 19 Nov. 9/2 The confusion of the times..the scumbling of boundary lines.

scummy, *a.* Delete *rare* and add: **1.** (Later examples.)
1936 A. Ransome *Pigeon Post* xxix. 312 'What'll it look like when we see it again?'..'All scummy on the top... The pure gold'll be underneath.' 1967 E. Chambers *Photolitho-Offset* xiv. 215 Under-exposure produces a weak, soft stencil, so that the image thickens-up in development and results in stencil breakdown and a scummy plate. 1979 K. M. Peyton *Marion's Angels* ii. 29 At high-water springs the river came right up over the saltings..and sent scummy fingers up the garden path.
3. *transf.* and *fig.* Filthy, dirty; despicable, disreputable. Also *Comb.* orig. and chiefly *U.S. colloq.*
1932 [see *pig *sb.*¹ 5]. 1952 B. Malamud *Natural* 67, I don't like the scummy tricks you play on people. 1973 E. Bullins *Theme is Blackness* 78 Hey, you white scummy-lookin' bourgeois bitch, take me to dinner? 1977 R. E. Harrington *Quintain* xvii. 202 Meeting scummy little men at..squalid Parisian restaurants. 1979 *Maledicta* III. 133 The scummy millionaire Marxist profs I know don't spend one *red* penny of their own: Let the workers & peasants pay!

scunge (skʌndʒ), *sb. colloq.* (orig. *Sc.*). Also **skunge.** [Origin unknown: cf. next.] **a.** A sly or vicious person. **b.** A scrounger or sponger. **c.** As a vague term of abuse.
1824 J. Mactaggart *Scottish Gallovidian Encycl.* 424 *Scun*, plan, craft. A scunge, a sly fellow; a maid seducer. 1900 in *Sc. Nat. Dict.* (1971) VIII. 104/1 A scunge has a crappin' for a' corns. 1912 G. Cunningham *Verse Maistly in Doric* 184 Jock, ye scunge! Come oot the dresser. 1948 *Football Times* 11 Sept., 'You great big skunge', meaning that you were always on the 'mootch'. 1967 *Comment* (N.Z.) June 14/1 He obviously thought I must be a bit of a scunge asking political questions. 1976 R. Barnard *Death on High C's* iv. 44 'Big joke, scunge,' said Gaylene, giving him the sort of shove that would have sent a lesser man through the wall.

scunge (skʌndʒ), *v. colloq.* (orig. and chiefly *Sc.*). Also **scundge, scunje,** etc. [Origin unknown: cf. prec. and *scrounge *v.*¹] *intr.* To prowl around looking for food, etc.; to scrounge, to sponge. So **scu·nging** *ppl. a.*
1843 J. B. Pratt *Life & Death Jamie Fleeman* (ed. 9) iv. 32 Hame wi you, ye scunging tyke, hame! 1844 W. Cross *Disruption* xxxi. 341 Neither will ye scunge after the gentry like McQuirkie. 1905 *Eng. Dial. Dict.* V. 299/2 *Scunge*, to slink about; to fawn like a dog for food. 1964 X. Herbert *Larger than Life* 243 Senile wrecks..coveting and scundging and bickering. 1966 *Huntly Express* 30 Sept. 2 It's maybe been a scunjin' dog.

|| **scungille** (skundʒi·le). Pl. **scungilli.** [ad. It. dial. *scunciglio* conch, seashell, prob. alteration of It. *conchiglia* seashell, shellfish.] A mollusc or conch, esp. the meat of a mollusc eaten as a delicacy.
1953 A. Boni *Talisman Italian Cookbk.* 65 Scungilli Marinara..½ pound of scungilli (pulp of conch)... Boil scungilli about 15 minutes and drain. 1963 T. Pynchon *V.* iii. 62 He tended each seashell on his submarine scungilli farm. 1972 *Village Voice* (N.Y.) 1 June 75/1 (Advt.), Italian–American fish specialties. Shrimp, calamari, scungilli, mussels, [etc.]. 1980 D. E. Westlake *Castle in Air* i. 12 Stuffing scungilli and spaghetti into her mouth.

scungy (skʌ·ndʒi), *a.* Chiefly *Austral.* Also **scungey.** [f. *scunge *v.* + -y¹.] Mean, dirty, disreputable. Also *Comb.*
1966 Baker *Austral. Lang.* (ed. 2) x. 215 *Scungey*, dirty, untidy, disreputable. 1969 *Coast to Coast 1967–8* 86 Y' hate me, don'tya? Don'tya? Don'tya, eh? Answer me, y' scungy bitch. Answer me! 1970 *Telegraph* (Brisbane) 10 Mar. 8/5 Nowadays people..talk about 'that scungy place, Bondi'. It's the truth, but it hurts. 1978 *Courier-Mail* (Brisbane) 19 May 5/5 Sometimes you see scungy-looking grapes that are worth a lot of money. 1980 F. Ansell *To fight Wild* 56 The dressing on her leg was getting very scungy but I wasn't game to touch it until the bone had had time to begin setting.

scunner, *sb.* For '*Sc.* and *north.*' read 'orig. *Sc.* and *north.*' and add: **1.** (Later examples.) Now freq. in a milder sense: a grudge, repugnance, dislike. Also *to take a scunner to.*
1900 R. J. Muir *Mystery Muncraig* ii. 21 He had never told his weakness to his brother, having had a 'scunner' against doing so. 1911 F. E. Crichton *Soundless Tide* ii. 20 He tuk some soort of a scunner til her, an' now he's just left her sittin'. 1927 J. Buchan *Witch Wood* i. 21 You'll give our young brother a scunner of the place. 1935 L. Kerr *Woman of Glenshiels* xiii. 207 It fair gies ye the scunner the way they all grumble. 1957 V. Palmer *Seedtime* ii. 15 He remembered he had taken a scunner against McCoy when he had flown down to her wedding. 1964 *Scotsman* 12 Nov. 5 Many of them have taken a scunner at religion because they took a scunner at it at school. 1974 P. De Vries *Glory of Hummingbird* ix. 123 He had taken a scunner to me... What had soured him on me..had been Jake's replacing him with me. 1977 L. Meynell *Hooky gets Wooden Spoon* xiii. 152 Thirty per cent of the calls..originated in personal spite, someone had taken a scunner against the next-door neighbours.
2. *Sc. dial.* **a.** Of persons: a nuisance, a pest, a good-for-nothing.
1796 J. Lauderdale *Poems* 91 Some poor waff detested scunner. 1899 *Shetland News* 11 Feb. 7/3 Von black pairts whaur som' o' da scunners o' boys is been makkin' slides. 1926 W. Queen *We're a' Coortin* iii. i. 69 Ye wee, bowly-leggit scunner ye. 1940 *Horizon* 11 Nov. 243 He was aye sittin' in ma road. A fair scunner! 1958 *Banffshire Jrnl.* 1 Apr. 7/1 A fraisie, meally-mou'd twa-faced scunner o' a lad.
b. Of things: a nuisance, a hardship, a plague, a vexatious matter.
1865 J. Horne *Poems* 24 Faigs, borrowed money is a sconner. 1917 A. S. Neill *Dominie Dismissed* xi. [illegible] 'Bairns is just a scunner,' said Sarah. 'Ye'll hae to stop yer typewriter or ye'll waken them.' 1926 W. Queen *We're a' Coortin* i. i. 10 It's a richt scunner walkin' up that long avenue tae the big hoose. 1947 H. W. Pryde *1st Bk. McFlannels* i. 4 Ah thocht the room floor was bad, but this is a fair scunner.

|| **scuola** (skwō·la). Pl. **scuole.** [It., = school.] In Venice, any of the buildings in which the medieval religious confraternities or guilds used to meet, a guild-hall; also *Hist.*, one of these guilds.
1851 J. Ruskin *Stones of Venice* I. 340 It is the most curious in conception of all the pictures in the Scuola. 1888 *Encycl. Brit.* XXIV. 155/1 Much of the splendour of Venice..was due to the wealth and religious zeal of the various trade guilds or confraternities, called *scuole* by the Venetians. 1902 R. Fry *Let.* 14 Oct. (1972) I. 198 Jacopo Bellini is known to have painted a large number of decorative pieces for the Venetian *Scuole* and these are, I believe, part of one of these series. 1936 A. B. Greene *Sunshine & Dust* xxiii. 397 In the Scuola, some wood-carvings of the life of St. Roche are worth study. 1961 L. Mumford *City in History* Note to plate 21, The architectural quality..is repeated..in the..scuola or guild hall. 1962 *Listener* 13 Sept. 386/2 The crucifix still exists..in the Church of S. Giovanni Evangelista in Venice; it was originally made for the *scuola* of that church, to house..a fragment of the True Cross. 1965 H. Honour *Companion Guide to Venice* iv. 53 The members of one *scuola* visited prisoners and paid for the last rites of those who were

executed. 1974 *Country Life* 25 Apr. 978/1 Most of the *Scuole* of Venice were..suppressed by the Napoleonic régime... Though teaching formed part of their function, the main aims of the *Scuole* were charitable.

scupper, *sb.* Add: **1. c.** *fig. coarse slang.* A depreciatory term for a woman, *esp.* a prostitute.
1935 A. J. Pollock *Underworld Speaks* 102/2 *Scupper*, a prostitute. 1970 G. Greer *Female Eunuch* 265 More familiar terms in current usage refer to women as receptacles for refuse,..as *tramp, scow, scupper.* 1972 F. Warner *Lying Figures* IV. 40 *Sapph* You were always firm... *Laz* Your limbs and trunk were in angles of contingency. *Sapph* I was your scupper.

scupper, *v.* Add: **b.** *colloq.* To defeat, ruin, destroy, put an end to.
a 1918 [see *knock *v.* 12 a]. 1948 [see *ditch *v.*¹ 6 c]. 1957 *Economist* 19 Oct. 235/1 The suspicion is still alive that there would have been secret rejoicing in Whitehall if the French Assembly had scuppered the common market. 1957 L. Durrell *Justine* III. 155 You can help us scupper them, old man. 1962 *Times* 2 Mar. 4/2 Underwood followed up his kick ahead and, when scuppered, found Rogers, as ever, there for a try at the post. 1974 *Times* 7 Feb. 14/8 If the Government wants to welsh on its promise, it will have to scupper Mr Money's Bill. 1981 W. Winward *Ball Bearing Run* iv. 51 'We're scuppered,' said Fallon... It was a crushing blow.

scuppernong. a. Substitute for def.: A cultivated grape-vine belonging to the variety of the southern muscadine, *Vitis rotundifolia*, so called, originally found in the region of the Scuppernong River; also, the fruit of a vine of this kind. Also *attrib.* (Earlier and later examples.)
1811 *Raleigh* (N. Carolina) *Star* 7 Mar. 40/2 Doctor James Mease..having seen Mr. Blount's account of the Scuppernong Grape..has requested of us to procure for him some specimens of the vine. 1829 *Free Press* (Tarboro, N. Carolina) 27 Feb. 3/3 Among them the Scuppernong, a native of North Carolina, growing in a swamp. 1857 *Harper's Mag.* May 746/1 The dwellings in the Piny Woods..almost always have..a trellis supporting an extensive scuppernong grape-vine. 1901 C. T. Mohr *Plant Life of Alabama* 136 The scuppernong grape yields its crops year after year with regular abundance. 1938 M. K. Rawlings *Yearling* ix. 74 The Scuppernong grapevine, a gift from his mother's kin in Carolina, was in bloom for the first time. 1944 *Clarke County Democrat* (Grove Hill, Alabama) 14 Dec. 1/5 The deer became entangled in a growth of scuppernong vines. 1949 B. A. Botkin *Treas. S. Folklore* II. i. 146 The poetic fable of the origin of the purple scuppernong grape in the seedling that sprouted on the edge of the pool stained with her blood from the silver arrow. 1972 J. Hewitt *N.Y. Times Cook Bk.* 308/1 Scuppernong Nectar South Carolina 12 pounds scuppernong grapes 1 cup white vinegar [etc.].
b. Wine made from the scuppernong grape. In full, **scuppernong wine.**
1825 *Catawba Jrnl.* (Charlotte, N. Carolina) 2 Aug. 3/1 The editor..having had a taste of the Scuppernong wine from North-Carolina, extols it in the highest terms. 1846 *Spirit of Times* 25 Apr. 97/1 A keg of 'Scuppernong' is on its way to us, having been shipped from Wilmington, N.C. 1862 'E. Kirke' *Among Pines* xvii. 280 [He] brought forth a box of Havanas, and a decanter of Scuppernong. 1887 [in Dict.]. 1936 M. Mitchell *Gone with Wind* xxxviii. 685 It never occurred to him that a decent woman would drink anything stronger than scuppernong wine.

scuppled (skʌ·p'ld), *a. nonce-wd.* [Cf. Scuffle *v.*²] Grooved, furrowed.
1873 G. M. Hopkins *Jrnls. & Papers* (1959) 235, I saw big smooth flinty waves, carved and scuppled in shallow grooves.

scur (skɔ̈ɹ). [Origin unknown.] A small horn found in polled cattle and sheep or their cross-bred offspring, not rooted in the skull but loosely attached by the skin. Hence **scurred** *a.*, having scurs.
1882 *Nat. Live Stock Jrnl.* (Chicago) Oct. 460/1 A heifer with only 'scurs', as the modified horns sometimes found in Polled cattle and in cross-bred offspring of Polled and horned breeds, are called in Scotland. 1902 W. Bateson *Mendel's Princ. Heredity: A Defence* 6 The offspring of the Polled Angus cow and the Shorthorn bull is almost invariably polled or with very small loose 'scurs'. 1919 J. Biggar *Galloway Cattle* 7 The Galloway increased the total of polled stock in the county, and knocked out the 'scurs' or abortive horns very considerably. 1960 *Farmer & Stockbreeder* 23 Feb. 81/1 Not every heterozygous bull showed scurs. *Ibid.* 22 Mar. 81/1 Some had small loose horns, some just had scurs. 1963 *Guardian* 3 Dec. 6 Only about half the ewes carry thin spiky horns, the others being polled or 'scurred'.

scurfer (skɔ̈·ɹfəɹ). [f. Scurf *v.* 3 + -er¹.] An operative who removes incrustations of dirt from boilers, metal plates, etc.
1881 *Instructions to Census Clerks* (1885) 36 Scurfer (Ships). 1921 *Dict. Occup. Terms* (1927) § 699 *Scurfer, retort scurfer* (gas works);..chips off deposited gas carbon from sides of retorts when it gets too thick. *Ibid.* § 952 *Boiler cleaner, boiler scaler, boiler scraper, boiler scurfer,..scurfer boy*..; removes 'scale' (incrusted deposit) from inner surface of boilers and from boiler tubes.

S-curve, -curved: see *S I. 2 c.

scurvy, *sb.* Add: **1.** Now recognized as due to insufficient ascorbic acid (vitamin C) in the diet. (Further examples.)
1966 Dunlop & Alstead *Textbk. Med. Treatment* (ed. 10) 390 Scurvy is a nutritional disease which results from prolonged subsistence on diets practically devoid of fresh fruits and vegetables. 1968 Passmore & Robson *Compan. Med. Stud.* I. v. 10/1 Five major diseases: scurvy, beriberi, pellagra, keratomalacia and rickets, arise as a result of a dietary lack of one of the vitamins.

scuse, *v.* Delete † and for '*Obs.* exc. in illiterate use' read 'Now chiefly in representations of colloq. speech, esp. in form **'scuse.**' (Further examples.)
1864 Dickens *Our Mutual Friend* (1865) I. 231 'Scuse *me*, Lawyer Lightwood, it's a part of the truth. 1902 [see *excuse *v.* 6 b]. 1922 Joyce *Ulysses* 418 All poppycock, you'll scuse me saying. 1971 G. Sims *Dead Hand* I. iv. 46 Scuse fingers.

scut, *sb.*¹ **2. b.** Delete † *Obs.* and add later examples.
1877 *Coursing Cal. 1876* 21 Handling her hare in grand style, never left the scut until she killed. *Ibid.* 78 Keeping well to the scut, she never gave her opponent a chance. 1921 Galsworthy *To Let* III. xi. 310 He sat there a long time dreaming his career, faithful to the scut of his possessive instinct.

scut (skʌt), *sb.*⁴ *dial.* or *slang.* Also **scutt, skut.** [? Var. of Scout *sb.*²] A term of contempt for a person.
1873 Trollope *Harry Heathcote* (1874) vi. 146, I thought you was ringing trees for that young scut at Gangoil? 1895 M. E. Francis *Frieze & Fustian* 63 I'll pinnance ye, ye little scut! 1901 *Longman's Mag.* Sept. 405 Moran, ye scut! don't be skirmishin'. 1916 J. B. Cooper *Coo-oo-ee* viii. 95 The likes of them skuts to find fault with my cookin'—'deed it's more than O'Callaghan himself would dare do. 1929 W. Deeping *Roper's Row* ix. 96 He always was a precocious little scut. 1936 M. Franklin *All that Swagger* xxvii. 254 You speak like a low-down scut. 1970 L. Sanders *Anderson Tapes* xii. 41 You bloody scut! Forget it! 1978 'M. M. Kaye' *Far Pavilions* lxiii. 863 You'd think those scutts could have given us a bit more notice... It's a shabby lot they are.

scut (skʌt), *sb.*⁵ *U.S. colloq.* [Origin unknown: cf. prec.] Tedious menial work. Freq. *attrib.* in *scut work.*
1960 Wentworth & Flexner *Dict. Amer. Slang* 454/1 *Scud, scut*,..hard, boring, or tedious tasks; minor details that are unrewarding and time-consuming. 1972 *Newsweek* 10 Jan. 37/2 Huber inflicted a fair amount of scut work on the boy—washing dishes, fetching books and journals from the library. 1976 *National Observer* (U.S.) 20 Mar. 14/2, I did all the scutwork: paid the bills, ran the houses, drove the children. 1978 L. Pryor *Viper* (1979) iii. 40 The servants..were..brought from Iran and Turkey to do the scut work. 1978 *Maledicta* II. 69 *Scut*, menial medical procedures that must be carried out, usually relegated to the least senior member of the medical team.

scutching, *vbl. sb.*¹ (Later example.)
1962 H. G. Green *Time to pass Over* xii. 142 I'll damn well have to give you a scutching for this.

scutching, *vbl. sb.*² Add: **1.** (Later examples.)
1875 [see *blowing *vbl. sb.*¹ 1 b]. 1902 [see *dropping *vbl. sb.* 5 b]. 1931 S. A. G. Caldwell *Preparation & Spinning of Flax Fibre* I. i. 15 The old method of scutching by hand has now given place to mill scutching. 1937 W. E. Morton *Introd. Study of Spinning* I. iv. 84 It is the essential purpose of scutching to remove only the boon and the bark. 1973 *Times* 7 May 11/5 One particularly alarming process, scutching, was always done by hand, and I am surprised that a whole generation of Ulstermen have any fingers uncrushed, for to scutch you feed a hank of flax under the karate-chop action of a wooden propeller!
2. *scutching blade, machine* (later examples); **scutching knife** (later *Hist.* example).
1973 L. Russell *Everyday Life Colonial Canada* ix. 111 A scutching blade, a wooden tool shaped like a butcher's knife. 1969 E. H. Pinto *Treen* 301 Old Irish, Scottish and English scutching knives are usually plain and straight bladed. 1839 Scutching machine [see *blowing-machine* s.v. *blowing *vbl. sb.*¹ 5]. 1901 T. Thornley *Cotton Spinning* I. iii. 75 The rollers..are seldom used on any scutching machines.

scute (skiūt), *sb.*² *Zool.* [f. Scute(llar *a.*] The name of any of a group of closely linked X-linked genes in *Drosophila* which act to reduce the number of scutellar bristles; also, a phenotype produced by these genes.
1923 Bridges & Morgan *Third-Chromosome Group Mutant Characters Drosophila Melanogaster* 160 Scute arose in the line selected for increased number of scutellar bristles. 1923 *Anat. Rec.* XXVI. 397 In three of these species the closely linked [*sc.* to yellow] character 'scute' or 'scutellar' is also known. 1940 *Genetics* XXV. 566 The great phenotypic similarity of the three scutes in question is an expression of the extreme similarity of their gene arrangements. 1974 Goodenough & Levine *Genetics* xi. 500 The *Basc* chromosome..carries the *Bar* eye gene *B*.., the *apricot* eye color gene *apr*, and a double inversion involving the *scute* (*sc*) region of the chromosome.

scutt, var. *scut *sb.*⁴

scutter, *sb.* (Later examples.)
1935 E. Pound *Let.* 23 May (1971) 274 The turn of the wave and the scutter of receding pebbles. 1961 H. R. F. Keating *Rush on Ultimate* i. 9 Clearly visible from their moderate height the violent scutter of human activity—figures running up and down on the beach in short sharp bursts. 1980 *Times* 16 Jan. 14/8 The Anglo-Egyptian treaty was settled up in 1936 in a sort of scutter.

scutter, *v.*² For '*colloq.* and *dial.*' read 'orig. *colloq.* and *dial.*' and add later examples. Also *fig.*
1916 A. Bennett *Lion's Share* xvii. 128 Miss Ingate scuttered to Audrey. 'Well,' she whispered. 'Here I am.' 1920 Wodehouse *Jill the Reckless* (1922) xxi. 313 All these people... Scuttering about and thinking they know all there is to know. 1934 E. Pound *ABC of Reading* II. 99 Inferior passages where he..has..scuttered over less interesting matter. 1947 A. Ransome *Great Northern?* vi. 82 A family of baby water-hens scuttered across. 1948 L. MacNeice *Holes in Sky* 38 A pebble Scutters from under the wheel. 1972 R. Adams *Watership Down* viii. 28 They watched him ..shake a shower of drops out of his fur and scutter into the alder bushes.

scuttle, *sb.*¹ Add: **3.** The part of a motor vehicle which connects the bonnet with the body. Also *attrib.*
1914 *Chambers's Jrnl.* Mar. 206/1 The scuttle-dash protects his body and chest from the wind. 1922 *Autocar* 10 Nov. 973 The coachwork, which now tapers from bonnet to scuttle in an unbroken sweep. 1925 *Morris Owner's Man.* 71 Under the butterfly nut at the back of the bonnet hinge (on top of the scuttle). 1963 *Times* 4 June 7/7 Above 65–70 m.p.h..., road noise was high and there was some scuttle shake. 1970 *Motoring Which?* July 109/2 Front wing to scuttle seams cracking, driver's seat had cut through carpet. 1973 'A. Hall' *Tango Briefing* vii. 94 A hole appeared in the scuttle three inches forward of the windscreen. 1980 *Daily Tel.* 5 Mar. 12/6 Even on some of the worst French road surfaces there was hardly any sign of scuttle shake.

scuttle, *sb.*⁵ Add: **1. a.** (Later example.)
1967 O. Wynd *Walk Softly, Men Praying* iv. 49, I stopped a scuttle back to his den with yet another request.
b. For (See Scuttle *v.*¹ 2.) in Dict. read (See Scuttle *v.*¹ 1 b.), and add later examples.
1906 *Daily News* 16 June 6/4 The word scuttle..in the Jingo Press, where the 'policy of scuttle' is used whenever we give up something to a small Power. 1954 *Economist* 7 Aug. 428 Mr. Attlee has always been particularly offended by Tory charges of 'scuttle' during his years in office. 1967 *Daily Express* 17 Feb. 2/2 The speed-up of the scuttle is inadvertently revealed in the Defence White Paper. 1969 *Guardian* 31 July 8/1 President Nixon has ruled out both a military solution and a 'scuttle'.

scuttle, *v.*² Add: **1. a.** Hence, by extension: to sink (one's own vessel) deliberately; to submerge; to destroy or smash (a car, aeroplane, etc.). Also *refl.*, of a vessel.
1779 J. Ramsay *Let.* 23 Nov. in *Parl. Papers 1910* (Cd. 5038) XXXV. 675 Captain Wilkinson is particularly celebrated for having said..he wished that all the English sugar islands were skuttled (sunk). 1939 *Times* 16 Dec. 7/3 The German steamer Adolf Leonhart (2,989 tons) was scuttled by her crew in the South Atlantic when she was intercepted by a British warship. *Ibid.* 18 Dec. 8/1 The Admiral Graf Spee the German 'pocket battleship' which was cornered by British warships, scuttled herself outside Montevideo harbour last night. 1941 *Collier's* 20 Dec. 50/1 Bruce, in a heavy sedan, had come banging out of his driveway to a collision. No one was hurt, but the flivver was scuttled. 1942 *R.A.F. Jrnl.* 13 June 10 The enemy would reveal herself by firing her guns or scuttling herself. 1955 *Times* 1 June 2/7 His Black Sea fleet had been scuttled and the war was confined to a narrowing circle round Sebastopol.
b. *fig.* (esp. in political contexts).
1888 [in Dict.]. 1940 *Star* 15 May 3/2 A favourite way of settling an argument is to exclaim, 'You go and scuttle yourself!' 1942 *Capital* 20 Jan. 1/3 The ill-concealed Axis maneuver, apparently part of a..scheme to scuttle the [Rio de] Janeiro conference. 1955 *Times* 16 May 5/1 'Now they scuttle us,' said Mr. Oatley, 'by slapping controls on what can be considered the cornerstone of every household—a domestic cooker.' 1965 Mrs L. B. Johnson *White House Diary* 21 Oct. (1970) 329 The day..began with bad news. The Rent Subsidy Bill had been scuttled without opportunity to work on it. 1976 *National Observer* (U.S.) 24 Apr. 18/3 Earlier this month the White House, in apparent agreement, scuttled a legislative proposal to deregulate cable TV. 1977 *Time* 7 Nov. 21/3 His effort to pay off a campaign promise to maritime unions by fixing the percentage of imported oil that must be carried in U.S. ships was scuttled.

scuttle-butt, scuttlebutt (skʌ·t'lbʌt). *Naut.* [See Scuttled *ppl. a.*] **a.** A cask of drinking-water on board ship; a drinking-fountain. Also *fig.*
1805 J. J. Moore *Midshipman's or Brit. Mariner's Vocab.*, *Scuttle-butt*, or cask, is a cask having a square piece sawn out of its bilge and lashed upon the deck. It is used to contain the fresh water for daily use. 1832 E. C. Wines *Two Years & Half in Navy* I. iii. 45 At sea the marines in succession all do duty as sentries in the following places;—one at the cabin door, one at the scuttle-butt, one at the brig. 1840 [see Scuttle *sb.*² 4]. 1844 J. F. Cooper *Afloat & Ashore* II. xi. 178 People never can tell so much of other person's affairs, without bailing out most of their ideas from their own scuttle-butts. 1846 [see Scuttled *ppl. a.*]. 1920 *U.S.S. Oklahoma Sea-Bag* 25 July 2/1 The Scuttle Butt has justified its existence as a source of prognostic rumor. The water is freezing cold—the Scuttle Butt is iced... Come down and get a drink of cold water. 1972 F. van W. Mason *Roads to Liberty* 104 Katie, on her way to draw water from the scuttlebutt, saw a flying fish.

b. orig. *U.S. Naut. slang.* Rumour, idle gossip, unfounded report.
1901 *Smoking Lamp* June 55/2 (*title of miscellany column*) Scuttle butt. 1933 *Leatherneck* July 18/1 We will endeavor to convey all of the scandal, scuttle-butt, dope and dopes to you through the..Editor. 1943 *Sun* (Baltimore) 20 Sept. 11/7 Also a cause for betting was the ultimate destination. In navy slang 'scuttlebutt' was rife and had the ship bound everywhere from China to Murmansk. 1950 'D. Divine' *King of Fassarai* iv. 20 I'd got the scuttle-butt about that from the Marine Corps boys. 1966 K. Giles *Big Greed* 76 He'd heard a rumour that the Frog drank.. but he dismissed it as a scuttlebutt. 1977 *Time* 11 Apr. 17/1 Rawlings Co. now makes the official major league baseball after a 101-year Spalding reign, and the scuttlebutt is that Rawlings is turning out a rabbit ball. 1980 *Rydge's* (Sydney) Jan. 23/1 They are privy to vast amounts of corporate intelligence (and scuttlebutt) that runs daily around the Rialto.
c. *attrib.*, as **scuttle-butt gossip,** (*a*) one who exchanges gossip at a scuttle-butt; (*b*) the gossip exchanged there; **scuttle-butt yarn,** a yarn originating from talk around a scuttle-butt.
1901 *Smoking Lamp* May 18/1 (*title*) Scuttle butt gossip. 1918 R. W. Kauffman *Our Navy at Work* xiii. 198 It's wilder than anything the scuttle-butt gossips could make up! *Ibid.* 199 Ships are full of..rumours..which originate in talk exchanged around the skuttle-butt, or drinking barrel, so that all wild stories are branded as 'scuttle-butt yarns'. 1923 *Our Navy* 1 May 15/2 While these rumors are branded as of the 'scuttle butt' variety they persist as rumors. 1930 P. Buranelli *Maggie of Suicide Fleet* ii. 57 There are stories and rumors, scuttle-butt gossip.
Hence **scu·ttlebutt** *v. intr.*, to gossip, to spread unfounded stories.
1945 H. I. Phillips *Private Purkey's Private Peace* xi. 62 There were many who felt..Oscar was just scuttle-butting.

scuzzy (skʌ·zi), *a. N. Amer. colloq.* [Perh. blend of Scummy *a.* + Fuzzy *a.*] Dirty, grimy; murky. So **scuz(z)** *sb.*, a dirty, messy person.
1968 *Sunday Sun* (Baltimore) 3 Nov. D1/5, I..did 'Midnight Cowboy' where I'm *Ratso Rizzo*, a complete scuzz. 1969 *Publ. Amer. Dial. Soc.* LI. 16 *Scuzzy, groady, skoady*, and *grungy* should probably be listed also under 'Blends'... *Scuzzy*, for example, seems to imply *fuzzy* and *scummy*: 'Your teeth are scuzzy.' 1972 J. Wambaugh *Blue Knight* vi. 78 One white, bearded scuz in a dirty buckskin vest and yellow headband. 1974 A. Fowles *Pastime* vii. 63 The scuzzy, grey, February days, neither cold nor clear. 1976 *Daily Colonist* (Victoria, B.C.) 14 Apr. 5/4 Perhaps Mr. Vander Kalm has good intentions about evicting scuzzy malingerers from the dole.

scyphozoan (səiːfozōᵘ·ăn), *sb.* and *a.* [f. mod. L. class name *Scyphozoa* (A. Goette 1898, in *Zeitschr. f. Wiss. Zool.* LXIII. 292), f. Scypho- + *-zoa* (see Zoon).] A jellyfish belonging to the class Scyphozoa; of or pertaining to an animal of this kind.
[1892 J. A. Thomson *Outlines Zool.* x. 137 (*caption*) Contrast between a Hydrozoon and a Scyphozoon medusoid.] 1915 Shipley & MacBride *Zool.* (ed. 3) iii. 75 The construction of the Scyphozoan organ is quite different. *Ibid.* 78 The original Scyphozoan was probably an organism like a polyp. 1938 R. Buchsbaum *Anim. without Backbones* viii. 95 *Aurelia* is one of the commonest of the scyphozoan jellyfish. 1967 P. A. Meglitsch *Invertebrate Zool.* vi. 148/1 Stauromedusae are the most unusual scyphozoans.

Scyth. a. For 'Now *rare*' in Dict., read '*Obs.* exc. *Hist.*' and add later examples.
1914 D. G. Hogarth *Ancient East* iii. 122 The predatory Scyth..probably lacked skill to inscribe them. 1950 [see *Kimmerian]. 1964 *Listener* 6 Feb. 238/2 So we came to the plain and the Scyths. 1973 R. L. Fox *Alexander* iv. 75 The barbarian Scyths and Thracians.
b. = *Scythian *sb.* 1 b. *rare.*
1972 B. Thomson *Premature Revolution* I. vii. 130 The 'Scyths' revived the old Slavophile faith in the Russian peasantry, as a bastion of spiritual values in an age of materialism... Usually the 'Scyths' were concerned with purely Russian questions, but Blok's poem *The Scyths* (*Skify*, 1918) reveals a chauvinistic and aggressive side to the movement.

scythe, *sb.* Add: **4. a.** *scythe-stroke* (also *fig.*).
1913 D. H. Lawrence *Mowers* in *Smart Set* Nov. 12 There's four men mowing down by the river; I can hear the sound of the scythe strokes, four Sharp breaths swishing. 1940 W. S. Churchill *Into Battle* (1941) 216 This armoured scythe-stroke almost reached Dunkirk—almost but not quite.

scythe, *v.* Add: **2. b.** *fig.* To cut down swiftly and drastically.
1970 *Daily Tel.* 12 Mar. 22/3 Net attributable profits are scythed from £602,000 to £210,000.
3. Delete *nonce-use* and add later examples.
1946 J. W. Day *Harvest Adventure* vii. 107 'Pleu-eu! Pleu-eu! Pleu!' and whimbrel went scything off low across the water, putting up a mixed lot of sandpipers. 1955 E. Pound *Classic Anthol.* II. 94 Feckless Huns town'd in Tsiao, seized Huo, lacking provisions, Scythed into Hao up to its border. 1966 Gillman & Haston *Eiger Direct* v. 107 John scythed up on his skis and stopped in a spray of snow. 1978 *Antiques & Art Monitor* 28 Oct. 23/3 High-rise aerial perspectives of a motorway scything through a city.

scythed, *ppl. a.* Add: **1.** (Later example.)
1922 Joyce *Ulysses* 649 Humanely his driver waited till he (or she) had ended, patient in his scythed car.
3. *fig.* Swept over as though by a scythe.
1952 Dylan Thomas *Coll. Poems* 172 Who knows the rocketing-wind will blow The bones out of the hills, And the scythed boulders bleed.

Scythian, *a.* and *sb.* Add: Now freq. with pronunc. (si·ðiăn). **B.** *sb.* **1. b.** *Russ. Lit.* An advocate of Scythism (sense *c).
1923 *Contemp. Rev.* Aug. 193 There is an immense wealth of pride in these *Scythians*, they heartily and sincerely despise the West. 1970 M. Ginsburg tr. *Zamyatin's Soviet Heretic* 22 Ivanov-Razumnik..leader of the Scythians, a literary group that included Blok and Bely. 1974 Moore & Parry *Twentieth-Cent. Russ. Lit.* ii. 18 He was now briefly one of the Scythians, who confronted the new social events with fervor and a sense of exultation.
2. (Later examples.)
1870 [see *Babylonian *sb.* 2]. 1894 [see *Medic *sb.*]. 1939 L. H. Gray *Foundations of Lang.* xiii. 425 Hesychius..cites words..from many non-Classical languages, such as..Scythian. 1972 W. B. Lockwood *Panorama Indo-European Lang.* xii. 235 The exiguous records of the Median language are of the same character as those of Scythian and Sarmatian.

Scythianism: see *Scythism c.

scything, *ppl. a.* Add: Also, having the sweeping action of a scythe. Also *fig.*
1960 E. Hamilton *Great Teresa* i. 37 The black, scything wings of swifts dipping and swerving. 1963 *Times* 17 May 5/4 Smith's scything forehand stroke. 1978 *Daily Tel.* 18 Feb. 1 Some victims were killed by collapsing masonry and scything debris.

scything (səi·ðiŋ), *vbl. sb.* [f. Scythe *v.* + -ing¹.] The action of the verb; the result of this, scythed grass, etc. Also *fig.*
1969 *Gloss. Landscape Work* (*B.S.I.*) v. 21 *Scything*. 1. The cutting of grass with a scythe. 2. Now usually the cutting of long grass, or other vegetation, either with a scythe or by a machine with reciprocating blades. 1969 M. Pocock in R. Blythe *Akenfield* xii. 196 Back she arrived later when I was lying on the scythings. 1978 *Maledicta* II. 232 In order to give you a taste of the 'vilest venom' of scholarly verbal aggression mentioned in our first issue..I wish to share with you excerpts from Zizi Quirk's scything in *Maledicta*.

Scythism. Restrict † *Obs.* to senses in Dict. and add pronunc. (si·þiz'm, si·ðiz'm). **c.** *Russ. Lit.* A movement among Russian men of letters soon after the Revolution of 1917 which favoured the peasant values of Asiatic Russia as against Western European civilization. Also **Scy·thianism.**
The term is a rendering of Russ. *skifstvo*.
1921 D. H. Lawrence *Sea & Sardinia* v. 164, I am glad that Russia flies back into savage Russianism, Scythism, savagely self-pivoting. 1923 *Contemp. Rev.* Aug. 193 National Bolshevism is of much more recent growth than either Bolshevism or Scythianism. 1926 *Encycl. Brit.* III. 1070/1 Moscow became the scene of a struggle between what may be called 'Europeanism' and what is known there as 'Scythism'. Scythism, which achieved the miracle of rallying to the Bolshevik standard the Russian reactionaries who were most violently opposed to it, inculcates scorn and hatred of western civilization, and aims at nothing less than Asia's vengeance on Europe. 1958 E. H. Carr *Socialism in One Country* I. ii. 60 After the publication of Blok's poem the name 'Scythism' (Skifstvo) came to be applied, not to a literary movement, but to a tendency which inspired many writers in the first years of the revolution. 1963 G. Struve in Hayward & Labedz *Lit. & Revol. in Soviet Russia* 4 R. Ivanov-Razumnik..the main ideologist, in the early days of the Revolution, of the so-called 'Scythianism'. 1972 B. Thomson *Premature Revol.* I. vii. 130 The peasant poets were..more sympathetic to the Social-Revolutionaries than to the Bolsheviks... Under the guidance of Ivanov-Razumnik they formed a movement called Scythianism ('Skifstvo').

S

(continued)

|| **se** (se). *Chinese Mus.* Also **che, she, tche.** [a. Chinese *sê*.] A twenty-five-stringed plucked musical instrument, somewhat similar to the zither.

This ancient instrument was already widely used during the Spring and Warring States period (770–476 B.C.). The number of strings was not fixed in ancient times. In quot. 1955, wrongly described as a lute.

1874 C. Engel *Catal. Mus. Instruments in S. Kensington Mus.* 53 The ancient stringed instruments, the *kin* and *chê*, were of the dulcimer kind. *Ibid.* 182 *Tche.* Wood, mounted with bone. Sixteen thin brass wires. The movable bridges belonging to the instrument are wanting. **1884** J. A. Van Aalst *Chinese Mus.* 62 The Sê..is said to have been invented by P'ao Hsi.., and to have had originally 50 strings... But the sê now in use has 25 strings..elevated on a movable bridge. **1908** A. C. Moule in *Jrnl. N. China Branch R. Asiatic Soc.* XXXIX. 108 Shê (Che)... A horizontal psaltery, curved above and flat below... There are twenty-five silk strings..stretched along the body. **1955** A. Fang in E. Pound *Classic Anthol.* p. xi, Of the two kinds of lute, the *k' in* has seven strings and the *se* twenty-five. **1977** Kwang-chih Chang *Archaeol. Anc. China* (ed. 2) ix. 402 Remains of wooden bases for the musical instrument *se* have been collected from the tombs.

sea, *sb.* Add: **I. 2. c.** [tr. L. *mare* (see *MARE⁴).] (Earlier and later examples.)

1667 R. Hooke *Micrographia* I. lx. 245 Those mountains, which are by Hevelius call'd the Apennine Mountains, and some other, which seem to border on the Seas of the Moon. **1698** C. Huygens *Celestial Worlds Discover'd* II. 130 Those vast countries which appear darker than the other, commonly taken for and call'd by the name seas, are discover'd with a good long telescope, to be full of little round cavities. **1833** J. F. W. Herschel *Treat. Astron.* vi. 229 What is, moreover, extremely singular in the geology of the moon is, that although nothing having the character of seas can be traced, (for the dusky spots which are commonly called seas, when closely examined, present appearances incompatible with the supposition of deep water,) yet there are large regions perfectly level, and apparently of a decided alluvial character. **1907** G. P. Serviss *Moon* iii. 146 This..does not invalidate what I have said about the lunar 'seas', or plains, darkening near sunset more rapidly than we should expect them to do, as a simple result of the low angle at which the sunlight strikes them. **1949** *Jrnl. Brit. Interplanetary Soc.* VIII. 185 The origin of the characteristic features of the lunar surface, craters, mountain ranges and 'seas', is far from being understood. **1974** *Times* 17 Apr. 16/3 It seems that the maria and the basins of the lunar 'seas' are of volcanic origin.

5. d. Roughness of the sea brought about by wind blowing at the time.

1927 G. Bradford *Gloss. Sea Terms* 152/1 The waves prevailing at any time are spoken of collectively as the sea, but they must be due to the wind then blowing. **1970** J. Verhoogen et al. *Earth* vii. 341/1 In the presence of the generating wind, waves have steep, sharp, asymmetric crests, and broad troughs, and the whole water surface is irregularly choppy. This condition is known as sea. **1977** *Offshore Engineer* July 35/1 In August 1975, the LWC began by using graphical methods to produce sea-swell forecast charts, combining 'sea', or wind-driven waves and 'swell', which is persistent wave movement continuing after the wind has dropped.

8. c. *Physics.* A (physical or mathematical) space filled with particles of a certain kind, *esp.* one in which only the particles near the boundary or surface are significant.

1955 [see *FERMI I]. **1965** Phillips & Williams *Inorg. Chem.* I. vi. 207 Because there are electron energy levels lying only very little above the surface of the calm Fermi sea, electrons can take up energy at normal temperatures in a metal and so make a contribution to the specific heat. **1972** *Sci. Amer.* Apr. 26/3 Once an atom has lost an electron it becomes a positive ion that finds itself in a deep electrostatic potential well created by the surrounding sea of negative electrons. **1979** *Ibid.* Sept. 76/3 These events are explained by interactions involving a 'sea' of quarks and anti-quarks that have a virtual existence in the vicinity of a proton.

II. 10. c. *worse things happen at sea* and varr.: a consolatory catch-phrase.

1829 P. Egan *Boxiana* 2nd Ser. II. 346 The Fancy were too game to complain.., contenting themselves with the old saying, 'that worse accidents occur at sea!' **1869** C. H. Spurgeon *John Ploughman's Talk* v. 41 To be poor is not always pleasant, but worse things than that happen at sea. **1948** 'N. Shute' *No Highway* xii. 297 Oh well, worse things happen at sea. I expect we shall get over it. **1978** M. Kenyon *Deep Pocket* viii. 97 Worse things 'ave 'appened at sea, he told himself, if 'e shoots..you'll 'ardly feel a thing.

16. b. In the Naut. proverbial phr. *he that would go to sea for pleasure, would go to hell for a pastime* and varr.

1899 A. J. Boyd *Shellback* viii. 110 Shentlemens vot goes to sea for pleasure vould go to hell for pastime. **1910** D. W. Bone *Brassbounder* xxvi. 289 He gave a half-laugh, and muttered the old formula about 'the man who would go to sea for pleasure, going to hell for a pastime!' **1924** R. Clements *Gipsy of Horn* iii. 50 'He who would go to sea for pleasure would go to hell for a pastime' is an attempt at heavy satire. **1933** M. Lowry *Ultramarine* i. 50 Well, a man who'd go to sea for fun'd go to hell for a pastime.

III. 18. Simple attributive: **a.** Of or belonging to the sea or a sea: *sea-bed* (later examples), *-marge* (later examples), *-romp*, *-surge*, *-swell*, *-swill*.

1937 *Discovery* Sept. 279/2 The sea-bed gave out a bluish light. **1975** *Offshore* Sept. 49-04/1 Other firms in this business include Heerema, with three special ships designed to drill seabed holes up to 1,200 ft in 700 ft of water. **1923** H. Belloc *Sonnets & Verse* 159 The rank sea-marge. **1976** *New Yorker* 8 Mar. 122/3 The bird has been watched on the sea-marge of Jamaica Bay. **1876** G. M. Hopkins *Wr. Deutschland* xvii, in *Poems* (1967) 57 They..rolled With the sea-romp over the wreck. **1912** E. Pound *Ripostes* 25 Known on my keel many a care's hold, And dire sea-surge. **1930** —— *XXX Cantos* vii. 25 Ear, ear for the sea-surge. **1880** W. Whitman *Daybks. & Notebks.* (1978) III. 628 A little sea-swell on the water. **1927** H. Crane *Let.* 12 Sept. (1965) 306 The movement of the verse..of the 'Ave Maria', with its sea-swell crescendo. **1878** G. M. Hopkins *Poems* (1967) 74 Till a lifebelt and God's will Lend him a lift from the sea-swill.

b. That is an attribute or quality of the sea: *sea-beat*, *-sound*, *-voice* (later examples), *-wash*.

a **1953** Dylan Thomas *Quite Early One Morning* (1954) 16 Sea captains..going down into a..cabin of sleep, rocked to the sea-beat of their ears. **1961** Sea-sound [see *HIGHVELD]. **1974** *BP Shield Internat.* Oct. 18/2 All sea-sounds were eclipsed by the noise of the drilling operation. **1930** T. S. Eliot *Ash Wednesday* 20 The lost lilac and the lost sea voices. **1955** C. Tomlinson *Necklace* 9 The sea-voice Tearing the silence from the silence. **1930** W. de la Mare *On Edge* 297 With the sea-wash in her ears. **1965** E. Richardson *Living Island* 123 There is no sigh of wind and scarcely a whisper of seawash.

c. Consisting of sea: *sea-approach.*

1913 J. Masefield *Mainsail Haul* (ed. 2) 139 The defences to the sea-approach were powerful. **1940** E. Colston Shepherd *Britain's Air Power* 9 The more usual work of these [coastal reconnaissance] aircraft is that of continuous patrol over all the sea approaches to Germany.

d. Phenomena occurring at sea: *sea-dusk*, *-gust*; also designating actions or events which take place at sea, as *sea-burial*, *-death*, *-rescue*.

1838 Poe *Narr. A. G. Pym of Nantucket* vii. 74 The mate..ordered the men to..allow it [*sc.* the body] the usual rites of sea-burial. **1922** Joyce *Ulysses* 51 Seadeath, mildest of all deaths known to man. **1916** —— *Portrait of Artist as Young Man* (1969) v. 226 Swallows flying through the seadusk over the flowing waters. **1970** T. Hughes *Crow* 31 The curlew trawled in seadusk through a chime of wineglasses. *c* **1866** G. M. Hopkins *Poems* (1930) 138 She listened how the sea-gust shook. **1874** *Trans. Highland Soc.* 245 Great loss and much misery is often caused by these destructive 'sea-gusts'. **1959** *Listener* 6 Aug. 217/3 A British sea-rescue plane. **1976** *Morecambe Guardian* 7 Dec. 25/6 A dramatic sea rescue during the early hours of July 5.

e. (*a*) Deposited by or in the sea: *sea-stone*; (*c*) proceeding from the sea: *sea-fog* (later examples), *-fret* (earlier and later examples), *-mist.*

1834 J. J. Audubon *Ornith. Biogr.* II. 201 The sea-fog began to approach the land so swiftly, that..we judged it prudent to return to our vessel. **1893** Kipling in *Pall Mall Budget* 14 Dec. 1950/2 West you'll turn and south again beyond the sea-fog's rim. **1972** *Gloss. Aeronaut. & Astronaut. Terms* (*B.S.I.*) xv. 6 *Sea fog*, fog formed at sea, usually by condensation of moisture in the lower layers of a warm air current passing over a relatively cold sea surface. **1842** C. Ridley *Let.* Feb. in *Cecilia* (1958) vii. 86 This evening everything was thawing but I imagine it was only what they call a sea fret. **1963** *Times* 13 June 4/6 Those who came yesterday out of the heat in the surrounding country were surprised to find Brighton enveloped in a sea fret, which..reduced visibility to a furlong or two. **1893** Kipling in *Pall Mall Budget* 14 Dec. 1947/2 The *Northern Light* drove into the bay and the sea-mist drove with her. **1974** L. Deighton *Spy Story* i. 11 A flurry of sea mist that rolled in upon us. **1918** D. H. Lawrence *New Poems* 27 Sad he sits on the white sea-stone. **1922** Joyce *Ulysses* 291 From his girdle hung a row of sea-stones. **1936** *Geogr. Jrnl.* LXXXVIII. 105 Bib Nambas..are very Melanesian,..with a frequent pig-moid strain, often with white seastones through their noses.

f. Situated in or by the sea: *sea-cave* (later examples), *garden*, *marsh* (later examples), *pen*, *-track.*

1917 D. H. Lawrence *Look! We have come Through!* 26 They dwelt in a huge, hoarse sea-cave. **1979** *Amer. Poetry Rev.* Mar.–Apr. 45/2 Three craft..negotiate intricate sharp turns and arcs through..narrow canals into sea-caves. **1881** W. D. Hay *Three Hundred Years Hence* vii. 135 With..sea-garden food, life in these deep-down Harbours is by no means unenjoyable. **1947** I. L. Idriess *Isles of Despair* xxxvii. 246 The sea gardens of the lagoons. **1977** *Times* 14 May 12/7 Snorkelling among the magnificent sea gardens which eddy among the rocks. **1835** J. J. Audubon *Ornith. Biogr.* III. 241 The Long-billed Curlew spends the day in the sea-marshes. **1982** 'J. Gash' *Firefly Gadroon* vi. 65 The sea-marshes show between the long runs of banks and dykes. **1976** *Daily Colonist* (Victoria, B.C.) 3 Oct. 32/3 They [*sc.* salmon fry] were transferred to sea pens on barges..and continued rapid growth in salt water. **1890** Kipling *Gipsy Trail* in *Poems 1886–1929* (1929) III. 284 Out on a clean sea-track. **1949** E. Muir *Coll. Poems* (1960) 164 The smooth sea-tracks that open and close again.

h. *sea-battle* (later example).

1940 N. Last *Diary* 9 Apr. in *Nella Last's War* (1981) 47, I kept..wondering if our sailors were winning in the reported sea battle.

j. Pertaining to life at sea: *sea-boot* (later examples; also in Naut. slang phr. *a face like a sea-boot*, a dejected or wry expression), *-booted* (later example); *-bread* (earlier example), *-clothes* (later example), *-language* (later example), *-rig*, *-stock* (later example).

1916 'Taffrail' *Pincher Martin* viii. 150 Wot's up wi' yer? You've got a face on yer like a sea-boot. **1946** *Nature* 14 Sept. 386/2 Land Army hose, sea-boot stockings,..and jungle-green pullovers also came under the scheme. **1971** G. M. Brown *Fishermen with Ploughs* 47 His sea boots filled, and Willag said no more. **1933** L. Luard *All Hands* 44 The decks..were alive with jovial sea-booted men. **1834** A. Underwood *Jrnl.* Dec. in *Southwestern Hist. Q.* (1928) Oct. 131, I in company with two of my fellow passengers started taking with us some sea bread water &c. determined to camp out that night. **1933** W. de la Mare *Fleeting* 45 His stiffening sea-clothes grey with salt. **1928** L. P. Smith *Words & Idioms* 20 The sea-languages of the world. **1840** R. H. Dana *Two Years before Mast* 4, I made my appearance on board at twelve o'clock, in full sea-rig. **1892** C. H. Fretwell *Anc. Mariner* 38, I..purchased my sea-stock of warm clothing, intending to join on the following day.

n. By sea; also, pertaining to navigation: *sea-crossing*, *-trader*, *-trading.*

1936 *British Birds* XXIX. 367 They undertake a 1,200 mile sea-crossing from Greenland to Ireland. **1962** H. R. Loyn *Anglo-Saxon England* i. 24 A sea-crossing is perilous to tribal institutions. **1899** C. J. Cutcliffe Hyne *Further Adv. Capt. Kettle* i. 4 Kettle had come across many types of sea-trader in his time. **1921** *Nineteenth Cent.* July 150 She failed..to become a great sea-trading nation.

o. In designations of persons: *sea-fellow.*

1909 E. Pound *Personae* 37 As Glaucus tasting the grass that made him sea-fellow with the other gods. **1918** D. H. Lawrence *New Poems* 27, I wish a wild sea-fellow would come down the glittering shingle.

p. In appellations of imaginary beings: *sea-girl.*

1917 T. S. Eliot *Prufrock* 16 Sea-girls wreathed with seaweed red and brown. **1923** E. P. Mathers tr. *J. C. Mardrus's Bk. of Thousand Nights & One Night* VII. 80 Suddenly they saw twelve sea girls..come up out of the water and dance a round upon the sand. **1939** Dylan Thomas *Map of Love* 5 The sea-girls' lineaments Glint in the staved and siren-printed caverns.

q. That lives or is found in the sea: *sea-gem*, *-harvest*, *-spawn*, *-vegetable* (later example).

1922 Joyce *Ulysses* 289 Golden ingots, silvery fishes,.. purple seagems and playful insects. **1888** L. A. Smith *Music of Waters* 341 The herring fishery in the Isle of Man is the staple industry of the place—the Manx sea-harvest it is called. **1916** Joyce *Portrait of Artist as Young Man* (1969) iv. 171 The sea harvest of shells and tangle. **1922** —— *Ulysses* 38 Signatures of all things I am here to read, seaspawn and seawrack. **1979** *Dallas* (Texas) *Times Herald* 30 May 8-E/1 Fans of the delicacy [*sc.* seaweed] believe that the term 'sea vegetables' would..enhance the image of native dishes.

19. Objective: *sea-rider.*

1939 W. B. Yeats *Last Poems* 29 That sea-rider Oisin led by the nose Through three enchanted islands.

20. Similative: *sea-cold* (hence *sea-coldly* adv.), *-grey* (later example), *-shot*, *-smiling* adjs.

1921 W. de la Mare *Veil* 78 In sea-cold Lyonesse. **1931** A. Huxley *Cicadas* 57, I reach for grapes, but from an inward vine Pluck sea-cold nipples, still bedewed with brine. **1955** E. Bowen *World of Love* vi. 105 'You mean, you were late at the sea?' 'Not at all,' said Antonia sea-coldly. **1970** T. Hughes *Crow* 34 Seeing sea-grey mash a mountain of itself. **1874** G. M. Hopkins *Jrnls. & Papers* (1959) 248 The sea-shot blue-and-green woollen gown our Lady wears. **1922** Joyce *Ulysses* 277 Charming, seasmiling and unanswering Lydia on Lidwell smiled.

21. Instrumental: *sea-blown*, *-fed*, *-lulled*, *-strewn*, *-sucked* adjs.

1857 J. G. Whittier in *National Era* 22 Oct. 170/5 So to us who walk in summer through the cool and sea-blown town. **1945** J. Betjeman *New Bats in Old Belfries* 27 Whose fantastic mausoleum Sings its own seablown Te Deum. **1922** Joyce *Ulysses* 238 A sailorman, rustbearded, ..eyes her. A long and seafed silent rut. **1847** J. R. Lowell *Poet. Wks.* (1912) 121 Fair Beatrice's spirit wandering now In some sea-lulled Hesperides. **1921** W. de la Mare *Veil* 85 The mild noon air of Spring again Lapped shimmering in that sea-lulled lane. **1892** W. B. Yeats *Countess Kathleen* 125 When fades the sea-strewn rose of day. **1934** T. S. Eliot *Rock* ii. 56 Many left their bodies to the kites of Syria Or sea-strewn along the routes. **1934** Dylan Thomas *18 Poems* 33 Half of the fellow father as he doubles His sea-sucked Adam in the hollow hulk. **1966** *New Statesman* 11 Feb. 196/1, I used To think of the soul As round and smooth Like a sea-sucked pebble.

23. a. Special combinations: **a. sea-affairs**: delete † and add further examples; **sea-air** *attrib.*, pertaining to or involving both the sea and the air; **sea-bag** *U.S.*, a seaman's travelling bag or trunk; also *transf.*, a heavy artillery shell; **sea bed,** † (*a*) a bed for use on board ship (*obs.*); (*b*) the floor of the sea: see sense 18 a in Dict.; **sea-beggar** (later examples); **sea-blessing** *Naut. slang* = *sailor's blessing* s.v. *SAILOR 5 c; **Sea Cadet,** a member of the Sea Cadet Association (see quot. 1976), a voluntary youth organization which seeks to foster and develop for public benefit a sea cadet corps and to provide sea training and promote education in maritime affairs; **sea-change** (later examples); now freq. *transf.* with or without allusion to Shakespeare's use (quot. 1610 in Dict.), an alteration or metamorphosis, a radical change; **sea chest,** (*b*) (see quot. 1909); **sea-cloth,** (*a*) (earlier example); (*b*) cloth used for making sailors' clothing; **sea clutter** = *sea return(s)* below; **sea-cook** (earlier and later examples of phr. *son of a sea-cook*); **sea-corpse** *poet.*, the corpse of a person drowned at sea; **sea-dingle** (later *arch.* example); **Sea Dyak**: see *DYAK; **sea-farm**: also **sea-farmer**; **sea-farming** *vbl. sb.*, mariculture; also as *ppl. a.*; **sea-fever,** longing or desire for the sea or sailing on it; **sea-fire** (later examples); **sea ivory** (earlier and later examples) (see also sense 23 f below); **sea-jockey** *N. Amer.*, a nimble sailor; the sailor of a small craft; occas. *derog.* (cf. *JOCKEY 5 b); **sea-keeping,** of a ship, hovercraft, etc.: the endurance of (rough) conditions at sea; **sea-kindly** *a.* (later examples); **sea-kindliness** (later examples); **sea-lane,** a route at sea for shipping; **sea-lift** *N. Amer.*, a large-scale transportation of troops, supplies, etc., by sea (cf. *AIRLIFT 2); hence as *v. trans.*, to transport by sea; **sea-loch** (later examples); **sea-lock** (later example); **sea-log** (earlier example); **sea-longing,** a yearning for the sea, sea-fever; **sea-mail,** mail conveyed by sea; a service for conveying letters, parcels, etc., by sea (not an official term); so as *v. trans.* (*rare*), to send by sea; cf. *AIRMAIL; **sea marker,** a device which can be dropped from an aircraft to produce a distinctive patch on water below it; **sea-mount,** a large natural elevation rising abruptly from the ocean floor, usu. entirely underwater; an underwater mountain; **Sea People(s)** = *Peoples of the Sea* s.v. *PEOPLE *sb.* 1 c; **sea-price** *Naut. colloq.*, an inflated price; **sea return(s),** unwanted radar images due to reflection from a rough sea; **sea-road** *rare*, a route by sea; **sea scout,** a member of the (Boy) Scout movement engaged in activities pertaining to the sea and seamanship; **sea seiche,** a seiche occurring in the open sea; **Sea Sled** (see quot. 1948) (a proprietary name in the U.S.); **sea-speed** (earlier example); **sea-stack** = Stack *sb.* 7; **sea-state,** the degree of turbulence at sea, esp. as measured according to a scale of average wave height; **sea-time,** (*a*) (later examples); (*c*) the duration of a journey at sea; **sea-toss** (example); **sea-train,** (*a*) a ship used for the transportation of railway cars; (*b*) a group of ships carrying supplies or equipment; **sea-valve** (examples); **sea-wise** *a.*, versed in the ways of the sea; also *absol.* as *sb.*; **sea-woman,** (*b*) a female sailor; a woman working at sea.

1726 Swift *Gulliver* III. i. 2 Having experienced my knowledge in sea-affairs to be at least equal to his, he would enter into any engagement to follow my advice. **1939** J. Masefield *Live & Kicking Ned* 29 He told me something of sea-affairs. **1945** L. E. O. Charlton *Roy. Air Force* 266 A strong Japanese battle fleet..delivered an attack..reminiscent of the sea-air battles of Midway and the Coral Sea. **1959** H. Barnes *Oceanogr. & Marine Biol.* iii. 106 Information about temperature and salinity conditions and their variation enables deductions to be made about physical processes taking place in the sea-air interface. **1918** M. Denig *Let.* July in K. Cowing *Dear Folks at Home* (1919) 250 A few big 'sea-bags' had hit near by. **1919** *Sea-Bag* 9 Feb. 3 Down in the bottom of a sea bag you may find the suit that a German Sub would have fired at if we had been lucky enough to really see a Fritz. **1926** J. W. Thomason *Fix Bayonets!* 148 If Brother Boche had kept flingin' them seabags around here, he'd a-hurt somebody. **1958** J. Kerouac *On Road* II. ii. 187 He grabbed his seabag and threw things into that. **1977** *N.Y. Rev. Bks.* 23 June 6/2 How easy it would have been at that point, one thinks, for the Marine Corps to have packed up its sea bags and departed. **1637** in *Archives of Maryland* (1887) IV. 76 The Inventary of the goods & chattells of mr John Baxter... 1. rugg & an old sea-bed. **1722** Defoe *Moll Flanders* 381 My Governess.. came down herself..bringing me in the first Place, a Sea Bed, as they call it, and all its Furniture. **1774** N. Cresswell *Jrnl.* 8 Apr. (1925) 9 Bought a Sea Bed; paid Captn. Parry my passage. **1922** P. S. Allen *Let.* 30 Mar. (1939) xvi. 183 We are now on our way to Rotterdam..to attend..the 350th anniversary of the recapture of Brill by the 'Sea-beggars' from the Spaniards on 1 April 1572. **1963** *Times* 22 Feb. 17/3 He spoke for an hour to a packed audience of intent undergraduates and history dons about ..the 'invasion' of the Netherlands by Prince William of Orange's sea beggars. **1883** W. C. Russell *Round Galley Fire* 109 The sea-blessings showered out by the cook as he chases his dishes and pans and burns his fingers. **1912** W. I. Downie *Reminisc. Blackwall Midshipman* ii. 19 Sea blessings galore descended on my unfortunate head. **1933** S. Bradford *Shell-Backs & Beachcombers* viii. 181 The mate..gave me his sea-blessing for having recommended such a man to him. **1976** *Times* 13 May 5/8 The Navy League, formed 81 years ago to press for more naval power for Britain, announced yesterday that it is changing its name to the Sea Cadet Association. **1977** *Navy News* June 32/5 Members are grateful to the Leicester Unit, Sea Cadet Corps, for the use of their H.Q. for branch meetings. **1917** E. Pound *Lustra* 193 Full many a fathomed sea-change in the eyes That sought with him the salt sea victories. **1923** J. M. Murry *Pencillings* 164 The characters which have suffered this sea-change, 'of whose bones are coral made', are the only unpleasant characters we remember. **1948** A. C. Baugh *Lit. Hist. England* II. ix. 173 An interesting paper suggesting that romance is transplanted epic, which has undergone a kind of sea-change in the passage. **1974** R. Helms *Tolkien's World* ii. 32 Even before *The Hobbit* was published he was at work on its sequel, a work in which Middle-earth has undergone a wondrous sea change. **1976** *Listener* 8 Apr. 450/3 The Messianic vision..has undergone some strange sea-changes outside Judaism. **1977** 'E. Crispin' *Glimpses of Moon* vii. 117 He..could, moreover..bring about a sea-change in the image of even the most bumbling police officers going about their duties, so that they emerged as prodigies of intelligence, zeal and kindness. **1909** *Cent. Dict.* Suppl., *Sea-chest*, in ship-building, a short open pipe extending from the outside plating to the interior just inside the inner bottom, the inner end of which is closed by a sea-valve placed in a position accessible from the interior of the vessel. **1942** G. C. Manning *Man. Ship Construction* (1943) iii. 76 Sea valves must be so placed as to be easily worked from the engine-room platforms. When they make connection with the sea through the double bottom or otherwise so that they would require a long neck if fastened directly to the shell they are attached to sea chests which are secured on the inside of the shell plating. **1972** L. M. Harris *Introd. Deepwater Floating Drilling Operations* 248 All sea chest strainers were removed. The sea valves were opened and examined. The sea chests were thoroughly examined. **1883** Stevenson *Treas. Isl.* VI. xxxi. 263 'He was a seaman,' said George Merry, who..was examining the rags of clothing. 'Leastways, this is good sea-cloth.' **1890** 'Biff' Hall *Turnover Club* xviii. 172 The wings are removed, and what is technically known as a 'sea cloth' takes their place. **1905** 'Q' *Shining Ferry* III. xviii. 218 A bustious, big fellow, with a round hat like a missionary's, and all the rest of him in sea-cloth. **1946** Sea clutter [see *CLUTTER *sb.* 2 c]. **1970** P. Clissold *Radar in Small Craft* ii. 26 Sea clutter is not likely to be of any consequence beyond three or four miles, but at short range it can obscure stronger targets. **1806** J. Davis *Post-Captain* v. 27 'A precious husband!' exclaimed captain Brilliant... 'A son of a sea-cook! If he was to fall overboard, I would not heave him a rope.' **1922** Joyce *Ulysses* 624 Boisterously trolling, like a veritable son of a seacook. **1977** A. Hunter *Gently Instrumental* iv. 59 You're a right son of a seacook, aren't you? **1878** G. M. Hopkins *Poems* (1967) 74 They say who saw one sea-corpse cold He was all of lovely manly mould. *c* **1931** Auden in M. Roberts *New Signatures* (1932) 30 Doom is dark and deeper than any sea-dingle. **1968** E. S. Iversen *Farming Edge of Sea* ii. 31 The most important group of animals to sea farmers are mollusks (clams, oysters, and mussels), crustaceans (shrimps, crabs, and lobsters), and fishes. **1962** *New Scientist* 18 Oct. 129 Sir Alister Hardy, of Oxford, the leading prophet of sea-farming. **1972** *Aquaculture* I. 232 Seafarming is feasible and it can be carried out with profit. **1975** *Times* 24 Apr. 3/2 By next year the group of sea-farmers expect to have 1,250,000 Pacific oysters..ready for the market. A company, Western Aquaculture, is one of the sea-farming organizations. **1902** J. Masefield (*title of poem*) Sea-fever. **1931** *Daily Express* 23 Sept. 9/4 Men with the sea-fever on them pottered about among the debris of the docks. **1980** P. Moyes *Angel Death* i. 9 The much smaller island..has been infected by the current sea-fever to the point of constructing a small yacht basin. **1903** Kipling *Five Nations* 74 Flying-fish about our bows, Flying sea-fires in our wake. **1947** K. Tennant *Lost Haven* ii. 30 About her the impersonal sea-fire broke and the strange lights vibrated and shone. **1851** H. Melville *Moby Dick* I. xvi. 111 Those thews ran not through base blocks of land wood, but deftly travelled through sheaves of sea-ivory. **1968** G. Jones *Hist. Vikings* I. i. 23 Southwards.. went skins and furs, amber, sea-ivory, and slaves. **1847** H. Melville *Omoo* xvi. 58 Jermin, sea-jockey that he was, sometimes stood in the fore-chains. **1897** *Outing* Dec. 234/1 Aboard one of these well-balanced and swift little vessels the sea jockey's art can easily be acquired. **1971** D. Conover *One Man's Island* 67 The sea jockeys have taken over the waterways... Outboard cruiser owners—sea jockeys, as we call them. **1963** *Times* 2 Mar. 8/4 Their employment in certain roles will depend largely on their sea-keeping qualities. **1972** C. Mudie *Motor Boats & Boating* 144 Maximum speeds have crept up from some forty knots to eighty knots in ten years and seakeeping has improved out of all recognition with rough water speeds nearly doubled. **1936** C. Winchester *Shipping Wonders of World* I. 690/3 The Livonia..proved her 'sea-kindliness' by crossing the Atlantic in the worst of weather. **1976** *Yachting World* Oct. 110/2 Of course, the boat doesn't usually match true wind speed in open sea conditions, but it does exhibit truly phenomenal sea-kindliness with the hydrofoils set in moderate and heavy conditions. **1958** J. L. Kent *Ships in Rough Water* xi. 157 A seakindly ship is one which rides the seas in rough weather without shipping green water and with little spray blown inboard. **1981** *Times* 2 Feb. 22 There is a possibility of building hulls which can achieve speeds in excess of 30 knots on a waterline length of only 75 metres. These..should be extremely sea-kindly. **1878** Sea-lane [in Dict., sense 18 c]. **1948** *British Birds* XLI. Suppl. 1 After some months on its sea-lanes one could not but feel that the true answer to any one question could only be known if it were possible to cover the whole area in a matter of a few days. **1978** J. A. Michener *Chesapeake* 346 Can we keep the sea lanes open? **1956** *Sun* (Baltimore) 19 Dec. 2/1 The General Eltinge will sail..tomorrow with the first of 5,500 Hungarian refugees to head toward the United States by sea... The sealift complements commercial air services and the airlift inaugurated by United States military planes. **1967** *Economist* 4 Mar. 802/2 [The United States] has the air-lift and sea-lift capacity to be on hand whenever a power vacuum develops. **1972** S. Burnford *One Woman's Arctic* i. 15 It had been brought in by the annual sea-lift the year before. **1974** *Greenville* (S. Carolina) *News* 23 Apr. 14/3 The Middle East fighting proved, he says, our capacity to airlift and sealift needed munitions and equipment over long distances. **1980** *N.Y. News* 11 May 14/2 Officials put at 30,598 the total number of Cubans sealifted to freedom across the Florida strait. **1934** Sea loch [see *FIARD, FJARD]. **1975** J. G. Evans *Environment Early Man Brit. Isles* iii. 67 The various long bays of south-west Ireland, some of which, like the sea lochs of western Scotland, have been glacially deepened as well. **1959** *Times* 8 Dec. 13/6 Down at the sea-locks..the tugs would be worrying like strange small sea animals. **1853** D. G. Rossetti *Let.* 16 Apr. (1965) I. 131 Your 'sea-log' gave me the greatest pleasure. **1955** J. R. R. Tolkien *Return of King* 149 Deep in the hearts of all my kindred lies the sea-longing. *a* **1973** —— *Silmarillion* (1977) xxiii. 244 The sea-longing woke in his heart. **1951** R. Macaulay *Let.* 12 Aug. in *Lett. to Friend* (1961) 173, I think I shall airmail this [letter]... But I really will seamail the next. **1971** *New Society* 14 Jan. 47/1 This [order coupon] gives inland and overseas rates (airmail and seamail). **1933** *Gloss. Aeronaut. Terms* (*B.S.I.*) 64 *Sea marker*, a device dropped from an aircraft on to water, providing a distinguishable patch for determining the drift-angle. **1944** 'N. Shute' *Pastoral* iii. 43 He began a chat with the Equipment Officer about sea-markers that did not mark. **1941** *Bull. Geol. Soc. Amer.* LII. 338 A number of remarkable submarine mountains, termed 'seamount' by the United States Board on Geographic Names, rise sharply to heights of 1 to more than 2 miles above the gulf floor. **1959** *New Scientist* 1 Jan. 14/1 The few Pacific seamounts whose summits do form islands are mostly coral atolls. **1962** [see *GUYOT]. **1977** *Dædalus* Summer 118 Both noticed that the magnetic field over some seamounts could be explained only if the seamount was reversely magnetized. **1928** C. Dawson *Age of Gods* xv. 358 It is extremely improbable that the Sea Peoples actually penetrated into the Hittite homelands. **1957** *Antiquity & Survival* II. 145/1 Sisera stood at the head of the Canaanite coalition, and perhaps belonged to the Sea People who invaded Palestine in the 12th century B.C. and gained control of the sea-coast. **1978** N. K. Sandars *Sea Peoples* iv. 83 Wild northerners..took ship to arrive on the borders of Egypt as those mysterious 'Sea Peoples' who so terrified Rameses III. **1910** D. W. Bone *Brassbounder* 64 'Good ol' "sea price"', said Martin. 'Many an 'appy 'ome, an' garden wit' a flagstaff, is built o' "sea price".' **1924** R. Clements *Gipsy of Horn* iv. 71 Sea-price is often a figure which a Maltee Jew would hesitate to ask. **1972** N. Ayland *Schooner Captain* xv. 134 All the bread he would let them have was a two pound loaf, for which he charged sea price. **1945** E. W. Cowan *Sea-Return Effects & their Elimination in AN/APS-6* (M.I.T. Radiation Lab. Rep. No. 707) 1 An airplane flying very close to the sea may be hidden by sea return. **1959** *Listener* 12 Feb. 277/1 It is almost impossible to pick up that iceberg with the radar equipment, because of what we call 'sea return' or 'sea clutter'. **1966** D. Taylor *Introd. Radar & Radar Techniques* iii. 40 The actual performance obtained with this form of A.S.V. equipment depended on aircraft height, state of sea (because of sea returns), operator's experience, etc. **1893** *Columbus* (Ohio) *Dispatch* 9 Nov., If fish disappeared from the sea-roads and fiords. **1906** *Outlook* 19 May 677/1 We hold the great sea-roads to the East. **1907** T. C. Middleton *Geogr. Knowl. Discov. Amer.* 25 The Vivaldi brothers of Genoa..in 1291 essayed a sea-road to India. **1911** R. S. S. Baden-Powell *Sea Scouting for Boys* 8 Sea Scouts are of two kinds, viz. (1) Coastguard Scouts; (2) Seamen Scouts. **1912** C. Beresford in W. Baden-Powell *Sea Scouting & Seamanship for Boys* p. vi, The Sea Scouts were formed as an auxiliary to..the Boy Scouts. The object of the Sea Scouts is to teach lads at or near the sea seamanship, navigation, pilotage, knotting and splicing, how to handle boats under oars and sail, [etc.]. **1950** *Oxf. Jun. Encycl.* IX. 87/1 The boy who is fond of the sea can become a Sea Scout... There are Sea Scout Troops on rivers and inland waterways as well as on the sea. **1977** *Listener* 24 Mar. 382/3, I was a sea scout and sailed and rowed boats. **1925** J. Proudman in *Monthly Notices R. Astron. Soc.* (*Geophysical Suppl.*) I. 247 By 'sea-seiches' we mean those oscillations of fairly definite period but of irregular amplitude and phase which are frequently observed on the sea coast. **1967** *Oceanogr. & Marine Biol.* V. 42 These modes have been called sea seiches..and are basically similar to the transient oscillations or seiches set up by wind and atmospheric pressure in closed basins. **1916** *Rudder* Apr. 175 One noticeable thing about..the sea sled..is the absence of bow-wave. **1948** R. de Kerchove *Internat. Maritime Dict.* 639/2 *Sea sled*, a type of construction adopted for small craft of high speed in which the ordinary V bottom is inverted in order to collect a layer of air under the bows of the boat. **1957** *Official Gaz.* (U.S. Patent Office) TM-10/2 Norman A. McDonald, Skokie, Ill. Filed Jan. 24, 1957. Sea Sled.. For Boats. First use Mar. 15, 1953. **1887** W. H. White *Mod. War Ships* 94 The 'sea-speeds' of all war-ships are always estimated on different assumptions. **1899** *Geogr. Jrnl.* Mar. 288 The isolated rock-masses and sea-stacks, which we are enabled to trace by means of the soundings. **1973** C. Bonington *Next Horizon* xxi. 292 A sea-stack on the north coast of Scotland. [**1963** *Meteorol. Gloss.* (Met. Office) (ed. 4) 222 The degree of sea disturbance is reported in a 'state of sea' code in which the scale number increases from 0 to 9 according to the average wave height.] **1967** *Jane's Surface Skimmer Systems 1967–68* 85/1 The journey has been covered successfully in Sea States 2–4, with wave heights up to 5 ft 0 in (1·5 m). **1977** *Offshore Engineer* Apr. 74/1 The calculator multiplies measured value of the load by a factor determined by sea-state, and compares result with safe-load for the particular crane luff angle specified in manufacturer's table. **1933** J. Masefield *Bird of Dawning* 16, I wished to get sea-time, sir, so as to be able to pass for master. **1930** *Times* 26 Mar. 17/4 Her sea-time beat the Bremen's best..by eighteen minutes. **1977** *Navy News* Aug. 22 (Advt.), Service includes normal roster sea-time in Leander and Type 12 frigates and small ships. **1847** H. Melville *Omoo* xxiv. 92 'Give him a sea-toss!' 'Overboard with him!' **1933** *Nat. Geogr. Mag.* May 581/1 Freight-car contents are transferred here into the holds of liners, and recently a terminal was established which places loaded cars themselves within huge vessels called 'seatrains'. **1942** W. S. Churchill *Second World War* (1951) IV. I. xxii. 349 These equipments will sail for Suez..in two sea-trains taken from the Havana sugar traffic, doing 15 and 13 knots respectively. **1947** *Sun* (Baltimore) 14 July 7/3 The ships which..should be started this year..are..two for Alaskan trade specifically, two sea trains and two 'mystery'. **1895** Kipling *Day's Work* (1898) 81 A sea-valve that communicated directly with the water outside. **1915** Chesterton *Poems* 16 On them the sea-valves cluster and the grey sea-forests curl. **1934** *Sun* (Baltimore) 17 Sept. 12/1 The sea-wise reason that in a strong breeze and an attendant unruly sea that elongated prow will come down and pound against the chop or plunge into a heavy swell. **1966** T. H. Raddall *Hangman's Beach* I. ii. 29 The sea-wise folk of Halifax awaited word from Europe. **1939** *Sun* (Baltimore) 1 July 20/2 Twenty-three seawomen sailed into Baltimore harbor yesterday afternoon aboard the ship William J. Stanford. **1963** *Punch* 21 Aug. 288/3 The endless queue of frustrated seawomen.

d. sea-attorney (examples); **sea-biscuit** = *sand dollar* s.v. Sand *sb.*[2] 10 b in Dict. and Suppl.; **sea butterfly** (examples); **sea-clam,** substitute for def.: one of several species of clam found on the Atlantic coast of North America, esp. the surf clam, *Spisula solidissima*; cf. Hen-clam; (earlier and later examples); **sea-moth,** a small fish of the family Pegasidæ, found in Indo-Pacific waters and having bony plates covering the body and enlarged pectoral fins; **sea-poacher** (later examples); **sea tiger** = Barracuda; **sea wasp,** a poisonous jellyfish belonging to the order Cubomedusæ, found in Indo-Pacific waters.

1849 H. Melville *Mardi* I. 55 There is the ordinary Brown Shark, or sea-attorney, so called by sailors. **1854** *Putnam's Mag.* Apr. 362/2 The dippers dip carefully, lest they get a stroke from the ray..or a rip from his cousin the 'sea-attorney'. **1949** G. E. & N. MacGinitie *Nat. Hist. Marine Anim.* xxvi. 236 The sand dollars, sea biscuits, or cake urchins..resemble very much flattened sea urchins. **1972** *Islander* (Victoria, B.C.) 12 Mar. 16/4 White people call them [*sc.* sea-urchins] sea-biscuits or sand-dollars. **1932** Borradaile & Potts *Invertebrata* 494 The Pteropoda (sea butterflies)..are modified for pelagic life. **1972** M. S. Gardiner *Biol. Invertebrates* v. 161/1 Planktonic 'pteropods'..or 'sea butterflies'. **1765** J. Bartram *Jrnl.* 29 July in *Trans. Amer. Philos. Soc.* (1942) XXXIII. 16/2 There is many clam shels of different sises..yᵉ very same with our sea clams. **1782** [see *HARD-SHELLED *a.* 1]. **1935** J. C. Lincoln *Cape Cod Yesterdays* 49 Along the outer bar, almost two miles from shore..were the large 'sea clams'. **1960** J. J. Rowlands *Spindrift* 83 Sea clams are from four to six inches long and about four inches wide. **1905** D. S. Jordan *Guide to Study of Fishes* II. xiii. 239 These 'sea-moths' are fantastic little fishes. **1947** K. H. Barnard *Pict. Guide S.Afr. Fishes* III. 76 The Dragon-fish or sea-moth..is also encased in bony plates like the sea-horses... It derives its name of sea-moth from its habit of skimming over the surface of the water. **1978** *Nature* 26 Oct. 693/1 The sea moths are a small family (Pegasidae) of marine fishes found only in the Indian and Western Pacific oceans, from East Africa to Hawaii. **1836** W. Yarrell *Hist. Brit. Fishes* I. 70 The Armed Bullhead, Pogge, Lyrie, Sea-poacher, Pluck, Noble. **1905** D. S. Jordan *Guide to Study of Fishes* II. xxv. 449 The sea-poachers or alligator-fishes, *Agonidæ*, are sculpins enclosed in a coat of mail. [**1924** L. L. Mowbray in J. O. La Gorce *Bk. Fishes* 143 Well deserving its nickname of 'The Tiger of the Sea', the carnivorous Barracuda..darts at its prey on sight.] **1937** M. N. Kaplan *Big Game Anglers' Paradise* iv. 180 Although ichthyologists gave the great barracuda the euphonious name *Sphyraena barracuda*, in common parlance it bears the *nom-de-guerre*, 'Sea-tiger'. **1963** Sea tiger [see *PICUDA]. **1910** A. G. Mayer *Medusæ of World* III. 504 The flexible part of the tentacles are [*sic*] armed with nematocysts, the stinging power of which is so great that the name 'Sea Wasp' is commonly given to these medusæ. **1966** J. H. Barnes in W. J. Rees *Cnidaria & their Evolution* 332 The origin of such stings was not known, but..it must have been a sea wasp. **1977** C. McCullough *Thorn Birds* xiii. 301 We're too far south here for sea wasps.

e. sea-bamboo *S. Afr.*, a large kelp, *Ecklonia maxima*; = Sea-trumpet 3.

1798 S. H. Wilcocke tr. *Stavorinus' Voy. East Indies* I. I. i. 25 On the 10th of November, we saw for the first time trumpets, or sea-bamboo, floating on the ocean. **1822** W. Burchell *Trav. Interior S. Afr.* I. ii. 28 The Dutch call this plant *Zee bambos* (sea-bamboo), and boys after cutting its stem to a convenient length when dry, sometimes amuse themselves in blowing it as a horn or trumpet. **1946** L. G. Green *So Few are Free* viii. 116 The place is called Bamboes Bay, because the sea bamboo is piled high on the beach after heavy gales. **1973** *Stand. Encycl. S. Afr.* IX. 562/2 The largest kelp of Southern Africa is the sea-trumpet or sea-bamboo..which commonly reaches lengths of over 6 metres.

f. sea arrow-grass, a marsh plant, *Triglochin maritima*, with fleshy grass-like leaves and spikes of green flowers; **sea-aster** (later examples); **sea convolvulus** = Sea-bell 1, *sea-bindweed*; **sea ivory,** a pale greyish lichen, *Ramalina siliquosa*, growing in flattened branches on sea-shore rocks; see also sense 23 a in Dict. and Suppl.; **sea lungwort** (later examples); **sea myrtle** = *groundsel-tree* s.v. Groundsel *sb.*[1] 2.

1770 J. Hill *Herbarium Britannicum* II. 215 (*heading*) Sea Arrow Grass. **1851** C. A. Johns *Flowers of Field* II. 245 Sea Arrow-grass... Salt marshes, common. **1975** J. G. Evans *Environment Early Man Brit. Isles* ii. 37 A number of species..today confined to coastal or estuarine situations, such as sea arrow grass..and sea thrift. **1925** G. Bonnier *Brit. Flora* 96 Sea Aster. Salt marshes: July–Sept. **1960** *Oxf. Bk. Wild Flowers* 136/2 Sea Aster.. is common in salt-marshes and on cliffs and rocks round the coasts. **1921** 'K. Mansfield' *Let.* 8 Aug. (1977) 227 Bathing dresses hanging over verandas, and sandshoes on window sills, and little pink 'sea' convolvulus. **1966** *Oxf. Bk. Flowerless Plants* p. viii/1 'Sea Ivory'..grows abundantly wherever there are rocks exposed at high-water mark. **1797** J. E. Smith *Eng. Bot.* VI. 368 (*heading*) Sea lungwort. **1960** Sea lungwort [see *oyster-plant* s.v. *OYSTER 5d]. **1883** G. O. Shields *Rustlings in Rockies* xxi. 195 Within the space of this five acres may be found.. sea myrtle, grape vine and ivy of several varieties. **1938** M. K. Rawlings *Yearling* xxiii. 279 The October blooming of dog-fennel and sea-myrtle had turned to a feathery fluff.

sea-bass. Restrict *U.S.* to senses in Dict. and substitute for defs. of senses a and c: **a.** A marine food fish of the family Serranidæ; cf. Jew-fish. **c.** The black sea-bass, *Centropristis striatus*, found along the Atlantic coast of North America (earlier and further examples). Add: **e.** = Bass *sb.*[1] 2; *Loup *sb.*[4]

1765, 1888 [see *BLACK FISH 1]. **1900** F. Norris *Blix* 129 There were..sheaves of fishing-rods, from the four-ounce wisp of the brook-trout up to the rigid eighteen-ounce lance of the king-salmon and sea-bass. **1961** E. S. Herald *Living Fishes of World* 177/2 The giant sea basses, sometimes called jewfishes,..are the largest American serranids. *Ibid.* 192/1 The genus *Cynoscion* includes..the California white seabass, *C. nobilis*. **1966, 1969** [see *LOUP *sb.*[4]]. **1973** J. Grigson *Fish Cookery* 104 To walk into a fishmonger's and see a tray of sea bass is a beautiful sight.

sea-bathe, *v.* Delete *rare* and add earlier and later examples. Hence **sea-bathing** *ppl. a.*

1792 Lady Templetown *Let.* 11 June in A. E. Newdigate-Newdegate *Cheverels* (1898) vii. 104 Eliza is sea-bathing at Ramsgate. *a* **1930** D. H. Lawrence *Last Poems* (1932) 156 These all-but-naked sea-bathing city people. **1951** N. M. Gunn *Well at World's End* xviii. 141 We did nothing but sun-bathe and sea-bathe.

sea-bathing. (Earlier examples.)

1749 S. Richardson *Let.* ?Nov.–Dec. (1964) 133 The sea-bathing I have not tried. *attrib.* **1781** J. Hare *Let.* 13 Feb. in *15th Rep. R. Comm. Hist. MSS.* (1897) App. VI. 457 We are to go in about six weeks' time to some sea-bathing place.

sea beach. (Earlier example.)

1742 *Rep. Comm. Houses Assembly S.-Carolina Exped. against St. Augustine* 21 He encamped that Night at Lacenela (the first Palmeta Hut on the Sea-Beach).

Seabees (sī·bīz), *sb. pl.* Also with small initial. [f. representation of initial letters of *construction battalion* + pl. *-s*.] **a.** (Members of) the Construction Battalions formed as a volunteer branch of the Civil Engineer Corps of the U.S. Navy.

1942 *Army & Navy Jrnl.* 21 Mar. 790/2 'Seabees' is the new name chosen to designate the new Naval Construction Regiments... With the name an insignia has been adopted —a flying bee, fighting mad. On its head it sports a sailor hat. In its fore hand or leg it clutches a spitting 'Tommy Gun'; in its amidship hand, a wrench, and in its aft hand, a carpenter's hammer. **1945** D. Dempsey in *Best One-Act Plays of 1944* 21 I'm in the Seabees, which means I'm in the Navy. **1950** 'D. Divine' *King of Fassarai* xi. 90 'The Seabees are starting in here—' he placed his pencil on the chart. **1960** S. W. Roskill *War at Sea* III. I. 339 The 'Seabees' also proved themselves first-class fighting troops when the need arose. **1980** G. V. Higgins *Kennedy for Defense* xix. 174 Some cops were in the Seabees, lotta cops was in the Army.

b. *attrib.* or as *adj.* (Sometimes in *sing.* form.)

1942 *Sun* (Baltimore) 17 Nov. 13/3 Brief ceremonies attended by naval officers marked the placing in operation of the third 'seabees' center constructed in Virginia. **1977** *Hongkong Standard* 12 Apr. 8/4 Of the 1,400 navy men here, more than 800 are seabee construction workers who will leave when the facility is completed. **1981** G. V. Higgins *Rat on Fire* v. 35 You were running that Seabees reunion.

seaboard, *sb.* **4.** (Earlier example.)

1788 F. Asbury *Jrnl.* 10 July (1821) II. 36 The Gnats are almost as troublesome here, as the moschetoes in the low-lands of the sea-board.

sea-boat. 1. a. Delete † *Obs.* and add later examples. **c.** *spec.* A small, manœuvrable craft sent out from a larger vessel, as in cases of emergency at sea.

1856 Ruskin *Harbours of England* 1 One object there is still, which I never pass without the renewed wonder of childhood; and that is the bow of a Boat... The blunt head of a common bluff undecked sea-boat. **1892** R. C. Leslie *Sea-Boat* 4 The building of a bluff-bowed, flat-floored sea-boat, is a much greater test of a boat-builder's skill than the construction of any form of longer and sharper craft. **1909** *Man. Seamanship* (Admiralty) II. viii. 142 Never call away a 'lifeboat' at sea except for the purpose of saving life; on all other occasions call away the 'sea boat'. **1914** 'Bartimeus' *Naval Occasions* xx. 179 The seaboat's crew had gone through an undress rehearsal of 'Man overboard!' **1963** H. C. de Mierre *Long Voyage* ix. 148 Port and starboard sea-boats had been put in the water. **1972** C. Mudie *Motor Boats & Boating* 76 Over twelve feet in length..the types of boat begin to diverge from the general purpose dinghy..into seaboats for family use and for fishing.

Seabright (sī·brəit). Also with small initial. The name of a small town on the coast of New Jersey, used *attrib.* and *absol.* to designate a variety of flat-keeled fishing-boat from the region, esp. adapted for landing on beaches in heavy seas.

1911 *Rudder* XXV. 83 (*caption*) Jersey Coast fishermen getting one of their power Seabright skiffs off and on the beach. **1930** G. Pinchot *To South Seas* i. 7 The bronze and mahogany launch was replaced by a Seabright dory twenty-two feet long by seven feet wide. *Ibid.* xii. 237 In the Seabright again I put it on the floor. **1971** P. J. Guthorn *Sea Bright Skiff* v. 74 The Sea Bright skiff rum boats were progressively, but never completely, replaced by boats of V-bottom design.

sea-cat. 2. Restrict † to sense in Dict. and add: **b.** *S. Afr.* One of several species of octopus, esp. *Octopus vulgaris*.

1785 G. Forster tr. *Sparrman's Voy. Cape of Good Hope* I. ii. 26 The *sepia loligo*, and the *sepia octopodia*,..are known to our sailors by the name of black-fish and sea-cats. **1882** *Cape Q. Rev.* Oct. 36 Even the sea cat responded to the hook. **1913** W. W. Thompson *Sea Fisheries Cape Colony* ii. 51 The octopus or sea-cat..appears to find a more congenial habitat on the rocky stretches of sea-board. **1957** S. Schoeman *Strike!* iii. 38 If seacat is used, a baby octopus put on whole will be most acceptable.

5. *Mil.* (In form *Seacat* or *Sea Cat*.) The name of a short-range, ship-to-air, guided missile (system).

1959 *Times* 5 Feb. 7/3 By early next year a considerable number of Britain's warships will be equipped with surface-to-air missiles, and these weapons..will almost certainly include the Short SX-A5 which was named the Seacat by the Admiralty and the Ministry of Supply yesterday. **1969** B. B. Schofield in P. Kemp *Hist. Royal Navy* xxvii. 294/1 In April 1958 a contract was awarded for the development of a close-range ship-to-air guided weapon to replace the 40-mm Bofors A/A guns in H.M. ships... This was achieved by the production of the Seacat missile. **1977** *Navy News* June 6 The 2,500 ton warship is armed with an automatic rapid fire 4·5 in. gun, Sea Cat and Exocet missile systems, and she operates a Wasp helicopter.

sea-conny. Add: Also **secunnie.** (Further examples.)

1810 M. Graham *Jrnl. Residence India* (1812) 85 The gunners and quarter-masters..are Indian-Portuguese; they are called *secunnies*. **1929** D. J. Munro *Roaring Forties* xxxii. 160 With a crash and lurch that sent the secunnie (helmsman), at the wheel flying over the top of it the ship struck.

sea-craft. Add: **3.** Sea-going craft; sea-vessels considered collectively.

1919 *Q. Rev.* Jan. 184 Vessels, whether sea-craft or air-craft. **1928** *Daily Express* 18 Jan. 1 Our oldest craft, seacraft.

sea-crafty, *a.* (Later example.)

1955 J. R. R. Tolkien *Return of King* v. ix. 153 Sea-crafty men of the Ethir gazing southward spoke of a change coming with a fresh wind from the sea.

sea-dog. Add: **5.** (Earlier U.S. example.)

1823 J. F. Cooper *Pilot* II. xi. 187 Ahead, heave ahead, sea-dogs!

Hence **sea-doggery,** behaviour or practice characteristic of a sea-dog or sea-dogs (sense 5); sailors collectively.

1928 *Daily Express* 9 Oct. 3/3 A little grey Dover full of small, sturdy ships..and a general air—assisted by a strong smell of oil, pea soup and roasting mutton—of waggish sea doggery. **1961** *John o' London's* 20 July 111/3 A background of adventure and sea-doggery.

seadrome (sī·drōᵘm). [f. Sea *sb.* + *-DROME, after *AERODROME.] A floating aerodrome, an offshore airport; *spec.* (in early use) one of a series of constructions on or at which a (sea) plane could alight (for refuelling) during a journey.

The *seadrome* exists solely as a concept. None has yet been built.

1923 *Daily Mail* 17 July 10 The feature of the project is that there should be eight 'seadromes'..moored permanently on the Atlantic air route. **1936** J. Grierson *High Failure* xiv. 293 A series of floating platforms (colloquially called 'sea-dromes') on which planes could alight, spaced at intervals of 5–600 miles across the Atlantic. **1947** *Jrnl. R. Aeronaut. Soc.* LI. 143/2 If alighting places for flying boats were more easy to provide than aerodromes, could Mr. Lipscomb offer any reason why they..suffered from lack of 'sea-dromes', whereas there were plenty of very costly aerodromes for landplanes? **1948** *Trinidad Guardian* 18 June 2/1 (*heading*) Seadrome site inspected. **1969** *Daily Tel.* 4 Sept. 24/2 The seadrome scheme for Foulness plans a floating airport complex measuring five million square yards. **1970** *New Scientist* 22 Jan. 156/1 The two firms of engineers..are already discussing their plans for a 'Seadrome' with one of the Thames estuary development companies.

sea-fishing. (Earlier example.)

1745 Ellis *Mod. Husbandman* VI. II. 66 Sea-fishing is the..healthiest Fishing of all others.

sea-floor. 1. The floor of the sea.

1855 [in Dict. s.v. Sea *sb.* 18 a]. **1922** [see *OUTWANDER *v.*]. **1946** *Proc. Prehist. Soc.* XII. 24 The seal must have sunk in deep water and come to rest on the sea-floor. **1967** *Oceanogr. & Marine Biol.* V. 317 Two principal dispersal mechanisms affect shelf benthos; these are, (1) shallow seafloor connections and, (2) surface currents. **1981** *London Mag.* July 9 A door slams, a heavy wave, a door, the sea-floor shudders.

2. sea-floor spreading *Geol.*, (the hypothesis of) the formation of fresh areas of oceanic crust, occurring through the upwelling of magma at mid-ocean ridges and its subsequent outward movement on either side.

1961 R. S. Dietz in *Nature* 3 June 854/1 [The concept proposed here, which can be termed the 'spreading sea-floor theory', is largely intuitive, having been derived through an attempt to interpret sea-floor bathymetry.] *Ibid.* 856/2 Sea floor spreading obviates this difficulty: continents never move through the sima. **1971** *Nature* 1 Jan. 9/2 The whole of geophysics has been transformed within a few years by the discovery of seafloor spreading. **1977** A. Hallam *Planet Earth* 42/3 If its magnetic signature allowed an age to be assigned to each piece of ocean floor, it was possible to calculate a velocity for sea-floor spreading.

sea-food. orig. and chiefly *U.S.* Food obtained from the sea; fish, crustacea, etc., used as food. Freq. *attrib.*

1836 *Knickerbocker* VIII. 423 She said that she had come to Screamy Point to get 'sea-food'. **1906** *N.Y. Even. Post* 10 Mar. 5 Up State residents are among the best customers of the sea food, fruit and produce dealers. **1927** *Weekly Dispatch* 1 May 1/2 The correct name of the Poydras levee [at New Orleans] is the Carnavon levee, so named by the Carnavon family in England, who built it in 1870 to protect the plantation and seafood packing plant they then owned. **1935** A. Baugh *Hist. Eng. Lang.* xi. 462 A writer in the London *Daily Mail* recently complained that an Englishman would find 'positively incomprehensible' the American words..*sea-food*,..and *hired-girl*. **1944** T. Barbour *That Vanishing Eden* 166, I was dining with some friends at a popular seafood restaurant in Miami. **1953** J. Hilton *Time & Time Again* I. 6 'I hope you like sea-food.'..'Sea-food?..Fish, that is? Oh yes, I do, indeed.' (..True enough, though this 'sea-food' set Charles thinking that he also enjoyed 'land-food'.) **1965** H. Gold *Man who was not with It* III. xxv. 234 Grack's contact man cooked in a diner on the seafood coast of Baltimore. **1978** K. Hudson *Jargon of Professions* iii. 82 Twenty or thirty years ago..fish was upgraded to seafood.

sea-god. (Later examples.)

1900 A. Nutt *Fairy Mythol. Shakes.* 31 Mongan, son of the Irish sea-god Manannan mac Lir. **1950** 'D. Divine' *King of Fassarai* xiii. 100 To-day the fish had been wary and the sea god not kind. **1978** R. Mitchison *Life in Scotland* iii. 60 The islanders of north Lewis sacrificed to a sea god at Hallowtide.

sea-going, *a.* Add: **1. b.** Capable of being used or suitable for use on a sea-going vessel; carried or conducted by sea.

1895 [see Sea-going *sb.*]. **1928** C. Dawson *Age of Gods* viii. 182 The sea-going trade of the Ægean world. **1962** *Listener* 29 Mar. 540/1 Land-base missiles and sea-going missiles.

sea-grass. Add: **2.** (*b*) (Later examples.) Also, the dried stems of this plant, used in ropes, chair-seats, etc.

1900 *Jrnl. Soc. Arts* 12 Jan. 170/1 Sea grass is a long thin grass that grows on the protected flats of the Lower St. Lawrence River. *Ibid.* (*heading*) Canadian sea grass for upholstering. **1911** [see *porch chair* s.v. *PORCH 7]. **1933** P. T. Tucker *Riding High Country* ii. 31 His las' rope was sea grass. **1979** *Dictionaries* I. 64 Sea grass is..a derived product with a local name.

4. *attrib.* (Later examples.)

1911 *Daily Colonist* (Victoria, B.C.) 11 Apr. 7/1 (Advt.), Special sale of sea grass chairs. **1967** *Southerly* XXVII. 152 Immemorial deck-chairs and seagrass tables.

sea-green, *a.* and *sb.* Add: **A.** *adj.* **2.** In phr. *sea-green incorruptible*, applied to Robespierre by Carlyle (see quot. 1837 in Dict.) and now commonly used allusively (often followed by some other word) to designate a person of rigid honesty or uncompromising idealism. Also in extended use and *absol.*, impervious to moral corruption.

1931 *Economist* 7 Mar. 492/1 Although Mr. Hu Han-min at Manking may be a 'sea-green incorruptible' of as pure a dye as Robespierre, the local representatives of the Party in the country districts are often oppressors of the poor. **1936** H. G. Wells *Anat. Frustration* ix. 94 His [*sc.* Philip Snowden's] early appearance as the 'sea green incorruptible' of the British Labour revolution. **1958** *Spectator* 1 Aug. 157/1 Utopia is to be attained only by sea-green incorruptibility. **1960** C. P. Snow *Affair* xl. 372 'I shall have to,' Skeffington replied, obdurate and sea-green. **1976** *Listener* 5 Aug. 143/3 He is, for all that, a man of sea-green integrity. **1977** *New Society* 17 Feb. 328/1 A former Chief Constable from the north west..a man of cast-iron integrity from a religious background and ..a sea-green incorruptible.

sea-gull. Add: **2.** A casual, non-Union, dock labourer. *N.Z. slang.*

c **1926** 'Mixer' *Transport Workers' Songbk.* 46 What a study! Let us paint it As the sea-gulls fly about, While the stringer birds are anxious For the meeting to come out. **1943** *2 N.Z.E.F. Times* 20 Sept. 5 As a result of a survey of non-Union labour on the wharves..500 out of 800 'sea-gulls' were interviewed at the Auckland District Manpower Office. **1959** G. Slatter *Gun in my Hand* xvii. 225 Ended up as a sea-gull on the Wellington wharves loading up the Home boats. **1966** G. W. Turner *Eng. Lang. Austral. & N.Z.* vii. 150 The watersiders have their own special language, but the only term I know from experience is *seagull*, the casual non-union labourer who picks up a job during busy times on the waterfront.

sea-island, *a.* Add to def.: *Gossypium barbadense*, distinguished by long silky fibres. (Earlier and later examples.)

1803 J. Davis *Travels in U.S.A.* 78 Of cotton there are two kinds; the sea-island and the island. The first is the most valuable. **1807** *Salmagundi* 15 Oct. 327 The lady of a Southern planter will..trail a bale of sea-island cotton at her heels. **1934** *Nat. Geogr. Mag.* Feb. 260/2 At Hampton Point and Retreat the first sea-island cotton was grown from seeds introduced from the island of Anguilla, in the West Indies. **1970** *Observer* 15 Mar. 48/6 (Advt.), Shirts and pyjamas made to measure, sea island poplins ..pure silk. **1977** P. Moyes *To kill Coconut* xiii. 180 Sea Island isn't grown only in the Caribbean.

B. *sb.* An offshore mooring station where oil tankers can discharge their cargo and from which the oil can be pumped ashore.

1975 *Offshore Engineer* Sept. 17 (*caption*) When linked to the 'sea islands' already installed, it will be more than 1 km long. **1979** F. Forsyth *Devil's Alternative* ix. 218 They berthed at 'sea islands', networks of pipes on stilts, well out to sea, from which their oil could be pumped ashore.

seal, *sb.*[1] Add: **4. a.** *seal-oil* (earlier and later examples).

1732 in *Calendar State Papers, Amer. & W. Indies* (1939) 227 Value of seal oil made last winter, £2478 10s. **1973** L. Russell *Everyday Life Colonial Canada* xii. 155 Until the 1860s, seal oil was an important lamp fuel in the eastern colonies, and was also used in food and as a lubricant.

c. *seal-brown* (earlier and later examples).

1881 [see *FAWN *sb.*[1] 3]. **1963** B. Vesey-Fitzgerald *Cat Owner's Encycl.* 36 The body should be a solid colour of rich dark seal-brown shading.

5. seal-hole, substitute for def.: a hole in ice kept open by seals coming to it for air and getting out of the water through it; (further example); **seal rookery** = Rookery 2 b.

1974 R. Adams *Shardik* iii. 22 He crouched and watched, vigilant as an Eskimo at a seal-hole. **1901** *Daily Colonist* (Victoria, B.C.) 9 Oct. 8/3 That the contention.. that the seal herds in the Behring Sea are decreasing is not being borne out by fact, is shown by recent advices from the seal rookeries in the Pribyloffs. **1974** G. Jenkins *Bridge of Magpies* iii. 46 Possession [Island] was as inviting as a seal rookery—and as smelly.

seal, *sb.*[2] Add: **7. d.** *gen.* Any means of preventing the passage of gas or liquid into or out of something, esp. at a place where two surfaces meet.

a **1884** Knight *Dict. Mech.* Suppl., *Seal*, an automatic valve closing a pipe. **1938** J. Strong *Mod. Physical Lab. Practice* iv. 152 A method of making a vacuum-tight seal between metal and porcelain. **1970** K. Ball *Fiat 600, 600D Autobook* x. 118/2 The seal..is compressed on the forward stroke to prevent leakage past the plunger. **1974** *Encycl. Brit. Macropædia* XIII. 857/2 Metal-can closures operate by..vacuum seals (which rely on atmospheric pressure on the lid).

8. a. *seal-engraver* (earlier example), *-engraving.*

1786 J. Wedgwood *Let.* 24 June (1965) 295 The material..is..nearly as hard as agate... It will bear to be cut..at the seal-engraver's lathe. **1948** D. Diringer *Alphabet* 73 From E. M. I. (thirtieth century B.C.) onwards, seal-engraving was practised.

seal, *v.*[1] Add: **II. 8. c.** To render (a surface of wood, etc.) impervious by the application of a special coating.

1940 H. T. Davey *Wood Finishing* xvii. 202 When dealing with inlays it is best to seal them with shellac before attempting to spray. **1958** *Listener* 18 Dec. 1055/2 If you still want a glossy finish, you must seal the hard-board first, and the easiest way of doing that is to apply a thin coat of plastic emulsion paint. **1977** *Reader's Digest Bk. Do-It-Yourself Skills & Techniques* iv. 140/3 Some porous timbers may require two light coats, but do not apply any more than is needed to seal the surface.

d. To prevent access to and egress from (an area or space); to close (entrances) for this purpose. Usu. with *off*.

1931 *Industr. & Engin. Chem.* (Analytical Ed.) 15 Oct. 349/2 The tube is immersed in liquid air, and when the gasoline is frozen, the area *B* is sealed off. **1946** A. Vagts *Landing Operations* IV. lix. 775 The Germans..concluded that they must try to seal off the Cotentin at its base. **1948** *Sci. News* VII. 44 For the moment, treated areas are being 'sealed' by the total removal of all bush within a two-miles belt, and a watch is being kept to see what happens to the small remaining fly population. **1974** 'M. Innes' *Mysterious Commission* xi. 99 This cellarage had..been boarded over and sealed off from the studio. **1981** E. Clark *Send in Lions* v. 53 The search..began immediately. Air and sea ports were sealed.

11. Also, to confine so as to prevent access or egress. Also with *off*.

1926 R. W. Lawson tr. *Hevesy & Paneth's Man. Radioactivity* i. 2 If we seal off 1·3138 gm. $RaCl_2$..in an evacuated tube. **1940** W. Faulkner *Hamlet* II. i. 140 We sealed it up in a asbestos matchbox. **1946** A. Vagts *Landing Operations* IV. lix. 774 The choice facing Rundstedt and Rommel was between an attempt to throw the Allies back into the sea.., or sealing them off in the Cotentin.

sealable, *a.* (Later example.)

1979 *Nature* 19 Apr. p. xiv/2 The processing of liquids in sealable glass containers.

sealant (sī·lănt). [f. Seal *v.*[1] + -ant[1].] A substance designed to seal a surface or container against the passage of a gas or liquid; a material used to fill up cracks.

1945 *Materials & Methods* XXII. 1750 (*caption*) A bronze flame thrower part which required pressure tightness and chemical resistance on the part of the sealant. **1954** *Wall St. Jrnl.* 13 July 8/2 Goodrich tires have a further sealant—an extra layer of butyl in a gummy form which flows into and seals a puncture after the puncturing object is removed. **1966** T. Pynchon *Crying of Lot 49* ii. 26 Barbed wire again gave way to the familiar parade of.. sealant makers, bottled gas works,..and whatever. **1971** *Good Motoring* Sept. 4/2 The Beetle, a 'standard' model which has done about 100,000 miles on the road, was fitted with..a magneto and sealant round doors and body. **1978** S. S. Peng *Coal Mine Ground Control* xii. 417 Mine sealant sprayed in a thin coating on the roof and rib not only strengthens..but also seals the surface from the wet-dry cycles of the ventilated air.

sealapack, var. *SILLAPACK.

sea-lawyer. 2. (Earlier and later examples.)

1829 W. N. Glascock *Sailors & Saints* I. ii. 31 What tack are we on now?—got hold of a sea-lawyer it seems. **1873** [see *pleasure-navy* s.v. *PLEASURE *sb.* 6 a]. **1953** Dylan Thomas *Under Milk Wood* (1954) 4 Alfred Pomeroy Jones, sealawyer, born in Mumbles. **1967** A. Dubus *Lieutenant* iii. 78 Forget about this sea lawyer business and face your Goddamn punishment like a man.

sealchie, -kie. Add: Also **silkie, silky.** Also, in folklore, a creature or spirit having the appearance of a seal; *spec.* one able to assume human form. (Further examples.)

1852 in *Proc. Soc. Antiquaries Scotl.* I. 88, I am a man, upo' the lan' An' I am a Silkie in the sea. **1856** E. Edmondston *Sk. & Tales Shetland Islands* vii. 80 The seal ..retired to the neighbourhood..where a mermaid had her abode. The latter..asked if she could help him. *Selkie* imagined she might, but only by regaining for him..the covering of which he had been so ruthlessly bereft. **1899** J. Spence *Shetland Folk-lore* 24 In old times there was an aversion to and superstitious dread of killing a *selkie* lest it should be a metamorphic Finn. **1933** J. Buchan *Prince of Captivity* 28 He has heard the silkies singing at dawn on farther islets than St. Kilda. **1976** K. Briggs *Dict. Fairies* 354 In Orkney..the great seal, the grey seal, the crested seal and others, are called 'the selkie folk' because it is believed that their natural form is human, that they live in an underwater world..and put on seal-skins and the appearance of seals to enable them to pass through the waters from one region of air to another.

sealed, *ppl. a.* Add: **1. h.** (Earlier example.)

1850 [see *bull's wool, bullswool* s.v. *BULL *sb.*[1] 11].

i. *Sealed Knot*, the name of an organization which re-enacts battles of the English Civil War for pleasure and charitable purposes.

The name is taken from a secret Royalist organization of the mid-seventeenth-century dedicated to the Restoration of the Stuart line.

1971 *Certificate of Incorporation* No. 1014700, I hereby certify that *The Sealed Knot* Limited is this day incorporated under the Companies Acts 1948 to 1967 and that the Company is limited. Given under my hand at London the 17th June 1971. **1976** A. Price *War Game* I. iii. 71 There are a number of these Civil War groups—the Sealed Knot was the first one. **1978** R. Westall *Devil on Road* ii. 7 I'd heard about the Sealed Knot. Guys..poncing around in Cavalier gear, losing the Civil War all over again.

2. a. Also of a railway train or carriage: closed to entry or exit, or admitting restricted movement, during its journey.

Freq. with allusion to Lenin's passage from Finland to Russia in 1917 in a train 'sealed' to prevent contact with German citizens.

1949 *Radio Times* 15 July 3/1 We took a sealed train to Avonmouth, a crowded troopship out into the Atlantic. **1975** M. Pearson *Sealed Train* vi. 90 The sealed carriage and its baggage wagon were..drawn backward toward Switzerland. *Ibid.* 94 As the Sealed Train steamed north, the Kaiser's troops were striving..to check the new Allied offensive. **1979** O. Sela *Petrograd Consignment* 159 Zinoviev..has suggested that they travel through Germany in a sealed carriage, without stops.

b. Also with *up*. *a sealed book* (earlier example).
[**1611** Bible *Isaiah* xxix. 11 And the vision of all is become vnto you, as the wordes of a booke that is sealed.] **1814** Lamb *Let.* 29 Aug. (1888) 278 My left arm reposes on the *Excursion*. I feel what it would be in quiet. It is now a sealed book. **1884** W. James in *Mind* IX. 201 No impression penetrates to the sealed-up sensibility. **1943** C. Day Lewis *Word over All* 17 Or a heightening At most of the sealed-up hour wherein we awaited What?

d. *sealed-off*: closed so that neither access nor egress is possible.
1926 R. W. Lawson tr. *Hevesy & Paneth's Man. Radioactivity* ii. 19 A wider evacuated tube with sealed-off ends and thick walls. **1938** *New Statesman* 19 Feb. 277/2 One is left wondering whether the town hall has sealed-off rooms, and whether the chief officials enter by different doors. **1963** B. Fozard *Instrumentation Nucl. Reactors* iv. 42 For beta-particle measurements sealed-off tubes with thin end-windows can be used. **1978** R. Ludlum *Holcroft Covenant* iii. 40 Are you telling me that two people got off that plane, walked through sealed-off corridors into the sealed-off, guarded customs area, and *vanished*?

e. *sealed room* = *locked room* s.v. *Locked *ppl. a.* e. Also *absol.* in related use. (Freq. used in detective novels.)
1939 'M. Innes' *Stop Press* ii. v. 286 Mr. Eliot's was distinctly not a mystery of the sealed-room type. **1944** J. D. Carr *Till Death do us Part* v. 52 If we have any clue to these sealed-room affairs, my guess is that there's the clue. **1971** A. Morice *Death of Gay Dog* xii. 135 One victim and five suspects, all together in a sealed room, as the saying goes.

f. *sealed-beam*: applied (usu. *attrib.*) to a motor-vehicle headlamp in which light source, reflector, and lens form a sealed, self-contained unit.
1939 *Daily Progress* (Charlottesville, Va.) 19 Aug. 4/1 The result of three years of cooperative effort by engineers.., the invention is described in a special story to The New York Times as 'a "sealed beam"', which at once provides greatly increased range and breadth of light with stronger intensity through its high, or "country", beam and reduction of glare with increased illumination of the right side of the road with its low or "traffic" beam'. **1954** [see *Prefocused *ppl. a.*]. **1965** *Economist* 23 Oct. p. x/1, Lucas did not adopt sealed-beam headlights..until well after the Americans. **1972** 'S. Abbey' *Bk. of Marina* xi. 97 Sealed-beam units form, in effect, large bulbs, each with either one or two filaments, an integral reflector and a front lens.

g. *sealed source*: a pellet of radioactive material in a sealed capsule, used in radiotherapy and radiography.
1962 *B.S.I. News* Mar. 17/2 Radiography sealed-sources—pellets of radioactive material contained in sealed capsules. **1971** *New Scientist* 1 Apr. 26/2 The US market is estimated to be worth over $80 million a year with some..$20 million on basic radioisotopes and sealed sources.

h. Of a road: surfaced with tar macadam, etc. Cf. *Tar-sealed *ppl. a.* Chiefly *Austral.* and *N.Z.*
1938 *Ann. Rep. Dept. Main Roads New South Wales 1937* 4 Generally, for country roads in New South Wales the sealed gravelled pavement has proved to be quite adequate. **1966** *Weekly News* (N.Z.) 5 Dec. 47/6 It was good to be back on the sealed highway. **1969** *Northern Territory News* (Darwin) *Focus '69* 11 (Advt.), Contractors for bitumen sealed surfaces across the Territory. **1977** *Weekly Times* (Melbourne) 19 Jan. 63/7 (Advt.), Situated on sealed road handy Frankston and Dandenong. **1979** B. L. C. Johnson *Pakistan* xii. 184/1 Roads of a high standard (sealed)..now total 27,152 km.

sea legs, *pl.* Add: Also *fig.*
1895 Lloyd George *Let.* 27 Mar. (1973) 83, I have got my sea legs in the House. They now listen to me with deference. **1977** *Rolling Stone* 7 Apr. 78/3 *Sailin'*, then, is a solid album by a promising singer/songwriter who is still finding her sea legs.
Hence **sea-legged** *a.*
1913 W. de la Mare *Peacock Pie* 109 Here is a sea-legged sailor. **1946** Dylan Thomas *Deaths & Entrances* 55 Goodbye to the man on the sea-legged deck.

sealer, *sb.*[1] Add: **5.** One who or that which seals (Seal *v.*[1] 8 in Dict. and Suppl.).
1881 *Instructions to Census Clerks* (1885) 58 Blacking Manufacture: Liquid:..Sealer. **1921** *Dict. Occup. Terms* (1927) §159 *Sealer, ink bottle*; seals filled bottles of ink with sealing wax or with paper seals. **1928** *Daily Express* 6 July 5/5 The machine seals the lids of special cans... Mr. A. Appleyard..and Mr. Hirst..have tested this British hand sealer scientifically. **1940** H. T. Davey *Wood Finishing* xvii. 199 The object of a sealer is to block the pores of the wood so as to stop the succeeding coats of lacquer sinking in. **1958** *Listener* 4 Sept. 359/2 Mix yourself some sanding sealer from starch and cellulose lacquer. **1971** *Home Preserv. Fruit & Veg.* (Min. Agric.) (ed. 13) 63 The can sealer must be cleaned thoroughly after use. **1977** *Reader's Digest Bk. Do-It-Yourself Skills & Techniques* iv. 140/2 Apply a light coat of the sealer by brush or cloth direct to the unfilled timber, working it well in.

6. *Canad.* A glass jar designed to preserve fruit, vegetables, etc.
1932 N. M. Jamieson *Cattle in Stall* 12 Just take in a great half-gallon sealer and get it filled—for about half a dollar! **1959** *Weekend Mag.* (Montreal) 15 Aug. 39/2, I went into the fruit cellar, turned on the light and was actually picking up two quart sealers of icicle pickles. **1970** *Canad. Antiques Collector* July–Aug. 23/2 The demand for these preserving jars dwindled with the beginning of the cheap production of glass 'sealers' and the discovery of the method of canning in tins.

sealer, *sb.*[2] **2.** (Earlier examples.)
1770 G. Cartwright *Jrnl.* 31 Dec. (1792) I. 76 After breakfast we set off homewards, being accompanied by two of the sealers. **1825** J. F. Cooper *Lionel Lincoln* II. vi. 138 The sealers of New-England have been able to discover Terra Australis.

sealing, *vbl. sb.*[1] Add: **5. b.** The action or process of rendering impervious; also *concr.*, material used for this.
1955 C. Jasper *Handbk. Decorating & Painting* ii. 51 Sealing. The best form of treatment..for making the plaster suitable to receive oil paint, is to apply..alkali resisting primer. **1960** *McGraw-Hill Encycl. Sci. & Technol.* XIV. 532/2 Sealing..prevents penetration of or redissolution in subsequent finish coats. **1962** *Which? Car Suppl.* Oct. 139/2 The rubber sealing round the ventilating panes was not effective.

sealing, *vbl. sb.*[2] (Earlier and later examples.)
1839 *Southern Lit. Messenger* V. 3/1 In a few years [they] made Stonington famed for sealing. *attrib.* **1786** G. Cartwright *Jrnl.* (1792) III. 237 All the sealing-posts now exhibit a very different appearance from what they originally did. **1911** *Daily Colonist* (Victoria, B.C.) 26 Apr. 14/2 The United States expressed a willingness to compensate the owners of the sealing fleet for the loss of the industry. **1933** J. Buchan *Prince of Captivity* i. iii. 91 A sealing sloop had crawled up the coast as soon as spring opened the shore waters. **1977** *Time* 10 Oct. 62/1 Before he was out of his teens he had..shipped on a sealing expedition to the Bering Sea.

sealing-wax. Add: **b.** Used *attrib.*, esp. as **sealing-wax red,** to designate a bright red colour, vermilion.
1907 *Yesterday's Shopping* (1969) 156/2 These Colours Kept in Stock..Sealing wax red..Sky..Stone. **1912** Mrs. P. Campbell *Let.* Aug. in *B. Shaw & Mrs. Campbell* (1952) 35 The scullery maids..with their bloody nails and sealing-wax lips make my hair stand on end. **1930** [see *post-office red* s.v. *Post Office 3]. **1930** A. P. Herbert *Water Gipsies* xxv. 380 Her lips were the hot sealing-wax lips which she knew he hated. **1939** *Burlington Mag.* Oct. 162/2 About 1550 the famous 'sealing-wax red' first appears on Turkish pottery. **1957** [see *Isnik].

sea-lion. Add: **3. b.** [tr. G. *seelöwe.*] In full, *Operation Sea-lion*, the code name for the German plan to invade the United Kingdom after the fall of France in the war of 1939–45. (Announced by Hitler in July 1940 and cancelled in October of that year.)
1949 W. S. Churchill *Their Finest Hour* xiv. 261 Our excellent Intelligence confirmed that the operation 'Sea Lion' had been definitely ordered by Hitler. **1951** F. H. Hinsley *Hitler's Strategy* iv. 83 In his [*sc.* Hitler's] mind 'Sea Lion' was never more than a colossal bluff. *Ibid.* x. 191 The final cancellation of operation 'Sea Lion'..was less important for itself than as an indication of the state of mind to which Hitler had already been reduced. **1978** D. Kyle *Black Camelot* xi. 178 Heydrich..must have obtained Raeder's copy after Sea Lion was cancelled.

seal point. [f. Seal *sb.*[1] + *Point *sb.*[1] A. 26 c.] One of the dark brown markings on the buff fur of one type of Siamese cat; also, a Siamese cat with markings of this colour. Hence **seal-pointed** *a.*
1934 P. Wade *Siamese Cat* xv. 101 Even the best Blue Pointed cannot..equal in beauty our Seal Pointed cats. *Ibid.* 102 Let us concentrate on breeding perfect Siamese with deep seal points. **1939** I. M. Mellen *Pract. Cat Bk.* i. 35 The adult body color of the Siamese is pale fawn..and the seal brown, almost black ears, muzzle, tail and feet are called collectively 'points', the color described being known as seal point. *Ibid.* 39 (*heading*) Qualities of a good seal point Siamese. **1958** *Listener* 18 Sept. 410/2 The first Siamese to become a champion arrived in 1896. He was a seal point. **1966** 'K. A. Saddler' *Gilt Edge* vi. 91 A seal point Siamese cat sneaked in. **1980** A. Scholefield *Berlin Blind* iii. 106 He..saw a cat. It was a beautiful sealpoint Siamese.

sealskin, *sb.* **2.** (Earlier example.)
1858 Queen Victoria *Let.* 27 Feb. in R. Fulford *Dearest Child* (1964) 62, I am so glad you find the sealskin comfortable; it is a nice warm thing.

Sealyham (sī·li,ăm). [The name of *Sealy Ham* (House) near Haverfordwest, the home of the Edwardes family which developed the breed in the 19th cent.] A small stocky white terrier, sometimes with markings of other colours, distinguished by a thick, rough coat, drooping ears, a small, erect tail, and a square, bearded muzzle. Also *attrib.*
1894 R. B. Lee *Hist. & Descr. Mod. Dogs: Terriers* xvi. 386 There is a strain of terrier much talked about of late known as the Sealy Ham. **1907** [see *Jack Russell (terrier)* s.v. *Jack *sb.*[1] 37]. **1917** W. J. Locke *Red Planet* xix. 250 You remember Jingo, the Sealyham. **1924** [see *Cairn 2]. **1930** R. H. Mottram *Europa's Beast* x. 241 The Sealyham..merely slept. **1954** M. K. Wilson tr. *Lorenz's Man meets Dog* viii. 79 A Sealyham's love of fun, and his fidelity to his master can prove a real moral support to a melancholy type of person. **1966** J. Betjeman *High & Low* 12 Where's Kathleen Stokes with her Sealyhams? **1979** R. Rendell *Means of Evil* 153 He emerged from the front door with two Sealyhams on a lead.

seam, *sb.*[1] Add: **I. 1. d.** Also in colloq. phrases, as *to burst* (*fall apart*, etc.) *at the seams*.
1962 Ross & Singer *Guilty Party* in *Plays of Year* XXIV. 74 You were bursting at the seams a little. Can we ease the strain for you? **1965** J. Von Sternberg *Fun in Chinese Laundry* (1966) ii. 43 In a few instances when I thought that I would come apart at the seams..I managed to make the director listen. **1969** 'V. Packer' *Don't rely on Gemini* (1970) xv. 131 He had begun to fall apart at the seams—to imagine..that the stars..were conspiring against him. **1977** *Times* 30 Apr. 9/6 My marriage..came apart at the seams.

f. *Cricket.* The raised band of stitching around the centre of a ball. Cf. *seam bowler*, sense 10 below.
1888 Steel & Lyttelton *Cricket* iii. 119 The ball is usually, by a slow bowler, grasped firmly with all the fingers resting on the seam, as this gives more purchase and resistance for the fingers to operate. **1906** F. R. Spofforth in Beldam & Fry *Great Bowlers & Fielders* 10 It is almost impossible to swerve unless the seam of the ball is up and down. **1948** C. Sly *How to bowl them Out* x. 75 The ball..resembles the planet Saturn in that it has a ring or projecting ridge round its waist..known as the seam. **1972** R. Illingworth *Young Cricketer* iii. 50 Grip the ball with the seam upright... Angle the seam to fine leg and not to third man.

g. *French seam* (Needlework), a double-stitched seam that is folded and sewn to resemble a plain seam on the right side.
1903 *Home Fashions* 12 Sept. 18/3 The sleeve is joined by French seam. **1964** *McCall's Sewing* 276/1 After cutting the panels, join them with tiny French seams.

6. b. *Metallurgy.* A superficial linear defect on worked metal usu. caused by closure of a blow hole.
c **1840** B. Legge *Guide to Iron Trade* 36 Having sems [*sic*] in a longitudinal direction. **1923** Glazebrook *Dict. Appl. Physics* V. 362/2 These surfaces become oxidised both during cooling of the ingot and during reheating for forging or rolling, and give rise to rokes or seams. **1924** Greaves & Wrighton *Pract. Microsc. Metallogr.* ix. 78 Seams show a similar microscopical appearance to rokes and are caused in rolling billets or bars by one portion of the metal folding over another. **1967** A. K. Osborne *Encycl. Iron & Steel Industry* (ed. 2) 373/2 Seams may also be caused by rippled surfaces or by recurrent teeming laps.

10. **seam allowance,** the amount of material in sewing which is calculated to be taken in by a seam; **seam bowler** *Cricket*, a medium or fast bowler who uses the seam to make the ball deviate in the air or off the pitch during delivery; hence **seam-bowling; seam-squirrel** *U.S. slang* (chiefly *Mil.*), a louse; **seam welding,** a form of resistance welding in which a linear weld is obtained by means of rolling disc-shaped electrodes which produce a line of overlapping welds (see also quot. 1964); so **seam-weld** *v. trans.*; also **seam weld** *sb.*; **seam welder,** a machine for **seam welding.**
1949 I. R. Duncan *Compl. Sewing Book* xv. 301 Plain seams may be used for every type of garment... The amount of seam allowance depends upon the material; generally from ⅜ to ½ an inch is required. **1964** *McCall's Sewing* 277/1 Press fold formed along edge of seam allowance. **1977** R. Richardson *Discovering Patchwork* 63/2 Window templates made of clear plastic with a shaded area round the edge... The shaded area is the seam allowance. **1948** J. Arlott *How to watch Cricket* iii. 14 The term 'seam-bowler' is almost identical [with 'pace bowler'] since it refers to those bowlers who use the seam to swing, or cut the ball. **1978** R. V. Jones *Most Secret War* xliv. 414, I had previously not believed such tales as the swinging of cricket balls by seam bowlers. **1956** R. Alston *Test Commentary* 136 Johnson persisted in a seam-bowling attack. **1976** Dexter & Makins *Testkill* 174 The steady England pressure which was now resting on the excellent seam bowling of Kirkstead. **1899** J. R. Skinner *Hist. Fourth Illinois Volunteers* 129 When it was first rumored that the old confederate seam squirrel had invaded our quarters, a small panic seized many. **1929** L. Thomas *Woodfill of Regulars* 240 The rest of the boys..stopped chasin' seam-squirrels in their undershirts. *a* **1956** C. J. Post *Little War of Private Post* (1960) 255 There is the gray-back, or seam-squirrel, from the days of our Civil War. **1920** *Whittaker's Electr. Engineer's Pocket-bk.* (ed. 4) 646 On light work, about 3 ft. of seam weld can be made per min. **1980** L. M. Gowd *Princ. Welding Technol.* xi. 164 After the first weld has been made,..the current must be raised to maintain the size of the welds. Accepting this limitation, satisfactory continuous seam welds can be made. **1921** Seam welder [see *seam welding* below]. **1959** Neumann & Bockhoff *Welding of Plastics* vii. 126 In using high-frequency seam welders, the breakdown strength of the plastic must be considered. **1976** *Western Mail* (Cardiff) 27 Nov. 6/5 Spot-welders, seam-welders and projection welding machines up to 20 KVA are also among the factory's equipment. **1917** Oberg & Jones *Machinery's Encycl.* VI. 496/2 By the seam-welding process, two sheets of wrought iron or steel are welded together along the edge by a continuous lap-weld... Plates can be successfully seam-welded at a rate of about one foot per minute. **1921** *Automobile Engineer* XI. 108/1 Seam welding..can be adopted with advantage when a tight joint is required.., a seam welder differing from a spot welder mainly in that roller electrodes are used instead of the pointed electrodes. **1964** Wordingham &

Reboul *Dict. Plastics* 158 *Seam welding*, with thermoplastic materials, the method of forming a welded seam, either by the use of welding rollers..for continuous welding, or by jig welding. **1975** Bram & Downs *Manuf. Technol.* ii. 62 Seam welding is similar to spot welding, the difference being that the spots overlap each other, making a continuous weld seam.

seam, *v.*[2] Add: **5.** *Cricket.* **a.** *intr.* Of a pitch: to aid a seam bowler; of a ball: to swing during delivery on account of the seam.
1960 E. W. Swanton *West Indies Revisited* 118 According to the players, the pitch was still 'seaming' a bit. **1974** *Reading Evening Post* 3 Sept. 14/7 Both opening bowlers made the ball seam considerably and Fletcher was next to go. **1976** *Liverpool Echo* 23 Nov. 18/7 The pitches out there have been known to seam.

b. *trans.* Of a bowler: to cause (a ball) to seam.
1963 T. Bailey *Improve Your Cricket* ii. 81 At medium pace it is possible to seam the ball..and naturally to vary pace, length and angle of flight. **1976** J. Snow *Cricket Rebel* 21 Thomson..and..Pountain..taking seven of the first eight wickets between them on a damp wicket which helped them seam the ball about.

seamanite (sī·mănait). *Min.* [See quot. 1930 and -ite[1].] A rare hydrated phosphate and borate of manganese,

$$Mn_3^{2+}(OH)_2[B(OH)_4][PO_4],$$

occurring as transparent, yellow, orthorhombic crystals.
1930 E. H. Kraus et al. in *Amer. Mineralogist* XV. 220 In recognition of Professor [A. E.] Seaman's long and influential service as Professor of Geology and Mineralogy at the Michigan College of Mining and Technology, and his valuable contributions to the geology and mineralogy of the Upper Peninsula of Michigan, the authors propose the name of seamanite for this new mineral. **1971** *Ibid.* LVI. 1531 Seamanite is certainly a candidate for one of the most exotic of mineral structures. The Mn-O octahedral arrangement is one of the most peculiar and unexpected on record.

sea-mark. **2. b.** (Later examples with allusion to Shakes. *Oth.* v. ii. 271.)
1932 *Proc. Brit. Acad.* XVII. 57 He [*sc.* Virgil] fixed for the imagination of the Roman race..the limit of its aspiration and achievement, the very sea-mark of its utmost sail. **1955** *Times* 11 May 11/4 Often they even marked, though not the very butt and seamark of his utmost sail, at least his objective for the time being.

seamer. Add: **2.** *Cricket.* A seam bowler; also, a delivery by a seam bowler.
1952 *Observer* 25 May 10/1 R. Smith, with his medium-paced 'seamers'—to use a modern and not too unmeaning term—and Bailey..began almost to persuade us that their names might be G. Lohmann and S. F. Barnes. **1955** *Times* 5 July 4/1 Silk was probably right in thinking that his seamers would get more out of it than his spinners. **1963** *Times* 13 June 3/1 It indicated that the intention of both captains was to rely entirely upon their seamers, which they duly did. **1976** Dexter & Makins *Testkill* 140 With the new ball due well before lunch he sensibly switched to Flinders' leg-spin and Lytton's seamers.

seamfree (sī·mfrī), *a.* (and *sb.*) [f. Seam *sb.*[1] + Free *a.*] Of stockings: = Seamless *a.* in Dict. and Suppl. Also *absol.* as *sb.*
1959 *Manch. Guardian* 27 July 4/2 Plaza..have a seamfree called 'Riviera' which wears well. **1960** *Sunday Express* 14 Feb. 14/4 The percentage of seam-free stockings sold today is steadily rising. **1976** *Leicester Trader* 24 Nov. 2/4 There are several styles [of stockings] to choose from at Plants including Wolsey Monte Carlo seamfree 15 denier at 59p a pair.

seaming, *vbl. sb.* **b.** *seaming machine* (earlier U.S. example).
1847 *Rep. Comm. Patents 1846* (U.S.) 214 My improved seaming machine for turning down and forming a seam of the flange surrounding the bottoms of the buckets.

seamless, *a.* Add: **1.** (Further examples.) Now *esp.* of tubing and women's stockings or tights.
1876 *Encycl. Brit.* IV. 218/2 Solid or seamless brass tubes..are made by drawing down short thick cast cylinders of brass till they reach the desired gauge or thinness. **1904** Goodchild & Tweney *Technol. & Sci. Dict.* 146/1 The [cycle] frame is generally formed of thin seamless steel tubing. **1921** *Daily Colonist* (Victoria, B.C.) 4 Oct. 5/1 Seamless (fashioned without seams). The women's hose knit to fit without a seam. **1959** *Observer* 22 Mar. 3/8 The seamless stocking, or 'circular' as the trade calls it, has always been made, but used to be thought of as an inferior thing. **1968** *Economist* 9 Nov. 94/1 Tubes is commissioning a new seamless tube mill, where profits ought to start this autumn. **1969** A. J. Hall *Stand. Handbk. Textiles* (ed. 7) iii. 154 The seamless hose are knitted on circular machines.

2. fig. *spec.* as *seamless web*, orig. with reference to the concept of history as an integral whole (see quot. 1898).
1862 E. Dickinson *Poems* (1955) I. 246 As if some Caravan of Sound Had parted Rank, Then knit, and swept—In Seamless Company. **1898** Pollock & Maitland *Hist. Eng. Law bef. Edw. I* (ed. 2) I. i. i. 1 Such is the unity of all history that any one who endeavours to tell a piece of it must feel that his first sentence tears a seamless web. **1929** *Oxf. Poetry* 12 A small patch of light on the seamless blank. **1952** Auden *Nones* 39 The three wise Maries come Sossing through seamless waters. **1964** [see *Integralism]. **1976** T. Eagleton *Crit. & Ideology* iii. 94 *The Prelude* is *formally* fissured by its ideological contradictions, unable to rise to the seamless impersonal epic it would wish itself to be. **1977** *Time* 10 Oct. 10/3 'Many of us find it intellectually difficult', he said, 'to penetrate the seamless web of the Japanese politico-economic system.'

sea monkey. **1.** A heraldic animal which is part monkey, part fish.
1909 A. C. Fox-Davies *Compl. Guide Heraldry* xiii. 230 It may be as well to allude to the asserted heraldic existence of the sea-monkey, though I am not aware of any instance in which it is borne.

2. A brine shrimp, *Artemia salina*, often used as food for fish in aquaria.
1973 *Sunday Mail* (Brisbane) 3 June 6/2 'Sea monkeys'..are currently booming in the United States. They come as a parcel of eggs that look like dried crystal. But..after the eggs are dropped into water they grow into what looks like sea shrimp. **1976** 'D. Halliday' *Dolly & Nanny Bird* ii. 29 'They're Sea Monkeys.'.. 'Brine shrimps... Fish eat them.'

sea-monster. **2.** (Later examples.)
1917 E. Pound *Lustra* 186 The sea-monster Bulges the squarish bronzes. **1952** C. Day Lewis tr. *Virgil's Aeneid* iii. 65 Below, she is a weird sea-monster With dolphin's tail and a belly of wolverine sort.

Seanad Eireann (ʃæ·năd ēə·rān). [Ir. *Seanad Éireann* the senate of Ireland.] The upper house of the parliament of the Republic of Ireland. Freq. *ellipt.* as **Seanad.**
1922 [see *Oireachtas 2]. **1923** W. B. Yeats in *Senate Speeches* (1961) 36, I hope I have the leave of the Seanad to use a different form of words. **1937** *Bunreacht na hÉireann (Constitution of Ireland)* 48 Article 18. 1. Seanad Eireann shall be composed of sixty members, of whom eleven shall be nominated members and forty-nine shall be elected members. **1962** M. Ameller *Parliaments* I. ii. 10 While the method of appointment of members of the Second Chamber is not altogether in keeping with ordinary democratic principles, that is offset by the fact that the powers of the Seanad are considerably curtailed. **1974** *Encycl. Brit. Macropædia* IX. 886/2 There are..60 members of the Seanad... The Seanad may delay..bills passed by the Dáil.

séance. Add: **1.** Also *séance royale*, a royal audience. (Earlier and later examples.) Also *loosely* (chiefly *U.S.*), a meeting or discussion.
1789 A. Young *Jrnl.* 20 June in *Trav. France* (1792) I. 115 A message from the King.., that he should meet them on Monday; and, under pretence of preparing the hall for the seance royale, the French guards were placed..to prevent any of the deputies entering the room. **1922** Joyce *Ulysses* 644 Come, he counselled, to close the *séance*. **1934** E. Pound *Eleven New Cantos* xxxiv. 16 At the Seance Royale last Thursday he had talked of His death in defence of the country. **1962** W. Schirra in *Into Orbit* 32 We would lock ourselves up in our office at Langley until we had a solution that satisfied us all... We called a session like this a 'seance'. **1977** *Time* 17 Jan. 41/1 For an hour most nights, he conducts a long-distance séance (at $3 a minute) with..his Australian proconsul, from the..desk in his study.

3. b. A 'sitting' for a portrait.
1877 Disraeli *Let.* 14 Apr. (1929) viii. 117 Now I am going to the Palace for my 3rd Seance [to Von Angeli]. **1919** R. Fry *Let.* 21 Oct. (1972) II. 460 All wanted me to draw their portraits so that every evening in the café I had to have a *séance*.

sea-otter. Add: **2.** The thick dark fur of *Enhydra lutris*.
1813 A. Henry *Jrnl.* 19 Nov. in E. Coues *New Light Early Hist. Greater Northwest* (1897) II. 753 His son had a robe of two sea otters, for which he demanded 48 beavers. **1915** *Chambers's Jrnl.* Jan. 48/2 Of sea-otter, too, perhaps one of the loveliest furs of all, the supply is failing. **1956** J. G. Links *Bk. Fur* iii. 139 A sea otter collar to his coat was to the great industrialist of the last century what a private stratocruiser is to his successor.

seaplane (sī·plēⁱn). [f. Sea *sb.* + *Plane *sb.*[5]] An aeroplane designed to be able to operate from water; *spec.* one with floats, in contrast to a flying boat.
1913 [see *Hydroplane *sb.* 3]. **1914** *Daily Express* 2 Nov. 5/2 The old cruiser Hermes..had been recently used as a seaplane-carrying ship. **1921** *Daily Colonist* (Victoria, B.C.) 5 Oct. 12/3 Seaplane stations have been established by the Air Board of Canada. **1938** *Sun* (Baltimore) 21 July 1/8 Hooked together—the seaplane Mercury above and the flying boat Maia below—the two planes rose from the river Shannon this evening. **1954** P. K. Kemp *Fleet Air Arm* 87 Three cross-Channel steamers were taken up to be transformed into seaplane carriers. **1973** [see *Refuel *v.* 2].

sea-purse. Add: **4. a.** [ad. the Algonquian language Munsee (Delaware) *sepoûs*, brook, small river: see Mathews *Dict. Americanisms* (1951).] (Earlier and later examples.)
1842 W. P. Hawes *Sporting Scenes* I. 102, I kept watch of him—when I came to a sea poose—I went in and to the east of it. **1904** *N.Y. Tribune* 29 May ii. 7/1 McDonald was a good swimmer, but, getting caught in a sea puss, was shot out to the deep sea with great velocity. **1932** *Sun* (Baltimore) 5 Sept. 6/3 The sea-purse swooped in and picked up a girl bather, who was suddenly seen to whirl about on the surface of the water like a cork.

seaquarium (sīkwēə·riŏm). Pl. **seaqua·ria.** [f. Sea *sb.* + A)quarium.] An aquarium for large marine animals.
1955 *Travel* Dec. 54/2 On the other side of the Sunshine State, at Miami Beach, a new $2,000,000 Seaquarium features giant fishbowls of what is probably the largest collection of tropical marine life in the world. **1959** *Listener* 30 July 186/1 A diver feeding dolphins in the Miami Seaquarium, Florida. **1962** *New Scientist* 15 Mar. 607/2 The great natural aquaria (or seaquaria or oceanaria) in Florida and California. **1969** *Daily Tel.* 24 Apr. 16/3 The seaquarium's Manatee tank is only five feet deep. **1978** *New York* 3 Apr. 74 Come bask in the warmth of exotic..Sonesta Beach Hotel & Tennis Club... Room with private balcony. Admission to the Seaquarium and Crandon Park Zoo. **1982** *Times* 21 Jan. 4/5 The three Clacton whales were kept in Iceland in a 'seaquarium'.

search, *sb.* Add: **2. d.** (Earlier U.S. example.)
1798 *Deb. Congress U.S.* 12 June (1851) 1907 Gentlemen appeared to confound the right of search with the right of capture.

6. *search area*, *-party* (later examples), *plane*, *team*; **search coil** *Electr.* = *exploring coil* s.v. *Exploring *vbl. sb.*; **search-warrant** (earlier examples).
1973 H. Nielsen *Severed Key* i. 13 Word of the airline crash had spread rapidly... Only the hardiest of the amateur sailors would be able to reach the search area. **1977** *New Yorker* 27 June 62/3 The search area could not be reduced much below forty thousand square miles. **1897** *Electrician* 30 July 439/2 You travel along the line of the main, carefully tracking the cable with the search coil, and listening intently at the telephone receiver. **1933** [see *fluxmeter*, *flux meter* s.v. *Flux *sb.* 13]. **1953** R. J. C. Atkinson *Field Archaeol.* (ed. 2) i. 39 The soil is usually removed in a series of shallow layers, and the surface of each layer can be swept with the search-coil. **1978** J. Irving *World according to Garp* ii. 30 She quickly organized a search party among the healthier and more mobile patients. **1981** A. Morice *Men in her Death* viii. 91 I'll be on my way. Robin will be sending out search parties if I'm not home soon. **1966** M. Woodhouse *Tree Frog* xxvi. 195 They would send out a search plane as soon as it got light. **1978** R. Ludlum *Holcroft Covenant* xxx. 348 A capsized craft fitting the description of the small boat was sighted by low-flying coastal search planes. **1976** L. Sanders *Hamlet Warning* xxvii. 239 Organize your men into search teams for the sectors they hold. **1739** W. Stephens *Jrnl.* 26 Mar. in *Colonial Rec. Georgia* (1906) IV. 306 It was thought proper to send out several Officers with a search Warrant. **1752** Fielding *Amelia* I. iii. xi. 269, I believe I should have applied to a Magistrate for a Search-warrant for that Picture.

search, *v.* Add: **I. 3. c.** *search me*: used (chiefly *imp.* in response to a question) to imply that the speaker has no knowledge of some fact or no idea what course to take. *colloq.* (orig. *N. Amer.*).
1901 Merwin & Webster *Calumet 'K'* iii. 37 'Search me,' said Denis. 'They've tied us up for these two weeks.' **1907** F. H. Burnett *Shuttle* xxvi. 262 If this ain't the limit! You may search *me*! **1916** 'B. M. Bower' *Phantom Herd* xi. 191 What ails that darned thing?.. You can search me. **1920** 'Sapper' *Bull-Dog Drummond* ii. 60 'Why did he send his confidential secretary..to Belfast?' 'Search me,' said Hugh. **1930** E. Waugh *Vile Bodies* i. 9 Word of eighteen letters meaning carnivorous mammal. Search me if I know how they do think of these things. **1949** G. Davenport *Family Fortunes* iii. i. 189 'How could the road be washed out—I went over it yesterday morning and it hasn't rained since!' 'Search me.' **1959** J. Thurber *Years with Ross* iv. 61 Faced with these formidable questions, any of his intimates..might easily murmur..'God Knows' or 'Search me'. **1965** D. Francis *For Kicks* viii. 110 'Where did he go for the summer?' I asked... 'Search me.' **1980** B. Bainbridge *Winter Garden* xiii. 102 'But where am I going?' asked Ashburner... 'Search me,' said Bernard. 'It's supposed to be a surprise.'

4. (Later example.)
1963 in H. W. Baade *Jurimetrics* 13 Western Reserve University established a Center..which has been engaged..in the investigation of methods of searching literature by electronic machines.

IV. 14. In special combs. with other vbs. forming verbal and substantival phrases (freq. used *attrib.*), as *search and destroy* (orig. *U.S.*), designating an uncompromising military strategy effected by the advance of troops through a given territory (first employed in the war in Vietnam); also *ellipt.*; *search and rescue* (orig. *U.S.*), designating a (chiefly *Mil.*) land or sea rescue service.
1966 *Economist* 22 Oct. 369/3 This is a different use of men from the highly mobile 'search and destroy' operations in which the American forces have been engaged up to now. **1970** [see *Green Beret* s.v. *Green *a.* 12]. **1973** D. Lees *Rape of Quiet Town* iv. 57 Sarrat's unit conducted search-and-destroy missions. **1977** 'E. McBain' *Long Time no See* xiii. 222, I was out there on a search-and-destroy. **1944** *Yank* 2 June 3/2 At last somebody thought of the Siberian Huskies assigned to a nearby Army search and rescue outfit. **1950** *Jrnl. R. United Service Inst.* XCV. 158 (*heading*) 'Search and rescue' radio watch. **1972** *Gloss. Aeronaut. & Astronaut. Terms* (B.S.I.) xiii. 3 *Search and rescue*, a service provided to notify the appropriate organizations of aircraft in need of search-and-rescue aid and to assist such organizations. **1977** *R.A.F. News* 11–24 May 9 (Advt.), Winching a capsized yachtsman to safety on board a Royal Air Force search-and-rescue Wessex.

searcher. Add: **3. d.** (Earlier example.)
1706 *Albany Fort Jrnl.* 1 July in G. Williams *Hudson's Bay Miscellany* (1975) 72/2 State of stores... 1 pair seachers for great guns.

4. b. A penetrating or embarrassing question. *colloq.*
1923 J. Manchon *Le Slang* 260 *Searcher*,..une question (ou un problème) embarrassante. **1951** N. M. Gunn *Well at World's End* xi. 81 He..asked me some questions, and searchers they were.

search-light. Add: **1.** Also, the beam of light thrown by such a lamp.
1919 Kipling in *Hutchinson's Story Mag.* July 12 And the crazy skies are lit By the searchlights of the Pit. **1929** S. Leslie *Anglo-Catholic* xvi. 225 He pointed..to the Northern sky, across which dropped a silver ribbon of attenuated search-light. **1943** L. B. Lyon *Evening in Stepney* 12 Across and across it move The search-lights, reckoning hate on a hidden clock.
Hence as *v. trans.*, to illuminate with a search-light; also *fig.*
1914 W. De Morgan *When Ghost meets Ghost* i. xxix. 344 That officer..was searchlighting areas for want of something to do. **1966** *Prison Service Jrnl.* VI. 21/2 We find our own failings..are searchlighted.

searlesite (sɜ̄·ɹlzəit). *Min.* [f. the name of John W. *Searles*, who put down the well in California from which the first specimen came: see -ite[1].] A hydrated borosilicate of sodium, $NaBSi_2O_6.H_2O$, occurring as white monoclinic crystals.
1914 Larsen & Hicks in *Jrnl. Washington Acad. Sci.* IV. 397 The mineral for which the name searlesite is proposed was found in samples from the deep well in Searles Lake, San Bernardino County, California. **1950** *Amer. Mineralogist* XXXV. 1017 In the stratum of trona 10 feet thick that is reached by the shaft at a depth of 1500 feet, subhedral crystals of searlesite up to 7 inches long, 4 inches wide, and one inch thick are not uncommon. **1974** *Mineral. Abstr.* XXV. 303/1 The phyllosilicate searlesite, $Na_2B_2(OH)_4Si_4O_{10}$, has been made from mixtures of Na_2O, B_2O_3, SiO_2 and excess H_2O heated under steam pressure at 250°C for up to 8 days.

sea-robin. **1.** (Earlier example.)
1814 S. L. Mitchill *Fishes N.Y.* 430 Gurnard, or Sea Robin, *Trigla lineata*.

Sears-Roebuck (sɪᵊɹz rōᵘ·bʊk), *a. U.S. slang.* Also **Sears and Roebuck.** The name of the American merchandising firm of *Sears, Roebuck* and Co., used allusively of an inexperienced, 'green', or amateurish person, or of something cheap or of little value.
1917 R. Lord *Captain Boyd's Battery, A.E.F.* (1919) ii. 23 Two tents of Shavetails (i.e. Reverse Officers, Ninety-Day Wonders, Sears & Roebuck Specials,..etc.) have been attached to us for instruction purposes. **1930** *Our Army* Jan. 4 He's a regular Sears-Roebuck product; three months of training, three tons of dignity and three ounces of horse sense. **1932** *Amer. Speech* VII. 270 *Sears-Roebuck driller*, an inexperienced and incapable driller. **1942** Berrey & Van den Bark *Amer. Thes. Slang* §21/14 Cheap; paltry... *Sears-Roebuck*. **1971** M. Tak *Truck Talk* 137 *Sears-Roebuck license*, the license said to be held by an inferior driver.

sea-salt, *sb.* Add: Hence **sea-salted** *a.*, impregnated or seasoned with sea-salt.
1925 V. Woolf *Mrs. Dalloway* 70 The brisk sea-salted air of their intimacy. *a* **1941** —— *Captain's Death Bed* (1950) 31 This gnarled and sea-salted man was no smug clergyman underneath.

sea-shell. **1. a.** (Earlier *attrib.* example.)
1871 Mrs. Stowe *Pink & White Tyranny* i. 4 He saw this distant vision of airy gauzes, of pearly whiteness, of sea-shell pink.

sea-sick, *a.* Add: **1. b.** *sea-sick medicine*, *pill*, *tablet*, a preparation taken to counter seasickness.
1925 E. H. Young *William* iii. 29 Mother's going to give us some sea-sick medicine before we start. **1951** N. Mitford *Blessing* ii. xii. 268 The stewardess..stood over her with a bottle of sea-sick tablets. **1959** 'M. M. Kaye' *House of Shade* iii. 38 He..had made her take several seasick pills. **1969** G. M. Brown *Orkney Tapestry* i. 17 In Scrabster they sip brandy or swallow sea-sick tablets. **1972** W. Ellis *Knife Edge* ii. 20 Seasick pills. We must get some before we set off.

sea-sickness. (Later *attrib.* examples.)
1955 E. Blishen *Roaring Boys* iii. 151 His sea-sickness pills he carried in a little phial. **1973** H. Nielsen *Severed Key* i. 12 Find the seasickness pills in the first aid kit in the forward locker.

sea-side, seaside. Add: **2.** (Earlier example.)
1782 W. Cowper *Let.* Nov. (1904) II. 20 Mr. Bull..is gone to the seaside with Mrs. Wilberforce, and will be absent six weeks.

4. a. (Earlier and later examples.) Freq. in *seaside café*, *holiday*, *resort*, *villa*.
1781 W. Cowper *Let.* 26 Sept. (1904) I. 358 The modern passion for seaside entertainments. **1873** *Young Englishwoman* June 281/1 The collars can be procured..of chemists at seaside resorts. **1909** *Sat. Even. Post* 5 June 17/2 As soon as the theaters or 'hall-shows', as the circus men call them, close, the summer-garden or seaside parks open. **1939** F. Thompson *Lark Rise* ii. 36 Children..went to parties and for sea-side holidays. **1967** N. Freeling *Strike out where not Applicable* 39 A messy, ugly building..that..had climbed beyond modest seaside-café beginnings. **1973** 'B. Mather' *Snowline* xix. 228 Looking like a seaside landlady who has just caught her daughter in bed with the star lodger. **1974** *Listener* 3 Jan. 27/1 The opera..went on, with success, at the seaside resort. **1976** P. R. White *Planning for Public Transport* vii. 141 The traditional seaside excursion demand has fallen, most weekend leisure trips now being made by car. **1981** 'W. Haggard' *Money Men* xi. 117 Dame Molly has rented a seaside villa.

b. **seaside finch** *U.S.*, a small sparrow, *Ammospiza maritima*, found on the Atlantic coast of North America; **seaside grape,** add: = Sea-grape 4; (further examples); **seaside sparrow** = *seaside finch* above.
1811 A. Wilson *Amer. Ornithol.* IV. 68 Sea-side Finch..derives its whole subsistence from the sea. **1872** E. Coues *Key to N. Amer. Birds* 137 Sea-side Finch. Olive-gray, obscurely streaked on the back and crown. **1756, 1792** Sea-side grape [see Grape *sb.*[1] 3 b]. **1837** J. L. Williams *Territory of Florida* 37 The seaside grape..and custard apples are frequently found in the hummocks. **1929** R. Hughes *High Wind in Jamaica* i. 16 The lane, or drive, was gorgeous: for the first few hundred yards it was entirely hedged with 'seaside grapes'. **1978** T. J. Wright in V. H. Heywood *Flowering Plants of World* 78/2 The purple berries of the West Indian seaside grape..are eaten. **1886** *Code Nomencl. & Check-list N. Amer. Birds* (Amer. Ornithologists' Union) 269 Seaside Sparrow..Salt marshes of the Atlantic coast. **1937** T. G. Pearson in Grosvenor & Wetmore *Bk. Birds* II. 272/2 The northern seaside sparrow inhabits..the vicinity of the ocean beaches. **1978** C. Harrison *Field Guide Nests, Eggs & Nestlings N. Amer. Birds* 394 Seaside Sparrow..breeds on salt marshes.

c. Special Combs., as **seaside (picture-)postcard,** a postcard of a type commonly sold at the sea-side, *spec.* one depicting a caricature of lewd or vulgar humour; **seaside rock (candy),** rock-candy in the form of a cylindrical stick, usu. with a cross-section displaying the name of a resort in coloured lettering, commonly sold at the seaside.
1955 M. Allingham *Beckoning Lady* ii. 22 They looked like a seaside picture-postcard..wedged in the tub cart together, with the donkey in front. **1957** R. Hoggart *Uses of Literacy* i. ii. 31 The fifty-year-old formality of seaside postcards: most of the year 'decent' working-class people would hardly approve of them, but on holiday they are likely to 'let up a bit' and send a few to friends—cards showing fat mothers-in-law and fat policemen, weedy little men with huge-bottomed wives, ubiquitous bottles of beer and chamber-pots. **1979** J. Wainwright *Duty Elsewhere* xv. 45 The exaggerated bawdiness of seaside postcards. **1963** *Listener* 3 Jan. 40/2 Seaside rock and candy-floss. **1968** W. Garner *Deep, Deep Freeze* vii. 96 Carnality ran through her like the letters through seaside rock candy. **1978** R. Hill *Pinch of Snuff* xxv. 262 He left them in a three-cornered trap..with a four-letter word burned on the lawn. Perhaps like sea-side rock it went all the way through.

sea-sider. [f. Sea-side + -er[2].] A frequenter of the seaside. (In quot. 1976 used as the nickname of an Association football team.)
1870 *Amer. Naturalist* III. 230 The Savannah Sparrow, though only occasionally found breeding so far south as Massachusetts, is evidently a sea-sider. **1892** *Pall Mall Gaz.* 23 June 1/3 The average lady seasider in this country frugally wears out her summer finery by the 'sad sea waves'. **1976** *West Lancs. Evening Gaz.* 15 Dec. 1. 18/7 Alan Ball comes to Blackpool tomorrow to talk terms with the Seasiders when he will be offered some of the most attractive terms ever offered by Blackpool.

season, *sb.* Add: **II. 13. c.** (*a man*, etc.) *for all seasons*: (one) ready for any situation or contingency, adaptable to any circumstance.
Orig. used with reference to Sir Thomas More (1478–1535); cf. Erasmus *Encomium Moriae* (1509) Pref., *omnium horarum hominem*.
1520 R. Whittinton *Vulgaria* fol. 14ʳ And as tyme requireth, [Thomas More is] a man of merueylous myrth & pastymes: and somtyme of sad grauite: as who say a man for all seasons. **1960** R. Bolt (*title*) A man for all seasons. **1968** *Listener* 26 Dec. 842/3 If there can really be no simple account of Mr Powell's general political position save that he is a man for all seasons, the same is true of his position on immigration. **1972** A. Price *Col. Butler's Wolf* xii. 131 She's a nice sort of girl... A girl for all seasons. **1973** *Listener* 29 Nov. 745/1 He [*sc.* John Kennedy] was a man for all seasons, a man for all people. **1976** M. Birmingham *Heat of Sun* viii. 114 The vulture..a bird for all seasons, I decided, ready to match one's every mood. **1981** M. Warner *Joan of Arc* xiii. 263 Joan, a heroine for all seasons.

III. 16. b. (Later examples.)
1781 H. Newdigate *Let.* 15 Oct. in A. E. Newdigate-Newdegate *Cheverels* (1898) iii. 46 A Hamper of good things,.. Pears & Pines in perfection, *Rouleaux* never out of season. **1847** A. Brontë *Agnes Grey* xviii. 282, I pretended to want to save it [*sc.* a hare]..as it was so glaringly out of season. **1960** *Mrs. Beeton's Bk. Househ. Managem.* vii. 93 When a fruit is out of season in one part of the globe it is usually in season in another.

VI. 21. *season-cracked* adj.; **season-check** *U.S.* (see quot. 1905); **season cracking,** the occurrence of longitudinal cracks in cold-worked brass or bronze; so **season crack** *sb.* and *v. intr.*; **season ticket** (earlier and later examples); **season-ticket holder** (earlier example).
1887 J. Kirkland *Zury* 32 Ye see that season-check in the butt-end [of a black-walnut log]. **1905** *Bull. Bureau of Forestry* (U.S. Dept. Agric.) No. 61. 33 *Check*, a longitudinal crack in timber caused by too rapid seasoning... Syn.: season check. **1909** Webster, Season cracks. **1915** *Trans. Amer. Inst. Metals* VIII. 322 It was found that all those cartridge cases which had season cracked had primers on. **1957** R. A. Higgins *Engin. Metall.* I. v. 87 A controlled low-temperature anneal..applied to hard-drawn 70/30 brass tube will effectively reduce its tendency to 'season-crack'..without reducing strength or hardness. *Ibid.* II. xiv. 292 Residual..stresses in cold-worked alloys often manifest themselves during service in the form of 'season cracks'. **1938** M. K. Rawlings *Yearling* xxvi. 347, I got a old dug-out right above here, is mighty sorry and season-cracked, but hit'd carry you acrost the creek. **1910** *Brass World* VI. 269/1 One of the most annoying things that happens to brass while in use is its season cracking. **1923** Glazebrook *Dict. Appl. Physics* V. 410/2 The essential condition for 'season cracking' is the presence in the finished article of internal stresses of considerable magnitude. **1967** A. H. Cottrell *Introd. Metall.* xxiii. 467 The season cracking of α-brass in ammoniacal environments may be due to local dezincification along the grain boundaries. **1820** *Columbian Centinel* (Boston, Mass.) 2 Dec. 1/1 For sale, a Boston Theatre Season Ticket, at a fair price. **1869** W. James *Coll. Ess. & Rev.* (1920) 1 People who are comfortably in possession of a season-ticket over the Stygian ferry. **1953** C. Day Lewis *Italian Visit* i. 14 The season tickets that rattle us back and forth in a groove from Centre to circumference. **1862** J. Simmonds *Railway Travellers Handy Bk.* 48 We may here remind season-ticket holders to renew their privilege.

seasonable, *a.* Add: ¶**4.** *erron.* used for Seasonal *a.* 2.
1923 *Glasgow Herald* 20 Mar. 9 Persons engaged in seasonable trades in which the duration of seasonable employment is too short to enable them to qualify for benefit. [**1980** *Listener* 10 Jan. 51/3 Will the BBC please note that the word they want is 'seasonable', not 'seasonal'. One has seasonable items like mince pies and carols; 'seasonal' is applied to rainfall and fluctuations in car sales, i.e., things that happen with the changing seasons.]

seasonably, *adv.* Add: ¶**2.** *erron.* used for Seasonally *adv.*
1928 *Britain's Industr. Future* (Liberal Industr. Inquiry) iv. xxvii. §7. 388 The hours worked in all the jobs concerned (omitting those in which hours varied weekly or seasonably).

seasonality (sīz'næ·lĭti). orig. *U.S.* [f. Seasonal *a.* + -ity.] **1.** The condition of being dependent on the seasons or other temporal cycle; the state of recurring at regular intervals.
1934 in Webster. **1936** *S.P.E. Tract* xlv. 191 University professors and other persons of literary or academic reputation have been responsible for such coinages as..*seasonality*. **1959** *Economist* 18 Apr. 218/2 The common seasonality of the tourist trade, horticulture and fishing makes these occupations awkward neighbours for year-round industrial enterprise. **1971** *Nature* 5 Feb. 406/2 The relatively sharp seasonality of fruiting and leaf formation would have effectively restricted exclusive arboreal feeding for an animal as large as a chimpanzee. **1976** J. S. Flemming *Inflation* xi. 109 Moreover baskets including foodstuffs will display seasonality which one would probably not want to reflect in the adjusted payments.

2. The degree to which a climate has distinct seasons.
1968 R. W. Fairbridge *Encycl. Geomorphol.* 721/2 No consideration has been given to seasonality of climate. **1972** *Sci. Amer.* June 64/1 A sharp reduction in coral diversity that began in late Eocene times and lasted throughout the Oligocene epoch seems to reflect a continued increase in seasonality of climate and a substantial lowering of mean temperatures.

seasonally, *adv.* Add: **2.** According to the season.
1937 Davies & Yoder *Business Statistics* v. 239 In January..the unadjusted index was 88 and in February it was 91. This was a rise, but not so much as would be seasonally expected, as is shown by adjusted figures for the same months, which are 91 and 89. **1942** Croxton & Cowden *Appl. Gen. Statistics* xviii. 524 There will be an enormous drop in the seasonally adjusted data between December and January. **1971** *Daily Tel.* 17 Dec. 15 Seasonally adjusted..the figures show an increase in borrowing by industry of about £50 million over the previous quarter. **1974** *Nature* 1 Feb. 269/2 Both parameters depend on the mesospheric circulation, whose main features at a particular latitude vary seasonally. **1982** *Guardian* 15 Apr. 14/2 The seasonally adjusted level of unemployment..remained at 11·8 per cent.

sea-song. Add: Also *fig.*
1944 Blunden *Shells by Stream* 31 The wind may sing his sea-song later In your review as he will in mine.

seasoning, *vbl. sb.* Add: **1. d.** (Earlier example.)
1807 *Salmagundi* 16 May 198 Strangers always..undergo a *seasoning* as europeans do in the West-Indies.

g. The process whereby a transported slave becomes inured to the conditions of slavery. *Obs.* exc. *Hist.*
1771 A. Benezit *Some Hist. Acct. Guinea* xiii. 130 At a moderate computation of the slaves who are purchased by our African merchants in a year, near thirty thousand die upon the voyage and in the seasoning. **1786** T. Clarkson *Ess. Slavery & Commerce Human Species* iii. iv. 139 This seasoning is said to expire, when the two first years of their servitude are completed; It is the time which an African must take to be so accustomed to the colony, as to be able to endure the common labour of a plantation, and to be put into the *gang*. **1804** R. Bisset *Defence of Slave Trade* 88 Instead of *thirty-three* in the hundred dying, as asserted by the author of the 'Concise Statement', *not three* in the hundred die of the seasoning. **1977** *Time* 7 Feb. 59/3 The passage took longer, with 'seasoning' camps at the beginning, usually on an island off the African coast.

h. The application of one of various finishes to leather after tanning.
1897 C. T. Davis *Manuf. Leather* (ed. 2) 358 A seasoning mixture is applied to the surface after tanning and before coloring. **1974** *Encycl. Brit. Macropædia* X. 763/2 In unpigmented seasoning [of leather], a simple glazing finish or seasoning may contain egg albumin, water, and glycerin.

sea-swallow. Add: **2. c.** An edible swiftlet of the genus *Collocalia*, found in south-east Asia.
1902 [in Dict., sense 3].

seat, *sb.* Add: **II. 6. d.** *U.S.* A place in the membership of the New York Stock Exchange.
1820 *Constitution N.Y. Stock & Exchange Board* in E. C. Stedman *N.Y. Stock Exchange* (1905) iv. 68 If two-thirds of the members present are for reinstating him, he shall again be entitled to his seat at the Board. **1882** J. D. McCabe *N.Y. by Sunlight & Gaslight* xxi. 337 A seat in the Board costs about $6000, and is the absolute personal property of its owner. **1948** *Time* 14 June 90/2 All who buy and sell on the floor must own Stock Exchange seats, which are currently worth about $65,000 apiece (1929 price: $625,000). **1972** *Times* 16 May (Wall Street Suppl.) p. viii/5 Among the brokers the numbers of 'seats' on the exchange remained unchanged at 1,366 (as it has done since December 1953).

7. *spec.* in a coach, train, aeroplane, etc.
1810 E. Weeton *Let.* 28 Dec. (1969) I. 318 Perhaps when Mr. and Mrs. P. go to Preston..I may get a seat with them..if they go in the chaise. **1976** *Daily Mirror* 16 July 9/6 The bosses are taking up too many first-class seats on main line commuter services. **1977** C. Forbes *Avalanche Express* vi. 68 Harry Wargrave occupied his normal seat..the gangway seat [in the aircraft].

b. Also *spec.* of a lavatory.
1907 *Yesterday's Shopping* (1969) 518/2 Seat Covers..for use on 'w.c.'s'. **1938** E. Bowen *Death of Heart* i. iv. 81 Portia..re-wound the gramophone on the shut seat, and Stravinsky filled the bathroom. **1979** M. Hastings *Bomber Command* vi. 155 The Elsan toilet which most crews had used with acute caution since a 50 Squadron gunner left most of the skin of his backside attached to the frozen seat one icy night over Germany.

10. Delete ? *Obs.* and add later *dial.* examples.
1962 *Sunday Express* 1 Apr. 21/5 These lairs [of hares] are usually called 'forms',..though in..Kent, the cosier word 'seats' is preferred. **1972** Evans & Thomson *Leaping Hare* iv. 52 They'll dig a little hole so they can cover, so they're level with the top of the land... A *seat* we call it.

IV. 22. b. The position of a horseshoe with respect to the hoof. Cf. *Seated *ppl. a.* 7.
1851 H. Stephens *Bk. Farm* (ed. 2) I. 338/1 The transparent shoe, showing the usual seat given to the shoe upon the forefoot.

23. b. The surface on which the head of a poppet-valve rests when the valve is closed.
1841 Valve-seat [see Valve *sb.*[1] 7 a]. **1916**, etc. [see *Grind *v.*[1] 5 b]. **1936** E. A. Phillipson *Steam Locomotive Design* x. 353 The springs provided to assist the valves to return to their seats are located in the steam spaces. **1963** R. F. Webb *Motorists' Dict.* 220 It is essential that the valve is accurately ground to match the seat so as to form an effective gas seal. **1970** K. Ball *Fiat 600, 600D Autobk.* i. 13 Refacing of the valve head seating area must be done on a special universal grinder enabling the angle of the seat to be accurately set.

V. 27. d. *the seat of one's pants*: see *Pants *sb. pl.* 1 e.

e. *to be on seat*: to be present in one's office. *W. Afr. pidgin.*
1971 J. Spencer *Eng. Lang. W. Afr.* 29 A very useful one which might be recommended to English-speaking communities elsewhere is the expression (to be) *on seat*, as in a sentence such as 'The Deputy Secretary is back on seat today'; meaning he is in the office, or generally available, as opposed to being absent. **1976** *Listener* 17 June 773/1 If you ask his servant where the district commissioner has gone, the servant tells you he is at the office with the impressive phrase: 'Master's on seat.'

VI. 28. *seat-back*, *-cover*, *reservation*.
1872 'Mark Twain' *Roughing It* iii. 30 The conductor bent all the seat-backs down. **1976** M. Birmingham *Heat of Sun* ii. 21, I..turned my head, half expecting to be able to see over the seat-back. **1881** C. C. Harrison *Woman's Handiwork* iii. 193 A seat-cover of slate-green plush. **1970** *Washington Post* 30 Sept. b13/4 (Advt.), Morris Katz & Sons Car Radio & Seatcover Center, Inc. **1973** W. McCarthy *Detail* iii. 181 He checked in for his seat reservation.

29. seat belt, a safety belt for a person in a moving conveyance, *spec.* one worn in an aircraft, esp. at take-off or landing, or one worn in a motor vehicle as a protection in an accident or in an emergency stop; also *fig.*; hence **seat-belted** *a.*, wearing a seat belt; **seat-board,** (*d*) a board forming a seat in a vehicle; **seat-mate**: for *U.S.* read *N. Amer.* and add earlier and later examples; **seat-mile,** a statistical unit denoting one mile travelled by one passenger, *spec.* in travel by air; **seat-pack,** a parachute carried in a pack worn over the posterior.
1932 *Luftfahrt* (Illustrierte technische Wörterbücher XVII) 128/3 Seat belt. **1933** *Aeroplane* 27 Dec. 1101/2 'Please fix seat-belts.' (Note! not *safety* belts.) **1959** *B.S.I. News* Apr. 18/2 Arising from the interest now being displayed in seat belts for motorists, a new technical committee of the B.S.I. recently held its first meeting, at which it was decided that a British Standard for these articles would serve a useful purpose. **1966** T. Pynchon *Crying of Lot 49* vi. 150 You're chicken, she told herself, snapping her seat belt... She drove savagely along the freeway. **1970** C. Hampton *Philanthropist* iii. 32 He..came and sat next to me on the sofa, and I thought this is it, fasten your seat belts. **1977** B. Freemantle *Charlie Muffin* xix. 192 They had cleared the airport and the seat-belt sign had been turned off. **1967** J. Redgate *Killing Season* (1968) i. xv. 65 The redhead sat, seat-belted, talking. **1873** J. H. Beadle *Undevel. West* iii. 70 The wagon made fearful lurches, and our seatboard rattled over it in every direction. **1859** *Ladies' Repository* Nov. 645/1 She will tickle the neck of her seat-mate with a bit of grass. **1968** *Globe & Mail* (Toronto) 17 Feb. 7/1 A television interview by my hon. friend's seatmate. **1976** L. Sanders *Hamlet Warning* (1977) xv. 124 On the night flight to Lisbon..his seatmate was a German auto parts specialist. **1953** *Wall St. Jrnl.* 24 Mar. 22/2 Mr. Cole predicted the combined airlines would have an annual capacity of 1,470,000,000 seat miles..by June of next year. **1961** P. W. Brooks *Mod. Airliner* i. 26 The most important non-stop stage lengths have been achieved—notably London–New York, 3,500 seat-miles. **1977** *Guernsey Weekly Press* 21 July 1/4 The 'plane is claimed to be economical with a fuel consumption per seat-mile lower than that of any other modern commercial transport aircraft. **1930** O. H. Kneen *Everyman's Bk. of Flying* xii. 217 For use in airplanes, the *seat pack* is generally used. **1946** W. F. Burbidge *From Balloon to Bomber* iii. 45 The 'seat pack' forms a cushion during the plane journey.

seat, *v.* Add: **4. a.** Delete † and add later examples of *intr.* for *refl.* use with other consts.
1916 Hirshfeld & Ulbricht *Steam Power* xi. 207 The valves are all double-seated.., that is, they seat at both ends. **1963** C. R. Cowell et al. *Inlays, Crowns, & Bridges* iv. 41 This must be done quickly otherwise the cement will begin to set and the restoration will not seat accurately. **1972** L. M. Harris *Introd. Deepwater Floating Drilling Operations* ix. 93 As the well-head seats on the previously set permanent guide structure and foundation-pile housing, it is latched and rigidly attached to the housing.

Seatainer (sī·tēⁱnəɹ). *Austral.* Also **seatainer.** [f. Sea *sb.* + Con)tainer.] A container for the transportation of freight by sea.
1964 *Economist* 25 July 361/1 (Advt.), Carriage of all cargo in 'Seatainers'. **1965** S. J. Baker *Ampol Bk. Australiana* (ed. 2) 112 World's first ship designed solely to carry 'seatainers'—large aluminium containers of uniform size inside which up to 17 tons of cargo may be packed. **1969** *Jane's Freight Containers 1968–69* 332/1 A new 'Seatainer' terminal will soon be brought into operation. **1974** *Australian* 12 Aug. 12/8 (*heading*) Seatainer important facility to Melbourne.

seated, *ppl. a.* Add: **7.** Of a horseshoe: hollowed out so that the bearing surface rests on the wall of the hoof.
1831 W. Youatt *Horse* xvii. 311 (*heading*) The concave-seated shoe. **1908** *Animal Managem.* (War Office) 227 Seating is the hollowing out of the bearing surface, opposed to the sole, so that a seated shoe bears on the wall alone.

seater. Restrict *rare* to sense 1 in Dict. and add: **2.** Substitute for entry: As second element, designating a vehicle or article of furniture having a specified seating capacity, as *two-* (*three-*, etc.) *seater*.
1906 [in Dict.]. **1916** H. Barber *Aeroplane Speaks* Pl. xvi, The familiar biplanes with 80 h.p. Gnomes, and 5-seater with 100 h.p. Anzani. **1923** Galsworthy *Captures* 217 Hubert Marsland..had occasion to stay the progress of his two-seater about ten miles from London for a minor repair. **1941** J. D. Carr *Case of Constant Suicides* iii. 44 A..five-seater car was drawn up before the tourist-office. **1966** 'A. Hall' *9th Directive* xx. 184 It was a massive black Lincoln sedan; a seven-seater executive-style transport. **1978** R. V. Jones *Most Secret War* iv. 39 It seemed that, if we developed the detector to the operational stage, it would have to be mounted in single seater fighters.

seater, var. *Saeter, setter.

sea-thistle. **1.** Delete ? *Obs.* and add later example.
1979 *Bull. Yorks. Dial. Soc.* Summer 7 The only kind of flowers were the dark blue sea thistle with very strong prickly leaves but no smell.

seating, *vbl. sb.* Add: **3.** (Earlier U.S. example.)
1790 *Pennsylvania Packet* 11 Dec. 1/2 A very choice Parcel of Hair Seatings, of various widths and patterns.

5. b. The raised outer part of a horseshoe which rests on the wall of the hoof; also, the hollowing out of a horseshoe so that the outer part rests on the wall of the hoof.
1831 W. Youatt *Horse* xvii. 319 A strip of felt or leather is sometimes placed between the seating of the shoe and the crust. **1908** *Animal Managem.* (War Office) 228 The object of seating is to take pressure off the sole.

5*. Of garments: the process of going out of shape at the seat.
1960 *Sunday Express* 11 Sept. 15/4 *Never* wear a slick, straight skirt without a tight-fitting slip beneath... That way there'll be no 'seating'. **1974** Lippman & Erskine *Dressmaking made Simple* iii. 53 Linings are essential in straight skirts to prevent seating.

6. a. *seating-face*.
1925 *Morris Owner's Man.* 77 These two photographs show a valve before..and after grinding-in. Note the different appearance of their seating faces.

b. *seating plan*.
1929 'E. Queen' *Roman Hat Mystery* xxii. 303 We had already borrowed a seating-plan. **1949** M. Mead *Male & Female* 406 Men holding the highest titles from each village were combined in a formal seating-plan. **1974** O. Manning *Rain Forest* i. vi. 75 Millman, seating-plan in hand, put Murodi at Lady Urquhart's right and Ogden on her left.

Seato (sī·to, sīᵊ·to). Also **S.E.A.T.O., SEATO.** [Acronym f. the initial letters of *South East Asia Treaty Organization* (after N.A.T.O.), set up in 1954.] A military alliance, lasting from 1954 to 1977, of Australia, New Zealand, France, Pakistan, the Philippines, Thailand, the United Kingdom, and the United States.
Pakistan withdrew from the alliance in 1972.
1954 *N.Y. Times* 23 July 16/1 Preparations are under way to call an international conference in August or September..to devise a SEATO for the East to match NATO in the West. **1955** *Times* 31 May 6/7 The programme proposes the strengthening of the links of friendship with France, and membership of the S.E.A.T.O. pact. **1957** *Observer* 6 Oct. 10/6 Pakistan has openly shown that she would expect Seato to assist her in the event of an attack from India. **1966** 'A. Hall' *9th Directive* i. 14 We have the SEATO headquarters here... Bangkok is a key city in the South-east Asian complex. **1970** *Ann. Reg. 1969* 149 The fourteenth Seato Council met in Bangkok on 20–21 May. **1977** *Times* 30 June 6/4 The South-East Asian Treaty (Seato) Organization..will fade into history tomorrow [*sc.* 30 June]..when the flags of its six remaining members, the United States, Britain, the Philippines, Thailand, Australia and New Zealand, are lowered from the organization's deserted headquarters in Bangkok for the last time.

Seatonian (sītōᵘ·niăn), *a.* [f. the name of the Eng. divine Thomas *Seaton* (1684–1741), the founder of the prize + -ian.] *Seatonian prize*: a prize awarded (since 1750) for religious poetry at the University of Cambridge. Also *ellipt.* (in quot., a poem to be submitted for this).
[**1773** (*title*) Musæ Seatonianæ. A complete collection of the Cambridge prize poems, from their first institution..to the present time.] **1795** A. W. Trollope (*title*) The destruction of Babylon, a Seatonian prize poem. **1864** A. J. Munby *Diary* 27 Sept. in D. Hudson *Munby* (1972) 203 Finished my first Seatonian; hurriedly, and with brain throbbing. **1908** H. Pentin *Judith* iv. 68 In the year 1865 'Judith' was the subject set for the Seatonian Prize Poem at Cambridge. **1961** *Listener* 31 Aug. 323/3 Very occasionally in the blank-verse wastes of the Seatonian prize-poems one comes across a single line..that adumbrates the future author of *A Song to David*. **1972** *Cambr. Univ. Reporter* 6 Dec. 391 The Examiners for the Seatonian Prize for the best English Poem on a sacred subject give notice that the subject for the year 1973 is '*Apocalypse*'.

sea-turtle[2]. Substitute for def.: A marine turtle belonging to the families Cheloniidæ or Dermochelyidæ. (Earlier and later examples.)
1612 W. Strachey *Trav. Virginia* (1849) 195/1 A sea turtle, tuwcuppewk. **1888** H. C. Bumpus in J. S. Kingsley *Riverside Nat. Hist.* III. 444 Sea-turtles are of considerable value as food. **1958** J. Carew *Black Midas* iii. 35 Mendoza had trapped a sea turtle and four men had to lift it into his donkey cart.

seau (sōᵘ). *Ceramics.* Also *erron.* † **sceau.** Pl. **seaux.** [Fr., lit. 'bucket'.] A vessel in the shape of a pail or bucket used for cooling wine, etc. (freq. forming part of dinner services of the eighteenth century).
1784 H. Walpole *Descr. Strawberry-Hill* 82 A sceau for liquors, of Seve. **1851** *Illustr. Catal. Gt. Exhib.* iii. 710/1 Porcelain Inkstands, Seaux, Card Trays. **1869** C. Schreiber *Jrnl.* (1911) I. 37 Some Bleu du Roy vases, small,..a pair of sceaux of the same colour. *Ibid.* 53 Picked up..a marked St. Claud sceau at Bencoux's. **1875** E. Meteyard *Wedgwood Handbk.* Gloss. 409 Seaux formed a part of all costly dinners and dessert services, particularly if intended for foreign countries... A choice pair of seaux in sea-green jasper is in the Marjoribanks Collection. **1974** Savage & Newman *Illustr. Dict. Ceramics* 259 *Seau à bouteille*,..a bucket-shaped receptacle for holding ice to chill a single bottle of wine.

sea-view. **2.** (Earlier examples.)

1790 J. Woodforde *Diary* 10 May (1927) III. 188 My Brother and Wife..very highly pleased with Yarmouth and the Sea View. **1844** A. W. Kinglake *Eothen* iv. 63 The reality of that very sea-view, which had bounded the sight of the Greeks.

sea-washed, *a.* Add: **b.** *sea-washed turf*, a dense turf found in coastal regions of northern England.

1931 R. Beale *Bk. Lawn* vi. 56 Turf can roughly be divided into three categories—Cumberland, Sea-washed or Marsh; Down, Heath or Moorland; and Meadow Turf. **1954** A. G. L. Hellyer *Encycl. Garden Work* 137/1 Sea-washed turf or Cumberland turf..owes its fineness to the fact that it is washed by salt water at high tides. **1962** I. Greenfield *Turf Culture* ii. 73 One [variety of creeping bent]..is a common constituent of the sea-washed turf used for bowling greens. **1977** *Lancs. Life* Nov. 82/4 Lancashire's only herd of wintering wild Bewick swans from Siberia..crop the sea-washed turf.

sea-way, seaway. Add: The form *seaway* (no hyphen) is now the more usual. **1. b.** An artificial or natural channel connecting two tracts of sea.

1866 [see sense 1 a in Dict.]. **1977** A. Hallam *Planet Earth* 222/1 Towards the close of the period the old seaway of Tethys was progressively eliminated as the African plate moved northwards to impinge upon the Asian plate.

c. An inland waterway with passage to the sea, esp. one capable of accommodating large ocean-going vessels. *N. Amer.* (chiefly in phr. *St. Lawrence Seaway*).

1921 A. M. Evans in *Chicago Daily Tribune* 4 Aug. 21/7 Coastwise trade between Chicago and Atlantic ports.. stands second only to the foreign commerce possibilities offered by the St. Lawrence seaway project. **1933** *Sun* (Baltimore) 23 June 3/1 (*heading*) Lakes-to-the-Gulf seaway dedicated... The joining of the Great Lakes with the Gulf of Mexico..by a $102,000,000 inland waterway was completed officially today. **1941** F. D. Roosevelt in *Great Lakes-St. Lawrence Basin: Hearings* (1942) I. 2, I recommend authorization of construction of the St. Lawrence seaway and power project, pursuant to the [illegible]

1902 H. Johnston *Uganda Protectorate* II. xix. 868 The Sabei men also hang to their locks of hair and to their ear-lobes rather striking ornaments. *Ibid.*, The dwellings of the Sabei are like those of the Masaba Bantu tribes. **1935** Thomas & Scott *Uganda* v. 86 Half-Hamitic languages are spoken by the Teso..and by the Sebei. **1953** Trowell & Wachsmann *Tribal Crafts of Uganda* i. 50 The Sebei are members of another branch of the Nilo-Hamites, being very closely related to the Suk-Nandi group in Kenya. *Ibid.* 51 Both Karamaja and Sebei fit their gourds with leather thongs for carrying. **1963** M. de K. Hemphill in Oliver & Mathew *Hist. E. Afr.* I. xi. 419 The Sapei were moving to the west, and the Kony to the north. **1967** W. Goldschmidt *Sebei Law* p. vii, The first and outstanding debt is owed to the Sebei themselves, the officials, the informants, and the many citizens who helped to shape my image of Sebei customs. **1968** P. Ladefoged et al. *Lang. in Uganda* ii. 81 Kupsabiny, the language spoken by the Sebei, has some likeness to its neighbour, Suk (or Pokot), but is otherwise distinct from any other members of the group. **1973** *Sci. Amer.* July 77/1 Brideprice is the largest single outlay of goods that an ordinary Sebei makes in his lifetime. **1976** *Drum* (E. Afr. ed.) Nov. 18/2 All the Kenya tribes of the Luo, Masai, Kisii, Baluhya, Turkana, Sebei, Teso, Suk, Pokot, Nandi, Marakwet, Samburu, and Elgeyo are my grandchildren. **1977** Savei [see *Sapiny].

Sebilian (sĭbi·liăn), *a.* (and *sb.*) *Archæol.* [ad. F. *Sébilien* (E. Vignard 1923, in *Bull. de l'Institut Franç. d'Archéol. Orientale* XXII. 3), f. the name of *Sebil*, a village in Upper Egypt: see -ian.] Of or pertaining to an Upper Palæolithic and Mesolithic culture of Upper Egypt; also *ellipt.* as *sb.*

1932 *Antiquity* VI. 193 In the Nile Valley some rather peculiar industries, probably allied to the Aurignacian, are called Sébilian. **1936** L. S. B. Leakey *Stone Age Afr.* x. 193 In North Africa at Sebil, the Early Sebilian gradually evolves by way of a Middle Sebilian into a true Microlithic stage known as the Upper Sebilian. **1952** V. G. Childe *New Light Most Anc. East* iii. 32 The stage intermediate between the food-gathering culture of Sebilian hunters and the settled agriculture of the oldest sedentary inhabitants of Egypt. **1969** Coles & Higgs *Archaeol. Early Man* ii. xiii. 196 The later stages of the Sebilian material contain microlithic forms, and one associated with aurochs, buffalo and shell middens. Suggested dates for the Sebilian industries range from c. 16,000 to 11,000 B.P. **1976** *Sci. Amer.* Aug. 34/1 A fourth industry, the Sebilian..is found at Kom Ombo.

sebk(h)a, varr. *Sabkha.

seborrhœa. Add: Also (chiefly *U.S.*) **seborrhea.** (Further examples.)

1940 Becker & Obermayer *Mod. Dermatol. & Syphilol.* v. 56/2 It is advisable to assume that seborrhea and the diseases to which it predisposes, namely acne vulgaris, seborrheic dermatitis and rosacea, do not appear before puberty. **1973** *Nature* 8 June 350/1 In Parkinsonism the associated seborrhoea is induced by excessive secretion of a pituitary sebotrophic hormone. **1978** J. Kilmartin tr. *R. Aron-Brunetière's Beauty & Medicine* ii. 22 Even people with fairly mild seborrhea or acne are self-conscious about it.

seborrhœic *a.* (also -rrheic) (further examples).

1955 *Sci. News Let.* 12 Feb. 104/3 The scientists have reported using the glands in cases of..seborrheic dermatitis. **1971** [see *liver spot* s.v. *Liver *sb.*[1] 7]. **1977** *Lancet* 27 Aug. 440/2 One troublesome feature.., the seborrhœic scalp rash, responds excellently to application of a tar shampoo.

sebotrophic (sebotrō^u·fik), *a.* *Physiol.* Also **-tropic** (-trō^u·pik, -trǫ·pik). [f. Seb(um + -o + *-trophic, *-tropic.] Tending to stimulate sebaceous activity.

1957 Lorincz & Lancaster in *Science* 19 July 124/1 Methods of preparation and assay of a crude extract of anterior pituitary glands of hogs showing tropic effects on sebaceous, preputial, and Harderian glands in the rat are described. These effects can be called sebotropic because the ectodermal glands affected produce secretions rich in lipids. **1968** Ebling & Rook in A. Rook et al. *Textbk. Dermatol.* II. xlv. 1335/2 Biological tests for sebotrophic activity..were carried out..on the preputial gland. **1973** [see *Seborrhœa].

sec. Add: Later examples as abbrev. of Secretary *sb.*[1] and (in colloq. phrases) of Second *sb.*[1]

1869 *Bradshaw's Railway Man.* XXI. 48 *Officers.*—Gen. Man., James Smithells; Sec., Archibald Gibson. **1953** Wodehouse *Performing Flea* 130 One of those tall, statuesque, frozen-faced secs who took his dictation in an aloof, revolted sort of way. **1956** A. Wilson *Anglo-Saxon Attitudes* ii. ii. 319 'I'm afraid I can't stop now. I'm late for an appointment.' 'It won't take a sec,' said Vin. *a* **1960** E. M. Forster *Maurice* (1971) vi. 37 Wait a sec, and I'll come too. **1962** A. Lurie *Love & Friendship* x. 199, I wonder if you could hold the baby for me, Missus Turner, please, just for a sec. **1979** M. Boyce *I was There!* 14/2 The Rugby Club's General Committee Banned 'Sine Die' their ticket Sec, my Uncle Will.

sec (sek), *a.* [Fr.] Of champagne and other wines: = Dry *a.* 8. Also *fig.* Cf. *extra sec* s.v. *Extra *adv.* a.

1863 T. G. Shaw *Wine, Vine & Cellar* xviii. 334 It is evident that the word 'sack' cannot be understood to have denoted sec (dry). **1899** *Judge* (N.Y.) XVI. 419/2 Berton 'sec' Champagne. **1891**, etc. [see *Extra *adv.* a]. **1931** *Morning Post* 10 Aug. 4/3 The Bayreuth Festival... Wagner sec. **1960** Wodehouse *Jeeves in Offing* xvi. 165, I was in my room, having shed the moistened outer crust and substituted something a bit more *sec* in pale flannel.

sécateur. Delete ‖ and add: Now usu. in pl. form *secateurs* and with anglicized pronunc. (sekătə̄ɪz) or (se·kătə̄ɪz). Further examples.

1924 H. H. Thomas *Compl. Amateur Gardener* vii. 67 The pruning outfit should consist of..a hone or sharpening stone,..and sécateurs. **1937** *Carter's Blue Bk. Gardening 1937* 365 (Advt.), The amazing manner in which this Secateur quickly and cleanly cuts off the thickest of branches, very soon becomes a fascination to the user. **1967** E. Mavor *Redoubt* x. 172 She dressed her enormous bulk in..a baize gardening apron with capacious pockets for her bass and secateurs. **1968** R. H. W. Brown *Gardening Complete* ii. 27 The tools which the gardener must have for a start are two spades, a fork, a rake, three hoes, a trowel, a dibber, a pair of secateurs, [etc.]. **1977** P. Thrower *Every Day Gardening* i. 16/1 (*caption*) If the garden includes numerous trees, shrubs and fruit trees a pair of long-handled secateurs is especially useful.

Secchi (se·ki). Also **secchi.** [The name of Angelo *Secchi* (1818–78), Italian astronomer.] *Secchi('s) disc*: a type of opaque white disc which is used in determining the transparency of water; the disc, maintained in a horizontal attitude, is allowed to sink and the depth at which it ceases to be visible from the surface is recorded.

Secchi and Cialdi first described the use of discs of this kind in 1865 (*Compt. Rend.* LXI. 101).

1913 *Science* 14 Nov. 703/1 The inland lakes of Wisconsin are not very transparent; the transparency, as shown by Secchi's disk, varying from less than 1 m. to about 7 m. **1931** *Trans. Wisconsin Acad.* XXVI. 337 The transparency as determined by the Secchi disc shows some correlation with the amount of organisms present in the water. *Ibid.* 419 Transparency..is measured by the visibility of Secchi's disc. **1963** G. L. Pickard *Descriptive Physical Oceanogr.* iii. 23 The colour of sea-water can be judged most conveniently against the white Secchi disc.. as it is lowered to determine the transparency of the water. **1975** *New Yorker* 12 May 80/3 They threw a secchi disc into the ocean to measure turbidity.

secco, *sb.* Add: **2.** Ellipt. for 'secco recitative'.

1960 *Times* 2 July 12/2 There are three tenor recitatives accompanied by piano in a modern equivalent of the old *secco*. **1969** *Daily Tel.* 18 Jan. 17/5 The beloved melody contains no fewer than 18 appoggiaturas actually written out by Mozart (he never does in seccos).

secede, *v.* Add: **3.** *trans.* To withdraw (a component territory) from a federal union or the like; to detach or cede (a piece of land). *rare.*

1946 W. Faulkner *Portable Faulkner* 739 A plot to secede the whole Mississippi Valley from the United States and join it to Spain. **1963** A. Smith *Throw out Two Hands* xv. 156 Many people feel it was wrong..for any park to secede part of itself for any reason.

secentismo, secentist, varr. *Seicentismo, *Seicentist.

secesh, *sb.* **b.** Delete ¶ *nonce-use* and add earlier example.

1863 Trollope in *Good Words* Dec. 858/2, I won't talk secesh to you out here in the cold.

secession. Add: **3. d.** Also with capital initial. [tr. G. *Sezession.*] A radical movement in art that began in Vienna and was contemporaneous with, and related to, art nouveau; the style of this movement. Freq. with *the*. Cf. *Sezessionstil, Secessionist b (quot. 1901).

[**1890** *Art Jrnl.* July 221/1 The important secession which..followed upon the recent retirement of M. Meissonier from all connection with the great annual exhibition of Paris. **1894** *Mag. of Art* XVII. 416/1 The secession of Munich is only one of the effects of the painter's shyness of regulation.] **1896** *Amer. Architect & Building News* 8 Feb. 63/1 'Secession' is a *nom de guerre* of the 'Verein Bildener Kuenstler Muenchens (A.V.)'. *a* **1935** F. Ponsonby *Recollections of Three Reigns* (1951) ix. 123 Inside it was composed of every style of mural decoration, but predominantly what was called 'art nouveau' or 'secession'. **1972** T. Walters *Art Nouveau* 6 The Secessionists..were determined to carefully build up a whole revolutionary way of life in which every object..whether a lavatory seat or an underground station, was designed to meet the ideals of the Secession.

5. c. In sense *3 d, as *secession exhibition, school, style* [tr. G. *Sezessionstil*].

1911 R. Fry *Let.* 13 Apr. (1972) I. 345, I thought that the Grafton might be used for a general secession exhibition of all non-academy art of any importance. **1949** *New Yorker* 5 Feb. 78/2 De Chirico..declares that the École de Paris..was really founded on the Munich Secession School. **1973** *Times* 18 Dec. (Hungary Today Suppl.) p. i/3 The *Vigszinhaz*..is in the turn of the century *Secession* style, the Vienna version of *art nouveau*.

secessionist. Add: **c.** (Further examples.) Also *pred.* (quasi-*adj.*).

1894 *Mag. of Art* XVII. 379 Salon of the Champ de Mars..the secessionist Salon. **1954** B. & R. North tr. *M. Duverger's Pol. Parties* ii. i. 294 If a party is clearly in a minority in the country as a whole but in a majority in

certain districts its attitude becomes autonomist or even secessionist. **1962** *Daily Tel.* 9 Feb. 19/1 Mr. Gizenga, arrested and accused of 'secessionist activities'. **1978** *Detroit Free Press* 2 Apr. 16c/3, I learned of the secessionist movement in the Upper Peninsula.

Sechuana (setʃuā·nă). Also **Sechoana, Sechwana, Secuana, Setswana,** † **Sichuana.** [a. Tswana *se-*, prefix meaning 'language' + *-chuana* *Tswana.] = *Tswana *sb.* b. Also *attrib.* or as *adj.*

Setswana is now the preferred term for the language.

1811 [see *Kori[1]]. **1815** A. Plumptre tr. *H. Lichtenstein's Trav. S. Afr.* II. li. 407 Under the name of *Beetjuana, Sihtjuana*, or *Muhtjuana* are to be included all the tribes that inhabit the country which extends from the river Kurahman. **1824** W. J. Burchell *Trav. S. Afr.* II. xi. 295 The Bachapins call this language the Sichuána. **1839** W. C. Harris *Wild Sports S. Afr.* ii. 11 This.. individual..possessed a fair smattering of the English and Sichuana languages. **1850** [see *Bechuana]. **1905** G. W. Stow *Native Races S. Afr.* xxi. 407 The Bachoana proper, including the Batlapin, Barolong, etc., speaking the Sechoana and Serolong. **1916** S. T. Plaatje *Sechuana Proverbs* 15/1 The present confusion..is apt to hamper missionary enterprise among Sechuana-speaking Natives. **1919** H. H. Johnston *Bantu & Semi-Bantu Langs.* iii. 317 *Secuana* dialects are spoken in Bechuanaland, north of the 28th degree of South latitude. **1935** A. L. James *Broadcast Word* i. 10 There is a story told of the Bechuana that..the early missionaries taught them the Lord's Prayer in the native language, Sechuana. **1961** L. van der Post *Heart of Hunter* i. vii. 110 A language they had evolved in the course of the journey out of a little Swahili, some Sechuana and a lot of onomatopœia. **1964** D. Varaday *Gara-Yaka* xii. 108 When at a loss for a word he helped himself in Afrikaans, Chivenda, Sechwana, Zulu, and Shangaan. **1977** 'J. McVean' *Bloodspoor* iii. 22 *Morena*. 'Chief.' The Setswana term for any white man of authority.

secko (se·ko). *Austral. slang.* [Shortened form of Sex *sb.* + *-o[2].] A sexual pervert; a sex offender.

1949 R. Park *Poor Man's Orange* v. 38 'Just look at that dirty ole secko, will you?' he said disgustedly. **1969** W. Dick *Naked Prodigal* i. 13 You look like you'd be the sorta bloke who'd take little kids down a lane and give 'em two bob, yuh bloody secko. **1974** *Bulletin* (Sydney) 6 Apr. 45/2 'I noticed Australians use a lot of diminutives, like Chrissie, pressie and journo.' 'In jails sex offenders are called seckos,' I told him.

sec-mod (sek mǫd), colloq. abbrev. of *secondary modern* s.v. *Secondary *a.* 5 f.

1968 *Listener* 28 Mar. 421/2 The wretched life-style on offer to most sec-mod school-leavers. **1973** *Times* 31 May 10/7 Black and white kids, in their sec-mod school uniform.

seco- (se·ko). *Chem.* [f. L. *sec-āre* to cut + -o.] A formative element used in naming derivatives, esp. of steroids, in which fission of a ring has occurred (see quot. 1951). Hence also as quasi-*adj.*

1951 R. S. Cahn et al. in *Jrnl. Chem. Soc.* 3535 Ring fission, with addition of a hydrogen atom at each terminal group thus created, shall be indicated by the prefix *seco*, the original steroid numbering being retained. **1959** L. F. & M. Fieser *Steroids* iv. 156 This was identified as the 2,3-secodicarboxylic acid..by saponification, oxidation to the 7-ketone, and Wolff-Kishner reduction to the known 2,3-secocholestane-2,3-dicarboxylic acid. **1961** I. E. Bush *Chromatogr. Steroids* ii. 102 The opening of rings to form seco-dicarboxylic acids. *Ibid.* vi. 342 The C_{19} triols..are oxidized to the D-seco-16,17-dialdehydes. **1977** *Lancet* 16 Apr. 841/1 The physiological regulation of secretion of this *seco*-steroid by the kidney.

secobarbital (sekobā·ɹbităl). *Pharm.* Chiefly *U.S.* [f. Seco(ndary *a.* + *Barbital.] = *Quinalbarbitone.

1952 *Analytical Chem.* XXIV. 1605/1 Considerable differences in the [optical density] ratios at various wave lengths are found with these barbiturates. For example, at 270 mμ butylallylonal has the highest ratio, +0·81, followed by secobarbital +0·69. **1962** *New Scientist* 22 Feb. 426/1 Some results obtained with secobarbital.. and other drugs are also mentioned. **1974** M. C. Gerald *Pharmacol.* xi. 205 Short-acting barbiturates such as.. secobarbital ('red devils'). **1976** *Billings* (Montana) *Gaz.* 11 July 9-A/1 In order on DAWN's list of drugs most frequently recorded in crisis situations..were heroin, marijuana, aspirin, LSD, secobarbital (..known as 'red devils').

secodont (se·kodǫnt), *a.* *Zool.* [f. L. *sec-āre* to cut + Gr. ὀδοντ-, ὀδούς tooth.] Of a tooth: adapted or suited for cutting. Of an animal: having such teeth.

1891 Flower & Lydekker *Introd. Mammals* 32 Trituberculism differentiating into a secodont and a bunodont series, according as to whether the dentition becomes of a cutting or a crushing type. **1968** R. Zangerl tr. *Peyer's Compar. Odontol.* 244 Both upper and lower carnassials [in modern carnivores] are secodont; that is, they have sharp cutting edges that run parallel to the edge of the jaw.

secohm (se·kō^um). *Electr. Obs.* exc. *Hist.* [f. Sec(ond *sb.*[1] + Ohm[2].] A name proposed for a unit of inductance.

1887 Ayrton & Perry in *Nature* 9 June 131/2 Hence we are driven to suggesting a temporary name for the unit, and as the first three letters in 'second' are common to the name in English, French, German, Italian, &c., and ohm is also common, we venture to suggest 'secohm' as a provisional name. **1948** *Atlantic Monthly* May 613/2 The motion to adopt 'henry' as the name of this unit.. was seconded by..Professor Ayrton, who had himself, a few years ago, proposed the word 'sec-ohm' as being a proper name for the unit of induction. **1963** Jerrard & McNeill *Dict. Sci. Units* 127 The secohm was equal to the product of one legal ohm and one second and its magnitude was about the same as a henry.

Seconal (se·kŏnæl, -ăl). *Pharm.* Also **seconal.** [f. Secon(dary *a.* + Al(lyl.] A proprietary term for *Secobarbital. Also, a tablet of this.

1935 *Official Gaz.* (U.S. Patent Office) 23 July 727/2 Eli Lilly and Company, Indianapolis... *Seconal.* For products of secondary amyl allyl barbituric acid and the sodium salts thereof..useful as hypnotics. Claims use since Apr. 11, 1935. **1937** *Jrnl. Amer. Pharmaceut. Assoc.* XXVI. 1248 It was concluded that 'Seconal' had a shorter duration of action and that its minimal anesthetic dose and minimal lethal dose were smaller than those of 'Sodium Amytal'. **1938** *Trade Marks Jrnl.* 23 Feb. 218/1 Seconal. **1950** E. Hemingway *Across River & into Trees* ii. 11 He was also anxious to lie down and take a seconal. **1958** 'A. Bridge' *Portuguese Escape* xiii. 217 She swallowed some Seconal with a gulp of water. **1959** N. Mailer *Advts. for Myself* (1961) 214 Drugging myself into sleep with an overload of seconal. **1965** G. Marx *Let.* 12 Oct. (1967) 68 At 8 we take two Seconals, three aspirin and a shot of LSD. **1973** M. Amis *Rachel Papers* 176 'What did you give him?' 'Half a Mandie, a Seconal—I can't remember—and two Mogadon, I think.'

second, *sb.*[1] Add: **2.** Now one of the base units of the International System of Units, and redefined in terms of the frequency of a spectral transition of an isotope of cæsium (see quot. 1968). (Further examples.)

1955 *Sci. Amer.* Mar. 52/2 Accordingly the International Committee is to define the second as: 1/31,556,925·975 of the tropical year 1900. **1968** *Nature* 16 Nov. 651/1 The basic unit of time in the International System of Units, formerly identical with the astronomical second of ephemeris time, is now based on a natural periodicity of the caesium atom and is defined in the following terms: 'the second is the duration of 9 192 631 770 periods of the radiation corresponding to the transition between the two hyperfine levels of the ground state of the caesium-133 atom'. **1975** *Oxf. Compan. Sports & Games* 975/1 This reduced the previous world records by half a second, the biggest single advance in the history of the events.

3. second-foot, a unit of the rate of flow of water, equal to one cubic foot per second.

1898 *U.S. Dept. Agric. Yearbk. 1897* 640 Where water is abundant, the duty has been known to be as low as 50 acres..to the second-foot. **1914** Ries & Watson *Engin. Geol.* v. 250 The height of the black lines illustrates the relative quantity of water expressed in cubic feet per second, or second feet, occurring throughout the year. **1928** *Manch. Guardian Weekly* 31 Aug. 178/3 The assumption is that the main river, suitably fortified with levees, can carry rather less than 2,000,000 cubic feet of water per second (or 2,000,000 second feet as it is briefly described).

second, *a.* and *sb.*[2] Add: **A.** *adj.* **2. a.** (Later examples of phr. *second to none*.)

1861 Geo. Eliot *Let.* 17 May (1954) III. 414, I doing little else but feel eminently uncomfortable, for which..I have a faculty 'second to none'. **1961** J. Heller *Catch-22* (1962) xi. 112 He would stand second to none in his devotion to country.

b. (Later examples.)

1880, etc. Second last [see *Last *a.* 1 b]. **1910** W. M. Raine *Bucky O'Connor* (1920) xx. 226 I'll agree to the second dearest in the world. **1959** J. Kirkup tr. *S. de Beauvoir's Memoirs of Dutiful Daughter* ii. 116 She was.. the second eldest daughter. **1977** *Word 1972* XXVIII. 104 The second-youngest of the fluent speakers. **1979** *Nature* 15 Feb. 561/2 *Secernosaurus* is the second most primitive hadrosaur known.

c. *second master* (examples); similarly *second mistress; second mate* (examples in naut. slang phrases referring to measures of liquor).

1843 J. F. Cooper *Ned Myers* II. ii. 61 Putting a second-mate's nip of brandy into my glass. **1853** 'C. Bede' *Adv. Mr. Verdant Green* i. 6 The second master..'licked a feller' for a false quantity. **1866** *School Life at Winchester Coll.* xiv. 177 The *Roll* which was published every November, giving a list of the entire establishment of the College, commencing with the Warden, Head master, (Informator,) Second master, (Hostiarius,) the ten fellows, three chaplains; the under masters [etc.]. **1923** L. Magnus *Jubilee Bk. Girls' Public Day School Trust* iv. 58 Her retirement coincided with that of her Second Mistress, Mrs Withiel. **1933** P. A. Eaddy *Hull Down* 99 He pulls a pint bottle out of the case, and drawing the cork pulls out a good Second Mate's four fingers. **1952** V. Noake *Hist. Alice Ottley School Worcester* xiii. 140 Miss Spurling's successor..was Miss Hilda M. Roden, second mistress of the Stamford High School, Stamford, Lincs. **1967** S. Waters *Indentures Indorsed* xxxv. 232 A couple of second mate's pegs was usually enough to set us all singing. **1976** C. Dexter *Last seen Wearing* xxx. 211 School masters, even experienced second masters, aren't all that highly recompensed.

d. (Earlier and later examples.)

1776 W. Howe *Let.* 7 June in *9th Rep. R. Comm. Hist. MSS.* App. iii. (rev. ed.) 35 in *Parl. Papers 1910* (Cd. 5038) XXXV. 675 The seniority of his rank..would have placed him second in command in Canada had he not been previously employed to the southward. **1939** C. S. Forester *Captain Hornblower, R.N.* xvi. 173 On his first commission his second-in-command had taken advantages of lapses on his part. **1982** *Observer* 16 Apr. 14/6 His second-in-command is a Sierra Leonean major.

4. b. Hence (without allusion to the proverb) *to be* (*come as*, etc.) *second nature* (*to* one): to be as if natural or instinctive. Occas. in ellipt. constructions without a vb.

1910 S. E. White in *Sunset* Apr. 421/1 Bob..rolled over twice with the rapid, vigorous twist second-nature to a seasoned half-back. **1944** *Sun* (Baltimore) 28 Nov. 8/2 Civilian air defense comes as second nature to them. **1954** T. S. Eliot *Confidential Clerk* iii. 96, I do feel more at ease when I'm behind a desk: It's second nature. **1967** Singha & Massey *Indian Dances* xviii. 159 They become second nature to her when she is dancing.

6. b. ellipt. for *second class* (in travelling by rail, etc.).

1912 R. Brooke *Let.* Jan. (1968) 334 The maids of the Ordinary Rich go second, with you and the normal me. **1937** W. H. Saumarez Smith *Let.* 10 July in *Young Man's Country* (1977) ii. 80 As I'm not getting Travel Allowance for this trip, I'd decided to travel 2nd. *a* **1976** A. Christie *Autobiogr.* (1977) vi. i. 289 Ladies travelling alone would never have travelled third class... Even ladies' maids always travelled second.

7. a. Further syntactical combs., as *second empire, generation, house, language, opinion, reading, secretary, sex, slip, string, subject* (see these words); **second Adam** (earlier examples); **second ballot,** a deciding ballot taken between the candidate who won a previous ballot without securing an absolute majority and the candidate with the next highest number of votes; also *attrib.* of an electoral system using this; **second banana** *slang* (orig. *U.S.*), a supporting comedian (cf. *top banana* s.v. *Top *sb.*[1] 32); **second base** (see *Base *sb.*[1] 15 c); **second blessing** orig. *U.S.*, an experience of God's grace subsequent to conversion, believed by some Christian groups to be the means of receiving the power to live a sanctified life; **second bottom,** (*a*) *U.S.*, the first terrace above the normal flood plain of a stream; (*b*) *Austral.*, a second stratum of gold-bearing material found by sinking below the bottom (*Bottom *sb.* 4 c); **second breakfast,** a light meal taken late in the morning or early in the afternoon; **second car,** an additional family car; **second chamber** (earlier example); **second channel** *Radio* = *Image *sb.* 7**; usu. *attrib.*; **second cut,** (*b*) *Austral.* and *N.Z.* (the mark of) a blow made to remove badly-cut fleece; a piece of short or inferior wool produced by this; † **second day,** in Quaker usage, Monday; **second-degree** *Med.*, used to designate burns that are sufficiently severe to cause blistering but not permanent scarring (see quot. 1972); see also Degree *sb.* 6 d; **second feature,** the supporting feature in a cinema programme; also *fig.* and *attrib.*; **second finger,** the finger next to the forefinger, the middle finger; **second front,** in the war of 1939–45, a front in Nazi-occupied Europe in addition to the Russian sector of fighting; also *fig.* and *attrib.*; **second fronting** *Philol.* [tr. G. *zweite Aufhellung* (K. Luick *Historische Grammatik der englischen Sprache* (1914) 164)], a sound-change in varieties of Old English by which the vowels *æ* (produced by an earlier fronting) and *a* became *e* and *æ* respectively; **second gear,** the gear next above the lowest or bottom gear on a motor vehicle or bicycle; **second greaser** *Naut. slang*, a second mate; **second growth,** (*a*) (earlier example); (*b*) the second category of growths (see *Growth[1] 1 d) or qualities into which wines are divided; also *attrib.*; **second home,** a second dwelling-place owned or supported in addition to the principal home; a home from home; also *fig.* and *attrib.*; hence **second homer,** the owner of a second home; **Second Isaiah** = *Deutero-Isaiah* s.v. Deutero-; **second line,** also *gen.*: any second row or series; freq. *attrib.* or as *adj.*, esp. designating persons or things that rank second in ability, value, etc.; hence **second-liner, -lining** *vbl. sb.*; **secondman,** an assistant driver on a diesel or electric train, replacing the fireman on a steam train; **second messenger** *Physiol.*, a substance whose release within a cell is promoted by a hormone or 'first messenger', and which brings about a response by the cell; **second mortgage,** a supplementary or puisne mortgage; † **second (motion) shaft** = *Layshaft; **second ranker,** a member of the second rank, a second-liner;

[illegible]

[illegible]

tant reaction... From secondary amines, nitrosamines precipitate as non-basic, yellowish oils.

(ii) Applied to organic compounds other than amines, etc. (see prec. sense) in which the characteristic functional group is located on a saturated carbon atom which is itself bonded to two other carbon atoms. [Applied orig. to alcohols by H. Kolbe, who used G. *secundär* (*Ann. der Chem. und Pharm.* (1864) CXXXII. 102).]

1864 *Chem. News* 26 Nov. 260/1 By a secondary alcohol the author [*sc.* Kolbe] means a body in which two of the typical hydrogen atoms in a typical alcohol are substituted by two atoms of some other alcohol radicals. **1876** *Phil. Mag.* II. 162 To so-called normal butylic alcohol is generally assigned the structural formula $CH_2(C_3H_7)OH$; to secondary butylic alcohol the formula $CH(CH_3)(C_2H_5)$-OH; [etc.]. **1876** [in Dict.]. **1900** Perkin & Kipping *Org. Chem.* vi. 107 Tertiary alcohols are, as a rule, more difficult to obtain than the primary or secondary compounds. **1932** I. D. Garard *Introd. Org. Chem.* iii. 34 Secondary butyl alcohol..is made from butylene..just as isopropyl alcohol is made from propylene. **1972** R. A. Jackson *Mechanism* v. 88 In general, primary compounds undergo S_N2 substitution more readily than do secondary compounds, and S_N2 reactions on tertiary compounds go with great difficulty if at all.

(iii) Applied to a saturated carbon atom which is bonded to two other carbon atoms; also, bonded to or involving such an atom. Of an ion or free radical: having (respectively) the electric charge or the unpaired electron located on a secondary carbon atom.

1903 A. J. Walker tr. *Holleman's Text-bk. Org. Chem.* I. 46 If it [*sc.* a carbon atom] is linked to two carbon atoms it is named secondary; if to three, tertiary; if to four, quaternary. **1926** H. G. Rule tr. *J. Schmidt's Text-bk. Org. Chem.* 70 If two, three or all four valencies are linked to carbon, the atom under consideration is termed secondary, tertiary or quaternary respectively. **1950** E. R. Alexander *Princ. Ionic Org. Reactions* iii. 42 We find..that a primary or secondary carbonium ion extracts a hydrogen atom with a pair of electrons from an alkane so as to form a secondary or tertiary carbonium ion. **1972** [see *primary *a.* 6 f (iii)]. **1972** Norman & Waddington *Mod. Org. Chem.* vi. 82 The order of stability of carbonium ions is tertiary > secondary > primary.

k. Applied to bodily characteristics which are peculiar to one sex but are not essential to reproduction; sometimes the sexual ducts and organs are also included. Cf. *primary *a.* 6 h.

1780 J. Hunter in *Phil. Trans. R. Soc.* LXX. 529 It is my intention at present to extend my inquiry on this subject no farther than as to what relates to that resemblance which one sex bears to that of another in those distinguishing properties which I term secondary... There is often a change of the secondary properties of one sex into another. *Ibid.* 530 The male..loses that resemblance which he before bore to the female in various secondary properties, exclusive of what relates to the organs of generation. *Ibid.* 531 A change of those secondary characters. **1859** C. Darwin *Orig. Species* v. 156, I think it will be admitted..that secondary sexual characters are very variable. **1871**, etc. [see *primary *a.* 6 h]. **1926** H. M. Kyle *Biol. Fishes* xii. 290 It is amongst the freshwater Teleosts..that the secondary sexual characters are most developed... Usually it is the pectoral fins that are longer in the male. **1977** Steen & Price *Human Sex & Sexuality* iv. 59 Androgens, of which testosterone is the principle one, control the development of secondary sex characteristics (distribution of hair, quality of voice, skeletal form, sebaceous gland activity).

l. *Geol.* Of a mineral: that is not an original constituent of the rock; formed by the alteration or replacement of primary constituents of the rock.

1886 [see *primary *a.* 6 i]. **1897** G. P. Merrill *Treat. Rocks* III. iii. 249 Those dikes containing so large a proportion of secondary epidote as to be of a dull greenish hue are almost invariably more enduring than the granites. **1931** A. Johannsen *Descr. Petrogr. Igneous Rocks* I. ii. 28 Secondary minerals may be introduced by the addition of material such as boron, fluorine, etc., to form tourmaline, topaz, fluorite, etc. **1974** Flint & Skinner *Physical Geol.* vi. 94/2 Water combines with the remaining aluminum silicate radical to create the clay mineral kaolinite... The resulting kaolinite we call a secondary mineral, because it was not present in the original rock.

m. *secondary shaft* = *layshaft.

1888 [see *layshaft]. **1902** A. C. Harmsworth et al. *Motors* x. 205 Causing the secondary shaft..to be rotated. **1926** H. T. Rutter *Mod. Motors* II. vii. 261 Parallel to the gear-shaft in the gear box is another shaft, which is called the 'lay' shaft, 'secondary', or countershaft.

n. *secondary spectrum*: a fringe of colours bordering an image formed by a lens corrected for two wavelengths and due to the non-coincidence of the foci of other wavelengths.

1893 W. E. Baxter tr. *H. van Heurck's Microscope* 370 The final upper lens, which is also a triplet, is used to destroy the secondary spectrum. **1932** Hardy & Perrin *Princ. Optics* vi. 115 This residual chromatism gives rise to a fringe of color surrounding the image of an extended object, which is known as the secondary spectrum. **1978** R. Kingslake *Lens Design Fund.* iv. 75 The fact that achromatizing a lens for two colors fails to unite the other colors is known as secondary spectrum; it should not be confused with the secondary chromatic aberration.

o. *Physics* and *Astr.* Of, pertaining to, or designating radiation that has been produced by the interaction of other (primary) radiation with matter. Of cosmic rays: produced in the earth's atmosphere by the impact of primary rays.

1898 *Sci. Abstr.* I. 128 The secondary rays emitted by the metal..pass some centimetres through the air. **1921** J. Scott-Taggart *Thermionic Tubes* i. 11 Under some conditions the electron bombardment liberates a number of secondary electrons attached to the atoms of the plate. **1938** [see *primary *a.* 6 k]. **1944** *Electronic Engin.* XVI. 372/1 In order to avoid or minimise secondary emission it is necessary that grid structures shall be maintained reasonably cool during the operating life of a valve. **1959** [see *primary *a.* 6 k]. **1964** M. Gowing *Britain & Atomic Energy 1939–1945* i. 39 When the uranium oxide was bombarded with fast neutrons the initial fission did not propagate itself because the secondary neutrons lost energy. **1974** *Encycl. Brit. Macropædia* V. 200/1 Secondary cosmic rays consist mainly of subatomic particles that are short-lived..; they cannot have come far and are thus known to have been produced within the atmosphere.

p. *secondary poverty*: effective poverty due to waste, inefficiency, or some other drain on resources, rather than to insufficiency of means.

1901 B. S. Rowntree *Poverty* p. viii, Families whose total earnings would be sufficient for merely physical efficiency were it not that some portion of it is absorbed by other expenditure... Poverty falling under this head is described as 'secondary' poverty. **1909** M. F. Davies *Life in Eng. Village* xii. 146 These people..appear to have a struggle to keep going, and their incomes do not probably exceed the limit of secondary poverty. **1970** M. Rein in P. Townsend *Concept of Poverty* ii. 60 If the diet is to..avoid building into its definition a confusion between primary and secondary poverty, then the standards of economy must be relaxed and a more realistic assumption of human error accepted.

q. Designating action taken by workers on strike to prevent other firms from doing business with the strikers' employers; *esp.* applied to a boycott or the picketing of the premises of firms not otherwise involved in the dispute. orig. *U.S.*

1909 *Pacific Reporter* XCVIII. 1083/1 This is the argument commonly advanced to establish the illegality of what has been called..a 'secondary' rather than a 'primary' boycott. **1916** L. Wolman *Boycott in Amer. Trade Unions* i. 142 The secondary boycott is distinctly different in effect from the simple strike; since..it inflicts injury upon an innocent third party. **1938** *Atlantic Reporter* CXCV. 379/2 The Legislature..never contemplated..'secondary picketing'. *Ibid.* 378/1 Secondary picketing is illegal. **1942** *Yale Law Jrnl.* May 1209 Secondary picketing against the employer's vendee, is the only effective means of publicizing the facts of a labor contest. **1979** *Daily Tel.* 13 Jan. 1/2 The Freight Transport Association said secondary picketing had been reduced in some areas, but expressed concern about the position in [the] Midlands where the dispute was unofficial. **1980** *Illustr. London News* Mar. 19/1 The Law Lords referred to their judgment in the case of McShane v Express Newspapers, in which they had decided that secondary blacking on the part of journalists, on the instruction of their union, fell within the immunity granted under section 13 of the 1974 Act.

r. Designating an earthquake S wave (see *S 6).

1919, etc. [see *primary *a.* 4 h].

s. *secondary industry*: industry that converts the materials provided by primary industry (see *primary *a.* 6 n) into commodities and products for the consumer.

1930 *Economist* 19 July 107/2 The design behind the former movement is clearly to enable nascent secondary industries to compete in the home market. **1944** [see *light *a.*[1] 4 c]. **1950** *N.Z. Jrnl. Agric.* Aug. 127/3 The tending of land, livestock, and crops has figured so prominently in the lives of New Zealanders—and will continue to do so despite the growth of secondary industries. **1977** D. M. Smith *Human Geogr.* viii. 232 The mineral or crop may be exported in its raw state for processing in Europe or North America, thus depriving the producing country of a possible basis for building up secondary industry.

t. *secondary air*: air supplied to a combustion zone where combustion with primary air is occurring.

1931 *Engineering* 9 Jan. 40/2 Complete combustion to CO_2 takes place at the end of the chamber, when an enveloping stream of secondary air meets the first stream. **1951** Cohen & Rogers *Gas Turbine Theory* vii. 195 If devices are used to increase the turbulence and so distribute the secondary air more uniformly throughout the burning gases, the combustion efficiency will be improved but at the expense of increased pressure loss.

u. Of radar: relying on signals transmitted automatically by aircraft in response to signals reaching it from the radar.

1945 R. Watson-Watt in *Nature* 15 Sept. 323 Radar in war fell into three convenient categories, each of which has come to stay in the peace... Secondary radar requires that small measure of co-operation which is involved in the fitting and switching on of an otherwise automatic responder. **1961** *Engineering* 6 Jan. 1/2 What secondary radar does for the controller on the ground is to give him identification of aircraft as they come within range. **1967** *New Scientist* 19 Oct. 151/2 Air traffic control is increasingly making use of secondary radar.

v. *secondary structure* (Biochem.), the three-dimensional form that the chain of a polynucleotide or polypeptide molecule assumes as a result of non-covalent bonds between neighbouring amino-acid residues.

1952 [see *primary *a.* 6 v]. **1960** *Nature* 8 Oct. 99/2 Ribonucleic acid is a single-stranded molecule the secondary structure of which arises from intramolecular interactions. **1974** [see *primary *a.* 6 v]. **1977** D. E. Metzler *Biochemistry* ii. 102/2 The value of β is always positive but that of τ can be negative, the secondary structure (Watson–Crick helix) being fully formed but with left-handed superhelical turns present.

4. a. (Further examples in *Bot.*)

1861 R. Bentley *Man. Bot.* 122 Adventitious or Secondary Root.—This name is applied to all roots which are not produced by the direct elongation of the radicle of the embryo. **1880** C. E. Bessey *Botany* 147 Where the secondary leaves (leaflets) grow from an extremely short axis. **1973** H. C. Bold *Morphol. Plants* (ed. 3) xxx. 570/2 Branches of the radicle are secondary roots; all other roots are adventitious.

c. *secondary road*: a road of a class lower than that of a main road; a minor road.

1903 in *Parl. Papers 1904* XXIV. 279 (Cd. 1793) p. vi, Roads of this class are known in different parts of the country as Secondary Roads, Contribution or Contributory Roads, 'Grant in aid' Roads, &c. **1929** A. Huxley *Let.* 1 Dec. (1969) 321 Even the secondary roads were tolerable. **1938** E. Ambler *Cause for Alarm* xiv. 228 The only roads we'll have to worry about are..secondary roads. **1959** T. S. Eliot *Elder Statesman* III. 88 It was late at night. A secondary road. I ran over an old man lying in the road. **1974** J. Thomson *Long Revenge* iv. 45 He turned off into the network of secondary roads.

5. f. *secondary instruction* (earlier example); *secondary school* (earlier and further examples); also *attrib.*; *secondary modern school*: a secondary school of a kind established by the Education Act of 1944, offering a general education to children not selected for grammar or technical schools (cf. *central school* s.v. *central *a.* 4 and *modern school* s.v. *modern *a.* 2 e); also (in colloq. use) *ellipt.* as *secondary modern* (freq. *attrib.*).

1809 R. L. Edgeworth *Ess. on Professional Educ.* i. 41 In the secondary schools for boys of nine or ten.., the principles of general grammar should be explained. **1835** *Southern Lit. Messenger* I. 275 Others classify them into 1st *primary* schools..2nd *secondary* schools, for the rudiments of Arithmetic, Geography, English Grammar, and further progress in reading and writing. **1863** M. Arnold in *Macm. Mag.* VIII. 355/1 The Royal Commissioners have thought themselves precluded..from making a thorough inquiry into the system of secondary instruction on the Continent. **1892** in *Parl. Papers* IX. 373 This Act may be cited as the Secondary School Teachers Registration Act, 1892. **1926** W. H. Hadow et al. *Rep. Consult. Comm. Educ. Adolescent* (Board of Educ.) 266 The expression 'secondary school' was borrowed from the French *'école secondaire'*, which was used apparently for the first time in the *Rapport et projet de décret sur l'organisation générale de l'instruction publique*, submitted to the Legislative Assembly by Condorcet in April, 1792. **1937** *Burlington Mag.* Sept. 107/2 No student..can possibly acquire more than a secondary-school smattering in the subject. **1943** C. Norwood et al. *Curriculum & Examinations in Secondary Schools* (Board of Educ.) I. iii. 15 At the age of 11+, or earlier in some cases, a child would pass into one of the three types of secondary education which we have postulated, secondary Grammar School, secondary Technical School, secondary Modern School. **1955** *Punch* 30 Mar. 404/2 'The thing that makes me nervous,' I said tentatively, 'is if they fail their 11-plus and land up in a Secondary Modern.' **1956** H. Loukes *Secondary Modern* i. 45 They are not to be regarded, these secondary modern children, as a backward group. **1961** M. Kelly *Spoilt Kill* II. 103 He taught maths in a secondary modern somewhere down south. **1976** *Yorkshire Evening Press* 9 Dec. 13/6 Derwent Secondary Modern School, York, was entered and £6.50 stolen. **1976** *Evening Post* (Nottingham) 14 Dec. 6/2 His early education finished at 14 when he left the Player Secondary School. **1982** *Guardian* 26 Apr. 3/2 Critics say it is a back door way of re-introducing grammar and secondary modern schools.

g. Archæol. *secondary burial* or *interment*: a burial of human remains in a site used for burial at an earlier time (see also quot. 1960); *Secondary Neolithic*: (of or pertaining to) that part of the Neolithic period in Britain marked by the fusion of native Mesolithic cultural elements with those of immigrant European agricultural peoples.

1865 J. Lubbock *Pre-Historic Times* iv. 110 It appears reasonable to conclude that these interments belong to the ante-metallic period; especially when..we find several secondary interments, plainly belonging to a later age. **1877** W. Greenwell *Brit. Barrows* 13 These secondary interments have been made either by placing the body on the surface of an existing barrow..or by making an excavation into it. Secondary burials occur in all parts of a barrow. **1954** S. Piggott *Neolithic Cultures* i. 15 These Secondary Neolithic cultures, as I have called them, were to form the basis of the ensuing British Bronze Age. **1960** K. M. Kenyon *Archæol. in Holy Land* iv. 86 The burials as we find them were secondary. That is to say, the bones were only placed in their present position after the flesh had largely decayed. **1963** E. S. Wood *Collins Field Guide Archaeol.* I. iv. 60 The secondary Neolithic is now appearing more complicated than it looked a few years ago. **1963** H. N. Savory in Foster & Alcock *Culture & Environment* iii. 26 It is therefore no longer necessary to envisage a narrow horizon on which Primary and Secondary Neolithic and 'Beaker' elements can scarcely be disentangled. **1977** Kwang-chih Chang *Archaeol. Anc. China* (ed. 3) viii. 406 Three ways to dispose of the dead were distinguished..: cremation and ash urns;

interment of the dorsal and stretched type; and probably secondary burials.

h. *secondary succession* (Ecol.): (see quot. 1905).

1905 F. E. Clements *Research Methods in Ecol.* iv. 247 Generally speaking, all successions on denuded soils are secondary... The great majority of secondary successions owe their origin to floods, animals, or the activities of man, and they agree in occurring upon decomposed soils of medium water-content. **1932** Fuller & Conard tr. *Braun-Blanquet's Plant Sociol.* xi. 279 Fires are always followed..by a secondary succession, which tends anew towards the climax. **1973** P. A. Colinvaux *Introd. Ecol.* vi. 77 Secondary succession is best understood by considering what happens to a farm when it is abandoned.

i. *secondary hardening* (Metallurgy): a further hardening which occurs in some previously hardened steels when they are tempered; so *secondary hardness*.

1915 Edwards & Kikkawa in *Jrnl. Iron & Steel Inst.* XCII. 12 The temperature at which this secondary hardening begins is progressively raised with increasing percentages of tungsten. *Ibid.*, As regards the temperature at which the maximum secondary hardness is obtained..for the steel with no tungsten this is 494°. **1937** *Discovery* May 155/2 The tempering of high speed steel is primarily undertaken to give maximum secondary hardness. **1949** P. C. Carman *Chem. Constitution & Properties of Engin. Materials* v. 192 On tempering, the hardness decreases slightly between 300° and 500°C., and then secondary hardening takes place between 500° and 600°C. **1967** A. H. Cottrell *Introd. Metall.* xxv. 517 The steel is tempered at 650°C to produce secondary hardening by precipitation of alloy carbides.

j. *Psychol.* In various phrases. *secondary conditioned reflex*: a reflex transferred from the original stimulus to one associated with it; similarly *secondary conditioned stimulus*; *secondary conditioning*: conditioning in which the response is transferred to a subsequent, associated stimulus; similarly *secondary reinforcement, reinforcer, reward*.

1927 G. V. Anrep tr. *Pavlov's Conditioned Reflexes* iii. 34 The appearance of a black square in the dog's line of vision is now used as yet a further stimulus, which is to be given the character of a secondary conditioned stimulus. **1938** B. F. Skinner *Behav. Organisms* ix. 245, I am inclined to doubt the reality of secondary conditioning of a respondent in general. **1940** Hilgard & Marquis *Conditioning & Learning* iii. 63 Secondary rewards such as approval, money, prestige and so forth. **1944** B. Malinowski *Sci. Theory of Culture* xii. 138 The secondary reinforcement becomes attached to the instrumental performance as a whole, and to all its component parts. **1957** E. R. Hilgard *Introd. Psychol.* (ed. 2) x. 242/1 A feature of secondary reinforcement that is very important for human social behavior is its wide application. *Ibid.*, There is also experimental evidence in support of the principle that secondary reinforcers have wide generality. **1976** *Howard Jrnl.* XV. I. 12 The relics of past experiences, surviving through the mechanism of secondary reinforcement. **1977** R. A. Rescorla in Davis & Hurwitz *Operant-Pavlovian Interactions* vi. 155 No increase in response rate was produced by this supposed secondary reinforcer.

k. *secondary recovery*, the recovery of oil by means of special techniques from reservoirs which have been substantially depleted; freq. *attrib.*

1940 P. D. Torrey in E. DeGolyer *Elements Petroleum Industry* xiii. 289 The two most commonly employed secondary recovery methods are water-flooding and gas-repressuring. **1945** L. M. Fanning *Our Oil Resources* iv. 96 In most instances secondary-recovery operations are more costly than primary operations. **1971** I. G. Gass et al. *Understanding Earth* xxiii. 330/2 Oil is obtained at this site by a technique of secondary recovery which involves the injection of water under pressure. **1973** C. J. May in Hobson & Pohl *Mod. Petroleum Technol.* (ed. 4) v. 174 Of perhaps more general interest is the application of secondary recovery methods to reservoirs which have been largely depleted by natural forces.

6. b. *secondary constriction* (Cytology): a chromosomal constriction not associated with the centromere.

1932 C. D. Darlington *Recent Adv. Cytol.* ii. 34 There are also found in many chromosomes 'secondary' constrictions which have no relationship with any present spindle attachment. **1957** C. P. Swanson *Cytol. & Cytogenetics* v. 131 The secondary constrictions seen in somatic metaphase chromosomes generally arise as the result of nucleolar formation. **1975** [see *satellite *sb.* 9].

B. *sb.* **9.** Also, a secondary circuit, current, etc. (Earlier and further examples.)

1837 M. Faraday in *Ann. Electr., Magnetism, & Chem.* I. 199 Why do secondaries almost annihilate the terminal effects of primitives? **1896** [see *primary *sb.* 5]. **1923** E. W. Marchant *Radio Telegr. & Teleph.* v. 67 If the ratio of transformation is made too great, the primary circuit may be tuned for quite a different wave-length from the secondary. **1947** R. Lee *Electronic Transformers & Circuits* i. 5 The right-hand winding is connected to a load and is called the secondary. **1967** [see *primary *sb.* 5].

11. *Physics* and *Astr.* A secondary ray or particle, esp. a secondary cosmic ray.

1921 *Proc. Nat. Acad. Sci.* VII. 17 Practically no secondaries have a velocity of more than 5 volts, even when the exciting primary electrons have velocities of 300 volts. **1932**, **1942** [see *primary *sb.* 8]. **1964** *Cambr. Rev.* 24 Oct. 48/2 A shower of secondaries of total energy ≥ ε. **1975** D. G. Fink *Electronics Engineers' Handbk.* vi. 113 The yield..drops at higher energies, since high-energy electrons penetrate deeper in the material and the secondaries generated there are unable to reach the material surface with enough energy to be emitted.

12. *Gram.* = *adjunct *sb.* 5 b.

[**1914**: see *adjunct *sb.* 5 b.] **1924** Jespersen *Philos. Gram.* xviii. 252 (*heading*) Secondaries and tertiaries. **1928** —— *Internat. Lang.* II. 97 The definite article is a secondary and therefore uninflected in number or gender. **1940** —— *S.P.E. Tract* LIV. 157 We thus distinguish between clause primaries, clause secondaries, and clause tertiaries. **1959** M. Schlauch *Eng. Lang. in Mod. Times* viii. 221 In this system a leading term..is a primary; its direct modifier (e.g. an adjective) is a secondary.

13. *Path.* An additional tumour arising from cells carried to the site from the initial tumour.

1952 Raven & Hancock *Cancer in Gen. Practice* xxii. 153 No treatment is effective except in the case of prostatic, and occasionally breast, secondaries which may respond to androgens or oestrogens. **1969** Bethell & Burg tr. *Solzhenitsyn's Cancer Ward* II. ii. 12 She could not come to terms with the possibility that radioactive gold might exist somewhere while her son's secondaries were penetrating his groin. **1977** *Proc. R. Soc. Med.* LXX. 199/2 Patients with hypercalcaemia and breast cancer usually have widespread osteolytic bone secondaries.

14. *U.S. Football.* The defensive backfield.

[**1912** *Collier's* 23 Nov. 11/2 He hears people about him rattling away about 'Minnesota shifts', 'secondary defense', and so on.] **1955** *Sports Illustr.* 12 Sept. 31/2 Four of them are ready to leap back into the secondary as linebackers. **1972** J. Mosedale *Football* ii. 18 Dutch is like a rabbit in a brush heap when he gets into the secondary. **1980** *Washington Star* 13 Aug. 65 The Redskins are confident their secondary is in fine shape without White... 'We're going to be fine in the secondary,' Beathard said. 'To hell with him.'

15. Short for *secondary school* or *secondary modern school*. *colloq.*

1962 L. Davidson *Rose of Tibet* 7 'Where does he teach?' 'He used to at the Edith Road Girls' Secondary in Fulham.' **1975** 'J. Bell' *Victim* xiv. 148 The passenger was a girl of twelve from a local comprehensive. Which led back to a London secondary in a northern suburb.

second class, *sb. phr.* [Cf. Second-class *a.* in Dict. and Suppl.] The second of a ranked series of classes in which things are grouped; esp. of university degrees, railway carriages, and mail.

1810 *Oxf. Univ. Calendar* 57 The second class is subdivided into two parts, according to the different degrees of merit. **1844** *Punch* VII. 258/2 In travelling by the second class, you will..be choked with dust and ashes from the engine. **1852** [see Class *sb.* 5]. **1863** *Statutes at Large U.S.A.* XII. 705 The second class embraces all mailable matter exclusively in print, and regularly issued at stated periods. **1883** *U.S. Official Postal Guide* Jan. 733 'Nixes' is a term used..to denote matter of domestic origin, chiefly of the first and second class, which is unmailable. **1931** H. Crane *Let.* 20 June (1965) 373 The post (for books, etc. 2nd class) is apt to be very slow to Mexico.

second-class, *a.* Add: **1. a.** *absol.*, also a second-class railway ticket or compartment.

1863 A. H. Clough *Poems* (ed. 2) 300 Punctual they met, a second-class he took. **1888** Kipling *Wee Willie Winkie* (1907) 205 There was only one Second-class on the train. **1901** —— *Kim* vii. 166 Father Victor..put him into an empty second-class next to Colonel Creighton's first. **1956** *Times* 4 June 10/7 In 1875 one of the railway companies abolished 'second class', and called it 'third'.

b. *second-class matter* (earlier example); now replaced by *second-class mail*: mail sent at the lower of two rates; so *second-class letter*, etc.

1873 *U.S. Postal Laws* II. viii. 176 Postage on second-class matter must be prepaid *in money*. **1968** *Times* 19 Mar. 2/1 Principal recommendations are: a two-tier system providing a first class letter service..and a slower second class service. *Ibid.* 9/1 The proposal for a first and second class mail service is a confession of failure. **1972** R. Hill *Fairly Dangerous Thing* I. i. 11 A confirmation had been sent off the previous day, second-class mail. **1976** *Cumberland & Westmorland Herald* 4 Dec. 9/6 A second-class letter posted after the last collection on a Friday night would not be dealt with until the following Monday. **1981** G. Hammond *Revenge Game* xiv. 155 Put a second-class stamp on it and it may *never* get there.

c. *second-class road*: a road of a second class; a B-road (see *B 2 (ii)).

1906 *Min. Evidence R. Comm. Motor Cars* 191/2 in *Parl. Papers* (Cd. 3081) XLVIII. 89 The roads within the country used for heavy traffic may be divided into three classes: first-class roads,..second-class roads, which require strengthening only, [etc.]. **1914** in *5th Ann. Rep. Road Board* 77 in *Parl. Papers 1914–16* XXXV. 127 The roads are to be divided into three classes—1. First Class; 2. Second Class; 3. All other roads. **1922** *Michelin Guide Gt. Brit.* (ed. 6) (Atlas) 31 Second-class roads are designated 'B' with its [*sic*] number. **1975** 'G. Black' *Big Wind for Summer* ii. 29, I drove..into a second-class road that became acutely third class.

d. *second-class citizen*: a person assigned to an inferior class of citizenship; one deprived of normal civic and legal rights; also in extended and *fig.* uses. Hence *second-class citizenship*. orig. *U.S.*

1942 *Time* 6 July 16/3 Finerty..argued that destitute 'second-class citizens' like Waller were barred from serving on Virginia juries. **1948** G. Orwell *Let.* 4 Jan. in *Coll. Ess.* (1968) IV. 401 Zilliacus wrote in demanding what amounts to Fascist legislation and creation of 2nd-class citizens. **1951** I. Shaw *Troubled Air* xv. 217 Cohen.. can't get into an hotel. He tells Levy, '..we're second-class citizens in this country.' **1958** J. K. Galbraith *Affluent Soc.* ix. 107 We relegate one important class of production to a second-class citizenship. **1965** *Austral. Encycl.* I. 322/1 These people [*sc.* convicts] were second-class citizens, well versed in the routines of unorthodox speech. **1972** D. Lessing *Golden Notebook* (ed. 2) p. viii, Women are second-class citizens, as they are saying energetically and competently in many countries. **1974** *Listener* 17 Jan. 95/3 To show what the graphic artist (a second-class citizen in the art world) is capable of. **1975** S. Lauder *Killing Time on Corvo* xvii. 160 Brazil.. offered more opportunities to an educated girl than the restrictions, the second-class citizenship, of old-fashioned Portugal.

2. (Earlier and later examples.)

1863 Mrs. Gaskell *Let.* 1 June (1966) 702 We shall have..to return—I was going to say 2nd class. **1877** Trollope *Amer. Senator* II. xxv. 268 Why could she not go by herself, second class, like any other young woman? **1974** *Times* 22 Oct. 14/4 I'll pay rail fares, of course. Second class. I always travel second class.

secondee (sekǫ·ndī·). [f. Second *v.*[2] + -ee[1].] A person temporarily transferred to a new unit, department, etc.

1980 *Old Lady of Threadneedle St.* 16 June 56/2 As a Bank secondee I found no difficulty in being accepted in the Treasury. **1981** *Times* 10 Mar. 2/5 He was described as a 'secondee' or trainee in the Prime Minister's press office. **1982** *Ann. Rep. Nat. Westminster Bank 1981* 23 Feb. 15/1 Following their period of secondment, the Bank will benefit on their return from the wide experience that has been gained by the secondees.

second-guess (se:kənd ge·s, se·kənd ges), *v. colloq.* (orig. and chiefly *N. Amer.*). [Prob. back-formation from next.] **1.** *trans.* To anticipate the action of (a person), to outguess; to predict or foresee (an event), to apprehend (simultaneously or beforehand) by guess-work.

1941 *Broadcasting* 22 Dec. 11/2 *Do not* try to second-guess or master-mind our military officials. Leave this for established military analysts and experts, who are experienced enough to await the facts before drawing conclusions. **1942** Berrey & Van den Bark *Amer. Thes. Slang* §646/2 Predict the outcome [of a sporting event].., *second-guess*. *Ibid.* §733/7 'Dope the races.' (To figure out or prophesy probable results from past performances &c.).., *second-guess*. **1963** 'R. L. Pike' *Mute Witness* viii. 137 Desperate people get panicky, and I never try to second-guess panicky people. **1974** *Globe & Mail* (Toronto) 22 Oct. 7/5 Any attempt to second-guess the economics of the situation to the end of the decade and beyond is a hazardous and probably futile task. **1976** *Publishers Weekly* 29 Mar. 49/2 Just when you think you've second-guessed [the author] WS, he turns the tables on you. **1980** *Sci. Amer.* Feb. 68/3 A mechanism by which his world-class backgammon program will develop a profile of an opponent's over-all playing style so that it can second-guess his moves and play accordingly.

2. To subject (a person or his action, esp. a decision) to criticism after the result of the action is known; to judge, question, or reconsider by hindsight. Also *refl.* and *absol.* or *intr.*

1946 [implied in *second-guessing* vbl. sb. below]. **1950** *Sun* (Baltimore) 27 Jan. 2/8 Second-guessing the conduct of Pacific war, Admiral Frederick C. Sherman points..to a whole series of lost chances. **1951** *Ibid.* 28 Sept. (B ed.) 17/7 We lost 11 in a row and I still get nightmares thinking about some of those games. I'm still second guessing myself on some of them. **1955** *Galaxy Sci. Fiction* Apr. 86/2 They say that a century or so ago..there were only about five billion [people]. But anyone can second-guess that. **1965** H. Waugh *End of Party* x. 72 'They should have called the police,' Avery growled... Fellows said with a shrug, 'It's always easy to second-guess.' **1974** M. Hoyt *Thirty Miles for Ice Cream* x. 126 We second-guessed that the spot where I went through [the ice] had been sheltered. **1978** J. Krantz *Scruples* viii. 223 Even Billy's New York lawyers approved, because Josh Hillman was an exceedingly brilliant lawyer... He protected Billy's interests without trying to second-guess their own, far more informed, decisions. **1980** *Daily Tel.* 4 Jan. 3/3 He had pointed out the defect [in the manufacture of a car] to Sir Michael, who had replied in a cavalier manner, saying it was not his job to 'second-guess' his designers.

Hence **se:cond-gue·ssing** *vbl. sb.* (usu. in sense 2).

1946 *Richmond* (Va.) *Times-Dispatch* 26 Dec. 12/1 Pate, in naming Jake Kramer and Ted Schroeder as the entire United States Davis Cup team,..left himself wide-open to what could be the greatest second-guessing attack of many years. **1963** *Life* 9 Aug. 44/3 The besetting sin of the theater is second-guessing, trying to anticipate what the public wants and what will be commercially sound. **1981** *Washington Post* 25 Feb. E 7/3 As Dennie walked back to the jocks' room, the Instructor couldn't resist a bit of second-guessing. 'I told you to stay on the rail,' he said.

second-guesser (se:kənd ge·səɹ). *colloq.* (orig. and chiefly *U.S.*). Also as two words. [f. Second *a.* + Guesser, poss. in slang sense 'umpire (in baseball)', the orig. meaning being 'one who acts as if he is a second umpire': cf. also prec.] **a.** In *Baseball*, a spectator who criticizes the playing of a team or the decisions of the umpire, usu. with the benefit of hindsight; hence *gen.*, one who criticizes (the actions or decisions of) another person after

the event. **b.** One who predicts the result of a horse-race. *rare.*

1937 *Sporting News Record Bk.* 65 [*Guesser*, an umpire.] *Ibid.* 66 *Secondguesser*, one who is continually criticizing moves of players and manager. **1939** *New Yorker* 13 May 80/2 He may not be quite the wonder horse the flushed and eager second-guessers insist he is. **1941** B. Schulberg *What makes Sammy Run?* xi. 192 After *Deadline* the second-guessers were saying I could only make mellers [*sc.* melodramas]. **1942** Berrey & Van den Bark *Amer. Thes. Slang* §637/3 *Second guesser*, a 'fan' who criticizes the umpire. **1950** R. Chandler *Let.* 9 Oct. (1966) 80, I suppose these primping second-guessers who call themselves critics think he shouldn't have written the book at all. **1953** Berrey & Van den Bark *Amer. Thes. Slang* (1954) §671/5 *Second guesser*, a fan ready with advice on how the game should be played—after it is over. *Ibid.* §730/7 *Second guesser*, the pest who always knew what horse would win—after it has won. **1972** R. K. Smith *Ransom* IV. 154 You're a professional second guesser, Stuart. Why didn't you speak up when we were discussing the question? **1978** *Times* 18 Apr. 16/5 President Johnson recognized the value of opposition and even appointed George Ball as his in-house second-guesser.

second hand, second-hand. Add: **B.** *adj.* **2. c.** *absol.* or quasi-*sb.* A second-hand book.

1905 *Westm. Gaz.* 21 Oct. 18/2 If a book..doesn't get a good sale within the first two months, it is no good as a second-hand. **1966** J. B. Priestley *Salt is Leaving* i. 6 He might be going off to some auction sale—y'know, to buy some more secondhands.

3. *second-hand shop* (examples). Also *second-hand bookstall, store.*

1795 J.-B. Lechevalier *Let.* in W. B. Stevens *Jrnl.* 24 Sept. (1965) III. 306 The old Plates, Plaisters, Sermons, Pieces of Iron, Brass and Copper that you purchased lately in that second hand Shop. **1875** L. Troubridge *Life amongst Troubridges* (1966) 115 Here there were all sorts of things..old iron stalls, second-hand book stalls. **1904** *Dialect Notes* II. 421 You can get most anything at the second-hand store. **1942** *Tee Emm* (Air Ministry) II. (recto rear cover), They may even end up on a second-hand bookstall. **1981** *Country Life* 16 July 205/1 A book very well worth looking out for in second-hand shops.

second-ha·ndedness. [f. Second-handed *a.* + -ness.] The quality or condition of being second-hand or hackneyed; secondhandness.

1905 G. B. Shaw *Let.* c Aug. (1972) II. 551, I have striven hard to open English eyes to the emptiness of Shakespeare's philosophy, to the superficiality and second-handedness of his morality. **1920** R. Macaulay *Potterism* II. iii. 90 Once you are tied up with a party, you can only avoid second-handedness, taking over views ready made. **1929** A. N. Whitehead *Aims Educ.* iv. 79 The second-handedness of the learned world is the secret of its mediocrity.

second-ha·nder. *colloq.* Also **secondhander.** [f. Second-hand *a.* + -er[1].] **a.** A second-hand commodity. **b.** A second-hand shop.

1896 'Mark Twain' in *Harper's Mag.* Aug. 350/2, I.. see him buy a red flannel shirt and some old ragged clothes... I seen our other pal lay in *his* stock of old rusty second-handers. **1969** *Daily Tel.* (Colour Suppl.) 12 Dec. 27/4, I went to the secondhander and got some little nylon dresses, and bows for her hair. **1977** *Drive* Mar.–Apr. 40/3 Beware these secondhanders—they were made for fleet-buyers only.

seconding (se·kəndiŋ), *ppl. a.* *rare*⁻¹. [f. Second *v.*[1] + -ing[2].] That acts as a second or supporter.

1748 Richardson *Clarissa* VII. xlix. 191 Curse upon my contriving genius! Curse upon my intriguing head, and upon my seconding heart!

secondment[2]. Add: Also in *transf.* sense (see *second *v.*[2]); freq. in phr. *on secondment*.

1955 *Times* 25 July 7/5 It should be possible for oversea universities to defray such losses..if they wish to attract teachers on secondment. **1964** M. Gowing *Britain & Atomic Energy 1939–1945* x. 288 Ways would have to be found of overcoming the staffing difficulty by various kinds of secondment. **1966** C. Sweeney *Scurrying Bush* i. 14 From East Africa.. I went to the Sudan on secondment to the Sudan Government. **1976** *Milton Keynes Express* 30 July 33/3 (Advt.), An experienced teacher required..to cover the secondment of the permanent teacher. **1982** *Guardian* 20 Apr. 1/4 Sir Michael was on secondment from Chloride, for which they received an undisclosed payment from BL.

secondness (se·kəndnės). [f. Second *a.* + -ness.] The quality or fact of being second; *spec.*, in the philosophy of C. S. Peirce (see *Peirce), the category of fact or reaction that gives to the category of idea or quality ('firstness') its actual existence or form.

c **1890** C. S. Peirce *Coll. Papers* (1931) I. §358. 184 When the second suffers some change from the action of the first, and is dependent upon it, the secondness is more genuine. **1903** —— *Ibid.* §24. 7, I think we have here a mode of being of one thing which consists in how a second object is. I call that Secondness. **1934** *Mind* XLIII. 490 Secondness comprises the actual facts of the world—the *hic et nunc* of things. **1966** F. J. Copleston *Hist. Philos.* VIII. xiv. 312 The meaning of an intellectual concept can be explicated in terms of the ideas of necessary relations between ideas of secondness and ideas of firstness, between..ideas of volition or action and ideas of perception. *Ibid.* 322 From one point of view secondness can be called 'fact', while from another..it is existence or actuality. **1979** *Trans. Philol. Soc.* 177 The affixation of *me* will in El[amitic] terms have turned the '2nd' which *II-um* represents into '2nd-ness' for the mere purpose of rendering meaningful the addition to it of the locatival affix *ma*, 'in secondness' amounting in effect to an adverbial 'secondly'.

secondo (sěkǫ·ndo). *Mus.* [It., = second.] In a pianoforte duet, the lower part; the pianist who plays this part.

1792 J. A. K. Colizzi *Three Duets for Two Performers on Harpsichord or Piano Forté* 2 Secondo. **1883** Grove *Dict. Mus.* III. 30/2 In pianoforte duets..*Secondo* or 2 do is put over that [*sc.* the part] for the 'bass' [player]. **1954**, **1965** [see *primo *sb.* 1]. **1976** *Gramophone* Jan. 1218/1 The technical level of accomplishment, particularly on the part of the *secondo* player, is not all it should be.

second-rateness. Add: (Further examples.) Now the usual form.

1905 G. B. Shaw *Irrational Knot* p. xiii, This consoles us for the undeniable secondrateness of the people we do know. **1945** R. Knox *God & Atom* vi. 84 We tacitly acknowledged in ourselves a kind of moral second-rateness which served as an excuse for low standards. **1976** *Times Lit. Suppl.* 17 Dec. 1576/4 The second-rateness of Douglas (but to be second-rate is to be next to first-rate).

second-rater. (Later examples.)

1916 E. Pound *Lett.* (1971) 87 Virgil is a second-rater, a Tennysonianized version of Homer. **1955** [see *beached *ppl. a.* 3]. **1977** *Times Lit. Suppl.* 22 Apr. 481/2 His adamant opposition to American participation in Hitler's war damned him conclusively, for me, as a mean-minded second-rater or worse.

second sight. 3. (Earlier example.)

1859 L. Wraxall tr. *Robert-Houdin's Mem.* II. i. 4 Chance led me straight to the invention of *second sight*.

secos, var. *sekos.

secousse. Add: **b.** *gen.* in Fr. sense. *poet. rare.*

1945 Auden *Coll. Poetry* 130 Blows a wind that whispers..Of hopes that will not survive the *secousse* of this spring Of blood and flames.

secrecy. Add: **5.** Special Comb.: **secrecy system,** a system for ensuring the secrecy of transmitted speech by scrambling it at the transmitter.

1940 [see *privacy 1 b]. **1949** [see *communication theory* s.v. *communication 12].

secret, *a.* and *sb.* Add: **A.** *adj.* **1. h.** Also *secret session*, a meeting of a legislative or deliberative body, conducted in secret. orig. *U.S.*

1872 W. Bagehot *Eng. Const.* (ed. 2) p. xlvii, This objection might be easily avoided by requiring that the discussion upon treaties in Parliament like that discussion in the American Senate should be 'in secret session', and that no report should be published of it. **1916** H. H. Asquith in *Hansard Commons* 27 Nov. 37, I think it would be premature to consider this question till it has been decided whether a Secret Session should be held. **1940** W. S. Churchill *Secret Session Speeches* (1946) 17 The reason why I asked the House to go into Secret Session was not because I had anything particularly secret or momentous to say. **1946** G. B. Shaw *Geneva* (ed. 2) 4 All threatening news was mentioned only in secret sessions of parliament, hidden under heavy penalties.

i. Also *secret dovetail* (Joinery): (see quot. 1972).

1882 W. J. Christy *Joints* 168 Mitred Dovetail Joint... It is also designated secret dovetail. **1963** F. Hilton *Adv. Carpentry & Joinery* x. 180/2 Two members are jointed using a secret dovetail and the third stub-tenoned, with the surfaces mitred. **1972** *Gloss. Terms Timber* (*B.S.I.*) 52 *Secret dovetail*, a dovetailed angle in which dovetails are used but do not show on the face of either member.

n. Also *secret-dimpling, -smiling.*

1925 Blunden *Eng. Poems* 83 Black was the secret-dimpling stream. **1928** —— *Retreat* 33 Thus the bright-templed rhyme Before the secret-smiling author came.

p. *secret life*: a private life of a nature concealed from the common observer; *spec.* one consisting of covert sexual dealings.

1880 (*title*) My secret life. **1927** E. M. Forster *Aspects of Novel* v. 113 Happiness and misery exist in the secret life, which each of us lives privately. **1928** Galsworthy *Swan Song* III. vii. 272 A secret life and Lippinghall! Long, long might that conjunction be deferred! **1973** L. Cooper *Tea on Sunday* xxvii. 207 Did you know that Holdsworth has a secret life?.. Lisa..saw him just going out of the bar with a glamour girl. **1976** C. Bermant *Coming Home* I. vii. 107 My secret life was now revealed to my parents.

4. c. (*a*) **secret service.** Also, an organization which performs this function; *spec.* (*U.S.*) a government department concerned with national security.

1859 *Atlantic Monthly* Feb. 163/2 The Secret Service was doubled..while half Paris must have been under arrest. **1867** L. C. Baker *Hist. U.S. Secret Service* 34 There is nothing in the Secret Service that demands a violation of honor, or a sacrifice of principle, beyond the ordinary rules of warfare. **1939** T. S. Eliot *Old Possum's Bk. Pract. Cats* 34 And when the loss has been disclosed, the Secret Service say: 'It *must* have been Macavity!' **1972** *Police Rev.* 10 Nov. 1445/1 The principal mission of the United States Secret Service today is safeguarding the lives of the President of the United States, the Vice-President, and many other important personalities in public life. **1981** A. Price *Soldier no More* vii. 93 I've got it... You're in the Secret Service.

(*b*) **secret agent,** a person engaged on secret service, esp. espionage.

a **1715** [see Guard *sb.* 7a]. **1837** J. P. Henderson *Let.* 5 Nov. in *Diplomatic Corresp. Texas* (Amer. Hist. Assoc.) (1911) III. 827 The Government of the United States.. [sent] a secret agent to Texas to enquire into her situation, power etc. **1893** S. Weyman *Gentleman of France* II. xviii. 136 You are here as the secret agent of the King of Navarre. **1907** Conrad (*title*) The secret agent. **1939** G. Greene *Confidential Agent* I. ii. 67 In melodrama a secret agent was never tired. **1973** W. Fairchild *Swiss Arrangement* ix. 114 Lisa laughed suddenly. 'I never thought I'd be going to bed with a secret agent,' she said.

(*c*) **secret police,** a police organization operating in secret, *spec.* one owing allegiance to the state or government and used for political purposes. Also *secret policeman*.

1823 F. Burney *Waterloo Jrnl.* in *Jrnls. & Lett.* (1980) VIII. 394 Buonaparte..trusted in the address of that mental diving machine, his secret police, for warding off any hazard. **1863** 'Ouida' *Held in Bondage* I. x. 233 The world has a trick of serving, like the Swiss Guard and the secret police, whichever side is uppermost and pays them best. **1910** A. Bennett *Clayhanger* II. xiv. 257 Some concealed emissary of the Russian secret police. **1938** E. Ambler *Cause for Alarm* vii. 119 The Ovra..has become a regularly constituted secret police force. **1973** D. Miller *Chinese Jade Affair* xviii. 176 The woes of being a secret policeman during the visits of V.I.P. personalities. **1981** G. Priestland *Priestland's Progress* ii. 38 Paul..had begun life as a religious secret policeman commissioned to stamp out the Church.

e. secret list chiefly *Mil.*, a register of research work or developments about which information may not be disclosed. Also *transf.*

1933 *Meccano Mag.* Feb. 109/1 As the aeroplane is on the Air Ministry Secret List, performance figures are not yet available for publication. **1949** Koestler *Promise & Fulfilment* II. v. 269 The war research which they are doing..is still on the secret list. **1955** E. Waugh *Officers & Gentlemen* I. x. 129 There's an agitation..to take you off the secret list. Heroes are urgently required to boost civilian morale. **1977** 'J. D. White' *Salzburg Affair* vii. 63 A missile projector, brand new..and still on the secret list.

f. secret weapon, a weapon (often of potentially decisive force) classified as secret. Also *fig.* and *transf.*

1936 E. Ambler *Dark Frontier* I. vi. 91 He once told me that in these days there was no such thing as a secret weapon. **1939** W. S. Churchill *Into Battle* (1941) 150 The magnetic mine..may perhaps be Herr Hitler's much vaunted secret weapon. **1953** E. Simon *Past Masters* II. 78 See the candid camera at work, that misnamed secret weapon. **1962** *Listener* 2 Aug. 160/2 The formidable Signor Mattei, who is Italy's anything but secret weapon. **1980** A. Scholefield *Berlin Blind* II. 75 Ah, the secret weapons... They are going to bring England to her knees.

B. *sb.* **I. 3. c.** (Earlier example.) Also *secret of Polichinelle* = *secret de polichinelle.

1828 Carlyle in *Foreign Rev.* II. III. 101 The 'open secret' is no longer a secret to him, and he knows that the Universe is *full* of goodness; that whatever has being has beauty. **1853** C. Brontë *Villette* III. xxiii. 336, I wanted to prove to Miss Lucy that I *could* keep a secret... How many times has she saucily insinuated that all my affairs are the secret of Polichinelle!

secretagogue (sĭkrī·tăgǫg), *sb.* and *a.* *Physiol.* Also (*erron.*) secreto-. [f. Secret(e *v.*[1] + Gr. ἀγωγός drawing forth.] **A.** *sb.* A substance which promotes secretion. **B.** *adj.* Tending to promote secretion.

1924 *Amer. Jrnl. Physiol.* LXVIII. 143 Secretin solutions prepared from different duodena..showed variations in their efficacy as pancreatic secretagogues. **1926** *Jrnl. Amer. Med. Assoc.* 28 Aug. 641/1 The acid washings of the upper portion of the small intestine collected from a living dog possess marked secretagogue action on the pancreas when intravenously injected. **1935** *Amer. Jrnl. Physiol.* CXII. 512 An extract of the duodenal mucosa which would possess certain properties, namely, that the intravenous injection of it would cause no secretion of pancreatic juice until acidified or that the secretagogue potency already present would be greatly increased by acidification. **1971** *Nature* 15 Oct. 497/1 We have tested this possibility in rats treated with a combination of two gastric secretogogues (substances which stimulate secretion of gastric juice), pentagastrin and carbachol.

secretaire[2]. Add: Also 8 **secretare. a.** (Earlier examples.)

1771 in *Maryland Hist. Mag.* (1919) XIV. 136 If you have moved it thence it may be in the old secretare in the Chappell. **1792** T. Blaikie *Diary Scotch Gardener* (1931) 235 He was forced by them to come to force or break open the Secretaire and drawers.

b. ‖ **secrétaire à abattant** (a abatañ), a variety of fall-front writing cabinet (see quot. 1977).

1920 F. M. Atkinson tr. *R. de Félice's French Furnit. under Louis XVI & Empire* II. ii. 56 The *secrétaire à abattant* is one of the favourite pieces of this epoch... That is the large drop-front escritoire, a serious, rather masculine piece. **1936** *Burlington Mag.* May p. xvii/1 A magnificent upright Louis Quinze *secrétaire à abattant*, by B.V.R.B. (Boucher, already mentioned). **1967** *Times*

14 Mar. 21/6 Today,..at 11 a.m. at Blenstock House... Bureau Bookcase.., a Dutch Marquetry Secretaire a abattant. **1977** Fleming & Honour *Penguin Dict. Decorative Arts* 718/1 A *secrétaire à abattant*..was first made in the C17 and was very popular in late-C18 France. It stands against a wall and looks like a cabinet or cupboard with the fall-front flap closed vertically.

secretarial, *a.* Add: **a.** (Later examples.) Also *spec.* designed for the training of office secretaries, as *secretarial college, course, school*.

1922 A. L. Church *Training of Secretary* 193 (Index) Secretarial schools. **1935** R. Strachey *Careers & Openings for Women* II. iv. 140 Short six or seven months' secretarial courses to make sure of an early start. **1935** L. H. Turner *Dict. Careers* 73 A few [scholarships are] given by the leading secretarial colleges. **1941** A. Huxley *Let.* 27 Nov. (1969) 473 Sophie has deserted acting for a secretarial school. **1953** B. Gordon-Cumming *Gentle Rain* 119, I went through secretarial courses and things like that. **1967** K. Giles *Death in Diamonds* i. 7, I finished secretarial school and I speak four languages. **1976** M. Maguire *Scratchproof* vii. 107 She'd left secretarial college bubbling with big job enthusiasm.

secretariat(e. Add: The form **secretariat** (also with initial capital) is now usual. Also, the administrative and executive department of a government or similar organization (as the United Nations), usu. directed by a Secretary(-General). Freq. in Communist use [cf. Russ. *sekretariát*].

1926 *Encycl. Brit.* III. 428/1 This 'plenum' elects..the Secretariat of the Central Committee of five members with two deputies. **1934** B. Maxwell *Soviet State* iii. 42 The Central Committee of the Union..is divided into three sections: (1) a Secretariat, which performs the current work of organization and execution. **1934** Webster, *Secretariat*,..the permanent organ of the League of Nations, comprising the Secretary-General, with officials and secretaries appointed by him. **1949** T. Lie *Road to Peace* 1 Next, I want to thank my staff, the members of the Secretariat. **1955** *Bull. Atomic Sci.* Mar. 85/2 The movement works through its secretariat in Amsterdam where the activities of its national groups are coordinated. **1965** A. Nove in B. Pearce tr. *Preobrazhensky's New Economics* p. viii, The party secretariat did not yet have the importance it acquired under Stalin. **1977** *Whitaker's Almanack 1978* 957 The real power in the Party [Communist] is vested..in the Politbureau, the Secretariat and the permanent Departments of the Central Committee.

secretary, *sb.*[1] (and *a.*). Add: **A.** *sb.* **2.** Also *spec.* in various civil service and parliamentary sub-ministerial posts: *parliamentary private secretary*: see *parliamentary *a.* 1; *permanent secretary*: see *permanent *a.* 1 d; *second* (or *third*) *secretary*: a senior civil servant in the Treasury immediately subordinate to the permanent (or second) secretary.

1932 *Whitaker's Almanack 1933* 329/1 Treasury... Permanent Secretary and Head of H.M. Civil Service, Sir Warren Fisher... Second Secretary, Sir Richard V.N. Hopkins. **1939** *Whitaker's Almanack 1940* 433/1 Second Sec., Sir Richard V. N. Hopkins... Joint Third Secretaries, Sir Frederick Phillips..; Sir Alan Barlow. **1964** Ld. Bridges *Treasury* xv. 145 The next rank in the Treasury is known as Second Secretary, which is the equivalent of a Permanent Secretary in other major departments. *Ibid.* 146 The next rank in the Treasury is Third Secretary which is the equivalent of a Deputy Secretary in a major department. **1976** in R. Crossman *Diaries* II. 200 Philip Allen (K.C.B. 1964) was Second Secretary to the Treasury 1963–6.

3. (Later examples in British and examples in U.S. use.)

In the United Kingdom, there have been numerous changes (too complex to set down here) in the nomenclature and duties of Secretaries of State since the nineteenth century. Since 1945, principal Secretaries of State have included the Secretary of State for the Home Department (Home Secretary), Foreign and Commonwealth Affairs (Foreign Secretary), Industry, Defence, Employment, and Northern Ireland. The title Chief Secretary (to the Treasury) was introduced in 1961: it is a ministerial appointment as opposed to the various civil servant Treasury Secretaries. The principal Secretary of State (usu. the Home Secretary) is sometimes referred to as the 'First Secretary'.

The nomenclature of senior U.S. cabinet ministers has also been subject to many changes.

1789 *Deb. Congr. U.S.* 26 Sept. (1834) 90, I likewise nominate Thomas Jefferson, for Secretary of State. **1846** J. K. Polk *Diary* 20 Mar. (1910) I. 293 Forty or fifty persons..called; among them the Russian Minister, the Secretary of State, [etc.]. **1906** 'Mark Twain' *Autobiogr.* (1924) I. 236 He had been ambassador, brilliant orator,.. admirable Secretary of State. **1940** W. Faulkner *Hamlet* I. iii. 74 A gold-filled diploma from the Secretary of State at Jackson saying for all men to know by these presents, greeting, that them twenty thousand goats..is goats. **1961** *Times* 10 Oct. 12/1 As Chief Secretary (a title used for the first time) Mr. Brooke will come under the general policy direction of the Chancellor. **1962** *Hansard Commons* 19 July 632 The Prime Minister: My right hon. Friend the First Secretary of State will act as Deputy Prime Minister. *a* **1974** R. Crossman *Diaries* (1975) I. 610 Oh dear, it is a panjandrum committee—the Prime Minister, First Secretary, Foreign Secretary, the Minister of Defence, the Minister of Labour for some reason, myself. **1976** *Billings* (Montana) *Gaz.* 20 June 8-A/4 Nixon, whom Bill Rogers (secretary of state from 1969 to 1973) referred to as the world's youngest elder statesman, had acquired enormous stature in world affairs.

4. (Later examples shortened from *secretary hand*.)

1969 M. B. Parkes *Eng. Cursive Book Hands 1250–1500* p. xx, One of the outstanding features of the history of English handwriting in the fifteenth century is the gradual infiltration of this new script, which in its English form we now call 'secretary', into all classes of books and documents, until by the sixteenth century it had become the principal script in use in this country. **1978** *Bodl. Libr. Rec.* IX. 324 The writing exercises..are confined to the rectos of the pages, except for practice alphabets in secretary and in a text hand on ff. 30b and 57b respectively.

5. (Earlier and further examples.) Now chiefly *U.S.*

1803 T. Sheraton *Cabinet Dict.* 303 *Secretary.* This term..among cabinet makers..is applied to certain pieces of furniture to write at. **1805** *Times* 7 Nov. 4/4 Genuine household furniture, and valuable Effects..consisting of..Excellent mahogany secretary and bookcase. **1819** A. Constable *Let.* 21 Mar. in J. Constable *Corr.* (1962) I. 178 The secretary in the White Room sold for 9 pounds or guineas, I forget which. **1865** G. W. Bagby *Writings* (1885) II. 27 When you come to open his 'secretary'..you will find his bonds, accounts..lying about loose. **1975** D. Ramsay *Descent into Dark* ii. 68 Anita.. was..stripping the finish from a maple secretary with a blowtorch. **1980** A. N. Wilson *Healing Art* xi. 129 There was a grandfather clock, and a roll-top secretary.

7. *secretary desk* (later example: for † read † exc. *U.S.*); (appositively) *secretary-treasurer, -typist*; **secretary-bird,** (*b*) [*bird *sb.* 1 d], a punning term for a young woman employed as a secretary; **secretary-general**: also *spec.* the title of the principal official of a Communist party or of some international organizations (as the United Nations); hence **secretary-generalship.**

1969 W. Douglas-Home (*title*) The secretary bird. **1974** I. Murdoch *Sacred & Profane Love Machine* 50 Since Pinn had become what she called a 'secretary bird' she had become much smarter. **1976** Deakin & Willis *Johnny go Home* xvi. 184 Even London's 'Secretary Birds'..have problems finding somewhere to live. **1967** Mrs. L. B. Johnson *White House Diary* 23 Apr. (1970) 509 Mr. Hoes showed me a secret drawer in the secretary desk. **1934** B. Maxwell *Soviet State* iii. 42 In theory the Political Bureau is appointed by the Central Committee; in reality the Secretary-General of the Party, if he is powerful enough, makes the selection. This is the case at present, since Stalin is the Secretary-General. **1949** T. Lie *Road to Peace* 1 (*heading*) Secretary-General of the United Nations. **1954** E. H. Carr *Interregnum* 336 Speculating what the secretary-general would report at the next party congress. **1968** *U.N. Security Council Proc.* 10 in *Parl. Papers 1967–8* (Cmnd. 3757) XLII. 229 The Secretary-General deplores any resort to force to settle international problems, wherever it may occur, in contravention of the Charter of the United Nations. **1959** *Economist* 9 May 506/1 According to one view of secretary-generalship. **1977** *Westindian World* 3–9 June 10/1 The whole trend of his Secretary-Generalship so far..is to place the Commonwealth firmly in its global setting. **1920** *Constitution of Santa Barbara Club* (Santa Barbara, Calif.), Officers..Secretary-Treasurer William Wyles. **1979** *Yale Alumni Mag.* Apr. (Suppl.) CN 11/2 He is a past president and secretary-treasurer of the American Association of Law Schools. [**1939** *Daily Tel.* 18 Dec. 13/2 (Advt.), Secretary-shorthand-typist, good correspondent, required immediately for engineer.] **1957** S. Smith *Not waving but Drowning* 34 Dark was the day for Childe Rolandine the artist When she went to work as a secretary-typist. **1976** *Milton Keynes Express* 2 July 4/4 His wife, a secretary-typist, had left for work.

se·cretary, *v.* [f. Secretary *sb.*[1]] **a.** *trans.* To assist (someone) secretarially. *nonce-use.* **b.** *intr.* To work as a secretary (esp. an office secretary). Also const. *to*. *colloq.* Hence **se·cretarying** *vbl. sb.*

1927 *Punch* 26 Oct. 450/1 Poor old Henry..is in the soup again... He secretaries my uncle,..and as a rule we lunch together. **1933** Wodehouse *Heavy Weather* v. 73 Fellow named Carmody, who has been secretarying there. **1958** *Times Lit. Suppl.* 26 Dec. 749/4 Dish-washing here, secretarying there, finally helping out as Bursar in a school dominated by the headmaster's demented wife. **1971** K. Dick *Ivy & Stevie* 55 Stevie..secretaried..to Sir Neville Pearson and Sir Frank Newnes. **1975** P. G. Winslow *Death of Angel* vi. 142 She got fed up with secretarying.

‖ **secret de Polichinelle** (səkre də poliʃinel). [Fr., secret of Polichinelle: see Punchinello.] A supposed secret which is generally known; an open secret.

1857 *Sat. Rev.* 14 Nov. 435/1 The accredited phrase in certain circles about the Court of Spain is, that there is a mystery about it; but the mystery is like the *secret de Polichinelle*. **1908** G. K. Chesterton *All Things Considered* 117 There is a..class of things which humanity does agree to hide... But..though they are, in one sense, a secret, they are also always a 'sécret [*sic*] de Polichinelle'. **1952** A. Christie *Mrs McGinty's Dead* xxiv. 171 A *secret de Polichinelle* is a secret that everyone can know. **1979** A. Buchan *Scrap Screen* vii. 103 It was a *secret de Polichinelle* in the Grosvenor family that the boy..was not the Dean's.

secretin (sĭkrī·tin). *Physiol.* [a. G. *secretin* (Bayliss & Starling 1902, in *Centralbl. f. Physiol.* XV. 682); cf. Secretion and -in[1].] A hormone that is released into the bloodstream from the gut, esp. in response to acidity, and stimulates pancreatic secretion.

1902 Bayliss & Starling in *Jrnl. Physiol.* XXVIII. 331 We have already suggested the name 'secretin' for this body, and as it has been accepted and made use of by subsequent workers it is as well to adhere to it. **1927** Haldane & Huxley *Animal Biol.* i. 18 Food..stimulates the intestine chemically, causing it to secrete a special substance from its lining; this passes into the blood, circulates through the whole body, but, though it exerts no effect on most organs, it stimulates the pancreas..to activity. This substance is called secretin. **1962** [see *prosecretin* s.v. *pro-[2] 1]. **1965** Lee & Knowles *Animal Hormones* viii. 121 Hydrochloric acid is not the only substance which induces the secretion of secretin; both digesting fat and bile salts are effective.

secretive, *a.* Add: Now usu. with pronunc. (sĭ·krĕtiv).

secretor (sĭkrī·tǫɹ). *Physiol.* [f. Secret(e *v.*[1] + -or, originally to render G. *ausscheider* (Schiff & Sasaki 1932, in *Klin. Wochenschr.* 20 Aug. 1428/2).] **a.** One who secretes appreciable amounts of blood-group antigens with his or her bodily fluids.

1941 *Amer. Jrnl. Obstetrics & Gynecol.* XLII. 933 This applies to about 80 per cent of all individuals (secretors) and if a fetus of Group A belongs to the class of non-secretors (20 per cent)..the maternal iso-agglutinin anti-A may serve as the source of the intrauterine hemolytic process. **1950** *Sci. News* XV. 111 Most people secrete the appropriate blood group substances (antigens) in bodily secretions such as saliva and tears. About one-seventh of the population, however, do not do this, and are called 'non-secretors'... Some cysts, in 'secretors', contain the A and B substances in very high concentration. **1962** R. Jeffries *Exhibit No. Thirteen* xi. 112 An examination of the semen enabled a typing to be made... The murderer is from group B and is what is known as a secretor. **1971** J. Z. Young *Introd. Study Man* xl. 585 The secretor genes are quite independent of the ABO genes.

b. *gen.* One who or that which secretes.

1972 *Sci. Amer.* Aug. 46/2 Recently Everitt and his colleagues have learned that when the female monkey's adrenals—the principal secretors of the male hormone androstenedione—are removed, her sexual receptivity is greatly reduced. **1977** *Lancet* 22 Oct. 841/2 Of these secretors, 50% secreted 50 µl of fluid and 10% secreted as much as 400 µl of fluid.

2. *attrib.* **secretor character, status,** the state of being or not being a secretor (sense a).

1956 *Brit. Med. Jrnl.* 29 Sept. 728/1 We have obtained a series of unrelated duodenal ulcer cases and compared their secretor character with controls taken from the general population. **1970** Gershowitz & Neel in D. Aminoff *Blood & Tissue Antigens* 39 The secretor status of stomach cancer patients should be investigated. **1976** *Proc. R. Soc. Med.* LXIX. 36/2 Attempts have been made to discover whether individuals are carrying the dystrophia myotonica gene by examining their secretor status.

sect, *v.*[2] (Later example.)

1882 G. A. Sala *Amer. Revisited* II. ii. 19 Almost every thoroughfare in the city being sected and intersected by lines for horse-cars.

-sect [ad. L. *sect-*, pa. ppl. stem of *secāre* to cut: cf. Sect *v.*[2]], a formative element of vbs. (as *hemisect, transect, trisect*) and adjs. (as *multisect*); *spec.* in *Bot.* in adjs. denoting forms of leaves (as *palmatisect, pedatisect, pennatisect*).

sectarianism. (Earlier example.)

1817 Coleridge *Biogr. Lit.* I. xii. 249 The spirit of sectarianism has been..the cause of our failures. We have imprisoned our own conceptions by the lines, which we have drawn, in order to exclude the conceptions of others.

section, *sb.* Add: **2. e.** (*b*) (Examples.) (*d*) *Austral.* and *N.Z.* An area of undeveloped land, variable in size. (*e*) *Austral.* and *N.Z.* A plot of land suitable for building on. (*f*) In various African countries, an administrative district (see quot. 1951).

(*b*) **1785** *Jrnls. Continental Congr. U.S.* (1933) XXVIII. 299 The plats of the townships..shall be marked by subdivisions into sections of 1 mile square. **1809** F. Cuming *Sk. Tour Western Country* (1810) 197 This Crouse is a wealthy man, having..a farm of two sections, containing thirteen hundred acres. **1849** E. Chamberlain *Indiana Gazetteer* (ed. 3) 420 North of Eel river are about 40 sections of barrens intermixed with small prairies. **1890** *Stock Grower & Farmer* 8 Mar. 5/3 The intervening sections of the Atlantic and Pacific railroad land grant [in Arizona] are owned by the cattle men and are not fenced. **1924** H. Croy *R.F.D. No. 3* 3 He had only one hundred and twenty acres of land, while most of the farmers had a quarter, or a half section, even a section. **1975** *New Yorker* 27 Oct. 114/2 We have six hundred and forty acres—what you call a section—and they wish to flood it to make recreation.

(*d*) **1836** *S. Austral. Gaz. & Colonial Register* 18 June 4/2 Surveyed land shall be divided, as nearly as may be, into sections of eighty acres each, with the exception of the site of the first town, which shall be divided into acre sections. **1841** W. Deans *Let.* 25 Mar. in J. Deans *Pioneers of Canterbury* (1937) i. 31 Some part of the rural sections may not just be what could have been wished. **1923** in J. Reid *Kiwi Laughs* (1961) His idea was that he and I should get the firewood rights on a thousand-acre section, up under the mountain reserve. **1950** *N.Z. Jrnl. Agric.* Jan. 26/2 Ten 10-acre sections have been allocated to returned servicemen.

(*e*) **1836** [see (*d*) above]. **1851** *Lyttelton* (N.Z.) *Times* 11 Jan. 5 The immediate choosing of the town acre sections has been a most important and useful measure. **1886** F. Hume *Myst. Hansom Cab* (1887) v. 19 She.. purchased a small section at St. Kilda, and built a house on it. **1935** J. Guthrie *Little Country* ii. 43 They would much rather have had an eighth-acre section. **1961** B. Crump *Hang on Minute Mate* 97 Tony..was paying off a section in Tokoroa and talking about putting in for one of them Government loans to build a house with. **1977** *N.Z. Herald* 8 Jan. 4-6/2 (Advt.), Waiheke Island, sections and batches urgently wanted.

(*f*) **1951** K. L. Little *Mende of Sierra Leone* v. 104 The overall picture..is one of small towns around each of which is spread a number of component villages. This combination of town and villages constitutes a social and political entity which, in the older sense, corresponds to what is officially termed, nowadays, the 'section' of a chiefdom. **1957** M. Banton *W. Afr. City* viii. 151 He is assisted by seven section chiefs and certain tribal officials. **1977** *Times of Zambia* 7 Sept. 7/7 He has received the reports from all governors in the province on the recent village and section elections.

g. (Later examples.) Also, the fourth part of a platoon; now used of various small tactical units.

1913 *Army Order 323* 1 Oct. 4 The non-commissioned officers and men of the machine-gun section..will be distributed for discipline and administration in peace amongst the four companies. **1915** D. O. Barnett *Let.* 24 Mar. in *Denis Oliver Barnett* (1915) 100 At first I thought the whole section was done in, as rifles and equipment flew in the air. **1939** J. T. Gorman *Army of To-Day* iii. 69 All the men in a section or platoon are taught to use the light (Bren) machine-gun *individually*. **1943** *Britain's Mod. Army* ix. 192/2 Columns of threes are now used, each column in a platoon representing a section with the commander at its head. Thus a section can 'peel off' quickly to a threatened flank, without leaving a gap in the column as used to occur when a section left the old column of fours. **1968** R. M. Barnes *Brit. Army of 1914* i. 35 The infantry advanced in small parties—probably sections in fours, spaced out at fairly wide intervals, or in a 'diamond formation' of sections or platoons.

n. *U.S. Railways.* (See quot. 1890.)

1872 *Newton Kansan* 3 Oct. 3/2 The caboose and the next three cars to it of the 1st section was badly smashed up. **1890** *Railways of Amer.* 162 But the more usual way of handling extra trains, when circumstances will permit, is to let them precede or follow a regular train upon the same schedule. The train is then said to be run in 'sections', and a ten minutes' interval is allowed between them. **1948** *Chicago Tribune* 11 Apr. 1 Ho! the second section! And no flagman out from the train we stopped.

o. *Mus.* A group of similar instruments forming part of a band or orchestra; also the players of such instruments. See also *rhythm section* s.v. *rhythm *sb.* 9 a.

1880 Grove *Dict. Mus.* II. 569/2 The Instrumental Band, as now constituted, naturally divides itself into certain sections, as distinct from each other as the Manuals of an Organ. **1944** W. Apel *Harvard Dict. Mus.* 520/1 It is only in the use of a relatively strong string section that Monteverdi's orchestra is progressive. **1955** Keepnews & Grauer *Pictorial Hist. Jazz* 103/2 Those two men added were both saxophone players; the total of three, instead of a single clarinetist, made a 'section'. That of course is one of the key words, one of the fundamentals of big-band music. **1977** J. Wainwright *Do Nothin'* viii. 124 The sax section—Ric..fills it out, with the tenor.

p. A metal bar, esp. one with a cross-section that is not a simple shape (see quots.).

1881 *Jrnl. Iron & Steel Inst.* 703 A book containing rules and measurements for the construction of various forms of sections of rolled iron, has been drawn up... It is full of formulae applicable to different sizes and forms of sections. **1902** *Ibid.* LXII. 499 Vollkommer suggests an arrangement of plant for the continuous casting and rolling of light sections from fluid metal. **1924** H. J. Skelton *Econ. Iron & Steel* 278 In Great Britain the product in bars or rods shaped in a rolling mill, when not round or square or flat in cross section, is called a 'section' or sectional material. **1956** A. K. Osborne *Encycl. Iron & Steel Industry* 412/2 Structural shapes. (Sections.) Hot rolled steel bars of various cross-sectional contours such as channels, angles, bulb angles, I and H beams, T and Z bars, joists and other complicated contours. **1965** M. H. T. Alford tr. *Tselikov & Smirnov's Rolling Mills* ii. 28/2 The second type [of mill] is used for lighter sections nearer in size to the products of medium section mills. **1971** W. K. V. Gale *Iron & Steel Industry* 181 *Section (shape) (profile)*, any rolled product which is not a round, square, or flat. This is British usage. In USA the term is often shape and in Europe, profile.

q. *Austral.* and *N.Z.* A fare stage on a bus or tram route.

1931 V. Palmer *Separate Lives* 285 He had travelled out the two sections to Aunt Rachel's dingy little house in the suburbs. **1948** *Landfall* June 112 He fingered the two pennies that remained from the half-crown... He'd have to walk to the end of the first section, catch the tram there.

7. a. (sense 2 e (*b*)) *section corner*; (sense 2 k) *section boss, crew, hand, man* (earlier and later examples), *master* (earlier example), *work*; (sense *2 o) *section man, work*.

1870 *Daily Territorial Enterprise* (Virginia City, Nevada) 22 Oct. 3/1 The clothes of the section boss caught upon the brake..as he was in the act of jumping off. **1947** K. D. Lumpkin *Making of Southerner* 163 The deacon was section boss on the railroad. **1817** *Niles' Weekly Reg.* XII. 97/2 At the distance of every mile..section corners are established. **1947** *Mich. Hist.* Sept. 319 He traced it up to the section corner and discovered that the cruiser had signed his name on the tree. **1884** Section-crew [see *push-car* s.v. *push-]. **1962** W. Stegner *Wolf Willow* i. ii. 33 Anonymously denounced in the *Leader* for nearly derailing the speeder of a section crew. **1976** *Columbus* (Montana) *News* 1 July 8/4 My sister and her friend thought I should go on the hand-car with the section crew. **1873** *Newton Kansan* 27 Feb. 3/2 A drunken section hand..laid down upon the railroad track to take a nap. **1904** F. Lynde *Grafters* xxiii. 284 When the section hands pelt stray dogs with new spikes from the stock keg. **1969** *Islander* (Victoria, B.C.) 2 Nov. 5/2 There are no sectionhands in the pass these days, nature finally triumphed in the thirties, closing the line. **1869** W. H. Jackson *Jrnl.* 1 Aug. in *Time Exposure* (1940) xi. 182 Got the section men to take us aboard their handcar. **1921** *Daily Colonist* (Victoria, B.C.) 23 Oct. 27/5 A section man..was killed last night while jumping from a moving train. **1936** D. McCowan *Animals Canad. Rockies* viii. 73 In spring..the section men burn the grass along the right-of-way of the railroad. **1955** L. Feather *Encycl. Jazz* vii. 194 Reluctant soloist but excellent sectionman. **1872** W. S. Huntington *Road-Master's Asst.* p. iii, The enormous expense of track repairs..may be greatly reduced by a reform in the every-day practice of the track-layer and section-master. **1891** C. Roberts *Adrift in Amer.* 71 Section work is track repairing. **1958** C. Fox in P. Gammond *Decca Bk. Jazz* vii. 91 Its attack and vitality ..made up for any roughness in the section-work. **1977** J. Wainwright *Do Nothin'* v. 66 Miller's secret? Size, tight section work and damn good arrangers.

b. Section Eight (also **8**) *U.S. Mil. slang*, discharge from the Army under section eight of Army Regulations 615–360 on the grounds of insanity or inability to adjust to Army life; hence **section-eight** *v.*, (usu. in *pass.*) to discharge from the Army on such grounds; **section head,** (*a*) the person in charge of a section of an organization; (*b*) the heading of a section of a newspaper or periodical; **section house,** (*b*) *U.S.* a house occupied by the men responsible for the maintenance of a section of a railway; **section-line,** (*a*) delete † and add later examples; now only *U.S.*; **section sergeant,** a police sergeant in charge of a section.

1943 *Yank* 23 July 15/2 If it weren't for Yank and its puzzles I'm sure there would be plenty of Section 8s in places like this. **1945** *Yank* 7 Dec. 8/1 Nobody knew whether he was getting section-eighted out of the Army. **1950** E. Hemingway *Across River & into Trees* xxxiii. 207 You stay in until you are hit badly or killed or go crazy and get section-eighted. **1971** J. Aiken *Nightly Deadshade* iii. 31 The place..becomes a reservoir of feebles, bullied by the section heads. **1973** W. H. Hallahan *Ross Forgery* ii. 14 Redhaired man with a Sandhurst accent..a former section head of British Military Intelligence. **1977** *Time* 15 Aug. 5/2 The new format includes different section heads, a new type face for headlines, hairline rules to set off columns. **1869** W. H. Jackson *Jrnl.* 24 Aug. in *Time Exposure* (1940) xi. 183 Decided to board at the section house rather than cook ourselves. **1903** *N.Y. Even. Post* 29 Aug. 1/2 Crowbars and tools.. were identified as having been taken from the railroad section house. **1976** *Columbus* (Montana) *News* 1 July 8/4, I accompanied my sister and her beau to a dance at a section house near the railroad. **1872** *Newton Kansan* 12 Sept. 2/4 The farmers..are leaving space for a road along the section lines. **1948** H. A. Jacobs *We chose Country* 24 We bowled along, climbing past snatches of woods and the straight section-line roads to a high plateau. **1956** *Police Jrnl.* XXIX. 1. 52 Sergeants are 'right in the picture'. Supervising each man on his patrol is still an important part of his duties but he is now the head of a team whose job is to police the section. **1964** M. Banton *Policeman in Community* ii. 15 The division is divided into four sections, each of which is in the charge of a sergeant. .. Their section sergeants use the station as their headquarters and go out from there to supervise their constables. **1973** J. Wainwright *Pride of Pigs* 12 Sergeant Crawley took some backhanders... Are you in favour of section sergeants taking nawpings?

section, *v.* Add: **1. c.** With *off*: to make (an area, part of a structure, etc.) into a separate section.

1960 'E. McBain' *Give Boys Great Big Hand* xiii. 156 The Carellas had sectioned off one corner of the house and disconnected the heating to it. **1976** J. Snow *Cricket Rebel* 72 The authorities sectioned off the stand next to the dressing room for Army marksmen.

sectional, *a.* Add: **1. a.** (Earlier U.S. example.)

1806 *Deb. Congr. U.S.* (1852) 15 Apr. 1042 Let a narrow, selfish, local, sectional policy prevail and struggles will commence.

3. (Examples.)

1901 *Furniture Rec.* 15 June Suppl. 4 The Gunn K.D. Sectional Book Cases. **1909** 'O. Henry' *Roads of Destiny* v. 74 A word from me was more to them than a whole deckle-edged library from East Aurora in sectional bookcases was from anybody else. **1937** *Discovery* Feb. 35/1 They had hoped to use a very light sectional building. **1957** *Times* 2 July (Agric. Suppl.) p. vi/2 A typical old barn now houses the grain intake pit, cleaner, pre-dry bin, continuous dryer, and sectional storage bins. **1960** *Which?* Jan. 5 Sectional boiler, coke-fired.

4. Of steel: rolled in the form of sections (sense *2 p).

1881 *Jrnl. Iron & Steel Inst.* 703 The sectional forms given..are intended to supply the requirements of engineers engaged in the construction of railways, warehouses, public buildings, and similar work. **1916** *Ibid.* XCIII. 354 All the sectional material rolled differs from the ideal by not giving equal distribution of stress in the material. **1924** H. J. Skelton *Econ. Iron & Steel* 279 Wherever practicable, it is desirable that rolled shapes or sectional material in mild steel should be ordered from the list of British Standard Sections.

B. *ellipt.* as *sb.* A piece of furniture composed of sections which can be used separately; *spec.* one which can be used either as a sofa or as a set of chairs. *U.S.*

1961 in Webster. **1972** *Village Voice* (N.Y.) 1 June 3/1 (Advt.), Top quality convertible sofas & sectionals. **1977** *Time* 28 Mar. 50/2 One day Lily and Richard decided that the living-room sofa would look better as a sectional. Practical kids, they picked up a saw and divided it into three pieces. **1980** *Redbook* Oct. 86 (Advt.), Display the sectionals alone or group them with matching modular pieces.

sectionalization (se:kʃənəlaizēɪ·ʃən). [f. Sectionalize *v.* + -ation.] The action or result of dividing into sections.

1904 in Webster. **1920** *Public Opinion* 17 Sept. 268/1 The very formula of 'nationalisation' was a sham, sectionalisation was the object. **1933** *Archit. Rev.* LXXIII. 110/2 What gives interest to London is the diversity of its quarters without extreme sectionalization. **1966** *McGraw-Hill Encycl. Sci. & Technol.* VIII. 490/1 Sectionalization is a method of distributing mine power so that power cables can be isolated..without shutting off the mains supply to several working sections.

sectionalize, *v.* Add: **1.** Chiefly as **se·ctionalized** *pa. pple.* and *ppl. a.* (Further examples.) Hence **se·ctionalizing** *vbl. sb.*

1937 *Discovery* Nov. 360/1 A sectionalised enquiry into the bases of photographic construction. **1957** *Practical Wireless* XXXIII. 357/3 (Advt.), Sectionalised windings ensure low leakage inductance and brilliant top note response. **1965** J. Bingham *Fragment of Fear* iv. 67 Luckily I have a sectionalised mind, and my thoughts were now on Juliet. **1971** *Nature* 7 May 2/1 In the event.. the foundation may have been lumbered with a number of vested interests which will sectionalize its proceedings. **1976** *Gramophone* Sept. 506/1 It will be seen from the accompanying sectionalized photo..that there are two magnetic gaps. **1977** *Ibid.* Feb. 1343/3 By intelligent sectionalizing, it has been possible to produce modules which can be tested and proved individually before incorporation.

sectionist. Add: †**2.** *N.Z.* The owner or occupier of a section (sense *2 e (*d*)) of land. *Obs.*

1841 W. Deans *Let.* 25 Mar. in J. Deans *Pioneers of Canterbury* (1937) i. 32 The secondary sectionists are to choose their land at Wanganui.

sectionize, *v.* Add: **a.** (Later U.S. examples.) Also *absol.*

1872 *Rep. Indian Affairs 1871* (U.S.) 185 [Various tribes] were induced either to sectionize, or in some way to admit white settlers. **1873** J. H. Beadle *Undeveloped West* 399 He is the only Choctaw in the district who is in favor of sectionizing and admitting white immigration. **1949** *Surveying & Mapping* Jan.–Mar. 31/2 Long before Florida was sectionized by ranges and townships, rulers of the Old World bestowed favors on a selected few supporters by granting them titles to vast parcels of land in America.

sector, *sb.* Add: **I. 2.** Further extended or specific senses.

g. (*a*) *Mil.* A part or section of a front, corresponding generally to a sector of a circle the centre of which is a headquarters.

1916 'Boyd Cable' *Action Front* 237 The Colonel was.. vainly trying to recall any sap-head within his sector of line. **1917** W. J. Locke *Red Planet* xiv. 161 Somewhere in this region—or sector, as we call it nowadays—there was a certain bit of ground that had been taken and retaken over and over again. **1930** S. Sassoon *Mem. Infantry Officer* iv. 61 Rose Trench..and Willow Avenue, were among the first objectives in our sector [of the Somme attack].

(*b*) A part or branch of an economy, or of a particular industry or activity. Freq. in phrases *private sector* (see *private *a.* 7 j), *public sector* (see *public *a.* 2 j).

1937 A. Huxley *Ends & Means* xii. 196 The accomplished intellectual understands the relations subsisting between many sectors of apprehended reality. **1950** *Hansard Commons* 7 Mar. 183 Every Member of this House ..could point to examples of gross feather bedding both in Government Service, in the socialised sector of the economy and in private industry. **1959** *Listener* 5 Nov. 767/2 Problems of a comparable nature in other sectors of industry. **1964** *Ann. Reg. 1963* 195 The Government's failure to carry out its declared aims—land reform..and planned development in all sectors of the economy—was to some extent caused by the President's readiness to yield too easily to pressure. **1980** *Sci. Amer.* Sept. 134/1 This development is the outcome of an explicit long-term policy to establish an adequate indigenous capacity in all the basic sectors, particularly metals and machinery, heavy chemicals.

(*c*) *gen.* One of the regions or districts into which a geographical area has been divided.

1943 H. A. Wallace *Century of Common Man* (1944) 82 The ignorance that clouds many communities in many sectors of our own nation. **1958** *Listener* 9 Oct. 547/1 It has recently become fashionable to divide the Middle East

into two major entities: the Arab sector and the non-Arab sector. **1971** *Daily Tel.* 12 Apr. 2/1 Experts believe nearly half of the country's daily oil consumption will be produced from the British sector of the sea by 1976.

h. *Computers.* A subdivision of a track on a magnetic drum or disc, or the block of data stored on it.

1958 *Computer Jrnl.* I. 128/1 Information is stored on 'sectors', each capable of containing 32 numbers... There are 1,024 such sectors, two to each track on a drum. **1962** *Gloss. Terms Automatic Data Processing* (*B.S.I.*) 68 *Sector*, a specified part of a track or band on a magnetic disc or drum store: hence, in programming, a deprecated alternative name for a block applied to the group of words stored on a sector. **1976** G. Wiederhold *Database Design* ii. 40 If tracks cannot be divided by hardware into sectors, system software may divide a track into smaller units.

i. *Gram.* The position in a sentence normally occupied by any one of the basic units of which the sentence is composed. Cf. *sector analysis*, sense 3 below.

[**1955** E. H. Jorden *Syntax Colloq. Jap.* v. 13 Evidence furnished by focus-classes indicates that minor sector boundaries should be observed even here—that the IC division should occur between the gerund of the copula and the following verb, where the sector boundary occurs.] **1966** R. L. Allen *Verb Syst. Present-Day Amer. Eng.* iii. 88 An examination of a large number of sequences suggests that in most non-literary sentences there is a kind of 'spectrum' of basic positions, which may be called 'sectors'. **1968** R. Crymes *Some Syst. Substitution Correlations in Mod. Amer. Eng.* ii. 36 The major positions in the major English sentence, which is a sentence having time orientation [,]..exist in fixed sequence, and they are called sectors. **1974** Chisholm & Milic *Eng. Lang.* vii. lii. 424 According to the grammatical description called Sector Analysis, the English sentence consists of ten sectors.

3. sector analysis *Gram.*, the analysis of sentences in terms of the positions occupied by the basic units of which they are composed (cf. sense 2 i above); **sector scanning,** scanning with radar, sonar, or the like in which the detector rotates to and fro through a fixed angle; so **sector scan** *sb.* (freq. *attrib.*, with hyphen).

1966 R. L. Allen *Verb Syst. Present-Day Amer. Eng.* iii. 88 The order..of the occupied sectors remains constant... Many of the details of this 'sector' analysis lie beyond the scope of the present study. **1971** D. T. Binh *Tagmemic Comparison of Structure of Eng. & Vietnamese Sentences* iii. 66 Sector analysis..primarily emphasizes the positions of units on the sentence, trunk, and predicate.. levels. **1977** *Amer. Speech 1975* L. 127 Only the concluding chapters give any attention to the problems of composition, and much of this is a discussion of sector analysis. **1946** *Radar: Summary Rep. & Harp Project* (U.S. Nat. Defense Res. Comm.) 143/2 *Sector scan*, motion of the scanner reflector back and forth through a limited angle, instead of through 360°. **1969** R. P. Selby in C. J. Richards *Mech. Engin. in Radar & Communications* ix. 387 Radar installations used for air-traffic control are sometimes required to operate on demand in sector-scan mode, the area of scan usually not exceeding 20° and the rate of scan approximately 20 scans per minute. **1978** *Nature* 9 Nov. 174/1 A simple sector scan mode is used, at a frequency of 1 Hz, with the target coupled acoustically to the transducer with water. **1946** *Princ. & Applic. of Underwater Sound* (U.S. Nat. Defense Res. Comm.) (1968) xi. 213/1 A plan position indicator is..required for sector scanning, the CRO spot..tracing a synchronous map of the motion of the active region. **1969** R. P. Selby in C. J. Richards *Mech. Engin. in Radar & Communications* ix. 386 It is sometimes required to move an antenna system about a vertical axis in an oscillatory mode (sector-scanning), thereby turning the antenna through a limited arc in either direction. **1974** Y. Kikuchi in G. W. Stroke et al. *Ultrasonic Imaging & Holography* 267 Asberg has been proposing a high speed sector scanning of a focusing mirror system receiver for obtaining an ultrasonic cinematogram of the living heart. **1977** *Navy News* July 18/2 Because the number of wrecks on our continental shelf is so high..modern equipment such as Hydrosearch—the British sector-scanning surveying sonar—is particularly needed.

sectoral, *a.* Add: **b.** (See *sector *sb.* 2 g.)

1969 P. Anderson in Cockburn & Blackburn *Student Power* 223 There were various sectoral attacks and refutations of Marx by marginalist economists like Böhm-Bawerk. **1971** *Seminar* Nov. 12/2 The second..category of English dailies in India is of the more 'professional' dailies... Lumped together, they certainly represent what may be called 'a sectoral monopoly'. **1981** *Times* 13 Aug. 16/4 In the sectoral breakdown, the statistics show that the banks have increased their support to manufacturing industry.

2. *sectoral horn* (Radio), a horn antenna having a rectangular cross-section and plane sides flared in one dimension only.

1939 W. L. Barrow in *Proc. IRE* XXVII. 41 A horn whose cross section is rectangular and whose sides flare in one direction only..will be termed a sectoral horn. **1959** K. Henney *Radio Engin. Handbk.* (ed. 5) xx. 88 The sectoral horn is flared in one dimension only, either in the plane parallel to the electric vector (*E* plane) or in the plane parallel to the magnetic vector (*H* plane).

sectored (se·ktəɹd), *a.* [f. Sector *sb.* + -ed², or Sector *v.* + -ed¹.] Divided into sectors; applied *spec.* to a disc divided into alternate black and white sectors of equal size.

1900 W. M. Stine *Photometrical Measurements* i. 21 Illumination..viewed through a sectored disk, rotating at a critical speed. **1972** *Nature* 13 Oct. 407/1 He demonstrated this phenomenon by using a black-and-white sectored disk rotated on a phonographic turntable.

sectorial, *a.*¹ Add: **c.** *Bot.* Applied to a type of chimæra (see quot. 1968). Also as *sb.* [ad. G. *sektorialchimäre* (E. Baur 1909, in *Zeitschr. f. induktive Abstammungs- u. Vererbungslehre* I. 342).]

1927 *Jrnl. Genetics* XVIII. 257 In a sectorial plant a sector (of the stem) is formed of tissue of another species or type. *Ibid.*, Sectorials of the type to which the term is restricted in this paper are of very rare occurrence. **1934** W. Neilson Jones *Plant Chimaeras & Graft Hybrids* ii. 18 True sectorial chimaeras are relatively uncommon; most of the plants recorded as 'sectorials' were probably mericlinal in structure. **1959** *New Biol.* XXX. 38 The first composite branch to arise had one side producing tomato leaves and the other producing nightshade leaves, and it was, therefore, concluded that the branch was composed half of tomato tissue and half of nightshade tissue. This type of structure is known as a Sectorial chimera. **1968** [see *periclinal *a.* 2 b].

Hence **secto·rially** *adv.*, in or into sectors.

1963 *Heredity* XVIII. 266 These plants have..a sectorially divided growing point.

sectorization (se:ktəraizēɪ·ʃən). *rare.* [f. Sector *sb.* + -ize + -ation.] Division into sectors; administration or operation on the basis of sectors or local divisions.

1962 *Flight International* LXXXII. 479/1 Traffic growth in recent years has increased the pressure on both pilots and controllers, especially in the sphere of voice communications, so that means have had to be found to reduce their workload. Various alleviating possibilities have been tested, such as 'sectorization', but they in turn usually introduce new difficulties. **1976** *Social Psychiatry* XI. 27 Sectorization of the psychiatric care, which means that all psychiatric care within a geographical area should be under a common management.

Secuana, var. *Sechuana.

secular, *a.* Add: **I. 3. b.** Also *secular-minded* adj.

1899 T. Veblen *Theory of Leisure Class* xii. 314 The sacerdotal scheme of life..does not hold good for the clergy of those denominations which have..diverged from the old established schedule of beliefs or observances... Their manner of life..does not differ in an extreme degree from that of secular-minded persons. **1930** A. Birrell *Et Cetera* 159 An equally veracious, though most secular-minded Presbyterian divine. **1957** N. Frye *Anat. Criticism* 265 In the Anglo-Saxon congregation of Wulfstan there must have been a few secular-minded highbrows who were thinking..of the preacher's mastery of alliterative rhythm.

II. 7. a. *secular equation*, also more widely, any equation of the form $| a_{ij} - b_{ij}\lambda | = 0$ $(i, j = 1, 2, \ldots, n)$, in which the left-hand side is a determinant and which arises in quantum mechanics; (further examples).

1937 E. C. Kemble *Fund. Princ. Quantum Mech.* x. 361 Its components must yield a nontrivial (*i.e.*, nonvanishing) solution of the set of *g* equations $\Sigma_n(A_{mn} - a\delta_{mn})x_n = 0$... Such a solution exists only if the determinant of the coefficients vanishes, *i.e.*, if *a* is a root of the so-called 'secular' equation det $(A - aI) = \ldots 0$. **1974** Gill & Willis *Pericyclic Reactions* i. 21 To obtain the wave functions corresponding to these energies it is necessary to solve the secular equations using the appropriate values of *E*.

8. *Econ.* and *Statistics.* Of a fluctuation or trend: occurring or persisting over an unlimited period; not periodic or short-term.

1895 A. Marshall *Princ. Econ.* (ed. 3) I. v. v. 470 There are secular movements of normal price, caused by the gradual growth of knowledge, of population and of capital, and the changing conditions of demand and supply from one generation to another. **1926** L. D. Edie *Econ.* II. iv. 49 Economic fluctuations fall into four major types: seasonal, secular, cyclical, and residual. **1971** H. S. Shryock et al. *Methods & Materials Demography* II. xiii. 377/2 If the observations are made at different times of the year, seasonal movements may also be apparent. When we are trying to describe the growth of a population over a relatively longer period of time (for example, India from 1872 to 1961) we are generally interested in the secular trend only. **1973** *Daily Tel.* 15 Jan. 17/6 This is the first time the Government has had to pay so much for money but the secular trend of interest rates will stop rising only if the rate of inflation is brought down. **1976** *Sci. Amer.* Sept. 107/1 The secular trend of workers migrating out of agricultural jobs as a result of technological change in agriculture has recently slackened.

secularist, *sb.* (and *a.*) **4.** (Earlier example.)

1888 Mrs. H. Ward *R. Elsmere* III. vi. xxxviii. 166 The most notorious secularist lecturers held forth.

secularly, *adv.* Add: **3.** *Astr.* Over a long period of time.

1971 *Nature* 24 Dec. 453/1 We might expect δ_f to be secularly dependent in the same way as H_e ($\propto P^{-\frac{1}{2}}$). **1979** *Ibid.* 20 Sept. 200/1 Even in the worst case of deviation from thermal equilibrium, that is when the fully convective star expands adiabatically,..the system would still be secularly stable against mass exchange.

secule (se·kiul). *Geol. Obsolescent.* [f. L. *sæculum* age, generation.] A period of geological time corresponding to a stratigraphical zone; = *moment *sb.* 2 d.

1903 A. J. Jukes-Browne in *Geol. Mag.* X. 37 The term hemera may, however, be occasionally convenient to signify the duration of a subzone, as age signifies the duration of a stage, but if we want to avoid confusion we must not speak of the hemera of a zone. For this another word should be coined..I would suggest that the Latin word *seculum* will furnish us with 'secule' which finds an actual French equivalent in *siècle*. **1933** W. J. Arkell *Jurassic Syst. Gt. Brit.* i. 21 Several hemeræ may be contained in one secule or zone-moment. **1956** *Amer. Jrnl. Sci.* CCLIV. 459 Schindewolf reasserts..his extreme view ..that the zone as conceived by Oppel was a purely chronological notion and that the term zone should be used only as a time term (= secule). **1969** [see *moment *sb.* 2 d].

secundigravida (sɪ̆kʊndigræ·vidă). *Obstetrics.* Pl. **idas-, -idæ.** Also **secundagravida.** [mod.L., f. as next + *gravida*, fem. of *gravidus* Gravid *a.*] A woman pregnant for the second time.

1904 in Stedman *Dunglison's Dict. Med. Sci.* (ed. 23) 1007/1. **1940** *Amer. Jrnl. Obstetr. & Gynecol.* XL. 988 Eighty-seven examinations were upon primigravidas, 61 upon secundigravidas, and 114 upon patients of greater parity. **1977** *Lancet* 22 Jan. 195/2 This policy should be restricted to secundigravidæ and those who have had two babies.

secundipara (sɪ̆kʊndi·pără). *Obstetrics.* Pl. **-paras,-paræ** [mod. L., f. *secundus, secundi-* second + *-para*, fem. of *-parus*, from *parere* to bring forth.] A woman who has twice been delivered of children.

1897 *Lippincott's Med. Dict.* 918/2 *Secundipara*, a woman bearing a child for the second time. **1938** *Nature* 24 Dec. 1121/2 Cases with metastatic tumours were associated with a significant increase in the amount of trimethylamine... In primiparæ, the average amount was 0·30 and in secundiparæ 0·29. **1972** E. C. Hughes *Obstetr.-Gynecol. Terminol.* vii. 332 A secundipara is a woman who has given birth for the second time to an infant or infants, alive or dead, weighing 500 gm or more.

secundum. Add: *secundum idem*, 'according to the same argument, calculation, etc.', in the same manner or respect.

1696 J. Sergeant *Method to Sci.* 390 To Affirm that the Atome is Chang'd according to One of those Different Regards or Notions, *viz* the Form, and Not-chang'd according to the Other, viz the Matter, has not the least show of Affirming and Denying *secundum idem*; nor, consequently, the least show of a Contradiction. **1865** S. Hodgson *Time & Space* ii. 140 The two objects are then not limited and unlimited secundum idem. **1882** W. James in *Mind* Apr. 187 The union and the division are not *secundum idem*: it divides them by keeping them out of the space between, it unites them by keeping them out of the space beyond.

secure, *a.* Add: **II. 3. f.** Of a telephone (line): free from the risk of being tapped (*tap *v.*¹ 2 c).

1961 in Webster. **1975** B. Meggs *Matter of Paradise* (1976) v. i. 106, I don't want to say anything more right now; this telephone isn't secure.

secure, *v.* Add: **7.** *Hort.* (See quot. 1928.)

1928 *Daily Express* 11 Aug. 4/2 The Japanese varieties of the chrysanthemum are now beginning to show their flower buds, and these should be 'secured', as it is called, at the earliest possible moment. This is done by pinching out with the thumb and finger the incipient side shoots or laterals that will be found in process of formation immediately beneath the buds and in the axils of the leaves. **1951** *Dict. Gardening* (R. Hort. Soc.) I. 476/1 It should be possible to secure the first crown bud of many varieties during the last week in July.

Securicor (sɪ̆kiūᵊ·rikɒɹ). [Invented name f. Securi(ty + Cor(ps.] A private security organization employed in the guarding and safe transport of money, goods, and property. Freq. *attrib.*, esp. as *Securicor man, van.* Also *fig.* (with small initial).

There is a large number of firms bearing this name owned by Security Group Ltd. (formerly Associated Hotels).

1953 *Change of Name Certificate* 3 Jan. in Dept. of Trade file (354883) Night Guards Limited... Securicor Ltd. **1961** *Security Gaz.* Feb. 64/3 (*caption*) Securicor guards are responsible for the safety of Ireland's greatest art treasure, the 1,000 year old 'Book of Kells'. **1962** *Daily Tel.* 6 Sept. 13/1 An executive of Securicor, the security organization, said..'They were stupid to try to change vehicles.' *Ibid.*, Mr. Norman Negus, 54, a Securicor guard, walked out of the bank..carrying the cash box chained to him. As he approached the armoured Securicor van the ambush was sprung. *Ibid.*, Another Securicor man, locked inside the armoured van, sounded the alarm siren. **1968** *Listener* 12 Dec. 804/3 Unless future student audiences can be screened in advance by the BBC's own securicor, one sees small hope for this series. **1970** *Guardian Weekly* 14 Mar. 9/1 Securicor, Security Express, and Factoryguards—the three main companies which account for about 90 per cent of the manned protection in Britain. **1977** D. Bagley *Enemy* xxx. 239 The auctioneer has Securicor men all over the place.

security. Add: **I. 1. b.** The safety or safeguarding of (the interests of) a state, organiza-

tion, person, etc., against danger, esp. from espionage or theft; the exercise of measures to this end; (the maintenance of) secrecy about military movements or diplomatic negotiations; in espionage, the maintenance of cover. Hence (with capital initial), a department (in government service, etc.) charged with ensuring this. (This sense tends towards 'the condition of making secure'.)

1941 *Times* 16 July 3/1 In order to ensure public security, the occupation of the principal localities in Syria and the Lebanon will be undertaken in accordance with the programme which will allow immediate replacement of French by the occupying forces. **1941** E. John *Lofoten Let.* 34 Major Talbot..prides himself..on the 'security' of this expedition... [*note*] That is the Army term for what normal people call 'secrecy'. **1945** [see *leakage 2]. **1955** *Bull. Atomic Sci.* Apr. 165/3 'Security', as it relates to the continuing struggle between the free world and the Soviet bloc, is an abundantly common yet widely misunderstood word. **1959** *Listener* 8 Oct. 558/1 You can call at offices, clubs, studios, and institutions—anywhere that does not verge on security—and usually they will tell you, foreigner though you are, the telephone numbers of their staff. **1961** R. Seth *Anat. Spying* v. 83 In the spy's vocabulary, Security means doing nothing that is likely to reveal his clandestine rôle... Each separate aspect of Security may be small..but any one aspect neglected is sufficient to cause the spy's downfall. **1965** M. Allingham *Mind Readers* vi. 59, I thought that might have been what Security told you when they sent for you. **1976** M. Delving *China Expert* iv. 44 Security persuaded him to leave the army, and a place was found for him in..MI5. **1976** *Daily Tel.* 20 July 2/3 Security at places like the airport is always under review. **1982** *Observer* 6 June 1/7 While Israeli reaction has been to praise the British police..there is some evidence that security outside the hotel was lax.

3. Delete *arch.* and add further examples.

1977 *Monitor* (McAllen, Texas) 7 June 16A, A feeling of security comes with owning your own home. **1979** R. Jaffe *Class Reunion* (1980) iii. i. 305 She knew now it was one of her weaknesses to look for total security.

10. (Further examples.) Also, in the U.S., such a document issued to investors to finance a business venture.

1848 [see Negotiable *a.* 1]. **1899** *Ann. Amer. Acad. Pol. & Soc. Sci.* XIV. 181 The term 'negotiable securities' is applicable in a general sense to many forms of commercial paper, including drafts and bills of exchange, but it is usually employed for the share-capital of corporations and for the bonds of such corporations and of local and state governments. **1925** R. H. Montgomery *Financial Handbk.* vii. 526 The financial executive..will naturally adapt the securities offered by his company so as best to fit in with the market he is trying to reach. **1937** J. I. Bogen *Corporation Finance* xiii. 223 Most..successful enterprises can raise funds..through the sale of securities... So broad has the ownership of stocks and bonds become that the United States has been described as a 'nation of investors'. **1962** *2001 Business Terms* (Alexander Hamilton Inst.) 246 *Securities*, stocks and bonds of business firms used to raise long-term capital. **1970** M. Greener *Penguin Dict. Commerce* 296 *Security*, a misused term often applied indiscriminately to shares, debentures, etc. In fact a security is something given or guaranteed by the borrower as a safeguard for a loan. The term is often applied to debentures and similar loan stock, and to negotiable instruments. Certificates of liability are known as securities, so sometimes are government stocks or any loans whose repayment is guaranteed. The term should not be applied to shares.

III. 11. security (also **securities**) **analyst** *U.S.*, a person who analyses the worth of securities, as by measuring the ratio of their cost to their dividends and earnings; **security blanket** orig. *U.S.* [idea popularized by the American cartoonist Charles M. Schulz (b. 1922) in the comic strip 'Peanuts' in which a boy named Linus carries a cot blanket for comfort], an object (esp. a blanket) given to a child to afford reassurance by its familiarity; also *fig.*

1934 Graham & Dodd *Security Analysis* vii. l. 586 There is a fundamental cleavage of viewpoint between the speculator and the securities analyst. **1937** *N.Y. Times* 18 May 40/4 Plans for the organization of the New York Society of Security Analysts. **1961** 'E. Lathen' *Banking on Death* iii. 22 One of Robichaux and Devane's security analysts was leaving the firm. **1979** A. Maling *Koberg Link* (1980) xx. 109 I'm a securities analyst. [**1956** C. M. Schulz *Good Grief, More Peanuts* 25 (*caption*) This is a 'security and happiness' blanket... *All* little kids carry them.] **1971** *Newsweek* 19 July 48/1 Deferred-admissions plans—a sort of security blanket that prospective students can carry with them during a year's sabbatical. **1973** *Ladies' Home Jrnl.* Dec. 102 A worn, torn, one-eyed teddy bear about a foot long was my 'security blanket'. **1975** *New Yorker* 21 Apr. 99/1 There's a security-blanket comfort in thinking you have something there in actuarial terms that you can rely on. **1976** *New Society* 28 Oct. 179/2 This [book], to give well deserved recognition to its usefulness..is already as dog-eared as a security blanket. **1978** *Maledicta* II. 81 The subjunctive mood is his security blanket.

12. *attrib.* with reference to (the maintenance of) security in military, penal, civil, and commercial contexts (see sense 1 b above).

a. Of devices which assist security, as *security fence, gate, lock*, etc. Also, of areas so protected, as *security wing*.

1904 Security-door [in Dict., sense 11]. **1963** *Security Gaz.* V. 187/2 The main innovation at Blundeston is the 12 ft. high security fence. **1968** *Rep. Work Prison Dept.* 6 in *Parl. Papers 1967–68* (Cmnd. 3774) XXXI. 57 The use of special security wings to house prisoners who require the strictest security is an expedient which poses severe problems. **1971** *Country Life* 10 June 1439/3, I gathered that they [*sc.* four stone figures] once guarded the four corners of Aldgate, one of London's well-known security gates. **1976** 'M. Albrand' *Taste of Terror* xviii. 107 I'm going to..put security locks on every door in the house. **1976** *Washington Post* 19 Apr. c21/2 (Advt.), Will throw in bumper rack, tie downs, security chain & lock. **1976** *Evening Times* (Glasgow) 1 Dec. 3/6 A key that was missing would let anyone escape from the block into the grounds—but it wouldn't let them out of the security gates.

b. Of measures, etc., intended to ensure security, as *security clearance, measure, pact, rating*, etc.

1925 *Times* 2 Sept. 11/4 The jurists..are discussing the technical details of the proposed Security Pact at the Foreign Office. **1945** *News Rev.* 10 May 9 The security black-out will be lifted to enable us to print some details about the nation's war effort. **1952** *Ann. Reg. 1951* 424 There were renewed expressions of disquiet from scientists about the encroachment of security measures on personal freedom of speech and action. **1955** M. Reifer *Dict. New Words* 185/1 *Security clearance*, the establishment, by means of investigation and executive determination, that a prospective federal employee or consultant is not a security risk, and may be hired. **1958** *New Statesman* 20 Sept. 365/2 With the off-shore islands and Formosa blanketed by a security screen, it is difficult to know exactly what is happening there. **1963** Mrs. L. B. Johnson *White House Diary* 2 Dec. (1970) 13 Dr. Henry Smyth, the lone dissenter in the 4-to-1 Commission decision when Dr. Oppenheimer lost his security clearance. **1963** L. Deighton *Horse under Water* xxi. 94 Act grown-up or I'll cut your security rating back. **1976** H. Tracy *Death in Reserve* xv. 118 It's your red area security pass. **1978** R. Ludlum *Holcroft Covenant* xxxi. 363 Suppose the guards were more alert, security measures more effective.

c. Of persons or organizations charged with the maintenance of security, as *security guard, man, officer, police, service*, etc. Also *security van*.

1940 *Hutchinson's Pictorial Hist. War* 4 Feb.–9 Apr. 180 It was quiet on the island at first, the Nazis believing the raiders to be just another security patrol. **1944** P. Gibbs *Battle Within* 135 If one of our Security Police had been in the church to-day he might have tapped you on the shoulder after the service and led you off to Brixton Gaol. **1945** *Daily Express* 4 June 1/1 A double check is being made by security officers on the three Belfast–Dublin trains that stop at Goraghwood daily for Customs examination. **1948** *Straits Times* 20 July 6/2 It is still true that 'the estates, mines and kampongs'—to echo Mr. MacDonald's broadcast again—do not feel that the security forces are doing enough for them. **1951** N. Brook in *Ld. Denning's Rep.* 79 in *Parl. Papers 1962–3* (Cmnd. 2152) XXIV. 349, I recommend that the Security Service should in future be responsible to the Home Secretary. **1955** Earl Winterton *Fifty Tumultuous Years* 74 He had no aide-de-camp with him, no 'security guard' and no police escort, not even a groom. **1958** *New Statesman* 7 June 716/2 But the Socialist Deputies of Nord and Pas de Calais..replied to these fears by saying openly to Guy Mollet: 'If you're afraid of the paratroops and you can't depend on the security police, arm the miners.' **1959** *Times Lit. Suppl.* 27 Mar. 182/2 Ministers, Civil servants and security men buzzing round the graceless tycoon who is producing a new bomber. **1963** A. Douglas-Home in *Hansard Commons* DCLXXXVI. 859 If we were to set up a Standing Security Commission, I think, first, that it should have a judicial chairman. **1965** D. Francis *Odds Against* ix. 135 [We] could arrange for some sort of guard on the course. Security patrols, that kind of thing. **1966** M. Woodhouse *Tree Frog* xxi. 153 There were no security guards..there wasn't anywhere to run to. **1970** G. F. Newman *Sir, You Bastard* vii. 198 An attack on a security van. **1973** *Times* 15 May 12/6 Britain's involvement in Northern Ireland was real and earnest, as the security forces knew. **1975** D. Lodge *Changing Places* v. 165 A solitary security man in his shelter lifted a lazy hand in salute. **1975** A. A. Thompson *Message from Absalom* iii. 17 Security police carrying rifles. **1976** H. Wilson *Governance of Britain* ix. 167 Until 1952 the Prime Minister was directly responsible for the security service. **1978** R. V. Jones *Most Secret War* liii. 520 One morning just before a weekend, the Security Officer rushed round the M.I.6 offices telling everyone to take down all maps off their walls. **1981** 'W. Haggard' *Money Men* xiii. 144 A security van drove up... It contained his loot.

d. With adjs. to form adjs., as *security-conscious, -minded* (hence *-mindedness*).

1943 J. H. Fullarton *Troop Target* 26 'If you ask me,' said Quigg, 'this Fifth Column palaver is mostly propaganda to make us security-minded.' **1955** I. Fleming *Moonraker* xv. 149 Drax..seemed to be meticulously security-conscious. **1955** M. Gilbert *Sky High* v. 69 The Inspector's tone implied exactly what he thought about the security-mindedness of County Councillors. **1960** 'R. East' *Kingston Black* xx. 185 In military intelligence it's second nature to be security-minded about public telephones. **1968** M. Jones *Survivor* i. 22 I've always heard they're madly careless in the Air Force. Not security-conscious like the Navy. **1972** M. Gilbert *Body of Girl* xxiii. 200 He decided to test the security-mindedness of the person chiefly concerned. **1976** 'M. Barak' *Secret List Heinrich Roehm* i. iv. 46 The Israelis are going to be much more security-minded now.

e. Special Combs.: **security blanket,** an official sanction introduced in order to maintain complete secrecy or safety from danger; **security check,** (*a*) a verification of identity or reliability, *spec.* of the loyalty of an official employee, for the purposes of security; (*b*) a phrase incorporated in a broadcast message from a spy to confirm his identity or to indicate that he is not operating under duress; hence **security-check** *v. trans.*, to subject to a security check; **Security Council,** a principal council of the United Nations consisting permanently of the Great Powers of 1945 and temporarily of certain others, charged with the settlement of disputes (and orig. with the threat of military action against aggressors); **security risk,** a person whose tenure of an official position constitutes a possible danger to the security of the state, etc.; also, a situation endangering security.

1955 *New Yorker* 5 Feb. 88/2 The size of the Regular Army is under a security blanket at present. **1972** *Times* 13 Sept. 8/8 At Heathrow..there was a tight security blanket. **1945** *Daily Express* 4 June 1/1 A security check on their identity cards and permits on the Ulster–Eire border showed that they were apparently in order. **1961** R. Seth *Anat. Spying* vi. 98 The operator is instructed to insert in all his messages what is known as a Security-check. **1966** 'A. Hall' *9th Directive* iii. 31 With the flap on..they were probably security-checking the Ambassador himself. **1978** R. Ludlum *Holcroft Covenant* xxxii. 374 Security check requested. **1979** F. Forsyth *Devil's Alternative* x. 232 The cipher clerks are..security-checked to the highest level. **1944** *Times* 10 Oct. 5/6 Tentative proposals have been made for the establishment of a general international organization under the title of The United Nations. The proposals..deal with..its principal organs, including a General Assembly, a Security Council, and an International Court of Justice. **1968** *Security Council Proc. Czechoslovakia* 3 in *Parl. Papers 1967–8* (Cmnd. 3757) XLII. 229 The Security Council met at 6.30 p.m. (New York Time) on 21st August. After protracted debate, the Council agreed to the inscription of the item on Czechoslovakia on its agenda. **1977** *Whitaker's Almanack 1978* 806/2 The Security Council consists of fifteen Members, each of which has one representative and one vote. **1948** Security risk [see *clearance 5 c]. **1951** *Ann. Reg. 1950* 186 Mr. Acheson was..asked whether he himself, in view of his friendship for Hiss, could be considered a security risk. **1965** M. Spark *Mandelbaum Gate* vii. 216 He disapproved of letting young chaps into the Foreign Service who openly professed to have no religion at all. A security risk, Freddy felt decidedly. **1975** *Radio Times* 2–8 Aug. 43/3 He was accused of being a 'security risk' because of his early Communist associations.

sedan. Add: **1. c.** = *saloon 4 c. Chiefly *N. Amer.* (Not used in the U.K.)

1912 *Motor World* 14 Nov. 18/1 In the new [Studebaker] cars, there is another coupe, and a 'Sedan', both mounted on a new four-cylinder chassis. **1915** *Literary Digest* (N.Y.) 21 Aug. (cover advt.), A touring car when the windows are down... With the windows raised, a luxurious sedan. **1922** *Short Stories* Feb. 98/1 The sedan had been equipped with an exhaust foot warmer or heater. **1935** M. M. Atwater *Murder in Midsummer* i. 6 A black sedan was drawn up on the shoulder of the road. **1966** 'A. Hall' *9th Directive* xx. 184 It was a massive black Lincoln sedan, a seven-seater executive-style transport. **1977** *Time* 8 Aug. 23/1 The two were surrounded by four pistol-carrying men and ordered into a nearby Peugeot sedan.

3. *sedan car, model*; **sedan clock** *Hist.* = *sedan-chair clock* s.v. *Sedan chair c.

1931 M. de la Roche *Finch's Fortune* xxv. 325 A sedan car stopped before the door. **1957** *N.Z. Listener* 22 Nov. 4/4 New Zealand English has diverged from the English of England more than is generally realised owing to the influence of American usage. 'Sedan' car often appears in the advertisements, where an English advertisement would print 'saloon'. **1950** D. de Carle *Watchmakers' & Clockmakers' Encycl. Dict.* 129/1 *Sedan clock*, a small hanging clock usually associated with the period of the Sedan Chair. **1968** —— *Clocks & their Value* 93 The value of a sedan clock depends on the case but can be anything from about £15 to £35. **1948** *Herald-Press* (St. Joseph, Mich.) 14 Aug. 5/1 Besides making some substantial changes in its present sedan models it plans to put a hard top convertible into production.

sedanca (sĕdæ·nkă). Also **Sedanca.** [f. Sedan + the name of Count Carlos de *Salaman*)*ca*, Spanish nobleman and Rolls-Royce agent, by whom the word was apparently coined.] In full, *Sedanca de Ville.* A sumptuously appointed cabriolet de ville mounted on a Rolls-Royce chassis. Cf. (U.S.) *town car* s.v. *town *sb.* 10.

1926 *Motor* 26 Oct. 615/3 The other body shown on a Rolls-Royce chassis is styled a Sedanca limousine. It can also be used as a cabriolet, inasmuch as the Barker patent de ville extension over the front seat can be removed completely. **1929** *Motor* 2 July (Suppl.) p. xl/1, 1927 (October) 40–50 hp Phantom Rolls-Royce enclosed-drive Sedanca de Ville. **1937** *Times* 20 Oct. 20/1 The Rolls-Royce Company show a..Phantom III sedanca at £3,040. **1963** D. Scott-Moncrieff *Thoroughbred Motor Car 1930–40* ii. 121 (*caption*) Rolls-Royce 20/25 with French-built sedanca-de-ville coachwork, 1935. **1978** *Times* 8 June 28 (Advt.), This magnificent Brewster Salamanca is one of 2 mint Silver Ghosts entered, along with..Rolls, 220/25 Sedanca, Alvises, Lagondas, [etc.].

sedan chair. Add: **c. sedan-chair clock, watch** *Hist.*, a large travelling watch of a type supposed to have been hung in sedan chairs.

1904 F. J. Britten *Old Clocks & Watches* (ed. 2) iv. 244 During the eighteenth century watch movements having plain silver dials from three inches to four inches in diameter were fixed in circular frames of wood, polished and with a moulded edge. They were called 'Sedan Chair Watches', though I cannot aver that they were as a rule carried in those useful, but obsolete conveyances. **1951** E. Wenham *Old Clocks for Mod. Use* vii. 47 It is still possible to obtain one of these..portable timepieces generally referred to as 'Sedan chair clocks' or, to give them another earlier name, 'post-chaise clocks'. **1960** *House & Garden* Apr. 99/3 Sedan-chair clocks..reproductions and an antique.

sedate, *a.* Add: **1. e.** Of inanimate objects: not unduly striking in colour or design; quiet and restful in tone.

1924 A. D. Sedgwick *Little French Girl* i. vi. 48 Sedate chairs with backs and seats embroidered in green and dove-colour were ranged along the wall. **1978** J. Carroll *Mortal Friends* iii. v. 310 Brady stood in the bridge of the window, looking out on the sedate front lawn.

3. *Comb.*, as *sedate-looking* adj.

1925 T. Dreiser *Amer. Tragedy* I. ii. xxxvii. 411 So clean, modest and sedate-looking a girl. **1977** *Rolling Stone* 19 May 90/5 The sedate-looking trio sings R and B with fervor.

sedate, *v.* Restrict † *Obs.* to sense in Dict. and add: **b.** *Med.* To make (a patient) sleepy or quiet by means of drugs; to administer a sedative to.

1945 *Richmond* (Va.) *Times-Dispatch* 21 Sept. 1/1 Two capsules are ordinarily considered enough to sedate a person—that is, produce a tendency to sleep. **1961** *Amer. Speech* XXXVI. 145 The informal speech of physicians embodies a great many technical colloquialisms that may be called the argot of medicine... 'He was very apprehensive, so I sedated him heavily.' **1977** *Proc. R. Soc. Med.* LXX. 549/1 He was sedated, intubated and ventilated and full supportive therapy was given with further blood transfusions.

Hence **seda·ted** *ppl. a.*, under the influence of a sedative drug.

1953 R. Lehmann *Echoing Grove* 129 'How is she?'.. 'Expecting you. Be careful, won't you? She's still sedated—mildly, of course.' **1974** L. Deighton *Spy Story* xi. 107, I was half inclined to give the sedated Miss Shaw a miss. **1976** J. Philips *Backlash* (1977) iii. i. 125 Elliot wasn't going to come to. He was heavily sedated.

sedation. Add: **1. a.** (Further example.) Now esp. with reference to the use of sedative drugs.

1979 *Guardian* 19 Feb. 22/6 The dead youth's parents were under sedation yesterday at their home in Dan-y-Cribyn. **1982** J. Penn *Notice of Death* vii. 65 We may get more in the morning from Mrs H., but at the moment she's under sedation.

‖ **sedekah** (sedekā·). [Malay, f. Arab. *ṣadaqa.*] In Malaysia: alms; a voluntary offering.

1839 T. J. Newbold *Straits of Malacca* I. ii. 88 When a boy has gone through the Koran..his parents give *Sedekah*, or alms. **1900** W. W. Skeat *Malay Magic* vi. 403 In the case of a Sultan as many as possible bear a hand in sending him to the grave..partly for the sake of the *sĕdĕkah* or alms given to the bearers. **1972** A. Amin tr. *Shahnon Ahmad's No Harvest but Thorn* ix. 96 Lahuma need not worry any more, thought Jeha. The feast would be a *sedekah* to all.

sedentarization (se:dentăraizēɪ·ʃən). [f. Sedentar(y *a.* + -ize: see -ation.] The settlement of a nomadic people in a permanent homeland or place of habitation.

1960 F. Barth in *Problems of Arid Zone* (UNESCO) 342 The solution..favoured by the state authorities..has been the simple and radical one of sedentarization: the total elimination of nomads by settling them on the land as agriculturalists. **1969** *Times Lit. Suppl.* 21 Aug. 929/2 Both the sedentarization of the tribes and the redistribution of land were thought..politically desirable. **1979** *West Africa* 13 Aug. 1459/2 Nor, they concluded..would the pastoralists themselves regret such a social and economic transformation once they realised that sedentarisation was good for them.

sedentary, *a.* and *sb.* Add: **A.** *adj.* **5.** *Geol.* Of a soil or sediment: = *residual *a.* 2 f.

1870 S. W. Johnson *How Crops Feed* ii. iii. 143 Sedentary soils, or Soils in place, are those which have not been transported by geological agencies. **1906** [see *residual *a.* 2 f]. **1929** *Daily Tel.* 22 Jan. 4/7 The soil being considered 'sedentary' in character. **1943** Millar & Turk *Soil Sci.* i. 4 Since they have not suffered the mixing that accompanies transportation by ice and water there are many variations in the characteristics, both physical and chemical, of sedentary materials. **1975** Flegmann & George *Soils* iv. 103 This visual gradation is particularly obvious when a soil has been formed *in situ* by the gradual weathering of parent rock, such soils being referred to as sedentary.

6. *Comb.*, as *sedentary-looking* adj.

1937 W. B. Yeats *Vision* (rev. ed.) 37 Aherne..was stout and sedentary-looking.

‖ **Seder** (sēɪ·dəɹ). Also with small initial. [Heb. *sēdher* order, procedure.] A Jewish ritual service and ceremonial dinner held on the first evening of the Passover and repeated on the second evening by Orthodox Jews outside Israel. Freq. *attrib.*

1865 *Chambers's Encycl.* VII. 312/1 At a later period, a certain number of cups of red wine were superadded to this meal, to which, as its special ceremonies and the order of its benedictions were fixed, the name *Seder* (arrangement) was given. **1891** [see *haggadah 2]. **1892** I. Zangwill *Childr. Ghetto* II. xxii. 166 She would have to.. see the sour faces of her little ones round a barren *Seder* table. *Ibid.* xxv. 205 Seder Night was a charmed time. The strange symbolic dishes..the special Hebrew melodies. **1909** *Cent. Dict.* Suppl. s.v., The celebrant, generally the head of the family, begins with the first of the thirteen functions in the seder service. **1932** C. Roth *Hist. Marranos* vii. 185 The traditional *Seder*-service. **1958** C. P. Snow *Conscience of Rich* i. iii. 24 We were as good as engaged after seder night in '96. **1970** I. Sieff *Memoirs* vi. 95 Herbert Samuel..spoke the holy words, the traditional conclusion to the *Seder* service on the eve of the Passover: 'Next year in Jerusalem.' **1978** J. Sacks in P. Moore *Man, Woman, & Priesthood* iii. 39 The best known of these events is the *Seder* on the first nights of Passover. **1982** *Listener* 7 Jan. 13/1, I sat round the Seder table during Passover.

sedge, *sb.*¹ Add: **1. f.** The characteristic greenish- (or reddish-)brown shade of sedge.

1927 *Daily Express* 12 Mar. 3/5 Sedge, a bright shade similar to the always popular almond, but with a tendency towards jade. **1938** J. W. Day *Dog in Sport* iv. 66 By 1885 the present type had largely evolved, the main differences being that the breed then possessed one colour only, a dark brown shading into a reddish sedge.

5. a. *sedge-ground, -peat.* **b.** *sedge-choked, -embattled* adjs.

1942 W. Faulkner *Go down, Moses* 92 The old worn-out brier- and sedge-choked fields spreading away. **1934** E. Blunden *Mind's Eye* 138 We shall see..the gilt-leaved beechwood and the sedge-embattled lake. **1667** in *Rec. Town Plymouth* (Mass.) (1889) I. 95 All that pte of the pond or sedge ground which lyeth between a place there called the Gurnett and the bounds of Samuell Ryders land. **1910** C. Harris *Eve's Husband* 30 The poor brown sedge-ground of an old field. **1943** G. Erdtman *Introd. Pollen Analysis* i. 6 The comparatively low pine pollen frequency of a sedge-peat was considered to be due to the fact that the pine sheds its pollen at a time when the sedges have attained full growth. **1952** *Chambers's Jrnl.* Jan. 61/2 We propose to..fork in ample horticultural sedge-peat in the spring. **1977** R. Davies *Pract. Gardening Encycl.* ii. 21/1 Sedge (or fen) peats are the remains of reeds and sedges and are dark coloured and well decayed.

6. sedge-cock *dial.* = Missel-thrush.

1886 R. Holland *Gloss. Words County Chester* 305 Sedcock,..the missel thrush... Sedgecock. **1955** E. Pound *Classic Anthol.* i. 75 June's green hopper moves a thigh, 'Sedge-cock' wings it in July. **1965** *Jrnl. Lancs. Dial. Soc.* Jan. 9 Mistle thrush... Sedgecock, Setcock, Sedcock: Oldham; nr. Stockport.

sediment, *sb.* Add: **4. sediment ring** *Astr.*, a ring of rock masses orbiting a planet, regarded as debris from the time of its formation.

1955 *Sci. News Let.* 22 Jan. 53/1 Dr. Kuiper said that the moon, as it sped away from the earth, plowed through a 'sediment ring', a swarm of small satellites moving around the earth. **1970** [see *planetismal *a.* and *sb.*].

sediment, *v.* Delete *rare* and add: **1.** (Later examples.)

1908 *Chambers's Jrnl.* May 396/1 Chemical precipitation was found essential to coagulate the suspended matter and thus enable the greater proportion of it to be sedimented in subsidence basins. **1976** *Nature* 19 Aug. 662/1 We then sedimented the eggs rapidly..in a hand centrifuge.

2. *intr.* **a.** To settle as sediment.

1927 *Brit. Jrnl. Exper. Path.* VIII. 122 In a typical rough culture of *enteritidis*..the bacteria rapidly sediment to the bottom. **1961** *Lancet* 5 Aug. 322/1 The erythrocytes being allowed to sediment within the syringe. **1971** *Nature* 25 June 527/2 Each preparation sedimented in the analytical ultracentrifuge as a single component with a sedimentation coefficient..of about 9·5S.

b. Of a liquid: to deposit a sediment.

1934 in Webster. **1962** Luntz & Wright in A. Pirie *Lens Metabolism Rel. Cataract* 319 Blood was collected..in a mixture of 1% sodium ethylenediamine tetra-acetic acid and 5% dextran..and allowed to sediment. **1978** *Nature* 10 Aug. 611/1 (*caption*) Heparinated blood was allowed to sediment at room temperature to separate red cells from plasma.

sedimented *ppl. a.* (*fig.* example); **sedimenting** *ppl. a.* (later example).

1962 H. Bloemendal et al. in A. Pirie *Lens Metabolism Rel. Cataract* 303 More rapidly sedimenting material is observed, but the shape of the corresponding boundary does not allow calculation of the sedimentation coefficient. **1977** D. L. Altheide in D. E. Johnson *Existential Sociol.* iv. 149 These tasks become taken for granted as sedimented knowledge for the members.

sedimentable (se·dimentăb'l), *a.* [f. prec. + -able.] That may be deposited or obtained as sediment. Hence **se:dimentabi·lity.**

1943 *Jrnl. Gen. Physiol.* XXVI. 352 As an elemental volume within the ascending layer is partially or completely cleared of sedimentable material, it has to move only a relatively short distance..before it reaches a region of comparable concentration. **1971** *Jrnl. Insect Physiol.* XVII. 865 Additions of 0·1% Triton..increase..enzymatic activity..without markedly altering the sedimentability of the enzyme. **1978** *Nature* 2 Mar. 55/2 After storing for 1 week at room temperature, at least 50% of the antimony was retained within sedimentable liposomes, the remainder having escaped into solution.

sedimentation. Add: **2.** *Comb.*, as **sedimentation coefficient, constant,** a measure of the size of a microscopic particle, equal to the terminal outward velocity of the particle when centrifuged in a fluid medium divided by the centrifugal force acting on it; (expressed in units of time: cf. *Svedberg); **sedimentation rate,** the rate of descent of particles suspended in a fluid; *spec.* in *Med.*, of the red cells in drawn blood; **sedimentation tank,** a tank in which sewage is allowed to stand so that the solid matter in suspension may have time to settle.

1962 H. Bloemendal et al. in A. Pirie *Lens Metabolism Rel. Cataract* 300, α-Crystallin, prepared by vertical starch block electrophoresis, has a sedimentation coefficient of 19 S (molecular weight 810,000). **1978** *Nature* 12 Jan. 170/2 Nervous tissue and muscle in rat and chicken contain several molecular forms of acetylcholinesterase.., distinguishable by their sedimentation coefficient in sucrose gradient. **1929** Svedberg & Katsurai in *Jrnl. Amer. Chem. Soc.* LI. 3577 The term 'Sedimentation Constant' has been adopted as a simplified means of expressing the old term 'Specific Sedimentation Velocity'. **1966** B. Pollara et al. in R. T. Smith et al. *Phylogeny of Immunity* ix. 94/2 The antibody produced by this animal is different from that of the other lower vertebrates, having a sedimentation constant of approximately 9S. **1946** *Nature* 30 Nov. 794/2 In accordance with Stokes's law, sedimentation-rates of such red-cell aggregates will be greater than those of single non-polarized blood cells. **1978** *Ibid.* 12 Oct. 532/1 To predict long-range trends in marsh stability, accurate measurements are needed of both subsidence and sedimentation rates. **1920** *Glasgow Herald* 3 Nov. 13 The sewage passes into a sedimentation tank designed so as to bring down in the form of sludge as much of the suspended solids as possible. **1966** *McGraw-Hill Encycl. Sci. & Technol.* XII. 221/2 In some [sewage] treatment plants screenings are passed through a grinder and returned to the flow so that they will settle out in the sedimentation tank.

sedimentology (se:dimĕntǫ·lŏdʒi). [f. Sediment *sb.* + -ology.] The branch of geology which deals with the nature and properties of sediments and sedimentary rocks. So **se:dimentolo·gical** *a.*, of or pertaining to sedimentology; **se:dimentolo·gically** *adv.*, by means of sedimentology; from the point of view of sedimentology; **se:dimento·logist,** an expert in sedimentology.

1932 H. Wadell in *Science* 1 Jan. 20/2 Sedimentology is here suggested as a term for the subject taught, retaining sedimentation for the act or process of deposition. The new term and its derivatives sedimentologist, sedimentologic and sedimentological will tend towards clearness. **1957** *Jrnl. Geol.* LXV. 485/1 It might be possible to subdivide further the material in this core from a sedimentological point of view into two components: (1) material of sand size..., and (2) lutite. **1958** *Times* 22 Mar. 8/3 Mr R. J. L. Allen..has been appointed George Martin Lees Research Fellow in Sedimentology. **1961** *Jrnl. Sedimentary Petrol.* XXXI. 207/1 The expression of kurtosis has not been used extensively by sedimentologists. **1966** *Palaeogeogr., Palaeoclimatol., Palaeoecol.* II. 113 The ironstones..mark, sedimentologically, a complete reversal of the trend towards increasing oxygenation. **1967** *New Scientist* 20 July 146/1 A number of grants for further research into sedimentology have been made to the University of Reading. **1970** *Nature* 2 May 425 Sedimentological studies of depositional environments of the modern Omo River delta and floodplain. **1973** *Ibid.* 8 June 342/1 Their [*sc.* marine grasses'] former presence can nevertheless be deduced by sedimentology and palaeontology. **1977** A. Hallam *Planet Earth* 245/1 The paleoecologist has to be paleobiologist, sedimentologist and biologist.

seditty, var. *saditty *a.*

Sedobrol (se·dobrǫl). A proprietary name for a medicinal preparation.

1913 *Trade Marks Jrnl.* 22 Jan. 105 *Sedobrol*... Chemical substances prepared for use in medicine and pharmacy. John Henry Land.., 10, Market Place, Coalville, Leicestershire; Chemist. **1921** R. Macaulay *Dangerous Ages* v. 102 There have been wonderful cures for insomnia lately... Which new thing? Sedobrol? Paraldehyd? **1934** E. Waugh *Handful of Dust* iii. 129 They had boiled water in an electric kettle and were drinking Sedobrol together. **1939–40** *Army & Navy Stores Catal.* 406/3 Sedobrol Tablets 10's 2/–, 30's 5/–.

sedoheptulose (sĭdohe·ptiŭlōᵘz, -s). *Chem.* [f. Sed(um + -o + hept)ose s.v. *hepta- + *-ulose².] A heptose that is found in the leaves of certain plants of the genus *Sedum* (notably *S. spectabile*), a phosphate of which is involved as an intermediate in carbohydrate metabolism in animals. Formerly called † **sedoheptose.**

1917 La Forge & Hudson in *Jrnl. Biol. Chem.* XXX. 68 Since the analyses of the three crystalline derivatives that have been described show conclusively that the sugar of *Sedum spectabile* contains seven carbon atoms, it will be named, with reference to its origin, sedoheptose. **1938** *Thorpe's Dict. Appl. Chem.* (ed. 4) II. 286/2 The natural ketoheptoses are *d*-mannoketoheptose..; perseulose.., and sedoheptose obtained from *Sedum spectabile.* **1939** *Jrnl. Amer. Chem. Soc.* LXI. 343/2 In the presence of mineral acids, sedoheptulose is transformed to a crystalline anhydride. **1959** [see *RIBULOSE]. **1970** R. W. McGilvery *Biochem.* xvi. 325 The only known function for the 7-carbon sedoheptulose-7-phosphate in animals is its participation as an intermediate in the pentose phosphate pathway.

Sedormid (sĭdǭ·ɪmid). *Pharm.* A proprietary term for *N*-(2-isopropylpent-4-enoyl)urea, $C_9H_{16}N_2O_2$, a white crystalline solid employed as a sedative and hypnotic.

1928 *Trade Marks Jrnl.* 9 May 736/2 *Sedormid*... The Hoffman-La Roche Chemical Works Limited..London. **1929** *Official Gaz.* (U.S. Patent Office) 14 May 299/2 *Sedormid.* For sedative and hypnotic. Claims use since Mar. 14, 1929. **1934** *Lancet* 21 Apr. 845/1 Two..cases of thrombopenic purpura were due to prolonged administration of the new hypnotic Sedormid (allylisopropyl-acetylurea). **1964** M. Hynes *Med. Bacteriol.* (ed. 8) vii. 79 In some patients the hypnotic Sedormid combines with the platelets and thrombocytopenic purpura follows. **1974** M. C. Gerald *Pharmacol.* iv. 80 Compare the ease in saying and remembering..allylisopropylacetylcarbamide vs. Sedormid.

Sedra (se·drǎ). Also **Sedrah, Sidra(h,** and with small initial. [Aram. (via Yiddish *sedre*): cf. *Seder.] In Jewish sabbatical liturgy, one of the fifty-four sections of the Pentateuch read in the Synagogue at the Sabbath morning service. Cf. Parashah.

Quot. 1909 does not accord with modern usage.

1907 I. Zangwill *Ghetto Comedies* 128 We are reading the *Sedrah* (weekly portion) about Joseph. **1909** *Cent. Dict.* Suppl., The Pentateuch is divided into fifty-four sedras or sections, which are subdivided into parashoth. The sedras must be read at the morning Sabbath services during the year, but as a year has only 52 weeks, in order to finish the sedras two of them are read on two special Sabbaths. **1962** *New Jewish Encycl.* 370/2 In common usage, the word 'Parashah' also refers to the entire *Sidrah* (or Sedrah), or to any of the various sections or paragraphs in the scroll of the Torah. The Torah is subdivided into *Sidrot* (sections), the reading of which is completed in a one year cycle; and each Sidrah is further subdivided into Parashiyot (sing. Parashah). **1973** *Jewish Chron.* 19 Jan. 18/2 Nehama Leibowitz's studies in the weekly *sidra* have justifiably become renowned as a key for the unlocking of the treasures found among the Jewish Biblical exegetes.

seducee. (Later example.)

1962 Gregor & Nicholas *Moral & Story* iv. 120 The 'innocent' seducee dies a depraved alcoholic, while her seducer is redeemed.

see, *sb.*[1] Add: **3.** (sense 2) *see-city.*

1937 *Daily Tel.* 28 Aug. 13/3 On this day, very nearly 1,500 years ago, Augustine lay dying in his see-city of Hippo. **1956** D. E. W. Harrison in D. L. Linton *Sheffield* 199 Meanwhile Sheffield became in 1914 the see-city of a diocese stretching to Goole on the Ouse.

see, *sb.*[2] Restrict *rare,* etc., to sense b in Dict. and add: **a.** (Later examples.) Used as a colloq. replacement for Look *sb.* (cf. also *look-see).

1927 J. Masefield *Midnight Folk* 251 The son took out a pocket telescope..and handed it to his father. 'Take a see for yourself, pop.' **1938** E. Williams *Corn is Green* I. i. 10 Ought to be 'ere by now, I'll 'ave a see.

see, *v.* Add: **A. 3. a.** *β.* (Earlier and U.S. examples of *seed.*)

1746 G. Murray *Let.* 17 Apr. in C. Petrie *Jacobite Movement: Last Phase* (1950) v. 113, I never seed him in time of action, neither at Gledsmoor, Falkirk, nor this last. **1777** P. Thicknesse *Year's Journey* II. xlix. 134 An English servant..told me..that he *seed* her very plain. **1833** J. Neal *Down-Easters* I. i. 14 Never seed sich a fellow since I breathed the breath of life. **1893** H. A. Shands *Some Peculiarities of Speech in Mississippi* 55 *Seed*, Negro and illiterate white for *saw.* **1945** in B. A. Botkin *Lay my Burden Down* 18 He was the ugliest man I ever seed.

d. *colloq.* and *dial.* **seen** (chiefly *1st pers. sing.*: see also *Eng. Dial. Dict.*).

1796 *Aurora* (Philadelphia) 30 Sept. 3/3 So fine a sight (says Yankee to his friend) I swear I never seen—you may depend. **1850** *Knickerbocker* July 87 We spoke of Major Andre. 'Oh,' said the old lady, 'I seen him more'n fifty times.' **1861** T. Hughes *Tom Brown at Oxf.* II. vii. 114 'Hev'ee seed aught o' my bees?'.. 'E'es, I seen 'em.' *c* **1915** in N. I. White *Amer. Negro Folk-Songs* (1928) iii. 146, I seen King Pharoe's daughter seeking Moses on the water. **1976** *Alyn & Deeside Observer* 10 Dec. 12/6 Richardson told Detective Constable Mahoney: 'I seen this Irish bloke and followed him to the Hawarden Castle.'

8. *γ.* (U.S. examples.)

1845 *Great Kalamazoo Hunt* in *Big Bear of Arkansas* 49 Well, after I had looked out for about fifteen minutes or so, and seed the boss begin to get desperately frightened, [etc.]. **1938** M. K. Rawlings *Yearling* xvi. 188 Seems to me I've seed it before.

δ. 9— *U.S. colloq.* and *dial.* **saw.**

1867 C. F. Browne *A. Ward in London* II. vii. 123 We have saw a entertainment as we never saw before. **1941** J. Faulkner *Men Working* ii. 33 How-some-ever, I've saw them.

B. I. 1. a. *Proverb.* (Later examples.)

1560 T. Becon *Catechism* in *Works* I. sig. Bbb2, This also must honest maids provide, that they be not full of tongue... A maid should be seen, and not heard. **1773** R. Graves *Spiritual Quix.* I. III. xviii. 179 It is a vulgar maxim, 'that a pretty woman should rather be seen than heard'. **1858** Geo. Eliot *Janet's Repentance* viii, in *Sc. Cler. Life* II. 167 Little gells must be seen and not heard. **1908** L. M. Montgomery *Anne of Green Gables* ii. 22 It's such a relief to talk..and not be told that children should be seen and not heard. **1980** L. Lewis *Private Life of Country House* v. 63 Two or three children..supposed to be seen and not heard and not to speak unless spoken to.

d. Also in phr. *to see things,* to suffer hallucinations or false imaginings. (Usu. *colloq.* as pres. pple.)

1922 M. A. von Arnim *Enchanted April* iii. 48 Mrs. Fisher..had no wish to find herself shut up..with somebody who saw things... It would be disagreeable..if Mrs. Wilkins were suddenly to assert that she saw Mr. Fisher. Mr. Fisher was dead; let him remain so. **1928** Kipling *Woman in his Life* in *Limits & Renewals* (1932) 47 After a drink or two..he told the tale of a friend who 'saw things'. **1935** A. Christie *Three Act Tragedy* III. vii. 180 Says I imagined it. Says I was 'seeing things'. **1953** B. Glemser *Dove on his Shoulder* ii. 13 'I must be seeing things', the major said... 'You alcoholic bastard.' **1977** 'D Rutherford' *Return Load* ii. 31 Was I seeing things or was that Sally driving your truck?

i. *to see* (a person) *coming,* to make out one who can be fooled or deceived. Also in proverbial phrase *when you've seen one, you've seen them all*; a conjuror's phrase *now you see him, now you don't,* and varr. Also, *to see red*: see *red *sb.*[1] 1 f.

c **1811** Blake *Public Address* in *Writings* (1978) II. 1046 When you have seen one of their Pictures you have seen all. **1869** 'Mark Twain' *Innocents Abr.* xxiii. 177 To me it seemed that when I had seen one of these martyrs I had seen them all. **1931** T. R. G. Lyell *Slang, Phrase & Idiom* 671 D'you mean to say you paid £100 for that car? My dear fellow, they must have seen you coming! As scrap iron, it's worth perhaps £10—not a penny more!! **1949** G. Davenport *Family Fortunes* II. iv. 145 'If you've seen one you've seen them all,' said Sam. **1967** T. Stoppard *Rosencrantz & Guildenstern are Dead* II. 62 It's just a man failing to reappear, that's all—now you see him, now you don't. **1973** *Illustr. London News* May 100/4, I know many people who feel that once you've seen one Jancso film, you've seen them all. **1980** G. M. Fraser *Mr American* xix. 370 If the American..had subsequently proved to be of moderate means, would he have been quite so welcome?.. Old Man Clayton had seen him coming. **1980** P. G. Winslow *Counsellor Heart* xx. 221 The only way is to..have them think it's something else. Now you see it, now you don't.

j. *transf.* or *fig.* of radar equipment, cameras, artificial satellites, etc.

1923 E. W. Marchant *Radio Telegr. & Teleph.* iv. 36 The method that was described by Hertz for detecting or 'seeing' radio waves was to use a spark gap in a circuit which was tuned to the frequency of the waves. **1945** *Rev. Sci. Instruments* XVI. 46/1 The photo-tube camera is mounted beneath the photofluorograph hood and 'sees' the object image on the screen which is 'seen' by the photographic camera at the apex of the hood. **1952** E. Larsen *Radar works like This* 9 Thus the picture of what the waves 'see' is built up from glowing blobs on the screen. **1957** T. Adler *Seeing Earth from Space* v. 126 The earth satellite Vanguard II..looks down on the earth from a much greater height than 86 miles. So it can see more of the earth than the camera that took this picture did. **1975** D. G. Fink *Electronics Engineers' Handbk.* xxv. 4 When connected to the antenna, the receiver sees a low-noise background of empty space, modified by surrounding terrain or sea surfaces and atmosphere, [etc.].

2. a. *Proverb.* (Later examples.)

1718 Mrs. Centlivre *Bold Stroke for Wife* II. 31, I am sorry such a well-invented Tale should do you no more Service. We old Fellows can see as far into a Mill-stone, as him that picks it. **1862** H. Kingsley *Ravenshoe* II. ix. 80 He could see through a brick wall as well as most men. **1885** C. M. Yonge *Two Sides of Shield* II. i. 16 He should defer his letter till he had..talked to his sister Jane, who could see through a milestone any day. **1920** 'Sapper' *Bull-Dog Drummond* v. 127 He could see farther into a brick wall than most of the people who called him a fool. **1978** A. Price *'44 Vintage* xviii. 204 I had a grandma could see clear through me and a brick wall both, so it's no surprise you can figure us.

3. a. Also, to foresee or forecast (an event, trend, etc.) (later examples); *U.S.*, to understand (a person). Also, *to see* (something) *coming*: to foresee or anticipate.

1872 E. Eggleston *End of World* xxiii. 158 '[I] see yer,' said Bill, trying in vain to draw his coat. **1873** J. H. Beadle *Undevel. West* xx. 369 'Marshal's got a good thing, though.' I see you; best place to make money in the United States. **1946** G. B. Shaw *Geneva* Pref. 7 Historians and newspaper editors can see revolutions three centuries off but not three years off. **1966** M. Woodhouse *Tree Frog* xxii. 161 'Let me guess... I know radar and guidance.'.. I nodded. I could see it coming a mile off. **1971** *Times* 16 June 21/5 Boost in gas reserves seen... The Soviet Union will expand production of oil.. and of natural gas..over 'the coming years'. **1974** 'E. Lathen' *Sweet & Low* xv. 149 Thatcher sympathized with him. This was one he had not seen coming, either.

f. Also *colloq.* with omission of the second person subj. pronoun, appended parenthetically to a statement, freq. implying refusal to tolerate dissent, or as a mere filler; also standing alone (= 'do you see?') as an interrogative, with similar force.

1952 J. Bingham *My Name is Michael Sibley* xv. 183 You and me have got to understand each other right, see? **1959** N. Mailer *Advts. for Myself* (1961) 39 Listen, bud, you ain't talkin' to Joe Crap, see; you watch what you say with me. **1968** *Listener* 19 Dec. 810/3, I believe in having a go, see, so long as there's some fun in it, see. **1976** T. Sharpe *Wilt* v. 45 There was this student all dressed up like a waiter see.

g. (Later examples.)

1892 Kipling in *Sun* (N.Y.) 28 Aug. II. 6/2 Each, in his separate star, Shall draw the Thing as he sees It for the God of Things as they are! **1934** E. O'Neill *Days without End* III. ii. 113 He sees it clearly as a throwback to boyhood experiences. **1976** M. Machlin *Pipeline* xi. 135 Dad's idea of an oil man is a financier in a starched collar... I see it more like the way you did things Gramps.

h. To perceive (good or attractive qualities) *in* a person or thing, often in an interrogative clause; to perceive (a certain characteristic or type) *in* a person or thing.

1832 Scott *Ct. Robert* in *Tales of my Landlord* 4th Ser. I. iv. 113 Hereward, though flattered by the unusual degree of attention which the Princess bestowed upon him, saw in her only the daughter of his Emperor. **1835** Browning *Paracelsus* III. 419 A professorship At Basil! Since you see so much in it. **1863** 'Ouida' *Held in Bondage* I. viii. 193 What could De Vigne possibly see in that woman? **1864** J. Bryce *Holy Rom. Emp.* 62 He put to death the rebel Crescentius, in whom modern enthusiasm has seen a patriotic republican. **1916** 'Taffrail' *Pincher Martin* vii. 114 Can't think what he sees in her. **1927** A. P. Herbert *Plain Jane* 95 I'm not a jealous woman, But I *can't* see what he sees in her. **1971** P. O'Donnell *Impossible Virgin* v. 107 She said quietly, 'Don't ask me what I see in him, please... Don't make judgments.'

i. To recognize the rightness or desirability of (an idea or thing); to give credence to, believe, accept; to consent to (a proposal). Usu. with *it. colloq.* (orig. *U.S.*).

1850 *California Courier* (San Francisco) 14 Nov. 2/2 This may be all right—but if it is, we cannot see it. **1860** R. Nicholson *Autobiogr.* 67 'Get up, my man, and let us go on,' said the stranger, almost throttling Cracroft. That worthy gentleman, however, 'could not see it', as we now say in modern slang. With a struggle he stammered that he had lost the wager. **1864** Hotten *Slang Dict.* 223 In street parlance, 'to see' is to know or believe; 'I don't see that,' i.e., 'I don't put faith in what you offer, or I know what you say to be untrue.' **1877** H. Ruede *Sod-House Days* (1937) 8 The hack driver wanted us to go with him to Osborne, but the fare was $3.50 (trunks extra) and we 'could not see it'. **1890** Kipling in *United Service Mag.* June 236, I said..'I don't keep a canteen up my sleeve.' They couldn't see it. **1934** G. B. Shaw *Too True to be Good* II. 60 But the old man never could be brought to see it. He said the proper profession for me was the bar. **1945** J. L. Marshall *Santa Fe* vii. 98 Fred then tried to interest the Burlington in his idea... But the Burlington couldn't see it. **1971** 'E. Lathen' *Longer the Thread* (1972) vi. 65, I know that's what it looks like... But, for the life of me, I can't see it.

j. Usu. in negative or interrogative context with personal obj. (esp. *refl.*) and compl. (pple. or phrase): to perceive in one's mind's eye; to envisage as possible or acceptable.

1875 L. Troubridge *Jrnl.* 2 June in J. Hope-Nicholson *Life amongst Troubridges* (1966) x. 117 My dreadful yellow that I don't see myself wearing at all. **1915** R. Brooke *Let.* 26 Jan. (1968) 657, I don't 'see' Viola [Tree] as the Lithuanian. **1926** Chesterton in W. R. Titterton *G. K. Chesterton* (1936) II. vii. 169, I do not quite see myself as the President of the League of Little People. **1955** R. Bannister *First Four Minutes* 16, I could not see myself in the winning place. **1962** M. Trevor *Newman* 441 He stuck to his opinion that Bayswater was not the place for them; he could not see Faber going there. **1976** M. Machlin *Pipeline* iv. 45, I can't see dying because of your feelings about conservation.

4. d. *intr.* To read music. *colloq.*

1955 L. Feather *Encycl. Jazz* 347 *See*, read (music). 'He doesn't see too well' refers to a performer who reads music slowly. **1970** C. Major *Dict. Afro-Amer. Slang* 101 *See*,..to read music.

5. a. Also *to see and* (*to*) *be seen* (later examples); hence *see-and-be-seen* attrib. phr.

1738 Swift *Polite Conv.* 41 Her Ladyship went to see, and to be seen. **1828** W. Scott *Jrnl.* 3 May (1972) 468 After the dinner I went to Mrs. Scott of Harden to see and be seen by her nieces. *a* **1911** W. S. Gilbert *Lost Bab Ballads* (1932) 31 To see and be seen is for what we pay At Islington on the half-crown day. **1960** *Times* 3 June 6/5 London audiences to which the social see-and-be-seen set attaches itself. **1961** *Economist* 25 Nov. 770/1 This mixing of 'blind' traffic with see-and-be-seen aircraft is particularly dangerous in overcrowded terminal areas.

f. Also *see here,* a brusque form of address used to preface an order, expostulation, reprimand, etc. Cf. *look here* s.v. Look *v.* 4 a.

1898 G. B. Shaw *Mrs. Warren's Profession* II. 185 Now see here, George: what are you up to about that girl? **1925** F. Scott Fitzgerald *Great Gatsby* vii. 152 'Now see here, Tom,' said Daisy, turning around from the mirror. **1941** J. D. Carr *Case of Constant Suicides* ii. 29 'See here,' pursued Alan... 'Let us get this straight.' **1974** G. Jenkins *Bridge of Magpies* ix. 148 'See here,' I said. 'There's been another death. I want you to signal the fisheries frigate.'

6. b. (Later examples in sense 'to ascertain by consideration'. Sometimes used as a formula for not giving a direct answer on the spot.)

1851 I. Spencer *Let.* 11 Jan. in U. Young *Life I. Spencer* (1933) III. iii. 181 About going to France, we shall see. **1898** G. B. Shaw *Arms & Man* III. 62 We shall see. And you shall wait my pleasure. **1925** F. Scott Fitzgerald *Great Gatsby* vii. 137 What he really said was: 'Yes... Yes... I'll see.' **1959** R. Matthews tr. *J. Steinmann's Saint Jerome* I. xi. 49 He would tell his friend about it, and later they would see.

7. b. *to see* (a person) *home* (earlier example); *to see* (a person) (*all*) *right*: to ensure (his) well-being or safety; *to see* (a person) *over, through*: of a thing, to be sufficient for (his) needs; also with prep. *over* or *through* (a period of time or difficulty).

1819 Keats *Let.* 16 Apr. (1958) II. 92 Do you..get groggy..so as to be obliged to be seen home with a Lantern. **1894** 'Mark Twain' in *St. Nicholas* Mar. 393/1 Thirty camel-loads of treasures was enough to see a dervish through, because they live very simple. *a* **1914** 'Saki' *Beasts & Super-Beasts* (1914) 217 If you'll lend me three pounds that ought to see me through comfortably. **1959** *Times* 19 Mar. 5/5 He said he would see me all right if I said I saw the two police strike the boy. **1965** *Listener* 25 Nov. 865/1 Although Louis MacNeice was a fluent and sometimes facile poet, his sense of fact generally saw him through. *a* **1966** 'M. na Gopaleen' *Best of Myles* (1968) 87 To be saddled with the task of 'seeing' an inebriated friend 'right'. **1966** M. Steen *Looking Glass* iii. 52 He.. wrote me a cheque for twenty pounds—'to see me over'. **1971** *N.Z. Listener* 22 Mar. 13/1 Tell yer, I'll see you right at a boardin' place until you get jacked up. **1974** S. B. Hough *Fear Fortune, Father* i. 15, I could remember Lawson saying to me, 'I'll see you all right.' **1976** M. Butterworth *Remains to be Seen* vi. 89, I stopped the milk till Monday... But if you want a couple of pints to see you over the weekend.

10. a. Also, to experience (a specific age in life): usu. in negative context; *to have seen better days* (examples of things); *to have seen everything, it all*: to have experienced all the possible events and situations of life (often used as an expression of resignation or boredom); *to see the New* (*Old*) *Year in* (*out*): see *year 7.

1806 To have seen better days [see Day *sb.* 13 a]. **1899** H. James *Awkward Age* I. i. 3 He had..doubled the Cape of the years—he would never again see fifty-five. **1925** F. Scott Fitzgerald *Great Gatsby* i. 21 I've been everywhere and seen everything and done everything. **1930** W. S. Maugham *Cakes & Ale* v. 72 'But she's not as old as you are,' I said... 'She'll never see thirty again.' **1941** F. Thompson *Over to Candleford* x. 144 Laura [wore] a green smock which had seen better days. **1957** 'M. M. Kaye' *Shadow of Moon* xiv. 216 'I escorted her out from England.' '*What!*.. Now I have seen everything!' **1959** N. Mailer *Advts. for Myself* (1961) 209 Pot gave me a sense of something new about the time I was convinced I had seen it all. **1973** G. Greene *Honorary Consul* II. iii. 80 She's not twenty, and, you know, I won't see sixty again. **1973** 'E. McBain' *Hail to Chief* i. 6 Men..with..eyes that had seen it all, seen it all: Monoghan and Monroe from Homicide. **1977** G. Tindall *Fields Beneath* vi. 90 The workhouse itself was a 'handsome brick edifice' that had seen better days. **1978** T. Willis *Buckingham Palace Connection* viii. 161 'A boat race,' said Tremayne... 'Now I've seen everything!' said Story.

12. a. Also, *to see a man* (*about a dog*) and varr.: a joc. form of excuse or explanation used to avoid giving the real reason for one's absence or departure; *spec.* (*euphem.*) to obtain an alcoholic drink; to go to the lavatory. orig. *U.S.*

1867 *Ball Players' Chron.* 12 Sept. 3/1 The rest of our nine having gone to see a man there was nobody to take the bat. **1890** Barrère & Leland *Dict. Slang* II. 216 *To see a man* (American), to go and have a drink at the bar. **1927** *Amer. Speech* III. 221 *See a man about a dog*, to go out and buy liquor. **1931** T. R. G. Lyell *Slang, Phrase & Idiom* 670 Excuse me a moment,—I shan't be long; I just have to go and see a man about something! **1945** *Richmond* (Va.) *Times-Dispatch* 25 Oct. 14/5 Greet the homecoming hero with a load of this and he will immediately find that he has to go somewhere else and see a man about a dog. **1969** *Private Eye* 28 Mar. 14, I got to see a man about a dog! **1977** A. C. H. Smith *Jericho Gun* v. 63 I've got to dash. Must see a man about a horse.

b. (Earlier example of U.S. colloq. sense 'to interview in order to bribe'.)

1867 *Ball Players' Chron.* 12 Dec. 4/2 This, that or the other 'professional' is 'seen'—that is the professional term for the act of bribery—and lo and behold! the second game between the rival clubs is marked by a signal defeat.

e. *see you*: colloq. formula of farewell, often in weakened sense without reference to an anticipated meeting (in full *I'll see you*). Also with advbs. and other extensions, as *around, soon,* etc. Also, (*I'll*) *be seeing you.* Cf. F. *au revoir*, G. *auf Wiedersehen.*

1891 S. Weyman *New Rector* II. i. 25 He waved an awkward farewell to Jack, muttered 'See you soon!' and went off. **1906** 'O. Henry' in *McClure's Mag.* Aug. 392/1 Now lift your hat and come away, while you receive Lou's cheery 'See you again'. **1932** J. W. Harris *Days of Endeavour* xiii. 228 The boys..follow it with no more than a cheery, 'So-long, old son; see you in Liverpool!' **1937** D. & H. Teilhet *Feather Cloak Murders* ii. 33 He waved cheerfully to the Baron, said, 'I'll be seeing you.' **1945**, etc. [see *hooray *int.*]. **1951** M. Kennedy *Lucy Carmichael* v. i. 239 'Well..be seeing you.'.. 'Be seeing you,' agreed Owen without enthusiasm. **1959** I. Fleming *Goldfinger* xix. 264 'See you around.' He grinned at Bond and moved off down the room. **1962** L. Deighton *Ipcress File* xi. 71 Thanks, chief. See you. **1970** J. Porter *Rather Common Sort of Crime* ii. 24 Well, ta ever so! Be seeing you! **1975** I. McEwan *First Love, Last Rites* 96 'See you tomorrow, then.'.. 'Yes, tomorrow.' **1978** J. Irving *World according to Garp* xiii. 253 'See ya,' she called, and drove off... 'See ya,' Garp mumbled after her.

13. a. (Earlier example in *Poker.*)

1880 [see *bet *sb.* 1 b].

15. c. (Later example.)

1921 G. B. Shaw *Back to Methuselah* II. 83 That would be—let me see—five times three hundred and sixty-five is—um.

II. Phraseological combinations.

17. see about—. (Earlier example.)

1839 Dickens *Let.* Feb. (1965) I. 510 Will you dine with us at 5 — and see about a box without loss of time?

22. see over—. (Earlier and later examples with reference to buildings.)

1793 J. Woodforde *Diary* 23 June (1929) IV. 36 We took a walk to Weston House and saw it all over. *c* **1869** Taylor & Dubourg in M. R. Booth *Eng. Plays of 19th Cent.* (1973) III. 245 Mr. Secker's card for a friend of his to see over the ruins. **1909** M. Beerbohm *Yet Again* (ed. 5) 3 If I were 'seeing over' a house. **1920** 'O. Douglas' *Penny Plain* v. 62, I was going to ask if I might see over the house. **1977** 'M. Yorke' *Cost of Silence* i. 9 The house ..was..up for sale... People saw over it and were dashed by the need to replumb and repair.

26*. a. see off. To put to flight, chase off (esp. of a dog). Also *trans.* and *fig.*, to get the better of, defeat, put down. *colloq.* (orig. *Mil.*). Cf. sense 7 b in Dict.

1915 H. Bruckshaw *Diary* Aug. in *Times* (1976) 7 Feb. 12/6 We had at last cleared the place except for sundry stragglers who would no doubt be seen off later. **1919** W. Lang *Sea Lawyer's Log* xi. 137 You may 'see off' a messmate by overwhelming him with violence, outpointing him in cunning or overcoming him with policy. **1929** *Times* 21 Feb. 11 When he and another detective went to arrest the men Hughes called to the Alsatian. 'See 'em off.' **1944** R. P. Fleming *Jrnl.* in D. Hart-Davis *Peter Fleming* (1974) xii. 293 An unusually well-found fighting patrol..perfectly capable of seeing-off the small parties from L. of C. **1961** *Sunday Express* 10 Dec. 5/2 (*caption*) Fast as I tell him to 'See 'em off' they tell him to 'Sit'. **1969** Y. Carter *Mr. Campion's Farthing* ix. 86, I know an audience of stuffed shirts when I see one. Besides—I just had to see that pompous bastard off. **1981** 'M. Yorke' *Hand of Death* x. 90 He'd..been rebuffed... She'd seen him off good and proper.

b. *Cricket.* *to see off the new ball*: to bat until the shine has been removed from the ball (esp. at the start of an innings).

1969 J. Arlott *Cricket: Great All-Rounders* vii. 108 Bailey was the intractable substance which..made a good innings better by seeing off a new ball down the order to give the tail-enders a chance. **1977** *Observer* 20 Mar. 1/4 The openers had batted for a while and had seen a lot of the new ball off.

27. see out. c. (Earlier example.)

1756 *Connoisseur* II. 555 Tom Buck..can *see out* the stoutest freeholder in England.

d. (Earlier examples.)

1782 *Let.* 12 Feb. in *Essex Inst. Hist. Coll.* (1859) I. 13/2, I am Detarmend as I have beene so long in the servis to se it out. **1783** H. Newdigate *Let.* 23 Mar in A. E. Newdigate-Newdegate *Cheverels* (1898) iv. 50 Y^e Opera.. is to be wonderfully shewy & the last Dance ye best, so we must see it out.

e. Of a thing (esp. one's personal property): to last (at least) as long as (a person or his lifetime); to outlast or suffice to the end.

1969 M. Pugh *Last Place Left* xix. 143 The suits I have will see me out. **1976** *Guardian* 10 Apr. 10/8, I imported a German car, a convertible Beetle... No more of these are being imported, but I expect it to see me out.

28. see through. (Earlier and later examples.)

1828 L. Hunt in *Companion* 6 Feb. 48 *William III.* The Dutchman, call'd to see our vessel through. **1913** J. Vaizey *College Girl* xviii. 250 Her thoughts flew off to Ralph Percival..recalling with pleasure his promise to 'see her through'. **1916** H. G. Wells (*title*) Mr. Britling sees it through. **1939** *War Illustr.* 14 Oct. 147 (*heading*) Mr. Briton'll see it through. **1977** G. Butler *Brides of Friedberg* v. 129 Don't worry..I'll see you through.

Seebeck (zē·bek). *Physics.* [The name of Thomas Johann *Seebeck* (1770–1831), Russian-born German physicist, who discovered the effect (*Abhandl. der K. Akad. der Wissensch. zu Berlin: Phys. Klasse* (1822–23) 265).] **Seebeck effect,** the phenomenon whereby an e.m.f. is generated in a circuit containing junctions between dissimilar metals if these junctions are at different temperatures; the phenomenon of thermoelectricity.

1903 *Whittaker's Electr. Engineer's Pocket Bk.* 99 If a junction between two dissimilar metals be heated or cooled, a flow of electricity will take place across the junction. If a current be passed through such a junction, a change of temperature will be produced. The former phenomenon is known as the Seebeck effect, the latter as the Peltier effect. **1906** *Sci. Abstr.* A. IX. 481 The Thomson-, Peltier-, and Seebeck-effects may be combined in one diagram, and the author does this for the case of iron and silver. **1964** S. H. Avner *Introd. Physical Metall.* i. 6 The total emf in a thermoelectric pyrometer, sometimes called the Seebeck effect, is therefore the algebraic sum of four emf's, two Peltier emf's at the hot and cold junctions and two Thomson emf's along each of the wires. **1973** J. G. Tweeddale *Materials Technol.* I. iv. 97 If an electric circuit is suitably completed a current will flow between the materials (the Seebeck effect).

seed, *sb.* Add: **1. a.** Also, *to go to seed* (Go *v.* 44 b): to cease flowering as seeds develop; *fig.*, to become habitually unkempt, ineffective, etc.; to deteriorate.

1817 J. K. Paulding *Lett. fr. South* I. xvii. 188 His white dimity could not last for ever, and he gradually went to seed. **1859** [see Go *v.* 44 b]. **1929** G. Ade *Let.* 8 Feb. (1973) 139 We have ridden for miles and miles [in Peking], visiting temples and palaces of incredible size and beauty, some of them slightly gone to seed and others filled with the most wonderful museum displays of Chinese art. **1951** E. Paul *Springtime in Paris* ii. 33 Clients, mostly young and disreputable, or old and gone to seed. **1967** G. F. Fiennes *I tried to run Railway* iv. 42 He seemed to be going to seed a bit; to be a bit slow.

d. Cf. *bird-seed* s.v. *bird *sb.* 9.

2. b. *Chem.* A small crystal of the desired substance introduced into a liquid in order to promote crystallization and to provide a nucleus for crystal growth. Orig. *spec.* in sugar manufacture. Also *fig.*

1909 in *Cent. Dict.* Suppl. **1915** H. C. Prinsen-Geerligs *Pract. White Sugar Manuf.* II. i. 80 White sugar destined for direct consumption should not only be white and brilliant, but should also possess a regular form and a rather large size; this latter desideratum makes it preferable to start the building up of the grain from a well-developed seed. **1959** *Engineering* 13 Feb. 219/2 Seeds can be cut to provide grown crystals that can be sawn in the most efficient manner. **1966** *McGraw-Hill Encycl. Sci. & Technol.* III. 601/1 A small crystal of the desired substance is added to the solution as a 'seed' to induce the formation of the first crystals. **1973** *Nature* 12 Oct. 294/1 Some cardinal topics in evolutionary biology were adopted as 'seeds' on which other constituents of the programme might crystallise: protein polymorphism, for example.

2*. *Sport*, esp. *Lawn Tennis.* [f. sense *11 of the vb.] One of a number of seeded players in a tournament.

1933 M. D. Lyon in *Aldin Bk. Outdoor Games* 509 'But why put my beloved lawners last?' wails the Thibetan 'seed'. **1954** *Sun* (Baltimore) 22 June 17/3 The remaining four men's seeds won just the way they were supposed to due to the sudden decision by Wimbledon to seed 12 instead of the traditional eight. **1958** *Times* 20 Mar. 16/5 (*heading*) Badminton seeds dislodged. **1963** *Times* 12 June 5/6 Mr. McKinley, the United States number one and runner-up at Wimbledon in 1961, is top seed in the men's event. **1977** *Western Morning News* 30 Aug. 12/8 Fiona Moffitt, the number five seed from Dawlish, is Devon's main hope for the title.

3. (Later examples.)

1914 J. London *Let.* 24 Feb. (1966) 415, I have never wantonly scattered my seed. **1973** K. A. Sey *Ghanaian English* vii. 84 *To take seed*, to become pregnant.

6. b. *Med.* A small container for holding a radioactive material such as radon when it is placed in body tissue in radiotherapy. Cf. *radon seed* s.v. *radon 2.

1925 *Glasgow Herald* 26 Mar. 9 To capture the gas emanating from radium, purify it, and bottle it in tiny tubes called from their shape seeds. **1974** R. M. Kirk et al. *Surgery* iv. 66 The gas radon ^{222}Rn, a product of radium decay, with a half-life of four days, can be sealed in gold tubes or 'seeds' and implanted into the tissues. **1980** *Daily Tel.* 4 Dec. 6/8 It can implant radioactive seeds by needle when surgery has failed to remove malignancy or patients can no longer sustain external radiation therapy.

7. a. Simple attrib. (*a*) *seed catalogue, -fall, -house, packet, rate, set, tray, weight*; (*b*) *seed-ball, -speck, -stalk* (earlier example); (*c*) *seed-potato(es)* (examples); (*d*) *seed drill* (earlier and later examples).

1917 D. H. Lawrence *Look! We have come Through!* 119 The seed-ball of the sun Is broken at last. [**1760** J. Webb (*title*) A catalogue of seeds and hardy plants.] **1901** L. H. Bailey *Princ. Veg. Gardening* v. 168 The differences..might be of such a character that they could not be definitely described in a seed catalogue. **1938** N. Marsh *Death in White Tie* xxv. 266 He hastily gathered up..parish magazines, *Church Times*, and seed catalogues. **1973** K. Giles *File on Death* v. 138 'Do you get much mail?'.. 'Today there were three letters and a seed catalogue.' **1792** W. Rutherford *Let.* 28 Feb. in *Trans. Soc. Promotion of Useful Arts* (N.Y.) I. 121 Some years ago a farmer in Somerset county, in New Jersey, first introduced a seed-drill of his invention. **1850** *Mary Wedlake's Priced List Farming Implements* 14 (*heading*) Improved corn and seed drill. **1941** [see *combine *sb.* c]. **1973** L. Russell *Colonial Canada* iii. 38 The seed drill was invented in the 1850s; in this the seeds were not just dropped into the furrows, but were inserted into the soil through flexible tubes with a cutting edge in front. **1968** *Jrnl. Forestry* LXVI. 422/2 The estimated seedfall.. averaged..almost 1 pound per acre. **1981** *Country Life* 16 July 184/1 My fear of the allium menace..prompts me to dead-head every species before seed-fall. **1912** 'C. F. Benton' *Fairs & Fêtes* 116 Give to every purchaser a catalogue, which will be donated by any seed-house on request. **1941** *Sun* (Baltimore) 6 Aug. 16/7 Gray's standard manual apparently *versus* a Philadelphia seed-house. **1935** A. G. L. Hellyer *Pract. Gardening* v. 43 Beginners are safe in following the directions printed on the seed packet. **1981** 'M. Yorke' *Hand of Death* xvi. 143 Ronald ..sorted through old seed packets. **1742** J. Savage in *New Hampshire Probate Rec.* (1916) III. 115, I Give to my Dear and Loving Wife..Ground..for to plant one bushel of Seed pertators. **1901** L. H. Bailey *Cycl. Amer. Hort.* III. 1419/2 The seed Potatoes are cut to one eye, and dropped about 12 to 15 in. apart. **1977** *Belfast Tel.* 22 Feb. 22/4 (Advt.), Foundation stock seed potatoes for sale. **1960** *Times* 28 Nov. 16/5 The recommended seed-rate [for maize] is 30 to 35 lb. an acre. **1977** J. L. Harper *Population Biol. Plants* viii. 250 Seed of *T*[*rifolium*] *repens* was sown together with varying seed rates of *Lolium perenne.* **1946** *Nature* 12 Oct. 519/2 The reactions based on pollen-tube growth and seed-set determinations of these two groups of plants are given in the accompanying table. **1978** *Ibid.* 7 Sept. 54/2 In artificial field bean pollination, manual stripping of open flowers is a recommended practice for increasing seed set in autosterile lines. **1917** D. H. Lawrence *Look! We have come Through!* 59 Rose-leaves that whirl in colour round a core Of seed-specks kindled lately. **1744** W. Ellis *Mod. Husbandman* Jan. ii. 32 A Turnep runs up a Seed-stalk sometimes near seven Feet high. **1953** E. R. Janes *Sweet Peas* vii. 53 Standard seed trays give little trouble. **1971** P. D. James *Shroud for Nightingale* vi. 183 A small stack of seed trays, pruning shears, a trowel and small fork. **1927** Haldane & Huxley *Animal Biol.* x. 219 They will produce plants each of which will have the same range of seed-weight as did its parent. **1977** J. L. Harper *Population Biol. Plants* vii. 203 The year to year variation in seed weight is quite large.

b. Objective and objective genitive. *seed planter, sower, -sowing* (example), *-testing; seed-eating* adj.

1927 Haldane & Huxley *Animal Biol.* x. 205 If..a tame sea-gull is fed on corn instead of fish, the whole lining of its stomach alters, becoming thicker and more like that of seed-eating birds. **1977** J. L. Harper *Population Biol. Plants* xx. 629 A plague of seed-eating mammals, birds or insects may prevent regeneration. **1850** *Rep. Comm. Patents 1849* (U.S.) I. 151 Having thus fully described my improved grain and seed planter. **1848** *Commerc. Rev.* VI. 133 Seed-Sowers, &c.—These machines are quite ingenious and labor-saving in their contrivance, [etc.]. **1977** *Young's Sporting Appliances* (S. Young & Sons Ltd.) 5 Seed sower. **1865** *Rep. Maine Board Agric.* X. 65 The use of the drill for general seed-sowing is at present considered too expensive. **1896** *U.S. Dept. Agric. Yearbk. 1895* 175 The importance of seed testing is recognized not only by professional seedmen, but also by intelligent farmers. **1910** *Chambers's Jrnl.* Oct. 685/1 Next comes seed-testing under the microscope, which shows the weed, seeds, and rubbish amongst them. **1950** *N.Z. Jrnl. Agric.* Mar. 243/1 Some 80 years ago the first seed-testing station was established in Saxony by Professor Nobbe. **1973** *Country Life* 15 Mar. 714/2 Work on onions at the Cambridge Official Seed Testing Station.

d. Instrumental, as *seed-borne* adj.

1931 *Bull. W. Virginia Agric. Exper. Station* No. 245. 5 The economic losses occasioned by these few seed-borne parasites..are enormous. **1968** *Times* 16 Dec. 7/1 A seed-borne fungus disease.

8. seed bank, (*a*) a place where seeds of different plant varieties and species are stored as a safeguard against their possible extinction; (*b*) the seeds that have accumulated naturally in a given area of ground; **seed bull,** a bull kept to serve cows; **seed-cotton** (earlier example); **seed crystal** = sense 2 b above; **seed dressing,** a preparation applied to seed in order to protect it against pests; the practice or an instance of employing this; **seed fat,** a fat obtained from seeds; **seed-leaf,** (*b*) used *attrib.* and *absol.* to designate a kind of tobacco grown in the northern United States and used chiefly for wrapping cigars (so called because it was first grown from imported seed); **seed metering,** automatic control of the numbers of seeds sown or planted by a machine; **seed money** *U.S.*, money allocated (esp. from public funds) for the initiation of a project and designed to stimulate the independent economic expansion of the project; **seed orchard,** a group of trees cultivated for the production of seed; **seed parent,** in hybridization, a plant whose seed is fertilized by pollen from a different plant; **seed stitch** *Needlework* = *seeding *vbl. sb.* 3*; **seed year,** one of the years in which a particular tree produces a good crop of seeds.

1958 *Economist* 25 Oct. 328/2 In America's first central seed bank, which has recently been opened at Colorado State University, there is space to store supplies of several hundred thousand basic seed stocks. **1974** *Nature* 24 May 303/2 As part of a worldwide effort to conserve this fast disappearing genetic resource, a seed bank had already been set up. **1977** J. L. Harper *Population Biol. Plants* iv. 83 The store of seeds buried in soil (the seed bank) is composed in part of seeds produced on the area and partly of seeds blown in from elsewhere. **1932** E. Hemingway *Death in Afternoon* xi. 118 The usual ranch has two hundred cows and four seed bulls. **1797** F. Baily *Jrnl. Tour N. Amer.* 285 The seed-cotton loses three-fourths of its weight by jenning. **1934** *Industr. & Engin. Chem.* Nov. 1201/1 The initial formation of crystal nuclei is profoundly influenced by the chance presence of very small seed crystals of the solute. **1962** [see *melt *sb.*[2] 2]. **1974** *Encycl. Brit. Macropædia* V. 337/1 If the structure and interatomic spacing of the surface..approximate that of the crystal, growth on the surface can resemble growth on a normal seed crystal. **1926–7** *Army & Navy Stores Catal.* p. lxxi/2 Seed Dressings, Liquid. **1955** E. Holmes *Pract. Plant Protection* iv. 31 By far the most important fungicidal seed-dressings are those based on chemicals known as the organo-mercury products. **1977** M. B. Green et al. *Chemicals for Crop Protection & Pest Control* xii. 103 The convenience and economy of seed-dressing makes it..a clear choice when the disease can be controlled this way. **1940** T. P. Hilditch *Chem. Constitution of Natural Fats* i. 18 However varied the fatty acids in seed fats may be, the resulting triglycerides are..fundamentally similar in type. **1949** *Thorpe's Dict. Appl. Chem.* (ed. 4) IX. 8/2 In many seed fats..the main component acids are confined to palmitic, oleic, and linoleic in varying proportions. **1963** *Times* 22 May (Suppl.) p. iv/2 The main sources of edible oils and fats are vegetable fats—particularly palm oil; seed fats, notably those of groundnut, coconut, soya bean, palm kernel and cotton seed; [etc.]. **1852** *Hunt's Merch. Mag.* XXVII. 555 The 'seed leaf' is raised on the Miami River. **1888** *Encycl. Brit.* XXIII. 425/2 The 'seed-leaf' tobacco of Pennsylvania, Connecticut, and Ohio, grown from Havana seed, is devoted to cigar-making in the United States. **1910** 'Mark Twain' *Speeches* 267, I bought what was called a seed-leaf cigar with a Connecticut wrapper. **1946** W. W. Garner *Production of Tobacco* ii. 35 The process of fermenting the leaf in cases in preparation for manufacture..began about 1845 and gave great impetus to the manufacture of the 'Seed and Havana' cigar, composed of Cuban filler and Connecticut Seedleaf wrapper. **1968** B. C. Akehurst *Tobacco* xi. 266 The current U.S. Department of Agriculture classification is as follows: Class 4..Type 41 Pennsylvania Seed Leaf. **1955** R. Bainer et al. *Princ. Farm Machinery* xi. 225 Most seed-metering devices may be classified as: (*a*) those having cells on a moving member, the cells being sized to accommodate single seeds or groups of a few seeds each, (*b*) the so-called 'force-feed' devices.., (*c*) stationary-opening units. **1971** *Power Farming* Mar. 29/4 The gearing..has been improved to give greater precision in seed metering. **1966** *N.Y. Times* 21 Aug. F15/2 The bonds would have enabled the state to gain $17-million, to be used as seed money to set up the loan guarantee rotary fund. **1970** *Sat. Rev.* (U.S.) 10 Jan. 27/1 This has been seed money in the best sense of the term. As President Nixon pointed out to Congress last month, every dollar of Foundation money has stimulated the donation of three dollars from other sources. **1977** *Time* 12 Dec. 33/2 It calls for $120 million in federal seed money to create 14,000 new jobs and rehabilitate four neighborhoods. **1951** H. McKusick in *Tree Planters' Notes* VIII. 8 It.. seems logical that we should explore the possibility of establishing seed orchards. **1979** *Beautiful British Columbia* Fall 40 Others are from seed orchards maintained by the province and the forest companies. **1902** *Jrnl. R. Hort. Soc.* XXVII. 209 We suggested..crossing our common Wood Anemone..with the scarlet *Anemone fulgens*, making the Wood Anemone the seed parent. **1970** R. Gorer *Devel. Garden Flowers* 19, I have tried to follow the convention in which the seed parent is the first named and the pollen parent the second. **1934** M. Thomas *Dict. Embroidery Stitches* 182 Tiny stitches taken at all angles and in any direction but of more or less even length produce a surprisingly effective filling, as the diagram of Seed Stitch shows. **1964** *McCall's Sewing* xiii. 246/2 *Seed stitch*,..a very tiny chain stitch tightly drawn and scattered in all directions to fill an open area. **1889** W. Schlich *Man. Forestry* I. II. ii. 173 The quantity of seed is governed by two things:—(1.) The average yield of each seed-year; and (2.) The frequency of seed-years. **1979** H. W. Hocker *Introd. Forest Biol.* ii. 55 Heavy seed years do not occur at frequent intervals, but are usually offset by succeeding years of light to very light crops.

seed, *v.* Add: **II. 6. a.** Also *absol.*

1888 *Vermont Agric. Rep.* X. 48, I..then seed broadcast with grass-seed. **1979** [see *plant *v.* 1].

b. *spec.* (i) To introduce a crystal or small particle into (a liquid or apparatus) so as to induce crystallization. Cf. *seed *sb.* 2 b.

1909 in *Cent. Dict.* Suppl. **1921** *Jrnl. Physical Chem.* XXV. 534 Points on the stable curve..were then easily determined by seeding the proper mixtures..with small crystals from this lot of hydrate. **1930** *Amer. Speech* VI. 14 Sometimes a crystallizer is seeded with a nest egg of sugar. **1936** H. L. Alling *Interpretative Petrol. Igneous Rocks* iv. 41 Supercooled water, left undisturbed, is stable if not in contact with ice. 'Seeding' it with the solid phase produces crystallization. **1964** G. H. Haggis et al. *Introd. Molecular Biol.* iv. 94 It has further been possible recently to study this process *in vitro* by 'seeding' near-saturated solutions of calcium phosphate with collagen fibres.

(ii) To introduce crystals of a substance such as silver iodide into (a cloud, etc.) in order to cause precipitation.

1947 *Sun* (Baltimore) 9 Sept. 3/3 Today's storm developed too close to the Florida West Coast to be seeded with dry ice or other crystals by Army and Navy planes. **1958** *Observer* 12 Jan. 6/3 The nearest thing to weather control so far is 'seeding' clouds with ice or silver iodide to persuade them to give up their rain. **1974** *Nature* 11 Oct. 461/3 Potential hail clouds are observed by radar, and then seeded by lead iodide from a rocket fired into the cloud's centre. **1975** *Daily Colonist* (Victoria, B.C.) 16 July 3/3 The United States government did not seed hurricane Fifi, nor was it ever contemplated.

(iii) *Biol.* To inoculate (a culture vessel) *with* cells from a culture which is to be propagated.

1960 *Virology* X. 387 One hundred-millimeter petri dishes were seeded with cells from primary cultures of normal mouse. **1978** *Nature* 23 Mar. 372/2 The procedure was repeated using cultures of Wi-38 cells prepared in.. eight-compartment chamber slides..that had been seeded three days earlier with 15,000 cells in each compartment.

7. a. (Earlier modern examples.)

1814 J. Taylor *Arator* (ed. 2) 154 When the wheat was seeded on high and narrow ridges. **1851** C. Cist *Cincinnati* xv. 317 The cotton crop is seeded in the spring.

d. *Biol.* To inoculate (cells from a culture) *into* a culture vessel or medium.

1965 *Proc. Nat. Acad. Sci.* LIV. 1585 The cells were.. seeded into 60-mm plastic dishes at a density of 2·5 × 10^4 cells per dish. **1973** *Nature* 22 June 450/1 Cells (10^5) were seeded into each 100-mm Petri dish (Falcon Co.) and incubated at 37°C in a humidified incubator with 5% CO_2 in air.

11. *Sport*, esp. *Lawn Tennis.* To assign (to several of the better competitors) a position in an ordered list, so that those most highly ranked do not meet until the later stages of an elimination competition; to arrange (a draw or event) to this end. Cf. *seed *sb.* 2*. orig. *U.S.*

1898 *Amer. Lawn Tennis* 13 Jan. 4/2 Several years ago, it was decided to 'seed' the best players through the championship draw, and this was done for two or three years. **1900** *Spalding's Lawn Tennis Ann.* 78 It is generally advisable to 'seed' the draw in handicap tournaments so that the players in each class shall be separated as far as possible one from another. **1911** *Spalding's Off. Lawn Tennis Guide* 55 Unlike many big events, Longwood is never seeded, and in consequence the possibility of an uneven draw materialized. **1924** *Times* 23 June 4/4 This year, for the first time, the draw has been 'seeded'; how little seeding accords with British notions may be gathered from there being no reference in the Oxford Dictionary—at any rate in the smaller one... In some countries the seeding is designed to keep the better players apart until the final stages. **1929** *Times* 29 June 4/4 Three of the women who had been 'seeded' for the draw were defeated during the day. **1953** *Sunday Graphic* 7 June 22/4 Rose, likely to be seeded in the first four at Wimbledon, did not play up to his reputation. **1955** *N.Y. Times* 10 May 33/5 Joe Burk's Red and Blue eight, which beat Navy and Harvard Saturday for the Adams Cup, was seeded first in the draw for the tenth annual regatta. **1972** D. Delman *Sudden Death* (1973) i. 17 Timmy was up against a big Australian kid who'd given me fits at Wimbledon... Timmy..would probably go into the tournament seeded second behind Cole. **1982** *Guardian* 8 June 22/8 She is seeded 14th and is not particularly worried that a seeding at Wimbledon is unlikely.

seed-corn. Add: **1.** (Later *fig.* examples.)

1962 *Economist* 19 May 688/1 The loans and grants from his agency are 'seed corn'; they stimulate local initiative and self-help. **1977** *Times* 20 Dec. 10/8 The IBA should be prepared to use the secondary rental income from these established stations which are already well into profit, as seedcorn money to help establish and run independent stations in smaller townships.

2. seed-corn maggot *U.S.*, the yellowish-white larva of a fly, *Hylemya platura,* which infests the seed of many vegetables and other crop plants, preventing sprouting or causing the seedlings to be weak and sickly; also, the adult fly.

1869 C. V. Riley *First Ann. Rep. Noxious Insects Missouri* 154 (*heading*) The seed-corn maggot. **1902** *Bull. Div. Entomol. U.S. Dept. Agric.* XXXIII. 84 The Seed-Corn Maggot..has received no less than seven Latin names. **1949** *Jrnl. Econ. Entomol.* XLII. 77/1 The seed corn maggot..injures bean, pea and corn seedlings before the plants emerge. **1975** *Nature* 7 Aug. 487/1 Larvae of the seedcorn maggot..may damage or kill young plants of many crop species by feeding on the cotyledons and plumules.

seeded, *ppl. a.* Add: **5.** Of fruit, esp. dried fruit: having the seeds removed.

1921 *Daily Colonist* (Victoria, B.C.) 1 Apr. 6/1 (Advt.), Sun-Maid seeded raisins, bulk, packets of 15 oz.

6. *Sport*, esp. *Lawn Tennis.* Of a competitor: assigned a position in a list of seeds (*seed *sb.* 2*) in an elimination competition. Also of a draw arranged in such a manner. Also *transf.*

1922 *Spalding's Tennis Ann.* 30 Commencing in 1922, all championships and other sanctioned tournaments except handicap events shall have a seeded draw. *Ibid.* 31 Seeded men in the top half of the draw. **1954** *New Yorker* 28 Aug. 57/1 The semifinals had the No. 1 and No. 2 seeded teams in each division. **1961** *Listener* 21 Sept. 437/1 One more look at the seeded players, a convincing warts-and-all likeness of Wingate, Stilwell an 'unmitigated disaster', and some memorable marginalia to the saga of Slim.

seeder. Add: **2.** (Earlier U.S. example.)

1868 *Rep. Iowa Agric. Soc. 1867* 226 The seeder can be adjusted in five minutes.

3. *cloud seeder*: one who or that which seeds clouds.

1953 *Jrnl. Amer. Water Works Assoc.* XLV. 1144/1 The skilled commercial cloud seeder is qualified by training and experience to undertake the large-scale operations that make cloud seeding economically worthwhile. **1958** *Ann. Reg. 1957* 485 Experiments with an electrostatic cloud seeder consisting of an aircraft trailing two 300 ft. cables carrying a 50,000 volt charge of electricity. **1975** *Nature* 28 Aug. 690/3 The responsibility..rests as much with the policy makers as it does with the plant breeders, the climatologists and the cloud seeders. **1981** *Economist* 20 June 55/1 Utilities that hire cloud-seeders, hoping to increase water-flow into their reservoirs, may cause the heavens to fall on the tourist industry.

seeding, *vbl. sb.* Add: **2. c.** *transf.* (Cf. *seed *v.* 6 b.)

1926 *Jrnl. Chem. Soc.* II. 2774 Crystallisation could be easily induced by 'seeding'. **1935** J. N. Friend *Physical Chem.* II. v. 128 This cannot be done even by careful seeding as the hydrate is too soluble to be reached in this way. **1939** E. Liljencrantz *Cancer Handbk. of Tumor Clinic* xi. 94 Medulloblastoma is distinguished by its tendency toward subarachnoid dissemination or 'seeding' along the entire cerebrospinal axis. **1947** *Sun* (Baltimore) 8 Sept. 3/2 The Miami plane will fly above the storm to photograph the effects of 'seeding' by the weather-science plane. **1958** S. M. Brooks *Basic Facts Med. Microbiol.* i. 29 Inoculation means the seeding of a culture medium with an organism..using a sterile platinum wire (straight or looped) or a sterile glass pipette. **1966** Wright & Symmers *Systemic Path.* II. xxxiv. 1248/1 The ability of a glioma to spread by seeding is of considerable practical importance. **1972** *Materials & Technol.* V. xx. 713 This process may be induced by the introduction of a few crystals of ephedrine oxalate—a process known as 'seeding'. **1977** *Jrnl. R. Soc. Arts* CXXV. 160/1 By far the largest and most sustained effort to modify weather deliberately has involved the artificial seeding of clouds in an attempt to increase the rainfall or suppress damaging hailstorms.

3*. A stitch in embroidery (see quot. 1960). Cf. Seed *v.* 9; *seed-stitch* s.v. *seed *sb.* 8.

c **1840** Lady Wilton *Art of Needlework* xx. 317 There is slabbing—veining—and button stitch; seeding—roping —and open stitch. **1960** B. Dean in G. Lewis *Handbk. Crafts* 24 Stem stitch..and seeding.., which consists of a small back stitch with another worked over it, to form an irregular filling, are also useful.

3.** *Sport*, esp. *Lawn Tennis.* The placing of competitors in a list of seeds (*seed *sb.* 2*); (also *pl.*) the order or ranking so produced.

1912 A. F. Wilding *On Court & Off* 140 Arranging or 'seeding' is a distinction without any material difference. **1937** P. B. Hawk *Off Racket* i. 13 'Seeding' was believed to be unfair to certain contestants and to make for a less interesting tournament by eliminating..the probability of thrilling matches in the early rounds. **1955** *N.Y. Times* 10 May 33/5 (*heading*) Penn crew tops sprint seedings. **1958** *Oxford Mail* 15 Jan. 8/1 For the first time, seeding is to be introduced into the Amateur Golf Championship being played at St. Andrews in the week beginning June 2. **1978** *Times* 4 July 19/3 Another clay court specialist, Miss Jausovec, upset the seedings by beating Wendy Turnbull.

4. (sense 2) *seeding rate*; (sense *3**) *seeding(s)-committee, system*; **seeding felling** (see quot. 1928).

1936 *Times* 22 June 7/4 If six of the chosen eight come through the seeding committee will have guessed well. **1960** *Times* 4 July 15/6 In spite of the skilful pruning of the seedings committee. **1976** *Liverpool Echo* 7 Dec. 17/7 Mrs. Marshall had a tough fight in her semi-final..as the girls battled to give the seeding committee for the finals an idea of current form. **1928** R. S. Troup *Silvicultural Systems* iv. 32 The seeding felling, under which the canopy is opened out in order to afford sufficient light to ensure the survival for a short time of seedlings springing from seed shed by the overhead trees. **1968** Champion & Seth *Gen. Silviculture for India* v. 269 The overwood may be removed in only two fellings, first the main seeding felling. **1930** L. S. Dickinson *Lawn* iv. 34 The seeding rate 3½ pounds per 1,000 square feet of lawn. **1949** G. H. Algren *Forage Crops* xxiv. 241 The seeding rates shown for certain crops..are too high. **1929** *Times* 24 June 7/1 The 'seeding' system undoubtedly has its merits.

seedling, *sb.* and *a.* Add: **A.** *sb.* **4. seedling blight,** a disease of seedlings, esp. a seed-borne, sometimes fatal disease of flax that affects esp. seedlings and is caused by the fungus *Colletotrichum lini.*

1919 Pethybridge & Lafferty in *Jrnl. Dept. Agric. & Technical Instr. Ireland* XX. 327 It is usually not recognised in the brairding crop until the seedlings are one or two inches high... It is during these early stages that the disease causes most damage, and it is for this reason that we have proposed the name 'Seedling-blight' for it, although the trouble is not entirely confined to plants in the seedling stage. **1980** F. Hope *Recognition & Control of Pests & Dis. Farm Crops* (ed. 2) 159/1 Seedling Blight *Colletotrichum lini* can be a destructive disease of flax seedlings.

seedy, *a.* Add: **2. b.** Also *spec.* as a result of excessive eating or drinking; = Crop-sick *a.* (Earlier examples.)

1729 R. Savage *Author to be Lett* 7 After an Evening's hard boozing, my brother Bards..have been what we call Seedy or Crop-sick. **1845** *Punch* IX. 40/2 Young Oxford eats a wondrous meal, And drinks a lot of beer, And in the morning oftentimes, Full seedy does appear.

seegar. Delete *obs.* and add: Representing a U.S. colloq. or dial. pronunciation, with stress on the first syllable. (Later examples.)

1935 Z. N. Hurston *Mules & Men* (1970) I. v. 119 Settin' by de fire smokin' uh seegar. **1976** W. Goldman *Magic* III. viii. 171 'Got any of them jazzy see-gars?' Fats wondered. **1976** *Time* 27 Sept. 46/1 Carter does not fit many Southern stereotypes. He is not a hard drinker, poker player, or profane and garrulous see-gar-chomping raconteur.

seeing, *vbl. sb.* Add: **1. a.** *seeing is believing* (later examples).

1712 J. Arbuthnot *Lewis Baboon* iv. 21 There's nothing like Matter of Fact; Seeing is Believing. **1848** J. C. & A. W. Hare *Guesses at Truth* 2nd Ser. 497 *Seeing is believing*, says the proverb... Though, of all our senses, the eyes are the most easily deceived, we believe them in preference to any other evidence. **1909** *Times Lit. Suppl.* 28 May 198/2 Seeing is believing;..only art can make history really credible, or a great name more than a label to an abstraction. **1975** A. Price *Our Man in Camelot* v. 84 'Show him the stuff.'.. 'Okay. Maybe you're right... Seeing is believing, I guess.'

2*. *Astr.* The quality of telescopic observation; the extent to which a stellar image remains steady and free from twinkling, or a planetary image sharp.

In modern usage seeing is quantified as the apparent angular diameter of a point source as seen in a powerful telescope.

1903 *Phil. Mag.* V. 674 Observatories are put even on high mountains to get rid of the disturbances in this atmosphere, which tend to make the image of every object tremulous.., and to prevent what the astronomer terms 'good seeing'. **1969** N. Calder *Violent Universe* 21 The best 'seeing' at any working observatory is said to be that at Cerro Tololo, in Chile. **1977** *Nature* 21 Apr. 693/2 The seeing during the observations was generally between 1″ and 2″.

seeing, *ppl. a.* Add: **2.** *seeing-eye*: in various senses of the vb. See, the faculty of seeing; *seeing-eye dog* (U.S.): a guide-dog trained to lead the blind.

Seeing eye is registered in the U.S. as a proprietary name for guide-dogs trained to lead the blind.

1921 P. Lubbock *Craft of Fiction* x. 146 The 'seeing eye' to which it is presented is not his, but the reader's own. **1930** *Official Gaz.* (U.S. Patent Off.) 26 Aug. 553/2 The Seeing Eye, Nashville, Tenn... *The Seeing Eye* for trained dogs. Claims use since Mar. 30, 1929. **1938** *Sun* (Baltimore) 18 May 8/7 Many people..will be much moved by the story of how the first of the now famous 'seeing-eye' dogs has been carried by plane back to the Seeing Eye School at Morristown, N.J., there to spend its last days quietly. **1948** J. Cannan *Little I Understood* viii. 101, I don't pretend to be able to draw. I've just got the seeing eye. **1950** P. Bottome *Under Skin* xvi. 137 He had bought her [*sc.* a blind woman] a 'Seeing Eye' dog, who took her wherever she wished. **1969** K. M. Wells *Owl Pen Reader* II. 210 You see, Grandpa had the seeing eye and grandma hadn't. **1979** 'A. Hailey' *Overload* III. xiii. 267 The interior [of the bar] was dark and smelled of mildew. 'Christ!' Nancy said. 'We need a seeing-eye dog.'

seeing, quasi-*conj.* Add: Also (*colloq.*) with *as* (*how*).

1833 Dickens *Let.* 18 Mar. (1965) I. 17 Seeing as I cannot fail to do that I have engaged in a pursuit. **1895** *Dialect Notes* I. 399 Seein' as how it's you, I'll do so-and-so. **1952** M. Laski *Village* vii. 121, I suppose—seeing as how we've both been let down—you wouldn't care to come in with me? **1974** S. Gulliver *Vulcan Bulletins* 29 Seeing as how you're always short of £sd, I thought you could maybe earn a bit.

seek, *sb.* Restrict † *Obs.* to sense in Dict. and add: **2.** *Computers.* The movement of a read/write head to a new position on a storage device; **seek-time,** the time taken by this, as part of the total access time.

1965 *IEEE Trans. Computers* XIV. 580/2 No more than three concurrent seek operations per data channel are justified for System I. **1967** *AFIPS Conf. Proc.* XXX. 11/1 Suppose that secondary memory is a disk... The operation of moving the arm is known as a seek; but the policy shortest seek time first..is unsatisfactory. **1974** *Communications Assoc. Computing Machinery* XVII. 139/2 The objective of optimally scheduling a sequence of requests on the DASD to minimize seektime or rotational delays. **1980** *Sci. Amer.* Aug. 118/3 First the head must be positioned over the proper track. This requires a 'seek time'.

seek, *v.* Add: **I. 12. b.** (Later examples.) Cf. Sick *v.*[2]

1928 Kipling *Limits & Renewals* (1932) 64 Go seek, boy! It's Dinah! Seek! **1968** P. N. Walker *Carnaby & Gaolbreakers* xix. 178 'Seek,' and the two police dogs were cast about the mini car.

seeker. Add: **1. a.** Also in phr. *seeker after truth.*

1840 [see Scientist]. *a* **1968** A. Farrer *Interpret. & Belief* (1976) 138 Madame Blavatsky, than whom few women have been more remarkable for the power of making solid objects fade into thin air among the mountains of India, and crystallize back to physical solidity in the middle of English drawing-room cushions, thence to be hacked out with scissors by delighted seekers after truth.

seelapak, var. *sillapak.

seem, *sb.* For † *Obs.* read ? *Obs.* or *dial.* (chiefly *Sc.*) and add later examples.

1730 A. Ramsay *Tea-Table Miscellany* (ed. 5) 213 His seim in Thrang of fiercest Stryfe, When Winner ay the same. **1812** Jane Austen *Let.* 29 Nov. (1952) 499 It [*sc.* a cloak] is to be Grey Woollen & cost ten shillings. I hope you like the *sim* of it. **1837** J. M. Wilson *Tales of Borders* III. 131/1 There comes slowly, as if frae the womb o' a cloud o' mountain mist, the seim o' a turreted abbey. **1913** H. P. Cameron tr. *T. à Kempis's Imit. Christ* II. vii. 63 Ye'se sune be begunkit, gin ye regaird allenarlie the ootrin seim o' men.

seem, *v.*[2] Add: **II. 4. a.** Also in weakened sense (chiefly interrogative).

1901 A. K. McClure *'Abe' Lincoln's Yarns & Stories* 65 'What seems to be the matter?' inquired Lincoln with all the calmness and self-possession he could muster. **1958** B. W. Aldiss *Non-Stop* IV. iii. 219 'What seems to be the trouble?' he asked. **1974** Wodehouse *Aunts aren't Gentlemen* xvi. 134 My voice shook a bit as I applied for further details. 'What seems to be the trouble?' I asked. **1977** G. Markstein *Chance Awakening* xix. 55 He dialled 100..'What seems to be the trouble?' asked the operator.

e. Also, in negative contexts (preceded by *can*): to seem unable. *colloq.*

1874 Hardy *Far from Madding Crowd* II. xiii. 168 Troy could hardly seem to believe her to be his proud wife Bathsheba. **1898** G. Gissing *Human Odds & Ends* 57 As a lad, I couldn't stick to anything—couldn't seem to put my heart into any sort of work. **1937** I. Baird *John* xix. 229 He couldn't seem to get the boy out of his head. **1969** M. Pugh *Last Place Left* vii. 44 Somehow I can't seem to get warm.

seep, *sb.* Restrict *dial.* to senses 1 and 3 in Dict. and add: **2.** (Later examples.)

1902 *Nature* 4 Dec. 113/1 At Comanche Spring, a small 'seep',..the limestone bluffs have been covered in a number of places with rude paintings of characteristic Indian design. **1903** *Bull. U.S. Geol. Survey* No. 212. 97 In this well small seeps or pockets of petroleum were found at several depths. **1966** *McGraw-Hill Encycl. Sci. & Technol.* X. 60/2 Almost without exception, seeps are at topographically low spots where water has accumulated. Oil..rises to the surface of the water, covering it with an iridescent film. **1972** *Science* 16 June 1257/2 The existence of submarine seeps is often mentioned in discussions of oil pollution.

seep, *v.* For *dial.* and *U.S.* read formerly *dial.* and *U.S.* and add: Also *fig.*

1922 H. Crane *Let.* 6 May (1965) 85 A new literary magazine.., *Secession*, which is (first number) just seeping into this country. **1931** W. G. McAdoo *Crowded Years* xviii. 284 The..emotions generated by the European struggle seeped into American thought. **1942** *R.A.F. Jrnl.* 13 June 13 The Commandos had been part of the British Army for about eight months before they seeped into print. **1955** A. L. Rowse *Expansion Elizabethan England* iv. 141 In 1585 a plan for peopling Munster was drawn up... Meanwhile the Irish came seeping back. *a* **1974** R. Crossman *Diaries* (1976) II. 453 One thing I learnt from my brief visit was how well-informed the Transport House staff are about internal Cabinet affairs and how they seep with anti-Government gossip.

Hence **see·ping** *ppl. a.*

1927 M. Eiker *Over Boat-Side* xi. 161 The nagging, monotonous, unessential, seeping harassments that sap a man of achievement.

seepage. For *Sc.* and *U.S.* read orig. *Sc.* and *U.S.* and add: **1.** *spec.* the slow movement of water into or out of the ground (as distinct from percolation through it); also, the slow movement of water through the ground under the action of gravity. (Further examples.)

1913 V. B. Lewes *Oil Fuel* 61 The surface indications, apart from seepages of oil, escape of natural gas from the soil, [etc.]..are practically nil. **1923** *U.S. Geol. Survey Water-Supply Paper* No. 494. 43 Seepage may be divided ..into influent seepage..and effluent seepage. **1950** *N.Z. Jrnl. Agric.* June 559/2 Placing concrete 'seep collars' round the pipe reduces the possibility of seepage. **1967** R. C. Ward *Princ. Hydrol.* viii. 311 Most of the rainfall which percolates through the soil layer to the underlying groundwater will eventually reach the main stream channels as groundwater runoff (sometimes referred to as ..effluent seepage). **1974** *Daily Tel.* 4 Apr. 17 An extensive seepage of fuel oil from a fractured feed pipe. **1976** Raudkivi & Callander *Analysis Groundwater Flow* i. 1 The motion of groundwater can be subdivided as follows: seepage; capillary rise and capillary flow; percolation, which occurs..under the action of a hydraulic gradient; turbulent groundwater flow.

fig. **1952** *Time* 14 Apr. 17/2 *Seepage* (n.), small net decrease in circulation, resulting from temporary suspension of subscriptions. (Opposite of *creepage.*) **1961** L. Mumford *City in History* x. 285 The 'law of cultural seepage': the making of innovations by a favored minority and their slow infiltration..into the lower economic ranks. **1976** *National Observer* (U.S.) 17 July 3/1 So far the seepage has had minimal effect on the multibillion-dollar Social Security system.

2. *attrib.*, as *seepage flow, loss, spring, water, well*; **seepage lake,** a lake that loses water chiefly by seepage into the ground containing it.

1939 A. K. Lobeck *Geomorphology* v. 159 Run-off.. occurs in two ways: (*a*) as surface run-off..; and (*b*) as ground-water run-off,..often called seepage flow. **1976** Raudkivi & Callander *Analysis Groundwater Flow* i. 2 This movement is complicated by the presence of ground air, most of which is expelled from the ground or dissolved in the seepage flow. **1934** *Ecol. Monogr.* IV. 441 These lakes may be classified into..those with outlets and those without them... The movement of water through lakes of the second type is entirely under ground and they are named seepage lakes. **1975** R. G. Wetzel *Limnology* iv. 40 In seepage lakes the lake seal is..likely to be effective over much of the deeper portions of the basin. **1902** *U.S. Geol. Survey Water-Supply & Irrigation Paper* No. 67. 42 C. E. Grunsky has measured the seepage loss in King River and the Fresno canal. **1937** C. F. Tolman *Ground Water* vii. 169 Seepage loss from this river is less than 5 per cent of the total flow. **1908** T. C. Hopkins *Elem. Physical Geogr.* ii. 60 Sometimes..the water..seeps or trickles out along the line of outcrop of the layer in sufficient quantities to keep the surface wet,..forming a swamp or bog on the hillside. This is called a seepage spring. **1964** G. B. Schaller *Year of Gorilla* (1965) x. 264 We came upon six elephants pawing the soil of a seepage spring. **1883** Seepage water [in Dict.]. **1876** Raudkivi & Callander *Analysis Groundwater Flow* i. 1 The capacity of the soil to hold suspended capillary water and water in the attached films is called field capacity and it is the excess over the field capacity which is free to travel downwards as gravity or seepage water. **1969** N. W. Parsons *Upon Sagebrush Harp* xviii. 103 Later that winter the farm papers began to tell of farm dugouts and seepage wells for the well-less prairie.

seer[1]. Add: **5.** Comb. *seerlike* adj.; **seercraft,** the prophetic art.

1883 R. C. Jebb *Sophocles* I. 139 Thus did the messages of seer-craft [*Oedipus Tyrannus* 723 φῆμαι μαντικαί] map out the future. **1913** G. Murray tr. *Euripides' Rhesus* 6 Sage and prophet, learned in the way of seercraft. **1849** J. G. Whittier *Proem* in *Poems* p. iv, Nor mine the seer-like power to show The secrets of the heart and mind. **1975** *New Yorker* 21 Apr. 96/2 One indication of the seer-like quality of these underwriting judgments can be glimpsed in the Continental Insurance underwriters' manual concerning the accident probabilities of drivers.

seer: see also Seir-fish, seer-fish in Dict. and Suppl.

seersucker. Add: Also 8 **sea sucker, see-sucker.** Also *attrib.* or as *adj.* (Earlier Amer. and later examples.) Also, a garment made of seersucker.

No longer restricted to the U.S. and India.
1722 C. Carroll in *Maryland Hist. Mag.* (1925) XX. 64 To Corded Dimothy..To I Sea Sucker D°. 1736 *Virginia Gaz.* 15 Oct. 4/2 Ran away..a Servant Woman..took..a Seersucker Gown. 1958 B. Malamud *Magic Barrel* 124 He dried himself and dressed. When he came forth in his seersucker, she offered salami. 1964 *Punch* 29 Apr. p. xiv/2 For men, striped cotton seersucker jackets. 1975 B. Garfield *Hopscotch* vii. 77 A trim sandy man..in the regulation seersucker. 1981 L. Stephan *Murder or Not* vi. 40 Her seersucker shift..with its thin blue and white stripes.

see-safe, *a.* and *adv.* [f. See *v.* + Safe *a.*] (See quot. 1960.)

1926 S. Unwin *Truth about Publishing* vi. 180 Some firms occasionally seek to protect themselves when buying books of doubtful saleability by marking their order..'see safe'. If this condition is passed..the publisher can be called upon..to exchange any surplus copies. *Ibid.*, If the firm has bought fifty-two copies of a new novel and marked the order 'twenty-six see safe', the publisher..may find himself obliged to take twenty-six back and to supply some more saleable work in their place. 1939 F. D. Sanders *Brit. Bk. Trade Organisation* 42 The Committee recommended..that books ordered as a result of such information be treated on the 'see safe' principle. 1959 *Bookseller* 24 Oct. 728/3 After say, 20 evening class students have called at a bookshop, all asking for the same title, the bookseller may be moved to telephone an order, three copies firm and three see-safe. 1960 G. A. Glaister *Gloss. Book* 372/2 *See safe*, said of books bought by a bookseller from a publisher..with the understanding that at some future date the publisher may be asked to exchange the bookseller's surplus for copies of another title. 1974 I. Norrie in Mumby & Norrie *Publishing & Bookselling* (ed. 5) II. 423 As the paperback boom gathered force it was increasingly difficult to sell new novels..by unknown writers... What was known as the 'see safe' system (titles which did not sell could be swopped for those which did) was adopted.

see-saw, *v.* **2.** (Earlier example.)

1754 Richardson *Grandison* VI. 285 Your nurse, in your infancy, *see-sawed* you.

seesee (sī·si). Also **see-see, sisi.** [Echoic: see quot. 1969.] In full, *seesee partridge*. A small sand partridge, *Ammoperdix griseogularis*, found in parts of western Asia.

1851 J. Gould *Birds of Asia* VII. Pl. 1 Sportsmen reckon it very easy to kill, and it is said to be delicious eating: the name Seesee expresses its call. 1858 *Proc. Zool. Soc.* XXVI. 503 It [sc. *Ammoperdix bonhami*] is known by the name of 'Sisi' in the Punjab. 1864 T. C. Jerdon *Birds of India* III. 567 (*heading*) The Seesee partridge. *Ibid.* 568 The Seesee is only found in the Punjab. 1923 *Blackw. Mag.* July 125/2 Game abounded—black partridge and sisi, hare and pigeon. 1928 *Ibid.* Apr. 544/2 We had an exciting and noisy hunt after a see-see partridge. 1969 Ali & Ripley *Handbk. Birds India & Pakistan* II. 8 Seesee Partridge... A small sandy grey-brown, concealingly desert-coloured partridge... When flushed on a hillside the birds invariably fly downhill, the wings producing the characteristic high-pitched squeaking *see-see* noise..which has given the bird its name.

see-through (sī·þru), *a.* and *sb.* Also (chiefly *U.S.*) **see-thru.** [f. vbl. phr. *to see through*: See *v.* 24 a.] **A.** *adj.* That can be seen through; transparent, diaphanous; having spaces allowing the passage of light. **a.** Of a fabric or (usu. woman's) garment.

1950 *Life* 10 Apr. 100 (*heading*) See-through fabrics bring undercovering to the surface. 1951 *Sunday Pictorial* 21 Jan. 2/1 'See-through' nighties..may be heavenly for women, but they have many disadvantages. 1960 J. Ironside *Fashion Alphabet* 30 The present trend towards see-through dresses and no-bras indicates that it [sc. the bare breast] is only one step away. 1968 B. Norman *Hounds of Sparta* xix. 143 Some slinky girl spy in a see-through nightdress. 1979 M. Hebden *Death set to Music* iii. 24 You don't normally come down dressed in a see-through robe to answer the door. 1980 *Quilt World* Sept./Oct. 63/1 'Mirror' or 'See-Thru' quilt, make in one day, instructions 50c.

b. in other contexts.

1956 *Sun* (Baltimore) 5 Sept. 19/3 The 40 by 80-foot building at the northwest corner of Light and Cross streets is of the 'open' or 'see-through' style of architecture that reveals the revolution that banking in all its phases has been undergoing. 1966 *Punch* 5 Oct. 506/2 Each resident is given a few square feet of privacy, enclosed by see-through fencing. 1967 'G. Bagby' *Corpse Candle* (1968) ix. 122 'With a glass house..there's no place a man could go there to take his pants off.' 'It's not all that see-through,' I assured him. 1975 G. Seymour *Harry's Game* ii. 22 The troops..with..the medieval Macron see-through shields. 1978 J. Irving *World according to Garp* xv. 314 He cleared a see-through spot on the dusted and caked rear window.

B. *sb.* **1.** The quality of allowing the passage of light; the extent to which it is possible to see clearly through something; unimpeded vision.

1954 *Sun* (Baltimore) 21 Dec. 5/6 (Advt.), In opaque nylon tricot for less see thru. 1957 *Jrnl. Optical Soc. Amer.* XLVII. 785/2 A material with good 'see-through' qualities may, in fact, be quite hazy. 1959 *Motor Man.* (ed. 36) xiii. 273 The driver has a clear 'see-through' if the towing angle is correct. 1969 C. O. Raspor in W. R. R. Park *Plastics Film Technol.* iv. 97 Transparency or 'see through' refers to the capability of seeing objects through a film without loss of detail caused by blurring or distortion. 1974 E. Castagna in P. F. Bruins *Packaging with Plastics* 126 Contact clarity, i.e., see-through to contained liquids, is excellent for copolymers.

2. A see-through fabric or garment.

1962 G. Callingford *Third Party Risk* iii. 42 Might buy 'erself..brushed nylon if she don't fancy the see-through. 1971 *Guardian* 1 June 9/2 Conditioned as we are to seeing hot pants and cool see-throughs worn in city streets. 1974 P. Haines *Tea at Gunter's* xiv. 149 You know—a rented place, Lu; and me in one of those flimsy see-throughs, lying about on the settee.

‖ **Sefer Torah** (sēⁱ·fəɹ tōᵊ·ro). Also **Sepher Torah;** *pl.* **Sifrei Torah** (sifrēⁱ·). [ad. Heb. *sēper tōrā* book of (the) Law; cf. Torah.] = *Scroll of the Law* s.v. *scroll *sb.* 1 e.

1650 E. Chilmead tr. *L. Modena's Hist. Rites, Customes, & Manners of Jews* x. 29 And there are in this Ark, or Chest, sometimes Two, sometimes Four..of these Books: and they are called..Sepher torah, The Book of the Law. 1893 I. Zangwill *Ghetto Tragedies* 9 The *Sepher Torah* is to the Jew at once the most precious and the most sacred of possessions. 1936 S. M. Lehrman *Jewish Festivals* 139 The Megillah is completely unrolled..to distinguish it from the 'Sefer Torah'. 1960 *Commentary* June 495/2 Anyone who inscribes even one letter on a Sefer Torah earns a *mitzvah*. 1973 *Jewish Chron.* 18 May 39/1 (Advt.), Wanted for immediate purchase..two Sifrei Torah. Must be first-class condition. 1976 B. Williams *Making of Manch. Jewry* x. 242 The warden..encouraged them to purchase a *Sepher Torah*.

seg (seg), *sb.*^{3} [Abbrev. of Segment *sb.*] A metal stud attached to the toe or heel of a shoe (or boot) to strengthen or protect from wear.

1958 *Shoe & Leather Trades Buyers Reg., Brands Directory & Diary* 100/1 Blakey's Boot Protectors Ltd.,..Boot protectors..all sorts and sizes, malleable cast hob nails, segs & studs, cricket spikes. 1970 *Guardian* 24 Dec. 9/3 The boys made indoor slides across the hall..scoring great weals in the polish with their segs and blakeys. 1976 A. Hill *Summer's End* viii. 123 Might be metal segs, them half-moons of metal you hammered into the heels and sole-tips of your boots to stop 'em wearing out quick.

seg (seg), *sb.*^{4} **a.** Also **seggie.** U.S. colloq. abbrev. of *segregationist *sb.* Cf. *outseg *v.* **b.** Slang (chiefly *U.S.*) abbrev. of *segregation 1 g.

1965 *Britannica Bk. of Year* 869/2 *Seg*,..a segregationist. Also *seggie*. 1970 *New Yorker* 12 Dec. 107 Fulbright for the first time openly appealed for black votes, because he believed that he couldn't win without them and that the 'seggies'..would vote against him no matter what he did. 1971 *Harper's Mag.* Jan. 35 When people wore the American flag then it was to show that they were not segs, because the segs of course wore the Confederate flag. 1974 *Guidelines to Volunteer Services* (N.Y. State Dept. Correctional Services) 43 *Seg*, segregation unit. 1977 *New Society* 23 June 616/2 He went straight into the segregation unit [at Wormwood Scrubs]... He continued his [hunger] strike simply in order to prevent an early return to 'seg'.

segashuate, var. *sagaciate.

Seger (zēⁱ·gəɹ). [The name of Hermann August *Seger* (1839–93), German ceramics technologist.] *Seger* (also *seger*) *cone*: each of a series of small numbered cones or pyramids made of different mixtures of refractory material and flux so that they melt at different known temperatures, used to indicate the temperature inside kilns, etc.

[1890 *Jrnl. Iron & Steel Inst.* II. 680 Professor Seger's 'normal' clay pyramids..should find extended application. 1894 *Ibid.* XLV. 432 Le Chatelier's pyrometer was used.., as Seger's cones did not appear to exactly meet all requirements.] 1895 *Ibid.* XLVII. 304 These Seger cones give the temperatures with quite sufficient accuracy. 1931 G. W. Tyrrell *Volcanoes* iv. 115 He used a stout iron pipe within which seger cones were fastened at regular intervals, and thrust it into the lava as far as he could. 1964 H. Hodges *Artifacts* i. 40 There are several series of pyrometric cones of which the Seger cones are, perhaps, the most commonly used. 1971 *Materials & Technol.* II. v. 271 The plaque..is heated above 1000°C at about 10°C per minute, and the end point is taken as the temperature (as indicated by the condition of the Seger cones) at which the tip of the test cone is on a level with the base.

seggie: see *seg *sb.*^{4}

segment, *sb.* Add: **4. b.** *Anthrop.* An autonomous sub-branch of a lineage group which remains within the larger tribal or clan structure.

1940 M. Fortes in Fortes & Evans-Pritchard *Afr. Polit. Systems* 243 A maximal lineage has an hierarchical structure. It consists of two or more major segments, each of a lesser span than the (inclusive) maximal lineage... Each major segment comprises lesser segments constituted on the same principle. 1950 M. Gluckman in A. R. Radcliffe-Brown *Afr. Systems of Kinship* 169 The lineages within the clan are usually residential units. Their segments are cores of villages. 1977 Hunter & Whitten *Stud. Cultural Anthrop.* xviii. 397/1 A hierarchical type of authority structure that rests on levels of increasingly inclusive tribal segments.

c. *Linguistics.* A unit forming part of a continuum of speech or (less commonly) text; an isolable unit in a phonological or syntactic system.

1943 K. L. Pike *Phonetics* vii. 107 A *segment* is a sound (or lack of sound) having indefinite borders but with a center that is produced by a crest or trough of stricture during the even motion or pressure of an initiator. 1946 B. Bloch in *Language* XXII. 237 A segment is a word or a sequence of words that does not occur alone as a pause-group in a major sentence. 1953 C. E. Bazell *Linguistic Form* 7 Morphemic segments may frequently be classed together under one morpheme in the American sense, and considered from this standpoint are regarded as allomorphs. 1960 E. Sivertsen *Cockney Phonol.* iv. 122 /ð/ may be manifested, not as a separate segment, but in the dental quality of a preceding apical segment whose phonemic norm is otherwise alveolar. 1964 E. Palmer tr. *Martinet's Elem. Gen. Linguistics* i. 26 The word *puerum*, adequately characterized by the segment *-um* as the object of the verb. 1972 W. Labov *Language in Inner City* iii. 99 As a rule the ordering of variable constraints within a segment is more regular than ordering across segments.

d. *Computers.* (See quot. 1954.)

1954 *Computers & Automation* May 18/2 *Segment*, a part of a complete specific routine, which can be entirely stored in the internal storage and contains the coding necessary to automatically call in and transfer control to other segments. 1963 *Communications ACM* VI. 391/2 Segments of a program could not be treated as independent entities. In general, a symbol name, if used in one segment, could not be used in another segment with a different meaning. 1969 [see *overlay *sb.* 6]. 1977 Hughes & Michtom *Structured Approach to Programming* v. 107 A segment is both a logical and physical subdivision of a module. Logically, it is a subfunction of the module's function. Physically, it is limited to the number of source-code lines that will fit on one printer page of source output (50 to 60 lines).

9. c. In *Linguistics* (see sense 4 c above).

1961 F. W. Householder in Saporta & Bastian *Psycholinguistics* 20/2 Exact boundaries were not as important as some early workers thought; location of segment-centers is in general adequate. 1969 *Language* XLV. 303 They would be stated in terms of component-sized entities (hypophonemes), however, not in terms of segment-sized entities (classical phonemes). 1971 *Archivum Linguisticum* II. 135 The features of articles and the features of suffixes are extracted from this complex and distributed in their proper places by 'segment transformations' or 'segmentalization'. 1978 *Language* LIV. 47 Cf. Eng. *svelte*, *sphere* etc., whose initial clusters violate the segment-sequence constraints of English.

segment, *v.* Add: Now usu. with pronunc. (segme·nt). **2. b.** *Anthrop.* Of a lineage group or clan: to divide into smaller autonomous branches within the larger social structure. Cf. sense *4 b of the sb.

1940 E. E. Evans-Pritchard in Fortes & Evans-Pritchard *Afr. Polit. Systems* 284 In the diagram below, A is a clan which is segmented into maximal lineages B and C. 1965 P. C. Lloyd in M. Banton *Polit. Systems & Distrib. Power* 66 The Ngoni have a lineage structure which continually segments. 1974 L. Mair *Afr. Societies* x. 127 All lineages segment in the course of generations.

3. (Examples in *Computing* and *Linguistics*.) Cf. senses *4 c, d of the sb.

1959 E. M. Grabbe et al. *Handbk. Automation, Computation, & Control* II. ii. 130 A discussion of an automatic system which faces the problem of segmenting a program, either data or instructions, into pieces is given by this group. 1962 C. O. Frake in Gladwin & Sturtevant *Anthrop. & Human Behavior* 75 How do we segment the stream of speech into category-designating units? 1969 P. B. Jordain *Condensed Computer Encycl.* 202 No matter how much central memory is provided for a computer, it will always be possible to write a program too large to fit. The most common way to handle this situation is to segment the program into overlays. 1972 W. Labov *Language in Inner City* ii. 50 Young black children frequently find it difficult to segment *I'm* into *I am*. 1977 Hughes & Michtom *Structured Approach to Programming* vi. 122 A module..is segmented in the same hierarchical fashion that a system or program is developed.

Hence **segme·nting** *ppl. a.* and *vbl. sb.*

1912 J. S. Huxley *Individual in Animal Kingdom* vi. 149 The blastomeres or separate cells of the segmenting egg. 1959 E. M. Grabbe et al. *Handbk. Automation, Computation, & Control* II. ii. 130 It is imperative that a programmer be allowed to override any automatic segmenting and allocation system in order to provide increased efficiency. 1977 Hughes & Michtom *Structured Approach to Programming* vi. 122 Segmenting is best handled at the stepwise refinement stage.

segmentable (segme·ntăb'l), *a.* [f. Segment *v.* + -able.] Capable of being divided into segments (*Linguistics* in quots.). Hence **segmentabi·lity.**

1957 Trager & Smith *Outl. Eng. Structure* 52 The phenomena that are segmentable were analyzed as phonemes of one kind or another. 1962 H. C. Conklin in Householder & Saporta *Probl. Lexicogr.* 122 Unitary lexemes may be either *simple* (unsegmentable) or *complex* (segmentable). 1964 *Language* XL. 207 The 'segmentability' and the quantum mechanics available to linguists strikes these other workers as very rigorous. 1979 *Trans. Philol. Soc.* 82 The 'weak' adjective endings are clearly segmentable suffixed formatives.

segmental, *a.* Add: **2. a.** *segmental apparatus*, the brain-stem of a vertebrate.

1917 [see *palæencephalon]. 1974 D. & M. Webster *Compar. Vertebr. Morphol.* xi. 240 Segmentation in the developing brain is neither as regular nor as apparent as it is in the spinal cord. However, during early development (at least), segmentation can be determined in the basal portion, which is, therefore, known as the 'segmental apparatus'.

c. *Linguistics.* Of, pertaining to, or designating the division of speech or (less commonly) text into segments (cf. *segment *sb.* 4 c). Freq. in phr. *segmental phoneme*, a consonant or vowel phoneme, which can occur as one of the units in a sequence of such phonemes.

1938 B. L. Whorf *Lang., Thought, & Reality* (1964) 126 Timbre phonemics (segmental): Table of phonemes. Pattern congruity. Allophones or probitional variants. Allophonic constellation. 1942 C. F. Hockett in *Language* XVIII. 8 Features..which clearly extend over a series of several segmental groupings are suprasegmental. 1950 R. A. Hall in *Ibid.* XXVI. 12 Forty-seven segmental phonemes is not an excessive number to posit for a language. 1958 *English Studies* XXXIX. 104 The segmental sounds are almost exclusively transferred from indigenous languages of the area. 1966 *Amer. Speech* XLI. 225 Pitch levels are not always suprasegmental features. When they are short enough, they can be considered one of the distinctive features of a segmental phoneme. 1976 *Word* 1971 XXVII. 57 The speech rhythms and patterns and intonations of the pregnant mother will [not] generate..'segmental phonemes', or anything so phonetically sophisticated relatedly. 1981 *Canad. Jrnl. Linguistics* XXVI. 74 The initial portions of the word have the same segmental values and the same relative prominence as the base word.

segmentalization (segme:ntălaizēⁱ·ʃən). [f. next + -ation.] Division into segments; *spec.* in *Linguistics*, transformation of a grammatical feature into a distinct segment of speech or text. Cf. *segment *sb.* 4 c.

1964 E. Becker in I. L. Horowitz *New Sociology* 115 There is another kind of confusion..in complex society, which Mills called attention to as a 'segmentalization of conduct'. 1969 *Language* XLV. 718 The auxiliary *be* is introduced by a segmentalization transformation from features of the following element. 1970 *Canad. Jrnl. Linguistics* XVI. 17 Their [sc. linguistic features'] eventual extraposition (or 'segmentalization') is blocked in the case of participles.

segme·ntalize, *v.* [f. Segmental *a.* + -ize.] *trans.* To divide into segments. Hence **segme·ntalized, segme·ntalizing** *ppl. adjs.*; **segme·ntalizer.**

1956 H. Whitehall et al. in *Kenyon Rev.* XVIII. 413 The constructive features must necessarily serve a descriptive grammar as *segmentalizers*—they are our chief scissors of linguistic perception. 1956 *Ibid.* 417 Both the rhythm-pointing and the segmentalizing functions of the three interrelated configurational features are of basic prosodic importance. 1968 *Language* XLIV. 774 Even when the feature is segmentalized, like *should* in Modern English, it remains an 'auxiliary' of the main verb. 1976 *Brit. Jrnl. Sociol.* XXVII. 317/1 The teaching practice of academic sociology..presents its subject matter in segmentalized courses on stratification, organization, politics, religion, etc.

segmentally, *adv.* Add: (Examples in *Linguistics*.) Cf. *segment *sb.* 4 c.

1957 S. Potter *Mod. Linguistics* v. 105 Sentences may be analyzed *segmentally* into phonological units called *phonemes* and *syllables*; into morphological units called *morphemes* and *words*; and into syntactic units called *phrases* and *clauses*. 1964 R. H. Robins *Gen. Linguistics* vi. 262 Stress and pitch are just as fit to serve as the markers or exponents of grammatical categories and of syntactic relations as are segmentally represented morphemes.

segmentary, *a.* Add: **2.** (Examples in *Anthrop.* use.) Cf. *segment *sb.* 4 b.

1940 E. E. Evans-Pritchard in Fortes & Evans-Pritchard *Afr. Polit. Systems* 284 Fission and fusion are two aspects of the same segmentary principle. 1957 V. W. Turner *Schism & Continuity in Afr. Soc.* x. 291 Overlapping of territorial and kinship groupings such as one finds in segmentary societies. 1977 Hunter & Whitten *Stud. Cultural Anthrop.* xviii. 396/1 Like other Bedouin tribes, Mutayr have what is called a segmentary tribal structure.

segmentation. Add: **2. a.** (Later examples in *Anthrop.* and examples in *Linguistics* and *Computing*.) Cf. *segment *sb.* 4.

1943 M. Schlauch *Gift of Tongues* 254 French has its own formulas of segmentation like *quant à*. 1950 A. R. Radcliffe-Brown *Afr. Systems Kinship* 40 For structures having successive segmentations the term 'polysegmentary' has been suggested. 1953 C. E. Bazell *Linguistic Form* p. i, Space has been found to touch on several aspects of linguistic form which pass unnoticed in more extensive treatments of the subject... The problem of segmentation..has for this reason been left in the background. 1962 C. O. Frake in Gladwin & Sturtevant *Anthrop. & Human Behavior* 76 The segmentation of speech into the grammatically functioning units revealed by linguistic analysis is a necessary, but not sufficient, condition for terminological analysis. 1962 *Spring Joint Computer Conf.* 307/1 Segmentation is the process of dividing a single program into pieces. This is done to permit the operation of programs that are too large to completely fit into memory. 1971 W. Wilder in R. Needham *Rethinking Kinship & Marriage* 213 It might well be possible to decide whether local segmentation among the Purum appeared to lead to the formation of alliance groups. 1973 C. W. Gear *Introd. Computer Sci.* vi. 243 The process of breaking a program into a number of smaller segments is called *segmentation*.

segmentative (segme·ntătiv), *a.* *Linguistics.* [f. Segment *sb.* + -ative.] = *segmental *a.* 2 c (see also quot. 1936).

1936 *Language* XII. 127 Punctual and segmentative aspects of verbs in Hopi... The segmentative aspect is formed by final reduplication of this root plus the durative suffix *-ta*. 1961 *Amer. Speech* XXXVI. 159 This and other such 'segmentative' analyses fail to explain a number of other grammatical facts about nominals.

segmented, *ppl. a.* Add: **3.** *Archæol.* Of a prehistoric gallery (grave): divided into sections or segments; having compartments.

1920 *Glasgow Herald* 9 Mar. 9 Vestiges of a segmented central gallery were disclosed. 1939 V. G. Childe *Dawn Europ. Civilization* (ed. 3) xii. 206 On the slopes of the Pyrenees, [etc.]..gallery graves are divided into a series of intercommunicating compartments by low, transverse slabs... Such tombs are known as *segmented cists*. 1954 S. Piggott *Neolithic Cultures* vi. 160 In the typical form of 'segmented gallery' a number of such slabs divide the chamber into a series of compartments.

4. *gen.*

1950 *Sci. News* XV. Pl. 5 (*caption*) The segmented appearance of the flame is due to the shock wave which is formed in the jet as it leaves the nozzle. 1967 *Jane's Surface Skimmer Systems 1967–68* 59/2 The pockets, based on the segmented skirts of a hovercraft, form a seal around the body.

segmenter (segme·ntəɹ). *Zool.* [f. Segment *v.* + -er^1.] A fully developed sporozoan schizont ready to divide into a number of merozoites.

1929 R. W. Hegner et al. *Animal Parasitol.* xi. 144 (*caption*) *Plasmodium vivax*. 1, Ring stage; 2, schizont; 3, segmenter; 4, gametocyte. 1946 P. F. Russell et al. *Practical Malariol.* ii. 31 (*caption*) A large segmenter in an endothelial cell in the brain. 1978 *Jrnl. Protozool.* XXV. 449/1 In schizonts and segmenters knobs often obscure the unit membrane structure.

sego. Add: In full, *sego lily*. (Earlier and later examples.)

1851 H. Howe *Hist. Coll. Gt. West* 432 Hogs fatten on a succulent bulb or tuber, called the Seacoe, or Seegose Root, which is highly esteemed as a table vegetable by the Mormons. 1852 H. Stansbury *Expl. & Survey Gt. Salt Lake* 160 Sego..is much used by the Indian tribes as an article of food. 1875 *Amer. Naturalist* IX. 18 The general Indian name of 'Sego' is applied indiscriminately to all the edible bulbs of this region [sc. Utah]. 1915 Armstrong & Thornber *Western Wild Flowers* 64 [The mariposa] is called Sego Lily..in Utah and is the 'State flower'. 1963 J. J. Craighead et al. *Field Guide to Rocky Mt. Wildflowers* 18 Sego Lily... A white tuliplike flower with a triangular cup-shaped appearance.

segregable, *a.* Delete *rare*$^{-1}$ and add earlier and later examples.

1905 *Rep. Evol. Comm. R. Soc.* II. 124 An extracted type..may carry on segregable determinants, whereby the individuals may, in reality, differ from each other, though outwardly alike. 1978 P. Sutcliffe *Oxf. Univ. Press* vi. 241 Milford tried to divide his business into 'segregable categories'.

segregant (se·grĭgănt), *a.* Restrict † *Obs.* *rare*$^{-1}$ to *gen.* sense in Dict. and add: **2.** *Genetics.* Having or being a genotype derived by segregation; usu., one different from that of either parent.

1936 *Discovery* May 161/1 The earlier investigators of heredity emphasised the discontinuous nature of inheritance... This is understandable when it is remembered that the segregant types, with which the earlier work was done, were of a very sharply contrasted nature. 1971 *New Scientist* 8 July 92/1 When certain chromosomes were eliminated, the malignant phenotype reappeared and the segregant cells were again able to produce tumours. 1974 *Nature* 27 Sept. 322/1 So called 'tritypes' involving any three of the four segregant genotypes should also be recovered frequently.

B. *sb.* *Genetics.* A segregant organism.

1955 *Genetics* XL. 894 No instances of adenineless segregants were obtained. 1976 *Nature* 29 Apr. 785/1 All 28 haploid segregants of strain DP62..were indistinguishable from strain NP73.

segregate, *v.* Add: **1. b.** To subject (people) to racial segregation; to enforce racial segregation in (a community, institution, etc.). Cf. *desegregate *v.*, *integrate *v.* 2 b.

1908 R. S. Baker *Following Color Line* III. xiv. 299 All through my former chapters I have been showing how the Negroes are being segregated. So are the Chinese segregated, and the blacks in South Africa. 1930 *Economist* 27 Sept. 563/1 It is not surprising that a South African Nationalist politician should..proclaim his preference for his own party's policy of 'segregating' the natives and safeguarding 'the natural superiority' of the whites. 1948 *Rep. Native Laws Commission 1946–48* (Dept. Native Affairs, South Africa) 33/1 This effect was accentuated by the policy which sought to segregate the Africans as far as possible in specially demarcated 'Reserves'. 1958 *N.Y. Post* 20 Apr. II. 7/3, I guess the DAR is not so much for segregating the colored as it is against doing you-know-what with them.

3. b. *Genetics.* To undergo or display segregation (sense *1 e).

1904 W. Bateson et al. in *Rep. Evol. Comm. R. Soc.* II. 120 The fern-leaved type is recessive to the palm-leaved, segregating from it perfectly. 1930 R. A. Fisher *Genetical Theory Nat. Selection* i. 9 Mendel also demonstrated what a theorist could scarcely have ventured to postulate, that the different factors examined by him in combination, segregated in the simplest possible manner, namely independently. 1974 *Encycl. Brit. Macropædia* XIV. 775/1 The once-separate genes have been brought together..to produce a tightly packed unit of several genes sufficiently near each other on the chromosome that they segregate together.

segregated, *ppl. a.* Add: Also *spec.* of institutions, groups, etc.: divided or separated on the basis of race. Cf. *segregate *v.* 1 b.

1948, 1956 [see *integrated *ppl. a.* c]. 1958 *Listener* 11 Dec. 982/1 Nine-tenths of the Negro children in the whole Southern region still go to segregated schools. 1960 *Guardian* 22 Mar. 13/7 San Antonio, Texas, launched its campaign against segregated lunch-counters. 1971 *Graphic* (Durban) 7 May 4/5 You are the future Black citizens of this segregated Republic.

segregation. Add: **1. e.** *Genetics.* The separation of pairs of homologous alleles or chromosomes, esp. as occurs at meiosis in the formation of gametes by a heterozygous organism, to whose progeny different traits may consequently be transmitted.

1902 W. F. R. Weldon in *Biometrika* I. 229 If the hybrids of the first generations [of two races of peas]..be allowed to fertilise themselves, all possible combinations of the ancestral race-characters will appear in the second generation with equal frequency... Characters intermediate between those of the ancestral races will not occur... This may be called the Law of Segregation. 1904 *Rep. Evol. Comm. R. Soc.* II. 128 The balance of evidence is in favour of the belief that gametic segregation takes place at the reduction-division. 1930 R. A. Fisher *Genetical Theory Nat. Selection* i. 8 The segregation of single pairs of genes, that is of single factors, was demonstrated by Mendel in his paper of 1865. 1954 *Genetics* XXXIX. 432 The..abnormal segregation of *s* has been observed whenever C 602/λ has been used as the F-parent in cross. 1970 *Watsonia* VIII. 48 We hope that the present investigation will show clearly that the concept of segregation following allopolyploidy (intergenomic segregation, Jones 1967) can offer another possible explanation of this phenomenon.

f. The enforced separation of different racial groups in a country, community, or institution. Cf. *apartheid.

1903 T. T. Fortune in B. T. Washington et al. *Negro Problem* vii. 215 The Afro-American people have been held together rather by the segregation decreed by law..than by ties of consanguinity. 1916 *Virginia Rep.* CXVII. 692 The cities and towns of this State have the power..to pass segregation ordinances separating the places of residence of white and colored citizens, respectively. 1927 [see *Eurafrican *a.* and *sb.* 3]. 1947 *Forum* (Johannesburg) 17 May 29/1 Political segregation is only possible where territorial segregation is in force. 1952 [see *desegregate *v.*]. 1957 *Times* 18 May 6/3 The ruling of the Supreme Court that racial segregation in public schools [in South Africa] was unconstitutional. 1974 *Spartanburg* (S. Carolina) *Herald* 25 Apr. A1/6 The black majority has the franchise in tribal homelands under South Africa's race segregation policy.

g. The isolation or separate confinement of dangerous or troublesome prisoners. Hence *concr.* (also *segregation unit*) a part of a prison designated for this purpose. Chiefly *U.S.*

1952 K. J. Scudder *Prisoners are People* 82 We set aside a few cells at Chino for segregation... If some man at Chino becomes defiant or refuses to work, he is placed in segregation. 1955 T. E. Gaddis *Birdman of Alcatraz* x. 91 The rear half of the structure held eighteen segregation cells... It was a prison within a prison. 1964 D. Glaser *Effectiveness of Prison & Parole System* viii. 174 The duration of disciplinary segregation is much briefer in federal prisons than in most state prisons... Men in segregation now receive the regular inmate food. *Ibid.* 175 Such units, usually called 'administrative segregation' in federal prisons, resemble the regular disciplinary section of a prison... The inmate may be restricted to quarters temporarily without being transferred to the segregation unit. 1974 P. W. Keve *Prison Life & Human Worth* ix. 158 The segregation unit needs intensive service. *Ibid.* 162 A man..cannot get out of segregation unless he meets certain good behavior standards. 1977 *New Yorker* 24 Oct. 114/3 Such sentences often included a certain number of days in segregation.

segregationist (segrĭgēⁱ·ʃənist), *a.* and *sb.* [f. Segregation + -ist.] **A.** *adj.* Of, pertaining to, or designating persons or policies advocating or supporting political or racial segregation. Cf. *integrationist *a.*

1954 W. K. Hancock *Country & Calling* vi. 171 Segregationist theory had partial relevance in that it emphasized South Africa's diversity of cultural inheritance. 1957 *Economist* 30 Nov. 774/2 One has to go as far to the right as Senator Ellender, the segregationist Democrat from Louisiana. 1961 *Encounter* XVI. 7/1 There is no segregationist legislation in Algeria. 1976 *National Observer* (U.S.) 21 Feb. 6/1 He could not accept its [sc. a club's] 'segregationist policy prohibiting any black guest'.

B. *sb.* An adherent or advocate of segregation. Cf. *integrationist *sb.* and *seg *sb.*^{4} a.

1955 [see *integrationist *sb.*]. 1957 P. Worsley *Trumpet shall Sound* x. 206 One of their spokesmen..posed the 'sixty-four dollar' question of the segregationist 'Would you let a native marry your daughter or sister?' 1962 *Daily Tel.* 2 Apr. 12/2 This is because Southern segregationists are yielding to the inevitable, or at least refraining from militant resistance. 1977 *Time* 24 Jan. 27/1 Joseph Rauh..charged that Bell had given 'aid and comfort to segregationists' while an Atlanta attorney.

† **segregator** (se·grĭgēⁱtəɹ). *Med. Obs.* [f. Segregate *v.* + -or.] An instrument for obtaining the urine from one kidney unmixed with that from the other.

1903 *Ann. Surg.* XXXVII. 30 The segregator could not possibly have attained the results given thus by catheterism and the strong aspiration. 1919 W. C. Pedersen *Urology* xiii. 704 The development of segregators corresponds with that of catheterization cystoscopes.

segue (se·gwēⁱ), *v.* *Mus.* [a. It. *segue*, 3rd pers. sing. pres. of *seguire* to follow.] *intr.* ‖ **1.** (See quot. 1959.)

1740 J. Grassineau *Mus. Dict.* 214 *Segue*, it follows, or comes after; this word is often found before *aria*, *alleluja*, *amen*. 1801 T. Busby *Compl. Dict. Mus.* p. xxxii, The Italian word *Segue*, set against any of these abbreviations, signifies a repetition of the same notes, or passage. *Ibid.* s.v. *Segue*, (Ital.) it follows: as *Segue Coro*, the chorus follows. 1876 Stainer & Barrett *Dict. Mus. Terms* 390/1 *Segue* (It.), follows, succeeds, comes after; as, *segue il coro*, the chorus follows; *segue l'aria*, the aria follows. 1959 *Collins Mus. Encycl.* 592/2 *Segue*,.. 'Follows'. Used as a direction (1) to proceed to the following movement without a break, and (2) to continue a formula which has been indicated, such as arpeggiating of chords or doubling in octaves.

2. *slang.* Of a person or music: to move without interruption from one song or melody to another. Freq. const. *into*. Also *transf.* and *fig.*

1958 D. Halperin in P. Gammond *Decca Bk. of Jazz* xx. 250 Then, without stopping, the guitarist and Ellington segued into *Body and soul*. 1962 'K. Orvis' *Damned & Destroyed* (1966) iii. 26 The Haydn selection reached the oboe part—melody segued and started to build. 1967 A. Arent *Gravedigger's Funeral* (1968) xii. 190 Just then the orchestra segued into something I recognised. 1970 *New Yorker* 12 Sept. 32/2 The first started off with some flourishes from a Bach organ toccata and segued into 'She Comes in Colors'. 1976 C. Larson *Muir's Blood* (1978) xxv. 135 The organist..*segued* resonantly from 'In the Garden' to 'Rock of Ages'.
transf. and *fig.* 1972 G. Baxt *Burning Sappho* v. 78 The crowds..let up a roar which soon segued into a mixture of cheers, jeers, jests, gibes. 1977 *Time* (Chicago) 17 Oct. 79/3 Bertolucci abruptly and wisely segues from the festivities to an epilogue, set in the present, that brings the enormous film full circle. 1978 Logan & Woffinden *Illustr. New Musical Express Encycl. Rock* 199/1 How do the world's most celebrated adolescents [sc. the Rolling Stones] segue into middle age?

segue (se·gwēⁱ), *sb.* *Mus. slang.* [f. prec.] An uninterrupted transition from one song or melody to another. (Used of both live and pre-recorded music.)

1937 *Printers' Ink Monthly* May 42/1 *Segue*, the transition from one musical number to another without break or announcements. 1952 B. Ulanov *Hist. Jazz in Amer.* (1958) xix. 240 We could be doing 'Limehouse Blues' way up in tempo, look at the clock, and do a direct segue into the theme. 1977 *Rolling Stone* 13 Jan. 19/1 The band plays an hour straight, moving fluidly from one number to another; indeed, Parker calls Edwards the 'King of Segue'. 1980 S. Brett *Dead Side of Mike* xv. 162 'He just does a long sequence of slow, sexy numbers, so they can dance real close:..Straight *segue*.' 'What's a *segue*?' 'Record to record, no chat.'

‖ **seguiriyas** (sēⁱgirī·as). Also **seguiriya,** etc. [Andalusian-Gypsy var. of Sp. *seguidilla* Seguidilla.] In full, *seguiriyas gitana* [Sp. *gitano*, fem. *gitana* gypsy]: a regional variety of flamenco music; the song or dance which accompanies this.

1922 I. Brown *Nights & Days on Gypsy Trail* iv. 86 Silverio asked the guitar player to strike up a Gypsy seguirya, and throwing back his head..he sang..the very song that one of the cantadores present had improvised years before. 1926 J. B. Trend *Mus. Sp. Hist. to 1600* 7 The more modern forms (*flamenco*) often sound more 'oriental' than the older, traditional *Cante hondo*; and the oldest, the *Siguiriya gitana*, seems..less tinged with superficial orientalism than any. 1936 W. Starkie *Don Gypsy* xix. 289 The *siguiriya gitana*..is full of tears and tragedy and for that reason is often called *playera* (the weeping poem). 1948 'La Meri' *Sp. Dancing* vii. 92 Of the flamenco estilos which are more distinctly song than dance, the *Seguidillas Gitano* (seguiriyas gitana) is the most typical. It is composed of four lines as against the seven which go to make up the Spanish Seguidillas. 1967 Chujoy & Manchester *Dance Encycl.* (rev. ed.) 856/1 Rhythm-forms of flamenco dances are: Alegrias, Soleares, Bulerias, Farruca, Zapateado, Tango, Zambra, and Seguiriyas. 1972 H. MacInnes *Message from Malaga* ii. 40 Once, she too had danced the *seguiriyas*.

‖ **Seguridad** (segū·riðað). [Sp., security.] The Spanish security service.

1937 F. Borkenau *Sp. Cockpit* ii. 120, I soon found out that these people belonged to the Seguridad, in other words the ordinary police of the old régime. 1938 E. Hemingway *Fifth Column* (1939) II. i. 47 A room in Seguridad headquarters. 1977 *Guardian Weekly* 3 Apr. 13/1 From time to time it's learned that this or that army officer has been replaced by a liberal, or that new police and Seguridad chiefs have been appointed.

Sehna (se·nă). Also **Sena, Senne,** etc. The name of a town (now Sanandaj) in Kurdistan, used *attrib.* and *absol.* to designate a variety of finely-woven Persian rug or a knot used in weaving some oriental carpets (see quot. 1910). Also *Comb.*, as **Sehna-Kurd** (see quot. 1931^2).

1901 J. K. Mumford *Oriental Rugs* xi. 183 In most of the Sehnas the diaper of small patterns covers the entire field. *Ibid.*, A few Sehna rugs have the pear pattern wrought upon a large scale. *Ibid.* 198 The genuine Mir Sarabands are tied in the Sehna knot. 1910 *Encycl. Brit.* V. 393/2 The second traditional knot is the Persian or Sehna knot, which..is tied so that from every space between the warp-threads one end of the knot protrudes. 1931 A. U. Dilley *Oriental Rugs & Carpets* iv. 102 Sehna (Senneh or Senna) rugs..are distinguished by masterful accomplishment in small pattern. *Ibid.* 103 Sehna-Kurd is the name applied to rugs woven by the Kurds in emulation of the Sehnas. 1957 [see *Ghiordes]. 1962 C. W. Jacobsen *Oriental Rugs* 289 The finest old Sena with silk warp will often have 500 to 600 knots to the square inch. A good average new Sena with cotton warp will have from 150 to 250 knots to the inch. *Ibid.* 291 We class many of the better rugs from Hamadan district (woven in small villages) and superior to the usual Hamadan, Sena-Kurds. 1975 'E. Lathen' *By Hook or by Crook* xviii. 167 A very nice, versatile Sena... This Sena is..twenty-six hundred dollars. 1975 *Oxf. Compan. Decorative Arts* 612/1 The Ghiordes knot is quite commonly used in western districts [of Persia] but a fine Sena knotting predominates. 1978 *Times* 13 Oct. 19/4 One expensive carpet, an unsold Senna at £40,000.

‖ **Sehnsucht** (zē·nzuχt). [Ger.] Yearning, wistful longing.

[1847 Thackeray *Van. Fair* (1848) iv. 28 It is no blame to them that after marriage this *Sehnsucht nach der Liebe* subsides.] *a* 1861 A. H. Clough *Poems* (1869) II. 193 (*title*) Sehnsucht. 1862 J. A. Symonds *Let.* 9 Dec. (1967) I. 372 Today I had a wondrous Sehnsucht to hear our choir once more. 1902 W. James *Varieties Relig. Exper.* xvi. 383 An excellent old German Lady..used to describe to me her *Sehnsucht* that she might yet visit 'Philadelphia', whose wondrous name had always haunted her imagination. 1911 G. Bell *Let.* 6 May (1927) I. xii. 303 When the 1st of May came I had a great 'Sehnsucht' for the daffodils and the opening beech leaves at Rounton. 1941 [see *radio *sb.* 3]. 1955 C. S. Lewis *Surprised by Joy* i. 14 The Castlereagh Hills..taught me longing—*Sehnsucht*; made me..a votary of the Blue Flower. 1972 J. I. M. Stewart *Palace of Art* xiv. 141 The stickiest romance of all attends, of course, upon gondolas by moonlight, and Gloria felt she must by no means counter or abridge this small enclave of adolescent *Sehnsucht* in her almost undeviatingly rational friend.

‖ **Seicentismo** (sēitʃenti·zmo). Also **Secentismo.** [It., f. *seicento* Seicento.] = Seicento; also, the character or quality of works produced during this period.

1881 *Encycl. Brit.* XIII. 511/1 This period [sc. the hundred and forty years from the treaty of Cateau Cambresis to the war of the Spanish succession] is known in the history of Italian literature as the Secentismo. Its writers..tried to produce effect with every kind of affectation, with bombast, with the strangest metaphors, in fact, with what in art is called mannerism, 'barocchism'. 1923 *Oxf. Broom* Apr. 17 Timid critics..excusing themselves on the ground that Seicentismo is analogous to the modern spirit, Fascismo, nay—Futurism even. 1926 R. Fry *Transformations* 95 A more guarded enthusiasm for these works than altogether suits recent converts to Seicentismo.

Seicentist. Add: Also *pl.* ‖ **Se(i)centisti.** Also *attrib.* or as *adj.*

1841 W. Spalding *Italy & Ital. Islands* II. III. iv. 414 The *seicentisti*, or artists of the seventeenth century. 1881 *Encycl. Brit.* XIII. 511/2 The 'Secentisti' erred by an overweening desire for novelty, which made them always go beyond the truth. 1931 A. Huxley *Music at Night* I. 38 The bright reassuring Heaven,..the stage immensities and stage mysteries, all the stock-in-trade of the *seicentisti*, are absent from his [sc. El Greco's] pictures. 1938 L. MacNeice *Mod. Poetry* i. 7 Theocritus is more escapist than Euripides..the Seicentist poets than Dante.

Seicento. Add: Also **seicento.** (Later examples.) Also *attrib.*

1926 E. Hutton tr. *A. Venturi's Short Hist. Ital. Art* vii. 339 The glory of the *Seicento* at Milan is summed up in the names of Cerano and Cairo. 1938 *Burlington Mag.* Aug. 82/2 To pass from the seicento to the settecento is inevitably to experience a slackening of tension which the impressive body of Solimenas..do little to check. 1947 D. Mahon *Studies in Seicento Art & Theory* 1 The writer was primarily interested in discovering why this occurred, and became involved..with certain aspects of Seicento art theory. 1972 *Listener* 7 Sept. 317/2 Festive scraps from the Seicento. 1979 *Now!* 21–27 Sept. 116/1 Two outstanding *seicento* marble busts..remind the viewer of J. Pierpont Morgan's greater collection of sculpture formed in America with the help of Henry Duveen.

seiche. Add: A short-lived standing oscillation of a lake or other body of water (as a bay or basin of the sea), somewhat resembling a tide, which may be caused by abrupt changes in atmospheric conditions or by small earth tremors. (Later examples.)

1905 *Geogr. Jrnl.* XXVI. 46 A seiche was observed..within the shelter of the pier at the east end of Loch a' Chroisg... The amplitude was a quarter of an inch, and the period about 11½ minutes. 1932 *Geogr. Rev.* XXII. 476 The strongest current caused by the seiches was found in the south-east corner of Great Bear Lake. 1957 G. E. Hutchinson *Treat. Limnol.* I. v. 299 The phenomenon has long been recognized locally in the Lake of Geneva, and the term seiche was recorded by Fatio de Duillier (1730) as applied to the oscillation in that lake. 1962 *New Scientist* 13 Sept. 560/1 Wind can also cause whole lakes and bays to oscillate..and these movements, called 'seiches', can communicate energy to the ground. 1971 *Nature* 4 June 306/2 The dominant internal seiche has a wavelength twice the length of the Loch, and a marked asymmetry.

Hence **sei·ching,** the occurrence of a seiche; the motion occurring in a seiche.

1955 *Sci. Amer.* Jan. 2/2 The extreme heights reached by the water in specific locations is due to another phenomenon known to oceanographers as 'seiching', which is similar to the sloshing of water in a bathtub. 1967 *Oceanogr. & Marine Biol.* V. 27 In contrast, Corkan..and Rossiter..found..little evidence of North Sea seiching. 1971 *Nature* 4 June 308/1 It is likely that the effect of the Earth's rotation is principally to cause a small alternating transverse tilt of the isotherms across the Loch during the seiching motion.

Seidel^1 (zai·d'l). *Ophthalm.* [The name of Erich *Seidel* (1882–1946), German ophthalmologist, who described the sign (see below) in 1914 (*Archiv f. Ophthalm.* LXXXVIII. 102).] Used in the possessive, as **Seidel's sign,** the occurrence of one or two hooked scotomata extending from the blind spot.

1918 R. H. Elliot *Glaucoma* iv. 220 Should the condition of increased intra-ocular pressure continue, Seidel's sign will pass on into Bjerrum's sign. 1932 *Optician* LXXXIII. 398/1 With 2/1,000 white definite Seidel's signs were evident as horns extending about 1½ in. both above and below the projected blind-spot. 1964 S. Duke-Elder *Parsons' Dis. Eye* (ed. 14) xxi. 305 Occasionally there is a sickle-shaped extension of the blind-spot above or below, or both, with the concavity of the sickle directed towards the fixation point (Seidel's sign); this is of more doubtful significance [as a symptom of glaucoma].

‖ **seidel**^{2} (zai·d'l). [Ger., orig. a liquid measure varying locally between about a third and a half litre.] A beer mug or glass (in quot. 1930 used *loosely*); the quantity that such a vessel will contain.

1922 E. E. Cummings *Enormous Room* iv. 92 Such..hands as might have grasped six seidels..on 13th street. 1930 D. Runyon in *Collier's* 1 Feb. 12/1 Wilbur Willard all mulled up to a million, what with him having been sitting out a few seidels of Scotch with a friend. 1957 M. Swan *British Guiana* xi. 180, I was welcomed with a seidel of Pilsener. 1980 G. V. Higgins *Kennedy for Defense* ii. 14 Knackwurst, German potato salad, couple seidels of Jake's Special Dark.

seif (sīf, sēⁱf). *Physical Geogr.* Also **sif.** [ad. Arab. *saif*, lit. 'sword'.] A sand dune having the form of a narrow ridge elongated in a direction parallel to that of the prevailing wind. Also *seif dune*.

1925 W. J. H. King *Myst. Libyan Desert* xxiii. 221 In front of us..was a high three-headed *sif*, or longitudinal sand dune. 1931 *Geogr. Jrnl.* LXXVIII. 16 In form it is a typical *sif* dune, a long straight ridge of sand with a single longitudinal chain of crests rising to billowy pyramids set at regular intervals, in silhouette something like huge saw-teeth. 1933 *Ibid.* LXXXII. 107 The Prince's map of this area shows a sudden change in the direction of the straight parallel lines of *seif* dunes which cross the country. 1941 R. A. Bagnold *Physics of Blown Sand & Desert Dunes* xv. 234 The early observers..held that it blew at right angles to the dune direction, whereas later observations over a longer period show that it blows parallel to the dunes... Both are right, since cross-winds are essential for seif formation. 1953 *Sci. News* XXVII. 16 The greatest height recorded for a *seif* dune is 210 metres... Their lengths vary from about 60 to 120 kilometres. 1975 *Nature* 20 Feb. 617/2 Until the recent drought, seif dunes were mainly active in this region north of the 150-mm isohyet. 1976 *Ibid.* 26 Feb. 654/1 Nearby, near the oasis of Bilma, a small dune field of seifs and barchans again indicates that the wind at right angles to the barchans would be to the left of the main trend.

Seigneur. Add: **a.** Also *transf.* in extended use.

1924 Wodehouse *Ukridge* vii. 156 'Yes, yes, yes,' said Ukridge, with testy impatience, quite the seigneur resenting interference from an underling. 1978 A. Morice *Murder by Proxy* i. 20 They used to own all the land... He still sees himself as the Seigneur of the neighbourhood.

c. In the Channel Islands, the lord of the manor; *spec.* the overlord of Sark.

1694 P. Falle *Acct. Isle of Jersey* iv. 114 The place of Bailly being lately become vacant by the..death of Sir Philip de Carteret,..Seigneur of St. Ouen,..the States of the Island..have..chosen for Judge Delegate Philip le Geyt. 1815 T. Quayle *Agric. Islands on Coast of Normandy* 298 Some small specimens of copper ore are said to have been discovered... Little hope is entertained by the Seigneur of the fief [sc. Sark], that the quantity of ore will be found considerable. 1835 H. D. Inglis *Channel Islands* 313 The lord of Serk is the sole lay impropriator of tithes. The tithe paid to the seigneur is the tenth sheaf of wheat, barley, oats, beans, &c. 1856 C. Le Quesne *Constitut. Hist. Jersey* ii. 31 At the Assize d'Héritage, or first day of sitting, the principal feudal seigneurs, or lords holding *in capite* from the Crown, are bound to appear. 1885 A. Edwardes *Girton Girl* I. v. 118 He had come to be Seigneur of Tintajeux through the inheritance of his Guernsey wife. 1935 E. Platt *Sark as I found It* ix. 67 Mr. Collings, the late Seigneur, had more than one offer during his lifetime to restart the mines. 1957 *Encycl. Brit.* V. 231/1 In Jersey, if a land-owner dies without a direct heir, the *seigneur* claims for a year and a day the income of his estate; the *seigneurs* of St. Ouen's and La Motte still license taverners. 1976 *Times* 10 Mar. (Channel Islands Suppl.) p. ii/3 You can cross to Sark where buying a freehold property depends simply on availability and the consent of the island's Seigneur.

seigneurial, *a.* Add: Also *fig.*, lordly; authoritative.

1970 *Times Lit. Suppl.* 23 July 787/2 In the United States, Linguistics has long derived authority from the presence there of the two most seigneurial of living linguists, Roman Jakobson and Noam Chomsky. 1972 A. Friedman in Cox & Dyson *20th-Cent. Mind* I. xii. 428 Conrad's irony is heavy and fuming, his seigneurial distance from his madmen..woefully great.

Seigneurie. Add: **2. b.** In the Channel Islands, the residence of a Seigneur (sense *c).

1935 E. Platt *Sark as I found It* ix. 65 The Seigneurie must always be the centre of entertaining in Sark. 1978 *Times Lit. Suppl.* 26 May 572/5 There is a nice photograph of him in a group of them outside the Seigneurie, looking as if there is a nasty smell under his nose.

‖ **Seilbahn** (zai·lbān). [Ger., f. *seil* cable, rope + *bahn* way, road.] A cable railway; an aerial cableway.

1963 I. Fleming *On Her Majesty's Secret Service* ix. 99 Together with the Gemeinde, the local authorities, he constructed the Seilbahn. 1972 *Daily Tel.* 12 Feb. 14/7 Sonntag, a delightful old village clustered round a tall-spired church has a *seilbahn* from which the venturesome may 'sail' across the deep ravine in a cable cab to Stein.

seine, *sb.*^{1} Add: **c.** *seine twine* (earlier example); **seine-net**: also as *v. intr.*, to fish with a seine; hence **seine-netter, seine-netting** *vbl. sb.*

1961 Seine-net *v.* [see *nephrops]. 1947 A. C. Hardy *Seafood Ships* vii. 99 The seine netter..is invariably under 100 ft. long and of wooden construction. 1970 *Cape Times* 28 Oct. (S.A. Fishing Rev.) 1/2 The days when trawlers and seine-netters could make maximum hauls virtually on their doorsteps are over. 1905 J. Johnstone *Brit. Fisheries* p. xxviii, This latter method of seine-netting brings us to the consideration of the methods of the inshore or longshore man. 1977 *Grimsby Even. Tel.* 14 May 7/7 A list of..stretches of water to be closed to trawling and seine netting. 1770 *Boston Gaz.* 12 Mar. 4/2 (Advt.), The right sort of three-threaded Sein Twine.

‖ **se ipse** (se ipse). [Eng. adaptation of L. *se ipsum*; cf. Ipse *pron.*] Himself: used emphatically with preceding sb.

1853 W. Jerdan *Autobiogr.* IV. xiii. 238 It was Wordsworth *se ipse*... The ideal was complete. 1943 G. Cope *Democracy within Church* 8 The then Archbishop of Canterbury did not hesitate to rebuke his sovereign..and so demonstrated that the clerical oath of allegiance is not to the monarch *se ipse*.

seir-fish, seer-fish. Add: Also *ellipt.* as **seer.**

1913 C. F. Holder *Game Fishes of World* viii. 83 The Seer..leaps eight feet and has a fighting weight of fifteen pounds. 1971 *Fashion Panorama* (Ceylon) Apr.–June (Advt., verso front cover), 12 lbs. comprising Seer, Crabs or Prawns.

‖ **seises** (se·isez), *sb. pl.* [Sp., (*los*: also used) *seises* (the) sixes, pl. of *seis* six.] The choristers (formerly six, now usu. ten) in certain Spanish cathedrals, esp. Seville, who perform a ritual dance with castanets before the altar during the octave of Corpus Christi and certain other festivals.

1845 R. Ford *Hand-bk. for Travellers Spain* I. ii. 255/2 At this Octave and at Corpus, the Quiresters or *Seises* (formerly they were six in number) dance before the high altar [of Seville Cathedral] with castanets and with plumed hats on their heads... They are dressed as pages of the time of Philip III. They wear blue and white for the Virgin, red and white for Corpus. 1885 E. de Amicis *Spain & Spaniards* ix. 324 The most curious privilege..of the Seville Cathedral, is the..dance of *los seises*, which takes place every evening.., for eight consecutive days, after the festival of *Corpus Domini*. 1903 A. Symons *Cities* 128, I returned to the Cathedral to see the dance of the Seises... The sixteen boys..came forward and knelt before the altar. 1926 J. B. Trend *Mus. Sp. Hist. to 1600* v. 85 The ten little *Seises* who dance, sing, and clack their castanets before the high altar of Seville Cathedral for the festivals of Corpus Christi and the Immaculate Conception. 1938 B. Schönberg tr. *C. Sachs's World Hist. Dance* vii. 337 The classical number of participants in the *Morris Dance*, six, was once the same for the Spanish cathedral dance..—and even today, although there are ten performers, they are still known as *los seises*, 'the sixes'. 1941 G. Chase *Music of Spain* xvi. 256 The *seises* (choirboys) of Seville Cathedral, who every year, during the octave of Corpus Christi, dressed in quaint costumes of the time of Carlos III.., dance before a special altar to the sound of their own castanets and the accompaniment of an orchestra. 1969 S. Sitwell *Gothic Europe* xii. 132 Dancing with castanets before the high altar..is performed by the *seises* or choristers.

seismal, *a.* (Example.)

1977 A. Hallam *Planet Earth* 78/3 Gravimetric, seismal and geothermal investigations of several rift zones.

seismic, *a.* Add: **1. a.** Also, pertaining to or involving earth vibrations produced artificially by explosions. *seismic survey*, † (*a*) a survey of an area in connection with its liability to earthquakes; (*b*) a survey (for oil and gas) employing seismic methods.

1887 *Trans. Seismol. Soc Japan* X. 36 The first method of avoiding earthquake motion in a given district is to make a seismic survey of that area. 1935 *Geol. Natural Gas* (Amer. Assoc. Petroleum Geologists) 664 Seismic surveys isolated a relatively small area with structural closure on the upthrown side of the fault. 1941 *Bull. Amer. Assoc. Petroleum Geologists* XXV. 1261 The unsolved problems in the field of seismic surveying require for their solution a more complete understanding of the properties of the earth strata. 1944 A. Holmes *Princ. Physical Geol.* xii. 206 Near the middle of the sheet the ice has been shown by seismic methods..to be over 8,000 feet thick. 1969 *Times* 16 July 4/2 The scientific equipment bay where the seismic experiment and the laser rangefinding device are stowed. 1972 *Daily Tel.* 1 May 2/7 Seismic studies show these areas could be almost as rich in oil and gas as the North Sea. 1979 *Jrnl. R. Soc. Arts* CXXVII. 405/1 The first stage in oil and gas exploration is usually in the form of seismic surveys.

b. *fig.*

1962 S. E. Finer *Man on Horseback* xii. 222 These bands of successor-states form so many seismic zones of preternatural military eruption. 1979 A. Draper *Fish* 10 Churchill's seismic decision to empty the national coffers in one bold operation..was only taken when the Germans were hammering at the door.

2. Special collocations: **seismic prospecting,** prospecting by investigating the propagation in rock of artificially created elastic waves; **seismic reflection,** the reflection of elastic waves at boundaries between different rock formations; usu. *attrib.*; **seismic refraction,** the refraction of elastic waves on passing between formations of rock having different seismic velocities; usu. *attrib.*; **seismic sea-wave** = *tsunami; **seismic velocity,** the velocity of propagation of elastic waves in a particular rock; **seismic wave,** an elastic wave in the earth produced by an earthquake or by artificial means.

1929 *Trans. Amer. Inst. Mining & Metall. Engineers* LXXXI. 626 In seismic prospecting in general, fewer data are employed than in pure seismology. 1963 C. A. Heiland *Geophysical Exploration* ix. 439 Uses of seismic prospecting in mining have been few. 1932 *Trans. Amer. Inst. Mining & Metall. Engineers* XCVII. 469 A second application of the method of least squares is the outlining of a buried structure by seismic reflections. 1940 *Bull. Amer. Assoc. Petroleum Geologists* XXIV. 1391 The sequence of methods of examination..is likely to be that of reconnaisance geology, detailed geologic mapping,..and core-drilling or seismic reflection examination. 1977 A. Hallam *Planet Earth* 29/2 In the search for oil-bearing strata, geophysicists make use of the seismic-reflection technique to the virtual exclusion of all other exploration methods. 1934 *Trans. Amer. Inst. Mining & Metall. Engineers* CX. 473 (*heading*) Seismic refraction methods as applied to shallow overburdens. 1979 *Nature* 3 May 56/2 In 1977 an 800-km long seismic refraction profile was shot across Iceland and along the south-eastern flank of the Reykjanes Ridge. 1905 C. Davison *Study of Recent Earthquakes* vi. 163 (*heading*) Seismic sea-waves. 1944 A. Holmes *Princ. Physical Geol.* xvii. 362 Strong submarine earthquakes are followed by seismic sea waves. 1966 *McGraw-Hill Encycl. Sci. & Technol.* XIV. 355/1 Explosions or collapses taking place under water may cause water waves, sometimes of great size. These fall within the class of tsunamis (seismic sea waves). 1935 *Bull. Amer. Assoc. Petroleum Geologists* XIX. 1 The idea that the longitudinal seismic velocity of sediments increases with the age of the sediments has been held rather generally for some time. 1977 A. Hallam *Planet Earth* 11/1 Observed seismic velocities..allow the variation of density with depth in the Earth to be determined. 1900 H. Nagaoka *Elastic Constants of Rocks & Velocity of Seismic Waves* (Publ. Earthquake Investigation Comm. in Foreign Languages No. 4) 65 The investigation of the seismic waves affords the best means of feeling the pulse of the interior of the earth. 1925 M. Mühlberg in I. A. Stigand *Outl. Occurrence & Geol. of Petroleum* 217 The transmissive velocity for seismic (as also acoustic) waves is different in various media. 1966 *McGraw-Hill Encycl. Sci. & Technol.* XII. 147/1 Seismic waves from explosions and earthquakes occur in the frequency range from about 100 to 1/3000 cps.

seismically *adv.*: also, by seismic methods.

1941 *Bull. Amer. Assoc. Petroleum Geologists* XXV. 1258 The average depth to which the Gulf Coast has been seismically surveyed. 1978 *Nature* 27 Apr. 791/2 The absence of a seismically detectable Tertiary–Quaternary sediment cover.

seismics (sai·zmiks), *sb. pl.* [f. Seismic *a.*: see -ic 2.] Seismic exploration techniques, esp. such techniques considered as a whole.

1934 *Trans. Amer. Inst. Mining & Metall. Engineers* CX. 462 Roman has developed the application of the method to seismics very satisfactorily. 1976 D. McLeod tr. *Beckmann's Geol. Petroleum* II. iv. 64 Reflexion seismics are the most frequently-used, but also the most expensive of all methods used in petroleum exploration.

seismocardiography (sai:zmokāɹdiǫ·grăfi). [f. Seismo- + Cardiography.] The analysis of movements of the chest as a means of studying those of the heart. So **seismoca·rdiogram,** the record made by this process; **sei·smocardiographic** *a.*, of or pertaining to seismocardiography.

1962 *Probl. Space Biol.* I. 461 The constructional details of a seismocardiographic transducer on dogs are given. *Ibid.* 463 Seismocardiography is a form of ballistocardiography. *Ibid.*, The recording of the seismocardiograms is impossible during movements. 1965 *Proc. 14th Internat. Astronaut. Congr., 1963* III. 38 During weightlessness the animal showed an absolute and relative increase of the first cycle of the seismocardiogram. 1970 L. I. Kakurin et al. in D. E. Busby *Rec. Adv. Aerospace Med.* 43 Biomedical monitoring was carried out by means of a unified sensor system. It comprised seismocardiographic (SCG) and pneumographic (PG) sensors. 1979 *Human Physiol.* V. 542 The use of seismocardiography in clinical, athletic, and space medicine,..and for mass examinations of the population is..a promising development.

seismograph. Add: Also as *v. trans.* and *intr.*, to study (a region) by means of seismography; to prospect for oil by seismic methods; **seismogra·phically** *adv.* (*fig.* in quot. 1964).

1925 M. Mühlberg in I. A. Stigand *Outl. Occurrence & Geol. of Petroleum* 218 The times of arrival are recorded seismographically. 1940 *Trans. Amer. Inst. Mining & Metall. Engineers* CXXXVIII. 305 (*heading*) Continuous profiling method of seismographing for oil structures. 1964 *New Statesman* 10 Apr. 577/1 In the quintessential Hanley lodging-house, every creak or whisper registered seismographically. 1968 C. Helmericks *Down Wild River North* xxvi. 423 They go 'seismographing' right overland through the wilderness. 1969 J. Mavor *Voyage to Atlantis* ix. 203 Dr Marinatos suggested we seismograph the whole field in order to determine whether the Minoan layer was relatively level and therefore a logical town site.

seismologically, *adv.* Add: (Examples.) Also *fig.*

1955 *Sci. Amer.* Sept. 59/1 Seismologically speaking, the crust differs from the underlying part of the mantle in the fact that P and S waves travel in it more slowly and with more variable speed. 1970 *Nature* 15 Aug. 649/2 This earthquake was a nice reminder that Britain is not as seismologically quiet as people believe. 1977 H. Greene *FSO—1* xix. 171 In an arm of government as seismologically sensitive as the Foreign Service,..Friburn's sudden resignation was a shock on the level of six or seven on the diplomatic Richter scale.

seismologist. (Earlier and later examples.)

1859 R. Mallet in *Rep. Brit. Assoc. 1858* 133 The subject appears to me worthy of more examination at the hands of Vulcanologists and Seismologists. 1955 *Sci. Amer.* Sept. 168/2 Our values agree well with those obtained by seismologists, who measure the thickness of the crust by the travel time of earthquake waves. 1973 *Nature* 27 July 195/2 This region of northern California, highly populated with seismologists and seismometers, is proving to be quite a testing ground for prediction.

seismonasty (sai·zmonæsti). *Bot.* [f. Seismo- + *nasty *sb.*^{2}] A nastic movement made in response to a mechanical shock. Hence **seismona·stic** *a.*, of or pertaining to a movement of this kind.

1912 W. H. Lang tr. *Strasburger's Text-bk. Bot.* I. 323 (*heading*) Seismonasty. *Ibid.*, Every disturbance resulting from a mechanical shock acts as a stimulus... These movements are termed seismonastic. 1951 [see *nastic *a.*]. 1965 Bell & Coombe tr. *Strasburger's Textbk. Bot.* (new ed.) 382 A whole series of plants displays movements of this kind after touching, a phenomenon termed seismonasty. *Ibid.* 383 Many petals, stamens and stigmata show seismonastic reactions.

seismotectonic (sai:zmotektǫ·nik), *a.* [f. Seismo- + Tectonic *a.*] Of, pertaining to, or designating features of the earth's crust, such as faults, which are associated with or revealed by earthquakes; † *seismotectonic line* (see quots. 1907, 1924).

1907 W. H. Hobbs in *Beiträge zur Geophysik* VIII. 224 A tendency of the damaged communes to be arranged in essentially right lines (seismotectonic lines) is noteworthy. 1907 —— *Earthquakes* ii. 32 Buried planes may, however, often be traced as lines of destruction especially marked out upon the surface of the ground. Such straight lanes of special damage from earthquake have been called seismotectonic lines, or structure lines revealed by earthquakes. 1924 *Bull. Seismol. Soc. Amer.* XIV. 31 Hobbs, 1907, describes as seismotectonic lines certain structure-lines which he believes to be revealed by earthquakes. 1974 *Nature* 19 Apr. 661/2 Here, we describe three additional earthquake faults found during recent seismotectonic field studies in Iran. 1980 *Ibid.* 19 June 529/2 An aftershock study was begun..in an effort to examine seismotectonic processes near Mt. St. Helens.

Seistan (sī·stān). Also **seistan.** [The name of a low-lying region of eastern Iran and south-western Afghanistan.] A strong north-

westerly wind prevalent in this region in the summer months. Also *attrib.*, as *Seistan wind*.
1906 *Geogr. Jrnl.* XXVIII. 224 The extraordinary frequency and violence of the Seistan wind, and the regularity with which it blows from the same quarter, are very remarkable. **1940** *Chambers's Techn. Dict.* 754/1 *Seistan*, the 120-day summer north wind in E. Persia. **1959** R. E. Huschke *Gloss. Meteorol.* 503 The seistan is associated.. with the deep summer-time low over northwest India. **1967** R. W. Fairbridge *Encycl. Atmospheric Sci. & Astrogeology* 1155/2 The 'seistan' or 'Wind of 120 days' is the characteristic wind of Seistan in eastern Iran, which blows from late May till the end of September, from north-northwest, sometimes with velocities from 70–120 mph.

Seitz (zəits). [See quot. 1944: a proprietary term in the U.S.] Used *attrib.* and in *Comb.* with reference to filtration, as **Seitz disc,** a small disc of compressed asbestos fibres, used for filtration; **Seitz filter,** a type of filter in which liquids are purified by passage through a readily replaceable Seitz disc; so as *v. trans.*; **Seitz-filtered** *ppl. a.*, **Seitz filtration; Seitz pad,** a Seitz disc.
1925 *1st Progress Rep. Foot-&-Mouth Dis. Res. Comm.* (Min. of Agric. & Fisheries) App. ii. 20 A routine form of asbestos filter of German manufacture, the Seitz filter, has been employed with success. **1928** T. M. Rivers *Filterable Viruses* ii. 59 Smaller filters with Seitz discs have been made for use in experimental work. [**1944** *Official Gaz.* (U.S. Patent Office) 4 Apr. 12/2 Republic Filters, Inc., Paterson, N.J... *Seitz.* The trade-mark is a facsimile of the signature of Mr. Seitz, the late inventor of the basic construction of the filtering apparatus manufactured and sold by applicant.] **1946** *Nature* 31 Aug. 293/2 After the fungus has grown in the liquid medium for about two weeks, the culture medium is Seitz-filtered and collected in autoclaved tubes and ampoules. **1947** *Ann. Rev. Microbiol.* I. 378 The second experiment.. involved thirty-six men who were divided into three groups of twelve each and sprayed, respectively, with untreated sputum and throat washings, with the material filtered through sintered glass or Seitz pads, or with the same material autoclaved. **1956** *Nature* 21 Jan. 134/1 Another 2 ml. of the same medium, containing 0·05 per cent aesculin sterilized by Seitz-filtration, [was] added. *Ibid.* 10 Mar. 481/2 (*caption*) Curve *B* shows the results in Seitz-filtered artificial duodenal contents alone. **1961** *Lancet* 29 July 227/2 The pH was adjusted to 7·4 and the medium sterilised by Seitz filtration. **1964** M. Hynes *Med. Bacteriol.* (ed. 8) ii. 25 The Seitz filter has almost entirely replaced all other types of filter... Filtration is carried out through an asbestos pad clamped to the bottom of a metal container.

sei whale (sē¹ wē¹l). Also **sejhval;** (*erron.*) **seivhal.** [Anglicization of Norw. *sejhval*, f. *sei* coal-fish + *hval* Whale *sb.*] A blue-grey rorqual, *Balænoptera borealis*.
1912 *Rep. Brit. Assoc.* 158 The Right whales and Sejhvals are said to appear only during the earlier part of the season. **1916** R. S. Andrews in *Mem. Amer. Mus. Nat. Hist.* I. 295 The vernacular term Sei Whale (*Sejhval*), adopted in this memoir, is the name by which *Balænoptera borealis* is known to the Norwegians. **1919** R. W. Clark in E. Shackleton *South* I. 364 The sperm-whale and the sei-whale have shown a good deal of seasonal variation. **1923** *Chambers's Jrnl.* Jan. 112/1 The lesser rorqual, the seivhal, and the humpback, another whalebone whale, take some small fish. **1939** *Nature* 16 Dec. 999/2 (*caption*) Species..Sei, Right, Sperm. **1958** *Times* 12 Nov. 11/6 A unit is defined as equalling one blue whale..or six sei whales. **1974** *Country Life* 11 Apr. 866/1 With the cessation of whaling in the Faeroes and Norway the numbers of fin and sei whales in those areas..will..increase.

seize, *v.* Add: **III. 11.** Also with *up*. Of a machine or mechanism: to stick, jam, or lock fast; to become unworkable, as by reason of undue heat or friction. Also *fig.*
1917 *Blackw. Mag.* May 807/1 Our engine recovered slightly now that its recovery was not so important, and it behaved well until it seized up for better or worse when we had landed. **1963** *Listener* 31 Jan. 198/2 As for the camera itself, at 40 below zero the wind-on mechanism jammed and the range-finder seized up. **1981** P. Audemars *Gone to her Death* ii. 44 Better we should find the trouble on our hoist rather than having something seize up on the main road.
fig. **1955** Cormack & McDougall in C. Morris *Social Case-Work in Gt. Britain* (ed. 2) i. 35 When the social service system was primitive it could do without case-work: the more elaborate modern machine would seize up. **1960** C. Day Lewis *Buried Day* ix. 182, I read the book; then, for hour after hour, I sat trying to think of something to say about it. I could not... My brain had seized up. **1976** Dexter & Makins *Testkill* 100 Any exercise.. might make me seize up.
Hence **seize-up** *Mech.*, the action or state of seizure (cf. sense *11 of the vb.).
1912 *Motor Man.* (ed. 14) vi. 232 Unless one makes quite certain that every bearing and cylinder be properly relubricated before starting up again, a 'seize-up'..is not improbable.

seized, *ppl. a.* (Later examples.)
1911 E. W. Walford *Maintenance of Motor Cars* ii. 30 The procedure is to allow the parts to cool down..and to inject a copious supply of oil to the seized part. **1977** F. Durbridge *Passenger* iii. 154 He walked the six miles into Reading..told a story about a seized engine.

seizing, *vbl. sb.* **1.** (Later examples in sense 11 of the vb.)
1911 E. W. Walford *Maintenance of Motor Cars* ii. 30 An engine may suddenly pull up, and it be found impossible to turn the crankshaft with the starting handle. The popular expression for this is 'seizing'. **1925** *Morris Owner's Man.* 29 Neglect of this results in harsh running and an overheated engine, loss of power, and finally 'seizing-up' of pistons or connecting rods.

‖ **Sejm** (sẽ¹m). Also **Seym.** [Pol.] In Poland: a general assembly or diet; a parliament; *spec.* (since 1921) the lower house of the Polish parliament.
[**1698** B. Connor *Hist. Poland* II. ii. iii. 83 The Grand Diet or Parliament of Poland, by the Natives call'd Seym Walny.] **1893** W. R. Morfill *Poland* v. 40 Ladislaus had assembled the first known *Seym*, or Diet, at Checiny... It consisted of the princes..prelates, barons and knights. **1916** G. E. Slocombe *Poland* ix. 85 Their [sc. the nobles'] first step was to obtain a revival of the *sejm* or general Diet of Poland. **1924** A. E. Tennant *Studies in Polish Life & Hist.* ix. 162 The new Sejm is a democratic assembly in which all classes are represented. **1934** E. J. Patterson *Poland* iii. 83 Under the constitution the powers of the Seym were very great... In case of a disagreement between the two houses the Senate could be over-ridden by a majority of 11/20 in the Seym, in ordinary session. **1959** Z. Pelczynski in D. E. Butler et al. *Elections Abroad* ii. 125 The Sejm had been evolving and adapting itself to the policy of 'democratisation'. **1981** *Financial Times* 13 Jan. 1/3 It will be discussed at a special Polish Communist Party congress this spring, then go for approval to the Sejm (Parliament).

sejunct, *a.* For *rare*⁻² read *rare* and add later nonce-use.
1922 Joyce *Ulysses* 378 Images, divine and human, the cogitation of which by sejunct females is to tumescence conducive.

‖ **sekere** (sekere·). Also **shekere.** [Yoruba.] A Yoruba gourd-rattle.
1921 S. & O. Johnson *Hist. Yorubas* i. viii. 121 The Calabash drum—ornamental with strings of cowries—is called *Sèkèrè*. **1937** I. O. Delano *Soul of Nigeria* xiii. 160 The acrobatic dancers..tour the Yoruba land... They have their beaters, and what they beat is called 'Sekere'. These are not drums, but gourds to which a number of cowries have been attached, so that they make a noise when shaken. **1963** W. Soyinka *Lion & Jewel* 61 Distant music. Light drums, flutes, box-guitars, 'sekere'. **1975** *New Yorker* 29 Sept. 41/2 The other instruments are.. and two shekeres—large West African calabashes strung with beads.

‖ **sekos** (sī·kǫs). *Egyptology.* Also **secos.** [a. Gr. σηκός pen, enclosure.] A sacred enclosure in an ancient Egyptian temple.
1820 G. Belzoni *Narr. Egypt & Nubia* iii. 291 Having observed, that the part where the sekos and cella must be was not touched, I set the men to work there. **1837** *Penny Cycl.* IX. 316/2 Facing its entrance was that leading into the *sekos* or shrine containing the figure of the deity. **1887** *Dict. Architecture* (Architectural Publ. Soc.) VII. 45/1 In the Egyptian temples the *secos* was the same as the *adytum* of the Greeks and Romans.

‖ **Sekt** (zekt). [Ger.; cf. Sack *sb.*³] A German sparkling white wine or champagne.
1920 G. Saintsbury *Notes on Cellar-bk.* vi. 84 The 'Cabinet *Sekt*'..was 'a very *German*' champagne. **1924** [see *flute *sb.*¹ 3 a]. **1951** R. Postgate *Plain Man's Guide to Wine* v. 96 A horrible German imitation [of champagne] called Sekt. **1962** P. Purser *Peregrination 22* xvi. 72 Do you remember the time in Berlin you were so sick after drinking too much sekt? **1971** 'M. Sinclair' *Sonntag* xxii. 129 The bottle of *Sekt* in front of them was good though.

selachyl (sĭlē¹-kəil, -kil). *Chem.* [a. G. *selachyl* (Tsujimoto & Toyama 1922, in *Chem. Umschau auf dem Gebiete der Fette*, etc. XXIX. 36): see Selachian *a.* and *sb.* and -yl.] *selachyl alcohol*: an oily liquid, α-glyceryl *cis*-9-octadecenyl ether, $C_{21}H_{42}O_3$, found in the liver oils of elasmobranch fishes.
1922 *Chem. Abstr.* XVI. 1513 The formula $C_{20}H_{40}O_3$ [sic] was selected to represent this unsatd. alc. and the name selachyl alcohol given to it. **1944** *Jrnl. Biol. Chem.* CLV. 448 Selachyl alcohol has been isolated from natural sources so far only as an oil. **1964** *Oceanogr. & Marine Biol.* II. 177 The most abundant alcoholic component of the alkoxydiglycerides is invariably selachyl alcohol, followed by chimyl alcohol and then by batyl alcohol.

seladang (sĕlā·dæŋ). Also **saladang, salandang, sladang.** [Malay, in Borneo *seladang*, in Sumatra *saladang*.] = Gaur; also, formerly, the Malayan tapir, *Tapirus indicus*.
1821 T. S. Raffles in *Trans. Linnean Soc.* XIII. 270 It [sc. *Tapirus Malayanus*] is known by different names in different parts of the country [sc. Sumatra]. By the people of Limun it is called Saladang. **1839** T. J. Newbold *Straits of Malacca* I. vii. 435 The Seladang is supposed by some zoologists to be identical with the Tapir. **1868** *Proc. Asiatic Soc. Bengal* 194 The Malayan bison, an animal well known to the Malays under the name of *Sladang*..[is] more formidable when wounded than the tiger. **1875** E. Blyth *Cat. Mammals & Birds of Burma* 47 In the Malayan peninsula, where it is known as the *Salandang*, this animal [sc. the Gaur] would appear to be becoming extremely rare. **1905** T. R. Hubback *Elephant & Seladang Hunting* 44 The seladang has no dewlap and no hump. *Ibid.* 47 Seladang are only found in little-inhabited districts. **1927** H. M. Tomlinson *Gallions Reach* xxxv. 263 A likely corner for sladang, the instantaneous bull which does not wait for trouble but makes it when you are not looking. **1933** L. Ainsworth *Confess. Planter in Malaya* 219 The Seladang or Malayan Bison is diminishing in numbers, and is now found almost exclusively in Negri Sembilan and Pahang. **1965** R. McKie *Company of Animals* viii. 130 Seladang (this spelling will be used for both singular and plural) are among the largest of the ox tribe and are remote relatives of domestic cattle. **1974** R. Butler *Buffalo Hook* iv. 28 The seladang is the wild buffalo of Malaysia.

Selah. (Later allusive examples.)
1906 J. London *Let.* 1 Dec. (1966) 229, I want an answer... I want it to the point... Selah. **1924** E. E. Cummings *Let.* 13 May (1969) 107 The Independants found me not incapable of a 40′ × 50′ '*abstract*' canvas which..hung very well (by itself)—this lidel effut cost me 9 days work and was dry on time, selah-sounds. **1947** W. Stevens in *Q. Rev. Lit.* III. 110 The dove in the belly builds his nest and coos, Selah, tempestuous bird. **1975** *New Yorker* 26 May 33/3 Then I can go back to doing what every really dedicated writer wants to do. Nothing. Selah, Phil.

‖ **selamlik** (selā·mlik). Also with capital initial. [Turk., lit. 'place of greeting', f. *selām* a. Arab. *salām* Salaam *sb.*) + *-lik* place.]
1. A room in a Muslim house set aside for business or the reception of male friends; the part of such a house reserved for men. Cf. *haremlik.
1838 D. Urquhart *Spirit of East* I. xxi. 392 The European arrives... Some of the attendants, in reply to his inquiries, point to the door of the Selamlik. **1854** R. Curzon *Armenia* 79 We went into the selamlik (or reception room) together. **1888** K. Bey *Evil of East* vii. 127 Every Turkish house is divided into two distinct parts:—the *selamlik* for men, and the *haremlik* for women. **1900** 'Odysseus' *Turkey in Europe* vii. 331 One feels that.. their existence is really divided into the departments of Selamlik and Harem, which means the difference, not only between the men's and the women's part of the house, but between formal and domestic life. **1936, 1941** [see *haremlik]. **1965** J. Fleming *Nothing is Number* i. i. 15 The *selamlik*, or male reception room, is inviolate.
2. The public procession of the Sultan to a mosque on Friday at noon. Now *Hist.*
1888 'Bey Kesnin' *Evil of East* vi. 125 On Fridays, for Selamlik, the soldier makes himself as smart as possible to escort his sovereign to the mosque. **1905** *Globe* (N.Y.) 21 July 1/2 It is reported here officially from Constantinople that during to-day's selamlik an attempt was made to assassinate the sultan. **1935** H. Edib *Clown & his Daughter* ix. 45 Selim Pasha decided to speak about Rabia's musical training to the Imam after the Selamlik—the Friday ceremony of His Majesty's going to the mosque. **1955** H. Luke *Old Turkey & New* (rev. ed.) vii. 166 On Friday, the 10th November, 1922 Vahid ed-Din attended his last *selamlik*. **1980** J. Lees-Milne *Harold Nicolson* 56 The First Balkan War had begun. That day [sc. 17 Oct. 1912] Harold witnessed the ceremony of the Selamlik.

Selbornian (selbǭ·ɹniăn), *a.* Also **Selburnian.** [f. *Selborne*, the name of a village in Hampshire + -ian.] **1.** Of, pertaining to, or connected with Selborne, the parish described in *The Natural History and Antiquities of Selborne* (1789) by Gilbert White (1720–93). Also as *sb.*, an inhabitant of Selborne or an admirer of the writings of Gilbert White.
1869 J. R. Lowell in *Atlantic Almanac* 32/1 How pleasant is his innocent vanity in adding to the list of the British, and still more of the Selbornian, *fauna!* **1928** *Observer* 26 Feb. 8/5 Another book about 'Gilbert White', ..a book full of pleasures for all Selbornians. **1938** H. J. Massingham *Writings of Gilbert White* I. p. x, He writes of the Selburnian echo. *Ibid.* p. xxii, How absorbed the greatest and most typical of Selburnians was with gardening. **1954** R. M. Lockley *Gilbert White* i. 9 Already he was able to convey his love of the Selbornian scene.
2. *Geol.* A name given by A. J. Jukes-Brown (see quot. 1900) to the Upper Greensand and Gault beds in the Albian stage of the Cretaceous in Southern England, from the prominent occurrence of these deposits near Selborne. Also *absol.* Now *rare* or *Obs.*
1900 A. J. Jukes-Brown *Cretaceous Rocks Brit.* I. 1 The Selbornian comprises the beds which are generally known as the Gault and the upper Greensand. *Ibid.* 31 Gault clay and greensand are only two of the different kinds of deposits that make up the group for which the name Selbornian is now proposed. **1910** [see *Albian *a.* and *sb.*]. **1922** J. C. Hughes *Geol. Story of Isle of Wight* vi. 37 Some prefer to call the Lower Greensand Vectian, from Vectis, the old name of the Isle of Wight, and the Upper Greensand Selbornian, a name generally adopted, because it forms a marked feature of the country about Selborne.

seldom, *a.* For '*Obs.*..action' read 'Now chiefly *U.S.*' and add further examples.
1883 'Mark Twain' *Life on Mississippi* xxi. 222 The seldomest spectacle on the Mississippi to-day is a wood-pile. **1959** W. Golding *Free Fall* i. 26 My seldom night terrors. **1961** E. Wilson in Webster s.v., With her small seldom smile.

seldomly, *adv.* (Later example.)
c **1864** E. Dickinson *Poems* (1955) II. 610 The ships.. That touch how seldomly Thy Shore?

seldseen, *a.* (Later *poet.* example.)
1916 E. Blunden *Pastorals* 30 Even as she flung the seld-seen gaud away.

select, *sb.* (Later N. Amer. examples.)
1961 S. Taylor in Webster s.v., It is possible to buy ware that is composed wholly of selects. **1968** *Globe & Mail* (Toronto) 15 Jan. 17/6 The All-Star game tomorrow night with the selects meeting the Stanley Cup defending champions from Toronto.

selectable, *a.* Delete *rare* and add later example in sense 'capable of being selected'.
1975 *Gramophone* Nov. 950/3 Low and high pass filters..with two selectable cut-off points for bass and treble are also featured.

Selectasine (sĭle·ktăsīn). The proprietary name of a colour-printing process which uses a single silk screen for each of the colours (see quot. 1934).
1918 *Official Gaz.* (U.S. Patent Office) 29 Jan. 1075/1 Method of delineating or reproducing pictures and designs... Assignors to Selectasine System, San Francisco, Cal., a Corporation of California. Filed Dec. 1, 1915. **1926** H. L. Hiett *Man. Stencil Screen Process Work* 79 Invented by Edward A. Owens, San Francisco, Cal., assignor to Selectasine System, Inc., San Francisco. **1934** H. Curwen *Processes of Graphic Reproduction in Printing* I. 40 Selectasine is best suited to showcards of very bold design... A sheet of fabric ready impregnated with wax is the basis, and one sheet is used for each colour of the design. The wax filling is removed from the fabric following the shape of the colour to be worked, and a stencil is thus formed. **1948** H. Missingham *Student's Guide in Commercial Art* II. 137 (*heading*) Silk screen or Selectasine process.

selectee (sĭlektī·). *U.S.* [f. Select *v.* + -ee.] A person selected for military service under the Selective Service system; a draftee. Now chiefly *transf.*
1940 *Army & Navy Register* (U.S.) 30 Nov. 7/2 The public considers the percentage of rejections of selectees high. **1942** *Newsweek* 27 Apr. 41/3 The Selectee is to be given a furlough up to ten days to arrange his personal affairs. **1958** *Optima* Mar. 47/2 The suspicion remains that the selectee is being favoured because he is regarded as the 'comer'—the man most likely to succeed in the immediately foreseeable future. **1979** *Tucson (Arizona) Citizen* 20 Sept. 10A/3 Most of Carter's selectees ..passed in a breeze.

selection. Add: **2. a.** Also, a musical passage or a sequence of selected musical extracts.
1857 W. Acton *Prostitution* vii. 103 [We]..enjoyed in a grim kind of way the 'selection' from some favourite opera. **1899** *Observer* 1 Oct. 5/3 The 'selection', which always begins the second part of the concert, was the familiar fantasia on English airs. **1929** *Radio Times* 8 Nov. 411/1 Selection of Songs by T. C. Sterndale Bennett. **1945** S. Hughes in C. Madge *Pilot Papers* I. 94 The brass band repertoire..consists principally of marches.. waltzes and light opera selections. **1968** M. Ball (*title*) Selections from the classics for chime bar music making.
5. a. (sense 1) *selection board, committee, panel, test*; **selection pressure** *Biol.*, differential mortality or fertility such as tends to make a population adapt genetically; **selection restriction** *Linguistics*, a syntactic or semantic restraint on the concurrence of dependent lexical items; **selection rule** *Physics*, any of a number of rules which describe, within certain limits, which particular quantum transitions can occur in an atom, molecule, etc., and which are 'forbidden'.
1940 R. S. Lambert *Ariel & all his Quality* xi. 302 Methods of appointing new staff..by advertising vacancies and setting up selection boards. **1976** L. Deighton *Twinkle, twinkle, Little Spy* xvi. 162 Douglas was sent to a swanky private school..but was still unable to pass the U.S. Army officers' selection board. **1909** *Daily Graphic* 26 July 6/3 It is safe to believe that the Selection Committee has done its best, the players will do their best, Australia will do its best. **1932** Q. D. Leavis *Fiction & Reading Public* I. ii. 22 The Book Society...was started..in 1927 ..with a Selection Committee of five novelists and journalists. **1978** J. Pudney *Thank Goodness for Cake* 97, I was short-listed and looked over by a selection committee. **1974** *BP Shield Internat.* Oct. 8/1 Conference members..were screened by a selection panel. **1944** J. S. Huxley *On living in Revolution* 79 [In the Australian area] there is less scope for variation,..so that general selection-pressure never became so intense. **1977** J. L. Harper *Population Biol. of Plants* ii. 46 Groups of species caught in different evolutionary pathways may face the same selective forces, e.g. a selection pressure to disperse. **1964** Katz & Postal *Integrated Theory of Linguistic Descriptions* ii. 15 Each reading in the dictionary entry for a lexical item must contain a *selection restriction*, i.e., a formally expressed necessary and sufficient condition for that reading to combine with others. **1976** *Word 1971* XXVII. 133 One reason is that the selection restrictions of the verb *throw* require an animate subject. **1931** H. P. Robertson tr. *H. Weyl's Theory of Groups & Quantum Mech.* iv. 198 The selection rule for the inner quantum number *j* is obtained in an analogous manner. **1977** I. M. Campbell *Energy & Atmosphere* viii. 218 The origin of the selection rules for radiative transitions between electronically excited states and the ground state lies in fundamental quantum theory. **1935** *Discovery* Jan. 17/2 The most anxious moment in the design of a selection test now comes. If we take a group of workers..can the new test sort them into their correct relative order? **1967** Wills & Yearsley *Handbk. Managem. Technol.* xi. 202 Are we right to ignore *selection tests*?
b. *Forestry.* Used *attrib.* with reference to a system of forest management under which there is a continuing selection of individual trees for felling over the whole area, on the basis of their saleability.
1891 W. Schlich *Man. Forestry* II. 133 The term *selection system* was introduced into India; it is perhaps not an ideal term, since a certain amount of selection is practised in all systems; it has been retained, as none better is at present available. **1911** H. S. Graves *Princ. handling Woodlands* ii. 72 The development of the individual trees in a selection stand is somewhat different than in an even-aged stand. **1935** N. C. Brown *Gen. Introd. Forestry U.S.* viii. 105 The selection method is likely to be best adapted to general silvicultural and economic conditions found in this country. **1950** *Q. Jrnl. Forestry* XLIV. 15 Rabbits are probably the greatest drawback to the selection system or any irregular system in this country. **1979** O. Kuthanová tr. *Jeník's Pict. Encycl. Forests* 451 (*caption*) Diagram showing selection felling in a high forest; four stages of forest with alternating generations of trees in man-made clearings.

selectional (sĭle·kʃənăl), *a.* [f. Selection + -al.] Of or pertaining to selection. Freq. in *Linguistics*; *selectional restriction* = *selection restriction* s.v. *selection 5 a.
1961 R. B. Long *Sentence & its Parts* ii. 40 Predeterminer modifiers generally are adverbial in function, and mensurant, selectional..or adjunct-like in force. **1964** *Language* XL. 38 There are, then, selectional restrictions between determiners and relative clauses. **1968** *Ibid.* XLIV. 579 On the other hand, there are cogent observations about the role of 'suffixoids' such as *-er* of *hammer* as selectional factors favoring the development of homophonous derivational affixes. **1971** *Nature* 9 Apr. 410/2, I identified all the maxima and minima of PDS for a given population and summed the successive absolute differences to give me an index that combined both selectional and random factors. **1977** *Word 1972* XXVIII. 89 The NP analysis of copular predicates can be retained for comparatives if *nios* is viewed as some sort of denominalizing particle which makes a comparative adjective compatible with the selectional restrictions of *tá*.
So **sele·ctionally** *adv.*
1958 P. Maas *Textual Crit.* 13 The task of proving the existence of the conjecturally (or 'selectionally'..) presupposed errors plays a considerable..role in textual criticism. **1965** N. Chomsky *Aspects of Theory of Syntax* ii. 116 We label as Noun the one that is selectionally dominant.

selectionism (sĭle·kʃəniz'm). [f. Selection + -ism.] The belief that evolution proceeds by natural selection; opp. to Lamarckism.
1912 A. Tridon tr. *Delage & Goldsmith's Theories Evol.* xi. 163 (*heading*) A discussion of Roux's theory; its merits. —Its relation to selectionism and Lamarckism. **1944** J. S. Huxley *On Living in Revolution* vi. 70 In the last twenty-five years..an enormous amount of new facts about evolution have been discovered, and the balance has now swung over heavily, and, I think, permanently, in favour of Darwinism or selectionism. **1982** *New Scientist* 15 Apr. 162/1 Neutralism never seriously claimed to do more than ride piggyback on selectionism.

selectionist, *sb.* and *a.* For **a, b** read **A, B** and add: **A.** *sb.* **a.** (Later example.) † **b.** One who believes that evolution proceeds primarily by natural selection for small differences (*Obs.*); opp. *mutationist; **c.** One who holds a selectionist view of genetic variation (cf. sense B. b below).
1909 W. Bateson *Mendel's Princ. Heredity* i. 3 If species had really arisen by the natural selection for impalpable differences,..the limits between species should be..indefinite... The selectionists believe..that it represents the facts of nature. **1911** [see *mutationist]. **1959** *Encounter* Sept. 62/2 The selectionist must assume..that consciousness is *useful*. **1979** *Sci. Amer.* Nov. 96/3 Selectionists maintain that for a mutant allele to spread through a species it must have some selective advantage.
B. *adj.* **a.** (Later example.) **b.** Of or pertaining to the belief that the majority of observed genetic variation is maintained by natural selection rather than by random effects.
1944 J. S. Huxley *On Living in Revolution* vi. 69 By Darwinism..was meant the selectionist theory of the method of evolution. **1971** *Nature* 13 Aug. 487/1 They maintain that a selectionist hypothesis would predict radical changes to be favoured over conservative changes. **1974** *Ibid.* 1 Nov. 62/2 A selectionist interpretation of the more rapid rate of molecular evolution in a living fossil is possible. **1978** *Sci. Amer.* Sept. 45/1 Selectionist evolution..is neither a chance phenomenon nor a deterministic phenomenon but a two-step tandem process combining the advantages of both.

selective, *a.* Add: **1. d.** *Psychol.* Applied to the capacity for, or process of, selection manifested by the mind or senses in reacting to certain stimuli and not to others, esp. *selective attention*.
1875 W. James in R. B. Perry *Thought & Char. W. James* (1935) I. 528 The *whole* mass of impressions falling on any individual are chaotic, and become orderly only by selective attention and recognition. **1935** G. Murphy *Briefer Gen. Psychol.* xii. 210 From the point of view..of what takes the most prominent place in consciousness, there is a further selective function to be considered. **1958** D. E. Broadbent *Perception & Communication* ii. 15 The performance of selective listeners seems to vary with information as defined by communication theory, rather than with amount of stimulation in the conventional sense. **1968** Trabasso & Bower *Attention in Learning* i. 18 Our primary goal is to offer new theory and new results regarding selective attention in discrimination learning. **1978** G. Underwood *Strategies of Inform. Processing* vii. 237 The selective rehearsal of some members of the memory set in preference to others.., the selective 'forgetting' of some items previously encoded.
e. *Educ.* (See quot. 1960–1.) Also applied to any school within the selective (as opposed to the comprehensive) system. Cf. *comprehensive *a.* 1 d.
1926 *Educ. of Adolescent* (Board of Educ. Consult. Comm.) iii. 79 As post-primary education develops, the schools dealing with the post-primary or secondary stage of education should include..schools of the type of the existing selective Central Schools, which give at least a four years' course from the age of 11+. **1955** O. Banks *Parity & Prestige in Eng. Secondary Educ.* viii. 97 Both the selective central school and the junior technical school originated in the period before the first world war. **1960–1** *Where?* Winter 17/1 *Selective school*, a school for which the pupils have been selected (usually at 11 plus) on the grounds that they can benefit from a more academic education. **1974** *Times* 17 Apr. 1/5 Over half of our secondary school pupils are still in selective schools. **1978** *Jrnl. R. Soc. Arts* CXXVI. 335/1 The decision to change our schooling from a selective to a comprehensive system.
f. *Philos.* Applied to a doctrine of realism put forward in the early 20th century which maintained that sense-data exist in material objects and that the senses of the perceiver select those which are appropriate to be registered.
1932 H. H. Price *Perception* ii. 40 Now let us turn to the *selective* interpretation. According to this, the somatic data..merely..enable us to be conscious of environmental data. **1944** W. T. Stace in P. A. Schilpp *Philos. B. Russell* 365 The view of selective realism is quite different... The sense-qualities actually exist in the object, whether it is perceived or unperceived, just as common sense supposes. What the sense-organ does is to *select* which of the sense-qualities we shall perceive. **1967** *Encycl. Philos.* VII. 78/2 Their..attempt..to deal with illusions..is a version of what is often called the selective theory.
2. Special collocations: *selective breeding*; *selective employment tax* (also with capital initials), a tax levied by the British Government between 3 May 1966 and 1 Apr. 1973, payable on all employees at varying rates and refunded to employers in selected branches of manufacturing industry; cf. S.E.T. s.v. *S 4 a; *selective service* (U.S.), a system of military service (from 1917 to 1973) under which draftees were selected from those persons required to enrol; *selective strike*, a form of industrial action in which union labour is not totally withdrawn but withheld in chosen sectors; *selective weedkiller*, a substance which kills some plant species without harming others.
1931 J. S. Huxley *What dare I Think?* i. 40 Selective breeding I have just touched upon. **1971** F. Hamilton *World Encycl. Dogs* 354 Selective breeding down through the last 150 years has produced the Foxhound of today. **1966** *Times* 4 May 1/1 A selective employment tax payable by employers only is to be introduced by the Government from September 5 to divert manpower from the service industry into manufacturing. **1973** *Guardian* 13 Apr. 13/5 Selective Employment Tax ended on 1 April 1973. As an employer, you may be eligible for a refund of S.E.T. **1978** *Jrnl. R. Soc. Arts* CXXVI. 405/1 The next phase was the indictment of the so-called 'candyfloss industries' and their punishment through the Selective Employment Tax. **1917** *New Republic* 9 June 148/2 The New Republic advocated the principle of selective service for this emergency. **1920** E. H. Crowder *Spirit of Selective Service* v. 119 There were to be two phases to Selective Service, the one, enrollment, the other selection. **1969** G. L. Wamsley *Selective Service & Changing Amer.* i. 1 A step can be taken toward understanding why Selective Service has come under attack after years of anonymity. **1976** N. Thornburg *Cutter & Bone* vii. 157 Nor did he meekly submit to his Selective Service draft notice. **1959** *Wall St. Jrnl.* 29 Dec. 2/2 Selective strikes would be expedient to the union, since they would concentrate pressure on individual companies. **1979** *Arizona Daily Star* 1 Apr. A1/5 The powerful Teamsters union early today called 'selective strikes' against a portion of the trucking industry. **1928** H. Martin *Sci. Princ. Plant Protection* ix. 170 The peculiar virtues of the metallic sulphates as selective weedkillers. **1965** *Listener* 22 July 142/1 Historically, selective weedkillers have been used since 1895 when solutions of copper sulphate were used to destroy charlock in cereal crops. **1976** *Field* 18 Nov. 1040/2 Base fertilizer for lawns should be applied in March or early April..and the selective weedkiller a week or two later.

selectivist (sĭle·ktivist). [f. Selective *a.* + -ist.] One who supports a selective theory or policy. Hence **sele·ctivism.**
1932 H. H. Price *Perception* ii. 44 The Selectivist is asked to say *which* colour belongs to the rose when nobody

is looking at it. **1944** W. T. Stace in P. A. Schilpp *Philos. B. Russell* 365 There are grave objections to selectivism, but that is not the present point. **1967** *Guardian* 4 Sept. 14/8 The selectivists are right in saying that much more should be done to channel social benefits to the people who need them most. **1969** *Physics Bull.* Mar. 108/1 They are apt to ignore work not published in their country's journals, a selectivism matched by the singlemindedness of the articles.

selectivity (se:lekti·vĭti). [f. Selective *a.* + -ity.] **1.** *Radio.* The ability of a receiver to tune separately to signals of adjacent frequencies, measured by the frequency difference between the half-power points of the pass-band of the receiver.
1903 *Electr. World & Engin.* 1 Aug. 173/2 The system of selectivity devised by Sir Oliver [Lodge] was the outcome of a series of classical experiments based upon his syntonic Leyden jars in which one jar is caused to discharge through a circuit by the sparks of another jar, provided the two circuits are of equal electrical dimensions. **1930** *Manch. Guardian* 20 Sept. 15/7 It is no exaggeration to say that..four-valve sets are more than equal in power and selectivity to a six or seven-valve set of the old type. **1943** C. L. Boltz *Basic Radio* viii. 131 The sharper is the resonance curve the better it will select, for supplies at frequencies quite near the resonant one will have very much less effect. The sharpness of the curve is therefore a measure of the selectivity. **1976** *Which?* Sept. 203/2 If you want to receive particularly weak signals, you need a set which has good selectivity.
2. *gen.* The quality of being selective.
1948 E. G. Boring et al. *Found. Psychol.* x. 218/1 This selectivity of perception amounts to giving one sense impression a clear track. **1951** Ruesch & Bateson *Communication* iii. 84 An alternate solution of the psychiatrist is to abstract and condense his observations before writing them down; selectivity, therefore, becomes an unavoidable issue. **1969** *Computers & Humanities* III. 278 The main program was the Formatting Program in which the two files were used as input to produce six indexes, four of which were selective. A 'selectivity module' had been introduced to make the selections possible. *a* **1974** R. Crossman *Diaries* (1976) II. 471 We were able to discuss one of the minor sensations of the recess—Ray Gunter's speech in August, in which he said we must be realistic and brave and face up to the need for selectivity and the means test. **1978** G. Underwood *Strategies of Information Processing* vii. 235 To information processing theory..attention and selectivity are central features.

selector. Add: **c.** (Further examples.) *spec.* (*a*) in a gearbox, the part that moves the gearwheels into and out of engagement; (*b*) *Teleph.*, a mechanism which automatically establishes electrical connection with one of a group of available contacts according to the number of impulses in the incoming signal; (*c*) in a motor vehicle with automatic transmission, the control by which the driver selects the mode of operation of the transmission.
1903 *Electr. Rev.* (Chicago) XLIII. 583/1 Each subscriber is connected to the exchange by two lines which end in what is called a first selector switch... The first selector consists of a couple of relays. **1926** H. T. Rutter *Mod. Motors* II. vii. 262 By moving the gear lever by hand it actuates what are termed selector rods, which are forked rods that slide the gear wheels on the gear-shaft one at a time into a position to mesh with the respective gear wheels on the lay-shaft. **1930** *Bell System Tech. Jrnl.* IX. 22 By placing several selectors in series a network of central offices may be built up, each office serving 10,000 telephones. **1961** *Listener* 16 Nov. 832/1, I would have been waiting, like some predetermined selector, for the necessary keys to fall into position. **1967** K. H. Brinkmann tr. *K. Trautman's Design of Automatic Telephone Exchanges* 16 Each selector has a relay set which controls the switching functions... Ten final selectors would suffice to serve 100 subscribers since each switch interconnects two subscribers. **1967** *Times* 31 Mar. 3/7 The coroner said the selector lever must have been in 'drive'. **1973** D. Barnes *See the Woman* (1974) i. 9 Conrad stopped, pushed the selector to park..and opened the car door. **1973** H. Fantel *True Sound of Music* (1974) vii. 105 To start with, any amplifier has a selector switch. As its name implies, it lets you pick the program source you want to hear: radio, record player, or tape. **1976** L. Thomas *Dangerous Davies* ii. 13 The record..swooped again onto the turntable at her touch of the selector button.
d. *Sport.* One of a number of officials appointed to select a team.
1928 *Daily Mail* 2 Apr. 14/1 The selectors could not find in England a team good enough to stop this very perfect Scottish machine. **1934** F. J. C. Gustard *Eng. v. Austral.* 9 His action may have saved the selectors a certain amount of embarrassment. **1953** B. Harris *Cricket Triumph* i. 19 Laker did not come into the side until the third test match, for in the selectors' minds he and Roy Tattersall..were of equal talent. **1976** J. Snow *Cricket Rebel* 138, I said somewhere in a newspaper article that 1974 summer that the selectors would have to be shot before I made a Test comeback. I think my assessment was right.
Hence **selecto·rial** *a.*, of or pertaining to a team selector (sense d above).
1959 *Times* 12 Oct. 15/4 The manner of its achievement cannot but raise some doubts in selectorial minds. **1963** A. Ross *Australia 63* 13 Selectorial opinion, for one reason or another, was hardening against Sheppard. **1975** *Cricketer* May 37/1 Mr Holloway..seems to feel (as indeed I do) that selectorial bias in favour of the southern and more particularly Home counties is an inescapable fact.

selectorate (sĭle·ktŏrĕt). [Blend of Selector and Electorate.] That section of a political party which has the effective power to choose a representative.
1967 P. Paterson *Selectorate* i. 27 Within the Conservative selectorate there are almost as many grades and ranks and classifications as in a masonic order. **1969** *Daily Tel.* 10 Dec. 14 Most [primaries], while broadening the selectorate, are still open to the criticism that the many are brought in only at the final stage, to exercise a choice over a few candidates chosen by a committee. **1980** *Times* 20 June 14/4, I favour confining the selectorate to MPs.

Selectric (sĭle·ktrik). Also **selectric.** [Blend of Select *v.* and Electric *a.* and *sb.*] A proprietary name for a kind of electric typewriter.
1964 *Official Gaz.* (U.S. Patent Office) 29 Sept. TM 211/2 International Business Machines Corporation, Armonk, N.Y. Filed May 15, 1964. *Selectric* for typewriters. First use July 21, 1961. **1967** *Trade Marks Jrnl.* 8 Nov. 1716/2 *Selectric*... Electrically operated typewriting machines; electrically operated machines... International Business Machines Corp... New York... 23rd February 1967. **1970** [see *golf ball* s.v. *golf *sb.* b]. **1971** *Computers & Humanities* V. 2/3 If a magnetic tape selectric typewriter (MT/ST) or a keypunch is used, the sorting and page-entering processes can be eliminated. **1977** J. Wambaugh *Black Marble* vi. 76 They can also..start operating *your* Selectric.

selectron (sĭle·ktrǫn). Also **Selectron.** [f. Select *v.* + *-tron.] A kind of cathode-ray tube formerly used in computers as a means of storing digital information.
1947 *Math. Tables & Other Aids to Computation* II. 229 'The selectron—a tube for selective electrostatic storage' by Dr. Jan Rajchman. **1947** [see *instruction 4 c]. **1950** W. W. Stifler *High-Speed Computing Devices* xiv. 370 The Selectron utilizes the fact that an insulated secondary-electron emitter can be made to 'float' at either of two stable potentials. **1957** R. K. Richards *Digital Computer Components & Circuits* vi. 265 Among the more important forms of electrostatic storage are the 'Williams tube', the 'barrier-grid tube', the 'holding-gun tube', and the Selectron, each of which has at one time or another been successfully used in a digital computer.

selen-. **1.** **selensulphur,** delete † and substitute for def.: a native variety of elemental sulphur containing a small proportion of selenium; (later examples).
1882 [see *hieratite]. **1944** C. Palache et al. *Dana's Syst. Min.* (ed. 7) I. 142 Old specimens of reddish brown selensulfur from Vulcano gave *n* [sc. refractive index] 2·544–2·675 indicating 83–90·5 per cent Se; but this high Se content lacks confirmation.

selendang, var. *slendang.

selenious, *a.* (Earlier example.)
1827 Mitscherlich & Nitzsch in *Q. Jrnl. Sci., Lit. & Art* II. 471 The acid is isomorphous with the sulphuric acid, and may with propriety be called selenic acid, that described by M. Berzelius being considered as the selenious acid.

selenium. **a.** For 'increased' read 'decreased' in small-type note.
b. **selenium cell,** a photoconductive or photovoltaic cell containing selenium (later examples).
1929 [see *photoconductive *a.*]. **1946** *Nature* 20 July 88/1 In the early days of his training he [sc. J. L. Baird] devised an improved pattern of selenium cell, which led him to develop a crude form of television. **1977** J. Hedgecoe *Photographer's Handbk.* 32 Your meter may use a selenium cell which generates a minute current of electricity, measured on a galvanometer.

seleniuretted, *a.* (Later example.)
1947 *Electronic Engin.* XIX. 363/1 On exposure to a moist atmosphere, sulphuretted, seleniuretted and telluretted hydrogen can be detected.

selenocentric, *a.* (Examples.)
1970 N. Armstrong et al. *First on Moon* v. 118 On July 17 at 1300 Moscow time the probe was placed in a selenocentric orbit. **1973** *Nature* 2 Nov. 7/2 [They] have detected increases of up to 100% in the density of picogram particles, in this case in selenocentric space, by using microphone sensors on board the satellites.

selenodesy (selinǫ·dĕsi). *Astr.* [f. Gr. σελήνη moon + -o, after Geodesy.] The study of the shape and features of the moon. Hence **selenode·tic** *a.*, of or pertaining to selenodesy.
1962 Z. Kopal *Moon* p. ix, The contents of the present volume..have been divided into five parts: 1. Rocket exploration of the moon... 11. Selenodesy and mapping of the moon. **1962** D. W. G. Arthur in *Ibid.* 102 The simplest of the two major selenodetic problems is the determination of the Moon's geometrical figure. **1962** *Flight International* LXXXII. 251/1 Selenodetic measurements of the Earth-Moon distance. **1967** J. Rösch in Kopal & Goudas *Measure of Moon* 71 Our task in starting up Selenodesy is not to determine the shape of a fictitious surface, but simply to establish reference points on the surface of the Moon to which we can refer the positions of other points. **1977** R. W. King et al. in J. D. Mulholland *Sci. Applications Lunar Laser Ranging* 51 (*heading*) Lunar dynamics and selenodesy.

self, *pron.*, *a.*, and *sb.* Add: **A.** *pronoun* and *pronominal adj.* **7.** *spec.* written on a cheque or counterfoil.
1866 [in Dict.]. **1873** D. G. Rossetti *Let.* 14 July (1967) III. 1192 If however you would prefer the cheque made out to *Self* as usual..I will send you another. **1935** G. Heyer *Death in Stocks* vi. 74 He drew a cheque for a hundred pounds to self on Friday. **1967** C. Watson *Lonely Heart 4122* ii. 23 He turned back the counterfoils... The uppermost..was marked 'Self', a withdrawal of four hundred pounds.
B. *adj.* **I. 3. b.** Also in wider application.
1930 R. A. Fisher *Genet. Theory Nat. Selection* vii. 165 In rats, the hooded (black and white) pattern is a simple recessive to the 'self' or 'solid' coloration. **1950** *Sun* (Baltimore) 7 Jan. 22/7 The 'self' pigeon..has a solid color while the 'barred' has a blue background with black bars across the wing.
II. 5. b. (Later examples.) *spec.* in *self belt, -fabric.*
1960 *Times* 21 Jan. 14/4 Jackets had either let-in or low-placed self belts. **1961** *Guardian* 28 Feb. 7/4 Jacket caught at the hips by a self-fabric belt. **1969** *Sears Catal.* Spring/Summer 10 Shower-resistant coat features self-fabric yoke for added protection. **1979** *Daily Tel.* 13 Feb. 15/1 (Advt.), Soft shirt dress..—no waist seams and self belt.
C. *sb.* **I. 4. a.** (*one's*) *old self*: (examples).
1841 C. Fox *Jrnl.* 30 Sept. (1972) 115 Dr. Calvert..was quite his old self, talking on his old subjects in his old way. **1975** H. Fleetwood *Picture of Innocence* x. 177 You seem quite like your old self again.
II. 6. b. Something (as an animal or garment) of a single colour. Cf. sense 3 b of the adj. in Dict. and Suppl.
1930 R. A. Fisher *Genet. Theory Nat. Selection* vii. 165 Rats of both selected lines were bred back to unselected selfs. **1978** *Lochaber News* 31 Mar. 2/1 (Advt.), 'Dereta' coats, superb collection of tweeds and selfs in a variety of fabrics, designs and colours.

self (self), *v.* [f. Self *pron.*, *a.*, and *sb.*] *trans.* To fertilize by means of pollen from the same plant; = Self-fertilize *v.*
1905 R. C. Punnett *Mendelism* 61 When this, the normal mode of fertilisation, takes place, the flower is said to be selfed. **1924** E. W. MacBride *Study of Heredity* vi. 143 When the recessive green peas are sown they give rise to plants which, when 'selfed', bear only green peas. **1949** H. L. Mencken in Kirby & Woolf *Philologica* 316 So far I have heard of no verb made of what appears to be a pronoun save *to self*. **1970** *Watsonia* VIII. 142 Rousi selfed a number of plants.
Hence **selfed** (selft) *ppl. a.*; **se·lfing** *vbl. sb.*, fertilization in this manner (in quot. 1924: a plant grown from seed so derived).
1924 *Genetics* IX. 30 Two such selfings of plants in family J were grown. **1927** *Technical Bull. N.Y. State Agric. Exper. Station, Geneva* No. 127. 4 Four selfed varieties of *P*[*runus*] *salicina* gave no seed. **1942** *Jrnl. Genetics* XLIII. 312 Their doubly heterozygous progeny will on selfing produce a range of genotypes with phenotypic differences. **1953** *Heredity* VII. 185 This mating system occurs in plants which have an imperfect selfing mechanism. **1977** M. Allan *Darwin & his Flowers* xv. 256 Nine of the original selfed plants had died. **1978** *Nature* 2 Feb. 441/1 Experimental selfings and crossings each lead to similar and relatively high levels of fruit set.

self-. Add: **I. a.** With nouns of action. *self-abandon, -abandonment* (later example), *-acquittal, -adjustment, -advertisement* (earlier example), *-affirmation, -aggrandisement* (later example), *-betterment, -castration, -censorship, -commiseration, -comprehension, -concealment, -confession, -confrontation, -constraint, -correction* (later example), *-critique, -crucifixion, -dedication* (later example), *-definition, -deification, -demolition, -deprecation, -description, -desertion* (earlier example), *-development, -discipline* (earlier and later examples), *-discovery, -display* (earlier example), *-disposal* (later example), *-dramatization, -duplication, -education* (earlier example), *-employment* (later example), *-enjoyment* (later examples), *-enrichment, -exhibition, -expansion, -exposure* (later examples), *-fulfilment, -generation, -immolation* (later examples), *-incrimination, -ingratiation, -interrogation* (earlier example), *-invitation, -laceration, -linkage, -management, -manipulation, -mockery, -mortification* (later example), *-mutilation, -objectification, -organization, -parody, -perception, -perpetuation, -portrayal, -prostration, -protection* (earlier example), *-punishment* (later example), *-purification, -rebuke, -recognition, -recrimination, -regeneration, -renewal, -repetition, -replication, -reproduction, -rule* (*Pol.* examples), *-scrutiny* (later examples), *-starvation, -stultification, -therapy, -transformation, -valuation* (later example), *-verification.*
1901 'L. Malet' *Hist. Sir R. Calmady* i. ix. 74 Her mother love..had none of the sweet self-abandon.. which that earlier passion had. **1958** R. Knox *Priestly Life* ix. 96 Self-abandonment has been preached by writers of undoubted orthodoxy. **1750** Richardson

Let. 29 Mar. (1964) 157 The place allows the latter [*sc.* conversationalists] to sit far enough from the [card-] tables not to be interrupted with remarks, blames, self-acquittals, of the engaged. **1892** 'Mark Twain' *Amer. Claimant* xiv. 132 You can't get the best of all verdicts, self-acquittal. **1848** J. S. Mill *Pol. Econ.* I. ii. xi. 406 Population had exercised its power of self-adjustment. **1962** A. Battersby *Guide to Stock Control* v. 51 This capacity for self-adjustment is a necessary part of *any* practical control system. **1840** J. S. Mill in *Edin. Rev.* LXXII. 27 What is saved in the workmanship may be more profitably expended in self-advertisement. **1854** Geo. Eliot tr. *Feuerbach's Essence Christianity* iii. 45 The exclusive self-affirmation of the human nature. **1924** W. B. Selbie *Psychol. Relig.* 53 That desire for..self-affirmation..is so characteristic of man at all stages of his development. **1979** *Dædalus* Summer 92 We are thrown back on our own truth and to the act of self-affirmation that constitutes our personal identity. **1937** *Discovery* July 225/2 A Board of Directors seeking only self-aggrandisement. **1931** G. F. Stout *Mind & Matter* 174 Each blindly strives towards its own self-maintenance and self-betterment. **1977** A. Giddens *Stud. in Social & Polit. Theory* i. 39 He criticizes Comte along with Spencer for..treating the impulse to self-betterment as if it were a general cause of the evolution of society. **1934** R. Campbell *Broken Record* vii. 161 Amongst the average English literary men, it is usual for them to go soft at thirty (the moral self-castration of the exoletus). **1950** R. Ames in *Science & Society* XIV. 195 The obvious need of the oppressed..for self-censorship of his social ideas. **1904** K. C. Thurston *John Chilcote* xi. 119 He had..a feeling of bitter self-commiseration that for the moment outweighed all other considerations. **1909** W. James *Pluralistic Universe* i. 36 It may be a supreme reaction of the universe upon itself by which it rises to self-comprehension. **1935** *Mind* XLIV. 94 The Delphic Γνῶθι σεαυτόν..counsels self-knowledge *versus* his [*sc.* Nietzsche's] and every Power-Man's need for self-concealment. **1963** Auden *Dyer's Hand* 109 A capacity for self-disclosure implies an equal capacity for self-concealment. **1885** W. James *Lit. Remains H. James* 19 Society is the same Creator, with the nothingness saved, determined to transparency and self-confession. **1923** F. G. Ellerton *Let.* 4 Nov. in *John Bailey* (1935) 236, I take it that the *pargoletta* passage is of the nature of a self-confession on Dante's part. **1961** *New Statesman* 23 June 1010/3 'Mellowness' becomes a means of avoiding the self-confrontations he says he has funked all his life. **1883** F. H. Bradley *Logic* III. i. vi. 450 Synthesis..would not force its parts into violent conjunctions, but, itself in each, by the loss of self-constraint would embrace its own fulfilment. **1953** D. F. Pocock tr. *Durkheim's Sociol. & Philos.* ii. 36 The act..cannot be accomplished without effort and self-constraint. **1965** *Math. in Biol. & Med.* (Med. Res. Council) iii. 85 The use of computers in diagnosis will need provision for..self-correction by new data, and for questioning unusual or missed signs. **1959** E. Pulgram *Introd. Spectrogr. Speech* 7 Switching from one linguistic code to another is conducive to self-critique. **1897** W. J. Locke *Derelicts* ii. xxiii. 413 The tragic futility of such self-crucifixion. **1943** J. S. Huxley *Evolutionary Ethics* 67 The desire for self-dedication..should be available to society's common pool. **1957** M. J. Huntingdon in R. K. Merton *Student-Physician* 181 First-year students ..think of each other primarily as students. This is reflected in their self-definitions. **1980** S. Brett *Dead Side of Mike* ii. 19 Definition, and particularly self-definition is very important to me as an artist. **1903** 'Mark Twain' in *N. Amer. Rev.* Apr. 510 She..has carried self-deification to a length which has not before been ventured in ages. **1962** E. Snow *Other Side of River* xvii. 122 Chiang made a fair start toward self-deification. **1935** T. S. Eliot *Murder in Cathedral* ii. 66 Dominated by the lust of self-demolition. **1924** W. Holtby *Crowded Street* xxxv. 260 Don't you think that this self-deprecation of yours was a little like cowardice? **1977** A. Giddens *Stud. in Social & Polit. Theory* ix. 307 Suicide represents an extreme on a range of possible forms of self-aggression, which extends from relatively minor forms such as verbal self-deprecation to actual self-destruction. **1902** W. James *Var. Relig. Exper.* xvi. 415 In spite of their repudiation of articulate self-description, mystical states in general assert a pretty distinct theoretic drift. **1978** J. Dunn in Hookway & Pettit *Action & Interpretation* 157 Two main difficulties follow from this centrality of the human capacity for self-description in specifying the field of the sciences of man. **1796** F. Burney *Camilla* V. 454 Thy afflicting, however blamable self-desertion. **1817** Coleridge *Biog. Lit.* I. xii. 275 Intelligence is a self-development. **1895** W. J. Locke *At Gate of Samaria* i. 7 She read books with the eagerness only exhibited by the young girl craving for self-development. **1977** Warren & Ponse in Douglas & Johnson *Existential Sociol.* x. 285 The Goffmanic masked self..comes into play when the audience is..to be manipulated rather than for such motives as self-development, [etc.]. **1796** Ld. Glenbervie *Diary* 15 Oct. (1928) I. 87 Every young person..should be conducted to form himself to such habits of self-discipline. **1946** *Nature* 27 July 113/1 The man who pursues truth for its own sake and studies science in an impersonal way, with rigorous self-discipline, is really showing his belief in a religion and is taking up a religious attitude. **1965** J. A. Michener *Source* 532 The Arabs arrived when they were strongest, in the throbbing flush of self-discovery and unification. **1838** J. S. Mill in *Westm. Rev.* Apr. 7 M. de Vigny,..a man..of a rare simplicity of heart, and freedom from egotism and self-display. *a* **1885** G. M. Hopkins *Poems* (1918) 78 One part, Reason, selfdisposal, choice of better or worse way, Is corpse now, cannot change. **1933** *Week-end Rev.* 11 Feb. 151/1 Like Hamlet, he had an incurable habit of self-dramatization. **1959** *Encounter* July 66/2 It was not so much insincerity as self-dramatisation. **1817** Coleridge *Biog. Lit.* I. xii. 268 A perpetual self-duplication of one and the same power into object and subject. **1953** J. S. Huxley *Evol. in Action* i. 16 They [*sc.* the chromosomes] have divided longitudinally after self-duplication. **1831** J. S. Mill *Let.* 22 Oct. (1910) I. 8 The only thing which I can usefully do at present,..is to work out.. principles of morals, government, law, education, above all self-education. **1976** *Times* 13 Aug. 13/7 In the poorest parts of the country..self employment is frequently the only manner in which a living income can be put together. **1870** D. G. Rossetti *Let.* 7 Nov. (1965) II. 914 That sense of the poet's self-enjoyment which is indispensable to the enjoyment of the reader. **1960** H. Read *Forms of Things Unknown* iii. ix. 149 No explanation of art as 'objectified self-enjoyment'..can account for the facts of art history. **1920** B. Russell *Pract. & Theory Bolshevism* ii. i. 127 Self-enrichment seemed the natural aim of a man's political actions. **1939** A. Huxley *After Many a Summer* i. ii. 24 Jeremy made his mannequin's gesture of apologetic self-exhibition. **1939** *Mind* XLVIII. 238 He sees that the root motive of mysticism is self-expansion—though he does not use this expression. **1921** D. H. Lawrence *Tortoises* 40 Doomed to partiality,.. Want, Self-exposure, hard humiliation, need to add himself on to her. **1979** J. Rathbone *Euro-Killers* ii. 22 Neither man was prepared to risk possible self-exposure by offering his ideas. **1936** *Mind* XLV. 242 He affirms the awareness of God to be 'rooted in' the human interest of self-fulfilment. **1981** R. Barnard *Sheer Torture* x. 109 Cultivating your ego, aiming at total self-fulfilment, doing your own thing. **1950** *Essays & Studies* III. 37 Everything in his poetry goes to suggest that it was created..by a largely spontaneous..process of self-generation. **1898** G. B. Shaw *Perfect Wagnerite* 95 Her self immolation on the pyre. **1961** D. G. James *Matt. Arnold* iii. 57 The omission of it [sc. *Empedocles*] from the 1853 volume..was a fine piece of self-immolation. **1911** *U.S. Reports* (U.S. Supreme Court) CCXXI. 388 The clear and simple directness of the privilege against self-incrimination. **1977** 'E. McBain' *Long Time no See* xv. 267 You are warned of your right to counsel and your privilege against self-incrimination. **1948** *Commentary* Nov. 417/2 Intelligence, humor, and charm is often humiliatingly exploited..as entertainment and self-ingratiation. **1963** *Times* 9 Feb. 4/2 The pallid self-ingratiation of Sally Logan's performance. *a* **1854** J. S. Mill *Draft Autobiogr.* (1961) 122 Let..your scrutiny, your self interrogation exhaust themselves on that. **1754** Richardson *Grandison* VI. 68 We then endeavoured to recollect the words of his self-invitation hither. **1805** Jane Austen *Let.* 24 Aug. (1952) 162, I defy her to accept this self-invitation of mine, unless it be really what perfectly suits her. **1963** *English Studies* XLIV. 144 Here, however, he will not allow himself to feel the self-laceration in the final stanzas of that canto. **1977** *Jrnl. Protozool.* XXIV. 9/2 Pattern formation by self-linkage. *a* **1866** J. Grote *Exam. Utilitarian Philos.* (1870) vi. 116 Good self-management his highest aim. **1977** Warren & Ponse in Douglas & Johnson *Existential Sociol.* x. 277 There is a difference in degree between the problems and self-management of gays and most other people. **1949** M. Mead *Male & Female* x. 216 The female child's genitals are less exposed..to self-manipulation. **1964** E. H. Mizruchi in I. L. Horowitz *New Sociol.* 257 Mental self-manipulation appears to be characteristic of the assembly-line work process. **1954** Koestler *Invisible Writing* iii. 40 Amiable Hans was a bald, moon-faced little man with a wealth of humour and self-mockery. **1931** J. S. Huxley *What dare I Think?* v. 169 He [*sc.* the humanist] finds the desire for a sacrifice and self-mortification just as natural..as the desire for achievement and self-assertion. **1864** G. O. Trevelyan *Compet. Wallah* 380 What is there in common between the faith of Heber and Swartz and a creed which enjoins suicide and self-mutilation. **1980** *Jrnl. R. Soc. Arts* Feb. 137/2 It represents self-mutilation, it can only lead to genocide and biocide. **1933** *Times Lit. Suppl.* 16 Mar. 174/1 His book is one long and infinitely various act of self-discovery, self-objectification, made possible only by self-forgetfulness. **1969** T. F. Torrance *Theol. Sci.* i. 42 We are frequently engaged in mythological self-objectifications of this sort. **1898** G. B. Shaw *Perfect Wagnerite* 11 The..self-organization of life. **1967** *Oceanogr. & Marine Biol.* V. 274 Ecological succession is a process of self-organization. **1958** S. Spender *Engaged in Writing* 89 He [spoke]..in the comic manner which was half-serious self-parody. **1978** *New York* 3 Apr. 64/3 The majesty of our landscape has its own built-in note of self-parody, and it has eluded these artists. **1678** Cudworth *Intell. Syst.* i. iii. 160 If the souls of Men and Animals be at any time without Consciousness and Self-perception, then..Clear and Express Consciousness is not Essential to Life. **1972** M. Argyle *Social Psychol. of Work* iv. 60 The self-image also includes self-perceptions such as being 'intelligent' or 'lively'. **1843** J. S. Mill *Logic* II. iii. xv. 32 A permanent effect..possessing the property of self-perpetuation. **1906** W. G. Sumner *Folkways* 473 They show how deep is the interest of human beings in the sex taboo, and in the self-perpetuation of society. **1962** F. I. Ordway et al. *Basic Astronautics* vi. 247 We think of life as a system or aggregate of chemical reactions that possess the inherent capability of self-perpetuation. **1914** G. B. Shaw *Dark Lady of Sonnets* 121 Self-betrayal is one thing; and self-portrayal, as in Hamlet and Mercutio, is another. **1872** Geo. Eliot *Middlem.* III. li. 354 The divine tribunal had changed its aspect for him; self-prostration was no longer enough. **1834** J. S. Mill in *Monthly Repos.* VIII. 838 This is.. the best sort of self-protection. **1974** R. H. S. Crossman *Diaries* (1975) I. 14 My own generalizations and predictions..sometimes provide an exercise in self-punishment since some of them read very foolishly today. **1924** W. B. Selbie *Psychol. Relig.* x. 202 A longing for self-purification. **1871** Geo. Eliot *Middlem.* I. iii. 41 Dorothea checked herself suddenly with self-rebuke for the presumptuous way in which she was reckoning on uncertain events. **1927** T. Wilder *Bridge of San Luis Rey* ii. 15 Each..was on the point of losing her mind under the alternations of self-rebuke and the outbursts of passion. **1955** *Essays & Studies* VIII. 75 Emma is shown as incapable of self-recognition. **1961** Webster, Self-recrimination. **1965** J. A. Michener *Source* 700 He was thrown into a world of self-recrimination and remorse. **1918** W. S. Churchill in M. Gilbert *Winston S. Churchill* (1977) IV. Compan. i. 366 They should be.. a real act of self-regeneration make a definite break with the system. **1962** E. Snow *Red China Today* (1963) xlviii. 373 In prison there is the omnipresent *threat* of force and humiliation, combined with the demand for self-regeneration. **1959** B. & R. North tr. *Duverger's Pol. Parties* i. ii. 87 One of the constant features of the French Communist party is its perpetual self-renewal. **1971** J. Z. Young *Introd. Study Man* xi. 147 They [*sc.* 'post-mitotic' cells] are provided with especially active mechanisms for self-renewal. **1848** Poe in *Graham's Mag.* Feb. 131/1 Apparent plagiarisms ..arise from an author's self-repetition. **1959** *New Biol.* XXVIII. 93 The most characteristic property [of life], that of self-replication, has been studied theoretically by certain mathematicians. **1926** J. S. Huxley *Essays of Biologist* vii. 256 The attributes of living matter which mark it off from inorganic matter become dominant—its capacity for self-reproduction, [etc.]. **1964** *Sci. Amer.* Sept. 154/2 Seen in this light, crystal growth is self-reproduction. **1900** G. B. Shaw *Let.* 14 Mar. (1972) II. 156 The definite intention to clear out of India as soon as the natives are capable of self rule is the most pious of superfluities. **1978** *Times* 5 May 6/8 There is apprehension..that Israel will impose its own peace plan which offers a measure of self-rule for the West Bank. **1865** D. G. Rossetti *Let.* 21 Nov. (1965) II. 581 This [feeling of rage]..leads not to envy in the least, but to self-scrutiny. **1979** *Dædalus* Summer 3 Genuine charity and a constant and severe self-scrutiny are psychologically unthinkable without the moral pride that is all but inseparable from spiritual energy. **1947** C. Gray *Contingencies* 37 The voluntary self-starvation of the rich. **1904** W. James in *Mind* XIII. 466 Can there be self-stultification in urging any account whatever of truth? Can the definition ever contradict the deed? **1959** *Spectator* 11 Sept. 336/3 He came to each [session] with notes on the successive steps in his 'self-therapy'. **1949** Koestler *Promise & Fulfilment* ii. iv. 253 This difference..was demonstrated by Irgun's..self-transformation into a *bona fide* democratic party. **1946** P. Bottome *Lifeline* xxxix. 297 Their self-valuation was threatened. **1854** Geo. Eliot tr. *Feuerbach's Essence Christianity* i. 6 Consciousness is self-verification. **1975** T. P. Whitney tr. *Solzhenitsyn's Gulag Archipelago* II. iii. i. 11 Prisoners, wherever possible, were to be brought into self-verification and self-supervision.
b. With vbl. sbs. *self-advertising* (earlier example), *-compounding, -doctoring, -doubting, -loathing* (later example), *-poisoning, -policing, -questioning* (earlier example), *-scourging* (later example), *-searching* (later example), *-teaching, -understanding.*
1762 Goldsmith *Life of Richard Nash* 13 Little Things..without merit..by self-advertizing, attract the attention of the day. **1890** W. James *Princ. Psychol.* I. vi. 158 Self-compounding of mental facts is inadmissible. **1817** Jane Austen *Sanditon* v, in *Minor Wks.* (1954) 388, I can be no Judge of what the habit of self-doctoring may do. **1951** C. R. Acton *Dog Annual* 143 The all-important power of..self-doctoring. **1754** Richardson *Grandison* VI. 115 She embraced me,..and cleared up all my self-doubtings. **1971** P. O'Donnell *Impossible Virgin* ix. 176 She..was filled with self-loathing because it was she who had trapped him. **1907** W. James in *McClure's Mag.* (1908) Feb. 420/2 Democracy as a whole may undergo self-poisoning. **1968** *Times* 28 Oct. 3/1 Professor Kessel..has found that 34 per cent of 511 patients admitted for self-poisoning gave warning. **1960** *Times* 13 May 18/5 Unless some self-policing is done in the American film industry. **1977** *Jrnl. R. Soc. Arts* CXXV. 450/1 Self-policing by the data storage industry is simply not an adequate safeguard. **1856** Bagehot in *National Rev.* Apr. 370 He has no passionate self-questionings, no indomitable fears, no asking perplexities. **1916** G. B. Shaw *Androcles & Lion* p. xx, Saints amazed the world with their austerities and self-scourgings. **1978** J. Sacks in P. Moore *Man, Woman, & Priesthood* iii. 28 It has led to considerable self-searching in the Jewish community in recent years. **1891** Self-teaching [see *nursery school* s.v. *nursery 8 c]. **1967** G. Jackson *Let.* 31 Jan. in *Soledad Brother* (1971) 107 A good self-teaching book on Arabic. **1903** G. B. Shaw *Man & Superman* iii. 129 Life's incessant aspiration to higher organization, wider, deeper, intenser, self-consciousness, and clearer self-understanding. **1977** A. Giddens *Stud. in Social & Polit. Theory* i. 57 In the philosophy of science, as contrasted to the methodological self-understanding of the social sciences, the 'orthodox model' has long since become subject to broad-ranging attack.
c. With agent-nouns. *self-dramatizer, -educator, -seeder.*
1936 A. Thirkell *August Folly* viii. 247 'She is a self-dramatizer,' said Charles. **1948** *Medium Ævum* XVII. 1 (*title*) A medieval self-educator. **1970** P. Dickinson *Seals* ii. 35 Great self-educators, these Pibbles. **1966** 'J. Berrisford' *Wild Garden* viii. 99 One of the most..accommodating self-seeders is a biennial evening primrose—*Oenothera lamarckiana*.
d. With nouns of state or condition. *self-disgust, -disrespect, -doubt* (later examples), *-hero-worship, -infatuation, -mastery* (later examples), *-mistrust* (later example), *-picture, -scorn, -tolerance* (Immunol.).
1921 E. M. Hull *Sheik* v. 160 A self-disgust seized him. He had been within an ace of betraying the man. **1977** *Listener* 28 Apr. 552/4 George Grosz..the master of self-disgust. **1853** J. Brown *Let.* (1912) 132 Dr. Scott's illness and death, and my own profound self-disrespect and indifferent health and overwork,—altogether I was in a sorry condition. **1938** Ld. Lymington *Famine in Eng.* 74 They are waiting to be led to salvation, and out of the slough of self-disrespect into which they have been driven. **1930** E. Waugh *Vile Bodies* vi. 106 Magically, self-doubt began to spread in the audience. **1980** J. McClure *Blood of Englishman* xv. 142 His hunch..drifted dangerously towards the jagged reef of self-doubt. **1927** J. S. Huxley *Relig. without Revelation* iv. 130 'My country, right or wrong'..

words which are immortal as the fittest inscription on the pedestal of the golden calf of self-hero-worship. **1948** L. Spitzer *Linguistics & Lit. Hist.* iv. 156 An enthusiasm inspired by self-infatuation. **1973** M. Amis *Rachel Papers* 16 Thinking back, actually, 'self-infatuation' strikes me as a rather ill-chosen word. It isn't so much that I like or love myself. Rather, I'm sentimental about myself. **1933** *Burlington Mag.* Dec. 260/1 The art of self-mastery, the grand style. **1979** *Dædalus* Summer 90 One must be..forever inaccessible to the seductions that distract or subtract from the forces necessary for self-mastery and self-defense. **1977** R. Holland *Self & Social Context* i. 20 From Freud they need clinical methods which give such primacy to the individual case but they dislike the 'pathological' self-picture. **1946** Dylan Thomas *Let.* 30 May (1966) 288 Inevitable moments of depression and self-mistrust. **1861** E. B. Browning *Mother & Poet* in *Last Poems* (1862) 95 Some women bear children in strength, And bite back the cry of their pain in self-scorn. **1964** M. Hynes *Med. Bacteriol.* (ed. 8) vii. 74 An animal will not normally form antibodies to its own tissues, even though these may be powerfully antigenic to a litter-mate. The mechanism of this self-tolerance and the consequences of its breakdown are discussed later. **1977** I. M. Roitt *Essent. Immunol.* iv. 109 Although we are uncertain of the mechanism, the idea that deletion of specific clones is responsible for tolerance induction is attractive. For example, it can account for the development of self-tolerance.

e. With adjs. *self-analytical, -aware, -corrective, -corroborative, -critical, -derisive, -expressive, -incriminatory, -infirmative, -mistrustful, -protective, -reproductive, -submersive, -sure.*

1943 H. Read *Educ. through Art* iii. 42 It may be that such eidetic individuals are more self-analytical than an eidetic person without creative gifts would be. **1924** R. Graves *Mock Beggar Hall* 28 Alert, with both eyes open, self-aware. **1979** A. Chisholm *Nancy Cunard* viii. 75 An intelligent, imaginative, self-aware young man. **1909** Webster, Self-corrective. **1934** *Mind* XLIII. 506 The inductive method is self-corrective. *a* **1812** J. Bentham *Rationale Judicial Evid.* (1827) III. v. xv. 224 We must..express the distinction between the two opposite kinds of evidentiary chains: styling the one, for example, the *self-infirmative* chain, we may style the other the *self-corroborative*. **1909** W. James *Meaning of Truth* xiii. 267 The hypothesis will, in short, have worked successfully all round the circle and proved self-corroborative. **1936** *Mind* XLV. 98 This suggestion that art, like religion, is self-critical in its activity. **1979** B. Malamud *Dubin's Lives* ix. 346 The tone of the book was self-critical. **1867** J. R. Lowell *Fitz Adam's Story* in *Poems* (1912) 575 He went on with a self-derisive sneer. **1928** D. H. Lawrence *Woman who rode Away* 172 His rather hooked nose self-derisive. **1934** *Essays & Studies* XIX. 28 The most seemingly self-expressive 'human document'. **1977** R. Holland *Self & Social Context* v. 102 He must develop an autonomous aspect of self to account for the more self-expressive or impulsive actions. **1949** Koestler *Promise & Fulfilment* I. xiii. 149 [He]..refused to testify..on the grounds that his testimony would be self-incriminatory. *a* **1812** Self-infirmative [see *self-corroborative* above]. **1875** A. Swinbourne *Picture Logic* (ed. 2) xxii. 142 Self-infirmative inference is where each fresh fact weakens the conclusion. **1897** G. B. Shaw *Let.* 10 Aug. (1965) I. 795 It is sheer want of practice that makes actors self-mistrustful when they are asked to..tell a story on the stage. **1890** W. James *Princ. Psychol.* II. xxiii. 381 The tendency to contract is the source of all the self-protective impulses and reactions which are later developed. **1979** B. Malamud *Dubin's Lives* ix. 337 Some people complicate their feelings in self-protective ways. **1943** J. S. Huxley *Evolutionary Ethics* v. 34 The capacity for self-reproduction, or better self-reproductive evolution. *a* **1914** Joyce *Stephen Hero* (1944) xvi. 36 Stephen, as he looked contemptuously at the laughing faces, thought of a self-submersive reptile. **1918** G. Frankau *One of Them* in *Poet. Works* (1923) II. 75 Goddess indeed! A self-sure, jade-eyed, slim puss, of life's each latest luxury impassioned. **1954** *Numbers* (N.Z.) July 1/5 Paul was lying on his bed, reading. Twenty-five; good-looker; self-sure.

f. With ppl. adjs. in *-ing*. *self-abbreviating, -adjusting* (earlier example), *-analysing, authenticating, -blanching, -correcting, -defeating, -defining, -delighting, -deprecating, -describing, -developing, -distributing, -doubting, -dramatizing, -enduring, -enhancing, -enjoying, -equilibrating, -executing, -explaining, -generating, -guaranteeing, -hating, -humbling* (later example), *-hurting, -immolating, -incriminating, -ingratiating, -interpreting* (later example), *-mocking, -negating, -observing, -optimizing, -ordering, -organizing, -parodying, -perpetuating* (later examples), *-policing, -prolonging, -propagating* (earlier example), *-punishing* (later example), *-recognizing, -reflecting, -reinforcing, -renewing, -replicating, -reproducing, -searching, -serving, -stultifying, -supplying, -teaching, -validating.*

1922 Joyce *Ulysses* 718 The selfprolonging tension of the thing proposed to be done and the selfabbreviating relaxation of the thing done. **1848** J. S. Mill *Pol. Econ.* II. III. xx. 154 There is a self-adjusting power in the variations of the exchange itself. **1922** W. B. Yeats *Trembling of Veil* III. 144 The subtle torture of self-analysing passion. **1980** A. Wilson *Setting World on Fire* III. i. 264 Find a play..which has a thoughtful, self-analysing part for an old actor like me. **1935** H. H. Farmer *World & God* I. ix. 158 Our interest is in the Christian experience of God as personal, which in the nature of the case must be self-authenticating. **1976** H. Montefiore in *Christian Believing* 151 His character is to me self-authenticating. **1907** A. French *Bk. Vegetables* 78 Varieties [of celery] are many, and classify chiefly as to height, color (red tint), and 'self-blanching' properties. **1976** 'D. Halliday' *Dolly & Nanny Bird* xi. 138 Simon had turned the colour of self-blanching celery. **1944** *Mind* LIII. 212 The last topic with which Hr. von Wright deals is Reichenbach's contention that induction is a self-correcting process. **1979** *Dædalus* Summer 150 The idea of a developing and self-correcting body of scientific inquirers offered what seemed to him a more satisfactory model. **1909** W. M. Urban *Valuation* xiv. 413 In this concept of the self-defeating process we have a negative test of validity. **1949** M. Mead *Male & Female* ix. 196 The Weimar Republic's self-defeating practice of giving..jobs..to older men. **1979** B. Malamud *Dubin's Lives* ix. 336 Their talk was self-defeating. **1968** Self-defining [see *literal *sb.* 3]. **1908** W. B. Yeats *Poetry & Ireland* 10 None other has a continual deliberate self-delighting happiness. **1958** *Times Lit. Suppl.* 17 Jan. 30/1 Mr Amis, seen by himself in a way in which ironical, self-deprecating people do see themselves. **1971** S. Hill *Strange Meeting* ii. 111 Barton turned his head and smiled, and then his face changed again, the old, self-deprecating expression over it. **1864** Bagehot in *National Rev.* Nov. 34 Such self-describing poets describe what is *in* them, but not *peculiar* to them. **1978** J. Dunn in Hookway & Pettit *Action & Interpretation* 156 Between a describer and a self-describing object there exist relations which are peculiar. **1865** Self-developing [see *self-changing* in Dict.]. **1980** N. Marsh *Photo-Finish* iii. 79 The lens cap..produced instant self-developing results. **1945** Koestler *Yogi & Commissar* III. iv. 242 Koehler assumes that there are self-distributing electromagnetic currents between the cortical projections of retinal points. **1966** *English Studies* XLVII. 201 Only the second look of recent criticism has shown how homosexual were those claims, and how self-doubting and despairing. *a* **1974** R. Crossman *Diaries* (1975) I. 339, I had noticed already that Ray Gunter was a dramatic, self-dramatizing kind of chairman. **1945** Auden *Coll. Poetry* 120 That through his self-annulment the real world Of self-enduring instants may endure Its final metamorphosis. **1965** F. Sargeson *Memoirs of Peon* viii. 265 The self-enhancing aim of endeavouring to ameliorate the lot of the family. **1854** Geo. Eliot tr. *Feuerbach's Essence Christianity* ii. 40 The understanding alone..is the self-enjoying, self-sufficing existence. **1951** Auden *Enchafèd Flood* iii. 102 All it desires is to be in equilibrium, a self-enjoying, self-sufficient self. **1947** *Mind* LVI. 58 The problem of a self-equilibrating physical system can now be attacked with both vigor and generality. **1974** W. Rees-Mogg *Reigning Error* iv. 81 A gold system is a self-equilibrating system operating on a constant world money supply. **1962** *Listener* 15 Mar. 455/1 Not all the provisions of the treaty [*sc.* Treaty of Rome] are of this so-called 'self-executing' character. **1903** J. Joyce *Let.* 9 Mar. (1966) II. 35, I enclose you self-explaining documents. **1865** H. Maudsley *Meth. of Study of Mind* 18 It is ridiculous to suppose that the man of genius is ever a fountain of self-generating energy. **1968** H. Harris *Nucleus & Cytoplasm* vi. 122 On any model..we are forced to consider how a set of conditions, initially produced in a cell by external stimuli, can become self-generating. **1887** A. Seth *Hegelianism* vi. 205 Hegel, like Plato, seeks reality not in the actual world, but in the eternal realm of an absolute and self-guaranteeing thought. **1961** J. Wilson *Reason & Morals* iii. 164 What is irrational is that such people take their feelings as self-guaranteeing, that they treat them as carrying their own justification with them. **1977** *New Yorker* 29 Aug. 82/3 Perhaps there was the same sort of self-hating or fearful motive behind Hitler's orders to murder the most helpless patients in the hospitals of the Third Reich. **1938** R. Graves *Coll. Poems* p. xiii, It is an exorcism of physical pretensions by self-humbling honesties. **1915** D. H. Lawrence *Rainbow* ii. 52 A burst of religious, self-hurting fervour had passed over the country. **1939** R. Campbell *Flowering Rifle* VI. 150 And better maimed Of a self-hurting member so inflamed. *a* **1922** T. S. Eliot *Waste Land Drafts* (1971) 101 line 26 One soul,..Self-immolating on the Mound. **1925** T. Dreiser *Amer. Tragedy* I. I. iii. 16 Any self-abnegating and self-immolating religious theory. *Ibid.* II. xxi. 303 He was not without the self-incriminating thought that in seeking this..he was driving toward a relationship which was not legitimate. **1964** *Harvard Law Rev.* Nov. 219 Implicit..is the proposition that in a pretrial police interrogation the accused has a right to remain silent, which he must waive intelligently before any self-incriminating statements will be admissible at trial. **1925** T. Dreiser *Amer. Tragedy* I. II. xxiii. 313 Clyde..now approached, his manner the epitome almost of a self-ingratiating..dog of high breeding and fine temperament. **1935** A. C. Baugh *Hist. Eng. Lang.* vii. 266 One further habit which was somewhat weakened [in Middle English], although by no means broken, was that of combining native words into self-interpreting compounds. **1966** *English Studies* XLVII. 204 The self-mocking,..witty, ironical, comic Whitman. **1963** L. Trilling in N. Frye *Romanticism Reconsidered* 83 One of Keats's boldest expressions of his sense that there is something perverse and self-negating in the erotic life. **1960** Koestler *Lotus & Robot* II. xi. 243 The split between the acting self and the self-observing self disappears. **1961** *Times* 23 Mar. 17/2 The anti-hero, oddly likable and self-observing. **1961** D. Gabor et al. in C. Cherry *Information Theory* xxvii. 348 (*heading*) A self-optimizing nonlinear filter, predictor and simulator. **1968** *Brit. Med. Bull.* XXIV. 251/1 Priban and Fincham..have been examining the adjustment of the rate and depth of breathing and the generation of individual breaths, using a 'self-optimizing' model. **1923** J. S. Huxley *Essays of Biologist* vi. 217 A universe which the march of knowledge is showing us ever more clearly as self-ordered and self-ordering. **1958** R. Williams *Culture & Society* III. 298 The development of an organized and self-organizing working class. **1946** P. Larkin *Jill* 13 Never having heard before this self-parodying Southern coo. **1958** S. Spender *Engaged in Writing* 14 Asphalt's ..blotched face with its self-parodying leer. **1938** Huxley & Haddon *We Europeans* iii. 8 The gene is then as self-perpetuating in its new type as it was in its old. **1971** N. Stacey *Who Cares?* viii. 139 It was quite obvious that few of our efforts looked like being self-generating and self-perpetuating. **1955** G. Gorer *Exploring Eng. Character* xv. 296 The English character became, to a very marked degree, 'self-policing'. **1922** Self-prolonging [see *self-abbreviating* above]. **1843** J. S. Mill *Logic* I. III. xiii. 565 It is..upon such substances that the self-propagating power of chemical action is likely to exert itself in the most marked manner. **1978** H. McLeave *Borderline Case* (1979) i. 14 A bloody-minded, self-punishing type. **1964** *Punch* 20 May 760/3 His.. reaction..is bitterly self-recognising. **1921** J. McTaggart *Nature of Existence* I. IV. xxxi. 299 Every substance which is a self-reflecting unity possesses two sorts of unity—organic unity and unity of self-reflection. **1953** J. S. Huxley *Evolution in Action* iv. 91 The organization..of large groups of nerve-cells and their connecting outgrowths into self-reinforcing circuits of excitation. **1875** J. R. Lowell in *Cambridge in Centennial Proc.* 30 The wingèd years, that winnow praise and blame, Blow many names out: they but fan to flame The self-renewing splendors of thy fame. **1977** J. L. Harper *Population Biol. Plants* 306 The food of an animal is self-renewing. **1946** *Nature* 21 Sept. 406/1 Alternatively ..some self-replicating cytoplasmic constituent of a complex cell, such as a plasmagene, may become capable of multiplying when transferred to a new environment. **1926** J. S. Huxley *Ess. Pop. Sci.* 230 The chromosomes are self-reproducing. **1964** *Sci. Amer.* Sept. 149 (*caption*) The little red and blue 'creatures' in the photograph.. are the two kinds of part of an elementary self-reproducing machine designed by..L. S. Penrose. **1838** W. E. Channing *Self-Culture* 12 There are two powers of the human soul which make self-culture possible, the self-searching and the self-forming power. **1978** P. Griffiths *Conc. Hist. Mod. Music* iii. 34 Both have a public rhetoric which was..foreign to Schoenberg in his profoundly self-searching form of Expressionism. **1827** J. Bentham *Rationale Judicial Evid.* III. vi. ii. 415 The probability, of mendacity, self-serving mendacity, and consequent deception, attached to the admission of the testimony of the party in his own behalf. **1896** *Southwestern Reporter* XXXIII. 791/1 On a criminal trial, self-serving acts of the defendants are properly excluded. **1904** J. H. Wigmore *Treat. Syst. Evidence* III. lvii. 2273 There is no principle of evidence especially excluding 'self-serving' statements by an accused or by any one else. **1958** *Listener* 18 Sept. 412/2 How far would the majority go in applying the general rule excluding what are sometimes called self-serving statements to conduct? **1972** J. Philips *Vanishing Senator* (1973) II. iii. 76, I always thought he was a kind of cheap, self-serving jerk. **1904** W. James in *Mind* XIII. 474 It seems to me to illustrate beautifully how self-stultifying the conception of a truth that shall merely register a standing fixture may become. **1955** J. L. Austin *How to do Things with Words* (1962) iv. 51 This commits you to it and refuses to commit you to it. It is a self-stultifying procedure. **1842** E. FitzGerald *Let.* 16 Aug. (1960) 37 When I read of your..riding into Naples with huge self-supplying beakers full of the warm South I am sure you had best stay where you are. **1946** *Nature* 2 Nov. 606/1 A permanent self-supplying community of professional miners. **1964** J. Z. Young *Model of Brain* v. 69 In particular we are interested in a self-teaching homeostat, that is to say, one whose information and instructions are not entirely built in by heredity. **1945** *Mind* LIV. 46 These beliefs were after all cases of immediate knowledge which would therefore be self-validating and so require no further explanation.

i. With advs. *self-deprecatingly, -organizingly, -understandingly.*

1924 W. Holtby *Crowded Street* XXXV. 259 She laughed self-deprecatingly. **1966** S. Beer *Decision & Control* xiv. 346 More typically, and more 'self-organizingly', we say that energy evens out. **1933** Auden *Dance of Death* 32 Self-understandingly I come.

2. a. With pa. pples. and ppl. adjs. in which *self-* denotes the agent or what is conceived as the agent. *self-acquitted* (earlier example), *-adjusted, -bound, -built, -caused* (earlier example), *-chosen* (later examples), *-compelled, -confessed, -constituted* (earlier example), *-defined, -developed, -educated* (later examples), *-employed, -fancied, -fed* (later examples), *-generated, -indexed, -instressed, -invited* (earlier and later examples), *-observed, -ordered, -organized, -outwitted, -paced, -paid* (earlier example), *-proclaimed, -produced* (later example), *-seeded, -selected, -sentenced, -set* (later example), *-sold, -tinted; self-clinging, -pronouncing.* Also with related advs., as *self-confessedly*, and occas. with other adjs., as *self-adhesive*.

1814 F. Burney *Wanderer* III. xlv. 107 Their honour was self-acquitted and their generosity was self-applauded. **1958** *Engineers' Digest* XIX. 244/1 (*heading*) Self-adhesive nameplates. **1977** *Lancashire Life* Dec. 80/1 Products range from a reproduction of a timbered English pub for a Paris shoe exhibition to self-adhesive stickers for confectionery packets. **1924** W. B. Selbie *Psychol. Relig.* 55 They [*sc.* primitive religions]..become self-adjusted to man's growing intellectual capacity and needs. **1838** E. B. Browning *Seraphim* 50 The winding, wandering music that returns Upon itself in starry course, self-bound To praise, and praise, and praise. **1935** T. S. Eliot *Murder in Cathedral* i. 29 Cabined in Canterbury, realmless ruler, Self-bound servant of a powerless Pope. **1845** J. R. Lowell in *Broadway Jrnl.* 8 Mar. 154/2 A guess At the spirit's deathlessness, Which ye entertain with fear In your self-built dungeon here. **1970** A. G. Frank in I. L. Horowitz *Masses in Lat. Amer.* vi. 220 In Mexico City 30 per cent of the population lives in self-

built housing. **1748** Richardson *Clarissa* IV. 347 Evils self-caused..admit not of palliation. **1856** Bagehot *Coll. Works* (1965) I. 408 Anything free refers to the people; anything elected seems self-chosen. **1972** *Guardian* 19 Jan. 2/7 Vienna's self-chosen image of 'Schlamperei', roughly translated as carefree sloppiness. **1952** A. G. L. Hellyer *Sanders' Encycl. Gardening* (ed. 22) 240 *H[ydrangea] petiolaris*, self-clinging climbing species, white. **1922** Joyce *Ulysses* 712 Ever would he wander, selfcompelled, to the extreme limit of his cometary orbit. *c* **1900** H. A. Jones *Mrs. Dane's Defence* IV. 83, I intend that Mrs. Bulsom-Porter shall stay..as a self-confessed scandal-monger. **1981** M. Underwood *Hand of Fate* II. 91 Even a self-confessed liar is capable of telling the truth. **1977** A. Giddens *Stud. in Social & Polit. Theory* i. 72 Lakatos's sophisticated methodological falsificationism is self-confessedly an attempt to reconcile a version of Popper's philosophy of science with some of the major difficulties created for the latter by the works of Kuhn and others. **1809** J. Orrok *Let.* 28 Sept. (1927) 110 Here, and at every other Station, they had self-constituted Committees and carried everything their own way. **1856** E. B. Browning *Aurora Leigh* I. 28 Books, that prove God's being so definitely, that man's doubt Grows self-defined the other side the line, Made atheist by suggestion. **1977** R. Holland *Self & Social Context* v. 95 There is a strong sense of tension between socially given role demands and self-defined role demands. **1846** J. D. Morell *Hist. Philos.* II. v. 117 We..come, at length, at the end of the process, to a self-produced, or rather a self-developed, *subject-object*. **1898** 'Mark Twain' in *Harper's Mag.* Mar. 533/2 When an Austrian is called Doctor it means that he is either a lawyer or a physician, and that he is not a self-educated man. **1977** *Listener* 15 Dec. 783/1 From his self-educated mother, Trilling inherited a love of English literature. **1947** *Hansard Commons* 15 Dec. 1441 Mr. Amory asked the Minister of Food why persons who would qualify otherwise for the extra cheese are ineligible if they are self-employed. **1978** *Jrnl. R. Soc. Med.* LXXI. 768 'Persons' include both salaried and self-employed, including those self-employed persons established in one country who provide services in another. **1859** J. S. Mill in *Fraser's Mag.* LX. 767/1 An enemy, or a self-fancied rival. **1881** O. Wilde *Poems* 123 That holocaust, That self-fed flame. **1960** *Farmer & Stockbreeder* 1 Mar. 72/2 He had heard a lot of farmers expressing disappointment in the..meat yield from self-fed silage. **1935** M. Lowenfeld *Play in Childhood* i. 37 Play ..is taken as applying to all activities in children that are spontaneous and self-generated. **1965** J. D. Chambers in Glass & Eversley *Population in Hist.* xiii. 333 The increase in the industrial population was partly self-generated. **1938** (*title*) The Oxford companion to music: self-indexed and with a pronouncing glossary. **1978** *Canad. N. & Q.* Dec. 1/2, I am in the process of preparing a bibliography of Canadian periodicals that are self-indexed, or which have been indexed externally. **1879** G. M. Hopkins *Poems* (1967) 81 What the heart is! which, like carriers let fly..To its own fine function, wild and self-instressed. **1748** Richardson *Clarissa* VII. 288 Some other more distant relations.., self-invited..attended. **1979** C. MacLeod *Family Vault* (1980) xviii. 162 The first of the self-invited guests slumped into a wooden chair. **1955** S. Spender *Making of Poem* I. iv. 67 The observer is self-observed. **1923** Self-ordered [see *self-ordering*, sense 1 f above]. **1959** I. & P. Opie *Lore & Lang. Schoolch.* xii. 252 A contributor..gives a vivid description of a particular party of self-organized jollyboys. *c* **1865** G. M. Hopkins *Poems* (1967) 120, I storm and shock you. So I fail. And like a self-outwitted blast Fling to the convent wicket fast. **1962** H. C. Weston *Sight, Light & Work* (ed. 2) iii. 76 Persons of average vision are able to perform self-paced tasks requiring the perception of detail. **1835** J. S. Mill in *Lond. Rev.* I. II. iv. 360 The unpaid is apt to become the self-paid. **1943** A. V. Barber *Let.* 8 Feb. in W. Temple *Lett.* (1963) 58 Their arguments reveal elementary, and sometimes even self-proclaimed, ignorance of the monetary mechanism. **1979** A. Chisholm *Nancy Cunard* xi. 103 John Banting..was to become England's sole self-proclaimed Surrealist painter. **1977** *Lancet* 30 July 207/2 A large proportion of the food consumed is self-produced. **1851** *Illustr. Catal. Gt. Exhib.* III. 542/1 Complete phonological English alphabet constructing self-pronouncing words with the proper orthography. **1931** (*title*) The Royal Webster dictionary for home and school, self-pronouncing. **1970** P. Y. Carter *Mr Campion's Falcon* ix. 68 A belt of self-seeded larches. **1924** G. B. Shaw *St. Joan* Pref. p. xlii, The saints and prophets..are always really self-selected, like Joan. **1977** J. D. Douglas in Douglas & Johnson *Existential Sociol.* i. 17 At first the recruits to this rationalized segment of society were self-selected. **1865** G. M. Hopkins *Poems* (1967) 169 Will no one show I argued ill? Because, although Self-sentenced, still I keep my trust. *a* **1957** R. Campbell tr. *Camões's Lusiads* in *Coll. Poems* (1960) III. 129 Like Canace, self-sentenced and undone. **1956** *Nature* 10 Mar. 490/1 The distribution of this self-set seed mainly on the more heterozygous members of the population restricts the rate of fixation of genetic variability under these conditions. **1977** *Times Lit. Suppl.* 28 Jan. 94/2 Downright despair over his herculean, self-set task. **1856** J. G. Whittier *Panorama* 23 The self-sold knaves of gain and place. **1924** G. B. Shaw *Saint Joan* Pref. p. lii, We do not..rush to the opposite extreme in a recoil from her as from a witch self-sold to the devil. **1922** Joyce *Ulysses* 344 A neat blouse of electric blue, self-tinted by dolly dyes.

b. With nouns of action, as *self-selection* (see also *self-assembly 1, etc.). Similarly with verbs (usu. forming adjs., occas. sbs.), as *self-build, -erect, -feed* (see also *self-drive *a.*).

1952 *Times* 5 Mar. 5/6 The first self-build society put up its first houses over three years ago. There are now several scores of societies with hundreds of houses built or building. **1976** *Eastern Even. News* (Norwich) 13 Dec. 13 (Advt.), Renta-tower lightweight self-build staging. **1924** *Motor* 28 Oct. 697/1 The open tourer, with its self-erect transparent side screens. **1960** *Farmer & Stockbreeder* 16 Feb. 20/2 (Advt.), Self-erect cattle shelter. **1897** *Sears, Roebuck Catal.* 127/1 Self-feed Base Burner... A neat, attractive stove. **1951** *Sun* (Baltimore) 31 May 9/5 Mixing salt with cottonseed meal enables livestock growers to self-feed controlled amounts of protein supplement to their beef cattle. **1958** *Times* 1 July (Suppl.) p. viii/7 For up to 20 cows the trailer can be a home-made self-feed rack standing on an ordinary farm trailer. **1969** *Times* 24 Feb. 12/2 Larger farmers have taken advantage of improvement grants to erect concrete-and-asbestos covered yards and silos, generally using a combination of self-feed and easy-feed. **1975** *N.Z. Jrnl. Agric.* Sept. 63/1 We self-fed our grass silage. **1962** E. Godfrey *Retail Selling & Organization* i. 7 Members..can..obtain.. financial assistance in the conversion of their shops to self-service or self-selection. **1979** *Guardian* 5 Nov. 22/2 A half of all Jobcentre placements are made as a result of the self-selection of a job by a job-seeker.

3. a. With adjs. and ppl. adjs. *self-absorbed* (later examples), *-addressed* (later examples), *-conflicting* (earlier example), *-embedded, -enclosed* (earlier example), *-grounded, -imprisoned, -nodding* (pres. pple.). With sbs. and vbl. sbs. *self-aggression, -communing, -communion, -discontent, -dissatisfaction* (later examples), *-embedding, -preoccupation, -relation.*

1903 Somerville & 'Ross' *All on Irish Shore* 132 His face was pale and strange and entirely self-absorbed. **1980** D. Newsome *On Edge of Paradise* 8 One can admit to being self-absorbed... Arthur was genuinely self-critical. **1904** *Delineator* Dec. 1084 If you will send a stamped self-addressed envelope, we will tell you where you can take a course. **1976** *Oxford Consumer* Mar. 11/1 Sending a stamped self-addressed envelope (6 × 3½) to the Chief Superintendent, St Aldate's Police Station. **1964** M. Argyle *Psychol. & Social Probl.* iv. 55 There is some evidence that self-aggression is a result of displaced aggression. **1927** E. M. Forster *Aspects of Novel* iii. 67 The..self-communings which politeness and shame prevent him from mentioning. **1963** *Times Lit. Suppl.* 11 Jan. 21/3 A seven-year-old girl's self-communings. **1818** Byron *Don Juan* I. xci. 48 His self-communion with his own high soul. **1916** Joyce *Portrait of Artist* (1969) 160 The impression which effaced his troubled selfcommunion was that of a mirthless mask reflecting a sunken day from the threshold of the college. **1819** Shelley *Prometheus Unbound* IV. 259 Its intense yet self-conflicting speed. **1872** Geo. Eliot *Middlem.* III. lviii. 290 His endurance was mingled with a self-discontent. **1891** W. James *Let.* 6 July (1920) I. 310 You've been saved many forms of self-dissatisfaction and misery. **1981** V. Canning *Boy on Platform One* ii. 30 She felt a rare mood of self-dissatisfaction. **1964** *Language* XL. 226 Syntactically complex self-embedded constructions in English. **1963** Self-embedding [see *push-down *a.* 1]. **1978** *Language* LIV. 171 Yngve is probably best known as the man who erroneously attributed to left-branching the effects of self-embedding. **1814** F. Burney *Wanderer* IV. lxxi. 240 Juliet passed three days, self-inclosed. **1817** Coleridge *Biog. Lit.* I. xii. 265 We are to seek..for some absolute truth..a truth self-grounded, unconditional and known by its own light. **1940** *Mind* XLIX. 171 All demonstration leads back to indemonstrable bases, and grounds must themselves be grounded on what is self-grounded. **1920** 'K. Mansfield' *Let.* 27 Oct. (1977) 190 Don't you feel that what English writers lack today is experience of Life. I don't mean that superficially. But they are self-imprisoned. **1922** Joyce *Ulysses* 212 Amused Buck Mulligan mused in pleasant murmur with himself, self-nodding. **1959** 'M. Ainsworth' *Murder is Catching* xv. 173 The sublime self-preoccupation of so many actors. **1883** F. H. Bradley *Logic* III. I. vi. 448 The analysis in the end is hence *not* synthesis, if that means self-relation. **1906** D. H. MacGregor in *Hibbert Jrnl.* July 800 The fact of self-distinction from the world is as ultimate as that of self-relation to it.

b. With adjs. and related sbs., vbs., pples. *self-complete* (earlier example), *-significant, -sterile* (also as *sb.*).

1857 E. B. Browning *Aurora Leigh* VII. 289 Both faces leaned together like a pair Of folded innocences, self-complete. **1882** G. M. Hopkins *Lett. to R. Bridges* (1955) 165 This seems in English a point..insisted on, that words shall be single and specific marks for things, whether self-significant or not. **1913** Self-sterile [see *incompatibility 4 b].

4. With pples. *self-actuating, -closing* (later examples), *-defrosting, -hardening, -igniting* (also *fig.*), *-levelling, -loading, -locking, -lubricating, -operating, -propelling* (earlier example), *-recording* (earlier example), *-setting, -stabilizing, -threading, -tuning, -unloading, -winding* (earlier example); *self-powered, -propelled* (later examples). With sbs. *self-demagnetization, -ignition, -registration, -stabilization.* With vbs. (forming adjs., sbs., and vbs.): *self-lock, -register.*

1959 H. Barnes *Oceanogr. & Marine Biol.* 168 Our attention will, therefore, be focussed on remotely controlled or self-actuating underwater cameras. **1931** *Times Educ. Suppl.* 27 June 249/4 Noise is minimized by the use of self-closing double doors. **1971** *Oxf. Univ. Gaz.* 18 Feb. 671/1 The barrier must be self-closing after a car has passed through. **1960** *Times* 7 Mar. 13/5 Refrigerators.. self defrosting. **1912** *Proc. Physical Soc.* XXIV. 342 (*heading*) The self-demagnetisation of annealed steel rods. **1906** Self-hardening [see *air-hardening a.* s.v. *air *sb.*[1] B. II]. **1865** *Naval & Mil. Gaz.* 16 Sept. 589/2 Self-igniting cartridges. **1948** L. Spitzer *Linguistics & Lit. Hist.* iv. 167 Diderot has experienced to the bitter end the self-annihilation of the self-igniting mind. **1927** *Automobile Engineer* XVII. 500/1 Compression ignition stands out clearly as the one factor controlling the onset of detonation in engine practice, this simultaneous activation of compressed combustion being what is understood by the 'self-ignition temperature' of a combustible. **1969** *Gloss. Terms assoc. with Fire* (*B.S.I.*) I. 7 *Self-ignition temperature*, the temperature at which a flammable gas/air mixture will ignite without an external source of ignition. **1960** *Farmer & Stockbreeder* 16 Feb. 47/3 (Advt.), Patent self-levelling linkage ensures uniform depth of cultivation despite ground variations. **1965** *Sun* 28 Sept. 6/2 Citroen..continue their anti-flash campaign. They are fitting 'self-levelling' headlights to a new..version of their Drophead Coupe. **1977** *Observer* 8 May 33/3 Low-roll cornering and self-levelling rear suspension keep the handling predictable under all loads. **1899** *Kynoch Jrnl.* Oct.–Nov. 2/2 Jones's self-loading cartridge case. **1973** *Country Life* 28 June 1907 Self-loading forage machines, used primarily for handling loose hay. **1977** *R.A.F. News* 8–21 June 5/3 Their equipment includes..self-loading rifles and Sterling sub-machine guns. **1976** V. Canning *Doomsday Carrier* i. 6 The door swung back to self-lock. *a* **1884** E. H. Knight *Pract. Dict. Mech.* Suppl. 795/2 *Self locking hook*, one which automatically closes. **1938** *Archit. Rev.* LXXXIII. p. lx/3 The glass is held by a continuous self-locking spring aluminium cover strip. **1980** D. Bloodworth *Trapdoor* xxx. 185 The door was self-locking and he could not force his way in. **1967** M. Chandler *Ceramics in Mod. World* v. 154 Among the many highly-specialized uses of graphite refractories is the making of self-lubricating piston rings. **1947** M. M. Lewis *Language in Society* 136 The machines of war are its self-operating weapons. **1962** A. Nisbett *Technique Sound Studio* 260 '*Midget*' *recorder*,..a self-powered recorder which is light enough to be carried about without too much difficulty. **1928** C. F. S. Gamble *Story N. Sea Air Station* xxii. 428 To clear out the stores and take them by sea to other stations,..a large, self-propelled concrete lighter (*R.A.F. 110*) was used. **1945** *Finito! Po Valley Campaign* 31 Tanks and self-propelled guns were captured intact. **1977** J. Stern in *Winter's Tales 23* 178 Of his last years several were spent..in a self-propelled wheelchair on the roads..of Dorset. **1862** *Times* 7 Apr. 9/4 A self-propelling bathing-machine. **1866** *Trans. Brit. Assoc.* 1865 20 (*heading*) Description of the magnetic storm of the beginning of August 1865, as recorded by the self-recording magnetographs at the Kew and Lisbon observatories. **1847** *Phil. Trans. R. Soc.* CXXXVII. 111 It would be superfluous to speak of those proposals..for self-registering, photographically, the variation of the declination magnet. **1885** *Arch. Ophthalmol.* XIV. 54 The advantages offered by a perimeter with self-register are too evident to be overlooked. **1847** *Rep. Brit. Assoc. Adv. Sci.* 1846 II. 11 In order to adapt it for self-registration, a light conical brass tube..was affixed to the lower side. **1908** *Sears, Roebuck Catal.* 37/2 The shuttle is the most perfect self threading cylindrical shuttle... The needle is self setting. **1948** W. E. Stephens et al. *Nuclear Fission & Atomic Energy* ix. 128 An interesting possible mechanism for self-stabilization of a chain reaction in the presence of a cadmium absorber was suggested by Adler and von Halban. **1953** *Jrnl. Inst. Electr. Engineers* C. I. 101/1 The self-stabilizing effect is so important in establishing an inherently safe and stable plant [*sc.* nuclear reactor]..that it should always be carefully studied. **1908** Self-threading [see *self-setting* above]. **1964** *Discovery* Oct. 67/2 The Rank Organisation has recently marketed a 16 mm projector which is self-threading. **1956** R. Sheckley in Aldiss & Harrison *Decade 1950s* (1976) 189 The portable sub-space set was self-tuning. **1960** *Farmer & Stockbreeder* 16 Feb. 72/3 He could load five tons of chopped silage from the barn silo into two self-unloading trailers to feed 180 cows in three-quarters of an hour. **1964** *New Scientist* 4 June 596/2 The self-unloading ship is not new but economic factors have until recently restricted its operation. **1825** *Mechanics' Mag.* III. 293/2 M. Recordon.. proceeded to England, where he obtained a patent for his invention of self-winding watches, which were then in great request.

5. b. Delete ? *Obs.* and add: *self-power, -value, -worth; selftaste, -yeast* (nonce-wds.).

a **1688** R. Cudworth *Treat. Freewill* (1838) 62 God Almighty could not make such a rational creature as this is..which had no self-power, no hegemonic or ruling principle. **1896** W. James *Let.* 24 July (1920) II. 41 Full of swelling and bursting *Weltschmerz* and religious melancholy, yet no more flexibility or self-power in his mind than in a boot-jack. **1964** E. Becker in I. L. Horowitz *New Sociol.* 123 This is the basic phenomenology of alienations: the failure to develop self-powers by transacting with the world of things. **1880** G. M. Hopkins *Sermons & Devotional Writings* (1959) 125 Above all my shame, my guilt, my fate are the very things in feeling, in tasting, which I most taste that selftaste which nothing in the world can match. **1959** *Guardian* 28 Aug. 4/3, I did meet Colin Wilson, and..I found that with him I received my sense of my self-value in its fullness. **1944** *Horizon* Feb. 104 The feeling of self-worth experienced by contented individualists. **1978** M. Puzo *Fools Die* xlvii. 501 The lack of self-worth, the desire to please someone that they thought really cared about them. *a* **1889** G. M. Hopkins *Poems* (1967) 101 Selfyeast of spirit a dull dough sours.

self-abnegation. Add: Also **self-a·bnegator** (*rare*).

1907 G. B. Shaw *John Bull's Other Island* p. xix, The Catholic is theoretically a Collectivist, a self-abnegator, a Tory.

self-abso·rption. 1. [Self- 3 a.] Absorption in one's own emotions, interests, or situation; self-preoccupation.

1862 [in Dict. s.v. Self- 3 a]. *a* **1930** D. H. Lawrence *Last Poems* (1932) 94 The lovely and self-absorbed girl Looked back at the handsome and self-absorbed young man And thrilled. And in that thrill he felt: Her self-absorption is even as strong as mine. **1979** V. S. Naipaul *Bend in River* xi. 188, I did not wish to lose myself in the ..self-absorption of that fantasy.

2. *Physics.* [*Self- 2 b.] Absorption of radiation by the material emitting it. Cf. *self-shielding *vbl. sb.*

1938 *Proc. Cambr. Philos. Soc.* XXXIV. 600 With radium E, strong sources may be obtained showing negligible self-absorption. **1950** *Atomics* Nov. 334/2 The self-absorption of β-rays in radioactive films..has been investigated. **1978** H. H. Bauer et al. *Instrumental Analysis* x. 265 At too high a current, Doppler broadening and self-absorption (absorption of part of the emitted radiation by the dense cloud of atoms in the source itself) will occur.

self-abuse. 3. For 'self-pollution' read 'masturbation'.

self-active, *a.* (Later examples.)

1906 G. H. Howison *Let.* 9 Feb. in R. B. Perry *Tht. & Char. W. James* (1935) I. 776 The self-active unity of consciousness. **1936** Allport & Odbert in *Psychol. Monogr.* XLVII. I. 17 It was customary for psychologists to conceive some 'power of the mind'..and by naming it to regard the power as fixed and self-active.

self-activity. (Later *Psychol.* examples.)

1932 C. S. Myers *Absurdity of Mind-Body Rel.* 4 All conscious mental activity is self-activity; only the self is conscious—conscious, at first, solely of self-activity (conation) and of modifications of that self-activity (affects). **1961** R. C. Tucker *Philos. & Myth in Marx* viii. 134 'Self-activity'.., by which Marx means free creativity in which a person feels thoroughly at home with himself..and experiences his energies as his own.

self-actualiza·tion. Chiefly *Psychol.* [Self- 1 a.] Realization or fulfilment of oneself, esp. considered as a drive or need.

1939 H. Ansbacher tr. *Goldstein's Organism* v. 197 Experiences..teach us that we have to assume only one drive, the drive of self-actualization. **1943** A. H. Maslow in *Psychol. Rev.* L. 382 What a man *can* be, he *must* be. This need we may call self-actualization. **1957** C. Argyris *Personality & Organization* vii. 181 The degree of self-actualization increases sharply for individuals as their dependence..and submissiveness are decreased. **1975** *Ecologist* V. 123/1 The evolving form of marriage seems to aim at what might be called 'self-actualization' of the persons.

Also **self-a·ctualize** *v. intr.*; **self-ac·tualized** *ppl. a.*, **-a·ctualizing** *ppl. a.* and *vbl. sb.*; **self-a·ctualizer.**

1874 Self-actualizing *ppl. a.* [in Dict. s.v. Self- 1 f]. **1943** A. M. Farrer *Finite & Infinite* xi. 119 Will is the self-actualising potency of (the process of) a project. **1954** A. H. Maslow *Motivation & Personality* xii. 223 The creativeness of the self-actualized man. *Ibid.* 230 A firm foundation for a value system is automatically furnished to the self-actualizer. **1959** H. B. Anderson *Creativity* xv. 242 To do this much would..appear to be a step toward self-actualizing. **1961** H. C. Smith *Personality Adjustment* xiv. 381 Somerset Maugham shows..the qualities of spontaneity..and self-acceptance that are found in self-actualizers. **1977** R. Holland *Self & Social Context* iv. 61 Arguing that there is a given 'instinctoid' tendency for the human being to self-actualise, that is to go beyond the satisfaction of the more basic needs to seek satisfaction of the higher needs. **1980** *Redbook* Oct. 58/1 Once I had seen close up what Friedan labeled the 'self-actualized woman', I was damned if I would take a chance with anything else.

self-admiring, *ppl. a.* (In Dict. s.v. Self-admiration.) (Later examples.)

1871 Geo. Eliot *Middlem.* (1872) I. I. i. 7 Dorothea.. was open, ardent, and not in the least self-admiring. **1951** M. McLuhan *Mech. Bride* (1967) 141/1 Is this little tough the twin brother of any self-patting, self-admiring deb? **1981** J. Carey *John Donne* iv. 99 His poems..though self-absorbed..are not self-admiring.

self-aliena·tion. *Philos.* and *Social Sci.* [Self- 1 a.] Alienation that takes place within the self, esp. in *Marxism*. Cf. *alienation 1 c.

1906 J. B. Baillie *Outl. to Idealistic Construct. Exper.* vii. 242 For this Self-alienation is itself regarded as necessary, as the very expression of free self-consciousness. **1926** H. J. Stenning tr. *Marx's Sel. Ess.* 13 The immediate task of philosophy, when enlisted in the service of history, is to unmask human self-alienation in its unholy shape. **1938** K. Korsch *Marx* II. xi. 158 The actual 'self-alienation' of the wage-labourer. **1964** S. M. Jourard *Transparent Self* ii. 11 It often comes to pass..that our public selves become so estranged from our real selves that the net consequence is self-alienation. **1977** A. Giddens *Stud. in Social & Polit. Theory* v. 199 For Marx ..money is the epitome of human self-alienation under capitalism, since it reduces all human qualities to quantitative values of exchange.

self-ali·gning, *ppl. a. Mech.* [Self- 4.] Capable of aligning itself automatically. Of a bearing or the like: having a degree of flexibility as regards alignment. Hence **self-ali·gnment.**

1904 *Science* 29 Jan. 164/1 Where the ends of the polar axis are supported on separate piers the bearings can be made self-aligning. **1905** [in Dict. s.v. Self- 4]. **1930** *Engineering* 26 Sept. 394/3 A self-aligning double roller feed for vertical double-spindle moulders. **1960** *Farmer & Stockbreeder* 15 Mar. 148/1 All-steel welded construction with self-aligning ball bearings throughout. **1962** G. A. T. Burdett *Automatic Control Handbk.* iv. 9 A method of maintaining self-alignment.

self-ana·lysis. [Self- 1 a.] Analysis by or of oneself; *spec.* Psychol., psychoanalysis of oneself undertaken by oneself.

1860 A. J. Munby *Diary* 12 Mar. in D. Hudson *Munby* (1972) 54 Self analysis helps one... I am sensible of a feeling of placid halfcontemptuous indifference. **1862** [in Dict. s.v. Self- 1 a]. **1911** E. Jones in *Amer. Jrnl. Psychol.* XXII. 520 The greatest value is to be attached to self-analysis, a fact to which attention cannot too often be called. *a* **1930** D. H. Lawrence *Last Poems* (1932) 28 Self-analysis Which goes further and further, and yet never finds an end. **1958** K. Menninger *Theory of Psychoanal. Technique* vii. 159 The process of self-analysis continues automatically..with increasing freeing and expanding of ego functions. **1977** M. Jahoda *Freud & Dilemmas of Psychol.* iii. 52 The self-analysis is important beyond this and..stakes the claim for psychoanalysis to be a reflexive psychology. **1981** B. Masters *Georgiana* viii. 204 With time on her hands, Georgiana gave way to self-analysis, introspection, regret.

Hence **self-a·nalyst.**

1929 Sheehan & Gaffney *Tristram Lloyd* III. i. 187 He was a self-analyst, and he carried the principle into every-day life, sorting, sifting, examining motives and principles. **1958** B. W. Aldiss *Non-Stop* III. iv. 184 He was not enough of a self-analyst to see it was a quality he had..himself.

self-asse·mbly. 1. [*Self- 2 b.] Subsequent assembly of something bought in the form of a kit; usu. *attrib.*, denoting items (e.g. furniture) sold in this form.

a **1966** in G. N. Leech *Eng. in Advertising* (1966) xv. 137 Peerless brings within your reach..the luxury of a Built-in Bedroom at a price you can really afford With Dovetail Self-assembly Units. **1978** *Dumfries Courier* 20 Oct. 27/1 (Advt.), Broom unit for self assembly £10. **1980** *Daily Tel.* 9 July 2/1 A subsidiary..which makes self-assembly garages and prefabricated home extensions.

2. *Biol.* [Self- 1 a.] The spontaneous formation of a sub-cellular particle from its components, e.g. that of a ribosome or of a virus in a medium containing the appropriate RNA and proteins. Hence **self-asse·mble** *v. intr.*

1969 *Jrnl. Molecular Biol.* XL. 412 We feel that the general principle of self-assembly revealed in the present *in vitro* system also operates *in vivo*. **1977** *Jrnl. Protozool.* XXIV. 9/1 Self-assembly of a restricted number of different macromolecular subunits to form comparatively simple structures such as a rhizoplast, a pelta, or a trichocyst. **1978** *Sci. Amer.* Nov. 53 (*caption*) The virus 'self-assembles' spontaneously in the test tube from its constituent RNA molecule and protein subunits, giving rise to infective virus particles indistinguishable from those found in nature.

self-asserting, *a.* (In Dict. s.v. Self-assertion.) (Earlier example.)

1837 J. S. Mill in *Westm. Rev.* XXVIII. 98 Carrel's manner was not of the self-asserting kind.

self-asse·ssment. [Self- 1 a.] Assessment or evaluation of oneself, one's actions or attitudes by oneself; an instance of this; *spec.* calculation of taxable liability by oneself.

1954 *Brit. Jrnl. Med. Psychol.* XXVII. 142/1 The psychiatric inventory described here (to be known as the Tavistock Self-assessment Inventory) is one which has been developed over the past few years in the Tavistock Clinic. **1960** *Spectator* 1 July 34 In the study of myth, Professor Kerényi explained in a 'self-assessment' written ten years ago, one must accept the 'axiom' of depth psychology. **1972** *Accountant* 12 Oct. 441/2 Special arrangements will secure for them the equivalent of tax credits, and self-assessment is waiting in the wings. **1979** *Internat. Rehabilit. Med.* I. 53/1 They were asked to assess the ability with which they performed a variety of activities during the preceding few weeks..self-assessments.

self-ba·lancing, *ppl. a.* [Self- 1 f, 4.] **1.** In technical use: capable of balancing itself automatically; automatically producing balance.

1912 *Chambers's Jrnl.* Sept. 638/1 An upright position is always assured when the car is stationary, a continuous self-balancing motion being given to the vehicle. **1954** D. E. Carritt in Isaacs & Iselin *Oceanogr. Instrumentation* 182 Snodgrass et al. have constructed a self-balancing photoelectric filter photometer.

2. *Accounting.* Designating that in which the debit side of the accountable items is equal to the credit side; *self-balancing ledger* (see quot. 1970).

1898 S. S. Dawson *Accountant's Compendium* 350/2 The Customers' Ledger..is self-balancing because of the operation of the Controlling Account. **1931** *Economist* 3 Jan. 7/2 The gross figures, including self-balancing revenue, show receipts from all sources amounting to £486·5 millions. **1953** *Chambers's Jrnl.* June 343/1 Catering generally is a profitable or at least self-balancing item in the seaside resort budget. **1964** R. B. Notestein in I. L. Horowitz *New Sociology* 52 The economy was self-balancing, with authority exerted by impersonal regional markets. **1970** M. Greener *Penguin Dict. Commerce* 297 *Self-balancing ledger*, an accounting term for a personal ledger containing a control account.

self-being. Restrict † *Obs.* to *concr.* use and add: (Later examples.) Now *rare*.

1880 G. M. Hopkins *Sermons & Devotional Writings* (1959) 122 And this whether I speak of human nature or of my individuality, my selfbeing. **1915** A. Vonier *Personality of Christ* v. 31 The scholastic maintains that self-being underlies self-consciousness, as the cause underlies the effect. **1977** R. Holland *Self & Social Context* ii. 41 He attributes the non-disclosure to role-playing, which is placed in sharp contrast with real self-being.

self-bi·as. *Electronics.* [Self- 1 d.] Bias applied to the grid of a valve by means of a resistor in the cathode circuit or the grid circuit. So **self-bi·ased** *ppl. a.*, **self-bi·asing** *vbl. sb.*

1931 Moyer & Wostrel *Radio Handbk.* xiii. 708 (*heading*) Self-biasing grid-glow tube. **1932** F. E. Terman *Radio Engin.* xi. 389 (*caption*) Self-bias circuits for obtaining a negative grid bias..by utilizing the voltage drop across a resistance in series with the cathode. *Ibid.*, When an anode detector is self-biased by the arrangement shown.., the detector characteristics are quite different from those obtained with a bias derived from a battery. **1939** H. J. Reich *Theory & Application Electron Tubes* vi. 161 (*caption*) Impedance-capacitance-coupled amplifier, showing the use of self-biasing resistors. **1945** [see *push-pull *a.* 2]. **1960** *Practical Wireless* XXXVI. 319/1 This condenser discharges through the grid leak.., building up the self bias for the valve.

self-binder. Add: **2.** (See quot.)

1950 *Times* 26 July 6/5 A self-binder for *The Times*, which takes one month's issues on sprung cords, is now available.

self-bio·graphy. *rare* or *Obs.* [Self- 1 d.] = Autobiography.

1796 I. D'Israeli *Miscellanies or Lit. Recreations* 95 (*heading*) Some observations on diaries, self-biography and self-characters. **1813** J. F. Stanfield *Essay on Study & Composition of Biography* I. i. 2 Our supply of genuine self-biography is but scanty.

self-bli·mped, *a.* [Self- 4; cf. *blimp 1 b.] Of a ciné camera: fitted with a sound-proof cover; insulated from sound by its own housing.

1961 in Webster. **1965** R. Fielding *Technique Special-Effects Photogr.* x. 269 The projector must be somehow silenced... Either the machine must be self-blimped or installed within a sound-proof projection room. **1969** *Focal Encycl. Film & Television* 74/2 Sound cameras of recent design are..self-blimped..so that mechanical noise is reduced to an absolute minimum. **1979** D. Cheshire *Bk. Movie Photogr.* 174 The camera must run as silently as possible to avoid disturbing the animals. You may need a 'self-blimped' camera.

self-ca·ncelling, *ppl. a.* [Self- 1 f, 4.] That cancels itself; that negate each other. **1.** In technical use: designed to stop working automatically when no longer required; applied esp. to traffic indicators of motor vehicles.

1935 [see *dipper 8]. **1945** *Autocar Handbk.* (Autocar Techn. Staff) (ed. 18) x. 196 Some form of 'self-cancelling' arrangement is..fitted, to avoid drivers unknowingly going along with an indicator extended. **1960** *News Chron.* 6 May 8/8 They have omitted..to make the traffic indicators self-cancelling. **1963** *Times* 15 Jan. 12/1 By flicking in the self-cancelling overdrive in third and top the motorway cruising speed went up to about 80. **1972** *Daily Tel.* 12 July 10/5 Not only is the stalk on the column very flimsy and mounted on the wrong side..but the indicators are not self-cancelling.

2. *gen.*

1943 *Mind* LII. 295 It has been made a *self-cancelling* expression like the expression 'ride a motor cycle' as used by someone who is determined to use 'ride' only where what you ride is something living. **1965** *Spectator* 15 Jan. 60/1 The concept of fighting for export orders has become largely self-cancelling.

self-capa·citance. *Electr.* [Self- 5 b.] The inherent capacitance of a circuit or component. Also **self-capa·city.**

1923 E. W. Marchant *Radio Telegr. & Teleph.* v. 66 Such coils have a considerable inductance, and 'self-capacity'. **1937** L. D. Weld *Gloss. Physics* 207/2 *Self-capacitance*, distributed capacitance of an electric circuit due to its containing closely wound coils of insulated wire, the adjacent turns of which have a condenser effect. **1958** *Practical Wireless* XXXIV. 22/2 This was found to be essential since the timebase was affecting the appearance of the trace due to the self-capacity of the X switch. **1960** *Ibid.* XXXVI. 397/2 A small capacitance which is in fact the self-capacitance of the crystal.

self-care (stress variable). [Self- 1 a, *2 b, 3 a.] Care for oneself; self-interested behaviour; freq. used *attrib.* to denote an institution in which patients and convalescents attend to many of their own needs.

1904 in Webster. **1932** Auden *Orators* II. 41 The second law of thermodynamics—self-care or minding one's own business. **1962** *Punch* 7 Nov. 658/1 Recovering your strength you're passed to the Self-Care Unit. **1979** *Internat. Rehabilit. Med.* I. 59/1 Assessment of the competence of patients in essential activities of daily living (A.D.L.), such as mobility, self care, and domestic activities.

self-ca·tering, *vbl. sb.* [Self- 3 a.] Catering for oneself; *spec.* providing one's own domestic services (as meals and cleaning) in rented holiday accommodation. Chiefly *attrib.* of holidays and vacation accommodation.

1970 *Country Life* 1 Oct. 802/1 There has also been a discernible movement towards self-catering holidays, in

farmhouses, chalets, caravans and cottages. **1972** *Guardian* 9 Dec. 13/2 Holiday camps are increasingly turning to self-catering. **1973** *Times* 25 Jan. (Channel Islands Suppl.) p. v/3 Self-catering accommodation is.. increasing rapidly. **1977** *Western Morning News* 1 Sept. 6/5 (Advt.), Comfortable all-electric self-catering Farmhouse; 2 double bedrooms week 3/10 September. **1981** *Daily Tel.* 27 Jan. 12/5 Two redundant stone-built barns ..have been converted into self-catering bunkhouses for walkers.

self-ce·ntring, *ppl. a.* Add: **2. b.** Of the steering of a motor vehicle: tending to return automatically to a central alignment. Also as *vbl. sb.*

1926 H. T. Rutter *Mod. Motors* II. ix. 331 The steering gear is arranged to be self-centring, so that little pressure is required to maintain the car in a straight line. **1963** *Times* 5 Feb. 7/5 At first, the powerful self-centring action of the steering makes the car feel heavy and slightly unstable. **1975** *Drive* New Year 102/3 The tail certainly wags the Beetle, especially in the wet when even the responsive, self-centring steering won't help to avoid trouble. **1977** *Times* 18 Aug. 23/3 The main criticisms of steering were..stiffness of operation, and lack of self-centring.

self-cha·nging, *ppl. a.* **1.** (In Dict. s.v. Self- 1 f.)

2. [Self- 4.] Of a gearbox: preselective.

1930 [see *fluid flywheel* s.v. *fluid *sb.* 3]. **1939** *Country Life* 11 Feb. p. xxx/3 The self-changing gear box and the special anti-snatch Armstrong Siddeley transmission.

self-clea·ning, *ppl. a.* (*vbl. sb.*) [Self- 1 b, f, 4.] Designed to keep itself clean automatically. Also as *vbl. sb.*, the action or capability of doing this.

1898 L. S. Robertson tr. *Bertin's Marine Boilers* iv. 91 Self-cleaning grates with movable bars have been used in America. **1930** *Daily Express* 6 Oct. 7/5 (Advt.), The Ewbank Success sweeper is fitted with ball bearings, a self-cleaning brush and rubber protectors. **1960** *Farmer & Stockbreeder* 15 Mar. 119/2 The self-cleaning action of the slats has not been particularly evident. **1971** *Engineering* Apr. 31/1 So far, cost and complexity have ruled out self-cleaning [for vending machines]. **1979** *Arizona Daily Star* 5 Aug. (Advt. Section) 17/7 Self-cleaning patio pool, landscaped, block wall.

self-clea·nsing, *ppl. a.* (*vbl. sb.*) [Self- 1 b, f, 4.] Tending automatically to keep itself clean. Also as *vbl. sb.*

1911 H. S. Watson *Sewerage Systems* iii. 41 All sewers should be self-cleansing. **1921** T. Goodhugh *Operative Dental Surg.* xviii. 262 Food in mastication cleans the teeth on the exposed surfaces, which are called 'self-cleansing surfaces'. **1963** J. Osborne *Dental Mech.* (ed. 5) viii. 147 The resin..may then be lightly polished to give a surface that is readily self-cleansing. **1978** *Jrnl. R. Soc. Med.* LXXI. 718 In these situations, self-cleansing is less common and keratin may accumulate.

self-co·cking, *ppl. a.* (*vbl. sb.*) [Self- 4.] Of a firearm: cocked and fired by pulling the trigger; double-action. Also *fig.*, and as *vbl. sb.*

1862 [see *back-action]. **1880** *News & Press* (Cimarron, New Mexico) 23 Dec. 3/1 Mr. T. O. Boggs of Tramperos, while in the act of drawing a self-cocking pistol from his pocket, accidentally discharged it. **1892** [in Dict. s.v. Self- 4]. **1902** W. D. Hulbert *Forest Neighbors* 161 Four shots, as fast as the self-cocking revolver could pour the lead into his body. **1964** H. L. Peterson *Encycl. Firearms* 17/1 The revolver patented by Robert Adams in 1851 was the English challenge to the Colt revolver, but unlike the latter it was made with a solid frame and a self-cocking action.

Hence **self-co·cker,** a firearm thus cocked and fired.

1863 [in Dict. s.v. Self- 4]. **1902** S. E. White in *McClure's Mag.* Apr. 552/2 It was wonderful work, rattling fire, quicker than a self-cocker even.

self-coi·ncidence. [Self- 3 a.] The fact or state of coinciding with its former position following a displacement.

1904 *Knowledge* June 110/1 Every molecule of the medium..is brought by certain so-called 'movements' into the position previously occupied by some other molecule of the medium (the medium is said to be 'brought to self-coincidence' by such a movement). **1966** [see *isometric *a.* 7].

self-colour. 1. (Later examples of clothing and embroidery.)

1927 T. Woodhouse *Artificial Silk* ix. 86 Knitted garment of a self-colour are made, in this simplest form, by the manipulation of a single continuous thread. **1950** *Sun* (Baltimore) 9 May 18 (Advt.), Barbara Lee leg-contour proportioned-fit sheer nylons with contrast or self-color seams. **1964** A. Butler *Teaching Children Embroidery* II. iv. 35 Free experimenting, with stitches used in some areas, in which self-colour is worked as a texture. **1976** *Morecambe Guardian* 7 Dec. 1/1 (Advt.), 100% wool skirt suit with self-colour embroidery.

self-compa·tible, *a.* and *sb. Bot.* [Self- 3 b.] **A.** *adj.* Able to be fertilized by means of its own pollen. **B.** *sb.* A self-compatible plant or species. Opp. *self-incompatible *a.* and *sb.*

1922 *Bot. Gaz.* LXXIII. 111 There are in the feebly self-compatible plants of these species no specially marked tendencies to self-compatibility at any definite phase of the blooming period. **1938** [see *self-incompatible *a.* and *sb.*]. **1955** *Heredity* IX. 70 The trees he tested could be divided on their setting capacity into those which were self-compatible and those which were self-incompatible. **1969** *Ibid.* XXIV. 165 We may suppose that dioecy here evolved recently from hermaphroditism in a *self-compatible* ancestor. **1979** *Nature* 20–27 Dec. 837/2 The usual explanation that dioecy arises in a self-compatible taxon as a way of ensuring outcrossing is certainly possible.

So **self-compatabi·lity,** the state of being self-compatible.

1917 *Jrnl. Genetics* VII. 78 The behaviour of the various series indicates an irregular inheritance of the characteristics of self-compatibility. **1969** *Heredity* XXIV. 613 Different self-compatibility alleles may occur localised in different habitats.

self-co·ncept. *Soc. Psychol.* [Self- 1 d.] A person's concept or idea of himself. Also **self-conce·ption.**

1934 in Webster. **1938** *Psychol. Abstr.* XII. 770/1 (Index) Self-concept, adolescence. **1947** *Amer. Psychologist* Sept. 365/2 Our observations of psychotherapeutic experience would seem to indicate that absence of any threat to the self-concept is an important item in the problem. **1957** *Amer. Sociol. Rev.* XXII. 218/2 The following analysis is limited to the executives' and the supervisors' self-conceptions and conceptions of each other. **1968** *Internat. Encycl. Social Sci.* XIV. 150/1 Since the 1940s, problems related to self concept have surged forth as indispensable..topics for scientific study in psychology and sociology. **1977** M. Edelman *Polit. Lang.* ii. 29 The self-conceptions that are a part of these contradictory cognitive structures.

self-concern, -concerned. Delete † *Obs.* and add later examples.

1941 *Mind* L. 179 The self is part 'psyche' and part 'spirit': the psyche or animal soul self-concerned and partial, the spirit disinterested and catholic. **1972** P. D. James *Unsuitable Job for a Woman* iv. 148 Even their compassion had been overlaid with self concern. **1982** J. O'Faolain *Obedient Wife* i. 23 It says something about the way you see me. Forgive my self-concern.

self-condensa·tion. *Chem.* [Self- 1 a.] A condensation reaction between two molecules of the same compound. Also **self-conde·nse** *v. intr.*, to undergo such a reaction.

1946 *Nature* 12 Oct. 514/1 The highly reactive bifunctional compound, γ-bromopropylamine,..would immediately undergo self-condensation. **1959** A. Albert *Heterocyclic Chem.* vii. 269 β-Keto-esters..self-condense to pyrones, e.g. ethyl acetoacetate gives 2-methyl-5-acetyl-6-hydroxy-γ-pyrone. **1964** N. G. Clark *Mod. Organic Chem.* xxiii. 478 The self-condensation of acetophenone provides one of the rare examples of the carbonyl group of an aromatic ketone participating in a condensation reaction. **1976** *Austral. Jrnl. Chem.* XXIX. 1039 Self-condensation of cyanoacetamides under the influence of phosphorus oxychloride leads to pyridines and pyridinium salts in high yield.

self-confident, *a.* (Earlier and later examples.)

a **1617** P. Baynes *Christian Lett.* (1620) sig. C5, O Lord..thou hast promised that thou wilt circumcise our selfe-confident hearts, so that we shal haue no trust in the flesh, but haue all our reioycing in thee. **1921** E. O'Neill *Diff'rent* I, in *Emperor Jones* 221 Her face..attracts the eye by a certain boldly-appealing vitality of self-confident youth. **1951** in M. McLuhan *Mech. Bride* (1967) 72/1 It's a self-confident look.

self-co·njugate, *a.* Add: **b.** *Math.* Of a subgroup: having the property that for any elements h in the subgroup and g in the group, the product ghg^{-1} is in the subgroup.

1888 G. G. Morrice tr. *Klein's Lectures on Ikosahedron* I. i. 7 Every group contains..two self-conjugate subgroups: viz., in the first place, the totality of all its operations, *i.e.*, the group itself, and, in the second place, that simplest group which consists of the identical operation alone. **1937** A. A. Albert *Mod. Higher Algebra* vi. 131 We call A a normal divisor (or invariant subgroup, or self-conjugate subgroup) of ∂. **1979** Page & Wilson *Introd. Computational Combinatorics* iv. 74 A partition whose Ferrers graph reads the same by rows and by columns is called self-conjugate.

Hence **self-co·njugately** *adv.*, as a self-conjugate subgroup.

1897 W. Burnside *Theory of Groups* x. 208 If two of these sub-groups have a common subgroup of order 4, it must be contained self-conjugately..in a sub-group of order 24 or 56. **1901** L. E. Dickson *Linear Groups* p. vii, Largest subgroup containing the Abelian group self-conjugately.

self-conscious, *a.* Add: **1 a.** (Earlier example.)

a **1688** Cudworth *Treatise of Freewill* (1838) 71 We are certain by inward sense that we can reflect upon ourselves and consider ourselves, which is a reduplication of life in a higher degree; for all cogitative beings as such are self-conscious.

2. (Earlier and later examples.)

1834 J. S. Mill *Let.* 12 Jan. in *Wks.* (1963) XII. 208, I begin to think that instead of being, as I once thought I was, the most self-conscious person living, I am much less self-conscious *now*..than almost anybody. **1937** 'G. Orwell' *Road to Wigan Pier* xiii. 255 Self-conscious Socialists dutifully addressing one another as 'Comrade'. **1974** J. Irving *158-Pound Marriage* ii. 42 She had never been self-conscious about what she wore.

self-consciousness. 4. (Earlier and later examples.)

1833 J. S. Mill *Let.* 25 Nov. in *Wks.* (1963) XII. 195 A man singularly free, if we may trust appearances, from self-consciousness. **1932** G. Greene *Stamboul Train* I. i. 5 Her body..even while stumbling..retained its self-consciousness.

self-consistent, *a.* Add: **b.** *Physics.* Of a trial solution of Schrödinger's equation for a nucleus with more than one electron: consistent with its own postulates (see quots.).

1928 D. R. Hartree in *Proc. Cambr. Philos. Soc.* XXIV. 113 If the final field is the same as the initial field, the field will be called 'self-consistent'. **1958** Sykes & Bell tr. *Landau & Lifshitz' Quantum Mechanics* x. 232 Schrödinger's equation for atoms containing more than one electron cannot be directly solved in practice, even by numerical methods. Approximate methods of calculating the energies and wave functions of the stationary states of the atoms are therefore important. The most important of these methods is what is called the self-consistent field method. The idea of this method consists in regarding each electron in the atom as being in motion in the 'self-consistent field' due to the nucleus together with all the other electrons. **1974** Eisberg & Resnick *Quantum Physics* ix. 347 It might seem that there is no way to find the net potential of an atom at intermediate distances from its center... But it can be taken care of by demanding that the net potential be self-consistent.

Hence **self-consi·stently** *adv.*

1953 F. J. Whitfield tr. *L. Hjelmslev's Prolegomena to Theory of Lang.* 9 We require of linguistic theory that it enable us to describe self-consistently and exhaustively all conceivable or possible Danish texts. **1973** *Physical Rev.* B. VII. 674 Energy bands, Fermi surfaces, and densities of states of calcium as a function of lattice constant have been calculated self-consistently by the augmented-plane-wave method.

self-contained, *a.* Add: **b.** Now more usually of a flat, or suite of rooms, within a house.

1910 *Bradshaw's Railway Guide* Apr. 1174 The only Hotel in the Town having self-contained Suites. **1928** E. A. Robertson *Cullum* xi. 206 We came back in the evening to what was practically a self-contained flat—no one else slept on that landing. **1977** *Wandsworth Borough News* 16 Sept. 15/1 Planning Proposals..7 Eckstein-road, Clapham Junction—conversion to form two self-contained flats.

self-contented, *ppl. a.* (Earlier example.)

1819 Shelley *Rosalind & Helen* 29 Men, women, children, slunk away, Whispering with self-contented pride.

self-contradiction. Add: Also **self-contradi·ctorily** *adv.*

1943 *Mind* LII. 313 We remember the man who never said, 'To see an event we should have to see it in an instant', thus self-contradictorily misdescribing our use of 'see an event'. **1959** P. F. Strawson *Individuals* i. 34 The standard..is set self-contradictorily high. **1981** G. MacBeth *Kind of Treason* ii. 17 The Clapham Center, as it was rather self-contradictorily called.

self-cri·ticism. [Self- 1 a.] **1.** Criticism of oneself.

1857 [in Dict. s.v. Self- 1 a]. **1926** J. S. Huxley *Essays in Popular Science* 162 The normal tendency of strongly-felt religious emotion to..set itself up as an absolute standard untempered by tolerance or by the self-criticism of reason. **1937** O. Jespersen in *S. P. E. Tract* XLVIII. 269 While one pronounces the beginning of an utterance the mind is busy preparing the rest, not only what to say, but how to say it. But most of this linguistic self-criticism is..lost to the world. **1956** H. Kurnitz *Invasion of Privacy* ii. 20 In this last phrase of good-natured self-criticism, Stradling's voice shifted into a glacial gear. **1977** *Lancet* 13 Aug. 357/1 Dr Bartsch reiterates our self-criticism that the two groups of patients were not comparable in all respects.

2. *Pol.* Criticism undertaken publicly by oneself of one's actions, attitudes, or policies, considered as a duty in order to ensure conformity with communist party doctrine. Also *transf.*

1933 E. & C. Paul tr. *Stalin's Leninism* II. 122 An abyss divides the self-criticism of the opposition..from Bolshevik self-criticism, the purpose of which is to *strengthen* the Party spirit. **1966** J. Bingham *Double Agent* vii. 101 Colonel Golchenko has performed an act of self-criticism in accordance with the principles of our party. **1976** M. Machlin *Pipeline* x. 116 Finally, after a scathing self-criticism session in the collective's skid row tenement in Seattle, Sonia decided she had had enough. **1978** D. Bloodworth *Crosstalk* x. 84 She was in trouble —public accusations that she was a counterrevolutionary, self-criticism sessions, denunciations,..the works.

self-cultiva·tion. [Self- 1 a.] = Self-culture.

1873 J. S. Mill *Autobiogr.* iv. 119, I did not neglect other modes of self-cultivation. **1964** R. Miliband in I. L. Horowitz *New Sociology* 80 A society..where man's social..setting would encourage self-cultivation and craftsmanship.

self-culture. (Earlier and later examples.)

1829 J. Sterling *Let.* 10 Nov. in Carlyle *Life John Sterling* (1851) I. vii. 56, I have no doubt that, by practice

SELF-DELIVERANCE

and self-culture, she will be a far finer actress. **1926** B. Webb *My Apprenticeship* ii. 60 A device of my own for self-culture—reading the books of my free choice.

self-deli·verance. [Self- 1 a.] Suicide by an incurable patient who finds his suffering intolerable.
This euphemism is not yet (1982) in wide currency. The related expression *self-deliveration* has also been used.
1975 M. R. Barrington in M. Kohl *Beneficent Euthanasia* 214 Taking one's own life..would be more readily comprehended as responsible behavior if it were expressed as 'self-deliveration'. *Ibid.* 245 *Self-deliverance*, accession to deliverance. *Self-deliveration*, suicide where deliveration has or might have been given. **1975** *Hansard Lords* 4 Dec. 47/2 Self-deliverance to be regarded as death by misadventure. **1980** *Daily Tel.* 12 Aug. 6/2 The decision not to publish [a guide to suicide] would mean 'tragedy and continued distress..for many who wish to bring about their own self-deliverance'.

self-deluded, *ppl. a.* Add: **self-deluder** (earlier example), **-deluding** *ppl. a.* (later examples).
1748 Richardson *Clarissa* VII. xv. 61 Poor mistaken creature!—Unhappy self-deluder! **1955** *Bull. Atomic Sci.* Apr. 168/1 It is shortsighted and self-deluding to ascribe more than a small part of Russian success to efficient espionage. **1980** D. Francis *Reflex* xii. 142 He was a pernicious self-deluding little egotist.

self-denying, *vbl. sb.* (Later example.)
1878 Trollope *Is he Popenjoy?* I. xiii. 174, I hate all kind of strictness and duty and self-denying.

self-destru·ct, *v.* orig. *N. Amer.* [f. Self- 1 a, h: cf. Destruct *v.* in Dict. & Suppl.] **1.** *intr.* Of a thing: to destroy itself automatically. Also *fig.*
1969 *Daily Colonist* (Victoria, B.C.) 30 Mar. 43/4 This message will self-destruct in 10 seconds but the printed message is the one that lives on. **1970** *New Yorker* 28 Nov. 58 Our definition of 'history' is going to change as we raise our consciousness. Our definition's going to—it's going to self-destruct. **1973** *Guardian* 18 June 4/6 Watergate came from within. The system itself has begun to self-destruct. **1977** D. Francis *Risk* xiv. 188 He's programmed to self-destruct before the end of the season... He'll go bust to the bookies. **1979** R. Perry *Bishop's Pawn* i. 14 The tape would automatically self-destruct after twenty minutes.
Hence as *sb.*; **self-destructing** *ppl. a.*
1970 *New Scientist* 27 Aug. 406 (*title*) Self-destructing proteins may tick away our years. **1972** *Village Voice* 1 June 25/3 These are finally not poems or plays or stories, but self-destructs. **1977** *Daily Tel.* 21 May 12/4 Built into the whole modernist adventure was a kind of self-destruct. **1978** J. McNeil *Consultant* xxxvi. 295 Alloway's program has done a convenient self-destruct.
2. *attrib.* as *adj. phr.*
1966 R. W. Taylor *Doomsday Sq.* iii. 36 There's a double safeguard in a self-destruct system that would operate automatically in case of navigational error. **1969** M. Crichton *Andromeda Strain* x. 108 At the lowest level of this laboratory is an automatic self-destruct device. **1975** J. Grady *Shadow of Condor* xvi. 250 He flicked the last strap holding him to the machine. He also punched the delayed self-destruct switch.

self-destructive, *a.* Add: **self-destru·ctiveness.**
1733 A. Baxter *Enq. Human Soul* vi. xlv. 267 Nothing is a mark of impossibility, but a self-destructiveness in the idea. **1977** J. D. Douglas in Douglas & Johnson *Existential Sociol.* i. 46 Self-hatred and the resulting self-destructiveness pervades the lives of the poor.

self-determination. Add: **2.** *Pol.* The action of a people in deciding its own form of government; free determination of statehood, postulated as a right (see quot. 1929).
1911 *Encycl. Brit.* XXIII. 653/1 The more enlightened of the emperors..made a genuine endeavour to give a due share in the work of government to the various subject races. But nothing could compensate for the lack of self-determination. **1917** *Times* 28 Dec. 8/1 According to the declarations of..the quadruple alliance, protection of the right of minorities forms an essential component part of the constitutional right of peoples to self-determination. **1918** Woodrow Wilson in *N.Y. Times* 12 Feb. 1/5 National aspirations must be respected; peoples may now be dominated and governed only by their own consent. 'Self-determination' is not a mere phrase. It is an imperative principle of action, which statesmen will henceforth ignore at their peril. **1929** W. S. Churchill *World Crisis* V. xi. 203 Although the expression 'Self-determination' will rightly be forever connected with the name of President Wilson, the idea was neither original nor new. The phrase itself is Fichte's 'Selbst bestimmung'. **1946** D. L. Sayers *Unpopular Opinions* 100 Eire demanded self-determination... Northern Ireland also wanted self-determination, and was determined to remain with England. **1959** E. H. Carr *Socialism in One Country* II. iv. xxi. 278 It was not only in Soviet Russia that a potential clash could be discerned between the claims of national self-determination and the claims of economic progress. **1968** 'J. le Carré' *Small Town in Germany* x. 162 The Yanks are going crazy about self-determination. Why don't they try it in East Germany? **1976** Ld. Home *Way Wind Blows* xii. 168 'Self-determination' was a slogan to which the Soviet leaders paid lip-service.

self-differentia·tion. [Self- 1 a.] Differentiation arising from within oneself or itself; *spec.* in *Biol.*, that of embryonic tissue occurring more or less independently of other parts of the embryo.
1891 W. J. Greenstreet tr. *Guyau's Educ. & Heredity* ix. 288 We are capable of self-imitation, self-differentiation, or self-modification. **1902** *Encycl. Brit.* XXVIII. 150/2 This partial independence has been called self-differentiation (*Selbstdifferenzierung*) by Roux, and is entirely a characteristic feature of ontogeny. **1926** [see *Differentiation 1 a]. **1972** *Jrnl. Embryol. & Exper. Morphol.* XXVIII. 547 The capacity of Henson's node for self-differentiation and induction.

self-diffu·sion. *Chem.* [Self- 3 a.] Migration of constituent atoms or molecules within the bulk of a substance, esp. in a crystalline solid.
1924 *Chem. Abstr.* XVIII. 2632 By mixing such comps. with AgI interchange of ions (self-diffusion) can be effected. **1938** R. W. Lawson tr. *Hevesy & Paneth's Man. Radioactivity* (ed. 2) xviii. 173 Not until the introduction of radioactive indicators was it possible to open up to observational study the phenomenon of self-diffusion. **1958** *New Scientist* 2 Jan. 24/2 Autoradiograph (*a*) shows the self-diffusion after treatment at 800°C—this has occurred mainly along the boundaries of the grains or crystals of metal that make up bulk iron. **1974** D. M. Adams *Inorg. Solids* ix. 286 Self-diffusion work with silver isotopes confirms that these ions move freely between the available sites in the iodine lattice.

self-dissocia·tion. *Chem.* [Self- 1 a, d.] = *Self-ionization. So **self-disso·ciated** *a.*, that undergoes self-ionization.
1905 *Jrnl. Physical Chem.* IX. 178 The conductivity of the pure solvents is explained by assuming 'self-dissociation' and considerable space is devoted to mere speculation as to what the composition of the ions might be in the various individual cases. **1973** Schmidt & Siebert in J. C. Bailar et al. *Comprehensive Inorg. Chem.* II. xxiii. 873 Sulphuric acid is also slightly self-dissociated into sulphur trioxide and water. *Ibid.*, The complete self-dissociation reaction in the sulphuric acid solvent system can be described..by the above equation.

se·lf-drive, *a.* [*Self- 2 b.] Designating a motor vehicle hired to be driven by oneself, not by a chauffeur, or an agency which supplies such vehicles. Also *ellipt.* as *sb.*, a self-drive car or van, and **se·lf-driven** *ppl. a.*
1929 *Star* 21 Aug. 13/3 (Advt.), Motor-cars for hire. Self-drive Sal[oon]s, tourers fr. 17/6 day. **1932** Kipling *Limits & Renewals* 80 A natty little grey and black self-driven coupé came from Brighton way. **1953** R. Macaulay *Last Lett. to Friend* (1962) 97, I have hired a self-drive car, as there seems no other way of getting about Cyprus. **1969** J. Leasor *They don't make them like that any More* vi. 194 We could hire a self-drive from somewhere. **1972** L. P. Bachmann *Ultimate Act* xvi. 125 At the international self-drive agency I handed over my credit card. **1978** N. Freeling *Night Lords* ii. 14 French show-offs in Jaguars..Americans in self-drives. **1982** 'W. Haggard' *Mischief Makers* xi. 132, I have hired a self-drive removal van... We will load those crates.

self-effacement. Add: Also **self-effa·cingness.**
1951 S. Spender *World within World* 167 Forster's strange mixture of qualities—his self-effacingness combined with a positive assertion of his views.

self-election. Add: Also **self-ele·cting** *ppl. a.*
1855 Bagehot in *National Rev.* Oct. 271 In the towns, the franchise belonged to a close and self-electing corporation. **1980** D. Francis *Reflex* vii. 76 That they [*sc.* the Jockey Club] were also self-electing meant in practice that the members were almost all..upper-class.

self-e·nergy. *Physics.* [Self- 5 b.] The energy possessed by a particle in isolation from other particles and fields; the energy of interaction of a particle, quasi-particle, or current with its own field.
1883 O. Heaviside in *Electrician* 10 Mar. 390/1 We have next..to consider the potential energy of a current system on itself, as distinguished from its energy with respect to another system. The last being called the mutual energy, we may for brevity term the former the self-energy. **1933** *Nature* 30 Dec. 1004/2 The use of the classical function *L* gives infinite values of self energy and other physical quantities which are, in fact, certainly finite. **1956** E. H. Hutten *Lang. Mod. Physics* iii. 97 The electron as a point-particle would possess infinite self-energy: this must occur when a finite charge is concentrated into a point—thus making the charge density, or charge per unit volume, infinite. **1962** Corson & Lorrain *Introd. Electromagn. Fields* vi. 243 The first two terms on the right are self-energies arising from the interaction of each current with its own field. **1977** *Dædalus* Fall 25 Just as the exchange of virtual photons between two electrons produces an energy of interaction between them, so also in quantum field theory the emission of virtual photons and their reabsorption by the same electron produces a self-energy.

self-estra·ngement. *Social Psychol.* [Self- 1 a; cf. G. *selbstentfremdung*.] Estrangement from one's natural self, esp. such as is thought to result from the alienating development of consciousness or from involvement in a complex industrialized culture. Hence **self-estra·nged** *ppl. a.*
1878 A. C. Brackett *Science of Educ.* 43 [The mind's] first stage of development is self-estrangement—it is absorbed in the observation of objects around it... This process of self-estrangement and its removal belongs to all culture. **1910** J. B. Baillie tr. *Hegel's Phenomenol.* II. vi. 488 (*heading*) Spirit in self-estrangement—the discipline of culture. *Ibid.* 490 The equilibrium of the whole..rests on the alienation of its opposite. The whole is, therefore, like each particular moment, a self-estranged reality. **1951** P. Tillich *Systematic Theol.* I. ii. 74 A second part of the system must give an analysis of man's existential self-estrangement..and the question implied in this situation. **1972** M. Argyle *Social Psychol. of Work* ix. 226 Self-estrangement—failure to regard the work as a central life interest or means of self-expression, experiencing a depersonalized detachment while at work. **1978** J. Updike *Coup* (1979) iv. 138 You can't talk about that without talking about the self-estrangement induced by forced labor.

self-evalua·tion. *Psychol.* [Self- 1 a.] Appraisal of one's actions or attitudes, esp. in relation to an objective standard. Hence **self-eva·luative, -eva·luatory** *adjs.*
1933 *Psychol. Abstr.* VII. 473/2 To attain self-evaluation, psycho-physiological and psychotechnical methods must be supplemented by introspection. **1957** N. Frye *Anat. Criticism* 179 The hero Denis comes to a point of self-evaluation. **1957** R. B. Cattell *Personality & Motivation* vi. 220 Self-evaluatory methods of gathering personality data include questionnaires, biographical inventories, [etc.]. **1966** J. Flescher *Dual Therapy* vii. 154 The self-evaluative recording seems to promote the therapist's training as well as the patient's progress. **1979** A. Chisholm *Nancy Cunard* xxx. 315 A nightmare crisis of self-evaluation.

self-exciting, *ppl. a.* Add: **a.** (Later examples.) Also *transf.*, with reference to the hypothesis that the earth's magnetic field is generated in the earth's core by a mechanism analogous to that of a dynamo.
1922 Glazebrook *Dict. Appl. Physics* II. 195/1 The great advantage of direct-current machines is that: (i) They can be made self-exciting. **1954** E. Bullard in G. P. Kuiper *Earth as Planet* iii. 129 The possibility that the motion of the material of the core could cause it to act as a self-exciting dynamo. **1971** *Sci. Amer.* Dec. 80/1 The influence of the Coriolis force on the motions of the outer core is thought to be critical to the operation of the self-exciting dynamo that generates the main magnetic field of the earth.
b. *Radio.* Self-oscillating; (see quot. 1943).
1922 *Proc. IRE* X. 251 The basis of super-regeneration was the discovery that a variation in the relation between the negative and positive resistances prevented a system which would normally oscillate violently from becoming self-exciting. **1943** *Gloss. Terms Telecomm.* (*B.S.I.*) 69 *Self-exciting sender*, a radio sender in which the oscillator determining the frequency also generates the radio-frequency power.
self-excitation (later examples); **self-exci·te** *v. intr.* and *trans.*; **self-exci·ted** *ppl. a.*, self-exciting.
1921 J. H. Morecroft *Princ. Radio Communication* vi. 562 A possible arrangement of self-excitation, in which the phase of the voltage impressed on the grid is adjustable. *Ibid.* 563 (*caption*) Conditions occurring in the self-excited tube. **1922** Glazebrook *Dict. Appl. Physics* II. 196/2 It is also necessary that the winding should have a sufficiently low resistance, otherwise self-excitation will not occur. *Ibid.*, At this speed..the machine will not self-excite. **1940** *Amateur Radio Handbk.* (ed. 2) xvii. 245/1 Developments during the few years prior to the war tended to relegate self-excited transmitters into the background. **1962** *Newnes Conc. Encycl. Electr. Engin.* 672/2 An induction generator can be self-excited by means of a capacitor. *Ibid.* 673/1 Unwanted self-excitation with dangerously-high voltages may occur in induction generators. **1977** *Jrnl. R. Soc. Arts* CXXV. 764/1 Before the appearance of the self-excited dynamo in 1867 it was by no means obvious that magneto-electric induction would power the world's industries.

self-fertile, *a.* Add: Also **self-ferti·lity.**
1917 *Genetics* II. 525 That some of his families arising from selfed seed behaved exactly as the families arising from crossed seed shows that he is often (at least) dealing with a pseudo self-fertility. **1924** *Ibid.* IX. 16 The self-fertile plants differed among themselves in the expression of their self-fertility. **1956** *Nature* 10 Mar. 490/1 On hybridization, spontaneous self-fertility was restored. **1970** *Bot. Gaz.* CXXXI. 139/2 Species were rated for self-fertility by their ability to set seed under isolated conditions.

se·lf-field. *Physics.* [Self- 5 b.] A field intrinsically associated with a charged particle, particle beam, or current, esp. as contrasted with any externally applied field that may be present.
1934 *Proc. R. Soc.* A. CXLIII. 437 The usual quantum mechanics is the limiting case in which the self-field is regarded as rigidly bound to the centre. **1970** *Particle Accelerators* I. 1/1 If the self-fields are large enough to trap the ions,..the cluster can be accelerated. **1976** *Nature* 19 Aug. 651/1 In the vicinity of the disk there is a small toroidal self-field $B\phi$ from the proton beams.

self-fina·ncing, *ppl. a.* [Self- 1 f.] That finances itself; *spec.* (of a programme of development, etc) that pays for its own implementation or continuation. Also as *vbl. sb.* and (as a back-formation) **self-fina·nce** *v. trans.*

1957 A. C. L. Day *Outl. Monetary Econ.* xxiii. 312 Such self-financing by large-scale industry has been very important in recent years. **1962** H. E. Beecheno *Introd. Bus. Stud.* xvi. 150 The whole body of these policies must therefore be set on a self-financing basis. **1962** E. Snow *Other Side of River* (1963) lix. 448 Cadres working in the brigade..'penetratingly explained' the necessity to 'self-finance' next year's development program. **1964** *Financial Times* 12 Mar. 5/8 It is becoming harder to rely principally on self-financing in order to raise capital for expansion. **1972** *Accountant* 19 Oct. 489/2 The Canadian company has maintained its contribution to group earnings and at the same time is self-financing its own expansion. **1979** *Jrnl. R. Soc. Arts* CXXVII. 656/1 They discuss the Society's miscellaneous activities, such as..the establishment in 1851–2 of self-financing public lavatories. **1980** I. St. James *Money Stories* ii. i. 42 Can't you sell some diamonds and plough the cash back into fresh equipment? Make the project self-financing?

self-flu·xing, *ppl. a.* [Self- 4.] Of iron ore: capable of being smelted without the addition of a flux, usu. because it contains a high proportion of lime. Also as *vbl. sb.*
1884 [in Dict. s.v. Self- 4]. **1923** Harbord & Hall *Metall. of Steel* (ed. 7) II. xxvi. 504 A few ores have an excess of lime, and if unable to secure self-fluxing ores the blast furnace manager endeavours to mix siliceous and calcareous ores when obtainable to form a self-fluxing mixture. **1973** *Times* 12 Feb. (Anchor Project Suppl.) p. ii/5 To have constructed an integrated plant..near the ore terminal would have entailed..the transport of a considerable amount of local ore in the reverse direction (or the loss of 'selffluxing' economies).

self-forgetful, *a.* Add: (Earlier and later examples.)
1848 J. G. Whittier *Pæan* in *Poet. Works* (1898) 336/1 They died, their brave hearts breaking slow, But self-forgetful to the last..Their breath upon the darkness passed. **1981** F. Inglis *Promise of Happiness* i. ii. 37 Many arts and crafts..offer an occasion for self-forgetful joy and peace.
self-forgetfulness (earlier and later examples.)
1805–6 Wordsworth *Prelude* (1959) iv. 294 The quiet stream Of self-forgetfulness. **1942** C. S. Lewis *Screwtape Lett.* xiv. 72 Let him think of it not as self-forgetfulness but as a certain kind of opinion..of his own talents and character.

self-fulfi·lling, *ppl. a.* orig. *Social Sci.* [Self- 1 f.] In phr. *self-fulfilling prophecy*: a prophecy or prediction which gives rise to actions that bring about its fulfilment (see quot. 1949).
1949 R. K. Merton *Soc. Theory & Soc. Structure* vii. 181 The self-fulfilling prophecy is, in the beginning, a *false* definition of the situation evoking a new behavior which makes the originally false conception come *true*. **1962** Blake & Mouton in M. Sherif *Intergroup Rel.* v. 113 Another dynamic, the self-fulfilling prophecy, also operates to cause one to misplace motivation. **1973** *Times* 7 Dec. 18/8 Panic buying of spirits..caused largely by forecasts of the shortage—a self-fulfilling prophecy. **1979** D. Gurr *Troika* ix. 60 '*Wonderful*..to meet again. Didn't I tell you that we should? *Vodka!*'.. 'A self-fulfilling prophecy, Alexey Ilyich.'

self-given, *a.* For † *Obs.* read *rare* and add later examples.
1909 R. Kane *Sermon of Sea* viii. 120 This object of intuitive consciousness must have its moral kind, its spiritual character, its self-given growth in evil or in good. **1934** T. S. Eliot *Rock* ii. 84 Those who prize the serpent's golden eyes, The worshippers, self-given sacrifice of the snake.

self-glorification. (Earlier and later examples.)
1838 J. S. Mill in *Westm. Rev.* Apr. 37 Those antique characters which, without self-glorification or hope of being appreciated, 'carry out..the sentiment of duty to its extremest consequences'. **1953** *Encounter* Oct. 49/1 It was..only Stalin who was thus hymned in the inevitable major key of Soviet self-glorification.

self-governed, *ppl. a.* Add: **2.** (Earlier and further examples.)
1709 Shaftesbury *Moralists* ii. 150, I suppose you will send your Disciple to seek for Deity in Mechanism; that is to say, in some exquisite System of self-govern'd Matter. **1848** J. S. Mill *Pol. Econ.* II. iv. vii. 320 They will require that their conduct and condition shall be essentially self-governed. **1859** J. Brown *Let.* 19 Aug. (1912) 171 Frankfort is a little Republic, self-governed, and a thriving, handsome, well-conditioned town.
self-governing *ppl. a.* (earlier and later examples).
1845 J. S. Mill in *Edin. Rev.* LXXXI. 511 For two centuries, the Scottish peasant..has been a reflecting, an observing, and therefore naturally a self-governing, a moral and a successful human being. **1933** *Discovery* Feb. 68/1 The greatest problem facing British Statesmen of modern times—the problem of transforming that India, in which society is organized on a religious basis, into a self-governing community on modern democratic lines. **1976** *Glasgow Herald* 26 Nov. 6/6 What business man in his right mind would prefer minimum lending rate of 14½% to one in the 5–7% range which a self-governing Scotland would..achieve.

self-gravita·tion. *Astr.* [Self- 3 a.] The gravitational forces acting among the components of a massive body.
1962 K. H. Prendergast in L. Woltjer *Distribution & Motion Interstellar Matter* 233 In our own galaxy, if you consider the terms in the equations of motion, other than the gravitational field of all the matter, then there are three: the random velocities, some sort of ordered magnetic field, and the self-gravitation of the gas. **1968** R. A. Lyttleton *Mysteries Solar Syst.* i. 36 Self-gravitation within the disk can far exceed the solar disruptive effect. **1977** *Dædalus* Summer 36 Roughly speaking we can attribute this change from Newtonian theory to the greater self-gravitation which matter possesses according to general relativity.
Hence **self-gravita·tional** *a.*; **self-gra·vitating** *ppl. a.*, influenced by self-gravitation; **self-gra·vity**, self-gravitation.
1962 P. O. Lindblad in L. Woltjer *Distribution & Motion Interstellar Matter* 223 A massive ring in a central force field..can carry two different kinds of bisymmetrical waves by self-gravitational action. **1962** G. B. Field in *Ibid.* 318 Is the system a self-gravitating one in the sense that the gas acts on the gas, or is it not? **1972** *Sci. Amer.* Apr. 43/2 The restoring forces that limit the height of the tides are due to the planet's elasticity and self-gravity. **1976** *Nature* 11 Nov. 114/1 Ultimately, self-gravity must drive the enormous contractions which convert diffuse gas into new stars. **1979** *New Scientist* 3 May 424/2 A self-gravitating mass of gas, cohering under its self-attraction, would adopt spherical symmetry.

self-hate. [Self- 1 d.] = next.
1947 F. Frenaye tr. *C. Levi's Christ stopped at Eboli* (1948) iv. 28 To the hates of the gentry he added self-hate. **1951** K. Horney *Neurosis & Human Growth* v. 110 Pride and self-hate are actually one entity. **1979** R. Blythe *View in Winter* ix. 289 Creeping indifference is a large factor in the self-hate of the aged.

self-ha·tred. [Self- 1 d.] Hatred of oneself, esp. of one's actual self when contrasted with one's imagined self. Cf. prec.
1865 D. G. Rossetti *Let.* 21 Nov. (1965) II. 581 Ambition, i.e. the feeling of pure rage and self-hatred when any one else does better than you do. **1896** A. Thorold tr. *Dialogue of St. Catherine of Siena* lxiii. 134 Digging up the root of self-love, with the knife of self-hatred and the love of virtue. **1942** C. S. Lewis *Screwtape Lett.* xiv. 72 All the abjection and self-hatred are designed, in the long run, solely for this end. **1975** G. St. George *Proteus Pact* (1976) iii. 95 He descended vertiginously into utter self-loathing..paralyzed by..self-hatred.

self-hea·ting, *vbl. sb.* (*ppl. a.*) [Self- 1 b, f, 4.] The action of becoming heated spontaneously or automatically. Also as *ppl. a.*, that is designed to heat itself automatically; (of food) held in a self-heating container.
1929 *Chem. Abstr.* XXIII. 1561 The effect..of self-heating of metals after previous chilling is attributed here to establishment of equil[ibrium] in a metastable system. **1952** Koestler *Arrow in Blue* v. xxxii. 272 Some new 'colossal' project like the radioactive soap or the self-heating bricks. **1959** P. Capon *Amongst Those Missing* 23 They found several battered cans of self-heating soup. *Ibid.* 24 The fuses on the self-heating cans did not burn too well. **1961** *Guardian* 29 May 2/3 Tins of self-heating cocoa fell from their appointed places into the bilges. **1965** Phillips & Williams *Inorg. Chem.* I. xvi. 576 As a result of self-heating effects, polonium and polonium compounds are always at temperatures markedly above their surroundings. **1981** J. R. L. Anderson *Death at High Latitude* ix. 135, I opened three more cans of our self-heating soup, the only means we had of getting a hot drink.

self-help. Add: Also **self-he·lper.**
1891 G. B. Shaw *Quintessence of Ibsenism* ii. 34 No one ever feels helpless by the side of the self-helper. **1976** W. Goldman *Magic* iii. vii. 165 'How come you have all these same kind of books?'.. 'That's my self-help collection. I'm an addict.'.. 'Why're you such a self-helper?'

self-he·terodyne, *a.* *Radio.* [Self- 1 e.] Being or employing a heterodyne receiver in which the same valve is used for the generation and rectification of local oscillations. Hence **self-he·terodyning** *ppl. a.* and *vbl. sb.*
1918 [see *Autodyne]. **1922** *Proc. IRE* X. 245 In the various forms of self-heterodyne circuits a free oscillation of constant amplitude is maintained in the system and the circuit may be considered as having zero resistance, but only for that particular amplitude of current. **1929** Duncan & Drew *Radio Telegr. & Teleph.* xii. 344 The circuit is commonly known as a self-heterodyning detector because the tube functions both as an oscillator and detector. *Ibid.* 351 (*heading*) The phenomenon of self-heterodyning.

self-hypno·sis. *Psychol.* [Self- 1 a, d.] = *autohypnosis* s.v. *Auto-[1] a. Also **self-hy·pnotism, -hypnotiza·tion, -hy·pnotized** *ppl. a.*, **-hy·pnotizer.**
1852 J. Braid *Magic* 20, I stated that I had found in the writings.., 'many statements corroborative of the fact, that the eastern saints are all self-hypnotizers.' *Ibid.* 108 Special gifts imparted to them in token of the great superiority of their religious system, of inducing a state of self-hypnotism, or ecstatic trance. **1891** G. C. Kingsbury *Practice Hypnotic Suggestion* vi. 91 It would be..interesting..to try if suggestions written out by ourselves, and used as the object of our fixed gaze for the purpose of self-hypnotization, would have any after influence. **1902** H. B. Woolston in *Amer. Jrnl. Psychol* XIII. 71 This sort of self-hypnosis may lead even to an identification of the individual with the ideal form that absorbs his mind. **1925** A. S. M. Hutchinson *One Increasing Purpose* iii. xvi. 333 He..gazed up, up, into cloudless perfect blue until, selfhypnotised, he seemed to himself to be rising up there. **1939** G. Greene *Lawless Roads* x. 252 He had listened to the pilgrims' tales. It couldn't be self-hypnotism. **1959** *House & Garden* July 32/1 The wondrous and covetable gift of self-hypnotism. **1960** *Times* 11 Feb. 13/2 He [*sc.* McCarthy] was not a self-hypnotized genius of hate like Hitler. **1976** M. V. Kline in E. Dengrove *Hypnosis & Behav. Therapy* x. 139 The use of self hypnosis and..of audio tape recordings.

self-identifica·tion. [Self- 1 a.] Identification with something outside oneself.
1941 *Manch. Guardian Weekly* 14 Mar. 214/3 Are we to sympathise with the young? Certainly... But need this be called 'Self-identification with the experience, interests, and problems of other young people?' **1959** *Twentieth Cent.* June 629 The great admiration which James..felt for the work of Sainte-Beuve..in many ways a matter of self-identification. **1973** C. Sagan *Cosmic Connection* (1975) i. 6 There is a serious question whether such a global self-identification of mankind can be achieved before we destroy ourselves with the technological forces our intelligence has unleashed.
Hence **self-ide·ntifying** *ppl. a.*
1963 N. Frye *Romanticism Reconsidered* 14 The self-identifying admiration which so many Romantics expressed for Napoleon. **1976** H. Wilson *Governance of Britain* iv. 92 His post-box will include hundreds of letters a week from self-identifying officials or members of local organizations of his party.

self-i·mage. [Self- 1 d.] An image or conception of oneself, esp. considered in relation to others.
1939 S. Spender *Still Centre* 101 The self-image Lifted in light against the lens Stares back with my dumb wall of eyes. **1951** M. McLuhan *Mech. Bride* (1967) 66/1 She embodies that self-image of a knight in shining armor. **1969** C. Fremlin *Possession* xiv. 117 The way into these cloying, stagnant emotional pools is smooth... You come out..with your self-image in shreds. **1977** *Church Times* 14 Jan. 10 It stresses..the Vatican's self-image as the centre of Christian prayer and charity and of work for justice and peace.

self-important, *a.* Add: So **self-impor·tantly** *adv.*
1961 in Webster. **1977** T. Heald *Just Desserts* vi. 134 A police sergeant who arrived self-importantly. **1981** *London Rev. Bks.* 19 Nov.–2 Dec. 7/1 Marilyn Butler..is importantly and not self-importantly a citizen of the world.

self-incompa·tible, *a.* and *sb.* *Bot.* [Self- 3 b.] **A.** *adj.* Unable to be fertilized by means of its own pollen. **B.** *sb.* A self-incompatible plant or species. Opp. *Self-compatible *a.* and *sb.*
1922 *Bot. Gaz.* LXXIII. 119 Plants may be completely self-incompatible throughout. **1938** Crane & Lawrence *Genetics of Garden Plants* ix. 188 Self-compatibles crossed with self-incompatibles will give both self-compatibles and self-incompatibles in the proportion 1:1 or 2:1 according to the constitution of the self-incompatible plants used. **1944** *Nature* 5 Aug. 164/2 The nuclear and cytoplasmic systems of self-incompatible plants are mutually and constructively related. **1968** *Times* 23 May 17/5 In some species more individuals are self-incompatible than in others. **1979** *Nature* 25 Oct. 671/2 This species is self-incompatible.
So **self-incompatibi·lity.**
1917 *Genetics* II. 506 The words self-incompatibility and self-impotence have been substituted for self-sterility by various writers. **1952** *Heredity* VI. 286 The primary function of self-incompatibility is the avoidance of self-fertilisation. **1970** *Watsonia* VIII. 142 The failure of fruit production following self-pollination results from self-incompatibility rather than sterility.

self-induced, *pa. pple.* and *ppl. a.* Add: **2.** *gen.* Induced by oneself or itself.
1949 Wellek & Warren *Theory of Lit.* iii. 21 What it articulates is superior to their own self-induced reverie or reflection. **1954** G. I. M. Swyer *Reprod. & Sex* 246 This often happens in clumsy attempts at self-induced criminal abortion. **1981** J. Brabazon *Dorothy L. Sayers* xx. 267 The pressures on Patrick McLaughlin, some of them self-induced.

self-induction. Add: (Earlier and later examples.) Also, the coefficient of self-induction.
1865 J. C. Maxwell in *Phil. Trans. R. Soc.* CLV. 472 The equation of the current x in a circuit whose resistance is R, and whose coefficient of self-induction is L, acted on by an external electromotive force ξ, is $\xi - Rx = d(Lx)/dt$. *Ibid.* 475 Hence the effect..is to increase the apparent resistance and diminish the apparent self-induction of the circuit. **1902** *Encycl. Brit.* XXVIII. 56/1 The circuit in the jar C is provided with a sliding piece, F, by means of which the self-induction of the discharging circuit..can be adjusted. **1958** Condon & Odishaw *Handbk. Physics* iv. i. 17/1 If the current in the circuit increases for any reason, the flux through the circuit must also increase, and this gives rise to an emf of self-induction which acts to oppose the increase in the current by Lenz's law.

self-instructed, *ppl. a.* (Example.)
1833 J. S. Mill in *Tait's Mag.* III. 348 Narrowness and self-conceit..are the..failings of the self-instructed.

self-insu·rance. [Self- 1 a.] Insurance of oneself or one's interests by maintaining a fund to cover possible losses. Hence **self-insu·rer, -insu·ring; self-insu·red.**
1905 *Ann. Amer. Acad. Polit. & Social Sci.* XXVI. 452, I am informed by the managers and officers of the largest steamship lines that self-insurance is practiced extensively by their companies in one form or another. **1909** Webster, Self-insurer. **1932** *Sun* (Baltimore) 8 Sept. 4/4 Self-insuring might be permitted to employers or industrial groups who can guarantee maintenance of a benefit system equal or superior to the standards of the State system. **1934** Webster, Self-insured. **1972** *Accountant* 5 Oct. 422/1 Premiums..are tax-deductible... This is an important advantage for the self-insurer... The company with a poor loss record has most to gain from self-insurance.

self-ioniza·tion. *Chem.* [Self- 1 a, d.] Spontaneous dissociation of a proportion of the molecules of a liquid into ions.
1931 *Chem. Rev.* VIII. 201 Bishop..used the hydrogen electrode for titrations in ethanol and emphasized the effect of the low self-ionization of the solvent on the titration curves. **1972** Cotton & Wilkinson *Adv. Inorg. Chem.* (ed. 3) v. 181 Pure H_2SO_4 shows extensive self-ionization resulting in high conductivity.

selfish, *a.* Add: **1. c.** *Genetics.* Of a gene or genetic material: tending to be perpetuated or to spread although of no effect on the phenotype.
1976 R. Dawkins *Selfish Gene* i. 3 Let us understand what our own selfish genes are up to, because we may then at least have the chance to upset their designs. **1979** *Human Genetics* (Ciba Symp.) 41 It seems to me that repetitive DNA is the only true selfish gene. **1981** *Nature* 13 Aug. 648/1 Selfish DNA, which contains no genetic information but which is perpetuated in eukaryote genomes, has attracted a lot of attention recently.

self-limitation. **1.** (Earlier example).
1847 J. D. Morell *Hist. View Philos.* (ed. 2) II. v. 105 The idea of the objective arises from the self-limitation of our own free activity.

self-limiting, *ppl. a.* Add: *spec.* in *Med.* = Self-limited *ppl. a.* 2.
1954 S. Duke-Elder *Parsons' Dis. Eye* (ed. 12) x. 125 The factor dominating the prognosis..is the recurrence of relapses on its cessation unless the malady is eradicable or self-limiting. **1965** J. Pollitt *Depression & its Treatment* v. 71 Premenstrual depression is a short-lived, self-limiting depressive illness. **1977** *Lancet* 19 Mar. 648/2 Whooping-cough is a self-limiting infection which should never be fatal with proper medical care.

self-li·quidating, *ppl. a.* *Comm.* (orig. *U.S.*). [Self- 1 f.] Of, pertaining to, or designating credit, or a loan, that repays itself with the money accruing within a certain period after its investment. Also of a premium similarly offered.
1915 *U.S. Fed. Reserve Board 1st Ann. Rep. 1914* 9 It is recommended that the Federal Reserve Banks confine themselves strictly to dealing in short-term, self-liquidating paper growing out of commercial, industrial, and agricultural operations. **1928** *Burroughs Clearing House* May 32/3 If the purpose is a constructive one, stimulating production and of a self-liquidating character..then the loan should be made. **1939** *Construction & Financing of Self-Liquidating Projects* (U.S. Congr. House Comm. on Banking & Currency) 2 In view of this splendid repayment record there can be little doubt that the tenancy program has been established on a sound, self-liquidating basis. **1942** W. B. Taylor *Financial Policies of Bus. Enterprise* iv. xiii. 288 Short-term loans, commonly made for less than a year, are usually self-liquidating and hence not adapted to the raising of fixed capital. **1967** *Economist* 10 June p. xxxi/2 The traditional role of providing short term, self-liquidating trade finance to a nation of shopkeepers is too narrow. **1971** R. L. Willsmer *Directing Marketing Effort* ii. xviii. 292 A self-liquidating premium looks good because it theoretically costs nothing.
Also **self-li·quidator,** a self-liquidating premium.
1944 *Premium Practice* Mar. 8/2 Mr. Cahill mentioned the following: (a) self-liquidators. **1948** *Ibid.* Oct. 30/2 The juvenile field was once the objective of the self-liquidator. Now it is the adult thing. **1962** G. Meredith *Effective Merchandising with Premiums* ii. iv. 37 It would be impossible to identify one promotion as the beginning of modern self-liquidator practice. **1970** R. Willsmer in G. Wills *New Ideas in Retail Managem.* xii. 170 Self-liquidating premium offers continue to lead the field... Only eight store-promoted self-liquidators were recorded.

self-liquida·tion. [Self- 1 a.] **1.** *Comm.* (orig. *U.S.*). The action or process of repaying a self-liquidating loan. Cf. *Self-liquidating *ppl. a.*
1932 *Burroughs Clearing House* Mar. 23/1 The idea of self-liquidation, of having a definite source of repayment in sight, of inquiring into the purpose of the loan with a view to finding out how far the money is to be sunk in fixed assets, all these relate to the liquidity of loans. **1949** H. V. Prochnow *Term Loans & Theories of Bank Liquidity* i. 5 The belief that commercial banks..should extend credit only for short periods and for purposes which result in the self-liquidation of the credit. **1951** *Banco Nazionale del Lavoro Q. Rev.* July–Sept. 135/2 In the last few decades the theory of 'self-liquidation' has been gradually set aside and replaced by the 'shiftability theory'.
2. The destruction or elimination of oneself by oneself. Also *fig.* Cf. *Liquidation 3 b.
1949 Koestler *Promise & Fulfilment* ii. iv. 253 This difference..was demonstrated by Irgun's voluntary self-liquidation..as opposed to the Stern Group's persistence in terrorism. **1964** I. L. Horowitz *New Sociology* 17 The recent work in some quarters..seems to point precisely in the direction of the self-liquidation of sociology. **1977** *Canadian Jrnl. Sociology* II. 106 The inevitable self-liquidation of science is another way of speaking of science's tie to life.

se·lf-loop. [Self- 3 a.] In a graph or network, a line that returns to the node it leaves.
1964 S. E. Salmaghraby in *Managem. Sci.* X. 499 We shall reserve the terms 'self-loop' to designate a branch that leads from a node to itself. **1980** *Sci. Amer.* Mar. 18/1 In graph theory a graph is defined as any set of points joined by lines, and a simple graph is defined as one that has no self-loops (lines that join a point to itself) and no parallel lines (two or more lines joining the same pair of points).

self-made, *ppl. a.* Add: *self-made man* (earlier examples).
1832 *Reg. Deb. Congr. U.S.* 2 Feb. 277 In Kentucky,..every manufactory..is in the hands of enterprising self-made men. **1854** Dickens *Hard Times* i. iv. 18 Mr. Bounderby..could never sufficiently vaunt himself a self-made man.

self-mai·ling, *a.* *U.S.* [Self- 1 f.] Designating postal matter that may be folded or otherwise secured, and sent by post without enclosure in an envelope. Also **self-mai·ler.**
1950 *Self-mailer, -mailing* in Webster Add. **1963** *Publishers' Weekly* 2 Sept. 28/1 Two types of mailing pieces have been prepared for the book trade: a self-mailer and a statement enclosure. For the self-mailer, which is a full-color, four-piece unit, the print order is 650,000.

self-maintai·ning, *ppl. a.* [Self- 1 f.] That maintains or sustains itself or (oneself); *spec.* = *Homœostatic *a.*
1879 [see *race-maintaining* s.v. Race *sb.*[2] 11]. **1890** W. James *Princ. Psychol.* II. xxvi. 582 If this were the entire nervous mechanism, the movement, once begun, would be self-maintaining. **1933** *Mind* XLII. 146 Immortality demands self-maintaining effort and formed character. **1959** G. D. Mitchell *Sociol.* vi. 101 It can thus be seen that witchcraft as a system of beliefs and practices is self-maintaining. **1971** J. Z. Young *Introd. Study Man* viii. 115 The whole mass constitutes one single self-maintaining system.
Also **self-mai·ntenance.**
1867 H. Spencer *Princ. Biol.* II. vi. ix. 454 Increased cost of self-maintenance entailed decreased power of propagation. **1909** W. James *Pluralistic Universe* iii. 121 The problem of understanding how the complete coherence of all things in the absolute should involve as a necessary moment in its self-maintenance the self-assertion of the finite minds. **1942** *R.A.F. Jrnl.* 13 June 18 Every man has to know something about cooking as a part of his self-maintenance in the field. **1971** J. Z. Young *Introd. Study Man* xi. 143 Continuous replacement is the absolutely necessary condition of self-maintenance.

self-ma·ss. *Physics.* [Self- 5 b.] The mass of a particle arising relativistically from its self-energy.
1955 O. Klein in W. Pauli *Niels Bohr* 112 Since we are neglecting the interaction between electrons and electromagnetic fields the question of the self-mass does not appear. **1977** *Dædalus* Fall 27 Thus the electron mass found in tables of physical data..would have to be identified with the bare mass *plus* the infinite 'self mass', produced by the interaction of the electron with its own virtual photon cloud.

self-mate, *sb.* (Earlier example.)
1867 *Chess Player's Mag.* III. 45 The self-mate, though not difficult, is very prettily conceived.

self-medica·tion. [Self- 1 a.] Medication carried out on oneself.
1942 *Horizon* June 371 *Le vice anglais* is most certainly and typically self-medication. **1961** *New Scientist* 5 Jan. 16/3 In spite of the National Health Service British pharmacists continue to sell large quantities of preparations for self-medication.

self-mo·tivated, *ppl. a.* [Self- 2.] Motivated by oneself; *spec.* stimulated to work by one's own enthusiasm and ambition. Also **self-mo·tivating** *ppl. a.*; **self-motiva·tion.**
1973 *Daily Tel.* 27 July 13 (Advt.), Applicants should be..self-motivating, enthusiastic and determined to succeed. **1974** *Spartanburg* (S. Carolina) *Herald Jrnl.* 20 Apr. B5/1 Secretary. Must be good typist, have good personality and be self-motivated. **1978** *Detroit Free Press* 16 Apr. F6/6 (Advt.), Nationwide meat wholesaler looking for self-motivated people. **1980** *West Lancs. Evening Gaz.* 6 Nov. 15 (Advt.), If you're over 21 with a car and possess the self-motivation and determination to succeed [etc.].

self-movement. (Earlier example.)
1883 F. H. Bradley *Princ. Logic* 448 Want of individuality in the *datum* that we began with, absence of self-movement and impossibility of self-development, this is the first defect.

self-no·ise. [Self- 5 b.] Noise generated directly by a particular object.
1953 *Jrnl. Acoustical Soc. Amer.* XXV. 314/2 The self-noise characteristics of this [wind]screen were measured at two wind speeds as a function of the frequency. **1954** L. L. Beranek *Acoustics* vi. 174 The self-noise produced by an electrostatic microphone is that produced by the d-c resistance of the crystal or dielectric shunted by the capacitance. **1960** *McGraw-Hill Encycl. Sci. & Technol.* XII. 504/1 Self-noise increases rapidly with the speed of the ship. **1975** D. G. Fink *Electronics Engineers' Handbk.* xxv. 125 Self-noise has many directional characteristics..; ambient noise generally has an omni-directional distribution.

self-noughting. Also **-naughting.** [Self- 1 b.] Depreciation or effacement of oneself, as manifested in the lives of saints and mystics. (Probably derived by E. Underhill from Hilton's *Scale of Perfection*: see quot. *a* 1400 s.v. Noughting *vbl. sb.*)
1911 E. Underhill *Mysticism* ix. 477 The 'self-naughting' or 'purification of the will', which here takes place, is the struggle to resolve that disharmony. **1937** *Mind* XLVI. 91 We remember what has been written by writers such as those cited above about self-surrender, submission and self-noughting or annihilation. **1959** *Month* Dec. 331 This is the true self-noughting, to know meself incapable of any good. **1976** H. A. Williams *Tensions* v. 89 Through this discovery of our true identity in God and the self-naughting which inevitably accompanies it, we become truly ourselves.

self-observation. [Self- 1 a.] Objective observation of one's attitudes, reactions, or thought processes.
1832 J. S. Mill in *Monthly Repos.* VI. 652 The knowledge of supersensual things, of man's mental and moral nature, where the appeal is to internal consciousness and self-observation. **1865** —— *Comte* 63 It is clear to him that we can learn very little about the feelings, and nothing at all about the intellect, by self-observation. **1948** *Mind* LVII. 511 [Social scientists]..making use of ..official statistics.., interviews of various types.., and new techniques of self-observation. **1977** Mettee & Smith in Suls & Miller *Social Comparison Processes* iv. 98 A theory..based on self-observation of one's own behavior rather than on observation and comparison with..others.

self-orienta·tion. [Self- 1 a.] **1.** The orientation or directing of one's actions or attitudes for oneself or by oneself.
a **1896** G. du Maurier *Martian* (1897) vi. 250 The feeling of self-orientation which was so necessary to him. **1936** Wirth & Shils tr. *Mannheim's Ideol. & Utopia* i. 22 To the extent that mechanistic psychology and..the social impulsion towards all-embracing mechanization, negated these values, they destroyed an important element in the self-orientation of human beings in their everyday life.
2. *Social Psychol.* Underlying motivation that orients one's behaviour primarily towards what concerns oneself.
1951 Parsons & Shils *Toward Gen. Theory of Action* i. 77 We maintain that there are only five basic pattern variables... They are..2. Self-orientation—Collective orientation. **1964** Gould & Kolb *Dict. Soc. Sci.* 489/1 In applying the moral mode of value-orientation, the actor must choose between action for private goals (self-orientation) and action on behalf of collective goals (collectivity-orientation).
Also **self-o·riented** *ppl. a.*
1936 *Mind* XLV. 72 Mr Leon holds, with Butler if not with Plato, that all the natural appetites, though self-oriented, are in themselves innocent. **1975** R. H. Rimmer *Premar Experiments* (1976) i. 49 At first the child is completely self-oriented, autistic, hedonistic.

self-oscilla·tion. *Electronics.* [Self- 3 b.] The generation of continuous oscillations in a circuit, amplifier, etc., in circumstances of excessive positive feedback.
1921 J. Scott-Taggart *Thermionic Tubes* vi. 204 The potentiometer..if suitably adjusted, will effectually prevent self-oscillation. **1943** K. R. Sturley *Radio Receiver Design* i. viii. 394 As the coupling between anode circuit and input is increased a point is reached where self-oscillation occurs. **1963** B. Fozard *Instrumentation Nuclear Reactors* x. 118 Feedback may be defined as positive, in which case it produces an increase in gain and may lead to instability and self-oscillation.
Hence **self-o·scillate** *v. intr.*, to display self-oscillation; **self-o·scillating** *ppl. a.*
1928 L. S. Palmer *Wireless Princ. & Pract.* ix. 328 Similar trouble arises when unstable receiving circuits self-oscillate accidentally. **1962** Simpson & Richards *Physical Princ. Junction Transistors* xviii. 459 Such a circuit..is called a self-oscillating mixer or a frequency converter stage.

self-pity. Add: Also **self-pi·tying** *ppl. a.*
1754 Richardson *Grandison* V. xxxiv. 217, I should have thought myself concerned,..to have expatiated on the self-pitying reflexion conveyed in these words. **1927** C. Connolly *Let.* 1 Jan. in *Romantic Friendship* (1975) 205 Propertius too self-pitying and conceited really to suffer. *a* **1974** R. Crossman *Diaries* (1976) II. 266 Over the weekend I'd been a bit self-pitying and not merely defeatist but defeated.

self-pleased, *ppl. a.* (Earlier example.)
1748 Richardson *Clarissa* VI. lxxxvi. 311 How, self-pleased, she could smile round.

self-pollina·tion. [Self- 1 a.] The deposition on a stigma of pollen from stamens within the same flower or another flower on the same plant.

1872 *Jrnl. Bot.* X. 25 Hildebrand has shown that sometimes, where at first sight self-pollenation [*sic*] (a better word, I think for '*Bestäubung*', than 'pollenization', suggested by Mr. Bennett) seems to be the intention of Nature, this is not followed by fertilization. **1876** [in Dict. s.v. Self- 1 a]. **1974** A. Huxley *Plant & Planet* xiii. 124 A fail-safe device, such as..self-pollination as a last resort.
Hence **self-po·llinate** *v. trans.*; **self-po·llinated** *ppl. a.*, **-po·llinating** *vbl. sb.* and *ppl. a*; **self-po·llinator,** a species which commonly shows self-pollination.
1890 J. R. A. Davis *Flowering Plant* ix. 130 Regularly self-pollinated flowers are characterized by inconspicuousness. *Ibid.*, All this [*sc.* absence of scent and nectar, inconspicuousness, etc.] is carried to the extreme in cleistogamous flowers, i.e., minute self-pollinating ones, which never open. **1933** *Jrnl. R. Hort. Soc.* LVIII. 283 At first the trees were too young to yield reliable results by crossing and self-pollinating. *Ibid.* 358 Hips developed freely, but even in those which were self-pollinated rarely more than one achene was formed. **1960** *McGraw-Hill Encycl. Sci. & Technol.* VIII. 223/1 A second generation, bred by self-pollinating the hybrid. **1961** Webster, Self-pollinator. **1977** *Nature* 2 June 402/2 It was suggested that some of the Mediterranean floras..evolved..from cross- into self-pollinators because insects could not have followed the new hardy plants into the hot dry environment. **1979** *Ibid.* 25 Oct. 671/2 This species is self-incompatible (1978 seed sets of self-pollinated flowers on five bagged plants compared with those of open-pollinated flowers on eight unbagged plants were 0.3 ± 0.08 ($n = 256$) compared with 7.1 ± 0.71 ($n = 43$)).

self-po·rtrait. [tr. G. *selbstbildnis, selbstporträt*; cf. Self- 1 d.] A self-made portrait of oneself. Cf. Portrait *sb.* 1 b and 3 b.
1831 *Fraser's Mag.* Mar. 129/1 The two self-portraits, so far as they are filled up, may be looked upon as real likenesses. **1896** *Academy* 25 Apr. 350/3 A self-portrait of the artist in the act of drawing. **1919** *Q. Rev.* Oct. 322 The high society of the 17th century had shown their taste for an analysis of this kind in their self-portraits. **1975** *Amer. N. & Q.* XIV. 58/2 G. C. Williamson, *English Conversation Pictures of the Eighteenth and Early Nineteenth Centuries*..provides a self-portrait of an era. **1977** R. L. Wolff *Gains & Losses* vii. 413 As an authentic portrait (and self-portrait) of the late 1840's..*Oakfield* has no equal.
Also **self-po·rtraiture.**
1847 Bagehot in *Prospective Rev.* III. 532 Nor without the will can the self-scrutinizing power show to men those startling self-portraitures. **1979** *Dædalus* Summer 89 The self-portraiture is but the further elaboration of a thought that was from the outset oriented toward personal life.

self-possessed, *ppl. a.* (Earlier example.)
1818 Scott *Heart Midl.* I. vii. 228 He..came forward to meet him, with a self-possessed, and even dignified air.

selfquai·ned, *ppl. a.* *nonce-wd.* [Self- 2; cf. *Quain *v.*[2]] Having natural angles or corners.
1888 G. M. Hopkins *Poems* (1967) 199 A coffer, burly all of blocks Built of chancequarrièd, selfquainèd, hoarhuskèd rocks.

self-que·nching, *ppl. a.* (*vbl. sb.*) [Self- 1 f, b.] Having within itself a cause of quenching. Also as *vbl. sb.*, quenching that occurs spontaneously. Also **self-que·nched** *ppl. a.*, self-quenching.
1936 *Rev. Sci. Instruments* VII. 31/1 When the discharge is self-quenched, the total charge passed through the counter is directly proportional to the primary ionization. **1940** *Physical Rev.* LVII. 1035/2 All counters are self-quenching if they overshoot, and if they do not, the inherent instability of the discharge will cause the discharge to be extinguished. **1946** *Ibid.* LXIX. 689/2 A study was made of the characteristics of self-quenched G-M counters filled with mixtures of argon and 28 different organic vapors. **1959** F. W. Wilson *Tool Engineer's Handbk.* (ed. 2) xv. 27 With extremely rapid heating rates and shallow heating, the mass of the part may conduct heat away fast enough to quench the part suitably. This is termed 'self-quenching'. **1963** B. Fozard *Instrumentation Nuclear Reactors* v. 51 If bromine..is added to the neon–argon mixture, satisfactory operation as a self-quenching counter is achieved. **1970** D. L. Horrocks in E. D. Bransome *Liquid Scintillation Counting* 36/2 Self-quenching seems to occur in molecules that have unprotected, coplanar chromophores so that there can be a complete mirror image overlap of the chromophores of the excited and unexcited solute molecules. **1971** *Gloss. Electrotechnical, Power Terms* (*B.S.I.*) iii. i. 32 *Self-quenching oscillator*, a generator of short trains of oscillations, the trains being separated by intervals of quiescence caused by the accumulation of electric charge in a capacitor in the circuit.

self-raising, *ppl. a.* (Earlier example.)
1854 *Daily Placer Times & Transcript* (San Francisco) 7 Mar. 1/4 Among the advantages of the Self-Raising Flour, are: The saving of time in preparing it for the oven, [etc.].

self-ra·ting, *vbl. sb.* *Psychol.* [Self- 1 b.] Rating of one's own attributes, feelings, or behaviour; an instance of this.
1938 G. W. Allport *Personality* xvi. 444 In self-rating there is a tendency to overestimate those qualities considered desirable and to underestimate those considered undesirable. **1945** *Jrnl. Clin. Psychol.* I. 297/1 The second approach to verbal self-ratings..consists simply in the..assertion that a 'self-rating' constitutes an intrinsically interesting and significant bit of verbal behavior. **1958** M. Argyle *Relig. Behaviour* x. 124–5 Several investigators have examined the factors associated with marital happiness or 'adjustment'—this being measured by means of self-ratings. **1972** *Jrnl. Social Psychol.* LXXXVI. 121 Some researchers have gone beyond this correlational approach to include self-ratings and peer ratings.

self-realization. (Earlier example.)
1874 W. Wallace *Logic of Hegel* xxi. p. clx, This process..may be called self-realisation (or development).

self-re·ference. [Self- 3 a.] Reference to oneself or itself; the direction of one's attention at oneself, esp. in forming a comparative assessment of one's characteristics or experience; *spec.* in *Philos.*, the characteristic or quality of referring to itself contained in certain paradoxes, propositions, or statements.
1910 Whitehead & Russell *Principia Math.* I. ii. 64 In all the above contradictions..there is a common characteristic, which we may describe as self-reference or reflexiveness. **1943** *Mind* LII. 20 Man is a microcosm of *Natura*, and his self-reference involves him therefore only in partial impotence. **1952** *Psychol. Abstr.* XXVI. 31/2 (*title*) The use of a new experimental autokinetic situation in the evaluation of self reference. **1960** E. H. Gombrich *Art & Illusion* vii. 239 The perplexing effect of this self-reference is very similar to the paradoxes beloved of philosophers. **1966** G. N. Leech *Eng. in Advertising* viii. 81 Only in a few stereotyped contexts..is third person address and self-reference still used. **1978** M. Hesse in Hookway & Pettit *Action & Interpretation* 7 Self-reference of social theorising as part of its own subject matter. **1981** *Sci. Amer.* Apr. 6/2 One has only to watch the Muppets or Monty Python on television to see dense and intricate webs of self-reference.
Also **self-re·ferent, -refere·ntial** (**-ially** *adv.*) *adjs.*; **self-refe·rring.**
1943 *Mind* LII. 20 Man's power over..himself as he appears in the self-referent perspective. **1946** *Mind* LV. 65 If a theory is included in its own subject-matter, we say that it is a *self-referential* theory. *Ibid.* 67 Complete doubt of everything led to a self-referentially inconsistent view and so had to be abandoned. **1958** A. Pap *Semantics & Necessary Truth* ix. 263 The kind of self-referential proposition forbidden by the theory of types. **1969** *Mind* LXXVIII. 9 A partly self-referring sentence, such as art. 88 of the Danish constitution. **1979** *Sci. Amer.* Nov. 29/3 These irreducible questions tend, however, to be rather artificial and self-referent. **1980** *Times Lit. Suppl.* 5 Sept. 957/2 The self-referential and profoundly paradoxical late novels.

self-refe·rral. [Self- 1 a.] The referring of oneself to an expert or (esp. medical) specialist for advice or treatment. Cf. *Referral b.
1971 *New Society* 14 Jan. 65/1 Fewer of its clients died in the high-risk first year after self-referral. **1975** *Language for Life* (Dept. Educ. & Sci.) xix. 281 There was an immediate and large increase in the number of self-referrals. **1979** *Internat. Rehabilit. Med.* I. 75/2 Some of the outcome measures to which numbers can be applied are..responses to stress (e.g. self-referral to social services).
Also **self-refe·rred** *ppl. a.*
1960 *Cambr. Rev.* 7 May 506/2 'Self-referred' patients generally turn out to be at least as ill as those who are referred by their tutors or general practitioners.

self-reflexive, *a.* Delete † and add: Also, characterized by reflexive action on itself; containing a reflection or image of itself. (Later examples.)
1933 A. Korzybski *Science & Sanity* xx. 323 All human 'knowledge' is structurally circular and self-reflexive, and so depends on some conscious or unconscious theory of knowledge and undefined terms. **1957** *Essays in Criticism* VII. 46 When everything is worked in to comprise part of a highly involute and self-reflexive symbolic pattern, there is just too much of it for poetry; it becomes mainly a complex intellectual parlour game. **1977** *Dædalus* Fall 105 The best way to illustrate the complex and self-reflexive progress of a semiotic enterprise is to consider what semiotics has done and promised to do for the study of the most complex of sign systems, literature. **1980** *San Francisco Bay Guardian* 16–23 Oct. 17/3 We all know about Vladimir Nabokov and self-reflexive writing.
So **self-refle·xiveness,** the quality or condition of being self-reflexive.
1933 A. Korzybski *Science & Sanity* iv. 58 A word *is not* the object it represents; and languages exhibit also this peculiar self-reflexiveness, that we can analyse languages by linguistic means.

self-regulating, *ppl. a.* (Examples in non-technical use.)
1846 J. D. Morell *Hist. Philos.* I. iv. 323 The human mind..is a spontaneous, self-regulating existence. **1975** *Listener* 11 Sept. 328/3 This new ideal of the self-regulating individual.

self-regulation. Add: *spec.* in *Biol.* = *Homœostasis.
1912 J. S. Huxley *Individual in Animal Kingdom* i. 18 Protoplasm has primitively a great power of self-regulation. **1957** —— *Relig. without Revelation* (rev. ed.) ix. 215 There has been an enormous rise in level of harmonious organisation—think of a bird or a mammal as against a flatworm or a jellyfish; inflexibility and the capacity for self-regulation.

self-reinfo·rcement. *Psychol.* [Self- 1 a.] The reinforcement or strengthening of one's own response to a stimulus or situation. Also *transf.* Also **self-reinfo·rcer; self-reinfo·rcing** *ppl. a.* Cf. *Reinforcement 3 c.
1963 *Jrnl. Experim. Psychol.* LXVI. 245/1 This procedure [*sc.* switching on a light for oneself] was called self-reinforcement. *Ibid.*, The frequency and accuracy of self-reinforcing responses..increased with more learning. **1973** *Jrnl. Genetic Psychol.* Sept. 86 Generally self-reinforcement studies have not attempted to deal with the classroom learning situation. **1977** P. F. Secord in T. Mischel *Self* ix. 261 Positive self-reinforcers are seen as increasing the frequency of desirable behavior, and aversive self-reinforcers are thought to reduce undesirable behavior. **1979** *Sci. Amer.* Sept. 52/1 The process is self-reinforcing: the flow of sodium ions through the membrane opens more channels and makes it easier for other ions to follow.

self-reliance. (Earlier example.)
1833 J. S. Mill *Let.* 25 Nov. in *Works* (1963) XII. 195 Combining perfect self-reliance with the most unaffected modesty.

self-repo·rt. *Psychol.* [Self- 3 a.] A report about oneself or aspects of one's behaviour made by oneself. Freq. *attrib.*
1970 *Jrnl. Gen. Psychol.* Oct. 169 Self-report statements of awareness are influenced by the same independent variables as other verbal behaviors. **1972** *Jrnl. Social Psychol.* LXXXVI. 124 The LSQ was designed to elicit a self-report of actual behaviour. **1977** J. D. Douglas in Douglas & Johnson *Existential Sociol.* 305 My theoretical variables..could not be directly related to the experience of the people killing themselves by anything except my own imagination and a few snatches of self-report data.

self-reproach. (Earlier example.)
1754 Richardson *Grandison* III. 3 Dear Miss Grandison, don't give me cause for self-reproach.

self-rescuer. *Coal-mining.* [Self- 1 c.] A safety device carried by coal-miners to give protection against noxious gases (see quot. 1977).
1961 in Webster. **1962** *Guardian* 31 Oct. 5/5 A new safety apparatus known as a 'self-rescuer', which gives a miner 30 minutes to reach a fresh-air area. **1977** *Guardian Weekly* 4 Dec. 19/3 Both the lamp..and the self-rescuer (a steel box containing a mask to refine dangerous gases, also hung from the waist) were heavy. **1979** *Daily Tel.* 13 July 6/8 The miners also carry safety equipment called a 'self-rescuer' and have automatic gas detectors.

self-revealed, *ppl. a.* (Examples.)
1923 J. W. Harvey tr. *Otto's Idea of Holy* xx. 175 For the abyss between creature and Creator.., sin and holiness, is..increased by that deeper knowledge that comes from the Gospel of Christ: and, as a result of the emotion ..stirred in the recognition of it, that in which 'the holy' stands self-revealed is taken here..both as the refuge from, and the means by which to approach, Holiness. **1978** *Church Times* 11 Aug. 6/3 Barth's God is God self-revealed not as a doormat but as Father, Son and Spirit.

self-reve·rsal. [Self- 3 b.] **1.** Reversal (of motion) by agency of the mover itself.
1886 R. Wormell *Electr. in Service of Man* 61 This [induction] machine is exceedingly powerful in favourable weather, but has an important defect, in a tendency to self-reversal, which is apt to occur at a stoppage.
2. *Physics.* The darkening of the middle of a bright spectral line as a result of radiation emitted by a hot gas being partly reabsorbed as it passes through parts of the gas that are cooler.
1905 E. C. C. Baly *Spectroscopy* xii. 384 The lines generally were sharper than in the case of the arc in air,..and there were fewer self-reversals. **1934** H. E. White *Introd. Atomic Spectra* xiv. 250 The centers of the potassium-doublet, the calcium-singlet, and the copper-doublet lines show self-reversal due to absorption. **1977** A. Corney *Atomic & Laser Spectrosc.* x. 313 Systematic errors are difficult to avoid in these experiments since the profile..is usually considerably distorted by self absorption or even self reversal.
3. *Geol.* The postulated reversal of the magnetization of some rocks by intrinsic means, rather than by reversal of the Earth's magnetic field (see quot. 1971).
1952 T. Nagata et al. in *Jrnl. Geomagnetism & Geoelectr.* IV. 22 An experimental proof of self-reversal of thermo-remanent magnetism of igneous rocks is obtained. **1971** *Nature* 5 Feb. 378/1 At this point [*sc.* the early 1950s] an alternative to field reversal began to receive serious consideration. This was self-reversal, the possibility that some rocks possessed an intrinsic property whereby they could acquire a magnetization antiparallel to the ambient field, or whereby an originally parallel magnetization could reverse spontaneously.

self-rising, *ppl. a.* *U.S.* [Self- 3 b.] = Self-raising *ppl. a.*
1865 *Chicago Tribune* 10 Apr. 1/6 Rogers' Self Rising Flour. The best, cheapest and most convenient Flour. **1930** *Randolph* (W. Va.) *Enterprise* 20 Nov. 4/3 Girls love

to make 'skillet biscuits' with a self-rising flour when out camping. **1976** *Encycl. Brit. Micropædia* IV. 195/1 The wide variety of wheat flours generally available includes.. self-rising flour, refined and bleached with added leavening and salt.

self-satisfaction. (Earlier example.)

1739 HUME *Treat. Hum. Nature* II. I. 32 We are not much satisfy'd with the thing itself; and are still less apt to feel any new degrees of self-satisfaction upon its account.

self-sea·ling, *ppl. a.* [SELF- 4.] Becoming gas- or liquid-tight automatically; used esp. of a type of fuel tank.

1895 *Montgomery Ward Catal.* Spring & Summer 545/3 Owing to frequent inquiries for a self-sealing jar..we decided to purchase the Globe Glass Top Jar. **1920** *Flight* XII. 605/2 A tank of his design was dropped from a height of approximately 400 ft... Although badly battered and bent, it is stated that no leakage of petrol occurred. The tank is a metal one and designed to be self-sealing. **1978** R. V. JONES *Most Secret War* x. 78 The single bomber that we had shot down was found to have self-sealing tanks.

self-seeker. Add: **2.** A push-button device on a radio for automatic tuning to the desired station. Cf. *SELF-SEEKING *ppl. a.* 2.

1960 *News Chron.* 22 Apr. 11/3 If the programme displeases you, press the button again and the self-seeker will move on to the next station.

self-seeking, *ppl. a.* Add: **2.** Of a radio: fitted with a self-seeker (sense *2).

1972 *Times* 14 Sept. 31/8 At 90 mph it is still quiet enough to enjoy the self-seeking radio. **1976** *Drive* Jan.–Feb. 82/2 Electric quarter lights, self-seeking radio with automatic electric aerial, two cigarette lighters.

self-se·rvice. [SELF- I a.] **1.** A system whereby customers in a shop, restaurant, etc., serve themselves instead of being attended to or waited on by the staff, usu. paying for all purchases in one place. Also, an establishment or department where this system obtains.

1919 *Ladies' Home Jrnl.* Jan. 65/2 The Duffy-Powers Company, operating a full-fledged department store in Rochester, New York, inaugurated self-service—that is, the customers, not the store, provide the service—in its grocery department. **1928** *Daily Mail* 7 Aug. 3/3 This same principle of self-service has..been successfully applied to food shops. **1944** M. PANETH *Branch Street* 6 The outing ended in a 'Lyons' with 'self service'. **1952** *Economist* 29 Nov. 636/1 Food shops..may soon be turning over to self-service. **1961** *Listener* 7 Dec. 983/1 Big stores, self-services, cinemas losing business. **1962** H. O. BEECHENO *Introd. Bus. Stud.* xi. 101 Self-service is..a very suitable form of retailing for pre-weighed or measured, pre-packed goods which the customer can select for himself and pay for at a cash-taking position as he leaves the shop. **1972** J. MANN *Mrs Knox's Profession* v. 31 She had pulled into the motorway café..and had gone into the self-service to buy a cup of tea.

2. *attrib.* or as *adj.* Designating: **a.** A place where customers may serve themselves.

1919 *Ladies' Home Jrnl.* Jan. 65/3 After several months .., not only are all the self-service departments reported on a self-supporting basis, but with sales increasing. **1950** 'N. SHUTE' *Town like Alice* xi. 330, I sat on the veranda with Jean, studying her drawing of the layout of the self-service grocery. **1952** D. RIESMAN in *Antioch Rev.* Dec. 430 The self-service supermarket, with its abundance of foods capably displayed. **1967** *Times* 16 May 21/1 The situation with self-service filling stations is a classic case. **1977** B. PYM *Quartet in Autumn* ii. 24 Letty did her shopping..at a small self-service store run by Uganda Asians that stayed open till eight o'clock in the evening.

b. A system of this nature.

1919 *Ladies' Home Jrnl.* Jan. 65/2 It [*sc.* a department store] followed this with an even greater inauguration—the self-service system in the sale of shoes..and women's wearing apparel. **1969** *Which?* Nov. 338/2 For clothes, some people liked self-service arrangements... Older women preferred the help..of an assistant.

c. A facility operated by the customer or user.

1952 S. KAUFFMANN *Philanderer* (1953) xiv. 240 Then he walked back into the house and went up in the self-service elevator. **1960** [see **coin-operated* adj. s.v. *COIN *sb.* 8].

d. Merchandise bought by self-service.

1958 *Spectator* 18 July 118/1 Packaging for 'self-service' retailed goods. **1967** M. DRABBLE *Jerusalem the Golden* viii. 194 They..collected themselves a large self-service meal.

Also **self-se·rve** *attrib.* (chiefly *N. Amer.*).

[**1916** *Amer. Mag.* Oct. 69/1 'We told the customers to make their own selection, which they did.' 'It was a sort of serve-self arrangement, like a restaurant,' I said. '*Serve self*..,' repeated Garner. '*Serve-Self Sale.* I have it.'] **1931** W. FAULKNER *Sanctuary* xxxi. 305 A self-serve place, where the customers moved slowly along a railing in single file. **1981** *Univ. Coll. London Bull.* Dec. 4/2 Self-serve counters would replace the current type.

self-shie·lding, *vbl. sb.* Chiefly *Nucl. Physics.* [SELF- I b.] Shielding of the interior of a body as a result of the outer parts' absorbing radiation coming from the exterior; also = *SELF-ABSORPTION 2.

1958 *Proc. 2nd U.N. Internat. Conf. Peaceful Uses Atomic Energy* XIII. 435/1 As the poison becomes depleted, its self-shielding decreases and the poison burns out faster than the fuel, causing a reactivity rise. **1963** B. FOZARD *Instrumentation Nucl. Reactors* i. 3 In many practical cases difficulties [in measuring the radioactivity of a source] arise due to such phenomena as self-shielding. **1977** J. MOTEFF in J. Weisman *Elem. Nucl. Reactor Design* v. 143 The receptor is shielded from each point source not only by interposed shielding material but also by materials within the source itself. This second contribution to the total shielding is called self-shielding or self-absorption.

So **self-shie·lded** *ppl. a.* (quot. 1964 refers to shielding from electromagnetic fields).

1958 *Proc. 2nd U.N. Internat. Conf. Peaceful Uses Atomic Energy* XIII. 433/2 The analytical treatment of self-shielded or lumped poisons. **1964** R. F. FICCHI *Electr. Interference* vii. 119 A self-shielded distributor cap is available commercially that makes use of the same physical configuration as a nonshielded cap. **1973** P. F. ZWEIFEL *Reactor Physics* ix. 248 The center of the resonance line is strongly self-shielded.

self-si·milar, *a.* [SELF- 3 a.] Similar to itself; having no variety within itself, uniform; *spec.* in *Math.* similar to itself at a different time, or to a copy of itself on a different scale.

1867 R. W. EMERSON *May-Day* 48 Stumbling on through vast self-similar woods. **1956** A. A. TOWNSEND *Struct. Turbulent Shear Flow* v. 101 The rate of strain in shear flow is self-similar everywhere. **1967** *Science* 5 May 636 Many [geographical curves] are statistically 'self-similar', meaning that each portion can be considered a reduced-scale image of the whole. **1978** *Sci. Amer.* Apr. 22/3 A coastline, for example, may be self-similar when viewed from a height of several miles down to several feet, but below that the fractal property is lost. **1978** *Amer. Scientist* LXVI. 713/1 The flow tends toward a self-similar solution, i.e. a flow in which the profiles of the physical quantities behind the shock wave remain constant in time.

Hence **self-simila·rity.**

1967 *Science* 5 May 636/1 Seacoast shapes are examples of highly involved curves such that each of their portion can—in a statistical sense—be considered as a reduced scale image of the whole. This property will be referred to as 'statistical self-similarity'. **1978** *Sci. Amer.* Apr. 22/2 Unlike these striking artificial curves the fractals that occur in nature—coastlines, rivers, trees, star clustering, clouds and so on—are so irregular that their self-similarity (scaling) must be treated statistically.

self-sown, *pa. pple.* and *ppl. a.* Add: Hence **self-sow** *v. intr.*, to propagate itself by seed.

1909 *Country Life in Amer.* Aug. 394/1 Annuals that 'self-sow' are welcome. **1980** *Country Life* 3 July 52/2 The milk thistle..might self-sow over-exuberantly.

self-sta·rter. [f. next.] **1.** An electrical device for starting the engine of a motor vehicle without the need to crank it. Also *fig.*

1894 *Work* 17 Feb. 73/3 The many improvements made in recent years in the use of ignition tubes, self-starters. **1902** A. C. HAWKSWORTH *Motors & Motor-Driving* vii. 137 In the case of cars with two or more cylinders various self-starters have also been introduced. **1927** *Sat. Even. Post* (N.Y.) 24 Dec. 15/3 Conversation with him was never a self-starter; you had to crank it. **1973** M. RUSSELL *Double Hit* xxv. 190 There was the squawk of a self-starter, a roar, a piping of tyres.

2. A person who acts on his own initiative (*spec.* at work). *colloq.*

1960 *Times* 1 Feb. 2/4 (Advt.), The Man: Must be a self-starter yet able to work within references given to him. **1964** MRS. L. B. JOHNSON *White House Diary* 15 Nov. (1970) 203 Luci describes herself as a 'self-starter'. **1971** *Sunday Times* (Johannesburg) 28 Mar. 30/4 (Advt.), Suitable applicants will be self-starters in the 25–35 year age group. **1979** *Arizona Daily Star* 5 Aug. (Advt. Section) 4/2 Must be self-starter with project management experience.

self-starting, *ppl. a.* [SELF- 4.] That comes into operation automatically or semi-automatically. Of a motor vehicle: fitted with a self-starter. Also *fig.*

1866 *Proc. Inst. Mech. Engineers* 268 An arrangement for rendering the self-adjusting injector also self-starting has been contrived at the writer's works. **1887** [in Dict. s.v. SELF- 4]. **1910** *Daily Mail Year Bk.* 157/1 A few non-dazzling headlights, and one or two dual ignition and self-starting devices. **1912** *Chambers's Jrnl.* Aug. 556/2 Self-starting motor-cars. **1927** M. ASQUITH *Lay Sermons* i. 22 After making elaborate notes on all his self-starting symptoms, he wrote a long prescription. **1965** *Wireless World* July 56/1 (Advt.), Synchronous electric clock movements... Self-starting.

self-stee·ring, *vbl. sb. Naut.* [SELF- 4.] The steering or directing of a vessel on a predetermined course by automatic means; **self-steering gear** (also **sails**), apparatus by which this is achieved; also *absol.*

1950 E. C. HISCOCK *Cruising under Sail* II. xix. 363 For a single-handed or short-handed ocean passage I consider that self-steering sails are most necessary. **1956** P. JOHNSON in J. Morwood *Sail Evolution* (A.Y.R.S. Publ. No. 3) 22 It is not only to the single-handed ocean voyager that self steering is useful. **1957** T. HERBERT *Self Steering* (A.Y.R.S. Publ. No. 13) 6 Self Steering may be defined as the ability of a sailing boat to stay on a set course relative to the wind... The various types of self steering gears fall into two broad categories: 1. Those gears which convert variations in wind pressure on the sails into rudder movements [etc.]. **1971** J. R. L. ANDERSON *Reckoning in Ice* viii. 153, I had no self-steering on Gudrid, but she was a beautifully balanced boat. **1976** *Southern Even. Echo* (Southampton) 1 Nov. 3/6 The 29-year-old blonde Lymington girl..expressed her admiration for..the winner—'one of the greatest sailors who have ever been, because he sailed without self-steering gear.'

self-stimula·tion. *Psychol.* [SELF- I a.] Stimulation of oneself for pleasure or excitement; *spec.* masturbation. Also, stimulation of its own pleasure centres effected in an animal by means of implanted electrodes (see quot. 1956).

1947 M. M. LEWIS *Lang. in Society* i. 18 There, at that level [*sc.* babbling]—as a form of play, of self-stimulation, of narcissism—it [*sc.* language] might remain. **1951** FORD & BEACH *Patterns of Sexual Behav.* ix. 155 American men who begin to masturbate..discontinue the practice.. because self-stimulation is replaced by heterosexual intercourse. **1956** J. OLDS in *Sci. Amer.* Oct. 114/3 In the rhinencephalon the effects were milder, producing self-stimulation at rates around 200 times per hour. **1964** J. Z. YOUNG *Model of Brain* xiv. 234 With stimulating electrodes implanted in them animals can be made either to seek self-stimulation from one centre or to avoid it if the electrode is in..one nearby. **1977** R. HOLLAND *Self & Social Context* iv. 62 Olds' study of self-stimulation by rats with implanted electrodes.

self-stu·dy. [SELF- I a, *2 b.] **1.** Study or contemplation of oneself.

1683 [in Dict. s.v. SELF- I a]. **1936** L. PEARSALL SMITH *Reperusals & Re-Collections* i. 6 He found moral profit also in the self-study, for how, he asked, can we correct our vices if we do not know them.

2. Study by oneself; private study.

1958 *Practical Wireless* XXXIV. 1 (Advt.), Designed and arranged for self-study at home. **1962** E. SNOW *Other Side of River* (1963) xviii. 130 Mao had already made wide-ranging contacts and done omnivorous reading in the Changsha library. In his self-study he was guided by his favorite professor. **1977** P. STREVENS *New Orientations Teaching of English* ii. 13 The adult who works on his own with a self-study course is still, though in an extended way, learning with a teacher.

self-sufficiency. Add: **1. a.** Now freq. used in sense *1 b of the adj.

1932 *World Agriculture* (R. Inst. Internat. Affairs Study Group) viii. 133 A measure of self-sufficiency in food is regarded as one of the primary elements in natural defence. **1953** G. E. & K. R. FUSSELL *Eng. Countrywoman* iii. 90 The canopy of self-sufficiency that covered all the needs of the manor house. **1973** J. & S. SEYMOUR *Self-Sufficiency* i. 9 What I am interested in is *post*-industrial self-sufficiency: that of the person who has gone through the big-city-industrial way of life..and wants to go on to something better. **1974** *News & Courier* (Charleston, S. Carolina) 7 Apr. A-12/2 They argue that the 1985 target date for self-sufficiency is unrealistic, and that..some oil and gas still must be imported. **1980** 'M. INNES' *Going it Alone* iv. 31 She had persuaded herself..that self-sufficiency must now be the prime concern of the small rural gentry.

self-sufficient, *a.* Add: **1. b.** *spec.* Of persons, groups, or nations: able to provide enough of a commodity (as food, oil) to supply one's own needs, without obtaining goods from elsewhere; self-reliant, self-supporting, independent. Freq. const. *in.*

1932 [implied at *SELF-SUFFICIENCY I a]. **1955** K. A. H. MURRAY *Agriculture* iii. 40 For countries that are practically self-sufficient in their food requirements.., war or the threat of war may necessitate little or no change in agricultural effort or output. **1973** J. & S. SEYMOUR *Self-Sufficiency* i. 10 More and more people, in all the highly industrial societies, are trying partially to opt out of the big-industry set-up and become less specialized and more self-sufficient. **1977** *Undercurrents* June–July 11/2 If we are to become self-sufficient in food production in Britain we are told that we must eat more grain and less meat. **1978** *Lancs. Life* July 54/2 He was soon back in the Lake District.., becoming self-sufficient through selling his paintings and wood sculptures, doing odd jobs and growing his own food.

Hence as *sb.* (*rare*), one that is sufficient in itself (oneself).

1909 W. JAMES *Meaning of Truth* xiv. 276 Mr. Russell, and also Mr. Hawtrey..seem to think that in our mouth also such terms..are self-sufficients with no context of varying relation that might be further asked about. **1926** I. BROWN *H. G. Wells* iii. 43 They have the Cockney's superb self-mastery, his power to rise to an occasion... Bert Smallways in *The War in the Air* is another of these admirable self-sufficients.

self-suggestion. 2. (Earlier example.)

1893 A. LANG in *Folk-Lore* IV. 433 But I am not my own dupe. Others may be more fortunate or more amenable to self-suggestion.

self-supporting, *ppl. a.* (Earlier and further examples.)

1829 *Lond. Med. & Surg. Jrnl.* II. 341 (*heading*) Atherstone self-supporting dispensary. **1845** J. S. MILL in *Edin. Rev.* LXXXI. 521 It is an indispensable condition that there be a reasonable prospect of their being at some future time self-supporting. **1893** G. B. SHAW *Widowers' Houses* II. 52 *Trench*: We must do the best we can with seven hundred. I think we ought to be self-supporting.

1936 N. STREATFEILD *Ballet Shoes* iii. 46 You will at least have taken a step towards trying to make them self-supporting. **1972** *Guardian* 1 Sept. 1/1 Britain took a giant stride towards becoming self-supporting in oil yesterday.

self-surrendering, *ppl. a.* (Earlier example.)

1878 J. G. WHITTIER *Vision Echard* 12 Its [*sc.* love's] self-surrendering freedom, Its loss that gaineth all.

self-sy·stem. *Psychol.* [SELF- 5 b.] The organized complex of drives and responses pertaining to (an aspect of) the self; the final choice of potentialities which the individual seeks to develop.

1897 W. JAMES *Let.* 1 Feb. (1920) II. 57 It is a case of the conflict of two *self-systems* in a personality up to that time heterogeneously divided, but in which..the higher loves and powers come definitively to gain the upper-hand. **1940** H. S. SULLIVAN *Psychiatry* III. 9/1 Along with the learning of language, the child is experiencing many restraints on the freedom which it had enjoyed.. and from these restraints there comes the evolution of the self system. **1977** R. HOLLAND *Self & Social Context* ii. 31 Anxiety is the special 'dynamism' that attaches to the development of a 'self-system'.

self-ta·pping, *ppl. a. Mech.* [SELF- 4.] Designating a hardened screw that will cut its own thread in a hole in metal that would otherwise need tapping. Also **self-ta·pper** (*colloq.*).

1936 R. T. KENT *Mech. Engineers' Handbk.* (ed. 11) III. ix. 36 Self-tapping screws are screws that may be driven into an untapped hole, forming the thread in the hole as they are driven. **1955** *Archit. Rev.* CXVII. 213/1 From the point of view of large-scale production of sheet metal components, the greatest design change has been brought about by the Parker-Kalon or self-tapping screw. **1970** K. BALL *Fiat 600, 600D Autobook* viii. 99/1 Slightly tighten the self-tapping bushes. **1978** *Hot Car* June 98 (Advt.), A spoiler which needs only six self tappers to secure it to withstand heavy impact.

self-ti·mer. *Photogr.* [SELF- I c or 4.] A mechanism that introduces a delay between the operation of the shutter release and the opening of the shutter, so that the photographer can photograph himself.

1951 L. A. MANNHEIM et al. *Rollei Way* 53 To make exposures with the self-timer..pull the button outwards. **1968** M. WOODHOUSE *Rock Baby* xvii. 173 He had..a self-timer on the camera and there were pictures of the two of us together. **1978** *Detroit Free Press* 5 Mar. A14 (Advt.), The Polaroid Pronto! RF Land Camera..comes with self timer and tripod.

self-transce·ndence. [SELF- I a, d.] Transcendence or surpassing of oneself or one's limitations; the achievement of or capacity to achieve a higher level of awareness, compassion, etc. Also **self-transce·ndency.**

1885 W. JAMES in *Mind* X. 27 We are not to ask, 'How is self-transcendence possible?' **1895** —— in *Psychol. Rev.* II. 110 Those mysterious notions of self-transcendency and presence in absence which are such essential parts of the ideas of knowledge, both of common men and of philosophers. **1946** J. S. HUXLEY *Unesco* ii. 62 Individuals..can only achieve fullest self-development by self-transcendence. **1956** *Essays in Criticism* VI. 17 An irony of a more darkly moral colouring, a sardonic self-transcendence, was known to Friedrich Schlegel. **1975** P. BAELZ *Forgotten Dream* iii. 46 To place such emphasis on the possibility of creative self-transcendence..at once suggests that we have escaped from sober reality into a world of our own make-believe.

So **self-transce·ndent** *a.*, **self-transce·nding** *ppl. a.*

1884 W. JAMES in R. B. Perry *Tht. & Char. of W. James* (1935) I. xxxvi. 579 A consciousness not self-transcendent in form is inconceivable. **1904** —— in *Jrnl. Philos.* I. 538 Common-sense theories left the gap untouched, declaring our mind able to clear it by a self-transcending leap. **1948** J. L. ADAMS tr. *Tillich's Protestant Era* v. 67 Self-transcending realism is a universal attitude toward reality..combines two elements, the emphasis on the real and the transcending power of faith. **1975** P. BAELZ *Forgotten Dream* ii. 25 Man also has the ability to distance himself from what is immediately presented to him as himself... He is self-transcending.

self twist. [SELF- 4.] A method of spinning in which the yarn is twisted by the lateral movement of a roller. Usu. (with hyphen) *attrib.*

1970 SELLING & LORD in P. R. Lord *Spinning in '70s* vii. 96 The second section deals with the so-called *self-twist yarns* such as those produced experimentally in Australia. **1972** *Sci. Amer.* Dec. 55/1 Self-twist spinning was developed in Australia, where a great deal of research is done on wools. **1974** [see *open end* adj. s.v. *OPEN *a.* 22 c]. **1980** *Jrnl. R. Soc. Arts* May 364/1 Perhaps our most notable success has been the development of the wool spinning method called 'Self-twist'.

selham, var. *SULHAM.

‖ **selicha** (sĕlī·χă, sĕliχā·). Also **seliha**; pl. **selichot, -oth,** etc. [a. Heb. *səlīḥā*, pl. *səlīḥōṯ* penitential prayer.] A Hebrew liturgical poem recited in penance on fast days, before Rosh Hashanah, and before and on Yom Kippur.

1864 *Chambers's Encycl.* VI. 155/1 The *Selichoth*, or Penitential Prayers. **1909** *Cent. Dict.* Suppl. 1194/2 The selihot are acrostically arranged, many containing not only the alphabet but also the names of the composer and his father. **1925** W. O. E. OESTERLEY *Jewish Background Christ. Liturgy* ii. 76 What are called the *Selichôth* (i.e. prayers for forgiveness; the word comes from the root *salach* 'to forgive') form an important element in the Jewish liturgy. **1960** S. BECKER tr. *A. Schwarz-Bart's Last of Just* (1961) I. 9 He is the author of the famous Seliha, Oh God, cover not our blood with thy silence. **1974** *Jewish Chron.* 13 Sept. 34/3, I was at selichot the other morning (none of your midnight revels for me) with my soul on high. **1976** B. WILLIAMS *Making of Manch. Jewry* xii. 304 In October 1870 the congregation agreed on minor ritual changes—including the abbreviation of the *Piyyutim* and *Selicoth*.

seligmannite (se·ligmănəit). *Min.* [ad. G. *seligmannit* (H. Baumhauer 1901, in *Sitzungsber. d. K. Preuss. Akad. d. Wissensch. zu Berlin* 110), f. the name of Gustav *Seligmann* (1849–1920), German mineral collector: see -ITE¹.] An arsenic sulphide of lead and copper, $PbCuAsS_3$, occurring as brittle, dark grey to black, orthorhombic crystals having a metallic lustre.

1902 *Mineral Mag.* XIII. 205 Seligmannite... Of this mineral,..only five crystals, varying in size from ½ to 4 mm., have been found. **1969** *Zeitschr. für Krist.* CXXX. 224 The structure of seligmannite is built up essentially of a framework of CuS_4 tetrahedra and AsS_3 pyramids, with lead atoms inserted in the large holes.

selion. Add: **2.** A furrow turned over by the plough. *nonce-use.*

1877 G. M. HOPKINS *Poems* (1967) 69 No wonder of it: sheer plód makes plough down sillion Shine.

Selish, var. *SALISH.

Seljuk, *a.* and *sb.* Add: **A.** *adj.* **2.** Designating a style of oriental carpet.

1931 A. U. DILLEY *Oriental Rugs & Carpets* xiii. 267 Seljuk, lifted out of Turkish history, and Dinar..are the names of oriental rugs produced in Greece. **1952** B. MIALL tr. *H. Jacoby's How to know Oriental Carpets & Rugs* 13 The oldest known Oriental carpets, the so-called Seljuk carpets of the thirteenth century, which came originally from the Ala-eddin Mosque at Konia, are now preserved, with other magnificent examples, in the Evkaf Museum at Istanbul. **1963** I. SCHLOSSER *European & Oriental Rugs & Carpets* 118 The Seljuk carpets..in the Evkaf Museum..have small but very interesting geometric patterns.

Selkup (se·lkŭp). Also **Sel'kup.** [Native name.] A member of a Samoyedic people of northern Siberia; the language of this people belonging to the Uralic family of languages. (Formerly known as 'Ostyak Samoyed': cf. *OSTYAK; SAMOYED *sb.* and *a.* in Dict. and Suppl.)

1962 [see *NENETS]. **1964** in S. P. Dunn *Levin & Potapov's Peoples of Siberia* 597 The summer dwelling of the Northern Sel'kups was the birchbark tent. **1967** D. S. PARLETT *Short Dict. Languages* 107 The Southern languages include Ostyak Samoyed (Sel'kup), and Kamassian. **1972** *Language* XLVIII. 206 Of the southern group, only Sel'kup ('Ostyak Samoyed') can be considered a living language. **1974** *Encycl. Brit. Macropædia* XVIII. 1022/2 The fourth language, Selkup, lies to the south in a region between the central Ob and central Yenisey. **1975** G. F. CUSHING tr. *Hajdu's Finno-Ugrian Languages & Peoples* iii. 215 The only representatives of the southern group of Samoyeds still extant are the Selkups.

sell, *sb.*² Add: **2. a.** (Earlier example.)

1838 *Actors by Daylight* 4 Aug. 179 (*heading*) Editorial consequence—Specimens of wit—A decided sell.

b. The technique of selling by advertising or persuasive salesmanship; the practice or fact of this. Usu. with qualifying word: cf. *hard sell* s.v. *HARD *a.* 22 a; *soft sell* s.v. *SOFT *a.* 27.

1952, etc. [see *hard sell* s.v. *HARD *a.* 22 a]. **1970** G. GREER *Female Eunuch* 11 The main force of their energy filtered away..through the sexual sell of the fifties. **1976** SCOTT & KOSKI *Walk-In* (1977) xiii. 81 He was like an encyclopedia salesman moving into the soft, memorized sell.

3. Delete ? *U.S.* and add later examples.

1981 *Times* 20 July 20/1 Cooke, Lumsden waver between a hold and sell recommendation for Dowty Group. *Ibid.* 27 July 20/1 Woodside Petroleum is a sell. **1981** *Sunday Times* 2 Aug. 43/6 We rated them a sell ahead of the disappointing figures last month.

4. For *U.S.* read orig. *U.S.* and add: **a.** (Earlier and later examples.) Also *gen.*, (one who makes) a sacrifice of principle or betrayal.

1862 M. B. CHESNUT *Diary* 6 May in C. V. Woodward *M. Chesnut's Civil War* (1981) 336 Another sellout to the devil. It is this giving up that kills me. **1883** J. HAY *Bread-Winners* 151 How much did the Captain give you for that sell-out? **1940** 'G. ORWELL' *Jrnl.* 24 June in *Coll. Essays* (1968) II. 354 High-up influences in England are preparing for a similar sell-out [to Pétain's]. **1953** *Landfall* (N.Z.) Dec. 283 This film could have ended with a punch; but this would have been running counter to the sacrosanct Hollywood tradition of the inevitable happy ending. So there is a sell-out. **1959** *Economist* 11 Apr. 134/2 Specially elected members (reference to whom the wilder parts of the audience had greeted with familiar African cries of 'stooges', 'sell-outs'). **1960** J. LEHMANN *I am My Brother* 4 A gigantic sell-out to the Nazis. *a* **1974** R. CROSSMAN *Diaries* (1975) I. 182 Then Maurice Edelman made an inflammatory half-hour attack on the Government, charging me and Frank Cousins with every kind of crime, including a sell-out to the Americans. **1980** *Times* 19 Feb. 6 Mr. Robert Mugabe's Zanla guerrillas infiltrated the region..maiming or murdering those who were considered sell-outs.

c. The disposal of a commodity because of great demand; also, a completely disposable commodity. Also *transf.*, an event for which all tickets have been sold; the occasion of such an event.

1859 *N.Y. Herald* 11 July 6/5 (Advt.), Our goods shall be sold cheap!! In this great sellout. **1923** *Variety* 11 Oct. 17/4 Business at 'Give and Take' at the Adelphi was generally big last week... Wednesday..was a virtual sell-out. **1933** *Sun* (Baltimore) 29 Aug. 8/3 The [actor's] interpretation may not be art, and it may not even be O'Neill, but it may easily be a sellout and it will almost assuredly be entertaining. **1945** *Ibid.* 27 Apr. 10-0/1 On a sharp sell-out of rails and steels after the opening, dealings were relatively heavy. **1945** S. LEWIS *C. Timberlane* (1947) xliii. 289, I have four tickets... They're absolutely impossible to get, show is a sell-out, but the agent is a friend of mine. **1950** *Sport* 22–28 Sept. 2/1 The Cup Final..is always a sell-out but crowds at other games are always below maximum. **1962** *Listener* 27 Dec. 1095/1 The shortage of shopping times for working people ..causes overcrowding, queuing, poor service and sell-outs on Saturday. **1968** *Globe & Mail* (Toronto) 15 Jan. 17/2 A sellout crowd of 75,546 watched..Vince Lombardi's National Football League champions. **1977** *Time* 19 Dec. 41/2 Such delicacies are instant sellouts.

5. sell-off. *Stock Exchange* (orig. and chiefly *U.S.*). A sale or disposal of bonds, shares, or commodities, usu. causing a fall in price.

1937 *Sun* (Baltimore) 6 Feb. 19/1 The sell-off was less pronounced than that of the share market and was slower in developing. **1941** *Ibid.* 16 Oct. 21/1 Adverse war news was blamed in most quarters for the selloff in stocks and commodities. **1958** *Time* 8 Dec. 98/3 The sell-off did not alarm most market experts. **1981** *Times* 21 May 24/1 The recent sell-off by a major institution has done much to cloud market sentiment.

sell, *v.* Add: **3. g.** *to sell down the river*: see *RIVER *sb.*¹ 3* a.

h. To advertise or publish the merits of (a commodity, idea, etc.); to persuade (a person) to accept or buy. Also, to convince (someone) of the worth of (something). (Variously with direct and indirect object.) *colloq.* (orig. *U.S.*).

1916 *Amer. Mag.* Mar. 50/1 I'd make my readers want to enlist. I'd 'sell' them the army. **1925** *Publishers' Weekly* 5 Dec. 1863 An advertising campaign to sell New York as the printing center of the world. **1931** W. G. MCADOO *Crowded Years* iii. 41, I had to 'sell' the idea to men like the elder J. P. Morgan. **1938** E. BOWEN *Death of Heart* III. iii. 362 They forget Major Brutt has come here to get a job... Oh dear, oh dear, I shall never sell him at all. **1942** *R.A.F. Jrnl.* 2 May 30 My work—in horrid modern commercial terms—is to 'sell' the R.A.F. to the Army. **1951** H. MACINNES *Neither Five nor Three* I. v. 77 The people of France were sold such ideas as.. 'Patriotism is for the rich'. **1956** B. HOLIDAY *Lady sings Blues* (1973) xix. 158 It seemed like a crazy idea, but he sold me. And what's more important, he sold a lot of other people. **1960** *Guardian* 9 Nov. 6/4 We have to sell to the public the idea that being a foster mother is a service..to the community. **1976** J. I. M. STEWART *Memorial Service* iv. 60 It's just no good your trying to sell me those rotten dons.

i. Const. *on.* To make (someone) enthusiastic about, or convinced of the worth of, something. Freq. *pass. colloq.* (orig. *U.S.*).

1918 *Maclean's Mag.* Mar. 52/2 The writer believes it is possible to finally 'sell' the Teutons on the advantages of peace as compared with war. **1928** WODEHOUSE *Money for Nothing* vii. 133 Come to think of it. I'm not too sold on this thing, anyway. **1932** E. WILSON *Devil take Hindmost* x. 112 He rarely mentions Communism, but..he is as much sold on it as any..party member. **1948** *Manch. Guardian Weekly* 9 Dec. 8 He is not selling anybody on America. **1950** *Newsweek* 1 May 45/1, I am going to..sell her on the idea of picking up the option for another 25 years. **1969** L. HELLMAN *Unfinished Woman* xii. 177 He had been doing his thesis on modern American novelists... I tried hard to sell him on Faulkner and Fitzgerald. **1970** J. EARL *Tuners & Amplifiers* iii. 72 If you are sold on a medium price ceramic cartridge it would pay to look for an amplifier with a 2MΩ ceramic (piezo) input of around 50mV sensitivity. **1978** A. PRICE *'44 Vintage* iv. 46 I've never been absolutely sold on the classics.

j. *refl.* use of sense 3 h above.

1938 L. BEMELMANS *Life Class* I. v. 73 He told me that what was most important in life was..the ability to 'sell' oneself, to call [hotel] guests by their correct names and to remember their faces. **1968** *Globe & Mail* (Toronto) 3 Feb. 11/3 Supersalesmanship is used to sell to adolescents, who must also learn to sell themselves. **1978** H. JOBSON *To die a Little* iii. 36 Sales gimmicks are out. We don't need them. The scheme sells itself.

4. b. *to sell short*: see SHORT *adv.* 11.

6. † **a.** Phr. *made to sell*: manufactured or contrived to secure a ready sale without regard to quality. *Obs.*

1827 P. CUNNINGHAM *Two Yrs. New South Wales* (ed. 2)

I. xvii. 297, I..seized a musket 'made to sell', and sallied out. **1855** P. H. DELAMOTTE *Pract. Photogr.* (ed. 2) 42 In some cameras, 'made to sell', no care is taken to adjust this plane. **1860** *Cassell's Illustr. Family Paper* Apr. 300/3 We shall hear fewer complaints of seed, unless it be from those who obtain packets which are only 'made to sell'.

b. *transf.* with personal subj.

1915 R. FRY *Let.* 21 Nov. (1972) II. 391 My show is turning out a great success..as far as attendance goes... Of course I don't sell—I never expected to. **1966** C. ACHEBE in *Black Orpheus* Mar. 45, I had a Raleigh bicycle, brand new, and everybody called me Jolly Ben. I was selling like hot bread.

12. sell out. d. Also *intr.* for *pass.*

1914 *Daily Mail* 31 Jan. 1/2 My first parcel from you sold out very quickly. **1974** M. BIRMINGHAM *You can help Me* ii. 39 The Friday flower stall was rapidly selling out.

e. For *U.S.* read orig. *U.S.* and add: To betray a person or cause for gain (cf. *sell-out* s.v. SELL *sb.*² in Dict. and Suppl.): (*a*) *intr.* (Later examples.)

1903 G. B. SHAW *Man & Superman* III. 78 He has sold out to the parliamentary humbugs and the bourgeoisie. **1946** KOESTLER *Thieves in Night* 112 The English are going to sell out on us. **1976** *Survey* Winter 86 Barbé called for tactics of disobedience to the colonial administrators and to the traditional chiefs who had 'sold out' to the French government.

(*b*) *trans.* (Examples.)

1857 *Lawrence* (Kansas) *Republican* 2 July 1 If the *Times* has not been 'sold out' to the Border Ruffian party, it looks very much as if it had been 'chartered'. **1867** *Oregon State Jrnl.* 19 Jan. 3/1 The writer thinks the officers were 'badly sold out'. **1936** M. MITCHELL *Gone with Wind* ix. 189 Why quibble about the Yankees earning an honest penny selling out the Union? **1940** J. B. PRIESTLEY *Postscripts* 45 It let the old hands, the experts,..speak for it, and they sold it out. **1967** *Times* 17 Nov. 8/6 With shouts of 'They sold us out, the bastards', the meeting moved to 'the moment of truth'. **1976** 'J. CHARLTON' *Remington Set* xxviii. 141 What happened is, Rog sold us out.

14. sell up. c. *intr.* To dispose of (a house, business, etc.) by sale.

1862 *Manch. Examiner & Times* 8 July 6/2 We hed a varra good heawse i' Stanley-street, once; but we hed to sell up an' creep hitherto. **1977** E. DEWHURST *Curtain Fall* xviii. 216 She would be coming home only in order to sell up.

sellaite (se·lă,əit). *Min.* [a. It. *sellaite* (Strüver 1868, in *Atti d. R. Accad. d. Sci. di Torino: Classe di Sci. fis.*, etc. IV. 35), f. the name of Q. *Sella* (1827–84), Italian mining engineer and mineralogist: see -ITE¹.] Native magnesium fluoride, MgF_2, occurring as fibrous aggregates of colourless tetragonal crystals.

1872 G. J. BRUSH *Dana's Syst. Min.* (ed. 5) App. I. 14 Sellaite... Found with anhydrite at Geibroula in Piedmont. **1923** J. W. MELLOR *Comprehensive Treat. Inorg. & Theoret. Chem.* IV. xxix. 296 Warm crystals of sellaite show a faint violet luminescence with cleavage. **1933** *Mineral. Abstr.* V. 238 Microscopical examination of the dolomite of bore-cores from the southern Harz showed the presence of sellaite as minute (0·04–0·075 mm.) tetragonal prisms. **1968** I. KOSTOV *Mineralogy* 189 Sellaite is isostructural with rutile.., the chain type of structure explaining its perfect {110} cleavage, prismatic to acicular habit, and optically positive character.

seller¹. Add: **1. c.** *Business.* In various phrases, as: (*a*) *seller four* (*ten, twenty, the year,* etc.): a form of contract in which the seller has the right to effect delivery within the specified number of days (four, ten, etc.); *seller's option*: the right of the seller to specify the number of days after which a sale is effected.

1849 *Merchants' Mag.* (U.S.) XX. 670 Fifty-six, buyer 20; 3rd broker—55⅞, seller 10. **1857** *Ibid.* XXXVII. 134 Sales at seller's option are generally a fraction below the current cash price. **1869** 'MARK TWAIN' *Innocents Abr.* xxxiv. 369 Sales of one lot Circassians, prime to good, 1852–1854..; one forty-niner—damaged—at £23, seller ten, no deposit.

(*b*) *seller's market*: a market in which there is excess demand at the going rate so that it is easy to effect additional sales. Also *fig.*

1934 in WEBSTER. **1948** G. CROWTHER *Outl. Money* (ed. 2) v. 163 The raw material markets may be transformed overnight from 'seller's markets' (i.e. where buying is insistent and the seller is in the strategic position) into 'buyer's markets', while the opposite movement is much slower. **1965** ZIGROSSER & GAEHDE *Guide to Collecting Orig. Prints* vi. 91 In boom times and a seller's market, almost anyone can set up as a dealer and make a success of it. **1979** R. JAFFE *Class Reunion* III. vii. 275 Everybody..here seems to be going to some shrink. They're all into self-help,..or screaming sessions, or group therapy... It's a seller's market.

3. Also used of other commodities. (Earlier and later examples.)

1895 *Montgomery Ward Catal.* Spring & Summer 6/2 Brocaded Brilliantines... This line we expect will be one of the greatest sellers of the season. **1925** *Daily Tel.* 13 May 20/7 (Advt.), Traveller Wanted.—We offer the latest new line. Big seller. Live men can earn £10 week. **1976** *Times* 1 May (Food Suppl.) p. ii/6 On tinned meat he said: 'My advice is to stock the major sellers, such as stewed steak.'

4. A selling race. *colloq.*

1922 *N. & Q.* 12th Ser. XI. 207/1 *Seller*,..a selling race —one in which the winner is bound to be offered by public auction. **1927** *Daily Express* 23 June 12/2 The game little Congou colt took another seller. **1928** *Daily Sketch* 7 Aug. 22/4 Another interesting proposition at the Midland meeting is the Loud Report filly in the juvenile seller.

selling, *vbl. sb.* Add: **c.** *selling job, -power*; **selling-point,** a place at which sales may be effected (cf. *point-of-sale* s.v. *POINT *sb.*¹ D.12*****), a retail outlet; **selling price** (earlier example); **selling race**: also *selling chase.*

1965 D. FRANCIS *For Kicks* iii. 44 The horses had all won selling chases—races where the winner was subsequently put up to auction. **1963** MRS. L. B. JOHNSON *White House Diary* 23 Nov. (1970) 8, I will have to..see about getting Lynda Bird to come back and live in Washington with us and go to school somewhere up here (and that will be a selling job!). **1976** I. LEVIN *Boys from Brazil* iii. 77 He's never been as sure as the rest of us that the project will work... The selling job we had to do! **1953** *Chambers's Jrnl.* June 343/1 In recent years some seaside resorts have entered the catering business... Margate, starting with nothing in 1946, now has fifty-two selling-points. **1960** R. WILLIAMS *Border Country* 158 Within a month..he would have all the selling-points he needed. **1904** J. LONDON *Let.* 11 July (1966) 161 My selling-power has increased. **1960** C. S. LEWIS *Studies in Words* 104 The literary innovators want to retain the prestige, almost the 'selling-power', of the consecrated word. **1815** Selling price [see *PRICE-CURRENT].

selling, *ppl. a.* **3.** That helps to effect a sale; esp. in phrr. *selling point, title.*

1875 TROLLOPE *Way We live Now* II. lxxxix. 248, I don't believe that anything like real selling praise is ever given to anybody, except to friends. **1959** *Times* 4 Mar. 11/7 The educational usefulness of television in backward areas depends on a firm grasp of the staffing problem. Otherwise it will be a selling-point, not a reality. **1963** P. PHILLIPS in Sissons & French *Age of Austerity* 148 The old selling phrase 'pre-war value' lost some of its attraction. **1965** W. HAGGARD *Hard Sell* ii. 12 'A delightful name for an aeroplane, isn't it?'..'I'd call it a selling title.' **1978** *Times* 27 Jan. 13/7 A French name is still a selling point for clothes.

sell-off: see *SELL *sb.*² 5.

Sellotape (se·lotēⁱp), *sb.* Also **sellotape, cellotape.** [f. CELL(ULOSE *sb.* + -O + TAPE *sb.*¹] The proprietary name of a cellulose or plastic self-adhesive tape, freq. dispensed from reels for domestic use. Also *gen.*

1949 *Trade Marks Jrnl.* 9 Nov. 1002/2 Sellotape... Adhesive tape (stationery). Adhesive Tapes Limited, 8, Brunel Road, Acton, London, W.3; Manufacturers and Merchants. **1957** *Landfall* (N.Z.) XI. II. 169 Does she want to come a cropper, dashed earthward by wings casually fastened with sellotape? **1960** P. A. BENNETT in J. Pudney *Pick of Today's Short Stories* XI. 16 A wind.. blew down all the pictures stuck with cellotape to the wall. **1960** H. PINTER *Birthday Party* 76 There's some Sellotape somewhere. We can stick them together. **1971** *Petticoat* 24 July 3/1 It's certainly a lot less painful than ordinary cellotape which we've been using up till now. **1974** J. COOPER *Women & Super Women* 19 They always talk about having 'stuck together because of the children', as though the little blighters have been using glue and sellotape on them. **1980** *Guardian Weekly* 3 Feb. 20/5 So far he has largely run on string and Sellotape.

Hence as *v. trans.*, to fasten with Sellotape; **se·llotaped** *ppl. a.*, **se·llotaping** *vbl. sb.*

1960 P. A. BENNETT in J. Pudney *Pick of Today's Short Stories* XI. 16 She..cellotaped them [*sc.* pictures] to the wall. **1964** A. WILSON *Late Call* iii. 113 He continued to sellotape Beth's caricatures to the walls. **1965** M. DRABBLE *Millstone* 178 Lydia really did have to rewrite two whole chapters as well as doing a lot of boring sellotaping. **1969** E. MCGIRR *Entry of Death* iii. 56 An assortment of cellotaped cartons. **1976** *Daily Tel.* (Colour Suppl.) 14 May 36/3 There are four pictures (one bought in Piccadilly, the others from *Elvis Monthly*) Sellotaped to the wall.

sell-out: see SELL *sb.*² 4 in Dict. and Suppl.

‖ **selon les règles** (sĕloṅ lĕ regl'), *phr.* [Fr.] According to the rules (of polite society). Also *fig.*

Victorian writers frequently wrote the third word with an acute accent.

1825 H. WILSON *Mem.* IV. 30 Our's bid fair to grow into a strong mutual fancy, if not to real true love, selon les régles. **1837** G. H. LEWES *Let.* in A. T. Kitchel *G. Lewes & G. Eliot* (1933) i. 12 In spite of its not being *selon les régles* of this most artificial of worlds..I take the shortest and easiest way I can think of for our better acquaintance. **1864** C. M. YONGE *Trial* II. iv. 74 He thinks he is proceeding *selon les régles.* **1893** —— & COLERIDGE *Strolling Players* xxv. 225 There was so much laughter around that George thought a jest quite *selon les règles.* **1921** L. STRACHEY *Q. Victoria* iii. 73 Was not such a course of conduct..simply *selon les règles?*

Selsdon man (se·lzdən). *Pol.* [The name of the *Selsdon* Park Hotel near Croydon, Surrey (see below) + MAN *sb.*¹: after *Piltdown man*, etc.] Used orig. and chiefly by political opponents to denote an imagined person or persons believed to be pursuing the policies outlined at a conference of Conservative Party leaders held at the Selsdon Park Hotel 30 Jan.–1 Feb. 1970.

1970 H. WILSON in *Labour Govt.* (1971) xxxvii. 759 Selsdon Man is designing a system of society for the ruthless and the pushing. **1971** BUTLER & PINTO-DUSCHINSKY *Brit. General Election of 1970* vi. 131 To Mr. Wilson, these were demonstrations of atavistic Conservative instincts, which he summed up in a phrase he repeated time and again: 'Selsdon man'. **1974** *Times* 31 Dec. 12/4 Selsdon man went wrong because it appeared to make the Conservative Party into a set of decimalized economic liberals. **1979** *Internat. Jrnl. Sociol. of Law* Feb. 102 'Selsdon Man' climbed into office..by exploiting the traditional staple stuff of postwar British electoral politics—prices, unemployment and speculation about the 'economy'.

selsyn (se·lsin). Also **Selsyn.** [f. SEL(F- + SYN(CHRONOUS *a.*] A kind of electric motor closely resembling a magslip and employed similarly in pairs in order esp. to transmit and receive information about the position or motion of mechanical equipment. Also *selsyn motor.*

Formerly a proprietary term in the U.S.

1926 *Official Gaz.* (U.S. Patent Office) 27 Apr. 802/1 General Electric Company, Shenectady, N.Y.... *Selsyn*... Electrical apparatus for transmitting and receiving motion. Claims use since April, 1921. **1930** *Gen. Electr. Rev.* July 378/2 A distinct possibility exists in the application of Selsyn motors to remote control. **1945** *Rev. Sci. Instruments* XVI. 132/2 The Central Scientific Company announces the availability of a limited supply of a.c. Selsyn motors. **1948** *Electronic Engin.* XX. 17 A recorder ..was modified..by removing the motor and replacing it by a gear train driven by a selsyn. **1962** J. BELL in G. A. T. Burdett *Automatic Control Handbk.* iv. 11 In the German Navy of the 1914–18 war an early form of a.c. synchronous transmission was used; this was copied by U.S.A. in the G.E. 'Selsyns', the forerunners of the present U.S.A. synchros. **1967** [see *MAGSLIP].

selvatic: see *SYLVATIC, SILVATIC *a.* 2.

selvage, selvedge, *sb.* Add: **5. b.** *Geol.* An alteration zone at the edge of a rock mass.

1934 [see *PARAGENESIS 2]. **1958** *Econ. Geol.* LIII. 292 Selvages of hydrothermal alteration of quartz monzonite along the contacts of some larger aplite-pegmatite bodies. **1979** *Nature* 5 Apr. 511/2 That some basic dykes have conspicuous charnockite selvedges might indicate metasomatic interchange between acid and basic rocks as a primary cause of the conversion, except that many dykes ..have no such marginal alteration. **1981** *Cambr. Encycl. Earth Sci.* 211/1 The majority of M[id-]O[cean]R[idge] B[asalt] lavas have a pillow form..and glass selvages—characteristics of rapid cooling in seawater.

selve (selv), *v. rare.* [f. SELF *sb.*] *intr.* (only G. M. Hopkins) and *trans.* (To cause) to become and act as a unique self. Hence **selved** *ppl. a.*, **se·lving** *vbl. sb.* Also *Comb.*, as *selved-up.*

1880 G. M. HOPKINS *Sermons & Devotional Writings* (1959) 122 Human nature, being more highly pitched, selved, and distinctive than anything in the world. *Ibid.* 123 Nothing else in nature comes near this unspeakable stress of pitch, distinctiveness, and selving, this selfbeing of my own. *Ibid.* 125 Nothing can..exercise function and determination before it has a nature to 'function' and determine, to selve and instress, with. *Ibid.*, I may treat the question from the side of my being, which is said to be compounded, selved-up, or identified with this universal mind. **1953** K. RAINE *Coll. Poems* (1956) 166 Ceasing to trouble the flowing of things with the fleeting Dream and hope and despair of this transient perilous selving. **1976** H. A. WILLIAMS *Tensions* v. 87 Around that dim and dull awareness of our identity with God we begin, gradually and instinctively, to centre and selve the rest of what we are. That centring and selving takes the whole of our life.

Sem (sem). *Egyptology.* Also **sam.** [Egyptian.] An Egyptian officiating priest who wore a distinctive robe made from a leopard's skin. Also *attrib.*, as *sem priest.* Cf. *SETEM.

1882 G. RAWLINSON *Hist. Anc. Egypt* I. xi. 438 The 'Sem',..or officiating high-priest, wore, as his costume of office, a complete leopard-skin, with head, claws, and tail. **1898** E. A. W. BUDGE tr. *Bk. Dead* cxvi. 181 The goddess Maāt is carried by the arm of him who eateth the Eye, and who is its divine judge, and the Sem priest carrieth me over upon it. **1910** *Hastings's Encycl. Relig. & Ethics* III. 538/1 Canonical part of the dress of the *sam* priest. **1927** H. CARTER *Tomb of Tut-ankh-amen* II. 28 Ay as king with royal insignia, clad in a leopard's skin of the *Sem* priest.

sema (sī·mă, sē·-). *Linguistics.* Pl. **se·mas, se·mata.** [a. G. *sema* (V. Skalička *Zur ungarischen Grammatik* (1935) 13), f. Gr. σῆμα sign.] = *SEME. Also *transf.*

1938 C. E. BAZELL in *Trans. Philol. Soc.* 112 It is only from the standpoint of their formatives that these morphemes behave like the 'cellules' (we should say the semata) of a single morpheme. *Ibid.* 113 There seems no reason to doubt that the smallest element in grammar, as Skalička holds, is the sema. **1973** D. OSMOND-SMITH tr. *Bettetini's Lang. & Technique of Film* i. 4 One must take into account..the so-called 'semas'—that is, signs whose signifieds correspond to a verbal proposition.

Semainean (sĕmai·niăn), *a. Archæol.* Also **Semainian.** [f. the name of *Semaine(h*, a village in Upper Egypt + -AN.] A term used by W. M. F. Petrie to designate the last period of predynastic culture in Egypt. Also *absol.* as *sb.*

? **1925** *Catal. Egyptian Antiquities found at Badari in 1925* 3 Approximate dates..9,000 B.C. ...Gerzean Age... 7,000 B.C. ... Semainean Age ... 5,500 B.C. ... First Dynasty. **1928** [see *GERZEAN *a.*]. **1939** W. M. F. PETRIE *Making of Egypt* vii. 55 We now reach the last of the ages, the Semainean, best represented at Semaineh, 17 miles west of Qena. **1958** V. G. CHILDE *New Light on Most Anc. East* v. 99 Petrie..postulated a 'dynastic race' and interpolated a Semainian period to contain the conquest and a Semainian culture to result from it. **1964** *Jrnl. Near Eastern Studies* XXIII. 274 Sir Flinders Petrie.. divided the Naqadian into three principal phases, named, with reference to the sites of El Amrah, Gerzeh, and Semaineh, the 'Amratian', the 'Gerzean', and the 'Semainean' and otherwise known as Naqada I, Naqada II, and Naqada III.

semal, semul, var. of SIMOOL in Dict. and Suppl.

‖ **Semana Santa** (semā·na sa·nta). [Sp.] In Spain: = HOLY WEEK.

1910 S. L. BENSUSAN *Home Life in Spain* vi. 73 The *feria* persists for a full fifteen days, starting on the Sunday before Easter and only beginning to pass when the last day of Semana Santa is a week old. **1966** E. MCGIRR *Funeral was in Spain* 56 That's Semana Santa in Spain. The week before Easter Sunday. **1979** A. SCHOLEFIELD *Point of Honour* 73 The flames and the smoke reminded me of a *semana santa* procession.

Semang (sĕma·ŋ), *sb.* and *a.* Also **Samang.** [Mal.] **A.** *sb.* (A member of) a Negrito people inhabiting the interior of the Malay peninsula. **B.** *adj.* Of or pertaining to this people.

1812 [see *POLYNESIAN *a.*]. **1814** [see *PAPUAN *sb.* 1]. **1839** T. J. NEWBOLD *Brit. Settlements in Straits of Malacca* II. xv. 377 It would appear that the Semang does not differ much in personal appearance from the Jakun. *Ibid.* 379 The Semang women..are said to be in common. **1860** MAYNE REID *Odd People* 415 The Samangs—a tribe inhabiting the mountainous parts of the Malayan peninsula —are also a negro or negrillo race. **1902** *Encycl. Brit.* XXVI. 485/2 The Vaalpens represent..a state of arrested development analogous to that of the Samangs and other Negritoes of the Malay Peninsula. **1920** R. J. WILKINSON *Hist. Peninsular Malays* (ed. 2) i. 2 The word Semang.. has come to be regarded as contemptuous. No negrito will answer to it. *Ibid.* i. 3 For practical purposes a Semang is a nomadic primitive Peninsular negrito whose numeral system stops at two. **1948** A. L. KROEBER *Anthropol.* (ed. 2) x. 424 Asia is particularly rich in tribal societies with 'internally marginal' cultures. Examples are..Palaung, Kachin, Moi, Semang, Sakai, and many others in the states of Farther India and Malaya. **1974** *Encycl. Brit. Micropædia* IX. 46/2 *Semang*, Negrito people of the Malay Peninsula... In the 1970s their population was estimated to be less than 3,000, with only about 100 in Thailand (where they are known as Ngok, or Ngo).

semanteme (simæ·ntīm). *Linguistics.* [a. F. *sémantème*, f. Gr. σημαντικός significant, after *morphème* *MORPHEME, *phonème* PHONEME.] A unit of meaning; a linguistic element which expresses a concept; = *SEMEME. Hence **semante·mic** *a.*

1925, etc. [see *MORPHEME]. **1933** *S.P.E. Tract* XXXVIII. 596 We already have plenty of synonyms for rumour, news, report, &c., to which semantemes the word *khubber* does not add any new shade of meaning. **1938** I. GOLDBERG *Wonder of Words* xvi. 324 If, in addition to its phonemic and semantemic elements, a word has another element, that element is morphemic. **1949** C. E. BAZELL in E. P. Hamp et al. *Readings in Linguistics II* (1966) 213 The inflections of tense..are naturally always determinants of the verbal semanteme. **1960** G. THOMSON *Greek Lang.* 1 The words composing a sentence may be divided into two kinds: those which express concepts ('full words' or semantemes) and those which express relationships between concepts ('empty words' or morphemes). **1977** A. SHERIDAN tr. *J. Lacan's Ecrits* iii. 63 The effects would no longer be produced, thus revealing that they do not depend even conditionally on the semanteme.

semantic, *a.* and *sb.* Add: **A.** *adj.* **2. a.** (Earlier and later examples.)

1894 E. W. FAY in *Amer. Jrnl. Philol.* XV. 433 Freedom of interchange between *r* and *l* is limited by semantic considerations. **1901** H. OERTEL *Lectures on Study of Lang.* i. 72 He was the first to distinguish clearly between the *formal* and the *semantic* side of a word. *Ibid.* v. 297 In the discussion of all semantic changes the logical aspect must be carefully kept separate from the psychological aspect. **1920** B. MALINOWSKI in *Bull. School Oriental Studies London Inst.* I. iv. 62 Sound semantic definitions valid for a wide range of linguistic types are needed before any grammatical analysis of native languages is possible. **1943** *Time* 22 Nov. 99/3 U.S. intellectuals in 1943 went out and ratified the Constitution all over again. But some of them had semantic reservations. **1968** *N.Y. Post* 15 Jan. 45/3 Each day passes with some new semantic quibble emanating from Washington. **1976** J. S. GRUBER *Lexical Structures in Syntax & Semantics* 1 We would acknowledge the necessity for interpretive semantics of some sort (e.g. a semantic calculus), but not one based on the interpretation of words and phrases.

b. In weakened uses.

1959 W. R. FISHEL in *New Leader* 2 Nov. 13/1 We do ourselves and our Asian neighbors a distinct disservice when we insist on stretching them or shrinking them to fit our particular semantic bed. **1971** L. KOPPETT *N.Y. Times Guide Spectator Sports* ii. 41 Lesson No. 1 must be clung to through all the semantic storms.

B. *sb. pl.* **1. a.** Also, (the study or analysis of) the relationships between linguistic symbols and their meanings. Const. as *sing.* and *pl.* (Earlier and further examples.)

Now the usual word in this sense.

[**1883** M. BRÉAL in *Études Grecques en France* XVII. 133 Cette étude..nous l'appellerons la *Sémantique*..c'est-à-dire la science des significations.] **1893** E. WILLIAMS tr. M. Bréal in *Trans. Amer. Philol. Assoc.* XXIV. 27 All, or almost all, the chapter of linguistics treating of Semantics, or the science of meanings, has yet to be written. **1895** C. R. LANMAN in *Ibid.* XXVI. p. xi, The doctrine of the principles that underlie the processes of the development of the meanings of words may be called semantics or semasiology. **1912** E. WEEKLEY *Romance of Words* 79 The convenient name semantics has been applied of late to the science of meanings, as distinguished from phonetics, the science of sound. **1920** B. MALINOWSKI in *Bull. School of Oriental Studies London Inst.* I. iv. 35 All these works..are résumés of the present state of linguistics, and they reflect the insufficient attention hitherto given to Semantics. **1933** L. BLOOMFIELD *Language* viii. 138 When the phonology of a language has been established, there remains the task of telling what meanings are attached to the several phonetic forms. This phase of the description is *semantics.* It is ordinarily divided into two parts, *grammar* and *lexicon.* **1941** J. RANSOM *New Criticism* i. 5 *The Meaning of Meaning* is in terms of the new philosophy of language; the authors refer to the latter as Symbolism, but since their book the name of it appears to have become standardized as Semantics. **1952** *Economist* 21 June 813/2 Professor Hayakawa says nothing about..the importance of semantics in the determination of word-origins and word-history. **1964** E. A. NIDA *Toward Sci. Transl.* iii. 35 While semantics deals with the relationship of symbols to referents, syntactics is concerned with the relationship of symbol to symbol. **1972** HARTMANN & STORK *Dict. Lang. & Linguistics* 204/2 *Linguistic semantics* has studied meaning more in terms of the connexions between speech acts and the physical and intellectual environment of the speaker. **1980** *Times Lit. Suppl.* 7 Mar. 268/1 Frege's goal was not to provide a semantics for natural language as he found it.

b. In weakened uses.

1944 M. RYSKIND in *Sat. Rev. Lit.* 23 Dec. 4/1 The technique of character-assassination instead of arguments is..standard totalitarian semantics. **1966** *N.Y. Post* 3 Aug. 6/4 Sen. Pastore said that everybody was engaged in semantics. 'It comes down to a very fine point,' he said, stating the obvious in a nutshell. **1978** K. HUDSON *Jargon of Professions* 16 Almost daily in the press briefing, whenever a newsman raises his hand to ask for clarification of some mealy-mouthed statement: 'I am not going to debate semantics with you,' the spokesman replies.

2. *transf.* The interpretation of signs in general.

1946 C. MORRIS *Signs, Lang. & Behavior* viii. 219 *Semantics* deals with the signification of signs in all modes of signifying... When so conceived, pragmatics, semantics, and syntactics are all interpretable within a behaviorally oriented semiotic. **1962** *Listener* 11 Jan. 70/2 Exposure to art, erroneous notions mixed with some accurate ones of history, the private struggle with semantics and meaning. **1970** G. GREER *Female Eunuch* 33 The notion of a curve is so closely connected to sexual semantics that some people cannot resist sniggering at road signs.

C. Special collocations. **semantic aphasia** *Path.*, disturbance in understanding the significance of any but the simplest forms of words or speech caused by disorder in the cerebral cortex; **semantic differential** *Psychol.*, a technique devised to measure the distribution of meaning that a person attaches to a concept, by rating descriptive words selected by him from an evaluated list; a scale or test to achieve this; **semantic paradox** *Logic*, a paradox caused by ambiguity of meaning in the language of a statement, rather than by its logical reasoning; **semantic poetry** (see quot. 1969).

1926 H. HEAD *Aphasia* II. 259 A case of Semantic Aphasia. **1958** *Lang. & Speech* I. 26 This symptom, appearing distinctly in cases where the affection damages the most complex and most recently formed zones of the parieto-occipital region at its border with the temporal region, constitutes a basic symptom of so-called 'semantic aphasia'. **1974** L. F. SIES *Aphasia Theory & Therapy* i. 51 Semantic aphasia produces an inability to perceive the complex relationships by which language classifies separate concepts. **1953** C. E. OSGOOD *Method & Theory in Exper. Psychol.* xvi. 713 The distribution of his [*sc.* the subject's] judgments on a standardized series of such scales serves to differentiate the meaning of this concept from others; for this reason this measuring instrument has been called a 'semantic differential'. **1962** *Listener* 11 Jan. 62/1 Identification with parents was measured by the similarity of children's description of their parents, and of the kind of person they would most like to be themselves, on a series of seven-point rating scales known as the 'semantic differential'. **1962** U. WEINREICH in Householder & Saporta *Probl. Lexicogr.* 26 Semantic-differential tests. **1979** T. SHAPIRO *Clinical Psycholinguistics* iii. 26 Procedures such as the semantic differential offer valid and important experimental properties to understand meaning. [**1939** *Mind* XLVIII. 358 This semantico-empirical paradox can easily be solved by the ramified theory of types without using the simple theory of types. **1948** H. C. BRODIE tr. *Chwistek's Limits of Sci.* ii. 40 Logical paradoxes must be distinguished from semantical paradoxes.] **1960** P. ZIFF *Semantic Anal.* iv. 134 The fact that the semantic paradoxes can be formulated in English has led some philosophers, primarily logicians, to the conclusion that English is in a muddled state. **1978** T. J. SMILEY in F. P. Ramsey *Foundations* 8 Ramsey transforms the problem by drawing the now standard distinction between the logical and the semantic paradoxes. **1949** S. THEMERSON (*title*) Bayamus and the Theatre of Semantic Poetry. **1951** *Times* 5 Apr. 5/1 Nothing could prevent Mr. Vyshinsky or Mr. Acheson from discussing Etruscan pottery or semantic poetry if they really wished to do so. **1969** *Poetry Rev.* LX. 274 Semantic Poetry is based on the idea that words such as *moon, night, heart, flower,* etc., having become clichés have become devalued and devoid of affective effect. SP avoids all form of rhetorical device and relies upon a text derived from traditional language by replacing each word by its dictionary definition... Semantic Poetry does not rarefy the verbal material to condense the meaning.

Hence **sema·ntical** (also **-ly** *adv.*) *a.*; **semanti·cian, sema·nticist,** a student of semantics; **semanti·city,** the quality of being semantic or possessing meaning derived from signs.

1895 M. BLOOMFIELD in *Amer. Jrnl. Philol.* XVI. 409 Every word, in so far as it is semantically expressive, may establish, by hap-hazard favoritism, a union between its meaning and any of its sounds. **1917** *Jrnl. Eng. & Gmc. Philol.* XVI. 472 The professional semanticist is visualized in his work-shop,..feverishly fingering the leaves of a host of lexical tomes, standard and dialect, old, middle and new. **1921** H. E. PALMER *Princ. Lang.-Study* 62 The lexicologist or semantician will study the meanings. **1926** C. M. DOKE *Phonetics of Zulu Lang.* 217 (*caption*) Words semantically alike but differing in tone. **1936** *Mind* XLV. 272 Chwistek, a semanticist and metamathematician noted for his work on the theory of types. **1941** J. RANSOM *New Criticism* iv. 282 All discourse consists in signs.., there is the *semantical* dimension proper, involving the reference of a sign to an object. **1960** *Sci. Amer.* Sept. 90/2 The dog's panting [does not] exhibit the design-feature of 'semanticity'... The calls of gibbons.. possess semanticity. **1960** *Economist* 15 Oct. 251/1 In Natal..there has been some talk of secession—semantically disguised because secession today bears a stigma it did not carry ten years ago. **1973** J. M. ANDERSON *Structural Aspects Lang. Change* 186 Many of the outstanding semanticists have been optimistic about finding some kind of regularity behind semantic processes. **1975** *Language* LI. 207 This distinction [between competence and performance]..is constantly attacked..by the 'semanticians'. **1975** I. ROBINSON *New Grammarians' Funeral* viii. 165 He says that music hasn't semanticity because 'Vivaldi's *Four Seasons* is stylized away from the reality it pretends to imitate'. **1978** C. HOOKWAY in Hookway & Pettit *Action & Interpretation* 26 The semantical and intentional discourse of the subjects provides an additional control.

semanticize (simæ·ntisəiz), *v. Linguistics.* [f. SEMANTIC *a.* + -IZE.] *trans.* To invest (something) with meaning; also, to analyse semantically. So **sema·nticized** *ppl. a.*, **sema·nticizing** *vbl. sb.*

1942 *Sat. Rev. Lit.* 10 Jan. 14 (*heading*) Semanticizing. **1961** WEBSTER, Semanticize. **1964** P. MEADOWS in I. L. Horowitz *New Sociol.* 450 The straitlaced purity of semanticized communication-theory. **1976** *Times Lit. Suppl.* 31 Dec. 1631/2 Lotman appears to be over-eager to semanticize, in his own words, 'any element on the parole level'.

semantico- (simæ·ntiko), also **sema·nto-,** combining form f. Gr. σημαντικός of SEMANTIC *a.*, used with adjs. and advbs. in sense 'semantic(ally) and ..'.

1932 W. L. GRAFF *Lang. & Languages* 277 All linguistic change is a process pertaining to..semanto-phonetic expression. *Ibid.* 420 Archaic Chinese seems to have had a system of..semanto-phonetically changeable radicals. **1939** Semantico-empirical [see *semantic paradox* s.v. *SEMANTIC *a.* C.]. **1971** *Language* XLVII. 80 These structures are semanto-syntactic, which means that the semantic properties or bundles of properties are arranged not in a linear order but in a hierarchical one. **1976** *Word* *1971* XXVII. 268 Semantico-intentionally there are two stages of development. **1977** P. STREVENS *New Orientations Teaching of English* ii. 25 *Semantico-grammatical categories*—expressing universal concepts of time, quantity, space and matter, as well as expressing grammatical concepts of *case* ('who did it, who it happened to, and what got changed').

semantron (sĭmæ·ntrən). *Gr. Orthodox Ch.* Also **simandro, simantron**; pl. **semantra.** [med. Gr. use of Gr. σήμαντρον sign, mark.] A wooden or metal bar struck by a mallet used to summon worshippers to service.

1849 R. CURZON *Visits to Monasteries of Levant* p. i, Interior of the Court of a Greek Monastery. A monk is calling the congregation to prayer, by beating a board called the simandro..which is generally used instead of bells. **1850** J. M. NEALE *Hist. Holy Eastern Ch.* I. II. ii. 217 The word *semantra*..properly signifies..the instruments ..by which the people were called together before bells were introduced into the east... They are of two kinds, wooden and iron. The wooden semantron is generally a long, well planed piece of timber. **1912** W. G. HOLMES *Age Justinian & Theodora* I. i. 110 At the boom of the great *semantron*..the various congregations issue forth to attend their respective places of worship. **1939** *Archit. Rev.* LXXXV. 289/1 Round the church..is a broad path along which a monk walks summoning the faithful to prayer by hammering on a simantron, a long piece of wood which he balances in one hand. **1958** *Times Lit.*

Suppl. 10 Oct. 581/5 The simantron..which summons the monks of Athos to prayer is a wooden not a brass instrument.

semaphore, *sb.* Add: **2.** This method of signalling; *spec.* a system for conveying messages by a code whereby the arms are moved through certain positions in a vertical plane relative to the body.

1904 [in Dict., sense 1 b]. **1918** E. S. Farrow *Dict. Mil. Terms* 544 *Semaphore*, a method of signalling in which the letters depend on the position of one or both arms in relation to the body. When sending semaphore, the signaller always faces the distant station. **1975** *Scout Handbk.* (1976) 124/1 Semaphore uses a different position of the arms for each letter.

semaphore *v.*: also *fig.*

1957 R. Campbell *Coll. Poems* II. 32 The lonely hamlets semaphore their loss. **1981** *Economist* 24 Jan. 88/1 American firms are nervously semaphoring price rises after a strong recovery at the end of 1980.

semasiology. (Earlier example.)

[*a* **1829** C. K. Reisig *Vorlesungen über Lateinische Sprachwissenschaft* (1839) II. 286 (*heading*) Semasiologie oder Bedeutungslehre.] **1847** J. W. Gibbs *Philol. Studies* (1857) 18 The development of intellectual and moral ideas from physical, constitutes an important part of *semasiology*, or that branch of grammar which treats of the development of the meanings of words.

semasiolo·gic *a.* = Semasiological *a. rare.*

1909 L. Bloomfield in *Mod. Philol.* VII. 248 A number of examples are here given of secondary Germanic ablaut forms exhibiting a..feature of semasiologic differentiation.

sematic, *a.* Restrict *Biol.* to sense in Dict. and add: † **2.** = Semantic *a. Obs. rare*$^{-1}$.

1855 J. W. Powell in *Trans. Anthropol. Soc. Washington* III. 189 While in the present state of knowledge it is perhaps not possible to set forth clearly the resultant sematic and structural effects upon any language, in linguistic arts important effects are discovered.

sematology. Add: Hence **sematolo·gical** *a. rare.*

1882 J. A. H. Murray *Let.* 27 Mar. in K. M. E. Murray *Caught in Web of Words* (1977) x. 190 All that you urge against phonetic statements, can be urged with far greater force against sematological ones.

semblable, *sb.* **2.** Delete † *Obs.* and add later examples in revived use.

1922 Joyce *Ulysses* 377 It behoves every most just citizen to become the exhortator and admonisher of his semblables. [**1923** T. S. Eliot *Waste Land* i. 8 You! hypocrite lecteur!—mon semblable,—mon frère!] **1941** V. Woolf *Between Acts* 242 There was Dodge, the lip-reader, her semblable, her conspirator. **1979** *Dædalus* Summer 30 These thoughts picture Othello as, in various ways, a semblable of yours.

semble, *a.* For † *Obs.* read *Obs.* (exc. *arch. poet.*) and add later example.

1965 Auden *About House* (1966) 40 Six lenient semble sieges, None of them perilous, Is now a Perfect Social Number.

sembling (se·mbliŋ), *vbl. sb. Ent.* Also symbolling. [See Semble *v.*[1]] The coming together of a male and a female moth; *spec.* a method of trapping male moths by using a captive female to attract them.

1748 J. Dutfield *New Nat. Hist. Engl. Moths & Butterflies* s.v. Emperor Moth, What is called Symbolling, or, the Coming together, is particularly observable of this Species. **1870** [see Semble *v.*[1] 1 b]. **1894** *Science* 23 Mar. 156/2 The sembling of a large native moth. **1924** *Contemp. Rev.* Sept. 364 Collectors of lepidoptera have long known the trick of 'sembling' to obtain a large series of males of certain moths.

seme (sīm). *Linguistics.* [ad. Gr. σῆμα sign: cf. *sema.] **a.** A sign. **b.** A unit of meaning; *spec.* the smallest unit of meaning. Cf. *pheme, *rheme.

a **1866** J. Grote in *Jrnl. Philol.* (1872) IV. 158 The noematism of it [*sc.* a language] might be..written..by symbols naturally suggesting themselves for the visual percepts, and by others agreed upon for abstract terms and those related to other senses than the eye. It might be well to call such quasi-writing or exhibition to the eye by other than vocal elements, *sematism*, and the symbols *semes*. **1906** C. S. S. Peirce in *Monist* XVI. 506 By a *Seme*, I shall mean anything which serves for any purpose as a substitute for an object of which it is, in some sense, a representative or Sign... The term 'The mortality of man' is a Seme. **1923**, **1931** [see *pheme]. **1951** E. A. Nida in *Word* VII. 5 The simple term *seme* identifies any minimal feature of meaning. **1966** D. G. Hays in *Automatic Transl. of Lang.* (NATO Summer School, Venice, 1962) 163 These semes are more nearly the units wanted in translation than the morphs or morphemes that comprise them. **1973** *Screen* Spring/Summer 18 [Eric Buyssens]..established a certain number of notions and distinctions (*seme* and *semic act, intrinsic* and *extrinsic semes, direct* and *substitutive semes*).

semeiological, *a.* (Later examples in sense 3 of *semeiology.)

1932 W. L. Graff *Lang. & Languages* 303 What is called change of meaning is essentially a semeiological phenomenon. **1968** Jacobson & Schoepf tr. *Lévi-Strauss's Structural Anthrop.* xvii. 364 Anthropology aims to be a semeiological science, and takes as a guiding principle that of 'meaning'. **1968** *Listener* 25 Jan. 122/1 The wide-awake analogue to this use of television is the employment of the medium for basic communication—infralinguistic and, if semiological at all, concerned with the most primitive human signals. **1973** D. Osmond-Smith tr. *Bettelini's Lang. & Technique of Film* i. 1 The basic material that the Barthes group take as a starting-point for their semiological and aesthetical observations is not their own. **1978** *Guardian Weekly* 5 Feb. 21/4 In semiological parlance, the film's lacunae often seem rather more important than what's actually happening on screen.

semeiology. Add: Sense 1 is now *obs.* The form **semiology** (sīmi,ǫ·lŏdʒi) is now usual. **3.** The branch of science concerned with the study of linguistic signs and symbols. Also in extended use.

[**1916** F. de Saussure *Cours de Linguistique Générale* iii. 34 On peut donc concevoir une science qui étudie la vie des signes au sein de la vie sociale; elle formerait une partie de la psychologie sociale, et par conséquent de la psychologie générale; nous le nommerons *sémiologie* (du grec *sēmeîon* 'signe').] **1923** Ogden & Richards *Meaning of Meaning* i. 8 The initial recognition of a general science of signs, 'semiology', of which linguistic would be a branch, was a very notable attempt in the right direction. **1932** W. L. Graff *Lang. & Languages* 72 Semeiology, the science of signs and symbols, is only in its infancy. **1947** *Word* III. 29 [According to de Saussure] there is a science of semiology, hitherto unrecognized... This semiology is differentiated by definition from semantics. **1959** *Times Lit. Suppl.* 20 Nov. 669/2 Joyce has become an inexhaustible hunting ground for hermeneutical exegesists to whom semeiology..is a science beside which plain criticism offers no excitements. **1967** *Economist* 14 Oct. 156/2 The tired businessman who refreshes himself with courses like 'Structural Linguistics, Semiology and Criticism'. **1972** *Times Lit. Suppl.* 21 July 833/1 He has written..about literature and about the semiology of the cinema. **1976** T. Eagleton *Crit. & Ideology* v. 166 Literature must indeed be re-situated within the field of general cultural production; but each mode of such production demands a semiology of its own.

Hence **sem(e)io·logist.**

1973 D. Osmond-Smith tr. *Bettelini's Lang. & Technique of Film* i. 3 There exists a certain confusion in the use of terms recently coined by semiologists. **1975** *Listener* 20 Mar. 367/3 Though the 'synchronic' approach of the semiologists is for the moment more fashionable, it is impossible not to be interested in the history of social myths. **1979** *Dædalus* Summer 111 It has proved much more elusive to disclose the overall intention underlying their visual assembly in any way that an anthropologist or semiologist would recognize as coherent.

semeiosis, var. *semiosis.

semeiotic, *a.* Add: The form **semiotic** (sīmi,ǫ·tik) is now usual. **3.** Of or pertaining to semiotics or the use of signs. Cf. *semeiotics 2.

1923 H. G. Baynes tr. *Jung's Psychol. Types* i. 82, I say 'semiotic' in contradistinction to 'symbolic'. What Freud terms symbols are no more than *signs* for elementary instinctive processes. **1957** *Publ. Amer. Dial. Soc.* xxviii. 4 It is an utterance that 'craves' a verbal or other semiotic (e.g., a nod) response. **1973** Akhmanova & Marčenko *Meaning Equivalence & Linguistic Expression* 7 The Morse code is a semiotic system *par excellence*, for in it every unit of content and every unit of expression are in regular one-to-one correspondence... The same applies..to all the other semiotic systems such as, for instance, notation in music, or chemical formulae, or mathematical signs. **1974** S. Morawski *Inquiries into Fundament. of Aesthetics* viii. 302 The fourth approach..considers the artistic communication itself and its semiotic connections. **1978** J. Updike *Coup* (1979) vii. 257 No doubt this semeiotic treasure-lode [*sc.* a wallet] enriches the arcana of some light-fingered ex-nomad.

B. *absol.* as *sb.* = *semeiotics 2.

[**1690** Locke *Hum. Und.* iv. xx. 361 The third Branch may be called σημιωτικὴ, or the *Doctrine of Signs*, the most usual whereof being words, it is aptly enough termed also λογικὴ, Logick.] *c* **1897** C. S. Peirce *Coll. Papers* (1932) II. ii. ii. 134 Logic, in its general sense, is..only another name for *semiotic* (σημειωτική), the quasi-necessary, or formal, doctrine of signs. **1937** C. Morris *Logical Positivism, Pragmatism, & Sci. Empiricism* 4 Semiotic being the general science which includes all of these [dimensions] and their interrelations. **1953** F. J. Whitfield tr. *Hjelmslev's Prolegomena to Theory of Lang.* 76 The so-called metalanguage (or, we should say, *metasemiotic*), by which is meant a semiotic that treats of a semiotic. **1973** R. Jakobson *Sci. of Lang.* ii. 32 The subject matter of semiotic is the communication of any messages whatever, whereas the field of linguistics is confined to the communication of verbal messages.

Hence **sem(e)ioti·cian.**

1946 C. Morris *Signs, Lang., & Behavior* i. i. 4 At some point the semiotician must say: 'Henceforth we will recognize that anything which fulfills certain conditions is a sign.' **1946** *Mind* LV. 46 Other groups of workers in the same field, as, for example..the Semioticians (*e.g.* Carnap, Morris). **1960** H. Read *Form of Things Unknown* i. ii. 34 In general, semioticians have confined themselves to the study of the various types of discourse which make use of language. **1976** *Visible Language* X. 68 It is possible, in the case of some 'auto-illustrations', to follow those semioticians who prefer to view iconic motivation as a special case of metonymic *pars pro toto*.

semeiotical, *a.* Add: **2.** = *semeiotic *a.* 3.

1938 C. Morris in *Internat. Encycl. Unified Sci.* I. ii. 29 'Rules of sign usage', like 'sign' itself, is a semiotical term and cannot be stated syntactically or semantically. **1946** *Mind* LX. 146 As a semiotical psychiatrist, a Therapeutic Positivist has to hand a technique for the resolution of philosophical problems and disputes.

Hence **sem(e)io·tically** *adv.*

1916 C. E. Long tr. *Jung's Analytical Psychol.* p. vii, The Vienna School interprets the psychological symbol semiotically, as a sign or token of certain primitive psychosexual processes. **1972** W. C. Stokoe *Semiotics & Human Sign Lang.* i. 15 Semiotically considered the difference between fingerspelling and a sign language could hardly be greater.

semeiotics. Add: The form **semiotics** (sīmi,ǫ·tiks) is now usual. **2.** The science of communication studied through the interpretation of signs and symbols as they operate in various fields, esp. language (see *semeiotic *sb.* for parallel form). Cf. *semeiology.

1880 G. Mallery *Introd. Study Sign Lang. among N. Amer. Indians* 4 Our native semiotics will surely help the archæologist in his study of native picture-writing. **1911** A. M. Ludovici tr. *F. Nietzsche's Antichrist* in *Twilight of Idols* xxxii. 169 One should guard against seeing anything more than a language of signs, semeiotics, an opportunity for parables in all this. **1955** A. Huxley *Genius & Goddess* 42 He kissed her—kissed her with an intensity of passion..for which the semiotics and the absent-mindedness had left her entirely unprepared. **1964** T. Sebeok et al. *Approaches to Semiotics* 5 Margaret Mead proposed semiotics..as a term which might aptly cover 'patterned communication in all modalities'. **1973** D. Osmond-Smith tr. *Bettelini's Lang. & Technique of Film* i. 2 Some talk of a universal semiotics, capable of including within itself all aspects of the film as sign-system. **1980** *Semiotica* XXIX. 185 (*heading*) A firework for the semiotics of visible human action.

sememe (sī·mīm). *Linguistics.* [f. Gr. σῆμ-α sign + *-eme.] A unit of meaning; *spec.* the smallest unit of meaning. Cf. *seme.

[**1904** A. Noreen *Vårt Språk* V. iii. 6 Detta sålunda definierade abstrakta begrepp 'betydelse' uppträder in concreto alltid i någon viss språklig dräkt, som ofta kan efter omständigheterna vara mycket varierande, och det kallas då semem, hvilken term alltså kan definieras som: ett visst bestämdt idéinnehåll, uttryckt i någon språklig form, likgiltigt hvilken. **1911** A. L. Elmquist in *Jrnl. Eng. & Gmc. Philol.* X. 321 Semology [in Noreen's *Vårt Språk*] deals with the 'semem' and Morphology with the 'morfem'.] **1913** *Jrnl. Eng. & Gmc. Philol.* XII. 87 This need not have been the only or even perhaps the usual sememe associated with the morpheme *flik* in O. Scand. **1926** L. Bloomfield in *Language* II. 157 The sememes..which stand in one-to-one correspondence with the morphemes, cannot be further analyzed by linguistic methods. **1933** —— *Language* x. 162 The meaning of a morpheme is a *sememe*. The linguist assumes that each sememe is a constant and definite unit of meaning, different from all other meanings, including all other sememes, in the language. **1949** [see *contrastiveness]. **1954** C. E. Bazell in E. P. Hamp et al. *Readings in Linguistics II* (1966) 329 At the European Semantic Conference, held at Nice in March, 1951..it was agreed to call the fundamental unit of content (in the linguistic norm) the *sememe*. **1965** N. Chomsky *Aspects of Theory of Syntax* 230 We could perfectly well state the facts..in terms of such new notions as 'semantic subject', 'semantic object', various kinds of 'sememes', etc., but such proliferation of terminology contributes nothing toward clarifying the..issues. **1973** *Screen* Spring/Summer 234 Greimas analyses the units of content (lexemes) into constant minimal semantic nuclei (semes) and contextual semantic units (sememes).

Hence **seme·mic** *a.*, **seme·mically** *adv.*

1953 C. E. Bazell *Linguistic Form* 81 The sememic level is approached when several congruent morphemes of this type are regarded as a single unit, for commutation is then usually possible. *Ibid.*, Sememically the unit may normally be taken as zero. **1968** P. M. Postal *Aspects Phonol. Theory* viii. 198 The four current properly linguistic strata are, in hierarchical order from 'top' to 'bottom', the sememic, the lexemic, the morphemic, and the phonemic. **1973** *Archivum Linguisticum* IV. 119 By a 'semantic base' Hutchins means a set of 'sememic graphs'.

semen. Add: **2.** *attrib.*, as **semen bank**, a store of semen which is kept available for artificial insemination.

1954 *Fertility & Sterility* V. 28 A semen bank could be organized and maintained, with the same scientific attitude and spirit of research as is found in all other therapies for helping the infertile. **1972** *Sat. Rev.* (U.S.) 10 June 40/3 Semen banks..are gaining popularity among vasectomy candidates as a solution to the irreversibility problem.

semi[2] (se·mi). Colloq. abbrev. of: **1.** Semi-detached house.

1912 R. Macaulay *Views & Vagabonds* xvi. 292 To settle down in the new semi and 'do for' her Ben,..what had heaven to offer more than that? **1939** R. Ferguson in *Queen's Bk. of Red Cross* 234, I wish my girl could see it, but..it might spoil her for a little semi on a housing estate. **1958** J. Symons *Gigantic Shadow* ii. 8 These were the moments when the watchers in their suburban semis wriggled most deliciously in their overstuffed armchairs. **1977** B. Pym *Quartet in Autumn* iii. 28 That house which, in the estate agents' language, was on its way to becoming a 'twenty thousand semi'.

2. *semi-evening dress* s.v. *semi- 8. Now *rare.*

1927 N. Coward *Sirocco* ii. i. 38 Miss Johnson is wearing what might technically be described as a 'semi'. Mrs Breeze is in a tea-gown. **1934** C. L. Anthony *Touch Wood* ii. ii. 57 Mrs McCrossan and Mr Berridge enter. Mrs McCrossan is in her semi.

3. *semifinished* (steel): see *semi- 8. Usu. in *pl. Industry.*

1931 *Economist* 5 Dec. 1054/2 If Continental exporters of 'semis'..feel that a high tariff is likely to be imposed in the..Budget, steel imports..are likely to increase. **1950** *Engineering* 21 July 59/2 Home produced steel semies are reaching the re-rolling mills in very large quantities. **1975** 'D. Jordan' *Black Account* xii. 62 Have you ever thought that the next stage would be a semis plant?

4. *semi-final* s.v. Semi- 8.

1942 Berrey & Van den Bark *Amer. Thes. Slang* § 643/6 Next to last event... *semi.* **1976** *Star* (Sheffield) 29 Oct. 2/5 Last heat of the TV sheepdog trials and it's no secret that Scotland meet England West in the first semi.

5. *semi-trailer; *semi-truck* s.v. *semi- 8. *U.S.* (with pronunc. se·mai) and *Austral.*

1942 Berrey & Van den Bark *Amer. Thes. Slang* § 81/16 *Semi*,..a two-wheeled truck. **1961** *Amer. Speech* XXXVI. 272 It is common to call..a trailer a *semi*. **1968** J. M. Ullman *Lady on Fire* (1969) viii. 102 They..rolled past a row of semis lined up at an unloading dock. **1969** *Northern Territory News* (Darwin) *Focus* '69 97/2 The Stuart Highway..has become increasingly busy with road trains and semis to meet the needs of the increased population. **1976** B. Bova *Multiple Man* (1977) iv. 49 The Beltway..was almost always jammed with..heavy semis delivering the daily bread.

6. *semi-submersible* (rig, barge, etc.): see *semi- 8. *Oil Industry.*

1975 *Offshore* Aug. 19/1 We will operate three semis of the Aker H-3... These rigs are well published in various publications. **1977** *Offshore Engineer* Apr. 51 (*caption*) Drillships closely follow semis in the oversupply stakes.

semi-, *prefix.* Add: The pronunc. (se·mai) is common in the U.S. **I. 1. a.** Compounded with adjs. and pples.: (further examples, including some used *ellipt.* as *sbs.*); *semi-rural* (earlier example), *-servile* (earlier example), *-social* (later example).

1871 Bagehot in *Fortn. Rev.* 1 Aug. 158 A semi-abstract discussion of practical topics. **1974** K. Clark *Another Part of Wood* ii. 52 From behind the semi-abstract clouds there appeared the same sexy girls. **1963** W. R. Rose in Wyndham Lewis *Lett.* 558 René Harding, the uncompromising, semi-autobiographical hero of *Self Condemned*. **1926** J. M. Keynes *End of Laissez-Faire* iv. 41, I suggest, therefore, that progress lies in the growth and the recognition of semi-autonomous bodies within the State. **1974** tr. *Wertheim's Evolution & Revolution* 371 This semi-autonomous activity by living beings..is increasingly determining the course of nature on earth. **1952** C. P. Blacker *Eugenics: Galton & After* 178 The countries of Asia are rapidly emerging from their colonial or semi-colonial status. **1974** tr. *Wertheim's Evolution & Revolution* 67 The colonial or semi-colonial backyards which capitalism..was exploiting in such a way as to provoke strong popular counter-forces. **1965** *Math. in Biol. & Med.* (Med. Res. Council) i. 7 A service for the collection and recording of this information is also being made available on a semi-commercial basis to any hospital in the United States. **1839** De Quincey in *Tait's Edin. Mag.* Sept. 573/2 This..semi-conscious feeling..taught them to feel the extremity of their danger. **1977** J. D. Douglas in Douglas & Johnson *Existential Sociol.* i. 22 Most of our experiences in everyday life are only semiconscious. **1937** *Discovery* Sept. 277/2 The comfort and interest with which the [television] pictures may be viewed in a semi-darkened room. **1680** T. Browne *Let.* 6 Sept. (1946) 181 Esquire Mildmay..a melancholy & semi-delirious person, yet fayre condition'd. **1975** M. Amis *Dead Babies* xiii. 67 His strangled shout had been a semi-delirious reply to Quentin's courtly knock. **1942** E. Partridge *Usage & Abusage* 91/2 *Conditional clauses* have always caused trouble to the semi-educated and the demi-reflective. **1954** Koestler *Invisible Writing* iv. xxxiv. 371 There is a character in *The Magic Mountain*, the semi-educated Frau Stoehr, who is always trying to be refined. **1920** T. S. Eliot *Sacred Wood* 49 You see..how completely any semi-ethical criterion of 'sublimity' misses the mark. **1943** V. Sackville-West *Eagle & Dove* ii. ix. 138 A semi-experimental perception of God, in very varying degrees of intensity and clarity. **1938** H. Nicolson *Diary* 22 Aug. (1966) 356 Russia has no sympathy for the semi-fascist systems established in the Balkans. *a* **1974** R. Crossman *Diaries* (1977) III. 299 This semi-fascist reaction has been accompanied by a general lack of credibility in the whole establishment. **1956** *Railway Mag.* Feb. 113/2 When travelling from Cologne to Wiesbaden in a semi-fast, we left Cologne 20 min. late. **1937** Koestler *Spanish Testament* iii. 63 Spain was still, when the People's Front came into power in 1936, a semi-feudal country, with sharp social contrasts. **1970** C. Furtado in I. L. Horowitz *Masses in Lat. Amer.* ii. 31 As an instrument for domination over a society where some forms of semi-feudal decentralization prevailed, the State emerged in colonial times as a strong bureaucracy. **1960** *Farmer & Stockbreeder* 26 Jan. 66/2 Any kind of fixed or semi-fixed equipment..will be considered. **1906** E. Johnston *Writing & Illuminat. & Lettering* xv. 317 The writing approaches the *stylographic*... It may conveniently be termed *Semi-formal.* **1977** *Stornoway Gaz.* 27 Aug. 1/7 The possibility of formal or semi-formal competitions could not be discounted. **1947** A. Koestler in *Partisan Rev.* XIV. 144 The unions become more and more absorbed into semigovernmental, managerial functions. **1950** B. Russell *Unpopular Essays* iv. 64 A rare moment of self-knowledge must have inspired the initial aphorism, which was made bearable to its author by its semi-humorous form. **1851** Mill *Logic* (ed. 3) II. v. v. 352 As the former is the error of sheer ignorance, so the latter is especially that of semi-instructed minds. **1955** J. Burnaby *Christian Words & Christian Meanings* iii. 62 The same tendency in modernised form is to be observed in much semi-instructed Christian thinking of our own time. **1921** F. Scott Fitzgerald *Let.* 25 Aug. (1964) 148 I'm sick of the flabby semi-intellectual softness in which I flounder with my generation. **1957** J. Kerouac *On Road* (1958) ii. 176 In Frisco eager crowds of young semi-intellectuals sat at his feet and listened to him. **1851** H. Melville *Moby Dick* III. xl. 230 Most mariners cherish a very superstitious feeling about seals, arising..from..the human look of their round heads and semi-intelligent faces. **1949** Koestler *Promise & Fulfilment* i. viii. 90 Haganah had a kind of semi-legal status which varied according to the political constellation. **1979** *Dædalus* Winter 157 The strength of the 'second' or 'parallel' market within the economy—in reality an entire spectrum of legal, semilegal, and illegal markets. **1909** O. Lodge *Survival of Man* vi. 86 Public performances..often tend to obscure a phenomenon by covering it with semi-legitimate contempt. **1927** *Mod. Philol.* Nov. 221 Even the semi-literate speaker actually obtains some of his speech material by linguistic borrowing from written records. **1957** E. Partridge *English gone Wrong* i. 5 The best English of the semi-literate. **1976** P. Cave *High Flying Birds* i. 10 Those who have perused my semi-literate journals masquerading as novels in the past may be familiar with some of my previous exploits. **1924** W. B. Selbie *Psychol. Relig.* 269 The persistence of the belief in such a semi-material ghost soul is a most interesting fact. **1924** *Econ. Jrnl.* XXXIV. 346 Meanwhile I got a good deal interested in the semi-mathematical side of pure Economics. **1902** W. James *Var. Relig. Exper.* iii. 69 Here is the abridged record of another mystical or semi-mystical experience. **1951** S. Ullmann *Princ. Semantics* iii. 158 The somewhat abstruse and occasionally semi-mystical jargon in which his views are often couched has given rise to a number of misinterpretations. **1962** H. R. Loyn *Anglo-Saxon England* i. 26 The semi-mythical island of Brittia, to which..the souls of the dead were ferried. **1962** Y. Malkiel in Householder & Saporta *Probl. Lexicogr.* 5 Perspective essentially involves the deliberate or semi-naive attitude of the collector toward the chosen slice of material. **1890** W. James *Princ. Psychol.* II. xxiv. 420 In ordinary fear, one may either run, or remain semi-paralyzed. **1922** Joyce *Ulysses* 302 The semiparalysed *doyen* of the party..had to be assisted to his seat by the aid of a powerful steam crane. **1973** *Morning Star* 16 Jan. 5/3 There [is] a much better and wider distribution of shops selling semi-prepared foods [in Moscow now]. **1931** E. H. Morris *Temple of Warriors* i. 5 A more than life-size sculpture of a semireclining human figure. *a* **1974** R. Crossman *Diaries* (1976) II. 742, I am semi-retired. **1835** Dickens *Let.* (1965) I. 56 You would prefer living in Chambers to remaining in your present semi-rural tranquility. **1930** Auden *Poems* 75 And I, stung by the sun, Think, semi-satisfied That, 'ere the smile is done, The eye deliberate May qualify the joy. **1954** S. Duke-Elder *Parsons' Dis. Eye* (ed. 12) vi. 86 (*caption*) Semi-schematic diagram showing the correspondence of the structures seen in the normal eye by gonioscopy with those of a microscopic section. **1955** D. W. Maurer in *Publ. Amer. Dial. Soc.* xxiv. 4 The language [of criminal subcultures] is usually secret or semisecret. **1939** *Daily Tel.* 18 Dec. 12/8 (Advt.), Audit clerk, semi-senior, able to work without supervision. **1951** *Sport* 16–22 Mar. 9/2 Alec Talbot..finished his career with another semi-senior club, Stourbridge. **1976** *Eastern Daily Press* (Norwich) 16 Dec. 5/4 (Advt.), Applicants from ambitious semi-seniors would be welcome. **1840** Poe *Autobiogr.* in *Graham's Mag.* Nov. 225/1 The design was never more than semi-serious. **1977** J. F. Fixx *Compl. Bk. Running* xiv. 167, I don't know of a single even semiserious runner who smokes. **1848** Mill *Pol. Econ.* 22 The immediate cultivators of the soil..ceased to be in a servile or semi-servile state. **1896** T. W. Sanders *Encycl. Gardening* (ed. 2) 35 Place plants in semi-shady position outdoors June to Aug. **1973** M. Amis *Rachel Papers* 67 Semi-shaven, dishcloth hair, duffle-coat, baggy brown Farmer-Giles corduroys. **1927** Carr-Saunders & Jones *Soc. Struct. Eng. & Wales* v. 50 It is not usual to think of those occupied in retail dealing as divided into skilled, semi-skilled, and unskilled. **1940** W. S. Churchill *Into Battle* (1941) 166 We have to make a huge expansion of our labour force, and especially of those capable of performing skilled or semi-skilled operations. **1976** E. Stewart *Launch!* (1977) 16 In 810—semiskilled—they were playing Tensor beams across eight-by-five-inch printed circuits. **1925** T. Dreiser *Amer. Trag.* (1926) I. ii. viii. 206 They were in the midst of one of those semi-religious, semi-social and semi-emotional church affairs, the object of which was to raise money for the church. **1922** Joyce *Ulysses* 659 A noggin and a quarter of soured adulterated milk, converted by heat into water, acidulous serum and semi-solidified curds. *Ibid.* 680 A series of static, semistatic and peripatetic intellectual dialogues. **1962** J. T. Marsh *Self-Smoothing Fabrics* xxiii. 382 Variations of the crumpling method have produced a few semi-subjective tests. **1943** *List Retail Controlled Prices* (Ministry of Food) (ed. 3) 6 Biscuits, sweet or semi-sweet. **1972** *Times* 15 Nov. (Ital. Wine Suppl.) iv./3 Est! Est!! Est!!!..comes in both dry and semi-sweet. **1954** W. K. Hancock *Country & Calling* viii. 227 There are..many words of technical or semi-technical origin which have lodged themselves..firmly in everyday speech. **1976** *Classical Q.* XXVI. 216 Both lines, as Nicolaus notes..allude to a semi-technical legal term. *a* **1974** R. Crossman *Diaries* (1976) II. 27 It was a reasoned, semi-theoretical speech and it produced an amazingly good response. **1851** H. Melville *Moby Dick* II. xliii. 294, I am convinced that from the heads of all ponderous profound beings, such as Plato,..Dante, and so on, there always goes up a certain semi-visible steam, while in the act of thinking deep thoughts.

b. Compounded with a sb. to form an adj. phr.: (further examples).

1955 Keepnews & Grauer *Pict. Hist. Jazz* xi. 117 They made much use of jazz and semi-jazz conceptions in their dance music. **1938** *New Statesman & Nation* 23 July 143/2 What is needed is an extended application of the Trade Board method over a wide range of growing trades which mass-produce luxury or semi-luxury goods. **1962** H. E. Beecheno *Introd. Bus. Stud.* xi. 98 The more expensive type of goods..which are not the subject of so much branding and national advertising—the luxury and semi-luxury goods. **1935** *Discovery* Aug. 220/1 Subsistence or semi-subsistence farming was the rule, the settlers eked out a bare existence from such poor land. **1975** *Sci. Amer.* May 76/3 Most of them work as semi-subsistence farmers and live according to Mayan cultural patterns.

2. Compounded with sbs.: (further examples).

1948 D. Diringer *Alphabet* ii. iv. 286 Groups of 'Assyrians'..developed a semi-autonomy, owing allegiance only to their *maliks*. **1874** J. S. Mill *Ess. Relig.* 70 An argument for the utility of religion is an appeal..to semi-believers to make them avert their eyes from what might possibly shake their unstable belief. **1964** L. MacNeice *Astrol.* vii. 230 The semi-charlatan who may not subscribe to any code of ethics. **1962** E. Snow *Other Side of River* iv. 36 China's major ocean and river ports fell under foreign control and she became a semicolony not of one nation but of all the major industrial and naval powers. **1865** J. S. Mill *Exam. Hamilton's Philos.* vii. 104 But such a semi-concession..cannot save him. **1929** *Evening News* 18 Nov. 4/5 His weariness was so heavy that it bore him into a state of semi-consciousness. **1958** T. Stanwell-Fletcher *Clear Lands* 89 That cold, beautiful semidark of the Arctic night. **1914** W. S. Churchill *Let.* 8 Oct. in M. Gilbert *W. S. Churchill* (1972) III. *Compan.* i. 182 We must not be led into frittering away resources by keeping half a dozen anchorages in a state of semi-defence. **1977** *Financial Times* 7 Oct. 23/3 Mr Wedgwood Benn's silky semi-defence of Government strategy from the platform. **1963** *Times* 7 Feb. 4/4 A Prime Minister with the support of the House of Commons may be in a stronger position than a President who has not got the support of Congress, but he is not the sole executive officer as a President is, or a dictator such as Khrushchev, or a semidictator such as de Gaulle. **1938** *New Statesman & Nation* 20 Aug. 288/2 They operated in a twilight of semi-illegality. **1949** Koestler *Insight & Outlook* vi. 84 The comic effects of *misspelling* in the letters of children or semi-illiterates. **1952** E. Grierson *Reputation for Song* vii. 58 She lived with an aunt, a semi-invalid. **1974** *Howard Jrnl.* XIV. 78 Sir Cyril Burt defined semi-literates as those 'who cannot make effective use of reading or writing'. **1920** J. M. Keynes *Econ. Consequences Peace* 95 Germany is threatened with a deluge of luxuries and semi-luxuries from abroad. **1961** tr. Zhou Enlai in *Look* 31 Jan. 104/2 A considerable period is needed before China can surpass the norm with regard to mechanization and semimechanization. **1906** E. Dyson *Fact'ry 'Ands* xiii. 169 He smote himself on the breast..leaving the packer in a condition of semi-paralysis. **1958** F. Newton in P. Gammond *Decca Bk. Jazz* v. 65 Variety songs, effect-catching numbers and calculated semi-pornography. *a* **1930** D. H. Lawrence *Phoenix* (1936) v. 597 Some sort of semi-profession, such as school-teaching. **1805–6** Wordsw. *Prelude* (1850) v. 113 This Arab phantom, which my Friend beheld, This semi-Quixote. **1835** J. S. Mill in *London Rev.* II. 272 A last desperate attempt of the Tories to creep back into power as semi-reformers. **1935** *Mind* XLIV. 524 Retirement (or shall we say semi-retirement) has given Dr. Schiller increased leisure. **1971** D. J. Smith *Discovering Railwayana* viii. 47 After the grouping..many of the relics were consigned to semi-retirement in backyard and basement. **1952** A. G. L. Hellyer *Sanders' Encycl. Gardening* (ed. 22) 40 Position, semi-shade or north aspect. **1900** W. S. Churchill in *Morning Post* 20 July 5/7 It was not possible for the conquering army to allow the capital..to be in a state of semi-siege. **1929** *New Yorker* 12 Oct. 25/3 It is like the semi-stupor of an habitual intoxication. **1977** *Proc. R. Soc. Med.* LXX. 689/1 High alcohol intake probably meets this need as well as providing them with an intermittent refuge in sedation and semistupor. **1863** J. S. Mill *Let.* 7 Jan. (1910) I. 273 He has triumphed wonderfully over the difficulty of rendering the thoughts or semi-thoughts of Plato. **1867** —— *Let.* 19 Oct. (1910) II. 90 Parliamentary semi-work or idleness.

3. Compounded with vbs.: (further examples).

1973 M. Amis *Rachel Papers* 102 Kiss and semi-lick throat and neck. **1962** *Times* 9 Apr. (Suppl.) p. iii/3 The dispensation of powdered lime with oxygen is used to semi-refine molten iron before final treatment.

4. Compounded with advs.: (further example).

1945 *Plastics* IX. 143/1 Practical mechanical [safety] devices capable of functioning either fully or semi-automatically.

4*. The prefix used *absol.* as an advb., in sense 'partly, to some extent'. *colloq.*

1979 K. M. Peyton *Marion's Angels* vi. 102 'I thought you were on holiday.' 'Semi.'

II. 5. a. semi-interquartile range *Statistics*, half of the interquartile range.

1911 G. U. Yule *Introd. Theory Statistics* viii. 134 There are three such measures in common use—the standard deviation, the mean deviation, and the quartile deviation or semi-interquartile range. **1971** T. R. Harshbarger *Introd. Statistics* v. 91 The semi-interquartile range is half the quartile range: $Q = \frac{1}{2}(Q_3 - Q_1)$. Q is also called the quartile deviation.

b. semi-monthly (earlier examples); **semi-weekly** (earlier and later examples).

1851 C. Cist *Sk. & Statistics Cincinnati* 75 These are semi-monthlies. **1860** *Ex. Doc. 36th U.S. Congr. 2 Sess. Senate* No. 1. 435 The present contract..provides for an additional monthly trip between New York and San Francisco, making the service tri-monthly instead of semi-monthly as heretofore. **1791** T. Jefferson *Let.* 21 July in A. A. Lipscomb *Writings T. Jefferson* (1903) XIX. 79 Besides this, Fenno's being the only weekly or semi-weekly newspaper. **1833** A. H. Tracy *Let.* 10 June in T. W. Barnes *Mem. T. Weed* (1884) iv. 49 Put Millard Fillmore on your list for the Semi-Weekly. **1926** *Jrnl. Biol. Chem.* LXIX. 92 The weights of the rats and of their food intake were recorded semiweekly.

6. b. *Math.* **semi-convergent** *a.*, applied to a series the sum of whose terms converges while the sum of the moduli of its terms diverges; hence **semi-convergence**; **semi-major (-minor) axis**, half of the length of the longest (shortest) diameter of an ellipse.

1902 E. T. Whittaker *Mod. Anal.* 12 Absolute convergence and semi-convergence. **1872** *Monthly Notices R. Astron. Soc.* XXXII. 262 Most functions are expansible in an ascending (convergent) series of the form $A_0 + A_1x + A_2x^2 + \ldots$, and a descending (semi-convergent) series of the form $B_1/x - B_2/x^2 + B_3/x^3 - \ldots$ **1959** A. Zygmund *Trigonometric Series* (ed. 2) II. iv. 175 If f is continuous, and $S[f]$ is uniformly semi-convergent to f from below, then $S[f]$ converges uniformly. [**1850** J. Haan *Analytical Geom. & Conic Sections* v. 45 The equation to the ellipse [is]..$y^2/b^2 + x^2/a^2 = 1$..., a and b being termed respectively the semi-axis major and the semi-axis minor.] **1899** Grace & Rosenberg *Coordinate Geom.* iv. 50 Find the length of the ordinates of each of the curves in Ex. 3 corresponding to the middle points of the semi-major axis. **1977** *Whitaker's Almanack 1978* 156/1 The squares of the periodic times vary as the cubes of the semi-major axes. **1909** C. N. Schmall *First Course in Analytical Geom.* viii. 191 Prove that the semi-minor axis is a mean proportional between the parts of tangent cut off. **1962** Corson & Lorrain *Introd. Electromagn. Fields* iv. 175 A charge Q is uniformly distributed throughout the volume of an ellipsoid of revolution whose semi-major axis is a and whose semi-minor axes are b.

7. a. *semi-liquid* (later example), *-lucent* (later example), *-matt, -moist.*

1963 D. W. & E. E. Humphries tr. *Termier's Erosion & Sedimentation* x. 193 Some marly deposits still contain 41·3% of water, which results in numerous landslips in this plastic, semi-liquid material. **1919** V. Woolf *Night & Day* xi. 138 Rodney's windows..were a semilucent red colour. **1967** Semi-matt [see *matt *a.*]. **1967** E. Chambers *Photolitho-Offset* xvii. 259 The soft, tacky gelatinous Everdamp paper makes for easier working, being semi-moist and ready for use.

b. *semi-Christianity, -socialism* (later example), *-socialist.*

1882 J. Snodgrass tr. *H. Heine's Relig. & Philos. in Germany* II. 67 Leibnitz..was well able to defend Christianity in its integrity. I say, in its integrity, for he defended it against semi-Christianity. **1961** C. S. Lewis *Let.* Feb. (1966) 297 For some people at some moments what I call semi-Christianity may be useful. **1930** W. K. Hancock *Australia* xi. 225 The Liberal party still continued to tread..the familiar path of semi-socialism. **1939** N. & S. Weyl *Reconq. Mexico* iv. 110 Mexico's long indoctrination with semisocialist theory..made leftist political candidates almost a political inevitability. **1976** N. O'Sullivan *Conservatism* i. 30 A great body of conservative thought..has itself tended to move in a semi-socialist or collectivist direction.

c. Gram. *semi-compound* (also as *sb.*), *-grammatical, -phonetic* adjs.; **semi-consonant** (earlier example); **semi-predicative** *a.*, quasi-predicative; forming part of a predicate; **semi-sentence,** a statement or utterance which possesses some of the features of a sentence.

1963 F. T. Visser *Hist. Syntax* I. iv. 389 An older semi-compound verb of the type *overpass*. **1964** *English Studies* XLV. 50 Some combinations..can therefore be considered as compounds or semi-compounds. **1820** F. Bopp in *Ann. Oriental Lit.* I. 6 The letter *y*..in Sanskrit or Gothic words..is always to be considered as a semi-consonant. **1964** P. Strevens *Papers in Lang. & Lang. Teaching* (1965) iv. 61 The body of data..includes grammatical, semi-grammatical and non-grammatical features. **1933** L. Bloomfield *Language* xvii. 290 Semi-phonetic devices, such as rising combinations of letters for a single phoneme. **1964** *Language* XL. 302 He had to depend mainly on the semiphonetic materials recorded by H. Paasonen. **1914** O. Jespersen *Mod. Eng. Gram.* II. 386 It is used in that sense as a semi-predicative post-adjunct. **1933** L. Bloomfield *Language* xii. 206 Some numeratives are used also in other syntactic positions, as..*all, both* as semi-predicative attributes (*the boys were both there*). **1954** *Acta Universitatis Carolinae* VII. 35 In the present article, the term 'semi-sentence' constructions covers both participial and infinitive phrases as well as those having neither of these, but which, owing to their binary character, have a distinct predicational form (e.g. *If lawful.*). **1969** *Word* XXV. 195 A sentence which contains no predicate is a semi-sentence. **1975** N. Chomsky *Logical Struct. Linguistic Theory* viii. 244 We are building a system of phrase structure only for first-order grammatical sentences, a category that presumably excludes such semisentences as 'sincerity appointed the table'.

d. Nat. Hist. *semi-erect, -sessile, -social, -terrestrial*; **semi-evergreen,** normally evergreen but shedding some leaves if conditions become severe.

1894 R. B. Lee *Hist. & Descr. Mod. Dogs of Gr. Brit. & Ireland (Terriers)* xv. 353 Ears..if not cut, to be small V-shaped and carried semi-erect. **1931** A. G. L. Hellyer *Sander's Encycl. Gardening* (ed. 21) 409 S[milax] *herbacea*, greenish, bluish-black fruits, twining or semi-erect, N. America. **1952** *Ibid.* (ed. 22) 10 Adenocarpus... Deciduous or semi-evergreen shrubs or small trees. **1978** *Detroit Free Press* 16 Apr. (Parade Suppl.) 32/2 (Advt.), The lush, semi-evergreen foliage erupts in massive clusters of star-shaped, fiery red blooms. **1962** D. Nichols *Echinoderms* xii. 156 These facts suggest that the animal remained semi-sessile on the sea-bottom and searched the surrounding area with its tentacles. **1962** *Sci. Survey* XVII. 285 Numbers of insects are affected [by sounds]..which are related, not to sexual activity, but to social or semi-social activities. **1932** J. S. Huxley *Probl. Relative Growth* i. 33 Twenty-four grams is a very small weight for many crabs, including forms of semi-terrestrial and burrowing habits. **1964** *Oceanogr. & Marine Biol.* II. 302 The semi-terrestrial hermit crab, *Coenobita perlatus*,..varies the frequency with which it visits water of different salinities.

e. *semi-ape* (example), *-dwarf* (also as *adj.*).

1886 Kipling *General Summary* in *Departmental Ditties*, We are very slightly changed From the semi-apes who ranged Pre-historic India. **1931** Semi-dwarf [see *Peter Pan 2]. **1959** *Sci. News Let.* 22 Aug. 120/3 Stiff-stemmed semidwarf wheat may be the answer for Pacific Northwest growers whose wheat suffers extensively from lodging. **1979** *Nature* 3 May 7/1 In India..rice research has tended to focus on light-yielding semi-dwarfs for irrigated land.

g. Path. *semi-narcosis.*

1937 Koestler *Spanish Testament* ii. 235 That merciful state of semi-narcosis induced by..spinning illusions. **1965** J. Pollitt *Depression & its Treatment* v. 67 In very severe cases, continuous sleep (continuous narcosis) or continuous semi-narcosis with chlorpromazine and small amounts of Sodium Amytal must be given.

h. Chem. *semi-reduced*; **semialdehyde,** a derivative of a compound containing two identical functional groups (e.g. a dicarboxylic acid) in which one of the groups has been converted into an aldehyde group; **semi-covalent** *a.*, having some covalent character.

1942 *Jrnl. Biol. Chem.* CXLV. 69 In the presence of this enzyme..α-ketoglutaric acid formed succinic semialdehyde and CO_2. **1976** *Nature* 16 Dec. 652/2 GABA concentrations were measured using GABA transferose and succinic semialdehyde dehydrogenase to generate NADH, which was measured spectrofluorometrically. **1965** Phillips & Williams *Inorg. Chem.* I. v. 156 Semi-covalent bonding involving d orbitals may also be expected to lead to high coordination numbers for early members of the transition series. *Ibid.* x. 386 The anaerobic system..tends to yield acetic acid, alcohol, and similar semi-oxidized and semi-reduced compounds.

j. semi-coke, a smokeless fuel that leaves little ash, made from coal by carbonization at a low temperature (usu. 500–600°C); **semi-coking** *a.*, designating a coal that is intermediate between a good coking coal and one that does not produce coke; **semi-steel,** substitute for def.: a low-carbon cast iron produced by melting mild steel with pig iron in a furnace (earlier and later examples).

1918 *Chem. & Metallurgical Engin.* XIX. 580/1 Charles Howard Smith..proposed to get an intermediate soft 16 to 20 per cent volatile semi-coke. **1972** Harker & Allen *Fuel Sci.* v. 70 The tar and most of the volatile matter are driven off leaving a material known as 'semi-coke'. **1915** *Iron & Coal Trades Rev.* XCI. 421/2 No serious attempts have hitherto been made to utilise in the low-temperature system non-coking and semi-coking coals for the production of oils. **1977** *Jrnl. R. Soc. Arts* CXXV. 64/2 High grade coke can be made from a coal mixture including a proportion of only semi-coking coal. **1858** *Q. Jrnl. Chem. Soc.* X. 145 Mr. Bessemer claims the production..of a particular product called semi-steel. **1929** *Iron & Steel Industry* III. 35 The metallurgist has long regarded the term 'semi-steel' as a misnomer, although the use of the term within the foundry business is harmless when it is simply made to imply the use of steel in a cast iron mixture... Instead, we have at present the innuendo that 'semi-steel' is more or less a definite iron-carbon alloy of superior properties compared to cast iron, whereas it is in reality an 'unknown quantity'. **1958** *Engineering* 21 Mar. 384/2 The bottom grate is of heat resistant 30 per cent chrome semi-steel and is reversible for burning coal or coke.

k. Archit. *semi-Gothic* (examples).

1768 *Acct. Denmark* 94 The royal palace of Rosenburg..is a handsome structure in the semi Gothic taste. **1925** F. Madan *Oxf. outside Guide-Books* (ed. 2) 186 The Firm's premises..were entirely rebuilt and included the present 'semi-Gothic' elevation.

8. semi-active *a.*, designating a method of missile guidance in which the missile responds to a signal transmitted from elsewhere and reflected by the target; **semi-antique** *a.*, of a rug or carpet: between fifty and one hundred years old; also *absol.* as *sb.*; **semi-Bantu** *a.*, of or pertaining to a number of languages in Central and West Africa closely related to the Bantu family; also as *sb.*; **semi-basement,** a basement room or rooms set only partially below ground level; **semi-broch** *Archæol.*, in the Hebrides and western mainland of Scotland: a hollow-walled fort-like structure representing a stage of development between the galleried dun and the broch; **semi-cell** *Bot.*, each of the two parts of a cell which is constricted in the middle, as in desmids; **semi-chemical** *a.*, applied to (wood pulp made by) a pulping process in which wood chips are subjected to mild chemical delignification followed by mechanical processing; **semi-classical** *a.*, (*a*) *gen.*, esp. in *Music*; (*b*) *Physics*, designating a theory that is intermediate in its assumptions and methods between the classical, or Newtonian, description and that of modern physics, esp. in quantum mechanics and relativity; hence **semi-classically** *adv.*; **semi-closed** *a.*, (*b*) *Med.*, applied to methods of administering anæsthetics employing a gas supply that is closed from the atmosphere and in excess of the patient's needs, the excess being vented to the atmosphere; (cf. *open *a.* 11 h, *semi-open *a.*); **semi-cursive** *a. Palæogr.*, of or pertaining to one of various scripts combining cursive features with elements of a more formal style; also *absol.* as *sb.*; **semi-diesel, -Diesel** *a.*, of an engine: (see quot. 1974); also as *sb.*, an engine of this type; **semi-display** *Typogr.*, a lay-out (for advertisements, etc.) intermediate between the run-on and displayed styles; hence **semi-displayed** *a.*; **semi-documentary** *a.*, of or pertaining to a film that presents factual or semi-factual material in fictional form; also as *sb.*; **semi-empirical** *a.*, that derives in part from theoretical considerations and in part from the results of experiment; so **semi-empirically** *adv.*; **semi-evening dress** (also **gown,** etc.), (a) fashionable dress (gown, etc.) of less than fully formal design suitable for both afternoon and evening wear; also *ellipt.* as **semi-evening** *sb.*; **semi-fabricated** *a.*, (of a material) formed into some standard shape for use in the making of finished articles; so **semifabricator,** a manufacturer of semifabricated goods; **semi-finalist,** a competitor in a semi-final; **semifinished** *a.*, (of a material, esp. steel) manufactured or treated for use in the making of finished articles; **semi-fitting** *a.*, designating clothing that partly fits the figure; **semi-gloss** *a.*, designating a finish intermediate between matt and glossy; **semi-intensive** *a. Agric.*, of or pertaining to a method of rearing livestock that includes features of intensive farming; **semileptonic** *a. Particle Physics*, involving both leptons and hadrons; hence **semileptonically** *adv.*; **semi-main** *U.S.* = *repêchage; **semi-manufacture,** a product made from raw materials and used in the manufacture of finished goods; **semi-portal** *a.*, applied to a crane mounted on a frame consisting essentially of a horizontal member supported at one end by an upright; **semi-proletariat** *Communism* [tr. Chinese *bànwú-chǎnjiējí* (1926 Mao Zedong)], the class of poor peasants and others intermediate between the proletariat and the petty bourgeoisie; **semi-psychic** *adv.* and *a. Bridge*, (see quot. 1964); **semi-quantitative** *a.*, partly quantitative; approximate; based on or yielding approximate figures; hence **semi-quantitatively** *adv.*; **semi-reflecting** *a.*, designating a material of low reflectivity (usu. a film deposited on a transparent base) which permits partial reflection and partial transmission; hence **semi-reflection; semi-reflector; semi-reflective** *a.*; **semi-scale,** half or some fraction of full-scale; usu. *attrib.*; **semi-sequitur** [after *non sequitur*], an inference or conclusion which is related only indirectly to the premisses; **semi-sub,** short for **semi-submersible* sb.; **semi-submersible** *a.* and *sb.*, (applied to) an offshore drilling platform or barge equipped with submerged hollow pontoons that can be flooded with water when the vessel is anchored on site in order to give it stability against waves and wind; **semi-synthetic** *a.*, that is a mixture of synthetic and natural materials, or has been prepared by artificial modification of a natural material; that is a combination of synthetic and natural processes; **semi-tropical** *a.* (earlier example); **semi-truck** *U.S.*, = *semi-trailer; **semi-variable** *a. Econ.*, of a cost (see quot. 1965); **semi-works** *U.S.*, a manufacturing plant used to develop and perfect a new product or process after testing in a pilot plant and before full-scale production; usu. *attrib.*

1954 K. W. Gatland *Devel. Guided Missile* (ed. 2) iii. 83 Final guidance is obtained by the technique known as 'semi-active homing', the missile responding to target reflection from the ground radar. **1945** G. G. Lewis *Pract. Bk. Oriental Rugs* (ed. 6) v. ii. 304 Antique or semi-antique Chinese rugs are as scarce as hen's teeth. **1970** J. Franses *European & Oriental Rugs* 3 Its wool is hand-spun on all the early carpets and semi-antiques. **1979** *Tucson Mag.* Apr. 23/1 The semi-antique Iranian prize rug was rolled up. **1919** H. H. Johnston *Compar. Study Bantu & Semi-Bantu Languages* ii. 17 The semi-Bantu languages on this north-west borderland have a vocabulary which contains a greater or smaller amount of Bantu roots. *Ibid.* 814 The Indiki language of the southern Maneñguba country [is an]..interesting form of Semi-Bantu. **1977** *Language* LIII. 291 Bantu and Semi-Bantu classifiers do have meaning. **1934** Semi-basement in Webster. **1963** *Punch* 20 Nov. 729/1 Desks crowded among filing cabinets in semi-basement. **1974** *Country Life* 7 Mar. (Suppl.) 32l/1 Semi-basement boiler/drying room, playroom. **1903** E. Beveridge *Coll & Tiree* x. 73

The Semi-broch type. It may be bold to introduce a new name in our description of the Tiree Duns, but a type seems to occur here which is..distinct from the ruder Hill-Forts. **1963** *Times* 16 Feb. 10/4 Tiree has some 20 dun sites, of which four appear to be hollow-walled semi-brochs. **1980** *Glasgow Archaeol. Jrnl.* VII. 73/1 There are only three semibrochs on the whole of the NW coast of Scotland. **1872** H. C. Wood *Contrib. Hist. Fresh-Water Algæ N. Amer.* 101 Divided into two symmetrical semi-cells. **1927** West & Fritsch *Treat. Brit. Freshwater Algæ* 259 The young semicells are clearly recognisable and nuclear division is complete before any change takes place in the chloroplasts of the parent. **1969** F. E. Round *Introd. Lower Plants* ii. 27 Each mature semicell is ornamented or produced into spines and looked at from the apices is constructed on a bilateral..or triangular basis. **1925** *Paper Trade Jrnl.* 15 Oct. 57/1 The various processes..for the production of semi-cellulose as above defined may conveniently be termed semi-chemical inking processes, as they consist essentially of some chemical treatment combined with a mechanical disintegration of the partially cooked wood. **1961** *Times* 2 June 26/2 A new semi-chemical pulp mill to use indigenous woods. **1974** *Sci. Amer.* Apr. 56/3 Semichemical processes are applied mainly to hardwoods because of the lower content of lignin in such wood. **1949** *Billboard* 2 Apr. 34/3 Albums listed are those classical and semi-classical albums selling best in the nation's retail record stores. **1964** J. W. Linnett *Electronic Struct. Molecules* i. 6 In quantum mechanics the orbits of the Bohr-Sommerfeld semi-classical methods are replaced by orbitals. **1970** G. K. Woodgate *Elem. Atomic Struct.* i. 4 Another important quantity corresponds semi-classically to the speed of the electron in the first Bohr orbit as a fraction of the speed of light. **1914** J. T. Gwathmey *Anesthesia* vi. 272 Semi-Closed method.—Martin Ware has reported one or two thousand cases without a fatality. **1977** *Proc. R. Soc. Med.* LXX. 784/2 Anaesthesia with spontaneous breathing usually employs a semi-closed circuit. **1927** *Bull. Bezan Club* IV. 11 These lines, on both the Greek and Latin sides, are written not in a stiff, formal book-script, but in an easy-going semi-cursive. **1948** D. Diringer *Alphabet* ii. x. 545 The Italian semi-cursive minuscule..developed from the Roman minuscule, was employed throughout Italy from the seventh to the ninth century, and continued to be used in Tuscany until the twelfth century. **1968** *PMLA* LXXXIII. 23/1 The script..may be described as an upright, bold but somewhat ungainly *cursiva* (*anglicana*) *formata*, or semi-cursive in older and looser nomenclature. **1911** *Engineer* 7 June 27/1 (*caption*) Semi-Diesel oil engine. **1920** R. A. McMillan *Guide to B.O.T. Exam. for Extra First-Class Engineers* xv. 396 A common type of Semi-Diesel is the Bolinder engine. **1960** G. Blanchet *Search in North* i. 19 From the engine exhaust came the hollow staccato of the semi-diesel. **1974** *McGraw-Hill Dict. Sci. & Techn. Terms* 1323/2 *Semidiesel engine.* 1. An internal combustion engine of a type resembling the diesel engine in using heavy oil as fuel but employing a lower compression pressure and spraying it under pressure, against a hot (uncooled) surface or spot, or igniting it by the precombustion or supercompression of a portion of the charge in a separate member or uncooled portion of the combustion chamber. 2. A true diesel engine that uses a means other than compressed air for fuel injection. **1971** *Cabinet Maker & Retail Furnisher* 24 Sept. 537 Run-on and Semi-display Announcements are set only in Times Bold and Times Roman. **1976** *Horse & Hound* 10 Dec. 65/1 (Advt.), Advertisers wishing to take a smaller space may use semi display with a minimum size of 2 cms but no blocks or illustrations are permissible in this style. **1972** *Daily Tel.* 5 July 6 *Run-on* (minimum setting) and *semi-displayed* (with lines of white space, indents, or double-line capitals)—£1.40 per line. **1939** L. Jacobs *Rise of Amer. Film* 413 Allied in spirit to the semi-documentary films were the realistic regional dramas. **1948** L. Levy *Music for Movies* xiii. 131 (*heading*) The semi-documentary. **1958** *New Statesman* 3 May 567/3 Granada's main achievement is to prove that serious semi-documentary or discussion programmes (*Under Fire, Youth wants to Know, What the Papers Say*) can be successful entertainment. **1935** Condon & Shortley *Theory of Atomic Spectra* i. 9 The results..were obtained in semi-empirical ways from consideration of a formulation of the theory that was only true in the limit of large quantum numbers. **1970** Semi-empirical [see *pre-exponential* s.v. *pre- B. 1 d]. **1976** *Physical Rev. Lett.* XXXVI. 375/1 We may now semiempirically incorporate the major factor neglected so far. **1917** *Vogue* Early Apr. 3/1 Semi-evening or Afternoon Gown in best quality Chiffon Taffetas and Ninon de Soie. **1923** A. Huxley *Antic Hay* x. 156 For semi-evening dress, shell rims with gold ear-pieces. **1923** B. Ruck *Dancing Star* i. vi. 117 Ripple's frock was the kind of garment dear to the compromise-loving British heart. It was '*a semi-evening*', which, to a purist, means that the dress is appropriate neither to evening or afternoon. **1938** E. Bowen *Death of Heart* i. ix. 155 She asked me to come to Peter Jones's with her to help her choose a semi-evening dress. *a* **1976** A. Christie *Autobiogr.* (1977) iii. iv. 155, I had a pale grey *crêpe de Chine* semi-evening dress. **1947** J. Newton *Introd. Metallurgy* (ed. 2) v. 238 Rolling not only produces finished shapes such as plates, sheets,..and rails, but many semifabricated shapes such as steel, copper, and brass rod for wire drawing; steel billets for forging and piercing. **1976** *Scotsman* 24 Dec. 4/5 Alcan (US), are also to raise prices for their semifabricated aluminium products. **1967** *Economist* 29 July 425/3 Hitherto most of this east European aluminium has been bought by the giant non-American producers... It has then either been passed on to their own semi-fabricators or sold to the independent semi-fabricators at the official producers' price. **1898** Semi-finalist [see -ist 4 b]. **1922** *Daily Mail* 24 Nov. 11 Beaten semi-finalists. **1972** D. Delman *Sudden Death* (1973) ii. 39 The year before he'd been champion at Wimbledon and a semi-finalist at Forest Hills. **1942** *R.A.F. Jrnl.* 16 May (*verso rear cover*), Germany would have to supply raw materials and semi-finished products. **1959** [see *forge-master* s.v. *forge *sb.* 6 b]. **1972** *Daily Tel.* 9 Feb. 2/4 The plastics industry expects that about 50 per cent. of finished and semi-finished products will be wholly or partially metricated by mid-1971. **1897** Semi-fitting [in Dict., sense 1 a]. **1930** *Daily Express* 8 Sept. 7/5 (Advt.), Semi-fitting bodice, with belt forming waist-line. **1963** *Times* 8 Jan. 11/2 A mid-tone matt or semi-gloss single colour. **1935** J. S. Hicks *Encycl. Poultry* II. 509 A semi-intensive house may be of any size from one, say, 6 ft. by 6 ft. to a capable of housing ten or a dozen birds, to a mammoth affair for the accommodation of 500 or more layers. **1966** *Economist* 1 Oct. 72/2 Beef growers are turning to what is termed 'semi-intensive' beef rearing instead. This combines intensive grazing with fattening on high protein compounds to produce a 15- to 18-month-old beast for market. **1965** *Physical Rev. Lett.* XIV. 51 (*heading*) SU(6) and semileptonic interactions. **1979** *Nature* 14 June 588/2 If the D decays semi-leptonically some of its energy is taken by the unobserved neutrino. **1968** *Surfer Mag.* Jan. 48/1 Overland finally got into the finals by winning the men's semi-main. **1979** *Tucson* (Arizona) *Citizen* 28 Apr. 2B/5 The top two finishers in the consolation and 25-lap semi-main will qualify for the final run for the lion's share of the $12,000 purse. **1935** *Economist* 16 Feb. 353/1 Japan is exporting more finished goods and manufactured foodstuffs, importing more raw materials and semi-manufactures. **1979** *Shell Trade in Eastern Europe* (Shell Internat. Petroleum Co.) 3 Historically, Eastern European exports to the West have been principally raw materials, semi-manufactures, fuels and agricultural produce. **1908** A. Tolhausen tr. *Böttcher's Cranes* vi. 256 (*heading*) Semi-portal travelling crane, with central steam supply station. **1958** *Times Rev. Industry* Oct. 20/3 Electric semi-portal cranes serving three transit sheds. **1951** tr. M. Litvinoff in J. Degras *Soviet Documents on Foreign Policy* I. 136 This new apparatus of power should embody the dictatorship of the working class (and in some places also the rural semi-proletariat, i.e. the poor peasants). **1965** J. Ch'ên *Mao & Chinese Revolution* (1967) i. v. 110 The semi-proletariat, according to Mao, consisted of the overwhelming majority of the semi-tenant peasants, poor peasants, handicraftsmen, shop assistants and pedlars. **1975** J. De Bres tr. *Mandel's Late Capitalism* xi. 362 Many of the producers in the export branch are recruited from the stratum of the semi-proletariat who engage in wage labour only to obtain a supplementary income to eke out their means of subsistence in agriculture. **1960** T. Reese *Play Bridge with Reese* 118 Unless he has opened semi-psychic he should have both the minor suit Queens. **1962** *Listener* 3 May 790/3 North opened with a semi-psychic One Club. **1964** *Official Encycl. Bridge* 493/2 *Semi-psychic*, a departure from normal bidding methods which is not a complete bluff but is still intended to deceive the opponents. The term usually refers to an opening bid well below minimum values, but lead-inhibiting bids belong in the same category. **1929** Parker & Crozier in C. Murchison *Found. Exper. Psychol.* viii. 362 The comparison of odors is possible, in an empirically semi-quantitative way, by the use of several such instruments. **1977** *Sci. Amer.* May 39/2 The first semiquantitative step in generalizing the theory of crystalline semiconductors to amorphous materials was taken by Sir Nevill Mott of the University of Cambridge. **1956** *Nature* 21 Jan. 127/1 The intermediate products formed in the hydrolysis of the cyclic oligomers of ε-caprolactam..have been identified and semi-quantitatively determined. **1927** *Jrnl. Sci. Instruments* IV. 491 (*heading*) Semi-reflecting surfaces. **1946** *Nature* 20 July 101/2 For glass surfaces that have not been made semi-reflecting these [interference patterns] do not have the contrast of reflected interference patterns. **1976** Z. Knittl *Optics of Thin Films* ix. 374 A common feature of many synthesis problems is the condition for $r(\omega^2)$ to be flat in a certain range about $\omega = 0$ and at a certain level. Depending on this level we have the anti-reflection or the semireflection problem. **1973** *Sci. Amer.* June 69/1 The devices can be made transmissive for rear-lighting applications,..or semireflective for both kinds of operation. **1945** *Jrnl. Sci. Instruments* XXII. 103/1 Before 1936, the majority of semi-reflectors were made by chemical deposition of silver. **1976** Z. Knittl *Optics of Thin Films* ix. 397 Chebyshev-type semireflectors..may be only one of the ways of achieving a broad-band maximum. **1946** *Nature* 7 Sept. 337/2 It is impossible in chemical engineering and many other branches of applied science to conduct research entirely in the laboratory; full-scale or semi-scale plant must be used. **1973** *Ibid.* 2 Feb. 319/1 In 'semi-scale' tests (about a tenth the size of a real reactor) water failed..to stay in the test vessel. **1965** *Punch* 24 Nov. 779/1 The discords between received Edwardian fiction and child-observed fact work as poignantly as ever, as do the semi-sequiturs: 'He was broad and stout and had a manful way of carelessly swinging his arms that gave him many friends.' **1975** *Offshore Engineer* Sept. 55/1 *Staflo*..and *Sea Quest*..along with the ill-fated *Ocean Prince*, are the only semi-subs built in the UK. **1962** *World Oil* Sept. 96/3 The rig is an all weather, semi-submersible which is submerged to the 40 foot level in normal drilling operations. **1963** *World Petroleum* Aug. 47/1 The semi-submersible vessels give a desired balance between cost, safety and performance. **1975** *North Sea Background Notes* (Brit. Petroleum Co.) 11 Semi-submersibles can be used for drilling when resting on the sea bottom, but they are generally employed in the floating position. **1980** *Christian Sci. Monitor* (Mid-western ed.) 4 Dec. 11/1 A third delineation well..was spudded on Nov. 14 five miles west of P-15 by the newly arrived semi-submersible drilling rig. **1946** *Nature* 7 Sept. 350/2 The method may give still better results if..semi-synthetic mediums are used for the toxin production. **1974** *Ibid.* 19 Apr. 706/2 Rats..were fed a semisynthetic diet for 3 months. **1856** J. C. Patteson *Let.* 18 June in C. M. Yonge *Life J. C. Patteson* (1874) I. 258 Many New Zealand and many semi-tropical plants. **1975** J. Grady *Shadow of Condor* viii. 130 The bus parked between two idling semi-trucks. **1965** H. K. Compton *Gloss. Purchasing & Supply Managem. Terms.* 123 *Semi-variable cost*, a cost which is partly fixed and partly variable, such as the cost of placing orders, carrying stock, etc., each of which has a fixed cost content, and a variable cost proportional to the volume of throughput. **1971** D. C. Hague *Managerial Econ.* ii. v. 104 Depreciation is a semi-variable cost. **1935** *Industr. & Eng. Chem.* XXVII. 863/2 The main purpose of the semi-works is the development to a financially profitable stage of those processes which have been initiated in the laboratory. **1945** H. D. Smyth *Gen. Acct. Devel. Atomic Energy Mil. Purposes* vii. 74 These include all aspects of the research, development and semi-works studies necessary for the design, construction, and operation of chain-reacting piles. **1956** A. H. Compton *Atomic Quest* 152 A 'semi-works' installation where they could train the men needed for the final operation and where they could make preliminary tests of their equipment and processes.

semi-a·rid, *a.* [Semi- 7 i.] Having slightly more precipitation than an arid climate, grasses being the characteristic vegetation.

1898 [in Dict., s.v. Semi- 7 i]. **1916** *Daily Colonist* (Victoria, B.C.) 22 July 4/2 Irrigation is being more and more applied to cultivated lands that are neither arid nor semi-arid. **1941** J. S. Huxley *Uniqueness of Man* ii. 61 The semi-arid bush country provides but scanty nutriment. **1976** *Sci. Amer.* Sept. 99/2 The wheat plant, which dominates the semiarid croplands of the world, fills the need in this area for a cultivated crop with a lower demand for water and a great tolerance of drought.

semi-automa·tic, *a.* and *sb.* [Semi- 1 a, 7 l.] **A.** *adj.* Partially automatic; *spec.* designating a system, device, or machine whose function is not completely automatic.

1890 W. James *Princ. Psychol.* I. 5 The performances of animal *instinct* seem semi-automatic, and the *reflex acts* of self-preservation certainly are so. **1937** [see *chrome *sb.* 1 b]. **1970** *Computers & Humanities* IV. 351 *Scope*: To make more specific and semiautomatic the testing technique of scientific works, intensive critique of books and other lengthy works.

2. Specialized uses. **a.** Of a type of lathe: that can perform a number of distinct operations on a given work-piece without intervention from an operator.

1903 T. R. Shaw *Lathes, Screw Machines, Boring & Turning Mills* xii. 505 A semi-automatic lathe, having four spindles, is illustrated. **1950** S. J. Gibson in A. W. Judge *Centre, Capstan & Automatic Lathes* II. iv. 181 The 'Maximatic'..is semi-automatic only in the sense that the loading is done by hand.

b. *Mil.* Of a firearm: that loads itself or performs part of the loading operation automatically, but does not fire continuously.

1911 H. A. Bethell *Mod. Artillery in Field* i. 15 Semi-Automatic Actions.—In these the breech opens automatically during the run up, and ejects the cartridge case. **1945** C. E. Balleisen *Princ. Firearms* i. 2 Most 'automatic' pistols are only *semiautomatic*. **1976** J. Wainwright *Walther P.38* 7 The Luger..was a *semi-automatic* pistol... Its trigger had to be squeezed for each shot.

c. Applied to a telephone system of which the operation is automatic except that dialling of the required number is done by an operator (see quots.).

1912 J. Poole *Pract. Telephone Handbk.* (ed. 5) xxxii. 536 Semi-automatic systems are in use on the Continent, and also in some towns in America, but not to any great extent. **1927** C. W. Wilman *Man. Automatic Telephony* xxii. 219 It is claimed for semi-automatic systems that many of the advantages of automatic systems, such as the rapidity of connection and instantaneous clearing, are retained, while the subscribers are relieved of the duty of dialling. **1976** T. H. Flowers *Introd. Exchange Systems* iii. 65 It was asserted that if a subscriber paid for service as he did, it was not right and was even dishonest to make him do his own operating, so semi-automatic exchanges with operators to do the dialling was the right thing.

B. *sb.* **1.** A semi-automatic lathe.

1902 *Lockwood's Dict. Mech. Engin.* (ed. 3) 444 *Semi-automatics*, these constitute a large class of machines which occupy a middle position between the..'full' automatic machines and those which involve the constant attendance of an operator. **1963** N. Weinstein tr. *Boguslavsky's Automatic & Semi-Automatic Lathes* i. 32 If internal surfaces are to be machined in addition to external surfaces, semi-automatics having a central end-working toolslide, or a turret are employed.

2. A semi-automatic firearm.

1964 H. L. Peterson *Encycl. Firearms* 31/2 In 1916, the Germans introduced a limited number of Mauser semi-automatic rifles. The French followed in 1917, with the Saint-Étienne gas-operated semi-automatic. **1978** S. Brill *Teamsters* vi. 239 The gunman reached down, held the High Standard semiautomatic against Mrs. Rand's chin and fired three more shots.

semi-barbarous, *a.* (Earlier example.)

1798 A. F. M. Willich *Elements Crit. Philos.* p. cxxii, The ancient Britons were as little acquainted with the art of writing, as any of the rude and semi-barbarous nations of those times.

se:mibrachia·tion. *Zool.* [Semi- 2 a.] A mode of progression exhibited by certain monkeys in which the forelimbs may be used both as legs in a quadrupedal gait and as arms by which to grasp and swing.

1961 J. R. Napier in *Symp. Zool. Soc. Lond.* V. 127 The twist [of the head] is generally seen in Primates that combine a quadrupedal gait with the specialized arboreal locomotor patterns of brachiation and semi-brachiation. [*Note*] The term semibrachiation has been devised by the author in association with Dr E. H. Ashton to describe the highly arboreal and acrobatic activities of certain of New and Old World monkeys. **1972** W. C. O. Hill *Evolutionary Biol. Primates* ix. 65 Old World exponents of semibrachiation include the six or more genera of the family Colobidae (the Asiatic leaf monkeys and the African *Colobus*).

Hence **semibra·chiating** *ppl. a.*; **semibra·chiator,** an animal that exhibits semibrachiation.
1963 J. R. Napier in *Symp. Zool. Soc. Lond.* X. 186 The mode of locomotion of semibrachiators is basically that of an arboreal quadruped but, in addition, a variable amount of time is spent in swinging by the arms and leaping with the forelimbs outstretched to grasp a handhold. *Ibid.* 188 The African semibrachiating *Colobus* shows a decided preference for the higher strata of the forest canopy. **1973** *Nature* 10 Aug. 373/2 With respect to locomotion, the first component separates quadrupedal cercopithecoids from both knuckle-walkers..and the quadrupedal arm-swinger (so-called 'semi-brachiator') *Ateles.*

semicarbazide (semikā·ɪbăzəid). *Chem.* [f. Semi- 7 h + Carb- + Az(o- + -ide.] A colourless, crystalline, basic solid, NH_2·CO·NH·NH_2, derived from urea by substitution of a hydrazino group for one of the amino groups, which reacts with carbonyl compounds to form semicarbazones. Also, a derivative of this.
1892 *Jrnl. Chem. Soc.* LXII. ii. 1297 When amidoguanidine is treated with dilute acids, or with caustic alkalis, it is first converted into semicarbazide with liberation of ammonia. **1938** G. H. Richter *Textbk. Org. Chem.* xiii. 224 Semicarbazide is a valuable reagent for..the Wolff-Kishner reduction of the carbonyl group. **1968** R. O. C. Norman *Princ. Org. Synthesis* x. 315 In order to obtain derivatives for the characterization of the carbonyl compound it is more satisfactory to use a monosubstituted hydrazine; semicarbazide, phenylhydrazine, and 2,4-dinitrophenylhydrazine are commonly chosen.

semicarbazone (semikā·ɪbăzōᵘn). *Chem.* [f. prec. + -one.] Any of a class of (usu. crystalline) compounds of general formula RR′C:N·NH·CO·NH_2 which are prepared by the condensation of semicarbazide with carbonyl compounds, in order to characterize the parent carbonyl or to protect the carbonyl group in synthesis.
1896 *Jrnl. Chem. Soc.* LXX. i. 343 Mesityl oxide semicarbazone melts at 156°. **1938** G. H. Richter *Textbk. Org. Chem.* vii. 102 The semicarbazones may be converted into hydrazones by the action of sodium ethylate. **1973** H. J. E. Loewenthal in J. F. W. McOmie *Protective Groups in Org. Chem.* ix. 340 In general, and in the steroid field in particular, ease of semicarbazone formation from different types of carbonyl groups follows the order observed with other protecting groups.

semicha (semī·χā). *Judaism.* Also **semichah, semikhah,** and with capital initial. [Heb. *sᵉmīkhāh,* lit. leaning.] The laying-on of hands by which a rabbi is ordained; the ordination of a rabbi. Also, a diploma of rabbinical ordination.
The laying-on of hands, practised only in antiquity, was later replaced by a proclamation (also called *semicha*).
1866 *Chambers's Encycl.* VIII. 70/1 Out of the number of the regular disciples (Talmidim) were chosen the Chaberim (Colleagues), who, again, were elected to the dignity of a rabbi by the 'Semichah', or imposition of hands by three members of the Sanhedrim. **1914** J. Hastings *Encycl. Relig. & Ethics* VII. 604/2 Among these Rabbis there grew up the desire to re-establish the old Rabbinic supremacy of Palestine. They desired to institute once more the *sᵉmīkhāh*, or ordination, and thus ordain a Sanhedrin which would be recognized throughout the world. **1962** *New Jewish Encycl.* 438/2 Technically, Semikhah ceased some two thousand years ago, and was not established anew until the 14th or 15th century. In modern times rabbinical students are granted Semikhah, that is, they receive a rabbinical diploma and become ordained as rabbis upon graduation from a rabbinical school or *Yeshivah.* **1973** *Jewish Chron.* 9 Feb. 22/1 The principal..accepts the fact that some of his best students go on to Gateshead for their semicha, not merely with resignation, but with positive approval. **1976** B. Williams *Making of Manch. Jewry* vii. 184 The possession of Rabbinical *semikhah* endorsed by European rabbis of unquestioned repute.

semicondu·cting, *a. Physics.* [Semi- 7 a.] Having the properties of a semiconductor.
1782 tr. A. Volta in *Phil. Trans. R. Soc.* LXXII. p. xii, The surface of those bodies does not contract any electricity, or if any electricity adheres to them, it vanishes soon, on account of their semi-conducting nature. **1787** [in Dict. s.v. Semi- *prefix* 8]. **1884** J. T. Sprague *Electricity* (ed. 2) xiii. 573 A semi-conducting incandescent material compounded of infusible earths and carbon or metals. **1975** D. G. Fink *Electronics Engineers' Handbk.* vii. 31 By far the most widely used semiconducting materials are germanium and silicon.
Also **semicondu·ctive** *a.*, in the same sense; **semicondu·ction, se:miconducti·vity.**
1931 *Proc. R. Soc.* A. CXXXIII. 469 To explain the general outlines of semi-conduction. **1953** *Jrnl. Inst. Electr. Engineers* C. i. 76/2 The value of semiconductive glaze lies in its use in situations where trouble is expected due to flashover of insulators under adverse weather..conditions. **1954** R. P. Turner *Transistors* i. 8 Many..elements and compounds have been found to possess semiconductivity in varying amounts. **1960** *Cambr. Rev.* 27 Feb. 394/1 Although a Cambridge physicist, A. H. Wilson, put forward the basic theory of their behaviour in 1931, it is really only in the last ten years that the phenomena of semiconductivity have been clearly understood. **1973** K. Seeger *Semiconductor Physics* i. 2 Semiconduction is specified by the following properties:[etc.].

semicondu·ctor. *Physics.* [Semi- 7 a.] **1. a.** A material whose capacity to conduct electricity is intermediate between that of a good conductor and an insulator. *Obs.* exc. as in next sense.
1838 *Ann. Electr., Magn., & Chem.* III. 316 Lichtenberg ..observes..'it is deserving of a trial also whether phosphorus would not become ignited at points whence a stream [of electricity] is issuing, on a semi-conductor being inserted between them'. **1863** E. Atkinson tr. *Ganot's Elem. Treat. Physics* ix. iv. 592 The retardation which electricity experiences in traversing a semi-conductor, such as a wet string. **1879** [in Dict. s.v. Semi- 8]. **1900** *Engineering* 28 Sept. 412/3 Semi-conductors like iron filings.
b. *spec.* Such a material in which there is a narrow gap between permitted energy bands, so that the only current carriers are electrons thermally excited from the valence band into the conduction band (*intrinsic semiconductor*: see *intrinsic *a.* 3 e) or into intermediate energy levels provided by impurity ions (*extrinsic semiconductor*).
1931 *Proc. R. Soc.* A. CXXXIII. 459 It is not possible to maintain that the difference between good and bad conductors is one of degree only... There is an essential difference between a semi-conductor, such as germanium, and a good conductor, such as silver, which must be accounted for by any theory which attempts to deal with semi-conductors. **1946** *Electronic Engin.* XVIII. 66/2 It is well known that 'semi-conductors', such as carbon, silicon.., etc., possess negative temperature coefficients of resistance at ordinary temperatures. **1961** G. R. Choppin *Exper. Nuclear Chem.* iii. 41 The semiconductor detectors..are made from thin (approximately 1mm) wafers of semiconductor silicon. **1970** J. Shepherd et al. *Higher Electrical Engin.* (ed. 2) xx. 623 In an extrinsic or doped semiconductor, impurities are added to the intrinsic material to give a predominance of either electrons (in *n*-type material) or holes (in *p*-type material) as charge carriers. **1979** *Jrnl. R. Soc. Arts* Oct. 692/2 Sometimes a significant advance in technology may itself create a new market, as did the advent of the semiconductors to the small 'transistor radio' market.
2. Special Combs.: **semiconductor diode,** a diode whose rectifying action depends on the properties of a junction between a semiconductor and either a metal or another type ef semiconductor; cf. *junction diode* (s.v. *junction *sb.* 4); **semiconductor junction** = *junction *sb.* 2 b; **semiconductor rectifier,** a semiconductor diode, usu. one intended for large currents; **semiconductor triode,** a junction transistor having two junctions.
1954 *Trans. IRE Prof. Group Broadcast & Television Receivers* July 34 (*heading*) Semiconductor diodes for TV receivers. **1975** Fink & McKenzie *Electronics Engineers' Handbk.* vii. 34 One of the highest-volume uses of the semiconductor diode is in computers. *Ibid.* 35 When a semiconductor junction is exposed to light, photons generate hole-electron pairs. **1946** *Physical Rev.* LXIX. 42/2 This effective contact e.m.f. is one important parameter in the theory and practice of semi-conductor rectifiers. **1962** *Times* 14 May 14/7 Semiconductor rectifiers on heavy-duty electric railway locomotives. **1948** *Physical Rev.* LXXIV. 230/1 (*heading*) A semi-conductor triode. **1970** D. F. Shaw *Introd. Electronics* (ed. 2) xii. 262 The transistor is a semiconductor triode possessing characteristics which are similar in many respects to those of thermionic triodes.

se:mi-conse·rvative, *a. Biochem.* [Semi- 1.] Of the replication of nucleic acid: such that one complete strand of each double helix is directly derived from its parent.
The term was originally proposed (see quot. 1957) to designate a class of models of the replication of DNA; it is now accepted that its true mechanism of replication falls in this class.
1957 Delbruck & Stent in McElroy & Glass *Chem. Basis of Heredity* 707 The considerable number of proposed schemes may be divided into three general classes as conservative, semi-conservative, and dispersive. *Ibid.* 709 Semi-conservative mechanisms are those which conserve the atomic identity of single chains of the parental DNA duplex, although effecting a permanent separation of the two chains from each other in the course of replication. **1970** *Nature* 7 Nov. 522/1 There is good evidence that DNA replication is semi-conservative and involves separation of the two strands which then act as templates. **1976** P. Collard *Devel. Microbiol.* viii. 106 Semi-conservative replication thus provided a possible answer to the riddle of the stability of the genome from generation to generation.
Hence **se:miconse·rvatively** *adv.*
1966 *Jrnl. Molecular Biol.* XV. 372 Since all the DNA is replicated semiconservatively..such a segregation pattern could arise most simply if every chromosome contained all its DNA within a single molecule. **1979** *Nature* 3 May 75/2, pBR322 DNA replicates semiconservatively and completely in a crude lysate of *E. coli.*

semi-cry·stalline, *a.* [Semi- 7 a.] Having or being a structure of crystals embedded in an amorphous groundmass; having or being a structure possessing crystalline character to some extent.
1816, etc. [in Dict., s.v. Semi- 7 a]. **1871** *Phil. Mag.* XLII. 404 This silica forms a series of semi-crystalline bands parallel with the walls of the fissure. **1927** *Proc. Physical Soc.* XXXIX. 370 In 1924 I [*sc.* Rutherford] put forward a suggestion that the central nucleus was a closely ordered arrangement of α particles and electrons in a semi-crystalline formation. **1975** *Sci. Amer.* Dec. 99/2 Representative values for the percentage by volume of crystals in a number of semicrystalline polymers are high-density polyethylene, 75 percent; low-density polyethylene, polypropylene and nylon, 50 percent.

semi-demi-. (Earlier examples in both senses.)
a **1661** Fuller *Worthies* (1662) ii. 246 [Rabbits'] wool is..used in making of hats, commonly..called half-beavers, though many of them hardly amount to the proportion of semi-demi-castors. **1826** J. F. Danneley *Mus. Gram.* 4 The Semidemisemiquaver has four hooks.

semi-det (semi,de·t), *sb.* Short for Semi-detached house.
1960 J. Stroud *Shorn Lamb* x. 110 An unexpected footpath in between a couple of semi-dets. **1978** J. McNeil *Consultant* xx. 174 A neat semidet with a bay window.

semi-detached, *a.* Add: **B.** *absol.* as *sb.* A semi-detached house. *colloq.*
1928 D. H. Lawrence *Lady Chatterley's Lover* xi. 188 New little streets of semi-detacheds were run up. **1957** M. & A. Potter *Interiors* 43/2 The garden city notions.. blotting the green countryside everywhere with varieties of the popular 'semi-detached'. **1979** C. Dexter *Service of All Dead* iii. 23 He'd found a quiet little semi-detached to rent.

semidine (se·midīn). *Chem.* [ad. G. *semidin* (P. Jacobson 1893, in *Ber. d. Deut. Chem. Ges.* XXVI. 700), f. *semi-* Semi- + *benzi-din* benzidine (s.v. Benzo-).] Any compound which is either (*a*) an *ortho*- anilino-derivative, or (*b*) an *N*-*para*-aminophenyl derivative, of a *para*-substituted aniline (distinguished as *ortho*- and *para*-semidines respectively); also *semidine base*; *semidine reaction, transformation,* etc., the rearrangement of *para*-substituted hydrazobenzenes in the presence of acid to yield *ortho*- and *para*-semidines (in proportions governed by the nature of the substituents).
1893 *Jrnl. Chem. Soc.* LXIV. i. 330 The hydrazo-compound..undergoes molecular change yielding two compounds, derivatives of ortho- and of para-amidodiphenylamine. It is proposed to call this reaction the semidine reaction, and to designate the orthamidodiphenylamine bases thus obtained by the name orthosemidines; the paramidodiphenylamine bases by the name parasemidines. **1898** *Ibid.* LXXIV. i. 441 Only 50 per cent. of the total semidines obtained. **1938** A. J. Mee tr. *P. Karrer's Org. Chem.* xxxiv. 498 If a *para*-position in hydrazobenzene is already occupied by a substituent, there are still further possibilities of isomerization. In addition to a diphenyline base and benzidine compounds .., two diphenylamine derivatives are formed in which only one of the benzene nuclei has rotated, the so-called *p*-semidine and *o*-semidine bases. The transformation is known as the semidine transformation. **1959** E. S. Gould *Mech. & Structure in Org. Chem.* xv. 658 The rearrangements of hydrazobenzenes to benzidines, to diphenylenes, and to *o*-semidines are third-order reactions, first order in substrate and second order in hydrogen ion. **1975** R. F. Brown *Org. Chem.* xxii. 772 A small amount of *o* shift occurs anyway, as well as a halfway shift to give semidines.

semi-dire·ct, *a.* [Semi- 1 a.] Not wholly direct; *spec.* (of lighting) so disposed that most but not all of the light reaches the illuminated area without first being reflected.
1914 S. C. Batstone *Electr. Light Fitting* viii. 166 If a frosted glass bowl or something of a similar nature be placed between the lamps and the surface or space to be illuminated, the result is 'semi-direct lighting'. **1926** J. S. Huxley *Essays of Biologist* i. 15 We can learn a great deal from the semi-direct methods of paleontology. **1957** *Encycl. Brit.* XIV. 107/2 Semidirect [lighting equipment] .. 10–40 [% upward]... 60–90 [% downward]. **1971** Powell & Higman *Finite Simple Groups* viii. 294 We form the semi-direct product *GI*.

se·mi-field. *Math.* [Semi- 7 a.] Used variously to denote a set, together with operations answering to addition and multiplication, that has certain specified properties of a field but not all of them.
1923 *Ann. Math.* XXIV. 240 A set *D* which satisfies these postulates will be called a semi-field. **1966** *Math. Rev.* XXXI. 39/2 A semifield is essentially an algebraic structure which satisfies all field axioms except perhaps associativity and commutativity of multiplication; the more customary terminologies are 'division ring' (not necessarily associative) and 'distributive quasifield'.

se·mi-group. *Math.* [ad. F. *semi-groupe* (J.-A. de Séguier *Élem. de la Théorie des Groupes Abstraits* (1904), i. 8); cf. Semi- 7.] A set together with an associative binary operation under which it is closed.
1904 *Bull. Amer. Math. Soc.* XI. 160 The author [*sc.* de Séguier] introduces..a semigroup *G* in connection with any subset *S* containing a system of generators of *G*. The postulates defining *G* are: (1) associativity; (2) for any *a* in *S* and *b* in *G*, there is at most one solution (*n* in *G*) of

an = *b*; (3) similarity for *na* = *b*. **1905** *Trans. Amer. Math. Soc.* VI. 205 The correct theorem involves the concept semi-group, which reduces to a group when there is a finite number of elements, but not in general for an infinitude of elements. **1968** P. A. P. Moran *Introd. Probability Theory* ii. 66 The convolution operation..has some of the properties of multiplication in that it is associative..and commutative,..but division is not in general possible. With this operation the set of all discrete distributions on (0, 1, ...) is therefore said to form a 'semi-group'. **1972** A. G. Howsow *Handbk. Terms Algebra & Anal.* v. 25 A semigroup..possessing..an identity element..is called a monoid. **1979** *Proc. London Math. Soc.* XXXVIII. 335 First we find exactly when the resolvent operators and the semigroup operators are strong Feller operators.

se:mi-indire·ct, *a.* [Semi- 1 a.] Of lighting: so disposed that most but not all of the light reaches the illuminated area indirectly, after having been reflected or scattered by some surface. Cf. *semi-direct *a.*
1914 J. Eck tr. *Högner's Light, Radiation & Illumination* (*table facing p.* 44), Ordinary direct-current arc lamp with standard arrangement of carbons and opal glass bowl below the arc producing semi-indirect illumination. **1964** S. Duke-Elder *Parsons' Dis. of Eye* (ed. 14) xxxvii. 559 In semi-indirect lighting the use of opalescent bowls permits a certain amount of direct illumination.

semi-invariant: see Seminvariant in Dict. and Suppl.

semikhah, var. *semicha.

semi-le·thal, *a.* and *sb. Genetics.* [Semi- 7 a.]
A. *adj.* Of an allele or a chromosomal abnormality: causing impaired viability of most of the individuals homozygous for it.
1917 *Proc. Nat. Acad. Sci.* III. 621 The viability of the three sex-linked dominants was .. already known; the remaining six dominants were tested. In all, it was found that three of the nine are not lethal, one..is semi-lethal, and five..are completely lethal when homozygous. **1927** *Brit. Jrnl. Exper. Biol.* V. 124 This mutation appears to be semi-lethal, as the animals homozygous for it are delicate and difficult to rear. **1937** *Genetics* XXII. 471 A group of nine changes obtained from male 8..proved to be semilethal. **1962, 1973** [see *lethal *a.* 1 d].
B. *sb.* A semi-lethal gene.
1919 *Proc. Soc. Exper. Biol. & Med.* XVII. 12 Four of the lethals (perhaps five) are more strictly speaking 'semi-lethals', as they occasionally allow the male possessing them to live. **1944** *Proc. Nat. Acad. Sci.* XXX. 174 *Fu* is a tail mutation closely resembling *Ki*. It acts as a semi-lethal, some *Fu Fu* being viable. **1974** *Nature* 2 Aug. 451/3 Most of such populations in North America and Europe proved to be polymorphic for variant genes, usually embryonic lethals or semilethals, belonging to the T series.

Sémillon (semiyoñ). Also **Semillon** and with small initial. [Fr. dial. (Midi), ult. ad. L. *semen* seed.] A white grape of France; also a similar one grown abroad; the white wine made from this grape.
1875, etc. [see *Sauvignon a]. **1926** [see *Monbazillac]. **1963** *Times* 12 Mar. (Austral. Suppl.) p. xv/7 The best dry whites are made from the riesling and the semillon. **1978** *Courier-Mail* (Brisbane) 16 Nov. 20/5 If you think you know semillon, you may be surprised by the flavour of Huntington Estate's Mudgee semillon, 1978.

semi-log (stress variable), *a.* [Shortened f. next: cf. *log *sb.*[3] and *a.*] = next.
1921 W. C. Marshall *Graphical Methods* ii. 16 It often happens that log-log or semi-log paper would be preferable to rectangular ruled paper. **1941** *Trans. Amer. Soc. Mech. Engineers* LXIII. 539/2 We may expect..information on just why actual log curves do not plot straight on semilog paper. **1978** *Sci. Amer.* Mar. 150/3 A plot of the data on a semilog graph is shown.

se:mi-logari·thmic, *a.* Also **semilogarithmic.** [Semi- 8.] Having or being a scale that is linear in one direction and logarithmic in the other. Cf. prec.
1919 A. C. Haskell *How to make & use Graphic Charts* iv. 20 The semi-logarithmic chart has a wide use for the plotting of comparative statistics of similar kind but dissimilar magnitude. **1930** R. Pearl *Med. Biometry & Statistics* (ed. 2) vi. 183 The scale of the ordinates is divided not in arithmetic progression but in proportion to the logarithms of numbers in arithmetic progression. Such a ruling is called an arithlog or semi-logarithmic grid. **1977** J. L. Harper *Population Biol. of Plants* xv. 458 Some of these are redrawn in Fig. 15/1 on a semi-logarithmic scale to show the change in range of dispersal if the seed crop is halved.
Hence **se:mi-logari·thmically** *adv.*
1919 A. C. Haskell *How to make & use Graphic Charts* iv. 18 This is shown very clearly by charting the same data on both arithmetically and semi-logarithmically ruled paper. **1976** *Nature* 11 Mar. 153/2 Figure 1 is an action spectrum obtained by plotting semi-logarithmically the reciprocal of retinal irradiance ($1/W\ cm^{-2}$) against wavelength (nm) for each exposure duration.

semi-mature, *a.* Restrict † *Obs. rare*⁻¹ to sense in Dict. and add: **b.** [Semi- 7 a.] Partially mature. (Examples in senses 1 b and *d of Mature *a.*)
1928 *Bull. Amer. Soil Survey Assoc.* IX. 39 Semimature soil. **1976** *Southern Even. Echo* (Southampton) 16 Nov. 2/8 Semi-mature trees are to be planted in Station Road.

semi-metal. Delete *Old Chem.* and add: Now usu. signifying incomplete metallic character in other physical properties, esp. electrical conductivity; *spec.* an element (as arsenic, antimony, bismuth) or other substance having properties intermediate between those of true metals and those of semiconductors. (Later examples.)
1912 W. E. Ford *Dana's Man. Mineral.* (ed. 13) iv. 115 The semimetals—tellurium, arsenic, antimony and bismuth—belong together in a crystal group, all of them showing rhombohedral crystals with closely agreeing fundamental angles. **1952** *Chem. Abstr.* XLVI. 1187 Semimetals with continuous properties between metals and ceramics. **1972** *Science* 19 May 753/1 There is a clear trend both with increasing pressure and with atomic number from a semiconductor to a semimetal with a distorted simple cubic structure to a metallic, simple cubic phase.

semi-metallic, *a.* Add: (Later example.) Cf. *semi-metal.
1974 D. M. Adams *Inorg. Solids* v. 88 In contrast, the heaviest elements of the same groups (Pb, Bi, Po) are metallic or semimetallic as revealed by a progressive change down the group in properties such as electrical conductivity and optical behaviour, as well as by changes of structure type.

semimicro- (semimɘi·kro), *prefix* and quasi-*adj. Chem.* [f. Semi- 1 a + Micro- 2 a, *8 b.]
a. Formative element denoting a scale of quantitative analysis between micro-scale and macro-scale (commonly 0·01–0·1 gramme), as in *semimicroanalysis, -analytical* adj., *-determination, -method.*
1951 A. I. Vogel *Text-Bk. Quantitative Inorg. Analysis* (ed. 2) xi. 814 Semi-micro-analysis is concerned with the manipulation of 10–100 mg. of material, whilst micro-analysis deals with 1–10 mg. of material. **1938** *Jrnl. Soc. Chem. Industry* Dec. 464/1 (*heading*) The determination of free and saline ammonia using a semi-microanalytical method. **1937** *Industr. & Engin. Chem.* (*Analytical Ed.*) 15 June 296 (*heading*) Micro-, semimicro-, and macro-determination of halogens in organic compounds. **1933** *Ibid.* Nov. 402/1 Northrop..has used a semi-micromethod with an accuracy of 0·2 per cent. **1963** tr. *Alexeyev's Quantitative Analysis* i. 16 The main advantages of the micro- and semimicromethods are their high speed and the need for only very small amounts of material.
b. Used without hyphen as an independent word.
1935 *Industr. & Engin. Chem.* (*Analytical Ed.*) Nov. 432/1 Semi-micro adaptations of the classical Kjeldahl method..have been reported. **1946** F. Schneider *Qualitative Organic Microanalysis* ii. 39 Bernhauer describes several types of micro or semi-micro receivers for vacuum distillations. *Ibid.* vii. 195 A test for the nitro group carried out on a semi-micro scale. **1974** [see *micro- 8 b]. **1976** *Lancet* 13 Nov. 1091/2 Plasma creatinine (semi-micro alkaline-picrate method) and urea (diacetylmonoxime method) concentrations were measured.

semi-mo·nocoque. [Semi- 7 a.] **a.** *Aeronaut.* A fuselage or other structure having a rigid outer skin and a framework of longerons or stringers, so that stresses are shared between the skin and the framework. Usu. *attrib.*
1918 *Flight* 28 Feb. 224/2 This semi-*monocoque* structure would retain its strength even after damaging some of the longitudinal members. **1931** Warner & Johnston *Aviation Handbk.* 653 The semimonocoque has the skin reinforced by longerons and vertical bulkheads but has no diagonal web members. *Ibid.,* For the semimonocoque type the verticals should be designed as for the true monocoque designs. **1948** [see *monocoque a]. **1960** C. H. Gibbs-Smith *Aeroplane* xii. 93 As many of the so-called 'monocoque' structures came to include internal stiffening members—the skin bearing most but not all of the loads —the term 'stressed-skin' is better, implying a 'semi-monocoque' rather than a wholly unaided 'shell' structure. **1980** R. C. Mikesh *Albatros D.Va* i. 13 Their semi-monocoque, smoothly contoured plywood fuselage was a radical change from the boxy, fabric-covered structures then in general use.
b. In a motor vehicle, a body or chassis combining features of the monocoque and space-frame types. Usu. *attrib.*
1966 *Publ. Amer. Dial. Soc.* 1964 xlii. 8 *Semi-monocoque, adj.*, applied to a chassis which combines constructions of both monocoque and space frame, such as that of the Jaguar. **1973** C. Campbell *Design Racing Sports Cars* viii. 134 The Formula 1 solution is usually called 'semi-monocoque' since the typical design has a centre section that is more like a bath-tub than an egg.

seminal, *a.* and *sb.* Add: **A.** *adj.* **4. a.** Also, freq. used of books, work, etc., which are highly original and influential; more loosely: important, central to the development or understanding of a subject.
1947 *Partisan Rev.* XIV. 409 To be sure, Engels' more specialized *Anti-Dühring* and *Feuerbach*, if less seminal are more systematized, more apposite to Lenin's immediate purposes. **1957** D. J. Enright *Apothecary's Shop* 233 The seminal works of modern literary criticism (such as Eliot's earlier essays and Leavis's *Revaluation*). **1960** *Guardian* 14 Oct. 8/6 Everything he says is of real value... This is what academics call a seminal book. **1960** *Twentieth Century* Nov. 438 Since the war there has been..no seminal poet in the younger generation. **1977** *New Yorker* 6 June 122/2 That the two pianists, each seminal, agreed to play together at all was startling.

seminar[2]**.** Delete ‖, transfer stress-dot to first syllable, and add: **b.** A conference of specialists; also, more generally, a course of instruction for managers, etc. orig. *U.S.*
1944 *Sun* (Baltimore) 21 Sept. 6/2 Problems of condemnation for the temporary war use of property will be featured at a seminar conducted by the American Institute of Real Estate Appraisers. **1969** *Listener* 10 Apr. 482/2 For the first time in its history, the Conservative Party has held a one-day seminar, not on Exports, or on Trade Unions, or on Defence, but on the Arts. **1982** *REMARC Database News* July 1 Key members of the Scottish Library Network..attended a REMARC seminar at the Carrollton Press data-entry facility at Irvine, Scotland.
2. *attrib.*
1948 *Sun* (Baltimore) 12 May 12/3 Two years ago he started the personal campaign that took him into large communities and small, and he developed a sort of '*seminar*' method of question-and-answer discussion at luncheons, meetings of students and similar gatherings. **1959** Seminar-paper [see *fence-sitting* s.v. *fence *sb.* 11]. **1977** J. D. Douglas in Douglas & Johnson *Existential Sociol.* i. 27 One can even see the feelings leading to important actions, such as separating from a lover or starting a seminar workshop to deal with the feelings. **1981** *Times Lit. Suppl.* 6 Feb. 136/4 The names of these daunting authors..make an occasional modest appearance on reading-lists and in seminar-rooms.

seminarist. 3. (Later example.)
1972 *Daily Tel.* 14 Jan. 17 For a fee of £60, seminarists will be impressed with the need for strategic plans for growth..over the next decade, fed but not accommodated.

Seminole (se·minōᵘl), *sb.* and *a.* Also 8 **Seminollie,** 8–9 **Siminole,** etc. [ad. Creek *simanó:li* wild, runaway, earlier and dial. *simaló:ni,* f. Amer. Sp. *cimarrón* (cf. *cimarron).] **A.** *sb.* **1.** A member of any of several groupings of North American Indians that comprise or comprised Creek Confederacy emigrants in Florida, or their descendants now resident in Florida and Oklahoma, esp. the present-day Florida Indians, who speak the Muskogee and Mikasuki languages of the Muskogean family. Also as *collect. sing.*
1763 W. G. de Brahm in *Amer. Indian Ethnohist.: S. & S.E. Indians: Florida Indians* (1974) I. 244 The Surveyor General..must have fallen in with many Hunting Ganges of Semiolilies. **1771** in *Ibid.* III. 17 The Seminolies or East Florida Creeks had frequent intercourse with Spaniards..by means of Fishing vessels. **1789** *Amer. State Papers: Indian Affairs* (1832) I. 15 Some of the most southern towns of the Lower Creeks, or Seminoles, are within the territory of Spain. **1866** 'F. Kirkland' *Pict. Bk. Anecdotes* 318/2 He fainted at the spectacle, and was soon after butchered by a Seminole. **1910** F. W. Hodge *Handbk. Amer. Indians* II. 500/2 While still under Spanish rule the Seminole became involved in hostility with the United States. **1946** *Nat. Geogr. Mag.* Jan. 53/2 The later Seminole, who were primarily an offshoot of the Creeks and Hitchiti, were also a Muskhogean people. **1972** *Listener* 28 Dec. 904/3 Hidden here in the Everglades..a race of forgotten, proud but degenerate Seminoles.
2. An Eastern Muskogean language of the Seminoles.
1848 *Southern Lit. Messenger* XIV. 482/2, I concluded at the time [that the opera] was written in Seminole, as the only word which I distinctly heard was *en ca.* **1933** [see *Creek *sb.*[2]]. **1972** J. L. Dillard *Black Eng.* iv. 153 The sentence is in fact Pidgin English with some relexification by Seminole.
B. *adj.* **1.** Of, pertaining to, or designating any of the Seminole groupings, or these peoples collectively.
[**1774**: see sense 2 below.] **1775** W. Bartram in *Trans. Amer. Philos. Soc.* (1943) XXXIII. 160 These were Seven likely Young Siminole Fellows all elligantly dresst & painted after the Indian fashions. **1797** J. Morse *Amer. Gazetteer* s.v. *Calos,* Not far from this is a considerable town of Seminole Indians. **1837** H. Martineau *Society in Amer.* II. 71 The Seminole fathers would not deliver them up. **1881** *Rep. Indian Affairs* (U.S.) p. lv, [They] were willing to incorporate the whole Seminole tribe into their nation. **1945** L. R. Tryon *Poor Man's Doctor* 3 The belt-like stone-crop, symbol of fertility to the original Seminole inhabitants of this paradise. **1973** A. H. Whiteford *North Amer. Indian Arts* 96 Seminole patchwork is unique among Indians of North America.
2. Special collocations: **Seminole horse,** a small horse belonging to a feral stock once found in south-eastern North America and locally domesticated by Indians and others; **Seminole war,** any of three wars waged by the U.S. against the Seminole Indians in 1817–18, 1835–42, and 1855–58.
1774 W. Bartram in *Trans. Amer. Philos. Soc.* (1943) XXXIII. 148 Here we saw herds of deer bounding before the chace of the naked active Floridian mounted on his fleet Siminole horse. **1806** P. Wakefield *Excursions N. Amer.* xvi. 107 Extensive savannahs..maintain

innumerable herds of deer, cattle, and Siminole horses, which are of a small breed. **1931** F. Harrison *John's Island Stud* (*South Carolina*) *1750–1788* 170 The Seminole (or Creek) horse, small in size and capricious in nature, having its origin in Florida. **1818** *Repub. Constellation* (Winchester, Va.) 11 July 2/1 Gen. Jackson..obtained full proof that the Spanish authorities at Pensacola had been active in fomenting the Seminole war. **1837** H. Martineau *Society in Amer.* II. ii. i. 71 Probably few of the United States troops who fell in the late Seminole war knew how the strife arose. **1948** *Florida Hist. Q.* July 35 Had not the Seminole war intervened, there is little question that the settlement of the present Taylor county would have begun ten years earlier than it did. **1973** D. Aaron *Unwritten War* iv. xi. 171 Sergeant Weber.. veteran of the Seminole War.

seminoma (seminōᵘ·mă). *Path.* [mod.L., ad. F. *seminome* (M. Chevassu *Tumeurs du Testicule* (1906) i. 15), f. L. *sēmin-, sēmen* Semen: see *-oma.] A malignant tumour of the testis, now acknowledged to derive from spermatogenic tissue.
[**1919** J. Ewing *Neoplastic Dis.* xl. 773 (*caption*) Embryonal carcinoma of testis. ('Seminome of Chevassu').] **1931** *Amer. Jrnl. Obstetrics & Gynecol.* XXII. 697 The term seminoma was chosen because this testicular new-growth in parts of its structure resembles seminal tubules. **1966** Wright & Symmers *Systemic Path.* I. xxvi. 820/1 Between 40 and 50 per cent of all testicular tumours are seminomas. **1974** J. D. Maynard in R. M. Kirk et al. *Surgery* viii. 190 The seminoma is a firm uniform mass of sheets of polyhedral cells with a tendency to undergo central necrosis.

semi-noma·dic, *a. Anthrop.* [Semi- 1 a.] Of a people, way of life, etc.: partially nomadic and partially settled. Freq. applied to a social group that depends largely on seasonal pasturing.
1843 J. C. Prichard *Nat. Hist. Man* xxx. 316 The Kafirs are associated together in large communities under chiefs, or kings... They are semi-nomadic, although living in towns of considerable size..which they occasionally move. **1918** G. Bell *Let.* 17 Mar. (1927) II. xvii. 450 They were men of the Ghazzi, a semi-nomadic tribe near Nasiriyeh. **1954** Koestler *Invisible Writing* x. 110 The natives were left to their semi-nomadic existence. **1960** J. Bright *Hist. Israel* ii. 73 The patriarchs..were semi-nomadic stockbreeders such as we know from the Tale of Sinuhe (twentieth century) or the Mari texts.
Hence **semi-no·mad** *a.* and *sb.*
1948 in D. Diringer *Alphabet* ii. iv. 277 Semi-nomad Arabs may have had their settlements round the natural wells from time immemorial. **1960** J. Bright *Hist. Israel* ii. 72 The patriarchs are portrayed as seminomads living in tents. **1972** *Catholic Bibl. Q.* Apr. 231 A seminomad sheep-farmer.

seminvariant. Add: **2.** *Statistics.* Any of a set of functions of a statistical distribution, each expressible as a polynomial in the moments.
Now usually spelt *semi-invariant.*
[**1903** T. N. Thiele *Theory of Observations* vi. 24 From the sums of powers we can easily compute also another serviceable collection of symmetrical functions, which for brevity we shall call the half-invariants.] **1922** A. Fisher *Math. Theory Probabilities* (ed. 2) xiv. 191 (*heading*) Semi-invariants of Thiele. **1930** *Biometrika* XXII. 225 Thiele, in 1889, after defining the semi-invariants, used symmetric functions of the observations of a sample which are the same functions of the sample moment coefficients as the population semi-invariants are of the population moment coefficients. He supplied an expression covering all the semi-invariants of the mean. **1968** P. A. P. Moran *Introd. Probability Theory* vi. 267 The κ_n ($n \geqslant 2$) are dependent only on μ_2, μ_3,..and are the same for all distributions of the form $F(x + d)$ ($-\infty < d < \infty$). They are therefore sometimes known as 'semi-invariants' since they are invariant under translation.

semi-official, *a.* (Earlier example.)
1806 *Deb. Congr. U.S.* 6 Mar. (1852) 597 Must we have semi-official authority, even for a title-page?

semiology. The usual spelling is now *semi-* not *semei-*: see Semeiology and derivs. in Dict. and Suppl.

semi-opacous, *a.* For '= next' read = Semi-opaque *a.*

semi-o·pen, *a.* [Semi- 1 a, 7 a.] Partially open; *spec.* in *Med.*, applied to methods of administering anæsthetics in which the inspired gas is atmospheric air partially restricted or controlled by some device. Cf. *open *a.* 11 h, *semi-closed* adj. (*b*) s.v. *semi- 8.
1914 J. T. Gwathmey *Anesthesia* vi. 276 If ethyl chlorid [*sic*] is administered to adults from an open or semi-open inhaler large quantities of the drug will be needed to bring about even partial anesthesia. **1972** [see *open *a.* 11 h].

semiosis (sīmi,ōᵘ·sis). Also **semeiosis.** [a. Gr. σημείωσις sign, inference from a sign.] The process whereby something functions as a sign (see also quots. 1971 and 1981).
c **1907** C. S. Peirce *Coll. Papers* (1934) V. iii. i. 332 It is important to understand what I mean by *semiosis.* All dynamical action..either takes place between two subjects..or..is a resultant of such actions between pairs. But by 'semiosis' I mean, on the contrary, an action, or influence, which is, or involves, a coöperation of *three* subjects, such as a sign, its object, and its interpretant. **1938** A. Huxley *Let.* 18 Nov. (1969) 438 It interests me a lot and has set me reading along a number of interesting lines—Carnap, Neurath, Morris and Korzybski on the problems of semiosis. **1963** J. Wiesenfarth *H. James* v. 97 The novel..is, rather, a study in logic and semiosis. **1971** Heath & Prendergast tr. J. Kristeva in *Signs of Times* 3 What we call semeiosis is not the signifying activity in all its complexity, but only one of the signifying acts such as the structure of judgement allows it to filter through. **1981** M. Warner *Joan of Arc* i. i. 28 Elements of mimesis..cling to..accounts of her..death, while..an accretion of semiosis, the search for inner meaning, covers their story.

semiotic, semiotics: see Semeiotic *a.*, Semeiotics in Dict. and Suppl.

semipe·rmeable, *a.* [tr. G. *halbdurchlässig* (J. H. van't Hoff 1887, in *Zeitschr. für physik. Chem.* I. 482), f. *halb-* Half, semi- + *durchlässig* pervious, permeable.] Of a membrane or other structure: selectively permeable to certain atoms and molecules; *spec.* permeable to molecules of water but not to those of any dissolved substance. Also in extended use.
1888 W. Ramsay tr. J. van't Hoff in *Phil. Mag.* XXVI. 82 The porous membrane..will be termed in the following pages a 'semipermeable membrane'. **1895** W. C. D. Whetham *Solution & Electrolysis* iv. 34 These semi-permeable membranes are made by filling a porous pot with the solution of a salt such as potassium ferrocyanide, and surrounding the outside with another solution..which gives an insoluble precipitate when in contact with the first. **1930** *Engineering* 28 Nov. 670/1 Each droplet [of sap] being subject to forces equivalent to the osmotic pressure acting on a plant cell within a semi-permeable membrane. **1974** L. Thomas *Lives of Cell* (1975) 162, I am glad to have a semipermeable memory after getting into this. **1978** P. W. Atkins *Physical Chem.* viii. 222 The chemical potential of the solvent on both sides of the semipermeable membrane must be equal.
Hence **se:mipermeabi·lity,** the property or condition of being semipermeable.
1900 W. H. Howell *Text-bk. Physiol.* (ed. 2) I. ii. 66 The semi-permeability is only approximately complete. **1974** L. Thomas *Lives of Cell* (1975) 173 Oxygen filters out the very bands of ultraviolet light that are most devastating for nucleic acids and proteins, while allowing full penetration of the visible light needed for photosynthesis. If it had not been for this semipermeability, we could never have come along.

semi-pro-, *a.* and *sb. colloq.* (orig. and chiefly *U.S.*). **A.** *adj.* **a.** = *semi-professional *a.* 1 b. *spec.* of sports.
1908 *Spalding's Official Base Ball Guide* 368 (Advt.), The 'Semi-Pro' League Ball; regulation size and weight.. Price, $1.00. **1910** *Baseball Mag.* Apr. 4/2 The semi-pro league flourishes. **1969** *Wall St. Jrnl.* 30 Sept. 28/2 Sam turned to umpiring..working in semipro and minor leagues. **1972** J. Mosedale *Football* viii. 116 He was playing semi-pro ball on Sundays. **1980** *Washington Post* 15 Aug. D2/1 The Metro Buccaneers, a first-year semipro football team, will play an exhibition game..Sunday.
b. = *semi-professional *a.* 1 a. Chiefly in *Music* and *Photography.*
1927 *Melody Maker* Aug. 757/3 This competition will be open to amateur, semi-pro. or full pro. combinations. **1934** S. R. Nelson *All about Jazz* iv. 82 There are thousands of 'semi-pro' dance bands, the members of which work at their ordinary occupations during the day and play in their leisure. **1965** *Melody Maker* 17 July 15, I am forming a semi-pro trio to play at weddings. **1977** *Time* 26 Sept. 43/1 She is a semipro photographer.
B. *sb.* = *semi-professional *sb.*
1910 *Baseball Mag.* Apr. 4/2 The despised semi-pros were drawing big crowds. **1912** *N.Y. Tribune* 7 Oct. 8/5 Enste, of the semi-pros, connected with his slow ball.. for a home run wallop. **1966** *Melody Maker* 15 Oct. 6 Semi-pros have always played an important role on the jazz scene.

semi-profe·ssional, *sb.* and *a.* [Semi- 1 a, 2 b.] **A.** *sb.* One who receives payment for an occupation or activity but does not rely on it for subsistence.
1897 *Sporting News* (St. Louis, Missouri) 27 Mar. 1 Doheny was only a semi-professional before he joined the New Yorks. **1971** *Daily Tel.* 4 Oct. 11/2 From the beginning..semi-professionals and professionals swooped down on the city's many basements..where they offered new drama.
B. *adj.* **1. a.** Designating a person or group receiving payment for an occupation or activity but not relying on it for subsistence.
1900 G. Bell *Let.* 21 Mar. (1927) I. v. 70 He is a photographer, semi-professional. **1951** Koestler *Age of Longing* i. 8 Gaston..had become a semiprofessional dancer and near-gigolo. **1972** *Daily Tel.* 22 May 1/8 The Storm are a semi-professional group formed about nine months ago.
b. Applied to an organization, activity, etc., involving semi-professionals.
1976 *Eastern Even. News* (Norwich) 9 Dec. 19/5 They would have to take drastic measures to keep the club within the semi-professional Magnet League any longer. **1978** *Homes & Gardens* Oct. 166 As each new instrument is made..it is tested in one of the semi-professional concerts given by the group.
2. Of or pertaining to an occupation considered intermediate in standing between a learned or skilled profession and a trade or handicraft.
1950 T. H. Marshall *Citizenship & Social Class* 150 They [*sc.* techniques] lend themselves in the same way to the establishment of semi-professional associations. **1965** *Word Study* Oct. 3/1, I wonder, however, if most of us have not observed an equally wanton use of the word *type* among various professional and semiprofessional groups in our country.
3. Of equipment: of a kind or quality close to that appropriate for professional use.
1953 E. T. Canby *Home Music Systems* xiii. 229 The semi-professional machines come with two or three speeds. **1975** *Gramophone* Sept. 531/2, 3M announced a new line..called the CTR series, featuring..a semi-professional 8-track recorder-player.

semiquinone (semikwi·nōᵘn). *Chem.* [Semi- 7 h.] A molecule or ion derived from a quinone and having one of the two oxygen atoms ionized or bonded to a hydrogen atom.
1931 L. Michaelis in *Jrnl. Biol. Chem.* XCII. 213 The difference between 1 molecule of each of the two successive steps will be proved to be only 1 electron (*i.e.* hydrogen atom) without any change in molecular size. The intermediary form will in this case be designated as a semiquinone. **1956** *Nature* 10 Mar. 483/1 Alcohols are attacked by photo-excited anthraquinones *A** with the production of semiquinones *A*H. **1973** B. J. Hazzard tr. *Organicum* vi. 380 Since..the semiquinone and *p*-benzoquinone are not resistant to alkali, the oxidation is carried out in acid solution and then takes place via quinhydrones.
Hence **semiqui·(no)noid** *a.*, having or being a structure resembling that of a semiquinone; also *absol.* as *sb.*
1932 *Jrnl. Biol. Chem.* XCVI. 704 Let us designate the three forms of the substance as *R* (the reduced form), *S* (the semiquinoid or intermediary form), and *T* (the totally oxidized, or quinoid, or holoquinoid form). **1935** *Chem. Rev.* XVI. 265 Among the three constituents of a two-step system it is not always the semiquinoid that exhibits the greatest instability. **1964** J. W. Linnett *Electronic Struct. Molecules* vii. 110 The two members are such that the transfer of one electron from one of the molecules to the other (e.g. from diamine to chloranil) will produce two semiquinonoid-type radicals.

semi-ri·gid, *a.* and *sb.* [Semi- 1, 7 l.] **A.** *adj.* **a.** Of an airship: having a flexible gas container to which is attached a stiffened keel or framework.
1908 [in Dict., s.v. Semi- 7 l]. **1919** H. Shaw *Text-bk. Aeronautics* xvii. 200 The semi-rigid type includes the Forlanini, Lebaudy, etc., of which the envelopes are not rigid, but are stiffened longitudinally by a framework running the whole length of the envelope. **1935** C. G. Burge *Compl. Bk. Aviation* 148/2 As with non-rigid and semi-rigid airships, a rigid at the start of a journey has its gas space about 90–95 per cent full of hydrogen. **1955, 1974** [see *keel *sb.*[1] 3 b].
b. *gen.* Somewhat rigid; having a certain amount of rigidity; *semi-rigid theory* (see quot. 1959).
1929 *Rep. & Mem. Aeronaut. Res. Comm.* No. 1155. 19 (*heading*) Specification of a simple type of semi-rigid wing. **1937** *Jrnl. R. Aeronaut. Soc.* XLI. 723 It was..possible to deduce from the semi-rigid theory design recommendations. **1946** *Nature* 30 Nov. 798/2 Rigid frames of triangular and square shapes, and semi-rigid frames. **1959** J. L. Nayler *Dict. Aeronaut. Engin.* 234 *Semi-rigid theory*, an approximate theory of elastic structure in which the theoretical infinite number of degrees of freedom is represented by a finite number, each being associated with an invariable mode. **1963** C. R. Cowell et al. *Inlays, Crowns, & Bridges* xi. 118 The retainer which is united to a pontic by a semi-rigid joint is called the 'minor' retainer.
B. *sb.* A semi-rigid airship.
1920 [see *rigid *sb.* 2]. **1925** E. H. Lewitt *Rigid Airship* i. 1 The airship may be divided into three classes: non-rigids, semi-rigids, and rigids. **1935** C. G. Burge *Compl. Bk. Aviation* 148/2 The semi-rigid is very similar in operation to the non-rigid.

‖ **semis**[1] (sē·mis). *Rom. Antiq.* [L., app. reduced f. *semi-* Semi- + *as* As *sb.*] A Roman coin, equivalent under the Republic and the early Empire to half an as, and under the later Empire to half a solidus. Cf. Tremis.
1853 H. N. Humphreys *Coin Collector's Man.* I. xxi. 260 The *Semis, Semisis,* or *Semi-as*, has an S upon it to denote its weight, as half that of the 'as'; it represented six ounces, and the type most usual in the Roman *series* is the head of Jupiter. **1949** *Oxf. Class. Dict.* 210/2 The coinage [of the early Empire] comprised *aureus* and half-piece in gold, *denarius* and half-piece in silver, *sestertius, dupondius,* and *semis* in orichalcum (brass), *as* and *quadrans* in copper. **1962** R. A. G. Carson *Coins* 180 Gold solidi were issued with some frequency, the semis or half-piece only rarely in the reign of the three brothers. *Ibid.* 197 In keeping with the earlier fifth-century tradition the obverse of the semis and the tremissis showed the diademed and cuirassed bust of the emperor in profile. **1979** *Nature* 5 July 46/2 The magnetisation of an orichalcum semis of the Roman Emperor Tiberius remained below the noise level up to the maximum fields available.

‖ **semis**[2] (sĕmī·). [F., lit. 'sowing'; cf. Semée *a.* and *sb.*] A form of decoration used in bookbinding, in which small ornaments are repeated regularly.
1926 R. Glazier *Man. Historic Ornament* 155 Some French bindings for Henry IV. are tooled with a *semis* of monograms on flowers. **1960** G. A. Glaister *Gloss. Book* 323/2 *Powdered,* the effect obtained on a book cover when small flower ornaments are repeated regularly in rows over it. Early examples date from 1560 on books bound for Charles IX of France. Also known as *semis* or *semé.* **1960** H. Hayward *Antique Coll.* 253/1 *Semis,*.. a repeating pattern of small ornaments used by bookbinders, 'sown' over the covers of books. **1977** Fleming & Honour *Penguin Dict. Decorative Arts* 719/1 *Semis,* an heraldic term also used in bookbinding for a diaper design made by the repetition of one or more small tooled ornaments.

se·mispecies. *Biol.* [Semi- 7 e.] A subdivision of a species regarded as having more individuality than does a subspecies.
1940 E. Mayr in *Amer. Naturalist* LXXIV. 256 The taxonomist finds it sometimes useful for practical purposes to treat as full species what should be regarded as subspecies on the basis of the definition. Groups of such 'semi-species' are called superspecies in the subsequent discussion. **1953** J. S. Huxley *Evolution in Action* iii. 71 The borderline of incipient species or semi-species. **1978** *Sci. Amer.* Sept. 60/1 Semispecies from the same locality will not crossbreed in the laboratory but semispecies from different localities will.

semi-spo·rts. [Semi- 1 b.] Used *attrib.*: **1.** Of articles of attire: somewhat informal or casual.
1929 *Footwear Organiser* July 25/1 There is no doubt that semi-sports shoes in two-colour effects will be largely worn. **1973** Schoeffler & Gale *Esquire's Encycl. 20th Cent. Men's Fashions* 194 By the early forties not only was a semisports outfit acceptable for town wear but, according to *Esquire,* it was 'smart to be sporty'.
2. Of a motor car: possessing some of the characteristics of a sports car.
1933 *Motor* 10 Oct. 513/1 The 20 h.p. Daimler semi-sports saloon. **1966** 'E. Peters' *Piper on Mountain* iii. 56 Half the world buys British when it comes to cars, especially semi-sports jobs like this.

semi-ste·rile, *a. Biol.* [Semi- 7 d.] Reduced in fertility by approximately 50 per cent.
1914 J. Belling in *Jrnl. Heredity* V. 71/1 The plants with perfect pollen have also perfect ovules; the plants with semi-sterile pollen have also half their ovules sterile. **1927** *Ibid.* XVIII. 267/1 Ears produced on semi-sterile individuals bear only 50 per cent of the normal number of seeds irrespective of the source of the pollen used, since the lethal operates prior to fertilization. **1956** *Nature* 10 Mar. 452/2 The cytological properties of a series of semi-sterile stocks. **1978** *Ibid.* 20 July 253/1 Using this method, dieldrin-resistant sterile or semisterile males would be released into the field.
Hence **semi-steri·lity.**
1914 J. Belling in *Jrnl. Heredity* V. 73/1 These plants are favorable for an investigation of semi-sterility since no complications arise from self-sterility, incompatibility, or intercrossing by insects. **1978** *Nature* 20 July 253/1 A release programme involving..males carrying genetic aberrations that give high semisterility.

Semite. (Earlier example.)
1848 C. Bunsen in *Rep. Brit. Assoc. Advancem. Sci. 1847* XVII. 266 The country which, according to the most ancient traditions of the Semites, was the cradle of mankind.

Semi:to-Hami·tic, *a.* and *sb.* = *Hamito-Semitic *a.* and *sb.* Also **Semitic-Hamitic** *a.* and *sb.*
1879 E. S. Roberts tr. *D. Pezzi's Aryan Philol.* i. i. 46 Hence the Southern and Central African dialects, the Erythræan (Semito-Hamitic) and the Aryan. **1910** *Encycl. Brit.* XII. 894 The development of a grammatical gender, this principal characteristic of Semito-Hamitic, in Bari and Masai, may be rather accidental than borrowed. **1928** *Language* IV. 129 If there is a Semitic-Hamitic language-group, then not primitive Semitic, but primitive Semitic-Hamitic must be correlated with primitive Indo-European. **1949** W. F. Albright *Archaeol. Palestine* iii. 61 Since very similar human skeletons have been found in the Badarian of Egypt as well as in late chalcolithic Gezer and Byblus, it seems to follow that these folk belonged to the ancestral Semito-Hamitic stock. **1964** *Jrnl. Semitic Studies* IX. 137 The Semito-Hamitic family, for which the very apt term Erythraic has been recently proposed by A. N. Tucker and M. A. Bryan. **1974** *Encycl. Brit. Macropædia* VIII. 589/2 Also known as the Semito-Hamitic, Erythraean, Afro-Asiatic, and Afrasian language group, it [*sc.* Hamito-Semitic] is the main language family of northern Africa.

semitone. 3. (Earlier example.)
1818 'T. Brown' *Brighton* I. i. 39 'Yes, my lord,' said Zephyr, and, in a semi-tone, muttered, 'he'll overlook many things.'

semi-trai·ler. [Semi- 7 a.] A road trailer that has a wheel system at the rear only and is coupled to a suitable tractor to form an articulated lorry. Freq. *transf.*, an articulated lorry made up in this way.
1919 *Engineering* 28 Nov. 718/2 With the semi-trailer the tractor partly carries and partly hauls the load, the front end of the two-wheeled vehicle being carried on a platform directly over the rear or driving wheels. **1926** *Encycl. Brit.* II. 987/2 This semi-trailer was a two-wheeled construction, the forward end of which was supported on the tractor frame by means of a swivelling fifth wheel. This end of the semi-trailer could be supported by means of jacks. **1949** *Automobile & Carriage Builders' Jrnl.* CIII. iii. 38/1 Probably the world's most curvaceous semi-trailer. **1956** 'N. Shute' *Beyond Black Stump* 44 The diesel semi-trailer..ground to a standstill in a swirl of red dust. **1962** *Coast to Coast 1961–62* 141 A loaded semi-trailer stood beside the kerb in front of the hotel.. and Keppler started towards it, hoping for a lift. **1978** O. White *Silent Reach* xv. 154 His car collided with a semi-trailer.

semi-transparency. (Earlier example.)
1787 J. Wedgwood *Let.* 16 June (1965) 308 The original artist availed himself of the semitransparency of the white glass.

semi-uncial, *a.*: Add: Cf. *half-uncial* s.v. *half- II. n.

semo- (sī·mo), combining form of Gr. σῆμα sign, used as the first element in **semolexe·mic** *a.*, of linguistic rules: governing the conversion of units of meaning into lexical items; **semolo·gical** *a.*, of or pertaining to semology; hence **semolo·gically** *adv.*; **semo·logy,** the study of meaning as an aspect of language; = Semantic *sb.* 1 a in Dict. and Suppl.; **semota·ctic** *a.*, of or pertaining to the ordering of units of meaning; hence **semota·ctically** *adv.* Cf. *semantico-.
1968 *Language* XLIV. 578 The last four pages of chapter 6 are devoted to a discussion of 'semolexemic rules', which map sememic networks onto lexemic trees. **1913** *Jrnl. Eng. & Gmc. Philol.* XII. 78 The change may be purely semological. **1928** C. Bergener *Contribution to Study of Conversion of Adjectives into Nouns* iv. 170 *Light, dark,* and *dusk* form a semological association-group. **1975** M. A. K. Halliday in S. Rogers *Children & Lang.* iv. 225 This is an analysis at the semological level in which the elements of structure are functional in character. **1913** *Jrnl. Eng. & Gmc. Philol.* XII. 78 Morphologically and semologically the stem *flik* exhibits types of word-formation and meaning development that are representative. **1911** A. L. Elmquist in *Ibid.* X. 318 The grammar [of Noreen's *Vårt Språk*] is in four parts.. (3), Semology, a new term for semasiology. *Ibid.,* The grammar offers a large number of new features, the Semology, perhaps, most of all. **1928** C. Bergener *Contribution to Study of Conversion of Adjectives into Nouns* i. 7 We are not only concerned with the semology but also with the morphology of converted adjectives. **1958** G. L. Trager in *Studies in Linguistics* XIII. 8 There seems to be no subdivision of either kinesics or paralinguistics exactly analogous to the phonology–morphology–semology of language. **1970** *Canadian Jrnl. Linguistics* XVI. 22 Below each disjunction each wire leads (though not necessarily immediately) to an inverse conjunction, the other wire of which leads up to the semology. **1966** S. M. Lamb *Outl. Stratif. Gram.* 14 Semotactic patterns differ from tactic patterns for lower strata in having considerable numbers of upwards ands. **1967** D. G. Hays *Introd. Computational Linguistics* xiii. 215 If parsing is performed after this dictionary lookup, its object is to determine semotactic well-formedness. **1969** *Language* XLV. 491 The principal area of stylistic concern is the discourse, not primarily the pleasing sound patterns or the juxtaposition of semotactically felicitous phrases.

semon (sī·mɒn). *Linguistics.* [Irreg. f. Gr. σῆμα: cf. -on[1].] In stratificational grammar: an element of meaning or one which combines with others to make up a sememe.
1965 S. M. Lamb in *Amer. Anthropol.* LXVII. ii. 46 Turning now to semology, the elementary unit is the semon, and a sememe is a unit composed of one or more semons. In a clause such as *he found his brace and bit* all the sememes are simple, but the following expressions contain or are complex sememes, i.e. sememes composed of multiple semons: *she put all her eggs in one basket.., don't give up the ship.* **1968** *Language* XLIV. 576 Hockett divides the 'semons' (the labels on the nodes in the sememic networks) into three classes, called 'links', 'kernels', and 'modifiers'. **1973** *Archivum Linguisticum* IV. 119 Within his general model the speaker makes a selection of 'semons' (elementary semantic features) on the basis of his 'awareness of a "cognitive experience"'.

sempervirent, *a.* Add: Also *absol.* as *sb.*, an evergreen plant (*rare*).
1957 J. D. Salinger in *New Yorker* 4 May 42/1 She nudged an unopened box of Sal Hepatica..to align it with the other sempervirents in its row.

sempiternity. Delete † *Obs.* and add later examples.
1933 *Mind* XLII. 309 Spinoza did not mean by 'eternity' either endless future duration or endless past and future duration ('sempiternity'). **1980** *Dædalus* Spring 255 The fraternal impulse of the heroic youth in the barque unsettles the moral ground of Hell's sempiternity.

Semple (semp'l). *Med.* The name of Lt.-Col. Sir David *Semple* (1856–1937), English bacteriologist, used *attrib.*, *absol.*, in *Comb.*, and in the possessive to designate a vaccine against rabies described by him in 1911 (*Sci. Mem. by Officers of Med. & Sanitary Depts. of Govt. of India* No. 44), and the techniques of preparing and administering it.
1934 *Q. Bull. Health Organisation* III. 615 Killed phenol vaccines... 4000 mg. Semple. *Ibid.* 618 Semple's modification of Fermi's original method (*i.e.*, the phenol vaccine is incubated at 37°C. for twenty-four hours) is employed. **1938** *Jrnl. Amer. Med. Assoc.* 20 Aug. 690/2 The Semple modification of the Pasteur treatment is commercially available and should be given twice daily. **1939** *Ibid.* 29 July 392/1 The Alabama State Health Department distributes Semple vaccine without charge to physicians for the treatment of those exposed to rabid animals. **1949** *New Gould Med. Dict.* 920/2 The vaccine used is made from 4% inoculated rabbit brain attenuated with 0·5% phenol; called *Semple's vaccine.* **1971** E. S. Tierkel in Nagano & Davenport *Rabies* 3 The Semple vaccine eventually became the biologic of choice in many countries. **1977** C. Kaplan *Rabies* vii. 108 A Semple-type vaccine was first used in 1921 for the successful mass vaccination of dogs in Japan.

‖ **semplice** (se·mplitʃe), *a.* and *adv. Mus.* [It., simple.] **A.** *adj.* Simple. **B.** *adv.* Simply. (Chiefly as a direction.)
1740 Grassineau *Mus. Dict.* 218 *Semplice,* simple, not doubled, compounded, or composed of any thing else, as *cadenza simplice,* is a cadence in which all the notes are equal in all the parts. **1801** Busby *Dict. Mus.* s.v., *Semplice,* (Ital.), a word implying that the movement before which it is placed is to be performed with chasteness and simplicity. **1883** Grove *Dict. Mus.* III. 461/1 *Semplice,* 'simple'; a direction denoting that the passage so marked is to be performed without any adornment or deviation from the time, used particularly in passages of which the character might possibly be misunderstood. **1976** *Gramophone* Nov. 786/2 The moment I suggest as a prime sample of Zukerman's profound insight is the third theme in that same movement with its rising phrases marked *pianissimo* and *semplice.*

‖ **sempre** (se·mpre), *adv. Mus.* [It.] Always, still: used to qualify an adj. or advb. Also *transf.*
1801 Busby *Dict. Mus.* s.v., *Sempre,* (Ital.), always, or throughout; as *sempre piano,* soft throughout. **1816** in G. Thomson *Sel. Collection Irish Melodies* 139 Judy Lovely Matchless Creature *Sempre dolce.* **1883** Grove *Dict. Mus.* III. 461/2 *Sempre,* 'always'; a word used in conjunction with some other mark of time or expression to signify that such mark is to remain in force until a new direction appears. **1959** *Collins' Mus. Encycl.* 594/1 *Sempre,* 'always, still', as in *sempre piano,* still softly. **1959** E. Pound *Thrones* xcvii. 29 Earth under Fortuna, each sphere hath its Lord, with ever-shifting change, sempre biasmata.

semseyite (se·msi,ɘit). *Min.* [ad. Hung. *semseyit* (J. S. Krenner 1881, in *Magyar tudományos Akad. Értes.* XV. 113), f. the name of Andor *Semsey* (1833–1923), Hung. nobleman and amateur mineralogist: see -ite[1].] A monoclinic sulphide of lead and antimony, $Pb_9Sb_8S_{21}$, found usu. as grey or black tabular or prismatic crystals with a metallic lustre.
1886 *Jrnl. Chem. Soc.* L. 313 Semseyite from Felsöbánya; small, grey tabular crystals. **1920** *Brit. Mus. Return* 142 in *Parl. Papers* XXXVI. 673 Semseyite from Dumfriesshire. **1955** [see *rosetted *a.* 1 a]. **1976** *Mineral. Abstr.* XXVII. 306/2 Genetic antagonism between semseyite..and boulangerite..may have been caused by variation in oxidizing power of sulphur between the two media.

sen, *sb.* (Earlier example.)
1727 J. Scheuchzer tr. *Kæmpfer's Hist. Japan* I. 17 The use of silver Money was forbid, and in its stead brass Sennis coin'd.

‖ **sen** (sen), *sb.*[2] [Indonesian, etc., repr. Cent[1].] In Indonesia, Malaysia, and other countries of the Far East: a coin or unit of currency valued at one hundredth of the principal measure.
1957 *Whitaker's Almanack 1958* 965 Indonesia... *Rupiah* of 100 *Sen.* **1959** *Ibid.* 1960 961 Cambodia... *Riel* of 100 *Sen.* **1962** R. A. G. Carson *Coins* 557 The islands became the independent republic of Indonesia in 1950. The unit of the new coinage is the rupiah of 100 sen.

Sen., sen., abbrev. of Senior *a.* 1 a. Cf. Jun. in Dict. and Suppl.
1676, etc. [see Senior *a.* 1 a]. **1837** [see *jun.]. **1955** *Times* 8 July 6/7 In two suits in another bedroom was found £115 10s., which Thomas Foote, sen., said was his. **1960** *Bedside 'Guardian'* IX. 140 It was an old tradition, Professor Arthur M. Schlesinger, sen., has said, [etc.].

Sen., abbrev. of Senator 1 f. *U.S.*
1857 Webster & Goodrich *N. Webster's Explanatory & Pronouncing Dict. Eng. Lang.* App. 462/1 *Sen.,* Senator. **1972** *Guardian* 3 July 1/2 (*heading*) Sen McGovern fights on. **1974** *Sumter* (S. Carolina) *Daily Item* 23 Apr. 1B/1 She has written Sen. Thomas O. Bowen..several times.

Sena, var. *Sehna; **Senacar,** var. *Seneca.

senaite (sē·i-, se·nā,ɘit). *Min.* [f. the name of Joachim da Costa *Sena,* 19th-cent. Brazilian mineralogist + -ite[1].] A rhombohedral titanate of iron, lead, and manganese, found as rough black crystals and rounded fragments in diamond-bearing sands in Brazil.

1898 Hussak & Prior in *Mineral. Mag.* XII. 30 (*heading*) Senaite, a new mineral belonging to the ilmenite group, from Brazil. **1976** *Acta Crystallographica* XXXII. B. 1509/1 Senaite, crichtonite, and davidite form a closely related series of minerals with similar morphologies and chemical compositions.

senate. Add: **1. f.** (Earlier U.S. example.)

1776 A. Adams *Let.* 15 Sept. (1875) 227 Whether you are in the American Senate or on board the British fleet, is a matter of uncertainty.

senatorial, *a.* **3.** (Examples.)

1785 T. Jefferson *Notes Virginia* App. 2 Let each county at the time of electing its delegates, chuse senatorial electors... Let the senatorial districts be divided into two classes. **1864** *Harper's Mag.* Mar. 568/2 Mrs X—..resides in our senatorial district. **1948** *Daily Ardmoreite* (Ardmore, Oklahoma) 23 July 14/6 This Senatorial district of Carter county is the birthplace and home of Joe.

send, *v.*[1] Add: **I. 2. c.** *slang* (orig. *U.S.*). To transport or arouse emotions in (a person); to enthral, delight (esp. of popular music). Also *absol.* Hence **se·nding** *ppl. a.*

1932 *Melody Maker* Oct. 836/1, I enclose the following wire which Louis (Musicmouth) Armstrong sent to Big John... 'My boy Earl was marvellous as ever yessir he sent me.' **1935** *Vanity Fair* (N.Y.) Nov. 71/3 Hot artists or bands that can put across their licks successfully are '*senders*'; they '*send*'. **1937** *Amer. Speech* XII. 47/1 The action of this trumpet really sends me and that's no jive. **1943** *N.Y. Times* 9 May II. 5/4 There has [*sic*] been some really solid trumpet players who can really send; some like Louis Armstrong who had a trumpet like heaven. *Ibid.*, Jimmy has a sending band and when he plays, brother, even the seats jump. **1955** V. Nabokov *Lolita* I. xxiv. 138, I do not know if in these tragic notes, I have sufficiently stressed the peculiar 'sending' effect that the writer's good looks.. had on women of every age and environment. **1956** B. Holiday *Lady sings Blues* (1973) ix. 86 Meade Lux Lewis knocked them out; Ammons and Johnson flipped them..Newton's band sent them. **1959** C. MacInnes *Absolute Beginners* 16 A film we went to ages ago that rather sent us. **1975** N. Mitchison *All change Here* iv. 39 So much modern poetry is ironic or deliberately held on a low note; that may be artistically admirable, but it doesn't send the reader.

8. a. (Examples of messages sent mechanically or electrically.)

1873 *Independent Defender* (San Francisco) 15 Nov. 3/1 The operator..excitedly telegraphed back, don't send so d—d fast. **1924** *Radio Times* 19 Dec. 585/3 This is only a receiving station. We can't send. We can only listen. **1929** *Amer. Speech* IV. 288 The sender's task is to 'move it',..—or simply 'send'. **1974** W. Garner *Big Enough Wreath* xi. 140 What if he'd asked to see the print-out? What if he'd gone over to see you send?

c. Of a shop: to deliver goods ordered.

1871 G. H. Lewes *Let.* 27 Aug. in *Geo. Eliot Lett.* (1956) V. 181 Take care the Stores people send on Thursday. **1968** *Observer* (Colour Suppl.) 22 Dec. 17/4 The shops won't send and now they've stopped the bus.

III. 23. send back. (Earlier example.)

1870 *Baily's Mag.* Aug. 359 A good catch..sent him back when only a few runs were wanted.

25. send down. c. (Earlier example.)

1871 *Baily's Mag.* Oct. 415 At times, no bowler in England sends down such utterly unplayable balls.

d. To dispatch or commit to prison by sentence. Freq. *pass. slang* (orig. *U.S.*).

1840 *Picayune* (New Orleans) 2 Aug. 2/5 She scorned to find surety in $500 to keep the peace, so she was sent down. **1880** G. A. Sala *Amer. Revisited* (1882) I. v. 85 They were 'sent down' for ten days. **1941** R. West *Black Lamb* II. 315 We caught the murderer..and he was sent down for a long sentence. **1960** G. Butler *Death lives Next Door* vi. 118 I'm Ted Springer's missus. Sent him down for three years, you did. **1976** 'P. B. Yuill' *Hazell & Menacing Jester* iii. 39 Is there any chance he could go to gaol?' 'You'd like him sent down, would you?'

e. To cause to accompany someone (to dinner).

1882 Mrs. H. Ward *R. Elsmere* II. II. xvii. 74 They would be sent down to dinner together to a certainty. **1892** 'A. Hope' *Mr. Witt's Widow* viii. 98 That lady.. sent Laura down to dinner with him.

f. *send her down, Davy* (also *Hughie*, etc.) and varr.: phr. expressing a wish for rain to fall. Cf. *Hughie. *slang* (chiefly *Austral.* and *N.Z.*).

1919 W. H. Downing *Digger Dial.* 44 *Send her down, Steve!*, let it rain on. **1925** Fraser & Gibbons *Soldier & Sailor Words & Phrases* 74 *David* (or *Davy*), *send it down*, a soldier's greeting to a shower of rain likely to postpone a parade. **1928** L. H. Nason *Sergeant Eadie* xi. 321 Hurray! Send her down, Davie; no drill today! **1937**, **1958** [see *Hughie]. **1975** *Panorama* (Austral.) Nov. 25 'Send 'er down, Hughie!' An expression in nationwide use since the turn of the century, which is..an invocation to Heaven..to send rain.

27. send in. c. Also, to send (the opposing side) in to bat first.

1912 P. F. Warner *England v. Australia* vi. 48 Trumper beat Douglas in the toss, and sent us in. **1930** C. G. Macartney *My Cricketing Days* iii. 18 They might have given us a good game had not our captain..won the toss and sent them in on a bad wicket. **1969** *Wisden Cricketers' Almanack* 478 Although Lancashire were without..their opening bowlers, Leicestershire sent them in on winning the toss.

28. send off. † **b.** = sense 25 b. *Obs.*

1843 [see *faculty 9 b].

c. *Sport.* To order (a player) to leave the pitch as a punishment.

1906 W. Pickford in Gibson & Pickford *Association Football* III. xvi. 6 A referee may send a player off at once and without any previous caution, if he is guilty of violent conduct. **1976** *Milton Keynes Express* 2 July 43/4 Newton..suffered a severe setback in the first half when they had a player sent off.

29. send on. c. To dispatch (a letter, etc.) forward from the place to which it was addressed.

1833 S. Smith *Life & Writings J. Downing* liii. 183 Dear sir, I want you to send this on to cousin Jack. **1879** Geo. Eliot *Let.* 7 July (1956) VII. 179, I have had 2 letters from him which Miss Gibson sent on. **1895** 'G. Mortimer' *Like Stars that Fall* xiv. 198 Didn't you get the letter sent on?

33. send up. d. (Earlier example.)

1821 *Salt-Bearer* (Eton Coll.) 129 He more than once had the honour of being 'sent up for good', *i.e.* having his verses read over by the head master as particularly worthy of commendation.

f. To mock, make fun of (a person or thing); to parody.

1931 T. R. G. Lyell *Slang, Phrase & Idiom in Colloq. Eng.* 673 The last time he came in, he was sent up unmercifully by half the room. **1957** 'N. Blake' *End of Chapter* 68 Who's Johnnie Ray? He's—go on! you're sending me up! **1962** *John o' London's* 29 Nov. 506/3 The effect..is as if he is attempting to 'send up' the whole picture. **1969** *Times* 13 Dec. p. v/2 These represented the British sense of humour, our genius for sending ourselves up, but they seem to me rather to be reinforcements of such attitudes. **1977** P. G. Winslow *Witch Hill Murder* II. xvii. 220, I wasn't sending you up the other night. I was *afraid*.

IV. 34. The infin. used: **a.** to describe the position of a switch for transmission.

1876 *Jrnl. Soc. Telegr. Engineers* V. 494 The switch has been put on 'send'. **1976** C. Egleton *State Visit* ix. 88 Because he kept the switch on 'send', they could just hear the band.

b. *attrib.* in the sense 'sending', as the name of a part.

1973 C. Bonington *Next Horizon* xiii. 194 He ended up by telling me to press the send switch of the radio three times as affirmative and twice for negative, in reply to his questions. **1976** K. Thackeray *Crownbird* ix. 198 He pulled his microphone towards him..and depressed the send button.

sendable, *a.* (Later examples.)

1901 *Chambers's Jrnl.* July 464/1 We have become so used to connecting the convenient little slips of sendable money with the Post-Office. **1965** P. Wylie *They both were Naked* I. iii. 150 She had been sendable because she wanted to be sent.

Sendai (se·ndəi). *Biol.* and *Med.* The name of a city in northern Honshu, Japan, used attrib. as **Sendai virus,** a paramyxovirus (first identified in Sendai) which causes disease of the upper respiratory tract in mammals, and is used in the laboratory to produce cell fusion.

[**1953** T. Sano et al. in *Yokohama Med. Bull.* IV. 215 We consider this disease a new form of virus pneumonitis and have termed it 'Newborn Virus Pneumonitis (Type Sendai)'.] **1958** *Ann. Rev. Microbiol.* XII. 66 Similar to the influenza virus, one of the new agents, Sendai virus, propagates sufficiently well in eggs to permit primary isolation by amniotic inoculation. **1970** *Nature* 25 July 339/2 The possibility of introducing alien genetic material into mammalian eggs by fusion with somatic cells using Sendai virus. **1976** *Ann. Rev. Microbiol.* XXX. 29 The transcriptive complex of a representative paramyxovirus, Sendai virus, contains a major structural polypeptide, NP..a less abundant polypeptide, P..and a high-molecular-weight polypeptide that is present in minute amounts.

sender. Add: **d.** One who or that which moves or enthrals, esp. a popular musician. Also in phr. *solid sender* (cf. *solid *a.* 18**). Cf. *send *v.*[1] 2 c. *slang* (orig. and chiefly *U.S.*).

1935 *Vanity Fair* (N.Y.) Nov. 38/1 None of these plates will be senders. *Ibid.* 71/3 Hot artists or bands that can put across their licks successfully are '*senders*'. **1938** *Metronome* Apr. 26/2 A really solid sender is the third record from the right in my collection. **1938** *Amer. Speech* XIII. 314/2 *Sender*, one who is extremely well-dressed or witty. **1954** [see *drag *sb.* 3 d]. **1960** *Spectator* 7 Oct. 523 Fabian, the teenagers' sender, indistinguishable from Cliff Richards. **1978** G. Vidal *Kalki* vi. 147 Arlene was addicted to the slang of her youth. 'A solid sender!' she added, nicely dating herself to World War II during which she entertained the troops.

sending, *vbl. sb.* Add: **1. b.** (Later examples.)

1938 *Times* 17 Feb. 15/6 The great expense of the traditional sending-off parties for soldiers called to the colours. **1978** *Rugby World* Apr. 7/3 The laws..make no provision for a foul committed *before* a try—unless it be sending off. **1982** 'W. Haggard' *Mischief-Makers* vi. 65 The referee, white, had sent off a black... There was a code which governed sendings-off, particularly in amateur [football] matches.

4. [ON., in same sense.] An unpleasant or evil thing supposed to be sent by a wizard, or through a wizard at the request of another party, as a punishment or act of revenge.

1864 Powell & Magnusson *Icel. Legends* 238 If he did not return to them by Christmas-day next, they would despatch a Sending to him who should kill him. **1888** Kipling *In Black & White* 68 A Sending..is a Thing sent by a wizard. **1915** *Hastings's Encycl. Relig. & Ethics* VIII. 218 A phenomenon analogous to that of the werwolf is that of the 'sending'—a thing or animal, sometimes animated or even created by the sorcerer, or some part of the sorcerer himself (his soul, etc.) and sent out by him to injure or kill his enemies. **1980** G. Household *Sending* v. 68 What your Norse shamans and the sagas called 'a sending'..a sort of portable ghost.

send-off. Add: **1.** (Earlier examples.) Also *transf.* and *fig.*

The earliest sense of the word appears to be 'a sending off or starting of contestants in a race'.

1841 *Spirit of Times* 18 Dec. 499/2 Sleepy John was the favorite against the field; but by bad management of the groom John got a miserable send off and lost at least fifty yards. **1867** *Harper's Mag.* Dec. 135/1 As they say at Jerome Park, a 'good send off' at life's outset is, four to one, better than a 'false start'. **1875** *Chicago Tribune* 15 July 1/3 ... send-off (in the boat-race) was not obtained. **1908** G. G. Greenwood *Shakespeare Problem Restated* xv. 483 If we could only get to the back of his [sc. Jonson's] mind, we should find that there was some different cause operating to induce him to give the best possible send-off to that celebrated venture. **1934** [see *faith 1 c]. **1977** J. McClure *Sunday Hangman* ii. 19 Every warder has to witness at least one little send-off.

se·nd-up. *colloq.* [f. vbl. phr. *to send up*: see *send *v.*[1] 33 f.] An act of mocking or teasing; a parody, a satire.

1958 A. Wilson *Middle Age of Mrs. Eliot* 355 If she teased him a bit maliciously, it was with such caressing malice that his gagging was more like a flirtation than a 'send up'. **1962** *Guardian* 24 Dec. 4/6 'Merry Christmas You Suckers'..., according to the record company, 'is a send-up of the whole commercialised business'. **1970** G. F. Newman *Sir, You Bastard* iii. 106 Gordon moved away, accepting his opinion as infallible, and never suspecting the blatant send-up. **1975** J. I. M. Stewart *Young Pattullo* xi. 256 What had taken place would have been described a decade later as a send-up.

Sendzimir (se·ndzimi³z). The name of Tadeuz or Thaddeus K. Sendzimir (b. 1894), Polish-born American engineer, used *attrib.* with reference to a type of rolling mill developed by him for cold rolling of steel, in which each of two working rolls is supported by two larger rolls, which are themselves backed by three still larger rolls (a further tier of four larger rolls is sometimes used).

1936 *Jrnl. Iron & Steel Inst.* CXXXIV. 166P, In the case of cold rolling mills for wide strip, simpler designs could be developed, like those described by Mr. Noll. In these forms, however—as in the Sendzimir mill which has been described—it was essential to avoid at all costs friction on the pulling drum, which might entail damage to the upper surface of the strip. **1952** W. D. Hargreaves in D. L. Linton *Sheffield* 254 A Sendzimir mill, which can work to within 0·0001 of an inch... **1978** C. R. Cottrell *Introd. Metallurgy* xxiii. 442 Long slender rolls are too elastically flexible to compress the metal unless they are supported by heavy backing rolls... This principle has been particularly developed in Rohn and Sendzimir mills for rolling thin foil. **1979** C. M. Black tr. *Thelning's Steel & its Heat Treatment* vi. 290 To a large extent the steels are used for tools for shaping and forming. The so-called Sendzimir rolls, i.e. rolls forming a cluster-roll mill, constitute an example of this.

Seneca. Add: Also † Senacar, Senecke, Sinecque, etc. Substitute for etym.: [ad. N. Amer. Du. *Sennecaas*, *Sennecaas*, collect. name for the Upper Iroquois tribes, perh. orig. a Mahican name for the Oneida or their village.] **A.** *sb.* (A member of) one of the five (later, six) tribes of the Iroquois Confederacy of North American Indians; their language. **B.** *adj.* Of or pertaining to this tribe.

Formerly used also to designate the four Upper Iroquois tribes, Seneca, and the Iroquois collectively.

[c**1616** in *Documents Colonial Hist. N.Y.* (1856) I. *facing* p. 11 Sennecas.] **1664** J. Winthrop *Let.* 6 Feb. in *Mass. Hist. Soc. Coll.* (1863) 4th Ser. VI. 531, 3000 of the Senecks, a people in league with the Mohawks beyond them, are gathered together. **1684** *New-Hampshire Hist. Soc. Coll.* (1827) II. 199 The 5 of Mohauck, Senacar, or other Indians, [shall] be paid out of such monies as shall be raised in the said Province. **1684** [see *Onondaga]. **1709** S. Sewall *Jrnl.* 11 Aug. in *Mass. Hist. Soc. Coll.* (1879) 5th Ser. VI. 262, 200 Eastern Indians..and 5th Senecks, 5 Nations to pray leave to dwell with them; and..others refusing them, they were gone to the Senecas. **1724** H. Jones *Pres. State Va.* I. 13, 5 The Senecas Indians in their War Dress may appear as terrible as any of the Sons of Anak. **1775** J. Adair *Hist. Amer. Indians* 393 A party of the Senekah Indians came to war against the Katabba. **1823** [see *Cayuga]. **1874** [see *Onondaga]. **1895** [see Six *a.* 1 d]. **1900** *Congress. Record* (U.S.) 26 Jan. 1232/2 Among the Seneca Indians a singularly beautiful belief prevailed. **1910** Kipling *Rewards & Fairies* 187 Senecas aren't Hurons, they're Iroquois, of course, and Toby told him so. **1933** L. Bloomfield *Language* iv. 72 The languages of the Iroquois type..*Seneca*. **1969** *Observer*

(Colour Suppl.) 25 May 61/1 The Iroquois leader Sagoyewatha, known as Red Jacket, was one of the Seneca tribe who fought alongside the British during the American War of Independence. **1976** T. A. Sebeok *Native Languages of Americas* I. 537 Wright at one time devised a unique set of letters to be used in printing Seneca, but they were never adopted.

C. *attrib.* **Seneca grass** (earlier example); **Seneca oil,** delete † and add: now only *Hist.* (earlier and later examples).

1814 J. Bigelow *Florula Bostoniensis* 145 Seneca grass... An erect, early grass, with a small panicle of short flowers. **1795** J. Scott *U.S. Gazetteer* s.v. *Allegany*, In this county is Oil creek. It flows from a spring much celebrated for a bitumen resembling Barbadoes tar, and is known by the name of Seneca Oil. **1910** Kipling *Rewards & Fairies* 161 He took orders for that famous Seneca Oil which he had the secret of from Red Jacket's Indians. **1959** *Chambers's Encycl.* X. 619/1 In North America crude oil was undoubtedly used by the Indians, and Seneca oil skimmed from the surface of water near Lake Seneca..was utilized for rheumatism, coughs, burns, sprains etc.

Senecal, *a.* For † *Obs.* read *Obs.* exc. in allusion to Chapman's use (see quot. 1612) and add later examples. Also as *sb.*: a member of a group of early seventeenth-century playwrights (see quot. 1926).

1926 T. S. Eliot in *Times Lit. Suppl.* 9 Dec. 906/3 In effect so to 'place' Davies, who appears anomalous, ... to the Senecals, to Chapman and Daniel and Greville... The type of his thought..separates him from the Senecals. **1934** H. Ellis *Chapman* 62 Chapman often refers to Seneca and the 'Senecal man', and he was also influenced by Epictetus. **1954** E. Rees *Tragedies of George Chapman* iv. 114 Chapman was too much of a humanist to forget that even a 'Senecal man' is human.

Senecan, *a.* Add: Hence **Se·necanism.**

1934 T. S. Eliot *Elizabethan Essays* 40 Much of Chapman's Senecanism was lately been shown..to be directly borrowed from Erasmus. **1978** *Studies in Eng. Lit.: Eng. Number* (Tokyo) 173 Jonson's use of Hoskyns' essay as a foundation for his own brand of Senecanism looks back to the past as well as to the future in the development of English prose.

senecio (senɪ·ʃio, -sio, -e·kio). [L. *senecio* old man, groundsel, in reference to the hairy pappus of the plant: adopted by Linnæus (*Hortus Cliffortianus* (1737) 406) as the name of a genus: cf. Sencion.] An annual or perennial herb or shrub of the large genus so called, which belongs to the family Compositæ and includes many cultivated plants and a few poisonous ones. Cf. Groundsel *sb.*[1] 1, *Jacobæa 1, Ragwort[1] 1.

1562, **1657** [see Groundsel *sb.*[1] 1]. **1784** J. Abercrombie *Propagation & Botanical Arrangement Plants & Trees* II. 676 (*heading*) Senecio longifolius. Long-leaved Cape Senecio. **1859** D. Bunce *Travels with Dr. Leichhardt* ii. 103, I obtained specimens of..a new species of *Stackhousia*, and a yellow, shewing flowers, [illegible] **1883** [illegible] **1902** L. H. Bailey *Cycl. Amer. Hort.* IV. 1656/1 A distinguishing mark of the Senecios lies in the character of the involucre. **1950** *Lancet* 23 Oct. 848/2 Senecio disease, or cirrhosis of the liver due to senecio poisoning. **1966** [illegible] Senecios (members of the same genus as common groundsel). **1975** P. Lively *Going Back* i. 4, I came out..past the buddleias and senecios past the iris garden hedge.

Senegal (se·nĕgǫl). The name of a river and a republic, formerly a French overseas territory, in western Africa.] **1.** Used *attrib.* in names of local animals and plants.

1783 J. Latham *Gen. Synopsis Birds* IV. 456 Senegal Warbler, *Motacilla Senegalensis*. Length five inches and a quarter... Inhabits Senegal. **1781** T. Pennant *Hist. Quadrupeds* I. 41 Antelope. Senegal. *Antilope Bubalis*... with horns almost close at the base, a little above bending over greatly... Inhabits Senegal. **1833** Penny Cycl. I. 183/1 The Senegal custard apple (*Anona Senegalensis*). **1896** H. A. Bryden *Tales S. Afr.* 60 Great spur-heeled Senegal cuckoos flapped heavily from one reed-bed to another. **1952** Mackworth-Praed & Grant *Birds E. & N.E. Afr.* I. 516 Senegal Coucal. *Centropus senegalensis*. **1966** C. Sweeney *Scurrying Bush* xi. 158 The Senegal galago is one of the commonest mammals in the Nuba Mountains in southern Kordofan.

2. Special Comb. **Senegal gum** = gum-senegal s.v. Gum *sb.*[2] 3 a.

1897 P. L. Simmonds *Dict. Trade* (rev. ed.) Suppl. 454/1 *Senegal gum*, an African gum obtained from *Acacia Senegalensis*. **1951** Kirk & Othmer *Encycl. Chem. Technol.* VII. 313 Kordofan gum is the finest gum obtainable.... Senegal gum is reported to be of lower quality and is much used for technical purposes although some of the better grades are used in pharmaceutical work.

Senegalese (se:nĕgǫlī·z), *sb.* and *a.* [f. prec. + -ese.] **A.** *sb.* A native or inhabitant of Senegal. **B.** *adj.* Of or pertaining to Senegal.

Formerly applied loosely to an inhabitant of any French colony in West or Central Africa.

1927 *19th Cent.* Feb. 313 The Senegalese were in the fight with us. **1919** H. H. Johnston *Compar. Stud. Bantu & Semi-Bantu Langs.* p. vi The facilities accorded to me in 1915 to visit the camps and hospitals of the 'Senegalese' soldiers in France. This assemblage of negroes from all parts of French West and West-Central Africa was a singularly fortunate circumstance. **1926** *Blackw. Mag.* Oct. 501/1 His French was fluent...The French of the Senegalese troops. **1928** E. Bergmann *Life Class* ii. 131 In the Ballroom kitchen, Kakabobé the Senegalese is dropping the huge casseroles up out of the elevator. **1978** *Black World* May 93/1 The struggle of the Senegalese against the invading French forces became a war of the [illegible] **1980** *Times Lit. Suppl.* 24 Oct. 1200/2 The belligerent exchange between the Senegalese writers at Frankfurt was symbolic of larger antagonisms.

Senegambian (senĕgæ·mbian), *sb.* and *a.* [f. *Senegambia* (see below) + -an.] **A.** *sb.* A native or inhabitant of Senegambia, former name of the region surrounding the Senegal and Gambia rivers in West Africa (cf. *Senegal, *Gambia); *U.S. colloq.*, applied to a Black American. **B.** *adj.* Of or pertaining to Senegambia; also used of Black Americans.

Senegambia is now the name of a confederation of Senegal and Gambia, formed for military and trade purposes in February 1982; both countries retain their sovereignty; the president of Senegal is president of the confederation.

1900 *Dialect Notes* II. 58 *Senegambian*, a negro or negress. **1902** *Encycl. Dict. Suppl.* 646/3 *Senegambian*, [illegible] **1911** *Encycl. Brit.* XXIV. 642/1 The conquest of the Senegambian region by the French followed. **1920** H. C. Witwer in *Collier's* 11 Dec. 21/1 A little bimbo [illegible] **1942** W. Faulkner *Go down, Moses* 49 What the hell Kingdom of Senegambian Montague and Capulet is this anyhow? **1949** [illegible] *Pittsburgh Courier* 11 Sept. 13/7 There are thousands of Negroes living in similar or better houses, despite the race hustling talk about the 'horrible houses' of Harlem. All the civilized Senegambians live in good houses. **1947** *Publ. Amer. Dial. Soc.* viii. 36 *Senegambian*. I am acquainted with this word as meaning a Negro from Senegambia, [illegible] **1976** P. Oliver *Savannah Syncopators* 10 The importance of the Senegambian slave trade lies in the accessibility of their ports to the Sudan savannah interior.

senesce, *v.* Delete *rare* and add later examples.

1909 G. B. Shaw *Let.* 22 June (1972) II. 847, I am not adolescing but senescing. **1955** *New Biology* XVIII. 12 If they [sc. experimental animals] do not senesce, they will not become old. [illegible] **1979** *Nature* 5 July 156 [illegible] deter grazers while the plants are alive do not disappear immediately when the plants senesce.

‖ **senex** (se·neks). Pl. **senes.** [L., old man.] In literary contexts, the stock figure of an old man. Also in various L. phrases. Cf. *old man f.

1898 E. P. Morris *Captives & Trinummus of Plautus* p. xxxvi, Hegio becomes in part the *comicus senex*, chiefly concerned with the humiliation of having been deceived. **1913** G. Norwood *Art of Terence* v. 76 How much better to collect the necessary funds permanently and elegantly from the senes! *Ibid.* vii. 128 Micio has gone back to his normal position of a [illegible] **1975** [illegible] victory of the young man over the *senex*... The emotions of a comic *senex*.

Senhouse (se·nhaus). *Naut.* [Origin unknown.] In full, *Senhouse slip*. A slip (Slip *sb.*[3] 3 e) designed to secure the end of a cable.

1923 *Man. Seamanship* (Admiralty) II. 114 (*caption*) Senhouse slip. **1948** R. de Kerchove *Internat. Maritime Dict.* 648/1 *Senhouse slip*, a short length of chain of the same strength as the anchor cable, with a slip hook at one end and a shackle at the other. It is shackled to the cable clench, its purpose being to allow the bitter end of the cable to be easily slipped in case of emergency. **1961** F. H. Burgess *Dict. Sailing* 183 *Senhouse slip*, a large slip in the cable locker near the clench bolt for holding the cable should it have to be unbolted and slipped. **1965** P. J. Abrahams *Last Honest* 187 He remembered knocking the Senhouse slip on one of the gripes away. **1976** *Oxf. Compan. Ships & Sea* 126/2 *Senhouse slip*... Its normal place in a ship used to be in the cable lockers where the inboard end of the cable is secured, but in several modern ships the end of the cable is shackled on to a deck bolt in the locker, no Senhouse slip being used. Smaller Senhouse slips are used in many smaller vessels and yachts to hold the ends of the guard-rails to the stanchions.

senicide. For *rare*[−1] read *rare* and add later example.

1931 R. R. Marett in W. Rose *Outl. Mod. Knowledge* 419 One must not make too much of the occasional cases of..senicide, namely, the putting away of the old.

senile, *a.* Add: **1. b.** *Path.* **senile dementia,** a severe form of senile deterioration, in which loss of memory, disorientation in time and space, and inability to cope with everyday life are strongly marked. Hence **senile dement,** one who suffers from this. Cf. Dementia.

1851 R. Dunglison *Dict. Med. Sci.* (ed. 8) 276/1 *Senile dementia, Insanity of the aged*, a form of moral insanity, in which the whole moral character of the individual is changed. **1902** A. R. Diefendorf *Kraepelin's Clin. Psychiatry* viii. 273 Senile dementia includes those forms of mental disease appearing in the period of involution. **1948** W. A. O'Connor *Psychiatry* xii. 269 The senile dement exhibits to a profound degree the characteristic failings of the deteriorated senile person. **1954** W. Mayer-Gross et al. *Clin. Psychiatry* xi. 482 The two pictures of senile dementia and normal ageing are qualitatively different and must be distinguished. **1976** *Scotsman* 20 Nov. (Weekend Suppl.) 3/4 Laing comments on this: 'I am sure he was not suffering from senile dementia, nor was it a slip of the pen.'

senile (sī·nail), *sb.* [f. the adj.] An aged person; one who exhibits the weakness or diseases of old age.

1938 N. Cameron in *Amer. Jrnl. Psychol.* LI. 664 The seniles..exhibited..the loose cluster type of organization (asyndesis). **1962** *Lancet* 8 Dec. 1221/2 Of every 100 potential long-stay male patients, 10 were schizophrenics, 21 seniles, and 12 manic-depressives. **1981** J. B. Hilton *Playground of Death* v. 67 The old girl was well into her eighties... You never know where you stand with these so-called seniles.

senior, *a.* and *sb.* Add: **A.** *adj.* **1. c.** *senior citizen*, a term for an elderly person, esp. one who is past the age of retirement, orig. *U.S.*

Freq. used in official communications and by the media as a euphemism for 'old-age pensioner'.

1938 *Time* 24 Oct. 12/2 Mr. Downey had an inspiration to do something on behalf of what he calls, for campaign purposes, 'our senior citizens'. **1952** *School & Society* (U.S.) 12 May 169 As a basis for their education, it [sc. pragmatism] is good for the young, the middle-aged, and our senior citizens. **1962** *British Advent Messenger* 28 Sept. 30/2 Owing to the extensive alterations to be done it was October 27, 1955, before we could welcome any of those dear senior citizens who were so anxiously waiting to enter the Home. **1966** T. Pynchon *Crying of Lot 49* iv. 90 Vesperhaven House, a home for senior citizens that Inverarity had put up. **1969** *Listener* 23 Jan. 101/2 More organised resistance came from the 'senior citizen'—or old age pensioners'—lobby. **1974** H. MacInnes *Climb to Lost World* iv. 56 We staggered up the bank to the village like senior citizens en route to the post office to collect their pensions. **1977** B. Pym *Quartet in Autumn* xvi. 152 Being retired, a senior citizen, you might say.

2. c. *Stock Exch.* Applied to securities the owners of which have first claim to be repaid by the issuing company. Cf. *junior stock* s.v. *Junior *a.* (*sb.*) 5.

1914 H. Halford *Dict. Stock Market Terms* (ed. 2) 79 Senior stocks. Debentures and Preference Stocks carrying a fixed rate of interest and ranking for dividend in priority to the Ordinary and Deferred Stocks. **1925** H. Parkinson *ABC of Stocks & Shares* 63 Among the 'senior' securities of the large railway companies the investor may roam at will. **1939** Mann & Grodinsky *Fin & Flow of Investment Values* I. 5 Graded these formal requirements in ratios, priorities, and margins of security, the senior securities of certain issuers secured by certain forms of property... are recommended. **1964** P. Wyckoff *Dict. Stock Market Terms* 238 Senior Securities. Bonds and preferred stock which receive prior consideration when a corporation fails or is being dissolved.

4. *senior common-room* (examples).

1774 J. Woodforde *Diary* 14 Jan. (1924) I. 122 We all went into the Senr Common Room. **1850** W. Bradbury *Eating People is Wrong* i. 28 As Treece was leaving the Senior Common Room, the Vice-Chancellor appeared in the doorway. **1981** E. North *Dames* iii. 41 They were in the Senior Common Room standing by the first school photograph.

5. Special collocations: **senior class** *U.S.*, a class in college or high school made up of students in their fourth year of academic study; **senior college** *U.S.*, a college in which the last two years' work for the bachelor's degree is done; **senior high (school)** *N. Amer.*, a secondary school comprising the three (or four) upper high school grades (cf. *junior high* (*school*) s.v. *Junior *a.* (*sb.*) 5); **senior school**, a high school or college course.

1707 T. Clap *Ann. or Hist. Yale College* 14 The Senior Class were removed to Milford. **1837** [in Dict., sense 2 b]. **1900** C. W. Winchester *Victories of Wesley Castle* 25 Wesley and Chester went to the city of Dorchester on some business for the senior class. **1980** *Redbook* Oct. 231/2, I couldn't go on the senior-class trip to Washington. **1899** *Univ. of Chicago Reg.* 37/1 The Faculties of the Schools of Arts, Literature, and Science have been organized as follows:..(2) The Faculty of the Senior Colleges [etc.]. **1942** *Bull. Vanderbilt Univ.* 15 May 69 The

College is divided, for certain purposes, into the Junior College and the Senior College. **1977** *Information Please Almanac for 1978* 826 (*heading*) Accredited U.S. Senior Colleges and Universities. **1909** *Ann. Rep. Bd. Educ.* (Columbus, Ohio) 29 You have established a Junior High School..leaving the tenth, eleventh and twelfth grades for the Senior High Schools. **1949** *Los Angeles Times* 23 June II. 5/1 Then they enter senior high school, and become 'juniors' and then seniors. **1955** [see *composite *a.* 6 c]. **1974** *Encycl. Brit. Macropædia* VI. 422/1 The elementary–secondary sequence overall is 12 years in length.., but the subdivisions of these years are various: ..six-three-three (elementary school, junior high school, and senior high school), [etc.]. **1871** Senior school [see *junior school* s.v. *Junior *a.* 5]. **1930** *Times* 26 Mar. 12/1 The first step in reorganization is to group all the senior children from 11 upwards in separate senior schools or departments or 'senior tops'. **1963** [see *djebba, djibbah]. **1963** Barnard & Lauwerys *Handbk. Brit. Educ. Terms* 175 *Senior School*, (1) An obsolete term used to describe the free non-selective post-primary schools (age-group 11–14) established under the pre-1944 elementary code. They provided a course of general studies with some vocational bias... (2) A term sometimes used to describe the top classes/forms of a grammar or public school. **1796** J. Morse *Amer. Univ. Geogr.* (ed. 3) I. 420 The undergraduates are not permitted to attend them [*sc.* medical lectures] till their senior year. **1924** S. S. Colvin *Introd. High School Teaching* 12 A number of high schools offer in their senior year a vocational course.

B. *sb.* **3*.** With *the*. The familiar name of the United Service and Royal Aero Club.

1906 G. W. E. Russell *Social Silhouettes* xxviii. 195 If he is an old soldier, he is eligible for 'The Senior', and may make free with the Duke of Wellington's dry sherry and Dugald Stewart's still drier library. **1974** *Financial Times* 29 June 15/3 Commander James Allen, secretary of the United Services and Royal Aero (which is widely known in club circles as the Senior), [etc.]. **1974** R. McDouall *Clubland Cooking* 11 Going west from Trafalgar Square we come first to the United Service Club, known as 'The Senior', because it was for senior officers of the Army and Navy. **1975** *Sunday Times* 25 May 24/1 The closure of the Senior will shake all the generals and admirals who have taken it for granted since 1815.

senium (sī·niəm). *Med.* [a. L. *senium* debility of age, f. *senēre* to be feeble, f. *senex* old.] The period of old age. Usu. with *the*.

1911 Stedman *Med. Dict.* 786/1 *Senium*, old age, especially the debility of the aged. **1932** F. M. Lipscomb *Dis. Old Age* p. v, I have included only..those maladies which have a high incidence in the senium. **1968** J. G. Howells *Family Psychiatry* ii. 23 Individual members of the family..may be of any age, from infancy to senium. **1977** T. R. Harrison *Princ. Internal Med.* (ed. 8) 8/2 In the absence of disease there is a steady falloff [of nerve and muscle cells] which begins at about the end of the growth period and continues at an accelerated pace into the senium.

Senna, etc., varr. *Sehna.

sennegrass (se·nĕgras). [a. Norw. *senegras*: cf. ON. *sina*, Sw. dial. *sena* withered grass.] An Arctic sedge, *Carex vesicaria*.

1897 tr. *Nansen's Farthest North* II. 95 Turn them [*sc.* Finn shoes] inside out, fill them with sennegraes [*sic*] or sedge,..and creep into the sleeping bag. **1919** E. Shackleton *South* xii. 229 Oil mixed with reindeer hair, bits of meat, sennegrass, and penguin feathers form a conglomeration which cements the stones together.

sennet[1]. For † *Obs.* read *Obs.* exc. *Hist.* and add later examples.

1922 Joyce *Ulysses* 471 Four buglers on foot blow a sennet. **1942** E. Blom *Music in Eng.* iii. 49 Shakespeare was much attached to music, and it would be rash to conclude from the mere evidence of printed texts that he contented himself in his plays with a few songs and an occasional tucket (*toccata*) or sennet (*sonata*) for trumpets behind the scenes.

Senni (se·ni). The name of a tributary of the River Usk, Powys, S. Wales, used *attrib.* as **Senni Bed,** any of a series of fossiliferous sandstones in the Lower Old Red Sandstone of S. Wales, well seen in the valley of the Senni. Usu. *pl.*

1904 A. Strahan *Geol. S. Wales Coalfield* V. i. 3 Nearly all the remainder of the Old Red [Sandstone] tract is occupied by the red sandstones of the Brownstone division, but a thick and persistent group of sandstones and marls with cornstones, generally characterised by a green colour, appears in some of the deeper valleys, among others that of the Senni, from which fact the name of Senni Beds has been applied to them. **1927** *Q. Jrnl. Geol. Soc.* LXXXIII. 197 The plant-remains are entombed in sage-green arenaceous shales, which are intercalated among the typical sage-green sandstones of the Senni Beds of this locality. **1970** R. M. Black *Elements Palaeont.* xix. 302 (*caption*) A psilophyte; a primitive vascular plant from the Lower Old Red Sandstone (Senni Beds), South Wales.

‖ **sennin** (se·n,nin). Also **sennen** [Jap., wizard, recluse, f. Chinese *hsien-jên* an immortal man.] In Oriental mythology: originally in Taoism, an elderly recluse who has acquired immortality through meditation and self-discipline; hence, a human being with supernatural powers, a recluse embodying the spirit of nature.

1875 Audsley & Bowes *Keramic Art of Japan* II (caption to plate X, division 1), Figure of a Buddhist *Sennen*, playing the *Koto*, and seated on the back of a fish. **1908** [see *kirin]. **1912** F. H. Davis *Myths & Legends of Japan* xxix. 356 The *Sennin* are mountain recluses, and many are the legends told in connection with them. Though they have human form, they are, at the same time, immortal, and adepts in the magical arts. **1915** E. Pound *Cathay* 19 In the storied houses of San-Ko they gave us more Sennin music. **1930** —— *XXX Cantos* iv. 17 Père Henri Jacques would speak with the Sennin, on Rokku.

Senoi (senoi·). [Native name, meaning 'man'.] **A.** *sb.* The name of a people inhabiting the provinces of Perak, Kelantan, and Pahang in West Malaysia; the language of this people. **B.** *adj.* Of or pertaining to this people. Cf. *Sakai *sb.* (and *a.*).

1891 *Jrnl. Straits Branch R. Asiatic Soc.* Dec. 14, I shall deal chiefly with the Sên-oi dialect. *Ibid.* 16 The country South of this line being inhabited by Sên-oi, and the northern division by Têm-be. *Ibid.* 22 There are two distinct *r*'s in Sên-oi. **1910** R. J. Wilkinson *Papers on Malay Subjects: Aboriginal Tribes* 21 The Northern Sakai of the Plus valley, a different race, also speak of themselves as *Senoi* and *Mai Darat*. *Ibid.* 23 The true 'Senoi' quiver is plain. **1923** I. H. N. Evans *Studies in Religion, Folk-Lore & Custom in Brit. N. Borneo & Malay Peninsula* II. 202 A Senoi man told me the following legend. **1958** J. Slimming *Temiar Jungle* 3 The largest of the three racial divisions, the Senoi, is divided linguistically into..the Semai-Senoi who are scattered across southern Perak & North Pahang and the Temiar-Senoi living to the north of the Semai. **1972** E. A. Nida *Bk. of Thousand Tongues* 386/2 Senoi is spoken by 15,000 to 20,000 people in the South Perak, Ipoh, Tanjong Malim, and Central Pahang states of Malaya. *Ibid.*, The Senoi moved into the higher inland areas during the early 19th century. *Ibid.*, Now speakers of different Senoi dialects can understand one another only with great difficulty.

señorita. 1. b. (Earlier example.)

1823 J. A. Quitman *Let.* 23 Aug. in J. F. H. Claiborne *Life & Corr. J. A. Quitman* (1860) I. iv. 85 The belles.. 'tote' their fans with the air of Spanish señoritas.

Senousian, Senoussi, etc., see *Senussi.

senr., abbrev. of Senior *a.* 1 a. Cf. *Sen., sen.

1763 J. Bell *Trav. from St. Petersburg* p. v/1 Peter Bell, senr. Esq. **1885** T. Hardy *Let.* 18 Mar. (1978) I. 131 The arrangement I made with Mr Macmillan Senr & Mr Craik. **1932** Blunden *Face of England* 73 John Bowers, senr., came through the clap-gate.

‖ **senryu** (se·nrĭ,u). The name of Karai *Senryu* (1718–90), a Japanese poet, used to denote a type of Japanese verse, similar in form to *haiku but more intentionally humorous or satirical in content and usually without seasonal references.

1938 T. Kunitomo *Jap. Lit. since 1868* II. i. 156 His submissive attitude which he likened to the spirit of *senryū* increased in his later writings. **1958** *Japan: its Land, People & Culture* xiii. 665/2 By applying the rule of 5.7.5. but disregarding other rigid rules *senryu* (satirical poems) were written in a freer spirit and with humour. **1977** G. Grigson *Faber Bk. Epigrams & Epitaphs* p. viii, Both *haiku* and *senryu* are epigrams—if epigram is taken to mean brevity; but a *haiku* has been defined as expressing a moment of vision into the nature of the world and a *senryu* as expressing a moment of satirical insight into the nature of ourselves... With us, rather unfortunately, 'epigram' has come only to suggest something like *senryu*, short and sharp.

sensal (se·nsăl), *a.* *Philos. rare.* [f. Sense *sb.* + -al.] Of or pertaining to sense or meaning (opp. *verbal*), or to the senses.

a **1866** J. Grote *Treat. Moral Ideals* (1876) xxi. 518 Part of our sensal organization. **1896** V. Welby in *Mind* V. 29 We might be allowed to coin a new derivative and speak of 'sensal' where we often now speak of 'verbal' questions. **1927** J. M. E. McTaggart *Nature of Existence* II. v. xxxviii. 116 We rejected the existence of matter and of sensa, because material and sensal qualities, as ordinarily defined, would not permit the determination, within the substances possessing them, of an infinite series of parts within parts. **1938** C. D. Broad *Exam. McTaggart's Philos.* II. vii. xxxiii. 249, I conclude then that McTaggart's argument against the possibility of extended particulars, whether material or sensal, breaks down at the fourth step in my synopsis of it.

sensate, *a.* Add: **5.** *Sociol.* In the theory of P. A. Sorokin, a type of culture in which the satisfaction of material needs and desires is the main objective. Cf. *idealistic *a.* 2 and *ideational *a.* 2.

1937 [see *ideational *a.* 2]. **1959** C. C. Zimmerman in J. S. Roucek *Contemp. Sociol.* 18 In sensate culture the main outlook for the individual is for extra-person stimuli, for articles which appeal to the ordinary untrained tastes, such as is seen in a quantity consumption culture. **1967** T. Parsons *Sociological Theory & Mod. Society* iv. xii. 388 The idealistic synthesis has then proceeded to break down into an increasingly sensate phase. **1977** J. D. Douglas in Douglas & Johnson *Existential Sociol.* i. 69 Most men have distinguished between such sacred thought and everyday, practical thought. (It is important, however, to note that rarely has this distinction been as sharp and important as in our increasingly sensate or secular culture.)

sensation. Add: **5. a.** *sensation drama* (earlier example), *novel* (earlier example), *novelist, scene* (earlier example), *story*; **b.** *sensation-mongering, -seeker, -seeking*; *sensation-giving, -hungry, -mongering* adjs.; **c.** Special comb.: † **sensation cell,** a sense-cell (*obs.*).

1892 Liew & Beyer tr. *Ziehen's Introd. Physiol. Psychol.* 160 He has lost the acoustic memory-cells, but retained the acoustic sensation-cells. **1904** E. B. Titchener tr. *Wundt's Physiol. Psychol.* I. 289 It thus becomes necessary to posit the existence of two sorts of cortical cells: sensation cells and idea cells. **1860** Mrs. S. Cowell *Jrnl.* 13 Mar. in M. Willson Disher *Cowells in Amer.* (1934) 36 We..saw Matilda Heron..in a 'new sensation Drama' called 'Mathilde'. **1865** Mill *Exam. Hamilton's Philos.* xxvii. 526 The knowledge-giving and the sensation-giving properties of an impression of sense. **1951** Koestler *Age of Longing* v. 86 She was sorry to disappoint the expectations of sensation-hungry journalists. **1925** W. Deeping *Sorrell & Son* xvi. 147 It wasn't..our hard work, Stephen, that saved us, but luck, and the noise made by a section of a sensation-mongering press. **1937** *Downside Rev.* LV. 402 The idea of his indulging in 'sensation-mongering' of any sort or kind was ridiculous. **1980** *Times Lit. Suppl.* 24 Oct. 1210/5 They have given a..sober account of..the trial, leaving all the sensation-mongering to frequent interspersions from newspaper headlines. **1863** *Q. Rev.* Apr. 486 A sensation novel, as a matter of course, abounds in incident. **1863** Sensation novelist [see *purpose *sb.* 3]. **1932** Q. D. Leavis *Fiction & Reading Public* II. iv. 154 Mrs. Radcliffe makes an appeal less to the nerves than to the imagination... The sensation novelists make a brute assault on the feelings and nerves in quite another way. **1861** H. Morley *Jrnl. London Play-goer* (1866) 282 Mr. Falconer's 'Peep o' Day'..deserves full houses..for what is called, according to the new term in theatrical slang, which Mr. Boucicault imported for us from the other side of the Atlantic, its 'sensation' scene. **1976** D. Francis *In France* i. 21 All day..cars..disgorged crowds of reporters, photographers and plain sensation-seekers. **1923** R. Macaulay *Told by Idiot* IV. 296 It was a queer affair, born of the emotionalism and sensation-seeking that beset many people at that time. **1869** L. M. Alcott *Little Women* II. xi. 157 She took to writing sensation stories—for..even all-perfect America read rubbish.

d. *Audiometry.* **sensation level,** the number of sensation units by which the loudness of a sound (supposedly proportional to its pressure amplitude) exceeds the loudness at which it would be barely perceptible; **sensation unit,** the unit of loudness by which two sounds differ if one is louder by a factor $10^{0\cdot05}$. (Both terms are disused.)

If loudness were truly proportional to pressure (or displacement) amplitude, and hence to the square root of power, a sensation unit would be equal to a decibel.

1925 J. C. Steinberg in *Physical Rev.* XXVI. 508 By sensation level is meant the number of units that the amplitude of any sound wave must be reduced in order to reach the threshold. **1927** *Ibid.* XXIX. 597 A unit of loudness somewhat better for our purpose would have been the least perceptible increment of loudness of a 700 cycle tone compared with its sensation level. **1929** H. Fletcher *Speech & Hearing* iii. 68 A change of the power level of a sound by one decibel is approximately the smallest that the ear can detect. When this unit is used in this connection the term 'sensation unit' has come into use. The sensation level of any sound reaching the ears is the number of sensation units it is above the threshold level for audition. **1931** Stewart & Lindsay *Acoustics* ix. 224 For the sake of convenience several authors are using the terms sensation unit and sensation level.

sensational, *a.* **2.** (Earlier example.)

1854 A. G. Henderson tr. *Cousin's Philos. of Kant* iii. 32 The sensational philosophy..pretends to deduce all knowledge from experience.

sensationalism. 1. (Earlier example.)

1846 J. D. Morell *Hist. Philos.* I. p. x, There are four expressions which occupy a very prominent place throughout the whole work, and those are—sensationalism, idealism, scepticism, and mysticism. Now of these four, the first, I believe, is a word entirely new, and, therefore, demands some apology for its introduction.

sensationalist. Add: **1.** (Earlier example.)

1847 J. D. Morell *Hist. Philos.* (ed. 2) I. i. 118 Sensationalists have attempted to contravene this view.

2. Also *attrib.* or as *adj.*

1979 *Guardian* 24 Aug. 8/6 A sensationalist and grossly misrepresentative newspaper story. **1980** *Times Lit. Suppl.* 18 Apr. 443/5 Compared with the style of Lady Falkender, Joe Haines *et toute cette galère*, so effectively convicted by their own sensationalist memoirs, the Garden Suburb was distinctly civilized in tone.

sensationalize, *v.* Add: Hence **sensa:tionaliza·tion.**

1955 *Times* 15 Aug. 7/5 By silence, and by mistrust of any publicity save that in the jargon of scientific journals, science has succeeded (in the words of Rutherford) in its own 'sensationalization'. **1977** *Lancet* 27 Aug. 449/1 It fell into disuse with..the sensationalisation of the 'opium vice' by writers such as De Quincey.

sensationism. Add: **2.** = Sensationalism 1.

1846 J. D. Morell *Hist. Philos.* I. p. xi, Next, I thought of sensism and sensationism, as being terms well adapted to describe the philosophy which builds itself up upon sense, or sensation; but these seemed to fail in respect to taste and euphony. **1936** *Brit. Jrnl. Psychol.* July 96 Stout saw the futility of associationism and sensa-

tionism forty years ago. **1948** W. McDougall *Social Psychol.* (ed. 29) 427 The essential novelty (for German psychology) of the teaching of this [sc. *Gestalt*] school is the repudiation of atomistic sensationism.

sensationist. Add: **2.** = Sensationalist 1. Also *attrib.*

1890 W. James *Princ. Psychol.* I. ix. 277 Now most believers in the ego make the same mistake as the associationists and sensationists whom they oppose. **1942** E. G. Boring *Sensation & Perception* i. 4 An empiricist is apt to be a sensationist, because it is by way of the senses that the mind has experience of the external world. **1953** K. Britton *J. S. Mill* vi. 192 From Locke to James Mill, there continued a complicated debate about the way in which we know physical objects; Reid and Hamilton providing an intelligent opposition to the sensationist school.

Hence **sensationi·stic** *a.*

1936 *Brit. Jrnl. Psychol.* July 97 The agreement, in principle, to discard the sensationistic hypothesis.

sense, *sb.* Add: **I. 1. f.** With defining word: the intuitive knowledge or appreciation of what action or judgement is appropriate to a given situation or sphere of activity. (Closely related to sense 1 d.)

1879, **1880** [see *colour-sense* s.v. Colour, color *sb.* 18]. **1923** G. Atherton *Black Oxen* vii. 23 The reportorial news-sense died painlessly. **1926**, etc. [see *dress sense* s.v. *dress *sb.* 4 a]. **1932**, etc. [see *clothes-sense* s.v. *clothes *sb. pl.* 4]. **1932** E. V. Lucas *Reading, Writing & Remembering* i. 29 Had he [*sc.* Dickens] been possessed of more prudence or money-sense..his last years would have been more leisurely and peaceful. **1957** H. Read *Tenth Muse* xxii. 182 The producer, and the actor, are firmly convinced that there is some sixth sense, a feeling for what is possible in the theatre, a 'stage-sense'.

9. (Further examples.)

1962 A. Nisbett *Technique Sound Studio* vi. 106 The live broadcast seems to have a greater sense of occasion. **1974** R. Adams *Shardik* lvi. 472 From natural awe and sense of occasion, they did not press forward.

III. 27. *to make sense*: also in extended use (freq. in neg. contexts).

1921 G. B. Shaw *Back to Methuselah* iv. 148 She spoke to me without any introduction, like any improper female... Improper female doesnt make sense. **1936** *Punch* 12 Feb. 170/2 It can't be right, it can't be. Spats and a bowler-hat, but no umbrella—it doesn't make sense.

29. b. Chiefly *Math.* That which distinguishes a pair of entities which differ only in that each is the reverse of the other.

1894 H. W. L. Hime *Outl. Quaternions* I. i. 2 No two vectors are equal unless they have, first, equal lengths, and, secondly, similar directions—the phrase 'similar directions' meaning 'parallel directions with the same sense'. **1947** Courant & Robbins *What is Math.?* (ed. 4) iii. 159 Although inversion preserves the *magnitude* of angles, it reverses their *sense*; i.e. if a ray through *P* sweeps out the angle *x*. in a counterclockwise direction, its image will sweep out angle *y*. in a clockwise direction. **1950** [see *oriented *ppl. a.* 1]. **1962** A. Nisbett *Technique Sound Studio* 251 The doublet [microphone] can become bi-directional, cardioid or omnidirectional, simply by varying the size and sense of the potential on one of the diaphragms. **1977** Holland & Treeby *Vectors* i. 10 The vector (1/*a*)a is a unit vector in the direction and sense of a.

IV. 30. *sense-appearance, -apprehension, -awareness, -cell, -consciousness* (earlier example), *-element, -feeling, -idea, -impulse, -life, -material, -modality, -object, -observation, -percept, -perception* (earlier and later examples), *-phenomenon, -picture, -presentation, -symbolism, -verification*; (senses 19 and 20) *sense-assimilation, -change, -development, -group, -link, -linkage, -loan, -unit, -word*; (instrumental) *sense-given* adj.; **sense aerial** = *sense-finder* below; **sense-content,** (*a*) *Philos.*, whatever is present to one of the senses; a sense-datum; (*b*) the sense or meaning contained in an idea or literary passage; **sense-experience,** experience that is derived from the senses; **sense-field** *Philos.* (see quot. 1925); **sense-finder,** an aerial designed for sense-finding; **sense-finding,** with some radio direction-finders: the operation of determining which of two indicated directions 180° apart is correct; **sense history,** (*a*) *Philos.* (see quot. 1923); (*b*) the history of the development of meaning attached to a word; **sense-quality** *Philos.* and *Psychol.*, the quality of the sensory properties inherent in an object; **sense-withdrawal** *Yoga* = *pratyahara; **sense-world,** the external world as it is known through the senses; the 'world' of experience that is derived from one of the senses.

1941 W. J. D. Allan *Radio Navigation* ii. 42 The third method of finding sense is by means of a sense aerial. **1970** Taylor & Parmar *Ground Studies for Pilots* vii. 245 By adjusting the phasing of the loop and sense aerials, the cardioid and its image are produced in rapid alterations. **1894** A. C. Fraser in *Locke's Essay Hum. Und.* II. iv. xi. 328 When ideas, or qualities of things, are..not merely revived in memory or imagination, in the absence of the actual sense-appearances. **1947** *Mind* LVI. 300 A chief begetter of the sense-datum theory was the problem set by illusory sense-appearances. a **1902** R. Adamson *Devel. Mod. Philos.* (1903) II. ii. i. 229 Leibniz..maintains that our sense-apprehension of the colour green is a confused sense-apprehension of the two colours blue and yellow. **1921** Hannay & Collingwood tr. *Ruggiero's Mod. Philos.* 206 The first 'something' is mere sense-apprehension. **1935** M. E. Houtzager *Unconscious Sound- & Sense-Assimilations* i. 26 Place-names..change according to sound-laws, but also..through unconscious sound- and sense-assimilations. **1922** A. N. Whitehead *Princ. Relativity* ii. 20 Divest consciousness of its ideality, such as its logical, emotional, aesthetic and moral apprehensions, and what is left is sense-awareness. **1978** *English Jrnl.* Dec. 57/2 The second group of students did get to try to drink from a water fountain which was truly a unique experience in sense-awareness. **1908** *Practitioner* Oct. 548 In the case of all our senses, the effects are produced by reponsive protoplasmic movements of the specially adapted sense-cells. **1953** N. Tinbergen *Herring Gull's World* iii. 19 The sense-cells in the retina are the units of vision. **1976** H. R. Schiffman *Sensation & Perception* ix. 125/2 (*caption*) The tips of the sense cells extend into a pit. **1931** G. Stern *Meaning & Change of Meaning* x. 261 Clippings seldom give rise to sense-changes. **1951** W. Empson *Structure Complex Words* 26 The cause of a sense-change need have nothing to do with the use made of it after it has been pushed through. **1858** A. C. Fraser *Rational Philos.* 94 So-called *sense*-consciousness can be analysed. **1896** L. T. Hobhouse *Theory of Knowl.* ii. 42 It is quite enough for our purpose that *some* sense-contents should be complex. **1902** W. James *Varieties Relig. Experience* iii. 55 The words 'soul', 'God', 'immortality', cover no distinctive sense-content. **1962** W. Nowottny *Lang. Poets Use* v. 111 The action of the poem..is not something agglomerated out of the successive sense-contents of each line. **1975** W. S. Robinson in H. N. Castaneda *Stud. in Sellars' Philos.* 105 An analogue..designed to apply specifically to sensing sense contents, is presupposed. **1882** J. A. H. Murray *Let.* in K. M. E. Murray *Caught in Web of Words* (1977) x. 190 Nobody exc[ept] my predecessors in specimens of the Dicty has yet *tried* to trace out historically the sense-development of English words. **1960** C. S. Lewis *Studies in Words* 29 Indeed the sense-development of the word *proper* itself..is a striking instance. **1889** J. Venn *Empirical Logic* vi. 150 The adhesive power between the sense-element and the notion is particularly strong in the case of..smell. **1862** A. C. Fraser in *Macm. Mag.* VI. 194/2 The steady reference to sense-experience ..distinguished Locke. **1871** Sense-experience [see Pre-perception]. **1923** T. P. Nunn *Educ.* xiii. 171 All that constitutes..the quantitative as distinguished from the qualitative aspects of sense-experience. **1968** *Listener* 30 May 685 Principles native to the mind which we utilise in grasping sense-experience. **1890** W. James *Princ. Psychol.* II. xx. 268 The *education* of our space-perception consists largely of two processes—reducing the various sense-feelings to a common *measure*, and *adding them together* into the single all-including space of the real world. **1925** C. D. Broad *Mind & its Place* iv. 195 A sensum is not something that exists in isolation; it is a differentiated part of a bigger and more enduring whole, viz., of a sense-*field*. **1971** A. J. Ayer *Russell & Moore* iii. 65 To obtain the equivalent of sensibilia, on the basis of our primitive percepts, all that is needed, I believe, is the projection of spatial and temporal relations beyond the sense-fields in which they are originally given. **1934** Webster, Sense finder. **1953** C. H. Cotter *Elem. Navigation* xlvii. 485 The principle of the sense finder is as follows. Depending on whether the transmitting station lies in a certain direction or the opposite direction, the e.m.f. in the loop aerial will be altered in phase by 180°. **1937** D. C. T. Bennett *Compl. Air Navigator* iv. 134 A sense-finding arrangement is usually incorporated in Fixed Loop Direction Finders. **1957** R. Watson-Watt *Three Steps to Victory* lviii. 361 Our Radio Research Station work..had included 'sense-finding', the removal of the direction-finding ambiguity between one compass bearing and its exact opposite. **1871** A. C. Fraser *Life of Berkeley* x. 369 We may even, with Berkeley, call these sense-given phenomena 'sensations'. **1933** *Mind* XLII. 292 A great variety of sense-given shapes—squares, parallelograms, trapezia, etc.—would then..all be either portions or distortions of the surface of this cube. **1928** C. Bergener *Contrib. to Study of Conversion of Adjectives into Nouns* 1, I have..made an attempt.., after arranging the material in sense-groups, to ascertain the productivity of this mode of word-formation during the different periods of the language. **1966** G. N. Leech *Eng. in Advertising* viii. 89 One of the skills of writing formal English consists in..arranging one's ideas so as to make the end of each sense-group..as far as possible the appropriate place for emphasis. **1923** C. D. Broad *Sci. Thought* x. 362 Let us call the whole series of sensible fields which an observer O senses in the course of his life, O's *sense-history*. **1933** *Oxf. Eng. Dict.* I. p. v, The aim of this Dictionary is to present..the words that have formed the English vocabulary..with all the relevant facts concerning their form, sense-history, pronunciation, and etymology. **1954** A. J. Ayer *Philos. Ess.* iv. 95 The occurrence, within a given sense-history, of a series of sense-fields. **1871** A. C. Fraser *Life of Berkeley* iii. 75 Sense-ideas are with Berkeley real and presentative; not representative images. **1900** B. Russell *Philos. Leibniz* xiv. 161 Sense-ideas must..be distinguished by their own nature. **1896** L. T. Hobhouse *Theory of Knowl.* ii. 56 A felt total impression resulting from the forty separate sense-impulses. **1894** A. C. Fraser in *Locke's Essay Hum. Und.* II. iv. xi. 332 For it is not metaphysically impossible that there may be a *dream*, continuous and orderly, like the actual sense-life of a man. **1964** M. McLuhan *Understanding Media* i. i. 19 Money has reorganized the sense life of peoples just because it is an *extension* of our sense lives. **1957** N. Frye *Anat. Criticism* iv. 272 The poetic creation..is an associative rhetorical process, most of it below the threshold of consciousness, a chaos of paronomasia, sound-links, ambiguous sense-links, and memory-links very like that of the dream. **1962** W. Nowottny *Lang. Poets Use* i. 15 The sense-linkage effected by rhyme is an effect of which Arnold was well aware. **1931** Sense-loan [see *cross-influence* s.v. *cross-B.]. a **1902** R. Adamson *Devel. Mod. Philos.* (1903) II. 68 Nor does inner sense-material lend itself even to the less complete theoretical form of natural science. **1894** Creighton & Titchener tr. *Wundt's Hum. & Anim. Psychol.* vii. 119 The individual sensation is estimated by the relation in which it stands to other sensations of the same sense-modality. **1977** P. Strevens *New Orientations Teaching of Eng.* ix. 115 The sense-modalities of vision and hearing. **1908** W. James *Meaning of Truth* (1909) xii. 239 Our private concepts represent the sense-objects to which they lead us, these being public realities independent of the individual. **1920** A. N. Whitehead *Concept Nat.* viii. 170 The appearance of sense-objects is conditioned by the adventures of material objects. **1949** *Mind* LVIII. 58 The position will be, as Mr. Russell in fact saw, that the constituents of physical constructs are not simply sense-*objects* but sensations or sense-experiences. **1909** W. James *Pluralistic Universe* iv. 145 Hypotheses, and deductions from these, controlled by sense-observations and analogies with what we know elsewhere, are to be thanked for all of science's results. **1956** E. L. Mascall *Christian Theol. & Nat. Sci.* ii. 48 Scientific theories need for their expression technical terms..whose definition in terms of sense-observations is extremely complicated and remote. **1907** W. James *Pragmatism* vi. 218 Some part of a system that dips at numerous points into sense-percepts. **1846** J. D. Morell *Hist. Philos.* I. ii. 205 In doing this, Kant took it for granted, as a thing lying altogether beyond the region of proof, the reality of our sense-perceptions. **1971** R. I. Aaron *Knowing & Function of Reason* iv. 79 This conclusion is reinforced by arguments from the change induced in sense-perception by drugs, such as mescaline. **1871** A. C. Fraser *Life of Berkeley* x. 371 There is no evidence that an unperceived sensation or sense-phenomenon exists. **1971** R. I. Aaron *Knowing & Function of Reason* iv. 79 What is sensed has been variously described as idea, impression,..sense-phenomenon, sensum, and so on. **1920** A. S. Eddington *Space, Time & Gravit.* ii. 32 It would be unreasonable to limit our thought of nature to what can be comprised in sense-pictures. **1884** S. H. Hodgson *Let.* 14 Feb. in R. B. Perry *Thought & Char. W. James* (1935) I. 625 The answer must, for me, be given by a 'what' which is at least a *possibility* of sense presentation. **1932** W. T. Stace *Theory of Knowl.* iii. 34 The images of hallucination and dream are just as much part of the given as are sense presentations. **1896** L. T. Hobhouse *Theory of Knowl.* ii. 38 The idea of sensation as giving these simple sense-qualities and nothing else. **1954** R. Wells in Saporta & Bastian *Psycholinguistics* (1961) 280/2, I refer to the fact that they are all names of sense-qualities. **1871** A. C. Fraser *Life of Berkeley* x. 375 The substantiality and causality of matter thus resolve into a Universal Sense-symbolism, the interpretation of which is the office of physical science. **1892** H. Sweet *New Eng. Gram.* I. 20 A word may be defined as an ultimate independent sense-unit. **1974** R. Quirk *Linguist & Eng. Lang.* vi. 97 We are dealing with languages whose structures differ so much that..translation..is possible only if we deal in large sense-units. **1907** W. James *Pragmatism* vi. 209 Their relations are perceptually obvious at a glance, and no sense-verification is necessary. **1937** K. T. Behanan *Yoga* xiii. 215 In *pratyahara* or the sense-withdrawal stage, a deliberate effort is made to diminish the impulses streaming in through the sense organs. **1960** J. Hewitt *Yoga* viii. 116 Sense-Withdrawal is something which you must do for yourself, your 'I' must be in complete control. c **1874** Sense-word [see *rhythm-word* s.v. *rhythm 9 a]. **1911** W. James *Some Probl. Philos.* viii. 139 Monism usually treats the sense-world as a mirage or illusion. **1932** W. T. Stace *Theory of Knowl.* ix. 212 Those characters of the 'thing'..are in different sense-worlds.

sense, *v.* Add: **4.** (Earlier examples.)

1841 A. M. Maxwell *Run through U.S.* I. 102 The noun *sense* they convert into a most comical verb—'I sense', or 'She sensed him to do it'. **1849** *Knickerbocker* XXXIII. 201 'Do you sense what you are doing, Jack?' said she. 'Sense it, Suzy?' replied B.,—'I do, to the letter.'

6. Of a machine, instrument, etc.: to detect (some circumstance or entity).

1946 *Ann. Computation Lab. Harvard Univ.* I. 22 In the event that one or both of the factors involved in a multiplication are negative numbers, this fact is sensed and stored by the multiply unit. **1962** F. I. Ordway et al. *Basic Astronomics* v. 197 After arrival on the Moon the fluid is vented, an operation sensed by a pressure switch. **1978** *Sci. Amer.* June 54/3 In general particle detectors operate by sensing the ionization of atoms caused by the passage of a charged particle.

sense-datum (se·ns,dē¹:tɒm). *Philos.* Pl. **-data.** [f. Sense *sb.* 1 + Datum.] Whatever is the immediate object of any of the senses, usually, but not always, with the implication that it is not a material object.

1882 J. Royce in *Mind* VII. 44 What relation does the external reality bear to the sense-datum? **1890** W. James *Princ. Psychol.* II. xx. 146 It is no wonder if some authors have gone so far as to think that the sense-data have no spatial worth at all. **1912** B. Russell *Probl. Philos.* i. 12 Let us give the name of 'sense-data' to the things that are immediately known in sensation: such things as colours, sounds, smells, hardnesses, roughnesses, and so on. **1938** W. S. Maugham *Summing Up* 260 The sense-datum, on which I thought all knowledge was based, seemed to me something given, which had to be accepted whether it suited the convenience or not. **1956** A. J. Ayer *Probl. Knowl.* 85 What..is immediately given in perception is an evanescent object called an idea, or an impression, or a presentation, or a sense-datum, which is not only private to a single observer but private to a single sense. **1980** *Dædalus* Spring 11 From the point of view of strict empiricism, the attempt to go beyond sense data..seems to fail.

‖ **sensei** (sensē¹·). [Jap.] In Japan: a teacher or instructor; a professor; a respectful title, occasionally with ironic connotations, for one skilled in an art. Also *transf.*

1884 tr. *J. J. Rein's Japan* II. i. 378 The Ban-i (foreign barbarian) of yesterday was the Ijin-san (foreign gentleman) of to-day, and in the mode of address even a sen-sei (worthy scholar). 1934 E. Blunden *Mind's Eye* 93 A copy of some newly acquired book, or a Japanese clay figure for the sensei's table. 1959 *Times Lit. Suppl.* 10 July 411/1 The ordinary reader begins to feel that he cannot get very far in poetry without a *sensei* or a *guru*. 1972 J. Ball *Five Pieces of Jade* iii. 35 My karate sensei tells me that I should learn Japanese. 1981 J. Melville *Sort of Samurai* i. 7 I'm afraid your colleague my father never really forgave me, Horiguchi-sensei.

Sen-Sen (se·nsen). *N. Amer.* Also **sen-sen.** The proprietary name of a breath-sweetener, freq. used to disguise the smell of drink or cigarettes.

1911 *Official Gaz.* (U.S. Patent Office) 30 May 1273/2 Sen-Sen Chiclet Company, New York, N.Y. Filed Mar. 23, 1911... Chewing-Gum and Cachous. Claims use since on or about the month of March, 1894. 1936 J. Dos Passos *Big Money* 497 He was waiting..in the lobby..chewing sensen to take the smell of three whiskeys..off his breath. 1947 J. Steinbeck *Wayward Bus* vii. 98 He took a few grains of sen-sen out of his inner shirt pocket and threw them in his mouth. 1951 P. Branch *Lion in Cellar* i. 8 He was eating sen-sen cachous. 1972 *Even. Telegram* (St. John's, Newfoundland) 28 June 12/1 Imbibers chewed Sen Sen to take the odor of liquor off their breath.

‖ **sensibile** (sensi·bile). *Philos.* Usu. in pl. **sensibilia** (sensibi·liă). [L., neut. of *sensibilis* Sensible.] A term popularized by Bertrand Russell to denote the kind of thing which, if sensed, is a sense-datum.

1856 J. Hinton *Sel. from MSS.* (1871) II. 159 The matter of the schools, the substratum that underlies and is to be distinguished from the 'properties' or 'sensibilia', must be the 'actualistic', eternal (or spiritual.) That is, it is the eternal *not seen*..i.e. it is the hypothesis. 1906 J. A. Stewart in *Mind* Oct. 521 The Ideas [of Plato]..are 'known'..only as performing their function of making *sensibilia* intelligible. 1918 B. Russell *Mysticism & Logic* viii. 148, I shall give the name *sensibilia* to those objects which have the same metaphysical and physical status as sense-data, without necessarily being data to any mind. Thus the relation of a *sensibile* to a sense-datum is like that of a man to a husband: a man becomes a husband by entering into the relation of marriage, and..a *sensibile* becomes a sense-datum by entering into the relation of acquaintance. 1921 tr. *Ruggiero's Mod. Philos.* 324 On this basic duality Varisco builds his theory. On the one side there exists the reality of sensibilia. 1940 A. J. Ayer *Found. Emp. Knowl.* ii. 71 We shall have to take as a criterion for the existence of a sensibile the truth of a single hypothetical proposition. 1962 J. L. Austin (*title*) Sense and sensibilia.

‖ **sensibilité** (sańsibilite). *rare.* [Fr.] = Sensibility.

1926 D. H. Lawrence *Plumed Serpent* v. 90 And in all the crowd, a sense of guardedness..a curious soft *sensibilité*. 1960 *Encounter* XV. II. 64 It was a matter of being formed alike, of having the same *sensibilité*.

sensibility. Add: **2. c.** *dissociation of sensibility*: T. S. Eliot's term for a separation of thought from feeling which he held to be first manifested in poetry of the later seventeenth century.

1921 T. S. Eliot in *Times Lit. Suppl.* 20 Oct. 669/4 The poets of the seventeenth century..possessed a mechanism of sensibility which could devour any kind of experience... In the seventeenth century a dissociation of sensibility set in, from which we have never recovered. 1930 E. M. W. Tillyard *Milton* 356 Some sort of dissociation of sensibility in Milton, not necessarily undesirable, has to be admitted; but that he was responsible for any such dissociation in others (at least till this general dissociation had inevitably set in) is untrue. 1943 L. C. Knights *Explorations* (1963) 93 It is as a contribution to our understanding of the seventeenth century 'dissociation of sensibility'—from which, as Mr. Eliot remarked,.. 'we have never recovered'—that I wish to consider some of the work of Francis Bacon. 1947 T. S. Eliot *Milton* 7, I wish first to mention another reproach against Milton, that represented by the phrase 'dissociation of sensibility'... I believe that the general affirmation represented by the phrase 'dissociation of sensibility'..retains some validity, but I now incline to agree with Dr. Tillyard that to lay the burden on the shoulders of Milton and Dryden was a mistake. 1957 F. Kermode *Romantic Image* viii. 143 The theory of the dissociation of sensibility is, in fact, the most successful version of a Symbolist attempt to explain why the modern world resists works of art that testify to the poet's special, anti-intellectual way of knowing truth.

sensible, *a.* (and *sb.*). Add: **IV. 14. c.** Of clothing, footwear, etc.: practical rather than attractive or fashionable.

1855 Mrs. Gaskell *North & South* xii. 146 Margaret was busy embroidering a small piece of cambric... Mrs. Thornton..liked Mrs. Dale's double knitting far better; that was sensible of its kind. 1888 Kipling *Under Deodars* 8 Nice, large, sensible shoes for all couples to stumble over as they go into the verandah! 1907 *Yesterday's Shopping* (1969) 339/3 The 'Sensible' carrier bag..is the only paper Bag with a firm bottom. 1924 A. Christie *Man in Brown Suit* xx. 169 Forty, if she's a day, wears pince-nez and sensible boots and an air of brisk efficiency that will be the death of me. 1944 Auden *For Time Being* ii. 36 The river on this side of which initiative and honesty stroll arm in arm wearing sensible clothes. 1959 *Observer* 22 Mar. 1/1 Chintz curtains and no-nonsense bundles of flowers in sensible pots. 1978 R. Hill *Pinch of Snuff* x. 100 Genuine English county..unobtrusively elegant..in simple twinset and sensible shoes.

‖ **sensiblerie** (sańsiblĕri). [Fr.] = Sentimentality a.

1931 *Times Lit. Suppl.* 7 May 559/1 The sentiment is obviously genuine, and never falls into excesses of *sensiblerie*. 1960 *Twentieth Cent.* Aug. 169 A piece of long-drawn-out, commercial *sensiblerie*, with no real tragic bite. 1974 *Financial Times* 4 Apr. 32/1 A brotherhood utterly devoid of sentimentality or *sensiblerie*.

sensibly, *adv.* **4.** (Earlier and later examples.)

1898 G. B. Shaw *Candida* I. 103 Do you think that the things people make fools of themselves about are any less real and true than the things they behave sensibly about? 1932 D. L. Sayers *Have his Carcase* i. 9 She was dressed sensibly in a short skirt and thin sweater. 1970 N. Marsh *When in Rome* iii. 51 They wore sensibly shady hats.

sensillum (sensi·lŏm). *Zool.* Pl. **sensilla.** Also **sensilla** (pl. **-æ**), **sensillium** (*rare*). [mod.L. (coined in Ger. by E. Haeckel in *Systematische Phylogenie* (1895) III. ii. 119), dim. of L. *sensus* perception.] A simple sensory receptor in invertebrates, esp. arthropods, consisting of a cell or small group of cells that is a modification of the cuticle or epidermis and is often hair- or rod-shaped.

1925 A. D. Imms *Gen. Textbk. Entomol.* 65 The tactile sensillæ of insects are often distributed over the entire integument. 1935 R. E. Snodgrass *Princ. Insect Morphol.* xvii. 525 The scolopophorous organs are usually compound sense organs, each consisting of a bundle of simple sensilla having a common point of attachment on the body wall. 1962 Gordon & Lavoipierre *Entomol. for Students of Med.* xxxiv. 211 On the 9th tergite of both the male and female flea, there is a small pincushion-like structure (the pygidium or sensillium) which is believed to have a sensory function. 1962 *Science Survey* III. 281 The last main class of hearing organs in insects comprises the various hair sensillae, scattered all over the body. 1963 R. P. Dales *Annelids* vii. 143 The sense cells on these tentacles or cirri are commonly fusiform in shape with a sensilla or sensillae projecting through the cuticle. 1971 *Nature* 24 Dec. 477/1 Each campaniform sensillum [on a cockroach leg] functions through a single primary sense cell. 1973 *Jrnl. Invertebrate Path.* XXII. 409 (*caption*) Portion of another integumentary growth with apparent sensillium.

sensing, *vbl. sb.* (in Dict. s.v. Sense *v.*). Add: *spec.* **a.** *U.S. Mil.* An observation of the point of impact of a shot with respect to the target.

1937 *Sun* (Baltimore) 16 Aug. 16/2 Transfers to the targets will be made by means of a high burst with shrapnel, the bursts being brought to a check point at a particular elevation and range by means of sensings from lateral observation points. 1944 H. F. Gregory *Anything Horse can Do* vii. 74 Conducting an artillery adjustment by communicating 'sensings' by changing the line of flight of the Autogiro. 1962 *Ordnance Techn. Terminol.* (U.S. Army Ordnance School, AD 660 112) 269/2 *Sensing*, the direction of a point of burst or impact, or centers of burst or impact with respect to the target; such as over, short, air or graze.

b. The action of an automatic device in detecting, observing, or measuring something. Freq. *attrib.*

1950 *Jrnl. Res. Nat. Bureau of Standards* (U.S.) XLV. 295/1 A sensing element in the gas stream normally attains a steady temperature below that of the gas itself, because of radiation and conduction from the sensing element. 1955 *Flight Test Man.* (Advisory Group on Aeronaut. Res. & Devel., NATO) I. IIA. x. 15/1 The collection methods of determining liquid water content below freezing..are essentially based upon the sensing of icing. 1962 F. I. Ordway et al. *Basic Astronautics* v. 185 (*caption*) Infrared vidicon tube... It could also be employed for horizon sensing. 1967 *Electronics* 6 Mar. 127/1 An infrared scanner will compete in the fast-developing market for thermal sensing and display devices. 1977 *Dædalus* Fall 38 All the rest of the electromagnetic spectrum became available to astronomers only when the development of sounding rockets and artificial satellites made it possible to send their sensing instruments above the atmosphere.

c. = *sense-finding* s.v. *Sense *sb.* 30.

1961 C. H. Cotter *Princ. & Pract. Radio Direction Finding* ii. 36 The device provided to resolve the 180° ambiguity is known as a sensefinder, and the process of doing so is known as sensing. 1976 W. H. P. Canner *Air Navigation* vi. 139 Sensing refers to the resolution of the ambiguity in bearings.

sensism. b. (Earlier example.)

1846 [see *sensationism 2].

sensitive, *a.* and *sb.* Add: **A.** *adj.* **3.** *sensitive plant* (earlier and later *fig.* examples: now freq. of persons).

1821 P. Egan *Boxiana* 1st Ser. III. 236 Martin went to work with both his hands so quickly, that his opponent's sensitive plant rolled about like a humming top, and he fell out of the ring covered with crimson. 1907 J. London *Let.* 27 Sept. (1966) 251 All 'sensitive plants' are egomaniacs; they are colossally stuck upon themselves. 1926 Galsworthy *Silver Spoon* I. vi. 41 Well, sir, the Press is a sensitive plant. I'm afraid you might make it curl up. 1974 'S. Woods' *Done to Death* 190 'So nice to be considered a sensitive plant,' said Emma.

5. e. Of a drilling machine: designed to give the operator continuous and sensitive control over the pressure and rate of drilling.

1895 C. J. Appleby *Illustr. Handbk. Machinery* IV. 53 Six speed sensitive drilling machine..is capable of drilling holes up to ⅝ in. diameter, and will swing 18 inches. 1942 [see *pillar drill* s.v. *pillar *sb.* 12]. 1971 C. R. Hine *Machine Tools & Processes for Engineers* xi. 261 This machine is slightly heavier than the sensitive drill press.

f. Involved with or likely to affect national security. Also with reference to other issues: that must be treated with care; likely to give offence if mishandled.

1953 *Manch. Guardian Weekly* 7 May 2 People in 'sensitive' jobs or departments—that is in positions having access to top secret or policy information. 1968 *Globe & Mail* (Toronto) 17 Feb. 7/1 Under the policies of the Government of Canada no one can buy arms and ship them from Canada to any sensitive area, whether that be Vietnam or anywhere else. 1973 P. Geddes *Ottawa Allegation* x. 138, I realise it's from a sensitive source, but could I see it for myself? 1977 T. Heald *Just Desserts* iv. 68 Probe gently. We are in what is known as a sensitive area.

g. Of a mathematical, statistical, or physical quantity: largely or appreciably influenced by changes in some other quantity, the choice of method or model, etc. Const. *to.*

1955 [see *robust *a.* 4]. 1966 A. Battersby *Math. in Managem.* ix. 231 The cost is not 'sensitive' to the batch size. 1968 Fox & Mayers *Computing Methods for Scientists & Engineers* iii. 31 The results show that y_r is extremely sensitive, for large *r*, to small changes in the initial condition. 1979 G. E. P. Box in Launer & Wilkinson *Robustness in Statistics* 211 How sensitive are inferences made about θ to these contemplated misspecifications of the model?

sensitivity. Add: **3. a.** The degree to which a device, test, or procedure responds to small amounts of or slight changes in that to which it is designed to respond; the ratio of the response of a device to the stimulus causing it; = Sensitiveness 3.

1918 E. S. Ferry *Handbk. Physics Measurements* II. iv. 179 (*heading*) Determination of the sensitivity of a galvanometer. 1937 H. Eagle *Lab. Diagnosis of Syphilis* vi. 117 The longer the incubation period, the greater was the sensitivity of the test. 1944 E. S. Smith *Automatic Control Engin.* iii. 17 Sensitivity is merely the ratio effect/cause. The over-all sensitivity is equal to the product of all the component sensitivities of the instrument. 1955 *Sci. Amer.* Mar. 68/2 The three procedures represent an ascending scale of sensitivity, and a descending scale of specificity. 1973 *Nature* 7 Dec. 343/2 An unsuccessful search for gravitational radiation was reported about a year ago by Braginskii *et al.*, with detectors of comparable sensitivity to those of Weber.

b. *spec.* in *Radio*, (a measure of) the ability of a receiver or other part of a radio system to pick up or respond to weak radio signals.

1928 L. S. Palmer *Wireless Princ. & Pract.* ix. 305 Most poor rectifying contacts can be improved by either the application of a small potential difference or by the application of heat, but with some of these contacts the increased sensitivity persists after the removal of the potential difference or heat. 1931 Moyer & Wostrel *Radio Handbk.* iii. 124 Many crystals do not have a uniform sensitivity over the entire surface. 1962 *Which?* Feb. 40/1 We tested the radios to see how well they would receive weak stations. Their ability to do this is called sensitivity. 1965 *Wireless World* Sept. 457/1 The sensitivity of an audio amplifier is nowadays (following the British Standard) often specified in terms of a 'sensitivity voltage', i.e. the e.m.f. applied in series with the stated source resistance, to the input terminals in order to obtain the rated output power or voltage. 1975 G. J. King *Audio Handbk.* v. 111 Each channel of a two- or four-channel amplifier should be measured for sensitivity independently.

4. *Psychol.* Used *attrib.*, esp. in *sensitivity group, training*, to denote training in small groups aimed at increasing a person's awareness of the behaviour, feelings, and motives of others and of himself. Cf. *T-group* s.v. *T 6*.

1954 *Personnel* XXX. 256/1 The suggested approach to leadership training combines these two features in order to focus sensitivity training on those interpersonal problems which intimately involve the members of the training group. 1964 M. Argyle *Psychol. & Social Probl.* x. 133 Many students could..benefit from sensitivity training, aimed at increasing the accuracy of perception of social situations. 1969 *Listener* 26 June 881/1 A sensitivity group of persons gets together in order to cultivate a heightened awareness of themselves and each other, in a sort of group therapy. 1971 *Harvest Years* Mar. 8/2 (*caption*) A few scenes from a sensitivity session. 1977 E. G. & N. C. Bormann *Speech Communication* (ed. 2) i. 12 Sensitivity groups have been used to train management personnel..and to institute individual and organizational change.

sensitization. Add: **1.** (Later examples in *Physiol.*)

1916 [see *sensitizin]. 1947 *Ann. Rev. Microbiol.* I. 298 Very striking is the effect of mycobacteria on sensitization to some simple compounds such as picryl chloride and 2:4 dinitrochlorobenzene... When these are injected intraperitoneally a state of anaphylaxis alone ensues, while the use of mycobacteria produces in addition

sensitization of the contact dermatitis type. 1969 *Daily Progress* (Charlottesville, Va.) 12 Jan. A2/4 It also brings its own problems of sensitization to the horse serum from which it is made.

2. *Psychol.* The fact or condition of responding in a sensitized or sensitive (as opp. to a repressed) manner, esp. to an emotional stimulus; the process of being sensitized to a particular stimulus.

1947 *Jrnl. Personality* XVI. 75 Such sensitization represents..the obverse of defense... We now find sensitization in the presence of 'dangerous' stimulus objects. 1959 *Ibid.* XXVII. 364 The opposite syndrome, composed of high Admission, low Denial, and high Anxiety scores describes the other end of the repression continuum, which has been labeled 'sensitization'. 1967 *Psychol. Reports* XX. 459 The word 'sensitization' is used because the purpose of the procedure is to build up an avoidance response to the undesirable stimulus. 1976 *Jrnl. Clin. Psychol.* XXXII. 321 Results were interpreted according to an approach-avoidance model of repression-sensitization.

sensitize, *v.* **2.** Transfer quot. 1904 to sense 3 below and add later example.

1978 *Dædalus* Spring 228 It is..reasonable to hope that the fraction of abuses, mistakes, surprises, and other alarming problems will drop as the professionals involved become more and more sensitized to the possibility of such problems.

3. *Physiol.* To render (an organ or organism) sensitive *to* the presence of some agent; *esp.* to render (the immune system) sensitive *to* the presence of antibody. Also *absol.*

1904 [in Dict., sense 2]. 1909 *Jrnl. Amer. Med. Assoc.* 30 Oct. 1473/1 The substance which is produced in the corpus luteum and which sensitizes the mucosa of the uterus has a specific affinity to the uterine tissue. 1922 *Jrnl. Physiol.* LVI. 143 The presence of free CO_2 'sensitises' the nerve cells to H ions. 1947 *Ann. Rev. Microbiol.* I. 305 It would be of great interest to identify the fraction or fractions which exert the adjuvant effect of mycobacteria and to know whether the effect could be produced without sensitizing to tuberculin. 1970 Passmore & Robson *Compan. Med. Stud.* II. xxii. 12/1 IgM antibodies are unable to sensitize tissues for anaphylactic reactions..although IgG and IgA molecules, are able to do so.

sensitized *ppl. a.*, **sensitizing** *vbl. sb.* and *ppl. a.* (later examples in *Physiol.*).

1909 *Jrnl. Amer. Med. Assoc.* 30 Oct. 1471/2, I have..been able to ascribe a definite function to the corpus luteum, namely, that of supplying a sensitizing substance to the uterus which prepares the latter to respond with the production of the maternal placenta, if an external stimulus of a mechanical nature is added. *Ibid.* 1472/1 We wished..to select the safest period for the egg to attach itself to a sensitized uterine mucosa. *Ibid.* 1473/1 This process of sensitizing enables the connective tissue of the uterine mucosa to proliferate periodically. 1941 *Nature* 26 July 116/1 The small, often-repeated doses of these drugs..provide a chance for the patient to become sensitized to the drug. 1947 *Ann. Rev. Microbiol.* I. 299 For the rapid production of allergic encephalitis in both the monkey and the guinea pig the presence of mycobacteria in the sensitizing injection seems to be essential. 1978 *Price's Textbk. Practice of Med.* (ed. 12) iv. 381/1 The signs and symptoms of generalized anaphylaxis that may follow..the parenteral injection of foreign serum, protein, or drugs, or sometimes even insect bites or stings in a sensitized individual, include a marked fall in blood pressure [etc.].

sensitizer. Restrict *Photogr.* to sense in Dict. and add: † **2.** *Immunol.* = *sensitizin. *Obs.*

1903 *Nature* 13 Aug. 360/2 The sensitizers of the tubercle bacillus. 1935 N. P. Sherwood *Immunol.* vi. 112 Anaphylactic sensitizers.—Many persons have considered these interesting antibodies as identical with the precipitins but more recently some doubt has been cast on this hypothesis.

3. *Psychol.* A person or thing that has a sensitizing effect; one who reacts by being sensitive to a stimulus (rather than repressing it).

1948 *Jrnl. Abnormal & Soc. Psychol.* XLIII. 151/2 Nature orientation acts as a sensitizer, lowering thresholds for acceptable stimulus objects. Let us call this mechanism selective sensitization. 1951 *Ibid.* XLVI. 557/1 Needs could act as sensitizers, lowering the recognition thresholds for need-related stimuli. 1972 S. R. Maddi *Personality Theories* v. 207 Psychiatric outpatients were..classified on the basis of their therapeutic interviews and other clinical tests as either sensitizers or repressers. 1976 *Psychol. Rep.* XXXIX. 189 The death message resulted in significantly more..anxiety than the neutral message for repressors as well as for sensitizers.

† **sensitizin** (se·nsitaizin). *Immunol. Obs.* [f. Sensitize *v.* + -in[1].] A substance which confers sensitivity on a species of antibody.

1916 R. Weil in *Jrnl. Immunol.* I. 1 In the following papers the term 'Anaphylactic antibody' has been replaced by the word 'Sensitizen'. This has the advantage of brevity. The word is formed on the analogy of the words precipitin and agglutinin, and carries its own significance—namely that substance which confers sensitization. 1920 *Ibid.* V. 319 The experiments furnish further evidence that precipitin and 'sensitizin' are identical.

sensitometric (se:nsitome·trik), *a.* *Photogr.* [f. Sensitometry: cf. *-metric.] Of or pertaining to sensitometry.

1881 *Brit. Jrnl. Photogr.* 25 Feb. 97/1 A committee was formed..to decide, if possible, the sensitometric question. 1949 *Electronic Engin.* XXI. 118/3 In such cases as motion picture processing where it is desirable to develop to constant gamma in the face of increasing exhaustion of developer, sensitometric strips are frequently included every few hundred feet so that the development may constantly be checked. 1967 E. Chambers *Photolitho-Offset* vii. 85 The starting point of sensitometry requires an understanding of sensitometric principles, involving the sensitometric curve.

Hence **sensitome·trically** *adv.*

1891 *Photographic Jrnl.* XVI. 65 The raw emulsion was mixed with the silver salts of various dyes dissolved in alcoholic solution of ammonia and tested sensitometrically, as well as spectroscopically. 1969 *Amat. Photographer* 9 Apr. 67/1 The integrating method of enlarging photometry is perhaps more difficult to explain sensitometrically.

sensitometry. (Earlier example.)

1881 *Brit. Jrnl. Photogr.* 25 Feb. 97/1 Photographic literature during the last six months has contained very numerous articles on the question of sensitometry.

sensize, *v.* Add: So **se·nsized** *ppl. a.*

This happens to pre-date the finite verb (in Dict.).

1846 J. D. Morell *Hist. View Philos.* II. v. 86 'The world,' says Fichte, 'is the sensized material of our practical life.'

sensor (se·nsəɹ, -ǫɹ). [f. the adj. or f. Sense *v.* + -or.] A device giving a signal for the detection or measurement of a physical property to which it responds.

1958 *New Scientist* 10 Apr. 22/2 The 'sensor' is a small cylinder enclosed in a bigger cylinder full of silicone fluid and set on bearings which allow it to turn. 1958 *Guided Missiles* (U.S. Dept. of the Air Force) vi. 273/1 Pickoffs include any of the devices that are used to transfer the energy received at the sensor to the following detecting and amplifying stage. 1963 *Ann. Reg.* 1962 401 Infra-red sensors designed to detect rocket launchings. 1969 *New Yorker* 12 Apr. 104/2 Inside his space suit, the astronaut has a number of sensors that report on the state of his health. 1975 *Sci. Amer.* July 108/2 A repellent acts in one way on the carbon dioxide sensor and in a different way on the moisture sensor. 1976 *Early Music* July 351/3 Its sensor passes over the object to be copied thousands of times in different directions, and passes this information to cutting heads. 1977 *Navy News* Aug. 32/1 Vast improvements in propulsion, sensor systems and weapons..have placed great demands on training facilities. 1980 *Sunday Express* 19 Oct. 27 Every Metro has brake pad wear sensors to tell you when to change the brake pads.

sensori-. Add: **sensori-motor** (later examples); also, that relates to activity involving both sensory and motor pathways; **sensorineu·ral** *a.*, applied to defective hearing that is due to a lesion of the inner ear or auditory nerve.

1908 W. McDougall *Introd. Soc. Psychol.* ii. 29 An innate or inherited psycho-physical disposition, which..probably has the form of a compound system of sensori-motor arcs. 1932 S. Zuckerman *Soc. Life Monkeys & Apes* ix. 147 The sensori-motor mechanisms of the primates differ from those of the lower mammals. 1977 *Language* LIII. 153 Piaget 1952 and Piaget and Inhelder 1971 view representation as an internalization of active sensori-motor imitation. [1960 *Laryngoscope* LXX. 885 A sudden unilateral or bilateral sensory-neural (perceptive) hearing loss.] 1964 *Arch. Otolaryngol.* LXXX. 382/1 In this type of slowly progressive sensorineural hearing loss the only finding is atrophy of the stria vascularis, the functional manifestation of which is hearing loss showing a flat audiometric curve. 1977 *Lancet* 12 Nov. 1003/2 Perforation of the round-window membrane was found in three children with severe sensorineural deafness.

sensorial, *a.* Add: Hence **senso·rially** *adv.*

1890 W. James *Princ. Psychol.* I. ii. 42 All of Munk's *birds* seemed totally blind (blind sensorially) after removal of the hemispheres by his operation. 1935 *Jrnl. Compar. Psychol.* XX. 10 A stimulus, a..property of which was its membership in a series which was only in part sensorially present on any one trial. 1962 Y. Malkiel in Householder & Saporta *Probl. Lexicogr.* 22 Multicolored plates..might be useful, especially to the sensorially perceptive etymologist.

sensory, *sb.* Add: **3.** *Psychol.* A person in whom sensation supposedly dominates over action. *rare.*

1902, 1929 [see *motor *sb.* 2 c].

sensory, *a.* Add: **2.** Special collocations: **sensory aphasia** *Path.*, aphasia evidenced by impaired speech, memory, writing, or reading which is due to cerebral defect or injury affecting comprehension or the ability to integrate incoming acoustic information, and freq. differentiated from incapacities deriving from motor defects; **sensory deprivation** *Psychol.*, the act or process whereby an organism is deprived of stimulation affecting one or more of the sense organs; the state or condition produced by such deprivation; **sensory-motor** *a.* = *sensori-motor a.* s.v. Sensori- in Dict. and Suppl.

1884 *Brain* VI. 401 The author [*sc.* Wernicke] also makes good use of the phenomena of the different forms of aphasia, which he divides into motor, conductive, sensory, and total aphasia. 1926 H. Head *Aphasia* I. II. iii. 202 'Sensory' aphasia, or amnesia, was divided into 'visual' and 'auditory'. 1959 *Psychol. Rev.* LXVI. 46/2 It is significant that with careful study of even a small number of patients, the traditional dichotomy between motor and sensory aphasia began to disappear. 1976 E. D. Mysak *Path. Speech Syst.* iii. 85 Sensory aphasia [is reported] with tumors in the left parietal region. 1948 D. Bakan *Investig. Effect of Sensory Deprivation on Stall Perception* i. 4 What is the effect of sensory deprivation on the accuracy with which the pilot can detect the edge of the stall proper. 1961 S. Cobb in P. Solomon et al. *Sensory Deprivation* p. xviii, The symptoms of the deprived child with 'atypical' and 'autistic' reactions are without doubt related to the phenomena seen in adults after experimental sensory deprivation. 1978 O'Connor & Hermelin *Seeing & Hearing & Space & Time* v. 65 Sensory deprivation, especially of audition, appears to decrease duration. 1957 Menon & Patel *Teaching of Eng. as Foreign Lang.* xi. 125 Spelling is a sensory-motor habit acquired by motor responses to certain sensory stimuli.

Hence **se·nsorily** *adv.*

1925 E. Sitwell *Poetry & Criticism* 20 Though it seems to us as though we heard them sensorily, yet the sound is unheard in reality. 1949 M. Mead *Male & Female* i. 17 Needs..for continuous contact with one sensorily identifiable human being throughout the first two years of life. 1954 *Essays in Criticism* IV. 313 Mill's attempt to define poetry as something not heard but *overheard*..is successful..in so far as it removes the sensorily ascertainable audience and replaces it with a mysterious audience.

sensual, *a.* Add: **4. a.** For Now *rare* or *Obs.* read Now only in phr. *the average sensual man* (see *average *a.* 2 b).

sensualization. (Earlier example.)

1798 A. F. M. Willich *Elem. Crit. Philos.* 141 The sensualization of an idea of reason.

sensu lato: see *sensu stricto.

sensum (se·nsŏm). *Philos.* Pl. **sensa.** [L., sensed, that which is sensed, neut. pa. pple. of *sentīre* to discern by the senses, to perceive.] = *sense-datum.

1868 A. Bain *Senses & Intellect* (ed. 3) 376 In Sensation, we seem to have the sentient mind, and the thing felt—*sentiens* and *sensum*. 1920 S. Alexander *Space, Time, & Deity* II. III. ii. 58 The non-mental external object which in this case is the *sensum* or *sensibile*. 1923 C. D. Broad *Sci. Thought* viii. 240 Such objects as *y* I am going to call Sensa. 1937 L. S. Stebbing *Philos. & Physicists* vi. 130 It is only because Russell and Joad *first* know that there are external objects that they are able to *infer* that there are private sensa. 1949 G. Ryle *Concept of Mind* vii. 213 The theory says that when a person has a visual sensation..his having this sensation consists in his finding or intuiting a sensum. 1967 *Encycl. Philos.* VII. 80/2 The essential point is that perceiving proper is the direct awareness of sensa. 1974 R. M. Yost in Carterette & Friedman *Handbk. Perception* I. ii. 33 One cannot locate a visual sensum in empty physical space without presupposing the Absolute Theory of Space.

Sensurround (se·nsŏraund). Also **sensurround.** [Blend of Sense *sb.* and Surround *v.*] The proprietary name of a special-effect technique whereby a cinema audience is apparently surrounded by low-frequency sound and air vibrations generated from the soundtrack of a film. Also *attrib.* and *transf.*

1974 *Newsweek* 2 Dec. 104/2 The quake under your seat is created by 'Sensurround', a system that hooks into the film's soundtrack and sets off low-level tremors at a cost of $500 a week to the theater owners. 1975 *New Yorker* 26 May 81/1 'But my dear,' said Mrs. Vreeland, her voice achieving in the small room much of the effect that Sensurround does in movie theaters, 'those wrapped heads are my *greatest achievement*.' *Ibid.* 82/2 'My *God*, Ferle,' said Mrs. Vreeland, employing her Sensurround voice. 1976 *Official Gaz.* (U.S. Patent Office) 6 Apr. 22/1 MCA Systems, Inc., Universal City, Calif.... *Sensurround* for electronic apparatus for generating special effects, including simulated earthquake effects, in motion picture theaters... First use Sept. 3, 1974. 1977 *Time* 11 July 51/3 *Rollercoaster* is the latest—and so far least—excuse to trot out Sensurround, that technology that is still in search of a character and, for that matter, a plot worthy of its woofers. 1980 *Spectator* 31 May 24/1 Every corner of the theatre is used to produce a sensurround, stereophonic effect, the noise and flashing all around us.

‖ **sensu stricto** (se·nsu stri·kto). [L., lit. 'in the restricted meaning'.] Strictly speaking; in the narrow sense (of a term, esp. in the natural sciences). Opp. to **sensu lato** (lā·to) [L. *lātus* broad], in the broad sense. Cf. *stricto sensu.

1941 J. S. Huxley *Uniqueness of Man* xi. 240 Human biology is but an extension of biology *sensu stricto*. 1942 W. B. Turnbull in *Bot. Rev.* VIII. 656 (*heading*) Algae (sensu lato). 1952 Sensu stricto [see *isochemical* adj. s.v. *iso- a]. 1954 [see *neocerebellum]. 1959 A. R. Clapham et al. *Excursion Flora Brit. Isles* 569 Sensu lato. In the broad sense. 1963 D. W. & E. E. Humphries tr. *Termier's Erosion & Sedimentation* xiv. 299 Griottes 'sensu stricto' composed of alternate beds of shale and limestones, irregularly corrugated. 1973 B. J. Williams *Evolution & Human Origins* xi. 176 In the material that follows I shall use the term Neandertal *sensu lato*, that is, in the broad sense. 1977 *Verbatim* Sept. 4/1 The trouble is caused not by length *sensu stricto* but by complexity.

sent, *ppl. a.* (Later examples.)

1825 H. C. Robinson *Diary* Dec. (1967) 86 *Irving.* He is a highly gifted man; he is a sent man, but they who are sent sometimes go further than they ought. 1940 *Amer. Speech* Oct. 337/1 *To be sent*, to be completely satisfied and in a stupor from the drug. 1951 *Manch. Guardian* 21 June 5/1 The slang of jazz addicts, which is full of phrases like 'hepsters', getting 'high', being 'sent' and other euphemisms for the delirium induced by improvised solos on the cornet and slide trombone. 1958 *Spectator* 25 July 133/3 The girls wore thick eye-makeup and 'sent' expressions.

sentence, *sb.* **4. a.** For † *Obs.* read *Obs.* exc. *Hist.* and add later example.

1962 T. P. Dunning in Davis & Wrenn *English & Medieval Studies* 178 That element of the *sentence* expressed by Pandarus in Book 1—Fortune as the way of the world—is here stated at some length by Criseyde.

b. For † *Obs.* read *Obs.* exc. *poet.* and add later example.

1917 T. S. Eliot *Prufrock & Other Observations* 15 Politic, cautious, and meticulous; Full of high sentence, but a bit obtuse.

6. c. *Logic.* A correctly ordered series of signs or symbols that expresses a proposition in an artificial or logical language.

1937 A. Smeaton tr. *Carnap's Logical Syntax Lang.* I. 26 We have already surveyed all the possible ways of constructing sentences and numerical expressions in Language I. 1957 P. Suppes *Introd. Logic* (1959) iii. 54 A sentence is a formula which has no free variables. 1976 Evans & McDowell *Truth & Meaning* p. viii, Conditions (1), (2), and (3) require that *L* be a logically perfect language, with sentences free from structural or lexical ambiguity.

9. (sense 6) *sentence-building, -completion, -construction, -form, -formation, -forming, -formula, -frame, -intonation, -meaning, -melody, -modifier, -pattern, -rhythm, -type*; *sentence-final, -forming, -initial* (also *-initially* adv.), *-modifying, -opening* adjs.; **sentence adverb** *Gram.*, an adverb used to qualify a complete sentence (see also quot. 1892); also **sentence adverbial**; **sentence diagram,** a schematic representation of the relationships between the constituent parts of a sentence; so **sentence diagramming**; **sentence-particle** *Gram.* (see quot. 1953); **sentence-token** *Logic* (see quot. 1936); **sentence-word** (earlier example).

1892 H. Sweet *New Eng. Gram.* I. 127 Sentence-adverbs. The answer to the question *is he here?* can be either the affirmative *yes* or the negative *no*. It is evident that *yes* and *no* are sentence-modifying adverbs and at the same time sentence-words like come! John!, alas! 1916 E. A. Sonnenschein *New Eng. Gram.* 33 Several..adverbs..may be used to qualify the sentence as a whole; when so used they are called sentence-adverbs. *Ibid.*, Some sentence-adverbs (especially 'too', 'else', 'only', 'even') may be used in such a way as to emphasize the word which stands next to them in the sentence. 1980 *Amer. Speech* 1976 LI. 168 Among the adverbs were some that are often classed together as sentence adverbs: *luckily, wisely, foolishly, rightly*. 1964 Katz & Postal *Integrated Theory of Linguistics Descriptions* iv. 95 The answers to yes–no questions are in fact sentence adverbials, i.e., *yes, no*, and perhaps by extension *maybe, of course, certainly*, etc. 1921 H. E. Palmer *Princ. Lang.-Study* 22 Exercises exist which ensure accuracy in..sentence-building. 1966 J. Derrick *Teaching Eng. to Immigrants* v. 205 Other sentence-completion exercises, in which there can be slightly more freedom of choice, can consist of 'half sentences' in which part or whole of the subject or predicate is missing and has to be filled in by the pupil. 1921 H. E. Palmer *Princ. Lang.-Study* iii. 58 The learner need know little about the sciences dealing with inflexions, sentence-construction, or meanings. 1937 Moffett & Johnson *Basic Writing* 509 A sentence diagram is merely a device by which the structure of a sentence can be..shown. *Ibid.* 632 (Index) Sentence diagramming. 1977 *Language* LIII. 493/1 The syntax section is fully and competently developed... L compares phrase-structure trees to traditional sentence diagrams, with which many students will be familiar. 1959 *College Composition & Communication* May 91 The question of just what ends we hope to attain by the use of any system of sentence diagraming. 1949 Sentence-final [see *intonationally* adv. s.v. *intonation[1]]. 1978 *Language* LIV. 79 Lehmann observes that the interrogative in Japanese is marked by placing *ka* after verbs in sentence-final position. 1930 T. Sasaki *On Lang. of R. Bridges' Poetry* 92 The language of poetry is in not a few points similar to primitive language, which usually favours 'gegenständliches Denken', and therefore prefers attributive sentence-form. 1965 *Language* XLI. 372 A form that differs from any elementary sentence-form of the language. 1935 G. K. Zipf *Psycho-Biol. of Lang.* v. 185 They [*sc.* substantive and verb] are not a *sine qua non* of sentence-formation. 1921 H. E. Palmer *Princ. Lang.-Study* 25 In choosing the units of our vocabulary we may be guided by..sentence-forming utility. 1936 J. Kantor *Objective Psychol. Gram.* III. xvii. 241 Those grammarians who call the verb a sentence-forming word, a phenomenon word, or an *Aussagewort*, also pay tribute to its action-referring character and save themselves from a too great stress of time. 1932 W. L. Graff *Lang.* I. iii. 132 Because it [*sc.* How do you do?] is itself a sentence, not merely sentence material, it may be termed a sentence formula or a formular sentence. 1962 G. A. Miller in *Amer. Psychologist* XVII. 756/1 One opinion is that we learn 'sentence frames' that we keep filed away in a sort of sentence-frame dictionary. The declarative, interrogative, affirmative, negative, active, passive, compound, complex, etc., sentence frames are all supposed to be learned separately and to have no intrinsic relation to one another. 1964 *Language* XL. 6 Mere sentence-initial position of *Wh* does not suffice to differentiate between inversion and noninversion. 1978 *Ibid.* LIV. 85 Sentence initial subjects in Japanese and English may be viewed as most distant from the verb. 1976 *Word* 1971 XXVII. 302 This word is then placed sentence-initially. 1934 J. J. Hogan *Outl. Eng. Philol.* I. iv. 25 It [*sc.* the sentence] has a musical tune, Sentence-Intonation. 1945 *Mind* LIV. 366 The vital question 'how separate words..can combine to yield sentence-meanings' is dealt with very summarily. 1922 O. Jespersen *Language* I. iv. 97 The heightened interest in everything concerning 'accent' (stress and pitch) has also led to investigations of sentence-stress and sentence-melody. 1928 H. Poutsma *Gram. Late Mod. Eng.* (ed. 2) I. I. v. 320 The question whether an adverbial adjunct is a sentence-modifier or a word-modifier, is of considerable importance. *Ibid.* i. 101 Weak *do* is used in connexion with sentence-modifying *not*. 1962 J. Söderlind in F. Behre *Contrib. Eng. Syntax* 117 This is a bold type, found in sentence-opening subject position. 1934 Priebsch & Collinson *German Lang.* II. xi. 445 German has one Greek characteristic which makes it neater and fuller of expressive shades than English, viz. the use of sentence-particles (*ja, doch*..etc.) and their cumulations (*ja doch*..etc.). 1953 *Trans. Philol. Soc.* 1952 6 The name 'sentence particle' (*satzpartikel*) was given by Kuhn to all unstressed and weak-stressed words which modify not one part of the sentence but the whole of it, and are therefore syntactically independent parts of the sentence. 1935 G. K. Zipf *Psycho-Biol. of Lang.* v. 201 The question of equilibrium which lies at the root of the development of all sentence-patterns. 1926 Fowler *Mod. Eng. Usage* 560/1 The separating adverb could have been placed outside the infinitive with little or in most cases no damage to the sentence-rhythm. 1957 R. W. Zandvoort *Handbk. Eng. Gram.* VI. 239 The different word order..may be due to a desire for variety, as much as to the requirements of sentence-rhythm. 1936 *Jrnl. Philos.* XXXIII. 703 A *sentence-token* is a particular set of particular symbolic marks (of a sort, let us say, to represent a complete assertion). 1976 A. N. Prior *Doctrine of Propositions* i. 35 We may say at once that the dominant tendency in Logic is for the term 'proposition' to be used not for a 'sentence-token' but for a 'sentence-type'. 1933 L. Bloomfield *Language* x. 169 The use of the secondary phoneme [!] gives us the sentence-type of exclamation. 1848 C. Bunsen in *Rep. Brit. Assoc. Advancem. Sci.* 282 The Egyptian root is not the unalterable particle, or rather sentence-word, of the Chinese.

Hence **se·ntencehood** [-hood], the condition of constituting a grammatically complete sentence.

1961 *Language Learning* XI. 175 These..are some of the requirements which one might reasonably set for an adequate theory of English 'sentencehood'. 1967 *Philos. Rev.* LXXVI. 151 In many sentences 'probable' can replace 'possible' without destroying sentencehood.

‖ **sententia** (sente·nʃiă). Pl. **-æ** (erron. **-a**). [L.] = Sentence *sb.* 4 a. Also in mod. use, a thought or reflection.

1917 E. E. Cummings *Let.* 5 May (1969) 19 The immemorially delightful sententia of a pocket-size sailor..i.e. 'submarines pooh-pooh'. 1926 C. Connolly *Let.* 25 Dec. in *Romantic Friendship* (1975) 199, I sent you one diary for quotations and the other was for sententiae. 1933 R. Tuve *Seasons & Months* i. 43 The Seasons motif found new uses... It became a vehicle for 'sententia', folk proverbs which appealed to generations fed upon Hending and Alfred. 1960 *Times* 29 Sept. 15/5 Those platitudinous *sententiae* that pass for conversational small change. 1964 C. S. Lewis *Discarded Image* vii. 191 Chaucer begins with a *Sententia* or maxim in the *Parlement*. 1971 *English Studies* LII. 456 An attempt is made to trace the process by which the popular proverb attained equal status with the literary *sententia*.

sentential, *a.* Restrict Now *rare* to other senses and add: **2. a.** (Later examples in mod. Linguistics.)

1956 J. H. Greenberg in Saporta & Bastian *Psycholinguistics* (1961) 471/2 *Sentential meaning.* The meaning of a maximal linguistic structure (i.e., of a sentence). 1965 N. Chomsky *Aspects of Theory of Syntax* ii. 100 Nouns with sentential Complements (such as 'the *idea* that he might succeed'). 1978 *Studies in Eng. Lit.: Eng. Number* (Tokyo) 96 They are excluded either by Ross's 'rightward bounding constraint' and 'sentential subject constraint', or by Chomsky's 'subjacency condition'.

b. *Logic.* In collocations denoting logical operations relating to sentences or propositions; esp. as *sentential calculus, connective, function.* Cf. *propositional *a.* b.

1937 A. Smeaton tr. *Carnap's Logical Syntax of Lang.* III. 91 Primitive sentences of the sentential calculus. *Ibid.* 138 Frege himself had already made a similar classification of all sentential functions into levels and kinds which also were arranged according to the kinds of their arguments. 1957 P. Suppes *Introd. Logic* iii. 43 We have developed the logic of the sentential connectives. 1966 *Math. Reviews* Jan. 7/1 The author presents a set of nine axiom schemes and two rules for the predicate calculus based on the infinite-valued sentential calculus of Łukasiewicz. 1976 G. Evans in Evans & McDowell *Truth & Meaning* viii. 215 Intuitionistic sentential connectives cannot be regarded as representing truth functions in any finite many-valued logic. 1976 A. R. Lacey *Dict. Philos.* 75 *x is red* can be called a propositional, statemental (rare) or sentential function, according as *blood is red* is regarded as a proposition, statement or sentence. Sentential functions are often called open sentences. The term *closed sentential function* is occasionally used of ordinary sentences.

sentiment. Add: **6. d.** *them's my sentiments*: a colloq. expression of agreement or approval. (In quot. 1847, a declaration of belief.)

1847 Thackeray *Van. Fair* (1848) xxi. 179 The sooner it is done the better, Mr. Osborne; them's my sentiments! 1886 J. Bailey *Let.* 28 Nov. (1935) 26, I was delighted, as I could have said to every word: 'Them's my sentiments!' 1937 A. Huxley *Let.* 16 Feb. (1969) 414, I ought to have written long since to thank you for your Sunrise Poem, about which I felt strongly that them was my sentiments. 1940 'B. M. Bower' *Spirit of Range* xiv. 162 'I'm willing to be just a boneheaded cow-puncher.' 'Accent on the bone,' Pink murmured. 'Them's my sentiments, old socks.'

sentimentalist. (Earlier examples.)

1783 *Scots Trimmer* III. 27 Dean Milles, who is ravished with the beauties of Sterne, would, in all probability, have given the admirers of this charming sentimentalist a huge quarto, price only one guinea. 1784 R. Bell (*title*) Illuminations for legislators and for sentimentalists.

sentimentali·stic, *a. rare.* [f. Sentimental *a.* + -istic.] Possessing sentimental characteristics; characterized by an exaggerated sentimentality.

1904 M. Beerbohm *Let.* 13 Apr. (1964) 161 All this sounds very 'literary' and sentimentalistic, but it is real enough to me. 1912 E. Pound *Prolegomena* in *Poetry Rev.* Feb. 75 As for the nineteenth century,.. I think we shall look back upon it as..a rather sentimentalistic, mannerish sort of a period.

sentimentalize, *v.* Add: **1. a.** (Earlier example.)

1788 W. Combe *Orig. Lett. Sterne* 14 In the mean time we will philosophize and sentimentalize;—the last word is a bright invention of the moment in which it was written, for yours or Dr. Johnson's service.

sentimentalizing *vbl. sb.* (earlier example).

1789 E. Sheridan *Jrnl.* 20 Aug. (1960) viii. 184 Tickell marries Miss Lee next week and so ends his sentimentalising.

sentinel, *sb.* Add: **6. a. sentinel pile** *Path.*, an external hæmorrhoid situated at the lower end of an anal fissure.

1910 *Practitioner* Apr. 520 It is probable that the fissure results from the tearing down of one of the anal valves, the free border of which eventually appears at the anus as a rounded œdematous tag—the so-called sentinel pile. 1974 R. M. Kirk et al. *Surgery* vi. 132 The oedematous skin at the lower end of the fissure protrudes as a 'sentinel' pile.

sentoku (se·ntoku). [Jap.] Originally, a Chinese bronze produced during the era (1426–35) of Emperor Hsüan of the Ming Dynasty; later, a golden-yellow Japanese bronze vessel made after the Chinese fashion; the bronze itself.

1902 *Encycl. Brit.* XXIX. 722/2 A golden yellow bronze, called *sentoku*. 1904 E. Dillon *Porcelain* vi. 92 Hsuan-te (1425–35)... This period gave its name to the famous pale bronze so admired in later days by the Japanese... The name *Sentoku* that they give to it is the Japanese reading of the characters forming this emperor's name. 1931 *Illustr. London News* 15 Aug. 268/3 That..characteristic Japanese alloy, *sentoku*, a sort of yellow bronze which is very soft and resembles brass rather than bronze. 1968 G. Savage *Conc. Hist. Bronzes* iv. 128 *Sentoku*, containing up to thirteen per cent of zinc, may have been used in the fifteenth century, and legend has it that vessels of this kind also contained gold.

sentry, *sb.*[1] Add: **6.** *sentry duty.*

1917 W. Owen *Let.* 16 Jan. (1967) 428 Servants don't do Sentry Duty. 1977 *Belfast Tel.* 28 Feb. 4/3 A soldier who shot dead a Derry man and injured two other people while on sentry duty in the city last May was jailed for five years.

sentry, *v.* Add: **a.** (Earlier example.) **c.** *intr.* and *refl.* To place as a sentry; also *fig.*

1820 J. S. Knowles *Virginius* v. II. 78 Though a legion Sentried that brothel, which he calls his palace, I'd tear her from him! 1922 Joyce *Ulysses* 221 Corny Kelleher..glanced..at a pine coffinlid sentried in a corner. 1979 G. Swarthout *Skeletons* 155 John and Paul had sentried themselves before the bank door, barring my re-entry.

sentry-go. Add: **b.** Also *transf.*

1922 [see *cave *int.*]. 1938 G. Greene *Brighton Rock* I. i. 4 This was Hale's job to do sentry go, until a challenger released him, in every seaside town in turn.

Senufo (sənū·fo). Also **Senoufo, Senufu.** [Akan.] A people of the Ivory Coast in western Africa; any of the sub-group of Gur (Niger–Congo) languages (or dialects) spoken by them. Also *attrib.* or as *adj.*

1911 F. W. H. Migeod *Languages of W. Afr.* I. i. 34 The Senufu..inhabit a big area of territory in the Ivory Coast hinterland. 1913 *Ibid.* II. xx. 317 Primeval languages not elsewhere classed:—*Berber*,.. *Bisogo*,.. *Senufo*. *Ibid.* xxi. 324 The Senufo group,.. in the Ivory Coast lagoons. 1928 *Africa* I. 220 The best examples of this Western Sudanese art, in the work of the Bammana, Senoufo, Itabe and Mossi, show a strong preponderance of the tendency to cover with ornament. 1935 *Chambers's Encycl.* VIII. 609/1 The Siena or Senufo hold that a man's soul passes into the body of his totem-animal, and conversely the spirit of the dead animal enters a new-born child of the clan. 1969 Morgan & Pugh *W. Afr.* I. i. 24 The Senoufo are a Voltaic community who have adopted

many Manding and Akan customs. They live well to the south, in sub-Sudanic and sub-Guinean environments. 1972 *Times* 28 Nov. 24/5 (Advt.), A Senufo carved-wood mask. 1972 *Language* XLVIII. 848, 7a is given by..Sedlak for..Baule, Gbeya, Senufo-Senadi.

Senussi (sĕnū·si). Also **Sanusi, Senoussi,** etc. [Arab. *sanūsī*, the name *Senussi* (see below).] (A member of) a Muslim religious fraternity founded in 1837 by Sidi Muhammad ibn Ali es-Senussi. Also *attrib.* or as *adj.*

1891 F. R. Wingate *Mahdiism of Egyptian Sudan* i. 2 The Senussi branch of the Shadli school, so called from the Senus mountain in Algiers, dates its inception about 1837. 1899 A. S. White *From Sphinx to Oracle* xii. 118 A man may become a Senussi without abandoning his Order. 1906 *Daily Chron.* 24 Apr. 7/6 The mysterious influence of the Senoussi. 1915 T. E. Lawrence *Let.* 22 Mar. (1938) 195 The Idrisi family, who are the Senussi and Assyr together. 1942 *R.A.F. Jrnl.* 27 June 15 The Senussi Arabs have a legend about the creation of the desert. 1949 E. E. Evans-Pritchard *Sanusi of Cyrenaica* i. 8 The Sanusi have never shown themselves more hostile than other Muslims to Christians and Jews. 1959 *Listener* 15 Jan. 100/2 The Sanusi religious family, settled among the Bedouin tribes of Cyrenaica. 1978 A. Melville-Ross *Blindfold* xxi. 123 King Idris, with his strong pro-British leanings and strict Senussi code of honour.

Also **Senu·ssian** *sb.* and *a.*; **Senu·ss(i)ism, Senu·ssist** *a.* and *sb.*; **Senu·ss(i)ite** *sb.* and *a.*

1884 *Science* 14 Nov. 457 A Mussulman confraternity known as the Senousians. *Ibid.* 459 Five hundred camels..ready at a moment's notice to convey to the interior the persons and property of the Senousian authorities. 1899 A. S. White *From Sphinx to Oracle* xii. 114 Absolutism and occultness are the two most potent powers in Senussi-ism. 1900 *Daily News* 15 Jan. 6/4 The only great religious organization of Moslem Protestants in Northern Africa are the Senoussi-ites, and they are harmless. *Ibid.*, So far, owing to the secrecy observed by Senoussist emissaries, no direct evidence regarding the movement is obtainable. 1934 Webster, Senusi... Also Senousi, Senussite. 1957 *Encycl. Brit.* XX. 331/2 The Darfur revolt of 1888–89..was nevertheless carried out in the name of the Senussites. *Ibid.* 332/2 In Cyrenaica Senussite resistance to the Italians was organized by Sheikh Rida. 1977 B. Lucas tr. *De Foucauld's Lett. from Desert* vii. 139 Our Tuaregs remain calm despite the capture of Djanet by the Senoussists.

Senussia (sĕnusī·ă). Also **Senusiya,** etc. [Arab. *sanūsiya*.] The fraternity founded by es-Senussi.

1888 *Encycl. Brit.* XXIII. 575/2 The sectaries of Senûsîya are found in all parts of North Africa. 1891 F. R. Wingate *Mahdiism Egyptian Sudan* i. 4 The Senussiyeh attacked the robbers. 1911 *Encycl. Brit.* XXIV. 649/1 Considerable diversity of opinion has prevailed among writers and travellers claiming knowledge of the *Senussia*. 1949 E. E. Evans-Pritchard *Sanusi of Cyrenaica* i. 8 The Grand Sanusi had himself been a member of a succession of orders before he started his own and he allowed..members of other Orders to belong to the Sanusiya at the same time. 1974 *Encycl. Brit. Micropædia* VIII. 888/1 *Sanūsīyah*, also spelled *Senussiya*, in a strict sense, a Muslim Ṣūfī (mystic) brotherhood established in 1837 by Sīdī Muḥammad ibn 'Alī as-Sanūsī.

‖ **senza** (se·ntsă), *prep. Mus.* [It.: see note s.v. Sans prep.] Without; in various phrases, as *senza bassi* without the basses, *senza tempo* in no definite time, etc.

1724 *Short Explic. Foreign Wds. in Mus. Bks.*, *Senza*, without. This word is used in the following Manner. *Senza d'aria*, without the Air... *Senza Violino*, without the Violins [etc.]. 1740 Grassineau *Mus. Dict.*, *Senza*, signifies without, as *senza stromenti*—without instruments. 1762 Sterne *Tr. Shandy* VI. xi. 50 *Con l'arco* upon this;—*senza l'arco* upon that.—All I know is, that they are musical terms. 1945 A. Huxley *Time must have a Stop* ii. 16 Forbidden themes, repulsively fascinating, disgustingly attractive! Sebastian would embark on them with a quiet casualness—*pianissimo*, so to speak, and *senza espressione*. 1959 *Listener* 4 June 1001/1 Aubrey Brain played this *senza sord*.

separate, *pa. pple.*, *a.*, and *sb.* Add: **B.** *adj.* **2. b.** In a hotel or boarding-house: *separate table*. Also *spec.* of rooms, etc., to which each of a married couple retires separately.

1756 *Old Maid* 21 Feb. 86, I have proposed separate beds, but he will never hear of it. 1817 Jane Austen *Let.* 20 Feb. (1952) 480, I w[d] recommend to her & Mr. D. the simple regimen of separate rooms. 1838 H. Martineau *Retrospect* I. 236 We..had..a separate table, at Mrs. Peyton's boarding-house. 1910 *Bradshaw's Railway Guide* 1008 (Advt.), *White Lion*... Coffee Room (separate tables), Billiard Room. 1971 J. Fleming *Grim Death & Barrow Boys* xi. 161 A Private Hotel on the sea-front where they had dinner at night and separate tables. 1977 C. Storr *Tales from Psychiatrist's Couch* 36 She sleeps in a twin bed in London, but in the cottage we have separate rooms.

e. Phr. *separate but equal*, asserting the equality of races under racial segregation. *U.S.*

[1776 T. Jefferson in *Dunlop's Pennsylvania Packet* 8 July 1/1 When in the course of human events, it becomes necessary for one People..to assume among the powers of the earth, the separate and equal station to which the laws of Nature and of Nature's God entitle them. 1890 *Louisiana Acts* CXI. 152 An Act..requiring all railway companies..to provide equal but separate accommodations for the white and colored races.] 1892 F. W. Gage *Negro Problem in U.S.* iii. 92 If railroad companies care to furnish separate but equal accommodations on equal terms to each race, no objection need be made. 1914 *U.S. Reports* (1915) CCXXXV. 160 It was not an infraction of the Fourteenth Amendment for a State to require separate, but equal, accommodations for the two races. 1948 *Time* 9 Feb. 75/1 In Missouri, where a 'separate but equal' law school has had its longest test. 1954 E. Warren in *U.S. Reports* CCCXLVII. 495 We conclude that in the field of public education the doctrine of 'separate but equal' has no place.

f. *separate school* Canad., a school receiving pupils from a racial or religious minority.

For detailed evidence and comment see *Dict. Canadianisms.*

1852 *Dundas Warden* (Canada West) 28 May 2/7 The law makes provision for Separate Schools, to meet an exigency—namely, the anticipated intrusion of the religious dogmas of a majority upon a minority. 1857 H. F. Douglass in *Ontario Hist.* (1963) June 88 Separate schools and churches are nuisances that should be abated as soon as possible, they are dark and hateful relics of Yankee Negrophobia. 1872 *Canadian Monthly* July 64/1 The Roman Catholics spoke frankly and sincerely for their separate schools, the New Brunswickers for their local liberties. 1911 *Daily Colonist* (Victoria, B.C.) 14 Apr. 5/2 Steps are being taken by the Roman Catholic authorities towards the establishment of separate schools in and near Vancouver. 1968 [see *junior high (school)* s.v. *junior *a.* (*sb.*) 5]. 1976 *Globe & Mail* (Toronto) 16 Jan. 29/8 That meant I was Roman Catholic,.. that my oppressed and persecuted parents had to pay for my separate school education as well as the education of all the heathens in the public schools.

g. *separate development*, the systematic development or regulation of a group or race by itself independently of other groups or races in a society; orig. and chiefly *S. Afr.*, = *apartheid.

1955 *Summary Rep. Comm. Socio-Econ. Devel. Bantu Areas S.Afr.* III. xxv. 105/1 (*heading*) Objections to the policy of separate development. 1962 [see *parallelism 2]. 1968 *Economist* 12 Oct. 17/1 A rigid, and openly acknowledged, form of 'separate development' operates there [*sc.* in Londonderry]. The most populous ward..is wholly Catholic..but skilful use of the 'property qualification' for local government elections [etc.]..ensure that these 12 councillors are Protestant Unionists. 1977 [see *plural democracy* s.v. *plural *a.* (*sb.*) 2]. 1979 E. Norman *Christianity & World Order* v. 61 The Dutch Reformed Church does not teach white racial superiority, nor is Separate Development an attempt to institutionalize *racial discrimination*.

C. *sb.* **5.** (Earlier example.)

1862 *Cornh. Mag.* Nov. 640 Professional thieves..form a net-work..by..which all criminal knowledge circulates. In prison and out of it, in the lowly village lodging-house and..'doing their separates' at Pentonville..they..spread criminal knowledge.

6. *Geol.* Any of the fractions into which constituents of a soil or other material can be separated according to a property such as particle size or mineral composition. Cf. *soil separate* s.v. *soil *sb.*[1] 10.

1909 A. G. McCall *Physical Properties of Soils* 84 The separates to be determined are as follows: Fine gravel 2·0–1·0 mm, Coarse sand 1·0–0·5,..Clay 0·005. 1924 F. E. Bear *Soil Managem.* vii. 56 In the Illinois soil survey, silt is defined as a separate the particles in which may vary from 0·03 to 0·001 millimeter in diameter. 1952 L. M. Thompson *Soils & Soil Fertility* ii. 10 The sand separate which occurs in an amount greater than any other separate is used to indicate the name; for example, fine sandy loam indicates a predominance of fine sand. 1977 *New Scientist* 21 Apr. 120/1 Isotopic abundance anomalies in mineral separates from meteorites.

7. *pl.* Articles of (esp. women's) dress which may be worn in various combinations and not only as parts of a matching outfit.

1945 *Britannica Bk. of Year* 276/2 These 'separates' were outfits of which the several parts could be interchanged to form many combinations. 1948 *Sun* (Baltimore) 3 Apr. 3/7 (Advt.), Tropical separates... Of crisp tropical rayon suiting nicely tailored... You can either 'mix 'em or match 'em'. 1958 *TV Times* 20 June 15/2 She finds 'separates' ideal for her type of performance. 1964 *McCall's Sewing* i. 13/1 Separates are the answer to the schoolgirl's needs. Skirts, sweaters, jackets and blouses that can mix and match are perfect. 1979 *Sunday Star* (Toronto) 30 Sept. D2/2 She's learned the knack of putting her own looks together with separates. She's off to school one day in gray dress pants, hot pink sweater and pale pink tam.

8. A self-contained, free-standing component of a sound reproduction system. Usu. *pl.*

1974 *Times* 8 Apr. 12/1 Demand for all kinds of audio systems—'separates' and otherwise. 1977 *Gramophone* Apr. 1625/2 Akai showed, along with five new receivers, that it too was getting into a wide line of separates and speakers.

separate, *v.* Add: **I. 1. a.** Also with *out*.

1962 H. E. Brecheno *Introd. Bus. Stud.* xi. 93 For the mass of smaller businesses these functions must be separated out. 1980 V. Cunningham *Sp. Civil War Verse* 64 The various elements of his poetry can't be separated out like this.

d. To discharge (a person) *from* the armed forces (*U.S. Mil.*); † to remove from employment.

1859 R. Thornton *Jrnl.* 25 June in E. C. Tabler *Zambezi Papers of Richard Thornton* (1963) I. 103 About 3 p.m. Dr. L. gave me an official letter separating me from the Expedition. 1888 *Civil Service* (U.S.) *Comm. 4th Rep.* 51 A statement of the number of persons who have been 'separated' from the classified service by removal, resignation, and death cannot be made. 1946 *Britannica Bk. of Year* 833/1 *Separate*, to discharge or release from active duty in the armed services. 1971 *Reader's Digest* (U.S. ed.) Oct. 13/1 This year one million veterans will be separated from the service.

separated, *ppl. a.* Add: **1.** (Further examples); also, *absol.* examples in sense 'one who has withdrawn from a conjugal relationship but is not divorced' (chiefly *pl.*). *U.S.*

1951 N. Mitford *Blessing* II. v. 192 It's always like that with separated couples... Each one is trying to give the child a better time than the other. 1960 *Time* 17 Oct. 112/2 A collection of psychologizing short stories about young separateds. 1975 *Observer* (Colour Suppl.) 5 Jan. 7/2 Feiffer is 45, separated, has a daughter, lives in New York. 1975 *Publishers Weekly* 27 Jan. 234/3 Explores the bisexual lifestyle through in-depth interviews with marrieds, singles and 'separateds'.

2. *Math.* (See quot. 1968.)

1956 J. M. H. Olmstead *Intermediate Analysis* iii. 77 The set of rational numbers and the set of irrational numbers are disjoint, but are about as far from being separated as two disjoint sets of real numbers can be. 1964 T. O. Moore *Elem. Gen. Topology* ii. 40 A space *X* is connected iff *X* is not the union of two separated sets. 1968 E. T. Copson *Metric Spaces* v. 62 Two sets *A* and *B* in a metric space *M* are said to be separated if neither has a point in common with the closure of the other.

separation. Add: **1. b.** *U.S.* Resignation or dismissal from employment, a university, etc.; discharge from the armed forces.

1779 T. Jefferson *Let.* 27 Mar. in *Writings* (1893) II. 179 The separation of these troops would be a breach of public faith. 1897 C. M. Flandrau *Harvard Episodes* 229 He would feel [sorrow] at what the official college gracefully terms the 'separation' of Billy from the University. 1923 J. D. Hackett *Labor Terms* in *Managem. Engin.* May, *Separation*, the termination of employment, either voluntary or involuntary, at the instance of the employer or worker. 1955 *Univ. of Va. News Letter* 15 June 1/2 Just as births exceed deaths to yield an expanding population, so new entrants exceed separations through deaths and retirements to yield an expanding labor force. 1976 *Washington Post* 19 Apr. C15/10 (Advt.), Excellent opportunity in proposal writing for former surface Naval officer who completed 1 or 2 tours prior to separation.

c. *separation of powers* Pol.: the vesting of the legislative, executive, and judiciary powers of government in separate bodies.

[1748 Montesquieu *De l'Esprit des Loix* I. xi. vi. 245 Il n'ya a point encore de Liberté, si la puissance de juger n'est pas séparée de la puissance législative & de l'exécutrice. 1788 A. Hamilton et al. *Federalist* II. xlvii. 92 The Meaning of the Maxim, which requires a Separation of the Departments of Power, examined and ascertained.] 1896 A. L. Lowell *Govts. & Parties in Continental Europe* I. i. 55 The Declaration of The Rights of Man proclaimed in 1789 that a community in which the separation of powers was not established had no constitution. 1921 J. Bryce *Mod. Democracies* II. II. xxxix. 23 No official of the Federal Government is eligible to sit in Congress, no official of the Government of a State to sit in its legislature. This provision, a tribute to the famous doctrine of the Separation of Powers, was meant to prevent the Executive from controlling the Legislature. 1973 *N.Y. Times* 15 Aug. 36/1 President Nixon's attorneys..asserted that the constitutional separation of powers precluded the courts from commanding him to make those tapes available to a grand jury.

13*. *Photogr.* and *Printing.* **a.** Each of three or more monochrome reproductions of a coloured picture, made in different colours in such a way that they combine to reproduce the full colour of the original.

1922 [see *colour separation* s.v. *colour *sb.*[1] 18]. 1933 T. S. Barber *Art & Pract. Printing* IV. xiv. 163 The Three-colour Process requires three half-tones, made from photographic colour separations. 1967 Karch & Buber *Offset Processes* v. 170 In making the separation from a colored original or transparency, the circular glass halftone screen..is used. 1972 *Physics Bull.* Sept. 533/3 An original colour picture must first be processed to obtain four continuous tone 'separations', that is images on film which present the red, green and blue content of the original together with a 'key'.

b. The process of obtaining a set of monochrome reproductions of a coloured picture in each of which the tones correspond to the proportions of a particular colour in the original.

1924, 1930 [see *colour separation* s.v. *colour *sb.*[1] 18]. 1931 F. R. Newens *Technique Colour Photogr.* iii. 25 This tri-pack gives extremely good colour separation. 1949 Melcher & Larrick *Printing & Promotion Handbk.* 48/2 Color separation is the technique by which the colors of the original art work..are sorted out so that all the reds appear in the red plate, the blue and the shades of blue in the blue plate, etc. *Ibid.* 49/1 Flat-color jobs present no problem... Full-color originals are more difficult. Here the printer may do his separations by the fake process method or by the process-color method. 1974 *Encycl. Brit. Macropædia* XIV. 304/2 In the direct method [of making colourplates], screen negatives are prepared directly from the copy through the colour-separation filters and a halftone screen.

13.** *Physics* and *Aeronaut.* The separation of the boundary layer from the surface of a body moving relative to the surrounding fluid.

1926 H. Glauert *Elem. Aerofoil & Airscrew Theory* viii. 100 When two parallel layers of fluid are moving in the same direction with different velocities, the surface of separation is a vortex sheet. 1935 K. D. Wood *Techn. Aerodynamics* ii. 46 At zero lift, there is commonly a certain amount of separation under the nose of the airfoil. 1949 O. G. Sutton *Sci. of Flight* ii. 40 The air stream has found it difficult to turn the corner... In technical language the flow separates. We shall see later that separation is of immense importance in all problems of aerodynamics. 1978 D. Küchemann *Aerodynamic Design of Aircraft* ii. 37 The most important boundary-layer phenomenon is flow separation.

13*.** Distinction or difference between the signals carried by the two channels of a stereophonic system; a measure of this.

1960 Markell & Stanton *Installing Hi-Fi Systems* i. 11 The portion of the room in which the maximum stereo effect is heard is fairly limited, and complete separation between the sound signals at the ears of the listener is impossible in a practical situation. 1962 *Times* 5 July 15/6 In general quality the discs were still preferable although on the tapes the stereo 'separation' was better. 1974 Harvey & Bohlman *Stereo F.M. Radio Handbk.* vi. 129 Some adjustment over the degree of cross-coupling may be provided by a preset control... labelled separation. 1975 G. J. King *Audio Handbk.* viii. 185 For good stereo image placement the separation should not be less than 20 dB over the important part of the spectrum.

14. separation anxiety *Psychol.*, anxiety provoked in a child by the threat or actuality of separation from its mother or mother substitute; also *transf.*; **separation factor** *Nucl. Engin.*, the ratio of the concentration of a particular isotope after a process of enrichment to the concentration before; also, the ratio of the concentrations in the two mixtures produced by the process; **separation negative**, a separation (sense 13* a above) in the form of a photographic negative; **separation plant**, an installation for the separation of isotopes of a chemical element; **separation point** *Physics* and *Aeronaut.*, a point on a surface at which boundary-layer separation begins.

1943 W. R. D. Fairbairn in *Brit. Jrnl. Med. Psychol.* XIX. 340/2 The problem of separation-anxiety in the soldier is anticipated under a totalitarian regime by a previous exploitation of infantile dependence. 1973 J. Bowlby *Attachment & Loss* II. vi. 59 Despite Freud's increasing insistence on the key role of separation anxiety in neuroses, there has been marked reluctance to adopt his ideas. 1977 *Sunday Mail* (Brisbane) 23 Oct. 24/7 Separation Anxiety: The child fears the parent will abandon him. Newson and Newson showed that very many parents used this as a threat. 1945 H. D. Smyth *Gen. Acct. Devel. Atomic Energy Mil. Purposes* ix. 94 In nearly every process a high separation factor means a low yield, a fact that calls for continual compromise. 1978 *Sci. Amer.* Aug. 31/3 A typical separation factor for an early machine was 1·25, which means that if the fraction of uranium 235 in the feed gas is 0·71 percent, as it is in natural uranium, the product contains ·794 uranium 235 and the waste contains ·635 percent. 1931 F. R. Newens *Technicolor Colour Photogr.* iii. 25 Although it is possible to produce the three-colour separation negatives by a single exposure, either in a one-exposure tricolour camera, or by means of a film tri-pack... neither method is at present readily available. 1957 P. Jenkins *Colour Separation Negatives* 30 A fundamental rule in separation-negative making is that any neutral (grey, white, or black) should be reproduced as an equal density on each of the three negatives. 1974 A. Sussman *Amateur Photographer's Handbk.* (ed. 8) xviii. 478 The problem of separation negatives, so far as the amateur is concerned, was overcome in 1935 when Eastman Kodak introduced Kodachrome film. 1945 H. D. Smyth *Gen. Acct. Devel. Atomic Energy Mil. Purposes* viii. 84 The principal installations constructed at the Clinton Laboratory site were the pile and the separation plant. 1974 *Encycl. Brit. Macropædia* XIII. 325/1 Groves arranged contracts for a gaseous diffusion separation plant, a plutonium production facility and a calutron pilot plant. 1946 A. W. Sherwood *Aerodynamics* vii. 101 If the velocity of flow over the sphere is increased, the local Reynolds number Rx for any point in the boundary layer is proportionately increased, with a maximum value at the separation points. 1978 D. Küchemann *Aerodynamic Design of Aircraft* ii. 37 The flow lifts off the wall at a separation point where the skin friction becomes zero and the air flows backwards behind it.

separationist. (Earlier example.)

1831 Ld. Holland *Let.* 3 Jan. in R. B. McDowell *Public Opinion in Ireland 1801–1846* (1952) vi. 143 Withdrawing from O'Connell and the separationists.

separatism. (Further examples.)

1957 P. Kemp *Mine were of Trouble* x. 190 Separatism and Anarchism were the strongest political forces in Catalonia. 1962 *Globe & Mail* (Toronto) 30 Jan. 4/1 Whoever is responsible..has done more for the advancement of separatism in this province [sc. Quebec] in 60 minutes than Dr. Chaput could in 60 years. 1971 *Daily Tel.* 19 July 7/4 The background that gave rise to Moslem separatism in British India, the emergence of the State of Pakistan, and the subsequent history. 1978 *Encounter* Feb. 12/2 Separatism is a problem in Quebec as well as in Scotland.

separatist, *sb.* and *a.* Add: **A.** *sb.* **1. f.** A critic who ascribes the *Iliad* and the *Odyssey* or any portions of them to separate authors. Cf. Separator 1.

1903 A. Platt *Iliad Bk. XVIII* p. xiv, Even among the ancients..there was a set of people called χωρίζοντες or Separatists, who held that the *Iliad* and *Odyssey* were by different authors. 1976 W. R. Johnson *Darkness Visible* 159 If I speak of things Homeric now as a separatist, now as a literal or oral unitarian, [etc.].

separatrix. Add: Pl. **separatrices** (-trisīz).

4. *Physics.* A boundary between regions having differing configurations of magnetic lines of force.

1958 *Proc. CERN Symp. High Energy Accelerators* I. 50/1 We see in fig. 3 the region of stability or 'bucket' within which particles execute stable phase oscillations... The bucket boundary, or separatrix is given by eq. (43). *Ibid.* 54/1 The region between the two separatrices. 1979 *Sci. Amer.* Aug. 44 (*caption*) The tokamak..is equipped with a magnetic diverter, an arrangement in which a magnetic-field separatrix is employed to prevent impurity ions from entering the main plasma column.

sepetir (sepĭtīə). [Mal.] A forest tree of the genus *Sindora* or *Pseudosindora*, belonging to the family Leguminosæ and native to south-east Asia; also, the hardwood timber produced by a tree of this kind.

1893 G. King in *Jrnl. Asiatic Soc. Bengal* LXII. II. 203 Mr. Curtis gives the Malay name of this Sindora intermedia [illegible] in Pangkor as 'Sepetir'. 1927 F. W. Foxworthy *Commercial Timber Trees Malay Peninsula* 91 It is doubtful if Sepetir produces more than one per cent of the timber in the forest. 1934 A. L. Howard *Man. Timbers of World* (ed. 2) 465 Sepetir... The colour of the wood is yellow-brown, with dark streaks. 1948 *Handbk. of Hardwoods* (Forest Prod. Res. Lab.) 210 Sepetir is the British Standard name for the timber of *Sindora* spp. and *Pseudosindora palustris*. 1971 *Country Life* 1 Apr. [illegible]

Sephadex (se·fădeks). *Chem.* A proprietary term for a preparation of dextran used for the separation and purification of chemicals on the basis of molecular size.

1959 *Trade Marks Jrnl.* 16 Sept. 827/1 Sephadex... Polymers being chemical compounds for industrial use in the purification of chemical substances. Aktiebolaget Pharmacia... Sweden... 11th March 1959. 1970 [see *Gel *sb.* 2]. 1975 Williams & Wilson *Biologist's Guide to Princ. & Techniques Pract. Biochem.* iii. 79 Cellulose derivatives (and Sephadex materials) have been used successfully in this way. 1978 *Nature* 12 Oct. 565/2 It was reduced with sodium dithionite and then isolated in 0·01 M phosphate buffer (pH 7) by Sephadex chromatography.

Sepher Torah, var. *Sefer Torah.

sepia. Add: **1. a.** Also **sepia print** (see quot. 1940).

1940 W. E. Woodbury *Encycl. Photogr.* 556 Black and sepia prints must not be washed together in the same dish. 1940 *Chambers's Techn. Dict.* 758/2 *Sepia print*, a release print in which the image is dyed sepia instead of being left black. 1977 *Spare Rib* July 62/4 A marvellous collection of sepia prints showed women at work in the hospital's wards.

B. Of American Blacks: *euphem.* for 'black'. *U.S.*

1943 Berrey & Van den Bark *Amer. Thes. Slang* §371 Negro color distinctions... sepia. 1944 H. L. Mencken in *Amer. Speech* XIX. 166 Some of them also use such terms as..*sepia* to get away from the..inaccurate *black*, and in 1944 there was a Sepia Miss America contest. 1947 S. Lewis *Kingsblood Royal* xxiii. 236 A certain number of sepia merchants get rich on the rest of us chosen people.

Sepik (se·pik). [Native name.] The name of a river and district in Papua New Guinea, used *attrib.* of the peoples of the district and of their languages and artefacts.

1949 M. Mead *Male & Female* viii. 178 The Sepik peoples—Iatmul, Tchambuli, and Mundugumor—make little of menstruation ceremonies. 1966 E. Lindall *Time and Sons* (1967) xii. 120 We've got two Sepik policemen here. 1971 *Current Trends in Linguistics* VIII. 516 The first modern study of Sepik languages was that undertaken by Laycock. 1973 *Sunday Times* 10 June (Colour Suppl.) 42/1 Certain Protestants were notorious for destroying all native works of art, especially Sepik sculptures.

‖ **seppuku** (sepū·ku). Also **Seppuku,** etc. [Jap., colloq. pronunc. of *setsu fuku*, f. Chinese *qie* to cut (with a sword or knife) + *fù* belly.] = Hara-kiri.

1871 A. B. Mitford *Tales Old Japan* II. 193 Seppuku (*hara-kiri*) is the mode of suicide adopted amongst Samurai when they have no alternative but to die. 1890 B. H. Chamberlain *Things Japanese* 141 The Japanese word *harakiri*, so well-known all over the world, is but little used by the Japanese themselves. The Japanese almost always prefer to employ the synonym *seppuku*, which they consider more elegant because it is derived from the Chinese. 1923 J. Street *Mysterious Japan* xvi. 198 At the sound of the guns he took his short sword and committed seppuku. 1947 R. Benedict *Chrysanthemum & Sword* x. 200 He could only come to terms with chu by killing himself according to the rules of seppuku. 1973 A. Broinowski *Take one Ambassador* xi. 178 You would at once..kill yourself. By *seppuku*, slitting of the stomach.

septal, *a.*¹ Add: **2.** *Bot.* Growing in hedges.

1847 H. C. Watson *Cybele Brit.* I. 66 The proposed series of terms run thus:—..Septal.—Plants of hedge-banks, and hedge-rows. 1926 J. J. Walker *Nat. Hist. Oxford Distr.* 114 *Cuscuta europæa* L. is very rare..; in Oxford it was associated with another septal species, *Humulus Lupulus*.

3. *Archæol.* Designating a stone or slab forming a barrier between compartments in a burial chamber.

1920 T. H. Bryce in J. A. Balfour *Book of Arran* 61 The chamber is divided by two septal slabs into three compartments. 1937 *Proc. Prehist. Soc.* New Ser. III. 167 The segmentation of the gallery is achieved by means of septal slabs, rather lower than the side slabs, and in the Irish cairns these septal slabs are set between pairs of vertical jamb stones. 1958 G. Daniel *Megalith Builders Western Europe* ii. 44 This segmentation or septalisation may be done by jambs projecting from each side, or by transverse stones or septal stones. Septal stones sometimes reach..half-way up the height of a chamber.

septanose (septănōᵘz, -s). *Chem.* [a. G. *septanose* (Micheel & Suckfüll 1933, in *Ann. d. Chem.* DII. 89), f. L. *sept-em* seven + *-anose* after *furanose, *pyranose.] A structure containing a seven-membered ring, adopted by some sugars; a sugar having this structure. Freq. *attrib.* Hence **se·ptanoside**, a glycoside in septanose form.

1933 *Chem. Abstr.* XXVII. 3453 An equil. mixt. of α- and β-1,3,4,5-tetraacetylgalactoseptanoses. 1934 *Ibid.* XXVIII. 1025 For an approx. 1% soln. in 0·01 N HCl of the septanoside..the half-time value of hydrolysis at 95° is about 25 min. *Ibid.*, The 7-membered septanose ring is not stable... 1948 R. J. McIlroy *Chem. of Polysaccharides* i. 4 A third type of ring, the seven-membered or septanose structure has been proposed. This ring is comparable to the furanose ring in stability. It has not been obtained from natural sources. 1973 *Carbohydrate Res.* XXVIII. 75 It is found that the pseudorotation of the septanose ring and the dioxolane rings are correlated to the position of attachment of the latter rings to the central septanose ring. 1974 *Jrnl. Chem. Soc.: Chem. Commun.* 190/1 Nitroethane condensed smoothly..with the hydrated dialdehyde (I) to give..the crystalline septanoside..in 38% yield.

septate, *a.* (Example other than in *Nat. Hist.*: cf. *Septum f.)

1947 G. H. Huxley *Survey Princ. & Practice Wave Guides* ii. 39 The field configuration in the septate coaxial system is very similar..to the pattern of the principal wave on a coaxial transmission line.

septcentenary (septsentī·nări). [f. L. *sept(em* seven + Centenary *sb.*, after *bicentenary*, etc.] A seven-hundredth anniversary.

1924 *Glasgow Herald* 21 Aug. 5/2 The present condition of Dornoch Cathedral, the septcentenary of which is to be celebrated on the 27th inst., presents a striking contrast to that of Elgin Cathedral. 1978 *Church Times* 29 Sept. 5/1 Septcentenary... The celebrations to mark the 700th anniversary of the establishment of a Christian church in Macclesfield is about to begin.

septectomy (septe·ktŏmi). *Surg.* [f. Sept(um + *-ectomy.] **a.** Resection of the nasal septum. *rare*⁻⁰.

1949 in *New Gould Med. Dict.* 922/2. 1961 in A. S. MacNalty *Brit. Med. Dict.* 1292/1.

b. Resection of the atrial septum; septostomy.

1974 *Lancet* 27 May 1140/1 Previous atrial septectomy or septostomy had been performed. 1977 *Ibid.* 18 June 1275/1 At the age of 4 weeks the pulmonary artery was banded,..and atrial septectomy was performed.

September. Add: **c. September massacres** *Fr. Hist.*, a mass killing of political prisoners in Paris on 2–6 September 1792.

1805–6 Wordsworth *Prelude* (1959) x. 370, I thought of those September Massacres, Divided from me by a little month. 1909 [illegible] *Scarlet Pimpernel* xi. 109 The news of the awful September massacres, and of the Reign of Terror and Anarchy. 1976 *Listener* 23–30 Dec. 837/1 There began to seem a fatal unsteadiness in the Revolution... There had been the atrocity of the September massacres.

septet(t, -ette. **a.** (Earlier example.)

1828 E. Holmes *Ramble among Musicians of Germany* 263 He [sc. Hummel] was pleased to hear that a lady..had repeatedly played in public his septett for the pianoforte.

septic, *a.* and *sb.* Add: **A.** *adj.* **2. septic tank,** a tank (associated either with a sewage works or with a residence that is not connected to a main sewer) in which the solid portions of sewage are allowed to settle and accumulate and are purified by the action of anaerobic bacteria.

1902 *Encycl. Brit.* XXIX. 379/1 The 'septic tank-system' was devised by Cameron of Exeter in 1896.

Ibid. XXXII. 526/1 At the present time the common cesspool is being resuscitated and improved under the name of a septic tank. 1909 *Chambers's Jrnl.* Feb. 87/1 The sewage system is of the latest character, with a septic tank. 1939 *Archit. Rev.* LXXXVI. 11/2 Bathrooms are installed on both ground and first floors, while other equipment includes an electric pumping plant, a deep well with electric pump and septic tank. 1951 *Good Housek. Encycl.* 315/1 An adequate supply of good water at a safe distance from the septic tank. 1976 *Eastern Even. News* (Norwich) 23 Dec. 27/1 His septic tank did not work, and nor did most others in the village.

3. In trivial use: unpleasant, nasty, 'rotten'. *slang.*

1914 '1 Hay' *Knight on Wheels* xviii. 172 Philip enquired after Mr. Brett, and learned that that 'septic blighter' (Timothy's description) had retired from the position of Housemaster. 1932 S. Gibbons *Cold Comfort Farm* xviii. 248 Rennett had had a pretty septic life. 1958 L. Durrell *Balthazar* 248 What septic weather to-day! 1976 G. Mitchell *Winking at Brim* vii. 89 Mummy and Daddy have had a row. Isn't it septic of them?

B. *sb.* Restrict † *Obs.* to sense in Dict. and add: **2.** *ellipt.* A septic tank. *Austral. colloq.*

1961 P. White *Riders in Chariot* iii. viii. 231 Rosetrees lived..in a texture-brick home—city water, no sewerage, but their own septic. 1977 *Weekly Times* (Melbourne) 19 Jan. 63/2 (Advt.), Lovely new home.., 2 bathrooms, 2 septics and large living area.

septo-². Add: **septo-stomy** *Surg.* [*-stomy], the surgical creation of a hole through the atrial septum; **septo-tomy** *Surg.* [-tomy] = prec.

1967 *Circulation* XXXVI. Suppl. 217/1 At the time of the diagnostic catheterization, atrial septostomy is performed by the balloon-catheter technique. 1977 *Lancet* 18 June 1276/1 The arterial oxygen saturation was 34% and did not improve after balloon septostomy. 1966 *Jrnl. Amer. Med. Assoc.* 13 June 992/1 Early clinical trials on infants with TGV [i.e. transposition of the great vessels] indicate that the procedure is as effective in prolonging life as surgical septostomy.

Septoria (septō·riă). Also **septoria**. [mod.L. (E. M. Fries *Novitiæ Floræ Suecicæ* (1819) v. 78, as *Septaria*; *Systema Orbis Vegetabilis* (1825) I. 119, as *Septoria*), f. L. *septum* Septum + *-ăria*, *-ōria*.] **a.** An imperfect fungus of the genus *Septoria*, which includes forms having spores borne in dark pycnidia and many of which cause disease in plants.

1891 *Bull. Torrey Bot. Club* XVIII. 372 The *Septoria* on celery is so very prevalent in this region. 1932 *Phytopathology* XXII. 795 Both Septorias were isolated from celery from the Kalamazoo marshes. 1946 K. S. Chester *Nature & Prevention Cereal Rusts* x. 129 Later in the spring, when abundant foliage is available to both fungi, *Septoria* sometimes destroys the lower leaves while the leaf rust attacks the upper ones. 1972 R. Gair et al. *Cereal Pests & Dis.* iii. 128 In Scotland the common *Septoria* on barley is *S. avenae* f. sp. *triticea*.

b. One of several leaf spot diseases caused by a fungus of this kind. Also *attrib.*

1903 *Special Bull. Michigan Agric. Exper. Station* No. 77. 2 (*title*) The Septoria leaf spot disease of celery, or celery blight. 1920 *Bull. W. Australia Dept. Agric.* No. 69. 11 Septoria, or Dry Blight, is a fungus disease caused by one of three species of fungus. 1926 Fawcett & Lee *Citrus Diseases & their Control* xviii. 478 (*caption*) Septoria spots on California lemon fruits. 1952 J. G. Dickson *Dis. Field Crops* xi. 217 Two Septoria blotches occur on wheat throughout the world. 1968 *Times* 16 Dec. 7/1 One disease, of which we had heard relatively little up to this year, seems to have had marked effects on the wheat crop. This was septoria, commonly known as glume blotch, although it also attacks seedlings and leaves earlier in the season. 1972 R. Gair et al. *Cereal Pests & Dis.* vii. 124 Like other *Septoria* diseases, speckled blotch is favoured by wet and humid weather. 1976 E. Scarrow *N.Z. Vegetable Gardening Guide* 36 Brown spots appearing on the leaves of celery are due to a fungus disease called septoria, which is seed-borne.

septum. Add: **b.** (Further examples.) *septum lucidum* or *pellucidum*, a thin double layer of tissue forming a partition between the two lateral ventricles of the brain.

1698 W. Cowper *Anat. Humane Bodies* App., 7th Table, fig. 5B The part of the corpus callosum by Vieussens, call'd fornix vera, between which, and the fornix..is plac'd the septum lucidum. 1713 W. Cheselden *Anat. Humane Body* III. xii. 135 Under the corpus callosum appear the two lateral or superior ventricles, which are divided into right and left by a very thin membrane named septum lucidum, which is extended between the corpus callosum and fornix. 1803 C. Bell *Anat. Human Body* III. I. i. 15 Those septa, or, as they are called, processes of the dura mater, being extended across from the internal surface of the cranium, support the brain in the sudden motions of the body. 1807 R. Morris et al. *Edin. Med. & Physical Dict.* II, Septum lucidum, or Septum pellucidum. 1899 F. H. Gerrish *Textbk. Anat.* [illegible] an anterior compartment and a posterior. 1942 F. A. Mettler *Neuroanatomy* iv. 97 Stretched between the copula, rostrum, genu and anterior half of the corpus callosum is a thin, vertically arranged membrane, the translucent septum (s. pellucidum). 1978 *Nature* 19 Jan. 209/1 Whether this input results in cellular discharge depends..on input through the second major afferent pathway to the hippocampus from the septum.

f. *Electronics.* A metal plate placed transversely across a waveguide and attached to the walls by conducting joints.

1947 G. H. Huxley *Survey Princ. & Practice of Wave Guides* ii. 39 The original electromagnetic field..transforms to that shown..where the magnetic loops are so disposed that their longitudinal portions run in opposite senses, one on each side of the septum. 1964 J. L. Altman *Microwave Circuits* iii. 87 Symmetrical obstacles, such as irises, septa, and posts of various cross-sections and positions within the waveguide, are of great practical importance as matching elements..or as elements of waveguide filters and periodic structures.

Sequanian (sekwā·niăn), *a.* *Geol.* Also ‖ **Séquanien**. [ad. F. *Séquanien*, f. L. *Sēquani* an ancient Celtic people of eastern Gaul: see -ian.] Name of a substage of the Upper Jurassic in north-western Europe, below the Kimmeridgian; of or pertaining to this substage and the rocks that characterize it, and the geological age during which they were deposited.

1851 [see *Kimmeridgian *a.*]. 1881 *Q. Jrnl. Geol. Soc.* XXXVII. II. 571 De Loriol..wishes to call the Astartian limestone by one name, Sequanian. 1882 A. Geikie *Text-bk. Geol.* 797 Corallien. Some authors take the upper part of this group into a separate section under the name of Sequanian, largely developed in the east of France, where it consists of massive limestones sometimes 400 feet thick. 1903 *Ibid.* (ed. 4) II. 1156 The Oxfordian and Corallian divisions of the Jurassic system, or Callovian, Oxfordian, and Sequanian formations, are in general feebly represented in the Alpine region. 1970 R. J. Small *Study of Landforms* iv. 145 The widely out-cropping Bathonian dolomite has been carved into a multitude of minor karstic forms which are overlooked by gently rounded hills of Sequanian limestone.

‖ **sequelula** (sĭkwe·liŭlă). *nonce-wd.* [mod.L. dim. of *sequēla* Sequela.] A little sequel (sense 7) or continuation of a literary work.

1912 M. Beerbohm *Christmas Garland* 61 (*title*) A Sequelula to 'The Dynasts'. 1941 E. Blunden *Thomas Hardy* 237 Soon after the completed publication of *The Dynasts*, a 'Sequelula' to it was delivered to the world.

sequenator (sĭ·kwĕnēᶦtəɹ). *Biochem.* [Irreg. f. *Sequence *v.* + *-ator.] = *Sequencer² 2.

1967 Edman & Begg in *European Jrnl. Biochem.* I. 81/1 We propose the term 'sequenator' for an instrument which determines the sequence of an ordered linear polymer by repeating a chemical process. 1973 *MTP Internat. Rev. Sci., Org. Chem.* VI. 48 The operation of the sequenator depends on the insolubility of the protein during the extraction procedure and short peptide chains are too soluble to be successfully degraded. 1978 *Nature* 19 Jan. 281/2 The sequence of the first 44 residues of this peptide has been determined on a protein sequenator (unpublished results).

sequence, *sb.* Add: **I. 2. c.** *Biochem.* The order of the constituent nucleotides in a nucleic acid molecule or of the amino-acids in a polypeptide or protein molecule.

1959 *Arch. Biochem. & Biophysics* LXXXV. 290 The sequence of these trinucleotides was determined by digestion with spleen monoesterase followed by snake venom diesterase with the resulting formation of a purine nucleotide, a purine nucleotide (Pu), and a pyrimidine nucleotide (Py). 1965 *Science* 19 Mar. 1462/1 During protein synthesis, the amino acid sequence of the polypeptide chain is determined by the interaction of a messenger RNA with transfer RNA's specific for a given amino acid. 1970 *Jrnl. Biochem.* CXVIII. 831/1 The recent determination of the sequences at the cohesive ends of DNA from bacteriophage λ..is an excellent example of the application of repair reactions with DNA polymerase..to nucleotide sequence studies. 1977 *Sci. Amer.* Dec. 55/1 The complete nucleotide sequence of the DNA of a small bacterial virus, φX174, has been established.

d. *Cinematogr.* and *Television.* A passage consisting of several shots unified about a single theme or event.

1929 *Morning Post* 24 May 12/7 Until recently, in all talking sequences, the actor has been compelled to be [illegible]. 1934 C. Lambert *Music Ho!* iv. 262 A famous sequence in the silent film *Mother*. 1941 B. Schulberg *What makes Sammy Run?* vi. 125 He stayed up..reading one script after another... The plan was for him and Sammy to write alternate sequences. 1958 *Daily Mail* 19 July 8/8 Parody of a French film sequence set in a sleepy bistro. 1976 D. Clark *Dread & Water* v. 105 He's got a movie shot of Silk climbing that mountain... The sequence is just one of Silk climbing.

e. *Geol.* (*a*) An ordered succession, esp. of strata in conformity.

1931 Gregory & Barrett *General Stratigr.* vi. 96 The fullest Russian sequence is in the Urals, where the Lower Devonian consists of marine slates, quartzites, and occasional limestones. 1975 A. E. Ringwood *Composition & Petrology of Earth's Mantle* vii. 243 In estimating the abundance of andesitic volcanism in Precambrian shield areas, allowance should be made for the andesitic component of associated geosynclinal sediments.

(*b*) In various specific usages (see quots.).

1933 R. C. Moore *Hist. Geol.* v. 54 No designation for the rocks of an era is in common use. The term 'sequence' will be used in this book. 1949 L. L. Sloss et al. in *Mem. Geol. Soc. Amer.* No. 39. 110 The writers term the assemblages of strata separated by the above-described objective horizons 'sequences'. Sequences should be considered as rock units, assemblages of formations and groups. 1962 Silberling & Roberts in *Geol. Soc. Amer. Special Paper* No. 72. 6 A different kind of subdivision..is required in northwestern Nevada for the upper Paleozoic and lower Mesozoic rocks. The subdivisions adopted are lithologically and geographically discrete units of major rank termed 'sequences' that are set apart from underlying or overlying sequences by unconformities.

III. 8. *attrib.* and *Comb.*, as **sequence control** *Computing*, a method of controlling the execution of distinct operations in a defined order; so **sequence-controlled** *ppl. a.*; **sequence dancing** (see quot. 1949); also **sequence dance**; **sequence date** *Archæol.*, a relative chronological date based upon comparison of a series of objects from an archæological site; hence **sequence dating**; **sequence shot** *Cinematogr.* (see quot. 1973); **sequence space** *Math.*, a space whose points are sequences.

1946 *Electr. Engin.* LXV. 387 (*caption*) Front view of calculator showing sequence control mechanism.., which tells machine what to do and when to do it. 1962 Sequence control [see *control register* s.v. *control *sb.* 5]. 1962 C. Dent *Quantity Surveying by Computer* iii. 24 Control.. ..directed to address No. 2 in the memory store for its next instruction, and so on, in numerical sequence. This mode of operation is called 'Automatic Sequence Control'. 1946 *Ann. Computation Lab. Harvard Univ.* I. p. ix, In May 1944, the Staff of the Computation Project began operations with the Automatic Sequence Controlled Calculator as an activity of the Bureau of Ships. 1950 W. W. Stiffler *High-Speed Computing Devices* v. 63 An automatic sequence-controlled calculator is a computing machine into which such a [sequencing] mechanism is built. 1927 *Melody Maker* Sept. 865/2 They are to a great extent sequence dances and based on what many consider to be old-fashioned steps and movements. 1978 *Abingdon Herald* 12 Jan. 1/9 The Wootton and Dry Sandford Sequence Dance Club. 1940 A. H. Franks *Ballroom Dancer's Handbk.* 109 Sequence dancing really has no place in the art of modern ballroom dancing and such dances are regarded as novelties. 1949 A. Chujoy *Dance Encycl.* 424/2 *Sequence Dancing*, a term used in England to describe those ballroom dances in which the steps have to be taken in a certain definite order, as a consequence of which all couples are always making the same movement at one time. 1980 *Radio Times* 29 Nov.–5 Dec. 86/3 This is Sequence Dancing... When one lady twirls 200 other ladies twirl. 1901 Sequence date [see S.D. s.v. *S 4 a]. 1920 W. M. F. Petrie *Prehist. Egypt* ii. 4 For permanent reference the whole 900 graves, when placed in their most probable order or sequence, were divided in 51 equal sections, and these were numbered 30 to 80, and such numbers termed Sequence Dates, marked as S.D. 1923 T. E. Peet in *Cambr. Anc. Hist.* I. vi. 247 Petrie, at Diospolis Parva, invented the now famous system of 'Sequence Dating'. 1958 L. Cottrell *Anvil of Civilisation* ii. 39 He [sc. Petrie] invented the system which we call 'sequence dating' which..enables archaeologists to establish the *comparative* age of a site by the type of pottery found on it, even when it lies below the 'historic horizon'. 1973 S. Heath in *Screen* Spring/Summer 11 A *sequence-shot*, a whole scene in one shot (e.g. *autonomous segment* 17 of *Adieu Philippine* showing Michel, the hero, and his friend Daniel working in the TV studio). 1974 M. Taylor tr. *Metz's Film Lang.* iii. 42 There was Jean Renoir with his many statements in favor of the sequence shot. 1940 H. S. Allen in *Proc. London Math. Soc.* XLVIII. 310 A set *S* of sequences containing the origin and such that for every *x* and *y* in *S* and every number *r*, *x*+*y* and *rx* are in *S* is called a sequence space. 1968 G. Ludwig *Wave Mech.* I. iii. 37 The formulation of a matrix in diagonal form and the solution of the eigenvalue problem are therefore equivalent problems, the first being defined only in the sequence space.

sequence (sī·kwĕns), *v.* [f. the sb.] **1.** *trans.* To arrange in a definite sequence or order.

1954 *Computers & Automation* Dec. 20/2 *Sequence*,..to select A if A is greater than or equal to B, and select B if A is less than B, or some variation of this operation. 1966 J. S. Bruner in *Beyond Information Given* (1974) xxv. 442 We..closed our eyes to the pedagogic problem of how to represent knowledge, how to sequence it in a form appropriate to young learners. 1974 M. B. Brown *Economics of Imperialism* ix. 226 Countries can be sequenced as markets for different products according to their standards of consumption. 1976 *Daily Tel.* 12 Aug. 2/3 To get the maximum use out of Heathrow's two main runways aircraft are carefully 'sequenced' from the four reporting points that serve the airport.

2. *Biochem.* To ascertain the sequence of monomers in (a biological polymer such as a polypeptide or a nucleic acid).

1970 S. Blackburn *Protein Sequence Determination* xx. 274 The future should see the increasing use of methods able to sequence very large molecules. 1977 *Sci. Amer.* Dec. 67/2 Now that DNA can be sequenced readily and rapidly we can expect that in the next few years the precise composition of many DNA's will be established.

Hence **se·quenced** *ppl. a.*, **se·quencing** *vbl. sb.*

1961 P. Siegel *Understanding Digital Computers* xv. 329 A sequencing unit to be used with a drum memory and a two-address instruction is shown. 1970 *Nature* 14 Mar. 1026/2 Data..on the patterns of change in amino acid substitutions in all the completely sequenced proteins show anything but a random pattern of substitution. 1971 *Archivum Linguisticum* II. 139 The realization rules accept as input specific pairs of such feature-sets and render them as sequenced strings of morphemes which are, in surface structure, simple *NPs*. 1977 *Sci. Amer.* Dec. 56/2 The smallest DNA molecules, those of certain viruses, are perhaps 70 times longer than the 75-nucleotide transfer-RNA molecules that were the subject of early RNA sequencing.

sequencer² (sī·kwĕnsəɹ). [f. Sequence *sb.* + -er¹.] **1. a.** (See quot.)

1954 *Computers & Automation* Dec. 21/1 *Sequencer*, in punch card machinery, a mechanism which will put items of information in sequence.

b. An apparatus for performing or initiating operations in the correct sequence; *spec.* one forming part of the control system of a computer.

1964 *New Scientist* 6 Feb. 319/1 A late decision to photograph the Moon in passing was defeated by a malfunction in the so-called central computer and sequencer. 1977 N. Freeling *Gadget* v. 222 We..make an automatic sequencer which takes over the work of sending the signals. 1978 D. J. Kuck *Struct. Computers & Computations* I. iv. 282 The sequencer drives the entire machine through a specified sequence of events which carry out (indeed are) whatever the instruction is defined to do.

2. *Biochem.* An apparatus for determining the sequence of monomers in a biological polymer.

1971 *European Jrnl. Biochem.* XX. 89 Using the sequencer and unpurified reagents, it was possible to degrade entirely a 0·24 μmole sample of the 21 amino acid insulin A-chain (oxidized) and to analyze the products. 1977 *Sci. Amer.* Oct. 103/1 The exact positions at which the labeled amino acid appears are determined with the aid of an automatic machine called a sequencer, which chemically removes one amino acid at a time from the polypeptide chain beginning at the amino terminal (NH_2) end.

sequential, *a.* Add: **1. d.** *sequential induction* (Biochem.): the formation in sequence of a group of related enzymes, consequent upon the induction of the first enzyme of the series (see quot. 1953).

1953 M. Cohn et al. in *Nature* 12 Dec. 1096/2 We..propose the following terms and designations... The exposure of an organism to a single inducer which is also a substrate may result in the induction of a sequence of enzymes, since the metabolism of the primary, exogenous inducer gives rise to the formation of a succession of intermediary metabolites each of which in turn serves as an inducer for the enzyme which converts it into the next member of the metabolic chain. This phenomenon is termed 'sequential induction' (simultaneous or successive adaptation). 1968 H. Harris *Nucleus & Cytoplasm* vi. 118 We have glimpses of this kind of organization in the phenomenon of 'sequential induction' (induction *en chaîne*) of enzymes in bacteria. 1971 *Bacteriol. Rev.* XXXV. 89/2 A sequential induction is characterized by a shift in the chemical nature of the inducer.

2. b. *Computers.* Of, pertaining to, or designating various aspects of a computer system and its control programming that operate or are utilized serially; *sequential search*, a search through a data list or file that is carried out serially.

1951 *Proc. IRE* XXXIX. 276/2 *Sequential control*, the manner in which instructions to a digital computer are set up in sequence and are fed consecutively to the computer during the solution of a problem. 1964 T. W. McRae *Impact of Computers on Accounting* iii. 53 A computer..carries out each of these operations in automatic sequence under the control of the computer programme. This particular characteristic is known as 'sequential processing'. 1965 *Information & Control* VIII. 159 (*heading*) Discrete sequential search. 1969 P. B. Jordain *Condensed Computer Encycl.* 447 Once written, a sequential file has to be read in the same order (or sometimes in the inverse order) in which it has been written. 1970 O. Dopping *Computers & Data Processing* x. 133 The magnetic tape can be called a sequential access memory, or serial access memory, because the records must be written and read in sequence. 1973 C. W. Gear *Introd. Computer Sci.* ii. 44 We refer to the set of input cards as a sequential data set because it is possible to read a card only after the preceding card has been read. 1979 J. E. Rowley *Mechanised In-House Information Syst.* i. 26 Any search must process the complete tape, from start to end, seeking matches between terms, i.e. a sequential search.

5. *Television.* **a.** Of or pertaining to the normal method of scanning a television image, in which all the lines are traversed in the same direction, with a rapid, blanked fly-back after each.

1940 *Chambers's Techn. Dict.* 759/1 *Sequential scanning*, scanning in which the spot traverses each line in the same direction, returning rapidly from the end of one line to the beginning of the next. 1942 *Electronics* Apr. 164/1 System 2 employs sequential scanning in order to eliminate interline flicker. 1966 G. H. Hotson *Television Receiver Theory* i. 5 A sequential raster..would be set up by drawing 625 lines one under the other. 1967 H. A. Cole *Basic Television* iii. 26 Provided that the rate of sequential scanning is high enough, the eye can be successfully 'tricked', by reason of the persistence of its vision, into believing that a very rapidly renewed image on the viewing screen has in fact been there all the time.

b. Of, pertaining to, or designating various systems of colour television in which picture information for the primary colours is transmitted successively in quantities corresponding to a dot, line, or field. Cf. *dot-* (also *field-*, *line-*) *sequential system* s.v. *dot *sb.*¹ 5 f, *field *sb.* 21, *line *sb.*² 32.

1947 *Electronics* Jan. 72/2 The sequential system is characterized by the fact that the transmitted signal contains information about one primary color only at any instant of time. 1951 *Proc. IRE* XXXIX. 1195/1 In the case of field-sequential or line-sequential presentation, step-wise switching from color to color is desired. In the case of dot-sequential presentation, sine-wave switching by circular deflection with uniform angular velocity is preferred. 1975 D. G. Fink *Electronics Engineers' Handbk.* xx. 12 The field sequential system employs a monochrome television camera, with a color-scanning disk mounting near the focal plane... The video signal derived from the camera tube thus consists of sequential color fields in the order that the primary light filters appear in front of the camera tube.

6. *sequential circuit* (Electronics), a logic circuit whose output depends on the order or timing of the inputs.

1954 D. A. Huffman in *Jrnl. Franklin Inst.* CCLVII. 165 In a circuit having secondary relays, the possibility of a 'memory' exists since the states of operation may not uniquely determine the output transmissions. A circuit having secondary relays will be called a sequential circuit. 1969 J. J. Sparkes *Transistor Switching* iv. 93 In combinational and sequential circuit diagrams it is usual to use special symbols. 1975 D. G. Fink *Electronics Engineers' Handbk.* xxiii. 41 In some cases problems exist in sequential circuits when a circuit action depends critically on which relay or logic element completes its operations first.

sequester, *v.* Add: **5.** *Chem.* To form a stable complex, esp. a chelate, with (an ion) so as effectively or actually to remove it from solution; to form a stable complex with (a biochemical molecule).

1934 R. E. Hall *U.S. Patent 1,956,515* 5/2 The water softening action of the sodium metaphosphate..is rather to sequester or lock up the calcium in a but extremely slightly [*sic*] ionizable condition in a soluble sodium-calcium-metaphosphate complex molecule. 1953 *Sci. Amer.* June 70/2 The iron..is tightly imprisoned and hidden away—'sequestered', in the poetic language of chelation technology—by EDTA's chelate rings. 1962 *Which?* Oct. 297/2 Instead of softening water by replacing the calcium and magnesium in hard water by sodium..you can 'wrap up' the calcium and magnesium—sequester them—and so isolate them from the soap during washing. 1973 *Nature* 13 July 103/2 Insect yolk proteins ..are synthesized and secreted by the fat body, and are sequestered from the haemolymph by the developing oocytes. 1977 *Sci. Amer.* July 92/1 When the cations are sequestered in an organic cage molecule, the resulting complex is so stable that the 'backsliding' reaction is prevented.

sequestering *vbl. sb.* (later examples).

1949 *Thorpe's Dict. Appl. Chem.* (ed. 4) IX. 512/2 The term 'sequestering' introduced by Hall Laboratories to designate the virtually complete elimination of Ca^{++} ions whilst retaining the calcium in solution in the form of a soluble complex. 1962 *Which?* Oct. 297/2 The best known sequestering agents for softening water in this way are the sodium metaphosphates. 1973 P. A. Allum *Politics & Society in Post-War Naples* ix. 316 Antonio Gava's manoeuvres to try to become Campanian Regional Chairman..included the sequestering of a DC regional councillor in a trunk.

sequestrant (sĭkwe·strănt), *sb.* (*a.*) *Chem.* [ad. L. *sequestrānt-em*, pr. pple. of *sequestrāre* (see Sequester *v.*).] A sequestering agent. Also *attrib.* or as *adj.*

1951 *Chem. Abstr.* XLV. 3756 (*heading*) Analytical applications of complexones (sequestrants). 1967 [see *lactobionic* adj. s.v. *lacto- 2]. 1972 *Sci. Amer.* Mar. 19/3 Sequestrants are added to food to bind trace metals and thus prevent any oxidative activity the metals in an ionized state might have on the food.

sequestrate, *v.* Add: **1. b.** *Physiol.* To render (a biochemical compound) metabolically unavailable without destroying it; to remove from the circulation.

1961 *Lancet* 29 July 258/1 The placenta, like the liver, can sequestrate and degrade insulin. 1977 *Proc. R. Soc. Med.* LXX. 521/1 They suggested that the increased titres might be due to failure of the cirrhotic liver to sequestrate gut-derived antigens, which then reached immunologically competent areas of the body.

sequestration. Add: **5*.** *Chem.* The action or state of being sequestered (sense *5).

1948 *Jrnl. Chem. Education* XXV. 483/1 In sequestration, the multivalent positive ion has practically disappeared from the solution without being evolved as a gas, removed as a precipitate or deposited as an element. 1959 R. L. Smith *Sequestration of Metals* iii. 26 Sequestration is most usually achieved by chelation, even although chelation itself covers many phenomena which would not be considered sequestration. 1973 D. F. Long tr. *Degrémont's Water Treatm. Handbk.* (ed. 4) ix. 293 The total sequestration of calcium requires about 50g of polyphosphate per degree TH.

sequestrectomy (sĭkwestre·ktŏmi). *Surg.* [f. Sequestr(um + *-ectomy.] The surgical excision of a sequestrum or sequestra; = Sequestrotomy.

1940 in *Chambers's Techn. Dict.* 759/1. 1954 E. L. Farquharson *Textbk. Operative Surg.* v. 136 Sequestrectomy.—Small loose sequestra may be discovered only during an operation to improve drainage, and are then removed as part of the operation. 1963 R. Warren *Surgery* xxxi. 1071/2 Sequestrectomy. Removal of sequestrum (or dead bone) is done primarily for osteomyelitis.

Sequestrene (sĭkwe·strīn). Also **sequestrene.** [f. Sequestr(ation + -ene.] A proprietary term for preparations of ethylenediamine tetra-acetic acid and its salts used as sequestering agents; *spec.* one containing sequestered iron for use on iron-deficient soils.

1949 *Official Gaz.* (U.S. Patent Office) 8 Feb. 320/2 Alrose Chemical Co., Inc., Cranston, R.I.... *Alro Sequestrene*... Claims use since Jan. 16, 1948. 1949 *Agric. Chemicals* IV. iv. 73/2 'Sequestrene', unlike the polyphosphates, is compatible with cationic surface-active agents. 1951 *Proc. Soc. Exper. Biol. & Med.* LXXVI. 619/1 Four Sequestrenes are available at present, Sequestrene AA (the free acid), SNA 2 (di-sodium salt), SNA 3 (trisodium salt) and SNA 4 (tetrasodium salt). 1958 *Times* 4 Oct. 9/5 The green leaves become pale and chlorotic. There may be several reasons for this condition but nearly always application of sequestrene compound.. will effect a quick..cure. 1965 *Listener* 1 July 25/3 Sequestrenes..are now widely sold for application to plants suffering from an iron deficiency. 1977 *Vole* No. 1. 34/3 Rugosas..may need sequestrene if there is an excess of lime.

sequined, *ppl. a.* Add: Also *fig.*

1918 E. Sitwell *Clown's Houses* 8 Beside the sea, metallic-bright And sequined with the noisy light. 1969 'E. Lathen' *When in Greece* xiii. 140 The sea spread a sequined carpet..below.

‖ **serab** (serā·b). Also **sirab.** [ad. Arab. *sarāb*.] A mirage.

a 1835 F. D. Hemans *Wks.* (1844) III. 87 Suns of blasting light perchance illume The glistening Serab which illudes his eye. 1839 *Penny Cycl.* XV. 261/2 This kind of mirage is not peculiar to Egypt; it is known in Persia also, where it is called *Serab* or *Sir-ab* (miraculous water). 1883 *Encyl. Brit.* XVI. 50/2 When the soil is parched up the appearance of the mirage (seráb) is very common.

Serabend, var. *Saraband².

serac. Add: ‖ **sérac.** (Later examples.)

1900 *Proc. Boston Soc. Nat. Hist.* XXIX. 295 Weathering occurs where variations of external temperature penetrate to the bed-rock, as is particularly the case between the séracs of glacial cascades. 1933 J. Buchan *Prince of Captivity* I. iii. 92 They came on ice-fields..and mountainous seracs which would have puzzled an Alpine climber. 1936 M. Roberts *Poems* 36 The snow falls, and the séracs; and the green glacier-ice Moves down. 1963 G. Carr *Lewker in Norway* vi. 124 On the further side of the right-hand ridge he could just see the upper *séracs* of the Bojumsbre. 1979 C. Kilian *Icequake* vi. 86 Huge fields of seracs—the topographical nightmare caused by intersecting crevasses.

seradeh, obs. var. Shraddha, sraddha.

seral (sīᵊ·răl), *a.*² *Ecol.* [f. *sere *sb.*² + -al.] Of or pertaining to a sere; being a member of a sere other than its climax.

1916 F. E. Clements *Plant Succession* ix. 184 In lowland and montane regions examples of priseres are often more numerous than those of subseres, and such regions are of the first importance for seral investigations. 1926 Tansley & Chipp *Study of Vegetation* ii. 18 We have a special technical term for the developmental series of communities... We apply the adjective *seral*, as opposed to *climax*, to such communities. 1932 *Forestry* VI. 190 The principal seral stages in natural succession from grassland or heath to beechwood are shortly described for certain soil types. 1955 P. A. Buxton *Nat. Hist. Tsetse Flies* ix. 278 It appears to be generally true that the grassland is seral and that it is prevented from developing into bush or woodland by annual fires. 1973 P. A. Colinvaux *Introd. Ecol.* vi. 77 The communities are.. classified into a number of subordinate communities, the seral stages, and the generic taxon, the *Beech-maple climax community*.

serape. Delete ‖ and add: Also † **zarape.** Now in wider use. (Earlier and later examples.)

1834 A. Pike *Prose Sk. & Poems* 138 The men with.. the zarape or blanket of striped red and white. 1836 [see *ranchero]. 1892 *Dial. Notes* I. 194 *Serápe*, a Mexican blanket, generally woven by hand by Indian women, with stripes of variegated colors. The *serape* has no opening or slit for the head, like the *poncho*, but is worn by men only, thrown across the shoulders. 1916 'B. M. Bower' *Phantom Herd* 68 He had finished with an old Mexican serape draped around his person for warmth. 1950 *Chicago Tribune* 1 Mar. 20/3 The feminine counterpart of the serape is the rebozo. 1979 *United States 1980/81* (Penguin Travel Guides) 49 Mexican, Indian, and 'Old West' items are especially good buys. Serapes..and wool rebozos.. make nice gifts.

Serapeum (serăpī·ŏm). *Egyptology* and *Anc. Hist.* Also **Serapeion** (-əi·ǫn), **Serapeium**; pl. **Serapeia.** [a. late L., ad. Gr. Σεραπεῖον, f. Σέραπις Serapis.] A temple of Serapis; *spec.* the great precinct near Memphis, where the sacred Apis bulls were buried, and a temple in Alexandria.

1841 *Penny Cycl.* XXI. 260/2 He had temples (Serapeia) in several parts of Egypt. 1847 J. Leitch tr. *Müller's Anc. Art* 243 The Serapeum was at the same time a sanatory institution. 1877 A. B. Edwards *Thousand Miles up Nile* iv. 86 According to one of these precious Serapeum tablets, the wounded bull did not die till the fourth year of the reign of Darius. 1927 Tarn & Griffith *Hellenistic Civiliz.* x. 294 The Serapeum at Delos has revealed that the triad who were so to influence Hellenism were..Isis, Sarapis, and Anubis. 1928 *Daily Tel.* 11 Dec.

13/4 It was suggested that the so-called Greek Serapeum was in truth nothing but the resting-place of the mother cows of Egyptian Apis. 1961 A. Gardiner *Egypt of Pharaohs* xii. 326 Not a single inscription of Dyn. XXI was found in the Serapeum [at Memphis]. 1972 P. M. Fraser *Ptolemaic Alexandria* I. v. 271 The Serapeum on Rhacotis Hill [in Alexandria] was within the Ptolemaic and Roman city-walls.

seraph¹. Add: **3.** *seraph-arrival*, *-cloud*, *-sense*; *seraph-bright*, *-haunted*, *-sent* adjs.

1876 G. M. Hopkins *Wr. Deutschland* xxiii, in *Poems* (1967) 59 With the gnarl of the nails in thee,..his Lovescape crucified And seal of his seraph-arrival. 1949 Blunden *After Bombing* 49 Marbles, mosaics, carvings, seraph-bright Paintings of wall and window. 1928 —— *Japanese Garland* 30 Their mysteries luring that young seraph-cloud Swan-like between the mountain and the moon. 1958 G. Barker *Two Plays* 52 Lie dreaming on that seraph-haunted shore. 1928 Blunden *Retreat* 65 Her touch is seraph sense. 1932 —— *Face of England* 126 They sparkled free In seraph-sent lucidity.

seraphim. Add: **2. c.** *β* (Later example.)

1924 E. Sitwell *Sleeping Beauty* iv. 23 From flowers as white as seraphims' breath.

d. Delete † *Obs.* and add later examples of *β*.

1920 'K. Mansfield' *Let.* Nov. (1928) II. 80 A cherubim and a seraphim come winging their way towards me. 1974 *Times Lit. Suppl.* 29 Mar. 314/3 Lamartine is content to be a seraphim.

Serax (se·ræks). *Pharm.* A proprietary name in Canada and the U.S. for *oxazepam.

1957 *Official Gaz.* (U.S. Patent Office) 27 Aug. TM 148/2 American Home Products Corporation... *Serax.* For ataractic. First use Feb. 4, 1957. 1968 *Jrnl. Pharmaceut. Sci.* LVII. 312/2 Oxazepam is a psychotropic agent. [*Note*] Marketed as Serax by Wyeth Laboratories, Philadelphia, Pa. 1974 M. C. Gerald *Pharmacol.* xvi. 309 The benzodiazepine derivatives include..oxazepam (Serax). 1977 *Rolling Stone* 30 June 81/3, I..reached into my shirt pocket, removed two ·30 mg. Serax capsules, popped them into my mouth, and washed them down with the drink.

seraya (serai·ă). [Mal.] A forest tree of the genus *Shorea* or *Parashorea*, belonging to the family Dipterocarpaceæ and native to south-east Asia; also, the hardwood timber produced by a tree of this kind. Cf. *lauan, *meranti.

1893 G. King in *Jrnl. Asiatic Soc. Bengal* LXII. II. 112 Its vernacular name in Penang is *Seraya*. 1916 *Bull. Dept. Forestry Brit. N. Borneo* No. 1. 19 The better coloured and figured pieces of Seriah make very acceptable substitutes for Mahogany, in panels, veneers, etc. 1920 A. L. Howard *Man. Timbers World* 256 The wood has been called by a variety of names such as East Indian mahogany and East Indian cedar, as well as by its proper name of serayah. 1940 E. J. H. Corner *Wayside Trees Malaya* I. 213 The *Seraya*..is perhaps the only forest-tree that can be identified from afar by its pale, outstanding crown. 1956 *Handbk. of Hardwoods* (Forest Prod. Res. Lab.) 153 Red seraya or light red seraya..may be described as the North Borneo equivalent of light red meranti. *Ibid.* 212 White seraya grows to an average height of 120 ft. 1962 J. C. S. Brough *Timbers for Woodwork* (rev. ed.) xvi. 174 Seraya..ranges from straw to reddish-brown. 1965 R. McKie *Company of Animals* i. 36 We stopped high on the ridge in light jungle topped by seraya trees. 1971 [see *lauan].

Serb, *sb.* **1. b.** (Earlier example.)

1861 Mill *Repr. Govt.* xvi. 292 The population of Hungary is composed of Magyars, Slovacks, Croats, Serbs, Roumans, and in some districts, Germans.

Serbian, *sb.* (Earlier example.)

1848 C. Bunsen in *Rep. Brit. Assoc. Advancem. Sci. 1847* 267 The sixth family is that of the Slavonic nations in their two great branches;..the western, the languages of the Tschechs (Bohemians), Slovaks, Poles, and Serbians.

Serbo-. Add: *Serbo-Croat*; *Serbo-Bulgarian*, *-Greek* adjs.

1923 G. Buchanan *My Mission to Russia* I. vi. 69 The so-called Serbo-Bulgarian Customs Union Treaty, negotiated in 1905, was never ratified by the Skuptschina. 1905 Serbo-Croat [in Dict.]. 1931 *Times Lit. Suppl.* 29 Jan. 82/3 Translations..from the Russian,..Judaeo-Spanish and Serbo-Croat. 1976 W. H. Canaway *Willow-Pattern War* vii. 78 Petar was bilingual in Serbo-Croat and Albanian. 1958 *Everyman's Encycl.* XI. 234/1 Later there were attempts to replace the decaying Byzantine empire by a Serbo-Greek empire. 1972 D. Dakin *Unification of Greece* ix. 126 The Serbo-Greek alliance of 1867.

sere (sīᵊɹ), *sb.*² *Ecol.* [f. L. *serĕ-re* to join in a series.] A series of plant communities, each naturally succeeding the previous one.

1916 F. E. Clements *Plant Succession* i. 4 A sere is a unit succession. It comprises the development of a formation from the appearance of the first pioneers through the final or climax stage. 1940 *Geogr. Jrnl.* XCVI. 8 The seres which follow the destruction of climax vegetation in the alpine region [of the Himalayas] vary. 1960 N. Polunin *Introd. Plant Geogr.* xi. 323 Such is succession, the developmental series of communities constituting a sere and leading up to a state of relative stability and permanence known as the climax.

sere, sear, *a.*¹ Add: **3. c. sereward** *adv.*, towards decay (*rare*⁻¹).

1902 T. Hardy *Poems of Past & Present* 142 The sun and shadows wheel, Season and season sereward steal.

serein. Delete 'after sunset'.

serendibite (sĕre·ndibəit, sĕrendi·bəit). *Min.* [f. *Serendib*, *Serendip*, a former name for Sri Lanka + -ite¹.] A borosilicate of aluminium, calcium, and magnesium, $(Ca,Mg)_5Al_5BSi_3O_{20}$, found as bluish triclinic crystals in which iron often replaces some of the aluminium and magnesium.

1902 *Nature* 20 Feb. 383/2 Messrs. G. T. Prior and A. K. Coomára-Swámy gave an account of the mode of occurrence and characters of 'serendibite', a new boro-silicate from Ceylon. 1978 W. A. Deer et al. *Rock-Forming Minerals* (ed. 2) IIA. 661 A number of serendibite.. occurrences in spinel-diopside skarns have been described from the Tayezhnoye iron ore localities of southern Yakutia... Serendibite is also found with sinhalite, warwickite and tourmaline..in the skarns of Handemi district, Tanzania.

serendipitous (serendi·pitəs), *a.* [f. Serendipit(y + -ous.] **a.** Of persons: having the faculty of making happy and unexpected discoveries by accident.

1958 *Times Lit. Suppl.* 22 Aug. 468/4 In the matter of adventure Miss de Banke was serendipitous to the *n*th degree. 1968 'E. McBain' *Fuzz* ix. 146 La Brisca seemed to be a serendipitous type who led them on a jolly excursion halfway across the city. 1975 *Reader's Digest* Oct. 150/2 And all for the best, too, as serendipitous San Diegans gladly tell you.

b. (The more usual sense.) Applied to discoveries, meetings, etc., of this kind.

1965 J. Wakefield *Death the Sure Physician* 50 It's rather fortunate that I should come across a chap with similar interests...distinctly serendipitous, in fact. 1971 *Nature* 20 Aug. 538/2 This suggestion was confirmed by the isolation of a stable tricarbonyliron complex of tetraphenylbutadiene by a serendipitous method (many of the best discoveries in the field have been made by chance). 1979 *Amer. Speech 1978* LIII. 272 As among these three systems, the girls couldn't have cared less, Yerke's suggestion was serendipitous.

Hence **serendi·pitously** *adv.*

1969 C. C. Winter *Pract. Urol.* vii. 211 Prostatitis is one of the most common of urologic disorders. It may be symptomless and discovered serendipitously in a routine, two glass urinalysis in which the first specimen shows some white blood cells or a few more than in the second glass. 1974 *Daily Tel.* (Colour Suppl.) 29 Nov. 16/3 We can imagine Hodder meeting Stoughton..and their discovering, serendipitously, a mutual interest in books. 1980 *Times Lit. Suppl.* 14 Nov. 1275/4 He had the knack of always being serendipitously on hand when a tenement caught fire.

serendipity. Add: (Later examples.) Also, the fact or an instance of such a discovery.

Formerly rare, this word and its derivatives have had wide currency in the 20th century.

1926 E. Meynell *Life of Francis Thompson* xiii. 221 To the Serendipity Shop—the venture of a friend in Westbourne Grove—he would often go. 1955 *Sci. Amer.* Apr. 92/1 Our story has as its critical episode one of those coincidences that show how discovery often depends on chance, or rather on what has been called 'serendipity' —the chance observation falling on a receptive eye. 1971 S. E. Morison *European Discovery Amer.: Northern Voy.* i. 3 Columbus and Cabot..(by the greatest serendipity of history) discovered America instead of reaching the Indies. 1980 *TWA Ambassador* Oct. 47/2 It becomes a glum bureaucracy, instead of the serendipity of 30 people putting out a magazine.

Hence **serendi·pitist.**

1939 Joyce *Finnegans Wake* 191 You..semisemitic serendipitist, you (thanks, I think that describes you) Europasianised Afferyank! 1968 *Punch* 13 Nov. 684/1 There are the financial serendipitists, the men blessed monetarily by a fortunate law.

serg, var. *sarge².

sergeant, serjeant, *sb.* Add: **9. b.** *sergeant-pilot.*

1919 J. T. B. Mccudden *Five Years in R.F.C.* III. iii. 86 About the end of August, 1915, a Serjeant-Pilot named Watts arrived for duty. *a* 1963 J. Lusby in 'B. James' *Austral. Short Stories* (1963) 221 The new boys comprised Australians, Englishmen, and Canadians... Most were sergeant-pilots, and in age retired school-boys.

11. sergeant-fish, delete *U.S.*, for Latin name substitute *Rachycentron canadum* and add: the cobia, a large game fish found in tropical and subtropical seas (earlier and later examples).

1873 *Forest & Stream* I. 258/1 Sergeant Fish..derives its trivial name from a black stripe running along its silvery sides..like that on the trowsers of a sergeant. 1947 K. H. Barnard *Dict. Guide S.Afr. Fishes* 112 Sergeant-fish... A somewhat rare fish, of elongate shape,..occasionally caught at Port Elizabeth and Natal. 1958 *Washington Post* 24 Sept. 1/2 An unusually fine run of cobia (the sergeant fish)..has caused big game fishermen to toss caution to the gods of Izaak Walton.

sergeant-major. Add: **4.** *Mil. slang.* Used *attrib.* to designate: (*a*) coffee with cream or milk and sugar (*U.S.*); (*b*) strong sweet tea; tea with rum; also in the possessive and *ellipt.*

1923 T. Boyd *Through Wheat* viii. 131 'Bring your canteen cups. Sergeant-major coffee.'..'Coffee, hot! And milk and sugar in it!' 1925 Fraser & Gibbons *Soldier & Sailor Words* 254 *Sergeant Major's Tea*, tea with sugar and milk, or a dash of rum, in it. 1929 J. L. Hodson *Grey Dawn—Red Night* II. v. 210 Two of them got up before the rest and made a fire and produced 'sergeant-major's tea' and bacon done to a turn. 1929 J. B. Priestley *Good Compan.* I. iv. 115 I'd like a drop o' tea with some rum in it, good old sergeant-major's. 1939 Joyce *Finnegans Wake* 331 Pointing up to skyless heaven like the spoon out of sergeantmajor's tay. 1948 Partridge *Dict. Forces' Slang 1939–1945* 164 *Sergeant-major's*, a Samson-strong, love-sweet brew of tea, popularly supposed to be the perquisite of holders of that rank. 1981 J. Wainwright *Urge for Justice* I. v. 30 This tea..it damn near dissolved the spoon. A real 'sergeant major' brew. The way tea *should* be made.

Hence as *v. trans.*, to order or shout in a brusque and stentorian manner; **sergeant-majorish, -majorly** *adjs.*, characteristic of or resembling a sergeant-major.

1925 G. W. Deeping *Sorrell & Son* viii. 77 Moreover, he might pocket a sergeant-majorly share of the tips. 1926 A. Bennett *Lord Raingo* xxxvi. 168 'Bow,' said the sergeant-majorish official behind him, in a no-nonsense voice. 1931 E. A. Robertson *Four Frightened People* ii. 77 Then we heard the voice of Mrs. Mardick sergeant-majoring the truant few. *a* 1935 T. E. Lawrence *Mint* (1955) II. iii. 108 Cursing fellows forbidden to look resentful..is a sergeant-majorish trick which good corporals would not allow themselves. 1962 M. Duffy *That's how It Was* iv. 43 'She'll soon learn,' the voice sergeant-majored high above me.

serial, *a.* and *sb.* Add: **A.** *adj.* **a.** (Earlier example.)

1840 A. Brisbane tr. *Fourier's Social Destiny* xxiv. 344 Industry was developed sufficiently..to admit of the application of the Serial mechanism to it.

b. Also of a radio play: broadcast in (usu. weekly) episodes. *serial rights* (later examples).

1890 [see *right *sb.*¹ 9 f]. 1903 J. London *Let.* 10 Mar. (1966) 150 The serial right has passed out of my hands. 1933 *B.B.C. Year-Bk. 1934* 213 Serial plays were a popular innovation: and their exciting episodes seemed to have appealed to..as many grown-ups as youngsters. 1944 *R.A.F. Jrnl.* Aug. 290 The American market..still offers big money for serial rights. 1955 *Radio Times* 22 Apr. 42/1 A new serial play in six parts written for broadcasting. 1960 *B.B.C. Handbk.* 68 An increased output of serial plays and characterized documentaries. 1970 [see *film rights* s.v. *film *sb.* 7 c].

c. (Further examples.) *serial section*, each of a series of sections through tissue made in successive parallel planes; hence *serial-section* vb. trans., *serial sectioning* vbl. sb. In Computing = *sequential *a.* 2 b.

1885 A. B. Lee *Microtomist's Vade-Mecum* xxiv. 203 (*heading*) Serial section mounting. 1908 *Q. Jrnl. Exper. Physiol.* I. 129 Where the epithelium persists..serial sections show that the cleft is completely closed by it. 1948 *Gloss. Computer Terms* (U.S. Office of Naval Res. Special Devices Center: M.I.T. Servomechanisms Lab. Rep. R-138) 10 *Serial programming*, execution of complete arithmetic operations one at a time. Coding is simpler and easier to organize where simultaneous arithmetic operations are avoided. Serial programming is possible with either parallel or serial digit transmission. 1960 Gregory & van Horn *Automatic Data-Processing Systems* viii. 248 Latency time for instructions stored in serial-access memories can increase program running time enough to warrant using other arrangements for storing instructions. 1961 *Lancet* 2 Sept. 523/1 The hypothalami ..were embedded in celloidin and serial-sectioned. 1964 G. H. Haggis et al. *Introd. Molecular Biol.* v. 113 Consideration of the confusion which would result from the examination of fifty serial-section electron microscope pictures placed on top of each other. 1969 P. B. Jordain *Condensed Computer Encycl.* 449 In character-oriented memory computers, serial addition permits forming sums with inexpensive hardware... In faster, word-organized computers, parallel addition is used. 1977 *Sci. Amer.* Sept. 130/1 Serial-access and block-access memories have access times that depend on the storage location selected. 1979 *Nature* 22 Feb. 596/2 Here was a man who had pioneered..the technique of serial sectioning, which enabled palaeontologists to examine the internal structures of fossils that would never have been accessible for study.

d. *Biol.* Involving or produced by the propagation of a micro-organism or tissue by means of a series of cultures, each grown from material derived from the previous one.

1904 *Proc. Amer. Acad. Arts & Sci.* XL. 277 In investigating the persistence of the (+) and (−) characters in the individual strains, the writer has begun a number of serial cultures. 1947 *Ann. Rev. Microbiol.* I. 26 During the early period of study the original culture on serial plating continued to produce colonies about 5 per cent of which contained only z_{38} and 95 per cent contained $z_{30}z_{31}$. 1970 L. T. Morton *Med. Bibliogr.* (ed. 3) 609 Laveran and Mesnil discovered that trypanosomes could be maintained indefinitely in rats and mice by serial passage.

e. *Educ.* and *Psychol.*: *serial learning*, the learning of words, numbers, etc., as a series so that each item acts as a stimulus for the next; *serial position*, the position of items in a serial test studied for its effect on learning; hence *attrib.* as *serial-position curve*, *effect*; *serial test*, a test of ability that makes use of items in serial arrangement; hence *serial testing*. Also *serial reproduction*.

1926 H. HEAD *Aphasia* I. II. i. 149 The order in which these serial tests are applied must be varied to suit the circumstances of the case. 1926 *Jrnl. Exper. Psychol.* IX. 195 (*title*) Specific serial learning; a study of backward association. 1926 *Amer. Jrnl. Psychol.* XXXVII. 538 It is apparent..that the effects of serial position upon memorization still constitute something of an issue. 1932 F. C. BARTLETT *Remembering* vii. 173 There is some suggestion that material treated by way of serial reproduction may gain a kind of group stamp or character. 1948 E. R. HILGARD *Theories of Learning* iv. 97 (*caption*) Serial position effect in the memorization of a list of 15 nonsense syllables. 1952 MCGEOCH & IRION *Psychol. Hum. Learning* iv. 115 (*heading*) Learning as a function of serial position. *Ibid.* x. 369 The results of one series of experiments by the method of serial reproduction..are important for their bearing upon the social diffusion of information. 1962 E. R. HILGARD *Introd. Psychol.* (ed. 3) ix. 273/2 Serial learning is easier than paired-associates learning. 1971 *Jrnl. Gen. Psychol.* LXXXV. 100 RFT performance was not found to be stable..but rather changed in the direction of greater field dependence on serial testing. 1972 *Jrnl. Social Psychol.* LXXXVI. 106 For both liked and disliked names the typical serial position curve was noted with most errors occurring in the middle of the lists. 1979 A. C. CATANIA *Learning* x. 243 Another variety of intraverbal relation occurs in serial learning, the learning of a list of items in a particular order.

f. In grammatical terminology: *spec.* in certain West African languages, designating a construction consisting of a series of verbs.

1933 L. BLOOMFIELD *Language* xii. 195 Endocentric constructions are of two kinds, *co-ordinative* (or *serial*) and *subordinative* (or *attributive*). 1957 S. POTTER *Mod. Linguistics* v. 114 It [*sc.* the phrase *good men*] is a subordinate or attributive construction as opposed to such a phrase as *men and women*, which is said to be co-ordinate or serial. 1963 *Jrnl. Afr. Languages* II. II. 145 One..feature of the syntax of Twi and many other West African languages which seems to have escaped the notice of the grammar-writers is that the only possible position for an object pronoun is immediately after a verb... It is necessary to introduce an extra verb to take the extra object pronoun... This introduction of an extra verb in this way results in a *serial verbal construction*. 1971 G. ANSRE in J. Spencer *Eng. Lang. W. Afr.* 157 Many of them [*sc.* the languages of West Africa] exhibit similarities in their grammatical patterning, such as the occurrence of a sequence of verbal forms within the same sentence which has come to be known as 'serial verbal construction'. 1977 E. A. GREGERSEN *Lang. in Afr.* v. 49 A distinctive feature of many West African languages is a multiple verb construction, known in the literature as serial verbs.

g. *serial number*, a number assigned to a person, item, etc., indicating position in a series; *spec.* a number printed on a banknote or manufactured article by which it can be identified.

1935 F. W. CROFT *Crime at Guildford* xiv. 201 All these high-class cameras bore a serial number. 1938 L. M. HARROD *Librarians' Gloss.* 135 *Serial Number*, the number indicating the order of publication in a series. 1959 *Ibid.* (ed. 2) 246 *Serial Number*... 2. One of the consecutive numbers appearing in front of an entry in a bibliography or catalogue. 1960 *Bedside 'Guardian'* IX. 135 It shows a willingness to surrender but a refusal to reveal one's serial number. 1962 L. DEIGHTON *Ipcress File* i. 11 People posted to him..were..given a new serial number from the batch..reserved for Civil Servants seconded to military duties. 1968 'R. SIMONS' *Death on Display* iv. 55 Crow..took himself off to check on the serial numbers of the five-pound notes. 1971 R. K. SMITH *Ransom* (1972) III. 121 Very good field glasses for a kid... Probably stolen. He typed the serial number on the form. 1976 J. CROSBY *Snake* (1977) xxiv. 129 She paid cash with bills that had been carefully laundered... Elf doubted whether the Feds had the serial numbers on her bills but she was taking no chances.

h. *Mus.* Applied to a type of composition which takes as its starting-point an arrangement of the twelve tones of the chromatic scale. Cf. *DODECAPHONIC *a.*, *SERIES 20; *twelve-note, -tone* s.v. *TWELVE *numeral a.* and *sb.* III. c.

1947 H. SEARLE in *Penguin Music Mag.* Dec. 22 Fartein Valen, whose *Sonetto di Michelangelo*..uses a serial technique derived from Berg. 1958 *Times* 6 June 4/4 Reti considers a number of alternatives to serial tonality, which is what dodecaphonists now practise in default of the milk of the word of Schönberg. 1963 *Times Lit. Suppl.* 3 May 320/4 Most of us reserve the term 'series' for an ordered succession of notes, as in the works of Schoenberg, but do not apply it to a collection of pitches such as are found in the works of Scriabine or Debussy. Mr. Perle extends 'serial composition' to both classes of music. 1978 P. GRIFFITHS *Conc. Hist. Mod. Music* vii. 88 The plate opposite shows the opening of his [*sc.* Webern's] Symphony (1928), arranged to display the serial structure. 1982 *Sunday Times* 25 July 41/6 In his [*sc.* Eisler's] film music he made bold use of the technique of montage, juxtaposing elements from jazz, cabaret and serial polyphony.

B. *sb.* **b.** A film shown in a number of episodes; a radio or television play broadcast in (usu.) weekly episodes.

1914 R. GRAU *Theatre of Science* xi. 245 The latter arranged with the late Thomas W. Hanshew..to prepare a serial. 1939 *BBC Handbk.* 20 An interesting aspect of the year's radio-dramatic work was the development of serial plays. The serial feature, which is the backbone of American radio, had made comparatively few appearances here before 1938... Publishers..found that the 'Monte Cristo' serial caused a great demand for the novel. 1950 G. WEBB *Inside Story of Dick Barton* i. 13 One certain way of arousing interest and gaining an audience was through the medium of the radio serial. 1955 *Radio Times* 22 Apr. 21/1 *Counterspy*, the six-part serial which begins in Children's Hour on Friday. 1964 K. C. LAHUE (*title*) Continued next week: a history of the moving picture serial. 1974 *Broadcast* 22 July 14/1 There is abundant evidence that the serial, or its twin brother the series, is a popular form of TV programming. People seem to like stories in which the same characters appear and reappear.

serialism (sīə·riăliz'm). [f. SERIAL *a.* + -ISM: cf. next.] **1.** The name given by J. W. Dunne (1875–1949) to a theory of the serial nature of time, which he evolved to account for the phenomenon of precognition, esp. in dreams (see quots.).

1927 J. W. DUNNE *Exper. with Time* xxi. 153 The serialism of the fields of presentation. *Ibid.* xxvi. 206 Serialism as a theory of the Universe. 1934 *Discovery* Aug. 239/1 His theory that in dreams the dreamer appears sometimes to move out of one dimension of time into another. Serialism, as Mr Dunne terms his main principle, is a fascinating idea. 1937 *Mind* XLVI. 165 The novelty of Serialism lies in this: in a Serial Universe it is permissible to rotate the geometrically mapped-out axis of a time-dimension (T_2) until its divisions coincide with those of a time (T_1) one dimension lower. 1974 *Country Life* 7 Feb. 233/1 J. W. Dunne's theory of Serialism..that we may discover the future in our dreams.

2. A belief or assumption that every process takes place in a regular succession.

1943 C. S. LEWIS *Abolition of Man* iii. 39 Such a reply springs from the fatal serialism of the modern imagination—the image of infinite unilinear progression which so haunts our minds... We tend to think of every process as if it must be like the numeral series.

3. *Mus.* The practice or principles of serial composition.

1958 [see *ATONAL *a.*]. 1967 *Spectator* 18 Aug. 200/1 If we are to assume that atonality refers to the idiom characterising Schoenberg's works prior to his adoption of serialism..then Penderecki's Passion..does not come into this category. 1977 P. JOHNSON *Enemies of Society* xvii. 228 Serialism does not provide a workable order, at least for most listeners, because the structure is mathematical rather than aural.

serialist (in Dict. s.v. SERIAL *a.* and *sb.*). Add: **1.** (Later example.)

1902 A. BENNETT *Truth about Author* xii, in *Academy* 5 July 44/2, I found an outlet..more remunerative than the concoction of serials; and I am a serialist no longer.

2. One who holds views that accord with a serial theory; *spec.* one who learns by studying items arranged in a series.

1936 *Mind* XLV. 31 The controversy between substrativists and serialists is one of long standing. 1975 G. PASK *Conversation, Cognition & Learning* 561 Serialists learn, remember and recapitulate a body of information in small, well-defined and sequentially-ordered segments.

3. *Mus.* A composer or advocate of serial music.

1959 *Atlantic Monthly* Feb. 88/2, I do believe, however, that success will not wholly pass by the rhythmic experimenters and the 100 per cent serialists. 1962 *Times* 26 Feb. 14/7 The two pieces..illustrated the difference of outlook between the newest generation of English serialists and their predecessors. 1980 *Early Music* Apr. 253/3 Many of these works are characterized by a degree of formal organization which would delight serialists.

4. *attrib.* or as *adj.*

1936 *Mind* XLV. 31 The serialist hypothesis..seems to me beset with difficulties and obscurities. 1959 *Times* 13 Feb. 13/4 The Institute of Contemporary Arts is presenting a whole serialist programme. 1975 G. PASK *Conversation, Cognition & Learning* iv. 108 The respondent is free to learn in any way and is found to adopt a holist or serialist approach.

serialization (in Dict. s.v. SERIAL *a.* and *sb.*). Add: **1.** Also, the broadcasting of a radio or television play in serial form.

1965 *Radio Times* 18 Feb. 15/1 *The Mill on the Floss*, of which a four-part serialisation..begins tonight. 1972 *Daily Tel.* 31 Jan. 7/2 This serialisation may well prove such compulsive viewing as to create new interest in this neglected German liberal.

2. *gen.* The action or state of forming a series.

1857 H. CLAPP tr. *Fourier's Social Destiny* I. iv. 37 The Administrative unity of the Globe is nothing more nor less than the Serialization of the general interests, operations and relations of the Human Race. 1962 *Listener* 22 Mar. 513/1 The fate of man is now 'serialization'. We lose our individuality and our capacity for action by being turned into merely one term in a series which could equally well be replaced by any other term. 1966 A. MANSER *Sartre* xiii. 214 Sartre, in demanding the abolition of serialisation, seems to be asking for an impossible Utopia.

3. *Mus.* The composition of serial music.

1959 *Observer* 23. Aug 7/3 This group [of composers] practices a technique of *total serialisation*, whereby not merely notes but all elements of music (pitch, instrumentation, rhythm, volume, etc.) are used in row formation, i.e., in regular patterns. 1966 F. HOYLE *October First is too Late* xi. 126 The style of this Greek music was more akin to the key system than to the modern serialization. 1976 P. STADLEN in D. Villiers *Next Year in Jerusalem* 328 Stravinsky..turned into a serial convert in his

the equivalent of the oolitic series. 1836 W. BUCKLAND *Geol. & Mineral.* I. ix. 76 The Tertiary Series introduces a system of new phenomena, presenting formations in which the remains of animal and vegetable life approach gradually nearer to species of our own epoch. 1882 A. GEIKIE *Text-bk. Geol.* 648 The rocks of the Cambrian series present great uniformity of lithological character over the globe.

(ii) *Stratigraphy.* The primary subdivision of a system, composed of a number of stages and corresponding to an epoch in time; the rocks deposited during any specific epoch.

At the 1881 meeting of the International Geological Congress, a scheme of nomenclature was adopted in which the stratigraphical terms *group, system, series, stage* in decreasing order of comprehensiveness correspond to the terms *era, period, epoch, age* for time intervals. The *system* and its subdivisions are now regarded as the primary time-stratigraphical terms, and the use of *group* in this sense is deprecated.

1881 *Geol. Mag.* Decade II. VIII. 558 The final result of the discussions was the adoption of terms in the following order, the most comprehensive being placed first:..Series..Epoch... As equivalents of *Series*, the terms *Section* or *Abtheilung* may be used... According to this scheme, we would speak of the Palæozoic Group or Era, the Silurian System or Period, the Ludlow Series or Epoch, and the Aymestry Stage or Age. 1898 *Jrnl. Geol.* VI. 355 The faunas of the Trenton limestone, the Utica and Hudson River shales are very intimately related, and that relation should be indicated by grouping the three together as stages of a single series. 1931 *Bull. Geol. Soc. Amer.* XLII. 426 The Pleistocene or Glacial Period will be divided into epochs and ages, and the Pleistocene or Glacial system into corresponding rock terms, series and stages. 1931 GREGORY & BARRETT *General Stratigr.* x. 155 In Scotland the Upper Estuarine Series includes the Brora Coal seam, of which the roof is Callovian. 1961 *Bull. Amer. Assoc. Petroleum Geologists* XLV. 658 The term 'series' is not restricted to stratified rocks, but may be applied to intrusive rocks in the same time-stratigraphic sense. 1976 H. D. HEDBERG *Internat. Stratigr. Guide* vii. 72 The series is a unit in the conventional chronostratigraphic hierarchy, ranking above a stage and below a system. The geochronologic equivalent of a series is an epoch. A series is always a subdivision of a system; it is usually but not always broken up into stages.

b. Any group of (usu. igneous) rocks having similar forms of occurrence and petrographical characteristics.

1844 C. DARWIN *Volcanic Islands* vi. 123 Is it not more probable, that these dikes have been formed by fissures penetrating into partially cooled rocks of the granitic and metamorphic series, and by their more fluid parts,..oozing out, and being sucked into such fissures? 1892–94 *Bull. Philos. Soc. Washington* XII. 178 Since neighboring centers may be erupting different phases of the rock series at one and the same time,..the same kinds of rock may occur in different parts of the whole complex series representing the order of eruption of the rocks in one region. 1909 J. P. IDDINGS *Igneous Rocks* I. II. iii. 408 The term series should be applied to groups of rocks characterized by similarity of certain chemical or mineral constituents and by variations in others; the rocks being members of one family. Series may traverse the general system of classification in various directions. 1975 A. E. KINGWOOD *Composition & Petrology of Earth's Mantle* vii. 243 The behaviour of the orogenic series is fundamentally different from that of the tholeiitic and alkalic series.

12. a. *in series* (earlier and later examples): in *Electr.* also said of circuit components connected together so as to form a single electrical path between two points (also *transf.*); const. *with*.

1884 *Jrnl. Soc. Telegr.-Engineers* XIII. 498 If you couple two such alternate-current machines in series, they will so control each others phase as to nullify each other. 1922 *Proc. IRE* X. 249 It was necessary to use a two-electrode tube in series with the auxiliary emf. 1943 C. L. BOLTZ *Basic Radio* viii. 132 When a condenser and resistor are in series in a circuit, the charging current when a D.C. supply is switched on causes a p.d. across the resistor for a fraction of a second. 1960 *Practical Wireless* XXXVI. 412/1 In series with the key jack is filled a potentiometer VR1 which provides a useful variation of the oscillator tone. 1968 *Brit. Med. Bull.* XXIV. 250/1 Each tissue consists of two compartments connected in series. 1975 G. J. KING *Audio Handbk.* iv. 84 The two transistors are connected in series across the supply.

b. (Later examples.) Also more generally, pertaining to or involving connection in series. Also *Comb.*

1920 *Whittaker's Electr. Engineer's Pocket-bk.* (ed. 4) 227 A motor has a series characteristic when the exciting or main flux is produced by the load current (or by part of it). 1926 R. W. HUTCHINSON *Wireless* vi. 101 In the above example of resonance the capacity and inductance were in series and such a case is often referred to as series resonance. 1950 *Engineering* 6 Jan. 8/2 The noise limiter ..employs a series-diode circuit. 1957 *Practical Wireless* XXXII. 379/1 (Advt.), It is essential to use mains primary types with T.V. receivers having series-connected heaters. 1961 *Amateur Radio Handbk.* (ed. 3) ix. 257/2 Valves such as the 807 can be used in both positions in a series-modulation system from a 1000 volts supply. 1962 G. A. T. BURDETT *Automatic Control Handbk.* i. 26 The outstanding characteristic of the d.c. series motor is powerful torque at starting and also at low speeds. 1970 J. SHEPHERD et al. *Higher Electr. Engin.* (ed. 2) ix. 265 In Fig. 9.5(*a*) two mutually coupled coils are connected in series. The connexion is called series aiding, since current enters the dotted ends of the coils, which thus produce aiding fluxes. 1974 HARVEY & BOHLMAN *Stereo F.M. Radio Handbk.* iii. 51 The attenuation produced by the series insertion of a crystal into a circuit operating at a variable frequency. 1975 D. G. FINK *Electronics Engineers' Handbk.* XIII. 80 The two series-connected windings in series with the load are called gate windings.

13. b. *Phonology.* (See quot. 1952.)

1952 A. MARTINET in *Word* VIII. 13 A number of consonantal phonemes characterized by one and the same articulation will be said to form a 'series' if their other characteristic articulations can be located at different points along the air channel. Thus in English /p/, /t/, /č/, /k/,..will form a series, and so will /b/, /d/, /ǧ/, /g/. 1956 E. STANKIEWICZ *Phonemic Patterns of Polish Dialects* in *For Roman Jakobson* 521 This reduction resulted in the fusion of alveolars and palatals into a single series (ś, ź, ć, ʒ́). 1969 C. A. M. BALTAXE tr. *Trubetzkoy's Princ. Phonol.* I. iv. 125 Many languages have two apical series, one characterized by the tip of the tongue pointed upward, the other by the tip of the tongue pointed downward, instead of a single series characterized by the participation of the tip of the tongue.

15. a. *Chem.* A set of related elements or compounds, esp. a group or period of the periodic table, or a number of compounds differing successively in composition by a fixed amount; a set of elements or compounds arranged in order of magnitude of some property.

1849 *Q. Jrnl. Chem. Soc.* II. 297 (*heading*) On a new series of organic bodies containing metals and phosphorus. 1869, 1876 [see HOMOLOGOUS *a.* 3]. 1922 [see *GROUP *sb.* 3 c (ii)]. 1943 [see *electro-chemical* adj. s.v. *ELECTRO-]. 1958, etc. [see *NEPHELAUXETIC *a.*]. 1962 COTTON & WILKINSON *Adv. Inorg. Chem.* xxiv. 495 For practical purposes..the third transition series begins with hafnium ..and embraces the elements Ta, W, Re, Os, Ir, Pt and Au. 1964 N. G. CLARK *Mod. Org. Chem.* ii. 12 Members of the series may be represented by a general molecular formula, and each member differs from the next by CH_2; the paraffins have the general formula C_nH_{2n+2}. 1972 COTTON & WILKINSON *Adv. Inorg. Chem.* (ed. 3) xiii. 373 The stability of these hydrides falls rapidly along the series, so that SbH_3 and BiH_3 are very unstable thermally.

b. = *radioactive series* s.v. *RADIOACTIVE *a.* 4.

1904 [see *DISINTEGRATION a]. 1926 R. W. LAWSON tr. *Hevesy & Paneth's Man. Radioactivity* xxiv. 180 The resulting end-product of the uranium-radium series does not emit rays, and is hence stable. 1949 F. SODDY *Story of Atomic Energy* v. 50/2 The RaE changes to Radium F,..the last radio-element in the main uranium series. 1974 *Encycl. Brit. Macropædia* I. 67/2 The mass numbers of all isotopes of the so-called thorium series..turn out to be multiples of four, and the series is known as the 4n series.

16. A set of alloys or minerals having the same chemical composition except for the relative proportions of two elements that can replace one another.

1855 *Phil. Mag.* X. 249 We..prepared a series of alloys in which copper predominated. 1859 *Phil. Trans. R. Soc.* CXLVIII. 357 The alloys of a series such as those of 2 equivalents of bismuth and 1 of lead, 3 Bi and 1 Pb, 4 Bi and 1 Pb, 5 Bi and 1 Pb, all conduct the same, viz. about 1·9, the various increasing quantities of lead exercising no influence on the conductibility of the alloys. 1911 *Encycl. Brit.* XVIII. 512/2 In other groups [of minerals] the replacement may be indefinite in extent, and between the ends of the series the different members may vary indefinitely in composition. 1914 C. H. DESCH *Intermediate Compounds* vi. 50 The compound Mg_3Bi_2 has a conductivity very near that of bismuth, and the two series Mg–Mg_3Bi_2 and Mg_3Bi_2–Bi are simple conglomerates. 1971 I. G. GASS et al. *Understanding Earth* i. 17/1 The plagioclase feldspars show a slightly more complex type of ionic replacement and form the series $NaAlSi_3O_8$ (albite)–$CaAl_2Si_2O_8$ (anorthite).

17. a. *Baseball.* A set of games played on successive days between two teams. Also *World Series*: see *WORLD *sb.*

1862 *Sunday Mercury* (N.Y.) 13 July 6/3 This last game ended the series, and the players were to return this.. morning. 1906 *World* (N.Y.) 26 July 8/4 To wind up their series with the Western teams, the hilltop boys gave the Michiganders a double drubbing. 1960 *Time* 3 Oct. 67/2 The Yankees have..individual stars who can rouse themselves to greatness and win a short series by themselves. 1973 *Internat. Herald Tribune* 15 June 15/4 It was the first time in almost a month that the Mets had won two straight. And it was the first time in exactly a month that they had captured a series.

b. *Cricket.* A set of Test matches between two sides on any one tour.

1912 A. A. LILLEY *Twenty-Four Years Cricket* xiv. 195 The only Test match of the tour that had a definite conclusion was the second of the series. 1935 *Wisden* II. 1 The Australian team of 1934 arrived in this country with the knowledge that during the previous series of Test Matches in Australia they had been beaten four times. 1966 J. ARLOTT in B. Johnston *Armchair Cricket 1966* 12 The fifth—Oval—Test of that series was the first scheduled for regular eye-witness accounts on each day. 1976 *0–10 Cricket Scene* (Austral.) 5/2 Ian Chappell and Ian Redpath both gave away Test cricket, and with Edwards leaving the scene on a series before, Australia had lost their three most consistent, fighting batsmen.

18. *Physics.* A set of lines in a spectrum whose frequencies are mathematically related in a fairly simple way.

1890 *Jrnl. Chem. Soc.* LVIII. II. 674 The corresponding components of the pairs form series whose wave-numbers are functions of the successive natural numbers. 1922 [see *LYMAN]. 1952 R. W. DITCHBURN *Light* xvii. 543 These formulæ suggest that the wave numbers of all these series may be expressed as differences of a set of wave numbers which are known as spectroscopic 'terms'. 1966 WILLIAMS & FLEMING *Spectrosc. Methods in Org. Chem.* ii. 21 When more than two triple bonds are conjugated, the spectrum shows a characteristic series of low intensity bands..at intervals of 2300 cm^{-1}..and high intensity bands..at intervals of 2600 cm^{-1}. 1978 E. P. BERTIN *Introd. X-Ray Spectrometric Analysis* i. 37 X-ray spectral lines are grouped in series *K, L, M, N*; all lines in a series result from electron transitions from various higher orbitals to the indicated shell.

19. *Soil Sci.* A group of soils which are derived from the same parent material and are similar in profile, though not necessarily in the texture of the surface horizon; = *soil series* s.v. *SOIL *sb.*[1] 10.

1904 *Ann. Rep. U.S. Dept. Agric.* 1903–4 269 These types have been arranged in 31 series, in which the soils are related in point of origin. 1913 *U.S. Bureau of Soils Bull.* No. 96. 8 A soil series is named from some town, village, county, or natural feature existing in the area when it was first encountered. 1917 MOSIER & GUSTAFSON *Soil Physics & Management* viii. 79 The Cecil series include the most important and widely distributed soils of the Piedmont Plateau. 1952 L. M. THOMPSON *Soils & Soil Fertility* vi. 86 Several of the great soil groups of the United States include hundreds of series. 1970 E. M. BRIDGES *World Soils* v. 34/2 The Ettrick Association derived from Silurian greywackes and shales has six component series in the Jedburgh and Morebattle district.

20. *Mus.* The arrangement of the twelve-tone chromatic scale which is used as the starting-point of a piece of *SERIAL music; = *tone-row* s.v. *TONE *sb.* 11.

1930 *Mod. Music* VII. IV. 5 The tonal material of a composition [by Schönberg] is a *series* of Twelve tones, borrowed from the chromatic scale and grouped in a special arrangement... The word 'series' is by no means identical with the idea of 'theme'... The series is to be considered rather as a tone-complex, whose successions and intervalic relations always recur. 1940 E. KŘENEK *Studies in Counterpoint* p. viii, The primary function of the series is that of a sort of 'store of motifs' out of which all the individual elements of the composition are to be developed. 1959 *Observer* 23 Aug. 7/3 According to this new system [of musical composition], a fixed *series* or succession of the twelve notes of the chromatic scale forms a framework which is the basis of the composition. 1978 P. GRIFFITHS *Conc. Hist. Mod. Music* vii. 88 The music is ..constructed as a four-part canon, each part of which begins with a statement of the series in a different form.

21. *Eccl.* With a specifying number: a designation of one of the alternative experimental forms of service used within the Church of England since 1965.

These rites were replaced in 1980 by those printed in the *Alternative Service Book*.

1965 (*title*) Alternative Services: First Series. 1965 (*title*) Alternative Services: Second Series. 1967 *Church Quarterly Rev.* CLXVIII. 442 It is undoubtedly the rite of Series 2 which points the way forward. *Ibid.* 449 It is appropriate here to look at the third form of Series 1 in which the self-oblation is omitted from the Canon. 1971 *Churchman* LXXXV. 212 The amended text..has now been published as *Holy Communion: Series 3*. 1973 *Franciscan* XV. 169 In our worship at S. Bene't's we have moved..from Series II to Series III, using John Rutter's setting. 1977 B. PYM *Quartet in Autumn* i. 15 What would be the reaction of the congregation if Father G. tried to introduce Series Three? 1981 BARTON & HALLIBURTON in *Believing in Church* iv. 107 The Durham book, which had in fact proposed that form of invitation which became the invitation to confession in Series 2 Communion.

22. Special Comb. **series-parallel** *Electr.*, used *attrib.* with reference to combinations of series and parallel connection, esp. to denote a method of control of sets of electric traction motors in which the motors work in series on starting and are switched to parallel working when a certain speed is reached; **series spectrum**, a spectrum consisting of a series (sense *18) of lines.

1894 K. HEDGES *Amer. Electr. Street Railways* vi. 68 In the series parallel method of control, the motors are first connected in 'series'. 1903 *Trans. Inst. Naval Archit.* XLV. 182 A voltage of 220, the motors to have series parallel control. 1957 *Railway Mag.* June 427/2 The operating voltage is 500 volts d.c., with orthodox series-parallel control for the four-motor equipments. 1968 *Radio Communications Handbk.* (ed. 4) i. 13/2 This is the value of the equivalent inductance of the four coils in this series-parallel arrangement. 1922 A. D. UDDEN tr. *Bohr's Theory of Spectra* II. ii. 29 Although the series spectra of the elements of higher atomic number have a more complicated structure than the hydrogen spectrum, simple laws have been discovered showing a remarkable analogy to the Balmer formula. 1974 G. REECE tr. *Hund's Hist. Quantum Theory* vii. 100 With the aid of the *n, l, j* scheme it was possible to understand the multiplicity of the terms in the optical series spectra for atoms with one, two or three external electrons.

serif. Add: Hence **se·rifed, seriffed** *a.*

1936 *Geogr. Jrnl.* LXXXVII. 571 Any criticism..should be directed towards the lettering, particularly to the large serifed capitals indicating principal settlements. 1957 S. MORISON *Aspects of Authority & Freedom* 12 It is not impossible that seriffed letters were considered more suitable for a revered text than unseriffed ones. 1980 B. CRUTCHLEY *To be Printer* 134 Hence the danger that seriffed forms will be passed over when they could do a job as well or better and at the same time enhance our aesthetic enjoyment.

serigala (sĕri·gală). [Mal.] The Malaysian

two directions. In the one the venter and dorsum became arched; the outer consequently embraced the inner whorls and the umbilicus became narrower and even disappeared. The shell thus produced was globose in shape and is described as a sphærocone... In the other modification the whorl increased in height..; the effect of this was to produce a shell with a widely open umbilicus and having the appearance of a coiled snake. The descriptive term serpenticone is therefore applied to this type of shell. **1970** R. M. Black *Elements Palaeont.* viii. 87 *Dactylioceras*..has an evolute shell with a wide, shallow umbilicus; the whorl section is oval. It is a typical serpenticone.

serpentine, *sb.* Add: **10. c.** A lake or canal of a winding shape, esp. the one constructed in Hyde Park in 1730.

1837 W. Tayler *Jrnl.* 2 Jan. in J. Burnett *Useful Toil* (1974) II. 176 Went round Hyde Park, saw some thousands of people sliding and skaiting on the Serpentine. **1853** Geo. Eliot *Let.* 22 Oct. (1954) II. 120, I am hoping for a row..on the Serpentine, which is really almost as good as a lake. **1885** C. M. Yonge *Nuttie's Father* II. xiii. 157 He said he'd take him to the Serpentine to sail his ship. **1971** *Country Life* 2 Sept. 546/2 By 1747 he [*sc.* George Anson of Shugborough] had erected a Chinese house on a little island formed by a new canal or serpentine. **1977** P. Willis *Charles Bridgeman* iv. 96 When Bridgeman's widow in her Petition to the Lords of the Treasury asks for money for making 'the lake in Hyde Park' she is obviously referring to the Serpentine.

serpentine, *a.* Add: **3.** Also, esp. in reference to canals or lakes.

1730 *London Jrnl.* 26 Sept. 2/3 Next Monday they begin upon the Serpentine River and Royal Mansion in Hyde-Park. **1824** J. C. Loudon *Encycl. Gardening* (ed. 2) III. iv. 1011 Those wavy serpentine canals..are never mistaken for natural scenes. **1948** C. Hussey in M. Jourdain *Work of William Kent* 23 The most famous Serpentine Lake, that in Hyde Park, was ordered by Queen Caroline and is probably due to Bridgeman.

11. a. serpentine superphosphate *N.Z.*, a mixture of superphosphate and crushed serpentinite, used as a fertilizer.

1941 Elliott & Lynch in *N.Z. Jrnl. Agric.* 15 Sept. 179/1 The name serpentine superphosphate will be used in future in place of 'silico superphosphate', as it is a more accurate description of the material. It is made by mixing three parts of hot, newly-made superphosphate with one part of ground serpentine and allowing the mixture to 'mature' in heaps for several days. **1965** G. J. Williams *Econ. Geol. N.Z.* x. 143/1 Serpentinite is quarried in considerable quantity in New Zealand for the manufacture of 'serpentine-superphosphate'.

serpentinite (sə̄·ɹpĕntinəit). *Petrogr.* [f. Serpentine *sb.* + -ite[1].] A rock consisting largely of serpentine or related minerals.

1936 G. C. Selfridge in *Amer. Mineralogist* XXI. 463 The term serpentinite is suggested for rocks composed of serpentine or antigorite or a mixture of both. **1956** *Amer. Jrnl. Sci.* CCLIV. 201 Bare *roches moutonées* of serpentinite stand out from the heather. **1965** [see *serpentine superphosphate* s.v. *serpentine *a.* 11 a]. **1977** A. Hallam *Planet Earth* 137 Serpentinites are rocks originally very rich in olivine that has been entirely converted to serpentine.

Hence **serpentini·tic** *a.*

1975 *Nature* 25 Dec. 701/2 The ubiquitous association of ultrabasic (mostly serpentinitic) rocks with glaucophane-schists may be represented here by the serpentinite-dunite of Gibbs Island and the glaucophane-schists of Smith Island.

serpentinize, *v.* **2.** (Earlier example.)

1886 [see *peridotitic *a.*].

serpierite (sə̄·ɹpiᵊrəit). *Min.* [a. F. *serpierite* (A. des Cloizeaux 1881, in *Bull. de la Soc. min. de France* IV. 92), f. the name of J. B. *Serpieri*, 19th-cent. Italian engineer: see -ite[1].] A hydrated basic sulphate of copper, zinc, and calcium found as crusts or aggregates of small pale blue orthorhombic crystals.

1892 E. S. Dana *Dana's Syst. Min.* (ed. 6) 963 Serpierite... Crystals minute... Occurs on smithsonite at the zinc mines of Laurium, Greece. **1927** *Mineral. Mag.* XXI. 387 Smithsonite ($ZnCO_3$) is also tolerably abundant ..and it is in cavities of this smithsonite that the serpierite occurs. **1964** *Amer. Mineralogist* XLIX. 1145 Devillite and serpierite were identified on the basis of morphology, optics, and microchemical tests.

serpulid. Add: Also *attrib.* or as *adj.*

1935 *Discovery* Apr. 98/2 The only growths..are..two species of serpulid worms. **1963** R. P. Dales *Annelids* 15 The most specialized tube-dwellers are the sabellid and serpulid fan-worms. **1980** *Nature* 29 May 323/1 The coarse and medium sand fractions consist of rock and pelecypod fragments, aragonitic and calcitic algae, serpulid tubes and peneroplid Foraminifera.

Serrano (serā·no). Also **serrano; † Serano.** [a. Sp. *serrano* of the mountains; a highlander.] (A member of) an Indian people of southern California; the Uzo-Aztecan language of these Indians, a component of the Takic branch.

1858 *San Francisco Bull.* 5 Nov., The true native Americans of the wild forests—such as the Yumas,.. Mohaves and Serranos—predominate. **1876** *Ann. Rep. U.S. Geogr. Surveys West of 100th Meridian* III. 553 Case-inflection is formed here..by adding to nouns post-positions as suffixes:..tumuet, in Serrano, mountain. **1896** F. Boas in *Proc. Amer. Assoc. Advancem. Sci. 1895* XLIV. 261 The Serano call themselves Mā'ringayam. *Ibid.* 262 The Serano is less closely related to the other Shoshonean dialects than these are among themselves. **1907** A. L. Kroeber *Shoshonean Dialects Calif.* I. 69 The Gitanemuk..vocabulary was..obtained at Tule river reservation... The vocabulary is the first that has been printed of this dialect, although it differs but little from Serrano, which has been known for years. **1921** *Glasgow Herald* 17 May 3/8 In due time he will become a 'Serrano' as the inhabitants of the highland regions are contemptuously styled by the dwellers in the cities of the coast. **1927** D. H. Lawrence *Mornings in Mexico* 84 The serranos, the Indians from the hills, wearing their little conical black felt hats. **1946** C. McWilliams *Southern Calif. Country* 26 The Serranos and the Gabrieleno were associated with the Mission San Gabriel. **1969** J. Mander *Static Soc.* vi. 158 The *serranos* are pure Indian. They are the direct descendants of the Incas, and of the tribes subjugated by the Incas; and they speak Quechua in preference to Spanish. **1974** *Encycl. Brit. Micropædia* IX. 73/2 In the early 1970s, there were fewer than 400 Serrano proper remaining. **1977** *Language* LIII. 459 The Serrano element which L gives as *-nuk*, base /-nowk(*i*·)/, shows up as *-nuk* only in 3sg. *pə-nuk*.

‖ **serré** (sere), *a.* [Fr., pa. pple. of *serrer* to close together.] Compact, logical; constricted by grief or emotion.

a **1854** J. S. Mill *Early Draft Autobiogr.* (1961) 115 Our debates..habitually consisted of the strongest arguments & most philosophic principles which either side was able to produce, thrown often into close & *serré* confutations of one another. **1908** D. H. Lawrence *Let.* 9 Oct. (1962) I. 28 My heart is '*serré*'—I shall soon have nothing inside my chest but the spent fragments of my organ of affection. **1931** T. S. Eliot *Sel. Essays* (1932) VII. 448 Whether the transition is cogent or not, is merely a question of whether the mind is *serré* or *délié*.

‖ **sertão** (ser·tɑuň). *Geogr.* Also † **Sertam; sertao, Sertão.** Pl. **sertãos, sertões** (se·rtoiňʃ). [Pg.] The name of an arid, barren region, characterized by caatinga, in the interior of Pernambuco and neighbouring states in NE Brazil; also applied to other areas in Brazil of similar character. Also, more widely, the remote interior or outback of Brazil.

1816 H. Koster *Trav. in Brazil* vi. 77 The trees had mostly lost their leaves. I had now entered upon the Sertam, and surely it deserves the name. **1851** *Illustr. Catal. Gt. Exhib.* v. 1429/2 The cap is made in the Sertaõ (the interior) of the province of Pernambuco. **1876** *Encycl. Brit.* IV. 226/1 Except on the loftiest mountains, and on the wide *sertãos*, the vegetation of Brazil is luxuriant beyond description. **1903** W. R. Fisher tr. *A. F. W. Schimper's Plant-Geogr.* III. iii. 275 In contrast with its southern portion, the middle part of Central Brazil, the so-called Sertão district, possesses a xerophilous woodland climate. **1926** R. Nash *Conquest of Brazil* iv. 144 This Negro-Indian cross was considerable only in the lawless sertões of Matto Grosso. **1930** C. F. Jones *S. Amer.* xxxiii. 471 The physical landscape of Northeast Brazil consists of three major divisions: marginal lowlands; *sertão*, parched uplands of brush-wood and grasses..; *serra*, elevated mountain zones. **1950** E. G. Ashton *S. Amer.* iv. 51 With a rather more generous rainfall on the *sertãos*..the characteristic vegetation resembles the savanna of the Matto Grosso. **1961** *Times* 8 Aug. 9/7 President Kubitschek's decision to move the capital from Rio to the heart of the undeveloped *sertao*. **1966** K. Webb tr. *Pohl & Zepp's Latin Amer.* ii. 43 A particular vegetation type of these drier lands is the *caatinga* of the north-east, the *sertão*. This region of semi-desert has a vegetation cover made up of sparse thorn forest... *Sertão* in this sense means a particular place—the dry interior of north-east Brazil; in a more general usage *Sertão* means the sparsely inhabited backlands of the interior of Brazil. It is the equivalent of the Australian 'outback'. **1971** P. C. C. Garnham *Progress in Parasitology* ix. 191 The scene is laid in the sertão and its wide empty spaces in northern Minas Gerais.

Hence **sertanista** (-ani·sta), one engaged in activity in the *sertão*; one knowledgeable about the *sertão* and its inhabitants.

1944 S. Putnam tr. *E. da Cunha's Rebellion in Backlands* i. 45 The northern *sertanistas*..were fully a match for the bandeirantes of the south. **1973** *Daily Colonist* (Victoria, B.C.) 22 June 5/2 Francisco Meireles and the two other sertanistas (Indian experts),..flew from Belem.

Sertoli (sə̄ɹtōᵘ·li). *Anat.* The name of Enrico *Sertoli* (1842–1910), Italian histologist, used *attrib.*, with *of*, and in the possessive, to designate a type of somatic cell described by him, found in the walls of the seminiferous tubules.

1880 E. Klein *Atlas of Histol.* xxxi. 270 Inside this *membrana propria* are several layers of epithelial cells, the seminal cells... They correspond to the germ-cells of Sertoli. **1888** *Buck's Handbk. Med. Sci.* VI. 522/2 Next to the tunica comes a layer... This layer contains two kinds of cells: First, the large Sertoli's cells, as they may be called after their discoverer. **1901** *Gray's Anat.* (ed. 15) 1004 The supporting cells, or cells of Sertoli. **1930** W. Bloom *A. A. Maximow's Text-bk. Histol.* xxxi. 595 The Sertoli cells in a seminiferous tubule with active spermatogenesis are very infrequent in comparison with the number of spermatogenic cells. *Ibid.*, The sustentacular cells or the cells of Sertoli. **1965** Lee & Knowles *Animal Hormones* iii. 65 In many vertebrates oestrogens are secreted by the testis, probably by the Sertoli cells, but whether this hormone has a physiological action is still in doubt.

serum. Add: **2. b.** *serum jaundice, rash, urticaria*; **serum broth** *Bacteriology*, a broth (*broth *sb.* 1 b) containing added serum.

1934 L. E. H. Whitby *Med. Bacteriol.* (ed. 2) ix. 109 Salicin serum broth..is an excellent selective medium for *Streptococcus pyogenes*. **1979** H. McLeave *Borderline Case* v. 53 The tubes of serum broth and tissue culture supplied by WHO. **1945** *Jrnl. Amer. Med. Assoc.* 28 July 911/1 (*heading*) Transmission experiments in serum jaundice and infectious hepatitis. **1908** *Glasgow Med. Jrnl.* LXIX. 277 The most obvious and constant features of the symptom-complex are the skin eruptions, or serum rash. **1905** *Practitioner* May 665 The duration of serum urticaria varies within somewhat wide limits.

c. Used *attrib.* (with or without a following hyphen) to denote (the concentration of) substances in the serum.

1958 J. B. Miale *Lab. Med.—Hematol.* viii. 372 When the iron deficiency is caused by inadequate utilization of iron a *decreased* serum iron concentration and *increased* storage of iron in the tissues are usually seen. **1959** *Jrnl Physiol.* CXLVI. 353 All the serum proteins are capable of binding thyroxine to some extent. **1960** Leavell & Thorup *Clin. Hematol.* iv. 187 The normal level of serum bilirubin on cord blood is considered to be from 0·8 to 2·6 mg. per 100 ml. **1961** *Lancet* 26 Aug. 492/2 Treatment of atherosclerosis..may best be directed towards improving fat tolerance as well as reducing specific serum-lipid fractions. *Ibid.* 2 Sept. 499/1 The associations of high serum-cholesterol levels with coronary heart-disease do not necessarily indicate any *causal* relationship. **1962** *Ibid.* 22 Dec. 1293/1 There was no correlation on admission between the degree of weight deficit and the serum-transaminase levels. **1977** J. F. Fixx *Compl. Bk. Running* i. 5 In Southern California not long ago, fifty-eight doctors were given physical exams... More than half had high serum lipid levels.

d. Special Combs.: **serum agglutination,** agglutination of antigens by components of serum; **serum disease,** serum sickness; **serum hepatitis,** a viral hepatitis transmitted by injections of blood serum; **serum reaction,** serum sickness; **serum sickness** [tr. G. *serumkrankheit* (C. von Pirquet 1903, in *Wiener klin. Wochenschr.* XVI. 1244/2)], anaphylactic reaction to injected foreign serum.

1914 *Jrnl. Hygiene* XIV. 264 In order to be able to observe serum agglutination and acid agglutination with the same bacterial extract the bacilli from a 24 to 48 hrs. agar slope were washed off with 10 c.c. of distilled water and the resulting emulsion was centrifuged. **1970** W. H. Parker *Health & Dis. in Farm Animals* xii. 161 An infected cow will give a positive reaction to the serum agglutination test. **1908** *Glasgow Med. Jrnl.* LXIX. 277 (*heading*) The serum disease in man after single and repeated doses. **1951** Serum disease [see *serum sickness* below]. **1943** *Lancet* 16 Jan. 83/1 (*heading*) Measles serum hepatitis. **1946** *Med. Clinics N. Amer.* XXX. 1408 'Virus hepatitis'..includes both infectious hepatitis..and homologous serum hepatitis. **1971** *New Scientist* 25 Mar. 676/1 This work promises a screening method which should help eliminate the danger of serum hepatitis developing after blood transfusions. **1905** *Practitioner* May 664 In cases of relapse, or of a second attack of diphtheria, the serum reaction may be very marked. **1916** *Arch. Internal Med.* XVIII. 497 Most of the cases of serum sickness occurred during convalescence from pneumonia. **1951** B. Schick tr. *von Pirquet & Schick's Serum Sickness* i. 5 We have abandoned the expression 'serum exanthema'... In its place we have proposed the name 'serum disease' or 'serum sickness'. **1970** W. H. Parker *Health & Dis. in Farm Animals* xxi. 289 The most important practical significance of anaphylaxis in farm animals is that in the treatment of disease 'serum sickness' may follow the repeated use of serum from another species.

servagerous, var. *savagerous *a.*

serval. Add: Also *serval cat.*

1919 F. W. Fitzsimons *Nat. Hist. S. Afr.: Mammals* I. 142 These Serval Cats often break cover. **1933** N. Douglas *Looking Back* II. 340 Somebody shot a serval cat.

servaline, *a.* Substitute for etym.: [a. mod.L. specific name of *Felis servalina* (W. Ogilby 1839, in *Proc. Zool. Soc.* 94), f. Serval + -ine[1].] and add: used to designate a serval in a darker colour phase than usual, with less obvious markings, once considered a separate species. Also as *sb.* (Later examples.)

1915 G. Aylmer in *Proc. Zool. Soc.* 154 The differences between the Servals and the Servaline Cats are of no systematic importance. **1964** L. S. Crandall *Managem. Wild Mammals in Captivity* 364 Specimens with markings reduced almost to dots were once distinguished under the name of servaline cat. **1970** Dorst & Dandelot *Field Guide Larger Mammals Afr.* 138 The servaline is a mere colour phase.

servant, *sb.* Add: **1.** *servant of all work* (earlier example); *servants' hall* (earlier examples).

c **1702** C. Fiennes, *Journeys* (1947) I. 55 Just behind the hall is the Servants hall. **1785** G. Washington *Diaries* 12 Mar. (1925) II. 349 In a line with the East end of my Kitchen, and Servants' Hall. **1819** *Morning Post* 14 Jan. 1 (Advt.), Wanted immediately, a servant of all-work, where others are kept for the children.

5. *servant-class, -problem, -trouble, -wife.*

1876 C. M. Yonge *Womankind* xxvii. 236 To raise the notions of the servant and factory-classes about marriage. **1920** 'O. Douglas' *Penny Plain* xiii. 134 Mawson..belonged to that fast disappearing body, the real servant class. **1916** A. Bennett *These Twain* III. xvii. 361 The servant problem had been growing acute. **1973** R. Lewis *Blood Money* viii. 117, I suppose you see it as somewhat anachronistic—a young girl 'in service'... But you won't be here to discuss the servant problem, Inspector Crow. **1859** Geo. Eliot *Let.* 26 Feb. (1954) III. 26 The servant-trouble seems less mountainous to me than it did the other day. **1977** A. Wilson *Strange Ride of Rudyard Kipling* vi. 272 Carrie Kipling['s]..diaries are entirely mundane..financial transactions..servant troubles. **1906** J. Joyce *Let.* 3 Dec. (1966) II. 199 Servant-wife blows her nose in the letter and lawyer confronts the mistress.

serve, *sb.*[2] Add: **3.** In fig. phr. *to give* (someone) *a serve*: to deal roughly with; to criticize or reprimand sharply. *Austral. slang.*

1974 Stackpole & Trengrove *Not just for Openers* 104, I continued to give Snow a bit of a serve. **1977** *Australian* 1 June 3 He was glad to be leaving and he would be giving the country a serve in an unnamed English newspaper if it was willing to pay enough for his views. **1980** *Daily Tel.* 25 June 17/3 He debunks the fashionable diets but also criticises doctors for not speaking out about bad eating habits. 'I give them a little serve,' he said, using an Australian expression.

serve, *v.*[1] Add: **III. 33.** (Later *fig.* examples.)

1884 L. A. Tollemache *Safe Studies* 359 The violent recoil against materialism which..has induced many good..persons to sell their scientific birthright and to serve tables. **1906** F. Pollock *Let.* 28 Dec. in *Pollock-Holmes Lett.* (1942) I. 136 Such men should not be put to serve the tables of university routine.

35. d. Phr. *serve-yourself*, used *attrib.* of a shop, restaurant, etc., where the customer serves himself.

1937 M. Hillis *Orchids on your Budget* iii. 50 Those serve-yourself emporiums..often have simple ones [*sc.* evening wraps]..which..won't look very different from the costly one on your neighbor. **1949** E. S. Gardner *Reluctant Witness* in *Case of Crying Swallow* (1974) 166 Mugs handed him a photograph. It showed a young woman standing in a serve-yourself grocery store. **1971** *Guardian* 2 Jan. 8/5 There are plans for..bold new excursions into what some experts see as the coming serve-yourself hotel era.

IV. 42. b. Also in phr. *serve-yourself*, used *attrib.* of foods or meals which one serves to oneself.

1971 M. Lee *Dying for Fun* ii. 19 His host, left to himself, would have taken him to a little piece of America in London, with serve-yourself fried chicken. **1976** *Newmarket Jrnl.* 16 Dec., A serve-yourself buffet with a choice of several hot and cold dishes is the ideal solution to entertaining over the festive season.

43. c. (Earlier example.)

1793 J. Macdonell *Jrnl.* 15 Aug. in *Five Fur-Traders of Northw.* (1933) 101 Our Bourgeois came up with us and ordered each man a dram, which I served out to them.

V. 47. b. *serve(s)* (*you, etc.*) *right*: also as *attrib. phr.*

1935 H. Straumann *Newspaper Headlines* i. 29 Lastly there is the *Daily Worker*, the Communist paper, with its serve-him-right attitude. **1946** N. E. Orchard in W. S. Knickerbocker *Twentieth Cent. Eng.* 164 Another little girl was sent off by herself when she needed punishment and made to read the family genealogy. Today when she visits the cemetery where her forebears lie,..she walks between the tombstones, a 'serves-you-right' expression on her face. **1977** *New Yorker* 15 Aug. 66/2 The widespread serves-them-right judgment that greeted New York's misfortune.

servery. Add: (Later examples.) Also, = *serving-hatch* s.v. Serving *vbl. sb.* 3.

1942 G. Mitchell *Laurels are Poison* vi. 54 Kitty returned..to get her own tea from the Servery. **1960** E. W. Hildick *Jim Starling & Colonel* ix. 75 They finally lined up in front of the servery. **1974** *Times* 3 May 11/3 The dividing unit acts as a servery, with drawers on the dining side for cutlery and linen.

Servian, *sb.* Add: **b.** The Serbian language.

1808 C. Stower *Printer's Gram.* 287 (*heading above the alphabet*) Servian. **1842** *Penny Cycl.* XXII. 127/2 The Servian was employed as a written language for the first time by Dositheus Obradovich. **1885** [see *Croatian *sb.* and *a.*]. **1900** H. H. Chadwick in *Indogerman. Forsch.* XI. 168 The *-à-* was probably accented, as in Servian.

Servian, *a.*[2] (Earlier example.)

1839 *Dublin Rev.* Aug. 87 This method of election was a manifest reaction and encroachment upon the Servian constitution.

service, *sb.*[1] Add: **I. 5. b.** Also, the Air Force and intelligence departments.

II. 12. a. *active service* (earlier example).

1799 *Times* 1 June 3/2 General Moreau..caused the administrations that were unfit for the active service of the war..to be removed back towards France.

IV. 19. b. Also in phr. *for services rendered* (orig. *Mil.*).

[**1916** *Times* 24 July 9/4 The King has approved the issue of a silver badge to..men..who on account of age or..wounds or sickness caused by military service have..been discharged from the Army... The badge is in the form of a circle... The circle bears the words 'For King and Empire—Services rendered', and circumscribes the Imperial cipher.] **1933** *Radio Times* 14 Apr. 75/2 The war was newly over... Everywhere you saw the little silver badge 'For Services Rendered'. **1938** M. Allingham *Fashion in Shrouds* xx. 378 That's where Mazarini used to pay his thugs for services rendered on the race-course. **1976** W. H. Canaway *Willow-Pattern War* xviii. 189 As a *quid pro quo* for services rendered in another context, the Americans made their information available to Bonn.

V. 27. d. In the restaurant-car of a railway train, on a ship, etc.: the serving of a meal at one of a number of separate sittings, as *first service*, etc.

1914 Kipling in *Nash's Mag.* July 484/1 Here is a fragment from the restaurant-car... 'I will give you the number, sar, at the time—for places at the first service.' **1926** E. Hemingway *Sun also Rises* 86 Leaving the dining-car I asked the conductor for tickets for the first service. **1932** G. Greene *Stamboul Train* III. ii. 140 Late for the last service Dr. Czinner came down the restaurant-car.

30. a. Also applied to other facilities, such as electricity, waste disposal, etc., esp. provided for domestic use. Freq. *pl.*

1963 *Ann. Reg. 1962* 450 Both make use of clusters of towers, mostly containing services, to give vertical emphasis and to free interior space. **1963** *Gloss. Gen. Building Terms* (*B.S.I.*) 24 *Services*, installations for (1) the introduction into and distribution within a building or structure of water, air, gas, liquid fuel, electricity, heat or other source of energy (2) the disposal of waste from a building or structure or (3) fire-fighting within a building or structure. **1979** *Nature* 1 Feb. p. xiii/2 The overhead service booms may provide such services as gas/electricity/water/vacuum/lighting.

31. b. Expert advice or assistance given by manufacturers and dealers to secure satisfactory results from goods made or supplied by them; *spec.*, the provision of maintenance or repair work to ensure the efficient running of a motor vehicle, etc.; a routine operation of examination and maintenance performed on a motor vehicle, etc.

1919 W. H. Berry *New Motoring* xxiv. 183 The need of a better service system for motorists has often been emphasised... There is ample room for a big development of a scheme for rendering practical car service. **1925** *Morris Owner's Man.* 93 (*caption*) Whenever you see this hanging sign you know that it denotes an establishment where Morris Service can be obtained. **1930** *Economist* 6 Sept. 454/2 There are obviously wide undeveloped markets, of which Britain should be able to secure at least a proportion..if English manufacturers can compete with those of America in the matter of 'service'. **1947** *E. African Ann. 1946–7* 24 'Service After Sales' is as much a Ford feature here as it is in other parts of the world. **1960** I. Jefferies *Dignity & Purity* iii. 44 The flat was paid up and the car never needed any service at all. **1974** J. le Carré *Tinker, Tailor, Soldier, Spy* xx. 168 Take your car in for a service at your local garage. **1977** *Western Morning News* 30 Aug. 10/4 (Advt.), Backed by our largest combine stores and skilled after-sales service.

c. *Broadcasting.* The supply of programmes by a particular broadcasting station.

1927 *B.B.C. Handbk. 1928* 32 Broadcasting Service. By 'service' is meant providing this public..with at least one programme a day. **1933** *B.B.C. Year-bk. 1934* 167 In August last the West Regional, the fourth of the new Regional Stations, began radiating a full programme service. **1949** *Radio Times* 15 July 9/1 *Variety Fanfare* returns to the air... Bowker Andrews first introduced this show in the North of England Home Service. **1957** [see *carry *v.* 40 b]. **1977** *Church Times* 9 Dec. 8/3 Now that the 7.30 half-hour on Radio Four has been taken over for other things, there is a gap from 10.15 a.m. until 11 p.m. on Sunday on this service without a word being said about religion.

d. pl. *Econ.* The section of the economy that supplies needs of the consumer but produces no tangible goods.

1936 *Discovery* Nov. 355/2 The distinction between capital goods and current goods is, of course, one of the most important in the whole of economics, but the remarkable growth in the number of those engaged in 'services', now estimated at 40 per cent. of the working population, is not so generally realised. **1941** *Economist* 22 Feb. 235/1 The British public spent almost £900 millions in 1937 on services, excluding entertainments, rent, rates and taxes. The largest constituents of the total were travel, domestic service, public utilities, hotels and restaurants. **1948** [see **current goods*]. **1965** *McGraw-Hill Dict. Mod. Econ.* 466 *Services*, the component of the gross national product that measures the output of intangible items. Services include such items as telephone service, railway, bus, and air transportation, private education, and radio and television repair. **1972** *Accountant* 17 Aug. 211/2 Manufacturing costs are of diminishing importance in an economy in which services are a major part of the whole.

e. *pl.* The provision of petrol, refreshments, etc., for motorists in buildings constructed near to or beside a motorway or other major road; the group of buildings themselves.

1967 *Autocar* 28 Dec. 6/2 It was a relief to see the 'Services 1 mile' sign. **1968** *Listener* 1 Aug. 134/3 We drove back on Sunday night along the endless M1, punctuated only by almost identical airport-international-style 'Services'. **1975** C. Storr *Chinese Egg* xxix. 193 She was passing the Heston Services, she'd be at the Henley exit in another quarter of an hour.

VI. 34. b. *service-side* (later examples).

1898 [see *half-court* s.v. *half- II. i]. **1963** *Times* 13 May 3/5 Lawrence's policy of holding the service-side at all costs, often using the side galleries to achieve the change of ends.

36. (Earlier example.)

1822 M. McSwiney *Let.* 6 Jan. (1972) II. 348 This debt I understand was due for the service of a *bull*.

VII. 37. a. (Further examples.) Also, belonging to the Air Force; military (opp. civilian). (ii) in *pl.*

1917 *Times* 1 Nov. 3/5 The Services Club, 19, Stratford-place, W., have just acquired an adjoining house. **1926** E. Hemingway *Sun also Rises* xvii. 212 Always slept with a loaded service revolver. **1929** *Star* 21 Aug. 9/1 Private and even Service pilots have appeared near to the station at prohibited times. **1933** *Radio Times* 14 Apr. 75/2 The slang of the hour was Service slang. **1937** A. Christie *Dumb Witness* i. 13 Emily Arundell's people ..were what is known as 'all service people'. **1945** *Manch. Guardian Weekly* 21 Dec. 323/3 There was an all-night Services canteen at Victoria station. **1945** *Tee Emm* (Air Ministry) V. 35 The Service scale of Tropical Kit, issued to airmen. **1954** *Economist* 11 Sept. 3/1 A trained test pilot is less likely to come to harm in a new and temperamental machine than an enthusiastic service flyer. **1958** *Listener* 6 Nov. 719/2 In 1947 Montgomery became Chief of the Imperial General Staff... He had to co-operate on equal terms with the Service Chiefs of the other two fighting services. **1973** J. Rossiter *Manipulators* iv. 48, I was so horribly humiliated—my service career finished. **1977** *R.A.F. News* 22 June–5 July 18 (*caption*) Alan won trophies for both rapid and snap shooting using the service-issue self-loading rifle. **1980** C. Smith *Cut-Out* xiii. 92 Photographs of her husband in service dress and holding a swagger cane.

b. (Further examples.)

1902 A. Bennett *Grand Babylon Hotel* i. 13 Jules walked to the service-door. **1909** H. G. Wells *Tono-Bungay* I. i. 28 One came down the main service stairs.. and..one went through a red baize door. **1919** *Chambers's Jrnl.* Jan. 57/2 What is known as the 'service' door..is likely to become popular for hotels, if not for private houses, all over the world. **1933** *Archit. Rev.* LXXIII. 24/2 (*caption*) The service-bay between the kitchen and the restaurant. **1950** T. S. Eliot *Cocktail Party* I. iii. 65, I shall take the precaution Of leaving by the service staircase. **1955** W. Gaddis *Recognitions* II. viii. 653 A transmitter? Mr. Inononu demanded at the head of the service stairs. **1956** H. Kurnitz *Invasion of Privacy* xiv. 92 She went out the back, through the service entrance. **1961** M. W. Barley *Eng. Farmhouse & Cottage* II. i. 64 His home consisted of a hall with a chamber over it, a ground floor chamber in which he slept, and four service-rooms, kitchen, buttery, milkhouse and cheese-house. **1976** H. Nielsen *Brink of Murder* xii. 106 The manager conceded that there was a service door and.. accompanied them upstairs on the service stairway. **1976** *Washington Post* 19 Apr. c20/1 (Advt.), Eat in kitchen with service bar. **1978** R. Ludlum *Holcroft Covenant* xxxix. 452, I could say that someone fitting his description was seen leaving by the service entrance. **1979** M. Soames *Clementine Churchill* xxiv. 395 A service-lift to the floor below.

c. Of or pertaining to services (sense *31 d), as *service industry, occupation, sector, trade.*

1941 *Economist* 18 Jan. 65/2 The very considerable increase in the standard of living..explains the growth of the 'service', as distinct from the 'productive' industries since the last war. **1959** *Times* 5 Sept. 10/3 The rapid expansion of service occupations—administration, the professions, retailing, entertainment and numerous health and welfare services. **1966** *Listener* 5 May 642/2 There is to be a selective employments tax to help get workers out of service industries into the factories. **1970** S. L. Barraclough in I. L. Horowitz *Masses in Lat. Amer.* iv. 150 This is reflected in the rapid increase of employment in the 'service' sector. **1970** G. Jackson *Let.* 17 Apr. in *Soledad Brother* (1971) 221 The new slavery..places the victim..in the case of most blacks in support roles inside and around the factory system (service trades), working for a wage. **1979** G. Wagner *Barnado* ii. 18 All the service industries took on extra labour at the beginning of the season.

38. service alley, a road or passage giving access to the back of a row of houses; **service area,** (*a*) the area in which broadcast transmissions can be received distinctly; (*b*) a space adjoining a house for the accommodation of dustbins, etc.; (*c*) an area providing petrol, refreshments, etc., for motorists; **service car** *Austral.* and *N.Z.*, a small motor-coach for public transport; **service ceiling** *Aeronaut.* (see *ceiling *vbl. sb.* 6 b); **service charge,** a charge made (additional to that for the food, etc.) for services rendered, esp. for service in a hotel or restaurant; **service club** *N. Amer.*, an association of business or professional people which seeks to promote community welfare and goodwill; **service contract,** a contract of employment; a business agreement between contractor and customer, normally one guaranteeing the maintenance and servicing of equipment; **service engineer,** an engineer engaged on the maintenance and servicing of equipment; **service flat,** a flat in which domestic service and other facilities are provided at a charge included in the rent; hence **service flatlet; service mark** orig. *U.S.*, a name or designation, protected by law, used by a commercial undertaking to distinguish a service offered by it from the services of competitors; **service module** *Astronautics*, a separable section of a spacecraft, esp. one in the U.S. Apollo series, containing the main

engine and other supporting equipment; **service plate** *U.S.*, a large ornate plate which marks a place at table and on which dining plates, etc., are set during the first courses; **service record,** the record of service of a soldier, employee, etc.; **service reservoir,** a (usu. small) reservoir filled from an impounding reservoir at times of low demand to supplement the supply to the local area at time of high demand, so as to reduce the necessary capacity of the conduits from the impounding reservoir; **service road,** a subsidiary road giving access to houses, shops, etc., away from a main road; **service routine** *Computers*, = *utility routine* s.v. *utility *sb.*; **service station,** an establishment providing service and maintenance for motor vehicles; more recently, merely = *filling-station* s.v. *filling *vbl. sb.* 4.

1922 J. Hergesheimer *Bright Shawl* 11 The street outside was narrow..once no more than a service alley for the larger dwellings back of which it ran. **1974** P. McCutchan *Call for Simon Shard* xi. 98 You take the back, Alan. There'll be a service alley. **1927** *B.B.C. Handbk. 1928* 62 Rival schemes can be compared as between those which base themselves upon very high power and believe that service areas can be over 100 miles radius, and those which cover the country from more centres and with therefore somewhat less power. **1956** *Good Housek. Home Encyl.* (ed. 4) 327/2 The service area behind or at the side of the house. **1958** *News Chron.* 25 Nov. 7/4 The great London–Birmingham Motorway... A road remarkably straight, soaring over 150 specially built bridges; with flyovers, flyunders; with service and eating areas every 12 miles. **1970** J. Earl *Tuners & Amplifiers* iii. 74 If you have in mind trying for more distant stations..outside the normal service area, then you will certainly need a tuner of top sensitivity. **1971** M. McCarthy *Birds of Amer.* 76 The yard in back of their house was all flagged, with..a 'service area' containing garbage cans. **1980** J. McNeil *Spy Game* xiv. 141 They careered into the narrow, curving entrance to a service area... He cruised the crowded car park. **1924** R. Rees *April's Sowing* ii. 19 I'd have gone in the service car. **1933** *Bulletin* (Sydney) 26 July 20/1 The drivers of service cars Outback. **1948** V. Palmer *Golconda* xxi. 173 He piled his traps into the dusty service-car and climbed up beside the driver. **1965** S. T. Ollivier *Petticoat Farm* xii. 166 The road was open, they knew, because the service cars were running through. **1920** *Flight* XII. 980/2 Principal characteristics of the Martin torpedo 'plane are:—Service ceiling..12,000 ft. **1944** H. F. Browne *Aeroplane Flight* vi. 97 At the service ceiling the aeroplane can fly only at speeds between 180 m.p.h. and 240 m.p.h. **1978** J. D. Anderson *Introd. to Flight* vi. 237 The service ceiling represents the practical upper limit of steady, level flight. **1929** *Post Office Guide* July 157 Any sum not exceeding £10 may be withdrawn by telegraph if the depositor pays the cost of the telegrams and a service charge of one shilling. **1955** R. Chandler *Let.* 7 Feb. (1981) 382, I paid a service charge on the bill... This service charge is supposed to take the place of tipping. **1977** B. Bainbridge *Injury Time* ii. 23 She looked at the bill and was astonished at the service charge. **1926** *Daily Colonist* (Victoria, B.C.) 16 July 2/5 He intended this time to take a party of executive officers of service clubs. **1978** J. L. Hensley *Killing in Gold* (1979) v. 65 He'd joined.. one of the service clubs, Lions, Rotary, or Kiwanis. **1948** *Rep. Native Laws Commission 1946–48* (Dept. Native Affairs, S. Afr.) 26/1 The service contract duplicate, which is issued to a Native in urban areas where registration of service contracts in terms of the Natives (Urban Areas) Consolidation Act is in force, is likewise regarded by them as a pass. **1958** *Listener* 20 Nov. 824/1 It would be wrong to abandon altogether the distinction between service contracts and other types of contract. **1975** *Petroleum Economist* Aug. 299/2 In 1968, a service contract was agreed with Elf/ERAP by which the French company would have a share of any production which resulted. **1958** *Practical Wireless* XXXIV. 67/1, I do feel pity for the poor service engineer. **1980** *West Lancs. Even. Gaz.* 28 Nov. 25 Sweda International..has vacancies for Service Engineers to service and maintain their range of electro-mechanical and electronic machines. **1922** W. J. Locke *Tale of Triona* iv. 38 She found herself the lucky tenant of a little suite in a set of service flats in Victoria Street. **1973** 'E. Ferrars' *Foot in Grave* iv. 70 Being solitary in my service flat suits me, and that's how I'm going on. **1960** M. Spark *Bachelors* x. 141 Those who were conducting love affairs in service flatlets found it convenient that the maids did not come in with their vacuum cleaners on Sundays. **1945** *Business Week* 30 June 86/2 A separate register would be authorized for 'service' marks to identify services rather than merchandise. This register would include 'names, symbols, titles, designations, slogans, character names, and distinctive features of radio or other advertising used in commerce'. **1949** H. Bennett *Trade-Marks* 130 Under the Lanham act service marks will be registrable for the first time. The new law thus gives great protection to a category of names and designations whose trade importance was not recognized under the old law. **1959** *Listener* 31 Dec. 1147/1 Trade marks, trade names, the so-called 'service-marks' of organizations such as transport undertakings and radio stations. **1973** S. A. Diamond *Trademark Problems & how to avoid Them* i. 2 Trademarks are not limited to goods; they may be used for services, like transportation, insurance, entertainments and advertising. Strictly speaking, they are then called 'service marks'. **1961** *Space Technol.* Oct. 41 (*caption*) Service module. **1962** *Manned Space Flight Program of N.A.S.A.* (U.S. Congress Senate Comm. on Aeronaut. & Space Sci.) 131 The service module..will contain the propulsion used for midcourse guidance corrections, for emergency abort situations, and for lunar takeoff. **1968** *Times* 16 Dec. 7/4 Before reentry, the command module holding the astronauts will separate from what is called the service module, the section of the spacecraft containing the main engine and power supplies. **1974** Service module [see *retro-fire *sb.*]. **1929** *Woman's Home Compan.* Apr. 67/2 The service plate has been removed and a..fish plate has taken its place. (The individual place..is never left without a plate before it.) **1934** J. B. Priestley *Eng. Journey* vii. 223 A very queer American custom, that of having what are known as 'service plates', which are never loaded with food but are placed before guests between courses, to be looked at and admired... These luxury plates were usually very ornate and would be specially made, here in the Potteries, for each customer. Some Americans liked to have a picture of their college on their plates. **1977** H. Fast *Immigrants* I. 49 A maid placed a plate of crab meat and mayonnaise on his service plate. **1918** E. S. Farrow *Dict. Mil. Terms* s.v., When an enlisted man is detached from his company, his service record will be forwarded by endorsement to his new commanding officer. **1923** J. D. Hackett *Labor Terms in Managem. Engin.* May, *Service Record*, a summary of all the facts necessary for appraising the worth of an individual to an employer. **1981** J. B. Hilton *Playground of Death* x. 123 There's his service record... He was a brave man. **1869** *Bradshaw's Railway Man.* XXI. 266 The service reservoir of the Ashton waterworks. **1967** J. H. Stephens *Water & Waste* iii. 49 The purpose of the service reservoir, and its close relative, the water tower, is to allow for the peaks in demand. **1921** H. Foston *At Front* 16 Unmetalled portions of the 'service' road. **1935** *Times* 30 Dec. 13/6 There are signs in this neighbourhood that the future development will be the right one of groups of houses set back and approached by service roads. **1976** *Southern Even. Echo* (Southampton) 12 Nov. (Advt. Suppl.) 7/6 In quiet and well screened service road, a most attractive and compact 3-bedroom bungalow. **1954** *Computers & Automation* Dec. 21/1 *Service routine*,..a routine designed to assist in the actual operation of the computer. **1969** P. B. Jordain *Condensed Computer Encycl.* 451 The most common service routines are the input/output programs... Other service routines perform such services as program loading, common calculations.., tracing, memory dumps, tape dumps, and the like. **1921** *Sci. Amer.* Dec. 135/3 Each..pipe at the service station is provided with a plug... Each automobile comes into the service station charged with static. **1925** *Morris Owner's Man.* 61 Do not forget that Messrs. Lucas have for your benefit, Service Stations in the following towns. **1935** *Economist* 7 Dec. 1132/2 The tendency..for manufacturing companies to transfer their service stations away from the centre of London..has.. lessened the convenience to the..motorist of running his car direct to the maker's service station for minor repairs. **1940** R. Chandler *Farewell, my Lovely* vii. 52 He gassed up there and the service station kid recognized him. **1977** *Lubricants Business* (Shell Internat. Petroleum Co.) 7 Supermarkets are increasingly competing with service stations for motor oil sales.

service, *sb.*[2] **4. service-berry,** (*b*) substitute for def.: a N. American tree or shrub of the genus *Amelanchier*, belonging to the family Rosaceæ and bearing clusters of white flowers followed by small, dark-coloured berries; also, the fruit of this tree or shrub; (earlier example).

1784 F. Asbury *Jrnl.* 31 July (1821) I. 370 The child he fed with..sawice berries.

service (sə̄·ɹvis), *v.* [f. Service *sb.*[1]] **1.** *trans.* To be of service to; to serve; to provide with a service.

1893 R. L. Stevenson *Catriona* I. xvi. 178 If I am to service ye the way that you propose, I'll lose my lifelihood. **1948** J. Steinbeck *Russian Jrnl.* (1949) 15 Airports are so far from the cities they supposedly service. **1955** *News of North* (Yellowknife, N.W.T.) 18 Nov. 1/5 A new town house, available to water and sewer service, would be assessed at a much higher rate than duplicate property in a part of the town not serviced this way. **1969** D. Widgery in Cockburn & Blackburn *Student Power* 139 It is unlikely that a radical Executive would be able to.. service the entirely different attitude of the apolitical small colleges. **1974** R. Adams *Shardik* lviii. 518 How many permanent camps or staging-forts would be needed to service a regular trade-route?

2. To perform routine maintenance or repair work on (a motor vehicle or other piece of equipment). orig. *U.S.*

1926 *Amer. Speech* II. 112/2 The automobile dealer says: 'Run the new car five hundred miles at twenty or less an hour, then have it thoroughly serviced with grease and oil.' **1930** *Bookman* Dec. 398 Probably the greatest cost in Television will be that expended for servicing the equipment. **1935** A. P. Herbert in *Punch* 27 Feb. 236/1, I denounce, Comrades, the foul new verb 'to service', an invention, I believe, of someone in the motor-trade. **1949** 'G. Orwell' *Nineteen Eighty-Four* II. 131 She enjoyed her work, which consisted chiefly in running and servicing a powerful but tricky electric motor. **1958** *Listener* 23 Oct. 655/2 Vehicles—whether moving, parked, unloading or being serviced—have already taken charge of the present ground level. **1978** R. Lewis *Uncertain Sound* vi. 154 Your car was serviced on the Thursday.

3. To pay interest on (a debt).

1942 *Sun* (Baltimore) 15 Jan. 2/1 Secretary of Interior Ickes announced today that interest payments on outstanding bonds of the Philippine Government would continue to be met and serviced through the United States Treasury. **1952** *Times* 1 Aug. 9/2 Nobody oversea will ever accept the idea that a company which has all the pesetas it could want should be bankrupted in Spain.. because the Spanish Government has not allowed it to buy sterling to service a sterling debt. **1975** *Daily Tel.* 21 Oct. 17/7 Unless there is a huge Federal loan guarantee by the end of next month the city will not be able to meet its payroll, let alone service its debts.

4. = Serve *v.*[1] 52; also of a man, to have sexual intercourse with (a woman).

1961 in Webster. **1966** [see *post coitum* s.v. *post Latin preposition*]. **1973** W. H. Canaway *Harry doing Good* II. v. 188, I knew a feller that married a twin. Identical. He was getting worn out..till he found out he was servicing both of these twins. **1976** T. Heald *Let Sleeping Dogs Die* iv. 78 One dog could presumably service several bitches in a day.

5. a. To supply (a person) *with* something. **b.** To process.

1969 *Daily Tel.* 6 Oct. 9/6 It is a proper function of local theatres to service their audiences with the latest in world taste. **1971** *Ibid.* 27 May 14/8 We'll be able to service the retailer with merchandise much faster than it can possibly be shipped in from overseas. **1971** D. Potter *Brit. Eliz. Stamps* vi. 70 A Post Office First-Day Cover service was available, and no charge was made if the envelope carried sufficient postage. The complete set was serviced for the cost of the envelope and the stamps. **1973** *Daily Tel.* 16 Apr. 2/2 This year building societies have been without funds to service all the mortgages required, and sales are being cancelled.

Hence **se·rviced** *ppl. a.*, provided with service or services; maintained.

1938 C. Himes *Headwaiter* in *Black on Black* (1973) 150 Over by the elevator where the room service was stationed, a waiter lounged indolently by a serviced table. **1968** *Globe & Mail* (Toronto) 13 Feb. 31/6 (Advt.), Serviced industrial land Northwest Metro bargain price. **1978** R. V. Jones *Most Secret War* xliv. 419 Each pilot could draw a serviced aircraft, probably a different one from that which he last flew, for each new operation.

serviceability. Add: (Later examples.) Also of machinery: the capacity to be maintained or repaired; reliability.

1913 *Chambers's Jrnl.* June 478/1 The claims of reliability, serviceability, and efficiency have been established. **1942** *R.A.F. Jrnl.* 3 Oct. 25 He looks at the Situation Map, the serviceability blackboard, and other items of visual intelligence. **1946** *Happy Landings* July 3/3 Drills connected with the serviceability of the engine ..are clearing the supercharger gears and exercising propellers in flight. **1961** B. Fergusson *Watery Maze* xiii. 319 It had been calculated at Quadrant that the 'serviceability rate'—that is, the number which one could actually count on for use on D-Day—would be 90 per cent for LSTs and 85 per cent for the rest. **1981** S. Dunmore *Ace* II. ii. 162 The shortages affected serviceability.. damaged aircraft had to be cannibalized to provide spares.

se·rviceman. Also with capital initial, hyphened, and as two words. [Service *sb.*[1]]

1. A man who serves, or has served, in the armed forces.

1899 (*title*) Constitution of the Service men of the Spanish war (Columbus, Ga.). **1917** *Hansard Commons* 4 May 608 (*heading*) Disabled service men. **1918** *Sheffield Daily Tel.* 18 Sept. 2/6 Discharged soldiers should be represented on all committees that have to deal with the training of discharged service men. **1945** *Daily Express* 14 Aug. 3/7 Duty-Free cigarettes sent to Service men in Europe may be officially rationed soon. **1959** *Listener* 23 July 119/2, I go about a good deal, talking to trade unionists and to work-people generally, to managers, to civil servants, to servicemen, and to students and teachers of all kinds. **1976** M. Green *Children of Sun* (1977) viii. 366 Only 750 Old Etonian servicemen died in this war... In the first one..1,150 died.

2. A man who maintains and repairs equipment (*service *sb.*[1] 31 b).

1961 in Webster. **1969** D. E. Westlake *Up your Banners* xxviii. 195 Our serviceman wouldn't make the installation at night, sir. **1970** *Which?* May 143/2 All except the Bendix needed at least one visit from a serviceman before they were working properly. **1974** *Sci. Amer.* Jan. 120/3 Beginners will find that neighboring radio hams and television servicemen can pass along useful construction tips.

Similarly **se·rvicewoman.**

1945 *Observer* 21 Oct. 6/1 (Advt.), For thousands of Service men and women the order is 'carry on'. **1972** J. Williams *Home Fronts* v. xv. 255 The historic Norfolk House in St James's Square was given over to servicewomen from overseas.

servicer (sə̄·ɹvisəɹ). [f. *service *v.* + -er[1].] One who services.

1973 *Houston* (Texas) *Chron.* (Suppl.) 14 Oct. 8/2 A better way to collect the oil from driveway auto servicers is being sought. *a* **1974** R. Crossman *Diaries* (1976) II. 406 Many of the owners are people of substance who have servicers for their vehicles, chauffeurs so to speak, who look after them as a whole-time occupation.

servicing (sə̄·ɹvisiŋ), *vbl. sb.* [-ing[1].] **1.** The action of maintaining or repairing a motor vehicle, etc. Also *fig.*

1935 *Economist* 7 Dec. 1131/2 Recognised dealers with facilities for the..'servicing' of the vehicles sold. **1956** *Good Housek. Home Encycl.* (ed. 4) 143/2 Servicing and repairs which can be carried out by the handyman of the house. **1959** I. Jefferies *Thirteen Days* x. 156 'Where's the bint, John?'..'I left her at the hospital for servicing and overhaul.' **1970** *Motoring Which?* July 91/1 All three needed routine servicing every 3,000 miles. **1978** D. Devine *Sunk without Trace* 253 There was the car... Why did she put it in for servicing that particular week?

2. The action of paying interest on a debt.

1939 *Times* 2 Mar. 13/2 Financial circles hope that an arrangement will be reached in the meantime for the continued servicing of foreign loans. **1946** *Sun* (Baltimore) 7 Oct. 5/2 Servicing of these two issues has required $647,507.50 a year.

3. The action of providing a service.

1944 *Ann. Reg. 1943* 90 Most of their reciprocal aid.. had therefore been 'servicing', taking the form of providing transportation, accommodation, airfields,..local supplies to American forces. **1946** J. S. Huxley *Unesco* ii. 58 The general educational servicing of the public in the fields of science and culture.

4. *attrib.* and *Comb.*

1930 *Economist* 31 May 1207/1 The unclassified group of trades—which includes road transport and most of what Americans term 'servicing' industries—contributed 20.08 per cent. **1944** *R.A.F. Jrnl.* Aug. 256, I joined the Servicing Commando unit at the Transit Camp. **1955** *Univ. of Virginia News Let.* 15 June 1/5 Growth of the servicing-type industries, such as retail and wholesale trade, utilities, business and repair services. **1968** [see *Poisson 1]. **1978** R. V. Jones *Most Secret War* xxxix. 360 At Zempin we hoped to have the aircraft and its servicing installations isolated.

servigrous (səɹvəi·grəs), *a.* *U.S. dial.* Also **savigrous, sevigrous, survigrous.** [Orig. unknown: freq. associated with a dial. pronunc. of *vigorous*, and cf. *savagerous *a.*] Fierce, severe, tough, vigorous. Hence **servi·grously** *adv.*

1835 A. B. Longstreet *Georgia Scenes* 227 'Pretty *sevigrous*, but nothing killing yet,' said Billy Curlew, as he learned the place of Spivey's ball. **1888** 'C. E. Craddock' *Keedon Bluffs* 88 He's a servigrous jumper, sure! *Ibid.* 215 The mos' servigrous singer they hed. **1890** T. N. Page in *Harper's New Monthly Mag.* Dec. 114/1 She so savigrous I tolt her I 'ain' nuver had nobody to prevaricate nuttin' 'bout me. **1901** *Nashville* (Tenn.) *Banner* 28 Oct. 4/3 The New York Sun..employs that expressive provincialism of the Southern mountaineer, 'servigrous'... The Sun says Mr. William Travis Jerome is 'servigrous' in the kind of oratory he uses. **1913** H. Kephart *Our Southern Highlanders* xiii. 294 Survigrous (ser*vi*-grus) is a superlative of vigorous (here pronounced *vi*-grus, with long *i*): as 'a survigrous baby', 'a most survigrous cusser'. **1928** M. Chapman *Happy Mountain* vii. 70 The Preacher must have been a rip-snorting and most survigrous young-un before he caught religion. **1938** C. H. Matschat *Suwannee River* 81 Ary fowkses knows as how onion juice, rubbed in servigerously, sprouts hair like weeds after a rain, iffen so be ye stand in the sun.

serving, *vbl. sb.* **3.** *serving-mallet* (earlier *transf.* example), *spoon*; **serving cart** *U.S.*, a small trolley from which food and drink may be served.

1969 *Sears Catal.* Spring/Summer 9/2 Serving cart... Removable tray. **1978** *Neiman Marcus Christmas Bk.* 98 'Le Gourmand' is a serving cart and more. Made of solid maple with butcher block top that slides back to reveal a stainless well for ice or storage. **1850** H. Melville *White Jacket* II. xliv. 296 Ten of us Waisters mean to club together, and buy a serving-mallet boat. **1960** M. Spark *Bachelors* x. 155 'What!' said Marlene, holding the cold peas in the serving-spoon suspended. **1978** M. Russell *Daylight Robbery* xix. 184 Mrs Braithwaite dropped the serving spoon upon her plate with a clatter.

servo (sə̄·ɹvo), *sb.* [The first element of Servo-motor (and *servo-mechanism) used substantively.] **1. a.** A servo-mechanism or servo-motor.

1910 *Engineering* 14 Jan. 56/2 A patent 'Servo' regulator valve is fitted in the dome [of the boiler]. **1924** *Motor* 7 Oct. 448/3 The servo actually trebles the braking effect usually obtainable for a given pedal pressure. **1948** I. A. Greenwood in I. A. Greenwood et al. *Electronic Instruments* viii. 220 A convenient classification of servos may be made in accordance with their uses, the principal examples of which are 'position servos' and 'velocity servos'. **1959** *Times* 1 Sept. 12/2 The..company have now introduced a vacuum-hydraulic brake servo which can be fitted easily to cars and light motor vehicles and is claimed to make a vast improvement in braking. **1966** *Electronics* 17 Oct. 108 To test a servo that controls a radar antenna, the antenna must be moved. **1971** *Sci. Amer.* July 120/3 The first truly automatic servo..was the fantail, a small windwheel mounted at right angles to the main sails and geared to turn the entire top of the windmill. **1976** B. Jackson *Flameout* (1977) iv. 28, I put electronic sensors on the wing tips and tail to respond to phugoid oscillations and dampen yaw. The control servos respond to the input from the sensors.

b. *transf.* = *servo-mechanism b.

1953 P. A. Merton in G. E. W. Wolstenholme *Spinal Cord* (Ciba Foundation Symp.) 249 The stretch reflex servo, which is so obviously concerned with maintaining length and not tension, must necessarily have receptors which record length. **1969** *Proc. Roy. Soc.* B. CLXXIII. 156 Voluntary movement could be initiated by the α route without sacrificing the advantages of a servo.

2. *attrib.* and *Comb.*, as *servo actuation, -actuator, assistance, loop*; *servo-actuated, -assisted, -driven, -operated* adjs.; **servo-a·mplifier,** the part of a servo-mechanism that responds to the small error signal and delivers a corresponding large signal to drive the servo-motor; **servo brake,** (*a*) a vehicle brake whose application is assisted by the momentum of the vehicle; (*b*) a brake that is operated by a servo-mechanism; hence **servo-braking** *vbl. sb.*; **servo flap** *Aeronaut.* = *servo tab* below; **servohydrau·lic** *a.*, both servo-controlled and hydraulic; **servo-mu·ltiplier,** a device for separately multiplying each of several voltages by a single voltage, the former being applied to a set of potentiometers on the same shaft as a potentiometer controlled by a servomotor receiving the latter voltage; **servo system** = *servo-mechanism a; also *transf.*; **servo tab** *Aeronaut.*, a tab, operated directly by the pilot, which gives rise to aerodynamic forces that assist in moving the main flap.

1961 *Hovering Craft & Hydrofoil* Oct. 18/2 The ultimate servo-actuated mechanical device may take ten years. **1959** K. Henney *Radio Engin. Handbk.* (ed. 5) xxv. 9 Deviations of the airplane about the yaw axis result in rotation of the heading stator with respect to the rotor and a signal to the rudder servoamplifier, causing servo-actuation of the rudder surface and the servo loop action described above. **1970** *Times* 4 Sept. (Aviation Suppl.) p. iv/4 Savings will also accrue from the replacement of the common mechanically coupled control systems in the aircraft by electronic systems with electro-hydraulic servo-actuators operating various aircraft control surfaces. **1946** *Radar: Summary Rep. & Harp Project* (U.S. Nat. Defense Res. Comm., Div. 14) 143/2 *Servo-amplifier*, the amplifier of power impulses in a servo system. **1947** *Electronic Engin.* July 215/1 In a characteristic arrangement, a valve circuit, known as the 'servo amplifier', controls the rotation of an electric motor. **1959** K. Henney *Radio Engin. Handbk.* (ed. 5) xxv. 8 The signal from the servoamplifier is applied to the elevator servomotor which actuates and drives the elevator surface. **1966** *McGraw-Hill Encycl. Sci. & Technol.* XII. 199/2 The servoamplifier is often electronic but may be a magnetic amplifier, a relay type of amplifier, or any combination of these types. **1929** *Times* 2 Nov. 4/7 A pedal applies the rearmost Euro [brakes] with vacuum servo assistance. **1972** *Guardian* 28 Jan. 3/4 (Advt.), To stop, there's a dual circuit four-disc system with servo assistance. **1951** *Automobile Engineer* May 199/1 (*heading*) Servo-assisted hydraulic brakes. **1976** *Field* 18 Nov. 1021 (Advt.), All four alloy wheels enjoy servo-assisted disc brakes. **1924** A. W. Judge et al. *Mod. Motor Cars* II. vii. 118 The Hispano-Suiza servo-brake controls are shown in perspective arrangement. **1926** *Encycl. Brit.* II. 980/2 This led to the development of so-called servo brakes, in which the momentum of the car was utilised to reinforce the effort of the driver in applying the brake. **1951** *Automobile Engineer* Feb. 60/1 (*heading*) Duo servo brakes. **1924** *Motor* 28 Oct. 707/1 We are now able..to give the first complete description of the servo-braking system adopted. **1947** *Proc. IRE* XXXV. 450/2 It is possible to divide one voltage by another using a servodriven potentiometer. **1966** M. Woodhouse *Tree Frog* xxi. 153 One small directional dish [antenna] which..I guessed was servo-driven. **1929** *Rep. & Mem. Aeronaut. Res. Comm.* No. 1262. 2 The control column is attached directly to the servo flap by means of wires. **1935** *Aircraft Engin.* VII. 303/1 During the past year or so there has been a growing tendency towards the use of small servo flaps on elevators and rudders either for trimming or for balancing. **1972** *Physics Bull.* Aug. 492/2 Fulmer Research Institute has recently installed a Mand servohydraulic testing facility to meet the increasing demand for testing larger specimens under widely varying rates of loading. **1980** *Jrnl. R. Soc. Arts* May 319/2 Facilities include a £350,000 Schenk servo hydraulic rig for testing structural fabrications to destruction. **1946** *Radar: Summary Rep. & Harp Project* (U.S. Nat. Defense Res. Comm., Div. 14) 143/2 *Servo loop*, that collection of elements in a servomechanism which measures the error in the quantity to be controlled and applies a correction tending to reduce that error to zero. **1953** P. A. Merton in G. E. W. Wolstenholme *Spinal Cord* (Ciba Foundation Symp.) 247 The stretch reflex is a feedback or servo loop, the feedback being negative. **1978** R. Jansson *News Caper* 10 Even with all the damping in the servo loops the controls leaped out of my hands... We flopped around the sky. **1952** G. A. & T. M. Korn *Electronic Analog Computers* vi. 245 In the case of servomultipliers, probably the best solution..is to make sure that the follow-up and multiplying potentiometers are all loaded by equal resistances. **1965** A. W. Langill *Automatic Control Systems Engin.* II. xviii. 503 Although the servomultiplier is normally employed to form the product of two analog voltages, the system is also used extensively in the generation of arbitrary functions. **1928** *Rep. & Mem. Aeronaut. Res. Comm.* No. 1171. 9 (*heading*) Theory for aerofoil R.A.F. 28 with servo operated flap. **1947** *Proc. IRE* XXXV. 444/1 Servo-operated torque amplifier. **1946** *Radar: Summary Rep. & Harp Project* (U.S. Nat. Defense Res. Comm., Div. 14) 143/2 *Servo system*, a mechanical, frequently electromechanical, system for transmitting accurate mechanical position from one point to another by electrical or other means. The position is corrected by feeding back an error signal. **1947** *Electronic Engin.* July 215/2 All quantities may be converted to a consistent electrical basis by employing a servo system. **1964** *Language* XL. 219 The fact that some sort of neurological servosystem does monitor encoding cannot be doubted. **1971** *Engineering* Apr. 45/2 A servo-system can be used to restore the balance after change of displacement. **1939** F. K. Teichmann *Airplane Design Man.* xiii. 207 If the tab can be controlled from the cockpit, it may be used to operate the larger surface and is then called a control-tab or a servo-tab. **1975** L. J. Clancy *Aerodynamics* xvi. 551 Another related device is the so-called servo-tab, which is designed to provide all the hinge movement required to deflect the control.

Hence as *v. trans.*, to control or operate by a servo-mechanism; **se·rvoed** *ppl. a.*

1971 *New Scientist* 12 Aug. 359/2 Ferguson solved this problem with their ingenious double-sided vacuum servo, which servos the brakes on and servos them (and the driver's foot) off. **1978** *Nature* 20 Apr. 704/2 The mapping operation was achieved by driving the telescope in declination at the maximum servoed rate of 130′ min⁻¹ back and forth across the galactic plane.

Servo-. Add: Also *Servo-Turkish* adj.

1897 E. A. Bartlett *Battlefields of Thessaly* iii. 72 The last Servo-Turkish campaign, that of 1876. **1914** W. G. Lawrence in *Home Lett. T. E. Lawrence* (1954) 509 The Servo-Turkish war.

servo control. [f. *servo *sb.*] **1.** † **a.** *Aeronaut.* An aircraft control using a servo tab (see *servo *sb.* 2). *Obs.* **b.** *gen.* = *servo-mechanism a.

1928 *Rep. & Mem. Advisory Comm. Aeronaut.* No. 1187.4 There is no spring force on the servo control. **1934** *Jrnl. Aeronaut. Sci.* I. 155/1 The device now known as the servo control flap is a development of the 'Flettner Rudder' which was invented by the designer of the rotor ships. **1947** *Jrnl. Inst. Electr. Engineers* XCIV. IIA. 184/2 No output is produced in the bearing channel, and the feed to the corresponding servo control is consequently zero. **1951** *Instruments* XXIV. 650/1 Servocontrols for the steel industry require..relatively high power level. **1977** *Proc. R. Soc. Med.* LXX. 208/2 As with all tools, the servo-control incubator is only as good as the persons who use it.

2. The use of a servo-mechanism to assist with the control of a system; the action or practice of controlling a system by means of a servo-mechanism.

1929 *Rep. & Mem. Aeronaut. Res. Comm.* No. 1262. 1 If no limit be set to the force which a pilot can exert then the maximum rolling moment possible with servo control is always less than that without it. **1953** P. A. Merton in G. E. W. Wolstenholme *Spinal Cord* (Ciba Foundation Symp.) 247 (*heading*) Speculations on the servo-control of movement. **1977** *Proc. R. Soc. Med.* LXX. 207/1 The principle of servo-control is that the instrument senses the baby's temperature and adjusts the thermal environment automatically.

Hence **se·rvo-control** *v. trans.*, **se·rvo-controlled** *ppl. a.*

1935 *Rep. & Mem. Aeronaut. Res. Comm.* No. 1652. 1 There have been several occurrences of flutter of servo-controlled rudders. **1948** I. A. Greenwood in I. A. Greenwood et al. *Electronic Instruments* viii. 217 Practically all heavy antiaircraft artillery..is servo-controlled by the output of some type of computer. **1972** *Physics Bull.* Apr. 232/2 The gratings are ruled in a very uniform layer of aluminium deposited on to an optically flat substrate.. using a large precision ruling engine servocontrolled by laser interferometers. **1977** *Proc. R. Soc. Med.* LXX. 207/2 The disadvantage of a servo-controlled incubator is that the temperature of the baby as a clinical guide to serious conditions..is lost.

se·rvo-me:chanism. Also without hyphen and as two words. [f. *servo *sb.* + Mechanism.] **a.** A powered mechanism in which a controlled motion is produced at a high energy or power level in response to an input motion at a lower energy level; *esp.* one in which feedback is employed to make the control automatic, and generally comprising a measuring device, a servo-amplifier, and a servo-motor.

1926 *Encycl. Brit.* II. 980/2 Some cars..were fitted with an hydraulic servo mechanism. **1935** *Jrnl. R. Aeronaut. Soc.* XXXIX. 794 The [de-icing] device..comprises two essentials, the ice detector and the servo-mechanism which controls the supply of alcohol. **1959** *Times Lit. Suppl.* 24 Apr. 245/2 All self-regulating machines are servo-mechanisms, in which one part of the machine regulates the operation of the other. **1960** *Times* 29 Jan. 3/1 (Advt.), The applicant should have a good knowledge of servomechanism theory. **1973** *Sci. Amer.* Nov. 129/1 Our seismographs and tiltmeters operate in tunnels several hundred feet underground and are provided with servomechanisms that automatically compensate for temperature changes. **1977** D. Fry *Homo Loquens* vii. 91 For centuries firearms..operated on the ballistic principle of the boy who throws the ball; in the last decades, feedback control has been introduced into this area by the employment of servo-mechanisms.

b. *transf.* A non-mechanical system that is characterized by self-regulating feedback, esp. in *Genetics* and *Physiol.* Also *fig.*

1953 J. S. Huxley *Evolution in Action* ii. 46 Natural selection..provided a genetic servo-mechanism to regulate the mutant back towards normality in its effects. **1958** *Times Lit. Suppl.* 17 Jan. 34/3 The human or animal body is a remarkable example of servo-mechanisms, in which the expenditure of a small amount of energy is used to release a large amount of energy, and in which a balance..is maintained by 'feed-back' processes. **1964** M. McLuhan *Understanding Media* I. iv. 41 This extension of himself by mirror numbed his perceptions until he became the servomechanism of his own extended or repeated image. **1971** *Nature* 10 Dec. 325/1 Natural selection is not a chance process but a feedback servo-mechanism between the gene pool and the environment. **1976** *New Yorker* 15 Nov. 157/1 Their basic assumption is that speech can best be understood as the output of a servomechanism—that is, a device with built-in feedback loops which continuously monitors its own performance and modifies its activity accordingly.

Hence **servo-mecha·nical** *a.*

1947 *Proc. IRE* XXXV. 770/2 The..servomechanical type of control may be preferable where slow drifts over long periods of time are encountered. **1966** S. Beer *Decision & Control* xvi. 418 In this servomechanical model there is a special facility for identifying precisely what is self-regulating about the system. **1977** *Lancet* 5 Nov. 955/2 The finding of only partial suppression of A.C.T.H. with normal levels of cortisol is difficult to understand in conventional servomechanical concepts.

servo-motor. Add: Also **servomotor, servo motor.** More widely, any device used as the motive element in a servo-mechanism. (Later examples.)

1932 *Jrnl. Iron & Steel Inst.* CXXV. 642 The apparatus ..makes use of the dilatations of the test bar to regulate the temperature of the furnace by means of a servo-motor. **1933** *Electrician* 16 June 796/2 The servo-motor for controlling the vanes is located between the turbine and generator shafts, so that the coupling flanges of the respective machines form the lower and upper servo-motor covers. **1951** *Automobile Engineer* May 199/1 A mechanically operated transmission brake functions as a servo-motor to apply pressure to the master cylinder of a hydraulic braking system. **1976** *Gramophone* Sept. 510/1 A DC servo motor powers the movement of the arm.

‖ **servus** (zeˑrvus), *int.* [Ger., a. L. *servus* servant.] An informal greeting or farewell used in Austria and southern Germany.

1893 E. R. Pennell *To Gipsyland* ii. 33 All..drank to me in the wine of their country, and cried aloud their '*Servus! Viva! Eljeu!*' **1966** R. E. Pickering *Himself Again* iv. 33 'Servus,' he said shortly. I ordered a coffee. **1978** H. McInnes *Prelude to Terror* viii. 89 As for her hotel in Vienna..the Sacher..everyone tugging a forelock and saying '*Servus!*'

sesame. Add: **c.** *sesame-seed* (later examples).

1972 *House & Garden* Feb. 98/2 Swiss fondue.. Sesame seed sticks..make a happy accompaniment. **1973** *Time Out* 2–8 Mar. 27/4 They also have..sesame seed rolls with really thick fillings for around 13p. **1978** *Nagel's Encycl.-Guide: China* 380 Rolls covered with sesame seeds are eaten at the same time.

Sesotho (sesūˑtu). Also **Sesuto, Sesutu.** [Bantu, = 'language of the Sotho'.] A south-eastern Bantu language spoken by members of the *Sotho people. Also *attrib.* or as *adj.*

1846 J. C. Brown tr. *Arbousset & Daumas's Narr. Tour N.-E. of Cape Good Hope* xxiii. 251, I spoke Sesuto, a dialect which the chief of the place.., also understood. **1871** J. Mackenzie *Ten Years North of Orange River* 492 They [*sc.* clicks] are found in other African languages, as in Zulu and Kaffir, and a few words in Sesuto. **1894**, etc. [see *kitchen Dutch* s.v. *kitchen *sb.* 7]. **1916** J. Buchan *Greenmantle* ix. 120, I spoke rapidly in Sesutu, for I was afraid the captain might know Dutch. **1939** tr. *E. N. Marais's My Friends the Baboons* vii. 81 From the struggling mass there arose constant laughter, mingled with Sesutu curses of the grossest kind. **1953** P. Lanham *Blanket Boy's Moon* iv. iii. 208, I am a Mosotho; I love Lesotho; I love my language Sesotho. **1975** J. McClure *Snake* x. 138 Marais made a point of thanking her in Sesotho, the only Bantu language he spoke. **1980** —— *Blood of Englishman* ix. 86 Zondi was fluent in Afrikaans ..and Sesuto.

Sesquesahamock, obs. var. *Susquehannock.

sesqui-. Add: **1. b. sesquiteˑrpene, -teˑrpenoid,** any terpene having the formula $C_{15}H_{24}$; any simple derivative of such a compound.

1888 Sesquiterpene [in Dict.]. **1922** *Nature* 16 Feb. 226/2 The principal constituents identified are safral, camphor, pinene, sesquiterpenes, eugenol, and alcoholic bodies. **1966** *New Scientist* 8 Dec. 576/2 The substance.. is derived from farnesol, one of the basic sesquiterpenes, vegetable substances..which imitate the function of the juvenile hormone. **1976** *Nature* 5 Aug. 487/2 The essential oils D-bornyl acetate, α- and β-santalol and several plant sesquiterpene hydrocarbons have been shown to induce sexual excitement in male American cockroaches. **1981** *Jrnl. Chem. Soc.* [illegible] (*heading*) Sesquiterpenoids. [illegible] Evidence for a nine-membered ring in caryophyllene. **1976** *Nature* 5 Aug. [illegible] Cadinol, a sesquiterpenoid biogenetically related to germacrene D, has been isolated from a plant and shown to be a stimulant for both male and female cockroaches.

sesquicentenary (se:skwisentīˑnări). [f. Sesqui- + Centenary *sb.*] A one-hundred-and-fiftieth anniversary; a festival celebrating this; = Sesquicentennial *sb.*

1961 in Webster. **1969** *New Scientist* 20 Nov. 389/1 The Cambridge Philosophical Society has just celebrated its sesquicentenary. **1978** *Times* 9 Jan. 9/7 This weekend's two 'Mainly Schubert' concerts..were of a kind to make one wish that his centenary celebrations, this year is the sesquicentenary of his death, came around more often.

sesquicentennial, *a.* and *sb.* For *U.S.* read chiefly *U.S.* and add: **B.** *sb.* (Later examples.)

1931 [see *Sesqui- 1 g]. **1976** *Times Lit. Suppl.* 12 Mar. 278/2 In 1926 the sesquicentennial [of Adam Smith's *Wealth of Nations*] was celebrated more modestly with a series of lectures given in the new home of economic orthodoxy, the London School of Economics. **1978** *Nature* 8 June 421/2 This year is the sesquicentennial of the announcement by Wöhler that marked the birth of synthetic organic chemistry.

sesquioxide. Add: Hence **se:squioxiˑdic** *a.*

1906 *Proc. R. Inst.* XVII. 102 Sesquioxidic mordants. **1932** [see *red *a.* 17 c].

sesquiplane (seˑskwiplēⁱn). Now *Hist.* Also ‖ **sesquiplan.** [ad. F. *sesquiplan*, f. *sesqui-* Sesqui- + *plan* Plane *sb.*³] A biplane having one wing of surface area not more than half that of the other.

1921 *Flight* 29 Sept. 650/2 The Nieuport-Delage 'Sesquiplans' are to all intents and purposes monoplanes, but with a small plane covering-in the wheel axle as in some of the German Fokkers. **1921** *Aeroplane* 5 Oct. 293/2 Why the machine is called the 'Sesquiplan' is even a greater mystery than the name of the 'Bamel'. *Ibid.* 19 Oct. 348/2 (*heading*) What is a sesquiplane? **1930** *Flight* 17 Jan. 115/2 Its most unusual feature was that, although a *sesquiplane*, its top wing was smaller in span and chord than the lower wing. **1939** C. H. L. Needham *Aircraft Design* I. vi. 74 The sesquiplane arrangement, in which one wing, generally the top, has roughly twice the area of the lower wing, is a compromise which enables the advantages of the biplane structure to be combined with somewhat improved aerodynamic characteristics. **1960** C. H. Gibbs-Smith *Aeroplane* II. 220 The Coanda sesquiplane of 1910. There has recently arisen some controversy about this machine... Until recently it has been accepted as an all-wood sesquiplane, with cantilever wings, powered by a 50-h.p. Clerget engine driving a 'turbo-propulseur' in the form of a large but simple ducted air-fan. **1981** Andrews & Morgan *Supermarine Aircraft since 1914* 174 The superstructure was of sesquiplane form, that is with the lower wing only one third the area of the top wing.

sessile, *a.* Add: **2.** Also in extended use.

1917 M. Webb *Gone to Earth* xiii. 118 People remained in a sessile state over tea for a long time. **1926** T. E. Lawrence *Seven Pillars* (1935) 7 The current of tribal movements..sessile or nomad. **1930** Auden *Poems* 56 No chattering valves of laughter emphasised Nor the swept gown ends of a gesture stirred The sessile hush. **1971** *Guinness Bk. Records* (ed. 18) 169/2 The longest recorded push of a normally sessile object is of 411 miles in the case of a wheeled hospital bed.

2*. *Cryst.* Of a dislocation in a crystal: unable to migrate through the lattice; fixed.

1949 F. C. Frank in *Proc. Physical Soc.* A. LXII. 202 Glide is prevented by a large restoring force... Such a dislocation will be called 'sessile', in contrast with 'glissile' dislocations—those which are capable of glide. **1966** C. R. Tottle *Sci. Engin. Materials* iv. 101 Frank described one form of sessile dislocation, in which an aggregate of vacant lattice sites collapses to form a loop of dislocation surrounding a disk of stacking fault. **1973** J. G. Tweeddale *Materials Technol.* I. v. 111 In the latter case they lock together (forming a sessile dislocation) and become very difficult to separate.

3. sessile oak, *Quercus petræa*, which has stalkless acorns; = Durmast.

[**1838** J. C. Loudon *Arboretum et Fruticetum Britannicum* III. 1736 (*heading*) The sessile-flowered Oak.] **1906** Elwes & Henry *Trees Gt. Brit.* II. 291 Sessile or Durmast Oak... More regular branching, resulting in a denser crown of foliage. **1971** *Country Life* 23 Dec. 1772/1 The lighter soils and hills were covered by the sessile oak, with acorns pressed against the twigs, and leaves on long stalks.

session, *sb.* Add: **2. a.** Also, a business period on the Stock Exchange and other commercial markets.

1928 *Daily Mail* 25 July 19/3 At second session Tin cash £217 15*s.* to £217 17*s.* 6*d.* **1981** *Times* 1 May 20/2 Leading industrials enjoyed one of the best sessions for some time.

3. d. (Earlier and later examples.) Now in general use, of both the school and the university year. Also, a portion of the day during which classes are given.

1714 J. Morice *Let.* 2 Aug. in W. C. Dickinson *Two Students at St. Andrews* (1952) 53 Alexander Sharp..being a Double Bajan with Mr Pringle last Session. **1862** G. Meredith *Let.* 23 Dec. (1970) I. 180, I presume that if I send to Bankers at Norwich, according to direction, before [illegible] Children go to school at seven o'clock in the morning and stay until eleven; then there is a break until two in the afternoon, the second session commencing at two and going on till five. **1923** *Leader* 7 July 1 The college is reopened [illegible]. **1976** *Billings* (Montana) *Gaz.* 27 June 2-D/6 Temporary shelter became a problem... Ricks College in Rexburg, a junior college on high ground, has opened its doors until its summer session starts.

4. Cf. *quarter-sessions 1.

6*. *transf.* [Senses not necessarily dependent upon the notion of 'sitting'.] A period of time given to or set aside for the pursuit of a particular activity: **a.** *gen.*

1920, etc. [see *bull *sb.*¹ 3 b]. **1970** [see *rap session* s.v. *rap *sb.*¹ 2 c]. **1976** *Cumberland News* 3 Dec. 24/5 A short session of dominoes followed.

b. in which musicians perform music, esp. for recording. Also, the music so recorded. Cf. *jam session* s.v. *jam *sb.*⁴ 3; *recording session* s.v. *recording *vbl. sb.* 5.

1927 [see *recording session* s.v. *recording *vbl. sb.* 5]. **1929** *Melody Maker* Apr. 381/1 The trouble is due to inferior musicians being engaged for this session. **1947** G. Beall in R. de Toledano *Frontiers of Jazz* vi. 87 He is present on most of the records, however, taking part in the recording session although by some means his part would not be directly apparent. **1962** *Radio Times* 17 May 43 The jazz musician..is merely inviting himself back to his friend's place for a beer after their session. **1969** R. A. Noblett *Season Chain* 7 This version has not been released on record... The session is, however, interesting.

c. A disturbance or argument. *colloq.* (chiefly *Austral.* and *N.Z.*).

1919 H. L. Wilson *Ma Pettengill* iv. 130 Then Ben came down and had a wholehearted session with me. He said I ought to have a talk with Ed and reason him out of his folly. **1930** L. W. Lower *Here's Luck* i. 5 We had a bit of a session—a 'go in' as they call it. I tried to reason with him. **1949** J. R. Cole *It was so Late* 10 'Don't shoot the barman, he's half shot already.'.. 'Bit of a session, eh?'

7. (sense 6) *session clerk* (later example); (sense 6*) *session fee, man, work*; **session musician,** one who is engaged to play music, usu. accompaniments, at a recording session.

1876 Session clerk [see *parochial *a.* (*sb.*) 1 a]. **1977** *Times* 1 Nov. 14/5 The orchestra had a choice: either to take a share of the royalties or settle for what the trade calls a session fee—a once-and-for-all payment. **1958** J. Asman in P. Gammond *Decca Bk. Jazz* xiv. 170 Men who worked in a number of musical fields, providing the recording studios with a reliable nucleus of session-men for every kind of date. **1980** *Oxford Times* 20 June 18 She is expertly backed by..fine sessionmen. **1968** *Guardian* 23 Feb. 10/6 The vast majority of pop records made rely to some extent on session musicians. **1980** P. Gosling *Loser's Blues* ii. 12 Separately as session musicians on other pop discs they were occasionally..in the charts. **1976** J. Wainwright *Walther P. 38* 82, I moved around from band to band..and sat in on my share of session work at the recording studios.

sessional, *a.* Add: **c.** Also Canad., *sessional indemnity*, the remuneration received by a member of a legislative assembly.

1900 E. B. Osborn *Greater Canada* 105 The average partisan gets to Ottawa for the sake of his sessional indemnity and what he can make by means of his position. **1963** *Globe & Mail* (Toronto) 12 July 8/2 Increases in sessional indemnities and expenses that the Quebec Legislature has just approved put it in a class by itself.

d. Pertaining to, or lasting a session (sense 3 d) in an educational institution.

1965 *Listener* 28 Jan. 137/2 Each student has to pursue two sessional courses, which last the whole year, designed to give an introduction to some subject not studied at sixth-form level. **1978** *Sci. Amer.* Jan. 12/2 Geesey is a sessional lecturer in the department of biochemistry and microbiology at the University of Victoria in British Columbia.

e. Pertaining to any period of activity of limited duration. Cf. *session *sb.* 6*.

1973 *Scotsman* 12 Jan. 17/4 (Advt.), Part-time medical officers for sessional work. **1977** *Times* 29 Aug. 6/2 Members of mixed card clubs paid a sessional fee according to the stake at their table.

sessioneer (seʃəniᵊˑɹ). [f. Session *sb.* + -eer.] = *session musician* s.v. *session *sb.* 7.

1958 T. Hall in P. Gammond *Decca Bk. Jazz* xix. 235 Joe Muddel is one of the busiest 'sessioneers' in Britain's radio, film and TV studios. **1977** *Sounds* 9 July 34/3 He's marshalled such star sessioneers as Richie Albright..and the Memphis Horns into fine order.

‖ **sestiere** (sestįę̄ˑre). Pl. **-ieri.** [It., f. L. *sextārius* the sixth part of a measure.] In Italy: one of six districts or areas of a city. Cf. *quartiere.

1599 L. Lewkenor tr. *Contarini's Commonwealth & Govt. Venice* 185 The Citie of Venice is divided into sixe parts, which they call Sestieri. **1673** J. Ray *Observations Journey Low-Countries* 151 This City [*sc.* Venice] is.. divided into six parts or regions, called thence *Sestieri*. **1832** S. de Sismondi *Hist. Ital. Republics* iv. 84 The town [of Florence] was divided into six parts, each *sestier*, as it was called, named two *anziani*. **1893** J. A. Symonds tr. [illegible] The most satisfactory way of tackling the surface of [Venetian] sights is to concentrate on one of the city's six *sestieri* (districts) at a time.

seston (seˑstǫn). *Biol.* and *Oceanogr.* [a. G. *seston* (R. Kolkwitz 1912, in *Ber. Deut. Bot. Ges.* XXX. 341), ad. Gr. σηστόν, neut. of σηστός that which is filtered, f. σήθειν to strain, filter; cf. Plankton.] Fine particulate matter suspended in water, esp. that which is organic or living.

1926 B. D. Jackson *Gloss. Bot. Terms* (ed. 3) 344/2 *Seston*, plankton material retained by very fine meshed sieves. **1941** *Ecol. Monogr.* XI. 58/1 The varieties in sestonic phosphorus are correlated with both the mass of organic seston and the quantity of phytoplankton, as measured by its chlorophyll content. **1957** G. E. Hutchinson *Treat. Limnol.* I. [illegible] **1967** *Ibid.* II. xix. 235 The seston consists of bioseston, or plankton and nekton, which latter is ordinarily quantitatively negligible, and abioseston or tripton. *Ibid.* 245 The entire mass of suspended matter in a volume of free water is called seston, the nonliving part, tripton. **1971** *New Scientist* 15 July 143/1 [illegible] Hence **sestoˑnic** *a.*, of, pertaining to, or being seston.

1941 [see above]. **1967** *Oceanogr. & Marine Biol.* V. 221 At a glance these ribbons appeared to be sestonic debris from coastal algae of phanerogams.

Sesuto, Sesutu, varr. *Sesotho.

set, *sb.*¹ Add: **I. 7. b.** A grudge. Chiefly in phr. *to have* (or *take*) *a set on* (a person), to have a grudge against. *Austral.* and *N.Z. colloq.* Cf. *set *v.*¹ 125 b.

1903 'T. Collins' *Such is Life* (1937) i. 36 'Hasn't Warrigal Alf got a set on you, too?' asked Thompson coldly. **1941** Baker *Dict. Austral. Slang* 64 *Set*, a grudge against (someone), e.g., 'have a set on someone'. **1946** K. Tennant *Lost Haven* (1947) xiv. 228 If the Old Man hadn't tried to give Mark Thorne such particular hell when he was starting his shop, perhaps Thorne wouldn't have taken a set on all the Sudermans... If he hadn't the set on the Sudermans.. he wouldn't have wanted to cut off his nose to spite his face. **1948** D. Ballantyne *Cunninghams* (1963) ii. 155 He had a bit of a set on Frank and Sydney and was always pinching their cheeks and telling them they were young roughnecks.

9. (Later examples.)

1923 *Rep. Progr. Appl. Chem.* VIII. 231 The time of set has been found to depend upon the proportion of combined water in the hydrated calcium aluminate. *Ibid.*, Removal of water..results in the time of set being reduced. **1957** V. J. Kehoe *Technique Film & Telev. Make-Up* xii. 249 Warm water hastens the set of the material, so chilling the bowl is advisable to slow down the set. **1963** D. Seton *Essent. Mod. Cookery* 156 The use of lemon juice or citric or tartaric acid is essential to ensure a good set [in marmalade].

b. *initial set* (Building), a condition attained by cement when it begins to stiffen, but before hardening commences.

1891 T. Potter *Concrete* (ed. 2) ii. iii. 104 If a plasterer finds his mortar too stiff [illegible] initial set has [illegible]. **1927** *Engineer* 5 Aug. 143/2 At the completion of the operation the cement shows but an initial set. **1953** Van den Branden & Knowles *Plastering* iv. 38 The initial set of Portland cement mortar occurs about two to three hours after the dry materials have been wetted.

II. 12. Also *spec.* in *Psychol.*, a predisposition or expectation that influences the response of a person or animal: used variously of conscious or unconscious, or of mental or physical, states. Cf. *set *v.*¹ 93 c.

1890 W. James *Princ. Psychol.* I. iv. 114 It is not in the moment of their actual presence [illegible] producing *motor effects*, that resolves and aspirations communicate the new 'set' to the brain. **1911** E. L. Thorndike *Animal Intell.* [illegible] button around with its nose [illegible] general 'set' [illegible] it will be less [illegible] it had been chiefly [illegible]. **1918** R. S. Woodworth *Dynamic Psychol.* iii. 56 Danger arouses a 'set' of the nervous system towards escape. **1931** *Brit. Jrnl. Psychol.* Apr. 379 The theory..that ability in proof-reading is largely a matter of experimental 'set'. **1953** J. B. Carroll *Study of Lang.* iii. 77 There are actually prelinguistic organismic events (sets, attitudes, etc.) which are the phenomena which expression theorists regard as 'thoughts' and 'ideas'. **1968** *Science* 13 Dec. 1252/1 'Set' refers to the subject's psychological expectations of what a drug will do to him in relation to his general personality structure. **1979** Forgus & Shulman *Personality* i. 9 We can measure the dominant perceptual sets..and these sets, in fact, direct perceptual selectivity.

18. b. The action or result of fixing the hair when damp so that it dries in the required style. Also with reference to fixing the hair by other means (with heat, a setting lotion, etc.). Also *hair-set*. Cf. *set *v.*¹ 81 b.

1933 C. A. Fox *Art & Craft Hairdressing Spec. Suppl.* iv. [illegible] **1938** H. Goodman *Princ. Professional Beauty Culture* [illegible] **1940** W. Peck *Bewildering Cares* iv. 110, I met her once at the hairdresser's bewailing that she couldn't afford a nice steak for Herbert on their income, and she had obviously spent the price of it on a 'set'. **1946** K. Tennant *Lost Haven* (1947) xii. 204 You can't get a hairset here and I don't know what to do about my hair. **1975** *Country Life* 27 Mar. [illegible]

19. d. *Bell-ringing.* The inverted position of a bell when it is set. Cf. Set *v.* 66.

1677 F. Stedman *Campanalogia* 23 A prospect of true ringing after a certain compass under the Sett, may thus be taken. *Ibid.* 59 The reason why one of them is said to move up, is, because he that rings that bell, in the making of the change must hold it up at the Sett a little longer than ordinary, to delay its striking, whereby 'tis made to follow the other note which before it preceded. **1901** H. E. Bulwer *Gloss. Techn. Terms Bells* (1904) 33 [illegible] the bell is brought to rest mouth upward a little beyond the balancing point [etc.].

e. *Carpentry.* The amount that the blade of a plane projects beyond the body.

1898 F. & H. P. Fletcher *Carpentry & Joinery* xxvi. 281 The set of the plane may be adjusted during use by tapping the iron of the nose. **1938** W. T. Telling *Carpentry & Joinery* ii. 116 All [planes] will do specially true work if properly set and sharpened and many of them have mechanical means of adjusting the cutting iron so to a fine set.

III. 23. d. An undeveloped or rudimentary fruit; *collect.*, flowers that have been fertilized and should develop into fruit. Also, the development of fruit following fertilization. Cf. Set *v.* 98.

1888 C. M. Doughty *Trav. in Arabia Deserta* II. xv. 436 Every cluster, which we had inclosed in it a spray of the male blossom, was lapped about with a wisp of dry forage; and this defended the sets from early flights of locusts. **1928** *Daily Tel.* 11 June 5/2 Of culinary apples the set appears good on the whole... Dessert cherries have had a fair set. **1930** Auchter & Knapp *Orchard & Small Fruit Culture* viii. 169 In such orchards, if the blossoms are properly pollinated, much better sets occur. **1964** H. B. Tukey *Dwarfed Fruit Trees* [illegible] **1976** H. Kingham *U.K. Tomato Man.* xvi. 126 For all crops overhead damping with a course spray helps to improve set.

26. *Lawn Tennis* (always spelt *set*). **a.** A group of games counting as a unit towards a match for the person or pair of persons who win the greater number of games in it.

1969 *Lawn Tennis* ('Know the Game' Ser.) 15 The first player or pair to win six games wins the set, except that should the score become five games each—'Five All'—one player or pair must become two games ahead to win the set. **1980** *Guardian* 14 July 18/5 Miss Jevans...had a bad patch in the second set before winning 6–1, 7–5.

b. *Comb.*, as **set point,** the state of a set when one side or player needs only one point to win the set; also, the point itself (cf. *match-point* (a) s.v. *match *sb.*¹).

1928 *Observer* 1 July 25/3 When that cunning player.. would, at set-point, send one up so high that it [illegible] straight down the centre line. **1946** *Times* 26 June 2/3 The Dutch pair, after missing a set point, squared 6 all by six games to five, finally secured the first set at 9–7. **1974** D. Delman *Sudden Death* vi. 152 Set point. I crouch, racket twirling.

28. (Earlier and later examples.) Also, more widely, the setting, stage furniture, etc., used on stage in a theatre. In *Film-making* and *Television*, the scenery (usu. built up rather than painted) and other properties used in the filming of an individual scene; the place or area in which filming takes place. Freq. in phr. *on* or *off* (*the*) *set*. Also *attrib.* and *Comb.* Cf. *film set* s.v. *film *sb.* 7 c.

1859 E. Fitzball *Thirty-Five Years of Dramatic Author's Life* II. vi. 91 The vast scenes were pushed into [illegible] **1918** H. Croy *How Motion Pictures are Made* 107 With the sets determined upon, preparations for the taking of the picture is begun. **1936** Wodehouse *Laughing Gas* iv. 35 She was supposed to be on the set, made up, at six on the following a.m. for some retakes. **1947** A. Huxley *Let.* 27 July (1969) 574 The ticklish situation on the set has made it impossible to come to New York for Claire's wedding. **1953** K. Reisz *Technique Film Editing* i. 26 Dialogue-writing, set-design and acting all become subjugated to this central purpose. **1956** C. McCullers in *Mademoiselle* Sept. 274/2 Mabel Goodley, the painter and set-designer. **1961** G. Millerson *Telev. Production* i. 15 The set designer, responsible for the scenic treatment. **1973** *Listener* 22 Nov. 727/3 The same people are very much less agreeable in *West Penwith* than they are 'off-set' in *Day for Night*. **1977** M. Babson *Murder, Murder, Little Star* xviii. 254 Had there been a further scene..in the dressing-room? Twinkle was being too good on set.

30. h. A trap or snare; a series of traps. *N. Amer. Trapping.*

1912 V. E. Roe *Maid of Whispering Hills* 74 [illegible] **1942** *Sun* (Baltimore) 2 Feb. 4/5 Each morning the trapper makes the rounds of his 'set'. [illegible] **1977** *Globe & Mail* (Toronto) 30 Mar. 33/3 We were still within 50 yards of the trap's position, when a 55-pound beaver, swimming unseen under the ice, hit the set.

33. (Later examples.) *esp.* a heavy punch or chisel for use on metal or stone. Cf. *sate *sb.*

1905 P. N. Hasluck *Handyman's Bk.* 134/1 For punching the nail head below the surface of the work, the set is used. **1920** A. H. Fay *Gloss. Mining & Mineral Industry* 605/1 *Sett*, a quarryman's term for a square-faced steel tool which is held in position and struck with a sledge to cause a fracture in a rock mass. **1924** [see *bolster *sb.*¹]. **1942** W. H. Atherton *Workshop Pract.* (ed. 2) v. 176 The Hot Sate or Sett..is in constant use for cutting away extraneous metal while hot. **1962** J. G. Robertson *Metalwork* viii. 95 The Hot Set (Sett or Sate)..is used for cutting off on the cutting face of the anvil. A smith holds the work and hot set whilst a striker wields the sledge hammer. The hot set is designed to cut hot metal. **1964** H. Hodges *Artifacts* iv. 77 The heads were either cast, or formed as the rivets were closed using sets (setts) or snaps.

set, *sb.*² **I. 2. d.** Substitute for def.: a subdivision of pupils or students (esp. in a single year) for instruction on a particular subject: usu. one of a number of such groupings and often constituted according to ability. (Earlier and later examples.)

1882 in R. S. Churchill *Winston S. Churchill* (1967) I. Compan. i. iii. 90 Place in 3rd Set of 14 boys for ½ Term—14th. **1914** 'I. Hay' *Lighter Side School Life* i. 15 He must know whether Mr. A. in the Senior Science Set is expounding the theories of inorganic chemistry which have been obsolete for ten years. **1963** M. Braine *These Ruins are Inhabited* vi. 86 Sets are ability groups. In each subject the boys had been divided into [illegible] each of these sets met as a class. **1977** P. D. James *Shroud for Nightingale* ii. 4 We haven't used the demonstration room since Nurse Pearce's death, but otherwise the set is continuing to work according to plan.

3. a. (Later examples.) Freq. with qualifying adj. or sb. indicating the location, affiliation, or characteristic activities of the group, as *the Bloomsbury* (*Chelsea, Cliveden*, etc.) *set*. Cf. *jet set* s.v. *jet *sb.*³ 11.

1914 [see *Bloomsbury]. **1922** M. Cowley in *Dial* LXXIII. 231 She [*sc.* K. Mansfield] has three backgrounds only: continental hotels, New Zealand upper-class society, and a certain artistic set in London. **1938** H. Nicolson *Diary* 19 Sept. (1966) 361 We talk of..how terrible has been the influence of the Cliveden set. **1942** N. Coward *Middle East Diary* 49 This place is the last refuge of the out-dated 'International Set'. **1960** J. Betjeman *Summoned by Bells* [illegible] **1977** *News of World* 17 Apr. 5/5 The Prince of the Beatniks abdicated... He said goodbye to the Chelsea Set.

b. A meeting of a street gang or group of 'street people', esp. a party; the place where such a group meets. Also, the group itself. *U.S. colloq.*

Freq. in Black English.

1959 *Esquire* Nov. 70 *Set*, a party. **1967** *Trans-Action* Apr. 5/2 The more or less organized component of the set—the meaning both the peer group and the place where it hangs out. **1969** R. L. Keiser *Vice Lords* iv. 40 A set had been planned. [illegible] **1972** J. Mills *Report to Commissioner* 200 [illegible] **1975** *Amer. Speech* 1972 XLVII. 151 Blue eyes, you are not on my set.

II. 5. b. = *pump-set, pumpset* s.v. *pump *sb.*¹ 11.

1889 W. Tate *Princ. Mining* xxi. 157 The lifting set delivers into a cistern from which the forcing set pumps the water to bank. [illegible] The installation comprises two vertical sets consisting of motor and pump only. **1977** [illegible] The circuits were modified to give a signal 'pump unprimed' but not to shut down the set.

c. A piece of electrical or electronic apparatus, as a telephone, a telegraph receiver or transmitter, a radio or television receiver, etc. Also, a radar transmitter and receiver. Cf. *handset.

1891 *Man. Instruct. Army Telegr. Field Telegraphs* Plate II (caption) Two single current sets. **1898** *Electrician* 4 Mar. 623/2 A diminutive telephone set..is now being put on the market. **1903** *Science Siftings* XXV. 49/1 The instruments of the portable military outfits are similar to those of the permanent station sets. **1915** A. Fage *Aeroplane* iv. 43 A wireless set driven by a motor-cycle engine is mounted in front of the passenger's seat. **1923** *Radio Broadcast* Jan. 181/1 Drug stores, music stores, cigar stores, even farm furnishing stores have radio sets for sale. **1931** B. Brown *Talking Pictures* vi. 146 [illegible] **1936** W. H. S. Smith *Let.* 13 Dec. in *Young Man's Country* (1977) ii. 46, I dropped in on Stansbury..to hear his wireless which is a very good set. **1943** J. L. Hornung *Radar Primer* v. 123 The electrical features of radar sets for use in airplanes are similar to those of radar sets that are for use in ships. **1955** *Radio Times* 22 Apr. 30/1 (Advt.), Here is a..table radiogram... Fine sets these Ferguson's. **1961** L. Mumford *City in Hist.* xvi. 496 Reality has been progressively reduced to what filters through to the television set. **1972** *Works Engineer* June 12 (*heading*) Standby electric generator sets. **1974** P. L. Wilson *Water Turbines* 17 (caption) [illegible] **1976** M. Gilbert *Night of Twelfth* iii. 88 We used to have that old set going all day. You'll be just in time for the six o'clock news.

8. f. A number of pieces of Jazz or popular music performed in sequence by a musician or group.

1946 B. Treadwell *Big Bk. Swing* 125/2 *Set*, group of musical selections. **1955** S. Whitmore *Solo* ii. v. 150 Between sets at Fack's Jasper found himself alone. **1967** *New Yorker* 21 Jan. 43, I played two sets and Marsala asked me to join the band. **1977** *Sounds* 1 Jan. [illegible] We all have hits but they're too disgusting to be included in the set.

10. † b. *Math.* Used variously, as defined by the individual author. *Obs.*

1837 W. R. Hamilton in *Trans. R. Irish Acad.* (Sci.) XVII. 422 The author hopes to publish hereafter..a Theory of Triplets and Sets of Moments. **1848** —— in *Ibid.* XXI. 201 When we have in any manner been led to form successively the separate conceptions of any number of moments of time, we may afterwards form the new conception of a system, or momental set, to which all these separate moments belong. **1886** *Phil. Trans. R. Soc.* CLXXVII. 23 If the collection be such that whatever undistinguished components *abcd..., pqrs...*we select, and

whatever other component *lmno...* we select, *w, x, y, z...* can always be selected from the collection, then the collection will be termed a set.

c. *Math.* and *Logic.* An assemblage of distinct entities, either individually specified or which satisfy certain specified conditions. Cf. *element *sb.* 5 d.

1857 *Phil. Trans. R. Soc.* CXLVII. 717 Any values $(z_1, y_1, z_1,...)$ satisfying the equations, are said to constitute a set of roots of the system. **1897** W. Burnside *Theory of Groups* i. 1 Let $a_1, a_2, ..., a_n$ be a set of *n* distinct letters. **1903** *Trans. Amer. Math. Soc.* IV. 27 A set of elements in which a rule of combination O is so defined as to satisfy the following three postulates shall be called an Abelian group with respect to O. **1937** *Jrnl. Symbolic Logic* II. 66 According to the leading idea of the von Neumann set theory we have to deal with two kinds of individuals, which we may distinguish as sets and classes. The distinction may be thought of in this way, that a set is a multitude forming a proper thing, whereas a class is a predicate regarded only with respect to its extension. **1965** Patterson & Rutherford *Elem. Abstr. Algebra* i. 3 If *x* is an element of a set *S*, we write $x \in S$. **1972** A. G. Howson *Handbk. Terms Algebra & Anal.* ii. 8 A set is a totality of certain definite, distinguishable objects of our intuition or thought—called the elements of the set. This classic definition of a set was given by Georg Cantor in 1874. Such attempts to give elementary definitions of a set are, however, doomed to failure, their being in the main based on the use of undefined synonyms, such as 'collection', and leading to logical inconsistencies (see *Russell paradox*..). For this reason, mathematicians now regard the notion of a set as an undefined, primitive concept. **1975** I. Stewart *Concepts Mod. Math.* iv. 47 There is only one empty set. All empty sets are equal.

d. *transf.* Used variously in *Linguistics* (see quots.).

1935 W. F. Twaddell *On Defining Phoneme* 60 A modification occurs only in phonetic fractions corresponding to forms, the relations of which constitute relations of sets of micro-phonemes. **1942** Bloch & Trager *Outl. Linguistic Anal.* iii. 45 A structural set is a group of all the phonemes which occur in a given phonetic environment and hence, in that position, directly contrast with each other. **1964** M. A. K. Halliday et al. *Linguistic Sci.* ii. 22 The range of possibilities in a closed choice is called technically a system, that in an open choice a set... We often talk of 'closed system' and 'open set'.

14. *running set*: see *running *ppl. a.* 17 f.

III. 15. Special Comb.: **set theory,** the branch of mathematics which deals with sets without regard to the nature of their individual constituents; an axiomatization which allows of the discussion of sets; **set-theoretic, -theoretical** *adjs.*, of or pertaining to set theory; hence **set-theoretically** *adv.*

1964 E. Mendelson *Introd. Math. Logic* p. vii, In the belief that beginners should be exposed to the most natural and easiest proofs, free-swinging set-theoretic methods have been used. **1957** P. Suppes *Introd. Logic* xi. 232 A function is a set-theoretical, not a linguistic, entity. **1952** S. C. Kleene *Introd. Metamath.* xiv. 424 B is a 'theorem' set-theoretically. **1936** W. V. Quine in *Jrnl. Symbolic Logic* June 45 Set-theoretic Foundations for Logic... In his set theory Zermelo uses the variables 'x', 'y', etc. for the representation of 'things' generally. **1937** *Jrnl. Symbolic Logic* II. 65 The system of axioms for set theory to be exhibited in this paper is a modification of the axiom system due to von Neumann. **1971** *Where* Nov. 332/1 Many would probably 'solve' it by using set theory and drawing a Venn diagram. **1975** N. Chomsky *Logical Struct. Linguistic Theory* iii. 107 We will assume ..that each level includes a full set theory, so that we can also form sets of strings, sequences of strings, etc.

set, *v.*¹ Add: **I. 5. a.** (Later U.S. examples.)

1884 C. H. Smith *Bill Arp's Scrap Bk.* vi. 74 Lawyers and doctors have to set about town. **1913** H. Kephart *Our Southern Highlanders* xiii. 298 'Come in and set.' 'Cain't stop long.' **1938** M. K. Rawlings *Yearling* i. 12 'If a feller'd light me a candle,' she said, 'I'd git shut o' the dishwashin' and mebbe have time to set and enjoy myself.' **1974** P. De Vries *Glory of Hummingbird* (1975) iii. 37 Lolly came almost every evening to set a spell.

III. 14. c. (Earlier example.)

1844 W. J. Pell *Treat. Game of Dominoes* 22 The largest count that can be made..is 129. To effect this, the winning hand must set.

IV. 25. a. Also, *to set afoot* or *on foot* (see Afoot 3, Foot *sb.* 32 c).

1615 R. Cocks *Jrnl.* 30 July (1883) I. 28 An other matter is now set on foote, which I never did heare of till this instant. **1638** [see Afoot 3]. **1736** T. Lediard *Life Marlborough* III. 364 A Treaty of Peace was again set on foot. **1829** Scott *Anne of Geierstein* III. ix. 263 He..has in a right godly manner tried to set afoot a treaty of peace with my own father. **1890** *Sunday Mag.* Aug. 531/2 Enquiries were at once set on foot.

34. Also *to set a fire* (without prepositional complement): to kindle or start a fire. *U.S.*

1906 *N.Y. Even. Post* 15 Nov. 3 Two fires in tenement house letter boxes were set to-day at an early hour. **1976** *Washington Post* 19 Apr. B1/7 The school had been broken into and the fire had been set.

VI. 65. b. *Computers.* To cause (a binary storage unit) to enter a prescribed state, *spec.* that representing 1. Also *intr.*, to enter a prescribed state.

1948 *Electronics* Apr. 127/1 The initial values can be set into the computer without too much time lag. **1957** R. K. Richards *Digital Computer Components & Circuits* vi. 263 The real problem in devising a large-capacity storage system is not so much in the storage elements themselves as in providing means to gain access to any specified individual storage element for the purpose of sensing..or setting its status. **1968** Maley & Heilweil *Introd. Digital Computers* vi. 82 The latch is simply a circuit whose output can be set to 1, or reset to 0, and it will remain at either one of these two values until another set or reset operation changes its value. **1971** J. H. Smith *Digital Logic* i. 12 A binary divider is a modified toggle which has only one input. If electrical pulses are applied to this input the unit will 'set'. The second pulse will 'reset' the circuit.

72. Delete † and add later *absol.* example.

1964 F. Bowers *Bibliogr. & Textual Criticism* vi. i. 161 The sole purpose of saving the printer the labour of setting from a difficult manuscript.

73. to put (words) *to* music (examples); to put (music) *to* words: no longer *Obs.*

1965 *Listener* 3 June 836/2 One does not make music 'colloquial' by using it to set colloquial words. **1966** Bennett & Smithers *Early M.E. Verse & Prose* 108 The music to which it [*sc.* a lyric] is set clearly shows that the words were composed to fit the tune. **1970** *Oxf. Compan. Music* (ed. 10) 498/1 The tunes set to these hymns were partly adaptations of the ancient plainsong, partly arrangements of folk song and partly original. **1979** *N.Y. Rev. Bks.* 17 May 32/4 Byrd set this notorious poem to music, and the setting certainly did not escape notice.

74. *to set the stage* (*for*): also *fig.*, to prepare the way or conditions for (an event, etc.).

1937 *Discovery* June 175/1 Given suitable conditions, the stage is always set for the transformation. **1972** *Review & Herald* 7 Dec. 12/2 However, it is first necessary to 'set the stage'. **1980** *Sci. Amer.* Jan. 122/1, I can best set the stage for describing Morelli's instrument by reviewing the two basic types of spectroscope and spectrophotometer.

76. b. To adjust (the blade of a plane in relation to the sole) in order to vary the depth of cut.

1677 J. Moxon *Mech. Exerc.* iv. 63 When you set the Iron of the Fore-Plain, consider the Stuff you are to work upon. **1857–9** E. L. Tarbuck *Encycl. Pract. Carpentry & Joinery* I. iii. 26 The projection of the plane iron may be very nicely regulated, or set, rank, or fine, that is projecting from the face in a greater or less degree. **1938** C. H. Hayward *Carpentry Bk.* i. 27 When a piece of wood with a difficult grain has to be planed, the back-iron is advanced and the plane set as fine as possible.

81. Restrict † *Obs.* to sense in Dict. and add: **b.** To arrange and fix (the hair) when damp so that it dries in the desired style; *occas.*, to fix a hair-style by other means.

1926 *Hairdressing* 10 Sept. 241/1 This can only be done by superior work; namely, excellent setting of the finished permanent. **1932** *Mod. Woman* Feb. 72/1 A perfectly easy method of keeping your hair perfectly waved, set and curled at home. **1957** V. J. Kehoe *Technique Film & Telev. Make-Up* xv. 214 Hair lacquer or spray..is used for setting the hair in place after it has been dressed. **1967** N. Freeling *Strike out where not Applicable* 10 Ash-blonde hair cut fairly short and set every week in Leiden. **1976** C. Bermant *Coming Home* i. vii. 105 Her hair was always smartly set.

86. Delete *Obs.* or *arch.*

1980 M. Boddy *Building Societies* iv. 46 The composite rate [of tax paid by building societies] was set at 79·3 per cent of the basic rate (then 35 per cent), i.e. 27·75 per cent.

VIII. 92. Delete 'Now *dial.*' Now usu. in sense 'likely, about (to)'. Also, in *Journalese*, const. *for* followed by *sb.*

1976 *Daily Tel.* 30 Nov. 1/6 Electricity prices are set to go up again on New Year's Day. **1978** *Sunday Tel.* 10 Dec. 1 (*heading*) Callaghan set for showdown with Benn. **1979** *Daily Tel.* 28 Feb. 2/6 The Inner City partnerships outside London seem set for increases above the average. **1982** *Times* 16 Oct. 9/6 The armchair moralists of Academe..are now set to carp about the sinking of the Belgrano.

93. c. *Psychol.* To predispose (a person or other organism) to a given response; usu. *pass.* Also *intr.* for *pass.* Cf. *set *sb.*¹ 12.

1909 *Amer. Jrnl. Psychol.* XX. 569 The psychophysical organism 'sets' to meet an imminent situation; and on the conscious side, this 'set' is expectation. **1938** *Mind* XLVII. 88 An observer in an experiment is said to be *set* towards an aspect of a situation if he is directed to it by the instructions. **1961** Lindgren & Byrne *Psychol.* vi. 143/2 Alterations in our familiar surroundings are often missed because we are 'set' to perceive certain stimuli.

95. c. Also *spec.* of an athlete poised to start a race (further *pass.* examples). In wider use: to be prepared for action; to be ready (*to* do something). Freq. in phr. (*to be*) *all set*. Cf. *on your mark*(*s*), (*get set, go*) s.v. *mark *sb.*¹ 12 e.

1844 J. Gregg *Commerce of Prairies* I. 51 Each teamster vies with his fellow..and it is a matter of boastful pride to be the first to cry out—'All's set!' **1882** W. A. Baillie-Grohman *Camps in Rockies* i. 3 'All set!' echoes from each of the horsemen in front. **1913** [see *crouch *sb.*² b]. **1930** *Amer. Speech* VI. 120 Set for big bout. **1935** *Encycl. Sports* 580/1 At the words 'Get set' you should let the weight come forward on to the finger tips and the leading foot, raising the left knee but lowering the back and head. **1949** N. Marsh *Swing, Brother, Swing* v. 84 All set, boys? Let's go. **1956** A. H. Compton *Atomic Quest* iii. 162 The du Pont Company was getting set to build the plutonium production plant. **1957** Duncan & Bone *Oxf. Pocket Bk. Athletic Training* (ed. 2) v. 62 On the command 'set' the body rises up smoothly with the body-weight on the hands and front foot. **1962** J. Heller *Catch-22* vi. 51 Just when I was all set to really start stashing it away, they had to manufacture fascism and start a war. **1979** *Daily Tel.* 26 Feb. 21/4 National Westminster is set to produce full year figures tomorrow.

97. a. (Later example.)

1974 M. Lindlaw *Super Sweets & Puddings* 9 *To set jelly quickly.* Dissolve the jelly tablet in ¼ pint (1½dl) hot water, then make up to 1 pint (6dl) with cold water or ice cubes. Stir until on the point of setting.

c. (Later examples.)

1963 D. Seton *Essent. Mod. Cookery* 151 The sugar is very important in jam-making. If too much or too little is used, the jam will not set. **1973** *Cooking for Today* (Good Housekeeping) 264/4 Pour one third of this jelly into a picnic jelly mould and put in a cool place to set.

X. 125. b. Phr. *to have* or *get* (a person) *set*: to have a score to settle with, 'have it in for' (that person). *Austral.* and *N.Z. slang.* Cf. *set *sb.*¹ 7 b.

1916 C. J. Dennis *Songs Sentimental Bloke* 40 This Romeo 'e's lurkin' wiv a crew—A dead tough crowd o' crooks—called Montague. 'Is cliner's push—wot's nick-named Capulet—They 'as 'em set. *c* **1926** 'Mixer' *Transport Workers' Song Bk.* 17 You growl and swear you can't get work Or the boss has got you set. **1945** Baker *Austral. Lang.* vi. 121 A man who has acquired a strong dislike of another person... He *gets someone set* and *words him*, rebukes him. **1959** —— *Drum* 112 *Get someone set*, to have a grudge against a person; to prepare to pay someone out.

XII. 141. set back. a. (Later examples.) Hence, with a sum of money as compl.: to cost (a person so much). Also *fig.*, to take aback, to disconcert.

1847 *Spirit of Times* 31 May 159/1 The captain used to boast that he could pack a gallon without its setting him back any. **1884** 'Mark Twain' *Huck. Finn* viii. 66 The nigger was set back considerable, because he reckoned it was all done with witchcraft. **1900** Ade *Fables in Slang* 131 Daughter was..seated under a Canopy that had set Father back thirty-two Dollars. **1922** S. Lewis *Babbitt* x. 142 How much'll it set me back? **1937** J. Steinbeck *Of Mice & Men* 79 'What's it set you back?' George asked. 'Two and a half [dollars].' **1940** H. L. Ickes *Secret Diary* (1954) III. 183 This set him back on his heels. **1966** 'J. Hackston' *Father clears Out* 53 'Goin' t' leave it?' the prince asked, a bit set back. **1974** *Country Life* 14 Nov. 1445/1 Even a moderately-sized piece of cheesecake sets you back 20p.

146. set in. d. Also *pass.*

1837 Dickens *Let.* ? Dec. (1965) I. 346, I was in the humour for writing last night—..was regularly set in—when there came a double knock.

149. set out. a*. To spread (leather) on a flat surface while wet, in order that it may dry free from wrinkles.

1885 C. T. Davis *Manuf. Leather* xxiii. 423 The side having been stuffed, and next 'set out'..the next step in the process of manufacturing upper leather is that of whitening. **1909** H. G. Bennett *Manuf. Leather* xxi. 261 The butts are now struck out, 'set out' or 'pinned'. **1946** J. W. Waterer *Leather in Life, Art & Industry* ii. ii. 147 The butts are piled up to drain... They are 'set out' to remove wrinkles and smooth the grain. **1969** T. C. Thorstensen *Pract. Leather Technol.* v. 70 After bleaching, the bends are wrung and sent to an oil wheel... This process usually takes about one hour. The leather is 'set out' to smooth and dry.

150. set over. g. To kill or murder. *U.S. Criminals' slang.*

1931 G. Irwin *Amer. Tramp & Underworld Slang* 166 *Set over*, to kill, probably since the victim is set over or apart. **1944** W. R. Burnett *Nobody lives Forever* xxii. 159 I've been trying to find you ever since you set Doc over. **1949** —— *Asphalt Jungle* xxxiii. 211 They have to set a guy over.

154. set up. m. (*f*) (Earlier and later examples.) Also in phr. *to set them up*, to provide free drinks.

1880 A. A. Hayes in *Harper's New Monthly Mag.* Jan. 209/1 You bet he lived high; always set up the drinks. **1883** Sweet & Knox *On Mexican Mustang* iii. 47 Then he swore, and cussed the 'demmed country, you know', but finally got into good humor, and set 'em up all round. **1906** C. de L. Canfield *Forty-Niner* ix. 83 Of course, it was drinks all around; you can't do anything in this country without setting 'em up first. **1949** [see *rock candy* s.v. *rock *sb.*¹ 9 a]. **1965** G. Melly *Owning Up* vi. 64 In exchange for a song or two from me, he was prepared to set them up all night.

p. (*c*) To make the necessary interconnections and initial settings in (a computer) for the performance of a particular calculation; to do this so that the computer will solve (an equation), perform (a calculation), etc.

1931 *Jrnl. Franklin Inst.* CCXII. 459 A bus shaft is assigned to each significant quantity appearing in the equation. The several relations existing between these are then set up by means of connections to the operating units. **1948** *Electronics* Apr. 124/1 When combining circuit elements to form an analog computer, the first step is to set up the differential equations to be solved. *Ibid.* 126/3 Consider setting up the computer for solving the differential equation $p^2y - 0{\cdot}2py - y = 0$. **1962** MacKay & Fisher *Analogue Computing at Ultra-High Speed* xiii. 171 The procedure for setting up a given equation on an analyser is not difficult. **1964** G. A. & T. M. Korn *Electronic Analog & Hybrid Computers* ii. 37 The computer is 'set up' for the given problem when a suitable arrangement of computing elements establishes the correct relationships between computer voltages.

aa. (*c*) To prepare, set in readiness (apparatus, machinery, etc.). (A more generalized application of prec. sense.)

1922 H. D. Burghardt *Machine Tool Operations* II. viii. 157 Sometimes an unskilled man or boy can operate several machines after they have been 'set up' by a skilled mechanic. **1962** A. Nisbett *Technique Sound Studio* viii. 142 To be able to put the disc on the turntable, locate the right groove,..and set the record up ready for playing in. **1977** P. Dickinson *Walking Dead* i. iii. 39 Foxe felt most fully alive..when he was setting up a new experiment.

(*d*) *gen.* To make preparations or arrangements for; to contrive, plot (a move, trick, etc.); to arrange (a social engagement).

1965 P. O'Donnell *Modesty Blaise* vii. 83 If Gabriel or anyone else has been setting up a job from here, Paco will know about it. **1968** [see *nim *sb.*²]. **1971** *Daily Tel.* 28 Oct. 3 (*heading*) Boy, 12, set up cripple's death jury is told. **1973** R. Busby *Pattern of Violence* vi. 104 Let's set this thing up. I'll get onto the divisional commander. **1973** *Houston* (Texas) *Chron.* (Suppl.) 14 Oct. 8/4 We set up a date and a couple of weeks later Agnew and I sat down in his suite in a Chicago hotel. **1978** R. Thomas *Chinaman's Chance* xxii. 228 'Could you set it up?' 'No problem.'

cc. (*c*) To bring (someone) to a position from which he may be knocked down, to make vulnerable (*lit.*, as in *Pugilism*); *fig.* (*colloq.* and *slang*), to lead on in order to fool, cheat, or incriminate (a person); to 'frame'. orig. *U.S.*

1950 J. Dempsey *Championship Fighting* x. 49 If you can land solidly with a straight left or with a left hook, you'll generally knock your opponent off balance, at least, and 'set him up' for a pot-shot with your right. **1956** B. Holiday *Lady sings Blues* xxi. 168 When I saw them running across the rooftops with my money, I knew I'd been had. Somebody had set me up. **1963** L. Deighton *Horse under Water* xxxi. 127 Either Mr Ivor Batcher was double-crossing his boss or I was being set up. **1964** S. Bellow *Herzog* 109 Of course he understood that Tennie was setting him up, and that he was a sucker for just the sort of appeal she made. **1979** A. Price *Tomorrow's Ghost* ii. 23 'You're deliberately using them for bait, for God's sake.' 'Oh no we're not... We didn't set them up.' **1981** 'E. V. Cunningham' *Case of Sliding Pool* ix. 101 He had a partner, whom he set up from the very beginning for the kill.

kk. (Later examples.)

1935 R. Bass in *Scribner's Mag.* Feb. 122/1 The body must never be left alone for an instant until it is left in the grave. It must be 'set up' with. **1968** E. R. Buckler *Ox Bells & Fireflies* ix. 127 Neighbors took turns 'setting up' with the patient night after night.

set, *v.*² [f. Set *sb.*²] *trans.* To group (pupils) into sets (see Set *sb.*² 2 d in Dict. and Suppl.); also *absol.* Hence **seˑtting** *vbl. sb.*²

1953 *Organ. Comprehensive Secondary Schools* (London County Council) 14 A practicable arrangement would be to re-set only across three adjacent forms. 'Setting' in this way would not determine the rate at which each set would work. **1957** B. Simon *New Trends in Eng. Educ.* ii. 46, I will not ask the reader to follow me in the intricacies of fifth-year setting. **1962** J. Vaizey *Britain in Sixties* v. 56 Some..feel that..children should be 'setted' for each subject. **1965** *Observer* 7 Nov. 4/8 Mathematics teachers consider it necessary to set after two terms. **1973** Morrison & McIntyre *Teachers & Teaching* (ed. 2) iii. 126 There is reason to believe that the practise of 'setting' —different streaming for each of several subjects—reduces these effects. **1975** *Language for Life* (Dept. Educ. & Sci.) xv. 224 Speaking purely for English, most of us have reservations about arrangements by which pupils are streamed or setted according to ability.

set, *ppl. a.* Add: **I. 1. a.** (Later examples of *set book*. Also *set text*.)

1966 N. Nicolson in H. Nicolson *Diaries & Lett.* (1966) 28 He read..the whole of Aeschylus' *Seven Against Thebes* because it was my set-book at school. **1968** *Listener* 22 Aug. 244/3 By the end of 1967, however, it had sold more than 15,000 copies, mainly because a few enterprising examining bodies had chosen it as a set book for A-level GCE. **1982** *Times* 12 Aug. 8/3 Mrs Thatcher's Family Policy Committee has been given a set text in the form of a paper by the recently appointed head of the Downing Street Policy Unit.

i. Of jelly: that has become firm. Cf. Set *v.*¹ 97 a, c.

1973 *Cooking for Today* (Good Housekeeping) 264/4 Pour half this vanilla jelly on to the set coffee jelly. **1974** M. Lindlaw *Super Sweets & Puddings* 58 Make up the Angel Delight..and pipe or swirl on to the set jelly.

2. e. Of a meal in a hotel, etc.: consisting of a predetermined collection of dishes or items of food at a fixed price.

1914 'Saki' *Beasts & Super-Beasts* 308 The one-and-sixpenny set dinner receded..to a Sunday extravagance. **1923** C. Stone *Let.* 30 June in C. Mackenzie *My Life & Times* (1966) V. 250, I fancy F. will get herself set teas, and other meals out. **1938** D. du Maurier *Rebecca* xxvi. 423 Colonel Julyan waded through the whole set lunch. **1957** W. Camp *Prospects of Love* iii. i. 148 She promptly chose the five shilling set meal. **1973** J. Pattinson *Search Warrant* vii. 105 If you have the set lunch, it comes cheaper. **1978** *Times* 3 June 11/4 There was an advertised set lunch at £5.50 plus VAT.

6. c. *set fair*: also *fig.* and in extended use.

1873 Browning *Red Cotton Night-Cap Country* ii. 108 Like some kindly weathercock..stuck fast at Set Fair. **1921** W. de la Mare *Mem. Midget* xxix. 197 Her mood, like our weather that April, was almost always 'set fair'. **1978** J. Pearson *Façades* xxiii. 399 Everything appeared set fair for the happiest of stays.

II. 8. set copper, a form of metallic copper containing about 6 per cent of cuprous oxide, produced by oxidation during refining; **set joint** *U.S. slang* (see quots. and *flat joint* (*b*) s.v. *flat *a.* 15); **set point,** the value of a physical quantity that an automatic controller or regulator is set to maintain; also *transf.*; **set scene** (earlier example); **set scrum(mage),** *Rugby Football*, an organized scrummage ordered by the referee during the course of play; opp. *loose scrum*(*mage*) s.v. *loose *a.* 9; **set shot** *Basketball* a shot at the basket made from a still position; **set stocking** *Agric.* (orig. *N.Z.*), the grazing of animals, esp. sheep, in the same pasture for a considerable period; so **set-stock** *v. trans.*

1904 *Trans. Amer. Inst. Mining Engineers* XXXIV. 671 Some of the copper is oxidized to cuprous oxide and dissolved by the metal bath. When the quantity of dissolved cuprous oxide has reached about 6 per cent, the metal is said to have been brought to 'set-copper'. **1959** J. Newton *Extractive Metallurgy* vi. 376 Usually it is not possible to take any short-cuts in refining copper—the metal must be carried to the set-copper stage and then poled. **1926** Maines & Grant *Wise-Crack Dict.* 14/1 *Set joint*, unbeatable game. **1931** *Amer. Speech* VI. 335 *Set-joint*,..a gambling device operated with a numbered wheel and arrow-spindle. These are always fitted with a *gimmick* which prevents the customer from winning too often, or which may be used by the operator to lead the customer on until he will place a large bet, when the operator applies the gimmick and the customer loses. **1941** T. J. Rhodes *Industr. Instruments for Measurem. & Control* ix. 419 Where it is not permissible for the process temperature to deviate for any appreciable period from the original set point, it is necessary to use a mode of control previously described as proportional and floating. **1972** *Science* 9 June 1125/1 One of the principal homeostatic 'set points', that for body temperature, seems to depend on the constant ratio of Na^+ to Ca^{2+} in the caudal hypothalamus. **1975** D. G. Fink *Electronics Engineers' Handbk.* xxiv. 14 Display of the measurement, set point, and output levels is normally provided. **1866** 'Old Stager' *Stage Reminisc.* ix. 122 The sizes and sets of yarns occupied his thoughts much less than theatrical 'lengths' and 'set scenes'. **1938** Macdonald & Rees *Rugger Practice & Tactics* ii. 34 Few tries are scored in good football by movements that start from a set scrum. **1960** E. S. & W. J. Higham *High Speed Rugby* xiv. 185 We deal with the set scrum first, because it forms the basis of loose scrums and loose rucks. **1977** *S. Wales Guardian* 27 Oct. 16/3 By this stage Llandovery's forwards were dominating the set scrums. **1925** R. M. Rayner *Man. Rugby Football for Public Schools* viii. 47 (*heading*) On getting possession in 'set' scrummages. **1971** *Times* 15 Feb. 9/4 Bryce, Miller and Moroney are an experienced front row and rubbed in the fact..at the set scrummages. **1940** *N.Y. Times* 21 Jan. v-1/3 The cadets, their set shots hitting the mark with a remarkable degree of accuracy, gained the upper hand at the outset. **1976** *Milton Keynes Express* 9 July 42/3 Wickham, Wynn and Waller were all desperately unlucky with set shots. **1956** *N.Z. Jrnl. Sci. & Technol.* A. XXXVII. 555 Most New Zealand farmers producing fat lambs..generally prefer to set-stock the ewes and lambs from lambing time onwards. **1964** *Weekly News* (Auckland) 21 Oct. 54/4 Under New Zealand conditions, ewes and lambs are usually set-stocked from lambing until weaning, although rotational grazing is practised on some farms. **1981–2** *Deer Farmer* (N.Z.) Summer 8/3 Hinds are set-stocked or mob-stocked over winter at about 10 to the acre. **1950** *N.Z. Jrnl. Agric.* Feb. 100/2 Make sure that the calves are rotated through the paddocks at intervals of a few days. Set stocking at this time of the year is bound to result in..some deaths. **1975** *Country Life* 26 June 1702/1 In the last 35 years we would seem to have gone full circle—from set-stocking, strip-grazing, paddock-grazing..and now the so called 'intensive' set-stocking.

10. set-in, (*a*) (later examples, of clothing).

1895 *Montgomery Ward Catal.* Spring & Summer 280/2 Men's overshirts... Set in bosom, yoke back, [etc.]. **1969** *Sears Catal.* Spring/Summer 37/3 Wing collar, set-in sleeves.

set-. Add: **set-ope** *dial.* [Ope *a.* and *sb.*], a device for holding open a gate, window, etc. (? *obs.*); **set-screw,** add: a screw that enables two contiguous parts to be brought into and held in their correct relative position, and is usually threaded the full length of the shank (earlier and later examples).

1823 E. Moor *Suffolk Words* 336 The gate oont keep back; the *set-ope* is gone. **1912** *Civ. Serv. Supply Assoc. Catal.* 1351 Melon frames..glazed with 21 oz. sheet glass, and fitted with improved set-opes. **1850** T. Tredgold *Steam Engine* (ed. 3) I. iv. 19 When the piston is taken to pieces.., the two set-screws *h* are taken out of the holes into which they are screwed in the piston cover, and two handles screwed into them. **1947** J. C. Rich *Materials & Methods of Sculpture* ix. 277 A setscrew is used [in proportional dividers] to hold the two arms together and to adjust the dividers. **1970** K. Ball *Fiat 600, 600D Autobook* vi. 69/1 Undo the setscrew securing the angled extension piece to the gear-rod.

setar, var. Sitar in Dict. and Suppl.

set-aside, *sb.* and *adj. phr. U.S.* [f. vbl. phr. *set aside*: cf. Set *v.* 139 a, Aside *adv.*] **A.** *sb. phr.* Something set aside; *spec.* a quantity of commodities, agricultural produce, etc., reserved by governmental order for a special purpose (*orig.* for supplying the military forces). Also, the action of securing thus.

1943 *Sun* (Baltimore) 16 June 28/8 'Set asides' for the armed forces have increased measurably. **1954** *Ibid.* 21 July B-12/3 One of the foggier sections of the pending overall farm bills would provide for 'set-asides' of surplus commodities now held by the Government. The purpose of the set-aside is clear enough, if the method of effecting it is not. **1966** *Wall St. Jrnl.* (Eastern ed.) 5 Dec. 12/2 The Government ordered drastic increases in defense set-asides on copper during the first half of 1967. **1975** *New Yorker* 26 May 66/2 A visitor asked Butz, who had become Secretary of Agriculture late in 1971, if as Secretary he would have preferred to limit the 'set-aside' of cropland withheld from production in 1972. **1980** *Outdoor Life* (U.S.) (Northeast ed.) Oct. 47/1 Another may be squeaky clean on wilderness set-aside, for example, but falter badly when it comes down to a sensible attitude on gun control.

B. *adj. phr.*

1943 *Daily Progress* (Charlottesville, Va.) 6 Oct. 3/5 A protest..to the War Food Administration against inclusion of smaller sizes of certain varieties of fruit in the government's 'set-aside' order to protect needs for the armed services. **1979** *Financial Times* 19 Jan. 37/7 There is in addition a 'set-aside' programme in the U.S. for both wheat and maize to restrict plantings.

set-back. Add: **2. b.** *N. Amer.* The setting back or recessing of a building from the edge of a roadway (as an element of environmental planning); the limit of this withdrawal or the open area created.

1916 *Ann. Rep. Planning Board of Brookline, Mass., 1915* 5 Disregard of the customary setback which has hitherto maintained a margin of cheerful green. **1923** *Stud. Building Height Limitations* (Zoning Comm. Chicago Real Estate Board) 163, I would like to ask in the case of a narrow street whether it would be automatically widened by the establishment of a fixed set-back. **1937** *Sun* (Baltimore) 30 Sept. 24/2 Claiming the building of a stone wall along a lot boundary on Wendover road 'is in total disregard of the theory of setbacks and open spaces in Guildford'. **1947** *Daily Progress* (Charlottesville, Va.) 13 Sept. 2/1 (*heading*) Staples approves rural setbacks. **1961** L. Mumford *City in Hist.* Note to plate 51, The architect is freed from arbitrarily uniform prescriptions as to garden allotments, setbacks from roads and obsolete street patterns. **1975** *Canadian Antiques Collector* Mar.–Apr. 27/1 One fence encloses a typical store and defines a deep setback for a distinctive house.

c. *N. Amer.* A feature of the design of a skyscraper by which higher storeys are successively set back a certain distance behind the line of lower storeys, leaving a horizontal area in front; the horizontal area so formed.

1923 *Stud. Building Height Limitations* (Zoning Comm. Chicago Real Estate Board) 23 Additional Heights of Buildings should be allowed, provided proper set-backs are required for such additional stories. **1926** *Daily Colonist* (Victoria, B.C.) 12 Jan. 3/6 It [*sc.* a proposed skyscraper] is to be of the type recently developed..with a reduction of area, or a 'set back' from the street line above the tenth floor and similar reductions taking place at intervals thereafter. **1934** *Sun* (Baltimore) 16 May 2/1 Former Governor Alfred E. Smith..announced that on June 1 the club would open a new terrace cafe on one of the setbacks of the Empire State Building. **1945** *Washington Post* 17 Aug. 1/8 His body cleared more than 100 feet of building setbacks in the seven tiers of parapets designed as obstacles to suicide plunges. **1964** 'J. H. Roberts' *Q Document* i. 12 The façade of the building was deceptive, with sculpted indentations and setbacks. **1977** *Guardian Weekly* 6 Nov. 18/4 One of the victims is hurled through a closed window, his body coming to rest on a setback seven floors below.

5. (sense *2 b) *setback line, space.*

1917 *Establishment of Setbacks* (City of N.Y. Comm. on City Plan) 4 The set-back line secures on certain streets a uniform set-back of buildings from the street line. **1948** *Daily Progress* (Charlottesville, Va.) 4 Nov. 1/4 An agreement..to permit the super market..to stay four feet inside the prescribed ten-foot setback line. *Ibid.*, If and when the city needs the full ten-foot setback space to widen Preston Avenue, [etc.].

set-down. Add: **1. b.** (Earlier example.)

1727 Swift *Let.* 29 Feb. in *Corr.* (1963) III. 268 Mr Schutz's coach that usd to give you so many a Set-down, is wheeled off to St. James's.

2. (Earlier example.)

1780 T. Pasley *Jrnl.* 1 Aug. in *Private Sea Jrnls.* (1931) 105 Gave John a set-down for impertinence.

3. (Earlier and later examples.)

1824 J. R. Anthony *Diary* 3 Feb. in Z. Pease *Life in New Bedford 100 Yrs. Ago* (1922) 76 After we had got our fill, returned back to our house with a bowl full [of fried oysters] for the girls. Had a good set down and parted at 12 o'clock. **1858** *Harvard Mag.* Sept. 281 We Americans think that we cannot live unless we have our three 'set-downs'..each day, which is absurd. **1907** J. London *Road* 28 At the very next house I was given a 'set-down'. Now a 'set-down' is the height of bliss. **1914** *Sat. Even. Post* 4 Apr. 10/2 That kid don't want no handouts. He gets setdowns. Yes, siree, bo; every time. Setdowns in the kitchen. **1941** J. Smiley *Hash House Lingo* 48 *Set down*, good meal.

4. The action or an instance of landing in or from an aircraft.

1951 *Sun* (Baltimore) (B ed.) 28 May 28/4 The pilot had called in to report his 'set down'. **1968** O. Wynd *Sumatra Seven Zero* xii. 183 Would it surprise him to know..that he had been under observation from the moment of set down?

‖ **setem** (seˑtem). *Egyptology.* Also **Setem.** [ad. Egyptian *sm, stm.*] = *Sem.

1963 C. M. Forde tr. *Desroches-Noblecourt's Tutankha-*

men vi. 181 He officiated as the last scion of the royal family in the role of the *Setem* priest at the funeral of his predecessor. **1972** I. E. S. Edwards *Treasures of Tutankhamun* 43 (*caption*) King Ay, wearing the leopard skin of a *setem*-priest performs the Opening of the Mouth ceremony.

‖ **se-tenant** (səˌtənaṅ) *a.* *Philately.* [Fr., 'holding together'.] Of postage stamps: joined together as when printed; usu. applied to two or more stamps of different denominations or designs. Also as quasi-*adv.*

1911 F. J. Melville *Chats on Postage Stamps* 49 Se tenant.—A French expression signifying that the stamps referred to have not been separated: usually employed in reference to an error, or variety, when still forming a pair with a normal stamp. **1938** D. B. Armstrong *Key to Stamp Collecting* s.v., *Se Tenant.*—French phrase meaning 'joined together', and applied to a pair of stamps, one of which differs from the other. **1957** *Encycl. Brit.* XVII. 715A/1 *Se-tenant* stamps are two or more unsevered from each other. The term is usually applied to unusual pairs or larger pieces in which one or more of the stamps differs from the other. Sometimes two denominations of stamps will be printed in one sheet and a pair of stamps of the different denominations would be described as 'a *se-tenant* pair'. **1971** [see *funny *a.* 1 b]. **1980** *Jrnl. R. Soc. Arts* July 535/2 The stamps have been printed as four se-tenant designs of one value, that is four different designs joined together on one sheet.

seter, var. *saeter, setter.

Seto (se·to). The name of a city 12 miles north-east of Nagoya in Japan used *attrib.* and *absol.* to designate the pottery and porcelain produced from the kilns established there in the 13th cent.

1881 Audsley & Bowes *Keramic Art of Japan* 125 The specimens..cost about four times as much as corresponding articles of Arita or Seto make. *Ibid.* 226 Other descriptions are called..Seto-Suke, Seko-Kuro, and..Ki-Seto, or yellow Seto after the colour of the glaze used. **1925** W. Weston *Wayfarer in Unfamiliar Japan* xii. 127 The region whose chief and oldest settlement, Seto, gives its name to the Japanese term *Seto-mono* (lit. 'Seto ware'), porcelain, just as we ourselves employ the word 'china' to connote articles of a similar nature. **1945** W. B. Honey *Ceramic Art of China & Other Countries of Far East* v. 181 The 'yellow Seto' (*ki-seto*) was possibly suggested by a variety of the *temmoku*, but more probably by late Corean ware. **1959** R. Kirkbride *Tamiko* ix. 67 The tea arrived in small porcelain cups which Ivan recognized as Seto, very old and rare, from the kiln at Nagoya. **1972** P. Murray tr. *Shoya Yoshida's Folk-Art* (ed. 2) 52 (*caption*) Seto water dropper with peony relief.

set-out. Add: **1. g.** (Earlier and later examples.)

1833 Dickens *Let.* ? Jan. (1965) I. 14, I am consequently unable to tell the story and to deliver a plain unvarnish'd tale of the set out. **1903** Somerville & 'Ross' *All on Irish Shore* i. 15 'I'm sure Fennessy wishes to hear no more of it,' said Barnet acridly to Mrs. Griffen, when Mrs. Alexander had passed swiftly out of hearing, 'after the way those girls have been worryin' on at him about it all the morning. Such a set out!' **1933** D. L. Sayers *Murder must Advertise* x. 166 'Coo! that was a set-out, that was.'

set piece. Also **set-piece, setpiece.** [Set *ppl. a.*] **1. a, b.** See Set *ppl. a.* 8.

c. A (passage of) formal composition in prose or verse; a discourse, narrative, etc., composed according to a set pattern.

1932 C. Brown *Eng. Lyrics of 13th Cent.* p. xiv, In the English romance *Arthour and Merlin* a series of lyrics on the various months—May, June, February, &c.—are introduced as set pieces to divide the romance into Fitts. **1954** *Essays in Criticism* IV. 1 Little reason to suppose that Menenius is as impartial or as wise as his famous set-piece, the fable of the belly and the members, might at first sight suggest. **1959** I. & P. Opie *Lore & Lang. Schoolch.* ix. 156 And there is the recurrent set-piece: 'What's your name?' 'Sarah Jane.' 'Where do you live?' 'Down the lane' [etc.]. **1968** *Listener* 10 Oct. 475/3 Amis's prose is very good, and some of his little set-pieces are brilliant, as well as modish. Thus, the American road-scene. **1977** *Broadcast* 7 Nov. 13/3 'Hard Times'..as a novel..has a few splendid set pieces and many incidental pleasures. **1980** *Times Lit. Suppl.* 19 Sept. 1012/1 The ceremony of the Holy Fire in the Holy Sepulchre in Jerusalem—the major set-piece of the novel.

2. *Theatr.* A piece of scenery, either flat or three-dimensional and usu. free-standing, that represents a single feature such as a tree, a gate, or the like.

1859 E. Fitzball *Thirty-Five Years of Dramatic Author's Life* I. vi. 140 His long costly robes, becoming entangled with a set piece, pulled down with it, the orange tree excepted, every morsel of scenery on the stage. **1884** [see *cloth *sb.* 7]. **1930** Selden & Sellman *Stage Scenery & Lighting* I. vi. 163 The character of an outdoor setting is better suggested by a few plastic 'set' and 'built' pieces, such as ground rows, silhouette hills, rocks, trees, and fences placed in front of the sky than by anything placed on the drop. **1970** H. Nelms *Scene Design* ii. 19/1 Except for the cyc and the ground row needed to mask its lower edge, the whole setting..consists of a single set piece. This is a flat piece of scenery placed by itself somewhere in the playing space.

3. An organized movement, action, or manœuvre; *spec.* in *Sport*, a prescribed (and usu. rehearsed) movement or feature of the game by which the ball is returned to play, as at a scrummage in Rugby or a free kick in Football.

1938 D. S. Milford *Hockey* vi. 122 We have now finished our survey of the full-back positions for what may be called 'set pieces'. **1947** *Jrnl. R. Aeronaut. Soc.* LI. 840/1 The fourth and last type of operation is the raid against an enemy shore base, usually referred to as a 'set-piece'. This is really rather similar to a Bomber Command operation at shorter range and with smaller aircraft. **1960** E. S. & W. J. Higham *High Speed Rugby* xx. 314 Set-piece Rugby. This is not a game in the strict sense, but a series of set pieces, following rapidly, one on another, at a word from the coach or referee. **1977** *Western Mail* (Cardiff) 5 Mar. 18/5 Modern rugby is a game of pressure and most of this is upon the halves, from set pieces and loose play.

4. (with hyphen) *attrib.* passing into *adj.* Having the attributes of a set-piece; formally or elaborately planned or composed; set (cf. Set *ppl. a.*).

1947 *Jrnl. R. Aeronaut. Soc.* LI. 840/1 (*heading*) The 'set-piece' attack. **1962** E. Godfrey *Retail Selling & Organization* ii. 11 The windows themselves may be enclosed, providing a background for set-piece displays. **1968** *Economist* 17 Feb. 43/1 The not quite universal tendency of reporters to compare the setpiece battle situation that has been building up at Khe Sanh, close to the boundary between South and North Vietnam, to Dienbienphu. **1976** H. Wilson *Governance of Britain* 9 The prime minister is not only required to make a set-piece ministerial or other broadcast on major occasions; he is constantly in the news.

Setswana, var. *Sechuana.

settable, *a.* Delete *rare* and add later examples.

1967 *Electronics* 6 Mar. 6/2 The synthesized center-frequency marker and side markers are accurate, stable, and precisely settable. **1981** *Times Lit. Suppl.* 22 May 573/2 So much of modern poetry has no sense of rhythm or rhyme, and therefore is neither memorable nor settable to music.

‖ **settecento** (setetʃe·nto). [It.: short for *mil settecento* one thousand seven hundred.] The eighteenth century considered as a period of Italian art, architecture, music, etc.

1926 E. Hutton tr. *A. Venturi's Short Hist. Ital. Art* vi. 286 The secret of the fascination Corregio exercised on the art of the *Seicento* and *Settecento*..must be sought in the extraordinary sensibility of his nature. **1936** A. Huxley *Olive Tree* 151 It is a scene from a *settecento* Earthly Paradise—before the Fall of 1789. **1941** *Burlington Mag.* Aug. 38/2 With our last selection, we pass from the Cinquecento to the Settecento. **1966** T. Pynchon *Crying of Lot 49* i. 10 The Fort Wayne Settecento Ensemble's variorum recording of the Vivaldi Kazoo Concerto.

setter, *sb.*[1] Add: **I. 1. a.** (Further examples.)

1964 *New Statesman* 20 Mar. 455/1 Mr Holloway is severe on other people's abuses of English—particularly the setters of examination questions. **1976** *Listener* 23 & 30 Dec. 820/3 The *Listener* setter [of crossword puzzles] caters..for the cognoscente.

II. a. *Irish setter* = *red setter* s.v. *red *a.* 17 a (examples). *English, Gordon setter*: see under first element in Suppl.

1866 J. Walker in *Field* 6 Jan. 3/2, I should feel obliged by your allowing me to say a word or two..on the colour and general characteristics of the Irish setter. **1912** A. Huxley *Let.* 13 May (1969) 42 He, the dog, is a beautiful Irish setter, the only one of his kind within a radius of miles, as the beast has only just been introduced into Germany. **1975** J. M. Brearley (*title*) This is the Irish Setter.

setter: see *saeter, setter.

setting, *vbl. sb.*[1] Add: **I. 1. a.** (Further examples.)

1921 H. G. Crockett *Pract. Leather Manuf.* ii. x. 141 When the goods have become sufficiently dry for setting, and in a fairly stiff condition, they are taken down and brushed over with water. **1953** D. Woodroffe *Leather Dressing* xiv. 144 Sometimes the setting is done by striking out a second time on the same machine. **1965** M. McIntyre *Place of Quiet Waters* ii. 23 There was a..bucksaw, badly in need of setting.

2. a. (Later examples in contexts other than Croquet.)

1948 'Duplex' *Sharpening Small Tools* iv. 71 To adjust the setting of the plane..the tension screw is slightly slackened and the set adjustment lever is moved upwards or downwards. **1979** *Homes & Gardens* June 154/1 Some recent models also have a thermometer and control dial so you are able to alter the setting.

e. A set of cutlery or crockery, or of both, sufficient for one place at table. Cf. *place-setting* s.v. *place *sb.*[1] 29.

1952 A. Vanderbilt *Compl. Bk. Etiquette* ix. 98 A young bride can do very well with four- or six-place settings consisting of dinner knife, dinner fork, salad fork, butter knife, teaspoon, and dessert spoon. **1961** *Times* 30 May 15/6 Veneered oak canteen containing settings for 8 people, including fish knives and forks, in Mappin Plate. **1975** M. Orr *Rich Girl, Poor Girl* xxi. 278 Maggie..was given..a twelve-piece setting of Danish flatware.

6. c. *Psychol.* and *Sociol.* A person's disposition or cast of mind formed by experience and colouring his behaviour. Also, the immediate environment considered as an influence upon behaviour; *spec.* an environment designed to create a particular atmosphere, esp. for experiments with mind-affecting drugs.

1914 M. Prince *Unconscious* x. 311 Antecedent experiences of life..conserved in the unconscious formed a setting that gave the point of view and attitude of mind. **1954** Barker & Wright *Midwest & its Children* iii. 45 A behavior setting has been defined as a standing pattern of behavior and as part of the milieu which are synomorphic and in which the milieu is circumjacent to the behavior. *Ibid.* vi. 223 A day from the life of a child in the settings of a community gives a sample of behavior and habitat that is..limited. **1963** E. Goffman *Behav. in Public Places* ii. 21 The same physical space can come to be used as a setting for more than one social occasion, and hence as a locus for more than one set of expectations. **1968** *Science* 13 Dec. 1236/1 It is necessary to control set and setting... The total environment in which the drug is taken is the setting. **1974** M. C. Gerald *Pharmacol.* xviii. 341 Among the variables that modify the marijuana response are dosage, route of administration, set, and setting.

7. (Earlier example.)

1871 D. G. Rossetti *Lett.* (1967) III. 923 Dr. Bennett..can publish his setting if he makes no alteration in the words.

9. b. A clutch (of eggs).

1902 E. Nesbit *Five Children & It* iii. 85 A setting of Buff Orpington eggs that had not turned out well. **1938** M. K. Rawlings *Yearling* xiv. 140 The setting was hatched. The young quail, each no bigger than the end of his thumb, scattered like small windblown leaves.

11. (Further examples.)

1963 D. Seton *Essent. Mod. Cookery* 150 The setting property of jam is due to the presence of pectin in the fruit. *Ibid.* 151 Setting-point is reached when the jam forms a flake and drops off the spoon cleanly or sharply when shaken. **1969** *Jams, Preserves & Homemade Sweets* (Good Housekeeping Libr. Cooking) i. 10 Lemon juice..aids the setting. *Ibid.* 11 After the sugar has been added, the jam should be watched carefully and tested for setting without undue delay.

III. 13. a. (Further examples.)

1894 J. E. Davis *Elem. Mod. Dressmaking* (1895) 46 The setting-in of the second sleeve [is] proceeded with. **1942** P. I. Smith *Princ. & Processes Light-Leather Manuf.* v. 161 After dyeing and fat-liquoring the next process is setting-out or striking out, which today is usually done by hand. **1942** W. S. Churchill *End of Beginning* (1943) 27 It was agreed that I should propose to those concerned the setting-up of a Pacific Council in London. **1953** D. Woodroffe *Leather Dressing* xiv. 144 Striking out forms an excellent treatment prior to the setting out process. **1959** *20th Cent.* Nov. 345 The setting-up of an alternative..television service. **1962** A. Nisbett *Technique Sound Studio* viii. 146 If wear is noticeably accelerated,..rather complicated ways of setting up have to be devised. **1975** Bram & Downs *Manuf. Technol.* vi. 168 To assist 'setting up', tenon blocks are provided to engage in table slots. **1979** A. B. Emary *Woodworking* viii. 37 The steps..are setting out, which means making certain full-size drawings—in the trade this is called making a workshop rod, compiling a list of timber required for the job, and placing the necessary marks on the timber from the information on the rod.

b. setting-out, (*b*) *U.S.* = Set-out 3; **setting-up** *U.S. dial.*, an all-night vigil, esp. one kept by relatives beside the body of a dead person (cf. Set *v.*[1] 154 kk in Dict. and Suppl.).

1848 *Ladies' Repository* VIII. 337, I think you can afford to give that to Hen and Kate as part of their 'setting-out'. **1900** J. de F. Shelton *Salt-Box House* xxi. 169 Despite the high prices caused by the war, her [bridal] setting-out was not inferior, having its full complement of silver, china..Irish-stitch (damask), [etc.]. **1835** C. Gilman *Recoll. Southern Matron* in *Southern Rose* 14 Nov. 41/2 This solemnity is usually styled by the negroes 'a setting up'. **1905** 'P. Pennington' *Jrnl.* 25 Dec. in *Woman Rice Planter* (1913) viii. 272 All the grown servants have gone to the 'setting up', which is one of the strongest articles of their creed..the feeling that they must not be found in their beds on this mysterious night when the King of the world was born. **1949** 'J. Nelson' *Backwoods Teacher* xv. 160 Most of them stayed a few minutes and departed, perhaps leaving one member of the family for the 'settin'-up'.

d. Phr. *setting-to-rights*, the action of putting things in their correct places (on a shelf, etc.). Cf. Set *v.*[1] 25 a.

1847 C. M. Yonge *Scenes & Characters* xxiii. 284 That wearisome operation, a complete setting-to-rights; Eleanor..extended her cares from the stores to every other household matter. **1911** K. D. Wiggin *Mother Carey's Chickens* xiv. 124 Dozens of shelves in odd spaces helped much in the tidy stowing away of household articles... In the midst of all this delightful and cheery setting-to-rights a letter arrived.

14. setting-board, (*b*) (later examples); **setting coat** (later examples); **setting lotion,** lotion that is applied to the hair in order to assist the process of setting; **setting-room** *N. Amer. dial.* = Sitting-room 1; **setting stuff,** the fine plaster from which a setting coat is made; **setting-up exercise** (examples).

1894 W. Furneaux *Butterflies & Moths* ix. 122 The most important requirement is the setting boards, of which several are necessary. **1976** V. Nabokov *Details of Sunset* 158 He would first pin the carefully killed insect in the cork-bottomed groove of the setting board. **1916** E. A. Doncaster *Limes & Cements* xiii. 144 The setting coat..is made of the pure lime as it runs from the basin.

1927 A. H. Telling *ABC of Plastering* 206 The setting coat should be about one-eighth of an inch thick. **1926** *Hairdressing* 27 Aug. 181/1 No setting lotion nor dressing of any kind was used. **1941** N. Marsh *Death & Dancing Footman* (1942) ii. 42 She was met by the..familiar smells of hot hair, setting lotion, and the sachets used in permanent waving. **1977** J. Wilson *Making Hate* xiii. 157 That thick sweetish smell you always get in ladies' hairdressers..setting lotion or hairspray. **1741** *Probate Rec. New Hampshire* (1915) III. 30, I give to my Beloved Wife ..ye furniture of ye Chamber over our Setting room. **1832** W. D. Williamson *Hist. Maine* II. xxviii. 703 Our indigenous cherry, black-birch, and curl maple,..were shoved from the parlour and setting-room, to admit articles of foreign mahogany. **1908** J. C. Lincoln *Cy Whittaker's Place* iii. 38 It's your dad's house come back alive, it is so! Look at this settin' room. **1911** *Encycl. Brit.* XXI. 785/1 Setting stuff should not be applied until the floating is quite firm and nearly dry, but it must not be too dry or the moisture will be drawn from the setting stuff. **1927** A. H. Telling *ABC of Plastering* 187 Setting or fine stuff consists of one part of plasterer's putty to two or three parts of sand. **1939** W. Verrall *Solid & Fibrous Plastering* v. 64 Setting or skimming stuff can be applied in its raw state or an addition of 10 to 15 per cent. of plaster of Paris added. **1935** O. Nash *Primrose Path* 37 A few setting-up exercises. **1970** *Soviet Weekly* 25 Apr. 2 He gets up at half past five, does a few setting-up exercises, takes a bath, has breakfast, goes for a walk before starting work.

setting, *vbl. sb.*[2]: see *set *v.*[2]

setting, *ppl. a.* (Later examples.)

1969 *Jams, Preserves & Homemade Sweets* (Good Housekeeping Libr. Cooking) ii. 38 Poor setting fruits can be combined with the better ones to give added colour or flavour. **1974** M. Lindlaw *Super Sweets & Puddings* 53 Make up the Dream Topping..or whisk the cream until it begins to thicken. Whisk into the setting jelly.

setting-pole. Add: Chiefly *N. Amer.* (Earlier and later examples.)

1763 J. Bell *Trav. from St. Petersburg* II. xiii. 140 The barques..run often a-ground..and the people were obliged to..heave them off..with levers and setting poles. **1931** G. L. Nute *Voyageur* 40 Up to this point they had used 'setting poles' as well as paddles whenever the current was too swift for the ordinary method of propelling the canoe. **1959** *Moosehead Gaz.* (Dexter, Maine) Feb. 18/3 Junior eased the canoe down through the rocky rips with a setting pole.

settle, *v.* Add: **I. 3. b.** Also, *rarely*, in *active*: to install (someone) in a residence.

1853 Geo. Eliot *Lett.* (1954) II. 97 What do you think of my going to Australia with Chrissey and all her family?—to settle them, and then come back.

4. g. *U.S. slang.* To sentence (a person) to imprisonment, put in prison.

1899 'J. Flynt' *Tramping with Tramps* 396 *Settled*, in prison. **1914** Jackson & Hellyer *Vocab. Criminal Slang* 75 *Settled*,..convicted of misdemeanor or statutory offence. Example: 'He's settled for a two spot.' **1916** *Literary Digest* 19 Aug. 425/1 Foley was 'pinched' and 'settled' in San Quentin. **1930** *Amer. Mercury* Dec. 457/2 He goes to the counter and gets settled for a nickel. **1955** D. W. Maurer in *Publ. Amer. Dial. Soc.* xxiv. 151 Maybe he will get *settled*, or sent to prison; among pick-pockets this term does not carry the implication of a long sentence or a life-term..; it usually means two years.

II. 11. a. Also with *in*, to become established in a new home. Hence, to become accustomed to a new abode or to new surroundings.

1929 *Star* 21 Aug. 15/1 The Jellicoes..are 'settling in' at their new London home this autumn. **1951** M. McLuhan *Mech. Bride* (1967) 67/2 It will want to 'settle in, and enjoy the sense of belonging in America. **1960** J. Stroud *Shorn Lamb* xiii. 151 He'll settle in, I feel sure. It'll be a long job though. **1977** 'A. York' *Tallant for Trouble* iii. 48 We met the Brices after we came. We threw a party, to settle in, and the Brices were top of the list.

IV. 18. b. Of food or a meal: to be digested.

1944 L. P. Hartley *Shrimp & Anemone* iv. 44 What about these toboggans? We've given our tea time to settle.

19. b. Also *spec.* of an infant or a child.

1972 P. D. James *Unsuitable Job* iii. 88 The Webbers couldn't bear the boy to go to hospital; they'd tried it once and he didn't settle. **1976** 'D. Halliday' *Dolly & Nanny Bird* viii. 106 You can let go now, he won't settle.

21. c. (Later example.)

1900 H. A. Jones *Mrs. Dane's Defence* ii. 39 We will very soon settle Mrs. Bulsom-Porter.

23. b. Also *to settle down* or *in*, of the weather, a season, etc.: to set in.

1889 J. K. Jerome *Three Men in Boat* v. 70 'Going to clear up, d'ye think?'.. 'Well, nó, sir; I'm afraid it's settled down for the day.' **1939** K. Pinkerton *Wilderness Wife* x. 110 By the first week in December the winter began to 'settle in', as they say in the North.

V. 26. b. Also without prepositional compl.

1976 *Cambridge Independent Press* 16 Dec. 1. 10/5 But Felixstowe settled and after 25 minutes they took the lead when Bailey touched home Goffin's cross.

28. b. Also *to settle down.*

1869 J. Greenwood *Seven Curses London* ii. 19 At sixteen..the pair embark in housekeeping and 'settle down'. **1891** G. Chamier *Philosopher Dick* v. 120 'Didn't I tell you he was a looking hout for some crib to settle down.' 'Settle down, indeed! What do you mean?' 'Why, he's about to get spliced.' **1911** G. B. Shaw *Getting Married* 118 They had all, as they called it, settled down, like balloons that had lost their lifting margin of gas. **1928** E. O'Neill *Strange Interlude* vi. 230 He looks pretty dissipated..too many women..ought to get married and settle down.

VI. 32. e. *to settle for*, to decide or agree on, to content oneself with.

1959 P. Bull *I know Face* i. 11 My father wanted me to be a chartered accountant, a profession which seemed to me to lack glamour. However, in order to show willing, I did settle for 'journalism'. **1963** H. Garner in R. Weaver *Canad. Short Stories* (1968) 2nd Ser. 27 There were plates of doughnuts..but I settled for a mug of coffee. **1972** C. Fremlin *Appointment with Yesterday* xi. 82 'You couldn't start straight away, could you?.. Or would you rather have some coffee?'.. Milly found the courage to settle for the coffee.

settleable, *a.* Restrict *nonce-wd.* to sense in Dict. and add: **2.** Having the property of settling or sinking to the bottom of a liquid.

1940 Imhoff & Fair *Sewage Treatment* ii. 30 It is important to know how much of the solid matter is in suspension, how much is settleable, and how much is in solution. **1947** [see *floc]. **1969** *Sci. Jrnl.* Mar. 81/2 In the activated sludge process the ability of the organisms to flocculate into settleable clumps is automatically inbred.

settlement. Add: **III. 11. c.** The sinking of floc and other solid particles in liquid sewage. Also *attrib.*, as *settlement tank.*

1912 H. Lemmoin-Cannon *Textbk. Sewage Disposal in U.K.* xviii. 62 Tanks of the same kind..are used for the purpose of attaining the settlement of the suspended organic solids by sedimentation. **1927** T. H. P. Veal *Disposal of Sewage* v. 54 Quiescent settlement tanks are operated on what is known as the fill-and-draw principle. *Ibid.* 55 The amount of clarification effected in a given time by quiescent settlement is greater than that effected by the continuous flow method. **1977** C. B. Capper in A. G. Callely et al. *Treatment Industrial Effluents* vi. 90 Probably the oldest method of removing suspended solids was by the use of horizontal-flow settlement tanks.

17. *attrib.* and *Comb.*, as (senses 6, 14) *settlement area, pattern*; (sense 9 b) *settlement price, terms*; **settlement day** = *settling day* s.v. Settling *vbl. sb.* 3 b; **settlement house** *U.S.*, an institution in an inner city area, usu. sponsored by a church or college, that gives educational, recreational, and other social services to the community (cf. sense 16).

1963 H. N. Savory in Foster & Alcock *Culture & Environment* iii. 31 The south Wales seaboard was a primary settlement-area of the continental colonists. **1977** *Word 1972* XXVIII. 72 Brittany is a dispersed settlement area, and farms are either isolated or in small clusters. **1896** W. H. S. Aubrey *Stock Exch. Investm.* 314/2 (Index), Settlement days. **1901** C. Duguid *How to read Money Article* xvii. 75 Directly one account is ended by the fortnightly settlement, another account begins. It commences at noon on the first settlement day. **1907** J. Strong *Challenge of City* 307 Your letter..was duly received and reply thereto delayed awaiting report from the inspection districts wherein are located the Settlement houses you mentioned. **1959** *New Statesman* 24 Oct. 534/2 In relation to the street gangs, most of these disquisitions regard the conventional 'agencies'—boys' clubs, mixed clubs, settlement houses, community centres—as ineffectual. **1978** G. Vidal *Kalki* v. 114 Of course, Amelia did work in settlement houses, helping the poor. **1958** G. Lienhardt *Tribes without Rulers* 98 Dinka settlement-patterns differ from each other according to the two broadly different kinds of country. **1928** *Daily Mail* 25 July 19/3 Tin: Standard cash quoted £217 10s. to £217 12s. 6d.; three months, £214 10s. to £214 12s. 6d.; settlement price, £217 10s. **1931** C. Maughan *Markets of London* 122 Rubber is also sold on 'settlement terms', which means that a buyer receives a profit or pays a loss every fortnight, in a similar way to settlements on the Stock Exchange.

settler. Add: **1. c.** A clerk in a betting shop who calculates the winnings.

1963 L. Meynell *Virgin Luck* vii. 172 'But what do you do..?' 'Settle. I'm a settler. I work out a bet. It goes down on a slip and after the race the settlers go through the slips and work them out.' **1966** P. Willmott *Adolescent Boys of East London* vi. 102 Non-manual jobs of a more 'routine' kind, mainly clerks and shop assistants, draughtsmen and betting-shop 'settlers'. **1977** *Evening Post* (Nottingham) 27 Jan. 14/1 (Advt.), Saturday settler required by independent licensed Betting Office.

2. a. (Earlier Amer. examples.) *old settler*: see *old *a.* D 4.

1696 *Rec. Early Hist. Boston* (1881) VI. 51 The first goers or first setlers of Woodstock. **1739** W. Stephens *Jrnl.* 15 Dec. in *Colonial Records State of Georgia* (1906) IV. 469 One Bunyon, a Builder of Boats and a Settler there.

5. settler's clock (earlier example); **settler's effects** *Canad.*, goods brought into the country by an immigrant for his personal use that are exempt from import duty; also *transf.*; **settler's matches** (earlier example).

1827 P. Cunningham *Two Years in N.S.W.* I. 232 The loud and discordant noise of the laughing jackass (or settler's clock, as he is called)..acquaints us that the sun has just dipped behind the hills. **1911** *Daily Colonist* (Victoria, B.C.) 2 Apr. 14/3 She landed 150 tons of cargo here, including three boilers..settler's effects, and general freight. **1939** K. Pinkerton *Wilderness Wife* i. 13 Beside us on the station platform that morning were our canoe, camp outfit and a few 'settler's effects'. **1965** I. Reekie *Along Old Melita Trail* ii. 12 Coming to Manitoba with two cars of settlers' effects. **1891** H. Lawson in *Bulletin* (Sydney) 19 Dec. 21/2 And we walked so very silent—being lost in reverie—that we heard the 'settlers'-matches' gently rustle on the tree.

settling, *vbl. sb.* Add: **5. a.** Also *settling down.*

1911 G. B. Shaw *Getting Married* 118 The process of settling down would go on until they settled into their graves.

b. *settling in*, the action of establishing one's residence in a new place or of becoming accustomed to new surroundings. Chiefly *attrib.*, as *settling-in allowance, grant.*

1955 *Times* 16 July 2/4 Generous settling-in allowance during first month, and housing assistance can be afforded to selected applicants. **1965** *Wireless World* Aug. 120 (Advt.), Free air travel will be provided for a successful applicant..and a 'settling in' allowance will also be paid on arrival in Salisbury. **1973** *Soviet Weekly* 17 Feb. 11/3 Some redundant teachers were willing to be sent to localities where there was a shortage of teachers. They were given settling in grants totalling two or three months average earnings, plus a quarter of that sum for each member of their family. **1974** *Economist* 31 Aug. 83/1 (Advt.), A settling-in allowance of up to £400 is payable in approved circumstances and assistance with housing for a short time is a possibility.

c. Special Comb.: **settling time,** the time taken for a measuring or control instrument to get within a certain distance of a new equilibrium value without subsequently deviating from it by that amount.

1951 Chestnut & Mayer *Servomechanisms & Regulating System Design* I. xiv. 410 (*heading*) Settling time t_s to reach 5 per cent of final value. **1974** J. W. Brewer *Control Systems* x. 285 Rise time is a measure of the speed of response, and the ratio of settling time to rise time is a measure of damping.

6. a. (Further *attrib.* examples.)

1867 J. A. Phillips *Mining & Metallurgy of Gold & Silver* ix. 183 When settling pits are used for the purpose of collecting the tailings for subsequent treatment, it is necessary that at least two of them should be provided, so that whilst one is being filled the other may be cleaned up. **1868** *Times* 4 Aug. 4/3 A complete churning and intimate union are effected by the sewage passing through a number of small apertures into cells, in each of which revolves a stirrer, and thence out of the cells into two very spacious settling tanks. **1894** G. E. Waring *Mod. Meth. Sewage Disposal* xvii. 234 The flocculent matter passing the screen clogged the 4-inch absorption-tiles after a time. This was obviated by constructing a settling-chamber. **1899** *Rep. Dept. Mines on Goldfields N.Z. 1898–99* 45 The crushed ore passes over amalgamated copper-plates, 12 ft. long, into four sets of spitzlutten or settling-boxes. **1915** G. B. Kershaw *Sewage Purification & Disposal* iv. 117 In most settling towers the bottom of the tank is formed in the shape of a cone, the sludge drain being placed at the apex. **1928** W. A. Mitchell *Civil Engineering* v. 572 The term settling basin is ordinarily applied to a reservoir which contains from one to four days' supply of water. **1940** Imhoff & Fair *Sewage Treatment* iv. 72 In these tanks, the lower story serves as a sludge chamber and the upper one as a settling chamber. **1941** C. A. Ward *Those Raw Materials* vi. 272 The separation of sludge [from oil] may take place in settling tanks or may be brought about by centrifugal means. **1955** Linsley & Franzini *Elements Hydraulic Engin.* x. 249 Settling basins..just below the point of diversion so that sediment can be collected and sluiced back into the river. **1964** *Grouts & Drilling Muds in Engin. Pract.* 29/2 Retention [of the mud] in settling pits should be not less than 12 min. **1978** *Sci. Amer.* July 99/1 Conventional sewage treatment employs a combination of large settling tanks, bacterial cultures and sludge-thickening devices to decontaminate waste water and to concentrate the solid residue.

set-to. Add: **2. b.** Also, in weakened sense: an argument, a heated debate.

1898 Lloyd George *Let.* 25 Dec. (1973) 116 We have had several sets to over church matters & female suffrage & we hit without sparing. **1912** Ld. Fisher *Let.* 2 Aug. in R. H. Bacon *Life Ld. Fisher* (1929) II. xvi. 155 At the Defence Committee yesterday.. we had a regular set-to with Lloyd George. **1976** Nichols & Armstrong *Workers Divided* ii. 135, I like nothing better than a good set-to with a good shop steward.

c. (Earlier example in the sense 'a drinking-bout'.)

1813 W. Dunlap *Mem. G. F. Cooke* I. iv. 63, I doubt not that his dinner with the Irish manager was a roaring set-to, a full and convincing proof of what is called Irish hospitality.

3. (Earlier U.S. example.)

1840 *Spirit of Times* 9 May 115 For two miles he trailed, keeping his horse in position for a set-to in the last mile.

se·t-up, *sb.* [f. vbl. phr. *to set up*: cf. Set *v.* 154 in Dict. and Suppl.] **1.** See *set-up* (a) s.v. Set-. Also *spec.* a stand or display at a carnival, etc.

1841, etc. [see *set-up* (a) s.v. Set-]. **1925** C. R. Cooper *Lions 'n' Tigers* v. 135 Mike..responded almost immediately,..returning to his various 'stands' and 'set-ups' as though he had never been away. Mike is still on the job with the circus. **1938** E. Caldwell *Southways* 53 'You win the set-up!' Bess cried, ducking under the railing. 'It's all yours! Go on in there and take it!'

2. a. The way in which something is organized, arranged, or constituted; an organization, arrangement, system, or situation. See also *set-up* (b) s.v. Set-.

Set-up occurs in a vague and indefinite sense in a large number of contexts, but several reasonably distinct areas of use can be isolated: (*a*) a business or administrative structure or organization: also, an economic, social, or political system (both with reference to the system or the persons involved); (*b*) a domestic situation, as determined either by lifestyle or personal relationships; (*c*) a team (esp. in *Sport*); (*d*) the layout of some mechanical apparatus or equipment.

1890 [see *set-up* (*b*) s.v. Set-]. **1922** *Proc. IRE* X. 249 Oscillograms of the essential current and voltage relations existing in the systems of the type illustrated by Figures 1 and 2 were obtained with the set up of apparatus illustrated in Figures 5 and 6, respectively. **1928** *Daily Express* 12 Mar. 13/2 The national set-up of motion picture chairmen. **1932** *Sun* (Baltimore) 31 Oct. 8/2 It develops that this Princeton 'set-up' actually outrushed Michigan, to produce what..ought to be classed..as at least a Class B 'upset'. **1932** D. Frome *By-Pass Murder* v. 39 Servants..belong to that..class of people whose public and private lives have no connection with each other... Publicly and privately their set-ups are as different as Fulham Road and Grosvenor Square. **1933** *Sun* (Baltimore) 29 July 1/1 They provided the set-up for the conspiracy. They helped organize the groups responsible for bombings, sluggings and strikes. **1939** [see *amputee]. **1942** *Archit. Rev.* XCII. 52/2 The reason for this must be found in the general economic set-up of these countries. **1943** F. L. Wright *Autobiogr.* (rev. ed.) v. 403 Other workmen.., learning of our set-up, asked us to take them on. **1946** E. G. Webber *Johnny Enzed in Italy* 9 Nice set up you've got here, Dig. **1949** E. L. Mascall *Existence & Analogy* iii. 48 Before we know where we are we shall have the best possible world and the whole Leibnitian set-up. **1951** *Sport* 27 Jan.–2 Feb. 7/1 The transfer had had full notice in the Press and yet he was shown included in the home set-up. **1952** H. Read *Philos. Mod. Art.* xiii. 233 What the work of art 'expresses', in an emotional sense, depends very largely on what the spectator brings, in the way of an emotional set-up, to the work of art. **1953** W. Burroughs *Junkie* 7 But these people were jerks for the most part and, after an initial period of fascination, I cooled off on the setup. **1958** *Times* 19 May 11/2 At the time of the battle of Marathon the Athenians had..a rather inefficient command set-up. **1962** A. Nisbett *Technique Sound Studio* xi. 191 In a full-scale studio 'telephone conversation' set-up it is usual to have two microphones, one normal and the other with a filter in the circuit. **1962** J. Ludwig in R. Weaver *Canad. Short Stories* (1968) 2nd Ser. 254 What a set-up for those two kids! Jimmy would spend a fortune on Shirley, but Maxie would spend the night. **1971** *Physics Bull.* Sept. 513/1 Multinational Data, a joint setup linking ICL with the French Compagnie International pour l'Informatique and the American Control Data Corporation. **1973** B. Broadfoot *Ten Lost Years* xi. 122 A fellow..offered us a job. This was the set-up and..we jumped at it. **a 1974** R. Crossman *Diaries* (1976) II. 295 Burke is now far the most powerful of the Prime Minister's confidants, and has got Harold tightly integrated into the Whitehall set-up. **1977** *Western Morning News* 30 Aug. 3/5 The family atmosphere certainly grips those who come from abroad, many of whom can find nothing to compare with the set-up in their own countries.

b. *Computers.* The arrangement of interconnections and setting of parameters in a computer, esp. an analogue one, necessary for the performance of a particular calculation.

1935 *Mem. & Proc. Manchester Lit. & Philos. Soc.* LXXIX. 67 A second diagram is then drawn up showing the various gear ratios, positive directions of rotation of the various shafts, [etc.]..., and from this second diagram the actual set-up of the machine is made. **1945** *Jrnl. Franklin Inst.* CCXL. 277 After a few days of operation the 'set-up' time per solution became tolerably small. **1952** G. A. & T. M. Korn *Electronic Analog Computers* ii. 32 A number of computer laboratories have successfully used so-called setup sheets as an intermediate step between the preparation of the block diagram and the actual computer setup. **1970** D. E. Hyndman *Analog & Hybrid Computing* iv. 82 A considerable amount of time is generally required to patch and check out a problem set-up.

c. The arrangement of guns, decoys, etc., for shooting wildfowl.

1939 *Sun* (Baltimore) 6 Nov. 7/4 Mr. Leichhardt has directed them to..examine the shooting 'set-ups' at the various lodges. **1957** R. Scharff *Compl. Duck Shooter's Handbk.* iv. 105 You'll get more ducks if you have some goose decoys in your setup. **1973** G. Gresham *Compl. Wildfowler* x. 150 Some days the birds just won't decoy properly. They..pitch in before they reach your setup.

d. *Cinematogr.* (See quot. 1959.)

1941 B. Schulberg *What makes Sammy Run?* xi. 284 A director..worrying..about the camera set-ups. **1957** V. J. Kehoe *Technique Film & Telev. Make-Up* i. 17 In motion pictures, where the scene is shot from each angle separately in different lighting set-ups and schemes, the lighting effect can be designed to be complementary to the make-up. **1959** W. S. Sharps *Dict. Cinematogr. & Sound Recording* 128/1 *Set-up*, the arrangement of the scenery, props, performers, lights, microphones and cameras for a particular shot. **1963** *Movie* Feb. 31/1 The open court scenes are presented within a framework of two medium close camera set-ups. **1976** C. Larson *Muir's Blood* xviii. 101 He got here at a quarter after nine. We won't get our first setup before noon.

e. The difference between the maximum and minimum heights of a water surface tilted by wind action.

1951 *Jrnl. Res. Nat. Bureau of Standards* (U.S.) XLVI. 373/1 An attempt was made to establish the relationship between set-up and the depth of water for the case where waves were present. **1968** R. W. Fairbridge *Encycl. Geomorphol.* 235/1 The intertidal zone is not a simple transition but rather a zone of maximum energy, whose peak level ranges through the tide cycle (plus storm-wave 'setup').

3. *colloq.* (orig. and chiefly *U.S.*). In *Boxing*, a fighter who can be easily defeated by his opponent (with the implication that he has been deliberately chosen on these grounds); *gen.*, an opponent who is easy to defeat; a thing that is easily overcome or accomplished, a 'push-over'. *spec.* in *Lawn Tennis*, a ball that is easy to hit or smash.

1926 R. Hughes in *Hearst's Internat.* Feb. 44/2 A guy was tellin' me that set-ups are has-beens or never-wases who get paid to stand up just long enough to be knocked out. **1926** *Clues* Nov. 162/2 *Set-up*, something easy; soft. **1932** *Sun* (Baltimore) 23 Sept. 1/2 Blaine himself did not take him [*sc.* Chapple] seriously, considered his opposition a 'set-up'. **1946** *College Topics* (Univ. of Va.) 7 Dec. 3/1 It must not be forgotten that Georgia had pretty much of a set-up as far as the schedule was concerned and it would have been a real surprise if Georgia had been beaten. **1950** *Sun* (Baltimore) 24 Mar. B 22/4 Marciano has fought a string of set-ups, and there are few known boxers on La Starza's list. **1957** M. Millar *Soft Talkers* vi. 63, I went after him anyway, tooth and nail. It was easy. Ron was a perfect set-up. **1961** *Times* 18 May 5/2 He is well aware that the best players never miss a 'set-up'. **1969** *New Yorker* 14 June 44/3 Graebner could now probably explode one. He has what is almost a set up on his power side.

4. *U.S. colloq.* The glass, ice, soda, etc., required for mixing a drink, which is served to customers, who supply their own spirits, in unlicensed premises.

1930 *Sun* (Baltimore) 24 Dec. 11/1 If the club did not ban liquor drinking on the premises and cease serving 'set-ups' its lease would be in danger. **1944** C. Himes *Cotton gonna kill me Yet* in *Black on Black* (1973) 197 You pays two bucks to get in this joint, fo' bucks for a half-pint grog, two bucks for a coke setup. **1954** F. P. Keyes *Royal Box* xiii. 186 But Herb asked did I want to turn him into a lone drinker, and where was the icebox, he could make his own setups, and the first thing I knew we were sitting at the table drinking ice-cold highballs. **1964** *Listener* 19 Nov. 803/3 The head waiter will probably relieve him of the bottle, place it on his table, and then serve him unlimited quantities of what are called 'set-ups' to make it palatable. These are the tonic-water, the ice, the soda, or whatever it is. **1973** W. McCarthy *Detail* i. 65 He looked over to the sideboard and saw a complete assortment of liquors, rums and set-ups.

5. *U.S. colloq.* A place- or table setting at a restaurant; the dishes, cutlery, etc., which make up this.

1934 in Webster. **1941** J. Smiley *Hash House Lingo* 48 *Set up*, table eating utensils. **1978** J. D. MacDonald *Empty Copper Sea* iii. 34 He led us to a corner booth set up for four, whipped away the extra setups.

6. *colloq.* (orig. *U.S.*). A scheme or trick whereby an innocent person is caused to incriminate himself or a criminal is caught red-handed; a 'frame-up'.

1968 *Sun Mag.* (Baltimore) 13 Oct. 28/3 That's how the narcs get most guys on possession of narcotics—through set-ups. **1970** W. Burroughs Jr. *Speed* iv. 78 A set-up crossed my mind, but Jesus, who cared if it was a set-up or not, we were still busted. **1973** *Black Panther* 7 Apr. 8/2 As he was bringing the food inside he noticed the driver of the truck had turned and was running away. Grady immediately realized it was all a set-up. **1978** J. Gardner *Dancing Dodo* xxix. 236 Arthur's clean... It was a set-up... I had him checked like you'd check a dodgy engine.

7. Special Comb.: **set-up man,** (*a*) *U.S.* (see quot. 1953); (*b*) *N. Amer.* a man who sets up machinery or equipment.

1953 W. Burroughs *Junkie* i. 18 They are always looking for a 'setup man', someone to plan jobs and tell them exactly what to do. **1954** [see *job enlargement* s.v. *job *sb.*[2] 7]. **1968** *Globe & Mail* (Toronto) 13 Jan. 48/2 (Advt.), Foreman and set-up man, polyethylene film extrusion, for afternoon shift. **1978** *Detroit Free Press* 16 Apr. F1/3 (Advt.), Willing to train experienced set-up man who has leadership potential.

sevagerous, var. *savagerous *a.*

seven, *a.* and *sb.* Add: **A.** *adj.* **1. e.** In *Naut.* slang phrs. *to knock seven bells out of* (someone): to beat (someone) severely; similarly, *to scare seven bells out of*: to terrify. Cf. Bell *sb.*[1] 3 b.

1929 F. C. Bowen *Sea Slang* 121 *To knock seven bells out of a man*, to give him a hiding or knock him out. **1932** J. W. Harris *Days of Endeavour* ix. 158 Three angry Norwegians..knocked seven bells out of him. **1933** M. Lowry *Ultramarine* iv. 206 Yis. He's knocked seven bells out of harder cases than you in his time. **1943** F. C. Hendry *True Tales of Sail & Steam* i. 11 She [*sc.* a ship] scared seven bells out of us and gave us the worst month I have ever known at sea.

2. g. *seven-a-side(s*, a form of Rugby Union football played with only seven men on each side; hence, a rugby match or tournament played with teams of this size. Cf. sense *2 e of the sb.

1900 *Scottish Sport* 13 Apr. 6/1 Melrose..will to-morrow resound with a Babel of tongues..eager and excited over the first of the Border seven-a-side tournaments. **a 1917** E. C. Smith *Braid Haaick* (1927) 24 The 'Greens' hev wun the seevin-a-sides at their ain spoarts. **1935** *Encycl. Sports* 529/1 *Seven-a-Side.* This variant of the Rugby Union game is much played by the better clubs towards the end of the season... There are, usually, three forwards and four backs on either side. **1961** *Times* 5 May 11/1 We would much rather have seen the seven-a-sides at Twickenham.

B. *sb.* **2. e.** Short for *seven-a-side(s* (see sense *2 g of the adj.).

1926 *Times* 26 Apr. 5/5 Cussen showed in the semi-final what pace means in the game of Sevens. **1977** *Daily Mirror* 15 Mar. 30/1 Rosslyn Park are angry that they cannot call their famous schools sevens tournament by the names of their sponsors.

3. c. Phr. *to throw a* (or *the*) *seven* and varr., to die; also, to faint or vomit. *Austral. slang.*

1894 H. Lawson *Martin Farrell* in *Coll. Verse* (1967) I. 269, I am pretty cronk and shaky—too far gone for hell or heaven, An' the chances are I'm goin'—that I'm goin' to 'do the seven'. **1899** W. T. Goodge *Hits! Skits! & Jingles!* 17 You could bet on me chuckin' the seven If she slung me for some other bloke! **1908** [see *marble *sb.* 4 b]. **1932** A. Upfield *Royal Abduction* xxvi. 201 If she sees the thing she won't scream and throw a seven. She'll shoot. **1966** T. Ronan *Once there was Bagman* x. 217 The partially digested fruit must have swollen inside me, for before long I was chucking sevens around the flat as I had done a few years before when I had that touch of ptomaine poisoning.

d. *seven-and-a-half*, a round game of cards in which the object is to make the number seven-and-a-half without exceeding it, by counting the pips on the cards.

Court cards (except for the King of Diamonds) are counted as worth half a point.

1937 *Sun* (Baltimore) 26 Jan. 5/5 All other games and machines—..twenty-one, blackjack, seven-and-a-half, big Injun, Klondyke and craps are mentioned specifically—require a $50-per-month license. **1964** A. Wykes *Gambling* vii. 178 Blackjack and seven-and-a-half are found more rarely in the casinos of Europe than those of America.

C. 1. a. *seven-maned, -stringed, -tongued, -towered.*

1949 S. Spender *Edge of Being* 29 The seven-maned Golden lions. **1965** J. A. Michener *Source* (1966) 251 He carried a small seven-stringed lyre made of fir wood trimmed with antique bronze and strung with twisted sheep's gut. **1913** E. Stock *Heroic Bishop* ii. 12 In later days he became known in India as 'the seven-tongued man'. **1959** E. Pound *Thrones* xcvi. 19 From the Palace, half-circle that street is, ending near the seven-towered castello.

b. *seven-leaved* (later example).

1927 E. Sitwell *Rustic Elegies* 81 The seven-leaved man-plant.

2. *seven-course, -cubit, -eighths, -feet* (earlier example), *-figure, -foot* (examples), *-octave* (earlier example), *-point, -sacrament.*

1933 C. Day Lewis *Magnetic Mountain* 12 The penny-a-liner, the seven-course diner. **1858** M. Arnold *Merope* 28 Agamemnon's unhappy,..world-fam'd, Seven-cubit-statur'd son. **1881** O. Wilde *Poems* 117 The seven-cubit spear. **1939** M. B. Picken *Lang. Fashion* 123/1 *Seven-eighths length*, length of coat that is shorter than dress or skirt by a little less than one-eighth of the length from shoulder to hem. **1963** Bird & Hutton-Stott *Veteran Motor Car* 186 A patent plate clutch, a new form of universal joint and seven-eighths-elliptic back springs of which the upper portions were cantilevered. **1979** *Tucson* (Arizona) *Citizen* 20 Sept. 2B/1, I was told that nothing will make my fall wardrobe look more up-to-date than a seven-eighths-length coat. **1782** 'J. H. St. John de Crèvecœur' *Lett. from Amer. Farmer* vi. 169 The right whale, or seven feet bone..about sixty feet long. **1842** *Penny Cycl.* XXIII. 498/1 Seven-figure numbers to 100 thousand. **1933** *Brit. Jrnl. Psychol.* XXIII. 358 Seven-figure logarithms were used in all calculational work. **1972** *Daily Tel.* 10 Apr. 3/8 Lord Salisbury..died before the Budget so his family cannot benefit from the top duty scale reduction..for seven-figure fortunes. **1935** *Discovery* Feb. 62/1 We read on p. 42 that Watt's 'Old Bess' had a single cylinder of 33 inches diameter—usually termed bore—and a 7 foot stroke. **1950** *Amer. Speech* XXV. 237/2 *Seven-foot line*, a line of sight seven feet off the property line from one seven-foot point to another. **1869** 'Mark Twain' *Innoc. Abr.* lix. 580 Any one can see a part of the unquestioned and undisputed Temple of Solomon, the same consisting of three or four stones lying one upon the other, each of which is about twice as long as a seven-octave piano. **1889** *Cent. Dict.*, *Seven-point*.., *a.*, related to seven points, as, the seven-point circle. **1939** A. Rodger in F. C. Bartlett et al. *Study of Society* ii. xi. 259 Causes of occupational failure..are..classifiable under seven headings. [*Note*] This seven-point plan, has, of course, other uses. **1977** *Word 1972* XXVIII. 310 A seven-point scale. **1935** *Burlington Mag.* Feb. 81/1 The well-known series of Seven Sacrament fonts, nearly all confined to..Norfolk and Suffolk. **1955** M. D. Anderson *Imagery Brit. Churches* iii. i. 54 There are about forty Seven Sacrament fonts... The bowls are carved with panels of figure sculpture, each representing one of the Sacraments.

3. *seven-shooter* (earlier examples).

1860 *Charleston* (S. Carolina) *Mercury* 6 Nov. 3/5 (Advt.), Allen & Wheelock's seven shooters. **1872** 'Mark Twain' *Roughing It* ii. 23, I was armed to the teeth with a pitiful little Smith & Wesson's seven-shooter..and it took the whole seven to make a dose for an adult.

seven-league(d, *a.* (Earlier example.)

1793 W. B. Stevens *Jrnl.* 27 Feb. (1966) i. 70 Wrote to Dewe that I would put on my seven league boots next weekend and stretch my course to Appleby.

sevenpence. Add: Also *transf.*, † transportation for seven years (earlier example).

1821 P. Egan *Life in London* ii. iii. 230 My Lord, if I am to stand *seven-pence*, my Lord, I hope you'll take it into your consideration.

seven sisters. Add: **5.** The seven international oil companies noted for their dominant influence on the production and marketing of petroleum (see quot. 1976).

1962 *Times* 29 Oct. 11/2 Mattei..liked to take the view that the 'Seven sisters'—as he called the majors—were making excessive profits out of both Governments and consumers. **1966** J. Aldridge *Statesman's Game* xviii. 137 Was this..a challenge to the Seven Sisters of the oil world? **1976** *N.Y. Rev. Bks.* 15 Apr. 20/1 The group of international oil companies often referred to as 'the Seven Sisters'—Exxon, Mobil, Gulf, Standard Oil of California, Texaco, British Petroleum, and Royal Dutch Shell.

6. The group of long-established colleges (originally for women only) which were formerly regarded as the most prestigious women's colleges in the U.S. Cf. *Ivy League.

1962 *Changing Times* Apr. 37/2 The most difficult women's colleges to get into are the so-called 'Seven Sisters'—Barnard, Bryn Mawr, Mount Holyoke, Radcliffe, Smith, Vassar and Wellesley. **1979** *N.Y. Rev. Bks.* 17 May 43/2 (Advt.), Yale grad..would like to meet a 'seven sisters' graduate. **1980** L. Birnbach et al. *Official Preppy Handbk.* 86/2 The feminine equivalent of the Ivy League is the group of colleges known as 'The Seven Sisters'.

seventeen, *a.* Add: **4. seventeen-year cicada** = next; **seventeen-year(s') locust** (earlier and later examples).

1870 *Amer. Naturalist* III. 106 The eggs and young of the seventeen-year Cicada. **1950** *Chicago Daily News* 13 Jan. 42 The periodic or 17-year cicada lives the longest of any known insect. **1817** *Columbian Centinel* (Boston) 14 May 1/4 The southern papers have announced that the present is the year for the appearance of what is called..the Seventeen Years Locust... The insect lives above ground about two months, and 17 years in it. **1843** H. D. Thoreau *Let.* 7 July (1958) 121 Have you seen the Seventeen year locust in Concord? **1975** *Islander* (Victoria, B.C.) 3 Aug. 3/1 The juice-sucking 17-year locusts—which are really cicadas—were also around.

seventh, *a.* Add: **2. c.** *seventh wave*, the wave traditionally thought to be the biggest in an increasing swell of the sea; also *fig.*, a culminating act or experience.

1891 Kipling *City of Dreadful Night* i. 3 Six moderately pure mouthfuls of [Calcutta] air may be drawn without offence. Then comes the seventh wave and the queaziness of an uncultured stomach. **1908** G. Meredith *Let.* 24 Jan. (1970) III. 1623 You know the seventh wave. There must be a gathering of the waters before a big surge is thrown on shore. **1940** L. MacNeice *Poems 1925–1940* 251 The Northern Lights and the Seventh Wave. **1976** E. Ward *Hanged Man* i. 3 A seventh wave pulled him backwards into undertow.

seventh-day. 1. b. *Seventh-day Adventist*: (earlier and later examples.)

1860 *Advent Rev. & Sabbath Herald* (Washington) 23 Oct. 179/2 *Resolved*, that we take the name of Seventh-day Adventists. **1931** *Times Lit. Suppl.* 19 Feb. 139/2 Seventh Day Adventist missions. **1956** R. Macaulay *Towers of Trebizond* v. 49 The local Anglican priest..would give him news of the Seventh-Day Adventist Mission. **1977** *Hongkong Standard* 12 Apr. 5/1 Seventh Day Adventist Church Conference.

seventy, *a.* and *sb.* Add: **A.** *adj.* **2. b.** *seventy-eight*, a gramophone record designed to be played at a speed of seventy-eight revolutions per minute (a speed that was standard until the introduction of the long-playing record); also written *78*; *seventy-five*, (*b*) a gun of 75 mm. calibre formerly used in the armies of the French Republic and the U.S.A.; also written *75*; *seventy-three(s* (U.S. slang), best wishes, good-bye; also written *73*.

1951 Sackville-West & Shawe-Taylor *Record Guide* 720 The wise collector will retain his 78 gramophone and the pick of his 78 recordings, continuing to buy good new 78s (especially single discs). **1967** R. Petrie *Foreign Bodies* xi. 156 Stina looked out her two old seventy-eight records and the four tunes on them sounded through the villa all day. **1971** G. Steiner *In Bluebeard's Castle* iv. 92 There is a science and market..in worn 'seventy-eights'. **1977** *Gramophone* Apr. 1527/1 The LP, at 33⅓ revolutions per minute, and the single, at 45 rpm, have retained their popularity with record-buyers, at the same time sounding the death-knell of the 78, which was finally abandoned in 1958. **1915** E. Wharton *Fighting France* 56 We begin to come more and more frequently on big colonies of 'Seventy-fives'. **1969** A. Horne *To lose a Battle* xxi. 510 Amongst the guns taken were some 7,000 French '75s' of First War vintage. **1941** *Traffic World* LXVIII. 198/1 Morse code operators..used many arbitrary numbers to shorten their work..4 meaning 'where', ..73 'best regards' and 22 'kisses'. **1976** *S9* (N.Y.) May/June 31/2 Seventy-threes, and 'bye.

B. *sb.* **2.** (Earlier and later examples of a decade in a particular century.)

1887 W. E. S. Fales *Brooklyn's Guardians* iv. 57 Before the 'Seventies', Brooklyn was essentially a city of homes. **1922** Joyce *Ulysses* 630 The tattoo..was all the go in the seventies or thereabouts, even in the House of Lords. **1977** *Rolling Stone* 21 Apr. 66/1, I was fascinated with the notion that Mike was what I might have become had I been a man, the lastborn instead of the first—, a child of the Seventies rather than the Sixties.

seven-up. Add: **1.** (Earlier example.)

1830 *N.Y. Constellation* 11 Sept. 2/5 Some tugged at the bottle,..and some played seven-up.

2. Seven up. Also **Seven-Up,** etc., and as **7-Up.** The proprietary name of a popular carbonated soft drink.

1928 *Official Gaz.* (U.S. Patent Office) 13 Nov. 304/2 'Seven Up.' For carbonated, nonalcoholic, noncereal, maltless beverages sold as soft drinks and syrups, extracts, and flavors used in making the same. **1953** *Trade Marks Jrnl.* 24 June 555/2 Seven-up... Non-alcoholic drinks and preparations for making such drinks, all included in Class 32. **1968** N. Fleming *Counter Paradise* v. 78 Crystal ordered a coffee for herself and a Seven-Up for Jake. **1978** *Washington Post* 5 Sept. 10A/1 Bottles of 7-Up, suddenly the hottest commodity in Egypt.

seven year(s, -years', *attrib. phr.* Add: *seven year(s') itch* (orig. *U.S.*), used to designate conditions supposed to last for, appear, or recur after, seven years; freq. applied *joc.* to an urge towards infidelity after seven years of marriage.

1854 H. D. Thoreau *Walden* 355 These may be but the spring months in the life of the race. If we have had the seven-years' itch, we hâve not seen the seventeen-year locust yet in Concord. **1899** C. W. Chesnutt *Conjure Woman* 154 Lawsuits wuz slow ez de seben-yeah eetch. **1936** C. Sandburg *People, Yes* 112 'May you have the sevenyear itch,' was answered, 'I hope your wife eats crackers in bed.' **1955** *Sun* (Baltimore) 16 May 12/7 When I was a boy we called the skin rash from poison ivy..'the seven-year-itch' and firmly believed that it would reappear every year for seven years. **1959** 'O. Mills' *Stairway to Murder* xiv. 149 Lapse understood and sympathised with. It's so common they've even got a name for it. It's the seven-year itch. **1980** P. Moyes *Angel Death* x. 132 There's something called the seven-year itch..middle-aged men quite suddenly cutting loose.

severalth, *a.* *U.S. dial.* Also **severaleth.** [f. Several *a.* + -th[2].] The ordinal form of Several *a.* 4 as an indefinite number.

1902 A. H. Lewis *Wolfville Days* xvi. 238 Re-fillin' his glass for the severaleth time. **1949** 'J. Nelson' *Backwoods Teacher* ii. 23 But presently, apologizing for the 'severalth' time—to use a good word we learned at Big Piney—..Mrs Helms did sit down.

several-fold, *a.*, *adv.*, and *sb.* Delete *rare* and add later examples.

1945 *Times* 7 Aug. 4/2 The Secretary of War..said later that an improved bomb would be forthcoming soon. It would increase by 'severalfold' the present effectiveness of the new weapon. **1979** *Nature* 29 Mar. 468/1 The rate of protein synthesis increases severalfold during the early cleavage stages. **1981** *Spectator* 7 Nov. 29 The reasons were severalfold.

severance. Add: **2. d.** Discharge from contractual employment. Also *ellipt.* = *severance pay* below.

1941 *North Western Reporter* CCXCVII. 652/1 There was a complete 'severance of employment' and compensation would be calculated on basis of $2.50 per day. **1945** *Monthly Labor Rev.* Jan. 48 The American Newspaper Guild..regards dismissal pay as an equity which the individual builds up on his job and for which he should be compensated regardless of the reason of severance. **1965** *Bull. U.S. Dept. Labor* No. 1425-2. ii. 9 Some agreements gave no details of the plan's characteristics—i.e., when severance would be paid, [etc.]. **1977** *Time* 5 Dec. 72/2 When CBS decided that Schorr must go, its lawyers in February 1976 agreed to pay Schorr more than two years' salary, and severance besides.

3. *attrib.* and *Comb.*, as (sense 2 d) *severance arrangement, money, payment*; **severance cutting** (see quot. 1928); **severance felling** = prec.; **severance pay,** money paid in compensation to one whose contractual employment is terminated; cf. *redundancy pay* s.v. *redundancy 3.

1971 *Guardian* 14 Jan. 13/3 Some kind of compulsory severance arrangement..will have to be negotiated. **1905** *Terms Forestry & Logging* (U.S. Dept. Agric.) 20 Severance cuttings are made to strengthen the trees on the edge of a stand. **1928** R. S. Troup *Silvicultural Systems* ii. 8 'Severance cuttings'..are cleared lines of varying breadth, usually 30–50 feet, cut through the wood while it is still comparatively young in order to induce low branching by the border trees. **1895** W. R. Fisher tr. *Schlich's Man. Forestry* 469 Severance-fellings should be forty to fifty feet broad. **1951** W. L. Taylor *Estate Forestry* xiv. 131 Gale and flood are countered by..correct orientation of forest rides and severance fellings. **1975** *N.Y. Times* 5 Nov. 23/3 The parent company closed down the edition [*sc.* the Scottish Daily Express] 18 months ago.., and the employees decided to keep it going as a cooperative. They put severance money of $1.3 million into the venture. **1953** P. C. Berg *Dict. New Words* 143/1 Severance pay. **1956** *Economist* 7 July 12/2 The unions now appear to be ready to lay rather more emphasis on bargaining for higher severance pay. **1979** *Now!* 21–27 Sept. 95/1 The £1,750 tax-free severance pay provides a cushion for defeated MPs. **1962** *Listener* 19 July 86/1, I believe a compensation scheme—severance payment—is important.

severe, *a.* Add: **III. 9. b.** Substitute for def.: *transf.* Of geographical terrain etc.: causing exertion or making great demands of endurance or skill; taxing, hard to 'negotiate'. Also *spec.* of a rock or mountain or the route by which it may be climbed. Hence as *sb.* (usu. with capital initial).

1881 *Sportsman's Year-bk.* 49 Twenty miles a day is often the work of a crack greyhound intended to run in a severe country. **1897** [in Dict.]. **1897** O. G. Jones *Rock-Climbing in Engl. Lake District* p. xxiii, Exceptionally Severe Courses: Screes Great Gully. **1935** D. Pilley *Climbing Days* iv. 66 It was bizarre that a *severe* should sometimes seem simple, when a *moderate* caused nerve storms of impotent despair. **1951** C. Coxhead *One Green Bottle* iii. 86 'Ah yes, the Amphitheatre Buttress... An easy Difficult, isn't it?' She..herself led Very Difficults, and occasionally..an easy Severe. **1958** E. Newby *Short Walk in Hindu Kush* iii. 34 Easy, moderate, difficult, very difficult, severe, very severe, exceptionally severe, and excessively severe. **1970** *Guardian* 28 Aug. 18/6 The Wen Slab, a broad expanse of smooth, sheer rock classed as 'very severe'. **1975** G. Moffat *Miss Pink* iii. 43 It's the big stack off the north headland: a hundred and fifty feet high... A good Severe, we thought. **1976** H. MacInnes *Death Reel* iii. 23 'Is there any climbing on Bidean?' 'I reckon Lilly's route is the best line on this side. About 600 feet, Severe.'

11. (Earlier example.)

1805 T. E. White *Jrnl.* (1904) 32, I got up this morning with the determination to have a severe nap before night.

13. (Later example.)

1802 J. Woodforde *Diary* 13 July (1931) V. 401 It was like to be a severe contested Election.

IV. *Comb.*, as *severe-faced, -looking* adjs.

1939 W. Fortescue *There's Rosemary* xxxv. 220 Together we hurried down the garden path—to meet Queen Alexandra, Princess Victoria, and a severe-faced lady coming out of the garden door. **a 1957** J. Cary *Captive & Free* (1959) xliii. 186 A very severe-looking young woman.

severization. For *rare*[−1] read *Obs. rare* and add earlier example.

1849 *Theatrical Programme* 20 Aug. 84 Brilliant Feats of Swordsmanship, including the Severisation of a Leg of Mutton.

Sévigné, sevigné. Add: Also, a jewel or ornament of the kind used to decorate a head-dress. Also *attrib.* Now *Hist.*

1826 M. Wilmot *Let.* 29 Feb. (1935) 239 Black velvet stomacher..fastened to the top with a sevigné, garnett and pearl..and another garnett sevigné on her forehead. **1840** M. Edgeworth *Let.* 30 Dec. (1971) 574 Sevigné headdress of black velvet with Sevigné jewels in front.

Sevillan, *a.* (Examples.) Also as *sb.*

1883 G. Meredith *Let.* 16 Mar. (1970) II. 690 We have just produced pots [of marmalade], which are Sevillan. **1930** H. Baerlein *Spain* x. 138 Three hundred Murillos of the best period used to be the quota of a Sevillan gentleman when Théophile Gautier travelled in these parts, and the Sevillan was willing to sell one or two of his treasures. **1971** D. Cory *Sunburst* iv. 50 The Sevillan summer was getting into its blazing stride.

Sevillano (sevilʸā·no), *sb.* and *a.* Also fem. **-a** and erron. **Sevilliana.** [Sp.] = Sevillian *a.* and *sb.*

1884 O. Patch *Sunny Spain* i. 14 The Sevillanas, as the ladies of Seville are called, are remarkable for their beauty. **1897** H. C. Chatfield-Taylor *Land of Castanet* iv. 93 The fair Sevilliana sits in her darkened chamber. **1904** B. Kennedy *Tramp in Spain* iv. 25, I like the Sevillanos. They are a fine, free and easy people. **1932** E. Hemingway *Death in Afternoon* 324 Now Ronda means sober and tragic in the Plaza with a limited repertoire and sevillano means light-hearted..with flowery style and a lengthy repertoire. **1970** R. A. H. Robinson *Origins of Franco's Spain* 325 Protest of 2,000 Sevillano farmers to Alcalá-Zamora. **1976** E. P. Benson *Bulls of Ronda* iii. 14, I see..that you are a Sevillano.

Sevillian, *a.* and *sb.* Add: (Earlier and later examples.) Also **Sevilian.**

1830 Disraeli *Let.* 26 July (1885) 30 The Sevillians say that *Cadiz es toda facada. Ibid.* 32 You see what a Sevillian écritoire is by this despatch. **1926** E. A. Peers *Royal Seville* i. 18 The ceramic art which is so noted a Sevilian industry. *Ibid.* ii. 29 The Sevilians like to say that the Guadalquivir salts the sea. **1957** A. MacNab *Bulls of Iberia* xv. 170 A Sevillian from the classic San Bernardo quarter. **1967** A. Rosin tr. *G. Pillement's Unknown Spain* 55 The charm of the Sevillians, who are known for their laughter.

sevruga. Add: **1.** (Later examples.)

1940 A. Simon *Conc. Encycl. Gastron.* ii. 14/2 The Ship caviar is light in colour and small; that from the Sevruga is smaller still. **1964** A. Launay *Caviare & After* i. 18 There are three varieties of acipenser used in the production of caviare, the Beluga, the Ocietrova or sturgeon and the Sevruga.

2. Caviare made from the roe of this fish.

1959 W. Heptinstall *Hors d'Œuvre & Cold Table* 29 Beluga Malossol caviar..has..the largest grain and..the highest price. Next comes Sevruga Malossol at about two-thirds of the price of the Beluga caviar. **1977** *Times* 16 Nov. 18/5, I have never been able to say 'when', whether it be a second helping of Sevruga or just another wee drop of the hard stuff.

sew, *v.*[1] Add: **1. e.** (Later examples.)

1880 J. W. Zaehnsdorf *Art of Bookbinding* v. 21 A third sheet having been sewn.., the needle brought out at the kettle-stitch, must be thrust between the two sheets first sewn. **1929** A. J. Vaughan *Mod. Bookbinding* i. 24 Before a book is sewn by hand the back is required to be marked..as a guide for the needle. **1968** I. Robinson *Introd. Bookbinding* 27 When the second section has been sewn the long and short ends of thread are drawn taut.

4. c. (*b*) Also, to bring about the conviction of (a person).

1927 *Dialect Notes* V. 462 *Sew up*, v., to convict on overwhelming evidence. 1929 D. Hammett *Red Harvest* vii. 80, I expected something like that. That's why I sewed you up. And you are sewed up. 1945 E. S. Gardner *Case of Gold-Digger's Purse* xv. 159 The police have sewed him up on a written statement.

d. *colloq.* To bring (something) to a desired conclusion or condition; to complete satisfactorily; to organize or gain control of (a person or thing); *spec.* to ensure the favourable outcome of a game or match. Freq. in phr. *all sewn* (or *sewed*) *up*.

1904 Ade *True Bills* 136 The Man with the Megaphone Voice cut no Ice whatsoever, for they had him sewed up. 1915 *Dialect Notes* IV. 235 *Sew up, v. phr.*, to make certain of (a place on a team, in a club, etc.). 1933 E. E. Cummings *eimi* 245 We glide to marriage 'they've got that all sewed up' blonde's mari affirms. 1936 'P. Quentin' *Puzzle for Fools* xxi. 198 He said that..he had Broadway sewed up—him and a few other fellows. 1942 E. S. Gardner *Case of Careless Kitten* (1944) xii. 100 By the time you get there, Lieutenant Tragg will have things sewed up so tight you'll have to pay admission to get within a block of the place. 1945 —— *Case of Gold-Digger's Purse* xv. 165 The police have all the witnesses sewed up tight. 1953 A. Upfield *Murder must Wait* xxi. 191 A Chinese I..played draughts with..let me win a man..and I'd think I had him well sewn up..and then he'd clean the board. 1960 T. McLean *Kings of Rugby* xi. 163 It was Henderson who sewed up the match a moment later. 1977 *News of World* 17 Apr. 23/3 Charlton appeared to have the game sewn up. 1979 *Quarto* Oct. 3/1 During this period the novelists had it all sewn up.

e. To enclose or seal off. *colloq.*

1962 *New Statesman* 21 Dec. 899/1 Knowing that it's only a matter of minutes before the Law would sew up the district with a cordon, we drop one of the team at the local railway station.

5. sew-on *a.*, attached by sewing.

1905–6 *T. Eaton & Co. Catal.* Fall & Winter 158/4 Sew-on Hose Supporters, which are stitched on to corset. 1977 *Evening Post* (Nottingham) 24 Jan. 8/1 (Advt.), Sew on Patches By Leomotif. Fantastic range of over 300 designs.

sewage, *sb.* Add: **3.** *sewage disposal*; **sewage lagoon** *N. Amer.* = *LAGOON[1] 2**.

1873 *Practitioner* XI. 381 The health-aspect of sewage disposal. 1939 *Country Life* 11 Feb. p. xxi/1 (Advt.), Sewage disposal for country houses, factories, farms, etc. 1978 J. Wainwright *Thief of Time* 29 The septic tank..forms part of the sewage-disposal system of the bungalow. 1930 S. H. Adams *Mod. Sewage Disposal & Hygienics* 472/2 (Index), Sewage lagoon. 1958 *Progress* (Preeceville, Sask.) 28 May 1/6 Northwest of town the earthwork for the sewage lagoon has been completed. 1976 *Billings* (Montana) *Gaz.* 30 June 4-D/3 Froid's water, sewer and sewage-lagoon systems.

sewelling, var. Shewelling *sb.*

sewer, *sb.*[1] Add: **2. b.** (Later example.)

1945 N. Mitford *Pursuit of Love* vii. 56 Who is that sewer with Linda?

5. *sewer-scent, -stench*; **sewer-gas** (earlier example); **sewer lagoon** *U.S.* = *sewage lagoon* s.v. *SEWAGE *sb.* 3; **sewer line** *U.S.*, a pipeline that is a sewer.

1849 in E. R. Pike *Human Docs. Victorian Golden Age* (1967) 276 These gases, which so many people are daily inhaling..are identically the same in nature with..sewer-gas. 1959 *Washington Post* 31 Oct. B1/5 To construct the outfall line to the location of the sewer lagoons would require trenches in excess of 25 feet. 1977 *It* May 6/3 (*caption*) Rosselli and his back-up man went down a man-hole behind the fence..and followed the sewer-line away from Dealey Plaza. 1929 D. H. Lawrence *Pansies* 54 And it's funny my dear young men, that you in your twenties should love the sewer scent Of obscenity. *Ibid.*, A vapour of rottenness out of their mouths, like sewer-stench wreathing.

sewing, *vbl. sb.*[1] Add: **1. a.** (Earlier and later examples in *Bookbinding*.)

1835 J. A. Arnett *Bibliopegia* 20 There are various ways of sewing, according to the size and thickness of the sheets of a book. 1880 J. W. Zaehnsdorf *Art of Bookbinding* v. 22 This is the strongest sewing executed at the present day. 1951 L. Town *Bookbinding by Hand* v. 99 If the sewing is done too tightly the book will be 'nipped in' at the kettle-stitches.

4. *sewing card, chair, -room; sewing girl; sewing-bird* (examples), *table* (examples); of a gathering for the purpose of sewing, as *sewing bee, circle*.

1862 M. Colt *Went to Kansas* 23 Have had two sewing bees; one for the old ladies, and one for the young. 1936 F. Clune *Roaming round Darling* xxi. 209 Funds are raised in various ways, such as dances, sewing-bees, jam and wood days. 1976 R. Barnard *Little Local Murder* iii. 35 Mrs Smith, a woman of no importance who had had a forlorn hope of starting a sewing-bee. 1857 *Spirit of Times* (N.Y.) 21 Nov. 192/3 (Advt.), Gold bracelets, gold pencils, sewing-birds. 1949 R. J. Sim *Pages from Past* 10 Who can say when the ancestor of the sewing bird made its appearance on the edge of the table? 1887 A. M. Sullivan *Let.* 20 Mar. in H. Keller *Story of my Life* (1903) III. iii. 312 Her father..sees her contentedly stringing her beads or making horizontal lines on her sewing-card. 1961 M. K. Ashby *Joseph Ashby* vii. 87 Tripping round in action songs..and the sewing cards..certainly made them [*sc.* the children] happier. 1868 *Ann. Rep. Secretary Michigan State Board Agric.* VII. 354 A. Dondero, Detroit..[exhibited] 1 willow ladies' sewing chair. 1978 D. Clark *Liberties* v. 99 He looked at Mrs Middleton on the sewing chair. 1846 *Knickerbocker* XXVII. 373 As if I too belonged to a sewing-circle, and read charity sermons. 1912 L. M. Montgomery *Chron. Avonlea* ii. 50 The minister's wife..asked her if she wouldn't come to their Sewing Circle. 1979 B. Parvin *Deadly Dyke* ix. 47 Find out if there's a local sewing circle..in the village. 1848 'N. Buntline' *Mysteries & Miseries N.Y.* 11 What, a little sewing girl, eh? 1870 O. Logan *Before Footlights* 576 Among the same number of sewing-girls of our great cities. 1852 E. E. Hale *If, Yes & Perhaps* (1868) 56, I always offered my services in the Sunday-schools and sewing-rooms. 1978 R. Hill *Pinch of Snuff* v. 50 We use this as a sewing-room... Alice..makes all our clothes in here. 1863 A. D. Whitney *Faith Gartney's Girlhood* xxi. 199 In her low chair by her sewing-table, sat the young sister. 1924 H. T. Lowe-Porter tr. *T. Mann's Buddenbrooks* I. i. 6 There was a sewing-table by the window. 1979 *Country Life* 27 Sept. Suppl. 59/3 Faded mahogany sewing table.

sewing-machine. Add: **1.** (Earlier and later *attrib.* examples.) Freq. as *sewing-machine oil*.

1863 in *Rebellion Rec.* V. I. 70 Elias Howe, Jr., the inventor of the sewing-machine needle, was a private in this regiment. 1895 *Montgomery Ward Catal.* 262/3 See Index for Sewing Machine Oil. 1977 A. Scholefield *Venom* v. 205 He found a can of..sewing-machine oil and squirted it into the lock.

3. *attrib.* **a.** Designating a musical instrument whose operation resembles the action of a sewing-machine. **b.** *fig.* Of rhythm: precise, regular, inexpressive.

1874 J. Codman *Mormon Country* viii. 80, I found the solitary musician seated at one of those sewing-machine 'melodeons', and grinding out the Missionary hymn. 1934 C. Lambert *Music Ho!* IV. 247 Bach's sewing-machine counterpoint. 1974 *Early Music* Apr. 119/2 His purpose is to counter the 'sewing-machine' style of playing Bach by the 'purists'.

sewn, *ppl. a.* Add: **b.** With advbs. forming adjs., as *sewn-in, -on*.

1961 *Guardian* 15 Sept. 10/1 Garments..will have sewn-in labels giving simple washing instructions. 1965 E. C. Hiscock *Cruising under Sail* (ed. 2) I. vii. 136 In way of the grommet the boltrope is usually protected against chafe by sewn-on leather or plastic. 1977 *Guardian* 10 Jan. 8/1 The pallid youth in the tie-die shirt with a sewn-on picture of Marx.

sex, *sb.* Add: **1. c.** *the weaker sex* (later example); *the second sex* (further examples); *the sterner sex* (example).

1849 C. Brontë *Shirley* III. xiv. 312 'Mama' is rather a misanthropist, is she not? Not the best opinion of the sterner sex? 1928 D. K. Parker tr. *Schopenhauer: Selections* 443 Women..form the *sexus sequior*—the second sex. 1953 H. M. Parshley tr. S. de Beauvoir (*title*) The second sex. 1961 'F. O'Brien' *Hard Life* v. 42 Decent people should look after women—isn't that right? The weaker sex. 1974 J. Mitchell *Psychoanalysis & Feminism* II. ii. 306 Woman is the archetype of the oppressed consciousness: the second sex.

e. Delete 'Now *rare*' and add further examples.

1892 'Mark Twain' *Amer. Claim.* xvii. 160 The customers applauded, the sex began to flock in. 1920 D. Lindsay *Voyage to Arcturus* i. 2 He was used to such receptions at the hands of the sex.

3. b. Sexual intercourse. Freq. in phr. *to have sex* (*with*).

1929 D. H. Lawrence *Pansies* 57 If you want to have sex, you've got to trust At the core of your heart, the other creature. 1952 S. Kauffmann *Philanderer* (1953) x. 174 Her arms went around his neck and his hand rested on her waist, and they had a brief moment of friendship before the sex began. 1960 R. East *Kingston Black* viii. 82 She refused to have sex with him. 1962 *Listener* 7 June 1006/2 Why wasn't Bond 'more tender' in his love-making? Why did he just 'have sex' and disappear? 1962 *Woman's Own* 18 Aug. 29/3 Those trends in our society that make sex before marriage so easy. 1971 *Petticoat* 17 July 6/2 The most conspicuous consequence of sex before marriage is the possibility of pregnancy. 1980 *Times* 6 Sept. 2/2 Michael was alongside me, and in due course, on top of me. We had not had sex, but we were contemplating it.

c. Genitalia; a penis. *slang*.

1938 D. Gascoyne *Hölderlin's Madness* 18 And the black cypresses strained upwards like the sex of a hanged man. 1956 H. Gold *Man who was not with It* (1965) xviii. 162 His eyes turned to his pants, gaping open, and his sex sick as an overhandled rattler gaping through. 1977 T. Allbeury *Man with President's Mind* i. 9 The narrow white briefs that barely captured her sex.

5. (Many combs. are paralleled by a collocation with Sexual *a.* 3.) *sex activity, affair* [Affair 3], *aid, -anger, antagonism, awareness, behaviour, circuit* [Circuit 7], *clinic, complex* [*COMPLEX *sb.* 3], *-compulsion, -consciousness, -contrast, -craving, difference, distribution, education, emotion, equality, -excitement, -experience, -exploitation, -feeling, fiend* [Fiend 4 a], *film, -flow, game, -hate, -hatred, hygiene, -inertia, -instruction, joke, life, -longing, -love, -machine* [Machine *sb.* 4 d], *magazine, mania, maniac, manifestation, manual* [Manual *sb.* 1 b], *-morality, novel, obsession, organ* [Organ *sb.*[1] 5], *orgy, partner, -party* [Party *sb.* 9], *-power, problem, question, repression, show, -specificity, starvation, stereotype, -stereotyping, story, -talk, -thrill, -union, war, warfare*; *sex-alive, -angry, -conscious* [*CONSCIOUS *a.* 12], *-crazed, -emancipated, -influenced, -mad* [Mad *a.* 4 c], *-obsessed, -segregated, -smelling, -specific, -starved* adjs.; **sex act,** the (or an) act of sexual intercourse; **sex-appeal,** sexual attractiveness; qualities which attract members of the opposite sex; also *fig.*; hence **sex-appeal** *v. trans.* and *intr.*, to attract sexually; **sex-appealing** *ppl. a.*, having or exerting sex appeal; **sex attractant,** a substance produced by one sex of a species that attracts members of the opposite sex, or a synthetic substance with the same property; also *attrib.* or as *adj.*; **sex-blind** *a.*, not discriminating between the sexes; **sexboat** *U.S. slang*, **sex-bomb** *slang* = *sexpot* below; **sex change,** a change of sex; *spec.* an apparent change of sex brought about by surgical means, treatment with hormones, etc.; also *attrib.*; **sex chromatin,** the material of a small heterochromatic body (believed to be an inactivated X chromosome) of which there is one in a normal (XX) female cell and in general one less than the number of X (or Z) chromosomes in a cell; also *attrib.*; **sex chromosome,** each of the chromosomes (normally two in number) in a cell's chromosomal complement the particular combination of which (as XX or XY) determines an individual's sex; opp. *AUTOSOME; hence *sex-chromosomal* adj.; **sex crime,** a crime involving sexual assault or with a sexual motive; a sexual act regarded as a crime; hence *sex criminal*; **sex-determinant,** that which determines an individual's sex; **sex determination,** the biological process that settles the sex of an individual; **sex determiner,** a gene which determines the sex of the individual bearing it; **sex-determining** *ppl. a.*, determining an individual's sex; **sex discrimination** orig. *U.S.*, unfavourable treatment motivated by prejudice against members of a particular sex; hence *sex-discriminating* ppl. adj., *sex-discriminatory* adj.; **sex drive,** the principle which motivates satisfaction of sexual needs; **sex factor,** (*a*) a sex-determining chromosome or gene; (*b*) [tr. F. *facteur sexuel* (Jacob & Wollman 1957, in *Compt. Rend.* CCXLV. 1840)], a bacterial plasmid which can promote the transfer of genetic material from its ('male') host to another ('female') bacterium in which recombination then takes place; **sex-free** *a.*, (*a*) having a liberated attitude towards sex; (*b*) not involving sex; **sex hormone,** any of the (natural or synthetic) hormones that affect sexual development or behaviour, esp. those produced by the gonads; **sex impulse,** the impulse towards satisfaction of sexual needs; **sex instinct,** the behaviour and feelings associated with sexual reproduction considered as an instinct for the survival of the species; **sex-interest,** concern with a sexual relationship, esp. as a theme or episode in a story, film, etc.; **sex-intergrade,** an intersex; **sex-killer,** a murderer who sexually assaults his victim; hence *sex-killing* vbl. sb.; **sex kitten** *colloq.*, a young woman who exploits her sex appeal; hence *sex-kittenish* adj.; **sex-mosaic** *Biol.*, an individual some of whose cells are of one sex and the rest of the other; **sex object,** a person towards whom or thing towards which the sexual impulse is directed; a person regarded only as the object of sexual desire; **sex offence,** (*U.S.*) **offense,** a breach of law or etiquette involving sex; **sex-offender,** a person guilty of a sex offence; **sex play,** (*a*) a play about sex or with sexual content; (*b*) sexual activity stopping short of intercourse; **sexpot, sex-pot** *colloq.*, a sexually exciting person, esp. a woman; also, a sexually very active or sex-obsessed person; also *attrib.* or as *adj.*; **sex ratio,** the ratio of the numbers of individuals of each sex; **sex relation,** (*a*) sexual relationship; (*b*) *pl.* sexual intercourse; also *sex relationship*; *sex-related* adj.; **sex-reversal,** adoption of a form or role characteristic of the opposite sex; hence **sex-reversed, -reversing** *ppl. adjs.*; **sex role** *Social Psychol.*, the culturally determined role or behaviour

which a person learns as appropriate to his or her sex; **sex shop,** a shop selling sex magazines, aids, etc.; **sex surrogate,** a person employed as a sexual partner for a person undergoing therapy for sexual problems; **sex symbol,** (*a*) a person who is for many the epitome of sexual attraction and glamour; (*b*) a symbol with a sexual signification; **sex therapy,** therapy that deals with a person's psychological impediments to sexual intercourse or with other sexual problems; so *sex therapist*; **sex-typed** *ppl. a. Sociol.* and *Psychol.*, typified as being characteristic of either the male or the female sex; so *sex-typing* ppl. adj. and vbl. sb.; **sex urge** = *sex drive* above.

1918 M. Stopes *Married Love* vi. 62 What must be taking place in the female system as a result of the completed sex act? 1958 *Listener* 21 Aug. 263/1 A consciousness of guilt in the sex act. 1972 J. Symons *Bloody Murder* xii. 160 Detailed accounts of sex acts is [*sic*] still less frequent in the crime story than in ordinary novels. 1898 C. P. Stetson *Women & Econ.* iii. 44 Woman..has developed in the lines of action to which she was confined; and those were always lines of sex-activity. 1949 M. Mead *Male & Female* x. 208 The burden of choosing between sex activity and other activities has been taken off the individual. 1933 Dylan Thomas *Let.* Nov. (1966) 51 She who lives serene..And drags her tea-time sex affair all fresh To the dinner table. 1949 M. Mead *Male & Female* v. 118 Two inexperienced adolescents had a first sex-affair. 1977 *Gay News* 7–20 Apr. 14/3, I should like to dismiss this neat little toy as a sex-aid for sadomasochists. 1923 Sex-alive [see *ranch dog* s.v. *RANCH *sb.*[1] 3 a]. 1923 D. H. Lawrence *Kangaroo* xiii. 294 And if for a time you *do* overcome her with reason, the sex-anger only arises more hideously. *Ibid.*, You can reason with a sex-angry woman till you are black in the face. 1909 E. Robins *Votes for Women* I. 44 This ferment of feminism..[is] likely to bring a very terrible thing in its train... Sex antagonism. 1952 A. Christie *Mrs McGinty's Dead* xii. 93 We really get a feeling of sex antagonism between the chap and the girl. 1924 *Amer. Mercury* II. 318/1 An actress with sex appeal is four times out of five a more effective actress. 1924 G. R. Chester *On Lot & Off* 25 She'd sex appeal me all right! 1937 G. Frankau *More of Us* v. 61 'Ergo et propter hoc festina lente', Remarked Athene from that smoke-blue ceiling. But Innocent continued sex-appealing. 1953 *Encounter* Nov. 32/2 For the frustrated and starved, it [*sc.* Communism] has all the sex-appeal of a strong, monolithic creed. 1979 *Guardian* 30 Mar. 9/6 Two conflicting general [education] policies have been put forwards by the Conservatives... Both have political sex appeal. 1980 N. Freeling *Castang's City* xi. 68 A very feminine charm, next door to sex appeal. 1928 *Daily Express* 24 Nov. 4/1 Sex-appealing women should cut out the hurt feelings, the dewy eyes, trembling lips, the 'Please, I'm just a woman' stuff. 1932 *News Chron.* 20 June 4/1 Both [stories] are glittering, glamorous, sex-appealing. 1964 *Listener* 27 Aug. 291/2 The study of sex-attractant odours in insects has shown that these may be so highly specific that a very slight change in their chemical structure may make them less effective or even quite ineffective. 1976 *Globe & Mail* (Toronto) 13 Sept. 5/1 Many species manufacture and use chemicals as sex attractants. 1925 *John o' London's Weekly* 5 Dec. 360/1 In many fishes..there seems little hint of sex-awareness. 1949 M. Mead *Male & Female* xiv. 284 The period between childhood sexuality focussed on the parent and the stirring of adolescent sex awareness. 1923 J. S. Huxley *Ess. of Biologist* iv. 144 Castrated animals fail to realize either possibility of normal sex-behaviour. 1949 M. Mead *Male & Female* xvii. 354 The exaggerated over-concern with the other..puts an extraordinary strain on sex behaviour. 1975 *New Yorker* 29 Sept. 29/1 Even in an ideal sex-blind situation you are going to encounter different kinds of blocksmanship between you and your goals. 1977 *N.Y. Post* 30 Mar. 3 Spokesmen for several liberal groups..declared yesterday that totally sex-blind job assignments are a violation of prisoners' rights. 1962 E. Lacy *Freeloaders* ii. 22, I don't buy the bit that every mademoiselle is automatically a sexboat because she's French. 1963 L. Deighton *Horse under Water* xxii. 98 I've got the photo of your secretarial sex-bomb. 1976 P. Cave *High Flying Birds* iii. 42 Sex-bomb, Sonya Stelling might be. Oscar contender she was not. 1946 *Nature* 3 Aug. 173/2 These results.., while providing virtual proof of sex-change from male to female in a section of the male population, point also to the probable occurrence of two types of males in *P*[*atella*] *vulgata*. 1960 *Twentieth Century* Mar. 258 Sex-change may well seem, as *The Times* said, 'unprepossessing' as a subject for comedy. 1970 *Daily Tel.* 21 Dec. 2/6 More people with transsexual problems are seeking National Health sex-change operations. 1952 Graham & Barr in *Anat. Rec.* CXII. 709 The term 'sex influenced chromatin', or simply 'sex chromatin', will be used in the description to follow for the nuclear structure characterized by a size relation to sex. 1962 *Lancet* 27 Jan. 216/2 In all cases the sex-chromatin pattern was shown to be consistent with the morphological sex. 1913 *Jrnl. Exper. Zool.* XV. 593 The two white females..came from the union of the two-X egg with the no-X sperm of the vermillion pink male, and should be entirely maternal and entirely non-paternal in sex chromosomal composition, i.e., they should be exact counterparts..of their mother. 1906 E. B. Wilson in *Jrnl. Exper. Zool.* III. 28 These chromosomes are the bearers of the male and female qualities (or the factors essential to the production of these qualities) respectively. They may also be designated (whenever it is desirable to avoid circumlocution) as sex-chromosomes. 1926 J. S. Huxley *Ess. Pop. Sci.* 46 From all the facts, we can, I think, be sure that all the higher animals possess special X or sex-chromosomes, two in one sex, one in the other, by whose agency sex is determined. 1974 Goodenough & Levine *Genetics* x. 466 The sex of a fly is determined, in part, by the number of euchromatic sex chromosomes, called X chromosomes, possessed by an individual. 1920 D. H. Lawrence *Women in Love* xvi. 221 Each acknowledges the perfection of the polarised sex-circuit. 1951 R. Campbell *Light on Dark Horse* 319 These Germans had money from the Komintern and they set up sex-clinics and communist-cells. 1972 *Newsweek* 27 Nov. 65 The sex clinic is fast becoming as vital a part of the modern hospital as the emergency room. 1921 R. Macaulay *Dangerous Ages* vi. 111 You prefer to avoid discussing certain aspects of your life. You obviously have a sex complex. 1928 D. H. Lawrence *Lady Chatterley's Lover* vi. 64 Be damned to the artificial sex-compulsion! refuse it! 1912 T. Dreiser *Financier* xx. 222 From the first she was somewhat sex-conscious. 1952 S. Kauffmann *Philanderer* (1953) ii. 30 But she's certainly a sex-conscious girl. It's in her voice, in the way she sits, the way she drinks. 1911 *Freewoman* 7 Dec. 56/1 The impression given..is that the editors and most of the contributors picture the average woman as an individual wallowing in sex-consciousness. 1953 D. A. Bannerman *Birds Brit. Isles* II. 193 'Sex consciousness'..remains for a long time, the male continuing to feed the female..even after the young are hatched. 1911 J. A. Thomson *Biol. Seasons* iii. 263 The male 'ends' are salmon red or dull pink; the female 'ends' are greenish-grey or drab—the sex-contrast eking itself out in colour. 1949 M. Mead *Male & Female* v. 110 The little boys will feel..the potential sex-contrast with the mother. 1921 D. H. Lawrence *Psychoanal. & Unconscious* i. 20 The incest-craving is not the result of inhibition of normal sex-craving. 1954 R. Bissell *High Water* xx. 218 Well I'm reading this here story about a sex-crazed maniac. 1925 *Amer. Mercury* Feb. 196 Sex crimes, which are commonly regarded as a natural result of drug taking, actually never occur among addicts. 1965 G. McInnes *Road to Gundagai* vi. 99 The murder of Alma Tirtschke by Colin Ross..was a sex crime which gripped Melbourne for weeks. 1977 *Gay News* 7–20 Apr. 10/1 Three gay people locked away in a private house constitutes a public meeting, and sexual acts that take place between any or all of them are sex crimes. 1972 J. Symons *Players & Game* xv. 100 Inadequacy. That's the mark of sex criminals. 1902 *Biol. Bull.* III. 73 If we accept the theory that chromatin is the bearer of hereditary qualities, there could be little doubt regarding the necessary chromosomic character of a sex determinant. 1889 Geddes & Thomson *Evolution of Sex* iii. 49 The temperature of the time, not of birth but of sex determination, must of course be noted. 1977 *Dædalus* CVI. IV. 137 Primitive wasps..evolved the sex determination mechanism of haplodiploidy, whereby unfertilized eggs yield males and fertilized eggs yield females. 1912 *Jrnl. Exper. Zool.* XII. 509 Bateson and Punnett ('11) describe certain exceptions occurring in their sex-linkage experiment with fowls, which they suggest may be due to a failure of the usual association between the sex-linked factor and the sex-determiner, i.e., to 'crossing over' in the female. 1960 E. J. Gardner *Princ. Genetics* vii. 123/2 When the parallelism was discovered between the chromosome cycle and gene behaviour it was generally assumed that genes other than sex determiners were also located in the sex chromosomes. 1901 Geddes & Thomson *Evol. Sex* (ed. 2) 51 In regard to rotifers (*Hydatina*), Maupas maintains that temperature is the sex-determining factor. 1966 *Lancet* 24 Dec. 1397/1 This discrepancy between mammals and birds..might reflect different evolutionary origins of chromosomal sex-determining mechanisms in the two classes. 1918 M. Stopes *Married Love* iii. 19 Vaguely, perhaps, men have realised that much of the charm of life lies in the sex-*differences* between men and women. 1979 *Bull. Amer. Acad. Arts & Sci.* Feb. 25 Clear sex differences in the probability of labor force participation and in the kinds of jobs held. 1964 C. Barber *Ling. Change in Present-Day Eng.* iv. 105 The sex-discriminating word used to indicate that the member of some profession or the holder of some office is a woman. 1976 *Listener* 4 Mar. 264/3, I do not care if it is unfair, or sex-discriminating, for a woman to get an old age pension five years before a man. 1916 *Campaign Text-Bk.* (National Woman's Party) 62 Enfranchised women in the United States regard the removal of sex discrimination from our national constitution as a political need of primary importance. 1965 *Financial Times* 24 Nov. 3/3 New guide-line interpretations on sex discrimination have just been issued by the Commission... So far the Commission has received over 400 complaints of sex discrimination. 1976 *Times* 6 Jan. 2/3 Employers who have cheated women of equal pay by job-grading schemes..may be taken to court for sex discrimination. 1976 A. Oakley in Mitchell & Oakley *Rights & Wrongs of Women* i. 27 This practical equality challenged the law which was sex-discriminatory. 1949 M. Mead *Male & Female* xiii. 266 The very simple accidents of sex distribution inside any family give a structure within which a child can feel unwanted. 1918 R. S. Woodworth *Dynamic Psychol.* vii. 173 The association is not entirely a spreading of the sex drive into the esthetic sphere. 1963 A. Heron *Towards Quaker View of Sex* 54 There may be a period..when the sex-drive is latent. 1979 J. Sherwood *Hour of Hyenas* ii. 23 You are very striking-looking, but..my power drive is far stronger nowadays than my sex drive. 1920 *Jrnl. Amer. Med. Assoc.* 25 Sept. 884/2 The conference will..consider the topic of venereal disease from these..aspects: (1) medical measures; (2) enforcement of repression and protection laws; and (3) sex education. 1969 *Guardian* 7 Aug. 7/6 A series of sex-education talks I'd been giving at a North-east London youth club. 1936 D. H. Lawrence *Pornography & So On* 46 The most high-flown sex-emancipated young people. 1911 O. Schreiner *Women & Labour* 232 The ignorant savage,..who violates and then clubs a female into submission, may be dominated, and is, by sex emotions of a certain class. 1967 A. Marshall in *Coast to Coast 1965–6* 108 He was sure the men he knew were incapable of any sex emotion other than an animal lust. 1907 E. Densmore (*title*) Sex equality. 1921 *Daily Mail Year Bk.* 42/1 The organization of the League of Nations has from the outset been founded on the principle of sex equality. 1977 'J. Gash' *Judas Pair* i. 14 She settled weeping while I found a coat. I'm all for sex equality. 1922 Ld. Dawson of Penn *Love–Marriage–Birth Control* 23 If this harmful restraint succeeds in preventing conception there eventuates the inevitable prevalence of sex excitement. 1936 D. H. Lawrence *Pornography & So On* 30 Sex-excitement of a secretive, furtive sort. 1919 M. Stopes *Married Love* (ed. 7) ix. 141 Women so harried by the undue drains of unregulated sex-experience. 1953 Sex experience [see *sex play* below]. 1914 J. London *Let.* 26 Mar. (1966) 419 The recent sex-exploitation in our magazines and books. 1911 R. C. Punnett *Mendelism* (ed. 3) xi. 107 The factor which repels the red-eye factor is in this case to be found in the male, and here consequently it is the male which must be regarded as heterozygous for a sex factor that is lacking in the female. 1931 E. B. Ford *Mendelism & Evolution* i. 14 Two doses of the sex factors carried by the X-chromosomes evoking the development of one of the sexes. 1955 *Science* 25 Feb. 305/1 The sex alleles of monosporoidal lines from a cross can be determined by matings in corn seedlings with lines representing each of the four haploid combinations of the parental sex factors. 1959 *Genetics* XLIV. 497 (*heading*) A variant sex factor in *Escherichia coli*. *Ibid.*, The wild type sex factor (F') of strain K-12 is characterized by its low affinity for the chromosome and its lack of any preferential site of attachment. 1968 W. Hayes *Genetics of Bacteria & their Viruses* (ed. 2) xiv. 799 If the sex factor happens to be inserted into the chromosome.., the replication initiated in the sex factor proceeds around the complete, integrated structure so that the chromosome is transferred as well. 1973 R. G. Krueger et al. *Introd. Microbiol.* xv. 423/2 The spread of a sex factor like Δ through successive populations of cells carrying different resistance genes on nontransmissible plasmids could result at each step in a strain carrying more and more genes for antibiotic resistance. 1918 M. Stopes *Married Love* v. 50 Even after a woman's dormant sex-feeling is aroused..it may even take as much as from ten to twenty minutes of actual physical union to consummate her feeling. 1937 *Discovery* May 162/2 Conjugal affection, as distinct from sex-feeling. 1970 *Women Speaking* Apr. 4/2 The nearest man ever comes to this reduction to a word is when he offends society sexually: he is *a sex fiend*. 1976 T. Heald *Let Sleeping Dogs Die* vi. 111 She was picked up by a sex fiend..and raped. 1970 Sex film [see *sex-instruction* below]. 1936 D. H. Lawrence *Pornography & So On* 44 The sex-flow is dying out of the young. 1929 —— *Pornography & Obscenity* 24 But the bohemian is 'sex free'. 1960 *Encounter* Sept. 72/1 Gerda is one of those sex-free affairs between tormented men and life-accepting women. a 1911 D. G. Phillips *Susan Lenox* (1917) II. x. 244 The favorite children's games, often played in the open street..were sex games. 1976 J. Crosby *Nightfall* xxvi. 148 'Don't you like sex games?' she said. a 1930 D. H. Lawrence *Last Poems* (1932) 140 All this talk of equality between the sexes is merely an expression of sex-hate. 1913 J. London *Let.* 11 Oct. (1966) 408 She cherishes a sex hatred for a woman who was bigger than she. 1917 *Jrnl. Exper. Zool.* XXIII. 371 (*heading*) The free-martin; a study of the action of sex hormones in the foetal life of cattle. 1965 *New Scientist* 22 Apr. 218/3 The materials contained in the various forms of The Pill are synthetic sex hormones. 1951 M. McLuhan *Mech. Bride* (1967) 23/2 The reader is treated as the sluggish male is treated by the sex-hungry cave woman in the shirt ads. 1912 G. F. Lydston (*title*) Sex hygiene for the male and what to say to the boy. 1949 M. Mead *Male & Female* i. 17 Proper diet, rest, and sex hygiene have taken care of that. a 1911 D. G. Phillips *Susan Lenox* (1917) II. xi. 281 In the streets the sex impulse shows stripped of all disguise. 1957 J. S. Huxley *Relig. without Revelation* ix. 223 The powerful emergence of the sex-impulse in adolescence. 1936 D. H. Lawrence *Pornography & So On* 45 You may even bring about a state of utter indifference and sex-inertia. 1951 H. A. Lindsay et al. in *Jrnl. Nat. Cancer Inst.* XII. 244 During accelerated synthesis of Nissl material the sex-influenced chromatin moves from its usual position, next to the nucleolus, toward the nuclear membrane. 1898 C. P. Stetson *Women & Econ.* iii. 56 What business has a little girl with the instincts of maternity?.. They are sex-instincts, and should not appear till the period of adolescence. 1976 A. Montagu *Nature of Human Aggression* (1978) iv. 64 Everyone 'knows' from his or her own experience that there is such a thing as a sex instinct. 1935 E. Bowen *House in Paris* I. iii. 42 We do not consider him ripe for direct sex-instruction yet, though my husband is working towards this through botany. 1970 *Guardian* 10 Dec. 10/4 From time to time we have to see, not sex films, but sex-instruction films. 1911 *Maclean's Mag.* Apr. 139/2 There is scarcely any 'sex interest' in it at all. 1940 'G. Orwell' in *Horizon* Mar. 189 Both of these papers admit a certain amount of sex-interest into their stories. 1917 Sex-intergrade [see *INTERSEX]. 1962 *Lancet* 27 Jan. 216/2 The hypothesis, advanced by Lang, that some male homosexuals are sex intergrades—i.e., morphologically male but genotypically female. 1941 'G. Orwell' in *Horizon* IV. 155 More than half, perhaps three-quarters, of the jokes are sex jokes. 1959 M. Gilbert *Blood & Judgement* ii. 26 You'd be surprised..how many sex-killers turn out to be their mothers' favourite sons. 1972 J. McClure *Caterpillar Cop* ii. 15 Murders..kept things going... But wanton sex killings involving the young were quite another matter. 1958 *Daily Sketch* 2 June 11/4 Clever film men have moulded her sex-kitten type. 1966 *Guardian* 7 Jan. 9/2 Brigitte Bardot..the original sex kitten with the French charm. 1977 D. Morris *Manwatching* 256 This is why we like 'sex kittens' more than females who are 'catty'. 1963 J. Fowles *Collector* II. 166 Antoinette was almost parodying herself, she was so sex-kittenish. 1898 C. P. Stetson *Women & Econ.* vii. 143 It should be..understood..that the higher development of social life following the economic independence of women makes possible a higher sex-life than has ever yet been known. 1922 S. Paton *Signs of Sanity* vi. 205 Some phases of our instinctive activities..we discuss frankly..; others, notably the sex-life, we treat in a..furtive manner. 1936 'P. Quentin' *Puzzle for Fools* xx. 181, I get no kick out of the sex-life of the white-tailed baboon. 1976 *Vogue* 15 Mar. 13/2 He is being reviled for apparently having absolutely no sex life, none at all. 1979 R. Jaffe *Class Reunion* I. v. 47 Richard was the only boy..who had a regular sex life with a girl, and so he was a celebrity. 1925 T. Dreiser *Amer. Tragedy* (1926) II. III. xiv. 183 All those..sex-longings

..had long since been covered with an easy manner. 1898 C. P. Stetson *Women & Econ.* xii. 260 The generous giving impulse of sex-love. 1976 R. Delmar in Mitchell & Oakley *Rights & Wrongs of Women* ix. 281 There are several conditions which, in Engels..make sex love the rule within the proletariat. 1935 H. Edib *Clown & his Daughter* vii. 43 Now she wants to burden me with another sex-machine. 1970 *Times* 9 Dec. 16/4 Lulu herself, from her first days as an insatiably successful sex-machine to her last days as an amateur prostitute, has no existence outside her appetite. 1943, 1974 Sex-mad [see *MAD *a.* 4 c]. 1931 Sex magazine [see *confession magazine]. 1980 P. Kinsley *Vatchman Smith* vii. 66 The customs officer..sold any sex-magazines he could confiscate from foreign tourists. 1895 *Fortn. Rev.* 1 Apr. 592 Sex mania in art and literature can be but a passing phase. 1895 H. Garland *Rose of Dutcher's Coolly* viii. 90 The brakeman came through and eyed her with the glare of a sex-maniac. 1971 *Daily Tel.* 14 Dec. 3 The Jersey sex maniac..was jailed for 30 years yesterday for 13 indecent and sexual offences against young girls and boys. 1975 'D. Rutherford' *Mystery Tour* i. 8 Wasn't that [*sc.* the murder] the work of some sadistic sex maniac? 1911 O. Schreiner *Women & Labour* v. 187 It is among certain orders of birds that sex manifestations appear to assume their most harmonious and poetical forms on earth. 1949 M. Mead *Male & Female* viii. 167 These [*sc.* seizures] are extremely violent, but without specific sex manifestations. 1975 H. McCloy *Minotaur Country* iii. 21 He would be a sober, industrious lover. He would read all the sex manuals. 1926 W. R. Inge *Lay Thoughts* III. 254 The pleasantest side of our civilization—the ease with which innocent friendships are made between men and women—stands or falls with that Christian sex-morality which is now being openly flouted. 1958 *Listener* 4 Dec. 933/2 Queen Victoria..instituted a new reign of sex-morality. 1903 *Bull. Museum Compar. Zool. Harvard Coll.* XL. 197 Unilateral and mixed hermaphrodites are an exceptional form of sex-mosaic. 1926 J. S. Huxley *Ess. Pop. Sci.* 296 It is possible in the case of abnormal distribution of the sex-chromosomes during development to obtain some parts of the body with the male-determining, others with the female-determining complement of chromosomes, with the result that a sex-mosaic or gynandromorph is the result [*sic*]. 1955 *Japanese Jrnl. Zool.* XI. 350 The 1st gynandromorph is a simple bilateral sex mosaic showing yellow Ww gene in the left male wings and white Ww gene in the right female wings. 1923 in C. D. Stelling *Yea & Nay* 33 (*heading*) Is there any alternative to the sex novel? 1951 M. McLuhan *Mech. Bride* (1967) 23/2 Amid the unmitigated torrent of sadistic sex novels works of reflection are tolerated only if they are..'warmly human'. 1911 *Amer. Jrnl. Psychol.* XXII. 423 Instead of sublimating the sex impulse, he [*sc.* Leonardo da Vinci] directed it towards the physical Jesus *in toto*. It was simply the substitution of one sex object for another. 1963 B. Friedan *Feminine Mystique* (1965) xi. 266 For the woman who lives according to the feminine mystique, there is no road to achievement, or status, or identity, except the sexual one: the achievement of sexual conquest, status as a desirable sex object, identity as a sexually successful wife and mother. 1980 G. Greene *Doctor Fischer of Geneva* ix. 59 Deane is not an actor: he is a sex object. Teenage girls worship him. 1914 *New Republic* 26 Dec. 27/1 An almost lyrical open-airness..saved 'The Garden Without Walls',..from being sex-obsessed. 1979 *London Rev. Bks.* 25 Oct. 13/1 Civilised human beings are remarkable among animal species for being sex-obsessed. 1920 F. M. Ford *Let.* 19 Sept. (1965) 127 The end would be the most horribly costive neurasthenic you can imagine, with incredible sex obsessions sedulously concealed. 1911 J. London *Let.* 8 Jan. (1966) 330 You are suffering from what you deem a sex-offence. 1977 J. Thomson *Case Closed* vii. 85 The offences..included fraud, burglary,..assault, sex offences of various sorts. 1939 *Columbia Law Rev.* Mar. 535 Prior to the enactment of this statute in Illinois, a criminal sex offender received no special treatment. 1976 K. Bonfiglioli *Something Nasty* iii. 33 We 'ave only two known sex-offenders worth the name in this Parish. 1902 W. James *Var. Relig. Exper.* i. 12 That without the chemical contributions which the sex-organs make to the blood, the brain would not be nourished so as to carry on religious activities..may be true or not true. 1978 *Times* 15 Mar. 6/3 Old symbols like mutilation of sex organs..are passé. 1962 J. Heller *Catch-22* xxi. 208 Officers' clubs everywhere pulsated with blurred but knowing accounts of lavish, hushed-up drinking and sex orgies there. 1949 M. Mead *Male & Female* ix. 195 But patterns that regulate competition in the choice of sex partners are learned patterns. 1970 Sex partner [see *sex therapy* below]. 1958 J. Kerouac *On Road* I. i. 8 He told him of..his innumerable girls and sex-parties and pornographic pictures. 1916 G. B. Shaw *Overruled* 63 Plays occupied wholly with the conventional results are..unsatisfying as sex plays. 1932 Wodehouse *Louder & Funnier* 270 When I write my daring sex-play, I have to submit it to Lord Cromer, who starts licking his blue pencil the moment he has opened the envelope. 1953 H. M. Parshley tr. *De Beauvoir's Second Sex* II. vii. 682 The excessive sentimentality,..and platonic crushes of adolescent girls,..are much more injurious than a little childish sex play and a few definite sex experiences. 1961 W. Brown *Bedeviled* 17 Gradually, however, the fondling developed into open sex play which frightened her. 1957 F. Morton *Art of Courtship* 156 How pitiful the American who cannot command the smile of a sexpot. 1961 *Harper's Bazaar* May 57/2 Ovid..the dirty old sexpot. 1963 J. T. Story *Something for Nothing* i. 17 'I like the hockey type,' Albert said. 'I can't stand these sex-pots.' 1975 *New Yorker* 5 May 115/1 Graham Chapman, John Cleese..with Connie Booth and Carol Cleveland as their sexpot aides. 1981 *London Mag.* July 89/2 Tough Games Mistress. Rebellious sexpot pupil (pregnant again). 1918 M. Stopes *Married Love* viii. 76 The periods of complete abstinence should be opportunities for transmuting the healthy sex-power into work of every sort. 1977 *London's Outrage* I. 7 The bully-boy sex-power of Nazism/fascism is very attractive and an easy solution to our complex moral and social dilemmas. 1900 J. X. Merriman *Let.* 1 July (1966) 222 The shrieking sisterhood who write on sex problems and scream out for votes. 1979 *Guardian* 10 July 9/1 Professional partners used in the treatment of sex problems. 1902 G. B. Shaw *Mrs. Warren's Profession* p. xx, Plays which treat sex questions as problems for thought instead of as aphrodisiacs will be freely performed. 1917 F. W. S. Browne in *Proc. Brit. Soc. for Study Sex Psychol.* 7 General early marriage, even if possible under present conditions, does not solve the sex question. 1906 *Biometrika* V. 79 The sex-ratio for the family of each individual is directly calculated and tabled. 1974 J. Burnett *Useful Toil* I. 48 Because of the unequal sex ratio one in three [Victorian] women were 'doomed' to spinsterhood. 1977 *Detroit Free Press* 11 Dec. 9-C/1 A high-ranking official of Planned Parenthood says television's standards on sex-related programming should be as much a public issue as its standards on televised violence. 1898 C. P. Stetson *Women & Econ.* i. 5 The economic status of the human female is relative to the sex-relation. 1911 O. Schreiner *Let.* in First & Scott *Olive Schreiner* (1980) vii. 291 [She] thinks it's wrong for people, even if married, to have sex relations with each other except just when they want to make a child. 1949 M. Mead *Male & Female* i. 14 The infant..before he can toddle has absorbed a particular style of sex relations. 1980 First & Scott *Olive Schreiner* viii. 307 A cutting about sex relations between Indians and white women. 1898 C. P. Stetson *Women & Econ.* iv. 74 Let us bear in mind that all the tender ties of family are ties of blood, of sex-relationship. 1963 A. Heron *Towards Quaker View of Sex* i. 8 There are certain historical characteristics of the Society of Friends that ought specially to lead to..understanding of the significance of the sex relationship. 1926 M. Leinster *Dew on Leaf* v. 29 He talked..about health, climate, and sex-repression. 1958 J. Cannan *And be Villain* i. 27 Richard used to treat me as a case of sex repression. 1916 *Amer. Naturalist* L. 388 In the generic crosses which give all, or nearly all, males at the beginning of the season and all, or nearly all, females in the autumn what is happening?—true sex reversal? or is it selective fertilization, differential maturation or a selective elimination of ova in the ovary? 1926 J. S. Huxley *Ess. Pop. Sci.* iv. 53 Hens which had undergone sex-reversal to cocks. 1949 M. Mead *Male & Female* vi. 129 Peoples may provide sex-reversal rôles for both sexes. 1926 J. S. Huxley *Ess. Pop. Sci.* iv. 53 Similar results have now been obtained..by breeding from sex-reversed moths. *Ibid.* 52 Some sexually abnormal human beings are the victims of this sex-reversing power. 1959 Sex-reversed [see *FEMINIZATION 2]. 1927 W. B. Wolfe tr. *Adler's Understanding Human Nature* I. vii. 135 There are only *two* sex rôles possible. One must orient oneself according to one of two models, either that of an ideal woman, or according to that of an ideal man. 1969 W. H. Sewell in E. F. Borgatta *Social Psychol.* 218/1 The sex-role concepts of children and..sex-role pressure in the socialization of the male child. 1977 *China Now* July/Aug. 3/3 The sex-roles are traditionally presented and the girl who helps Mummy to hang out the washing is rewarded. 1955 T. H. Pear *English Social Differences* 200 This class-segregated, sex-segregated regime. 1970 *Guardian* 26 Nov. 4/2 'Sex-shops'..are now established in many West German cities. 1974 K. Millett *Flying* (1975) III. 307 Sarah stops by and gives us an account of the new sex shop in Tottenham Court Road. 1981 *Observer* 4 Jan. 3/2 How do you designate a sex shop anyway? Does selling contraceptives make Boots a sex shop? 1959 P. Bull *I know Face* x. 187 A city almost entirely devoted to sex-shows. 1922 Joyce *Ulysses* 535 The dark sexsmelling theatre unbridles vice. 1961 *Giornale di Microbiologia* IX. 149 Two phages have been isolated which are sex-specific on *E*[*scherichia*] *coli* K 12. 1976 P. Marks in Mitchell & Oakley *Rights & Wrongs of Women* v. 179 They may differ among themselves about how far vocations, and thus education, should be sex-specific. 1961 *Giornale di Microbiologia* IX. 147 The same strains used to test the sex-specificity of phages μ_2 and ϕ_1 were challenged with phages T_1, .., T_7. 1912 M. Hastings *New Sin* II. 57 Sex-starvation, they call it. It's awful, but it can be done because it must be done. 1977 'D. Rutherford' *Return Load* i. 12 He wondered whether sex starvation had..started to provoke hallucinations... He had seen..a gorgeous red-head. 1927 A. Huxley *Proper Studies* 292 St. Anthony and the unwashed, underfed, sex-starved monks of the Thebaid. 1978 M. Birmingham *Sleep in Ditch* 164 Their first guess is that I'm a sex-starved grass-widow, glimpsing seducers behind every door. 1949 M. Mead *Male & Female* vi. 137 A sex stereotype that decries the interests..of each sex is usually not completely without a basis. 1977 *N.Z. Herald* 8 Jan. 2-2/7 The society has new projects planned for its second decade, including a survey into sex-stereotyping. 1936 Sex-story [see *crime-story]. 1979 *Guardian* 10 July 9/1 Dr. Martin Cole's Institute for Sex Education and Research..has been supplying sex surrogate therapy for ten years. a 1911 D. G. Phillips *Susan Lenox* (1917) II. xx. 442 Men..might regard her as nothing but sex symbol; she regarded herself as an intelligence. 1951 M. McLuhan *Mech. Bride* (1967) 94/2 The 'line' [of chorus girls] itself..is even more basic than the sex symbol of the flower. 1976 Botham & Donnelly *Valentino* i. 12 The olive skin of the man who would..become the world's first and most enduring sex symbol. 1931 G. T. Renier *English: Are they Human?* iv. 85 The wide-spread ignorance of the technique of love among average Englishmen..results from the taboo on direct sex-talk. 1977 C. Fremlin *Spider-Orchid* xiii. 94 They have these sex talks..at school from the age of nine. 1978 J. Pudney *Thank Goodness for Cake* 80 At the football club sex talk proliferated, constantly spiced with dirty stories. 1976 *National Observer* (U.S.) 13 Mar. 6/3 American life-styles: a TV anchorman,..a rock star,..a groupie, a sex therapist. 1977 DeLora & Warren *Understanding Sexual Interaction* iv. 90 Sex therapists, who direct their creative energy toward the specifically sexual difficulties of individuals or couples. 1961 R. A. & F. R. Harper in *Encycl. Sexual Behaviour* I. 348/2 Individuals and groups vary..in the quality and quantity of sex therapy they need. 1970 Belliveau & Richter *Understanding Human Sexual Inadequacy* (1971) III. vii. 77 Success in sex therapy is dependent upon communication between the sex-partners. 1928 D. H. Lawrence *Lady Chatterley's Lover* i. 6 In the actual sex-thrill within the body, the sisters nearly succumbed to the strange male power. 1941 Miller & Dollard *Social Learning & Imitation* xii. 198 By the time a child reaches his second year sex-typing has already begun. *Ibid.*, The punishments meted out to adults who actually exhibit such tendencies tend to maintain and strengthen the sex-typed habits acquired. 1979 *Bull. Amer. Acad. Arts & Sci.* Feb. 31 The modal pattern for each sex is, nonetheless, conventionally sex-typed. 1976 *New Society* 4 Mar. 509/1 The 'blue/gun for a boy: pink/dolly for a girl', sex-typing syndrome. 1898 C. P. Stetson *Women & Econ.* x. 213 We confuse the natural result of marriage in children, common to all forms of sex-union, with the family,—a purely social phenomenon. 1923 M. Stopes *Contraception* iii. 39 The type of woman who..has acquired the view that all sex union after the procreation of the desired number of children has been accomplished, is wrong. 1920 M. Sanger *Woman & New Race* ix. 111 This man is not concerned with his wife's sex urge, save as it responds to his own at times of his choosing. 1966 D. Francis *Flying Finish* vi. 74 Bravery is built in... You can't stamp it out any more than the sex urge. 1912 L. Housman (*title*) Sex-war and woman's suffrage. 1978 J. Irving *World according to Garp* xvii. 362 'Fucking *women*,' the cabby said. 'Fucking *men*,' said Garp, feeling..that he had done his duty to ensure that the sex war went on. 1911 'I. Hay' *Safety Match* I. i. 4 The sides of the house are equally balanced both for purposes of companionship and in the event of sex-warfare.

sex, *v.* Add: **2.** *to sex up* (slang), to give a sexual flavour to, to increase the sexual content of.

Examples of the adj. or quasi-adj. *sexed-up* 'sexually aroused' are included here for convenience.

1942 Berrey & Van den Bark *Amer. Thes. Slang* § 361/5 *Make sexy, sex it up*, to introduce sex into, as a story. *Ibid.* § 361/9 Passionate; amative; lustful..(*all*) *sexed up*. 1958 *Observer* 24 Aug. 5/7 The business of 'sexing up' the titles of foreign films is..a trick well known in both France and Britain. 1959 *Ibid.* 11 Oct. 21/4 Reads rather like an old-time boy's book sexed up and sadistified for the 1950s. 1969 J. Gardner *Compl. State of Death* vi. 97 What do you do when you get sexed up? Mortify the flesh? 1976 *Nature* 15 July 177/3 Erickson and Zenone tested the reaction of 35 males to two groups of females... The males..showed more aggression and less courtship towards the 'sexed up' females.

3. *intr.* To have sexual intercourse. *slang*.

1966 'G. Black' *You want to die, Johnny?* ix. 172 The surprising thing isn't the number of teenagers who like sex and dope, but rather..that there are so many that don't. 1980 J. Barnett *Palmprint* i. 6 Maybe we sex together at yo' place.

sexational (seksē[1]·ʃənăl), *a. slang* (orig. *U.S.*). Also **sexsational.** [Blend of Sex *sb.* and Sensational *a.*] Sexually sensational. So **sexa·tionalism.**

1927 *Time* 16 May 39 Newspaper sensationalism has developed into sex-ationalism. 1928 *Daily Express* 30 July 3/3 You send us films of sexational novels and pornographic plays, which our censors cut to pieces. 1937 *Time* 27 Dec. 30/2 Sexational, robustious Cinemactress Mae West appeared on a commercial broadcast for the first time in four years. 1957 R. Hoggart *Uses of Literacy* viii. 213 Their authors' favourite descriptive epithet for them is 'sexational'. 1968 *Punch* 29 May 789/3 These are the qualities that save the film from being what *Variety* often calls 'sexsational'. 1976 *West Lancs. Evening Gaz.* 8 Dec. 3/2 (Advt.), 1st Blackpool showing of the Sexsational *Highway through the Bedroom* (X).

sexcapade (se·kskăpē[1]d). *colloq.* [Blend of Sex *sb.* and Escapade.] A sexual escapade.

1965 F. Raphael *Darling* xxvii. 134 We are not complicating our holiday with disgusting sexcapades. 1976 *Honolulu Star-Bull.* 21 Dec. A-11/1 A generally less swinging group than the lone men off on sexcapades who helped give tourism a bad name.

sexduction (seksdʌ·kʃən). *Microbiology.* Also **sex-duction.** [Blend of Sex *sb.* and Transduction.] The transfer of part of a bacterial genome from one bacterium to another by a sex factor.

1960 [see *F III. 1 l]. 1964 G. H. Haggis et al. *Introd. Molecular Biol.* x. 277 It is possible by exploiting a special phenomenon, known as sex-duction, to bring about a stable state of diploidy for a short length of the chromosome containing all the genes involved in the lactose system. 1974 *Molecular & Gen. Genetics* CXXX. 99 The number of sexductants in a λ-immune recipient is not significantly affected while sexduction into a non-immune recipient is increased by a factor 10–20.

sexed, *a.* Add: **3.** With prefixed adv.: having the sexual desires, emotions, or functions developed in a specified way.

1898, etc. [see *OVER-SEXED *a.*]. 1921 *Outward Bound* Mar. 13/2 Had Elizabeth been as strongly sexed as she. 1949 M. Mead *Male & Female* vi. 139 The tall, fiery, infinitely proud, specifically nervously sexed man and woman. 1974 M. Taylor tr. *Metz's Film Lang.* i. 9 Each one is disguised by a whole rigorously sexed body.

4. Of poultry: divided into the two sexes.

1960 *Farmer & Stockbreeder* 1 Mar. 148/3 (Advt.), Day old sexed pullets. 1971 *Farmers Weekly* 19 Mar. 92/2 (Advt.), Sexed poults.

sexennium (sekse·niŏm). [L.: see Sexennarian.] A period of six years.

1959 *Amer. Jrnl. Phys. Anthropol.* XVII. 132/1 Over the sexennium from age 5 years to age 11 years, some children augment their skeletal face depth more than

twice as much as other children. 1970 D. M. Walker *Princ. Scottish Private Law* I. li. 882 When the sexennium has run the constitution and resting—owing of the debt must be proved by the writ or oath of the debtor.

sexful (se·ksfŭl), *a. rare.* Also **sexfull.** [f. Sex *sb.* + -ful.] Conveying sexual emotions; sexy.

1898 G. Moore *Evelyn Innes* xvii. 232 Soprano voices of a rarer and more radiant timbre than any woman's sexful voice. 1959 D. Barton *Loving Cup* 144 A sexless voice..was singing a sexfull dance lyric.

sexi-. Add: **sexivalent** *a.* (earlier example).

1872 *Phil. Mag.* XLIII. 260 Carbon is quadrivalent; two carbon atoms may unite by mutually saturating 1, 2, or 3 pairs of affinities, thus giving rise to a radical C^2, which may be sexi-, quadri-, or bivalent accordingly: thus, C^2H^6, C^2H^4, C^2H^2.

sexing, *vbl. sb.* Add: **1.** Delete *rare* and add later examples.

1970 *Nature* 11 July 190/1 None of these specimens was sufficiently mature to allow accurate sexing. 1973 *Ibid.* 14 Dec. 423/2 Final proof of an increase in frequency of Y sperm must await sexing of offspring conceived of spermatozoa exposed to the isolation process.

2. With *up*: see *SEX *v.* 2.

1954 Koestler *Invisible Writing* xxxi. 330 [Otto] thought that a little sexing-up of the war could do no harm.

sexism (se·ksiz'm). [f. Sex *sb.* + -ism after *RACISM.] The assumption that one sex is superior to the other and the resultant discrimination practised against members of the supposed inferior sex, esp. by men against women; also conformity with the traditional stereotyping of social roles on the basis of sex.

1968 C. Bird in *Vital Speeches* (U.S.) 15 Nov. 90 Sexism is judging people by their sex where sex doesn't matter. 1968 S. Vanauken *Freedom for Movement Girls—Now* 7 The parallels between *sexism* and *racism* are sharp and clear. Each embodies false assumption in a myth. 1971 *Guardian* 15 Jan. 11/4 The concept of a 'woman's page'..perpetuates sexism by stressing the 'special' domestic interests supposedly adhering to women. 1971 *Publishers' Weekly* 22 Mar. 14 The Women's National Book Association panel during NBA Week on 'sexism' in children's books. 1973 *Ms.* Nov. 39/2 An insidious form of sexism pervades most biographies of famous women, a tendency to treat women's work as peripheral to their lives. 1981 *Amer. Speech* LVI. 84 Although they recognize the inherent sexism of the generic masculine, the Fowlers see no real alternative.

sexist (se·ksist), *sb.* and *a.* [f. Sex *sb.* + -ist after *RACIST *sb.* and *a.*] **A.** *sb.* One who advocates, practises, or conforms to sexism. **B.** *adj.* Of, pertaining to, or characteristic of sexism or sexists.

1965 P. M. Leet *Speech* 18 Nov. (mimeographed), When you argue..that since fewer women write good poetry this justifies their total exclusion, you are taking a position analogous to that of the racist—I might call you in this case a 'sexist'—who says that since so few Negroes have held positions of importance..their exclusion from history books is a matter of good judgment rather than discrimination. 1968 C. Bird in *Vital Speeches* (U.S.) 15 Nov. 90 There is recognition abroad that we are in many ways a sexist country. *Ibid.*, Women are sexists as often as men. 1968 S. Vanauken *Freedom for Movement Girls—Now* 2 The sexist myth is the greatest and most pervasive myth the world has ever told itself. *Ibid.* 7 *A sexist* is one who proclaims or justifies or assumes the supremacy of one sex (guess which) over the other. 1971 *Publishers' Weekly* 22 Mar. 20/2 We live in a sexist society. By sexist we mean predetermining social roles on the basis of sex alone. 1976 *New Yorker* 5 Apr. 57/1 He was very stern and disagreeable and a gross sexist. 1977 Miller & Swift *Words & Women* ix. 143 The language of American school books mirrors the sexist assumptions of society. 1978 J. Irving *World according to Garp* xvi. 344 The sexist notion that women are..the acceptable prey of predatory males.

sexlessly, *adv.* (Later examples.)

1965 F. Raphael *Darling* ix. 38 The West End was sexlessly bent on commerce. 1978 *Daily Tel.* 6 July 13/1 He is frequently bidden to the neat little house..where Leonora flirts with him, mothers him, indeed sexlessly gobbles him up.

sex-limited (stress variable), *a.* [f. Sex *sb.* + Limited *ppl. a.*] †**1.** = *SEX-LINKED *a.* 1. *Obs.*

1909 W. Bateson in A. C. Seward *Darwin & Mod. Science* v. 94 A study of the sex-limited descent of certain features in other animals. 1919 R. C. Punnett *Mendelism* (ed. 5) 95 Sex-limited inheritance..has been demonstrated in other birds besides poultry.

2. Of a genetic character or a phenotype: capable of occurring only in individuals of one sex. Of an individual: having such a character or phenotype.

[1871 C. R. Darwin *Descent of Man* viii. 282 (*heading*) Inheritance as limited by sex. *Ibid.* 285 If any variation appeared in a female pigeon, which was from the first sexually limited in its development to the females, it would be easy to make a breed with the females alone thus characterized. 1905 *Brit. Med. Jrnl.* 28 Oct. 1094/1 The abnormality, though affecting both boys and girls, is mainly limited to one sex in each family.] 1923 *Jrnl. Genetics* XIII. 215 As a contribution towards the adoption..of uniform terminology for the various modes of inheritance (at present, such terms as 'sex-limited' and 'sex-linked' are often used at random), I would point out that the denomination '*sex-linked*' (geschlechtsgebunden) ought to be applied only to conditions of inheritance explainable by the presence of the factor in question in those sex-chromosomes which are normally found in individuals of both sexes. *Ibid.* 216 The expression 'sex-limited inheritance' is not very judicious... 'Sex-limited manifestation' would be better. 1944 *Genetics* XXIX. 520 The mutant is strongly sex-limited in phenotypic expression. 1975 *Jrnl. Zool.* CLXXVII. 330 In butterflies, most sex-limited morphs are controlled by autosomal genes, their expression being dependent on the sex of the developing individual.

3. *transf.* Occurring only in one sex.

1949 M. Mead *Male & Female* xi. 237 To the phrasing of any piece of human behaviour, however sex-limited it may seem to be, both sexes contribute their imaginations.

Hence **sex-limita·tion.**

1911 *Jrnl. Genetics* I. 189 The operation of the system of sex-limitation is similar in all these examples, the only difference being that in the one group the repulsion is from the factor *F*, in the other from the factor *M*. 1922 *Encycl. Brit.* XXXII. 419/2 (*heading*) Sex-limitation and sex-linkage. 1975 *Jrnl. Zool.* CLXXVII. 329 Polymorphism in butterflies is complicated by various factors, notably sexual dimorphism and sex-limitation.

sex-linked (stress variable), *a.* [f. as prec. + Linked *ppl. a.*] **1.** Being or determined by a gene that is carried on a sex chromosome.

[1905 *Brit. Med. Jrnl.* 28 Oct. 1095/1 Thus, in haemophilia, the abnormality is habitually linked with one sex.] 1912 *Jrnl. Exper. Zool.* XIII. 80 In the male-producing sperm, where no X is present, the sex-linked characters are always absent. 1923 [see *SEX-LIMITED *a.* 2]. 1974 *Encycl. Brit. Micropædia* III. 23/2 Colour blindness, which affects about 20 times as many males as females, is a sex-linked recessive character.

2. Occurring only or characteristically in one sex.

1932 Auden *Orators* II. 42 Self-regard, in origin a mere accident of overcrowding, like haemophilia is a sex-linked disease. 1956 H. W. Papashvily *All Happy Endings* v. 75 In the writing of *Uncle Tom's Cabin*, Mrs. Stowe had exhibited many of those qualities considered by her contemporaries to be sex-linked to females—sympathy, tact, sensitivity, sensibility.

So **sex-linkage,** the state or condition of being sex-linked.

1912 *Jrnl. Exper. Zool.* XII. 512 In this case it would seem that complete sex-linkage, such as that found..in barred fowls, would not occur at all. 1975 *Zool. Jrnl.* CLXXVII. 330 Such sex-linkage..may give rise to a situation where a morph is common in one sex, and very rare in the other.

sexly, *a.* Delete † *Obs.*, for *rare*[−1] read *arch. rare*, and add later example.

a 1945 E. R. Eddison *Mezentian Gate* (1958) xxxviii. 203 You [*sc.* a woman] are hampered by no sexly weakness: as fit as any man living to undertake it.

sexology (seksǫ·lŏdʒi). orig. *U.S.* [f. Sex *sb.* + -ology.] The scientific study of sex and of the relations between the sexes.

1902 W. H. Walling (*title*) Sexology. 1912 *Amer. Jrnl. Urology* VIII. 441 The author believes that..there is practically no individual who can claim to have entirely avoided sexual activities of any kind, an opinion in which Lowenfeld, whose experience in sexology is so extensive, concurs. 1936 C. S. Lewis *Allegory of Love* I. 13 The second factor is the medieval theory of marriage—what may be called, by a convenient modern barbarism, the 'sexology' of the medieval church. 1977 E. J. Trimmer et al. *Visual Dict. Sex* (1978) i. 18 Reich is..in all the history of sexology, perhaps the most single-minded believer in the centrality of sex to human lives.

Hence **sexolo·gical** *a.*, of or pertaining to sexology; **sexo·logist,** one who studies sexology.

1914 *Amer. Jrnl. Clinical Med.* Aug. 687/1 There may be some homosexuals who are reconciled to or even proud of their abnormality, as some sexologists claim, but I must confess that I have not met such types. 1920 *Contemp. Rev.* July 93 A point of some sexological interest. 1949 Koestler *Insight & Outlook* xiv. 196 The distinction made by some sexologists between the detumescent and contractile components of the sexual drive. 1973 I. Robinson *Survival of English* v. 178 The only happy aspect of the sexological misapplication of the jargon of science is that the common language is strong enough to make it appear at times what it is, howlingly funny. 1980 *Times Lit. Suppl.* 28 Nov. 1355/1 The sexual inclinations of a sexologist presumably have to be investigated in some detail, even if the whole truth is not intrinsically interesting.

sexophone (se·ksŏfōun). *rare.* [Blend of Sex *sb.* + Saxophone.] An imaginary musical instrument resembling a saxophone and producing sexual sensations. Also *attrib.* So **sexo·phonist.**

1932 A. Huxley *Brave New World* v. 88 The Sixteen Sexophonists were playing an old favourite. *Ibid.* 89 The sexophones wailed like melodious cats under the moon. 1945 V. Nabokov *Real Life Sebastian Knight* xi. 93 Physical love is but another way of saying the same thing and not a special sexophone note, which once heard is echoed in every other region of the soul.

sexpert (se·kspəɹt). *slang* (orig. *U.S.*). [Blend of Sex *sb.* and Expert *sb.*] An expert on sexual matters.

1924 [see *DIM *a.* 4 b]. 1979 *Radio Times* 9–15 June 78/1 Every other interviewed sexpert seemed to come from California where..you can graduate in any old spurious subject.

sexploitation (seksploitē[1]·ʃən). [Blend of Sex *sb.* and Exploitation.] The exploitation of sex, esp. in films. Also *attrib.*

1942 in Berrey & Van den Bark *Amer. Thes. Slang* § 361/1. 1948 G. V. Desani *All about Mr. Hatterr* ii. 73 Damme, make the eternal triangle pay out a dividend! First that Portuguese feller victimised, then self! Damme, sexploitation! 1967 *Time* 17 Feb. 99 Jack Smith's four-year-old *Flaming Creatures*, an incredible tedious parody of a sexploitation picture, demonstrates how easy it is to fall asleep in the steamy midst of an hour-long transvestite orgy. 1977 E. J. Trimmer et al. *Visual Dict. Sex* (1978) xxiv. 272 Sexploitation in advertisements seeks to tell the public that the problems of human sexuality are easily solved by the purchase of manufactured goods. 1981 *Times* 29 Jan. 4/4 It will become the 164th 'sexploitation' establishment in an area of less than one square mile.

So **sexploi·t** *v. trans.* to exploit sexually; **sexploi·tative** *a.*

1970 M. Pei *Words in Sheep's Clothing* App. A 218 Lastly, there is the entrancing possibility of a vast extension of *sex*-portmanteau forms, for which a precedent already exists in 'sexploit' and 'sexploitation'. 1977 *Gay News* 24 Mar. 29/2 *Inserts* advertises itself as..simultaneously serious and sexploitative. 1977 M. Sokolinsky tr. *Merle's Virility Factor* ix. 180 That sociologist broad..she never stopped telling me: 'Bess, you've been sexploited by men.'

sexsational, var. *SEXATIONAL *a.*

sextal (se·kstăl), *a.* [f. L. *sext-us* sixth + -al.] Pertaining to a system of numerical notation with 6 as base.

1943 *Trans. Philol. Soc. 1941* 10 A purely native six-system; that is what we have in Finno-Ugrian... The sextal system, though not common among the languages of the world, is quite well-attested. 1962 *Punch* 4 Apr. 554/3 Age-grouping would be standardised (on the sextal scale, of course). 1971 *Nature* 12 Mar. 133/3 Like the SI prefixes, it perpetuates the sextal (and ternary) system for counting indices, while the rest of our number system is decimal.

sextary. 2. For 'sixth' read 'sixteenth'.

sextuplicate, *a.* Restrict *Obs. rare*[−1] to sense in Dict. and add: **B.** *sb.* (See quot. 1934.) Also in phr. *in sextuplicate*, in sixfold quantity, in six copies.

1934 Webster, *Sextuplicate, n.*, a sixth thing corresponding to five others of the same kind; an exact copy of something of which five other exact copies exist. 1975 *Daily Tel.* 1 Aug. 12 Contestants. have been asked to submit their entries in sextuplicate. 1978 *Nature* 31 Aug. 897/2 (*caption*) Each column represents the result for one patient, averaging 4 to 10 separate assays (in sextuplicate) per tissue. 1979 *Ibid.* 29 Mar. 465/1 After pre-incubation, the tubes were returned to ice and ^{3}H-5-HT was added in sextuplicate.

sextuplication. (Later example.)

1935 *Antiquity* IX. 298 The sextuplication of the King's Chamber roof.

sextupole (se·kstiupōul), *a.* and *sb. Physics.* [f. *sextu-*, in Sextuple *a.* and *sb.* and related words + Pole *sb.*[2]] **A.** *adj.* Having six magnetic (or electric) poles, three of each polarity. **B.** *sb.* A sextupole device.

1961 in *Progr. Nucl. Physics* (1964) IX. 114 A Sextupole Magnet Design. 1969 [see *QUADRUPOLE *sb.* b]. 1976 *Physics Bull.* Nov. 499/2 Extra dipoles, sextupoles and octupoles act as correcting magnets in case of defects in the magnet lattice and for superfine control. 1979 *Sci. Amer.* May 64/1 The atoms emerge from a nozzle as a low-energy beam, then pass into the strongly nonuniform field of a sextupole, or six-pole, magnet... It discriminates strongly between spin-up and spin-down electrons.

sexual, *a.* Add: Many collocations are paralleled by and equivalent to combs. with *sex* (see *SEX *sb.* 5). These are not individually listed or defined below.

The division of senses in the Dict. was satisfactory at the time of publication of this part (1913). In the sections below, our aim has been to compress as best we could into the original framework selected examples from our large collection of printed evidence.—Ed.

1. (Further examples.) *sexual politics*, the principles determining the relationship of the sexes; so *sexual-political* adj., *sexual politician*.

1968 S. Hynes *Edwardian Turn of Mind* vi. 201 It was an easy grammatical step..to shift the blame from sexual discrimination to sex itself. 1970 K. Millett *Sexual Politics* p. xii, The prospect of radical change in sexual politics. *Ibid.* II. iii. 110 The sexual-political predilections of each faction. *Ibid.* iv. 233 So we proceed to the counterrevolutionary sexual politicians themselves—Lawrence, Miller and Mailer. 1981 N. Tucker *Child & Book* vii. 212 Children's perceptions of their sexual roles are built up from many different sources.

2. (Further examples.) *sexual dimorphism*, the condition in which there exist marked differences in form or appearance between the sexes of a species in addition to differences in the sexual organs themselves; *sexual interference* (euphem.), sexual assault or molestation; *sexual selection*: see Selection 3 b.

1800 Wordsworth *Lyrical Ballads* I. p. xxxii, From this principle the direction of the sexual appetite, and all the passions connected with it take their origin. **1849** W. M. Thackeray *Pendennis* I. xxv. 239 That anxiety with which brooding women watch over their sons' affections... I have no doubt there is a sexual jealousy on the mother's part, and a secret pang. **1868** H. Maudsley *Physiol. & Pathol. of Mind* (ed. 2) II. iii. 405 Acute dementia..connected, he believes, with the effect produced on the nervous system by sexual intercourse. **1888** Sexual dimorphism [see Dimorphism b]. **1898** *Alienist & Neurologist* Oct. 613 From whatever side and from whatever symptoms we start, we always unfailingly reach the region of the sexual life. **1898** 'S. Grand' *Beth Bk.* xliv. 417 The sex question..is the stock-in-trade of every author, as if there were nothing..in the lives of men and women but their sexual relations. **1902** *Encycl. Brit.* XXVII. 625/2 *Bonellia* and *Hamingia* are very interesting examples of sexual dimorphism... The male is reduced to a minute..organism, which passes its life..in a special recess of the nephridia of the female. **1903** G. B. Shaw *Man & Superman* p. xvi, The Don Juan play..is to deal with sexual attraction. **1928** W. S. Maugham *Sacred Flame* III. 142 You can't go without food... But you can go without the satisfaction of your sexual appetites. **1929** D. H. Lawrence *Pornography & Obscenity* 18 The young man and the young woman went and had sexual intercourse together. **1932** S. Zuckerman *Soc. Life Monkeys & Apes* xiii. 212 It is possible that sexual dimorphism plays some part in determining the monogamy or polygyny of a species. **1934** Blunden *Choice or Chance* 58 You saw the mystical idea of course—..roughly, my Tristram is the Sexual Force. **1951** N. Mitford *Blessing* xi. 118 If you don't empty your mind and heart of sexual jealousy..you will never be happy with me. **1957** R. Chandler *Let.* 28 Apr. (1981) 441, I had not filled myself with sexual fantasies. **1960** C. Day Lewis *Buried Day* vi. 107 The conflict and guilt set up by masturbation.. and the sexual fantasies attached to it. **1968** S. Hynes *Edwardian Turn of Mind* vi. 195 The biological facts of sexual attraction and the urge to reproduce. **1968** 'A. Gilbert' *Night Encounter* iv. 45 Quite a young girl... No attempt at sexual interference, no signs of pregnancy. **1970** *Cambr. Anc. Hist.* (ed. 3) I. I. v. 156 Even allowing for marked sexual dimorphism, it is still obvious that more than one species [of Australopithecine] demands recognition. **1971** P. D. James *Shroud for Nightingale* vii. 246 The late Mr. Dettinger hadn't understood [many things], his wife's sexual needs among them. **1973** B. A. Tonkin tr. *Lorenz & Leyhausen's Motivation Human & Animal Behaviour* iii. 53 The sexual drives bring forces of a new kind pulling in a new direction. **1974** H. R. F. Keating *Underside* xi. 108 She must know..that men had sexual urges, that they could not live without any sexual experience of any sort. **1978** P. G. Winslow *Coppergold* 163 Post-mortem had shown that the girl had had sexual relations with a man just before her death. **1979** *Times Lit. Suppl.* 23 Nov. 20/1 Edel..suggests.. that after it was clear that a sexual relationship with Virginia was impossible he [*sc.* Leonard Woolf] sublimated his sexual drives in work. **1980** D. Newsome *On Edge of Paradise* 382 He had no sexual life; all his sexual instincts had to be sublimated. **1981** A. Edwards *Sonya* xix. 321 Her diary..is filled with musings on her sexual needs, frustrations, and fantasies. **1981** G. Markstein *Ultimate Issue* 261 'And at these places sexual intimacy took place—' Christ almighty..how many more times is he going to say it. Sexual misconduct. Sexual intercourse.

3. (Earlier and further examples.) *sexual athlete*, a sexually vigorous person, a skilled performer in sexual intercourse; so *sexual athleticism, athletics*; *sexual inversion*: see *Inversion 10; *sexual psychopathy*: see Psychopathy 1; hence *sexual psychopath, psychopathic* adj.; *sexual revolution*, the liberalization of established social and moral attitudes to sex.

1792 M. Wollstonecraft *Rights of Woman* vii. 273 (*heading*) Modesty.—comprehensively considered, and not as a sexual virtue. **1888** K. Pearson *Ethics of Freethought* v. 120 Another good example is that of sexual morality; here the most difficult questions arise. **1897** C. Lombroso in T. L. Stedman *Twentieth Century Practice* XII. 402 We have seen that in the most extraordinary tendencies of sexual psychopathy..the somatic and hereditary signs are epileptoid in character. **1908** G. B. Shaw *Getting Married* 182 Whilst the subject is considered shameful..we shall have no systematic instruction in sexual hygiene. **1910** tr. *Freud's Lectures* in *Amer. Jrnl. Psychol.* XXI. 218 The claims of our civilization make life too hard for the greater part of humanity..without producing an excess of cultural gain by this excess of sexual repression. **1924** *Internat. Jrnl. Psycho-analysis* V. 95 (*heading*) The sexual offender. **1934** A. Huxley *Beyond Mexique Bay* 44 Places where people..obey other sexual taboos. **1936** 'R. West' *Thinking Reed* viii. 265 It sounded insincere, as if she were merely obeying a sexual convention. **1938** *John Marshall Law Q.* III. 407 The proposed law is as follows: An Act to Provide for the Commitment and Detention of Criminal Sexual Psychopathic Persons. **1939** R. Pearl *Nat. Hist. Population* 293 Some present-day examples of sexual athletes who make Casanova, the traditional star, seem a somewhat puny performer. *Ibid.*, I thought you might be interested in some cases of—as it seems to me—prodigious sexual athleticism. **1941** *Horizon* Sept. 161 All societies, as the price of survival, have to insist on a fairly high standard of sexual morality. **1945** T. P. Wolfe tr. *Reich's Sexual Revolution* ix. 64 Soviet sexual legislation was the clearest expression of the first attack of the sexual revolution on the reactionary sexual order. **1950** *Amer. Jrnl. Sociol.* LVI. 142/1 Since 1937 twelve states and the District of Columbia have enacted sexual psychopath laws. *Ibid.* 142/2 The concept of the 'sexual psychopath' is so vague that it cannot be used for judicial and administrative purposes. **1954** B. Karpman *Sexual Offender* xii. 224 Determination of the question of sexual psychopathy is by a superior court and commitment is for an indefinite period. **1957** H. M. Hacker in *Marriage & Family Living* XIX. 232/1 In societies which differentiate strongly between masculine and feminine social roles, individuals who manifest personality traits ascribed to the opposite sex or who feel inadequate in fulfilling their part of the sexual division of labor may become confused in their sexual identification, and feel that they must also change their sexual object. **1958** J. Byrom *Or be He Dead* iii. 37 Essays about sexual perverts. **1961** R. F. C. Hull tr. *Jung's Freud & Psychoanal.* in *Coll. Works* IV. iv. 321 As soon as we enter the field of neurosis, this antithesis is stretched to the limit. God becomes the symbol of the most complete sexual repression. **1963** B. Friedan *Feminine Mystique* (1965) xi. 266 A woman who is herself only a sexual object, lives finally in a world of objects, unable to touch in others the individual identity she lacks herself. **1967** B. W. Tuchman in *Sat. Rev.* (U.S.) 25 Feb. 28/3 Sexual perversion and hallucinatory drugs..'are not what human history is about.' **1968** S. Hynes *Edwardian Turn of Mind* vi. 201 The [suffrage] movement never made sexual freedom a goal. **1968** A. Diment *Bang Bang Birds* II. vi. 81 You'd think his life work was spreading American sexual mores around the world. **1969** *Daily Tel.* (Colour Suppl.) 24 Oct. 35/2 The family is not composed of sexual athletes and cannot make love 20 times a day and so cannot try everything. **1970** *Guardian* 12 Nov. 10/3 Would he have to do anything awful in the way of sexual athleticism? **1970** K. Millett *Sexual Politics* iii. 62 A sexual revolution would require.. an end of traditional sexual inhibitions and taboos. **1971** G. Steiner *In Bluebeard's Castle* I. 15 Bourgeois sexual ethics were a veneer, masking a great area of turbulent hypocrisy. **1971** H. Lee *Surrogate Wife* (1972) 27 Nice girls..were less likely to be able to copulate with strangers, even if those strangers were sexual therapists. **1972** P. D. James *Unsuitable Job* iv. 144 One of the more innocuous of sexual deviations. **1972** P. Cousins (*title*) Christianity and sexual liberation. **1975** G. Howell *In Vogue* 62 Sexual deviation was becoming respectable. *Ibid.*, The sexual education of the jazz age. **1975** L. Farley in *N.Y. Times* 19 Aug. 38/1 Sexual harassment of women in their place of employment is extremely widespread. **1976** *Globe & Mail* (Toronto) 16 Feb. 1/6 Police said the report of a pathologist who examined the bodies makes no mention of a sexual assault. **1976** *Jrnl. R. Soc. Arts* June 351/2 The 4,500 magazines dealing in specialized trades or tastes from ironmongery to sexual athletics. **1977** *Evening Gaz.* (Middlesbrough) 11 Jan. 3/4 Sexual offences, mainly indecent assault on females, increased by 17. **1977** R. Green in W. H. Masters et al. *Ethical Issues in Sex Therapy & Research* vii. 198 There are clinicians of considerable sophistication utilizing sexual surrogates. **1978** *Times* 7 Aug. 2/4 Contraception..is producing..a profound alteration in sexual behaviour. **1979** J. Crosby *Party of Year* xvi. 99 Decadent bourgeois sexual perversions. **1977** C. Fremlin *Spider-Orchid* viii. 61 In spite of Permissiveness and the Sexual Revolution..nothing had changed! **1981** J. B. Hilton *Surrender Value* xiv. 104 He was a mild man... There was no record..of sexual aberration. **1982** R. Grayson *Montmartre Murders* xi. 91, I have been told that Suji's sexual appetite is formidable.

sexualization (seːksiuăləizēⁱ·ʃən). [f. Sexualize *v.* + -ation.] The act or process of sexualizing; the state of being sexualized; adaptation to a sexual role.

1889 *Classical Rev.* III. 391 We are inclined to doubt Pott's confident assumption that sexualization is a necessary consequence of personification. **1977** A. H. Williams in A. W. Franklin *Challenge of Child Abuse* xii. 168 In sadism, it is important to recognize the sexualization of the pleasure in the cruel transaction. **1979** *Country Life* 29 Mar. 947/4 The lives available to a woman..education, sexualisation, part-time prostitution, suburbanisation.

sexy (se·ksi), *a.* [f. Sex *sb.* + -y¹.] **a.** Concerned with or engrossed in sex. **b.** Sexually attractive or provocative, sexually exciting; also *fig.* Also *Comb.*, as *sexy-looking* adj.

[**1925** *La Nouvelle Revue Française* Jan. 313 Depuis que Joyce a publié un livre qu'ils croient 'sexy'—cet état d'esprit n'a pas d'équivalent français—on s'en empare.. que sa méthode sert de modèle à des gens qui..se disent surréalistes.] **1928** *Sunday Dispatch* 2 Dec. 19/2 Australian audiences..like sex plays, but they mustn't be too sexy. **1934** Dylan Thomas *Let.* 9 May (1966) 120, I shall now attempt to..look all sexy at the mantel-piece. **1935** J. T. Farrell *Judgment Day* ii. 22 He watched the dark sexy-looking waitress scurry with a large tray of food. **1940** N. Mitford *Pigeon Pie* ii. 29 He had been most famous as a singer of those sexy ballads which were adored by our grandparents. **1942** [see *glamorize *v.*]. **1947** J. Steinbeck *Wayward Bus* 84 Her voice was throaty and sexy. **1959** *Housewife* Oct. 131/2 Did she keep her figure beautiful and wear sexy nightgowns? **1963** A. Ross *Australia 63* 23 A sexy-looking torch singer breathed huskily into her microphone. **1970** *Wall St. Jrnl.* (Eastern ed.) 22 Sept. 1/6 Corn and soybeans may not sound as sexy as electronics or aerospace. **1975** R. Player *Let's talk of Graves* iii. 80 The seminarists were..casting perplexed eyes to Heaven, the Immaculate Conception being far too sexy to explain to adolescents. **1977** *New Scientist* 17 Mar. 638/1 Expensive, high-technology, politically sexy.. small wonder the big dam is popular all round. **1978** *Rolling Stone* 12 Jan. 44 A 'sexy' (TV for a good story) news idea. **1981** V. Glendinning *Edith Sitwell* iv. 57 Nancy Cunard—sexy, vital, unpredictable.

Hence **se·xily** *adv.*, **se·xiness.**

1925 *Glasgow Herald* 28 May 4/2 The stallion seems to vanish altogether near the end of the story, and the Welsh groom is put into prominence, with mere 'sexiness' thus supplanting magnificent vitality. **1947** J. Steinbeck *Wayward Bus* 83 Wide-set eyes meant sexiness. **1953** R. Lehmann *Echoing Grove* 240 Late party, the latest place, place of lugubrious eroticism, sexily spot-lit. **1971** *Ink* 12 June 6/3 So-called permissiveness, sexiness, moral relaxation have gone as far as they can without beginning to alter *radically* our civilization. **1978** P. Bryers *Cat Trapper* ix. 61 She curled sexily on a cushion.

Seychellois (sēⁱʃelwa·), *sb.* and *a.* Fem. **Seychelloise** (-wa·z). Pl. **Seychellois,** fem. **-oises** (-wa·z). [Fr.] **A.** *sb.* A native or inhabitant of the Seychelles, a group of islands in the Indian Ocean. **B.** *adj.* Of or pertaining to the Seychelles or their inhabitants.

1898 F. A. Barkly *From Tropics to North Sea* viii. 34 The Seychellois have a method of walling houses with the bark of the lathe palm. **1936** J. A. F. Ozanne *Coconuts & Créoles* xii. 184 The young Seychellois is quite safe from European competition. *Ibid.* 186 Most of these young Seychellois boys are very keen sportsmen. **1960** I. Fleming *For Your Eyes Only* 207 The only conceivable security hazard in the Seychelles lay in the beauty and ready availability of the Seychelloises. **1971** *Vogue* Dec. 73, I took passage on the launch with a crew of Seychellois police. **1978** T. Allbeury *Lantern Network* iii. 34 She was a Seychelloise from Mahé. **1980** *Illustr. London News* Mar. 67/2 As a tourist you will find both English and French spoken and understood; the Seychellois speak their own patois.

Seyfert (sēⁱ·fɔɪt). *Astr.* The name of Carl K. *Seyfert* (1911–60), U.S. astronomer, who first described such galaxies in 1943 (*Astrophysical Jrnl.* XCVII. 28), used *attrib.* with reference to a class of galaxies characterized by bright compact cores that show strong infra-red emission.

1959 *Astrophysical Jrnl.* CXXX. 26 Another characteristic feature of many of the Seyfert galaxies is the apparently small size of their nuclei in which the broad emission features arise. **1963** *Ibid.* CXXXVII. 1032 It is possible that an event of the Seyfert type is experienced by all spiral galaxies and not only by the Sa and Sb systems. We believe this to be a reasonable assumption, but it has the effect only of reducing the frequency of Seyfert objects among spirals by a factor of 2 or less. **1969** *Ibid.* CLVIII. 859 The detection of such [nonstellar] absorption features in Seyfert spectra is rather difficult. **1970** *Nature* 31 Jan. 410/2 Seyfert galaxies with their extraordinarily bright and unusual nuclei could easily be intermediaries between quasars and normal galaxies. **1970** *Sci. Jrnl.* Feb. 57/1 Many of the phenomena observed in Seyfert galaxies are similar to those seen in quasi-stellar objects. **1971** *New Scientist* 17 June 695/1 A compact infrared source with a spectrum similar to that of the Seyfert nuclei, but with a smaller energy output.., appears to be located at the centre of our Galaxy.

b. *absol.* A Seyfert galaxy.

1968 *Physical Rev. Lett.* XXI. 1540/1 NGC1275 and 3C120..are a hundred times more luminous in the radio than most of the Seyferts. **1977** *Dædalus* Fall 56 The association of X-ray emitters with active compact galaxies..has been confirmed by the discovery of an X-ray emission by the Seyfert galaxy 3C120..and of a dozen more Seyferts by Pounds and his associates working with Ariel-5.

Seym, var. *Sejm.

seymouriamorph (sīmōᵊ·riămǭɪf). [a. mod.L. name of suborder *Seymouriamorpha*, f. generic name *Seymouria* (F. Broili 1904, in *Palæontographica* LI. 81), f. *Seymour* name of a town in Baylor County, Texas + -ia¹.] A fossil tetrapod belonging to the suborder Seymouriamorpha, considered to include transitional forms between amphibians and reptiles.

1945 A. S. Romer *Vertebr. Paleontol.* (ed. 2) xxvii. 527 The degenerate genus *Kotlassia* represented the seymouriamorphs in the late Permian. **1977** A. Hallam *Planet Earth* 269 The terrestrial anthracosaurs, known as seymouriamorphs, include the only large terrestrial amphibians known.

Seyssel (sēⁱ·sel). The name of two villages on the upper Rhône, used *attrib.* or *absol.* to designate various white wines made there.

1926 P. M. Shand *Book of Wine* v. 171 The best-known Savoy wine is Seyssel... Much of the wine sold as Seyssel only enjoys the type-name by courtesy. **1968** V. & M. Pettitt *Len Deighton's Travel Dossier* 46 *Savoy*... Seyssel wines (much is made into 'vin mousseux'). **1968** [see *Roussette 3].

sez (sez). ¶. Jocular spelling of *says*, 3rd person sing. pres. of Say *v.*¹, esp. in representations of uneducated speech, and in phrase *sez you* (see *say *v.*¹ 3 b).

1844 'J. Slick' *High Life in N.Y.* I. i. 8 Cousin John took out his watch..and, sez he,—'Come, Mr. Slick.' **1886** F. H. Burnett *Little Lord Fauntleroy* i. 9 Sez he to me: 'Mary,' sez he, 'I'm very much int'rusted in the 'lection,' sez he. **1904** Wodehouse *William Tell* vii. 45 'What I sez,' said Friesshardt, 'is, wot's the use of us wasting our time here?' **1930** *Outlook* (N.Y.) 12 Nov. 417/3, I am so tired of hearing sap, oh boy, and how, sez you, guts and dirty bum, that I could almost leave for the Fiji Islands to escape them. **1931** *Week-end Rev.* 24 Oct. 513/1 Mr. Lowe is well known for his invention of the famous catch-

phrase 'Sez you!' **1933** Dylan Thomas *Let.* Nov. (1966) 53 May I borrow that foul expression of yours—it isn't yours, really—and whisper Sez You into his ear. **1940** H. G. Wells *Babes in Darkling Wood* II. i. 125 They must not attempt either to monopolise or possess. (Sez we.) **1960** J. Stroud *Shorn Lamb* xi. 128 If I make a movement, he sez: 'Oh, don't be disgusting!' he sez. **1973** B. Graeme *Two & Two make Five* xiii. 123 'He's..not nearly so useful in a rough house.' 'Sez you!' Sanders growled. **1977** [see *say *v.*¹ 3 b].

‖ **Sezession** (zetsesiō·n). *Art.* Pl. **Sezessionen.** [G.] = *secession 3 d.

1905 *Burlington Mag.* VI. 422/2 The Austrian Government..did not look with a favourable eye upon this 'Sezession', and withheld all support which the 'Sezession' claimed as well as other societies of artists. **1959** P. & L. Murray *Dict. Art & Artists* 296 The *Sezessionen* were groups of artists in Germany and Austria who resigned from established academic bodies and exhibiting societies in order to forward the aims of various modern (usually Impressionist) movements. **1962** *Listener* 1 Mar. 384/3, I said nothing about..the artists of the *Sezession*. **1970** *Oxf. Compan. Art* 106/1 When in 1910 a number of young painters were rejected by the Sezession—among them members of Die Brücke—they started the *Neue Sezession*.

Also **Sezessioni·st,** an artist belonging to the *Sezession*; **Sezessionstil** (-ʃtī·l) [G. *stil* style], the style of the *Sezession*.

1958 M. L. Wolf *Dict. Painting* 264 *Sezessionists*, an art group in Vienna associated with the *art nouveau* movement, popular between 1890 and 1905. **1967** J. N. Barron *Lang. of Painting* 19 In Belgium, it [*sc. Art Nouveau*] was associated with *Les Vingt* (The XX); in Vienna, with the Sezessionists. **1970** *Oxf. Compan. Art* 80/1 In Germany the style was called *Jugendstil*..; in Austria it was called *Sezessionstil*. **1978** *Country Life* 10 Aug. 394/1 (*caption*) The Bull Inn, Paisley. Built in 1900–1901..it combines Scottish Baronial features with a curved gable akin to the contemporary Viennese *Sezessionstil*.

sferics (sfe·riks), *sb. pl.* orig. *U.S.* Also with capital initial; (*rarely*) **spherics.** [Contraction and respelling of *atmospherics *sb. pl.*] Atmospherics; sometimes used to denote a radio direction-finding system used to locate storms by means of the atmospherics they produce. Hence **sfe·ric** *a.*, of or pertaining to sferics.

1945 in *U.S. Army Signal Res. & Development Lab., Techn. Rep. 2199* (AD 266–795) (1961) 97 (*heading*) Military characteristics for automatic atmospherics (sferics) equipment. **1949** *Marine Observer* XIX. 199 'Sferic' is the code word which has been used for some years now to designate reports of positions of areas in which thunderstorms are taking place. **1951** R. C. Wanta in T. F. Malone *Compendium Meteorol.* 1297/1 Sferics (less commonly spherics) is a contraction of the word atmospherics meaning natural electrical phenomena detected by radio methods. **1963** T. Pynchon *V.* ix. 230 As it turned out, the whistler was only the first of a family of sferics whose taxonomy was to include clicks, hooks, risers, nose-whistlers and one like a warbling of birds called the dawn chorus. **1968** B. W. Atkinson *Weather Business* ii. 41 Sferic fixes depend on the radiation of electromagnetic waves caused by lightning flashes in the clouds. **1974** *Nature* 10 May 134/2 Observers under the balloon reported no thunder or lightning; thus we attribute the spherics to an intense thunderstorm system that was over the eastern United States at the time. *Ibid.*, Our local v.l.f. monitor recorded strong spheric activity.

sforzando. Delete sense b and substitute: **b.** As *sb.* Pl. **sforzandi, sforzandos.** A note or group of notes specially emphasized or rendered louder than the rest; an increase in loudness and emphasis; also *transf.* Also *attrib.* or as *adj.*

1849 *Belfast News-Letter* 6 Feb. 4/4 Prima donna instrumentalists of the sforzando order. **1890** G. B. Shaw in *Scots Observer* 28 June 143/2 The smoothing-out of all the old jerks and jigs and *sforzandos* from the surface of the stream of melody. **1902** [in Dict., sense b]. **1947** Auden in *Amer. Scholar* Autumn 405 Fill a Dwarf's ears with sforzandos and the dwarf will Believe he's a giant. **1956** T. Beecham 26 Apr. in H. Procter Gregg *Beecham Remembered* (1976) II. 180 Well, you know what dramatic effects are in my experience? It's making loud thumps on a percussion instrument, *sforzandi* in the strings—constant *sforzandi*—a few pathetic whinings on the wind, and sudden blasts on trumpets..and trombones. **1965** *New Statesman* 7 May 736/3 The woodwind skirls and *sforzando* brass exclamations still evoke the native dance-song. **1976** *Gramophone* Mar. 1463/1 He is meticulous in observing the jab of sudden sforzandos.

sforzato. Add: (Later examples.) Pl. **sforzati, sforzatos.**

1969 *Listener* 17 July 89/2 The hard sound of the first 'g' in the word is reflected in *sforzato* harp notes. **1976** *Daily Tel.* 8 Mar. 8/3 Presumably, if the lost autograph of the 32 in C minor had meanwhile surfaced,..with all *sforzatos* abolished, this fact would have been mentioned. **1977** 'E. Crispin' *Glimpses of Moon* iii. 41 '*Derngh!*' he exclaimed in his nose, imitating sforzato stopped horns. **1977** *Gramophone* Dec. 1098/2 No. 2 in E minor has as its first movement a funeral march that alternates between minor and major and is full of suspensions and *sforzati*. **1978** *Jrnl. R. Soc. Arts* CXXVI. 354/1 It is impossible for us to say definitely how soft is soft, or how loud is loud or how much emphasis to put on a sforzato.

sfumato, *a.* Add: Also as *sb.*, the technique of softening outlines and allowing tones and colours to shade gradually into one another; a softened outline or hazy form produced in this way.

1909 R. Fry *Let.* 16 Feb. (1972) I. 312 The black is a wonderful *sfumato* with scarcely any modelling. **1935** *Burlington Mag.* Feb. 72/2 The fine 'sfumato' and the delicacy of the surface. **1936** *Ibid.* Oct. 192/1 The influence of Venice is said to have succeeded the abandonment of Leonardo's *sfumato*. **1965** *New Statesman* 7 May 732/2 In all encounters with Leonardo there is never a definite answer to any of the artistic, aesthetic, historical or psychological questions which he raises. The *sfumato* —the suggestive smoke-like contour—is as much a characteristic of his personality as it is a feature of his drawing.

‖ **sgraffiato** (sgraf,fiā·to). *Pottery.* Pl. **sgraffiati** (-tī). [It., pa. pple. of *sgraffiare* to scratch, to produce sgraffito: see Sgraffito.] = Sgraffito b.

1862 J. C. Robinson in *Catal. Special Exhibition Wks. of Art S. Kensington Mus.* (1863, rev. ed.) Section xxi. 400 (*heading*) Sgraffiato or Incised Wares. **1900** F. Litchfield *Pott. & Porc.* ii. 13 This ware has been termed *sgraffiati, sgraffiato*, or incised ware. **1957** Mankowitz & Haggar *Conc. Encycl. Eng. Pott. & Porc.* 71/2 Several important slipware potteries existed in North Devon... Jugs and other presentation wares decorated in *sgraffiato* style with mariner's compasses, Royal arms, country scenes, or birds and foliage are typical. **1973** *Times Lit. Suppl.* 28 Dec. 1590/3 The medieval sgraffiato pottery of north and north-west Iran was based on Sasanian metalwork. **1980** *Catal. Fine Chinese Ceramics* (Sotheby, Hong Kong) 24 A large sgraffiato jar with shouldered ovoid body.

sgraffito. b. (Earlier example.)

1878 [see *Doulton].

Sgt., abbrev. of Sergeant *sb.*

1899 *Morning Post* 6 Dec. 5/2 No. 3732 Squad.-Sgt.-Maj. should be Shoeing-Smith John Hobbs. **1909** *Army & Navy Gaz.* 10 Apr. 342/2 Sgt. Huntley beat Sgt. Kilvert. **1948** *R.A.F. Rev.* June 22/3 Sgt. A. Roff, R.A.F., Selitar, Singapore. **1977** *Belfast Tel.* 28 Feb. 1/7 'Sgt. Joe', as he was affectionately known, was considered to be the model village policeman.

sh (ʃ), *v.* Also **sh-sh-sh.** [f. Sh *int.*; cf. Shshsh, *shush *v.*] **a.** *trans.* To reduce to silence or tranquillity with the sound of 'sh!', or attempt to do so.

1887 A. J. Evans *At Mercy of Tiberius* vii. 125, I patted and, 'she-e-d' her [*sc.* a baby], but she got her head above cover..and..set up a squall. **1916** 'B. Cable' *Action Front* 62 Ainsley 'sh-sh-sh-ed' him to silence.

b. *intr.* To become quiet in response to an order 'sh!'.

1925 'R. Crompton' *Still—William* iii. 53 'Sh!' William said fiercely. Violet Elizabeth 'Sh'd' obediently. **1972** Wodehouse *Pearls, Girls, & Monty Bodkin* ix. 136 She uttered a 'Sh!' of such significance that Grayce instinctively lowered her voice. 'Mrs. Molloy!' 'Sh!' Grayce might have retorted that she *had* sh-ed and that if she sh-ed any further she could become inaudible.

Shabas, var. *Shabbos.

‖ **shabash** (ʃaba·ʃ), *int.* [Hindi or Urdu.] Well done! Also as *sb.*, an exclamation of *Shabash!*

1843 C. I. C. Davidson *Travels in Upper India* I. 209, I was awakened at night from a sound sleep by the repeated *savâshes! wâh! wâhs!* from the residence of the thannadar. **1886** Yule & Burnell *Hobson-Jobson* 618/1 *Shabash!* interj. 'Well done!' 'Bravo!' Pers. *Shāh-bāsh*. 'Rex fias!' **1901** Kipling *Kim* iv. 105 'Oh, *shabash!*' murmured Kim..'Well done, indeed?' **1907** P. S. Allen *Let.* 8 Mar. (1939) 58 However in 10 days I shall be clear for Erasmus for the rest of the year. Shabash! **1974** H. R. F. Keating *Bats fly up for Inspector Ghote* iii. 24 Cries of 'Jolly good show' and 'Shabash, Inspector, shabash.'

‖ **Shabbat** (ʃaba·t). Also **Shabat, Shabbath.** [Heb. *šabbāṯ* Sabbath. Cf. *Shabbos.] Among Sephardic Jews and in the State of Israel: the Sabbath. Also in phr. *Shabbat shalom* [*shalom *int.* and *sb.*], 'peaceful Sabbath', a form of salutation used on the Sabbath.

1934 Webster, Shabbath. **1965** J. A. Michener *Source* 42 After we see it, why don't we go on to Zefat? It'll be Shabbat and we can attend the Vodzher Rebbe's synagogue. **1966** L. Davidson *Long Way to Shiloh* viii. 116 'All right. Thanks. Shalom.' 'Shabat Shalom.' **1971** D. Meiring *Wall of Glass* ii. 21 What would Arabs be doing on Tel Aviv beach?.. Above all on a Shabat? *Ibid.* v. 40 I'd been waiting for the sound of the ram's horn, to light my Shabat candles. **1976** C. Bermant *Coming Home* I. iii. 47 There is a traditional Jewish dish called *cholent*.. normally eaten for Shabbat lunch. **1980** *Encounter* Oct. 56/2 We bade each other *shabbat shalom*.

b. *Comb.*, as **Shabbat-goy** = *Sabbath goy* s.v. *Sabbath 4.

1859 [see *remnant *sb.* 2 b].

shabbify, *v.* Delete *nonce-wd.* and add later *fig.* example.

1961 E. Williams *George* xi. 149 Thomas became, for life, Tom,..and Biblical Job..was overnight shabbified into Joe.

‖ **Shabbos** (ʃa·bəs). Also **Shabas, Shabbes, Shabbuss, Shabes, Shabus.** Pl. **-im.** [Yiddish *shabes* ad. Heb. *šabbāṯ* Sabbath. See also *Shabbat.] Among Ashkenazi Jews: the Sabbath. Also *attrib.* Phr. to *make Shabbos*, to prepare for the Sabbath.

1876 Geo. Eliot *Dan. Der.* II. iv. xxxiii. 348 'Shlav'm Shabbes fyock on,' said Adelaide Rebekah. 'Her Sabbath frock... She'll have her Sabbath frock on this evening.' **1892** I. Zangwill *Childr. Ghetto* I. 151, I shall have no fish for Shabbos. **1959** H. Pinter *Birthday Party* I. 14 After lunch on Shabbuss we'd go and sit in a couple of deck-chairs. **1960** F. Raphael *Limits of Love* III. iv. 317 The chauffeur always carried a box of good things..which the inmates came to enjoy on the following *shabas* eve. **1968** M. Richler in R. Weaver *Canad. Short Stories* 2nd Ser. 150 We're orthodox here. Today is *shabus*. **1968** L. Rosten *Joys of Yiddish* 318 *Shabbes* begins just before sunset on Friday. The wife and mother, dressed in her very best, lights the *Shabbes* candles and offers a benediction. **1975** C. Potok *In Beginning* ii. 126 Yesterday was such a nice day. It was a pleasure to make Shabbos yesterday. *Ibid.* iii. 185 I'll read it [*sc.* the Torah]..in the synagogue, for Shabbosim and holidays. **1979** H. Howard *Sealed Envelope* xiv. 181 Every week she expects you for shabbos dinner but always you have excuses.

b. *Comb.*, as **Shabbos-goy** [*goy] = *Sabbath goy* s.v. *Sabbath 4; also fem. **Shabbos-goya(h**; also *fig.*

1892 I. Zangwill *Childr. Ghetto* I. 158 Poor women, frequently Irish, known as *Shabbos-goyahs* or *fire-goyahs*, acted as stokers to the Ghetto at twopence a hearth. **1959** M. Wilson *They came as Strangers* III. iii. 194 Victor Gollancz's parents were strictly orthodox,..never took a bus on the Sabbath and had a *Shabbes goy* to light the fires on that day. **1962** B. Abrahams tr. *Life of Glückel of Hameln* IV. 102 Send the *shabbos-goya* to me. I want to send her somewhere. **1969** A. Laski *Dominant Fifth* v. 187 Perhaps..we..have deliberately assimilated him, made him a tame cat, or *Shabbos-goy*. **1978** *Maledicta 1977* I. 323 Reinhold Aman, president, director, chief-lucubrator and *shabes-goy* of Maledicta, edits, typesets and publishes this journal considered by the repressed as a Gross Encounter of the Worst Kind.

shabby, *a.* Add: **1. c.** (Earlier example.)

1805 T. F. Fremantle *Let.* 1 Aug. in P. Fremantle *Wynne Diaries* (1940) III. vii. 195, I shall not apologise.. for the very shabby letters I have been..writing to you.

3. b. For *dial.* read '*dial.* and *colloq.*' and add earlier and later examples.

1853 D. G. Rossetti *Let.* 12 July (1965) I. 148 The weather had been generally very shabby. **1950** W. Stevens *Let.* 1 Feb. (1967) 663 While we have had an occasional day of proper winter, mostly it has been pretty shabby.

shabby (ʃæ·bi), *v.* [f. the adj.] **1.** *intr.* To act shabbily. *rare.*

1898 M. Deland *Old Chester Tales* 213 'They'll be shabbying on me,' said Katy.

2. *trans.* and *intr.* To make or become shabby.

1912 *Daily Chron.* 5 Mar. 9/2 She will probably find that a good deal of the..'shabbying' of her clothes is caused by..throwing her dresses carelessly down on chairs. **1920** H. Begbie *Life of W. Booth* I. iii. 56 The shadow of poverty deepening every day upon the shabbying walls of his unhappy home. **1962** D. Lessing *Golden Notebk.* IV. 536 You'll be one of those tough, square, solid middle-aged men, like a shabbying brown bear, your golden crew-cut greying judiciously at the temples.

shabby-gentility. (Earlier and later examples.)

1829 Carlyle *German Playwrights* in *Foreign Rev.* III. 115 Old threadbare material, scoured up into a state of shabby-gentility. **1898** G. B. Shaw *Plays Pleasant & Unpleasant* II. 156 The shifts of impecunious shabby-gentility.

Shabes, var. *Shabbos.

‖ **shabti** (ʃa·bti). *Egyptology.* [a. Egyptian *šbty*.] = *ushabti. Cf. *shawabti.

1864 *Zeitschrift für Ägyptische Sprache und Alterhumskunde* Oct./Nov. 90 The principal variants of the names of these figures have been given by M. Chabas as..*šuabti*, ..*šabt*,..*šabti* and..*šebti*; to these..may be added many additional forms. **1935** W. M. F. Petrie (*title*) Shabtis. **1936** *Times Lit. Suppl.* 18 July 601/1 Next to the scarab the *shabti* is perhaps the object most readily acquired by visitors to Egypt... This book is of first importance for the study of *shabti*-figures. **1953** *Flinders Petrie Centenary Exhib.* 21 The *shabti* is a figure placed in tombs to answer by magic for the deceased when called to labour in the Fields of the Blessed. **1964** I. E. S. Edwards et al. *Introd. Guide Egyptian Collections in Brit. Mus.* vi. 155 *Shabti*-figures are exhibited in the Third Egyptian Room. The name *shabti* is of uncertain meaning. The same figures were called *shawabtis* in the New Kingdom and *ushabtis* in the Late Period.

shabunder. Add: Also **shahbandar** (now the usual spelling), **shahbender.** Also, the title of an officer with wider responsibilities; *spec.* one of three chief local officials who administered Sarawak under the Sultan of Brunei.

1922 O. Rutter *Brit. North Borneo* v. 135 The Padas Damit district..was taken over..as a result of the operations against Pengiran Shabandar Hassan. **1960** S. Runciman *White Rajahs* III. ii. 93 It [*sc.* the land] be-

longed to a Brunei Princess, whose brother, the Pangiran Shabandar, administered it. **1964** D. K. Bassett in Wang Gungwu *Malaysia* II. vii. 122 The chief *shahbandar* of Kedah. **1969** J. M. Gullick *Malaysia* ii. 40 Control of the port, supervision of the merchants and collection of customs duties were divided among four harbour-masters, called *shahbandars*, who looked after the traders in one region. **1971** N. Tarling *Britain, Brookes & Brunei* I. i. 4 The system of offices, of a mixed Hindu-Muslim character, included..men of noble blood or *pengirans*, the most important being the *shahbandar*. **1974** S. E. Morison *European Discovery of America: Southern Voyages* xviii. 443 The Shahbender (title of the ruler of Brunei) sent out..a beautiful prao with gilt work on bow and stern.

Shabus, var. *Shabbos.

‖ **shabu-shabu** (ʃa·buʃa·bu). [Jap.] A Japanese dish of thinly sliced beef or pork cooked with vegetables in boiling soup.

1970 T. Egami *Oriental Cookery* 130 Shabu-Shabu. The word *shabu-shabu* derives from the sound of thin slices of succulent beef being gently swirled with chopsticks..in hot broth. **1973** *Times* 9 June 11/2 The speciality dish shabu-shabu (£1.80); raw beef and geometrically cut vegetables, briefly cooked by the waitress in a pagoda-shaped pot of broth, and served with rice and savoury dip. **1979** *United States 1980/81* (Penguin Travel Guides) 179 You can sample delicious shabu-shabu, Japan's answer to Swiss fondue.

Shabzieger, var. *Schabzieger.

shack, *sb.*³ Add: **1.** (Earlier and later examples.) Also applied to other similar structures, and used outside U.S. and Canada.

1878 *Rep. Indian Affairs* (U.S.) 42 Too much praise cannot be given to these homesteaders for..the erection of this building, while they, themselves, were living in shacks. **1932** A. Christie *Peril at End House* v. 70 We saw a lot of messy-looking shacks, and then by good luck we found this. **1936** D. Glover *Home Thoughts* 18 A mountain shack Where blankets, candles, frying-pan Bespeak the only needs of man. **1939** *Denver Post* 2 Jan. 16-B/6 Other work will include the building of a ski shack. **1950** J. Baxter in *Landfall* (N.Z.) XIII. 10 There in a corrugated iron shack Behind a brushwood fence, he lives alone. **1960** *Daily Mail* 11 Apr. 4/4 In Durban..Bren guns and heavy machine-guns covered the hillsides spotted with native shacks.

b. *shack town.*

1923 H. Steele *Spirit-of-Iron* 105 Where little shack-towns rose, it knew there should be cities. **1962** G. MacEwan *Blazing Old Cattle Trail* i. 4 The residents of what had been an unprepossessing shack-town found their community overrun with rip-roaring cowboys, gamblers, gunmen, even women.

2. *U.S. slang.* A house.

1910 C. E. Mulford *Hopalong Cassidy* xiii. 128 You stay in that shack. Don't leave it for a second, understand? **1930** *Living Age* 1 Apr. 188 I've gotta tote this outfit of waffles and candy to grandmomma's shack.

3. *U.S. slang.* = *radio shack* s.v. *radio *sb.* 7.

1929 *Amer. Speech* V. 49 *Shack*, wireless room or office. **1947** *Christian Science Monitor* 15 Jan. 9/1 Al's [ham radio] station, like most of the other 75,000 American amateurs, is a bedroom converted into what they call a 'shack'. **1960** [see *rig *sb.*⁴ 3 d].

shack (ʃæk), *sb.*⁵ *U.S.* [prob. f. Shack *sb.*² or Shack *v.*²] A slow trot. Also *attrib.*

1881 *Harper's Mag.* Feb. 375/2 [He] walked with a peculiar shack gait. **1900** H. Garland *Eagle's Heart* 144 He continued his steady onward 'shack' toward the West. **1938** G. Butler *Running & Runners* iii. 85 Probably the best exercise of the whole lot is the 'shack' a word derived from the ponderous movement of a cart-horse. This is a movement mid-way between running and walking.

shack (ʃæk), *sb.*⁶ *N. Amer. slang.* Also **shacks.** [Origin obscure.] The brakeman or guard on a train.

1899 'J. Flynt' *Tramping with Tramps* 397 *Shack*, a brakeman. **1907** J. London *Road* 213 As the freight got out of Philadelphia she began to hit up speed. Then I understood what the shack had meant by suicide. **1926** *Amer. Speech* I. 652/2 *Shacks*, brakeman on train. **1931** 'D. Stiff' *Milk & Honey Route* ii. 27 A great many hobo writers..are full ready to tell the novice how to outwit the brakemen, or shacks. **1947** L. M. Beebe *Mixed Train Daily* 313 The stock was valuable and a roundup was imperative, but, as the shacks and hoggers of the S.V. were unaccustomed to the saddle, a score of professional cowpokes were engaged for the task. **1976** Lieberman & Rhodes *Compl. CB Handbk.* vi. 136 *Shack*, railroad conductor.

shack, *v.*² Add: **1. b.** To move with a slow ambling gait, to go at a slow trot. *U.S.*

1833 in B. F. Hallett *Full Rep. Trial E. K. Avery* 61, I *shacked* down some of the hills, (partly run). **1916** H. Titus *I Conquered* ii. 31 Yonder [was] a man shacking along on a rough little horse, head down, listless. **1947** *Sat. Even. Post* 8 Mar. 53/1 Each winter Steve shacked in to Barry's camp a couple of times, sat in the log office a day and shacked out.

shack (ʃæk), *v.*⁴ *slang* (orig. *N. Amer.*). [f. Shack *sb.*³; cf. Shack *v.*² 2.] **1.** *intr.* To live in a shack.

1895 *Dialect Notes* I. 393 *Shack*..(v.) to live in a *shack* or keep a bachelor's hall in general. 'They sent away their wives and *shacked* for a time.' **1935** Z. N. Hurston *Mules & Men* I. vi. 127 You ain't de Everglades Cypress Lumber Comp'ny sho nuff. Youse just shacking in one of their shanties. **1954** C. Bruce *Channel Shore* 16 Men had sailed east from here to the Cape Breton coast, to shack on the beaches and fish the waters off Petit de Grat. **1975** *Maclean's Mag.* (Toronto) May 43/3 We used to shack there, camp ourselves where the mine was.

2. a. *intr.* Usu. with *up*. To obtain temporary accommodation, to shelter for the night; to lodge *with* (esp. as a sexual partner), to set up house *with*, to cohabit (*with*); hence, to have sexual intercourse *with*.

1935 Z. N. Hurston *Mules & Men* I. vii. 161 Ah..was doin' fine till Ah shacked up with a woman dat had a great big ole black cat. **1942** Berrey & Van den Bark *Amer. Thes. Slang* § 62/9 *Shack up*, to stay in a camp for the night. **1945** *Sun* (Baltimore) 1 Mar. 6/6 More wanderlust grips the sow and she shacks up with half a dozen families before the original owner gets wind of her again. **1946** *Time* 14 Oct. 40/3 The medicine man..had shacked up with a halfbreed cook. **1947** L. Waller *Show me Way* III. xxii. 191 She wanted me to shack with her tonight. **1949** R. Chandler *Little Sister* xviii. 120 I'm not talking about her love life... She doesn't have to shack up with a red-hot. **1951** J. D. Salinger *Catcher in Rye* ix. 73, I was going to shack up in a hotel for a couple of days and not go home till vacation started. **1959** H. Hobson *Mission House Murder* iii. 22 Besides appearing at performances she has to shack up with Johnny. **1965** S. T. Ollivier *Petticoat Farm* ix. 128 A man's got to have something to offer a girl before he asks her to shack up with him. **1968** *Listener* 15 Feb. 210/1 Some [trusted prisoners] even had their own cars to go up town and shack in some motel with a woman. **1972** P. Lively *Driftway* x. 136 We'll shack up for the night. There's a field farther on where the farmer's not one of those choosy fellows as'll turn me off after half an hour. **1976** W. Greatorex *Crossover* 193 Galina's not my wife... We shack up, that's all. **1981** A. Morice *Men in her Death* viii. 80 This must have been..before they become friendly enough to shack up together.

b. *trans.* Usu. with *up*. To provide with accommodation or lodging, esp. as a sexual partner. Chiefly in pass. *to be shacked (up)*, to be staying or lodging, to be cohabiting (*with*).

1927 *Dialect Notes* V. 462 *Shack up*, v., to put up for the night. **1946** *Amer. Speech* XXI. 252 'I'm shacked up around here' means that the speaker has found a friendly *fräulein* who in substance maintains a home for him. The *fräulein* herself is a 'shack job'. **1953** P. Frankau *Winged Horse* IV. 242 He's shacked up with Celia. **1957** *Economist* 30 Nov. 787/1 Private Girard's marriage to the Japanese girl with whom he had been 'shacked up'. **1958** 'E. McBain' *Killer's Payoff* (1960) ii. 19 'Where is this Newton?' 'He's shacked in a hotel..downtown.' **1967** [see *girl *sb.* 2 g]. **1973** *Globe & Mail* (Toronto) 29 Sept. 1/2 Even the mayor was shacked up and everybody knew. **1975** D. Lodge *Changing Places* iii. 125 Philip Swallow is shacked up with Melanie at that address.

So **sha·cking** *vbl. sb.*

1884 *Prince Albert* (Saskatchewan) *Times* 13 June 3/2 Of all the enjoyments Prince Albert can number, there's none equals shacking on a *pre-emption* claim. **1945** *Yank* 8 June 14 Must be. I'm sure not crackin' up from shackin' up. **1980** M. Underwood *Clear Case of Suicide* xiii. 96 Casual shacking up was quite different from holy matrimony.

shack-job (ʃæ·kdʒǫb). *U.S. slang.* Also **shack job.** [f. *shack *v.*⁴ + *job *sb.*² 4 f.] = *shack-up 2.

1946 [see *shack *v.*⁴ 2 b]. **1951** *New Yorker* 10 Mar. 112/2 Allowing him to sleep with their daughter (this was an early shack-job, not the girl mentioned above). **1955** W. Gaddis *Recognitions* I. iv. 158 Look, rabbit, I'm looking for a shack-job, see? **1966** *Sunday Times* (Colour Suppl.) 4 Dec. 73/3 *Shack job*, easy-woman.

shackle, *sb.*³ Delete *U.S.* and add earlier example.

1835 D. Webster *Original Scottish Rhymes* 194 There'll be gude tents an' shachels For drinkers to roar an' to rift.

shackles (ʃæk'lz). *dial.* and *slang.* [Prob. f. Shackle-bone.] Broth, soup, or stew.

1886 F. T. Elworthy *West Somerset Word-Bk.* 658 *Shackles*.., broth. Every mornin' my old 'ummun makth me a basin o' *shackles*, and her knowth how to make 'em too, mind, way a plenty o' liks (leeks) in 'em. **1909** W. H. Davies *Beggars* xiii. 104 The following are a few slang words used by beggars... soup—shackles. **1931** 'G. Orwell' *Coll. Essays* (1968) I. 70 New words (i.e. words new to me)... *Shackles*, broth or gravy. **1969** *Tel.* (Brisbane) 29 July 2/4 Mr. Coppard records how one night he stumbled on a field kitchen and enjoyed a wonderful meal of schackels, a soup made up from leftovers.

shackle-up (ʃæ·k'lɒp). *slang.* Also **shackle up.** [Origin uncertain: cf. prec.] An act of preparing food in a pot.

1935 H. Neville *Sneak Thief on Road* 347 Shackle-up —a great cooking of food in a pot. **1936** J. Curtis *Gilt Kid* xx. 202 A spare shirt and a couple o' tins in case they want to have a shackle up.

shackling, *ppl. a.* **2.** (Earlier examples.)

1790 W. Maclay *Jrnl.* 24 May (1890) vii. 272 His whole figure has a loose, shackling air. **1793** J. Lindley in *Friends' Misc.* (1836) (ed. 2) II. 63 And the wagon very shackling, made the tour very disagreeable. **1846** J. K. Polk *Diary* 9 June (1929) iii. 114 Mr. Bancroft reminded Mr. Buchanan of a remark which he had made in the Cabinet some months ago, that the title of the United States north of 49° was a shackling one.

shackly, *a.* Add: (Earlier examples.) Also **shackley.**

1843 *Indiana Q. Mag. Hist.* III. 121, I stopped at a small poverty-stricken little town called Mt Meridian; shackly houses, huts and hovels..gave no great expectation of refinements. **1843** *New Mirror* 18 Nov. 116/2 Hitched with oakum before a shackley go-cart, the rocking evolution of whose wheels showed that it was long since they had firmly revolved in their own proper axis.

shack-shack, shac-shac (ʃæ·kʃæk), vars. Chac-chac. Cf. Shak-shak.

1848 in *Caribbean Q.* (1956) IV. III & IV. 184 Bands of music (soi-disant) including those inelegant instruments, the tin kettle and salt box, the bangee and shack shack. **1953** P. L. Fermor *Violins of Saint-Jacques* 58 The leaders wielded shackshacks: cylinders of bamboo filled with rattling seeds. **1955** *Caribbean Q.* IV. II. 101 The band includes..a cuatro..a set of home-made drums with cymbal and triangle, and shac-shacs. **1959** P. Capon *Amongst Those Missing* 243 The music..was strongly rhythmic and the rhythm was marked by the rattling of shack-shacks and maracas.

shack-up (ʃæ·kɒp). *slang* (chiefly *U.S.*). Also **shack up, shackup.** [f. *shack *v.*⁴ 2.] **1.** Cohabitation. Also *attrib.*

1935 Z. N. Hurston *Mules & Men* I. ii. 54 'Oh, you kin be had,' Gold retorted... 'Yeah? But not wid de trace chains. Never no shack up. Ah want dis tip-in love and tip yo' hat and walk out.' **1974** *Times Lit. Suppl.* 18 Oct. 1155/2 An affair with David,..a shack-up with Colin. **1977** *Toronto Star* 21 May B5 One down-to-earth mother referred to 'my child's shack-up partner'.

2. A partner in cohabitation or sexual intercourse.

1969 E. R. Johnson *Mongo's back in Town* ii. 20 That's not like Angel. She was still Mike's shackup. **1972** J. Gores *Dead Skip* (1973) xxiii. 163 He didn't even know if the guy was married or single. He might have a shack-up there for the night.

shac-shac: see *shack-shack.

shad, *sb.* Add: **3.** For † *Obs. rare* ⁻¹ read *rare* and add later example.

1894 'Mark Twain' in *St. Nicholas* Jan. 252/2 Spiders in a desert, you shad?.. You don't ever reflect, Huck Finn, and I reckon you really haven't got anything to reflect with.

4. a. *shadbone, -fisher* (earlier example), *roe.*

1962 Auden *Dyer's Hand* (1963) 303 Thus, she describes a tomcat's face: the shadbones regularly set about the mouth. **1860** *Harper's Mag.* Nov. 795/1 A party of shad-fishers, pulling in their seine. **1888** *All about Alaska* (Pacific Coast Steamship Co.) 54 Herring roe is to the native Sitkans what the shad roe is to the dwellers on the Susquehanna and the Potomac. **1976** *National Observer* (U.S.) 23 Oct. 19/4 And, this is the place for exotic fish eating, with surprises like..shad roe, wolf fish.

b. **shad-blow** = *shad-bush*; **shad-bush** (earlier example); **shad-flower,** (*a*) (earlier example); **shad-fly** (earlier example); **shad-worm** (earlier example).

1846 D. J. Browne *Trees Amer.* 282 The Canadian Amelanchier..[also called] June Berry, Shad-blow, Shad-flower. **1890** *Harper's Mag.* Apr. 710/2 Shadblow, with leaves of bluish green, white flowers or green berries waiting for the sun to make them red. **1960** *Washington Post* 25 Jan. B1/4 Trees considered to be worth only 60 per cent..are..common horse-chestnut, shadblow serviceberry, [etc.]. **1818** A. Eaton *Man. Bot. N. & Middle States* 145 *Aronia..botryapium* (shad-bush). **1817** —— *Man. Bot. Northern States* 55 *Aronia..botryapium*, (shad-flower). **1825** *Canad. Mag.* IV. 474 The ephemeral Spring Fly, called..by the English the Shad Fly, as they are supposed to indicate the approach of the fish. **1851** M. H. Perley *Rep. Fisheries Bay of Fundy* 88 At Windsor, the 'Shad-worm' is found upon the mud flats.

shad-bellied, *a.* **a, b.** (Earlier examples.)

1832 J. P. Kennedy *Swallow Barn* III. i. 3 A shad-bellied blue bobtail coat. **1847** [see *night-owl 2].

‖ **shadchan** (ʃa·dχăn). Also **schadchen, schatchen, shadchen, shadkhan, shadkin, shatchen.** [Yiddish *shadkhn*, ad. Heb. *šaddᵊḵān*, f. *šiddēḵ* to arrange a marriage; cf. *shidduch.] A Jewish professional matchmaker or marriage-broker. Also *fig.*

a **1890** in Barrère & Leland *Dict. Slang* (1890) II. 219/2 Ten per cent of the dowry goes to the shadkin when the others become kin. **1892** I. Zangwill *Childr. Ghetto* I. 60 He sent a *Shadchan* to propose to her, and they were affianced, Chayah's father undertaking to give a dowry of two hundred gulden. **1897** F. Moss *Amer. Metropolis* iii. 216 A man named J— H—..acted as schatchen (match-maker). **1950** D. Riesman in *Psychiatry* May 177/1 The lies and sales talk of the schadchen, the Jewish marriage brokers. **1957** L. Stern *Midas Touch* I. viii. 68 'Shatchen,' Israel muttered into his beard, 'Matchmaker.' **1959** 'W. Haggard' *Venetian Blind* vii. 98, I married her..because my parents, on the advice of a reliable *shadchan*..considered the match suitable. **1968** [see *marriage broker* s.v. *marriage 8]. **1976** *Publishers Weekly* 11 Oct. 94/2 Taking the role of *shadkhan*, the authors bring together seeker and supplier.

‖ **shadda** (ʃa·dda). *Gram.* Also **śèdda, shaddah.** [a. Arab. *šadda*, lit. strengthening.] In Arabic, a sign, also called *tašdīd*, written or printed above a consonant to indicate that it is doubled.

1896 W. Wright tr. *Caspari's Gram. Arabic Lang.* (ed. 3) I. I. iii. 14 In African Mss. the vowel is not always written with the šèdda. **1925** W. H. T. Gairdner *Phonetics of Arabic* ix. 58 The sign written over the consonant-letter in Arabic writing is called aʃ ʃadda ('force'). **1958** D. Cowan *Introd. Mod. Literary Arabic* 5 If two identical consonants come together and are not separated by a vowel only one is written with the mark ّ above it. This mark is called..*shadda* or 'strengthening'. **1962** Haywood & Nahmad *New Arabic Gram.* i. 10 A doubled letter is not written twice, unless separated by an intermediate vowel. Instead, the sign ّ (called.. tashdid or..shadda) is written over the letter. **1969** A. G. Chesne *Arabic Lang.* ii. 28 Other diacritical marks were..introduced. Among these are..the *shaddah* for doubling a consonant. **1971** R. A. Wisbey *Computer in Lit. & Ling. Research* 228 Though the vowels are not to be included, as they are usually omitted in the script, shadda, the sign which denotes the doubling of a letter, and the various combinations with hamza are to be plotted.

‖ **Shaddai.** (ʃa·dəi). *Judaism.* Also **Shadai.** [Heb., of uncertain meaning; in the English versions of the Bible usu. translated 'Almighty'.] One of the names of God in the Bible and cabbala, inscribed on certain ritual objects and on talismans.

1620 J. Donne *Sermon* (1957) III. 191 Shaddai is the name of God, and yet Shaddai is spoyle, violence and depredation. **1881** *Encycl. Brit.* XIII. 812/1 The angel Metatron inhabits this world. He alone constitutes the world of pure spirit, and is the garment of Shaddai, *i.e.*, the visible manifestation of the Deity. **1892** I. Zangwill *Childr. Ghetto* I. v. 133 The doorposts twinkled with *Mezuzahs*—cases or cylinders containing sacred script, with the word *Shaddai* (Almighty) peering out of a little glass eye at the centre. **1926** W. & E. Muir tr. *Feuchtwanger's Jew Süss* I. 37 Three furrows, sharp, deep, short and almost vertical above his nose cleft his forehead; and they formed the sacred letter Shin, the first letter of God's name, Shaddai. **1962** I. B. Singer *Slave* ii. 219 The incantation a scribe had written out for her:.. *Yuhah* will guard me! *Shaddai* will save me! *Taftifiah* will be a wall for me. **1969** E. Stewart *Heads* 91 There was an amulet..the little golden heart with the letters that spelled *Shadai*..*Shadai*, he thought: *eternal*.

shaddup (ʃɒdɒ·p), repr. a colloq. or vulg. pronunc. of imper. *shut up!* (see Shut *v.* 19 m).

1959 R. Condon *Manchurian Candidate* iv. 78 Shaddup! You hear? Shaddup! **1977** *Daily News* (Perth, Austral.) 19 Jan. 3 (*caption*) 'Forget it, mother it doesn't bother me.' 'It bothers me! People'll think I'm gettin' old!' 'Shaddup.' **1977** *Daily Mirror* 6 Apr. 24 (*caption*) 'Snooker isn't a trifle!' 'Aw, Shaddup!!'

shade, *sb.* Add: **II. 6. c.** (Further examples.) Now usu. in *pl.* and no longer exclusively in humorous use. Also *loosely*, with reference to some person or thing in the past of which a present event is reminiscent.

1899 R. Whiteing *No. 5 John Street* xviii. 183 Shade of Tilda! not a bud but would outvalue your entire stock. **1928** H. Crane *Let.* 22 Feb. (1965) 317 A paean from Venusberg! Oy-oy-oy! I have just had my ninth snifter of Scotch. O shades of Bert Savoy! **1968** *Listener* 25 July 98/2 The persistent..demand..for a major change in the relationship between a free people and the state, for an end to arbitrary, secretive and alien government and for the restoration and maintenance of free institutions. Shades of Disraeli, maybe. **1977** *Times* 26 Nov. 4/3 Colleges..were..conducting campaigns to ban Jewish societies... Shades of Nazi Germany (he said). **1978** H. Wouk *War & Remembrance* xxiv. 238 There's a fridge, but it doesn't work. Shades of Singapore.

7. For † *Obs.* read 'Now chiefly *Hist.*' and add later examples.

1960 H. Hayward *Antique Coll.* 255/1 Edward Foster of Derby often painted faces in brown, blue or some other colour, and unless details are shown in the faces, such may also be termed shades. **1970** *Oxf. Compan. Art* 1065/1 The great vogue of silhouette portraits (more often known in England as 'shades') came between 1750 and 1850. **1979** *Jrnl. R. Soc. Arts* July 513/1 Anything but an average shade, it is, none-the-less, a competent head-and-shoulders in strict profile.

III. 11. e. Also (*pl.*), sunglasses, tinted glasses. *colloq.* (chiefly *U.S.*).

1958 *Amer. Speech* XXXIII. 225 Less frequently used among nonmusicians (primarily for lack of an occasion) are *shades* (dark glasses). **1965** *N.Y. Times* 11 Apr. E14/6 Your teen-age daughter asks what you think of her 'shades', which you are canny enough to know are her sunglasses. **1976** *National Observer* (U.S.) 10 Apr. 1/4 'Hiya, Rog,' says somebody else, popping up clip-on shades. **1980** G. V. Higgins *Kennedy for Defense* vi. 68, I looked at Emerson, hiding behind his shades and his imported-cigarette smoke.

h. (See quot. 1894.)

1894 T. Elliston *Organs & Tuning* 127 *Shade*, a flap of metal at the top of a reed pipe to regulate the power, at the top of a flue pipe to tune by—also applied when the tone, pitch, or power of a pipe is affected through being shaded or shadowed by an obstruction. **1925** H. F. Milne *How to build Small Two-Manual Chamber Pipe Organ* 127 The pipes in many reed stops are of an inverted conical shape, and the regulating device may take the form of either a cap or shade. **1951** R. Whitworth *Organ Stops & their Uses* i. 13 The pipe represented at letter L is the much over-used swell oboe for 4 ft C. The bell and its shade on the top should be noticed.

IV. 12. c. Instrumental, as *shade-softened* adj.

1866 G. M. Hopkins *Jrnls. & Papers* (1959) 138 Very level clouds, long pelletted sticks of shade-softened grey in the West.

13. shade-bearer, a plant which is shade-tolerant; **shade-bearing** *a.*, = *shade-tolerant* adj. below; **shade-card,** a card illustrating the range of colours in which goods are supplied; also *fig.*; **shade-lover, plant,** a plant which thrives in shady conditions; **shadepull** *U.S.*, a cord for pulling down a window-shade; **shade-tolerant** *a.*, able to grow normally in the shade of taller plants.

1891 Shade-bearer [see *light-demander* s.v. *light *sb.* 16]. **1959** *Times* 7 Dec. (Agric. Suppl.) p. viii/4 Beech can be the underplant used for amenity work, because it is a shade-bearer. **1889** W. Schlich *Man. Forestry* I. ii. 117 Certain species cannot thrive unless they enjoy a large measure of light throughout life, while others will bear a certain amount of shade. Accordingly, the former are called 'light demanding', and the latter shade bearing species. **1975** T. C. Whitmore *Tropical Rain Forests of Far East* vi. 71/2 The population structure of light-demanding and shade-bearing species in a stand of high forest is markedly different. **1895** *British Warehouseman* Feb. 38/2 A new and very attractive shade-card, comprising all the newest tints. **1955** *Radio Times* 22 Apr. 22/2 A free illustrated colour booklet about Snowcem and a shade card. **1961** P. Mason *Common Sense about Race* IV. i. 120 There are words conveying ..subtle nuances of skin-colour..a kind of verbal shade-card. **1960** *Farmer & Stockbreeder* 9 Feb. (Suppl.) 4/2 Other worthwhile shade-lovers—the climbing fig..and the sweetheart plant. **1926** H. A. Spoehr *Photosynthesis* ii. 103 It would be interesting to determine whether shade plants such as the *Oxalis*..do not utilize a greater proportion of the light absorbed than plants growing in the direct sunlight. **1974** Shade plant [see *solarization 1 c]. **1955** W. Gaddis *Recognitions* I. vi. 202 The housefly..drawn to a new destination the instant it halted, from the shade-pull to the floor, from there to the lampshade. **1973** *Philadelphia Inquirer* (Today Suppl.) 7 Oct. 41/3 Meg is replacing such geegaws with tasteful black shadepulls. **1952** J. D. U. Ward *Woodman's Diary* 310 Some other species, mostly in the shade-tolerant category..allow a wide latitude for neglect and error. **1964** V. J. Chapman *Coastal Veg.* ix. 214 On Fair Isle, Red campion (*Melandrium rubrum*) grows well in the fescue swards, probably because being a woodland species it is shade-tolerant.

shade, *v.*¹ Add: **5. a.** Delete '(now *dial.*)' and add further examples.

1928 S. Lewis *Man who knew Coolidge* I. 45 And I got to admit that Walt's radio shades mine just the least little bit. **1972** *Sydney Morning Herald* 26 Aug. 31/7 University slightly shade Gordon in points scored for and against. **1973** *Observer* 3 June 28/7 Denness, whose 534 runs in first-class matches this season have him shading even Boycott. **1975** *Cork Examiner* 30 May 15/1 Womble survived a bad last hurdle mistake to shade strongly challenging Glenicmurrin by a short head.

9. c. *trans.* To make a slight or gradual reduction in (a price, value, etc.). Also *intr.* of shares, prices, etc.: to decline slightly in value, cost, etc.

1875 *Chicago Tribune* 27 Oct. 6/4 Prices are not strong, the quotations being shaded on fair orders. **1899** *Pitman's Commercial Correspondence & Commercial English* xii. 119 Please, therefore, do your best to deliver the finest quality you possibly can at the figure named, or, if you can shade the price a little, it would be advisable to do so. **1903** *Boston Transcript* 24 Oct. 22 To spur his freight traffic manager to get business without shading rates. **1928** *New Statesman* 28 July (Finance Suppl.) p. x, The newsprint and pulp industry..has..been developed rather faster than the demand, with the result that prices have been shaded. **1966** *Times* 17 June 16/4 If a favourite or near-favourite was being quoted at two-to-one as its price in the ring and on the rails, and if a certain bookie had not got it in his book, far from 'shading the odds', he had to increase them to attract money on that horse for his book. **1973** 'R. MacLeod' *Burial in Portugal* iv. 90 When he'd bought, Consolidated had already been shading at 130 and Maltsters had been on the upswing at 146. **1978** *Daily Tel.* 29 Mar. 21 Banks may be invited..to shade the margin over base rate which they charge private customers and small businesses. **1981** *Times* 7 May 24 General Accident..shaded 10p to 334p.

d. *trans.* To modify the pitch of (an open organ stop) by placing something near the top of the pipe. *rare.*

1876 Stainer & Barrett *Dict. Mus. Terms* 395/1 *Shading of pipes*, the placing of anything so near the top of an organ pipe as to affect the vibrating column of air which it contains. **1894** [see *shade *sb.* 11 h].

shaded, *ppl. a.* Add: **1. b.** (Earlier and later examples.) Also as second element with a specifying colour.

1836 Dickens *Sk. Boz* (1837) 2nd Ser. 352 A shaded lamp by the bed-side. **1903** H. James *Ambassadors* xvi. 222 His dinner with Maria Gostrey, between the pink-shaded candles. **1956** E. Grierson *Second Man* i. 32 The secluded corner table with the shaded lights. **1973** I. Drummond *Jaws of Watchdog* xii. 157 A red-shaded lamp.

6. Of prices, values, etc.: see *shade *v.*¹ 9 c.

1960 *Farmer & Stockbroker* 26 Jan. 4/1 Oilcakes in limited demand at shaded rates. **1976** *Birmingham Post* 16 Dec. 9/11 R. and A. G. Crossland at 16½p and Moss Engineering 44p shaded, brighter contrasts being provided by Willmot Breeden at 49p.

‖ **shadi** (ʃā·dī). [Hind. *shādī*.] In the Indian subcontinent: a wedding, marriage.

1893 Kipling *Many Inventions* 223 Make the *shadi* swiftly, and the girl will make him a Mussalman. **1897** F. A. Steel *Potter's Thumb* 299 There's going to be a big shâdi (wedding)..to-morrow morning. **1978** 'M. M. Kaye' *Far Pavilions* xxx. 443 A canopy..beneath which the sacred fire would be lighted and the officiating priests perform the *shadi*, the marriage ceremony.

shading, *vbl. sb.* Add: **4. c.** A spurious variation in brightness over parts of a televised image. Freq. *attrib.*

1940 D. G. Fink *Princ. Television Engin.* ix. 414 The remaining item of equipment necessary to produce a composite video signal of adequate quality is the shading correction generator required with camera tubes of the iconoscope (storage-mosaic) type. **1961** G. Millerson *Technique Television Production* iii. 50 Shading is reduced manually, by adjusting electronic correction circuits. **1969** G. L. Hansen *Introd. Solid-State Television Systems* xi. 269 If the red, green, and blue channels were called upon to reproduce a white scene, the unbalance caused by the corner shading in the red channel would produce a red hue in the corner. *Ibid.*, Shading generators..supply waveforms to the cathodes of the camera tubes to offset the variations that are present because of shading irregularities.

shadkhan, shadkin, vars. *shadchan.

shadoof. Add: Also **schaduf, shaduf.**

1937 *Times Lit. Suppl.* 10 Apr. 268/2 The sounds which run as an undercurrent beneath the activities of Egyptian life..are the creak of the water-wheel and of the *shaduf*. **1961** G. Clark *World Prehist.* v. 109 For any great extension of the fertile zone it was necessary to cut channels and lift the Nile waters into them by some such device as the *schaduf*. **1974** *Encycl. Brit. Micropaedia* IX. 100/3 (*caption*) Villagers in the state of Tamil Nadu, India, using a shaduf to raise water from a stream into irrigation channels.

shadow, *sb.* Add: **I. 1. d.** *Psychol.* In the theory of C. G. Jung (1875–1961), the dark aspect of personality formed by those fears and unpleasant emotions which, being rejected by the self or persona of which an individual is conscious, exist in the personal unconscious; an archetype in which this aspect is concentrated.

1923 H. G. Baynes tr. *Jung's Psychological Types* iv. 203 For the sake of understanding, it is, I think, a good thing to detach the man from his shadow, the unconscious... One sees much in another man which does not belong to his conscious psychology, but which gleams out from his unconscious. **1940** S. Dell tr. *Jung's Integration of Personality* (1941) iii. 70 To take his [*sc.* the devil's] place there are human beings to whom we gratefully resign our shadows. With what pleasure..we read newspaper reports of crime. *Ibid.* 88 The three archetypes so far mentioned—the shadow, the anima, and the wise old man—are of the kind immediately experienced in personified form. **1959** *Listener* 29 Oct. 723/2 Jung defined an archetypal image which he called the shadow... The shadow actually became, in his designation, a term which covered a wide variety of impulses and wishes, most of which were felt to be evil or at least inadmissible. **1973** *Jrnl. Genetic Psychol.* Mar. 165 The shadow is described as the dark side of the personality or representing the original conception of evil in the world. The latter conception places the shadow in the collective unconscious.

3. b. = *eye-shadow* s.v. *eye *sb.*¹ 28.

1936 *Time* 26 Oct. 39/2 Make-up Man Senz 'deepened' Miss Phillips' bulgy eyes with dark brown 'shadow'. **1966** *Vogue* Dec. 84/3 Soft liquid shadows in browns, greys and seaweed greens to put near the curve. **1976** 'E. McBain' *Guns* vii. 198 She wears orange lipstick... There is green shadow on her eyelids.

II. 4. a. Phr. *under* or *in the shadow of*: within the purlieus of, close up against, in proximity to.

1853 C. Brontë *Villette* I. v. 85, I lie in the shadow of St. Paul's. **1931** *Times Lit. Suppl.* 20 Aug. 625/4 The gradual rise of Innsbruck from a little village lying under the shadow of the great castle of the Dukes of Andechs to the..capital city of Tyrol.

e. (Earlier example.)

1873 *Proc. London Math. Soc.* IV. 271 Immediately in the rear of a sufficiently large sphere there will be a sound shadow.

f. A dark area in a (positive) radiograph (appearing as a light area in a negative).

1903 Pusey & Caldwell *Practical Application Röntgen Rays* I. v. 120 (*caption*) Apparatus for orthographic projection of *x*-ray shadows on fluorescent screen. **1928** A. Turnbull tr. *Köhler's Röntgenol.* 187 Dense bean-like shadows lateral to the upper opening of the hip-joint.. have been observed. **1964** le Roux & Dodds *Portfolio Chest Radiographs* i. 17 (*caption*) A normal P.A. chest radiograph of a young adult female with dense mammary shadows.

6. e. *spec.* The Opposition counterpart of a cabinet minister; a member of the shadow cabinet (see sense 16 b below).

1912 Ld. Lansdowne *Let.* 23 Feb. in R. Blake *Unknown Prime Minister* (1955) v. 103 But if the House of Commons 'shadows' are to number 11, I don't see how I can leave out Londonderry. **1961** *Daily Tel.* 1 Dec. 14 The five members of the Labour front bench who have exchanged 'shadows'. **1975** R. Lewis *Margaret Thatcher* i. 4 When he resigned from the leadership, out of all the Shadows, only Lord Carrington, one of nature's gentlemen, went round to his old chief to express his consolation and regrets. **1980** *Times* 8 Dec. 2/4 Mr Denis Healey..has continued as shadow on Treasury affairs.

f. Freq. in phr. *to wear* (oneself or another) *to a shadow*.
1840 Dickens *Lett.* (1969) II. 51 Commend me to him though he *does* wear me to a shadow. **1847** C. M. Yonge *Scenes & Characters* xviii. 236 And poor Lily wearing herself to a shadow, in vain attempts to mend matters. **1887** [in Dict.]. **1977** *Grimsby Even. Tel.* 14 May 1/6 He was wearing himself to a shadow touring the country and Holland and Sweden trying to get new contracts.

8. d. *Football.* A player who marks (Mark *v.* 15 c) another player in the opposing team.
1976 *Southern Even. Echo* (Southampton) 15 Nov. 13/7 The rare occasions he outwitted his experienced close-marking shadow, Billy Tucker. **1976** *Times* 2 Dec. 12/2 The ability of Everton's forwards to escape from their marking shadows had been apparent throughout.

IV. 15. a. *shadow-leaf, -pattern, -tackle, -train, -wife, -word, -world* (earlier and later examples).
1957 C. Day Lewis *Pegasus* 55 Frecklings of sunlight and flickerings of shadowleaf. **1943** Koestler *Arrival & Departure* III. 86 He stared at the ceiling of the dim room on which the shutters projected a streaky shadow-pattern of grey and white ribs. **1967** E. Short *Embroidery & Fabric Collage* i. 32 *Shadow patterns.* If any three-dimensional object is suspended between a bright light and a sheet of white card or paper, and the object revolved, a series of patterns will be made by the shadow of the object on the card. **1888** G. M. Hopkins *Poems* (1967) 72 Shivelights and shadowtackle in long lashes lace, lance, and pair. **1932** Auden in *Rev. Eng. Stud.* (1978) Aug. 282 A shadow-train flitted foreshortened through fields. **1939** —— & Isherwood *Journey to War* 279 Loss is their shadow-wife. **1932** D. H. Lawrence *Etruscan Places* ii. 42 Pelasgian is but a shadow-word. **1957** E. Partridge *English gone Wrong* II. 38 In the U.S.S.R., *right* is a shadow-word; and rights, something one possesses only theoretically. **1853** Shadow-world [see *live *v.*[1] 5]. **1953** S. Spender *Creative Element* 93 What Lawrence protested against was not intellect but the kind of intellectualization whereby men create a shadow-world for themselves.

b. *shadow-coloured, -hung, -stroked* adjs.; also similative, as *shadow-white* adj.
1947 K. Tennant *Lost Haven* x. 147 In these sweeps of land were shadow-coloured birds and the beautiful mid-night blue of the wild pigeons. **1952** R. Campbell tr. *Baudelaire's Poems* 43 When you're asleep, dear shadow-coloured wench. **1913** 'Saki' *When William Came* (1914) xviii. 288 A grey shadow-hung land which seemed to have been emptied of all things that belonged to the daytime. **1866** G. M. Hopkins *Jrnls. & Papers* (1959) 144 Prettily shadow-stroked spikes of pale green grain. **1918** D. H. Lawrence *New Poems* 54 Into the shadow-white chamber silts the white Flux of another dawn.

16. shadow box, a case with a protective transparent front in which is displayed a painting, jewel, etc.; also *attrib.*; **shadow-box** *v. intr.* and *trans.*, to box (against) an imaginary opponent, as a form of training; also *fig.*; so **shadow-boxing** *vbl. sb.*; **shadow-check** (see quot. 1957); chiefly *attrib.*; **shadow cretonne, -print, -tissue,** a reversible material having a woven-in pattern which gives a shadowy, blurred effect; **shadow embroidery** = *shadow work* below; **shadow figure** = *shadow puppet* below; **shadow-grey** *a.* and *sb.*, (a) dark grey; **shadow lace,** a lace with an indistinct pattern; **shadow mask** *Television*, a perforated metal screen situated directly behind the phosphor screen in certain types of colour television tube, and having a pattern of precisely located holes through which the electron beams pass so as to strike the correct dots on the phosphor screen; freq. *attrib.*, as *shadow-mask tube*; **shadow-picture,** (*b*) a picture formed by a shadow (usu. of a person's hand or hands) thrown upon a screen or other surface (cf. Shadowgraph 1); **shadow-play** (earlier and later examples); also, a puppet play of the shadow theatre; also *attrib.* and *fig.*; **shadow-price** (see quot. 1965); also *transf.*; hence **shadow-pricing** *vbl. sb.*; **shadow print**: see *shadow cretonne* above; **shadow puppet,** a puppet used in a shadow play; **shadow-site,** an archaeological site revealed by shadowing on the ground; **shadow-stitch,** (*b*) a criss-cross embroidery stitch used on sheer materials for filling in spaces, and which, being worked on the wrong side, shows through on the right side in a shadowy way with an outline resembling a back-stitch; **shadow stripe** (see quots. 1940, 1947); so **shadow-striped** *a.*; **shadow tag** *N. Amer.* (see quot. 1977); **shadow test** (examples of sense (*a*)); **shadow theatre,** a form of puppetry in which flat figures are passed between a strong light and a translucent screen, the audience watching on the other side of the screen; also, a place where such puppet shows are performed; **shadow tissue**: see *shadow cretonne* above; **shadow work,** embroidery done in shadow-stitch; also *attrib.* and *fig.*
1909 *Cent. Dict.* Suppl., Shadow-box..*n.* **1969** [see *optical *a.* 2 c]. **1973** *Houston Chron.* 21 Oct. 18 (Advt.), Giant hutch mirror with shadow-box frame and shelves. **1976** *National Observer* (U.S.) 17 Jan. 14/1 (Advt.), This stunning golden shadow box pendant. **1924** S. Lewis *Free Air* i. 8 She fought the steering-wheel as though she were shadow-boxing. **1927** [see *punch-bag* s.v. *punch *sb.*[4] 3]. **1932** H. S. Drago *Champ* i. 15 Andy protested that it wasn't necessary as he shadow-boxed an imaginary opponent. **1951** *Scott. Jrnl. Theol.* IV. 321 Unlike many Fundamentalists he is aware that the battle has passed into new phases and he is not satisfied to shadow box on deserted fields. **1971** *Nature* 22 Oct. 510/1 These representatives of European governments are still shadow-boxing with each other. **1977** *Time* 19 Dec. 68/2 It was O.K. to shadowbox at a professional gym. **1919** E. Corri *Refereeing 1,000 Fights* 69 The mascot stripped to the waist to do some shadow boxing. **1939** *Sun* (Baltimore) 17 Feb. 10/1 Shadow boxing over the selection of a site for the Leakin Memorial Park will continue next week. **1966** *Illustr. London News* 10 Sept. 10/3 But in any case, the gnomes know that a good deal of what is going on is 'shadow-boxing'. **1978** A. Garve *Counterstroke* I. 60 He did a little shadow boxing and some skipping. **1908** *Sears, Roebuck Catal.* 1058/2 The background is a fairly dark shadow check effect. **1957** *Terms & Definitions* (Textile Inst.) (ed. 3) 89 *Shadow stripe*.., an effect, due to different reflections of light, produced in woven fabrics by employing yarns of different physical properties, usually of 'S' and 'Z' twist, in warp or weft (or in both, when it becomes a shadow check). **1960** *Woman* 23 Apr. 9/1 Dainty shadow-check shirt-waisters. **1932** *Sale Catal.*, Made of good quality Shadow Cretonne. **1943** E. Bowen *Seven Winters* 48 Pink-and-cream 'shadow' cretonne. **1973** 'D. Halliday' *Dolly & Starry Bird* xviii. 284 His bruises stood out like shadow cretonne on a chesterfield. **1920** J. Hergesheimer *Linda Condon* ii. 11 Shadow embroidery and fine shell edges. **1935** H. Edib *Clown & his Daughter* xliii. 240 It meant that she could easily buy a leather set of shadow figures for Tewfik. **1976** *Jrnl. R. Soc. Arts* Apr. 254/2 Flat Figures and Shadow Figures are a distinct type of puppet... In the Shadow Theatre the figures are placed between a light and a translucent screen. **1918** W. Beebe *Jungle Peace* (1919) ii. 26 The shadow-grey sea. **1932** *Sale Catal.*, A beautiful quality plain silk... Shades:..shadow grey and gunmetal. **1914–15** *T. Eaton Catal.* Fall & Winter 32 All White Evening Dress of Paillette Silk and Allover Shadow Lace. **1977** C. McCullough *Thorn Birds* iii. 61 His mother clad in a long bustled gown of palest pink shadow lace. **1951** *Proc. IRE* XXXIX. 1187/1 The first public demonstration of.. shadow-mask color tubes..was made in March, 1950. *Ibid.* 1188/2 The triangular pattern [of holes] was chosen for the shadow mask in experimental tubes primarily because of its mechanical properties. **1965** *Wireless World* July 354/2 The Mullard colour-selection shadow mask with graded holes. **1975** K. Wicks *Television* 54 The most common type of picture tube in use today is the shadow-mask tube. **1889** J. Pollard *Plays & Games for Little Folks* 32 *Shadow Pictures.* In order to make these pictures show well on the wall, there must be but one lamp in the room, and that must stand back of the performer. **1977** O. Schell *China* (1978) III. 244 At break we sit on the freshly turned earth and make shadow pictures with our hands. **1890** Champlin & Bostick *Young Folks' Cycl. Games & Sports* 625/2 *Shadow plays*, plays in which not the actors, but their shadows, are seen by the audience. **1900** W. W. Skeat *Malay Magic* vi. 514 Another very characteristic performance is the Shadow-Play. **1932** E. Waugh *Black Mischief* iii. 92 He liked..to appear in society..to survey the shadow-play of fashion. **1938** *Burlington Mag.* Aug. 87/2 Shadow-play puppets. **1964** *Catal. National Mus. Kuala Lumpur* 3/1 The shadow play exhibit is arranged so that the visitor can see backstage and learn how the figures are manipulated during the drama. **1971** *Country Life* 17 June 1544/1 A shadow play, the Wayang Kulit of parchment puppet figures manipulated from behind a lamplit screen. **1965** A. Waterston *Development Planning* ix. 322 If the true economic cost of a project is to be determined in situations where market prices are out of line..it may be necessary to 'adjust' the prevailing prices by estimating the extent to which they deviate from 'equilibrium' prices. The adjusted prices, variously known as 'shadow' or 'accounting' prices, are then substituted for prevailing prices and used to determine real costs and benefits to an economy and to compare the project under consideration with other projects on a comparable basis. **1970** S. L. Barraclough in I. L. Horowitz *Masses in Lat. Amer.* iv. 157 Should labor be counted as a cost valued at current wage rates when there are no alternative job opportunities? If not, what 'shadow prices' should be used? **1981** *Sci. Amer.* June 116/3 Marginal values are sometimes called shadow prices or imputed prices. **1965** A. Waterston *Development Planning* ix. 323 Shadow-pricing can also permit valid comparisons to be made of a public sector project with a private sector project. **1976** *Nature* 8 July 84/1 Does this justify the attachment of a money-tag to all values, even though this means what economists call 'shadow pricing' (for example, the 'value' of a view of the South Downs is the extra cost of not defacing the view if a road or a line of electric pylons has to be built in the neighbourhood)? **1926** G. G. Denny *Fabrics* (ed. 2) I. 111 *Warp print or shadow print.* Silks, ribbons and cretonnes woven with plain filling on a printed warp which gives a faint and shadowy design. **1968** J. Ironside *Fashion Alphabet* 246 *Shadow print*, the warp yarns are printed with the design before weaving, giving a shadowy print effect. **1923** H. W. Whanslow *Everybody's Theatre, & How to make It* iv. 42 A fine collection of these Javanese shadow puppets. **1971** H. Trevelyan *Worlds Apart* iii. 43 There were the ingenious hand-made toys, the shadow-puppets manipulated on sticks. **1929** O. G. S. Crawford *Air-Photography for Archaeologists* 3/1 Inequalities in the surface of the ground produce shadows. All sites where remains are visible on the ground fall into this class. They may be called *shadow-sites*. **1946** R. J. C. Atkinson *Field Archaeol.* I. 47 Shadow-sites are those whose surface is irregular, consisting of banks, mounds, ditches and terraces whose presence is revealed by the shadows they cast when seen in the low light of the rising or setting sun. **1932** *Mod. Woman* Feb. 56/1 This shadow stitch is just like herring-boning worked rather closely together... It gives you the shape of the leaf outlined in back-stitch on the right side and padded with long, crossed stitches on the wrong. **1932** *Pontings Whitsun Sales Catal.* 11 Morning Washing Frock for the larger than stock size in shadow stripe art. silk. **1940** *Chambers's Techn. Dict.* 762/2 *Shadow stripes*,..cotton cloths, of plain or satin weave, in which stripes are produced by using warp yarns of different directions of twist. The shadow effect is due to light being reflected in different directions by the different twists. **1947** J. Stevenson-Hamilton *Wild Life S. Afr.* vi. 52 Burchell's zebra.. 'Shadow stripes', that is to say light brown bands impinged upon the white ground which separates the black markings. **1930** *Economist* 18 Oct. 713/1 As a result a substantial amount of business was booked, principally in shadow striped poplins. **1969** I. & P. Opie *Children's Games* ii. 86 The game [*sc.* Shadow Touch] is also played in Canada and the United States ('Shadow Tag'). **1977** *Hartford* (Conn.) *Courant* 6 June 15/3 There was 'Shadow Tag' on sunny days—the 'It' player ran after the others, trying to jump on a shadow with a foot. **1884** H. E. Juler *Ophthalmic Sci. & Pract.* xiv. 363 The two following [methods] are very useful in estimating refraction; in both the ophthalmoscopic mirror alone is employed, and is held at a considerable distance from the eye. The first of these may be called the 'Fundus-Image' test; the other has been called 'Retinoscopy', but would be more appropriately designated by some such term as 'Shadow Test'. **1889** G. A. Berry *Dis. of Eye* xiv. 462 It has been called the shadow test because attention is directed perhaps more to the dark shadow which borders the illuminated area than to the area itself. **1964** S. Duke-Elder *Parsons' Dis. of Eye* (ed. 14) vii. 69 Retinoscopy, or, more correctly, skiascopy or the shadow test, is the most practicable method of estimating the conditions of the refraction objectively. **1923** H. W. Whanslow *Everybody's Theatre, & How to make It* iv. 41 China.. has had its shadow-theatres for many centuries. **1932** J. Nicoll tr. *Van Boehn's Dolls & Puppets* viii. 35 The Chinese shadow theatre..has no public, and the educated classes pay no attention to it now. **1970** *Guardian* 22 July 20 Mr P. L. Amin Sweeney, who has just gained a Ph.D. for a thesis on the Malay shadow theatre, yesterday demonstrated the art with a lamp, a screen, and 40 flat hide puppets. **1920** *Queen* 3 Apr. 17 (Advt.), Shadow Tissue. **1939–40** *Army & Navy Stores Catal.* 1073/2 Shadow tissues at 1/3 per yard. **1919** 'C. Dane' *Legend* 94, I possess that underlying shadow-work (I admit it's no more) of fact to guide me in deciphering her method in the first book. **1932** D. C. Minter *Mod. Needlecraft* 25/1 If the material is very transparent, a white thread on a white ground is..effective. This 'shadow work', as it is called.., can be prettily used on collars and cuffs and small articles. **1932** *Mod. Woman* Feb. 56 The shadow work tea cloth and cosy. **1967** Shadow work [see *pattern darning* s.v. *pattern *sb.* 13 b].

b. Designating members of an opposition party nominated as counterparts of members of the government in power holding cabinet or other offices, or the offices held, as *shadow cabinet, minister, ministry,* etc.
1906 A. J. Balfour *Let.* in Ld. Newton *Ld. Lansdowne* (1929) 354 If we are to have, as you suggest, a Committee consisting of members selected from the Front Bench in both Houses,..what we should really have would be a shadow Cabinet once a week. **1925** J. O'Connor *Hist. Ireland 1798–1924* II. xxiii. 302 The Dail might go on to the crack of doom passing secret resolutions, appointing shadow ministers, [etc.]. **1953** Earl Winterton *Orders of Day* p. xi, I was in Mr. Churchill's 'Shadow Cabinet' from 1945 to 1950. **1958** *Spectator* 20 June 799/2 The Chancellors and Shadow-Chancellors. **1965** *New Statesman* 19 Mar. 436/2 Mr Ernest Marples, 'Shadow' Minister of Technology, will start work today at the English Electric Leo-Marconi works at Kidsgrove, Staffs. **1970** C. Hampton *Philanthropist* ii. 18 The Shadow Minister of Health..was hit in the ankle by a ricochet. **1973** *Ottawa Jrnl.* 21 Feb. 29/2 Opposition Leader Stanfield and his shadow cabinet have been using it to try and discredit Liberal economic policies in advance of the budget. **1976** H. Wilson *Governance of Britain* vii. 150 As shadow chancellor, I had..made some strong comments on some of the projects. *Ibid.* viii. 158 The Conservative leader.. also nominates the members of the so-called Shadow Cabinet and allocates the shadow 'portfolios'. **1977** M. Walker *National Front* iii. 57 The Shadow Home Secretary..supported the motion. **1980** *Austral. Financial Rev.* 11 Apr. 13/2 Labor's energy policy for the next Federal election, which was unveiled..by the Leader of the Opposition..and the Shadow Minister for Minerals and Energy.

c. Designating organizations, structures, etc., built or instituted to substitute for or duplicate those existing in an emergency or to fulfil special needs, esp. before and during the war of 1939–45. Also as *adj.*
1936 *Economist* 31 Oct. 195/2 There was the scheme for the 'shadow' industry... This..was to consist of a set of new factories built at the expense of the Government, but supplied with skilled labour and management by the private companies. **1937** *Sunday Express* 24 Jan. 14/2 Experts other than Lord Nuffield have doubts whether the Government's shadow factory system for air-craft production is wise or workable in war time. **1938** *Times* 16 Mar. 7/2 Both in the regular industry..and in the shadow scheme, which was designed as a reinforcement and an insurance, engines were somewhat ahead of air-frames. **1939** *Sun* (Baltimore) 6 July 1/5 The factories themselves, conventional and 'shadow', are turning out a certain number of aircraft and engines each month—the actual number could not be learned. **1939** *Air Ann. Brit. Empire* 3 The Standard Motor Company is also concerned in shadow manufacture of new engine components. **1940** *Ann. Reg. 1939* 20 The whole 'shadow' organisation should be in a position to function as soon as an emergency arose. **1944** *Jrnl. R. Aeronaut. Soc.* XLVIII. 370 Considerable experience had been gained by the Bristol Co., in their licence manufacture all over the world, which had already taught them the method of laying out drawings and preparing data remote from the parent factory, and this was of the utmost help in getting going on the 'shadow' production. **1946** *R.A.F. Jrnl.* May 160 He may have spent his last few years before donning a uniform on a war job in a shadow factory—a shadow factory which, with the coming of peace, was closed. **1980** J. Dutton *Copley's Hunch* ii. iv. 178 The war came..then they put up one of those shadow factories at—well, I'd best not say where.

shadow, *v.* Add: **12. b.** *Speech Therapy. trans.* and *intr.* To repeat (another's words) with the minimum of delay, as a treatment for stuttering.
1955 *Nature* 5 Nov. 874/2 The subject 'shadows' an unseen message read by the operator steadily and continuously. [illegible] **1973** C. Van Riper *Treatment of Stuttering* iii. 69 When stutterers 'shadowed' the speech of a model speaker almost complete 'suppression' of stuttering occurred. **1977** D. Fry *Homo Loquens* x. 149 A stammerer who is shadowing will hear the appropriate sequence of sounds in advance and this should cancel out any built-in delay in his system.

c. *trans.* To act as a shadow (see *shadow *sb.* 6 e) in respect to (a parliamentary minister, ministry, etc.). Also *absol.*
1969 *Daily Tel.* 26 Oct. 1 An unusual trio of Tory political partners is associated with the..gallery... One is Geoffrey Rippon, who 'shadows' Defence. **1971** R. E. Leavis in *Human World* Aug. 8 The politician..was at that time 'shadowing' Education. **1974** *Times* 11 Mar. 1/1 Mr Carr shadows Mr Healey at the Treasury... Sir Alec Douglas Home maintains..foreign affairs and Mr Rippon will shadow Europe. **1977** *Times* 5 Nov. 1/5 The new spokesman on Treasury and economic affairs..will be Mr Peter Tapsell, who formerly helped to shadow the Foreign Office.

13. *Microscopy.* To subject (a specimen) to the process of *shadow-casting *vbl. sb.* 1.
1945 *Proc. Soc. Exper. Biol. & Med.* LVIII. 267 (*caption*) A micrograph of a similar preparation after it has been shadowed by the oblique deposition upon it of a thin layer of chromium. **1966** D. G. Brandon *Mod. Techniques Metallogr.* 48 By shadowing the surface of the replica with a heavy metal from a carefully collimated source at a known angle, the intensity differences from point to point on the surface can be related directly to the surface topography of the specimen. **1978** *Nature* 19 Jan. 213/2 Increased ammoniation is indicated principally by the change in morphology of particles collected (during ascent) on a carbon surface and 'shadowed' with silicon oxide later in the laboratory.

shadow-casting, *ppl. a.* and *vbl. sb.* **A.** *ppl. a.* That casts a shadow or shadows.
1859, 1882 [in Dict. s.v. shadow *sb.* 1 c]. **1904** W. de la Mare *Henry Brocken* 145 [illegible] shadow-casting moonlight [illegible]. **1954** *Jrnl. Exper. Psychol.* XLV. 206/1 The shadow-casting object is placed as close to the screen as possible, whereas the distance between the light source and the object is made large.

B. *vbl. sb.* **1.** *Microscopy.* A technique for enhancing an electron-microscope image by projecting a beam of small particles or atoms (usu. of a heavy metal) on to the sample at a small angle to the horizontal, so as to form a deposit giving the appearance of shadows cast by sideways illumination.
1944 Williams & Wyckoff in *Jrnl. Appl. Physics* XV. 712/2 Information was needed [illegible] stereoscopic photography. A procedure for measuring heights based on shadow casting was accordingly developed to meet this need. **1947** *Ann. Rev. Microbiol.* I. 11 With the addition of the shadow casting technique, about eleven fibrils were clearly seen to compose the cilium of *Paramecium*. **1971** V. A. Phillips *Mod. Metallogr. Techniques & Applications* v. 183 It is customary to enhance the contrast of replicas by shadow casting, that is, by evaporating a heavy metal from an angle to give shadow effects in the final electron micrograph, equivalent to oblique light illumination.

2. *lit.* The casting of shadows, esp. in *Psychol.* as a technique of perceptual research. Also *attrib.*
1957 *Psychol. Rev.* LXIV. 291/2 Previous shadow-casting devices..have not been constructed for this systematic purpose. **1971** *Nature* 3 Sept. 55/2 The geometric principles of shadow casting and the use of chromatic filters for dichoptic stimulation are established techniques.

Hence (as a back-formation) **sha·dow-cast** *v. trans.*, to enhance (a microscopic image) by shadow-casting; to subject (a microscopic specimen) to shadow-casting; also as *ppl. a.*; **shadow caster,** a device employed in perceptual research in order to cast shadows, esp. one for producing seemingly three-dimensional shadows.
1944 *Jrnl. Appl. Physics* XV. 715/2 (*caption*) Example of a shadow-cast electron micrograph for the determination of the heights of objects. **1957** *Psychol. Rev.* LXIV. 291/2 Considering the mount and the screen as two geometrical planes, changes in the position of the mount will yield all possible perspective transformations of the shadow relative to the shadow caster. **1969** *Vision Res.* IX. 154 The principal advantage of the stereoscopic shadow-caster over other stereoscopic projection methods lies in its versatility in the study of binocular kinetic space perception. **1971** V. A. Phillips *Mod. Metallogr. Techniques & Applications* v. 187 After being allowed to harden, it is dry-stripped and shadow cast. **1971** *Nature* 3 Sept. 55/2 To produce this effect, a stereoscopic shadow caster was used, in which two point sources slightly separated horizontally cast the shadows of three vertical rods on a transparent screen. **1979** *Ibid.* 27 Sept. 287/2 Film samples were transferred to electron microscope grids... Transferred films were shadowed once [illegible] with platinum-palladium at an angle of 10° before examination in a Siemens Elmiskop.

shadowed, *ppl. a.* Add: **10.** *Microscopy.* Subjected to the process of *shadow-casting *vbl. sb.* 1.
1944 *Jrnl. Appl. Physics* XV. 714/2 Photographing and measuring the lengths of the shadowed areas thus formed on the preparation. **1949** *Proc. Soc. Exper. Biol. & Med.* LXXI. 80/1 (*caption*) Chromium shadowed preparation of the same slide. **1973** P. J. Goodhew *Specimen Preparation in Materials Sci.* v. 151 If a high resolution is required then the grain size of the shadowed replica is important.

shadowgraph, *sb.* Add: **1. a.** Also *fig.* Also, = Shadowgraphist.
1886 *St. Stephen's Rev.* 27 Mar. 5/2 At the New Club..on Saturday next..Mason and Titus, the American shadowgraphs, who nightly provoke so much laughter at the Oxford, will appear at 11.30 p.m. **1928** A. S. Eddington *Nature of Physical World* p. xvi, In the world of physics we watch a shadowgraph performance of the drama of familiar life. **1965** J. Van Strenberg [illegible] i. 2 [illegible] shadowgraphs of my concepts of them [illegible].

b. An image formed by light which has passed through a fluid and been refracted differently by regions of different density (used esp. in the study of fluid flow).
1926 *Proc. R. Soc.* A. CXI. 336 Shadowgraphs of the jets emerging into the atmosphere from nozzles of different forms and at different initial pressures were obtained by the method described above. **1945** *Jrnl. Optical Soc. Amer.* XXXV. 595/1 [illegible] **1955** F. J. Weyl in *High Speed Aerodynamics & Jet Propulsion* IX. 21 By far the most extensive use of shadowgraph techniques..concerns the recording of shock waves and slip discontinuities. **1974** W. Merzkirch *Flow Visualization* iii. 85 [illegible] **1978** *Nature* 5 Jan. 41/2 (*caption*) Shadowgraph photograph showing the tilted layers and interfaces evolved in inserting a block of ice into salt-stratified water at room temperature.

2. (Further examples.)
1975 *Nature* 25 Sept. 276/2 The X-ray shadowgraph image is converted into a charge image on a dielectric by the ionisation of a gas or liquid in a chamber. **1978** *Sci. Amer.* Nov. 62/1 The simplest and most successful way to produce an image with X rays is with contact X-ray microscopy. This technique, which achieves a resolution substantially better than that of the light microscope, creates a shadowgraph of the specimen.

shadowing, *vbl. sb.* Add: **2.** Also *transf.* (cf. *shadow *sb.* 4 i).
1977 *Lancet* 3 Sept. 512/1 A chest radiograph showed slight inflammatory shadowing in the right upper zone.
4. For † *Obs.* read *rare* and add later example.
1827 J. F. Cooper *Prairie* I. 17 The martin's fur..was of a fineness and shadowing that a queen might covet.
6. b. *Speech Therapy.* Repetition of another's words with the minimum of delay, as a treatment for stuttering.
1955 *Nature* 5 Nov. 874/1 'Shadowing' means repeating concurrently. **1975** in J. Eisenson *Stuttering: 2nd Symposium* 54 It would appear that speech behavior during shadowing shares something in common with singing.
6*. *Microscopy.* = *shadow-casting *vbl. sb.* 1.
1945 *Science* 8 June 596/1 Their orientation with respect to the direction of shadowing. **1973** P. J. Goodhew *Specimen Preparation in Materials Sci.* v. 150 Shadowing is normally carried out with heavy elements, a wide variety of which have been used.

shadow-land. Add: **2.** A place in shadow; a gloomy, unhappy place; an indeterminate borderland between other places, states, etc.
1923 *Daily Mail* 13 July 13 To lead Ireland out of the shadow-land of much unnecessary suffering and turmoil into a brighter and happier state. **1949** Brooks & Warren *Fundamentals of Good Writing* vi. 241 If we understand the extremes..we can use common sense to discriminate among the examples in the shadowland in between. **1960** Auden *Homage to Clio* 18 [illegible] shadowland of trees. **1966** *New Statesman* 27 May 775/1 [illegible] in the sexual power-game of the race war's shadowlands.

Shadrach. Add: **2.** (See quots.) *local.*
1827 in S. Holland *Mem. Sydney Smith* (1855) I. 259 His fires are blown into brightness by *Shadracks*, tubes furnished with air from without, opening into the centre of the fire. **1954** D. Hartley *Food in England* iv. 46 In some places, a draught is obtained by shadrack, an underground arrangement like a small blast furnace through which the blast of a rotary fan is carried under the [peat] fire.

shaduf, var. Shadoof in Dict. and Suppl.

SHAEF (ʃeɪf). Also **S.H.A.E.F., Shaef.** [Acronym f. the initials of Supreme Headquarters Allied Expeditionary Force.] The operational headquarters of the allied expeditionary force that invaded occupied Europe in 1944–5.
1944 *N.Y. Times* 21 May iv. 8/2 'SHAEF' stands for Supreme Headquarters Allied Expeditionary Force. **1944** C. Milburn *Diary* 8 June (1979) 215 The great assault is known as S.H.A.E.F., pronounced 'shafe'—Supreme Headquarters Allied Expeditionary Force. **1945** *Times* 28 May 3/3 Shaef announced yesterday that it is expected that the port of Hamburg will be open on June 1 to allied shipping and supplies for our armies in the liberated countries. **1958** *Listener* 21 Aug. 272/1 The Cossac spirit carried on into Shaef and through to triumph. **1977** P. Ustinov *Dear Me* xi. 145 SHAEF, the Allied Supreme Headquarters, wished an official film to be made about the war in the West.

shaft, *sb.*[1] Add: **9*.** In various slang uses. **a.** The penis. Also *to shaft of delight*.
[**1719** T. D'Urfey *Wit & Mirth* IV. 72 It is a Shaft of Cupid's cut, 'Twill serve to Rove, to Prick, to Butt.] **1772** G. A. Stevens *Songs, Comic, & Satyrical* 11 For Cupid's Pantheon, the Shaft of Delight Must spring from the Masculine Base. **1971** B. W. Aldiss *Soldier Erect* 43 It was never enough merely to lower your trousers—they had to come off,..so that you could crouch there naked but for your shirt, frantically rubbing your shaft.
b. A human leg. *U.S.*
1935 A. J. Pollock *Underworld Speaks* 103/2 *Shaft*, a woman's leg. **1939** C. Morley *Kitty Foyle* 95 If anyone showed a good shaft Pop would wink at me.
c. *U.S.* An act or instance of unfair or harsh treatment; slighting, rejection, 'the push'; esp. in *to give* or *get the shaft*.
1959 *Amer. Speech* XXXIV. 155 A girl or boy who makes a play for another's date is shafting... If he succeeds, the loser gets the shaft (sometimes with barbs), the *purple shaft*, or the maroon macaroon, depending upon the degree of injury to his pride. **1960** Wentworth & Flexner *Dict. Amer. Slang* 461/1 *Shaft*..an act or instance of being taken advantage of, unfairly treated, deceived, tricked, cheated, or victimized; a raw deal. Usu. in 'to get the (or a) shaft'. Fig. the image is the taboo one of the final insult, having someone insert something, as a barbed shaft, up one's rectum. **1964** *Mad Mag.* July 14 Looks like somebody gave him the shaft! **1977** *Amer. Speech 1975* L. 66 She gave him the shaft after he broke their date last weekend. **1979** *Mod. Photography* Dec. 86/2, I would give more of my business to Minolta but for the company's uncooperative, anti-consumer thinking. Doubtless there are many such as myself who have gotten the shaft.
10. c. shaft-hole (*attrib.* examples).
1928 [see *core-casting s.v. *core *sb.*[1] 15]. **1958** W. Willetts *Chinese Art* I. ii. 75 (*heading*) Objects derived from the shaft-hole adze. **1971** *Listener* 7 Jan. 14/1 (*caption*) Copper shaft-hole axe from the Balkan late neolithic.
e. shaft-tailed whidah, widow bird, a dark-coloured African weaver-bird, *Vidua regia*, having long tail-feathers with bare shafts.
1881 F. & C. G. Gates *Handbk. Land & Water Birds of Natal* 64 (*pl. caption*) Shaft-tailed Whydah Bird. **1900** A. C. Stark *Birds S. Afr.* I. 148 Shaft-tailed Widow Bird... The four central, elongated tail-feathers are webbed at their ends... [illegible] **1948** C. D. Priest *Eggs of Birds breeding S. Afr.* 135 Shaft-tailed Whydah..undoubtedly parasitic. **1974** *Sci. Amer.* Oct. 96/2 The shaft-tailed widow bird of South Africa..mimics the repertory of its host, the violet-eared waxbill.
f. shaft-alley *Naut.* (see quot. 1884); also used *attrib.* to designate unofficial or unreliable information or its source, attributed to gossip in shaft-alley; **shaft horsepower,** brake horse-power, *spec.* power delivered to a propeller shaft or the shaft of a turbine; **shaft turbine** (see quot. 1958).
1884 *Naval Encycl.* 732/1 *Shaft-alley*, a passage extending from the engine-room to the stern..in which is contained the propeller-shaft and its bearings. **1922** L. Hersey *Sea Grist* 155 It was rumored by shaft alley wireless that we would reach Antwerp, Belgium, in two days. **1941** R. G. M. Ehlers *Diary of Ship's Surgeon* (1944) 67 A 'shaft alley' rumor brought word that all ships had been ordered out of Hong Kong. **1945** *Sun* (Baltimore) 30 Aug. 7-0/5 It's the job of these six men to go down to the nethermost portion of this ship in 'Shaft Alley', where the big propeller shafts whirl. **1908** A. E. Tompkins *Marine Engin.* (ed. 3) v. 61 The torsion-meter is used to measure this angular twist between two points of a shaft, and from this angle the shaft horse-power is calculated. **1974** *Petroleum Rev.* XXVIII. 490/1 The high shaft horsepower was the conditioning factor for this proportion of pilot fuel. **1958** *Chambers's Techn. Dict.* Add. 1013/1 *Shaft turbine*, any gas turbine aero-engine wherein the major part of the energy in the combustion gases is extracted by a turbine and delivered, through appropriate gearing, to a shaft. **1970** Lambermont & Pirie *Helicopters & Autogyros of World* (ed. 2) 147 It had two shaft-turbine engines mounted on the cabin top instead of two Pratt and Whitney piston engines.

shaft, *sb.*[3] Add: **4. a.** Also objective, as *shaft-sinker*.
1922 D. H. Lawrence *Aaron's Rod* vii. 70 His father had been a shaft-sinker.
b. shaft-house (earlier example); **shaft pillar** *Mining*, a body of coal or rock unworked in order to provide support for an adjacent shaft.

1872 *Statistics of Mines & Mining 1870* (U.S. Treasury Dept.) 344 The quartz is brought from the mine, unless the mill is in or near the shaft-house, in wagons. **1855** G. C. Greenwell *Pract. Treat. Mine Engin.* vi. 155 The situation of coal pits varies so much, together with the position of the seams of coal, dykes and slips, that no rule can be laid down for the form of the pillars of coal, left near the shaft, which are called the shaft pillars. **1929** I. C. F. Statham *Winning & Working* xxx. 499 This subsidence was not..wholly due to the removal of the shaft pillar, but was partly accounted for by crushing of the shaft pillar in an upper seam. **1977** *Irish Press* 29 Sept. 8/4 A third semipermanent pillar, known as the shaft pillar, cuts across the orebody from north to south.

shaft, *v.*[2] Add: **4.** *trans.* To treat unfairly or harshly; to cheat, deceive; to take advantage of; to slight, reject. *slang* (orig. and chiefly *N. Amer.*).
1959 *Amer. Speech* XXXIV. 155 A raw deal from any other source may also be referred to in this way; for example, one may be *shafted* or *jabbed* by the opposite sex, a professor, a policeman, parents, or anyone else for any real or imagined injury. **1966** 'E. Lathen' *Murder makes Wheels go Round* xiii. 108 He was a menace to Wahl... He'd railroaded Orin Dunn into jail... He was shafting Buck Holsinger! **1970** *Deb. Senate Canada* 1 June 7551/2 As I have told my constituents in Hamilton, Ontario, which seems to have been continually shafted by this government. **1971** B. Malamud *Tenants* 19 Rent control..is an immoral situation. The innocent landlord gets shafted. **1976** M. Machlin *Pipeline* xxxv. 397, I think how they're shafting us with this whole deal.
5. = *fuck *v.* 1 *trans. coarse slang.*
1970 G. Lord *Marshmallow Pie* xxi. 185 There was this young girl among them, not even sixteen yet..like as not being shafted by every dirty long-haired crud in town. **1971** B. W. Aldiss *Soldier Erect* 82 How sinful he looked, squatting there by the water while his wife was being shafted by some dirty big Mendip only a few feet away! **1971** J. Wainwright *Last Buccaneer* II. 228 He was Jimmy Needler—that's all..and the rest of the world could go shaft itself.
Hence **sha·fting** *vbl. sb.*
1971 B. W. Aldiss *Soldier Erect* 124 Hello there, gran! What do you do? Gobble? Where are the birds? We want three as are fit enough to stand a gude shafting. **1972** J. Wainwright *Requiem for Loser* iii. 50 What a monumental shafting he'd deliver to some lucky bint. **1973** *Farm & Country* 20 Nov. 23/3 Hugh Blaine charged that farmers 'suffered a shafting at the hands of feed dealers last year'. **1975** R. H. Rimmer *Premar Experiments* i. 94 After double-dealing with his own people and selling them to the slavers, some slaver gave the king and his family a shafting and enslaved them too.

shag, *sb.*[1] Add: **3. c.** *ellipt.* A shag carpet or rug; shag pile. See sense 6 c below.
1951 K. R. Gillespie *Home Furnishings* v. 164 A few cotton floor coverings woven on standard carpet backs have come into the market, in addition to the bouclé weaves, shags, [etc.]. **1974** *Anderson* (S. Carolina) *Independent* 18 Apr. (Sears Advts. Suppl.) 5 Nylon pile shag. Long shag that's slow to show soil! **1976** H. Nielsen *Brink of Murder* xii. 108 The floor was carpeted with soft yellow shag.
6. a. shag end *N. Amer. colloq.* = Fag-end 2.
1972 J. Mosher *Adultery* iv. xxi. 176 It was the shag end of winter and there were scarcely any victuals to be had. **1977** G. V. Higgins *Dreamland* i. 13 The years that came between that night..and the shag end of 1971.
c. Of carpets, rugs, etc.: having a long, rough, pile. Also *shag pile*. Cf. Shaggy *a.* 1 c.
1946 *House Beautiful* Oct. 199 (Advt.), Charm Tred Shag Cotton Rugs. **1947** *Sun* (Baltimore) 1 Dec. 13/4 (Advt.), Heavy Loop Pile Shag Rugs. **1969** D. E. Westlake *Up your Banners* (1970) xlii. 309 The silence had the texture of a shag rug. **1974** *Times* 3 May 11/4 Kosset Panorama, the cheapest shag carpet I have seen. *Ibid.* 12 Aug. 22/8 (Advt.), *Carpets*..Shag Piles and Berber Weaves.

shag, *sb.*[2] Add: **b.** Also *a shag on a rock*: used in various Austral. colloq. phrases as a type of the isolated or exposed.
1845 R. Howitt *Impressions of Australia Felix* 233 'Poor as a bandicoot', 'miserable as a shag on a rock', &c.; these and others I very frequently heard them make. **1929** J. Raeside *Golden Days* 16 The flood waters did not subside, and we were there like three shags on a rock. **1971** D. Ireland *Unknown Industrial Prisoner* 275 It's easy enough to curse England. Leaving us out here like a shag on a rock.

shag, *sb.*[6] *coarse slang.* [f. *shag *v.*[3]] **a.** An act of copulation.
1937 Partridge *Dict. Slang* 748/2 *Shag*, a copulation; also, copulation generically. **1971** B. W. Aldiss *Soldier Erect* 114 It was not just a good shag I needed. It was romance.
b. One who copulates. *rare.*
1971 K. Amis *Girl, 20* ii. 76 Ageing shag tries to stimulate jaded appetite by re-creating situation of days of first discovery of sex. **1978** —— *Jake's Thing* ix. 94 The moustached shag and the flat-chested bint..had moved away from the bar with their drinks.

shag, *sb.*[7] [Perh. f. Shag *v.*[2] or *v.*[3]] A dance popular esp. in the U.S. in the 1930s and 1940s, and characterized by vigorous hopping from one foot to the other. Hence as *v. intr.*, to dance the shag; **sha·gger**[1], one who dances the shag.
1932 (*title of jazz tune*) Shag. **1937** [see *Big Apple* s.v. *big *a.* B. 2]. **1938** *Sun* (Baltimore) 24 June 4/3 The Virginia reel, the shag, the sugarfoot and trucking predominate on the dance program. **1939** Ramsey & Smith *Jazzmen* xiii. 271 The Crescent Billiard Hall.. frequented by the best shaggers in town. *Ibid.*, Usually when Brunions reaches the third chorus..the kids have stopped shagging. **1940** *Time* 29 Jan. 17/1 A citizenry shagging to the tune of *Oh Johnny!* refused to take the 1940 Campaign seriously. **1954** S. G. L. Dannett et al. *Down Memory Lane* 131 The shag is a fast, nervous, hopping dance, performed in time to a strongly accentuated rhythm. **1963** *N.Y. Times Mag.* 27 Oct. 104/2 [The Negroes'] body rhythm and frank sensuality turned the formal European waltz into..the shag, the Susie Q. and the big Apple.

shag, *a.* **2.** For † *Obs.* read 'Now *rare*' and add later example.
1975 R. L. Duncan *Dragons at Gate* (1976) i. 12 Jenkins feigned the appearance of his early thirties..sandy-colored shag hair.

shag (ʃæg), *v.*[3] *coarse slang.* [Origin unknown, perh. f. Shag *v.*[1]] **1.** *trans.* and *intr.* To copulate (with).
1788 Grose *Dict. Vulgar Tongue* (ed. 2), *Shag*, to copulate. **1879–80** *Pearl* (1970) 258 A fellow who's had the mishap, To forget, when he shagged her, to button his flap. **1958** N. Levine *Canada made Me* iv. 102 You know what they're talking about? If they got shagged last night. **1969** J. Wood *Three Blind Mice* iii. 32 We..go and shag ourselves half-stupid all night..and *pay* them for it in the morning! **1973** *Nation Rev.* (Melbourne) 24–30 Aug. 1417/4 The credo of the new fashioned mammy is if you shag, I shag. **1977** C. McCullough *Thorn Birds* xvii. 413 There are plenty of men who will shag anything if it's a virgin. **1980** R. Adams *Girl in Swing* xxi. 279 'He's never absent.' And the corporal next to Jack muttered, 'Well, I 'ope 'e ain't 'angin' around when I'm shaggin' my missus.'
2. Used profanely in imprecations and exclamations. Cf. *fuck *v.* 2 and sense 1 of next.
1933 M. Lowry *Ultramarine* ii. 88 'Paddy—give us Paddy McGulligan's daughter, Mary Ann.' 'Oh shag off!' **1971** R. Ludlum *Scarlatti Inheritance* iv. 43 'Get four men and get out there.' 'Go *shag*, Captain.' 'Are you disobeying your superior officer?' **1973** G. Pinsent *Rowdyman* 135 'Then shag you!' I shouted, as he swaggered away.
Hence **sha·gger**[2]; **sha·gging** *vbl. sb.*[2] and *ppl. a.*
1970 G. Greer *Female Eunuch* 41 All the vulgar linguistic emphasis is placed upon the *poking* element;.. *rooting, shagging* are..acts performed upon the passive female. **1971** B. W. Aldiss *Soldier Erect* 9 God, what sodding, shagging, scab-devouring misery it all was! *Ibid.* 12, I could watch my reflection in the mirror of the wardrobe... Now *there* was a born shagger, if ever I saw one, given the chance. **1977** *Zigzag* Aug. 4/3 Plus the fact it gets hotter than shagging in the back of a car during the summer of '76.

shag (ʃæg), *v.*[4] [Origin unknown. Possibly connected with Shag *v.*[1] or (esp. in sense 1) *v.*[3]; but it is not even certain that senses 1 and 2 belong to the same verb.] **1.** *intr.* To make off; to wander aimlessly; to traipse. Freq. with advbs. *slang.*
When followed by *off* there is some overlap with sense 2 of *shag *v.*[3]
1851 *Gloss. Provinc. Words Gloucs.* 11 *Shag*, to steal away. **1932** J. T. Farrell *Young Lonigan* iv. 192 He watched a familiar looking airedale dog shag about. **1938** D. Runyon *Furthermore* xiv. 278 The Princess is getting too grown-up to be shagging around Broadway, and.. she is now going to public school. **1968** 'B. Mather' *Springers* xv. 162 I'll take you into Russia with me—or you can shag off on your own. **1976** W. H. Canaway *Willow-Pattern War* xiv. 140 We'd been shagging around over these mountains for four days now, and we hadn't seen one single musk deer.
2. *trans.* To chase. Also const. *up. spec.* in *Baseball*, to go for or catch (fly balls). *U.S. colloq.*
1913 C. H. Claudy *Battle of Baseball* 318, I was allowed to 'shag' foul balls. **1932** J. T. Farrell *Young Lonigan* ii. 66 Demons..would come and lean over his bed..until his old man came and shagged them away. **1955** *Sun* (Baltimore) 3 Mar. 19/8 Coan..shagged flies under the tutelage of Coach Tom Oliver. **1979** *Navajo Times* (Window Rock, Arizona) 24 May 15/1, I was originally picked as an outfielder, so I played two years shagging balls in the outfield. **1981** G. V. Higgins *Rat on Fire* v. 37 Every so often..I got to shag up a couple of guys who haven't told a clean joke in years and give the guy free entertainment.

shaganappi, *sb.* and *a.* For *U.S.* read *N. Amer.* and add: **a.** *sb.* (Earlier examples in both senses.) **b.** *adj.* Also, cheap, inferior, makeshift.
1743 J. Isham *Obs. Hudsons Bay* (1949) 46 Shag, a nap, pee or a string of Leather tauk' a miss. **1820** G. Simpson *Jrnl.* 10 Sept. (1938) 58 Any dressed skins, sinews, and Shaganapy lines that can be spared, you will send here for New Caledonia. **1895** W. Elkington *Five Years in Canada* xi. 101 A few years ago the only horses in the country were bronchos and Indian 'shaganappies' as they are called. **1900** W. A. Fraser *Mooswa* 35 Have patience, little shaganappi (cheap) Bird. **1961** W. O. Mitchell *Jake & Kid* i. 2 At the Rabbit Hill school concert last night, folks heard a shaganappy speech. *Ibid.* viii. 106 Thinka anybuddy havin' a shaganappy thing like that in their house!

shagged, *a.*[1] **2. b.** (Earlier and later examples.)
1784 J. Byng *Jrnl.* 11 July in *Torrington Diaries* (1934) I. 161 A most fearful mountain, call'd Drwsycood, of awful, and shagged front. **1927** F. B. Young *Portrait of Clare* III. viii. 350 The twisted apple-trees stood shagged with a silvery blight.

shagged (ʃægd), *a.*[2] *slang.* [Origin uncertain: perh. rel. to Shagged *a.* or *shag *v.*[3] Cf. also Fagged *ppl. a.*] Weary, exhausted. Also with *out*.
1932 Auden *Orators* III. 99 Wakeful at night, in the morning fagged; They feel like angels, but they look just shagged. **1947** D. M. Davin *Gorse blooms Pale* 178 They're all in pretty good nick, considering. Shagged, of course. **1950** Dylan Thomas in *Circus* Apr. 8/2 He is..thin, not to say of a shagged-out appearance. **1960** *Observer* 20 Mar. 10/3 Oh cut it out, Sarge—let up! I'm shagged. **1971** *Peace News* 10 Sept. 8/2 The haggard and shagged-out end products of a lifetime spent in the pursuit of materialism. **1975** G. W. Target *Strike Strikers* iii. 51 The two other-rankers were now sitting in the back of the jeep, with all of 'em looking shagged out.

shaggery (ʃæ·gəri). *N.Z.* [f. Shag *sb.*[2] + -ery.] A breeding colony of shags.
1882 W. D. Hay *Brighter Britain!* II. vi. 222 They [*sc.* Kawau] build in trees, in large 'shaggeries'. **1921** H. Guthrie-Smith *Tutira* xxii. 207 Fish are attracted to the vicinity of shaggeries.

shaggily, *adv.* (Examples.)
1859 Geo. Eliot *Adam Bede* I. x. 196 His hair was tossed shaggily about his forehead. **1921** D. H. Lawrence *Birds, Beasts & Flowers* (1923) 82 It was a fiery fortress frowning shaggily on the world. **1977** T. Heald *Just Deserts* i. 11 A dark, shaggily suave person.

shaggy, *a.* Add: **3.** *shaggy-chested, -haired* (later examples), *-legged, -throated*; **shaggy dog story,** a lengthy tediously detailed story of an inconsequential series of events, more amusing to the teller than to his audience, or amusing only by its pointlessness; also *shaggy dog yarn*, etc.; **shaggy (ink-)cap** = next; **shaggy mane** (examples).
1894 M. C. Cooke *Edible & Poisonous Mushrooms* 57 Shaggy Caps... This is one of the best of edibles, and common enough everywhere. **1979** *Guardian* 31 Oct. 14/1 The delightful pleasures of Shaggy Cap soup or Lawyer's Wig stew. **1922** Joyce *Ulysses* 510 Ben Jumbo Dollard, rubicund,..shaggychested, shockmaned,..stands forth. [**1945** D. Low in *N.Y. Times Mag.* 4 Feb. 40/1 The logical lunacy of 'Shaggy Dog'.] **1946** *Coll. Shaggy Dog Stories* facing p. 1 Stories of the Shaggy Dog variety are essentially tales to be told rather than read. **1947** *Beat* Apr. 6/3 Here's one of my favourite 'shaggy dog' stories. **1952** A. R. K. Barnard in A. Redman *Somewhat 'Shaggy'* 4 The comparatively recent type of story—the 'Shaggy Dog' yarn. **1952** Koestler *Arrow in Blue* I. viii. 68 The people of Budapest have a peculiar shaggy-dog kind of humour. **1958** *Listener* 16 Oct. 623/1 It was a shaggy-dog story about a small-town worthy who shams madness to avoid paying bills. **1972** P. Ruell *Red Christmas* xi. 102 He seemed to be in the middle of an autobiographical shaggy-dog story. **1866** Geo. Eliot *Felix Holt* II. xvi. 15 The shaggy-haired, cravatless image of Felix Holt. **1974** L. Deighton *Spy Story* i. 14 A shaggy-haired giant, complete with kilt. **1953** J. Ramsbottom *Mushrooms & Toadstools* Pl. 22 (*caption*) Shaggy Ink-Cap..often in enormous numbers on made-up ground. **1970** J. Webster *Introd. Fungi* II. iv. 311 *Coprinus comatus* is a large terrestrial species (the shaggy ink-cap or lawyer's wig) which is edible. **1927** D. H. Lawrence *Etruscan Places* (1932) i. 16 He grins and drinks wine, and immediately one sees again the shaggy-legged faun. [**1885** J. A. Palmer *Mushrooms Amer.* Pl. II (*caption*) Shaggy-Maned Mushroom.] **1895** W. H. Gibson *Our Edible Toadstools & Mushrooms* 28 The Shaggy-mane..is conspicuously even-gilled, and is a decided delicacy. **1976** *National Observer* (U.S.) 13 Mar. 19/2 Now is the time of the shaggy manes and field mushrooms. **1946** R. S. Thomas *Stones of Field* 17 Thunder-browed and shaggy-throated All the men were there.

shagroon (ʃăgrū·n). *N.Z. slang* (now *Hist.*). [Perh. ad. Ir. *seachrán* wandering.] An early settler in Canterbury, New Zealand, from anywhere except Britain, esp. one from Australia.
1851 W. Lyon (*title*) Dream of a shagroon. **1851** E. Ward *Jrnl.* 20 Feb. (1951) 132 Started with Henry and a 'shagroon' cattle-driver. **1898** E. E. Morris *Austral English* 410/2 The men who came from England were called *Pilgrims*, all others *Shagroons*; probably a modification of the Irish word *Shaughraun*. **1930** L. G. D. Acland *Early Canterbury Runs* 1st Ser. i. 3 The Australians were known as 'Prophets' or 'Shagroons'. **1966** G. W. Turner *Eng. Lang. Austral. & N.Z.* i. 16 In Canterbury, immigrants from Victoria, locally called *shagroons*, set up sheep stations on the plains and were contemptuous of the agricultural enterprises of the *pilgrims* as the Canterbury Association's settlers were called.

Shah Abbas (ʃā æ·bās). The name of a Shah of Persia (1558–1628) used *attrib.* and *absol.* to designate Persian rugs and carpets like those made for him or their characteristic design.

1901 J. K. Mumford *Oriental Rugs* vi. 74, I have known a Persian who paid thirty dollars..for a fragment of one of these old Shah Abbas rugs. *Ibid.*, The Shah Abbas pattern is still made in rug factories. **1913** W. A. Hawley *Oriental Rugs* ix. 105 Only a few Persian rugs have the formal repetitive patterns, such as the Herati, Guli Hinnai, Mina Khani, and Shah Abbas. **1973** P. O'Donnell *Silver Mistress* i. 16 The sumptuous silken glow of the Shah Abbas carpet.

‖ **shahāda** (ʃahā·da). Also **shahādah.** [Arab. *šahāda* testimony, evidence.] The Muslim profession of faith, 'Lā ilāha illā Allāh, Muḥammad rasūl Allāh' ('there is only one God, and Muḥammad is his prophet').
1885 T. P. Hughes *Dict. Islam* 571/1 *Shahādah*.., 'evidence'. **1929** E. D. Ross tr. *Lammens's Islam* iii. 56 The customary offering of prayer, of which the *shahāda* forms an integral part, takes the place of this obligation. **1970** *New Yorker* 29 Aug. 45/1 A European who repented the error of his faith and proclaimed the *shahada*—'There is but one God and Mohammed is His Prophet'—before dying would always go directly to Heaven. **1981** *Daily Tel.* 19 June 15/8 Everything in Islamic art and thought should be seen in the light of the *shahada*, or profession of faith.

‖ **Shahanshah** (ʃā·ənʃā·). Also **Shahenshah, Shahinshah, Shah-in-Shah,** etc. [Pers. *šāhan-šāh* king of kings: see Shah.] 'King of kings': a title given to the Shah of Iran (Persia).
1815 J. Malcolm *Hist. Persia* I. vi. 92 The son of Babek was hailed..with the proud title of Shahan Shah, or King of Kings. **1824** J. Morier *Hajji Baba* II. iii. 63 The *Shah-in-Shah* speaks like an angel. **1877** E. S. Dallas *Kettner's Bk. of Table* 478 It is to be hoped that the land of vegetable marrows sometimes makes a dish of them for the Shah en Shah—king of kings. **1892** Ld. Curzon *Persia & Persian Question* I. xiv. 434 He remains the Shahinshah, or King of Kings. **1938** 'M. Essad-Bey' et al. *Reza Shah* iii. 44 The trim appearance of the Cossacks..impressed the Shahanshah tremendously. **1953** J. H. Iliffe in A. J. Arberry *Legacy of Persia* i. 31 The Shāhanshāh (King of kings) is..always aloof from his subjects. **1972** N. Gordimer *Livingstone's Companions* 5 He had been sent to Iran for the coronation of the Shahanshah. **1980** *Listener* 14 Feb. 199/3 The set book was the historic speeches of the Shahenshah published in English translation.

Shahaptan, var. *Sahaptin. **shahbandar,** now the usual spelling of Shabunder in Dict. and Suppl.

‖ **Shahbanu** (ʃā·bānu). Also **Shahbanou, Shahr-banu.** [Pers., f. *šāh* Shah + *bānū* lady.] The title of the wife of the Shah of Iran (Persia).
1915 P. M. Sykes *Hist. Persia* II. xlvii. 44 It is universally believed by Persians, that Husayn married the daughter of Yezdigird, who is known throughout Persia as 'Shahr-bánu' or the 'Queen'. **1975** *Vogue* Dec. 149 The Shahanshah and Shahbanou and the Imperial Family. **1976** *Times* 6 Apr. 9/1 (*caption*) H. I. M. Farah Pahlavi, Shahbanu of Iran. **1980** J. Cartwright *Horse of Darius* xi. 155 A stiff family group..with the Shah.., the Shahbanu and the younger children.

shaheed, var. *shahid. **Shahenshah,** var. *Shahanshah.

‖ **shahid** (ʃā·īd). Also **shaheed.** [Arab. *šahīd* witness, martyr.] A Muslim martyr.
1881 *Calcutta Rev.* LXXVII. 74 The martyrs of the new Indian religion, known by the Musalman name *shahid*, were to have their exceeding great reward in a future state. **1934** *Encycl. Islam* IV. 259/1 The Muslim who falls on the battlefield is called *Shahid*..'witness, martyr'. **1967** P. M. Hubbard *Custom of Country* (1969) vii. 87 All Pakistanis killed fighting the Indians were *shaheeds*, martyrs of Islam. **1977** *Bangladesh Times* 20 Jan. 8/3 Shaheed Asaduzzaman, a student of Dacca University, was one of the many shaheeds who laid down their lives during the mass movement of 1969.

Shahinshah, Shah-in-Shah, vars. *Shahanshah.

‖ **shahnai** (ʃa·nai). Also **shannai, shehnai.** Pl. same. [ad. Hindi and Urdu *šahnāī*, f. Pers. *šāhnāy*.] An Indian wind instrument of the oboe class.
1914 A. H. Fox Strangways *Music of Hindostan* i. 46 In the temple at Madura I heard the *nāgasāram* (N. India *shahnai*), a kind of oboe with a very loud tone. **1957** *New Oxf. Hist. Music* I. iv. 223 Among the imported wind-instruments we may count the different varieties of the oboe class, such as *shannai* (*surnahi*) which spread from the Near East across continents and to the far islands of the Indonesian archipelago. **1967** *Evening Standard* 20 Sept. 10/2 There has also been this extraordinary blossoming of Indian music... You are now expected to know about Bismillah Khan and his shehnai. **1969** R. Shankar *My Music* i. 40/2 The oboelike *shahnai* ..is thought to be an auspicious instrument and is often played..to celebrate..marriage. **1971** *Illustr. Weekly India* 18 Apr. 27 No Mendelssohn's Wedding March for her. She must have *shehnai* and Gujarati songs! **1981** Ld. Harewood *Tongs & Bones* xii. 195 The two greatest woodwind players I ever heard—Heinz Holliger, the Swiss oboist, and Bismillah Khan, the Indian *shahnai* player.

Shahr-banu, var. *Shahbanu. **shaikh,** var. Sheikh. **shaikha(h,** var. *sheikhah. **shaikhdom,** var. Sheikhdom. **shaitel,** var. *sheitel. **Shaivism, Shaivite,** vars. *Saivism, *Saivite.

shake, *sb.*[1] Add: **II. 2. h.** (Later examples.) Also, *in three* (or *two*) *shakes of a sheep's* (or *lamb's*) *tail*, (*in*) *half a shake*.
1858 S. A. Hammett *Piney Woods Tavern* xxiv. 260 Out come my mare, and in a couple of shakes of a sheep's tail we was a doin' our three minits jest as fine as silk. *Ibid.* xxvi. 283 In hafe a shake Bingham broke through 'em. **1867** G. W. Harris *Sut Lovingood* 113 Pat tuck me at my word, an' wer outen site in the shake ove a lamb's tail. **1884** 'Mark Twain' *Huck. Finn* xli. 414, I says to myself spos'n he can't fix that leg just in three shakes of a sheep's tail, as the saying is? spos'n it takes him three or four days? **1902** E. Nesbit *Five Children & It* ii. 51 He'll be ready in a brace of shakes, he says. **1934** N. Scanlan *Tides of Youth* 117 Half a shake—any more beer? **1936** W. Greene *Death in Deep South* II. 93 If you boys will just hold your horses, I'll have a statement for you. Harmon's typing it now. It'll be ready in a shake. **1958** J. Wain *Contenders* xii. 265 In two shakes he's solved the problem. Or shaken it anyway. **1966** *Guardian* 29 July 8/7 Then they are off again... I nearly wrote 'in two shakes of a lamb's tail'. **1973** E. Lemarchand *Let or Hindrance* xii. 140 I'll knock you up bacon and eggs in a brace of shakes.
i. *a fair shake*: a fair deal. Also, *an even shake, a good shake*, and opp., *an unfair shake*. U.S. slang.
1830 *Central Watchtower & Farmer's Jrnl.* (Harrodsburg, Kentucky) 22 May 1/3 Says I, any way that will be a fair shake. **1902** S. E. White *Blazed Trail* xxxi. 218 'That ain't a fair shake,' cried the man excitedly. **1949** E. B. White *Let.* 20 Nov. (1976) 315 *The New Yorker* disagrees with practically everything Boyer believes in... Nevertheless, it has given Boyer a fair shake. **1969** L. G. Arthur in A. E. Wilkerson *Rights of Children* (1973) x. 124 What does the child receive in return:..just psychiatric screen, shiny tiled walls, and electronic listening. It doesn't seem an even shake! **1972** *Time* 17 Apr. 33/1 The Administration took office..expecting an unfair shake. **1976** M. Machlin *Pipeline* xix. 243 What about the natives? They're not getting such a good shake. **1980** in S. Terkel *Amer. Dreams* 341 I'd like to see an America where so much power was not in the hands of the few. Where everybody'd get a fair shake.
j. orig. *Naut.* An act of shaking a sleeper to rouse him. Also *fig.*, a morning call.
1933 P. A. Eaddy *Hull Down* 49 If I'm asleep give me a shake at eight bells. **1945** 'Tackline' *Holiday Sailor* iv. 47 A shake. Another shake. I peer muzzily down at Gordon's upturned face. 'Quarter to four, Smiler, if you want any tea.' 'Uh? Oh, righto—thanks. I'll be up.' **1979** D. Gurr *Troika* vii. 43 The knocking intruded slowly into consciousness. The room was dark, although my shake was for six-thirty.
k. A party, *esp.* a rent party. *U.S. slang.*
1946 [see *percolator c]. **1956** S. Longstreet *Real Jazz Old & New* xvi. 126 Depression came... You could always wrassel up a piano and get together to listen and charge a few coins and have a skiffle. Or, as some said, a rent party, or a shake. **1977** *Amer. Speech 1975* L. 65 *Shake*.., party. 'There's a shake at Jim's house.'
3. b. For 'only *U.S.*' read '*U.S.* and *N.Z.*' and add further examples.
1845 E. J. Wakefield *Adventure in N.Z.* II. xv. 368 The most severe earthquake occurred that I had yet felt... The natives..acknowledged that they had never experienced so bad a *ru*, or 'shake'. **1929** 'E. Milton' *Love & Chiffon* 219 In good old New Zealand, you'd realize these shakes are mere nothings. **1948** J. Courage in *Landfall* II. 298 The earthquake happened late..but the shake woke Mr Blakiston immediately. **1949** *Los Angeles Times* 14 May 1/4 Newspaper and police switchboards were flooded immediately with requests for information on the shake.
4. a. *the shakes* (later examples).
1966 M. Woodhouse *Tree Frog* xii. 93 It was like getting the shakes on an exposed pitch of rock. **1976** B. Bova *Multiple Man* xiii. 135 The sliding glass doors.. were locked... So I sat around and waited, trying not to get the shakes.
b. *the shakes* (further examples of sense 'delirium tremens').
1927 *New Republic* 9 Mar. 72/1 The following is a partial list of words denoting drunkenness now in common use in the United States..to have the shakes. **1947** A. Marshall in *Coast to Coast 1946* 177 The longest bender I ever had was eight months. It took me three years to get over the shakes it gave me. **1977** *New Yorker* 3 Oct. 40/1 Have you ever had the D.T.s? The shakes?
d. A shaking movement in a dance; *the Shake*, a dance characterized by such movements.
1946 [see *bump *sb.*[1] 1 f]. **1956** B. Holiday *Lady sings Blues* (1973) iv. 41 The Cotton Club—a place Negroes never saw inside unless they played music or did the shakes or shimmies. **1962** *Guardian* 31 Dec. 5/1 The Madison threatens to become compulsory dancing; creeping up behind it come the Slop, the Shake, the Waddle,.. the Bossa Nova. **1966** [see *jerk *sb.*[1] 2 f].
6. b. *to give* (someone) *a* or *the shake* or *the cold shake*: to cold-shoulder, rebuff; evade, escape. *U.S.*
1875 E. Eddy *Let.* 29 Oct. in J. F. Daly *Life A. Daly* (1917) xxv. 215, I desire to give the 'Two Orphans' a shake. **1883** 'Mark Twain' *Life on Miss.* iii. 33 None of them herded with Dick Allbright. They all give him the cold shake. **1930** D. Runyon in *Collier's* 1 Feb. 13/3 Although I give her..all my affection, she will probably give me the shake. **1970** N. Fleming *Czech Point* (1971) viii. 106 If these jokers want to tail us, they've damn well got to do it properly from behind. Overtake and give them the shake.
7. (Later examples.)
1913 D. H. Lawrence *Sons & Lovers* x. 266 You think you're terrific great shakes, and that you live under the eternal insult of working in a factory. **1939** *Sun* (Baltimore) 8 Nov. 6/8 Women feel..that, no matter what poor shakes of wives they are, their husbands are blessed beyond their deserts in getting them. **1948** G. H. Johnston *Death takes small Bites* vii. 159 He couldn't have been any great shakes as a driver because he didn't beat you by much. **1970** H. McLeave *Question of Negligence* xiii. 191 I'm no great shakes at this modern dancing. **1976** *Daily Mirror* 18 Mar. 2/3 Sir Richard may not have been particularly great shakes. But he was never given much chance to show his paces.
III. 10. b. Also *sing.* when *attrib.* or *comb.* (see sense 13 below). Occas. also used outside the U.S. (Earlier and later examples.)
1772 Tillinghast & Holroyd *Let.* 23 Nov. in *Commerce of Rhode Island* (1914) I. 420 We herewith send you all the Shakes we can yet get in. **1939** I. Baird *Waste Heritage* xviii. 240 Weathered barns with the lichen growing on the shakes. **1964** L. Linton *Of Days & Driftwood* ix. 51 The first place of worship was a very small building of shakes. **1977** *Tel.* (Brisbane) 20 Dec. 36/5 Shakes are hand split and have a rustic appearance. **1982** *Times* 26 Jan. 11/3 The design of these shakes was identical to that of the traditional oak shakes used in England for many centuries.
12. (Examples.)
1948 D. Ballantyne *Cunninghams* I. xxix. 146 You sat..sucking raspberry shakes through straws. **1953** [see *parfait]. **1966** B. H. Deal *Fancy's Knell* iii. 44 'I'll have a burger too,' the redhaired boy said. 'And a shake.' **1981** J. D. MacDonald *Free Fall in Crimson* x. 114 She sucked up the shake.
IV. 13. (sense 10 b) *shake cabin, house, roof, shanty*; *shake-maker*; *shake-sided* adj.; **shake dancer** *slang* (see quot. 1968); so (as a back-formation) **shake dance**; **shake music** (see quot. 1942); **shake wave** = *S wave* s.v. *S 6.
1885 L. W. Spring *Kansas* v. 64 Big Springs in the autumn of 1855 was a place of four or five shake-cabins and log-huts. **1967** M. Craven *I heard Owl* (1968) v. 36 Old Marta was there and the girl called Keetah, and the two small children.., come from the shake cabins to pick blueberries. **1956** B. Holiday *Lady sings Blues* (1973) x. 98 A shake dancer with her pimp. **1968** J. Lock *Lady Policeman* xi. 102, I was..assigned..to the women entertainers. They were known as 'shake dancers'... The art consisted of shaking bare or almost bare breasts to music. *Ibid.*, Her daughter..did bare-breasted shake dances. **1976** *National Observer* (U.S.) 2 Oct. 21/5 She becomes a shake-dancer and B-girl. **1857** *Lawrence* (Kansas) *Republican* 9 July 3 You are always welcome to his log or shake house. **1901** J. Muir *Our National Parks* ix. 298, I found many shake-makers at work in it, access to these magnificent woods having been made easy by the old mill wagon road. **1935** Shake music [see *jungle *sb.* 3 c]. **1942** Berrey & Van den Bark *Amer. Thes. Slang* § 579/1 *Syncopated music; jazz*... Shake music, a savage style similar to 'jungle music'. **1947** *Michigan Hist.* June 178 It was a small log cabin with a shake roof. **1978** J. Hyams *Pool* vi. 68 It had..a steep moss-covered shake roof that turned green in the rain. **1879** *Atlantic Monthly* Aug. 154/1 Every one of the frail shake shanties is a centre of destruction. **1970** J. Hansen *Fadeout* i. 8 The overhang of a shake-sided cabana. [**1929** H. Jeffreys *Earth* (ed. 2) vi. 86 The type (9) are called the longitudinal, irrotational, condensational, primary, or *P* waves; (10) and (11) the transverse, distortional, equivoluminal, secondary, or *S* waves. Prof. H. H. Turner has very appropriately called them the 'push' and the 'shake'.] **1944** A. Holmes *Princ. Physical Geol.* xvii. 369 The S or 'shake' waves are distortional waves, in which each particle vibrates at right angles to the direction of propagation. **1969** *Daily Tel.* 2 Sept. 12 Seismic records obtained so far from the landing site show the absence of the so-called 'shake-waves'.

shake, *sb.*[2] (Earlier example.)
1846 *Swell's Night Guide* 36 Many of the Haymarket shakes frequent this lumber.

shake, *v.* Add: **III. 5. b.** *more than you can shake a stick at* (and vars.): more than one can count, a considerable amount or number. *colloq.* (orig. and chiefly *U.S.*).
1818 *Lancaster* (Pa.) *Jrnl.* 5 Aug. 3/1 We have in Lancaster as many Taverns as you can shake a stick at. **1843** R. Carlton *New Purchase* I. xii. 86 Our queen snake was..retiring, attended by more of her subjects than we even dared to shake a stick at. **1904** J. C. Lincoln *Cap'n Eri* iv. 56 There's more Snows in Nantucket than you can shake a stick at. **1939** L. M. Montgomery *Anne of Ingleside* xxi. 137, I had more beaux than you could shake a stick at. **1960** 'E. McBain' *Give Boys* (1962) iv. 32 We get more damn cancellations than you can shake a stick at. **1982** *Folio* Spring 4 More consuls and dictators hanging from her family tree than a prudent man would shake a bundle of twigs at.
6. f. *to shake a foot* (earlier example), *hoof* (U.S.), *leg* (*b*) to hurry; *to shake it*: for † read *obs.* exc. *U.S. Blacks'*; also, *to shake that thing*.
1848 *Buffalo Gals* (song) 3, I ax'd her would she hab a dance... I taught dat I might get a chance, To shake a foot wid her. **1904** *N.Y. World Mag.* 1 May 6/3 Shake a leg..meaning to 'hurry up'. **1927** *Jrnl. Abnormal & Social Psychol.* XXII. 16 'Shake it', 'shake that thing', etc. Such expressions are very frequent in the blues.

Ostensibly they refer to dancing, but they are really Negro vulgar expressions relating to coitus. 1927 S. Lewis *Elmer Gantry* xxv. 333 Come on, Reverend. I bet you can shake a hoof as good as anybody! The wife says she's gotta dance with you! 1935 F. M. Davis *Black Man's Verse* 34 Strut it in Harlem, let Fifth Avenue shake it slow Plink plank plink a plink. 1952 Wodehouse *Barmy in Wonderland* viii. 82 'Clean the place up.' 'Yes, sir.' 'And shake a leg.' 1967 M. C. Melnick in A. Dundes *Mother Wit* (1973) 273/1 If you shake it, I'll buy you a diamond ring.

9. a. (a) fig. *to shake hands with an old friend, the wife's best friend* (colloq.), of men: to urinate.
1952 M. Tripp *Faith is Windsock* iii. 44 'I'm going out for a crafty smoke; anyone coming?' 'Sure, I'll come.. I want to shake hands with an old friend, anyway.' 1965 *Times Lit. Suppl.* 16 Sept. 812/2 Expressive Australianisms to describe this prosaic function.. pointing Percy at the porcelain, shaking hands with the wife's best friend, [etc.].

c. For 'Now only *U.S. slang*' read 'Now chiefly *U.S.*' and add further examples.
1911 M. Beerbohm *Zuleika Dobson* xv. 218 'Are you going to die to-day, or not?' 'As a matter of fact, I am, but—' 'Shake!'.. Oover wrung the Duke's hand. 1927 *Punch* 20 Apr. 444/3 'Long may it flourish!' said Roger, shaking vigorously. 1938 M. K. Rawlings *Yearling* vii. 61 'You got to promise.. not to beat the very puddin' outen me after you've hunted him.' 'Shake.' A hairy paw closed over Penny's hand. 1966 'J. Hackston' *Father clears Out* 109 Tom.. said, almost benignly, 'Now shake!' and they shook,.. in the true spirit of eternal friendship. 1972 J. Gores *Dead Skip* viii. 55 He stood up, stuck out his hand.. They shook.

11. b. Also, to upset the composure or complacency of (someone). *colloq.*
1943 C. H. Ward-Jackson *It's a Piece of Cake* 54 'That'll shake him,' as the transport officer said when he refused to provide a vehicle for the Group Captain without written authority. 1966 *New Yorker* 25 June 52 It shook me some when I looked at the label.

d. (Further examples.) Also *N.Z.* Also, less strongly, to be keen on, to be impressed by, to admire, and const. *after*.
1907 H. Lawson in *Austral. Short Stories* (1951) 84 The trouble is that I'm so long, and I always seem to get shook after little girls. 1926 K. S. Prichard *Working Bullocks* 301 Didn't know she was so shook on Mark Smith. 1934 L. G. D. Acland in *Press* (Christchurch) 27 Jan. 15/7 *Shook on*, to be, to admire; to be keen on; e.g., 'I'm not so shook on his horse.' 1940 F. Sargeson *Man & his Wife* (1944) 22 Mother wasn't too shook on our doing it at first, but afterwards she didn't mind. 1947 D. M. Davin *Gorse Blooms Pale* 78, I wasn't as shook on Phyllis as all that. 1965 M. Shadbolt *Among Cinders* xiii. 209 The bush. Still not too bloody shook on it, are you? 1975 *Sunday Tel.* (Sydney) 29 June 49 I'm not all that shook on cocktail parties myself.

12. c. *to shake down* intr. and refl., to settle down, to accommodate oneself to circumstances, a condition, position, etc.
1864 C. M. Yonge *Trial* II. x. 178 Mr. Cheviot, as the family shook down together, became less afraid of Ethel. 1875 Trollope *Prime Minister* (1876) I. vii. 109 You'll find they'll shake down after the usual amount of resistance and compliance. 1916 'Taffrail' *Pincher Martin* i. 4 You needn't look so scared. You'll soon shake down. Is this your first ship? 1959 *Times* 31 May 11/2 By the time a new American President has.. shaken himself down in the White House, the West German federal elections.. will be approaching. 1973 *Times* 26 Apr. 17/1 Agricultural prices.. produced.. several sharp clashes of interest. That is an integral part of the process of shaking down. 1980 R. Adams *Girl in Swing* (1981) xvii. 190 And how is the beautiful Karin? Is she shaking down nicely in England?

13. b. (Later example.)
1931 Galsworthy *To Let* vii. 181 Impressions of the United States, whose dust he had just shaken from off his feet—a country.. so barbarous in every way.

c. For *U.S.* read 'orig. *U.S.*' and add to def.: Also, to give (a person) the slip; to jilt; occas. to abandon (a place); to shake off (an illness, feeling, etc.). (Earlier and further examples.)
1872 'Mark Twain' *Roughing It* xlvi. 336 He never shook his mother... No indeedy.. he looked after her and took care of her. 1874 (song) *I'll Never Get Drunk Any More*, Chorus: The pledge I will take, the whiskey I'll shake, Oh I'll never get drunk any more. 1884 'Mark Twain' *Huck. Finn* xxxi. 323 That little rascal has stole our raft and shook us, and run off. 1893 Kipling *Seven Seas* 96 We've shaken the Clubs and the Messes To go and find out and be damned. 1907 R. W. Service *Songs of Sourdough* 13, I was all caked in on a dance-hall jade, but she shook me in the end. 1934 in J. A. & A. Lomax *Amer. Ballads & Folk Songs* xx. 459 She shook me for the driver. 1935 M. de la Roche *Young Renny* iv. 28 'He was paying us a visit and the time went on and—he just came with us.' 'You mean you couldn't shake him?' 1949 R. Chandler *Let.* 21 Mar. (1981) 157, I'm going down to Palm Springs for a week to try to shake this cough. 1953 'S. Ransome' *Drag Dark* (1954) xiii. 131 We deliberately shook you that night, then tailed you back here. 1958 J. Kerouac *On Road* v. 34 'I have a date with my boy friend.' 'Can't you shake him?' 1965 V. Canning *Whip Hand* v. 51 If anyone was following you must have shaken them. 1972 'T. Coe' *Don't lie to Me* (1974) vii. 75 The picture of the murderer stayed in my head... Trying to shake it, trying to shake the mood it was giving me, I searched for other things to think about. 1974 'J. Ross' *Burning of Billy Toober* x. 97 If you don't shake it [sc. heroin], it'll kill you in the end. 1977 *Rolling Stone* 16 June 34/4 Blauer had admitted himself to a New York state hospital hoping to shake a debilitating depression. 1979 'S. Woods' *This Fatal Writ* 129 If you know you're being followed, it isn't too difficult to shake a tail.

16. c. *to shake down*, to extort money from, to blackmail or otherwise pressurize (a person) for (occas. of) money, etc. *slang* (orig. and chiefly *U.S.*).
1872 G. P. Burnham *Mem. U.S. Secret Service* p. viii, Sharps, 'come down', 'shake down'; to extort money from individuals. 1916 J. London *Let.* 12 Oct. (1966) 473 'Uncle Charley'.. then proceeded to shake you down in proper money-lender.. fashion. 1922 J. Barbican *Confess. Rum-Runner* xiv. 148 For only last week they were shook down for five hundred by a stray fellow from the Department. 1949 *Los Angeles Times* 3 Jan. 1/3 Ferguson.. accused them of trying to 'shakedown' Mickey Cohen of $5000. 1958 H. Kurnitz *Invasion of Privacy* vii. 54 'You weren't by any chance trying to shake him down?'.. 'No, sir. Not a penny.' 1966 T. Pynchon *Crying of Lot 49* ii. 28 He left after shaking her down for four bits for carrying the bags. 1976 'J. Ross' *I know what it's like to Die* xxii. 144 Sickert had been shaken down for protection money.

d. *to shake down*, (esp. of police, etc.) to search (a person or place). *slang* (orig. and chiefly *U.S.*).
1915 *N.Y. World* 9 May (Suppl.) 14/1 *Frisk*, to shake down or search. 1955 D. W. Maurer in *Publ. Amer. Dialect Soc.* xxiv. 46 They.. shook down my hotel. 1968 *Listener* 15 Feb. 210/1 Inmate guards have been in complete control of the prison. They.. shook down incoming prisoners to take radios and watches and so on. 1977 D. Bagley *Enemy* xvii. 142 Once Mayberry had been shaken down, the guards were taken from Ryan and Gilliam. 1979 D. Anthony *Long Hard Cure* xiv. 198 The Sony had been in plain sight.. Billy Combs was shaking down the rest of the house.

21. g. (Earlier example.)
1884 'Mark Twain' *Huck. Finn* vi. 40 People allowed there'd be another trial to.. give me to the widow for my guardeen... This shook me up considerable, because I didn't want to go back to the widow's.

IV. 22. shake-hands (earlier example).
1800 F. Burney *Let.* 18 July (1973) IV. 425 William will be much pleased by a private congratulatory *shake hands* from you in his own Apartment.

shake-down. Add: **2. b.** A forced contribution; an instance of extortion. Cf. *Shake v. 16 c. orig. and chiefly *U.S.*
1902 in *Dict. Americanisms* (1951) s.v. *shake*, To the historic phrase 'blackmail'.. have been added, as words of similar evil omen, the new and expressive terms shake-down and rake-off. 1903 A. H. Hodder *Fight for City* iv. He [sc. a New York policeman] was fined 30 days' pay because he would not stand for a 'shake-down', which means that he had refused to give from time to time upon demand $5 or 10 dollars.. to his superiors to be used for purposes unknown. 1916 J. London *Let.* 12 Oct. (1966) 473 A usurer.. stuck out because.. he saw the shake-down of me would not go through. 1942 *Sun* (Baltimore) 30 Mar. 11/7 Jack Pollack was named.. as the man behind a demand for a $2,500 'shakedown' to kill a liquor license bill in the Legislature. 1978 S. Brill *Teamsters* xix. 329 While the shakedown was proved, it was never shown that the money went to Presser personally.

c. A search of a person or a place. Cf. *Shake v. 16 d. orig. and chiefly *U.S.*
1914 Jackson & Hellyer *Vocab. Criminal Slang* 75 *Shake down*, noun. General currency. A personal search; a deprivation of one's personal belongings... Example: 'If this dick nails you you'll have to stand a shake down.' 1936 *Sun* (Baltimore) 31 Jan. 3/5 A sudden 'shakedown' of the Stateville Penitentiary resulted in the seizure of several knives. 1958 *Landfall* (N.Z.) XII. 18 Just about nine o'clock, without any warning, there was a shake-down (of prisoners). 1977 D. Bagley *Enemy* vii. 53, I really wanted.. to give Ashton's study a good shake-down. But.. it's bad form.. to be found searching through your host's private papers.

3. *attrib.* (sense 2 a) *shakedown test*; (sense *2 b), as *shake-down racket, scheme*; (sense *2 c), as *shake-down party*; **shake-down cruise** *Naut.* (orig. *U.S.*), a cruise designed to test a newly-launched ship and its equipment and to train its crew; also *fig.*; similarly **shake-down flight** *Aeronaut.*
1927 C. Bradford *Gloss. Sea Terms* 156/1 *Shakedown cruise*, one for.. the purpose of adjusting machinery and instruments, and familiarizing a crew with a new vessel. 1933 *Sun* (Baltimore) 21 Jan. 16/7 The newest addition to the navy.. is on its 'shake-down' cruise. 1968 *Wall St. Jrnl.* (Eastern ed.) 13 Sept. 1/1 The first week on the road.. has been a shakedown cruise for the Senator and his staff—enabling the candidate to test stump themes,.. giving aides experience at writing speeches and organizing motorcades, [etc.]. 1978 P. O'Donnell *Dragon's Claw* ii. 27 The boat's had a shakedown cruise and.. it 'andles beautifully. 1934 D. Runyon in *Collier's* 24 Nov. 52/2 He is only going back to his old shake-down dodge, so all you have to do is to buy him off. 1939 *Sun* (Baltimore) 25 Mar. 20/3 The 10,000-mile shakedown flight will require about two weeks. 1952 *Here & Now* (N.Z.) II. iv. 32 The shake-down parties have been through the cells pretty often. 1976 *Honolulu Star-Bull.* 21 Dec. A-2/4 The mayor's own press secretary was convicted in a shake-down scheme. 1942 *R.A.F. Jrnl.* 2 May (recto rear cover), To send them into actual service.. without modification or further shakedown test.

sha·ke-out. [f. vbl. phr. *to shake out*: see Shake v. 20.] **1.** *Stock Exch.* (See Shake sb.[1] 8 a.) Also, a sudden fall in prices, a sudden general disposal of particular stocks, etc.
1928 *Sun* (Baltimore) 7 Dec. 1/3 Measured by the Associated Press averages of twenty leading industrials and twenty leading rails, which dropped $9.45 and $4.38, respectively, it was one of the quickest and most drastic shakeouts in recent market history. 1981 *Times* 14 Aug. 18/3 Properties came in for a small shake out with Stock Conversion a weak market 10p lower at 379p.

2. An upheaval or reorganization, esp. one involving contraction, streamlining, shedding of personnel, closure of some businesses, etc.
1969 *Times* 5 Mar. 8/1 There had been what was sometimes called a 'shake-out' in the film industry during the past year. A number of those elements which did not raise the repute of the film industry had been removed. 1956 *Sun* (Baltimore) 20 Jan. 13/1 New claims for jobless pay increased about 20 per cent in Maryland last week under the impetus of the usual year-end economic 'shake-out'. 1957 *Time* 2 Sept. 59/1 In downtown Washington, D.C., eight, or about half, of the city's big discount houses went out of business in the past year. The shake-out is almost as severe in Los Angeles, Boston and Dallas. 1963 *Listener* 21 Feb. 329/2 Public-house people is perhaps most busy about the need for a shake-out of the party system. 1964 *Financial Times* 3 Mar. 12/7 A shake-out in the business world, with pressure on profits and profit margins forcing the inefficient producer, or.. retailer, out of business or into the arms of the stronger. 1967 *Listener* 19 Jan. 80/1 A nation-wide witch-hunt and counterbalancing resistance movement which could well make the upheaval of the past year [in China] seem like the mildest of shake-outs. 1974 *Howard Jrnl.* XIV. 19 Successive recessions and mechanization have meant a 'shake-out' of labour in traditionally labour-intensive industries. 1981 *Economist* 28 Nov. 26/1 Workers left in droves, because they were laid off. The worst of that shake-out is over.

shaker. Add: **1. a.** *mover and shaker, shaker and mover* (U.S.), a person who influences events, a person who gets things done.
1874 A. O'Shaughnessy *Music & Moonlight* 1 Yet we are the movers and shakers Of the world for ever, it seems. 1972 F. Knebel *Dark Horse* (1973) ii. 14 The rich movers and shakers.. always manage to manipulate the Congress for their own benefit and screw the rest of us. 1975 J. F. Burke *Death Trick* iv. 61 Beniamino Tucci was.. known as the Little Godfather of the Upper West Side.. He was a mover and shaker with many interests. 1977 *Time* 10 Oct. 13/2 Perish the thought that a shaker and mover should work for the Government.

2. b. A simple percussion instrument that is shaken; *spec.* = Cha-chac. Cf. *shack-shack, etc.
[1837 I. M. Belisario *Sk. Negro Pop. in Jamaica* (caption to Plate 7), Shaka, a rattle used by the French Set Dancers.] 1943 *Penguin New Writing* XVIII. 96 He finished with a flourish of the shakers and threw his drumstick into the air. 1958 E. Borneman in P. Gammond *Decca Bk. Jazz* xxi. 275 A dale leader and a small group.. who accompanied themselves on.. shakers and guitar. 1978 A. Matthew *Play with a Purpose for Under-Sevens* ix. 145 Shakers can be made from a wide variety of empty containers to make a noisy drumstick. 1978 S. Dickinson *Mother's Help* iii. 43 Shakers, or rattles, are probably the easiest instrument of all to make.

4. b. *attrib.* (Earlier and further examples, esp. of artefacts produced by or of a type produced by Shakers.)
1817 *Niles' Reg.* XII. 371/1 At Enfield, Vermont, he visited the 'Habitation of the Shakers [sic] community,' to see their own phraseology, and to mimic familiar language the Shaking Quakers. 1856 in C. C. Richards *Village Life in Amer.* (1912) 77 We went down town this morning and bought us some shaker bonnets to wear to school. 1863 *Trans. Illinois Agric. Soc.* (1865) V. 456 Your committee would.. suggest that they put on their.. shaker bonnet. [etc.] 1947 T. Morris *Amer. Angler's Bk.* xiii. 371 Two or three pairs of stout yarn socks ('Shaker' socks are best). 1866 A. D. Whitney *Summer in Leslie Goldthwaite's Life* vi. 92 Do you think this good Shaker rocking-chair. 1869 *Montgomery Ward Catal.* 561/3 Baby Carriage Robes. Shaker flannel, pinked edge and embroidered corners. 1928 *Antiques* XIV. 124 A study of the characteristic forms of Shaker furniture suggests the hypothesis that the early craftsmen adapted to their own designs existing Colonial models. 1967 D. Sheridan *I was following this Girl* iii. 20 Early American Engraving up this way. Early Shaker Woodwork in the Brook Street Foyer. 1975 J. Gores *Hammett* i. 11 He wore a maroon worsted Shaker coat over a wool shirt. 1978 *Jrnl. R. Soc. Arts* CXXVI. 305/2 The Fraktur and Shaker artists, many of whose works.. seem to have anticipated Paul Klee.

5. d. Substitute for def.: A container in which cocktails or other mixed drinks are blended by shaking. Freq. as the second element of a comb. (Earlier and later examples.)
1868 [see *cocktail shaker]. 1895 *Montgomery Ward Catal.* 435/1 Liquor Mixers or Lemonade Shakers of tin. 1922 S. Lewis *Babbitt* viii. 110 He did not possess a cocktail-shaker. A shaker was proof of dissipation. 1929 E. Linklater *Poet's Pub* ii. 39 Holly poured his chosen liquors into a long silver shaker, added broken fragments of ice, screwed down the top, and, like a man with the palsy, shook. 1946 'P. Quentin' *Puzzle for Fiends* v. 35 She carried a small shaker of Manhattans. 1959 A. W. Sherring *Tip Off* xiii. 135 Big Boy Gale watched them in the bar mirror as he poured drinks from a shaker. 1971 *Scope* (S.Afr.) 19 Mar. 77/2 Many people consider the shaker and mixer as being the same thing, but there is a considerable difference: cocktails with clear ingredients are prepared by stirring in a mixer, cloudy liquids are agitated in a shaker.

f. *U.S.* = Castor[2] 1.
1910 J. W. Tompkins *Mothers & Fathers* 29 Miss Elsie would be terribly shocked at this shaker. 1969 J. A. McPherson in A. Chapman *New Black Voices* (1972) 153

Shouldn't you polish the shakers or clean out the Pantry or squeeze oranges? 1978 S. Brill *Teamsters* vii. 284 The small formica booth table with the mini-juke box built into the wall just above the salt and pepper shakers.

9. Short for *Shaker bonnet*: see sense 4 b above. *U.S. local*.
1858 M. D. Colt *Let.* in *Went to Kansas* (1862) xiii. 238, I did not wear the green silk calash, but a shaker, made of brown muslin smoothed over a pasteboard frame; it was very fashionable; besides it kept the sun out of my face, and was very genteel for a school-ma'am. 1881 *Harper's Mag.* May 854/2 The bonnet is far too fine. I had a shaker she got at the store. 1905 K. D. Wiggin *Rose o' River* 9 Young Shaker Sisters, in close-fitting 'Shakers', mourning bonnets for aged dames, [etc.]. 1909 *Dialect Notes* III. 413 *Shaker*, a palm leaf sunbonnet.

shakerism (earlier example).
1828 *Catholic Vindicator* 5 Dec. 41 Anabaptism or independentism, quakerism, or shakerism.

shakerful (ʃēi·kəfʊl). [f. Shaker 8 d + -ful.] The contents of a (cocktail) shaker; the amount that a shaker will hold.
1928 H. E. Hodgins *Mr. Blandings builds his Dream House* vii. 93 A second shakerful of still paler Martinis. 1966 T. Pynchon *Crying of Lot 49* i. 15 Gliding like a large bird he dispensed a shakerful of sweating booze. 1977 [see *screwdriver 3].

Shakespeare (ʃēi·kspīə). [The name of William *Shakespeare*: see Shaksperian a. (and sb.).] **1.** A person (occas. a thing) comparable to Shakespeare, esp. as being pre-eminent in a particular sphere.
1828 M. Edgeworth *Let.* 23 Oct. (1971) 243 Humboldt is the Shakespeare of travellers—as much superior in genius to other travellers as Shakespeare to other poets. 1859 A. J. Munby *Diary* 17 Mar. (1972) 28 When.. the poetic soul.. has learnt.. to see the poetic side of all such things, then we may have a Homer of the railway and a Shakespeare of the Ballot. 1905 'Mark Twain' in *N. Amer. Rev.* Jan. 3 The telegraph, the telephone.. the Pullman car.. the Shakespeares of the inventor-tribe, so to speak. 1931 R. Campbell *Georgiad* ii. 36 A Fabian Shakespeare of the Summer Schools To other poets laying down my rules.

2. *attrib.*, as **Shakespeare collar**, (a) = *polo collar* s.v. *polo[1] 3; (b) (see quot. 1960); **Shakespeare country**, the part of Warwickshire around Stratford-on-Avon, birthplace of Shakespeare; **Shakespeare industry**, the large-scale production of writings about Shakespeare, items commemorating Shakespeare, etc.; the commercial exploitation of objects, places, etc., associated with Shakespeare.
1907 *Yesterday's Shopping* (1969) 873/1 Cotton Football Shirts.. Shakespeare collar, and three buttons. 1913 [see *polo collar* s.v. *polo sb.[1] 3]. 1960 C. W. Cunnington et al. *Dict. Eng. Costume* 192/1 *Shakespeare collar*. 1860's on. A shallow turn-over collar, the points projecting downwards onto the shirt-front. 1900 J. Leyland *Shakespeare Country* 92 This survey of Shakespeare Country has traversed a rich district of middle England that was familiar to the great poet in his boyhood. 1966 J. Wainwright *Crystallised Carbon Pig* iv. 20 The plan worked.. as smoothly.. as an American tourist's trip through the Shakespeare country. 1972 *Times* 4 Aug. 4/2 The 'Shakespeare country' around Stratford-upon-Avon, Oxford and Cambridge are the main non-metropolitan attractions. 1939 Brown & Fearon (*title*) Amazing monument: a short history of the Shakespeare industry. 1958 *Listener* 2 Oct. 523/1 Is not much of this book-making on the Bard another branch of the Shakespeare industry? 1962 *Observer* 4 Mar. 13/6 The 1864 affair.. marked the dawn of the Shakespeare industry. At Stratford.. 'streets were adorned with flags and banners; the townsfolk and visitors wore the.. Shakespeare badge', [etc.].

Hence **Sha·kespeare** *v. intr.* (*nonce-wd.*), to act in a Shakespeare play.
1896 G. B. Shaw *Our Theatres in Nineties* (1932) II. 90 Madame de Navarro has declaimed, spouted, statuesqued, Shakespeared, and all the rest of it.

Shakespearian, *a.* (and *sb.*) Now the usual form of Shaksperian *a.* (and *sb.*). Add: **a.** *adj.* (Earlier examples.)
1755 H. Fielding *Voy. to Lisbon* 100 A poetic, if not a Shakespearian genius. 1805 C. Wilmot *Let.* 4 Aug. in *Russ. Jrnls.* (1934) II. 164, I rooted out Hamlet's Garden.. & got into a *Shakespearian tantrum* at finding myself in the place.

b. *sb.* (Further examples.) Also, one who believes that Shakespeare wrote the plays usually attributed to him; an imitator of Shakespeare's style, one of his school; an admirer of Shakespeare's works.
1874 [see *Baconian a. and sb. 2]. 1912 E. Nesbit *Let.* in D. L. Moore *E. Nesbit* (1933) xv. 268 Are you a Baconian or a Shakespearean? 1930 N. Streatfeild *Ballet Shoes* iii. 38 We'll read some more one day. I'll make a Shakespearean of you. 1964 *English Studies* XLV. 353 It also establishes the negative method of praising Heywood, as a minor Shakespearian. 1971 *Daily Tel.* 8 Mar. 10/4 One of those devoted Shakespeareans who knows his author backwards. 1979 F. Kermode *Genesis of Secrecy* iv. 79 Shakespearians may find explanations of the mysteriousness.. of *Hamlet*, by considering instead the *ur-Hamlet*.

Shakespeariana (further examples); **Shakespea·rianizing** *vbl. sb.*, the action or instances of imitating passages from the works of Shakespeare; **Shakespea·rianly** *adv.*, in a Shakespearian manner; **Shakespearism** (further example); **Sha·kespearite**, one who believes that Shakespeare wrote the plays traditionally attributed to him.
1861 *Sat. Rev.* 30 Nov. 557/2 'The national pulse beats Shakespearianly.' So at least says Mr. James Orchard Halliwell in one of two circulars.. which have lately reached us about the National Shakespearian Fund. 1865 F. Timm (*title*) Shakespeare's Genius from 1564 to 1864. 1890 *Merry England* July 242 A fine, Shakespeareanly virile bit of poetry. 1903 G. B. Shaw *Let.* 12 Jan. (1972) II. 303 Ben Jonson never could quite get over the absurdity of the Shakespearisms which he knew so well at the Mermaid passing off in cold ink as literature. 1909 'Mark Twain' *Is Shakespeare Dead?* v. 50 Two of these cults are known as the Shakespearites and the Baconians... The Shakespearite knows that Shakespeare wrote Shakespeare's Works. 1921 G. B. Shaw in *John Keats Memorial Volume* 176 The lines beginning (Shakespearianly) with How fever'd is the man who cannot look Upon his mortal days with temperate blood! 1936 F. R. Leavis *Revaluation* vi. 243 *The Cenci*.. is full of particular echoes of Shakespeare... This Shakespearianizing.. is.. quite damning. 1953 *John o' London's* 12 June 520/4 A musical piece, with Arthur Askey.. Shakespeareanly disguised. 1963 J. Kerr *The Kid from Stratford*. 1955 *Times* 13 Aug. 9/5 One such foundation has for many years placed a standing order for the purchase of Shakespeareana offered at our leading sale rooms. 1964 *Economist* 11 Apr. 144/3 The appetite for Shakespeareana.

sha·ke-up. (In Dict. s.v. Shake sb.[1] 8 b.) Substitute for def.: An act of shaking up or being shaken up, or the result of this; a thorough or drastic change or rearrangement; a disturbing or unsettling experience. (Earlier and further examples.)
1847 J. S. Mill *Let.* (1910) I. 131 To give that general shake-up to the torpid mind of the nation which the French Revolution gave to Continental Europe. 1880 'Mark Twain' *Tramp Abr.* xxxviii. 438 My nerves had hardly grown quiet after this affair when they got another shake-up,—one which utterly unmanned me for a moment. 1882 *National Police Gaz.* 18 Nov. 7/2 (*heading*) The Union Square Company has a matrimonial shake-up all around. 1899 R. H. Barbour *Half-back* vi. 59 There'll be a shake-up to-morrow... He's going to put Greer on the scrub to-morrow. 1916 'Taffrail' *Pincher Martin* xvi. 312 But, orl the same, Tubby boy, I reckons it's done us orl good ter 'ave a bit of a shake up like this 'ere [sc. a naval engagement]. 1938 E. Bowen *Death of Heart* III. i. 325, I can't see that this change has done you harm. Nor the shake-up either; you were getting too quiet. 1962 E. Snow *Red China Today* (1963) xxii. 165 'Rectification', self-criticism, retraining and restudy among party and nonparty cadres are followed by shake-ups which affect millions. 1967 D. Pinner *Ritual* viii. 85 You didn't half give me a shake-up. 1969 *Listener* 27 Mar. 410/2, I read that BBC radio is due for a programme shake-up. 1970 'D. Halliday' *Dolly & Cookie Bird* iii. 35 It was sporting of your father to ask me. I can imagine what a shake-up it must have been, without taking me on as well. 1980 *Christian Sci. Monitor* (Midwestern ed.) 4 Dec. 4/1 As a result of the latest shake-up, which occurred at the party meeting, the political balance has shifted away from the conservatives.

shaking, *vbl. sb.* **1. c.** (Further examples.)
1926 J. S. Huxley *Ess. in Pop. Sci.* ix. 118 Taking them [sc. Infusoria] on a railway journey to give them a good shaking-up. 1928 *Daily Mail* 9 Aug. 12/4 The best opinion is that a thorough shaking-out will do much good, but there is no need for alarm. 1958 L. Durrell *Mountolive* vi. 134 Only Pursewarden had not put in an appearance... Mountolive planned to give him a shaking-up at the first opportunity.

shaking, *ppl. a.* **b.** (Further examples.)
1888, 1955 [see *Parkinson[1]].

Shaksperian, etc.: see also *Shakespearian *a.* (and *sb.*).

Shakti, var. *Sakti.

‖ **shaku** (ʃa·ku). Also 8 **sackf, sak, saku.** Pl. same. [Jap., ad. Chinese *chĭ* a foot.] **1.** A Japanese measure of length, equal to 11·9 inches (30·3 cm.); (see also quot. 1974).
1727 J. G. Scheuchzer tr. *Kæmpfer's Hist. Japan* I. xi. 136 One *Sackf* and a half long. *Ibid.* II. iv. 180 Snow.. to the height of four *Sak* and five Suns, that is about four foot and a half. *Ibid.* III. vi. 246 His Stature.. of nine *Saku*, and nine *Suns*, proportionable to the greatness of his Genius. 1878 *Trans. Asiatic Soc. Japan* VI. II. 249 The seismograph consisted of a copper vessel, whose diameter was 8 *shaku* or feet. 1884 tr. *J. J. Rein's Japan* ii. 415 The interval of three shaku (1 metre). 1893 E. Arnold *Adzuma* I. i. 2 He could.. run so fleetly that a cord of thirty *shaku*, tied to his waist, would stream in a straight line behind him. 1974 *Encycl. Brit. Micropædia* IX. 106/3 *Shaku*, a unit of length, area, and volume in Japan, equivalent to 10/33 metre, 3·306 square decimetres, and 18·039 cubic centimetres.

2. A flat baton made of wood or horn, a little over a foot in length, upon which a Japanese court noble formerly would note memoranda, but later carried as a mark of honour in the presence of the emperor, or by the emperor himself.
1875 F. V. Dickins tr. *Chiushingura* (1876) 208 'Twas the Emperor's whim That the tree should from him Have a *shaku* with Ta-iu writ on. 1880 *Trans. Asiatic Soc. Japan* VIII. 351 A short staff called the *Shaku*, which was generally held vertical in the right hand. 1894 C. M. Salwey *Fans of Japan* 6 The *shaku*.. was a stick in the shape of the outside frame of a folding fan, about two feet in length, about an inch and a half to two inches at the top, decreasing at the base to about one inch. 1928 *Daily Express* 12 Nov. 3/7 The Emperor, after seating himself on the throne, was presented with the small wooden baton (shaku) which is a traditional symbol of authority found in many Shinto rites.

shakudo. Add: (Earlier example.) Also *attrib.*
1860 S. B. Kemish *Jap. Empire* 114 The beautiful work called *syakfdo*, in which various metals are partly blended, partly combined, producing an effect much resembling fine enamel. 1911 *Encycl. Brit.* XV. 179/2 To apply a lining of silver to a shakudo box. 1981 G. MacBeth *Kind of Treason* ix. 91 The little black cups of the *nanako* on the *shakudo* grounds.

‖ **shakuhachi** (ʃa:kuha·tʃi). [Jap., f. *shaku + *hachi* eight (tenths).] An end-blown Japanese flute, made of bamboo.
1893 F. T. Piggott *Music & Musical Instruments of Japan* I. 43 The Shakuhachi, introduced into Japan from China by Prince Tsuneyoshi as far back as.. 1335, seems to have been treated from the first as a solo instrument. 1949 *Western Folklore* July 202 Their melodies are found to be very pleasing, especially if the *shakuhachi*, or bamboo flute, is used as accompaniment. 1965 W. Swaan *Jap. Lantern* ii. 22 Two itinerant beggar-priests playing the shakuhachi, an archaic type of bamboo flute. 1981 *Daily Tel.* 17 Mar. 11/7 The first half was devoted to traditional pieces, performed on the shakuhachi, a simple bamboo flute.

shaky, *a.* Add: **2. b.** *the Shaky Isles* (colloq.), New Zealand (from the frequency of earthquakes).
1933 *Bulletin* (Sydney) 2 Aug. 20/2 The widespread notion that they're peculiar to the Shaky Isles. 1941 S. J. Baker *N.Z. Slang* vi. 49 New Zealand was no longer merely a colony; it became *the Dominion, the Shakey Isles*,.. and so on. 1971 *Sunday Tel.* (Sydney) 16 May 37/2 He came over from the Shaky Isles in his early 30s.

5. e. *shaky do* [*do sb.[1] 2 b], a difficult or risky situation, a close shave. *slang* (orig. *R.A.F.*).
1942 T. Rattigan *Flare Path* III. 150 They had rather a shaky do last night. 1943 C. H. Ward-Jackson *It's a Piece of Cake* 54 *A shaky-do*, any occurrence that has serious consequences or just escapes them. This may vary from a pilot temporarily losing control of his Whitley as a result of being hit by flak, to an erk who is out of station bounds without a pass and only just avoids a Service policeman. 1944 T. H. Wisdom *Triumph over Tunisia* ix. 79 No. 18 Squadron, which had been involved in many similar 'Shaky do's', was asked to lay on the raid. 1949 F. Maclean *Eastern Approaches* II. vi. 244 The earth all round was kicked up by a burst from the plane's tail-gunner... 'This,' said the Australian, 'is going to be a shaky do.'

shale, *sb.*[2] Add: **2. b. shale oil** (further examples); **shale shaker,** a vibrating screen used in oil and gas drilling to remove drill cuttings from the circulating drilling mud that is passed through it.
1945 Heald & Ayres in L. M. Fanning *Our Oil Resources* vi. 185 Crude shale oil is produced from oil shale by retorting. 1976 *Time* 20 Dec. 41/1 Prices for getting shale oil or using wet-steam deposits in the earth to generate electricity are also far from commercially acceptable. 1959 *Petroleum Handbk.* (Shell) (ed. 4) 85 On reaching the well head it is diverted via a horizontal flow line to a vibrating screen or 'shale shaker'. 1974 G. S. Ormsby in P. L. Moore et al. *Drilling Practices Manual* vi. 152 The term 'shale shaker' is used in drilling mud work to cover all the devices that in another industry might be differentiated as 'shaking' screen, 'vibrating screens', and 'oscillating screens'.

shaler, var. *sheila.

shall, *v.* Add: **II.** ** *The past tense* should *with temporal function.*

14. d. Also in noun-clause dependent on expressions of willing, etc., (esp. the verb *want*) in the pres. tense. *colloq.* (orig. and chiefly *U.S.* and in representations of Jewish speech).
1852, 1903 [see Want v. 5 b]. 1920 W. D. Howells *Vacation of Kelwyns* 188 Want I should drive ye home? 1960 F. Raphael *Limits of Love* 3 You want we should go bankrupt? 1970 R. Millar *Abelard & Heloise* I. iii. 11 He asks you should go to him. 1978 J. Rosenthal *Evacuees* iv. 89 They want they should take you away.

17. (Later example).
1945 R. Gibbings *Lovely is Lee* xxvii. 133 On the 23rd of March 1889 who should be born in Cork but myself?

*** *The past tense* should *with modal function.*

18. a. Also, in statements of expectation, likelihood, prediction, etc.
1922 Galsworthy *Loyalties* III. i. 82 'Mr. Twisden's not in, then?'.. 'No. He's at the Courts. They're just up; he should be in directly.' 1954 Wodehouse & Bolton *Bring on Girls* viii. 101 'It will run a bit short, I suppose, but it should have a wide appeal.' 'Very wide,' said Guy. 'You've got a winner.' 1961 E. F. Schumacher in *Small is Beautiful* (1973) II. iii. 117 Proved oil reserves should be enough for forty years. 1963 'J. le Carré' *Spy who came in from Cold* x. 94 A couple of weeks should see you through. 1966 T. Frisby *There's a Girl in my Soup* I. 2 That blanket should be warm by now. 1970 R. Millar *Abelard & Heloise* I. ii. 27 A makeshift effort, but it should serve. *Ibid.* xiv. 35 Master Simon says he should be up in a week.

Phrase. *this is as it should be.*
1829 Carlyle in *Edin. Rev.* June 458 This is as it should be; for not in turning back,.. but only in resolutely struggling forward, does our life consist. 1860 J. W. Palmer tr. *M. J. Michelet's Love* IV. i. 184 The mother lives entirely in that cradle; the world is as nothing to her. This is as it should be, for it is the saving of the babe.

b. Delete ? *Obs.* and add further examples. Also, ought according to expectation to be, presumably will be (cf. sense *18 a).
1821 Byron *Cain* I. i. 365, I have heard it said, That seraphs *love most*—cherubim *know most*—And this should be a cherub—since he loves not. 1855 Kingsley *Westw. Ho!* II. ix. 249 That should be Barbados.. unless my reckoning is far out.

c. (Earlier and later examples.)
1811 Lady Granville *Let.* 6 Oct. (1894) I. 21 You should have heard the shout when he said by mistake, [etc.]. 1908 Belloc *Cautionary Tales for Children* 26 That Night a Fire *did* break out—You should have heard Matilda Shout! 1971 S. Gray *Butley* II. 58 But you should see our flat. Even Joey's room is like a pigsty.

d. Used ironically, expressing the inappropriateness or unlikeliness of the action advocated or state envisaged, as *I should worry*, there is no reason for me to worry, I am not worried. *colloq.* (orig. a Yiddishism).
1892 Kipling *Naulahka* xiii. 154 [Amer. loq.] I should murmur!.. It makes me feel good all over. 1906 F. H. Burnett *Shuttle* (1907) xxxviii. 381 'Hope you had a fine time, Mr. Selden?' 'Fine! I should smile! Fine wasn't in it.' 1914 'High Jinks Jr.' *Choice Slang* 13 *I should worry*, I do not care. 1929 'E. Queen' *Roman Hat Myst.* vi. 80 'Well,' grinned the District Attorney, 'I carry a lot of insurance, so I should worry.' 1937 D. L. Sayers *Busman's Honeymoon* x. 224 'You watch your step, Polly. Maybe 'e's married three times a'ready.' 'I *should* worry,' said the girl, with a toss of the head. 1945 A. Kober *Parm Me* 155 'The cilling you think he's going to fix?' 'You should live so long!' he'll say. 1957 *N.Y. World-Telegram* 13 Sept. 22/5 All I ask of these scientists is that they put in writing their guarantee that insects will get us yet. We should be so lucky. 1967 V. C. Welburn *Johnny So Long* II. iii. 76 Don't try to digest everything at once. Hell, I should talk. 1970 M. O'Brine *Crambo* lxiii. 170 If that's the best their gunners can do, we should worry. *Ibid.* lvii. 230 'It's your life,' said Waterhouse. 'I should live that long,' said Gesing. 1975 R. Rendell *Shake Hands for Ever* iii. 29 They both came in at about ten—my God, I should be so lucky!

shallot, shalot. 2. Delete †*Obs.* and add later examples.
1938 *Oxf. Compan. Mus.* 658/1 A certain number [sc. of organ pipes] have a tongue of metal like that of a toy trumpet, vibrating against the open (or partially open) side of a little brass tube (called *eschallot* or *shallot*) at the bottom of the pipe.., and these are called *Reed Stops*. 1969 J. Curnutt tr. *Andersen's Organ Building & Design* iii. 51 The *shallot*.. consists of a tube which is closed at the lower end and planed flat on one side so that a throat-shaped opening is produced.

shallow, *a.*[1] and *sb.*[3] Add: **A.** *adj.* **1. a.** *shallow end*, spec. of a swimming-pool; also *fig.*; *shallow well*, a well that is not deep; *spec.* (see quot. 1972[2]).
1877 J. T. Fanning *Practical Treat. Water Supply Engin.* I. vii. 104 Shallow well and spring supplies are, usually, yields of water from the drift formation alone. 1924 'I. Hay' (*title*) The shallow end. 1929 R. Hughes *High Wind in Jamaica* i. 9 The little ones, of course, only splashed about the shallow end: but John and Emily dived. 1943 *Bull. Amer. Assoc. Petroleum Geologists* XXVII. 838 Of the shallow wells drilled for gas, 80 per cent were producers and 20 per cent were dry. 1959 Ackerman & Löf *Technol. in Amer. Water Development* x. 281 The first irrigation was from shallow wells within the suction lift of centrifugal and piston pumps. 1972 J. Rossiter *Rope for General Dietz* vi. 78 She broke away, swimming to the shallow end and climbing out. 1972 *Gloss. Geol.* (Amer. Geol. Inst.) 650/2 *Shallow well*. (a) A water well.. that taps the shallowest aquifer in the vicinity. The water is generally unconfined ground water. (b) A well whose water level is shallow enough to permit use of a shallow-well (suction) pump, the practical lift of which is taken as 22 ft.

shallow, *a.*[2] (Earlier example.)
1839 H. Brandon in W. A. Miles *Poverty, Mendicity & Crime* 165/1 *Shallow Coves*, or *Shallow Fellows*,—fellows who go about the country, half-naked, with a Guernsey jacket, but no hat, shoes, nor stockings.

shalloway (ʃæ·lowēⁱ). *Canad. Obs.* exc. *Hist.* [Origin unknown.] A small coastal sailing craft.
1676 in D. W. Prowse *Hist. Newfoundland from Eng., Colonial, & Foreign Rec.* (1895) viii. 206 No Indians come [to Placentia] but some Canida Indians from forts of Canida in french shallowayes. 1774 G. Cartwright *Jrnl. Residence Coast of Labrador* (1792) II. 14 Hooper's shalloway having sprung her foremast,.. I sent the boat-builder to make her a new one. 1971 E. R. Seary *Place Names* v. 86 The fishing boats in the cod and seal fishery were formerly called shallops and shalloways... The shalloways were open boats, what are now called punts.

shallow-pate. Add: Now *arch.* (Further examples.)
1930 [see *dowd sb.[1]]. 1964 *Listener* 24 Dec. 1002/1 That complicated shallow-pate, Jean-Paul Sartre,.. makes of his subjective and personal passions a sort of objective truth.

‖ **shalom** (ʃalǫ·m), *int.* and *sb.* [Heb., lit. 'peace'.] In Jewish society, a word used as a salutation at meeting or parting. See also *Shabbat shalom* s.v. *Shabbat. So **shalom aleichem** (alēⁱχe·m) *int.* [Heb. *ʿalēḵem*], peace be with you.
1881 E. B. Tylor *Anthropology* i. 11 The Arab still salutes the stranger with *salâm alaikum*, 'peace upon you', nearly as the ancient Hebrew would have said shâlôm lâchem, that is, 'peace to you'. 1898 I. Zangwill *Dreamers of Ghetto* p. vi, I saw two Jews that met by chance... *Shalom Aleichem*, mournfully each said. 1959 I. Jefferies *Thirteen Days* iv. 49 Shalom. Cigarette, Sarge? 1962 L. R. Banks *End to Running* II. xiii. 265 There would be cries of 'Shalom, Shalom! Come in! Have a drink!' 1972 O. Sela *Bearer Plot* xxxix. 219 With a whispered '*Shalom*' she was gone. 1977 *Rolling Stone* 19 May 6/4, I was ready to say goodbye America, shalom Israel.

‖ **shalwar** (ʃɒ·lvār). Also **salvar, salwar, shalvar.** [Urdu *šalwār*, Hindi *salvār*, ad. Pers. *šalwār* Shulwar.] **a.** Loose trousers worn by both sexes in some South Asian countries, esp. those worn by women together with a *kameez*.
1951 P. Miles *They came to Mountain* ix. 91 Over her *salvar* she wore stockings tied up with bootlaces. 1955 R. Prawer Jhabvala *To whom she Will* xv. 102 She was very fine now in a pink silk kamiz with blue roses on it and a pink salwar. *Ibid.* 295 The salwar is 104 inches wide at the waist and tied with a cord: when the cord is pulled the material falls in a bunch of folds over the stomach. The legs taper towards the ankles. 1957 *Geogr. Mag.* Aug. 198/1 A sophisticated taste will prefer the fashions of Paris and Bombay to the unbecoming *qamis* and *shalwar* of the North. 1967 *Times* 28 Sept. 2 (*caption*) Muslim children wearing their traditional baggy trousers outside Moat Girls' School, Leicester, yesterday. The shalwars will be allowed by the education authority on religious grounds. 1972 H. R. F. Keating *Inspector Ghote trusts Heart* iii. 34 She was wearing not a sari but a bright, cherry-red kameez and salwar. 1973 *Observer* (Colour Suppl.) 28 Oct. 72/1 Semi-transparent trousers (shalvar) approx £30.

b. *Comb.*, as **shalwar-, salwar-kameez(e** [see *kameez(e], a woman's outfit consisting of *shalwar* and *kameez* or loose tunic.
1955 R. Prawer Jhabvala *To whom she Will* xxx. 222 A lively sturdy girl in coloured salwar-kamiz. 1966 J. & R. Godden *Two under Indian Sun* iv. 104 She in salwar-kameeze, the loose tunic and trousers.. worn by Muslim and up-country girls. 1977 A. Desai *Fire on Mountain* III. xi. 136 A bunch of schoolgirls in bright indigo *salwar-kameez*.

sham, *sb.*[1] and *a.* Add: **A.** *sb.* **3. c.** (Earlier example.)
1884 *Cottage Hearth* (Boston) Aug. 254/1 Large shams made of four very small handkerchiefs.. are elegant in appearance over blue or pink under covers.

B. *adj.* **1. b.** *sham operation* (Biol.), an operation in which an incision is made but nothing is removed, performed on animals of an experimental control group so that they suffer the same incidental effects of the operation as the animals on which a true operation is performed. Hence *sham-o·perate* vb. trans., to perform a sham operation on; *sham-o·perated* ppl. adj.
1963 *Life Sci.* II. 475 Rats.. were thymectomized within the first eighteen hours after birth. Approximately half of the litters were sham-operated. The polyoma virus was injected subcutaneously either immediately after thymectomy or sham-operation or two to three weeks afterwards. *Ibid.* 477 Rats thymectomized at birth seem to be much more sensitive to the oncogenic action of the polyoma virus than are normal or sham-operated rats. 1970 *Physiol. Zool.* XLIII. 91/1 Matched animals in an approximately 1:1 ratio were 'parietalectomized'.. or sham-operated according to standard procedures. 1975 *Nature* 27 Mar. 349/1 Pinealectomy and sham operation were performed as described previously, and 10 d later a 2-mm semicircular wound was made in the right ear of each animal, including the controls. *Ibid.* 349/2 The result indicated that control, sham-operated, and melatonin-treated animals form one group.

5. *sham-dead, -Tudor.*
1934 *Discovery* Oct. 304/1 It cannot be long before the incongruity of the sham-Tudor house with the 1934 interior is generally recognised. 1945 [see *Jacobethan a.]. 1970 T. Hughes *Crow* 53 So in one hand he held a sham-dead spider.

shama, var. *shamma.

shamal. Add: Also **shamaal, shemmal.**
1900 S. M. Zwemer *Arabia* x. 107 The prevailing wind at Bahrein.. is the *shemmāl* or Northwester. 1980 D. Creed *Scarab* xiii. 129 The Shamaal blows strong from the north.

shaman, *sb.* (and *a.*) Add: **A.** *sb.* Delete † from sense 'an adherent of Shamanism' and add later example. Also more recently, with recognition of the widespread similarity of primitive beliefs, the term denotes esp. a man or woman who is regarded as having direct access to, and influence in, the spirit world which is usu. manifested during a trance and empowers them to guide souls, cure illnesses, etc. Also *fig.* Hence **sha·manka, sha·maness, sha·manin,** terms sometimes applied to a female shaman.
1921 R. H. Lowie *Primitive Society* xii. 328 It was indeed through the shaman, who revealed the will of the spirits, that the chief was chosen. 1925 G. Róheim *Austral. Totemism* vii. 350 This rite.. is based on the scheme of death and rebirth and.. the vocation of a shaman is often chosen at puberty. 1936 *Jrnl. R. Anthropol. Inst.* LXVI. 80 The term *shamanka* is used by travellers and anthropologists for all female shamans. This usage is unscientific and misleading... For the sake of convenience, however, I shall follow current usage. 1938 in F. Boas *Gen. Anthropol.* ix. 469 Because the North Californian woman happens to be a shaman does not mean that she treats her family and friends differently. 1952 Koestler *Arrow in Blue* xiii. 106 Vladimir Jabotinsky.. became the first political shaman in my life. 1955 H. V. Elwin *Relig. Indian Tribe* v. 146 A shamanin who has done the wrong things is regarded rather as a nun who has broken her vows. 1964 W. R. Trask tr. *Eliade's Shamanism* i. 4 The shaman is also a magician and medicine man... But beyond this, he is a psychopomp, and he may also be priest, mystic, and poet. *Ibid.*, Through this whole region in which the ecstatic experience is considered the religious experience par excellence, the shaman, and he alone, is the great master of ecstasy. *Ibid.* vii. 241 A shamaness.. resolves to bring back his soul and goes down to the 'world of the dead'. 1968 N. K. Sandars *Prehist. Art of Europe* i. 26 In Siberia there were also women who were shamankas. 1971 I. M. Lewis *Ecstatic Relig.* ii. 56 We are perfectly justified in applying the term shaman to mean.. a 'master of spirits'. 1971 *Times Lit. Suppl.* 19 Nov. 1453/3 The Maori shaman clasps in his arms the tree on which his people rely for food, clothing, shelter and transport. 1972 P. M. Bartz *South Korea* 42/1 Primitive spirit worship (shamanism) was followed by Buddhism... Today, there are said to be 27,000 shamans, 10,000 of them women. 1977 D. R. McCann *Black Crane* p. i, These oracles to Chesŏk.. were recited by a *mudang*, or shamaness. 1979 *London Rev. Bks.* 25 Oct. 1/1 America lacks this type of magician—the shamans there are grander, more worldly, more pretentious.

shamanic, *a.* Add: Also, of or connected with a shaman.
1964 *Listener* 29 Oct. 677/2 The initiation dreams, the general schema of shamanic flight.. are not a shaman monopoly.

shamanism. Add: Also the beliefs, rituals, techniques, etc., associated with a shaman, the general pattern of which is found almost universally in primitive cultures at the food-gathering stage of social development.
1930 G. Róheim *Animism, Magic, & Divine King* III. iv. 166 The sexual organs play a large part in Chukchee shamanism. 1947 H. C. E. Zacharias *Protohistory* iv. 109 Shamanism.. by which term I mean not merely the forms of hysteria and of falling into a trance. 1963 in H. N. Michael *Stud. in Siberian Shamanism* 120 Evenk shamanism was characterized by such phenomena as.. a special shamanistic language, numerous and extremely effective pieces of equipment, [etc.]. 1964 W. R. Trask tr. *Eliade's Shamanism* i. 4 A first definition of this complex phenomenon, and perhaps the least hazardous, will be: shamanism = *technique of ecstasy*. 1972 G. Jones *Kings, Beasts, & Heroes* III. i. 129 They nourished him with foaming hornfuls drawn from the deep casks of wonder, myth, shamanism, make-believe, wish-fulfilment, unreason.

shamanize *v.*, *intr.*: hence **sha·manizing** *vbl. sb.* and *ppl. a.*
1949 W. Howells *Heathens* viii. 126 Evans-Pritchard has the same thing to say about Zande witch doctors, who do shamanizing of a less distinct type. 1963 in H. N. Michael *Stud. in Siberian Shamanism* 8 The people were beating the drum ('shamanizing'). 1964 *Listener* 29 Oct. 677/2 The vital function shamanizing can take on.. may be seen in the *Bardo Thodol*, the Tibetan 'Book of the Dead'. *Ibid.*, In a shamanizing society, 'Venus and Adonis'.., 'The Wanderings of Oisin', 'Ash Wednesday', would all qualify their authors for the magic drum.

shamas(h (ʃa·məs). Also **shammas(h, shammes, shammos,** etc. Pl. **-im.** [Yiddish *shames*, Heb. *šammāš* attendant, f. *šimmēš* to serve.] **1.** A beadle or sexton in a Jewish synagogue.
1650 [see *chazzan]. 1675 L. Addison *Present State of Jews* xi. 90 First the Summas, or Sacristan. 1862 *Once a Week* VII. 191/2 The shamas, a kind of curate and clerk combined, brings a glass of wine. 1892 I. Zangwill *Childr. Ghetto* I. 6 Many of the worshippers were tempted to give beyond their means for fear of losing the esteem of the *Shammos*, or beadle, a potent personage only next in influence to the President. 1896 I. Abrahams *Jewish Life Mid. Ages* 8 It was an ancient custom in several places for the Shamash or verger to announce every Saturday the result of law-suits, and to inform the congregation that certain properties were on the market. 1903 *Standard* 27 Apr., There is a 'shammas' acting as beadle, door-keeper, collector, cook, and utility-man in emergencies. 1909 *Cent. Dict.* Suppl., Shammash. 1946 Koestler *Thieves in Night* 288 It was opened by the *shamash* or door-keeper. 1960 L. P. Gartner *Jewish Immigrant in England 1870–1914* vii. 189 The observant

immigrant in the East End.. could not look to.. the rabbi nor to the more commonplace *shammash* (sexton) for edification. 1967 C. Potok *Chosen* i. 27 You should see his father. He's one of Reb Saunders' shamashim. 1968 L. Rosten *Joys of Yiddish* 329 The visiting rabbi stopped in the middle of his sermon and signaled to the *shammes*. 1973 *Jewish Chron.* 2 Feb. 26/2 The death of Mr Harry Goldman in his 74th year has cast a shadow over the members of the Singers Hill Synagogue, where he served as shammas for 40 years. 1976 *New Yorker* 29 Mar. 64/2 The divinely inspired shammes, or custodian, of the synagogue.

2. An extra candle used for lighting the Chanukah candles.
1961 in Webster. 1976 Y. L. Bialer *Jewish Life* 161 The Hanukkah lights are.. not to be used for ordinary purposes... Alongside each Hanukkah lamp the sages instituted the use of a special candle for normal household purposes, calling it the 'caretaker' (*shamash*).

shamateur (ʃæ·mătəɹ, ʃæmătə̄·ɹ). [f. Sham *a.* + Am)ateur.] A sportsman who is classed as an amateur but behaves like a professional, esp. one who makes money out of his performances.
1896 *Badminton Mag.* II. 533 For frank and open professionalism there may be a good deal to be said, but nothing can make the 'promateur' and the 'shamateur' attractive. 1928 *Sunday Dispatch* 8 July 22/3 The Football Association do not regard their clean-up of non-professional Soccer as completed by the sensational exposure and punishment of four hundred 'shamateurs' and their accomplices in the North. 1955 T. H. Pear *Eng. Social Differences* 247 An Oxbridge college which trained its 'eight' so rigidly that they did not take a reasonable part in university life, was criticised as exposing the University to the suspicion of encouraging 'shamateurs'. 1962 *Punch* 18 Apr. 627/2 The bitchery of high shamateur tennis. 1973 Riessen & Evans *Match Point* I. x. facing p. 92 (*caption*) It was hard work being a 'shamateur'—lugging all that booty around.

Hence **sha·mateurism.**
1928 *Sat. Rev.* 4 Feb. 126 Where the interests of amphitheatre and arena come first, 'shamateurism' must pass eventually into an honest professionalism. 1964 *Punch* 6 May 668/1 Shamateurism has grown steadily in lawn tennis. 1979 *Financial Rev.* (Sydney) 27 Aug. 25/2 Shamateurism is common enough and their activities probably do not inflict any real damage.

shamba. Add: Also, a farm or plantation.
1942 *E. Afr. Ann.* 1941–2 17/2 They are to be found in every kind of work.. from shamba labourer to school teacher. 1952 *Chambers's Jrnl.* Apr. 247/1 There would be rain soon and the ground must be ready for planting before it came. There were two shambas to be cultivated, one for millet and one for beans. 1973 *Reader's Digest* Feb. 200/1 Most Kenyans, for example, have *shambas* (small farms) back in their tribal homelands. 1980 *Times* 23 Feb. 6/2 Jos had pegged out several acres to turn into a coffee shamba.

shamble, *sb.*[1] Add: **5. b.** *pl.* In more general use, a scene of disorder or devastation; a ruin; a mess. orig. *U.S.*
1926 P. H. de Kruif *Microbe Hunters* III. iv. 83 Once more his laboratory became a shambles of cluttered flasks and hurrying assistants. 1942 E. Waugh *Put out More Flags* ii. 150 Alastair learned, too, that all schemes ended in a 'shambles' which did not mean, as he feared, a slaughter, but a brief restoration of individual freedom of movement. 1966 M. R. D. Foot *SOE in France* viii. 184 Helped the commandos to make a thorough shambles of the main dockyard. 1979 *Daily Tel.* 5 Sept. 6/6 Haiti remains a dictatorship, its economy in a shambles.

shambled, *ppl. a.* Add: **2.** *U.S.* Wrecked, ruined. Cf. *shamble sb.[1] 5 b.
1940 *Newsweek* 17 June 21/2 (*caption*) Nazis photographed the shambled Dunkerque's water front. 1952 *Time* 11 Aug. 25/1 (*caption*) Reconstruction of the shambled town.. is expected to take at least five years.

shambolic (ʃæmbǫ·lik), *a. colloq.* [f. *shamble sb.[1] 5 b, perh. after Symbolic *a.*] Chaotic, disorderly, undisciplined.
Reported to be 'in common use' in 1958.—Ed.
1970 *Times* 18 June 9 His office in Printing House Square is so impeccably tidy that it is.. a standing reproach to the standard image of shambolic newspaper offices. 1975 *Times* 14 June 8/5 The average listener is in the position of anybody who encounters an organization at work for the first time. It may appear shambolic but how much is that because he hasn't yet made sense of it. 1978 R. Jansson *News Caper* xiii. 110 We may have a shambolic landing, Jean. I want you to go right through the aircraft reminding people about the emergency drill. 1980 *Jrnl. R. Soc. Arts* July 509/1 It will continue in a much more shambolic manner than the urbanization that has occurred in the Western World.

shame, *sb.* Add: **I. 7.** (Later example.)
1922 Joyce *Ulysses* 533 And with loving pencil you shaded my eyes, my bosom and my shame.

II. 16. a. (*d*) Also, in S. Afr., as an expression of sympathy or pleasure.
1932 *Grocott's Mail* (Grahamstown, S. Afr.) 9 Jan. 3 During the address of our local dairy representatives.. I heard several murmurs of Oh! and Shame! and grant the statements were given in a manner that commanded much sympathy. 1952 N. Gordimer *Soft Voice of Serpent* (1953) i. 6 'Shame, isn't he a funny old man,' she said. 1976 *Sunday Times* (Johannesburg) 14 Nov., Oh, look, look!.. those foals. Oh, shame, aren't they sweet.

III. 17. (sense 7) *shame-cloth, -rag; shame-closing, -making, -wounded* adjs.; **shame culture,** a culture in which conformity of behaviour is maintained through the individual's fear of being shamed.
1922 Joyce *Ulysses* 55 She blinked up out of her avid shameclosing eyes. 1963 M. Laurence *Tomorrow-Tamer* 226 Not yet five years old, she wore only a shamecloth, a mere flutter of red and beaded rag around her middle and between her legs. 1947 R. Benedict *Chrysanthemum & Sword* x. 223 True shame cultures rely on external sanctions for good behaviour. 1953 M. B. Singer in Piers & Singer *Shame & Guilt* II. iii. 56, I shall consider whether the test data support the conclusion that American Indian cultures are predominantly shame cultures. 1977 A. Giddens *Stud. in Social & Polit. Theory* 393 Some anthropologists have sought to contrast the 'guilt cultures' of Western Europe with 'shame cultures'. 1934 R. Macaulay *Going Abroad* xxxv. 297 He adoring some one else, that was shame-making and humbling too. 1977 D. Ramsay *You can't call it Murder* I. 26 Nothing to do with her, thank God. Offering such thanks was shame-making. 1938 R. Graves *Coll. Poems* 166 Those froward hermits.. Wore but a shame-rag, dusk or dawn, And rolled in thorny places. 1922 Joyce *Ulysses* 49 Our souls, shamewounded by our sins, cling to us yet more.

shamedly, *adv.* For *nonce-wd.* read *rare* and add later example.
1913 D. H. Lawrence *Sons & Lovers* x. 260 Then she herself took her place on the sofa, shamedly.

shaming, *ppl. a.* Add: Hence **sha·mingly** *adv.*
1970 D. Neville-Rolfe *Power without Glory* II. 247 Shorthand and typing... I originally took a shamingly long time to learn both. 1979 *Homes & Gardens* June 169/1 Eventually my bus arrived; a collection of shamingly healthy-looking people in breeches and clumpy boots gathered round.

shamisen, var. Samisen in Dict. and Suppl.

‖ **shamma** (ʃæ·mă). Also **chamma, shama.** [Amharic.] A long loose robe resembling a toga, worn by both men and women in Ethiopia.
1862 H. A. Stern *Wanderings among Falashas in Abyssinia* xxi. 311 The costume of the Abyssinian is exceedingly simple. Men of all ranks.. wear a *shama*, or loose dress of white cotton. 1893 J. T. Bent *Sacred City of Ethiopians* ii. 35 A young stripling in cotton drawers and the red-striped *shamma* of everyday wear. 1930 H. Norden *Africa's Last Empire* iii. 62 She was swathed in the usual *chamma*. 1969 *Daily Tel.* 18 Oct. 10/8 The streets are full of the gentle flutter of the white muslin *shammas* worn by the shy, slim Ethiopian women.

Shammar (ʃæ·măɹ). [Native name.] (A member of) a bedouin tribe originating in the Nafud desert of Saudi Arabia. Also *attrib.* or as *adj.*
1911 G. Bell *Let.* 14 Apr. (1927) I. xii. 300 Now the Shammar are *Beda*; only the Shammar and the Anazeh are real Bedawin, the others are just Arabs. 1916 T. E. Lawrence *Home Lett.* (1954) 311 There is a very heady old Shammar, and an Aneyze. 1929 F. Stark *Let.* 24 Nov. (1974) I. 215 He.. has now suggested a visit to one of the Shammar chiefs up the Tigris. 1959 W. Thesiger *Arabian Sands* iii. 54 In Syria I had seen the Shammar migrating, a whole people on the move, covering the desert with their herds.

shammas(h, shammes, -os, vars. *shamas(h.

shampoo, *sb.* Add: **1.** (Later examples in sense *2 b of the vb.)
1951 *Good Housek. Home Encycl.* 172/1 Soiled silk lampshades can be given a soap-and-water shampoo. 1970 *Which?* Aug. 251/2 Which carpet shampoo you choose depends on how much carpet you have to clean and how often you want to clean it.

2. *shampoo and set* (cf. *set sb.[1] 18 b).
1935 A. Christie *Death in Clouds* xiii. 136 Is it a shampoo and set, or are you having a tint to-day? 1977 *Belfast Tel.* 22 Feb. 1/4 Customers will have to shell out an extra 10–15 pc for a snip of the scissors or a shampoo and set in many cases.

3. *slang.* Arbitrary alteration of Champagne. Cf. Sham *sb.*[2]
1957 R. Longrigg *Switchboard* 192 'You'd better have some shampoo, darling.'.. 'Shampoo?' 'Champagne.' 1959 A. Sinclair *Breaking of Bumbo* ii. 31 The waiter brings a bottle of champagne... Shampoo, Sheila dear?

shampoo, *v.* Add: **2. a.** Also *absol.*
1976 *Glasgow Herald* 26 Nov. 17/4 This conditioning treatment is used before you shampoo, and so resembles the sort of reconditioning treatments available at a salon.

b. *transf.* To wash (a carpet, upholstery, etc.) with a cleansing agent. Also *absol.*
1954 A. C. Moore *How to clean Everything* I. 138/1 Synthetic (soapless) detergents.. are available in liquid, paste and powder form to shampoo hair, rugs and upholstery. 1969 *Sears Catal.* Spring/Summer 1359 Outdoor carpets... They'll take whatever comes in stride.. resist spots, stains.. just vacuum or shampoo clean. 1970 *Which?* Aug. 252/1 If you are going to shampoo your carpet only now and then.. a cheap aerosol would be most economical.

shampooer. Add: **b.** A device for applying carpet shampoo.
1960 *Farmer & Stockbreeder* 22 Mar. (Suppl.) 11/2 The new mechanical shampooers simplify the job greatly. 1974 *Spartanburg* (S. Carolina) *Herald* 25 Apr. A5 (Advt.), Rent a twin brush carpet shampooer.

shamshin, var. Samisen in Dict. and Suppl.

shamus (ʃā·mɒs, ʃēⁱ·mɒs). *U.S. slang.* Also **sharmus, shommus.** [Orig. uncertain: perh. f. *shamas(h or the Irish proper name *Seamus*.] A police officer; a private detective.
1925 H. Leverage *Dict. Underworld* in *Flynn's Mag.* 28 Mar. 660 *Sharmus*, a detective; a cop. 1928 J. O'Connor *Broadway Racketeers* xvii. 186 Every Shommus on the beat knew we were going South with the stuff, but they couldn't prove it. 1930 [see *kosher a. (sb.) c]. 1950 'S. Ransome' *Deadly Miss Ashley* ii. 19 Men in my profession don't call each other shamuses... We private operatives call ourselves private operatives. 1960 Wodehouse *Jeeves in Offing* iv. 49 You mean that I'm to be a sort of private eye or shamus, tailing them up? 1977 *New Yorker* 2 May 38/3, I think my wife is having me tailed by a private shamus.

Shamvaian (ʃæmvai·ăn), *a. Geol.* [f. *Shamva* + -ian.] Of or pertaining to rocks in the vicinity of Shamva, a town in northern Zimbabwe; *spec.* the epithet of a Pre-Cambrian mountain-building episode in southern Africa when these were affected.
1947 A. M. Macgregor in *Bull. S. Rhodesian Geol. Surv.* No. 38. 8 Although the Shamva Grits of the Mazoe Valley were not the first to be described they occupy the largest continuous area... The name 'Shamvaian' is proposed. *Ibid.* 15 The Shamvaian rocks of the Bembesi valley are generally similar to the upper series at Que Que. 1951 *Trans. Geol. Soc. S. Afr.* LIV. p. xxix, The largest area of these rocks.. is that of the Shamva grits in the Mazoe Valley which is selected as the type area of the Shamvaian system. 1964 A. E. Phaup in S. H. Haughton *Geol. Ore Deposits S. Afr.* II. 2 Some reefs may be pre-Bulawayan and pre-Shamvaian in age, but the majority are post-Shamvaian. 1971 I. G. Gass et al. *Understanding Earth* xxii. 317/1 At least seven major mountain-building events (periods of orogenesis) are recorded in Africa and are.. (ii) 2500–2800 million years ago (Shamvaian orogeny); [etc.].

Shan (ʃān), *sb.*[4] and *a.*[2] Also **Sciam, Shaan.** [Burmese.] **A.** *sb.* (A member of) a group of Mongoloid peoples of the Tai family, inhabiting parts of Burma, south China, and Indo-China. Also, the Tai language spoken by these peoples. **B.** *adj.* Of or pertaining to the Shan or their language.
1800 M. Symes *Acc. Embassy Ava* xi. 274 An intelligent man.. informed him, that.. the first Shaan town was called Thangdat. *Ibid.*, Shaan, or Shan, is a very comprehensive term given to different nations, some independent, others the subjects of the greater states. 1833 W. Tandy tr. *Sangermano's Descr. Burmese Empire* ix. 57 The zaboà or petty princes of the Sciam, subject to the Burmese. 1898 [see *Lolo]. 1920 *Blackw. Mag.* June 839/1 You sell liquor and opium to Burmans and Shans. 1927 *Ibid.* June 819/1 Men armed with.. long Shan knives were patrolling the approaches. 1951 R. Firth *Elem. Social Organiz.* iv. 145 The Pai of the Chinese Shans; the *anga* of the Tikopia.. are.. examples of the allocation of large-scale resources in goods and labour with primary reference to status yields. 1968 O. Wynd *Sumatra Seven Zero* ii. 25 Your companion knows Shan. She learned it to pass the time. 1977 *Whitaker's Almanack* 1978 828 Burmese is the official language, but minority languages include Shan, Karen, Chin, and the various Kachin dialects.

shananacking, var. *shenanigan.

shandite (ʃæ·ndəit). *Min.* [ad. G. *shandit* (P. Ramdohr in *Sitzungsber. d. Deutsch. Akad. d. Wissensch. zu Berlin* (*Math.-naturwissensch. Kl.*) 1949 (1950) vi. 26), f. the name of S. J. Shand (1882–1957), Scottish geologist: see -ite[1].] A sulphide of nickel and lead, $Ni_3Pb_2S_2$, found as yellow rhombohedral crystals.
1950 *Amer. Mineralogist* XXXV. 450 On receiving this result Dr. Ramdohr intimated in a private communication (Nov., 1949) that he proposed to name the new mineral shandite, after Professor S. J. Shand,.. who had suggested.. the study of the Insizwa nickel ores which led to the discovery of parkerite. *Ibid.*, Two lots of artificial shandite were prepared by fusing charges of 1 gm. and 2 gm. of the powdered elements in the proportions 3Ni:2Bi:2S. 1968 I. Kostov *Mineralogy* 117 Shandite has a distorted spinel structure, the nickel atoms being arranged along the pseudocubic diagonals; this explains its ferromagnetic properties.

shandy, *sb.* Add: (Later examples.) Also, a mixture of beer and fizzy lemonade.
1919 *Chambers's Jrnl.* Sept. 593/1 Staff-Sergeant Jack Dorley, R.E., finished off his shandy with a long draught. 1947 K. Tennant *Lost Haven* xvii. 267 Miss O'Shea was drinking ginger-beer and her escort had a shandy. 1969 A. Christie *Hallowe'en Party* v. 39 For me, I think a shandy. The ginger beer and the beer? 1976 *Milton Keynes Express* 30 July 2/7 He said Mr Westley had drunk four pints of shandy during the evening but his driving had not appeared to be affected.

Shang (ʃæŋ). Also Chang, Xanga. [Chinese *shāng*.] The name of a dynasty which ruled China during part of the second millennium B.C., probably from the 16th to the 11th century B.C. Also *attrib.* or as *adj.* Also called **Shang-Yin** [f. the place-name *Yin* in Honan Province, the dynasty's final capital].

1669 J. Ogilby tr. *Nieuhoff's Embassy Grand Tartar Cham* xviii. 281 [Then] arose the family of Xanga, whereof the Emperor Tangus, in the year before Christ's Birth 1766, was the first who called it Xanga, from a Lordship of the same name he possessed. **1736** R. Brookes tr. *Du Halde's Gen. Hist. China* I. 298 (*heading*) The Second Dynasty, called *Chang*, which comprehends the Lives of Twenty Eight Emperors in the Space of 644 Years. **1797** *Encycl. Brit.* IV. 653/1 The whole of their (*sc.* of the Chinese) emperors..are comprehended in 22 dynasties, mentioned in the following table..2. Shang, or Ing. **1877** *Ibid.* VI. 259/1 Confucius's own ancestry is traced up, through the sovereigns of the previous dynasty of Shang, to Hwang-ti. **1933** R. Fry *Let.* 15 Dec. (1972) II. 686 Some..accomplished pre-Cheou bronzes: Shang they say but anyhow earlier than anything I knew. **1939** *Burlington Mag.* Feb. 85/1 Practically everything is problematical... This applies to the use of Shang weapons, though their date seems fairly safe. **1958** W. Willetts *Chinese Art* I. 108 Writing in Shang-Yin times may have been more widespread through society than was at first believed. **1972** S. H. Hansford *Gloss. Chinese Art & Archaeol.* Ku..is applied especially to objects attributed to the..Hsia, Shang-Yin and Chou, and the Han Dynasty. **1978** *New Archaeol. Finds in China* II. 2 Shang relics have been newly discovered in regions other than the Central Plains.

Shangaan (ʃaŋgā·n). [Native name.] (A member of) a Bantu people inhabiting Zimbabwe, Mozambique, and South Africa; the language of the south-east Bantu group spoken by this people. Also *attrib.* or as *adj.*

1887 J. W. Matthews *Incwadi Yami* 185 The native servants in the diamond fields..includes nearly twenty different tribes such as Zulus, Swazees, Basutos, Shangaans, [etc.]. **1911** *Encycl. Brit.* XXVII. 189/2 The Shangaan are members of a Bantu tribe from the Delagoa Bay region who took refuge in the Transvaal between 1860 and 1862 to escape Zulu raids. **1932** C. Fuller *Louis Trigardt's Trek* 39 Trigardt adopted a version from other than local natives, probably his Shangaan guides. **1948** *Rep. Native Laws Comm.* 1946–48 (Dept. Native Affairs, S. Afr.) 38/1, I have had a most valuable document, prepared by a young Shangaan. **1968** C. Burke *Elephant across Border* ix. 128 An elderly man..got out, speaking quietly in Shangaan to an African. *Ibid.* 128 They suddenly saw Murray and his Shangaan tracker. **1973** *Standard Encycl. S. Afr.* IX. 607/1 The Transvaal Tsonga—called Machangana (Shangaans),..—have been recognized as one of the eight Bantu peoples living in the Republic of South Africa.

shanghai, *sb.* Add: **2.** Also in *N.Z.* and as **shanghye.**

1940 F. Sargeson *Man & Wife* (1944) 7, I was out in the backyard with my shanghai, and..I took a shot at a thrush. **1947** D. M. Davin *Gorse blooms Pale* 57 Bits of shangeye as they called their catapults. **1972** M. Gee *In my Father's Den* 44 He made me shanghais and bows and arrows.

shanghai, *v.* Add: **1. b.** *transf.* To transfer forcibly or abduct; to constrain or compel. *colloq.* (orig. *U.S. Mil. slang*).

1919 in *Amer. Speech* (1975) XLVII. 97 The conscripts..were shanghaied out to..on duty. **1934** *Sun* (Baltimore) 21 May 7/5 Arguments will be heard..on Insull's plea that his jurisdiction over him because he was 'shanghaied' from Istanbul. **1958** *People* 4 May 15/6 We began to wonder if she'd got herself shanghaied. **1974** *Sunday Times* 15 Dec. 3/1 Hunt..thought he was being 'Shanghai-ed'—prison jargon for a transfer to another prison as a punishment. **1976** J. Gibson *As I saw It* xxviii. 491 Most of my guests got shanghaied into giving a general knowledge talk to the boys.

Hence **shanghai-er,** one who shanghais.

1917 *Chambers's Jrnl.* Jan. 19/2 Once..a shanghaier had been shanghaied by a rival shanghaier. **1926** J. Black *You can't Win* xii. 152 Here I learned to beware the crafty shanghaier with his knockout drops.

Shanghailander (ʃæŋhai·lændəɹ). [f. Shanghai, after Highlander, Islander, etc.] A native or inhabitant of Shanghai.

1917 *China Press* (Shanghai) 20 June 6/3 The Shanghailanders are still working to get a pipe band together and yesterday there were two leathery-lunged Scots extracting sounds from their machines in the Town Hall. **1937** E. Linklater *Juan in China* vii. 123 They were Shanghailanders. **1959** *Times* 8 June 36/3 The impertinent, self-assured Shanghailander of the past has disappeared.

Shanghainese (ʃæŋhainī·z). [f. Shanghai + -n- + -ese, after Chinese, etc.] The Chinese dialect of the Wu group spoken in Shanghai; a native or inhabitant of Shanghai (also *collect.*). Also *attrib.* or as *adj.*

1964 *Asia Mag.* 12 July 22/3 The Chinese [in Hong Kong]..speak no less than seven tongues—Cantonese, Hokk..Shanghainese, Chiuchow and Fukienese. **1965** M. West *Ambassador* vi. 116 He looked more like a Shanghainese than a Viet. **1970** T. Lilly *Projects Section* iv. 36 The Shanghainese must surely be among the most handsome people in the world. **1971** K. Hopkins *Hong Kong* 225 Shanghainese firms were mainly established in Hong Kong, in the late 1940s and early 1950s, by capitalists from Shanghai who fled from Communism. **1977** H. Fast *Immigrants* ii. 114 'You see, my mother is Shanghainese—.' 'Not really,' Feng Wo interrupted. 'She comes from a tiny village to the south of the city.' *Ibid.* 115, I was speaking Shanghainese and Mandarin before I ever knew a word of English.

Shango (ʃæ·ŋgo). Also **shango.** [f. the name *Shango* of the god of thunder and lightning in the Yoruba religion of W. Nigeria.] **1.** A syncretistic cult practised in the Caribbean. Freq. *attrib.*

1953 *Caribbean Q.* III. 1. 16 In 1916 I had the first experience of the Shango. **1958** J. Carew *Wild Coast* xi. 155 The shango gods, Dumbhalla, Legba, Moko. **1963** G. J. McCall *Social Problems* x. 364/2 'Hoodoo'..corresponding to..shango in Trinidad. **1974** *Trinidad Guardian* 2 Nov. 7/6 Sometimes they used the heady chants heard at 'wakes', shango meetings, and other such ceremonies.

2. A dance associated with the Shango cult.

1948 E. Leaf *Isles of Rhythm* vi. 142 It had a more pronounced sex base than the Afro-Caribbean religious dances such as *Voodoo, Shango* and *Obeah*. **1971** *Advocate-News* (Barbados) 17 Sept. (Guyana Suppl.) p. i/1 *Cumfa*, a spirit dance like the Trinidad 'Shango'.

Shangri-La (ʃæŋgrilā·). Also **Shangrila, shangri-la,** etc. The name of *Shangri-La* [f. Tibetan *la* mountain pass], a Tibetan utopia in *Lost Horizon* (1933), a novel by James Hilton, used *transf.* to designate an earthly paradise, a place of retreat from the worries of modern civilization. (In quot. 1945 as quasi-*adj.*)

[**1933** J. Hilton *Lost Horizon* ix. 212 When the High Lama asked him whether Shangri-La was not unique in his experience..he answered..'To be quite frank, it reminds me very slightly of Oxford, where I used to lecture.'] **1938** 'E. Queen' *Four of Hearts* xv. 197 'It's a simply hideous place.'..'It's not exactly another Shangri-La.'] **1941** *Time* 23 June 53/1 The Captain operates an insular Shangri-La in the South Pacific. **1945** L. Durrell *Let.* 15 Dec. in *Spirit of Place* (1969) 81, I was afraid I would sound so heartlessly healthy and the country so Shangri la that you would write me a stinker. **1960** D. Lessing *In Pursuit of English* i. 15 Their Shangri-La would be populated..with nice professional people. **1971** *Sat. Rev.* (U.S.) 11 Dec. 53/1 The Windward Islands..excel any of the shangri-las of the South Pacific. **1977** *China Now* July/Aug. 19/1 The lamasery, for all its association with the Shangrila myth of eternal youth and joy, brought misery.

shank, *sb.* Add: **5. w.** Substitute for def.: A straight piece of metal tubing fitted to a brass instrument to lower its pitch. (Examples.)

1885 G. B. Shaw in *Our Corner* Nov. 313 Brass instruments have resources in shanks and tuning-slides for flattening. **1938** *Oxf. Compan. Mus.* 114/1 By the addition (to a horn or trumpet) of a crook (a curved additional length of tubing) or a shank (a straight additional length), the fundamental note could be altered. **1977** *Early Music* V. 221/2 Every hand horn that we found..had a C alto shank among the set of crooks.

6. a. (Later examples.)

1873 *Catal. Loan Exhib. Anc. & Mod. Jewellery* (South Kensington Museum) 72 Gold ring, the shank formed of leaves; in centre a transparent stone. **1928** *Daily Express* 18 June 5/2 The middle stone may be placed into a knife-edged shank with a 'coronet' setting, producing a solitaire ring. **1978** *Morecambe Guardian* 14 Mar. 16/5 Mrs Tyson was fined a further £20 for stating orally..that a ring was solid gold when it had a hollow shank filled with wax.

7. e. *Fishing.* (*a*) A line of pots attached to a rope, used to catch crabs, whelks, etc. (*b*) = *shank-net* (see sense 11 in Dict.).

1962 *Listener* 28 June 1105/2 Not far short of 600 fathoms of rope go to one shank of pots (a shank holds thirty-six pots). **1971** *Country Life* 29 Apr. 1000 (*caption*) Fishermen shooting out their shank of pots. Between 24 and 70 pots are attached at intervals along the rope to form the shank. **1973** W. Elmer *Terminol. Fishing* ii. 72 Shanks are designed to be dragged in shallow waters.

9. Esp. in phr. *shank of the evening* (earlier and later U.S. examples).

1829 *Virginia Lit. Museum* I. 418/2 'Won't you spend the *balance* of the evening with me?' In some places, shank is quaintly used with the same signification. **1856** P. Thompson *Hist. Boston* 722 *Shank of the evening*, the twilight or dusk of the evening, and in some cases the latter part of it. **1972** Wodehouse *Pearls, Girls, & Monty Bodkin* vii. 97 'It's very late.' 'Shank of the evening.' **1973** *Publishers Weekly* 19 Nov. 56/2 The stuff that makes the antennae of music pros and music-lovers twitch during cultural quarrels in the shank of a Lincoln Center evening.

10*. *Golf.* An act of striking the ball with the heel of the club.

1942 *Sun* (Baltimore) 8 July 12/7 It should be stated here that a slice isn't a shank. **1960** *Times* 31 May 4/1 Miss Price had a shank at the 13th.

shank, *v.* Add: **4.** *Golf.* To strike (the ball) with the heel of the club.

1927 *Daily Express* 26 Oct. 3/4 Of all the awful things a man may do to a golf ball the most demoralising and the most mystifying is to 'shank' it. **1942** *Sun* (Baltimore) 8 July 12/7 Try to shank one. That's about the best cure I know after thirty years of golf. **1976** *Par Golf* Aug. 39/2 He had shanked his second and bunkered his third.

shanking, *vbl. sb.* Add: **4.** *Golf.* The action of striking (the ball) with the heel of the club.

1924 C. J. H. Tolley *Mod. Golfer* 247 Shanking is a fault which is frequently occurring. **1942** *Sun* (Baltimore) 8 July 12/6 Shanking, in golf, is hitting the ball deep in the heel of the club, thereby causing the ball to fly away at a right angle. **1976** *Sunday Mail* (Glasgow) 26 Dec. 34/2 Norman..had a comparatively poor season last year..(mainly due to a bout of shanking which he is convinced has cleared).

shannai, var. *Shahnai.

Shannon[2] (ʃæ·nən). *Information Theory.* The name of Claude Elwood *Shannon* (b. 1916), U.S. mathematician, used *attrib.* and in the possessive to designate various concepts arising from his work, esp. *Shannon's* (*second* or *capacity*) *theorem*, a theorem regarding the ability of a noisy channel to carry information with no more than an arbitrarily small frequency of errors (see quot. 1970).

1956 L. Brillouin *Sci. & Information Theory* i. 7 This is exactly Shannon's formula..for a problem with just two signals. **1956** *IRE Trans. Information Theory* II. 102/1 In the discrete case this quantity [of information] is evaluated correctly according to the well-known Shannon formula. **1963** N. Abramson *Information Theory & Coding* vi. 173 Shannon's second theorem can..be characterized as a little more than an existence proof. **1970** H. A. Rodgers *Dict. Data Processing Terms* 98/1 *Shannon's capacity theorem*, in information theory, a theorem stating that it is possible to encode a source of messages having an information rate *H* bits/sec so that its information can be transmitted through a noisy channel with an arbitrarily small frequency of errors, provided that H ⩽ C bits/sec, where C is called the limiting capacity of the channel. **1972** L. L. Gatlin *Information Theory & Living System* iv. 98 A fundamental condition under which the Shannon theorem is valid is that the rate of emission from the source..must not exceed the channel capacity.

shant. Add: Hence *loosely*, a drink.

1960 *News Chron.* 5 Aug. 7/8 We did not want to roll anybody but we had a few shants and I always get a bit garritty then. *Ibid.*, We all like a fight when we have had a few shants. **1970** A. Draper *Swansong for Rare Bird* vii. 52 'So I had a few shants,' I said.

shanty, *sb.*[1] Add: Also **chanty. 1. c.** (Earlier examples.)

1824 *Canadian Mag.* III. 201 They commence by building a log cabin called a *Chanty* to shelter them from the weather, and hence another appellation they are known by, namely *Chanty Men*. **1829** J. MacTaggart *Three Years in Canada* I. 242 In these shanties they pass the time pretty well, considering them to be made up of Highlandmen, Irishmen, and Yankees.

2. Also in *N.Z.* (Earlier and further examples.)

1862 *Otago Goldfields & Resources* 28 These accommodation houses are not mere 'shanties' and the traveller, with ordinary precautions, is always safe. **1880** H. Lapham in D. M. Davin *N.Z. Short Stories* (1953) 57 When I first saw it..nearly every second house was a 'shanty' or a store.

3. (sense 1) *shanty-dweller, home, shop, slum*; (sense 2) *shanty-keeper* (earlier example); **shanty Irish** *a. U.S.*, belonging to the Irish lower-classes; also *ellipt.*; so **shanty Irish** *sb. pl.*, **shanty Irishman; shantyman** (earlier example); **shanty town,** a suburb consisting of shanties, *spec.* a poor or depressed area of a city or town.

1970 *E. Afr. Standard* (Nairobi) 23 Jan. 1/3 Rich-quick land racketeers who leased small plots to shanty-dwellers in return for 'rent'. *Ibid.* 2 Jan. 15/4 The fire..destroyed some 98 shanty homes. **1928** J. Tully *Shanty Irish* xi. 117 I'm just plain Shanty Irish an' I'll go to hell when I die. **1966** [see *Residency 1 a]. **1975** J. F. Burke *Death Trick* (1976) v. 79 That shanty Irish bitch!..She hit me..and got away. **1934** J. T. Farrell *Young Manhood Studs Lonigan* xx. 334 The Irish made a shanty Irishman out of Christ. **1874** V. Pyke *Adventures G. W. Pratt* i. iii. 6 The shanty-keeper interposed. **1824** Shanty man [see sense 1 c above]. **1891** H. Melville *Timoleon* 63 And here and there a shanty-shop Where Fez-caps, swords, tobacco, shawls Lay orderless. **1969** *Cultural News from India* Nov. 20 Shanty shops on pavements, packed buses and tram cars..mark the biggest annual festival of Bengal. **1969** A. G. Frank *Latin Amer.* (1970) xix. 300 The wealth and elegance of downtown Mexico City dazzle the visitor..but equally do the miles of Mexico City's shanty slums depress. **1876** *Potter's American Monthly* Oct. 400/2 (*caption*) Shanty town. **1880** *New York Daily Graphic* 4 Mar. 38 (*caption*) A scene in shantytown, New York. **1917** U. Sinclair *King Coal* 36 There's lots of people have boarders in shanty-town. **1946** [see *Hooverville]. **1954** H. Gibbs *Background to Bitterness* ii. vii. 121 By the end of 1871 over 10,000 diggers occupied the hot, corrugated iron shanty-town of Kimberley. **1980** *Times* 4 Jan. 6/5 The overpopulated [Turkish] cities are girdled with slummy shanty towns.

shanty, *sb.*[2] (Earlier examples.)

1856, 1867 [see *chantey, chanty].

shantying, *vbl. sb.* (Earlier and later examples.)

1824 *Canadian Mag.* III. 202 Such is the usual routine of what is called *Shantying* in Canada. **1926** F. Rickaby *Ballads & Songs of Shanty-Boy* 47 Shantying I'll give o'er when I'm landed safe on shore, And I'll lead a different life.

Shaoshing (ʃauʃi·ŋ). Also **shao hsing, shaohsing, shao shing.** The name of a town (Pinyin *Shaoxing*) in the Zhejiang province of China, used *attrib.* to designate the rice wine produced there. Also *ellipt.*

1961 *Sunday Express* 26 Nov. 19/4 Everything I wanted to cook called for 'shao shing' or yellow rice wine. **1965** O. A. Mendelsohn *Dict. Drink & Drinking* 306 There are three types of shaohsing: Shan Niang is full-bodied [etc.]. **1969** *Times* 9 Dec. (Taiwan Suppl.) p. viii/7 I would praise Shanghai because the best *Shaoshing* wine once came from spring-water on a hill on the outskirts. **1980** E. Behr *Getting Even* vii. 90 They drank the ritual toast, in hot shao hsing wine.

shape, *sb.* Add: **I. 9. c.** Phr. *in all shapes and sizes*: in a great variety of forms.

1958 J. Townsend *Young Devils* xxi. 196 Parents come in all shapes and sizes. **1967** 'A. Cordell' *Bright Cantonese* xvii. 189 You've got competition..in all shapes and sizes. **1980** P. Moyes *Angel Death* v. 60 Tourists come in all shapes and sizes.

10. a. (Later examples.)

1968 *Listener* 28 Mar. 400/3 The BBC in the shape of Harman Grisewood referred him to the government. **1976** *Eastern Even. News* (Norwich) 22 Dec. 14/4 Gothic had mixed luck, falling foul of County Council in the shape of David Simpson.

11. *to take shape* (later examples).

1939 *Daily Tel.* 18 Dec. 1/2 The great grey hulk of Germany's pocket-battleship..began to take shape. **1982** *New Scientist* 2 Sept. 609/1 The idea that nuclear armaments could be used..had taken firm shape in the minds of the technical people.

13. Delete *Sporting* and substitute for def.: Condition, state of health, repair, or fitness. orig. *U.S.*

1865 O. W. Norton *Army Lett.* (1903) 249, I got through it all in good shape. **1901** [in Dict.]. **1924** J. Galsworthy *Forest* ii. ii. 52 With only nine Soudanese..and less than thirty carriers—all in bad shape; it's precious long odds against our getting through. **1976** *National Observer* (U.S.) 24 July 3/5 Most of the corn-growing areas..were in pretty good shape for moisture.

14. b. (Earlier example.)

1880 *Girl's Own Paper* 20 Mar. 191/3 You might also buy a shape, and make a little hat to match.

e. pl. *Gambling.* (See quot. 1936.) *U.S. slang.*

1928 [see *miss-out]. **1936** *Flynn's Mag.* 21 Mar. 139/2 'Shapes' are dice which have beveled faces on some sides of the cube. These cause the dice to trip faster when these surfaces strike the playing table.

f. *Bridge.* The distribution of suits in a hand of cards.

1954 G. S. Coffin *Bridge Play from A–Z* i. 17 There are in bridge three dominant Playing Shapes: I. Notrump Shape. II. Trump/No-trump Shape. III. Ruff Shape. **1958** *Listener* 27 Nov. 901/3 The shape is unsuitable for a double—the bidding might go too high. **1961** A. Truscott *Bridge* ii. 22 The shape of a hand is the way the cards are divided between the four suits.

IV. 17. *shape-shifting* sb. (earlier example), *-changer, -changing* vbl. sb.; **shape elastic** *adj. phr. Physics*, pertaining to or designating a component of the scattering cross-section of an atomic nucleus that is regarded as independent of the formation of a compound nucleus; **shape factor** *Physics*, an algebraic factor in the expression predicting the profile of a spectral line; **shape memory,** a property exhibited by certain alloys of recovering their initial shape when they are heated after having been plastically deformed; **shape-note** (chiefly *U.S.*), one of a series of notes having heads of different shapes, used to represent the degrees of a scale.

1906 W. B. Yeats *Poems, 1899–1905* 63 Shadows, illusions, That the shape-changers..have cast into his mind. **1978** H. R. E. Davidson in Porter & Russell *Animals in Folklore* 141 The tales of shape-changers in the Sagas are not told 'for true'. *Ibid.* 127 (*heading*) Shape-changing in the Old Norse Sagas. **1954** H. Feshbach et al. in *Physical Rev.* XCVI. 449/2 It will be practical..to subdivide the elastic cross section into two parts... We call the second part..the 'compound elastic' cross section... The first part we call 'shape elastic' cross section; this is the part of the elastic scattering which occurs without the formation of a compound [nucleus]. **1971** P. E. Hodgson *Nuclear Reactions & Nuclear Structure* vii. 142 The method of analysis described..applies only to the shape elastic part of elastic-scattering cross-sections. **1955** E. Konopinski in K. Siegbahn *Beta- & Gamma-Ray Spectrosc.* x. 301, S_n will be called the 'shape factor'... Whenever S_n happens to be independent of the energy W,..the spectrum has the 'statistical shape' $\sim pW(W_0-W)^2$, modified only by the Coulomb effect. **1970** *Physical Rev.: C* I. 644/1 It is important that the correlation coefficients and the energy dependence of the shape factor be known to an accuracy of a few percent to provide a meaningful test of nuclear models. **1968** de Lange & Zijderveld in *Jrnl. Appl. Physics* XXXIX. 2195/1 On heating above 90°C the reverse transformation takes place. The sudden change of configuration into the original shape, which occurs then, is called here the shape-memory effect. **1975** *Nature* 22 May 281/2 The spring has to be made of one of the alloys—a select band—which exhibit shape-memory. **1932** V. Randolph *Ozark Mountain Folks* 248 Right hyar is whar I get in some good licks for shape-notes, too. **1980** P. M. Young *George Grove* vii. 146 The congregational singing in the enthusiastic manner derived from *Sacred Harp*,..and the 'shape-note' books. **1884** A. Lang in M. Hunt *Grimm's Household Tales* I. p. lxvii, He escapes with her..by her magical gift of shape-shifting.

shape, *v.* Add: **I. 11. c.** *intr.* To assume a shape or form; to develop or progress. Freq. const. *up.*

1865 O. W. Norton *Army Lett.* (1903) 278 As things are shaping I do not much think I shall try till after Congress meets. **1903** *N.Y. Times* 10 Sept. 6/3 Matters are shaping for an effort on the part of the organized teamsters to reproduce in this city the..conditions which exist in Chicago. **1921** R. D. Paine *Comr. Rolling Ocean* xvii. 293 Here is how it shapes up to me. **1941** B. Schulberg *What makes Sammy Run?* vii. 153 It's shaping up something terrific... It looks like the biggest opening this town ever had. **1951** *Sport* 27 Apr.–3 May 12/1 How will Yorkshire shape up this summer? **1965** *Listener* 25 Nov. 871/1 The autumn output has shaped up most satisfactorily, far better than its schedules might suggest. **1980** N. Marsh *Photo-Finish* vi. 156 He pulled out... He didn't fancy the way things shaped up.

II. 16. d. Also *absol.* or *intr. rare.*

1848 J. F. Cooper *Oak Openings* I. iv. 49 Perhaps it would be best for me to shape at once for Ohio.

19. b. Also in *Golf*: to get into the proper attitude or stance for a stroke.

1930 Wodehouse *Very Good, Jeeves!* vi. 160 It was while I was shaping for a rather tricky shot that the front-door bell went.

d. *to shape up*: to pull oneself together or meet a required standard; to show one's capabilities. Also, to get oneself into good physical condition.

1938 E. Bowen *Death of Heart* i. v. 98 There seemed no reason why he should not shape up. **1951** *Chambers's Jrnl.* Nov. 645/1 He shaped up awkwardly against a man who was not only champion but twelve years his junior. **1963** *Time* 8 Nov. 10/3 You stated that an icosahedron is a two-sided solid figure... Shape up, sir! It's really a 20-sided solid figure. **1976** *National Observer* (U.S.) 10 July 8/2 After that [*sc.* adolescence] one is expected to shape up, get a job, get married. *Ibid.* 14 Aug. 11/1, I have gained 5 more pounds, and so once again am embarked on a semiserious effort to shape up. **1977** N. Marsh *Last Ditch* vi. 154 He taught her to ride and was uncommon proud of the way she shaped up.

e. Phr. *to shape up or ship out*: used as a threat of transference or dismissal if a satisfactory performance is not achieved. *slang* (orig. *U.S. Mil.*).

1956 *Amer. Speech* XXXI. 108 *Shape up or ship out!*, start soldiering or be sent to a combat zone. **1968** *Review & Herald* 19 Sept. 24/2 We ought to tell them to 'shape up or ship out'. **1977** *Guardian Weekly* 30 Oct. 15/1 If the International Labor Organization didn't shape up within two years, the U.S. government would ship out.

20. (Earlier example.) Also with *out* or *up.*

1855 R. Carboni *Eureka Stockade* 9 By this time two covies..had stripped to their middle, and were 'shaping' for a round or two. **1899** S. MacManus *In Chimney Corners* 12 'I'll fight you,' says Billy, shaping out and winding the bit of stick three times over his head. **1927** *Daily Express* 31 May 7 He shaped up to Murphy, when he punched the watchman on the jaw with his fist and knocked him insensible. **1977** N. Marsh *Last Ditch* ii. 47 If you feel like a fight you've only to say so and we'll shape up and make fools of ourselves.

SHAPE (ʃēɪp). Also **S.H.A.P.E., Shape.** [Acronym f. the initials of *Supreme Headquarters Allied Powers in Europe*, set up in 1951.] An organization established by the N.A.T.O. Council embodying a structure of command for the defence of western Europe.

1950 *Sun* (Baltimore) 20 Dec. 4/3 The 60-year-old five-star general..will create another international staff at SHAPE (Supreme Headquarters, Allied Powers in Europe). **1951** *N.Y. Times* 3 Apr. 8/3 General of the Army Dwight D. Eisenhower formally assumed command and activated the Supreme Headquarters of the Allied Powers in Europe (Shape) into an operational headquarters this morning [2 Apr.]. **1955** *Times* 20 July 8/2 He had accepted the leadership of the Supreme Headquarters in Europe (S.H.A.P.E.) as being a true agency for peace. **1958** *Listener* 25 Sept. 453/1 The signatures of the four Shape Supreme Commanders. **1976** H. Wilson *Governance of Britain* vii. 136 On 8 March the list concludes with southern Africa, Staffordshire, the United Nations and SHAPE.

shaped, *ppl. a.* Add: **3.** Special collocations: **shaped charge,** an explosive charge having a cavity which causes the blast to be concentrated into a small area; **shaped note** = *shape note* s.v. *Shape *sb.* 17.

1889 F. H. Gilson *Hist. Shaped or Character Notes* 4 The great variety of systems embodying the same idea,—that is, a separate shape for each syllable,—came to be so confusing that the majority of the advocates of shaped notes finally agreed to adopt one system. **1945** *Chicago Tribune* 18 Nov. vii. 1/5 We sang from song books printed with old time shaped notes. **1948** *Sun* (Baltimore) 2 Jan. 1/3 A light but potent recoilless gun of the new type, using a 'shaped charge' like that of the bazooka. **1979** A. Hailey *Overload* i. ii. 14 What the saboteur used, they decided, was a 'shaped charge'—a cone of dynamite which, when detonated, had a forward velocity similar to that of a bullet.

shaper. Add: **5.** *Electronics.* A device which modifies an input to produce an output having a specific waveform.

1967 [see *keyer]. **1971** J. H. Smith *Digital Logic* vi. 109 When the count of ten is reached..a gate will produce a signal which is shaped by the shaper unit.

sha·pe-up. *U.S.* Also **shapeup.** [f. vbl. phr. *to shape up*: see Shape *v.* in Dict. and Suppl.] **1.** A system of hiring dock workers for the day or half-day by arbitrary selection from a gathering of men on site. Also *transf.*

1940 *Sun* (Baltimore) 8 Nov. 22/7 Under the shape-up system..longshoremen are forced to gather on the docks..every morning from 5 o'clock on..for the sake of a half day's pay. **1948** *Ibid.* 26 Nov. 2/2 Retention of the traditional twice-a-day 'shapeup' or work call, with the guarantee of four hours work for men called to work only once in a single day. **1954** *Ibid.* 9 Apr. 20/4 There are now from 22 to 24 cities being considered for the major leagues... This is the opinion of Ford Frick, baseball's high commissioner... 'Of course,' Frick said, 'I don't know when their shape-up will take place.' **1967** *Boston Sunday Herald* 30 Apr. 1. 7/1 Boston's union longshoremen have sounded the death knell of their traditional but unwieldy dock shape-up. **1977** *Time* 17 Oct. 57/2 When Marlon Brando starred in *On the Waterfront* (1954), the morning shape-ups of New York Dock workers were pretty much as the movie portrayed them—noisy, brawling scenes of men fighting for the jobs available.

2. The action or an instance of shaping up.

1963 *Washington Post* 2 Oct. D2 The United States Olympic track and field team will have a final 'shapeup' meet in the Los Angeles Coliseum. **1977** *Time* 7 Feb. 20/2 He [*sc.* Lipshutz] presides at the daily 8 a.m. staff meetings... She is the only..outspoken liberal at Lipshutz's daily shape-ups. *Ibid.* 4 Apr. 60/1 Then, as the Central Intelligence Agency became mired in inefficiency, Schlesinger was tapped for the shape-up operation. As CIA director he immediately began to demythologize the agency.

shaping, *vbl. sb.* Add: **1. b.** *Electronics.* The process of modifying the waveform of an electrical signal.

[**1902** *Electr. Rev.* 10 Oct. 641/2 On adjusting the inductance and resistance of the shunt, along with the receiving condenser and its shunt, the signals are effectively curbed and shaped at the receiving end without reference to the sending station.] **1924** *Jrnl. Inst. Electr. Engineers* LXII. 192/1 The relation of damping to the reception of wireless signals, viz. its relation to the 'shaping' of the received dots and dashes. **1949** H. E. Penrose *Princ. & Pract. Radar* x. 152 A diode may..be employed with the primary object of exercising amplitude control, or..with the primary object of shaping. **1971** J. H. Smith *Digital Logic* iv. 68 The output of the shaping circuit is fed to the S_3 trigger pulse input and a 20 μF capacitor is used to control the pulse length.

c. *Radar.* Modification of a radar beam so as to obtain a desired spatial configuration.

1945 C. S. Pao *Shaping Primary Pattern of Horn Feed* (M. I. T. Radiation Lab. Rep. No. 655) 2 Such beam shaping cannot usually be achieved by merely changing the geometrical dimensions of the horn. **1975** D. G. Fink *Electronics Engineers' Handbk.* xxv. 66 In the horn-fed reflector the shaping can be achieved by either the reflector or the feed.

‖ **shapka** (ʃæ·pka). [Russ., = hat.] A brimless Russian hat of fur or sheepskin. (See also quot. 1945.)

1945 *Richmond* (Va.) *News-Leader* 2 Aug. 14/3 Newest thing in casual headgear is called a 'shapka' (meaning a small informal hat in Hungarian)... This headpiece is a cross between a 'babushka' and a snood, and can be worn over any kind of coiffure. **1958** *Philadelphia Sunday Bull.* 30 Mar. v. 1/3 Wearing one of those 'shapka', or tall Russian hats, the wisecracking Hope descended on a Moscow shivering in six-degree-below-zero temperature. **1963** V. Nabokov *Gift* iv. 271 He never removed either his fur-lined dressing gown or his lambskin shapka. **1977** *Time* 14 Feb. 33/2 Every [Russian] man wears a *shapka*, a fur..hat with ear flaps.

sharable, var. *Shareable *a.* **Shararat,** var. *Sherarat.

‖ **sharav** (ʃara·v). Also **Sharav.** [ad. Heb. *šārāḇ* parching heat.] A hot desert wind occurring in the Middle East in April and May; = Khamsin.

1968 *Listener* 27 June 827/2 In Crete they really cannot help it if the killingly hot *sharav* has blown across from the Levant. **1969** O. Hesky *Sequin Syndicate* ix. 96 Mornings when the *sharav* would blow. **1973** *New Scientist* 14 June 670 Ill winds such as the Föhn in Germany and the Sharav in the Near East can produce a malaise in humans. **1980** *Times* 4 Sept. 2/6 Seasonal 'ill-winds' such as the Mediterranean föhn and sirocco and the Middle Eastern sharav.

Sharawaggi. Delete ‖ and *Obs.* and add: (ʃæ:rawæ·dʒi). Also with small initial. Revived in the twentieth century with particular application to landscape gardening and architecture.

For a discussion of etymological hypotheses see **1949** *Archit. Rev.* CVI. 391/2.

1933 *Times Lit. Suppl.* 28 Dec. 913/4 Gothicism, *Chinoiserie*, all that may be summed up in the magic word *Sharawadgi* that was imposed by ignorance or mystification upon the connoisseurs of the time as a genuine Oriental art term. **1937** A. R. Humphreys *William Shenstone* ii. 41 Sharawadgi, as understood in England, has the main ingredients... It has no faith in mathematics and defies irregularity... It finds beauty in infinite variety... It treats natural material according to that material's own potential of grace. **1944** *Archit. Rev.* XCV. 3 (*heading*) Exterior furnishing or sharawaggi: the art of making urban landscape. **1966** Nairn & Pevsner *Buildings of England: Sussex* 294 What Petworth shows more than anything else is Sharawaggi..good buildings of all dates mixing perfectly at a hint of man.

shard, sherd, *sb.*[1] Add: **II. 4. a.** *spec. Archæol.*, a piece of broken pottery. (Later examples.)

Sherd is now established as the normal *Archaeol.* spelling.

1937 *Jrnl. R. Anthrop. Inst.* LXVII. 233, I could find no bronze-age sherds. **1955** *Sci. Amer.* July 46/3 We came upon a few fragmentary sherds of Aegean painted pottery. **1971** *World Archaeol.* III. 203 Many historic Amphlett sherds were recovered.

Shardana, Sherden (ʃa·rdāna, ʃə·rdən), *collect. pl.* [ad. Egyptian *Šrdn*.] One of the Sea Peoples, tentatively identified with the later Sardinians, who fought against the Egyptians in the 13th century B.C. and afterwards served them as mercenaries. Hence **Sha·rdan,** a member of this people. Also *attrib.*

1877 *Encycl. Brit.* VII. 739/2 The king of the Rebu (Libyans), with the warriors of several tribes joined the Shardana (Sardones), the Shakalasha (Sikels), [etc.]. **1910** *Ibid.* IX. 85/2 The Sherden had been in the armies of Rameses II, and are distinguished by their remarkable helmets and apparently body armour of metal. **1948** C. Dawson *Age of Gods* xiii. 300 The Shardana..were especially important and formed the royal bodyguard. **1952** O. R. Gurney *Hittites* i. 35 The Sherden..appear frequently in Egyptian inscriptions. **1960** K. M. Kenyon *Archaeol. in Holy Land* ix. 224 For centuries Shardans..had been in the habit of serving the Egyptians as mercenaries. *Ibid.* 227 It is known on literary evidence that Shardan mercenaries were employed by the Egyptians.

share, *sb.*[2] Add: **5. d.** Also, *upon shares, on the shares.*

1817 *Massachusetts Spy* 29 Jan. 1/2 To be let, upon Shares or Hire, a Farm. **1882** [see *renter *sb.*[1] 4 b].

6. *share bonus, broker* (earlier example), *capital, -dealing, index, -mart, -owner(ship), premium, price*; **share-farmer** chiefly *Austral.*, one who works on a farm for an agreed portion of the profits; so **share-farming** *vbl. sb.*; **share-hand** *U.S.*, a farm-worker or tenant who raises crops on shares; **shareman** (earlier and later N. Amer. examples); **share-milker** *N.Z.*, one who works on a dairy farm for an agreed portion of the profits (cf. *share-farmer* above); hence **share-milk** *v. trans.*, **share-milking** *vbl. sb.*; **share-pusher** (see quot. 1914); hence **share-pushing** *vbl. sb.* and *ppl. a.*; **shareswoman,** (*b*) (earlier and later examples).

1928 *Daily Chron.* 9 Aug. 8/4 A share bonus of 50 per cent. was provided on account of the year 1917–18. **1845** (*title*) Railway Mania; or, The Irish sharebroker. **1848** *Bradshaw's Railway Almanack* 57 Guaranteed 5 per cent. in perpetuity upon £3,000,000 (the authorized Share Capital). **1974** *Terminol. Management & Financial Accountancy* (Inst. Cost & Managem. Accountants) 62 *Equity share capital*, the issued share capital of a company which carries an unrestricted right to participate beyond a specified amount in a distribution. **1955** *Times* 17 May 18/4 Not inconsiderable profits have been made from time to time in sharedealing. **1969** *Times* 2 May (Suppl.) p. viii/4 The finance houses do not distribute as dividend the profits made on the realization of investments, colloquially known as share-dealing profits. **1928** R. G. Stapledon *Tour in Austral. & N.Z.* iv. 28 Many successful men have started as share-farmers... The system of old share-farming was probably unfair. **1927** *Southerly* XXVI. 203 Sharemilker..is a New Zealand term, which is certainly not in use in New South Wales, where the popular term is share-farmer. **1927** *Austral. Encycl.* I. 46 The details of 'share-farming' contracts varied with the district: thus in some cases the landowner provided everything but the labour and took two-thirds of the crop in return, in others the farmer provided plant, labour, and half the bags required and took half the crop. **1932** A. Jose *Australia Human & Economic* 262 Thus there came into favour a system of 'share-farming' (in Europe better known as metayage), which gave the actual cultivators an interest in good tillage while retaining in the owner's hands full control of his property. **1960** *Farmer & Stockbreeder* 12 Jan. 51/1 Several new variations of share-farming were propounded. **1911** Jenks & Lauck *Immigration Problem* 83 How much value careful cultivation, kitchen gardens and small store accounts may be due to the cotton 'share hand' and tenant. **1930** *Financial News* 13 Aug. 5/5 (*heading*) Industrial Share Index. **1982** *Financial Times* 1 May 24/2 The FT-Actuaries 500 share index eased only 0.9 per cent from Thursday's record high. **1687** *Connecticut Colony Public Rec.* (1859) III. 425 Fishermen..shall not presume to break off their voyage..without the consent of the owner, master and share-men. **1820** in C. R. Fay *Life & Labour in Newfoundland* (1956) vii. 159 Sharemen are frequently indigent planters who have fallen into debt with their merchant and who cannot afford to use their own boats. **1966** A. R. Scammell *My Newfoundland* 26, I was shareman with his father..three summers on the lower Labrador. **1870** J. K. Medbery *Men & Mysteries Wall St.* 19 In all the great European share-marts there is a general executive organization. **1937** Gordon & Bennett *Gentlemen of Jury* ii. 67 Two months later the mother is 'out in the sheds' helping to share-milk 100 cows. **1950** J. Guthrie *Little Country* xii. 205 Advertisements..for a farm-hand or a share-milker. **1977** *N.Z. Herald* 8 Jan. 4-7/1 (Advt.), A position has become available for a 50–50 sharemilker on a 130-acre dairy farm. The present sharemilker has purchased his own property. **1937** H. G. Philpott *Hist. N.Z. Dairy Industry* ii. 65 [He] adopted the system of farm labour now commonly known as 'share milking'. **1958** *Times* 16 June 12/7 In New Zealand 'share milking', as it is called, is controlled by a wages board award. **1973** *Massey Ferguson Rev.* (N.Z.) Mar.–Apr. 3/3 Sharemilking is an important contributor to the dairy industry in New Zealand. **1968** *Sci. Jrnl.* Nov. 85/2 Competitors, suppliers, customers, shareowners, bankers and the government. **1978** *Detroit Free Press* 5 Mar. B 13/3 [They] have made lifetime careers of trying to give shareowners a voice in the running of publicly held corporations. **1969** *Economist* 24 May 119/2 (*heading*) Wider shareownership. **1930** *Daily Express* 6 Oct. 14/2 The discount on this issue has been entirely written off from share premium and capital reserve account. **1950** *Economist* 29 Apr. 896/2 Bonus issues..would not be allowed to have fully discounted in advance the retention of the dividends. **1980** W. Ash *Incorporated* xiii. 156 You can imagine what could've happened to share prices if that got out beforehand. **1914** H. Halford *Dict. Stock Market Terms* 79 *Share pusher*, one who endeavours to dispose of Shares to the public by circular or advertisement, instead of selling them on the market. **1928** 'N. Shute' *Ruined City* xii. 247 We're a pension part..Couple of bloody share-pushers, if you ask me. **1965** B. Sweet-Escott *Baker Street Irregular* i. 34 A look which suggested that I was..somewhere between a share-pusher and a black marketeer. **1928** *Daily Mail* 3 Aug. 13/3 The day on which the sections penalising share-pushing shall come into force. **1928** *Evening News* 18 Aug. 11/3 Shares of this sort are among those favoured by the share-pusher... the share-pushing circulars of the 'bucket-shop' fraternity. **1972** *Times* 28 Dec. 17/4 The City section..has developed considerable expertise in cases involving prospectuses, share-pushing and market rigging operations. **1977** N. Faulks *No Mitigating Circumstances* xi. 152 A number of other persons were charged with taking part in the share-pushing transactions. **1867** G. E. Clark *Seven Years of Sailor's Life* xxvi. 272 The sharesmen were all looking at the mate. **1932** J. Robertson *Alaska* 12/1 *Oysterman & Fisherman* Mar. 14/1 The crew wage and crew feeding system practised by the 'sharesmen'-outfitters.

share, *v.*[2] Add: **4. f.** *Chem.* Of an atom, orbital, etc.: to hold (one or more electrons) in common with another atom or orbital, so as to form a covalent bond. (See also *shared *ppl. a.*)

1919 *Jrnl. Amer. Chem. Soc.* XLI. 888 An octet may share an even number of its electrons with 1, 2, 3, or 4 other octets. **1923** *Trans. Faraday Soc.* XIX. 461 In chemically stable molecules we have only to consider each atom sharing pairs of electrons. It is well known that such structures do not exhibit any signs of electrical polarity. One must therefore suppose that the net charge on both atoms is zero, i.e. that the two shared electrons are in general so distributed that when one is in one atom the other is in the other. **1964** J. W. Linnett *Electronic Structure of Molecules* ii. 29 The two electrons may be regarded as being shared between the 1s orbital of the hydrogen and one of the 2p orbitals of the fluorine.

6*. *intr.* and *trans.* In the language of Moral Rearmament: to confess one's sins openly; to impart to others one's spiritual experiences. Also const. *with*. Also in wider use.

1932 [implied at *sharing *vbl. sb.*[2] 2]. **1933** S. A. King *Challenge to Oxford Groups* v. 45 What does the Bishop think a man feels when he has 'shared' to 'witness' and finds that God has used that 'sharing' to bring a brother out of..bondage? **1934** R. Macaulay *Going Abroad* xiv. 151 She would, she thought, be able to share with another girl in a way she could not with Hugo. *Ibid.* xvii. 197, I must say, I did annoy my father a bit by trying to make him share with me... **1940** Graves & Hodge *Long Week-End* xi. 200 One of the practices was to 'share' confessions of their sins. **1949** A. Wilson *Wrong Set* 9, I do believe you're trying to get me to 'share'. And I never even guessed that you were a groupist. **1980** B. Pym *Few Green Leaves* (1980) xix. 140 Sometimes she had been a good listener. **1982** J. Heath *Murder of Plumed* (1982) xix. 112 The way she'd shared her..right to tell me about her..problems.

7. *share-out*: for 'Chiefly' in Dict. read 'occas.' Also that which is distributed; a portion or share (of profits, interest, etc.), a 'cut'. (Later examples.)

1941 *Sun* (Baltimore) 27 Jan. 4/1 It could be 'well in at the head of the table for the shareout' when the war ended. **1951** A. L. Rowse *England of Elizabeth* viii. 325 The new nobility around the young king helped themselves to a vast share-out of Crown and Church lands. **1963** *Times* 7 May 18/2 No, a share-out it may be—and each shareholder may have his own private, affectionate name for it—but in..the businesslike print of the Birmingham Post, 'dividend' is the better term. **1976** *Scottish Daily Express* 23 Dec. 6/2 Kilkerr was forced to accept his share-out from Soho vice bosses on Friday nights.

shared *ppl. a.* (later examples in *Chem.*).

1923 [see sense 4 f above]. **1939** L. Pauling *Nature Chem. Bond* ii. 6 In methane the carbon atom, with its two inner electrons and its outer shell of eight shared electrons, has assumed the stable two-shell configuration of neon. **1977** H. S. Pickering *Covalent Bond* ii. 17 A variant of the covalent bond occurs in which the two shared electrons of the bond come originally from one of the two atoms.

shareable (ʃɛə·rāb'l), *a.* Also **sharable.** [f. Share *v.*[2] + -able.] That may be shared. Hence **shareabi·lity.**

1920 *Q. Rev.* July 161 It must be shared or at least must be shareable; otherwise it were nothing at all. **1932** W. T. Stace *Theory of Knowl. & Existence* 332 Shareability of perception is what distinguishes the real from the unreal. **1935** W. de la Mare *Early One Morning* i. 19 In the telling of these vividly sharable experiences no direct hint is given that the author is actually seeing himself..as a child. **1977** Fontana & Van de Water in Douglas & Johnson *Existential Sociol.* iii. 102 The potential shareability of human understanding..is so great that complex systems of exchange can be constructed by groups who share no language. **1980** *Dædalus* Spring 20 Scientific understanding is essentially and eminently shareable.

share-crop (ʃɛ̄ə·ɹkrɒp), *v.* Chiefly *U.S.* Also **sharecrop, share crop.** [f. Share *sb.*[2] + Crop *v.*] **a.** *intr.* To farm on shares (see Share *sb.*[2] 5 d). **b.** *trans.* To grow (a crop) on this system; also *transf.* So **sha·re-crop** *sb.*, a crop raised on shares; also *attrib.*; **sha·re-cropper,** one who share-crops; also *attrib.*; and hence **sha·re-cropped** *ppl. a.*; **sha·re-cropping** *vbl. sb.* and *ppl. a.*

[**1867** in J. H. Easterby *S. Carolina Rice Plantation* (1945) 231 This will be cheaper in the end than the contract or share of the crop system.] **1907** *Springfield* (Mass.) *Weekly Republ.* 25 Apr. 1 The 'share crops' system is what its name implies, the immigrant being housed and fitted with all the necessaries and then sharing the proceeds of the harvest with the landowner. **1925** *Annals Amer. Acad.* Jan. 61/1 Number of farmers operated by tenants of various kinds (cash, share and share-cash 'croppers', standing tenants, etc.) increased about 100,000. **1928** J. Peterkin *Scarlet Sister Mary* xxi. 235 The..thing would cost as much as his whole share-crop of cotton would make in five years. **1929** L. R. Gottschalk *Era French Revolution* 33 Most of them had become métayers, who, like our share-croppers, farmed a piece of land for a stipulated portion..of the harvest. **1930** *Dialect Notes* VI. ii. 83 *Share-crop*, v., to farm 'on shares'... The word, like the practice, is very common. **1936** *Daily Progress* (Charlottesville, Va.) 3 Nov. 1/5 Bridges, who sharecrops for W. G. Gray, Senath attorney, sat in a big chair and poked wood in the fire. **1944** *Chicago Daily News* 2 Dec. 4/6 The Capone gang owned several locals of the hotel and restaurant workers in Chicago, and had Lou Romano share-cropping them. **1945** *Reader's Digest* Nov. 26/1 He found hundreds of blacks who share-cropped cotton. **1945** B. A. Botkin *Lay My Burden Down* 225 That was the beginning of the sharecropping system. **1947** *Social Forces* Dec. 202/1 One hundred Negro sharecropper families on the King and Anderson plantation. **1960** H. E. Bates *Aspidistra in Babylon* 215 A toothless Chinaman..share-cropped vanilla farther up the hill. **1962** E. Snow *Red China Today* (1963) lvi. 427 The 'well-to-do' peasants (living at a prosperous share-crop level) resented anything that looked like loss of control over their savings, homes and land. **1970** S. L. Barraclough in I. L. Horowitz *Masses in Lat. Amer.* iv. 104 When legalized slavery and forced labor were abolished..various forms of share-cropping took their place. **1973** W. McCarthy *Detail* i. 62 Paul..came from a large, poor, sharecropper family, in the heart of the delta land of Mississippi. **1973** *Advocate-News* (Barbados) 17 Feb. 8/2 A depression time share-cropping family. **1974** *Evening Herald* (Rock Hill, S. Carolina) 18 Apr. 13/2 He was born in Greenville County, S. C., and spent his youth helping his father sharecrop. **1976** *Jrnl. Devel. Econ.* III. 345 In this exercise, we limit ourselves in focus to a single sharecropped plot on the fazenda.

Sharia (ʃarī·ā). Also **Shariah, Shariat,** † **Sharieh,** † **Sheriat,** and with small initial. [Arab. *šarī'a*.] The Islamic religious law, including the teachings of the Koran and the traditional sayings of Muhammad.

1855 R. F. Burton *Personal Narr. Pilgrimage to El-Medinah* II. xxi. 281 In fact, justice at El Medinah is administered in perfect conformity with the Shariat or Holy Law. **1877** *Encycl. Brit.* VII. 113/2 Shi'at or Sher'iat, i.e. legal religion under the supervision of a murshid. **1920** *19th Cent.* Sept. 500 Questions of divorce and inheritance are decided by the religious or Sharieh judge, from whom there is an appeal to the Sharieh Court of Appeal in Jerusalem. **1921** *Glasgow Herald* 16 July 7 A code based on the Shariat and prepared upon his orders would come into force. **1927** *Ibid.* 21 Oct. 11 The Sheriat (Moslem) canonical law has practically been abolished, a civil code borrowed from Switzerland being substituted for all questions of marriage and inheritance. **1936** F. Stark *Southern Gates Arabia* iii. 39 The law is the Muhammedan shari'a. **1965** *Mod. Law Rev.* XXVIII. v. 543 The marriage was dissolved, according to Mohammedan law by the unilateral declaration of *talak* divorce in the appropriate Sharia Court. **1971** *Illustr. Weekly India* 4 Apr. 47/1 For long *Sharia* has not been administered by a Caliph or Imam. **1979** *Guardian* 28 Mar. 18/4 At the basis of a way of life which was remarkable for its homogeneity, is the Shariah—meaning simply the way or path.

sharif, Sharifian, varr. Shereef, Shereefian *a.* in Dict. and Suppl.

sharing, *vbl. sb.*[2] Add: **2.** *spec.* The action of *share *v.*[2] 6*.

1932 A. J. Russell *For Sinners Only* ii. 25 They [*sc.* The Oxford Group] defined Sharing as meaning two distinct things—further definable as Confession and Witness. **1945** N. L. McClung *Stream runs Fast* xxx. 298 They have class meetings, too, but they call them 'quiet times', and they tell their spiritual experiences, but they call that 'sharing', and they care nothing about money. **1964** T. Driberg *Mystery of Moral Re-Armament* ii. 38 An ill-judged attempt by Buchman to promote, if only vicariously, the practice of Sharing.

shark, *sb.*[1] Add: **1. d.** *transf. Naut. slang.* A sardine.

1916 'Taffrail' *Pincher Martin* viii. 144 There was a peculiar tang in the air... He found out afterwards that it emanated from various sardine-preserving factories, and the discovery put him off canteen 'sharks' for quite a week.

2. c. (Further example of sense 'the press-gang'.)

1866 'Mark Twain' *Lett. from Hawaii* (1967) 81 The professional 'sharks' in New Bedford and New London who furnish crews to ships.

d. (Earlier example.)

1806 *Port Folio* 17 May 304/1, I got plenty of promises, Latin, and jaw, And who ever got more from a lawyer? Of the sport I got sick, so threw up the game, For my cash by the sharks had got eaten.

e. *U.S. College slang.* A highly intelligent or able student. ? *Obs.*

1895 W. C. Gore in *Inlander* Dec. 111 *Shark*,..a person who is very bright either in a general way or (more often) in some particular line of work. **1903** *Williams College Class Book* 29 'Dido' is a Math. shark of the first water. **1909** *Springfield* (Mass.) *Weekly Republ.* 8 July 12 The 'shark' does well in his lessons, but recognizes that study is the first thing in college. 'Sharks' play games. **1920** [see *elocute v.].

4. a. *shark-fisher, -steak* (earlier example); *shark-fishing* (later examples); *shark-infested, -proof* adjs.

1897 'Mark Twain' *Following Equator* xiii. 142 He was passing by a nodding shark-fisher. **1914** *Chambers's Jrnl.* Feb. 89/1 Shark-fishing is regarded as being as much a trade as a sport. **1976** L. Deighton *Twinkle, twinkle, Little Spy* xi. 117 Is she interested in stud farms or shark fishing? **1978** *Detroit Free Press* 5 Mar. A 17/1 Rescue crews Saturday searched shark-infested waters for the bodies of..crewmen killed in the crash of a..domestic airliner. **1923** 'R. Daly' *Enchanted Isl.* x. 92 She had been bathing in the shark-proof palisade below. **1967** *Coast to Coast 1965–66* 162 Some evenings the Roebourne mob..would..swim in..our shark-proof pool beside the wharf. **1847** H. Melville *Omoo* xiv. 65 A shark-steak and be hanged to you!

b. **shark-bait,** *Austral. colloq.*, a lone or daring swimmer well out from shore; hence **shark-baiter, -baiting; shark net** *S. Afr. local*, a length of netting positioned off-shore to protect bathers from sharks; also *shark netting*; **shark's fin** = *shark-fin*; also in **shark's fin soup; sharkskin,** (*a*) also *attrib.*; (*b*) (i) a woven or warp-knitted fabric of wool, silk, or rayon with a smooth, slightly lustrous, finish; freq. *attrib.*; (ii) an outfit made of this fabric; **shark-toothed** *a.* (earlier and later examples).

1920 A. H. Adams *Australians* 177 Farther out in the deep water swam the venturous line of experts, technically known as 'shark-bait'. **1937** K. S. Prichard *Intimate Stranger* i. 16 'Shark-bait', boys and girls on the beaches called her, she was so daring. Always swimming out there beyond the reef. **1924** A. Wright *Rung In* iii. 31 It might be only some foolhardy 'shark baiter', as he heard the more venturesome of the bathers termed. **1965** *Austral. Encycl.* VIII. 82/2 Solitary bathers are more often attacked than groups, but the 'shark-baiter' farthest off shore is not necessarily the victim. **1951** Cusack & James *Come in Spinner* 221 I've given up shark-baiting. Mug's game. **1967** K. S. Prichard *Subtle Flame* 99 I'm no good at shark baiting. **1970** *Studies in English* (Univ. Cape Town) I. 33 These bracelets were originally made out of shark netting. The surfer would dig his way out to the shark nets, cut himself a piece and tie it around his wrist. **1977** J. McClure *Sunday Hangman* ix. 95 The shark nets protecting the bathers off its [*sc.* Durban's] beaches. **1933** *Gourmet's Bk. Food & Drink* iii. 49 In his own country the Chinaman's evening meal is a somewhat variegated affair..and includes..shark's fins, cucumber, fish brawn. *c* **1938** *Fortnum & Mason Price List* 58/1 Soups..Sharks' Fins per bot. 7/6. **1966** *Guardian* 30 July 7/3 In the heart of Chinatown, shark's fin soup with crab sauce. **1978** *Nagel's Encycl.-Guide: China* 379 Sharks' fins need lengthy preparation, because they are bought dried. **1851** H. Melville *Moby Dick* III. xxvii. 174 With matted beard, and swathed in a bristling shark-skin apron..Perth was standing between his forge and anvil. **1932** C. Beaton *Diary* Mar. in *Wandering Years* (1961) xiii. 255, I bought vast quantities, at almost negligible cost, of football vests, exotic footgear, the scantiest shorts in all colours and in white sharkskin. **1944** R. Chandler *Lady in Lake* ii. 11 The man wore trunks and the woman what looked like a very daring white sharkskin bathing suit. **1957** L. Durrell *Justine* iii. 183 Now in his ice-smooth sharkskin with the scarlet cummerbund he seemed..the richest and most handsome of the city's bankers. **1974** D. Ramsay *No Cause to Kill* ii. 110 Ivy Eastbrook..in silk shirt and sharkskin trousers. **1979** E. Koch *Goodnight, Little Spy* ii. 6 During the winter he wore..five serge suits and two sharkskins. **1794** T. Dwight *Greenfield Hill* 79 What stretches Avarice's gulphy maw, And opens wide her shark-tooth'd jaw. **1935** C. Day Lewis *Time to Dance* 42 Over the shark-toothed Timor sea Lost their bearings.

shark, *v.*[3] Delete *U.S. local.* and add: Only as **sha·rking** [formed after *fishing*, etc.], shark-fishing; also *attrib.*

1881 A. J. Northrup *'Sconset Cottage Life* xi. 100 No summer experience at 'Sconset is complete without..one 'sharking' expedition. **1882** E. K. Godfrey *Island of Nantucket* 329 A visit can be made to the 'sharking grounds'. **1937** J. W. Day *Sporting Adventure* 219 The Isle of Arran, off the Scottish coast, is the centre from which the new sport is being followed. A fishing-smack has been fitted out specially there for parties who wish to go out 'sharking'. **1960** *Sunday Express* 24 July 13/5 Good sharking!

sharka (ʃā·ɹkā). Also (*rare*) **sarka.** [f. Bulg. *sharka na slivite* pox of plums.] = *plum pox* s.v. *Plum *sb.* 6 d.

1961 *Tidsskrift for Planteavl* LXV. Saernummer 138 The sarka virus disease (plum pox) is the most significant virus of fruit trees in Yugoslavia. **1973** *Plant Virol.* 167 Sharka has become one of the very serious problems for the plum production in Europe. **1974** K. M. Smith *Plant Viruses* (ed. 5) ii. 13 In the parenchyma cells of fruit from plum trees infected with the 'Sharka' virus, 'plum-pox', cytoplasmic and intranuclear inclusion bodies have been observed. **1976** [see *plum pox* s.v. *plum *sb.* 6d].

sharp, *a.* and *sb.*[1] Add: **A.** *adj.* **1. d.** (Further N. Amer. examples.)

1836 *Col. Crockett's Exploits in Texas* i. 20 A fellow..who in those parts was considered as sharp as a steel trap. **1912** *Dialect Notes* III. 589 They won't fool him; he's a *sharp as tacks*. **1976** *National Observer* (U.S.) 10 Apr. 13/4 Mrs. Owen..is not only as sharp as a tack but is perhaps the best-looking school principal in Texas or elsewhere.

3. b. Substitute for def.: Of reasoning or discourse: acute, sagacious. Also, of remarks: pointed, apt, witty. (Later example.)

1968 *Observer* 14 Apr. 24/7 It was a sharp idea of the BBC's Religious Department, letting Malcolm Muggeridge wander round the Holy Land.

f. In colloq. phr. *you're so sharp you'll cut yourself* and varr.: variously used as an observation, reproof, or warning implying over-cleverness.

[**1903** 'T. Collins' *Such is Life* (1944) 278 Gosh! you've been on the turkey; you'll be cutting yourself some of these times.] **1910** 'H. H. Richardson' *Getting of Wisdom* xiv. 142 If you're so sharp, you'll cut yourself! **1930** W. S. Maugham *Cakes & Ale* x. 116 You're so sharp you'll cut yourself if you don't look out. **1968** J. Fleming *Kill or Cure* xiv. 189 He was as sharp as a bag of monkeys, that sharp he'd cut himself.

4. k. Of vehicular transport: ahead of schedule, early; hence of a time-table, etc.: tight, demanding. *colloq.*

1942 Berrey & Van den Bark *Amer. Thes. Slang* § 768 *Hot, sharp*, ahead of schedule. **1945** *Transit News* (Capital Transit Co., Washington, D.C.) 15 June, When a car or bus is ahead of schedule, it's 'Hot or Sharp', while when late it's 'Dragging'. **1977** *Modern Railways* Dec. 480/2 Certain of the intermediate schedules are quite sharp.

6*. As a general term of approbation. orig. *U.S. slang.* **a.** Excellent, fine.

1940 J. O'Hara *Pal Joey* 97, I sound like everything was sharp. **1963** in C. Booker *Neophiliacs* (1969) viii. 186 WIP'S opens late february london's sharpest nightclub. **1979** *Arizona Daily Star* 5 Aug. (Advt. Section) 20/4 The home is sharp with four bedrooms. *Ibid.* 20/3 Sharp and roomy 4 bdrm split plan with spacious modern kitchen.

b. Of clothes: stylish, fashionable, smart, 'snappy'. Hence of the wearer: well-dressed, attractive.

1944 C. Calloway *Hepsters Dict.*, *Sharp*.., neat, smart, tricky. Ex., 'That hat is sharp as a tack.' **1956** B. Holiday *Lady sings Blues* i. 20, I was always the sharpest kid in the block when I was dressed up. **1962** *Observer* 18 Feb. 23/2 It's more a desire for things you haven't got but feel you've a right to, because other people have them—a sharp suit, good things, neat things, flashy things. **1969** W. Ash *Take-off* iv. 57 When Jacques turned up, he was looking pretty sharp..in the sort of dark suit which..looks expensive. **1977** N. Marsh *Last Ditch* iii. 55 Louis..looked almost embarrassingly smooth in breeches, boots, sharp hacking-jacket and gloves.

c. Of a motor vehicle: smart, well-equipped; in good condition. Cf. *Sharp *sb.*[1] 13.

1970 *Globe & Mail* (Toronto) 28 Sept. 27/4 (Advt.), Chevrolet convertible, fully equipped, a real sharp car. **1974** *Anderson* (S. Carolina) *Independent* 19 Apr. 10B/6 (Advt.), V-8, automatic power steering, electric seats.. one of the sharpest around. **1977** *Drive* Sept.–Oct. 16/1 The Saab interior, however, is drab—not sharp at all.

11. a. *spec.* of the definition of a photographic image; also *transf.* of a lens producing a sharp image. (Further examples.)

1883 J. H. T. Ellerbeck *Amateur's First Handbk.* iv. 22 Screw out the whole until, having taken the cap off the lens, you find the image, upside down, coming up sharp, then take a magnifier and see that it is perfectly sharp. **1921** *Daily Colonist* (Victoria, B.C.) 6 Apr. 5/1 (Advt.), Negatives which are exceptionally clear and 'sharp' make splendid enlargements. **1961** G. Millerson *Technique Television Production* iii. 34 Many simple photographic and motion-picture cameras have no focusing mechanism. And yet, at the push of a button, they produce acceptably sharp pictures. **1979** *SLR Camera* Jan. 42/3 Although we did not shoot our optical test target we can say that this is a very sharp lens.

b. *Physical Sci.* Of a phenomenon, condition, or state, esp. resonance: having, or occurring over, a narrow range of values of energy; capable of graphical representation by a curve showing a sharp peak; clearly defined.

1906 G. Eichhorn *Wireless Telegr.* vi. 40 The slighter the damping, the sharper the resonance. **1936** R. S. Glasgow *Princ. Radio Engin.* ix. 248 The effect of resistance predominates and the tuning is sharpest at the low-frequency end. **1960** Dicke & Wittke *Introd. Quantum Mech.* xvi. 308 The longer a particle can stay trapped before escaping, the sharper the energy level is. **1971** P. E. Hodgson *Nuclear Reactions & Nuclear Structure* xiv. 414 If the resonance is sharp..the cross-section due to the resonating partial waves greatly exceeds that due to all the other partial waves.

B. *sb.* **8. b.** (Earlier example.)

1840 *Spirit of Times* 12 Sept. 330/2 This race completely took in the 'sharps', who brought the bay filly as a 'bite' on purpose to beat the chesnut, who won the race.

10. (Earlier example.)

1834 *Chambers's Edin. Jrnl.* III. 129/2 The traveller, knowing the fondness of the Africans for needles, had brought..a great quantity of *Whitechapel sharps*.

12. *Diamond-cutting.* **b.** A sharp piece of diamond used to mark the point of intended cleavage; a pencil-like tool to which such a diamond is attached.

1903 W. R. Cattelle *Precious Stones* 67 To cleave, the crystal is fastened to the end of a stick and a V-shaped incision made in the grain with a sharp piece of diamond, called a 'sharp'. **1973** G. Jenkins *Cleft of Stars* iii. 36 Called technically a 'sharp', my diamond pencil looked like an ordinary pencil made of metal.

13. *N. Amer. slang.* A second-hand car in excellent condition (see quot.). Cf. sense 6* c of the adj. above.

1960 Wentworth & Flexner *Dict. Amer. Slang* 463/2 *Sharp*,..a used but well-cared-for automobile having extra accessories.

C. *Comb.* **1.** *sharp-bladed, -boned, -bowed, -ribbed, -scented, -textured, -thorned.*

1913 J. London *Let.* 5 Sept. (1966) 397 You must in your dealings be..as straight as the edge of the sharp-bladed sword. **1933** W. de la Mare *Fleeting* 119 A homelier music than this bleaching wind's In these sharp-bladed grasses. **1794** T. Dwight *Greenfield Hill* 44 His sharp-bon'd horse..Tied, many an hour, in yonder tavern-shed. **1976** W. Trevor *Children of Dynmouth* i. 14 Timothy Gedge was..a boy with a sharp-boned face and wide, thin shoulders. **1865** W. Whitman *Drum-Taps* 41 O the beautiful, sharp bow'd steam-ships. **1946** I. Irving *Royal Navalese* 155 *Sharp-bowed*, the description of a man who has had a very close-cropped hair-cut. **1844** J. R. Lowell *Poems* 220 Grim Boaz, who, sharp-ribbed and gaunt, yet feared A thing more wild and starving than himself. **1910** W. de la Mare *Three Mulla-Mulgars* xvii. 224 Thimble lay in a sleep so quiet..it seemed to Nod the heart beneath the sharp-ribbed chest was scarcely stirring. **1927** E. Sitwell *Rustic Elegies* 81 The sharp-scented rose-boughs. **1864** G. M. Hopkins *Let.* 20 July (1956) 213 *Roughed it*; I believe it means irritating the skin on sharp-textured blankets. **1967** *Coast to Coast 1965–66* 193 You lie down in the sharp-textured air of the desert night. **1912** W. de la Mare *Listeners* 92 Wreathed shall with incense be Thy sharp-thorned may. **1965** J. A. Michener *Source* (1966) 76 Sharp-thorned vines clutched at them and sucking mud tried to grasp their ankles.

2. sharp end *Naut. slang*, the bows of a ship; also *transf.*, the front line, the centre of activity; esp. in phr. *at the sharp end*; **sharp eyespot,** a fungal disease of cereals similar to eyespot but caused by *Corticium solani* (*Thanetophorus cucumeris*) and characterized by more clearly defined markings; **sharpshins** *dial.*, (*a*) a fleet-footed person; (*b*) a sharp-witted person; an intelligent child; **sharp-toothed** *a.* (later examples).

1948 Partridge *Dict. Forces' Slang* 166 *Sharp end, the*, the bows of the ship... (Navy.) (2) Hence, *at the sharp end*..at the front, well forward. **1973** D. Francis *Slay-Ride* i. 9 Arne pointed the sharp end back... The dinghy slapped busily through the little waves. **1976** *New Scientist* 28 Oct. 230/2 Within a few months I was appointed financial controller... But I still wasn't at the sharp end. **1980** A. Price *Hour of Donkey* ii. 36 The distant sound of bombing indicated that he was very close to the sharp end of the war. **1943** *Nature* 7 Aug. 161/1 In the first wheat crop after grass, eyespot is generally absent or rare, but sharp eyespot is found just as commonly as on old arable land. **1980** F. Hope in E. Gram et al. *Recognition & Control of Pests & Dis. Farm Crops* (ed. 2) 136/1 Sharp Eyespot (*Rhizoctonia cerealis/Corticium solani*) is similar in appearance to Eyespot, the main difference being that the lesions are more defined and angular, whilst the dark borders are easily distinguished from the linear areas. **1883–6** C. S. Burne *Shropshire Folk-Lore* xxxv. 581 'Sharpshins' is still applied in Shropshire, 1st, to light heels, 2nd, to sharp wits, *e.g.* 'Be off, sharpshins!'=run away, make haste... 'Now then, sharpshins! taking me up as usual!'..said in rebuke to some smart speech, display of cleverness, or captious criticism. **1915** D. H. Lawrence *Rainbow* iii. 76 'What does she say, that I'm a fawce little thing?' the small girl asked afterwards. 'She means you're a sharp-shins.' **1855** W. Whitman *Leaves of Grass* 62 Blind loving wrestling touch! Sheathed hooded sharptoothed touch! **1938** M. K. Rawlings *Yearling* xxi. 270 They had found..the weak and the strong brought together to earth, the sharp-toothed and the dull.

sharp, *adv.* Add: **1. a.** Also, smartly, nattily (after *Sharp *a.* 6* b).

1951 J. H. Smyth *I, Mobster* xiii. 142 He was dressed sharp, like the wise guys on Broadway. **1981** 'D. Shannon' *Murder most Strange* i. 15 He was..dressed real sharp, a gray suit, not just sports clothes.

sharp, *v.* **1. a.** For 'Now only *dial.*' read 'Now *dial.* or *arch.*' and add later example.
a **1945** E. R. Eddison *Mezentian Gate* (1972) xxviii. 137 A ready means lay to hand in converse with brother: a merry war, sharping and training up the claws of her wit.
5. c. (Later example used *intr.* with personal subject.)
1895 [see *flat *v.*² 7].
f. *intr.* for *refl.* To dress *up*, to dress smartly. *U.S. colloq.* Cf. *sharp *a.* 6* b.
1957 J. Kerouac *On Road* (1958) i. ix. 53 Tim, Rawlins, and I decided to sharp up for the big night.

Sharpa, obs. var. *Sherpa.

sharpen, *v.* Add: **1. c.** In fig. phr. *to sharpen one's pencil*: to prepare to work; to revise or improve one's work.
1957 *Times Lit. Suppl.* 15 Nov. 689/3 This is where the post-historic Ph.D. men will sharpen their pencils. **1965** *Daily Progress* (Charlottesville, Va.) 17 June 33/5 He suggested that Stahr and Hovde go home and suggest to local power officials that they 'sharpen their pencils' and figure out lower rates. **1969** *Listener* 2 Jan. 10/2 We.. published our findings. Radio telescopes all over the world were trained on the sources, while theoreticians sharpened their pencils.
2. Now freq. const. *up*. **a.** (Later examples.)
1947 'L. Starr' *Corrie* xii. 161 Just a little cocktail..to sharpen up our appetites for lunch. **1953** S. Plath *Let.* 15 May (1978) i. 114 Sharpening up writing again, once it's rusty, is very painful.
c. Also, of a political measure.
1962 *Listener* 2 Aug. 160/1 French anti-cartel policy was sharpened up by the act of 1958.
g. *refl.* To improve one's appearance; to smarten oneself up.
1952 S. Selvon *Brighter Sun* ii. 24 Since the Americans came the girls sharpened themselves up and wouldn't be had for less than five Yankee dollars.

Sharpey's fibre (ʃā·ɹpi). *Anat.* [Named after William *Sharpey* (1802–80), Scottish anatomist, who described such fibres in 1856 (J. Quain *Anat.* (ed. 6) I. p. cxx).] A fibre of connective tissue passing from the periosteum through the lamellæ of a bone or tooth.
[**1878** *Q. Jrnl. Microsc. Sci.* XVIII. 132 The lamellæ when stripped off from a bone that has been softened in acid but subsequently freed from all traces of the acid by long steeping in water or spirit exhibit under the microscope the appearance of intercrossing fibres (the reticulating fibres of Sharpey).] **1890** G. M. Gould *New Med. Dict.* 400/2 *Sharpey's fibers*, calcified fibers of white, fibrous tissue bolting together the peripheric lamellæ of bone. **1896** A. Clarkson *Text-bk. Histol.* vi. 151 In addition to fibres forming the lamellæ themselves, there are others—perforating, or Sharpey's fibres—to be found in bone. **1946** *Nature* 24 Aug. 269/1 Superficially placed osteocytes with their processes, and Sharpey's fibres, stand out black against the colourless matrix. **1962** Blake & Trott *Periodontology* ii. 12 The parts of the fibres which lie embedded within cementum or bone are known as Sharpey's fibres.

sharpie. Restrict *U.S.* to sense in Dict. and add: **1.** (Earlier example.)
1860 *Diary* 10 Dec. in *Outing* (1913) Mar. 688/2, I took some of the skiffs and sharpies behind the Emma S...and we went down to Whig inlet.
2. *colloq.* (orig. *U.S.*). **a.** = Sharper¹ 2; **b.** = Sharp *sb.*¹ 8 b.
1942 Berrey & Van den Bark *Amer. Thes. Slang* §461 *Clever Crook*,..sharper, sharpie, sharpshooter, slicker. *Ibid.* §637 *Sports predictor*,..dopester, sharp, sharpie. **1944** *Chicago Daily News* 4 Nov. 6/1 Central characters of both plays are engaging highbinders and sharpies who are not exactly thieves, but more than slightly overoptimistic in their use of..other people's money. **1949** W. R. Burnett *Asphalt Jungle* xiv. 92 He..couldn't make up his mind whether he'd been a chump or a sharpie. **1964** S. Bellow *Herzog* 3 He had chosen to be dreamy..and the sharpies cleaned him out. **1974** *Times Lit. Suppl.* 3 May 465/4 The same wise-cracking, classless sharpie..can make throwaway remarks about Ingrid Haebler in his own tatty old flat. **1979** T. Gifford *Hollywood Gothic* vi. 71, I had drunks.. directors, producers, New York sharpies of every kind.
3. *Austral. colloq.* A young person who adopts certain extreme or provocative styles of hair, dress, etc. (see quot. 1975); the Australian counterpart of the *skinhead.
1965 W. Dick *Bunch of Ratbags* 202 The more a sharpie protested he was not a bodgie, the more they [*sc.* the police] laughed and belted him. **1972** *Sydney Morning Herald* 20 Jan. 2 It was alleged in evidence that Still died during an incident involving 'sharpies' and 'long-hairs' outside Greystanes Progress Hall. **1975** *Sun-Herald* (Sydney) 13 Apr. 7 A sharpie is usually aged between 14 and 19 years. The boys wear their hair cropped short on the top and sides and longer at the back. The girls often wear 'dolly' makeup and have their ears pierced. Tattoos are often worn by both sexes. The sharpies wear blue jeans or high-waisted slacks supported by old-fashioned braces, matched with a tee shirt and sometimes a woollen cardigan. **1977** *Sunday Mail* (Brisbane) 21 Aug. 37/3 Carmel says her mother accepted her being a sharpie—even a punk—till she shaved her hair off.
4. *N. Amer. colloq.* That which is smart or in good condition. Used esp. of cars. Cf. *sharp *a.* 6* a, c; *sharp *sb.*¹ 13.
1970 *Globe & Mail* (Toronto) 26 Sept. 47/7 (Advt.), Chevrolet Malibu 2-door hardtop, fully equipped, a real sharpie. **1979** *Tucson* (Arizona) *Citizen* (Advt. Suppl.) 28 Apr. 17/1 Starter home..carpeting, drapes and remodeled kitchen. Call..to see this little sharpie.
5. *attrib.* or as *adj.* in above senses.
1961 W. Brown *Bedeviled* 19 He was a tall, slender youth with the sharpie clothes and the long sideburns of the juvenile delinquent. **1975** *Sunday Mail* (Brisbane) 13 Apr. 30/1 Police will mount an all-out campaign against Sydney's sharpie gangs. **1980** *Times Lit. Suppl.* 7 Nov. 1270/4 Higgins..tells his latest story from the point of view of a hustling Irish-American sharpy lawyer.

sharping, *vbl. sb.* **1. a.** (*U.S.* and *transf.* examples of *sharp *v.* 5 c.)
1895, 1956 [see *flatting *vbl. sb.* 3].

sharpish, *a.* Add: **2.** quasi-*adv.* = Sharply *adv.* (esp. sense 4); somewhat sharply. *colloq.*
1886 Baumann *Londinismen* 174/1 He looks sharpish for his rents. **1899** *Longman's Mag.* July 273 When a barge does come, Dorcas 'bustles her about sharpish', and there is a great to-do. **1952** *Chambers's Jrnl.* Feb. 119/1 We're late ourselves. Better be off sharpish if we're going to be home for tea. **1975** T. Allbeury *Special Collection* viii. 54 They shuffled him back to Moscow pretty sharpish.

sharpite (ʃā·ɹpəit). *Min.* [a. F. *sharpite* (J. Mélon 1938, in *Bull. Séances Inst. R. Colonial Belge* IX. 333), f. the name of Major R. R. *Sharp*, who discovered the uranium deposit where it was first found: see -ite¹.] A hydrated carbonate of uranium found as greenish yellow crusts of thin radiating fibres.
1939 *Mineral. Abstr.* VII. 225 Sharpite forms yellowish-green, radially fibrous crusts on curite and becquerelite from Shinkolobwe, Katanga. **1971** *Ibid.* XXII. 92/1 Secondary uranium mineralization occurs among secondary Fe, Cu, and Pb minerals in Kletno; it is represented by nasturan, uranium black, gummite, fourmarierite, and traces of sharpite.

sharply, *adv.* Add: **7*.** Smartly, fashionably.
1965 V. Canning *Whip Hand* ii. 15, I was sharply dressed for the part, young man on holiday, well-heeled. **1981** P. Inchbald *Tondo for Short* vii. 73 He was dressed as sharply as he could manage.

sharpness. Add: **9. b.** *Physical Sci.* The extent to which a phenomenon, condition, etc., is sharp (sense *11 b).
1906 G. Eichhorn *Wireless Telegr.* vi. 40 Wien clearly demonstrated the greater sharpness of resonance in loosely-coupled systems than in the simple system. **1921** [see *peakiness]. **1966** Cotton & Wilkinson *Adv. Inorg. Chem.* (ed. 2) xxvii. 733 (*caption*) Note the greater sharpness of the solution spectra.

sharp practice. Add: **2. a.** (Earlier example.) **b.** (Earlier and later examples.)
The second instance in quot. 1836 may also belong with 2 b.
1836 Dickens *Pickw.* (1837) xx. 209 'Dodson and Fogg—sharp practice their's—capital men of business is Dodson and Fogg, Sir.' Mr. Pickwick admitted the sharp practice of Dodson and Fogg. **1845** C. M. Kirkland *Western Clearings* 42 His 'law studies'..were comprised in six months' 'sharp practice', as clerk to a gentleman who had quitted the shoe-maker's bench for the law. **1914** Wodehouse *Man Upstairs* 182 He could not say exactly that it was sharp practice on Owen's part. **1944** 'Brahms' & 'Simon' *Titania has Mother* ix. 95 'Sharp practice?' said the Fairy Peaseblossom. 'Oh no, ma'am. Just diplomacy.'

Sharps (ʃāɹps). *U.S.* Also with small initial. The name of Christian *Sharps* (1811–74), American gunsmith, applied *absol.* to any of a number of firearms invented and manufactured by him; esp. a celebrated variety of breech-loading, single-shot rifle. Also in the possessive and *attrib.*
1850 *Sci. Amer.* 9 Mar. 193/2 (*heading*) Sharps' breech-loading patent rifle. **1873** *Forest & Stream* XXIX. 2 As long as the Indian agents send cases of breech-loading Sharps marked 'hardware'..as Government annuity, we must expect these murders, robberies, etc., on the part of the Indians. **1929** B. Davis *Truth about Geronimo* 161 His gun, an antiquated Sharps rifle, had gone off accidentally. **1958** 'W. Henry' *Seven Men at Mimbres Springs* xv. 176 Doc Harnaday and his old Sharps could..'hit a yearling bull..at 900 yards'.

sharp-set, *a.* **1. b.** (Later example.)
1918 V. Woolf *Diary* 5 Feb. (1977) I. 119 She [*sc.* a dog]..wags her tail as hard as she can, & snatches at any scrap of talk as if she were sharp set.

sharp-shin. Add: **2.** The small North American sharp-shinned hawk, *Accipiter striatus*; cf. *sharp-shinned* s.v. Sharp *a.* C. 2.
1912 W. B. Barrows *Michigan Bird Life* 264 There were Sharp-shins everywhere—sweeping about through the woods. **1937** *Nat. Geogr. Mag.* July 132/2 We have watched the sharp-shin dodge through the thickest brush after its quarry. **1960** R. T. Peterson *Field Guide Birds Texas* 58 Sharp-shin's square-tipped tail can look slightly rounded when spread.

sharp-shod, *a.* Chiefly *N. Amer.* [f. Sharp *adv.* + Shod *ppl. a.*] Of a horse: provided with calked shoes; rough-shod. Also [as a back-formation] **sharp-shoe** *v. trans. rare.*
1889 *Cent. Dict.*, Sharp-shod. **1904** in *Eng. Dial. Dict.* **1906** *Daily Colonist* (Victoria, B.C.) 20 Jan. 8/2 It was quite apparent from the horses' tracks in the snow that one saddle horse was sharp shod. **1935** H. Davis *Honey in Horn* xviii. 305 A man run down by a bunch of sharp-shod horses has a tendency to scatter around badly. **1962** J. Onslow *Bowler-Hatted Cowboy* xi. 100 We sharp-shod our horses in front with shoes that had sharp calkins at the heels and one on each toe.

sharpshooter. Add: **1. b.** Restricted to *fig.* and add later examples.
1933 *Sun* (Baltimore) 26 May 1/3 They were set apart as a shining target upon which all political sharpshooters practiced. **1960** *Twentieth Cent.* May 447 Sociology is not at present and is not likely to become in the near future a subject for intellectual sharpshooters.
c. *U.S. Mil.* A rifleman of a particular grade; *spec.* one who has attained the level of proficiency between that of marksman and expert.
1880 *N.Y. Times* 15 Sept. 12/2 Until within a comparatively recent period all officers and members of organizations armed with the Remington rifle scoring 25 points or better at 200 and 500 yards, 5 shots at each distance..were accounted 'marksmen'... Later on a distinction was made in respect of men making 42 points or better, by classing them as 'sharpshooters'. **1918** E. S. Farrow *Dict. Mil. Terms* 551 *Sharpshooter*,..in small-arms firing, a grade of rifleman just below that of expert rifleman. **1974** M. Allen *Super Tour* i. 6 The entrance was guarded..by the chief gatekeeper or one of his three sons, all armed with rifles and all rating as sharpshooters when they had done their military service. **1977** *Time* 22 Aug. 34/3 He flunked his first rifle-shooting test but eventually qualified as an infantry sharpshooter (the middle ranking between marksman and expert) with the M-16 rifle.
2. Transferred senses. **† a.** *Cricket.* = Shooter 9. *Obs.*
1863 *Boy's Own Vol.* Christmas 218 Among the best men who played cricket were Wenman, among the old players, Carpenter and Daft, and last, not least, Mr. E. Grace.
b. *Sport.* A player whose aim is particularly accurate. *U.S.*
1913 *N.Y. Tribune* 21 Sept. 12/1 Devore..was the principal sharpshooter of the second encounter, as he answered for all four runs. **1974** *State* (Columbia, S. Carolina) 27 Feb. 3-B/1 Virginia Military..builds its offense around freshman sharpshooter John Krovic.
c. *U.S. colloq.* = Sharp *sb.*¹ 8 b.
1942 *Sun* (Baltimore) 7 Mar. 10/5 The sharpshooters have been saying that Market Wise should romp in. **1944** *Ibid.* 21 Sept. 17/5 The real romance of the race track..is the betting crowd. First you get the experts or the sharpshooters.
3. One of several leafhoppers of the family Cicadellidæ, feeding on grasses, grape-vines, and other plants.
1902 *Yearbk. U.S. Dept. Agric. 1901* 377 Early cotton..avoids to a great extent damage to the plant by the boll-worm, cotton worm, and sharp-shooter. **1959** *Washington Post* 23 July 42/1 Priemsborder identified them as leafhoppers, commonly called sharpshooters.
sharpshooting *vbl. sb.* and *ppl. a.* (later *transf.* examples).
1948 *Sporting Mirror* 19 Nov. 5/1 First Division clubs are queuing up to bid for Vic Lambden, Bristol Rovers' sharpshooting centre forward. **1959** *Norwich Mercury* 10 Dec., It was the same Carver player who had been the first goal, and he was certainly giving Old Lakenham the first lessons in sharpshooting. **1978** M. Puzo *Fools Die* xvi. 172 The client was a sharpshooting Wasp Wall Street broker named Buddy Stove.

sharpster (ʃā·ɹpstəɹ). *colloq.* (chiefly *U.S.*). [f. Sharp *sb.*¹ + -ster.] **1.** = Sharper¹ 2.
1942 Berrey & Van den Bark *Amer. Thes. Slang* §436 *Cheat, swindler*,..sharpster. **1953** *Archivum Linguisticum* VII. 153 An adjectival derivative..with the pejorative sense of 'sharpster'. **1965** *English Studies* XLVI. 464 A boxer, among a dozen other appellations, may be a fighter..a discounted gambler, a sharpster.
2. A 'sharp' or stylish dresser. (In quot. *attrib.*)
1957 J. Kerouac *On Road* (1958) ii. vi. 144 Once I knocked on his door..and he opened it wearing..a vest with nothing underneath, and long striped sharpster pants.

sharrer (ʃæ·rəɹ). *colloq.* Also **sharra.** Short for Char-à-banc. Cf. *chara².
1934 D. L. Sayers *Nine Tailors* 114 There's a regular party comin'..in Jack Brownlow's sharrer. **1966** 'L. Lane' *ABZ of Scouse* 102 She fell outer ther sharrer. **1977** *Listener* 15 Dec. 801/2 And charabanc: it went from bang to bong and back again to bang... 'Meanwhile,' says Mr Ferris, 'the nation went about its business and called it either a coach or a sharra.'

sharry (ʃæ·ri). *colloq.* Also **sharrie.** As prec. Cf. *charry.
1923 *Chambers's Jrnl.* Dec. 7/1 Many who travel by 'sharry' do so because they want the pleasures of the road. **1974** P. Wright *Lang. Brit. Industry* ix. 79 Part of many a market square..was daily reserved as a parking area for motor-coaches (then called *sharries* or *charry-bangs*).

‖ **shashlik** (ʃæ·ʃlɪk). [ad. Russ. *shashlýk*, ult. f. Turk. *šiš* a spit, skewer; cf. *shish kebab.] An Eastern European and Asian kebab of mutton and garnishings often served on a skewer. Also *attrib.*
1925 R. F. Wilson *Paris on Parade* vi. 163 He gives them..shashlik, alternately spitted morsels of mutton and bacon grilled over charcoal. **1951** 'A. Garve' *Murder in Moscow* xv. 144, I was giving Watson luncheon..I had had the shashlik..I found almost uncooked. **1960** *Guardian* 29 Aug. 5/6, I ate..shashlik with retsina. **1977** *N.Y. Times* 9 May 14/2 An outdoor shashlik stand just off Ashkhabad's Marx Prospekt was pulling in passers-by.

Shasta (ʃæ·stă, *locally* ʃæ·sti), *a.* and *sb.* Also **Saste, Shaste, Shasty.** [Native name.] **A.** *adj.* **1.** Designating a tribe of American Indians living in the highlands of northern California or the language of the Hokan group spoken by this people.
1843 T. J. Farnham *Trav. Gt. Western Prairies* II. 208 The doctor..had his face very much slashed in a contest with the Shasty Indians near the southern border of Oregon. **1851** G. Gibbs *Jrnl.* in H. R. Schoolcraft *Indian Tribes* (1853) III. iv. 161 This man was afterwards dispatched..to make another attempt to assemble the Shasté tribes. **1963** *Language* XXXIX. 40 The Shasta language of northern California and southern Oregon has a long history of relevance for Hokan studies.
2. Special collocations in *Nat. Hist.*: **Shasta cypress,** the Macnab cypress, *Cupressus macnabiana*, a small tree native to California; **Shasta daisy,** a perennial herb, *Chrysanthemum* × *superbum*, of the family Compositæ, a hybrid developed by Luther Burbank (1849–1926) and bearing large white flowers; **Shasta (red) fir,** a large red fir, *Abies magnifica* var. *shastensis*, native to western North America; **Shasta lily,** a yellow-flowered fragrant lily, *Lilium kelleyanum*, native to the mountains of central and northern California.
1897 G. B. Sudworth *Nomencl. Arborescent Flora U.S.* 76 Cupressus macnabiana..Macnab Cypress..[also called] Shasta Cypress. **1908** N. L. Britton *N. Amer. Trees* 101 The tree..is also called Fragrant cypress, Shasta cypress, Macnab's cypress. **1901** L. Burbank *New Creations* (Suppl.) 8 When 'Shasta Daisies' were being bred and educated up to their present state more admiration has been bestowed upon them by visitors than upon any other flowering plant growing on my grounds. **1977** *Monitor* (McAllen, Texas) 19 June 1c/3 The church was decorated with pedestal urns of white Shasta daisies. **1897** G. B. Sudworth *Nomencl. Arborescent Flora U.S.* 58 Abies magnifica..Shasta Fir..[also called] Shasta Red Fir. **1949** Collingwood & Brush *Knowing your Trees* 106 Red Fir, also known as red-barked fir, Shasta fir and golden fir, is found on high mountain slopes and meadows. **1915** Armstrong & Thornber *Western Wild Flowers* 34 Shasta Lily is a variety with a small bulb. **1937** J. H. McFarland et al. *Garden Bulbs in Color* 155 There are several varieties of this Washington or Shasta Lily.
B. *sb.* **1.** (A member of) the Shasta people.
1843 T. J. Farnham *Trav. Gt. Western Prairies* II. 311 (*population table*) Shastys 500. **1846** H. Hale *Ethnogr. & Philol.* (U.S. Exploring Exped.) 218 The women of the Saste..are tattooed in lines from the mouth to the chin. **1855** *Crescent City* (California) *Herald* 3 Oct., Some of the Shastas..are represented as having been the last to retreat. **1900** G. W. James *Indian Basketry* 79 The fine white grass, used by the Shastas in the manufacture of their baskets is gained from great elevations in the mountains. **1935** R. Benedict *Patterns of Culture* ii. 43 Among the Shasta it was the convention that only women were so blessed. **1973** A. H. Whiteford *N. Amer. Indian Arts* 47 The..Shasta..had..food baskets.
2. Their language. Also *Comb.*, as **Shasta-Achomawi.**
1851 G. Gibbs *Jrnl.* in H. R. Schoolcraft *Indian Tribes* (1853) III. iv. 151 Higher on the main river, the prevailing language is the Shastè. **1903** *Amer. Anthropologist* V. 213 (*heading*) The Shasta-Achomawi: a new linguistic stock, with four new dialects. **1913** [see *Hokan]. **1963** *Language* XXXIX. 43 Dixon's Shasta-Achomawi group..was placed in Dixon and Kroeber's Hokan stock.
Hence **Sha·stan** *a.* and *sb.* [-an], (designating) a linguistic grouping of the Shasta and certain other peoples (formerly including the Achomawi-Atsugewi), or the peoples so comprised.
1920 *Bull. U.S. Bureau Amer. Ethnol.* XXX. ii. 528/2 Shastan Family..a linguistic stock comprising two principal groups, the Sastean and the Palaihnihan of Powell. **1963** *Language* XXXIX. 43 This neatly validates the theory of contraction with the Shastan subfamily. *Ibid.*, There may have been as many as five such families a little over a century ago, viz. *Shastan*..; *Palaihnihan*..; *Pomoan*..; *Chumashan*..; and *Yuman*. **1974** *Encycl. Brit. Micropædia* IX. 118/2 *Shastan Indians*, a group of Indian peoples speaking related languages of the Shastan family of Hokan stock... The culture of the Shastans was imitative of the Yurok and Karok.

Shaster, Shastra. Add: The form *sastra* (or the scholarly transliteration *śāstra*) is now freq. used. (Further examples.) Also, a body of teaching, a science; a treatise. Hence **Shastra·ic, Sastra·ic** *a.*
1845 *Encycl. Metrop.* XXI. 673/1 Those who study the Vêdas..are styled *Waidik*,..when learned in the six Sâstras they are called *Sâstri*. **1932** S. Dasgupta *Hist. Indian Philos.* II. xiv. 445 At the first stage a man performs his duties in accordance with the injunctions of the *śāstras*. **1956** V. Raghavan *Indian Heritage* p. lvii, These systems later came to be called *Sastras*, meaning thereby merely that they are authoritative and systematic schools of thought. **1960** Koestler *Lotus & Robot* i. 59 It is significant that every science in India is called a Sastra—a system of thought with a spiritual purpose. **1961** Webster, Shastraic. **1967** Singha & Massey *Indian Dances* i. 33 The shastraic literature which kept it alive. **1968** *Indian Mus. Jrnl.* V. 46 Its gigantic volume, comprehensive subject-matter, *śāstraic* treatment of the subject-matter. **1974** *Encycl. Brit. Macropædia* III. 433/1 From the 2nd century AD onward, Mahāyāna authors wrote 'treatises' (*śāstras*) in their own names.

shat, *pa. pple.* of Shit *v.* in Dict. and Suppl.; also var. Shott in Dict. and Suppl.

shatchen, var. *shadchan. **shatranji,** var. *sitringee.

shatter, *sb.* **1.** (Further example.)
1872 *Argosy* XIII. 199 It's a sad thing..for men like you to be obliged to work yourselves to shatters to keep them.

shatter, *v.* Add: **7. shatter belt** *Geol.*, a belt of fractured or brecciated rock formed as a result of faulting; **shatter cone** *Geol.*, a fluted conical structure produced in rock by intense mechanical shock, esp. by that associated with meteoritic impact; hence **shatter-coned** *a.*, characterized by the presence of shatter cones; **shatter-coning** *vbl. sb.*, the formation or presence of shatter cones; **shatter crack** *Metallurgy* (see quot. 1958); **shatter-pate** (later example); **shatter-pated** *a.* (later examples); **shatter-proof** *a.*, proof against shattering; also *fig.*
1910 Peach & Horne in Murray & Pullar *Bathymetr. Survey Scottish Fresh-Water Lochs* I. 459 Shatter belts situated along lines of fault or dislocations of the strata have exercised a considerable influence in producing the isolation of these individual masses. **1970** R. J. Small *Study of Landforms* iii. 102 Where crustal movement leads to the creation of crushed or brecciated zones ('shatter-belts')..., the development of river valleys may be closely guided. **1933** W. H. Bucher in *Rep. 16th Internat. Geol. Congr.* 1070 In the same vicinity more convincing evidence of the action of an explosive force is seen in the local development of 'shatter cones', innumerable incipient cracks traversing beds of rather coarsely crystalline dolomite in the shape of interpenetrating cones, not unlike cone-in-cone. **1979** *Sci. Amer.* Mar. 43/3 Macroscopic evidence includes 'shatter cones', structures of quartzite that flare outward and downward, away from the direction of impact. **1967** *McGraw-Hill Yearbk. Sci. & Technol.* 110 The Steinheim Basin..is the prototype shatter-coned structure. **1968** *New Scientist* 28 Nov. 501/2 A really definitive theory of shatter coning is still lacking. **1975** *Nature* 29 May 394/1 Shatter coning and intense microtwinning of calcite..are indicators of shock metamorphism. **1930** *Jrnl. Iron & Steel Inst.* CXXI. 703 Rapid cooling through the secondary brittle range is believed to cause the formation of shatter cracks in rails showing secondary brittleness. **1958** A. D. Merriman *Dict. Metallurgy* 315/2 *Shatter cracks*, a name used in reference to fine internal fissures, particularly when found in the heads of steel rails. The cracks lie at random in all directions and occur most frequently in large steel forgings. **1976** *Times* 7 Dec. 14/5 This poor shatterpate's condition. **1917** J. B. Cabell *Cream of Jest* (1927) xxi. 115 Everywhere, in every age,..men stumbled amiable and shatter-pated through a jungle of miracles, blind to its wonderfulness. **1938** *Sun* (Baltimore) 18 Feb. 12/2 Ah reckon you're a mite shatter-pated in yore wit-box. **1936** *Ibid.* 6 Feb. 6/2 Two men who locked the doors of their automobile from the inside..watched State police vainly try to break through shatterproof glass. **1953** M. McCarthy in *Reporter* 3 Mar. 38/2 These people live in shatterproof hierarchical structures. **1978** M. Dewis *Law Health & Safety at Work* v. 245 The plaintiff relied on a statement that the windscreens were shatterproof and bought one of their cars.

shattered, *ppl. a.* Add: **c.** *colloq.* Extremely distressed or exhausted; upset, overcome.
1930 A. Christie *Murder at Vicarage* ii. 14 How tiresome everyone is. I feel shattered... If only I had some money I'd go away. **1968** *Listener* 12 Sept. 337/2, I came in at tea-time, I sat down and I was absolutely shattered. **1978** S. Radley *Death & Maiden* viii. 75, I honestly can't tell you anything about Mary. I'm shattered, that's all I know.

shatterer. (Later example.)
1923 *Weekly Dispatch* 25 Feb. 2 Stravinsky..is a disturber of our peace, a shatterer of illusions.

shattering, *vbl. sb.* (Examples in sense 3 of the vb.)
1960 *Farmer & Stockbreeder* 16 Feb. 77/1 Some plants produced seed heads which were less susceptible to shattering than others. **1974** E. Stacey *Peace Country* ii. 112 The report said that he lost considerable of the crop from shattering.

shattering, *ppl. a.* Add: **2. c.** In trivial use, astounding, upsetting; tiresome.
1924 Wodehouse *Bill the Conqueror* v. 97 Any ordinary disaster she might have coped with, but this was too shattering. **1948** R. Lehmann *Note in Music* (ed. 2) 114 We don't converse much. But now and then she lets fall a shattering remark. **1958** [see *lavatorial *a.* 2]. **1967** *Listener* 16 Nov. 637/3 The hundreds of quotations.. about..murders, the savage punishments, and slave life in the New World, are shattering.
shatteringly *adv.* (later examples).
1911 G. K. Chesterton *Ballad White Horse* v. 112 On the helm of a high chief Fell shatteringly his brand. **1939** H. J. Massingham *Countryman's Jrnl.* xxx. 132 The argument applies far more shatteringly to the Purbeck limestone.

shattuckite (ʃæ·tə̆kəit). *Min.* [f. the name *Shattuck* (see quot. 1915) + -ite¹.] A hydrated silicate of copper found as pale blue orthorhombic crystals.
1915 W. T. Schaller in *Jrnl. Washington Acad. Sci.* V. 7 Shattuckite is a blue hydrous copper silicate from the Shattuck Arizona Copper Company's mine at Bisbee, Arizona... Shattuckite forms pseudomorphs after malachite and also occurs as small spherulites. **1928** *Mineral. Abstr.* III. 485 Scalenohedra..of calcite from Tantara mine, Katanga, are completely changed to dioptase or to a mixture of deep-blue pleochroic shattuckite and pale-blue non-pleochroic plancheite. **1977** *Amer. Mineralogist* LXII. 491 The crystal structure of shattuckite, $Cu_5(SiO_3)_4(OH)_2$, has been refined in the orthorhombic space group *Pcab*.

shaughraun (ʃa·xrā, ʃaxrā·n). *Anglo-Ir.* and *Newfoundland.* Also **shaughran,** (Newfoundland) **shaugraun** (-g-). [ad. Ir. *seachrán* a wandering, a straying, an error.] **a.** In phrs. *to go a shaughraun*: to go wrong; *on* (or *in*) *a* (or *the*) *shaughraun*: in a vagrant or drifting state. **b.** A vagabond.
1843 W. Carleton *Traits & Stories Irish Peasantry* I. 5 His speculation was gone a shaughran, as he termed it. *c* **1874** D. Boucicault in M. R. Booth *Eng. Plays of 19th Cent.* (1969) II. 165 (*title*) The Shaughraun. *Ibid.*, Conn, the shaughraun, the soul of every fair, the life of every funeral, the first fiddle at all weddings and patterns. **1892** E. Lawless *Grania* I. ii. ii. 153 'Tis eight days in the week she'll find herself working..yes, and going a shaughraun most like at the tail of it all. **1922** Joyce *Ulysses* 134 We'll paralyse Europe as Ignatius Gallaher used to say when he was on the shaughraun, doing billiardmarking in the Clarence. **1955** L. E. F. English *Historic Newfoundland* (St John's Newfoundland Tourist Devel. Div.) 36 *Shaugraun*, a vagabond state. **1961** 'F. O'Brien' *Hard Life* v. 34 Well the dear knows I think you are trying to destroy my temper, Father, and put me out of my wits and make an unfortunate shaughraun out of me. **1963** *Amer. Speech* XXXVIII. 300 *Shaugraun*,..(1) A vagabond state, (2) a person in a vagabond state, a bum. 'He spent his youth in a shaugraun.' 'He was a shaugraun.' [Newfoundland].

‖ **shauri** (ʃau·ri). Pl. **shauries, shauris.** [a. Swahili, f. Arab. *šūrā*.] Counsel, debate, problem.
1925 N. K. Strange *Wife in Kenya* xvii. 123 Sometimes such strangers came on business intent, to have a grand shauri, or a remunerative deal in posho. **1938** E. Hemingway *Fifth Column* (1939) 121 The gun-bearers..go with us. It's their *shauri*. You see, they signed on for it. **1970** *Kenya Farmer* Feb. 8/5 Often he can solve a problem by calling a meeting of the staff and obtaining their views and suggestions, not only on their personal *shauris*, but also on improvements in sales and service. **1975** T. Dinesen *My Sister, Isak Dinesen* vi. 79 Sometime in the future, in which we shall remember all the shauries (especially difficult and unpleasant things) as shadows and smile at them.

shave, *sb.*² **3. a.** (Earlier examples.)
1834 C. A. Davis *Lett. J. Downing* 39 I've got some real shaves myself in that way. **1855** J. R. Planché *Discreet Princess* in *Extravaganzas of Planché* (1879) V. 130, I much suspect this is some barbarous 'shave'.

shave, *v.* Add: **12. a.** To cut down in amount, to reduce. orig. and chiefly *U.S.*
1898 *Boston Herald* 23 Jan. 14/3 There are indications that tariff sheets are being secretly shaved. **1941** B. Schulberg *What makes Sammy Run?* vii. 141 The studio was having one of its periodic drives to cut overheads—which seemed to mean shaving stenographers' wages first. **1962** Wodehouse *Service with Smile* xi. 180 In the hope of making him shave his price a bit? **1972** *Times* 16 Feb. 13/8 This coin enables traders to shave their prices and if it did not exist there would be a considerable rounding up.
b. To deduct (a small amount) *from* or *off* (a quantity, time, etc.).
a **1961** in *Webster* (1961) s.v., New procedures shave minutes from the unloading process. **1982** *Times* 26 Aug. 15/2 The Bank shaved another 1/8 of a percentage point off the rate at which it buys bills from the discount houses.
13. *Comb.*, as **shavecoat,** a man's casual garment resembling a housecoat; **shavetail** orig. *U.S. Mil. slang*, (*a*) an untrained pack animal, identified by a shaven tail; also *attrib.*; (*b*) *fig.* a newly commissioned officer, *spec.* a second lieutenant; also gen., an inexperienced person; also *attrib.*
1964 *N.Y. Times Mag.* 29 Nov. 73 Weldon makes something like a shavecoat. It's extra comfortable and convenient for shaving. **1970** *Sunday Mail* (Brisbane) 27 Dec. 6D The bridegroom wore a mustard colored shavecoat. But it was a very nice shavecoat. **1846** *New Orleans Delta* 31 Aug. 366/2 [This mule] was followed by Shavetail Kicky, Esq., who, in a few pertinent remarks, expressed his ass-ent to the proceedings. **1891** J. G. Bourke *On Border with Crook* 153 Officers..are known as 'bell-sharps' and 'shave-tails'..the former being the old captain..and the latter the youngster fresh from his studies. **1908** R. E. Beach *Barrier* 283 The first shavetail desperado that meets him will spit in his eye. **1948** F. Blake *Johnny Christmas* i. 26 He had..six shavetail Indian ponies packed with articles he had brought down to sell. **1976** L. Deighton *Twinkle, Twinkle, Little Spy* vii. 70, I was a shavetail, just out of pilot training. **1980** *Blair & Ketchum's Country Jrnl.* Oct. 43/1 That last practice started in the mines, where a 'shavetail' was a snaky mule, not to be trusted.

shaved, *ppl. a.* Add: **1. a.** *spec.* of ice cut in thin slices or shavings for chilling drinks.
1927 E. Hemingway *Fiesta* iii. xix. 272, I..had a glass of lemon juice and shaved ice. **1962** *Listener* 16 Aug. 248/2 Snowball, which consists of shaved ice covered with scarlet syrup. **1971** J. Philips *Escape a Killer* (1972) i. ii. 20 The martini shaker in its bed of shaved ice.

shaver. **5. a.** Delete † *Obs.* and add later examples. Now esp. a small electrical appliance with a set of blades working against a perforated guard.
1897 *Sears, Roebuck Catal.* 111/3 A little gem and a dandy shaver, no better steel put in a razor. **1924** *Punch's Almanack for 1925* 3 Nov. p. xxviii/1 (Advt.), The Wilkinson safety shaver with hollow-ground blades is the ideal..gift for men. **1960** N. Mitford *Don't tell Alfred* xviii. 193 'What do you pack?' 'Shavers—you know, razors.' **1980** R. Moody *Devil you Don't* iii. 33 He..took out his shaver and toothbrush.
6. *colloq.* = *shavecoat* s.v. *shave *v.* 13.
1926 H. Pearson *Whispering Gallery* vii. 106 At a big meeting of hospital directors he turned up..in..unbusiness-like colours. One of the directors whispered to another: 'He'll be coming in his 'shaver' next.'
7. *attrib.*, as **shaver point, socket,** a power point for an electric shaver.
1971 *Fremdsprachen* XV. 145/1 *Shaver point*, Trockenrasiersteckdose. **1977** *Western Morning News* 30 Aug. 10/4 (Advt.), Good-sized kitchen; radios, shaver points, &c. all bedrooms. **1965** *Newnes Practical Householder Encycl.* III. 1329/1 The shaver is operated from a special shaver socket unit. **1978** *Cornish Guardian* 27 Apr. 19/2 (Advt.), Strip lights with shaver sockets.

Shavian (ʃēi·viăn), *a.* and *sb.* [f. *Shavi(us*, the latinized surname of George Bernard Shaw (1856–1950), playwright and critic + -an.] **A.** *adj.* Pertaining to, characteristic of, or resembling G. B. Shaw or his works or opinions.
1904 *Times* 2 Nov. 6/2 Not a play but a thoroughly characteristic 'Shavian' farago. **1935** G. K. Chesterton *G. B. Shaw* 292 The Shavian evolutionist does really want to cast the whole body of man into Chaos. **1960** [see *metabiological* adj. s.v. *metabiology]. **1977** M. T. Bloom *13th Man* vi. 101 Your mother..named you for a Shavian heroine and..you've turned into one.
B. *sb.* An admirer or follower of G. B. Shaw. (In quot. 1921, a character of Shaw's.)
1905 G. B. Shaw *Let.* 10 Feb. (1972) II. 512 Though he is an admirer of Shaw, he is no Shavian. *Ibid.* 18 Feb. 515 Are you going to write a natural history, like a true Shavian? **1921** *Spectator* 28 May 680/2 The poet Eugene Marchbanks is perhaps as near to primary emotion as any Shavian. **1967** O. Lancaster *Eye to Future* i. 8 My mother..always remained a devoted Shavian. **1978** P. Boardman *Worlds of Patrick Geddes* xi. 405 P. G. uses modern socialism as his next example... He names Marx ..the Fabians..and the Shavians with income-equalisations.
Hence **Shavia·na,** objects or texts relating to G. B. Shaw; **Sha·vianism,** the tenets or a characteristic saying of G. B. Shaw; also *nonce-wds.* in Shaw's writings, as **Shaviani·smus** = prec.; **Shavia·nity,** the quality or state of being Shavian; **Sha·vianized** *a.*, that has been rendered Shavian in character.
1903 G. B. Shaw *Let.* 2 Sept. (1972) II. 357 Your theme ..is that the book is a mere rechauffée of stale Shavianized Nietzsche. **1905** —— *Let.* 3 Jan. (1972) II. 496 The dawning of Ibsenism & Nietzscheanism & 'Shavianism' seemed to him the coming of chaos. **1920** —— *Shaw on Theatre* (1958) 133 The younger generation, Shavians to a man, demonstrated their Shavianity by scoffing at me as a Back Number. **1927** *Observer* 12 June 15/1 It [*sc.* a foreword to a volume of plays] will be prized by collectors of Shaviana for its friendly and human qualities. **1927** G. B. Shaw *Pen Portraits & Reviews* (1932) 2 The Shavianismus tickled him enormously; and he was never tired of quoting..my jokes. **1958** *Sunday Times* 27 Apr. 8/2 This slender item of Shaviana. **1975** *Listener* 14 Aug. 218/4 A revival of Shavianism.

shaving, *vbl. sb.* Add: **4. b.** (Earlier example.)
1813 T. Eaton *Rev. N.Y.* 127 Again the broker claims per cent. For he on shaving is attent.
5. *shaving cream, -cup* (earlier example), *dish, foam, -glass* (earlier and later examples), *mirror*, † *rag, soap* (earlier example), *-tackle* (earlier and later examples), *water* (earlier example); **shaving box** (earlier example); **shaving brush** (earlier example); **shaving horse** (earlier and later examples); **shaving-knife,** (*a*) (later example); **shaving stick,** a stick of shaving soap.
1774 J. Woodforde *Diary* 14 Jan. (1924) I. 122 For a Shaving Box of one Darcy..pd 0.1.0. **1792** —— *Diary* 26 July (1927) III. 364 At D[itt]o for Shaving Brush and Powder. **1851** C. Cist *Sk. Cincinnati in 1851* 226 A. E. Wetherill, manufacturer of..soaps and shaving creams. **1922** S. Lewis *Babbitt* i. 5 He snatched up his tube of shaving-cream, furiously he lathered. **1976** 'Z. Stone' *Modigliani Scandal* iii. iv. 136 Peter Usher put down his safety razor..and washed the remains of the shaving cream off his face. **1871** 'Mark Twain' *Sketches* 258 Noted the numbers on the private shaving-cups in the pigeon-holes. **1879** C. Schreiber *Jrnl.* 17 Sept. (1911) II. 205 An Oriental shaving dish with the arms of Groningen. **1961** L. G. G. Ramsey *Connoisseur New Guide Antique Eng. Pott. Porc. & Glass* 28 One-handled bleeding bowls, and shaving-dishes with a segment out of the rim. **1974** *Harrod's Christmas Catal.* 11 'His' contains..deodorant, shaving foam, soap and talc. **1979** M. Page *Pilate Plot* i. 10 A can of aerosol shaving foam. *a* **1817** Jane Austen *Persuasion* (1818) IV. i. 18 Now I am quite snug, with my little shaving glass in one corner. **1967** S. Beckett *Stories & Texts for Nothing* vi. 99 The glass, a round shaving-glass, double-faced. **1841** *Southern Lit. Messenger* VII. 527/2 A receptacle for spinning-wheels, wash-tubs, pitchforks, shaving horses, and sundry other implements of domestic industry. **1930** *Times Educ. Suppl.* 18 Oct. p. iv/1 A few tools for the lathe..a saw, a shaving-horse. **1974** P. Blandford *Country Craft Tools* ix. 128 Most shaving horses still in existence show that all sizes were made. **1843** *Knickerbocker* XXII. 386 The rub-a-dub of the cooper's mallet, the creak of his shaving-knife were still. **1911** *Daily Colonist* (Victoria, B.C.) 30 Apr. 10/2 In the surrounding staterooms, the rest began to hang up shaving mirrors and get into deck shoes. **1958** 'R. Crompton' *William's Television Show* vii. 187 The double reflection of the dressing-table mirror and the shaving mirror over the hand basin gave her a perfect view of her profile. **1796** T. Wale in H. J. Wale *My Grandfather's Pocket-Bk.* (1883) xvii. 341 Wash given out to Mrs. Wheeler... 1 pair under stockings, thread. 3 shaving wrages. **1814** Jane Austen *Let.* 18 Nov. (1952) 412 The dirty Shaving Rag was exquisite! **1790** *Pennsylvania Packet* 19 Apr. 4/2 He has likewise for Sale..Shaving soap. **1886** Kipling *Departmental Ditties* (ed. 2) 14 Pears's shaving-sticks will give you little taste and lots of lather. **1915** S. Lewis *Trail of Hawk* 328, I must go in and get a shaving-stick. **1978** D. Bagley *Flyaway* xxviii. 268 I'm old-fashioned enough to use a soap shaving-stick. **1842** Dickens *Let.* 3 Apr. (1974) III. 181 My shaving tackle, dressing case, brushes, books, and papers. **1980** *Daily Tel.* 16 Feb. 3/1 In some units..patients are expected to share toothbrushes and shaving tackle. **1837** Dickens *Pickw.* xxxix. 426 'Shaving water, Sam,' said Mr. Pickwick.

shaving, *ppl. a.* (Further examples.)
1789 E. Butler *Diary* 12 Jan. in G. H. Bell *Hamwood Papers Ladies Llangollen* (1930) 169 The most bitter Cold I ever remember, cutting shaving wind. **1810** B. Hunt *Diary* 20 Jan. in *Chester County* (Pennsylvania) *Hist. Soc. Bull.* (1898) 17/2 Most shaveing Cold weather.

Shavuoth (ʃāvū·əs, ʃavuo·t). Also **Shavuot, Shevuos, Shevuoth.** [a. Heb. *šābūʿōṯ*, pl. of *šābūaʿ* week.] = Pentecost 1. Cf. *Feast of weeks* s.v. Week *sb.* 2 e.
1892 I. Zangwill *Childr. Ghetto* III. xii. 194 He died four years ago come next *Shevuos*. **1941** G. G. Scholem *Major Trends Jewish Mysticism* viii. 297 An incident during the night before the feast of *Shevuoth* in 1665. **1944** M. Samuel in M. W. Weisgal *Chaim Weizmann* i. 88 The Jews were more transfigured by *their* celebration of Shavuoth and Sukkoth than the Russian peasants by *their* thanksgiving celebrations. **1962** *New Jewish Encycl.* 442/2 Since the early 1800's Shavuot has been generally the time when annual confirmation exercises are held. **1968** L. Rosten *Joys of Yiddish* 338 Synagogues are decorated with greens on *Shevuoth*. **1975** C. Potok *In the Beginning* i. 63 The last time I had been in the synagogue was about a week ago during the Festival of Shavuoth. **1981** *Amer. Speech* LVI. 13 The word spelled *Shavuot* 'Pentecost' will almost always be read..with the traditional Yiddish or Ashkenazic pronunciation as if it were spelled *Shevuos*.

‖ **shawabti** (ʃawæ·bti). *Egyptology.* Pl. **shawabtiu,** (anglicized) **shawabtis.** [ad. Egyptian *šw3bt(y)*, prob. f. *šw3b* persea-wood, from which the figurines may originally have been made. In later Egyptian this was replaced by *wšbty* *ushabti.] = *ushabti. Cf. *shabti.
1922 Ld. Carnarvon in *Daily Mail* 18 Dec. 10 Propped against the wall is a most beautiful portrait shawabti of the King. **1923** Carter & Mace *Tomb of Tut. Ankh. Amen* I. 120 Beside this shrine there was a large *shawabti* statuette of the king. **1960** *Oxf. Univ. Gaz.* 4 Mar. 805/2 A XXth Dynasty polychrome terracotta shawabti figure of the chantress of Amun, Inhay. **1961** *Ibid.* 10 Mar. 832/1 Eight fragments of royal shawabtiu in various stones inscribed with cartouches of King Akhnaten. **1970** P. R. S. Moorey *Anc. Egypt* iii. 68 Among a wide variety of other funerary equipment the magical servant figures or *shawabtis* are the most common.

Shawanese, *a.* Substitute for entry: Obs. var. *Shawnee *a.* and *sb.*

Shawian (ʃǭ·iăn), *a. nonce-wd.* [f. the name of George Bernard *Shaw* (1856–1950), playwright and critic, + -ian.] = *Shavian *a.* Also in other *rare* or *nonce* formations, esp. in Shaw's correspondence, as **Sha·wism, Sha·wite: Sha·wesque, Sha·wist** adjs., etc.
1894 G. B. Shaw *Let.* 2 Dec. (1965) I. 466 He resolved to follow up the vein of comedy opened by Henry Arthur Jones..before venturing upon the Shawian quicksand. **1895** —— *Let.* 27 Aug. (1965) I. 551, I have finished the draft of a one act piece about Napoleon, a very Shawesque curtain raiser. *Ibid.*, I want to write a big book of devotion for modern people..—a gospel of Shawianity. **1897** E. Terry *Let.* 9 Jan. in *Ellen Terry & Shaw* (1931) 144 His are my sentiments, but how comes it it's all put in a Shawesque manner? **1897** G. B. Shaw *Let.* 14 July (1965) I. 782 If I steal the plot my version will be so Shawified that nobody will recognize it. **1898** *Westm. Gaz.* 15 Jan. 8/1 Mr. Raymond Blathwayt's 'Talk' with Mr. George Bernard Shaw..is full of characteristic Shawisms. **1899** G. B. Shaw *Let.* *c* 26–30 Dec. (1972) II. 125, I again urge you to make that Marxian column of yours an intelligently Socialist (that is, Shawist) one. **1900** —— *Let.* 27 Jan. (1972) II. 142 The Shawish quality of my characters. **1903** —— *Let.* 26 Jan. (1972) II. 305 [Dr. L. Kellner]..was one of the first German Shawites. **1904** *Times* 2 Nov. 6/2 A rivulet of 'story' meanders through a meadow of 'Shawisms'. **1928** *Weekly Dispatch* 13 May 12/7 Here was a golden opportunity for an exhibition of Shawishness that was instantly forthcoming.

shawl, *sb.* Add: **2*.** [Anglo-Irish.] A common prostitute. Cf. *shawlie. *slang.*
1922 Joyce *Ulysses* 308 Blind to the world up in a shebeen in Bride street after closing time, fornicating with two shawls.
3. b. shawl collar (see quot. 1960); also **shawl-collared** *a.*
1913 *T. Eaton & Co. Catal.* Spring & Summer 4/3 The graceful shawl collar is edged with whipcord silk. **1960** C. W. Cunnington et al. *Dict. Eng. Costume* 192/2 *Shawl collar*, 1820's on..a term denoting a broad turn-over collar of a coat or waistcoat, continuous with the lapels, i.e. without a notch between. **1974** *Country Life* 17 Jan. 106/3 Shawl-collar cardigans are the thing to look for. *Ibid.* 107/1 Shawl-collared, kimono-style cardigan.

shawl, *v.* Add: Also *absol.* Hence **shaw·ling** *ppl. a.* (In quots. *fig.*, of snow.)
1930 R. Campbell *Adamastor* 72 Around your rocks you furl the shawling snow. *a* **1953** Dylan Thomas *Prospect of Sea* (1955) 97 Our snow was not only shaken in whitewash buckets down the sky, I think it came shawling out of the ground.

shawlie (ʃǭ·li). *colloq.* (chiefly *Irish* and *northern*). Also **shawly.** [f. Shawl *sb.* + -ie.] A woman (usu. poor or working-class) who wears a shawl over her head.
1914 F. Niven *Justice of Peace* ii. iii. 258 'Shawlies'—as the girls who are to be seen in the neighbourhood of the Trongate of Glasgow, wearing shawls over their heads, are locally called. **1928** F. T. Jesse *Many Latitudes* 233, I will put a shawl over my head. I will go along the way I was one of them Shawlies from the quays. **1934** S. Beckett *More Pricks than Kicks* 63 A lowly house dear to shawlies. **1947** P. Doncaster *Sigh for Drum Beat* ii. 12 He knew they were the footsteps of a woman. A shawly, with her shawl wound about her body and a baby snuggled into it. **1966** 'L. Lane' *ABZ of Scouse* II. 72 A working-class Liverpool woman usually elderly and of scruffy appearance. Once known as a *shawlie* from the local habit of wearing a shawl. **1980** J. Masters *Heart of War* v. 68 She had been transformed into a typical Dublin shawlie, only more ragged and dirtier than most.

shawmist (ʃǭ·mist). [f. Shawm *sb.* + -ist.] One who plays the shawm. (In earlier centuries called a *shawmer* or *shawm-player*.)
1961 A. Baines *Musical Instr. through Ages* ix. 233 The European shawmist presses the lips to a wooden 'pirouette'. **1977** *Early Music* July 333/1 The shawmist, perhaps accompanied by the tambourine player, can only have performed monophonic melodies.

Shawnee. Substitute for entry: (ʃǭnī·), *a.* and *sb.* Also † **Savan(n)a, Shawanee, Shaweno, Shawnese, Shonee,** etc. [ad. Munsee *šā:wano:w*, f. Shawnee *ša:wano:ki* people of the south. Early forms in *-ese* prob. represent a hybrid formation with -ese, later interpreted as the pl. of a stem in *-ee*.] **A.** *adj.* **1.** Of or pertaining to a tribe of Algonquian Indians, formerly resident in the eastern U.S. and now in Oklahoma; designating a member of this tribe or its language.
1674 H. Woodward *Let.* 31 Dec. in *S. Carolina Hist. Soc. Coll.* (1897) V. 460 Two days before my departure arrived two Savana Indians. **1714** J. Lawson *Hist. Carolina* 171 The Savanna Indians, who formerly lived on the Banks of the Messiasippi, and remov'd thence to the Head of one of the Rivers of South-Carolina. **1728** in C. S. R. Hildeburn *Century of Printing* (1885) 94 Two Indian Treaties..Between the Honourable..Lieut. Governour of the Province of Pennsylvania..And The Chiefs of the Conestogoe, Delaware, Shawanese and Canawese Indians. **1737** *Documents Colonial Hist. New-York* (1855) VI. 106, I am very Sorry that there has been any Misunderstanding betwixt Governour Pen & the Shaweno Indians. **1748** C. Weiser *Jrnl.* 10 Sept. in R. G. Thwaites *Early Western Travels* (1904) I. 31, I made a Present to the old Shawonese Chief. **1788** [see Mohican, Mohegan *a.*]. **1817** *State Papers & Publick Documents U.S.* (1819) (ed. 3) XII. 450 The balance of our men (except five Shonee Indians who had left us several days before),..came on this side of the first chain of mountains. **1821** T. Nuttall *Jrnl. Trav. Arkansa* 54 We stopped awhile at a Shawnee camp. **1826** T. Flint *Recoll.* 231 A rich commandant..married a Shawnese wife. **1855** A. W. Whipple *Explor. Railway Route* III. 51 [Figure D illustrates] one of a pair of Shawnee ear-drops..made by a native artist. **1949** *Democrat* 30 June 7/4 Several Shawnee towns were set up in various places over the state [of Alabama]. **1973** A. H. Whiteford *N. Amer. Indian Arts* 17 Creek and Shawnee bowls have designs incised around the rims.

2. Shawnee salad, a perennial herb, *Hydrophyllum virginianum*, native to eastern North America and bearing clusters of white or purple flowers; **Shawnee wood** (examples).
1780 J. Donelson *Jrnl.* 29 Mar. in *Three Pioneer Tennessee Documents* (1964) 9 Gathered some herbs in the bottoms of Cumberland, which some of the Company called Shawanee Sallad. **1822** *Trans. Hort. Soc. Lond.* 1st Ser. IV. 445 The *Hydrophyllum Virginicum* is called by the Americans of the Western States, Indian Sallad, or Shawanee Sallad, because these Indians eat it as such, when tender. **1829, 1866** [see Shawanese *a.*]. **1931** W. N. Clute *Common Names Plants* 31 At least three plants commemorate them: the Shawnee Law.., the Shawnee salad.., and the Shawnee tree, also called Shawnee wood. **1818** W. P. C. Barton *Compendium Floræ Philadelphicæ* I. 9 Catalpa-Tree, Catawba-Tree, Schawnes-wood. **1931** Shawnee wood [see *shawnee salad* above].
B. *sb.* **1.** The name of a tribe of Algonquian Indians; a member of this tribe.
1693 *Documents Colonial Hist. New-York* (1854) IV. 43 Wee are glad that the Showannoes..did make their application to you last fall for protection. **1737** *Ibid.* (1855) VI. 107, I Recommend to you to keep the Shawenoes among your Selves as you have Done the Tuskierores to prevent their Going to the French. **1755** R. Dinwiddie *Let.* 14 Dec. in S. M. Hamilton *Lett. to Washington* (1898) I. 149 The Cherokees have taken up the Hatchet against the French & Shawnese. **1786** [see *Huron]. **1828** M. Neville *Last of Boatmen* in *Western Souvenir 1829* 118 The scout thought it as praiseworthy to bring in the scalp of a Shawnee, as the skin of a panther. **1837** R. M. Bird *Nick of Woods* I. 15 The Shawanee and the Wyandot still hunted the bear and buffalo in the cane-brake. **1854** *Southern Lit. Messenger* XX. 401 The Shawnese still deserved their chastisement. **1891** H. F. O'Beirne *Leaders Indian Territory* p. viii/1 The Shawnees awaited their opportunity until the Tonkaway braves had departed on a big hunt. **1964** B. G. Hoffman *Anthrop. Papers of Smithsonian Inst. No. 70* 223 The Shawnee have what may be considered the best claim to having resided in..the Fort Ancient territory.
2. The language of this tribe.
1792 H. H. Brackenridge *Mod. Chivalry* I. v. iii. 61 It would be necessary for him only to talk Irish, which he might pass for the Shawanee. **1940** [see *lexeme]. **1972** *Language* XLVIII. 846 [It] is supposed to be found in..Tacana, Shawnee..and Wichita.

Shay² (ʃēi). *N. Amer.* The name of Ephraim *Shay* (1813–1916), Amer. engineer, applied to a geared locomotive designed by him in 1874 for hauling timber.
1894 J. Dredge *Rec. Transportation Exhibits at World's Columbian Exposition of 1893* 308/1 (*heading*) The Shay Locomotive. *Ibid.* 308/2 The Shay Locomotive Engine No. 75, on the Montana Union Railway, on December 6th, 1890, hauled a train of fifteen loaded cars up the Anaconda high line, a distance of 1½ mile on a 3½ per cent. grade..at a rate of speed equal to 4 miles per hour. **1936** *ABC Brit. Columbia Lumber Trade Directory* 74 Elk River Timber Co., Ltd...75 Miles of Track; 5 Locos; 4 Shays. **1942** R. L. Haig-Brown *Timber* 253 *Shay*, a geared locomotive, slow but powerful and adapted to heavy hauling. Shays of about 70 tons are commonly used in hauling logs from the landing to the mainline. **1953** *Harmac News* (Vancouver, B.C.) Oct. 8/1 A 42-ton Shay was hauling logs for Wiest logging in 1914. **1962** *Amer. Speech* XXXVII. 135 *Shay*, the most popular of the geared locomotives used in the woods. The several cylinders are upright on the engineer's side of the frame, and the boiler is offset to balance them. A propeller shaft driven by cranks runs along one side of the engine and is geared to the axles. **1974** *Islander* (Victoria, B.C.) 7 July 3/1 The 45-ton Shay (named for designer Ephraim Shay) will be replaced by a 120-ton Baldwin diesel-electric locomotive.

shazam (ʃăzæ·m), *int. Children's slang.* [Invented word: see quots. 1940 and 1976.] A 'magic' word used like 'abracadabra' or 'presto' to introduce an extraordinary deed or story.
1940 *Whiz Comics No. 2* 5 'Speak my name!' 'Shazam!' ..As Billy speaks the magic word he becomes Captain Marvel. **1964** *Playboy* May 63 Shazam! **1976** in M. Horn *World Encycl. Comics* 157/1 Captain Marvel was really a homeless orphan named Billy Batson who was taken to see the old wizard Shazam. When Billy spoke his name, he was magically transformed... The lure of simply yelling 'Shazam'—which stood for Solomon's wisdom, Hercules' strength, Atlas' stamina, Zeus' power, Achilles' courage, and Mercury's speed. **1978** *Amer. Poetry Rev.* July/Aug. 44/4 Incantatory rituals are often marked by the repetition of magic formulas. Abbra Cadabra. Enny Meeny Minny Mo. Shazam. A Tisket a Tasket or Hi do Hi do hi de ho.

shcherbakovite (ʃəɹbā·kǫvəit). *Min.* [ad. Russ. *shcherbakovit* (Es'kova & Kazakova 1954, in *Doklady Akad. Nauk SSSR* XCIX. 837), f. the name of D. I. *Shcherbakov* (1893–1966), Russian mineralogist: see -ite¹.] A silicate of potassium, sodium, barium, titanium, and niobium, $(K,Na,Ba)_2(Ti,Nb)_2(Si_2O_7)_2$, found as brittle, brown, orthorhombic crystals.
1955 *Mineral. Abstr.* XII. 569 (*heading*) Shcherbakovite—a new mineral. **1964** *Doklady Acad. Sci. U.S.S.R.: Earth Sci. Sect.* CLI. 129/1 The goniometric measurements were made on small long prismatic crystals of shcherbakovite from an arfvedsonite-feldspar vein in the Khibiny alkalic massif.

shchi, var. Stchi in Dict. and Suppl.

shd, shd., abbrev. or contraction of *should* s.v. Shall *v.*
1780 J. Woodforde *Diary* 9 Dec. (1924) I. 297 Edm[d] also mentions..that his Father shd say that he had left some shirts behind. **1811** Shelley *Let.* 12 Jan. (1964) I. 45, I see no reason why they shd. always continue so. **1885** Hardy *Let.* 13 Mar. (1978) I. 131, I shd say that a married daughter..who is here, strikes me as a particularly sensible woman. **1930** E. Pound *XXX Cantos* viii. 29 And I shd. like to be party to it, as was promised me. **1962** L. Davidson *Rose of Tibet* 6, I feel strongly we shd go out for a fictional kind of title. **1973** *Black World* Sept. 84 Ever get tired of other people telling you what you shd be doing for yr self?

she, *pers. pron.* Add: **I. 2. e.** Applied *colloq.* to things (both material and immaterial) to which the female sex is not conventionally attributed (esp. in *Austral.* and *N.Z.*). Freq. in idiomatic phrases *she's jake* (or *right*): see *jake *a.*, *right *a.* 15 d.
1903 'T. Collins' *Such is Life* (1944) i. 5 You dunno what you're doin' when you're foolin' with this run. She's hair-trigger at the best o' times, an' she's on full cock this year. **1938** N. Marsh *Artists in Crime* vi. 81, I went down to the studio... She was locked, but the key's left on a nail. **1941** J. Street *Father's House* x. 190 A wet spell would ruin us and she was coming rain before long. **1958** *N.Z. Listener* 18 July 6/2 If you tear the hamstring, in the back of the leg just above the knee, you're gone a million—she's nasty. **1969** *Private Eye* 25 Apr. 12 She's apples Eric—I don't reckon I feel like brekkie! **1973** P. Wilson *N.Z. Jack* xxi. 187 'Fix it all right?' I asked. 'She's jake now, mate,' he said. 'There she is, good as new'.
V. 9. *she-wolf* (examples); also *fig.*; **she-stock, -stuff** *U.S.*, female cattle.
1868 J. S. Northcote *Celebrated Sanctuaries of Madonna* v. 297 The marriage, doomed to so unhappy an issue, between Edward II. of England, and Isabella, the she-wolf of France. **1923** H. G. Evarts *Tumbleweeds* 87 The herd would have been worked on the spot,..the she-stuff ..being allowed to scatter. *Ibid.* 88 There were..no she-stock on the range. **1936** *Discovery* Nov. 333/2 A figure of the Roman she-wolf is being erected in Addis Ababa. **1937** *Sun* (Baltimore) 15 Nov. 14/6 She stock gathered enough strength to finish around 25c. higher than a week earlier. **1961** *Wranglin' Notes* (Eaton's Ranch, Wyoming) Nov., Our hay crop was cut, baled and stacked early this year. It is fed to the 200 'she stuff' and calves that winter on the lower ranch.
10. a. *she-being.*
1881 G. M. Hopkins *Sermons & Spiritual Writings* (1959) 170 The woman, that is she-being, not she-man, of the Apocalypse.
b. *she-fiend, -goblin, -guardian, -man* (later example).
1846 Dickens *Pictures fr. Italy* 25 A gate, which this She-Goblin unlocked. **1881** She-man [see *she-being*, sense 10 a above]. **1922** Joyce *Ulysses* 48 A shefiend's whiteness under her rancid rags. **1937** E. Pound *Fifth Decad Cantos* xlii. 12 By Della Rena and M. Magdalene the She Guardian, *tutrice*.
d. *she-woman.*
1931 *Sunday Dispatch* 31 May 12/4, I am one hundred per cent. SHE-woman. **1951** B. Russell *New Hopes for changing World* 162 Correlative to the he-man is the she-woman, who is equally undesirable.
12. *she-tragedy* (later example).
1978 *N. & Q.* Feb. 85/1 As a 'she-tragedy' and a history play (of sorts), the choice of Banks's play is an interesting move.
13. *she-male* (*U.S. colloq.* and *dial.*) = Female *sb.*
18.. in B. A. Botkin *Treas. Amer. Folklore* (1944) i. 8 Davy Crockett's hand would be sure to shake if his iron was pointed within a hundred mile of a shemale. **1917** S. Lewis *Job* xv. 246 Course you high-strung virgin kind of shemales take some time to learn to get over your choosey, finicky ways. **1941** E. P. O'Donnell *Great Big Doorstep* 125 How can that one inside be cruel to dogs like that? If they were she-males always dropping pups, I wouldn't say.
c. In catch-phr. *who's she—the cat's mother?* and varr., said to one (esp. a child) who uses the pronoun of the third person singular impolitely or with inadequate reference.
1897 'S. Grand' *Beth Bk.* xx. 204 Don't call your mamma 'she'. 'She' is the cat. **1913** C. Mackenzie *Sinister St.* I. i. 9 'Who's She?' demanded Nurse. 'She's the cat's mother.' **1949** N. Streatfeild *Painted Garden* ix. 105 'She said so.' Jane looked superior. 'She, my boy, is the cat's mother.' **1959** I. & P. Opie *Lore & Lang. Schoolch.* iii. 52 To one who keeps saying 'she' in an impolite manner the reproof is: 'Who's *she*, the cat's mother?' **1972** Casson & Grenfell *Nanny Says* 21 Who's she? The cat's grandmother?

she, var. *se.

s/he, written representation of 'he or she', used as nom. sing. third person pron. to include both genders.
1977 *Gay News* 24 Mar. 6/3 The questionnaire asks congregations whether they would call a minister to their pulpit if s/he were gay. **1978** *Amer. Educator* Winter 65 A child's sexual orientation is determined before s/he enters school. **1982** *Benedicta!* Fall 13 Can s/he figure out your address?

sheaf, *sb.* Add: **6. b.** *Math.* A topological space each point of which is associated with a structure having all the properties of an Abelian group (e.g. a vector space or a ring) in such a way that there is an isomorphism between the structures on neighbouring points. [The sense is due to J. Leray, who used F. *faisceau* (*Jrnl. de Math.* (1950) XXIX. 5).]
1955 *Ann. Math.* LXII. 56 The French word 'faisceau' has been translated into English as 'sheaf' or 'stack'. In this paper we use the word 'stack', since 'sheaf' has been used before in mathematics. **1958** R. G. Swan *Theory of Sheaves* 1 Sheaves are very useful in proving theorems. **1973** R. O. Wells *Differential Anal. Complex Manifolds* ii. 43 A sheaf..on a space *X* is a carrier of localized information about the space.
7. sheaf catalogue *Librarianship* (see quot. 1976); **sheaf oats** *U.S.* (now *rare*), oats bound in sheaves.
1902 *Library World* V. 129 Some librarians I know..are hesitating in their adoption of a ms. catalogue between the card and a new rival—the sheaf catalogue. **1913** J. H. Quinn *Library Cataloguing* iv. 33 The book-form of catalogues with separate leaves, known as 'sheaf-catalogues'. **1976** *Gloss. Documentation Terms* (*B.S.I.*) 60 *Sheaf catalogue*, a catalogue recorded on slips of paper of uniform size filed in loose-leaf binders. **1765** G. Croghan *Jrnl.* 10 June in R. G. Thwaites *Early Western Travels* (1904) I. ii. 140 The young reeds being preferable to sheaf oats. **1894** *Outing* XXIV. 337/1 Ten minutes later the horses were quietly eating their corn and sheaf oats.

shear, *sb.*¹ Add: **1. j.** In phr. *off* (*the*) *shears*: of sheep, just shorn. *Austral.* and *N.Z.*
1888 J. Bradshaw *N.Z. of Today* vi. 110 The hoggett.. in 1882 could be readily sold 'off the shears' at twelve shillings. **1896** T. W. Heney *Girl at Birrell's* 69 Now and again a buyer visited the stations to get cheap sheep 'off shears'. **1930** L. G. D. Acland *Early Canterbury Runs* 1st Ser. viii. 216 He drove them over Porter's Pass off the shears. **1964** T. Ronan *Packhorse & Pearling Boat* 147 The sheep had arrived off-shears.
6. *shear-blade* (earlier example), *-mark.*
1812 *Niles' Weekly Reg.* 25 Jan. 390/1 The subscriber at short notice can furnish clothier's shear blades. **1934** Dylan Thomas *Let.* 25 Apr. (1966) 111 There were no shear-marks visible in my last letter for the reason that I had cut out nothing.

shear, *sb.*² Add: **II. 8. shear centre,** the point in the plane of a section of a structural member through which a shear force can be applied without producing torsion; **shear flow,** flow which is accompanied by or occurs under the influence of a shearing force; **shear modulus** = *modulus of rigidity* s.v. *rigidity 1 b; **shear plane** *Geol.*, a boundary surface between bodies of rock or ice which have experienced relative motion parallel to the surface; **shear strength** = *shearing strength* s.v. Shearing *vbl. sb.* 8 b; **shear stress,** stress tending to produce shear; **shear-thickening** *sb.* and *a.*, (the property of) becoming more viscous when subjected to shear; similarly, **shear-thinning** *sb.* and *a.*; **shear wave,** an elastic wave which vibrates transversely to the direction of propagation; an S-wave.
1937 A. P. Poorman *Strength of Materials* (ed. 3) vii. 142 This point of application of the load, in order that there shall be no twist of the beam as it deflects, is called the shear centre. **1972** T. H. G. Megson *Aircraft Struct.* vii. 247 For cruciform or angle sections..the shear centre is located at the intersection of the sides. **1950** *Phil. Mag.* XLI. 890 (*heading*) The eddy viscosity in turbulent shear flow. **1975** Raudkivi & Callander *Adv. Fluid Mech.* iv. 155 A shear flow has non-zero gradients of mean velocity and the fluid is being sheared by the mean motion. **1937** Dodge & Thompson *Fluid Mech.* viii. 165 An analogy is often drawn between the coefficient *μ* and the shear modulus of elastic materials. **1973** J. G. Tweeddale *Materials Technol.* I. iv. 86 The shear modulus may be obtained from experimental values of torsional load and torsional strain measured on a cylindrical test segment. **1888, 1903** Shear plane [in Dict., sense 7]. **1969** Bennison & Wright *Geol. Hist. Brit. Isles* xvi. 356 Recent work on the mechanism of ice flow shows that shear planes occur and that these carry material from the sole of the ice sheet to the surface. **1931** Laurson & Cox *Properties & Mechanics of Materials* i. 13 For most brittle materials..the tensile strength is least, the shear strength next, and the compressive strength greatest. **1978** *Sci. Amer.* Apr. 122/3 Under some circumstances molten silicates may not behave like ordinary fluids; they may have a shear strength greater than zero. **1937** Dodge & Thompson *Fluid Mech.* viii. 165 There is..an important distinction between the effects of shear stress on solids and on liquids. **1971** J. W. Ireland *Mech. Fluids* viii. 242 Determine the shear stress at the pipe walls when

water flows at the rate of 300 litres/min. through a 7·5 cm diameter pipe 150 m long. **1963** A. J. de Vries in P. Sherman *Rheol. Emulsions* 146 (Index), Shear thickening, activation energy. **1966** *Jrnl. Colloid & Interface Sci.* XXI. 554/1 This shear thickening leads to a maximum viscosity beyond which shear thinning occurs as the shear rate is increased. **1978** *Sci. Amer.* Nov. 143/2 The easiest example of a shear-thickening fluid that you can whip up in the kitchen is a simple mixture of water and cornstarch (or any common starch). **1966** Shear-thinning [see *shear-thickening* above]. **1974** P. L. Moore et al. *Drilling Practices Manual* ii. 25 Most [drilling] muds are shear thinning. **1978** *Sci. Amer.* Nov. 142/3 The advantage of shear-thinning is perhaps most apparent in ink. You want the ink in your ball-point to flow freely (by being sheared) as you write, but you do not want it to flow when the pen is in your pocket. **1936** J. B. Macelwane *Introd. Theoret. Seismol.* vii. 147 An isotropic elastic solid can transmit two types of waves, compressional and shear waves. **1977** A. Hallam *Planet Earth* 17 The outer core does not transmit shear waves and so must be liquid.

shear, *v.* Add: **5. d.** (Earlier example in mod. sense.) Now *rare*.

1852 *Trans. Mich. Agric. Soc.* III. 139 An article upon Sheep, describing tracts that shear the big fleeces.

e. *Austral.* and *N.Z.* To own or keep (sheep).

1930 L. G. D. Acland *Early Canterbury Runs* 1st Ser. ii. 35 The homestead.. belongs to J. E. Scott who still shears about 5000 sheep there. **1965** J. S. Gunn *Terminol. Shearing Industry* ii. 18 *Shear*.. this word is.. used by woolgrowers to indicate the size of their flocks, for example, 'I shear about 5000'.

sheared, *ppl. a.* Add: **2.** Subjected to shear; strained or distorted by shearing stress.

1930 J. S. Flett in Peach & Horne *Geol. Scotl.* ii. 56 Sheared Granite-Gneiss. This rock belongs to the group of muscovite-biotite-granites, and is rich in alkali felspar.

shearer. Add: **6. b.** A coal-cutting machine that cuts in a vertical plane parallel to the coal face.

1956 E. Mason *Deputy's Man.* I. xviii. 258 Most of the cutter-loading machines use vertical shearers which make a cut perpendicular to the plane of the seam.. Disc shearers have picks attached to a vertical wheel rotating on a horizontal shaft. **1972** *New Scientist* 29 July 260/1 As the shearer cuts its way along the face, coal was automatically loaded on to a conveyor. **1979** *Jrnl. R. Soc. Arts* Jan. 90/2 By far the most common coal-getting machine in use to-day is known as the shearer.. which removes a slice of coal of about two-thirds of a metre in depth from the face by a revolving drum fitted with tungsten carbide tipped picks.

shearing, *vbl. sb.* Add: **1.** (Later *concr.* examples.)

1848 J. R. Lowell *Poet. Works* (1896) 136/2 Your goddess of freedom, a tight, buxom girl,.. who can sing at a husking or romp at shearing. **1901** M. Franklin *My Brilliant Career* xvi. 136, I was looking forward to the shearing.

7. *shearing time* (delete † and add later examples); *shearing-floor*, *paddock*, *-shed* (earlier example); *shearing-machine* (later example).

1863 R. Henning *Let.* 26 Nov. (1966) 146 The shearing floor is made to accommodate twelve shearers. **a 1914** A. B. Paterson in *Penguin Bk. Austral. Ballads* (1964) 178 Round the shearing-floor the listening shearers gape. **1977** Yin Ming *United & Equal* 50 The commune has gradually mechanized its operations. It now has over 80 items of mechanized equipment—trucks, tractors, diesel engines, fodder-processors, mowers and shearing machines. **1933** L. G. D. Acland in *Press* (Christchurch, N.Z.) 25 Nov., *Shearing paddock*, handy paddock to hold sheep during shearing. **1851** H. W. Haygarth *Recoll. Bush Life Austral.* vii. 76 The shearing-sheds. **1941** L. Hauber in *New* ... **1924** ... *Sheep* ... **1900** G. Comm. *Patents 1861: Agric.* (U.S.) 138 Shearing time.. is the month of June. **1953** O. E. Middleton in C. K. Stead *N.Z. Short Stories* (1966) 188 Shearing-time was always a worry for Charlie.

shearling. 2. Delete † *Obs.* and add: In recent use, *spec.* the woollen lining or body of a coat, etc. Chiefly *U.S.*

1971 *New Yorker* 16 Oct. 143/2 (Advt.), Suede on the outside, the white shearling on the inside. **1977** L. O'Donnell *Aftershock* (1979) ii. 31 She shrugged out of her shearling-lined coat and hung it on the rack.

sheat-fish: see Sheath-fish, sheat-fish in Dict. and Suppl.

sheath[1]. Add: **2. k.** A contraceptive made of thin rubber worn on the penis; a condom.

1861 Cf. Drysdale *Elem. Soc. Science* (ed. 4) ii. 349 The accessory and sensational part of the venereal act is obtained.. by the use of the sheath (which is.. very frequent), but more so on the continent than in this country. **1897** *Science of Generation* xx. 235 The use of various mechanical contrivances, such as French Safes, Condom Sheaths, etc. **1929** M. C. Stopes *Let. to Working Mothers* 24 Some men like to use a sheath, and this is quite a safe method. **1962** *Lancet* 2 June 1194/1 The survey shows that the sheath and coitus interruptus are still the methods most commonly used for contraception in this country. **1977** J. E. Trimmer et al. *Visual Dict. Sex* (1978) xiv. 134 Careful attention must be taken in the rolling on of the sheath so as not to damage it.

1. A long close-fitting dress or skirt, usu. with a slit or pleat on one side. Cf. *sheath dress, gown*, etc., sense 7 a in Dict. and Suppl.

1904 H. O. Sturgis *Belchamber* xx. 273 A sinuous young lady, clad in a sheath of some glittering, shimmering blackness. **1932** in C. W. Cunnington *Eng. Women's Clothes in Present Cent.* (1952) vii. 222 The moulded sheath glorifies the body beautifully. **1956** *Observer* 20 July 9/3 Dresses for day and evening are cut as figure-gripping sheaths to which are added built-out and stiffened hip-pockets of a vast size. **1976** G. Moffat *Over Sea to Death* ii. 24 She wore a bright green jersey sheath which emphasised her large angles.

3. b. *Electronics.* The anode of a thermionic valve. [Disused.]

1919 R. D. Bangay *Oscillation Valve* 57 It [sc. the Fleming Valve] consists of a metal or carbon filament (F) and a metal cylinder (S) (usually called the sheath) surrounding the filament.. the sheath [thus forming] the anode of the valve. **1922** Glazebrook *Dict. Appl. Physics* II. 880/2 A battery.. has its positive terminal connected to the sheath.. and its negative one to the filament.

c. *Physics.* A region of charged particles or plasma surrounding an object.

1923 *Science* 12 Oct. 290/2 Around each negative electrode there is thus a sheath of definite thickness containing only positive ions and neutral atoms. **1955** A. von Engel *Ionized Gases* vii. 194 The anode is therefore covered with a luminous sheath—the anode glow which is sometimes divided into several luminous spots. **1973** Hellund & Tryvelpiece *Princ. Plasma Physics* i. 46 If the potential of the probe is much larger than the local potential of the plasma, the probe attracts electrons and repels ions, forming a sheath region around the probe, of which..

7. a. *sheath dress.*

1924 in C. W. Cunnington *Eng. Women's Clothes in Present Cent.* (1952) vi. 171 The beltless sheath dress. **1965** N. L. McClung *Stream runs Fast* xv. 132 She was a sweet-faced little violet of a woman, in a sheath dress of silver cloth. **1980** 'M. Herden' *Pel under Pressure* xiii. 139 She had one of those sheath dresses on. Fitted like a skin.

b. *sheath cell Anat.*, a Schwann cell (*Schwann b).

1906 Sheath cell [see *Schwann b*]. **1967** D. Bodian in G. C. Quarton et al. *Neurosciences* 13/1 Axons that are enclosed by a single fold of a sheath cell are referred to as unmyelinated axons.

sheath-fish, sheat-fish. Add: The latter spelling is now usual.

sheave, *sb.*[1] Add: **2. c.** *sheave-wheel.*

1906 M. T. Towne *Her Majesty Montana* 118 A Butte miner.. is lowered to his labors in a steel cage suspended from a heavy wire cable passing over sheave-wheels on a head-frame. **1971** *Financial Mail* (Johannesburg) 26 Feb. 648/1 They depend upon our equipment—like high speed man cage and skip guide rollers, sheave wheels.

sheave, *sb.*[2] Add: **3.** *Paper-making* = *Shive* *sb.*[2] ? *Obs.*

1880 J. Dunbar *Pract. Papermaker* 15 All rags.. contain sheive, which nothing but judicious boiling will remove. **1888** Cross & Bevan *Text-bk. Paper-Making* vi. 90 Such impurities as weeds.. if not removed would be liable to appear in the finished paper as dark-coloured specks, technically known as 'sheave'. **1964** G. Clapperton *Pract. Paper-Making* xi. 133 The pressure applied in the super-calender is often such as to cause all the sheave and gritty matters to show up.

shebang. Also † chebang, shee-bang. For *U.S.* read *N. Amer.* and add: **1. a.** (Earlier examples.)

1862 W. Whitman *Jrnl.* 23–31 Dec. in *Specimen Days & Collect* (1882–3) 27 Their shebang enclosures of bushes. **1863** — *Jrnl.* Jan. in *Ibid.*, The soldiers guarding the road came out from their tents or shebangs of bushes.

b. (Example in phr. *the whole shebang*.)

1879 B. Harte *Story of Mine* 85 That.. don't fetch me even ef he'd chartered the whole shebang.

c. A low drinking establishment, a tavern.

1878 C. Hallock *Hallock's Amer. Club List & Sportsman's Glossary* 131 Chebang, any sort of structure from a shanty to a hotel. **1901** H. G. Parker *Right of Way* vii. 90 There were people who called this place a 'shebang'. **1908** H. W. Sinclair *Raw Gold* vi. 68 There was a sort of shebang.. you couldn't call it a hotel if you had any regard for the truth—on the outskirts of Walsh for the accommodation of wayfarers without a camp-outfit. **1963** E. C. Guillet *Pioneer Farmer & Backwoodsman* I. xix. 314 Less picturesque were the shebangs dotted along the rivers, where squaws and whiskey awaited the shanty boys and their winter pay.

2. (Examples.) Freq. in phr. *the whole shebang.*

1869 'Mark Twain' *Let. to Publishers* (1967) 26, I like the book, I like you and your style and your business vim, and believe the chebang will be a success. **1903** W. N. Harben *Georgians* ix. 88, I sold out my shebang, put the money in my pocket. **1924** H. Crane *Let.* 5 Dec. (1965) 196, I am growing more and more sick of factions, gangs, jealousies, recriminations, excoriations and the whole literary shee-bang. **1933** E. E. Cummings *Let.* 13 Sept. (1969) 124 Camels placidly nibble the whole shebang, not merely the smallish but the spike thorns. **1946** F. Palmer *Golconda* xiv. 109 I've.. seen him standing up there on one of those outcrops overlooking the company's buildings as if he'd like to call down fire from heaven on the whole shebang. **1967** *Boston Sunday Herald* 7 May iv. 4/3 You can't get rid of the feeling.. that the people in charge of the shebang are far more pessimistic and confused than they were the last time you were here. **1977** R. E. Megill *Introd. Risk Analysis* iii. 28 The standard deviation is then calculated by dividing the total number of wells, *N*, into the sum of all the group deviations.. and then taking the square root of the whole shebang.

shebeen. Add: **a.** Also in South Africa, a (usu. Black) illicit establishment where liquor is sold or consumed. In extended sense: a drinking-party, esp. among West Indians.

1900 *Kruger's Secret Service* vi. 135 In Fordsburg there was a shebeen kept by a certain Pulinski.. Pulinski took me inside, where I found the place to be full of Kaffirs. **1931** J. Mockford *Khama* x. 77 So zealous was Khama in enforcing his ban on beer, that he attacked with his fist on shebeens himself and fired the huts of the brewers with his own hand. **1949** *Cape Argus* 13 Jan. 2 Carousals that cannot be described as shebeens, but that are almost as mischievous. **1944** L. du Plessis *Cape Malays* v. 81 Shebeens have sprung up in clusters, wine is brought in from Monday to Saturday by 'runners'. **1958** *Times Lit. Suppl.* 22 Aug. 472/3 The flashy, teeming, squalid world of the urban localities.. laughter, gramophones, shebeens, and 'tsotsi' thugs making night excursions dangerous for Africans and police alike. **1962** D. Lessing *Golden Notebk.* i. 57 Peter.. spends his last night in the Colony drunk, and by chance encounters his dark love in some shabby Shebeen. **1975** *New Society* 10 July 71/2 The West Indians (in Southampton), who are more free and easy, and tend to have noisy all night parties and shebeens. **1980** *Times* 11 May 5/1 The South African Government, after years of battling to control illicit drinking dens, known as shebeens, in black townships, has conceded defeat and legalized them.

b. *shebeen-keeper*; **shebeen queen** *S. Afr.*, a woman who runs a shebeen.

1922 Joyce *Ulysses* 443 In the shadow a shebeenkeeper haggles with the navvy and the two redcoats. **1954** R. St. John *Through Malan's Afr.* xxx. 236 Shebeen queens were warned to stop their liquor supply. **1970** *Times* 10 Jan. 24/1 After fire-bombing of a few that stayed open, the shebeen queens (women that operate most speakeasies) duly shut up shop, and Sowetoans did their Christmas drinking.

shebeener (later *S. Afr.* examples).

1942 [see *beer-hall* s.v. *beer sb.*[1] 4]. **1950** [see *dop sb.*[2] 2].

shebo, var. *Shiveau.

‖ **shechita** (ʃəˈxiːtɑ). Also shecheta, shechitah, shehita(h). [Heb. *šᵉḥîṭā*, f. *šāḥaṭ* to slaughter.] The Jewish method of slaughtering animals. Cf. Shochet.

1876 J. Picciotto *Sk. Anglo-Jewish Hist.* xxvii. 217 The Portuguese would willingly join in the formation of a Board for.. the management of the Shechita arrangements for slaughtering and preparing cattle for food more Judaico. **1891** M. Friedlander *Jewish Relig.* 463 Any deviation from these rules in the act of killing the animal renders the *Shechitah* invalid. **1920** *Jewish Chron.* 14 Jan. 34/1 (*heading*) The Shechita Board, the Public, and the Butchers. *Ibid.* 2 Apr. 13/2 Pinkey von Konow.. constantly describes Shechita as a religious observance, but when it suits her purpose she speaks of it as an act 'practised under the cloak of religion'. **1921** *Dict. Occup. Terms* (1927) § 448 *Slaughterer, Jewish*; .. *shochim*.. is employed by Jewish Board of Shechita. **1939** *Contemp. Jewish Record* II. iii. 32 Charges of alleged brutality have been made the basis for the prohibition of *Shehitah* in several European countries. **1947** *Ibid.* IV. 539 The Iron Guards.. bestially murdered them in parody of *shehita*. **1966** E. Huxley *Back Street New Worlds* ii. 25 Animals must be killed by a method known as *shechita*, the severing of all veins and arteries in the neck at one cut. **1976** *Jewish Chron.* 16 Apr. 4/5 In America, after examining the subject thoroughly, the.. Institute of.. said that.. found shechita to be humane. **1976** C. Bermant *Coming Home* i. i. 18 Insofar as slaughter can be humane, shechita is humane—to the animal. It is, however, inhumane to the shochet, for.. he has to face the.. almost pleading eyes of the animal.

shed, *sb.*[1] Add: **6. b.** *Comb.*, as **shed-rod**, **-stick**, a device by which the warp is opened.

1908 W. Bray *Everyday Life of Aztecs* vii. 144 This passage is easy to achieve by means of a roller or shed-rod inserted across the warp under every alternate thread. **1930** L. Hooper *Hand-Loom Weaving* i. vi. 84 After the weft has been drawn straight, the shed-stick being again in a flat position, its edge may be brought down smartly upon the whole weft in order to beat it together. **1960** G. Lewis *Handbk. Crafts* 98 Shed sticks.. hold the cross.. in place.

shed, *sb.*[2] Add: **1. b.** For *Austral.* read *Austral.* and *N.Z.* and add examples.

1857 F. Cooper *Wild Adventures in Austral.* 105 He was bound for the shearing through New England. By this time, most likely, he has set in at some of the sheds on the Namoi. **1893** H. Lawson *Coll. Prose* (1972) II. 24 Men tramping in search of a 'shed' are not called 'sundowners' or 'swaggies'; they are 'trav'lers'. **1911** C. E. W. Bean *'Dreadnought' of Darling* v. 50 They were rich men—shearers—probably making from one big shed to another on their horses. **1940** F. Sargeson *Man & his Wife* (1944) 47 We got a job picking up fleeces in a big shed. **1955** G. Bowen *Wool Away!* (1956) 2 Good wool-handling, good shed, good sheep, good boss, and a good gang can create an atmosphere of work and action.

c. = *Hangar b.* ? *Obs.*

1909 *Daily Chron.* 5 Nov. 1/6 They have been watching the great shed gradually nearing completion.., and have been eagerly awaiting the advent of the airship. **1916** H. Barber *Aeroplane Speaks* 17 The Aeroplane is wheeled out of its shed on to the greensward of the Military Aerodrome.

3*. [By analogy with *barn sb.*[1] 1 d.] In nuclear physics, a proposed unit of area of

nuclear cross-section equal to 10^{-24} barn (10^{-48} cm.²).

The unit is impracticably small and appears to have had minimal use.

1956 W. C. Michels et al. *Internat. Dict. Physics & Electronics* 820/2 *Shed*, a unit of nuclear cross section equal to 10^{-24} barn or 10^{-48} square centimetre. **1965** *Guinness Bk. Records* (ed. 12) 80 The smallest unit of area is a 'shed', used in sub-atomic physics. **1968** F. Kertesz *Lang. Nuclear Sci.* (Oak Ridge Nat. Lab. TM 2367) 20 During the study of the neutrino, a much smaller surface was used in theoretical studies and the area 10^{-48} cm[2] was quite logically named the shed; however, this latter name did not receive general acceptance. **1979** *New Scientist* 12 July 168/2 The shed.. seems to me to be less witty [than the barn].

4. (sense 1 b) *shed-boss*, *-hand*; *shed-shaped* adj.; **shed master**, one in charge of a locomotive shed (see quot. 1921); **shed roof** (earlier and later *Amer.* examples); **shed-room** *U.S.*, a shed attached to a house and serving as a room.

1894 W. E. Swanton *Notes on N.Z.* ii. 97 Then there is the 'shed boss', who looks after everything, sees the sheep are shorn properly, takes the tally, looks after pressing, etc. **1946** E. C. Stoddart *Te Waimate* 110 In the early days the manager acted as shed-boss. **1905** W. Baucke *Where White Man Treads* 219 The shed-hands and shearers were mostly Maoris. **1961** *N.Z. Listener* 26 May 8/1 The two shedhands played their unending game of poker. **1921** *Dict. Occup. Terms* (1927) § 700 *Shed master*, in charge of smaller locomotive shed than that supervised by shed superintendent. **1960** *Times* 2 Sept. 6/6 He should have sent a message to the shed-master. **1978** J. Blackburn *Dead Man's Handle* ii How the shed-master had cursed when he heard the old steamer come clanking home with a fractured bearing. **1786** in *Maryland Hist. Mag.* (1908) III. 45 The Shead Ruff of Capt Cressap's house. **1878** *Essex Inst. Hist. Coll.* (1866) VIII. 235 These [carriages] are.. built like our mud-scows, with a shed roof over your head, looking like a floating creepshole. **1976** 'D. Baxter' *Ex Quesitus Asked* xiii. 129 Above the wall I caught a glimpse of a white-gravelled, sloping shed roof. **1843** *Knickerbocker* XXI. 304, I had in the morning secured a bed in a shed-room. **1932** H. Clarke *Let.* 11 Jan. (1965) 39 We've relegated him to the shed-room in back of the kitchen. **1857** Thomas *Mains Woods* (1864) 246 A shed-shaped tent will catch and reflect the heat like a Yankee-baker.

shed, *v.*[1] Add: **1. a.** (Further examples.)

1921 *Kelso Chron.* 26 Aug. 2 A better never lifted paw, To shed or wear off a stell. **1942** R. B. Kelley *Animal Breeding* xv. 140 The shepherd has to shed or separate these [marked sheep] from the flock of 20. **1949** *Scots Magazine* Sept. 463 Wicket-gates for 'shedding' the sheep into various pens. **1951** N. M. Gunn *Well at World's End* xvii. 132 Some evenings ago, I fell in with a shepherd. I had shed one of his ewes and ultimately run her into a corner. **1977** *Field* 13 Jan. 56/2 The shepherd guided the dog to cut out, or 'shed' the marked sheep. **1981** I. A. Gordon in *N.Z. Listener* 27 June 86 When you shed your sheep they are out in the open.

d. Also *absol.*

1958 *St. Nicholas* Nov. 84/2 He still grows till he is called a 'Buster', and then sheds. Then he is called a 'Shell-Crab'. **1973** M. G. Eberhart *Danger Money* (1975) xii. 139 Her jacket still bore some stray beige hairs from Toby [sc. a cat]; did he never stop shedding? she thought.

f. *transf.* To take off (a garment); to doff, divest oneself of. Also *fig.*

1858 *Lawrence* (Kansas) *Republican* 28 Oct. 1/6 She was compelled to 'shed' her woman's 'fixin's', and put on a man's breeches and hickory shirt. **1884** 'Mark Twain' *Huck. Finn* xx. 196 The duke shed his coat and said he was ready. **1922** Joyce *Ulysses* 134 You will shed your bankrupt ideas, you understand, Ruby Cohen? **1969** *Times* 18 Mar. 1 Cabinet members had been advised to shed their ministerial cloaks and campaign for their own beliefs.

g. Of a share: to fall in price by (an amount). *Financial colloq.*

1947 *Financial Times* 29 Jan. 1/7 Preferences remained comparatively steady, although B.A.G.S. Sixes shed ⅞ to 22¼. **1981** *Times* 11 Apr. 19/5 Hawker Siddeley added 4p to 384p and GKN shed a similar figure at 322p.

h. *to shed* (*a*, *the*, etc.) *load*: temporarily to curtail the electricity supply to an area in order to prevent excessive demand on the generating plant. Cf. *load-shedding* s.v. *Load sb.* 8.

1967 *Times* 27 Feb. 7/3 The alternative was to go on running every day, and to shed the load because they could not carry the peak load. **1952** *Blackw. Mag.* Dec. 483/1 And if the local electrician chose the middle of your party to shed a load—well, where were you then? **1975** *IEEE Trans. Power Apparatus & Systems* XCIV. 360/1, 65% of the companies shed 25% or 30% of their load on underfrequency. *Ibid.* 360/2 Most companies shed load in two or three steps.

shed, *v.*[2] Add: **b.** Also with *up* (*N.Z. colloq.*).

1960 *N.Z. Jrnl. Agric.* Oct. 310 Sheep brought in for shedding were either to be shed up, otherwise the pens in the shed get very dirty and much wool is stained. **1981** I. A. Gordon in *N.Z. Listener* 27 June 86 When you shed-up sheep you put them under cover to prevent their fleeces from getting wet before shearing.

shed: see *She'd.

shedding, *vbl. sb.* Add: **1.** (Examples spec. with reference to sheep.)

1832 *Trans. Highl. Soc.* 295 When gathered to the same fold, or shedding-place together. **1921** *Kelso Chron.* 26 Aug. 2 Test—Hauld between two poles, drive round trainer, and proceed between two other sets of poles before penning, shedding, and wearing. **1942** R. B. Kelley *Animal Breeding* xv. 140 In Scotland 'shedding' is a special feature of the dog's training. **1966** J. Murray *Rural Rhymes* 27 An' them they had a sheep tae pairt Frae a' the rest—ca'ed sheddin'. **1957** *Dumfries & Galloway Standard* 19 Oct. 5/6 Midge had a good run out and the most perfect lift in the competition, perfect fetch, driving and shedding but lost one mark in the penning to total 49 points. **1959** *Times* 18 Sept. 7/1 He appeared to be well in the lead when he reached the shedding ring. **1977** *Field* 13 Jan. 55/2 Each dog guided a bunch of sheep round the field, obstacle course and finished by bringing them to his master in the 'shedding' ring.

3. d. The premature falling of the young bolls of cotton plants.

1899 *Yearbk. U.S. Dept. Agric.* 728 Over the eastern portion shedding was reported, with complaints of drought in portions of the Carolinas. **1974** J. W. Purseglove *Trop. Crops* 348 The first flowering period requires relative dryness, otherwise excessive boll shedding ensues.

4. *Electr.* = *load-shedding* s.v. *Load sb.* 8. Also *fig.*

1945 *Electrician* 25 May 457/1 Some shedding of the load may still be unavoidable during next winter. **1958** *Listener* 12 June 980/1 In one year there was a shedding of load of serious drama. **1975** *IEEE Trans. Power Apparatus & Systems* XC. 1460/2 The load selected for shedding will be different for various systems because of geographical, historical, political and reliability factors.

Sheehan (ʃiːən). *Path.* [The name of H. L. Sheehan (b. 1900), English pathologist.] *Sheehan's syndrome*: pituitary insufficiency (cf. *Simmonds' disease*) caused by necrosis of the gland as a result of post-partum hæmorrhage and shock.

1950 R. F. Escamilla in S. Soskin *Progr. Clinical Endocrinol.* x. 125 Case II represents an instance of most of the disease following a complicated delivery, and is therefore an example of the subclassification of 'Simmonds' disease described by Sheehan, and called 'Sheehan's syndrome' by some authors. **1973** *Daily Colonist* (Victoria, B.C.) 18 July 2/3 Since the pituitary controls the ovaries, the thyroid, and the adrenal glands, in Sheehan's syndrome produces symptoms resulting from failure of other glands.

sheelah, var. *Sheila.

‖ **sheela-na-gig** (ʃiːlə,nəˈgɪg). Also **sheila-**, **shiela-**, **shela-**, **-gyg**. [ad. Ir. *Síle na gcíoch* Julia of the breasts.] A medieval carved stone female figure sometimes found on churches or castles in Britain and Ireland (see quot. 1934). Also *ellipt.*, as **sheela**, **sheila**.

1846 *Proc. R. Irish Acad.* 1843–4 III. vii. 575 In the church at Dowth there is a sheila-na-gig, carved in stone quite different to that which composes the walls of the church. **1861** *Jrnl. Kilkenny Archæol. Soc.* 1860–61 VI. 69 This effigy.. belongs to that class of sculpturings which in Ireland have extended down to the middle of the sixteenth century.. They are known amongst the peasantry of the southern counties by the name of 'Sheela-na-gig'. **1894** *Jrnl. R. Hist & Archæol. Soc. Irel.* 1870–84 XV. ii. 285 The name by which [such figures] are generally known is 'Sheelanagig'. Our 'Sheela' here measures two feet in length. **1929** *Man* XXIX. 134 A stone.. hollowed out to form in relief the rudely carved figure of a woman of the kind known in Ireland as Sheela-na-gig. **1934** *Jrnl. R. Anthropol. Inst.* XXXVII. 99 There are.. some of the figures are known as Sheila-na-gig. These are always nude and are represented in a frontal aspect, the legs usually wide apart, and the hands so posed as to call attention to the genitalia. *Ibid.* 98 The Sheila from Blackhall Castle.. is represented with breasts and long hair. **1977** *Times* 14 Sept. 5/1 The figures, called Sheela-na-Gig, are not found only in Ireland.. one of the best preserved is on the corbel of a church in Kilpeck [Hereford and Worcester]. *Ibid.* 5/2 The British Sheelas are slightly older than the Irish.

sheen, *sb.*[1] Add: **3.** A very thin film or slick of oil (esp. on water).

1970 *Daily Colonist* (Victoria, B.C.) 29 Apr. 1/1 Investigators reported four large 'sheens' of oil.. The term being used instead of [oil] slicks because the material being spread so quite lightly on the water—was sighted from aircraft. **1978** *Daily Tel.* 28 Mar. 1/4 There was 'an extremely thin sheen of oil six to seven miles off Jersey' which was of no danger to marine life or holidaymakers.

sheen, *sb.*[2] (Earlier example.)

1839 H. Brandon in W. A. Miles *Poverty, Mendicity & Crime* 165/1 *Sheen*, bad money.

sheen (ʃiːn), *sb.*[3] *U.S. slang.* [Prob. abbrev. of *machine sb.* 1 c.] A car; an automobile.

1968–70 *Current Slang* (Univ. S. Dakota) III–IV. 107 *Sheen*, car. **1975** *Amer. Speech* 1972 XLVII. 153 Hey, look down the street pas' that sheen double-parked. **1980** in S. Terkel *Amer. Dreams* 125 My friends are always talkin' about havin' a nice sheen. That's a nice car or van, something set up real nice on the inside.

sheened, (ʃiːnd), *ppl. a.* [f. Sheen *v.* + -ed[1].] Having a sheen, shining. Freq. in predic. use coupled with *with*.

1920 E. Sitwell *Wooden Pegasus* 100 Beneath umbrellas I can see Pink faces sheened with stupidity. **1942** [see *Sheeting*]. **1958** *Listener* 3 July 24/2 Their huge sheeks, 'all silver white', or sheened with lilac.

sheeny, *sb.*[1] Add: (Earlier and later examples.) Now only as a term of vulgar abuse.

1816 J. H. Lewis *Lectures on Art of Writing* (ed. 7) vi. 84 A motley-fool the *thing* I mean is, One of the common puffing sheenies. **1918** G. Frankau *One of Them* x. 75 Yid, and Hebrew, and Sheeny, .. Porn Russ, Yank, Dago, Teuton, Gaul, or Sheeney? **1946** [see *Lib sb.* 1 f]. **1976** *Honolulu Star-Bull.* 21 Dec. a-3/1 Her mom, there's a couple of sheenies at our door with a turkey. **1977** H. Fast *Immigrants* ii. 88 Maybe we didn't do so bad for a Dago fisherman and a sheeny storekeeper.

sheep, *sb.* Add: **2. a.** (*c*) (Later examples.) Also *attrib.*, as *sheep-and-goat.*

1923 [see *one-nighter* s.v. *one B.* 33]. **1943** J. S. Huxley *Evolutionary Ethics* iii. 19 Our ethics will be unrealistic if, after dividing our impulses into sheep and goats, we.. transform the goats into scapegoats. **1954** N. Coward *Future Indefinite* iii. 138, I.. tried repeatedly to analyse my emotions coldly and objectively and to still my anxieties by segregating them, by separating the sheep from the goats. **1962** *Listener* 15 Mar. 469/2 This 'sheep and goats' view, though it may appear plausible, is not to be taken for granted. **1978** K. Hudson *Jargon of Professions* 15 In the author is sheep and goats among the countries who tried to defy all reasonable precautions.

c. (Later examples.)

1922 Joyce *Ulysses* 433 He was down and out but, though branded as a black sheep,.. he meant to reform. **1932** R. Aldington *Soft Answers* 76 Every privileged class tries its best to whitewash its black sheep. **1944** W. S. Maugham *Razor's Edge* v. 176 There was a time when the black sheep of the family was sent from my country to America. **1958** 'J. Byrom' *Or be he Dead* viii. 125 I'm the Black Sheep of the family, so if you ever meet any of my relatives, you'd better not tell them you've met me. **1979** *Jrnl. R. Soc. Arts* CXXVII. 650/1, I should like to think that they would play Black Sheep of Britain [sc. as the] in the propaganda deliberately as a means of sorting out the sheep from the goats?

3. a. Also, *one might as well be hanged for a sheep as a lamb* and varr. (earlier and later examples).

1678 J. Ray *Eng. Proverbs* (ed. 2) 350 As good be hang'd for an old sheep as a young lamb. **1836** F. Marryat *Mr. Midshipman Easy* I. xi. 181 We might as well be hanged for a sheep as a lamb,.. I vote that we do not go on board. **1913** D. H. Lawrence *Sons & Lovers* x. 259 It seemed as if she did not like being discovered in her home circumstances.. But she might as well be hung for a sheep as for a lamb. She would have to pay hell for the sheep. **1977** R. Pym *Quartet in Autumn* xv. 143 Letty.. decided that she might as well be hung for a sheep as a lamb and make the most of her meal.

d. *to count sheep*: as a soporific, to count imaginary sheep jumping over an obstacle one by one.

[**1854** S. Smith *'Way down East* xi. 273 He shut his eyes with all his might, and tried to think of sheep jumping over a wall.] **1920** E. O'Neill *Beyond Horizon* iii. i. 128, I couldn't get to sleep to save my soul. I counted ten million sheep if I counted one. **1952** J. Tey *Singing Sands* iv. 46 'Count sheep.' **1974** T. S. Eliot *Waste Land Drafts* (1971) 27 When restless nights distract her brain from sleep She may as well spend them counting sheep. **1977** H. Pitcher *When Miss Emmie was in Russia* x. 75 Nanny.. was trying her hardest to persuade Irina to go to sleep. Did you know that if you count sheep, it is watching the sheep *jump* that sends you off?

5. a. (Later examples.)

1914 G. B. Shaw *Misalliance Pref.* p. lxviii, Bullied and ordered about, the Englishman obeys like a sheep. **1930** — *Apple Cart* ii. 72 The way you fellows scuttle backward and forward from one mind to another whenever Joe holds up his finger is disgusting. This is a Cabinet of sheep. **1946** Wodehouse *Uncle Dynamite* xiii. 226 She looks on you as a.. poor, spineless sheep who can't say boo to a goose.

7. b. *sheep-flock* (later *attrib.* example).

1876 G. M. Hopkins *Poems* (1967) 65 And sheep-flock clouds like worlds of wool.

c. *sheep crib, -down* (earlier example), *paddock, ranch, shed, station* (later examples), *-wagon.*

1921 K. S. Woods *Rural Industries round Oxf.* ii. i. 80 Hazel.. is made into wattle or 'flake' hurdles and sheep cribs. **1798** W. Wyrick *Eng. Country Crafts* (ed. 2) 40 Other specialities are the making of such articles as hen-coops, pump-buckets, sheep-cribs. **1789** G. White *Selborne* i. 2 A vast hill of chalk.. divided into a sheep down, a high wood, and a long hanging wood. **1930** L. G. D. Acland *Early Canterbury Runs* 1st Ser. v. 103 This part of the station is still called the 'sheep paddocks'. **1874** J. G. McCoy *Hist. Sk. Cattle Trade* i. 1 This it is common to hear of a corn ranch, a wheat ranch, a sheep ranch. **1981** G. McDonald *Fletch & Widow Bradley* xvii. 72 She worked six months on a sheep ranch. **1946** J. W. Day *Harvest Adventure* vii. 110 Allus came up to my sheep-shed, an' if I ai' people a-watchin' ner at work—tourists an' lokie—would say, 'Ah! company I see.' **1911** C. E. W. Bean *'Dreadnought' of Darling* i. 8 The long blue-grey galvanised-iron wool-shed of some sheep station. **1967** F. Clune *Red Heart* 99 They get out into the last outpost of civilisation, at Mount Abundance sheep station. **1909** E. Rupert *Let.* 14 May in *Atlantic Monthly* (1913) Oct. 434/2 About noon the first day out we came near a sheep-wagon. **1962** G. MacEwan *Blazing Old Cattle Trail* xx. 134 The canvas-roofed sheep wagon was the ultimate in household compaction, combining the essentials of kitchen, dining-room, bedroom and shop quarters.

d. *sheep-rancher.*

1904 *Country Life* July 287/1 The Montana sheep-rancher figures that the wool will pay all expenses, leaving the increase for his profit. **1976** A. J. Russell *Pour*

Hemlock (1979) vii. 61 A sheep rancher who owned vast lands on the Colorado Plateau, in northeastern Arizona.

e. *sheep-bitten*, *-grazed*, *-proof*, *-scattered*, *-white* (later examples).

1917 J. Masefield *Lollingdon Downs* 31 Night is on the downland, on the lonely moorland, On the hills where the wind goes over sheep-bitten turf. **1925** W. de la Mare *Three Sleeping Boys in Broomsticks* 256 The bird-haunted, sheep-grazed meadows. **1976** *Southern Even. Echo* (Southampton) 2 Nov. (Advt. Suppl.) 3/8 Turfs, good quality, sheep grazed and weed treated, machine cut. **1882** Armstrong & Campbell *Austral. Sheep Husbandry* xvii. 186 This fence can be made still more sheep-proof.. by leaving out the bottom wire.. and having.. a light embankment thrown up. **1905** 'T. Collins' *Such is Life* iv. 134 The fence, much damaged by floods, was repaired merely to the sheep-proof standard. **1950** *N.Z. Jrnl. Agric.* July 74/2, 20 paddocks, all sheep-proof fenced. **1978** I. Murdoch *Sea* 401 After the dog there was ordinary farm land, sheep-scattered hillsides. **1945** Dylan Thomas in *Poetry* (Chicago) July 175 The frosts hold Flocked with the sheep white smoke of the farm house cowl.

8. sheep blowfly, a large greenish blowfly belonging to the genus *Lucilia*, esp. *L. coprina*, the larva of which is a pest of sheep in Australia; **sheep-bush** *Austral.*, either of two species of *Geijera*, *G. parviflora* or *G. linearifolia*, of the family Rutaceæ, a small evergreen tree sometimes used as fodder for sheep; **sheep-camp**, (*a*) *N. Amer.*, a camp for sheep herders; (*b*) *Austral.* and *N.Z.*, a resting or assembly place of sheep (cf. *camp sb.*[2] 4 c); (*c*) *S. Afr.*, a fenced-in enclosure for sheep (cf. *camp sb.*[2] 15); **sheep cocky** *Austral.* and *N.Z. colloq.*, a sheep-farmer on a small scale (cf. Cocky *sb.*[2] 2); **sheep-crook** (later example); **sheep-dip**, (*a*) (later examples); also *fig.* (see quots. 1945, 1976); **sheep-dipping** (later examples); **sheep-dog**: also as *v. trans.*, to urge (someone) on in the manner of a sheep-dog; to direct or 'herd'; also **sheep-dogging** *vbl. sb.*; **sheep-farm** (earlier example); **sheep-kill** = *sheep-laurel*; **sheep-laurel** (earlier and later examples); **sheep-meat**, (*b*) in mod. trading use: meat obtained from sheep; mutton and lamb; (also written as one word); **sheepnose**, a small cider apple (see quots.); **sheep-poison**, (*a*) (earlier examples); **sheep-run** (further examples); **sheep-sick** *a.* (later example); **sheep-sorrel** (earlier U.S. example); **sheep trot** *nonce-wd.*, a dance as of satyrs.

1932 *Discovery* July 210/2 The sheep blowfly.. is reliably estimated to do £4,000,000 worth of damage every year [in Australia]. **1974** R. D. Hughes *Living Insects* v. 128 Cool temperatures in autumn can induce a pupal diapause in the prepupal larvæ of the sheep blowfly. **1889** Sheep-bush [see *Wilga*]. **1933** *Bulletin* (Sydney) 7 June 25/2 Sheep bush.. is quite as good as that in long narrow leaves. **1965** *Austral. Encycl.* IX. 220/2 The smaller related (*Geijera*) *linearifolia*, which extends into Western Australia, is called sheep-bush. **1869** J. Muir *Jrnl.* 29 June in *My First Summer in Sierra* (1911) 85 Though only a sheep camp, this grand mountain hollow is home. **1921** H. Guthrie-Smith *Tutira* xx. 180 Before the establishment of sheep-camps growing grass and clover, there was nothing to tempt the pig from the low grounds. **1931** *Amer. Speech* VII. 120 A *sheep camp*, or the migratory home of a pair of shepherders, consists of a canvas-topped wagon with a stove in it and a bunk or bed at the back. **1959** P. A. Rollins *Gone Haywire* v. 114 He had stopped at a sheep camp and played casino. **1947** [see *camp sb.*[2] 15]. **1950** *N.Z. Jrnl. Agric.* May 483/1 The paddock was a sheep-camp paddock of similar shape where considerable numbers of sheep were frequently concentrated. **1973** *Kingston* (Ontario) *Whig-Standard* 11 Aug. 7/2 A few miles down-river there was a sheep-camp. **1949** F. Sargeson *I Saw in My Dream* ii. xiv. 206, I never can touch my hat to a sheep-cocky's dogs aren't pets. ? **1873** Hardy *Let.* (1978) I. 25, I have sketched in my note-book during the past summer a few current instances of smockfrocks, gaiters, sheep-crooks, rick-'staddles'.. and some other out-of-the-way things that might have to be shown. **1911** W. H. Koebel *In Maoriland Bush* v. 93 He was selling a new species of sheep-dip. **1915** *N.Z. Jrnl. Agric.* 20 Nov. 121 Do not overtake in the purchase of sheep-dip. **1945** Mencken *Amer. Lang. Suppl.* I. 262 Many generic names for alcoholic stimulants.. *sheep-dip*, [etc.]. **1968** K. Weatherly *Roo Shooter* 128 His seat was an old five-gallon petrol tin that had once held sheep dip. **1976** *New Yorker* 3 May 65/1 'Sheep dip' was what the lumberjacks called their tea. **1915** J. R. Macdonald *N.Z. Sheepfarming* xxvii. 71 It is needless to set forth all the conditions for complete success in sheep dipping, seeing that.. it is the custom to attach directions for use on every packet of drum. **1968** J. Arnold *Shell Bk. Country Crafts* 228 The primary use, for a coracle, now, is for fly-fishing and sheep-dipping. **1975** *Times Lit. Suppl.* 13 Apr. 418/4 Working with Thomas Jones and sheep-dogged by vigilant helpers, I entered a new dimension of scholarship. **1981** S. Jackman *Game of Soldiers* i. 15 The Group Senior Signals officer.. had done his time.. on Coastal Command Sunderlands, sheep-dogging convoys in the Western Approaches. **1969** R. Blishen *This Right Soft Lot* i. ii. 40 A surprising number of boys seemed never to have seen the Thames before.. So I did a little quick sheep-dogging, and at last we reached the gallery. **1775** T. Pennant *Tour Scotl.* II. 200 A letter from Mr. George Malcolm, concerning Sheep-farms, &c. **1968** R. Buckler *Ox Bells & Fireflies* vii. 106 The purple loops of the sheepkill. **1810** F. A. Michaux *Hist. Arbres Forestiers Amér. Sept.* I. 39 Mountain laurel.. sheep laurel,.. noms secondaires. **1954** *Sheep Laurel* [see *Lambkill*]. **1975** *Austral. Outlook* XXIX. 298 New Zealand supplies 80 per cent of the British sheepmeat imports. **1978** *Times* 30 June 17/3 The word 'sheepmeat' with which Brussels refers to mutton and lamb, is translated from the official French term, *Viande ovine*. **1979** *Times* 13 Nov. 17/6 The recent use of the term 'sheepmeat' in place of mutton and lamb is depressing in the extreme and will, I fear, make many people put off buying what is one of our most important farm products. **1817** W. Coxe *View Cultivation of Fruit Trees* 115 Bullocks Pippin.. is more generally known by the vulgar name of Sheep-nose, from a supposed resemblance between the form of the apple and that part of a sheep. **1925** C. Morley *Safety Pins* 178 We have seen apples of a strange shape, something like a pear (sheepnoses, they call them). **1943** B. Damon *Sense of Humus* 234 The Sheepnose, for example, had an interesting shape and a name just right. **1790,1814** *Sheep poison* [see *Lambkill*]. **1889** W. Fox *Six Colonies of New Zealand* i. 27 The.. plain.. is surrounded by hills which afford excellent sheep-runs. **1911** C. E. W. Bean *'Dreadnought' of Darling* i. 8 The names painted in big letters on the railway stations were the names of the sheep runs. **1936** A. Russell *Gone Nomad* iv. 23, I even learnt to operate on the lambs myself, and to perform the many other jobs that combine to make up the yard work on a sheep run. **1962** *Times* 6 June 15/6 Most of it goes [sheep-sick]. **1967** N. Wyt. *Law Jrnl.* 14 Mar. (1827) 191/1 Sheep-sorrel, **1978** [see *Rumex*]. **1926** D. H. Lawrence *Sun* (1960) 136 ... **1960** Neighbourhood sheep trot.. of the naïads.. **1781** ... 'Sheep trot'.. **1975** *Fashion* 10 The satyrs danced the sheep-trot all the day.

9. *sheep's trotters* (later *sing.* example); **sheep's foot**, (*c*) *sheep's foot roller*, a kind of tamping roller consisting of a steel drum studded with projecting feet; **sheep's grey**, material composed of a mixture of black and white wool; also *attrib.* or as adj.; **sheep's nose** = *sheepnose* sense 8 above; **sheep's wool**, (*a*) (later example).

1934 J. L. Harman *Highway Engin.* (ed. 2) xiii. 224 Various types of tamping rollers have been developed.. sheep's-foot and sectional rollers. **1973** G. E. Bateman in Hirschfeld & Poulos *Embankment-Dam Engin.* 13/1 The recent development of heavy vibratory rollers capable of compacting rockfill has produced the most significant change in placement procedures in the construction of earth and rockfill dams since the introduction of the sheepsfoot roller for the compaction of earthen core materials. **1852** *Trans. Mich. Agric. Soc.* III. 483 Ten yards or over of sheeps grey cloth. **1877** *Rep. Vermont Board Agric.* IV. 92 The men and boys wore homemade coats of sheep's gray. **1976** *National Observer* (U.S.) 24 July 21/3 The cold winter. **1919** ... 14 Apr. 15/4 (Advt.), Pullover or Cardigan. Colors: Blue Heather, Natural White, Sheep's Grey. **1936** *Ag & Cole* CLXX. 183/2 Sheep's Nose, an old-time variety of apple whose name is almost forgotten. **1969** *Sheep's trotter* [see *Hand-down* s.v. *Hand sb.* 23]. **1922** Joyce *Ulysses* 427 A cold sheep's trotter, sprinkled with wolf's bane. **1978** S. Sheldon *Bloodline* viii. 105 Samuel huddled into his threadbare sheep's-wool coat.

sheep-hook. Add: Also *fig.*

1866 C. M. Hopkins *Jrnls. & Papers* (1959) 143 Beeches.. scatter their tops in charming tufted sheephooks drooping towards each other and tilted.

sheepman. 1. For *U.S.* and 'in Canada' read orig. *N. Amer.* and add later examples.

1936 W. M. Raine *Troubled Water* xx. 174 If one sheepman was permitted to invade the range, dozens of others would drive across into the forbidden territory. **1950** L. G. D. Acland *Early Canterbury Runs* 1st Ser. v. 106 D. Oliver, the head shepherd, became manager. He was a fine sheepman and a good rider. **1937** *Times Lit. Suppl.* 6 Feb. 86/3 There has never been in Britain the hostility between the cattlemen and the sheepmen. **1962** H. J. Massingham *Curious Traveller* ix. 163 The commonwealth: sheep-men of the mountain parishes only paid half a crown a year to the Lady Marcher if they drove a hundred cattle.. **1965** 'J. Hackston' *Father Clears Out* 37 A new volume on sheep-raising,.. proclaiming to the world that Father was a sheep-man. **1977** F. Ormsby *Store of Candles* 13 Even the hardlies move to corner tables, Mouthing 'Sheepman'.

sheep-o(h) (ʃiːpəʊ), *int.* and *sb. Austral.* and *N.Z.* [f. Sheep *sb.* + *-o*[1].] **A.** *int.* A shearer's call for a sheep to shear. **B.** *sb.* (Chiefly *N.Z.*) = *penner-up* s.v. *Penner*[2].

1900 H. Lawson *On the Track* 132 'Go it, you—tigers!' yell's a tarboy. 'Wool away!' 'Tar!' 'Sheep Ho!' We rush through with a whirring noise till breakfast time. **1911** W. H. Koebel *In Maoriland Bush* viii. 134 The 'sheep-oh' sets to work to fill the nearly emptied pens. **1925** R. Rees *Lake of Enchantment* vii. 111 The [shearing] gang [included].. some boys to act as 'sheep-os'—that is to keep the pens in the shed filled up from the yards outside. **1940** Coo. [see *Penner*]. **1949** P. Newton *High Country Days* 5 The cry of 'Sheep-o!' would rouse the 'penner-up'. **1955** G. Bowen *Wool Away!* (1956) vii. 96 The 'sheepo' is the man who fills the catching pens.. The shearer has to call for sheep; he gives the call of 'sheepo'.

sheep's head. Add: **2. b.** *Cards.* A simple form of skat.

1886 E. E. Lemcke *Skat* 4 Skat is of quite recent origin. It bears a great resemblance to the Wendish game of 'Schafskopf' (Sheepshead) and to 'Solo'. **1913** *Official Rules Card Games* (U.S. Playing Card Co.) (ed. 17) 205 Schafkopf, (Sheepshead).. Object of the Game—To win in tricks certain cards of counting value as follows: A's, 11, 10, K's 4, Q's 3, and J's 2. **1939** C. Isherwood *Goodbye to Berlin* 186 A group of youths.. were playing Sheep's Head. **1951** F. Brown *Murder can be Fun* iv. 57, I know a feller plays schafskopf—sheepshead. **1976** *Warren* (Ohio) *Tribune* 27 June 1/2 National *Skat & Sheepshead Q. Dec.* 12 Fred Suter always wanted to prove that Sheepshead is just as good as a Classic Tournament.

3. a. Also applied to similar fishes, esp. the marine *Archosargus rhomboidalis* of Florida and the fresh-water drumfish, *Aplodinatus grunniens.*

1884 J. O. L. Goode *Bk. Fishes* 73/2 The Sheepshead belongs to the Porgy family. **1935** E. F. Lacler et al. *Ichthyol.* v. 147 In some fishes such as the sheepshead and the sea bream (*Archosargus*) they [sc. the cutting teeth] look almost human. **1979** *South Padre Parade* (South Padre Island, Texas) Dec. 6/3 These visiting fishermen caught their limit of reds, trout and sheepshead.

sheepskin. Add: **1. c.** A coat made of sheepskin.

1917 W. J. Locke *Red Planet* xxi. 270 From the shapeless lean-to sheepskin he had the huge boots.. [he] was clad in wool. Over a filthy sheepskin were slung all kinds of paraphernalia. **1977** A. Schofield *Venom* ii. 83 The policeman made no reply, hunching down again in his sheepskin.

3. a. *sheepskin-lined* adj.

1936 E. Kirwan in *Blackw. Mag.* Mar. 68/3 A pair of sheepskin-lined.. boots. **1980** *Country Life* 3 July 46/1 The ad is the colourful Mongolian dress.. Winter dels are sheepskin-lined (a man's del can require up to 40 lambskins).

b. *sheepskin rug.*

1917 *Harrods Catal.* 763/1 Sheepskin rugs 35/to 120/0. **1963** 'J. Tey' *Shilling for Candles* xxvi. 275 An atmosphere of marble mantlepieces and sheepskin rugs. **1976** J. Bingham *God's Defector* vi. 73 In front of the dressing-table was a white sheepskin rug.

sheepy, *a.* Add: **b.** Full of sheep.

1934 Dylan Thomas *Let.* 25 Apr. (1966) 110, I even go without a coat (sometimes) in this cold weather, & tread the sheepy-bumped over the sheepy fields. **1979** E. Barnard *Posthumous Papers* viii. 139 The.. hills beyond.. the lonely, sheepy lands.

sheer, *a.* and *adv.* **A.** *adj.* **6. a.** For Now *U.S.* read Chiefly *U.S.* (exc. of stockings) and add later examples.

1928 E. M. Clowes *On Wallaby* viii. 225 America seems to have taken to fashioning her literature with a crimping-iron and 'sheer-lawn', while Australia hacks hers out with a lily-hook from bush-block and bush. **1934** A. Woollcott *While Rome Burns* 284 Ravishing French peasant girls with sheer, shadowy stockings, and a disposition to say 'Ooh, la la' at appropriate intervals. **1951** in M. McLuhan *Mech. Bride* (1967) 90 Ivory Flakes care helps safeguard sheerer nylons. **1977** D. Clarke *Gimmel Flask* iii. 56 She was.. dressed in a tan moygashel suit, sheer nylons on excellent legs.

b. *absol.* as *sb.*

1934 in *Webster*. **1937** *Jrnl.* (Lincoln, Nebraska) 25 Apr. (Advt.), Saucy sheers for budgeteers. **1943** *Amer. Speech* XVIII. 94 [New Zealand] Sheer (dress material) (English 'kirby-grip'), *sheer* (dress material), *tubbles* (washable frocks), are taken from American, not English, advertisements. **1952** C. W. Cunnington *Eng. Women's Clothes in Present Cent.* vii. 297 Stockings.. in service sheers, 47½ to 61/11. **1966** *Daily Tel.* 28 Oct. 15/4 Tiny curtain sheers.. are better than the German Dralon sheers we saw in the shops last winter. **1978** *Lancs. Life* Apr. 85/1 There is.. a new range of Swiss sheers and prints with louvred blinds to match.

sheerly, *adv.* **4.** (Later examples.)

1920 G. M. Crewe *Northern Numbers* 69 The barrier.. Lifts sheerly. To the unknowledgeable skies. **1947** E. Mannell *Sussex* i. 15 Across the road,.. the chalk downs again sheerly, down to the water-levels and the river valley. **1964** G. B. Schaller *Year of Gorilla* (1965) xiii. 24 The volcano's.. Its slopes dropping sheerly.. The cliff gorge.. the rugged slopes.. and the base of the huge rock wall that rose sheerly to the peaked summit.

sheesh kabab, var. *Shish kebab.

sheet, *sb.*[1] Add: **3. c.** In proverbial phr. *as white* (*or pale*) *as a sheet*. Cf. White *a.* 5 b.

1751 Fielding *Amelia* III. vii. viii. 84 She entered.. with a face as white as a sheet. **1830** M. T. Thompson *Chron. Pilgrim* ix. 4 He turned as pale as a sheet. **1872** T. Hardy *Under Greenwood Tree* I. 1. viii. 125 You'll be white as a sheet to-morrow. **1929** E. Rice *Street Scene* i. 72 Well, there was three of 'em.. Mr. Maurrant looked in' at Sankey as if he was ready to kill him, an' Mrs. Maurrant as white as a sheet, an' Sankey.. as innocent as the babe unborn. **1953** A. J. Cronin *Adv. in Two Worlds* xxxix. 278 He was as pale as a sheet. **1960** J. Wain *Strike the Father Dead* 226 The sheets seemed little larger than a shrimp, pale as a sheet, with.. big dark eyes.

5. e. A dollar bill (*U.S.*) or pound note; the monetary value of this. *slang.*

1937 *Research Stud. State Coll. Washington* V. 19 What a fellow gets for one sheet from an officer he can sell to the boys.. for five and ten sheets. **1958** F. Norman *Bang to Rights* i. 48 Which if it did happen would cost some geezer sixty sheets at least. **1963** *New Society* 29 Aug. half a sheet. **1969** M. Pugh *Last Place Left* xxvi. 191 A sheet the night. Five quid if you last a week. **1978** *Red Car Jrnl.* Gone.. .. is, 3.. Buying a house.. a double high-pitched, screaming noise.. It cost.. you. Pair, two sheets. *Ibid.* Tis cash..

f. *U.S. slang.* = *rap sheet* s.v. *Rap sb.*[1] 7.

1958 *N.Y. Times Mag.* 16 Mar. 88/3 *Sheet*, a criminal

record. **1976** C. Weston *Rouse Demon* (1977) xxvi. 125 Somebody scared him into it. Let's take a look at his sheet. I want to know who.

6. b. (Later examples.)

1930 [see *Ashburne* ... *Art of Bookbinding* i. 1 Should the amateur wish to have his books in sheets, he may get them by asking his bookseller for them. **1979** P. Gaskell *New Introd. Bibliogr.* 144 Long books were divided in quires of 12–24 sheets before this folding took place; hence 'books in quires' as a synonym for 'books in sheets'.

c. For *rare* read chiefly *U.S.* and add later examples.

1926 R. Hughes in *Hearst's Internat.* Feb. 44/1 'How come the newspapers keep saying your fights are all fixed?'.. 'Ah, who cares what the dirty sheets say?' **1958** *Spectator* 20 June 807/2 A mass-circulation London Sunday 'sheet'. **1970** G. Dutton in Boyd & McLeod *Wanders to Newspapers* iii. 220 Rivington made it clear that he intended no partisan sheet.

12. a. *sheet-whiteness*; *sheet-pale* adj.

1906 Hardy *Dynasts* III. iv. v. 225 Sir David Baird, still helpless from his wound, is carried past on a cot, sheet-pale and thin. **1940** H. Gold *Man who was not with it* xxvii. 250 This.. creature who was Pauline's dark daughter,.. so her nose.. with the nose slightly.. .. pale. **1967** B. Ward *Nov.*.. blue nor sheet-whiteness, sheet-whiteness,.. in her still cold beauty.

b. sheet band *Printing* (see quot.); **sheet erosion**, the erosion of soil by rain-water acting more or less uniformly over a wide area; **sheet-fed** *a. Printing*, using paper in the form of cut sheets; **sheet-flood**, a short-lived expanse of running water that spreads as a continuous film over a large area following sudden heavy rain; **sheet-flow** *Geomorphol.*, a flow that covers a wide expanse of a surface instead of being confined in a channel; **sheet glass**, add: (*a*) in mod. use, a kind of flat glass made by a vertical drawing process (cf. *Fourcault*); **sheet ice**, ice formed in a thin, smooth layer on water; **sheet lightning** (earlier example); **Sheetrock**, the proprietary name of a plaster-board made of gypsum between heavy paper (also with small initial); **sheet-wash**, sheet erosion; (erosion caused by) a sheet-flood.

1940 V. S. Gaskin *Print Mss. of Printing & Binding* x. 120/2 Carefully set, the sheet bands hold the sheet up to the cylinder and help to expel air from between the sheet and the cylinder, and thus minimize the slurring. **1928** Mosier & Gustafson *Soil Physics & Management* xvii. 494 'Sheet erosion' is the removal of soil from the surface in thin layers. **1961** *Daily Tel.* 10 Nov. 32/1 Without sheets of stonework, the greatest loss than gullying. **1978** W. W. Emmett in M. J. Kirkby *Hillslope Hydrol.* v. 171 ... 'sheet erosion' is obliged to be the evidence of more accelerated erosion than sheet erosion. **1966** *Sheet-fed* [see *Rotary a.* 3 b]. **1977** A. Hellard *Fleet Bk.* 49 After storms, flows in the form of sheet-floods, comparatively shallow floods running over the.. large bodies.. which spreads over the land. **1973** W. H. Hallahan *Ross Forgery* iv. 52 The paper McLellan used.. .. stamps made on sheet-fed presses. **1975** ... floods.. sheet-floods of.. formed in.. bodies of.. **1932** D. W. Johnson in *Jrnl. Geomorphol.* i. 1933.. *Ibid.* It was.. and *Trans. Amer. Geophys. Union* ... The deposition of.. loose materials.. by sheet-flow.. **1954** ... XXXIX. 481 The deposit was obviously not a sheet-flow; it was a stream [of the typical material] of unknown flow. **1977** A. Hallam *Planet Earth* 85/2 This leads to preferential weathering at the break in slope, the weathering products being removed to the adjacent lowland by unchannelled sheet-flow. **1950** ... 1969 Rock.. by.. process.. **1974** *Encycl. Brit. Macropædia* VIII. 202/1 Sheet glass.. ... the pouring of.. the drawing process is.. **1912** ... floats on a.. the production of.. sheet ice.. **1934** ... Regional Lang. Stud.—*Newfoundland* (1978) viii. 24 Sheet ice.. thin ice.. .. 1916.. *Sheet lightning* .. **1794** J. B. S. Morritt *Let.* 24 June (1914) 44 We had.. sheet lightning.. every evening, and have.. **1921** *Official Gaz.* (U.S. Patent Office) 22 Mar. 765/1 Sheetrock.. Plaster Wall-Board. Claims use since Aug. 28, 1917. U.S. Gypsum Co., Chicago. **1924** *Trade Marks Jrnl.* 5 Nov. 2475 Sheetrock.. Plaster.. used in building. **1966** *Times* 23 Mar. .. .. Chicago. **1973** R. B. Parker *Godwulf Manuscript* (1974) ix. 71 I've.. only.. a.. the windows, sheetrock partitions painted green. **1936** Finch & Trewartha *Elem. Geogr.* xxv. 559 One of the most widespread and least noticed kinds of erosion is sheet-wash. **1939** *Geogr. Jrnl.* XCIII. 305 If Tibu accounts of the nature of the rainfall are even partially credited, some form of sheet-wash can readily be imagined covering the whole floor of a huge wadi, and undercutting its sides. **1964** A. Holmes *Princ. Physical Geol.* (ed. 2) xx. 613 A sudden change of slope seems to be favoured by torrential seasonal rainfalls and by the liberation of only minute amounts of fine debris which can be readily swept away by sheet-wash over the pediment. **1973** J. G. Cruickshank *Soil Geogr.* ii. 39 Fluvial erosion by rivers or sheet wash is the most important process further south in the tundra.

13. a. *sheet metal.*

1933 *Rep. & Mem. Aeronaut. Res. Comm.* No. 1553. 18 Constructions in thin sheet metal (*e.g.* monocoque fuselage) normally consist of a large area of sheet divided into a number of small panels by a system of stiffeners. **1959** *Motor Man.* (ed. 36) i. 3 The sheet metal forming the front wings and the sides of the bonnet. **1976** Lieberman & Rhodes *Compl. CB Handbk.* v. 97 I left it to the inside of the sheet-metal screws that actually screw into the rain-gutter groove of the trunk.

d. *sheet music*, music published in sheet form (as opp. to bound form).

1857 *Lawrence* (Kansas) *Republican* 11 June 3 (Advt.), Sheet music. **1880** ... **1891** .. [*Dict.*]. **1920** J. B. Priestley *Good Companions* iii. iii. 254 Performing rights, sheet music, gramophone records. **1981** J. Wainwright *Urge for Justice* i. ii. 14 The top of the shop was crammed with sheet music.

sheet, *v.*[1] Add: **5.** Also of rain: to fall in a sheet or sheets (sense 7 b). Freq. with *down.*

1971 D. Beaty *Temple Tree* 5 The monsoon rain was sheeting down. **1978** *Detroit Free Press* 16 Apr. 19/1 Bumping over the high moon thunderheads, with the sheeting rain lashing the windows, the air passenger from the South Pacific grips the seat arms.

sheeted, *ppl. a.* Add: **5.** *Geol.* Of rock (esp. granite) or a rock formation: having been divided into thin laminæ; *sheeted zone*, a belt of highly fissured rock associated with a fault, the fissures freq. being occupied by veins of minerals.

1893 *Bull. U.S. Geol. Survey* No. 213. 99 The granite is sheeted near the veins, the planes of sheeting being parallel to the veins themselves. **1904** H. Ries *Econ. Geol.* xvii. 339 (*caption*) Ore along sheeted zone. *Ibid.* 340 Composite veins in sheeted basalt dikes. **1939** W. H. Emmons et al. *Geol.* (ed. 2) xvii. 425 Some veins fill single openings.. others fill closely spaced parallel openings, which are sheeted zones. **1953** *Amer. Jrnl. Sci.* CCLI. ... During the glacial epoch these sheeted granites.. must have been easily plucked and quarried by the advancing ice. **1974** *Nature* 29 Nov. 375/2 Shattered pebbles and sheeted bedrock are common weathering phenomena in most desert regions.

sheeting, *vbl. sb.* Add: **3*.** *Geol.* The occurrence or development of closely spaced, approximately parallel fractures or joints in rock. Also *fig.*

1899 *Trans. Inst. Mining & Metallurgy* VIII. 67 The multiple fracturing, parallel to the walls of the dyke, is a characteristic feature of such lodes.. This class of lode can be described as a sheeting of the rock. **1903** [see *Sheeted ppl. a.* 5]. **1942** E. Eckel *Building Stones & Clays* iii. 59 Geologists.. ascribe some or all sheeting structure to the effects of late external stresses. **1942** ... That sheets ... **1964** ... sheeting on the ... **1975** D. Bowles *Stone Industries* viii. 116 Widely separated sheeting planes occur at a depth of 250 feet at Quincy, Mass. **1965** A. Holmes *Princ. Physical Geol.* (ed. 2) vii. 161 The sheeting or sheeted structure that is often seen in exposures of granite and other plutonic intrusions.

sheetlet (ʃiːtlɛt). [f. Sheet *sb.*[1] + -let.] A small sheet; *spec.* in *Philately* (cf. Sheet *sb.*[1] 5 d).

1930 in *Webster*. **1971** D. Potter *Brit. Eliz. Stamps* xi. 123 Three separate editions, 25, 43 and 65, contained sheetlets of one stamp. **1978** *Sunday Telegraph* (Colour Suppl.) 29 Jan. 39 (Advt.), A beautifully designed pack containing the unique series of stamps (in sheetlets) to be issued by the Crown Agents.

sheevo, var. *Shiveau.

Sheffer (ʃɛfə(r)). *Logic.* [The name of Henry M. Sheffer (1883–1964), U.S. philosopher.] *Sheffer's stroke*: the symbol |; also, the logical function of non-conjunction (and sometimes non-disjunction) that it represents, described by Sheffer in 1913 (*Trans. Amer. Math. Soc.* XIV. 481–8) (so *Sheffer stroke function*).

1932 Lewis & Langford *Symbolic Logic* ix. 306 The prefixes are analogues of Sheffer's stroke-function *p*|*q* in its two interpretations. **1950** R. M. Hammond et al. tr. *Hilbert & Ackermann's Princ. Math. Logic* i. 11 Since Sheffer's stroke.. can be expressed by Sheffer's stroke, the same holds for the other fundamental connectives. **1957** [see *non-conjunction* s.v. *Non- 2*]. **1960** N. R. Scott *Analog & Digital Computer Technol.* x. 392 If the low-potential and high-potential signal levels are identified respectively with logical 0 and 1, the circuit performs the Sheffer stroke function. **1972** A. G. Howson *Handbk. Terms Algebra & Anal.* i. 2 Here basic connective... *ibid.* The Sheffer stroke... the connective |.. alone. All of the other connectives can be defined in terms of | alone.

Sheffield. Add: **2.** The name of Henry North Holroyd, third Earl of *Sheffield* (1832–1909), used *attrib.* in **Sheffield Shield**, a trophy presented by him in 1892 and contested annually by Australian state cricket teams (first played in 1892–3).

1901 *Wisden's Cricketers' Almanack* (ed. 38) 462 The result determined the possession of the Sheffield Shield for the season. **1912** *Dict. National Biogr.* 2nd Suppl. II. 290/1 In 1891–2 Lord Sheffield, at his sole expense, took to Australia a team.. under the captaincy of.. Alfred Shaw. This enterprise greatly stimulated Australian cricket; the 'Sheffield Shield', which he presented to the Australian Cricket Association, is still competed for annually by cricketers of Victoria, New South Wales, and South Australia. **1930** C. G. Macartney *My Cricketing Days* iii. 20 My first season with the new club coincided with my first in interstate and Sheffield Shield cricket. **1977** *World of Cricket Monthly* June 19/1 The Australian cricketers who are signed are available to play Test, Sheffield Shield and club cricket when they are not playing in the super Test and other series.

‖ **shefstvo** (ʃɛfstvo). Also **chefstvo**. [Russ.] Patronage, sponsorship: variously used (see quots.).

1937 S. N. Harper *Govt. Soviet Union* viii. 134 The institution of so-called 'patronage' (*chefstvo*) is extended also to the Red Army and aims to keep those in military service in the everyday life of the community. **1948** J. Towster *Political Power in U.S.S.R.* xii. 322 Another form of tutelage was the so-called *shefstvo* or patronage by industrial regions over agricultural regions. **1950** B. Moore *Soviet Politics* viii. 173 Another method, which evidently sprang up in the early thirties and then was permitted to die a natural death, was called 'patronage' (*shefstvo*), usually the patronage of a specific factory or of a specific group of workers over some section of the state administration. **1955** H. Hodgkinson *Doubletalk* 120 *Shefstvo*, patronage exercised by a *shef* or 'chief' is the Soviet equivalent of 'empire building' in Western bureaucracy.

‖ **shehecheyanu** (ʃeheçeyāˈnuː). Also **shehechyoni**, etc. [Heb., lit. 'that has sustained us'.] A Jewish benediction pronounced on the evening of a principal holy day and on new occasions of thanksgiving.

1892 I. Zangwill *Childr. Ghetto* II. xvi. 53 He was wondering whether he ought not to say *Shehecheyanu*—the prayer over a new pleasure. **1959** E. S. Hoddes *Duct. Judaism* 213 *She'hecheyanu*, a blessing recited on the first evening of festivals, on eating the first fruits of the season and on donning a new raiment [etc.] for the first time. **1974** L. Davidson *Long Way to Shiloh* xv. 227 When they.. told me that the possibility existed, that by some miracle of God's grace, the holy Menorah was still in Erets Yisroel I said a *shecheyanu*. **1974** C. Raphael *Feast of History* i. 23 The Seder ceremony was an occasion for joy.. We thanked God—in the famous *She'hecheyanu* prayer—that he had 'kept us alive and sustained us and brought us to this moment'. **1980** *Jewish Chron.* 28 Mar. (Passover Suppl.) p. iii/2 He'd done the kiddush and the *shehecheyanu*.

shehita(h), var. *Shechita. **shehnai**, var. *Shahnai.

sheikh. Delete ‖ and add: **1. b.** (Chiefly in spelling *sheik*.) A type of a strong, romantic lover; a lady-killer. [f. *The Sheik*, a novel by E. M. Hull (1919), and its cinematic adaptation *The Sheikh*, 1921, starring Rudolph Valentino.]

1925 *Lit. Digest* 14 Feb. 28/2 We hear almost nothing more of the matinee idol any more.. The 'sheik' has taken his place. **1927** *Amer. Speech* II. 209/2 The girl calls the young man.. 'my sheik'. **1933** S. Gibbons *Cold Comfort Farm* xviii.245 The new shiek appears clad in a silver screen: 'Seth Starkadder is "Small-Town Sheik"'. **1939** 'G. Orwell' *Coming up for Air* i. i. 10 When your last natural tooth goes, the time when you can kid yourself that you're a Hollywood sheik is definitely at an end. **1956** S. Longstreet *Real Jazz Old & New* xii. 99 John Held Jr. drew the jazz-flapper and her sheik best.. The lad was apple-headed, his hair buttered tight down. He wore bell-bottomed trousers, a racoon coat, drove a Stutz Bearcat and played or danced hot jazz a lot. **1958** 'L. Egan' *Motive in Shadow* v. 87 He's sure a handsome sheik.

shei-khling, a petty sheikh.

1914 G. Bell *Let.* 19 Jan. (1927) I. xiii. 316 A young Sheikhling from Jauf. **1970** W. Garner *Big enough Wreath* iv. 40 Mini-presidents and rich sheikhlings seeking advice.

‖ **Sheikha** (ʃeɪkə, ʃiːkə). Also **Shaikha(h)**, **Sheika**, and with small initial. [Arab. *šaiḵa*.] An Arab lady or matron of good family; hence, the (chief) wife of a sheikh. (Also as a title of respect.) Also *transf.*, the consort of a 'sheikh' (cf. *Sheikh* 1 b above).

1853 J. Richardson *Narr. Mission Cent. Afr.* II. xiv. 247, I had a visit from a great sister of the Sarkee, a woman who is a Sheikha (female Sheikh), and receives the revenues of fifty villages for her own private use. **1926** Maines & Grant *Wise-Crack Dict.* 11/2 Sheikha, sheik's sweetie. **1949** R. P. Dickson *Arab of Desert* viii. 144 Al Jazi, sister of Khalid al Hithlain.. had a will of her own, and was a veritable 'Shaikhah'. **1967** M. Childers *Taint of Innocence* iii. 189 The Emir, the Sheika, and a party.. had left the palace. **1976** *Sheika* [see *shaikh* s.v.]. **1977** *Sunday Times* (Colour Suppl.) 3 Apr. 23/3 Shaikha.. a Saudi Arabian noblewoman.. who went.. to 50 Chepstow Villas, Notting Hill. **1976** L. Blandford *Oil Sheikhs* xxii. 256 Sheikha Osheh is.. married to Sheikh Mubarak. **1980** D. Creed *Scarab* xv 180 They were Arabs—and so are you, sheikha.

sheila (ʃiːlə). Now *Austral.* and *N.Z. colloq.* Also 9 shaler; **sheelah**, etc. [Orig. uncertain. Early *shaler* is not formally explained. It may represent a generic use of the (originally Irish) personal name *Sheila*, the counterpart of Paddy *sb.*[2] 1 a (see quot. 1828); in any case it was assimilated to this name at a later stage.] A girl or young woman; a girl-

friend. (Playfully affectionate and predominantly in male use.)
[**1828** *Monitor* (Sydney) 22 Mar. 1053/3 Many a piteous Sheila stood wiping the gory locks of her Paddy, until released from that duty by the officious interference of the knight of the baton.] **1839** H. Brandon in W. A. Miles *Poverty, Mendicity & Crime* 165/1 *Shaler*, girls (country phrase). **1847** G. W. M. Reynolds *Myst. London* III. xxv. 71/1 Cop that young shaler unto thee. **1864** Hotten *Slang Dict.* 225 *Shaler*, a girl. **1895** C. Crowe *Austral. Slang Dict.* 72 *Shaler*, a girl. **1918** *N.Z.E.F. Chrons.* 2 July 252/1, I goes and stays out at Ngaire with my shieler's people. **1919** W. H. Downing *Digger Dialects* 44 *Sheila*, a girl. **1928** A. Wright *Good Recovery* 117 Leave the sheilas alone, they're sure to pool a man sooner or later. **1930** V. Palmer *Men are Human* xxvii. 251 There was a sheilah he had working for him once, a lively piece with black eyes. **1940** F. Sargeson *Man & his Wife* (1944) 66 I've got a job in a grocer's shop and I'm trotting a sheila. She's a pearl of a sheila too. **1951** F. Hardy *Power without Glory* (ed. 3) ii. 21 'What d'yer expect us to do, just sit around and starve ourselves?'..'Please yer bloody self, but you've got to think of yer mother, and that sheila of yours.' **1959** *Woman's Own* 4 July 11/3 They have a daughter—a nice-looking sheila, too. **1962** *Coast to Coast 1961–62* 11 'I know a sheila,' Sonny began. 'A real trimmer.' **1969** *Private Eye* 21 Nov. 14 Don't be shit-scared of these sheilas who work for me. **1976** *Times Lit. Suppl.* 9 Apr. 418/3 His past would be re-mythologized by a host of pre-war piano accompanists, sepia footballers and nude sheilahs from the master brush of Norman Lindsay. **1977** D. Seaman *Committee* 63 They made the usual jokes about the local Sheilas.

sheila-na-gig, var. *SHEELA-NA-GIG. **sheirut**, var. *SHERUT.

‖ **sheitel** (ʃēɪ·t'l). Also **shaitel**. [ad. Yiddish *sheytl*, f. MHG. *scheitel* crown of the head.] Among strictly Orthodox Ashkenazi Jews: a wig worn by a married woman.
1892 I. Zangwill *Childr. Ghetto* I. ii. 41 A small, sickly-looking woman, with..the wig without which no virtuous wife is complete... A lower stratum of smatchel brown peeped out in front of the Shaitel. **1957** L. Stern *Midas Touch* i. vii. 63 His mother..no longer wore her sheitel, or traditional wig. **1973** *Jewish Chron.* 19 Jan. 22/3 A comely young rebbetzin in a glamorous sheitel and eyelashes long enough to brush dandruff from lapels has been touring American universities to attract straying Jewish youths back to their faith.

shekel. Add: **1. c.** An Israeli unit of currency introduced in February 1980, equivalent to ten former Israeli pounds; a note of this value.
1980 *Times* 23 Feb. 1/2 From next week the Israeli pound is to be replaced by a new currency named after the Biblical shekel... Each shekel will be purchased with 20 present Israeli pounds. **1980** *Whitaker's Almanack 1981* 979 Israel..Israeli Shekel of 100 New Agora..[Notes] Shekels 10, 50, 5, 1.
2. Also in phr. *to rake in the shekels*, to make money rapidly or 'hand over fist' (from a venture). Cf. *RAKE *sb.*[1] 2 a (*a*).
1887 [see *HIGH *a.* 20 b]. **1915** J. Buchan *39 Steps* i. 18 The capitalists would rake in the shekels, and make fortunes by buying up wreckage.

shekere, var. *SEKERE. **shekest(h)eh**, obs. vars. *SHIKASTA.

sheld-duck. Add: Now usu. **shelduck**.
1923 C. W. R. Knight *Aristocrats of Air* x. 124, I did not succeed in locating a Sheldduck's nest. **1939** J. Fisher *Birds as Animals* ii. 190 In..black skimmer and shelduck, both birds incubate. **1966** E. Palmer *Plains of Camdeboo* xi. 190 Shelduck, well-known for making their nests in holes, often choose ant-bear burrows. **1972** *Country Life* 3 Feb. 274/1 The shelduck is the largest and most striking bird of the English shores.

shelf, *sb.*[1] Add: **I. c.** *off the shelf*: from a supply of ready-made goods. Also (with hyphens) as *adj. phr.*
1936 *Industr. & Engin. Chem.* Feb. 150/2 The individual customer must generally have his material fabricated to his order and cannot obtain material [aluminium] 'off the shelf'. **1958** *Engineering* 11 Apr. 455/1 Using sets built to a standard pattern which would be available 'off the shelf' and made up of interchangeable parts. **1962** J. Glenn in *Into Orbit* 37 They had to use 'off-the-shelf' items in order to save time. **1978** *Nature* 26 Oct. 784/1 As CAMAC equipment is often sold as individual products to system builders, items are often available either 'off the shelf'.
3*. A police informer. *Austral. slang.*
1926 J. Doone *Timely Tips for New Australians* (Gloss.), *Shelf*, a slang word denoting an informer. **1952** *People* (Austral.) 5 Dec. 8/2 The jail authorities knew such trafficking went on and often set traps for the warders through the good offices of 'shelfs' or 'trusties' (prisoners who were informers).
II. 4. b. Also unqualified (freq. *attrib.*).
1913 *Jrnl. Geol.* XXI. 525 The mode of formation keeps the face of the shelf within a certain distance from the sea-surface. *Ibid.*, The shelf zone is..a biologic horizon of the first importance. **1934** C. R. Longwell et al. *Outl. Physical Geol.* vii. 125 The shelf beyond the long-quiescent Atlantic coast of North America is 60 to 80 miles wide off the Carolinas. **1964** *Oceanogr. & Marine Biol.* II. 61 The temperature minimum at 150m may be formed, in part, by mixing of winter Bering Sea water with shelf water. **1978** Friedman & Sanders *Princ.*

Sedimentol. xii. 360/2 (*caption*) Shelf lagoon between margin of a continental block and a reef tract that is situated at the edge of the continental mass.
III. 6. *shelf-load*, *-space*; *shelf-like* adj.; **shelf appeal**, the attractiveness to a customer of packaged goods displayed in a shop; **shelf back** *U.S.*, the spine of a book; **shelf cod**, cod found in inshore waters above the continental shelf; **shelf ice** [tr. G. *schelfeis* (O. Nordenskjöld 1908, in *Zeitschr. der Ges. für Erdkunde zu Berlin* XLIII. 628, following suggestion of A. Penck)], ice which forms a thick level layer on water (usu. the sea) but is attached to land; **shelf life**, the length of time that a commodity may be stored without becoming unfit for use or consumption; **shelf-list** (examples); so **shelf-lister**, one who compiles shelf-lists; **shelf paper**, paper used for lining shelves; **shelf sea**, an expanse of sea overlying continental shelf.
1933 *Shelf Appeal* July 3 (*heading*) Shelf Appeal. A monthly publication devoted to the planning, designing, manufacturing & display of the package. **1963** *B.S.I. News* Apr. 20/1 The 'shelf-appeal' pack designed to catch the eye of the ordinary shopper. **1964** *Times Rev. Industry* Feb. 23/1 With the post-war swing to more branded goods, self-service, enhanced hygiene and the need for 'shelf-appeal', the demand for more and better packaging has expanded fast. **1925** J. A. Holden *Bookman's Gloss.* 97 *Shelf-back*, the back of a book, on which the title is lettered. **1931** *Publishers' Weekly* 9 May 2322 In stamping the name of the author and the name of the periodical stamped in gold on cover and shelfback. **1960** G. A. Glaister *Gloss. Book* 375/1 *Shelf back*, the spine of a book. **1935** L. Luard *Conquering Seas* 37 Cod... Shelf Cod. **1976** *Eastern Even. News* (Norwich) 9 Dec. 12/5 Grimsby fish. Poor supply, good demand; six boats landed 1317 kit. Shelf cod (4 to 5). **1910** *Geogr. Jrnl.* XXXV. 726 To this ice formed *in situ* out of snow accumulations in the sea Nordenskiöld gives the name 'shelf-ice'. **1938** *Ibid.* XCI. 521 Of all the glacial features in this region, perhaps the greatest interest attaches to the shelf-ice filling King George VI Sound. **1977** *New Yorker* 20 June 91/1 The river's edges are lined with ice that is stationary—'shelf ice', 'shore ice', the first to freeze at the start of winter and the last to go in spring. **1927** *Manufacturing Confectioner* Jan. 21 (*heading*) What is the shelf life of your hard candy? **1933** R. A. Watson Watt et al. *Applications of Cathode Ray Oscillograph in Radio Res.* i. 27 The batteries may..be of very small size; 'shelf-life' and loss through casual leakage..are more important than their actual load current. **1940** *Austral. Pat. Dentistry* XLIV. 39 Either copper or copper and zinc must be present in the alloy if it is to possess a reasonable 'shelf life'. **1956** *Visible Packaging of Flour Confectionery* (British Cellophane Ltd.) 3 For small fancy cakes a moistureproof heat-sealing wrap..should give a shelf-life of several days. **1969** *Observer* 26 Jan. 5/5 The shelf-life of donated blood is about 21 days. **1980** D. Francis *Reflex* viii. 90 Some photographic chemicals lose their power with age. Shelf life, and so on. **1851** H. Melville *Moby Dick* I. iii. 18 On one side stood a long, low, shelf-like table covered with cracked glass cases, filled with dusty rarities. **1962** *Science Survey* XI. 178 The inner membrane whose shelf-like folds protrude into the interior of the organelle. **1920** A. E. Bostwick *Amer. Public Library* 171 The name 'shelf list' is somewhat unfortunate. **1979** *Amer. N. & Q.* June 166/2 Each volume of the shelflist is in four parts, viz., the classification schedule, the shelflist proper (showing shelf mark, author, title, place of publication, date of publication, and the same items (excluding serials) in chronological order), and an author and title index. **1927** W. W. Bishop *Pract. Handbk. Mod. Libr. Catal.* 21 The catalog room..should be..on the same floor with the order clerks, classifiers and shelf-listers. **1950** G. Greene in *Dickens's Oliver Twist* p. vii, We must forget that long ... *Arts* Mar. 177/2 ... ten shelf loads of regulations, there is no escaping that there is no risk of failure. **1895** *Montgomery Ward Catal.* 113/3 Shelf Paper, pinked in fancy designs, each sheet 8½ inches wide, 33 inches long. **1968** *Listener* 27 June 841/1 Kerouac types *On the Road* on a 120-foot roll of shelf paper but cannot get it published. **1913** T. C. Chamberlin in *Jrnl. Geol.* XXI. 523 The waters that rest upon the continental platforms may be known conveniently as shelf-seas. **1966** Bennison & Wright *Geol. Hist. Brit. Isles* ii. 25 Sediments laid down in deeper water..have a quite different faunal content from the shelf-sea deposits. ... iii. 95 It became clear that the shelf sea added to the sedimentary mass that ... **1978** J. Irving *World according to Garp* ii. 28 Her books...outgrew the shelf space.

shelf, *v.*[1] Add: **2.** *Austral. slang.* To inform upon. Cf. *SHELF *sb.*[1] 3*.
1953 K. Tennant *Joyful Condemned* xi. 96 Central has only to lamp you coming in here, and we all go up. Jimmy here shelfed me before. **1958** V. Kelly *Greedy Ones* 104 We were mates in the affair and you don't shelf your mates. And anyone who does shelf a mate has got to take what's coming to him.

shell, *sb.* Add: **I. 8. b.** The empty case of a fruit.
1902 H. L. Wilson *Spenders* xxvii. 313 Mr. Milbrey glanced at the two shells of the orange which the butler was then removing. **1974** *Times* 20 Apr. 10/8 Grapefruit mixes well with cottage cheese, and you can use the shell to hold the salad.
9. b. (Earlier example.)
1815 C'tess Granville *Lett.* (1894) I. 73 Madame de Coigny has difficulty in re-uniting people *chez elle*, and

if one meets a Frenchman there, he draws into his shell and sits in gloomy silence.
II. 12. b. A thin body bounded by two closely spaced curved surfaces: (*a*) as a concept in *Statics*; (*b*) in *Civil Engin.*, a structural member of this form that has strength by virtue of its shape.
1877 G. M. Minchin *Treat. Statics* xiv. 432 Hence..every shell of uniform density and small thickness bounded by similar, similarly situated, and concentric surfaces produces a constant potential at all points in its interior. **1892** A. E. H. Love *Treat. Math. Theory Elasticity* I. vii. 221 Consider the case of a spherical shell whose outer and inner surfaces are subjected to hydrostatic pressure. **1952** O. Faber *Reinforced Concrete* xiii. 192 For clear widths of about 150 ft. and over it is found to be economical to provide arched shells spanning direct, with stiffening ribs at about 25 ft. to 35 ft. centres. **1967** H. Kraus *Thin Elastic Shells* p. vii, Sophisticated uses of shells are currently being made in missiles and space vehicles, submarines, nuclear reactor vessels, refinery equipment, and the like. **1972** R. E. Owen *Roofs* vi. 81 A cylindrical shell transmits direct load to the columns.
c. *U.S.* A concave structure designed to accommodate a band or orchestra.
1938 D. Baker *Young Man with Horn* I. vi. 71 At the rear of the room was the orchestra shell, very shell-like, fluted along the upper edge. **1978** *Chicago* June 22/2 Each concert will be given on two evenings, and performances will take place..in the new James C. Petrillo Music Shell at Jackson and Columbus.
18. a. (Earlier example.)
1867 *Harper's Mag.* Oct. 654/2 Look at these beautiful 'shells', resting one above the other on the brackets on either wall.
19. b. *Physics.* (A set of electrons forming) one of a number of concentric structures around the nucleus of an atom; *spec.* a set of electrons each having the same principal quantum number. Also, (a set of nucleons forming) a corresponding structure within a nucleus.
1904 J. J. Thomson in *Phil. Mag.* VII. 255 When the corpuscles [*sc.* electrons] are not constrained to one plane, but can move about in all directions, they will arrange themselves in a series of concentric shells. **1919** *Proc. Nat. Acad. Sci.* V. 252 The electrons in any given atom are distributed through a series of concentric (nearly) spherical shells, all of equal thickness. **1933** *Physical Rev.* XLI. 370/1 For some time, there has been speculation as to whether or not the atomic nucleus can be regarded as consisting of shells of protons, just as the external structure is known to consist of shells of electrons. **1952** *Sci. News* XXIII. 36 Neon has 10 electrons in two complete shells. **1961** G. R. Choppin *Exper. Nuclear Chem.* iii. 30 Frequently, rather than emit a gamma ray, a nucleus will interact with its extranuclear electronic shells and cause emission of an electron. **1972** *Sci. Amer.* Oct. 101/1 In nuclei there also is a periodic recurrence of certain properties as nucleons are added to fill successive shells of quantum states. **1974** G. Reece tr. *Hund's Hist. Quantum Theory* viii. 106 The formation of molecules was thus a problem of atomic structure, namely the tendency of atoms to form ions with complete shells.
23. b. Also *N. Amer.*, the unlined body of a coat; *U.S.*, an article of clothing for the upper body, *spec.* a woman's (usu. sleeveless) overblouse or a light all-weather jacket.
1913 *T. Eaton & Co. Catal.* Semi-Annual Sale 17/3 Men's muskrat-lined coat..Collar is genuine Canadian otter; the shell is cut from standard quality black beaver cloth. **1962** *Mademoiselle* Aug. 276/2 A white cardigan to show a matching sleeveless shell. **1967** *Boston Sunday Globe* 23 Apr. 5/2 (Advt.), 3-pc. acetate double knit with rayon metallic shell. **1976** *National Observer* (U.S.) 1 May 7/4 (Advt.), Nylon shell for men and women. Ultra light, all weather sports jacket of tough two-ply coated nylon. **1976** U. Curtiss *Dig Little Deeper* x. 89 Paula came in, wearing a topaz-colored pants suit over a ruffled cream shell.
e. The body of a car.
1937 *Times* 13 Apr. (Brit. Motor Suppl.) p. xiii/1 The various stages through which the car body shell can pass, therefore, are as follows. **1972** *Oxford Mail* 13 Oct. 1/5 Output of body shells for the Marina range was halted for a time.
IV. 25. b. A company which has ceased to trade but which is still quoted on the stock exchange.
1964 *Economist* 19 Dec. 1378/2 A company had ceased normal trading and was a pure shell. **1969** 'D. Rutherford' *Gilt-Edged Cockpit* iv. 68 It's called buying a shell. A tax loss company. When you set the Hackforth loss against our profit we're left with practically no tax to pay. **1981** *Times* 27 May 20/6 Mr Alastair Milne..headed a consortium bidding for former cash shell Phoenix Mining two years ago.
VI. 34. d. *shell-grey*, *-pink* (later examples).
1963 *Times* 8 June 12/3 Short dresses of shell-grey silk with flared skirts. **1932** J. C. Powys *Glastonbury Romance* II. xxi. 701 The new silk lining of her ottoman had turned itself..into an incredible shell-pink. **1951** E. Paul *Springtime in Paris* xii. 229 Anatole turned shell pink, then a kind of raspberry shade. **1977** 'E. Anthony' *Silver Falcon* (1978) 133 The house was..painted shell-pink.
35. b. *shell-gravel*, *-grit*, *-sand* (later example).
1882 W. D. Hay *Brighter Britain!* I. xi. 307 A straight, broad path, smooth and white with shell-gravel. **1922**

Joyce *Ulysses* 48 Loose sand and shellgrit crusted her bare feet. **1938** Shell-grit [see *ROUGHETTED *pa. pple.* b]. **1964** *Oceanogr. & Marine Biol.* II. 418 *Spatangus purpureus*..lives in shell-gravel. **1977** *Stornoway Gaz.* 27 Aug. 4/6 These are well worth looking for in June or July in the grassland behind our many shell-sand beaches.
c. *shell-box*, *-road* (earlier example).
1836 T. Power *Impressions of Amer.* II. 99 We soon gained the shell road however, and found it as good as the streets of Mobile. **1853** C. Brontë *Villette* III. xxxi. 35 Slipping into his hand the ruddy little shell box. ... J. Fleming *To make Underworld* ii. 21 She makes these shell boxes, y'know..all stuck over with shells.
e. *shell-pattern* (earlier and later examples).
1780 J. Wedgwood *Let.* 21 Oct. (1965) 260, I now expect to sell a good deal of this green shell pattern. **1897** *Private Life of Queen* xxii. 180 The enormous 'shell pattern' service of knives, forks and spoons. **1967** E. Short *Embroidery & Fabric Collage* iii. 68 Aluminium templates can be bought in a number of geometric shapes and also a shell pattern.
36. a. *shell-burst*, *crater*, *-fire* (examples), *-gun* (later example), *-hole*, *-madness*, *-splinter*, *-storm*, *-trap*; **b.** *shell-dodging*; *shell-firing* adj.; **c.** *shell-pitted*, *-pocked*, *-smitten*, *-torn* adjs.
1917 W. Owen *Let.* 2 Mar. (1967) 440 Did you see any shell-bursts? **1980** G. M. Fraser *Mr American* xxi. 556 It was a shell-burst that Ah stopped. **1916** 'Boyd Cable' *Action Front* 49 The neutral ground..was a sea of mud, broken by heaped earth and yawning shell craters. **1977** J. Cleary *High Road to China* i. 35, I..was trapped in a shell crater with three dead men. **1917** 'Contact' *Airman's Outings* 245 Freed from the immediate necessity of shell-dodging. **1900** W. S. Churchill in *Morning Post* 25 June 5/7 In spite of an accurate shellfire they continued to advance boldly against the highest part of the hill. **1917** *Lancet* 28 Apr. 559/2 They had had barely a hundred shell-fire casualties. **1915** 'Boyd Cable' *Between the Lines* 56 Bullet and shell-firing attacks had failed to drive them from their trenches. **1918** W. S. Churchill *Let.* 15 July, There is a shell-firing Maxim gun. **1943** *R.A.F. Jrnl.* 15 May 15 These include..constant-speed 3-blade propeller; shell-firing guns; wing-flaps. **1940** *Flight* 12 Dec. 522/2 The shell-gun or 'cannon' has been in action recently on the machines of Fighter Command. **1916** 'Boyd Cable' *Action Front* 141 The stretcher-bearers who lifted him from the shell-hole. **1917** S. Hill *Strange Meeting* iii. 203 Then suddenly they came between the stumps of some trees, dropped down into a shell hole. **1923** Kipling *Irish Guards in Gt. War* I. 293 A stand of dead 'shell-madness', where all ears and eyes were intolerably overburdened with echoes and pictures. **1918** W. S. Churchill *Let.* 15 July, in Gilbert *W. S. Churchill* (1977) IV. I. vii. 147 For an hour we ran through devastated, shell pitted facias—scraggy shreds of woods. **1925** *Scribner's Mag.* Sept. 234/2 Only the 4th bay before us in the open, on a bleak, shell-pocked slope. **1917** J. Masefield *Old Front Line* 76 It has been more shell-smitten than most parts of the lines. **1920** W. S. Churchill *Let.* 11 Nov. in R. S. Churchill *W. S. Churchill* (1967) I. Compan. II. 1071 The driver..was wounded severely in the scalp by a shell-splinter almost immediately. **1945** H. Fielding *Dressing of Diamond* 137 He had been ripped by a shell splinter and sewn up casually. **1903, 1914** Shell-storm [see *RAFALE]. **1918** W. Owen *Let.* 4 Jan. (1967) 525 He was badly wounded, and..still wears the shell-torn boots. **1940** S. Spender *Edge of Being* 24 Moving in death through shell-torn streets. **1930** 'Ludwig Renn' *Krieg* ... A shell-trap..so unsuspiciously avoided by modern [military] engineers. **1923** Kipling *Irish Guards in Gt. War* I. 97 Annequin..had become more than ever a shell-trap.
38*. In sense *25 b: *shell company, corporation, game, operation, transaction*.
1968 *Economist* 25 Mar. 557/1 Shell companies have nothing to do with oil. They are corporate entities empty of their trading assets; they hold only cash or near cash assets in their balance sheets and otherwise have nothing but a stock exchange quotation—which is essential. **1977** *Irish Press* 29 Sept. 7/7 It was in April 1972 that Fitzwilliam Resources, of the same stock that formed Fitzwilliam Securities and Fitzwilton Ltd., (the Irish 'shell' company, which is now a shadow of its former self), took a 66 per cent stake in Tara. **1979** *Wall St. Jrnl.* 5 July 4/4 The Securities and Exchange Commission said it's disturbed by the increasing use of inactive 'shell' corporations as vehicles for distributing unregistered stock to the public. **1974** A. A. Thompson *Swiss Legacy* xx. They are shell corporations... They have no assets, no activities, nothing. They are merely conduits for money going elsewhere. **1969** *N.Y. Rev. Books* 2 Jan. 42/3 What becomes almost obscene about such a reactionary shell game..is that these very same corporate chiefs right now planning an increase in unemployment. **1977** P. Branston *Up & Coming Man* xi. 102 A shell operation, where you buy a dormant or nearly defunct company and inject assets into it. **1958** *Spectator* 11 July 68/3 This shell operation could be good value.
39. *shell-back*, (*a*) (earlier U.S. example); also *transf.*; hence *shell-backed a.*; **shell beach**, a beach composed wholly or predominantly of sea-shells; *spec.* the name of one such on the Channel Island of Herm; **shell-bird**, (*a*) *Canada*, the red-breasted merganser, *Mergus serrator*; (*b*) *nonce-use*, a tortoise; **shell-briar** *a.*, designating a type of tobacco-pipe with a rough, dark-stained stem and bowl; **shell button** (earlier U.S. example); **shell cocoa**, the husks of cocoa-beans or the drink made from an infusion of these; **shell concrete** *Building*, concrete used in shell construction; **shell construction** *Building*, the use of thin curved shells (sense 12 b above) to roof areas

having wide spans; **shell egg**, an egg in its natural state in the shell (opp. to *dried egg*: cf. *DRIED *ppl. a.* 1); **shell-game** (earlier and later examples); also *fig.*; **shell-gritted** *a.* *Archæol.*, denoting a ware made of a paste mixed with particles of shell; **shell ice** *Canad.* = *cat-ice* s.v. CAT *sb.*[1] 18; **shell midden** *Archæol.* = *shell-heap*; **shell model** *Nuclear Physics*, a theoretical description of nuclear structure in which the nucleus is considered to consist of nucleons arranged in shells (sense *19 b); **shell-moulding** *vbl. sb.*, in *Founding*, a method of making moulds and cores in which a shell of resin-bonded sand is formed in parts around a heated metal pattern, the parts being joined together after removal of the pattern; so **shell-mould**, a mould made in this way; **shell rock** *N. Amer.*, hard rock consisting largely of compacted sea-shells; **shell roof**, a roof consisting of a shell (sense 12 b above); **shell-shock**, (*a*) a name given, esp. during the war of 1914–18, to certain psychological disturbances occurring in conditions of active warfare and supposed to result primarily from exposure to shell-fire; also *fig.*; (*b*) *slang*, cocoa; hence **shell-shock** *v. trans.*, to affect with shell shock; **shell-shocked** *a.*, suffering from shell shock; **shell-shocker**, a sufferer from shell shock; **shell steak**, a steak cut from the short loin; **shell-stitch**, one of various knitting or sewing stitches producing shell-like patterns; **shell structure** *Physics*, the structure of the atom envisaged as consisting of a number of electron shells (sense 19 b above); **shell transformer**, a shell-type transformer (see below); **shell-type** *sb.* and *a.*, (applied to) something having or resembling a shell in any sense; *shell-type transformer*, a transformer having its windings wholly or largely enclosed within the metal 'core'.
1853 J. T. Downey *Filings from Old Saw* (1956) vi. 30 Both the nerve of ... shell-backs, and the occasional disbursement of an extra *tot* of whiskey ... **1943** A. Ransome *Picts & Martyrs* xi. 103 He felt as if he was going to sit for an examination and he wanted to make no mistakes with those two old shellbacks, Nancy and Peggy, as examiners. **1959** J. Cary *Captive & Free* 207 The old hulk was full of crabs—there doesn't seem to be anything else in the sub-editor's room. Old shellbacks that have been chewing on Fowler for forty years. **1963** *Listener* 21 Feb. 330/1, I have no doubt a lot of right-wing shell-backs are now conceding, with blimpish magnanimity, that there's really something to be said for these young fellows after all. **1974** *Times* ... In both division lobbies right-wingers rubbed shoulders with ... parliamentary ... **1930** R. Campbell *Adamastor* 30 A shell-backed saint ... **1972** *Daily Tel.* 29 Dec. ... shell-backed experience. **1835** H. D. Inglis *Channel Islands* 323 Herm possesses another attraction,..the shell beach. **1925** E. R. Lankester *Diversions of Naturalist* 144 The shells which are accumulated on shell-beaches have come from animals which lived in quantity at depths of ten or twenty fathoms. **1964** H. Mitchell *Introducing Channel Islands* v. 114 It is the accumulation there of countless thousands of shells of great variety. There are said to be over five hundred distinct species represented, and they have given it the name of the Shell Beach. **1776** G. Cartwright *Jrnl.* 2 Oct. (1792) I. 40 They returned with three shellbirds and a saddleback. **1973** E. Goudie *Woman of Labrador* 32 ..so Shell birds are not very good eating because they taste very fishy. **1921** D. H. Lawrence *Tortoises* 14 Nay, tiny shell-bird, What a huge vast inanimate it is, that you must row against. **1972** N. J. Bosse *Incident at Naha* i. 17 He..lit a pipe, his largest shell-briar Apple. **1977** A. Scholefield *Venom* v. 203 The chubby face, the merry eyes, the small rounded shell-briar emerged. **1789** *Deb. Congr. U.S.* 29 Aug. (1834) 796 An exclusive privilege of making and selling buttons of different dimensions. **1874** T. J. Law *Law's Grocer's Man.* (ed. 2) 1170/2 Shell cocoa. **1906** J. Joyce *Let.* 21 Aug. (1966) II. 358, I sent Nora a tin of shell cocoa. Pay the duty on it which cannot be high and see that Nora takes it every morning and evening. **1922** W. B. Yeats *Trembling of Veil* II. xiii. 119 She had lived for many weeks upon bread and shell-cocoa, so that her food never cost her more than a penny a day. **1949** *Archit. Rev.* CVI. 302/2 The boiler house, which has a shell-concrete roof. **1958** *Times* 23 Sept. 16/3 The structures he [*sc.* Candela] has designed there—mostly in shell-concrete—have begun to attract attention far outside Mexico. **1946** *Archit. Rev.* C. 8 The roofs of the canteen and the concert studio are of shell construction of 4 in. thickness. **1974** *Encycl. Brit. Macropædia* IV. 1078/1 Shell construction, where the strength of a thin curved concrete membrane is used advantageously to produce a light and aesthetic roof capable of bridging wide spaces without appreciable bending. [**1942** *Sun* (Baltimore) 18 Feb. 24/7 There are, according to experts, three kinds of markets for eggs—shell (direct-to-consumer variety), frozen and dried.] **1943** E. Oliver *Night Thoughts of Country Landlady* viii. 60 Before buying the very small but essential allowance of grain required to make these hens lay, you must hand over your coupons for 'Shell Eggs'. **1949** S. Gibbons *Matchmaker* i. 11 On Tuesday we have bacon and egg pie,

Father, and on Wednesday boiled shell eggs. **1972** *Guardian* 24 Mar. 10/8 Present minimum import prices for shell eggs and for..dried whole egg are to continue unchanged. **1890** B. Hall *Turnover Club* 169 Would endeavour to make a collection of Japanese coins, with their cards and a shell game. **1942** *Sun* (Baltimore) 29 Mar. 19/6 The defendant pleaded innocent to charges of operating a shell game. **1972** *Times Lit. Suppl.* 29 Dec. 1570/1 Both memory and history are shell games. **1977** *Rolling Stone* 21 Apr. 88/2 Both of them create with the sleight of hand of a shell-game swindler. **1954** S. Piggott *Neolithic Cultures* iii. 108 A bowl of typically 'Abingdon' shell-gritted ware from Great Ponton in south Lincolnshire. **1965** I. F. Smith *Windmill Hill & Avebury* v. 50 The heavier rims are decorated more often than the simple and rolled rims, and shell-gritted ware more often than flint-gritted. **1875** *United Service Mag.* CXXXIX. 41 [It] is brittle and bad for skating, 'shell-ice' as it is called. **1977** *Globe & Mail* (Toronto) 9 Mar. 36/7 Travel isn't very good. There's shell ice with pockets of water underneath and flooding around the cracks and heaves, but no actual danger yet. **1924** *Proc. Prehist. Soc.* 1923–24 IV. 11. 206 Directly resting upon the brickearth was a shell midden. **1971** *Nature* 11 June 397/2 Between 1881 and 1913 four Mesolithic 'shell midden' sites were excavated on the Island of Oronsay in the Inner Hebrides. **1946** *Physical Rev.* LXIX. 538 On the shell model the radius should be equal to Gamow's radius plus the radius of the alpha particle. **1970** I. E. McCarthy *Nuclear Reactions* i. iv. 83 The independent particle model for finite nuclei is the shell model. [**1947** *F.I.A.T. Final Rep. No. 1168* (Brit. Intelligence Objectives Sub-Comm.) 2 Such a bed lined by the thin mold shell resist the hydrostatic pressure of the influent liquid metal.] **1950** *Materials & Methods* Aug. 45/3 For the investigation of the metallurgical characteristics of the tin bronze alloys as affected by plastic bonded shell molds, a master pattern plate is being utilized. **1953** J. G. Tweeddale *Materials Technol.* II. ii. 39 Since, for simplicity, a shell mould is made up from two, outer mould parts, it is not always possible to build in the best pouring channel system. **1951** *Iron Age* 15 Nov. 111/1 The Builders Iron Foundry has been working with the Croning Process, or shell molding method of producing castings. **1979** J. Neely *Pract. Metall. & Materials of Industry* xxiv. 325/2 The advantages of shell molding over other forms of sand casting are that high precision, better finishes, and more complex shapes are possible, and less machining is needed. **1807** J. Barlow *Columbiad* ix. 521 And mark thy native orb!..What an age her shell-rock ribs attest! **1837** J. L. Williams *Territory of Florida* 56 The bank is formed of concrete shell rock. **1935** H. Davis *Honey in Horn* xvi. 261 The road under the horses' feet was black shellrock. **1954** G. Magnel *Prestressed Concrete* (ed. 3) x. 303 (*caption*) Prestressed beams for shell roof. **1972** R. E. Owen *Roofs* vi. 91 Most shell roofs are easy to drain to their edges or ends. **1915** *Brit. Med. Jrnl.* 11 Dec. 848/2 The necessity of investigating cases of 'shell shock' very carefully in order to differentiate those that are functional from those that are due to organic lesions. **1918** E. A. Mackintosh *War, the Liberator* 146 The Corporal..collapsed suddenly with twitching hands and staring, frightened eyes, proclaiming the shell-shock he had held off while the work was to be done. **1925** Fraser & Gibbons *Sailor & Soldier Words* 255 *Shell shock*... Since the war, the term has been officially abolished, in favour of the technical term 'Psychoneurosis'. **1933** J. F. C. Fuller *Generalship* 20 The most rapid way to shell-shock an army is to shell-proof its generals; for once the heart of an army is severed from its head the result is paralysis. **1935** M. Harrison *Spring in Tartarus* III. 300 The cocoa which Jim sold at a penny the cup, was called 'shell-shock'. **1943** G. Greene *Ministry of Fear* II. i. 111 There's not a finer shell-shock clinic in the country. **1952** S. Kauffmann *Philanderer* (1953) vii. 108 An unfortunate rambling man, supposed to have been shell-shocked in the war. **1959** *Listener* 5 Mar. 406/1 A mug of 'shell-shock'—that is what we call cocoa. **1976** *Ibid.* 9 Feb. 168/2 Seeking relief from this shell-shock, I phone a screenwriter friend. **1978** *Maledicta* 1977 I. 121 The student was shell-shocked by the letter. **1918** E. A. Mackintosh *War, the Liberator* 146 The man rejected the offer with scorn, as badly shell-shocked men will. **1958** P. Dickinson *Green Gene* ix. 180 'How are you?' he said. 'Burnt out,' said Mr. Leary. 'Shell-shocked.' **1926** Kipling *Debits & Credits* (1926) 65 It appeared that the silent Brother was a 'shell-shocker'. **1968** *Funk & Wagnalls Cook's & Diner's Dict.* 213/2 *Shell steak*, another name for club steak. **1969** R. Lockridge *Murder in False Face* v. 67 You can watch a tall stranger cut the fat from a shell steak. **1973** *Listener* 19 Apr. 501/2 A landscape of luscious rib roasts, lamb chops, shell steaks, T-bone steaks, sirloin steaks, fillet mignon,..and so on. **1895** *Montgomery Ward Catal.* 291/1 Fascinators, hand made, shell stitch, made of Shetland floss. **1976** *Woman's Day* (U.S.) Nov. 128/1 Crocheted rainbow afghan in shell-stitch pattern fairly glows with its twelve different colors. **1955** Friedman & Weisskopf in W. Pauli *Niels Bohr* 146 Some years ago when the evidence for the shell structure was accumulating and some of the inadequacies of the compound nucleus picture were becoming more apparent. **1974** G. Reece tr. *Hund's Hist. Quantum Theory* vii. 92 During this period new facts were discovered which made it possible to understand..the shell structure. **1888** G. Kapp in *Jrnl. Soc. Telegr.-Engineers & Electricians* XVII. 96 We may divide transformers broadly into two classes—one in which the copper coils are spread over the surface of the iron core, enveloping the latter more or less completely; and the other in which the core is spread over the surface of the copper coils, forming a shell over the winding. I propose to call the former 'core transformers', and the latter 'shell transformers'. **1902** *Encycl. Brit.* XXXIII. 418/1 Shell transformers have the disadvantage generally of poor ventilation for the copper circuits. **1888** *Jrnl. Soc. Telegr.-Engineers & Electricians* XVII. 113 These figures show that even in stout rings..the core type [of transformer] is better than the shell type. **1922** Glazebrook *Dict. Appl. Physics* II. 911/2 The three-phase shell-type transformer is a development of the single phase, having three individual sets of coils and the three cores arranged to form one composite core. **1935** *Discovery* Nov. 333/2 The early

pottery lamps of the Ægean, Phœnicia, etc. (known to the British Museum as the 'cocked-hat' type, though 'shell-type' seems much more expressive, both as to shape and origin). **1947** R. Lee *Electronic Transformers & Circuits* ix. 239 Lower capacitance obtains with two coils than with a shell-type transformer of the same interleaving. **1964** W. L. Goodman *Hist. Woodworking Tools* 179 In 1864 the first shell-type chuck with adjustable jaws was patented by Barber.
b. *shell-bark*, for Latin names substitute *Carya ovata* or *C. laciniosa*; also, the nut produced by one of these trees; (earlier and later examples); **shell-cracker** *U.S.*, the red-ear sunfish, *Lepomis megalotis*; **shell parrot** = *BUDGERIGAR.
1769 R. Smith *Jrnl.* 11 May in F. W. Halsey *Tour of Four Great Rivers* (1906) 21 The Timber in these Parts..consists of..red Oak Hazel Bushes, Ash and Gum together with Butternut and Shellbark, Hiccory in plenty. **1785** G. Washington *Jrnl.* 15 Apr. (1925) II. 362, I planted..a row of the Shell bark hickory Nutt from New York. **1885** *Harper's Mag.* Dec. 78/2 The chipmunk..[has] his hoard of hazel-nuts and shell-barks. **1948** *N.W. Ohio Q.* Winter 13 Two or three did not get in until dark bearing the big loads of fine shellbarks. **1969** T. H. Everett *Living Trees of World* xii. 98/2 The big shellbark hickory..chiefly inhabits rich, deep, fairly moist soils. **1889** Shell-cracker in *Cent. Dict.* **1947** B. W. Dalrymple *Panfish* 180 The name 'Shellcracker' comes from his habit of feeding on small crustaceans. **1975** *Southern Living* Aug. 18/3 Fishing is good for bass, crappie, bream, bluegill and shellcracker. **1890** 'Lyth' *Golden South* xiv. 127 The tiny budgerigar, sometimes called the shell parrot. **1954** *Coast to Coast 1953–4* 88 The shell-parrots, in glittering, swerving flights, were shrill over the reaches of the river.

shellac. Add: **1.** (Earlier *attrib.* example and later examples in the manufacture of gramophone records.)
1913 *Jrnl. Franklin Inst.* CLXXVI. 192 These modern composition disk records are in reality seals of the human voice, because the substance they are made of is a modified sealing wax,..containing shellac as a basic substance. *Ibid.*, Shellac is much adulterated, and the mineral and fibrous substances which are added require careful selection. **1933** *Amer. Speech* VIII. 13/1 Professor Jones..has them [*sc.* cardinal vowels] recorded on shellac. **1962** A. Nisbett *Technique Sound Studio* 266 Formerly, shellac was used for pressing records, and having much greater elasticity was suitable for record materials when only very heavy..pick-up heads were available.
attrib. **1765** T. H. Croker et al. *Compl. Dict. Arts & Sciences* II. s.v. *Japanning*, The proper japan ground..is much the best formed of shell-lac varnish.
2. A gramophone record made of shellac.
1954 *Billboard* 21 Aug. 18/2 Unless a publisher could get hold of an acetate of his song, he had to wait until the shellacs were ready. **1977** G. V. Higgins *Dreamland* xvi. 180, I remember the phonograph playing... There was one tune... He played it all the time, scratchy and noisy as the old shellacs were.

shellac (ʃelæ·k), *v.* orig. and chiefly *U.S.* Also **shellack**; *pa. pple.* **shellacked**. [f. the sb.]
1. *trans.* To coat or varnish with shellac.
1876 *Scribner's Monthly* Feb. 488/1 It is made of plain white pine, brought to a good surface and shellacked. **1917** C. Mathewson *Second Base Sloan* 90 Wayne threatened to varnish or shellac the paper so that it would turn the rain. **1969** *Sunday Times* 9 Feb. 58/1 They tend to wear dark blue silken suits or little black dresses and look as if they had been shellacked or sprayed with fixative just before they left home.
2. *slang.* To beat, thrash, punish.
1930 C. F. Coe *Gunman* iv. 53 These two bums that Lefty shellacked were members of Red Karfola's gang. **1935** J. Hargan *Gloss. Prison Lang.* 7 *Shellack*, to punish or beat. **1977** *Time* 8 Aug. 28/2 Pitcher McArdle was shellacked for..six runs in the first inning.
Hence **shella·c(k)ed** *ppl. a.*, (*a*) coated, varnished, or fastened with shellac; (*b*) *U.S. slang*, intoxicated, 'plastered'.
1881 C. C. Harrison *Woman's Handiwork* III. 139 Book-shelves have been made of pine, painted in flat color or stained and shellacked. **1884, 1902** [see Shellacked *pa. pple.*]. **1922** *Dialect Notes* V. 148 *Shellacked*, stewed, bunned, etc. **1935** J. T. Farrell *Judgment Day* I. iv. 85 You know, when I first found out about how you'd get shellacked, I thought it was pretty terrible. **1941** Wyndham Lewis *Let.* 17 Oct. (1963) 300 The silly 'toughness' of the Irish immigrant mass, shellacked into a sly, bluff, servility. **1948** H. L. Mencken *Amer. Lang. Suppl.* II. 644 When a novelty is obvious it seldom lasts very long, *e.g.*, *shellacked* for drunk.

shella·c(k)ing, *vbl. sb.* Chiefly *U.S.* [f. *SHELLAC *v.* + -ING[1].] **1.** In Dict.
2. A beating or thrashing, a 'pasting'; a defeat. *slang.*
1931 E. H. Lavine *Third Degree* x. 121 When this method failed, as it invariably did, he would leave the room and the *shellacking* continued. **1941** *Sun* (Baltimore) 18 Sept. 13/1 The main bulk of the Twenty-ninth Division handed the One Hundred and Fifteenth Regimental Combat Team a shellacking on the combat range today. **1956** W. H. Whyte *Organization Man* ii. 22 By the time of the First World War the Protestant Ethic had taken a shellacking from which it would not recover. **1960** T. McLean *Kings of Rugby* 56 'Ronnie' left the field to run into the greatest shellacking he had ever had. **1978** H. Wouk *War & Remembrance* xxxiii. 351 The Japs can't recover from the shellacking they took at Midway.

shelled, *ppl. a.* Add: *shelled corn*: Indian corn removed from the cob. *U.S.*
1676 in *Maryland Archives* (1884) II. 560 A Peck of Indian, shell'd Corn or Oates. **1714** J. Green *Diary* in *Essex Inst. Hist. Coll.* (1869) X. 104, I agreed to give Mr. Ganson five bushels of shelled corn at harvest, for ye damage my oxen did ye last night. **1828** W. Cobbett *Treat. Cobbett's Corn* § 136 The Americans call it, and..we must call it, 'shelled corn'. **1950** *Chicago Tribune* 20 Mar. IV. 1/3 The class of Illinois shippers primarily affected would be those who consign shelled corn to far western states.

sheller. **1. b.** (Earlier U.S. example.)
1859 *Rep. Comm. Patents 1858* (U.S.) I. 361 The nature of this invention relates..to the form and arrangements of the shellers.

Shelleyan, *a.* (and *sb.*) Add: (Examples of variant form *Shelleian*.)
a 1907 F. Thompson in *Dublin Rev.* (1908) July 36 Perhaps none of his poems is more purely and typically Shelleian than *The Cloud*. **1930** Blunden *De Bello Germanico* 26 Such a bed is not surpassed by the Shelleian shakedown of roses.
Hence **Shelleya·na** [ANA *suff.*], books or items relating to Shelley; **She·lleyist** = SHELLEYITE.
1886 *Academy* Mar. 218/2 'Shelleyana' of all kinds. **1924** P. Creswick *Beaten Path* xxxvii. 203 'Had I any books about Shelley?'..'We have considerable Shelleyana,' said I. 'Here is the special shelf.' **1934** G. B. Shaw *Prefaces* 506/1 They were Shelleyists, but not atheists.

shell-fishing. (Earlier example.)
1868 *Harper's Mag.* Dec. 12/2 Our men, some of whom were practiced in the business of shell-fishing,..very soon dug up a very fine feast of clams.

Shelluh, Shelook, obs. varr. *SHILLUK.

shelt (ʃelt). *Sc.* Also **shalt, shilt, sholt**. [Abbrev. of SHELTIE.] A Shetland pony.
1774 *Aberdeen Jrnl.* 27 June 4/2 (Advt.), There are Three Shalts, and several Year-olds of very fine Kinds. **1777** J. Duff *Let.* 1 May in A. & H. Tayler *Lord Fife & his Factor* (1925) iv. 103, I would wish my Horses, My black Shelt, the roan Poney and Smith..to be at Mar Lodge. **1817** J. Christie *Instructions* 29 We on with our sholts, a jogging and budging. **1880** J. Watt *Poet. Sk.* 59 To wirk some orra beast, or drive a milk shilt. **1920** *Glasgow Herald* 21 Oct. 8 It was the custom of the priest to go about his pastoral duties on a favourite pony..; indeed, he and the 'shelt' were inseparable. **1951** *Scots Mag.* July 278 They have their work cut out for them, those sturdy shelts. **1980** D. K. Cameron *Willie Gavin* xx. 199 Mettlesome black *shelts* they had been, high-stepping, jingling their bright harness.

shelter, *sb.* Add: **1. b.** Also, an enclosed shelter from air-raids, nuclear fall-out, etc., usu. underground. Cf. *ANDERSON, *MORRISON.
1918 *Ann. Reg. 1917* I. 175 Much greater public attention was paid to the question of air-raids... Arrangements were made to provide shelters throughout London. **1938** *Times* (Weekly ed.) 29 Dec. 3/1 Sir John Anderson outlined his plans for the provision of shelters against high explosive bombs. **1943** G. Greene *Ministry of Fear* I. i. 17 The sirens began their nightly wail... She was making for her favourite shelter down the street. **1961** E. S. Turner *Phoney War* v. 50 The authorities had no intention of allowing the Underground to be used as a shelter.., but when the bombing began the people simply bought tickets and took possession of it. **1961** [see *FALL-OUT *sb.*]. **1978** L. Deighton *SS-GB* ii. 18 She was killed..during the air attacks... He was in the shelter that day.
f. (Earlier and later examples.)
1890 W. Booth *In Darkest Eng.* II. ii. 97 You come along to one of our Shelters. On entering you pay fourpence, and are free of the establishment for the night. **1934** *Changing Men* 25 There has been no homelier, happier place than the Woman's Shelter in High Street. **1976** *New Society* 17 June 633/2 All lodging houses, hostels and night shelters in Glasgow were visited on two nights last winter.
g. A (temporary) home for animals. *U.S.*
1971 *New Yorker* 30 Oct. 41/1 The Bide-A-Wee animal shelter in Westhampton. **1979** *Arizona Daily Star* 22 July J 4/4 Lillian Schaaf willed her $1 million estate for a new animal shelter to be built on land she owned outside Worthington, Ohio.
3. (sense 1 b) *shelter-life*, *marshal*, *warden*; designating conditions and ailments attributed to time spent in air-raid shelters, as *shelter cough*, *paralysis*, *rash*, *throat* (all *temporary*); **shelter foot, leg**, a painful, swollen foot or leg after a person has slept in a sitting position (*temporary*); **shelter half**, one half of a shelter tent; **shelter tent** (earlier example); **shelter tree**, also, any tree grown to provide shelter; **shelter wood**, trees left standing to provide shelter in which saplings can grow; freq. *attrib.*
1940 *New Statesman* 19 Oct. 372/1 In every shelter I have been in during the past six weeks I have heard that hacking 'shelter cough' and the wheezy sleep of the bronchial cases. **1941** *Lancet* 6 Dec. 690/1 In shelter-foot the most potent causative factors are venous stagnation and increased capillary permeability. **1942** *Sun*

(Baltimore) 23 Apr. 22/2 When the Doaks family goes into a bomb shelter.., Mr. Doaks is likely to say, '..I'll just doze off here in this deck chair'. Next morning Mr. Doaks has a pair of painfully swollen legs and feet—shelter foot. **1911** F. Funston *Mem. Two Wars* 354 Not even shelter halves, popularly known as 'dog tents', were carried, and many..a night we stretched out in the rain. **1966** *Sunday Times* (Colour Suppl.) 4 Dec. 73/3 *Shelter half*, half a pup tent, and carried by every GI. **1940** *Lancet* 7 Dec. 722/1 (*heading*) Shelter legs. *Ibid.* 722/2 The elderly and infirm are loth to make use of bunks in the shelters..and since they are apt to sit about all day as well as all night they readily acquire shelter leg. **1944** *Newsweek* 20 Mar. 90/1 Renewed Nazi raids brought an old ailment back to London—'shelter legs'. **1943** Wyndham Lewis *Let.* 26 Jan. (1963) 342 A sculptor of course cannot exactly be commissioned to do marbles of shelter-life. **1974** *Times* 21 Jan. 4/3 The Medical Officer of Health reported in 1940 that the increase in deaths from respiratory diseases was at least partially due to shelter life. **1940** *New Statesman* 19 Oct. 372/2 Lord Horder's Committee attaches much importance to the shelter marshals, who should be paid, whole-time officials, chosen from the wardens' service for their proper human understanding and given a status which would entitle them to the help of the police and authority over the shelterers. **1942** *Sun* (Baltimore) 23 Apr. 22/2 A similar ailment is called 'shelter paralysis'. **1943** *Our Towns* (Women's Group on Public Welfare) iii. 78 *Scabies* or 'the Itch' has now almost achieved respectability under the name of 'Shelter Rash'. **1862** O. W. Norton *Let.* 16 June in *Army Lett.* (1903) 88 Anyone who has lived in these shelter tents any length of time can appreciate the difficulties of writing in a hard shower. **1940** *New Statesman* 19 Oct. 372/1 Equally common are the complaints of 'shelter throats', which may mean anything from tonsilitis to diphtheria. **1889** W. Schlich *Man. Forestry* II. ii. 133 They in their turn become mother and shelter trees. **1980** *Garden* CV. 198/1 The Corot-like view of immense weeping willows fortifying the lake's head must surely have gained by the recent deaths of elms and other shelter trees, though the garden is now more exposed to north-east winds. **1974** C. Franklin *By Horror Haunted* 12 The kids larking about..and the Shelter Wardens shouting at them. **1889** W. Schlich *Man. Forestry* I. ii. 208 The wood is created, or regenerated, under the shelter of the whole or part of the old crop, which forms the shelter-wood. **1928** R. S. Troup *Silvicultural Systems* vii. 82 The original idea underlying the adoption of the shelter-wood strip system was the necessity for working against the prevailing westerly winds, so that the newly exposed edges of mature woods should always be protected from it. **1979** *Sci. Amer.* Feb. 84/1 The final system is called shelterwood cutting, because the mature stand is removed in two or more partial cuttings so that the new stand can become established under the shelter of a partial canopy of remaining trees.

shelter, *v.* Add: **1. e.** To protect (invested income) from taxation; to invest with this purpose.
1963 *Vital Speeches* XXIX. 357/2 A tightening of the personal holding company rules, to end the escapes from individual taxation now available through the use of these devices to shelter investment income. **1972** P. C. Reid *Corporate & Executive Tax, Sheltered Investments* ix. 125 The prospective investor should always keep in mind that the major objective is to shelter his income. **1973** *Times* 10 Dec. 7/5 False rumours..that I sheltered the income on which my daughter, Tricia, should have paid taxes. **1975** *Tax Shelter Investments* (U.S. Congress Jt. Comm. on Internal Revenue Taxation) III. 22 He sheltered $13,000 with a cattle feeding operation.

sheltered, *ppl. a.* Add: **b.** *sheltered life*, a life protected from the ordinary hazards and hardships of living. Also *sheltered existence*.
1888 Kipling *Plain Tales* 14 There was a Boy once who had been brought up under the 'sheltered life' theory; and the theory killed him dead. **1920** *Ladies' Home Jrnl.* Feb. 185/1 Three thousand dollars and no business experience, thirty-three years of sheltered life, and two children under ten years of age—this was my problem when I was left a widow. **1937** Wodehouse *Summer Moonshine* (1938) xix. 221 Hers had been a sheltered life... She had never been brought face to face with tragedy. **1959** J. Kirkup tr. *S. de Beauvoir's Mem. Dutiful Daughter* II. 117 She led a very sheltered existence in the Basque country, where there were not many eligible young men. **1977** W. H. Saumarez Smith *Young Man's Country* ii. 35, I had lived a fairly sheltered life in England... Now for the first time in a position of responsibility and power, I was learning how people behave.
c. *Econ.* Designating trades, industries, etc., which are not exposed to competition, and the commodities in which they deal.
1924 *Westm. Gaz.* 18 Aug. 4/5 In the so-called sheltered trades real weekly wages have generally been maintained at at least their pre-war level. **1930** *Economist* 4 Jan. 9/1 The higher Japanese price-level is accounted for largely by such 'sheltered' goods as red beans, miso. **1972** *Wall Street Jrnl.* 9 Aug. 1/5 A number of tax shelter plans are designed to allow investments in sheltered industries like oil.
d. Affording relief or exemption from tax; untaxed. *U.S.*
1955 W. J. Casey *Tax Sheltered Investments* (ed. 2) xix. 205 Partnership operation allows you a write off of losses, a sheltered return on a quick success by sale of your partnership interest. **1970** *Tax Sheltered Investments* (A. Anderson & Co.) i. 1 All sheltered investments generate one or more of these advantages. **1974** *Los Angeles Times* 13 Oct. III. 9/2 The second $50,000, which was to be 'sheltered income', was to be sent..directly to an insurance company as a payment for an annuity purchased by Hunter.

e. Designating places for living or working (or suitable work) provided for the mentally or physically infirm, where special assistance and facilities are available.
1961 *Oxford Mail* 16 Mar. 4/6 The three-storey house [for patients of a mental hospital in the final stages of readjustment to community life]..was opened last month and is known officially as a 'sheltered hostel'. **1971** *Rand Daily Mail* 3 Apr. 11/4 One of these tenants is a humble man who earns R14 a week in a sheltered employment factory. **1973** *Howard Jrnl.* XIII. 276 Sheltered workshop facilities, industrial training or punitive labour. **1976** *Ilkeston Advertiser* 10 Dec. 16/2 The county council had already decided to stop the grants—towards general improvement areas,..council house adaptation for the handicapped and sheltered housing. **1977** *New Society* 3 Mar. 441/2 Patients leaving Herrison [Mental Hospital] have been 'graded' according to the kind of home they can cope with outside. Those least able to run their own lives are in sheltered accommodation. **1980** *Times* 23 July 12/2 If you are still undecided about sheltered housing or a home, you might investigate the various facilities.

shelterer. Add: **1. b.** One who takes shelter from an air-raid.
1940 R. Macaulay *Lett. to Sister* (1964) 116 The Central London tube was so crammed with thousands of shelterers that I couldn't get near the platform. **1944** *Ourselves in Wartime* v. 71 Tiers of bunks, canteens, adequate sanitation and first-aid posts were provided. These amenities helped the shelterers to make the best of a bad job. **1957** R. W. Zandvoort in *Wiener Beiträge zur englischen Philologie* LXV. 270 Air raid shelters were a necessary refuge... Books..were the best means to divert the attention of *shelterers* (as they were called) and others from the pressures and anxieties of life in wartime. **1980** P. Fitzgerald *Human Voices* xi. 163 The LPTB's bunks occupied the walls... Other shelterers had arrived.

Sheltie, shelty. **2.** For *nonce-use* read *rare* and add later example.
1967 H. W. Sutherland *Magnie* vii. 94 They were queer, the Shelties, came from nothing... England was heaven to them.
3. = *Shetland sheep-dog* s.v. *SHETLAND 1 d. Also *attrib.*
1911 *Our Dogs* 27 Oct. 984/3 (Advt.), Two promising Sheltie pups for sale. **1916** B. Thynne *Shetl. Sheep-Dog* 5 The origin of the Sheltie is 'wrapt in mystery'. **1950** A. C. Smith *Dogs since 1900* xii. 248 After the interruption caused by the first war, the Shelties began to assert their charms. **1972** *Times* 19 Sept. 24/6 (Advt.), King Charles and Sheltie Collie, both 2 years old..free to loving family. **1980** 'T. Hinde' *Daymare* iv. 26 He had been council dust-cart driver till he ran down one Sheltie too many.

shelving, *vbl. sb.*[1] **2.** (Earlier example.)
1817 M. Austin *Let.* 30 June in E. C. Barker *Austin Papers* (1924) I. 316 He has the Plank..selected,..as many as is wanted for shelving.

shelvy, *a.*[2] (Earlier example.)
1811 D. Buchan in J. Barrow *Chron. Hist. Voyages* (1818) App. I. 8 At noon several difficulties presented themselves in crossing a tract of shelvy ice, intersected with deep and wide rents.

Shema (ʃema·). Also **Shemah, Shemang**. [Heb. *šema'* hear, imper. of *šāma'* to hear.] The first word of the verse Deut. vi. 4 used as a name for three portions of the Scriptures, Deut. vi. 4–9, xi. 13–21, Numbers xv. 37–41, to be repeated twice daily by all adult Jewish males, and used as a Jewish confession of faith.
1706 I. Abendana *Discourse Eccl. & Civil Polity of Jews* iv. 106 Every day we use an office containing three sections of the law, commonly styled *Shemang*, i.e. *Hear*. **1816** J. Allen *Mod. Judaism* xix. 331 Another essential part of the daily service..is the reading of three portions of scripture. The first of these portions beginning with the word *Shema*... This term is applied to all the portions taken together, and the recital of them is called *Kiriath Shema*, the *Reading of the Shema*. To recite these passages twice every day they maintain to be expressly enjoined. **1864** *Chambers's Encycl.* VI. 155/1 The first additions to the *Shemah* formed the introductory thanksgiving for the renewed day. **1876** Geo. Eliot *Daniel Deronda* IV. VII. lxi. 221 The *Shemah*, wherein we briefly confess the divine Unity, is the chief devotional exercise of the Hebrew. **1926** *Brit. Weekly* 5 Aug. 367/3 Esther..joins in that immemorial declaration of her people the Shemang or Confession of Unity. **1936** E. Underhill *Worship* (ed. 2) I. ii. 28 The Jews' daily repetition of the *Shema*, the Christians' ritual use of the Lord's Prayer, and the Moslems' of the First *Sura*, are all justified by psychology no less than by religion. **1949** W. F. Albright *Archaeol. of Palestine* x. 221 In 1902 the Nash fragment containing the Ten Commandments and the Shema ('Hear, O Israel..') was discovered in the Faiyum. **1972** J. Caine *Hamlet, my Boy* vi. 86 He reminded himself that a Jew was supposed to say the *Shema* in the hour of death. **1977** H. Kaplan *Damascus Cover* (1978) xii. 123 'Don't you know the Shema?' She was either a Jewess or a highly trained agent.

Shemite, *a.* Add: (Earlier example.)
1791 *Gentl. Mag.* Feb. 107/1 Arabic..contains a good deal of Persian and Gothic and other Shemite dialects.
Shemitic *a.* (earlier example); **Shemitish** *a.* (earlier example).

1822 *Malte-Brun's Universal Geogr.* I. XXIII. 570 As most of the nations that speak these languages descend, according to Moses, from Shem, this stock has been distinguished under the general name of the Shemitic languages. **1838** W. B. Winning *Man. Compar. Philol.* 277 The assumptions..that the Hebrew tongue must necessarily be of Shemitish derivation.

shemmal, var. SHAMAL in Dict. and Suppl.
shemmi, var. *SHIMMY *v.*[1]

shemozzle (ʃėmǫ·z'l). *slang.* Also † **chimozzle**; **s(c)h(e)-**, **s(c)hi-**, **s(c)hlemozzle**. [Of uncertain origin. In early use (*shlemozzle*, etc.) apparently East End slang: perh. ad. Yiddish *shlimazl* misfortune, unlucky person (see *SCHLIMAZEL), with subsequent reduction of *schle-* to *sche-*.] A muddle or complication; a quarrel, row, rumpus, mêlée.
1899 A. M. Binstead *Houndsditch Day by Day* 23 It was through no recklessness or extravagance that he was in this shlemozzle. **1900** *From Front* xiv. 183 We might look upon this little chimozzle as a kind o' misunderstanding. **1901** J. M. Cobban *Golden Tooth* xvii. 170 If Will comes out of this shemozzle. **1916** 'Peter' *Trench Yarns* ii. 16 In the ensuing shemmozle Samuel got laid out with the butt-end of a rifle. **1916** 'Taffrail' *Pincher Martin* vii. 120 'We ain't the best o' friends, 'cos me an' 'im 'ad a bit of a shimozzle—' 'Shimozzle!..What on earth's that?' 'Bit o' a dust-up, sir.' **1928** *Sunday Dispatch* 29 July 15/2 Those..who saw so little of war that they still think it to be a gloriously romatic shlemozzle. **1930** S. Sassoon *Mem. Infantry Officer* vi. 156 When I showed the battle-plan to the Sergeant-Major, all he said was 'We'll have a rough house from Ale Alley.' But no one had any idea it was going to be such a schimozzle as it was! **1936** J. G. Brandon *Pawnshop Murder* i. 9 Has a *schlemozzle*, and takes 'is stuff over to Paris. **1937** G. Frankau *More of Us* xv. 160 Then Sophie called; and, brooding, 'Nice schemozzle If that lot stays to feed as well as sozzle.' **1943** T. Worling *White Ensigns* viii. 146 'How many more shimozzles like yesterday are we going to get?' Mr. Hebard queried. **1949** 'N. Blake' *Head of Traveller* xi. 191 Even if it turns out that none of us is implicated in the murder, there's still going to be a shocking shemozzle, I'm afraid, about his original disappearance. **1951** S. Kaye-Smith *Mrs. Gailey* v. 292 It was in the papers. What a schemozzle! **1955** E. Waugh *Officers & Gentlemen* 276 There was something of a schemozzle last night but we weren't in that. **1960** I. Cross *Backward Sex* 167 The whole schmozzle is over. **1966** *Telegraph* (Austral.) 27 Dec. 8/4 The next day saw one of the greatest shemozzles in Australian sport. **1978** L. Meynell *Papersnake* ii. 22 There was going to be one hell of an uncomfortable shemozzle in his life if he didn't get his priorities right.
Hence **shemo·zzle** *v. intr.*, to decamp, make off, 'scarper'.
1903 Farmer & Henley *Slang* VI. 172/2 *Shemozzle*..Verb (East End). To be off; to decamp. **1925** Fraser & Gibbons *Soldier & Sailor Words* 256 *Shemozzle, to*, to make off: to get out of the way—e.g., 'We saw the M.P.'s (Military Police) coming, so we shemozzled.' **1944** Auden *For Time Being* (1945) 118 He was caught by a common cold and condemned to the whiskey mines, But schemozzled back to the Army.

‖ **shen** (ʃən). Also **Shen, Shin**, etc. [Chinese *shén*.] In Chinese philosophy: a god, person of supernatural power, or the spirit of a dead person.
1847 W. H. Medhurst *Dissert. Theol. Chinese* 5 It would appear, that..Shin and..Kwei, are terms equivalent to spirit and anima, in the human system. **1901** J. J. M. de Groot *Relig. Syst. China* IV. II. 4 Confucius means to say, the shen manifests itself in its full development in man by his khi or 'breath'; indeed, only animated man lives and breathes. **1905** E. H. Parker *China & Religion* i. 21 The spirits of men were called *kwei*, and those of Heaven and Nature *shên*. **1934** A. D. Waley *Way & its Power* 29 'If the monarch loses his *shên*,' says *Han Fei Tzŭ*, 'the tigers will soon be on his tracks.' **1955** E. Herbert *Taoist Notebk.* 43 Thus Mencius in his moral hierarchy posited two grades—the 'spirit-like' (*Shên*), the saint beyond man's power of comprehension. **1978** F. Mann *Acupuncture* (ed. 3) iv. 32 Shen is usually translated as 'spirit', a word which to the Western mind more often than not suggests the supernatural. But Shen..is a down-to-earth word.

shenanigan (ʃėnæ·nigăn). orig. *U.S.* Also **shenan(n)egan, -igan, -gin; shin-; -(c)kin**; and other vars. [Origin obscure.] Trickery, skulduggery, machination, intrigue; teasing, 'kidding', nonsense; (usu. *pl.*) a trick, a prank, an exhibition of high spirits, a carry-on. Hence **shena·niganning, shena·nigin(g)**, *pres. pple.* and *vbl. sb.*
1855 *Town Talk* (San Francisco) 25 Apr. 2 Are you quite sure? No shenanigan? **1856** *Spirit of Age* (Sacramento) 30 Apr. 2 These facts indicate that there is some *shenanegan* going on. **1857** C. E. De Long *Jrnl.* 15 Aug. in *Calif. Hist. Soc. Q.* (1930) IX. 156 Race came off Whiskey Bill winner, the Mare's rider held in, and Smith pronouncing it shenanigan. **1862** 'Mark Twain' *Let.* May (1917) I. iii. 77 Consider them all..guilty (of 'shenanigan') until they are proved innocent. **1894** M. J. Jaques *Texan Ranch Life* xiii. 115 He assured me that he was not 'shenan-neganning' me, and that the dish would prove a delicacy. **1897** *Outing* (U.S.) XXIX. 483/1 A man who is firmly kind, but who will stand no shinanigan. **1901** W. S. Walker *In Blood* xxxi. 332 We're mates all round, an' no more shenannikin. **1902** R. Barr *Victors* v. 81 If

I were to pay them they might think there was some shenanigan about it. **1924** J. Masefield *Lord Harker* III. 146 Now, brother, answer me and no damned shinanniking. **1926** E. Ferber *Show Boat* xiv. 305 I never had a fight on my boat and wasn't going to have any such low life shenanigans now. **1928** *Saturday Even. Post* 10 Mar. 11/2 The renunciation of Mr Coolidge was a distinct disappointment to the great mass of the Republican Party,..and this left the way open for some astute shenanigan in various states. **1930** 'S. S. Van Dine' *Scarab Murder Case* xv. 212 There's too much shenanigan going on around here to suit me. I want action. **1932** S. O'Casey *Let.* c 10 Feb. (1975) I. 540 It is really hateful that I should be compelled to turn my thoughts to the pious shinanachin of a few Jesuits & a group of Methodist Preachers. **1936** 'R. Hyde' *Passport to Hell* x. 152 The more followed him to take care of him and see he did not lose his pay-roll shenannigin in his Irish way with the mademoiselles. **1938** J. I. Rodale *King's English on Horseback* 137/1 *Mischief*,..Shenanigins. **1940** Dylan Thomas *Portrait of Artist as Young Dog* 122 No shananigins in the old moonlight. **1947** W. Greenwood *Only for Love* I. 23 'Here, listen, you two. No shinanigin'...' 'Shinanigin'? What's that?'..'Shinanigin'? Well-er-just ..Well-er-just shinanigin'... Messing about.' **1948** D. Ballantyne *Cunninghams* I. vi. 34 You see you do tomorrow, then... No shinnanicking now. **1960** T. Griffith *Waist-High Culture* II. 81 Readers [of newspapers] do not recognize every shenanigan inflicted upon them. **1969** in Halpert & Story *Christmas Mumming in Newfoundland* 90 The entire household looks on, laughing at the girls' shenanigans. **1973** B. Bainbridge *Dressmaker* 11 They'd all catch their death of cold shenannigging about in the middle of the night. **1973** *Nation Rev.* (Melbourne) 31 Aug. 1448/2 (*heading*) Shennanigans behind the silver screen. **1974** *Ridge Citizen* (Johnston, S. Carolina) 18 Apr. 2/1 We don't condone whatever wrongdoing or shenanigans that may have taken place at Watergate or elsewhere. **1976** R. Lewis *Witness my Death* ii. 56 There might have been a certain amount of shenanigans going on behind the scenes. **1978** *New Yorker* 3 Apr. 16/2 House Calls—Glenda Jackson, Art Carney, Richard Benjamin, and Charlie Matthau in a film directed by Howard Zieff, about medical shenanigans in and out of a hospital. **1980** *Times* 3 Jan. 10/4 Doubtful political shenanigans..in the Central Pacific.

shending, *vbl. sb.* For † *Obs.* read *Obs.* exc. *arch.* and add later example.
1935 J. D. Wilson *What happens in Hamlet* vi. 179 The King at his prayers, the shending of Gertrude, the slaying of Polonius, Hamlet's departure to England.

‖ **sheng**[1] (ʃʊŋ). *Mus.* Also 8– **cheng**; 9 **sang, sing**. [Chinese *shēng*.] A Chinese wind instrument consisting of a set of reed pipes.
1795 W. Winterbotham *Hist., Geogr., & Philos. View Chinese Empire* ix. 428 The ancient *cheng* differed in the number of their pipes; those used at present have only thirteen: this instrument appears to have some affinity with our organs. **1839** *Chinese Repository* VIII. 52 The *sāng*..is a collection of tubes varying in length so as to utter sounds at harmonic intervals from each other. **1845** *Encycl. Metropol.* XVI. 579/2 The *Sheng* or *Sing*; the lower half of a gourd, in which a row of pipes is fixed, with a curved and lateral one on which the performer blows. **1937** *Times Lit. Suppl.* 16 Jan. 41/3 The *sheng* is assigned by the notation to a certain level of the voice. **1961** J. Howarth in A. Baines *Musical Instruments through Ages* xiii. 321 In 1777,..Père Amiot..sent the present of a sheng from China to Paris. **1972** Liu Jung-en *Six Yüan Plays* 15 The *shêng*, consisting of thirteen pipes of different lengths forming a circle each with a finger hole and having a mouthpiece through which the musician blows and sucks. **1973** *Times* 11 June 14/2 The concertina was inspired by the Cheng (Chinese mouth organ). **1980** *Early Music* July 355/1 These, the sho of Japan and the sheng of China, are forms of free-reed mouth organ with a rigid wind chest held in the hands, the fingers remaining free to open and close the reeds in the cane pipes.

‖ **sheng**[2] (ʃʊŋ). [Chinese *shēng*.] The principal male character in a Chinese opera. Also *attrib.*
1886 *Jrnl. R. Asiatic Soc.* (*North-China Branch*) XX. 208 The characters in Chinese plays are arranged under five denominations... The hero is *shêng*. He wears a black beard, but his face is not concealed. **1937** Arlington & Acton *Famous Chinese Plays* p. xxiii, *Shêng* are divided into *Wên*, Civil, and *Wu*, Military. These are the leading actors. **1972** C. P. Mackerras *Rise of Peking Opera* i. 7 There were seven types of actors in the *nan-hsi*... Some of these terms can still be found..in Peking Opera today. For example, the *sheng* and *tan* were—and still are—the principal male and female characters respectively. **1973** R. F. S. Yang in Yuan-li Wu *China* 74/1 Yen was the founder of the Yen school of the 'bearded *sheng*' voice.

shenzi (ʃe·nzi). [Swahili.] In East Africa, an uncivilized tribesman. In extended sense, a barbarian, a person outside the person's cultural group.
1910 T. Roosevelt *Afr. Game Trails* x. 258 The 'shenzis'—wild natives called in Swahili..'wa-shenzi'. **1921** *Blackw. Mag.* Jan. 121/2 He, an askari of G company, ran away from a lot of miserable shouting shenzis. **1926** *Spectator* 3 July 10/2 Local *shenzis*..had gathered round the gramophone. **1976** K. Thackeray *Crownbird* vi. 111 The Chinese face was impassive... That he should be..involved with this bunch of shenzis..was very confusing.

she-oak. Add: **2.** (Earlier example.)
1873 J. C. F. Johnson *Christmas on Carringa* i. Able to put away at a sitting a larger quantity of colonial 'sheoak' than any man of his inches.

b. *attrib.* and *Comb.*, as *she-oak beer*; **she-oak net**, a safety net for sailors boarding ship (see quot. 1898).

1927 F. H. SHAW *Knocking Around* 106 She-oak beer, the common Australian beverage, was a potent tipple. **1898** MORRIS *Austral Eng.* 415/1 *She-oak nets*, nets placed on each side of a gangway from a ship to the pier, to prevent sailors who have been indulging in *she-oak* (beer) falling into the water. **1925** R. CLEMENTS *Gipsy of Horn* 111 Old Australian traders used to spread a net under the gangway, called therefrom the sheoke net, whose office it was to save mariners who 'missed stays' when coming aboard from falling into the dock. **1938** W. E. DEXTER *Rope Yarns* 234 A dog..followed every drunken sailor—never by any chance a sober one—down the pier to his ship. If he managed to get on board safely the dog returned, but if he fell into the she-oak net it would howl until the man was rescued.

sheogue (ʃī·ōg). Also **shee-og.** [ad. Ir. *sióg* fairy.] In Ireland: a fairy.

1852 W. WILDE *Irish Pop. Superstitions* 14 The sheeogue [*sic*] is the true fairy. *Ibid.* 52 The mystic pipers of the sheogues..are said..to favour mortals with their melodies. **1892** W. B. YEATS *Countess Kathleen* 26 You poured out wine as the wood sheogues do When they'd entice a soul out of the world. *Ibid.* 50 Brother, where wander all these dwarfish folk, Hostile to men, the sheogues of the tides? **1893** —— *Celtic Twilight* 117 He would not hear of ghosts or sheogues. **1959** D. A. MAC MANUS *Middle Kingdom* i. 20 There are the stories of human encounters with the small spirits, the 'wee folk' of Ulster, the 'little people' of the South, the 'Shee-og' of Gaeldom.

sheol. (Examples of *slang* use.)

1889 [see *RAH *int.* and *sb.*]. **1903** A. M. BINSTEAD *Pitcher in Paradise* v. 117 Briefly, Dickie, I have been out all night and there's sheol to pay. **1920** B. CRONIN *Timber Wolves* vii. 116 Them big bugs are the meanest things this side sheol.

shepherd, *sb.* Add: **4.** (Earlier example.)

1855 R. CARBONI *Eureka Stockade* 9 The faithful shepherds..were sure to snore in peace a foot and a half under ground from the surface and six score feet from 'bang on the gutter'.

5*. = *German shepherd (dog)* s.v. *GERMAN *a.*[2] and *sb.*[2] 4 a.

1938 J. STEINBECK *Long Valley* 13 The rangy dog darted from between the wheels and ran ahead. Instantly the two ranch shepherds flew out at him. **1978** R. LUDLUM *Holcroft Covenant* ix. 104 Suddenly, the menacing faces of enormous long-haired black shepherds lunged at the windows on both sides of the car.

6. d. shepherd plaid = *shepherd's plaid* (SHEPHERD *sb.* 7 c).

1940 [see *KILLER 7]. **1970** *Globe & Mail* (Toronto) 25 Sept. 3/1 (Advt.), New multiple colored striped worsteds, shepherd plaids and a host of plains are now ready for your inspection.

7. b. shepherd's crook arm, a chair-arm shaped somewhat like a shepherd's crook; **shepherd's pie** (earlier and later examples).

1960 H. HAYWARD *Antique Coll.* 256/2 *Shepherd's crook arm*, chair or settee arm of elegantly curving shape, the end in the form of a shepherd's crook, fashionable during the first three decades of the 18th cent. **1973** *Country Life* 30 Aug. (Suppl.) 74/2 Walnut Queen Anne armchair..has a spoon-back, shepherds' crook arms and graceful cabriole legs. **1877** E. S. DALLAS *Kettner's Bk. of Table* 256 In Scotland they produce..such a stew, cover it over with a crust, and call it shepherd's pie... The shepherd's pie of Scotland is..too farinaceous—potatoes within and paste without. **1969** R. WOLLHEIM *Family Romance* 228 What I couldn't face was ordering shepherd's pie. **1977** B. PYM *Quartet in Autumn* xviii. 164 Put a shepherd's pie in the oven.

shepherd, *v.* Add: **3.** (Earlier example.) Also *N.Z.*

1855 R. CARBONI *Eureka Stockade* 8 Here begins as a profession the precious game of 'shepherding', or keeping claims in reserve; that is the digger turned squatter. **1864** *Append. Jrnls. House Reps. N.Z.* C. IV. 8 Shepherding forbidden.

Shararat (ʃerərā·t). Also **Shararat, Shererat.** [Arab.] (A member of) a nomadic tribe of northern Saudi Arabia. Also *attrib.*

1830 J. L. BURCKHARDT *Notes on Bedouins & Wahábys* 17 *El Sherárát*, in the sandy plain S. of the Akabe el Shamye and eastward of the Hadj route. Their numbers are considerable, and all are Wahabys. *Ibid.*, South of the Sherárát on the E. of the Hadj road, as far as the vicinity of Mekka, the whole country is inhabited by Aenezes. **1875** *Encycl. Brit.* II. 247/2 Thirdly, in the northern desert, the Howeytat and Sherarat, comparatively small and savage tribes. **1888** C. M. DOUGHTY *Travels in Arabia Deserta* I. iii. 57 Upon the other hand are..low heights much more distant, in the Sherarát nomad country. **1917** T. E. LAWRENCE *Lett.* (1938) 234 The Arab losses in the fight came to two killed (a Rualla and a Sherarat) and several wounded. **1955** H. ST. J. PHILBY *Sa'udi Arabia* iii. 90 'Abdul-'Aziz now [in 1798] switched his offensive to the north: sending..the Amin of Buraida, to raid the Shararat on the Syrian border with telling effect.

sherardize. Substitute for entry:

sherardize (ʃe·rāɪdəiz), *v.* Formerly also **Sherardize.** [f. the name of *Sherard* O. Cowper-Coles (1867–1936), English chemist + -IZE.] *trans.* To coat (iron or steel articles) with zinc by heating in contact with zinc dust at a temperature below the melting point of zinc. So **she·rardized** *ppl. a.*, **she·rardizing** *vbl. sb.*

1901 *Brit. Pat.* 9927 1 By our process, which we name 'Sherardizing' to distinguish it from galvanizing, the thickness and evenness of the deposit can be regulated..and any description or shape of iron or steel can be..dealt with. **1904** [in Dict.]. **1904** *Engin. Rev.* XI. 107/2 In practice Sherardised iron and steel are found to withstand the ordinary corrosive agents galvanised iron is exposed to, to a remarkable degree. *Ibid.*, Bolts which had not been Sherardised. **1935** H. R. SIMONDS *Finishing Metal Products* ii. 12 It is possible to sherardize sash of large size only before it is assembled, as sherardizing is destroyed by welding. **1963** *Times* 22 Apr. (Zinc Suppl.) p. v/1 Sherardizing is excellent for small things like screws, for which only regular and small changes in dimensions can be tolerated. **1977** *Offshore Engineer* June 67 (Advt.), We still sherardize every nut, washer and bolt to cut maintenance to virtually zero.

Sherari (ʃerā·ri), *a.* and *sb.* Also **Sherary.** [Arab.] **A.** *adj.* Of or pertaining to the Sherarat. **B.** *sb.* A member of the Sherarat; also, a dromedary bred by the Sherarat.

1888 C. M. DOUGHTY *Travels in Arabia Deserta* II. ii. 32 'It is he,' said the Emir, 'Sheràry hound! how durst thou do this violence?' Meṭaab bade the stranger take the Sheràry's lance. *Ibid.* ix. 239 The thelûls [riding camels] of the Sherarát..are praised above other in Western Arabia: Ibn Rashîd's armed band are mounted upon the light and fleet *Sheràries*. **1926** T. E. LAWRENCE *Seven Pillars* IV. lii. 273 At last the Sherari boy said if we gave him scope he would settle his account and leave him living. *Ibid.* liv. 284 My camel, the sherari racer, Naama, stretched herself out. **1927** —— *Revolt in Desert* xi. 154 It had killed only two of us, one Rueli and one Sherari.

sherbet. Add: **3. b.** (Later Austral. examples.)

1917 H. LAWSON *Coll. Verse* (1969) III. 214 Beer that *we* called 'sherbet'. **1974** F. ARCHER *Treasure House* i. 18 He had a strident voice and with a few sherbets under his belt you knew he was about.

4. *sherbet cup* (later examples); **sherbet dabs** (see quot. 1957); **sherbet fountain**, a confection consisting of a bag of sherbet with a liquorice 'straw' through which it is sucked up.

1805 M. WILMOT *Jrnl.* 9 Apr. in Londonderry & Hyde *Russian Jrnls.* (1934) I. 140 Some Sherbet Cups in silver stands. **1966** MRS L. B. JOHNSON *White House Diary* 2 Aug. (1970) 403 My brother Tony Taylor had sent a set of sherbet cups that had belonged to my mother. **1957** J. KIRKUP *Only Child* 118 Another treat was Sherbet Dabs: we got a caramel-flavoured lollipop which we dipped into a bag of sherbet. **1958** *Listener* 23 Oct. 649/2 The Bonds' shop was not the place for liquorice root, tiger nuts, or sherbet dabs. **1957** R. HOGGART *Uses of Literacy* I. ii. 57 The boy's odder pleasures of taste, not so much..the sherbet-fountains, monkey nuts and aniseed balls, but..a penny stick of licorice or some cinnamon root from the chemist.

Sherbro (ʃɜ·ɪbro). [ad. a native name.] (A member of) a people of the southern coast of Sierra Leone; also, their language. Also *attrib.* or as *adj.*

1836 F. H. RANKIN *White Man's Grave* I. ii. 33 The Bulloms of Sherbro, or the Sherbros, pronounced 'Say-bras', are to the south of the colony. **1887** *Encycl. Brit.* XXII. 44/2 The following are the more important races that can be distinctly classified:—Mandingos, 1190;..Sherbros, 2882 [; etc.]. **1944** M. GORVIE *Our People of Sierra Leone Protectorate* iii. 20 There is today to be found in the Temne language a large number of Sherbro words. **1957** M. BANTON *W. Afr. City* v. 95 The Sherbro specialize in fishing and water-front work. **1970** J. R. CARTWRIGHT *Politics in Sierra Leone 1947–67* i. 13 It is likely that all these tribes with the possible exception of the Limbas and the Sherbros along the coast, entered Sierra Leone within the past 600 years.

Sherden: see *SHARDANA.

shereef. Add: Forms: **sharif** (later examples).

1959 *Economist* 3 July 11/2 Even the assorted amirs, sultans and sharifs of south Arabia. **1974** C. MAJUL in Gowing & McAmis *Muslim Filipinos* I. 4 The sultans of Sulu have all claimed descent from this *sharif*.

Hence **Sheree·fate**, the office of Shereef; **Sheree·fial** *a.* = SHEREEFIAN *a.* in Dict. and Suppl.

1917 R. STORRS *Orientations* (1937) x. 244 He [*sc.* the Sultan of Muscat] professes entire satisfaction..with the Sharifial movement. **1920** *19th Cent.* Aug. 233 It was to the British Government..that the Sherifial family of Mecca addressed their communications during the war. **1924** *Glasgow Herald* 8 Mar. 9 He began his official career by assisting successive Sherifs, and in 1908..was appointed to the Sherifate by the Porte.

Shereefian, *a.* Add: **2.** Of or pertaining to the Shereef of Mecca. Also as *sb.*, a supporter of the Shereef.

1921 G. BELL *Let.* 17 Apr. (1927) II. xx. 590, I am therefore identified as a Sharifian. **1926** T. E. LAWRENCE *Seven Pillars* I. xvi. 82 Blood feuds were nominally healed, and really suspended in the Sherifian area. *Ibid.* II. xviii. 93 Garland single-handed was teaching the Sherifians how to blow up railways with dynamite. **1929** W. S. CHURCHILL *World Crisis* V. 462 He came of the Sherifian family which, as guardians of the Holy Places at Mecca, commanded wide veneration throughout the Islamic world. **1935** R. H. KIERNAN *Lawrence of Arabia* viii. 174 He was defeated by the fanatical Wahabis in a fight which resulted in the destruction of a Sherifian force of four thousand men. **1976** *Times Lit. Suppl.* 30 Apr. 519/2 Correspondence which on the Sharifian side was a continuation of the attempts made before the outbreak of the 1914–18 War to enlist the services of Great Britain in Sharif Husayn's resistance to the centralizing policy of the Turkish government. *Ibid.*, To these proposals the Sharifians demanded an answer within thirty days.

3. Of, pertaining to, or designating descent from Muhammad.

1936 E. WAUGH *Waugh in Abyssinia* 86 The Emir's family claimed high, Sheriffian descent. **1976** *Times Lit. Suppl.* 13 Feb. 164/2 No doubt sultans did issue certificates of 'sherifian' descent. *Ibid.* 164/3 The conversion to the specifically 'sherifian' idiom of prestigious ancestry. **1977** P. RAYMOND *Matter of Assassination* vii. 75 His Sharifian Majesty, exalted of God, received us in his audience chamber.

Shererat, var. *SHERARAT. **Sheriat,** var. *SHARIA.

Sheridanesque (ʃe:ridăn·esk), *a.* [f. the name of Richard Brinsley *Sheridan* (1751–1816), British dramatist + -ESQUE.] Of, pertaining to, or characteristic of Sheridan or his plays.

1931 [see *RACINIAN *a.* and *sb.*]. **1933** BLUNDEN *Charles Lamb* ii. 61 He [*sc.* Lamb] ventured to show what he could do independent of the Sheridanesque cleverness around him.

So **She·ridania:na** [*-IANA], anecdotes about Sheridan.

1826 F. REYNOLDS *Sheridaniana* p. iii, The..volume..is intended to comprehend all that is most interesting..about..Sheridan—a person so eminently qualified to form the subject of such a work, that it seems somewhat singular that the *present* should be the first collection of Sheridaniana. **1931** *Times Lit. Suppl.* 16 Apr. 303/3 Sheridaniana? There is 'at the bottom of this stagnant well' some trace of truth, the misery of a reputation for wit.

sheridanite (ʃe·ridănəit). *Min.* [See quot. 1912 and -ITE[1].] A chlorite, $(Mg,Al,Fe^{2+})_6(Si,Al)_4O_{10}(OH)_8$, chemically similar to clinochlore but containing less silicon.

1912 J. E. WOLFF in *Amer. Jrnl. Sci.* CCXXXIV. 476 Although it seems hardly permissible to add a new name to the forty or fifty now found under the chlorite group, yet the purity of this material, its peculiar chemical composition and the certainty that it will be available in large quantity, perhaps justifies the name of 'Sheridanite', from the county [in Wyoming] in which it occurs. **1967** *Canad. Mineralogist* IX. 30 The fine white material in the core of the specimen from Zlatoust, Russia... This material gave an extremely sharp chlorite *x*-ray pattern. The structural formula..classes it as a sheridanite, as defined by Foster (1962).

sheriff. Add: **1. a.** (*a*). (Later examples.) The boroughs and cities which had their own sheriffs are no longer styled 'counties of themselves', but many of them, e.g. Canterbury, Chester, Gloucester, still have sheriffs. The Sheriffs of the City of London ceased to be also Sheriffs of Middlesex in 1888.

1888 *Act 51 & 52 Vict.* c. 41 s. 41(8) The sheriffs of the city of London shall not have any authority except in the city. *Ibid.* s. 46(6) The right of the mayor, commonalty, and citizens of the city of London to elect the sheriff of Middlesex shall cease, and it shall be lawful for Her Majesty the Queen to appoint a sheriff of the county of Middlesex. **1972** *Local Govt. Act* c. 70 s. 219(1) Sheriffs appointed for a county or Greater London shall be known as high sheriffs. **1974** C. A. CROSS *Principles Local Govt. Law* (ed. 5) xxvii. 488 The Queen in granting a charter to preserve the privileges of a city or borough existing prior to April 1, 1974, may confer on the city or borough the power to appoint a sheriff, as opposed to a high sheriff.

(*b*). Short for *sheriff's officer* (sense 4 a in Dict.); = BAILIFF 2.

1928 *Daily Mail* 30 July 7/1 You have had the sheriff in your house? *Ibid.*, Who put the sheriff into your house?

4. a. sheriff court (later examples); **sheriff officer** (later example); **sheriff's sale** *N. Amer.*, a public sale conducted by a sheriff following a court order for seizure and sale of property to satisfy a judgment.

1894 *Scots Law Times* I. 701 Except in Sheriff Court Cases, the figures refer to the number of the Case, and not to the number of the Page. **1962** T. B. SMITH *Scotland* iv. 102 The sheriff court has jurisdiction to try all crimes committed within the sheriffdom, except treason, murder, attempt to murder, rape, incest, certain offences against the Official Secrets Acts, deforcement of messengers, and breach of duty by magistrates. **1978** *Dumfries Courier* 13 Oct. 2/5 The sequel took place at Dumfries Sheriff Court on Friday when Savage pleaded guilty to the offence. **1932** *Encycl. Laws Scotland* XIII. 527 Sheriff Officers are the persons by whom writs are served and executions carried out in the Sheriff Courts. **1798** *Pittsburgh Gaz.* 6 Oct. 1/2 (Advt.), Sheriff's Sales. **1883** *Brandon* (Manitoba) *Daily Mail* 9 Jan. 4/1 There will be a sheriff's sale of goods and chattels belonging to the late firm of Hambly & Miller, barbers, etc., on Saturday next. **1947** *Steamboat* (Colorado) *Pilot* 30 Jan. 2/8 The electric light plant..was sold at sheriff's sale. **1966**

Globe & Mail (Toronto) 6 Sept. 30/1 Sheriff's Sale of Lands—Under and by Virtue of an Execution issued out of the Ninth Division Court of the County of York to me directed against the lands and tenements of [etc.].

Sherlock (ʃē·ɪlɒk), *sb.* [See *HOLMESIAN *a.* and *sb.*, and next.] A person who investigates mysteries or shows great perceptiveness; a private detective.

1903 G. V. HOBART *Back to Woods* iii. 57 'Down there, eh?' snorted the country Sherlock. **1928** D. L. SAYERS *Ld. Peter views Body* 42 I'm riding with Freddy Arbuthnot,..as you might see by my legs, if you were really as big a Sherlock as you make out. **1932** KIPLING *Limits & Renewals* 178 We aren't exactly first-class Sherlocks. **1967** N. FREELING *Strike out where not Applicable* 27 Mr. van der Valk, my dear, our police Sherlock. **1972** 'L. EGAN' *Paper Chase* xii. 191 You'll have to turn Sherlock and solve the case yourself.

So **She·rlock** *v. intr.* and *trans.*, to engage in detective work; to investigate (something), to make deductions about; **She·rlockian**, *sb.* and *a.* (*a*) *sb.* = *SHERLOCK *sb.*; (*b*) *adj.* pertaining to or characteristic of Sherlock Holmes; = *HOLMESIAN *a.*; **She·rlockia:na** [*-IANA], things connected with Sherlock Holmes, writings about Sherlock Holmes; **She·rlocking** *vbl. sb.*, detective work.

1903 *Bookman* XVII. 5/2 If you decipher this you are a real Sherlockian. **1913** *Manch. Guardian* 15 Jan., Any man with a bundle or package was suspicious, so we 'sherlocked' around for a bit and watched him go into a barber's shop to get disguised by having his hair cut. There we 'pinched' him. **1920** J. GALSWORTHY *Foundations* I, in *Plays* (1929) 468 Don't call in the police!..Let me do the Sherlocking for you. **1934** *Discovery* Sept. 273/2 It is now close on four hundred years since that door was used and sandals trod those steps, but we are able to 'Sherlock' this detail of old times. **1937** STEVENS & SHORTEN *How to watch Football Game* vi. 51 No use in trying to Sherlock the next play. Anything can happen. **1942** H. HAYCRAFT *Murder for Pleasure* xiii. 276 Vincent Starrett's *Private Life of Sherlock Holmes* with its valuable appended bibliography of Sherlockiana. **1957** J. KEROUAC *On Road* (1958) 135 They tried some amateur Sherlocking by asking the same questions twice. **1959** *Listener* 3 Dec. 993/1 A startling piece of detective work, followed up with exact, devoted, Sherlockian tenacity. **1962** W. S. BARING-GOULD *Sherlock Holmes* 263 Late 1895–late 1896 called by many Sherlockian commentators 'The Missing Year', and the subject of much learned speculation. **1963** 'G. CARR' *Lewker in Norway* i. 25 You're really disappointed because you can't go Sherlocking after that young man. **1972** E. ROUTLEY *Puritan Pleasures of Detective Story* ii. 27 This is not a contribution to Sherlockiana, but the..cult of Sherlock Holmes is itself..in the field of our enquiry. **1975** *Daily Tel.* 27 May 14/3 He..had built up an outstanding collection of Sherlockiana, including such rarities as a copy of the Beeton's Christmas Annual of 1887 in which the first Holmes adventure..was originally published. **1977** *New Yorker* 20 June 71/1 He lights his pipe. It is long and low and looks somewhat Sherlockian.

Sherlock Holmes (ʃɜ·ɪlɒk hōᵘmz). [See *HOLMESIAN *a.* and *sb.*] A person resembling Sherlock Holmes; = *SHERLOCK *sb.*

1896 E. TURNER *Little Larrikin* x. 108 It took her nearly five minutes to wonder sufficiently at him..and call him a Sherlock Holmes. **1914** T. A. BAGGS *Back from Front* xxiv. 118 It needed no Sherlock Holmes to discover where English cavalry had bivouacked for the night. **1957** A. MACNAB *Bulls of Iberia* xvi. 237 The press critics kept very quiet about it, and one need not be a Sherlock Holmes to guess why. **1981** *Times* 22 Apr. 6/5 The doctor becomes a medical Sherlock Holmes.

So **Sherlock Holmes** *v. trans.*, to make deductions about, to assess, to deduce (cf. *SHERLOCK *v.*); **Sherlock Ho·lmesian** *a.* = *SHERLOCKIAN *a.*; **Sherlock Ho·lmesing** *vbl. sb.* = *SHERLOCKING *vbl. sb.*

1922 JOYCE *Ulysses* 620 He had been meantime taking stock of the individual in front of him and Sherlockholmesing him up. **1929** C. I. LEWIS *Mind & World-Order* ix. 287 All the Sherlock Holmesing in the world would not help him a bit because he would not be able to recognize evidence when he found it. **1954** E. EAGER *Half Magic* 35 'She just as good as said so,' said Jane, 'and I Sherlock Holmesed the rest.' **1958** *Observer* 16 Mar. 14/3 The stern Sherlock Holmesian K.C. **1958** A. WILSON *Middle Age of Mrs. Eliot* I. 89 All this woman's intuition is just a lot of Sherlock Holmesing. **1972** *Sci. Amer.* Mar. 106/3 He proceeded indeed by a kind of Sherlock Holmesian logic, which presumes that in eliminating the impossible and the more implausible it has arrived at the true.

Sherman[1] (ʃɜ·ɪmăn). The name of John *Sherman* (1823–1900), U.S. senator, used *attrib.* to designate either of two acts passed by Congress in 1890, one to prohibit combinations in restraint of inter-state or foreign trade, the other to maintain the price of silver by government purchase of silver bullion; also, to designate treasury notes issued under the provisions of the latter act.

1892 *Dem. Platform* in K. Porter *Nat. Party Platforms* (1924) 162 We denounce the Republican legislation known as the Sherman Act of 1890. **1894** *Harper's Mag.* Jan. 318/1 Mr. Voorhees's substitute repealing the Sherman law was passed by a vote of 43 to 32. **1897** *Money* May 23 Government notes called Treasury notes of 1890, sometimes *Sherman notes*, sometimes *Coin notes*..were issued in payment of purchases of silver bullion from 1890–93. **1947** *Atlantic Monthly* June 73/1 In 1890 the Sherman Act was passed to 'appease the restive masses', but it would be fifty years before Thurman Arnold would demonstrate, briefly, that the Act could be made to work. **1948** *Duncan* (Oklahoma) *Daily Banner* 1 July 1/3 The U.S. Supreme Court..held that officers of the company and the firm itself had violated the Sherman anti-trust law. **1974** *Encycl. Brit. Micropædia* IV. 303/1 The Bland-Allison Act was superseded in 1890 by the Sherman Silver Purchase Act, which increased the government's monthly silver purchases by 50 percent. Fear that the U.S. was about to abandon the gold standard precipitated the Panic of 1893, causing the Sherman Act to be hastily repealed the same year.

Sherman[2] (ʃɜ·ɪmăn). The name of W. T. *Sherman* (1820–91), U.S. general, used *attrib.* and *absol.* to designate an American type of medium tank, much used during the war of 1939–45. Also *General Sherman (tank)*.

1942 *Times* 6 Nov. 6/3 (*caption*) A picture just received from America of General Sherman tanks. **1942** W. S. CHURCHILL *End of Beginning* (1943) 229 This very powerful force.., including all the best tanks, the Grants and the Shermans, was withdrawn from the battle front. **1944** *Sun* (Baltimore) 15 Nov. 11/2 The Sherman tanks jumped off with orders to cross the river. **1965** A. J. P. TAYLOR *Eng. Hist. 1914–45* xv. 554 To aid Auchinleck, the Americans diverted 300 Sherman tanks to Suez. **1969** STUBBS & CONNOR *Armor-Cavalry* I. 64 A much improved M3 medium was standardized in 1941 as the M4, better known throughout the war by its British designation, the General Sherman. **1971** E. LUTTWAK *Dict. Mod. War* 176/1 The Israeli army uses a large number of converted Shermans fitted with a new 105-mm gun.

Sherpa (ʃɜ·ɪpă). Also 9 **Serpa, Sharpa.** [ad. Tibetan *sharpa*, inhabitant of an eastern country.] **1.** (A member of) a Tibetan people living on the southern slopes of the Himalayas. Also *attrib.* or as *adj.*

1847 B. H. HODGSON in *Jrnl. Asiatic Soc. Bengal* XVI. 1237 Cis-Himálayan Bhotias vel Tibetans, called..Serpa, &c. *Ibid.* 1238 The sub-Himálayan races..inhabit all the central and temperate parts of these mountains, the juxta nivean or nethermost tracts being left to the Rongbo vel Sérpá. **1874** —— *Ess. on Nepál & Tibet* II. ii. 30 Cis-Himálayan Bhotias vel Tibetans, called Rongbo, ..Sérpa or Sharpa etc. **1922** G. H. LEIGH-MALLORY in C. K. Howard-Bury *Mount Everest* xiv. 224 The Tibetan coolies..were notably less strong than our Sherpas. **1924** *Glasgow Herald* 16 Apr. 10/6 Two of them after a merry-making battered each other in a terrific manner, but the following morning..they were nursing each other with the greatest care and mutual pity. Such is a Sherpa porter. **1950** T. LONGSTAFF *This my Voyage* iv. 67 He..had not yet shown us that Sherpas could be the mainstay of any Himalayan expedition. **1970** *Daily Colonist* (Victoria, B.C.) 11 Nov. 38/4 Sen Tensing, the sherpa bearer, shuddered in fright. **1979** *Daily Tel.* 17 May 1/4 One of mountaineering's best known Sherpas, Ang Phu, was feared killed yesterday in a fall on Everest after helping a Yugoslav expedition put a second team on the summit.

2. *transf.* and *fig.* A mountain guide or porter; a guide; an official who makes the preparations for a summit conference.

1959 M. PUGH *Chancer* 137 What was the idea of trying that cliff? Did you fancy your chance as a sherpa? **1976** P. CAVE *High Flying Birds* iii. 35 'O.K.,' I said. 'Lead the way, sherpa.' **1980** *Times* 23 June 1/2 The seven leaders inevitably based much of their comment on the draft communique drawn up for publication after the meeting by the seven government officials—known as the 'sherpas'—who have been charged with preparing this summit.

sherri-varrie: see *SHIVAREE.

sherry, *sb.*[1] Add: **1. a.** Now a fortified wine. (Further examples.)

1958 A. L. SIMON *Dict. Wines, Spirits & Liqueurs* 146/1 Sherry is made from the best wine of each vintage, to which some Brandy is added, after which it is kept for many years with the best wine of other vintages. **1967** *Times* 1 Aug. 6/5 'Sherry' means a wine coming from the Jerez district of Spain. The Court, giving judgment,..decided that it would be unjust now to restrain Vine Products Ltd...from using the expressions 'British sherry', 'English sherry', 'Cyprus sherry', 'South African sherry', and 'Australian sherry', used for certain wines in England.

c. A glass or drink of sherry.

1924 GALSWORTHY *White Monkey* II. ix. 192 Will you have a sherry? **1979** M. MCCARTHY *Cannibals & Missionaries* i. 19 He had been counting on picking Gus's brains..over a sherry or a bourbon.

2. (Examples.)

1907 [see *PORT *sb.*[5] c]. **1925** [see *LIQUEUR *sb.* 2]. **1974** [see *PORT *sb.*[5] c].

3. a. *sherry decanter, trifle* [TRIFLE *sb.* 6 b]; *sherry-style, -type* adjs.; **sherry-bar**, a bar at which sherry is the principal drink sold; **sherry morning**, a morning sherry party; **sherry party**, a party at which sherry is the principal drink served.

1951 G. GREENE *End of Affair* v. ii. 181 Waterbury was waiting in a sherry-bar off Tottenham Court Road. **1950** J. CANNAN *Murder Included* vii. 164 A tray with two glasses and a sherry decanter. **1977** G. MCDONALD *Confess, Fletch* xxiii. 104 There were Scotch bottles, bourbon bottles, gin bottles, sherry and port decanters. **1976** *Milton Keynes Express* 16 July 2/3 A sherry morning on Sunday brought in £98.32 for Olney Town Cricket Club funds. **1936** *Cherwell* 7 Mar. 158/2 A tendency to throw sherry parties and get a little drunk. **1977** 'J. LE CARRÉ' *Honourable Schoolboy* xi. 239 Christmas was hardly noticed apart from a rather battered sherry party. **1960** *Times* 3 Oct. (Wine Trade Suppl.) p. iv/5 In the United States there is a domestic wine business which last year turned out..20 million gallons of sherry-style wine. **1951** H. SMITH *Master Bk. Dessert Pies & Sweets* viii. 279 *Sherry Trifle*, arrange slices of Swiss roll at the bottom of a round glass dish and sprinkle liberally with sherry. **1979** K. BONFIGLIOLI *After you with Pistol* xxii. 176 Two helpings of sherry-trifle. **1962** *Times* 3 Feb. 9/4 A glass of sherry-type wine.

c. sherry-negus (earlier example).

c **1863** T. TAYLOR *Ticket-of-Leave Man* I. 7 Two sherry negus, two shillings.

sherry, *v.*[2] Restrict *nonce-wd.* to sense in Dict. and add: **2.** To add sherry to. Chiefly as **she·rried** *ppl. a.*

1970 *Guardian* 15 Apr. 10/2 A heavily sherried trifle. **1977** D. J. ELLIOTT in D. Marcus *Best Irish Short Stories* 2 156 They ate prawns in aspic, sherried.

‖ **sherut** (ʃē[ɪ]rū·t). Also **sheirut.** [Heb., lit. 'service'.] In Israel, a large taxi shared by several passengers.

1950 G. MIKES *Milk & Honey* 91 Between the two [bus and taxi] there is the communal taxi, the *sherut*. **1968** P. DURST *Badge of Infamy* xiv. 133 He flagged down a *sherut*, one of the ramshackle taxi-cum-buses which plied between Haifa and Tel Aviv. **1977** H. KAPLAN *Damascus Cover* (1978) iii. 21 The *sheirut* drivers, loudly beckoning arriving passengers to share a taxi to Tel Aviv or Jerusalem. **1981** *Daily Tel.* 24 June 11 Jitneys and sheruts should be introduced on the roads to make public transport more efficient.

‖ **sherwani** (ʃɜɪwā·ni). [Hindi.] In the Indian sub-continent, a knee-length coat, buttoning to the neck, worn by men.

1911 [see *ACHKAN]. **1964** D. N. WILBUR *Pakistan* viii. 168 Among well-to-do Moslems generally, a long coat known as a *sherwani* is worn. **1971** R. RUSSELL tr. *A. Ahmad's Shore & Wave* vi. 54 He gravely took off his *sherwani* and hung it on the coat-hanger.

sheshbesh (ʃe·ʃbeʃ). Also **shesh-besh.** [Turk., f. Pers. *shash* six + Turk. *beş* five.] A variety of backgammon played in the Middle East. Also *attrib.*

1971 L. DAVIDSON *Smith's Gazelle* ii. 37 The sheshbesh players pressed his arm in sympathy. **1975** C. A. HADDAD *Moroccan* (1977) iv. 44 My friends would come over..to play sheshbesh. **1977** H. KAPLAN *Damascus Cover* (1978) vii. 67 In this part of the world we have been playing sheshbesh, what you call backgammon, for nearly three thousand years.

shet (ʃet), **shed** (ʃed), repr. a U.S. dial. and colloq. pronunc. of SHUT *v.* and *ppl. a.*, esp. in phrase *to get (be, stay) shet of* (see SHUT *v.* 11 a).

1837 A. SHERWOOD *Gazetteer of State of Georgia* (ed. 3) 70/1 *Get shet of*, for get rid of. **1848**, etc. [see *open-and-shut* s.v. *OPEN *a.* 22 c]. **1871** E. EGGLESTON *Hoosier School-Master* (1872) xxxii. 162 I'm glad to be shed of you! **1930** G. B. JOHNSON in B. A. Botkin *Folk-Say* vii. 357 The Negro raises..'great big' hogs, and tries to 'get shet of' his enemies, just as poor white folk have done for hundreds of years. **1935** Z. N. HURSTON *Mules & Men* I. iii. 77 Throw mah trunk out befo' you shet up dat place! **1943** E. CALDWELL *Georgia Boy* 80, I thought you was trying to get shed of it. **1974** *State* (Columbia, S. Carolina) 15 Feb. 1-B/1 A gentleman at the coffee counter began munching his hamburger a whole lot slower yesterday when the waitress admitted she just couldn't 'get shet' of her strep throat. **1976** *Verbatim* Sept. 8/1 The range of his scholarship and the deftness of his intuition..are bound in an engaging prose style that makes a reader grateful to be shet for a while of the orthodox taxonomist. **1978** J. A. MICHENER *Chesapeake* 726 'Turlocks hate us colored,' he warned his daughters, 'so the smart thing, stay shed of 'em... Just you stay clear, like me, an' you find no trouble.'

Shetland. Add: **1. c.** Also with small initial. (Earlier and later examples of things made of this wool.) *Shetland floss*, Shetland wool. *Shetland knitting*, a traditional style of knitting characterized by a distinctive technique and by the following of Scandinavian patterns in 'natural' colours.

1854 *Morning Post* 7 July 1/5 (Advt.), Shetland shawls and veils in great variety. **1895** Shetland floss [see *double-knit* adj. s.v. *DOUBLE *a.* A. 6]. **1934** W. MOFFATT *Shetland: Isles of Nightless Summer* i. 23 The wool is invariably short, curly and silky... This wool is famous all over the world, and is so valued that unscrupulous people in other lands sell wool called 'Shetland Floss' that was never within a thousand miles of Shetland. **1935** in *Scottish Woollens* (Nat. Assoc. Sc. Woollen Manufacturers) (1956) 55 Shetlanders..still knit a great variety of articles, and the finest Shetland knitting is still unrivalled. **1966** N. FREELING *Dresden Green* I. 17 A beige shetland pullover. **1974** *People's Jrnl.* (Inverness & Northern Counties ed.) 7 Sept. 9/1 They studied island

crafts such as silverwork, handloom weaving, carding and spinning, and Shetland knitting. **1979** R. JAFFE *Class Reunion* 5 Plain dark shetland cardigans. **1980** L. BIRNBACH et al. *Official Preppy Handbk.* 107/1 You should be comfortable with a turtleneck,..a button-down shirt and a Shetland sweater.

d. *Shetland sheep-dog*, a small long-coated collie belonging to the breed so called; also *Shetland collie*.

1908 *Our Dogs* 27 Mar. 681/2 Shetland Collies.—This variety is at present in the hands of a few fanciers only. **1909** *Ibid.* 15 Oct. 937/1 Shetland Collies or Sheepdogs, the original type and colour. **1960** E. MILLER *Shetland Sheepdog* 5 So much experimenting has gone on with this breed, that the Shetland Sheepdog is still far from being 'typed' like a Poodle or Boxer. **1971** F. HAMILTON *World Encycl. Dogs* 97 The name Shetland Collie was the first choice but the Collie Clubs objected so strongly that the Kennel Club would not allow it. **1977** *Belfast Tel.* 14 Feb. 12/9 (Advt.), Shetland sheepdog puppies.

2. *absol.* **a.** A shetland pony.

1836 S. C. STEVENSON *Let.* Sept. in E. Boykin *Victoria, Albert, & Mrs. Stevenson* (1957) 30, I cannot tell you how I covet one of these little shetlands for my little nephews. **1975** *Country Life* 6 Feb. 327/1 The ponies shown included Connemaras,..Shetlands and Welsh.

b. A Shetland shawl, sweater, etc.

1870 CROWN PRINCESS OF PRUSSIA *Let.* 18 Mar. in R. Fulford *Your Dear Letter* (1971) 207 In my condition..I..never am in my own room without a lace shawl or a shetland or black mantilla. **1972** *Vogue* 1 Mar. 57/1 Classic Shetland and cashmere sweaters... Shetlands in 79 colours. **1979** A. V. BADGLEY *Rembrandt Decisions* xi. 156 Stout oxfords, ribbed wool stockings, itchy shetlands.

c. A Shetland sheep-dog.

1945 C. L. B. HUBBARD *Observer's Bk. Dogs* 143 A perfect Collie in miniature, the Shetland is efficient and nimble. **1958** O. GWYNNE-JONES *Shetland Sheepdog Handbk.* i. 8 By 1912 Shetlands were classified at..sixteen shows.

‖ **sheva.** Add: Also **shewa** (ʃəwā·), **shva.** **1.** (Earlier and later examples.)

1582 MULCASTER *Elementarie* xvii. 113 Like to a silent Hebrew Scheua. **1818** P. S. DUPONCEAU in *Trans. Amer. Philos. Soc.* I. 241 A small vacant space, as it were, between the consonants, like the *Sheva* of the Hebrews. **1914** DAVIDSON & MCFADYEN *Introd. Hebrew Gram.* (ed. 19) 23 The place of shᵉwa vocal, simple or composite, is under the first of two consonants that begin a syllable. **1939** J. WEINGREEN *Pract. Gram. for Classical Hebrew* 9 The shewa is not a vowel. The quick vowel-like sound is like the 'e' in 'because'. **1965** *Language* XLI. 543 The shva is a masoretic grapheme.

2. (Earlier and later examples.) See also *SCHWA.

1818 [see *phonologist* s.v. *PHONOLOGY]. **1939** E. PROKOSCH *Compar. Germanic Gram.* 94 IE ə and ь are distinguished as 'shva primum' and 'shva secundum', but the term 'shva' alone always refers to ə.

Shevuos, Shevuoth, varr. *SHAVUOTH.

Shiah. (Later examples of form *Shia*.)

1926 T. E. LAWRENCE *Seven Pillars* II. xxvii. 140 They were Shias, and had been since the days of Kerbela. **1955** *Times* 31 Aug. 6/6 The Shia mourning day, Ashura, commemorating the death of Imam Hussein, grandson of the prophet, passed off without incident. **1969** *Pioneer* (Lucknow) 13 Aug. 4/8 Syeds are mostly Shias in Lucknow. **1974** *Educ. & Community Relations* Jan. 5 Celebrated with processions mostly by the Shia sect.

shiatsu (ʃi,æ·tsu). Also **Shiatsu, shiatzu.** [Jap., lit. finger-pressure.] A kind of therapy, of Japanese origin, in which pressure is applied with the thumbs and palms to certain points on the body. Also *attrib.*

1967 *Tel.* (Brisbane) 4 Dec. 12 A Japanese physiotherapist..believes that his shiatsu finger massage is good for treating high blood pressure, insomnia and hernia. **1969** T. NAMIKOSHI *Shiatsu: Health & Vitality at your Fingertips* i. 10 Widely practiced in Japan today, shiatsu is described by the Ministry of Welfare as follows: Shiatsu is a treatment in which the thumbs and palms of the hands are used to apply pressure to certain points in order to correct irregularities of the living body. *Ibid.* 13 The thumbs are often used in shiatsu treatment. *Ibid.* iv. 81 Apply strong shiatsu pressure to the eight points on the calf. **1975** *Publishers Weekly* 18 Aug. 29 (Advt.), Shiatzu was developed centuries ago in Japan as a refinement of the acupuncture treatment from China. **1980** *Daily Tel.* 21 June 12/4 There are 13 pressure-points in *shia-tsu*.

Shibayama (ʃibāyā·mā). Also **shibayama.** The name of a Japanese family of carvers, used *absol.* and *attrib.* to denote a distinctive style of inlay work which they originated.

1928 F. M. JONAS *Netsuké* 174 Shibayama is the name by which all encrusted work is known. **1956** F. MEINERTZHAGEN *Art of Netsuke Carver* III. 45 'Shibayama' work, mostly of poor quality, in the form of vases, jars, trays, etc., was made from the late 19th century up to recent times. **1971** *Country Life* 22 Apr. 941/1 The design is composed of gold lacquer and *shibayama*, that is pieces of ivory, tortoiseshell, mother-of-pearl used to build up the picture. **1982** 'J. GASH' *Firefly Gadroon* iv. 36 A set of Shibayama knife-handles.

‖ **shibui** (ʃi·bu,i), *a.* and *sb.* Also **shibu.** [Jap., f. *shibu* an astringent substance.] **A.** *adj.* Tasteful in a quiet, profound, or unostentatious way. **B.** *sb.* Tastefulness, refinement, appreciation of elegant simplicity.

In Japanese the substantival form is *shibu* (the substance) or *shibumi* (its quality); the adj. is *shibui*.

1947 J. MORRIS *Phoenix Cup* II. 28 A picture, a piece of pottery, a kimono, or what you will, may be, from the Japanese point of view, in exquisite taste, and yet not *shibui*; to be thus described there must be also some invocation of quietude and austerity. **1958** *Japan: its Land, People & Culture* xxix. 1019/2 As beauty approaches the highest level it becomes a subtle beauty represented by what is known as the *shibu* taste. **1960** E. MANNIN *Flowery Sword* x. 166 Vulgarity and *shibui* side by side. **1965** *This is Japan 1966* 121/2 The sense of appreciation known as *shibui*, which enables Japanese to derive such satisfaction from the drinking of the tea, might seem to the foreigner a type of high, super-refined, even affected taste. **1970** J. KIRKUP *Japan behind Fan* 27 That ghastly good taste, common to all modern hotels, that turns the most *shibui* atmosphere into something expensive and pretentious.

‖ **shibuichi** (ʃibui·tʃi). [Jap., f. *shi* four + *bu* part(s) + *ichi* one.] An alloy consisting of three parts of copper to one of silver, extensively used by the Japanese on account of its beautiful silver-grey patina. Also *attrib.*

1880 T. W. CUTLER *Gram. Jap. Ornament* 19 The *shibu-ichi* is inlaid with gold and *shakudo*. **1902** *Encycl. Brit.* XXIX. 720/1 Neither metal, when it emerges from the furnace, has any beauty, *shakudo* being simply dark-coloured, and *shibuichi* pale gun-metal. **1911** *Ibid.* XV. 179 On the surface of a shibuichi box-lid we see the backs of a flock of geese. **1977** *Times* 14 Oct. 17/5 The top priced *tsuba*..is..made from a Japanese alloy called shibuichi.

shice (ʃais), *sb.* and *a.* Also **shise.** *slang* (? *Obs.*). [G. *scheiss*; cf. SHICER.] **A.** *sb.* Nothing; base money; something worthless. **B.** *adj.* Worthless, counterfeit, spurious.

1859 HOTTEN *Dict. Slang* 91 *Shice*, nothing; 'to do anything for *shice*', to get no payment. **1877** *Five Years' Penal Servitude* iii. 240, I ascertained while at Dartmoor that a very large 'business' is done in 'shise'. *c* **1890** *Five Years of Prison Life* ii. 62 Seeing how the fellow was acting he sent him two 'shise' notes, which gave him a dose that 'corked him'. **1939** J. B. PRIESTLEY *Let People Sing* x. 256 'I keep tellin' Knocker it's a shice,' said Micky earnestly.

shicer. Add: **1.** (Earlier example.) Also, something worthless, a failure.

1846 *Swell's Night Guide* 61 The shiser thinks to bounce us by flashing a shofel quid. **1874** A. BATHGATE *Colonial Experiences* viii. 97 There are, of course, many what may be called technical terms in connection with the pursuit of mining... Such, for example, as 'duffer' or 'shiser', anything that is useless. **1916** J. B. COOPER *Coo-oo-ee* xiv. 203 'The case is a "shicer" already,' replied Jack. 'Hawley has given it up.'

shick (ʃik), Austral. and N.Z. abbrev. of *SHICKER *a.*, *sb.* 1.

1916 C. J. DENNIS *Moods of Ginger Mick* 19 The toff's too shick or silly fer to 'eave 'is carkis ect. **1941** BAKER *Dict. Austral. Slang* 65 *Shick*, drunk. Also (n.) a drunken person. **1966** B. BEAVER *You can't come Back* 144 The wonder was we got that far without falling over and breaking them, because we were both pretty shick by then.

shicker (ʃi·kəɪ), *a.* and *sb.* *Austral.* and *N.Z. colloq.* and in Jewish speech. Also **shiker, shikker, shikkur.** [ad. Yiddish *shiker* ad. Heb. *šikkōr*, f. *šākar* to be drunk. Perh. infl. by *SHICKERY *a.* and *adv.*] **A.** *adj.* Drunk, intoxicated.

1892 I. ZANGWILL *Childr. Ghetto* I. ii. 53 'But I'll get drunk on gingerbeer,' Pesach laughed back. 'You can't,' Fanny said... 'Ha! ha! ha! Can't even get *shikkur* on it. What a liquor!' **1898** *Bulletin* (Sydney) 17 Dec. Red Page, *Shiker*, drunk. **1899** A. BINSTEAD *Houndsditch Day by Day* 46 She comes over shikkur an vants to go to shleeb. **1916** J. B. COOPER *Coo-oo-ee* v. 60 Hickford said the deceased was 'shicker' on the night before his death. **1949** *Hilltop* (N.Z.) I. II. 20 Sid should have some chance to get shicker as everybody else. **1963** 'M. CORRIGAN' *Why do Women—?* xv. 106 She was shikker and she took it into her head to come here. **1970** *N.Z. Listener* 12 Oct. 12/5 After midnight, Jerry got so shicker that he was quarrelling with everyone.

B. *sb.* **1.** A drunk.

1906 [see *on one's ear* s.v. *EAR *sb.*[1] 1 c]. **1938** X. HERBERT *Capricornia* xviii. 234 He's the biggest shikker in Town. Now nick off, you old sponge. **1945** A. KOBER *Parm Me* 156 Right away they are coming here, the whole bunch!.. Will be Uncle Henry and Uncle Philip—two foist-cless *shikkers*! **1949** D. M. DAVIN *Roads from Home* I. iii. 41 A shicker. Broke too, by the sound of it. **1964** S. BELLOW *Herzog* 135 He drank his pay—a *shicker*. **1970** S. ELLIN *Bind* xlvii. 234, I was hunting in the swamps with my *shikker* friend... I dumped my *shikker* off at Flagler Street.

2. Liquor; esp. in phr. *on the shicker.* (Only *Austral.* and *N.Z.*)

1916 C. J. DENNIS *Moods of Ginger Mick* 154 *Shicker*, intoxicating liquor. **1918** *N.Z.E.F. Chron.* 27 Feb. 33/1 He saved his strength for shicker. **1928** A. WRIGHT *Good Recovery* 85 'Yes, I've been on the shikker,' he answered huskily. **1945** R. RENE *Mo's Memoirs* 49 One night the magician had been on the shicker, and with a fine disregard for life and limb he let the lion out. **1958** H. D. WILLIAMSON *Sunlit Plain* 58 He was on the shicker when I was there last week.

shicker (ʃi·kəɪ), *v.* *Austral.* and *N.Z. colloq.* [f. as prec.] *intr.* To take alcoholic drink, to get drunk.

a **1922** H. LAWSON *Benno & his Old 'Uns* in *Prose* (1964) III. 230 Her Old 'Un 'shickered' till he got 'mucked' every pay day. **1951** CUSACK & JAMES *Come in Spinner* 33 He'd gamble his shirt off on any damn thing that's got a leg to run on, but he doesn't shicker.

shickered (ʃi·kəɪd), *a.* *Austral.* and *N.Z. colloq.* Also **shikkered.** [f. *SHICKER *a.* and *sb.*, or *v.*] Drunk, intoxicated.

1911 L. STONE *Jonah* I. ix. 124 'E's bin shickered since last We'n'sday. **1911** H. FOSTON *In Bellbird's Lair* vii. 34 He thinks we're 'shickered'. **1916** C. J. DENNIS *Moods of Ginger Mick* 19 There's a shickered toff slings Rosie goo-goo eyes. **1930** A. W. GROOM *Merry Christmas* xvi. 123 Every time I get shikkered I sober up after an' say 'Never again, Sandy'. **1934** T. WOOD *Cobbers* xvi. 209 He *tole* me he was goin' to get sh-shickered, an' he has... He'sh as drunk'sh forty catsh! **1947** D. M. DAVIN *Gorse blooms Pale* 212 When you're shickered you know..things aren't really as good as they seem. **1952** J. CLEARY *Sundowners* i. 42 He's drunk! Shickered to the eyeballs! **1961** P. WHITE *Riders in Chariot* 261 'I'm gunna get out of this!' he announced at last. 'I'm gunna get shickered stiff!' **1972** C. DRUMMOND *Death at Bar* v. 114 'You look half shickered, love,' said Charlie. 'I'll be right when I've been sick,' said Mrs. Gaukroger.

shickery (ʃi·kəɪi), *a.* and *adv.* *slang* (? *obs.*). [Origin unknown.] **A.** *adj.* Shabby, rickety, shaky; also, drunk. **B.** *adv.* Shabbily, badly.

1851 H. MAYHEW *London Labour* I. 424/1 There's another sort who carry on the crocussing business, but on a small scale; they..are called hedge crocusses... But as the hedge crocus is shickery togged, he makes poorly out. **1859** HOTTEN *Dict. Slang* 91 *Shickery*, shabby, badly. **1878** 'R. BOLDREWOOD' in *Town & Country Jrnl.* 26 Jan. 170/1 Old Tom had a goodish cheque this time, and was at it a week afore I come in. He *was* rayther shickerry. **1888** —— *Robbery under Arms* III. xv. 220 Four panels of shickery two-rail fence.

shicksa, var. *SHIKSA.

shickster (ʃi·kstəɪ). *slang* (? *obs.*). [f. *SHIKSA: see -STER.] (See quots. 1937, 1965.)

1839 H. BRANDON in W. A. Miles *Poverty, Mendicity & Crime* 165/1 *Shickster*, a lady. **1846** *Swell's Night Guide* p. iii, The..bully, or cracksman, who would screw a drum,..collar the shiksters denarly, or paste a greenhorn. **1882** *Sydney Slang Dict.* 10/2 He got the cant of togs from a *shickster* whose husband's in a bone-box. **1937** PARTRIDGE *Dict. Slang* 755/2 *Shickster*, any (Gentile) woman or girl..a none too respectable girl or woman. **1965** *English Studies* XLVI. 454 The way for *shyster* in American English is paved by *shickster*..of Yiddish origin; in criminals' cant it designated a respectable girl or lady..but it soon deteriorated to 'prostitute'.

‖ **shidduch** (ʃi·dŭχ). Also **shiddach.** [Yiddish, ad. Heb. *šiddūḵ* courtship, arranged marriage.] An arranged marriage, a (good) match. See also *SHADCHAN.

1892 I. ZANGWILL *Childr. Ghetto* I. 108 Every match is a grand *Shidduch* before the marriage; after, we hear another tale. **1968** L. ROSTEN *Joys of Yiddish* 339, I am considering getting married. But I warn you I'll accept nothing but a remarkable *shiddach*! **1976** C. BERMANT *Coming Home* I. i. 16 A *shidduch* was arranged for him and he married the pretty..daughter of a Vitebsk..merchant.

shiela-na-gig, var. *SHEELA-NA-GIG.

shield, *sb.* Add: **I. 1. d.** *two sides of a shield*: two ways of looking at something, two sides to a question; *the other side* (or *reverse*) *of the shield*: the other side of a question or consideration, the side which is less obvious or which has not been presented (cf. *the reverse of the medal* s.v. MEDAL *sb.* 3 b, Fr. *le revers de la médaille*).

? **1855** A. W. CHAPMAN *Let.* in R. K. Webb *Harriet Martineau* (1960) i. 27, I consider it a great misfortune, in one sense,..a blessing, often times, in another. There are two sides to every shield. **1885** C. M. YONGE (*title*) The two sides of the shield. **1909** P. COLLIER *England & English* i. 16 These beef-eating, port-drinking fellows in Piccadilly..are well enough..but this other side of the shield is distressing to look at. Poor, stunted..denizens of the East End. **1911** H. S. WALPOLE *Mr. Perrin & Mr. Traill* ii. 24 The reverse of the shield is..given in that first letter to his mother.

II. 8. b. A protective device in clothing, as a *dress-shield*.

1884, etc. [see *dress-shield* s.v. *DRESS *sb.* 4 a]. **1897** *Sears, Roebuck Catal.* 321/3 Kleinert pays for the dress if it is ruined by perspiration, if his shields are used in it.

c. *Mech.* (See quots.)

1888 *Lockwood's Dict. Mech. Engin.* 316 *Shield.* (1) A covering employed to protect the bearings and spindles of emery-grinding machines from the action of the gritty dust. (2) A guard placed over or in front of band and circular saws and portions of machinery to protect the workmen from accidents. **1967** J. L. & G. H. F. NAYLER *Dict. Mech. Engin.* 323 *Shield*, a contrivance or covering, protective plate, or screen, to protect machinery or the operator, from damage or accident.

d. *Physics.* An electrically conducting cover of a device or apparatus intended to protect it from external electric or magnetic fields or to reduce or eliminate interference radiated by the device or apparatus itself. Cf. SCREEN *sb.*[1] 6.

1919 J. A. FLEMING *Thermionic Valve* ii. 66 (*caption*) Fig. 34 shows the valve..with copper-gauze shield for protecting from external electric fields. **1947** R. LEE *Electronic Transformers & Circuits* vi. 174 Multiple shields increase the action..because eddy currents induced in the shields set up fluxes opposing the stray field. **1975** D. G. FINK *Electronics Engineers' Handbk.* VI. 32 It is often desirable to shield part of the circuit from electromagnetic fields. The shields can absorb, reflect, or degrade (by multiple internal reflections) the electromagnetic energy. The most commonly used shields are braided copper.

e. *Physics.* A mass of material, usu. lead or concrete, intended to absorb neutrons and other ionizing radiation emitted by a reactor or accelerator. See also *biological shield* s.v. *BIOLOGICAL *a.*

1933 *Proc. R. Soc.* A. CXLI. 262 The steel shield S prevents the impact of secondary electrons upon the glass walls. **1947** M. D. KAMEN *Radioactive Tracers in Biol.* iv. 93 Protection against γ radiation is best afforded by working with remote control devices behind heavy lead shields at least 2–3 inches thick. **1962** *Newnes Conc. Encycl. Nuclear Energy* 756/2 Such a shield is an iron or barium loaded concrete. The main part of the shield in most reactors, called the bulk or biological shield, is made of such material. **1974** *Encycl. Brit. Macropædia* XIII. 319/1 Typically, a 'core barrel'..is enclosed in a thermal shield, a pressure vessel, a water shield against neutrons, and a blanket of reinforced concrete for gamma-ray absorption.

III. 15. d. A policeman's badge of office. *U.S.*

1903 *N.Y. Evening Post* 29 Oct. 3 The ex-policeman who turned in his shield in September. **1956** 'E. MCBAIN' *87th Precinct* (1959) 19 Why don't you turn in your shield? Become a hackie or something? **1970** E. R. JOHNSON *God Keepers* xv. 167 Nobody said that a detective rating..and a shield provided detectives with all the right answers.

16. b. A shield-shaped centre of a chair-back.

1897 [see *shield-back* adj., sense 19a below].

17. (Earlier example.)

1855 *Poultry Chron.* III. 140/1 (*table*) Toys, or any other Variety [of pigeon] not mentioned above; such as Suabians.., Shields, Swiss, &c.

17*. *Physical Geogr.* **a.** A large, seismically stable mass of Archaean basement rock having the form of a flat or gently convex peneplained platform and usu. forming the nucleus of a continent. Freq. with capital initial in proper names, as *Baltic, Canadian Shield.* [tr. G. *schild* (introduced in this sense by E. Suess *Das Antlitz der Erde* (1888) II. III. ii. 42).]

In quot. 1968 *ellipt.* for *Canadian Shield.*

1906 H. B. C. SOLLAS tr. *Suess's Face of Earth* II. III. ii. 30 The whole of the north-east of America, from the mouth of the St. Lawrence to that of the Mackenzie..belongs to a broad table-land of horizontal Palaeozoic beds, from beneath which the Archaean foundation crops out in the middle of the table-land not unlike a flat shield. This Archaean shield is thus surrounded by a ring of horizontally stratified sediments... It is to the exposed Archaean surface that we give the name of the Canadian shield. **1906** [see *BALTIC *a.* 3]. **1915** C. SCHUCHERT *Text-bk. Geol.* II. xxi. 461 Most of the present continents have been formed around ancient protuberances of the lithosphere, the nuclear lands or shields. **1939** A. K. LOBECK *Geomorphol.* i. 4 Suess showed that certain substantial areas of the earth have always been rigid and unyielding, as, for example, the Canadian and Baltic Shields of America and Europe, the eastern Siberia Shield of Asia, [etc.]. **1963** D. W. & E. E. HUMPHRIES tr. *Termier's Erosion & Sedimentation* ii. 40 The ancient shields, which are also called 'old platforms', are none other than peneplains which have almost attained a final form. **1968** *Beaver* Autumn 14/1 He was with the voyageurs somewhere out in the bush covering some historic route over the Shield. **1969** BENNISON & WRIGHT *Geol. Hist. Brit. Isles* iii. 41 Before the break up of the continents the Lewisian rocks were probably contiguous with the Canadian-Greenland shield, one of the original continental cores. **1971** I. G. GASS et al. *Understanding Earth* iii. 55/1 The oldest continental regions, the Precambrian Shields, are often dominated by igneous rocks such as granite or by highly metamorphosed rocks such as gneiss.

b. The dome of a shield volcano.

1937 *Bull. Volcanologique* I. 94 Composite structures resulting from the accumulations of a series of shifting vents of shield type. **1943** *Amer. Jrnl. Sci.* CCXLI. 241 On the north and northwest its great lava shield abuts against the dormant or extinct volcanoes of Mauna Kea and Hualalai, and on its southeastern slope rests the smaller, younger shield of Kilauea volcano. **1976** *Sci. Amer.* Jan. 33/2 Olympus Mons is enormous by terrestrial standards. Its shield is between 500 and 600 kilometers across, some five times larger than the largest shield on the earth.

IV. 18. a. (sense 15 d) *shield number.*

1972 J. GORES *Dead Skip* i. 8 The policeman..repeated his name, adding his shield number.

c. *shield-backed* adj. (earlier example).

1880 [see *REGENCY 7 b].

19. shield-back *a.*, having a shield-shaped back; **shield volcano** [tr. G. *schildvulkan* (H. Reck 1910, in *Geol. und palæont. Abhandl.* IX. 84)], a volcano having the form of a very broad dome with gently sloping sides, characteristic of the eruption of basic lavas of low viscosity.

1897 K. W. CLOUSTON *Chippendale Period Eng. Furnit.* 65 In the 'shield back' chair, which is Hepplewhite's favourite shape, the shield and its interior ornament making the splat never touch the seat of the chair at all. **1939** *Country Life* 11 Feb. p. xxvii (Advt.), Fine Antique Hepplewhite Mahogany Shield-back Arm Chair with attractively carved splats. **1978** *Morecambe Guardian* 14 Mar. 9/7 Other new items are shield-back dining chairs and matching carver chairs with loose drop-in seat. [**1911** *Geol. Mag.* VIII. 59 The so-called 'Schild' volcanoes in Iceland.] **1911** *Geogr. Jrnl.* XXXVII. 666 These shield-volcanoes rise from a roundish base, with a gently convex surface surmounted by an elevated ring surrounding the crater. **1944** A. HOLMES *Princ. Physical Geol.* xx. 457 Hawaii..has been built up from the sea floor by the coalescence of several shield volcanoes. **1977** *Whitaker's Almanack 1978* 1037/1 The Isla Fernandina is the summit of a very large shield volcano rising from the sea floor to 1,495 metres above sea-level.

b. shield snake, a venomous southern African snake, *Aspidelaps scutatus*, distinguished by a large scale on its head.

1910 F. W. FITZSIMONS *Snakes S. Afr.* iv. 84 This snake is known as the Shield Snake, because it has a large scale on its nose which is partly detached at the sides. **1973** *Stand. Encycl. S. Afr.* IX. 613/1 The shield-snake is egg-laying.

shield, *v.* Add: **1. b.** *Electr.* = *SCREEN *v.* 1 c. Const. *from, against.* Also *absol.*

1922 *Wireless World* 1 July 416/1 The high amplifications possible with multi-stage valve receivers..lead to unexpected results when endeavouring to shield instruments from radio frequency fields. **1935** F. E. TERMAN *Measurements in Radio Engin.* xiv. 342 There are circumstances when it is desirable to shield against electrostatic fields without interfering in any way with the magnetic fields which are present. **1970** J. SHEPHERD et al. *Higher Electr. Engin.* (ed. 2) vii. 226 Sometimes sufficient shielding can be obtained by a few short-circuited copper turns, placed round the object to be shielded in such a direction that the axis of the turns is in the direction of the magnetic field.

shielded, *ppl. a.* **4.** (Further examples.)

1927 *Morning Post* 8 Sept. 10/1 The high frequency stage should be screened, and if one of the new shielded valves is used the circuit will be a simple one, easily handled, and cheaply built. **1970** J. SHEPHERD et al. *Higher Electr. Engin.* (ed. 2) vii. 225 For steady (or static) fields, the only method of achieving shielding is to provide a low-reluctance magnetic path for the stray flux, in such a way that this flux bypasses the shielded point.

shielding, *vbl. sb.* Add: **2.** *Physics.* Material which protects or shields: **a.** against electric and magnetic fields (cf. *SHIELD *sb.* 8 d); also, a shield.

1930 *Proc. IRE* XVIII. 435 The complete exciter unit is enclosed with a metal grill in addition to the individual shielding compartment for the various stages. **1933** *Practical Wireless* 4 Feb. 962/2 A heavy flexible metal shielding fits tightly over these bakelite shells, and through the shells ordinary rubber-covered lead-in wire is threaded. **1975** D. G. FINK *Electronics Engineers' Handbk.* XVII. 31 The Wagner ground connection..can be used in place of shielding at lower frequencies if the utmost precision is not required.

b. against radiation (cf. *SHIELD *sb.* 8 e).

1945 HAWLEY & LEIFSON *Atomic Energy* 185 There was not sufficient radioactive emanation to be dangerous within a radius of fifty feet of the pile; notwithstanding this, precautionary shielding would be necessary in a locomotive unit. **1950** *Chemical Engin. Progress* XLVI. 109/1 To critical size must be added enough shielding to make reactors safe, and the amount of shielding required..is of considerable volume and weight. **1958** W. K. MANSFIELD *Elem. Nucl. Physics* v. 45 The shielding of a reactor will be designed to absorb the γ-rays rather than α and β-rays. **1974** *Encycl. Brit. Macropædia* XIII. 319/1 The shielding of the reactor must keep heat losses and radiation levels external to the reactor down to acceptable levels.

shift, *sb.* Add: **IV. 10. a.** For 'Now *rare*' read 'Now chiefly *N. Amer.*' and add further examples.

1927 M. DE LA ROCHE *Jalna* xix. 250 He pictured her in a fine embroidered shift, curled softly beneath the silk eiderdown. **1929** W. FAULKNER *Sartoris* 177 The flowers you know are all there, in their shifts and with their hair combed out for the night. **1936** M. DE LA ROCHE *Whiteoak Harvest* xxii. 301 She is such a slack creature that I dare say the poor child doesn't own a clean shift.

b. A straight loose dress.

1957 M. B. PICKEN *Fashion Dict.* 293/1 *Shift*,..loose dress hanging straight from shoulders, with fulness closely belted at waistline. **1965** H. L. BROCKMAN *Theory of Fashion Design* v. 95/2 The shift automatically lengthens the figure at the expense of widening it at the waistline. **1975** D. LODGE *Changing Places* v. 177 Girls in kaftans, saris, skinny sweaters, bloomers, shifts, muu-muus.

V. 14. a. *to get a shift on* (colloq.), to get a move on (see MOVE *sb.* 6 in Dict. and Suppl.).

1906 [see *POLE *v.*[1] 8]. **1977** *Times Educ. Suppl.* 21 Oct. 9/2 We could have started certainly a year earlier, even two years earlier if we had got a shift on.

c. *Physics.* A displacement of a spectral line from the expected position or from some reference position; hence, a change of an energy level in an atom, molecule, etc.; *chemical shift*, in nuclear magnetic resonance or Mössbauer spectroscopy, the position of a resonance in the spectrum measured relative to some standard signal, the separation being characteristic of the chemical environment of the resonating nucleus. See also *RED SHIFT.

1884 *Phil. Mag.* XVIII. 161 A shift of the lines towards the more refrangible side of the spectrum. **1897** *Astrophysical Jrnl.* V. 210 Here is certainly a *vera causa* for some shift towards the red in molecules causing light. **1932** *Physical Rev.* XLII. 350 The direction of the shift is again such that Hg^{204} has the highest energy. **1945** R. A. SAWYER *Exper. Spectroscopy* v. 118 Changes in temperature and pressure may lead to serious difficulties in prism spectrographs through broadening and shifts of spectral lines. **1952** *Physical Rev.* LXXXVIII. 1070/1 A shift in the nuclear resonance, known as the chemical shift, is due to the effects of diamagnetism and induced paramagnetism in a molecule. **1961** A. D. THACKERAY *Astron. Spectroscopy* xiii. 186 Interpreted as a radial velocity this shift implies that the nebula in question is running away from us at a speed of slightly over 60,000 km/sec. **1966** *McGraw-Hill Encycl. Sci. & Technol.* VIII. 600a/1 Chemists have become interested in using the Mössbauer effect because of the isomer shift (also called isomeric or chemical shift); this results from the interaction of the electron density..at the nucleus with the nuclear charge. **1970** G. K. WOODGATE *Elem. Atomic Struct.* viii. 154 Since the perturbing states of opposite parity lie a long way away,..one expects the Stark shift of the ground state to be small. **1978** P. W. ATKINS *Physical Chem.* xix. 625 The two methylene protons are in a different part of the molecule; they therefore have a different chemical shift, and come into resonance at another magnetic field.

d. *Philol.* A phonetic change. See also *accent-shift, *consonant-shift, *sound-shift, *stress-shift, *vowel-shift.

[**1875**: in Dict., sense 14 b.] **1894** O. F. EMERSON *Hist. Eng. Lang.* xiv. 241 § 271 The shift from voiceless to voiced in certain positions has taken place since Teutonic times. **1909** O. JESPERSEN *Mod. Eng. Gram.* (1949) I. viii. 231 In most cases the spelling had become fixed before the shift, which..is one of the chief reasons of the divergence between spelling and sound in English... The shift may be represented graphically. **1934** PRIEBSCH & COLLINSON *German Lang.* II. i. 86 A clean cut was made between those dialects which underwent the shift and those which remained unaffected. *Ibid.* 88 The shift from stop to spirant was carried out over the whole High German area.

e. A change of gear in a motor vehicle. *N. Amer.*

1915 V. W. PAGÉ *Questions & Answers* (rev. ed.) xxvii. 446 The clutch must be disengaged before a shift can be made. **1947** R. F. KUNS in Kuns & Plumridge *Automotive Fundamentals: Chassis & Power Transmission* 164 The overdrive shift is made automatically, by simply lifting the foot from the accelerator for about 1¼ seconds.

f. *Chem.* A migration of an atom or group, or of electrons, from one point in a molecule to another, or occas. between molecules, in a chemical reaction.

1932 *Jrnl. Amer. Chem. Soc.* LIV. 3278 The shift of the electron pair includes the atom or group which it holds. **1947** *Ibid.* LXIX. 290/2 On the other hand, the hydrogen atom with its pair of electrons might be transferred by an *inter* rather than an *intra* molecular shift. **1953** C. K. INGOLD *Structure & Mechanism in Org. Chem.* ix. 482 Other rearrangements involve only the shift of a methyl group to an adjoining position. **1968** R. O. C. NORMAN *Princ. Org. Synthesis* xiv. 435 A typical example of a hydride shift occurs in the reaction of a primary aliphatic amine with nitrous acid; e.g. n-propylamine gives isopropanol, together with propylene, and only a trace of n-propanol. **1975** C. J. COLLINS in R. F. Brown *Org. Chem.* xvi. 535*b* Prior to our explanation it was commonly held that all 1,2-shifts—for example, of hydrogen, alkyl, or aryl during Wagner-Meerwein, pinacol, Demjanov rearrangements and the like—took place with inversion of configuration at the migration terminus.

g. *Computers.* The movement of the digits of a word in a register one or more places to left or right, equivalent to multiplying or dividing the corresponding number by a power of whatever number is the base.

1946 *Ann. Computation Lab. Harvard Univ.* I. 73 The first molding is..used for reset and the second to read out the tens digit of the amount of shift in conjunction with the proper molding of the first column... The shift is counted to the right. **1966** *IFIP-ICC Vocab. of Information Processing* 70 Digits shifted beyond the end of the word or register may simply be dropped, or in a cyclic shift (or end-around shift) they may be returned to the opposite end of the word or register in a circular fashion. **1970** O. DOPPING *Computers & Data Processing* v. 80 Sometimes it is necessary to analyze the individual characters of a word. The computer can do this by means of shift instructions. These are instructions for left shift and right shift.

15. b. *Pianoforte.* The mechanism for or act of shifting the keyboard action by means of the soft pedal.

1896 A. J. HIPKINS *Pianoforte* 41 Unless these are directly opposite the strings by a decided shift or return, a snarling quality of tone will be heard. *Ibid.* Up to about 1830 there was a further shift permissible to one string only, the *Una Corda* of Beethoven. **1944** W. APEL

Harvard Dict. Mus. 778/2 Beethoven...not only calls for a gradation in three steps..but even for a gradual execution of the shift: *poco a poco due corde*.

17*. Something which effects a shift. **a.** A mechanism for changing gear in a motor vehicle; a gear-lever. Cf. *gear-shift* s.v. *GEAR *sb.* IV. *N. Amer.*

1914 *Automobile* 9 Apr. 771/2 (Advt.), New electric shift. **1926** F. Scott Fitzgerald *Great Gatsby* vii. 144 'Shall we all go in my car?'..suggested Gatsby... 'Is it standard shift?' demanded Tom. **1968** *Globe & Mail* (Toronto) 13 Jan. 26/1 (Advt.), Mercury Parklane Marauder...radio, bucket seats, floor shift. **1978** J. Irving *World according to Garp* xii. 224 The gear knob of the Volvo's stick shift came off in her hand.

b. = *shift key*, sense 18 below.

1919 H. Etheridge *Dict. Typewriting* 208 It is usual to provide duplicate keys on each side of the keyboard, so that the shift may be operated with either hand. **1936** A. Dvorak et al. *Typewriting Behavior* 2. 260 Really you strike the shift just a tiny fraction of a second before you strike the capital letter. **1957** A. C. Lloyd et al. *Gregg Typewriting for Colleges* 10 A-finger reaches over, to Shift.

c. = *shift code*, sense 18 below.

1957 *Encycl. Brit.* XXI. 886/2 With such a code [as the Baudot code] it is possible to obtain 32 different combinations, 26 of which are assigned to letters of the alphabet, leaving 1 for the idle condition, and 5 for functions such as space, figure shift, letter shift, etc. **1972** *Computers & Humanities* VI. 149 The tape punch would consequently have fewer possibilities than the card punch, if this number of 44 were not doubled by a shift giving an extra punch code to change from lower to upper case, or from upper to lower case. **1980** L. Moore *Foundations of Programming with Pascal* ii. 38 The 5-bit code commonly used by Creed teleprinters had two shift-codes, a 'letter shift' and a 'figure shift'. Each of the remaining thirty codes was mapped to two characters, one belonging to the 'letter' set and the other to the 'figure' set.

17.** *Telegr.* and *Computers.* A change from one set of characters to another; also, a set of characters indicated by any particular shift code.

1913 H. W. Pendry *Baudôt Printing Telegraph System* 2 He adapts therein several elements of the earlier Hughes system—namely, the type-wheel and printing arrangement as well as a similar figure shift device. **1928** A. Williams *Telegr. & Teleph.* ii. 33 The possible number of permutations is thirty-one, but each of these can be made to signify either of two characters by a 'shift' at the receiving end corresponding to the shift key of an ordinary typewriter. **1960** M. G. Say et al. *Analogue & Digital Computers* ix. 265 Such an arrangement is very appropriate in telegraphy, where changes from one shift to the other are not common. **1967** D. G. Hays *Introd. Computational Linguistics* iv. 75 Some of the shifts are capitalization, boldface, superscript, and large. Most alphabets require shifts and diacritics. **1970** O. Dopping *Computers & Data Processing* ii. 41 We say that the characters are in two shifts, a letter shift and digit shift, in the same way as the characters on a typewriter are in two shifts or cases. **1971** T. C. Collocott *Dict. Sci. & Technol.* 1064/1 In teleprinters, one shift is capital letters, the other figures and special signs.

VI. 18. *shift-worker*, *-working* (sense 12); *shift-strap* (sense 10); **shift character, code,** *Telegr.* and *Computers*, a character in a code that indicates that subsequent characters are to be interpreted in terms of a different fount or coding scheme; **shift dress** = sense *10 b; **shift-key,** a key for adjusting the mechanism in a typewriter when characters in a different position on the keys, such as capitals, are to be used; **shift lever** *N. Amer.*, a gear-lever in a motor vehicle; **shift-lock,** a device for holding the shift-key of a typewriter continuously depressed; also *attrib.*; **shift register** *Computers*, a register specifically intended for subjecting data to a shift (sense 14 g above); **shift-round** *colloq.*, reallocation of positions, a move to another position; **shift-sign** *Phonetics* (see quot. 1939); **shiftsman** (see quot. 1921); **shift-stick** *colloq.*, a gear-lever in a motor vehicle; **shift terminator** *Computers*, a character introduced into a string of text to cancel the effect of a preceding shift code; **shift valve,** a valve that moves to produce automatic gear-changes in a motor vehicle.

1967 D. G. Hays *Introd. Computational Linguistics* iv. 75 But there are also 8 shift characters, that influence the style or position of following graphic characters, and a shift terminator. **1970** O. Dopping *Computers & Data Processing* ii. 41 After the letter shift character in the teleprinter code, all the following characters are interpreted as belonging to the letter case until there is a digit shift character, and vice versa. **1967** D. G. Hays *Introd. Computational Linguistics* iv. 70 When we read a shift code, we must remember what shift we are in until receiving another. **1972** *Computers & Humanities* VI. 149 We get 44 characters which may be preceded by either the upper-case or the lower-case shift code. **1980** Shift code [see sense 17* c above]. **1966** Shift-dress [see *Ming *sb.*² c]. **1970** 'D. Halliday' *Dolly & Cookie Bird* iii. 30, I was wearing a high-necked shift dress. **1893** *Manual of Typewriter* i. 15 When the machine in use is one with a single keyboard,—that is to say, one with a shift-key by the depression of which the upper-case characters are brought into play,—the shift-key should be governed by the little finger. **1940** M. Crooks *Home Instruction Course in Touch Typewriting* 56 You may like to note, whilst on the subject of the shift key, that there is an additional key—usually above one of the shift keys—called the 'Shift Lock'. **1980** *Daily Tel.* 4 Nov. 13/4 Beth Porter as shelfmate (archie couldn't work the shift lock) in the roach and the pussycat. **1920** F. B. Scholl *Automobile Owner's Guide* 7 Place the shift-lever into the first-speed slot and let up on the clutch pedal. **1973** R. Naylor *Hungarian Game* xviii. 266 When the engine turned over he jammed the shift lever into reverse and pressed the accelerator. **1899** J. Wardle *Universal Typewriter Man.* 21 Shift lock.—When it is desired to write a large number of capital letters or signs, the Cylinder may be brought forward by means of the Lock Handle, and this action will lock the Cylinder in that position. **1936** M. Crooks *Bk. of Remington Typewriter* iii. 37 The action of the shift lock key is quite simple. **1977** E. Mackay *Typewriting Dict.* 195 The shift key should be depressed by the little finger... If a whole word, heading, sentence, etc., is required in capital letters, the typist should depress the shift lock, which 'locks' the typewriter mechanism. **1950** W. W. Stifler *High-Speed Computing Devices* viii. 209 A multiplier might be designed using the parallel adder and the shift register... The product accumulator is twice the length of the operand registers and is also a shift register. **1975** *Nature* 27 Mar. 366/3 A bubble device consists simply of an assembly of a number of integrated circuits each of which carries magnetically activated tracks, that is, shift registers, along which are driven patterns of bubbles and gaps representing binary data. **1940** J. Reith *Diary* 5 Apr. (1975) v. 244 Cabinet changes tonight...a weird shift-round. a **1974** R. Crossman *Diaries* (1975) I. 611 Thursday, the day of my shift-round. **1939** B. Bloch in H. Kurath et al. *Handbk. Linguistic Geogr. New England* iv. 139 Shift Signs. In order to avoid the necessity of using special symbols for the innumerable shades of sound intermediate between any two of the vowels shown in the diagram..., the phonetic alphabet of the Linguistic Atlas provides shift signs in the form of small arrowheads, which are placed after a vowel symbol to indicate varieties heard as articulated in a higher, a lower, a more advanced or a more retracted position than the vowel denoted by the unmodified letter. **1970** *Publ. Amer. Dial. Soc.* 1968 L. 5 Shift sign, a raised, V lowered, ..are used to show modification of the vowels. **1921** *Dict. Occup. Terms* (1927) § 044 Shifter, shiftsman,..general term for labourers assisting repairers, timberers, etc., in building stoppings and clearing falls of stone. *Ibid.* § 054 Shifter, shiftsman, shiftman..works at night, when mine workers are absent, repairing road-ways, etc. **1924** *Public Opinion* 8 Feb. 127/1 Machinery shall be in charge of a competent shiftsman. **1968** *Autocar* 14 Mar. 25/1 (Advt.), Aussies have better things to do with their arms than give them to a shift-stick. **1975** *Publishers Weekly* 17 Mar. 53/1 Even readers who don't know a shiftstick from a lollipop may find themselves caught up in the pace of this exciting inside-story of a veteran Indy 500 racing-car mechanic. **1922** Joyce *Ulysses* 222 A white petticoat/bodice and taut shiftstraps. **1967** Shift terminator [see *shift character]. **1965** D. G. Hays *Introd. Computational Linguistics* iv. 76 If a whole sentence is in italics, the italic-shift character occurs just once in continuous text, followed by the sentence... shift valve for direct drive, having a modulator valve at one end. **1955** W. H. Crouse *Automotive Transmission & Power Trains* vii. 23 The throttle pressure is applied to the spring end of the shift valve. **1970** *AA Bk. Car* 110/3 A system of brake bands and clutches selected by hydraulic shift valves. **1942** T. K. Djamo *Factory Inspection in Gt. Brit.* vii. 142 The Home Secretary may require certain conditions for the safe-guarding of shift workers. **1973** *Rep. Comm. Future of Broadcasting* (Cmnd. 6753) iii. 25 Shift workers wanting more entertainment during daytime hours. **1937** M. L. Yates *Wages & Labour Conditions in Brit. Engineering* iv. 54 Shift-working was the subject of a separate Agreement between the Employers' Federation and the Amalgamated Engineering Union in 1920. **1963** *Times* 9 May (Suppl. Elect. Power Brit.) p. iii/7 Because our tempo of life is geared to what we regard as orthodox hours, shift working is a burden and one disrupts family life.

f. *Computers.* To move (data) to right or left in a register. Also *absol.*

1946 *Ann. Computation Lab. Harvard Univ.* I. 72 The quotient shift counter..is used to calculate the number of columns the quotient must be shifted to the right upon reading out to the buss in order to conform with the operating decimal position. **1947** A. W. Burks et al. in *Coll. Wks. J. von Neumann* (1963) V. 44 We do not consider multiplication by 2 as a true product since we will have a facility for shifting right or left in one or two pulse times. **1966** *IFIP-ICC Vocab. of Information Processing* 70 Digits shifted beyond the end of the word or register may simply be dropped. **1968** Fox & Mayers *Computing Methods for Scientists & Engineers* ii. 21 The first operation 'shifts' a_2 to the right by b_1-b_2 places.

16. b. Also, to spend (money).

1923 E. P. Oppenheim *Inevitable Millionaires* xiv. 148, I should trip it to Monte. That's the place to shift the shekels.

23. To move, to travel, esp. quickly; to get a move on. Cf. *SHIFT *sb.* 14 a.

1946 H. Arlen '*Piracy*' i. 11, I am..going to leap on my motor-bike and shift like hell to London. **1968** A. Diment *Bang Bang Birds* ix. 170 The speedo needle clawed its way up..to finally flicker over the 180 mark. Kilometres an hour of course but we were still shifting. **1970** M. Kenyon *Whoop-for Welcome* v. 37 You'll have time for a bite at Murphy's if you shift.

shifta (ʃi·ftä). Pl. **shifta, shiftas.** Also with capital initial. [Somali *shifto* bandit, ad. Amharic.] A Somali bandit or guerrilla, operating mainly in northern Kenya. Also *attrib.*

1950 *Times* 18 Feb. 5/3 Mr. H. V. Rose, who is in charge of the anti-Shifta operations, has reported a steady decline in the number of incidents this month, as the number of contacts made by 'ferret' forces with the Shifta bands has increased. **1959** *Times* 29 June 11/7 The shifta, although they may relieve the traveller of his goods, seldom harm his person. **1967** *Economist* 27 May 906/1 The Kenyan government warned its Somali neighbours that if they went on supporting the ethnic Somali shifta terrorists (freedom-fighters, the Somalis call them) in Kenya's north-eastern province, Kenya might hit back. **1980** *Daily Tel.* 8 Jan. 1/8 She might have been killed by marauding Somali bandits called 'Shifta', who are blamed for much of the game poaching in Northern Kenya.

shiftable, *a.* **2.** Delete † *nonce-use* and add earlier example.

1832 Let. 17 Apr. in J. Constable's *Corr.* (1962) I. 269 Abram would have accompanied her on Monday, but as business required his going this day, we thought it quite unnecessary to alter his plans for so *shiftable* a young lady.

Hence **shiftabi·lity,** the ability to be shifted.

1951 [see *SELF-LIQUIDATION]. **1972** *Linguistic Inquiry* III. 377 (*heading*) On the shiftability of past participles. **1976** *Language* LII. 39 The shiftability of the NP has nothing to do with whether the containing prepositional phrase has or has not originated in a relative clause.

shifter. Add: **6. b.** The gear-change mechanism or control in a motor vehicle. Freq. *attrib. N. Amer.*

1910 J. E. Homans *Self-Propelled Vehicles* (ed. 7) xxix. 378 The arm of the sliding gear shifter meets a raised portion of the reverse shaft. **1915** Hobbs & Elliott *Gasoline Automobile* iii. 67 These two sets [of sliding gears] are operated by two shifter yokes which lead to the gear control lever in the car. **1920** V. W. Pagé *Model T Ford Car* (new ed.) vii. 341 To stop the speeder you reverse gear and then shifter to the horizontal position. **1946** W. H. Crouse *Automotive Mech.* xvii. 390 A close-up view of the gear selector and shifter rods. **1973** G. V. Higgins *Friends of Eddie Coyle* xiii. 78 'How come you got the automatic?...I'd want the Hurst shifter in it.' 'You wouldn't want it once you started buying clutches for it.'

shift, *v.* Add: **III. 12. c.** To change (gear), move (a gear lever). Also *intr.*, to change *from* one gear *into* another; *to shift up* or *down*, to engage a higher or a lower gear. Also *fig.*

1910 J. E. Homans *Self-Propelled Vehicles* (ed. 7) xxix. 381 In shifting from high to low gears, all intermediate speeds were engaged. *Ibid.* xlv. 624 To shift the transmission lever for the speed changes, if the transmission is of the selective type, the two movements may offer some difficulty to the beginner. **1946** W. H. Crouse *Automotive Mech.* xvii. 388 Let us shift into second and note the actions that take place. *Ibid.* 389 Synchromesh devices come into use when gears are shifted into second and third. **1961** Webster, *Shift gears*, to make a change from one method, tempo, or approach to another. **1962** J. D. MacDonald *Girl* viii. 99 She drove with her brown hands high on the wheel... She shifted up and shifted down. **1963** A. Miller *Incident at Vichy* 32 For some of us it's difficult to shift gears and go into reverse. **1969** *New Yorker* 6 Sept. 105/2 The Rumanians, having barely paused to shift their ideological gears, began building the...congress. **1970** D. MacKenzie *Kiss Combust* (1971) 13 He drove out of Palamos... He shifted into drive and settled back. **1975** R. Hayes *Hungarian Game* iii. 319, I..shifted from second to third and..let my hand linger a while on the gear lever's mahogany knob. **1973** *Sci. Amer.* Apr. 11/2 (Advt.), Once on the valley floor I shifted up into fourth. **1973** *Time* 16 Apr. 53/1 'We're shifting gears,' says Vail, 'and hiring guys with a track record of seven, eight, nine years' experience.' **1976** H. Nielsen *Brink of Murder* xv. 132 Simon switched on the ignition and shifted into reverse.

shifting, *vbl. sb.* Add: **2. c.** (Earlier example.)

1833 Poe in *Southern Lit. Messenger* (1835) Dec. 35/2, I therefore thought proper to contrive a hiding-place in the hold. This I did by removing a small portion of the shifting boards. **1885** F. Gordon *Pyotshaw* 51 Ma guid shiftin' claes clean spilt. **1957** *Scotland's Mag.* June

46 The first of the noisy band of workers came 'skailing' out of the pithead baths, in their shifting clothes. **1913** D. H. Lawrence *Sons & Lovers* viii. 199 She wiped him in a desultory fashion, and went upstairs, returning immediately with his shifting-trousers. When he was dried he struggled into his shirt.

shifting, *ppl. a.* Add: **1. b. shifting agriculture** = **shifting cultivation*; **shifting cultivation,** any of several forms of agriculture in which an area of ground is cleared of vegetation and cultivated for a (usu. small) number of years and then abandoned because of nomadic habits or deliberate fallowing or because the yield of crop has become uneconomic, when cultivation is begun elsewhere; hence **shifting cultivator**; **shifting keyboard** *Pianoforte*, a keyboard action of a grand piano, etc., which is moved by the use of the soft pedal; so **shifting pedal**; **shifting spanner,** an adjustable spanner.

1934 W. Fitzgerald *Africa* iii. v. 354 Cocoa-planting necessitated the abandonment of the old system of shifting agriculture. **1973** W. T. W. Morgan *E. Africa* iv. 51 In general absence of fertilizers and lacking any complete rotation system, a system of fallowing was necessary. In the tropics this system has come to be referred to as 'shifting agriculture'. **1922** Shifting cultivation [see *CHENA]. **1952** P. W. Richards *Tropical Rain Forest* xvii. 378 The destruction of the primary forest which gives rise to secondary successions may take place in different ways and for various reasons. By far the most important cause of destruction up to the present has been the system of shifting cultivation (the ladang system of Malaysia and the taungya system of Burma) which is practised by nearly all the native peoples of the tropics. **1962** J. H. Galloway in Blakemore & Smith *Latin America* viii. 382 Much of the agriculture of the Andes...is shifting and still primitive, unproductive shifting cultivation. **1963** K. J. Pelzer *Pioneer Settlement in Asiatic Tropics* ii. 16 The shifting cultivator does not use the same piece of land every year; instead, he kills or cuts down at regular intervals—every year, every other year, or every third year—the trees of a small forest patch. **1979** *Nature* 16 Aug. 533/1 Shifting cultivators fell and burn forest land, then plant food crops and raise animals and later move on when soil fertility drops. **1896** A. J. Hipkins *Pianoforte* 20 In many upright pianos, however, although some are made with shifting keyboards..a soft pedal is contrived by mechanically dropping a strip of cloth...between the hammers and the strings. **1922** A. H. Lindo *Pedalling in Pianoforte Music* ii. x. 143 The mechanism of the shifting keyboard, which is fitted to many Uprights and to nearly all Grands, is preferable. **1880** Grove's *Dict. Mus.* II. 668/1 The shifting pedal ('Verschiebung')..because the whole mechanism, keyboard and hammer, is shifted slightly to the right when the left pedal is pressed down. **1829** *Mechanics' Mag.* 31 Jan. 413/1, I send you..a plan of a new shifting-spanner, which answers better than the wedge-spanner. **1936** J. Gutteridge *Little Country* ii. 38 I'll lend you the shifting spanner.

shifty, *a.* Add: **1.** (Earlier U.S. examples.)

1783 *Maryland Jrnl.* 18 Feb. 3/3 Ran away...a Negro Man, named Pompey,..very artful and shifty. **1838** 'Texan' *Mexico v. Texas* 217 Flambeau, who was extremely shifty, soon built up a nice little booth.

2. b. Also *Comb.*, as *shifty-eyed* adj.

1922 H. Jenkins *John Dene of Toronto* i. 11 A shifty-eyed little man. **1977** M. Babson *Lord Mayor of Death* xiv. 92 He must have looked a right shifty-eyed bastard.

4. Delete *rare* and add further examples.

1907 J. G. Millais *Newfoundland* xii. 244 The wind had dropped, and showed signs of being shifty. **1976** *Yachts & Yachting* 20 Aug. 177/1 Weatherwise the 116-strong fleet had mostly light and shifty winds.

Shiga (ʃi·gä). *Bacteriol.* The name of Kiyoshi *Shiga* (1870–1957), Japanese bacteriologist, who discovered this bacterium in 1898, used *attrib.* and in the possessive to denote the Gram-negative bacterium *Shigella dysenteriæ*, which causes dysentery in man, and the toxin produced by it.

1900 *Philad. Med. Jrnl.* VI. 425/1 They proceeded upon the false assumption that Shiga's microorganism was a variety of B[acillus] coli communis. *Ibid.*, Comparison of the Eberth-Gaffky and Shiga bacilli show the criterions of difference are by no means numerous. **1946** *Nature* 10 Aug. 207/2 It supervised the production of typhus vaccine and Shiga toxoid, and made suitable recommendations to the Department of National Defence concerning their use. **1947** *Ann. Rev. Microbiol.* I. 313 The Shiga bacillus is rare except in the Middle East and in India. *Ibid.* 316 There is no evidence that the Shiga toxin plays a significant part in producing the pathology of dysentery. **1976** Higginson & Gillies *Path. in Tropics* (ed. 2) vi. 318 Shigella...Subgroup A: Ten antigenically distinct serotypes, including Shiga's bacillus.

Shigella (ʃige·lä). *Bacteriol.* Also **shigella.** Pl. **-ellæ, -ellas.** [mod.L., f. *SHIG(A + L. *-ella* (see -EL²).] **1.** A member of the genus of Gram-negative, rod-shaped bacteria so called, which includes some causing dysentery in man and other animals.

[**1919** Castellani & Chalmers *Man. Trop. Med.* (ed. 3) xxxvi. 934 Non-motile—Genus 3, *Shigella* Castellani and Chalmers, 1918.] **1937** M. Frobisher *Fund. Bacteriol.* xxx. 316 A number of the shigellas..(also commonly known as dysentery bacilli) cause intestinal disturbances. **1969** *New Yorker* 11 Oct. 123/3 A shigella infection produces a rather distinctive kind of damage that can be detected by microscopic examination. **1977** *Lancet* 19 Mar. 609/2 Faeces from 11 children and 3 adults who were examined for salmonellæ, shigellæ, and *Staphylococcus aureus* with negative results.

2. = *SHIGELLOSIS. Also *Shigella dysentery*.

1963 G. Maxwell *Rocks Remain* ii. 21 Immediate veterinary tests showed the presence of liver fluke... Even further tests showed the presence of Shigella dysentery [*sic*], we decided to abandon interest in the fishes altogether. **1973** *Daily Colonist* (Victoria, B.C.) 20 May 25/1 Belinda Manybears, three months old, was suffering from shigella, a type of dysentery, at the time of her death. *Ibid.* 1 July 32/2 An intestinal disease doctors have tentatively diagnosed as Shigella swept through the luxury ship. **1977** Shigella dysentery [see *SHIGELLOSIS].

shigellosis (ʃigelǭu·sis). *Path.* [f. prec. + -OSIS.] Infection with or a disease caused by shigellæ.

1944 Hardy & Watt in *Jrnl. Amer. Med. Assoc.* 22 Apr. 1179/1 In line with the accepted use of 'brucellosis' for all Brucella infections, we recommend the adoption of 'shigellosis' for all infections due to pathogenic varieties of shigella. **1977** *Proc. R. Soc. Med.* LXX. 374/1 Shigellosis in monkeys closely resembles Shigella dysentery in humans.

Shiho, var. *SAHO *sb.* and *a.*

shih-tzu (ʃī·tsu). Also **shitzu.** [ad. Chin. *shīzigǒu*, f. *shī* lion + *zǐ* son + *gǒu* dog, formerly transliterated *shih-tzu kou*.] A small long-coated dog of the breed so called, originally developed in China, often tan or grey and white in colour, with long ears and a tail curling over the back. Also *attrib.*

1921 V. W. F. Collier *Dogs China & Japan* iii. 52 These books [sc. the imperial dog-books]..portray dogs closely resembling the 'Pekingese' type, as also the 'Shih-tzu' dog and the 'Pug'. **1934** *Kennel Gaz.* May 385/3 It was decided that dogs might be registered under the heading of A.O.V. Shih-Tzu, and that those now registered as Apso could be altered without charge. **1948** C. L. B. Hubbard *Dogs in Britain* xx. 328 The Shih Tzu ..is a close relative of the Lhasa Apso with which it has often been confused. **1960** *Guardian* 9 Nov. 10/6 Who, ever heard of a shih-tzu? **1969** *Queen* 17 Jan. 46/1 She sits..softly scratching the tummy of one of her Shitzu dogs. **1979** A. Chisholm *Nancy Cunard* xxii. 332 Perhaps she had tripped..trying to save one of the Shih Tzu puppies.

Shiism. (Further examples.)

1919 P. M. Sykes *Hist. Persia* II. xlvii. 43 Persia, where Shiism is the official religion. **1964** W. Thesiger *Marsh Arabs* vi. 43 Shiism had started as a political movement, designed to advance the claims of Ali and his descendants to the Caliphate... In time, Shiism split Islam as decisively as the Reformation divided the Catholic Church. **1982** *Times* 31 May 6/1 From the Gulf to the Levant, there are 2,000 miles of Shiism, broken only by the irritation of Iraq's survival.

shiitake (ʃi,ita·ke). [Jap., f. *shii* a name used for several evergreen trees + *take* mushroom.] A mushroom, *Lentinus edodes*, of the family Agaricaceæ, cultivated in Japan and China on logs from various trees of the family Fagaceæ, esp. *Castanopsis cuspidata* or *Quercus* species. Also *attrib.*

1877 *Grevillea* V. 103 The Shii-take species..have this peculiar excellence, that though they are all but tasteless in their raw state, when they are dried they have an extremely fine flavour. **1925** *Bot. Mag. Tokyo* XXXIX. 319 The best and most common mushrooms in Japan are shii-take and Matsu-take. **1936** *Nature* 31 Oct. 746/2 Production of the shiitake has little horticultural similarity to mushroom growing. **1953** J. Ramsbottom *Mushrooms & Toadstools* vii. 74 The cultivation of shii-take in Japan is believed to date back over two thousand years. **1962** R. Singer *Mushrooms & Truffles* v. 139 Logs with Shiitake mycelium were dragged to a suitable site. **1975** J. Grigson *Mushroom Feast* 262 On account of the...laborious method of cultivation, shiitake are more expensive even on their home ground.

Shijō (ʃi·dʒǭu). [The name of the street in Kyōto, Japan, where the founder lived.] Used *attrib.* to designate a school of Japanese painting.

1884 Satow & Hawes *Murray's Handbk. Japan* (ed. 2) 97 The Shijō art was a compromise, retaining Chinese perspective and ignoring the laws of chiaroscuro, but copying details of form in flowers and animals with remarkable fidelity. **1902** *Encycl. Brit.* XXIX. 717/1 Amongst the associates of the Shijō master was the celebrated Ganku. **1970** *Oxf. Compan. Art* 696/2 A subdivision of the Maruyama School called the Shijo School.

shikar, *sb.* Add: Phr. *on shikar*, on a hunting expedition, out hunting.

1944 J. Corbett *Man-Eaters of Kumaon* 186 Of all the men I have been on shikar with Ibbotson is by far and away the best. **1955** *Times* 14 May 10/5 His service in the Indian Army and his periods of leave, which were almost invariably spent on *shikar*, developed his eye for country and his taste for natural history.

‖ **shikara** (ʃikā·rā). Also **shikarah, shikari.** [Hind.] A long, swift boat used in Kashmir. Also *attrib.*

1875 F. Drew *Jummoo & Kashmir Territories* viii. 181 A *shikāri* is the sort of boat that is in daily use with the English visitors; a light boat, manned..by six men, it goes at a fast pace. **1893** E. F. Knight *Where Three Empires Meet* iii. 35 Leaving our slow doongahs to follow us, we hailed some of the gondolas of Srinagur; long, swift canoes known as *shikarahs*, in which we reclined luxuriously on soft cushions. **1933** *Discovery* Nov. 348/2 A *shikara* or river taxi. **1964** V. S. Naipaul *Area of Darkness* iv. 103 Their *shikara* boats were a cluster of red and orange awnings and cushions; and in *shikaras* we were ferried over to the houseboats. **1981** S. Rushdie *Midnight's Children* i. 14 The thaw had come rapidly... Many of the small boats, the shikaras, had been caught napping. *Ibid.* 16 He floated past the shikara moorings.

shikara, var. *SHIKHARA.

shikari. Now the usual form of SHIKAREE.

1881 *Encycl. Brit.* XII. 741/1 Rewards are given by government to native *shikāris* for the heads of tigers. **1885** [see SHIKAREE a]. **1907** [see SHIKAREE b]. **1936** W. H. Saumarez Smith *Let.* 21 Nov. in *Young Man's Country* (1977) ii. 43 When they find that I am neither a Blue, nor a brilliant horseman, nor an experienced shikari, nor a bridge-player, they may think twice about offering the appointment to me. **1964** R. Perry *World of Tiger* i. 8 Peacocks and tigers live together, say Indian shikaris, because the peacock is a thirsty bird and the tiger is a thirsty animal. **1971** *Illustr. Weekly India* 11 Apr. 45/1, I had become a keen shikari; there were few better areas than the country 80 miles round Delhi for small game. **1977** *New Yorker* 27 June 28/2 We're going to use you for bait, as a decoy. The way shikaris stake out a goat in India.

‖ **shikasta** (ʃikæ·stā). Also † **shekest(h)eh, shikast, shikasteh,** etc. [Pers., lit. 'broken'.] A late cursive Persian script.

1771 W. Jones *Gram. Persian Lang.* 15 As to the Shekesteh, it is very irregular and inelegant, and is chiefly used by the idle Indians, who will not take time to form their letters perfectly. **1849** F. Madden tr. *Silvestre's Universal Palæogr.* I. 52 For private affairs and official papers, the writing is generally careless and inelegant, destitute of its diacritical marks, and thence named *shekestheh*, or broken. **1889** Sachau & Ethé *Catal. Persian Manuscripts in Bodl. Libr.* 186 Ff. 51–67 is partly written in Shikasta.., partly in Nasta'līk. **1901** Kipling *Kim* xiv. 364 He..tore a leaf from a note-book, and.. wrote in gross Shikast—the script that bad little boys use when they write dirt on walls. **1954** A. F. L. Beeston *Catal. Additional Persian Manuscripts in Bodl. Libr.* 4/2 Ff. 604–5 are pieces of a letter in shikasta. **1966** Hosking & Meredith-Owens *Handbk. Asian Scripts* ii. 20 The last phase in Persian writing was the evolution of the *Shikasteh* (literally 'broken') script from Nasta'līk. **1970** G. Unwin tr. *Jensen's Sign, Symbol & Script* xi. 331 For correspondence the Persians invariably employ the Sīkästä script.

shikepoke, var. SHITEPOKE in Dict. and Suppl. **shiker,** var. *SHICKER *a.* and *sb.*

‖ **shikhara** (ʃi·khärā). Also **shikara, śikhara, sikr(a).** [Skr. *śikhara*, point, peak, spire.] A pyramidal tower on a Hindu temple, sometimes having convexly curved sides.

1829 J. Tod *Ann. & Antiquities Rajast'han* I. 670 The pinnacle or *sikra* rises, like the crown of the Hindú Cybele. **1838** *Penny Cycl.* XII. 238/1 The body of the temple, or sanctuary,..over which rises a pyramidical *sikr*, or roof. **1891** J. Fergusson *Hist. Indian & Eastern Archit.* II. ii. 221 The towers or spires called Sikras.., which invariably surmount the cells in which the images are placed. **1927** A. Coomaraswamy *Hist. Indian & Indonesian Art* III. 80 In more characteristic examples in the Ganges valley the *śikhara* and cella together form a tower. **1930** K. T. Shah *Splendour that was 'Ind* viii. 151 Prof. Macdonell is of the view that the *Shikhara* is a natural evolution of the Buddhist Stupa. Later critics of India's art treasures are inclined, however, to hold that the *Shikhara*, with all the symbolism of which it was the concrete and complex expression, was introduced in India by the Vedic Aryans. **1959** Hooykaas & Christie tr. *Frederic's Indian Temples & Sculpture* 297 The temple itself has a *sikhara* of Khajuraho type. **1972** 'E. Peters' *Death to Landlords* xi. 158 On the outermost platform of rock, its *shikhara* tapering into the air..stood the modern white memorial. **1974** *Daily Colonist* (Victoria, B.C.) 8 Sept. 13/6 But the most noteworthy aspect of the Khajuraho temple is the shikara, or spire. **1977** *Jrnl. R. Soc. Arts* CXXV. 579/1 In the larger temples the *śikhara* tends to be..surrounded by smaller subsidiary *śikharas*.

shikimi (ʃikī·mi). Also **skim(m)i.** [Jap.] A small evergreen tree, *Illicium anisatum*, the Japanese anise, belonging to the family Illiciaceæ, native to Japan and Korea, and bearing aromatic leaves and fragrant white or yellow flowers followed by star-shaped fruits; often associated with funeral rites. Also *attrib.*

1727 J. G. Scheuchzer tr. *Kæmpfer's Hist. Japan* v. xv. 598 He had..a large tub of water standing by him..and some skimmi branches lying by it. **1881** *Jrnl. Chem. Soc.* XL. 918 To this substance..the author [sc.

J. F. Eijkman] gives the name of 'shikimine', from the Japanese name of the fruit 'shikimi'. **1889** J. J. Rein *Industries Japan* 136 The fruits of the Skimi, which is consecrated to Buddha and therefore much grown about Buddhist temples and cloisters, made a great stir some time ago. They came to market as a spice, instead of the Star anis, which they closely resemble, and turned out to be poisonous. **1896** L. Hearn *Kokoro* iii. 43 A vase containing *shikimi*—that sacred plant used in the Buddhist ceremony of making offerings to the dead. **1976** E. H. Walker *Flora of Okinawa & Southern Ryukyu Islands* 472 Shikimi, a variant of *ashiki-mi*, bad (i.e. poisonous) fruits. In Buddhist ceremonies the leaves are burned as incense.

shikimic (ʃiki·mik), *a. Biochem.* [ad. F. *shikimique* (J. F. Eijkman 1885, in *Recueil des Travaux chim. des Pays-Bas* IV. 49), f. Jap. *shikimi* Japanese anise (from which it was first isolated): see prec., -IC.] *shikimic acid*: a hydro-aromatic acid, $C_6H_6(OH)_3(COOH)$, which is formed in many bacteria and higher plants as an intermediate in the synthesis of phenylalanine, tyrosine, and other aromatic compounds from aliphatic precursors.

1886 *Jrnl. Chem. Soc.* L. 95 Shikimic acid, $C_7H_{10}O_5$,..is a white, crystalline compound, insoluble in alcohol, ether, and chloroform, but readily soluble in water. **1953** Fruton & Simmonds *Gen. Biochem.* xxxi. 742 It would appear..that at least four aromatic compounds can be formed from carbohydrate by mechanisms that involve shikimic acid as an intermediate. **1978** *Nature* 20 July 216/2 The exact nature of the bracken fern 'carcinogen' remains elusive and although various chemicals, such as shikimic acid, which have been extracted from bracken show some oncogenic activity, the major compound responsible has not yet been identified.

shikker, -ur, varr. *SHICKER *a.* and *sb.*

shikkered, var. *SHICKERED *a.*

‖ **shiksa** (ʃi·ksä). Also **shicksa, shiksah, shikse(h).** [Yiddish *shikse*, ad. Heb. *šiqṣâ*, f. *šeqeṣ* a detested thing + *-â* fem. suff.] In Jewish speech, a gentile girl. Also *attrib.* or as *adj.* Cf. *SHICKSTER.

1892 I. Zangwill *Childr. Ghetto* I. I. vi. 158 We must keep a *Shiksah* to attend to the Shabbos fire. **1928** *Daily Express* 21 Feb. 9/2 There is a suggestion that he has fallen in love with a 'shiksa' (a Christian girl), played by May McAvoy, the 'Ben Hur' star. **1930** E. Ferber *Cimarron* xi. 183 His deep-sunk eyes looked at them. *Shicksas.* **1959** M. Levin *Eva* 11 How often as children had we..watched our *shikseh* maids at their prayers, or in moments of closeness with the *shiksehs*..fingered their crosses. **1963** M. McCarthy *Group* xiv. 318 An Orthodox Jew can't marry a shiksah... They frown on exogamy. **1964** D. Gray *Devil wore Scarlet* iii. 26 Daisy isn't Jewish, of course. She's a Shicksa. **1969** L. Michaels *Going Places* 87 Shikse blonde or purple eggplant, she was his wife. **1978** J. Krantz *Scruples* viii. 226 His mother, a lady of the old school, had repeatedly and solemnly warned him that there is a yellow-haired, blue-eyed shiksa lying in wait for every good Jewish boy. **1979** R. Jaffe *Class Reunion* II. iv. 156 She had heard stories that young doctors had affairs with shiksa nurses.

shilajatu, var. *SILAJIT.

Shilha (ʃi·lhā). Also **Shilh, Shilhah, Shleuh, Shluh.** [Native name.] (A member of) a Berber people of southern Morocco; also, the language of this people. Also *attrib.* or as *adj.*

1713 S. Ockley *Acct. S. W. Barbary* i. 28 It is about 30 days journey distant from Macquanes... They also differ from the other Moors in their language, and have a peculiar dialect to themselves, which they call *Shilhah*. **1841** G. Borrow *Zincali* I. I. vii. 113 The tribes who speak the Shilha language, and who are the descendants of the ancient Numidians. *Ibid.* 118 The sect of Sidi Hamed au Muza... Their language is the Shilhah, or a dialect thereof. *Ibid.* II. III. 117 The coast of Northern Africa, where only Arabic and Shilhah are spoken. **1882** [see *KABYLE]. **1921** E. Sapir *Language* 73 The Hamitic languages of Northern Africa, e.g. Shilh. **1951** W. Blunt *Black Sunrise* iv. 42 The Berber-speaking inhabitants can be further subdivided into Berbers proper, Shluhs, etc. **1972** E. A. Nida *Bk. Thousand Tongues* (ed. 2) 389/2 Shilha is a Berber language spoken in several dialects throughout southern Morocco. Most Shilha men are bilingual in Arabic and their mother tongue. **1973** A. Adam in Gellner & Micaud *Arabs & Berbers* III. xvii. 325 *Tachelhit*, the language of the Shleuh, who inhabit the western High Atlas, the Anti-Atlas and the plains or valleys of southwestern Morocco. **1976** K. L. Brown *People of Salé* III. viii. 153 Generally the term used in Salé for those speakers of Berber languages, whether they came from the Rif, Middle Atlas, High Atlas, Anti-Atlas or the valleys of southern Morocco, was Shluh.

shill (ʃil), *sb. slang* (chiefly *N. Amer.*). [Perh. abbrev. of *SHILLABER.] A decoy or accomplice, esp. one posing as an enthusiastic or successful customer to encourage other buyers, gamblers, etc.

1916 *Editor* 2 Dec. 518/2 *Shill, copper*:—One who leads the others by patronizing a show or game. **1926** *Amer. Mercury* Dec. 466/1 A wrestler..offered to throw anyone for 500 smacks and a couple of shills accepted his defy. **1935** H. Davis *Honey in Horn* xv. 231 She had often thought of renting him out as a shill for some tent-show evangelist. **1955** T. Sterling *Evil of Day* xxi. 208, I used to be a shill in a Reno gambling club. **1971** J. Gray *Red Lights* vi. 136 The commonest trap was for a shill to haunt Ninth Avenue disguised as a farm hand. **1978** M. Puzo *Fools Die* ii. 19 As a shill she played with casino money... She was subject not to fate but to the fixed weekly salary she received from the casino.

shill (ʃil), *v.*³ *slang* (chiefly *N. Amer.*). [f. *SHILL *sb.*] **1.** *intr.* To act as a shill.

1914 L. E. Jackson *Vocab. Criminal Slang* 75 To 'shill' is to act in the capacity of a hired criminal. **1928** *Amer. Speech* III. 376 *Shill*, to boost for the auctioneer. **1948** F. Brown *Dead Ringer* 156 She was going to shill on Walter's wheel. **1965** H. Gold *Man who was not with It* xxv. 236 It's how to get the audience... I shilled for my wife. **1975** *Weekend Mag.* (Montreal) 11 Jan. 9 Canadian advertisers are confined mainly to hockey players when they're looking for an athlete to shill for them.

2. *trans.* To entice (a person) as a shill; to act as a shill for (a gambling game, etc.).

1974 R. B. Parker *God save Child* xxii. 150 Doctor Croft was the one who shilled old Fraser Robinson onto Vicki's scam. **1978** M. Puzo *Fools Die* x. 96 Diane, the blonde that shills baccarat.

shillaber (ʃi·lābəɹ). *slang* (chiefly *N. Amer.*). [Origin unknown.] = *SHILL *sb.*

1913 *Collier's* 6 Dec. 29/2 The business men turned out to be 'shillabers', if you know what 'shillabers' are. **1924** G. Bronson-Howard *Devil's Chaplain* vii. 111 One time 'ballyhoo' and 'shillaber', proprietor of 'Chief Bigspoon's..' medicine show. **1940** *Amer. Speech* XV. 122/1 *Shill* or *shillaber*, an accomplice who plays a confidence game so that the *mark* sees him win.

shilling. Add: **1. a, 2. a.** No longer in official use after the introduction of decimal coinage in 1971, but still occas. used to denote five new pence or the five-pence piece.

The coin itself was allowed to circulate for some time after decimalization, alongside the new (and equivalent) five-pence piece, which it resembled in shape, size, weight, and composition.

1974 'J. le Carré' *Tinker, Tailor, Soldier, Spy* i. 18 Spikely discovered..a draft of the next day's examination paper, and rented it to candidates at five new pence a time. Several boys paid their shilling.

b. Also used as a unit of currency (representing variously 12 pence and 100 cents) in other countries, as Kenya, Uganda, Malta, etc.; *freq.* preceded by the name of the issuing country; also, the coin itself.

1921 W. S. Churchill in *Hansard Commons* 30 May 596 As recommended by the Currency Committee appointed in Kenya in February, 1921, the standard coin will be, not a florin, but a shilling... Rupee contracts.. will be construed at the rate of two shillings to one rupee. **1927** W. McG. Ross *Kenya from Within* xii. 208 The new scheme..was that both florin and rupee should disappear, the shilling be introduced and all existing cental coins be degraded, by edict, to half their value. **1969** *Times* 16 Sept. (Somali Republic Suppl.) p. v/3 The internal value of the Somali shilling has..been relatively stable. **1977** *Times* 24 June 14/8 On the free market in Kenya.. 100 Uganda shillings usually bring no more than 20 Kenya shillings.

6. (sense 1 b) *shilling bill*; *shilling gallery* (earlier example); **shilling mark** *Typogr.* = SOLIDUS¹ 2.

1976 K. Thackeray *Crownbird* viii. 161 Priest..tucked some hundred shilling bills into his pocket. **1801** *Monthly Mirror* June 421 He grins and looks broad nonsense with a stare, to the vast delight of the shilling gallery. **1888** C. T. Jacobi *Printers' Vocabulary* 123 *Shilling mark*, the sign thus / which was used in old books as a 'scratch comma'. **1904** Murray & Bradley *Hart's Rules for Compositors* (ed. 15) 29 The diagonal sign / or 'shilling-mark'.

Shilluk (ʃilu·k). Also † **Chillouk,** † **Shelluh,** † **Shelook, Shillook.** [Native name.] The name of a Sudanese people dwelling mainly on the west bank of the Nile: a member of this people; also the Nilotic language of this people. Also *attrib.* or as *adj.*

1790 J. Bruce *Travels to discover Source of Nile* IV. VIII. ix. 458 This race of negroes is, in their own country, called Shillook. They founded Sennaar. **1799** W. G. Browne *Travels in Africa, Egypt, & Syria* App. II. 453 Shillûk is a town of idolaters... The name *Shillûk* is not Arabic, and its meaning is unknown.— When asked concerning their name or country, the people reply *Shillûk*. **1832** W. M. Leake in *Jrnl. R. Geogr. Soc.* II. 26 On the twelfth day they reached the first island of the Shillúks. **1835** *Jrnl. R. Geogr. Soc.* V. 42 Twelve hours farther brought them to the first island of the Shelooks. **1840** *Penny Cycl.* XVI. 232/1 The first island of the Shilluks.. is not far from Aleis. **1873** E. E. Frewer tr. *Schweinfurth's Hrt. Africa* I. 261 The jet-black Shillooks, Nueir, and Dinka, native of the dark alluvial flats, stand out in marked distinction to the dwellers upon the iron-red rocks. **1894** [see *NUER]. **1913** *Rep. Brit. Assoc.* 633 Dr. Seligmann's discoveries among the Shilluks of the Nile Valley. **1921** E. Sapir *Language* 80 *Shilluk*, one of the languages of the headwaters of the Nile. **1927** *World Domin.* Oct. 319 All travellers note the Shilluk style of hairdressing. **1927** *Times* (Weekly ed.) 29 Dec. 30/1 Across on the east bank you will see the Shilluk. **1949** [see *NUBA]. **1964** E. A. Nida *Toward Sci. Transl.* iii. 51 Shilluk uses 'break' only with objects such as wood. **1973** *Times* 27 Mar. (Sudan Suppl.) p. viii/3 (*caption*) A Shilluk tribesman with a traditional musical instrument in the Southern Sudan. **1976** D. Topolski *Muzungu* iv. 56 The Shilluk repeat the cutting process from quite an early age.

shim, *sb.*² Restrict *local* to senses 1–4 in Dict. and add: **5. a.** (Earlier and later examples.)

1860 Clark & Colburn *Recent Practice Locomotive Engine* 62/2 Where no gibs are employed in the crosshead blocks, 'shims' or thickness pieces of sheet-tin or copper are interposed under the ends of the guide-bars. **1916** R. T. Nicholson *Bk. of Ford* viii. 118 Paper 'shims' —that is, slips of paper shaped to the flats—between cap and socket. These 'shims' will prevent your tightening the nuts up too far. **1953** J. Lawrence *Questions & Answers on Automobile Transmission & Steering* IV. 90 To rectify the pre-load, adjust the shim pack between the outer bearing cone and the pinion shank or spacer. **1977** *New Yorker* 4 July 33/2 He took out the shims and adjusted them for proper clearance.

b. *Criminals' slang* (chiefly *U.S.*). = *LOID.

1968 L. O'Donnell *Face of Crime* i. 12 The lock was of the deadbolt type that doesn't yield to the opportunist's plastic shim. **1973** R. Parkes *Guardians* i. 8 Had the door fitted flush to the frame, the old perspex shim wouldn't have slipped in. **1977** 'L. Egan' *Blind Search* x. 172 Denny and I went to Nonie's place, and he used a shim to get us in.

shim, *v.*² Add: **2. a.** *N. Amer.* Also const. *out*. (Examples.)

1937 H. E. Stafford *Troubles of Electr. Equip.* iii. 46 The only permanent way is..by decreasing the air gap by shimming the pole shoes. **1967** E. B. Nickerson *Kayaks to Arctic* xiv. 126 Each cabin had a single door, well shimmed but still hanging out of plumb. **1974** R. M. Pirsig *Zen & Art of Motorcycle Maintenance* v. 57 You're going to have to shim those out.

b. *Criminals' slang* (chiefly *U.S.*). To open (a lock or door) with a shim. Cf. *SHIM *sb.*² 5 b.

1972 J. Wambaugh *Blue Knight* (1973) ii. 25 The burglar..would shim doors which isn't too hard to do in any hotel.

‖ **shimada** (ʃimā·dā). The name of a town in Honshu, central Japan, applied *absol.* and *attrib.* to a young unmarried ladies' formal hairstyle in which the hair is drawn into a queue and fastened at the top of the head.

1910 Jukichi Inouye *Home Life in Tokyo* ix. 113 Both the *shimada* and the *marumage* are heavy as they require false hair. **1936** K. Nohara *True Face of Japan* v. 173 The unmarried girl wears the *shimada* coiffure. **1959** R. Kirkbride *Tamiko* vii. 53 A Geisha girl, gorgeous in kimono and shimada hairstyle.

‖ **shime-waza** (ʃī·me,wā·zā). *Judo.* Also **shime waza.** [Jap., f. *shimeru* to tighten, constrict + *waza* art, deed, work.] The art of strangulation; a strangle-hold. Also *attrib.*

1954 E. Dominy *Teach Yourself Judo* 191 *Shime Waza*, the art of Strangulation. **1956** K. Tomiki *Judo* iii. 92 There are two methods of strangling, namely necklock and cheeklock. But the latter is excluded from the practice in *judo*, only the former being referred to as *shime-waza*. **1957** *Judo* (Know the Game Series) 26/2 Lastly there is the shimewaza group (strangling and choking techniques)... The definition is as follows: the strangle aims at compressing the common carotid artery just behind the sternomastoid muscle which runs up both sides of the neck... If this pressure is maintained it is only a matter of a few seconds before the man becomes unconscious. **1978** D. Starbrook *Judo Starbrook Style* vi. 98 A shimewaza is a stranglehold.

shimiyana (ʃimiyā·nä). *S. Afr.* Also **shimiaan, shimiyane, shimya(a)n.** [ad. Zulu *isi)shimeyana*.] An intoxicating home-brewed drink made from treacle or sugar and water.

1870 A. F. Lindley *After Ophir* xix. 306 *Shimyan* and *jwarlar* were produced for our consumption, and we were invited to witness the usual dancing performances at the *kraal* after dark. **1900** J. Robinson *Lifetime S. Afr.* 307 'Shimyaan', a concoction of treacle and water allowed to ferment in the sun. This beverage was maddening in its effects, and the parent of much crime. **1934** R. Campbell *Broken Record* 68 You loaf, you drink *shimiaan*. **1946** *Cape Times* 29 Oct. 5 The presiding magistrate called in a number of native spectators to decide whether a drum of brew before the court was shimiyana. **1949** *Cape Argus* 6 Aug. 1/4 Malinga pleaded guilty to being in possession of three gallons of shimiyane which, he said, he brewed for his own consumption. **1961** T. Matshikiza *Chocolates for My Wife* 76 They plug you cockfull of Shimiyana. Some randy home brew.

shimmer, *v.*¹ Add: **2.** *intr.* To move effortlessly; to glide, drift (*by*, *off*, etc.).

1904 in *Eng. Dial. Dict.* V. 385 [Yorks.] He shimmered by, a piece o' way off. **1923** Wodehouse *Inimitable Jeeves* x. 102 Jeeves shimmered off, and Cyril blew in, full of good cheer and blitheringness. **1930** C. Williams *War in Heaven* xi. 176 'I just want to shimmer up, like Jeeves, not walk,' she said. **1973** M. Amis *Rachel Papers* 151 Move my hand over her bronze tights, tracing her hip-bone, circling beneath the overhang of her buttock, shimmer flat-palmed down the back of her legs.

shimmeriness (ʃi·mərinės). [f. SHIMMERY *a.*¹ + -NESS.] The condition of being shimmery; a flickering or insubstantial quality.

1913 D. H. Lawrence *Sons & Lovers* II. vii. 153 Only this shimmeriness is the real living. The shape is a dead

crust. **1948** M. Schorer in *Hudson Rev.* I. 76 His [sc. Lawrence's] belief in a..poetry in which nothing is fixed, static, or final, where all is shimmeriness and impermanence.

shimmy, *sb.*¹ (Earlier and later examples.)

1837 F. Marryat *Snarleyyow* xliii. 267 We have nothing but petticoats here and shimmeys. **1889** *Macmillan's Mag.* Sept. 360, I did count on gettin' myself a new shimmy. **1952** *New Yorker* 20 Sept. 35/1 To persuade the young matron to doff her wet shimmy.

shimmy (ʃi·mi), *sb.*² Also **shimi.** [App. a use of SHIMMY *sb.*¹] **1.** A lively modern dance resembling a foxtrot accompanied by simulated quivering or shaking of the body which first achieved wide popularity in the early nineteen-twenties; a performance of this dance. Also in phr. *to shake a shimmy*. orig. *U.S.*

[**1917** *Variety* 30 Nov. 19/1 The opening number was programed as a combination of 'Strutter's Ball', 'Shimme-Sha-Wabble' and 'Walking the Dog'.] **1918** *Dancing Times* Nov. 35 It is still very, very crude—and it is called 'Shaking the Shimmy'... It's a nigger dance, of course, and it appears to be a slow walk with a frequent twitching of the shoulders. **1920** C. Sandburg *Smoke & Steel* 223 Shimmying the fast shimmy to the Livery Stable Blues. **1922** *Weekly Dispatch* 31 Dec. 9 'Shimmy' banned in New York... The Chicago camel-walk, scandal, balconnades, and shimmy dances must cease. **1924** P. Marks *Plastic Age* 275 That music was enough to make a saint shed his halo and shake a shimmy. **1935** J. T. Farrell *Judgment Day* I. xvi. 387 The building began to waver and dance before his eyes. Funny. The building was doing the shimmy. **1947** M. Berger in R. de Toledano *Frontiers of Jazz* viii. 96 They did the Virginia reel, slow and fast quadrilles and the shimmy. **1956** B. Holiday *Lady sings Blues* (1973) iv. 41 White people.. came to the Cotton Club—a place Negroes never saw inside unless they played music or did the shakes or shimmies. **1975** P. G. Winslow *Death of Angel* xii. 232 Frayne..held the towel behind his hips and did what.. used to be called the shimmy. **1977** *New Hampshire Times* 27 July 12/2 Glasses in New Hampshire cupboards began to rattle as houses started modest shimmies.

2. *transf.* An oscillation or vibration of the wheels, etc., of a motor vehicle or of an aircraft undercarriage; *spec.* = **wheel wobble*.

1925 *Proc. Inst. Automobile Engineers* XIX. 822 This phenomenon..is variously termed 'wheel flap', 'shimmy', 'goldfishing', 'tramping', 'wobble', according to the nationality and imagination of the writers. **1936** *Aircraft Engin.* July 199/1 With the use of the front castorable wheel another difficulty develops..in the form of wheel shimmy. **1940** G. Frankau *Self-Portrait* lxii. 385 'Frankie'..developed a shimmy in her full elliptic springing that made her solid steering column feel like india-rubber. **1943** F. L. Wright *Autobiogr.* (rev. ed.) v. 411 At high speed it would settle down and shake itself almost to pieces in a perfect frenzy (the garage-doctors called it a shimmy). **1958** H. G. Conway *Landing Gear Design* viii. 150 Shimmy can be divided into two basic types: large angle and small angle (or kinematic) shimmy. **1968** K. J. Bunker in J. G. Giles *Steering, Suspension & Tyres* vii. 132 Shimmy..is usually started by road irregularities. **1977** *Grimsby Even. Tel.* 5 May 5/1 (Advt.), Terrific tyre bargains!.. Wheel balancing. Got the 'shimmy'—Got the 'shakes'? Expert correction.

3. *attrib.* and *Comb.*, as (sense 1) *shimmy dance, dancer, dress*; **shimmy damper,** a device fitted to aircraft undercarriages and motor vehicles in order to prevent or reduce shimmy; **shimmy-fox(trot)** = next; also, (a piece of) music to accompany this dance; **shimmy shake, shiver** = *SHIMMY *sb.*² 1; hence *shimmy shaker*.

1928 *Proc. Inst. Automobile Engineers* XXII. 741 It is important when using a shimmy damper to avoid the use of spring connections..in the steerage linkage. **1946** *Jrnl. R. Aeronaut. Soc.* L. 533/2 In only one aircraft had a hydraulic shimmy damper been used, and that was a direct copy of the damper used in Douglas aircraft. **1958** H. G. Conway *Landing Gear Design* viii. 153 (*caption*) A well-known type of American shimmy damper on the rotating vane principle. **1919** *N.Y. World* 17 Jan. 7/5 (*heading*) Shimmy dance is banned in greater New York. **1922** Joyce *Ulysses* 533 You found me in evil company, highkickers, coster picnic makers, pugilists..and the nifty shimmy dancers. **1967** *Boston Globe* 5 Apr. 59/3 A largely nude shimmy dancer put in all the bumps and grinds with a gyrating G.I. **1919** *Honey Pot* I. I. 8 The Eton collar which, in addition to her plain blue 'shimmy' dress.. made her resemble a school-girl of sweet sixteen. **1968** P. Oliver *Screening Blues* vi. 206 The women dressed exotically and were supported by scantily clad chorus girls wearing the shimmy dresses of the period. **1934** C. Lambert *Music Ho!* III. 224 Jannings going to the dogs is not a more melancholy spectacle than some worthy Teutonic fiddler putting a little pep into a 'shimmy-fox'. **1926–7** *T. Eaton & Co. Catal.* 305/1 *Collegiate*—Shimmy Fox Trot. **1928** *Observer* 15 Apr. 12 Instead of a scherzo she has written a shimmy-foxtrot. **1925** Infanta Eulalia of Spain *Courts & Countries after War* i. 18 The history of dancing during the Revolution repeated itself, with the differences that the Carmagnole of '93 was the Shimmy Shake or the Bunny Hug of 1914. **1920** *Sat. Even. Post* 27 Nov. 42/4 Then they was a pair of young shimmy shakers. **1919** *N.Y. Sun* 16 Jan. 14/4, I was dancing the shimi shiver.

shimmy, *v.*¹ orig. *U.S.* Also **shemmi.** [f. prec.] **1. a.** *intr.* To dance the shimmy.

1919 J. R. Pickell *Twenty-Four Days on Troopship* xiv. 74 O, boys, we don't care, If we never get home, If mother will shimmy So long as we roam. **1919** A. J. Piron (*song-title*), I wish I could shemmi like my sister Kate. **1932** J. Laver *Nymph Errant* viii. 199 Constantine shimmied until beads of perspiration gathered on his shiny forehead. **1977** *Zigzag* Apr. 28/1 He gyrates, shimmies, shakes his ass.

b. *trans.* To dance (the shimmy); to shake (part of the body) as in the shimmy.

1920 [see *SHIMMY *sb.*² 1]. **1956** H. Gold *Man who was not with It* xi. 85 Pauline used to tell me..shimmying her loose bare arms of which she was so proud for their milky flesh, 'like this, like that, and ziggety-zaggety'. **1974** J. Irving *158-Pound Marriage* vii. 148 She shimmied her fingers the way Tyrone Williams did before the whistle.

2. *intr. fig.* and *transf.* To shake, quiver, vibrate; to progress hastily or irregularly.

1925 C. R. Cooper *Lions 'n' Tigers* ix. 235 Leader Mary was beginning to shimmy slightly with increased flight. **1941** *Picture Post* 3 May 9/2 The gunfire came surging back... The floor of the basement shimmied underneath me and the whole house shook like a Chinese lantern in a breeze. **1942** *Jrnl. Aeronaut. Sci.* IX. 400/1 It is impossible for any side force to build up on the tire to cause it to shimmy. **1958** H. G. Conway *Landing Gear Design* viii. 152 Aircraft..with less than a certain amount of castoring friction shimmied and those with more did not. **1969** L. Michaels *Going Places* 135 She..shimmied up my arm and hung from my shoulder like a bunch of bananas. **1976** *Times Lit. Suppl.* 2 July 814/3 When his wife was asleep, he would shimmy down a pillar to the ground floor. **1980** *Daily Tel.* 29 Nov. 17/3 Palm, shimmying in the warm breezes all along the coasts.

Hence **shi·mmying** *vbl. sb.* and *ppl. a.*

1919 J. R. Pickell *Twenty-Four Days on Troopship* 73 The star in the heavens, Looked down with a frown, To see mother so shimmied In her shimmeying gown. **1928** Galsworthy *Swan Song* II. xiii. 217 He..watched the dancing on deck—funny business nowadays, shimmying, bunnyhugging, didn't they call it. **1942** R. H. Bound in R. A. Beaumont *Aeronaut. Engin.* xv. 412/2 Main tail wheels have been subject to one very serious defect, namely, shimmying; this consists of violent oscillations of the tail-wheel from side to side when the aircraft is running over the ground. **1972** *Sci. Amer.* Oct. 100/3 Most of the behavior of a nucleus undergoing nuclear fission can be understood as the splitting of a shimmying electrically charged drop. **1977** *Gay News* 24 Mar. 23/1 A mere suggestion of a shimmying hip and you were lectured by a bartender.

shimmy (ʃi·mi), *v.*² [Alteration of SHIMMER *v.*¹ under infl. of *SHIMMY *v.*¹] *intr.* and *trans.* To 'dance' *in*; to transport (a person) quickly.

1923 Wodehouse *Inimitable Jeeves* vii. 76, I bounded into the sitting-room, but it was empty. Jeeves shimmied in. **1930** G. MacMunn *Behind Scenes in Many Wars* x. 187 A small destroyer..would shimmy us over to the beaches from Imbros. **1980** G. V. Higgins *Kennedy for Defense* x. 104, I just love seeing fat fees shimmy out the door to go elsewhere.

† **shimose** (ʃimǭu·se). *Mil. Obs.* [The name of Masachika *Shimose* (1859–1911), Japanese engineer.] A form of lyddite made in Japan. Also † **shi·mosite.**

1904 *Amer. Inventor* 1 June 256/2 An explosive used by the Japanese, and called Shimose, after its inventor,.. is said to be more powerful than either dynamite or guncotton, and to possess features found in no other high-power explosive. **1915** A. Marshall *Explosives* 322 The first satisfactory solution of the problem was the adoption of picric acid by France. This was quickly followed by similar measures taken by practically all the other Powers, each of whom, however, gave the substance a different name; France..Mélinite, England..Lyddite,.. Japan..Shimosite. **1917** *Chambers's Jrnl.* Apr. 258/2 The picric acid compound known as lyddite in England, melinite in France, and *shimose* in Japan.

shimozzle, var. *SHEMOZZLE. **shimya(a)n,** vars. *SHIMIYANA.

shin, *sb.*¹ Add: **5.** *shin-guard* (earlier example); **shin-cracker** *Austral.* (see quots.); **shin-oak** (earlier example); **shin-plaster**: for *U.S.* read *Hist.* (orig. *U.S.*) and add: (*b*) (further examples); (*c*) *Canad.*, a twenty-five dollar bill; **shin-splint,** (*a*) (*dial.* (see quot. 1893–4); (*b*) *pl.* (const. as *sing.* and *pl.*), any of a number of painful conditions of the lower leg that may be caused by running on hard surfaces; **shin-tangle** *Canad.* (see quot. 1905).

1928 *Wentworth Mag.* (Sydney) June 33 'Shin-crakers', that is, blows on the shin owing to the rock suddenly breaking off. **1945** S. J. Baker *Austral. Lang.* 99 *Shincracker*, a subsoil of close-grained, brittle sandstone where the potch or silica runs. **1969** E. Waller *There's Opal out There* 20 Ailments common to the Lightning Ridge, such as shincracker shin. **1971** J. S. Gunn *Opal Terminol.* 42 *Shin cracker*. Also *shincracker*, common name for the fine-grained Coocoran claystone which on exposure at the surface becomes a hard, brittle, siliceous rock that usually has to be dug through to get to the opal ground. Its name is appropriate because, when worked with a sinking pick or jack hammer, pieces shatter or fly off to strike the digger's shins, hence the injury called 'shin-cracker shin'. **1884–5** *Derbyshire Football Guide* 97 (Advt.), *Shin-guards* ..2s. 6d. **1844** J. Gregg *Commerce of Prairies* II. 200 Black-jacks..[are] intermixed with a very diminutive dwarf oak, called by the hunters 'shin-oak'. **1887** *Grip* (Toronto) 2 Apr. 10/2, I will give further particulars on receipt of a shin-plaister. **1929** H. Colebatch *Story of Hundred Years* xxxvii. 458 The 'shin-plasters' of Connor, Doherty, and Durack, and of many hotel and store keepers, form the regularly accepted currency. **1936** M. Mitchell *Gone with Wind* xvii. 308, I haven't a cent. Rhett, give me a few shin plasters. Here, Big Sam, buy some tobacco for yourself. **1962** H. Green *Time to pass Over* v. 77 Old Josh felt disposed to part with a few of his mouldy shinplasters. **1972** *Tel.* (Brisbane) 10 Nov. 40/1 Some years ago I was working in Boulia, where there wasn't a bank. Shin plasters were issued by Mr. J. P. Howard who owned the hotel at Boulia. **1812** in J. Bell *Rhymes of Northern Bards* 35, I lost a' my shin-splints among the great stanes. **1893–4** R. O. Heslop *Northumberland Words* II. 632 *Shin-splints*, a kind of greave or leg armour worn on the shins by trimmers, etc., to protect the legs in working. **1930** Stedman *Med. Dict.* (ed. 11) 951/1 *Shin splints*, myositis and periostitis affecting chiefly the extensor muscles of the lower lateral aspect of the legs. **1938** A. Thorndike *Athletic Injuries* xxi. 180 Shin splints in track, cross country and other sports are a very definite injury—a tearing of the origin of the tibialis posticus muscle from the tibia in its lower third. **1977** J. F. Fixx *Compl. Bk. Running* v. 71 Shin splints—pains in the front of the leg that are common in beginning runners. **1905** J. Outram *In Heart Canad. Rockies* 176 A dense undergrowth..is often designated by the expressive term 'shin-tangle'. **1973** P. Geddes *Ottawa Allegation* xiv. 181 Nothing else was about except for the birds, making for cover under the shintangle.

Shin (ʃīn), *sb.*² [Native name.] One of the Dardic peoples inhabiting the Gilgit agency of Kashmir; a member of this people.

1875 F. Drew *Jummoo & Kashmir Territories* xviii. 428 The table..shows in what countries the Shīn caste is found... The Shīn occur, mixed with Yashkun, along the Indus Valley. **1879** *Encycl. Brit.* X. 598/1 The middle castes, Shin and Yashkun, form the body of the Dard people. The pure Shin looks more like a European than any high-caste Brahman of India. **1910** *Ibid.* XII. 20/1 The dominant race is that of the Shins, whose language is universally spoken. **1913** A. Neve *Thirty Years in Kashmir* ix. 84 My own impression is that the Rajah families were originally Shins... The Shin races (*i.e.*, the Dards) first occupied the eastern Hindu Kush. **1938** R. C. F. Schomberg *Kafirs & Glaciers* xii. 206 The people of Ashret are not Chitralis but Dangariks who speak Palula, a language allied to the Shina spoken at Gilgit... They are probably Shins who came from Chilas.

Shin (ʃin), *sb.*³ [Jap., = genuine, authentic.] The name of a major Japanese Buddhist sect which teaches salvation by faith in the Buddha Amida and emphasizes morality rather than orthodoxy. Usu. *attrib.* or as *adj.*

1877 W. E. Griffis *Mikado's Empire* (ed. 2) I. xvi. 173 The Shin sect hold a form of the Protestant doctrine of justification by faith, believing in Buddha instead of Jesus. **1904** L. Hearn *Japan: Attempt at Interpretation* xiii. 302 Nobunaga agreed to spare the lives of the Shin priests. **1960** B. Leach *Potter in Japan* ii. 49 Dr. Suzuki is the leading writer on both Zen and Shin Buddhism, both in English and Japanese. **1976** *Education & Community Relations* July/Aug. 8/1 Talks by the Chief Abbot of the Nishi Hongwanji (Shin Sect) of Japan.

shin, *v.* Add: **2.** (Earlier examples.)

1838 J. C. Neal *Charcoal Sketches* 106 Shin it, good man..shin it as well as you know how! **1840** G. T. Strong *Diary* 8 May (1952) 138 One banner in particular —representing Matty shinning away from the White House.

3. Also, to shoot in the shins.

1819 E. Evans *Pedestrious Tour* 214 Soldiers are apt to fire too high. He was often heard to say to his troops in battle: 'Shin them, my brave boys!'

Shina (ʃī·na). [Native name: cf. *SHIN *sb.*²] The Indo-Aryan language spoken by the Shin. Also *attrib.*

1854 A. Cunningham *Ladāk* ii. 37 The Persian character, which all the Dards make use of in writing their own language, of which there are three distinct dialects,— the Shiná, the Khajunah, and the Arniya. The *Shiná* dialect is spoken by the people of Astor, Gilgit, Chelas, Darél, Kohli, and Pálas. **1903** Risley & Gait *Census India 1901* I. I. 310 Shina, one of the non-Sanskrit Indo-Aryan forms of speech. **1936** R. C. F. Schomberg *Unknown Karakoram* I. iii. 39 Shum, which means dog in the Shina tongue. **1977** D. Murphy *Where Indus is Young* vii. 147 Talking rapidly in Shina, the Gilgit language.

shinanigan, -gin, etc., vars. *SHENANIGAN.

Shin Bet (ʃin bet). Also **Shin Beth** and as one word. [mod.Heb., f. *šīn* + *bēṯ*, the names of the initial letters of *šērûṯ biṭaḥôn kelālī* (general) security service.] The principal security service of Israel. Also *attrib.*

1964 L. Deighton *Funeral in Berlin* xlii. 264 Samantha was a Shinbet agent after him for war crimes. **1968** C. Leader *Angry Darkness* xii. 111 Schererzade was now in the hands of the Shinbeth, the coldly efficient and unemotional Israeli Intelligence Service. **1969** A. Marin *Rise with Wind* vi. 76 The statement..went directly to the Shin Beth. **1972** *Guardian* 27 Jan. 3/4 The street always becomes stiff with Shin Bet (Secret Service) men looking studiously unobtrusive. **1981** A. Winch *Blood Money* xiii. 134 A man..who has been positively identified by the Israeli Shin Beth.

shindig. Substitute for entry:

shindig (ʃi·ndig). Also **shin-dig.** [Of uncertain origin: perh. f. SHIN *sb.* + DIG *sb.*¹, but infl. by SHINDY in later senses.] † **1.** *U.S.* (See quot. 1859.) *obs.*

[**1849**: see *HOE-DOWN.] **1859** [in Dict.].

2. A country dance; a party, ball, 'knees-up'; a lively gathering of any kind. Also *fig.* orig. *U.S.*

1871 B. HARTE in *Atlantic Monthly* Sept. 373/1 'Is this a dashed Puritan meeting?'.. 'It's no Pike County shindig.' **1892, 1899** [in Dict.]. **1935** C. W. PARMENTER *Kings of Beacon Hill* I. xv. 98 Does everyone attend those shindigs, Sandy, or is a girl invited by some special boy? **1946** [see *FURORE 2]. **1956** WALLIS & BLAIR *Thunder Above* (1959) ix. 98 He was killed in an air defence exercise. One of those NATO shindigs. **1959** *New Statesman* 27 June 883/2 The competition among the 'old nobility' to attend what they termed 'Aspers' little shindig' was so fierce that five private detectives were hired to keep out the unwelcome. **1962** E. LUCIA *Klondike Kate* viii. 172 Kate never lacked a date for such shindigs. **1977** C. McCULLOUGH *Thorn Birds* xi. 267 'What's a ceilidh anyway?'..'It's Gaelic for a gathering, a shindig.'

3. = SHINDY 3.

1961 PARTRIDGE *Dict. Slang* Suppl. 1268/2 *Shindig*, an altercation, a violent quarrel, a tremendous fuss. **1966** *Listener* 17 Feb. 255/1 A classic row developed over the half birth of Peter Watkins's film *The War Game*, and this did not wholly distract from the shindig on the other side of the fence about a programme on the police. **1977** 'E. CRISPIN' *Glimpses of Moon* vi. 87 They'd kick up a shindig, naturally, but it was always their husbands they were furious with.

shindy. 3. (Earlier and later examples.)

1829 B. HALL *Trav. N. Amer.* III. 325, I never saw a more complete row, or as a fellow near me called it, 'a more regular shindy'. **1841** *Sporting Rev.* July 52 The docket of bankruptcy..created, as our polite continental neighbours call it, 'a sensation', or, in downright English, 'a shindy'. **1916** 'TAFFRAIL' *Pincher Martin* vii. 115 If you want to kick up a shindy, Mister Parkin, you'd best do it ourside. **1903** SOMERVILLE & 'ROSS' *All on Irish Shore* vii. 177 There was a frightful shindy, Carew wanting to have his blood, and all the rest of us trying to prevent a row. **1962** L. DEIGHTON *Ipcress File* x. 61 We're not having another Burgess and Maclean shindy, questions in the House and all that. **1976** J. I. M. STEWART *Memorial Service* iv. 53 There was quite a shindy, and there might have been more of it.

shine, *sb.*[1] Add: **2. f.** *spec.* The shiny surface of a new cricket ball.

1950 [see *OPENER 1 e]. **1976** J. SNOW *Cricket Rebel* 30 Mike Smith didn't call upon me until Fred Trueman and Rumsey had seen the shine off the new ball.

3. a. (Further examples opposed to *rain.*) Also fig. phr. (*come*) *rain or shine*, in any circumstances, come what may.

1905 H. A. VACHELL *Hill* vi. 138 With me you're first, rain or shine. **1908** *Sears, Roebuck & Co. Catal.* 1076/2 These overcoats do double service, being adapted for all kinds of chilly weather, rain or shine. **1952** M. R. RINEHART *Pool* vii. 54 She walks everywhere here in the city, rain or shine. **1978** M. BIRMINGHAM *Sleep in Ditch* 115 He said..we'd got to put up with each other for ten years, 'come rain or shine'.

b. Abbrev. of *MOONSHINE 4.

1933 *Sun* (Baltimore) 22 July 8/7 You should take a julep made from the wonderful 'shine' made in the hills of Western Maryland. **1938** M. K. RAWLINGS *Yearling* i. 12 'Goin' to Grahamsville allus do make me hongry.' 'You git a snort o' 'shine there, is the reason,' she said. **1969** P. KAVANAGH *Such Men are Dangerous* (1971) iii. 42 Whiskey? A quart of shine, which the Lord loves, it being a natural product. **1977** E. LEONARD *Hunted* (1978) ix. 90 'Hundred-proof pure Kentucky bourbon. How about that.' Like it was a treat and all Davis drank was some kind of piss-poor shine.

5. An abusive term for a Black. Also *attrib.* *U.S. slang.*

1908 J. M. SULLIVAN *Criminal Slang* 24 *Shine*, a colored person. **1929** D. HAMMETT *Dain Curse* iv. 34 How'd you make out with the shine? **1934** J. T. FARRELL *Young Manhood* III. xv. 227 They saw one beautiful blonde girl with a coal-black, sweating nigger, and they said nothing, only because there were too many shines in the place. **1940** R. CHANDLER *Farewell, My Lovely* iii. 13 His voice said bitterly: 'Shines. Another shine killing. That's what I rate after eighteen years in this man's police department.' **1953** W. BURROUGHS *Junkie* v. 51 A Negro sitting opposite us smiled. 'The shine is wise,' said Roy in my ear. 'He is O.K.' **1969** S. GREENLEE *Spook who sat by Door* xiii. 116 He's a shine detective lieutenant.

shine, *sb.*[2] **4.** For U.S. read *colloq.*, orig. *U.S.* and add earlier and later examples.

1839 *Crockett Almanac 1840* 14, I wonst had an old flame I took sumthin of a shine to. **1934** F. H. BUSHICK *Glamorous Days* xxiii. 278 Nobody wanted the old corn cutter except this Irishman, who took a shine to it. **1956** P. SCOTT *Male Child* II. iv. 152, I suppose I oughtn't to blame you if Marion's taken a shine to you... You must have a way with women. **1961** *Guardian* 23 Mar. 10/6 Shaw evidently took a shine to the young hero-worshipping woman. **1978** L. MEYNELL *Papersnake* x. 132 He took to you... He took a shine to you. **1980** *Times Lit. Suppl.* 18 July 799/1 If her [*sc.* Barbara Pym's] heroines were married, they were not unfaithful to their husbands, although they might take a shine to the curate.

shine, *v.* Add: Forms: *Pa. t.* and *pa. pple.* **shined** (further U.S. examples).

1929 [see *CONCORDANCY]. **1929** W. FAULKNER *Sound & Fury* 105 He wore a derby and shined shoes. **1948** *Sun* (Baltimore) 18 Oct. 12/5 It was full of adept and memorable phrases... It shined with wit and humor. **1950,** etc. [see sense 9 c below]. **1974** *Black World* Jan. 57/1 Her shiny black paint shined in the sun. **1982** *Chicago Sun-Times* 12 July 65 But Red Smith was a beacon who shined for half a century.

9. c. To direct the rays of (a light) *on, on to, under*, etc.

1889 *Cent. Dict.* 5573/3 The policeman *shone* his lantern up the alley. **1950** *Sun* (Baltimore) 14 July 8/4 Two men in the office shined a flashlight under the platform. **1967** P. SHAFFER *Black Comedy* 48 The Colonel takes the torch from Harold and shines it pitilessly in Schuppanzigh's face. **1978** J. IRVING *World according to Garp* xi. 210 The policeman shined his light over Garp. **1979** *Sci. Amer.* Mar. 85/2 The intense light from this source was shined on a crystal that served as a frequency doubler.

10. c. *to shine up to*: to try to please; to make oneself pleasant to. *U.S.*

1882 *Century Mag.* Oct. 827 It was then that David first set out to shine up to her. **1902** S. E. WHITE *Blazed Trail* xlii. 204 You might shine up to Hilda Farrand and join the rest of the fortune-hunters. **1971** C. FICK *Danziger Transcript* (1973) 143, I never saw him sell a single secret..or shine up to a Kraut PW.

shiner. Add: **1. f.** A diamond or other jewel. Usu. *pl. slang.*

1884 *Queenstown Free Press* 15 Jan. (Pettman), When they dug it up they at once came to the conclusion it was a real shiner. **1928** M. C. SHARPE *Chicago May* 287 *Sparklers, sparks, shiners*.., etc.—diamonds. **1934** D. L. SAYERS *Nine Tailors* 274, I never had those shiners. **1959** [see *ICE *sb.* 4 c].

g. A black eye. *slang.*

1904 'No. 1500' *Life in Sing Sing* 253/1 *Shiner*, a discolored eye. **1932** [see *HECK *sb.* and *int.*]. **1934** A. MERRITT *Burn Witch Burn!* vii. 89 All I can do is..be dignified an' maybe hand out a shiner or two if they get too rough. **1943** C. S. FORESTER *Ship* xviii. 114 That's a rare shiner you've got there, Grant. **1958** C. WILLIAMS *Man in Motion* (1959) iii. 29 At first I thought it was because of the shiner and the bruises on my face, but then I began to wonder. **1967** [see *LEFT *sb.* 2 a]. **1977** *Daily Mirror* 16 Mar. 3/5 Annie Walker, Coronation Street's snooty landlady, is about to show up in the snug.. sporting a real shiner. But her black eye is not the result of a well-rehearsed punch-up in the taproom.

h. *Paper-making.* A glistening particle of a mineral impurity on the surface of finished paper.

1922 *Manufacture of Pulp & Paper* III. VIII. 3 In colored papers, shiners will not take the dye. **1963** R. A. HIGHAM *Handbk. Papermaking* vii. 197 When supercalendering papers with a high percentage of straw in the furnish, there is a tendency towards shiners and windows.

2. d. A window-cleaner. *slang.*

1958 *Listener* 20 Nov. 818/2 His fellow shiners disregarded the L.C.C. by-law, because very few windows are equipped with metal hooks for holding on a safety belt. **1967** *Sunday Times* 8 Jan. 3/7 Len is widely regarded as London's top 'shiner' (window cleaner). **1977** *Centuryan* (Office Cleaning Services) Christmas 1/1 There we were, shiners and cleaning ladies, surrounding Fred and Dora on the float by London Wall.

shingle, *sb.*[1] Add: **1. b.** (Further examples.)

1957 J. FRAME *Owls do Cry* 26 Francie Withers has a brother who's a shingle short. **1966** P. WHITE *Solid Mandala* 82 He accepted Arthur his twin brother, who was, as they put it, a shingle short. **1968** *Southerly* XXVIII. 3 Royal said: 'I reckon we're a shingle short to 'uv ended up on the Parramatta Road.'

d. *to hang out* (or *set up*) *one's shingle*, to begin to practise a profession.

1879 [in Dict.]. **1944** V. W. BROOKS *World of Washington Irving* xvi. 308 Catlin hung out his shingle as a portrait-painter and made a little money for his next trip. **1963** J. N. HARRIS *Weird World Wes Beattie* i. 8 He had hung up his shingle and commenced the practice of criminal law in the lower courts. **1977** *Time* 22 Aug. 48/2 Any academic can set up his shingle and be a literary critic.

e. A style of cutting women's hair short, as in the bob, but with the back hair shingled (cf. SHINGLE *v.*[1] 2 a). Also, hair cut in this way.

1924 *Hairdressing* Feb. (*caption*), Based on the 'shingle'. **1927** F. E. BAILY *Golden Vanity* xvii. 265 Doris powdered her face, combed her dark shingle, lit a cigarette, and picked up her beef cubes. **1945** N. MITFORD *Pursuit of Love* xx. 172 She had a short canary-coloured shingle (windswept) and wore trousers. **1975** G. HOWELL *In Vogue* 13/1 The small pitted cloche brought in the bob, which became the 'shingle' or the 'bingle' of the twenties.

2. a. *shingle effect* (sense 1 e); **shingle cap, net,** a cap-shaped hair-net for preserving the hair-style in bed; **shingle wig,** a short-haired wig cut in a shingle.

1926 *Vogue* Late Nov. 85 A charming little shingle cap for night wear. **1934** A. CHRISTIE *Murder on Orient Express* II. xi. 146 She had on a shingle cap and I only saw the back of her head. **1977** 'E. McBAIN' *Long Time no See* x. 152 Her blond hair was cut in..bangs on the forehead, a shingle effect at the back of her head. **1928** R. MACAULAY *Keeping up Appearances* ix. 89 She had bought..three shingle nets. **1928** *Times* 19 Dec. 15/7 After bathing the shingle-wig was slipped over the dishevelled head.

b. (*b*) *shingle saw.*

1882 R. GRIMSHAW *Suppl. Grimshaw on Saws* 235 One we know of is running a 42-inch shingle saw in heading 1500 revolutions per minute. **1974** D. SEARS *Lark in Clear Air* i. 19 Snoring with a shrill gutter like a shingle-saw slicing knotty cedar.

shingle, *sb.*[2] Add: **1.** In New Zealand also loose angular stones in mountain country.

1900, etc. [see *shingle-slip* below]. **1944,** etc. [see *shingle-slide* below]. **1959** *Tararua* XIII. 46 The word *shingle* itself is given an unusual meaning in New Zealand. In standard usage it refers only to the small roundish water-worn stones of the seashore or rivers. We use it also of moderately-sized, angular stones, such as in fact are found in shingle slides.

3. *shingle-bed* (earlier example); **shingle slide** *N.Z.* (see quot. 1944); **shingle-slip** = **shingle slide.*

1861 C. C. BOWEN *Poems* 76 Ghastly white beneath, Lay stretched the rough, drear shingle-bed. **1944** *Mod. Junior Dict.* (Whitcombe & Tombs) (ed. 7) 365 *Shingle-slide* or *-slip*,..a term used in New Zealand for (steep) mountain-sides covered with loose, sliding stones, in England called 'screes'. **1959** A. McLINTOCK *Descr. Atlas N.Z.* 32 With the baring of the ground between the tussocks, sheet and wind erosion have taken place and there has been a speeding up of the creep of the mantle of rock waste, resulting in the formation of new shingle slides and an increase in area of old ones. **1900** *Canterbury Old & New* 190 One of the most characteristic features of our Canterbury Alps is afforded by the numerous 'shingle-slips' formed by the weathering of rocks. **1971** *N.Z. Listener* 19 Apr. 56/4 The creek beside the shingle slip just below the confluence.

shingle, *v.*[1] **2.** Restrict *U.S.* to senses in Dict. and add: **a.** Also, to cut (women's hair) so that it tapers from the back of the head to the nape of the neck; also *absol.*, to have the hair so cut.

1924 *Punch* 17 Sept. 319 It moves me not if Araminta shingles Her locks, or Evelina has them bobbed. **1926** GALSWORTHY *Silver Spoon* iv. 25 Fully dressed for the evening, she had but little on, and her hair was shingled. *Ibid.*, She had been one of the first twelve to shingle. **1976** M. GREEN *Children of Sun* (1977) v. 207 Women began to bob their hair immediately after the war, were shingling it by 1925.

shingled, *ppl. a.*[1] Add: **3.** Of hair: cut with the ends exposed all over the head or in a shingle. Of persons: having hair so cut.

1889 KIPLING *From Sea to Sea* (1899) I. xxi. 414 The American missionary teaches the Japanese girl to wear bangs—'shingled bangs'—on her forehead. **1924** M. ARLEN *Green Hat* i. 42 Iris Storm was the first Englishwoman I ever saw with 'shingled hair'. This was in 1922. **1926** R. MACAULAY *Crewe Train* I. iii. 19 She looked..like a Beardsley woman shingled. **1930** W. S. MAUGHAM *Cakes & Ale* xxiii. 231 'Very quiet,' I said to the shingled barmaid. **1953** 'N. SHUTE' *In Wet* vii. 210 He stroked the soft, shingled hair at the back of her head. **1978** *Church Times* 1 Sept. 5/4 Their hair is shingled and rigidly marcel-waved. They stand in Junoesque poses like overblown flappers.

shingler[1]. Add: **2.** A woman who has her hair shingled. *transitory.*

1926 *Glasgow Herald* 11 Nov. 3/4 Was the first shingler a suffragette? **1929** D. MACKAIL *How Amusing!* 337 Though..Duval has done a certain amount of shingling,.. his shop..has no separate entrance for shinglers.

shingling, *vbl. sb.*[1] Add: **2.** (See *SHINGLE *v.*[1] 2 a.)

1924 *Chambers's Jrnl.* June 483/2 You do not care for the shingling and bobbed hair styles? **1926** GALSWORTHY *Silver Spoon* iv. 25 'My dear girl,' Michael had said, when shingling came in, 'to please me, don't! Your *nuque* will be too bristly for kisses.' **1939–40** *Army & Navy Stores Catal.* 901 Hair Clipper... Specially designed for Shingling, Bobbing, and for removing any superfluous hair.

shingly, *a.*[2] Add: (Austral. and N.Z. examples: see *SHINGLE *sb.*[2] 1.)

1857 J. T. THOMSON in N. M. Taylor *Early Travellers in N.Z.* (1959) 336 The plains are alluvial and shingly. **1878** E. S. ELWELL *Boy Colonists* 182 After a long.. climb, they reached the top. It was bare and shingly. **1926** K. S. PRITCHARD *Working Bullocks* v. 52 They rode for hours..along the shingly ledges of steep hill-sides. **1949** A. E. WOODHOUSE in A. E. Currie *Centennial Treasury Otago Verse* 87 The shingly rivers seaward swirling.

Shingon (ʃi·ŋgǫn). Also 9 **Singon.** [Jap., = true word, mantra, f. *shin* true + *gon* word.] The name of a Buddhist sect founded in Japan in the eighth century and devoted to esoteric Buddhism. Also *attrib.*

1727 J. G. SCHEUCHZER tr. *Kæmpfer's Hist. Japan* I. II. v. 199 In the 1850 streets of this city, there were ..10 070 of the sect *Singon*. **1834** *Chinese Repository* Nov. 323 There are now in Japan the following sects which are tolerated by government. 1. Zen... 5. Singon... *Singon* means to repeat true psalms. **1880** E. J. REED *Japan* I. iv. 81 The learned Kobo Daishi..was likewise the founder of the *Shingon* ('True Words') sect of Buddhists in Japan. **1894** *Trans. Asiatic Soc. Japan* XXII. 382 (*heading*) The history of the Shingon sect. **1908** A. LLOYD *Wheat among Tares* iv. 40 Kōbō's faith—the so-called Mantra or Shingon Buddhism—so much resembles Manichaeism that it may be said to be practically the same system. **1931** G. B. SANSOM *Japan* III. xii. 222 The Shingon doctrines are mystical, and not to be explained in words. **1961** *Listener* 31 Aug. 316/1 A Buddhist sect, called Shingon,..is one of the most flourishing sects in Japan today. **1977** T. KASHIMA *Buddhism in America* i. 4 Shingon Buddhism..is based on the *Dainichi* Sutra (the Great Sun Sutra).

shining, *ppl. a.* Add: **1. d. shining cuckoo,** a

copper-coloured cuckoo, *Chalcites lucidus*, found in New Zealand and other parts of the Pacific.

1782 J. LATHAM *Gen. Synopsis Birds* I. II. 528 Shining C[uckow]..Size of a small Thrush..inhabits New Zealand. **1888** W. BULLER *Birds N.Z.* (ed. 2) I. 133 A peculiar whistling cry..announces the arrival in our country of the shining cuckoo. **1965** F. SARGESON *Memoirs of Peon* vi. 155 It was..something like the call of the shining cuckoo, a sound just in the air.

2. c. *shining armour* (see quot. 1533 in Dict. under 1 a, and *knight in shining armour* s.v. *KNIGHT *sb.* 4 e) (freq. ironic): a sign of preparedness to fight nobly in a good cause, esp. in defence of the weak.

1910 *Times* 22 Sept. 5/1 The action of an ally in taking his stand in shining armour at a grave moment by the side of your most gracious Sovereign. **1913** S. SHAW *William of Germany* xi. 249 The Emperor's soldiers and his Dreadnoughts, his mailed fist and shining armour, are built and put on in the spirit of precaution and defence. **1919** G. B. SHAW *Inca of Perusalem* 220 What other defence have we poor common people against your shining armor, your mailed fist, your pomp and parade? **1940** E. F. BENSON *Final Edition* xiii. 268 His Field Marshalls took his sabre-rattlings and his stupendous announcements that he was the chosen instrument of the Lord of Hosts at their face value,..and made him keep polishing up the shining armour which he had donned for his secret reassurance. **1960** J. STROUD *Shorn Lamb* xxii. 242 When I first came galloping out of the University, in shining armour..it was To the Rescue of the Deprived Child. **1968** 'M. UNDERWOOD' *Man who killed too Soon* vi. 61, I could tempt him to don his shining armour and try a rescue operation.

5. (Further examples.)

1887 G. M. HOPKINS *Poems* (1967) 71 Raced With, along them, cragiron under and cold furls—With-a-fountain's shining-shot furls. **1915** D. H. LAWRENCE *Rainbow* ii. 52 Sometimes, all shining-eyed, she was back at her own home. **1923** KIPLING *Irish Guards in Great War* I. 289 Everything was as shining-new as death.

shinnanicking, var. *SHENANIGAN.

Shinner[2] (ʃi·nəɹ). Colloq. abbrev. of *SINN FEINER (ʃin fēi·nəɹ).

1921 *Glasgow Herald* 9 Apr. 12 The sands are running out as Ireland will shortly be ruined. 'Shinners' may be killed daily. **1974** J. JOHNSTON *How Many Miles to Babylon?* 51, I thought I'd heard it about that you were with the Shinners.

shinnery (ʃi·nəri). *U.S.* [f. SHIN *sb.*[1] + -ERY.] An area of scrub in which shin-oak predominates.

1901 *Rev. Reviews* XXIV. 310/1 It [*sc.* 'creeps'] is due mainly to an insufficiency of nourishment in the grass, particularly in pastures where 'shinnery' or dwarf oak trees abound. **1913** W. C. BARNES *Western Grazing Grounds* 268 The scrub oak of the western ranges..forms ..great areas called 'shinneries'. **1946** *Oklahoma Game & Fish News* Mar. 4/1 The located crow roosts in the shinnery motts west of Elk City.

shinning, *vbl. sb.* **2.** (Earlier example.)

1834 A. GREENE *Perils of Pearl Street* i. 16 The exercise of *shinning*.

shinny (ʃi·ni), *sb.*[2] *Southern U.S.* [Alteration of *SHINE *sb.*[1] 3 b: see -Y[6].] = *MOONSHINE 4.

1934 in WEBSTER. **1944** D. VAN DE VOORT in B. A. Botkin *Treas. Amer. Folklore* v. 686 Wiley went over to the safe and got out his pappy's jug of shinny. **1960** H. LEE *To kill Mockingbird* xiii. 139 Miss Maudie Atkinson baked a Lane cake so loaded with shinny it made me tight. **1972** J. CARR *Second Oldest Profession* xi. 166 There are often regional names by which the illicit distillate is recognized. Some of these are 'cannonball swig'.. 'preacher's lye'..'shinny'..'kickapoo joy juice'.

shinny, *v.* Add: (Later examples.) Also with *down, absol.*, and with advb. acc., as *to shinny one's way.*

1936 J. STEINBECK *In Dubious Battle* vi. 86 Jim shinnied down the tree. **1937** *Sun* (Baltimore) 23 Oct. 12/7 It is difficult to imagine a man over 60 shinnying up a porch post. **1967** 'E. QUEEN' *Face to Face* xiii. 61 Somehow he's managed to shinny his way back into her good graces. **1976** *Daily Tel.* 29 Sept. 15/2 They must.. shinny up ropes, and slide down vines. **1977** *Time* 4 Apr. 42/2 Coming on fast is Robert Shaw, Israeli counterterrorist, who must shinny down a rope from a helicopter.

Hence **shi·nnying** *vbl. sb.*

1906 *Washington Post* 22 May 2 As its girth precluded 'shinnying', Gladden procured a ladder.

Shinshū (ʃi·nʃū). [Jap., f. *shin* *SHIN *sb.*[3] + *shū* sect = *SHIN *sb.*[3]

1727 J. G. SCHEUCHZER tr. *Kæmpfer's Hist. Jap.* I. IV. i. 264 The monks of the Chinese and other *Sensju* monasteries send also some of the fraternity to go a begging six times a month. **1896** L. HEARN *Kokoro* x. 193 Wealthier sects had established Buddhist schools on the Western plan: and the Shinshū could already boast of its scholars.

Shinwari (ʃinwā·rī). [Native name.] (A member of) a nomadic people inhabiting areas of Afghanistan around the Khyber Pass. Also *attrib.*

1875 *Encycl. Brit.* I. 232/1 Lead is found..in the Shinwari country. **1888** KIPLING *Phantom 'Rickshaw* 85 Would they could have foretold that my *kafila* would have been cut up by the Shinwaris almost within shadow of the Pass! **1958** O. CAROE *Pathans* xv. 234 The Mohmands and Safis formed a tribal confederacy with large numbers of Afridis and Shinwaris to oppose the passage of the royal troops. **1978** 'M. M. KAYE' *Far Pavilions* v. 87 We have a proverb in the country beyond the Khyber, that says 'A snake, a scorpion and a Shinwari have no heart to tame.'

shiny, *a.* Add: **A.** *adj.* **b.** Also, apparently excellent.

1915 KIPLING *Fringes of Fleet* 40 'Why didn't you then?' I asked. There were loads of shiny reasons. **1970** *Guardian* 14 Dec. 4/1 The Andean Pact is Latin America's ..shiniest attempt at creating a regional common market, but its gleaming paintwork is likely to take some hard knocks.

ship, *sb.*[1] Add: **2.** *ship in a bottle*, a model ship inside a bottle the neck of which is smaller than the ship.

1949 N. MITFORD *Love in Cold Climate* I. xii. 128 The safes..were full of treasures..a carved nut; a ship in a bottle; [etc.]. **1976** *Times* 2 Feb. 16/4 Construction kits are popular..including a ship-in-a-bottle outfit.

3. a. *to burn one's ships* (earlier example).

1871 HARDY *Desperate Remedies* II. i. 39 He saw the strokes plainly, instantly resolving to burn his ships and hazard all on an advance.

b. *ship of fools* [after the title of Sebastian Brant's satirical work *Das Narrenschiff* (1494), translated into English by Alexander Barclay as *The shyp of folys of the worlde* (1509)], a ship whose passengers represent various types of vice or folly.

1609 DEKKER *Guls Horne-Booke* 3 Any person aforesaid, longing to make a voyage in the Ship of Fools. **1807** W. H. IRELAND (*title*) Stultifera navis; qua omnium mortalium narratur stultia. The modern ship of fools. **1864** TENNYSON *Voyage* x, in *Enoch Arden* 149 'A ship of fools' he shriek'd in spite. **1919** KIPLING *Debits & Credits* (1926) 358 He Who launched our Ship of Fools many anchors gave us. **1975** *Times Lit. Suppl.* 7 Feb. 126/4 The Apocalypse as depicted by Bosch, the upside-down world of Goya, the Ship of Fools having landed its cargo.

c. *ships that pass in the night* [after the phrase by Longfellow: see quot. 1873], used of people whose acquaintance is necessarily transitory.

1873 LONGFELLOW *Aftermath in Tales of Wayside Inn* III. iv. 59 Ships that pass in the night, and speak each other in passing... So on the ocean of life we pass and speak one another, Only a look and a voice, then darkness again. **1893** B. HARRADEN (*title*) Ships that pass in the night. **1939** WODEHOUSE *Uncle Fred in Springtime* xiv. 198 The thought that they had met and parted like ships that pass in the night was very bitter to him. **1978** D. BAGLEY *Flyaway* xxv. 235 'Inquisitive, isn't he?' 'Not abnormally so. Chit-chat between ships that pass in the night.'

d. *a tight ship*, a ship in which ropes, etc., are tight; hence a strictly run ship; usu. *transf.* and *fig.*

1971 'H. CALVIN' *Poison Chasers* i. 6 Dai liked a tight mainsheet... 'Pull in tighter, boy... I want a tight ship.' **1972** *Sat. Rev.* (U.S.) 24 June 42/1 The two student judges ..ran a tight ship. Firm commands—'There will be no knitting in my courtroom.' **1977** *Times Lit. Suppl.* 13 May 593/3 Dr Kelly runs a tighter ship altogether than Dr Sheeran: her bibliography is a model of both inclusion and exclusion.

4. b. Substitute for def.: A balloon, aircraft, or powered spacecraft. (Examples.)

1679 R. HOOKE *Philos. Collections* No. 1. 18 A demonstration, how it is practicably possible to make a ship which shall be sustained by the air, and may be moved either by sails or oars. **1709** *Evening Post* 20–22 Dec. 2 The description of a flying ship, lately invented. **1784,** etc. [see *aerial ship* s.v. *AERIAL *a.* 5]. **1860** *Brit. Patent 1598* 1 An improved navigable balloon or aerostatic ship. **1908** H. G. WELLS *War in Air* v. 151 The ships of the German air-fleet rising one by one. **1930** *Sci. Wonder Q.* Spring 352 Both men ran toward the ship..for if the rocket were destroyed, they would be lost in the icy wastes of Venus. **1980** J. CARTWRIGHT *Horse of Darius* xvi. 251 'O.K. Let's get in the ship.'.. As soon as they were airborne, Teymour told him what had happened.

5. d. (Later example.) Now *Hist.*

1981 *Times* 14 July 3/2 Two hollow log 'ships' for boiling brine were found, dating to the sixteenth century.

8. a. *ship-canal* (earlier and later examples), *-channel* (earlier example), *-crane, -pump* (earlier example).

1798 I. ALLEN *Hist. Vermont* 268 A ship canal would be the means of importing salt, and exporting the preceding articles cheap. **1959** *Chambers's Encycl.* III. 38/1 The great ship canals of modern times have been built to carry large ocean-going vessels; but the earlier ship canals..can take only small ships and barges. **1775** J. QUINCY *Let.* 31 Oct. in J. Sparks *Corr. Amer. Revolution* (1853) I. 73 The ship-channel..runs between the east head of Long Island and the south point of Deer Island. **1932** AUDEN *Orators* I. 16 Like those ship-cranes along Clydebank. **1742** W. ELLIS *Timber-Tree Improved* II. xxxvii. 181 The Timber is..of especial Use..for Ship-pumps.

b. For 'Scandinavian' read 'Scandinavian and Anglo-Saxon'. (Later examples.)

1907 H. M. CHADWICK *Origin Eng. Nation* xi. 288 The launching of the funeral ship really was an ancient custom ..from which both ship-cremation (on land) and ship-burial were derived. **1940** *Burlington Mag.* Dec. 174/1 The great Anglo-Saxon ship-burial at Sutton Hoo..was excavated in the summer of 1939. **1963** C. GREEN *Sutton Hoo* ii. 33 (*heading*) The ship-barrow excavation.

c. *ship-jumper, -jumping.*

1964 *Punch* 4 Mar. 336/3 Except for a few ship-jumpers, most come by air. **1959** P. McCUTCHAN *Storm South* xv. 213 Genuine cases of ship-jumping by men who had had enough of sail.

d. *ship-based, -borne* (later examples) adjs.

1973 J. D. R. RAWLINGS *Pictorial Hist. Fleet Air Arm* vi. 69 The Navy..could see a use for the helicopter as a ship-based submarine spotter. **1932** *19th Cent.* Feb. 206 The second [method] is the limitation by agreement of numbers of ship-borne aircraft. **1978** *Navy News* May 5/2 The Phoebe..came out of her two-and-a-half-year refit with..shipborne torpedoes fitted.

9. a. ship('s) biscuit (earlier example with the possessive); **ship-broking** *vbl. sb.* = *ship-brokerage* in Dict.; **ship('s) company** (later examples); **ship('s) decanter,** a decanter with a base of greater width than the shoulder; **ship-lap,** (*b*) boards interlocked by rebates, used esp. for cladding; **ship-lap** *v.*: hence **ship-lapped** *ppl. a.*; **ship-mate**: also in phr. *to be ship-mates with*, to sail in the same vessel with; hence *transf.*, to be acquainted with, to have knowledge of (*colloq.*); **ship plane,** an aeroplane specially adapted for operating from an aircraft carrier; **shipside,** (*a*) (see sense 8 a in Dict.); also *spec.*, the outside of the hull of a ship; (*b*) the dock adjacent to a moored ship; **ship('s) stores,** (*a*) (earlier example); (*b*) sing. (*U.S.*) a shop on board ship; **ship('s) time,** (*a*) (earlier example); (*b*) *Canad. local*, (the time of) the arrival of an annual supply ship; **ship-to-air,** used *attrib.* to designate a missile fired from a ship at an aerial target; **ship-to-ship,** used *attrib.* to designate communications, missiles, etc., directed from one ship to another; **ship-to-shore,** used *attrib.* to designate communications, missiles, etc., directed from a ship to land; also *ellipt.* as *sb.*, a radio-telephone operating in this manner.

1855 E. ACTON *Mod. Cookery* (rev. ed.) xxxi. 603 The residents are then compelled to have recourse..to ship's biscuit. **1955** *Times* 29 June 14/5 Our shipbroking department had an active 12 months and profited during the latter part of the year from the substantial rise in tramp freights. **1969** *Daily Tel.* 24 Jan. 5/3 Wigham-Richardson is largely concerned with marine insurance, shipbroking and chartering. **1891** Ship's company [see *ship's writer*, sense 9 c in Dict.]. **1978** *Cornish Guardian* 27 Apr. 13/5 Shore-based organisations..will be involved..as well as 25 to 35 members of the ship's company. **1929** W. A. THORPE *Hist. Eng. & Irish Glass* II. Plate cxxix (*caption*), Ship's decanter, four angular rings round the neck. **1976** J. CARROLL *Madonna Red* (1977) iii. 93 The ambassador was holding a crystal ship's decanter. **1979** P. ALEXANDER *Show me Hero* vii. 90 A ship decanter and two wine glasses. **1939** W. FAULKNER *Wild Palms* 15 The flimsy walls (they were not even tongue-and-groove..but were of ship-lap). **1977** *Cornish Times* 19 Aug. 13/2 (Advt.), Our..plumbing, shiplap claddings and drainage systems are always in stock. **1958** *Archit. Rev.* CXXIII. 327 (*caption*) Northern elevation with ship-lapped pine used as facing for the first floor. **1876** W. LAMONT *Yachting in Arctic Seas* 13 These six Tromsönians were, in seagoing phrase, the *hardest bargains* I was ever shipmates with. **1880** W. CLARK RUSSELL *Sailor's Sweetheart* III. ii. 60, I had never been shipmates with an island of this kind before. **1961** G. FOULSER *Seaman's Voice* i. 13, I was never shipmates with a boom mainsail. **1919** in C. G. Grey *All World's Aircraft* 1. 96A The *Beardmore* W.B. III. was evolved from the *Sopwith* 'Pup' in an effort to turn this machine into a ship-plane. **1922** *Flight* XIV. 126/2 Landplanes designed so as to facilitate their landing on a ship's deck will ordinarily be known as Ship Planes. **1942** *Ark Royal* Aug. 13/2 A ship-plane represents certain constructional problems which entail a sacrifice of speed. **1439–1887** Shipside [in Dict., sense 8 a]. **1937** *Sun* (Baltimore) 26 Mar. 26/3 He parked his car in a garage, left orders with an automobile company to have a new machine at shipside [etc.]. **1969** *Jane's Freight Containers 1968–69* 83/3 General cargo facilities..have two shipside tracks. **1972** C. MUDIE *Motor Boats & Boating* 93 Most sports fishermen therefore incorporate a section of the cockpit coamings, shipside, or transom which can be removed to help loading. **1785** *Daily Universal Register* 1 Jan. 4/3 Sundry ships stores, consisting of sails, cables, anchors. **1943** *U.S. Navy Bluejacket's Man.* (ed. 11) 1143 The ship's store, perhaps better known as the 'Canteen', is also under the jurisdiction of the supply officer. **1969** A. R. BOSWORTH *My Love Affair with Navy* ii. 44 He had been to the ship's store, and he came into the ward with several candy bars. **1771** A GRAHAM *Observations on Hudson's Bay* (1969) ix. 282 How affairs went on last shiptime I know not. **1869** 'MARK TWAIN' *Innoc. Abr.* v. 47 Young Mr. Blucher..was a good deal worried by the constantly changing 'ship time'. **1956** *Beaver* Winter 52/1 Time to plan the spring work—but why do that; shiptime is far away and now is really the time for that rest. **1957** *Times Survey Brit. Aviation* Sept. 2/4 A ship-to-air weapon. **1972** *Times* 29 Sept. 4/8 The through-deck carriers will carry..the ship-to-air missile, Sea Dart. **1904** H. W. WILSON in *Cambr. Mod. Hist.* VIII. xv. 482 The battle of Camperdown..was not, as had been planned, a mere ship-to-ship encounter. **1944** *Proc. IRE* XXXII. 326/2 Ship-to-ship telephone communication. **1977** *Navy News* Aug. 19 (*caption*) H.M.S. Fife, one of four guided missile destroyers in Portsmouth Navy Days, with her new Exocet ship-to-ship missiles mounted just below her

bridge. **1923** *Monthly Weather Rev.* LI. 5/1 The cost of radio ship-to-shore tolls. **1962** K. C. HUTCHIN *How not to kill your Husband* xlvii. 221 The worst invention of recent years connected with sailing is 'ship-to-shore' radio-telephone. **1971** N. FREELING *Over High Side* III. 197 We've got the ship-to-shore. Couldn't we phone someone? **1977** B. GARFIELD *Recoil* xiv. 148 'Why the hell don't you ever turn on your ship-to-shore?'..'I go on this boat to get away from telephones.' **1979** *Daily Tel.* 22 Sept. 36/3 Makers of ship-to-shore oil pipes.

c. *ship's doctor* (later example); **ship's writer** (earlier and later examples); **ship's yeoman** (earlier example).

1974 L. DEIGHTON *Spy Story* xviii. 190 There was the ship's doctor. **1881** *Naval Encycl.* 745/2 *Ship's writer*, a petty officer who, under the directions of the executive-officer, does the writing and keeps the watch-, muster-, conduct-, and other books of the ship. **1969** T. PARKER *Twisting Lane* 43 He wasn't a sailor, he was a kind of a clerk on board ship... I believe he was called a ship's writer. **1850** Ship's yeoman [see YEOMAN 2 c].

ship, *v.* **6. a.** Delete † *Obs.* and add later *U.S.* examples. Cf. sense 6 c below.

1904 H. JAMES *Golden Bowl* II. xxxvii. 279 You regularly make me wish that I had shipped back to American City. **1978** M. PUZO *Fools Die* xvi. 171 By the time Frank and his units left the armory and shipped to Fort Lee there was a lot of bad blood.

c. U.S. Mil. slang. *to ship out*: to depart, to be transported; also *fig.* (cf. *shape up or ship out* s.v. *SHAPE *v.* 19 e); *to ship over*: to re-enlist, to volunteer for a tour of duty.

1908 L. G. TISDALE *Three Years behind Guns* xxiii. 259 Do you want to ship over? **1924** ANDERSON & STALLINGS *What Price Glory?* I. i. 7 When I left China the Yangtse was full of the bodies of virgins that drowned their beautiful selves because I was shipping over. **1948** [see *CASH *v.*[2] 2 b]. **1953** *CEC Bull.* Jan. 31/1 This outfit shipped out of Davisville 12 September 1943. **1964** G. L. COON *Short End* 223, I wouldn't ship over in Korea, and especially in Pankari. **1978** M. PUZO *Fools Die* xvi. 171 At the end of the month, when everybody shipped out, I bought Frank a present.

7. c. (Earlier example.)

1857 *Harper's Mag.* Sept. 459/2 A few of the more enterprising operators..thought nothing of shipping two or three thousand tons per annum.

e. *intr.* Of perishable goods: to admit of being transported.

1867 *Trans. Illinois State Agric. Soc.* VII. 510 It ships well, and is a very good peach. **1927** *Daily Express* 9 Nov. 5/5 Persimmons..will probably be as plentiful and popular as the banana, because it ships well and grows.. freely.

10. b. (Earlier U.S. example.)

1833 B. SILLIMAN *Man. Sugar Cane* 80 [The pan] is made to ship and unship.

11. a. (Further example.)

1851 H. MELVILLE *Moby Dick* I. xxxiii. 236 He pauses, ships a new face altogether.

b. *to ship a stripe*: to gain promotion in the navy or air-force. *colloq.*

1915 H. ROSHER *In R.N.A.S.* (1916) 38, I see in this morning's paper that I have shipped another stripe (Flight Lieutenant). **1924** *Blackw. Mag.* Mar. 333/2 For once his clothes were more interesting than mine for he had 'shipped' his half-stripe, and was a whole degree more important in the world!

shipboard. Add: **3.** Esp. in phrs. *shipboard acquaintance, romance*, etc., to denote casual or ephemeral relationships.

1916 G. B. SHAW *Overruled* 78 Was it the usual aimless man's lark: a mere shipboard flirtation? **1933** F. BALDWIN *Innocent Bystander* vi. 107 A shipboard romance will do a lot for his ego. **1963** 'W. HAGGARD' *High Wire* iv. 37 I'm a casual pick-up in the snow, a sort of shipboard acquaintance. **1978** 'M. M. KAYE' *Far Pavilions* ix. 148 It had been possible for Mrs Harlowe to introduce both young men as shipboard acquaintances. **1980** J. GARDNER *Garden of Weapons* II. viii. 194 He should be able to treat the business with Miriam like some shipboard romance. But his growing bewitchment would not allow that.

ship-breaker. Add: Also, a firm or company engaged in the business of breaking up old vessels.

1935 *Sun* (Baltimore) 21 May 12/6 The sale of some of the big outdated ocean liners to shipbreakers (that's the name for companies that take old ships apart with hammers and acetylene torches).

ship-breaking, *vbl. sb.* **2.** (Later examples.)

1931 A. HUXLEY *Let.* 6 Aug. (1969) 351 There are bits of Toulon harbour—ship-building and ship-breaking yards..I have always longed to paint. **1976** *S. Wales Echo* 26 Nov. 38/8 The Welshman..has business interests which include farming..and ship-breaking.

ship-load. (Earlier Amer. example.)

1639 *Portsmouth* (Rhode Island) *Rec.* (1910) 10 For men to gett a shipp lood of..pipe stauffes.

shipoo, var. *SHYPOO.

shippable, *a.* Restrict *rare*⁻⁰ to sense in Dict. and add: **2.** That can be shipped.

1920 *Glasgow Herald* 17 Nov. 11 The Southern Hemisphere promises for 1921 a shippable surplus of 40,000,000 qr. **1979** *Sci. Amer.* Apr. 31/1 This magazine page is a coated stock, several layers of clay filler and white pigment having been rolled onto the moving web during its passage from a wet slurry to a shippable roll.

shipper. 3. b. For *U.S.* read orig. *U.S.* and add earlier and later examples.

1840 *Niles' Register* 4 Apr. 80/2 Principal transportation lines have resolved to give the shipper or owner the full advantage of the reduction of twenty cents per barrel. **1950** *Times* 28 Feb. 4/5 Charter aircraft have been carrying bulk cargoes... By carrying full loads in each direction low rates have been available to shippers.

4. (Earlier example.)

1852 *Trans. Michigan State Agric. Soc.* III. 160 By the shipper the logs may be geared deeper or shallower.

shipping, *vbl. sb.* Add: **6. a.** *shipping company, house, lane, line.*

1897 *Whitaker's Almanack* 709/1 The New Zealand Shipping Company was established to run steamers direct to New Zealand, Tasmania, and Australia. **1924** *Times Trade & Engin. Suppl.* 29 Nov. 239/1 Shipping companies are complaining that apple shipments are not as heavy as they desire. **1919** *Brit. Manuf.* Nov. 40/2 The attempt to do away with the shipping house. **1931** W. G. CARR *By Guess & by God* xii. 194 German submarines which.. lurked about the shipping lanes waiting to attack the Allied merchantmen. **1974** L. DEIGHTON *Spy Story* xii. 122 Ice-breakers keeping two shipping lanes clear all through the winter. **1908** J. R. SMITH *Ocean Carrier* II. iii. 275 The starting of rival shipping lines is deterred by the certainty of fierce competition. **1981** A. GRAHAM-YOOLL *Forgotten Colony* xviii. 244 The Houlder Brothers shipping line..had been prominent in the meat trade for almost one century.

b. shipping fever *Vet. Sci.* (orig. *U.S.*), any of several diseases typically contracted by cattle while being shipped from place to place, esp. one caused by bacteria of the genus *Pasteurella.*

1932 *Jrnl. Amer. Vet. Med. Assoc.* LXXX. 165 The incidence of shipping fever was greatest during wet, cold weather. **1955** *Sci. News Let.* 15 Oct. 249/3 Shipping fever, the costly cattle disease that strikes like human influenza, makes the movement of cattle from range to feedlot one of the most dangerous activities in the livestock industry. **1970** T. G. HUNGERFORD *Dis. Livestock* (ed. 7) 332/1 Smeal recorded a case which clinically suggested shipping fever.

ship-repair (ʃip,ripēə·ɹ). [f. SHIP *sb.*[1] + REPAIR *v.*[2]] The business or craft of restoring a ship to a sound condition. Usu. *attrib.* So **shi·p-repairing** *vbl. sb.*; hence **shi·p-repairer,** a firm engaged in the business of repairing ships.

1941 W. S. CHURCHILL *Secret Session Speeches* (1946) 32 At least another 40,000 men must be drawn into ship-repairing. **1969** *Jane's Freight Containers 1968–69* 198/2 An adjacent yard also provides a ship-repair service. **1976** *Eastern Evening Press* (Norwich) 19 Nov. 1 The Government continued on a collision course..after winning a Commons vote..to retain ship-repairing yards. **1976** *Western Mail* (Cardiff) 22 Nov. 5/2 The ship-repairers don't want to be nationalised. **1976** *Jrnl.* (Newcastle) 26 Nov., The House of Lords refused to back down over 12 ship-repair firms in the old session of Parliament.

shipwreck, *v.* **2.** (Later *fig.* example.)

1932 J. BUCHAN *Gap in Curtain* iii. 153 His only success was with me, for I..could talk to him about..the inaccuracies of the Greville Memoirs. But the real rock on which the thing shipwrecked was Protection.

shipyard. Add: **2.** *attrib.* as **shipyard eye,** an epidemic form of keratoconjunctivitis caused by a virus.

1943 *Sun* (Baltimore) 17 June 13/6 The Health Department mentioned the new eye infection, shipyard eye, which appeared last winter. **1974** *Jrnl. Hygiene* LXXIII. 158 Because of this frequent occurrence of the disease among workers in shipyards, the term 'shipyard eye' was coined.

shiralee (ʃirălī·). *Austral. slang.* Also **shirallee.** [Origin unknown.] A bundle of blankets or personal belongings, a swag (sense 10).

1892 G. PARKER *Round Compass in Australia* 49 Let him down easy and slow... Drop in his shirallee and water-bag by him. **1945** BAKER *Austral. Lang.* v. 102 A drum..is the equivalent of *swag,..shiralee,*..or *bluey* as the tramp's rolled blanket is variously called. **1955** D. NILAND (*title*) The shiralee. **1955** *Times* 4 Aug. 9/2 The shiralee is the swag, the burden, the Australian swagman carries. **1974** *Sunday Sun* (Brisbane) 5 May 4/2 The fences, the barns, the houses—they're all gone and I'm out on the road with my shiralee.

Shiraz. Add: **1. b.** (Earlier example.) **d.** Denoting a rug or carpet made in the district of Shiraz; also *ellipt.*

b. 1840 N. P. WILLIS *Loiterings of Trav.* II. 224 He politely begged pardon for smoking in the Countess's presence, and filled the enamelled bowl with Shiraz tobacco.

d. 1900 J. K. MUMFORD *Oriental Rugs* xi. 213 The Shiraz displays unusual features of finish. *Ibid.* 214 The true Shiraz rugs may be known almost invariably by the small checked selvage. **1920** T. E. LAWRENCE *Let.* 16 Feb. (1938) 299 As for the rugs, please take any that seem worthy to you. There were two Afghans in the Arab Bureau, & a big (and not bad but thin) Shiraz, in the Savoy. **1932** P. SELVER tr. *Čapek's Tales from Two Pockets* 189 I've got piles and piles of Shiraz, Shirvan.. and other common-or-garden carpets. **1968** L. DURRELL *Tunc* III. 153 A small and lovely carpet—an authentic Shiraz according to the label. **1980** G. THOMPSON *Murder Mystery* (1981) xxviii. 215 A large brandy snifter now lay on the floor on the Shiraz rug.

2. The name of a variety of grape from which red wine is made, grown orig. in the Rhône valley of France; the wine made from this grape.

The French name for the grape is *syrah* (*scyras, sirrah* are also found). The Eng. form is app. an alteration of this, influenced by the belief that the vine was brought (by Crusaders) from Iran and is therefore to be identified with that from which *Shiraz* (sense a in Dict.) is made.

[**1908** E. & A. VIZETELLY *Wines of France* 140 For red Hermitage the vine..is the Ciras, Scyras, or Sirrah, a corruption, it is alleged, of Shiraz, the tradition being that the hermit of the mount brought some vine cuttings with him from the East. The Ciras is, at any rate, a distinct variety.] **1927** A. I. PEROLD *Treat. Viticulture* v. 271 Shiraz, the grape of Hermitage,..produces a fine, famous red wine. **1966** *Courier-Mail* (Brisbane) 25 Oct. 2/10 He thought his 1952 shiraz was of such vast quality there was no bottle in a restaurant cellar to equal it. **1973** 'E. FERRARS' *Small World Murder* viii. 102 They drank one of the Lyndon's own wines, a Shiraz. **1977** A. SCHOLEFIELD *Venom* I. 38 Replanting areas with *shiraz* vines.

shire, *sb.* Add: **3. f.** A rural administrative district in some states of Australia. Freq. *attrib.*

1909 BRIERLEY & IRISH *Crown Lands Acts New South Wales* (ed. 2) (Advt.), Ordinances for Municipalities or Shires. **1947** K. TENNANT *Lost Haven* iii. 58 A man with any push would form a progress association and devil the shire council about the roads. **1977** *Bulletin* (Sydney) 22 Jan. 46/2 The town also has a shire community centre.

6. a. Also = *shire counties* (see sense 8 b below).

1977 *Daily Tel.* 14 Mar. 2/7 In the shires Labour are defending a rump of seven non-Metropolitan counties they still hold out of 39: Cleveland, Derbyshire, Durham, [etc.].

8. b. shire county, a non-metropolitan county of the U.K., as instituted by the local government reorganization of 1974; **shire-town,** (*b*) *U.S.* = *county seat* s.v. COUNTY[1] 8 b.

1972 *Times* 21 Sept. 4/2 The AMC received no guidance whether metropolitan county councils would want to be grouped with 'shire' counties or with district councils. **1977** *Daily Tel.* 25 Apr. 6/8 Britain's great conurbations and the shire counties are preparing for a..tussle. **1648** *New Hampshire Provincial & State Papers* (1867) I. 189 The Court doth think fitt that the shire town of Norfolke be referred to further consideration. **1717** S. SEWALL *Diary* 13 Jan. (1882) III. 132 Cambridge is the Shire-Town for Middlesex. **1881** *Century Mag.* Dec. 251/1 It was the central town in the county, and yet not the shire-town. **1969** *Bangor* (Maine) *Daily News* 10 July 1/5 (*caption*) This particular sign in Whiting..has omitted an 'a' from the shiretown of Washington County.

shirk, *sb.*[2] **2.** (Earlier example.)

1863 *Sat. Rev.* 29 Aug. 278/1 Small shirks may be apples of Sodom, but they clearly constitute with some people one of the main pleasures of life.

shirk, *v.* Add: **1. c.** *intr.* To shift or fend *for* oneself. *U.S.*

1843 C. MATHEWS *Various Writings* 71/1 As for Harvest, let him shirk for himself. **1874** *Rep. Vermont Board Agric.* II. 422 They are then turned into the pasture to shirk for themselves.

Shirley[2] (ʃɜ̄·ɹli). The name of a district of Croydon, Surrey, used *attrib.* in **Shirley poppy** to designate an annual poppy bearing single or double flowers, usu. red, pink, or white, and belonging to a variety of *Papaver rhœas* developed by William Wilks (1843–1923), vicar of Shirley and secretary of the Royal Horticultural Society. Also *absol.*

1886 W. WILKS in *Jrnl. Hort.* 21 Oct. 367/1, I call them 'Shirley Poppies', as there seems so much doubt as to their specific name. **1889** *Ibid.* 15 Aug. 126/2, I sent a pinch of seed to a leading firm of London nurserymen last spring for comparison with the Shirley. **1904** *Nature* 25 Aug. 408/1 Many a white-edged poppy may have germinated and perished before Mr. Wilks saved the individual which in a few generations gave rise to the Shirleys. **1932** E. G. WHEELWRIGHT *Garden of Pleasant Flowers* iii. 82 Other common annuals easily raised are godetias,..Shirley poppies, and love-in-a-mist. **1978** R. GORER *Growing Plants from Seed* iv. 53 Shirley and opium poppies transplant very badly.

shirred, *a.* Add: **2.** Also *fig.*

1929 S. JAMESON in *Legion Bk.* 105 The waters..in sunshine ruffled and shirred with living light. **1946** C. McCULLERS *Member of Wedding* I. 29 The March winds banged on the window-panes, and clouds were shirred and white on the blue sky. **1973** P. WHITE *Eye of Storm* viii. 363 A brisk day: the harbour waters slightly shirred, newspaper rising and flapping in gutters.

shirt, *sb.* Add: **1. h.** A shirt of a particular colour worn as the emblem or uniform of a political party or movement. Also *transf.*, the wearer of such a shirt. Cf. *BLACKSHIRT, *RED SHIRT.

1864 [see *RED SHIRT, REDSHIRT]. **1922** [see *BLACKSHIRT, BLACK SHIRT]. **1934** *Times* 28 Feb. 15/5, I beg leave to point out that our election law requires to be brought up to date, since it was framed at a time when the political 'shirt' parties were undreamt of. **1939** H. G. WELLS *Holy Terror* II. i. 114 Two purple shirts who had visited his rooms in his second year. **1940** E. A. WALKER *South Africa* 23 Latterly more than one anti-Semitic 'shirt' movement has arisen owing a good deal to German encouragement and example. **1975** *Times Lit. Suppl.* 11 Apr. 392/1 *Antifalange*, a commentary on an apologia for the old shirts of the Spanish fascist movement.

2. d. (Later examples of *the shirt off one's back.*)

1925 W. N. BURNS *Saga Billy the Kid* 67 He was a free-hearted, generous boy. He'd give a friend the shirt off his back. **1980** *Times* 7 Oct. 10/5 One day this industry will have the shirt off my back.

f. *to put one's shirt on* (a horse, etc.) (examples); *to get* (a person's) *shirt out* (later example); *to keep one's shirt on*: to remain calm (orig. *U.S.*); *to lose one's shirt*: to lose all one's possessions.

1854 *Spirit of Times* (N.Y.) 4 Nov. 447/3, I say, you durned ash cats, just keep yer shirts on, will ye? **1904** W. H. SMITH *Promoters* i. 15 I'll tell you how, if you'll keep your shirt on. **1932** WODEHOUSE *Louder & Funnier* 113 Sure he knows you know, Bill... Don't get your shirt out. **1935** E. B. MANN *Thirsty Range* xi. 144 He hit the market..about the time the bottom dropped out of it. He lost his shirt! **1938** E. BOWEN *Death of Heart* I. i. 25 He had not foreseen ever having to put his shirt on either [woman]. **1945** *Chambers's Jrnl.* Oct. 554/1 Okay, okay—keep your shirt on. Let's see what can be done. **1954** T. S. ELIOT *Confidential Clerk* II. 63 Marriage is a gamble. But I'm a born gambler And I've put my shirt—no, not quite the right expression—Lucasta's the most exciting speculation I've ever thought of investing in. **1981** P. THEROUX *Mosquito Coast* xi. 131 'Keep your shirt on,' Father shouted.

5. a. *shirt pocket, -tail* (earlier example).

1962 L. DEIGHTON *Ipcress File* xxiii. 150 The very young soldier reached into his shirt pocket. **1977** D. AITKIN *Second Chair* i. 3 My hand in my shirt pocket, tugging at the little diary. **1846** J. W. WEBB *Altowan* I. vi. 174 He..leaped into the river,..and made a shirt-tail across the prairie on the other side.

c. shirt-bosom (earlier example); **shirt-button** (earlier and later examples); **shirt-dress,** a dress having a bodice styled like a shirt; **shirt front**: also *transf.* of the breasts of other animals; also *attrib.* of a cricket pitch: very smooth and even (*colloq.*); **shirt-jac(ket)** chiefly *U.S.*, a garment resembling a shirt but worn as a jacket; **shirtlifter** *Austral. slang*, a male homosexual; **shirtmaker** = *shirt-waist dress* below; freq. *attrib.* (a proprietary term in the U.S.); **shirt-tail,** (*a*) see sense 5 a in Dict.; (*b*) *U.S.*, used *attrib.* or as *adj.* to designate something small and insignificant, or a remote relationship; freq. as **shirt-tail boy,** a very young boy; **shirt-waist**: for *U.S.* read 'orig. *U.S.*' and add earlier example; also formerly a garment worn by men and boys; freq. *attrib.* as **shirt-waist dress,** a dress having a shirt-waist bodice; **shirtwaister** = prec.

1833 J. NEAL *Down-Easters* I. 3 His collar turned back, and his shirt-bosom all open to the waist. **1651** R. VERNEY in M. M. Verney *Memoirs* (1894) III. ii. 38 Blew Thread, Shirt Buttons and old White..Buttons. **1742** C. CARROLL *Let.* 24 Nov. in *Maryland Hist. Mag.* (1925) XX. 178 Three or four Papers good shirt Buttons but not made on Wire. **1978** P. NIESEWAND *Underground Connection* 123 He undid his shirt buttons and stripped to the waist. **1943** in C. W. Cunnington *Eng. Women's Clothing Present Cent.* (1952) viii. 273 Necklines avoid the shirt-dress look which has been so widespread in recent years. **1973** *Country Life* 22 Feb. 492/1 If ever there was a right season to wear a shirt-dress, this is it. **1978** *Detroit Free Press* 5 Mar. D 12/1 A delicate mini-floral two-piece shirtdress. **1877** E. S. DALLAS *Kettner's Bk. of Table* 104 The carp is to be stuffed... The skin may be left on his shirtfront. **1920** P. F. WARNER *Cricket* 212 The result of all this work is that the pitch literally shines—and looks as if it had been ironed. 'Shirt-front wickets' they call them. **1963** *Times* 18 May 4/5 By merely bowling accurately, with seam upright, they presented problems which West Indian batsmen, reared on the shirt-front surfaces of their own grounds, found too complicated. **1967** L. EGAN *Nameless Ones* xiii. 158 He [*sc.* the cat] was a handsome sight, his gray tiger stripes smooth and his white shirt front immaculate. **1964** *Playboy* Nov. 173 The cool, crisp and comfortable shirt-jac which looks like a shirt, but is worn outside the trousers. **1977** *Guardian Weekly* 10 July 9/1 'Shirt-jacs', as they call tropical suits in Trinidad. **1975** *Daily News* (N.Y.) 26 July 12 Many leisure suits have shirt-jackets rather than the traditional jacket. **1966** BAKER *Austral. Lang.* (ed. 2) x. 216 *Shirt lifter*, a sodomite. **1974** B. HUMPHRIES in *Bulletin* (Sydney) 19 Jan. 13 When I first seen them photos of him in his 'Riverina Rig' I took him for an out-of-work ballet dancer or some kind of shirtlifter. **1926** *Official Gaz.* (U.S. Patent Office) 15 June 584/2 Best & Co... Shirtmaker frock..women's and children's dresses. **1960** *Guardian* 27 July 7/3 For this summer they have chosen a shirtmaker in drip-dry cotton. **1976** 'R. ROYLE' *Cry Rape* xx. 91, I chose a simple navy shirtmaker dress. **1845** J. HOOPER *Adventures Capt. Simon Suggs* 13 From the time he was a 'shirt-tail boy', [his wits] were always too sharp for his father's. **1846, 1873** Shirt-tail [in Dict. and Suppl., sense 5 a]. **1878** J. C. GUILD *Old Times in Tennessee* 411, I traversed these granite hills and beautiful vales as a shirt-tail boy. **1929** W. FAULKNER *Sound & Fury* 256 My people owned slaves here when you all were running little shirt tail country stores. **1938** M. K. RAWLINGS *Yearling* xxxiii. 421 Nobody but your folks'll bother with a little ol' shirt-tail boy like you. **1941** *Amer. Speech* XVI. 24/2 *Shirt-tail kin*, a remote relationship. **1975** *Publishers Weekly* 8 Sept. 57/2 A shirttail relation of the hotel-owning branch of the family. **1879** *Harper's Bazaar* 14 June 377 Kilt suits made here have the pleats stitched to a belt at the waist, and are then buttoned to a white shirt waist. **1902** *Sears Catal.* 819/3 Three hundred dozen men's regular $1·50 shirtwaists to go at 50 cents. **1957** *Observer* 1 Dec. 11/2 This gives the many lovely, tight-belted shirt-waist dresses a heavy look. **1980** *Times* 2 Sept. 10/1 The shirt-waist dress..is still a basic article of apparel. **1957** *Observer* 1 Dec. 11/2 These shirt-waisters are lovely, bodices luxuriously bloused [etc.]. **1973** *Country Life* 8 Mar. 633/2 The longer cardigan jacket..is worn in the daytime over shirtwaisters.

shirtless, *a.* (Later examples.)

1962 J. F. POWERS *Morte d'Urban* 218 Some shirtless youths in an old car rolled up from behind him. **1971** *Daily Tel.* (Colour Suppl.) 16 July 36/1 It had been a long slog up, shirtless, sweating, rucksacks bumping.

shirtsleeve. Add: **1. a.** (Earlier example of *in one's shirtsleeves.*) Also used *loosely* in pl. with reference to the absence of a coat.

1789 J. WOODFORDE *Diary* 30 July (1927) III. 126 The latter was..working in his garden in his Shirt Sleeves. **1832** F. TROLLOPE *Dom. Manners Amer.* II. xxv. 56, I saw one man..take off his coat that he might enjoy the refreshing coolness of shirt sleeves. **1942** D. M. CROOK *Spitfire Pilot* 45 We were flying in shirt sleeves. **1977** *Times* 29 Oct. 5 (*caption*) A Panama hat and shirtsleeves for the Duke of Edinburgh in the tropical sun.

2. *attrib.* (in sing. or pl.) **a.** That is in shirt-sleeves; usu. *fig.*, hard-working, workman-like; down-to-earth, informal; (see also quot. 1959).

1864, 1908 [in Dict.]. **1924** LAWRENCE & SKINNER *Boy in Bush* 11 The shirt-sleeves familiarity, the shabby clothes. **1959** *Times* 7 Sept. 16/1 Shirtsleeves weather. **1967** *Boston Sunday Herald* 26 Mar. I. 44/2 (Advt.), A 'self-starter' and a guy who will fit into a small shirtsleeve agency. **1979** G. SEYMOUR *Red Fox* iii. 44 [He] would have given much to have exchanged the brilliance of the surroundings for a shirtsleeves working area.

b. *shirtsleeve(s) diplomacy*, management of political affairs which is characterized by lack of formality or sophistication; *shirtsleeve order* (Mil.), the wearing of uniform without a jacket.

1931 W. F. SANDS *Undiplomatic Mem.* 22, I admit some pride in believing that they acted like gentlemen in the matter, though perhaps it was only shirt-sleeves diplomacy. **1959** *Times* 22 Aug. 9/5 Maximum cartridges are an asset for long-range shots, but perhaps hardly ideal for shirt-sleeve order. **1977** *Listener* 11 Aug. 164/3 Accessibility ..and, in the best sense, shirtsleeve diplomacy are on. **1979** 'J. D. WHITE' *Brandenburg Affair* iv. 37 Colonel Petrov..was another huge man... Even in shirt-sleeve order..his massive arms..gave him an air of permanence.

shirty, *a.* Restrict *slang* to sense in Dict. and add: **1.** (Earlier and later examples.)

1846 *Swell's Night Guide* 54 'I am exceedingly obliged,' grunted Tomkins, in rather a shirty tone, and continued reading. **1856** H. PHILLIPS *Jrnl.* 18 Apr. (typescript) 30 Jem and Mrs R. Shirty. **1911** [see *ROUSE *v.*[4]]. **1916** *Chambers's Jrnl.* June 404/2 Please don't get shirty, old chap. **1927** [see *RAG *sb.*[1] 3 c]. **1934** WODEHOUSE *Right Ho, Jeeves* vii. 73 But don't tell me that when he saw how shirty she was about it, the chump didn't back down? **1960** J. RAE *Custard Boys* I. v. 54 All right; all right; there's no need to get shirty about it.

2. [f. SHIRT *sb.* 1.] Resembling or modelled on a shirt.

1958 J. KEROUAC *On Road* IV. 288 A sixteen-year-old colored girl..in her short shirty dress. **1973** *Guardian* 10 Apr. 13/2 Fox-trimmed parka with shirred waist, shirty cuffs.

Hence **shi·rtily** *adv.*

1974 P. CAVE *Mama* (new ed.) xiv. 113 'Seems a bloody con to me,' he mumbled shirtily. **1978** D. FRANCIS *Trial Run* vii. 105 They told us pretty shirtily just now not to bother them.

Shirvan (ʃɜ̄ɹvā·n). The name of a region in the Soviet republic of Azerbaijan, used *attrib.* or *ellipt.* to denote a short-napped rug or carpet made in that area and similar to those of Daghestan.

1892 CARDINAL & HARFORD *Oriental Carpets & Rugs* 21 Daghestan. This title includes those kinds known as Kazaos, Karabaghs, Cubas, Shirvans, &c. **1913** [see *KABISTAN]. **1931** [see *DAGHESTAN, DAGESTAN]. **1970** L. DEIGHTON *Bomber* xxiii. 350 Modern Shirvans. They won't interest you, you know too much about carpets.

shise, var. *SHICE *sb.* and *a.*

‖ **shishi** (ʃī·ʃi). [Jap.] A lion, *spec.* as a decorative motif on Japanese porcelain.

1970 *Ashmolean Mus. Rep. Visitors 1969* 48 Saucer, with two *shishi* in Kakiemon enamels. **1976** *Daily Tel.* 20 July 12/3 The interior is enamelled in iron-red, blue, turquoise and yellow with a shishi beneath a spray of peony.

shish kebab (ʃi·ʃ kĭbæb). Also **sheesh kabab, shish-kebab, shishkebab, shushkabab.** [a. Turkish *şişkebap*, f. *şiş* skewer + *kebap* roast meat.] **1.** A dish consisting of pieces of meat (usu. lamb) grilled on skewers. Cf. CABOB 1, *KEBAB.

1914 S. LEWIS *Our Mr. Wrenn* ii. 26 I'm sure you'll like shish kebab. **1921** A. C. TRAIN *By Advice of Counsel* 73 Sardi had ordered *sheesh kabab*. **1951** KOESTLER *Age of Longing* I. vii. 153 'What kind of dishes do you really like?' ..'Shashlik. And shushkabab.' **1960** *Times* 4 June 7/6 Stands of appetizing *shish-kebab* on bamboo skewers. **1976** *Outdoor Living* (N.Z.) I. II. 63 Perfectly suited to barbecue cooking are shishkebabs and the variations are infinite. **1980** P. WAY *Icarus* xli. 180 The cluster of shish kebab stalls.

2. *Physical Chem.* A fibrous crystalline structure formed in some flowing or agitated polymer solutions, consisting of many plate-like crystallites (*kebabs*) growing outwards from a long ribbon or rod (a *shish*).

1966 A. J. PENNINGS in H. S. Peiser *Crystal Growth* 391/2 Most of the fibres exhibit lamellar overgrowth..and helical structures can also be observed. These structures will be referred to as Shish-kebabs. **1974** J. SCHULTZ *Polymer Materials Sci.* ii. 111 Electron and x-ray diffraction experiments have shown that the polymer chains in both shish and kebab are aligned parallel to the fiber axis. **1975** *Nature* 15 May 195/3 The shish kebab, of overall diameter about 1 μm and length up to several mm, apparently has a central core, the 'shish', about 20 nm diameter skewering lamellar crystals, the 'kebabs'. **1979** *Ibid.* 29 Mar. 440/1 The formation of fibrous precipitates, almost entirely of the shish kebab structure, on the stirrer by stirring supercooled solutions of polyethylene and isotactic polystyrene respectively.

shit, *sb.* Add: The form **shite** now chiefly occurs as an occasional jocular or quasi-euphemistic variant.

1971 B. W. ALDISS *Soldier Erect* 137 Do you think Churchill gives a shite for the Fourteenth Army? **1976** *New Musical Express* 17 Apr. 11/4 If you have to spend a lot of time with people who are interested in their chess boards and little card games and shite like that, it can drive you *nuts*.

1. a. (Later examples.)

1961 F. KING *Custom House* xix. 275 Leave that shit alone! Filthy dog. **1967** P. ROTH *Portnoy's Complaint* (1969) 47 Trying to clear my feet of my undershorts before anybody can peek inside, where..I always discover in the bottommost seam a pale and wispy brushstroke of my shit. **1973** E. JONG *Fear of Flying* (1974) ii. 25 In general the toilets run swift here and the shit disappears long before you can leap up and turn around to admire it. **1980** K. DOVER *Greeks* ii. 38 We might pick on his revelation of what Greek warfare was like... Blood and shit and pus are the same..in all ages.

b. (Later examples.)

1921 D. H. LAWRENCE *Let.* 10 Nov. (1962) II. 673 They are both such abject shits it is a pity they can't be flushed down a sewer. **1922** E. M. FORSTER *Let.* 27 Sept. in P. N. Furbank *E. M. Forster* (1978) II. v. 106, I think that most Indians, like most English people, are shits. **1926** C. CONNOLLY *Let.* 3 Aug. in *Romantic Friendship* (1975) 157 Her son is a complete little shit though.. witty and humorous beyond his years. **1941** J. REITH *Diary* 20 Oct. (1975) v. 281 Beaverbrook—to no one is the vulgar designation shit more appropriately applied—telephoned about park railings. **1956** I. MURDOCH *Flight from Enchanter* xii. 176 'You beastly contemptible shit of a crook,' said Hunter. **1968** *Observer* 29 Sept. (Colour Suppl.) 25/3 We hate the staff here. Keep away from them as much as possible. The shits. **1975** D. LODGE *Changing Places* ii. 67 Is that little shit still shooting his mouth off in there? **1976** J. I. M. STEWART *Young Pattullo* viii. 165 She was a third-class harlot who made up for it by being a first-class shit. **1978** J. IRVING *World according to Garp* xii. 217 Oh, I never knew what *shits* men were until I became a woman.

c. In negative contexts: Anything. Phr. *not to give a shit*: not to care at all.

1922 JOYCE *Ulysses* II. 587 He's a whitearsed bugger. I don't give a shit for him. **1969** W. LABOV in J. E. Alatis *Teaching Standard Eng. to Speakers of Other Languages* (1970) 15 The average whitey out here got everything, you dig? And the nigger ain't got shit, y'know. **1970** *Landfall* (N.Z.) Sept. 218 Nobody gives a shit for nobody. **1973** D. BARNES *See Woman* (1974) I. 19 Don't tell them shit. The skipper is on his way, and he'll decide what to tell them. **1978** K. AMIS *Jake's Thing* iii. 30 An interviewer..being very rude to a politician..and the politician not giving a shit.

d. *transf.* Rubbish, trash.

1930 A. HUXLEY *Let.* 7 Jan. (1969) 326 In every case something precious and lovely had been taken away and replaced by a mound of shit. **1957** I. CROSS *God Boy* (1958) xxii. 192 They just tell me she's in a hospital and that God knows best and all that shit. **1966** L. COHEN *Beautiful Losers* I. 8 Listen, F., don't give me any of your mystical shit. **1976** M. SPARK *Takeover* x. 149 Even if it's shit it gets people thinking about religion. **1977** *Rolling Stone* 5 May 6/2, I enjoyed Simmons' logic that Shakespeare is 'shit' simply because he can't understand it.

e. *fig.* Misfortune, unpleasantness. Esp. in phr. *to be in the shit*: to be in trouble or difficulty.

1937 PARTRIDGE *Dict. Slang* 758/2 *sh**, in the*, in trouble. **1958** S. BECKETT *Malone Dies* 98 In any case, here I am back in the shit. **1971** B. W. ALDISS *Soldier Erect* 162 We were all in the shit together and it was madness to try and escape it. **1977** *Rolling Stone* 24 Mar. 55/5, I feel really lucky that I've had the opportunity to go through some of the heartaches and shit we've been through the past year.

f. An intoxicating or euphoriant drug, *spec.* cannabis, heroin, or marijuana.

1950 L. Rivers in *Neurotica* Autumn 45 Senor! You want some shit? How much? Senor, I have great stuff. **1960** J. Gelber *Connection* II. 88 At that time shit was relatively scarce and I had to go out of the city to score. **1972** *Daily Tel.* 3 Apr. 8 Acid (LSD) and 'shit' (cannabis), were on open sale, and..a notice was pinned to a tent stating: 'Anybody with some black shit for sale, ask for Irish Mick.' **1980** S. Wilson *Dealer's War* III. ix. 229 'Hope it's good shit,' I whispered as he swabbed my arm.

g. In phrases *up shit creek*: in an unpleasant situation or awkward predicament (cf. *up the creek* s.v. *creek *sb.*¹ 5); *shit out of luck*: (see quot. 1942); (*when*) *the shit flies* or *hits the fan*: alluding to a moment of crisis or its disastrous consequences; *to beat, kick,* or *knock the shit out of* (someone): to thrash or beat severely; *to get one's shit together* (U.S.): to collect oneself, to manage one's affairs.

1937 J. Dos Passos *U.S.A.* I. 70 We're up shit creek now for fair. **1942** Berrey & Van den Bark *Amer. Thes. Slang* § 219/10 Unlucky..shit out of luck, (all) washed up. **1966** P. O'Donnell *Sabre-Tooth* iv. 62 We're all going to be there, where the shit's flying. **1966** L. Cohen *Beautiful Losers* I. 122 Let's beat the shit out of him. **1967** Partridge *Dict. Slang* Suppl. 1355/2 Wait till the major hears that! Then the shit'll hit the fan! **1968** A. Diment *Bang bang Birds* ix. 172 Should the shit hit the fan and the Swedes come over stroppy, he could say..'weren't nothing to do with us, son!' **1969** *Win* 15 May 31/2 We sense the government and its agents daily becoming more ineffective as we get our own shit together. **1971** B. W. Aldiss *Soldier Erect* 260 The Japs..were meek and respectful... The shit had been knocked out of them. **1973** *Black World* June 62 He sure didn't want a family..to support..just when he was '..gettin' my shit together to finish school'. **1977** H. Fast *Immigrants* III. 171 It's been too quiet. Tomorrow, the shit hits the fan. **1978** M. Puzo *Fools Die* i. 10, I will show you the artist getting the shit kicked out of him for the sake of his art. *Ibid.* xl. 450 So you see, my dear, you're shit out of luck. **1981** *Private Eye* 31 July 11/2 If they'd followed her this far up shit creek it's a long way to walk back.

2. b. *the shits*, diarrhœa (in persons). Also *fig.*

1947 *Amer. Speech* XXII. 305 I'd rather die with the screaming shits. **1967** *Coast to Coast 1965–6* 200 Women have always given me the shits. **1977** *Zigzag* Mar. 8/1 'I've had the shits,' he cried. 'You want to avoid the food.'

3. a. In terms of abuse, as *shit-ass, -bag, -breeches, -face, -head, -heel, -pot*; **b. shit-hole** (see quot. 1937); usu. *fig.*; **shit-hot** *a.* (see quot. 1961); also used *loosely* as a term of approbation; **shit-house** (later examples); also in *gen.* use as a term of disgust or contempt (freq. *attrib.*); **shit-kicker** *U.S.*, a rustic; **shit-list** (see quots. 1942, 1945); **shit-scared** *a.*, extremely frightened; **shitwork**, (esp. in the language of feminists) work considered to be menial or routine, esp. housework.

1942 Berrey & Van den Bark *Amer. Thes. Slang* § 396 Terms of disparagement..shit-ass. **1971** B. Malamud *Tenants* 165 He then cried out, 'Oh what a hypocrite shitass I am to ask a Jew ofay for advice how to express my soul work.' **1937** Partridge *Dict. Slang* 758/2 *Sh**-bag*, the belly; in pl., the guts. **1961** *Ibid.* Suppl. II. 1269/1 *Shit-bag*,..an unpleasant person. **1968** Bethell & Burg tr. *Solzhenitsyn's Cancer Ward* I. viii. 121 All he could see was this shitbag wolfing a chicken bone. **1973** Shit-bag [see *mental *a.*¹ 1 c]. **1922** Joyce *Ulysses* 428 Hey, shitbreeches, are you doing the hattrick? **1937** W. L. G. Cowan *Loud Report* II. 97 'Hallo, s— face.' **1973** M. Amis *Rachel Papers* 115 'Why,' I wondered, 'did old shitface come round? What was he after?' **1961** Partridge *Dict. Slang* Suppl. 1269/2 *Shit-head*, an objectionable person. **1971** J. Michener in *Reader's Digest* Apr. 240 Again the girls were particularly abusive, taunting the guards, calling them 'shit-heads', 'half-ass pigs'. **1979** P. Niesewand *Member of Club* viii. 56 You lying shithead! **1935** J. Hargan *Gloss. Prison Lang.* 7 *Shitheel*, an inmate who considers himself superior to all the others. **1939** J. Steinbeck *Grapes of Wrath* 212 And Mae, when she is alone with Al, has a name for them. She calls them shitheels. **1977** H. Fast *Immigrants* VI. 359 You could have sent a registered letter, or that little shitheel of an errand boy, Clancy? **1937** Partridge *Dict. Slang* 758/2 *Sh**-hole*, the rectum. **1969** A. Cornelisen *Torregreca* v. 176, I made up my mind early I wasn't going to..spend my life..in one of those shit-holes. **1977** *Zigzag* June 28/3 John went to a Catholic school in Caledonian Road—'a right shit-hole'. **1961** Partridge *Dict. Slang* Suppl. 1269/2 *Shit-hot*, unpleasantly enthusiastic,..very skilful, cunning, knowledgeable. **1973** M. Amis *Rachel Papers* 199 They've elected a new guy... I don't know anything about him. Except that he's shit-hot. **1976** *Sounds* 11 Dec. 29/2 Chuck Leavell's pretty damn good all the time, and the rhythm section's still shit-hot. **1922** Joyce *Ulysses* 335 Cute as a shithouse rat. **1972** G. Morley *Jockey rides Honest Race* 173 You're probably right..but I still feel shithouse about it. **1973** J. Wainwright *Devil you Don't* 46 Have you explained all this shithouse philosophy to the rate-payers? **1976** P. Cave *High flying Birds* ii. 19 'Nothing wrong with it—safe as a brick-built shithouse,' I assured her. **1977** *Zigzag* Aug. 5/2 If you're banned in town A and then banned in town B, well then town C has just got to ban you or it's, 'well what kind of shithouse place are you running there, councillor?' **1966** *Publ. Amer. Dial. Soc.* 1964 XLII. 29 The commonplace generic term for any rustic, *shit kicker*. **1969** L. Michaels *Going Places* 23, I was a city boy. No innocent shitkicker from Jersey. **1969** *Rolling Stone* 28 June 14/1 Saturday nights the avid shitkicker can whoop it up. **1942** Berrey & Van den Bark *Amer. Thes. Slang* § 336/2 *Blacklist*,..shit or stink list. **1945** *Amer. Speech* XX. 263 In the vulgar talk of the barracks, soldiers uninhibitedly use the phrase *shit list*, for a list of men whom one dislikes and is anxious to see embarrassed or inconvenienced. **1965** *Liberator* Aug. 23/1 Sweet Mac is on my shit list. **1970** R. D. Abrahams *Positively Black* i. 8 Moynihan had made it onto the black shit-list in spite of his obvious sympathies. **1937** Partridge *Dict. Slang* 758/2 *Sh**-pot*, a thorough or worthless humbug (person); a sneak. **1971** B. Malamud *Tenants* 132 Lesser, don't think you so hot, You got the look of a shit-pot. **1958** P. Scott *Mark of Warrior* 169 I'm shit-scared stuck up there with all my men gone. **1977** *Rolling Stone* 13 Jan. 12/4 Stewart was 'shit scared' about opening night. **1968** *No More Fun & Games* Oct. 43 Along with their equal integrated position they can equally misuse their less political sophisticated sisters to do their shitwork. **1972** *Guardian* 30 Mar. 13/3 They call it..'shit work' and they equate it with emptying dustbins and crawling on your belly in a coal mine... The resentment against housework came up like a great surging wave. **1980** D. Spender *Man Made Lang.* i. 48 Because of its parallels with housework, Fishman argues that women do the shitwork in conversation.

c. *attrib.* or as *adj.*

1968 H. Davies *Beatles* ix. 66, I think it [*sc.* jazz] is shit music, even more stupid than rock and roll. **1971** B. Malamud *Tenants* 104 He sat on the bed with a shit smile on his mouth. **1973** Boyd & Parkes *Dark Number* v. 55 'Look, so you've got a crippled leg.'.. He winced at that and turned away. 'That was a shit thing to say.'

shit, *v.* Add: *Pa. t.* and *pa. pple* **shat.** Forms: *Pa. t.* 7—shit; shat, shitted. *Pa. pple* shat, shit. **1.** (Further examples.)

c **1784** W. Blake *Island in Moon* in *Compl. Writings* (1966) 46 'The trumpeter shit in his hat,' said the Epicurean. **1929** C. Connolly *Let.* Nov. in *Romantic Friendship* (1975) 329 It [*sc.* a kinkajou]..had a genius for sitting where it shat. **1952** M. Lowry *Let.* 24 Nov. (1967) 323 A seagull has shat on the roof of the convent. **1975** R. Hoban *Turtle Diary* xlix. 199 Their dogs shitted on the paths. **1976** *Listener* 22 Jan. 92/1 The titillation of finally finding out that great men and women spat and shat and had piles. **1979** *Guardian Weekly* 4 Nov. 21/2 Shat in his pants with fear.

2. (Later example.)

1978 T. L. Smith *Money War* (1979) I. 149 The planes..had shit a neat stream of Day-Glo orange bricks.

3. a. Esp. in phr. *to shit oneself*: (*a*) to defile oneself with excrement; (*b*) *fig.*, to be afraid.

1914 Ld. Fisher *Let.* 14 Aug. in M. Gilbert *Winston S. Churchill* (1972) III. *Compan.* I. 35 The French Admiral shot himself. The English Admiral 'shit' himself. **1968** A. Diment *Bang Bang Birds* iv. 51 You can shit yourself. I'm not working for you. **1977** *Spare Rib* May 8/3, I was shitting myself before I came, looking for all kinds of excuses. **1980** *Sunday Times* 27 Apr. 42/3, I can easily arrange not to be diverted by knowing when a sorry old man shat himself.

b. In slang phrases *to shit* (*someone*): to tease or attempt to deceive; *to shit a brick*: (see quot. 1961); also as *int.*

1934 H. Miller *Tropic of Cancer* 61 Carl looks at me in despair. 'Is he shitting me, that bastard?' **1961** Partridge *Dict. Slang* Suppl. 1269/1 *Shit*, esp. *be shitting, bricks*, to be really worried,..to be thoroughly frightened. **1965** M. Shadbolt *Among Cinders* x. 83 'A queer thing. Something psychological.' 'Shit a brick,' he said. **1971** B. W. Aldiss *Soldier Erect* 187 You're shitting me, Jock! You never did it! **1976** H. Ferguson *Confessions Long Distance Acid Head* 48 By the time I got back to the hospital they were all shitting bricks. **1978** J. Krantz *Scruples* vii. 216 'Prince will shit a brick,' Billy said with a giggle. **1979** C. Kilian *Icequake* iii. 47 Didja see the wave comin' across the Shelf?.. There was a wave. I'm not shittin' you.

shitting *ppl. a.* (later *fig.* examples); **shitter,** (*a*) (later example); (*b*) a privy; a lavatory pan.

1952 *Amer. Speech* XXXIII. 270 Here I bring together all the current terms [for 'horse']. The most common are *pony, cayuse, shitter*, and *scate*. **1967** M. Shulman *Kill 3* II. iv. 77 You bastard! You shitting little bastard! **1969** R. Fernandez in A. Chapman *New Black Voices* (1972) 380 Markings on a shitter wall. **1971** B. W. Aldiss *Soldier Erect* 168 The blokes say Calcutta's got more whore-houses than it has shitters. **1971** *Black Scholar* Sept. 46/2 He lit a square and sat down on the shitter and tried to collect his thoughts. **1980** L. Cooper *Desirable Residence* xlii. 160 That shitting girl looks at me as if I was dirt.

shit (ʃit), *int.* Also **shee-y-it, she-it** (emphatic forms), **shite.** [f. the sb. and vb.] **1.** A coarse exclamation of annoyance or disgust.

1920 Joyce *Let.* 3 Jan. (1957) I. 134 O shite and onions! When is this bloody state of affairs going to end? **1922** E. E. Cummings *Enormous Room* xi. 219 My father is dead! Shit! Oh, well. The war is over. Good. **1925** D. H. Lawrence *Let.* 7 Nov. (1962) II. 865 Why doesn't somebody finally and loudly say Shit! to it all! **1959** 'E. McBain' *87th Precinct* xix. 138 'When I came back, he was gone.' 'Shit,' Willis said. **1969** 'J. Morris' *Fever Grass* xvii. 150 'Shit!' Scully said disgustedly. 'Oh, shee-y-it!' **1977** *Time* 14 Nov. 66/3 Aw, she-it, as the street kids say.

2. In trivial use.

1937 J. Dos Passos *U.S.A.* I. 73 Shit, let's try pick 'em up. **1969** *Private Eye* 23 May 14 Shit! What a lovely scene. **1976** P. Cave *High flying Birds* iii. 42 'Aw, shit. It was nothing,' she muttered, writing the matter off casually.

shite-hawk (ʃəi·thǭk). [f. Shit(e) *sb.* + Hawk *sb.*¹] **1.** In India, a name used for a kite of the genus *Milvus*. (This sense probably much earlier in oral use.)

1967 Partridge *Dict. Slang.* Suppl. 1356/2 Shite-hawk... A vulture: British Army in India: ca. 1870–1947. **1971** B. W. Aldiss *Soldier Erect* 75 The universal kite-hawks—universally known as 'shite-hawks'—had been plentiful in Kanchapur; in Vadikhasundi, they were two a penny. Like the fly, the shite-hawk was one of India's essential scavengers.

2. *transf.* = Shit *sb.* 1 b.

1958 M. K. Joseph *I'll soldier no More* iv. 95 That shitehawk... Why can't a man like that be shot like a sick horse. **1981** J. B. Hilton *Playground of Death* v. 64, I liked the man... And yet he was a shite-hawk. He was a journalist.

shitepoke. Also **shikepoke. 1.** Add: also, the black-crowned night heron, *Nycticorax nycticorax*, or the bittern, *Botaurus lentiginosus*. (Earlier and later examples.)

1775 *First Bk. Amer. Chron.* ii. 19 They drummed with their drums,..running too and fro like shite-pokes on the muddy shore. **1913** [see *hop-toad]. **1942** W. Faulkner in *Sat. Even. Post* 28 Mar. 11/3, I..went to the barn and got the slingshot and the shikepoke egg. **1966** *Publ. Amer. Dial. Soc.* XLII. 17 The green heron... Another widely-used name is shitepoke (or shikepoke). **1972** G. Beine *Land of Coyote* 88 A shitepoke waded about searching for frogs.

2. (See quot. 1926.)

1926 in H. Wentworth *Amer. Dial. Dict.* (1944) 550/2 *Shitepoke*,..applied opprobriously to a person. **1936** D. Lutes *Country Kitchen* 19 I'll return it—when they've returned all the molasses and sugar and eggs and everything else they've borrowed in the last year—the old shitepoke.

shitless (ʃi·tlės), *a. coarse slang.* Alluding to a state of extreme fear or physical distress. Esp. in phr. *to be scared shitless.*

1936 L. Durrell *Spirit of Place* (1969) 42 We're scared shitless because if there's any place Benito wants more than Ethiopia it's Corfu. **1964** L. Auchincloss *Rector of Justin* x. 158 You're scared shitless this little affair will do you out of old Tanager's dough, aren't you? **1971** B. Malamud *Tenants* 62 On my first solo gig I was..beaten shitless, and dumped in jail. **1976** *New Musical Express* 12 Feb. 24/1 The self-appointed custodians of public morality who campaign against pornography because they're simply scared shitless by it.

shitten, *a.* **b.** (Later examples.)

1846 *Swell's Night Guide* 49 Which of us had hold of the crappy (sh-ten) end of the stick? **1931** [see *lord *sb.* 12 d].

shitticism (ʃi·tisiz'm). *joc.* [f. *shitty *a.*, after *witticism.*] A scatological figure of speech.

1936 R. Frost *Let.* 9 May (1964) 277 My contribution was the witticisms: yours the shitticisms. **1977** *N.Y. Rev. Bks.* 10 Nov. 21/1 On his deathbed Frost forgave Pound his shitticisms.

shitty (ʃi·ti), *a.* [f. Shit *sb.* + -y¹.] **a.** = Shitten *a.* b.

1924 E. Hemingway *Let.* 19 July (1981) 119 In all other arts the more meazly and shitty the guy, i.e. Joyce, the greater the success in his art. **1952** S. Kauffmann *Philanderer* (1953) iv. 65 The first thing we do is change the looks of *Hearts Today*. We keep the same size but we get rid of that shitty confession look. **1970** *Guardian* 18 Nov. 11/6 You keep asking shitty questions that are irrelevant. **1977** *Spare Rib* Sept. 4/3 All the shitty jobs that most women..do every day of their lives.

b. = Shitten *a.* a.

1935 in A. W. Read *Lexical Evidence from Folk Epigr. in W.N. Amer.* 75 This Shithouse stinks like shit Because it is so shitty. **1977** C. McCullough *Thorn Birds* i. 7 If I catch you flaming little twerps touching that doll again I'll brand your shitty little arses!

shitzu, var. *shih-tzu.

shiur (ʃī·uɹ). [Heb. *šiʿūr* measure, portion.] A lesson in Jewish traditional sources.

1959 D. D. Runes *Conc. Dict. Judaism* 206/2 *Shiur*, fixed measure; generally used to designate Talmudic study hour. **1967** [see *rav]. **1973** *Jewish Chron.* 9 Feb. 21/4 Like other fortnightly meetings..it began with a shiur on the sidra of the week given by Rabbi Rosin.

shiv (ʃiv), **shive³,** *sb.* = *chiv *sb.*¹, Chive³. Also *attrib.*

1915 *N.Y. World* 9 May (Suppl.) 14/3 *Shive*, a razor. **1926** J. Black *You can't Win* (1927) vii. 87 'Better get busy with your 'shive', kid.' I started cutting on the side opposite the boarding house. **1926** *Clues* Nov. 162/2 *Shiv*, a knife. **1930** *Sat. Even. Post* 5 Apr. 46/1 Those he does not shoot he sticks with his shiv—which is a knife. **1951** F. Brown *Murder can be Fun* viii. 120, I won't use no shiv. I'll take him apart with my bare hands. **1959** 'M. Ainsworth' *Murder is Catching* vi. 72 Has he any criminal connections..could he get hold of shiv-boys, for instance? **1959** H. Hobson *Mission House Murder* xiii. 88 He held a white-handled open razor... "Strewth!" he said, 'a shiv!' **1972** *Daily Progress* (Charlottesville, Va.) 17 Feb. B. 11/5 The guards even planted a 'shive'—a tableknife—in McKinney's cell. **1976** L. Deighton *Twinkle, twinkle, Little Spy* ix. 89, I never heard of the K.G.B. using a shiv artist who hit the wrong target, and..let them grab the knife.

shiv (ʃiv), **shive³,** *v.* = *chiv *v.*, Chive *v.*

1926 [see *jam *sb.*¹ 1 b]. **1959** 'M. Ainsworth' *Murder is Catching* xv. 163 The boys been shiving you? **1967** 'E. Queen' *Face to Face* xxx. 136 'When was that?' 'The night they found Spotty shivved.' **1980** J. Wainwright *Venus Fly-Trap* 165 Two guys. So much hatred—eh? One shivs the other.

shiva(h (ʃi·vā). [Heb. *šiḇʿā* seven.] A period of seven days' mourning for the dead, beginning immediately after the funeral; *to sit shiva*, to observe this period. Also *attrib.*

1892 I. Zangwill *Childr. Ghetto* I. 177 If you had come round when he was sitting *Shivah* for Benjamin—peace be upon him!—you would have known. **1910** *Jewish Chron.* 29 Apr. 15/2 As an orthodox family we 'sat shiva'; and we took to the low seats with great heaviness of heart and in utter distress. **1938** *Ibid.* 10 June 15/2 The mourners requested that three 'shiva' chairs be sent to the house. **1959** B. Kops *Hamlet of Stepney Green* II. i. 33 The Shiva is starting soon. Oh, those seven days of mourning. **1964** D. Gray *Devil wore Scarlet* vi. 41 For seven days from the day of the funeral onwards a Jewish family sits Shiva. They sit on low stools in the drawing-room..and they sort of receive their friends and relations and get their sympathy. **1976** B. Williams *Making of Manchester Jewry* xi. 279 Members paid..4*d* a week, which entitled them..to *shiva* benefits, and to free burial. **1977** *New Yorker* 23 May 38/3 Elka's husband, Yontche, died, but Elka didn't observe shivah.

shivaree, *sb.* (Earlier examples.) Also in Cornwall. Also *fig.*

1805 J. F. Watson in *Amer. Pioneer* (1843) May 229 When a *sherrie-varrie* is announced, it is done by a running cry through the streets. **1843** 'R. Carlton' *New Purchase* II. lv. 231 The musicians..letting off at each repetition of the demand peals of shiver-ree. **1876** 'Mark Twain' *Old Times* 55, I started such a rattling 'shivaree' down below as never had astounded an engineer in this world before. **1926** Galsworthy *Silver Spoon* III. v. 251 And now came the usual 'shivaree' about such and such a case. **1942** A. L. Rowse *Cornish Childhood* 9 The most splendid of 'shivarees' given at Tregonissey was for the wedding of Eliza Dyer to Eneas (pronounced Enas) Kellow. **1966** G. E. Evans *Pattern under Plough* xi. 115 It was called the *Kiddly Band* in Cornwall..and it took part in the *shivaree* or wedding junketings there until the 'twenties. **1977** *Western Morning News* 30 Aug. 6/4 There was another word for a shallal: perhaps imported, or re-imported, from America. This is a shivaree, probably a corruption of charivari.

shivaree, *v.* (Earlier examples.)

1805 J. F. Watson in *Amer. Pioneer* (1843) May 229 Edward Livingston, esq., was *sherri-varried* here; on which occasion the parties came out promptly to the balcony and thanked the populace for their attention. **1872** E. Eggleston *End of World* xlvi. 294 Among the manly recreations which they have proposed to themselves is that of shivereeing 'that Dutchman, Gus Wehle'.

shive, *sb.*² Restrict '*Obs.* exc. *dial.*' to sense in Dict. and add: **2.** With pronunc. (ʃəiv). *Paper-making.* A dark particle in finished paper resulting from incomplete digestion of impurities in the raw material; such particles collectively. Cf. *sheave *sb.*² 3.

1922 *Manuf. Pulp & Paper* III. VIII. 1 The difference between shives and slivers should be clearly understood. **1937** E. J. Labarre *Dict. Paper & Paper-Making Terms* 207/2 *Sheave*, also spelt *shive*..are dark specks in the finished paper, due to impurities in raw materials, rags and esparto grass, [etc.].., hence 'shivery' or 'shivey' paper. **1952** F. H. Norris *Paper & Paper Making* xx. 291 Shives are usually light in colour..and..shaped like a minute splinter. **1968** R. R. A. Higham *Handbk. Papermaking* (ed. 2) ii. 79 Shive, knots..and similar impurities, cause breaks in the web.

3. *Comb.*: **shivelight** *nonce-wd.*, a sliver of light.

1888 G. M. Hopkins *Poems* (1967) 105 Shivelights and shadowtackle in long lashes lace, lance, and pair.

shive, *sb.*³, *v.*³: see *shiv *sb.*, *v.*

shiveau (ʃivǝ͞u·). *dial.* Now *rare* or *Obs.* exc. as *shivoo. Also 8 **chevaux,** 9 **chiveau, shebo, sheevo.** [Origin unknown.] = *shivoo.

1798 C. Cathcart *Let.* 6 May in Fraser & Gibbons *Soldier & Sailor Words* (1925) 255 We have just left Gibraltar... Sir John Orde gave a grand chevaux to which he was so good as to invite me. **1828** 'Jon Bee' *Living Pict. Lond.* ii. 83 A pressing invitation to what he calls a '*Chiveau*', or merry dinner. **1849** A. Harris *Emigrant Family* I. vii. 114 A 'shiveau' at the hut. **1862** C. C. Robinson *Dial. Leeds* 402 *Sheevo*, a shindy. 'A bonny sheevo thuh wor.' **1877** F. Ross et al. *Gloss. Words Holderness* 123/2 We'd a meetin I vesthry las' neet aboot a new cess, and them at didn't want yan kick'd up a riglar *shebo*. **1880** M. A. Courtney *Gloss. W. Cornwall* 50/2 *Shee-vo*, a disturbance; a row. 'There was such a grand *shee-vo*.'

shivering, *vbl. sb.*¹ Add: **1. b.** *Pottery.* Peeling and splitting of the glaze.

1921 A. B. Searle *Clayworkers' Hand-bk.* (ed. 3) xi. 208 Shivering is a variety of 'peeling' which may be produced by adding flint which has been too finely ground or an excess of fine silica to a body. **1947** J. C. Rich *Materials & Methods of Sculpture* ii. 51 Another cause of shivering is firing at too low a temperature. **1964** H. Hodges *Artifacts* ii. 52 Peeling..or shivering of a glaze usually results from a failure to fit the glaze to the body, the contraction of the glaze being less than that of the body.

shivering, *vbl. sb.*² Add: **2.** *Vet. Sci.* A pathological condition of horses in which certain muscles undergo rapid spasms, most commonly those in the hindquarters.

1847 *Vet. Record* III. 4 'Shivering', an affection of the stiffle-joint, associated with luxation of the patella. **1907** J. W. Axe *Horse* VIII. 374 During this test it will be noticed whether the action is close, ..and whether there are any indications of stringhalt or shivering. **1978** Geddes *Horse* 218 The most common conditions are:.. Shivering, a condition of the hind limbs, characterized by shaking movements of the hind limb and tail when the leg is flexed and lowered to the ground.

shiversome (ʃi·vəɹsəm), *a.* [f. Shiver *sb.*² + -some¹.] Causing shivers.

1939 *Observer* 16 Sept. 8 An ... [illegible] ... **1948** I. Brown *No Idle Words* 33 Fever is a hideous paradox, being cold heat and shiversome ardour. **1970** *Daily Tel.* 17 Nov. 12/4 An brilliantly shiversome a film as I have seen for years.

shivery, *a.*² Add: **2.** Also *shivery-shaky.*

1864 *Derby Day* iv. 50 He's all shivery shakey as if he'd got the staggers or the cold shivers. **1934** J. Joyce *Let.* 15 Aug. (1966) III. 317, I have a fit of ague for the past three days... O Lord, the one day I feel so shivery-shaky!

shivey, *a.* Delete *dial.* and add a further example from *Paper-making.*

1937 [see *shive *sb.*² 2].

shivoo (ʃivū·). *Austral. colloq.* Also **chivoo.** [Var. *shiveau.] A celebration, a party, a spree. **b.** A disturbance, a row.

1889 J. H. Hunt *Hunt's Bk. Bonanzas* 52 [illegible] ... **1919** W. H. Downing *Digger Dialects* 16 *Chivoo*, a celebration. **1924** Lawrence & Skinner *Boy in Bush* I. 15 There was a chivoo. ... [illegible] ... **1961** P. White *Riders in Chariot* ix. 314 ... [illegible] ... **1979** *Sunday Mail Mag.* (Brisbane) 8 Apr. 3/1 ... [illegible]

Shkyipetar, Shkypetar, varr. *Shqipetar.

shl-, var. *schl-* in many words (mostly German or Yiddish), qq.v. under this latter spelling. **shlanter,** var. *schlenter *sb.* and *a.* **shlemozzle,** var. *shemozzle. **Shl(e)uh,** var. *Shilha. **shlinter,** var. *schlenter.

shm-, var. *schm-* in many words (mostly German or Yiddish), qq.v. under this latter spelling (cf. *schm-). **shmear,** var. *schmeer. **shmock,** var. *schmuck.

shmoo (ʃmū). *U.S.* [Invented word.] A fabulous animal invented by the U.S. cartoonist Al Capp in 1948. It is small and round, and ready to fulfil immediately any of man's material wants. Also, a model or toy version of this animal. Also *Comb.*, as *shmoo-like* adj.

1948 *Newsweek* 12 July 56 ... [illegible] ... **1948** *Times-Herald* (Washington, D.C.) ... [illegible] ... **1948** *Sun* (Baltimore) ... [illegible] ... **1949** ... [illegible] ... **1962** L. Hearn *Kill* ... [illegible]

shmoos, var. *schmooze *v.*, *sb.*

‖ **sho** (ʃō), *sb.*¹ Pl. **sho.** [Jap.] A Japanese unit of capacity equal to ten *go*; equivalent to approximately 3·18 pints (1·80 litres).

1876 W. E. Griffis *Mikado's Empire* (1877) II. Notes & App. 609 Measures of capacity.. ... [illegible] ... **1938** D. T. Suzuki *Zen Buddhism & its Influence on Jap. Culture* ... [illegible]

‖ **shō** (ʃō), *sb.*² [Jap.] A small Japanese organ, made from seventeen vertical bamboo pipes, which is held in the hand and blown into.

1888 L. A. Smith *Music of Waters* 281 The 'shō'..seems to correspond to our organ, but only in so far as it has pipes. **1936** K. Sunaga *Jap. Music* 27 The *shō* is an instrument of which there seems to be no equivalent outside oriental music, though it is sometimes spoken of as a mouth organ. **1972** *Times* 18 Sept. 5/4 The head priest showed the instruments to Mr Heath, who took special interest in the sho pipes... He watched closely as a musician demonstrated the sho.

‖ **sho** (ʃō), *sb.*³ [Tibetan.] A former Tibetan unit of currency; a coin of this value.

1902 S. C. Das *Journey to Lhasa & Central Tibet* vii. 182 Shopkeepers and pedlars pay five *sho* ... annually. **1970** R. D. Taring *Daughter of Tibet* ix. 107 Ten *sho* made one *sang*.

sho, sho' (ʃō), repr. U.S. Blacks' pronunc. of Sure *adv.*

1893 H. A. Shands *Some Peculiarities of Speech in Mississippi* 56 *Sho*.. ... [illegible] ... **1942** S. Kennedy *Palmetto Country* ... [illegible] ... **1973** E. Bullins *Theme is Blackness* ... [illegible]

shoal, *a.* (and *adv.*). Add: **1. c.** The phr. *shoal water* used *fig.*

1884 'Mark Twain' *Huck. Finn* xxviii. 284 It jolted her up like everything, of course; but I was over the shoal water now, so I went right along, and told her every blame thing... ... [illegible]

shochet. (Later examples.)

1907 I. Zangwill *Ghetto Comedies* 386 The *Shomer* and the *Shochet* are the official twain of ritual butchery. **1961** A. W. Moss *Valiant Crusade* vi. 81 In Shechita, the Shochet, or official who actually does the killing, is specially trained. **1972** *Daily Tel.* 8 June 18 Before the Shochet can use his knife the animal has to be got into the correct position.

‖ **shochu** (ʃōu·tʃu). [Jap.] A rough Japanese spirit distilled from various ingredients, including sake dregs. Also *attrib.*

1938 Bush & Kagami *Japanalia* 143/1 ... [illegible] ... **1964** I. Fleming *You only live Twice* x. 124 The herdsman..handed Bond a bottle of what appeared to be water. Tiger said, 'This is shochu. It is a very raw gin.' **1970** J. Kirkup *Japan behind the Fan* ... [illegible] ... **1980** 'J. Melville' *Chrysanthemum Chain* 88 A bottle of cheap shochu ... [illegible]

shock, *sb.*¹ Add: **3.** *shock corn, fodder.*

1865 *Trans. Illinois Agric. Soc.* ... [illegible] ... **1949** H. Hornsby *Lonesome Valley* ... [illegible]

shock, *sb.*³ Add: **2.** (Further examples.) Also, a sudden large application of energy other than mechanical energy, esp. thermal energy (cf. **thermal shock*); a shock wave.

1904, 1907 [see *shock wave a]. **1932** *Trans. Amer. Soc. Mech. Engineers* ... [illegible] ... **1955** K. F. Herzfeld in F. D. Rossini *Thermodynamics & Physics of Matter* ... [illegible] ... **1966** *McGraw-Hill Encycl. Sci. & Technol.* ... [illegible] ... **1974** *Acustica* ... [illegible]

4. a. Also *ironically* in apposition with *horror*; so *shock-horror* used *attrib.*

1977 *Gay News* ... [illegible] ... **1981** *Brit. Jrnl. Med.* ... [illegible]

d. (*a*) *culture* (or *cultural*) *shock*: a state of distress or disorientation brought about by sudden subjection to an unfamiliar culture; (*b*) *future shock*: an analogous state brought about by too rapid a pace of social or technological change.

a. 1940 J. B. Holt in *Amer. Sociol. Rev.* Oct. 744 ... [illegible] ... **1952** *Human Organisation* Spring ... [illegible] ... and its aftermath. **1960** [see *culture shock* s.v. *culture *sb.* 5d]. **1970** A. Toffler *Future Shock* i. 12 Culture shock is the effect that, immersion in a strange culture has on the unprepared visitor... Culture shock is what happens when a traveler suddenly finds himself in a place where yes may mean no, where a 'fixed price' is negotiable, where to be kept waiting in an outer office is no cause for insult, where laughter may signify anger. **1978** *Lancashire Life* July 43/3 To see a defender in a protective helmet playing at cover point can cause culture shock in newcomers to the game.

b. 1965 A. Toffler in *Horizon* Summer 109/1 Culture shock is relatively mild in comparison with a much more serious malady that might be called 'future shock'. Future shock is the dizzying disorientation brought on by the premature arrival of the future. It may well be the most important disease of tomorrow. **1972** *Newsweek* 17 July 34 Fortunately, Japan's innate self discipline and sense of harmony have cushioned much of the 'future shock'..and so far Japan is relatively free of the..disorders so prevalent in other industrialized societies. **1975** *Whig-Standard* (Kingston, Ontario) 7 Jan. 6/3 Canada's foreign policy is in poor health these days—suffering from a severe case of future shock complicated by a crippling inferiority complex.

5. a. (Further examples.) Now used more precisely in *Med.* for a condition whose principal characteristic is low blood volume (see quot. 1968); also *ellipt.* for *shell shock* s.v. *shell *sb.* 39; *in shock*, in a state of shock; incapacitated by very low blood pressure and associated symptoms, or debilitated by ill-treatment or bad news; so *into shock, out of shock.*

1889 *Amer. Jrnl. Med. Sci.* XCVII. 282 For the purpose of stimulating a patient in shock it is more rational to give ether than to give alcohol. **1917** W. Owen *Let.* 23 May (1967) 463 Sorrel was mentioned for Shock [in the Casualty List]. **1928** *Jrnl. Amer. Med. Assoc.* 9 June 1859/2 The blood pressure..suddenly fell to 52 systolic and 44 diastolic as he went into shock. **1958** J. Cannan *And be Villain* vi. 137 Mrs Hallow was subjected to quite an ordeal this morning and is practically in shock. **1959** *Woman's Own* 27 June 44/2 She's in shock, of course, but we're giving her a transfusion. **1968** Passmore & Robson *Compan. Med. Stud.* I. xxviii. 41/1 Loss of blood volume is an important but not the only cause of shock; a similar state of shock occurs in acute heart failure and in severe infections in which the responsible haemodynamic mechanisms are different. **1975** *Publishers Weekly* 21 July 67/2 When Joe signed his first pro contract, Rose..went into shock. **1980** M. Russell *Death Fuse* xvi. 149, I think she's able to talk. She's out of shock.

b. A paralytic seizure or stroke. Chiefly *Sc.* and *U.S. dial.*

1794 J. Woodforde *Diary* 13 Oct. (1929) IV. 143 Mr. Whitmell had a kind of Paralytic Shock this last Spring. **1896** H. Johnston *Dr. Congalton's Legacy* xxiv. 315 The mistress of Windy-yett had taken 'a terrible turn—a shock or something'. **1903** K. D. Wiggin *Rebecca of Sunnybrook Farm* xxvi. 279 We had three o' the worst shocks in our family that there ever was..and I know every symptom of 'em better 'n the doctors. **1951** E. Graham *My Window looks down East* xi. 96 That was when he came the nearest to the stroke. (In Maine we call them 'shocks'.) **1955** W. P. Milne *Eppie Elrick* xxx. 284 Haein teen a shock an' soocht awa in 'is sleep. **1981** M. Cantwell in *N.Y. Times Mag* 9 Aug. 6/4 One cause of my future demise might be that I 'took a shock.' In other places, a person in that condition would be said to have suffered a paralytic stroke.

c. *anaphylactic shock*: see s.v. *anaphylaxis.

6. (Earlier example of sense 'electric shock'.)

1746 W. Watson *Sequel to Experiments in Electr.* 10 He receives a violent shock through both his arms.

6*. ellipt. for *shock-absorber*. Chiefly *U.S.*

1961 in Webster. **1968** *Hot Car* Nov. 14/2 (Advt.), These are the shocks to end all shocks. **1977** *Transatlantic Rev.* LX. 22 He told her that the car would also need new shocks, brakes, a muffler, a starter and an engine job. **1979** *Guardian* 4 June 6/7 Shock absorber people who will swop your worn-out shocks.

7. a. *shock effect, -value*; *shock-resistant* adj.; also, of things that startle or shock, as *shock headline, language, news, story*; **shock cone** *Aeronaut.*, a nose cone or other conical fairing which serves to streamline an aircraft for supersonic flight; **shock cord,** heavy elasticated cord designed to absorb or resist mechanical shock; a length of this; **shock excitation,** the excitation of natural oscillations in a system by a sudden impulse of energy from an external source; so **shock-excited** *ppl. a.*; **shock front,** the wave front of a shock wave; **shock measure,** a severe or exceptional measure taken usu. to deal with an emergency; **shock-mount** *sb.*, a mounting designed to absorb or resist mechanical shock; also as *v. trans.*, to attach by means of such a mounting; **shock-mounted** *ppl. a.*, **shock-mounting** *vbl. sb.*; **shock police,** in Spain, a republican force of specially armed police for use in assault operations; **shock-proof** *a.*, proof against damage by mechanical shock or by a surge of electrical power; also *fig.*; hence **shock-proofing** *vbl. sb.*, the process of rendering shock-proof; **shock stall** *Aeronaut.*, a stalling condition undergone by an aircraft at a speed close to that of sound, involving increased air resistance and loss of lift and control; also **shock stalling** *vbl. sb.*; **shock strut** *Aeronaut.*, a strut containing a shock absorber in the landing gear of an aircraft; **shock-tactics,** add: also *transf.* and *fig.*; also occas. **shock-tactic; shock test,** a test in which an object is subjected to mechanical shock; hence **shock-testing** *vbl. sb.*; **shock therapy, treatment,** treatment by means of artificially induced shock, whether anaphylactic, electrical, or drug-induced; *spec.* electro-convulsive therapy; also *fig.*; **shock tube,** an apparatus for producing shock waves by making a gas at high pressure expand suddenly into a low-pressure tube or cavity.

1947 *Shell Aviation News* No. 112. 6/3 Republic is building a fighter aircraft with a needle-nose shock cone to permit supersonic speed. **1961** *Aeroplane* CI. 548/1 The 'shock cones' are merely fairings over the boundary-layer bleed, which brings the intakes proud of the fuselage. **1930** P. White *How to fly Airplane* xx. 279 (*caption*) Two boys at the rear are holding against the pull of the shock cord. **1980** *TWA Ambassador* Oct. 32/3 Secure all gear, such as the coffee-cans, in the canoe with rope or shock cord. **1959** *N. & Q.* CCIV. 36/2 The most intense shock effects come from the imposition of war's horrible destruction upon familiar places usually associated with quiet and peace. **1977** A. Giddens *Stud. in Social & Polit. Theory* ix. 317 Suicide attempts do appear to have a 'shock' effect on relatives and friends. **1920** E. W. Stone *Elem. Radiotelegr.* iii. 54 Exciting a vibratory circuit into oscillation is variously termed impulse excitation, shock excitation, and whip-crack excitation. **1930** A. B. Wood *Textbk. Sound* 213 A single explosion impulse is often sufficient to set a resonator into vibration, thereby producing a musical note or a noise by shock excitation. **1967** R. F. Graf *Mod. Dict. Electronics* 138/2 *Free oscillations*, commonly referred to as shock-excited oscillations. **1975** D. G. Fink *Electronics Engineers' Handbk.* XIX. 17 The sound levels of shock-excited tones are more difficult to specify because they vary so much during decay and can be excited over a very wide range. **1949** L. M. Milne-Thomson *Theoret. Hydrodynamics* (ed. 2) xx. 577 If the shock front is sufficiently oblique to the oncoming air, the conditions behind may still be supersonic. **1969** *New Scientist* 28 Aug. 434/2 The boom signature appears to coalesce at a relatively short distance from the flightpath into two shockfronts. **1977** M. Walker *National Front* iii. 60 Shock headlines like 'Five Million Coloured Asians now in Britain?' (*RPS News*). **1959** *Listener* 13 Aug. 253/1 One is reminded more of the shock-language of the *avant-garde* Russian poets of the time. **1962** *Daily Tel.* 13 Dec. 10/2 Mr. Marples's shock measures..will be more than justified if this savage record [of death and injury on the roads] can be substantially improved. **1939** *Interavia* 5 Dec. 10 Two sets of shock mounts on either side of the blades are employed. **1942** P. C. Sandretto *Princ. Aeronaut. Radio Engin.* i. 10 Shock mounting in the past consisted of certain rubber shock absorbers mounted to the radio units; however, the trend is toward the use of shock-mounted shelves permanently installed in the airplane. **1947** B. W. Pike et al. in A. Roberts *Radar Beacons* xvi. 361 The beacon should be either internally shock-mounted or carried in a padded case. **1964** R. F. Ficchi *Electr. Interference* iv. 40 Since many equipments are presently being installed on shock mountings so they will not be affected by vibration, it is important to bond adequately across shock mounts. **1973** *Times* 31 Oct. (Suppl.) p. viii, His company believe they can avoid fatigue failure..by overdesign of various components and by shock-mounting valves, pipes and other equipment. **1976** *National Observer* (U.S.) 1 May 3/4 (Advt.), Precision, shock-mounted lucite level. **1974** *Times* 3 Apr. 1/1 (*heading*) Shock news is broken to EEC ministers. **1937** *Ann. Reg. 1936* 249 The murder, on July 12, of Lieutenant Castillo of the Shock Police (known to have Socialist sympathies) by gangsters of the Right. **1957** P. Kemp *Mine were of Trouble* i. 16 Shock Police (*Guardias de Asalto*), posted in side streets on motor cycles. **1911** T. Dreiser *Jennie Gerhardt* xviii. 138 He was..curiously elated beneath a sturdy, shock-proof exterior. **1925** *Wireless World* 8 Apr. 280/3 (*heading*) Shock-proof valve holder. **1930** R. Macaulay *Staying with Relations* xiv. 203 They needed a shock-proof screen between them, to deaden the assaults of each on the other's strained nerves. **1952** H. R. Clauser *Pract. Radiogr. for Industry* iii. 37 Protection is obtained by immersing the [X-ray] tube in oil inside a shockproof casing. **1971** J. Wainwright *Dig Grave & let him Lie* 20 He was shockproof—the ultimate example of what bobbying can do to a man. **1978** R. Good *Watches in Colour* viii. 116 Shock-proof bearings for the balance pivots. **1952** H. R. Clauser *Pract. Radiogr. for Industry* iii. 37 Shockproofing of x-ray tubes can be done by enclosing the tube in a casing which is at ground potential. **1946** D. de Carle *Practical Watch Repairing* (1947) xx. 262 The shock-resistant watch cannot be looked upon as a novelty, in the sense that it is a passing phase. **1963** *New Yorker* 7 Dec. 136/1 (Advt.), Self-winding, thin, water- and shock-resistant. **1938** *Proc. R. Soc.* A. CLXIX. 188 This critical speed at which the sound wave condenses into a single pressure discontinuity is probably the shock stall or compressibility stall. **1948** *Sci. News* VII. 30 The use of swept-back wings raises the critical Mach number for the aircraft and allows the higher speeds to be reached without the danger of shock stall. **1966** D. Stinton *Anat. of Aeroplane* vi. 103 The buffeting and sharp loss of lift (Shockstall) caused by compressibility gave rise to the early misconception of a 'sound-barrier', beyond which man might not fly. **1937** *Flight* 4 Nov. 450/1 The elliptic cylinder..has a lower maximum velocity..and therefore presumably a higher shock stalling speed. **1952** W. J. Duncan *Princ. Control & Stability of Aircraft* xiii. 310 The separation of the shock stalling and critical Mach numbers may vary from almost zero to about 0·2. **1946** *Sun* (Baltimore) 5 Nov. 9/6 (Advt.), You'll see the shock story of the year. **1931** F. D. Bradbrooke *Light Aeroplane Man.* vi. 92 When the advantages of the divided undercarriage began to compel attention the shock-absorbing gear had to be incorporated in the compression or shock strut. This was done by having a telescopic strut, generally with projections in each portion which were lashed with rubber cord in such a way that a shortening of the strut stretched the cord. **1949** J. W. Vale *Aviation Mechanic's Aircraft Man.* xii. 343 The fluid used in the Pneudraulic shock struts..may be used for initial filling or for partial refilling. **1919** *Manch. Guardian* 4 Feb. 5/1 It was the 'shock Tactics' of labour warfare. **1954** *Essays & Studies* VII. 84 To judge from the vehement reactions of the critics, the revolutionary shock-tactics seem in this aim to have succeeded remarkably well. **1959** *Listener* 19 Nov. 895/2 This definition of Zen as a kind of natural mysticism whose shock-tactic and other techniques are designed to stimulate the intuitive and..religious awareness. **1960** G. E. Evans *Horse in Furrow* xx. 239 His initial shock tactics were successful: he gained control of the horse. **1961** *Encounter* Apr. 56/2 This shock-tactic had by no means the same general appeal as his witch being burnt at the stake. **1977** R. Barnard *Blood Brotherhood* ix. 92 There was a hidden design behind..Chief Inspector Plunkett's questions, or perhaps..he was trying shock tactics. **1904** *Proc. Inst. Mech. Engineers* 1135 Considerable movement has been made of late towards the establishment of a shock test for steel. **1949** J. F. Blackburn *Components Handbk.* xiii. 516 In certain cases the shock tests resulted in permanent mechanical damage to the relays, often without opening the contacts. **1974** *Jrnl. Soc. Environmental Engineers* XIII. 17/1 Maximum displacement during a shock test can be obtained in several ways, one of the simplest is to measure the deformation of a piece of plasticine. **1917** *Jrnl. Iron & Steel Inst.* XCVI. 65 The regularity of the results which can be obtained with carefully manipulated shock-testing machines. *Ibid.*, A system of calibrating apparatus for the shock-testing of metals. **1963** C. T. Morrow *Shock & Vibration Engin.* I. v. 114 The advantages of an asymmetrical pulse shape in shock testing. **1917** *Jrnl. Exper. Med.* XXVI. 699 The mechanism of recovery following the so called 'protein shock therapy'. *Ibid.* 705 Bacterial infection not confined to the lymph spaces will not be influenced by shock therapy to the same extent. **1939** *Jrnl. Amer. Med. Assoc.* 16 Sept. 1170/2 Dislocations and fractures occur in insulin and metrazol shock therapy. *Ibid.* 2 Dec. 2100/2 The author recommends the use of electric shock therapy for old cases of depressive psychosis in which other therapeutic methods have failed. **1953** H. Read *True Voice of Feeling* I. viii. 149 There may be more sense in [Ezra] Pound's shock-therapy. **1973** W. Barlow *Alexander Principle* i. 13 The psychiatrists..treated her with shock therapy and anti-depressant drugs. **1979** *Tucson Mag.* Jan. 24/1 The shock therapy that was Simon's murder induced a spirited public outcry. **1938** M. Sakel (*title*) The pharmacological shock treatment of schizophrenia. **1939** *Jrnl. Amer. Med. Assoc.* 16 Sept. 1170/2 (*heading*) Faradic shock treatment of 'functional' psychoses. **1945** Koestler *Yogi & Commissar* III. iii. 202 That diplomatic shock-treatment of which they have so far only had a faint foretaste. **1947** *Sun* (Baltimore) 20 Jan. 7/1 We believe that the 'shock treatment' of prompt action is needed to halt the insane spiral of mounting costs and rising prices. **1974** *Listener* 24 Jan. 124/3 Munch suffered a complete breakdown, received shock treatment at a clinic in Copenhagen. **1949** *Sci. Amer.* Nov. 18/2 These pictures illustrate the great potentiality of the shock tube in observing and recording supersonic flow patterns. **1977** I. M. Campbell *Energy & Atmosphere* v. 105 In the shock tube these conditions are produced by the rapid adiabatic compression of air or N_2-O_2 mixtures across the shock front created by the sudden release of high-pressure driver gas at one end of the tube. **1933** *Archit. Rev.* LXXIV. 68/1 The *Daily Mail*, drawn naturally to the new medium by its tradition of public-spirited interest in aviation and its awareness of the shock-value of surprise. **1959** *Encounter* Nov. 59/2 He had never dreamt that his [ballet] company would have such shock-value.

b. shock troops [tr. G. *stosstruppen*], forces of selected and specially armed men trained for deployment in assault operations, especially against strong positions or large numbers; (rarely) *sing.*, such a force. Also *fig.* and *attrib.* (in *sing.*). Hence **shock trooper.**

1917 *Times* 20 June 5/5 The second shock-troop battalion of the Third Army. **1918** E. S. Farrow *Dict. Mil. Terms* 553 *Shock Troops*, troops especially selected for assault work. They usually wear steel breastplates and other protection strong enough to turn a bullet at 50 or 60 feet. **1927** *Daily Express* 14 July 9/4 The G.P.U. troops..in..a war against Poland and Britain..would be used as shock troops. **1928** *Dict. Amer. Biogr.* I. 382/2 A master of strategy usually stays behind the lines. Not so Bishop Asbury. He asked no more of his skirmishers and shock troops than he was himself prepared to undergo. **1934** W. A. Edwards in F. R. Leavis *Determinations* 155 Swinburne, like some Soviet shock-trooper exhorting feeble comrades, batters and bullies us into thinking every playwright a demi-god. **1938** Auden & Isherwood *On Frontier* I. i. 30 As for the Shock Troops..the whole organization's rotten from top to bottom. *Ibid.* II. i. 61 Such a nice boy! And quite high up in his shock-troop already. *Ibid.* ii. 84 My Peace Speech. I shall stand before my shock troopers and I shall tell them [etc.]. **1940** *Ann. Reg. 1939* 225, 54 front-line divisions..of which 14 divisions represented shock troops. **1940** *War Illustr.* 12 Apr. 353 Those who man the submarines may well be called the 'shock troops' of the sea, for of all the seafaring men their task is surely the most arduous and the most dangerous. **1952** C. Day Lewis tr. *Virgil's Aeneid* II. 48 At the Scaean gate, panoplied Juno Heads the shock-troops. **1955** M. Banks *Commando Climber* i. 6 As the war developed, the general trend of commando operations had been..towards the amphibious shock-troop variety of fighting, usually on a brigade or at least on a commando scale. **1955** A. Galland *First & Last* xii. 91 The fighter supporting land operations of the army is exclusively a tactical arm, a kind of 'flying shock-troop', for attacking at low level enemy positions and troop movements in the front line, even with bombs, rockets and small arms. **1959** *Listener* 10 Dec. 1031/1 The main line of Chartist stock in the north is poorly represented. Where are those shock troops of Feargus O'Connor, from whose enduring loyalty his influence in the movement was derived? **1962** *Daily Tel.* 14 Sept. 24/5 While Mr. Macmillan was having longish talks with Mr. Menzies of Australia and Mr. Diefenbaker of Canada and others, his shocktroop Ministers, Mr. Sandys and Mr. Heath were standing up to all comers.

1968 *Economist* 16 Nov. 67/1 The National Union of Mineworkers' leaders voted to accept exactly the same percentage pay rise for their men, once the shocktroops of the wage struggle. **1973** R. L. Fox *Alexander* iv. 78 The Shield Bearers..served as shock troops on night raids, hill climbs and forced marches. **1975** *Guardian* 21 Jan. 14/1 The cliché about the Jesuits as the 'Pope's shock troops'. **1977** *Times* 21 Nov. 15/6 Many members of the Waffen SS..were used simply as shock troops during Hitler's War.

c. Applied to a worker in the U.S.S.R. who voluntarily exceeds the production quotas and is regarded as exemplary, and to a brigade formed by such workers and used for the achievement of arduous or urgent tasks; also to such methods of work.

1931 S. N. Harper *Making Bolsheviks* iii. 52 The so-called 'shock-brigade movement' among workmen, which is interpreted by the Communists as an illustration of a new attitude and as a new and vital force produced by the Revolution. *Ibid.* 53 The shock-brigade of a given factory or mine is a voluntary organization of workmen who have come together to fulfil and, if possible, exceed the quota of production assigned to the factory under the Five-Year Plan of expansion. **1931** *Times Lit. Suppl.* 9 July 536/4 'Cultural work' is still in effect a kind of propagandist 'shock work'. **1931** *Morning Post* 11 Aug. 11/7 Three hundred and fifty Russian 'shock workers' landed in London from the S.S. Ukraine yesterday for a two days' 'holiday'. **1934** *Spectator* 26 Oct. 615/2 The 'shock-brigaders' and star 'go-getters'. **1938** *Times* 25 July 13/6 The other projects undertaken by the groups include organization of labour to help in times of special agricultural pressure (like the Shock Brigades in Russia). **1939** R. Campbell *Flowering Rifle* v. 124 Each grim, shock-working Stakhanov. **1946** —— *Talking Bronco* 16 For us mere Shock-workers of the Camp and City Whose sweat, and life-blood, is their beer. **1949** F. Maclean *Eastern Approaches* I. xi. 157 From what I had heard of Soviet 'shock' methods a group of Stakhanovites or shock-workers should be able to put any paddle boat in order in an hour or two. **1962** E. Snow *Red China Today* (1963) lviii. 441 By intensive cultivation, luck and shock-brigade methods, we brought in a rich harvest in record time. **1981** I. Boland tr. *Ginzburg's Within Whirlwind* I. x. 84 The management did not want to use force... These girls were shock workers.

shock, *a.* Add: **b.** *shock-maned* adj.

1922 [see *shaggy-chested* s.v. *shaggy *a* 3].

shock, *v.*² Add: **3. c.** To subject to or transform by mechanical shock. Cf. *shocked *ppl. a.* 3.

1950 D. Q. Kern *Process Heat Transfer* xiv. 382 Hard scale which can be shocked from the tube. **1965** *Ann. N.Y. Acad. Sci.* CXXIII. 602 Salt shocked by a chemical explosion adjusted mainly by plastic glide along close-spaced slip planes. **1968** Ahrens & Rosenberg in French & Short *Shock Metamorphism of Natural Materials* 59/2 The Hugoniot curve is defined as the locus of pressure-volume-energy states that may be achieved within the material by shocking it from a given initial state.

4. b. (Later example.) Also *intr.* for *pass.*, to suffer shock.

1959 N. Clad *Love & Money* (1960) 15 She liked to say things to shock Clarence, partly because he shocked so easily. **1967** *Listener* 5 Oct. 448/3 Triana's play does shock.

shockability (ʃǫkăbi·līti). [f. Shockable *a.*: see -ity.] The capacity for being shocked. Also *attrib.*

1929 G. Gould *Democritus* 89 What we want is to preserve the precious gift of shockability while remaining too intelligent to be shocked. **1963** O. Stewart in C. Irving *Scandal '63* xix. 213 In England shockability is primarily sexual. **1969** J. Elliot *Duel* III. iii. 259 She had a low shockability threshold when it came to language. **1977** *Zigzag* Aug. 5/2 The new wave audiences have been beaten into applauding what they can get while ignoring any new trends that don't quite fit the Sunday People blueprint for shockability.

shockable, *a.* Delete *nonce-wd.* and add later examples. Also *absol.* in pl. sense (with *the*). Hence **sho·ckableness** = *shockability.

1929 M. Arlen *Babes in Wood* 254 How..nice I think you are—to be so shockable. How different from your books! **1939** H. G. Wells *Holy Terror* III. i. 207 Shockableness is still high. Reverence is something that survives belief. **1965** *Listener* 1 July 12/1 Experimental plays, full of loose living and talk..should be put on at a rather late hour at regular intervals, so that the shockable know not to look. **1976** S. Barstow *Right True End* III. xii. 183, I think she's shockable. I'm not certain how her loyalty to me would stand the strain if she knew I was knocking off a colleague's wife.

shock-absorber. [Shock *sb.*³] **1.** A device, esp. on a motor vehicle, aircraft undercarriage, etc., which serves to absorb mechanical shock and to damp vibration.

1906 [in Dict. s.v. Shock *sb.*³ 7]. **1929** *Times* 31 Oct. 21/3 The considerable number of cars which were fitted with the Luvax hydraulic shock absorbers. **1931** *Henley's ABC of Gliding & Sailflying* 193 If the landing gear is not provided with shock absorbers, it is desirable to pad the pilot's seat well. **1945** Crocker & King *Piping Handbk.* (ed. 4) xviii. 1319 Hydraulic piping can be supported in box guides so as to permit free longitudinal movement except as restrained by the shock absorbers. This arrangement is..effective in cushioning shocks which are due to water hammer. **1951** *Engineering* 26 Oct. 533/2 Long semi-elliptical reverse-camber springs of conventional design are used for the rear suspension, which is..controlled by double-acting hydraulic shock absorbers. **1973** *Times* 15 Oct. 21/8 British Leyland have admitted that fires have occurred in 18 of the new buses and shock-absorber brackets have broken off.

2. *fig.* Something which (or someone who) reduces or mitigates the worst effects of a new and unpleasant occurrence or experience. Also *attrib.*

1924 *Foundry* 15 Feb. 105 (Advt.), Where lower production costs are necessary our products are real shock-absorbers. **1954** Koestler *Invisible Writing* iv. 52 The elastic shock-absorbers of my Party training began to operate at once. **1954** B. & R. North tr. *M. Duverger's Pol. Parties* I. ii. 83 Stable and stabilizing communities, playing some part as political shock-absorbers. **1957** *Economist* 28 Sept. 1003/1 The process..would close when the promised elections set up an all-Korean government—perhaps with a degree of regional autonomy, and with shock-absorber clauses written into the constitution to cushion the losers. **1969** *Listener* 15 May 666/1 They cannot rely on help from friends or families in emergency—they are people without shock-absorbers. Often a single misfortune gives the push towards disintegration. **1977** *Time* 15 Aug. 13/1 Black labor acts as a shock absorber enabling Italy to survive economic crisis.

Hence **shock-absorbent** *a.*, **shock-absorbing** *ppl. a.* and *vbl. sb.*

1909 Shock-absorbing [in Dict. s.v. Shock *sb.*³ 7]. **1946** D. de Carle *Pract. Watch Repairing* xx. 262 Had the balance been fitted with some shock absorbent device the damage would not have occurred. **1958** *Listener* 30 Oct. 683/1, I think you want a car that has adequate shock absorbing. **1973** *Times* 13 July 9/1 Each [crash-helmet] has a shock-absorbent liner as well as ear and cheek pads. **1973** J. G. Tweeddale *Materials Technol.* II. iii. 56 Drums..may be lined with shock-absorbing material, to minimise wear of the drum and to reduce noise.

shocked, *ppl. a.*¹ Add: **2.** (Earlier example.)

1840 Queen Victoria *Let.* 21 Jan. in B. Connell *Regina v. Palmerston* (1962) i. 20 A letter..which she has kept near three years, she is shocked to say.

3. Subjected to mechanical shock, esp. by the passage of a shock wave.

1962 E. M. Shoemaker in Z. Kopol *Physics & Astron. Moon* viii. 317 Part of the kinetic energy of the meteorite engulfed by shock is converted to internal energy in the meteorite, and part is transferred as kinetic and internal energy to the shocked rock ahead of the meteorite. **1973** *Nature* 27 July 211/2 Hypervelocity impact cratering studies have shown that about ninety-nine volumes of crushed and shocked rock are formed and ejected for each volume of liquid melt (glass) that is formed and ejected. **1976** *Ibid.* 11 Nov. 114/3 This ionisation front is preceded by a shock front, and after a time there is a shocked layer of neutral gas between these two fronts.

Hence **shockedly** (ʃǫ·kėdli, ʃǫ·ktli), *adv.*

1926 Fowler *Mod. Eng. Usage* 531/2 *Shockedly*. A bad form. **1963** D. Hughes in Sissons & French *Age of Austerity* 93 Then, rather shockedly, a knight's name figured in court.

shocker¹. Add: **1.** (Further examples.) Also, something or someone shockingly bad.

1954 M. Croft *Spare Rod* I. ix. 62 A lot of people believe that to be a good writer a man must have a well-nigh perfect character... On the contrary, many of them have been perfect shockers. **1958** *Manch. Guardian* 7 June 4/7 The 'musical' is still in the ascendant. I caught one real shocker—'Jamaica'. But the others..were fun. **1960** J. Stroud *Shorn Lamb* xxiii. 248, I remember a girl we had..an absolute shocker, never settled anywhere. **1965** Mrs. L. B. Johnson *White House Diary* 17 June (1970) 291 Then Lyndon delivered the shocker of the evening. **1973** *Express* (Trinidad & Tobago) 9 Apr. 1/3 The series will be the sort of shocker that should cast out smug complacency. **1976** *Daily Times* (Lagos) 12 Oct. 1/3 This is the shocker for teachers in Bomo State. **1977** *Time* 17 Jan. 9/1 What began as a shocker killing has grown steadily more sensational. **1977** *Horse & Hound* 10 June 8/2 Lucky Sovereign ran a shocker, presumably either unable or unwilling to give his true running on this firm ground and/or the Epsom course. **1977** *Time* 8 Aug. 39/1 Then on Wednesday came a shocker from Bethlehem Steel, which reported an operating loss of $75·4 million for the first half. **1978** *Chicago* June 119/2 The jury found the guy guilty on only one count. A real shocker, and entirely due to Sullivan's work.

2. A shock-absorber. *colloq.*

1949 Partridge *Dict. Slang* (ed. 3) 1168/1 *Shocker*,..a shock-absorber: motorists': since ca. 1925. **1969** *Guardian* 23 Aug. 5/8 The models with worn shockers were thrown from side to side. **1977** *Caravan World* (Austral.) Jan. 19/2 Incorrect loading on the tow-ball..results in overloaded tyres, springs and shockers if the weight is forward.

shocker². (Earlier U.S. example.)

1786 G. Washington *Diaries* 15 July (1925) III. 91 For every two Cradlers to allow 4 rakes, 1 shocker, and two Carriers.

shock-headed, *a.* Add: *shock-headed Peter* = *Struwwelpeter.

1848 tr. *H. Hoffmann's English Struwwelpeter* (ed. 4) 2 Any thing to me is sweeter Than to see Shock-headed Peter. **1895** R. Stephens *Cruciform Mark* ix. 47 'Shock-headed Peter', as he was familiarly dubbed..was a very interesting personality... Red-haired and red-bearded, his head was a perfect burning bush. **1905** C. Mackenzie *Diary* 23 May in *My Life & Times* (1964) III. 224 Went to see Martin Harvey as Hamlet... He looked like..shock-headed Peter. **1926** D. L. Sayers *Clouds of Witness* vii. 159 Shaking her head so angrily that she looked like shock-headed Peter.

shocking, *ppl. a.* Add: **3. d.** *shocking pink*: a vivid, garish shade of pink.

1938 *Encycl. Brit. Bk. of Year* 248/2 Only one new colour arrived; it is 'Shocking Pink', introduced by Schiaparelli in Feb. 1937, then taken up by other designers, with the result that the vanguard of fashionable women everywhere are now seen wearing this crude, cruel shade of rose. **1939** *Archit. Rev.* LXXXV. 305/3 There it is, in its gay 'shocking pink' cover, compact, comprehensive and invaluable. **1954** E. Schiaparelli *Shocking Life* ix. 97 My friends and executives..began to say that I was crazy and that nobody would want it because it was really 'nigger pink'... The colour 'shocking' established itself for ever as a classic. Even Dali dyed an enormous stuffed bear in shocking pink. **1960** M. Stewart *My Brother Michael* x. 136 They [*sc.* socks] were luminous, and of a startling shade of shocking pink. **1977** C. McCullough *Thorn Birds* xix. 520 An involuntary grimace at first sight of her shocking-pink slipper satin.

Shockley (ʃǫ·kli). The name of William B. *Shockley* (b. 1910), British-born U.S. physicist, used *attrib.* to designate concepts and devices he invented, as **Shockley diode,** a semiconductor diode consisting of four regions of alternate conductivity types (*n* and *p*), with the anode and the cathode connected to the end ones; **Shockley partial (dislocation),** a partial dislocation in which the lattice displacement, as represented by the Burgers vector, lies in the fault plane, so that the dislocation is capable of gliding.

1962 *Instrument Pract.* XVI. 1466/2 The arrangement will then have the characteristics of the Shockley or four layer diode which has only two terminals. **1965** *Wireless World* Aug. 397/2 Transistors, unijunction transistors or Shockley diodes can be used. **1975** T. D. Towers *Semiconductor Circuit Elements* xii. 146 A Shockley diode represents a switch with an off resistance of megohms and an on resistance of ohms. **1953** W. T. Read *Dislocations in Crystals* vii. 98 The Shockley partial is one of the three types of partial dislocations associated with plane faults in f.c.c. **1969** [see *partial *sb.* 2]. **1976** M. T. Sprackling *Plastic Deformation Simple Ionic Crystals* iv. 51 Shockley partial dislocations can glide in the fault plane; Frank partial dislocations cannot.

shock wave. Also **sho·ck-wave, sho·ckwave.** [f. Shock *sb.*³; cf. F. *onde de choc* (common in early 20th-cent. Fr. writings on fluid dynamics).] **a.** A disturbance that travels through a fluid as a narrow region in which there is a large, abrupt change in pressure and related quantities, esp. one separating regions of subsonic and supersonic fluid flow such as is created by an object moving faster than sound or by an explosion; *loosely*, any pressure wave of large amplitude.

[**1904** *Sci. Abstr.* A. VII. 646 Quasi-waves of shock propagating a dilatation in accordance with Hugoniot's law.] **1907** *Chem. Abstr.* I. 1470 The explosive wave is a shock wave accompanied by a decided combustion. **1931** *Proc. Nat. Acad. Sci.* XVII. 534 Let us regard a plane discontinuity or 'shock wave' from the point of view of the observer moving with it, so that the shock wave will appear to us stationary. **1945** *Times* 8 Nov. 2/2 When the oncoming aeroplane is travelling at, say, 600 m.p.h. or more the steadiness of the air-flow over the wing then breaks down with the result that what is known as a shock-wave occurs at the thickest part of the wing. **1950** *Sci. News* XV. Plate 5 (*caption*) The segmented appearance of the flame is due to the shock wave which is formed in the jet as it leaves the nozzle and which is reflected to and fro within the jet. **1958** *N. & Q.* CCIII. 139/1 Aside from other factors in nuclear explosions, the shock-wave, or blast, is possibly the greatest source of danger. **1973** C. Mason *Hostage* v. 65 The building was meanwhile hit by a shockwave so severe it felt like an earthquake. **1979** *Jrnl. R. Soc. Arts* CXXVII. 403/1 A small explosive charge is detonated in the ground and the resulting shock waves picked up by sensitive microphones called geophones.

b. *fig.*

1969 *Harper's Mag.* Dec. 125 The last writer who sent shock waves through Western literature. **1974** I. Murdoch *Sacred & Profane Love Machine* 145 The shock wave had not yet really come... Awful grief and pain hovered somewhere near to her. **1979** P. Harcourt *Sleep of Spies* I. iv. 56 The question caused shock waves round the room.

shod, *ppl. a.* Add: **2. a.** (Earlier mod. example.)

1840 J. Rowand *Let.* 8 July in G. P. de T. Glazebrook *Hargrave Corr.* (1938) 317 A couple of blood Indians got afighting..which ended by driving two shod arrows through one fellows body.

b. Of a motor vehicle: having tyres of a certain quality, as *well shod*. Cf. *shoe *v.* 2.

1967 Partridge *Dict. Slang* Suppl. 1356/1 *Shod*. 'Colloquially applied to motor vehicles. A car with good tyres is described as *'well shod'* (B.P.): Australian: since ca. 1945. **1977** *Horse & Hound* 14 Jan. 44/2 (Advt.), Bedford T.K. diesel, 1964, horse/cattle box... Well shod.

shodden, *a.* (Earlier example.)

1829 J. F. Cooper *Borderers* vi. 74 His beast hath had a shodden hoof.

shoddy, *sb.* Add: **4*.** = **reclaimed rubber.* Now *rare.*

1892 *Sci. Amer.* 7 May 293/2 Mould work of the lower

grades is often made of shoddy with no addition of pure rubber. **1898** *India-Rubber & Gutta-Percha & Electr. Trades' Jrnl.* XVI. 190/2 It is not surprising..that the volume of mechanical 'shoddy' should be placed by the best estimates at not above one-sixth of the total production of reclaimed rubber in the United States. **1974** K. F. Heinisch *Dict. Rubber* 428/2 *Shoddy*, jargon formerly used for reclaim; now rarely used.

5. shoddy dropper *Austral.* and *N.Z. slang*, a pedlar of cheap or falsely described clothing; a hawker.

1941 Baker *N.Z. Slang* vi. 52 We have [this century]..acquired some underworld slang of our own..[e.g.] *shoddy dropper*, a seller of cheap serge; [etc.]. **1964** *Australasian Post* 28 May 38/3 The 'shoddy droppers' (Indian hawkers) always carried a supply of patent medicines..with them on their rounds. **1972** *Telegraph* (Brisbane) 30 Aug. 24/6 The operators were known as dudders and professional shoddy-droppers.

shoddy, *a.* Add: **2.** Also, cheap, inferior; displaying signs of use, shabby, dilapidated.

1927 *Melody Maker* Apr. 305/1 The great majority of dance bands have settled down to a very stereotyped and shoddy sort of music. **1929** V. Woolf *Room of one's Own* v. 133 She will still wear the shoddy old fetters of class on her feet. **1932** E. Waugh *Black Mischief* vii. 255 The royal box was still there, shoddy sort of affair, but it provided a platform. **1952** *Manch. Guardian Weekly* 31 July 7/2 Because Stevenson was the man to beat, and Kefauver was their man, they had to fall back on the shoddy pretence that Stevenson was the tool of the big city machines.

shoddyism. (In Dict. s.v. Shoddy *sb.*) (Earlier example.)

1865 *Three Years among Working Classes in U.S.* vi. 124 Shoddyism among a large class of the people, corruption in official stations, an absorbing passion for making money..are the prevailing characteristics of the day.

shoe, *sb.* Add: **2. b.** Also, *the shoe is on the other foot*: the facts are otherwise, the position is reversed. Cf. *the boot is on the other leg* s.v. Boot *sb.*³ 1 b.

1933 *Mystery* May 122/1 Inspector Queen has not been able to discover our man-about-town's source of income. A gigolo? Gigolos do not pay for ladies' apartments; the shoe is rather on the other foot. **1939** B. K. Harris *Purslane* 179, I tell him if he had the waitin' on him to do the shoes would be on the other foot. **1953** J. S. Huxley *Evolution in Action* ii. 49 All the objections to a selectionist explanation of evolution that are based on the improbability of its results, simply fall to the ground. In fact the shoe is now on the other foot. Improbability is to be *expected* as the result of natural selection. **1976** *Times Lit. Suppl.* 2 Jan. 13/4 Most obviously, the shoe is on the other foot. Anyone who knows what rigour is will find it lacking in most of Hegel's transitions.

5. c. (Later examples.)

1972 L. M. Harris *Introd. Deepwater Floating Drilling Operations* ix. 90 The shoe of the foundation pile is equipped with a breakaway guide frame. **1976** *Offshore Platforms & Pipelining* 8/2 Deviation will begin about 100 ft below the conductor shoe.

q. A tyre. *slang.*

1917 E. E. Cummings *Let.* 2 Aug. (1969) 32 The rear axle looked like a mosquito's beak, and there were 2 shoes blown. **1934** R. Blaker *Night-Shift* vii. 87 If the tyres were worn to the fabric smooth as an egg, he could dismiss this outlay as 'a set of new shoes'.

r. A box for dealing the cards in baccarat or *chemin de fer*. Also, a game of baccarat.

1923 W. J. Locke *Moordius & Co.* xi. 149 Moordius dealt from the shoe. One card to the right, one to the left, one to himself. **1930** D. Byrne *Golden Goat* i. 8 They were only interested in the passing of the 'shoe' as the *chemin-de-fer* box is called. **1960** O. Manning *Great Fortune* i. 54 Hadjimoscos took his place before the shoe. As soon as he had drawn cards, he became serious and businesslike. **1964** A. Wykes *Gambling* xiii. 324 Zographos could remember every card that was played throughout a game (or 'shoe') of baccarat. **1965** D. Francis *Odds Against* iv. 52 He kept his side of the bargain by digging out the chemmy shoe. **1976** 'J. Welcome' *Grand National* iv. 54 The bank passed and the polished walnut and silver shoe slid along the table to a gaunt, henna-haired woman.

s. On a camera, a socket or other mounting for the temporary attachment of an accessory.

1953 A. Matheson *Leica Way* 47 The Leica IF and IC models carry a detachable brilliant viewfinder..in one of the two accessory shoes. **1971**, etc. [see *hot shoe* s.v. *Hot *a.* 12 c]. **1979** *SLR Camera* Feb. 74/1 The modification consists of adding an extra contact to the shoe of the sensor lead.

6. a. *shoe-bag*, *-factory*, *-store* (for U.S. read orig. U.S. and add earlier and later examples).

1873 S. Coolidge *What Katy did at School* vii. 139 Hang your dresses up..and put your shoes in the shoe-bag. **1972** J. Wilson *Hide & Seek* i. 7 The children began to pour out into the playground. Alice was..dragging her shoe bag along after her. **1855** J. Holbrook *Ten Years among Mail Bags* 276 How many persons are employed in that shoe factory? **1960** M. Spark *Ballad Peckham Rye* vi. 111 She told him all of her life in the shoe factory. **1789** *Boston Directory* 175 Bond and Bryant, shoe-store. **1976** *Milton Keynes Express* 11 June 12/2 The shoe store of Leslie Wheeler at New Bradwell.

b. *shoe-repairer.*

1933 *Radio Times* 14 Apr. 127/1 The well-known Kensington Shoe Repairers. **1976** M. Hinxman *End of Good Woman* vii. 94 Shoe repairer, now that was 'poshe' if you like!

c. shoe-bench *U.S.*, a shoemaker's bench; **shoe-bill,** (*a*) (earlier example); so **shoe-bill(ed) stork**; **shoe-brush** (earlier example); also *attrib.* of an object shaped like a shoe-brush; **shoe-button,** a button used for fastening a boot or a shoe; freq. *attrib.* of a small expressionless eye; **shoe-deep** *a. U.S.*, deep enough to cover a person's shoes; **shoe-flower** (earlier example); **shoe-last,** also *fig.*; also used *attrib.* in *Archæol.* to designate or with reference to polished stone implements, flat on one side and curved on the other, found in the area of neolithic Danubian culture; **shoe-piece,** (*a*) (in Dict., sense 6 a); (*b*) a piece of wood at the back of a chair, supporting the splat; **shoe-shine, shoeshine** (orig. and chiefly *U.S.*), a polish given to shoes, esp. by a shoe-shiner; freq. *attrib.*; **shoe-shiner,** one who polishes shoes for money; **shoe-tree** (earlier example).

1841 *Knickerbocker* May 362 A few weeks' rumination on the shoe-bench, or cogitation on the tailor's board. **1891** *Harper's Mag.* June 57/1 An express wagon was..loaded with the old shoe bench. **1861** Geo. Eliot *Let.* 20 Feb. (1954) III. 381 There is a shoe-bill, a great bird of grotesque ugliness. **1957** *Bull. Brit. Mus. (Nat. Hist.): Zool.* V. 111 (*heading*) The Pelecaniform characters of the skeleton of the Shoe-bill Stork. **1964** E. A. Nida *Towards Sci. Transl.* iii. 40 A person points to the beak of a shoe-billed stork and says *That's a big bill.* **1975** *Daily Colonist* (Victoria, B.C.) 17 Dec. 6/3 Shoebill stork..is one of only eight. **1740** E. Purefoy *Let.* 8 Mar. (1931) II. x. 248, I received Mr Robotham's letter..with half a dozen of oranges, a dozen of Delft plates, & 2 shoe Brushes. **1968** J. Arnold *Shell Bk. Country Crafts* 185 For use in difficult angles an extra leggat is used, having a shoe-brush handle to make it easy to use in those awkward places. **1895** *Montgomery Ward Catal.* Spring & Summer 525/2 Button Machine... No family should be without this machine..for putting on their own shoe buttons. **1928** E. O'Neill *Strange Interlude* v. 170 Lust ogling me for a dollar with oily shoe-button Italian eyes! **1973** 'A. Gilbert' *Is she Dead Too?* (rev. ed.) ii. 29 Her eyes round and hard as shoe buttons. **1773** P. V. Fithian *Jrnl.* 28 Dec. (1900) I. 75 Last night there fell a snow, which is about half Shoe-deep. **1891** M. E. Wilkins *New England Nun* 174 There had been a light fall of snow..but it was not shoe-deep. **1814** J. Lunan *Hortus Jamaicensis* I. 176 They are also put to a use which seems little consistent with their elegance and beauty, that of blacking shoes, whence their names of *rosæ calceolariæ* and shoe-flower. **1879** G. M. Hopkins *Lett. to R. Bridges* (1955) 76 Look upon them [*sc.* suggestions] as shoelasts on which to shape your final handiwork. **1927** Peake & Fleure *Priests & Kings* vi. 126 They had..stone celts or hoes, among which is a type, not unlike that found in the Danube basin, known as the shoe-last celt. **1961** G. Clark *World Prehist.* vi. 126 Boian pioneers..used a variety of stone tools, including adze-blades of bevelled and shoe-last form. **1970** Bray & Trump *Dict. Archaeol.* 210/2 *Shoe-last adze* or *celt*, a long thin stone adze employed by the Danubian farmers of the Early Neolithic, possibly as a hoe for cultivating their fields. **1923** J. C. Rogers *Eng. Furnit.* II. ii. 63 The splat..rose from a shaped shoe-piece planted on the rear seat rail. **1969** J. Gloag *Short Dict. Furnit.* (rev. ed.) 607 *Shoe-piece*, the shaped projection that rises from the back rail of a chair seat, into which the base of the splat is socketed. **1911** H. P. Fairchild *Greek Immigration to U.S.* vii. 127 In 1904 there were but three shoe-shine parlors in the hands of Greeks in the city. **1931** *Kansas City Times* 29 Oct., Cecil, the Negro shoe shine boy at the City barber shop, has organized a band. **1957** *New Yorker* 5 Oct. 35/1 President Romano,..resting tensely in his shirtsleeves, getting a shoeshine. **1958** X. Fielding *Corsair Country* i. 20 Where's the nobility in..these importunate shoe-shine boys? **1976** *National Observer* (U.S.) 21 Feb. 4/2 Complaints about the all-female shoeshine parlors in Salt Lake City may diminish... The commissioners said that the parlors..must not allow the shoe shiners to mingle with the customers or sit on their laps. **1910** *Chambers's Jrnl.* July 431/1 The hotel will not be bothered with boot-cleaning, that service being performed by the 'shoe-shiner' in the basement. **1976** Shoe shiner [see *shoe-shine* above]. **1827** Drake & Mansfield *Cincinnati* viii. 60 In the third story the manufacture of shoe trees is carried on.

shoe, *v.* Add: **2.** Also, to provide (a motor vehicle) with tyres of a specified type or quality. Cf. *Shoe *sb.* 5 q.

1925 *Morris Owner's Man.* p. lxx (Advt.), Every car is turned out in sound order and condition, shod with good tyres. **1971** *Drive* Summer 121/1 The test car was shod with radial tyres. **1976** *Southern Even. Echo* (Southampton) 2 Nov. 15/3 Braking is by servo-assisted discs at the front and rear drums, with radial tyres as standard shoeing equipment.

shoe (ʃū), *a. U.S. slang.* [Origin obscure.] Conforming to the dress, behaviour, or attitudes of students at exclusive educational establishments; acceptable to or commended by such people.

1962 *Punch* 13 June 895/3 A girl at these institutions [*sc.* schools] must prove herself to be 'shoe'—and woe betide her if her dress and manner don't manifest..'shoeness'. **1973** *N.Y. Times Mag.* 17 June 38/3 Perhaps it is significant that one favourite mode of protest in the fifties was satire. We—a lot of us—were cool, ironic, 'shoe'. **1980** L. Birnbach et al. *Official Preppy Handbk.* 222/2 *Shoe, adj.* Top-drawer. Very acceptable.

shoeblack. Add: **b.** Also simply *shoeblack.* (Earlier and later examples.)

1837 J. Macfadyen *Flora Jamaica* I. 66 The flowers, from the mucilaginous juice they contain, are employed to give a polish to the leather of shoes; and hence the plant has received the name of the shoe-black. **1965** *Harper's Bazaar* Feb. 18/3 The scarlet blossoms of the bush they [*sc.* Jamaicans] call the Shoe-Black.

shoe-box, shoebox (ʃū·bɒks). [Shoe *sb.*]

1. A box in which a pair of shoes is packed.

1860, 1897 [in Dict. s.v. Shoe *sb.*¹ 6 a]. **1930** J. Dos Passos *42nd Parallel* i. 101 A small house like a shoebox. **1970** J. Earl *Tuners & Amplifiers* i. 14 A modern stereo amplifier capable of yielding 20 watts..is nowadays barely larger than a shoe box.

2. *fig.* A building or part of a building resembling a shoe-box. Also *attrib.*

1968 *Listener* 1 Aug. 134/2 Leeds hasn't changed much. There are a few changes. Some of those glass shoe-boxes have been plonked down at random in the city centre. **1972** *Times* 8 June (Birmingham Suppl.) p. v/4 The mandatory shoe-box buildings. **1978** N. Freeling *Night Lords* xxxiii. 153 Bianchi..waved casually at the building; the usual pile of open shoe-boxes. **1979** M. A. Sharp *Sunflower* iii. 29 Shoebox buildings, nestled together like children's blocks.

shoe-horn, *v.* Add: **2. b.** To manœuvre or compress (someone or something) *into* (*in, on to*) an inadequate space (occas. *into* an inadequate period of time).

1927 D. L. Sayers *Unnatural Death* vi. 64 He shoe-horned himself into his seat [in a motor car]. **1954** *Archit. Rev.* CXVI. 212/1 For the second edition he shoehorned in material on Eiffel, Maillart and Alto, thus playing havoc with his illustration numbers. **1968** *Economist* 18 May 69/1 Big aircraft are shoe-horned on to small landing strips with only elementary navigation aids. **1969** J. Wainwright *Take-over Men* iii. 37 Lewis..was shoe-horning himself behind the steering wheel. **1972** *Real Estate Rev.* Winter 107/1 The logical-thought input might be assumed to have been forcibly pressed and shoe-horned into a preconceived formal framework. **1974** *Daily Tel.* (Colour Suppl.) 23 Aug. 7/2 In order to shoe-horn in more people airlines and aircraft manufacturers have given up any interest at all in comfort. **1978** *Science* 17 Mar. 1161/2 Nearly 140 symposia and almost 1000 speakers shoehorned into five days and nights. **1979** R. L. Simon *Peking Duck* ii. 19 The houses were..shoe-horned onto thirty feet of beachfront land.

shoepack (ʃū·pæk). *N. Amer.* Also **shoepac, shupac.** [ad. Delaware Jargon *seppock, sippack* shoes, f. Unami Delaware *čipahko* moccasins, infl. by Shoe *sb.*] Orig., and still locally, a moccasin with an extra sole; more recently, a commercially manufactured oiled leather boot, usually with a rubber sole. Cf. *Pac.

1755 in S. M. Hamilton *Lett. to Washington* (1898) I. 99 It would be a good thing to have Shoe-packs or Moccosons for the Scouts. **1824** [in Dict. s.v. Shoe *sb.* 6 c]. **1853** S. Strickland *Twenty-Seven Years in Canada West* II. 286 Shoe-packs, a species of mocassin peculiar to the Lower Province, cow-hide boots, and a *bonnet rouge* for the head, complete the costume of the Canadian lumberman. **1882** J. M. LeMoine *Picturesque Quebec* 201 He came pounding along Notre Dame street, in Montreal, in his red shirt and tan-colored *shupac* boots, all dripping wet. **1903** S. E. White *Forest* x. 120 He brought to light..oil-tanned shoepacs, with and without the flexible sole. **1940** R. Marshall *Arctic Village* 101 It is only in the fall and the spring that the snow is soggy, and in those seasons shoepacks with rubber bottoms and leather uppers replace the moccasins. **1977** *New Yorker* 20 June 69/2 After the cast comes off, I can walk with a shoepac and a cane.

shoe--string, shoestring. 1. (In Dict. s.v. Shoe *sb.* 6 c.)

2. A small or inadequate amount of money; a very little capital; a small margin. Chiefly in phr. *on a shoe-string. colloq.* (orig. *U.S.*).

[**1882** *Century Mag.* Apr. 884/2 [He] could draw to a shoe-string, as the saying went, and obtain a tan-yard!] **1904** *Cosmopolitan* May 89 He..speculated 'on a shoe-string'—an exceedingly slim margin. **1926** J. Black *You can't Win* viii. 90 The new owners had no bankroll, just opened up on a shoestring. **1932** *Atlantic Monthly* Mar. 310/1 Every business man who has made a big success of himself started on a shoestring. **1957** *Listener* 28 Nov. 893/3 Reformative efforts have to be..as they say, 'run on a shoe-string'. **1977** C. McCullough *Thorn Birds* xvii. 441 Australians in England, youth-hostelling on a shoestring.

3. = *shoe-string potato*, sense 4 c below.

1931 B. Starke *Touch & Go* x. 156, I..found that the word 'shoe-strings' on the menu really meant Julienne potatoes. I ate every last shoe-string.

4. *attrib.* and *Comb.* **a.** *attrib.* Narrow (*lit.* and *fig.*).

1878 *Congress. Rec.* App. 13 June 478/2, I will promise to meet him on the northern border of 'the shoe-string district'. **1897** *Pop. Sci. Monthly* L. 309 Bad roads and shoestring paths..fringe them. **1953** *Times* 30 July 7 A shoe-string majority.

b. *attrib.* Operating on a shoe-string, costing or spending little; cheap, informal; petty.

1890 J. P. Quinn *Fools of Fortune* 494 The gamblers, aside from a lot of 'hangers on', known as 'shoestring' or 'tin horn' gamblers, do not figure in the criminal records. **1923** 'B. M. Bower' *Parowan Bonanza* xi. 137 The little shoestring propositions that go broke and leave empty

houses behind them. **1936** *Sun* (Baltimore) 20 Nov. 1/5 The governors of the Federal Reserve System today proposed steps to plug up loopholes through which 'shoe-string' and other operators have been able to trade extensively without even posting margins. **1941** B. Schulberg *What makes Sammy Run?* vi. 123 A shoe-string producer..had bought the stock shots from *Hell's Angels* and *Wings* and needed an airplane story. **1958** *Vogue* Oct. 203 Winter after winter ski-crazy students flock to the snow for shoestring holidays which have been planned to the last farthing. **1959** 'M. Neville' *Sweet Night for Murder* i. 17 She talked, thought, dreamed clothes. On a shoe-string allowance, however, she could do little about them. **1977** *S. Wales Guardian* 27 Oct. 1/3 The Education Committee was being penny wise and pound foolish by giving some contracts to small private contractors running on a shoe-string budget and using non-union labour. **1978** J. Krantz *Scruples* iii. 78 The ad that launched Spider was for a new type of fingernail hardener, put out by a shoe-string company.

c. Special comb.: **shoe-string catch** *Baseball*, a running catch made close to the ground; **shoe-string fungus** = *honey fungus* s.v. *Honey *sb.* (*a*) 7 b; **shoe-string potato** *U.S.*, a julienne potato (see *Julienne 2) (chiefly *pl.*); **shoe-string (root) rot**, the disease caused by shoe-string fungus; **shoe-string tie**, a very narrow neck-tie.

1926 *N.Y. Times* 11 Oct. 25/2 Haley ran up on it and tried to make a shoestring catch of it. **1957** *New Yorker* 13 July 17/1 Like a shoestring catch in center field, make it and you're the hero. Muff it and you're a dope. **1926** F. D. Heald *Man. Plant Diseases* xxvi. 794 The causal fungus is generally referred to as the 'honey agaric'..or the 'shoestring fungus'. **1978** T. A. Tattar *Diseases of Shade Trees* xxiii. 315 The shoestring-fungus..and the two-lined chestnut borer..are the most common organisms of secondary action. **1906** 'H. McHugh' *Skiddoo!* ii. 30 The next course was French fried potatoes with some shoe-string potatoes on the side. **1940** *Amer. Mercury* Sept. 72 Old Fred Harvey started turning a shoestring potato into a 2500-mile of railroad eating-places. **1976** U. Curtiss *Birthday Gift* xv. 141 As fruitless as looking for shoestring potatoes in strange supermarkets. **1978** T. A. Tattar *Diseases of Shade Trees* x. 140 Shoestring root rot affects a wide range of shrubs and trees. **1931** E. E. Hubert *Outl. Forest Pathol.* ix. 417 Shoestring rot is a disease well known to the forest pathologist. **1903** F. Norris *Pit* 337 His shoestring tie straggled over his frayed shirt front. **1961** *Sunday Express* 29 Jan. 15/6 A short, dark Frenchman in a shoe-string tie.

shoey (ʃū·i). *slang.* [f. Shoe *v.* + -y⁶.] A shoeing-smith in a cavalry regiment.

1919 in *War Terms* in *Athenæum* 1 Aug. 695/2. **1925** in Fraser & Gibbons *Soldier & Sailor Words* 256. **1969** S. Mays *Fall out Officers* xii. 93 Shoey..Slap some shoes on my new horse.

shoful. Add: Also **shofel. 1.** (Earlier examples.)

1828 *Sessions' Papers Old Bailey 1827–28* 602/2 The twenty counterfeit shillings were found on me; the sister came to me and asked if I had any *shofle* about me, if I had to put it away. **1846** *Swell's Night Guide* 61 The shiser thinks to bounce us by passing a shofel quid.

shog, *v.* Add: **1. a.** Also, to shake *off* a load. *rare.*

1949 D. L. Sayers tr. *Dante's Divine Comedy* I. xvii. 178 Having shogged our burden off..away he bounded.

3. b. (Later *Lit.* examples.)

1884 C. M. Yonge *Armourer's Prentices* I. x. 192 Bolt..bade him shog off, and not come sneaking after other folks' shoes. **1929** J. C. Powys *Wolf Solent* vii. 154 Lob began to swagger slowly away. 'I knows why you wants me to shog off,' he called back. *Ibid.* ix. 208 Wolf shogged off by himself. **1962** L. R. Banks *End to Running* i. v. 71 I'll just say to hell with her, to hell with the money and the house and everything else—I'll just shog off.

‖ **shogi** (ʃōu·gi). Also **Sho-gi, † Sho-Ho-Ye, Shongi.** [Jap.] A Japanese board game resembling chess.

1858 *Japan Opened* (Relig. Tract Soc.) vii. 267 The game is called *Sho-Ho-Ye*, and is a great favourite among the Japanese. **1884** tr. *J. J. Rein's Japan* II. ii. 430 Among those of which adults of all classes..are very fond, the most conspicuous at present are Shōgi, or chess, and Go. **1890** B. H. Chamberlain *Things Japanese* 66 Japanese chess (*shōgi*) was introduced from China centuries ago. **1905** Cho-Yo *Japanese Chess (Shō-ngi)* 27 The governing class of people valued the chessological Art or Science of struggles, commonly known as *Shōngi* (Chess). **1969** R. C. Bell *Board & Table Games* II. ii. 38 There are about ten million Sho-gi players in Japan. **1975** *Way to Play* 50/1 There have been many forms of shogi since its introduction in about the eighth century.

shogunate. (Earlier example.)

1871 A. B. Mitford *Tales Old Japan* I. 99 After..the abolition of the Shogunate, he accompanied the last of the Shoguns in his retirement.

Shoho, var. *Saho *sb.* and *a.*

‖ **shoji** (ʃōu·dʒi). [Jap.] **1.** In Japanese architecture, a sliding outer or inner door made of a latticed screen covered usu. with white paper.

1880 I. L. Bird *Unbeaten Tracks in Japan* I. 90, I closed the sliding windows, with translucent paper for window panes, called *shōji*. **1922** J. Street *Mysterious Japan* ii. 24 Children glimpsed through the open wood and paper *shoji* of their matchbox houses. **1959** R. Kirkbride *Tamiko* iv. 28 He swung himself over the sill and dropped into her room, closing the shoji behind him. **1979** H. McCloy *Smoking Mirror* Inside the house there were..sliding partitions like the panels on a Japanese shoji.

2. *attrib.* or as *adj.*

1886 J. La Farge *Let.* 1 Sept. in *Artist's Lett. Japan* (1897) 27 To look out of the *shoji* screens into the garden. **1896** L. Hearn *Kokoro* ii. 19 The light *shōji* frames serving at once for windows and walls, and repapered twice a year. **1958** R. Gaskin *Few Ways with Dried Flowers* 126 (*caption*) Shoji type screen decorated with a variety of pressed leaves. **1967** M. M. Pegler *Dict. Interior Design* 407 The shoji panels are used as screens, dividers, doors that slide behind one another on a track (Japanese style), or as window coverings. **1977** *Time* 24 Jan. 17/1 The hero tears his way through the hard paper covering of a shoji screen.

‖ **shokku** (ʃɔ·kū). [Jap., f. Shock *sb.*³] Used *joc.* to denote a shock or surprise in political or economic affairs concerning Japan.

1971 *Time* 4 Oct. 36/1 The President had convulsed Japan..with the 'Nixon shokku'—his spectacular policy shifts on China and the economy. **1973** *Times* 3 Sept. 18 Indeed, the Nixon Administration's diplomatic *shokku* in 1971 did lasting damage to Japan's relations with the U.S. **1978** *Encounter* Sept. 56/2 The Japanese were able to cope with their frightful 'oil shokku' with far more self-restraint than marked the response of other countries to the oil embargo.

sholy (ʃō·li). Also **sholey.** Representation of U.S. Black and Southern dial. pronunc. of Surely *adv.*

1919 H. W. Odum in *Amer. Mercury* Sept. 48/1 Camp sholy was roughish place. **1940** W. Faulkner *Hamlet* i. ii. 56 'You brought it to me?' 'Sholy,' Ratliff said. 'Here it.' **1966** M. Treadwell in C. Chapman *New Black Voices* (1972) 140 Sholey. She, hit'll be jes fine.

‖ **shomer** (ʃō·mɑɪ). Pl. **shomrim.** [ad. Heb. *šōmēr* watchman.] **1.** (See quot. 1971.)

1909 *Cent. Dict.* Suppl., Shomer. **1923** H. Trager *Pioneers in Palestine* ii. iii. 147 The other passengers talk of the killing of one of the Shomrim by an Arab. **1935** *Zionist Review* Mar. 10/2 At 4:30 a.m...pounded the shomer (night watch) on our door. **1938** *Ibid.* Feb. 35/1 The renowned old Jewish watchmen..have been in the country for twenty-five years and more, having been members of the 'Shomrim' organisation. **1944** M. Samuel *Harvest in Desert* … **1971** *Encycl. Zionism & Israel* … *Hashomer* … Self-defence organization of Jewish pioneers..established in Palestine in 1909 to protect Jewish settlements there. Each..colony..was supplied with Shomrim (watchmen).

2. An inspector who verifies that food is prepared in accordance with Jewish religious laws.

1909 *Cent. Dict.* Suppl., Shomer. **1923** H. M. Lazarus *Ways of her Household* … **1954** A. H. Wyndham *London Board for Shechita* … The shomer visited the community of Paris..and was given every facility to inspect the methods..of shechita. **1972** *Times* … I am quite happy to accept supervision of our food supplies by a shomer appointed by the local Jewish authorities.

Shona (ʃō·nă). See *Mashona *sb.* and *a.*

1930 C. M. Doke *Rep. Unification Shona Dialects* 61 The Language Committee..hope to unify..the dialects of the area..known as Mashonaland and the name for the language should..be one which indicates this width of range. The name 'Shona' has been suggested which the Committee now recommends. **1930** C. G. Seligman *Races of Africa* viii. 187 The Shona peoples of Southern Rhodesia and of Portuguese East Africa immediately south of the Zambesi as far as the Sabi River. **1936** *Discovery* June 179/1 The Shona..were the later or perhaps the earlier occupiers of the Hill which we call Mapungubwe. … **1979** *Financial Times* … there is still time to negotiate a three tier federal solution which would avoid domination of the minority Ndebele and the minority Whites by the majority Shonas.

Shonee, var. *Shawnee *a.* and *sb.*

shoneen. Add: Used (esp. *attrib.*) to indicate a person's inclination towards English rather than Irish standards and attitudes in cultural life, sport, etc. (Later examples.) Hence **shonee·nism.**

1918 F. Hackett *Ireland* iii. 65 West Britonism makes us what we are, shoneenism and toadyism, so it is, they're the curse of Ireland. **1920** B. MacNamara *Clanking of Chains* iv. 44 But the shoneenism of Ambrose was in more perfect keeping with the shoneen heart of Ballycullen... The songs which he sang were out of the English music halls, the books which he read were English drivel. **1922** Joyce *Ulysses* 311 Irish sport and shoneen games. **1927** [see *pro-Britisher* s.v. *Pro-¹ 5 b]. **1958** B. Behan *Borstal Boy* iii. 326 Now, there was a lot of shoneen writing and playing up to the herrenvolk by Rugby writers. **1960** *20th Cent.* Oct. 324 This aunt practically invented the whole concept of shoneenism... She believed that God was Anglo-Saxon, Protestant.

shonicker (ʃɒ·nikəɹ). *U.S. slang.* Also **shoniker, shonniker.** [Orig. uncertain: see quots. 1966, 1970.] An offensive name for a Jew (see also quot. 1914).

1914 Jackson & Hellyer *Vocab. Criminal Slang* 75 *Shoniker*, current amongst pawnbrokers and thieves, especially Jews. A neophyte or inexperienced hand at the game. **1927** *Dialect Notes* V. 461 *Shonnicker*, n., a Jewish pawnbroker. **1932** J. T. Farrell *Young Lonigan* vi. 169 Two booknoses..did come along. Andy and Johnny O'Brien..stopped the shonickers. **1966** *Publ. Amer. Dial. Soc.* 1964 XLI. 45 This folk etymology derives *shonicker* from Yiddish *schmoozle*... My colleague..suggests a derivation from Hanukkah. **1970** L. M. Feinsilver *Taste of Yiddish* 318 Shon, shonik, shonnicker, shonniker. … opprobrious terms for a Jew in England are supposed to have come from Yiddish *shoniker* (petty trader or peddler).

shonk (ʃɒŋk). *slang.* (Shortened form of *Shonicker.) An offensive name for a Jew. Hence **sho·nky** *a.*¹ (see quot. 1951).

1938 W. Matthews *Cockney Past & Present* v. 153, I diffidently suggest the following words as the most familiar slang terms rarely used except by cockneys.. *shonk*, nose, Jew. **1940** R. Postgate *Verdict of Twelve* i. v. 75 Let's have a bit of fun with the shonks. **1951** E. Partridge *Dict. Slang* (ed. 4) Add. 1168/1 *Shonky*, adj., mean; money-grubbing: late C. 19–20. **1961** 'W. Haggard' … 'Brighton!' It's full of shonks.' 'Which means there are hotels with night life.'

shonkinite (ʃɒ·ŋkinəit). *Geol.* [f. *Shonkin*, Indian name for the Highwood Mountains, Montana + -ite¹.] A dark granular form of syenite consisting largely of augite and orthoclase. Hence **shonkini·tic** *a.*, having the character or consisting of shonkinite.

1895 Weed & Pirsson in *Bull. Geol. Soc. Amer.* VI. 415 For this type of rock, then, we propose the name of shonkinite,..and shonkinite as a name for a granular plutonic rock consisting of essential augite and orthoclase, and thereby related to the syenite family. **1900** *Bull. Geol. Soc. Amer.* XI. 395 The relations of the 'fine grained' (shonkinitic) syenite to the leucite-porphyry are uncertain. **1928** Peach & Horne *Chapters Geol. Scotl.* (1930) iv. 112 A discontinuous zone of fine-grained shonkinite and pulaskite. **1943** S. J. Shand *Eruptive Rocks* (ed. 2) xiv. 255 There are no lavas of shonkinite composition. **1955** Turner & Verhoogen *Igneous & Metamorphic Petrology* viii. 196 The syenite undoubtedly is a differentiate derived from a parent shonkinite magma. **1978** S. R. Nockolds et al. *Petrology for Students* v. 61 Other shonkinites from the Highwood Mountains and from the Bearpaw Mountains, Montana, may have a little plagioclase.

shonky (ʃɒ·ŋki), *a.*² *Austral. slang.* Also **shonkie.** [Perh. f. *Shonky *a.*¹: see *Shonk.] Unreliable, dishonest, 'crooked'. Hence as *sb.*, one engaged in irregular or illegal business activities.

1970 R. Beilby *No Medals for Aphrodite* 116 You shonkie sod! **1976** *Sunday Mail* (Brisbane) 2 May 8/4 Queensland has some 'shonky' charities. I cannot name them because of libel. **1979** *Financial Rev.* (Sydney) 25 July 11/6 Mr Groom is right when he refers to the building industry as being characterized by initiative and drive, but unless something is done to eliminate these 'shonkies' quickly, then such qualities will be characteristics of the past. **1981** *Australian* 2 Feb. 7/7 The woman..was forthright about the cut-price air fares... 'We call these tickets shonky,' she said.

sho' nuff: see *Nuff *b.*

shoo, *v.* Add: **1. a.** Also, to do this by means of a movement or gesture. Also *trans.*

1919 Conrad *Arrow of Gold* i. i. 9 Shells were falling all round till a tiny French gunboat..shooed the *Numancia* away out of territorial waters. **1938** W. de la Mare *Memory* 14 She shoo'd it away with her gloves. **1959** D. Beatty *Cone of Silence* iii. 35 Then she shooed cups and plate away from her. **1959** *Listener* 15 Jan. 113/2 He shakes or nods his head to shoo the flies away. **1973** 'H. Carmichael' *Too Late for Tears* viii. 108 'I hope you won't mind if I shoo you out now. I've got work to do.' **1977** *Times* 22 Aug. 10/1 Israeli artillery regularly fires into south Lebanon to shoo away Palestinian guerrillas from Lebanese Christian enclaves in the border area.

b. To drive or urge (a person, animal, etc.) in a desired direction.

1903 *N.Y. Sun* 17 Nov. 12 The police shoo everybody to the south side of the loops. **1923** 'B. M. Bower' *Parowan Bonanza* xiii. 151 You're supposed to shoo a lady gently before you down the aisle. **1946** M. Dickens *Happy Prisoner* xi. 267 The first pony had already been shooed into the ring. **1973** M. Amis *Rachel Papers* 150, I do not churlishly flatten her on to the sofa nor shoo her downstairs.

4. *trans.* With *in*, to allow a racehorse to win easily. *U.S. slang.* Cf. *Shoo-in.

1908 G. E. Smith *Racing Maxims & Methods of 'Pittsburgh Phil'* ix. 123 There were many times presumably that 'Tod' would win through such manipulations, being 'shooed in', as it were. **1935** D. Runyon *Money from Home* 128 They are going to shoo in Never Despair. **1976**

New Yorker 22 Mar. 85/2 To be sure, Shoemaker's confreres could have shooed him in long before this, but jockeys never, never do such things.

shoo-fly (ʃū·flai), *vbl. phr.* and *sb. U.S.* Also **shoofly, shoo fly.** [f. Shoo *int.*¹ + Fly *v.*¹ and *sb.*¹]

A. *vbl. phr.* A catch phrase, popularized by a song, used as an exclamation of annoyance. *Obs.*

1867 *Chicago Republican* 24 July 8/1 [Baseball] players invariably say 'Shoo-fly', when they make a miss. **1889** Farmer *Americanisms* 484/2 *Shoo! fly! don't bother me!* … The full phrase is now familiarly colloquial. **1919** Mencken *Amer. Lang.* 311 Shoo-fly afflicted the American people for at least two years, and 'I don't think' and other are quite as long.

B. *sb.* **† 1.** A device or structure intended to afford protection from flies. *Obs.*

1879 *Glendale* (Montana) *Atlantis* 28 Dec. 4/4 A Dutchman drove rapidly along Main Street, with a new shoo-fly attached to his wagon, making forty flips a second and striking back and forth with the vigor of a steam engine. **1925** J. Ralph *Dixie* iv. 126 In many cases they order great pavilion like giant tents built over their houses, and they call them 'shoo-flies', a name utterly without significance in that connection.

2. A policeman, usu. in plain clothes, whose duty is to watch and report on other policemen. *slang.*

1877 *Daily Graphic* May 1 A 'shoofly' is the term applied by a policeman to another officer who is detailed to watch him. **1903** H. Hapgood *Autobiogr. Thief* (1904) xii. 265, I was gathered in to make a reputation for those two shoo-flies. **1932** *Detective Fiction Weekly* 27 June 790/2 A force of 'shoo flies'—roundsmen in civilian clothes—were sent out regularly from headquarters to sweep into a precinct and look over the men. **1952** *Sun* (Baltimore) 10 May 2/2 Evans said he spent eighteen months on 'shoo-fly duty'—a special tour of duty that he worked out of headquarters keeping other policemen in line and honest. **1980** 'E. McBain' *Ghosts* viii. 127 'You want a beer?'..'Officially I'm still on duty, but fuck it.' 'Shooflies are heavy around the holidays.'

3. A rocking horse in which the seat is placed between two rockers representing the animal. Freq. *attrib.* as **shoo-fly rocker.**

1887 *Chicago Tribune* 27 Nov. 16/7 (Advt.) Shoo fly hobby horse, 75 c. **1895** *Montgomery Ward Catal.* Spring & Summer 563/2 Shoo-fly rockers... **1947** *Chicago Tribune* 30 Dec. 13/1 The Teaneck, N.J., library has installed something called a 'shoo-fly', a metal rocker in which little Elmer..can rock his head off.

4. A temporary railway track constructed for use while the main track is obstructed or under repair. Also *transf.* (see quot. 1907).

1902 *N.Y. Even. Post* 29 July 1 The Southern Pacific Company's shoo-fly around the tracks now submerged will be completed in a few days. **1907** *Dialect Notes* III. 249 Shoo-fly, n., suburban railway train. … **1933** *Highway Mag.* Jan. 9/1 Beginning in the spring of 1936 the railroad built two temporary 'shoo fly' tracks about 75 feet west of the existing tracks. **1961** *Washington Post* 17 Feb. A4 (*caption*) Workers on top of a construction train adjust the overhead wires for a bypass—or 'Shoo-fly' in railroad parlance—of the main railroad tracks serving the area to the south of Washington.

5. *Printing.* In some flat-bed presses, a set of narrow strips which lift the edge of the sheet off the cylinder ready for delivery. Also *shoofly finger.*

1888 *Inland Printer* XI. 551/1 Where the delivery construction uses shoo-fly fingers to give the forward edge of the sheet a lift as it passes out onto the delivery, the proper setting of these fingers or shoo-flies, as they are usually called, is of prime importance. … *Theory & Practice of Presswork* (U.S. Govt. Printing Office) (rev. ed.) xxvii. The shooflies, stripper fingers, and tape delivery … are to be found …

6. In needlework, a traditional patchwork design.

1931 R. S. McKim *One Hundred & One Patchwork Patterns* 84/1 As shoo fly is one of the simplest of old-fashioned patchworks, both to cut and to piece, it would be a good choice for one on which a little girl could learn sewing. **1977** E. Y. Wood *Amer. Patchwork Quilts* 15 Four-patch designs are here, such as the classic 'Shoo Fly'.

7. *attrib.* and *Comb.* **a.** (Examples in various unspecified senses.)

1870 *North Alabamian & Times* (Tuscumbia, Alabama) 21 Apr. 2/7 The gentlemen can gratify their taste by ornamenting themselves with the latest style of 'Shoo Fly' Hats. **1886** M. B. Buckley *Diary Tour in Amer.* viii. 213 There were 'Shoo fly neckties' … O. Wister *Jrnl.* 17 June (1958) 106 That's a terrible plain woman Hank's got. All driven and fixed up..Looks like a picture on one of these shoo-fly boxes. **1897** *Sears, Roebuck Catal.* … Shoo-fly … pints.. quarts. **1946** C. Richter *Fields* 278 Huldah had gone with Amy MacMahon, a red shoofly ribbon low on both their necks.

b. shoo-fly pie, a rich tart made of molasses baked in a pastry case with a crumble topping; **shoo-fly plant,** a large annual herb, *Nicandra physaloides*, belonging to the family Solanaceæ, native to Peru, and bearing pale blue flowers followed by berries enclosed in the enlarged calyx.

1935 *Esquire* Dec. 200/1 'Shoo-fly pie'—a brown-and-white crumb-cake, faintly spiced. **1971** *Daily Colonist* (Victoria, B.C.) 9 July 34 The pie is nothing but … shoo-fly pie and other Pennsylvania items. **1979** *United States 1980/81* (Penguin Travel Guides) 48 Vermont cheese and maple syrup..and shoofly pie and pretzels in the Pennsylvania Dutch country—are typical of their respective regions. [**1902** L. H. Bailey *Cycl. Amer. Hort.* IV. 1664/1 Shoo-fly plant. A name proposed by one seedsman for Physalis.] **1949** L. H. Bailey *Man. Cultivated Plants* (ed. 2) 871 *Nicandra physalodes*..Apple-of-Peru, Shoo-fly Plant. **1955** *Times* 10 Oct. 16/7 Left-over seeds thrown out for the birds from special mixtures..are responsible for the appearances of casual weeds like..the Apple of Peru, or 'shoo-fly plant', and its pale blue flowers …

shoo–in. *N. Amer.* [f. vbl. phr. *to shoo in*: see *Shoo *v.* 4.] **1.** In horseracing, a predetermined or 'fixed' race, or the winner of it. Hence *loosely*, a horse which is a certain winner.

1928 *National Turf Digest* (Baltimore) Dec. 409/2 A 'skate' is a horse having no class whatever, and rarely wins only in case of a 'fluke' or 'shoo in'. **1937** *Collier's* 11 Sept. 11/3 … **1950** *Sun* (Baltimore) 2 June 22/4 Some horsemen wondered whether Chris Chenery's Virginia flyer would be such a shoo-in for the Belmont Stakes..after all. **1969** R. Lockridge *Risky Way to Kill* xi. 132 … Figuring him to be pretty much of a shoo-in for the..hunter championship.

2. *transf.* (esp. *Pol.*). A certain or easy winner; a certainty, a 'walk-over'.

1939 … 30 Jan. 15/5 (*heading*) Bear cagers appear shoo-in for southern division title. **1948** *Tuscaloosa* (Alabama) *News* 30 July 4/2 This type of registration might endanger their balloting in the local elections, where Democratic candidates are usually chosen on a 'shoo-in' basis. … **1968** *Economist* 20 Jan. 29/1 … The press thought him a shoo-in for the Republican nomination. **1976** *Islander* (Victoria, B.C.) 10 Oct. 15/3 From then on, … it was a shoo-in. …

shook, *ppl. a.* **2.** *colloq.* **a.** *to be shook on*: to be enamoured of or enthusiastic about. *Austral.* and *N.Z.*

1888 'R. Boldrewood' *Robbery under Arms* II. iii. 46 He was awful shook on Mad; but she wouldn't look at him. *Ibid.* xix. 291 I'm regular shook on the polka. **1926** J. Devanny *Butcher Shop* … **1965** … *Political Farm* vii. 100 Tom said he wasn't too shook on makin' a public fool of himself. **1975** *Sunday Tel.* (Sydney) 29 June 80 Like Chappell, I'm not all that shook on cricket parties myself.

b. Emotionally or physically disturbed, discomposed, upset. Usu. const. *up*. orig. and chiefly *U.S.*

1891 Kipling in *Macmillan's Mag.* Oct. 473/2 He took my 'and an' pulled me up, an' I was pretty shook. **1897** … *Captains Courageous* … You was shook up and silly. **1914** *Dialect Notes* IV. 79 Shook, *p.a.* startled, shocked, grieved. … **1959** H. Salisbury (*title*) The shook-up generation. **1960** *Washington Post* 25 May 1 'Elva behaved all the way through just like a good policeman,' … said of his wife's role in the chase. 'But she was pretty shook up after it was all over.' **1976** *Daily Mirror* 11 Mar. 1/1 Rock superstar Rod Stewart was all shook up last night after a 'violent' fight with..Britt Ekland.

shool, var. *Shul.

shoot, *sb.*¹ Add: **1. i.** *transf.* The action of shooting a film. Cf. *Shoot *v.* 22 f.

1929 *Morning Post* 24 May 12/7 A Wembley 'Shoot' Described… A 'talkie' sequence is being 'shot' in … and then suddenly you found yourself part of a commercial shoot?

j. *Mil.* An act of bombardment; esp., an exercise in which anti-aircraft drill is practised.

1941 *Hutchinson's Pictorial Hist. War* July–Sept. 219 (*caption*) … Bofors guns, of proved efficiency against low-flying aircraft, are included in Malta's defence against bombing... A practice shoot is in progress. **1961** … xiv. 344 The prize shoot was … **1977** *R.A.F. News* 8–21 June 11/2 Such black days, and other times when 'shoots' are cancelled, can be disappointing and frustrating to the range team as to the fliers and their units.

3. f. In slang phr. *to give* (a person) *the shoot*: to dismiss from employment, sack; also *transf.* So *to get the shoot.* Cf. *Boot *sb.*³ 1 c.

1846 *Swell's Night Guide* 50 'You nasty old man,' said she, 'and your doss gorger cracked a wid about you to me, and said she must give you the shoot.' **1906** [see *Move *sb.* 6].

g. = *Shot *sb.*¹ 7 h. *U.S.*

1959 *Time* 5 Jan. 24/2 Another 20 or 30 Atlas shoots must be made. **1961** *N.Y. Times Mag.* 5 May 28/2 (*caption*) In a recent 'shoot' the capsule was picked up at sea fifty-six minutes after take-off.

6. c. (Earlier example.) Also *Austral.*, an opening and ramp leading from one pen to another in a sheep-shearing shed. Cf. *Chute *sb.* 3 b.

1873 J. H. Beadle *Undevel. West* xxii. 432 About a quarter section of cattle-yards and 'shoots' extend around the depot. **1900, 1905** [see *Pen *sb.*¹ 1 c]. **1955** Stewart & Keesing *Austral. Bush Ballads* 239 The shearers squint along the pens, they squint along the shoots; The shearers squint along the board to catch the Boss's boots.

shoot, *v.* Add: **I. 1. i.** (Earlier example.)

1816 W. Lambert *Instructions & Rules of Cricket* 29 When a ball is pitched short of its usual and proper length ..it may cut or shoot on the ground.

2. a. (Further examples with *off*.)

1828 W. Carr *Dial. Craven* (ed. 2) II. 121 *To shoot off*, to go off precipitately. **1930** J. Devanny *Bushman Burke* xii. 72 Whatja shoot off for? Had great time after you left. **1946** [see *Coast *sb.* 4 c].

d. *colloq.* To depart, go away. Freq. *int.*

1897 *Leeds Mercury* 19 June Suppl., Nah, then, shooit, or ah'll mak yo! **1970** G. F. Newman *Sir, You Bastard* viii. 230 I'll shoot then, if it's all right. Nothing else? **1974** H. L. Foster *Ribbin', Jivin', & Playin' Dozens* v. 203 Two of them said, 'Shoot.' They all turned and walked away.

e. *to shoot through*: to escape, abscond; to depart, leave. *Austral.* and *N.Z. slang.* Cf. *Go *v.* 89 d.

1947 *Pix* 20 Sept. 15 *Shoot through*, escape, abscond. **1951** S. Mackenzie *Dead Men Rising* 37 I'm shooting through—my woman's sick and I've waited longer than I should have. **1965** M. Shadbolt *Among Cinders* xviii. 168 'Well,' I said... 'I guess I'd better be shooting through. Thanks for the sausage.' **1978** *Telegraph* (Brisbane) 11 Jan. 28/1 I've been advised to shoot through and forget about the debts.

4. f. U.S. *to shoot the chute(s*: = *to chute the chute(s* s.v. *Chute *v.* 1. Also *shoot-the-chute* used as *sb.*

1895 *N.Y. Dramatic News* 30 Nov. 17/4 Shooting the Chutes, the latest craze that has struck the town, is.. drawing large crowds. **1920** R. Frost *Let.* 23 July (1964) 116 This man's island..will be full of divers and entertainment dives such as movies, con games, and shoot-the-chutes. **1946** E. O'Neill *Iceman Cometh* iii. 165 We're goin' to beat it down to Coney Island and shoot the chutes. **1977** *Time* 4 July 26/2 They are the not-so-spiritual descendants..of the Parisians who in 1817 rode the original shoot-the-chute.

8. a. Also of prices, sales, etc.: to rise sharply.

1968 *Listener* 27 June 826/3 The sales of vodka in Moscow shot up by 25 per cent. **1977** *Evening Gaz.* (Middlesbrough) 11 Jan. 1/6 The pound shot up two cents against the dollar.

II. 11. h. Also, to dispatch (a thing) rapidly.

1926 *S.P.E. Tract* xxiv. 126 If the article is ready, shoot it in. **1942** *Tee Emm* (Air Ministry) II. 129 You have grumbled at the amount of bumph the Group has shot at you. **1971** *Black Scholar* June 54/2, I don't have a picture at this time, but when I get one I will shoot it to you.

j. To discard, get rid of; orig. in *shoot that hat*, etc., as a mild imprecation. Also *to shoot trouble* = *Trouble-shoot *v. slang* (orig. *U.S.*).

1877 in Bartlett *Dict. Amer.* (ed. 4) 586 One lady.. with derisive scorn..observed in the language of the day, 'Oh, shoot that hat!'..The slang the gang is using now, You'll hear from every lip; It's *shoot the hat!* and get it boiled; And don't you lose your grip. **1884** J. Hay *Bread-Winners* xvi. 249 If I had all the cash he takes in to-night, I'd buy an island and shoot the machine business. **1902** Farmer & Henley *Slang* VI. 188/1 *Shoot that* (*hat, man*—anything)!..a mild imprecation, 'Bother!' **1928** *Sunday Express* 18 Mar. 9/2 'Tell him to shoot that song-and-dance outfit'.., and jump into some plain overalls. **1965** 'W. Haggard' *Hard Sell* xiv. 154 When there was trouble..then Murco Monti shot it. Political trouble especially. **1980** P. Harcourt *Tomorrow's Treason* i. v. 72, I want you to shoot trouble for me till the conference is over.

k. *Cricket.* To bowl (a side or part of one) *out* quickly and cheaply.

1900 P. F. Warner *Cricket in Many Climes* iv. ii. 159 Next day Ainsworth and 'Bos' shot the opposing side out for 30. **1976** J. Snow *Cricket Rebel* 112 Half the Warwickshire side had been shot out for 46.

l. Slang phr. *to shoot a card*: to leave a visiting card. ? *Obs.*

1901 *Captain* V. 7/2 The second-year man is always careful to 'shoot' his card when the fresher is abroad. **1924** 'Sapper' *Third Round* vi. 139 He may have heard that Mrs. Goodman is here, and has come to shoot a card.

12. a. Also *spec.* in *Angling*, to allow a quantity of (line) to run out through the hand at the forward motion of the rod in casting.

1931 *Hardy's Anglers' Guide* 13 In Figs. 1 and 2 the left hand gathers slack line, while in Fig. 3 this line is released during the forward stroke. This is termed 'shooting' line. **1977** *Chicago Tribune* 2 Oct. iii. 10/2 The current will pull on the line, making casting and 'shooting' of fly line on the next cast extremely difficult.

15. Also *colloq.*, to convey or transfer (a person) with speed.

1919 F. Hurst *Humoresque* 200 Come; I'll shoot you to the club. **1921** E. B. White *Let.* 15 Sept. (1976) 24, I go to one person and he says hello and shoots me on to another. **1972** T. Stoppard *Jumpers* i. 50 I'll shoot him in here... You can try your charms on him.

17. d. Now usu. *to shoot one's cuffs.* Also *fig.*

1909 M. Beerbohm *Yet Again* 230 The large young man, shooting his cuffs, strode forward. **1929** W. Deeping *Roper's Row* iii. 25 You felt that you had shot your cuffs and scored a point when you wiped Moorhouse's stately eye. **1942** L. A. G. Strong *Unpractised Heart* 27 The millionaire pulled out a gold pencil and shot his cuff. **1974** S. Coulter *Château* i. xvii. 133 He shot his cuffs and walked resolutely towards Mademoiselle Aurélie. **1977** J. Cheever *Falconer* 129 He shot his cuff to check the time.

18. Restrict † to first three senses in Dict. and add: **c.** *intr.* To ejaculate; orig. in phr. *to shoot one's roe. slang.*

1879–80 *Pearl* (1970) 217 And the little creatures found, When they dragged him to the ground, That, while lecturing, he'd shot his noble roe, roe, roe. **1922** Joyce *Ulysses* 553 Bloom: (*His eyes wildly dilated, clasps himself*). Show! Hide! Show! Plough her! More! Shoot! **1969** [see *Come *v.* 16*]. **1972** H. C. Rae *Shooting Gallery* iii. 191, I wanted him to shoot and get it over.

III. 22. c. Also used in sports involving the scoring of goals. Cf. sense 27 in Dict. and Suppl.

1981 E. North *Dames* xiii. 255 'Well shot! Well shot!' Some feeble cheering on the touchline.

d. (*e*) In phr. *to shoot to kill* (cf. *Kill *v.* 6 c), implying the desire to kill, rather than frighten or wound, a living target. Also *shoot-to-kill* adj. phr.

1867 *Harper's Mag.* Feb. 274/2 Wild Bill with his own hands has killed hundreds of men... 'He shoots to kill', as they say on the border. **1949** N. Marsh *Swing, Brother, Swing* iii. 48 Plays like he shoots an' he shoots to kill. **1956** 'J. Christopher' *Death of Grass* v. 79 'Must you shoot to kill?' He began to say: 'It's a matter of safety.' **1973** *Black Panther* 7 Apr. 10/3 LEAA was the liberal establishment's attempt to modernize police techniques as a substitute for 'shoot to kill' repression. **1977** *New Yorker* 15 Aug. 67/1 A Cuban businessman.. recalled favorably Mayor Richard Daley's calling for the National Guard and giving the police shoot-to-kill orders because of the Chicago riots.

(*f*) *to shoot it out*: to settle (a dispute) by shooting or by the exchange of military fire. Cf. Fight *v.* 8.

1912 W. M. Raine *Brand Blotters* xii. 327 Had he shown any sign of indecision, they would have taken a chance and shot it out. **1939** *War Illustr.* 18 Dec. 458/3 Two gangsters have temporarily joined hands to plunder the wealthier citizens, on the understanding that they will 'shoot it out' between them later. **1949** F. Maclean *Eastern Approaches* ii. iv. 221 To try and shoot it out with them would bring the whole place about our ears.

f. Now usu. in extended sense: *trans.* To photograph (a scene, action, person, etc.) with a cinematographic camera; to take (cinematographic film), to film; occas. with the actor as subject. Also *intr.*

1916 'B. M. Bower' *Phantom Herd* ii. 22 He..debated whether it should be 'shot' with two cameras or three. **1919** *Conquest* Dec. 70/2 First, the camera man 'shoots' on the tank containing the fishes with one half of the lens open. **1930** E. Waugh *Vile Bodies* ix. 156 'All right,' said one of the men with megaphones... 'We'll shoot the duel now.' **1953** *Manch. Guardian Weekly* 27 Aug. 7/1 While the big-city exhibitors were pondering this expensive outlay a small studio in Hollywood shot a poor film with two interlocking lenses. **1962** Montagu & Leyda tr. *Nizhny's Lessons with Eisenstein* iii. 66 Could the set-up be so changed as to shoot past Dessalines' back? **1976** *Observer* (Colour Suppl.) 9 May 10/2 She has also shot the odd film here. **1978** J. Krantz *Scruples* iii. 77 If anyone was going to go down to the Virgin Islands and shoot three models in next year's monokinis..it was Hank.

g. *to shoot a profile*: see *Profile *sb.* 4 e.

23. g. For '*U.S.* slang' read '*slang* (orig. *U.S.*)' and add: Also, to talk unrestrainedly or at length, to assert one's opinions; to boast or brag. Also *to shoot one's mouth off* and similar phrases.

1864 *Rocky Mountain News* (Denver, Colorado) 3 Aug. 4/2 A Dutch married woman..was taxed $17·80 for 'shooting off her mouth' against the virtue and morality of a neighbouring maiden. **1880** *News & Press* (Cimarron, New Mexico) 8 Apr. 1/5 Nobby, you've..never yit shot off yer mouth on the marryin' biz. **1919** O. W. Holmes *Let.* 5 Apr. (1964) iii. 184 They make me want to write a letter to ease my mind and shoot off my mouth; but of course I keep a judicial silence. **1933** D. L. Sayers *Murder must Advertise* x. 165 That don't prove nothing... Not without you know 'ow long it took Mr. Tompkin to shoot 'is mouth off. **1946** E. O. O'Neill *Iceman Cometh* ii. 99 He ought to do it, and not just shook off his old bazoo about it. **1954** Wodehouse *Jeeves & Feudal Spirit* xiii. 123 So this was how the woman was accustomed to shoot off her bally head about me in my absence. **1967** *Boston Globe* 5 Apr. 51/5 The only way we can keep Red..quiet is to beat Boston... I'm tired of hearing him shoot off his mouth. **1973** W. J. Burley *Death in Salubrious Place* i. 24 With Matthew Eva shooting his mouth off about Peters it could turn ugly.

h. *trans.* and *intr.* To inject by means of a hypodermic syringe (used esp. with reference to the taking of addictive drugs). Also *refl.* Freq. const. *up. slang* (orig. *U.S.*).

1914 Jackson & Hellyer *Vocab. Criminal Slang* 75 *Shoot*, verb, current amongst hypodermic habitues. To inject morphine or other drug with a syringe. Example; 'How many times do you shoot a day?' **1926** J. Black *You can't Win* xii. 161 They grew so despondent over their plight..they decided to 'shoot up' the small portion of white stuff they had left. **1951** *Life* 25 June 120/1 But furnishes hypodermics and other paraphernalia so he may 'shoot himself' on the spot. **1953** W. Burroughs *Junkie* i. 23 He shot another syrette. *Ibid.* iii. 36, I began shooting in the main line to save stuff and because the immediate kick was better. **1969** H. Waugh *Young Prey* iv. 65 The junkies..need a place to go where they can shoot themselves and where an expert can shoot the novices. **1971** *Oz* May 5/2 They were using those needles man, they were shooting up. **1979** R. Jaffe *Class Reunion* (1980) ii. iv. 204 He [*sc.* a dermatologist] would be the magician who would..peel off wrinkles, shoot silicone into laugh lines.

i. *intr.* To proceed, go ahead (with a speech, question, etc.), to 'fire away'. Usu. *imp.*, as an invitation to introduce a topic. Also *trans.* to direct (words); to say, speak; occas. *imp.* with *it*, and *to shoot back*, to riposte, retort. *colloq.* (orig. *U.S.*)

1915 *Dialect Notes* IV. 235 *Shoot, imper.*, continue; go ahead. **1917** Ade *Let.* 12 June (1973) 64 The other day a harelipped man working for me stopped me and said he wished to ask a question. I told him to shoot. **1920** S. Lewis *Main Street* 121 'Shall we try 'The Idylls of the King? They're so full of color.' 'Go to it. Shoot.' **1934** J. M. Cain *Postman always rings Twice* ii. 9, I shot it right close to her ear, almost in a whisper. 'How come you married this Greek, anyway?' **1935** W. D. Hubbard *Thousandth Frog* 157 'Well,' said Gratton. 'Shoot it.' **1942** *Amer. Mercury* July 90 Dat what you shooting ain't worth a damn! **1951** J. G. Fennessy *Sonnet in Bottle* vii. iii. 259 'I want to ask your advice about something.'.. Rupert said, 'Well, shoot.' **1974** 'E. Lathen' *Sweet & Low* xi. 112 'Where's Amory?'... 'How do I know?' Yeoman shot back. **1978** H. Wouk *War & Remembrance* xxxvii. 388 'Can I pick your brain on one more point?' 'Shoot.'

j. Chiefly U.S. slang. *to shoot the works*: to effect something to the fullest extent; *spec.* to discharge the necessary business; to tell the truth, reveal all; *to shoot the bull*: to talk nonsense (cf. *Bull *sb.*⁴ 3); *to shoot a line*: see *Line *sb.*² 13 g.

1922 E. O'Neill *Hairy Ape* vii. 74 Can't youse see I belong? Sure! I'm reg'lar. I'll stick, get me? I'll shoot de woiks for youse. **1930** *Amer. Speech* V. 197 (*heading*) Shooting the bull. **1946** Mezzrow & Wolfe *Really Blues* i. 5 We had a yen..to strut and act biggity and shoot the works. **1951** W. Stevens *Let.* 27 Dec. (1967) 735 We have people who seem to hand a list of names to a stenographer and tell her to shoot the works. **1972** *Maclean's Mag.* Mar. 41/2 'Writes all my speeches,' he'd say and slap me on the back. 'Smart boy! He can sure shoot the bull.'

k. *Bridge.* To play abnormally in a tournament in order to achieve a high score, e.g. one needed to win a tournament.

1957 M. Miles *How to win at Duplicate Bridge* vii. 318 Why do people invariably overbid when they are shooting? There are many better opportunities to shoot by underbidding. **1972** *Times* 3 June 8/6 It was the 98th deal of a 100 board match and North South were 'shooting' for top scores.

l. To strive *for*, to aim at. *U.S. colloq.*

1967 *Technology Week* 23 Jan. 15/3 The space agency is currently shooting for a nuclear engine that would operate for 1,000 hours before refueling. **1976** *Billings* (Montana) *Gaz.* 20 June 6-E/2 Mississippi College long jumper Larry Myricks is shooting for a jump of 27 feet.

26. a. Also *U.S.*, to throw a die or dice; to play at dice. *colloq.*

1909 Webster 526/1 The caster throws or 'shoots' the dice, and wins if the throw is 7 or 11. **1929** *Amer. Mercury* Sept. 49/2 We got to stop shootin' dices. **1932** W. Faulkner *Light in August* 34, I would have thought that maybe shooting dice would be the one thing he could do. **1940** —— *Hamlet* 258 Vagrancy or razor fights or shooting dice for ten or fifteen cents.

b. *U.S. colloq.* To play (a game), as in *to shoot pool, casino*, etc. Cf. *to shoot crap(s)* s.v. *Crap *sb.*⁴ 1, *Craps.

1926 E. Hemingway *In our Time* 183 They talk and tell stories and shoot pool. **1935** A. Squire *Sing Sing Doctor* xiv. 207 Perhaps he goes for a walk, or attends a movie, or shoots a practice game of pool. **1949** A. Miller *Death of Salesman* i. 48 Come in later, we'll shoot a little casino. **1979** R. Jaffe *Class Reunion* (1980) ii. ii. 187 He liked to dance, play golf, drink, shoot skeet, and laugh.

27. Now widely used in other sports, as *Basketball, Netball*, etc. **a.** (Later examples.) Also with *basket*, etc., as object.

1900 *Springfield Daily Republican* 4 Dec. 3/1 Both teams warmed up slowly and played open polo. Curtis shot the first goal in from the side. **1901** A. Farrell *Ice Hockey & Polo Guide* 54 This sudden movement surprises the man and he is liable to shoot the puck inaccurately. **1908** in H. A. Fisher *Basket Ball Guide 1908–9* 67 It was he who shot Williams' two baskets in the first overtime period. **1935** F. Hewitt *Down Ice* (rev. ed.) iv. 88 The best play is to shoot the puck at the boards and jump ahead to receive the carom. **1975** *New Yorker* 7 Apr. 112/2 He went over to shoot some baskets by himself, and while he was doing this it suddenly hit him that the game meant too much to him to give up.

b. (Later examples.)

1917 *N.Y. Times* 4 Feb. viii. 1/5 The Aggies won the game in the final period when D. Ross shot at random from the centre of the rink. **1929** J. G. Buss *Basketball* vi. 71 *Shoot high*—and do not aim at the rim of the basket. **1951** *Netball* ('Know the Game' Series) 28/2 The player.. must..aim directly at the goal... If she shoots and then catches the ball again, she must not shoot again directly. **1978** *Washington Post* 26 Oct. A 19/5 The 'problem' is that the girls' game has only forwards and guards, and the three guards do not shoot.

c. *Golf* (orig. *U.S.*). *trans.* To record (a score) for a round or part of one; also *to shoot a birdie*, to achieve a score of one under par for a hole. Also *intr.*

1922 *Golfers Mag.* Aug. 28/1 This class of golfers, the fellows who shoot from 90 up, make golf possible in this country. **1923** [see *Birdie 2]. **1933** *Amer. Golfer* July 34/1 How about birdies and eagles?..In order to shoot a birdie, you'd have to play the hole in 3.3 strokes. **1941** *Sat. Even. Post* 19 Apr. 118/3 They shot a twelve-under-par score in winning their first match. **1977** *Rolling Stone* 5 May 47/5 He played a round every day, shooting in the low 100s.

IV. 30. a. (Later *fig.* examples of *to shoot to pieces.*) Also in similar phrases. Cf. Blow *v.*¹ 24 in Dict. and Suppl.

1939 R. G. Collingwood *Autobiogr.* vi. 49 A philosophical doctrine was stuck up and shot to pieces by the 'realistic' criticism. **1946** W. H. Auden in *Harvard Alumni Bull.* 15 June 707 And nerves that never flinched at slaughter Are shot to pieces by the shorter Poems of Donne. **1955** E. Pound *Section: Rock-Drill* xcii. 81 Semele's personality shot to atoms. **1973** *Black Panther* 31 Mar. 2/1 These same pilot officers..are now being encouraged to avenge their injured pride which was shot to hell because many of the anti-war enlisted men were Black. **1979** *Homes & Gardens* June 81/1 The Season has been shot to pieces this century.

c. Now freq. to bring down (an aircraft, hence a pilot, etc.) by shooting; freq. in phr. *to shoot down in flames.* Also *fig.*, to overwhelm (a person) in argument, to destroy (an argument or theory); to assail with objections; to bring down to size.

1918 W. A. Bishop *Winged Warfare* xii. 160, I turned on the nearer of the two-seaters and..managed to shoot him down. **1928** E. Wallace *Tam* iii. 29 Captain Muller shot down his twenty-seventh aeroplane. **1943** N. Balchin *Small Back Room* 5 'Just army conservatism,' he said wearily. 'Just the army's usual trick of shooting things down.' **1943** C. H. Ward-Jackson *It's a Piece of Cake* 54 *Shot down in flames*, hopelessly beaten at anything. **1948** *Daily Tel.* 26 May 4/5 Another enemy aircraft was shot down in flames. **1958** *Times* 18 Dec. 11/4 To be shot down in flames may be an exaggerated description of getting the worse of an argument. **1959** *Listener* 3 Sept. 351/1 This is the way in which we shoot down cosmological theories. **1962** J. F. Powers *Morte d'Urban* iv. 98 He had no choice but to shoot the woman down. **1969** M. O'Brine *Mills* xi. 43 She, herself, had been a little shocked by his answer, but had secretly enjoyed seeing Eileen shot down in flames. **1977** *R.A.F. News* 11–24 May 8/6 He saw..Baron von Richtofen, shot down. **1981** J. B. Hilton *Playground of Death* vi. 81 Please shoot me down in flames if you think I'm making a bloody idiot of myself.

e. *colloq.* (orig. *U.S.*). *to shoot up*: to assail (a person, thing) by shooting; to terrorize or rampage around (a place). Also *R.A.F. slang* of an aircraft or its pilot: to dive over (a person, thing) as if or in order to attack.

1890 *Stock Grower & Farmer* 18 Jan. 5/2 This so enraged the boys that they began shooting up the town. *Ibid.* 21 June 3/1 Three cowboys shot each other up. **1909** *Chambers's Jrnl.* Feb. 104/1 Armed and masked men..took entire possession of it [*sc.* Princeton], 'shot it up' until opposition subsided. **1926** *Daily Colonist* (Victoria, B.C.) 22 July 15/7 Paul Davis..wounded after 'shooting up' rooming houses here early yesterday morning, died in hospital today. **1937** *New Statesman & Nation* 20 Feb. 288/2 Mr. Partridge is not quite so strong with regard to the slang of flying... I think..to *shoot-up*, to dive onto, and the *wind-sock* should all find a place. **1946** D. Hamson *We fell among Greeks* xvii. 180 When he had finished unloading his parachutes and parcels, he would take a long circle round and 'shoot us up', i.e. dive and roar over us at less than fifty feet, and, as he zoomed up, dip his wings left and right in salute. **1973** L. M. Boston *Memory in House* iv. 40 A squadron would roar over the house from which one plane swooped down to shoot us up. **1976** *Daily Tel.* 5 July 1, The Air France airbus which was skyjacked..a week ago, was also 'shot up'.

f. With *out*: to render (something) useless with a shot; *spec.* to puncture (a tyre), extinguish (a lamp) by shooting.

1972 *Daily Tel.* 9 Feb. 4/5 The vehicle was halted when police shot its tyres out. **1976** D. Barnes *Yesterday is Dead* (1977) ii. 267 Couple of cut-'em-up family disputes and a guy that shot out fourteen street lights. **1977** J. Cartwright *Fighting Men* viii. 102 He was sorely tempted to shoot out the plane's tyres.

31. (Earlier examples.)

1761 Sterne *Tr. Shandy* IV. xiv. 85 Can'st thou carry *Trismegistus* in thy head... If she can, I'll be shot, said my father. **1803** J. Kenney *Raising Wind* ii. i. 24 What a fine seal; and I'll be shot if it [*sc.* a letter] don't feel like a bank note.

32. d. *shot in the neck* (earlier example); *to shoot the breeze* (U.S.): to chat, talk idly; *to* *shoot the crow* (Sc.): to steal away without paying one's bill; to depart hurriedly, abscond, 'do a bunk' (see *S.N.D.*).

1830 *Cherokee Phoenix* (New Echota, Georgia) 21 Apr. 4/3 *Counsel.* What do you mean by *corned*? *Witness.* I mean, pretty well *shot in the neck*. 1887 *Fun* 8 June 246/2 A canny Scot was recently sentenced to ten days' hard for shooting the crow—*i.e.* ordering half-a-quarter of whiskey, drinking it rapidly, and neglecting to pay. 1941 *Guide to U.S. Naval Academy* 149 *Breeze, shoot the*, to refight the Civil War, etc. 1943 *Sun* (Baltimore) 4 June 30/2 He can..walk across the camp to meet some friend in another outfit, and 'shoot the breeze'. 1971 R. K. Smith *Ransom* (1972) III. 113 There were other negative signs, too. No one had come by to shoot the breeze, to have a cup of coffee. 1973 'J. Patrick' *Glasgow Gang Observed* xi. 97 He had been serving a sentence of twenty-eight days detention in the last week of which he had 'shot the crow' and 'jolted', i.e. absconded. 1977 W. McIlvanney *Laidlaw* xliv. 206 There'll only be his mother in the house. His father shot the crow years ago.

34. Also *Oil Industry*, to detonate an explosive charge in (a well) in order to increase the flow of oil or gas.

c 1870 [in Dict.]. 1903 *Dialect Notes* II. 344 *Shoot (the well)*, to cause an explosion of several quarts of nitroglycerine at the depth of the *pay-streak*..so as to break and crack the oil rock, enabling the oil to flow faster from the pores. 1921 *Daily Colonist* (Victoria, B.C.) 11 Mar. 12/3 The report states that in Ironville No. 1 a good showing of thick oil was obtained at various depths... It was decided to shoot this well, but owing to water it was not yet known what result this would have. 1949 *Our Industry* (Anglo-Iranian Oil Co.) (ed. 2) ii. 52 Some rocks..containing oil are..compact and 'tight'... In such cases the well is often 'shot' in order to shatter the rock.

37. b. For *dial.* read 'Now *colloq.*' and add later examples.

1952 'R. Gordon' *Doctor in House* i 9 His love for his old hospital, like one's affection for the youthful homestead, increased steadily with the length of time he had been shot of it. 1976 *Daily Tel.* 22 Sept. 16/1 Advising its members to make haste to get shot of unsuitable employees.

shoot, *int.* *U.S. slang.* An arbitrary alteration of *SHIT *int.*

In some instances this may perh. be regarded as an imp. use of *SHOOT *v.* 11 j.

1934 Webster 2319/2 *Shoot..*, *interj.* Pshaw! Bother!—often with *it*. 1941 E. Welty *Curtain of Green* 42 Oh, shoot, that was about three and a half years ago. 1950 R. Moore *Candlemas Bay* 301 'Oh shoot,' she told Jen, when Jen suggested they'd better write the next batch of boarders not to come. 1979 *Tucson Mag.* Feb. 42/1 Back in high school, I tried other so-called sports, but I always went back to rodeo. Shoot, that's the only sport there is.

shoot-'em-up (ʃū·təmʌp). *slang* (orig. *U.S.*). Also **shoot-em-up, shootemup.** [f. vbl. phr. *to shoot them up*: see *SHOOT *v.* 30 e.] A fast-moving story or film, esp. a Western, of which gun-play is a dominant feature. Also *attrib.* (or as *adj.*) and *Comb.*

1953 *Variety* 11 Feb. 6/2 A standard outdoor action plot is unfolded in 'Gunsmoke' to make it a Western feature for the shoot-'em-up market. 1958 *Washington Post* 30 July A25/7 That doesn't leave much room for anything except shoot-'em-ups. 1968 *Listener* 18 July 86/1 A racetrack-gang, shoot-em-up-bang plot and one of Mr Hawkes's disturbingly discontinuous surfaces of experience are yoked by violence together. 1973 *N.Y. Times* 10 June VII. 28/2 The new or free-form Western has several choice entries..'Oklahoma Crude', a splendid shootemup about a lady wildcatter in the oil fields. 1976 *Publishers Weekly* 15 Mar. 50/2 Her decision to put expediency ahead of love nearly costs Corey his life when the shoot-'em-up finale swings the action back to North Africa.

shooter. Add: **I. 2. b.** *Sport.* One who kicks or drives a ball at goal; also in extended use in *Basketball*, *Netball*, etc.

1901 A. Farrell *Ice Hockey & Ice Polo Guide* 54 The goaler..may skate out to meet him, being careful that he is directly in line between shooter and goal. 1922 W. E. Meanwell *Basket Ball for Men* vii. 68 Line 'throwing is the chief fault with most shooters. 1957 *Encycl. Brit.* III. 181/1 Previously, a star shooter could attempt all free throws for his team. 1963 C. Glyn *Don't knock Corners Off* xxi. 178, I stood and shivered as Miss Pratt picked people for the netball teams... 'You can be Shooter.' 1978 T. L. Smith *Money War* (1979) III. 218 The only gambit he could think of was darts. He was the fifth best shooter in the very active St. Louis league.

c. One who throws a die or marble. Cf. also *crap-shooter* s.v. *CRAP *sb.*[4] 2.

1910 A. Bennett *Clayhanger* I. i. 9 The bearded shooter, pleased by this tribute..twisted his white apron. 1926 [see *little Joe* s.v. *LITTLE *a.* 13]. 1969 R. C. Bell *Board & Table Games* II. v. 84 When any pair is thrown and the third die is 2, 3, 4, or 5, the number of the third die becomes the shooter's point.

II. 5. (Later example.)

1981 *Country Life* 18 June 1772/1 Sappy shooters, from which to take cuttings now.

8. b. (Further examples.)

1931 E. O'Neill *Hunted* IV, in *Mourning becomes Electra* 156 Easy goes, shipmate! Stow that pistol!.. Not that I'm skeered o' you or your shooter! 1970 G. F. Newman *Sir, you Bastard* v. 138 Why did you pull the shooter on the two detectives? 1972 L. Henderson *Cage until Tame* ix. 77 We'll need a shooter, one barrel into the ceiling straight off.

9. (Earlier example.)

1843 'Wykehamist' *Pract. Hints Cricket* 7 Another advantage of this mode of holding the Bat close to the ground, is the greater facility the player has in stopping 'shooters'.

10. *Public School slang.* A black morning coat. ? *Obs.*

1870 *Harrovian* 9 Apr. 134/2 Although the use of slang words and phrases has now become almost universal among the greater portion of the community, we generally find that each University or School possesses a dialect peculiar to itself... *Shooter*, a shooting coat. 1920 Galsworthy *In Chancery* II. vii. 181 'I suppose I'd better change into a "shooter",' he muttered, escaping to his room. He put on the 'shooter', a high collar, a pearl pin, and his neatest grey spats.

11. *U.S. slang.* A measure or drink of spirit, esp. whisky.

1971 *Car & Driver* Jan. 75/1 He made his famous call for 'shooters'. Now in case you haven't heard, a 'shooter' is a Turner variation of the word 'shot', as in a 'shot of likker', and..refers to a shot of Canadian Club mixed into a few fingers of 'Co-cola'. 1973 W. McCarthy *Detail* i. 61 Let's have a shooter and a beer. 1981 W. Safire in *N.Y. Times Mag.* 2 May 16/4 The word coming up fast for a *straight shot* is a *shooter*. 'A shooter is a shot of liquor swallowed in one quick gulp,' says Jeff Dee.

shooting, *vbl. sb.* Add: **1. a.** Also, an incident in which a person is shot with a firearm.

1873 'Mark Twain' *Gilded Age* xlvi. 425 What some of the journals lacked in suitable length..they made up in encyclopaedic information about other similar murders and shootings. 1977 *Whitaker's Almanack 1978* 590/2 During the election campaign 50 people were reported killed in shootings and bombings.

e. *Oil Industry.* Detonation of an explosive charge in a well to increase the flow of oil or gas. Cf. *SHOOT *v.* 34.

1914 F. A. Talbot *Oil Conquest of World* v. 64 'Shooting' is undertaken only when the limestone or sandstone is of such a nature that it restricts the flow of oil. 1937 *Amer. Speech* XII. 154/1 *Shooting a well*, using nitroglycerine to make oil flow. 1946 [see *OIL WELL]. 1969 *Times* 2 May 25/1 The international oil companies are stepping up their interest in the Irish Sea in search for oil and gas... The area involved covers at least 15,000 miles and although the 'shooting' will be selective, the cost will..be..high.

5. Also, in extended use in other sports, as *Basketball*, *Netball*, *Hockey*, etc.

1897 *Encycl. Sport* I. 518/2 The goalkeeper should run forward..so as to attempt to tackle him [*sc.* the hockey player] before he can get within shooting range. 1935 *Encycl. Sports* 436/1 It [*sc.* the game of netball] proceeds when..the ball..is received by one standing within the shooting circle. 1961 *Netball* ('Know the Game' Series) (ed. 5) 20 (*heading*) Footwork for throwing and shooting. 1974 *Plain Dealer* (Cleveland, Ohio) 26 Oct. 4-D/1 This system paid off in the team's shooting this week.

6. b. The action or process of taking film with a cinematographic camera.

1920 I. P. Gore in *Stage Year Book* 56 Many companies are paying trips to the Continent for the 'shooting' of certain scenes in the actual 'locations'. 1941 B. Schulberg *What makes Sammy Run?* xi. 284 A director exhausted from the day's shooting. 1955 *Times* 31 May 10/3 Mr. Orson Welles, for one, has shown..the methods of 'shooting' which lay emphasis on rehearsals. 1979 *Beautiful Brit. Columbia* Spring 4 Victoria was one of the shooting locations for Harry in Your Pocket.

c. The action or process of injecting an (addictive) drug intravenously. *slang* (orig. *U.S.*).

1951 *Evening Sun* (Baltimore) 27 Mar. 4/1 A powerful combination of 'bernice snorting' and heroin 'shooting' was called 'blowing speed balls'. 1953 W. Burroughs *Junkie* 8 You don't wake up one morning and decide to be a drug addict. It takes at least three months' shooting twice a day to get any habit at all. 1971 *Black Scholar* Apr.–May 46 Mugging, theft, pimping and shooting dope are not themselves political actions.

8. a. *shooting-season* (earlier and later examples); (sense *6 b) *shooting schedule, script*. Also *shooting-boot* (later examples; also (*fig.*) in sense 5), *dress, jacket* (earlier example), *shoe, suit* (example).

1894 *Country Gentleman's Catal.* 154 Fagg Brothers,.. makers of shooting boots to *H.R.H. The Duke of Saxe-Coburg*. 1947 *Sporting Mirror* 7 Nov. 11/3 Grimsby were having a sad and sorry season until the unexpected revival at Manchester United when Cairns found his shooting boots. 1948 C. Day Lewis *Otterbury Incident* 26 Everyone knows he's a deadly shot when his shooting-boots are on. 1978 *Cornish Guardian* 27 Apr. 5/5 Tintagel found their shooting boots in the second half of their game. 1794 J. Woodforde *Diary* 27 Oct. (1929) IV. 149, I met Mr. Stoughton..in a Shooting Dress. 1852 J. R. Planché *Day of Reckoning* III. i. 30 *Claude*..in a shooting dress, is seated on the steps of the terrace, examining the lock of his gun. 1796 Jane Austen *Let.* 5 Sept. (1952) 11 Let me know..how many of the Gentlemen, Musicians & Waiters, he will have persuaded to come in their Shooting Jackets. 1950 'E. Crispin' *Frequent Hearses* i. 36 'It would be possible for me to meet her?'..'That depends on the shooting schedules. The film's on the floor.' 1976 M. Maguire *Scratchproof* i. 11 Shooting schedules were being delayed and people were beginning to say the film was jinxed. 1929 I. Montagu tr. *Pudovkin's On Film Technique* vi. 176 The *Shooting-script* is the scenario in its final cinematographic form. 1933 A. Brunel *Filmcraft* 141 Here follow two sequences of the actual shooting script of 'A Light Woman'. 1976 H. Orel in M. Drabble *Genius of T. Hardy* 103 Perhaps John Wain exaggerates by describing the entire work [*sc.* *The Dynasts*] as a shooting script. 1781 G. Selwyn *Let.* 19 May in *15th Rep. R. Comm. Hist. Manuscripts* App. VI. 484 in *Parl. Papers 1897* (C. 8551) LI. 1 Boothby proposes to go to you in the shooting season, that is near Christmas. 1981 C. Miller *Childhood in Scotland* 54 The opening of the shooting seasons varied with the type of game. 1839 A. Mathews *Mem. Charles Mathews* III. vii. 162, I had them made after a plan of my own, for shooting-shoes. 1976 *Shooting Times & Country Mag.* 18–24 Nov. (Advt.), The golden boot—our famous shooting shoe. 1893 Kipling *Day's Work* (1898) 43 The Rao Sahib, in tweed shooting-suit and a seven-hued turban.

b. **shooting booth,** a booth at a fair in which shooting for prizes is carried out (cf. *shooting gallery* (*a*)); **shooting brake,** an estate car, now *rare*; orig. a light, horse-drawn wagonette designed to accommodate passengers and goods (cf. Break *sb.*[2]); **shooting-gallery,** (*a*) also *fig.* in colloq. phr. *the whole shooting gallery* = *the whole shoot* s.v. Shoot *sb.*[1] 8; (*b*) *U.S. slang*, a place where addictive drugs may be obtained and 'shot' or taken by injection; **shooting-ground,** (*a*) (earlier and later examples); **shooting-iron** (earlier example); **shooting match** (later examples); also *fig.* in colloq. phr. *the whole shooting match* = *the whole shoot* s.v. Shoot *sb.*[1] 8; **shooting phaeton** = *shooting brake* (orig. sense) above; **shooting seat** = *shooting stick* (*c*) below, now *rare*; **shooting-stick,** (*b*) *slang* = *shooting iron* (obs.); (*c*) a walking-stick with a handle that may be opened to form an impromptu seat, first used by shooters; **shooting war,** hostilities involving armed conflict, as opposed to *cold war* s.v. *COLD *a.* 19; first used with reference to U.S. involvement in the war of 1939–45.

1900 *Times* 7 July 10/1 We may soon expect swings erected in the practice-ground, shooting booths under.. the big stand. 1970 R. Lowell *Notebk.* 202 The shags Flying in straight lines like duck in a shooting booth. 1912 H. J. Butler *Motor Bodies & Chassis* iv. 48 Wagonettes, Shooting Brakes, and Luggage Cars.—This type of body fulfils the requirements of the sporting dogcart, and generally has sufficient capacity to replace two of these horsed vehicles. 1934 A. G. Street *Endless Furrow* xvii. 301 After a few minutes occupied with introductions and drinking a glass of sherry James found himself in the shooting brake, and soon the four-in-hand swept through the gates into the town. 1948 H. McCausland *Eng. Carriage* iv. 77 A very neat, very sporting little brake.. intended for private use in the country with a team or pair, was the Shooting Brake, which had, behind its high box, a strong suggestion of the dog-cart in its bodywork. 1958 *Times* 13 Aug. 4/5 One man was killed and 11 people were injured when a shooting brake and a motor coach were in collision at Holcombe Brook, Bury, to-night. 1951 *Life* 11 June 120/1 Sometimes he runs a 'shooting gallery', an establishment which not only sells the addict dope but furnishes hypodermics. a 1966 'M. na Gopaleen' *Best of Myles* (1968) 323 Put the whole shooting gallery into a saucepan of cold water. 1972 J. Wambaugh *Blue Knight* (1973) ii. 36 He knows this boss dyke, a real mean bull dagger. Her pad's a shooting gallery for some of us. 1973 R. Busby *Pattern of Violence* v. 79 'Did you call in?'.. 'Yeah... The whole shooting gallery 'll be here in a few minutes.' 1835 J. J. Audubon *Ornith. Biogr.* III. 37 There is no lack of shooting grounds, for every creek of salt-water swarms with Marsh Hens. 1897 *Outing* Mar. 536/2 A shooting friend..and myself were staying at a farmhouse, near the shooting-grounds. 1775 S. Adams *Let.* 31 Jan. in *Writings* (1907) III. 172 It puts me in mind of what I remember to have heard you observe, that we may all be soon under the necessity of keeping *Shooting Irons*. 1813 *Niles' Weekly Register* IV. 35/1, I..gained their applause for my activity at our shooting matches. 1896 [see *CON b]. 1922 D. H. Lawrence in *N.Y. Times* 24 Dec. 9/4 What a lively shooting match will go on between all the Jacks and the Juans! 1953 K. Reisz *Technique Film Editing* II. 76 The final chase..was best presented as a 'battle of wits', instead of a wild action-packed shooting match. 1974 *BP Shield Internat.* Oct. 2/4 This had the effect of tilting up the whole shooting match. 1890 *Coach Builders' Jrnl.* 15 Nov. 181/2 Another of this firm's exhibits was a shooting phaeton... It was furnished with luncheon basket. 1898 *Carriage Builders' Jrnl.* Dec. p. ix/2 (Advt.), Four-wheel shooting phaeton; varnished walnut; pigskin cushions, brass mounts and lamps, mat, and gun-box complete. 1895 *Army & Navy Co-op Soc. Price List* 15 Sept. 954 Cane shooting seat. *Ibid.*, Wood, folding Shooting Seat, can be used as a Walking Stick. 1917 *Harrods Gen. Catal.* 1089/2 Mills' Patent Shooting Seats. Strongly recommended as being the lightest and best seat, it is also telescopic. 1845 E. J. Wakefield *Adventure N.Z.* I. xi. 319 Every article of trade with the natives has its slang term,—in order that they may converse with each other respecting a purchase without initiating the natives into their calculations, thus pigs and potatoes were respectively represented by '*grunters*' and '*spuds*', guns..by '*shooting-sticks*'. 1866 'F. Kirkland' *Pictorial Bk. Anecdotes* 237/2 Sambo.. fell back in confusion when the 'shooting stick' was brandished toward his own breast. 1926 E. P. Oppenheim *Golden Beast* I. xvii. 163 Judith had already disappeared, swinging her shooting stick in her hand. 1967 *Guardian* 23 May 2/6 The shooting sticks will prod the roots of every stately garden. 1941 *Time* 4 Aug. 15/3, 55%..are ready to risk some kind of shooting war at once. 1956 F. Castle *Violent Hours* vi. 51, I got into the real shooting war towards the close, at Okinawa. 1978 L. Heren *Growing up on The Times* iii. 86 Pat had joined me before the end of the shooting war, and was almost killed in Jerusalem.

shootist. (Earlier and later U.S. examples.)

1864 *Gold Hill* (Nevada) *News* 15 Jan. 3/1 (*heading*) A Shootist. 1976 *National Observer* (U.S.) 4 Sept. 18/2 J. B. Books, the protagonist of Wayne's new movie, *The Shootist*,..not only restores the legend but expands it, giving the man and his memory grace and dignity. A shootist is a man good with a gun, and J. B. Books is a retired marshal who was good enough to kill 30 men.

shoo·t-out. orig. *U.S.* Also **shootout.** [f. vbl. phr. *to shoot it out*: see *SHOOT *v.* 22 d (*f*).]

1. A sustained exchange of shooting, a gunfight. Also *fig.*, a dispute or competition.

1953 *N.Y. Times* 5 July VII. 13/2 The justly famous shoot-out between the Earps and the Clantons in the O-K Corral. 1968 'R. Macdonald' *Instant Enemy* xxx. 188 The last thing needed was the kind of shoot-out in which innocent people could get hurt. 1969 *Daily Tel.* (Colour Suppl.) 17 Jan. 15/3 In one 'shoot-out' that developed, Oakland police killed a 17-year-old Panther, Bobby Hutton, while he was unarmed. 1975 *Atlanta Jrnl.* 20 Jan. 1/2 In the ensuing shootout Sunday, the man who fired the shot was killed. 1976 *Washington Post* 19 Apr. A4/1 Church's strategy is to have the decisive shoot-out on the Senate floor in a major public debate. 1978 *Fortune* 31 Dec. 59 In the quick-draw tradition of the Old Wild West stagecoach, the two major lines have responded to adversity..by taking on one another in a fierce and profitless shootout over passenger fares. 1981 *Economist* 8 Aug. 34/1 The normal run of muggings, burglaries and rapes has been exacerbated by shootouts.. among 'cocaine cowboys'.

2. *transf.* In *Football*, a tie-breaker (see quot. 1978).

1978 *Guardian Weekly* 16 June 19/4 If the match is still tied..the teams resort to a shoot-out. Five different members of each team take a free shot at goal, starting from the 25 yard line and having five seconds to dribble the ball before shooting. If even these ten shots fail to produce a decision, they play a sudden death shoot-out until one side wins. 1979 *Globe & Mail* (Toronto) 1 May 48/9 Johann Scharmann converted the deciding shot in a shootout to win it for Detroit Express.

shoo·t-up. [f. vbl. phr. *to shoot up*: see *SHOOT *v.* 30 e.] **1.** A furious exchange of shooting, a gun-battle, a shoot-out; also, an assault by gun-fire.

1922 *Blackw. Mag.* Oct. 441/2 A favourite form of amusement of the I.R.A. used to consist in what was commonly called 'shooting up' a district: these outrages took the form of shooting at every Loyalist who appeared within range for a whole evening... Before one of these shoot-ups you might search every house in the district to be shot-up till dusk, and not find any arms..; but soon after dusk..men would..distribute arms to the gunmen. 1942 Berrey & Van den Bark *Amer. Thes. Slang* §348/4 *Gun battle*,..shoot-up. 1962 *Listener* 20 Sept. 438/3 A shoot-up in Alcatraz. 1972 P. Dickinson *Lizard in Cup* ix. 122 The cops in New Jersey got him in a shoot-up with the Black Panthers. 1978 *New York* 3 Apr. 10/2 Were the Egyptian commandos killed by Cypriot soldiers—as the Cyprus government claims—or by the PLO? Was the shoot-up at Larnaca airport a 'misunderstanding' or a well-conceived plan?

2. The act of flying low over a target as if to or actually to attack. *R.A.F. slang.*

1942 I. Gleed *Arise to Conquer* vi. 62 We do a gentle shoot-up of our billets.

shop, *sb.* Add: **2. a.** *Shop!* (earlier example).

1888 Kipling *Plain Tales from Hills* 242 A little wife to call 'shorp!' when the door-bell rung.

e. [Back-formation f. Shop *v.* 4.] An act of shopping for purchases. *colloq.*

1960 *Housewife* May 121/2 You should find it possible to have one big 'shop' a week with a small mid-week 'shop' for perishables. 1978 D. Murphy *Place Apart* ix. 198 It was a Saturday morning, when many go into the city centre to do their weekly 'big shop'.

f. *shop-within-a-shop*, a shop which functions independently within the premises of a larger store, usu. dealing in the goods of one manufacturer.

1962 E. Godfrey *Retail Selling & Organ.* i. 5 Another practice..is that of opening a shop-within-a-shop, selling the manufacturer's goods and staffed by his employees. 1978 *Country Life* 5 Oct. 1054/1 Mulberry Company.. makers of..high fashion accessories are opening shops-within-shops at nine Nieman Marcus stores.

3. b. (Earlier *absol.* example.)

a 1779 'J. H. St. John de Crèvecœur' *Sk. 18th-Cent. Amer.* (1925) 143 The truly economical farmer has always what we call a shop, that is, a house big enough to contain a loom. There..our wives can..weave.

d. *Glass-making.* A team or gang of workers (see quots.).

1889 *Harper's New Monthly Mag.* July 259/1 Generally four [glass factory workers] constitute a shop, the most skilful workman (the blower) at the head, the gatherer (a young fellow) next, and two boys, one handling moulds or tools, and the other carrying the products to the annealing oven. 1905 *28th Ann. Rep. New Jersey Bureau Statistics of Labor* III. 201 A case in point..is the change from single blower method of doing work, which prevailed previous to 1870, to what is now known as the 'shop system'; that is to say, three men now work together, two of them gathering glass and blowing the ware, while the third makes the neck smooth. 1949 P. Davis *Devel. Amer. Glass Industry* x. 230 The operation was performed by a three-worker shop composed of a gatherer, a blower, and a crimper. 1970 *Awake* (Austral.) 8 Jan. 23/1 The glassblowers function as a 'shop' of six or seven men.

e. *N. Amer.* A schoolroom equipped for teaching the arts of the workshop; this study as a classroom discipline. Cf. *shop class*, sense 9 c below.

1914 J. S. Taylor *Handbook of Vocational Education* iii. 54 The school shop now resembles the abode of the cabinet maker. *Ibid.* v. 65 The student learns much of what industrial life is like..in the successful operations of..the school shop. 1941 *School Shop* Oct. 2/2 *School Shop* has been established to serve shop teachers. 1948 G. O. Wilbur *Industrial Arts in General Educ.* ix. 127 There is some evidence which seems to indicate a close correlation between the atmosphere of the school shop and the type of learning which takes place there. 1974 J. Heller *Something Happened* 224 The new teachers, the old teachers,..the shop teacher, and the science teacher (he has always been leery of shop teachers and science teachers. Perhaps because they are men.) 1978 *Detroit Free Press* 2 Apr. 3D/1 A school..cannot prevent a girl from taking shop or a boy from taking home economics.

4. (Earlier examples.) *The Shop*: (later example); also (*Austral. slang*), the University of Melbourne.

1779 E. Gibbon *Let.* 15 May (1956) II. 215 So much remains to be done, that I can hardly spare a single day from the Shop. 1827 T. Surr *Richmond* II. i. 5, I hurried off with Bucks to the office, or *shop*, as he called it. 1889 *Centennial Mag.* II. III. 218 It related how 'a medical student came up to the Shop' as a freshman, and 'thought through exams. he would speedily pop'. 1918 G. Wall *Lett. of Airman* 15, I would be quite glad to get the Shop exam results. 1964 G. Johnston *My Brother Jack* 260 The years at the Shop gave me nothing except a worthless B.A. and the privilege of being thrown into the University lake. 1978 G. M. Fraser *Flashman at Charge* 110 We treated each other decently, and weren't one jot more incompetent than this Sandhurst-and-Shop crowd.

b. (Earlier and later examples.) Also in gen. use (*rare*).

1885 J. K. Jerome *On Stage & Off* 126 After that it was next to impossible for him to get a shop (this expression is not slang, it is a bit of local colour). 1922 E. Wallace *Flying Fifty-Five* xxx. 178 Fired, are you?..Well, what are you going to do? Get another shop? 1978 G. Mitchell *Wraiths & Changelings* xii. 128 He was an out-of-work actor and was very anxious to get a shop, as he called it.

5. *to talk shop* (earlier example).

1860 C. Fox *Jrnl.* 28 Sept. (1972) 232 Holman Hunt.. does not talk 'shop', but is perfectly willing to tell you anything you really wish to know of his painting.

8. b. *to set up shop* (*fig.* example); *to live over the shop*, to live on the premises where one works; *to mind the shop*: see *MIND *v.* 11.

1930 D. L. Sayers *Strong Poison* i. 23 'He's put her into a house somewhere round about, I fancy,' said Freddy, 'with a typewriting office to look after and live over the shop and run those comic charity stunts of his.' 1963 A. Huxley *Let.* 27 Mar. (1969) 952, I..heard of his plans for an LSD institute... He may be more successful in setting up shop within the US. 1976 H. Wilson *Governance of Britain* iv. 83 In 1964–70 I lived in No. 10. In 1974 I decided that I did not want to live over the shop again, and I slept each night in my home in Lord North Street.

c. Also, in a state of confusion. (Earlier and later examples.)

1874 Hotten *Slang Dict.* 288 In pugilistic slang, to punish a man severely is 'to knock him all over the shop'. 1916 'Taffrail' *Pincher Martin* xiv. 267 'Wagglin' about a bit,' the coxswain answered, gazing at his compass-card. .. 'She's all over the shop. Up to sou'-east one minute, an' back to south-eighty the next.' 1926 G. B. Shaw *Intell. Woman's Guide* lxxi. 345 The unconventional ones are all over the shop with all sorts of opinions. 1935 F. M. Ford *Let.* 15 Oct. (1965) 245 He is in the greatest danger of going slack all over the shop. 1978 J. I. M. Stewart *Full Term* ix. 93 At one of Anthea Gender's [parties] one was substantially although not too obtrusively in the presence of grandees drawn from all over the shop.

9. a. (*a*) *shop-bell, -counter, -door* (later examples), *-front* (earlier example: also *attrib.* and *fig.*), *sign*; (*c*) *shop-goods* (later examples); (*d*) *shop hours*; (*e*) *shop-boy* (earlier and later examples), *-clerk, -girl* (earlier and later examples), *-worker*.

1853 Mrs. Gaskell *Cranford* xv. 299 She..was only extricated from her dilemma by the sound of the shop-bell. 1972 J. Thomson *Not One of Us* viii. 90 The tinkle of the shop bell severed the conversation and she went through to serve. 1813 Jane Austen *Pride & Prejudice* I. xv. 166 Mr. Jones's shop boy..had told her that they were not to send any more draughts to Netherfield. 1903 G. B. Shaw *Man & Superman* III. 132, I breathe an atmosphere of sweetness, like a confectioner's shopboy. 1977 *Daily Times* (Lagos) 25 Dec. 22/4 (Advt.), Drivers—Houseboys, Shopboys, Shopgirls, Housegirls. 1911 H. S. Harrison *Queed* xiii. 151 There is your public..shop-clerks, stenographers [etc.]. 1921 *Dict. Occup. Terms* (1927) § 939 *Shopclerk*.., keeps record of amount of work done by piece workers for purpose of calculating cost and wages. 1822 D. Wordsworth *Jrnl.* 21 Sept. (1941) II. viii. 361 One a gentlemanly, middle aged man; the other rather younger, with a dash of the shop-counter. 1972 *Listener* 23 Nov. 690/1 'Voluntary price control'..has certainly not worked over the shop counter where it was most needed. a 1745 Swift *Works* (1766) XIII. 47 Our shop-doors will be no longer crowded with so many thieves and pick-pockets. 1832 *Chambers's Edin. Jrnl.* I. 277/1 Transported, he through the shop-door pops his head. 1977 A. Hunter *Gently Instrumental* v. 60 The hour of the lunchtime siesta when every shop door was closed. 1835 Dickens in *Evening Chron.* 14 July 3/3 He..got his butcher to skewer them up on conspicuous joints in his shop-front. 1934 *Times Lit. Suppl.* 25 Jan. 61/1 (*title*) Modern shopfront construction. 1961 D. Holbrook *Eng. for Maturity* 15, I never knew how much shop-front is behind—or perhaps in front of—teaching. 1975 *Sunday Times* 3 Aug. 24/6 The keening nature of the sounds resembles so often the seizing music one has heard in shop-front gospel churches all over America. 1820 M. Edgeworth *Let.* 21 May (1979) 134 The fishwomen, criers and shopgirls whose manners to customers are in general a curious mixture of the affected indifference.. and of the real anxiety for your custom. c 1855 Geo. Eliot in J. W. Cross *George Eliot's Life* (1885) I. vi. 364 She looked like a shop-girl who has donned a masquerade dress impromptu. 1951 A. Baron *Rosie Hogarth* 176 She forced herself to speak calmly, in her precise shopgirl's voice. 1796 J. Woodforde *Diary* 2 Apr. (1929) IV. 268 Betty Cary went wth. him, to bring home some Shop Goods. 1875 [see *ABUSEFULLY *adv.*]. 1972 *Morning Star* 11 Oct. 1 Wage earners might receive more in their pay packets than they do at present but might pay more than they gained in the extra tax paid on shop goods. 1892 *Act* 55 & 56 *Vict.* c. 62 § 1 This Act may be cited as the Shop Hours Act, 1892. 1967 *Observer* 14 May 28/7 Shop hours are 10–7 p.m. 1930 *Daily Express* 6 Oct. 3/5 A great flame which lit up the whole sky..and clearly illuminated the shop signs. 1969 E. H. Pinto *Treen* 410/2 Trade labels on London goods sold between 1765 and 1770, are sometimes printed with the old shop sign. 1896 *Shop Assistant* Aug. 11/2 Manchester may again be counted as a stronghold of unionism amongst shop workers, eager and ready for the fray..which shall emancipate the shop slaves from slavery. 1966 *Listener* 25 Aug. 264/2 Shop-workers' union is to join the opposition to the Government's wage freeze.

b. *shop-done* adj. (nonce use).

1888 G. M. Hopkins *Let.* 1 May (1956) 291, I may be able to send you one [*sc.* a photograph] of me, not shop-done but artistically better.

b*. Applied to food, goods, etc., produced commercially for sale, as (often unfavourably) opposed to home-made or made to order, as *shop-bread, cake*, etc.; shop-bought.

1859 Geo. Eliot *Adam Bede* II. I. xx. 95 A cloth made of homespun linen... None of your bleached 'shop-rag' that would wear into holes in no time. 1876 C. M. Yonge *Three Brides* II. i. 304 I'm sent for one of Herbert's shirts... I believe their hearts would break outright if he took to shop ones. 1928 E. G. Millar *Eng. Illuminated MSS. of XIVth & XVth Cent.* iii. 38 Sarum Horae... These are seldom of more than mediocre quality, and are merely 'shop' copies. 1949 D. Smith *I capture Castle* xii. 203, I had..two slices of cake (real shop cake) and milk. 1957 J. Braine *Room at Top* xxi. 180 It must have seemed that she was offering me a good home-cooked dinner and that I was rejecting it in favour of a slice of chalky shop bread spread with factory-made meat paste. 1975 *Times* 22 Nov. 11/6 As late as the 1930s the better-off continued to look down on those who..spread 'shop' jam on their bread. 1978 D. Murphy *Place Apart* xi. 229 She brought out a slice of Christmas cake... 'It's only shop,' she apologised.

c. **shop assistant,** a salesman or saleswoman in a retail shop or store; **shop class** *N. Amer.*, a class in which the arts of the workshop are taught (cf. sense 3 e above); **shop committee** *U.S.* (see quot. 1923); **shopcraft** *N. Amer.*, an association of railway employees working in repair shops, etc.; **shop-dropper** *local Austral. colloq.* (see quot. 1957) (cf. *DROPPER 1 d); **shop-finish,** the professional finish of an article produced in a commercial workshop (sometimes depreciatory); also *transf.*; hence **shop-finished** *ppl. a.*; **shop-fitting,** (*a*) *pl.*, the fitments (as counters and shelves) with which a shop is equipped; (*b*) the action or process of fitting out a shop with these; hence **shop-fitter; shop-gaze** *v. intr.*, to window-shop; **shop-house,** in S.E. Asia, a shop opening on to the pavement and also used as the residence of the proprietor; **shop-soiled** *a.*, also *fig.*; **shop steward,** a person elected by his or her fellow-workers in a factory, etc., or a branch of it as their spokesman on conditions of work, etc.; **shop-talk** (later examples); **shop-walker** (earlier example); so **shop-walk** *v. intr.*, to act as a shop-walker; **shop-work,** work done in a shop or workshop; **shop-worn** *a.* orig. *U.S.* (earlier examples).

1880 *Girl's Own Paper* 25 Sept. 612/1 There are two great enemies for the shop assistant—the severe shop-walker..and the inconsiderate lady-customer. 1921 *Dict. Occup. Terms* (1927) § 775 *Shop assistant*.., serves customers with goods in retail shop or store, makes out bill or docket. 1977 D. James *Spy at Evening* xii. 86 They were mostly school kids..or young shop assistants and working boys. 1948 G. O. Wilbur *Industrial Arts in General Educ.* xiv. 212 If students go home enthusiastic about the work in their shop classes, a general approval of the whole school program by the parents is apt to follow. 1962 A. Lurie *Love & Friendship* iv. 70 On the last day of school he would take home the present he had made for his mother in shop class. 1978 M. Puzo *Fools Die* xxxix. 435 In the shop class of the asylum school I made myself such a hat. 1908 *Mod. Business* Aug. 69/1 With a good shop committee the men will not be afraid to ventilate their grievances. 1923 J. D. Hackett *Labor Terms in Managem. Engin.* May 344/2 *Shop Committee*, a committee appointed by members of a works committee for the consideration of some special labor problem. 1954 C. E. Dankert *Introd. Labor* x. 187 In many labor organizations there are structural units smaller than, and subordinate to, the locals. These are the so-called shop committees, which are under the leadership of shop stewards. 1973 S. Aronowitz in G. Hunnius et al. *Workers' Control* I. 105 The impulse to dual forms of struggle—shop committees, wild cat strikes, steward movements—may become important in the labor movements of the future. 1919 W. Hines *Let.* 10 Nov. in *Official Proc. 5th Biennial Convention Railway Employees Dept. Amer. Fed. Labor* (1920) 133 The fullest cooperation of..the national officers of the Shop Crafts organizations. 1942 H. E. Jones *Wages & Labor Relations in Railroad Industry 1900–1941* 14 For shop craft employees, annual compensation stood at $1,754 in 1922. 1973 *Daily Colonist* (Victoria, B.C.) 7 Sept. 1/4 Latest union flareup occurred ..as members of Canadian National Railways shopcraft unions walked off the job. 1957 *Courier-Mail* (Brisbane) 26 Nov. 2 'Shop-droppers' are truck owners who buy large quantities of fruit and vegetables at the market and sell them to shopkeepers in and around Brisbane. 1967 *Sunday Mail* (Brisbane) 12 Feb. 18 The suppliers—known as 'shop-droppers'—have been operating for several years. 1923 *New Statesman* 6 Oct. 738/1 They [*sc.* early plays by Somerset Maugham] had the handy compactness, shop-finish and alluring shinyness of a new dressing-case. 1931 R. Fry in W. Rose *Outl. Mod. Knowl.* 914 This last perfection of finish, for which craftsmen have adopted the excellent term 'shop-finish'. 1938 R. G. Collingwood *Princ. Art* xv. 329 The slick shop-finish of a ready-made article. 1932 R. Fry *Characteristics French Art* II. 43 Elsheimer's pictures are so tight, so horribly shop-finished and over-polished. 1885 *List of Subscribers, Classified* (United Telephone Co.) (ed. 6) 188 (*heading*) Shop fitters. 1921 *Dict. Occup. Terms* (1927) § 483 *Shop fitter*, receives wooden parts or sections of counters, desks,..and other shop-fittings..fits and joins these parts or sections together. 1951 A. Baron *Rosie Hogarth* 13 Fred was an engineer and Jack a shopfitter by trade. 1978 *Detroit Free Press* 16 Apr. (Parade Suppl.) 21/1 The 33 indicted..included..a shopfitter. 1858 P. L. Simmonds *Dict. Trade Products* 342 *Shop-fittings*, the counters, desks, shelves, gas-burners, and other fixtures of a shop. 1911 *Rep. Labour & Soc. Cond. Germany* III. VI–VII. 29 The building and shopfitting trade. 1939 C. Vernon *Sweet Shop* XII. xlix. 178 We give in this chapter some general hints on shop fitting and decoration. 1959 R. Buckner *Design for Selling* ii. 14 So many bakers are altering their shop fittings to comply with the hygiene regulations. 1977 *Centuryan* (Office Cleaning Services) Christmas 2/4 The firm was concentrating too much on shopfitting for one client. 1876 L. Troubridge *Life amongst Troubridges* (1966) 143 Shopped the whole morning—*flanéed* down Regent Street, shop-gazing with true country zeal. 1946 S. Spender *European Witness* 21 Crowds who a few years ago were shop-gazing in their city. 1949 *Malayan Pictorial Observer* Aug. 9 Shop-houses line the main street. 1957 G. W. Skinner *Chinese Society in Thailand* iii. 107 By the 1880's..the junk bazaar was..a thing of the past... The former floating population of Chinese tradesmen moved to the two-story shop-houses built in rows along the new streets. 1966 'A. Hall' *9th Directive* i. 7 Where the trishaw had dropped me..was a narrow street of shop-houses. 1978 L. Heren *Growing up on The Times* v. 182 The shophouse had four small rooms. The front room, or shop, was given over to a dispensary... Behind were two small bedrooms and a kitchen. 1926 T. E. Lawrence *Seven Pillars* (1935) v. lix. 333 Beyrout was the door of Syria, a chromatic Levantine screen through which cheap or shop-soiled foreign influences entered. 1927 M. Arlen *Young Men in Love* II. 137 Always together... That shop-soiled man and the tall girl with the curly gleaming hair. 1977 J. Wainwright *Day of Peppercorn Kill* 33 Not the love of a wife... A shop-soiled love—which..he'd reject. 1904 *Rules Amalg. Soc. Engineers* 46 Committees may also appoint shop-stewards to..keep the committee posted with all events occurring in the various shops. 1928 *Britain's Industr. Future* (Liberal Industr. Inquiry) III. xviii. 226 The shop-steward movement, which reached such magnitude during the War, was essentially an attempt to base the struggle for better conditions upon the natural unit of the factory. 1950 A. P. Herbert *Independent Member* 251 At Short's works at Rochester..the shop-stewards threatened a strike. a 1974 R. Crossman *Diaries* (1975) I. 478 This is the first big event of this election campaign—the revelation of a so-called kangaroo trial by shop stewards at the B.M.C. works. 1922 S. Lewis *Babbitt* x. 143 The shop-talk roused Paul Riesling... He was..a very able salesman. 1971 D. E. Westlake *I gave at Office* 76, I must have given her my complete life story..and virtually tons of shoptalk about my job. 1905 H. G. Wells *Kipps* I. vi. 135 Buggins, whose place it was to shopwalk while Carshot served, shopwalked with quite unparallelled dignity. 1825 in A. Nicoll *Hist. Eng. Drama 1660–1900* (1959) VI. 459 (*title of play*) The shop-walker. 1899 W. James *Talks to Teachers* v. 35 Laboratory work and shop work engender a habit of observation,..a knowledge of the difference between accuracy and vagueness. 1932 O. E. Saunders *Hist. Eng. Art in Middle Ages* xiii. 157 Countless lesser Books of Hours were turned out all through the fifteenth and sixteenth centuries for private patrons, but they represent mere shop-work. 1974 J. Burnett *Useful Toil* II. 141 Girls could now go into shop work, into the new light factory trades and into.. clerical work. 1838 *Amer. Comic All-I-Make for 1839* 7 The piece of goods got kinder shop worn, and the old man thought he'd never get her off his hands. 1849 Thoreau *Week Concord Riv.* 220 He is even envied by his shop-worn neighbours. 1871 P. T. Barnum *Struggles & Triumphs* 40 A large quantity of tin ware which had been in the shop for years and was considerably 'shop-worn'.

shop, *v.* Add: **1. b.** To dismiss (a person) from a position or post. *rare.*

1864 Hotten *Slang Dict.* 228 *Shop*, to discharge a shopman. 1915 H. L. Wilson *Ruggles of Red Gap* xvii. 308, I would have shopped the fellow in an instant,..had it been at any other time. He was most impertinent.

4. a. Also *transf.*

1951 M. McCarthy in *Holiday* May 47/2 He determined to attach his name to some lasting benevolent enterprise and settled on woman's education after cautious shopping and advice-seeking. 1973 *Times* 27 Feb. 16/3 The National Portrait Gallery went shopping at Phillips sale room yesterday.

b. With *around*. To visit different shops examining the prices of comparable goods offered for sale before making a purchase; to make purchases at different shops according to which offers the best price. Freq. *transf.* and *fig.*

1922 *Management Engineering* Feb. 89/1 During the war, although orders greatly exceeded production, absenteeism increased. Men took days off to 'shop around', knowing that if unsuccessful they would be welcomed back. 1936 D. Powell *Turn, Magic Wheel* II. 195 Can't you just see those little embryos shopping around for security. 1948 *Economist* 31 July 171/2 It is impossible to shop around for cheaper raw materials. 1952 A. Huxley *Let.* c 20 July (1969) 647 Since success depends on a satisfactory relation between the hypnotised person and the operator you must be prepared to 'shop around' until you find someone sympathetic as well as skilful. 1960 W. Taplin *Advertising* iv. 83 We have.. noticed the..case..of the people who buy advertised products and in effect accept the advertiser's persuasion rather than spend time 'shopping around'. 1976 J. I. M. Stewart *Memorial Service* ii. 35 It's usual to shop around a little. To send in a list of three or four colleges.

c. *trans.* To shop at (a store); to examine goods on sale in (a shop). *N. Amer.*

1955 in H. Galinsky *Amerikanisches und Britisches Englisch* (1957) 49 Shop the store that gives you more. 1961 *Ford Times* Mar. 28 (*heading*) Shopping the southern roadside. 1974 S. Marcus *Minding Store* iv. 85 One man who had shopped the entire store complained that he hadn't found what he was looking for. 1980 'E. McBain' *Ghosts* ii. 18 Maybe all the burglars..were out shopping the department stores.

5. (Earlier example.)

1808 *Rules of Journeymen, Hat-Makers & Finishers of Stockport* in A. Aspinall *Early Eng. Trade Unions* (1949) iv. 110 And when any person comes wishing to be asked for, the person that goes and asks for him, to take his ticket, and in case that man is shopped, he must leave his ticket at the place he is shopped.

Shope (ʃōᵘp). *Biol.* The name of Richard Edwin *Shope* (1902–66), U.S. physician, used *attrib.* to designate a transmissible papilloma of rabbits described by him in 1932 (*Jrnl. Exper. Med.* LVI. 793, 803), and the DNA virus which causes it.

1934 *Jrnl. Exper. Med.* LX. 756 The Shope papilloma, as occurring in nature, manifestly falls into the group of infectious warts, condylomas, and papillomas. 1935 *Proc. Soc. Exper. Biol. & Med.* XXXIII. 193 Shope virus from some sources gives rise to progressively enlarging papillomas, and that from others to growths which tend to disappear. 1938 *Ann. Reg. 1937* 349 A high molecular weight protein apparently associated with the viral activity was isolated from the Shope rabbit papilloma. 1961 R. D. Baker *Essent. Path.* xiii. 311 Carcinogenic viruses are known to produce the Rous sarcoma in chickens and the Shope papilloma of rabbits. 1970 *New Scientist* 29 Jan. 194/1 The Shope virus..induces warts on the skins of rabbits.

shop floor. [f. Shop *sb.* + Floor *sb.*[1]] **1.** The floor of a workshop or factory, where the operatives work; the part of a factory or workshop concerned with productive as opp. to administrative work. Freq. (with hyphen) *attrib.* Also in phr. *on the shop floor*.

1951 E. Jacques *Changing Culture of Factory* III. xii. 316 The workers' representatives..were regarded as a more reliable source of information about shop floor matters. 1956 *Nature* 18 Feb. 300/2 It is difficult to see how, in a Handbook of this kind, the real challenge and opportunity which shop-floor industry presents could be presented. 1962 *B.S.I. News* Apr. 17/2 Polymethylmethacrylate might be understood by a chemist but 'Perspex' (a trade name) would be understood at shop-floor level. 1962 *Listener* 12 July 44/2 The lack of participation on the shop floor may be one of the underlying causes of malaise. 1967 C. Margerison in Wills & Yearsley *Handbk. Management Technol.* 31 The worker on the shop floor does not tend to identify himself either with the values or the goals of the manager. 1970 *Guardian* 11 July 11/6 The unanimous decision of a dockers' delegate conference—shop floor democracy at its most democratic. 1977 M. Walker *National Front* vi. 156 The new trade union policy did not focus solely on the exploitation of racialism on the shop floor.

2. By extension, the workers on the shop floor considered *collect.*

1958 *Listener* 30 Oct. 680/2 A dispute between management and shop floor about the number of men to be employed upon a machine. 1977 *Times* 20 Apr. 4/7 They have decided to cancel... The shop floor is upset about it.

shophar. Add: The standard form is now **shofar.** (Earlier and further examples.)

1833 *Children's Mag.* V. 113 The crooked trumpet, or *shophar*, was appointed by the Law of Moses to be blown ..when the year of jubilee was proclaimed. 1891 M. Friedlander *Jewish Relig.* 403 The *shofar* is intended to awaken us. 1931 *Times Lit. Suppl.* 24 Sept. 722/2 The blowing of the Shofar turns out to have been originally the imitation of the voice of the dying God. 1973 *Synagogue Light* Sept. 41/2 Every morning at the conclusion of the service a blast of the shofar reminds us of the approaching Day of Judgment.

shoplift, *v.* [Back-formation f. Shoplifting *vbl. sb.*] To steal from a shop while pretending to be a customer. **a.** *intr.*

1820 [see Sheep-steal *v.*]. 1843 *Punch* 8 Apr. 150/1 Policeman, here's a wench Shoplifting, take the customer to jail. 1959 *Times* 9 Mar. (Britain's Food Suppl.) p. ix/3 The temptation to shop-lift is one facet of the principle on which every self-service store depends. 1971 *Guardian* 22 Feb. 9/1 When Mrs Brown gets depressed, she starts to shoplift.

b. *trans.*

1922 Joyce *Ulysses* 741 A whore always shoplifting anything she could. 1979 K. Conlon *Move in Game* I. iii. 39 She'd shoplifted a bottle of nail varnish remover.

shopman. Add: **3.** A man employed in a railway workshop.

1926 *Times* 8 Mar. 9/4 He was glad to be able to state that the case of the railway shopmen had been advanced. 1960 *Listener* 18 Aug. 250/1 They were trainmen, shopmen, telegraph operators, and so on, all highly specialized in their respective callings.

shoppe (ʃɒp, ʃɒ·pi), an archaic form of Shop *sb.* now used affectedly (as in the names of tea-shops, etc.) to suggest quaint, old-world charm. Cf. *OLDE *a.*

1933 J. Betjeman *Ghastly Good Taste* 138 Arts and Crafts. Gentle folk weaving and spinning; Modern Church Furnishing; Old Tea Shoppes. 1948 [see *gift shop* s.v. *GIFT *sb.* 9 b]. 1957 E. Pound tr. *Rimbaud* 15 What lures the aintient truss-maker from his shoppe whose luxury Sucked in the passers-by. 1979 L. Kallen *Introducing C. B. Greenfield* i. 10 This is a business office, not a tea shoppe.

shopper. Add: **1. a.** (Earlier example.)

1860 Mrs. Gaskell *Let.* 27 Aug. (1966) 632 She is very dainty-fingered, a beautiful ready workwoman, a capital shopper &c.

b. An advertising sheet or newspaper.

1958 Palmer & Gilmore in W. C. Clark *Journalism Tomorrow* iii. 32 The little advertising sheet, often called a shopper... In a few cases, the shopper can be converted into a fine-looking, profitable weekly newspaper. 1976 *National Observer* (U.S.) 16 Oct., He would junk a lot of the expensive radio and TV commercials, spend the money instead in shoppers and small town weeklies, where you can buy the whole back page for $65.

c. A shopping bag or trolley.

1968 H. C. Rae *Few Small Bones* II. v. 113 He bought enough tinned food..to last him a full week, then lugging the laden shopper, set out along the main street. 1970 *Kay & Co.* (Worcester) *Catal.* 1970–71 Autumn/Winter 934 (*caption*) Giant size trolley shoppers. *Ibid.*, Tartan shopper. Novelty-shaped large capacity zip-top Shopping bag in Black Watch Tartan. 1978 H. R. F. Keating *Long Walk to Wimbledon* IX. 137 The wheels of Marigold's shopper squeaked.

2. *slang.* An informer.

1924 E. Wallace *Room 13* xxxi. 300 Jeffrey's going to shop you sooner or later, because he's a natural born shopper. 1955 P. Wildeblood *Against Law* 105 'Shoppers'...people who go to the cop-shop and squeal on their friends.

shoppie (ʃɒ·pi). *Sc.* [f. Shop *sb.* + -ie.] A little shop.

1872 Queen Victoria *Jrnl.* 13 June in D. Duff *Victoria in Highlands* (1968) 260 At half-past ten drove out in the waggonette..and drove beyond Mrs. Patterson's 'shoppie' a little way. 1887 W. Carnie *Waifs of Rhyme* 17, I ken a winsome wifikie that keeps a snug bit shoppie.

shoppie, var. *SHOPPY *sb.*

shopping, *vbl. sb.* Add: **1. a.** Also with *around*. Freq. *transf.*

1940 *Sun* (Baltimore) 16 Feb. 24/2 By the simple expedient of 'shopping around' before making a deal, Walter N. Kirkman..has succeeded in saving the State $13,300 a year in office rent here. 1971 *Nature* 10 Dec. 368/1 An author does much better by bargaining hard and honestly with almost any single good publishing house than by 'shopping around'.

b. *transf.* The goods that have been purchased (in quot. 1934, 'something that has been purchased').

1934 *Punch* 2 May 489/1 Thank you so much, darling, for those marvellous Shoppings... The pyjamas are divine. 1948 P. Wentworth *Traveller Returns* i. 6 There were three people in front of her—a very stout woman with a basket full of shopping..., and a stooping elderly man. 1975 *Oxford Consumer* June 4 There are those, i.e. the elderly, the infirm, people with shopping,..for whom cycling is not always possible.

c. = *SHOP *sb.* 2 e. *colloq.*

1934 *Punch* 2 May 489/1 Would you think me the most terrific pest if I asked you to do yet another shopping for me? 1980 M. Forster *Bride of Lowther Hall* xviii. 272, I could perhaps go to Wigton and do a mammoth shopping.

2. *attrib.* and *Comb.* Simple attrib.: *shopping bag, basket, day, expedition, hours, spree, tour, trip*; in the names of places where shopping is performed, as *shopping arcade, area, centre, complex, mall* (chiefly *N. Amer.*), *parade, plaza* (chiefly *N. Amer.*), *precinct, street*; **shopping-bag lady** *U.S.*, a vagrant woman

carrying her possessions in shopping bags; **shopping cart** orig. *N. Amer.*, a large wire basket on wheels provided for the use of supermarket customers; **shopping net,** a shopping bag made of string or plastic net; **shopping service,** a department or organization offering advice or assistance with shopping; **shopping tray,** an open wire receptacle for shopping designed to fit over the chassis of a pram; **shopping trolley,** (*a*) a shopping bag set on a wheeled frame; (*b*) = *shopping cart* above. See also *SHOPPING LIST.

1933 *Radio Times* 14 Apr. 72/2 Dear Covent Garden.. we hear, alas, that you are soon to..make way for shopping-arcades. 1977 J. Bingham *Marriage Bureau Murder* xvi. 184 There was a shopping arcade... He often shopped there. 1959 *Manch. Guardian* 11 Aug. 6/5 More shopping areas for pedestrians only. 1974 A. Morice *Killing with Kindness* ii. 18 You have to pass through the shopping area to get to the main road. 1886 S. Coolidge *What Katy did Next* ix. 247 In her shopping-bag one or two of the Carnival bonbons still remained. 1973 J. Stranger *Walk Lonely Road* xiii. 97 Millie came in with a brimming shopping-bag. [1972 S. R. Curtin *Nobody ever died of Old Age* vi. 85 Letty the Bag Lady..would pack all her valuables in two large shopping bags and carry them with her.] 1976 *N.Y. Times* 30 Sept. 43/5 The chief thing about Sally, the former 'shopping bag lady', is that she has personality. 1978 *Harper's Mag.* Mar. 104 On a nearby bench, apparently keeping an intermittent vigil on the vigil, were two shopping-bag ladies. 1979 *N.Y. Times* 10 Jan. B 7 An elderly 'shopping bag lady', ..one of the legion of homeless, independent, often eccentric women who live on the city's streets. 1923 E. Bowen *Encounters* 166 She had forgotten her shopping basket and her purse. 1977 *Times* 29 June 5/7 She arrived at the court by bus carrying a shopping basket containing her dressing gown. 1956 *Sun* (Baltimore) (B ed.) 19 Jan. 14/1 A feature of the supermarket is the now familiar shopping cart, a contraption of recent invention. 1958 M. Dickens *Man Overboard* v. 83 A woman with a shopping cart bumped into him. 1976 G. A. Browne *Slide* (1977) ix. 74 The shopping cart somehow getting fuller than intended. 1898 Shopping centre [see *CENTRE *sb.* 6 a]. 1933 L. E. Neal *Retailing & Public* v. 25 There is..great competition between multiple branch organisations..to acquire the best positions in the popular shopping centres. 1944 H. G. Wells *'42 to '44* 136 In Welwyn Garden City to-day there is a single 'shopping centre' associated with one bazaar. 1971 P. Gresswell *Environment* 183 Very few shopping centres are imaginatively designed. 1977 B. Garfield *Recoil* v. 62 Caruso..drove him..to a shopping center in Santa Monica. 1970 *Times* 27 Feb. 19/7 Fram Gerrard, of Manchester, has won the building contract for the £1·5m. shopping complex in Leicester. 1861 C. M. Yonge *Young Step-Mother* xxix. 439 This was a grand shopping day, an endless business. 1973 D. Miller *Chinese Jade Affair* xxiii. 220 'Only thirty Shopping Days to Christmas!' advised the big department stores. 1847 A. Brontë *Agnes Grey* xxv. 358, I took a little more pains with my attire than if I had merely been going on some shopping expedition alone. 1885 [in Dict.]. 1979 L. Meynell *Hooky & Villainous Chauffeur* xi. 139 What with cooking..and the necessary shopping expeditions she didn't have much time over. 1964 A. Adburgham *Shops & Shopping* i. 7 There were no stated shopping hours. 1967 *Canad. Ann. Rev. 1966* 137 Its proposals for..restriction of billboards, abolition of overhanging signs and overhead wires, closed pedestrian shopping malls, [etc.]. 1972 *N.Z. News* 26 Jan. 7 (*caption*) This scene is taken from the Cuba Street shopping mall, Wellington. 1979 *Jrnl. R. Soc. Arts* July 505/1 Shopping malls are the nearest thing to the market place which you could find in North America. 1955 T. Sterling *Evil of Day* i. 1 Within the cabin, women with shopping nets and men with folded newspapers awaited their turns. 1969 P. Highsmith *Tremor of Forgery* xxi. 195 Adams had a shopping net in his hand. He was putting things away in the kitchen. 1969 *Morning Star* 1 Dec. 4/1 There will be various amenities when Stage 1 and Stage 2 are completed, [including] a shopping parade. 1957 *Ottawa Jrnl.* 6 Aug. 1/1 The break-ins were compared with the noisy June 3 'tow-truck' robbery at another grocery store in the shopping plaza. 1981 P. Theroux *Mosquito Coast* vi. 44 We came to a shopping plaza, where we parked. 1958 Shopping precinct [see *PRECINCT *sb.* 4]. 1980 A. Auswaks *Trick of Diamonds* iii. 75 At the end of the High Street stood a new shipping precinct. 1925 *Eaton's News Weekly* 26 Sept. 18 When Ordering by Mail..Address Letter to *Shopping Service*. 1972 S. Ellin *Mirror, Mirror on Wall* 79 The lady happens to work for a shopping service... If you want to do some shopping for your wife ..these people take you to the right places and pick the right stuff. 1962 M. Summerton *Nightingale at Noon* (1963) iv. 55 A shopping spree to purge a memory. 1976 'S. Woods' *My Life is Done* 69 Ana had been on a shopping spree. 1914 A. Bennett *Price of Love* xiv. 283 Her first apparition in the shopping streets of the town.. as Mrs. Louis Fores, married woman. 1975 P. Moyes *Black Widower* ii. 19 The great shopping street that bisects the area. 1878 Shopping tour [in Dict.]. 1970 *Kay & Co.* (Worcester) *Catal.* 1970–71 Autumn/Winter 227 Leeway shopping tray provides secure storage space and is readily adaptable to most prams. 1977 *Cornish Times* 19 Aug. 7/3 (Advt.), Silver Cross pram, detachable body..with shopping tray. 1969 G. Lyall *Venus with Pistol* xviii. 116 Henri has been *killed*... Doesn't that mean more than this—shopping trip? 1976 *Economist* 17 Apr. 79/3 A motorised pram-cum-shopping-trolley would arguably be the most egalitarian form of transport. 1978 *Green Shield Stamp Catal.* Feb. 113 *Shopping trolley*. Adjustable telescopic handle. Detachable shopping bag. 1979 *Criminal Appeal Reports (Sentencing)* (1980) I. 255 On the morning of January 25, 1979 at Brent Cross Shopping Centre she went into a Waitrose Supermarket and loaded up a shopping trolley with groceries.

shopping list. [f. Shopping *vbl. sb.* + List *sb.*[6]] **1.** A list of purchases to be made or shops to be visited.

1913 *Vanity Fair* Sept. 7/2 You can easily clip half a dozen and attach them to your own shopping list. 1921 *Daily Colonist* (Victoria, B.C.) 31 Mar. 6/5 We have a beautiful selection of Oriental Rugs and Jewels. Put our name on your shopping list. 1947 W. Stevens *Let.* 5 Sept. (1967) 567 But now that the weather is growing cooler, I begin to look at my shopping list. 1977 'E. Crispin' *Glimpses of Moon* xii. 256 He went to investigate, finding a message..scribbled on a sheet torn from a Shopping List pad.

2. *transf.* and *fig.* A list of items to be considered, acted upon, etc.; *spec.* a list of weaponry sought for purchase.

1959 *Manch. Guardian* 8 July 3/3 He..did not want to put down a shopping-list of industries to be nationalised. 1963 *Times* 21 Feb. 11/3 Ground-to-air missiles were on the 'shopping list' of defence equipment taken to Moscow by the Secretary-General of the External Affairs Ministry. 1969 *Nature* 2 Aug. 436/2 The council should say which fields of medical and biological research come at the top of its shopping-list—biological organization, arterial disease, population control and drug dependence. 1970 *Cape Times* 28 Oct. 26/5 His shopping list will include a full-back..and a creative mid-field player. 1975 *N.Y. Times* 8 Sept. 1/3 The Israeli sources said negotiations had resumed on a shopping list that includes..F-15 fighter planes to match the MIG-23's deployed by Egypt and Syria. 1977 *Times* 10 June 17/1 The [French] left have indeed a fearsome shopping list for nationalization.

shoppy, *a.* **3.** (Later example used disparagingly.)

1914 W. Owen *Let.* 10 May (1967) 249 Miss H— who, you confess, is 'shoppy'. Now that is distressing. The fact of being employed in a shop does not matter; but shoppiness does matter.

shoppy (ʃɒ·pi), *sb. slang.* Also **shoppie.** [f. Shop *sb.* + -y[6].] A shop assistant.

1909 P. Webling *Story of Virginia Perfect* i. 6 Her manner towards him..had none of the affectation of the ordinary 'young lady in business', or the vulgar intimacy of a poorer class of 'shoppie'. 1916 'Taffrail' *Pincher Martin* vii. 114 She's in Skeets the draper's... Never could stand them shoppies; they give themselves such airs. 1934 H. A. Vachell *Disappearance of Martha Penny* i. 20 Her sparkling eyes, her fine figure, were gifts rarely bestowed upon urban 'shoppies'.

shop-window. Add: **2.** (Later examples.)

1929 C. Connolly *Let.* Nov. in *Romantic Friendship* (1975) 328 [In America] a wife is a man's shop window. 1933 P. Godfrey *Back-Stage* iii. 34 A 'shop-window' is a part which carries a low salary, but which, nevertheless, is likely to enhance his acting reputation. Certain small theatres are also 'shop-windows', like the Everyman Theatre in its early days. 1954 [see *END *sb.* 7c]. 1961 *Radio Times* 6 Apr. 4/2 Once a year the BBC..stages a couple of all-star concerts as a shop-window for the kind of 'pop' programmes that it broadcasts..during the rest of the year. 1977 *Listener* 17 Mar. 347/3 London was to be a city of individuals..rather than a shop-window of state, monarchy and empire.

Hence **shop-wi·ndowful,** as much or as many as a shop-window will hold.

1898 G. B. Shaw *Mrs. Warren's Prof.* IV. 231, I shouldn't enjoy..being bored at the opera to shew off a shop windowful of diamonds.

shor (ʃǭɹ). [Turki *shōr*.] In Turkistan, an elongated saline depression in desert sand.

1888 *Encycl. Brit.* XXIII. 512/1 A feature distinctive of the Turcoman desert is seen in the very numerous *shors*, or elongated depressions, the lower portions of which are occupied mostly with sand impregnated with brackish water. 1898 *Geogr. Jrnl.* XII. 308 The *shors*.. are seen on the southern border of the Kara-kum sands. 1951 N. T. Mirov *Geogr. of Russia* xv. 155 Lower depressions, so-called sor or shor, equally extensive, are occupied with alkaline soil of the solonchak type.

shoran (ʃǭᵊ-, ʃǭ·ræn). orig. *U.S.* Also **Shoran.** [f. the initial letters of *short-range navigation*.] A secondary radar system, used for precision navigation and for distance measurement, in which an aircraft or ship determines its distance from two widely-spaced ground stations which it interrogates alternately with radio pulses. Freq. *attrib.* Cf. *LORAN.

1946 *N.Y. Times* 28 Apr. 1/3 Those who have..tested shoran call it 'perhaps the greatest single invention of its type for long-range mapping to come from the family of radar in the war'. 1946 *Trans. Amer. Geophys. Union* XXVII. 459 Basically, the shoran equipment consists of three main units, the airborne transmitter-receiver-indicator and the two ground station transponders. 1949 *Sun* (Baltimore) 2 Mar. 13/2 What prompted the use of the Superforts was the belief that the development of Shoran (Short Range Navigation) equipment had reached the point where it could be put to work for the astronomer. 1966 *McGraw-Hill Encycl. Sci. & Technol.* IX. 18/2 The single-path round-trip system is the basis of distance determination in all radars. It is the system employed by Benito, Condar, Oboe, shoran, and the distance-measuring portion of Tacan equipment. 1970 *Canad. Cartographer* VII. 23/1, 12 shoran ground stations. 1977 *Sci. Amer.* Oct. 92/3 At each one a transponder was set up for a shoran radio-positioning system.

shore, *sb.*[1] Add: **1. b.** Similarly in *Geomorphol.*

1919 D. W. Johnson *Shore Processes & Shoreline Devel.* iv. 160 The most important of the four zones extends from low water mark to the base of the cliff,.. which usually marks the landward limit of effective wave action. This is the zone over which the water line, the line of contact between land and sea, migrates; and it will here be called the shore. 1968 R. W. Fairbridge *Encycl. Geomorphol.* 62/1 The shore is technically the coastal zone extending from the low tide limit to the maximum swash line. 1978 A. L. Bloom *Geomorphology* xix. 437 The shore zone, or simply shore, is the zone affected by wave action.

5. a. *shore clothes, duty, -ice* (earlier and later examples), *suit*; *shore-based* adj.; **shore break** *Surfing* (see quot. 1962); **shore dinner** *U.S.*, a dinner consisting mainly of sea-food; **shore face** (see quots.); **shore fishery** *N. Amer.* (see quot. 1948); **shore leave,** leave of absence granted to a sailor to go on shore; **shore liberty** = prec.; **shore-loafer** *Naut. slang*, a civilian; **shore party,** (*a*) *N.Z.*, a body of whalers using a land-based station (*obs.*); (*b*) a body of persons going ashore from a ship; *spec.* a body of soldiers sent ashore; **shore patrol** *U.S.*, a naval police organization responsible for the conduct of sailors on land; hence **shore patrolman; shore platform,** a horizontal or gently sloping platform cut at about sea level in a cliff by wave action; **shore seine,** a seine used near the shore; **shore-shooting** (earlier example); **shore station,** a base on land used for shore-whaling; **shore-whaling** (earlier examples); also *spec.* = *bay whaling* s.v. *BAY *sb.*[2] 5; hence **shore whaler,** a person engaged in shore-whaling; **shore zone,** the intertidal zone, or the zone affected by wave action; = Shore *sb.*[1] 1 b in Dict. and Suppl.

1927 *Daily Tel.* 22 Mar. 10/7 The limitation of air armaments shall be effected by limiting the number of shore-based aircraft of service type maintained in commission. 1950 A. Lee *Soviet Air Force* 34 Its naval force was shore-based except for a few reconnaissance machines on cruisers. 1980 *Jrnl. R. Soc. Arts* July 521/2 The professional mariner currently gives cautious approval to shore-based information services. 1962 T. Masters *Surfing made Easy* 65 *Shore break*, waves which break close to the beach. 1965 J. Pollard *Surfrider* ii. 20 The next one you might take right to the 'shore break', the waves breaking on the very edge of the beach. 1972 Y. Maley in G. W. Turner *Good Austral. Eng.* iv. 77 Drouyn gets it on,..hangs five, re-enters the shore break, then steps off onto the sand. 1862 E. Hodder *Mem. N.Z. Life* 24 Shore clothes were unpacked, the ship was made tidy. 1922 E. O'Neill *Anna Christie* I. 100 He is dressed in a wrinkled, ill-fitting dark suit of shore clothes. 1895 *Outing* XXVI. 408/2 Happy-Go-Lucky Beach is proud of their achievements..in the ordering of and presiding at a good shore-dinner. 1947 E. H. Paul *Linden on Saugus Branch* 267 It was arranged for the party to eat at the Massasoit a shore dinner cooked by Jeff. 1881 *Library Universal Knowl.* XI. 408 Paymasters..on shore-duty are employed in the naval purchasing agencies. 1901 *Chambers's Jrnl.* Aug. 551/2 Four keepers are employed in connection with the lighthouse, three being in constant attendance while the fourth is on shore-duty. 1912 J. Barrell in *Bull. Geol. Soc. Amer.* XXIII. 385 The shore face is the relatively narrow slope developed by the breaking waves, a slope which separates the subaerial plain above from the subaqueous below. 1944 A. Holmes *Princ. Physical Geol.* xiv. 291 In appropriate circumstances some of the sediment in transit across the wave-cut platform accumulates in the deeper water beyond, to form a shoreface terrace which grows forward like a broad embankment with its upper surface in smooth continuity with the platform. 1972 *Gloss. Geol.* (Amer. Geol. Inst.) 654/2 *Shoreface*,..the narrow, rather steeply sloping zone seaward or lakeward from the low-water shoreline, permanently covered by water, and over which beach sands and gravels actively oscillate with changing wave conditions. 1767 T. Hutchinson *Hist. Province Mass.-Bay* II. iv. 445 In what they call a sedentaire and we a shore fishery we shall always outdo them. 1948 R. de Kerchove *Internat. Maritime Dict.* 671/1 *Shore fisheries*. Under this head are included all those fisheries prosecuted from small boats or from the shore without the aid or use of vessels. 1971 E. R. Seary *Place Names Avalon Peninsula* iv. 65 [The Killigrews] either settled permanently or had a summer plantation at Killigrews for the shore fishery. 1752 J. Robson *Acct. Six Years Residence in Hudson's Bay* 58 At York-fort and Churchill-river I have observed that the ice did not break off close at the shore, but gradually; the first field leaving the shore-ice two or three miles broad, the second less, and so on till it was cleared away. 1953 *Beaver* June 22 They walked across the shore ice, perhaps as far as two miles. 1977 *New Yorker* 10 June 55/1 The river's edges are lined with ice that is stationary —'shelf ice', 'shore ice', the first to freeze at the start of winter and the last to go in spring. 1888 E. L. Dorsey *Midshipman Bob* 205 They set about making the most of their shore-leave. 1941 C. S. Forrester *Captain from Connecticut* xv. 216 Shore leave..meant rum and women. 1974 M. Hastings *Dragon Island* iv. 37 Darley was leaning on the rail. 'Shore leave?' he asked. 1906 J. London *Let.* 17 Nov. (1966) 220 You can depend upon me giving good opportunities for shore-liberty. 1971 S. E. Morison *European Discovery Amer.: Northern Voy.* ix. 287 *La Dauphine* almost always anchored in an uncomfortable roadstead, and they had shore liberty but once in the entire voyage. 1916 'Taffrail' *Carry On!* 25 If an ordinary 'shore-loafer', as a bluejacket sometimes calls a civilian, were suddenly transported to one of His Majesty's battleships he would probably spend his first few days on board in a state of hopeless bewilderment. 1841 H. W. Petre *Acct. Settlements N.Z. Co.* iv. 77 System of 'shore-parties'..is much more economical than the pursuit of the whale by ships equipped for the purpose. 1901 G. B. Shaw *Caesar & Cleopatra* III. 160 My men at the barricades are between the sea party and the shore party. 1974 M. Hastings *Dragon Island* v. 42 Tallander's concern regarding any shore parties from our ship. 1917 *Blue Jacket's Man.* (ed. 5) 644 Perhaps the establishment of the Shore Patrol has done more than any other one institution to make petty officers realize their duty as a class. 1973 H. Gruppe *Truxton Cipher* xvii. 176 The phone rang urgently in Shore Patrol headquarters at the fleet landing. 1944 *Bull. Bur. Naval Personnel Information* (U.S.) Sept. 12/1 The shore patrolman could have barged in to break up the argument. 1973 H. Gruppe *Truxton Cipher* xx. 208 Dieter leaped..straight into the arms of two waiting Shore Patrolmen. 1895 J. D. Dana *Man. Geol.* (ed. 4) 220 Besides battering and degrading cliffs, wave-action makes shore-platforms, by shearing away the rocks of coasts down to a horizontal surface near low-tide level. 1978 A. L. Bloom *Geomorphology* xix. 448 Shore platforms are developed by water-level weathering at various heights, relative to tide level, depending on structural factors..and also on wave energy, tidal range, and climate. 1884 G. B. Goode et al. *Fisheries & Fishery Industries of U.S.* I. 289 It seems..absurd that the Massachusetts people should have supposed that the use of shore-seines was exterminating the Mackerel on the coast of Massachusetts. 1973 W. Elmer *Terminol. Fishing* ii. 69 The shore seine is worked with a boat and a shore party. 1829 G. Griffin *Collegians* III. xxxi. 2 He had gone down to the Dairy farm, for the purpose of shore-shooting. 1966 *Austral. Encycl.* IX. 276/2 In 1947 a small chaser fed a shore station at Albany. 1966 *Encycl. N.Z.* III. 639/1 Hunting, therefore, occurred from vessels ranging considerable distances off shore, from others at bay anchorages, and also from a large number of open boats based on shore stations. 1885 W. D. Howells *Rise S. Lapham* xxiii. 415 A young fellow in the shabby shore-suit of a sailor. 1924 J. Masefield *Sard Harker* I. 30 Steward, will you have the goodness to set out my shore-suit presently? 1872 *Trans. & Proc. N.Z. Inst.* V. 156 The females visit the bays and inlets round the coast to calve..where they are captured by the shore whalers. 1966 *Encycl. N.Z.* III. 640/1 Hundreds of right whales killed by pelagic whalers off shore and in the bays where ships' boats were often in direct competition with those of the shore whalers. 1841 S. Revans *Lett. to H. S. Chapman* (typescript) II. 163 If no shore whaling were allowed the cow would rear the calf and get fat. 1851 H. Melville *Moby Dick* II. xxxix. 273 In the Shore Whaling.., when a Right Whale gives token of sinking, they fasten buoys to him. 1922 E. C. Starks *Hist. Calif. Shore Whaling* 6 Whaling may be classified under three heads:.. Third.—For want of a better term we may call the third form modern shore whaling. The whales are not taken from small boats, but from a seaworthy steam whaler... The whaler stays out until it has secured one or more whales, which it tows to a whaling station on shore. 1959 A. H. McLintock *Descr. Atlas N.Z.* p. xvii, Today there is a shore whaling station at the seaward entrance to Tory Channel. 1921 A. W. Grabau *Textbk. Geol.* I. xvii. 518 This [littoral] district naturally falls into two zones, (*a*) that of the shore between high and low tide (shore zone) and (*b*) that permanently submerged..(neritic zone). 1978 A. L. Bloom *Geomorphology* xix. 444 Where the postglacial rise of sea level has created a shoreline on a former hill slope, shore-zone processes cut a cliff and bench.

b. **shore-bug**, a bug belonging to the family Saldidæ; **shore fly**, a small black fly of the family Ephydridæ, found in damp or marshy places.

1895 J. H. Comstock *Man. Study Insects* 134 Some of the Shore-bugs dig burrows, and live for a part of the time beneath the ground. 1968 *Oxf. Bk. Insects* 28/2 The most common and widespread British shore bug..lives around the margins of ponds, ditches, and semi-stagnant streams and lakes. 1942 E. O. Essig *College Entomology* xxxv. 743 (Shore Flies, Ephydrid Flies.) Ephydridæ. 1954 Borror & DeLong *Introd. Study Insects* xxvii. 633 The shore flies are small to very small; most of them are dark coloured. 1979 *Nature* 29 Nov. 501/2 Eighty per cent of their diet comprises three insect species, the shore fly, *Ephydra riparia*, the waterboatman, *Trichocorixa reticulata*, and the mosquito, *Aedes dorsalis*.

Shore (ʃōəɹ), *sb.*[5] *Metallurgy.* The name of Albert F. *Shore* (fl. 1907), U.S. manufacturer, used *attrib.* with reference to the *Scleroscope he invented and to a scale of relative hardness associated with the use of this instrument, as *Shore hardness, Scleroscope, test*, etc.

1908 *Iron Age* LXXXII. 555 (*heading*) The Shore Scleroscope. 1908 *Jrnl. Iron & Steel Inst.* LXXVIII. 639 Maurer also gives an account of his investigations on the Shore hardness test. 1924 Jeffries & Archer *Sci. of Metals* i. 21 The Shore numbers are more representative of the yield point than of the tensile strength. 1937 R. T. Rolfe *Steels for User* vi. 110 Thus a Brinell hardness of 131 should be equivalent to a Shore figure of..22, which is the same as the Shore hardness determined. 1967 E. Chambers *Photolitho-Offset* xv. 226 The Shore hardness of the rubber stock should be between 8 to 16. 1979 J. Neely *Pract. Metall. & Materials of Industry* vi. 71/2 Elastic hardness is measured by an instrument called a Shore Scleroscope.

shore (ʃōəɹ), repr. *colloq.* or (in *U.S.*) *dial.* pronunc. of Sure *a.* and *adv.*

1890 *Dialect Notes* I. 71 *Shore*, sure. 1898 G. B. Shaw *Candida* I. 97 Glad to meet you, I'm shore. 1932 V. Randolph *Ozark Mountain Folks* ix. 163 Hit shore was a bad night at our place. Yas, sir, hit *shore* was! 1938 M. K. Rawlings *Yearling* vii. 62 Well, stay, then, if these folks is shore you're welcome. 1973 R. Hoban *Lion of Boaz-Jachin & Jachin-Boaz* xviii. 100 It's a proper thing for a man to do—not like running a restaurant or some shore thing like that. 1979–80 *Verbatim* Winter 14/1 My cousin Sharon, a University of Missouri homecoming queen, shore did look pretty, and her mother shore could fry chicken.

shore, *v.*[1] **1.** Delete (now *rarely*) and add later *fig.* examples with *up*.

1892 'Mark Twain' *Amer. Claimant* xxii. 219 This prop shored him up and kept him from floundering back into democracy and re-renouncing aristocracy. 1959 *Listener* 10 Dec. 1021/1 These are all signs that local authorities are likely to shore up their position for the time being. 1978 *N.Y. Times* 30 Mar. D9/4 The Carter Administration was not contemplating any emergency measures to shore up the dollar.

shore-going, *vbl. sb.* (Earlier *attrib.* examples.)

1846 *Knickerbocker* XXVIII. 64 He rigged himself in his 'shore-going togs' after supper. 1887 B. Harte *Crusade of 'Excelsior'* 63 Captain Bunker..wore a shore-going suit of black broadcloth.

shoreside. Add: *attrib.* passing into *adj.* (later examples).

1937 *Sun* (Baltimore) 4 Sept. 3/1 Harry Bridges, leader of the longshoremen, proclaimed his union's aim of a 'march inland'—to organize all shoreside transportation and commodity handling under the stevedores. 1966 *Economist* 25 June 1439/3 Norway makes sure of a flow of young men into the merchant service. Bachelors pay half the income tax of workers ashore; married men pay a little more, though still less than the shoreside worker. 1979 D. Lowden *Boudapesti 3* xxx. 160 Buildings going up...Shoreside villages, without a fishing boat in sight.

Hence as *adv.*, to the shore, to land (*rare*).

1948 Partridge *Dict. Forces' Slang* 168 Are you coming shore-side this afternoon? 1949 *Sun* (Baltimore) 6 July 10/2 A desolate peninsula in Venezuela, expected to become the Western Hemisphere's largest oil port, needed facilities for seamen going 'shoreside'.

shoreward, *adv.* Add: **3.** *shoreward of*: towards the shore in respect of; on the shoreward side of.

1941 *Sun* (Baltimore) 12 July 7/3 All interested parties are urged to be present, particularly those engaged in fishing or operating boats shoreward of or adjacent to existing authorized fishing areas. 1974 R. Adams *Shardik* xvi. 110 They came ashore..close to..the cluster of storage huts and servants' quarters lying shoreward of the Sindrad.

shorn, *ppl. a.* Add: **4.** *shorn lamb*: also applied to the dressed fur of the sheep used in garment-making.

1945 N. Mitford *Pursuit of Love* xxi. 183 Linda..comes back covered with rich furs, while you and I..get ..three-quarter-length shorn lamb. 1968 J. Ironside *Fashion Alphabet* 153 *American broadtail.* This was the trade-name for the fur of the very young Argentine lamb... The name is now prohibited in the U.S.A. and the fur is sold under its true name of 'shorn lamb'. 1978 *Lancashire Life* Oct. 101/1 (*caption*) The 'Rambler' jacket in shorn lamb with leather piping all round and on the pockets, plus leather belt, is about £130.

short, *a.*, *sb.* and *adv.* Add: **A.** *adj.* **I. 1. b.** *to get by the short hairs*: see *Hair *sb.* 8 p. Also, in same sense, *to get* or *have* (a person) *by the short and curlies. slang* (orig. *Mil.*).

1948 Partridge *Dict. Forces' Slang* 168 *Short and curlies*, the short hairs, in the phrase 'He got me by the short and curlies'—he caught me out properly. 1956 P. Scott *Male Child* III. i. 191 'Is there any point in going on, if you can't fool yourself?'..'Of course not. That's where we've got by the short and curlies. We cling on.' 1969 J. Gardner *Founder Member* iv. 72 'Stalemate?'..'Looks like it... Got us hard by the short and curlies. I wouldn't try arguing.' 1971 D. Francis *Bonecrack* xii. 153 Suppose..that I abducted Alessandro. ..I would then have Enso by the short and curlies. 1976 P. Hill *Hunters* xi. 164 There is no need for kid gloves now, we've got him by the short and curlies.

II. 5. a. *in short order*: see *Order *sb.* 27 d.

8. a. Colloq. phr. *the short answer to* (something) *is* (and variants), used to introduce a straightforward, immediate, or peremptory response or solution. Also *transf.* and *fig.*

1955 *New Statesman* 24 Dec. 851/1, I suppose the short answer is money. 1962 *Times* 24 Nov. 4/6 The short answer is that they seem to indicate a far less cavalier attitude to viewers' wishes on the part of the television companies than has previously been manifest. 1966 W. Cooper *Memoirs of New Man* I. iv. 47 'The short answer to that, my dear, is No.' 'But what about the *long* answer?' 1968 *Guardian* 10 June 7/6 There's no short answer. We have tried..to persuade our dealers that giving service increases sales. 1978 A. Price *'44 Vintage* vi. 69 We don't get captured, Jack—that's the short answer. 1980 *N.Y. Times Book Rev.* 23 Mar. 11/3 The short answer is: yes, it is ill will.

b. *short story* (earlier and later examples). Hence *short-storyist.* Also *short short story*, a very short story; also *ellipt.* as *short-short.*

1877 *Independent* 17 May 9/2 His various books have been eminently readable, in the highest sense of the adjective, and some of his short stories have been almost without a flaw in their glittering beauty. *a* 1882 Trollope *Autobiogr.* (1883) I. viii. 182, I had..written from time to time certain short stories, which had been published in different periodicals. 1902 H. Belloc *Path to Rome* 140 Terror..is half the plot of their insane 'short stories'. 1923 J. M. Murry *Pencillings* 82 Mr H. G. Wells's definition of the short story as a fiction that can be read in a quarter of an hour. 1929 *Science Wonder Stories* Nov. 485/1 A few years ago, a short story was anywhere from ten thousand to twenty thousand words. Of late the short, *short* story has gained ascendency in a number of magazines. A short, *short* story is one that runs to not more than fifteen hundred words. 1936 E. Bowen *Faber Bk. Mod. Short Stories* 17 H. E. Bates has, as a shortstoryist, already a substantial body of work to his name. 1940 G. V. Martin *For our Vines have Tender Grapes* iv. 32 Unemployed unemployables..typing endlessly the Great American Saga..cannot sell a short-short to the *Chicago Daily News.* 1957 R. Hoggart *Uses of Literacy* vi. 166 The magazines go beyond the stories to the 'short short stories' or the 'one-minute stories'. 1962 E. Lacy *Freeloaders* vi. 106 An airmail letter from my agent telling me he'd sold a short-short of mine. 1972 J. Symons *Bloody Murder* xiii. 164 The 'short short story' of 2,000 words or less. 1977 V. S. Pritchett *Gentle Barbarian* vi. 90 From a short-story writer's point of view, the timing..is perfect... Turgenev is a master of his craft.

c. *short and sweet* (further examples).

1866 *Harper's Mag.* Oct. 674/2 The letter-book was consulted, and there stood, short and sweet, and right to the point: 'Dear Sir' [etc.]. 1882 H. Munby *Let.* 9 June in D. Hudson *Munby* (1972) 408, I don't like burning your letters & I don't like to *keep* them either—short & sweet is what I like from you. 1970 W. Smith *Gold Mine* xxxv. 92 Reasons first. I'll make it short and sweet, right?

13. (Further examples of *short vowel.*)

1952 [see *checked *ppl. a.*[1] 1 b]. 1962 [see *duration 1 c].

14. a. *short drink*: a small measure of liquor; a drink which is relatively strong in alcohol and hence drunk in small measures; a dram of spirits or the like.

1883 *Daily Tel.* 2 July 5/3 All these are short drinks—that is to say, drams. 1937 A. J. Cronin *Citadel* II. vii. 155 Challis..was successfully and cheerfully despatching his third short drink. 1957 M. Spark *Comforters* iv. 85 Caroline and Laurence had been on short drinks, and both were rather lit up. 1973 J. Aiken in V. Whitaker *Winter's Crimes 5* 22 A large Whisky Mac—his favourite short drink.

III. 15. a. Also *short change* (Change *sb.* 7 b); *short commons*: also *fig.* Phr. *in short supply*: inadequate to demand.

1852 Disraeli in *Hansard Commons* 9 Feb. 303, I cannot, however, help congratulating Parliamentary reformers on the content with which they have accepted the repast provided for them; the voracity of their appetites seems to me satisfied with very short commons. 1874 Short change [see Fluff *sb.*[1] 3 a]. 1908 U. B. Sinclair *Metropolis* 351 Three times in a single day in another of these great caravanseries, Montague was offered short change. 1928 Foy & Harlow *Clowning through Life* 81 Our Peanut and juice vendors were all short change artists. 1942 *Times Rev. 1941* 3 Jan. p. v/4 There has been an exemplary pooling of machine tools and of other requisites in short supply. 1943 *Sun* (Baltimore) 7 Jan. 24/2 The State's lawmakers are going to be on 'short commons' and the employes are going to be comparatively flush. 1951 D. Howarth *Shetland Bus* xi. 150 Larsen..now found himself in urgent need of bootlaces, so he tried to take the opportunity to buy some; but they were 'in short supply', and he had to make do with string. 1970 *Observer* 13 Sept. 38/5 It's shortcommons for tourists unless they're white.

b. *short ton*: see Ton[1] 4.

18. f. (Earlier and later examples.) Also followed by a sb. or an expression of quantity. *a shingle short*: see Shingle *sb.*[1] 1 b.

1873 Trollope *Phineas Redux* in *Graphic* 22 Nov. 486/2 He did take the key with him... We were a key short at the time he was away. 1923 Kipling *Irish Guards in Great War* I. 1 They were short one officer. 1944 M. Paneth *Branch Street* 97 Our last warden left us... Now we were two people short. 1976 J. Snow *Cricket Rebel* 43 We were more than 150 short when Derek joined me at the wicket. 1977 *Cork Examiner* 6 June 10/1 Cork.. were short three of their regulars.

g. *U.S. Stock Exch.* Having sold as yet unacquired stock which the seller hopes can be bought at a lower price before the time fixed for delivery. Also *short of* (such stock). Cf. Short *adv.* 11.

1849 *Merchants' Mag.* (U.S.) XXI. 118 If he does not own the stock he is 'short', or what is the same thing, a 'bear'. 1865 *Harper's Mag.* Apr. 616/2 If he has sold 500 Hudson for future delivery, expecting it to fall, he is pronounced 'short of Hudson'. 1884 A. Daly *Big Bonanza* 20 The market opened lively with a demand for speculative shares by those who have been 'short' of the leading stocks.

h. *short on*: having an insufficient quantity of, deficient in respect of.

1922 P. A. Rollins *Cowboy* iii. 54 The actual 'bad man' was 'short on conversation'. 1942 E. Waugh *Put out More Flags* ii. 113 It's just this kind of influence these children need... They're rather short on culture at the moment. 1959 *Omaha World-Herald* 20 Dec. D 18/3 Many a gypsy, short on funds and long on nerve, is a master at avoiding inspections. 1977 *Lancet* 13 Aug. 357/1 Dr Bartsch's comment, though interesting, is short on facts.

i. *U.S.* Of a race-horse, not in top form. Also in *attrib.* use.

1942 *Sun* (Baltimore) 1 May 17/1 Ben would prefer to pass the Derby altogether and point his guns at the Preakners' $50,000 the following week, rather than take a chance on setting a 'short' horse back farther in the Derby. 1960 *Washington Post* 23 May A 22 He said..that Venetian Way was 'short', meaning the horse wasn't quite ready for the 1¼ mile race. 1977 *Time* 20 June 51/2 Horsemen were quick to point out that he was slightly 'short'—not in peak form—for the Kentucky Derby.

V. 23. *short-date, -form, -grain* (see also sense 26 below), *-haul* (*Haul *sb.* 1 c), *-leaf* (earlier and later examples), *-life, -line, -persistence, -stroke* (later examples), *-take-off, -vowel.*

1909 *Q. Rev.* Oct. 358 British railways offer facilities for cheap, short-date bookings. 1947 *Sun* (Baltimore) 15 May 2/8 The bill carries a new set of tables for the so-called short-form taxpayers—those with adjusted gross incomes of $5,000 or less. 1961 R. B. Long *Sentence & its Parts* x. 234 The Appalachians, the Rockies..are short-form phrasal proper names. 1972 *Accountant* 19 Oct. 487/2 The position [should] be clarified by an appropriate reference in the short-form report. 1947 L. P. De Gouy *Gold Cookery Bk.* xi. 766 There are many varieties of rice produced in the United States. They are ..of three general types, long grain, medium grain, and short grain. 1970 Simon & Howe *Dict. Gastronomy* 326/2 Rice generally can be divided into long, medium and short grain types. 1895 *Funk's Stand. Dict.* II. 1658/1 The interstate commerce law forbids a greater charge in the aggregate for a shorter than for a longer distance over the same line..and is commonly called the long and short haul clause. 1939 *Jrnl. R. Aeronaut. Soc.* XLIII. 900 This figure represents about 3 lbs. per rated h.p. of the engines, which is appreciably better than the figure for most modern short-haul air liners. 1960 *Times* 14 Nov. 13/6 The most popular aircraft in use for short-haul work is the Bristol Freighter. 1976 P. R. White *Planning for Public Transport* ix. 190 Increased fuel costs affect short-haul operations more radically than long-haul. 1979 *Arizona Daily Star* 5 Aug. 1/1 It's mainly on short-haul trains. 1796 B. Hawkins *Let.* 4 Dec. in *Coll. Georgia Hist. Soc.* (1916) IX. 24, I..came..to oak and short leaf pine. 1969 T. H. Everett *Living Trees of World* iv. 51/1 The shortleaf pine..is found from New York to Florida and Texas and has dark bluish-green foliage. 1966 M. Woodhouse *Tree Frog* xix. 144 A self-contained short-life powerplant. 1973 *Guardian* 16 Feb. 6/1 Foods classified by the Ministry as 'short-life'—that is with a storage capacity after packing of less than three months—would eventually all be stamped with a 'sell by' date. 1977 *Spare Rib* July 40/3 (Advt.), Willing to expand the project into emergency and shortlife housing and building coops. 1941 L. MacNeice *Poetry of W. B. Yeats* vi. 114 Yeats.. is outstanding among modern poets for his mastery of the short-line poem with three or four stresses to a line. 1974 *Aiken* (S. Carolina) *Standard* 24 Apr. 1-B/1 (*caption*) Train enthusiast Larry Raid, Denmark, Iowa, stands on the tracks of a shortline railroad that he has rebuilt from Keithburg, Ill., to Oakville, Iowa. 1965 *Math. in Biol. & Med.* (Med. Res. Council) iv. 192 The 1·5-mil-spot cathode ray tube..produces the short-persistence flying spot, with a sweep diameter of 10 cm. 1973 *Gloss. Electrotechnical, Power Terms* (*B.S.I.*) I. vi. 16 *Short-persistence screen*, of a cathode ray tube. A screen whose luminance decays rapidly after the stimulus has been reduced or removed. 1921 A. W. Judge *Automobile & Aircraft Engines* iv. 167 Tests..show that at low piston speeds the short stroke engine has a somewhat better thermal efficiency. 1977 D. Bastow *W. O. Bentley—Engineer* xx. 337 The short stroke engine would be at least no worse off in terms of specific weight per horsepower. 1959 *Times Lit. Suppl.* 27 Feb. 115/2 The Army pressed for ample supplies of vertical or short take-off aircraft to move troops on the battlefield. 1973 *Lebende Sprachen* XVIII. 69/1 Britain's aircraft manufacturers also have in mind a new medium size short take-off aircraft to replace the Hawker Siddeley 748. 1965 *Language* XLI. 25 We can posit for Proto-Germanic a short-vowel system consisting of four phonemes.

24. *short-barrelled, -frocked, -handled, -leaved* (earlier example), *-necked* (later examples), *-nosed* (later examples), *-vowelled.*

1978 R. Ludlum *Holcroft Covenant* xxxi. 365 He withdrew a small, short-barreled revolver and handed it to Tennyson. 1915 G. Frankau *Tid'apa* ii. 17 Perched sideways, short-frocked, on the mattress, he thought her a child in the gloom. 1933 Dylan Thomas *Let.* Nov. (1966) 52 My only sister passed through the stages of..short-frocked flappery and social snobbery into a comfortable married life. 1851 H. Melville *Moby Dick* II. xxxvi. 241 A short-handled sharp spade being sent up to him, he diligently searches for the proper place to begin breaking into the Tun. 1964 W. L. Goodman *Hist. Woodworking Tools* 29 The T- axe was gradually superseded from the middle of the 14th century onwards by the short-handled version of type 3. 1748 M. Catesby *Nat. Hist. Carolina* App. p. xxii, The short-leav'd Pine is usually a small tree. 1955 *Archaeol. News Let.* VI. 15 Professor Piggott proposed the replacement of the alphabetical classification by descriptive terms... Classes A and C should be known as 'necked beakers'... Abercromby..suggested that Class C—short necked beakers—was derived from Class A. 1970 Bray & Trump *Dict. Archaeol.* 36/2 The international bell-beakers are uncommon in Britain, where they are replaced by local variants, the long-necked (formerly A) beakers of eastern England and the short-necked (formerly C) beakers of Scotland. 1910 *Blackw. Mag.* Feb. 287/1 The Jumna, like all Indian rivers, is full of crocodiles both of the short- and long-nosed descriptions. 1948 C. L. B. Hubbard *Dogs in Britain* 3 Such breeds as the Pug and short-nosed dogs. 1935 G. O. Curme *Gram. Eng. Lang.* II. xii. 307 In early Modern English there was alongside of the long-voweled *bete* or *beat* the short-voweled *bett.*

26. **short-arc** *a.* (see quot. 1972); **short-arm** *a.*, (*a*) designating a punch thrown with the arm not fully extended; also *ellipt.* as *sb.*; (*b*) *slang* (orig. and chiefly *Mil.*), designating an inspection of the penis for venereal disease or other infection; also *ellipt.* as *sb.*; **short-arse, -ass** *slang*, a person of small stature; a person of little account; hence **short-arsed, -assed** *a.*, of small stature; **short back and sides**, a haircut in which the hair is cut short at the back and sides of the head; also *attrib.*; **short ballot** *U.S.*, a ballot in which only the more important offices are held up for election, the minor offices being filled by appointment; *spec.* (see quot. 1940); also, a form for such a ballot; also *attrib.*; **short cards** *U.S.*, one of various card games played for money (see quots.); also *attrib.* in *sing.*; **short chain** *Chem.* [*Chain *sb.* 5 g], a relatively small number of atoms (usu. of carbon) linked together in a line; usu. *attrib.* (with hyphen); **short clothes**, an infant's short-coats (see Short-coat *sb.* 2); also *fig.*; **short con** *U.S. slang*, a small-scale confidence racket; also (with hyphen) *attrib.*; **short corner** *Hockey*, a penalty hit taken from a spot on the goal-line up to within ten yards of the goal-posts, a penalty corner; **short-cycle(d** *adjs. Bot.*, (of a rust fungus) not having a complete life cycle; **short-day** *a.*, (of a plant) not flowering until the period of light each day falls below some limit; **short-eat** *Sri Lanka*, a snack; **short end**, (*c*) a remnant of cloth; (*d*) *U.S. slang*, the inferior part or share (*of* something), the losing end, a bad deal; (*e*) *Comm.* that part of a stock market which deals in short-term stocks; **short focus**, a focal point that is near to the lens; chiefly *attrib.*, esp. in *short-focus lens*, spec., a photographic lens whose focal length is less than the length of the diagonal of the negative or plate with which it is used; **short-frock** (earlier example); **short fuse** *U.S. slang*, a quick temper; hence **short-fused** *a.*; **short game** *Golf*, the style of golf played at the approach to and on the green; **short grain**, a condition of the fibres which gives rise to brittleness in wood (cf. Short-grained *a.*); **short-grass**, (*b*) used, usu. *attrib.*, to designate the vegetation of certain prairies; **short-hairs** (earlier example); also *attrib.* in *sing.*; **short head** *Racing*, a distance less than the length of a horse's head; a horse that has lost by a short head; also *attrib.* and *fig.*; hence **short-head** *v. trans.*, to defeat by a short head; also *transf.* and *fig.*; **short horse** *U.S.*, (*a*) = *quarter-horse; also *attrib.*; (*b*) (see sense 18 i above); **short octave** (further examples); also in keyboard instruments other than the organ; **short order** *U.S.*, an order for food to be prepared and served up quickly; a dish so served; also *attrib.*; **short-period** *a.*, extending over or lasting for a brief period of time; recurring at short intervals; **short-punt** *v. intr. Rugby Football*, to punt the ball a short distance; **short sauce**: see Sauce *sb.* 4 a; **short score** *Mus.* (see quots. 1876, 1954); **short-sea(s)** *a.*, of or pertaining to short sea crossings; **short-service** (earlier example); **short shorts** *U.S.*, very short drawers or trousers; briefs; † **short-six**, (*a*) = Six *sb.* 3 h; (*b*) *U.S.*, a type of cigar (cf. *long nine); **short sleeve**, a sleeve which does not reach below the elbow; also *attrib.*; hence **short-sleeved** *a.*, having short sleeves; **short-snorter** *U.S. Mil. slang*, (see quot. 1954); also, a person who collects a short-snorter; also *attrib.*; **short spoon** *Golf*, a short wooden club (see Spoon *sb.* 4 c); **short-staffed** *a.*, not adequately provided with staff, understaffed; **short-stage** *a.*, with short distances between stopping places; also *ellipt.* as *sb.*, a coach travelling in this way (*obs.* exc. *Hist.*); **short staple** *a.* (earlier example); **Short Street**, an imaginary street where people in financial difficulty are supposed to reside; **short suit** *Cards*, a suit of which a player has few cards; also *attrib.*; hence **short-suited** *a.*, having a short suit; also *fig.*; **short sweetening** *U.S. dial.*, (*a*) cane sugar (as opposed to molasses); (*b*) maple sugar (as opposed to cane sugar); **short-title**, also, an abbreviated form of a full title of any work; also *attrib.*; **short-weight** *v. trans.*, to give short weight to (see sense 15 a in Dict.) (*U.S.*); **short whist**: see Whist *sb.*[3] a.

1955 *Sci. News Let.* 27 Aug. 136/2 Use of xenon and platinum eliminates the usual warm-up period required before today's signaling searchlights can be used. The bulb, called a short-arc mercury-xenon lamp, was developed by Westinghouse Electric Corporation. 1972 *Gloss. Electrotechnical, Power Terms* (*B.S.I.*) IV. iii. 14 *Short-arc lamp*, discharge lamp in which the distance between the electrodes is small (of the order 1 mm to 10 mm). 1906 'H. McHugh' *Skiddoo!* 10 To the Bury Little Bunch of Newspaper knockers who have so assiduously plied hammer and harpoon since this series began, I want to say that 575,000 John Henry books were sold up to March 1st, 1906. There is your answer, O Beloved of the Short Arm Jab! 1911 J. Masefield *Everlasting Mercy* 11 Billy bats Some stinging short-arms in my slats. 1919 in *Wine, Women & War* (1926) 307 Short arm inspection between vomits. Doctor sicker than patients. 1953 *Sun* (Baltimore) 5 Jan. (B ed.) 12/5 At the Gay street station you are taken to the second floor for a brief physical check, the army 'short-arm', given in this case to see if anything has turned up since your preinduction physical. 1975 C. Allen *Plain Tales from Raj* xv. 159 Periodical medical checks, known as 'short arm inspections', ensured that any man who availed himself of the 'tree rats' or 'grass *bidis*' was properly dealt with. 1978 M. Puzo *Fools Die* xl. 450 Before you go to bed with a guy, give him a short arm... You strip down his penis, you know, like you're masturbating him, and if there's a yellow fluid coming out like a drippage, you know he's infected. 1706 Short-arse [see Spud *sb.* 4]. 1949 D. M. Davin *Roads from Home* 212 That little shortarse tried to report him. 1973 M. Amis *Rachel Papers* 54 'What's her real name?' I implored. 'Jean.' 'Oh. The short-arse? Yeah, she's all right. Boring dress.' 1951 Partridge *Dict. Slang* (ed. 4) 1168/2 *Short-arsed*, (of a person) that is short. 1973 M. Amis *Rachel Papers* 114 He was a short-arsed little bastard—about five-five. 1962 H. Hood in R. Weaver *Canad. Short Stories* (1968) 2nd Ser. 203 'Now this Pearson,' said one of the revellers, 'he's just a little short-ass. He's just a little fellow without any brains.' 1962 *Canadian Jrnl. Linguistics* Autumn 49 Short-ass(ed). 1965 M. Allingham *Mind Readers* xxii. 241 A bony young man who..wore 'short back and sides'. 1972 N. Bentley *Events of that Week* 72 The gents, in tweed caps or with their silvery short-backs-and-sides exposed to the unrelenting sun. 1974 'G. Black' *Golden Cockatrice* v. 81 Mr Long favoured short back and sides hair-cuts for his personnel. 1982 *Observer* 16 May 4/3 Bejeaned teenagers seemed to outnumber the elderly short-back-and-sides 'sweats'. 1909 R. S. Childs in *Outlook* 17 July 638/2 On such a short ballot basis the entry of our best men into public life becomes possible. 1914 *Cycl. Amer. Government* I. 104/2 A short ballot is any voting paper which requires the selection of only a few important candidates. 1940 *Amer. Pol. Sci. Rev.* Oct. 955 The term 'presidential short ballot' is applied to the ballot form in which the names of candidates for presidential electors are omitted, and only the names of the candidates for president and vice-president appear. 1952 R. Rienow *Introd. to Government* IV. xix. 362 A suggested reform would limit the offices upon which people vote to those which..have a broad policy-making function. The plan is called the short ballot. 1968 *Economist* 27 July 33/3 As for the electors themselves, they could scarcely be more anonymous. In fact, thirty-five states have abandoned any attempt to tell the voters who they are. Instead these states only use what is called 'the short ballot'. 1845 J. J. Hooper *Some Adventures Simon Suggs* 134 Thar never were a *peaceabler* or more *gentlemanlier* game o' short cards played. 1876 *Scribner's Monthly* May 45 It is worthy of a short-card sharp and a keno flopper. 1935 A. J. Pollock *Underworld Speaks* 105/2 *Short card player*, a gambler who plays all card games well except draw poker. 1938 H. Asbury *Sucker's Progress* 286 Short card games predominated, the favorites being Brag, Poker, Seven-Up and Whist. 1942 Berrey & Van den Bark *Amer. Thes. Slang* § 743/1 *Short-card player*, a poker player, esp. a cardsharp. 1940 *Jrnl. Dairy Sci.* XXIII. 1054 The short chain fatty acids are by products of this synthesis. 1961 *Lancet* 12 Aug. 343/1 There was a high content of short-chain acids and unidentified long-chain..acids in the cholesterol esters. 1972 *Jrnl. Chromatogr.* LXXIV. 335 Measurement of short-chain fatty acids in various biological materials is becoming increasingly important in the physiology and taxonomy of microorganisms and in the dairy, food, and beverage industries. 1816 E. Weeton *Jrnl. of Governess* (1969) II. 140, I have been..making short-clothes for Mary... I think of having her weaned in a few weeks. 1843 C. Ridley *Let.* Nov. in *Cecilia* (1958) xii. 138, I believe it is much better for children to be put early into short clothes if they are well. 1921 J. Buchan *Path of King* xii. 246 He held that the country had grown up and couldn't be kept much longer in short clothes. 1932 *Detective Fiction Weekly* 6 Feb. 126/1 Little tricks known as the 'short con'. 1948 Mencken *Amer. Lang.* Suppl. II. 667 Short-con workers operate on a modest scale, and are usually content with whatever money the victim has on him at the time he is rooked. 1965 H. Gold *Man who was not with It* viii. 67 This was better than..any of the other short-con moments of which Grack had told. 1967 J. Potter *Foul Play* i. 10 His team had failed to score from the resulting short corner. 1976 *Southern Even. Echo* (Southampton) 2 Nov. 22/5 Yateley..reduced the arrears from a short corner. 1915 H. C. Travelbee in *Proc. Indiana Acad. Sci. 1914* 231 We note the teliospores of a short-cycled rust appearing on the æcial host of a long-cycled heterœcious rust. 1926 Short-cycle [see *macrocyclic *a.* 1]. 1950 E. A. Bessey *Morphol. & Taxon. Fungi* xii. 396 Another anomaly in the life cycle of a short-cycle rust. *Ibid.* 397 In most of the short-cycled rusts studied..it has been shown that the mycelium is of monocaryon type until the telium or aecium is formed, when dicaryon cells appear. 1920 Garner & Allard in *Jrnl. Agric. Res.* XVIII. 559 It will be convenient to use the expressions 'long day' as meaning exposure to light for more than 12 hours and 'short day' as referring to an exposure of 12 hours or less. *Ibid.* 576 It [*sc. Aster linearifolius*] is a typical short-day' flowering perennial.

1947 *Sci. News* IV. 129 By and large, short day plants flower if they receive 8-9 hours of light a day, and long day plants flower if they receive 14-16 hours of light a day. 1980 *Sci. Amer.* May 105 (*caption*) The cocklebier is a short-day plant and will flower only if it receives at least 8·5 hours of continuous total darkness each day. 1962 *Housewife* (Ceylon) Feb. 25 (Advt.), Order your:—Short Eats..Cakes & Pastries From Grosvenor Caterers. 1971 *Times Weekender* (Ceylon) 3 Oct. 4/7 She wanted to go to a creamery and after looking at the short-eats on display, ordered a special bun. 1860 Short end [see *balk *sb.*[1]]. 1904 Ade *True Bills* 14 Each Partner naturally believed that he was getting the Short End of the Arrangement. 1942 Berrey & Van den Bark *Amer. Thes. Slang* § 371/2 Lose,..get the short end. 1960 *Lebende Sprachen* V. 35/1 Left-overs, remnants, *short ends.* 1964 *Financial Times* 10 Feb. 9/1 A fair business was done in temporary funds in the Local Authorities loans market last week. Rates tended to move erratically at the short end. 1976 *Scottish Daily Express* 24 Dec. 12/1 At the short end of the market prices were clipped by £1/8. 1977 *Time* 8 Aug. 28/2 *Annie* went back to Broadway on the short end of a 6–2 score. 1979 E. Newman *Sunday Punch* xxiii. 205 You're getting the very short end of the purse. 1845 *Encycl. Metrop.* IV. 408 Another.. method is to substitute for the sun its image formed in the focus of a convex lens of short focus. 1862 *Illustr. Catal. Internat. Exhib., Industr. Dept., Brit. Div.* II. No. 3154 The same camera can be used for either short-focus portrait or long-focus landscape lenses. 1882 *Encycl. Brit.* XIV. 580/2 The sun's image formed by a lens or burning glass of short focus is our best mode of attempting to realize the conception of a luminous point. 1935 *Discovery* Jan. 25/1 The picture [*sc.* the oldest existing photograph] was taken on sensitised paper, probably with a small short-focus camera having a large-aperture lens. 1973 *Focal Dict. Photogr. Technol.* 559 Short focus lenses of normal angle are used in macrophotography at natural size or larger scales of reproduction. 1885 Kipling in *Pioneer* 27 Jan. 5/2 Clad in short frocks in the West, Are you growing the charms that shall capture and ravish the heart from my breast? 1968 *N.Y. Times* 13 Oct. IV. 10 Tully, a fellow notorious around Sausalito for his short fuse. 1980 G. Thompson *Murder Mystery* xix. 149 Postel's first-rate but he's got a short fuse. You lie to him..and he'll walk off your case. 1979 *Observer* 16 Dec. 9/1 He's quite short-fused, but he knows how to control his temper. 1858 *Chambers's Jrnl.* 4 Sept. 157/2 The 'short game'—coming into play when the ball lies from a hundred to one hundred and fifty yards from the hole. 1903 H. G. Hutchinson et al. *Bk. Golf* IV. ii. 238 Treat the combination of mashies (sometimes irons) and putting together, calling it the short game. 1976 *Webster's Sports Dict.* 386/2 *Short game*.., the aspect of play in which control of relatively short shots (as approach shots or putts) is of primary importance. 1947 J. C. S. Brough *Timbers for Woodwork* iv. 30 Short grain means that the fibres lie in such a direction that the timber may snap or fracture with practically no splintering. 1956 F. W. Jane *Struct. Wood* xi. 254 Where this type of grain occurs it is clearly impossible to produce lumber which is straight grained—the wood must, inevitably, have short grain and suffer from the defects associated with such a structural peculiarity. 1916 *Jrnl. Ecol.* IV. 49 This *Bouteloua* mixed consocies, representing the most radical departure from the typical short-grass, really differs mainly in the possession of a derived element. 1929 Weaver & Clements *Plant Ecol.* xvii. 401 The short-grass plains extend over areas in western Nebraska and include much of the western half of Kansas, eastern Colorado, [etc.]. 1961 *Listener* 7 Sept. 346/2 The short-grass plains of the Serengeti. 1867 *Ball Players' Chron.* 4 July 2/1 Being assisted by their brutish followers of the short-hair grade, they generally manage to make large hauls of plunder. 1875 *Nation* 1 Apr. 218 A very real division of the Democratic party in this city into two sets of politicians known familiarly as 'Short Hairs' and 'Swallow Tails'—the former comprising the rank and file of voters, and the latter 'the property owners and substantial men'. 1883 J. Greenwood *Odd People in Odd Places* 107 Fancy him having that horribly anathematized 'short head' all his own and in hand to do as he likes with—to revile it, and punch it. 1898 A. E. T. Watson *Turf* 133 There are legends of judges having made mistakes in short head verdicts. 1922 *Weekly Dispatch* 12 Nov. 7 Danny caught him napping and shortheaded him on the post. 1932 A. J. Worrall *Eng. Idioms* VII. 55 Teckla took the lead on the straight and won by a short-head from Bomba II. 1935 N. Mitchison *We have been Warned* I. 55 She successfully short-headed a wool merchant for the bath. 1963 J. Prescot *Case for Hearing* iii. 44 The favourite was beaten on the post... I was short-headed out of a fiver. 1976 Ld. Home *Way Wind Blows* xiii. 186 Had these two pulled their weight, I have no doubt at all that our short-head defeat would have been converted into a narrow victory, and a win at that time for the Conservative Party could well have smashed the Socialists. 1977 *Irish Times* 8 June 2/1 Frozen Tiger beat the flying outsider, Poppy Fields, by a diminishing short head. 1922 *Breeders' Gaz.* 24 Aug. 212/4 There is little doubt he came from Southern Illinois, the home of many well-known 'short-horses'. 1971 *Amer. N. & Q.* Apr. 127/2 They..have had many and strong infusions of thoroughbred blood through the years, but not such that the prized short horse characteristics were lost. 1880 Grove *Dict. Mus.* II. 588/1 In the short octave two of the natural keys were omitted, and the succession stood thus:— CC (on the EE key), FF, G, A, B, C. 1961 A. Baines *Musical Instruments through Ages* iv. 77 The earlier instruments [of the harpsichord family] very generally had a 'short' or 'broken' octave in the bass. 1980 *Early Music* Apr. 215/1 In my own field of keyboard instruments, the important questions used to be how the jack mechanism worked and how the short octave was tuned. 1906 'O. Henry' *Four Million* 103 The clatter of steel, the screaming of 'short orders', the cries of the hungering and all the horrid tumult of feeding men. 1927 *Amer. Speech* II. 414/1 The nomenclature of the short-order restaurant. 1928 S. Lewis *Man who knew Coolidge* I. 31 We plan to have a restaurant there serving short-orders twenty-four hours a day. 1956 J. Potts *Death of Stray Cat* vii. 75 Working..as a short-order cook in his diner. 1978 J. Updike *Coup* (1979) iv. 137, I..worked as waiter and short-order cook in various eating establishments. 1887 *Encycl. Brit.* XXII. 509/1 Short-period fluctuations between a maximum and minimum, within the limits of each single stroke [of a piston]. 1895 *Knowledge* 1 May 111/2 There is no longer any reason to doubt that all 'short-period variables' are really close binaries. 1900 *Ibid.* 1 Dec. 285/2 Brorsen's Comet... This interesting short-period comet. 1923 P. B. Ballard *New Examiner* 107 For..an overwhelming majority, short-period testing, when properly carried out, is as sound and as valid as long-period testing. 1962 L. S. Sasieni *Optical Dispensing* viii. 195 A second pair of lenses for special purposes, or for short-period use. 1967 *Oceanogr. & Marine Biol.* V. 128 Such a rise is a normal aspect of the short-period low-amplitude, climatic oscillations of the Holocene epoch. 1937 C. Day Lewis *Starting Point* I. iii. 48 The next time he received the ball, he short-punted ahead. 1954 J. B. G. Thomas *On Tour* 184 Morgan was half through before he short-punted. 1876 Stainer & Barrett *Dict. Mus. Terms* 388/2 A *short* or compressed score is when all the parts are arranged or transcribed so that they shall appear in two staves... In transcribing four-part music into short score, the two upper parts are arranged in the treble stave. 1946 A. Hutchings in A. L. Bacharach *Brit. Music* xvi. 207, I do not think that, even now, Rubbra finds it easy to bring off the orchestration conceived in his mind's ear while writing his 'short score'. 1954 *Grove's Dict. Music* (ed. 5) VII. 765/1 *Short score*, a term meaning either (*a*) a condensation of a vocal or instrumental full score for pianoforte or organ for use at rehearsal or (*b*) a composer's first draft of a full score in which a large orchestral lay-out is reduced to a few staves. 1980 *Early Music* July 414/3 The collection is most unusual in that it..has..a fully notated 'short-score' keyboard accompaniment. 1952 J. W. Day *New Yeomen of England* iii. 40 The bargemen were a race apart, born and brought up to the short-seas trade. 1966 *Guardian* 28 Feb. 16/7 The coastal and short-sea container services. 1976 *Daily Tel.* 9 Sept. 1/2 All cross-Channel and short-sea vessels crewed by union members will be affected immediately the strike starts. 1882 E. W. Hamilton *Diary* 19 Nov. (1972) I. 361 What struck him [*sc.* Mr. Gladstone] most was the magnificent appearance of the line regiments—an unanswerable proof against the charges of short service. 1946 *Sun* (Baltimore) 24 Apr. 7/2 'What are briefs,' asked Senator Millikin... Cheney dug into his satchel, came up with a pair, and waved them at Millikin. 'Oh,' said the senator. 'Short shorts.' 1964 [see *Jamaica b]. 1976 *Billings* (Montana) *Gaz.* 2 July 2-A (Advt.), Coordinate your tops with shorts from the great selection of short shorts cuffed or uncuffed and jamaica length shorts. 1831 H. J. Finn *Amer. Comic Ann.* 219, I was drest all in white, and lookt like a short-six goin to be dipt. 1838 W. E. Burton *Burton's Comic Songster* 188 Give me some short six's. 1843 [see Six *sb.* 3]. 1865 C. F. Browne *A. Ward: his Trav.* 57 Tom Slink, who used to smoke short-sixes and get acquainted with the little circus-boys. 1885 *Pioneer* 19 Aug. 5/1, I buy me not twelve-button gloves, 'short sixes' eke, or rings. 1890 J. Jefferson *Autobiogr.* 146 The very cornerstone of Juliet's balcony contained twenty pounds of the best 'short sixes'. 1639 in *Rec. Governor & Co. Massachusetts Bay* (1853) I. 274 No garment shalbee made wth short sleeves, whereby the nakedness of the arme may be discovered. 1976 *Lady's Mag.* Dec. 564/2 Short sleeves in small plaits. 1847 Dickens *Dombey* (1848) xxxv. 354 Mrs Skewton..in a very youthful costume, with short sleeves. 1931 *Mod. Woman* Feb. 46 Length of short sleeve seam, 4½ ins. 1969 *Sears Catal.* Spring/Summer 11/1 Short sleeve tops in windowpane printed plaid. 1839 C. Brontë *Caroline Vernon* in W. Gérin *C. Brontë* (1967) viii. 133 [A] short-sleeved frock, worked trousers and streaming sash that would better have suited the age of 9 or 10 than that of 15. 1973 D. E. Westlake *Cops & Robbers* 7 With the heat the way it was, I was glad the Police Department let its people wear a short-sleeved shirt in the summer. 1944 *Sun* (Baltimore) 1 May 13/5 A 'short-snorter' made up of 18 pieces of currency. *Ibid.*, The short-snorter fellowship consists of persons who have flown over salt water. Its 'membership card' is a piece of currency signed by other short-snorters. 1954 Berrey & Van den Bark *Amer. Ther. Slang* (ed. 2) § 896/1 *Short snorter*, an autograph list on a string of attached bills, usually of the currency in the countries visited. 1976 R. M. Stern *Will* ii. 15 On the study wall in a plain wood frame were the connected, signature-scrawled dollar bills..called short-snorters, or some such silly name. 'They were the in thing,' his father had told him... 'You carried them with you..and got as many signatures as you could... If there ever was a reason, I've forgotten it.' 1858 *Chambers's Jrnl.* 4 Sept. 157/1 The names of the wooden-headed clubs..short-spoon. 1901 *Encycl. Sport* II. 459/2 The long spoon, mid spoon, short spoon and baffing spoon..are now rarely seen, having been supplanted by the brassy, and the modern irons and mashies. 1970 F. C. Avis *Golf Dict.* 166 Long spoon, the No. 3 wood; short spoon, the No. 4 wood. 1953 K. Tennant *Joyful Condemned* xxxiv. 334, I should get my patients to bed. They're very short-staffed. 1968 'M. Carroll' *Dead Trouble* ii. 27 They've got a room but they're short-staffed. I said we'd take our cases up. 1977 J. Sherwood *Honesty will get you Nowhere* i. 16 Matron.. was short-staffed, at her wits' end to keep the place going. 1837 Dickens *Pickwick Papers* xxxii. 339 Numerous cads and drivers of short stages. 1903 W. Gilbey *Early Carriages & Roads* 56 It seems..certain that the year 1662 saw a great increase in the number of 'short stages'—that is to say, coaches running between London and towns twenty, thirty, forty miles distant. 1963 *Times* 24 May (London Underground Suppl.) p. vi/4 Local transport in and around the Metropolis was by short-stage carriages, which ran from point to point (usually an inn). 1969 J. E. Tuffs *Essex Coaching Days* vii. 63 Spreading out from London in all directions was the net-work of short-stage coach routes. 1802 J. Simons *Let.* 15 Dec. in *Steele Papers* (1924) I. 341 Short Stable [*sic*], or Green seed Cotton if the best Quality, 16 cents. 1920 in *Further Lett. from Man of No Importance* (1932) 79 Soldiering is an honourable but not a profitable profession, and landed property is apt to land the owner in 'Short Street'. 1938 *Daily Tel.* 25 July 9/1 She was finally unable to help Wilde, then penniless, for his defence: and having lived in 'Short Street' myself, I quite understood. 1876 A. Campbell-Walker *Correct Card* p. xiii, *Short suit*, one of which you hold originally not more than three cards. 1893 'L. Hoffman' tr. *Hertefeld's Game of Skat* 8 As..a suit consists of seven cards only, three or more constitute a long suit; and two or less a short suit. 1931 E. Culbertson *Contract Bridge at Glance* 43 With an Ace, a King, or honours not in sequence in one or two suits, and a *worthless* doubleton or a singleton in the third suit, lead the short suit. 1964 Frey & Truscott *Official Encycl. Bridge* 499/1 The short-suit lead is also indicated when there is a bidding inference that this is partner's suit. 1935 Auden & Isherwood *Dog beneath Skin* II. ii. 82 Hullo, you short-suited? Here, let's see your hand. 1940 C. Woolrich in *Ellery Queen's Magicians of Mystery* (1976) 281 Did I say she was beautiful? Double it in spades, and you're still short-suited. 1850 *Quincy* (Illinois) *Whig* 19 Nov. 2/2 He put.. all the money she had in short sweetening, and left her without a cent. 1883 [see *long sweetening* s.v. Long *a.*[1] 18]. 1914 B. T. Washington *Selected Speeches* 218 This good lady asked whether we wanted long or short sweetening in our coffee. 1948 E. N. Dick *Dixie Frontier* 291 'Short sweetening', or maple sugar, was also obtained in its raw state from the trees. 1896 in A. H. Chester *Dict. Names Minerals* 1 (Advt.), Short-Title Catalogue of..Publications... Arranged under subjects. 1945–51 D. Wing (*title*) Short-title catalogue of books printed in England, Scotland, Ireland, Wales, and British America.., 1641–1700. 1978 *Amer. N. & Q.* XVI. 151/1 The short-title lists.. ought to be based on meticulous bibliographical descriptions. 1932 T. S. Stribling *Store* vii. 73 It is much mo' dangerous to accuse a white man of shawt-weightin' you when he ain't 'an when he is. 1952 *Sun* (Baltimore) 8 Apr. (B ed.) 30/5 Many Harford county coal dealers apparently are short-weighting their customers. 1977 *Time* 7 Nov. 72/3 A leading Catholic contractor short-weights the church.

b. **short-eared owl**, a light-coloured owl flecked with brown or black, *Asio flammeus flammeus*, distinguished by short ear-tufts and found in Europe, northern Africa, and North America.

1766 T. Pennant *Brit. Zool.* I. 71 The Short eared Owl ..is found in the mountanous [*sic*] wooded parts of our island. 1833 [in Dict., sense 24]. 1974 *Times* 20 Mar. 18/4 Twenty short-eared owls have settled on a Humber nature reserve.

d. In Cricket: **short leg** (examples); **short slip**: see *Slip *sb.*[3] 14; **short square (leg)**, a square leg standing close in to the wicket; **short stop** = *short slip* (see also in Baseball, below). In Baseball: **short field**, that part of the field in which the short stop plays; also, = *short fielder*; **short fielder** (earlier example); **short stop** (earlier and later examples); also *fig.*; see also in Cricket, above.

1856 *Spirit of Times* 6 Dec. 229/1 Adams, as short field has for many years, been deservedly distinguished. 1948 *N.Y. Times* 25 Apr. 51/6 Jack Conway was shifted to the short field. 1961 J. S. Salak *Dict. Amer. Sports* 397 *Shortfield*,..area around shortstop position, between second and third bases. 1857 *Spirit of Times* 18 July 309/3 He is a splendid short fielder. 1843 'Wykehamist' *Pract. Hints Cricket*, (*caption*) Short leg or Middle On. 1851 J. Pycroft *Cricket Field* x. 191 Short-leg is often a very hardly used personage, expected to save runs that seem easy, but are actual impossibilities. 1877, 1894 [see Leg *sb.* 6 c]. 1963 *Times* 14 Jan. 8/3 Jarman overbalanced in pushing the first ball he received to forward short leg. 1860 *Baily's Mag.* Aug. 364 Willsher, too, made a rare catch at short square leg. 1963 *Times* 1 May 4/5 (*caption*) M. J. K. Smith (M.C.C.), at short square leg, ducks as P. J. Sharpe (Yorkshire) hooks a ball from J. A. Flavell during the match at Lord's yesterday. 1977 *Sunday Times* 30 Jan. 30/4 He made it strike like a snake and Fletcher was caught at short square. *c* 1837 W. Martin *Bk. Sports* vi. 104 *Order of the Players*... 4 Long stop. 5 Short stop. 1857 *Spirit of Times* 25 July 324/3 Second Nine Fahys, pitcher;..Smith, short stop. 1860 in H. T. Peters *Currier & Ives* (1942) Pl. 162, I thought our fusion would be a 'short stop' to his career. 1877 *London Soc.* XXXI. 533/1 Dorrington was almost as good at cover as Hillyer was at short-stop. 1912 *Australasian* 6 Jan. 21/2 Hobbs..was caught at short-stop by Carter. 1950 *Nature Mag.* Mar. 131/2 A sudden lunge with the net will often cut off its escape. If the net misses, a lucky shortstop may nab the lizard in passing. 1977 *Time* 8 Aug. 28/3 Andrea played shortstop and first base. 1978 *Verbatim* Feb. 2/2 If a batter hits safely between the shortstop and second, or second and first, the announcer will call that 'a seeing-eye base hit'.

e. Designating or pertaining to transactions in which a seller sells stock or goods that he does not at the time possess (cf. sense *A. 18 g and C. 11), as *short operation, position, sale, selling, side*; **short covering**, the buying in of stock or goods to cover a short sale; **short interest** (see quot. 1900); **short market** (see quot.).

1930 *Daily Express* 22 May 2/6 In late dealings the downward trend was more pronounced, but finally sporadic short covering resulted in some improvement from the day's levels. 1937 *Sun* (Baltimore) 17 Feb. 15/5 Steels, after putting on a draggy performance during the morning, whittled down part of the early losses with the aid of what brokers described as short covering. 1973 *Times* 13 July 21/8 Prices by now were really moving. Record after record was smashed. Massive short covering helped to keep prices on the boil. 1866 *Comm. & Fin. Chron.* III. 75/2 During the week a moderate short interest has been drawn out by the dullness of the market. 1900 S. A. Nelson *ABC of Wall St.* 159 *Short interest*, that

interest in the market whichi s represented by the aggregate sales of men who have sold at a price with the expectation of buying in at a cheaper price. 1949 *Time* 30 May 73/1 By mid-May, the short interest had risen 130,058 in a month to 1,628,551 shares. 1900 S. A. Nelson *ABC of Wall St.* 159 *Short market*, an oversold market, with the aggregate contracts for the delivery of stocks exceeding the supply at a certain range of prices. 1870 J. K. Medbery *Men & Mysteries Wall St.* 175 With ..realizations upon short sales, Jerome felt rich enough to dissolve partnership. *Ibid.* 202 A sale profit either in a 'long' or 'short' operation. 1931 *Daily Express* 22 Sept. 2/1 Short selling in the present circumstances would demoralise the market. Consequently all short positions carried must be reported in detail each day. 1911 *Amer. Year Bk. 1910* 385/2 All of these bills were directed against the use of 'options', 'short sales', and transactions in 'futures'. 1930 *Daily Express* 23 May 2/6 Short-selling in Case Threshing Machine, which declined more than 18 points, was an unsettling factor. 1966 'H. MacDiarmid' *Company I've Kept* iii. 71 We find..fantastic spectacles which, like short-selling, are, as the late Otto Kahn stigmatised that operation, 'inherently repellent to a right-thinking man'. 1902 A. D. McFaul *Ike Glidden* xviii. 139 He bought and sold on the short side for cash and sold on the long side on credit.

B. II. *sb.* **4. f.** (Further examples.) Also more widely = *short drink* (see *A. 14 a).

1953 *Word for Word* (Whitbread & Co.) 32/1 *Short*, a colloquial name for a gin or whisky drink, usually taken before a meal. 1973 J. Wainwright *Touch of Malice* 89, I wouldn't have thought..you were a beer man. I'd have said shorts. 1978 R. Barnard *Unruly Son* xvi. 176 There was a man and his wife... Didn't talk much, just sat and drank shorts. 1980 G. Mitchell *Whispering Knights* i. 7 They only drink shorts... Gin, and doubles at that.

g. Delete ? *nonce-use* and add earlier and later examples. Also, a contraction of a phrase.

1873 L. Troubridge *Life amongst Troubridges* (1966) vi. 53 We..have *names* for some of our relations... Uncle Tum and Aunt Kitty are Tumbo and Kitginx. *Of course* these names we never tell *anybody*..but only use them as *shorts.* 1879 Grove *Dict. Mus.* I. 332/2 *Change*..the word used as the short for change of key or *Modulation.* 1914 H. Stretton *Alone in London* iii. 28 Dolly was the short for Dorothy, and in early times he had often called his wife by that name. 1920 *Black's Domestic Dict.* 28/2 Bouquet is the short for 'Bouquet garni'.

h. In the Morse code, a dot (opp. 'long'); a short buzz, etc., sounded as a signal.

1875, etc. [see *long *a.*[1] B. 4 b]. 1978 P. Niesewand *Underground Connection* 91 Ziad pressed the bell for flat 23, two shorts, a long and a short, and waited for the entrance buzzer to sound. 1978 J. H. Bentley in *Islands* (N.Z.) Aug. 79 'I was waiting for the proper knock,' I said. Three shorts, one long.

i. A short story or article.

1912 E. A. Parry *What Judge Saw* xiv. 245 For many years I wrote dramatic criticism and reviewed books, and wrote 'shorts' and occasionally full-dress leaders for the *Manchester Guardian.* 1937 D. L. Sayers *Busman's Honeymoon* xiv. 297 That was a special effort. Three five-thousand-word shorts at forty guineas each for the *Thrill Magazine.* 1965 *Listener* 10 June 873/1 How many 'shorts' appear in comparable circumstances in this country today? Two dozen a year? The others have to make their bows between hard covers.

j. *U.S. slang.* A street-car; a car.

1914 Jackson & Hellyer *Vocab. Criminal Slang* 76 *Short*,..a street car. Derived from the limited extent of a street car ride compared with the distances negotiable by railroad transportation. 1932 *Literary Digest* 9 Apr. 36 *Short*, an automobile, used especially in the phrase 'hot short', for a stolen car. 1961 'D. Shannon' *Ace of Spades* vi. 70 This perfectly good almost brand new Caddy I got for him, a *present*, an' he says he can't handle it..comes back with this piece of old junk, my God, pickin' up a thing like—stickin' me with a hot short to get rid of! 1975 W. McCarthy *Fourth Man* I. 26 Everybody brings him hot cars..shorts, we get up north, he fixes 'em up and then sells 'em.

k. *Mil.* A shot that falls short of its target.

1922 [see *ladder *sb.* 3 c]. 1969 I. Kemp *Brit. G.I. in Vietnam* vii. 150 We were watching the shells bursting among them when a 'short' exploded right in front of us.

l. A short film for cinema or television.

1929 *N.Y. Times* 20 Oct. IX. 8 *Shorts*, short, audible pictures. 1930 *Times* 26 Mar. 14/2 The production of several multilingual talking pictures and a series of talking comedy 'shorts'. 1935 *Life & Letters* Sept. 195 This film is actually an advertising short. 1961 *Sunday Express* 2 Apr. 19/6 A half-hour 'short'..made by a brand-new director. 1980 A. Cornelisen *Flight from Torregreca* i. 19 A young actor, who was in town making a television short.

m. *U.S.* A pair of shorts (see B. 6 d in Dict. and Suppl.).

1936 *Institute News* (Underwear Institute, N.Y.) 15 Dec. 11/2 The knitted trunk short has done very well indeed. 1956 *Amer. Speech* XXXI. 109 *A short* (a pair of drawers). 1974 *State* (Columbia, S. Carolina) 28 Mar. 2-A (Advt.), Western jean short with pockets and belt loops in navy cotton denim.

5. (Earlier example.)

1849 G. G. Foster *N.Y. in Slices* 19 Some wild-looking 'short'..rushes down and hysterically inquires of his obliging neighbour, Mr. Smith, whether he hasn't a few hundred over.

5*. *Baseball.* = *short stop* (see A. 26 d).

1856 *Spirit of Times* 4 Oct. 86/1 The Eagle Club now made a very judicious change by placing..Mr. Place as short, which effectually prevented their opponents from making any more such scores as was done in the first innings. 1897 *Outing* May 203/1 Chandler at short is being very hard pushed. 1967 C. Potok *Chosen* I. i. 34 The first one hit a single, and the second one sent a high fly to short, which Sidney Goldberg caught without having to move a step. 1976 *National Observer* (U.S.) 14 Aug. 6/5, 'I try to imagine what I'd throw to get the guy out,' says Stone. 'You know, to get him to ground out to short or something.'

6. d. Substitute for def.: Trousers reaching only to the knees or higher. In the U.S. also *spec.* underpants. (Further examples.)

1927 *Amer. Speech* II. 278/1 *Shorts*, athletic trousers. 1933 *Sun* (Baltimore) 29 Sept. 14/3 Too long has man.. allowed himself to be made miserable by a summer garb which is anything but summery. We thrill to the bold challenge issued by A. Van Dyke.., 'Shorts for men!' 1941 B. Schulberg *What makes Sammy Run?* x. 253 He was stripped down to his silk shorts. 1965 H. Gold *Man who was not with It* i. 6 He..plucked a tricksie in shorts as she wiggled by. 1973 G. Robyns *Wimbledon* xx. 137 In the thirties Wimbledon abounded in beauty... Eileen Bennett..was the girl who dared to wear shorts for the first time in public. 1974 *Caribbean Contact* Mar. 10/5 Bishop's wife came to my hotel and reported that her husband had spent the night in his 'shorts' (or underwear) ..after being stripped.

g. (*b*) (Earlier example.)

1868 *Territorial Enterprise* (Virginia, Nevada) 11 Feb. 3/2 We believe..this rise is attributable to 'cornering' of the 'shorts' below.

h. *Comm.* Short-dated securities.

1932 *Manch. Guardian* 28 Jan. 15/1 The 'shorts' are all due for repayment at par at various dates between 1933 and 1936. 1940 *Economist* 13 July 48/2 If the above sequence of interest rates reveals an artificially wide gap, it is between the yields on medium shorts and the irredeemables. 1963 H. D. Berman *Stock Exch.* (ed. 4) v. 40 U.K. Government loans with less than five years to go to the final redemption date (known as 'shorts') are always dealt in plus accrued interest. 1980 *Times* 15 Jan. 14 Gains of up to £2 were seen in long gilts and of up to ⅛ in shorts.

C. *adv.* **5. b.** *to take* (a person) *short.* (*a*) In passive freq. *spec.* to have an urgent need to urinate or defecate. *colloq.*

1890 [in Dict.]. 1928 R. Campbell *Wayzgoose* i. 30 'Tis Nature's whim that dogs, when taken short, Still to the loftiest monument resort. 1967 'J. Ashford' *Forget what you Saw* xx. 180 Simon was in such a terrible state of nerves that he had already been taken short twice and had to rush for the lavatory. 1977 *Private Eye* 11 Nov. 10/2 Taken badly short when on his way to work, and finding that both of the public lavatories in Putney were closed, Mr. Peter Herring entered a police station and asked if he could use their convenience.

11. *to sell short*: (*a*) (earlier examples); (*b*) *fig.* to undervalue; to belittle.

1852 *Hunt's Merch. Mag.* XXVI. 738 The writer of the *Aurora* phillipic complains of the practice of 'selling short'. 1861 in *Rebellion Rec.* (1862) I. III. 27 When one of the members of the Board offered to sell Government Stock 'short' on time, he was instantly hissed down. 1936 B. & S. Spewack *Boy meets Girl* I. 35 Larry: You can't act with a baby. They steal every scene—Law: Are you selling motherhood short? 1959 *Times* 1 Sept. 3/5 Brown, the man who knocked out the present British featherweight champion..has been sold short before. 1972 *Guardian* 21 June 1/2 Who, throughout the Labour Government, spent his time 'selling sterling short' in speeches both at home and abroad? 1974 J. Cleary *Peter's Pence* x. 281 *Domine, non sum dignus*... Martin would always sell himself short. 1980 *Times Lit. Suppl.* 24 Oct. 1194/5 It is poetry that is being sold short by such determined efforts to be funny and clever.

12. *short-pitched* adj.; **short-acting** *a. Pharm.*, relatively transient in effect.

1951 A. Grollman *Pharmacol. & Therapeutics* vi. 143 In insomnia, where there is difficulty in falling asleep, the short acting drugs are indicated. 1978 *Price's Textbk. Practice of Med.* (ed. 12) III. 260/1 The short-acting barbiturates are apt to cause profound depression. 1867 G. H. Selkirk *Guide to Cricket Ground* ii. 36 A ball which grounds nearer the bowler than a length ball is 'short-pitched'. 1977 *World of Cricket Monthly* June 32/2 Some pointless, short-pitched bowling by Imran and Sarfraz.

short, *v.*[2] Add: (Earlier example.) Also with *out* and *fig.* Hence **sho·rting** *vbl. sb.*[2] and *ppl. a.*

1904 *Electr. Rev.* 3 Sept. 341 Should any line become 'shorted' or 'grounded'. 1912 *Motor Manual* Advt. facing p. iii, The separators allow the plates to be placed closer together and yet make 'shorting' impossible. 1957 *Practical Wireless* XXXIII. 734/1 A 150Ω resistor with a shorting switch. 1971 B. W. Aldiss *Soldier Erect* 87 Both men were immediately sympathetic, and Di made a lot of clicking noises like a shorting Morse key. 1971 P. O'Donnell *Impossible Virgin* x. 198 I'll take the Land-Rover and short out the ignition. 1974 *Sumter* (S. Carolina) *Daily Item* 24 Apr. 2A/6 The fire was started by an electric fence shorting out. 1976 *National Observer* (U.S.) 14 Aug. 9/2 A calcium-lead battery requires an inorganic sack around the lead grids to prevent materials of erosion from shorting out the battery. 1979 *N.Y. Rev. Bks.* 8 Feb. 12/2, I cannot see for the life of me why Miss Renault, that dedicated Hellenophile, should choose to live anywhere rather than the Aegean—unless she is anxious to avoid shorting out the overloaded circuits of fantasy by the insistent presence of the real world.

shortall (ʃǭ·ɹtǭl). *U.S.* [f. Short *a.* + Over)all *sb.*] Freq. *pl.* A child's one-piece suit with short sleeves and short trouser legs.

1966 *N.Y. Times* 17 Apr. 1-97/2 (Advt.), Both dress and shortalls have the fresh-from-the-laundry look of striped cotton seersucker that survives even after a hard day's play. 1969 *Sears Catal.* Spring/Summer 23 Shortall with soil release. 1976 *Billings* (Montana) *Gaz.* 5 July 3-B (Advt.), Infants & Toddler Shortalls, Orig. 4.00.

short-cha·nge, *v.* orig. *U.S.* [f. *short change*: see *Short *a.* 15 a.] *trans.* To rob by giving insufficient change. Also *fig.*, to deprive (a person) of his due; to cheat, deceive. So **short-cha·nged** *ppl. a.*; **short-cha·nger**; **short-cha·nging** *vbl. sb.*

1903 Ade *People you Know* 30 Brad was out in the back Townships short-changing the Farmers. 1914 [see *gyp *sb.*[1] 3]. 1920 C. R. Cooper *Under Big Top* 205 The gambling and the graft of the side shows, the short-changers in the 'connection', the constant form of Temptation ever beckoning! 1928 L. North *Parasites* 304 The girl at the cash-register short-changed him. 1946 *Richmond* (Va.) *Times-Dispatch* 14 Feb. 4/1 Henry C. Clausen..told Pearl Harbor investigators tonight that the Navy in the South Pacific was 'short-changing' the Army right up to the latter months of the war on the information it received from decoding Japanese messages. 1958 *Photoplay* Oct. 33/2 As a child, I felt that life had short-changed me. 1959 J. Braine *Vodi* vii. 108 He'd never bought a drink for a barmaid in his life; he said that..the bitches always short-changed you anyway. 1962 A. Lurie *Love & Friendship* xiv. 278 Short-changing in the stores. 1964 D. Francis *Nerve* i. 9 He probably shot himself because that whey-faced bitch short-changed him in bed. 1976 'D. Halliday' *Dolly & Nanny Bird* i. 14 The cultural circuit for short-changed minorities. 1978 J. A. Michener *Chesapeake* 788 My mom and pop have worked fourteen hours a day, six days a week for more than fifty years... From birth to death they've been short-changed.

short circuit, *sb.* Add: (Earlier example.) Also *attrib.* and *fig.*

1854 *Q. Rev.* XCV. 146 If the insulator should happen to get wet, the electric fluid will sometimes..run down the post to the earth, and make a short circuit home again to its battery. 1920 *Whittaker's Electr. Engineer's Pocket-bk.* (ed. 4) 253 The short-circuit test consists in measuring the voltage and power required to send current through the windings of the machine, and results in ascertaining.. the power lost in heat during the passage of the current through them. 1937 Koestler *Spanish Testament* II. 311 True, at least once a day there is a short-circuit in my consciousness. 1970 J. Shepherd et al. *Higher Electr. Engin.* (ed. 2) xvii. 547 The circuit-breakers..must be capable of dealing with the maximum possible short-circuit current that can occur at their points of connexion. 1972 D. Bloodworth *Any Number can Play* xvi. 149 You could both only be relied upon to play it straight if you didn't know everything about each other? As soon as you get a security short-circuit and double agents giggling..someone is bound to give the game away.

short-circuit, *v.* Add: **1. a.** (Earlier example.)

1867 R. S. Culley *Handbk. Practical Telegr.* (ed. 2) viii. 166 *To short circuit a battery*, to connect the poles by a wire.

d. *intr.* Of electrical apparatus: to fail or cease working as a result of a short circuit occurring in it.

1902 *Electr. Rev.* 31 Oct. 732/1 Many a motor..condemned for short-circuiting when it is really the fault of the brakes. 1975 *New Yorker* 21 Apr. 34/2 It was a gutsy performance,..recalling Margo's near-rendezvous with Rock 'n Roll Heaven last September, when another jump-suit-cum-guitar short-circuited during an impromptu hailstorm in Louisville. 1976 *Evening Post* (Nottingham) 15 Dec. 5/3 The machine short-circuited with a bang when it was switched on.

3. *fig.* To interrupt, to cut short; to bypass by taking more direct action. Also *absol.*

1899 [in Dict., sense 1 c]. 1924 J. Buchan *Three Hostages* iii. 48 If you had happened to look at that rag you might have short-circuited your inquiry. 1938 *Ann. Reg. 1937* 159 The Council decision naturally short-circuited much of the debate which might have been expected in the Assembly. 1953 E. M. Forster *Hill of Devi* 40 Dewas and King-Emperor! In Dewas it often seemed that they might have much in common. Could one but short-circuit, all might yet be well. 1978 D. Bloodworth *Crosstalk* v. 40 It was..essential to have a secure means of short-circuiting the usual channels. 1979 D. Cupitt in M. Goulder *Incarnation & Myth* iii. 32 Here is matter for a great deal of controversy. I propose to short-circuit it by simply stipulating that [etc.].

Hence **short-ci·rcuited** *ppl. a.*; **short-ci·rcuiting** *vbl. sb.*

1896 [see *entero-enterostomy* s.v. *entero-]. 1919 H. E. Penrose *Wireless Telegr.* IV. 75 Examine the band of the magnetic detector, the magnets, and the short-circuiting contacts of the manipulating key. 1949 Koestler *Insight & Outlook* iv. 39 The concept of bisociation implies a short-circuiting of two separate mental patterns. 1951 M. McLuhan *Mech. Bride* 145/2 A kind of streamlined or short-circuited version of the usual success pattern. 1972 *Jrnl. Social Psychol.* LXXXVIII. 247 It may be postulated that the more empathic two people are with one another, the more short-circuited or 'efficient' their communication.

short crust. Sense b also **shortcrust.** [Short *a.* 20 a.] **a.** A crust of pastry made short with butter or other fat. **b.** (Also *shortcrust pastry.*) A type of short pastry.

1747 [see Short *a.* 20 a]. 1868 M. Jewry *Warne's Model Cookery* 411/1 (*heading*) To make a short crust with dripping. 1951 *Good Housek. Home Encycl.* 586/1 Lard is suitable for shortcrust and flaky pastry. 1970 Simon & Howe *Dict. Gastronomy* 290/2 Shortcrust (basically 1 lb. flour to ½ lb. fat) is crumbly when cooked.

sho·rt-cut, *v.* Also **shortcut.** [f. Short cut

sb.[1]] **1.** *trans.* **a.** To overtake by taking a short cut. **b.** To traverse by a short cut.
1915 J. London *Let.* 26 Jan. (1966) 443 If I could short-cut men to such success, I'd quit writing for a living. *a* **1951** H. G. Lamond in *Austral. Short Stories* (1951) 211 But he short-cut her on one circle. **1960** *Times* 19 Sept. 3/5 An attempt to short-cut the way to success.

2. *intr.* To take a short cut. Also const. *it.*
1925 M. Arlen *May Fair* 210 You'll go short-cutting alone... I've heard enough tales about Carmion Wood to last me a life-time. **1933** J. Steinbeck *Red Pony* in *N. Amer. Rev.* Nov. 425/1 They crossed a stubble-field to shortcut to the barn. **1960** E. Bowen *Time in Rome* v. 137 He rushed up..the Janiculum hill, short-cutting from level to level. **1970** K. Giles *Death in Church* i. 11 He knows London... He's short-cutting it towards the South Bank all right. **1977** *S. Wales Echo* 18 Jan., 'This way,' says Dai Dogs. 'We'll short cut.' The short cut is through Abertridwr's dead pit.

So **sho·rt-cutting** *ppl. a.*
1901 Kipling *Kim* xiii. 330 Though low-lying clouds might be a hindrance to a short-cutting stranger, they made no earthly difference to a thoughtful man.

shortening, *vbl. sb.* Add: **2.** Substitute for def.: A fat or oil used to make pastry, etc., short. (Earlier and later examples.)
1796 A. Simmons *Amer. Cookery* 34 *Loaf Cakes* No. 2 Rub 4 pound of sugar, 3 and a half pound of shortning, (half butter and half lard) into 9 pound of flour. **1970** Simon & Howe *Dict. Gastronomy* 347/2 *Shortening*, a culinary term used more in the United States than in Britain and it applies to fats used in making breads, cakes, pastry etc. All fats, even oils, come under this nomenclature and are used because they make mixtures 'short' or tender. **1980** *Blair & Ketchum's Country Jrnl.* Oct. 34/3, 2 tablespoons shortening.

short-fall. Add: Also **short fall, shortfall.** (Further examples.) Also, a decline; a short-coming, a fault; a deficit, a gap; a loss. Also *attrib.*
1928 *London Mercury* May 4 We notice in Mr. Churchill's Budget Speech the repeated use of the word 'shortfall'. The word is used to indicate the difference between the amount estimated and the amount received, when the difference is on the wrong side. **1941** W. S. Churchill *2nd World War* (1950) III. 748 We cannot afford losses on that scale in view of the short fall of the American bomber programme. **1953** *Economist* 3 Jan. 35/1 The Exchequer return as a whole..showed a deficit of £951 million, in contrast with one of £602 million in the first nine months of 1951/52. These comparisons, moreover, understate the extent of the shortfall. **1960** *Times Rev. Industry* Mar. 99/3 The report estimates that the changes in American shares of these markets alone cost just over $500m. in 1958. This total is called a shortfall, for want of a better name, but is the equivalent value in 1958 of the net change in the United States share of these six markets for a list of 45 classes of manufactured goods. **1960** *Catholic Herald* 1 Apr. 3/3 He has..no illusions about her [*sc.* France's] immediate shortfall from that high level of being. **1960** *Washington Post* 16 Nov. A 16 West Germany is now running a foreign payment surplus about half the size of the country's deficit, which in view of the disparity in gross product suggests a considerable shortfall by the Federal republic in the assumption of common obligations. **1966** *Economist* 28 May 970/3 Lord Moran's apology for his [*sc.* Churchill's] shortfalls in world and domestic decisions is based too often on a too simple fallacy. **1969** *Times* 14 Feb. 25/4 The shortfall between exports and imports in January was $216m. **1971** 'D. Halliday' *Dolly & Doctor Bird* ii. 23 A scene of continuous short-fall pandemonium. **1976** *Ann. Rep. Manpower Services Comm. 1975–76* iii. 25/2 To make good the shortfall in employers' recruitment into long-term training occupations ITBs made available 'training awards' where the level of recruitment overall into long-term training was likely to fall below the particular industry's long-term needs. **1981** *Times* 4 Aug. 16/5 Half year figures from Standard Telephone & Cables revealed a profits shortfall.

short-grained, *a.* Add: **2.** Of rice: having a relatively short grain.
1955 N. Heaton *Cassell's Cooking Dict.* 422 Rice may be short-grained.., or long-grained and brightly polished. ..The former is used chiefly for puddings. **1978** J. Passmore *All Asian Cookbk.* (1979) 8 Chinese like their rice reasonably dry, short-grained.

shorthand. Add: **b.** (Further examples.)
1931 E. A. Robertson *Four Frightened People* ii. 57 Arnold Ainger spoke exactly the same language as we did... It is possible to talk from the first in mental shorthand with such an acquaintance. **1960** N. Mitford *Don't tell Alfred* xiv. 158 Americans..need the dissertation; the kind of shorthand that we talk would be useless to them. **1970** *Guardian* 10 July 10/6 'Oxfash' is undergraduate shorthand for the establishment of councillors, planners, shopkeepers, and dons.

c. *shorthand notebook, pad, reporter* (earlier example), *-writer*; **shorthand typist,** one who takes down dictation in shorthand and then types out the text.
1903 G. B. Shaw *Man & Superman* i. 2 He has no secretary with a shorthand notebook. **1977** A. Morice *Scared to Death* xxiii. 154 A single page of lined paper, perforated at the top and obviously torn from a shorthand notebook. **1960** C. Morris *Unloved* 463 Janet is making notes on a shorthand pad. **1977** M. Babson *Lord Mayor of Death* iv. 36 The recording officer began noting the words in his shorthand pad. **1881** 'Mark Twain' *Lett. to Publishers* (1967) 147 A short-hand reporter to travel with us in the spring. **1901** *Phonetic Jrnl.* 24 Aug. 541 To a large extent the occupation of the shorthand-typist has hitherto been synonymous with the lady typist. **1926** I. Pitman *Dictation Practice in Bus. Corr.* Pref. p. i, In addition to business correspondence, the shorthand-typist is frequently called upon to complete forms. **1973** J. R. L. Anderson *Death on Rocks* i. 21 She had worked as a shorthand typist. **1922** Joyce *Ulysses* 140 There was not even one shorthandwriter in the hall. **1964** T. L. Kinsey *Audio-Typing & Electric Typewriters* ii. 7 Any audio-typist can transcribe audio-dictation—the dictation is never lost, as can be the case with written notes recorded in an absent shorthand-writer's notebook.

Hence **sho·rthand** *v.*, (*a*) *trans.*, to transcribe in shorthand; also *fig.*; (*b*) *intr.*, to use shorthand.
1928 *John Blunt* 11 Aug. 1/3 (*heading*) Shorthand it! **1936** *Punch* 15 Jan. 60/1 All I wanted..was a secretary—just an ordinary girl who could shorthand and typewrite and answer a telephone and so on. **1975** *Times Lit. Suppl.* 28 Feb. 210/4 The urbanized cluster of inlets and coastal fringe the Venetians shorthanded as *La Dominante.* **1975** I. K. Martin *Regan & Manhattan File* 37 One of Broughton's silent colleagues had a notebook out, shorthanding Regan's words. **1977** *Monitor* (McAllen, Texas) 17 July 5F/6 'Shorthand' has also become quite popular in what is jocularly referred to in Hollywood as the 'creative community'. Ask a producer what his show is all about and he'll never say, 'I don't know.' He says instead, 'There's no way to shorthand that.'

short-handed, *a.*[2] Add: **2.** *spec.* in *Ice Hockey*, having fewer players on the ice than the opposing team because a penalty has been imposed; also, of a goal: scored while a team is short-handed.
1939 R. F. Vaughan *Hockey* 364 *Short handed*, a team with one or more players in the penalty box. **1951** L. Percival *Hockey Handbk.* iv. 126/2 The fundamental weapons are..aggressive use of the body, getting the puck into the opposing defensive zone and keeping it there with five men up, forcing the game (even though shorthanded), and generally keeping the pace of the play high. **1969** *Official Rule Bk. & Schedule of National Hockey League 1969–70* 30 'Short-handed' means that the team must be below the numerical strength of its opponents on the ice at the time the goal is scored... Thus coincident minor penalties to both teams do *not* cause either side to be 'short-handed'. **1970** B. Orr *Orr on Ice* 96 Our men attempt to get the puck down to the other end and keep it, for even a possible shorthanded score. **1976** *Washington Post* 19 Apr. D3/1 Serge Savard scored a short-handed goal..to lead the Montreal Canadians to a 4–1 victory over the Chicago Black Hawks.

shorthorn. Add: **2.** A small round carrot belonging to the variety so called. Also *attrib.*
1873 *Young Englishwoman* Oct. 499/1 There are several sorts of carrots: the earliest of which are generally termed short-horn. **1885** W. Robinson tr. *Vilmorin-Andrieux's Veg. Garden* 162 French Horn, or Early Short Horn, Carrot... Root almost globular, or slightly top-shaped. **1930** *Times Educ. Suppl.* 26 Apr. (Home & Classroom Suppl.) p. iv/3 Attend to the successional sowing of such crops as peas, spinach,..short-horn carrots, and lettuces. **1976** *Publishers Weekly* 19 Apr. 77/1 Such virtually unknown and unavailable items as perennial broccoli and short horn carrots.

3. *U.S. slang.* A new arrival, a greenhorn.
1888 *Outing* Nov. 129/2 Besides a few snipe killed at a swamp called by Shorthorns 'cineky', from the Spanish *sienica*, we still depended upon Uncle Sam's subsistence stores for our daily bread. **1905** A. H. Lewis *Sunset Trail* ii. 34 Don't let no shorthorn have my room. **1907** J. London *Road* 173 Gay-cats are short-horns, *chechaquos*, new chums, or tenderfeet. A gay-cat is a new-comer on The Road who is man-grown, or, at least, youth-grown. **1942** Berrey & Van den Bark *Amer. Thes. Slang* § 456/1 Inexperienced person;..*shorthorn.*

shortia (ʃɔ̄ᵊ·ɹtiă). [mod.L., f. the name of Charles W. *Short* (1794–1863), American botanist + -ia[1].] A small stemless evergreen herb of the genus so called, belonging to the family Diapensiaceæ, native to eastern North America and temperate parts of Asia, and bearing glossy leaves and white, pink, or blue flowers.
Shortia galacifolia was first described by Asa Gray from a dried specimen seen in Paris in 1839, which had been collected by André Michaux in 1798. The plant was not rediscovered in the wild until 1878.
[**1839** A. Gray *Jrnl.* 8 Apr. in *Lett. A. Gray* (1893) I. iii. 178 As this is a good North American genus and comes from near Kentucky, it shall be christened Shortia, to which we will stand as godfathers.] **1877** *Field & Forest* Sept. 40 More than once I was greeted with the query 'Found "Shortey" yet?' By which I suppose was meant the mythical Shortia of Michaux, for which any enthusiastic young or old botanist is at liberty to hunt. **1948** *Hyde Park Shopper* (Chicago) 29 Apr. 8/5 The rare flower, shortia, is found only in the mountains of North Carolina and Japan. **1962** *Times* 8 Dec. 11/3 The shortias, too, do well in a moist..position. **1974** *Country Life* 12 Dec. 1896/1 American woodlanders, such as shortias, schizocodons,..and erythroniums especially revel in it [*sc.* beech leaf-mould].

shortie, var. *shorty *sb.* and *a.*

shortite (ʃɔ̄·ɹtəit). *Min.* [f. the name of Maxwell N. *Short* (1889–1952), U.S. mineralogist + -ite[1].] A double carbonate of sodium and calcium, $Na_2Ca_2(CO_3)_3$, found as colourless to pale yellow wedge-shaped, orthorhombic crystals that are strongly pyroelectric.
1939 J. J. Fahey in *Amer. Mineralogist* XXIV. 514 Crystals of shortite..were found in cores of clay shale from Sweetwater County, Wyoming, at depths between 1258 and 1805 feet. *Ibid.*, Shortite is named in honor of Dr. Maxwell N. Short, Professor of optical mineralogy at the University of Arizona..and widely known for his contributions to the study of opaque minerals. **1973** *Jrnl. Geol.* LXXXI. 229 Shortite..occurs in the groundmass of micaceous kimberlite dikes from the Upper Canada Gold Mine, Ontario.

short list. [f. Short *a.* + List *sb.*[6]] A list of selected names, *esp.* of candidates for a post, from which the final selection is to be made. Hence **sho·rt-list** *v. trans.*, to put on a short list; **sho·rt-listing** *vbl. sb.*; **sho·rt-listed** *ppl. a.*
1927 W. E. Collinson *Contemp. Eng.* 124 Selection committees to University posts first familiarised me with the meaning of the short list. **1929** J. Van Druten *Young Woodley* ii. 28 He dreamed of a Headship elsewhere, always pulling wires and toadying to achieve it. He had applied for six, and lost them all; in only one case had he ever been on the Short List. **1955** M. Gilbert *Sky High* ii. 26 The police have got a short list of suspects. **1955** *Times* 12 July 2/3 Further particulars may be obtained from the Registrar, with whom applications (five copies) and the names of two referees, must be lodged not later than 23rd July, 1955. Candidates short-listed will be interviewed on 27th July, 1955. **1958** *New Statesman* 3 May 559/3 It is said that there is a Catholic majority on the party executive which did the short-listing. **1961** *Guardian* 4 Feb. 6/4 A documentary film..has been shortlisted for a 1961 British Academy award. **1962** *Listener* 4 Jan. 23/2 The mass of short-listed poems was taken to Farringford. **1974** *Country Life* 17 Jan. (Suppl.) 25 Discussions..in England will 'short list' the properties that should interest you. **1977** *Wandsworth Borough News* 7 Oct. 2/2 The local party's General Management Committee will vote for their choice next Monday evening after hearing each of the short-listed candidates give a 10-minute talk. **1979** *Financial Rev.* (Austral.) 3 July 26/1 A two-ship shortlist for the Melbourne aircraft carrier replacement.

short-lived, *a.* Add: Now usu. with pronunc. (-livd). **2. c.** Of a radioisotope or sub-atomic particle: having a relatively short half-life.
1926 R. W. Lawson tr. *Hevesy & Paneth's Man. Radioactivity* xxiii. 170 Short-lived thorium isotopes like uranium X, radiothorium, etc. **1947** *Radiology* XLIX. 286/2 The tracer and therapeutic studies with 'short-lived' artificially produced radioactive isotopes were not directed toward the study of the more general biologic effects. **1973** L. J. Tassie *Physics of Elem. Particles* ix. 91 Since the weak interaction causing the decay of the neutral kaon does not conserve *CP*, the short-lived neutral kaon and the long-lived neutral kaon are not necessarily eigenstates of *CP*.

short run, *sb.* and (with hyphen) *a.* [f. Short *a.* + Run *sb.*[1]] **A.** *sb.* **1.** (See Short *a.* 26 d.)
1830 [in Dict. s.v. Short *a.* 26 d]. **1921** P. F. Warner *My Cricketing Life* xi. 203 One of the features of his [*sc.* J. B. Hobbs's] long partnerships with Rhodes was the number of short runs they ran.

2. A relatively brief passage of time within a sequence of events (opp. Long run, long-run), usu. in phr. *in the short run.*
1879 Geo. Eliot *Let.* 18 Oct. (1956) VII. 212 Mrs. Healy's marriage is surely what you expected in the long or short run. **1928** *Britain's Industr. Future* (Liberal Industr. Inquiry) iv. xxvii. 396 But we do not believe that in anything but the very short run industry benefits from displacing adult workers by successive relays of young boys and girls. **1940** F. Pickersgill *Let.* 19 Jan. (1948) 151 Chamberlain is of course succeeding in the short run in getting rid of all the energetic and intelligent people in the Party. **1952** Isard & Whitney *Atomic Power* ix. 185 The question arises whether the economic benefits of atomic power may be greater for such a country as Russia in the short-run and middle-run as well. **1965** *New Statesman* 7 May 718/3 In the long run you will obtain substantial benefit from professional business schools like the one at Harvard... In the short run—well, I don't envy George Brown. **1971** D. C. Hague *Managerial Econ.* ii. v. 112 In economic theory, the short run is defined as that period of time during which the physical capacity of the firm is fixed.

3. *Theatr.* A short period of being represented on the stage. Also *attrib.*
1922 H. Granville-Barker *Exemplary Theatre* vi. 260 To replace the long run by the short run..or the hastily concocted 'repertory' season, is no remedy. *Ibid.* 262 Most of the 'short-run' theatres..by misplaced courtesy are dubbed 'repertory'. **1961** *Twentieth Century* Feb. 101 The short-run system of provincial repertory. **1967** *Oxf. Compan. Theatre* (ed. 3) 797/1 In Stratford-upon-Avon plays are now introduced into the bill one at a time, and given a short run before being merged in a changing bill.

4. A class or line of goods produced in limited quantity. Also *attrib.*
1957 *Times* 4 Nov. 13/1 The blot on the British copy-book is the complaint from a number of stores that most manufacturers [of furniture] will not consider what are known in the trade as short runs. **1959** *Times* 14 Jan. 12/4 Printing short-run quality bookwork. **1967** V. Strauss *Printing Industry* x. 660/1 Foot-operated wire stitchers.. are used for short runs or for jobs that cannot be handled on more automatic machinery. **1970** *Publishers' Weekly* 8 June 152 Many speakers complained about printing short-run jobs: price is too high, quality is not good, and.. deliveries are too slow. **1976** *Scotsman* 25 Nov. 4/3 (Advt.), Craft Bookbinding. Contemporary style bindings

in leather and cloth: short-runs re-backing and general restoration.

B. *adj.* Occurring in or relating to the short run; = *short-term *a.*
1947 *Partisan Rev.* XIV. 240 There is certainly a short-run tendency in critical situations toward reliance on reactionaries as counter to communists. **1958** *Listener* 9 Oct. 548/1 A combination of short-run and long-term programming on two different political and geographical levels. **1966** *Philos.* XLI. 294 Any cyclical view of history can harbour only a short-run optimism. **1979** *Internat. Jrnl. Sociol. of Law* VII. 308 Larceny, some of it responsive to short-run price changes.

short-stay, *a.* [f. Short *a.* + Stay *sb.*[3]] **1.** (See Short *a.* 26 a.)
2. That makes a short stay, *spec.* in a hospital or other institution providing care or treatment; providing accommodation for a short stay; of or pertaining to those who make a short stay.
1946 *Nature* 26 Oct. 578/2 It is proposed to establish a diagnostic and research centre at the teaching hospital.. to deal with short-stay in-patients and out-patients. **1952** C. P. Blacker *Eugenics: Galton & After* xi. 316 These mainly consist of nursery services (long- and short-stay residential nurseries; part and full-time day nurseries; nursery schools). **1960** J. Stroud *Shorn Lamb* i. 14 A short-stay case..[is] a case where..the children are only going to be In Care for a very short time. **1965** J. Pollitt *Depression & its Treatment* vi. 85 Out-patient treatment and short-stay admission is now widely practised for all psychiatrically ill patients. **1976** *Howard Jrnl.* XV. i. 45 Readers working with homeless alcoholics will appreciate the accounts of the birth and growth of two long-stay hostels and a short-stay house. **1981** D. Kavanagh *Fiddle City* ii. 31 Short-stay car parks, long-stay car parks.

short-term, *a.* [f. Short *a.* + Term *sb.*] Lasting for, pertaining to, or involving a relatively short period of time; maturing or becoming effective after a short period. Also quasi-*advb.*
1901 [in Dict. s.v. Short *a.* 23]. **1932** *Ann. Reg. 1931* 194 The withdrawal of French short-term credits has given the impetus to the original crisis, and other foreign creditors had followed suit. **1943** J. S. Huxley *Evolutionary Ethics* iv. 31 All existing societies manifest considerable ethical disunity, and..this is an expression of the conflicts and contradictions inherent in the situation —conflicts between classes and groups, between long-term and short-term good, [etc.]. **1948** 'N. Shute' *No Highway* 2 Short-term *ad hoc* experiments to solve a particular problem. **1956** *Planning* XXII. 41 The short-term forecasts, up to one year, are the basis for current production and buying and sales planning. **1959** *Times* 20 Jan. 9/3 It will not be enough, however, for the Government to solve the problems short-term. **1970** *Money Which?* Mar. 56/3 A gain counts as a short-term gain if, in general, you had held the asset for a year or less. **1972** *New Scientist* 24 Feb. 428 This type of memory is referred to as short-term memory because the number of events we can hold in this fashion is strictly limited, and forgetting is extremely rapid once our attention is diverted. **1980** *Daily Tel.* 28 June 16 The Minister of the Environment is seeking ways to control local high-spending by councils such as Manchester. Short-term, his powers are fairly limited.

short time. Also (attrib.) short-time. [f. Short *a.* + Time *sb.*] **1.** The state or condition of working less than the regular number of hours per day or of days per week. Also *attrib.* and quasi-*advb.*
1848 W. Fagan *O'Connell* II. 561 A supporter of 'short time' work for children in the Factories. **1861** J. S. Mill *Let.* 1 Mar. (1910) I. 245 The equality, if not superiority,.. of the short-time pupils. **1861** J. Ward *Diary* 16 Nov. in J. Burnett *Useful Toil* (1974) i. 80 The manufacturing districts..are all running short time through the scarcity of cotton. **1864** R. A. Arnold *Hist. Cotton Famine* iii. 84 Short time means short wages. **1906** *Daily Colonist* (Victoria, B.C.) 1 Jan. 13/2 Disorderly scenes were witnessed recently in Newcastle-on-Tyne..owing to an intimation that fifty [employees] would be put on short time. **1930** *Economist* 22 Feb. 411/1 Several watchmaking factories are working short time. **1955** *Times* 14 May 5/6 Three years ago more than 80,000 textile and clothing workers were flung out of work, and thousands more were on short time. **1961** *Ann. Reg. 1960* p. xix, Unemployment was at a low level but in the latter half of the year there was much short-time working particularly in the motor-car industry. **1978** P. Bailey *Leisure & Class in Victorian England* iv. 80 In the early 1870s...the success of the Short Time agitation encouraged further campaigns to cut working hours.

2. *slang.* A brief visit to a prostitute; a brief sojourn in a hotel for sexual purposes. Also *attrib.*
1937 Partridge *Dict. Slang* 763/1 *Short time*, a visit to a prostitute for one copulation only. **1939** G. Greene *Confidential Agent* i. ii. 48 It's no bother. It's the 'short times' that are the bother. In and out three times in a night. **1961** R. Seth *Anat. Spying* ii. 30 Ninety-five per cent of any ship's company, as soon as they are able to get ashore,..make straight for the nearest bar for a 'quick one', and then on to the nearest brothel for what is known in the jargon as 'a short time'. **1971** *Guardian* 8 July 3/1 Miles of girlie bars, short time hotels. **1979** J. Wainwright *Take Murder* i. 23 Three hours. Not the proverbial 'short time'... It means a bed somewhere.

short-ti·mer. [f. Short *a.* + Timer 4.] **1.** (See Short *a.* 26 a.) *Obs.* exc. *Hist.*
1863 Dickens *Uncommercial Traveller* in *All Year Round* 20 June 400/2 The Short-Timers, in a writing competition, beat the Long-Timers of a first-class National School. **1883** [in Dict. s.v. Short *a.* 26 a].

2. *U.S. Mil. slang.* One nearing the end of his period of military service.
1906 T. Beyer *Amer. Battleship in Commission* 73 In most cases..'short-timers' are sent home before their enlistment expires. **1918** L. E. Ruggles *Navy Explained* 132 Any man who has less than six months to do on his enlistment is called a 'short timer'. **1952** M. Russ *Last Parallel* (1957) 16 Being what is known as a short-timer..I'm at peace with service life. **1977** *Chicago Tribune* 2 Oct. ii. 4/6 An inspecting officer discovered a 'short-timer's calendar' centrally positioned in a foot-locker display.

3. *slang.* One serving a short prison sentence.
1915 J. London *Jacket* iv. 25 He was a pallid-faced, little dope-fiend of a short-timer.

4. *slang.* One who makes a brief sojourn in a hotel for sexual purposes; one who visits a prostitute. Cf. *short time 2.
1923 J. Manchon *Le Slang* 267 *Short-timers*,..un couple (amoureux) qui loue en meublé pour peu de temps. **1939** G. Greene *Confidential Agent* i. ii. 87 The shabby hotel to which 'short timers' come. **1960** *Amer. Speech* XXXV. 120 When the GI visited a *movie star*, it was usually for a *short-time*. *Short-time*..also..had an agent noun, a *short-timer.*

short-toed, *a.* [Short *a.* 24.] Having short toes; esp. in *short-toed lark*, any of various larks of the genera *Calandrella* and *Spizocorys*, which comprise small gregarious birds widely distributed in Europe, Africa, and Asia, esp. in deserts and steppes.
1837 [in Dict. s.v. Short *a.* 24]. **1863** T. C. Jerdon *Birds of India* II. i. 425 The short-toed Lark appears in India in October and November, in flocks. **1869–73** [in Dict. s.v. Short *a.* 24]. **1933** *Discovery* July 224/1 The short-toed lark, whose protective colouring makes him almost invisible on sandy soil. **1963** *Times* 27 Feb. 11/6 Overhead range the 14 species of birds of prey, which breed here or within everyday range, including the imperial, short-toed and booted eagles. **1972** Ali & Ripley *Handbk. Birds India* V. 21 Short-toed Larks become excessively fat before emigration and are then netted everywhere in vast quantities. **1980** Howard & Moore *Compl. Checklist Birds of World* 347/1 *Spizocorys starki* (Stark's Short-toed Lark). Angola to W. Transvaal.

sho·rtwall. *Mining.* [f. Short *a.* + Wall *sb.*[1]] A short coalface. Freq. *attrib.*, denoting (equipment designed for) a method of mining in which shortwalls are worked. Cf. *longwall* s.v. Long *a.*[1] 18 a.
1912 F. D. Power *Coalfields & Collieries Austral.* 409 *Shortwall machine*, a coal cutter for use in bords, which, when once the cutting part has made the sumping cut, is drawn across the face automatically by ropes, undercutting as it proceeds. **1931** *Trans. Inst. Mining Engineers* LXXXI. 474 The thin seams are extracted on a 'shortwall' (retreating) method. **1942** *Ibid.* CII. 42 The shortwall is antagonistic to the basic principle that the maximum area of seam should be won with the minimum of roadway consistent with roadway costs. **1958** I. C. F. Statham *Coal Mining Pract.* I. vii. 298 Mechanised methods are now generally adopted for the coal-cutting in these narrow opening-up places, using either a shortwall coal-cutter, a Universal coal-cutting machine or in some cases a long-wall machine. **1964** A. Nelson *Dict. Mining* 406 A shortwall face may be any length between about 5 and 30 yd and is generally employed in pillar methods of working. **1973** L. J. Thomas *Introd. Mining* vii. 258 A group of shortwalls are mined adjacent to each other with rib-pillars left between.

shortward(s (ʃɔ̄·ɹtwǭɹd(z), *adv.* [f. Short *a.*, *sb.*, and *adv.* + -ward, -wards.] Towards shorter wavelengths; on the short-wavelength side *of.*
1972 *Science* 1 Sept. 789/3 Figure 4c..shows features shortward of Lyman-α, which are attributed to the Birge-Hopfield bands of N_2. **1974** [see *longward(s *adv.*]. **1978** *Nature* 5 Oct. 414/1 IUE provides an excellent way for acquiring UV spectra in the wavelength region shortwards of the Earth's ozone cutoff (about 3,100 Å).

short wave. Also **sho·rt-wave, shortwave.** [f. Short *a.* + Wave *sb.*] **a.** An electromagnetic wave of relatively short wavelength, *spec.* a radio wavelength of less than about 100 metres, corresponding to a frequency of more than 3000 kilohertz; radio communication or broadcasting employing such waves. Usu. *attrib.* or as *adj.*; also in adverbial use.
1839 [see *long wave* s.v. *long *a.*[1] 18]. **1902** *Encycl. Brit.* XXVIII. 68/2 With very fine wires the condition *na* small can be fulfilled with quite short waves. **1907** J. Erskine-Murray *Handbk. Wireless Telegr.* i. 5 Short-wave Hertzian telegraphy has never been successful at distances beyond a mile or two. **1928** D. Brunt *Meteorology* v. 40 We have thus to picture a beam of short-wave light from the sun reaching the outer boundary of the earth's atmosphere. **1928** *Daily Mail Year Bk.* 240/2 The B.B.C.'s slowness to recognise the importance of short-wave Empire broadcasting. **1941** J. Steinbeck *Sea of Cortez* i. 7 In all the crackle and noise of the short-wave one of our men made contact with another boat. **1943** D. Powell *Time to be Born* xiii. 308 They..drove to N.B.C. studios where he broadcast short-wave to London. **1961** *Ann. Reg. 1960* 447 The mass-produced short-wave transistor. **1972** *Sci. Amer.* Sept. 109/1 Finally it was discovered that 'shortwave' frequencies (from three to 30 megahertz) can travel halfway around the earth and more by being repeatedly reflected between the F layer and the earth. **1978** W. F. Buckley *Stained Glass* xviii. 182 The Director rose and turned off the impressive short-wave speaker that had brought in the press conference.

b. *Med.* Used *attrib.* or as *adj.* to denote diathermy in which energy is applied to the tissues by means of oscillating electric fields having frequencies within the short-wave radio range.
1935 R. Kovács *Electrotherapy & Light Therapy* (ed. 2) xviii. 306 Short-wave diathermy comprises wave lengths from 12 to 30 meters... Ultra short-wave diathermy comprises wave lengths below 12 meters. **1935** Wilson & Dowse tr. *Holzer & Weissenberg's Found. Short Wave Therapy* 160 Short wave therapy has been used for about eight years, and..not a single case of irreparable injury has been published. **1965** E. D. R. Campbell in C. W. H. Havard *Fund. Current Med. Treatment* xvi. 596 Short-wave diathermy causes heat to be more evenly distributed through the bulk of a limb by two distinct processes. **1965** N. Freeling *Criminal Conversation* i. ii. 18 Dr Hubert van der Post, neurologist, specialist in short-wave and other electrical treatments.

short-winded, *a.* **b.** (Later examples.)
1934 C. Lambert *Music Ho!* i. 36 His [*sc.* Debussy's] melodies may be a little short-winded. **1976** *Gramophone* May 1772/3 The first movements, if short-winded for Bach, have an easy fluency.

short-windedness. Add: Also *fig.*
1934 C. Lambert *Music Ho!* ii. 102 His [*sc.* Stravinsky's] melodic style has always been marked by extreme shortwindedness.

shorty (ʃɔ̄·ɹti), *sb.* and *a.* *colloq.* or *slang.* Also shortie. [f. Short *a.* + -y[6], -ie.] **A.** *sb.*
1. *Sc.* (spelt *shortie.*) Shortbread; a piece of shortbread.
1882 in Jamieson, s.v. Short-Bread. **1919** C. Orr *Glorious Thing* i. 11 Home wouldn't be home to Minnie without your shorties, Jenny. **1974** *Sunday Post* (Glasgow) 27 Oct. 17/5 Shortie made with butter and browned real good is sold in St. Andrews.

2. A person of short stature; freq. with capital initial as a nickname or form of address.
1888 *Texas Siftings* 7 Jan. 10/3 One boy yelled out: 'Go it, Shorty!' **1908** S. Ford *Side-Stepping with Shorty* xix. 306 'Hello, Shorty!' says he, in that little squeak of his. **1914** 'Bartimeus' *Naval Occasions* xxiii. 213 Your middle watch, Shortie? *a* **1935** T. E. Lawrence *Mint* (1950) 152 The figure gives a large surplus of shorties. **1960** V. Nabokov *Invitation to Beheading* vi. 70 The door of this cell was wide open, and inside, the likable shorty whom he had seen before..was standing on a chair. **1978** *Oxford Times* 20 Oct. 13/2 A reader who is 5ft 1½in tall has sent me a delightful document entitled 'Reflections of a Suicidal Shortie'.

3. A drink of spirits; a short drink. orig. *U.S.*
1931 *Amer. Speech* VII. 83 *Shorty*, and *apple-jack* are illicit drinks. **1942** Berrey & Van den Bark *Amer. Thes. Slang* § 99/8 Strong liquor,..*shorty* (esp. hard, illicit liquor). *Ibid.* § 101/2 *Shorty*, a straight gin with a ginger ale chaser. **1963** *Freedomways* III. 523 Yarborough ..yelled, 'Bartender. Give the professor another shorty of gin there.' **1971** *Scope* (S. Afr.) 19 Mar. 77/2 Also popular are the tall cylindrical tumblers for long drinks, and the chunky tumbler for 'shorties'.

4. A short story, article, film, etc.
1934 M. H. Weseen *Dict. Amer. Slang* 151 *Shorties*, short motion picture films. **1968** *Listener* 4 July 22/1 Half the time we wanted to kill its crippling name and start on a new formula, with nudes and shorties to rival the money-making *Lilliput* and *Men Only*. **1976** 'K. Royce' *Bustillo* x. 137, I read rather an interesting article. Just a shortie. About your man Warton.

5. A short article of clothing; pl. *spec.* shorts.
1942 Berrey & Van den Bark *Amer. Thes. Slang* § 87/37 Underwear,..*shorties*, shorts. **1945** *Richmond* (Va.) *News Leader* 2 Mar. 19 (Advt.), Shorties: More popular than ever this spring! 100% wool Shetlands, in a belted all-around style. **1957** M. B. Picken *Fashion Dict.* 151/1 *Shortie*, glove, either slip-on or having one button; extending to wrist or a little beyond. **1958** *Vogue* Jan. 2 Jackets..just right for spring..like the jaunty shortie. **1959** 'D. Buckingham' *Wind Tunnel* xix. 157 She was already in her nightdress—a silly little nylon shorty. **1962** *Times* 19 Nov. 14/4 Among modern 'shorties' and light-weight garments in my wardrobe. **1963** [see *peep-hole].

B. *adj.* Designating products which are shorter than the norm.
1949 *Sun* (Baltimore) 19 Mar. 7/7 (Advt.), Short and sugary..cool and comfy..these delicious Rayon Jersey Shortie Pajamas in pastel shades. **1952** *N.Y. Times* 6 May 33/8 A black suede shorty glove. **1954** *Wall St. Jrnl.* 16 Apr. 1/1 Promoters of Florida tourist attractions are energetically courting a new, inexpensive and very rewarding publicity medium. It's the 'shortie' TV feature—films which Floridians make and which the networks admit they're eager to show without charge. **1956** J. Potts *Diehard* x. 165 Her bright-green shortie coat was spotted with rain. **1960** *News Chron.* 5 July 1/7 They were sacked..after a party in which they staged their own version of Florence Nightingale in the nurses' home, dancing about in 'shortie' nighties waving lighted candles. **1971** *Scope* (S. Afr.) 19 Mar. 64/1 The girls changed into shortie cat-suits. **1972** D. Haston *In High Places* v. 69 To go anywhere in the winter you have to be able to ski, so I had borrowed a pair of shorty skis from Grahame Tiso. **1980** *Dirt Bike* Oct. 64/3 Both bikes are using Oakley II grips and DeHandler shorty levers.

‖**shosagoto** (ʃosāgǭᵘ·to). [Jap., f. *shosa* acting, conduct + *koto* matter, affair.] In Japanese Kabuki drama: a dance play; a mime performed to music.
1911 *Encycl. Brit.* XV. 170/1 Mimetic posture-dances (*Shosagoto*) were always introduced as interludes. **1957** *Oxf. Compan. Theatre* (ed. 2) 412/2 The Japanese theatre recognizes three main classes, the *jidaimono* or histories..; the *sewamono* or melodramas; and the *shosagoto* or dances. **1967** 'J. H. Roberts' *February Plan* i. iii. 92 The play today was a *shosagoto*, a Kabuki drama adapted from a Noh play. **1975** J. R. Brandon *Kabuki* 5 In the late seventeenth century, three major divisions of kabuki drama were recognized: *sewamono..jidaimono*; and dance pieces, called *shosagoto.*

Shoshone (ʃoʃǭᵘ·ni), *sb.* and *a.* Also 9 Shoshonee, -ie; 9– Shoshoni. [From an unidentified American Indian language; the folk-etymology given in quot. 1918 is rejected by scholars.] **A.** *sb.* **1.** (A member of) a North American Indian people of Wyoming, Idaho, Nevada, and neighbouring states.
1805 M. Lewis *Jrnl.* 19 Aug. in *Orig. Jrnls. Lewis & Clark Exped.* (1904) II. xv. 370 The Shoshonees may be estimated at about 100 warriors. **1830** *Western Monthly Rev.* III. 562 The Shoshonee are a numerous and powerful tribe of Indians. **1834** A. Pike *Prose Sketches & Poems* 200 The Shoshones are the Snakes. **1836** W. Irving *Astoria* II. xi. 132 The Shoshonies are a branch of the once powerful and prosperous tribe of the Snakes. **1884** W. Shepherd *Prairie Exper.* 59 The Crows..came down to visit the Rapahoes, Shoshones, and other tribes. **1918** J. E. Rees *Idaho Chronol., Nomenclature, Bibliogr.* 111 The name comes from two Indian words, 'Shawnt', meaning 'abundance', and 'shaw-nip', 'grass', which was etymologically changed to the euphonious name 'Shoshoni' and in English conveys the thought of 'abundance of grass'. **1938** *Bull. U.S. Bureau Amer. Ethnol.* No. 120. 238 Shoshoni and Ute were periodically at grips. **1959** E. Tunis *Indians* 91/1 In time, the Shoshone in the far west were following social patterns that had been folkways along the Missouri. **1977** J. Gunn in Hill & Gunn *Individual in Prehistory* ix. 190 If it is assumed that Shoshoni ware is truly diagnostic of Shoshoni population movement, these data are incongruent with the proposed 1000-year-later migration of the Shoshoni into the Great Basin.

2. The language of this people, a member of the Uto-Aztecan family (formerly also applied to a grouping of languages including Shoshone).
1843 F. Marryat *Travels & Adventures of Monsieur Violet* xiv. 33/2, I addressed him in Shoshone, which beautiful dialect is common to the Comanches, Apaches, and Arrapahoes. **1933** [see *Paiute *sb.* b]. **1977** *Language* LIII. 459 The correlations with Cupan *yax* 'to be', Shoshoni *yikʷɨ* 'to sit (pl.)', and a present-tense suffix in Southern Paiute are improbable.

B. *adj.* Of or pertaining to the Shoshone or their language or a former grouping of languages to which this language was assigned.
1805 M. Lewis *Jrnl.* 17 Aug. in *Orig. Jrnls. Lewis & Clark Exped.* (1904) II. xv. 364, I was to bring on the party and baggage to the Shoshone Camp. **1837** W. Irving *Adventures Capt. Bonneville* I. xvi. 260 There was but little chance of meeting the Shoshonie bands. **1886** *Outing* Dec. 198/2 Dick had..a Shoshone woman for his wife. **1926** D. Branch in J. F. Dobie *Rainbow in Morning* (1965) 128 The precarious, abject living of the Shoshone Diggers. **1956** J. Whatmough *Lang.* xii. 221 A Shoshoni dialect spoken in southwestern Utah. **1976** *Billings* (Montana) *Gaz.* 16 June 1-A/6 What more fitting tribute to those who fought and died in this struggle, U.S. soldiers, Sioux, Cheyenne, Crow and Shoshone warriors, than to protect it for those who follow us?

Shoshonean (ʃoʃǭᵘ·niăn), *a.* and *sb.* Also -ian. [f. prec. + -an.] (Designating) a branch of the Uto-Aztecan languages including Shoshone; of or pertaining to speakers of these languages.
1891 J. W. Powell in *Ann. Rep. U.S. Bureau Amer. Ethnol.* VII. 109 Very likely much of the area occupied by the Atsina was formerly Shoshonean territory. **1893** A. F. Chamberlain in *Rep. Brit. Assoc. Adv. Sci. 1892* 589 A seeming similarity in a few points of general structure to the Shoshonian and to the Siouan tongues. **1904** *Rep. U.S. Nat. Museum 1901–2* 472 All the Shoshonean types of weaving, all their forms of baskets, and most of the patterns on them are ancient. **1921** A. F. Hall *Handbk. Yosemite Nat. Park* 51 The Mono..are an offshoot from the Paiutes and other Shoshoneans of Nevada and the Great Basin country. **1929** E. Sapir in *Encycl. Brit.* V. 138/1 *Shoshonean*..occupies the greater part of the Great Basin area and contiguous territory in southern California and the southwestern Plains. **1932** [see *Pima]. **1940** M. J. Rogers in E. C. Jaeger *California Deserts* xi. 117 In historic times the Mohave Desert had fallen almost entirely into the hands of the Shoshoneans. **1950** F. Eggan *Social Organization of Western Pueblos* i. 13 By comparing the Hopi system with other Shoshonean systems, he came to the conclusion that 'linguistic conservatism has been of slight importance in the history of the present Hopi nomenclature'. **1974** *Amer. Anthropologist* LXXVI. 11 (*heading*) An archaeological perspective on Shoshonean bands.

shoshonite (ʃoʃǭᵘ·nəit). *Petrogr.* [f. the name of the *Shoshone* River, Wyoming (cf. *Shoshone) + -ite[1].] A type of basaltic rock varying quite widely in composition and distinguished by containing, in addition to augite, labradorite, and usu. olivine, significant amounts of potassium feldspar.
1895 J. P. Iddings in *Jrnl. Geol.* III. 938 The classes will be described in the order just given under the names: *Absarokite, Shoshonite* and *Banakite*. *Ibid.* 943 The rocks classed as shoshonites are more numerous than the absarokites, and embrace a somewhat wider range of composition. They occupy the middle ground, as it were, in the series. **1937** Holmes & Harwood *Volcanic Area of Bufumbira* (Mem. Geol. Survey Uganda No. 3) xii. 269 The xenoliths of the shoshonites of Sabinyo include aggregates referable to kentallenite and olivine-monzonite. **1976** *Nature* 12 Feb. 472/1 The calc-alkaline lavas.. range from 'normal' basalts, basaltic-andesites and andesites to highly potassic varieties having all the compositional and modal characteristics of shoshonites.

Hence **shoshoni·tic** *a.*, resembling or consisting of shoshonite.
1937 Holmes & Harwood *Volcanic Area of Bufumbira* (Mem. Geol. Survey Uganda No. 3) ii. 16 These rocks are more feldspathic than the shoshonitic absarokites of Muhavura and Mgahinga. **1964** G. A. Joplin *Petrogr. Austral. Igneous Rocks* vii. 90 The different shoshonitic rocks show many common characteristics.

shot, *sb.*[1] Add: **I. 7. e.** (*b*) Also, a picture (or sequence of pictures) continuously shot by a single film or television camera; the action or process of taking such a picture.
1923 'B. M. Bower' *Parowan Bonanza* xxvi. 303 Bill and Tommy were both below examining the effect of their 'shots' of the evening before. **1937** *Discovery* Nov. 330/2 For each unit of programme transmission, called a shot, on account of similarity with sound-film technique, several electron cameras may be in use. **1957** W. Alwyn in Manvell & Huntley *Film Music* 9 Chapter 4 shows the various dramatic forms film music takes, and analyses in each case shot-by-shot and phase-by-phase a particular sequence [etc.]. **1963** *Movie* Jan. 8/1 The sequence in the Albert Hall auditorium..lasts about twelve minutes, 124 shots without any dialogue. **1972** *Listener* 21 Dec. 852/1 Sequence of calls before a shot. Production Assistant: 'Quiet. Going for a take. Standing by.' Director: 'Right.' **1979** D. Gurr *Troika* i. 2 The first picture is on the screen... He never told me they had that shot.

g. (*a*) A hypodermic injection of a narcotic, hallucinogen, or the like, or of a vaccine; a measure of a substance for injection. Also *fig.* *colloq.* (orig. *U.S.*).
1904 *San Francisco Chron. Suppl.* 30 Oct. 4/1, I varied hardly a minute each day in the time of taking my injection. My first shot was when I awoke in the morning. **1921** S. Lewis *Let.* 12 July in C. Mackenzie *My Life & Times* (1966) V. 199 Your book..was..at once a Social Document, and an opiate—or, as we say in the States, a shot of dope! **1936** L. C. Douglas *White Banners* xviii. 373 That reminds me—I've to take some typhoid shots. **1948** G. H. Johnston *Death takes Small Bites* iv. 81 If you've never had a plague shot and you've been here for five hours you might have contracted the disease. **1953** W. Burroughs *Junkie* viii. 74 About fifteen minutes later the attendant called, 'Shot line!' Everyone in the ward lined up. **1957** *London Mag.* Sept. 40 They were persons of a kind needing shots of the notion of art as others need shots of insulin. **1969** A. Lurie *Real People* 154 A doctor had come and given Charlie a shot and put him to bed. **1978** G. A. Sheehan *Running & Being* x. 136 We begin to hear about Butazolidine and cortisone shots.

(*b*) In *fig.* phr. *a shot in the arm*, a much needed stimulant or encouragement. *colloq.* (orig. *U.S.*).
1922 S. Lewis *Babbitt* viii. 108 All afternoon he snorted and chuckled and gurgled over his ability to 'give the Boys a real shot in the arm tonight'. **1939** I. Baird *Waste Heritage* xii. 157 He saw the thing because he recognized it and knew how the shot-in-the-arm worked. **1949** *Hansard Commons* 27 Sept. 82 The brake..will lead rapidly to that dollar competition..in which we..and the Belgians will use this 'shot-in-the-arm' only for the purpose of making our positions worse. **1951** M. McLuhan *Mech. Bride* (1967) 47/1 Their masters, who then decide what sort of shot in the arm the public needs. **1961** *Daily Tel.* 11 May 20/6 A 'shot in the arm' will be given to Minehead if Mr. Billy Butlin is allowed to build a holiday camp there. **1976** 'A. Garve' *Home to Roost* i. 21 Everyone felt better for seeing her. She was a shot in the arm.

(*c*) A measure of lubricant injected into the petrol tank of a motor vehicle.
1965 L. Sands *Something to Hide* v. 82 'Four gallons, Will, and four shots.'..The shots were squirted in. **1978** *Reader's Digest* Sept. 130 (Advt.), Regular shots of Redex can save you well over £30 a year at current petrol prices... Add one shot of Redex for every gallon of petrol you buy.

h. (See quots.) Cf. *moon-shot* s.v. *moon *sb.* 16; *space shot* s.v. *space *sb.*[1] 19.
1934 *Scoops* 19 May 456/1 Shot..a rocket flight. **1957** Williams & Epstein *Rocket Pioneers* viii. 188 The last moments before a rocket shot are always tense. **1959** F. D. Adams *Aeronaut. Dict.* 152/2 *Shot*, an act or instance of firing a rocket, esp. from the earth's surface.

i. *to call the shots*, to make the decisions; to exercise control over events. *colloq.* (orig. and chiefly *U.S.*).
1967 E. Liebow *Tally's Corner* v. 157 Sea Cat made no secret of the fact that Gloria was calling the shots in this relationship. **1972** *N.Y. Times* 3 Nov. 38/3 It is the majority party which calls the 'shots' on the rules and legislative policies affecting our city. **1978** S. Brill *Teamsters* v. 164 They stand off in a corner as if to say, 'I'm calling the shots here.' **1981** *Sunday Tel.* 5 July 8/5 They felt that an anti-Old Etonian cabal was calling the shots.

8. a. *in(to), out of shot*: also *Photogr.*, *Cinemat.*, and *Television*, in(to) or out of view of the camera.
1958 *Spectator* 18 July 87/1 One Coco-Cola-clutching teenager..darting little glances at the camera to see if he was still in shot. **1960** N. Kneale *Mrs. Wickens in Fall* 174 The maid Cecile hurried into Shot with a tray heaped with cut bread. **1960** I. MacCormick *Small Victory* 69 Thompson looks at each of them disgustedly, then he turns away and moves out of shot. **1969** J. Elliot *Duel* iii. iii. 248 You'll have to move the mike up... Unless you want it in shot. **1976** M. Maguire *Scratchproof* iii. 40 The camera pulled back as she dashed into shot. **1980** D. Francis *Reflex* x. 120 [He] told me it was important that he should be included in my photographs..prominently in shot.

9. b. Also *cheap shot* (*N. Amer. colloq.*).
1973 W. Just *Congressman who loved Flaubert* 97 He tells me it's going to be a sympathetic show... No cheap shots. **1979** R. Jaffe *Class Reunion* (1980) ii. xi. 288 'Every time you come back from those faggots you hang around with in New York you act like a bitch.'..A cheap shot.

d. *Not...by a long shot.* Substitute for entry: *by a long shot*, by a considerable amount, by far; freq. negative in emphatic use. *colloq.* (orig. *U.S.*). (Earlier and later examples.)
1848 in Bartlett *Dict. Americanisms* 215 Mr. Divver offered a resolution summarily removing the superintendent, and was quickly told..that he was going too fast by a long shot. **1861** M. B. Chesnut *Diary* 26 Aug. in C. V. Woodward *M. Chesnut's Civil War* (1981) 163 'They dont pay the soldiers every week.' 'Not by a long shot,' cried a soldier laddie. **1884** in I. M. Tarbell *Hist. Standard Oil Co.* (1904) II. xiii. 114 They are not the Democracy of Ohio by a long shot. **1931** Wyndham Lewis *Apes of God* 17 If those were my last wishes as they are not by a long shot, would they be perfectly clear or not? **1957** W. Saroyan *Whole Voyald* 17 It wasn't only to have pretty women swarm around that I hustled my first book into print. It wasn't that alone by a long shot.

10. a. Also, *a shot in the dark*, a guess, a random attempt. Cf. Dark *sb.* 5.
1895 G. B. Shaw in *Sat. Rev.* 9 Feb. 183/1 Never did man make a worse shot in the dark. **1935** C. Isherwood *Mr. Norris changes Trains* xii. 184, I could no longer resist trying a shot in the dark. 'But you get paid from Paris?' I had scored a bull. **1950** G. Greene *Third Man* iv. 39 'There's something queer about Harry's death.' It was a shot in the dark, but already he had this firm instinctive sense that there was something wrong. **1963** *Listener* 7 Mar. 420/2 It can have been nothing more than a 'shot in dark' [*sic*], but it was a strange prediction none the less.

c. Something which has a chance to succeed (as a racehorse, etc.): usu. preceded by the odds. *colloq.*
1923 Wodehouse *Inimitable Jeeves* iv. 49, It was one of those occasions about which I shall prattle to my grandchildren—if I ever have any, which at the moment of going to press seems more or less of a hundred-to-one shot. **1931** *Daily Express* 23 Sept. 17/6 Mick..will be a neat shot for anybody when the St. Leger is run. **1936** Wodehouse *Laughing Gas* i. 9 As far as my chances of ever copping the title went, I don't suppose I was originally more than about a hundred-to-eight shot, if that. **1941** *Sun* (Baltimore) 14 Aug. 13/6 I've seen 10-to-1 shots that I knew were better horses in certain races than 2-to-1 shots. **1977** *New Yorker* 10 Oct. 174/1 Proud Birdie, a lightly weighted, 4-1 shot in the betting, was next to last going down the back-stretch.

d. *U.S. Billiards. to call one's shot*, to announce which ball one intends to shoot into which pocket; also *fig.* (Properly distinguished from sense 7 i above.)
1953 *Official Rule Bk. Pocket & Carom Billiard Games* 27 Player does not have to 'call his shot' on opening stroke. **1954** Berrey & Van den Bark *Amer. Thesaurus Slang* (ed. 2) § 179/4 Call one's shot,..*to guess or predict rightly.* **1959** N. Mailer *Advts. for Myself* 22 It will be fine if I can write so well and so strongly as to call my shot. **1962** Wodehouse *Service with Smile* vi. 83 In making this statement, he called his shots correctly. **1976** *Billings* (Montana) *Sunday Gaz.* 20 June 1-A/3 There was no question in my mind that Nixon was calling his own shots.

11. f. In Cricket, tennis, golf, etc.: (*oh,*) *shot!*, an applauding exclamation used when a player makes a good stroke, or on an accurate throw; also used when a boxer delivers an effective blow. Freq. *good shot!*
1906 Wodehouse *Love among Chickens* 311 Oh, shot, sir! Shot, indeed! **1907** 'I. Hay' *Pip* x. 309 Here are two young men worth watching. Number One is addressing his ball for an approach shot... 'Good shot!' remarks Number Two. **1922** Wodehouse *Clicking of Cuthbert* ix. 218 He drove a perfect ball, hard and low with a lot of roll. Even Eunice was impressed. 'Good shot, partner!' she said. **1933** D. L. Sayers *Murder must Advertise* xviii. 310 He always hits out. I like to see a batsman hitting out, you know. There! Good shot! Good shot! **1940** E. F. Benson *Final Edition* iii. 52 Everybody chorused 'Good shot, my lord' on the smallest excuse. **1972** J. Burmeister *Running Scared* vii. 95 The resonant bonk of a tennis racket..a distant cry of 'Oh, shot!'

22. c. *big shot* (formerly also *great* or *high shot*), an important person; a prominent member of a profession, organization, etc. Also *attrib.* Chiefly *U.S.*

1861 G. Meredith *Let.* 9 July (1970) I. 91 The great 'shots' of Stanz parade the town with their prizes in their hats. **1929** *Cincinnati* (Ohio) *Enquirer* 5 Oct. 10/3 One of them is just as likely to win the series as one of the 'big shots'. **1933** D. Runyon in *Collier's* 28 Jan. 41/1 Many of these guys are very high shots during the gold rush. **1935** C. Odets *Waiting for Lefty* in *3 Plays* (1936) iii. 135 Sure, the big shot money men want us like that. **1941** Auden *New Year Letter* ii. 33 Unlike the big-shots of the day. **1957** H. Roosenburg *Walls came tumbling Down* iii. 74 Who are these new prisoners? Are they all the big-shot Nazis? **1960** *New Statesman* 9 Jan. 31/1 On arrival I was asked to dine with Thomas Lamont, along with a number of big-shots in the American newspaper world, including..Henry Luce of *Time-Life.* **1974** K. Millet *Flying* (1975) iii. 300 He would still go for his man in an interview. Used the program to get the big shots.

IV. 23. g. A dram of spirits.
1928 Wodehouse *Good Morning, Bill* ii. 72, I think *I'll* take a shot in a glass. **1935** J. T. Farrell *Judgment Day* ii. xviii. 449 Near White City he stopped in front of a speakeasy, deciding that one good, stiff shot would jack him up. **1955** 'A. Gilbert' *Is She dead Too?* vi. 112 Edwin..produced a very little whisky in a bottle. Lamb ..gave himself a generous shot in the cup of tea. **1979** R. Jaffe *Class Reunion* (1980) ii. v. 235 He poured two shots of vodka, one for her and one for himself.

V. 28*. *that's the shot!* and varr.: in expressions of approval, that's a good idea, or the 'very thing'. *Austral. colloq.*
1953 T. A. G. Hungerford *Riverslake* 142 That's the shot. Buy a bit of land and grow things. **1958** R. Stow *To Islands* ii. 46 Cattle's the shot... They worked it before. **1963** J. Cleary *Flight of Chariots* 370, I think a good strong cuppa brew would be the shot. **1976** D. Ireland *Glass Canoe* 227 'That's the shot,' said Mick. 'Stick around and guard the place.'

30. shot drill (later example); **shot effect** [tr. G. *schroteffekt* (W. Schottky 1918, in *Ann. der Physik* LVII. 547), f. *schrot* small shot], the fluctuation in the magnitude of the anode current in a thermionic valve due to the random character of electron emission; also *transf.*, any fluctuation having a similar stochastic character; **shot-firer,** (*a*) a man employed to fire the shot (sense 7 f) in blasting; (*b*) an electrical device for detonating the shot; hence **shot-firing**; **shot glass,** (*b*) *U.S.* a glass for holding a short drink; **shot gold** orig. *U.S.*, gold occurring in the form of small spheres like lead shot; **shot line** = *shot rope* below; **shot list** *Cinematogr.* and *Television*, a list of shots made by a camera; **shotmaking** *U.S.*, in golf, tennis, etc.: the playing of (esp. successful or attacking) strokes; also **shot-maker; shot noise** = *shot effect* above; **shot-peening** [see Peen, pene *v.* in Dict. and Suppl.], the use of a stream of hard metal particles directed against a metal part to harden and strengthen its surface; so **shot-peen** *v. trans.*; **shot-peened** *ppl. a.*; **shot rope,** a weighted rope hung over the side of a boat and used to guide the descent and ascent of divers.
1936 'R. Hyde' *Passport to Hell* xv. 232 Either he didn't know I was supposed to be in Le Havre doing shot drill, or he'd forgotten. [**1921** *Sci. Abstr.* A. XXIV. 759 The object is the measurement of the spontaneous current variations in high-vacuum discharge tubes, a subject which has been previously dealt with theoretically by Schottky and called by him the 'Schrot effect' (literally, the small shot effect).] **1923** *Chem. Abstr.* XVII. 924 (*heading*) The present state of the shot effect problem. **1930** *Proc. IRE* XVIII. 243 In the absence of space charge the noise has been termed by Schottky the 'schroteffekt', or 'small shot effect', from the analogy which the flight of electrons from the filament to the plate of a vacuum tube bears to the spattering of small shot fired from a shot gun. The simple term 'shot effect' will be used in this paper to denote this noise either with or without space charge. **1947** *Electronic Engin.* XIX. 82/1 Shot effect is more pronounced when the negative grid bias is greater than the usual value. **1964** N. Wiener *God & Golem* 41 There are..cases..where these irregularities are just what we wish to produce, and there are commercial devices for producing them. These are known as shot-effect generators. **1968** P. A. P. Moran *Introd. Probability Theory* ix. 423 Campbell's theorem originally arose in the study of the 'shot effect' in thermionic vacuum tubes. **1883** W. S. Gresley *Gloss. Terms Coal Mining* 219 *Shot firer*, a man specially appointed by the manager of a mine to fire off every shot in a certain number of stalls or heads during the shift. **1886** [in Dict., sense 29]. **1891** C. Pamely *Colliery Manager's Handbk.* xiv. 472 Shots are fired by the aid of litter straws, paper squibs.., safety fuze or by an electric shot-firer. **1939** G. Heyer *No Wind of Blame* xvi. 315 'What's that thing called that they use in mines when they want to blast? Electrical thing they touch off the dynamite with?' 'A shot-firer, do you mean? **1973** 'J. Patrick' *Glasgow Gang Observed* xvii. 140 He began work as a shot firer in a factory close to the approved school. **1884** *Engineering* 31 Oct. 420/2 (*heading*) Shot-firing in mines. **1959** *Times Rev. Industry* Feb. 14/1 For many years the infusion of coal *in situ* by water under pressure has been employed to reduce the dust hazard. Recently this procedure has been combined with that of shot-firing. **1955** A. Miller *Mem. Two Mondays* in *View from Bridge* 43 Enter Bert, carefully carrying a shotglass of whisky. **1970** A. Maling *Lambert's Son* (1972) xxxviii. 157, I put the lemon twists in a shot glass. **1858** *Pike's Peak Guide Book* 222 Those who have prospected over in the parks..say that they find the shot gold there. **1929** E. J. Dunn *Geol. Gold* xvii. 185 Spherical grains and small pieces of gold are found in the alluvial wash at Creswick... They are often quite spherical, generally of small size, but occasionally up to several dwts. in weight, and are known as 'shot gold'. **1971** A. P. McInnes *Dunlevy* 113 Shot gold..is always considered a coarse gold prospect, indicating coarser gold lower down. **1968** A. P. Balder *Compl. Man. Skin Diving* xiii. 248 A shot line..should be used from a boat when diving in bad visibility. **1976** Zanelli & Skuse *Sub-Aqua Illustr. Dict.* 84/2 *Shot line*, a line to which a very heavy weight (or 'shot') is fixed. It is used to guide the descent and ascent of divers. It must not be used as an anchor, because shot lines should be hung vertically. **1969** J. Elliot *Duel* iii. ii. 233 She..learned how to make production breakdowns, set out commentary scripts, type shot lists. **1971** P. Purser *Holy Father's Navy* xxi. 101 The film was back from the labs, the shot list neatly typed up, the editing facilities booked. **1974** *Union* (S. Carolina) *Daily Times* 23 Apr. 9/5 The best five shot-makers in each team. **1969** *New Yorker* 14 June 45/1 My style is playmaking—consistent, percentage tennis—and his style is shotmaking. **1977** *Ibid.* 8 Aug. 48/3 These statistics..give no indication of the absolutely superb shotmaking that Bolt produced on an exceedingly narrow, fast, and exacting course. **1930** *Proc. IRE* XVIII. 255 The solid line curve *D* is the sum of the calculated shot and thermal noises. **1978** *Nature* 8 June 432/1 Individual QSO continuum magnitudes, which are also affected by the added uncertainty introduced by shot noise were generally accurate to ±7%. **1944** H. F. Moore *Shot Peening & Fatigue of Metals* (Amer. Foundry Equipment Co.) 5/1 The metal just below the shot-peened layer is somewhat affected by the shot peening. **1944** *Proc. Soc. Exper. Stress Analysis* II. 172/2 These pieces..were shot-peened on both flat faces. **1956** F. H. Keating *Chromium-Nickel Austenitic Steels* v. 70 Well substantiated claims have been made for improvement in fatigue-resistance by shot-peening, which introduces compressive stresses in the surface layers. **1962** *Engineering* 23 Mar. 403/2, 85 per cent of the metal is machined away on electronically controlled milling machines before being curved by press or shot-peening. **1972** L. M. Harris *Introd. Deepwater Floating Drilling Operations* xii. 138 Each weld should be ground inside and out, and the transition area and weld shotpeened. **1972** H. T. Jensen in Mann & Milligan *Aircraft Fatigue* 156 We established an allowable strength of two-thirds of the strength of the machined and shot-peened strength for components that retain their as-forged surfaces. **1909** *Man. Seamanship* (Admiralty) II. vi. 116 As soon as the diver sees anything he can signal for shot rope to be lowered. **1940** 'N. Shute' *Landfall* 257 Then in slow motion he [*sc.* a diver] reached out and grasped the shot-rope, stepped off the ladder and was gone. **1960** Brookes & Broadhurst *Diving Manual* (ed. 2) 105 A shot rope should be very heavily weighted so that it will hang vertically in the water, uninfluenced by tides and currents.

shot, *ppl. a.* Add: **4. a.** Also with *down*: of an aircraft or its crew.
1943 'M. Coles' *Without Lawful Authority* ix. 115 'The 'plane..crashed in flames just this side of the Polish frontier... One of our fellows..says there were bullet-holes in the wings.' 'Shot down, eh?' **1957** H. Roosenburg *Walls came tumbling Down* 7 A local resistance group ..concentrated on picking up shot-down Allied pilots. **1968** *Listener* 26 Dec. 858/2 He was also the source of her story..of the shot-down pilot who told his captors that his sister in the States was a rich '*industrielle*'. **1980** E. Behr *Getting Even* x. 120 Organising escape routes for shot-down R.A.F. and U.S. air crews on the run.

b. Drunk. *slang* (chiefly *U.S.*, *Austral.*, and *N.Z.*). Cf. Shoot *v.* 32 d.
1864 *Harper's Mag.* May 856/2 He again sat down by the fire..by which time he was pretty well 'shot'. **1896** W. C. Gore in *Inlander* Jan. 146 *Shot*, a. Intoxicated. **1930** *Sat. Even. Post* 26 July 145/2 'I'm half shot,' he said... 'An' so are you. You're just as drunk as I am.' **1943** N. Marsh *Colour Scheme* x. 187 The chap was half-shot... He smelt of booze. **1957** *Nelson* (N.Z.) *Even. Mail* 18 May 7 He asked the man: 'Are you shot?' The man said: 'Yes.'..It was ascertained that he had fallen down while intoxicated. **1972** T. Lilley *K Section* ix. 43 He was well shot last night. Staggering.

c. In fig. phr. *shot through* (also *to hell* or *pieces*), in a state of ruin or collapse. *colloq.* (chiefly *U.S.*).
1926 E. Hemingway *Fiesta* (1927) iii. xix. 277 That meant San Sebastian all shot to hell. **1932** L. Golding *Magnolia St.* i. iv. 67 The old man was all shot to pieces... He had fallen into a sort of torpor. **1937** H. G. Wells *Brynhild* xi. 243 To-day I feel shot through. I feel shot to pieces. **1977** M. Babson *Murder, Murder, Little Star* vii. 50 Look at the price I pay. My private life is shot to hell.

d. Of things: worn out, ruined, used up, spent. *slang* (chiefly *U.S.*).
1933 C. K. Stewart *Speech Amer. Airman* (Univ. Akron thesis) 89 *Shot*, an adjective meaning 'useless' 'gone', or 'worn out'. **1960** *Analog Science Fact/Fiction* Oct. 136/1 With him gone, the interstellar drive project would've been shot. **1970** I. Petite *Meander to Alaska* i. vii. 66 At that point they discovered that the transmission bearings were 'shot'. **1981** G. V. Higgins *Rat on Fire* vii. 54 Your boiler is one of those old things... I think it's about shot.

e. With *up*, severely wounded or damaged by shooting. Also *fig.* (*colloq.*) and *transf.* drugged (*U.S. slang*).
1934 V. M. Yeates *Winged Victory* i. xviii. 146 If he tried to do anything on his own he would probably get himself shot down, or at least shot up. **1938** [see *polluted *ppl. a.* b]. **1945** *Richmond* (Va.) *Times-Dispatch* 22 May 2/8 The Forty-fifth Division's 'most shot up soldier to return alive' is back in the States. **1964** L. Nkosi *Rhythm of Violence* 63 Stop Shouting!..Are you all shot up or something! **1978** *Detroit Free Press* 2 Apr. (Detroit Suppl.) 8/1 Starting up the stairs, she steps around a recently shot-up addict who is just nodding off.

f. Of people: exhausted. *slang* (chiefly *U.S.*).
1939 [see *cosy *v.* 2]. **1945** G. Casey *Downhill is Easier* iv. 183 Late at night you could easily walk the twelve miles..without seeing a vehicle. I realized I was shot. **1951** E. B. White *Let.* 11 Dec. (1976) 346 Ross died last week and we have been in something of a scramble here, as well as feeling quite shot. **1967** 'V. Siller' *Biltmore Call* 120, I thought she was shot and her nerves had given out. **1972** J. Gores *Dead Skip* (1973) xii. 83 He..[was] literally too tired to move... Shot. Utterly shot.

sho·t-bag. Chiefly *U.S.* [Shot *sb.*[1]] A bag for carrying shot; a shot-pouch. Also *transf.* a purse.
1638 in *Archives of Maryland* (1887) IV. 32 It[em] one fowling piece & shott bagge. **1750** P. Hog *Let.* 14 May in S. M. Hamilton *Lett. to Washington* (1898) I. 260 Going to Load he missed his Shot Bag which had been Carried away by one of the Shots. **1800** A. Henry *Jrnl.* 4 Oct. (1897) I. iii. 111 The Indians were standing in the fort with nothing on but their breech-clouts, powder-horns, shot bags, and guns in their hands. **1872** 'Mark Twain' *Roughing It* ii. 5 We also took with us a little shot-bag of silver coin. **1946** *Aircraft Engineering* XVIII. 109/2 The time-honoured method of loading with shot bags the structure to be tested in an inverted position is still current practice across the Atlantic.

sho·t-blast. [Shot *sb.*[1]] A high-speed stream of steel particles directed at a surface to clean and roughen it.
1923 *Foundry* 1 Feb. 13 Have you noticed how the modern shot-blast and grit-blast are replacing the old dust breeding sand-blast? **1934** *Foundry Trade Jrnl.* LI. 340 The barrel..is essentially a contrivance for mechanically exposing every part of the casting or forging to be cleaned to the effect of the shot-blast. **1958** *Engineering* 21 Mar. 382/3 An airless type shot-blast plant has been installed, to deal with 35 tons of castings a day.

Hence **shot-blasting,** the use of a shot-blast; also **shot-blast** *v. trans.*, to subject to shot-blasting; **shot-blasted** *ppl. a.*; **shot-blaster,** a person using a shot-blast.
1934 F. W. Partington *G. F. Charnock's Mech. Technol.* (ed. 2) xxxi. 415 Steel castings are frequently cleaned of both sand and scale by subjecting them to shot-blasting. **1937** *Jrnl. Iron & Steel Inst.* CXXXV. 160a (*heading*) Buick shotblasts large castings automatically. **1941** W. K. Wilson *Pract. Solution Torsional Vibration Problems* (ed. 2) II. vii. 146 Shot-blasted springs are very susceptible to rusting. **1946** *Engineer* 15 Feb. 152/1 A large-capacity dust extractor, coupled to the shot-blasting chamber, exhausts all the fine dust and sand. **1959** *Times* 27 Apr. (Suppl.) p. xi/4 It has first to be shot-blasted to provide an absolutely clean surface. **1961** *Evening Standard* 26 July 18/5 (Advt.), Shot blasters & metal sprayers..reqd. **1964** S. Crawford *Basic Engin. Processes* iii. 87 Prior to bronze welding the joint faces or edges are cleaned by shot blasting. **1975** *Offshore Engineer* Sept. 143/4 A new version of their shot-blaster's helmet has been introduced by Martindale Protection. **1977** *Exchange & Mart* (South) 24 Feb. 19 M/3 One bodiless chassis, stripped, shotblasted, etc. partly reassembled.

shot-bush. (Earlier example.)
1785 [see *pigeon-weed* s.v. *pigeon *sb.* 6].

shot-gun, shotgun. For **a, b** read **1, 2** and add: **1. b.** *ellipt.* for (*a*) a shotgun building (sense 3 b below); (*b*) = *shotgun formation* (U.S. Football), sense 2 below.
1945 B. A. Botkin *Lay my Burden Down* 98 They had to go out and live in sod houses and little old boxed shotguns and turn their Negroes loose. **1966** [see *shotgun formation* below]. **1976** *Time* 19 Jan. 43 Staubach's talent is throwing from the shotgun. **1977** *New Yorker* 10 Oct. 178/2 Buckley, operating from the shotgun, threw some forty, of which some twenty were valid.

2. shotgun formation *U.S. Football* (see quot. 1966); also *fig.*; **shotgun marriage** orig. *U.S.* = *shotgun wedding* below; also *fig.*; **shotgun microphone** (or *colloq.* **mike**), a highly directional microphone with a long barrel that is pointed at a distant source of sound; **shotgun wedding** orig. *U.S.*, a wedding made in haste or under duress by reason of the bride's pregnancy; also *fig.*
[**1966** Rote & Winter *Lang. of Pro Football* iii. 137 *Shotgun*, offensive formation where quarterback sets four or five yards behind center with other backfield men split out as flankers or slot backs; an offensive formation designed to facilitate sending out as many pass receivers as possible while passer is in safer position to throw.] **1967** *Wall St. Jrnl.* (Eastern ed.) 30 Jan. 8/5 Mr. Hornung's argument [that his accomplishments constituted educational, artistic, scientific and civic achievements] constitutes 'a shotgun formation', the court said. **1972** J. Mosedale *Football* vi. 91 Hickey installed the shotgun formation, putting the quarterback at tailback where he could pass or run. **1929** E. W. Howe *Plain People* xxix. 267 Two people cordially disliked me for years because I thought it best to mention very briefly and respectfully their shot gun marriage. **1958** *Manch. Guardian* 22 Mar. 4/6 There were references to..the possible shotgun marriage of the 'Daily Herald' and the 'News Chronicle'. **1973** 'H. Carmichael' *Too Late for Tears* xv. 179 Shot-gun marriages went out with the advent of the Pill. **1968** J. M. Ullman *Lady on Fire* (1969) viii. 111 Even if they walked around Curley..could pick up the conversation

with a shotgun microphone. The device had a range of several hundred feet. **1972** *Jrnl. Social Psychol.* LXXXVI. 30 A shotgun microphone was located to one side of the group to record the verbal interactions. **1978** T. Gifford *Glendower Legacy* (1979) 187 We used that miserable shotgun mike and believe me, Brennan was watching television and sneezing. **1927** S. Lewis *Elmer Gantry* ix. 134 There were, in those parts and those days, not infrequent ceremonies known as 'shotgun weddings'. **1946** *Sun* (Baltimore) 29 Oct. 1/1 Charges that the veto system was a 'shotgun wedding' forced upon the small nations. **1974** E. Huxley *Gallipot Eyes* (1976) 61 She can't have been more than fourteen when she married. A shot-gun wedding, clearly.

3. a. Passing into *adj.* Made or done hastily or under pressure of necessity.

1937 *Sun* (Baltimore) 18 Aug. 8/2 Shotgun legislation... Measures pushed through in a last-hour rush. **1962** *Economist* 7 Apr. 73/1 Mr Sandys's shot-gun reorganisation of the aircraft industry. **1977** *Observer* 13 Mar. 13/2 By the end of last year, 464 men had been forced to quit, 74 after formal proceedings, the remainder in shotgun resignations.

b. Designating a house or other building with rooms set in a line on either side of a long central hallway. *U.S.*

1938 J. Stuart *Beyond Dark Hills* vi. 156 Their faces wore the blank expression of the Armco plant's shotgun dwelling houses. **1944** T. D. Clark *Pills, Petticoats & Plows* iii. 56 There was no wiser spot on earth than the porches which jutted out from the long shotgun buildings. **1950** *Penguin New Writing* XL. 67 Your Riverbottom Nigras lived in little shotgun houses. **1964** *Amer. Folk Music Occasional* 1. 92 De Ole souf unroll... Dog-trots and shotgun shantys. **1974** *Times* 14 Jan. 12/3 The American South is still unmistakably southern..grits at breakfast, blacks living in shotgun shacks.

c. quasi-adv. *to ride shotgun*: see *ride *v.* 1 m.

shot-hole. Add: **1. a.** (Earlier example.)

1745 Capt. Durrell *Jrnl.* 20 May in J. S. McLennan *Louisbourg* (1918) 177 We had several Shott holes in all our sides.

b. (Examples.)

[**1889** E. A. Ormerod *Rep. Observations Injurious Insects* 94, I found that the injury [to the tree] began by a small hole like a shot-hole in the side of the attacked stem.] **1946** *Nature* 13 July 52/2 Dry rot..is the result of the operations of fungi, and not of insects—the attacks of the latter being usually discernible by the presence of small pin or 'shot' holes in the wood. **1972** *Gloss. Terms Timber (B.S.I.)* 17 *Shothole*, a worm-hole usually more than 1·5 mm and not more than 3 mm in diameter.

c. A small round hole made in a leaf by a fungus or bacterium; also, a condition in which such holes occur.

1897 [implied in sense 4 below]. **1902** D. McAlpine *Fungus Dis.* iii. 33 There is a very familiar appearance in the leaves of many of our stone-fruit trees, where they are more or less punctured with round holes, as if riddled with shot; hence the name 'shot-hole' applied to the injury. **1926** F. D. Heald *Man. Plant Dis.* ii. 33 Some varieties are prone to shot hole whenever localized areas of leaf tissue are killed. **1946** H. Wormald *Dis. Fruits & Hops* 143 The chief agent of leaf spotting in plums is the organism which causes Bacterial Canker... Eventually the infected parts are killed and drop out, leaving 'shot holes'. **1976** A. Hellyer *Collingridge Encycl. Gardening* 259 Small round holes appear in the leaves, a symptom which is sometimes known as shot-hole and was once believed to be a separate disease [from bacterial canker caused by *Pseudomonas morsprunorum*].

4. *Comb.*, as **shot-hole borer**, a small bark beetle of the family Scolytidæ, esp. *Anisandrus dispar* (cf. *scolytid); **shot-hole disease**, a plant disease characterized by shot-holes in the leaves; **shot-hole fungus**, a fungus which causes shot-holes, esp. in certain fruit trees.

[**1890** E. A. Ormerod *Man. Injur. Insects* (ed. 2) 331, I found that the cause of the injury was the 'Shot-borer' Beetle (as it is called in America).] **1916** *Farmers' Bull.* (U.S. Dept. Agric.) No. 763. 2 The shot-hole borers or barkbeetles burrow into the bark. **1968** *Oxf. Bk. Insects* 190/2 The Shot-hole Borer.., one of the Ambrosia Beetles, is found locally in southern England. **1926** F. D. Heald *Man. Plant Dis.* xx. 511 The disease [*sc.* cherry leaf spot] is known by various common names, such as 'leaf blight', 'leaf spot', 'yellows', 'yellow leaf', and the 'shot-hole disease'. **1946** Shot-hole disease [see *abscission 3]. **1897** W. G. Smith tr. *von Tubeuf's Dis. Plant* II. 463 *Phyllosticta persicae*... The name 'shot-hole fungus' has sometimes been applied to this and allied forms. **1906** M. C. Cook *Fungoid Pests* 131 A shot-hole fungus has been found lately, several times affecting Peach leaves in this country.

shott. Add: (Later examples.) Also **shat.** Cf. *sabkha.

1898 *Geogr. Jrnl.* June 604 The shats, or salt lakes, of the south of Tunis are rather a disappointment to the traveller. **1902** *Encycl. Brit.* XXXIII. 482/1 These shats ..are, strictly speaking, not lakes at all at the present day. **1957** G. E. Hutchinson *Treat. Limnol.* I. i. 136 In north Africa there is a series of inland drainage basins or shotts between the Greater and Lesser Atlas Mountains. **1969** J. Mavor *Voyage to Atlantis* ii. 46 Paul Borchardt. in 1927..placed the lesser island of Atlantis..a few miles inland of the gulf of Gabès in Tunisia in the region of 'shotts'.

shotty, *a.* Add: *spec.* of gold: in the form of small, roundish lumps. (Earlier and later examples.)

1860 *Mining Surveyors' Rep.* (Mining Dept., Victoria) Aug. 236 There were also some very good patches of shotty gold and small nuggets found in the vicinity of this nugget. **1929** E. J. Dunn *Geol. Gold* xix. 222 Pounds weight of 'shotty' gold..were washed from a dish of gravel. **1959** *Observer* 17 May 8/3 Odd old prospectors still fossick in the hills in search of shotty gold.

shoulda, shouldda (ʃu·də), repr. colloq. or vulgar pronunc. of *should have* (see Shall *v.* 18).

1933 [see *junk *sb.*[2] 1 e]. **1943** C. Himes *Black on Black* (1973) 194 Man, you shoulda seen them cats. **1956** 'E. McBain' *Cop Hater in 87th Precinct* (1959) 77 The kid was a Junior... Nobody shoulda given him a reefer. **1967** V. S. Naipaul *Mystic Masseur* iv. 55, I shoulda get married long before now. **1978** J. Irving *World according to Garp* xviii. 393, I guess I shoulda *knocked*.

should-be, *sb.* and *a.* (Later examples.)

1878 W. James in *Jrnl. Speculative Philos.* XII. 14 The interest of survival..has hitherto been treated as an ideal *should-be*. **1885** G. M. Hopkins *Let.* 24 Apr. (1938) 107 God grant it may not be..that the should-be receiver was dead. **1951** [see *is *sb.*].

shoulder, *sb.* Add: **2. f.** (*straight*) *from the shoulder*: also *fig.*

1904 W. H. Smith *Promoters* v. 103 You'll..be in a shape to talk business, right from the shoulder. **1911** R. D. Saunders *Col. Todhunter* ix. 118 A man that talks old-fashioned American Democracy straight from the shoulder. **1926** N. Coward *Easy Virtue* I. 10, I must.. have a talk to her... A straight-from-the-shoulder chat might make her see things in a better light. **1947** L. P. Hartley *Eustace & Hilda* xi. 182 That letter had been written straight from the shoulder, or the heart. **1963** V. Nabokov *Gift* iv. 214 He subsequently wrote it right down, straight from the shoulder, in three nights. **1977** *Gramophone* Aug. 291/3 As to the power and authority, he takes Beethoven at his word when he sees an *ff* mark and lets you have it right from the shoulder.

h. *to weep* (or *cry*) *on* (a person's) *shoulder*: to pour out one's troubles to a person; also in phr. *a shoulder to cry on*, a sympathetic and consoling listener to a person's troubles.

1935 H. L. Ickes *Diary* 10 Feb. (1953) I. 292, I called Tugwell yesterday afternoon to tell him that if he wanted any shoulder to weep on, mine was a broad one. **1942** T. Bailey *Pink Camellia* iii. 19 She likes to talk of her troubles and weep on people's shoulders. **1966** L. Deighton *Billion Dollar Brain* xvi. 163 I'm always weeping on your shoulder. **1974** 'J. le Carré' *Tinker, Tailor, Soldier, Spy* xviii. 158, I asked 'What did he want?' And Ann said 'A shoulder to cry on.' Bill..wanted to pour out his heart, she said.

i. *off-the-shoulder* (attrib. phr.): of a dress, blouse, etc., that leaves the shoulders bare.

[**1813** Jane Austen *Let.* 15 Sept. (1932) 322 Stays now are not made to force the bosom up at all... I was really glad to hear that they are not to be so much off the shoulders as they were.] **1952** S. Kauffmann *Philanderer* (1953) xiv. 232 She was wearing an off-the-shoulder white blouse. **1960** *Guardian* 5 Jan. 6/7 A golden off-the-shoulder evening dress of only eight years ago. **1974** *Country Life* 17 Jan. 106/3 Off-the-shoulder and square necklines.

4. c. (*b*) *shoulder of mutton sail* (later examples).

1880 *Harper's Mag.* LXI. 350/2 But the Hampton boat —a modified pink-stern, with shoulder-of-mutton sails on its small masts—was the 'abler'..to stand the exigencies of all sorts of weather. **1961** F. H. Burgess *Dict. Sailing* 187 *Shoulder-of-mutton sail*, name given to the triangular Bermudan sail.

6. j. The edge of a road; *spec.* a strip at the side of the main carriageway on which vehicles may stop in an emergency. Cf. *hard shoulder* s.v. *hard *a.* 22 b; *soft shoulder* s.v. *soft *a.* 27. orig. *U.S.*

1933 *Sun* (Baltimore) 27 Dec. 8/7, I..stayed well over on the shoulder. But..only one of the numerous cars.. bothered to move nearer the middle of the road. Repeatedly, I stepped back into the bushes and mud. **1942** *Short Guide Gr. Brit.* (U.S. War Dept.) 32 *Shoulder*, (of road)—verge. **1965** 'E. McBain' *Doll* (1966) x. 127 The road was winding and narrow... The shoulders were muddy and soft. **1979** G. Seymour *Red Fox* xii. 185 The engine coughed and died, barren of petrol... They were about to stop on the hard shoulder.

k. A poorly resolved subsidiary maximum interrupting a part of a graph otherwise having a fairly uniform or smoothly varying slope.

1956 *Jrnl. Exper. Med.* CIII. 657 The existence of the shoulder in the survival curve..is unequivocal and constitutes evidence for a multiple hit killing mechanism. **1964** *Physics in Med. & Biol.* IX. 167 If the log of the surviving fraction is plotted against dose on a linear scale, after an initial shoulder, a straight-line graph is obtained. **1977** *Nature* 17 Feb. 660/2 The asymmetry evident in the low resolution scan is revealed to be a shoulder at ~10 cm⁻¹ to lower energy than the main band which occurs at 1,528 cm⁻¹.

l. *Surfing.* (See quots.)

1962 T. Masters *Surfing made Easy* 65 *Shoulder*, the unbroken section to the side of a breaking wave. **1965** J. Pollard *Surfrider* ii. 20 Take this one near the 'shoulder', the unbroken part of the wave reached at the end of a slide. **1968** *Surfer Mag.* Jan. 65/1 Positively the shoulder-hoppers paradise.

9. a. *shoulder belt, blanket, harness, pad, sack, -socket.*

1968 *N.Y. Times* 15 Sept. I. 46 A new safety seat, with built-in shoulder belts, is being developed by the General Motors Corporation. **1976** *Billings* (Montana) *Gaz.* 24 June 7-A/5 Ontario on Jan. 1 became the first jurisdiction on the North American continent to require the wearing of available lap or shoulder belts. **1973** A. H. Whiteford *North Amer. Indian Arts* 69 Shoulder blankets, worn by males, have checkered or tartan patterns. **1968** *Time* 5 Apr. 38 Padded roll bars and shoulder harnesses are standard on the Shelby Cobra. **1974** Hawkey & Bingham *Wild Card* xxii. 180 Wallcroft unfastened his seatbelt and shoulder harness and got out [of the car]. **1868** C. L. Eastlake *Hints on Household Taste* iii. 80 The 'Cromwell' chair..is..copied from examples of the seventeenth century... Both the seat and shoulder-pad are stuffed.. with horsehair. **1904** *Sci. Amer.* 21 May 406/1 Every coat has a shoulder-pad of various thicknesses made of wadding. **1951** *Sport* 16–22 Mar. 14/3 A slight 'teacup storm' occurred in Yorkshire Rugby Union circles because a Leeds team were alleged to be wearing shoulder pads. **1979** R. Perry *Bishop's Pawn* iii. 51 He ripped seams, split shoulder pads and carved up shoes. **1923** D. H. Lawrence *Captain's Doll* 232 Alexander was putting the bread back into his shoulder-sack. **1953** *Scrutiny* XIX. 289 He pulls out the picture from his shoulder-sack. **1921** D. H. Lawrence *Birds, Beasts & Flowers* (1923) 81 Shall great wings flame from his shoulder-sockets Assyrian-wise?

b. *shoulder-fired, -launched* adjs.

1967 J. S. Tompkins *Weapons of World War III* viii. 105 There is also a shoulder-fired descendant of the bazooka called the LAW, or Light Antitank weapon. **1974** *Times* 5 Mar. 6/8 Guards now are equipped with the General Dynamics Redeye infra-red-guided, shoulder-launched anti-aircraft missile. **1977** *Belfast Tel.* 22 Feb. 17/4 Our new Blowpipe shoulder-launched missile which is in service with the armed forces of both the United Kingdom and Canada.

c. shoulder-bag, a bag carried by a strap or straps slung over the shoulder; **shoulder board** chiefly *U.S.*, each of the two stiffened pieces of material worn at the shoulders of military uniform and bearing the insignia of rank; **shoulder charge**, a charge in which the shoulder is directed at the target; hence as *v. trans.*; **shoulder holster**, a holster suspended from a shoulder-strap; **shoulder-length** *a.*: of hair, etc., that reaches down to the shoulders; **shoulder line**, (*a*) a line drawn on the shoulder (of an object); (*b*) the line of a woman's garment over the shoulders; **shoulder patch**, a patch attached to the shoulder of a garment and bearing an emblem or insignia; **shoulder plane** *Woodworking* (see quot. 1954); **shoulder pod** [cf. Tripod *sb.*], a support for a camera that rests against the shoulder; **shoulder stand**, a position in which the body and legs are held up in the air and supported on the shoulders; **shoulder tab**, each of the two pieces of material worn at the shoulders of military or other uniform and bearing insignia of rank; **shoulder throw** *Judo* (see quot. 1968); **shoulder wing**, a monoplane wing mounted high on the fuselage but not in the highest position; usu. *attrib.*

1912 D. H. Lawrence *Let.* 19 Aug. (1932) 49 We walked from the Isarthal down here—F. and I—with our German shoulder-bags on our backs. **1960** L. Davidson *Night of Wenceslas* ii. 32 She was wearing a gaily coloured cotton frock and a shoulder bag. **1977** P. Theroux *Consul's File* 48 She sat down and threw her shoulder-bag on a side-table. **1949** J. Steinbeck *Russian Jrnl.* 20 The uniforms were without insignia and without shoulder boards. **1980** 'J. le Carré' *Smiley's People* xxv. 295, I saw no shoulder-boards, the guards wore plain clothes. **1930** *Daily Express* 6 Oct. 16/2 They exchanged good shoulder charges, and honours were about even. **1971** *Sunday Australian* 8 Aug. 3/4 Twice outside the motel where the Springboks were staying they were shoulder-charged by police. **1973** *Weekly News* (Glasgow) 11 Aug. 7/2 He shoulder-charged the door pushing it open. **1895** *Montgomery Ward Catal.* 481/2 Shoulder Holster, with breast and shoulder strap to wear under coat on left side. **1935** M. M. Atwater *Murder in Midsummer* xxi. 193 Mr. Henry Smith..buckled on his shoulder-holster, weighted by his old six-shooter. **1973** 'I. Drummond' *Jaws of Watchdog* xiii. 166 Sandro's own gun was in its shoulder-holster. **1951** Shoulder-length [see *cut *sb.*[2] 16 a]. **1976** C. Dexter *Last seen Wearing* xxix. 202 Long shoulder-length hair..brushed forward over her face. **1916–17** *T. Eaton & Co. Catal.* Fall & Winter 414/2 Semi-porcelain dinner set has..gold edges and green shoulder line. **1931** *McCall's Mag.* Sept. 74 A significant self-fabric cuff and a very notable shoulder line. **1979** *Guardian* 13 June 12/4 The best of this year's T-shirts..are loose with a dropped shoulderline. **1947** A. P. Gaskell *Big Game* 82 He recalled their first issue of shoulder-patches. **1970** N. Armstrong et al. *First on Moon* v. 101 An Apollo 1 shoulder patch..would be left on the moon. **1935** N. R. Rogers *Technol. Woodwork & Metalwork* I. 56 The Shoulder Plane is intended, as its name implies, for trueing tenon shoulders (end grain). **1954** W. E. Kelsey *Carpentry, Joinery & Woodcutting Machinery* i. 16 Shoulder planes..are metal rebate planes with a narrow mouth and a low-pitched cutter... They are used chiefly for planing against the end grain and are specially suitable for hardwoods. **1979** A. B. Emary *Woodworking* xxviii. 121 The bevelled portion of the mouldings can be made with a shoulder plane or a badger plane. **1963** D. Botting in A. Smith *Throw out Two Hands* 263 It was possible to

make hand-held movie shots (using pistol-grip or shoulder pod) with lenses of longer focus than usual. **1981** *Birds* Autumn 18/3 With miniaturisation and the wide use of telephoto lenses,..the stalking technique evolved, using shoulderpods rather than tripods. **1956** Kunzle & Thomas *Freestanding* iv. 44 *Inverted shoulder stand.* Start from back lying and raise the legs and hips until vertical. **1977** 'M. Yorke' *Cost of Silence* iv. 32 Sarah was practising the shoulder stand upstairs. **1966** D. Francis *Flying Finish* v. 66 Gold-braided shoulder tabs on his navy uniform jacket. **1956** K. Tomiki *Judo* iii. 73 *Seoi-Nage* (Shoulder throw). **1960** *Oxford Mail* 10 Mar. 8/3 Milsom scored a half-point for a shoulder throw then full points for a hip throw and a stranglehold. **1968** K. Smith *Judo Dict.* 186 *Shoulder throws*, those made from a standing position and using principally the action of the hands and arms. **1941** R. A. Saville-Sneath *Aircraft Recognition* I. ii. 15 Variants of the *high-wing* type are..*Shoulder-wing*, a type in which..the wing-roots join the fuselage at the 'shoulder', i.e. lower than the normal high-wing but appreciably higher than the mid-wing position. **1962** L. Deighton *Ipcress File* v. 33, I noticed a twin-engined shoulder wing Grumman S2F-3. **1969** K. Munson *Pioneer Aircraft 1903–14* 149/1 The Type A was a single-seat, warp-controlled, shoulder-wing monoplane with a 50 h.p. gnome rotary.

shoulder, *v.* Add: **7. a.** Also *spec.* of a racehorse, to carry (a specified weight) on the back.

1939 *Country Life* 11 Feb. 156/2 Last year, when shouldering 10st. 2lb., he fell at Becher's Brook on the second circuit. **1977** *Western Morning News* 30 Aug. 11/7 The six-year-old was returning to the course of his previous success this season, and for that win was shouldering a 7lb. penalty.

8. b. *to shoulder arms*: also *fig.* in *Cricket* (see quot. 1966).

1966 *Armchair Cricket 1966* 111 *Shoulder arms*, an expression used to describe a batsman's action when he holds the bat aloft over his shoulder as he allows the ball to go by on the off-side without attempting a stroke. **1975** *Daily Mirror* 16 Aug. 28/2 Ross Edwards immediately walked into the next ball, shouldered arms and was leg before. **1977** *Sunday Times* 30 Jan. 30/3 The next ball hit Gaekwad on the pad as he shouldered arms.

shou·lderless, *a.* [f. Shoulder *sb.*] Without a shoulder or shoulders, esp. of garments.

1928 Peake & Fleure *Steppe & Sown* vi. 78 Keller figures two shoulderless sleeves from Meilen, which Schenk placed in the third or Morgienne period. **1963** C. R. Cowell et al. *Inlays, Crowns, & Bridges* vi. 62 Shoulderless crowns present fewer difficulties in fitting than those with shoulders. **1979** *N.Y. Times Mag.* 30 Sept. 68/3 Giorgio Armani..has put together a take-off on classic men's wear that may be the most significant look to come out of Italy this season: the 'shoulderless' suit.

shouse (ʃaus). *Austral. slang.* [Syncopated form of *shit-house* (Shit *sb.* 3).] A privy.

1941 S. J. Baker *Dict. Austral. Slang* 66 *Shouse*, a privy. **1951** D. Stivens *Jimmy Brockett* 214, I seen that now as plain as a country shouse. **1957** 'N. Culotta' *They're a Weird Mob* (1958) 43 Yeah, chuck 'em ter the shouse. **1968** T. Keneally *Three Cheers for Paraclete* 84 I'd like some trees on it, pines and gums, so you don't have to see your neighbour's shouse first thing each morning. **1975** L. Ryan *Shearers* 98 Dewlap, who had been standing at the back of the ring, all alone like a country s'house, now sidled up.

shout, *sb.*[2] Add: **1. d.** *U.S.* Among American Blacks, a form of dancing accompanied by much loud singing, of religious origin (cf. *ring-shout* s.v. *ring *sb.*[1] 18); a song of the type sung during such a performance. Also *attrib.*

1862 in E. W. Pearson *Lett. from Port Royal* (1906) 27 As we walked home we asked Cuffy if they considered the 'shout' as part of their religious worship. **1908** *Sears, Roebuck Catal.* 199/3 *Negro Shouts.* Songs with laughing and whistling choruses. **1937** [see *praying band* s.v. *praying *ppl. a.* b]. **1938** *Mississippi* (Amer. Guide Ser.) 24 Soon a woman leaps out into the aisle. She is 'moved by the spirit', she cries, and slowly, rigidly, she begins 'the shout', or if it is a Holiness meeting, the 'Holy Dance'. It is shuffling, intricate; her heels thud on the floor. **1955** Keepnews & Grauer *Pictorial Hist. Jazz* xii. 127 Stomping variations of rags, known as 'shouts', were the show-pieces most often used in competition. **1972** *Listener* 10 Aug. 187/1 A musical innovator with tremendous vocal power, he brings gospel and shout singing to the blues.

e. *shout-up*, a noisy argument. *colloq.*

1965 G. Melly *Owning Up* ix. 107 Whenever one of us was describing to the other some drunken shout-up with a third party. **1973** *Times* 3 Nov. 11, I didn't mention it until it seemed to become a pattern and then we had a good old shout up.

2. For *Colonial slang* read *slang* (orig. *Colonial*) and add earlier and later examples.

1854 F. Fyans *Let.* in T. F. Bride *Lett. fr. Victorian Pioneers* (1898) 127 Do you forget the shout you stood—the shout for all hands? **1886** H. Baumann *Londinismen* 177/1 *It's my shout*, jetzt will ich euch traktieren. **1911** C. E. W. Bean *'Dreadnought' of Darling* xxxii. 282 'Boys,' he says, 'help yourselves. This is my shout.' **1914** *Bulletin* (Sydney) 17 Dec. 44/2 The..bloke..ses t' me: 'Your shout mate.' **1954** S. Mackenzie *Refuge* 16 Come up and have a cup of coffee—my shout. **1977** D. Bagley *Enemy* xxvii. 212 Honnister addressed the landlord. 'Hi, Monte: a large scotch and a pint of Director's.' 'My shout,' I said.

shout, *v.* Add: **1. g.** (Earlier example.)

1876 *Scribner's Monthly* Nov. 142/1 'Then why prevaricate?' Said he perversely, 'Now yer shoutin'!'

† i. To be loud in support of a candidate. Cf. Root *v.*[2] 1 d. *U.S. Obs.*

1875 [implied in *shouter[2] 1 b]. **1907** *N.Y. Evening Post* (semi-weekly ed.) 21 Nov. 4 Federal office-holders in various Southern States have been dutifully shouting for Roosevelt.

2. c. *fig.* To indicate plainly.

1931 E. F. Benson *Mapp & Lucia* ii. 30 Red-brick houses with tiled roofs, that shouted Queen Anne and George I in Lucia's enraptured ears. **1976** D. Francis *In Frame* iii. 48 From laquered hair via crocodile handbag to gold-trimmed shoes she shouted money.

4. Restrict *dial.* to senses in Dict. **f.** To howl *down* or reduce to silence by shouts of disapproval. Also *fig.*

1924 G. B. Shaw *St. Joan* vi. 89, I know that there is no faith in a Frenchman. [*Tumult, which he shouts down.*] **1965** M. Spark *Mandelbaum Gate* iii. 59 Freddy's thoughts whispered on, refusing to be shouted down by any other voice that might arise in his brain to hush them up. **1967** N. Freeling *Strike out where not Applicable* 36 Francis forces things sometimes by simply shouting her down. **1978** P. Moore *Man, Woman, & Priesthood* xi. 171 This challenge may be not only right, but vindicated; it cannot, however, be ignored and shouted down.

5. Also *N.Z.* **a.** (Earlier and further examples.)

1855 R. Carboni *Eureka Stockade* 68 You shouted nobblers round for all hands—that's all right; it's no more than fair and square now for the boys to shout for you. **1856** H. W. Harper *Lett. from N.Z.* (1914) 10 The first person in New Zealand to 'shout' for me, which here means to ask you into a house of call and stand treat. **1873** J. H. H. St. John *Pakeha Rambles through Maori Lands* v. 82 Our friend set to work pumping him, and 'shouted' liberally till the old fellow's tongue was unloosed. **1916** J. B. Cooper *Coo-oo-ee* iii. 36 Passing that stage of drunkenness, they started to quarrel over the question as to whose turn it was 'to shout'. **1963** N. Hilliard *Piece of Land* 32 'Going to shout, Horace?' Clarrie pulled out some change. **1981** *National Times* (Austral.) 25–31 Jan. 23/1 The tightwad..wouldn't shout if a shark bit him.

b. (Earlier example.) Also *to shout* (a person) *to* (a treat) and with indirect obj.

1855 [see sense a above]. **1906** E. Dyson *Fact'ry 'Ands* xiv. 185 He gave up beer..in order to have it in his power to shout the young lady to 2 s. seats at the Royal. **1940** F. Sargeson *Man & his Wife* (1944) 64 If he had a win he'd shouts us plenty of beer and cigarettes. **1964** V. M. Grayland *Grave-Digger's Apprentice* xx. 119 If my luck's in..I'll shout you to the pictures tonight for helping me out. **1965** S. T. Ollivier *Petticoat Farm* vii. 98 'Mingy old skinflints!' hissed Jane. 'They could have shouted us an ice-cream!' **1977** *Caravan World* (Austral.) Jan. 35/1 On meeting an old friend a miner would shout him, not a drink as in other places, but a bath.

shouter[2]**.** Add: **1. † b.** One who loudly supports a particular candidate. Cf. *shout *v.* 1 i. *U.S. Obs.*

1875 *Weekly New Mexican* 13 Oct. 2/1 The Carleton and Perea 'shouters', got up a procession with banners, transparencies and noise. **1904** *Rochester* (N.Y.) *Post-Express* 26 May 4 The canvass..was very thorough, Hearst shouters being busily engaged in every city.

2. b. In the West Indies, a member of a Baptist sect influenced by African religious practices.

1950 *Caribbean Q.* II. ii. 17 The Shouters and Shakers ..may practise a pseudo Christianity strongly influenced by African cult practices. **1956** M. Stearns *Story of Jazz* (1957) iii. 30 The northern religion did not make much headway except with a small group of converts to the Baptist faith in Toco, a village in the northeastern part of the island. They are called Shouters with some accuracy, for they generated enough excitement and noise to be officially banned. **1974** *Encycl. Brit. Macropædia* III. 906/1 Charismatic leaders frequently organize distinctive local variants of Christianity, such as the Baptist sects graphically termed 'Shouters' or 'Jumpers'.

3. Also *N.Z.* (Earlier example.)

1863 *Lyttelton* (N.Z.) *Times* 31 Dec. 4/1 Perhaps some of our readers do not know the extent to which the practice of 'shouting', or of inviting to drink at the 'shouter's' expense, is carried even here.

4. One who participates in a shout (*shout *sb.*[2] 1 d); a gospel-singer; a type of blues-singer. *U.S.*

1867 *Nation* 30 May 433/1 A band, composed of the best singers and of tired shouters stand at the side of the room to 'base' the others. **1931** R. W. Gordon in A. T. Smythe et al. *Carolina Low-Country* 199 The shouters form a circle and proceed around and around in a sort of slow processional. **1946** R. Blesh *Shining Trumpets* xiii. 311 In barrel-house vein are his records..with vocals by Joe Turner, then an unspoiled shouter. **1976** A. Murray *Stomping Blues* ix. 169 Joe Turner..has long been considered the Big Daddy of traditional blues shouters.

shouting, *vbl. sb.* Add: **1. d.** (Later examples.) Now usu. in form *bar the shouting.* No longer restricted to contests.

1869 A. L. Gordon *How We beat Favourite* in *Poems* (1912) 140 The race is all over, bar shouting. **1897** *Nat. Police Gaz.* (U.S.) 26 May 7/4 It was all over 'bar' the shouting, but the youngster refused emphatically to give way. **1909** A. Bennett *What Public Wants* IV. 54 If I wasn't sure that it's all over except the shouting, I wouldn't touch it. **1959** *Times* 12 June 5/3 In the absence of rain or miracles it was all over bar the shouting at Romford last evening. **1976** *Western Morning News* 25 Sept. 8/2 But if the Rhodesia affair is all over bar the shouting, can the same be said about South Africa?

† e. Loud support for a particular candidate. *U.S. Obs.*

1904 *Minneapolis Times* 29 May 6 Thus far most of the enthusiastic shouting for Gorman can be traced to the Gorman press bureau.

2. (Earlier and further examples.)

1862 E. Hodder *Memories N.Z. Life* 123 Among this class, going to these [public houses] and 'shouting'..is considered the acmé of pleasure. **1874** A. Bathgate *Colonial Experiences* viii. 99 One of the greatest social evils in the gold-fields is the system of 'shouting'. **1911** E. M. Clowes *On Wallaby* iv. 106 Of course, men still go 'on the bust', cheques are planked down, and 'shouting'—the Australian equivalent for 'treating'—indulged in till all the money is finished. **1963** *Evening Post* (Wellington, N.Z.) 10 July 13/5 Costs incurred by licensing trusts in dispensing free liquor or 'shouting' ostensibly for the purpose of encouraging patronage are under fire.

3. The performing of a shout (*shout *sb.*[2] 1 d); a declamatory style of singing among American Blacks.

1871 in *Rep. 42nd U.S. Congress 2 Sess. Joint Select Comm. Condition of Affairs Late Insurrectionary States: Georgia* (1872) I. 306, I have attended what they call their religious meetings; and they have what they call 'shouting'. **1927** *N.Y. Times Mag.* 24 Apr. 4/1 The type of song used in shouting is peculiar and has had much to do with molding and changing spirituals. **1946** R. Blesh *Shining Trumpets* (1949) v. 109 The rhythmic style of singing which we shall call *shouting*, a style clearly derived from, or related to, the declamatory sermons of the rural preacher.

4. *Comb.*, as **shouting distance** = *hailing distance* s.v. Hailing *vbl. sb.* b; chiefly in phr. *to be within shouting distance* (*of*) (also *fig.*); **shouting match**, a loud altercation.

1930 E. H. Young *Miss Mole* iii. 29 She must be within shouting distance of the rich old gentleman who was going to leave her a fortune. **1958** L. A. G. Strong *Light above Lake* 11 This is not to say that..O'Hara was an angel, or within shouting distance of one. **1961** *Guardian* 20 Jan. 22/7 Different ways of making..thermo-nuclear weapons cheaply are already within shouting distance. **1977** R. Gadney *Champagne Marxist* xiii. 83 I'll station two men outside... One will be within shouting distance. **1970** M. Braithwaite *Never sleep Three in Bed* vi. 68 We would begin a full-scale debate on which way we should have turned. Soon it would develop into a shouting match. **1981** V. Glendinning *Edith Sitwell* xv. 189 Edith was able to field, in this shouting match, one impressive new ally—John Sparrow.

shouting, *ppl. a.* Add: **2.** *U.S.* Denoting religious sects whose congregations express themselves by shouting, esp. in phr. *shouting Methodist.*

1851 T. A. Burke *Polly Peablossom's Wedding* 87 Forgeron was from that time 'a shouting Methodist'. **1876** J. Burroughs *Winter Sunshine* I. 23 About the only genuine shouting Methodists that remain are to be found in the coloured churches. **1941** W. C. Handy *Father of Blues* (1957) xi. 158 My mother was a 'shouting Methodist'. **1959** [see *religio-musical* s.v. *religio-].

shoutingly, *adv.* Add: Also *fig.*

1894 'Mark Twain' *Those Extraordinary Twins* ii. 335 The new lodger, rather shoutingly dressed.

shove, *v.* Add: **3. c.** *intr.* Of persons: to depart, go away. Const. with advbs., as *off*, *† out*, etc. Cf. *push *v.* 1 h. *colloq.* (orig. *U.S.*).

1844 *Spirit of Times* 24 Aug. 302/2 As we shoved off from Fort P. our boys made the welkin ring, and away we dashed down the Apalachicola. **1856** 'Mark Twain' *Adv. T. J. Snodgrass* (1928) 31, I shoved out for the Massasawit House. **1904** 'O. Henry' in *McClure's Mag.* Apr. 612/1 When dark came we fagged 'em a batch of bullets and shoved out the back door for the rocks. **1909** J. R. Ware *Passing Eng. Victorian Era* 223/1 *Shove off* (Navy), to quit, go, flee, depart—from shoving off a boat from land or ship. **1916** 'Taffrail' *Pincher Martin* vii. 105 'Ere, 'arf a mo'!..Don't shove orf. **1922** Joyce *Ulysses* 591 Well, I'll shove along. **1936** J. Steinbeck *In Dubious Battle* viii. 133 Them deputies knew we was goin' to shove off before daylight. **1956** P. Scott *Male Child* III. ii. 206, I wouldn't dream of telling you to shove off. You're there by Alan's invitation. **1979** D. Anthony *Long Hard Cure* xiv. 113 My, look at the hour. I'd better shove off.

d. Similarly without adv.

1866 'Mark Twain' *Lett. from Hawaii* (1967) 43, I then took what small change he had and 'shoved'. **1884** —— *Huckleberry Finn* xl. 409 We just unfurled our heels and shoved. **1944** *Sat. Even. Post* 9 Dec. 82/3 Well, I guess I'll shove. Good-by. **1954** C. Williams *Touch of Death* vii. 61 I'm going to shove. I can get away. **1975** N. Freeling *What are Bugles blowing For?* iv. 17, I have to ferry you down to the office... Let's shove, shall we?

4. e. To put or place. (In *colloq.* and casual use without notion of effort.) Also with *up*, *down.*

1902 Wodehouse *Pothunters* v. 93 You might shove up the list to-night. **1927** W. E. Collinson *Contemp. Eng.* 23 At Dulwich..we plunked things down, we shoved down notes or we shoved up lists. **1938** N. Streatfeild *Circus is Coming* vi. 76 He threw an envelope across to Santa. 'Shove yours to Mr Stibbings in there, and lick it up.' **1974** A. Fowles *Pastime* ii. 12 Shove your coat on the chair.

10. a. *to shove* (*the*) *queer* (examples). Now *Obs.* or *rare.*

1859 [implied in *shover[1] b]. **1873** G. W. Perrie *Buckskin Mose* ii. 36 If I had been detected in 'shoving the queer',..they wouldn't have cared one red cent. **1915** A. Conan Doyle *Valley of Fear* II. ii. 189 This man Pinto helped me to shove the queer... It means to pass the dollars out into circulation.

c. *to shove it*: to depart; to desist from a course of action. Usu. in *imp.*, as an expression of contemptuous dismissal. Cf. *stick *v.*[1] 18 d.

1941 Baker *Dict. Austral. Slang* 71 *Stick it!*, a contemptuous ejaculation. Also, '*shove it!*' **1956** B. Holiday *Lady sings Blues* (1973) viii. 84 It wasn't long after I left that he told them to shove it like I had. **1973** J. Wainwright *Devil You Don't* 18 'What say we pick one?'.. McGuire said: 'Shove it. It's not why we're here.' **1978** L. Stewart *Same Time, Next Year* (1979) xiii. 145 If he doesn't like it he can shove it, but don't worry—he won't.

11. (Earlier example.)

1836 *Montreal Transcript* 29 Dec. 2/2 About one it [*sc.* the ice] shoved for the second time, when it remained stationary till dark.

12. shove-halfpenny: delete *slang* and 'gambling', and add later examples; now usu. in form **-ha'penny.**

1915 T. Burke *Nights in Town* 126 She shot knife, fork, and spoon across the table with a neat shove-ha'p'ny stroke. **1942** *R.A.F. Jrnl.* 3 Oct. 5 Moving from bar to shove ha'penny table. **1969** *Listener* 20 Mar. 381/3 The Camley Arms sounded such a nice pub, with piano on most nights and darts and shove-ha'penny in the public bar.

shovel, *sb.* Add: **1. g.** *shovel and broom*: rhyming slang for 'room'. Chiefly *U.S.*

1928 M. C. Sharpe *Chicago May* 288/2 *Shovel and broom*, room. **1929** [see *line *sb.*[2] 13 f]. **1938** *Detective Fiction Weekly* 23 Apr. 75/1 In Australian slang...a house is a 'rat and mouse' and a room a 'shovel and broom'.

7. a. *shovel-footed* (later example).

1836 E. A. Poe in *Southern Lit. Messenger* Aug. 595/2 Not a shovel-footed negro waddles across the stage

b. shovel pass *U.S. Sports*, an underarm, forward pass made with a shovelling movement of the arms; so as *v. trans.* and *intr.*

1948 *News-Age-Herald* (Birmingham, Alabama) 31 Oct. c-5/2 Frank Tripucka..then shovel-passed to Sitko. *Ibid.*, Frank Tripucka..faked a handoff to Bill Gay then shovel-passed it to Sitko. **1976** *Honolulu Star Bull.* 21 Dec. H-1/1 Alabama trotted out such gimmicks as..an underhand shovel pass for a key 19-yard gain and an unbalanced line.

shovel-board, shuffleboard. Add: **2.** (Earlier and later examples.) Also, a variant of this game played on a court not necessarily on shipboard.

The usual form is now *shuffleboard.*

1836 T. Power *Impressions of Amer.* I. 14 Shuffle-board, chess, and backgammon, with exercise and pleasant converse, will while away the intervening hours. **1851** J. D. Lewis *Across Atlantic* 6 That ignominious game called shovel-board, which consists in stooping down and projecting flat slabs of wood at figures chalked on the deck. **1932** E. Waugh *Cruise* in *Work Suspended* (1949) Papa is very good at the deck games especially one called shuffle board. **1967** *Boston Sunday Herald* 26 Mar. VI. 7/7 (Advt.), 500-ft. sandy beach, 3 pools, putting green, tennis, shuffleboard, supervised Kiddie Playground. **1977** 'J. le Carré' *Hon. Schoolboy* iv. 90 Sometimes she stayed for old-tyme dancing or a game of shuffleboard.

shover[1]**. b.** (Earlier example.)

1859 *National Police Gaz.* (U.S.) 14 May 4/3 A 'shover' named Flynn,..obtained a quantity of 'queer' and went with it to Mrs. Beemer's house and left it on her table.

shover[2] (ʃɒ·vəɹ). Also **shovver, shuv(v)er.** Jocular alteration of *Chauffeur. Hence (as back-formation) **shove** *v.*[2] *intr.*, **sho·ving** *vbl. sb.*[2]

1905 S. A. Barnett *Let.* 17 Aug. in H. Barnett *Canon S. A. Barnett* (1918) II. xli. 192 Dick Batston had to leave us and we are waiting here till a 'shover' comes from London... Dick drives splendidly. **1912** *World* 5 Nov. 707/2 She can drive as well as any 'shuver'. **1914** *Professional Chauffeurs' Club Jrnl.* May 15/2 When Tommy, the bad boy of the family, has received the Order of the Boot..his fond mama..says, 'Let us make him a shuvver.' **1925** *Chambers's Jrnl.* Feb. 146/1 Joseph—our 'shover'—was in the back of the car. **1932** E. M. Keate *Mimic* vi. 77 'Paulett come and shove for me on Monday... My shover's ill.' Timothy enjoyed 'shoving'. **1953** J. Trench *Docken Dead* ix. 127 Sir R. went out first to tell the shuvver to bring the car round to the front. **1974** E. Lemarchand *Buried in Past* v. 95 It was class, for one thing. His Dad had been shovver to old Mrs P's mother and father. **1976** 'J. Charlton' *Remington Set* xiii. 63 Rabbiting on with the Pritchards' shover.

show, *sb.*[1] Add: **I. 3. a.** Also *to put up a* (good, etc.) *show*: to give (such) an account of oneself.

1934 A. P. Herbert *Holy Deadlock* 265 They've got to run the thing as they find it; and I think they put up a jolly good show. **1941** 'G. Orwell' in *World Rev.* (1950) June 41 Impossible to guess what kind of show the Russians can put up.

c. (Earlier modern examples.) Also *N.Z.*

1864 'Mark Twain' in Harte & 'Twain' *Sk. Sixties* (1926) iv. 141 Give him another show. **1866** H. W. Harper *Lett. from N.Z.* (1914) 102 There was a general response: 'We'll be there, and give you a show', a bit of diggers' slang, which I found meant—'Here's your chance, can you use it?' **1876–7** 'R. Boldrewood' *Colonial Reformer* (1890) II. xvi. 42 As he's a gentleman, he's bound to give you a show.

5. c. (*a*) (Earlier and later examples.)

1756 W. Fairfax *Let.* 26 Apr. in S. M. Hamilton *Lett. to Washington* (1898) I. 231 Prospect of great Wealth from his Share of a Copper Mine..which has a Show of much rich Ore. **1864** *Harper's Mag.* Dec. 60/2 A young farmer.. was eloquent upon the 'show' the new well had made. **1916** *Daily Colonist* (Victoria, B.C.) 29 July 4/3 The copper deposits which have been examined have not advanced beyond the stage of prospectors' shows. **1949** *Amer. Speech* XXIV. 34 *Black gold, crude, crude oil*,.. *show*, and *showing* are synonyms. **1975** *North Sea Background Notes* (Brit. Petroleum Co.) 19 Ten days later a more encouraging show was found at a deeper level, but it was two months before the company was able to announce that it had indeed made the first commercial discovery of gas in the North Sea.

(*b*) *Austral.* A mine.

1931 V. Palmer *Separate Lives* 186 He's sitting on twelve thousand since he got rid of his cobalt show, and he can't spend that here. **1942** [see *never *adv.* 9]. **1948** V. Palmer *Golconda* ii. 9 During the war they had rehabilitated themselves by carting wolfram on camels from a show they had discovered on the Western Australian border.

7. d. *to put on* (or *up*) *a show*: to present a good appearance that conceals the reality; to put a good or brave face on something.

1953 B. Gordon-Cumming *Gentle Rain* 140 'It was encouraging to see her looking so nice, wasn't it?'..'I expect she was putting on a show for our sakes.' **1960** O. Manning *Great Fortune* III. xviii. 215 He had learnt to 'put up a show'. He had hidden his fears and uncertainties.

II. 12. a. (Earlier example.)

1816 Jane Austen *Emma* III. xviii. 338 You were both talking of other things; of business, shows of cattle, or new drills.

15. a. Restrict *colloq.* or *jocularly* to senses other than those of an exhibition of works of art or a dramatic performance, and add earlier and further examples.

1830 J. Constable *Let.* (1965) III. 26, I have laid by my Wood, to enable me to pay some old and just debts, to Smith, to Woodburn, & others—which I shall have ready for the 'show' I hope. **1844** J. Cowell *Thirty Years passed among Players* xvii. 42/2, I consented to become his guest for a week, and 'show my show' in the town-hall. **1896** *N.Y. Dramatic News* 4 July 12/1 [The circus] well merited its title of the 'greatest show on earth'. **1912** W. Owen *Let.* 23 June (1967) 142 Methought I was treating you to a Show at a Picture Palace! **1935** J. Reith *Diary* 4 July (1975) 121 This was a very good show, interesting people, plenty of room, and plenty to eat. **1963** *Listener* 31 Jan. 210/1 That these painters were not a handful of isolated figures was made plain by the Whitechapel show devoted to recent Australian painting. This show..was, however eclectic it may have been, still an intelligent reflection of one man's taste. **1978** *Lancashire Life* Sept. 89/2 My favourite Blackpool show? Unhesitatingly, I nominate *No, No, Nanette.*

b. *Mil. slang.* An engagement, battle, or raid; a war. Also *big show*, a major campaign. Cf. *do *sb.*[1] 2 b.

1892 Kipling *Barrack-Room Ballads* 59 What was the end of all the show, Johnnie, Johnnie? Ask my Colonel, for I don't know. **1914** R. Brooke *Let.* Dec. (1968) 644, I entered this show (Sub-lieutenant R. Brooke R.N.D. at your service) in September and by the end of the month was in a trench. **1915** T. E. Lawrence *Let.* 29 Aug. (1938) 199 This Dardanelles show lags all the time... The big show must go wrong or go right first. **1918** E. A. Mackintosh *War, the Liberator* IV. 129 'Oh, God,' he whispered, 'don't let us get casualties before we start the show.' **1924** J. Galsworthy *White Monkey* II. xi. 205, I should very much dislike being blown up..but I should still more dislike missing the next show. **1939** A. B. Callaway *With Packs & Rifles* ii. 20 With less than a month's training I was on my way across to the big show. **1942** *R.A.F. Jrnl.* 18 Apr. 10 At the end of the show I flew to Mosul. **1944** V. G. Garvin tr. *Gary's Forest of Anger* xxvi. 110 The convoy was strongly guarded... Scenting 'a big show', the three Zborowski brothers spent their nights prowling round the lorries. **1977** *Daily Tel.* 15 Mar. 12/6 He was there in '98, I think, for the Malakand show. **1980** P. Fitzgerald *Human Voices* ix. 135 The pale pink smoke of London's fires..reminded him of a quiet sector of the line in the last show.

c. In radio and television, a light entertainment programme; more generally in the U.S., any kind of broadcast.

1932, etc. [see *radio show* s.v. *radio *sb.* 7]. **1937** *Amer. Speech* XII. 101 *Show* is used to designate nearly every type of broadcast. **1949** [see *service *sb.*[1] 31 c]. **1956** B. Holiday *Lady sings Blues* (1973) xv. 124 If you're an American citizen and unless you go to bed early these nights, you're liable to see me on the late-late show. **1964** Mrs L. B. Johnson *White House Diary* 22 Apr. (1970) 116 In a few minutes I got a call from Lyndon's office saying, 'Turn on TV and watch the David Brinkley Show.' **1972** *Listener* 22 Jan. 124/3 If we do get breakfast television, the chat show will flourish.

16. a. (Earlier U.S. examples); also with qualifying word (esp. *good, bad*); *to run the show* (examples); also *to run one's own show*: to be independent; *to give the show away* (earlier example).

1797 H. W. Foster *Coquette* 138 The show is over, as we yankees say; and the girl is my own. **1851** N. Kingsley *Diary* 29 Jan. (1914) 171 He got 500 dollars for his share, which takeing the show as we now have it I think was a plenty. **1879** L. Troubridge *Life amongst Troubridges* (1966) 152 Tanner [the maid] was tweaked up in a cap as our 'Aunt'. She rather gave the show away by dashing forward to open the cab door! **1900** W. S. Churchill in *Morning Post* 12 Apr. 5/7 'Yes,' replied the subaltern laconically, 'shoulder smashed up.' We expressed our sympathy. 'Oh, that's all right; good show wasn't it? The men are awfully pleased.' **1922** 'R. Crompton' *More William* ix. 156 Life was a rotten show. **1927** E. M. Forster *Aspects of Novel* viii. 205 James..has..a very short list of characters... For so fine a novelist it is a poor show. **1932** *Sun* (Baltimore) 23 Sept. 1/5 Farm representatives viewed the decision as another victory for the banker element which they fear intends to 'run the show'. **1946** L. P. Hartley *Sixth Heaven* iv. 95 You made a mistake..to absent yourself from the 'rag'—it was a really good show. **1955** *Times* 15 Aug. 6/1 They wanted to be independent... The Sudanese people wanted to 'run their own show'. **1957** J. Braine *Room at Top* vii. 61 The accountants and the engineers run the show no matter who's in charge. **1959** *Listener* 5 Nov. 766/1 He..came back with the bowl empty, which was taken by his mother as a jolly good show. **1969** *New Yorker* 12 Apr. 115 The astronauts..like to think they are running their own show. **1974** D. Seaman *Bomb that could Lip-Read* xi. 92 It's a bad show, Dickie... Why didn't you go straight in to clobber those terrorists?

b. *bad* (also *poor*) *show!*: an expression of dismay or disapprobation. Opp. *good show!* = an excellent performance or production! fine! splendid!

1916 H. Yoxall *Jrnl.* 18 Sept. in *Fashion of Life* (1966) iv. 37 They intend to do nothing to provide us with battle hdqrs. Bad show. **1936** 'M. Innes' *Death at President's Lodgings* x. 187 'Poor show,' said Horace. 'Distinctly where we step off,' said Mike. **1940** 'Gun Buster' *Return via Dunkirk* II. xvi. 195 The Battle of All Time is about to commence. Probably shan't get back. Jolly good show, chaps. Jolly good show. **1956** J. Symons *Paper Chase* iv. 12 Transport definitely not laid on to time. Bad show. **1963** N. Marsh *Dead Water* (1964) v. 123 'Bad show,' he said. 'Apologise. Not myself.'

c. *the show must go on*: things (orig. a circus or theatrical performance) must carry on as planned despite difficulty, calamity, etc.; *to get the show on the road*: to get started (*colloq.*); *to steal the show*: see *steal *v.*[1] 4 h; *to stop the show*: see *stop *v.* 23 c.

1941 E. Holding *Speak of Devil* xvii. 281 The hotel business is like the theatre. No matter what happens, the show must go on. **1943** *Amer. N. & Q.* Jan. 159/1 The Show Must Go On..is still primarily a circus slogan, although it can certainly be regarded as an axiom, in a lesser degree, of any form of show business, including the theatre. **1957** J. Blish *Fallen Star* II. vii. 88 They came trooping into the thawing shack... 'That's enough,' Jayne said at last. 'Let's get this show on the road.' **1957** 'Gypsy Rose Lee' *Gypsy* xxxiv. 309 Gertrude Lawrence, with a true show-must-go-on attitude had accepted my degree *in absentia*. **1961** L. Mumford *City in History* viii. 231 For the Roman the whole routine of the spectacle became a compulsive one: *The show must go on!* **1973** M. Russell *Double Hit* ix. 67 Now we can start to make plans... There's no harm in getting the show on the road. **1978** R. Hill *Pinch of Snuff* i. 10 How'd she look at the end of the film? I've heard that the show must go on, but this is ridiculous.

d. *all over the show* = *all over the shop* s.v. Shop *sb.* 8 c in Dict. and Suppl.

1947 'A. P. Gaskell' *Big Game* 24 Is he very shickered? Yes; he's all over the show. **1980** M. Drabble *Middle Ground* 171 The district's not what it was... Arabs, all over the show. Shocking, isn't it?

III. 18. b. *U.S. horse-racing.* The third place in a race. Freq. *attrib.*

1925 W. L. Comfort in *Sat. Even. Post* 11 July 13/3 He had broken into show money this afternoon. Yesterday he had won and placed. *Ibid.* 124/4, I wouldn't back Black Ball right now for any better than place or show. **1944** *Sun* (Baltimore) 14 Apr. 14/1 Rougemont won by three lengths... So Bluesteel took the place and Sea fight, the favorite, got the show award. **1964** A. Wykes *Gambling* viii. 197 There is no 'show' betting in England. **1975** *Cleveland* (Ohio) *Plain Dealer* 23 Mar. 13-c/2 Hail to Springtime was moved into the show spot and Top of the Morn dropped to fourth.

IV. 20. *show field, -ground, jump, -ring, saddle.*

1843 W. Dyott *Diary* Sept. (1907) II. 374 Both attended in the show field, and afterwards at the dinner. **1846** *Amer. Agriculturist* V. 333/1 The show-ground was located in the heart of the village. **1870** H. H. Dixon *Saddle & Sirloin* iv. 89 When they met in the show-ring. **1930** J. L. M. Barrett *Pract. Jumping* ix. 94 Blinks.. admitted to having had a whole set of show-jumps..made by the local carpenter. **1931** *Times Lit. Suppl.* 29 Oct. 840/4 The chapter on The Hound, with its warning against over-estimating the importance of success in the show-ring. **1955** *Times* 4 July 5/7 The showground, extending over 166 acres,..will be fully occupied by the trade stands..and the livestock lines, judging rings, and the pavilions. **1963** E. H. Edwards *Saddlery* xvii. 126 The English show saddle, confined in its use purely to the show ring, approximates more nearly to the dressage saddle than any other type, but its design is directed at displaying the horse's conformation rather than giving the rider any assistance in showing off the action of the horse. **1973** *Country Life* 8 Mar. 654/2 Large assortment of B.S.J.A. Show Jumps. All brightly painted and bolted construction. **1976** T. Heald *Let Sleeping Dogs Die* v. 100 'Judging of the Dog-lovers' League Dog of the Year will commence..in the main ring.'.. The crowds were already gathered round the main show-ring. **1977** *Western Morning News* 1 Sept. 8/5 This year a vast number of traders throughout the area will be displaying their goods in the large marquee on the showfield.

21. a. (Further examples.)

a **1855** C. Brontë *Emma* in *Cornhill Mag.* (1860) I. 480 To judge whether the acquisition now offered was likely to answer well as a show-pupil. **1869** S. R. Hole *Bk. about Roses* ix. 138 *Paul Ricaut*..is not reliable as a show Rose, expanding rapidly..on his arrival at the exhibition. **1915** W. B. Yeats *Reveries* 173 My friend, now in his last year at school, was a show boy, and had beaten all Ireland again and again. **1941** F. Thompson *Over to Candleford* xiii. 200 She was the show pupil of the school; good at every subject. **1950** Blunden *John Keats* 33 His friends..cultivated him with vague praise, as though they were raising the most scented, gorgeous show-rose ever exhibited. **1952** A. G. L. Hellyer *Sanders' Encycl. Gardening* (ed. 22) 159 Sweet Williams, Show-type—smooth-edged petals with dark centres. **1961** C. H. D. Todd *Popular Whippet* iv. 54 The exact type of brood bitch for which I was seeking in preference to the most brilliant show-type bitch.

22. show band, a jazz band which performs with verve and theatrical extravagance; **show flat**, a flat decorated and furnished for exhibition as an advertisement, usu. for others of similar construction (cf. *show house 2 b); **show-folk** (earlier example); **show-people** = *show-folk*; **show piece**, an item of work presented for exhibition or display; freq. *transf.*; **show-stopper**, an item (esp. a song or other performance) in a show that wins so much applause as to bring the show to a temporary stop; also *fig.*; hence **show-stopping** *a.*; **show trial**, a judicial trial attended by great publicity: usu. used with specific reference to a prejudged trial of political dissidents by a Communist government; **show tune**, a popular tune from a light musical entertainment; **show-window**: for *U.S.* read orig. *U.S.* and add earlier and *fig.* examples; examples; **show-woman** (earlier example in second sense); also, a female guide in a show-place; **show wood**, the exposed wood of the frame of an upholstered chair (chiefly *attrib.*). See also *show biz, *-boat, *-business, etc.

1927 *Melody Maker* Aug. 759/2 (*heading*) South African show band in England. **1933** *Fortune* Aug. 48/1 Ellington has never compromised with the public taste for..'show bands' combining music with scenic effects, low comedy, and flag drills. **1952** B. Ulanov *Hist. Jazz in Amer.* (1958) xiv. 162 The Cotton Pickers were best known as a show band. **1970** J. Wainwright *Freeze thy Blood less Coldly* 55 He could be corner-man in any showband in the country. **1962** *Guardian* 7 Feb. 8/2 Those show flats where colour seems to have been interpreted on a liquorice all-sorts plan. **1974** *Country Life* 14 Mar. 602/1 The first five-storey block of 20 flats is expected to be completed at the end of the year... A show flat is expected to be ready in the summer. **1755** C. Charke *Life* vii. 252 The very chairmen had something to say, by way of exultation, on the misfortunes of the poor show-folk, as they impudently and ignorantly termed them. **1853** *Diogenes* 9 Apr. 159/2 Here is a description of the President's inauguration costume, to which we call the attention of Tussaud, Springthorpe, and other waxwork show-people. **1954** *Encounter* Feb. 63/2 Arthur Helliwell..treats it [*sc.* society] as a kind of *demi-monde*, dominated by show-people and speculators. **1962** 'K. Orvis' *Damned & Destroyed* xii. 82 I'm showpeople. Well, anyway, I was once. I headlined in vaudeville for fifteen years. **1838** Dickens *Sk. Young Gentlemen* 52 He likes to place implicit reliance upon the play-bills when he goes to see a show-piece. **1885** [in Dict., sense 21 a]. **1928** W. de la Mare *Come Hither* (ed. 2) 773 The sampler..the show-pieces on canvas or linen of a little girl (aged six or upwards) to prove her skill and diligence with the needle. **1941** Blunden *Thomas Hardy* iv. 84 The poet John Clare..had become a sort of show-piece in a lunatic asylum. **1978** G. Household *Last Two Weeks Georges Rivac* ii. 23 He looked suspiciously international..a show piece to impress the foreigner. **1926** *Variety* 18 Aug. 63/1 The first half [of the programme] held two show-stoppers in the Dixie Four..who stopped the show..with their 'itch' dance finish, and Dave Apollon and Co., who stopped it, closing the first half. **1945** [see *gasser 2]. **1953** N. Coward *Noël Coward Song Bk.* 76 'The Stately Homes of England' was what is colloquially known as a 'show stopper'. **1960** *Sunday Express* 18 Dec. 14/3 A show-stopper of shimmering silver lamé. **1967** T. Stoppard *Rosencrantz & Guildenstern are Dead* II. 50 *Rosencrantz*: I can't remember how I did it. *Guildenstern*: It probably comes natural to you. *Rosencrantz*: Yes, I've got a show-stopper there. **1981** *Times Lit. Suppl.* 20 Feb. 202/3 Despite the inclusion of..other operatic items, and the Handel and Mendelssohn oratio arias which Lind loved to sing, the real show-stoppers were the 'Bird Song', [etc.]. **1962** *Times* 7 Dec. 8/7 There are few show-stopping solos of any length. **1975** *Listener* 2 Jan. 25/3 The song.. is one of the few heart-seizing, if not show-stopping, moments in Western theatre. **1937** E. Lyons *Assignment in Utopia* (1938) III. x. 370 The Ramzin affair..in my mind..figures as the classic example of the demonstration trial—those attributes which set the show trial off from ordinary trials..seem..sharply delineated. *Ibid.* 373 Sitmin..was the one whose young son had demanded his death... This was by now a standardized piece of business in important show trials. **1949** Koestler *Promise & Fulfilment* I. vii. 87 The Administration resorted to the eccentric and rather un-British expedient of staging several show-trials. **1966** M. Woodhouse *Tree Frog* viii. 63 The Russians had U2 sorted out... They were able to go to town on it, show trial, the lot. **1978** P. P. Read *Train Robbers* viii. 145 Because it was a show trial..many of the best barristers in England were retained. **1981** *Ann. Reg. 1980* 302 Jiang Qing constantly harangued witnesses and shouted defiance at the court—to such effect that she was forcibly removed on two occasions. Thus the trial was not equivalent to the Stalinist show trials of the 1950s. **1962** J. Heller *Catch-22* vii. 59 He was a short-legged, wide-shouldered, smiling young soul who whistled bouncy show tunes continuously. **1977** *New Yorker* 10 Oct. 177/1 The Brown band..opened up its offensive with one of those breezy, quickstep middle-period Jule Styne show tunes. **1826** 'N. Nondescript' *The —* I. III. 30 Have you not noticed a *shew window* full of pictures, and a sign with the words 'visiting cards for sale'? **1855** 'Q. K. P. Doesticks' *Doesticks, what he Says* xiv. 118 Those..ladies who used to perform their perpetual gyrations in the show-windows. **1914** 'B. M. Bower' *Flying U Ranch* 33 Where do you keep him when he ain't in the show window? **1958** S. Spender *Engaged in Writing* 125 The glass-making was a show-window of Venice. **1965** *Navy News* Apr. Suppl. 2/5 Yeovilton is the most suitable and convenient venue for show window air displays throughout the country. **1820** M. Edgeworth *Let.* 26 Dec. (1971) 231 A holy family by Leonardo da Vinci which the woman who shewed the pictures told us was by *Vincy* Ma'am... A cabinet brought from Italy by the 3d Duke of Beaufort as said shew-woman always carefully told us. **1848** Mrs. Gaskell *Mary Barton* I. iii. 37 Her beauty would have made her desirable as a show-woman. **1977** G. M. Hopkins *Let.* 15 Aug. in *Further Lett.* (1956) 147 Mary's photographs are of Snowdon, Cadair Idris, and Valley Crucis Abbey: in the foreground of the last is the show-woman Miss Lloyd in her green shade, a quaint old character. **1919** V. Woolf *Night & Day* xxiv. 335 While Katharine went on steadily with her duties as show-woman, Rodney examined intently a row of little drawings. **1909** Wells & Hooper *Mod. Cabinet Wk.* 368 *Show wood*, applied to stuffed chairs with part of the frame showing. **1921** F. Palmer *Pract. Upholstering* xii. 39 In some show-wood frames the whole surface is polished. **1958** *House & Garden* Mar. 5 The showwood legs can be polished to any shade. **1976** *Canadian Collector* (Toronto) Sept.–Oct. 16/2 The show-wood in this piece [*sc.* a sofa] is mahogany and the frame is birch.

show, *v.* Add: **II. 2. h.** *to show the flag*: see *flag *sb.*[4] 1 c.

l. To exhibit (an animal) in a show or display.

1854 *Poultry Chron.* I. 572/2 The best [birds] should never be shown more than once a month. **1976** T. Heald *Let Sleeping Dogs Die* i. 25 The best dog I ever had..went to a Count in Florence. He paid £3000 and never showed him.

m. To display (a slide, film, etc.) on a screen by projection for public viewing. Also *absol.* for *pass.*

1911 [see *film *sb.* 3 c]. **1931** B. Brown *Talking Pictures* xii. 290 Some of the first talking pictures to be shown in this country were from Fox Movietone News. **1942** E. Waugh *Put out More Flags* iii. 219 A film was showing in the Ministry theatre: it dealt with otter-hunting. **1964** G. McDonald *Running Scared* (1977) I. ii. 22 It was Sunday and there would be a new film showing. **1976** *Oxf. Compan. Film* 118/2 Local councils.. gradually accepted the principle that a film passed by the Board could be shown without fear of prosecution under the 1909 Act.

n. *absol.* Of an artist, fashion designer, etc.: to hold an exhibition of one's work.

1912 R. Fry *Let.* (1972) I. 357 I'm delighted that [Augustus] John wants to show. **1958** *Listener* 28 Aug. 310/3 Mr. Pooley, a young painter showing at the Hammersmith Gallery..is worth watching. **1960** *Guardian* 21 July 7/7 The autumn collections of the Incorporated Society of London Fashion Designers began yesterday... John Cavanagh and Lachasse showed on this the first of the three days. **1972** E. Lucie-Smith in Cox & Dyson *20th-Cent. Mind* II. xiv. 486 Other American artists who showed there included Robert Motherwell. **1977** *Times* 24 Feb. 8/1 Emanuel Ungaro was showing in the hotel.

4. b. (Later examples.)

1853 Lytton *My Novel* III. x. xxiv. 217 The Great Commoner had, indeed, 'something to show' for the money he had disdained and squandered. **1976** J. R. L. Anderson *Redundancy Pay* i. 11 He had been earning quite a lot of money, but had nothing real to show for it... They spent his salary as it came in.

7. Also in film.

1963 *Movie* June 29/1 The early part of the film shows their reactions to the death of Toni's wife.

9. f. *to show a leg*: see *leg *sb.* 2 a.

V. 22. d. (Later example.)

1916 'B. M. Bower' *Phantom Herd* vii. 112 As to the break I made in getting those boys out here, you'll have to show me—that's all. **1976** J. E. Taylor in L. Wing *Early Childhood Autism* (ed. 2) viii. 209 He [*sc.* the autistic child] should not be shown how to perform, since this strengthens his dependence on other people.

g. Used as an expression of defiance or self-assertion. With ellipsis of obj. clause.

1894 Mrs. H. Ward *Marcella* I. I. ii. 28 'They shall see—I will show them!' she said to herself with angry energy. **1910** A. Bennett *Clayhanger* I. ii. 16 'I'll show 'em!' he muttered. And he meant that he would show the world. **1921** H. Williamson *Beautiful Years* 170 My aunt, what a riff-raff of new bugs, Spotty. We'll show 'em, eh? **1935** C. S. Forester *African Queen* vi. 116 They hadn't believed anyone would try to get down those gorges... Well, this'd show 'em. **1952** M. Laski *Village* xvi. 215 Never did I think I'd live to hear my friends making excuses for me... I'll show them, she thought, I'll show them. **1974** M. Babson *Stalking Lamb* iv. 32 Perhaps she wouldn't come back and collect her at all—that would show her!

25. d. *to go to show*: see *go *v.* 42; freq. absol. in colloq. phr. *it just* (or *only*) *goes to show.* Also simply *it just* (or *only*) *shows.* Occas. with indirect obj.

1926 F. Scott Fitzgerald *Great Gatsby* ix. 209 It just shows you, don't it?..Jimmy was bound to get ahead. **1937** M. Sharp *Nutmeg Tree* xiv. 182 'Do you care for Galsworthy?' asked Julia... Sir William replied that he did. Which just showed. **1945** E. Waugh *Brideshead Revisited* II. i. 222 My wife's in a terrible way. She's an experienced sailor. Only shows, doesn't it. **1952** M. Laski *Village* iv. 68 An elementary school-teacher, that's what she was... Well, it just shows. **1977** J. Bingham *Marriage Bureau Murders* xi. 140, I was a little worried about you..but..here you are safe and sound, well, well, it just shows!

VI. 28. a. Also *transf.*, of a woman: to manifest visible signs of pregnancy. *colloq.*

1936 M. Mitchell *Gone with Wind* xxxviii. 681 Comforting herself..with the belief that she did not show at all when thus covered. **1957** [see *mistake *sb.* 1 d]. **1966** B. Askwith *Step out of Time* vi. 95 How the old lady knew, miss, we have no idea. Rose certainly hadn't begun to show. **1979** R. Rendell *Make Death love Me* iii. 25 In that field Christopher was conceived... Pam would marry before she began to 'show'.

f. Of an oil well: to give an indication of the presence of oil. Cf. Show *sb.*[1] 5 c.

1904 *Dialect Notes* II. 389 *Show*, v., to promise oil. **1977** *Times* 2 Nov. 3 Almost five years ago to the day, Beryl [*sc.* an oil well] showed.

g. In catch phr. *your slip* (etc.) *is showing* (cf. Slip *sb.*[3] 10, *4 c), addressed to someone thought to be unwittingly exposing a fault. Also in similar contexts.

1943 D. Powell *Time to be Born* v. 103 Pardon me, lady, your slip is showing. **1958** *Spectator* 1 Aug. 174/2 There were still standards, and hypocrisy, the homage of vice and virtue, was the duty of all public figures. But in a gratifying number of cases the slip showed and the shocked and gratified public learned that 'Anything Goes'. **1968** [see *credibility gap* s.v. *credibility c]. **1971** C. Fick *Danziger Transcript* (1973) 33 Your defense mechanisms are showing, Mr. Danziger. **1976** A. Miller *Inside Outside* vii. 83, I see in you a new broom, though your slip is still showing in places.

29. a. Also in weakened sense, an *ellipt.* use of sense 35 c, to put in an appearance, 'turn up'.

1864 Dickens *Mut. Fr.* (1865) I. I. xiii. 127 What if I.. take a look round?..None of you need show. **1907** B. M. Croker *Company's Servant* xxxiii. 335 'Gojar never shows by day,' explained Talbot. **1951** T. Sterling *House without Door* vii. 81 Big-shots like that didn't come to the police... The guy would never show. **1969** W. Garner *Us or Them War* xxxvii. 276 Jagger said flatly, 'I'm staying until she shows.' **1974** 'J. le Carré' *Tinker, Tailor, Soldier, Spy* vii. 56 She didn't show... It was the first time she'd broken a date.

d. (Later examples.)

1982 *Times* 13 July 16/1 Glaxo showed strongly, moving up 12p on revised profit forecasts. *Ibid.* 17 July 12/1 Mixconcrete showed very firm.

e. *N. Amer.* *Horse-racing.* To finish third or in the first three in a race.

1903 J. Ullman *What's the Odds?* 129 The customary limit of the handbooks around Chicago was twenty, eight and four, to win, place and show. **1936** [see *place *v.* 5 f]. **1968** *Globe & Mail* (Toronto) 13 Feb. 27/1, I myself..usually bet $200 to show, or $50-$100-$200 across the board. **1977** *New Yorker* 16 May 130/1 In the OTB betting shops..he paid three dollars straight, three dollars to place, and three dollars to show.

30. f. With ppl. adj. *to show willing*, to display readiness to please or satisfy. *colloq.*

Willing is sometimes construed as a noun, in which case the verb belongs to branch IV.

1957 *Loneliness* (Women's Group on Public Welfare) iii. 29 The older woman..must 'show willing' and be adaptable. **1959** P. Bull *I know Face* i. 11 My father wanted me to be a chartered accountant... However, in order to show willing, I did settle for 'journalism'. **1964** A. Wilson *Late Call* iii. 121 Luckily the poor creature was very willing, and there was one rule Sylvia always made— never turn off anyone who shows willing. **1973** J. Wilson *Truth or Dare* x. 121, I feel I've got to go, Claire, just to show willing.

34. show out. a. (Earlier *fig.* example.)

1839 M. Walker *Diary* 9 Feb. in C. M. Drury *First White Women over Rockies* (1963) II. 143 Mrs. W. on first reaching us seemed in good humor & I hoped she had made her a better heart. But at supper table & even before she began to show out.

VII. 35*. show-and-tell. *N. Amer.* A method used in teaching young children, who are encouraged to bring objects to school and describe them to their classmates. Usu. as *attrib. phr.*

[**1948** *Q. Jrnl. Speech* XXXIV. 361/1 Those who volunteered to speak during the regular classroom share-and-tell period.] **1950** *Amer. Childhood* Sept. 18/1 Woodridge Elementary School in Austin, Texas, along with many other schools in the nation, began solving this problem three years ago through a 'Show and Tell' period, conducted almost every day in the first, second, and third grades. *Ibid.* 18/3 Pictures cause a great sensation at 'Show and Tell'. **1958** J. E. Leavitt *Nursery-Kindergarten Educ.* xi. 235 The virtues of 'show and tell' are too often dissipated in either agonies of shyness, or frantic last minute panics at home to 'find something for show and tell'. **1962** P. Bracken *I hate to housekeep Bk.* (1963) vi. 62 The children might, at their Show-and-Tell sessions at school. **1980** in S. Terkel *Amer. Dreams* 112 No courses in show-and-tell and personality adjustment.

37, 38. See also as main entries in Suppl.

38*. show-through. *Printing.* The fact of print on one side of a sheet of paper being

visible from the other side. Cf. *PRINT-THROUGH 2, *strike-through* s.v. *STRIKE *v.* 88.
1947 *New Book Faces* (Lanston Monotype Machine Co., Philadelphia) 3/2 The 'show through' which is the result of printing in heavy color on thin or semi-transparent papers results in a loss of visibility and thereby affects readability. **1961** *N. & Q.* Apr. 160/2 A comparison of the facsimile with the original has shown that the facsimile is not so clear, but it has an adequate definition and the show-through, which occurs on several pages, does not seriously interfere with the reproduction. **1971** *British Printer* Jan. 80/2 The show-through is slight enough to permit printing both sides.

39. show up. a. Also, an instance of this; an exposé.
a **1854** MILL *Early Draft Autobiogr.* (1961) ii. 93 In my father's article the detailed shew-up of the Edinburgh Review had been left unfinished. **1937** W. H. S. SMITH *Let.* 30 Oct. in *Young Man's Country* (1977) ii. 97, I am beguiling myself..by reading Sinclair Lewis's *Elmer Gantry*, a show-up of American Nonconformists. **1949** 'G. ORWELL' *Let.* 16 June in *Coll. Essays* (1968) IV. 502 My recent novel is *not* intended as an attack on Socialism ..but as a show-up of the perversions to which a centralised economy is liable. **1961** *Guardian* 9 June 9/2 One of those tough, sexy, ostensibly moralistic show-ups.

b. A police identification parade. *U.S. slang.*
1929 M. A. GILL *Underworld Slang* 11/1 *Show-up*, where suspects are viewed by the police. **1932** *Sun* (Baltimore) 12 Dec. 1/3 The authorities conducted a mysterious 'show up' today for De Larm..police endeavoring to learn whether he had been seen in the vicinity..at the time of the killing. **1949** *Penguin New Writing* XXXVI. 96 They were real cops though. I had.. to stand the showup..and to put in twenty days at Juvenile. **1955** *Sun* (Baltimore) 22 Nov. 3/1 Lyman Brown..picked Graham out of a 'showup' of seven jail inmates.

showable, *a.* **3.** (Earlier example.)
1813 F. BURNEY 15 July *Let.* (1978) VII. 157 It was most fortunate that bit was shewable, for it gave propriety to the Measure.

show biz. Also show-biz and as one word. **a.** Colloq. (orig. *U.S.*) abbrev. of *SHOW BUSINESS.
1945 *Variety* 13 June 25/4 Big-league baseball already had rearranged its team travel schedules to a minimum. However show biz has done nothing about this yet. **1948** *N.Y. Times* 18 July 2E/7 Jack Pulaski... helped coin much of *Variety's* lingo, such as 'show biz'. **1953** [see *IMPRESSION *sb.* 6 d]. **1959** R. LONGRIGG *Wrong Number* iv. 60 Got to extemporise in show biz. **1960** M. T. WILLIAMS *Art of Jazz* p. i, A strange branch of big-time show biz. **1960** *New Left Rev.* May–June 33/2 They have their 'Pop Page'..their key to Showbiz. **1971** 'A. BURGESS' *MF* xv. 162 A lot of science gets turned into showbiz as they call it. **1976** *Liverpool Echo* 23 Nov. 6/4 Blackpool remains..the heartland of Northern showbiz.

b. *attrib.*
1945 *Variety* 30 May 28/3 (*heading*) Cantor's Showbiz tribute. **1946** J. B. PRIESTLEY *Bright Day* viii. 243 Wouldn't it be heavenly if you could mix them up —.. the retired majors..and the Show Biz boys. **1959** *News Chron.* 17 Oct. 4/2 Skirmishes with showbiz brigands. **1976** A. DAVIS *Television: First Forty Years* 42 Many of the public had misgivings, for the word 'newscaster' suggested a show-biz American approach.

Hence **show-bizzy** *a.*
1969 *Listener* 9 Jan. 62/2 The first night..was a show-bizzy occasion. **1973** S. COHEN *Diane Game* (1974) xvii. 149 After the welcoming speeches and..show-bizzy, bouquet-throwing speeches, the stage was cleared. **1981** *Times* 20 May 14/5 Even more controversial, however, is Cardin's idea to make Maxim's more showbizzy with a series of monthly soirees starring international artists.

show-boat. *U.S.* [SHOW *sb.*[1]] A river steamer on which theatrical performances are given.
1869 *Atlantic Monthly* July 85/1 The little steamer Banjo, a show-boat belonging to Dr. Spaulding, the manager of the Floating Palace, was advertised to be at Cape Girardeau. **1909** A. C. RICE *Mr. Opp* vii. 98 A new and handsome Show Boat will tie up at the Cove. **1926** E. FERBER *Show Boat* i. 13 Eager for entertainment as the dwellers were along the little Illinois and Missouri town,..they came sparsely to the show boat. **1952** M. ALLINGHAM *Tiger in Smoke* v. 88 Lighted buses crawling by looked as big as showboats. **1977** N. ADAM *Triplehip Cracksman* xii. 124 They were a mixed bunch.. drinker..showboat gambler.

show-boater. *U.S.* [f. prec. + -ER[1].] An actor on or manager of a showboat. Also *fig.*, one who performs (in other contexts) in the theatrical style characteristic of a showboat player.
1951 P. GRAHAM *Showboats* ii. 15 Green River..was in later years to become showboaters' paradise. **1952** *Times Lit. Suppl.* 4 Apr. 239/4 The show that Mr. Chapman, the first 'showboater', presented to the Mississippi Valley. **1968** 'G. BAGBY' *Another Day* iii. 45 'What was wrong with him?' 'He was a clown. He was a show-boater. He had a big mouth.' **1969** *Wall St. Jrnl.* 10 Oct. 14/1 Some of the best umpires in baseball today are almost unknown to the fans... They're not the show-boaters. **1977** *Time* Jan. 56/2 Streisand is a showboater, a sort of one-woman Hippodrome whose roots are in the brassiest tradition of the American musical theatre.

Also **show-boating.**
1951 P. GRAHAM *Showboats* vii. 66 The years 1884–1889 were important..to the growing institution of showboating. **1972** D. RAMSAY *Little Murder Music* 71 Maybe there's some other reason behind the recent tendency towards showboating. The fact is, there have been a lot of gimmicks lately. **1975** W. SAFIRE *Before Fall* vi. 384 The proposal was what Nixon would call..'showboating', presented primarily for its political impact in the States.. with little chance of its acceptance by the North Vietnamese.

show business. orig. *U.S.* [SHOW *sb.*[1]] **1. a.** The entertainment industry, esp. light entertainment (formerly, always with *the*). Occas., people engaged in show business collectively. Cf. *SHOW BIZ.
1850 T. FORD *Peep behind Curtain* vii. 26 This gentleman has been engaged..in the show-business, and is, beyond all question, the best. **1870** D. J. KIRWAN *Palace & Hovel* v. 65 I've been in the show business for sixteen years... I fust began jumpin', as a hacrobat in the penny gaffs. **1886** [see PERFORMING *ppl. a.* 2]. **1903** *Century Mag.* Apr. 819/1 General Grant..declined to remain in town for the occasion, saying that he had had enough of 'show business'. **1911** G. K. CHESTERTON *Innoc. Father Brown* v. 126 He happened to be picked up by some travelling show, and..got on quite well in the show business. **1936** 'P. QUENTIN' *Puzzle for Fools* iv. 26 He was crazy to get into show business; stunts and strong-man acts. **1956** H. GRISEWOOD in A. Pryce-Jones *New Outl. Mod. Knowl.* 426 The cinema and broadcasting to a large extent are part of what is called show-business. **1962** J. MCCABE *Mr. Laurel & Mr. Hardy* i. 28 He had a great opening. All show business on the British Isles came to it. **1978** M. MUGGERIDGE in R. Trevivian *So you're Lonely* 11 The Religious Broadcasting Department of the BBC..itself a curious no-man's-land lying between show business and evangelism.

b. In catch-phr. *there's no business like show business.*
1946 I. BERLIN (*song-title*) There's no business like show business. **1956** B. HOLIDAY *Lady sings Blues* (1973) vi. 62 It's like they say, there's no damn business like show business. You have to smile to keep from throwing up. **1972** *Guardian* 24 June 11/7 There is, as they say, no business like show business.

2. *attrib.*
1958 *Punch* 26 Feb. 278/1, I was delighted to see that Frankie Vaughan had been selected as Show Business Personality of the Year. **1961** *Radio Times* 6 Apr. 53/2 No..illusions about the permanence of show-business reputations. **1976** M. BUTTERWORTH *Festival!* xi. 187 Feature stories on the show business stars.

showcase, *sb.* (*a.*) (In Dict. s.v. SHOW *sb.*[1] 22.). Add: **1.** (Earlier and later examples.)
1835 F. LIEBER *Stranger in Amer.* II. ii. 64, I..found there..all the companions of my earliest youth, the show-cases..with their old Nuremberg prints. **1879** [see SHOW *sb*[1] 22]. **1897** R. M. STUART *In Simpkinsville* 121 Old Dr. Jenkins stood behind the showcase in his drug-store. **1935** D. L. SAYERS *Gaudy Night* vi. 124 What a blessing I hadn't put the Folio Chaucer and the other valuables in the show-cases. **1952** P. WENTWORTH *Brading Collection* xii. 69 Round the sides there were glass-topped show-cases. **1970** J. S. HARDMAN tr. *R. Boulanger's Turkey* 201 In the showcases against the wall..are various glass objects.

2. *fig.* A place or medium for presenting (esp. attractively) to general attention (freq. in Theatr. contexts). Chiefly *U.S.*
1937 'M. INNES' *Hamlet, Revenge!* II. iii. 131 Scamnum is..simply a Crispin show-case, dukedom and all. **1958** D. EWEN *Compl. Bk. Amer. Musical Theater* 79 It [sc. *Walk a little Faster*]..deserves to be remembered if only because it was the showcase for..Duke's most popular song, 'April in Paris'. **1967** N. FREELING *Strike out where not Applicable* i. 7 Its park has been turned into a landscaped garden which is a showcase for the bulb industry. **1976** M. BUTTERWORTH *Festival* xi. 173 The entertainment world fell over itself to get into this, the biggest 'showcase' on earth. **1978** S. BRILL *Teamsters* x. 364 Gibbons' local was becoming a showcase of progressive unionism.

3. *attrib.* or as *adj.* Freq. *fig.*
1903 A. H. LEWIS *Boss* 189 He's no show-case proposition!..To look at him folks might take him for a fool. **1934** *Architect. Record* Sept. 189 Show case height for standard selling. **1937** 'M. INNES' *Hamlet, Revenge!* II. iii. 131 The Duke has a show-case role. He's an Elder Statesman. **1955** *Times* 10 May 3/5 The London Philharmonic Orchestra has broken out for the summer season with some showcase programmes. **1975** B. GARFIELD *Death Sentence* (1976) vi. 36 The First Ward [of Chicago].. included the showcase hoop.

Hence as *v. trans.* (orig. *U.S.*), to place in or as in a showcase (chiefly *fig.*).
1945 H. L. MENCKEN *Amer. Lang.* Suppl. I. v. 387 A few of its [sc. *Variety's*] characteristic inventions will suffice: *to ash-can, to angel, to showcase* [etc.]. **1949** *Jrnl.* (Baltimore) 20 July 2/1 They showcase new acts, who want to be on TV. **1959** *Spectator* 14 Aug. 192/1 It [sc. the Festival Ballet] has show-cased any number of notable performers. **1961** *Times* 4 Feb. 11/5 An album which introduces a new verb to the English language. It has been recorded 'to showcase' 10 different bands at a *Jazz Festival in Hi-Fi.* **1982** *Time Out* 16 July 39/5 The film..showcases both sides of Young's music.

show-down (in Dict. s.v. SHOW *v.* 37). Add: **1.** *Card-playing.* (Earlier example.) Also **show-down poker** (see quot. 1901).
1892 W. J. FLORENCE *Handbk. Poker* 42 If a player miscalls his hands, innocently, and on the show-down has enough to win the pot, it remains his. **1901** R. F. FOSTER *Poker* 16 If the limit is ten cents only, and the blind is one, the game becomes 'showdown Poker' because a player will bet the limit on anything and everything. **1973** D. WESTHEIMER *Going Public* v. 79 How about one hand of showdown poker?..Low man does the job?

2. *fig.* Delete Chiefly *U.S.* and add: also, a declaration or trial of one's strength or position; a final confrontation, a reckoning (intended to be conclusive).
1904 F. LYNDE *Grafters* ix. 131 'You don't mean to say there is any doubt about our ability to do it?' 'Oh, no; I suppose not, if it comes to a show-down.' **1916** *Daily Colonist* (Victoria, B.C.) 9 July 4/4 It is quite time that there should be what is vulgarly known as a 'show down', and this 'show down' is what I have been endeavoring to bring about. **1927** *Glasgow Herald* 19 Aug. 10 An opportunity of 'forcing a show-down' concerning British flights in China. **1936** M. ALLINGHAM *Flowers for Judge* ix. 143 Salley has been stewing up for a row with his critics for some time and is spoiling for a show-down. **1945** E. WAUGH *Brideshead Revisited* 241 If he has a show down with the old gang, they'll just disappear. **1950** T. S. ELIOT *Cocktail Party* III. 161, I believe, Henry, if I may put it vulgarly, That Lavinia has forced you to a show-down. **1962** A. LURIE *Love & Friendship* xvi. 310 When it comes to a showdown, women stick to their husbands. **1977** J. CROSBY *Company of Friends* xvii. 113 Never had she told him her code name. Some things a woman needed—for the final showdown.

3. *Comb.*, as **showdown inspection** *U.S. Mil.*, a surprise inspection of kit. Also *ellipt.*
1920 in H. S. Duell *Hist. 306th Field Artillery* 4/1 First of numerous 'show-down' inspections for the missing 'Laces, shoe, russet, pair, extra'. **1928** L. H. NASON *Top Kick* 27 When we get back to billets they'll have a show-down, an' anyone that's shy, finds it on the payroll. **1943** *Yank* 2 Apr. 19 The Army custom of show-down inspection is unnerving. **1963** J. O. KILLENS *And then We heard Thunder* ii. 14 Showdown Inspection with all of your equipment spread out before you to be checked by the pink-cheeked officers.

shower, *sb.*[1] Add: **1. d.** A dust-storm: freq. qualified by a place-name. *Austral.*
1898 E. E. MORRIS *Austral Eng.* 115/1 *Darling Shower*, a local name in the interior of Australia, and especially on the River Darling, for a dust storm, caused by cyclonic winds. **1903** 'T. COLLINS' *Such is Life* (1944) 329 (*heading*) Wilcannia shower. *Ibid.* 331 The steady intensity of the shower augmented as I went on... The increasing broadside pressure, with the sand and dust, was becoming too much for the horses. **1933** A. B. PATERSON *Animals Noah Forgot* 36 The Bogan shower, that is mostly dust. **1949** *Geogr. Mag.* Feb. 373 Duststorms are called, in various parts of the country as indicated by the change in operative names, *Darling shower, Cobar shower, Bedourie shower* and *Wilcannia shower*.

e. In prov. phr. *I didn't come down in the last shower*, indicating that one is not inexperienced, or not so raw as to be easily fooled. Chiefly *Austral.*
[**1906** 'T. COLLINS' *Rigby's Romance* (1946) 256 *He* didn't come down with the las' rain. Pity that sort of bloke ever dies.] **1944** L. GLASSOP *We were Rats* 51 I'm awake-up, I am... I didden come down in the last shower. **1951** F. HARDY *Power without Glory* 259 'It's no use lying to me, Arty,' John West said. 'I didn't come down in the last shower.' *a* **1966** 'M. NA GOPALEEN' *Best of Myles* (1968) 277 No damn fear. I didn't come down in the last shower. **1971** B. VERNON *Big Day at Bellbird* 135, I didn't come down in the last shower, and neither did you.

f. A group or crowd (of people). Usu. *derog.*, a pitiful collection or rabble. *slang.*
1942 G. KERSH *Nine Lives Bill Nelson* ii. 13 I've seen him with some of the lousiest showers of rooks you ever saw in your life. **1958** A. HACKNEY *Private Life* xiv. 139 That unit was an absolute shower. **1962** 'H. CALVIN' *System* xiii. 176, I bet none of your shower ever even looked at Challen and his mob. **1967** M. PROCTER *Exercise Hoodwink* xiv. 100 'Have you still got the same shower in your lot?' 'Mainly. One or two new faces.' **1973** *Observer* 1 Apr. 13/3 Some of the people who go out with the hounds these days are a shower... We can't have people turning up as if they have been wearing the same pyjamas for a month. **1978** L. DAVIDSON *Chelsea Murders* I. v. 29 It's a group. What I was thinking..the Manson shower.

g. A term of mild abuse used of one person as opp. to a group. *slang.*
1949 M. LEIGH *Cross of Fire* iv. 71 You bat-eyed, buttock-brained..shower! **1959** S. GIBBONS *Pink Front Door* ii. 23 Hasn't he [sc. a baby] lain down *yet*? He is a little shower. **1966** O. NORTON *School of Liars* iv. 63 I'm of finer stuff, personally. I don't know why I waste my time playing with a shower like you. **1973** 'P. ALDING' *Field of Fire* xv. 123 'You're a right shower,' said Weiland.

2. c. (Earlier and later examples.) (Now the more usual term.)
1873 'MARK TWAIN' & WARNER *Gilded Age* xxxiii. 308 He has fell back on hot foot-baths at night and cold showers in the morning. **1930** P. MACDONALD *Link* xi. 218, I had a shower and rammed on some clothes. **1953** R. LEHMANN *Echoing Grove* 48 Must have a shower. I've been in a muck sweat all day. **1973** J. WAINWRIGHT *Pride of Pigs* 59 The bathroom..was small, but lush..with..a bidet and a corner shower.

e. *N.Z.* (See quot. 1943.)
1943 J. A. W. BENNETT in *Amer. Speech* XVIII. 86 A *shower* is..a light decorated covering spread over cups and saucers set out on a tray or table. **1957** J. FRAME *Owls do Cry* I. vi. 24 Parcels from the handwork sale, tablerunners and tea-showers in lazy-daisy and chain and shadow stitch. **1967** F. SARGESON *Hangover* xiv. 108 All was out of sight beneath a large and snowy fabric..—the kind of gossamer thing he could remember his mother had coveted many years ago in a shop window and described as a shower

3. b. An abundance of gifts of a similar kind presented by guests at a party to celebrate esp. a wedding or birth; a party given for this purpose. Also *attrib.* Also as second element in *kitchen shower* s.v. *KITCHEN *sb.* 7, *linen shower* s.v. *LINEN *sb.* 5, *wedding shower*, etc. *U.S.*
1904 *Grand Rapids* (Michigan) *Even. Press* 22 June 4 The 'shower parties' that through mistaken hospitality the wedded couple are forced to attend. **1926** *Publishers' Weekly* 26 June 2031/1 First comes June, then the showers, the wedding and after the honeymoon the settling down to a home life. **1949** *Los Angeles Times Home Mag.* 8 May 14/3 Wedding showers..are a particularly warmhearted American custom. **1958** *Even. Standard* 10 Apr. 8/2 A shower..is really a gift-giving party centred round a luncheon, tea or supper party. **1978** J. CARROLL *Mortal Friends* IV. ii. 394 She didn't explain, but implied that she wanted to discuss her trousseau and her silver and china patterns and the sort of showers she would want.

4. b. *Physics.* A number of high-energy particles appearing together; *spec.* a group generated in the atmosphere by cosmic radiation.
1933 *Proc. R. Soc.* A. CXXXIX. 702 Particles of great energy are thrown backwards in a direction nearly opposite to that of the incident shower. **1947** *Sci. News* IV. 125 Sixty-five Geiger-Muller counters have been used to follow the direction of some of the rays and establish the width of one shower. **1966** *McGraw-Hill Encycl. Sci. & Technol.* III. 498/1 The electrons and photons of such showers are referred to as the soft component of the atmospheric (secondary) cosmic rays, reaching a maximum intensity at an atmospheric depth of 150–200 g/cm². **1977** J. NARLIKAR *Struct. Universe* iii. 99 These showers contain particles produced after the cosmic rays have interacted with the atoms of the atmosphere.

6. *shower-cap, -coat, shoe*; **shower box** *N.Z.*, = next; **shower cubicle,** a cubicle containing a shower bath; **shower curtain,** a waterproof curtain separating the shower-bath from the rest of the room; **shower head,** a rose or nozzle from which the water issues in a shower-bath; **shower-room,** a room housing one or more shower-baths; **shower stall** *U.S.* = *shower cubicle* above; **shower unit,** a shower-bath or the principal apparatus of a shower-bath.
1965 F. SARGESON *Memoirs of Peon* vii. 201 Beyond that a shower-box of such microscopic dimensions. **1977** *N.Z. Herald* 5 Jan. 2–16/8 (Advt.), Must be good value, with..shower box, basement garage plus carport. **1964** *Punch* 19 Feb. 289/1 Shower-cap and bath-towel. **1972** R. K. SMITH *Ransom* III. 125 She..stepped into the shower, remembering only at the last minute to slip on a shower cap. **1964** *Punch* 14 Oct. p. xiii, Rainwear shop ..includes velvet showercoats. **1976** *Evening Standard* 29 Dec. (Advt.), Quelrayn showercoats, fleecey lined. **1966** P. O'DONNELL *Sabre-Tooth* iii. 56 She..padded into the big shower-cubicle. **1938** L. BEMELMANS *Life Class* II. iv. 158 He was inside his shower-curtain, whistling. **1974** HAWKEY & BINGHAM *Wild Card* xviii. 145 She pulled aside the shower curtains and leaned out. **1967** *Gloss. Sanitation Terms* (*B.S.I.*) 44 *Shower head*, a water fitting, for use in a shower bath, from which water issues as a film or spray. **1978** R. NIXON *Mem.* 369 The shower in the President's private bathroom in the Residence.. consisted of half a dozen different jets and showerheads. **1939** 'E. QUEEN' *Mind Over Matter* in *Blue Book* Oct. 21/2 The crowd was so dense it overflowed into the adjoining shower-room. **1951** *Good Housek. Home Encycl.* 20/2 The ideal is of course to have a separate 'shower room'. **1965** F. SARGESON *Memoirs of Peon* iv. 90 He..was washing his hands at a basin in the far corner of the shower-room. **1975** W. CRAIG *Strasbourg Legacy* (1976) i. 8 Hoess, the commandant at Auschwitz,..refused to take credit for killing three million inmates. He had been on leave of absence while a portion of them died in the shower rooms. **1960** *Amer. Speech* XXV. 264 From the verb *edewa* [sc. 'come here'], sandals or shower shoes were generally called *edewa shoes*. **1978** *Sat. Even. Post* July/Aug. 105/1 (Advt.), Swim sneaks. All-rubber bathing and shower shoes. Sure-footed protection on rocks, decks, or shower floors. **1956** 'E. MCBAIN' *Cop Hater* (1958) ix. 81 The café still served as a sort of no-man's-land..served the same purpose as the shower stall does in a honeymoon suite. **1975** J. F. BURKE *Death Trick* (1976) ii. 12 He took a quick look in the bathroom, noted that the shower stall was wet. **1973** *Times* 15 Dec. 3/1 (Advt.), Shower Units, Taps, Mixers, Splashbacks.

shower, *v.* Add: **4. b.** *intr.* To have a shower-bath. Rarely *refl.* or *trans.*, to give (oneself or someone) a shower-bath.
1930 U. PARROTT *Strangers may Kiss* 196 He..asked if he could do anything helpful about dinner, and when she said, 'no', went to shower himself. **1939** R. CHANDLER *Big Sleep* xxii. 213, I shaved and showered and dressed. **1948** *Sun* (Baltimore) 1 Oct. 15/4 Jockeys..had to shower before donning silks for their next riding assignment. **1956** K. HULME *Nun's Story* viii. 128 Time to shower and put on a fresh guimpe. **1966** T. PYNCHON *Crying of Lot 49* v. 115 The executive undressed, showered and hung his suit out on the line to dry. **1977** *Listener* 7 Apr. 446/2 The condemned man..will have been showered and then been dressed in his brown burial suit. **1978** R. LUDLUM *Holcroft Covenant* xiii. 153 Holcroft showered, shaved, put the soiled clothes in a hamper outside the door, and called the car-rental agency.

showered *ppl. a.*, also, having had a shower-bath.
1953 A. UPFIELD *Murder must Wait* xxiv. 209 Shaved showered and dressed, Boney sat at his desk. **1971** D. E. WESTLAKE *I gave at Office* (1972) 139 Dressed in..new clothes..showered, fed beef stew and beer, my basic optimism..began slowly to rise.

shower-bath. Add: **b.** Also in *fig.* phr. *to pull the string of the shower-bath*, to cause (something concealed) to be released or made known suddenly. *colloq. rare.*
1928 KIPLING *Limits & Renewals* (1932) 20 If I pull the string of the shower-bath in the papers..Castorley might go off his veray parfit gentil nut. **1937** V. WOOLF *Years* 441 Why can't he flow? Why can't he pull the string of the shower bath? Why's it all locked up, refrigerated? Because he's a priest, a mystery monger.

show·erproof, *a.* (*sb.*) [f. SHOWER *sb.*[1] + PROOF *a.*, after *rainproof*, etc.] **1.** Resistant to light rain.
1895 [in Dict. s.v. SHOWER *sb.* 6]. **1907** *Army & Navy Stores Catal.* 751 A Large Variety of Showerproof Garments in all the Newest Shapes. **1923** *T. Eaton & Co. Catal.* Spring & Summer 214/4 There is no coat more popular than the English Gabardine. It is stylish, light in weight, yet firm in the weave, making it showerproof. **1960** *Farmer & Stockbreeder* 9 Feb. Suppl. 5/2 (Advt), Showerproof coat..is in a simulated leather material. **1969** A. J. HALL *Stand. Handbk. Textiles* (ed. 7) v. 333 Silicones ..are now much used for making textile fabrics showerproof. **1973** *Times* 9 Apr. 6/1 Within the trade, 'showerproof' apparently means only that the garment will not soak through at the first hint of drizzle.

2. As *sb.*, a showerproof garment, esp. a raincoat.
1972 *Guardian* 6 June 13/3 Showerproofs will stand up to short sharp showers... But they are not for the drenching downpour. **1974** M. BIRMINGHAM *You can help Me* i. 14 A slight girl in a short off-white shower-proof, a crimson scarf at her neck.

Hence as *v. trans.*, to render showerproof; **show·erproofed** *ppl. a.*, **show·erproofing** *vbl. sb.* and *ppl. a.*
1933 *Dyer* 28 Apr. 450/3 Where a shower proofing effect is desired, many of the difficulties..are less in evidence. **1951** *Good Housek. Home Encycl.* 314/2 It is sometimes advisable to showerproof outer garments. **1958** *Vogue* Apr. 104 Reversible coat in showerproofed poplin. **1962** J. T. MARSH *Self-Smoothing Fabrics* ix. 123 With urea-formaldehyde and Velan..the latter can act as a catalyst as well as the shower-proofing agent or softener. **1970** *UK Trade Names* 73/3 Marriner Hand-knitting Wools..Halyard showerproofed, nylonised, chunky. **1973** *Country Life* 23 Aug. 529/1 Alligator are still shower-proofing melton cloth.

show·girl. [SHOW *sb.*[1]] **1. a.** An actress whose role is decorative rather than histrionic.
1836 G. SOANE (*title*) Lilian, the show girl. *Ibid.* (Duncombe ed.) I. ix. 8 The tinsel dress of the poor show girl. **1903** 'C. E. MERRIMAN' *Lett. from Son* xv. 208 Not quite all the modern Venuses have been corralled for the 'show-girl' department of musical comedy. **1923** A. TRAIN *Children's Children* xv. 192 He found it hard to believe that she was an actress, and a show-girl at that. **1936** J. BEYNON *Planet Plane* viii. 73 Give a show-girl smile, and everyone is only too glad to have you along. **1960** 'N. SHUTE' *Trustee from Toolroom* i. 5 She had gone with a party of show girls..to the Queen's Hotel after the performance. **1977** W. M. SPACKMAN *Armful of Warm Girl* 38 A leathery cousin of his own who when divorced had set up this show girl in a little flat.

b. *transf.* A mannequin.
1929 *Punch* 17 Apr. 444/3 Not the least interesting thing about these parades is that among the show-girls are many well-known titled heiresses, doing it purely for cigarette-money. **1936** J. B. PRIESTLEY *They Walk in City* i. 16 Manniekin. One o' them show girls in the big shops.

2. A girl in charge of a booth at a fair. *rare.*
1912 A. BENNETT *Matador of Five Towns* 91 The showmen and the showgirls and the showboys were titivating their booths.

show house. Add: **2. b.** A house specially finished for exhibition as an advertisement, usu. for others of similar construction (on a housing estate). Also *fig.* Cf. *show flat* s.v. *SHOW *sb.*[1] 22.
1962 *Listener* 15 Nov. 799/1 It [sc. the Labour Party] wished, as it were, to paddle its own canoe—to build in Britain a show-house of democratic socialism which the rest of Europe might inspect and then draw the lesson. **1963** *Times* 9 May 9/4 The first showhouse of a new private development which will ultimately provide 800 low-cost homes was opened here today. **1970** *Times* 4 Mar. 15/1 (Advt.), Don't miss the 'House and Garden' showhouse. **1974** *Country Life* 19 Dec. Suppl. 19 Family houses to be constructed... Showhouse available for visiting. **1978** J. SHERWOOD *Limericks of Lachasse* iii. 34 Frau Hoffmann's got the place looking like a show house in an exhibition.

3. A building used for staging theatrical performances; a travelling theatre.
1920 D. H. LAWRENCE *Lost Girl* vi. 105 A certain wooden show-house..an old travelling theatre. **1930** I. GOLDBERG *Tin Pan Alley* 217 The showhouses, too.. are taxed for this privilege, on the basis of their seating capacity.

showing, *vbl. sb.* Add: **1. b.** *showing-off* (later examples). Also with *up.*
1887 'F. ANSTEY' in *Graphic* 31 Dec. 727/3 He was annoyed with them for what he considered was 'showing off'. **1923** 'K. MANSFIELD' *Doves' Nest* 143 She detected that morning just the very faintest boyish showing off. **1962** N. STREATFEILD *Apple Bough* xviii. 255 Wolfgang put on his showing-off voice... 'I'm starring in this new picture.' **1973** J. PATRICK *Glasgow Gang Observed* vi. 64 Tim summed up..the boy's loss of face: 'Whit a showin' up in front o' aw the boays.'

d. *Cinemat.* and *Television.* The projection of a film on to a screen; the exhibition or broadcasting of a film. Also, an instance of this.
1947 *Ann. Reg. 1946* 376 British films..have brought $8,000,000 back to this country in 1946 from their American showings. **1967** *Listener* 6 July 15/1 The first London showings of..*New Faces* filled a cinema in Tooting with a stream of unaccustomed patrons. **1972** 'E. FERRARS' *Breath of Suspicion* iii. 45 They had arrived just as the earlier showing of the film was ending.

e. A public exhibition of the work of an artist or fashion designer; an art or fashion show.
1967 'T. WELLS' *Dead by Light of Moon* (1968) i. 8 Wouldn't you know something like this would happen at my first important showing? All the big art critics here, and somebody has to jinx the lights! **1969** 'H. PENTECOST' *Girl Watcher's Funeral* (1970) III. i. 136 It's about my showing on Friday... The fashion writers and the trade journals haven't given me much of a play. **1982** *Times* 3 Aug. 6/1 A decade or so ago, the couture salons of Paris were filled with American store buyers. At the showings this week, there weren't any store buyers.

8. a. Delete † and read *Obs.* exc. *Hist.* (Later examples.)
1859 GEO ELIOT *Adam Bede* III. l. 224 It is a vain thought to flee from the work that God appoints us... But now, I believe, I have a clear showing that my work lies elsewhere. **1978** F. BEER *Julian of Norwich's Revelations* 28 These four showings seem without question to fall into the first of Julian's 'thre partyes'.

c. = SHOW *sb.*[1] 5 c.
1926 *Daily Colonist* (Victoria, B.C.) 28 July 7/2 The work now in progress at the property consists in the stripping downwards of the showings in these workings, which will afford..an idea of the character of the ore showings over a depth of 900 feet. **1977** J. B. HILTON *Dead-Nettle* ii. 20 'I'll drive deep. Happen there's another seam.' 'Then your first job is to collect your showing.'

showing, *ppl. a.* Add: **4.** With advbs., as *showing-up*. *rare.*
a **1941** V. WOOLF *New Dress* in *Haunted House* (1943) 49 The looking-glass..that dreadfully showing-up blue pool.

show-·jumping, *vbl. sb.* [SHOW *sb.*[1]] Competitive jumping on horseback over a prepared course of hurdles or show jumps (also with horse as subj.).
1929 G. BROOKE *Way of Man with Horse* xvii. 235 Show jumping is a thing apart from steeplechasing... Riding over the ordinary English show-jumping course,.. necessitates a horse that will not rush. **1936** [see *show-jumper* below]. **1958** S. WILLCOX *Three Days Running* xii. 141 Chips led after the dressage and did a clear round of show-jumping. **1963** E. H. EDWARDS *Saddlery* xvi. 121 Recently there has been much correspondence as to whether..the National Hunt jockeys, should employ a saddle incorporating the features of a show-jumping saddle and so conform to something approaching a show-jumping seat. **1973** *Country Life* 13 Sept. 697/1 The second fence, to him, was simply a show jumping fence.

Hence [as a back-formation] **show-·jump** *v. intr.*, to compete in show-jumping; **show jumper,** a horse or rider that competes in show-jumping competitions.
1929 G. BROOKE *Way of Man with Horse* xvii. 236 A large percentage of show jumpers are half-bred horses. **1936** P. RODZIANKO *Mod. Horsemanship* xi. 193 A horse that has hunted is usually more successful as a show jumper than a horse that has never hunted... Hunting helps show jumping. **1943** R. S. SUMMERHAYS *Elements of Riding* (ed. 2) xix. 91 The method I have described is universally accepted as being the true way to show-jump. **1954** I. TOPTANI *Mod. Show Jumping* ii. 41 No normally trained show jumper will make for every single jump it encounters on its way. **1966** M. CATTO *Bird on Wing* iii. 58 I've show-jumped at the Tivoli. **1974** *Radio Times* 14 Mar. 4/2 Olympic silver medallist show jumper Ann Moore. **1977** N. MARSH *Last Ditch* v. 145 Dulce fancied it, though. Thought she'd make a show-jumper of it. **1980** 'E. ANTHONY' *Defector* iii. 69 She used to show-jump as a teenager.

showman. Add: **1. b.** Also *transf.* (in sports, politics, etc.), one who performs with a display of style or panache.
1964 *Guardian* 2 Mar. 7/6 There's two kinds of workers in the game, shooters and showmen. The shooter's the real wrestler... The showman's in there to make a splash, do the theatricals. **1967** C. SETON-WATSON *Italy from Liberalism to Fascism* xii. 547 In politics he [sc. D'Annunzio] was never more than a dilettante and a theatrical showman. **1977** *Irish Press* 29 Sept. 13/2 Meanwhile, showman Brian Barnes, was giving the small crowd their money's worth by nominating to them the type of shots he was going to play.

showmanship: restrict *nonce-wds.* to other derivative forms and add later (chiefly *transf.* and *fig.*) examples.
1874 G. H. LEWES *Let.* 19 Feb. in *Geo. Eliot Lett.* (1956) VI. 20 The story is very well told... But there is a chorus of objection against the excessive *showmanship* of the commentary. **1926** A. BENNETT *Ld. Raingo* II. lxiv. 287 He had made a fine display of courage and wit on the doctor's declaration of his malady, but it was only a display, a proud piece of showmanship perhaps unworthy of so solemn an occasion. **1927** *Daily Express* 12 Aug. 9/3 She held a great reception yesterday, sitting in her red caravan, and chatted of the olden times and the present conditions of showmanship. **1906** *People* 13 May 4/5 But Tony, no slouch when it comes to showmanship, helped it along by wearing..a rose brocade dinner jacket. **1972** *Newsweek* 10 Jan. 8/1 Pragmatism—coupled with some of the showmanship for which he has long been known—stood as the hallmark of Bhutto's first fortnight as President.

show-me, *attrib. phr. U.S.* [SHOW *v.*] That demands demonstration; believing only on clear evidence, extremely sceptical. (Orig. used to describe the people of Missouri: see *MISSOURI 2.)
1909 *N.Y. Even. Post* (semi-weekly ed.) 19 Apr. 88 Everything indicative of the 'show-me' State of Missouri. **1909** R. A. WASON *Happy Hawkins* 283 He belonged to the show-me club. **1933** *Sun* (Baltimore) 29 July 6/1 These administrators are going to spend a great deal of their time..trying to get money from Secretary Ickes, who appears to be a show-me man. **1949** *Manch. Guardian Weekly* 20 Oct. 2 A typically materialistic 'show-me' American. **1978** J. UPDIKE *Coup* (1979) i. 37 The premature gray and show-me squint of these Yankees is muddled in with their something eternally puerile awkward, winning, and hopeful.

show-off, *sb.* (*a.*) (in Dict. s.v. SHOW *v.* 38). Add: **c.** (Earlier example.)
1843 S. BAMFORD *Passages in Life of Radical* II. iii. 18 After some show off, by Mr Hunt, without which indeed, he scarcely knew how to get out of any matter, we left the dock.

c*. A person given to showing off. (The principal sense.)
1924 G. E. KELLY (*title*) The show-off. **1925** [see *GO *v.* 58 g]. **1932** W. S. MAUGHAM *For Services Rendered* II. 42 Lois: Well, I've always looked upon you as rather a show-off. **1943** D. POWELL *Time to be Born* ii. 52 We.. don't like to show off. Our women aren't show-offs. **1954** C. ARMSTRONG *Better to eat You* iv. 34 He knew how to deflate the show-offs and encourage the shy. **1960** T. GRIFFITH *Waist-high Culture* I. ii. 24 Grateful for an adult audience..we must have been unbearable showoffs. **1972** 'E. FERRARS' *Breath of Suspicion* vii. 102 That honeymoon couple... What show-offs they both are, acting as if no one else had ever had a love affair before them.

d. (Earlier and later examples.) Also, ostentatious, showy.
1818 S. FERRIER *Marriage* II. xxii. 298 Colonel Lennox was evidently not a shew-off character. **1843** *Punch* Feb. 79/1 We never see this show-off style of living. **1934** T. N. WILDER *Heaven's my Destination* iii. 48 He was one of that loud show-off kind. **1954** KOESTLER *Trail of Dinosaur* (1955) II. 89 The perfect symbol of it all is the show-off TV aerial on the roof of suburban houses. **1977** H. INNES *Big Footprints* IV. i. 317 You stupid show-off bastard.

Hence **show-offish, show-o·ffy** *adjs.*
1942 H. HAYCRAFT *Murder for Pleasure* ix. 189 He has too often stooped to merely show-offish quotation-spouting. **1952** S. KAUFFMAN *Philanderer* (1953) iii. 54 My sweet, dear husband. My darling, show-offy, gentle husband. **1971** LAVER & COLLINS *Educ. Tennis Player* xxv. 292. I hadn't jumped a net in a dozen years. I thought it was a bit show-offish for one thing. **1978** A. MALING *Lucky Devil* xxix. 155 He just came along to write the check. And he was kind of show-offy about it.

show-place. 2. (Earlier example.)
1794 J. B. S. MORRITT *Let.* 19 Mar. (1914) i. 10 This [sc. Dresden] I hear is a very fine show-place, particularly for pictures.

showroom. Add: **3.** (Earlier examples.) Occas. in *sing.*
1820 T. CREEVEY *Let.* 23 Jan. in J. Gore *Creevey* (1948) xiii. 179 She is like one of her numerous gold and silver musical dickey birds, that are in all the show rooms of this house. **1820** D. WORDSWORTH *Jrnl.* 29 Sept. (1941) II. 322 Having paced through every show-room of the palace, we surveyed again the exterior.

shox (ʃɒks), *sb. pl.* [Re-spelling of *shocks* (chiefly in advertising and trade journalism).] Shock absorbers.
1976 *Hot Car* Mar. 20/2 (Advt.), Full range of automatic adjusting shox for American and European cars. **1977** *Custom Car* Nov. 13/1 The Fiesta 1·3S feels like a dry-sprung Mini on competition shox.

‖ **shoyu** (ʃōᵘ·yu). Also **sho-yu, † soeju.** [Jap.: see SOY[1].] = SOY[1] 1. Freq. *attrib.* as *shoyu sauce.*
1727 J. G. SCHEUCHZER tr. *Kæmpfer's Hist. Japan* I. ix. 121 What they call Soeju, is also made of it, which is a sort of an *Embamma*, as they call it, which they eat at meals to get a good stomach. **1880** I. L. BIRD *Unbeaten Tracks in Japan* I. 232 Eels and other dainties are served with soy (*shō-yu*), the great Japanese sauce. **1920** *Japan Advertiser* 22 Aug. 5/1 The eel is laid out flat and broiled over a charcoal fire with a special shoyu sauce. **1936** K. W. COLEGROVE *Militarism in Japan* iv. 45 He came from a most humble home, and he worked as a boy for a manufacturer of *shoyu* (a Japanese sauce). **1960** B. LEACH *Potter in Japan* vi. 132 We ate it first with shoyu sauce. **1976** *Sci. Amer.* Sept. 172/1 Tempeh, ragi, sufu, shoyu, ang-kak, tea fungus and mizo are among those [fermentation products] eaten in Asian countries.

Shqip (ʃkyip). Also **Shqup, Shqyp.** [Alb.] = *ALBANIAN *sb.*[2] 2. Cf. next.
1969 'R. STARK' *Blackbird* xxv. 169 In Albania, they speak Shqyp... And in Shqyp,..Albania is Shqipenija. That means eagle country. **1974** *Encycl. Brit. Macropædia* I. 422/1 The origins of the general name Albanian.. and of the current official name Shqip or Shqipëri, which may well be derived from a term meaning 'pronounce clearly, intelligibly', are still disputed. **1976** W. H. CANAWAY *Willow-Pattern War* iii. 25 'From now on we speak only Shqyp'—which is what Albanians call Albanian.

Shqipetar (ʃkyi·pətɑːɪ). Also **Shkyipetar, Skipitar,** etc. [Alb.] = *ALBANIAN *sb.*[2] 1. Cf. prec.
1833 *Penny Cycl.* I. 256/2 The Albanian..calls himself Skipitar, and his native land Skiperi. **1860** *Chambers's Encycl.* I. 104/2 The *inhabitants*, estimated at 1,900,000, form a peculiar people, the Albanians or Arnauts; they call themselves Skypetars. **1876** [see *LATIN *sb.* 3 c]. **1902** *Encycl. Brit.* XXV. 246/2 The Albanians, both Ghegs and Tosks, call themselves *Shkyipetar*, and their land *Shkyipenia* or *Shkyiperia*, the former being the Gheg, the latter the Tosk form of the word. **1935** *Chambers's Encycl.* I. 122/2 The name Shkypetar is derived by some from *shkip*, a rock, and thus signifies Hillmen; while others derive it from *shkyup*, an eagle, signifying Sons of the Eagle. **1971** *Guardian* 10 Aug. 4/5 Shquiptars prefer folk songs..rather than the Internationale.

‖ **shraddha, sraddha.** Add: Also † **seradeh.** (Earlier example.) Now freq. with spelling *shradh.*
1776 N. B. HALHED tr. *Code Gentoo Laws* ii. 73 Who, after his Father's Death, performs not the *Serâdeh* (religious Offices to his Father's Memory). **1969** *Hindu Weekly Mag.* (Madras) 3 Aug. p. ii/4 Nimai performed the necessary rituals and ceremonies—tarpan, shradh, ablution, offering of oblations. **1971** *Illustr. Weekly India* 4 Apr. 13/2 Females and children whose *yagyopavit* has not been done are not permitted to perform the *shradh* ceremony.

shrap (ʃræp). Colloq. abbrev. of SHRAPNEL.
1918 in Hamilton & Corbin *Echoes from Over There* (1919) xi. 125 They come fast at times too... I mean shrap and heavies. **1920** *Amer. Legion Weekly* 4 June 15 Top kicks whose skulls no shrap could dent. **1944** *Yank* 7 Apr. 3 The shrap is flying like rain.

shrapnel, *sb.* Add: **1.** (Later *attrib.* and *Comb.* examples.)
1918 G. FRANKAU *Judgement of Valhalla* 7 And floundered, torn and bleeding, Over trenches, through the wire, With the shrapnel-barrage leading To the prey of our desire. **1923** KIPLING *Irish Guards in Gt. War* I. 222 A shrapnel-barrage fell also on the supports. **1939** AUDEN & ISHERWOOD *Journey to War* 71 If you looked closely you could see dull red shrapnel bursts. **1940** [see sense 3 below]. **1944** *Daily Tel.* 22 Feb. 4/6 Most people know what a shrapnel shell (now temporarily obsolete) used to be.

2. *Austral.* and *N.Z. Mil. slang.* Small change, notes, or coins of low denominations.
1919 W. H. DOWNING *Digger Dialects* 44 *Shrapnel*,.. tattered French bank notes of small denominations. **1977** *Camera & Ciné* Nov. 24 'I don't suppose you'd have a bit of shrapnel?'..I shook fifty cents out of my purse and handed it to him.

3. Fragments from shells or bombs (see quot. 1940[1]).
1940 *N. & Q.* CLXXIX. 278/1 The public has chosen to ignore the facts that shrapnel shell has become obsolete and that anti-aircraft guns fire high-explosive only. In consequence the shell fragments which are at present descending upon its devoted head are unhesitatingly referred to by the public as 'shrapnel' and the correct expression, 'shell fragments', has begun to verge on pedantry. **1940** W. S. CHURCHILL *Secret Session Speeches* (1946) 20 Our barrage will be firing, and..great numbers of shell splinters usually described most erroneously as shrapnel, will be falling in the streets. **1946** *Chambers's Jrnl.* May 228/2 A viciously singing piece of shrapnel put his helmet straight for him. **1976** *Times* 18 Aug. 12/5 What journalists and other non-gunners call shrapnel are in fact fragments from high explosive bombs or shells. **1982** *Times* 21 June 4/5 One found a piece of shrapnel from the bomb in the pocket of his overalls.

shraum (ʃrɔ̄m). *rare*[−1]. [ad. Ir. *sream* corrupt matter, phlegm, running from the eyes.] A mucous deposit.
1922 JOYCE *Ulysses* 241 He wiped away the heavy shraums that clogged his eyes.

shrdlu (ʃru·dlū). Also **shrdlu etaoin** (-eta,oin). A sequence of letters appearing on the keyboard of a type-setting machine, used as an example of an absurd or unintelligible utterance. Cf. *QWERT, QWERTY.
1943 D. POWELL *Time to be Born* xii. 290 She read..the words... For all the sense they made to Vicky they might have been a trail of 'shrdlu etaoin's'. **1970** *New Statesman* 25 Dec. 863/2 Those witty literals, exotic printers' pies and secret messages in the *shrdlu* code.

shred, *v.* Add: **4. b.** Also *spec.* to reduce (documents) to unreadable strips or fragments by means of a shredder. Cf. *SHREDDING *vbl. sb.* 3*.
1950 *Paper-Maker* Aug. 151 (Advt.), The 'Watford' Shredder and Duster..gives most excellent results in shredding and dusting waste papers. **1974** BERNSTEIN & WOODWARD *All President's Men* xiii. 267 People became afraid that the newspapers might be discovered, so someone said, 'Shred them.' **1980** *Daily Tel.* 3 Nov. 1/4 One of his jobs was to check and shred hundreds of bundles of secret and confidential papers.

shredded, *ppl. a.* Add: *shredded wheat*: freq. eaten as a breakfast cereal (earlier and later examples).
Often written with capital initials as if a proprietary term.
1899 *T. Eaton & Co. Catal.* Spring & Summer 171/3 Shredded wheat drink, 1-lb. package. **1906** *Mrs. Beeton's Bk. Househ. Managem.* lxvii. 1721 Menus for simple breakfasts..Buttered shredded wheat. **1944** [see *PUFFED, PUFT *ppl. a.* c]. **1980** *Times* 6 Feb. 13/8 The arrival of a truck loaded with packets of Shredded Wheat.

shredder. Add: **3.** Also *spec.*, a machine for reducing documents to small unreadable fragments.
1950 [see *SHRED *v.* 4 b]. **1962, 1973** [see *paper shredder* s.v. *PAPER *sb.* 12]. **1977** *New Yorker* 27 June 23/1 Papers were discussed behind sealed doors..and tossed into shredders.

shredding, *vbl. sb.* Add: **3*.** Reducing to shreds; *spec.* the reducing of documents to small strips or fragments by a machine esp. for reasons of security.
1954 *Paper-Maker* Dec. 94 (Advt.), Watford Shredder and Duster. A speedy and efficient method of shredding, dusting and cleaning. **1966** *Punch* 15 June 864/3 The special executive type shredder involved the user in actually removing pins from documents before shredding. **1973** *Times* 18 May 11/4 Mr Odle said he did not know what documents Mr Liddy was destroying in the paper shredder but in retrospect he conceded that the shredding had been very significant.

4. *shredding machine.*
1975 R. L. SIMON *Wild Turkey* xii. 77, I was in my office..when I heard someone at the shredding machine. **1980** P. KINSLEY *Vatchman Switch* xvii. 115 He switched on the shredding machine.

shredding, *ppl. a.* (Earlier example.)
1883 'MARK TWAIN' *Life on Mississippi* li. 501 Spectral trees, dimly glimpsed through the shredding fog.

shrew, *sb.*[1] Add: **2.** *shrew-faced* adj.
1913 D. H. LAWRENCE *Sons & Lovers* iv. 97 A tall, thin shrew-faced woman.

shrew, *sb.*[2] Add: **4.** *shrew-ridden* adj.
1922 JOYCE *Ulysses* 425 We have shrewridden Shakespeare and henpecked Socrates.

shrewd, *a.* Add: **15.** *shrewd-eyed* adj.; **shrewd-head** *Austral.* and *N.Z. slang*, a cunning person.
1856 J. G. WHITTIER *Panorama* 9 The shrewd-eyed salesman, garrulous and loud. **1959** *Daily Tel.* 20 May 17/1 A smiling, shrewd-eyed woman. **1916** C. J. DENNIS *Songs Sentimental Bloke* 43 Now, this 'ere gorspil bloke's a fair shrewd 'ead. **1946** J. MORRISON in *Coast to Coast* 163 Some shrewd-head overseas will get the blame for that pillaged case. **1960** N. HILLIARD *Maori Girl* III. i. 177 Only the shrewd-heads go for that hard stuff: the shysters the takes.

shrewd (ʃrūd), *sb. rare.* [f. the adj.] **1.** A shrewd or cunning person (see also quot. 1954).
c **1858** E. DICKINSON *Poems* (1955) I. 14 Could a shrewd advise me We might e'en divide—Should a shrewd betray me—Atropos decide! **1954** *Picture Post* 2 Jan. 34 The word 'Spiv', it seems, is out of date. The new word, we are reliably informed, is 'Shrewd'—and it is used as a noun, adjective and verb... The 'shrewd' is not an American by-product. He is home-bred and thoroughly English, in style and slang.

2. Shrewdness, sagacity, cunning.
1977 F. BRANSTON *Up & Coming Man* xii. 126 All you needed was some capital and a lot of shrewd and you couldn't go wrong.

shrewdie (ʃrū·di). *colloq.* (orig. *Austral.* and *N.Z.*). Also **shrewdy.** [f. SHREWD *a.* + -IE.]
1. A shrewd or cunning person.
1916 A. WRIGHT *Under Cloud* 35 Look here, Wilson, you're not such a shrewdie as you imagine. **1933** E. PARTRIDGE *Words, Words, Words!* III. 199 *Flankenheinrich*, to build a 'flank(ing) Henry', to show oneself a 'shrewdy'. **1949** F. SARGESON *I saw it in my Dream* 116 Johnny's a shrewdy. **1956** D. M. DAVIN *Sullen Bell* II. viii. 161 There'd been none of the shrewdies who dug themselves into good hospital jobs. **1967** 'E. QUEEN' *Face to Face* xx. 95 She couldn't have kept it under wraps indefinitely. Not with a shrewdie like Gegee Guild to account to. **1970** *Daily Tel.* 14 May 13/5 In the train of each social upheaval in America, like jackals at a carcase, come the shrewdies and hucksters in search of a quick dollar. **1979** J. DRUMMOND *I saw him Die* ii. 16 'A shrewdie, would you say?' 'Very shrewd.'

2. A cunning trick. *rare.*
1961 B. CRUMP *Hang on a Minute* 27 Everything was as good as he said it was, but I knew he was pulling a shrewdie on me somewhere.

Shrewsbury. Add: **2.** In phrases. *by Shrewsbury clock*: added (in allusion to quot. 1596) to statements of duration as a proverbial phr., esp. to indicate exaggeration; *as exact* (or *regular*) *as Shrewsbury clock*: very exact or regular.
[**1596**, SHAKES. *1 Hen. IV* v. iv. 151 We rose both at an instant, and fought a long houre by Shrewsburie clocke.] **1681** *Poor Robin Almanack* Mar. 12 A great many people shall feed (according as Sir John Falstaffe used to sight) three hours together by Shrewsbury Clock. **1784** H. COWLEY *More Ways than One* I. i. 1 My master is as exact as Shrewsbury clock. **1796** J. WHITE *Orig. Lett. Sir J. Falstaff* 17 Fifteen minutes, as thou say'st, by Shrewsbury clock. **1835** J. KINCAID *Random Shots from Rifleman* xiii. 332 Sir Arthur, in all his movements for twenty years, had been as regular as Shrewsbury clock. **1891** R. L. STEVENSON *Let.* in *Wks.* (1923) XXII. xi. 386, I remember, when I first saw this, laughing for an hour by Shrewsbury clock. **1901** G. B. SHAW *Admirable Bashville* III. 324 On the impenetrable sarcolobe That holds his seedling brain these fists have pounded By Shrewsb'ry clock an hour.

‖ **Shri, Sri** (ʃrī). [Skr. *Śrī*, the name of Lakshmī (goddess of prosperity or beauty and the wife of Vishnu); hence, as an honorific title.] In India, a title prefixed to the names of deities and distinguished persons (and to the titles of sacred books, etc.) as a mark of respect; also, more recently, used as the Indian equivalent of Mr.
1799 *Asiatick Researches* VI. 475 *Sitá* was an incarnation of *Devi*: for *Sri-De'vi* the wife of Daisha, and daughter of *Adima* and *Iva*, entreated the Goddess, to give her one daughter exactly like herself. **1832** *Ibid.* XVII. 178 The *Kerala Utpatti*..calls him [sc. Sankara] the offspring of adultery, for which his mother *Sri Mahádevi* was expelled her caste. **1924** E. M. FORSTER *Passage to India* xxxiii. 289 Infinite love took upon itself the form of Shri Krishna, and saved the world. **1938** M. K. GANDHI in D. G. Tendulkar *Mahatma* (1952) IV. 348 The readers will remember that in the heyday of non-co-operation, the terms 'Mr.' and 'Esquire', were dropped by Congressmen and the nationalist press, and 'Shri' was the title largely used for all, irrespective of religion. **1969** 'E. PETERS' *Mourning Raga* iii. 44 'We are looking for the house of Shri Satyavan Kumar.'..'Yes, this is house of Mr Kumar.' **1981** S. RUSHDIE *Midnight's Children* I. 77 My cousin, Shri Ramram Seth, is a great seer.

shriek, *sb.* Add: **d.** Also, an outcry of alarm, surprise, or reproof. *colloq.*
1929 'SEAMARK' *Down River* i. 22 'Yet this man is getting through?' 'Yes—and with bags of it, too, judging from the shriek we got from the Yard.'

e. (Later examples.) Also *shriek-mark.*
1864 H. ALFORD *Queen's Eng.* 93 Our friend the compositor is sure to write '*Oh*' with a shriek (!) and to put another shriek after 'Sir'. **1933** BLUNDEN & NORMAN *We'll shift our Ground* 16 It remained only to add the shriek-marks and to discover a heroine. **1969** A. GLYN *Dragon Variation* i. 9 In her mind's eye she saw the printed score-sheet, 'N × P!!', shriek-mark, shriek-mark. **1977** *Time Lit. Suppl.* 29 Apr. 521/1 He reviewed *Principia Mathematica*... He was the only man at the college who could read its curlicues, shrieks, and hooks.

shriek, *v.* Add: **4.** *fig.* **a.** *trans.* To indicate clearly or blatantly. **b.** *intr.* To provide a clear or blatant indication *of.*
1920 'O. DOUGLAS' *Penny Plain* xii. 127 The ospreys in her hat seem to shriek money. **1938** E. AMBLER *Cause for Alarm* xiv. 225 That hat of yours..shrieks English to high Heaven. **1944** M. LASKI *Love on Supertax* v. 57 In that gathering her dress, she felt with a hot rush of shame, simply shrieked of Grosvenor Street. **1972** K. BENTON *Spy in Chancery* xi. 106 The furniture was old, well-worn and miscellaneous, fairly shrieking of 'furnished let'.

shrieking, *ppl. a.* Add: **2.** *fig.* **a.** Great, excellent, splendid. **b.** Of colours: excessively bright; lurid, glaring.
1926 N. COWARD *Queen was in Parlour* I. i. 15 My first experience was such a shrieking success. **1958** P. POLLACK *Pict. Hist. Photogr.* xii. 155/1 Satin blouses of shrieking colors. **1966** [see *CAPRI 2].

shrill, *a.* and *adv.* Add: **4.** Also of colours: bright, glaring.
1973 D. LESSING *Coll. Afr. Stories* II. 117 She wore a tight shrill green dress.

5. *shrill-sounding, -voiced* (later *transf.* and *fig.* examples) adjs.
1838 POE *Narr. Arthur Gordon Pym* xxiii. 188 A shrill-sounding and phantom voice screamed within my ears. **1920** A. HUXLEY *Leda* 15 The sky Was full of strange tumult suddenly—Beating of mighty wings and shrill-voiced fear. **1960** R. CAMPBELL tr. *Paço d'Arcos's Nostalgia* 44 The voiceless city of the shrill-voiced lights.

shrill, *v.* Add: **3.** Also with *out* (later examples).
1947 A. RANSOME *Great Northern?* i. 16 Roger's voice shrilled out, 'Sail HO!' **1975** *New Yorker* 16 June 97/3 It was a lapse on Miss Sills' part to shrill out a high E flat at the end of the first finale, but otherwise she was tender, touching, and sensitive.

shrimp, *sb.* Add: **1. c.** A shrimp or prawn used as a bait in angling.
1856 'STONEHENGE' *Man. Brit. Rural Sports* 236/2 *Shrimps* are used for angling in docks and canals, and are good baits for perch, if used alive. **1910** *Encycl. Brit.* II. 29/1 Odd attractions such as boiled shrimps, caddis-grubs, small frogs, maggots, wasp-grubs, &c. are sometimes successful. **1924** *Blackw. Mag.* Apr. 489/1, I would not trust the most experienced salmon with Michael Lydon and a Galway 'shrimp'. **1931** *Hardy's Anglers' Guide* 31 The shrimp will wake the lazy dozer, and he'll take it or your fly with a rush. *Ibid.* 180 Prawn and Shrimp Tackles. **1962** *L. L. Bean Catal.* 52 Bean's shrimp fly is an excellent imitation of the natural food for trout.

d. A colour resembling that of a cooked shrimp, a bright shade of pink.
1895 *Montgomery Ward Catal.* Spring & Summer 12/1 Colored Surah silks..in the following colors: green.. shrimp, wine. **1927** T. WOODHOUSE *Artificial Silk* 81 The particulars of the colours and patterns are as under— No. 1. A shrimp ordinary crochet pattern. **1975** *New Yorker* 26 May 81 (Advt.), This plain white steerhide belt reverses to cool summer shades including seafoam green, shrimp, bone.

3. a. *shrimp net* (earlier example), *paste, -pink* (earlier and later examples), *-red*; *shrimp-brown, -coloured* adjs.
1932 A. HUXLEY *Brave New World* iii. 38 Two shrimp-brown children emerged from a neighbouring shrubbery. **1889** 'MARK TWAIN' *Connecticut Yankee* ii. 34 An airy slim boy in shrimp-colored tights. **1859** A. J. MUNBY *Diary* 18 July in D. Hudson *Munby* (1972) 39 She stood there leaning on her shrimp net (for she had been fishing). **1918** Shrimp paste [see *Gentleman's Relish* s.v. *GENTLEMAN 7 c]. **1976** *Western Living* (Vancouver, B.C.) June 50/2 A fishing village, processing salt fish, making shrimp paste, and doing a bit of duck-farming. **1882** *Cassell's Fam. Mag.* 236/1 Shrimp-pink with white is one of the happiest and latest combinations. **1973** J. ROSSITER *Manipulators* viii. 90 A shrimp-pink shirt. **1923** D. H. LAWRENCE *Kangaroo* vii. 137 The different shells, their sea-colours of pink and brown and rainbow and..shrimp-red.

b. shrimp-boat a boat engaged in fishing for shrimps; **shrimp cocktail** [COCKTAIL *sb.* 4], a dish of boiled shrimps served cold in a sauce; **shrimp cracker,** a light, crisp cracker flavoured with shrimp and served as an accompaniment to Oriental food; **shrimp gumbo** *U.S.*, a shrimp soup thickened with okra pods; **shrimp plant,** an evergreen shrub, *Justicia brandegeana* (formerly *Beloperone guttata*), belonging to the family Acanthaceæ, native to Mexico, and bearing small white flowers hidden in clusters of pinkish-brown bracts.
1872 B. JERROLD *London* p. viii, Smacks, barges shrimp-boats. **1979** *Guardian* 22 Oct. 26/7 The size of mesh permitted on shrimp boats in British waters. **1937** *America's Cook Bk.* 180 Lobster or shrimp cocktail... Chill thoroughly and serve in cocktail glasses. **1977** J. WAINWRIGHT *Nest of Rats* I. vi. 38 It was a nice meal. Shrimp cocktails, followed by a good mixed grill. **1969** *Listener* 12 June 814/1 The village chief himself asked us to a dinner of dried deer and shrimp crackers, chicken and lettuce. **1975** J. VAN DE WETERING *Outsider in Amsterdam* (1976) vi. 79 He..broke a piece of shrimp-crackers and grabbed the noodles. [**1805** Shrimp gumbo: see *GUMBO 1 b.]. **1885** L. HEARN *La Cuisine Creole* 21 (*heading*) Maigre shrimp gombo for Lent. **1889** J. WHITEHEAD *Steward's Handbk.* IV. 337/2 Shrimp-gumbo..not boiled after gumbo is in. **1938** C. H. MATSCHAT *Suwanee River* 255 The supper was fresh shrimp gumbo, hot and highly spiced. **1941** L. H. & E. Z. BAILEY *Hortus Second* 101 [*Beloperone*] *guttata*. Shrimp-plant. **1946** M. FREE *All about House Plants* xvii. 153 The Shrimp Plant..is a comparatively new introduction to the house-plant scene. **1956** X. FIELD *Housewife Bk. House Plants* III. 74 The Shrimp Plant has always been a favourite of mine. I delight in its prawn-like flowers. **1975** J. VAN DE WETERING *Tumbleweed* (1976) ii. 18 There were plants on all window-sills..the shrimp plant with a pink growth at the end of each stalk.

shrimp (ʃrimp), *v.* [f. the sb.] **a.** *intr.* To fish for shrimps. **b.** *trans.* To fish (a pool, etc.) with shrimp as a bait.
1844 M. HOLE *Diary* 26 Aug. in B. Massingham *Turn on Fountains* (1974) ii. 39 Making the most of our last day at old Blackpool... Loafed. Shrimped. **1926** R. MACAULAY *Crewe Train* II. v. 119 Torquay wasn't bad. One could shrimp and prawn and fish. **1931** *Hardy's Anglers' Guide* 31 And remember too that you may shrimp a pool in this manner and revert to fly without any fear of your pool being disturbed. **1938** *Mississippi* (U.S. Works Progress Admin.) 169 In many instances, however, boats are oystering at one season and shrimping at another.

shrimping, *vbl. sb.* Add: **2.** Fishing with shrimp as a bait.
1931 *Hardy's Anglers' Guide* 31, I have had splendid sport, owing to this way of shrimping.

shrine, *sb.* Add: **6. b.** With reference to the Order of Nobles of the Mystic Shrine. Cf. *SHRINER.
1968 *Chicago Tribune* 7 July I. 1/1 More than 2,500 Shriners gathered..to kick off the 94th annual Shrine convention. **1974** *Sunday Advocate-News* (Barbados) 3 Mar. 11/4 [The] Prime Minister..received members of the Manito Shrine Club. **1976** *Columbus* (Montana) *News* 27 May 6/3 Tom participated in the clown unit of the Shrine Ceremonial parade.

Shriner (ʃrai·nəɹ). orig. and chiefly *U.S.* [f. SHRINE *sb.* + -ER[1].] A member of the Order of Nobles of the Mystic Shrine, established in the U.S. in 1872. Also **Shrinite.**
1884 *Proc. Imperial Council Anc. Arabic Order Nobles of Mystic Shrine* 54 His brother..was a Noble of the Shrine, and well up in all that makes an efficient and sincere Shrinite. **1886** *Ibid.* 85 The Walee..was made a Shriner in 1883. **1927** E. O'NEILL *Marco Millions* II. i. 95 This costume is a queer jumble of stunning effects that recall the parade uniforms of our modern Knights Templar, of Columbus, of Pythias, Mystic Shriners, the Klan, etc. **1966** *Economist* 24 Dec. 1332/2 It appears that the celebrated Shriners are a sub-class of masons. **1979** P. THEROUX *Old Patagonian Express* xii. 193 It was Club-going Hour. At the officers mess and the VFW,..the Shriners Club, the Masons..the day's work was done.

shrink, *sb.* Add: Also *spec.* in *Textiles*, the reduction in dimension of a fibre or fabric, usu. caused by treatment with water.
1947 J. T. MARSH *Introd. Textile Finishing* ix. 244 Modern anti-shrink treatments are based on..two methods. **1954** A. J. HALL *Standard Handbk. Textiles* (ed. 4) v. 280 The well-known London shrink treatment is widely used to remove residual stretch in a wool fabric. **1967** SHAW & ECKERSLEY *Cotton* xv. 131 (*heading*) Dimensional stability (shrink resistance).

2. A psychiatrist. Cf. *head-shrinker* s.v. *HEAD *sb.* 66. *slang* (orig. *U.S.*).
1966 T. PYNCHON *Crying of Lot 49* i. 16 It was Dr Hilarius, her shrink or psychotherapist. **1969** C. YOUNG *Todd Dossier* 78 What you've written may prove helpful. That's what the man said, the shrink. **1973** *Nation Rev.* (Melbourne) 31 Aug.–6 Sept. 1434/1 A number of value judgments were offered..by a couple of the shrinks. **1978** M. WALKER *Infiltrator* iii. 39 He could have gone to a pricey shrink who would have certified him too delicate for the Army. **1980** *Times Lit. Suppl.* 3 Oct. 117/2 It does not take a shrink to see that a man so humanly flawed and artistically inept has got to be a loser.

3. *attrib.* and *Comb.*, as (sense 1) *shrink-proofing, -resistance*; *shrink-controlled, -proof, -resist, -resistant*, adjs.
1967 KARCH & BUBER *Offset Processes* ii. 31 Controlled sheets of rubber with shrink-controlled material in the centre of the plate. **1969** *Sears, Roebuck Catal.* Spring/Summer 44/2 Shrink-controlled..cotton. **1928** Shrink-proof [see *colour-fast* s.v. *COLOUR *sb.*[1] 18]. **1965** A. J. HALL *Stand. Handbk. Textiles* (ed. 6) v. 307 In recent years three important shrinkproof treatments have come into use. **1962** J. T. MARSH *Self-Smoothing Fabrics* ii. 11 The work is based fundamentally on the shrinkproofing of wool. **1963** A. J. HALL *Textile Sci.* v. 236 (*heading*) Shrink-resist finishes for cellulose fibre fabrics. **1958** *Times* 20 Oct. 13/2 This tweed has been woven for the school..and tested for washability, shrink-resistance. **1946** A. J. HALL *Stand. Handbk. Textiles* v. 275 (*caption*) Sanforising machine for making fabric shrink resistant in washing. **1973** *Times* 9 Apr. 6/3 The main terms covered by the new standard will be..'shrink resistant', 'crease shedding'.

shrink, *v.* Add: **17. shrink film** = *SHRINK-WRAP *sb.*; **shrink fit** = *shrinkage fit* s.v. SHRINKAGE 4.
1967 *Times Rev. Industry* May 76/3 *Shrink film*: as a replacement for fibre-board cartons in containing canned and bottled goods during distribution. **1969** L. S. MOUNTS in W. R. R. Park *Plastics Film Technol.* v. 124 Shrink films are sealed by special point sealers, hot wire.. or impulse. **1882** *Amer. Machinist* 8 Apr. 9/1 How much should be allowed in making a shrink fit of a wrought iron crank to the shaft? **1941** L. S. MARKS *Mech. Engineer's Handbk.* (ed. 4) 923 Shrink fits are used in places where a force fit would be difficult to assemble, as for example, locomotive wheel tires. **1970** K. BALL *Fiat 600, 600D Autobook* i. 16/1 The starter ring is a shrink-fit on the flywheel.

shrinkable, *a.* Add: (Example.)
1969 W. R. R. PARK *Plastics Film Technol.* viii. 181 Heat shrinkable polyvinyl chloride tubes in a range of diameters are becoming available.

Hence **shrinkabi·lity.**
1946 A. J. HALL *Stand. Handbk. Textiles* 273 Shrinkability in washing is a feature of textile materials to which a great deal of attention has been given. **1975** I. STEWART *Concepts Mod. Math.* xiii. 189 The 'shrinkability' of a closed curve is clearly a topological property.

shrinkage. Add: **3. a.** (Earlier example.)
1873 'MARK TWAIN' *Gilded Age* xliv. 397 They invariably allowed a half for shrinkage in his statements.

b. *spec.* in *Comm.*, an allowance made for the reduction in takings due to wastage, theft, etc.
1961 *Times* 6 Jan. 6/3 An allowance of up to 1 per cent is made for pilfering, the euphemistic word for it being 'shrinkage'. **1972** *Guardian* 14 July 12/6 Around £300 is lost each year..through..shrinkage. Shrinkage is not just customer pilferage. It includes errors, incompetence and inexperience. **1981** *Times* 4 Mar. 16/1 For some time supermarkets and department stores have referred to shoplifting euphemistically as shrinkage on their balance sheets.

4. *shrinkage-resistant* adj.; **shrinkage cavity,** a cavity in metal caused by shrinkage; **shrinkage crack,** also, a crack similarly formed in other materials; **shrinkage fit** (examples).
1923 GLAZEBROOK *Dict. Appl. Physics* V. 358 (*caption*) Shrinkage cavities at surface of aluminium alloy ingot. **1973** G. J. DAVIES *Solidification & Casting* ix. 180 (*caption*) A large shrinkage cavity in the interior of an aluminium-bronze sand casting. **1930** *New Statesman* 27 Dec. 357/2 And even as regards telegraph poles and the like a preference is given to those with long shrinkage cracks. **1895** W. KENT *Mech. Engineer's Pocket-bk.* 973 (*heading*) Shrinkage fits. **1928** F. D. JONES *Handbk. Encycl. Engin.*

977 A cylindrical part which is to be held in position by a shrinkage fit is first turned a few thousandths of an inch larger than the hole in which it is to fit; the diameter of the latter is increased by heating, and after the port is inserted, the heated outer member is cooled, causing it to grip the pin or shaft with tremendous pressure. **1946** *Nature* 14 Sept. 386/1 The greatly increased demand for shrinkage-resistant garments by the Forces.

shrinker. Add: **2.** A person employed in shrinking materials in various manufacturing processes.

1921 *Dict. Occup. Terms* (1927) § 190 Tyre shrinker. *Ibid.* § 384 *Crabber; cloth shrinker, potter, shrinker,..*tends crabbing machine, which passes cloth, at a suitable tension, through weighted rollers in a succession of troughs of water,..where it is expanded or reduced to a specified width. **1960** *Classification of Occupations* (Gen. Register Office) 99/2 Shrinker and dyer.

shrinking, *ppl. a.* Add: **2. d.** *shrinking violet*, a shy or modest person.

1915 N. L. McClung *In Times like These* vi. 83 Voting will not be compulsory; the shrinking violets will not be torn from their shady fence-corner; the 'home bodies' will be able to still sit in rapt contemplation of their own fireside. **1949** E. Coxhead *Wind in West* iii. 83 You believe in the shrinking-violet technique, do you? **1966** R. Severn *Desperate Rendezvous* viii. 70 'Time will convince you both how indispensable I *have* been...' Wilson scowled. 'A real shrinking violet, aren't you?' he said. **1976** *Listener* 22 July 90/1 Frayn has not forgotten the underdog... The shrinking violet (as he rightly recognizes) is the most dangerous plant in the glades of privilege.

shrink-wrapping (ʃri·ŋk,ræ:piŋ), *vbl. sb.* [f. Shrink *v.* + Wrapping *vbl. sb.*] The process of packaging an article by causing a thin plastic film to contract around it so as to cling tightly to its surface.

1959 *Packaging & Display Encycl.* (ed. 5) iv. 352 Polyvinylidene films..are still mostly connected with the shrink-wrapping of meats. **1968** *Packaging* Feb. 61/2 The..machine..represents a major forward step in the shrink-wrapping field. **1971** *British Printer* Jan. 99/2 Müller can now produce machinery for..shrink-wrapping.

Hence **shri·nk-wrap** *sb.*, the plastic film used in shrink-wrapping; also (as a back-formation) as *v. trans.*; hence **shri·nk-wrapped** *ppl. a.*; **shri·nk-wrapper,** a machine used for shrink-wrapping.

1961 *Packaging Rev. Data Bk.* 283/2 Oriented film..is under development as a shrink-wrap for poultry, hams, cheese and preserve pot covers. **1968** *Packaging* Feb. 61/2 One unit..replaces the tray-erector, product-loader, and shrink-wrapper. **1969** *Ibid.* Apr. 33 (*caption*) A..principle ..of..shrinkwrapping the entire load—in one continuous sequence. **1970** *Register of Packaging* 19 Up to 30 shrink wrapped packs per minute. **1978** *Listener* 20 July 73/3 Apples that..keep well..shrink-wrapped on the supermarket shelves. **1981** *Times Lit. Suppl.* 30 Jan. 108/3 *The Writers' and Poets' Yearbook*..is cased in boards coated with a shrinkwrap, transparent laminate that looks like a badly fitting condom.

shroud, *sb.*¹ Add: **7. d.** *Engin.* A circular band attached to the circumference of the rotor of a turbine; a flange on the tip of a turbine rotor blade (flanges on adjacent blades usu. interlocking so as to form a continuous band).

1906 J. W. Sothern *Marine Steam Turbine* (ed. 2) II. 54 At the outer ends..the blades fit..into a channel-shaped brass ring, or 'shroud', as it is called. **1951** Cohen & Rogers *Gas Turbine Theory* v. 106 Although shrouds have been used on superchargers, they have not come into general use so far on impellers for gas turbines. **1967** N. E. Borden *Jet-Engine Fundamentals* 93 The shrouds form a band around the perimeter of the turbine rotor which interlocks the blades at their tips and reduces vibration. **1971** P. J. McMahon *Aircraft Propulsion* v. 162 What may show the difference between a turbine and a compressor stage would be the fitting of shrouds on turbine blades.

e. A temporary covering for part of a spacecraft, esp. one which protects and streamlines the payload of a rocket during launching.

1965 W. R. Corliss *Space Probes & Planetary Exploration* x. 235 The shroud gives probes that customary conical appearance before deployment of the articulated sections. **1966** [see *dock *v.*³ 5]. **1975** K. W. Gatland *Missiles & Rockets* viii. 184 Above that, enclosed in a shroud, were the Airlock Module, Multiple Docking Adapter and Apollo Telescope Mount.

8. (sense 2) *shroud-plait* (poet.).

c **1864** G. M. Hopkins *Poems* (1967) 117, I desire They swathe and lace the shroud-plaits o'er my face.

shroud, *sb.*² Add: **2. b.** = *shroud line* below.

1942 F. H. Colvin *Aircraft Handbk.* (ed. 5) 679 (*caption*) Drawing the shrouds into their pockets in the seat pack. **1957** L. L. Beckford *A.B.C. of Aeronaut.* 74/1 Fastened firmly between the gores are strong cords, called Shrouds, which distribute the load evenly over the Canopy. **1973** 'A. Hall' *Tango Briefing* x. 119 Watch the ground. The whisper of wind in the shrouds... He'd had to give me five seconds..so that the 'chutes wouldn't foul each other.

3. shroud line, any of the straps joining the canopy of a parachute to the harness. Usu. *pl.* Cf. *rigging lines* s.v. *rigging (*vbl.*) *sb.*² 2 c.

1929 A. F. Collins *Aviation* xii. 184 The pilot chute and big parachute, together with its shroud lines that hold it to the harness, are made so that they fold up in a very small pack. **1973** 'A. Hall' *Tango Briefing* xix. 240 The supply 'chute was draped across a spur of rock... The shroud lines were badly twisted.

shroud, *v.*² Add: **9.** Also in gen. use with reference to the provision of a shroud in var. technical senses. Cf. Shroud *sb.*¹ 7 in Dict. and Suppl.

1913 S. J. Reed *Turbines applied to Marine Propulsion* iii. 41 In both of the above systems the tips of the blades are shrouded with a steel strip, a projecting piece being left on the blade tip which passes through a hole in the shroud and which is eventually riveted over. **1948** *Chambers's Jrnl.* July 392/2 Mica is used to support the input sockets, thus preventing breakdowns owing to heat, and the sockets are also shrouded for safety and the prevention of shock. **1966** *McGraw-Hill Encycl. Sci. & Technol.* IV. 292/2 Shrouding a propeller may be used on a ship to decrease interference of propeller and hull.

shrub, *sb.*¹ Add: **4. a.** *shrub border*.

1933 A. Osborn *Shrubs & Trees* v. 38 Endeavour to visualize the shrub-border when its inmates are fully grown. **1978** R. E. Heath *Miniature Shrubs* p. vii, A shrub border once planted needs very little after care.

b. shrub rose, a rose, esp. a species or variety long in cultivation, allowed to follow its natural pattern of growth. Cf. *old rose a.

1948 G. S. Thomas in *Jrnl. R. Hort. Soc.* LXXIII. 170 (*heading*) Shrub roses for the modern garden. **1980** A. Wilson *Setting World on Fire* II. iv. 117 She runs a shrub rose nursery garden in Sussex.

shrubbery. Add: **3.** *transf.* A beard or whiskers. *joc.*

1937 Wodehouse *Lord Emsworth & Others* ii. 101 Something has eaten off Sir Preston's moustache... I met him outside, and the shrubbery had completely disappeared. **1966** J. S. Cox *Illustr. Dict. Hairdressing* 137/2 *Shrubbery*, a false beard; also the natural beard.

shrug, *sb.* Add: **5.** A short, close-fitting woman's jacket or shoulder stole with sleeves, orig. knitted or crocheted. Also *attrib.*, as *shrug jacket*. orig. *U.S.*

1957 *Knitted Outerwear Times* 12 Aug. 23/2 Women are buying the Orlon shrug for wear over sleeveless dresses and in air conditioned rooms. **1962** *Guardian* 5 Oct. 8/5 (*caption*) This ensemble..is made up of skirt and shrug in royal blue loose weave mockknit. **1973** *Country Life* 25 Jan. 250/2 Lucca lamb shrug. **1980** *Times* 12 Feb. 7/6 Camisole top and shrug jacket takes 7 50g balls of 3 Suisses Barbara.

shrug, *v.* Add: **3. b.** *fig. to shrug* (something) *off* or *aside*: to dismiss or reject (something) in an offhand manner; to be unaffected by.

1909 Webster, Shrug off. **1932** *Now & Then* Spring 15/2 He might shrug aside or be bored or even disgusted by *Strange Interlude* or *Mourning Becomes Electra*. **1949** *Catholic Times* 4 Mar. 5/1 The disgrace of Yalta can never be shrugged off. **1963** *Observer* 17 Feb. 23/5 Some houseflies and mosquitoes can now shrug off not one but several of the most lethal poisons. **1981** *Times* 18 Apr. 21/5 The stockmarket has also shrugged aside the collapse of Hedderwick Stirling Grumbar.

10. *intr.* To manœuvre one's arms and shoulders *into* a garment. Also *refl.*

1930 H. Ashton *Dr. Serocold* I. iii. 43 She shrugged herself into her stiff overcoat and began to button it all the way up to her chin. **1937** D. Aldis *Time at her Heels* ii. 33 And shrugging into her brown suit coat she followed her sons and daughter downstairs. **1974** J. Dowell *Look-off Bear* 6, I shrugged into my warm red-and-black-checked mackinaw jacket.

shrugged *ppl. a.* Also *fig.*

1874 G. M. Hopkins *Jrnl.* 23 May (1937) 195 Trees, clouds, and mountaintops 'seized' or 'shrugged' as in Turner.

shtchee, shtchi, varr. Stchi in Dict. and Suppl.

‖ **shtetl** (ʃte·t'l, ʃtēⁱ·t'l). Also **shtetel.** Pl. **shtetlach, shtetlakh, shtetls.** [Yiddish, 'little town', f. G. *stadt*.] A small Jewish town or village in Eastern Europe. Now *Hist.* Also *transf.*

1949 *Yivo Ann. Jewish Soc. Sci.* IV. 87 Swistocz..was considered one of the larger towns (*shtetl*) in the district of Grodno. **1963** M. Samuel *Little did I Know* ix. 137 The *Shtetlach!* Those forlorn little settlements in a vast and hostile wilderness, isolated alike from Jewish and non-Jewish centers of civilization. **1968** L. Rosten *Joys of Yiddish* 370 The world of the Jews in Germany..was vastly different from the world of *shtetl* Jews in eastern Europe. **1972** *New Society* 3 Aug. 228/1 People who, in former reincarnations, were clerks and schoolmarms in the *shtetls* of Minsk, Pinsk and Milwaukee. **1973** *Times* 3 Feb. 13/4 Jewish food is the diet of the *shtetel*—the small village. **1976** *Nat. Observer* (U.S.) 6 Mar. 19/1 In the *shtetlakh*, the small towns, they lived lives of grinding poverty amid the constant expectation of external attack. **1977** *Listener* 31 Mar. 422/4 The little Jewish townships of Russia, the *shtetlach*. **1978** *Times Lit. Suppl.* 10 Nov. 1314/2 Hitler swept away the *shtetls* of Eastern Europe.

‖ **shtibl** (ʃtī·b'l). Also **shtiebel, shtieble, stiebel.** Pl. **shtibblach.** [Yiddish; cf. Ger. dial. *stüberl* small room.] A small synagogue or Jewish house of worship.

1929 I. Goller *Five Bks. Mr. Moses* I. i. 10 Reb Zalman, Rabbi of the *Shtiebel* off Commerical Street. *Ibid.* ii. 15 A single, large room the *Shtiebel*, combination of synagogue, house of study, library and club-room for the disciples of Reb Zalman. **1960** H. Rabinowicz *Guide to Hassidism* x. 120 The *stiebel* (literally 'room')..played a vital rôle. It served both as place of worship and as house of study... In the *stiebel* the Hassidim communed with God and with themselves. **1967** C. Porok *Chosen* vii. 116 Each Hasidic sect had its own house of worship—shtibblach, they were called. **1973** *Jewish Chron.* 14 Sept. 22/1 The new cathedral synagogue of Jerusalem..will make Herod's temple look like a shtibl. **1976** C. Bermant *Coming Home* II. iv. 173, I belong to two synagogues, a large one..and a gilded little bethel with about two dozen members called a *shtieble*.

‖ **shtik** (ʃtik). *U.S. slang.* Also **schtick, schtik, shtick.** [Yiddish, f. G. *stück* piece, play.] **1.** An act or stage routine; a joke, a 'gag'. Hence *transf.* (freq. slightly *derog.*), a patter, a 'line'; a gimmick or characteristic style. orig. *Theatr.*

1961 A. Berkman *Singers' Gloss. Show Business* 78 A piece of business; a gag or joke..*shtik*. **1962** 'E. McBain' *Empty Hours* 129 The girl didn't say a word. She didn't have to. The effect was almost comic, akin to the cocktail-party scene... The word '*shtik*' crossed Meyer's mind. **1964** S. Bellow *Herzog* 60 'Let's cut out all the *shtick*,' said Gersbach. 'Let's say you're a crumb.' **1968** P. Tamony *Americanisms* (typescript) No. 19. 2 Consistency is not the schtick of protestors. **1973** *Times* 30 July 5/5 Emotionally controlled..and minus the usual Streisand shtick, it [*sc.* a film] is arguably the best performance of her career. **1974** P. Gzowski *Bk. about this Country* 46/1 People who are professionally funny—guys who do schticks—are less funny than he is. **1977** *Time* 19 Dec. 12/2 The former Prime Minister is not at all apologetic about his Yuletide shtik, pointing out that he has chosen to write books and sell records rather than go the David Frost route.

2. A particular area of activity or interest, a sphere or 'scene'.

1968 *Atlantic Monthly* Sept. 50/1 My first assignment was to a gentle middle-aged Jewish household, hardly my schtik. **1972** *New York* 1 May 13/2 This unfortunately overlong satire on the sexology schtick of our times. **1976** *Publishers Weekly* 15 Mar. 55/2 A husband trying to puzzle out his woman, women-God-bless-them in general, and the whole female *shtick*.

Shtokavian, Shtokavski: see *Štokavian.

‖ **shtook** (ʃtuk) *slang.* Also **schtook, schtuck, shtuck,** etc. [Origin unknown: app. not a Yiddish word.] Trouble; esp. in phr. *in* (*dead*) *shtook*.

1936 G. Ingram *Muffled Man* iv. 68, I to come to you if I'm in 'stook'. **1960** *News Chron.* 16 Feb. 6 In fact, I'm in shtook. **1970** G. F. Newman *Sir, You Bastard* 280 The filth, who was supposed to be in dead shtook, was sitting there like he owned the gaff. **1971** F. Norman *Dodger-Greaser* vii. 136 It'd be a terrible fing if none or us could 'elp each uvva aht nar an' then wen we're in dead schtook. **1975** *Observer* 27 Apr. 33/1 The scheme went awry, landing David in shtuck with the Law and jeopardising his lucrative future. **1978** J. Gardner *Dancing Dodo* xxxi. 246 You know I'm in schtuck with my bosses.

‖ **shtoom** (ʃtum), *a.* and *v. slang.* Also **schtoom, shtum(m),** etc. [Yiddish, f. G. *stumm* silent, mute.] **A.** *adj.* Silent, speechless, dumb. Esp. in phr. *to keep* (or *stay*) *shtoom*. Occas. also as *sb.*

1958 F. Norman *Bang to Rights* 15, I think it's much better to keep shtoom. **1972** J. Caine *Hamlet, my Boy* x. 141 Keep *shtumm* and you won't get plugged. **1981** J. Barnett *Firing Squad* ii. 14 Stay schtoom. If the law does come round you don't go near... It's got to be total schtoom on this.

B. *v. intr.* To be quiet, to shut *up*. Also *trans.*: cf. *shut it* s.v. *Shut *v.* 4 b(*c*).

1958 F. Norman *Bang to Rights* 72 You can always shtoomup if any screws are earholeing. **1973** Boyd & Parkes *Dark Number* ii. 21 You start something, and then just when it's getting interesting you shtoom up. **1982** 'J. Gash' *Firefly Gadroon* xvii. 185 Shtum it. Sounds carry in this.

shtshi, var. Stchi in Dict. and Suppl.

shtuck, var. *shtook.

‖ **shtup** (ʃtup), *v. slang.* [Yiddish; cf. Ger. dial. *stupfen* to nudge, jog.] **1.** *trans.* To push. Hence also as *sb.*

1968 L. Rosten *Joys of Yiddish* 374 'Don't *shtup*', means 'Don't push'—both literally and figuratively... A man who *shtups* himself in 'di hoykhe fenster' (the high windows) is a man who is a social climber. **1977** M. T. Bloom *13th Man* vii. 136 What the fuck hell is now holding up the works?..Give them a *shtup*. *Ibid.* xi. 182 I'm trying to give fate a *shtup*, a push, to hurry up and resolve matters.

2. *trans.* and *intr.* To copulate (with).

1969 P. Roth *Portnoy's Complaint* 83 Why, *of course* he was *shtupping* her. **1974** D. Westlake *Help* v. 35 He'd go on home..shtup the wife..then shlep on back here. **1977** *Custom Car* Nov. 67/2 Italian men can actually murder their wives if they find 'em shtupping around.

Shuboth, var. *Shabbat.

shubunkin (ʃubʊ·ŋkin). Also **Shubunkin.** [Jap., f. *shu* vermilion + *bun* portion, division + *kin* gold.] A goldfish, *Carassius auratus*, that is multicoloured with black spots and red patches and has elongated fins and tail.

1917 W. T. Innes *Goldfish Varieties* ii. 26 One of the more recent introductions is the Shubunkin. **1928** *Daily Express* 5 July 8/4 'I'll just bring these Shubunkins to the front for you.' And he coaxed into view a magnificent creature with a blue body covered with red, black, yellow and brown spots. **1971** *Country Life* 1 Apr. 774/3 For common goldfish and shubunkins this [depth of water] can be reduced to 9 or 12 inches. **1980** *West Lancs. Evening Gaz.* 25 Aug. 17 First in the shubunkins..came second in the any variety of goldfish section.

shuck, *sb.*² Add: **1. e.** Phr. *to light a shuck*: to leave in a hurry, to hurry away.

1905 *Dial. Notes* III. 86 *Light a shuck*, to go in a hurry, to move on, to keep away from danger. **1938** in B. A. Botkin *Treas. S. Folklore* (1949) III. i. 459 He jumped outen the water and lit a shuck for camp. **1947** *True* Nov. 108/2 But the Espinosas lit a shuck for the mountains. **1971** J. V. Allen *Cowboy Lore* iv. 71 So he saddled up old Chaw one night and lit a shuck this way.

2. e. Nonsense, deception, sham.

1958 G. Lea *Somewhere there's Music* 163, I know about double negative too, but that's a lot of shuck. **1959** *Encounter* June 43 Despite his rejection of marriage as middle-class 'shuck' (phoney), the Beatnik's Wedding is an important event. **1972** *Islander* (Victoria, B.C.) 25 June 14/4 This is a good book and as they say in..Texas: 'I'm not putting the shuck on you' so get it and read it. **1980** A. Toffler *Third Wave* xix. 261 The recently graduated son..proclaims the nine-to-five job a degrading sham and a shuck.

3. (Earlier example.)

1847 J. M. Field *Drama in Pokerville* 68 And Mr. Bagly was there...[to shoot] any gentleman who might say 'shucks!'.

4. *shuck-mattress* (earlier example); (sense 2 e) *shuck-shark*.

1843 W. Frazier *Jrnl.* July (1930) 27 Our cargo..was a motley pile..from broken skillets, up to rickety bedsteads and *shuck*-mattresses. **1952** A. Lomax *Mister Jelly Roll* 136 I want you shuck-sharks and crooks to get out of town.

shuck, *v.*² Add: **2. a.** (Later examples.) (Now also occas. used outside N. Amer.)

1966 *Listener* 3 Nov. 650/2, I regard it as a great fortune to have shucked off this amount of remorse about intellectual achievement. **1968** *N.Y. Times Book Rev.* 23 June vii. 1/1 That an actively practicing attorney.. should ever be able to shuck them off long enough to produce a book..struck me as a most unlikely miracle. **1969** G. Macbeth *War Quartet* 28 The deflector bag Filled with loose cases, shucked out. **1969** *New Yorker* 12 Apr. 86/2 Then the astronaut shucks the box from the tube, which he discards as a doctor might throw away the protective part of a syringe. **1975** *Times Lit. Suppl.* 4 July 725/1 The work of a British historian that shucks off the weight of this ponderous tradition. **1976** *New Yorker* 19 Apr. 98/2 Odd thing: Joanne, now living in Connecticut, has hung on to her Southern accent; the two others, both New Yorkers, have shucked theirs. **1978** *Guardian Weekly* 1 Jan. 18/4 Spanish boys and girls have shucked the race for money... Marriage and children are not a goal. Neither is wealth.

3. a. *trans.* and *intr.* To deceive, fool or 'kid' (someone). *slang*.

1959 L. Lipton *Holy Barbarians* 25, I didn't shuck the customers enough to please the crook who was running the car lot. **1966** [see sense b below]. **1969** S. Greenlee *Spook who sat by Door* xiii. 114 He soothed them and told them to go home..and he did not shuck. You either work at a cover or forget it. **1976** C. Weston *Rouse Demon* xviii. 88 You shucking me, man, I didn't get rid of nobody! **1979** *Maclean's Mag.* 4 June 6/3 The petulant Keith Jarrett is an example: 'He's shucking.'

b. *shucking and jiving*: fooling. Cf. *jive *v.* 1 a. *U.S. Blacks*.

[**1966** E. Bullins *Theme is Blackness* (1973) 27 Yawhl jivin'...yawhl shuckin'.] **1969** H. R. Brown *Die Nigger Die!* ii. 25, I told him he should think about it, but I knew I was schuckin' and jivin'. **1974** H. L. Foster *Ribbin', Jivin', & Playin' Dozens* v. 195 For many blacks, shuckin' and jivin' is a survival technique to avoid and stay out of trouble.

shucker (earlier example).

1872 *Golden Hours* iv. 397/1 The colored shuckers are considered the best because they will throw down a small oyster, and only open the large or medium-sized ones.

shudder, *v.* Add: **1. d.** Esp. in colloq. phr. *I shudder to think* with obj. clause.

1872 Geo. Eliot *Let.* 4 Aug. (1956) V. 297, I shudder a little to think what a long book it will be. **1952** M. Laski *Village* iii. 53 What they're going to think of us abroad, I shudder to think. **1970** A. Price *Labyrinth Makers* xii. 161 What he'll make of you I shudder to think!

4. For *nonce-use* read *rare* and add earlier and later examples.

1639 Chapman & Shirly *Tragedie of Chabot Admirall of France* i. sig. B, Loud conscience has a voyce to shudder greatnesse. **1925** E. Blunden *English Poems* 58 A drowned sheep lodged In a black holt of alders, Its poor fleece brown and vile, To shudder young beholders.

shuddersome, *a.* Add: (Later example.) Also *transf.*, inclined to shudder. Hence **shu·dder-somely** *adv.*

1903 A. M. Binstead *Pitcher in Paradise* ii. 52 One of those shuddersomely refined affairs that are supposed to be meat and drink to the giddy suburbanite. **1941** I. L. Idriess *Great Boomerang* xxii. 168 Eastern daggers of shuddersome shape, weapons of many nations. **1969** O. Blakiston *For crying out Shroud* iii. 29, I don't.. really feel so shuddersome about the smallest of bald patches.

shuddup (ʃɒdɒ·p), vulgar corruption of imp. *shut up* (see Shut *v.* 19 m). Cf. *shaddup, *shurrup.

1940 C. McCullers *Heart is Lonely Hunter* ii. xii. 217 'You dumb dumb dumb dumb—' 'Shuddup! Shuddup!' **1978** F. Mullally *Deadly Payoff* xiii. 180 'Shuddup,' Macdonald snorted.

shuffle, *sb.* Add: **5.** (Earlier example of *double shuffle*.) Also *spec.* a modern popular dance to jazz or rock and roll music, evolved orig. from Negro folk-dance; the music to which this is danced.

1821 P. Egan *Life in London* ii. v. 287 The *kidwys* and *kiddiesses* were footing the double shuffle against each other. **1925** (*jazz music title*) Shanghai Shuffle. [**1935** K. Burchill *Step Dancing* iv. 14 Swing the leg forward from the knee, so that the ball of your foot strikes the ground as it comes through... From this position swing the lower part of the leg back to its original position, striking the ground at the same time with the ball of the foot... These two movements..done in this order, constitute what is known as the 'Shuffle'.] **1955** Keepnews & Grauer *Pictorial Hist. Jazz.* ix 97 Such slightly later recordings as *Riverboat Shuffle*. **1956** M. Stearns *Story of Jazz* xvii. 203 The arrangements now sound heavy and cluttered and the rhythm was almost of the 'shuffle' variety. **1976** J. van de Wetering *Corpse on Dike* (1977) xvi. 159 The combo..played a slow shuffle, very easy to get into.

6. b. phr. *lost in the shuffle*: overlooked or missed in the mêlée or multitude. *U.S.*

1930 D. Runyon in *Collier's* 22 Mar. 21/4, I find we are about lost in the shuffle of guys with little mustaches. **1955** *New Yorker* 11 June 74/3 Mr. Ewell's efforts to be quietly funny are lost in the shuffle. **1981** W. Safire in *N.Y. Times Mag.* 8 Feb. 12/2 The book itself would then get lost in the shuffle.

c. *transf.* A redistribution of ministerial posts within a government or cabinet. Cf. Reshuffle *sb.* in Dict. and Suppl.

1941 C. Mackenzie *Red Tapeworm* xxii. 296, I hope we shall have no more of these Cabinet shuffles for the time being! **1966** —— *Paper Lives* xiv. 184 Mr Williamson, who was hoping like Mr Upjohn to find himself in the Cabinet at the next shuffle, ceased to argue. **1976** J. I. M. Stewart *Memorial Service* xiii. 204 The government was judged likely soon to undergo one of those 'shuffles' that English political mythology declares to be periodically essential.

shuffle, *v.* Add: **1. a.** Also *fig.* with *off* to die. (In playful allusion to *Hamlet*: see sense 5 d, quot. 1602.) *colloq.*

1922 A. Huxley *Mortal Coils* 124 One has to bring them [*sc.* obituary notices] up to date every year or so for fear of being caught napping if one of these old birds chooses to shuffle off suddenly. **1977** *N.Y. Rev. Bks.* 4 Aug. 29/3 She thought—if one had to 'shuffle off'—it would be terrific to be electrocuted while playing a bass guitar in a rock group.

12. shuffle beat = next; **shuffle rhythm,** a slow strongly syncopated rhythm (see quot. 1940); **shuffle-wing,** for *Accentor* substitute *Prunella*; (later examples).

1955 Shapiro & Hentoff *Hear me talkin' to Ya* 21 They played the shuffle beat on the snare drum. **1977** *National Observer* (U.S.) 22 Jan. 22/4 A lot of it's the old southern shuffle beat. Music you could get up and dance to. **1940** *Swing* June 13/3 The typifying characteristic of the Savitt band is its 'shuffle rhythm', which is distinguished..by its..4/4 jazz time. **1967** *Boston Sunday Herald* 30 Apr. 14/4 From shuffle rhythm to rock he waxed them all. **1909** W. H. Hudson *Afoot in England* xxiv. 289, I also love the smaller vocalists—the modest shufflewing and the lesser whitethroat. **1977** *Sunday Tel.* 6 Feb. 15/7 This is a day on which to..watch truly aggressive chaffinches competing with shufflewings.

shuffly (ʃɒ·fli), *a.* [f. Shuffle *v.* + -y¹.] Characterized by shuffling; inclined to shuffle.

1926 *Blackw. Mag.* Oct. 539/2 The step slower and, if possible, more shuffly. **1952** J. Masters *Deceivers* xxi. 240 That shuffly, shy old fool. **1974** N. Freeling *Dressing of Diamond* 66 He banged on the counter, brought the shuffly man out.

shufti, shufty (ʃu·fti, ʃɒ-), *sb. slang* (orig. *Mil.*). [f. next.]

1. A look or glance. Esp. in phr. *to take* (or *have*) *a shufti*.

1943 *Gen* 16 Jan. 11/2 Take a shufti at that. That's Stephanie. **1947** D. M. Davin *Gorse blooms Pale* 201 She took another good shufty at us. **1965** J. Porter *Dover Two* x. 133 Take a shufty at this lot! **1971** Whillans & Ormerod *Don Whillans* x. 81 'Let's have a shufti at that description then,' said Joe, and we both examined the piece of paper covered with Snell's handwriting. **1980** R. Adams *Girl in Swing* xix. 245 Good idea, old boy. I'm game. Let's 'ave a crafty shufti round with that in mind, shall we?

2. *Comb.*, as **shufti-kite,** a reconnaissance aircraft; **shufti-scope,** a probe (see quots.).

1944 T. H. Wisdom *Triumph over Tunisia* 172 When the Hun shufti-kites were kept away..the U.S. Second Corps moved..to the coastal stretch in the north. **1948** Partridge *Dict. Forces' Slang* 169 *Shuftiscope*, instrument used by doctors for exploring the interior of a dysentery case. A telescope or periscope. **1962** *Times* 12 Apr. 9/4 When officers were making a routine check of the vehicle with an instrument they call a 'shufti-scope' they found the watches. Mr. Eaton explained that the 'shufti-scope' is an instrument consisting of a probe with a light through which it is possible to see into cavities.

shufty (ʃu·fti, ʃɒ-), *v. slang* (chiefly *Mil.*). Now *rare*. [f. Arab. *šufti* have you seen?, f. *šāf* to see.] *intr.* To look or watch; to glance. Freq. in *imp.*

1943 C. H. Ward-Jackson *It's Piece of Cake* 54 *Shufty*, look, watch. Thus, the orderly room sergeant to a Waaf clerk as he sights a squadron of Liberators through the headquarters' window, 'Shufty!' **1947** *Amer. Speech* XXII. 267 The word *shufti* was among the most commonly used, comprehending in its meaning both the noun and the verb 'look'..'to shufti around' [etc.].

‖ **shugo** (ʃū·go). Also **Shugo.** [Jap.] In the Japanese feudal system: a military governor.

1893 F. Brinkley tr. *Hist. Empire Japan* iv. 171 His Majesty sanctioned the appointment of High Constables (*Shugo*) in the various provinces. **1933** F. C. Jones *Japan* i. 13 As is usual under feudal conditions, the *Shugo*..made their posts hereditary. **1974** *Encycl. Brit. Macropædia* X. 64/2 It was the job of the *shugo* to recruit metropolitan guards and keep strict control over subversives and criminals.

‖ **shuka** (ʃū·kā). Also **Shukkah.** [Swahili.] A long piece of fabric, now commonly worn as a shawl-like garment in East Africa.

1856 R. F. Burton *First Footsteps E. Africa* ii. 29 As regards the word Tobe, it signifies, in Arabic, a garment generally: the Somal call it 'Maro', and the half Tobe a 'Shukkah'. **1936** F. Stark *Southern Gates Arabia* xi. 123 The visitors from outside came wrapped in their shuka, a square black shawl with two fringed sides which they wore draped over their heads and bodies, the two lower corners knotted together and the upper ones thrown over one arm. **1971** *Standard* (Dar es Salaam) 7 Apr. 6/3 (Advt.), Tasini leads the way in the latest dress print designs. Shukas, Vitenges etc.

‖ **shul** (ʃūl). Also **shool.** [Yiddish, f. G. *schule* school.] The synagogue.

[**1804** M. Wilmot *Let.* 17 July in *Russian Jrnls.* (1934) i. 114 They walked with down cast eyes and penitent countenances to the School (as the Synagogue is call'd).] **1874** Hotten *Slang Dict.* 288 *Shool*, Jews' term for their synagogue. **1876** Geo. Eliot *D. Deronda* IV. xxxiii. 136/1 This evening is the Sabbath..and I go to the *Shool*. **1892** [see *hesped]. **1932** L. Golding *Magnolia Street* i. xiii. 235 The presentation will be at night, after *shool*. **1957** L. Stern *Midas Touch* i. i. 19 The Kosher food, the ceremonial prayers, coming here to *shul*. **1977** *New Yorker* 9 May 41/3 Her great-uncle Zindel, a former shammes in a shul that had been torn down.

‖ **Shulchan Aruch** (ʃu·lχān ā·ruχ). Also **Shulh(h)an Arukh.** [Heb., lit. table prepared.] The name of a work on Jewish religious practice by Joseph ben Ephraim Caro (1488–1575) now accepted as the standard guide to Orthodox observance.

1901 M. Gaster *Hist. Anc. Synagogue Spanish & Portuguese Jews* 47 It is an abstract of the *Shulkhan Arukh of Caro* containing all the laws and ceremonies of a practical utility, condensed into a very small form. **1922** *Aspects Jewish Life & Thought* 43 This popular edition of the greater work is the *Shulchan Aruch*..which was first published in the year 1565. **1957** *Encycl. Brit.* X. 18/2 The codes, especially Joseph Qaro's *Shulḥan Arukh*, became the halachic authority in Jewry. **1978** I. B. Singer *Shosha* ii. 22 He breaks every law of the Shulchan Aruch, yet at the same time he preaches Jewishness.

shulwar. Add: Also † **shalwar.** (Earlier example.)

1824 J. Morier *Adventures Hajji Baba* II. ix. 144 Can I offer him five tomâuns, and a pair of crimson *shalwars*?

shun, *v.* Add: **9. shun-pike** (earlier example); hence **shunpike** *v. intr.*, to drive along minor roads, avoiding the toll on turnpikes, or for pleasure; **shunpiker; shun-piking** *vbl. sb.*

1853 *Albany Even. Atlas* 9 Apr. 4 The Oswego Canal.. has been called a 'shun pike'. Produce sent by Lake Ontario and the Oswego Canal, avoids tolls on the canals west of Syracuse. **1964** *Collier's Encycl. Yearbk.* 70 Besides making long trips at high speed, motorists could take part in sports car rallies, chug about in antiques, 'shunpike' on quiet back roads. **1967** *Sat. Rev.* (U.S.) 22 Apr. 55 (Advt.), Smooth roads, beautiful scenery—what more could a shunpiker want? **1961** M. Beadle *These Ruins are Inhabited* iv. 41 George's reaction was to avoid all main roads—which is a good idea anyway if you're not in a hurry, the virtues of shun-piking being self-evident. **1972** *Alberta Motorist* (Edmonton) Apr. 6/3 There are, of course, many times when shunpiking is preferable for the traveller.

'shun (ʃɒn). Colloq. abbrev. of Attention 5, representing the verbal distortion and stress when used as a military command.

1888 Kipling *Plain Tales from Hills* 242 Stan'at—*hease*, 'Shun. **1928** *Granta* 2 Nov. 76/1 Prisoner! 'Shun.

Move to the right in fours. **1955** W. Faulkner *Fable* 108 'Bridesman,' he said but at that moment the major said ''Shun!'

‖ **shunga** (ʃu·ngā). Also **shun-ga.** [Jap., f. *shun* spring + *ga* picture.] An example of Japanese erotic art; a painting or print of a pornographic nature.

1964 *New Society* 17 Dec. 20/2 These paintings were *shunga*, 'Spring Pictures', which is the delicate Japanese euphemism for..erotic art. **1970** *Oxf. Compan. Art* 1172 With the exception of Sharaku almost all the 600 to 700 artists in the movement produced pornographic prints, known as *shun-ga* or 'spring pictures'. **1981** G. Macbeth *Kind of Treason* ii. 23 Quite a good collection of [Japanese] prints. A Hiroshige. Some *shunga*.

shungite (ʃu·ŋgəit). *Min.* Also **schungite.** [ad. G. *schungit* (A. von Inostranzeff 1886, in *Neues Jahrb. für Mineral.* i. 92), f. *Schunga* (Russ. *Shunga*), name of a village in Russia close to the Finnish border: see -ite¹.] (See quote. 1972.)

1892 E. S. Dana *Dana's Syst. Min.* (ed. 6) 8 Schungite from the Olonets Government, Russia, is a similar amorphous form of carbon intermediate between anthracite and graphite, occurring in phyllite. **1916** *Trans. & Proc. Geol. Soc. S. Africa* XVIII. 130 Schungite and graphitoid merge into graphite, but the difference is not crystalline merely, for the latter yields graphitic acid on oxidation, while the former do not. **1941** *Compt. Rend.* (*Doklady*) *de l'Acad. des Sci. de l'URSS* XXXIII. 358 Other authors arrived at an opposite conclusion, to the effect that the crystalline phase existing in anthracite and shungite is graphite, but..in a highly dispersed state. **1946** *Mineral. Abstr.* IX. 202 Eleven analyses of shungite from Shunga, Lake Onega, and other localities in eastern Karelia show C 46–99, ash 1–46, V_2O_5 0·016–0·77%, &c. **1968** *Ibid.* XIX. 102/1 The use of schungite to obtain combustible gases. **1972** *Gloss. Geol.* (Amer. Geol. Inst.) 656/2 *Shungite*, a hard, black, amorphous, coal-like material containing over 98% carbon, found interbedded among Precambrian schists. It is probably the metamorphic equivalent of bitumen, but it may represent merely impure graphite.

shunt, *sb.* Add: **1. b.** (Earlier example.)

1862 J. Simmons *Railway Traveller's Handy Bk.* 12 A thin line in the middle of trains..represents a shunt.

c. *slang.* A motor accident, a crash: esp. a nose-to-tail collision. Also **shunt-up** (on analogy with *pile-up 1), a multiple crash.

1959 in *Chambers's Twentieth Cent. Dict.* Add. **1967** *Economist* 11 Feb. 544/3 The characteristic American accident is a multivehicle shunt on a freeway. **1976** J. Wainwright *Bastard* vii. 91 Mist..happens on motorways, and is the cause of multi-car shunt-ups, where radiators kiss bumper bars. **1978** G. Vaughan *Belgrade Drop* vi. 41 'Another bloody shunt,' Yardley groaned. The Zagreb trunk was notorious for accidents.

2. c. *Med.* A natural or artificial route, esp. from a vein to an artery, whereby blood may bypass a capillary bed; the passage of blood along such a route. Also, the surgical construction of such a route.

1923 *Medicine* II. 20 Deficient oxygenation of the arterial blood..caused..by an unaerated shunt of venous blood into arterial. *Ibid.* 33 The readiness with which cyanosis develops or increases, during exercise, in a patient with an unaerated shunt. **1937** Best & Taylor *Physiol. Basis Med. Pract.* xxxv. 581 (*caption*) A portion of the blood passes through unaerated channels (shunt) from the venous to the arterial system. **1961** *Lancet* 30 Sept. 728/2 2 patients..had both undergone splenectomy some time before portocaval shunt was performed. **1980** *Amer. Speech* LV. 47 Clamps used to stop blood flow through a shunt when kidney dialysis is initiated.

d. *Biochem.* An alternative metabolic pathway; *spec.* (freq. as *hexose monophosphate shunt*) the pentose phosphate cycle.

1953 *Jrnl. Bacteriol.* LXVI. 17 It has been possible to infer..the existence of a sequence of reactions, the hexosemonophosphate shunt, which may serve in nature as a major pathway for the aerobic breakdown of carbohydrate by filamentous microorganisms. **1964** [see pentose phosphate shunt s.v. *pentose 2]. **1967** M. E. Hale *Biol. Lichens* viii. 118 Insoluble metabolic shunt products often serve as reserve food. **1970** [see *ribose].

5. a. shunt machine, motor, a direct-current motor in which the field and armature windings are connected in parallel with respect to the supply.

1888 *Rep. Brit. Assoc. Adv. Sci.* 1887 616 In a shunt machine the current through the coils of the second magnet may be controlled by the addition of a resistance in series with it. **1953** E. Molloy *Small Motors & Transformers* ii. 17 [Differential compounding] is sometimes employed to secure further improvement of the speed characteristic of a shunt machine, in cases where the load is variable. **1883** *Jrnl. Soc. Telegr. Engineers & Electricians* XII. 310 A coiled magneto-machine.., in which we have a magneto-generator acting as a brake, and a shunt motor. **1977** *N.Z. Herald* 5 Jan. 3/7 And once in Auckland, tied up by traffic lights and other vehicles, he switches over completely to his shunt motor for a quiet, anxiety-free tour of the town.

shunt, *v.* Add: **4. a.** (Earlier example.)

1845 *Min. Proc. Inst. Civil Engin.* IV. 252 At intermediate stations, the waggons are now 'shunted' to their proper places, in a siding, by the engine which has propelled them along the main line.

6. b. *Med.* To pass (blood) through a shunt. Cf. *shunt *sb.* 2 c.

1923 *Medicine* II. 18 A condition..in patients with congenital perforate septum of the heart, where a fraction of the blood is shunted directly from the right heart to the left without passing through the lungs. **1950** Best & Taylor *Physiol. Basis of Med. Pract.* (ed. 5) xxxv. 435/1 Most of the blood is shunted to the arterial side through channels which normally close at, or shortly after, birth.

shunting, *vbl. sb.* (Later examples in *Med.*)

1961 *Lancet* 23 Sept. 693/2 Arteriovenous shunting may be..applicable to more mammalian species..than has been previously considered. **1974** *Nature* 9 Aug. 489/1 Because of the incompletely divided ventricle, shunting of venous blood from the right atrium to the systemic circuit..may occur.

shupac, var. *shoepack.

shurrup (ʃɒrɒ·p, ʃuru·p), vulgar corruption of imp. *shut up* (see Shut *v.* 19 m). Also **shurr up.** Cf. *shaddup, *shuddup.

1893 S. Grand *Heavenly Twins* I. i. iv. 27 Barbara politely requested her to 'shurrup!' a word of the boys which she permitted herself to borrow. **1929** J. B. Priestley *Good Companions* iii. iv. 553 'Shurr up!' the voice jeered, before anybody else could make a sound. **1960** C. Ray *Merry England* 173 You shurrup, shurrup: I've just about had enough of you. **1978** J. Wainwright *Ripple of Murders* 40 Shurrup, Richard, don't show your ignorance.

shush (ʃuʃ, ʃɒʃ), *v.* [Echoic, representing a repetition of Sh *int.* Cf. Shshsh.] **1.** *trans.* To call or reduce (someone) to silence by uttering the sounds denoted by *sh-sh*. Also const. advs.

1925 P. Gibbs *Unchanging Quest* xvii. 127 She would.. 'shush' away any intruders who came to interrupt her private conversation. **1930** E. Ferber *Cimarron* i. 6 The woman listened... They shushed their children when they moved or whimpered. **1949** 'J. Tey' *Brat Farrar* xxv. 225 'The Ashbys cleaning up as usual,' the voice said, and was instantly shushed. **1961** J. Heller *Catch 22* (1962) xxii. 223 'Stop shouting, will you?' Yossarian shushed him. **1971** Wodehouse *Much Obliged, Jeeves* xv. 152 The ancestor..would, I think, have developed the theme had I not shushed her down with a raised hand. **1980** *Christian Sci. Monitor* (Midwestern ed.) 4 Dec. B3/2 He shushes them or storms at them or coaches them.

2. a. *intr.* To call for silence in this manner. Freq. *imp.* **b.** To become or remain silent.

1924 'O. Douglas' *Pink Sugar* v. 52, I stood patiently while Nellie 'shushed' under her breath as she brushed, directing me at intervals to 'Stand still, will ye!' **1929** J. B. Priestley *Good Companions* ii. i. 273 He gets no further, being fiercely requested by several of his colleagues to 'shush'. *Ibid.* iii. iv. 553 Some people laughed. The remainder indignantly shushed again and then clapped. **1969** M. Pugh *Last Place Left* xix. 140 'Shush, you bloody pig,' he said, for the collie was growing impatient. **1972** J. Wilson *Hide & Seek* v. 81 'Sh, now, there's a good girl.' 'I won't shush. I want to go home.' **1977** 'J. Bell' *Such a Nice Client* vi. 63 'That's the idea. Shush!' The manageress was back.

shush (ʃuʃ, ʃɒʃ), *sb. colloq.* [f. the vb.] An utterance of the exclamation 'Sh!' Also *transf.*, quiet. Cf. Hush *sb.*² 1 a.

1954 R. Fuller *Fantasy & Fugue* v. 115, I got him out among the shushes and tut-tuts before the attendants could reach him. **1959** I. & P. Opie *Lore & Lang. Schoolch.* x. 193 Let's 'ave a bit of shush. **1971** 'A. Burgess' *MF* vi. 71 There was a response of frightened shushes. **1982** *Observer* (Colour Suppl.) 25 Apr. 11/2 Can we have a little shush please?

shushing (ʃu·ʃiŋ, ʃɒ·ʃiŋ), *vbl. sb.* [f. *shush *v.* + -ing¹.] The action of the verb. Also *attrib.*, as *shushing, noise, sound*.

1929 J. B. Priestley *Good Companions* iii. iv. 552 It raised a loud and jeering laugh from that quarter, though the rest of the audience immediately made shushing noises. **1937** *Richmond* (Va.) *News-Leader* 9 Sept. 3/1 (*heading*) No 'shushing' at Nazi parley. **1972** *Listener* 15 June 806 The sound of a gourd rattle to make a 'shushing' sound. **1981** A. Edwards *Sonya* i. 27 There was much shushing in the front parlor if the children came in.

shush-shush, *a. rare.* [Reduplicated form of *shush *v.* or *sb.*] **a.** Echoic, designating something that makes a repeated soft sound like a shush. **b.** = *hush-hush.

1954 *New World Writing* vi. 44, I measured..his shush-shush tread in my presence. **1963** *Listener* 24 Jan. 182/1 It is his [*sc.* the defence correspondent's] job to know what is going on in the shush-shush world behind the 'D' notices.

Shuswap (ʃu·swɒp), *sb.* and *a.* Also † **Shooshap, Shouswap, Shushwap.** [ad. Shuswap *sxʷepmx* Shuswap Indians.] **A.** *sb.* **a.** A North American Indian tribe inhabiting southern British Columbia. **b.** Their language. **B.** *adj.* Of or pertaining to this tribe or their language.

1838 S. Parker *Jrnl. Exploring Tour* (1842) 313 West of these [*sc.* the Kettle Falls Indians] are the Sinpauēlish ..; and below these are the Shooshaps, having a population of five hundred and seventy-five. **1845** [see *carrier 2 b]. **1904** A. G. Morice *Hist. Northern Interior Brit. Columbia* 6 In case of death the bodies were buried among the Chilcotins and the Shushwaps. *Ibid.* iv. 51 This so-called Carrier vocabulary is made up of Shushwap words. **1927** *Internat. Jrnl. Amer. Linguistics* IV. 120 Salisham—Interior dialects: Shuswap; Lillooet. **1959** E. Tunis *Indians* 130/2 The Shushwap were a typical tribe of the interior Salish. **1965** *Canad. Jrnl. Linguistics* Spring 86 Terms found in those Salisham languages which are geographically the closest, particularly Kalispel..and Shuswap where available. **1976** *Shuswap Visitor's Guide* 10/3 According to a Shuswap Indian legend, you will find a place of perpetual summer.

shut, *sb.* Add: **6.** *Comb.*, as **shut-knife** *dial.*, a clasp-knife, a pocket-knife.

1879 J. Spilling *Johnny's Jaunt* i. 8, I took out my shet-knife and cut her a..huncheon off the loaf. **1913** D. H. Lawrence *Sons & Lovers* vii. 210 But they managed to procure a loaf and a currant-loaf, which they hacked into pieces with shut-knives, and ate sitting on the wall near the bridge. **1979** in R. Blythe *View in Winter* i. 63 He'd whittle away at things... He was that cliver [*sic*] with his shutknife.

shut, *v.* Add: **I. 3.** Also in pa. pple. with verbs of movement, as *draw, push*, etc., denoting completion of an action; equivalent to To *adv.* 4. orig. *U.S.*

1884 *Century Mag.* Nov. 13 He..pushed the ground-glass door shut. **1902** O. Wister *Virginian* xiv. 163 Our wheels clucked over the main-line switch. A train-hand threw it shut after. **1911** H. S. Harrison *Queed* ii. 23 The last boarder rising drew shut the folding-doors into the parlor. **1933** E. O'Neill *Ah, Wilderness!* i. 18 She slams the door shut. **1957** T. Slessor *First Overland* 256 We slam shut the windows, as the car slides down through the rocks.

4. b. (*a*) *to shut* (one's) *face*: see *face *sb.* 2 a for examples.

(*c*) *shut it* (in imperative): close one's mouth, hold one's tongue.

1886–96 in Farmer & Henley *Slang* (1903) VI. 202/1 Oh, shut it! Close your mouth until I tell you when. **1908** G. Sanger *Seventy Years Showman* x. 33 'Shut it!' said one of the showmen roughly; 'save your breath for the next scene.' **1945** G. Millar *Maquis* viii. 163 'Enough,' cried Boulaya. 'Shut it, Frisé... You know nothing.'

6. (Later examples.)

1949 K. S. Woods *Rural Crafts England* ii. ii. 33 The tyres have to be tightened by cutting out a piece and rejoining or shutting them with a smaller circumference to grip the wheel. **1964** H. Hodges *Artifacts* v. 86 For nearly all purposes the most effective way of joining iron was by welding or shutting.

11. a. Also (*dial.*) *shut on* and *ellipt.*

1848 Mrs. Gaskell *Mary Barton* I. v. 68 As for a bad man, one's glad enough to get shut on him. **1914** D. H. Lawrence *Widowing of Mrs. Holroyd* iii. 84 Who dost think wor goin' ter stop when we knowed 'e on'y kep on so's to get shut on us. **1976** S. Barstow *Right True End* i. iv. 65 'I haven't *got* her.' 'You're well shut, from all I hear.'

II. 13. shut down. a. (Earlier example of night.)

1880 'Mark Twain' *Tramp Abroad* xix. 182 We got to ..Heidelberg before the night shut down.

d. *Mech.* To stop or switch off (a device or machine, esp. an engine); to cause to stop working or running. Also *absol.*

1895 G. W. Lummis-Paterson *Management of Dynamos* x. 148 When shutting down a machine, the load should first be gradually reduced..by easing down the engine. **1911** Marshall & Sankey *Gas Engines* vi. 175 [Filling the reservoir] is done when shutting down the engine so soon as the gas is turned off. **1948** H. Constant *Gas Turbines* xi. 137 The best that can be done is to shut down as many engines as possible and operate the remainder at a power output giving reasonable efficiency. **1969** I. Kemp *Brit. G.I. in Vietnam* viii. 163 We all 'shut down'—switched off our engines and made fast the rotor blades to the tails of the helicopters. **1976** *Physics Bull.* Aug. 339/2 Two samples were taken from the low-sulphur plant: one at ambient temperatures 30 days after the boiler had been shut down. **1980** *Daily Tel.* 10 Mar. 3/2 A nagging 'oil migration' problem..could eventually have forced a pilot to shut down the engine.

e. *Physics.* To stop (the chain reaction) in a nuclear reactor; to stop (a nuclear reactor) from producing useful power by making the fuel assembly subcritical.

1945 Hawley & Leifson *Atomic Energy* 157 Knoowing just when to shut down the chain-reaction..would be quite a problem. **1951** *Nucleonics* Jan. 5/2 If the temperature of the uranium exceeds 60°C, the pile is automatically shut down. **1963** B. Fozard *Instrumentation Nuclear Reactors* xiii. 164 Unless a heavy-water reactor is shut down for such a long period that there is a significant fall in the activity of the high-energy gamma emitters, it has a built-in neutron source. **1976** *Sci. Amer.* Jan. 27/2 The level of radioactivity in a standard 1,000-megawatt reactor is very high: about 10 billion curies half an hour after the reactor is shut down.

f. *intr.* Of a device, machine, or installation, esp. a nuclear reactor: to cease to operate.

1945 H. D. Smyth *Gen. Account Devel. Atomic Energy for Military Purposes* 1940–45 viii. 81 The half-life of the U-239 is so short that its concentration becomes negligible soon after the pile shuts down. **1960** *Engineering Index 1959* 246/2 During power failures of up to 1·5 sec duration..synchronous motors usually shut down. **1976** *Sci. Amer.* July 36/1 After the reactor had shut down, the evidence of its activity was preserved virtually undisturbed through the succeeding ages of geological activity. **1978** *Nature* 19 Oct. 576/2 As the voltage dropped rapidly

various sensors began to shut down, and within a few minutes the satellite ceased to respond to commands sent up from ground control stations. **1979** *Daily Tel.* 15 Aug. 32/3 The Sea Kings will operate throughout the night but the others do not have night capability and will have to shut down for the night.

15. *shut in.* **c*.** *Oil Industry.* To cease drawing oil or gas from (a well).

1931 W. H. Emmons *Geol. Petroleum* (ed. 2) xvi. 527 In 1923, during a period of overproduction, certain groups of wells were shut in while others near by were pumped. **1962** T. C. Frick *Petroleum Production Handbk.* II. xxx. 8 Pressure readings obtained while the well is being cleaned before the well is shut in. **1971** *West Indian World* 5 Nov. 16/2 All 17 wells had encountered satisfactory oil sand sections and were at present 'shut in' and awaiting the installation of an eight-inch pipeline.

d. For † *Obs.* read 'Now *rare*' and add later example.

1924 [implied at *shutting *vbl. sb.* 3].

16. shut off. c. *intr.* To come to a halt; to cease talking or writing. *U.S.*

1896 'Mark Twain' in *Harper's Mag.* Sept. 526/1 'Now who—' He shut off sudden. **1902** J. London *Let.* 12 July (1966) 136 Someone is going down town, so I'll shut off and give them a chance to mail this. **1938** V. Woolf *Let.* 18 June (1980) VI. 241 He rang me up late on Wednesday... He said he had travelled post haste from Prague to see Leonard. I said, A misunderstanding. Then we shut off.

17. shut out. d. (Earlier examples.) Also *transf.* in other games and *fig.* *N. Amer.*

1881 *N.Y. Herald* 17 July 10/3 The Domestics were shut out in every inning up to the eighth, when by bunching their hits they scored two earned runs on a single by Mahny. **1894** *Spalding's Base Ball Guide* 40 Nichols..shut out the St. Louis team without a game to their credit out of four games played. **1952** in Wentworth & Flexner *Amer. Dict. Slang* (1960) 475/2 The last time [the Princeton football players] were shut out Penn did it on Nov. 3, 1946. **1957** *Northland News* (Uranium City, Sask.) 7 Jan. 7/2 [Ice hockey] The Flyers shut-out the blue and gold for the second time by a 2–0 count. **1976** *National Observer* (U.S.) 15 May 2/1 Reagan shut out Ford in Texas, winning all 96 delegates to the National Convention.

19. shut up. l. (Earlier example.)

1814 Jane Austen *Mansfield Park* III. xvi. 305 Her son, who was always guided by the last speaker, by the person who could get hold of and shut him up.

m. (Earlier example.)

1840 *Picayune* (New Orleans) 10 Oct. 2/4 The Dutchman got a hint to 'shut up' from one of the officers.

shut, *ppl. a.* Add: **2.** *shut-away* adj. Also with adjs., as *shut-eyed, -minded* (so *-mindedness*), *-mouthed*.

1911 Galsworthy *Patrician* ii. xvi. 253 Her face had a strange, brooding, shut-away look, as though he had frightened her. **1913** 'Saki' *When William Came* (1914) xvi. 272 He looked round again at the rolling stretches of brown hills; before he had regarded them merely as the background to this little shut-away world. **1959** *Listener* 12 Mar. 473/1 The sensitive, shut-away man. **1934** J. A. Lee *Children of Poor* (1949) iii. 68 Prayer was tiring in the extreme, in an atmosphere of tense, sweaty, shut-eyed sanctity. **1956** H. Gold *Man who was not with It* (1965) xxx. 283. I watched Belle's shut-eyed face compose with fatigue. **1960** T. Hughes *Lupercal* 37 And look in at the byre's Blaze of darkness: a sudden shut-eyed look Backward into the head. **1977** *Sunday Mail* (Brisbane) 20 Nov. 27/4, I don't want to get shut-minded as I get older. **1981** *Times Lit. Suppl.* 16 Jan. 60/4 At this stage in such a review it is a common *topos* to remark that thanks are due to the editor or author for raising weighty questions. It may seem churlish or shut-minded, but for £45 one might also expect a few answers. **1933** C. C. Martindale in M. Leahy *Conversions to Catholic Church* ix. 91 A priestly work of incredible shut-mindedness, but..homage-worthiness. **1936** C. Sandburg *People, Yes* 113 In Vermont a shut-mouthed husband finally broke forth to his wife. **1959** W. R. Bird *These are Maritimes* viii. 217 That made him awful mad but he wouldn't say anything. He's what you'd call 'shut-mouthed'.

shut, var. *shutt.

shut-down. Add: **1. a.** (Earlier and later examples.)

The sense in quot. 1857 is uncertain.

1857 *Knickerbocker* XLIX. 35 'I'll be just exactly shot if you *don't*!' he added with a patent diabolical shut-down. **1884** *Boston Jrnl.* 16 Oct. 7 The Acushnet papermill at New Bedford has started up after a shut-down of two months. **1901** *Chambers's Jrnl.* Jan. 29/1 In the old days when we had a grievance we could talk it over with the boss; but today if there is a shut-down nobody knows anything about it except that it has been ordered from headquarters in New York. **1931** *Economist* 15 Aug. 308/2 The Ford shutdown will restrict August production. **1959** *News Chron.* 2 July 4/3 What do you propose..to prevent a shutdown of the national papers? **1967** *N.Y. Times* (Internat. Ed.) 11 Feb. 2/2 Classes resumed quietly at Madrid University today after a 10-day shutdown because of campus battles between students and the police. **1972** M. Jones *Life on Dole* vi. 51 The Dowlais shut-down had several consequences. It took away the one industrial enterprise that kept going all through the week.

b. The cessation of broadcasting for the day on any particular channel or station; the time at which this occurs.

1959 *Listener* 19 Mar. 499/1 The peak viewing-hours from 7.0 p.m. to shut-down. **1973** *Times* 15 Dec. 2/7 (*heading*) TV staggers shutdown.

2. a. The cessation of operation of a machine, device, or installation, esp. as a result of a fault.

1911 J. F. C. Snell *Power House Design* x. 352 The.. security against a shut-down arising from the duplicate sets of bus bars. **1916** *Standard. Rules Amer. Inst. Electr. Engin.* 35 This method consists in the measurement of the temperature of windings by their increase in resistance, corrected to the instant of shut-down when necessary. **1936** *Discovery* Apr. 113/2 In the event of a shut-down of the power unit, an alternative supply is available from the storage battery. **1953** [see *input *v.* 3]. **1971** *New Scientist* 7 Jan. 19/1 An abrupt shutdown is a very different affair from the computer operator pushing the computer 'off' buttons. **1976** M. Machlin *Pipeline* liv. 542 The breakdown at Pump Station One and Three..resulted in an automatic shutdown of the entire pipeline.

b. *Physics.* The stopping of the chain reaction in a nuclear reactor when the fuel assembly is made subcritical.

1945 H. D. Smyth *Gen. Acct. Devel. Atomic Energy for Military Purposes* 1940–45 vii. 71 A shut-down for servicing could be effected. **1951** *Nucleonics* Jan. 5/1 There are two sets..of emergency shut-down rods. **1963** B. Fozard *Instrumentation Nuclear Reactors* xiii. 162 The delay due to xenon poisoning in restarting after a shut down cannot be reduced or eliminated. **1978** N. L. Franklin in Foley & Van Buren *Nuclear or Not?* 120 [Boron steel control rods] drop down into the core ensuring a rapid shutdown.

shute². Add: **2.** A variety of raw silk; tram silk.

1839 Ure *Dict. Arts* 1102 There are three denominations of raw silk; viz., organzine, *trame* (shute or tram), and floss. **1868** [see Tram *sb.*¹ 1].

shute³. Add: **5.** = *chute *sb.* 3 b.

1961 R. P. Hobson *Rancher takes Wife* iii. 47 Each had a corral system with shute and squeeze, and a log horse-pasture. **1971** J. S. Gunn *Distrib. Shearing Terms N.S.W.* 10 *Shute*, the opening through which a shorn sheep is pushed. *Ibid.*, *Shute*, the ramp outside down which a shorn sheep is pushed to the counting-out pen.

shu·t-eye. *colloq.* [f. Shut *v.* + Eye *sb.*¹] Sleep; sleeping.

[**1896** Kipling *Seven Seas* 217 (*heading*) The shut-eye sentry.] **1899** *Navy & Army Illustr.* 9 Dec. 307 The remainder of the dinner hour..is spent in smoking and perhaps dozing (a little shut-eye). **1923** C. E. Montague *Fiery Particles* 174 'We'll go on to-morrow,' I said. 'A bit of shut-eye for me now.' **1936** A. Huxley *Eyeless in Gaza* xxiv. 346 Time for your spot of shut-eye. **1942** 'M. Innes' *Daffodil Affair* ii. 68 Or game of cards... After a long party nothing like a hand or two before shut-eye. **1958** 'Castle' & 'Hailey' *Flight into Danger* i. 25 Good-night... I can sure use some shut-eye. **1977** *Time* 31 Jan. 6/1 Air Force Two indeed came equipped with a place to catch up on shut-eye between stops. 'Can you sleep on a plane?' asked Ford. Said Mondale, 'I'm going to find out.'

shu·t-in, *a.* and *sb.* orig. *U.S.* [Shut *ppl. a.* 2: cf. Shut *v.* 15.] **A.** *adj.* **1.** Enclosed, hemmed in; *esp.* of a person: confined by severe weather or by physical or mental disability; isolated by self-absorption; abstracted.

1849 [in Dict. s.v. Shut *ppl. a.* 2]. **1909** *Sunday School Times* (Philadelphia) 27 Feb. 110 She had brought a handful of flowers and a heart full of sunshine to the shut-in mother. **1912** [see *autism]. **1932** *Brit. Jrnl. Psychol.* Apr. 301 Subject 6 has fewer friends..because she is so 'shut in' and difficult to get to know. **1943** J. B. Priestley *Daylight on Saturday* xviii. 133 You could tell by the shut-in look on their faces as they worked that they were busy thinking about these things. **1957** [see *claustrophobic *a.* b]. **1975** *Budget* (Sugarcreek, Ohio) 20 Mar. 7/4 Neal C. Troyer..and Dan L. Schwartz's spent Sat. at Nappanee visiting shut-in relatives, Bis. John L. Schwartz and Mrs. Lizzie Borkholder who is blind.

2. *Oil Industry.* Of or pertaining to oil and gas wells that are shut in; applied *esp.* to production capacity that is available but not being utilized.

1931 *Economist* 28 Mar. 671/2 Production from the new East Texas fields is weakening prices, and the vast amount of shut-in production is a constant menace. **1960** *Fortnightly Rev.* (Anderson & Strudwick, Richmond, Va.) 19 Aug., The long struggle to obtain markets for Canada's large shut-in natural gas reserves was finally concluded last week. **1962** *Listener* 10 May 796/2 There remains a huge quantity of shut-in oil capacity overhanging the market and depressing prices. **1974** P. L. Moore et al. *Drilling Practices Manual* xii. 314 The shut-in drill pipe pressure is easy to obtain if there is no back-pressure valve in the drill string.

B. *sb.* **1.** A person who is confined by severe weather or by a physical or mental disability; a withdrawn person, one who is isolated from normal social interaction.

1904 [in Dict. s.v. Shut *ppl. a.* 2]. **1949** *Richmond* (Va.) *Times-Dispatch* 1 Feb. 9/2 (*heading*) Salad dressing diet palls on storm shut-ins. **1952** *Catholic Times* 25 July 7/4 It is the 'shut-ins' and sufferers who give the greatest of all contributions—their sufferings. **1966** *Daily Tel.* 25 May 19/5 The first Sunday in June is Shut-In's Day, a special day set aside each year to remember the sick and housebound. **1975** C. Potok *In Beginning* ii. 118 Better a man..who is a man of the world and can also learn than a bearded shut-in with the brain of a genius and the soul of a calf. **1979** *Arizona Daily Star* 22 July J6/4 Volunteers are needed to grocery shop for elderly shut-ins.

2. *Oil Industry.* A state or period of being shut in.

1962 T. C. Frick *Petroleum Production Handbk.* II. xxx. 7 Subsurface-pressure gauges are very useful in wells where liquids accumulate in the wellbore during shut-in. **1977** R. D. Langenkamp *Handbk. Oil Industry Terms & Phrases* (ed. 2) 152 There is a great difference between a shut-down and a shut-in... A well is shut in when its wellhead valves are closed, shutting off production.

Hence **shut-i·n-ness,** the quality of being confined, secluded, or withdrawn into oneself.

1913 D. H. Lawrence *Let.* ?10 June (1962) I. 210 The world gets a queer feeling of shut-in-ness, as if it stifled one, the horizon being too near, the sky too low. **1920** *Chambers's Jrnl.* Jan. 23/2 Leaving the bowl of the crater with its strange sense of shut-in-ness. **1952** *Jrnl. Mental Sci.* XCVIII. 310 There is evidence that shy, shut-in people are more liable to schizophrenia than outgoing folk, but this is disputed, and some psychiatrists maintain that the 'shut-inness' is just an early stage of the illness.

shut-off. Add: Also **shutoff. 1.** (Further examples.) Also, something used for stopping the operation of anything.

1951 *Nucleonics* Nov. 16/2 Two..plates are used in controlling the reactor; the other two act as emergency shut-offs. **1970** N. Armstrong et al. *First on Moon* xiii. 313 Okay, I'm going to open up the main shutoffs. Ascent feed closed. **1973** R. Ludlum *Matlock Paper* xxii. 188 He found the shut-off button and pressed it. **1978** *N.Y. Times* 30 Mar. B 8 (Advt.), A pocket-size recorder that's perfect for school, office or home. Has end-of-tape shut-off.

2. A cessation of flow, supply, or activity.

1889 *Cent. Dict.* 5606/1 *Shut-off*, stoppage of anything. **1919** *Summary of Operations Calif. Oil Fields* (Calif. State Mining Bur.) V. i. 8 *Collar shut-off*, an accidental 'shut-of' supposed to be occasioned by the accumulation of material between the walls of a well and the casing at, or just above, a collar. **1942** *Sun* (Baltimore) 28 Dec. 18/4 A threatened shutoff of tin containers. **1974** *Spartanburg* (S. Carolina) *Herald-Jrnl.* 21 Apr. A2 (Advt.), Easily adjusts from powerful stream to fine spray to complete water shut-off. **1977** *Nature* 6 Jan. 29/1 Neither *alt* nor *mod* mutants affect the shutoff of host protein or rRNA synthesis *in vivo*.

B. *adj.* [Shut *ppl. a.* 2.] Disconnected, stopped; isolated, withdrawn.

1913 D. H. Lawrence *Sons & Lovers* v. 96 That peculiar shut-off look of the poor who have to depend on the favour of others. **1933** S. Spender *Poems* 45 The airliner with shut-off engines Glides over suburbs. **1939** C. Isherwood *Goodbye to Berlin* 91, I felt a most marvellous sort of shut-off feeling from all the rest of the world. **1982** 'J. Ross' *Death's Head* iv. 30 She said nothing more, waiting in a shut-off silence.

shut-o·ut, *a.* and *sb.* [Shut *ppl. a.* 2: cf. Shut *v.* 17.] **A.** *adj.* **1.** That is shut out or excluded; isolated, remote.

1853 [in Dict. s.v. Shut *ppl. a.* 2]. **1853** Mrs. Gaskell *Let.* Sept. (1966) 245 He..was rather intimate with Lord Palmerston at Cambridge, a pleasant soothing reflection now, in his shut-out life. **1860** Geo. Eliot *Let.* 5 Sept. (1954) III. 342 It is a better house than I care to have, but as it is more shut out than anything we have seen..I accept the luxury.

2. In Bridge, of a bid: pre-emptive or otherwise intended to discourage the opposition from bidding. Also *transf.* of a financial bid.

1916 [see *pre-emptive *a.* 2]. **1921** A. E. M. Foster *Auction Bridge* 52 Pre-emptive bids are, in my opinion, a mistake, unless they are of the nature of necessary shut-out bids. **1932** [see *pre-emptive *a.* 2]. **1959** Reese & Dormer *Bridge Players' Dict.* 167 An opening bid of three or four in a suit is in nearly all systems a weak shutout bid, based on a long suit with little or no outside strength. **1969** *Observer* 12 Jan. 11/1 Joe Hyman, chairman of Viyella International, ought to be worried by Courtaulds' 15s. 6d. a share shut-out bid for English Calico. **1982** *Times* 22 Jan. 13/5 The Council for the Securities Industry moved yesterday to ban 'shut out' takeover bids. It has become popular for one company to take control of another by buying or obtaining promises which give it 50 per cent control before anyone else can make counter proposals.

3. In Baseball and other games: characterized by the failure of the losers to score; that prevents the opponents from scoring. Chiefly *N. Amer.*

1949 *Minot* (North Dakota) *Daily News* 22 July 8/8 He led his Grand Forks team to a one-hit shutout victory over Duluth. **1974** *News & Press* (Darlington, S. Carolina) 25 Apr. 11/1 Riding to victory on the two hit shutout pitching of Reece Ammons and the strong bat of Darrell Lloyd, the St. John's Blue Devils blitzed the Manning Monarchs 6–0. **1978** *Monitor* (McAllen, Texas) 21 May 3B/5 The Purple team scoring a 9–0 shutout win over the White unit in a controlled scrimmage.

B. *sb.* **1.** In Baseball and other games: a match or innings in which one side does not score; prevention from scoring. Also *fig.* Chiefly *N. Amer.*

1889 *Pueblo* (Colorado) *Opinion* 21 July 4/5 The Springs were 'fated' from the start, and narrowly escaped a shut out. **1897** [in Dict. s.v. Shut *ppl. a.* 2]. **1936** *N.Y. Herald Tribune* 2 Oct. 10/2 The national scoreboard looked pretty bad. In fact it looked so much like a shutout for the team that you voted a change of management in order to give the country a chance to win the game. **1937** *Evening Standard* 25 Feb. 31 The Swiss may fully extend Britain, and maybe even break Foster's proud run of 'shut-outs'. **1955** *Edmonton Jrnl.* 4 Jan. 10/4 A few hours later he secured that elusive shutout that had escaped him so far this season. 1972 *Newsweek* 31 July 43/3 Fischer routed Mark Taimanov..6 to o, for the first shutout in the history of grandmaster chess. 1972 *N.Y. Times* 3 Nov. 39/1 Senator McGovern may lose every state in the Union, and..his only chance of avoiding a shut-out lies with the people..in California and the District of Columbia. 1977 *Wandsworth Borough News* 16 Sept. 10/3 Putney St. Mary's Senior 'D' and U/12 'A' teams were in good scoring form, both winning their matches after scoring 10 goals apiece. For the 'D' team it was a complete shut-out. 1978 J. Irving *World according to Garp* xix. 426 Roberta pitched a shutout.

2. A shut-out bid: see sense 2 of the adj. above.

1936 E. Culbertson *Contract Bridge Complete* 17 *Shut-out*, an unnecessarily high bid, designed to make it difficult for the other side to enter the auction. 1982 *Observer* 17 Jan. 17/5 The Takeover Panel's consent to the 'shut-out'.

shutt (ʃɒt). *local.* Also **shut.** [See Shoat^1.] = Grayling 1 a.

1939, 1952 [see *pink sb.^{4} 2 b].

shutter, *sb.* Add: **2. a.** *to put up the shutters* (earlier example).

[1837 Dickens *O. Twist* (1838) I. iv. 62 The undertaker had just put up the shutters of his shop, and was making some entries in his day-book.] 1877 Trollope *Amer. Senator* I. iii. 27 If..you won't have any client that isn't a gentleman, you might as well put up your shutters at once.

j. A device for regulating the supply of cooling air to the radiator of an internal-combustion engine.

1918 *Aëronaut. Jrnl.* XXII. 119 A metal ring is in position on the underside of the radiators within reach of the pilot. This is apparently intended to carry a semi-circular disc to act as a radiator shutter. 1935 *Times* 2 Oct. 6/5 The temperature of the water is controlled by thermostatically-operated radiator shutters. 1957 Frazee & Eshelman *Tractors & Crawlers* v. 173 Shutters..are used on some tractors to reduce the air flow through the radiator. 1971 M. Tak *Truck Talk* 142 *Shutters*, louvers that are located between the tractor's grill and the radiator of the engine and that open and close like venetian blinds.

3. a. shutter eye (see quot. *a* 1884); **shutter weir,** a type of movable weir consisting of one or more leaves pivoted about a horizontal axis at or towards the bottom and held nearly vertical until released.

a 1884 Knight *Dict. Mech.* Suppl. 807/2 *Shutter eye*, an eye for hanging a shutter to, having a projecting flange or support, which is built into the wall. 1930 *Gen. Catal. Tools & Supplies* (Buck & Hickman Ltd.) 1081 (*heading*) Pointed gate and shutter hooks and eyes. 1880 *Engineering* 30 Jan. 102/1 The needle and shutter weirs had been the most extensively adopted, and were the best types of movable weirs... The shutter weir was the most suitable for high weirs across rapid rivers, and where the navigation was conducted by flushing. 1928 F. Johnstone-Taylor *River Engin.* iii. 69 The shutter weir, as a form of barrage, is not a new introduction, having been used..principally in Central Europe, for a number of years.

b. shutter-bug *slang* (orig. *U.S.*), an enthusiastic photographer; also *attrib.*; **shutter priority** *Photogr.*, used *attrib.* and *absol.* to designate automatic working in which the user sets the shutter speed and leaves the appropriate aperture to be set by the camera when the exposure is made; **shutter release,** the button on a camera that is pressed to cause the shutter to open; **shutter speed,** the nominal time for which a shutter is open at a given setting.

1940 *Amer. Speech* XV. 357 The amateur..is known as a clicker, a snapper,..a shutterbug. 1972 *New Nation* (Singapore) 25 Nov. 9/8 It is unspoilt by shutter-bug tourists. 1979 *Arizona Daily Star* 8 Apr. 1. 3/6 (Advt.), 4 canyon tours... Our best for shutterbugs includes round trip transportation & lodgings plus photo stop at Casa Grande Ruins. 1974 L. Gaunt *Canon Reflex Way* 24 The system can be converted to shutter-priority automatic working. 1978 *SLR Camera* Aug. 31/1 (Advt.), With the new Minolta XD7 a flick of a switch allows you to change the system to suit the subject—shutter priority *or* aperture priority *or* fully manual. 1958 *Newnes Compl. Amat. Photogr.* 37 At the B setting, the shutter remains open as long as the shutter release is kept pressed. 1979 *Amat. Photographer* 30 May 97/3 Initial pressure on the shutter release lights the viewfinder LEDs which indicate the speed to be chosen by the camera. 1889 Shutter speed [in Dict.]. 1906 R. C. Bayley *Compl. Photographer* xiii. 158 In the Frena camera a method of regulating the shutter speed by regulating the size of the opening in a rotating disc was employed. 1977 J. Hedgecoe *Photographer's Handbk.* 162 Fast shutter speeds—say 1/250 sec or shorter—usually eliminate problems of camera shake.

shuttered, *ppl. a.* Add: Also *fig.*

1927 J. R. Theobald in *Oxf. Poetry* 34 A clothèd woman all alone Beneath her shuttered sky. 1930 R. Campbell *Adamastor* 76 The day burns through their blood Like a white candle through a shuttered hand. 1957 M. Stewart *Thunder on Right* i. 15 Her heart began to beat lightly and fast, but her face was shuttered, and she gave no sign.

shuttering, *vbl. sb.* Add: **2. a.** = *formwork* s.v. *form *sb.* 22. **b.** = *form *sb.* 18 b.

1895 E. de V. Buckingham *Contractors' Price-Bk.* 371 (*heading*) Concrete building casing and shuttering (per yard super). 1902 *Min. Proc. Inst. Civil Engin.* CXLIX. 298 If the shutterings or drums are well made and kept well greased, a fairly good and even face can be obtained by mixing the concrete rather wet, and well chopping down behind the shutterings or drums. 1919 *Spectator* 25 Oct. 539/2 The soil..to be used without admixture..and to be thrown direct into the shutterings. 1939 *Archit. Rev.* LXXXVI. 35 The facing slabs..can be used, when braced, as external shuttering to a reinforced concrete wall. 1959 *Listener* 27 Aug. 316/2 Exposed concrete shows the least incidents of the shuttering, the joints of the planks, the fibres and knots of the wood. 1976 *Star* (Sheffield) 20 Nov. (Advt.), Shuttering joiners required.

shutting, *vbl. sb.* Add: **3.** (Later example.)

1924 C. Mackenzie *Heavenly Ladder* xix. 247 They finished decorating the church just before the shutting in of a still and humid dusk.

5. *attrib.*, as **shutting joint, post,** the joint or post against which a door or gate closes; **shutting stile** (see quot. 1955).

1823 *Practical Builder* iii. 182 (*heading*) On the formation of the shutting-joints of doors. 1929 T. Corkhill in R. Greenhalgh *Joinery & Carpentry* I. iii. 171 The hinge is on the concave and the result is an extremely awkward shutting joint. 1944 N. W. Kay *Pract. Carpenter & Joiner* vi. 140 As a curved rebate is not always possible the shutting joint is set out to a straight surface tangential to the greatest radius on the framing. *a* 1877 Knight *Dict. Mech.* III. 2170/1 *Shutting post*, the post or joint against which a gate or door is closed. 1909 *Chambers's Jrnl.* Nov. 764/1 On the under side of the shutting post is a small roller which runs on to a bracket on the shutting post itself, thus taking up the whole weight of the gate when it is closed. 1909 Webster, Shutting stile. 1955 N. W. Kay *Mod. Building Encycl.* 613 *Shutting stile*, door or window stile opposite the hanging stile.

shuttle, *sb.*^{1} Add: **8. b.** A shuttle service of aircraft; *esp.* one operated by an airline for which reservation of seats is not a requirement; an aircraft flying on such a service.

1942 [see *shuttle route*, sense 9 a below]. 1944, 1961 [see *shuttle service*, sense 9 b below]. 1964 Mrs. L. B. Johnson *White House Diary* 9 Apr. (1970) 104, I could have caught a much later plane if I could only have ridden the shuttle. 1971 R. Thomas *Backup Men* x. 84, I got in line for the Eastern shuttle... It's rumored that if Eastern doesn't have a seat for you on its regular shuttle to New York, it will roll out a special plane just for you. 1973 *Daily Tel.* 11 Sept. 6/4 British Caledonian is to extend its low fare 'Moonjet Service'—Britain's first no-reservation walk-on, walk-off shuttle—to Belfast. 1977 *Time* 10 Oct. 4/1 Freddie Laker's bargain-basement transatlantic shuttle, the no-frills, no-reservations Skytrain, was finally aloft, carrying passengers between London and New York at the rock-bottom round-trip fare of $236. 1978 R. Ludlum *Holcroft Covenant* xiii. 153 The shuttles to Paris were frequent, the customs procedures lax.

c. More fully, *space shuttle.* A space rocket with wings enabling it to land like an aircraft and be used repeatedly.

Quot. 1960 is fictional.

1960 'J. Wyndham' in *New Worlds* Nov. 41 The acceleration in that shuttle would spread you all over the floor. 1969 *New Scientist* 5 June 513/2 NASA has announced the formation of task groups to look into..a re-usable low-cost 'Space Shuttle' to relay men and materials to and from the [space] station... The space shuttle..would be fired off vertically, shed its fuel tanks and, upon return, land horizontally at an airport. *Ibid.* 2 Oct. 7/1 Another shuttle plying on a regular basis between Cape Kennedy and this large space laboratory. 1972 *National Observer* (U.S.) 27 May 6/3 The shuttle's primary mission is to carry satellites into earth orbit and release them, at a cost below that of the expendable rockets now used to launch satellites. 1981 *Daily Tel.* 15 Apr. 1 The American space shuttle landed on a dry lake bed in California's Mojave Desert yesterday to complete the maiden flight of the first re-usable rocketship.

d. A series of journeys for the purpose of shuttle diplomacy (see sense 9 b below).

1975 *Daily Tel.* 29 Aug. 24/3 (*heading*) Raid as peace shuttle nears end. *Ibid.*, Dr Kissinger completed the last legs of his Middle East shuttle yesterday. 1977 *Time* 17 Jan. 30 It was a diplomatic shuttle, but not exactly in the Kissinger mode. *Ibid.*, Thus [Ivor] Richard's shuttle has been dubbed by some officials and journalists in southern Africa a safari of salvation.

9. a. (sense 3) *shuttle-winder*; (sense 8) *shuttle bus, flight, plane, raid, rocket, route, ship.*

1951 *Sun* (Baltimore) 18 May 3/1 The cars—some are called 'shuttle busses' because they operate from the West (executive offices) Wing to the East Wing [of the White House]—carry messengers too. 1972 *Times* 8 June 7/2 Traffic jams..that officials hoped would be averted by the bicycles and shuttle buses. 1979 *United States 1980/81* (Penguin Travel Guides) 259 During the summer, a shuttlebus runs from the lakefront to the courthouse. 1944 *News* (Tuscaloosa, Ala.) 25 June 1/3 Three crewmen also were lost as a result of the attack on the fields, apparently those used by Italian based and Britain-based bombers in the shuttle flights over Axis targets. 1961 *N.Y. Times* 10 May 90/5 The shuttle flights between the two pairs of cities carried 6,147 passengers in their first week... Passengers arriving by bus, cab or car would be able to step out at the terminal door within 150 feet of the shuttle planes. 1977 *New Yorker* 19 Sept. 40/1 She had to drive home alone, while he took a shuttle flight in the opposite direction. 1944 *Britannica Bk. Year* 770/1 *Shuttle*, combining form. Involving vehicles, especially aircraft, making repeated trips between fixed points, as ..'shuttle raid', 'shuttle plane'. 1961 [see *shuttle flight* above]. 1976 J. Crosby *Nightfall* xii. 68 [He] left for the shuttle plane to New York. 1943 *Time* 18 Oct. 85/1 The..pilot flew on his first mission eight weeks ago, joined the first shuttle raid on Germany, flew safely to Africa, [etc.]. 1953 J. N. Leonard *Flight into Space* 87 They say that von Braun's great shuttle rockets—to say nothing of his space station—would surely fail. 1942 *R.A.F. Jrnl.* 30 May 22 The danger zone, which is the shuttle route of the German Fock-Wulf Condors. 1959 *Amazing Stories* June 12/1 Hubbard visited the spaceport ..and watched the shuttle-ships come and go. *a* 1877 Knight *Dict. Mech.* III. 2171/2 *Shuttle-winder*, a device for winding a shuttle, such as the round shuttle of the Wheeler and Wilson sewing-machine, or a tatting-shuttle.

b. † **shuttle armature** *Electr.*, an armature having a single coil wound upon an elongated iron former shaped like a shuttle (*obs.*); **shuttle bombing,** bombing carried out by planes taking off from one base and landing at another; so **shuttle bomber; shuttle car,** a vehicle for making frequent short journeys, *spec.* one for the underground haulage of coal; **shuttle diplomacy,** diplomatic activity involving a series of journeys to and fro, *esp.* by a mediator travelling between disputing parties; hence **shuttle diplomat; shuttle service** (earlier and later examples); more widely, any transport service in which vehicles or aircraft travel to and fro between fixed points at frequent intervals; **shuttle-train** (examples); † **shuttle-wound armature** *Electr.* = *shuttle armature* above (*obs.*).

1890 Slingo & Brooker *Electr. Engin.* viii. 241 That the design of the shuttle armature is faulty may easily be proved, for, after being rotated for a little time, the iron shuttle or core gets quite warm. 1924 S. R. Roget *Dict. Elect. Terms* 226/2 *Shuttle armature*, a simple form of armature now rarely used, except in very small machines, with a single coil connected to a two-part commutator and lying in the two broad slots in an elongated core built up of stampings in the shape of an H with rounded sides. Also called *Siemens 'H' armature.* 1944 *Yank* 28 July 7 They are not shuttle bombers, and they did not fly from Italy to Russia intentionally. 1944 *Newsweek* 10 Jan. 27 Last summer the RAF and the Eighth both tried shuttle bombing. 1954 *Times* 10 Aug. 4/1 The city may be important for another reason—as one end of a shuttle-bombing route similar to those which worked so effectively in Europe. 1905 Calkins & Holden *Mod. Advertising* v. 89 They also have many shuttle cars, or [street]cars that make short runs. 1956 Atkinson & White in D. L. Linton *Sheffield* 276 The shuttle cars transport the ore to the main-road conveyors which discharge the ironstone at the surface into wagons. 1979 *Jrnl. R. Soc. Arts* CXXVII. 89/2 Rubber tyred shuttle cars can be used from the continuous miner to the main transport system if the floor is hard enough. 1974 *Between Lines* (Newtown, Pa.) 15 Feb. 2/3 So beware of an over-celebration of Kissinger's shuttle diplomacy, heroic as it's been. 1976 *Birmingham Post* 16 Dec. 2/5 Mr. Richard plans a round of 'shuttle diplomacy' in Southern Africa seeking support for more direct British involvement in the decolonisation of Rhodesia. 1979 H. Kissinger *White House Years* p. xxi, The October 1973 Middle East war and the 'shuttle diplomacy' that followed. 1977 *Time* 13 June 80 Or consider Henry Kissinger. Understandably, Citizen K's style has changed perceptibly from that of the shuttle diplomat. 1892 *Q. Rev.* Oct. 486 The South-Eastern used, twenty years back, to run a 'shuttle' service every ten minutes between Charing Cross and Cannon St. 1933 *Times* 28 Feb. 9/4 Shuttle services from the outer districts connecting with the trunk and City routes can be substituted for through services from the suburbs to the City. 1944 A. Jacob *Traveller's War* xxviii. 419 It is the same kind of non-stop bombing shuttle service with which Conyngham, the A.O.C. Western Desert, achieved such great results in Africa. 1961 *Wall St. Jrnl.* 20 Mar. 2/3 Eastern Airlines said it wants to start a low-cost air 'shuttle' service between Boston, New York and Washington. 1966 'H. MacDiarmid' *Company I've Kept* viii. 189 About..the ..date of my birthday, Biggar Post Office had to run what was virtually a shuttle-service several times a day to deliver the masses of mail. 1969 *Guardian* 18 Jan. 1/4 We can expect to see a permanent Russian space station in orbit.., probably with a shuttle service of Russian scientists from earth. 1978 *Detroit Free Press* 5 Mar. D. 16/3 It is more a commercial than a resort hotel, but it has a pool and runs a daily shuttle service to nearby public beaches for guests. 1888 A. R. Diehl *Two Thousand Words* 190 *Shuttle-train*, one that takes short runs back and forth. 1923 *World Almanac* 503/2 A shuttle train runs between 50th Street and 59th Street on Sixth Avenue. 1942 *Sun* (Baltimore) 7 Mar. 20 When loss of tires has forced the automobile from use, shuttle trains, supplemented by busses, will be the most practical ..means for the transportation of workers in this area. 1974 *Encycl. Brit. Macropædia* XV. 494/2 In 1965 the first Freightliner container..shuttle trains began running on British Railways. 1893 G. Kapp *Dynamos, Alternators, & Transformers* ix. 209 The simplest example of an open-coil armature is the so-called shuttle-wound armature. 1902 *Encycl. Brit.* XXVII. 577/1 The second or drum method was used in the original 'shuttle-wound' armatures invented by Dr. Werner von Siemens in 1856.

shuttle, *v.* Restrict *Obs.* exc. *dial.* to senses 1 and 2 in Dict. and add: **1. b.** To transport in a vehicle or craft operating a shuttle service. Also *transf.*

1930 E. Ferber *Cimarron* xxi. 334 With his geological knowledge..and his familiarity with the region, he was shuttled back and forth from one end of the state to the other. 1945 *Times* 13 Sept. 5/7 There has been no difficulty about shuttling prisoners resident in the British and American zones. 1965 *Listener* 30 Sept. 482/2 So what happens to the old patient? Does he or she get shuttled

around to one hospital after another? 1971 *Nature* 27 Aug. 632/1 That malate may serve to 'shuttle' reducing equivalents from cytoplasm to mitochondria. 1975 *Daily Tel.* 1 May 1 Scores of transport aircraft shuttle Vietnamese evacuees from Guam, Wake Island, and the Philippines to America at the rate of up to 5,000 a day. 1977 *Offshore Engineer* May 49/2 Two 15,000t tankers shuttle the oil to Spanish refineries. 1978 H. Wouk *War & Remembrance* xiv. 148 Trains devoted to shuttling the Jews rolled eastward jam-packed and went back empty.

2. (Further examples.) Also, to travel in one direction using a shuttle service. Also *transf.*

1935 M. M. Atwater *Murder in Midsummer* i. 6 A few automobiles, like overgrown beetles, shuttled back and forth along the concrete highways. 1966 *Aviation Week & Space Technol.* 5 Dec. 95/1 Analyses could be made automatically or by astronauts shuttling from earth to the satellite laboratory, staying one month or more and then returning to earth. 1971 'A. Burgess' *MF* iv. 42 He was not to be seen: perhaps he had shuttled off to Boston or somewhere. 1973 *Internat. Relations Dict.* (U.S. Dept. State Library) 38/1 Henry Kissinger personally shuttled back and forth between Jerusalem and Cairo. 1975 *Sci. Amer.* Jan. 13/3, I moved 'temporarily' to the University of Liverpool in 1965 and have shuttled between the departments of genetics and zoology ever since. 1977 *Time* 15 Aug. 19/3 Although it was not on his original schedule, Vance decided to shuttle back to Amman, Damascus and Alexandria to convey Israeli views to Sadat and Assad. 1978 R. Ludlum *Holcroft Covenant* xiii. 157 France's domestic airline shuttled about the country with splendid irregularity. 1979 *Sci. Amer.* Jan. 29/1 (*caption*) The trains shuttle back and forth without being uncoupled, acting much like a conveyor belt.

shuttlecock, *v.* Add: **1.** (Further *fig.* example.)

1955 *Times* 10 Aug. 5/2 What Rostand lightly shuttlecocks to and fro across this barrier M. Anouilh might nowadays have said more sadly.

2. (Later example.)

1960 *Times* 23 Feb. 4/5 Miss Maxine Audley..played a part shuttle-cocking between lover, son, and husband.

shuttleless, *a.* Delete *rare* and add further examples.

1961 *Textile World* July 30 Shuttleless looms. 1967 *Economist* 29 Apr. 459/3 Popularly known as 'the green box', the BD 200, together with the shuttleless jet loom, could revolutionise the Czech industry. 1975 *Guardian* 27 Jan. 15/1 Most loom-builders were thinking in terms of shuttleless machines.

shuttling, *vbl. sb.* Add: **2.** Travelling to and fro.

1937 R. S. Morton *Woman Surgeon* i. 15 This periodical shuttling between Virginia and New Jersey entailed as much preparation and inconvenience as a modern voyage to the Antarctic.

shu·t-up, *a.* [Shut *ppl. a.* 2: cf. Shut *v.* 19.] **1.** That has been shut up: see Shut *ppl. a.* 2.

2. That can be shut up; foldable.

1799 Jane Austen *Let.* 8 Jan. (1952) 50 Martha kindly made room for me in her bed, which was the shut-up one in the new nursery. 1911 R. Nevill *Floreat Etona* x. 307 The furniture of the rooms... For the most part..consisted of a shut-up bed, a 'burry' (bureau) washstand, which also closed up, and sock cupboard.

shuv(v)er, var. *shover. **shva,** var. *sheva. **shvartze, shvartzer,** varr. *schvartze, schvartzer. **shwa,** var. *schwa.

shy, *sb.*^{2} Add: **2. b.** (Earlier example.)

1824 P. Egan *Boxiana* IV. 149 'I am sure you are too generous to let a brave man want; and I never knew an appeal made here in vain.' 'Well then, go it' echoed one of the East-enders; 'I like to have a shy for my money.' Half crowns, shillings and sixpences were instantly thrown upon the stage.

3. b. *coconut shy*: see *coco, cocoa 4 d.

shy, *a.* Add: **2. d.** As last element in Combs.: frightened (of), averse or reluctant (to).

1884, etc. [see Gun-shy *a.*]. 1928, etc. [see *work-shy* adj. s.v. *work *sb.* 34]. 1934, etc. [see *book-shy a.* s.v. *book *sb.* 18]. 1938 *Amer. Speech* XIII. 188/1 *Needle-shy*, a phobia..which manifests itself in a revulsion against using the hypodermic needle or seeing it used. 1972 *Sat. Rev.* (U.S.) 28 Oct. 33/3 The extent to which rather sophisticated people remain telephone-shy is remarkable. 1972 *Guardian* 24 Nov. 32/1 Traffic shy commuters.

3. (Later example without const.)

1940 *Ann. Reg. 1939* 362 So much money had already been lost..that investors were shy.

6. b. Also const. *on.*

1903 [see *long green* s.v. *long *a.*^{1} A. 18]. 1975 R. Stout *Family Affair* (1976) iv. 46, I merely thought *some* women were a little shy on brains, present company not excepted.

7. b. (Earlier example.)

1821 D. Haggart *Life* 39 Although I had not been idle during these three months, I found my blunt getting shy.

9. *shy-eyed, -footed* adjs.; *shy-making* ppl. adj.; *shy-brightly* adv.

1922 Joyce *Ulysses* 205 Eglintoneyes..looked up shybrightly. 1910 J. Masefield *Ballads & Poems* 80 The shy-eyed delicate deer. 1952 R. Finlayson *Schooner came to Atia* 118 Shy-eyed kiddies ran out to look. 1917 J. Masefield *Lollingdon Downs* 86 Shy-footed beauty dear. 1930 E. Waugh *Vile Bodies* ii. 19, I shall just ring up every Cabinet Minister and *all* the newspapers and give them all the most shy-making details. 1940 M. Allingham *Black Plumes* xv. 175 Great heroism, like great cowardice, is shy-making, and they were all.. embarrassed. 1974 *Listener* 21 Nov. 677/1 Dr Ray rightly quotes enough of their shy-making exchanges.

shy, *v.*^{2} **2. a.** (Earlier example.)

1793 W. B. Stevens *Jrnl.* 26 Mar. (1965) 1. 74 It was but the other day he thought that every man ought to *shy* Jack Dawson from their Houses and Lo now he is his *dear friend.*

Shylock. Add: (Further examples.) Also, a Jew, a pawnbroker; in *U.S.* (with small initial), an abusive term for a moneylender; = *loan-shark* s.v. *loan *sb.*^{1} 5. (These uses are considered offensive.)

1786 R. Cumberland *Observer* (ed. 2) III. No. 64. 30 Smoke the Jew!..Out with Shylock. 1898 W. J. Locke *Idols* vi. 81 He could raise the money, cry quits with the urbane and gentle-mannered Shylock. 1930 *Sat. Even. Post* 5 Apr. 48/1 There are also guys present who are called Shylocks, because they will lend you dough when you go broke at the table, on watches or rings..at very good interest. 1935 *Sun* (Baltimore) 28 Dec. 1/3 The jury held the backbone of the 'shylock' or usury racket had been broken in the city by Dewey. 1951 Turkus & Feder *Murder, Inc.* v. 121 'Sometimes it's as good as 3,000 per cent,' one of the shylocks..explained. 1959 I. & P. Opie *Lore & Lang. Schoolch.* xvi. 346 Today, children colloquially refer to a Jew as a Yid, Shylock, or Hooknose. 1972 J. Mills *Report to Commissioner* 121 The Panthers are worse than the shylocks... They keep bleeding you. 1976 *Sunday Times* (Lagos) 1 Aug. 21/4 They are expected to alleviate the suffering of the common man who for long has been a victim of Shylock landlords. 1978 S. Brill *Teamsters* iv. 150 A member who couldn't meet his shylock payments often found that the union people he 'elected' became enforcers against him.

Hence **Shy·lock** *v. trans.*, to force (a person) to repay a debt, esp. at an exorbitant rate of interest; **shy·locker** *U.S.*, one who charges an exorbitant rate of interest; **shy·locking** *vbl. sb.* (*U.S.*)

1930 H. G. Wells *Autocracy of Mr. Parham* iv. v. 308 They bullied and quarrelled when we were only too ready for acquiescent action. They Shylocked Europe. 1933 A. G. Macdonell *England, their England* xiii. 242 One or two of them have had a bit of hard luck lately, and one can't Shylock the poor devils. 1951 Turkus & Feder *Murder, Inc.* v. 120 Loan-sharking. 'Shylocking,' it is called. 1961 *Brooklyn Law Rev.* XXVIII. 41 Such activities are the 'shakedown racket', 'shylocking' (where interest of 20% per week is charged..) and labor extortion. 1973 *Listener* 27 Jan. 878/3 Today's Dillinger.. controls vice, dope and shylocking (usury) throughout his State. 1973 J. Gores *Final Notice* xii. 73 He's an enforcer for a shylocker..with lots of sweet loans to push the vigorish up.

shypoo (ʃəi·pū, ʃəipū·). *Austral. slang.* Also **shipoo.** [Origin unknown.] Inferior liquor; a public house that sells this. Also *attrib.*

1901 *Bulletin Reciter* 30 We drank the shypoo deeply, till the lateness of the night. 1908 E. G. Murphy *Jarrahland Jingles* 108 The swell exclusive club, Have swept the shypoo shanty from its lair amid the scrub. 1917 [see *Aussie *sb.* and *a.* 1]. 1934 *Bulletin* (Sydney) 22 Aug. 20/1 That bubbling drink and spuming We used to call shypoo? 1936 H. Drake-Brockman *Sheba Lane* 237 How about managing that shipoo for me? 1962 T. Ronan *Deep of Sky* 218 A hostelry...restricted to the sale of beer and wine. Locally this was known as the 'Shypoo Shop'. I'm not sure of the derivation of 'Shypoo'. I think it is bastard Chinese for soft drink. To the sturdy second wave of pioneers of West Kimberley, beer and wine were soft drinks.

shyster. Add: **1.** For *U.S. slang* read 'orig. and chiefly *U.S. slang*' and add earlier U.S. and later non-U.S. examples.

1844 G. Wilkes *Mysteries of Tombs* 44/1 He is consulted by the magistrates on all important points of law, and the inferior shysters look upon him with a reverence approaching veneration. 1849 G. G. Foster *New York in Slices* 20 He must..wait next day for the visits of the 'shyster' lawyers—a set of turkey-buzzards whose touch is pollution and whose breath is pestilence. 1943 M. H. Harris *Vegetative Eye* 15 Not to Memory, with its shyster lackey, Association. 1952 *Manch. Guardian Weekly* 19 June 3 They call Taft's 'shyster methods' so necessary. 1961 *Listener* 14 Dec. 1046/1 A solicitor's chief clerk who persuades his shyster employer to leave the country to avoid embezzlement charges. 1981 J. Wainwright *All on Summer's Day* 31 The shyster lawyers..swear blind the client's been manhandled while in police custody.

2. *Austral.* Alteration of Shicer.

1938 X. Herbert *Capricornia* (1939) xxi. 306 You lousy sweatin' old shyster you. 1941 Baker *Dict. Austral. Slang* 66 *Shyster*, a worthless mine.

siafu (sī,ā·fu). [Swahili.] = *safari ant* s.v. *safari 3 d. Also *attrib.*

[1920 G. D. H. Carpenter *Naturalist on Lake Victoria* xii. 276 The ant that most obtrusively calls for notice is the well known *Dorylus* or 'Safari ant',..in Kiswahili called 'Siafu'.] 1959 E. Huxley *Flame Trees of Thika* xv. 154 The rain brought out the *siafu*. Those fearful black rivers of implacable insects poured between their low mud banks over the garden. 1976 [see *safari ant* s.v. *safari 3]. 1977 Reader & Croze *Pyramids of Life* 212/1 The most effective and thorough carnivore hunter in the forest is a colony of siafu (*Dorylus*), the so-called army ant.

sial (səi·al). *Geol.* Also **Sial.** [a. G. *sial*, f. *si-licium* Silicon + *al-uminium* Aluminium.

The name was altered from *sal* (see *sal^4 2) by G. Pfeffer in order to avoid confusion with other meanings of that word (see A. Wegener *Die Entstehung der Kontinente und Ozeane* (1920) iii. 22 and quot. 1924 below).]

The discontinuous upper layer of the earth's crust represented by the continental masses, which are composed of relatively light rocks rich in silica and alumina and may be regarded as floating on a lower crustal layer of sima; the material of which these masses are composed.

1922 [see *sal^4 2]. 1923 *Rep. Brit. Assoc. Adv. Sci. 1922* 364 Sial masses drift through the Sima like icebergs through the sea. 1924 J. G. A. Skerl tr. *Wegener's Orig. Continents & Oceans* iii. 36 Following a short communication from Pfeffer I should like to write 'Sial' instead of 'Sal' in order that there may be no confusion with the Latin word for Salt. 1927 Peake & Fleure *Apes & Men* ii. 18 Various crises have led to the splitting of the Sial, and..its fragments have wandered off in many directions. 1950 P. H. Kuenen *Marine Geol.* ii. 118 The isostatic equilibrium of the crust requires a considerable thickness of granitic sial in the continents and a rock of greater density in the oceanic sections. 1954 [see *sal^4 2]. 1978 D. Bridgwater et al. in D. H. Tarling *Evol. Earth's Crust* ii. 63 In the type of Archaean craton displayed in southern Africa break up and partial destruction of earlier sial probably took place in a different tectonic regime.

Hence **sia·lic** *a.*^{1}, of or pertaining to the continental crust or the material of which it is made.

1924 J. G. A. Skerl tr. *Wegener's Orig. Continents & Oceans* iv. 59 Molten sialic masses (granite) from the under side of the South American block.. have emerged on its posterior edge. 1944 A. Holmes *Princ. Physical Geol.* ii. 14 There should be bulges..where gravity is relatively low; that is to say, wherever the outer part of the crust is composed of light sialic rocks. Such places are the continents. 1955 *Sci. Amer.* Sept. 62/1 Because granite is the chief sialic rock and basalt the chief simatic one, the continents are most commonly described as granitic and the ocean basins as basaltic. 1970 *Ibid.* Feb. 32/3 Stresses in the earth's crust may crack the sialic layer, producing faults and fissures that can be as much as 20 meters wide. 1975 *Nature* 3 Apr. 397/2 Localised partial melting of ancient, sialic crust may..occur in the zone of heating above large bodies of basic igneous magma introduced into high crustal levels.

sialadenitis: see s.v. *sialo-.

sialectasis (səi,ăle·ktăsis). *Path.* [f. *sialo- + Gr. ἔκτασις dilatation.] Dilatation of the ducts of the salivary glands, usu. the parotids.

1940 *Brit. Jrnl. Surg.* XXVII. 713 The term 'sialectasis' which has been used to describe the condition is not strictly correct etymologically, translating as 'a stretching out or dilatation of the saliva'. 1953 *Brit. Med. Jrnl.* 19 Dec. 1359/2 The underlying pathology in..sialectasis, sialangiectasis, or chronic recurrent parotitis, is obscure, but the essential feature is obstruction of the parotid ducts. 1974 J. D. Maynard in R. M. Kirk et al. *Surgery* ix. 201 The parotid architecture may be normal or reveal the curious appearance reminiscent of radiographs of bronchiectasis, called sialectasis. 1977 [see *sicca*].

sialic (səi·ălik), *a.*^{2} *Biochem.* [f. Gr. σίαλ-ον saliva + -ic.] *sialic acid*: any acyl or related derivative of a neuraminic acid.

Orig. applied *spec.* to a substance of this type (then of unknown structure) isolated from the salivary glands of cattle.

1952 G. Blix et al. in *Acta Chem. Scand.* VI. 359 Besides this substance, for which we propose the provisional name sialic acid, the mucin contains small amounts of carbohydrate of the dihexose-hexosamine type. 1957 —— in *Nature* 25 May 1088/2 Sialic acid is suggested as group name for the acylated neuraminic acids. 1967 *New Scientist* 16 Feb. 414/1 Twenty years elapsed before the structure of this complex sugar was determined, and by then it was clear that it was only one of a family of sugars called sialic acids, all of which had an unusually long backbone of nine carbon atoms. 1978 *Nature* 16 Feb. 674/2 Sialic acids..of the red blood cell (RBC) membrane are considered to play an important part in the physiology of the RBC.

sialidase (səi,æ·lidē^{1}z). *Biochem.* [f. prec. + -id^4 + *-ase.] = *neuraminidase.

1956 Heimer & Meyer in *Proc. Nat. Acad. Sci.* XLII. 731 The bond split must be O-glycosidic rather than N-glycosidic... The name 'sialidase' is proposed for the enzyme responsible for this action. 1963 Barka & Anderson *Histochemistry* iii. 94 Sialidase, an enzyme that selectively removes sialic acid..from mucopolysaccharides, eliminates the metachromatic staining and Alcian blue affinity of some acid mucopolysaccharides. 1978 A. White et al. *Princ. Biochem.* (ed. 6) xxix. 916 It remains unclear how sialic acid in mammals or galactose in birds and reptiles is removed in vivo from plasma glycoproteins, although sialidase and β-galactosidases are present in many tissues.

siallite (səi·ăləit). *Geol.* [ad. G. *siallit* (H. Harrassowitz 1926, in *Fortschr. Geol. Palæont.* IV. 263), f. as *sial: see -ite^1.] Weathered rock that is largely composed of hydrous aluminium silicates and is highly leached of alkalis. So **sialli·tic** *a.*

1933 H. Greene tr. *Vageler's Introd. Tropical Soils* v. 127 Where the amount of water is small and the period

of action is short, as in the case of arid and semi-arid districts, siallitic and allitic bodies are formed in small amount. *Ibid.* 140 The soil becomes podsolised. The long-held opinion that this produces not merely siallites in general but a quite special kaolin and kaolinite can no longer be maintained. 1965 B. T. Bunting *Geogr. of Soil* xiv. 168 Other forms of brown earth are formed on siallitic parent materials which may overlie chalk.

sialo- (səi,ălo), before a vowel also **sial-**, comb. form of Gr. σίαλον saliva, used as a formative element in *Med.*, as **si:al(o)adeni·tis** [Gr. ἀδήν gland], inflammation of a salivary gland; **si·alolith** [-lith], a calculus in a salivary gland; hence **si:alolithi·asis,** the presence of such a calculus; **sialorrhœa** (-rī·ă) [*-rrhœa, -rrhea], excessive flow of saliva.

1859 Mayne *Expos. Lex.* (1860) 1159/2 Sialadenitis. 1925 Marshall & Piney *Textbk. Surg. Path.* vi. 123 Chronic sialo-adenitis may affect either the parotid or submaxillary and may be associated with calculi. 1977 *Arch. Virol.* LIV. 352 Sialoadenitis produced by PRCV.. was morphologically compatible with SDAV-induced lesions. [1855 Dunglison *Dict. Med. Sci.* (ed. 12) 787/2 *Sialolithi*, calculi, salivary.] 1862 Mayne *Med. Vocab.* (ed. 2) 373/2 *Sialolithus*, a salival calculus; a sialolith. 1910 Adami & Nicholls *Princ. Path.* II. xvii. 396 Concretions composed of phosphate or carbonate of lime are not uncommonly found within the [salivary] duct (sialoliths), sometimes causing or associated with cystic dilatation of the ducts and acini. 1973 *Brit. Dental Jrnl.* CXXXV. 292/2 The sialolith was club-shaped, yellow in colour and measured 14 mm in length. 1859 Mayne *Expos. Lex.* (1860) 1160/1 Sialolithiasis. 1916 L. F. Barker *Monographic Med.* III. viii. 264 Stone is sometimes palpable in a salivary duct (sialolithiasis). 1973 *Brit. Dental Jrnl.* CXXXV. 291/1 Sialolithiasis, the formation of a calculus in the duct or gland substance of a major or minor salivary gland, occurs most commonly in the middle-aged adult. 1846 *Lond. Med. Gaz.* XXXVII. 379/1 In the idiopathic sialorrhœa..the flow takes place as well in the night as in the day time. 1888 *Buck's Handbk. Med. Sci.* VI. 251/1 The term sialorrhea, while indicating an abnormal flow, does not necessarily imply abnormal secretion. 1933 *Ann. Rep. London County Council* IV. iii. 135 At the same time the face may be distorted into a spasmodic grimace, accompanied by flushing, lachrymation and sialorrhœa. 1978 *Jrnl. R. Soc. Med.* LXXI. 346 Abnormal conditions, such as excessive diarrhoea, sialorrhoea or sweating.

sialoglycoprotein (səi:ălogləikoprōᵘ·tīn). *Biochem.* [f. *sial(ic *a.*^{2} + -o + *glycoprotein* (s.v. *glyco-).] Any glycoprotein in which sialic acid residues form a major constituent of the side chains.

1963 A. Gottschalk in Florkin & Stotz *Comprehensive Biochem.* VIII. i. 29 All sialo-glycoproteins have some features in common. Invariably their sialic acid residues occupy a terminal position and are linked through a neuraminidase susceptible α-ketosidic linkage to another sugar residue. 1965 *Biochimica & Biophysica Acta* CI. 166 A homogeneous 7·1 S sialoglycoprotein..composed of 5·9% neutral carbohydrates, 4·8% aminosugars and 4·4% sialic acid. 1976 *Nature* 20 May 236/2 Liposomes bearing sialoglycoproteins can participate in agglutination reactions mediated by appropriate lectins.

sialography (səi,ălǫ·grăfi). *Med.* [f. *sialo- + -graphy.] Radiography of the ducts of a salivary gland after they have been injected with a radio-opaque fluid.

1931 *Brit. Jrnl. Surg.* XIX. 142 Sialography will demonstrate abnormalities, dilatations, and obstructions of the larger and smaller parotid ducts. 1960 *Radiology* LXXIV. 138/1 Secretory sialography has been a useful aid in determining whether tumors in the submaxillary and parotid regions are within or adjacent to the major salivary glands. 1975 *Ibid.* CXVII. 220/1 Sialography, the demonstration of ducts and parenchyma of the major salivary glands by injection of contrast material in the gland's main secretory duct, has been practiced since its first description by Arcelin in 1913. 1977 [see *sicca*].

Hence **si·alogram,** an X-ray photograph made by this technique; **sialogra·phic** *a.*

1931 *Brit. Jrnl. Surg.* XIX. 144 The sialogram shows the main duct of the gland, the larger branches, and the ductules, all well filled with lipiodol. 1938 *Surg., Gynecol. & Obstetr.* LXVII. 777 A sialographic study of 23 normal parotid glands and 76 cases of neoplastic disease of the parotid. 1974 J. D. Maynard in R. M. Kirk et al. *Surgery* ix. 201 If symptoms are sufficiently worrying to the patient and a sialogram reveals gross dilatation of the main duct, ablative..surgery is indicated. 1976 *Radiology* CXXI. 747/1 A..curved side-hole cannula for use in sialographic procedures.

sialomucin (səi,ălomiū·sin). *Biochem.* [f. *sial(ic *a.*^{2} + -o + Mucin in Dict. and Suppl.] Any mucin containing sialic acid residues in the molecule.

1958 L. Odin in Wolstenholme & O'Connor *Ciba Found. Symp. Chem. & Biol. Mucopolysaccharides* 235 The second main type of epithelial glycoprotein contains sialic acid as its characteristic carbohydrate constituent. .. This type of glycoprotein will for convenience be called sialomucin. 1976 *Path. Ann.* XI. 172 A recent study of the mucin..suggests that it is a sialomucin with minor sulphomucin and neutral mucin components.

sialon (səi·ălǫn). *Chem.* [f. the chemical symbols for silicon (*Si*), aluminium (*Al*), oxygen (*O*), and nitrogen (*N*).] Any of a large class of refractory materials which have crystal structures similar to those of silica and the silicates but contain aluminium and nitrogen in the polymeric framework in addition to silicon and oxygen.

1973 K. H. Jack in *Trans. & Jrnl. Brit. Ceramic Soc.* LXXII. 380/1 The structural unit of β-silicon nitride is SiN_4; of α it is on average $SiN_{3.5}O_{0.5}$; of oxynitride it is SiN_3O; and of the new 'sialons' it is $(Si, Al)(O, N)_4$. 1975 P. Popper *Special Ceramics* VI. p. ix, Several papers deal with materials in the Si-Al-O-N or Mg-Si-Al-O-N system. Colloquially, workers in this field have become accustomed to refer to them as 'sialons'. 1981 *Economist* 11 Apr. 98/3 The new materials mentioned at last week's conference included..very strong ceramics called sialons.

Siamese, *a.* and *sb.* Add: **A.** *adj.* **1. b.** *Siamese cat*, a lightly built shorthaired cat belonging to a breed originally found in Thailand (formerly Siam), distinguished by buff-coloured fur with points of brown, blue, or other colours, and a narrow head with large ears and slanting blue eyes; so *Siamese kitten.*

1871 *Illustr. London News* 22 July 63/2 It [sc. the variety class at the Crystal Palace cat show] contained.. a singular Siamese cat, coloured precisely like a black-faced pug-dog. 1889 H. Weir *Our Cats* 73 Siamese Cat. Among the beautiful varieties of the domestic cat brought into notice by cat shows, none deserve more attention than 'The Royal Cat of Siam'. 1942 Siamese kitten [see *pet-shop* s.v. *pet *sb.*^{1} 3 d]. 1958 *Listener* 18 Sept. 410/1 There are many legends and stories about the origin of Siamese cats but few facts. 1972 Ing & Pond *Champion Cats of World* 109 The Siamese Cat Club was founded in 1901.

c. *Siamese fighter, fighting fish*, a brightly coloured, often red, tropical fish, *Betta splendens*, native to Malaysia and Thailand, and distinguished by enlarged fins and tail.

1929 W. T. Innes *Goldfish Varieties* (ed. 11) 256 (*caption*) *Betta splendens*, Cambodia variety, or Veiltail Siamese Albino Fighting Fish. 1968 R. Clapperton *No News on Monday* i. 9 Fifty bucks' worth of aquarium: two purple Siamese fighters hovering motionless, separated by glass. 1971 *Ceylon Observer* (Mag. ed.) 19 Sept. 2/6 (Advt.), Goldfish..Siamese Fighters. 1977 D. J. Coffey *Encycl. Aquarium Fish* 67/1 Siamese Fighting-fish are aggressive.

2. *Siamese twins*: For '1814' read '1811' and add: Also *gen.*, any pair of twins physically united by their tissues; *sing.* one of a pair of such twins. Also *attrib.* (in sing.) and *fig.* (Further examples.)

1835 Dickens in *Evening Chron.* 18 June 4 They were three long graces in drapery, with..another..the three fates with another sister—the Siamese twins multiplied by two. 1859 Geo. Eliot *Let.* 27 Feb. (1954) III. 27 People who have been inseparable and found *all* their happiness in each other for five years are in a sort of Siamese-twin condition that other people are not likely to regard with tolerance or even with belief. 1879 *Mind* IV. 331 Should the empiricists succeed in their attempt to resolve such Siamese-twin elements into habitual juxtapositions. 1883 E. W. Hamilton *Diary* 18 Dec. (1972) II. 526 Chamberlain also spoke briefly. He passed a high eulogy on Dilke—they are the Siamese twins of politicians. 1922 Joyce *Ulysses* 404 Heated argument..regarding the juridical and theological dilemma in the event of one Siamese twin predeceasing the other. 1926 J. S. Huxley *Ess. Pop. Sci.* 235 Partial constriction [of a newt embryo] produces partial doubling or 'Siamese twins'. 1937 H. H. Newman et al. *Twins* xiii. 355 Conjoined twins (Siamese twins) show marked differences in height, weight, features, and intelligence. 1957 Mankowicz & Haggar *Encycl. Eng. Pott. & Porc.* 202/1 The Kentish Siamese twins, Eliza and Mary Chalkhurst (who died in 1734 at the age of 34) were apparently made in redware. 1957 E. H. Shepard *Drawn from Memory* ix. 172 She was a queer experiment in ship design, a sort of Siamese twin of a ship; two complete hulls joined together in the middle. She had two sets of engines and two funnels. 1961 R. B. Long *Sentence & its Parts* iv. 106 Here *on Thursdays* is followed by two complete nucleuses, tied together in Siamese-twin fashion by joint possession of the introductory adjunct. 1965 E. Gowers *Fowler's Mod. Eng. Usage* (ed. 2) 554/1 *Siamese twins.* This seems a suitable term for the many words which, linked in pairs by *and* or *or*, are used to convey a single meaning. *Ibid.* 554/2 Whenever a Siamese twin suggests itself to a writer, he should be on his guard; it may be just the phrase he wants, but it is more likely to be one of those clichés that are always lying in wait to fill a vacuum in the brain. 1970 G. Greer *Female Eunuch* 245 The bitter animosity..of divorce is unknown where individuals have not become Siamese twins. 1972 Miller & Keane *Encycl. & Dict. Med. & Nursing* 876/1 New techniques in surgery..are making it possible to separate some Siamese twins whose physical links are highly complex. 1981 *Birds* Spring 63/1 My neighbour found Siamese twin starlings caught on some old wire... They had two legs each, but only three feet. The inner leg of each bird was joined together with one foot between them. 1981 *N.Y. Times Mag.* 19 July 6/4 The Kennedy prose style was a product of him and his Siamese twin, Theodore Sorensen, writing freely with his free arm.

b. Also, pertaining to or characteristic of Siamese twins.

1851 H. Melville *Moby Dick* II. xxx. 206 So, then, an elongated Siamese ligature [sc. monkey-rope] united us. 1955 E. Bowen *World of Love* v. 87 In step, in Siamese closeness, they paced towards it. 1969 N. J. Berrill *Person in Womb* xiii. 158 Siamese conditions meant death of mother and offspring during the agony of delivery.

c. Also *Siamese connection.*

1914 J. Kenlon *Fires & Firefighters* xxii. 322 A length of three-inch hose is attached to the pipe and strapped to the ladder with a siamese connection on the ground.

B. *sb.* **2.** (Earlier example.)

1759 *Universal Hist., Mod.* VII. 238 The Siamese resembles the Chinese in several respects: it consists mostly in monosyllables, and has neither declensions nor conjugations.

3. A Siamese cat.

1893 J. Jennings *Domestic or Fancy Cats* ii. 17 A pure-bred Siamese is a valuable cat. 1939 T. S. Eliot *Old Possum's Practical Cats* 15 It was a Siamese had mauled his missing ear. 1950 W. de la Mare *Inward Companion* 70 That crafty cat, a buff-black Siamese, Sniffing through wild wood. 1973 'E. McBain' *Let's hear it for Deaf Man* iii. 37 Janik himself resembled a cross-eyed Siamese, blue eyes magnified behind bifocals..a tuft of black hair behind each ear.

4. A Siamese coupling or connection.

1914 J. Kenlon *Fires & Firefighting* xxii. 322 Run in two lines, connect to the siamese, raise the bed ladder to the desired position and the stream is controlled from the street by guys. 1969 *Publ. Amer. Dial. Soc.* lii. 55 *Siamese*,..a connector joint used for reducing two lines into one line. 'A Siamese will allow more pressure than a single line.'

Siamesed, *ppl. a.* (Later examples.)

1914 J. Kenlon *Fires & Firefighting* 377 For siamesed lines, an allowance was made for the loss in the siamese connection and for 20 feet of 3½-inch lead hose. 1942 Potts & Harriss *Fire Pumps & Hydraulics* viii. 59 Where series pumping has to be resorted to it is desirable to employ 'Siamesed' lines, i.e. to duplicate the deliveries from the pump or pumps in the series up to the final pump. 1970 B. Knox *Children of Mist* i. 20 The Jaguar ..had twin carburetters plus a siamesed exhaust.

sib, *a.* and *sb.*^{2} Add: **1. a.** (Earlier and later examples in *spec.* sense.)

1875 R. L. Wallace *Canary Bk.* iii. 51 He said, very gravely, 'Weel, sib bred is sib bred, an I thocht that anybody kenned what that was.' However, I found that the meaning of the words was consanguinity. 1902 Bateson & Saunders *Rep. Evolution Comm. R. Soc.* I. 4 The possibility..should not be forgotten that the prepotency of the sib-bred hens may have been an original character of their particular strain. 1961 R. B. Bennett *Budgerigars, Canaries & Foreign Finches* xxi. 155 *Sib-bred*, the young from related birds bred in and in for many years. In-bred.

3. a. *spec.* in *Anthrop.* A group of people of either sex that are recognized by an individual as his kindred, and who freq. form an exogamic unit. (Further examples.)

1919 *Amer. Anthropologist* XXI. 28 The sib, like the family, is a kinship group... On the one hand, it excludes one half of the blood-kindred—the father's side of the family in matronymic, the mother's side in patronymic societies. On the other hand, it admits on equal terms all kindred of the favored side regardless of degree and even individuals considered blood-relatives merely through legal fiction. 1950 A. R. Radcliffe-Brown in Radcliffe-Brown & Forde *African Systems of Kinship & Marriage* 15 The arrangement of kin by degrees of nearness or distance was based on sib-ship... A man's sib were all his cognates within a certain degree. *Ibid.*, It is evident that no two persons can have the same sib, though for two unmarried full brothers, *A* and *B*, every person who was sib to *A* was sib to *B*, and *A* and *B* were sib-kinsmen of one another. A person cannot be said to 'belong' to a sib or be a member of a sib in the sense in which he can be said to belong to a lineage or a clan or a village community. 1967 R. Fox *Kinship & Marriage* vi. 167 The kindred..was known as the sib... Amongst the Teutonic peoples, the sib was the exogamic unit. 1968 G. D. Mitchell *Dict. Sociol.* 160 The British usage confines *sib* to an ego-centred group of cognates within a certain degree; it is thus synonymous with some meanings of kindred.

c. *Anthrop.* A kinship sub-group whose members claim unilineal descent from a single ancestor of either sex; a clan. Cf. sense 3 a in Dict. and Suppl. orig. and chiefly *U.S.*

1890 F. B. Jevons tr. *Schrader's Prehist. Antiq. of Aryan Peoples* iv. xii. 398 The Teutonic sib.., as long as it was an agrarian and military unit, is to be conceived as having been purely agnatic. 1901 *Jrnl. R. Anthrop. Inst.* IV. 68 Two brothers, and still more, a father and son, cannot fall into two different sibs. 1918 F. S. Philbrick tr. *Huebner's Hist. Gmc. Private Law* i. iii. 114 The primitive Germans lived in..sibs that based their kinship solely upon descent from a common tribal male ancestor. 1949 G. P. Murdock *Social Structure* iii. 47 If all persons born with the name Smith in our society regarded themselves as related, they would constitute a patri-sib. Some unilinear societies lack true sibs, possessing only lineages. The great majority, however, possess sibs... A sib normally includes several lineages. 1967 R. Fox *Kinship & Marriage* i. 50 There are other usages [for *clan* and *lineage*] which cause considerable confusion. .. American writers often use *sib* as the generic term with *patri-sib* and *matri-sib* as the sub varieties. This is quite wrong... The Anglo-Saxon *sib*..was not a descent group at all.

d. Chiefly *Genetics.* A brother or sister, another individual of the same parentage (see also quot. 1933).

1919 *Genetics* IV. 496 The observations..show themselves to be a strictly homogeneous population, with correlation much larger than that between sibs. 1931 E. & C. Paul tr. E. Baur et al. *Human Heredity* xi. 508 The methods..known as the brother-and-sister method or as the sib method. [*Note*] The word 'sib' or 'sibling' is coming into use in genetics in the English-speaking

world, as an equivalent of the convenient German term 'Geschwister'. 1933 *Proc. R. Soc. Edin.* LIII. 106 A comparison of the resemblance between ordinary sibs and fraternal twins may be used. [Note] Here and throughout sib is used to denote brothers and sisters of different birth-rank. 1937 *Nature* 2 Oct. 573/1 Genetical research shows that height has little value as an indicator of relationships and specific distinctness, since a plant two inches high may be a sib to a plant twenty inches high. 1958 *Times* 18 Jan. 4/3 Among other problems possibly was the danger of sibs—offspring of the same parent—marrying. 1974 *Nature* 12 Apr. 594/2 The foundation stock of the highly resistant flock had parents and sibs that did not develop scrapie.

e. *attrib.*, as *sib bond, group, -mate, -mating, -pair, selection, system.*

1938 *Jrnl. R. Anthrop. Inst.* LXVIII. 301 With Trobriands and Hopi the successive sib bond exists through females. 1901 *Ibid.* XXXI. 68 The size of the sib group has always been determined by economic facts. 1920 R. H. Lowie *Primitive Society* vi. 107 Sib-mates of the same generation usually call one another siblings, and from this..it is but a step to feeling that marriage between sib-mates would be incestuous. Hence we find as one of the most common traits of the sib the law of exogamy. 1949 G. P. Murdock *Social Structure* iv. 73 They are wives and clansmen of one group of disputant men, sisters and sibmates of the other. 1949 R. A. Fisher *Inbreeding* iii. 47 A mating of parent and offspring interpolated in a series of sib-matings does not advance the inbreeding process so much as a sib-mating would have done. 1971 *Brit. Med. Bull.* XXVII. 45/2 Sib-pair comparisons have also been used to separate the effects of family size..upon intelligence. 1956 *Genetics* XLI. 367 (*heading*) Isolation of preadaptive mutants in bacteria by sib selection. 1934 *Nature* 19 May 743/1 The sib system, and its attendant naming habits, is the most flourishing part of the old thought system.

sib (sib), *sb.*^{3} Also **Sib.** Colloq. abbrev. of *sibilant *sb.* 2.

1957 D. E. Walker *Lunch with Stranger* 142 British memoirs published since the war have confirmed both the speed with which the Sibs reached the German High Command and the disruptive influence they had there. 1958 *Punch* 8 Jan. 96/1 A Sib, short for sibilant, is a story or rumour concocted to promote a set purpose, generally the sales of a particular product. 1965 B. Sweet-Escott *Baker Street Irregular* iii. 98 The rumours or 'sibs', as they were called, were devised by a high-powered committee in London... Success or failure was judged by the degree to which the sibs were repeated by enemy or neutral newspapers and broadcasting transmissions. 1975 P. Fussell *Gt. War & Mod. Memory* ix. 328 One department [of the Special Operations Executive] did nothing but contrive 'sibs'—bizarre and hair-raising rumours to be spread over the Continent.

Sibbald (si·băld). The name of Sir Robert Sibbald (1641–1722), Scottish naturalist, used in the possessive to designate the blue whale, *Balænoptera musculus.*

1897 R. Lydekker *Conc. Knowledge Nat. Hist.* 173 The largest..of all whales, is the blue, or Sibbald's rorqual.., commonly known to the American whalers by the name of 'sulphur-bottom'. 1937 [see *blue whale* s.v. *blue *a.* 12 a]. 1972 *Oxf. Bk. Vertebrates* 190/1 Sibbald's Rorqual..has a world-wide distribution and migrates between high and low latitudes.

Sibelian (sibē^1·liăn), *a.* and *sb.* [f. *Sibelius* (see below) + -an.] **A.** *adj.* Of, pertaining to, or characteristic of the Finnish composer Jean Sibelius (1865–1957) or his music. **B.** *sb.* An admirer or adherent of Sibelius; an interpreter of his works.

1935 C. Gray *Sibelius* 64 The instrumental roles are reversed, with the strings constituting a background of a typically Sibelian kind. 1937 *Sunday Times* 29 Aug. 5/2 The No. 4..is of exceptional interest to the Sibelian student now in the light thrown on it by the No. 7 and 'Tapiola'. 1943 *Scrutiny* XI. 171 Some passages of characteristically Sibelian cross-rhythmed moto perpetuo. 1966 *Punch* 23 Feb. 291/2 The Sibelian atmosphere of magic, melancholy and old solitude. 1975 *Gramophone* Oct. 598/2 Shostakovich's view of the symphony is Mahlerian rather than Sibelian. 1976 *Ibid.* June 52/2 She has put all Sibelians in her debt by including the inspired Serenades on her record. *Ibid.* Oct. 608/2 Last month Decca released a version of the *Four Legends* conducted by no less a Sibelian than Jussi Jalas, the composer's son-in-law.

Siberia (səibīᵊ·riă). The name of a region of the U.S.S.R. in Asia used as a type of a cold, inhospitable place, or a place of exile, banishment, or imprisonment. Also *fig.*

1841 Geo. Eliot *Let.* 17 Feb. (1954) I. 81 Probably this projected transportation may be to a Cape of Good Hope instead of a Siberia. 1876 C. M. Yonge *Three Brides* I. x. 159, I used to be Camilla to all the neighbourhood, and here I find myself..banished to Siberia. 1926 C. Plumb in *Oxf. Poetry 1925* 40 The seas shall not seem vast Siberias of Time. 1952 R. Campbell tr. *Baudelaire's Poems* 80 In my Siberia a bright explosion as of tropic heat. 1972 *Listener* 21 Dec. 857/3 In 1830 Andrew Jackson..[ordered] that all the Indian tribes..be removed to the west of the Mississippi..to the 'Siberia' of the Far West. 1974 'J. le Carré' *Tinker, Tailor, Soldier, Spy* xi. 82 Guillam departed for the siberias of Brixton.

Siberian, *a.* and *sb.* Add: **A.** *adj.* **2. a.** *Siberian husky, thrush, tiger.*

1930 *Amer. Kennel Gaz.* Jan. 26/1 The Siberian huskies are recognized as the ideal dogs for driving. *Ibid.* Nov. 73/1 The breed of dog known as the Siberian Husky has been recognized by the Stud Book Committee. 1950 J. Hambleton *Abitibi Adventure* 74 In their equipment was included a two-dog team, made up of Siberian huskies, blue-eyed beauties trained for generations to pull their hearts out and to survive the utmost rigors of the north. 1972 *Even. Telegram* (St. John's, Newfoundland) 24 June 14/1 Various types of Northern dogs were used including Siberian Huskies. 1901 *Ibis* 417 Siberian Thrush... The nest was found on the Yenesei River by Mr. Popham in 1895. 1954 D. A. Bannerman *Birds Brit. Isles* III. 166 There is some reason to suppose that the Siberian thrush..has occurred in Britain. 1895 R. Lydekker *Hand-bk. Carnivora* I. 150 A specimen of the Siberian Tiger, apparently the first brought alive to Europe, was exhibited recently in Hagenbeck's menagerie in Amsterdam. 1956 M. L. Taylor *Tiger's Claw* vii. 59 'Can you read the Chinese ideogram on his head?' he inquired of me. 'It is *wang* meaning king. In the north it is believed that only Siberian tigers carry this mark.' 1978 *Times* 27 Oct. 32/8 (Advt.), Large Siberian tiger skin mounted on red satin, £600.

b. *Siberian cedar, elm, larch, pine* (later example), (*stone*) *pine* (earlier and later examples), *wallflower.*

1763 J. Bell *Trav. St. Petersburg* I. 250 Towards the Baykall lake, are high hills..covered with tall trees; among which are many..Siberian cedars. 1838 J. C. Loudon *Arboretum & Fruticetum Britannicum* IV. 2275 The Siberian Stone Pine, or Siberian Cedar..—The cones are said to be longer, and the scales larger, than in the Swiss variety. 1967 M. T. Mirov *Genus Pinus* iii. 233 Russians call *P. sibirica* 'Siberian cedar', which causes a great deal of confusion. 1974 *Nomencl. Commerc. Timbers (B.S.I.)* 69 'Siberian cedar' (UK)... This name is liable to be confusing and its use should be discontinued. 1904 *Outing* Oct. 84/2 The English elm and the cork-bark and Siberian elms are also desirable. 1981 *Sci. Amer.* Aug. 40/3 Many Asian species, such as the Chinese elm..and the Siberian elm.., are comparatively resistant to infection. 1838 J. C. Loudon *Arboretum & Fruticetum Britannicum* IV. 2352 The Siberian larch was introduced into England by Messrs. Loddiges..about the end of the last century. 1969 T. H. Everett *Living Trees of World* iv. 41/1 The Siberian larch.., a close relative of the European larch, grows to 120 feet tall. 1958 *N.Z. Timber Jrnl.* Sept. 87/1 Siberian pine: *Pinus cembra* var. *sibirica* ..and *P. koraiensis*... Shipped from Vladivostok. 1838 Siberian stone pine [see *Siberian cedar* above]. 1923 A. Rehder in L. H. Bailey *Cultivated Evergreens* v. 303 Siberian Stone P[ine]. A form with shorter leaves and larger cones [than *Pinus cembra*]. 1967 N. T. Mirov *Genus Pinus* iii. 233 Its 'common' English name, 'Siberian stone pine', is rather inappropriate. 1925 G. W. Deeping *Sorrell & Son* xxxvii. 375 A half wild patch of..purple and gold tulips, and burning orange siberian wallflowers. 1933 *Jrnl. R. Hort. Soc.* LVIII. 172 The Siberian Wallflower..probably merits that name as little as the one by which it is commonly known. 1979 C. E. L. Phillips *New Small Garden* x. 137 The slightly later Siberian wallflower..is a glowing orange ball of fire.

B. *sb.* **3.** A Siberian husky.

1928 *N.Y. Times* 29 Jan. ii. 1/5 Seppala left the line without a word to his furry, sharp-eared Siberians. 1944 C. Clay *Phantom Fur Thieves* 137 The Siberian..was originally bred and raised in Siberia, is smaller in size than a huskie, more stockily built, more heavily furred.

sibilance. Delete *rare* and add: (Further examples.) Also, an undue prominence of sibilants, *esp.* in reproduced sound.

1939 A. Clarke *Coll. Plays* (1963) 92 The words of the Chorus become a mere sibilance. 1943 A. L. Rowse *Cornish Childhood* 87 Her voluminous skirts which made such a lovely sibillance [*sic*] whenever she moved. 1960 G. A. Briggs *A to Z in Audio* 183 *Sibilance*, a fault in reproduction in which consonants and, in particular, 's' sounds are given unnatural prominence. 1962 A. Nisbett *Technique Sound Studio* iv. 82 'Tilting' such a ribbon was at one time commonly used as a means of discriminating against sibilance (a voice defect which seems to be less of a nuisance than it has been in the past). 1974 —— *Use of Microphones* 24 Sometimes intelligibility is improved by using a microphone with a peak..in the 6000–8000 Hz frequency range... But this may also enhance the natural sibilance of some voices. 1979 *Amer. Poetry Rev.* Mar./Apr. 26/1 Now if we take out the first one we risk sibilance by having the *s* of 'Ignatius' run into the *s* of 'swallows'.

sibilant, *sb.* Add: **1.** (Earlier example.)

1788 W. Jones in *Asiatick Researches* I. 11 Next come different classes of *dentals*, and among the first of them should be placed the *sibilants.*

2. A rumour started and spread for propaganda or advertising purposes. Cf. Whisper *sb.* 2.

1957 *Observer* 27 Oct. 18/7 The sib-spreader, that fortunate extrovert with a cast-iron digestion who is employed to dash about bars and cocktail parties spreading carefully composed sibilants or selling rumours. 1958 [see *sib *sb.*^{3}].

sibilate, *v.* Add: **2. a.** More generally, to say with a hissing sound.

1903 K. D. Wiggin *Rebecca of Sunnybrook Farm* xix. 199 'How about cookies?' 'Do you think it's worth while?' sibilated Miss Miranda. 1910 O. Browning *Mem. Sixty Yrs.* xx. 318 Two portly gentlemen..turned round towards me, hissed violently and sibilated the word 'Poet' thinking, I suppose, that I was Robert.

Sibiriak (sibi·ryæk). Also **Sibiryak.** [Russ. *Sibiryák* Siberian.] A Siberian descended from European Russian settlers. Also *attrib.* or as *adj.*

1903 W. Gerrare *Greater-Russia* viii. 105 The Siberiaks, as the descendants of the early settlers in the west province are called, are of average height. 1916 M. A. Czaplicka *My Siberian Year* xii. 243 The Sibiriak—that is, broadly speaking, the colonial whose ancestors have been settling in Siberia, voluntarily or involuntarily, since, say, the end of the Middle Ages. *Ibid.* 245 An anthology of Sibiriak songs..would provide an illuminating document for the student of Siberian life. 1974 *Encycl. Brit. Macropædia* XVI. 726/1 There were long-established Russian peasant societies in certain parts of Siberia, in many cases the descendants of exiled religious dissidents. Such people, known as Sibiryaks or 'local Russians', have a culture and an outlook differing markedly from those of the people of European Russia.

sibling. Restrict † *Obs.* to sense in Dict. and add: Pronunc. (si·bliŋ). **2.** = *sib *sb.*^{2} 3 d. Also (chiefly *Anthrop.*), each of two or more children belonging to a family or kinship group having at least one parent in common. Also *fig.*

1903 K. Pearson in *Biometrika* II. 369 These [calculations] will enable us..to predict the probable character in any individual from a knowledge of one or more parents or brethren ('siblings', = brothers or sisters). 1930 *Nature* 15 Nov. 766 A few were odd twins who had a brother or sister at school, and the remainder were either siblings of twins, or pairs of siblings unconnected with twins. 1933 L. Spier *Yuman Tribes* vii. 209 She needed to know no more than the sex and relative ages of the siblings from whom the lines were traced to give..the terms used between any pair in the succeeding generations. 1941 *Jrnl. R. Anthrop. Inst.* LXXI. 7/2 A group of siblings is constituted by the sons and daughters of a man and his wife in monogamous societies, or of a man and his wives where there is polygyny, or of a woman and her husbands in polyandrous communities. 1950 M. Fortes in Radcliffe-Brown & Forde *Afr. Systems of Kinship* 273 Next to the bond between mother and child none is so strong as that between siblings by the same mother. 1957 V. W. Turner *Schism & Continuity in Afr. Soc.* iii. 68 Thus uterine siblings and matrilineal parallel cousins remained together throughout life. 1957 *Observer* 24 Nov. 15/1 There are..echoes of 'Death of a Salesman'—the loyal, harassed wife, the two disenchanted siblings, and the salesman father with an unfulfilled dream. 1970 *Nature* 19 Dec. 1221/2 In the herring gull.., the chick hatching from this third egg suffers a much higher mortality than either of its siblings. 1972 *Daily Tel.* 20 Mar. 2/7 The line dividing the Kevin Street Sinn Fein organisation and its terrorist sibling, the Provisional IRA, is almost invisible. 1974 'J. Melville' *Nun's Castle* vii. 153 Siblings and kinsfolk did not have to be friends.

3. *attrib.* and *Comb.*, as *sibling group*; **sibling rivalry,** rivalry arising from the jealousy that can exist between siblings; **sibling species** *Biol.*, one of a pair or group of reproductively isolated populations whose members are morphologically very similar; cf. *twin species.*

1957 V. W. Turner *Schism & Continuity in Afr. Soc.* p. xx, The headman's uterine sibling group. 1975 H. & C. Geertz *Kinship in Bali* ii. 57 One brother may farm the plot of land as sharecropper to the sibling group as a whole. 1930 *Smith Coll. Stud. in Social Wk.* I. 6 (*title*) Two studies in sibling rivalry. 1959 Martin & Stendler *Child Behav. & Devel.* (rev. ed.) xvi. 553 The most common form of jealousy in young children is sibling rivalry. 1972 C. Raphael *Feast of History* iv. 118 Moses..shows more than a hint of sibling rivalry in his attitude to his brother Aaron. 1979 Watson & Lindgren *Psychol. of Child* (ed. 4) iv. 125 Adler was also keenly aware of the conflicts and tensions among siblings..the intense competition—sibling rivalry—for status, power, and parental attention. 1940 E. Mayr in *Amer. Naturalist* LXXIV. 258 Sibling species: The opposite condition exists where pairs or larger groups of related species are so similar that they are generally considered as one species. 1942 —— *Systematics & Origin of Species* vii. 151, I call..morphologically similar and closely related, but sympatric species, sibling species. This corresponds to the 'dual species' of Pryer and of Hering (1935). 1979 *Nature* 7 June 557/2 The eastern meadowlark (*Sturnella magna*) and western meadowlark (*S. neglecta*) are sibling species of songbirds in secondary contact throughout a narrow zone of sympatry in central North America.

Hence **si·blingship** = *sibship 1.

1950 M. Fortes in Radcliffe-Brown & Forde *Afr. Syst. of Kinship & Marriage* 273 The range within which the bond of siblingship is accepted as automatically binding. 1957 V. W. Turner *Schism & Continuity Afr. Soc.* vii. 226 These are all means..of reducing the strength of uterine siblingship. 1970 [see *kinship 1 b].

sibship (si·bʃip). [f. Sib *sb.*^{2} + -ship.]

1. *Anthrop.* The state of belonging to a sib, or to the same sib (*sib *a.* and *sb.*^{2} 3 a, c).

1908 *Rep. Brit. Assoc. Adv. Sci. 1907* 654 *Sib and sibship*, the old word *sib* may be used for the relationship set up by membership of the sept. 1924 W. H. R. Rivers *Soc. Organ.* ii. 22 All the people of a village or district.. believe themselves to be related to one another, and thus form a characteristic example of sibship. 1950 [see *sib *a.* and *sb.*^{2} 3 a].

2. *Biol.* and *Med.* A group comprising all the individuals born to a particular pair of parents.

1919 *Genetics* IV. 489 Fraternal resemblance is usually not far from ·54, so that 46 percent of the variance of the population occurs within the sibship. 1925 *Jrnl. Genetics* XV. 259 An acre of F_2 plants and an acre of F_3 sibships. 1939 *Nature* 18 Mar. 484/2 The fall in the birth-rate may be due to diminution in family size... Thus the average sibship in Great Britain now is said to be half of what it was fifty years ago. 1949 C. Stern *Princ. Human Genetics* 389 In sibships of 10, chance alone will give rise, on the average, to 1 sibship in 1024 of all boys. 1958 *Immunology* I. 49 Eight goldfish from the same sibship as before. 1978

Brit. Med. Jrnl. 14 Jan. 72/1 There is a high incidence of spontaneous abortion in sibships in which a case of anencephaly or spina bifida has occurred.

sic, *adv.* Add: Also as *sb.*, an instance of 'sic'.
1910 [used s.v. Sic *v.*]. **1937** *Scrutiny* Sept. 131 As for what Miss Lynch calls 'his really serious affair with Harriet' (I feel this deserves a *sic*), it is purely theatrical. **1963** J. Mitford *Amer. Way Death* ii. 27, I do not like repeated use of *sic*... The reader who is fastidious about usage will hereafter have to supply his own *sics*. **1973** E. Taylor *Serpent under It* (1974) xiv. 224 He called the librarian..and asked him to check... Hence the 'sic'.

sic, var. Sick *v.*[2] in Dict. and Suppl.

Sican (si·kăn). Also **Sikan.** [ad. L. *Sicānus*, pl. *-ī*: see Sicanian *a.*] A member of an ancient people inhabiting Sicily at the time of the coming of the Sicels (see *Sicel *sb.* and *a.*).
1887 *Encycl. Brit.* XXII. 15/1 It is possible..that..the Sikans..belonged to the earlier non-Aryan population of western Europe. **1911** *Ibid.* XXV. 24/2 They [*sc.* the Sicels] found in the island a people called Sicans. **1968** M. Finley *Hist. Sicily* ii. 23 The Sicans who were apparently thinner on the ground than the Sicels, seem also to have been more resistant to Hellenization. **1970** *Oxf. Classical Dict.* (ed. 2) 985/1 Thucydides..attributes an Iberian origin to the Sicans.

Sicanian, *a.* Restrict *poet.* to sense in Dict. and add: Pronunc. (sikē̄·niăn). **A.** *adj.* **b.** *Archæol.* Denoting the Neolithic period in Sicily. **c.** Of or pertaining to the Sicans. **B.** *sb.* = *Sican.
1629 T. Hobbes tr. *Thucydides' Peloponnesian War* vi. 350 After them, the first that appear to have dwelt therein, are the Sicanians. **1876** F. Tozer *Class. Geogr.* x. 117 The original inhabitants of Sicily were two tribes, the Sicanians in the west, and the Sicels in the east, both of whom belonged to the same Graeco-Italian stock as the Greeks themselves. **1909** T. E. Peet *Stone & Bronze Ages Italy & Sicily* v. 123 By Professor Orsi..the pre-Hellenic period in the island, excluding the palaeolithic, is divided into five divisions. To the first of these he gives the name Sicanian; the other four are called respectively First, Second, Third and Fourth Siculan periods. **1911** *Encycl. Brit.* XXV. 20/1 The most important of the towns to which a Sicanian origin can be with certainty assigned..are: Hyccara..; Omphakē..; and Camicus. **1957** *Ibid.* XX. 603/1 The term Sicanian is applied to that period of the Stone Age which followed the palaeolithic period exemplified in the remarkable rock engravings of the cave near Palermo.

sicca[2] (si·kă). *Path.* [ellipt. for mod.L. *keratoconjunctivitis sicca*.] The symptom of reduced or no lachrymation, with consequent dryness and inflammation of the conjunctiva, characteristic of Sjögren's syndrome; used *attrib.* as *sicca syndrome* to designate the occurrence of this symptom in the absence of rheumatoid arthritis.
1938 *Acta Ophthalmologica* XVI. 176 (*heading*) The initial stage of glandular changes in the sicca syndrome. **1949** *Ibid.* XXVII. Suppl. 33. 1 (*heading*) Keratoconjunctivitis sicca and the sicca syndrome. *Ibid.* 9 The bulk of the sicca patients in my material are rheumatics. **1967** *Amer. Jrnl. Med.* XLIII. 50/1 Sjögren's syndrome is characterized clinically by dryness of the conjunctiva and mucous membranes (sicca syndrome), and by frequent episodes of salivary and lacrimal gland enlargement. **1977** *Proc. R. Soc. Med.* LXX. 483/2 In the 'sicca' or Sjogren syndrome, Patey & Thackray (1955) have described lymphoid tissue in and around intralobar ducts which breaches the walls and allows radiopaque dye to escape during sialography, causing 'punctate sialectasis'.

Sicel (si·sĕl, si·kĕl), *sb.* and *a.* Also **Sikel.** [ad. Gr. Σικελός.] **A.** *sb.* **a.** A member of an ancient people of Sicily. **b.** The language of this people. **B.** *adj.* **a.** Of or pertaining to the Sicels or their language. **b.** *poet.* Sicilian.
1838 C. Thirlwall *Hist. Greece* II. xii. 92 The Sicels and the Phœnicians gradually retreated before the Greeks. **1881** B. Jowett tr. *Thucydides* I. 409 The Sicels were originally inhabitants of Italy,..there are Sicels still in Italy, and the country itself was so called from Italus a Sicel king. **1887** *Encycl. Brit.* XXII. 15/1 Some Sikel elements made their way into the Greek life of Sicily. **1895** L. Johnson *Poems* 37 Oh! Hellas lies far hence, Far the blue Sicel sea. **1911** *Encycl. Brit.* XXV. 24/2 That the Sicels spoke a tongue closely akin to Latin is plain from several Sicel words which crept into Sicilian Greek, and from the Siceliot system of weights and measures. **1939** L. H. Gray *Foundations of Lang.* 335 Besides the Italic dialects proper, mention must also be made of *Sicel*, of which a few glosses and an inscription of three lines have been preserved, and which seems to have belonged either to this group or to Ligurian. **1948** T. J. Dunbabin *Western Greeks* i. 40 It appears that the Sikels moved from Sicily to Italy, not vice versa. **1974** *Encycl. Brit. Micropædia* IX. 182/2 The most important Sicel gods were the Palici..; Adranus..; and the goddess Hybla. **1977** *Canad. Jrnl. Linguistics* Spring 31 Messapic and Sicel in the south take on new significance vis à vis Iberian.

Siceliot (sis-, sike·liǫt), *sb.* and *a.* Also **Sikeliot, -ote.** [ad. Gr. Σικελιώτης, f. Σικελία Sicily: see -ot[2], -ote.] **A.** *sb.* One of the ancient Greeks who colonized Sicily, distinguished from the Sicels who had settled in Sicily before their coming. **B.** *adj.* Of or pertaining to the Siceliots.
1836 C. Thirlwall *Hist. Greece* III. xxii. 263 [They were] linked together by the common name of Siceliots. **1842** *Penny Cycl.* XXIV. 407/1 The intimate knowledge which he [*sc.* Thucydides] shows respecting the history of the Italiotes and Siceliots. **1887** *Encycl. Brit.* XXII. 16/1 The ancient kingship was perhaps kept on or renewed in some of the Sikeliot and Italiot towns. **1892** *Athenæum* 7 May 597/3 In Syracuse then lay the last hopes of rescue for the Siceliot Greeks. **1931** D. MacIver *Greek Cities in Italy & Sicily* xii. 172 Gela was chosen for the Pan-Sikeliot conference of Greek Cities in 424 B.C.

‖ **Sicherheitsdienst** (zi·χ^ərhaitsdīnst). [G., f. *sicherheit* security + *dienst* service.] The security branch of the Schutzstaffel (SS) of the Nazi party, set up in 1931–2. Also *attrib.*
1947 [see S.D. s.v. *S 4 a.]. **1958** P. Kemp *No Colours or Crest* i. 9 His murder at the hands of Gestapo or *Sicherheitsdienst*. **1966** N. Freeling *Dresden Green* i. 44 The highly successful delivery of five of the area Sicherheitsdienst group, trousers round their ankles. **1976** T. Allbeury *Only Good German* iv. 23 The SD, the Sicherheitsdienst, were an SS outfit under Walter Schellenberg.

Sichuana, obs. var. *Sechuana.

Sicilian, *a.* and *sb.* Add: **A.** *adj.* **1. b.** Of or pertaining to the Italian dialect of Sicily.
1842 [in Dict.]. **1881** *Encycl. Brit.* XIII. 495/1 The Sicilian vocalism is conspicuously etymological. **1975** *Times Lit. Suppl.* 31 Oct. 1296/3 Dialect seems diminishing and parochial to the serious writer... We can meet good Sicilian or Venetian or Roman dialogue, but not good Italian dialogue.
2. a. (Chess) *Sicilian defence, game, opening* (earlier example).
1847 H. Staunton *Chess-Player's Handbk.* v. ii. 371 The Sicilian Game... In the opinion of Jaenisch..this is the best possible reply to the move of I.P. to K's 4th. **1852** —— *Chess Tournament* 29, I have before taken occasion to remark that in this position of the Sicilian Opening, the first player may gain time..by taking off the Kt. at once. **1875** G. H. D. Gossip *Chess-Player's Manual* iv. xxx. 799 The 'Sicilian' is now considered by most modern authorities to be a comparatively weak mode of play... We are of the opinion that the Sicilian defence is not so bad as it has been represented. **1900** *Knowledge* 1 Aug. 192/1 The success attending the Sicilian defence is especially noteworthy. **1975** *Amer. Speech 1971* XLVI. 232 One can hear heated arguments on the virtues of the Maroczy Variation of the Scheveningen System in the Sicilian Defence to the King's Pawn Opening.
B. *sb.* **4.** A language or dialect spoken in Sicily, *spec.* a dialect of modern Italian.
1818 Keats *Let.* 3 May in R. M. Milnes *Life, Lett., &c. J. Keats* (1848) I. 135 Or may I woo thee In earlier Sicilian? **1859** B. W. Dwight *Mod. Philol.* i. 187 Italian: (Dialects, Lombard; Genoese; Florentine; Neapolitan, Sicilian,..&c.). **1880** A. H. Sayce *Introd. Sci. Lang.* II. vii. 119 Sicilian, for instance, reads like a new language. **1933** L. Bloomfield *Language* iv. 64 *Ligurian* (round the present Riviera) and *Sicilian* in Sicily, may have been close to Italic. **1968** D. Mack Smith *Medieval Sicily* v. 63 Giacomo of Lentini, author of a Provençal-type lyric which is the first poem in true Sicilian that has survived. **1978** *Language* LIV. 184 Sicilian reflects the seven-vowel Southern Romance vocalism.

Siciliana. Add: (Further examples.) Also **Siciliano, siciliano**; pl. **sicilianos, siciliani.** Also, a piece in $\frac{6}{8}$ or $\frac{12}{8}$ time resembling this music. Also *attrib.*
1883 Grove *Dict. Mus.* III. 491/2 *Siciliana, Siciliano, Sicilienne*, a dance rhythm closely allied to the Pastorale. **1947** C. Gray *Contingencies & Other Essays* v. 118 The frequent recurrence in Bellini's music of 12–8 rhythms, called in seventeenth- and eighteenth-century music *Siciliani*, seems always to have been a feature of the popular music of his countrymen. **1959** D. Cooke *Lang. of Music* ii. 100 The siciliano movement in Brahms's St. Anthony Variations. **1968** *Listener* 20 June 814/1 Their variety was astonishing, from the lyrical sicilianos which the history books praise, to vigorous virtuoso pieces. **1970** W. Apel *Harvard Dict. Music* (ed. 2) 774/2 The *siciliana* occurs as a slow movement in early sonatas..as well as in vocal music..whenever gentle pastoral scenes are to be represented musically. **1974** *Early Music* July 197/1 A delightful siciliano aria for alto to the words 'Qui sedes ad dexteram Patris'. **1979** *Ibid.* Oct. 545/2 A profusion of characteristic ideas, charming *Sicilianos*, bubbling *Allegros*, idiomatic and elegant writing for the instrument.

Sicilienne. Add: **1.** (Earlier example.)
1873 *Young Englishwoman* Jan. 24/1 Sicilienne, a new kind of silk, both soft and glossy.
2. *Mus.* = Siciliana in Dict. and Suppl.
1883 [see *Siciliana]. **1927** *Daily Tel.* 12 Feb. 5/2 The Sicilienne was quiet and restful.

sick, *a.* and *sb.* Add: **A.** *adj.* **I. 1. a.** Delete 'Now chiefly literary and *U.S.*' and add further examples. Also, *to go sick*, to become ill, to report sick.
1879 [see sense 1 f below]. **1902** W. B. Yeats *Where there is Nothing* (1903) iv. i. 77 No fear, they won't refuse a sick man. **1915** D. O. Barnett *Lett.* 53 He's lots better this morning,..and he is not 'going sick' at all. **1927** E. J. Thompson *These Men thy Friends* 12 Filthy climate. No fun. But she just carries on. Hasn't gone sick once in six months. **1936** G. B. Shaw *Millionairess* ii. 164 You are my doctor: do you hear? I am a sick woman: you cannot abandon me to die. **1945** *Chambers's Jrnl.* Sept. 452/1 'And you're telling me that you've never had a few days off?.. Not even for sick-leave?' 'I was never sick, sir.' **1952** [see *Lead *sb.*[1] 6 b]. **1956** D. Jacobson *Dance in Sun* ii. ix. 91 'Hey,' he said rudely to Fletcher, 'are you sick?' **1959** V. Watkins *Cypress & Acacia* 23, I found him feeble and sick. And cold. **1962** G. Lawton *John Wesley's English* iii. 57 When Wesley is sick he is 'laid-up.' **1976** *Evening Post* (Nottingham) 15 Dec. 24/4 Willis went sick during the opening match in Poona.
d. (Further *transf.* examples.) Also *fig.*, orig. applied to Turkey and hence to other countries, regions, etc., and in extended uses.
1860 S. S. Cox *Eight Years in Congress* (1865) 129 'Mexico is our "sick man".' 'Yes; she is to America what Turkey is to Europe.' **1868** C. Schurz *Speeches, Corr. & Pol. Papers* (1913) I. 456 The South is our 'sick man'... The 'sick man' has been operated upon by Democratic doctors once more. **1888** S. Lane-Pool *Turkey* xvii. 343 The Powers have always acted on the principle that somebody must serve as a dyke between Russia and the Bosphorus, and that Turkey, being there, had better be maintained in her position. The 'Sick Man' of the morbid mind of Nicholas must be galvanized into sufficient vitality to sit up and pretend to be well. **1897** *Japan Times* 30 Mar. 3/4 Mr Valentine Chirol, who shortly after the war published in the London *Times* a series of remarkable articles exposing the rottenness of China..has recently been in the East again..and has commenced a second series of equally striking articles on the 'Sick Man of Asia'. **1918** *Times* 3 Jan. 5/1 The Sick Man of Europe has changed his doctors, and the new doctors..have prescribed participation in the European war. *Ibid.* 5/2 The Sick Man finds himself less sick than his neighbour, and Russia defenceless offers her flanks to Turkey's sharpest blades. **1929** H. M. Kallen *Frontiers of Hope* 451 Under the terms of the Peace the Jew has simply been made to replace the Turk as the Sick Man of Europe. **1959** *Listener* 30 July 168/2 It was Italy which turned the Austrian empire into a second 'sick man'. **1961** N. Smart in I. Ramsey *Prospect for Metaphysics* v. 80 Natural theology is the Sick Man of Europe. **1963** *Times* 31 Jan. 11/1 There is no imminent threat to it, but once that is passed India would be on the way to becoming in economic terms the sick man of Asia. **1967** *Listener* 26 Jan. 116/2 In December 1958 France was the sick man of Europe; it had no exchange reserves and was incapable of facing the Common Market. **1970** R. Lowell *Notebook* 205 The movie's not always the sick man of the arts. **1974** *Times* 4 May 8/4, I have been wondering who now qualifies for the title of Sick Man of Sound Broadcasting. **1979** G. St. Aubyn *Edward VII* vii. 319 China was the sick man of the Orient over whose corpse the vultures hovered.
† **f.** Said of pigeons which have lost their young and so have no recipient for the soft food that they regurgitate. *Obs.*
1765 *Treat. Domestic Pigeons* 21 If your Pigeons do not hatch, because their eggs are addle, or otherwise, you should give them a pair, or at least one young one, to feed off their soft meat, which would be apt to make them sick. **1854** L. A. Meall *Moubray's Poultry* viii. 455 We have never observed the old birds 'sick' (as most books assert they are) when the young have died. **1879** L. Wright *Pract. Pigeon Keeper* iii. 37 In order that another young one from some other pair..may be given the parents to feed off their soft meat, and save them from 'going sick' with it.
g. *slang* (orig. *U.S.*). Of a drug addict, craving for a dose of a drug, suffering from withdrawal symptoms.
1951 *N.Y. Times* 15 June 14/6, I..would walk up and..ask the bartender: 'Say, have you seen so-and-so yet?' I says: 'Man, I'm sick.' **1953** W. Burroughs *Junkie* vii. 69 The usual routine is to grab someone with junk on him, and let him stew in jail until he is good and sick. **1967** M. M. Glatt et al. *Drug Scene* vii. 91 Even now, more than two years after leaving hospital, I still feel sometimes sick in the morning when I am tense or upset, and I feel sick whenever I see syringes or 'addicts' in TV plays.
2. b. Delete ? *Obs.* and add later examples. Also *sick to the stomach*.
1863 Trollope *Small House at Allington* xxxvi, in *Cornh. Mag.* Aug. 228 How well can I remember the terror created within me by..a certain fine old gentleman. ..I would become sick in my stomach. **1923** [see *Here *adv.* 5 b]. **1947** A. Huxley *Let.* 9 Apr. (1969) 570, I heard a bit of the *Parsifal* Good Friday music at Easter-time.., and it made me feel even more 'sick to my stomach', as the Americans say, than in the past. **1948** 'J. Tey' *Franchise Affair* xiii. 139 You make me sick—Cat-sick. Sick to my stomach. **1955** *Jrnl. Canad. Linguistic Assoc.* Mar. 17 Another expression which has some striking variants depending on the choice of preposition is *sick at the stomach*. In the Northern speech area of the United States the usual equivalent is *sick to the stomach*; in the Midland and Southern areas, *at* is the usual preposition... In New England..and most of the Yankee settlement areas, *to* enjoys a virtual monopoly. In northwest New York State, however, *sick at the stomach* is unusually common. **1975** *Times* 30 June 17/5 If all the factories are nationalized I shall walk out of here sick to the stomach.
c. (Further examples: sense sometimes merging with 4.) Also *sick as a parrot* (a fanciful catch-phrase, chiefly used *joc.*)
1861 Hughes *Tom Brown at Oxf.* III. xi. 207 It turned me as sick as a dog. *a* **1906** 'T. Collins' *Rigby's Romance* (1946) xli. 221 Well, by-and-by I woke up, sick as a dog, with my face all scorched, and I lay down again. **1915** J. Buchan *39 Steps* vii. 161, I had a crushing headache, and felt as sick as a cat. **1947** A. Ransome *Great Northern?* xix. 238 'Sick as cats with himself,' said Nancy. **1979** *Private Eye* 16 Feb. 12/1 The Moggatollah admitted frankly that he was 'sick as a parrot' at the way events

had been unfolding. **1982** *Daily Star* 5 Feb. 5/6 Peter the budgie was sick as a parrot until a vet diagnosed his problem yesterday. Peter..has got gout!
d. In phrases *to worry* (oneself), *be worried, sick*.
1952 M. Laski *Village* v. 89 Edith Wilson had heard about Wendy's illness, and worried herself sick, not knowing what to do for the best. **1961** 'J. le Carré' *Call for Dead* iv. 37 You look worried sick. **1977** R. Ludlum *Chancellor Ms.* xxx. 320 She hasn't been able to sleep. She's worried sick.
II. 4. a. (*d*) Freq. in phr. *to make* (a person) *sick*.
1819, 1860 [in Dict.]. **1911** G. B. Shaw *Blanco Posnet* 30 A man like you makes me sick. **1937** 'G. Orwell' *Road to Wigan Pier* xii. 228 It makes one sick to see..men sweating their guts out to dig a trench.., when some easily devised machine would scoop the earth out in a couple of minutes. **1944** M. Laski *Love on Supertax* i. 18 The Duchess lost her temper. 'You make me sick!' she shouted. **1978** T. Allbeury *Lantern Network* iii. 34 He talks like a schoolboy. All that 'knocking the Germans for six' stuff, it makes me sick.
5. b. In phrases *sick and tired of* (cf. *sick-tired*, sense 11 in Dict.), *sick to death of*.
1783 'J. H. St. John de Crèvecœur' *Sk. 18th-Cent. Amer.* (1925) 298, I am quite sick and tired of these pretended conscientious non-fighting mortals. **1884** E. W. Nye *Baled Hay* 124 We are sick and tired of pointing out different avenues of wealth to be laughed at and ridiculed. **1890** E. Dowson *Let.* c 11 Sept. (1967) 166, I am sick to death of this place. **1925** F. S. Fitzgerald *Great Gatsby* ix. 205 'You young men think you can force your way in any time,' she scolded. 'We're getting sick in tired of it.' **1953** R. Lehmann *Echoing Grove* 117 He was sick to death of the sound of these three crass monosyllables which he seemed always to be reiterating. **1976** *Milton Keynes Express* 9 July 2/6, I believe people are sick and tired of half-truths and evasions.
III. 6. Also, morbid, enjoying sick humour (see sense 7 f below).
1959, 1960 [see sense 7 f below]. **1961** *Times* 17 July 14/5 Mr. Sahl is disapproving of the so-called 'sick' comedians of America. **1961** Webster s.v. [1]*sick*, A sick personality. **1962** *Listener* 25 Oct. 692/3 From Korea James Mossman reported on the Panmunjom truce-line (a raree-show for tourists these days, I gather: how sick can people get?). **1964** L. Nkosi *Rhythm of Violence* 45 Don't mind them, honey! They're the sickest bunch of people you ever saw.
7. a. (Further examples.)
1817 in *Trans. Ill. State Hist. Soc. 1910* (1912) 147 Sick Milk, Sick Wheat, a plenty of Ague near the large streams. **1847** H. Howe *Hist. Collect. Ohio* 274 Those lands were too sick for wheat, making 'sick' wheat, so termed, because when made into bread, it had the effect of an emetic. **1915** *Rep. Brit. Assoc. Adv. Sci. 1914* 672 The fertility of this 'sick' soil can be restored by merely heating it for an hour or two to a temperature approaching that of boiling water. **1921** *Brit. Mus. Return* 74 in *Parl. Papers* XXVII. 651 The treatment and cleaning of sick and dirty coins. **1930** *N. & Q.* 16 Aug. 124/2 A cheese..is sick when it has been over soured or over acidulated, and in time 'weeps', gradually becoming soft inside. **1947** I. L. Idriess *Isles of Despair* xvi. 106 Some roots are 'sick', eaten through and through by boring insects. **1965** *Listener* 2 Sept. 358/1 Soils can be said to be 'potato sick', 'rose sick', 'flax sick', etc.
fig. **1931** H. Crane *Let.* 13 June (1965) 371 As for Mexico..you were right, it's a sick country. **1959** *Washington Post* 18 Nov. A 16/2 Some czars in the labor movement will scream over this resort to the courts to straighten out so-called internal affairs of sick unions, but for racket-harassed workers it is an event of the first importance. **1960** *Wall St. Jrnl.* 2 Feb. 4 He has taken other sick businesses and has done a marvelous job with them. **1973** *Black Panther* 14 July 6/2, I was basically looking at myself..people of my complexion struggling for their liberation. I saw how sick conditions were. **1976** Smythies & Corbett *Psychiatry* iii. 29 Concepts like a 'sick' society have become commonplace.
f. *colloq.* Of humour, a joke, etc.: macabre, providing amusement by reference to something that is thoroughly unpleasant.
1959 *Punch* 2 Sept. 106/1 The prototype of sick jokes is one that goes 'But apart from that, Mrs. Lincoln, how did you enjoy the play?' **1959** *Guardian* 16 Oct. 10/3 Feiffer..belongs..to the new American fashion of sick humour... Like those gifted sick comedians Mort Sahl and Lenny Bruce..he is able to go straight to the springs of derision and aggression where so much humour begins. **1959** *Washington Post* 26 Nov. D 22/1 'Sick comedy,' defines Berman carefully, 'is comic material which violates what we regard as the limits of sensitivity—poking fun at a cripple..or kidding a typhoon that killed thousands.' **1960** *Guardian* 7 Oct. 15/3 Jules Feiffer, regarded as one of the 'sick' school of cartoonists, is not as sick as all that... No one is sicker than Charles Addams. **1961** *Harper's Bazaar* Feb. 84/2 To enjoy..the sick joke..you have to..swallow jokes about cancer, corruption, homosexuality, third degree, race prejudice and insanity. **1965** *Times Lit. Suppl.* 25 Nov. 1035/4 This has been a time of sick laughter. **1968** M. Woodhouse *Rock Baby* xvii. 164 How long exactly does it take to become a bomb-disposal expert? And don't tell me that you learn by your mistakes because I'm not in the mood for sick jokes. **1975** P. Fussell *Gt. War & Mod. Memory* vi. 228 There is extant a postwar version of such a record [of battle],..aimed at what today might be called the Sick Nostalgia Market. **1978** D. Devine *Sunk without Trace* xxv. 226 'How does it feel..to be back in the bosom of your family?' Judy said sharply: 'I'm not in the mood for sick jokes.'
10. a. *sick bay* (later examples).
1919 W. Lang *Sea Lawyer's Log* i. 6 Then our guide, a Leading Seaman,..conducted us to the doctor's quarters—or 'sick bay', as he expressed it. **1971** P. D. James *Shroud for Nightingale* iii. 60 She's in the sick bay... It's part of the private wing.
b. *sick-benefit, -leave* (later examples).
1909 *Chambers's Jrnl.* 26 Dec. 56/2 Members who..may have received sick-benefit. **1952** Sick benefit [see *Approved *ppl. a.* 6]. **1943** J. B. Priestley *Daylight on Saturday* xxix. 231 He'll be home on sick leave.., the doctor says. **1976** *Times* 8 Mar. 12/8 Staff are entitled to paid sick leave only if there is a reasonable prospect of their return to duty.
11. *sick-making* ppl. adj. and vbl. sb.; *sick-sweet* adj.
1930 E. Waugh *Vile Bodies* i. 7 Sometimes the ship pitched and sometimes she rolled... 'Too, too sick-making,' said Miss Runcible, with one of her rare flashes of accuracy. **1938** Dylan Thomas *Let.* c 6 July (1966) 203 There will be speechmaking, drunkmaking, sickmaking and we must all dress up. **1949** N. Mitford *Love in Cold Climate* i. vi. 59 I'm in a terrible do about my [stolen] bracelet of lucky charms—no value to anybody else—really—too too sick-making. **1976** I. Illich *Limits to Medicine* 7 What has turned health care into a sick-making enterprise is the very intensity of an engineering endeavour. **1978** *Times* 5 Oct. 2/4 What is sickmaking is the IBA..trying to make the BBC out as the monster and them the viewers' guardian. **1922** Joyce *Ulysses* 444 The odour of the sicksweet weed floats towards him.
12. Special combs., as **sick-bag,** a bag provided in aircraft, ships, etc., as a receptacle for vomit; **sick call,** (*a*) (orig. and chiefly *Mil.*), a call sounded to summon those reporting sick to a place of treatment; an assembly for medical treatment; (*b*) a visit made to a sick person; (*c*) a summons to visit a sick person; **sick headache** = Migraine; also in phrases as a type of something useless or unhelpful; **sick parade** *Mil.*, an inspection of those who are ill; the people on sick parade; **sick visiting** (see quot. 1933).
1962 W. Schirra in *Into Orbit* 33 On the plane, John Glenn and Al Shepard took one of the brown paper 'sick-bags' and scribbled on it: 'Here is the answer to the air sickness problem.' **1976** 'D. Halliday' *Dolly & Nanny Bird* vi. 71 The accustomed routine with Kleenex and sick bags..and barley sugar. **1836** J. Hildreth *Dragoon Campaigns to Rocky Mountains* 114 Every morning..'sick call' blows. **1850** E. Price (*title*) Sick calls: from the diary of a missionary priest. **1883** Lady Herbert tr. *Life St. John Baptist de Rossi* iii. iv. 147 The servants never again dared to fail to warn him of any sick call. **1930** F. A. Pottle *Stretchers* 31 Sick call is blown before the dispensary door. **1931** P. J. Joyce *John Healy* ii. 37 That imperious, unmistakable sick-call knock. **1945** *Yank* 13 July 19/2 A punitive measure to discourage falling out for sick call. **1976** *Billings* (Montana) *Gaz.* 27 June 1-B/4 There is a daily sick call by a local doctor for the inmates. **1977** *New Yorker* 24 Oct. 106/2 A third way for an inmate to see a doctor is to go to sick call, which is held each weekday morning at Green Haven on the first floor of the Hospital-Segregation Building. **1978** J. Carroll *Mortal Friends* iv. ii. 389, I was a young priest at the time, see, and I get this sick-call. **1778** Fothergill in *Med. Observ.* (1784) VI. 103 Remarks on that Complaint commonly known under the Name of the Sick Head-ach. **1799** [in Dict., sense 9]. **1857** M. O. Colt *Jrnl.* 18 May (1862) xii. 218, I..was obliged..to..stay two nights and one day, suffering with a sick headache. **1915** D. O. Barnett *Lett.* 153 Shrapnel is for defenders, to stop an advance of infantry, but no more use against prepared positions than a sick headache. **1977** 'E. Crispin' *Glimpses of Moon* xii. 252 That pair in the back, between them, are about as much use as a sick headache. **1915** 'I. Hay' *First Hundred Thousand* xi. 137 M'Splae departs, grumbling, and reappears on sick parade a few days later. **1927** R. H. Mottram *Spanish Farm Trilogy* 258 Do you know what a sick parade I've got? Eighty! Yes, I have. **1966** *Times* 9 July 9/7 Command Orders say... Sick Parade has now become 'sick list'. **1933** *O.E.D. Suppl.* s.v. Sick *a.* and *sb.* 10 b, *Sick visiting*, the visiting of the sick, esp. by a minister of religion. **1960** N. Nicholson *William Cowper* 13 A most exacting life of piety, prayer-meetings, self-denial, and sick-visiting. **1977** *West Briton* 25 Aug. 11/1 He paid tribute to Mr. Clay's work, especially his sick visiting.
B. *sb.* **3. a.** For *rare* read '*rare* exc. in phr. *to give* (a person) *the sick*, to nauseate, to disgust'.
1849 *Sessions Paper* 26 Nov. 5 If I have many such markets as this, it will give me the sick. **1897** [in Dict.]. **1939** 'G. Orwell' *Coming up for Air* IV. v. 257 As for the picturesqueness,..it merely gives me the sick. **1960** *Spectator* 11 Nov. 751 Rackham and all give me the sick.
b. Colloq. phr. *on the sick*, incapacitated by illness, receiving sickness benefit.
1976 *News of World* 14 Mar. 11/2 My Dad used to be on the sick for a long time and couldn't work. **1976** *Par Golf* Aug. 39/3, I didn't realise this would get in the papers. It could cost me my job. I'm on the sick. **1976** L. Thomas *Dangerous Davies* vii. 68, I took it [*sc.* an allotment] on..but then I was on the sick for months and the council..takes it off me.
4. Vomited matter.
1959 I. & P. Opie *Lore & Lang. Schoolch.* ix. 162 Spread it on the butty nice and thick, Swallow it down with a bucket of sick. **1966** *Listener* 3 Nov. 651/3 Middle-aged Chelsea ladies are crawling about in each others' sick. **1977** *Ibid.* 3 Mar. 282/4 There's blood on the windscreen, sick on the trousers.

sick, *v.*[1] Add: **2.** For † *Obs.* read 'Now *rare*' and add later example.
1909 J. Masefield *Tragedy of Nan* iii. 64 You talk rude to the quality... Talk as'd sick a savage.
4. *trans.* and *intr.* To vomit, to spew *up*. Also *fig.*
1924 C. Mackenzie *Old Men of Sea* xix. 333 The volcano started in sicking up red-hot pitch and all. **1930** Kipling *Thy Servant a Dog* 25, I have ate grass and sicked up. **1930** *Dial. Notes* VI. 83 [Child loq.] I sicked all over my yew dress. **1937** L. A. G. Strong *Swift Shadow* 209 But the snow do turn my stomach and I sicked in the hedge. **1948** 'N. Shute' *No Highway* 162 It can't do me any good if I sick it all up. **1954** 'N. Blake' *Whisper in Gloom* vii. 100, I can't go sicking it all up to the police. **1966** C. Sweeney *Scurrying Bush* xiii. 188 On the way the reptile sicked up another hen, and half-way it regurgitated a third hen on the floor of my vehicle. **1975** *Times* 16 Jan. 18/3 A planeload of passengers sicking their breakfast. **1980** *Sunday Tel.* (Colour Suppl.) 21 Dec. 11/3 She sings *Away in a Manger*..and drinks lots of drinks and then she sicks up.

sick, *v.*[2] Add: **1.** (Earlier and later examples.)
1845 J. J. Hooper *Some Adv. S. Suggs* 154 Sick him Pomp,..sick, sick, si-c-k him Bull. **1908** *Westm. Gaz.* 19 Sept. 16/1 'Sick' un then.' Now 'sicking' a hedgehog is a job which few dogs care to tackle. **1933** 'R. Crompton' *William—the Rebel* i. 14 The small white dog, evidently mistaking William's contemptuous 'Huh!' for a new form of 'Sick him!' gave a low growl and sprang forthwith upon the astonished Wotan. **1952** Wodehouse *Barmy in Wonderland* v. 53 'Sic 'em, Tulip,' he said. **1977** *Globe & Mail* (Toronto) 2 Mar. 5/2 All my dogs are attack-trained ..but they won't respond to English commands... It's so little kids can't tell him to sic someone.
2. a. (Earlier example.) Also, to set (a dog or other animal) *on* or *at*.
1845 J. J. Hooper *Some Adv. S. Suggs* 151 If I was to sick them on your old hoss yonder, they'd eat him up afore you could say Jack Robinson. **1899** B. Tarkington *Gentl. Indiana* viii. 131 Seems some of the boys..sicked the dogs on him. **1909** J. Masefield *Tragedy of Nan* ii. 28 Hope they'll catch 'im and 'ang 'im. I'd like to sick the dogs at 'em. **1932** W. Faulkner *Light in August* xiii. 286 They couldn't run him away if they was to sick them bloodhounds on him. **1977** J. Hodgins *Invention of World* iii. 75 He threatened to turn the stones into slobbering wolves and sic them on her.
b. *fig.* To set (a person) to work *on*; to set (a person) to pursue, observe, accompany, etc. (const. *on* or *on to*).
1923 E. B. White *Let.* 2 Jan. (1976) 62 The Times sicks me on feature stuff because the city editor discovered early in the game that city politics appear only in humorous light to me. **1929** R. Lardner *Round Up* xxvi. 327 All I told him was that he'd have to let me pick my own roommate after this and not sic no wild man on to me. **1939** Wodehouse *Uncle Fred in Springtime* i. 18 Why should you barge in here, gnashing your bally teeth, just because Horace sicked Claude Polt, private investigator, on to you? **1958** 'E. Dundy' *Dud Avocado* ii. iv. 221 I'll never forgive you for..sic-ing the sort of person the Contessa is on him. **1958** R. Stout *Champagne for One* (1959) xiv. 172 He had cleared away some underbrush, for instance who had sicked the cops on Laidlaw. **1972** R. Thomas *Porkchoppers* (1974) xxviii. 240 Penry works for me. If you need something done..then I'll sic him on it.

sicken, *v.* Add: **1. d.** *to sicken for*: to be in the early stages of (a disease which is not yet manifest); to be 'coming down' with.
1883 F. Montgomery *Blue Veil* ii. vii. 218, I was sickening for the mumps. **1977** *Sunday Times* 16 Jan. 30/5 Amiss, sickening for the flu which prevented him fielding, got his bat caught in his pads.

sickening, *ppl. a.* Add: **2.** Also in weakened sense.
1922 [see *Fed *pa. pple.*]. **1924** M. Arlen *Green Hat* ii. 67 In ten years' time..Hilary will be the only Liberal left in Parliament, looking happier and younger and more sickening than ever. **1925** D. Mackail *Greenery Street* viii. 179 But, Ian, Daphne's thing is a subscription dance. ..*Please* don't be so sickening. **1937** W. H. Saumarez Smith *Let.* 8 Feb. in *Young Man's Country* (1977) ii. 56 Isn't it too sickening that I shall get to Singapore just after Margaret's and Ronald's departure.

Sickertian (sikɔ̄·ɪtiăn), *a.* [f. *Sickert* (see below) + -ian.] Reminiscent of, or in the manner of, the works of the English painter W. R. Sickert (1860–1942).
1959 *Times* 27 May 7/3 He would seem to be an unwilling draughtsman—although a few delightful drawings in a Sickertian manner are shown. **1964** L. Deighton *Funeral in Berlin* xiv. 85 The Sickertian backwaters of Camden Town. **1976** *Times* 1 May 7/1 Sylvia Gosse..became a competent artist, practising in a Sickertian style.

sickie (si·ki). Also **sicky.** [f. Sick *a.* and *sb.* + -ie.] **1.** *Austral.* and *N.Z. colloq.* A day's sick leave, usu. taken without sufficient medical reason.
1953 T. A. G. Hungerford *Riverslake* 11 Now and then there would be one or more off on a sickie—they changed their jobs so frequently that they never let their sick leave accumulate. **1959** D. Hewett *Bobbin Up* 81 She wished she could take a sickie tomorrer, but it was payday. **1969** *Telegraph* (Brisbane) 28 Oct. 8/2 If we don't feel just right we don't go up. We have a 'sickie' and everyone understands. **1974** *N.Z. Woman's Weekly* 8 July 5/3 Because of the nature of the work it was impossible to plan time off for social functions at short notice. As a result a small percentage of cabin crew staff resorted to 'sickies'. **1981** *Courier-Mail* (Brisbane) 5 Sept. 5/2 A part-time fireman's sense of duty cost him his job after he answered an emergency call when he was taking a 'sickie' from work.
2. *N. Amer. slang.* One who is mentally ill or perverted. Also *attrib.* or as *adj.*
1973 *Ottawa Jrnl.* 1 Aug. 34/2, I hope she gets professional help because these sickies usually get worse, not better. **1974** P. De Vries *Glory of Hummingbird* (1975) x. 135 'Shall I..make it clear..I'm a sickie?' 'No!..this—ailment of yours..it's an expression of some deep-seated conflict.' **1975** *Chronicle-Herald* (Halifax, N.S.) 2 Aug. 26/2 Dade County's entire homicide squad was mobilized..to search for a 'sickie' who murdered two attractive young women. **1977** *Chicago Tribune* 2 Oct. vii. 3/2 A sickie Army lieutenant who tries to run off with the reporter's daughter.

sick-in (si·kˌin). *U.S.* [f. Sick *a.* + *-in[3].] Industrial action in which a group of workers absent themselves from work on the pretext of sickness. Cf. *sickout, sick-out.
1974 *Spartanburg* (S. Carolina) *Herald* 25 Apr. B 8/9 The 'sick-in' forced cancellation of eight of the 11 scheduled U.N. meetings Tuesday. **1976** *N.Y. Times* 7 Sept. 35/4 New Orleans police stage sick-in. **1977** *Monitor* (McAllen, Texas) 12 June 54/1 Millard Holden,..president of the Independent Produce Haulers of America, said he had heard 'sick-in' by independent truckers would be called over the July 4 holiday.

sickish, *a.* **2.** (Earlier example.)
1727 S. J. *Vineyard* 36 Your Grapes..must not be over Ripe, for..the Wine will be Sickish and Ropey.

sickle, *sb.* Add: **4. sickle scaler** *Dentistry*, an instrument with a curved blade for removing scale from teeth.
1930 W. H. O. McGehee *Text-bk. Operative Dentistry* 930/2 (Index), Sickle scalers. **1956** H. M. Goldman et al. *Periodontal Therapy* v. 94 The sickle scaler has a blade with two or four cutting edges. **1962** Blake & Trott *Periodontology* x. 97 Fine sickle scalers are used for subgingival scaling.

sickle (si·k'l), *v.*[2] [f. Sickle *sb.*: cf. Sickled *ppl. a.* b, Sickling[1].] **1.** *trans.* To cut with a sickle. Also *absol.*
1922 J. Masefield *Dream* 13 All golden ripe and ready to be shorn By sickling sunburnt reapers singing staves. **1927** H. E. Fosdick *Pilgrimage to Palestine* i. 4 The harvesters were sickling golden grain on the Shephelah hills. **1971** *Country Life* 2 Dec. 1501/1 The English labourer sickles his corn in August, the French labourer has it in by that time.
2. *Path.* **a.** *intr.* Of red blood cells: to become crescent- or sickle-shaped. Of blood: to exhibit sickling.
1923 *Amer. Jrnl. Dis. Children* XXVI. 133 The blood of the father..was normal on being drawn, but 'sickled' after standing for variable periods of time at room temperature. **1946** *Lancet* 10 Aug. 204/1 Severely anæmic blood always sickles far more readily than blood which is not anæmic. **1970** R. W. McGilvery *Biochem.* iv. 75 Even the cells of heterozygotes will sickle if the oxygen tension is low enough. **1981** *Sci. Amer.* Mar. 117/1 After cyanate treatment and washing, the *AS* cells remained competent as hosts for *P. falciparum*, but now they did not sickle as readily.
b. *trans.* To cause to sickle.
1977 *Lancet* 20 Aug. 411/1 The desickling agent..reacts with red cells which had been deoxygenated and sickled with sodium metabisulphite.

sickle cell. *Path.* [f. Sickle *sb.* + Cell *sb.*[1]] **1.** One of the characteristic crescent-shaped red cells found in the blood of people with sickle cell anæmia.
[**1910** *Arch. Internal Med.* VI. 517 (*heading*) Peculiar elongated and sickle-shaped red blood corpuscles in a case of severe anemia.] **1923** *Johns Hopkins Hosp. Bull.* XXXIV. 42/2 It may..be appropriate to mention here the relationships of blood grouping to another phenomenon occurring in this family, namely, the presence of the so-called 'sickle cells' or crescentic red blood corpuscles. **1946** *Lancet* 10 Aug. 204/1 As long as the diagnosis rested on the presence of sickle cells in stained films, the condition was rarely identified. **1968** *Times* 13 Nov. 16/1 The membranes of sickle cells took longer to remove than those of normal cells, a difference that could be related to the impaired ability of sickle cells to carry oxygen.
2. *attrib.*, as *sickle cell count, family, individual, phenomenon*; **sickle cell anæmia, disease,** a frequently fatal form of anæmia, characterized by the presence of red blood cells that are rich in sickle cell hæmoglobin and sickle readily, and occurring in individuals homozygous for the sickle cell gene; **sickle cell gene,** an autosomal gene found in man, which when heterozygous produces the sickle cell trait and when homozygous sickle cell anæmia, and which is especially common in tropical Africa; **sickle cell hæmoglobin,** an abnormal hæmoglobin which tends to produce a characteristic crescent shape in red cells containing it; **sickle cell trait,** a relatively harmless condition, characterized by the presence of red blood cells containing some sickle cell hæmoglobin and conferring some resistance to malaria; (formerly applied to the characteristic sickling of the red cells which is seen *a fortiori* in sickle cell anæmia).
1922 *Jrnl. Amer. Med. Assoc.* 14 Oct. 1318/2 (*heading*) Sickle cell anemia. **1924** *Ibid.* 5 July 16/1 Several new phases have been brought out since the time, fourteen years ago, when I described what for want of a better term, I [*sc.* J. B. Herrick] called sickle cell anemia. **1969** *Times* 28 Aug. 8/2 The gene for sickle cell anaemia..persists in certain African populations because it confers resistance against malaria. **1977** *Rolling Stone* 21 Apr. 8/1 He missed..the appearance of a frightening new phenomenon in the American black community: the replacement of sickle cell anemia with Potomac Fever as *the* crippling disease among blacks. **1970** P. Oliver *Savannah Syncopators* 41 Those [tribes] on the coast include a large proportion with high counts of the sickle-cell gene making them resistant to malaria and able to withstand heat and high humidity; those in the savannah belt include large numbers of tribes with low sickle-cell counts and lower resistance to malaria. **1949** *Science* 25 Nov. 547/2 (*heading*) On the genetics of sickle cell disease. **1961** R. D. Baker *Essent. Path.* xviii. 492 In sickle cell disease the spleen is often enlarged with intensely red cut surface. **1981** *Westindian World* 28 Aug. 13/2 Sickle cell disease is a blood disorder. It only affects African, Afro-Caribbean and Asian people. **1972** *Science* 13 Oct. 138/3 Physicians who treat sickle cell families emphasize the difference between sickle cell anaemia..and sickle cell trait. **1946** *Lancet* 10 Aug. 204/1 It seems clear that the sickle-cell gene originated in Africa and was carried by slaves to North and South America. **1961** *Times* 21 July 9/5 The population has a genetic resistance to malaria... They possess what is known as a sickle-cell gene, which results in their haemoglobin being abnormal. **1950** *Nature* 21 Oct. 677/1 Dr. F. Eirich, who had first directed our attention to this peculiar disease, gave us a sample of sickle-cell hæmoglobin solution. **1971** *New Scientist* 24 June 762/2 He found that in sickle cell haemoglobin one negatively charged glutamic acid in each of the two normal β chains was replaced by an electrically neutral valine. **1958** *Oxf. Univ. Gaz.* 23 Apr. 892 A study of the physique, growth, and fertility of sickle cell and normal individuals in malarious and non-malarious areas of Nigeria. **1926** *Amer. Jrnl. Dis. Children* XXXII. 334 (*heading*) The sickle cell phenomenon. **1928** *Lancet* 24 Mar. 614/1 Sickle cell or drepanocytic anæmia is found only in negroes with the sickle cell trait—that is, whose red cells become distorted into sickle-shaped cells under certain conditions. **1959** *Listener* 26 Nov. 919/2 Sickle cell trait can be found in parts of Africa, in some Mediterranean countries, and in parts of India. **1978** R. B. Scott *Price's Textbk. Pract. of Med.* (ed. 12) xiv. 1156/1 In sickle-cell trait where there is only 30–40 per cent haemoglobin S, symptoms are rare.

sickled, *ppl. a.* Add: **2.** *Path.* [or f. *sickle *v.*[2]] Of a red blood cell: sickle-shaped.
1923 *Johns Hopkins Hosp. Bull.* XXXIV. 339/1 All the cells in the preparation kept in the dark had reversed within two days, whereas the controls remained sickled for a period of three days to one month. **1949** *Science* 25 Nov. 543/1 Under sufficiently low oxygen pressure..all the cells of both types assume the sickled form. **1971** *New Scientist* 24 June 762/1 The sickled cells tend to clog the circulation and to break up before they reach the life span of normal red cells.

sicklemia (sik'lī·miă). *Path.* [contraction of *sickle (cell anæ)mia.] Sickle cell anæmia or the sickle cell trait. Hence **sickle·mic** *a.*
1932 *Amer. Jrnl. Med. Sci.* CLXXXIII. 386 The term sicklemia has been suggested instead of sickle-cell anemia because a grave anemia is not constantly found in these patients. It is perhaps best reserved for those cases that show sickling without anemia. **1949** *Science* 25 Nov. 543/1 Less than 1 percent of those [erythrocytes] in the venous circulation of sicklemic individuals..are normally sickled. *Ibid.* 545/2 We have assumed that one of the two components of sicklemia hemoglobin is identical with sickle cell anemia hemoglobin and the other is identical with the normal compound. **1961** R. D. Baker *Essent. Path.* xviii. 492 The hereditary trait of sickling of red corpuscles is usually found in apparently normal individuals. However, 2 to 4 per cent of Negroes with sicklemia have sickle cell anemia because of hemolysis of the abnormal red cells. **1973** B. J. Williams *Evolution & Human Origins* iv. 61/1 In thalassemia and, to a less marked extent, sicklemia, there are changes in bone marrow.

sickler. Add: **2.** *Path.* A person with sickle cell anæmia or the sickle cell trait.
1932 *Amer. Jrnl. Med. Sci.* CLXXXIII. 388 Normal gall bladders have been observed in 'sicklers' at autopsy who have had attacks of severe pain and rigidity in the upper right quadrant of the abdomen. **1954** *Brit. Med. Jrnl.* 6 Feb. 291/2 In a group of 102 sicklers from the Balovale district of Northern Rhodesia only 10 (9·8%) had blood slides showing malaria parasites, whereas in a comparable group of 491 non-sicklers 75 (15·3%) had malaria parasites. **1978** *Observer* (Colour Suppl.) 26 Nov. 58/1 In the old days, 'sicklers' rarely reached adulthood.

sicklerite (si·klərəit). *Min.* [f. the name *Sickler* (see quot. 1912) + -ite[1].] A phosphate of lithium, bivalent manganese, and ferric iron, $Li(Mn,Fe)(PO_4)$, found as masses of brown orthorhombic crystals.
1912 W. T. Schaller in *Jrnl. Washington Acad. Sci.* II. 145 Sicklerite. Named after the Sickler family, formerly of Pala [in San Diego county, California]. Found in cleavable masses..on Hiriart Hill near Pala... Results from the alteration of lithiophilite. **1975** *Fortschritte der Mineral.* LII. Special Issue 299 A large suite of secondary phosphates including sicklerite,..heterosite and others were derived from the primary minerals by hydrothermal alterations which leached alkali ions from the structures with concomitant oxidation of bivalent iron and manganese.

sickling[1]. Add: **2.** *Path.* The adoption of a crescent shape by red blood cells.
1923 *Johns Hopkins Hosp. Bull.* XXXIV. 339/1 From these experiments it may be concluded that the sickling is an inherent property of the red blood cells and that the patients' sera are without effect on normal erythrocytes. **1961** R. D. Baker *Essent. Path.* xviii. 491 In the absence of oxygen the hemoglobin S forms crystals which cause the erythrocytes to assume peculiar sickled or oat forms. The sickling can be demonstrated in Negroes with sicklemia by allowing moist preparations of blood to stand, thus using up the oxygen. **1977** *Time* 28 Nov. 56/3 The Rockefeller scientists realized that any treatment for this genetic disease..had to be directed at stopping the characteristic sickling, or distortion, of the red blood cells that occurs after they unload their cargo of oxygen.

sick-list. (Earlier and later examples.)
1748 Smollett *R. Random* I. xxvii. 246 After the captain came on board, our first mate..went to wait on him with a sick list. **1883** 'Mark Twain' *Life on Miss.* 309 Which one numbered the biggest sick-list? **1951** *Sport* 27 Jan.–2 Feb. 4/2 With Flewin, Thompson and Spence all cluttering up the Fratton sick-list, Jack volunteered his services in the pivotal position.

sickly, *a.* Add: **8.** *sickly-scented, -sweet.*
1932 D. Gascoyne *Roman Balcony* 85 The courtesan in the sickly-scented secrecy of her thick-curtained chamber. **1951** S. Spender *World within World* 267 We had to distinguish between those which smelt like pear-drops, carnations and sickly-scented hay. **1912** R. A. Freeman *Singing Bone* 136 The same idea having occurred to me, I applied the handle of the knife to my nose and instantly detected the sickly-sweet odour of musk. **1965** G. McInnes *Road to Gundagai* xiii. 222 The room was..sickly-sweet.

sickness. Add: **6.** *sickness insurance.*
1911 *Q. Rev.* July 209 Sickness-insurance. *a* **1974** R. Crossman *Diaries* (1976) II. 745 This sickness insurance provides that if you are sick for ten days or more you are paid at the end of the tenth day and the three waiting days at the beginning are included.

sicknik (si·knik). *U.S. slang.* Also **sicknic.** [f. Sick *a.* + *-nik.] **a.** One who is mentally ill. **b.** One who indulges in sick humour.
1959 R. Bloch *Blood Runs Cold* (1963) 161 This is a real sicknik. A masochist, like. **1961–2** *Amer. Scholar* XXXI. 108/1 Then all at once (seemingly), in 1959, the sicknics were at hand. **1966** *New Statesman* 19 Aug. 261/1 The silliest definition of the art of Lenny Bruce is that of a 'sick' comedian (or 'sicknik', according to *Time* magazine), since it is so manifestly the sicknesses of society he was getting at. **1968** [see *-nik].

sick-nurse, *sb.* (Earlier example.)
1816 Scott *Antiquary* III. xiii. 284 He has had an infernal tumble..and..I have sent your friend, Sweep-clean..to act as his sick-nurse.

sicko (si·ko). *U.S. slang.* [f. Sick *a.* + *-o[2].] = *sickie 2.
1977 J. Wambaugh *Black Marble* (1978) xi. 245 But, Clarence, listen! She's a sicko. Some kinda fruitcake or somethin. **1982** *Chicago Sun-Times* 25 Nov. 7/2 Is it asking too much for these sickos to stop bothering decent women?

sickout, sick-out (si·kˌaut). orig. and chiefly *U.S.* [f. Sick *a.* + Out *adv.*, on the analogy of 'walk-out'.] = *sick-in. Also *attrib.*
1970 *Wall St. Jrnl.* 13 Apr. 1/3 The air controllers' 'sickout' is showing 'continuing improvement', the Government reported yesterday. On Saturday a Federal district judge ordered leaders of the Professional Air Traffic Controllers Organization to tell their members to go back to work or produce medical certificates of illness. **1974** *Trinidad Guardian* (Port-of-Spain) 16 Oct. 4/1 (Advt.), After all these strikes, sickouts, go-slows, and not a penny increase yet, now these merciless taxis up by 25% plus. **1977** *N.Y. Post* 18 June 1 The Yonkers police sickout crisis. **1978** *Telegraph* (Brisbane) 30 Oct. 12/1 Flights worldwide were delayed by a 'sick-out' of [Pan-American] flight attendants who had voted against a strike at the weekend.

sicky, var. *sickie.

‖ **sic transit** (sĭk træ·nzit). Also erron. **transitur.** [L.] A catch-phrase expressing the impermanence of things, in full *sic transit gloria mundi*, 'thus passes the glory of the world'.
The phrase is possibly an adaptation of a passage of Thomas à Kempis (see quot. *a* 1471). For the use of these words in papal coronations see *King's Classical & Foreign Quotations* (ed. 3, 1904) 319.
[*a* **1471** T. à Kempis *Imitatio Christi* (1827) i. iii. 7 O quam cito transit gloria mundi.] **1601** Jonson *Every Man in his Humour* v. i. sig. M^v, See, see, how our Poets glory shines brighter and brighter, still, still it increaseth, oh now its at the highest, and now it declines as fast: you may see gallants, *sic transit gloria mundi*. **1777** H. Walpole *Let.* 5 Dec. (1904) X. 163 General Howe must probably return to defend New York. *Sic transit gloria mundi!* **1833** Lord Lytton *Godolph.* III. xxiii. 250 His breathing..died away as insensibly as an infant. *Sic transit gloria mundi!* **1851** Geo. Eliot *Let.* 4 Oct. (1954) I. 364 Sic transitur—i.e. the money from my pocket. **1915** D. H. Lawrence *Crown* in *Reflections on Death of Porcupine* (1925) 92 Despair comes over us when it [*sc.* the body] passes away. '*Sic transit*', we say, in agony. **1951** 'J. Wyndham' *Day of Triffids* v. 104 'Never—never again now will you see a sight like that,' I told myself. '*Sic transit*—' **1965** A. Nicol *Truly Married Woman* 16 They looked at it [*sc.* a Roman ruin] silently... 'Sic transit,' he'd said as they drove away. **1971** S. Jepson *Let. to Dead Girl* v. 50 What was once my dressing room is dismantled..(*Sic transit gloria mundi*, and so on).

sicula (si·kiulă). *Palæont.* [a. L. *sīcula*, dim. of *sīca* curved dagger.] The small conical or dagger-shaped structure secreted by and containing the initial member of a colony of graptolites.
1893 *Lunds Univ. Årsskrift* XXIX. xii. 3 The sicula is seen to project from the right side obliquely upwards towards the median groove, thus separating the rounded bases of the two moieties of the rhabdosoma. **1910** [see *rhabdosome* s.v. *Rhabdo-]. **1970** R. M. Black *Elements Palaeont.* xiv. 211 The sicula was secreted by the first members (a zooid) of the colony and the thecae were formed by the subsequent zooids by a process of budding.

Siculan (si·kiulăn), *a.* and *sb.* [f. L. *Sicul-us*, ad. Gr. Σικελ-ός (see *Sicel *sb.* and *a.*) + -an.] **A.** *adj.* **a.** *Archæol.* Denoting the Chalcolithic, Bronze, and Iron Ages in Sicily. **b.** *poet.* Sicilian. **B.** *sb.* *Archæol.* A member of a people of Siculan culture. Also **Siculian, Siculic.**
1896 *Jrnl. Hellenic Studies* XVI. 134 Orsi's investigations during the last seven years have included nearly every period, from the pure stone age with its pre-Siculic indigenous population, to the time of the Christian catacombs... Orsi has been able to follow down to the fifth century B.C. traces of this strange Siculian civilization. **1901** G. Seigi *Mediterranean Race* xiv. 283 The presence..of objects of Mycenean character in the first Siculic period. **1909** [see *sicanian *a.*]. **1928** *Antiquity* II. 145 The veil has been half lifted from the romantic history of the Siculans who lived near Locri Epizephyrii. **1948** T. J. Dunbabin *Western Greeks* i. 40 The Siculan culture of prehellenic Lokroi and kindred sites in Bruttium is not older than the tenth century. **1957** *Encycl. Brit.* XX. 603/2 A few rock-hewn tombs of the first Siculan period have been discovered near Palermo. **1975** G. Ewart *Be my Guest!* ii. 45 Not Siculan feastings sweet will bring forth smell.

Siculo-. Add: *Siculo-American, -Arabic.*
1939 *Burlington Mag.* Oct. 171/1 Twelfth-century Siculo-Arabic combs and croziers. **1966** *Economist* 2 July 51/1 It [*sc.* the Mafia] revived with the help of the Siculo-American gangsters employed to ease the Allied landings in Sicily. **1974** K. Clark *Another Part of Wood* iii. 115 Leigh Ashton..had a good eye for art of all kinds, but what he really loved was a small fragment of Sassanian silk, a Persian pot or a Siculo-Arabic ivory.

Sicyonian (sikiōᵘ·niăn), *sb.* and *a.* Also 7 **Sicionian; Sikyonian.** [f. L. *Sicyōnius*, f. *Sicyōn*, a. Gr. Σικυών: see -ian.] **A.** *sb.* A native or inhabitant of Sicyon, an ancient Greek city in the northern Peloponnese. **B.** *adj.* Of or pertaining to Sicyon.
1642 J. Howell *Forraine Travell* xvi. 206 Among the Sicionians there were admitted..[no] Physitians. **1841** *Penny Cycl.* XXI. 126/2 They [*sc.* Dipœnus and Scyllis] were employed by the Sicyonians to make for them certain statues of their gods. **1887** *Encycl. Brit.* XXII. 32/2 Clisthenes was the most powerful and famous of the Sicyonian despots. **1958** R. Liddell *Morea* ii. i. 41 There is also a small sanctuary of the nymphs: here Sicyonian brides made offerings. **1976** R. J. Hopper *Early Greeks* viii. 214 His victorious chariot was later preserved at Delphi under a form of *baldacchino*, the 'Sikyonian *monopteros*' decorated with archaic sculpture.

sidalcea (sidæ·lsiă). [mod.L. (A. Gray in G. Bentham *Plantas Hartwegianas* (1848) 300), f. Sida + *Alcea*, the names of related genera.] An annual or perennial herb of the genus so called, belonging to the family Malvaceæ, native to western North America, and bearing racemes of white, pink, or purple flowers.
1882 *Gardener's Chron.* XVIII. 439 The remark applies with full force to the Sidalcea. **1922** A. J. Macself *Hardy Perennials* ii. 190 The Sidalceas grow erect with spikes of bloom resembling small Hollyhocks. **1962** R. Page *Educ. Gardener* vii. 213 One of the sidalceas..has spikes of soft rose-pink flowers and rather dull green foliage. **1976** *Country Life* 11 Mar. 619/1 Gardeners are..dividing treasured perennials..sidalceas, ligularias and chelone among them.

Sidamo (sidā·mo). Also **Sidama, Sydama.** [Native name.] A group of Cushitic-speaking peoples in southwestern Ethiopia; a member of one of these peoples. Also, the language of these peoples. Also *attrib.* or as *adj.*
1834 S. Gobat *Jrnl. Three Years' Residence in Abyssinia* ii. 166 A month's journey to the w.s.w. of Shoa he had found a little Christian empire, who have a particular language, called Sidama, and books. *Ibid.* iv. 305 Kidam Mariam had with him a slave..of Sidama... He said that the Sidamas are Christians. **1868** J. C. Hotten *Abyssinia & its People* ii. v. 200 Kaffa is a wealthy and fertile kingdom; the inhabitants are proud and handsome... They still retain a recollection and traces of the Christian faith... The Gallas, with whom they are incessantly at war, call them, in derision, 'Sydama', originally signifying Christian. **1910** *Encycl. Brit.* XII. 894/1 Similar dialects are those of the Sid(d)āma tribes, south of Abyssinia. **1939** L. H. Gray *Foundations of Language* 366 Mixed with Ethiopic-speakers are the Agaw dialects..; and in the south-west are the Sidama (Gudella, Kaffa, Kullo-Walamo, Bambala, etc.). **1972** *Bk. of Thousand Tongues* (ed. 2) 392/2 Sidamo is spoken by an estimated 100,000 people in highland areas north and east of Lake Abaya, southwestern Ethiopia. **1974** *Encycl. Brit. Macropædia* VI. 1013/1 Those Galla and Sidamo who retain their traditional cosmologies..are more firmly anchored in the present.

Sidcot (si·dkǫt). Also **sidcot,** (*erron.*) **Sidcott.** [f. the name of the Australian-born aviator *Sidney Cotton* (1894–1969), who designed it.] In full *Sidcot* (*flying*) *suit*. A warm one-piece suit worn by aviators or the like (see quot. 1969). Hence **Si·dcotted** *a.*, dressed in a Sidcot.
1921 *Flight* XIII. 635/2 When Mr. Courtney was testing the 'D.H.6' fuselage fitted with the experimental 'Alula' wing at Brough..the weather was rather cold, and he was wearing his 'Sidcot' suit. **1927** T. E. Lawrence *Let.* 30 Dec. (1938) 563, I often think of you, and always as a rather shapeless Sidcotted bundle, peering over the rim..of a Virgin in mid air. **1928** *Motor-Cycling* 11 Jan. 240/1 An objection frequently raised against it, is that the Sidcot leaves the ankles entirely unprotected. **1940** *War Illustr.* 26 Jan. 19 Here are the two pilots of the Meteorological Flight, wearing warm Sidcot flying suits and oxygen masks, before taking to the air. **1942** R. Hillary *Last Enemy* ii 45, I pulled on my sidcot and gloves and slipped my feet into the comforting warmth of my fur-lined boots. **1954** W. Faulkner *Fable* 113 He took the pistol from the Sidcott's knee-pocket. **1964** G. Lyall *Most Dangerous Game* xxii. 186 Climbing into the Sidcot suit and slinging a Sten gun round my neck. **1969** Barker & Cotton *Aviator Extraordinary* iii. 33 The suit had a warm lining of thin fur, then a layer of airproof silk, then an outside layer of light Burberry material, the whole being made in one piece just like a set of overalls... For a name I took the first three letters of my Christian and surnames—'Sidcot'.

‖ **siddha** (si·dā). Also **Siddha.** [Skr., f. *sidh-* to be fulfilled.] In Indian religions, one who has attained perfection, a saint, a semi-divine being; *spec.* in Jainism, a perfected, bodyless being, freed from the cycle of rebirths.
1846 [see *Gandharva]. **1883** M. M. Williams *Relig. Thought & Life in India* vii. 191 All who are uninitiated into this system [*sc.* Śāktism] are styled 'beasts' (pasu), the initiated being called Siddha, 'the perfect ones'. **1971** *Illustr. Weekly India* 11 Apr. 11/3 At the end of a period of thirty years, the venerable ascetic Mahavir became a *siddha*, freed from the cycle of birth.

‖ **siddhi** (si·di). Also **Siddhi.** [Skr., lit. 'fulfilment', f. as prec.] In Indian and Tibetan religion, a collective name for supernatural or magical powers acquired by meditation or other practices. Also in *pl.* in same sense.
[**1863** E. Schlagintweit *Buddhism in Tibet* vi. 56 The reciting of Dhāranis, if combined with the practise of magical rites and supported by morality and contemplation, leads to superhuman faculties (in Sanskrit Siddhi).] **1882** [see *Mandala]. **1921** C. Eliot *Hinduism & Buddhism* II. xxxii. 282 The religious life prescribed in the Tantras commences with the initiation and requires the supervision of the Guru. The object of it is *Siddhi* or success, the highest form of which is spiritual perfection. **1941** A. Huxley *Grey Eminence* ix. 218 He was deeply impressed by any manifestation of the *siddhis*, as the Indians call them, the psychic powers which may be aroused by meditation and to which the wiser mystics pay as little attention as possible. **1960** [see *Mudra]. **1970** *Man, Myth & Magic* v. 146/3 The idea of an astral body is very old. Ancient Indian writings describe the eight *siddhis* or supernormal powers which can be acquired through a type of yoga called *Pranayama*. **1977** *Time* 8 Aug. 44/2 The teachers have now brought those wares to the American market: lessons that will lead trainees to the Siddhis, or supernatural powers.

Siddonian (sidōᵘ·niăn), *a.* [f. *Siddon(s* (see below) + -ian.] In the style of, or typical of, the acting of the English tragic actress Mrs. S. Siddons (1755–1831).
1795 T. Wilkinson *Wandering Patentee* III. 120, I have not observed the audience ever thought of heat in the Dog-days, when the Siddonian queen was followed. **1894** W. Archer *Theatrical World of 1893* 192 She is an imposing artist of the Siddonian school. **1931** C. St. John *Ellen Terry & Bernard Shaw* 64 The concealment of her ears, and her Siddonian deportment, made an indelible impression on him [*sc.* Shaw]. **1946** G. B. Shaw in *Drama* Winter 8 [She] burst in on me and demanded a Siddonian part.

siddown (sidau·n), repr. a colloq. pronunc. of the imperative *sit down!* (see Sit *v.* 21).
1936 J. Curtis *Gilt Kid* iv. 44 'Siddown,' he retorted waving an imperious hand. **1953** K. Tennant *Joyful Condemned* iii. 22 'Siddown,' he advised Jake Fletcher, who took the straight-backed chair. **1967** G. Greene *May we borrow your Husband?* 110 'Siddown,' he said, 'make yourself comfortable.' **1975** D. Beaty *Electric Train* 165 What do you know about it, sonny?..Siddown! Siddown!

Siddur (si·duᵊɹ). Also **Sidoor, Sidur,** and with small initial. Pl. **Siddurim.** [Heb. *siddūr*, lit. 'order'.] A Jewish prayer book containing prayers and other information relevant to the daily liturgy. Cf. *Machzor.
1864 *Chambers's Encycl.* VI. 155/1 These early collections of prayers generally contained also compositions from the hand of the compiler, and minor additions, such as ethical tracts, almanacs, &c., and were called *Siddurim* (Orders, Rituals), embracing the whole calendar year, week-days and new moons, fasts and festivals. **1905** *Jewish Encycl.* X. 171/1 The collection, in one book, of the year's prayers for week-days, Sabbaths, holy days and fast-days is generally known as the 'Seder Tefillot', or simply the 'Siddur'. **1911** Z. Hodes (*title*) Studies in Sidoor. **1925** W. O. E. Oesterley *Jewish Background of Christian Liturgy* i. 35 The next [extant collection of prayers] in date is the Siddur ('Order' of prayers..) of Sa'adya (A.D. 892–942). **1949** S. Gaon *Devel. Jewish Prayer* 6 Collections of prayers..called..*sidur*..from the noun 'seder', which..originally..meant 'collection'. **1976** H. Kemelman *Wednesday Rabbi got Wet* li. 297 He had forgotten to take the prayerbook with him... She picked up the *siddur* and followed him.

side, *sb.*[1] Add: **I. 2. b.** *side by side*: also (hyphened) as *attrib. phr.*
1908 *Daily Chron.* 29 Nov. 9/5 The side-car..has the advantages of ready convertibility, low cost, and high speed..together with the far greater sociability afforded by the side-by-side accommodation. **1930** *Times Educ. Suppl.* 27 Dec. p. i/3 Side-by-side valves instead of the overhead valves. **1956** R. Redfield *Peasant Society & Culture* i. 20 Those early comparisons were side-by-side comparisons of societies unaffected by cities and civilization. **1970** *Gloss. Aeronaut. & Astronaut. Terms* (*B.S.I.*) viii. 9 *Side-by-side assembly*, an assembly of connecting rods in which a number of similar plain connecting rods are arranged successively side-by-side with narrow big-ends usually carrying roller bearings.
c. Designating a double-barrelled shotgun with barrels set side by side (cf. *over-and-under *a.*). Also *absol.* and *ellipt.* as *sb.*
1950 R. Shaughnessy *Skeet & Trapshooting* iii. 30 Double-barreled guns are manufactured in two styles, one known as the *over-and-under*, the other as the *side-by-side*. **1961** Side-by-side *sb.* in Webster. **1964** H. L. Peterson *Encycl. Firearms* 139/1 Henry Nock's patent breech of 1787... It was possible to shorten barrels..for a sporting gun... Shortening the barrels made the guns still lighter, and this in turn made the side-by-side double-barreled fowling piece practical. **1979** G. Hammond *Dead Game* ii. 32, I saw all the guns... Two..were over-and-unders, the rest side-by-sides. **1980** *Outdoor Life* (U.S.) (Northeast ed.) Oct. 53/1, I had this double-barreled side-by-side 20-gauge Savage-Fox, the ugliest thing ever made.
II. 8. c. In prov. phr. *the other side of the coin* (*penny*, etc.) = *the reverse of the medal* s.v. Medal *sb.* 3 b; *the other side of the shield*: see *shield *sb.* 1 d.
1904 Yeats *Let.* ? 20 Jan. (1954) iii. 425 *The Shadowy Waters*..is more of a ritual than a human story... *Cuchulain* or *The King's Threshold* are the other side of the halfpenny. *Ibid.* Apr. 433, I am reckless in mere speech that is not written. You are the other side of the penny, for you are admirably careful in speech. **1966** *Listener* 19 May 713/1 The social and psychological pressures are not different things but often just different sides of the same penny. **1975** M. Russell *Murder by Mile* ix. 92 Angus Hamilton's..to address members and answer questions. He thought it might present an opportunity to put across the other side of the coin.
d. Each of the two grooved faces of a gramophone record. Also *slang*, a recording made on this; a record. In extended use, of tape recording. Cf. *flip side* s.v. *flip *sb.*[2] 7.
1936 *Rhythm* Apr. 28/1 American Brunswick, Columbia and Vocalion have a blanket contract with Irving Mills for so many sides a year. **1948** *N.Y. Age* 18 Dec. 2/6 We expect 'Skiffle Blues' to be one of our big sides in the coming weeks. **1950** *Down Beat* 14 July 11 (*heading*) Will the Louis sides on cylinder ever turn up? **1960** J. Baldwin *Another Country* (1962) ii. iii. 310 'How about some sides?'.. Lorenzo put on something..by the Modern Jazz Quartet. **1971** D. E. Westlake *I gave at Office* 133 There was some tape left. Tape three, side two, the one just before this. **1979** *Guardian* 9 June 12/7, I had to wait until side two for any bloom of Schubertian joy.
9. c. *Theatr.* A page of typescript containing an actor's part and cue words (usu. *pl.*).
1933 P. Godfrey *Back-Stage* iii. 37 An experienced actor, being offered a part, is unimpressed by the number of 'sides' it contains, a 'side' being a half-quarto sheet of typescript. **1963** 'E. McBain' *Ten plus One* (1964) vii. 73 'She had memorized all of her sides—' Richardson paused here to see whether or not anyone had caught his use of the professional term 'sides'..—'in the first two nights of rehearsal.' **1976** R. James *House is Dark* xiii. 135 'Don't see why actors ever gave up sides... They're so much easier.'.. 'Haidee, it wasn't for the sake of the actors that they used sides. It was supposed to prevent actors from duplicating a hit script and going off with it on their own.'
10. Also *to look on* (or *to*) *the bright* (or *worst*, etc.) *side*: see *look *v.* 18 d.
III. 12. a. *fig. the other side*: see *other *a.* 2 d (*a*).
c. *fig. on the* (adjective) *side*, tending towards the condition or aspect described. Cf. *on the safe side* s.v. Safe *a.* 9 c in Dict. and Suppl.
1713 [see Right *a.* 10 c]. *c* **1805** G. Colman in M. R.

Booth *Eng. Plays of 19th Cent.* (1973) III. 69 It's prophesying on the sure side, to foretell a thing when it has happened. **1864** TROLLOPE *Can You forgive Her?* I. xi. 90 He is just a shade too good... But it's a fault on the right side. **1923** A. J. ANDERSON *Soul Sifters* xxiv. 252 'Michelmore was always on the rough side!' he remarked aloud. **1952** A. J. CRONIN *Adventures in Two Worlds* xii. 97 She was on the thin side..and her liquid, brownish eyes were too large. **1974** A. MORICE *Killing with Kindness* ii. 14 He was a bit on the tired side, but..he's accustomed to long hours.

d. *on this* (or *the other*) *side*: with reference to the Atlantic Ocean. Cf. *this* (etc.) *side of the puddle* s.v. *PUDDLE *sb.* 1 c. *colloq.* (chiefly *U.S.*).

1884 *Naturalist's World* Sept. 155/2 Canadian Postal Science College..is a society which has grown up very rapidly 'on the other side'. **1928** WODEHOUSE *Money for Nothing* vii. 129 There's dozens of people on the other side who'll buy it.

13. e. In *fig.* phr. *the other side of the hill*, those aspects of a situation which are unknown at present. Also *transf.*, the latter part of life, and in *Mil.* contexts, the enemy position or activities.

1852 DUKE OF WELLINGTON in *Croker Papers* (1884) III. xxviii. 275 We amused ourselves by guessing what sort of a country we should find at the other side of the hills we drove up... When I expressed surprise at some extraordinary good guesses he [sc. Croker] had made, he said, 'Why, I have spent all my life in trying *to guess what was at the other side of the hill*.' **1926** C. B. WATERLOW in H. Golding *Wonder Bk. of Motors* 12 It is not only what is on the other side of the hill that matters, but everything along the road. **1948** B. H. L. HART (*title*) The other side of the hill. **1957** C. SMITH *Case of Torches* i. 5, I had to go through a lot of badinage..about..how old I was getting and what it was like on the other side of the hill. **1960** G. MARTELLI *Agent Extraordinary* 15, I..wish..to express my gratitude..to [the]..technical director of the flying bomb sites..for allowing me a glimpse of the 'other side of the hill'. **1978** *Times* 30 Jan. 13/2 Mr Peyton..began to argue for a revalued green pound... He..correctly read what was on the other side of the hill (that is, the Government itself would soon have to revalue).

14. e. *on the side* (orig. *U.S.*). (*a*) Served separately from the main dish.

1884 *Bad Lands Cow Boy* (Little Missouri, Dakota Terr.) 7 Feb. 1/5 'Gimme that snake rare—milk gravy on the side,' was hallooed to the cook. **1916** *Literary Digest* 18 Mar. 766/3 'Beef stew and a cup of tea for me,' the new arrival said. 'Bossy in a bowl—boiled leaves on the side,' sang the waiter. **1975** D. LODGE *Changing Places* ii. 95 A club sandwich with french fries on the side.

(*b*) In addition; surreptitiously, without acknowledgement. (Freq. with connotation of dishonesty: illicitly; outside wedlock.)

1893 *Congress. Rec.* 18 Dec. 360/1 He will have no pension attorney, for a silent partner, no relative doing business 'on the side' with that bureau. **1904** *N.Y. Times* 22 June 3 To attend the big fair and receive the entertainment of St. Louis on the side. **1927** *Daily News* 11 Mar. 2/2 Y' see, Bill's in the rag-and-bone trade and he does a bit [of receiving] on the side. Just anythink he can pick up. **1937** D. L. SAYERS *Zeal of thy House* ii. 44 Pocketing commissions and that sort of thing? Doing little deals on the side? **1953** R. LEHMANN *Echoing Grove* iv. 226 An independent career-woman with a successful love life on the side. **1968** R. L. HUDSON *Grace is not Blue-eyed Blond* xi. 145 What would some of you say if I told you that I, as a married man, have had three women on the side? **1977** *Gay News* 24 Mar. 14/4 They may..gear their expectations to include sexual contacts on the side.

(*c*) *spec.* in addition to one's regular or ordinary occupation; as a subsidiary source of income (also with occas. implication of irregularity).

1898 *N.Y. Jrnl.* 26 Aug. 9/3 Samuel..started an ice cream parlor, with cigars, tobacco and delicatessen on the side. **1915** WODEHOUSE *Something Fresh* iv. 107 'I'm not asking you to be a valet and nothing else.' 'You would want me to do some cooking and plain sewing on the side, perhaps?' **1928** S. LEWIS *Man who knew Coolidge* i. 13 I'd never made a peep about how maybe it'd be a good stunt for him to go out and maybe earn a little money on the side. **1945** *Reader's Digest* July 22/1 There is a good job teaching music theory..lined up for him. And he is composing on the side. **1960** N. MARSH *False Scent* i. 29 I'm trying, on the side, to break out in a rash of serious writing. **1977** *Navy News* Dec. 1/3 We do not have information about how many people do jobs on the side, but I suspect that that practice is not confined to the Armed Forces.

15. b. Also succeeding or suffixed to the names of places or regions to form adj. or advb. phrases in the sense 'in (or on, towards) the area specified', esp. as *STATESIDE *a.* and *adv.*

1726 [see SOUTH-SIDE b]. **1924** E. M. FORSTER *Passage to India* xxxvii. 323 Jolly good poems, I'm getting published Bombay side. **1966** K. GILES *Provenance of Death* ii. 58 An Italian industrialist who does a lot of business Moscow-side.

e. *side-of-the-mouth* adj. phr., spoken aside or (as) from the side of the mouth; delivered in a rough drawling manner; pungently demotic. Also, of the style of such utterances.

1958 *Listener* 7 Aug. 203/1 Hoarse, side-of-the-mouth cracks of quite shattering pessimism. **1960** G. COULTER in M. T. Williams *Art of Jazz* 170 A racy, side-of-the-mouth idiom. **1974** *Publishers Weekly* 25 Mar. 52/3 Describes in blunt, side-of-the-mouth prose how he was given the 'contract' [to kill someone].

f. Also *to laugh on the other, wrong side* (*of one's face, mouth*): see LAUGH *v.* 1 b in Dict. and Suppl.

16. b. *side-to-side* adj. phr., characterized by movement from one side to the other.

1934 in WEBSTER. **1950** J. DEMPSEY *Championship Fighting* xxii. 157 Motions that made my head an elusive side-to-side target. **1955** W. W. DENLINGER *Compl. Boston* i. 132 This formation [sc. bandy legs] results in a side-to-side gait. **1962** 'K. ORVIS' *Damned & Destroyed* ix. 59 Her head began a loose, disjointed, side-to-side swaying.

IV. 18. a. Also in prov. phr. *on the side of the angels*: in favour of a spiritual interpretation (of human nature); more loosely, on the side of right despite the risk of unpopularity.

1864 DISRAELI *Church Policy* 26 Is man an ape or an angel?.. I am on the side of the angels. **1926** *Punch* 22 Dec. 700/1 Miss Marguerite Williams..is so firmly posted on the side of the angels that I can forgive her if she occasionally seems rather to force the note. **1941** A. L. ROWSE *Tudor Cornwall* ii. 52 Mr. Tawney tells us.. that 'their silence was the taciturnity of men, not the speechlessness of dumb beasts'. Though that may a little be questioned, no doubt he is on the side of the angels. **1956** G. H. VALLINS *Pattern of Eng.* vii. 171 'Different from' reminds the reader that whatever other men have done.., Fowler himself is on the side of the angels. **1979** 'C. AIRD' *Some die Eloquent* vii. 99 He had always in any case been on the side of the angels anyway. Apes were less appealing.

20. b. (Later examples.)

1898 J. A. GIBBS *Cotswold Village* xi. 230 The rest of our team included the jovial miller;..the village curate, who captained the side..; one or two farmers; [etc.]. **1947** N. CARDUS *Autobiogr.* ii. ii. 194 A boy fixes figures painted on square bits of tin—just the total of the batting side, the fall of the wickets, and the score of the last man out. **1977** C. MARTIN-JENKINS *Jubilee Tests* ii. iv. 87 England's only difference from their Centenary Test side was the replacement of Fletcher by Barlow.

c. (so many)-*a-side*, indicating the number of players that may compose a team on the field of play. Usu. *attrib.* of a sport or match, as *five-a-side football*, etc. Cf. *SEVEN *a.* 2 g.

1900 [see *SEVEN *a.* 2 g]. **1926** *Times* 12 Apr. 6/5 The following are the results of yesterday's matches in the Seven-a-side Tournament. **1932** *Times Lit. Suppl.* 30 June 484/2 It is now over a century since it [sc. shinty] was played, seventy-five or so a-side. **1951** *Sport* 27 Apr.–3 May 4/3 A five-a-side football match is being played between Glasgow's Celtic and Rangers. **1973** J. M. WHITE *Garden Game* 104 We do sanction two-a-side encounters from time to time, or even three-a-side. **1978** *Cornish Guardian* 27 Apr. 13 At a Cub Scout's six-a-side football competition..Mrs. M. Dean..presented the winners' shield.

d. *to let the side down*: see *LET *v.*[1] 29 b.

V. 23. a. *side-entrance, outlet, -path* (earlier and *fig.* examples), *turning, ward, -window* (examples of motor vehicles), *-yard.*

1907 *Daily Mail Year Bk.* 74/2 In turn the prevailing form of body has been the..side-entrance phaeton, and the landaulet and limousine. **1926** W. W. JACOBS *Sea Whispers* v. 113 To leave by the side-entrance was the best way of avoiding trouble. **1976** *Northumberland Gaz.* 26 Nov. 18/1 (Advt.), Hall, sitting room,..side entrance, porch. **1967** *Gloss. Sanitation Terms* (B.S.I.) 10 *Side outlet tee*, a tee which incorporates an additional branch at 90° both to the main pipe and to the leg of the tee. **1972** L. M. HARRIS *Introd. Deepwater Floating Drilling Operations* x. 98 The subsea blowout-preventer stack..should have two side outlets for the choke and kill-line connections. **1854** DICKENS *Hard Times* ii. xii. 250 Indifferent to the rain,..she struck into a side-path parallel with the ride. **1924** R. S. HICHENS *After Verdict* ii. vii. 181 She's always in the wholesome centre... No false steps into side-paths for her. **1946** *Law Rep.* 5 Oct. 334 She was executing a manoeuvre of turning from the near side into a side turning on her off-side. **1965** M. SPARK *Mandelbaum Gate* iii. 54 He dodged down a side-turning into the shop of an Arab dealer. **1968** M. ALLINGHAM *Cargo of Eagles* iii. 42 She shared a side ward..with two other old ladies. **1959** M. SUMMERTON *Small Wilderness* x. 131, I rolled down the side window. Instantly the car was hazed with incoming fog. **1976** *Derbyshire Times* (Peak ed.) 3 Sept. 6/2 The youth admitted kicking the side window of the car twice. **1879** W. WHITMAN *Daybooks & Notebooks* (1978) I. 139 The window where I sit ..opens on a spacious side-yard. **1979** *Kingston* (Ontario) *Whig-Standard* 29 Mar. 21/2 Narrow side yards could block fire trucks.

b. *side-curtain* (examples of use on early open motor vehicles), *parting, -whiskers* (earlier example).

1912 *Motor Manual* (ed. 14) iv. 161 The only car for such weather conditions is a covered one, either one with a Cape cart hood with side curtains well down,..or the more complete enclosure of limousine or landaulet. **1980** L. LEWIS *Private Life of Country House* iii. 35 We bought a secondhand T model Ford... It..had a hood and talc side-curtains. **1942** Side parting [see *PIPE *sb.*[1] 1 c]. **1982** D. PHILLIPS *Coconut Kiss* v. 43 Grace asked if she could do her hair with a side parting and a slide. **1867** *Amer. Naturalist* Aug. 287 Their ears are often tufted, and one species, at least, has 'side-whiskers' formed by the true fur, in addition to the labial bristles which ordinarily receive this name.

24. a. *side-beam, -eye, -glimpse.*

1935 *Discovery* Nov. 341 (*caption*) The advantages of a side-beam:..the beam can be followed as it strikes the objects bordering a country road. **1978** R. V. JONES *Most Secret War* xx. 169 If another aircraft flew down one of the side beams, the result should be a second line intersecting with the first at the exact location of the station. **1922** JOYCE *Ulysses* 48 A side-eye at my Hamlet hat. **1958** J. KEROUAC *On Road* 189 Looking at me with the same wary insolent side-eye. **1890** 'MARK TWAIN' *Let.* 11 Feb. in J. Brown *Lett.* (1912) 452 The charm of the painter is so strong that one can't keep his entire attention on the developing portrait, but must steal side-glimpses of the artist.

c. *side-remark, -talk.*

a **1910** 'MARK TWAIN' *What is Man?* in *Harper's Mag.* Oct. 673/1 He was treated to many side remarks by his fellows. **1968** *Economist* 3 Feb. 13/1 Formal statistical tables.., so that everybody can tell what each Wilsonian side-remark is meant to mean. **1917** G. S. GORDON *Let.* 14 Feb. (1943) 70 This is all side-talk compared with the great thing that has happened to you. **1931** B. BROWN *Talking Pictures* x. 250 The need for silence in the studio is increased, since side talk, coughs, etc., are liable to be picked up.

26. *side-lock.*

1907 *Yesterday's Shopping* (1969) 629/2 Plain quality side-lock hammerless guns. **1955** R. CHURCHILL *Game Shooting* iv. ii. 193 Apart from the action of fully-automatic guns, all game guns are built either with what is called the sidelock action or the box lock.

27. side action *Pharm.* = *SIDE EFFECT 2; **side band** *Telecommunication*, a band of frequencies above or below a carrier frequency, within which lie the frequencies produced by modulation of the carrier; **side boy** (earlier and later examples); **side-burn** [alteration of *BURNSIDE, after *side-hair*, etc.], for *U.S.* read orig. *U.S.* and add: usu. *pl.* (later examples); hence **side-burned** *a.*; **side chair**, an upright wooden chair without arms; **side circuit** *Teleph.* (see quot. 1916); **side cut** *Oil Industry* = *side stream* (*b*) below; **side drift** *Mining*, etc., a horizontal tunnel leading off the main passage (cf. DRIFT *sb.* 15); **side-drum**, also, in *Jazz*, etc.: a drum (usu. part of a set) placed on a stand beside the performer; **side-entry**, (*a*) a side-entrance; an area outside the side-door of a house; (*b*) *Bridge*, a card providing access to a hand in a suit other than trumps (cf. *ENTRY 1 f); **side frequency** *Telecommunication*, a particular frequency in a side band, equal, in the case of amplitude modulation, to the carrier frequency plus or minus a particular modulating frequency; **side gallery**, either of the two galleries along the side of the debating chamber of the House of Commons, divided to seat Members and others; **side-head(ing)** *Journalism* (see quot. 1889); **side-hold** *Mountaineering*, a hold in which the rock is gripped from the side; **side lamp**, a lamp placed at the side (see also quot. 1885); *spec.* of a motor vehicle = *SIDE-LIGHT 3 c; **side-lay** (later example in *Printing*); **side lever** *Mech.*, each of two beams located on the sides of some forms of steam engine, which transmit motion from the cross-head of the pistons to the connecting rods; **side-loader**, a fork-lift truck in which the fork is located at the side of the vehicle; **side lobe**, any lobe in the response or radiation pattern of a radio aerial other than the central, or main, lobe; **side-lock**, † (*a*) *pl.* that part of a wig that covers the ears and neck; (*b*) a lock of hair worn at the side of the head (also *fig.*); **side mill** *Engin.*, a circular milling cutter with teeth on its face, so that it cuts in the direction of its axis of rotation; also, a cylindrical cutter used with its axis parallel to the surface of the workpiece, so that the cutting action occurs along its length; hence **side milling** *vbl. sb.*; **side-note** (earlier example); **side-partner** *U.S. colloq.*, a close associate at work; hence, a colleague or 'side kick'; **side-piece** (*gen.* examples); **side play** *Mech.*, freedom of movement from side to side; **side reaction** *Chem.*, a subsidiary reaction taking place in a chemical system at the same time as a more important reaction; **side road**, (*a*) a minor or subsidiary road; a road leading from or to a main road; (*b*) *spec.* (*Canad.*) in Ontario, a road which passes along the side boundary of a concession; **side salad**, a salad served as a side dish; **side scraper** *Archæol.*, a broad flint implement with a scraping edge on one of the longer sides of the flake (cf. *RACLOIR); **side-screen**, † (*a*) in landscape, a secondary feature set on both sides of the principal to show perspective; (*b*) one of the side-curtains of an open motor vehicle (in quot. 1970, of the cab of a railway locomotive); **side-seat**, (*b*) a seat facing or placed at the side in any form of transport;

side-split *Canad.*, a split-level house with fewer storeys on one side than the other; **side-splitter** (examples); **side-splitting** *a.* (earlier example); hence **side-splittingly** *adv.*; **side-stream**, (*a*) a tributary stream or subsidiary current; also *fig.*; (*b*) *Oil Industry*, a fraction drawn off at an intermediate tray in a distillation column; **side-stroke** *Swimming*, a stroke employed in swimming on the side; also as advb.; **side suit**, in *Cards*, a suit other than trumps, esp. (in *Bridge*) a long suit; **side-sway**, (*a*) a rolling motion from side to side in a moving vehicle; (*b*) a sideways movement or displacement of the upper part of a building or structure as a result of wind pressure; **side tone**, *Teleph.*, the reproduction of the user's voice in a telephone receiver; a sound so heard; **side-trawler**, a trawler in which the nets are set and hauled over the sides; **side trip**, a detour or deviation (also *fig.*); a voyage or excursion aside from the main journey, esp. for sightseeing; **side valve** *Mech.*, a valve that is mounted alongside the cylinder in an internal-combustion engine and opens into a sideways extension of the combustion chamber; freq. *attrib.*; **side-wheeler**, (*b*) *U.S. Baseball*, a side-arm or left-handed pitcher; (*c*) *U.S.*, a pacing horse with a rolling gait (see quots.); **side-wing** *Theatr.* = WING *sb.* 9 c; also *transf.* in *pl.*, side-whiskers (*slang*).

1933 M. B. MUSE *Pharmacol. & Therapeutics* i. 42 Morphine sulphate..when administered as an analgesic has numerous side actions some of which are harmless. **1922** *Proc. I.R.E.* X. 363 A modulated radio telephone wave consists of two components, one, the carrier frequency itself and the other, the so-called side bands, which are the actual modulated components. **1943** *Electronic Engin.* XV. 339/2 If only one side-band is received, a local oscillator must supply the suppressed carrier frequency.. before the detector stage. **1974** *Physics Bull.* Mar. 91/2 Attention was drawn to..the use of side bands generated in the visible region by laser mixing techniques as an alternative to direct comparison of infrared and visible wavelengths. **1823** J. F. COOPER *Pilot* I. iii. 31 The shrill whistle of the boat-swains mate, as he recalled the side-boys. **1916** 'TAFFRAIL' *Pincher Martin* i. 2 Eyed critically by the grinning side-boy and the messenger. **1977** *Navy News* Sept. 25/2 Shipmate Fred Talbot was nominated 'side boy' for the evening and he piped the Mayor aboard. **1936** G. GREENE *Journey without Maps* ii. iv. 197 He was ..handsome in his native robe and his sideburns. **1977** 'J. LE CARRÉ' *Hon. Schoolboy* xii. 264 Cy..had sideburns and..looked like a Mormon missionary. **1941** B. SCHULBERG *What makes Sammy Run?* iv. 75 A swarthy, sideburned Latin. **1976** *Listener* 29 July 103/1 The headmaster, the richly sideburned Mr Terry Ellis. **1905** Side chair [see *ROCKER[1] 4 b]. **1968** *Canadian Antiques Collector* July 32 (Advt.), Sheraton side chairs in mahogany with red damask upholstery. Circa 1810. **1916** *Standard. Rules Amer. Inst. Electr. Engineers* 95 A side circuit is a two-wire circuit forming one side of a phantom circuit. **1957** W. FRASER *Telecommunications* v. 122 It is possible to transmit speech on the phantom circuit without interference to either side-circuit. **1949** *Our Industry* (Anglo-Iranian Oil Co. Ltd.) (ed. 2) iii. 95 Reverting now to the question of drawing off 'side-cuts', the product on any one tray must always be contaminated by some traces of lighter components. **1970** W. G. ROBERTS *Quest for Oil* viii. 85 At the point where we wish to draw off side-cuts or intermediate fractions, special trays are put in. **1872** 'MARK TWAIN' *Roughing It* 279 He disappeared in the gloom of a 'side drift' just as a head appeared in the mouth of the shaft. **1940** *Chambers's Techn. Dict.* 768/1 *Side drift*, an adit. **1926–7** Side-drum [see *gong-drum* s.v. *GONG[1] 1 c]. **1956** Side-drum [see *BOOMY *a.*]. **1885** S. O. JEWETT *Marsh Island* 195 The old farmer and his crony moved their chairs into the square side-entry. **1901** W. CHURCHILL *Crisis* 13 He did not discuss his ambitions at dinner with the other clerks in the side entry. **1958** *Listener* 23 Oct. 669/2 North is unlikely to have the heart suit and a side entry. **1977** *Homes & Gardens* Feb. 14 The hand with the long suit has few side-entries. **1925** *Proc. I.R.E.* XIII. 295 In the case of the side frequencies produced by the modulator tube and delivered through the circuit..there will be no emfs. to balance them out and they will be impressed upon the amplifier. **1978** P. H. SMALE *Telecommunication Systems* ii. 18 The sum of carrier and modulating signal frequencies is called the upper sidefrequency. **1883** T. E. MAY *Treat. on Law, Privileges, Proc. & Usage of Parliament* (ed. 9) xi. 341 A member may speak from the side galleries, appropriated to members, but not from below the bar. **1930** B. FELL *Palace of Westminster* 41 The seating on the floor of the House accommodates 368 members and there is room in the side galleries for another 82. *a* **1974** R. CROSSMAN *Diaries* (1976) II. 402 During the all-night sitting on the Abortion Bill the side galleries would have to be closed because otherwise there would be no door-keepers. **1889** *Cent. Dict.*, *Side-head*... In *printing*, a heading or a sub-head run in at the beginning of a paragraph, instead of being made a separate line. **1964** *New Statesman* 21 Feb. 303/3 We shall present these pieces in a way which makes their character as editorial opinion less equivocal, by prefixing a generic side-head. **1971** D. AYERST *Guardian* xxiv. 347 The *Guardian* gave Churchill only those two or three lines towards the end of the story below a modest side-head: 'The Home Secretary'. **1968** *Heidelberg News* Sept. 4/1 Send him a picture and supply a caption as a side heading. **1920** G. W. YOUNG *Mountain Craft* iv. 162 'Side'-holds, where the edge or point of rock projects and is grasped sideways. **1977** D. LAW *Starting Mountaineering & Rock Climbing* vi. 68 Most of the holds for jamming are side holds, but some can be used for a vertical pull-up. **1826** J. O'KEEFE *Recoll. Life* I. x. 376 His shaggy dress took fire from the side-lamps. **1885** E. B. IVATTS *Railway Management at Stations* 550 *Lamp* (*side*), lamps for showing red lights at the two sides of guards' vans at the end of a train, as signals to an approaching train..and white lights towards the engine driver to enable him looking back to see that no portion of his train has broken away. **1912** *Motor Manual* (ed. 14) iii. 124 It is possible..to adapt electric lighting very successfully to any car, both for interior lighting of limousines and landaulets, and for head, side and rear lamps. **1963** *Times* 30 Apr. 13/4 The many heavy lorries which..only have one pin-head size nearside sidelamp in the town. **1946** A. MONKMAN in H. Whetton *Pract. Printing & Binding* v. 64/2 Assuming the ordinary half-sheet of sixteens is on the machine, and the side-lay for printing the first side is at the foot of page 1, there would be a regular and accurate margin from page 1. **1839** Side lever [see *sway-beam* s.v. SWAY-]. **1846** [in Dict., sense 23 b]. **1882** SENNETT & ORAM *Marine Steam Engine* I. i. 3 The side-lever type of engine, though very heavy and occupying a large space for the power developed, was safe and reliable. **1939** H. W. DICKINSON *Short Hist. Steam Engine* vi. 108 This type was taken up by other makers under the name of the side-lever engine, and remained for about forty years the established type for marine-engine practice. **1960** *Times* 16 Mar. (Canberra Suppl.) p. iv/6 The forward bulkhead.. must move one frame aft to accommodate the mechanism of the side-loader. **1973** *Scotsman* 19 Feb. 3/1 (Advt.), We produce the largest and most versatile range of frontlift and sideloader trucks in the UK. **1946** *Proc. I.R.E.* XXXIV. 335/2 After either the side-lobe level or the position of the first null is specified, the position of the other nulls and of the side lobes can be found by simple calculation. **1975** D. G. FINK *Electronics Engineers' Handbk.* xxv. 25 In angle, the response function $\chi(\theta,\phi)$ is simply the antenna pattern... It has a main lobe in the direction to which it is matched, and side lobes extending over all visible space. **1688** R. HOLME *Academy of Armory* ii. xviii. 463/2 The *side Locks*, are those as cover and keep warm the ears and neck, being a degree shorter than the former [sc. 'The *Bottom Locks*']. **1848** [in Dict., sense 23 b]. **1889** *Century Mag.* Sept. 710/1 The monuments represent him as a prince and nothing more, still wearing the side-lock of juniority. **1944** S. BELLOW *Dangling Man* 23 The street..presented itself in one of its winter aspects, creased and with thin sidelocks of snow. **1978** I. B. SINGER *Shosha* xi. 196 Old women in bonnets of beads and ribbons, men with white beards and sidelocks. **1898** H. S. WILSON *Pract. Tool-Maker & Designer* i. 15 Select some good-sized straddle or side mills. **1954** H. W. PORTER et al. *Machine Shop Operations & Setups* ix. 312 This makes it possible to finish two or more parallel surfaces at the same time by using two or more side mills. **1910** D. DE VRIES *Milling Machines* xv. 449 The sharpening of the left side teeth of a side milling cutter with a cup wheel. **1954** H. W. PORTER et al. *Machine Shop Operations & Setups* ix. 312 Straddle milling requires two side-milling cutters. **1973** J. G. TWEEDDALE *Materials Technol.* II. vi. 146 Side-milling uses a cylindrical multifluted cutter (usually spirally fluted) which rotates on an axis parallel to the workpiece and is traversed across it to cut, progressively, tangentially into it. **1776** W. ROBERTSON *Let.* 8 Apr. in *Corr. Adam Smith* (1977) 193, I should wish that in the 2d Edition you would give..what the Book-sellers call Side-notes. **1890** *N.Y. Even. Post* 23 May 8/2 The arrest was made by the witness's side partner, it being his night off. **1921** R. D. PAINE *Comrades Rolling Ocean* ix. 159 We shall have to consult my side-partner, Briscoe. **1849** Side-piece [see MOB-CAP]. **1928** *Daily Express* 16 Aug. 5/2 A car (with a left-hand drive and a hood but no side-pieces). **1861** J. BOURNE *Treat. Steam-Engine* (ed. 5) viii. 352/1 The guide blocks are of brass, and in wearing down they maintain their position in the groove. This mode of construction prevents side play. **1905** E. C. C. BALY *Spectroscopy* iii. 50 It is..necessary that the jaws move smoothly in their grooves without any trace of side play. **1934** WEBSTER, Side reaction. **1936** *Jrnl. Amer. Chem. Soc.* LVIII. 2210/2 The small deviations observed..may be due to the fact that the side reactions..are more prominent in one case than in the other. **1973** *Sci. Amer.* Oct. 60/3 High-energy intermediates are frequently formed in chemical syntheses, but if they are not isolated from water or other reactive substances, they decompose in side reactions that lower the yield of the reaction. **1854** T. C. KEEFER *Ottawa* 72 The municipalities have taxed themselves too heavily for the main road..to be able to build also the side roads. **1873** *Woodstock* (Ontario) *Sentinel* 5 Dec. 3/4 To Joseph Whaley for pine timber for culvert on first side road, $1.00. **1958** R. LIDDELL *Morea* ii. viii. 190 South of Argos, a side-road, off the main road to Tripolis, leads to Kefaläri. **1968** *Globe & Mail* (Toronto) 13 Feb. 3/6 The partly frozen body..was found Sunday beside the Nixon Side-road. **1976** *Western Mail* (Cardiff) 22 Nov. 1/2 The other man was seen standing..near a vehicle on a little-used side road. [**1951** F. BROBECK *Good Salad Bk.* 9 A small salad served with the main course, or after it, on a decorative plate is the old-fashioned side-dish salad.] **1972** D. SALE *Love Bite* ii. xvii. 212 She..helped herself to a side salad of avocado in French dressing. **1980** P. ABLEMAN *Shoestring's Finest Hour* iv. 73, I queued for a hefty portion of shepherd's pie and a side salad. **1872** J. EVANS *Anc. Stone Implements* xiii. 272 When the instrument is broader than it is long, it has been termed a side scraper. **1921** R. A. S. MACALISTER *Text-Bk. European Archæol.* I. vii. 321 The side-scraper.., a flake with secondary chipping along one edge, making it fit to scrape the interior of hides in preparing garments. **1977** *Sci. Amer.* Nov. 126/2 Among the Upper Industry tools at Hoxne were a small number of the flake implements that are traditionally called 'side scrapers' and are presumed to have played a role in the dressing of hides. **1782** W. GILPIN *Observ. on River Wye* ii. 8 Every view on a river, thus circumstanced [with steep banks], is composed of four grand parts; the area, which is the river itself; the two side-screens, which are the opposite banks, and mark the perspective; and the front-screen, which points out the winding of the river. *a* **1817** JANE AUSTEN *Northanger Abbey* (1818) I. xiv. 263 He talked of fore-grounds, distances, and second distances—side-screens and perspectives. **1932** *News Chron.* 6 Aug. 3/5 The assailant thrust a six-chambered revolver through a side-screen and fired. **1958** L. DURRELL *Mountolive* xv. 276 He drove up..in his pennoned car, rejoicing in..the whickering of wind at the side-screens. **1970** N. FLEMING *Czech Point* xiv. 191 The canvas sidescreens to the cab flapped in the wind. **1978** A. PRICE *'44 Vintage* xiii. 160 The staff car..with a closed canvas hood and side-screens. **1889** *Cent. Dict.*, *Side-seat*,..in a vehicle of any kind, a seat with the back against the side of the vehicle, as usually in a horse-car or omnibus. **1901** *Daily News* 5 July 4 In the stern with the side-seats out there is room for 3 or 4 drift-nets. **1922** JOYCE *Ulysses* 591 The car and horse..turn. Corny Kelleher on the sideseat sways his head to and fro in sign of mirth at Bloom's plight. **1968** *Globe & Mail* (Toronto) 15 Jan. 23/1 (Advt.), Why not see this spacious side-split..in Etobicoke? **1864** *Harper's Mag.* Feb. 422/1, I send you three samples [of letters].., hoping thereby to reciprocate some of your side-splitters. **1903** A. M. BINSTEAD *Pitcher in Paradise* v. 133 As regards poetry I have already had a sidesplitter entitled 'Don't chalk your cue before a lady' accepted by Mr Arthur Roberts. **1860** S. MORDECAI *Virginia* xiv. 188 These among other side-splitting tales, which he told and acted with the skill of a Matthews. **1907** A. BENNETT *Grim Smile Five Towns* 7 Something side-splittingly funny—one of the best jokes that ever occurred. **1970** *Daily Tel.* 6 Aug. 6/2 The reader is..given an at times side-splittingly funny account of the eccentricities of Frazer and his wife. **1900** *Knowledge* 1 Dec. 273/1 The rotten condition of the surface is seen when one of the larger side-streams cuts its way down to the Durance. **1935** *Petroleum* (Inst. Petroleum Technologists) 84 The provision made for taking side-streams has greatly extended the usefulness of the fractionating tower. **1939** *John o' London's Weekly* (Suppl.) 9 June p. ii/2 The cliffs of Norfolk and Suffolk are menaced by the sidestream of a tidal current that flows westwards into the Wash. **1960** *Times* 29 Sept. 15/7 His writings are in the stimulating side-stream of scholarly diaries. **1973** D. ANDERSEN *Ways Harsh & Wild* i. 42 You'll see the waves splashing up the sides of the canyon and falling back to form a hogsback in midstream that's a yard higher than the sidestreams. **1973** R. PRIESTLEY in G. D. Hobson *Mod. Petroleum Technol.* (ed. 4) viii. 271 Products of a volatility intermediate between the overhead and bottoms products are withdrawn as sidestreams. **1867** C. STEEDMAN *Man. Swimming* 105 The five movements—three for the legs and two for the arms—required for the performance of the side stroke. *a* **1936** KIPLING *Something of Myself* (1937) ii. 34 One set of verses which exactly set the time for my side-stroke when I bathed in the big rollers. **1962** A. SEXTON *All my Pretty Ones* 55 The old-fashioned side stroke. **1976** J. MCCLURE *Rogue Eagle* xi. 185 Buchanan ..waded in and swam sidestroke downstream of the gelding. **1952** I. MACLEOD *Bridge is Easy Game* xiv. 181 The commonest case is when dummy has a long solid side suit. **1960** C. H. GOREN *New Contract Bridge in Nutshel* 13 In a side suit, the fourth card is considered a long card. **1974** *National Skat & Sheepshead Q.* Mar./Apr. 29 This decision is based upon the official rule of calling a side suit ace. **1930** Sidesway [see *ROLLING *vbl. sb.*[2] 6 b]. **1932** CROSS & MORGAN *Continuous Frames of Reinforced Concrete* iv. 108 In many problems in the analysis of rigid frames..a solution assuming no movement of the joints does not satisfy statistics because the shear in all columns of any one story is not equal to the known shear in that story. This indicates sidesway..of the frame sufficient to make $\Sigma H = O$. **1961** E. LIGHTFOOT *Moment Distribution* v. 123 After moments have been apportioned to the frame,..a single or a double cycle of joint balance and carry-over is performed (with the frame restrained against sidesway). **1980** *Daily Tel.* 11 June 14/5 [In the Princess] fast cornering produces rather a lot of side-sway. **1917** G. D. SHEPARDSON *Telephone Apparatus* xiv. 234 Some arrangements seem to give more trouble than others from 'side-tones', whereby the speaker hears his own words too strongly. *a* **1944** K. DOUGLAS *Alamein to Zem Zem* (1946) 38 The operator..breathed and hummed into the microphone, listening for sidetone. **1978** *Sci. Amer.* Mar. 59/1 Too little side tone gives the telephone an unnatural 'dead' sound and tends to cause the user to talk too loudly. **1962** J. TUNSTALL *Fishermen* ii. 46 The whole-freezer side-trawler is a..compromise solution in which conventional side-trawling is retained. *a* **1911** D. G. PHILLIPS *Susan Lenox* (1917) II. viii. 212 He's got a nasty streak in him... He put me on the Island once for a little side trip I made. **1929** L. F. CARR *Amer. Challenged* 3 He was forced to borrow money for a little side trip to New York. **1966** *Guardian* 24 Dec. 4/3 Another family..spent three weeks in Jutland, with a side trip of a couple of days in Copenhagen. **1979** N. & I. LYONS *Champagne Blues* 78, I..make arrangements for special side trips, room supplements, extended tour packages. **1928** *Daily Mail Year Bk.* p. lxiv, 3·49 H.P. side valve 'sports' model. **1946** [see *L 3]. **1973** J. LEASOR *Host of Extras* ix. 179 My heart was pounding like an old side-valve engine on a long hill when the ignition's too retarded. **1911** *Spalding's Official Base Ball Guide* 277 Redfern, side-wheeler, with Flowers, McFarlin and Reis with the other end up, made the pitching department a clever one. **1926** *Amer. Speech* I. 369/2 [Baseball] They are 'south-paws' or 'port-siders' or 'side-wheelers' when they are left-handed. **1936** *Literary Digest* 1 Aug. 35/2 Pacers have a rolling, lunging movement that has earned them the nickname of 'wigglers' or 'side-wheelers'. Left front and left rear legs come up together, followed by a counterbalancing lunge to the right. **1948** *Times Digest* (Richmond, Va.) 15 Mar. 17/4 The lanky Cincinnati Reds' sidewheeler has added a new pitch to his repertoire. **1953** *Sun* (Baltimore) 10 Aug. B15/2 Mac Hayman [will]..handle the speedy sidewheeler in her three stake engagements at the Ocean Downs Raceway. **1707** E. SETTLE *Siege of Troy* iii. 23 The Scene opens and discovers a Grove..over a Tarras Walk, is seen a Beautiful Garden of six side Wings. **1811** L. SUMBEL *Mem. Life* III. 220 And a fourth, with locks *bushed* out on his temples, burlesquing the side-wings of your noble head. **1814** JANE AUSTEN *Mansfield Park* I. xiii. 257 Just a side wing or two run up..and three or four scenes to be let down. **1881** *Atlantic Monthly* Sept. 402/1 It seems as if certain actors in some preceding comedy of his were standing at the side-wings, and critically watching the progress of the after-piece.

side, *v.*[2] *slang* (now *rare*). [f. SIDE *sb.*[2]] *intr.* To be conceited or boastful; to 'put on side'. Also with *about.*

1906 R. BROOKE *Let.* 4 June (1968) 54 This school-life.. calls to me... I play my part with zest, alternately 'siding' and ragging. **1909** WODEHOUSE *Mike* v. 26 There's just a chance you might try to side about a bit soon.

si·de-arm, *a.* (and *adv.*) [SIDE *sb.*[1] 26.]

1. Performed or delivered with a swing of the arm extended sideways, esp. in *Baseball*. Cf. ROUND-ARM *a.* and *adv.*

1908 *Baseball Mag.* June 32/1 The spit ball..when pitched with a side arm movement will go out. **1909** *Amer. Mag.* Aug. 402/2 He pitched..two fast side arm balls, high and outside. **1939** 'N. BLAKE' *Smiler with Knife* xii. 188 The unerring side-arm flick..had surprised many confident run-stealers before now. **1948** *Sun* (Baltimore) 15 Mar. 17/8 Blackwell said his best pitch was a fast sidearm ball. **1978** *Detroit Free Press* 5 Mar. c8/3 'He struck me out three times one day,' said Williams. 'He threw that sidearm slider and it would come into your face and explode.'

2. as *adv.* With a sweep of the arm extended sideways; in a side-arm manner.

1958 I. CROSS *God Boy* vii. 55 She swam sidearm alongside me. **1973** C. SAGAN *Cosmic Connection* (1975) xv. 112 As pitcher, he could throw the ball sidearm—at the horizon at between twenty and thirty miles per hour.

side-bar. Add: **5. side-bar whiskers** *U.S. local*, side-whiskers, side-burns.

1882 G. W. PECK *Peck's Sunshine* 55 He was a red-faced man, with these side-bar whiskers. **1975** *Amer. Speech 1972* XLVII. 155 Some of my earliest memories of childhood in the Up Country of South Carolina..center upon the hearing of the lively and colorful word *tea-hounds*... Invariably the expression was employed humorously..when a person referred to side-bar whiskers, rather bushy projections extending from the hairline to below the ears and worn with an unbearded chin.

6. A secondary newspaper article featuring some aspect of a main story in the same publication. *U.S. Journalism.*

1948 C. D. MACDOUGALL *Interpretive Reporting* (rev. ed.) 695 *Sidebar*, a complete article on one phase of a longer story, run separately. **1967** R. J. SERLING *President's Plane is Missing* iv. 70 Bat us out a good sidebar on anything that's ever happened to a presidential plane. **1977** *Time* 17 Jan. 5/2 The cover story was written by Morrison, with a sidebar on Felker by Michael Demarest.

sideboard. Add: **3. a.** (Example.)

1874 'UNCLE BOB' *Lett. to Children* (1875) xiv. 87 Starting with our standing collars on, we managed to get to the church... Some mischievous boy would cry out, 'Come out of those sideboards.'

b. (Examples.)

1907 *Daily Chron.* 7 Dec. 5/7 You have described the duke as having small whiskers?—Yes, they were sideboards. Where did you get that name?—I have been in America... You call them sideboards?—Yes, or sideburns. **1956** D. M. DAVIN *Sullen Bell* ii. iv. 136 He was a miserable little sod, with sideboards and an American tie. **1961** H. S. TURNER *Something Extraordinary* i. 9 The boys are dressed in the Teddy style, with tight trousers and sideboards. **1975** M. BRADBURY *History Man* vi. 97 He takes his razor..clipping at the line of the sideboards.

si·decar. Also **side-car, side car.** [SIDE *sb.*[1]]

1. a. (See SIDE *sb.*[1] 27.) Now *Hist.*

1881 [see SIDE *sb.*[1] 27]. **1963** 'A. BRIDGE' *Dangerous Islands* ix. 140 The Irish side-car, now almost extinct, is one of the least comfortable vehicles imaginable. The driver sits easily foursquare behind the horse, but the passengers perch on two long seats with a high back between them, parallel with the direction of the vehicle —and since side-cars are now never used except on very rough roads, they are jerked about, clinging to the back between the seats. **1974** *Encycl. Brit. Micropædia* V. 529/2 *Jaunting car*, also called *jaunty car* or *side-car*, two-wheeled, open vehicle, popular in Ireland from the 19th century, was unusual in having lengthwise passenger seats, either facing each other (inside the car) or back to back (outside).

b. A vehicle designed to be attached to the (near-)side of a motor-cycle to accommodate one or more passengers. Occas. attached to a bicycle.

1903 *Hardwareman* 13 June 520/1 The side car is..the most sociable attachment for a motor bicycle. **1927** KIPLING *Limits & Renewals* (1932) 177 'What did he do afterwards?' ' "Bought a side-car to his bike, to hold more vegetables.' **1935** H. MOORE *Compl. Cyclist* v. 44 Until recently the cycling of a married couple was seriously restricted by the birth of a child... Now, the tandem-cum-sidecar, bearing a family of three (or more), is a fairly common sight. **1951** T. STERLING *House without Door* ii. 12 He sat on a bicycle which had a large white sidecar. **1978** J. IRVING *World according to Garp* i. 17 The pilot hurried to have Garp transferred to the sidecar of a medic's motorcycle.

2. A cocktail made of brandy and lemon juice with a dash of an orange liqueur.

1928 S. LEWIS *Man who knew Coolidge* i. 61 Mame took a Bronx, and Delmerine took a side-car. **1930** AUDEN *Poems* 12 I'll have a sidecar, thanks. **1952** S. KAUFFMANN *Philanderer* (1953) v. 77 They ordered sidecars and Suzy said..'Not too strong... I've got a lot of drinking to do to-night.' **1978** M. DICKENS *Open Book* vi. 49 Once or twice at cocktail parties, I saw someone I knew and had to..keep my head down as I cruised the crowd with my tray of sidecars and white ladies.

Hence as *v. intr.*, to drive a motor-cycle with a side-car attached; to travel in a side-car; also **si·de-car(r)ing** *vbl. sb.*; **si·decarist**, one who drives or travels in a motor-cycle combination. Now *rare*.

1911 *Motor Cycle* 19 Jan. 21 (Advt.), Best after tests for sidecaring,..for economy. **1914** *Motor Cycling* 12 May 8, 3½ h.p. is insufficient for sidecarring at a satisfactory speed. **1920** *Motor Cycle* 29 Apr. 487/2, I make this request because of its importance to sidecarists. **1923** *Ibid.* 25 Oct. 665/2 No doubt she would feel small and lonely when sidecarring by herself in one of these ample-looking bodies.

side chain. [SIDE *sb.*[1]] † **1.** A chain mounted at the side of a vehicle for any purpose (see quots.). *Obs.*

1849–50 J. WEALE *Dict. Terms*, *Side Chains*, chains and hooks fixed to the sides of the tender and engine for safety, should the central drag-bar give way. **1883** W. S. GRESLEY *Gloss. Coal-Mining* 221 *Side chain*, a chain hooked on to the sides of tubs when running upon an engine-plane or jig, to keep all the tubs together in case a coupling breaks. **1886** *Encycl. Brit.* XX. 247/2 Some [railway] companies have gone further and placed the guard or side chains upon springs.

2. a. *Chem.* A chain of atoms attached to the principal part of a molecule.

1886 ROSCOE & SCHORLEMMER *Treat. Chem.* III. iii. 7 If two atoms of hydrogen [in benzene] be replaced by elements or radicals, termed 'side chains', three isomeric compounds may be formed whether the entering element or side chain be identical or different. **1927** HALDANE & HUXLEY *Animal Biol.* ii. 73 The whole [chromosome] is like a gigantic single organic chemical compound, since the molecules of such a chemical compound are all made up of smaller parts—the side-chains and radicals and single atoms. **1974** *Sci. Amer.* June 59/3 They [sc. the catecholamines] have in common a chemical structure that consists of a benzene ring on which there are two adjacent hydroxyl groups and an ethylamine side chain.

b. *Physiol.* A structure postulated to project from the surface of a cell and to constitute a receptor (RECEPTOR 3 a in Dict. and Suppl.) in Ehrlich's theory of immunological action; so *side-chain theory*.

1900 tr. P. Ehrlich in *Proc. R. Soc.* LXVI. 433 We may assume that the protoplasm consists of a special executive centre..in connection with which are nutritive side-chains, which possess a certain degree of independence, and which may differ from one another according to the requirements of the different cells. *Ibid.* 440, I have now laid before you the fundamental facts which up to the present constitute our knowledge in the field pertaining to immunity, and which can be most easily and successfully explained through the agency of 'side-chain theory'. **1911** *Rep. Brit. Assoc. Adv. Sci. 1910* 635 Ehrlich has given a graphic representation of this process in his side-chain theory. **1935** N. P. SHERWOOD *Immunol.* vi. 121 In Ehrlich's theory these free chemical entities or cast-off side chains constitute the antibodies found in the circulation. **1974** *Encycl. Brit. Macropædia* VI. 510/1 Only if the haptophore group of a toxic molecule combines with the side chain of the cell can a bacterial toxin act upon a cell. The affected organism then produces great quantities of the side chains, all of them 'gauged' to the disease-producing agent. These immune bodies prevent a renewed infection.

si·de-comb. orig. *U.S.* [SIDE *sb.*[1] 23 b.] A comb used to secure a woman's hair, esp. at the side of the head.

1824 *Missouri* (Columbia) *Intelligencer* 8 May 3/3 (Advt.), Tortoise shell, tuck and side combs. **1851** H. MAYHEW *London Labour* I. 346/1 A few earrings and ear-drops, and sometimes a few side-combs. **1897** [see SIDE *sb.*[1] 23 b]. *c* **1909** D. H. LAWRENCE *Collier's Friday Night* (1934) ii. 28 She finishes stroking her hair up with her side-combs. **1966** *Olney Amsden & Sons Ltd. Price List* 35 *Side comb*..boxed 3 dozen singles 10/- per box.

sided, *ppl. a.* Add: **1.** (Later examples.)

1904 *N.Y. Sun* 7 Aug. 20 The yard is sided with cabins. **1952** DYLAN THOMAS *Coll. Poems* 21 The boy she dropped from darkness at her side Into the sided lap of light grew strong.

sidedness, (*b*) one-sidedness; lack of symmetry in a superficially symmetrical structure or system.

1970 A. L. LEVNINGER *Biochem.* xxvii. 617 (*heading*) Sidedness of the transport process. **1972** *Sci. Amer.* Feb. 32/2 Such proteins are evidently located exclusively on only one side of the membrane. This information lends credence to the concept of sidedness in membranes. **1976** *Word 1971* XXVII. 240 The above contention finds support from another aspect of human evolution, that is, the sidedness or dominancy in either of the two hemispheres of our brain.

side-door. Add: **1. b.** *fig.*

1930 *Times* 21 Mar. 15/5 We maintain that these 'side door' credits are endangering our chances of securing recognition of our bonds. **1965** *New Statesman* 30 Apr. 676/1 The proposed conference on Cambodia—an attempt to settle Vietnam through a side-door—may prove abortive.

2. side-door Pullman *N. Amer. slang* (chiefly *Tramps'*), a railway goods wagon with sliding doors in the sides, a box-car; a freight car.

1887 M. ROBERTS *Western Avernus* xvii. 237 When the engine..started out, I lighted a match and took a look at my..travelling carriage, or 'side-door Pullman', as the 'tramps'..facetiously call them. **1918** R. W. LARDNER *Treat 'em Rough* 10 If they didn't have all the luck in the world they would be rideing [*sic*] around the country in a side door Pullman with all their baggage on. **1927** L. F. RANLETT *Let's Go* 34 A train of 'side-door Pullmans' ..drew up at the railway station.

si·de-effect. [SIDE *sb.*[1] 24 d.] **1.** *gen.* A subsidiary consequence of an action, occurrence, or state of affairs; an unintended secondary result.

1884 [see SIDE *sb.*[1] 24 d]. **1933** E. BLUNDEN *Mind's Eye* (1934) i. 59 Here the side-effects of the quarrels south of the Canal..were felt and paid for in some casualties. **1959** B. WOOTTON *Social Sci. & Social Pathol.* xi. 336 The problem is, moreover, made more difficult still by the second of the potentially unfortunate side-effects of reformative penal treatment. **1974** 'D. CRAIG' *Whose Little Girl are You?* ii. 23 A side effect of sobriety he had found to be deadly fatigue.

2. *Med.* An effect (usu. for the worse) of a drug or other chemical other than that for which it is administered; usu. *pl.*

1939 WRIGHT & MONTAG *Textbk. Materia Med. Pharmacol. & Therapeutics* x. 112 The effects which are not desired in any particular case are referred to as 'side effects' or 'side actions' and, in some instances, these may be so powerful as to limit seriously the therapeutic usefulness of the drug. **1952** *Chambers's Jrnl.* June 384 A new alkaloid, built to the same molecular pattern, but with slight modifications, has been synthesised; it is equally anti-malarial but its side-effects are less toxic. **1961** *Lancet* 19 Aug. 390/1 The commonest complication of anticoagulant therapy is hæmorrhage. Numerous reports describe the frequency and nature of this side-effect. **1978** E. HARTMANN *Sleeping Pill* ii. 16 In the usual doses, paraldehyde does not produce severe side effects.

si·de-foot, *v.* *Association Football.* [SIDE *sb.*[1]] *trans.* To strike (the ball) with the (out)side of the foot.

1950 *Sport* 22–28 Sept. 10/2, I like to see the player who ..sidefoots the ball to a nearby colleague and starts a quick-passing movement. **1960** *Times* 19 Sept. 17/4 Robson side-footing the ball into the net. **1977** *Irish Press* 29 Sept. 18/5 He sent a cross to Jovanovic, who side-footed the ball into the net.

Hence as *adv.*, with the side of the foot; **si·de-footed** *ppl. a.*

1968 *Listener* 23 May 682/3 It's the way I kick, side-foot. **1974** *Observer* 1 Sept. 18/5 Latchford..steered a side-footed volley to Rimmer's left.

si·de-issue. [SIDE *sb.*[1] 24 d.] A subsidiary issue.

1873 [see SIDE *sb.*[1] 24 d]. **1928** *Sat. Rev.* 20 Oct. 515/2 The secret is well kept and the various side issues are interesting and bewildering. **1938** [see *CREATIVE *a.* 1 b]. **1958** *Times* 29 Sept. 11/2 The floor will go down fighting on the side-issue of the platform's acceptance of the *status quo* for the public schools. **1982** F. DAVIES *Death of Hit-Man* ix. 151 Zio must be here for something far more important. Not just to kill Denton. That must have been a side-issue.

si·de-kick. *slang* (orig. *U.S.*). [Back-formation from next.] **1.** A companion or close associate; *spec.* an accomplice or partner in crime; a subordinate member of a pair or group. More *loosely*, a friend, a colleague.

1906 H. GREEN *At Actors' Boarding House* 85 The Red Swede..sat over a pint of champagne with Dopey Polly.. and his side kick, the Runt. *a* **1911** D. G. PHILLIPS *Susan Lenox* (1917) II. xvii. 394 'Now, what d'ye think of that?' said Black Mustache to his 'side-kick.'.. 'Guess we'd better run her in, Pete.' **1927** J. M. SAUNDERS *Wings* iv. 173 'I want two of you,' the Major said, 'who is your side-kick?' 'Armstrong,' Johnny admitted unwillingly. **1934** *Bulletin* (Sydney) 25 July 47/1 Snowy was good at soft things; as a rule, you could trust him as a sidekick to help you to a clear getaway. **1956** M. PROCTER *Pub Crawler* viii. 102 He's Frank McGeen's sidekick. They team up together in jobs of this sort. **1960** *Times* 14 Oct. 18/7 Miss Moira Redmond, as an ex-wife.., made a takingly crisp and sub-acid side-kick. **1976** *New Yorker* 23 Feb. 82/3 Christopher Lloyd was funny as a drug-ridden sidekick of the defunct singer. **1981** 'J. MCVEAN' *Seabird Nine* xiii. 154 It was the White House... And not just some little cotton-tail sidekick either, but counsel to the President.

2. *Criminals' slang.* = SIDE-POCKET 1. *U.S.*

1916 *Lit. Digest* 19 Aug. 424/2 Pockets range from 'side kicks' to 'double insiders'. **1935** *Amer. Speech* X. 20/2 *Side-kick* [formerly] one's pal. In modern argot a side pocket in the coat; it is doubtful if there is any connection. **1955** D. W. MAURER in *Publ. Amer. Dial. Soc.* xxiv. 125 The outside pockets in an overcoat are called *side kicks* (from which we get a venerable American idiom).

3. An incidental criticism; a passing or indirect attack, a 'side-swipe'.

1958 *Economist* 1 Feb. 384/2 Two intriguing passages in Mr. Thorneycroft's speech were, first, what could be taken as a sidekick at the Government's insistence on Britain having the H-bomb. **1971** *Scotsman* 20 May 10/1 Parents came in for a good many side-kicks these days, he said. 'Blame the parents' was a recurring cry.

si·de-kicker. *slang* (orig. *U.S.*). Now *rare*. [SIDE *sb.*[1]] = *SIDE-KICK 1.

1903 'O. HENRY' in *McClure's Mag.* Feb. 432 Billy was my side-kicker in New York. **1926** J. BLACK *You can't Win* xiv. 189, I cast about for a 'sidekicker'. **1929** *Papers Mich. Acad. Sci., Arts & Lett.* X. 322/1 *Side-kicker*, sleeping companion; bosom chum. **1933** *Bulletin* (Sydney) 6 Sept. 41/2 'One's known as Yargus,' George says, 'and his sidekicker as the Snake.'

side-light. Add: **1. b.** (Earlier example.)

1862 J. BROWN *Let.* 4 Feb. (1912) 193, I like so much your saying that about the breeding of my mind, and all the side-lights and sub-suggestions.

3. c. One of the small warning lights on either side of the front (or rear) of a motor vehicle, which when lit show the position and width of the vehicle, esp. at night; a side-lamp.

1912 *Motor Manual* (ed. 14) iii. 120 Much better side lights have been provided than ever existed before. **1955** *Times* 16 Aug. 2/7 British motorists are singularly reluctant to use the lights of their cars—motorists abroad invariably switch on their sidelights much earlier. **1973** A. BEHREND *Samarai Affair* xi. 108 One headlamp and both sidelights had been broken.

side-line, *sb.* Add: **1. a.** For '(see quot. 1862)' read 'and other sports: a line marking the edge of the playing area at the side; a touch-line. Also, the area immediately outside this.' (Further examples.) Also *fig.* with allusion to the position of a spectator observing but removed from the action of a game, esp. in phr. *on* or *from the sidelines*.

1886 J. DWIGHT *Lawn-Tennis* ii. i. 41 He may play down the side-line or he may lob. **1899** H. A. QUINN *Pennsylvania Stories* 24 The coaches on the side lines were not so jubilant. **1962** [see *BYE *sb.* 1 c]. **1977** *Cleethorpes News* 27 May 15/2 It was a close game, with plenty of support on the sidelines. *fig.* **1934** WEBSTER, *Side-line*... The standpoint of those not immediately participating. **1939** *Times* 2 Nov. 8/3 The Russian Government were well satisfied with the policy announced by the Supreme Soviet two months ago —standing on the sidelines and watching Germany, England, and France fight out the war. **1954** T. S. ELIOT *Confidential Clerk* i. 18 But as you're here, Eggers, I can just relax. I'm going to enjoy the game from the side-lines. **1974** J. MANN *Sticking Place* x. 151, I can't sit on the sidelines all my life, producing academic dissertations.

b. (Earlier example.)

1890 KIPLING *From Sea to Sea* (1899) II. xxix. 62 Livingstone is..the junction for the little side-line that takes you to the Yellowstone National Park.

c. *Canad.* = *side road* (*b*) s.v. *SIDE *sb.*[1] 27.

1834 in W. A. Langton *Early Days in Upper Canada* (1926) 91 The concession lines run N.17 1/2W. and the side lines I am told are not exactly perpendicular. **1896** J. L. GOURLAY *Hist. Ottawa Valley* 34 The concessions and sidelines in these townships were 66 feet wide.

3. (Further examples.)

1937 'G. ORWELL' *Road to Wigan Pier* i. 8 Brooker.. was a miner by trade, but he and his wife had been keeping shops of various kinds as a side-line all their lives. **1966** *Listener* 13 Jan. 67/3 Few of them managed to make a good living out of their art alone, without running a side-line such as a brewery or an insurance office. **1977** *New Yorker* 29 Aug. 47/2 His sideline computer-service business thrived. **1979** D. GAGEBY in J. J. Lee *Ireland 1945–70* 130 It caused suspicion among journalists whose sideline earnings on correspondence for English or American papers seemed to be threatened.

side-line, *v.* Add: **I.** *trans.* **1.** (Earlier U.S. example.)

1837 W. IRVING *Rocky Mountains* I. ii. 36 The horses were 'side lined', as it is termed: that is to say, the fore and hind foot on the same side of the animal were tied together, so as to be within eighteen inches of each other.

2. *pass.* (or as *ppl. a.*) Of a sportsman: to be forced to remain out of competition on the side-lines, esp. through injury. Also *fig.* Occas. *actively*, of an injury, etc. orig. and chiefly *U.S.*

1945 *Sun* (Baltimore) 30 June 8/2 Snead is once more sidelined with his back ailment. **1947** *News–Age–Herald* (Birmingham, Alabama) 20 July 1B/7 Charley Keller..is recovering..from an operation to relieve the back ailment that has sidelined him for several weeks. **1949** *Cavalier Daily* (Univ. of Virginia) 22 Oct. 1/5 Gene Schroeder, still side-lined with a shoulder injury. **1966** *N.Y. Times* (Internat. ed.) 22 Apr. 12/8 Buckpasser..is sidelined with a hoof injury. **1970** [see *defenceman* s.v. *DEFENCE *sb.* 9]. **1975** *Amer. Speech 1972* XLVII. 143 Although polio cruelly limited her for many years to the use of one arm, Betty Adler was never sidelined. **1977** *Daily Express* 29 Jan. 38/4 Ian Wallace and Mick Ferguson, their first-choice front men who have been side-lined for the last six weeks.

3. To mark (a passage of text) for special attention by a line or lines drawn in the margin; *spec.* (of confidential matter) to indicate that it should not be printed or published.

1968 *Guardian* 24 July 8/1 The..witnesses had been encouraged to speak freely with the assurance that they would be allowed to 'sideline' those parts of the evidence that they did not wish to see published. **1978** *Observer* 10 Dec. 1/7 The all-party committee..is expected to exercise traditional discretion in 'sidelining' or censoring Cabinet minutes.

4. *fig.* To remove from the centre of activity or attention; to place in an inferior position.

1973 H. GRUPPE *Truxton Cipher* (1974) xx. 218 He even persuaded the brass to put him in charge of the special project... The Navy was glad to do it. Sideline him for a bit. **1976** *National Observer* (U.S.) 22 May 10/2 President Nixon vetoed the legislation in 1971, and since then, a lack of congressional and Administration support has sidelined the Mondale approach.

II. 5. *intr.* To engage *in* as a subsidiary occupation or sport.

1944 *College Topics* (Univ. of Virginia) 30 Mar. 3 Captain Nat Boyd is specializing in the hurdles and broad jumping and sidelining in the high jump. **1975** B. GARFIELD *Hopscotch* xv. 147 The kingpin in town was a back-porch country lawyer who..sidelined in real estate.

si·de-looking, *a.* [SIDE *sb.*[1]] **1.** Characterized by looking sideways.

1829 J. F. COOPER *Wept of Wish-ton-Wish* I. iii. 37 A demure, side-looking young woman kept her great wheel in motion. **1956** H. GOLD *Man who was not with It* (1965) iv. 36 Those others turned me back..to the side-looking cast in Phyl's eyes, her black hair short-cropped with a calculated wildness.

2. Producing or being a radar or sonar beam transmitted sideways and downwards, usu. from an aircraft for the mapping of relief.

1961 B. L. CORDRY et al. in G. Merrill *Airborne Radar* xiv. 777 In side-looking systems..the orientation of the antenna may not be changed easily. **1964** *Jrnl. Geophysics Res.* LXIX. 3824/1 Narrow-beam lateral echo sounders are similar to standard echo sounders except that the transducer is mounted to give a side-looking beam that is a few degrees below horizontal. **1971** P. O'DONNELL *Impossible Virgin* i. 6 From film taken by side-looking radar he could penetrate..the earth's skin, to..the bed-rock below. **1977** *Sci. Amer.* Oct. 93/2 The acute grazing angle of the microwave illumination of side-looking radar emphasizes the form of the land, and the large areas that can be surveyed under constant conditions favor the recognition of extensive features.

sideman. Restrict † *Obs.* to senses in Dict. and add: **5.** A supporting musician in a jazz or dance band. Cf. *front man* (ii) s.v. *FRONT *sb.* 14. orig. *U.S.*

1936 *Amer. Mercury* XXXVIII. p. x/2 *Side man*, any musician in the band except the leader. **1943** P. E. MILLER *Yearbk. Pop. Music* 7/2 He began playing in bands just a few years later, and was soon accepted as a desirable sideman. **1961** *Radio Times* 21 Dec. 53/4 *Jazz Club*. Humphrey Lyttelton..welcomes as his guests two former Lyttelton sidemen Jimmy Skidmore on tenor sax and Johnny Picard on trombone. **1977** J. WAINWRIGHT *Do Nothin' till You hear from Me* v. 67 Goodman and Dorsey..had tight section work, when needed—but they let their sidemen cut loose, and weave their own patterns.

side meat. *N. Amer.* (chiefly *Southern* and *Western U.S.*). [SIDE *sb.*[1]] Salt pork or bacon, usu. cut from the side of the pig.

1868 *Overland Monthly* Nov. 468/1 But they do not thrive after transplanting any better than do the corn pone of Virginia..and the 'side-meat' of Missouri. **1873** J. H. BEADLE *Undevel. West* xxiv. 482 Two bright-eyed, graceful, copper-colored *señoritas* bring me a supper of coffee, side meat, eggs, and *tortillas de mais*. **1939** J. STEINBECK *Grapes of Wrath* v. 43 But—you see, a bank or a company can't do that, because those creatures don't breathe air, don't eat side-meat. **1957** M. SHULMAN *Rally round Flag, Boys!* (1958) xiv. 158 Accents that recalled hominy grits and sidemeat. **1975** G. V. HIGGINS *City on Hill* viii. 208 None of these intellectual-emotional phenomena is an adequate substitute for side meat and greens.

side-on, *adv.* and *a.* [SIDE *sb.*[1] + ON *adv.* 7 b; cf. *HEAD-ON *adv.* and *a.*] **A.** *adv.* (*si·de-o·n*). With one side directed towards the point of reference; from the side. **B.** *adj.* (*si·de-on*) Directed from or towards one side; indirect. Of a collision: involving the meeting of one side of a vehicle with an object.

1909 L. M. MONTGOMERY *Anne of Avonlea* xiv. 154 His farm is side-on to the Newbridge road. **1928** *Daily Mail* 16 Aug. 13/7 Side-on collisions frequently occur owing to blurred side curtains. **1960** *Times* 10 June 19/1 He may have been a little more side-on at the moment of delivery. **1976** LD. HOME *Way Wind Blows* xi. 156 One of our delegation was sitting with a side-on view of the rostrum. **1977** J. CLEARY *Vortex* viii. 204 Wind was hitting the car side-on.

sidero-[1]. Add: **1. siderona·trite** [ad. Sp. *sideronatrita* (A. Raimondi *Minerales del Perú* (1878) 209)], an orthorhombic hydrated basic sulphate of ferric iron and sodium, $Na_2Fe(OH)(SO_4)_2.3H_2O$, which is a secondary mineral found in very arid regions as yellow masses or crusts, and can be prepared artificially as needle-shaped crystals; **siderophy·llite** [Gr. φύλλον leaf], a variety of biotite containing a high proportion of ferrous iron and aluminium but little ferric iron and little or no magnesium; **si·derotil** (-til, -təil) [a. G. *siderotil* (A. Schrauf 1891, in *Jahrb. d. k.-k. geol. Reichanstalt* XLI. 381), f. Gr. τίλ-ος anything plucked (f. τίλλειν to pluck)], a hydrated ferrous sulphate, $FeSO_4.5H_2O$, found as triclinic fibrous crusts and needles of a white or pale yellow or green colour (see also quot. 1964).

1890 *Amer. Jrnl. Sci.* CXL. 202 Associated with the sideronatrite..is a grayish white laminated mineral, ferronatrite, which is also often intermixed through the whole mass of the sideronatrite. **1935** J. W. MELLOR *Comprehensive Treat. Inorg. & Theoret. Chem.* XIV. 345 Sideronatrite occurs in orange-yellow or straw-yellow, crystalline masses of fine, fibrous structure which separate into thin splinters. **1975** *Nature* 5 June 472/1 Alteration has produced haematite and goethite on the outer parts of the concretions, whereas the inner parts contain jarosite, gypsum, baryte, celestine and sideronatrite. **1880** H. C. LEWIS in *Proc. Acad. Nat. Sci. Philadelphia* XXXII. 255 The name of *Siderophyllite*..has been given in allusion to the large percentage of iron which it contains. **1967** *Mineral. Abstr.* XVIII. 142/1 High temperature metasomatically altered granites of Jurassic and Cretaceous age [in Upper Kolyma, Russian S.F.S.R.] locally include fayalite and siderophyllite greisens in association with Sn deposits. **1897** *Mineral. Mag.* XI. 335 Siderotil... $FeSO_4 + 5H_2O$. Idria, Carniola. **1920** *Amer. Jrnl. Sci.* CCL. 229 Melanterite, either natural or artificial, is commonly coated with a white powder of siderotil and the fine powder [of the former] will dehydrate after standing for some months to the pentahydrate. **1964** *Amer. Mineralogist* XLIX. 820 From a study of natural and synthetic iron sulfates, it is concluded that the name siderotil should be applied to $(Fe,X)SO_4.5H_2O$, where X is any cation or group of cations individually less abundant than Fe. The pure compound probably does not exist in nature, but several examples of the cuprian variety are known.

2. si:deroachrestic (-ākre·stik) *a.* *Path.* [ad. mod.L. *sideroachrestica* (coined in Ger. by L. Heilmeyer et al. 1957, in *Schweiz. med. Wochenschr.* LXXXVII. 1237/2), f. Gr. ἄχρηστ-ος useless (f. ἀ- A- 14 + χρῆσθαι to use)], designating a form of hypochromic anæmia in which impaired synthesis of hæmoglobin renders treatment with iron of no avail; **si·derochrome** *Biochem.* [ad. G. *siderochrom* (H. Bickel et al. 1960, in *Experientia* XVI. 131/2), f. Gr. χρῶμα colour], any of various compounds concerned with the transport of iron in bacteria; **sideropeni·a** *Med.* [*-PENIA], an abnormally low concentration of iron in the blood; hence **sideropeni·c** *a.*; **si·derophage** *Med.* [Gr. φαγεῖν to eat, devour] (see quot. 1970); **si·derophil(e, siderophi·lic** *adjs.* *Geol.* and *Chem.* [ad. G. *siderophil* (V. M. Goldschmidt 1923, in *Skrifter utgit av Videnskapsselsk. I: Mat.-nat. Kl.* iii. 5); see -PHIL, -PHILE], applied to elements which are commonly found in metallic phases (sometimes *spec.* in association with iron) rather than combined as silicates or sulphides, and are supposed to have become concentrated in the earth's core; **siderophi·lin** *Biochem.* [-PHIL, -PHILE + -IN[1]] = *TRANSFERRIN; **si·derosome** *Med.* [*-SOME[4]], a particle of non-hæmoglobin iron in a cell.

1961 *Amer. Jrnl. Clinical Path.* XXXV. 338/1 The patients suffered from an anemia which was resistant to all forms of therapy except blood transfusion and had an erythroid hyperplasia of the bone marrow with a conspicuous accumulation of iron-staining granules in the developing erythrocytes. Similar instances have been described..under the term 'sideroachrestic anemia'. **1970** A. E. LEWIS *Princ. Hematol.* xiii. 205 Sideroachrestic anemia is a very rare, hereditary anemia, refractory to treatment with iron, vitamins, or folic acid. **1961** *Chem. Abstr.* LV. 23684 Sideromycins, sideramines, and other unidentified Fe-contg. biol. active substances are taken together as a group called siderochromes. **1976** *Nature* 19 Aug. 722/2 Microbial iron-transport compounds, or siderochromes are of two general structural types, the phenolates and the hydroxamates. **1938** J. WALDENSTRÖM in *Acta Med. Scand.* Suppl. XC. 395 All these factors may lead to the same result, most suitably called sideropenia. **1946** M. M. WINTROBE *Clinical Hematol.* (ed. 2) xii. 533 Anemia was not present in all of his cases but low plasma iron (sideropenia) was consistently found. **1971** J. H. DAGG et al. in Goldberg & Brain *Rec. Adv. Haematol.* ii. 107 Sideropenia causes well-defined chemical and biochemical changes, and may be associated with the clinical tissue signs found in iron deficiency states. **1939** *Acta Radiol.* XX. (*heading*) 618 The roentgenological diagnosis of sideropenic dysphagia. **1971** J. H. DAGG et al. in Goldberg & Brain *Rec. Adv. Haematol.* ii. 105 An erythrocyte protoporphyrin level above 40·0 μg per 100 ml. erythrocytes and a transferrin saturation of less than 16 per cent taken together, allow a firm diagnosis of the sideropenic state without the necessity for marrow biopsy. **1970** PASSMORE & ROBSON *Compan. Med. Stud.* II. xxi. 11/1 In lesions where there has been much haemorrhage, phagocytosis of iron pigment results in a pigmented stippling of the cytoplasm [of macrophages], and such cells are termed siderophages. **1977** *Lancet* 30 July 244/1 No siderophages were found in the cerebrospinal fluid on the 10th day. **1923** Siderophil [see *lithophil(e* adj. s.v. *LITHO-]. **1950** RANKAMA & SAHAMA *Geochem.* iv. 93 It may be assumed that the typically siderophile elements..are enriched in the nickel-iron core of the Earth. **1954** A. MUIR *Goldschmidt's Geochem.* 680 The scarcity of all the platinum metals in the lithosphere is due to their extremely siderophil nature. **1977** *Nature* 20 Jan. 197/3 The siderophile and volatile elements on the Moon are depleted relative to the Earth and meteorites. **1971** C. B. MOORE in B. Mason *Handbk. Elemental Abundances in Meteorites* xiv. 127 Although by definition silicon is a lithophilic element, evidence also shows that under highly reducing conditions in meteorites it may also be siderophilic. **1949** A. L. SCHADE et al. in *Arch. Biochem.* XX. 170 (*heading*) Carbon dioxide in complex formation with iron and siderophilin, the iron-binding component of human plasma. **1971** *Nature* 28 May 250/1 Transferrin (siderophilin) is a beta-globulin found universally in vertebrate serum. **1970** *Haematologia* IV. 301 In spite of the active rhopheocytosis, ferritin aggregates (siderosomes) were found in the erythroblasts only exceptionally. **1972** W. J. Williams et al. *Hematology* viii. 80/2 Cells containing siderosomes or 'iron bodies' are usually reticulocytes. **1979** *Experientia* XXXV. 256/1 The hepatic increase of ferric deposits (ferritin, siderosomes and lipofuscin aggregates) more or less overloaded in iron in relatives of idiopathic hemochromatosis is well-known.

sideroblast (si·dĕroblast). *Med.* [f. Sidero-^{1} + -blast.] A normoblast containing one or more granules of ferritin.
1954 E. Kaplan et al. in *Blood* IX. 204 For the purpose of this report these cells were designated as 'sideroblasts' in analogy to the term siderocyte already generally accepted for erythrocytes with iron-staining inclusions. *Ibid.* 212 The term sideroblast was proposed for normoblasts with nonstaining inclusions. **1962** Bothwell & Finch *Iron Metabolism* 289 Sideroblasts are absent in iron-deficiency anemia. **1972** Passmore & Robson *Compan. Med. Stud.* III. xxi. 29/2 Ring sideroblasts were first observed in 1949 in guinea-pigs suffering from lead poisoning. They were first described in man in 1953. **1976** *Lancet* 18 Dec. 1364/1 The patient's bone-marrow exhibited a variable number of normal sideroblasts ranging from 20 to 70%.
Hence **siderobla·stic** *a.*, of, pertaining to, or characterized by the presence of sideroblasts; *esp.* in *sideroblastic anæmia*.
1956 *Blood* XI. 250 (*heading*) Chronic refractory anemia with sideroblastic bone marrow. **1972** *Nature* 10 Mar. 73/1 A similar although lesser degree of defective synthesis is seen in two other conditions due to haem deficiency, namely sideroblastic anaemia, and iron deficiency anaemia. **1977** *Blood* L. 165/2 Sideroblastic anemia..is a heterogeneous group of disorders with different basic biochemical abnormalities.

siderocyte (si·dĕrosəit). *Med.* [f. Sidero-^{1} + -cyte.] An erythrocyte containing one or more granules of non-hæmoglobin iron.
1941 H. Grüneberg in *Nature* 26 July 114/2 The 'iron cells' or 'siderocytes' do not stain diffusely, but show blue granules which vary in number from one to a dozen or more and in size from fairly large blobs down to the finest dust-like stipples. **1944** Whitby & Britton *Disorders of Blood* (ed. 4) ii. 36 In some conditions, especially after splenectomy, iron granules may be found in red cells by the prussian blue reaction. The significance of these so-called siderocytes..is unknown. **1966** [see *Pappenheimer]. **1977** *Blood* L. 165 The excessive nonheme iron accumulated in *f/f* fetal siderocytes is located within mitochondria.
Hence **siderocy·tic** *a.*, of, pertaining to, or characterized by the presence of siderocytes.
1957 *Blood* XII. 168 Neither red cells nor inclusion bodies were lost when siderocytic blood was transfused into a recipient without a spleen.

siderosis. For etym. read: [mod.L. (F. A. Zenker 1866, in *Deutsch. Arch. f. klin. Med.* II. 70), f. Gr. σίδηρος iron; see -osis.] and add: **2.** The accumulation in the tissues of siderotic material.
1890 G. M. Gould *New Med. Dict.* 401/1 *Siderosis*, the pigmentation of the lymphatic glands, liver and kidneys, so called from the presence of iron in the pigment. **1906** J. L. Salinger tr. Ehrlich & Lazarus in R. C. Cabot *Dis. of Metabolism of Blood, Animal Parasites, Toxicology* 314 Siderosis is *always* found [in progressive pernicious anemia], that is, an abnormally increased amount of iron in the internal organs, especially in the liver, in the spleen, in the bone marrow, and in the lymph-glands. **1966** Wright & Symmers *Systemic Path.* I. xxi. 661/1 The Bantu people of South Africa show a very high incidence of cirrhosis of the Laënnec type in association with extensive siderosis.
3. The condition in which the lens of the eye is stained with rust derived from an embedded particle of iron. [f. mod.L. *siderosis bulbi* (G. von Bunge 1891, in *Verhandl. des X. Internat. Med. Congr., 1900* IV. x. 151).]
1895 H. R. Swanzy *Handbk. Dis. of Eye* (ed. 5) vi. 190 A discoloration occurs in cases where particles of iron have been imbedded in the eye. Siderosis..is the name given to this..condition. **1926** *Trans. Ophthalm. Soc.* XLV. 281 Particles [of iron] embedded deeply in the sclera do not give rise to siderosis. **1962** D. G. Cogan in A. Pirie *Lens Metabolism Rel. Cataract* 292 One curious property of the epithelium is its capacity to bind and accumulate iron. The entity, called siderosis lentis, may result..from an intraocular foreign body of iron.
Hence **sidero·tic** *a*, formed from or rich in insoluble iron compounds derived from the breakdown of hæmoglobin; of or pertaining to siderosis.
1932 W. Boyd *Text-Bk. Path.* xxviii. 719 All the cases of splenic anemia do not show siderotic nodules, nor are the nodules confined to this disease. **1941** *Nature* 18 Oct. 470/1 The amount of siderotic material per siderocyte is generally small. **1978** J. Batten in R. B. Scott *Price's Textbk. Pract. Med.* ix. 909/2 Siderotic lung disease.

'sides, *prep.* and *adv.* For 'Now *dial.*' read 'Now *dial.* and *colloq.*' and add later examples.
1918 C. Mackenzie *Sylvia Scarlett* ii. iv. 331, I didn't seem to want them no more. 'Sides, I've got seven already. **1936** M. Mitchell *Gone with Wind* v. 79 It's too late den. Dey's already mahied[*sc.* married]. 'Sides, gempmums specs dey wives ter have sense. **1968** J. Wheeler Smith in W. King *Black Short Story Anthol.* (1972) 37 'He didn't hurt you none, did he?' 'No, he didn't bother me, sides looking mean.' **1975** 'J. Lymington' *Spider in Bath* i. 11 It's not like that. Sides, the job helps his pension.

si·de-scan, *a.* Also **side scan, sidescan.** [Side *sb.*^{1} 24.] Applied to side-looking sonar (and radar), esp. on a ship.
1967 *Undersea Technol.* Apr. 24/2 The most important feature of side scan sonar..is its ability to produce a permanent, continuous, graphic record of what it 'sees'. **1969** *New Scientist* 20 Feb. 393/2 A range of sidescan sonars is now available. **1978** *Nature* 5 Jan. 49/2 Side-scan sonar has been used previously to profile icebergs, by lowering a sonar transducer vertically, at a known rate, from the side of a boat. **1979** *Ibid.* 29 Mar. 399/1 These surfaces were revealed by side-scan radar from aircraft in surveys made by the Brazilian government.
So **si·de-scanner; si·descanning** *a.*
1960 *Guardian* 20 July 2/6 In view of the development of side-scanning photography these flights were provocative while not violating international air-space. **1968** *Observer* 4 Feb. 3/3 A standard sonar device is used to sweep a large area and echoes of all possible wrecks are picked up and charted. The side-scanner then moves in to check each echo in detail. **1968** *Proc. 5th Symp. Remote Sensing of Environment* 534 The resolution of a side-scanning sonar instrument depends primarily on the width of the beam in the horizontal plane..and on the duration of the pulses. **1978** *Navy News* Oct. 5/1 She is equipped with echo sounders to measure the depth of water as well as side-scanning sonar which gives a view of the sea bed when locating wrecks.

side-show. Add: (Earlier U.S. and later *fig.* examples, esp. in *Mil.* contexts: cf. *show *sb.*^{1} 15 b.)
1866 C. H. Smith (*title*) Bill Arp, so called. A side show of the Southern side of the War. **1869** 'Mark Twain' *Innoc. Abr.* liii. 573 And so I close my chapter on the Church of the Holy Sepulchre... With all its clap-trap side-shows and unseemly impostures of every kind. **1915** D. H. Lawrence *Rainbow* xiii. 381 But Ursula never told about Winifred Inger. That was a sort of secret side-show to her life. **1919** *Daily Express* (Dublin) 18 Mar. 3/6 (*heading*) Side shows. Sir Chas. Monro's despatch. Important minor operations. **1931** T. E. Lawrence *Let.* 13 Apr. (1938) 718 Your war-history has become one of my constant reference books, for the main war; and that its chapters on the side-shows are so crisply black-and-white as to make exciting stories of them. **1959** *Listener* 12 Mar. 444/1 Not even sideshows in the Yemen or police operations by the army in a riotous colony will save them. **1977** P. Moyes *To kill Coconut* xiii. 178 'What in hell is going on up in the forest.'.. 'I told you, that's a side-show.'

side-slip, *sb.* Add: **5. b.** *Aeronaut.* = *slip *sb.*^{3} 9 j; also, a manœuvre in which this is deliberately produced.
1910 *Flight* 25 June 493/1 Is the banking..sufficient to overcome centrifugal movement (which is of course a sort of side-slip) or not? **1915** [see *overbank *v.* 2 a]. **1928** *Observer* 1 July 17/3 With the greatest of ease they performed side-slips, vertical dives, and loops. **1969** K. Munson *Pioneer Aircraft 1903–14* 108/1 He eventually made a safe landing after the machine was put into a side-slip—a manoeuvre which, as Dallas Brett later recorded, was then 'popularly regarded as being in the nature of a preliminary funeral rite'. **1978** *Sci. Amer.* Nov. 137/1 The fixed rear fin had tended to correct this condition by causing the machine to turn in the direction of the sideslip.
c. *Skiing* and *Surfing.* The action of descending (a slope or wave).
1913 [see *side-slip *v.* c]. **1959** P. Moyes *Dead Men don't Ski* iv. 51 Now they were tackling the sideslip—skidding sideways down icy slopes, their skis flat against the mountain-side. **1968** *Surfer Mag.* Jan. 24/3 Sliding sideways in a controlled sideslip until you reach the bottom of the wave. **1970** N. Fleming *Czech Point* i. 10, I..went into a twenty foot long sideslip and stopped.
d. *fig.*
1916 H. Barber *Aeroplane Speaks* p. v, The dreadful haltings, the many side-slips, the irregular speed, and, in short, the altogether disconcerting ways of a pen. **1921** Galsworthy *To Let* ii. p. vi, He therefore confined himself to discussing with Dumetrius whether Monticellis would come again..and the future of Johns, with a side-slip into Buxton Knights.

side-slip, *v.* Add: **b.** *Aeronaut.* Of an aeroplane: to move sideways, *esp.* towards the centre of curvature while turning (cf. *skid *v.*^{1} 3 c). Also *trans.*, to cause to do this. Cf. *slip *v.*^{1} 9 c.
1911 *Flight* 23 Sept. 830/2 He turned sharply to the left, permitting the machine to bank up too much, whereupon it side-slipped to the ground. **1928** *Rep. & Mem. Aeronaut. Res. Comm.* No. 1187. 6 The aeroplane is side-slipped with the rudder bar central. **1935** C. Day Lewis *Time to Dance* 36 From three thousand feet they tilted Over, side-slipped away. **1941** Pope & Otis *Elem. Aeronaut.* ix. 84 It is possible..to sideslip the plane even during a turn of the 'S'. **1966** M. Woodhouse *Tree Frog* xxv. 183 The whole plane felt dead and we sideslipped fast. **1978** *Sci. Amer.* Nov. 139/1 Every so often the machine would lose control and simply sideslip into the ground.
c. *Skiing.* To descend sideways. Cf. sense 5 c of the *sb.*
1913 A. Lunn *Ski-ing* iii. 79 The expert makes great use of the side-slip for getting down difficult ground... Run with your ski in the normal position for traversing a slope... Flatten them against the slope and slip sideways. Then run a little way in the normal position and again side-slip. **1952** Iselin & Spectorsky *Invitation to Skiing* vi. 101 When you can side slip and stop at will, you are ready for the next exercise. **1972** 'M. Yorke' *Silent Witness* ii. 13 Knees flexed..he had side-slipped down the sheer drop from the shoulder of the mountain.
d. *transf.* and *fig.* Also occas. *trans.*
1917 'Contact' *Airman's Outings* v. 139 Snatches of familiar flying-talk..side-slipped away from Archie. **1921** Galsworthy *To Let* i. iii. 36 His heart moved in a disconcerting manner, as if it had side-slipped within his chest. **1930** *London Mercury* Feb. 319 Here he was suddenly realising that they controlled another mode of clutching. Better to side-slip that too—if he could. **1931** *Technol. Rev.* Nov. 67/1, I was cruising along 43d Street when along came Bill, Sam, and Charlie in formation, and we all side-slipped into a speakeasy and did a lot of barroom flying. **1960** T. McLean *Kings of Rugby* xi. 120 He offered dummy passes, sidestepped, sideslipped. **1964** D. Macarthur *Reminiscences* v. 125 The problem was to sideslip my troops westward..before their path would be cut off from the north.
So **side-slipping** *vbl. sb.* and *ppl. a.*
1887 [in Dict.]. **1904** *Motor Cycle* 18 Apr. 361/3 Side-slipping is a great bugbear to motor cyclists. **1916** 'Boyd Cable' *Action Front* 1 A wet night, a greasy road, and a side-slipping motor-bike. **1930** [see *nose-dive *v.* 1 a]. **1949** A. Fawcus *Skiing Simplified* ii. 49 Sideslipping is the easiest way to get down a steep slope without having to turn. **1950** *Gloss. Aeronaut. Terms* (*B.S.I.*) 1. 10 *Side-slipping*, motion of an aircraft relative to the air such that the air flow has a component along the lateral axis. **1975** *Oxf. Compan. Sports & Games* 957/1 The secret of giant slalom is good control of the ski edges to prevent wasteful sideslipping.

side-step, *sb.* Add: **1.** (Earlier example.) Also *spec.* in *Rugby Football*, a step to the side, as to avoid a tackle, made while running with the ball. Also *fig.*
1789 *Rules & Reg. Field Exercises Army in Ireland* i. 8 The *Side-step*, or march, is very necessary on many occasions when halted. **1927** [see *jink *v.*^{1}]. **1940** 'Gun Buster' *Return via Dunkirk* ii. viii. 148 We've got to make a side-step, giving up part of the ridge we're now holding. **1960** V. Jenkins *Lions down Under* xi. 161 Risman did the rest with two perfect side-steps.

side-step, *v.* Add: **1. a.** *intr.* (Later examples.) Also *transf.*, to go aside from the direct route; to make a side-trip.
1927 *Ladies' Home Jrnl.* Jan. 62, I was glad I side stepped, for the journey..was worth a deviation. **1951** *Sport* 6–12 Apr. 15/2 He twists, turns, side-steps and swerves beautifully. **1973** 'D. Halliday' *Dolly & Starry Bird* xv. 230 Sophia sidestepped and walked sharply past Charles.
b. *fig.* To practise evasion; to avoid an issue or prevaricate.
a **1911** D. G. Phillips *Susan Lenox* (1917) II. xi. 276 What do you think of that, Terry? I offered her a twenty and she sidestepped. **1930** P. Macdonald *Link* vii. 114 I've even asked Dinwater that flat out on two occasions. But he's always side-stepped.
c. *Skiing.* (See quots.)
1924 *Ski Terms* in *Tourist* Winter Sports No. 12/2 *Side stepping*, climbing by lifting the skis horizontally. **1976** *Webster's Sports Dict.* 390/2 *Sidestep*, to climb a slope by employing a sidestep.
2. a. *trans.* To avoid by stepping to the side; *spec.* in *Football.*
1905 'O. Henry' in *N.Y. World* (Mag. section) 12 Mar. 5/4 I've lost two inches of my tail trying to sidestep those swinging doors. **1920** W. Camp *Football without Coach* 93 The man who catches the ball will not be able to sidestep him or pass him. **1931** *Times* 16 Feb. 5/1 Once, Arigho.. managed to side-step Reeve. **1976** *Sunday Mail* (Glasgow) 28 Nov. 46/6 The striker took his time, sidestepped a tackle and neatly shot low past Rennie.
b. *fig.*
1900 Ade *Fables in Slang* 56 The Parents decided to give Clarence a large Measure of Liberty, that he might become Acquainted with the Snares and Temptations of the World when he was Young, and thus be Prepared to side-step the Pitfalls when he was Older. **1906** *Springfield* (Mass.) *Weekly Republ.* 27 Sept. 8 The Idaho republicans are deftly side-stepping the anti-Mormon issue. **1915** Wodehouse *Psmith Journalist* xxiii. 173 If I can put him away, it gets me into line with Jimmy, and he can't side-step me. **1932** E. Wallace *When Gangs came to London* xxviii. 287 One of the best gunmen that ever sidestepped the chair. **1947** E. O. Schlunke in *Coast to Coast 1946* 51 But to make sure he wasn't 'side-stepped', Krantz turned up at the railway station. **1964** *English Studies* XLV. 21 Many people side-step the recognition of a plurality of Englishes by such judgments as: 'Oh, that's not English, that's American.' **1977** *Time* 8 Aug. 25/1 If the British government continues on the line it appears to be following, we will be able to sidestep them and arrive at an internal settlement.
3. *trans.* To cause (a person or thing) to move sideways; to transfer to the side.
1969 A. Glyn *Dragon Variation* viii. 245 Jeff's Queen was *en prise*, and he side-stepped her one square where she would still be able to defend the Rook's Pawn. **1974** *Daily Tel.* 26 Mar. 16 Peter Dimmock, general manager of BBC Enterprises... Mr Dimmock, formerly general manager of BBC Outside Broadcasts, was sidestepped to his present job.
Hence **si·de-stepper**, one who steps sideways or avoids a direct course or issue; **si·de-stepping** *vbl. sb.*, stepping sideways, evasiveness, avoidance; also as *ppl. a.*

1901 G. V. Hobart *Down Line with John Henry* 100, I was the likeliest side-stepper that ever did a grass-chopping speciality. **1909** *N.Y. Even. Post* (semi-weekly ed.) 1 Mar. 1 Had not Mr. Fairbanks been a really wonderful side-stepper, their essential differences might have long ago become public property. **1912** J. Sandilands *Western Canad. Dict. & Phrase-Bk.*, *Sidestepping*, wandering from the argument, evading the question. **1932** *Blue Valley Farmer* (Oklahoma City) 17 Mar. 6/5 When America is grappling with things fundamental, tired and disgusted with side-stepping, buck-passing and plain lying..the country must content itself with a stone when it asked for bread. **1949** 'J. Tey' *Brat Farrar* xii. 100 The conversational ground he moved on was firm. There was need for neither side-stepping nor manoeuvre. **1960** *Times* 16 June 16/2 'Autumn' with its evocative side-stepping harmonies, was delightfully sung. **1970** *Times* 28 Sept. 12/7 The sidestepping at blinding speed of Gerald Davies on the right wing. **1980** N. Freeling *Castang's City* xvi. 105 A dodgy rapid sidestepper, a clever elusive runner.

sides to mi·ddle, *adv.* (and *adj.*) *phr.* Also -into-, side to middle. [Side *sb.*^{1}] Of a sheet of bed-linen: with the sides and middle changing places, as a worn sheet cut down the centre and resewn thus to prolong its useful life. Also (hyphened) as *adj. phr.*
1861 Mrs. Beeton *Bk. Househ. Managem.* ii. 24 Sheets should be turned 'sides to middle' before they are allowed to get very thin. **1884** E. Nesbit *Let.* in D. Langley Moore *E. Nesbit* (1933) v. 71, I have done 2 sheets 'sides into middle'. **1949** H. Ashton *Parson Austen's Daughter* viii. 309 She..had to sleep on sheets which had been turned side to middle. **1950** B. Pym *Some Tame Gazelle* xv. 161 Although the Archdeacon had not personally made the bed, he knew that there were sides-to-middle sheets on it. **1963** 'J. M. Berrisford' *Gardening in Lime* ii. 24 After six to eight weeks the whole [compost] heap may be turned, sides to middle, with the top material becoming the bottom. **1972** 'S. Woods' *They love not Poison* viii. 112 Jenny [was] diligently hemming a sheet that Mrs. Dibb wanted turned 'sides to middle'.

si·de-swipe, *sb.* [f. the vb.; cf. also Side-wipe.] **1.** A glancing blow from or on the side (esp. of a motor vehicle). Chiefly *U.S.*
1917 *Dialect Notes* IV. 400 He struck it with a *side-swipe*. **1935** *Evening Sun* (Baltimore) 7 Mar. 25/3 A few days ago we said there is no such word as *sideswipe* in the dictionary... *Side-wiping* puts the shine on the car—*side-swiping* takes it off. **1964** D. Solomon *Accidents on Main Rural Highways* 1. 2 Nearly half of all accident involvements were either rear-end collisions or same-direction sideswipes. **1968** *Autocar* 14 Mar. 24/3 With adjustable front seats there is weakness in sideswipe. **1977** *Time* 24 Jan. 2/3 (Advt.), The ESV's have proved their life-saving value in head-on and rear-end collisions, side-swipes and roll-overs.
2. A passing jibe or verbal attack; an indirect rebuke or criticism.
1924 Kipling *Prophet & Country* in *Debits & Credits* (1926) 193 A side-swipe at the practically non-existent birth-rate. **1959** *Sunday Times* 29 Nov. 25/6 Wolf Mankowitz has reshaped his satire on show business... A few side-swipes are allowed. **1964** *Ann. Reg. 1963* 26 He allowed himself one side-swipe at the security services, declaring that 'the £60 million spent on these services under the right hon. gentleman's premiership have been less productive..than the security services of the *News of the World*.' **1977** M. Walker *National Front* vii. 179 Martin Webster..again warned of the enemy within in an article..which included a series of side-swipes against 'the trimmers, the popularity-seekers and moderates'.

si·de-swipe, *v.* Chiefly *U.S.* [Side *sb.*^{1}] *trans.* To strike (something) a glancing blow on or with the side (esp. of a motor vehicle).
1904 *Philadelphia Even. Tel.* 12 Nov. 16 The west-bound St. Louis Express, while pulling on to a siding, was sideswiped by the east-bound Pittsburg Limited. **1916** *Daily Colonist* (Victoria, B.C.) 15 July 2/4 Five persons were injured..when a motor car..overturned on the boulevard, after sideswiping a stalled car. **1938** D. Baker *Young Man with Horn* i. v. 41 No sidewalks there, and one out of ten cars doing its best to sideswipe them. **1960** I. Cross *Backward Sex* iii. 66 The cup slid from one end of the tray to the other, sideswiping the teapot on the way. **1973** K. Giles *File on Death* ii. 29 A grey car.. came up fast... It side-swiped her and drove on.

side-track, *sb.* Add: **a.** (Earlier examples.)
1835 *Maine Farmer* 24 July 198/1 One of the principle [*sic*] dealers here has offered to lay a side track from the road to his own storehouse. **1876** G. A. Crofutt *Trans-Continental Tourist* 41 Waterloo is a small side-track station.
b. A side-path.
1892 [see *middle-of-the-road]. **1966** 'J. Hackston' *Father clears Out* 113 Trooper Caldecott riding along the soft, silent side-tracks that led to the Sunday trading.
c. *transf.* and *fig.*
1901 *Congress. Rec.* XXXIV. iii. 2476/1, I do not propose to be side tracked by any Senator from the other side of the Chamber. I myself will decide when I will go on the side track. **1935** B. Malinowski *Coral Gardens & their Magic* II. vi. v. 239 Man never runs on the sidetrack of magical verbiage. **1972** J. Philips *Vanishing Senator* (1973) iii. ii. 136 Couldn't you try rental agencies? It wouldn't be a side track for you, Inspector.

side-track, *v.* Add: **1.** *fig.* (Earlier and later examples.) Also *spec.* to divert or lead (a person) from the main course; to turn (something) aside from prominence.
1880 *News & Press* (Cimarron, New Mexico) 19 Feb. 4/3 Short skirts are now worn for dancing dresses, and the gentlemen are no longer obliged to wait for the ladies to side-track their trains before they can pass. **1887** *Scribner's Mag.* in Farmer *Americanisms* (1889) 487/1 Mebbe them thar lieyers side-tracked him with their everlastin' quashtuns, an' ef so, he warn't so pow'ful much ter blame. **1891** F. H. Smith *Col. Carter* 139 Yancey broke away again, but Fitz side-tracked him with a gesture, and asked the colonel to repeat Klutchem's exact words. **1918** W. S. Churchill *Let.* 12 Jan. in M. Gilbert *Winston S. Churchill* (1977) IV. Compan. i. 231 Munitions are everywhere being side-tracked to the claims of food, of civil imports, of Allies, and of dollars. **1929** C. Connolly *Let.* in *Romantic Friendship* (1975) 325 Delicious people not so young are all somehow sidetracked. **1931** L. A. G. Strong *Garden* xxxv. 324 He'd be all the worse to deal with, if he saw he'd been sidetracked. **1970** C. James in Rubinstein & Stoneman *Education for Democracy* 157 Acknowledgement of special interests is vitally important in adolescence and should not be side-tracked into extra-curricular events. **1978** T. Allbeury *Lantern Network* xi. 160 Bailey disliked the fruity voice..and..the attempt to side-track him.
side-tracking *vbl. sb.* (*fig.* examples).
1947 *Mind* LVI. 291 The abnormal violence of the reaction against Idealism must, I think, be held primarily responsible for the side-tracking of its doctrine of judgment. **1981** A. Price *I'll soldier no More* 54 Through all the verbiage and side-tracking he held to his primary objectives.

sidewalk. 2. For 'Now *U.S.*' read 'Now chiefly *U.S.*' and add further examples outside the U.S.
1824 F. Burney in *Jrnls. & Lett.* (1980) VIII. 525 The streets [of Trier] were dreadfully ill paved, without any side Walk. **1826** *United Empire Loyalist* (Toronto) 1 July 39/4 Some regulation with respect to the improvement of the sidewalks, may be considered as necessary. **1891** G. Meredith *Let.* 27 May (1970) II. 1030 The way to propitiate them [*sc.* reviewers] is to keep along the side-walks, out of the sun. **1936** D. Gascoyne *Man's Life is this Meat* 29 On the sidewalks houses eat the afternoon. **1951** R. Campbell *Light on Dark Horse* ii. 27 The sidewalk..was still unpaved in those days. **1966** G. W. Turner *Eng. Lang. in Austral. & N.Z.* viii 172 Cities have footpaths, though pavement is also used and Aucklanders seem to be able to use side-walk as well. **1971** *Rand Daily Mail* 27 Mar. 6/4 Wait on the sidewalk after getting off a bus. **1977** 'J. D. White' *Salzburg Affair* xiii. 111 The sidewalks with their cheerful.. tourists, street musicians, girls.
3. Comb., as *sidewalk cafe, skate*; **sidewalk superintendent** *joc.* (chiefly *U.S.*), an idler who watches and gives unsolicited advice at construction works, road repairs, etc.; **sidewalk surfing** *slang* (orig. *U.S.*) = *skateboarding *vbl. sb.*; hence **sidewalk surfer.**
1940 R. Chandler *Farewell, my Lovely* viii. 54 The sidewalk café..was bright and cheerful inside, but the.. tables outside under the striped awning were empty. **1979** P. Harcourt *Sleep of Spies* i. vii. 96 The week in Paris passed very quickly... We spent a lot of time..idling in sidewalk cafés. **1925** *Sears Catal.* (ed. 150) 751 Improved extension sidewalk skates. **1977** *Montgomery Ward Catal.* Spring–Summer 509/1 Junior rink-style sidewalk skates. **1940** *Sun* (Baltimore) 30 Mar. 20/7 The walk..is covered so that the sidewalk superintendents can meet in rainy weather. **1970** R. P. Warren *Incarnations* 46 Sidewalk superintendents turn now From their duties and at you stare. **1976** A. Cassorla *Skateboarder's Bible* 9 Weird-wheeling sidewalk surfers can be seen whipping over the blacktop from Reno to Rio de Janeiro. **1965** *National Observer* (U.S.) 5 Apr. 12/1 Remember the hula hoop? The popularity attained by that hip-swinging fad in the late 1950s is fast being dwarfed by a new teen-age craze known as sidewalk surfing or skateboarding.

sidewall. Add: **2.** *Sport* (esp. *Squash Rackets*). A wall forming one side of a court.
1902 [see *boast *v.*^{2}, *sb.*^{2}]. **1935** *Encycl. Sports* 582/2 [Squash] If..the server makes his ball hit the side wall or roof or floor before it hits the front wall,..he loses his innings. **1963** *Times* 4 Feb. 4/3 This service was nothing more than an astutely placed lob, but which, when hit with reverse spin, clung to the sidewall. **1975** *Oxf. Compan. Sports & Games* 823/1 [Real tennis] The side walls are in play up to a height of about 18 ft. (5·5 m.), where the windows begin.
3. a. The side of a vehicle's tyre (Tyre *sb.*^{5} 2), usu. untreaded and freq. marked or coloured distinctively. Freq. *attrib.*, as *side-wall tyre* (also *absol.* as sb.).
1922 *Encycl. Brit.* XXXII. 729/2 The outside of the [tyre's] carcass is entirely covered with rubber: the sides with a 'sidewall' layer. *c* **1949** in M. McLuhan *Mech. Bride* (1967) 83 White sidewall tires..available at extra cost. **1972** *Daily Tel.* (Colour Suppl.) 20 Oct. 10/4 Hans Galli, their chief test-pilot, was..convinced that radial-ply tyres were unsuitable for aircraft because of their flexible sidewalls. **1976** J. Lee *Ninth Man* 70 The car... White sidewalls. Radio and heater
b. A surface at either side of a hovercraft that projects downwards underneath it and helps to contain the air-cushion. Freq. *attrib.*
1960 *Aeroplane* XCIX. 149/1 This is a type of Hovercraft with side walls which are partially immersed in water. **1960** *Economist* 6 Aug. 585/3 William Denny..is to enter a new field for British companies and develop what are called side-wall craft. **1968** *Economist* 18 May 76/3 The car ferry versions of the sidewall hovercraft now being planned..should provide direct competition with the biggest hydrofoil..now building. **1975** *Nature* 6 Feb. 391/1 The development of a 200-passenger sidewall hover-ferry.

sidewards, *adv.* **1.** (Later example.)
1912 J. Stephens *Crock of Gold* vi. 57 He was not looking at her but far away sidewards across the spreading hill.

sideways, *adv.* Add: **2. c.** Const. *on* (On *adv.* 7 b): = *side-on adv.; also (hyphened) as *adj.*
1972 *Guardian* 28 Jan. 6/1 Methods of reducing injuries when a car is hit sideways-on by another car. **1973** W. Barlow *Alexander Principle* ix. 141 We need to look once more at our spines and our stance sideways on. **1976** J. Snow *Cricket Rebel* 37 By the time I returned to Sussex..I was very much more a 'sideways-on' bowler.
3. b. *to look sideways*: also, to regard something in a furtive or improper manner; *spec.* to glance amorously or suspiciously *at*.
1844 'J. Slick' *High Life in N.Y.* I. xiv. 217 If he dared look sideways at his [i.e. another's] wife or sister. **1895** Conrad *Almayer's Folly* 160 This thought caused him to pluck up heart and look at Nina sideways. **1921** B. Gilbert *Old England* 70 But he was known to be paying to three different women for a child each, And his housekeeper beginning to look sideways. **1974** N. Marsh *Black as he's Painted* vi. 162 It wouldn't..be anything out of the way if they got round to looking sideways at each other.
c. In colloq. phr. *to knock sideways*: to astound, as with pleasure or shock; to dumbfound, to amaze.
1925 B. Travers *Mischief* iv. 60 She could have engaged in a viva voce competition with the editor of *The Dog World* and knocked him sideways. **1942** J. B. Priestley *Black-Out in Gretley* vii. 169 When anybody ..does something or has something that suddenly knocks me sideways, I feel I ought to mention it... It's like paying a debt. **1957** R. Mason *World of Suzie Wong* ii. iv. 149 Their attitude is basically commercial... But my guess is that this stuff will knock them sideways. **1960** M. Stewart *My Brother Michael* xvi. 203, I can't seem to think straight... I feel knocked kind of sideways.
d. *Comb.*, as *sideways-looking* adj. = *side-looking *a.*
1962 *Daily Tel.* 29 Oct. 22/3 A new method of obtaining aerial pictures is by 'sideways looking' radar. **1966** *Geo-Marine Technol.* Oct. 18 Sideways-looking sonars produce acoustic pictures of the surface of the sea-bed. **1979** 'M. M. Kaye' *Shadow of Moon* (ed. 2) xxxiv. 408 The dark, secretive, sideways-looking eyes.

side-winder^1. (Earlier examples.)
1840 *Daily Pennant* (St. Louis) 14 May (Th. s.v. Sockdologer), Tim gives him a sockdologer and two side-winders, and leaves him for dead on the spot. **1846** J. Codman *Sailor's Life & Sailor's Yarns* 31 'Take that then, for want of a shillaleh!' said the lumper, giving him a side-winder with his fist.

side-winder^2. Add: **1.** (Earlier and later examples.) Also, one of several other small rattlesnakes.
1875 H. C. Yarrow in *Rep. U.S. Geogr. Surveys West of 100th Meridian* V. 535 They were also seen in Arizona, and are called 'side-winders' by the settlers, owing to their peculiar lateral progressive motion. **1906** *Out West* Feb. 136 It is..a land of the side-winder, the Gila monster, the scorpion and the centipede. **1949** *Nat. Hist.* May 212/2 Among other interesting tracks are the slanting 'ladder-rung' trails of the sidewinder rattlesnakes. **1971** 'D. Shannon' *Ringer* ix. 149 Royce shied back at the sight of him as if he'd been a sidewinder.
2. *transf.* and *fig.*
1906 *McClure's Mag.* XXVI. 414 You never could tell where Texas Pete was goin' to jump next. He was a sidewinder and a diamond-back and a little black rattlesnake all rolled into one. **1936** G. Rounds *Ol' Paul* 21 It was known as the orneriest river... It was an old 'side winder' for fair. **1940** W. Faulkner *Hamlet* iv. i. 244 Hup, you broom-tailed hay-burning sidewinders. **1964** R. Murphy *Pond* ix. 129 'They grew up with her and knew what a sidewinder she was.' 'What's a sidewinder?' Joey asked. 'It's a rattlesnake, but I meant it as a sort of troublemaker.'
Hence as a back-formation, **si·de-winding** *ppl. a.*, moving like a side-winder (also *fig.*); also as *vbl. sb.*; **si·de-wind** *v. intr.*
1902 H. Day *Pine Tree Ballads* 150 That was a side-windin' answer for him. **1954** J. A. Pringle *Common Snakes* p. vi, Some of the adders have a peculiar side-winding movement for crawling over loose sand. **1969** A. Bellairs *Life of Reptiles* i. iii. 105 It is generally believed that sidewinding is the most efficient type of locomotion which a snake can use over a smooth sandy surface... *Crotalus cerastes* can sidewind at about 2 mph. **1972** *World of Wild Wheels* (Custom Car) 65/1. His side-winding Anglia symbolises everything that Hot Rod racing is all about. **1977** 'J. le Carré' *Hon. Schoolboy* xvii. 403 Occasionally a yellow bus came sidewinding down the hill toward them.

sidewise, *adv.* and *a.* Add: **6.** as *adj.* (*fig.* example).
a **1914** Joyce *Stephen Hero* (1944) 48 You're a 'cute fellow, said Stephen in a sidewise fashion.

sidey: see Sidy *a.*^{2} in Dict. and Suppl.

Sidhe (ʃī), *sb. pl. Ir. Mythol.* Also **sidhe** and (*sing.*) **Sidh.** [Ellipsis (not found in Irish) of Ir. (*aos*) *sidhe* people of the fairy mound: cf. *folk of peace* s.v. *folk 3 c and Banshee.] The hills of the fairies; fairyland, faerie. Hence (esp. in the writings of W. B. Yeats), the fairy folk, fairies, freq. regarded as the mythical gods of ancient Ireland. Cf. *sheogue.
1793 J. Hely tr. *O'Flaherty's Ogygia* II. iii. xxii. 55 When the princesses saw these venerable gentlemen.. they looked upon them to be the people of the Sidhe. The Irish call these Sidhe, aërial spirits or phantoms; because they are seen to come out of pleasant hills, where the common people imagine they reside: which fictitious habitations are called by us Sidhe or Siodha. **1880** S. Ferguson *Conary* in *Poems* 95 These wicked sprites,.. men of the *Sidhs*..Who played their pipes before us, led us on Into..the night. **1899** W. B. Yeats *Wind among Reeds* 1 (*title*) The hosting of the Sidhe. **1899** —— *Let.* 21 June (1954) 321, I myself try to avoid the word 'fairy'. ..*Sidhe* or 'gentry' or 'the others' is better. **1906** S. Gwynn *Fair Hills Irel.* ii. 34 The heroes of the mysterious Tuatha de Danann who after their defeat by the Milesians withdrew from daylight into the recesses of the earth—and who are still there, fairy folk, the people of the *Sidhe*. **1919** W. B. Yeats *Only Jealousy of Emer* in *Two Plays for Dancers* 25 What one among the Sidhe has dared to lie Upon Cuchulain's bed and take his image? **1941** L. MacNeice *Poetry of W. B. Yeats* iv. 79 Yeats's world of the Sidhe and curlews. **1977** N. Arrowsmith *Field Guide to Little People* 20 The Sidhe live a very domestic life if undisturbed, caring for their animals, drinking whisky, borrowing milk and meal. *Ibid.* 21 The Sidhe are thin, up to six feet in height, handsome and young-looking despite their great age.

sidi. Add: *sidi-boy* (earlier example).
1867 G. E. Clark *Seven Years of Sailor's Life* viii. 86, I wandered off alone to the 'Seide' boys village of fishermen.

siding, *vbl. sb.* Add: **II. 5. b.** For *U.S.* read orig. and chiefly *U.S.* and add: (Earlier and later examples.) Also made of materials other than timber.
1829 J. F. Cooper *Wept of Wish-ton-wish* I. xvii. 246 [Dwellings] constructed of a firm frame-work, neatly covered with sidings of boards. **1946** E. Hodgins *Mr Blandings builds his Dream House* xiii. 191 The mason subcontractor was stalled in his tracks since..no finished siding could even begin to be nailed to the sheathing. **1958** *N.Z. Timber Jrnl.* Sept. 87/1 *Sidings*, weatherboards for vertical surfaces. Varieties incude: feather-edge, novelty, rebated, ship-lap. **1968** *Globe & Mail* (Toronto) 17 Feb. 1/2 Outside the house, a black hole in the white siding showed where one bullet had driven through the living room wall. **1970** *Washington Post* 30 Sept. b. 13/7 (Advt.), A sparkling white home is yours with Reynolds aluminum siding installed by Hechinger.
c. = Sideling *sb.* 2. *Austral.* and *N.Z.*
1891 G. Chamier *Philosopher Dick* i. xiii. 360 He told him to mind the siding by the shoot. **1902** H. Lawson *Joe Wilson* in *Prose Works* (1946) 350 The dark box-scrub-covered ridges ended in steep 'sidings' coming down to the creek-bank. **1904** G. B. Lancaster *Sons o'Men* 28 Must have gone over the siding. **1931** F. D. Davison *Man-Shy* (1934) xii. 165 The scrubbers were grazing along an ironbark siding. **1975** *N.Z. Jrnl. Agric.* Sept. 27/2 The animal which grazes mostly on non-treated areas—such as gullies or sidings—..will not be fully protected.

sidle, *v.* Add: **1. b.** To make one's way in a horizontal or transverse direction about an incline; *spec.* in *Mountaineering* = Traverse *v.* 21. *N.Z.*
1950 *N.Z. Jrnl. Agric.* Oct. 295/1 Sowing was done following the contours and from higher to lower altitudes, as a man tends to climb when sidling. **1958** *Tararua* XII. 29 To *sidle*, to go around the side or across the face of a hill, is a characteristic New Zealand expression, strange to the Englishman or Australian. **1971** *N.Z. Listener* 19 Apr. 56/5 They got up the lower scree, sidled across the first face into a couloir, but they were getting bombed so they cramponed up to just below a gendarme.

Sidneian (si·dniăn, sidnī·ăn), *a.* Also 7 **Sydnæan, Sydnian; Sidneyan.** [f. the name of Sir Philip *Sidney* (1554–86), English statesman and man of letters, + -an.] Of, pertaining to, or characteristic of the life and works of Sidney.
c **1610** Chapman *Sonnet* [to the Earl of Montgomery] in *Homer Prince of Poets* sig. Ee3^v, There runs a blood, faire Earle, through your cleare vains,.. Which still the liuing Sydnian soule maintaines. **1646** Crashaw *Steps to Temple* 137 Sydnæan showers Of sweet discourse, whose powers Can Crowne old Winters head with flowers.
1931 E. Blunden in *Mind's Eye* (1934) i. 53 It was Sidneian virtue in our colonel to invite even F. into the log cabin for a drink and a tune. **1952** D. Davie *Purity of Diction in Eng. Verse* App. B. 199 The 'sprezzatura' of Castiglione and the Sidneyan ideal. **1965** K. Graham *Eng. Criticism of Novel* ii. 39 Hardy's..eventual abandonment of the novel for poetry can perhaps be foreseen in the high, Sidneyan aims he set for it.

Sidoor, Sidur, varr. *Siddur. **Sidra(h,** varr. *Sedra. **Sidur,** var. *Siddur.

sidy, *a.*^{2} (Later examples.)
1913 C. Mackenzie *Sinister St.* I. ii. iii. 179 The porter was frightfully sick at having to give me a telegram. He is a sidy swine. **1935** 'N. Blake' *Question of Proof* viii. 154 'It is possible that he could have thought he had a chance of being elected?' 'Oh, I should think so; he was sidey enough.' **1946** B. Marshall *George Brown's Schooldays* xxviii. 115 He couldn't very well put himself in first because people might think it rather sidey.

Siebel (sī·b'l), the name of *Siebel* Flugzeugwerke K.G., a German aircraft manufacturing company, used *attrib.* in *Siebel ferry*, a power-driven troop and freight landing-craft developed by them during the war of 1939–45. Also *absol.*
1942 *R.A.F. Jrnl.* 18 Apr. 4 (*caption*) This is the largest scale photograph yet taken of..a Siebel Ferry. Gun positions are visible at the four extremities. **1946** R. Capell *Simiomata* i. 36 A Siebel ferry was moving out of the harbour. *Ibid.*, The Siebel..made things unpleasant with air-bursts. These ferries, accommodating about 150 troops, carry two 88 mm. guns. **1973** D. Hamilton-Hill *SOE Assignment* vii. 100 Siebel Ferries—the special invention of the German Kriegsmarine—double (in parallel) enlarged canoe-type gun boats.

siege, *sb.* Add: **I. 1. e.** *Siege Perilous*: the vacant seat at King Arthur's Round Table which could be occupied without peril only by the Knight destined to achieve the Grail. Also *fig.*
[*c* **1230** *La Queste del Saint Graal* (1967) 4 Et einsi alerent tant qu'il vindrent au grant siege que len apeloit le Siege Perilleux. *Ibid.* 7 Tuit li compaignon de la Table Reonde furent venu et li siege aempli, fors seulement cil que len apeloit le Siege Perilleus.] *c* **1470** Malory *Works* (1967) I. 102 But in the Sege Perelous there shall nevir man sitte but one, and yf there be ony so hardy to do hit he shall be destroyed, and he that shall sitte therein shall have no felowe. **1870** Tennyson *Holy Grail & Other Poems* 43 There stood a vacant chair... And Merlin call'd it 'The Siege perilous', Perilous for good and ill. **1922** J. Buchan *Huntingtower* xiii. 256 There in a coign of the old battlements he would prove an ugly customer to the pursuit. Only one at a time could reach that siege perilous. **1959** P. Le Gentil in R. S. Loomis *Arthurian Lit. in Middle Ages* xix. 261 Three scenes, the fateful occupation of the Siege Perilous and the two visits to the Grail castle, constitute the main pattern.
2. c. (Later examples in *collect.* sense.)
1801 J. Strutt *Sports & Pastimes* i. ii. 28 A *sege* of herons, and of bitterns. **1937** J. W. Day *Sporting Adventure* 106 They [*sc.* herons] are about in pairs instead of the 'sieges' of half a dozen or more which one met only a month ago fishing on the tide line. **1977** *Islander* (Victoria, B.C.) 5 June 3/2 A siege of herons flying home against a sunset sky.
II. 6. a. Also *transf.*
1911 *Times* 5 Jan. 6/2 (*heading*) Foreign opinion on the Stepney siege. **1980** *Daily Tel.* 5 June 8/6 Police forced their way into a flat..after a man had barricaded himself in... During the two-hour siege the man's wife sustained a broken nose.
c. A period of illness, struggle, or difficulty. *U.S.*
1840 [in Dict., sense 6 a]. **1898** E. C. Hall *Aunt Jane* 9 She was as pale and peaked as if she had been through a siege of typhoid. **1929** *Randolph* (W. Va.) *Enterprise* 11 Apr. 1/1 The..Literary Society had another heavy siege Tuesday night of this week. **1952** R. Chandler *Let.* 31 July (1966) 27 She is weakened by a long siege of bronchitis. **1975** *Publishers Weekly* 11 Aug. 113/1 After her own siege with breast cancer, the author consulted with other victims.
7. a. (In *transf.* senses) *siege action, tactics.*
1977 *Evening Post* (Nottingham) 24 Jan. 5/8 The threat to car jobs in the Midlands grew today as delivery drivers began another week of 'siege action' at three big Leyland factories. **1977** P. Hill *Fanatics* 109 Those two have been trained in siege tactics.
b. siege economy, an economic situation in which the availability of imported goods is severely restricted by import controls and the export of capital is curtailed; **siege mentality**, a defensive or paranoid attitude of mind based on an assumption of hostility in others.
1962 S. E. Finer *Man on Horseback* vii. 92 By 1940 the parties had been dissolved, the zaibitsu harnessed to a siege economy. **1979** H. S. Kent *In on Act* xii. 131 The phrase 'siege economy' is sometimes used today to conjure up a last desperate plight in which, under the protection of high tariff walls, we would try to grow our own food, labour grimly in our mines and make the things we needed most; and so control our foreign trade as to bring in the additional supplies that we could not do without. **1969** J. L. McKenzie *Roman Catholic Church* iii. iv. 222 This revival could not have come about without relaxation of the 'siege mentality'. **1976** *Deb. Senate Canada* 8 Mar. 1159/2 With the growing siege mentality in the suburbs of our major urban areas, the people know that crime is not under control.

Siegenian (sīge·niăn), *a. Geol.* [f. *Siegen*, name of a town and region in North Rhine-Westphalia, W. Germany, + -ian.] Pertaining to or designating a stage of the Lower Devonian in N.W. Europe, immediately above the Gedinnian, or the epoch or age during which it was deposited. Also *absol.*
1922 *Proc. Geologists' Assoc.* XXXIII. 12 The higher horizon also includes many Siegenian species. **1928** E. Neaverson *Stratigraphical Palaeont.* xi. 255 In Britain.. the Siegenian stage is best represented by the Meadfoot beds of South Devon. **1931** [see *Famennian *a.*] **1967** M. R. House in W. B. Harland et al. *Fossil Record* i. 47 The goniatites appear in the mid-Siegenian as simple primitive types which soon diversify rapidly. **1979** R. Anderton et al. *Dynamic Stratigr. Brit. Isles* x. 130/2 Non-marine faunas of ostracoderm and placoderm fish fragments and plant remains..indicate a lower Devonian (Siegenian) age for the group.

Siegfried Line (sī·gfrīd ləin). [tr. G. *Siegfriedlinie*, f. the name of the hero *Siegfried* of Wagner's *Ring* cycle (and of the MHG epic poem the *Nibelungenlied*).] The line of fortifications occupied by the Germans in France during the war of 1914–18. Similarly, the line of defence constructed along Germany's western frontier before the war of 1939–45.
[**1923** Kipling *Irish Guards in Great War* I. 204 The Hindenburg line, known to the Germans as 'Siegfried'.] **1936** H. A. L. Fisher *Hist. Europe* iii. xxxiii. 1144 A position which had been fortified with elaborate care, and was known by the Germans as the Siegfried and by the English as the Hindenburg line. **1938** *Times* 25 Oct. 14/4 The evening newspapers [in Berlin] published to-day the first photographs of the so-called 'Siegfried Line', the massive fortifications which are being erected on the western frontiers of the Reich. **1939** *Times* 22 Sept. 6/7 What song is to be the 'Tipperary' of this war? The first candidate would seem to be 'The Washing on the Siegfried Line'. Its chorus is sufficiently simple and singable:—We're gonna hang out the washing on the Siegfried Line. Have you any dirty washing, mother dear? We're gonna hang out the washing on the Siegfried Line... If the Siegfried Line's still there. **1946** *R.A.F. Jrnl.* May 157, I rode straight through the Siegfried Line with dug-outs and then a line of anti-tank traps strung out across the countryside. **1978** E. Malpass *Wind brings up Rain* xxiv. 221 Cholera..had broken out..in the Siegfried Line.

|| **Sieg Heil** (zīg hail), *int.* Also **Sieg-heil, sieg heil,** etc. [Ger., lit. 'Hail victory'.] The victory salute used by the Germans during the Nazi regime, esp. at political rallies, etc. Also as *sb.* and *v. intr.* Hence **sieg-hei·ling** *ppl. a.* Cf. *heil *int.*
1940 'N. Blake' *Malice in Wonderland* ii. 31 The hysterical pitch of the Sieg Heils at a Nazi congress. **1944** V. G. Garvin tr. *R. Gary's Forest of Anger* xxii. 89 He has only done his duty. Nothing else. Sieg-heil! **1967** R. M. Stern *Kessler Legacy* iii. 29 Your newspaper character probably..marched and *Sieg Heiled* with the rest of the boys. **1968** *Guardian* 25 Apr. 1/1, 200 dockers arrived..to shout 'Enoch! Enoch! Enoch!' in 'Sieg heil' tempo. **1968** *Listener* 26 Sept. 403/3 Thus, by 1935, her pictorial records of militarised youngsters, marching young men and sieg-heiling fathers of families were worth uncountable battalions to the Fuehrer. **1976** *Scotsman* 20 Nov. (Weekend Suppl.) 2/6 The film, with its 'Sieg Heils' and hysterical atmosphere, is still trotted out in documentaries about the 1930s. **1978** A. Neave *Nuremberg* xxii. 257, I half expected them to rise, salute and cry '*Sieg Heil*'!

Siemens (sī·mĕnz, || zī·mĕns). The name of four German-born brothers, Ernst Werner (1816–92), Karl Wilhelm or Charles William (1823–83), Friedrich (1826–1904), and Karl (1829–1906) (von) *Siemens*, used *attrib.* and in the possessive: **1.** To denote processes or devices discovered, invented, or developed by one or more of the brothers.
The brothers were closely associated in invention and manufacturing, and it is frequently not possible to attribute a given invention or process to any one of them. Their name is often combined with that of another inventor.
a. *Steel-making.* Sometimes in Comb. with the name of Pierre Blaise Emile *Martin* (1824–1915), French engineer, as *Siemens pyrometer, regenerator*; **Siemens('s) furnace,** an open-hearth furnace; **Siemens-Martin furnace** = *Siemens furnace*; **Siemens-Martin process,** the process, invented by Martin, of melting pig iron and scrap steel together in a Siemens furnace, usu. in alkaline conditions; **Siemens process,** a process similar to the Siemens–Martin process, but usu. carried out in acidic conditions; **Siemens producer,** a form of gas producer developed by the Siemens brothers (see Producer 3).
1866 *Chambers's Jrnl.* 25 Aug. 543/2 For..any..process in which an intense heat is required, the Siemens furnace is eminently suitable. **1875** *Ure's Dict. Arts* (ed. 7) III. 909 Another modification of the Siemens process consists in the use of finely-divided iron in the spongy state..instead of bars or other manufactured forms of malleable iron. *Ibid.* 910 Two processes are employed at the Landore works: the Siemens-Martin process, which consists..in dissolving scrap-metal or steel in a bath of pig-metal, to which spiegeleisen is finally added; and the ore-reducing process. *a* **1877** Knight *Dict. Mech.* III. 2365/2 In the Martin-Siemens reverberatory furnace the decarbonization of the pig-iron is effected by the reactions, upon the molten bath, of wrought-iron or ore and of the furnace-flame. **1877** Siemens producer [see Regenerator 2 a]. **1879** *Encycl. Brit.* IX. 846/1 The most perfect method of utilizing the waste heat hitherto applied is that of the Siemens regenerator, in which the spent gases are made to travel through chambers, known as regenerators or recuperators of heat. **1881** *Ibid.* XIII. 294/2 The calorific value of a unit of weight of gas from a Siemens producer is about 650. *Ibid.* 304/1 The other Siemens pyrometer depends on the alteration of the electrical resistance of a platinum wire when heated. *Ibid.* 305/1 The ball of a Siemens pyrometer can be introduced into the tuyere through the orifice. *Ibid.* 348/2 The Pernot furnace..is substantially a Siemens-Martin furnace with a rotating bed. **1923** Glazebrook *Dict. Appl. Physics* V. 515/1 The development of the mass production of steel by the Bessemer, Siemens, and subsequent processes rendered modern engineering possible on the present scale. **1973** R. D. Pehlke *Unit Processes Extractive Metallurgy* iv. 88 The Siemens-Martin process, commonly referred to as the open hearth process, was developed at about the same time as the Thomas process. **1974** *Encycl. Brit. Macropædia* XVII. 640/1 The Siemens furnace used a grate for burning solid fuel,..and the duct-work necessary to convey the gases from one end of the furnace to the regenerator chamber and the hot air to the grate at the other end was complex and inefficient. *Ibid.*, Pierre and Emile Martin in France in 1864 built a furnace that was fired by gas and placed a set of two Siemens regenerator chambers at each end of the furnace. *Ibid.*, This furnace became known as the Siemens-Martin furnace, or, more commonly, as the open-hearth furnace.

b. In similar technical applications, as *Siemens('s) direction finder, dynamo, (electro-)dynamometer, relay, wattmeter*, and in the names of various forms of lamp.
1867 R. S. Culley *Handbk. Pract. Telegr.* (ed. 2) ix. 184 Siemens' relay consists of an electromagnet of the usual horse-shoe form. **1879** *Telegr. Jrnl.* VII. 318/2 (*heading*) Siemens' electric lamp. *Ibid.* 412/2 (*heading*) Siemens' differential electric lamp. **1882** *Encycl. Brit.* XIV. 633/1 In the Siemens differential lamp,..a potential or shunt coil and a current coil oppose each other; as the arc lengthens the current becomes less, and the potential greater, each acting to cause the carbons to approach. **1884** H. R. Kempe *Handbk. Electr. Testing* (ed. 3) xii. 284 Like galvanometers, the Siemens electro-dynamometer is not susceptible of great accuracy when readings are very low. **1886** J. Maier *Arc & Glow Lamps* v. 40 A large Siemens' dynamo at 450 revolutions after two hours' work became so hot that the electro-magnets began to fire. **1892** W. P. Maycock *Electr. Lighting & Power Distribution* I. v. 122 Siemens' wattmeter is very similar in appearance to the electro-dynamometer. **1912** *Motor Manual* (ed. 14) iii. 125 The Siemens metallic tungsten lamps are practically unaffected by shock. **1922** Glazebrook *Dict. Appl. Physics* II. 4/1 If the current measured passes through both coils the scale division will approximately follow a 'square law', as in the case of the Siemens dynamometer. **1927** S. H. Long *Navigational Wireless* v. 86 (*heading*) Instructions for operating the Siemens direction finder. **1930** T. E. Herbert *Telegraphy* (ed. 5) viii. 258 Siemens-Halske Relay. This relay is developed from the original Siemen's [*sic*] relay. **1966** *McGraw-Hill Encycl. Sci. & Technol.* IV. 476/2 If the same current flows through all coils in series, as in the early Siemen's [*sic*] electrodynamometer.., the instrument can be calibrated as an ammeter.
2. *Electr.* † **a.** A unit of resistance, used esp. in Germany, slightly smaller than the ohm. Usu. in the possessive, as *Siemens'(s) unit. Obs.*
1867 R. S. Culley *Handbk. Pract. Telegr.* (ed. 2) ii. 30 'Siemens' Unit' is one metre of pure mercury, of one square millimetre section, at a temperature of 32°. Thus the Ohm and the Siemens are really alike. **1899** J. E. Young *Electr. Testing for Telegraph Engineers* iv. 48 A third standard, Siemens's unit, equals 0·954 B.A. ohm.
b. (Usu. written **siemens.**) [Named after Charles William *Siemens.*] A unit of conductance, equivalent to the mho.
1935 *Proc. Nat. Acad. Sci.* XXI. 579 (*table*) Siemens. **1936** *Jrnl. Inst. Electr. Engineers* LXXVIII. 238/2 The names 'hertz' and 'siemens' likewise were voted for the names of the practical units of frequency and conductance respectively. **1963** Jerrard & McNeill *Dict. Sci. Units* 128 The siemens is the practical unit of conductance and is equivalent to the mho... The unit, although approved by the I.E.C. in 1933 has not yet replaced the mho. **1972** [see *Pascal 2]. **1978** *Nature* 27 July 379/1 Unit ion conductances of a few picosiemens or greater have been reported.

Siena: see *Sienna.

Sienese, *sb.* and *a.* Add: Now usu. with pronunc. (sīĕnī·z). **A.** *sb.* **2.** An artist belonging to the Italian school of painting developed at Siena during the 14th and 15th centuries. Also *transf.*, a painting produced by such an artist.
1888 H. Attwell *Italian Masters* 6 The Sienese lack the robust simplicity of Giotto, and were little influenced by the austere spirit of Dante. **1921** A. Huxley *Crome Yellow* ii. 12 Henry Wimbush was forced to sell some of his Primitives..four or five nameless Sienese—to the Americans. **1959** *Listener* 26 Nov. 940/2 Minor Sienese are thick on the ground and so are lesser painters of the Dutch school. **1977** 'R. West' *Celebration* 541 Oh, your Florentines, your Sienese, your Umbrians!
B. *adj.* (Earlier example.)
1814 J. Mayne *Jrnl.* 20 Oct. (1909) ix. 154 We walked through the principal parts of the town..and enjoyed the pure Siennese tongue.

Sienna. Substitute for entry:
Siena (si,e·nă). Also **Sienna, † Syenna.** [The name of a city and province in the Tuscany region of central Italy. In sense 1, *ellipt.* for Terra Sienna.]
1. a, b. (In Dict.)
2. *Siena marble*, a reddish mottled stone obtained from the neighbourhood of Siena.
1774 *Builder's Magazine* 129 The body of the work may be of statuary, and the columns of Syenna marble. **1802** Mrs. Edgeworth *Let.* 6 Dec. in C. Colvin *Maria Edgeworth in France & Switzerland* (1979) 46 The Salle for public Lectures is..30 feet high supported by 4 pillars of Sienna Marble. **1848** E. Ruskin *Let.* 20 July in M. Lutyens *Ruskins & Grays* (1972) xiv. 126 The pillars of polished Sienna marble... John said..it made him quite sick. **1894** A. Heaton *Record of Work* Pl. 8 (*caption*) Grate with brass mouldings, Sienna marble slips, and plain..tiles. **1947** J. C. Rich *Materials & Methods of Sculpture* viii. 228 Most of the Siena marbles are veined. Siena Unie, a bright yellow variety, appears to be the only one with little or no marking. **1975** *Country Life* May (*Suppl.*) 35 George III painted console table, with..fine Siena marble tops.

sierozem (sī^ə·rozem). *Soil Sci.* Also **serozem.** [ad. Russ. *serozëm*, f. *séryĭ* grey + *zemlyá* earth, soil.] A type of soil, usu. calcareous and poor in organic material, that is characterized by a brownish-grey surface horizon grading into harder, carbonate-rich lower layers, and is developed typically under mixed shrub vegetation in arid climates.
1934 *Soil Sci.* XXXVIII. 485 On the steeper, and consequently drier, slopes sierozems (gray soils) are developed. **1965** B. T. Bunting *Geogr. Soil* xii. 142 The cooler Russian and Argentine deserts have light grey serozem. **1976** H. E. Dregne *Soil of Arid Regions* 79 A typical Serozem..from near Isfahan in Iran..had a 1 cm. thick desert pavement of fairly angular volcanic rocks overlying a loose, light brownish-gray, coarse sandy loam 4 cm. thick. **1977** J. C. F. Tedrow *Soils of Polar Landscapes* viii. 138 Traditionally, pedologists have focused their taxonomic investigations on the mature terrestric soils—Podzol, Chernozem, Sierozem, and so forth.

sierra. Add: **1. b.** (Earlier example.)
1807 R. Southey *Lett. from England* II. xxxiv. 95 A range of mountains standing in the three provinces of Worcester, Gloucester, and Hereford... This sierra is justly admired for the beauty of its form.
3. = *cero.
1889 in *Cent. Dict.* **1905** D. S. Jordan *Guide Study of Fishes* II. xvi. 266 Almost exactly like it [*sc.* the Spanish mackerel] in appearance is the pintado, or sierra. **1965** A. J. McClane *Stand. Fishing Encycl.* 793/2 Sierra are found along the Pacific coast of America from San Diego to Peru.

Sierra Leone (si,e·ră lī^ou·n), the name of a republic of West Africa, used *attrib.* in **Sierra Leone peach**, a shrub or small tree, *Nauclea latifolia* (formerly *Sarcocephalus esculentus*) of the family Rubiaceæ, native to tropical Africa, or its edible reddish fruit; = *Guinea peach* s.v. Guinea 1.
1866 Lindley & Moore *Treas. Bot.* II. 1020/1 S[*arcocephalus*] *esculentus* has pink flowers and an edible fruit, of the size of a peach, whence it has been called the Sierra Leone Peach. **1965** M. S. Nielsen *Introd. Flowering Plants W. Afr.* ix. 173 Sierra Leone or Guinea Peach..is a bush or small tree in savanna with dark red, juicy fruit balls.
Hence **Sierra Leo·nean, -ian**, an inhabitant or native of Sierra Leone; also as *adj.*, of or pertaining to Sierra Leone.
1791 A. M. Falconbridge *Let.* 8 June in *Narr. Two Voyages to River Sierra Leone* (1802) iv. 83 The people appear more inclined to industry than the Sierra Leonians. **1897** M. H. Kingsley *Travels W. Afr.* 680 Bishop Ingram would have been able to write a more cheerful and hopeful book..if the Sierra Leonians had had a thorough grounding in technical culture. **1910** T. J. Alldridge *Transformed Colony* ix. 75 The Sierra Leonean trader feels all this pretty badly, especially as every Sierra Leonean wants to trade. **1926** *Chambers's Jrnl.* Nov. 660/1 Sierra-Leonean solicitors were hurrying to final interviews with retained advocates. **1957** M. Banton *W. Afr. City* vi. 104 At the [1931] census the term 'Sierra Leonian' was adopted for persons previously classified as 'Liberated Africans and their descendants'. **1974** *Times* 4 May (Sierra Leone Suppl.) p. i/1 Rhythm plays a particularly important role in the way of life of all Sierra Leoneans. **1976** *Sunday Times* (Lagos) 31 Oct. 5/4 The Sierra Leonian repatriates..played a much greater role.

sies, var. *sis int.

siesta. Delete || and add: **a.** Also *transf.*
1946 D. C. Peattie *Road of Naturalist* i. 19 Its crepuscular flowers are large as those of a wild rose when they open..great mothlike petals languidly expanding, as if still oppressed with the long siesta of the day. **1947** J. Stevenson-Hamilton *Wild Life S. Afr.* iii. 27 The rhinoceros, startled suddenly from his midday siesta, rushes blindly at the intruder.
Hence as *v. intr.*, to take a siesta.
1899 C. J. C. Hyne *Further Adventures Capt. Kettle* ii. 30 'Right,' said Kettle. 'I'll siesta too.' **1956** G. Durrell *My Family & Other Animals* xv. 201 The rest of the family, finding they could not siesta with the argument going on, assembled to find out the trouble. **1976** L. Deighton *Twinkle, twinkle Little Spy* i. 7 The staff have noisy arguments about who should siesta on the cold stone floor.

Siesta^2 (si,e·stă). *Chess.* The name of a town in Italy, used *attrib.* in *Siesta gambit, variation*, a continuation of the *Ruy Lopez opening popularized in a tournament held there (see quot. 1965).
1935 Smith & Dash tr. *Znosko-Borovsky's How to play Chess Openings* 54 This line is known as the Siesta Gambit and has been much played in recent years. **1948** G. Abrahams *Teach yourself Chess* iii. v. 195 P to KB4 gives the Siesta variation which is playable. **1965** *Listener* 29 Apr. 651/3, 1 P–K4 P–K4 2 N–KB3 N–QB3 3 B–N5 P–QR3 4 B–R4 P–Q3 5 P–B3 P–B4: the so-called 'Siesta' variation, which, despite its name, leads to lively play and provides an early counter to White's positional grip on the centre.

|| **sieur^1** (syö̆r). Now *arch.* [Fr.: cf. Monsieur.] Used as a courtesy title or form of address.
1772 in D. Arundell *Sadler's Wells* (1965) iii. 28 Principal dancers. Sieur Daigueville and his pupils. **1893** S. Weyman *Gentleman of France* I. iii. 66 You have not told me yet, sieur, where we stay to-night. **1901** G. B. Shaw *Admirable Bashville* iii. 124 Bumpkin Fitz Algernon de Courcy Cashel Byron, sieur of Park Lane and overlord of Dorset. **1981** P. Vansittart *Death of Robin Hood* ii. ii. 40 John's dwarf, Sieur Marc,..now discarded winter as he might a cloak.

|| **sieur^2** (syö̆r). *S. Afr.* [ad. Afrikaans *seur*, f. Du. *sinjeur* lord, master: ultimately related and assimilated to prec.] A respectful form of address or reference to a superior; master, 'sir'.
1812 A. Plumptre tr. *Lichtenstein's Trav. S. Afr.* I. 118 The former [*sc.* the Hottentots] only address their master by the title of *Baas* (Master), while the slaves address him as *Sieur* (lord). **1886** G. A. Farini *Through Kalahari Desert* 312 You ought not to have stayed here last night; the klein Sieur was very anxious. **1942** 'B. Knight' *Sun climbs Slowly* xviii. 154 'Missis, *Sieur*,' she shrilled excitedly. 'The veld is on fire.' **1968** K. McMagh *Dinner of Herbs* 56 The maid who took the early morning coffee reported this to the sieur.

sieva (sī·vă). [Origin unknown.] A kidney bean belonging to an American variety of *Phaseolus lunatus*, or its edible seed.
1888 G. D. Merrill *Hist. Coös County* v. 42 The Indians had..a kind of bean called now 'seiva [*sic*] bean'. **1949** *Nat. Geogr. Mag.* Aug. 159/1 The lima beans grown by the various Indian tribes..varied from the present small types used by the Hopi Indians in the Southwest to the Sieva type found in the East. **1972** Y. Lovelock *Veg. Bk.* i. 55 The scimitar-podded kidney bean..., also known as Hibbert, sieva or sugar bean, grows wild in tropical America and is also cultivated.

sieve, *sb.* Add: **4*.** *Math.* **a.** In full *sieve of Eratosthenes* [tr. Gr. κόσκινον Ἐρατοσθένους, f. the name of the Greek scientific writer of the 3rd c. B.C. who devised it]. A method of finding the prime numbers in a (usu. consecutive) list of numbers by deleting in turn all the multiples of all possible prime factors.
1803 tr. *Bossut's Gen. Hist. Math.* 18 The famous sieve of Eratosthenes..affords an easy and commodious method of finding prime numbers. **1857** *Proc. Ashmolean Soc.* III. 128 To Eratosthenes of Alexandria..is attributed the invention of the method by which the primes may successively be determined in order of magnitude. It is termed..'the sieve of Eratosthenes'. **1945** E. T. Bell *Development of Math.* (ed. 2) iv. 89 Boethius reproduced the sieve of Eratosthenes and offered some amusing trifles on figurate numbers. **1966** Ogilvy & Anderson *Excurs. Number Theory* viii. 97 There is no known formula that turns out the prime numbers. Essentially the only way to find them is by the use of the 'sieve' devised by Eratosthenes.
b. A method of estimating or finding upper and lower limits for the number of primes, or of numbers not having factors within a stated set, that fall within a stated interval.
1897 *Nature* 6 May 10/2 (*heading*) Sieve for primes. **1952** *Proc. Internat. Congr. Mathematicians* I. 286 Ever since Viggo Brun introduced his ingenious sieve-method, it has been a very important tool in connection with problems in the theory of primes. **1972** M. N. Huxley (*title*) The distribution of prime numbers: large sieves and zero-density theorems.
6. a. sieve analysis, a particle-size analysis of a powdered or granular material made by passing it through sieves of increasing fineness; **sieve map**, a map upon which the distribution of a number of features is depicted by means of transparent overlays.
1928 C. C. Wiley *Princ. Highway Engin.* ii. 25 Gravel should be well graded from fine to coarse. This is determined by a sieve analysis. The sieve analysis curve for a high-grade gravel should approximate a straight line. **1971** R. Hardbottle tr. *Grassmann's Physical Princ. Chem. Engin.* v. 302 Sieve analysis, in which the grains are passed in succession through sieves of various finenesses, gives directly a cumulative curve, in which..the masses or weights of the different fractions are given. [**1938** E. G. R. Taylor in *Geogr. Jrnl.* XCII. 25 The last map..is constructed on what I have termed the sieve method.] **1952** Monkhouse & Wilkinson *Maps & Diagrams* iv. 190 E. G. R. Taylor produced a map upon which all areas in Great Britain unsuitable for industrial location were indicated in solid black. These areas were determined by superimposing isopleths representing certain specific factors... This process was termed 'sieving out' and the resultant maps are sometimes referred to as 'sieve-maps'. **1965** *Listener* 27 May 774/2 When to these were added those areas of real natural beauty within the conurbation..and areas of the highest agricultural value..the result (which we called the regional sieve map) was an exceedingly complex jig-saw puzzle.

sieved (sivd), *ppl. a.* [f. Sieve *v.* + -ed[1].] Passed through a sieve.

1949 *Nat. Geogr. Mag.* Aug. 172/2 [Kale and collards] ..in a finely chopped or 'sieved' form as food for babies. **1971** *Nature* 30 Apr. 559/2 Sanidines were extracted by elutriation of crushed and sieved pumice breccia.

Sievers (sī·vəɹz, ‖ zī·fərs), the name of Eduard *Sievers* (1850–1932), German philologist, used *attrib.* of the rule formulated by him (*Beitr. z. Gesch. d. deutschen Sprache u. Lit.* (1878) V. 129) that in Indo-European (post-consonantal) unaccented *i* and *u* before a vowel were consonantal after a short syllable and vocalic after a long syllable; also, this rule as modified by later scholars, or as applied by them to particular early Indo-European languages.

F. Edgerton (*Language* X (1934) 235 f. and elsewhere) played an important part in developing and modifying the rule, which is now sometimes called 'Sievers-Edgerton's Law'.

1934 *Language* X. 235 (*title*) Sievers's law and IE. weak-grade vocalism. **1939** E. Prokosch *Compar. Gmc. Gram.* 92 The difference in treatment according to the character of the preceding syllable..may have been Indo-European; this view was first expressed by Sievers... It is frequently referred to as 'Sievers' Law'. **1953** *Jrnl. Eng. & Gmc. Philol.* LII. 149 The essential characteristic of the Proto-[Indo-]European resonants was their threefold function as vowel, e.g. [i], consonant, [y], or vowel plus consonant [iy], as conditioned by the preceding phonemes; the description of this variation is generally known as Sievers' Law. It is most clearly apparent in our Vedic Sanskrit documents. **1959** A. Campbell *Old Eng. Gram.* 164 This interchange of *i* and *ij* is called Sievers' law. **1966** M. R. Emeneau in Birnbaum & Puhvel *Anc. Indo-Europ. Dialects* 126 While Sievers-Edgerton's law was in full working order, something other than phrase..must have been the phonemic form. **1975** Lass & Anderson *Old Eng. Phonol.* 273 The earliest instance in Germanic of an important role being played by the long/short syllable dichotomy, is the set of phenomena associated with the *sonalischen Auslautgesetz* ('law of vocalic finals'), now usually called 'Sievers's Law', or the 'Sievers-Edgerton Law'.

Sievert (sī·vəɹt). The name of R. M. *Sievert*, 20th-c. Swedish radiologist, used to denote either of two units of dose of ionizing radiation. †**a.** *Sievert unit* (see quot. 1955). *Obs.* **b.** (Written *sievert*.) (See quot. 1977.)

1945 C. W. Wilson *Radium Therapy* iii. 70 For the practical determination of the Sievert unit dose by this method, the measurement is made..with a chamber having walls of a finite thickness. **1955** *Gloss. Terms Radiology (B.S.I.)* 19 *Sievert unit*, a unit of gamma-ray dose, being the dose of radiation delivered in one hour at a distance of 1 cm from a point source of 1 mg of radium element enclosed in platinum 0·5 mm thick. It is numerically equal to about 8·4 röntgens. **1977** *Ann. ICRP* I. III. 4/1 The special name for the unit of dose equivalent is the sievert (Sv): 1 Sv = 1 Jkg^{-1} (= 100 rem). **1982** *Sci. Amer.* Feb. 34/2 The data come from fairly high doses of radiation (·5 sievert to two sieverts).

sif, var. *seif.

sifaka (sɪfæ·kă). Also **sifac.** [Malagasy.] A small arboreal primate belonging to the genus *Propithecus* of the family Indriidæ, native to Madagascar and distinguished by whitish silky fur with darker patches on the head and limbs, a hairless black face, and a long tail.

1845 *Encycl. Metrop.* XVI. 429/1 It [sc. *Cercopithecus Memæus* Cuv.] is a native of Cochin China and Madagascar, where it is called *Sifac*. **1901** W. Rice *Animals* 37 A smaller indri with a long tail is the sifaka, the native name of a pretty animal which..goes about by day in bands of six or eight. **1930** *Times Educ. Suppl.* 1 Mar. (Home & Classroom Suppl.) p. ix/2 The monkey-lemurs, or sifakas, are purely arboreal. **1961** *Listener* 2 Nov. 709/1 A sifaka lemur seen in David Attenborough's 'Zoo Quest to Madagascar'. **1978** *Nature* 19 Oct. 587/1 White sifaka and ringtailed lemur populations have remained stable since 1963 in the privately protected 100-hectare reserve at Berenty.

siff, var. *syph.

1947 C. Willingham *End as Man* ii. 18 Why don't you tell us about that time you got siff from your nigger maid? **1971** B. W. Aldiss *Soldier Erect* 157 Them mankey whores in yon knocking-shop'll give you a dose as soon as look at you. There's no' a one of them as isn't rotten with siff.

siffleur (siflör). Also †*erron. pl.* **siffleux.** [Fr., lit. a whistler.] **1.** *Canad.* One of several animals that make a whistling noise, esp. the hoary marmot, *Marmota caligata*, or its flesh used as food. Cf. Whistler 2 b.

1703 tr. *La Hontan's New Voy. N. Amer.* I. 62 Certain little Beasts, call'd Siffleurs, or Whistlers. **1808** S. Fraser *Jrnl.* 18 June (1960) 86 They gave us a *siffleur* (marmot) which is the first fresh meat we tasted since our departure. **1858** J. Palliser *Jrnl.* 22 Aug. in *Palliser Papers* (1968) 274 The only animal which we have seen is the siffleur, whose shrill whistle we heard for the first time close to our encampment. **1898** F. Russell *Expl. Far North* 249 Ground squirrels or 'siffleux' as they are known to the Company's people, are a characteristic feature of the barren portions of Arctic America. **1949** *Canad. Alpine Jrnl.* May 32 They dined on 'delicious siffleur' which tasted on the tongue like 'very delicate mutton or the fat of sucking pig'. **1968** R. M. Patterson *Finlay's River* 209 La Guarde shot a couple of siffleurs—marmots or whistlers—lean, miserable things at that season.

2. (with fem. *-euse*). One who entertains professionally by whistling.

1827 T. Dibdin *Reminiscences* II. vi. 122 The following is from a celebrated *siffleur*:..'I Take the Liberty of Inclosing a few Lines, to Inform you that I am a Beautyfull Whistler If you Please to Give Me one Trial on the Stage.' **1912** *Music Hall & Theatre Review* 11 Jan. 31/1 'Who is He', a clever siffleur, charms everyone with his mimicry of birds. **1981** *Times* 26 Feb. 13/1 Melba's friend Adeline Murrelli was not a singer but that almost forgotten artist, a siffleuse, or whistler.

sift, *sb.* 3. (Examples.)

1876 G. M. Hopkins *Wr. Deutschland* iv, in *Poems* (1967) 52, I am soft sift In an hourglass. **1962** M. E. Murie *Far North* II. ix. 192 The little sift of snow on the ice was marked only by tracks of ox.

sig, *sb.* Add: **2.** A solution applied to the grain side of leather before it is stained black. *Obs.*

1897 C. T. Davis *Manuf. Leather* (ed. 2) 623 In the making of 'sig' stains, blacks and pastes for leather, borax is the currier's friend. **1900** H. C. Standage *Leather Worker's Manual* iv. 69 *Seasoning for Oil-Grain Leathers*—This is sometimes called 'sig'. It is a fluid which is put onto the skin, so as to dye or stain it, or otherwise prepare it for being blackened on the grain side. **1903** L. A. Flemming *Pract. Tanning* 51 A good 'sig' is made of forty gallons of water, twelve pounds of salts of tartar, five pounds of bichromate of potash and one quart of ammonia.

sig., abbrev. of Signature. (Examples.)

1866 G. Simpson *Let.* 30 May in *Geo. Eliot Lett.* (1956) IV. 263 Vol. II. is on the Machines up to Sig. K. 9 and the whole of it will be on to morrow. **1959** *N. & Q.* Dec. 461/1 In the Preface to the Lay-Reader of Richard Baxter's *Gildas Salvianus: The Reformed Pastor* (1656), sig. C. 8, the following passage occurs. **1972** P. Gaskell *New Introd. Bibliogr.* 99 (*caption*) Sheet of duodecimo.. with two signatures..(12° in 8s and 4s, 2 sigs.).

Sigatoka (siŋgătōᵘ·kă). Also **Sing-.** [Name of the district of Fiji where the disease was first observed.] Used *attrib.* and *absol.* to designate a disease of banana plants caused by the fungus *Cercospora musæ*, characterized by the appearance on the leaves of elongated spots, followed by rotting of the entire leaves.

1925 *Agric. Circular Dept. Agric. Fiji* V. 68 Sigatoka Disease—This is by far the most important of the banana diseases of Fiji. **1958** *New Scientist* 26 June 258/1 Sigatoka has spread to virtually all the banana growing areas since it was discovered in 1903. **1963** A. Burns *Fiji* II. iii. 197 Leaf-spot disease, which affects bananas in various parts of the world, is often referred to as Singatoka Disease, because it was first identified in the Singatoka area of Viti Levu. **1972** C. W. Wardlaw *Banana Dis.* xi. 314 In Fiji, it became notorious in 1913 in the Sigatoka valley —hence the name 'Sigatoka Disease' or simply 'Sigatoka'.

sigh, *sb.* **3.** *sigh-like* (later examples).

1911 J. A. Thomson *Biol. of Seasons* II. 193 When the young bird appears to be contented and very comfortable, it utters a plaintive, almost sigh-like cheep. **1964** J. C. Catford in D. Abercrombie et al. *Daniel Jones* 32 Auditory effect, 'sigh-like' mixture of breath and voice: one form of voiced [h].

sigh, *v.* Add: **3. c.** With cognate obj.

1789 Blake *Songs Innoc.* f. 6, Think not, thou canst sigh a sigh, And thy maker is not by. **1847** C. Bronte *J. Eyre* II. ii. 37 She sighed a sigh of ineffable satisfaction, as if her cup of happiness were now full. **1888** Mrs. H. Ward *R. Elsmere* III. xli. 225 Robert sighed a long sigh.

sighfully, *adv.* (Later examples.)

1905 W. J. Locke *Morals Marcus Ordeyne* x. 213 My aunt sighfully acquiesced, and for a while we discussed the depravity of human nature. **1925** T. Dreiser *Amer. Trag.* (1926) II. III. xxv. 321 And here he wearily and sighfully drew forth his large white handkerchief once more.

sight, *sb.*[1] Add: **I. 1. d.** In colloq. phr. *a sight for sore eyes*: a person or thing one is glad to see, esp. a welcome visitor.

[**1738** Swift *Polite Conv.* I. 7 The Sight of you is good for sore Eyes.] **1826** Hazlitt in *New Monthly Mag.* XVI. 38 Garrick's name was..proposed..on condition he should act in tragedy and comedy... What a *sight for sore eyes* that would be! **1871** *Monthly Pkt.* Christmas 108 You're a sight for sair ee'n the now! I'm just aboot in the awfuest swither ever a body was. **1897** R. Marsh *Crime & Criminal* xxiii. 192 He was a sight for sore eyes... I like to see a man that is a man. **1931** E. O'Neill *Mourning becomes Electra* (1932) 125 You certainly are a sight for sore eyes, Vinnie! **1973** *People's Jrnl.* (Inverness) 28 July 4/5 Elizabeth..and Sheena had done wonders,.. and the buffet was a sight for sore eyes.

e. Something which calls forth contemptuous, horrified, or amused glances; a shocking, repulsive, or ridiculous spectacle. *colloq.*

[**1694** W. Penn *Rise of Quakers* ii. 53 It was not very easie to our Primitive Friends, to make themselves Sights and Spectacles, and the Scorn and Derision of the World.] **1862** F. W. Robinson *Owen* II. v. iv. 288 I'm getting better now,..I was a sight last week. **1911** M. Beerbohm *Zuleika Dobson* xx. 296 Clarence curbed the brotherly intention of telling her she looked 'a sight' in them. **1940** W. Faulkner *Hamlet* III. i. 229 'Aint he a sight now,' Snopes cackled.

f. *sight unseen*: without previous inspection, without seeing the object to be acquired. orig. *U.S.*

1892 *Dialect Notes* I. 231 To trade knives *sight unseen* is to swap without seeing each other's knife. **1898** *Yearbk. U.S. Dept. Agric. 1897* 427 The intelligent farmer of today has got beyond trading 'sight unseen' or 'buying a cat in a bag' when it comes to fertilizers. **1940** F. D. Davison *Woman at Mill* 94, I learned that he had selected sight-unseen, that he had now come to look over his property for the first time. **1962** V. Nabokov *Pale Fire* 82, I have had occasion to say something about the amenities of my habitation. The charming, charmingly vague lady ..who secured it for me, sight unseen, meant well, no doubt. **1968** *Listener* 7 Mar. 303, I said you were mad to advertise our modest needs—sight unseen—in the New Statesman. **1979** *Daily Tel.* 3 Feb. 34/2, I am recommending this [TV film], sight unseen, because the first offering in the series..was so good.

2. b. (Further examples.)

1931 R. Campbell *Georgiad* i. 12 He could be..heterosexual with either, too—A damn sight more than you or I could do! **1958** *Times* 16 Oct. 17/1 Surrey..will have to do a sight better than they did yesterday if they are to make their presence fittingly felt in the current Rugby Union county championship. **1977** 'E. Crispin' *Glimpses of Moon* xii. 241 Be a sight cooler there than it is here, I reckon. **1979** C. P. Snow *Coat of Varnish* xvi. 133 Money might be fun, but if I had to choose I'd a damn sight rather try for the top jobs.

c. Phr. *by a long, damn*, etc., *sight*, by a long way, by a good deal (usu. in negative contexts). *U.S. colloq.*

1834 C. A. Davis *Lett. J. Downing* 41 'Gineral, do you want another report?' 'Not by a darn'd sight.' **1840** *Niles' Register* 9 May 149/2 He asked him if he was not going for Harrison and the whigs. 'No,' said he, 'not by a d—d sight.' **1844** *Republican Sentinel* (Richmond, Va.) 22 June 1/2 These animals begin to venture out a little of nights, since the Baltimore Convention, but are slyer by a long sight than foxes. **1884** 'Mark Twain' *Huck. Finn* i. 5, I asked her if she reckoned Tom Sawyer would go there, and she said not by a considerable sight. **1894** —— in *Century Mag.* Mar. 779/2 It ain't on'y jist Essex blood dat's in you, not by a long sight. **1931** E. O'Neill *Mourning becomes Electra* (1932) 248 But I don't wish to convey that he approves of all I've set down—not by a damned sight! **1959** E. Pound *Thrones* ciii. 88 But not his fault by a damn sight. **1976** M. Machlin *Pipeline* xxii. 272 The excitement ain't through here by a damn sight.

II. 5. e. *Poker.* A show of hands; *spec.* one called for by a player who has insufficient chips or money to equal another's bet, but bets as much as he can. *U.S.*

1821 G. Long *Hoyle's Games Improved* 162 The youngest hand..may *call a sight*... If he *calls a sight* the cards must be shown in rotation, the player who calls showing last, and the best hand shown wins the pool. **1850** H. G. Bohn *Bohn's New Hand-bk. Games* 381 Should one of the party over-reach the amount that is in possession of an adversary, a '*sight*' may be demanded. **1887** *Courier-Jrnl.* (Louisville, Kentucky) 23 Jan. 15/7 Then a rule sprang up that a man should be allowed a sight for his money. **1940** O. Jacoby *On Poker* 150 Even though a player's hand is beaten in sight, he should make no move to fold it except in his proper turn. **1964** E. Sinclair *Poker* v. 142 If a player who is beaten in sight bets against the cinch hand, he will be allowed to withdraw his bet from the pot after his attention had been drawn to the fact that he is beaten in sight.

f. A sale of packets of uncut diamonds.

1940 *Economist* 2 Mar. 385/1 A significant hint on the likely evolution of the British exchange control technique was provided last week on the occasion of the latest diamond 'sights' held in London. **1966** J. Wainwright *Crystallised Carbon Pig* xv. 74 The Diamond Corporation hold monthly 'sights'—auctions, I suppose you could call them. They sell anything between three million and five million pounds' worth of stones at each 'sight'. **1978** *Times* 9 Mar. 23/6 The rough gems are sold by the CSO at 10 'sights' (sales) a year... London holds the most important sights.

7. a. For 'Now *rare*' read 'Now *rare* except in slang phr. *to take sights*, to observe, to watch.' (Further examples.)

1934 P. Allingham *Cheapjack* xiv. 175 He pointed out a Rolls-Royce which stopped at the entrance of the fair... 'Take sights at that Rolls,' he said. 'There'll be some right mugs turn up in a minute.' **1950** R. M. Howe *Gross's Criminal Investigation* (ed. 4) viii. 163/1 *Take sights*, looking out (especially for suitable house to break into). **1962** *New Statesman* 21 Dec. 897/3 Once we have all this information, we start 'taking sights', and this means watching the house, from the grounds, for the best part of a week.

b. (Earlier example.)

1834 *Reg. Deb. Congr. U.S.* 25 Feb. 691, I supposed for once in my life I saw gentlemen in the open field, and might be able to draw a fine sight upon them.

d. (Example.)

1848 Bartlett *Dict. Amer.* 303 In North Carolina the distance that can be seen on a road is called a *sight*.

III. 8. b. With definite article, *spec.* = Second sight. Chiefly *Sc.*

1924 W. Holtby *Crowded Street* xxix. 195 They say she's got the 'sight'—you know, second sight. **1925** W. Duke *Scotland's Heir* x. 223 At the last the Sight came upon him, and he reared upright, crying with outflung arms that he saw bloody claymores. **1959** E. H. Clements *High Tension* v. 80 The factor stared at him. 'Why, Kilmorrin ,you have the sight!' **1977** C. McCarry *Secret Lovers* x. 129, I do believe you see me in everything I do. If you haven't the sight, then what is the explanation?

c. Contrasted with *faith*.

c **1382** Bible (Wyclif) *2 Cor.* v. 7 For we walken bi feith, and not bi cleer siȝt. **1611** *Ibid.* (A.V.), For we walke by faith, not by sight. **1834** J. H. Newman *Parochial Sermons* I. xvii. 258 And all these inducements to live by sight and not by faith are greatly increased, when men are engaged in any pursuit which properly *belongs* to the intellect. **1858** W. Brock *Sir H. Havelock* xiii. 216 He might well have doubted of success had he walked by sight. **1871** H. Alford in *Hymns Anc. & Mod.* (1875) 285/1 Forward, marching eastward Where the heaven is bright, Till the veil be lifted, Till our faith be sight. **1981** M. Green *I believe in Satan's Downfall* vii. 201 To walk by sight would be the very antithesis of the trusting walk of faith to which God's Messiah, along with all men, was called.

10. b. See also *out-of-sight *adj. phr.* (*sb.*).

11. c. *line of sight* (further examples); also *transf.* with reference to the transmission of radio waves, etc.; freq. *attrib.* (with hyphens); *line-of-sight velocity* = **radial velocity*.

1920 A. S. Eddington *Space, Time & Gravitation* viii. 135 In the case of the sun we know by other evidence exactly what the line-of-sight velocity should be; but we have not this knowledge for other stars. **1955** *Times* 18 July 8/2 The others [*sc.* ways of transmitting radio waves], employing in one case 'very high', and in the other 'ultra high' frequencies, are extremely reliable but until now they have been limited to line-of-sight transmission. **1956** H. S. Jones in A. Pryce-Jones *New Outl. Mod. Knowledge* 129 When the first measurements of the line-of-sight velocities of some of these objects [*sc.* spiral nebulae] were made, they were found to be surprisingly large. **1963** G. Troup *Masers & Lasers* (ed. 2) ix. 158 Infra-red masers might be applied to line-of-sight terrestrial communications. **1972** *Sci. Amer.* Feb. 76/1 Microwave radio links are limited to line-of-sight operation.

IV. 14. b. Also, a telescopic device or other optical aid designed for this purpose; *in one's sights*, visible through the sights of one's gun; also *fig.*, esp. in phr. *to raise one's sights*, to adopt a more ambitious objective.

1942 T. Rattigan *Flare Path* I. 20 I've got 'im in my sights, and 'e's getting bigger all the time. **1950** *Economist* 9 Dec. 1002/2 The United States must now raise its sights, in terms of both manpower and production. **1956** A. H. Compton *Atomic Quest* 151 Colonel Marshall..had helped greatly in raising our sights as to the magnitude of the production task. *Ibid.* 339 It is in part the competition between societies that is forcing us to readjust our educational sights. **1959** *N.Z. Listener* 10 July 4/2, I did, for many months, seeing apprehensively that the Army would be raising its sights on compulsory marshalling of our manpower little by little. **1962** *Times* 26 Apr. 7/3 Set your sights a little higher than the kitchen and try to trim your appearance to the job. *Ibid.* 5 Dec. 4/3 Lawry setting his sights on a century. **1967** Mrs. L. B. Johnson *White House Diary* 5 Dec. (1970) 596 First, he said, we have raised our sights. We have set our national goals to have a clean country. **1971** *Nature* 31 Dec. 499/2 Two years ago, the Government Actuary was estimating that the population would have grown from 56 million at present to..68 million by the turn of the century, but he has since been forced by more recent trends to lower his sights. **1976** J. Snow *Cricket Rebel* 35 They were not Gloucestershire batsmen at the other end of my sights that day but the England selectors.

c. Any of a number of nails in the sides and ends of a billiard table, used in marking out the table for some forms of carom billiards.

1864 W. B. Dick *Amer. Hoyle* 419 A line is drawn down the centre of the table, from the centre nails or sights in the *head* and *lower* cushions. **1890** Champlin & Bostwick *Cycl. Games & Sports* 81/1 Each carom table has on it two spots, along an imaginary line drawn lengthways through the centre from the middle rails or 'sights' in the head and lower cushions: the first, opposite the second 'sight', is sometimes called the light red spot, the second, opposite the sixth 'sight', the dark red spot. **1910** *Encycl. Brit.* III. 939/1 In the case of the Triangular Baulk-line, lines are drawn at the four corners from the second 'sight' on the side-rails to the first sight on the end-rails, forming four triangles within which only a limited number of caroms may be made.

V. 16. a. *sight-seeker* (earlier example).

1814 F. Burney *Let.* 24 Aug. (1978) VII. 438 But for Heaven's sake send him no more sight-seekers, who expect 'The Hero' to give dinners, & shew Lyons!

b. *sight-piece* (earlier example).

1835 C. F. Hoffman *Winter in West* II. 171 The long western rifle has three sight-pieces on the barrel.

c. In terms relating to the practice of watching the keys of a typewriter while typing, as *sight method, system, technique, typing, typist, writer, writing*. Cf. *touch-typing*, etc. s.v. *touch-.

1904 A. E. Morton *Mod. Typewriting* (ed. 2) 12 There are two methods of manipulation, one the 'touch', and the other the 'sight' system. **1918** M. B. Owen *Typewriting Speed* 145 The constant shifting of the eyes in sight writing. *Ibid.* 147 The sight typist writes spasmodically. *Ibid.* 153 Many sight writers use all the fingers. **1928** M. Crooks *Touch Typewriting for Teachers* ii. 10 A typist writing by the Sight method expends about six times as much..energy..as that expended by the Touch typist. *Ibid.* 11 The properly trained Touch typist is capable of greater speed than the Sight typist. **1935** A. C. Marshall *Princ. Teaching Typewriting* i. 1 It is..hardly necessary now to advocate the 'touch' system as against 'sight-typing'. *Ibid.* 2 The maximum speed ever attained by a sight-typist has never exceeded 60 per cent of that of equivalent touch experts. **1969** L. J. West *Acquisition of Typewriting Skills* viii. 183 Will not early sight typists form a habit of sight typing?..How does one wean learners away from sight techniques?

17. sight bar, a metal bar forming part of the breech-sight of a gun; **sight bill** *U.S.*, a bill of exchange payable on presentation; **sight-board** = *sight-screen* below; **sight cheque, draft** *U.S.*, a cheque or draft payable on presentation; **sight edge** *Naut.*, (see quot. 1948); **sight feed,** a device through which the feeding of lubricant or fuel may be seen; also (with hyphen) *attrib.*; **sight gag**, a joke which achieves its effect visually; **sight-holder,** a diamond merchant entitled to buy diamonds at a sight (see sense 5 f above); **sight liability,** an obligation to pay money on presentation of a cheque or bill of exchange; **sight-line,** (*a*) (see sense 16 b in Dict.); (*b*) a straight line extending from the eye of a spectator to an object or area being watched; *spec.* a line from the eye of a spectator in a theatre to the edge of the part of the stage which that spectator can see; **sight-player,** one who is able to play music at sight; so **sight-playing; sight-read** *v. intr.* and *trans.*, to read (a piece of music) at sight; **sight record** *Ornith.*, a record of the sighting (not the capture) of a bird; **sight-screen** = *screen *sb.*[1] 1 h; **sight-setter,** on a warship, a member of a gun-crew whose duty is to keep the gun-sight at the correct elevation as shown by the range indicator (see also quot. 1973); **sight tube,** (*a*) a tube through which observations are made; (*b*) a transparent tube connected to a tank or cistern so as to display the level of the liquid inside it.

1884 *Naval Encycl.* 751/2 *Sight-bar*, a metal bar on which the range in yards, or in degrees, is marked. It is a part of the breech-sight, and, by raising or lowering it, different ranges are obtained. **1920** Carter & Arnold *Field Artillery Instruction* iii. 47 The rocking bar sight consists of a rocking bar..and a sight bar. **1853** *Southern Literary Messenger* XIX. 89/2 Mr. Thompson agreed to accommodate him with a sight bill on his correspondent in Raleigh. **1887** *Courier-Jrnl.* (Louisville, Kentucky) 5 May 7/3 Eastern exchange was firm, and there were more buyers than sellers of New York sight bills at 80c per $1,000 premium. **1898** K. S. Ranjitsinhji *With Stoddart's Team* (ed. 4) iii. 49 [At Adelaide] the sight-boards behind the bowler's arm appeared to be but reminders of the existence of such things for a better purpose in England. **1955** Miller & Whittington *Cricket Typhoon* I. i. 13 Tiny white pavilion and tinier white sight-boards. **1975** N. Nicholson *Wednesday Early Closing* vi. 129 Every..excuse for hindrance and delay was..tried—asking for the sight-boards to be moved,..looking round at the fielders, testing the bat. **1863** 'E. Kirke' *My Southern Friends* xxii. 232, I enclose you sight check of Branch Bank of Cape Fear on Bank of Republic, for $10,820. **1850** G. N. Jones *Florida Plantation Rec.* (1927) 60 Your favor of the 22nd ult. enclosing sight draft on Messrs Habersham for $200. **1904** 'O. Henry' *Cabbages & Kings* xiv. 254 It's a gold mine. It's a sight-draft on your president man for twenty thousand dollars. **1979** O. Sela *Petrograd Consignment* 34 At the bank..letters of authority were presented, mandates altered and instructions given for the preparation of sight drafts. **1911** *Encycl. Brit.* XXIV. 971/1 The projections of the plate and longitudinal sight edges are drawn in the body plan on the floor. **1948** R. de Kerchove *Internat. Maritime Dict.* 676/2 *Sight edge*, the edges of the plates, in clinker-built plating, which are visible on the outside of the shell, on the top of decks and tank top, and on the opposite side from the stiffeners on bulkheads. **1888** *Lockwood's Dict. Mech. Engin.* 319 *Sight feed lubricator*, a lubricator..in which the flowing or non-flowing of the oil is always apparent at sight, being enclosed in, or having to pass through a glass vessel. **1902** A. C. Harmsworth et al. *Motors* ix. 172 If a Dubrulle mechanical lubricator is used, examine the ball valves sometimes, and do not trust entirely to the sight feed. **1928** *Daily Tel.* 16 Oct. 7 Non-crushable back-lamps and sight-feed fuel gauges on the dashboard are in demand for the new cars. **1957** *N.Y. Herald Tribune* 7 Nov. 24/4 The line gags are like the sight gags: they're not quite sturdy enough to be up and around yet. **1977** *Time* 2 May 49/3 *I Love My Wife*..is dotted with paralyzingly funny sight gags. **1973** *Times* 19 June (Bombay Suppl.) p. xii/3 Bombay has at least 2,000 diamond businesses, of which about 1,400 are members of the Diamond Merchants' Association. Of these 43 are 'sight-holders' of the Diamond Trading Company of London, which means that they are notified of the 10 'sightings' which the DTC holds every year. The sight-holders are the only people in India to whom the DTC will sell. **1930** *Economist* 27 Sept. 556/2 It may be desirable to modify present standards as regards the ratio of gold cover to notes and sight liabilities. **1958** *Spectator* 24 Jan. 97/2 The proportion of sight-liabilities covered by reserves is no better than in 1945. **1917** E. B. Kinsila *Mod. Theatre Construction* iv. 60 One of the most important requisites in designing an auditorium is the establishment of correct sight lines. **1957** J. Osborne *Entertainer* 11 The sight-lines are preserved by swagging. **1958** *Archit. Rev.* CXXIII. 352/2 The second case [for the substitution of wire fence for hedgerow] is to provide sightlines at corners. **1971** P. Gresswell *Environment* 264 Sight lines have to be kept open at bends and corners. **1975** I. Melchior *Sleeper Agent* (1976) II. 65 He positioned himself so that he had optimum sight lines down the side street. **1977** *Time Out* 28 Jan.–3 Feb. 43/2 Check seating plan before buying tickets as many seats have restricted sight lines. **1909** *Chambers's Jrnl.* May 334/2 Ask an accomplished sight-player how he is able to translate so readily the symbols he reads with the eye into their relative notes. *Ibid.* 334/1 He maintains that sight-playing does not depend upon an accurate knowledge of the relationship between notes and keys. **1944** W. Apel *Harvard Dict. Music* 680/1 The greatest enemy of sight-playing is playing by heart. **1903** A. W. Patterson *Schumann* 181 We want more than a facility to 'sight read' in order to fully comprehend. **1959** 'F. Newton' *Jazz Scene* ii. 30 Jazz cannot at present be adequately noted down on paper, and if it could, would almost certainly be far too complex for players to sight-read. **1974** *Guardian* 22 Mar. 14/4 Paul Beard, the [orchestra] leader, asked him whether he would like to stay on, making him sight-read part of Vaughan Williams's Fourth Symphony as an audition. **1934** *Brit. Birds* XXVIII. 31 All but one of these are 'sight-records', but in some cases the writer had already made the acquaintance of the species in other lands. **1959** D. A. Bannerman *Birds Brit. Isles* VIII. 35 A sight-record of a frigate bird observed off the south-west coast of Ireland on 25th May 1953 by W. K. Richmond, was published in the *Fair Isle Bulletin*. **1956** N. Cardus *Close of Play* 20 The sixth ball..was fielded on the boundary's edge at the sight-screen behind MacDonald's arm. **1977** T. Heald *Just Desserts* v. 92 At either end of the ground were white sightscreens on wheels. **1909** *Cent. Dict.* Suppl., Sight-setter. **1916** 'Taffrail' *Pincher Martin* xvi. 307 Some order came through a voice-pipe to the gun; whereupon the sight-setter twiddled a small wheel and peered anxiously at a graduated dial. **1920** *Blackw. Mag.* Mar. 332/2 Dully from the concealed gun positions echoed the calls of the sight-setters. **1973** J. Quick *Dict. Weapons & Military Terms* 400/1 *Sight setter*, the gun-crew member who sets the range and deflection data ordered by the officer controlling the fire. **1851** H. Melville *Moby Dick* III. xxxviii. 221 The crushed copper sight-tubes of the quadrant. **1859** *Times* 7 Jan. 8/4 He can enter an enemy's harbour under water and make surveys, only showing above the surface a sight tube, no more than one half inch in diameter, and retire still under water. **1900** W. M. Stine *Photometrical Measurements* iii. 77 Adjust the telescopic sight tube until the different portions of the field are sharply outlined. **1905** *Motor Manual* (ed. 7) iv. 78 The oil..enters a series of sight tubes. **1951** *Proc. Physical Soc.* B. LXIV. 49 The level of the liquid in the annular gap can be deduced from observations of its level in a vertical sight-tube attached to the filling apparatus.

sight, *v.*[1] Add: **2. c.** To take aim at (an object); to level or aim (a fire-arm, etc.) at a target.

1871 *Harper's Mag.* Dec. 48/2 No sooner, however, did he 'sight', or try to sight, the horseman in question,.. than the thumping against the ribs again began. **1901** F. Norris *Octopus* II. vi. 521 With the words, he dropped to one knee, and sighting his rifle carefully, fired into the group of armed men. **1976** D. Storey *Saville* I. iv. 36 Take out the bullets, and sight it at various objects outside the window.

3. a. (Earlier example.)

1787 in *Maryland Hist. Mag.* (1924) XIX. 265 The mother of the complainants wife sighted with a compass from the tree.

b. With *in*. To correct the sights of (a fire-arm, etc.) by testing and adjustment. *N. Amer.*

1958 *Washington Post* 31 Oct. D6/3 The Berwyn Rod & Gun Club invites deer hunters to sight-in their rifles during all-day open house sessions on Nov. 2 and Nov. 9. **1971** W. Hillen *Blackwater River* x. 91, I started him off right by sighting-in his new rifle, and soon he was hitting the apple box every time. **1972** *Islander* (Victoria, B.C.) 24 Sept. 13/1 Heading into the woods with a rifle that hasn't been sighted-in makes no more sense than driving an automobile without a gasoline gauge. **1980** *Outdoor Life* (U.S.) (Northeast ed.) Oct. 94/3 A Leupold 4X compact scope (made specially for the Kimber rifle) mounted and sighted in at the factory.

sighter. Add: **3.** Also *transf.* and *fig.*

1920 G. S. Gordon *Let.* 17 Sept. (1943) 139 You were charming about my article... Richmond, in a brief post-card, called the thing a 'bull's-eye'. I call it a sighter; and some day I shall have the second shot. **1960** *Times* 11 Apr. 3/7 An early sighter by Albaladejo gave some inkling of what was in store in the matter of dropped goals.

sighting, *vbl. sb.*[1] Add: **2.** Also, an instance of catching sight of (esp. something rare or unusual).

1955 W. Girvan *Flying Saucers & Common Sense* i. 13 It was not long before sightings were being reported elsewhere than in Scandinavia. **1968** *Listener* 4 July 18/2 Our objection is against the extraterrestrial origin of the phenomena, and most of us base our disbelief on the very large number of 'sightings' that have been reported. **1976** *Scotsman* 15 Dec. 9/4 The experiment to reintroduce the white-tailed eagle to the Isle of Rhum National Nature Reserve is reported to be going well, with sightings of the released birds.

3. b. *sighting-in*, the action of correcting the sights of a fire-arm, etc. Cf. *sight *v.*[1] 3 b. *N. Amer.*

1958 *Washington Post* 31 Oct. D6/3 (*heading*) 2 Days are Set for Sighting-In. **1962** *Wildlife Rev.* Dec. 9 The opening of a sighting-in range at Maiden Creek some 12 miles north of Cache Creek has been welcomed with enthusiasm by sportsmen. **1970** R. A. Steindler *Firearms Dict.* 229/1 *Sighting-in*, process of adjusting the sights, usually the rear one, or the elevation..& windage.. adjustments of a scope, so that the bullet will hit a predetermined point of aim on the target at a specific distance. Much of the trial & error method of sighting-in a rifle can be eliminated by the use of a collimator.

4. *sighting-hood, -tube.*

1909 *Cent. Dict.* Suppl. 1217/1 Modern turrets usually have three sighting-hoods, one in the center line for the turret training-pointer and one on each side for the two gun-pointers. **1973** J. Quick *Dict. Weapons & Military Terms* 400/1 *Sighting hood*, an armored hood with viewing slits in the sides, as on the top of a turret, a submarine, etc. **1946** *Nature* 12 Oct. 518/2 Radiation from a 5 sq. mm. area of tyre falls on to the cell via an arrangement which comprises a water-cooled copper sighting-tube and a rotating slotted disk which serves as the radiation chopper. **1958** J. Needham in *Aspects of Translation* 86 And there are holes which take the place of the sighting-tube for looking up (at the heavenly bodies).

sight-see, *v.* Add: **1.** (Earlier and later examples.)

1824 R. Heber *Narr. Journey Upper Provinces India* (1828) I. xii. 302, I had been sight-seeing from five till nearly ten o'clock. **1913** E. Wharton *Custom of Country* xxx. 412 'I suppose you've been to that old church over there?'.. 'Oh, of course; when I used to sightsee. Have you never been to Paris before?' **1925** C. Connolly *Let.* 28 Feb. in *Romantic Friendship* (1975) 61 Nor is there anywhere near to tempt one to sightsee. **1976** *Church Times* 30 July 7/3 It seems to me infinitely more absurd —if art and architecture mean anything to you at all, and if you can sight-see in reasonably unhurried and uncongested comfort—*not* to want to see the best in this line that civilisation has to offer. **1979** R. Jaffe *Class Reunion* II. v. 168 'And then I thought I'd just sightsee.' Dutifully she told him everything she had seen.

2. *trans.* To visit the principal sights of (a place).

1968 J. Wainwright *Web of Silence* 109, I spent the day sight-seeing Berlin. **1976** *New Yorker* 29 Mar. 95/2 (Advt.), Meet the chefs, inspect kitchens, plus sightsee the highlights from Cortina to Rome. **1977** *Daily Colonist* (Victoria, B.C.) 19 June 25/3 Macao is only $2\frac{1}{2}$ miles long. Has 300,000 people. You sightsee it in an hour.

sight-seeing, *vbl. sb.* (Earlier example.)

1824 R. Heber *Narr. Journey Upper Provinces India* (1828) I. xv. 380 Morning rides, evening sight-seeing.

attrib. (Earlier and later examples.)

1827 Mrs. B. Hall *Let.* 13 Dec. in *Aristocratic Journey* (1931) 146 This has been another regular sight-seeing day. **1916** *Daily Colonist* (Victoria, B.C.) 9 July 6/3 The picnic party left the Gorge terminus during the early forenoon in the special sight-seeing car of the B.C. Electric Railway. **1925** C. Morley *Thunder on Left* xiii. 170 People were always driving up in crowds to visit his secrets. Like sight-seeing busses loaded with excursionists. **1976** *National Observer* (U.S.) 30 Oct. 5/1, I spotted an empty sightseeing bus moving slowly up Collins Avenue.

sight-seer. (Earlier example.)

1834 G. Crabbe Jun. in *Poet. Wks. G. Crabbe* I. viii. 207 A friend in town procured us those very eligible rooms for sight-seers, in Osborne's Hotel, Adelphi.

‖ **sigillata** (sidʒilē·tă, sigilā·tă). Also **Sigillata.** [L., = sealed: see Terra sigillata.] = *terra sigillata 3. Also *attrib.*

1903 *Amer. Jrnl. Archaeol.* VII. 485 This study of the origin of the Gallo-Roman *sigillata* is preliminary to a complete publication. **1936** J. H. Iliffe (*title*) Sigillata wares in the Near East. **1938** *Burlington Mag.* Jan. p. xiv/2 Southern Spain has already been known as one of the chief pottery centres of the Roman Empire... But that 'sigillata-ware' was actually made there is not so clear. **1948** *Proc. Prehistoric Soc.* XIV. 79 Sea-borne trade with the Province is shown by a sherd of Sigillata. **1966** G. Simpson in Oswald & Pryce *Introd. Study of Terra Sigillata* (ed. 2) p. iii, No attempt was made originally to give an exhaustive description of varieties of Sigillata which had no definite chronological value.

sigillum. For *rare*⁻¹ read *rare* and add: **2.** *R. C. Ch.* The seal of confession. Cf. Seal *sb.*[2] 2 b.

1927 F. A. Marks tr. *Kurtscheid's Hist. Seal of Confession* 1 Since the Middle Ages we have for this obligation the technical term Seal of Confession (*sigillum, signaculum confessionis*). **1937** S. O'Faoláin *Purse of Coppers* 49 To add to his difficulty—for it was no help to know what, under the *sigillum*, he must pretend not to know—he had just been told in the sacristy by her employer that a pair of her best boots was missing.

3. A sign or symbol; an abbreviation; an identifying character.

1966 B. Malamud *Fixer* iii. 83 Please note, if you will, the sigillum on my coat lapel. **1978** *Language* LIV. 5 The true picture of dialect classification, either regionally or socially, is not known; nor is it implied by the sigilla of *DED*.

SIGINT, Sigint (si·gint), abbrev. of *signal(s) intelligence* s.v. *signal *sb.* 5 e.

1969 Thomas & Crowley *New Acronyms & Initialisms 1969* 376 SIGINT, Signal Intelligence (Military). **1972** *New Scientist* 2 Mar. 467 Generally speaking the larger part of the staff of all Sigint headquarters consists of scientists and engineers. Apart from actual cryptanalysis, there is a continuing need to improve intercept equipment. **1976** *Time Out* 21 May 8/2 Each country's signals intelligence (SIGINT) agency has authority to monitor communications in one area. **1979** J. Barnett *Backfire is Hostile!* ii. 29 She was in charge of Sigint... Signals intelligence. **1980** J. McNeil *Spy Game* xviii. 182 What do you know about SIGINT?.. Signals Intelligence?.. You know that's the Agency's main role over here?

sigla. Add: **a.** Also *transf.* and *fig.*

1963 V. Nabokov *Gift* ii. 86 On the soft red sand one could make out the sigla of a summer day: the imprints of a dog's paws, the beaded tracks of a wagtail. **1973** D. Osmond-Smith tr. *Bettetini's Lang. & Technique of Film* i. 61 If this image comes to be interpreted as the *sigla* of the preconceptual schema..it will only be of value for its semantic content.

b. Editorial designations of versions of an early literary text, esp. those used in the preparation of an edition. Also *sing.* **siglum.**

1939 R. B. McKerrow *Prolegomena for Oxf. Shakespeare* iii. 83 Round brackets enclosing a siglum are also used as a warning that the edition thus indicated has a reading which differs. **1950** *Classica et Medievalia* XI. 7, I prefer to use the same *siglum* for the New Testament and the Patrick section [of a manuscript]. **1962** E. J. Dobson in Davis & Wrenn *Eng. & Medieval Studies* 128, I use the following sigla: A = MS. C.C.C.C. 402 (ed. J. R. R. Tolkien, E.E.T.S. 249, 1962); C = MS. Cotton Cleopatra C VI; [etc.]. *Ibid.*, References to any text other than A are preceded by the siglum of that text. **1975** *N. & Q.* Feb. 53/2 Twenty texts of the *Epistola Cuthberti* are not recorded by Dobbie. We give them below; for each of those that Brotanek enumerates we give his siglum..; we also give some versions not listed by Brotanek, with our sigla.

‖ **siglos** (si·glǫs). Pl. **sigli, sigloi.** [Gr.] **a.** A unit of weight (see quot. 1911). Also *attrib.* **b.** A silver coin of ancient Persia.

1911 *Encycl. Brit.* XIX. 871/2 The unit of weight in the East was the shekel (*siglos*)... Starting from the siglos as unit, they [*sc.* the Greeks] invented a money-mina of 50 sigli... The siglos-units..chiefly employed in Asia Minor were the following [etc.]. *Ibid.* 903/2 Darius chose two weights, the gold shekel of 8·4 grammes and the silver drachm of 5·58 grammes... The gold coin was called the daric, the silver the siglos. *Ibid.* The regal coinage is of darics..and subdivisions in gold and of sigli and subdivisions in silver. **1962** R. A. G. Carson *Coins* 82 Both the gold darics and the silver sigloi are bean-shaped pieces with a type on obverse only and an oblong incuse on the reverse. **1962** D. Harden *Phoenicians* xii. 166 The Persians themselves..minted their *darics* and *sigloi* primarily for use in their Greek dominions in Asia Minor. **1972** *Oxf. Univ. Gaz.* CII. Suppl. No. 3. 46 Perhaps the most important aspect of this hoard is the inclusion of a number of worn (and countermarked) Persian sigloi.

siglum: see *sigla b.

sigma. Add: **3.** *Physics* and *Chem.* **a.** [After *S 5.] Used to designate electrons, orbitals, molecular states, etc., possessing zero angular momentum about an internuclear axis; *sigma-* (or *σ-*) *bond*, a bond formed by a σ-orbital.

Usu. written σ when it refers to one electron and Σ when it refers to a molecule as a whole.

1929, etc. [see *pi *sb.* 3]. **1939** J. W. T. Spinks tr. *Herzberg's Diatomic Molecules* v. 260 For multiplet Σ states the character positive-negative depends on whether *K*..is even or odd. **1952** L. N. Ferguson *Electron Structures of Org. Molecules* ii. 21 In butadiene, each carbon atom forms three σ bonds. **1963** W. J. Moore *Physical Chem.* (ed. 4) xiv. 600 In the electronic excitation an electron is removed from a π orbital and placed in a σ orbital. **1964** R. G. Parr *Quantum Theory of Molec. Electronic Structure* iii. 41 It is supposed that somehow the effect of the others, the sigma electrons, can be lumped into the Hamiltonian for the pi electrons. **1966** Phillips & Williams *Inorg. Chem.* II. xxviii. 341 The vinyl anion, $C_2H_3^-$, binds directly to cobalt (III) of vitamin B_{12} as a simple σ-bonded ligand. **1972** DePuy & Chapman *Molec. Reactions & Photochem.* vi. 103 The numbers refer to the atoms at either end of the sigma bond which is thought of as moving. **1978** P. W. Atkins *Physical Chem.* xv. 471 Thus an oxygen molecule, which.. has two π^*-electrons, could be in either a Σ state (the electrons orbiting in opposite directions) or in a Δ state (the electrons orbiting in the same sense around the bond).

b. *Particle Physics.* Used, usu. *attrib.*, to denote any of a triplet of hyperons (and their antiparticles) having an average mass of approximately 1190 MeV (2340 times that of the electron), a spin of ½, zero hypercharge, and unit isospin, and which on decaying usu. produce a nucleon and a pion (if charged) or a lambda particle and a photon (if neutral). Freq. written as Σ.

1954 Gell-Mann & Pais in *Proc. Glasgow Conf. Nuclear & Meson Physics* (1955) 347 The..apparent existence of both a positive and a negative hyperon which we shall call $\Sigma^{\pm}$ with the decay schemes $\Sigma^{\pm} \rightarrow N + \pi^{\pm} + (\sim 115\text{ MeV})$, $\Sigma^{\pm} \rightarrow P + \pi^0 + (\sim 115\text{ MeV})$. **1955** *Nuovo Cimento* II. 824 An event, interpreted as the disappearance of a charged Σ-hyperon in flight, has been observed ..in the Brookhaven cosmotron. **1961** W. S. C. Williams *Introd. Elementary Particles* xii. 298 Any subsequent Σ decays in flight. **1963** K. W. Ford *World of Elementary Particles* vi. 179 (*caption*) A negative pion..collides with a proton and produces two strange particles, a neutral sigma and a neutral kaon... The sigma particle lives too short a time to move a measurable distance.., decaying almost at once into a lambda and a photon. **1976** *Sci. Amer.* Jan. 46/2 The baryons with the lowest mass are those with a spin of 1/2. There are eight of them: two nucleons (*N*).., a lambda particle (Λ).., three sigma particles (Σ)..and two cascade particles (Ξ).

4. Biochem. *ellipt.* for *sigma factor*, sense 6 below.

1970 *New Scientist* 23 July 176/1 Soon after infection.. the phage makes its *own* sigma, which redirects the host core enzyme to start making a different set of phage proteins.

5. *Statistics.* A standard deviation: used in the singular as if the name of a unit.

1978 N. R. Ullman *Elem. Statistics* iii. 68 The difference is 0·5 inches, or, since σ = 0·29..1·72 standard deviations apart or 1·72 σ (1·72 sigma) apart. *Ibid.* 69 You can express values in terms of sigma or in terms of 'so many sigma units from the mean'. **1979** *Nature* 29 Mar. 411/1 On each radiocarbon analysis, one sigma counting errors are given.

6. *Comb.*, as **sigma factor** *Biochem.*, a component of RNA polymerase which determines where transcription begins.

[**1969** R. R. Burgess et al. in *Nature* 4 Jan. 44/2 GG enzyme contains, in addition, two extra bands which we shall designate σ and τ. *Ibid.* 46/1 The presence of the stimulating factor, σ, greatly enhances the amount of RNA synthesis. *Ibid.* 46/2 σ and similar factors could.. act as positive control elements regulating the amount of synthesis of different classes of RNA.] **1969** *Times* 9 May 12/6 The sigma factor helps to specify which genes are expressed. **1970** *Nature* 29 Aug. 886/1 Today sigma factor proteins..are central to all attempts to explain the positive control of gene expression. **1976** *Proc. Nat. Acad. Sci.* LXXIII. 3961 (*heading*) Purification and characterization of a putative sigma factor from *Chlamydomonas reinhardi*. *Ibid.* 4405 (*heading*) Induction of sigma factor synthesis in *Escherichia coli* by the N gene product of bacteriophage lambda.

sigmatism. (In Dict. s.v. Sigma.) Add: **b.** Defective articulation of sibilants.

1888 A. H. Buck *Reference Handbk. Med. Sci.* VI. 617/1 Lisping, or Sigmatism, is the most common form of stammering. It consists of giving *s* a wrong sound, usually that of *th*, by carrying the tip of the tongue too far forward, so as to touch the upper teeth. **1933** S. M. Stinchfield *Speech Disorders* iv. 76 The boys led in the number of cases of sigmatism, stuttering.., deafness and speech defects as a whole. **1957** *Dental Practitioner* VII. 220/2 Speech may be 'thick', which I tend to associate with the large tongue and the lateral sigmatism which suggests tongue behaviour. **1965** W. R. Brain *Speech Disorders* 154 Various disorders of the production of s (sigmatism) have been recognized.

sigmatropic (sigmătrǫ·pik), *a. Chem.* [f. Sigma + *-tropic.] Involving the movement of a sigma bond to a new pair of atoms within a molecule.

1965 Woodward & Hoffmann in *Jrnl. Amer. Chem. Soc.* LXXXVII. 2511/2 We define as a sigmatropic change of order [*i, j*] the migration of a σ-bond, flanked by one or more π-electron systems, to a new position whose termini are *i* − 1 and *j* − 1 atoms removed from the original bonded loci, in an uncatalyzed intramolecular process. **1974** Gill & Willis *Pericyclic Reactions* iii. 84 The transition state of a sigmatropic change is..reminiscent of the transition state of a cyclo-addition reaction.

sigmoid, *a.* and *sb.* Add: **A.** *adj.* **2.** *sigmoid colon* = *sigmoid flexure* in Dict.

1896 *Quain's Elem. Anat.* (ed. 10) III. IV. 113 The sigmoid colon may be defined as that part of the colon which is attached to the left iliac fossa from the iliac crest to the brim of the true pelvis. **1962** *Lancet* 5 May 951/2 Anterior resection of the rectum with primary sigmoid-rectal anastomosis for neoplasm of the rectosigmoid region and sigmoid colon.

Hence **sigmoide·ctomy** [*-ectomy], surgical excision of the sigmoid flexure; **sigmoidi·city,** the extent to which a curve is sigmoid (S-shaped); **sigmoidi·tis,** inflammation of the sigmoid flexure.

1906 P. L. Mummery *Sigmoidoscope* 55 In the cases of acute proctitis or sigmoiditis an examination with the sigmoidoscope may afford useful information. **1915** B. G. A. Moynihan *Abdominal Operations* (ed. 3) II. 490/1 (Index), Sigmoidectomy. **1938** H. E. Bacon *Anus, Rectum, Sigmoid Colon* xix. 695 (*heading*) Abdominal resection (sigmoidectomy)—one-stage procedure. **1968** A. White et al. *Princ. Biochem.* (ed. 4) xi. 243 For hemoglobin, a similar plot yields a sigmoidal curve.. which obeys the relationship known as the Hill equation $y = Kx^n/(1 + Kx^n)$ where..the exponent n gives a measure of the sigmoidicity of the curve. **1977** *Arch. Biochem. & Biophys.* CLXXXIV. 300/1 With the enzyme from *Am[aranthus] edulis*, the response to increasing malate was qualitatively similar to that recorded for the *At[riplex] spongiosa* enzyme but sigmoidicity was more pronounced.

sigmoidoscope (sigmoi·dǒskōᵘp). *Med.* [f. prec. + -o + -scope.] A speculum for examining the lower bowel and for assisting in minor operations therein. Cf. Sigmoidscope.

1900 in Dorland *Med. Dict.* 599/2. **1906** P. L. Mummery *Sigmoidoscope* 7 The introduction of the electric pneumatic sigmoidoscope represents a great advance in our powers of accurate diagnosis. **1974** R. M. Kirk et al. *Surgery* vi. 123 Polyps can be removed through a sigmoidoscope using a snare.

Hence as *v. trans.*; also **sigmoidosco·pic** *a.*, performed or ascertained by means of a sigmoidoscope; **sigmoidosco·pically** *adv.*; **sigmoido·scopist,** one who uses a sigmoidoscope; **sigmoido·scopy,** examination by means of a sigmoidoscope.

1900 Dorland *Med. Dict.* 599/2 Sigmoidoscopy. **1906** P. L. Mummery *Sigmoidoscope* 37 A sigmoidoscopic examination will give valuable information if there is any question of being able to remove the growth. **1961** L. Martin *Clinical Endocrinol.* (ed. 3) vi. 163 Intestinal symptoms and sigmoidoscopy or barium enema appearances will be diagnostic. **1962** *Lancet* 26 May 1095/1 Of the 20 patients receiving control treatment only 7

improved symptomatically and 8 sigmoidoscopically. **1966** *Ibid.* 25 June 1420/2 Heaven forbid that some future 'Chief of Service' should licence me to sigmoidoscope my cases of earache in this surgery. **1976** *Path. Ann.* XI. 28 Rectal examination may reveal blood, as may sigmoidoscopy, when the blood can often be seen coming from higher in the bowel. **1977** *Proc. R. Soc. Med.* LXX. 273/2 Sigmoidoscopic biopsy of the stricture at 14 cm showed chronically inflamed mucosa with no tumour tissue. **1977** *Lancet* 29 Oct. 893/2 The sigmoidoscopist was not able to recognise the type of enema a patient was receiving by the sigmoidoscopic appearances on the morning after an overnight retention enema.

sign, *sb.* Add: **I. 2. e.** *Math.* That aspect of a quantity which may be either positive or negative.

1820 G. Peacock *Differential & Integral Calculus* 112 The sign of d^2u may be easily determined. **1836** A. de Morgan *Differential & Integral Calculus* xiv. 369 When there is a change of sign, *y* is a maximum (M), or a minimum (*m*), according as the change is from + to − or from − to + (*x* increasing). **1924** G. F. Swain *Structural Engin.* xiii. 350 It is obvious that n_1 will have the same sign as f_1, and n_2 the opposite sign. **1957** G. E. Hutchinson *Treat. Limnol.* I. ix. 597 Where biochemical oxygen uptake or production occurs, no general rule as to the sign of the divergence from saturation will be possible. **1978** C. P. McKeague *Elem. Algebra* i. 23 To multiply any two real numbers simply multiply their absolute values, the sign of the answer is 1. positive if both numbers had the same sign..2. negative if the numbers had opposite signs.

6. Also, a board giving information, directions, etc.

1904, etc. [see *road sign* s.v. *road *sb.* 9 b].

II. 7. a. Also *the signs of the times*, indications of current trends. Now freq. as *sing.* phr. with leading indef. article.

1525 Bible (Tyndale) (1526) *Matt.* xvi. 3 Can ye not discerne the sygnes of the tymes? **1833** *Daily Nat. Intelligencer* 17 July 3/3 We have stood upon our '*reserved rights*' of neutrality, to watch the signs of the times. **1907** *Nature* 14 Mar. 459/1 This book is an interesting sign of the times. **1921** J. Galsworthy *To Let* II. xi. 214 '*He's* a sign of the times,' muttered Soames, 'if you like.' **1953** A. J. Toynbee *World & West* vi. 93 The people who have read the signs of the times and have taken action in the light of these indications are the obscure missionaries of half-a-dozen Oriental religions. **1977** *Gay News* 24 Mar. 19/3 Last year, perhaps as a sign of the times, Take Six notched up over 80 mentions in everything from the *Daily Mirror* to the Italian glossioso *L'Uomo*.

b. (*a*) (Examples in *Linguistics* and *Semiotics*.)

1890 W. James *Princ. Psychol.* II. xxii. 356 Language is a system of signs, different from the things signified, but able to suggest them. *c* **1902** C. S. Peirce *Coll. Papers* (1932) II. § 92 Genuine mediation is the character of a *Sign*. **1922** tr. *Wittgenstein's Tractatus* 53 The sign is the part of the symbol perceptible by the senses. **1938** C. W. Morris (*title*) Foundations of the theory of signs. **1947, 1949** [see *signifiant]. **1954** [see *signifier b]. **1964** Gould & Kolb *Dict. Soc. Sci.* 641/2 Sign denotes any stimulus which, because of association with another stimulus, elicits a response appropriate to but in the absence of the original stimulus. **1978** *Incorporated Linguist* Summer 60/3 Modern society's haste to read inadvertently into signs (in the Barthesian sense) rather than decipher the simple message. **1979** S. G. J. Hervey *Axiomatic Semantics* vii. 61 By the law of excluded middle, any given sign is either simple or complex, but not both.

(*b*) *Theol.* Phr. *outward visible sign* and varr., in sacramental ordinances, the outward and visible aspect which symbolizes the inward and spiritual aspect. Also *transf.*

1553 J. Bradford in Coverdale *Lett. Martyrs* (1564) 293 There is Idolatry in worshipping the outwarde signe of breade and wyne. **1604** *Bk. Com. Prayer, Catechism*, *Q.* How many partes be there in a Sacrament? *A.* Two: the Outward visible signe, and the Inward spirituall Grace. *c* **1816** J. Marriott *Hymn*, Grant to this child the inward grace, While we the outward sign impart. **1861** tr. *O Food that Weary Pilgrims Love!* in *Hymns, Anc. & Mod.* (Introits & Anthems) p. xvii, O Jesu, Whom, by power divine Now hidden 'neath the outward sign, We worship and adore. **1898** A. G. Mortimer *Cath. Faith & Practice* I. 124 The matter [of a sacrament] is the outward sign; the form that which determines the matter to its special use or purpose. **1921** J. Galsworthy *To Let* III. x. 288 In the union of the great-granddaughter..with the heir of a ninth baronet was the outward and visible sign of that merger of class in class which buttresses the political stability of a realm. **1931** V. Dixon *Sebastian Wile* II. ii. § 1 Her governess had said farewell, outward and visible sign that Martha's days of childish servitude were over. **1938** *Doctrine in Church of England* II. 127 The ordinary scholastic use is to employ the word [*sc.* sacrament] as meaning the outward and visible sign. **1951** A. Powell *Question of Upbringing* iii. 157 Monsieur Dubuisson accepted the brandy as the outward and visible sign of reconciliation. **1962** Wilson & Templeton *Anglican Teaching* ix. 180 The Catechism..defines a Sacrament as 'an outward and visible sign of an inward and spiritual grace..ordained by Christ Himself'.

d. (Earlier examples.)

1692 *Cal. Virginia St. Papers* (1875) I. 44 We Ranged about to see if we could find ye tract of any Indians, but we could not see any fresh signe. **1746** *New Hampsh. Hist. Soc. Coll.* (1834) IV. 208 By the sign of this ambush, and by the sign of their going off, in a single file, it was supposed there could not be less than 50 or 60 Indians. **1821** J. Fowler *Jrnl.* 3 Nov. (1898) 33 Heare We find the first fresh Sign of bever. *Ibid.* 7 Nov. 36 We see old sign of Indeans... We again See the Sign of White men a Head of us.

e. *Med.* An objective evidence or indication of disease (as opposed to a subjective one, or *symptom*); often used with the name of one who associated an indication with a disease characterized by it, to designate the former.

1842 W. A. Guy *Hooper's Physician's Vademecum* (new ed.) I. iii. 16 The word sign has not precisely the same meaning as the term symptom, though the two terms are sometimes used without much discrimination... Cough, expectoration, dyspnœa, hectic fever, night sweats, and emaciation, are *symptoms* of pulmonary consumption, but they are not *signs*, for each of them may occur in other diseases; but cavernous respiration and pectoriloquy are signs. *Ibid.*, The term *physical* sign is in common use among medical men: it means a sign which is an object of sense. Thus heat, redness, and swelling are physical signs of inflammation, pectoriloquy of phthisis, coagulable urine of disease of the kidney. **1851** R. P. Cotton *Phthisis & Stethoscope* i. 12 Physical signs by themselves, as a general rule, determine nothing more than physical conditions..; hence it is, that we require the use of other rules, as well as a knowledge of the patient's history and general symptoms. *Ibid.* ii. 24 Diminished resonance is one of the earliest and most characteristic signs of phthisis. **1872** W. Williams *Princ. & Pract. Vet. Surg.* xiii. 244 The diagnostic signs of elbow-joint lameness are, first, the semi-flexed position of the limb..whilst standing still; and the dropping of the head and anterior parts of the body during action. **1886** J. Finlayson *Clin. Manual for Study Med. Cases* (ed. 2) II. 51 A pain is a 'Symptom' (subjective); a bulging chest, to which it may be due, is a 'Sign' (objective): giddiness is a 'Symptom' (subjective); the staggering resulting from it is a 'Sign' (objective). **1908** *Practitioner* Jan. 10 We do not obtain ankle clonus, or Babinsky's, or Oppenheim's sign. **1927** G. W. Deeping *Kitty* xv. 193 Mr. St. George had an undoubted paraplegia. There was definite spasticity of the lower limbs... Babinski's sign was present. **1956** A. I. Littlejohn tr. *D. Wirth's Vet. Clin. Diagnosis* 1 Symptoms in the medical sense are not available to the veterinary diagnostician, but the substitution of the term 'symptom' for 'sign' in veterinary usage is widespread. **1971** S. Magalini *Dict. Med. Syndromes* 148/1 Dercum's [syndrome]... *Symptoms.* Prevalent in women 40 to 60 years of age. Pain in part of body where localized accumulation of fat occurs. Asthenia, headache... *Signs.* Subcutaneous accumulation of fat elevated, dry, reddish, or bluish, anesthesia and diminished cutaneous sensibility. **1974** T. McGinnis *Well Dog Bk.* (1979) 95 Because your dog cannot describe their feelings in words, they technically have no symptoms, only *signs* which are any objective evidence of disease or injury you can detect.

III. 12. a. (sense 1) *sign-language* (earlier and later examples; also *fig.*); (sense 6) *sign-painter* (earlier and later examples), *-writer* (Writer 1 b), *-writing*; (sense 7) *sign-situation, -system, -using* vbl. sb. and ppl. adj.

1847 T. H. Gallaudet in *Amer. Ann. Deaf & Dumb* I. 59 They originate from elements of this sign-language which nature furnishes to man wherever he is found, whether barbarous or civilized. **1960** S. Plath *Colossus* 39 These..sheets..Speak in sign language of a lost other-world. **1981** *Amer. Speech* LVI. 130 Sign language is as adequate for the deaf as any vocal-auditory language is for a hearing person. **1725** *New-Eng. Courant* 15 Feb. 1/2, I would oblige every Sign-Painter to serve seven Years at College, before he presum'd to handle Pencil or Paint-Box. **1942** *Burlington Mag.* Jan. 9/1 Ireland takes this sketch as a proof that Hogarth contemplated setting up as a sign-painter. **1923** Ogden & Richards *Meaning of Meaning* i. 15 There may be a very long chain of sign-situations intervening between the act and its referent. **1977** *Dædalus* Fall 105 Literature..though it is..a form of communication..is cut off from the immediate pragmatic purposes which simplify other sign situations. **1924** R. H. Bell *Mystery of Words* 101 A study of the general principles of language has brought out the nature of the linguistic sign-system. **1977** R. H. Brown in Douglas & Johnson *Existential Sociol.* ii. 90 These norms and rules form a sign system that is itself subject to the feedback of experience. **1890** W. James *Princ. Psychol.* II. xxii. 357 In the human child..these ruptures of contiguous association are very soon made; far off cases of sign-using arise when we make a sign now; and soon language is launched. **1938** C. W. Morris *Found. of Theory of Signs* i. 1 Men are the dominant sign-using animals. **1957** C. E. Osgood et al. *Measurement of Meaning* i. 3 The behavior of the sign-using organism. **1871** J. Callingham *Sign Writing* i. 1 It is curious that the term 'sign-writer' is not to be found in any encyclopædia or dictionary, ancient or modern... Even Kelly's ponderous 'Post Office London Directory' does not deem the sign-writer worthy of separate enumeration in its list of trades. **1977** J. McClure *Sunday Hangman* xiii. 151 A family of losers trying to find the right words for the signwriter. **1871** J. Callingham (*title*) Sign writing. **1954** 'J. Wyndham' *Jizzle* 49 Elmer was a house-painter who doubled in the less spacious art of sign-writing. **1978** *Dumfries & Galloway Standard* 21 Oct. 21/2 (Advt.), All types of signwriting undertaken.

b. Special combs., as **sign-behaviour,** behaviour that is dependent on a sign (sense 7); **sign bit** *Computers*, a sign digit located in a sequence of binary digits; **sign-design** (see quot. 1942); **sign digit** *Computers*, a digit, located in a sequence of digits, whose value depends on the algebraic sign of the number represented; **sign-event,** a particular occurrence of the use of a sign (sense 7); **sign-process,** the process whereby a token or indication becomes operative or functions as a sign; **sign stimulus** *Biol.*, the component or characteristic of an external stimulus which is effective in initiating a particular innate behavioural response in an animal perceiving it, regardless of the presence or absence of the remainder of the stimulus; **sign-vehicle,** the token or indication that acts as a sign.

1946 C. W. Morris *Signs, Lang. & Behav.* i. 7 And goal-seeking behavior in which signs exercise control may be called sign-behavior. **1964** Gould & Kolb *Dict. Soc. Sci.* 641/2 Sign-behaviour is found in all levels of animal life. **1962** *Gloss. Terms Automatic Data Proc.* (*B.S.I.*) 19 Where the sign digit is a binary digit it is often known as a sign bit. **1975** T. Bartee *Introd. Computer Sci.* ii. 47 The sign bit is set apart from the magnitude bits by a . in each word... An alternate technique uses a box for the sign bit. **1942** R. Carnap *Introd. Semantics* § 3.5 The word 'sign' is ambiguous. It means sometimes a single object or event, sometimes a kind to which many objects belong. Whenever necessary, we shall use 'sign-*event*' in the first case, 'sign-*design*' in the second. **1944** *Mind* LIII. 36 The sign-design is what is usually meant when we use such words as 'symbol', 'word', 'sentence'. It is the form or structure common to a set of actual occurrences (sounds, marks, gestures) whereby they function symbolically. **1974** M. Taylor tr. *Metz's Film Lang.* iii. 90 Between words—pure 'sign events' as they are called in American semiotics, events that never occur twice..and language.. there is room for the study of 'sign designs', sentence patterns. **1947** A. W. Burks et al. in J. von Neumann *Coll. Wks.* (1963) V. 46 Our numbers are 40 digit aggregates, the left-most digit being the sign digit. **1950** *Proc. R. Soc.* A. CCII. 574 The first digit is regarded as a sign digit and a 'binary point' supposed to exist before the second digit. **1969** J. J. Sparkes *Transistor Switching* viii. 194 The sign digit is normally '0' for positive numbers. **1942** Sign-event [see *sign-design* above]. **1973** *Screen* Spring/Summer 164 *Spoken words*..are pure 'sign-events' incapable of being reproduced twice over and therefore impossible to study scientifically. **1946** C. W. Morris *Signs, Lang. & Behav.* i. 3 Terms which are commonly used in describing sign-processes. **1957** C. E. Osgood et al. *Measurement of Meaning* i. 5 A first step toward a behavioral interpretation of the sign-process. **1934** E. S. Russell *Behaviour of Animals* ii. 33 The principle of representative stimuli, or sign stimuli as we may call them for short, is illustrated not only in the flight reactions of animals..but even more clearly in..food-finding behaviour. **1967** A. Manning *Introd. Animal Behaviour* iii. 39 There are many examples of auditory and chemical sign-stimuli too. Turkey hens which are breeding for the first time will accept as chicks any object which makes the typical cheeping call. On the other hand..deaf turkey hens kill most of their chicks because they never receive the auditory sign-stimulus for parental behaviour. **1975**, **1980** Sign stimulus [see *releaser c]. **1938** C. W. Morris *Found. of Theory of Signs* i. 4 In such cases *S* is the sign vehicle.., *D* the designatum, and *I* the interpretant of the interpreter. **1955** T. H. Pear *Eng. Soc. Differences* i. 33 Status symbols are sign-vehicles, cues which determine the status to be imputed to a person.

sign, *v.*[1] Add: **I. 4. c.** With *in*. To secure the admittance of (a person) to a hotel, club, etc., by signing a register; to record the entrance of (a person) into a building, etc.

1930 A. P. Herbert *Water Gipsies* xxv. 368 Isn't he sleeping in the hotel himself?.. Didn't want to sign you in as his wife, I shouldn't wonder. **1957** C. MacInnes *City of Spades* I. xi. 79, I shall sign you in till Johnny come, and check with him later. **1971** R. Hill *Advancement of Learning* xvi. 222 'Have you been signed in?'.. Of course, it was a club. 'Then you can't buy a drink, can you?' **1977** J. P. Anderson in Douglas & Johnson *Existential Sociol.* vi. 191 His face fell a foot when the social worker told him that Viejas Rehabilitation Center was the only place he could get in, that he would have to sign himself in for from three to six months. **1978** M. Z. Lewin *Silent Salesman* xviii. 107, I know of at least one person who was in Research [Laboratory] Three on the twenty-seventh who isn't signed in or out.

d. With *out*. To secure the release of (a person or thing) by signing; to record the removal of (a thing) or the departure of (a person) from a building, etc.

1963 V. Nabokov *Gift* iii. 187 He signed out the complete works of Chernyshevski from the state library. **1968** *Globe & Mail* (Toronto) 13 Feb. 11/6 The nurse replied that he had been signed out by the doctor. **1972** D. E. Westlake *Cops & Robbers* (1973) x. 137 Why don't you shlep on back to the [police] station and sign us both out? **1978** [see sense 4 c above].

5. a. Also const. *for*, as authorization or acknowledgement of receipt. Also, to make a written contract *with*. *to sign on the dotted line*: see *dotted *ppl. a.* 1 c.

1938 L. Bemelmans *Life Class* II. iv. 160, I won't pay for anything that isn't properly ordered... I pay only for things I sign for. **1956** B. Holiday *Lady sings Blues* (1973) xxii. 181 The only royalties I get are on my records made after I signed with Decca. **1957** C. Smith *Case of Torches* i. 10 'Some of the boys in the laboratory ..think they compromise their independence if they sign for something.' 'All the other boxes have been signed for.' **1966** J. B. Priestley *Salt is Leaving* v. 61 If a Miss Tiller asks for me, tell her we've gone in. I've already signed for her. **1967** E. S. Gardner *Case of Queenly Contestant* xvi. 206 He said he would take care of all my expenses... I.. sign for meals in the hotel restaurant. **1974** *Times* 5 Feb. 11/7 John Alderton and Pauline Collins..have signed with London Weekend Television to appear as husband and wife in a new comedy series. **1977** P. D. James *Death of Expert Witness* II. 100 We let them borrow the key and they sign for it in a book in the office.

b. (*a*) With *off*. (Further examples.) *gen.*, to record that one is bringing something to an end, to stop doing something; *spec.* (i) *Broadcasting*, to cease broadcasting, to announce the end of a broadcast; (ii) to fall silent, to withdraw one's attention; (iii) to record leaving one's work, to stop work; (iv) *Bridge*, to indicate by a conventional bid that one is ending the bidding.

1923 *Sci. Amer.* Nov. 310/3 The local broadcasting stations have 'signed off' for the night. **1929** WODEHOUSE *Mr. Mulliner Speaking* vi. 206 If you're trying to propose to me, sign off. There is nothing doing. **1933** A. MCCABE *Contract without Tears* 165 Had North wished to sign-off at this point he would have bid five diamonds. **1937** *Speculum* Apr. 268 Tired copyists expressed their relief at signing off from their labors. **1948** *Times* 2 Sept. 2/7 Reluctance to sign off with no additional values has led to many [Contract Bridge] players getting out of their depth. **1953** W. R. BURNETT *Vanity Row* xxi. 188 Lynch was..listening to a comedy programme... 'Be with you in a minute... They're just about to sign off.' **1954** M. PROCTER *Hell is City* I. v. 30 What time did you sign off? ..Since then you've been in some pub... You've been working on that murder, I suppose. **1957** F. HOYLE *Black Cloud* xi. 210 If the politicians started..arguing.. the Cloud would sign off altogether. It's not going to waste its time talking to gibbering idiots. **1962** *Listener* 1 Mar. 394/3 He bid 5 N.T., which by convention asked his partner to bid Six Diamonds if he held the King of the agreed suit, hearts, and otherwise to sign off in Six Hearts. **1965** 'J. LE CARRÉ' *Looking-Glass War* xxiii. 241 'The transmission's stopped.'..'Did he sign off?' **1971** H. TREVELYAN *Worlds Apart* xvii. 193 By the summer of 1964 Khrushchev had decided to have nothing to do with Vietnam either and virtually signed off. **1974** R. M. PIRSIG *Zen & Art of Motorcycle Maintenance* i. 23 John signs off every time the subject of cycle repair comes up. **1976** *Times* 1 May 12/7 North can hardly be blamed for seeking a slam when his partner could have 'signed off' by responding Five Diamonds to Five Clubs. **1976** *Milton Keynes Express* 30 July 13/1 In a statement Hawkins said he did not sign off because the Works job was only temporary and he was afraid he would not be able to sign on again. **1979** *Irish Times* 28 Sept. 3/1 A decision will be made later as to whether this progressive three-year-old will sign off for the season in the St Simon Stakes or the Champion Stakes.

(*b*) With *on*. (Earlier and later examples.) *spec.* (i) to record one's arrival at work, to begin work; (ii) to sign a contract to join an organization, etc.; (iii) to register at the Department of Social Security (formerly Labour or Employment Exchange) in order to obtain unemployment benefit.

1862 *Railway Traveller's Handy Bk.* 8 In most Government offices the *employés* are compelled to 'sign on', as it is called, when they arrive in the morning. **1930** E. POUND *XXX Cantos* ix. 37 Until he signed on with Siena. **1936** N. MITCHISON *Fourth Pig* 29 If I didn't keep it up, there'd be a dozen knocking themselves over to get my job. And then it would be signing on again at the Labour. **1941** *Illustrated* 6 Sept. 21 She hands him the emergency slip. It says that he must sign on at 8 a.m. for the 9.30 special. **1955** *Times* 18 Aug. 5/1 Some of our men there had signed on for three or five years because they had been told they would learn a trade, but they were just batmen and doing no training at all. **1960** C. MACINNES *Mr. Love & Justice* 45 Frankie had paid his last visit to the Labour because..he wasn't going through the comedy of 'signing on' any more. **1974** P. WRIGHT *Lang. Brit. Industry* ii. 31 Bus drivers and conductors have instead to *sign on*... They may have to make a personal appearance before the traffic inspector to show that they are not drunk or otherwise unfit. **1976** *Yorkshire Evening Press* 9 Dec. 3/4 If you gave up work voluntarily then you could be disqualified from receipt of unemployment benefit for up to six weeks, and you would have to 'sign on' and hold yourself available for employment every week. **1981** B. HINES *Looks & Smiles* 18 You take this [card] up to the Social Security office and sign on at the time it says here. *Ibid.* 44 Miserable bunch of bastards, the sergeant said... Anybody'd think they'd been forced to sign on.

c. With *up*. To enrol, to enlist; to give support *to*.

1903 A. H. LEWIS *Boss* 186 You can tell by th'way they go to bat, whether th' Blackberry has signed up to them to kill our franchise. **1926** *Ladies' Home Jrnl.* Apr. 25 So she signed up for evening classes. **1942** E. PAUL *Narrow St.* xxxiv. 306 It was generally accepted in our street after that that France was eager to sign up with Russia against Hitler. **1942** E. WAUGH *Put out More Flags* i. 69 What I thought of doing was to sign up with you... It's a great help to start in a decent regiment. **1951** *Listener* 31 Jan. 172/2 Inducing other governments to sign up to professions of high moral and legal principles. **1975** M. BRADBURY *History Man* vi. 99 I've signed up for an evening class. **1977** T. HEALD *Just Desserts* i. 11 Collingdale had had to sign up as a novice friar.

d. With *out, in.* To record one's departure from, arrival at, a hotel, club, etc., by signing a register; also *fig.*

1951 G. GREENE *End of Affair* II. ii. 65 It was..as though I had signed out of the war. **1966** G. BURNETT *Dead Account* xii. 97 And my name's Brook. Where do we sign in? **1968** 'G. BAGBY' *Another Day—Another Death* vii. 142 It seemed impossible that..all the police who'd been poring over the book could have missed someone who signed in and hadn't signed out. **1978** M. Z. LEWIN *Silent Salesman* xviii. 108 I'd like a list of all the people who signed in or out of Research Three.

6. c. Also with *up*; also *fig.*

1927 WODEHOUSE *Meet Mr Mulliner* i. 29 If George had been a member of the Olympic Games Selection Committee, he would have signed this woman up immediately. **1932** *Radio Times* 1 Apr. 5/2 Seversky immediately signed the violinist up for his broadcast. **1956** B. HOLIDAY *Lady sings Blues* (1973) iii. 35 Joe Glaser, the big agent and manager..signed me up on the spot. **1963** WODEHOUSE *Stiff Upper Lip, Jeeves* iii. 26 While I personally..would run a mile in tight shoes to avoid marrying Stiffy, I knew him to be strongly in favour of signing her up. **1980** G. M. FRASER *Mr American* xxiii. 442 Your friend Pip is to be one of the top turns in the cabaret—I suppose they signed her up as soon as they saw the early editions.

II. 9. a. Also *spec.* to use a sign language.

1909 in WEBSTER. **1977** *Rolling Stone* 16 June 46/1 Washoe used to sign to the others quite a bit, but of course the chimps she was signing to didn't respond. **1978** *Detroit Free Press* 5 Mar. 10/4 Strangely, many educators of deaf students don't sign (use sign language). **1980** *Nat. Geographic* June 849 Bin was picking up sign language... He didn't talk to Princess; he signed to her as he also did with other non-signing orangutans.

b. Also *spec.* to communicate or express (something) in a sign language.

1909 in WEBSTER. **1975** J. GOULET *Oh's Profit* i. 4 Liedlich and his wife, Nancy, had signed the month, the hand-dance that was the month April. **1975** *Church Times* 15 Aug. 2/2 The lessons will be signed by deaf readers and the Lord's Prayer by one who is also blind. **1978** *Oxford Times* 16 June 2 The British Deaf Association Choir..'signed' the hymns.

III. 10. Comb., as **sign-in,** the action of signing in (see senses *4 c and *5 d); also used *attrib.* and *absol.* of a register in which people sign in; **sign-off,** the action of signing off (see sense *5 b (*a*)); *Broadcasting*, the end of transmission, an announcement of this; also *attrib.*; **sign-on,** the action of signing on (see senses *5 b (*b*) and 6 c); *Broadcasting*, the start of transmission; **sign-out,** the action of signing out (see senses *4 d and *5 d); the signature of one who has signed out; **sign-up,** the action of signing up or the state of having signed up (see senses *5 c and *6 c); also, a person who has signed up; also *attrib.*

1968 'G. BAGBY' *Another Day—Another Death* vii. 142 He showed me the porter's sign-in. The man had come in quite early. **1972** 'J. LANGE' *Binary* 8 A guard with a sign-in book stood in front of the elevator. Graves..took the pen and wrote his name, his authorization, and the time. **1978** S. BRILL *Teamsters* iii. 115 Their names were the first entered every morning in the sign-in register at the Fund's reception desk. **1942** E. CULBERTSON *Official Bk. Contract Bridge* xv. 187 Finally, there is the sign-off bid... The sign-off may be made even if the responder hold one Ace. **1949** *Cavalier Daily* (Univ. Virginia) 23 Sept. 1/3 A non-affiliated station..will be on the air only during the daylight hours. Sign-on and sign-off times will vary from month to month. **1958** *Listener* 30 Oct. 709/2 The sign-off for his partner would clearly be Five Hearts. **1960** *News Chron.* 27 June 3/1 One [question]..was used by the producer as a rather abrupt sign-off. **1961** [see *sign-on* below]. **1962** H. T. MOORE *Coll. Lett. D. H. Lawrence* I. p. xxi, Lawrence's sign-off line was often a foreign phrase. **1971** H. TREVELYAN *Worlds Apart* xvii. 194 So now they could no longer stay silent and issued the expected 'sign-off' statement on Vietnam, evidence that they were powerless to take any diplomatic initiative. **1976** *Time* 20 Dec. 47/2 Remember his sign-off as he was being expelled from a Democratic Convention: 'This is John Chancellor, somewhere in custody.' **1948** *Seafarers' Log* 9 Jan. 5/2 One thing about the pay-offs and sign-ons we had: All the beefs were settled aboard ship to everybody's satisfaction. **1949** [see *sign-off* above]. **1961** *Time* 19 May 53/1 The toughest TV critic..dared the station and network operators and owners to sit down in front of their sets from sign-on to sign-off. **1968** 'G. BAGBY' *Another Day—Another Death* vii. 142, I looked at the later sign-outs. Those covered the mob of cleaning women. **1940** *Sun* (Baltimore) 17 Sept. 9/7 Talbot, with a 'sign up' of 391,..led all the counties. **1941** *Ibid.* 14 Feb. 7/1 'There is a direct need for immediate sign-up' of nurses for army duty. **1945** *National Legionnaire* (U.S.) Sept. 1 (*heading*) Legion speeds sign-up of 12,000,000 War II victors. **1951** *Daily Progress* (Charlottesville, Va.) 5 Mar. 3/2 Sign-ups through the end of last month totaled 1,033. **1972** *Jrnl. Social Psychol.* LXXXVII. 118 A sign-up sheet was then distributed and students were asked to indicate whether..they would be willing to volunteer for the experiment. **1974** *News & Reporter* (Chester, S. Carolina) 22 Apr. 10-A/1 This will be a singles tournament and sign up will start Saturday morning at 10 a.m. **1980** *Dirt Bike* Oct. 5/1 The little gray-haired lady at the sign-up booth is your wife, or your girlfriend.

signal, *sb.* Add: **4. c.** A modulation of an electric current, electromagnetic wave, or the like by means of which information is conveyed from one place to another; the current or wave itself; also, a current or wave whose presence is regarded as conveying information about the source from which it comes. Also = *signal strength*, sense 5 e below.

1855 D. LARDNER *Electric Telegraph* v. § 121 The signals transmitted appear upon the telegraphic instrument informing the agent whence the dispatch will come. **1873** *Trans. Inst. Engineers & Shipbuilders in Scotland* XVI. 119 If several thousand Leyden Jars were distributed along an aerial line of telegraph..the signals through the line would exhibit exactly the same inductive retardation as those sent through the actual submarine line. **1902** *Proc. R. Soc.* LXX. 256 For transmitting signals, an aerial wire or wires were attached to one of two spark balls fitted to an induction coil, the other ball being earthed. **1923** *Radio Times* 28 Sept. 2/2 After sunset signals may increase very considerably. **1958** *Times* 18 Jan. 7/3 The problem of how to reach Iraq and the Persian Gulf area with an adequate signal has yet to be solved. **1961** *New Scientist* 26 Jan. 199/3 The picture signal which indicates by amplitude modulation of a carrier wave how bright each point on a line should be, is interrupted at the ends of each line by synchronizing pulses. **1961** G. MILLERSON *Technique Telev. Production* ii. 19 The current, known as the video or picture signal, is subsequently amplified and passed to the video switching console. **1965** *New Statesman* 30 Apr. 674/2 The signals received from it [*sc.* Early Bird satellite] on the ground are extremely weak—about one-millionth of a normal TV signal in a fringe area. **1970** J. EARL *Tuners & Amplifiers* iv. 78 Signal delivered by the control section..is just right for feeding into the power amplifier section. **1978** PASACHOFF & KUTNER *University Astron.* xi. 302 Various objects in space emit electromagnetic signals in the radio part of the spectrum.

5. a. *signal arm* (ARM *sb.*[1] 6 c), *code, gun* (earlier and later examples), *lamp* (further examples), *light* (further examples), *strip*; also, denoting something used in receiving a signal or signals, as *signal pad* (PAD *sb.*[3] 4). Also objective, as *signal-processing*.

1901 *Railway Mag.* May 463/2 A neighbouring signal-arm falls. **1949** M. TAYLOR *Railway Signalling* i. 6 Near the top of the post is the signal arm which is always on the left of the post when the viewer is facing the signal. **1962** A. LURIE *Love & Friendship* xv. 293 Coarse grass grew along the track, and the signal arm was rusted at all clear. **1877** J. HABBERTON *Jericho Road* 94 Between the societies of neighboring counties there often existed signal-codes, and unwritten extradition and reciprocity treaties. **1952** M. K. WILSON tr. *Lorenz's King Solomon's Ring* viii. 82 The whole complicated 'signal code' of the jackdaw. **1758** in J. S. McLennan *Louisbourg* (1918) 414 Light gales and fair weather, later thick fog. Fired signal gun. **1930** R. CAMPBELL *Adamastor* 74 The rocks, spray-clouded, are your signal guns. **1902** *Chambers's Jrnl.* July 479/2 The apparatus is simply a new glass for the signal-lamp, facing along the same way and throwing a powerful beam of light over the whole length of the arm. **1932** G. GREENE *Stamboul Train* I. i. 5 A signal lamp turned from red to green. **1881** Signal-light [see *running light* s.v. *RUNNING *vbl. sb.* 17 a]. **1936** *Discovery* Sept. 289/2 Motor-car headlights, signal lights and searchlights. **1976** *Billings* (Montana) *Gaz.* 5 July 8-A/6 The Utah Highway Patrol requested drivers to stay off the roads as signal lights were off through much of the state. **1958** P. KEMP *No Colours or Crest* iv. 54 By the operator's stool..were some signal pads. **1975** T. ALLBEURY *Palomino Blonde* x. 63 The Morse came and he was getting it down on his signal pad. **1964** R. F. FICCHI *Electr. Interference* vi. 99 Another way to reduce susceptibility to unwanted signals in cables is to use various signal-processing methods to improve the signal-to-noise ratio. **1917** 'CONTACT' *Airman's Outings* 261 The whole party circles round the aerodrome until the signal strips for 'Carry on' are laid out on the ground. **1954** W. FAULKNER *Fable* 87 He reached the aerodrome and saw the ground signal-strip laid out on it;..not until he saw the other aeroplanes on the ground or landing or coming into land did he recognise it to be the peremptory emergency signal to all aircraft to come down.

b. *signal gantry*.

1927 A. MEE *Children's Treasure House* III. 1819/1 A signal gantry (one of those large bridges covered with signals which stand near important junctions or great termini). **1939** [see *GANTRY 2 b]. **1976** *Physics Bull.* Dec. 556/1 A total of 15 signal gantries span the roadway for a distance of about 8 km.

c. *signal corps* (earlier example).

a **1885** G. B. MCCLELLAN *Own Story* (1887) 135 The weak point in the signal corps..was that its officers were not trained soldiers.

d. With other parts of speech, as *signal-like* adj.

1935 *Amer. Speech* X. 250/1 The result obtained..are functionally independent groups of sounds, each of the groups reflecting one of the basic, signal-like values in the given language. **1961** *Brno Studies in English* III. 51 Its survival may be satisfactorily explained by the signal-like character of *pi* in foreign and emotionally coloured words.

e. Special combs., as **signal anxiety** *Psychol.*, anxiety which, according to the theory put forward by Freud in 1926, acts as a signal of danger to the ego; **signal-caller** *N. Amer. Football*, a player who signals to other members of his team what the next move and formation should be; **signal detection** *Psychol.*, the detection of signals, esp. with regard to the observer's vigilance and sensitivity; also *attrib.*; **signal generator** *Electronics*, an instrument that can generate modulated or unmodulated electrical waveforms of known amplitude and frequency, used in adjusting and testing electronic apparatus; **signal intelligence**: see *signals intelligence* below; **signal-noise ratio** = *signal-to-noise ratio* below; **signal plate** *Television*, in some types of camera tube, a plate electrode whose capacitance relative to the adjacent photo-electric surface is used to provide the picture signal; **signal reaction** (see quot. 1976); **signal red,** a vermilion colour; **signal(s) intelligence,** intelligence derived from the monitoring, interception, and interpretation of radio signals and similar transmissions (cf. *SIGINT, SIGINT); **signal strength,** the amplitude or power of a signal, esp. of a broadcast signal as it reaches a given location or is received by a given aerial; **signal-to-noise ratio** *Radio*

and *Electronics*, the ratio of the strength of a desired signal to that of unwanted noise interference, usu. expressed in decibels; also *transf.* to non-electrical systems.

[**1928** J. RICKMAN in *Internat. Jrnl. Psychoanal.* (Suppl. 2) 63 The situation which conditions the anticipation is the danger situation which gives, so to speak, the 'Anxiety Signal' of impending helplessness.] **1948** E. JONES *Papers on Psycho-Anal.* (ed. 5) xiv. 315 Primary anxiety, no less than the later 'signal' anxiety, belongs essentially to these defensive measures. **1968** C. RYCROFT *Crit. Dict. Psychoanal.* 154 Signal anxiety..in Freud's formulation is the response of the ego to internal danger and the stimulus to the formation and use of defence-mechanisms. **1975** in S. Arieti *Amer. Handbk. Psychiatry* (ed. 2) IV. xxi. 485/1 The finely modulated, discrete and homeostatically balanced responses of the adult, such as 'signal anxiety'. **1971** L. KOPPETT *Guide to Spectator Sports* ii. 48 The quarterback now handles the ball on every play... He must be the signal-caller, too, because the attack is now concentrated in his hands. **1979** *Arizona Daily Star* 5 Aug. C 9/1 John Banaszak jarred the ball loose from Bills signal caller David Mays. **1954** TANNER & SWETS in *Psychol. Rev.* LXI. 409/2 The mathematical model of signal detection is applicable to problems of visual detection. **1971** D. E. BROADBENT *Decision & Stress* iii. 75 We cannot..take the success of signal detection theory in psychophysics as if it guaranteed its adequacy in the case of vigilance. **1979** in Hamilton & Warburton *Hum. Stress & Cognition* v. 148 Signal detection methodology is now the more widely used approach in the study of vigilance. **1929** K. HENNEY *Princ. Radio* xv. 376 The circuit diagram is that of the General Radio Signal Generator, a device which consists of a radio-frequency oscillator, a means of measuring and controlling its output, and a means of using any desired part of this output for purposes of measuring receivers. **1950** J. H. REYNER *Encycl. Radio & Television* 564/1 For the testing and alignment of receivers, signal generators giving modulated waves are available. **1972** *Jrnl. Social Psychol.* LXXXVII. 119 These tapes were made by recording square wave pulses produced by a Hewlet-Packard low-frequency signal generator. **1958** *Guardian* 22 May 3/1 Thompson and Miller trained for 'a specialised and secret duty in Signal Intelligence', Mr Jones continued. **1969** Signal intelligence [see *SIGINT, SIGINT]. **1972** *New Scientist* 2 Mar. 466/1 The generic term for the business today is Signal Intelligence (Sigint). **1934** A. L. ALBERT *Electr. Communication* xv. 412 Just as in transmission over wires, it is necessary that the signal-noise ratio is high, and that fading is not excessive. **1962** *Daily Tel.* 6 July 21/7 However efficient the detectors and however good the signal-noise ratio, there will always be the possibility of doubt. **1934** V. K. ZWORYKIN in *Jrnl. Franklin Inst.* CCXVII. 10 Consider the circuit of a single photo-electric element in the mosaic... Here *P* represents such an element, and *C* its capacity to a plate common to all the elements, which hereafter will be called the 'signal plate'. **1975** K. WICKS *Television* 19 After striking the signal plate, the electron beam returns along the tube, attracted by the positive charge on a series of five electrodes called dynodes. **1946** F. P. CHISHOLM in W. S. Knickerbocker *Twentieth Century English* 183 Signal-reactions..and other neuro-semantic disorders are often combined with brilliant verbal facility. **1976** N. POSTMAN *Crazy Talk* 195 A signal reaction is what happens when words have lost their referential or symbolic aspect and instead assume the character of religious icons. **1936** *Times Educ. Suppl.* 21 Nov. p. iv/1 Gules..is not included in the range, but signal-red, similar but not quite so yellow, takes its place. **1977** *Western Morning News* 30 Aug. 2/7 Jaguar XJS; brand new;..signal red; beige trim. **1976**, etc. Signals intelligence [see *SIGINT, SIGINT]. **1912** *Marconigraph* II. 269/2 Observations..showed that during totality the signal strength was increased. **1935** *Discovery* Sept. 278/1 The successive signal strengths are proportioned to the corresponding light and shade areas of the image focused by the lens upon the mosaic. **1968** M. WOODHOUSE *Rock Baby* xiii. 130 He..showed me..the tiny signal-strength meter on the side. **1978** *Broadcast* 20 Nov. 16/1 The report of the meeting..referred only to discussion of [Radio] Hallam's signal strength. **1935** *Proc. IRE* XXIII. 713 A single tone was used to modulate the transmitter when measuring signal-to-noise ratios. **1966** D. G. BRANDON *Mod. Techniques Metallogr.* 239 The overall quantum efficiency for high-energy electrons can be of the order of 0·2 and is only limited by the necessity to achieve a reasonable signal-to-noise ratio in the emulsion. **1974** HARVEY & BOHLMAN *Stereo F.M. Radio Handbk.* ii. 32 A reduction in carrier deviation means that a smaller signal is available at the output of the receiver demodulator and consequently the signal-to-noise ratio is lowered.

signal, *v.* Add: **2. b.** Also, to indicate, esp. unofficially or indirectly. Cf. *SIGNALIZE *v.* 5.

1962 *Amer. Speech* XXXVII. 214 Since one member of the pair [of pronominal forms] is functionally redundant, the usage in different dialects may be expected to vary far more than for forms which signal significant differences. **1963** *Ibid.* XXXVIII. 52 The slots, or, more accurately, the units that can fill the slots, signal their meanings by their positions relative to each other... Some units signal their meaning without regard to position. **1978** *Times* 24 Apr. 2/1 The government seemed yesterday to be signalling its willingness to concede a reduction in the highest rates of income tax. **1979** *Tucson* (Arizona) *Citizen* 20 Sept. 1A/5 U.S. economic indicators signal the dollar should be lower than current rates. **1981** *Times* 25 Sept. 19/1 Sterling is now lower than when the Bank of England signalled higher interest rates a week last Monday.

signalization (signălaizē̆ɪ·ʃən). *Psychol.* [f. SIGNALIZ(E *v.* 4 + -ATION.] A term derived from the work of I. P. Pavlov (1849–1936) for the process whereby a signal comes to elicit the same response as the original stimulus.

1927 G. V. ANREP tr. *Pavlov's Conditioned Reflexes* ii. 22 The underlying principle of this activity is signalization. The sound of the metronome is the signal for food, and the animal reacts to the signal in the same way as if it were food. **1969** in K. H. Pribram *Memory Mechanisms* 137 The conditioned reflex..enables the animal to adjust itself to the essential factors of the external and internal world on the basis of the principle of signalization.

signalize, *v.* Add: **5.** Also, to indicate.

1961 *Texas Stud. Lit. & Lang.* III. 283 Pip's abject leave-taking of Miss Haversham..signalizes his homage to a supposed patroness. **1964** W. H. DRAY *Philos. Hist.* 53 The willingness of the revisionists to apportion blame, however, does not signalize a return to the sectional type of partisanship. **1966** R. S. RUDNER *Philos. Social Sci.* 15 A definition signalizes the redundancy or eliminability of a term. **1976** *Network* (Brit. Social. Assoc.) No. 6. 2/1 When people are trying to change their status..they seek to change their name, both to signalize the change that has taken place and to give form to the identity they are trying to shape.

6. [SIGNAL *sb.*] *U.S.* and *Austral.* To provide (an intersection, etc.) with traffic signals.

1961 in WEBSTER. **1977** *Sunday Mail* (Brisbane) 29 June 2/2 We had 158 signalised intersections at June 30 last year, 65 signalised pedestrian crossings..and 281 floodlit ones.

signans (si·gnænz). *Linguistics* and *Semiotics.* [L., pr. pple. of *signāre* to signify.] = *SIGNIFIANT. Opp. *SIGNATUM.

1953 C. E. BAZELL *Linguistic Form* 29 Multiple oppositions, rare or unknown in the signans, are normal in the signatum. **1954** *Litera* (Istanbul) I. 31 In a code, each discrete signans has a discrete signatum; for instance in the Morse code a certain combination of dots and dashes signifies a certain combination of strokes and points. **1956** JAKOBSON & HALLE *Fundamentals of Lang.* I. ii. 15 The 'expression plane' of language, as he christened the aspect named *signans* in Stoic and Scholastic tradition,..is to be studied without any recourse to phonetic premises. **1959** *Jrnl. Individual Psychol.* XV. 62 The *signans* is perceptible, the *signatum* intelligible... Thus we perceive the sound-shape of the word *tree* and, on the other hand, we may translate this word by other verbal signs with more or less equivalent *signata* but each with a different *signans*.

signary (si·gnări). [f. L. *signum* sign + -ARY[1], after *syllabary*.] An arrangement of signs; the signs which constitute the syllabic or alphabetic symbols of a language.

1902 *Encycl. Brit.* XXVII. 730/2 Probably all the signs in the hieroglyphic signary can be employed in their primary sense. **1909** A. J. EVANS *Scripta Minoa* I. p. v, I have endeavoured to supply a preliminary apparatus criticus in the form of tables and explanatory catalogues of the different signaries. **1924** L. ECKENSTEIN *Tutankhaten* v. 42 There were other scribes of other signaries and languages who inscribed soft clay tablets with a copper stilus. **1932** *Antiquity* VI. 375 Nearly all the Phoenician signs were already in use in Egypt as far back as the 1st dynasty, part of the larger series of the Mediterranean signary.

signatum (signā·təm). *Linguistics* and *Semiotics.* [L., neut. sing. pa. pple. of *signāre* to signify.] = *SIGNIFIÉ. Opp. *SIGNANS.

1953, etc. [see *SIGNANS]. **1962** R. JAKOBSON *Sel. Writings* I. 658 This criticism has been repeatedly mistaken for an attempt to withdraw the *signatum* from the scope of any phonemic analysis.

signature, *sb.* Add: **4. b.** (Further examples.)

1940 R. MORRISH *Police & Crime-Detection* xii. 114 All these marks, however minute in themselves, form the specific 'signature' or identity of any fire-arm. **1952** M. ALLINGHAM *Tiger in Smoke* xi. 185 You knifed three people..and..you went and left your signature all over the shop. **1960** T. HUGHES *Lupercal* 27 No Signature but this threshold-held hollow Remained of some vigorous souls That had Englished for Elizabeth. **1966** *Listener* 17 Nov. 746/1 An obvious pointer..is the prevalence of Shostakovich's musical 'signature'—D-S-C-H (D-E flat-C-B)—and its derivatives. **1971** *Daily Tel.* 22 Jan. 13/2 Signature of his collection: the V-necked pullover on every outfit. **1979** F. KERMODE *Genesis of Secrecy* iii. 56 The episode..is a sort of reticent signature, like Alfred Hitchcock's appearances in his own films. **1979** *Studies in Eng. Lit.: Eng. Number* (Tokyo) 109 Indeed, this prosodic 'signature' is written on page after page of the *Pisan Cantos*.

d. *spec.* = *signature tune*, sense 9 below.

1932 [see *signature tune*, sense 9 below]. **1937** *Printers' Ink Monthly* May 42/2 *Signature*, the musical number or sound effect which regularly identifies a program. **1962** A. NISBETT *Technique Sound Studio* ix. 160 It is not so common for the start of a record to provide a good crisp opening signature.

e. Any pattern or characteristic in the physical properties or behaviour of a particular object, substance, etc., by which it can be identified; freq. a characteristic response which it gives to a test.

1960 *Jrnl. Histochem. & Cytochem.* VIII. 288/1 Failure to demonstrate striking differences in 'pH signature' between analogous components of normal and carcinoma cells did not particularly surprise us. **1967** *Electronics* 6 Mar. 50/3 The new radar will gather 'signatures' of orbital vehicles as well as reentering missiles. **1969** *New Scientist* 2 Oct. 21/2 The Concorde's signature from a given height will be less ponderous than that of the Boeing. **1971** *Sci. Amer.* July 77/3 The characteristic signature of a supernova remnant is the emission of radio waves whose distribution of energy with wavelength is nonthermal. **1973** D. KYLE *Raft of Swords* (1974) iii. 21 Super-sensitive acoustic receivers..which can instantly identify a vessel's 'signature'; in other words, identify the distinctive sounds made by a particular vessel. **1977** A. HALLAM *Planet Earth* 41/3 Its magnetic signature allowed an age to be assigned to each piece of ocean floor. **1980** *Globe & Mail* (Toronto) 23 Aug. 2/5 The concept includes means of reducing the infra-red 'signature' given off by engine heat, thus defeating the ability of sensors to 'see' aircraft or vehicles even in darkness.

6. b. (Later examples.)

1901 D. COCKERELL *Bookbinding* i. 34 The sheets of a newly printed book are arranged in piles in the printer's warehouse, each pile being made up of the same sheet or 'signature'. **1965** *Times Lit. Suppl.* 14 Oct. 928/3 This word 'signature' is..often used when the user appears to mean either 'leaf' or 'quire'. **1981** *Printing World* 28 Jan. 13/3 Signatures can be perforated down the back for 'slotted' (also known as 'notched' or 'burst') binding.

8. *Pharm.* (See quot. 1951.)

1856 E. PARRISH *Introd. Pract. Pharmacy* v. ii. 418 The prescription may be divided, for the purpose of study, into the following parts..: 1. The superscription. 2. The inscription. 3. The subscription. 4. The signature. **1901** T. SOLLMANN *Textbk. Pharmacol.* vi. 105 The directions to the patient (signature) are always written in English, so that the patient can read them. **1951** A. GROLLMAN *Pharmacol. & Therapeutics* 753 A prescription traditionally includes the following parts: 1. The superscription.. 2. The inscription or body of the prescription... 3. The subscription... 4. The signature.., which includes the directions for the patient.

9. Special Comb.: **signature tune,** a piece of music that always precedes or follows a particular programme or a performance by a particular entertainer or band; also *transf.* and *fig.*

1932 *Daily Mail* 4 Mar. 11/4 B.B.C. Band's 'Signature'. 'Just the Time for Dancing' and 'Till Next Time' are the titles of the 'signature' tunes selected by Mr. Henry Hall for his new B.B.C. Dance Band, to be used every time the band begins or concludes a broadcast. **1934** *Punch* 8 Aug. 164/2 My dearest memory of the place [*sc.* Bilgesea] is that there was never a moment at which wailing could not be heard. It is..the 'signature-tune'..of English holiday-makers throughout the country. **1938** O. SITWELL *Those were Days* IV. iii. 462 Diminutive moonstones and giant chrysanthemums were her signature-tune, her speciality almost, you might say, what she lived for. **1950** E. PARTRIDGE *Here, There & Everywhere* 181 With Lewis Carroll the verse-form often serves the same purpose as a signature-tune. **1958** P. GAMMOND *Duke Ellington* II. 72 The Duke and Bubber wrote *East St Louis toodle-oo*..and this became..the band's signature tune. **1962** A. NISBETT *Technique Sound Studio* ix. 161 Vocal music is not usually suitable for prefading, and is therefore rarely used for a closing signature tune. **1972** T. LILLEY *'K' Section* xl. 192 The six o'clock news signature tune. The familiar introductory pictures. **1977** *N.Y. Rev. Bks.* 26 May 3/3 'And trod so sweetly proud' and 'In this blind bitter land' come from Yeats's well-recognized signature tune rather than from an imagination strenuously engaged with its experience.

sign-board. Add: Also *fig.*

1934 V. WOOLF *Oliver Goldsmith* in *Captain's Death Bed* (1950) 15 Bodies and hearts are attached to these sign-board faces.

2. Chiefly *U.S.* A board on a guide-post to direct travellers, etc.

1829 A. ROYALL *Mrs. Royall's Pennsylvania* II. 38 You scarcely go a mile in Pennsylvania but you see a *Preacher*—as signboards are called. They point out the road but never travel it. **1883** *Wheelman* I. 298 They found a sign-board pointing to Swampscott and Lynn. **1972** *Straits Times* 25 Nov. 18/4 Although there are sign boards on either side of the estate indicating that it is closed to lorry traffic, these lorries continue to rumble through the estate.

signed[1], *ppl. a.* Add: **2.** (Later examples.) Also *fig.*

c **1893** H. G. WELLS *Let.* in *Experiment in Autobiog.* (1934) I. vi. 392 They have let me sign an article in the *Pall Mall Gazette*,..and signed articles in dailies is a distinct advance for a poor wretch like me. **1930** G. B. SHAW *Apple Cart* I. 30 The ultimatum is here..and I shall not leave this room until I have His Majesty's signed pledge that its conditions will be observed. **1942** *World Rev.* Apr. 38/1 Key features of the newspaper contents were signed articles, and dramatic and literary *critiques* contributed by the greatest writers. **1959** *Times Lit. Suppl.* 5 June 334/2 It is a signed portrait. It is also a highly appealing, fair and convincing one. **1973** *Times* 28 Apr. 11/3 Both recordings contain versions of the 'signed' motet *Illibata Dei*, so called because the composer's name is spelled out in the text as an acrostic.

signed[3] (saind), *a. Arith.* [f. SIGN *sb.* + -ED[2].] Having a (plus or minus) sign; esp. as *signed number*, a positive or negative integer.

1873 *Proc. Lond. Math. Soc.* IV. 111 A signed magnitude. **1905** *Proc. R. Soc. Edin.* XXV. 372 The sum of the signed primary minors of a determinant. **1950** *Math. Tables & Other Aids to Computation* IV. 103 A word may represent a signed-binary number lying somewhere between -2^{40} and $+2^{40}$. **1966** J. H. CADWELL *Topics in Recreational Math.* xiv. 160 We shall prove that all derivatives at these points are signed integers or zero. **1969** J. J. SPARKES *Transistor Switching* viii. 193 When writing 'signed' numbers the sign digit is identified with a bar over it.

signee (səinī·). [f. SIGN *v.*[1] + -EE[1].] One who has signed a contract or register.

1953 *Sun* (Baltimore) 24 Apr. B20/2 Ken Jackson.. became the thirty-third Colt signee yesterday. **1970** G. F. NEWMAN *Sir, You Bastard* i. 13 Generally the officials were never so subtle, nor the signees so sensitive, as to be influenced by surroundings. **1977** *Times of Zambia* 12 Sept. 8/7 Montgomery, a recent signee from Celtic of the Scottish Premier League in UK.

signer. Add: **1. b.** *U.S.* (Usu. with capital initial.) *spec.* One of the signatories to the Declaration of Independence.

1865 M. B. CHESNUT *Diary* 29 Mar. in C. V. Woodward *M. Chesnut's Civil War* (1981) xxxiii. 772 What is the use of being the grandson of a signer if one is not a loyal gentleman? **1913** E. WHARTON *Custom of Country* vii. 90 The high dark dining-room with..dim portraits of 'Signers' and their females. **1928** W. A. WHITE *Masks in Pageant* 67 The grandson of a President and the great-grandson of a Signer. **1973** A. POWELL *Temporary Kings* i. 48 He is descended..from what is known as a 'Signer', one Bulton Gwinnett, who set his name to the Declaration of Independence.

signet-ring. Add: **2.** *Path.* Used, usu. *attrib.*, to describe cells and organisms that resemble signet rings in appearance.

1901 J. EWING in *Jrnl. Exper. Med.* V. 446 At a very early period of its development the æstivo-autumnal parasite in the present cases assumed a very characteristic ring shape. Many of these rings early developed a thickening of one segment, and to these bodies of various sizes the term 'signet-ring' very aptly applies. *Ibid.* 448 In six cases taking quinine typical signet-ring forms were seen in the peripheral blood 60 to 72 hours after the beginning of the paroxysm. **1928** L. E. H. WHITBY *Med. Bacteriol.* xxi. 209 The most characteristic form of the trophozoite is the so-called 'signet-ring' form, in which the chromatin granule or 'dot' is at one side. **1961** [see *KRUKENBERG]. **1966** WRIGHT & SYMMERS *Systemic Path.* I. xv. 511/1 When the mucus-secreting cells are anaplastic they appear as rounded cells, with the nucleus displaced to one side by a globule of mucus in the cytoplasm ('signet-ring cells'); 'signet-ring cell' carcinomas are highly malignant.

signifiable, *a.* (Earlier example.)

1857 *Encycl. Brit.* XIII. 610/2 It might then be determinable, and signifiable through the form of expression.

‖ **signifiant** (sinʸifiaṅ). *Linguistics* and *Semiotics.* [Fr., pres. pple. of *signifier* to signify.] A sound, symbol, or image, or a sequence of sounds, etc., as opposed to the meaning expressed; the physical element of a sign. Opp. *SIGNIFIÉ.

[**1916** F. DE SAUSSURE *Cours de Linguistique Générale* I. i. 101 Nous appelons *signe* la combinaison du concept et de l'image acoustique; mais dans l'usage courant ce terme désigne généralement l'image acoustique seule... Nous proposons de conserver le mot *signe* pour désigner le total, et de remplacer *concept* et *image acoustique* respectivement par *signifié* et *signifiant*; ces derniers termes ont l'avantage de marquer l'opposition qui les sépare soit entre eux, soit du total dont ils font partie.] **1939** [see *SAUSSUREAN *a.*] **1947** *Word* III. 8 Signs are the primary objects of linguistic study. Words, word-groups, and sentences are all signs—signifiants linked with signifiés. **1949** *Archivum Linguisticum* I. 1 But the morpheme is a sign in the sense of de Saussure, an association of a *signifiant* and a *signifié* upon equal terms. **1964** *Language* XL. 307 He [*sc.* Firth] rejected all distinction between langue and parole, and signifiant and signifié. **1973** [see *MENTALISM 2]. **1973** *Screen* Spring/Summer 220 A *signifiant* (the pattern of alternating images) and a *signifié* (the indication of a simultaneity between the corresponding actions).

signific (signi·fik). *Linguistics.* [f. SIGNIFIC(ANT *sb.*] = *RADICAL *sb.* 1 c. Also *attrib.* Cf. *PHONETIC *sb.*

1923 B. KARLGREN *Analytic Dict. Chinese & Sino-Japanese* 1, I never use the term 'radical', as it wrongly conveys the idea of 'radix, racine, root' which is quite a different notion in general philology from that of signific: the meaning indicating part in the Chinese character. *Ibid.* 4 Nine tenths of all Chinese characters consist of one 'signific' and one 'phonetic'. **1948** R. A. D. FORREST *Chinese Lang.* ii. 38 The signific element in each case [*sc.* a character] denotes, or, more commonly, merely suggests, an order of ideas to which the meaning of the whole belongs. **1951** SHAU WING CHAN *Elementary Chinese* p. xvi, The last category is known as Chuafichü... It includes characters having identical significs and somewhat similar meanings but different phonetics. **1964** *Language* XL. 104 What sometimes appeared..to be a graphic element totally unrelated to the phonemic shape of the morph in question and hence by default a semantic 'key' or 'signific' was often in reality a phonetic element in the script. **1973** *Sci. Amer.* Feb. 54/2 When the 'horse' phonetic is combined with the signific for 'jade', we have *mă*, which means 'agate'.

‖ **significacio** (signifikā·sio). Also **significatio.** [med.L. :— L. *significātio* significance.] An allegorical meaning; an innuendo.

1933 R. TUVE *Seasons & Months* iii. 85 The commonplace *significatio* of Phoebus is to be explained..as a literary inheritance rather than the last faint rumbling of wheels in a pagan festival of the Sun. **1936** C. S. LEWIS *Allegory of Love* i. 1 It is essential to this form that the iteral narrative and the *significacio* should be separable. **1968** J. A. W. BENNETT *Chaucer's Book of Fame* iii. 104 The emphatic negative is not a mere trope; it underlines the double *significacio*. **1969** R. A. LANHAM *Handlist Rhetorical Terms* 92 *Significatio*,..an innuendo.

significance. Add: **3.** *Statistics.* The level at or extent to which a result is statistically significant; freq. *attrib.*, as *significance level*; **significance test,** a method used to calculate the significance of a result; hence *significance testing* vbl. sb.

1888 J. VENN *Logic of Chance* (ed. 3) xix. 486 As before, common sense would feel little doubt that such a difference was significant, but it could give no numerical estimate of the significance. **1907** *Biometrika* V. 183 Let it be reasonable to suppose a quantity significant when it is *β* times its standard deviation, or *β*/·67449 times its probable error, then we have for significance test: *m* — *M* > [etc.]. Several other cases of probable error tests of significance deserve reconsideration. **1947** *Ibid.* XLVII. 139 The problem of testing the significance of difference between two proportions..receives early attention in textbooks on mathematical statistics. *Ibid.*, Such a difference in levels of significance in the solution of an everyday problem is obviously puzzling. **1960** *Amer. Sociol. Rev.* XXV. 202/2 In the test of this hypothesis a Chi-square of 34·34 was obtained, considerably lower than Chi-square at the ten per cent significance level for 34 degrees of freedom. **1970** *Nature* 25 July 384/2 Calculations of significance are based on the significance of the difference between paired observations using Student's *t* test. **1972** A. W. F. EDWARDS *Likelihood* i. 2 The rejection of the theory led to the flowering of alternative methods of inference, particularly significance-testing and estimation, to which we are heirs today. **1977** P. JOHNSON *Enemies of Society* xi. 157 In psychology, for example, it is notorious that 'results' used to confirm hypotheses are often no better than random data because significance tests would validate almost anything.

significans (signi·fikænz). *Linguistics* and *Semiotics.* [L., pres. pple. of *significāre* to signify.] = *SIGNIFIANT. Opp. *SIGNIFICATUM.

1964 E. PALMER tr. *Martinet's Elem. General Linguistics* i. 24 Every linguistic sign comprises a significatum, its meaning or value..and a significans through which the sign is made manifest. **1972** HARTMANN & STORK *Dict. Lang. & Linguistics* 209/1 *Signifier*... Alternative term: significans,..signifiant.

significant, *a.* Add: **2. b.** Substitute for leading phrase and def.: Conveying information about the value of a quantity; esp. in *significant digit, figure*, a digit which has its precise numerical meaning in the number containing it, and is not a zero used simply to fill a vacant place at the beginning or end. (Further examples.)

1938 A. E. WAUGH *Elem. Statistical Method* ii. 8 If we are told that the distance is 1000·00 miles, there are six significant figures, since it was not necessary to put in the zeros to locate the decimal point. **1957** R. A. BUCKINGHAM *Numerical Methods* i. 6 The numbers 0·000101 and 0·000999, both of which have the same absolute accuracy afforded by 6 decimals, and 3 significant figures, may yet have relative errors differing by an order of magnitude. **1962** C. BELL et al. *Fund. Arith. for Teachers* xi. 192 The value 3·1416 is said to be accurate to five significant figures. *Ibid.* xv. 231 If a measurement is expressed as a natural number, it is not always possible to determine the number of significant digits. **1965** I. ADLER *New Look at Arith.* iv. 228 Zeros which are not significant can always be replaced by words which serve the same function. *Ibid.* 231 When two approximate numbers are multiplied, the product has at most as many significant digits as there are in that one of the two numbers that has the fewer significant digits. **1968** *Brit. Med. Bull.* XXIV. 216/2 This [*sc.* a mathematical value] is not 'significant' but is printed out routinely. **1971** *Physics Bull.* Oct. 597/3 Does any analyst doing routine tests have the right to quote his result to four figures, and pretend that the fourth is also significant?

c. *Significant Form, significant form* Aesthetics, a hypothetical quality, thought to be common to all great works of art, that evokes an aesthetic response and is considered to be more significant than the subject-matter.

1914 C. BELL *Art* II. i. 8 What quality is shared by all objects that provoke our aesthetic emotions?.. In each, lines and colours combined in a particular way, certain forms and relations of forms, stir our aesthetic emotions. These relations and combinations..I call 'Significant Form'; and 'Significant Form' is the one quality common to all works of visual art. **1914** R. FRY in *Nation* 7 Mar. 938/2 Why must the potter who is to make a superbly beautiful pot not think only of its significant form, but think first and most passionately about its functions as a pot? **1929** D. H. LAWRENCE *Paintings* 20 The critics stepped forth and abstracted his good apple into Significant Form, and henceforth Cézanne was Saved. **1959** H. B. ALLSOPP *Future of Arts* xiv. 120 Some abstractionists persist in seeking what Clive Bell called 'significant form' which is significant without being significant of anything. I suspect that this idea is nonsense. **1965** *Brit. Jrnl. Aesthetics* V. 113 Significant form cannot be attributed primarily to works of art on the ground that aesthetic emotion obtained from works of art is more intense than that felt in the contemplation of natural objects and pure forms.

5. *Statistics.* Of an observed or calculated result, such as the difference between the means of two samples: having a low probability of occurrence if the null hypothesis is true; *statistically significant*, significant at some conventionally chosen level, freq. five per cent.

A result is said to be significant at a specified level of probability if it will be obtained or exceeded with not more than that probability when the null hypothesis is true.

1885 *Jrnl. R. Statistical Soc.* (Jubilee Vol.) 187 In order to determine whether the observed difference between the mean stature of 2,315 criminals and the mean stature of 8,585 British adult males belonging to the general population is significant [etc.]. **1907** *Biometrika* V. 318 Relative local differences falling beyond +2 and −2 may be regarded as probably significant since the number of asylums is small (22). **1925** R. A. FISHER *Stat. Methods Res. Workers* iii. 47 Deviations exceeding twice the standard deviation are thus formally regarded as significant. **1931** L. H. C. TIPPETT *Methods Statistics* iii. 48 It is conventional to regard all deviations greater than those with probabilities of 0·05 as real, or statistically significant. **1969** *Sci. Jrnl.* Nov. 57/1 The attitude scores differed only slightly and the differences were not statistically significant. **1970** *Nature* 25 July 376/2 Analysis of variance gave highly significant population and fertilizer effects. **1971** *Daily Tel.* (Colour Suppl.) 29 Oct. 35/4 The result, although occurring more often than other conjunctions, did not occur often enough to be statistically significant; i.e. a statistician would have said it was a chance occurrence. **1971** *Nature* 26 Nov. 231/2 If.. fifteen experiments are performed to detect a relationship which is not present, the probability that one or more experiments will give a result significant at the 0·05 level is 0·54.

significatio, var. *SIGNIFICACIO.

signification. Add: **1. c.** *Semiotics.* The process of signifying; the production of signs.

1946 C. MORRIS *Signs, Language & Behavior* iv. 111 Since a sign can denote without its interpreter knowing whether or not it denotes..it is evident that signification and knowledge are not limited to that portion of the world which acts as a direct stimulus to an interpreter's behavior. **1957** J. MARITAIN in R. N. Anshen *Language* I. v. 88 Animals make use of signs without perceiving the relation of signification. To perceive the relation of signification is to have an *idea*, i.e., a spiritual sign. **1973** *Screen* Spring/Summer 94 Each of them is in other contexts the signifying substance of one or more signification systems other than the cinema. *Ibid.* 109 In cinema signification (the process of the production of signs) is described by Metz as always more or less motivated, that is non-arbitrary.

2. (Examples in *Semiotics*.)

1964 C. MORRIS *Signification & Significance* i. 3 'Significations'..are not 'entities' in any objectionable sense, but certain describable aspects of complex behavioral processes in the natural world. **1976** T. EAGLETON *Crit. & Ideology* iii. 72 The text takes as its object, not the real, but certain significations by which the real lives itself—significations which are themselves the product of its partial abolition.

Hence **significa·tional** *a.*

1953 H. H. PRICE *Thinking & Exper.* iv. 117 Our question concerning the 'tied' (as opposed to 'free') character of significational thinking.

significatum (signifikā·təm). *Linguistics* and *Semiotics.* [L., neut. sing. pa. pple. of *significāre* to signify.] That which is signified or denoted; *spec.* = *SIGNIFIÉ. Opp. *SIGNIFICANS.

1865 S. HODGSON *Time & Space* ii. 44 It is of the utmost importance in reasoning to distinguish which kind of object or significatum it is which is expressed, or concealed, by a word or set of words. **1946** C. MORRIS *Signs, Lang. & Behavior* 17 Those conditions which are such that whatever fulfills them is a denotatum will be called the significatum of the sign. **1964** [see *SIGNIFICANS]. **1974** G. LEECH *Semantics* v. 73 The set of conditions (e.g. the qualities of being edible, tasty, nourishing) which make the bone a denotatum of S_1 [*sc.* a buzzer sound] constitute the *significatum* of the sign.

‖ **signifié** (sinʸifie). *Linguistics* and *Semiotics.* [Fr., pa. pple. of *signifier* to signify.] A concept or meaning as opposed to its expression in a physical medium (phonetic, graphic, etc.); the semantic element of a sign. Opp. *SIGNIFIANT.

[**1916**: see *SIGNIFIANT.] **1939** [see *SAUSSUREAN *a.*]. **1947**, etc. [see *SIGNIFIANT]. **1963** J. LYONS *Structural Semantics* iii. 41 Here it may be noted that there is no need whatsoever to posit a common conceptual 'signifié', or 'meaning', to account for identity of application between different languages.

signified, *ppl. a.* Add: **B.** *absol.* as *sb.* That which is indicated; *Linguistics* and *Semiotics* = *SIGNIFIÉ (opp. *SIGNIFIER b).

1939 L. H. GRAY *Foundations of Lang.* 16 In the speaker's mind a concept arises... This concept is termed the *signified*. **1954**, etc. [see *SIGNIFIER b].

signifier. Add: **b.** *Linguistics* and *Semiotics.* = *SIGNIFIANT. Opp. *SIGNIFIED *sb.*

1954 U. WEINREICH *Languages in Contact* ii. 9 It becomes possible for the bilingual to interpret two signs whose semantemes, or signifieds, he has identified as a compound sign with a single signified and two signifiers, one in each language. **1960** W. BASKIN tr. *de Saussure's Course in Gen. Linguistics* I. i. 67 The bond between the signifier and the signified is arbitrary. **1967** LAVERS & SMITH tr. *Barthes's Elements of Semiology* III. 65 The commutation test consists of artificially introducing a change in the plane of expression (signifiers) and in observing whether this change brings about a correlative modification on the plane of content (signifieds). **1973** *Screen*

Spring/Summer 89 In a great many cases the term 'form' designates the film *signifier*..and the term 'content' its *signified*. **1976** T. EAGLETON *Crit. & Ideology* iii. 72 History, one might say, is the *ultimate* signifier of literature, as it is the ultimate signified. **1977** A. SHERIDAN tr. *Lacan's Écrits* iii. 69 The symptom is here the signifier of a signified repressed from the consciousness of the subject. **1979** *Dædalus* Summer 72 He has only to offer the Signifier, 'Lie'—which could mean 'tell an untruth'—for Othello to snatch the Signified, 'lie with'.

c. *U.S. slang* (chiefly *Blacks'*). One who boasts or makes insulting remarks or insinuations, esp. in an attempt to exceed others in exaggeration.

1962 *Jrnl. Amer. Folklore* July–Sept. 212 The monkey is a 'signifier', and one of the methods he uses for inflaming the lion is to indicate that the elephant has been 'sounding' on the lion. **1965** H. GOLD *Man who was not with It* xxii. 204 When he bragged like any carnie signifier, then I wondered where and why I was going. **1972** J. MARYLAND in T. Kochman *Rappin' & Stylin' Out* 209 The following verbal play is indicative of the type that might be found in any of a number of shine parlors..or street corners where the signifiers can be found congregating each day.

signify, *v.* Add: **7. b.** (Later examples.)

1903 SOMERVILLE & 'ROSS' *All on Irish Shore* iii. 75 'Did many people say it?' asked Mr Gunning... 'Oh, no one whose opinion signified!' retorted Fanny Fitz. **1930** A. P. HERBERT *Water Gipsies* ii. 16 Don't worry, Fred. It don't signify.

8. *intr.* *U.S. slang* (chiefly *Blacks'*). To boast or brag; to make insulting remarks or insinuations.

1932 *Evening Sun* (Baltimore) 9 Dec. 31/5 *Signify*, to pretend to have knowledge of a matter or subject in which one is poorly informed. **1935** Z. N. HURSTON *Mules & Men* I. vii. 161 'Aw, woman, quit tryin' to signify.' 'Ah kin signify all Ah please, Mr. Nappy-chin.' **1948** *Common Ground* Summer 42/2 He was signifying and getting his revenge through songs. **1968** *Down Beat* 7 Mar. 38/3 One night Billie brought the personal element into focus by 'signifying', which in Harlemese means making a series of pointed but oblique remarks apparently addressed to no one in particular, but unmistakable in intention in such a close-knit circle. **1969** C. MITCHELL *Lang. Behavior in Black Urban Community* iii. 96, I wasn't signifying at her, but..if the shoe fits, wear it. **1973** A. DUNDES *Mother Wit* 141/2 A sample of some of the special techniques and forms of extended word play should convince even the most adamant sceptic that no black child who can signify or play the dozens can rightly be called lacking in verbal skills.

signifying, *vbl. sb.* Add: **b.** *U.S. slang* (chiefly *Blacks'*). The act of boasting, baiting, insulting, or making insinuations. Also *attrib.*

1959 A. ANDERSON *Lover Man* 21 'Y'all hush your signifying,' I said. 'That there's a *lady*, and I won't have y'all signifying 'bout her like that.' **1964** *Amer. Folk Music Occasional* I. 75 'Signifying' is a children's device, and is severely 'put down' by adults. **1970** [see *JONING *vbl. sb.*]. **1974** H. L. FOSTER *Ribbin', Jivin', & Playin' Dozens* v. 206 Mezzrow describes a signifying scene in Big John's bar in Harlem. **1977** *Maledicta* Summer 15 Young blacks have verbal contests like this in anti-family insults, called *sounding* or *signifying*.

signifying, *ppl. a.* Add: **b.** *U.S. slang* (chiefly *Blacks'*). That boasts, insults, or makes insinuations.

1956 M. STEARNS *Story of Jazz* (1957) i. 11 The language is Creole French and the New Orleans Creoles call it a 'signifying song'. In spite of its gaiety and rhumba-like rhythms, this song cuts two ways and the *sali dame* (dirty lady) to which it is addressed is about to have her reputation shredded. **1969** *Negro Digest* Sept. 14 Signifying poetry holds a special fascination for me. **1974** H. L. FOSTER *Ribbin', Jivin', & Playin' Dozens* v. 207 Deep down in the jungle where the coconuts grow/Lived the signifyingest motherfucker that the world ever know.

signing, *vbl. sb.* Add: **1. b.** With adverbs, as *signing-in, -off, -on, -out, -up*. Also *attrib.*

1925 PATERSON & WEBSTER *Man. Locomotive Running Shed Management* viii. 103 The signing-on times of cleaners are arranged with due regard to the finishing times of the engines they are required to clean. **1948** H. INNES *Blue Ice* iii. 77 He gave the signing-off whistle.. then our attention was called back to the radio. **1950** *Sport* 7–11 Apr. 14/3 He is Allenby Chilton,..to whom Liverpool gave a trial before the United engaged him, for a £10 signing-on fee, in 1938. **1965** 'T. HINDE' *Games of Chance* II. iv. 203 The purpose of the signing-in book..is to ascertain that all staff are arriving punctually. **1968** *Brit. Med. Bull.* XXIV. 222/1 The computer equivalent of the signing out inspection of the report by the laboratory staff. **1973** *Guardian* 23 May 6/8 There is no signing-on at the new job centre. **1974** HAWKEY & BINGHAM *Wild Card* xiv. 124 A laxness among certain members of staff regarding signing-out procedures for items drawn from stock. **1976** E. DUNPHY *Only a Game?* v. 159 The signing-on fee is crucial in football these days. **1981** J. SCOTT *Distant View of Death* xii. 162 Signing in and signing out meant nothing to Rosher... He simply worked.

3. *transf.* One who has signed a contract.

1974 *Motor Cycle* 23 Mar. 10/6 (*caption*) Hackney's new signing, Norwegian Dag Lovaas (left) chats with Kings Lynn international Malcolm Simmons. **1977** *Times* 4 Aug. 6/6 Macdonald, a £330,000 buy from Newcastle..and Hudson, a £200,000 signing from Stoke..were sent home ..because of alleged misconduct on the tour.

sign-post, *sb.* Add: **2.** Also *fig.* and *attrib.*

1961 *Atlanta Constitution* 17 Aug. 5 In the..breathless state of being in love the usual signposts that guide you to lasting and satisfying relationships are sometimes obscured. **1962** *Listener* 6 Dec. 958/2 He therefore studies talk more closely and discovers that..by selecting signpost phrases, he can convey what is being felt by what is unsaid. **1969** I. & P. OPIE *Children's Games* p. viii, Full use has been made of the signposts, clearly marked for those willing to look for them, that are provided by *The Oxford English Dictionary* and *The English Dialect Dictionary*.

So **si·gnpostless** *a.*, not equipped with or marked by sign-posts.

1962 *Times* 22 Feb. 14/4 The signpost-less lanes of Surrey. **1968** *Economist* 16 Mar. 12/2 A preposterous, unanalysable, ignorant, signpostless, meandering mess.

si·gnpost, *v.* Also **sign-post.** [f. prec.] *trans.* To direct or indicate by means of or in the manner of a sign-post; to equip or provide with sign-posts. Also *fig.* Hence **si·gnposted** *ppl. a.*, **si·gnposting** *ppl. a.* and *vbl. sb.*

1895 A. A. GRACE *Maoriland Stories* 105, I shall just sign-post 'em up to the station when they come. **1922** W. J. LOCKE *Tale of Triona* xxiii. 265 The road undulated ..with a steeply sloping drop of thirty feet to the valley. Such spots were grimly sign-posted for motorists. **1923** *Daily Mail* 19 May 5 Where the road is not so good and badly needs proper signposting. *Ibid.* 21 May 4 Dartmoor is moderately well signposted. **1930** *Aberdeen Press & Jrnl.* 8 May 5 The R.A.C. propose only to signpost the main arteries. **1938** *Sun* (Baltimore) 16 May 8/1 Even more marked improvements have been made. There are what are called 'sign-posted' reviews. **1946** E. LINKLATER *Private Angelo* viii. 81 They had reconnoitred, by routes that camel-ribs signposted, the farthest Libyan oases. **1953** X. FIELDING *Stronghold* 57 A forbidden area metaphorically signposted 'noli me tangere'. **1955** *Times* 6 Jan. 6/1 They were also signposting a ring road for through traffic avoiding the central London area. **1961** *N. & Q.* Nov. 440/1 There are textual notes at the foot of each page, ludicrously signposted by a system of expanding alphabetical cycles. **1971** P. GRESSWELL *Environment* 104 Some councils are prepared to spend money on signposting. **1975** J. B. HARLEY *O.S. Maps* p. xiii, A principal objective is to indicate what the map user is likely to find on a particular map series and..to signpost what is omitted. **1976** *Howard Jrnl.* XV. I. 39 The different themes and possible groupings of the essays could perhaps have been signposted better to make the reader's task easier. **1978** A. & G. RITCHIE *Anc. Monuments Orkney* 22 This cairn is approached by a signposted path on the N side of the Trumland to Westness road. **1981** *Times* 2 Sept. 21/1 It was as recently as June that the board [of Trusthouse Forte] reported on the six months to last April. This signposted a dreadful year for hotels.

‖ **sigri** (si·grī). [Gujarati *sagḍī*.] A fire or stove used for cooking.

1949 J. R. LAWRENCE *Indian Embers* 56 All day our bedding is spread over wicker cages enclosing a charcoal sigri. **1954** J. MASTERS *Bhowani Junction* 127 The invisible charcoal fumes from the sigri tingled in my nostrils. **1964** M. MALGONKAR *Bend in Ganges* xxxiv. 334 Sundari was bending over the sigri..turning the toast. **1970** 'B. MATHER' *Break in Line* xx. 245 Out here, away from the charcoal *sigri*, the cold was intense.

‖ **sijo** (sī·dʒo). [Korean.] **a.** A type of Korean vocal music. **b.** A Korean lyric poem usu. consisting of twenty-four syllables divided into three lines. Also *attrib.*

1898 I. L. BIRD *Korea & her Neighbours* xii. 191 There are three classes of Korean vocal music, the first being the *Si-jo* or 'classical' style, *andante tremuloso*, and 'punctuated with drums'. **1954** W. STEVENS *Let.* 9 July (1967) 840 Here are your sijos... The poems are charming, at least to me. But Korean poetry..is a delicacy, like bees' knees and apple hips. **1960** P. HYUN *Voices of Dawn* 28 The metaphysical *sijo* poets. **1971** *Korean Folklore & Classics* III. 86 The satto read sijo silently shaking himself..to forget his tediousness. **1972** *Korea Past & Present* xiv. 319 Kagok is a five-stanza form which is accompanied by an orchestra, whereas Sijo is a three-stanza form without orchestra. But they are similar in the sense that both use Sijo poems as words. **1977** *Korea* (Korean Overseas Information Service, Seoul) 57/3 The most popular Korean poetic form, the *Sijo*.

sika[1]. Substitute for def.: A small red deer, *Sika nippon*, native to Japan and eastern China and widely naturalized elsewhere. Also *attrib.* (Later examples.)

1909 E. PROTHEROE *Handy Nat. Hist.: Mammals* x. 365 The Sika..is a beautiful brilliant chestnut, thickly spotted with white. **1957** O. BRELAND *Animal Friends & Foes* i. 38 Important deer..include..the sika deer of eastern Asia. **1966** *Punch* 19 Oct. 596/2 The sika-deer of Brownsea Island left its restricted space some years ago for the Dorset mainland. **1978** *Lancashire Life* Nov. 70/1 The last Lord Ribblesdale..also introduced the Sika deer which roam the district to this day. **1981–2** *Deer Farmer* (N.Z.) Summer 3 But we have farmed other breeds, such as fallow and sika.

‖ **sika**[2] (sī·kā). [Bengali, ad. Skr. *śikyā* sling.] A rope hanger for suspending baskets, etc.

1974 *Observer* (Colour Suppl.) 15 Sept. 40 (Advt.), Sikas are traditionally used in Bangladesh village homes instead of kitchen cupboards. **1979** *Church Times* 26 Oct. 13/3 Jute *sikas* (basket hangers), bags and place-mats from Bangladesh form an attractive part of Traidcraft's range.

Sikan, var. *SICAN. **Sikel,** var. *SICEL *sb.* and *a.* **Sikeliot(e,** varr. *SICELIOT *sb.* and *a.*

Sikh. 1. For definition read: A member of a monotheistic religious group, originally established in India (chiefly in the Punjab) by Guru Nanak in the early part of the 16th century.

The majority of Sikhs are still located in the Punjab in northern India, but many are now living in other parts of India, and in Africa, Europe, the United States, and elsewhere. Sikhs became famous for their military prowess in the 19th century during the period of British imperialism. The Sikh religion requires its members, among other institutionalized customs, to wear a turban, and this has brought Sikhs into conflict with authorities in some countries outside India in the second half of the present century, mainly because of local regulations about the wearing of crash helmets on motor cycles.

Sikhara, var. *SHIKHARA.

Sikhism. (Earlier example.)

1849 J. D. CUNNINGHAM *Hist. Sikhs* iii. 96/1 There are also elements of change within Sikhism itself.

Sikkimese (sikimī·z), *a.* and *sb.* [f. *Sikkim* (see below) + -ESE.] **A.** *adj.* Of or pertaining to Sikkim, a country in the eastern Himalayas. **B.** *sb.* A native or inhabitant of Sikkim; *collect.* the people of Sikkim.

1861 J. C. GAWLER *Dispatch* 15 Feb. in H. St. G. M. Mcrea *Regimental Hist. 45th Rattray's Sikhs* (1933) I. 149 Captain Impey's column surprised the Sikkimese camp at Temi. **1938** G. GORER *Himalayan Village* i. 36 During the eighteenth and early nineteenth centuries the Lepchas fought with the Sikkimese against the continued invasions of the Nepali and Bhutanese. **1955** *Times* 3 June 7/3 The devout Sikkimese who asked that no human foot should profane the final high places that their faith holds sacred. **1960** 'S. HARVESTER' *Chinese Hammer* i. 10 Sikkimese metalwork and Kashmiri brocades. **1973** *Times* 12 Apr. 8/7 The Bhutia-Lepchas and the Sikkimese Nepalis have, however, parity in the State Council. **1978** C. HUMPHREYS *Both Sides Circle* xx. 212 The inhabitants were a blend of Lepcha, Bhutia and other races besides the native Sikkimese.

sikr(a, varr. *SHIKHARA.

Siksika (si·ksikā). Also † **Seksikai, Siksikai.** [Blackfoot, f. *siksi-* black + *-ka* foot.] The Blackfoot Indians, *esp.* those of the northernmost of the three tribes which comprise the Blackfoot.

1843 tr. *Wied-Neuwied's Trav. Interior N. Amer.* xix. 245 The Blackfeet form a numerous nation, which is divided into three tribes, speaking one and the same language. These tribes are—1. The Siksekai or Seksekai, the Blackfeet properly so called. *Ibid.*, The Siksekai signifies, in their language, Blackfoot. **1902** *Encycl. Brit.* XXIX. 466/1 Blackfeet (Siksika). **1910** in F. Hodge *Handbk. Amer. Indians* II. 570/1 A band of the Kainah division of the Siksika. **1923** M. BARBEAU *Indian Days in Canadian Rockies* Addenda, 197 The Blackfoot..were the Bedouins of the Plains... Their three subdivisions, consisting of the Blackfoot proper (or Siksika), the Blood, and the Piegan..considered themselves as of one family. **1942** O. LEWIS *Effects of White Contact upon Blackfoot Culture* ii. 7 The Blackfoot, Blood and Piegan tribes are at present time located on four reserves, of which three are in Alberta, Canada, and one in Montana... To avoid confusion we shall use the Blackfoot term 'Siksika' to refer to the Northern Blackfoot, and the term Blackfoot to refer to all three tribes. **1952** J. R. SWANTON *Indian Tribes N. Amer.* 396 The Siksika are divided into the following subtribes: The Siksika or Blackfeet proper, [etc.].

Sikyonian, var. *SICYONIAN *sb.* and *a.*

silage. Add: **2.** *attrib.* and *Comb.*, as *silage clamp* (CLAMP *sb.*[3]), *loader, -maker, stack*; *silage-feeding, -making* vbl. sbs.; *silage-fed* ppl. adj.; **silage cutter,** a stationary machine for chopping a crop into short lengths for silage and elevating it into a silo; also, a silage harvester; **silage harvester,** a machine for cutting a standing crop as it travels, chopping it into short lengths for silage, and elevating it into another vehicle.

1961 *Farmers' Weekly* 6 Oct. 117/1 The loose housing and the self-feed silage clamps. **1978** *Cornish Guardian* 27 Apr. 15/7 (Advt.), 131 acre accredited dairy farm..milking parlour, silage clamp. **1962** *Trans. Amer. Soc. Agric. Engin.* XIX. 117 (*heading*) Code for testing silage cutters. **1967** MARTIN & LEONARD *Princ. Field Crop Production* (ed. 2) viii. 208 Corn and sorghum row crops for silage are harvested with a field silage cutter or a forage harvester. **1972** *Country Life* 15 June 1580/1 Silage feeding to the flock..usually begins just before Christmas. *Ibid.* 1580/3 It is usually the practice in silage-fed flocks to introduce a cereal supplement just before lambing. **1931** J. B. DAVIDSON *Agric. Machinery* xxv. 266 To reduce the labor of handling green fodder and to dispense with the use of binding twine, the operations of cutting and harvesting silage are combined in the silage harvester, and the cut silage is delivered to a wagon as it is drawn beside the harvester through the field. **1977** *Cork Examiner* 8 June 15/5 (Advt.), Used Kidd silage harvester, good working condition, £525. **1971** *Power Farming* Mar. 8/3 The operation of the American silage loader..is explained in the captions to the accompanying photographs. **1924** W. J. Malden *Grassland Farming* xi. 152 The silage-maker has two points to consider. **1960** *Farmer & Stockbreeder* 2 Feb. 55/2 Most of the people I have talked to have been experienced silage-makers. **1924** W. J. Malden *Grassland Farming* xi. 149 Heating is really an aid in silage-making, because under the moist heat the stiffest stems yield, and compression becomes easy. **1960** *Farmer & Stockbreeder* 16 Feb. 73/2 In spite of the great progress in silage-making in the last few years, the amount of permanent grass and temporary leys made into hay in Britain still far exceeded the amount conserved by all the other processes put together. **1888** Silage stack [in Dict.].

silajit (si·lădʒit). Also **shilajatu, sillajeet,** etc. [a. Hind. *shila-jit*, Skr. *śilājit, śilājatu* bitumen, f. Skr. *śilā* rock + *jit* conquering or *jatu* essence.] A name given to various solid or viscous substances found on rock in India and Nepal (see quot. 1903), esp. a usu. dark-brown odoriferous substance which is used in traditional Indian medicine and probably consists principally of dried animal urine.

1811–12 F. Buchanan *Acct. Bihar & Patna* (1936) II. iii. iii. 467 About three miles farther in the same direction..a very peculiar substance called silajit exudes. **1833** G. Playfair tr. *Taleef Shereef* 96 *Sillajeet* is the urine of the hill wild goat, which when the animal is rutting, is discharged on the stones and evaporated by the sun's heat. **1903** *Jrnl. R. Asiatic Soc. Bengal* LXII. ii. 98 One of the most peculiar medicinal substances of the East is that called Silajit or Shilajatu. It is known by the former name in Hindi and Persian, and by the latter in Bengali and Sanskrit. **1964** *New Statesman* 3 Apr. 517/2 He was extolling the miraculous properties of his ointment, guaranteed to contain the purest silajit brought from Tibet.

silane (sai·lēⁱn). *Chem.* [ad. G. *silan* (A. Stock 1916, in *Ber. d. Deut. Chem. Ges.* XLIX. 108): see Silicon and -ane.] Any of the large class of hydrides of silicon analogous to the alkanes; *spec.* silicon tetrahydride, SiH_4, a colourless gas which has strong reducing properties and is spontaneously flammable in air.

1916 *Jrnl. Chem. Soc.* CX. ii. 319 The suggestion is made that the term 'silanes' be accepted generally for the saturated compounds of silicon and hydrogen, the various members being distinguished thus: SiH_4, monosilane; Si_2H_6, disilane; [etc.]. **1935** *Nature* 9 Mar. 397/2 The oxidation of silane resembles very closely that of phosphine rather than that of methane. **1958** *Times Rev. Industry* June 26/3 The gaseous compound silane, SiH_4, is produced by reacting silicon tetrachloride with lithium aluminium hydride. **1977** *Whitaker's Almanack 1978* 1034/1 Work has now begun on cleaning and repairing these figures [on Wells Cathedral] and..it is hoped that the use of lime and silanes will prevent further decay of the stone surfaces.

silanize (sai·lănaiz), *v.* [f. *silan(e + -ize.] *trans.* To treat (silica-based material, esp. support material for chromatography) with reagents which render the surface more inert by converting reactive groups to organo-silicon groups. Hence **si·lanized** *ppl. a.*, **si·lanizing** *vbl. sb.* Also **silaniza·tion,** treatment of this kind.

1962 *Analytical Chem.* XXXIV. 891/3 To eliminate possible adsorption effects, the material was silanized with Siliclad. *Ibid.*, Silanizing reduced the specific surface to about 2 sq. meters per gram. *Ibid.* 892/1 A weighed amount of silanized Chromosorb-R. **1968** *Jrnl. Chromatogr.* XXXIV. 305 By silanizing the support surface we obtained changes in the chemical character of the surface. *Ibid.* 308 The effect of surface silanization on the structure of the internal support pores is obvious. **1973** *Nature* 30 Mar. 339/1 A 5·75 inch disposable pipette plugged with silanized glass-wool. **1975** Williams & Wilson *Biologist's Guide to Princ. & Techniques Pract. Biochem.* iii. 70 This is normally achieved by silanization of the support with such compounds as hexamethyl-disilazane.

Silastic (silæ·stik). *Chem.* Also **silastic.** [f. Si(licon + E)lastic *a.* and *sb.*] A proprietary name for silicone rubber. Freq. *attrib.*

1946 *Rubber Age* Feb. 580/1 Silastic is an almost entirely inorganic synthetic elastomer. **1947** *Official Gaz.* (U.S. Patent Office) 2 Dec. 26/1 Dow Corning Corporation... *Silastic.* For compositions..comparable to..rubber prior to vulcanization but containing organosilicon polymers... Claims use since July, 1945. **1953** *Electronic Engin.* XXV. 309/1 Also recently introduced is a range of tropical suppressor filter units which are hermetically sealed in a metal container with silastic terminals. **1965** *Trade Marks Jrnl.* 12 May 632/1 Silastic. **1980** *Recent Advances in Surgery* X. 66 A silastic strain gauge.

‖ **silat** (silæ·t). Also **Silat.** [Mal.] The Malay art of self-defence, practised in a series of exercises as a martial art or accompanied by drums as a ceremonial display or dance.

[**1900** W. W. Skeat *Malay Magic* vi. 381 Those who have any skill amuse the company with exhibitions of Malay fencing (*main silat*).] **1910** R. J. Wilkinson *Papers on Malay Subjects: Life & Customs* iii. 27 Even if it does drag on..as a game for Malay boys, the *main silat* will never preserve in its new form the curious wealth of technicalities associated with the national weapon of the country. **1972** M. Sheppard *Taman Indera* vi. 85 The first henna dance..took place in the palace garden... The only accompaniment to this dance was supplied by *Silat* drums—double-headed barrel drums. *Ibid.* x. 142 There are a great many varieties of Silat and different teachers specialize in different types. **1978** T. Williamson *Technicians of Death* xiii. 112 Silat, the Malayan martial art where the hands and arms are used for defence and the knees and feet for attack.

‖ **silbador** (silbădōᵊ·ɪ). Pl. **silbadores, silbadors.** [Sp., = whistler.] One who uses the whistled language *silbo* (see next).

1957 *Archivum Linguisticum* IX. i. 44 The melodies whistled to silbadores proved completely unintelligible. **1965** *Sun* 24 July 5/5 A top silbador can be heard up to nine miles away... The silbadors have a voice range of two octaves.

‖ **silbo** (si·lbo). Also **Silbo.** [Sp., = a whistle, whistling.] A form of whistled Spanish used by the inhabitants of Gomera in the Canary Islands, in order to communicate across long distances. Also known as *Silbo Gomero.*

1957 *Archivum Linguisticum* IX. i. 44 There are abundant references to the silbo and at least two full-length studies. **1964** A. Classe in D. Abercrombie et al. *Daniel Jones* 43 The investigation of a whistled form of Spanish (Silbo Gomero)... We tested one group of consonants which are invariably confused in the Silbo. **1977** *Sci. Amer.* May 141/2 In the Spanish Civil War military signals were on occasion cast in Silbo, but there were Gomerans on both sides and so the measure-countermeasure drama soon ran its course. **1978** *Verbatim* Sept. 13/1 A phonetic analysis of the Silbo Gomero of Canarian Spanish is extended feature by feature to the articulated whistle.

silcott (si·lkɒt). ? *Obs.* Also **silcot, Sillcott.** [f. Sil(k *sb.* + Cott(on *sb.*¹] A material made of cotton finished to resemble silk, chiefly used for underskirts. Cf. Silkette.

1894 J. E. Davis *Elem. Mod. Dressmaking* (1895) 93 Varieties of silkette or silcot (cotton finished to look and feel like silk). **1923** *Daily Mail* 17 Feb. 10 (Advt.), Petticoat of Sillcott edged with pleated and hemstitched flounce. *Ibid.* 28 Feb. 5 Silcott Petticoat with small crystal frill.

silcrete (si·lkrīt). *Geol.* [f. Sil(ica + Con)crete *sb.*] A quartzite formed of sand grains or pebbles cemented together by silica; a siliceous duricrust.

1902 G. W. Lamplugh in *Geol. Mag.* IX. 575, I have the hardihood to suggest that the term might be complemented by equivalents,—'silcrete', for sporadic masses in loose material of the 'greywether' type, indurated by a siliceous cement. **1950** *Antiquity* XXIV. 209 More than 90 per cent of these implements are of 'ferricrete sandstone', the others being of 'silcrete sandstone'. **1977** A. Hallam *Planet Earth* 85 Bedrock became deeply weathered to give striking weathering crusts (duricrusts) composed of iron and aluminum oxides (in the case of ferricretes) or silica (in the case of silcretes).

sild (silt). [a. Da., Norw. *sild* herring: cf. Sile *sb.*³] A small immature herring, *Clupea harengus*, esp. one caught in northern European seas.

1921 *Handbk. Norway & Sweden* (Admiralty) viii. 157 The true sardine..does not occur in northern waters. Norwegian 'sardines' are now known in commerce as 'sild'. **1962** E. M. Cruickshank in G. Borgstrom *Fish as Food* II. iv. 195 With small fish like the sild, the entire fish is processed.

silence, *sb.* Add: **1. a.** (Later pl. and *personif.* examples.)

a **1875** G. M. Hopkins *Poems* (1967) 31 Elected Silence, sing to me. **1967** G. Steiner *Lang. & Silence* 415 Dickens, Hopkins, Kipling are examples of modern writers whose root sensibility was oral, and who tried to adapt essentially oral means to the silences of print.

e. The renunciation of speech chosen or vowed by certain religious or monastic orders, *esp.* the Trappists. Also in extended sense. Freq. in phr. *the rule of silence.*

1387, *c* **1450** [see sense 7 in Dict.]. *a* **1631** Donne *Poems* (1633) 69 Harmelesse fish monastique silence keepe. **1884** Addis & Arnold *Cath. Dict.* 804/1 Probably the most trying part of all the discipline is the silence, no monk being allowed to speak to his brother on any occasion. **1921** G. O'Donovan *Vocations* xxii. 305 Hush, Sister. The rule of silence is no joke. **1957** P. L. Fermor *Time to keep Silence* 67 There is a special dispensation from the rule of silence for the monks who deal with the abbey livestock when they are actually addressing their dumb charges. **1978** *Oxford Diocesan Mag.* Dec. 17/1 Then there was the two days' retreat... At no time..was the sense of fellowship more apparent than during the silence.

f. Proverbial phr. *silence is golden.*

1834 Carlyle *Sart. Res.* iii. iii, in *Fraser's Mag.* June 668/1 As the Swiss Inscription says: *Sprechen ist silbern, Schweigen ist golden* (Speech is silvern, Silence is golden). **1865** W. White *Eastern England* ii. ix. 129 Silence is golden, says the proverb. We apprehended the full significance thereof when far away from busy thoroughfares. **1935** M. V. Hughes *Vivians* vii. 138 'Did you tell him about that?' 'No, and I'm wondering whether I ought to?' 'I shouldn't if I were you. Silence is golden.' **1980** J. O'Neill *Spy Game* xxv. 239 'I'll tell you the rest.. on the way back.' He sealed her lips with a finger. 'Meanwhile, silence is golden.'

2. d. Phr. *the rest is silence* and varr., in allusion to the last words of the dying Hamlet (Shaks. *Ham.* v. ii. 368).

1910 Galsworthy *Justice* ii. 49 Once this cheque was altered and presented, the work of four minutes—four mad minutes—the rest has been silence. **1939** A. Huxley *After Many a Summer* ii. i. 187 If only the rest were silence!.. What joy if the rest of Wordsworth had been silence, the rest of Coleridge, the rest of Shelley! **1982** *Daily Tel.* 2 June 16/4 In most of the countries involved the eternal tug-of-war between Government and news media has long since ended. The curtain has fallen. The rest is silence.

e. A period of silence observed in memory of the dead, *esp.* the two minutes' silence kept on the anniversary of Armistice Day (11 Nov. 1918) or, since 1946, on Remembrance Sunday.

1919 *Times* 12 Nov. 15/6 The Great Silence... At 11 o'clock yesterday morning the nation, in response to the King's invitation, paid homage to the Glorious Dead by keeping a two minutes' silence for prayer and remembrance. *Ibid.* 16/1 On the Stock Exchange, after the silence, a gong was sounded. **1926** A. Topham *Chron. Prussian Court* xx. 245 We discussed among other things the *Titanic* disaster [1912] which had recently happened, and I remember referring to 'the silence' of two minutes by which the Canadian railways and churches had honoured the memory of the Canadians who had perished. **1929** *B.B.C. Year-bk. 1930* 78 Broadcasting the Silence November 11th, 1928. **1972** 'E. Lathen' *Murder without Icing* (1973) xxii. 188 The game was preceded by a two-minute silence in memory of Billy Sicagusa. **1982** D. Phillips *Coconut Kiss* vi. 52 You march once round the playground and salute the flag... Then you go in for the two minutes' silence.

7. *silence-box* (earlier example), *room* (examples); *silence-loving* adj.; **silence cabinet,** (*a*) = *silence-box*; (*b*) (see quot. 1929).

1889 *Telephone* I. 471/1 The public is also admitted to a silence-box at the Nottingham Post Office. **1893** Preece & Stubbs *Man. Teleph.* 227 At most telephone exchanges a 'silence cabinet' is provided in the public office. **1929** *B.B.C. Year-bk. 1930* 309 In Savoy Hill there are nine studios, six of which are equipped with silence cabinets (these are small rooms adjacent to the studios from which the announcer can speak before switching over to the studio itself). **1855** F. W. Faber *Growth in Holiness* ix. 147 It wrung a cry even from the silence-loving Heart of our ever-blessed Saviour. **1912** W. Owen *Let.* 23 June (1967) 142 The firm Superintendent of their Sunday School, the silence-loving, and the melancholy-voiced, on that day capered about the lawn among them. **1958** S. Hyland *Who goes Hang?* xviii. 77 They were in the Silence Room of the Library, a room in which conversation..is..a tabu. **1959** T. S. Eliot *Elder Statesman* ii. 47 And remember, when you want to be *very* quiet There's the Silence Room. With a television set.

silenced, *ppl. a.* Add: **b.** Of a gun: fitted with a silencer.

1965 [see *gunsel 2]. **1974** 'I. Drummond' *Power of Bug* vii. 106 Why *did* the chap poke a silenced pistol through the window? **1980** *Daily Tel.* 15 Oct. 3 He would not shoot Henry MacKenny with a silenced firearm.

silencer. Add: **2.** For 'or rifle' read 'rifle, etc.' and add later examples. Hence **si·lencered** *a.*, of guns: fitted with a silencer (cf. *silenced *ppl. a.* b.).

1905 *Engineering* 20 Oct. 529/3 This form of silencer is not necessarily confined to marine motors. **1926** G. Hunting *Vicarion* vi. 98 He must have had a silencer on his gun. **1950** G. Brenan *Face of Spain* vii. 149 Nine hours in a bus without a silencer..over mountain roads full of pot-holes. **1958** *Economist* 25 Oct. 349/3 Silencers on the engines reduce their efficiency by 4 to 5 per cent, and the total weight of silencing equipment is 1,600 lb, equivalent to 8 passengers. **1967** J. Wainwright *Worms must Wait* lxxxii. 214 They heard the tiny crack of the silencered Luger. **1978** R. Ludlum *Holcroft Covenant* iii. 44 He opened the door, pulled out his revolver and fired, the gunshot muted by a silencer.

silent, *a.* and *sb.* Add: **A.** *adj.* **1. a.** Also in phr. *strong silent man* (or *person, type,* etc.): a man who conceals and controls his feelings.

1840 Carlyle *Chartism* iv. 30 With this strong silent people have the noisy vehement Irish now at length got common cause made. **1905** M. Beerbohm in *Sat. Rev.* 23 Sept. 401/1 He is going to cry? No, the hero is one of those strong, silent men. **1913** C. Mackenzie *Sinister Street* I. ii. xv. 407 She said I must be careful not to grow up into a strong silent Englishman, because their day was done. **1919** A. A. Milne *Not that it Matters* 142 It is useless to model ourselves now on the strong, silent man of the novel whose face is a shutter to hide his emotions. **1936** W. S. Maugham *Cosmopolitans* 260 These for the most part are strong silent men who waste no words. **1978** L. Charteris *Saint & Templar Treasure* (1979) i. 25 I've always fancied myself as the strong silent type.

d. *as silent as the grave*: of a place, hushed, containing no natural noise; of a person, secretive, discreet.

1823 J. F. Cooper *Pilot* I. vi. 78 'Does he keep silent?' 'As the grave.' **1829** W. Scott *Jrnl.* 1 July (1946) 89 The house..became silent as the grave. **1889** R. L. Stevenson *Master of Ball.* iii. 62 We..lowered ourselves softly into a skiff, and left that ship behind us as silent as the

grave. **1936** W. S. Maugham *Cosmopolitans* 269, I will be as silent as the grave, but honestly I don't understand. What does it all mean?

3. c. (Later example.)

1981 Macniven & Moore *Literary Lifelines* p. v, Silent corrections have been limited to restoring transposed letters.

e. Of a cinema film: unaccompanied by sound recording. Also in extended use to designate that which is related to or concerned with the silent film industry.

1914 [see *filmdom]. **1918** *N.Y. Times* 25 Nov. 11/3 (*heading*) Two opera stars in silent films. **1927** *Melody Maker* Sept. 933/3, I can see very little difference between the music appropriate to the spoken drama and that for the silent screen. **1941** B. Schulberg *What makes Sammy Run?* iv. 59 He was married to one of the big silent stars. **1967** *Listener* 30 Nov. 712/2 Rooming houses full of stars of silent pictures whom nobody remembers. **1977** R. Barnard *Death on High C's* ii. 19 In Owen's production.. you will be the silent-film heroine, and I will be the silent-film handsome seducer.

4. c. Of machinery, etc.: operating with or causing a minimum of noise.

1887 *Encycl. Brit.* XXII. 524/1 Dr. Otto's 'silent' engine, introduced in 1876, was the first successful motor of the modern type. **1904** A. B. F. Young *Compl. Motorist* (ed. 2) iv. 103 The silent working of the Lanchester car makes it also an extremely useful carriage for town use. *a* **1943** H. A. Whitcombe in J. Joyce *Trams of Past* (1979) 25 The citizens of Birmingham were proud of their steam trams and acclaimed them before all others for their..smooth and silent running.

5. b. (Later examples.)

1974 *Nature* 1 Feb. 295/2 Enhancing serum as used in series *b* was then absorbed to remove Ag-B antibodies but possibly not antibodies against (? serologically silent) products of other genes in the MHC, if such exist. **1979** *Ibid.* 5 July 12/3 The recessive scrapie allele is likely to be widespread but clinically 'silent' in these breeds.

d. *Med.* Not giving rise to or showing readily apparent signs or symptoms.

1928 W. Overend *Radiogr. of Chest* II. iv. 49 There are two forms of silent pneumonia: a hilar which does not reach the pleura; and a cortical which does not reach the hilum. **1951** [see *latent *a.* d]. **1979** *Jrnl. R. Soc. Arts* CXXVII. 171/2 We have had no great disease problem associated with them as yet but I have a feeling that the natural host is often a silent carrier.

6. b. *silent-footed* (earlier example).

1845 J. R. Lowell *To Future* in *Graham's Mag.* XXVIII. 52 And he can see the grim-eyed Doom From out the trembling gloom Its silent-footed steeds toward his palace goading.

7. Special collocations: *silent band* = *silent majority* (*b*) below; *silent cop* (*Austral.*) (see quot. 1934); *silent* (*dog*) *whistle,* a high-frequency whistle producing a note audible to a dog but scarcely audible to a human being; *silent heat* (*Vet.*), ovulation occurring without the signs of œstrus; † *silent highway,* a river or canal (*obs.*); *Silent Land,* used allusively to denote the state beyond this life; *silent majority,* (*a*) the dead; (*b*) the mass of people whose views remain unexpressed, esp. in political contexts; those who are usu. overlooked because of their moderation; *silent partner* (*U.S.*) = *sleeping partner* s.v. Sleeping *ppl. a.* 5 a; *silent policeman* (*N.Z.*) = *silent cop* above; *silent service* (see quot. 1929); *silent spring,* in allusion to the title of the work by R. Carson (see quot. 1962), which drew attention to the danger to the natural environment inherent in the use of toxic chemicals; *silent vote* (*U.S.*), the vote of those whose political leanings are not known in advance of their vote being cast; so *silent voter.*

1866 G. Meredith *Let.* 15 Jan. (1970) I. 326 Will bawlings in the street avail?.. They irritate the slumbering dominant party, without strengthening the insurgent. What is being done in the *Fortnightly*, for instance..does strengthen, while it increases the silent band. **1934** T. Wood *Cobbers* x. 122 A circle in the middle of cross-roads, for example, round which all traffic changing direction must swing; a round yellow blob, known here [*sc.* in Adelaide] as the Silent Cop, or the Poached Egg. **1959** D. Hewitt *Bobbin Up* 2 This was the corner, by the silent cop, where she and Roy had come to grief. **1961** C. Willock *Death in Covert* iv. 64 Attached..to the lapel of Gumbe-Howard's coat was a silent dog whistle, and attached..to his heels was a silent dog. **1965** D. Francis *For Kicks* xiii. 173 That's a silent whistle... For dogs... You can't hear it very well..but of course a dog can. **1980** J. W. Hill *Intermediate Physics* xvi. 150 The 'silent' dog whistle produces a note too high for the human ear but heard by a dog. **1950** N. Barron *Dairy Farmer's Vet. Bk.* vi. 65 Cows sometimes have short and possibly 'silent' heats that pass unnoticed, when the ovary produces the egg but the cow does not show any outward sign of being in season. **1970** W. H. Parker *Health & Dis. Farm Animals* vi. 58 The cow tends to be more generous with her signs of oestrus unless she is on a low plane of nutrition, when she may have a 'silent heat' like the ewe. **1848** *Punch* XV. 158/1 New towns have lately sprung up..on each side of the Thames. If the population manages to keep pace with the mania for building, the 'silent highway' will soon become as noisy as the New Cut. **1875** *Birmingham Daily Mail* 5 Mar. 2/5, I speculate on the incalculable good that a 'Home' would do in rescuing girl-babies of the silent highway from the unwomanly scenes of their wasted young lives. **1935** *Times* 28 Feb. 14/2 This Cinderella of the bridges that span London's silent highway. **1853** *Working Man's Way in World* xiv. 320 (*heading*) Parents and friends in the Silent Land. **1939** L. M. Montgomery *Anne of Ingleside* xxvii. 186 Ah well, Anne dearie, they've both passed long since into the Silent Land. **1874** *Harper's New Monthly Mag.* Sept. 468 (*heading*) The silent majority. **1910** *Motor World* 31 Mar. 851/1 (*heading*) Two join the 'silent majority'. Death calls at Detroit and Buffalo and claims well known men. **1955** C. V. Wedgwood *Great Rebellion* I. ii. iv. 256 The King in his natural optimism still believed that a silent majority in Scotland were in his favour. **1970** *Guardian* 11 May 10/2 The Midwest..is 'silent majority' country. **1976** *National Observer* (U.S.) 1 May B4/4 Two-thirds of all marriages in the United States still succeed. So often silent majorities of this kind are too readily forgotten. **1828–32** Silent partner [in Dict., sense 5 b]. **1894** S. Leavitt *Our Money Wars* 221 His Wall St. concern.. came to grief in 1890; also a concern in Buffalo in which he was a silent partner. **1974** R. L. Simon *Wild Turkey* xii. 82 They're only the directors. There's someone behind them. Another investor... It's a silent partner. **1965** F. Sargeson *Memoirs of Peon* iv. 88 A silent policeman had been prised from its street-moorings. **1929** *Papers Mich. Acad. Sci., Arts & Lett.* X. 323/2 *Silent Service,* the Navy. This is a reference to the long silent vigil of the British Fleet. **1937** T. Rattigan *French without Tears* i. 40 You naval people never talk about yourselves, do you? *Rogers.* Well, you know, silent service and all that. **1982** *Daily Tel.* 2 June 16/3 The Army learned certain lessons in Northern Ireland. The silent service did not share that experience. **1962** R. Carson (*title*) Silent spring. **1970** *N.Y. Times* 12 June 38 The Caspian Sea is probably the most dramatic battle-ground of Soviet Russia's looming silent spring and to date this battle is being lost to oil, petroleum products, industrial and city sewage, ballast and waste from ships. *Ibid.* 27 June 28 If we don't develop suitable pesticides—and use them—we really will have Silent Spring because there won't be any trees left for the birds to sing in. **1981** J. Sutherland *Bestsellers* x. 112 Guilt about man's depredation of his and other species' environment, at the 'silent spring' which he has brought. **1936** *Durant* (Okla.) *Daily Democrat* 2 Nov. 2/4 The regents and police pensions amendments have the best chances of carrying, but even they are endangered by the 'silent vote'. **1952** *Economist* 6 Sept. 556/1 It is believed that there will be a substantial 'silent' vote against Mr McCarthy by other citizens unwilling or unable to take a public stand against him. **1884** *Judge* 12 Nov. 140/2 To the Silent Voter, who was to make himself Felt for Cleveland: Tell me where you are and all will be forgiven.

B. *sb.* **3.** A silent film (see sense 3 e of the adj.).

1929 *Morning Post* 24 May 12/7 Every recognised tradition of the 'silents' seems to have gone by the board. **1977** 'J. le Carré' *Hon. Schoolboy* xvii. 404 Even the latest films up here are silents.

silentish, *a.* (Later example.)

1948 E. Blunden *Shakespeare to Hardy* (1956) xii. 168 Among those who were present..at Dickens's public readings, a 'silentish young man'.

Silesian, *a.* and *sb.* Add: **A.** *adj.* **b.** *Silesian stem,* a shouldered stem of a goblet or candle-stick, supposed to have been so named in honour of George I for whom a goblet with such a stem was first made. Hence *Silesian-stemmed* adj.

[**1925** F. Buckley *Hist. Old Eng. Glass* ix. 72 The waisted bowl,..and the pediment, or Silesian-shouldered stem, were both novelties at this time.] **1929** W. A. Thorpe *Hist. Eng. & Irish Glass* I. v. 171 The shouldered stem is commonly known as 'Silesian', but it is certainly not exclusive to that part of Germany. *Ibid.* vi. 204 For about twenty years..the Silesian stem remained constant and English stem growth was at a standstill. **1961** E. M. Elville *Collector's Dict. Glass* 166 (*caption*) Silesian-stemmed glass moulded on the four shoulders with GR in relief. **1978** *Country Life* 19 Oct. 1170/2 It is a very pretty glass (I always fall for the high-shouldered, so-called Silesian stem which is apparently not Silesian at all). **1979** 'J. Gash' *Grail Tree* ii. 22 An engraved lead-glass cordial glass..among some Silesian-stemmed glasses.

Silex² (sai·leks). The proprietary name of a coffee-making machine in which boiling water is drawn through ground coffee in a filter by the creation of a vacuum.

1914 *Official Gaz.* (U.S. Patent Office) 31 Mar. 1370/2 Silex. Particular description of goods.—Coffee-percolators. **1934** *Trade Marks Jrnl.* 12 Sept. 1187/2 *Silex*... Coffee making appliances, kettles and kettle stands, milk heating and pasteurizing utensils, urns and urn stands, coffee sets [etc.]... The Silex Company.., State of Connecticut, United States of America; manufacturers. **1949** H. MacLennan *Precipice* iv. 267 The coffee had bubbled into the top of the Silex. **1971** J. Henderson *Copperhead* xv. 191 A girl appeared with a Silex of coffee and the usual paper cups.

silhouette, *sb.* Add: **2. a.** (Earlier example.)

1843 Thackeray *Irish Sketch-Bk.* II. xiii. 233 Ghostly looking *silhouettes*.

b. The contour or outline of a garment.

1920 *Glasgow Herald* 27 Nov. 4 The silhouette of this season is..much more attractive than that last year approved by Dame Fashion. *Ibid.* 4 Dec. 4 See that you preserve the silhouette of the gown. **1978** *Country Life* 17 Aug. 472/1 The new silhouette..is straight, narrow and short, with well defined, padded shoulders.

silica. Add: **b.** *silica dust*; **silica glass** = *quartz glass* s.v. *quartz 2 a; **silica gel,** hydrated silica in a hard granular form which is very hygroscopic and is used as a desiccant; **silica wool** = *slag wool* s.v. Slag *sb.* 5.

1918 *Act* 8 & 9 *Geo. V* c. 14 § 1 (3) Any industry.. involving exposure to silica dust. **1920** *Jrnl. Amer. Chem. Soc.* XLII. 971 The mere fact that a chemically inert substance like silica gel is found exhibiting such marked absorptive properties is sufficient in itself to indicate that the cause of adsorption does not lie in the interaction of adsorbent and adsorbed substance. **1956** *Nature* 18 Feb. 329/1 The acid mixture was separated on a silica gel column into acetic and propionic acids. **1977** J. Hedgecoe *Photographer's Handbk.* 36 If you are storing a camera in an unfavorable climate put it in a plastic bag with the packet of silica gel which is supplied with most new cameras. **1916** *Chem. Abstr.* X. 103 (*heading*) Silica glass. **1919** *Nature* 23 Oct. 153/1, I have recently observed that 'silica glass' possesses a remarkable crystalline or quasi-crystalline structure when examined in the polariscope. **1965** B. J. Moody *Compar. Inorg. Chem.* xviii. 268/2 Silica may be fused in the oxy-hydrogen blowpipe flame, softening at 1500–1600° C and fusing above 1700° C... The amorphous vitreous product is a supercooled liquid, silica glass, quartz glass, fused quartz or just silica. **1906** *Chambers's Jrnl.* Aug. 599/1 Slag-wool, or silica-wool, is in appearance and properties similar to asbestos.

silicate. a. Substitute for def.: Any salt, ester, or anion of a silicic acid; any substance (e.g. very many minerals and rocks) which is regarded as being formed from silica together with other oxides, and has an extended polymeric anionic structure built up from linked (SiO_4) tetrahedra. (Later examples.)

1891 *Jrnl. Chem. Soc.* LX. ii. 814 (*heading*) Action of phosphorus oxychloride on ethereal silicates and their chloro-derivatives. **1937** W. L. Bragg *Atomic Structure of Minerals* ix. 146 The classification of the silicates by their silicon-oxygen structures..corresponds to a large extent to the usual mineralogical classification. The structure is so rigid in the pyroxenes and amphiboles, micas, felspars and zeolites that it decides the crystal form. **1959** Berry & Mason *Mineralogy* xv. 462 Silicate classification is based on the types of linkages, which are as follows: 1. Independent tetrahedral groups... 2. Double tetrahedra structures [etc.]. **1962** P. J. & B. Durrant *Introd. Adv. Inorg. Chem.* xviii. 605 In most silicates the SiO_4 tetrahedra are not discrete, but are parts of larger systems in which certain oxygen atoms are held in common by two tetrahedra.

silication. (Later examples.)

1904 *Monogr. U.S. Geol. Survey* XLVII. viii. 667 Silication of carbonates, forming silicates and releasing carbon dioxide, is one of the chief reactions of the zone of anamorphism. **1923** *Jrnl. Geol.* XXXI. 176 Magnetic end-stage emanations..produce a bewildering series of silication..silicification, gametization and metallic-mineralization effects.

silici-. Add: **silicicla·stic** *a.* *Petrol.*, applied to clastic rocks and deposits that are not carbonates; **si·licicole, silici·colous** *adjs.* *Bot.* [L. *colĕre* to inhabit], growing best in siliceous soil.

1961 J. Braunstein in *Bull. Amer. Assoc. Petroleum Geologists* XLV. 2017/2, I dare to submit two new terms... These are, 'calciclastic'.., referring to clastic carbonate rocks, and, 'siliciclastic'.., referring to clastic non-carbonate rocks (which are almost exclusively silicon-bearing, either as forms of quartz or as silicates). **1976** *Nature* 3 June 440/1 Two major subdivisions are considered—the siliciclastic and carbonate tidal deposits. Recent siliciclastic examples include the classic tidal flats of the North Sea, [etc.]. **1965** Bell & Coombe tr. *Strasburger's Textbk. Bot.* 753 Species characteristic of particular soils have long been known,..e.g. halophytes.., or calcicole or silicicole (calcifuge) plants. **1901** *Jrnl. R. Microsc. Soc.* 53 J. A. Cl. Roux has grown a number of silicicolous (arenaceous) plants in calcareous soils, and finds that..the seedling plants develope tardily and imperfectly. **1932** Fuller & Conrad tr. *Braun-Blanquet's Plant Sociol.* vi. 187 In humid regions the leaching of carbonates sets in upon dolomite.., and acidophilous plants follow in the train of the progressive acidification of the dolomite soil. This occurrence of silicicolous plants on dolomite substrata was first noticed by Sendtner.

silicic, *a.* Add: *silicic acid,* a very weakly acidic substance obtained esp. by the action of acids on solutions of silicates and usu. existing in the form of colloidal solutions which contain H_4SiO_4 (*orthosilicic acid*) but consist mainly of polymeric oxyacids derived from this; any of these constituent acids. (Later examples.)

1898 [see *lactonitrile* s.v. *lacto- 2]. **1913** E. Hatschek *Physics & Chem. of Colloids* v. 41 If a solution of sodium silicate..is decomposed by a slight excess of hydrochloric acid, and the mixture is dialysed until the free acid and the sodium chloride have been removed, there remains in the dialyser a perfectly clear colourless sol of silicic acid. **1955** *Sci. Amer.* Sept. 65/3 The more volatile or acidic or silicic constituents of the fluid rock beneath the surface may have concentrated in the original uplifts. **1973** E. G. Rochow in J. C. Bailar et al. *Comprehensive Inorg. Chem.* I. xv. 1409 The relations between silicon dioxide and water are so close..that it often is difficult to distinguish between solutions of silica, solutions of distinct molecular silicic acids, colloidal dispersions of silica, the definite hydrates of SiO_2, hydrated silica in general, and silica gel.

silicium. Add: Now *Obs.*

silico-. Add: **b. silicoca·rnotite** [cf. *carnotite, an unrelated mineral named after the same person], a silicate and phosphate of calcium, $Ca_5(PO_4)_2SiO_4$, found as orthorhombic crystals (coloured blue by impurities) in basic slag from steelmaking processes; **silico-ma·nganese,** a ferro-alloy containing relatively high proportions of manganese and silicon, *spec.* one containing 65 to 70 per cent. manganese and 12 to 25 per cent. silicon which is used in steel-making as a manganese-containing additive and as a deoxidizer; **si:licomoly·bdic acid,** any of a class of polyanionic oxyacids obtained when mixed solutions containing a molybdate and a silicate are acidified; *esp.* a yellow crystalline solid of this kind whose formation is the basis of a colorimetric determination of silicon; so **silicomo·lybdate,** an anion or a salt of such an acid; **silicopho·sphate** = *phosphosilicate* s.v. *phospho-.

1911 V. A. Kroll in *Jrnl. Iron & Steel Inst.* LXXXIV. 126 These crystals, which are of a beautiful dichroic-blue colour, will be distinguished in the present paper by the name Silico-Carnotite. **1949** *Mineral. Mag.* XXVIII. 496 Silicocarnotite is found in basic slags rich in phosphorus. **1971** *Tschermaks Mineral. und Petrogr. Mitteilungen* XVI. 19 The nature of the PO_4/SiO_4 substitution in silicocarnotite should be resolved by collecting X-ray intensity data from a crystal known to be free from impurities. **1895** E. L. Rhead *Metallurgy* ix. 112 Siliconeisen and silico-manganese are irons containing silicon, or silicon and manganese... They are employed in steel manufacture. **1941** *Trans. Amer. Soc. Mech. Engineers* LXIII. 367/2 Both high-manganese and silicomanganese steels equal the chrome-vanadium steel tested and may have commercial advantages. **1956** W. D. Hargreaves in D. L. Linton *Sheffield* 280 The city therefore uses quite large tonnages of pig-iron and of alloying metals..such as ..silico-manganese. **1881** *Jrnl. Chem. Soc.* XL. 880 Ammonium silico-molybdate is obtained in small yellow octahedrons, by mixing nitric acid solutions of ammonium molybdate and an alkaline silicate. **1928** J. H. Yoe *Photometric Chem. Analysis* I. xxxi. 366 Silicates and phosphates form yellow silico- and phosphomolybdates with ammonium molybdate in acid solution. On treatment with sodium sulfite the silico- and phosphomolybdates give a blue reduction product. **1973** E. G. Rochow in J. C. Bailar et al. *Comprehensive Inorg. Chem.* I. xv. 1466 Silicon is one of 36 elements which have been reported as central atoms of heteropoly acid aggregates (borotungstates, phosphovanadates, silicomolybdates, etc.). **1871** *Jrnl. Chem. Soc.* XXIV. 157 (*heading*) A few remarks on the yellow precipitate containing silicomolybdic acid. **1956** *Nature* 3 Mar. 435/1 Analysis was carried out by measuring absorptiometrically the colour produced by the reduction of the silico-molybdic acid complex with ascorbic acid. **1927** *Jrnl. Agric. Sci.* XVII. 143 Stead..concluded that the most soluble phosphate was a silico-phosphate represented by $5CaO.P_2O_5.SiO_2$. **1963** C. R. Cowell et al. *Inlays, Crowns, & Bridges* vii. 77 A fine-grain silicophosphate cement of the appropriate shade is most suitable for cementation because it possesses some degree of translucency.

silicoflagellate (si:likoflæ·dʒĕlēⁱt). [f. Silico- + *flagellate *sb.*] A marine flagellate of the family Silicoflagellidæ, distinguished by a siliceous skeleton and radiating spines.

1906 M. Hartog in Harmer & Shipley *Camb. Nat. Hist.* I. iii. 86 The Silicoflagellate family Dictyochidae..have a skeleton of a similar nature. **1961** R. D. Manwell *Introd. Protozool.* iii. 29 Living silicoflagellates are almost exclusively marine and pelagic in habit. **1978** *Nature* 16 Mar. 244/2 Apart from being an important structural component of diatoms and silicoflagellates..it [*sc.* silicon] is necessary for DNA polymerase activity in diatoms.

silicon. Add: **2.** Special comb.: **silicon carbide,** a hard refractory compound of silicon and carbon SiC: see *carborundum; **silicon chip,** a chip (*chip *sb.*¹ 2 f) of silicon; **silicon ester,** any ester of silicic acid, *spec.* tetraethyl orthosilicate, $Si(OC_2H_5)_4$, a colourless flammable liquid which is readily hydrolysed to silica and is used in paints, weatherproof coatings for masonry, etc., and as a binding agent for moulds; **silicon iron, steel,** cast iron, or steel (respectively) containing a relatively high proportion of silicon, added to increase the magnetic permeability and/or the resistance to corrosion and heat; **Silicon Valley** orig. *U.S.* [from the use made of silicon chips], the Santa Clara valley, S.E. of San Francisco, where many leading U.S. microelectronic firms are located; **silicon wafer,** a wafer of silicon from which individual silicon chips can be separated (cf. *wafer *sb.*).

1893 *Chem. News* LXVIII. 3A/2 (*heading*) The analysis of silicon carbide. **1982** Jackson & Day *Better than New* 138/2 Silicon carbide abrasives are used extensively for furniture renovation. **1965** *Sci. Amer.* Nov. 66/3 Engineers..saw the posibility of producing complete circuits within a silicon chip by forming all the circuit elements by diffusion. **1979** *Daily Tel.* 3 Oct. 14/4 Perhaps..the biggest gains will be possible from the sophisticated electronic 'engine management' control systems now being developed with the aid of the silicon chip and mini computer. **1923** A. P. Laurie *Brit. Pat. 221, 342* 1/1 According to my earlier application silicon esters are applied to the stone and allowed gradually to hydrolise in position. **1969** *Kirk-Othmer Encycl. Chem. Technol.* (ed. 2) XVIII. 217 The silicon esters of organic acids, or silicon carboxylates, are also known and are prepared by the reaction of the acid with silicon halides. **1878** *Chem. News* 27 Dec. 299/2 This metallic mass is a silicon iron, remarkably rich in silicon, and evidently the product of a blast-furnace. **1970** *Materials & Technol.* III. x 769 These principles have been incorporated in the manufacture of conventional materials such as silicon-iron for electrical transformers. **1882** *Jrnl. Iron & Steel Inst.* 376 Peculiarities of silicon steel are: the adhesive scale which covers it; a low degree of weldability; and a very fine grain. **1975** D. G. Fink *Electronics Engineers' Handbk.* vi. 80 Silicon steels, known as electrical steels, are very widely used for low- and intermediate-frequency applications. **1974** *Fortune* June 135/2 They have turned part of Santa Clara County into 'Silicon Valley', the world capital of semiconductor technology. **1980** *N.Y. Times* 22 June iv. 8E In more recent years 'Silicon Valley' has grown up along the peninsula from San Francisco through Stanford University to San Jose. **1956** *Bell Syst. Technical Jrnl.* XXXV. 3 After diffusion the entire surface of the silicon wafer is covered with the diffused n- and p-type layers. **1977** *Sci. Amer.* Sept. 111/3 The pure, single-crystal silicon wafers that bear the circuits are much larger: currently three or four inches in diameter.

silicone (si·likōᵘn). *Chem.* [f. Silic(o- + -one.] † **1.** Also **-on.** [ad. G. *silicon* (F. Wöhler 1863, in *Ann. d. Chem. u. Pharm.* CXXVII. 263).] A yellow solid obtained by the action of concentrated hydrochloric acid on calcium silicide and said to be a compound of silicon, hydrogen, and oxygen. *Obs.*

1863 F. Wöhler in *Chem. News* 10 Oct. 172 The relative proportions of the combined silicium and calcium are in general those necessary to form the compound $CaSi_2$... It is only by means of this compound that we can..explain the composition and mode of formation of the yellow body which is obtained by the action [on it] of hydrochloric acid, and which I shall now describe under the name of *Silicon.* **1909** *Jrnl. Chem. Soc.* XCVI. ii. 806 The colourless substance, leucone, which is produced by exposing silicone to light and air, represents an intermediate step in the oxidation of silicone to silicon dioxide. **1946** J. R. Partington *Gen. & Inorg. Chem.* xviii. 506 By the action of concentrated hydrochloric acid on calcium silicide Wöhler (1863) obtained a yellow solid which he called silicone and formulated $Si_4H_4O_3$.

2. a. Formerly, the name given to any supposed compound of silicon analogous to the ketones, having a formula RR′SiO (R, R′ being organic radicals); in mod. use, any of a large group of synthetic organosilicon polymers (siloxanes) based on chains or networks of alternating silicon and oxygen atoms, many of these being good electrical insulators with high durability, and finding uses as liquids, greases, rubbers (notably in cosmetic surgery), or resins.

[**1906** *Jrnl. Chem. Soc.* XC. i. 128 The gelatinous diphenylsilicone (diphenyl silicoketone).] *Ibid.* 563 The silicones present in the residues..are mixtures, in variable proportions, of siliciformic anhydride and silico-oxalic acid. **1912** *Ibid.* CI. ii. 2106 The term *silicone* has already been used to denote the analogues of the ketones, and may be advantageously retained for this purpose. **1948** *Q. Rev. Chem. Soc.* II. 26 The word 'silicone' was originally used by Kipping to denote the silicon analogues of ketones (RR′Si:O, cf. RR′C:O) but he soon recognised that monomeric silicones are incapable of existence; in fact, no compound containing the group Si:O is yet known. **1957** *Listener* 31 Oct. 719/2 The manufacturers use silicones partly because they give some resistance to spilled liquids and because dust does not readily adhere to the film of polish. **1979** [see *shoot *v.* 23 h].

b. *attrib.*, as *silicone polish, resin, rubber,* etc.

1944 *Chem. & Engin. News* 10 July 1134/3 Silicone resins, used as insulating varnishes and bonds, definitely fill the void between organic and inorganic insulating materials. *Ibid.* 25 Nov. 2016/1 Silicone rubber..has been developed by the General Electric Co., Schenectady, and was demonstrated at the Engineers' Club in New York on November 14. **1955** *Radio Times* 22 Apr. 52/1 (Advt.), New Goddard's Silicone Wax has revolutionized polishing. It's put wonder-working silicones into hard wax. **1958** *Times Rev. Industry* May 26/3 The core is clamped with steel clamps, and while still under pressure is bound with silicone varnish. **1960** *Farmer & Stockbreeder* 22 Mar. (Suppl.) 10/2 First, put the pointing in good repair, then treat the wall with a silicone waterproofing liquid. **1969** T. C. Thorstensen *Pract. Leather Technol.* xiv. 235 As a result of silicone treatment, shoes can be made that are for all practical purposes completely water-repellent. **1971** A. Diment *Think Inc.* vii. 129 Her breasts were.. very firm... Maybe a few silicone injections I thought cynically. **1973** *Materials & Technol.* VI. viii. 602 Silicone resins, being thermally stable, are used as impregnating varnishes, moulding compounds, encapsulants, and in laminates. **1974** *Sci. Amer.* Mar. 71/1 Artificial heart valves and experimental heart-bypass pumps are often fabricated from silicone rubber because the polymer has a lower tendency than most organic polymers to trigger the clotting of blood. **1976** *Gramophone* Nov. 755/3, I recommend the use..of silicone polish, applied thinly and evenly with a soft cloth, starting from the label to take up the surplus, and working outwards.

c. *Comb.*, as *silicone-impregnated, -proofed, -treated* adjs.; *silicone-treat* vb. trans.

1956 *Nature* 25 Feb. 365/2 Reversed-phase chromatography using silicone-impregnated kieselguhr columns. **1976** *Shooting Times & Country Mag.* 9–15 Dec. 7/1 (Advt.), The GARCIA gun and reel silicone impregnated cloth. Protects and cleans all metal. **1956** *Good Housek. Home Encycl.* (ed. 4) 251/2 Silicone-proofed rainwear and ski-clothes. **1958** *Times* 20 Jan. 11/5 Where traditional upholstery material is used on chairs..they should be silicone-treated. **1946** *Canad. Med. Assoc. Jrnl.* LV. 29/1 The difference between clotting times in untreated..and silicone-treated..tubes is equally marked with plasma. **1956** *Good Housek. Home Encycl.* (ed. 4) 99/2 Silicone-treated furnishing fabrics.

Hence **si·liconed** *ppl. a.*, coated, impregnated, or otherwise treated with a silicone or silicone-based material; **si·liconing** *vbl. sb.* Also (as a back-formation) **si·licone** *v. trans.*

1950 *Jrnl. Appl. Physiol.* III. 366 Venous blood was collected in a 10-ml. siliconed syringe. *Ibid.* 375 Siliconing of the syringe appears a worth-while precaution. **1959** *Listener* 1 Jan. 43/1 Siliconing really does prevent the food from sticking [to the frying pan]. **1973** 'E. McBain' *Let's hear it for Deaf Man* viii. 124 A woman..with bleached blond hair and siliconed breasts. **1977** *Listener* 10 Nov. 611/2 The motel on the dunes, its beams shiny and siliconed in the clear light. **1978** *Morecambe Guardian* 14 Mar. 29/1 (Advt.), Sandblasting and siliconing. **1980** 'D. Kavanagh' *Duffy* iv. 68 She was naked, thinnish, with..breasts which had probably been siliconed.

siliconize, *v.* (In Dict. s.v. Silicon.) **1.** *trans.* Substitute for def.: To cause to combine with silicon or its compounds; esp. to subject (a metal) to a process in which the surface is impregnated with silicon so as to form a protective coating.

1880 [implied in *siliconizing* below]. **1948** N. E. Woldman *Metal Process Engin.* viii. 222 The iron or steel to be siliconized is subjected to the action of silicon carbide or ferrosilicon, or a mixture of the two and chlorine at temperatures of 1700 to 1850° F. **1977** R. B. Ross *Handbk. Metal Treatments & Testing* 349 It is generally more economical to produce components in high-silicon cast iron rather than cast irons which are subsequently Siliconized.

2. [f. *silicone.] To impregnate, coat, or otherwise treat with silicones or silicone-based material. Usu. in ppl. adj. (see below).

1957 [implied in *siliconized* below]. **1963** *Obstetr. & Gynecol.* XXI. 47/1 The glassware was siliconized to prevent adsorption of the isotopes.

So **si·liconized** *ppl. a.*, **si·liconizing** *vbl. sb.*; also **si:liconiza·tion,** treatment or combination with silicon or a silicone.

1880 Siliconizing [in Dict. s.v. Siliconize *v.*]. **1920** Webster, Siliconized. **1924** *Industr. & Engin. Chem.* Nov. 1112/1 An oxidizing atmosphere would tend to oxidize both the silicon and the iron simultaneously and prevent true siliconization. **1946** *Iron Age* 4 Apr. 75/2 These siliconized shafts replaced shafts made of stainless and nitrided steel. **1947** Kirk & Othmer *Encycl. Chem. Technol.* I. 579 Other diffusion treatments for mild and other steels are calorizing, chromizing, and siliconizing. **1950** *Chem. Abstr.* XLIV. 4856 This method of siliconization permits a further increase of corrosion resistance of 18Cr-8Ni stainless steel, of Monel metal, and of pure Ni. **1957** *Archit. Rev.* CXXII. 354/2 T. & W. Farmiloe will demonstrate the qualities of their siliconized paints, put on the market since the last exhibition. **1977** *Hot Car* Oct. 97/3 (Advt.), Tough, resilient, siliconised enamel cuts out labour. **1978** *Nature* 28 Sept. 322/2 A double-barrelled glass micro-pipette, pretreated with a siliconising agent.

silicosis. Substitute for etym.: [ad. It. *silicosi* (A. Visconti: see C. L. Rovida in *Annali di Chim. applicata alla Med.* (1871) LIII. 103), ult. f. L. *silic-* Silex + -osis.] and add: (Earlier example.)

1881 *Jrnl. Anat. & Physiol.* XV. 395 According to the nature and character of the irritant, certain forms [of pneumokoniosis] are distinguished, the chief of which are anthracosis, chalicosis or silicosis, and siderosis.

Hence **silico·tic** *a.*, affected by silicosis. Also as *sb.*

1913 *Mem. S. Afr. Inst. Med. Res.* No 3. 120 The amount of silica in the ash is greater in that from the silicotic lung than in that from the normal lung. **1938** *Jrnl. Amer. Med. Assoc.* 19 Nov. 1928/1 Clinicians, who see only the disabled silicotic patient with advanced lesions, often complicated by tuberculosis, are insisting that only such conditions represent the true picture of silicosis. **1948** [see *pneumoconiotic* s.v. *pneumo-]. **1980** D. Pownall *Between Ribble & Lune* i. 20 Aaron's disease, a form of silicotic lung infection caused by the dust.

siliqua. Add: Pl. **siliquae. 4.** A Roman silver coin of the 4th and 5th centuries A.D., of the value of $\frac{1}{24}$ th of a solidus.

1889 *Jrnl. Hellenic Stud.* X. 95 All the mediaeval standards were based upon the gold solidus of Constantine the Great..divided into 24 *siliquae* or *κεράτια* (from whence comes *carat*). **1927** A. R. Burns *Money & Monetary Policy Early Times* x. 243 From about the middle of the 4th century A.D. the *siliqua*, the half of the *miliarense*, became increasingly important. **1940** [see *aureus]. **1962** R. A. G. Carson *Coins* 185 Siliquae were struck mainly by Italian mints for Arcadius and Honorius and by the Gallic mints for the usurpers in control there.

silk, *sb.* and *a.* Add: **I. 1. e.** Silk sold in the form of thread or twist for sewing; freq. with defining word, as *embroidery, sewing silk,* etc.

1480–1826 Sewing silk [see Sewing *vbl. sb.*¹ 4]. **1851** *Illustr. Catal. Gt. Exhib.* III. iii. 506/1 Veil, vest, and shawl embroidering silk. **1920** A. K. Arthur *Embroidery Bk.* ii. 10 Silks of different makes, embroidery or knitting, filosel,..and..'Tyrian', are all good for various purposes. **1951** L. Town *Bookbinding by Hand* xiv. 175 This is necessary to prevent fraying of the sewing silk as it passes round the headband. **1973** C. Gavin *Snow Mountain* xxiii. 392 The drawn-thread work they were doing, on coarse linen with silks brought from their home.

f. *artificial silk* [cf. F. *soie artificielle* (de Chardonnet 1884, in *French Pat. 165,349*)] = *rayon¹ 3. Also shortened to *art silk* (also *artsilk*).

1885 *Jrnl. Soc. Chem. Industry* 29 Jan. 34 Mr. J. B. Payne exhibited..some samples of 'artificial silk', a new filament produced by pressure through a die, from pyroxylin, the invention of Mr. J. W. Swan. **1922** *Daily Mail* 2 Dec. 1 (Advt.), Three charming designs in silcot, cotton, satin and art silk stockinette. **1924** [see *rayon¹ 3 a]. **1928** *Lancet* 24 Mar. 631/2 Mr. Kelly asked the Home Secretary whether the Home Office had received any reports as to the conditions of health of workpeople employed in artificial silk factories. **1928** *Daily Mail* 3 Aug. 18/2 Snias and British Enkas were firmer among Artsilks. **1935** *Economist* 2 Nov. 854/1 Swedish exports to Italy consist to a large extent of chemical pulp..most of it rayon cellulose for artificial silk. **1944** A. L. Bowley *Stud. Nat. Income 1924–1938* 170 The excise on Artificial Silk has been charged since July 1925. **1957** H. Croome *Forgotten Place* 15 A brilliant orange artsilk coverlet on a double bed..artsilk curtains of a different shade framing the windows.

2. b. (Earlier and later examples of *to take silk.*) Also (*rare*), *to have silk.*

1866 A. J. Munby *Diary* 2 Nov. (1972) 229 Dined in Hall..the talk was of who is to have silk presently and make way for us rising juniors. **1875** Trollope *Prime Minister* (1876) I. iii. 36 He had..worked in a stuff gown till he was nearly sixty... He would take his silk as an honour for his declining years. **1925** W. S. Maugham *Painted Veil* vii. 25 He was still a junior and many younger men than he had already taken silk. **1979** G. Wagner *Barnardo* viii. 130 Thesiger..had become a QC... He had two juniors..who later took silk.

c. Esp. in phr. *to sport, don,* or *wear silk*: to ride (in a race). (Earlier and later examples.)

1884 H. Smart *From Post to Finish* I. xv. 243 Next week Gerald would 'don silk'..and be embarked on the career she had marked out... Had she done right?.. And yet, with his aptitude for riding..what better path.. was open to him? **1898** A. E. T. Watson *Turf* x. 189 A gentleman, when this misfortune happens to him..can cease to wear silk, or at any rate need not ride over hurdles or fences.

d. A parachute; chiefly in phr. *to take to* or *hit the silk,* to bale out of an aircraft by parachute. *U.S. Air Force slang.*

1933 *Jrnl. R. Aeronaut. Soc.* XXXVII. 828 The American pilot..remarked that if he had engine trouble over England he would 'take to the silk', in other words abandon his machine and come down by parachute. **1943** R. Whelan *Flying Tigers* 100 After gaining altitude Mott's plane burst into flames and he 'hit the silk'. **1956** N. Marsh *Off with his Head* (1957) viii. 177 Over Germany ..we got clobbered and I hit the silk.

3. b. Also *spec.* a jockey's cap and jacket carrying the horse-owner's colours. Cf. sense 2 c above.

1946 *Sun* (Baltimore) 31 May 15/1 Lovely Imp, carrying the Maryland silks of R. Bruce Livie's..Stable, won on a disqualification. **1955** *Radio Times* 22 Apr. 9/3 The jockeys..in gaudy silks. **1977** *New Yorker* 4 July 71/1 Last Thursday, in his first appearance in silks since the accident, he won with his first mount..by a length and a quarter.

c. (Earlier examples.)

1793 F. Burney *Let.* 4 Feb. (1972) II. 12 My love & thanks to my dear Sarah; though she ought to send my black silk. **1819** M. Edgeworth *Let.* ? 10 Mar. (1971) 181 Fanny wore her green silk and it looked beautiful.

e. A silk hat.

1906 Joyce *Let.* 12 Aug. (1966) II. 148, I am curious to know how he looked in a tall silk. **1930** D. H. Lawrence *Love among Haystacks* 87, I assured her her hat was adorable, and, much to my relief, I got rid of my silk and into a dressing gown.

4. a For *U.S.* read 'orig. and chiefly *U.S.*' (Earlier and later examples.) Phr. *in silk,* at that stage when the silk is prominent.

[*c* **1662** in *New England Q.* (1937) X. 126 There groweth within the Huske upon the Corne a matter like small threads which appeare out of the top of the Eare like a tuft of haire or Silke.] **1770** G. Washington *Diary* 25 Aug. (1925) I. 395 Many Stalks were putting out entire new Shoots with young and tender Silk. **1774** P. V. Fithian *Jrnl.* 19 July (1900) 212 The Corn is beginning pretty generally to tassel, & I saw one hill in Silk, and in Blossom. **1847** D. Drake *Pioneer Life Kentucky* (1870) 52 By the month of August the corn is in silk. **1914** J. Burdett-Davy *Maize* v. 233 In some cases, and in the same breed, the silks appear before the tassels. **1950** *New Biol.* VIII. 46 Pollen from each plant is then poured on to the silks of the same plant. **1980** *Sci. Amer.* Jan. 101 Primitive corn and teosinte, with their tiny ears, have small pollen grains that cannot fertilize the kernels of large modern ears with their long silks.

b. A silky lustre in some rubies and sapphires, due to microscopic crystals, and considered a defect.

1886 *Jrnl. Franklin Inst.* CXXII. 380 In many genuine rubies we find a silky structure (called *silk* by jewellers). **1903** W. R. Cattelle *Precious Stones* 47 Rubies generally contain clusters of light or dark-colored spots... White, glistening streaks in the grain of the stone, called silk, are of frequent occurrence... If silk shows plainly when the stone is faced up, it is one of the most serious defects. **1929** M. Weinstein *Precious & Semi-Precious Stones* i. 6 The peculiar optical effect shown by many natural rubies and sapphires, known as 'silk', is never seen in synthetic stones. **1976** B. W. Anderson *Gemstones for Everyman* xii. 152 The Burma rubies..usually..show small patches of 'silk' consisting of fine needles of rutile intersecting at 60°. These have a silky appearance by reflected light.

II. 6. a. (Later examples of proverbial use.) Also used *allusively.*

1907 E. Gosse *Father & Son* ix. 239 'Even the Lord can't make a silk purse out of a sow's ear,' said Miss Marks. **1929** D. H. Lawrence *Pansies* 129 Women.. want to change the man himself And turn the poor silk glove into a lusty sow's ear. **1932** R. Aldington *Soft Answers* 47 Too late Julia realised that the best and most self-sacrificing of wives cannot make a silk purse out of a sow's ear, an Arnold Bennett out of an Oswald. **1959** M. Bradbury *Eating People is Wrong* ii. 55 For the mass of men there is not too much to be said or done; you can't make a silk purse out of a sow's ear. **1978** *Jrnl. R. Soc. Arts* CXXVI. 339/2 She and her colleagues in the teaching profession are expected to turn children like that into silk purses, able to count, to spell, to read, to write, to understand, and so on.

b. *silk chiffon, gauze* (later example), *jersey, velvet* (later examples).

1965 *Which?* Mar. 95/1 *Silk chiffon,* a light, open mesh fabric, soft and smooth. For scarves, lingerie, evening dresses, millinery. **1976** *Times* 9 Mar. 9 (Advt.), Hand made silk chiffon blouse. **1965** T. R. Tregear *Geogr. China* ii. 81 It was not the finely woven Chinese brocades and damasks that were wanted in Rome, for when they arrived they were unravelled and re-woven into lighter, flimsier silk gauzes. **1897** *Sears, Roebuck Catal.* 231/1 Ladies' black pure silk jersey mitts. **1925** *Eaton's News Weekly* 26 Sept. 17 Very new, the Gossard Dancelette girdle of silk jersey. **1980** D. Creed *Scarab* xviii. 173 Her camel-coloured silk-jersey dress. **1880** L. Troubridge *Life amongst Troubridges* (1966) xi. 156 It was silk velvet and ten shillings a yard. **1966** P. O'Donnell *Sabre-Tooth* vi. 91 Ilse..put on a white silk-velvet dressing gown.

d. (Later examples.)

1879 G. M. Hopkins *Poems* (1967) 84 The vault and scope and schooling..In silk-ash kept from cooling. **1888** in *Ibid.* 198 Silk-beech, scrolled ash.

8. a. *silk grower* (earlier example).

1842 in *Proc. Vermont Hist. Soc.* (1940) VIII. 156 Called on Mr. Dexter the silk grower.

9. *silk-hatted, -hung, -lined, -socked.*

1903 W. Le Queux *Seven Secrets* (ed. 2) xxii. 219 The silk-hatted, frock-coated existence of the fashionable physician. **1976** L. St. Clair *Fortune in Death* iii. 27 A bank messenger, very properly silk-hatted and frock-coated. **1947** Auden *Age of Anxiety* (1948) ii. 35 And mother wrote Swift and sure in the silk-hung saloon Her large round letters. **1901** *Westm. Gaz.* 29 Nov. 7/2 A romantic American,..after living as a hermit for fifteen years in consequence of an unhappy love affair, has been buried in a silk-lined grave. **1979** *Country Life* 1 Feb. 309/1 (Advt.), Silk-lined mohair coats. **1918** G. Frankau *One of Them* in *Poet. Wks.* (1923) II. xix. 123 Silk-socked; bright monocled; gallant. **1922** Joyce *Ulysses* 517 Bella ..lifts..a plump buskined hoof and a full pastern, silk-socked.

10. a. silk embroidery, embroidery worked with silk threads; **silk road, route** (freq. with initial capitals), a trade route from China through India to the West, used in ancient times by traders in silk; **silk waste,** the fibres which remain after the reeling of silk yarn, or those obtained from damaged cocoons; † **silk wool,** a mixed yarn made of wool and either silk or staple fibre (*obs.*).

1837 *Penny Cycl.* VII. 77/1 A piece of silk embroidery. **1889** J. J. Rein *Industries of Japan* iii. iv. 389 Oftentimes this silk embroidery is connected very skilfully with the painting or printing of the material. **1982** E. North *Ancient Enemies* iii. 32 I've never yet told her about the private [American] college..where you can major in any subject... I met a girl who'd been there majoring in something like silk embroidery. **1931** J. W. Gregory *Story of Road* i. iii. 43 The northern silk road in Asia crossed Persia and Kashgar to the Tarim Basin in Chinese Turkestan. **1936** P. Fleming *News from Tartary* i. iv. 29 The Silk Road takes..you through Sinkiang to Kashgar and the Himalayan passes by one of two alternative routes. **1982** *Times* 25 Feb. 10/5 A community of some 200 Chinese-Jewish descendants of Silk Road traders in the ancient capital of Kaifeng, who no longer identify with Judaism. **1913** J. Buchan *Divus Johnson* in *Runagates Club* (1928) vi. 152 Russian geographers were interesting themselves in the line of the old silk route to Cathay. **1949** D. Carruthers *Beyond Caspian* iv. 95 The Silk Route was not a disjointed affair, built up in sections, linking likely markets. **1981** *Daily Tel.* 30 Mar. 18/5 Tartar hats..recall the Silk Route of Marco Polo. **1842** *Encycl. Brit.* (ed. 7) XX. 350/2 To introduce such alterations in the spinning of silk waste as will supersede the cutting, carding, and scutching processes... The art of silk waste spinning..is still in its infancy. **1965** A. Brearley *Woollen Industry* v. 27 Silk wastes are used in woollen blends for their own distinctive merits. **1859** L. Oliphant *Narr. Earl of Elgin's Mission China & Japan* (1860) II. 255 The Japanese wear in winter garments thickly padded either with cotton or silk wool. **1908** *Practitioner* Nov. 760 Silcool is a form of vegetable silk-wool. **1928** F. M. Rowe tr. *Reinthaler's Artificial Silk* vii. 128 Staple fibre yarn..cannot belie its cellulose nature; it lacks..the tenacity of wool. This can be remedied..by spinning staple fibre in admixture with wool or recovered wool... Such mixed yarns (carded or combed) termed 'silk-wool',..are still used for needlework and machine knitting.

b. silk-bark, a small evergreen tree, *Maytenus acuminata,* belonging to the family Celastraceæ and native to southern Africa; **silk-tassel (bush, tree)** = *garrya; **silk wood,** (*b*) = Calabur tree; (*c*) = *Queensland maple* s.v. *Queensland.

1894 T. R. Sim *Flora of Kaffraria* 28 (*heading*) Silkbark. **1907** —— *Forests & Forest Flora Cape of Good Hope* xiv. 184 Silk-bark... A small branched unarmed tree. **1912** *Cape Times* 12 Oct. 9/8 In the gorge beneath the fall an indigenous thicket, yellow-wood, Hottentot cherry, silk-bark, has been allowed to remain. **1972** Palmer & Pitman *Trees S. Afr.* II. 1285 The silkbark or sybas has a wide distribution, occurring from eastern tropical Africa to the Cape. **1897** M. E. Parsons *Wild Flowers of California* 370 (*heading*) Silk-tassel tree. Quinine-bush. **1949** J. T. Howell *Marin Flora* 211 The graceful catkins of the staminate plants make the silk tassel bush one of the most beautiful shrubs in the chaparral. **1976** *Hortus Third* (L. H. Bailey Hortorium) 495/2 *Garrya* Dougl[as]. Silk-tassel, silk-tassel bush... The garryas are ornamentals flowering in late winter and early spring. **1888** Silk wood: used in def. of *Calabur tree.* **1891** [in Dict.]. **1909** F. M. Bailey *Comprehensive Catal. Queensland Plants* 91 *Flindersia*..*Brayleyana*.. Wood has been cut under the name of 'Silkwood'. **1948** A. L. Howard *Man. Timbers World* (ed. 3) 354 Maple silk-wood is moderately elastic.

c. silk-spinner, a spider or a silk-moth.

1868 C. M. Yonge *Chaplet of Pearls* II. xliii. 246 A colony of silk-spinners, attracted by the mulberry-leaves of the old abbey garden. **1869** 'Mark Twain' *Innoc. Abr.* xxxiii. 365 This old dried-up reservoir is occupied by a few ghostly silk-spinners now. **1896** R. Lydekker *Royal Nat. Hist.* VI. iii. 95 (*heading*) The silk-spinners,—Family *Bombycidæ.*

silk, *v.* [f. the sb.] **a.** *trans.* To remove the silk from (maize). *U.S.*

1847 D. Drake *Pioneer Life Kentucky* (1870) 52 My first business in the morning was to pull, and husk and silk enough [corn] for breakfast. **1892** *Hist. Rev. Industr. & Commercial Growth York County* (Pa., U.S.) 59 [They] make a specialty of..'silkers' for silking corn. **1972** E. Wigginton *Foxfire Bk.* 177 Shuck and silk corn that is in roasting ear.

b. *intr.* Of maize: to produce the silk. *U.S.*

1878 J. H. Beadle *Western Wilds* xv. 245 The summers are short and the nights cool. Corn will not silk. **1939** *Sun* (Baltimore) 21 July 13/4 The corn in the county is later..as the farmers have planted it so it will silk after August 10 when the danger from beetles is over. **1948** *Clarke County Democrat* (Grove Hill, Alabama) 3 June 1/3 This worm usually waits until corn bunches for tasseling or begins to silk before they attack.

silked, *a.* Add: **1.** (Earlier example.) Also, clothed in or covered with silk.

1837 *Annals Electricity, Magnetism & Chem.* I. 112 The bar is covered with several coils of silked copper wire. **1864** C. Rossetti *Farm Walk* in *Prince's Progress* (1866) 152 I've seen grand ladies plumed and silked. **1909** M. B. Saunders *Litany Lane* iii. 33 Gorgeously furred and laced and scented and silked.

2. Of (the pages of) a book, etc.: having been strengthened by silking (sense *2).

1943 *Amer. Archivist* VI. 153 Upon subjection to accelerated aging tests..it was found that the silked papers had lost 52 per cent of their folding endurance. **1971** *Catal. Mildred C. Esty Coll. MSS Rob. Burns* (Christie's) 7, 1 p., 4to., with conjoint address leaf (considerably repaired, silked, a few words rubbed along fold).

silken, *a.* **II. 7. a.** (Later example.)

1955 E. Pound *Classic Anthol.* i. 20 Lady of silken word.

silker (si·lkəɹ). [f. Silk *sb.* + -er¹.] One who works in or with silk; in various technical uses.

1881 *Instr. Census Clerks* (1885) 50 Piano Manufacturing... Silker. *Ibid.* 69 Cotton.. Silker. **1903** *Sci. Amer. Suppl.* 24 Jan. 22629/3 From the cutters' room the leather, which has assumed the shape of the glove, is sent to the 'silkers', who embroider the back. **1921** *Dict. Occup. Terms* (1927) 200/1 *Taper,..silker*; stitches tape by machine, down seam, in closing upper of a boot or shoe. *Ibid.* 203/1 *Silker*; (i) finishes cloth piece..by stitching folds together, by hand, with a silk thread to hold them in position; (ii) sews selvedges of cloth pieces with different coloured threads of silk or of mercerised cotton.

silkette. (Later example.)

1922 Joyce *Ulysses* 735 The second pair of silkette stockings is laddered.

silkie (si·lki). Also **Silkie.** [f. Silk *sb.* + -ie.] A small chicken belonging to the variety so called, distinguished by long, soft plumage.

1885 [in Dict. s.v. Silky *a.* 6 b]. **1937** W. W. Broomhead *Poultry Breeding & Management* ix. 325/1 Silkies are quaint little fowls. *Ibid.* 326/1 (*caption*) A White Silkie Cock. **1978** *Country Life* 11 May 1285/1, I have a clutch of pure-bred Silkies being incubated now... I am not sure where the Silkie stands in bantam genealogy.

silkie, var. Sealchie, -kie in Dict. and Suppl.

silkily, *adv.* Add: **a.** (Later examples.)
1947 A. P. Gaskell *Big Game* 91 Flash young things with lipstick, long-legged in high-heeled shoes, stood silkily, smoking tailormades. 1980 K. Follett *Key to Rebecca* iii. 37 'Thank heaven you're back,' he said silkily.
b. Smoothly, quietly; used esp. of the running of an engine or machine.
1923 *Daily Mail* 7 Aug. 3 (Advt.), I was much impressed with the vehicle..beautifully suspended and runs very silkily. 1962 I. Murdoch *Unofficial Rose* xxix. 280 He drew it [*sc.* the dagger] silkily out of its sheath. 1978 *Lancashire Life* Apr. 141/1 The 132-2000..would slip silkily into top at anything from 25 m.p.h. upwards.

silking, *vbl. sb.* Add: **2.** The attachment of a piece of silk or other fine material to one or both sides of a sheet of paper in order to strengthen or preserve it.
1943 *Amer. Archivist* VI. 152 The two principal methods of restoration employed at the present time, silking and lamination with cellulose acetate foil, are described below. 1980 S. G. Swartzburg *Preserving Library Materials* vii. 74 The sheets can be strengthened by a covering of a thin sheet of japan tissue, pasted over the original page... This process is often called 'silking' because originally a fine chiffon fabric was used.
3. Development of the silk in maize.
1976 R. W. Jugenheimer *Corn* xiii. 206 The most desirable strains of corn over a period of years often have been those in which the individual plants varied considerably in date of silking and tasseling. 1977 *N.Z. Jrnl. Agric.* Jan. 13/4 The optimum time for spraying is a fortnight either side of silking.

silkoline (si·lkolīn). Also **silkaline, silkolene, S-.** [f. Silk *sb.*, after Crinoline.] A soft cotton fabric with a smooth finish resembling that of silk.
1896 *Proc. Internat. Typogr. Union* 64/1, 12 yds. silkaline, $1.80. 1907 *Yesterday's Shopping* (1969) 742/2 Dress Linings..Silkoline, black..yd. 0/10. 1911 *Everybody's Mag.* XXV. 795/2 The last wrinkle and darn of their blue silkolene cotton tights had vanished from the stage. 1918 *Sears Catal.* 1164/1 A beautiful rose design in a border Silkoline. 1921 *Daily Colonist* (Victoria, B.C.) 8 Apr. 20/3 (Advt.), Silkolene, 36 inches wide, in a full selection of colorings and designs. 1950 '*Mercury*' *Dict. of Textile Terms* 465/1 Silkaline, a very light printed, plain weave, glossy cotton fabric, made in the grey and calendered. 1970 *Kay & Co.* (Worcester) *Catal.* 1970–71 Autumn/Winter 448/2 Bonsoir 'Silkaline' Pyjamas are fashionably styled in 100% cotton.

silk screen, *sb.* [f. Silk *sb.* + Screen *sb.*[1]] **1.** A screen (*screen *sb.*[1] 6 *b) made of silk for use in screen printing. Usu. *attrib.*, esp. in *silk screen printing, process* (also *absol.*).
1930 B. Zahn *Silk Screen Methods of Reproduction* 9 There is no other phase of the graphic arts which presents so many possibilities as the Silk Screen Process. *Ibid.* 10 A silk screen has the advantage of making a perfect imprint at low cost. 1934 F. A. Baker *Silk Screen Practice* i. 13 The Silk Screen Process has become a necessity to all businesses that deal in colour reproduction work. 1950 *Atomics* Jan. 22/2 The silk screen press consists simply of a piece of silk stretched wet on a frame which taughtens on drying. 1952 *Print* (U.S.) July 1/1 André Girard..has been the first to apply silk screen to the art of making books. Silk screen has gone far in the scant thirty years of its existence. 1959 *Daily Mail* 14 Aug. 1/3 Robert, a silk screen printer, and Rosemary, a typist, first met about eighteen months ago. 1967 M. Chandler *Ceramics in Mod. World* iii. 111 The ceramist may make use of silk-screen printing. 1981 *West Lancs. Even. Gaz.* 20 Feb. 21 (Advt.), Wanted: Silk screen printer with artistic and modelling abilities.
2. A print made by the silk screen process.
1977 J. Didion *Bk. Common Prayer* ii. ii. 59 The thin FBI man gazed over Charlotte's head at the 10' by 16' silk screen of Mao Tse-tung. 1979 *Farmington* (New Mexico) *Daily Times* 27 May (Entertainment Suppl.) 22/5 'Mural-sized graphics', which will be reproduced as signed, limited-edition silk-screens.
Hence **silk-screen** *v. trans.*, to print, decorate, or reproduce by the silk screen process; **silk-screened** *ppl. a.*, **silk-screening** *vbl. sb.*; also **silk-screener**, a silk-screen printer.
1961 M. Jones *Potbank* xxv. 110 Joan..was silk-screening coffee-pots. 1967 *Listener* 21 Dec. 829/3 Andy Warhol's carrier bag beautifully silk-screened with a Campbell's soup can. 1976 Woodward & Bernstein *Final Days* xi. 138 As the sheer mechanics of preparing the transcripts for public consumption intensified, David Hoopes..summoned the State Department silk-screener from a baseball game. 1976 *National Observer* (U.S.) 28 Aug. 9/2 She worked in the farm's silk-screening shop, helping to print Christmas cards. 1978 *Detroit Free Press* 16 Apr. (Detroit Suppl.) 34 (Advt.), Each is silk-screened several times to produce the most subtle of shading on multi-colored cotton prints.

silk stocking. Add: **2. a.** Hence in extended sense, a member of the wealthy or upper class. (Later examples.) *U.S.*
1896 [see sense 2 b in Dict.]. 1903 *Independent* 12 Nov. 2663/1 The mass of voters look upon him as a 'silk stocking'—as one who neither understands nor sympathizes with their life.
b. (Earlier example.)
1840 *Niles' Nat. Reg.* (Baltimore) 14 Mar. 22/1 They cried out in derision of locofoco slang—'Here go the silk stockings.'
3. *attrib.* and *Comb.* **a.** Simple *attrib.*, as (sense 2 a) *silk-stocking company, gentry*, etc. *U.S.*
1798 *Deb. Congress U.S.* 15 June (1851) 1948 If they wished to place them in a ridiculous point of view, or to produce for them the name of the *Silk Stocking Company*, or any other term of derision, they could not take a more effectual course to obtain it. 1812 T. Jefferson *Writings* (1904) XIII. 163, I trust..the Gores and Pickerings will find their levees crowded with silk stocking gentry, but no yeomanry. 1836 *Col. Crockett's Exploits & Adventures Texas* iv. 58 You may be called a drunken dog by some of the clean shirt and silk stocking gentry. 1874 'H. Churton' *Toinette* xiii. 154 She had managed to pick up..'a tolerable English education',..[possibly] through the charity of some teacher at the 'Silk-Stocking Academy', on 'Gentleman Ridge'. 1903 *N.Y. Sun* 28 Nov. 4 He is the representative of the wealthy intellectual, the cultured, the 'silk stocking' element, for which the people in general have no abiding affection. 1980 *Verbatim* Autumn 1/2 Next after *hill* the commonest generic is *row* (several examples already given). The most frequent response of this type was *Silk Stocking Row*.
b. Special combs.: **silk-stocking district** (or **quarter**), a district inhabited mainly by supporters of the (Whig or) Republican party.
1893 *World's Fair Puck* 18 Sept. 231/2 Mr. Astorbilt (*of the silk-stocking district*)—No; I thought I was a thousand miles away from Tammany! 1903 *N.Y. Even. Post* 30 Oct. 2 Political conditions change even in the 'silk-stocking' quarter—the middle reaches of Manhattan, between 14th Street and 96th Street. 1964 *Economist* 2 May 486/2 Mr Wallace may do well, both in the silk stocking districts, and on the waterfront.

silkweed. **1.** (Earlier and later examples.)
1784 *Mem. Amer. Acad.* I. 424 The seeds are contained in large pods, and are crowned with white down,..resembling silk, which has occasioned the name of Silk-weed. 1814 [see *milkweed 2]. 1940 J. Stuart *Trees of Heaven* 48 There is the musty smell of ironweeds, milkweeds, silkweeds,..and bull grass on the lazy wind.

silky, *a.* (*sb.*) Add: **1. b.** As *sb.* (See quot. 1976.)
1822 T. Bewick *Memoir* (1975) ii. 16, I..was only to walk along the dark passage to the back Door and to repeat something (rather ominous indeed) about 'Silkey & Hedley Kow'. 1866 W. Henderson *Folk Lore Northern Counties* vii. 230 Black Heddon..was greatly disturbed by a supernatural being, popularly called Silky, from the nature of her robes. 1912 in R. Tongue *Forgotten Folk-Tales English Counties* (1970) iii. v. 202 Gilsland's lord had a silky who cleaned the house-place, devilled and punched lazy serving-wenches and kept all shining clean. 1967 *Tablet* 16 Dec. 1307/2 Nearly all peoples, the world over, have believed in beings they called elves, silkies, trolls, elementals or fairies. 1976 K. M. Briggs *Dict. Fairies* 365 The Northumbrian and Border silky..is always female... She is a spirit dressed in rustling silk, who does domestic chores about the house and is a terror to idle servants.
3. c. Of a machine, mechanism, etc. or its motion: smooth.
1935 *Times* 23 Apr. 17/6 The engine is silky and quiet throughout its range. 1977 *Gramophone* Nov. 960/2 To provide a silky movement of the tuning control, and gearing down of the knob, the Tandberg engineers have made use of a two-gang variable capacitor with its inbuilt gearing.
5. b. silky cornel = next; **silky dogwood,** a large shrub, *Cornus amomum*, native to eastern North America, whose leaves have silky hairs on their lower sides; **silky oak,** one of several Australian trees of the family Proteaceæ, esp. *Grevillea robusta* or *Cardwellia sublimis*, or the oak-like timber produced by them.
1848 A. Gray *Man. Botany Northern U.S.* 168 (*heading*) Silky cornel. 1891 J. M. Coulter *Bot. W. Texas* i. 150 Silky cornel... Common in the Atlantic States and extending into eastern and northern Texas. 1900 B. B. Smyth *Plants & Flowers Kansas* ii. 25 It is the Silky Dogwood; grows in clumps; and had blue berries when ripe, in broad cymose clusters. 1957 W. C. Grimm *Bk. Shrubs* 351 The Silky Dogwood is sometimes called the Silky Cornel, Swamp Dogwood, or Kinnikinnick. 1866 Silky oak [in Dict.]. 1888 F. M. Bailey *Queensland Woods* 104 *S[tenocarpus] salignus*... One of the woods called Silky Oak. 1889 Silky oak [in Dict.]. 1965 *Austral. Encycl.* II. 180/2 The northern silky oak, *Cardwellia sublimis* of Queensland, is also sometimes known as bull oak.

silky, var. Sealchie, -kie in Dict. and Suppl.

sill, *sb.*[1] Add: **1. c.** The lower horizontal members of the frame of a motor vehicle.
1959 *Motor Man.* (ed. 36) i. 17 In the case of the Austin, a normal pressed-steel body was used, the channel-section sills of which were joined to the open faces of the channel section side-members to form substantial box sections. 1976 *Drive* Sept.–Oct. 75/1 The high boxed sills were a necessary structural link between the front and rear of the car. 1980 *Daily Tel.* 11 Sept. 7 (Advt.), Full underbody sealing and wax injection of sills and cross-members.
2. a. (Earlier and later *Comb.* examples.)
1885 C. M. Yonge *Nuttie's Father* I. i. 6 Lovely sill boxes full of flowers in the windows. 1955 *Archit. Rev.* CXVIII. 126/1 Panels of woven cane hanging from the sillboards cover the radiators.
3. a. Also *Comb.*
1870 *Jrnl. Ethnological Soc.* II. 417 At each end of this passage, and at right angles to it, are two square or somewhat oblong chambers. The first..was about 3 feet in width. 1981 *Glasgow Archaeol. Jrnl.* VIII. 52/2 The main uprights were set in newly dug postholes, linked by sill-beam trenches.
e. A high ridge on the sea bed that effectively separates the bodies of water on either side.
1933 *Geogr. Jrnl.* LXXXI. 571 Hamish island, situated on the shallow sill of the Red Sea. 1942 O. von Engeln *Geomorphology* xxx. 468 They [*sc.* fiords] are closed at the seaward end by a distinct rock sill at shallow depth, beyond which the descent to the deeper ocean waters begins. 1978 *Nature* 14 Dec. 680/2 Outflowing Mediterranean subsurface waters... They spill over the sill at Gibraltar (330 m).
4. b. In mod. use, a tabular igneous intrusion lying parallel to the surrounding strata. (Later examples.)
1914 J. P. Iddings *Problem of Volcanism* vii. 222 Intrusions along bedding planes of stratified rocks are commonly called sills at whatever angle they may be tilted, and intrusions in fractures that transgress stratified beds are commonly called dikes. 1977 A. Hallam *Planet Earth* 68 Fine examples of sills are the Carboniferous dolerite sill that forms Salisbury Crags in Edinburgh, Scotland, and the Palisades sill, up to 300m (1000ft) thick, along the west bank of the Hudson River near New York.

sillabub. Add: **1. a.** (Later examples.)
1911 M. A. Fairclough *Ideal Cookery Book* 722 Syllabubs... Fill some custard glasses rather more than half full with the mixture. 1976 *Sat. Even. Post All American Cookbk.* 259/2 A syllabub is a ladylike version of eggnog.
b. *syllabub glass, jug, pot* (later examples).
1677 in S. Young *Hist. Worshipful Company Glass Sellers of London* (1913) 69 All covers for drinking or 'Sullibub' glasses ribbed and plain shall be delivered at 3s. per lb. 1723 J. Nott *Cook's & Confectioner's Dict.* sig. Ll[1], Scum off the Froth, and put it into Syllabub Glasses. 1897 A. Hartshorne *Old Eng. Glasses* xix. 308 In Mr. Cuming's collection is an open-mouthed glass tumbler, a family relic,..said to be of the first part of the last century, and called from time immemorial 'a syllabub or whip glass'. 1970 G. Savage *Dict. Antiques* 418/2 In the 1770s the old syllabub glass, which was always on a stem and a foot, became unfashionable and was replaced by a stemless glass. 1975 *Country Life* 2 Jan. 11/3 A Syllabub Jug in Ravenscroft Glass. 1723 J. Nott *Cook's & Confectioner's Dict.* No. 188 S, Fill your Syllabub-Pot with Cyder..Sugar..Cream. 1910 *Queen* 9 July 65/1 This syllabub or posset pot is very interesting and..the date of it has been fixed at about 1700.

silladar. Add: Also **silladari, sillahdari.** (Later examples.)
1931 E. Howell *Mizh* i. 9 The men were on a silladari basis. 1960 Ld. Ismay *Memoirs* i. 7 The 21st Cavalry, like nearly all other Indian Cavalry regiments, was organised on what was known as the 'silladar' system. 1974 P. Mason *Matter of Honour* i. 26 Until 1924 many regiments of cavalry were still silladari—that is to say, the regiment was a kind of joint-stock company in which the trooper paid for his horse and equipment when he joined and sold them back when he left.

sillapak (sɪ·lăpăk). [ad. Eskimo *silapak*.] Also **sealapack, seelapak.** A white outer garment worn as camouflage by Eskimo hunters of Labrador.
[1916 E. W. Hawkes *Labrador Eskimo* 39 Over the fur or duffle dickie a cotton slip (ci'tapaq) is drawn.] 1942 *Beaver* (Winnipeg) Dec. 3/2 The chief preacher pulled his thin sillapak so that it might..prevent the fatal drowsiness which precedes death from creeping upon him. 1952 *Ibid.* Dec. 10/2 Sometimes the hunter dons a white 'seelapak' to make himself even less conspicuous. 1959 *Weekend Mag.* (Montreal) 22 Aug. 9 A tightly-woven poplin of cotton and nylon takes the place of the wind-blocking fabric Eskimos first used for the outer garment known as sealapack.

sillar (sīlyā·i). *Geol.* [Sp.] An ignimbrite or volcanic tuff that has not become indurated by welding.
1948 C. N. Fenner in *Bull. Geol. Soc. Amer.* LIX. 883 There has been a tendency to call all such deposits 'welded tuffs'... For those in which induration is primarily the result of recrystallization, and for those in which the fragments have little cohesion, another term is desirable. The local term 'sillar'.., commonly used in the Arequipa region [in Peru], has been applied in the present paper. *Ibid.* 887 These occurrences of both white and salmon sillar contain scattered inclusions of white pumice. 1965 F. H. Hatch et al. *Petrol. Sedimentary Rocks* (ed. 4) xv. 322 Both sillars and welded tuffs frequently exhibit columnar jointing, somewhat akin to that in lava flows proper. 1978 *Nature* 24 Aug. 750/1 In the outcrops visited, two distinct flow units, each up to 30 m thick, were consistently present, both exhibiting the properties of classic sillar.

Sillonist (si·lŏnist). *Ch. Hist.* Also ‖**Silloniste.** [ad. F. *silloniste*, f. the name of the review *Le Sillon*, founded in 1894 by Paul Renaudin.] A member of a French Catholic movement for social reform led from *c* 1902

to 1910 by Marc Sangnier. Hence **Si·llonism,** the principles and policies of the Sillonists.
1910 *Amer. Cath. Q. Rev.* XXXV. 707 The result..can only be a democracy which will be neither Catholic nor Protestant, nor Jewish; a religion (for Sillonism, its chief state, is a religion) more universal than the Catholic Church. 1910 *Daily News* 2 Sept. 5 The Sillonists have developed..in certain forms of private property. 1957 *Church Hist.* XXVI. 29 Certain theologians..pretended to find errors in Sillonist doctrine. *Ibid.* 241 The Sillonistes appeared to contravene in their actions the two basic principles of the papal policy in France. 1978 J. Sondheimer tr. *Aubert's Church in Secularised Soc.* I. iii. 49 This was an aspect..in which Sillonism marked itself out as being the best tradition of the liberal Catholicism of the nineteenth century. *Ibid.* 50 The Sillonists had envisaged forming a spiritual polity.

silly, *a.*, *sb.*, and *adv.* Add: **A.** *adj.* **5. c.** *silly season* (earlier example); also *transf.* and *attrib.*
1861 *Sat. Rev.* 13 July 37/2 We have, however, observed this year very strong symptoms of the Silly Season of 1861 setting in a month or two before its time. 1910 H. G. Wells in *Eng. Rev.* Sept. 308, I got..Burkett of the *Dial* to try some silly-season discussion of State Help for Mothers. 1930 *Forum* Dec. 375/2 The silly season was formally launched and the Big Parade began. 1952 M. Tripp *Faith is Windsock* i. 20 Fat daily newspapers, silly-season follies, cries of 'Give Chamberlain a peerage!' after Munich. 1971 *Jrnl. Gen. Psychol.* Jan. 151 (*heading*) The psychobiological silly season—or—what happens when neurophysiological data become psychological theories. 1976 T. Heald *Let Sleeping Dogs Die* vii. 129 The reporters were..embarrassed at having to attend such a..silly-season event.
d. (Earlier example.)
1888 R. H. Lyttelton in Steel & Lyttelton *Cricket* vi. 287 The English captain acceded to W. G. Grace's wish and allowed him to go forward point, or, as it is familiarly called, 'silly point'.
e. *to play silly buggers* (also *bleeders, b-s*), to fool about, to mess around. Cf. *to play buggers* s.v. *Bugger *sb.* 2. *slang.*
1960 Partridge *Dict. Slang Suppl.* 1174/1 *Silly buggers, play*, to indulge in provocative horse-play; hence, to feign stupidity: low: since ca. 1920. 1968 M. Woodhouse *Rock Baby* ix. 93 If they want to play silly buggers, let them. We're technicians. 1969 M. Pugh *Last Place Left* ii. 13 Don't play silly bugger, Rab. 1972 J. McClure *Caterpillar Cop* iii. 43 It was too easy..and too like what happened when the gods played silly buggers. 1972 'K. Royce' *Miniatures Frame* v. 50, I have to pin something on him to stop him playing silly b's. 1976 'N. Waterhouse' *Mondays, Thursdays* 45 I'm sure none of this had anything to do with the supposed threat to our privacy. It was our God-given right to play silly buggers that was threatened, and the nation responded magnificently. 1979 *Guardian* 19 Aug. 2/1 ..what you must people jeopardising our position by playing silly b's.
d.** Proverbial phr. *ask a silly question* (*and you get a silly answer*).
1960 'A. Gilbert' *Missing from her Home* v. 73 No, don't bother to answer that. 'Ask a silly question and you get a silly answer.' 1970 M. Pereira *Pigeon's Blood* xi. 122 'John? Tell me straight: do you or don't you?' John Raze looked at his friend. 'Ask a silly question..' he said. Then after a pause: 'No.' 1974 *Guardian* 16 Mar. 24/6 Questionnaires..coming under the heading of 'Ask a silly question, and you get a silly answer' get their just deserts.
e. *silly-clever.*
1896 G. B. Shaw in *Sat. Rev.* 11 July 36/2 Greene was really amusing, Marston spirited and silly-clever. 1946 'G. Orwell' in *Polemic* Sept.–Oct. 8 Innumerable silly-clever Conservatives..like Sir Alan Herbert, Professor G. M. Young, Lord Elton. 1963 *Economist* 11 May 538/1 Mr Khrushchev's silly-clever forward pass in Cuba.
7. Special collocations: **silly ass,** a foolish or stupid person (cf. Ass 2); *spec.* an amiable upper-class idiot; freq. *attrib.*; **silly billy,** a foolish or feeble-minded person; used *spec.* as a nickname of William Frederick, Duke of Gloucester (1776–1834), and of William IV (1765–1837); **silly house** *slang*, a mental hospital (cf. *funny farm* s.v. *Funny *a.* 4); † **sillypop** *slang*, a foolish or light-headed woman (cf. Pop *sb.*[3]); also *attrib.* (*obs.*); **Silly Putty** orig. *U.S.*, the proprietary name of an elastic putty-like substance with the remarkable properties of stretching, shattering, and bouncing sharply when appropriately handled, sold chiefly as a plaything; also *fig.* and with small initials; **Silly Symphony,** any of a series of animated cartoons (see quot. 1976) produced by the American cartoonist Walter Elias ('Walt') Disney (1901–66).
1901 G. B. Shaw *Captain Brassbound's Conversion* III. 290 You silly ass, you. 1905 *Punch* 22 Mar. 214/2 He inquired if Phyllis 'had done the Academy yet'? Which, as it didn't open for some days, was a silly-ass thing to say. 1945 'G. Orwell' in *Windmill* No. 2. 18 The silly-ass Englishman with his spats and his monocle. 1973 [see *knut]. 1978 R. V. Jones *Most Secret War* vii. 60 In the best manner of the silly-ass Englishman he blundered into one door after another in an apparent search for the lavatory. 1834 J. Romilly *Diary* 13 Apr. (1967) 55 He was in a towering passion for a minute but soon got into a good humour by laughing at the D. of Gloster. 'Did you see silly Billy squirted on last night? it was worth 5£.' 1872 B. Jerrold *London* xv. 124 The silly-Billy of the neighbourhood—on whom the neighbourhood is merciless. 1908 L. H. Dawson *Nicknames & Pseudonyms* 269 *Silly Billy*, a nickname of William Frederick (1776–1834), Duke of Gloucester; also of William iv (1765–1837). 1934 R. Nichols *Fisbo* 48 Come, come, don't be a silly-billy. 1958 N. Marsh *Singing in Shrouds* 173 You'll think me a frightful silly-billy. 1969 A. Christie *Hallowe'en Party* xvi. 173 The King what had a head like a pear was on the throne—Silly Billy, wasn't it, William IVth. 1977 in Lewis & Baker *Wordpower* II. vi. 15 Mr Healey is a Silly Billy to have waited so long before doing so little of what everyone knew was necessary. 1969 K. Giles *Death cracks Bottle* x. 116 They used to allow me my *News of the World* in the silly house. 1894 M. Beerbohm *Defence of Cosmetics* in *Yellow Bk.* Apr. 70 She is the veriest little sillypop. 1895 *Punch* 18 May 230/3 On stygie or on cinder-path, sillypop things As want to play Man and *be* Woman are trying to fly without wings. 1950 *New Yorker* 26 Aug. 20 After absorbing the elementary facts about Silly Putty..we sought out Mr. Lee Weber, the manager of the bookshop... He told us that Silly Putty is the most terrific item the Doubleday shops have been privileged to handle since 'Forever Amber'. 1952 *Official Gaz.* (U.S. Patent Office) 341 Peter Hodgson, New Haven, Conn... Silly Putty... For the Plastic Known as Organo Silicone Designed and Sold for Use as a Modeling Clay and Amusement..by Children. Claims used since July 1949. 1954 'E. Box' *Death in Fifth Position* v. 111 Silly putty is a pink substance which, if rolled in a ball, will bounce better than rubber, which will shatter if you hit it with a hammer and which will stretch to an unbelievable length. 1963 *Punch* 2 Oct. 495/1 What children today call 'silly putty', which can be pinched and stretched into any shape or length, like toffee. 1964 *Trade Marks Jrnl.* 19 Aug. 1364/2 *Silly Putty*... Playthings made of mouldable plastics. Peter Hodgson,..New Haven, State of Connecticut, United States of America; manufacturer. 1974 P. de Vries *Glory of Hummingbird* (1975) v. 79 It's only your hands I'm putty in... Silly putty I'm afraid. 1929 *Exhibitors Herald World* 16 Nov. 53/3 Booked into the Tivoli at Toronto for a week's run, 'The Skeleton Dance', one of the Disney Silly Symphonies which are being released by Columbia Pictures, has already made three weeks there. 1936 G. Greene *Journey without Maps* i. ii. 67 Natives..looked like grasshoppers in a Silly Symphony. 1976 *Oxf. Compan. Film* 22/1 He [*sc.* Disney] pioneered the precise integration of the animated image with sound—particularly music—in the Silly Symphony series which began in 1928.

silly, *v.* **1. b.** (Earlier example.)
1859 *Sessions Papers Central Criminal Court* 10 May 17, I felt great pain from the blows... It half *sillied* me at the time.

silo, *sb.* Add: **1, 2.** Also, a cylindrical tower or other structure erected above ground for this purpose. Cf. *pit silo* s.v. *pit *sb.*[1] 14.
1886 Stallmaier & Fux tr. *Luther's Constr. & Equipment of Grain Magazines* 11 A silo is erected with outside walls, and sometimes covered with slates. 1904 Wilcox & Smith *Farmer's Cycl. Agric.* 377/1 The first silos were simply pits dug in the ground... Since about 1875 silos of stone, brick and wood have come into use. 1948 *Coast to Coast 1947* 240 The silos stood up tall and straight, grey against the dazzling sky. A line of wheat-laden vehicles moved slowly up towards the hopper. 1950 *Amer. Speech* XXV. 165 Wherever it is possible to find ground that will be dry all seasons of the year, farmers build 'pit silos' and 'trench silos' rather than the cylindrical silos entirely aboveground. 1977 *Daily Tel.* 18 Mar. 8/3 The Norfolk agricultural engineering firm, Rowlands Engineers,..has started a three-shift system..to cope with export orders of more than £400,000 for grain, coffee and cocoa storage silos.
3. = Silage. *rare.*
1889 M. S. van de Velde *Cosmopolitan Recoll.* II. ii. 44 Near the spot where Mademoiselle de Montpensier, the daughter of Gaston d'Orléans, held her little court, rise the new constructions for the storage of silo. 1898 F. P. Dunne *Mr. Dooley in Peace & War* 17 If they'd put blinders on th' mules, they wudden't be scared back be wan iv thim Spanish fleets that a jackass sees whin he's been up all night, secretly stuffing himsilf with silo.
4. *transf.* A large bin used for the storage of loose materials, as cement, etc.
1920 *Glasgow Herald* 2 Sept. 3/8 The coal silos are of sufficient capacity to maintain the supply for about 20 hours in the event of the stoppage of the conveyors. 1958 *Times Rev. Industry* May 64/3 Sulphur will be imported at wharves nearby and stored in a concrete silo with a capacity of 5,000 tons. 1961 *Engineering* 9 June 794 Cement now comes in tankers and is stored in silos. 1973 *Daily Tel.* 25 July 2/3 Another [boy]..was trapped up to his waist in a cement silo for nearly three hours.
5. a. An underground structure in which a guided missile is stored and from which it may be fired. Also *attrib.* Cf. *hard *a.* 14 f, *soft *a.* 19 d.
1958 *N.Y. Times* 15 June 24/4 The system will be protected against neutralization in an enemy attack because the missiles will be installed in concrete-lined underground silos. 1960 *Aeroplane* XLIX. 18/1 For these 'silo' squadrons each missile will be emplaced vertically in a reinforced concrete-lined hole, 52 ft. in diameter and about 175 ft. deep. An elevator raises the missile to the surface a few minutes before launching. 1962 *Engineering* 5 Jan. 13 The Atlas-F [rocket] variant is to be housed within a 'silo' 174 feet deep and 52 feet in diameter... The first silos for Atlas-F are already under construction. 1968 *Economist* 31 Aug. 14/1 On present plans Minuteman III is scheduled gradually to replace Minuteman I and Minuteman II in the silos that dot the prairies and mountains of the western United States. 1975 'A. Hall' *Mandarin Cypher* xiii. 196 The Chinese Republic had silos all over the mainland for reaction-take-off missiles. 1978 *Daily Tel.* 27 July 1/2 Loading the Revenge's 16 ballistic missiles from their hillside silos at Coulport started..yesterday.
b. *Comb.* **silo buster** *slang*, a missile which can destroy an enemy missile in its silo; so **silo-busting** *ppl. a.*
1970 *Nature* 3 Oct. 11/1 The smaller but more accurate multiple warheads..are, like the SS-9, silo-busting weapons. 1977 *Time* 3 Oct. 22/3 The U.S. has also become increasingly concerned..about existing rockets that may become 'silo busters', with the explosive force and pinpoint accuracy to destroy U.S. missiles in their underground launchers.

siloxane (silǫ·ksēⁱn). *Chem.* [ad. G. *siloxan* (A. Stock 1917, in *Ber. d. Deut. Chem. Ges.* L. 170): see Silicon, Oxygen, and -ane.] Any compound having a molecular structure consisting of a chain of alternate silicon and oxygen atoms, the silicon atoms being bonded to hydrogen atoms or to organic radicals. Cf. *silicone 2 a. Freq. *attrib.*
1917 *Jrnl. Chem. Soc.* CXII. ii. 204 Perhaps the greatest confusion in the nomenclature of silicon compounds is to be found in the case of substances containing the —Si—O—Si— system. It is proposed to call the parent hydrogen compounds of this type 'siloxanes', and to specify the number of silicon and oxygen atoms in such terms as 'disiloxane', 'disildioxane', etc. 1941 *Jrnl. Amer. Chem. Soc.* LXIII. 800/2 A cross-linked structure of siloxane chains. 1948 *Electronic Engin.* XX. 82/3 The silicone oils are semi-inorganic polymers in which carbon atoms of organic radicals are linked to the silicon atoms of the so-called siloxane chain. 1960 *New Scientist* 12 May 1205/2 Silicone rubber..is built up from purified siloxanes. 1974 *Encycl. Brit. Macropædia* IV. 104/2 The most useful of man-made inorganic polymers are the siloxanes (also known as silicones).

Silsbee (si·lzbi). *Physics.* The name of Francis Briggs *Silsbee* (b. 1889), American physicist, used *attrib.* and in the possessive with reference to the phenomenon (discovered by Kamerlingh Onnes and explained by Silsbee) of the destruction of the superconducting properties of a superconductor when a current exceeding a certain critical value (the *Silsbee current*) is passed through it.
1926 *Jrnl. Franklin Inst.* CCI. 407 As is demanded for the correctness of Silsbee's hypothesis the supra-conductivity, which was disturbed by exceeding the threshold current, was restored by applying an external magnetic field. 1932 *Proc. R. Soc.* A. CXXXVI. 65 There remained the possibility..that the effect was a manifestation of the Silsbee effect, in which the current through the superconductor produces, by its own magnetic field, a depression of the critical temperature. 1968 C. G. Kuper *Introd. Theory Superconductivity* v. 89 If a current *I* in excess of the Silsbee current..is passed through a cylindrical wire.., some destruction of superconductivity must occur. 1975 M. Tinkham *Introd. Superconductivity* iii. 99 Silsbee's rule that the critical current cannot exceed that which produces a critical magnetic field at the superconductor.

silt, *sb.* Add: **1. c.** *Soil Sci.* Applied *spec.* to particles whose sizes fall within a specified size range between those of sand and clay and to soils having a specified proportion of such particles (see quots.). Hence *silt-grade*; *silt-size* sb. (adj.).
1873 *Amer. Jrnl. Sci.* CVI. 288 It makes a material difference whether the grains of sand contained in a soil or clay are prevalently half a millimeter in diameter, or the tenth or twentieth part of that amount. Sand (or more properly silt) of the latter size is by no means impalpable. 1909 A. G. McCall *Physical Properties of Soils* 88 Stir up the soil remaining in the centrifugal tube and allow to stand for about one minute, or until all particles larger than silt (0·05 [mm.]) have settled. 1920 Silt grade [see *siltstone]. 1958 I. W. Cornwall *Soils for Archaeologist* xi. 125 Once the silt-grade is reached (below 0·06 mm.) a moderate wind is able to transport the grains..for long distances. 1967 *Gloss. Highway Engin. Terms* (B.S.I.) 23 *Silt.* (1) A natural sediment of grading finer than sand consisting of granular products of rock weathering: it is gritty to the touch. (2) In soil analysis it comprises the fraction between 0·06 mm and 0·002 mm. 1968 R. W. Fairbridge *Encycl. Geomorphol.* 675/1 Primary forms of lime include minute grains, incrustations on silt-size grain aggregates and snail shells. 1971 *Gloss. Soil Sci. Terms* (Soil Sci. Soc. Amer.) 15/1 *Silt*, a soil separate consisting of particles between 0·05 and 0·002 mm in equivalent diameter. 1972 [see *sand *sb.*[1] 1 h]. 1976 L. F. Curtis et al. *Soils in Brit. Isles* i. 3 The mineral matter [of soil] includes particles of clay (less than 2 μm diameter), silt (2–50 μm diameter) and sand (50 μm–2 mm diameter).
4. *silt-land, -trap*; **silt loam,** a soil composed at least half of silt; **siltstone** *Petrol.* (see quot. 1920).
1927 *Daily Express* 11 July 11/3 Both fenland and silt-land in these counties boast rich alluvial soil. 1963 *Times* 1 Feb. 13/6 Being essentially a study of the silt lands that border the Wash, it did not attempt to deal in detail with the equally interesting Roman occupation of the southern part of the region. 1917 Mosier & Gustafson *Soil Physics & Management* x. 138 The silt loam soils cover extensive areas in the middle west of the United States and owe their origin to loess. 1957 H. B. Vanderford *Managing Southern Soils* iv. 93 Surface soils which have medium to coarse textures (sandy loams, silt loams, and

clay loams) are suitable for cultivation and relatively easy to keep in good tilth. 1920 A. Holmes *Nomencl. Petrol.* 211 *Siltstone*, a very fine-grained sandstone, the particles of which are predominantly of silt grade. 1946 L. D. Stamp *Britain's Struct. & Scenery* xii. 116 The lower Carboniferous deposits in the Central Lowlands of Scotland are sandstones, shales, cementstones and siltstones with only occasional bands of limestones. 1977 A. Hallam *Planet Earth* 264 The graptolites..are often common in offshore black shales and siltstones. 1946 F. D. Davison *Dusty* viii. 84 They..were going to clear out the silt-trap of one of the tanks. 1966 E. Palmer *Plains of Camdeboo* xviii. 301 Dams, water-troughs, silt traps.

silt, *v.* Add: **1. a.** Also *fig.*
1955 *Times* 10 June 7/3 [The] streets of London silt up with the swelling torrent of motor traffic.
silted *ppl. a.*: also with *up*.
1960 *Archaeologia Cambrensis* CIX. 56 The excavations..revealed a massive stone platform built partly on the peat filling of a silted-up cistern.

siltation (siltēⁱ·ʃən). [f. Silt *v.* + -ation.] The action or result of silting. Freq. *attrib.*
1932 *Min. Proc. Inst. Civil Engin.* CCXXXII. 70 Siltation Records.—As there is a considerable quantity of silt in circulation in Bombay harbour, a detailed investigation was carried out. 1949 *Radio Times* 15 July 7/3 On the foreshore the cunning Dutch made 'siltation fields'... The incoming tides deposit mud in these fields which gradually rise until they are above sea level. Then a dyke is built. 1967 L. E. Cronin in G. H. Lauff *Estuaries* 671/1 The Potomac is a very flashy river... Droughts, floods, heavy siltation,..all present problems in its control. 1977 *Offshore Engineer* Aug. 48/4 The Port needed accurate and quick surveying to assess the siltation characteristics.

silumin (si·liūmin). *Metallurgy.* [a. G. *silumin*: see Silicon and Aluminium.] Any of a series of casting alloys of aluminium containing about 9 to 13 per cent silicon.
1922 *Chem. Abstr.* XVI. 3861 Silumin... A new Al-Si alloy has been prepd. at the Metallbank in Frankfurt. 1930 *Engineering* 18 Apr. 508/1 This difficulty was partly overcome by resting the tent on slabs of Silumin. 1967 *Chem. Abstr.* LXVI. 21573 Effect of 0·0016–0·00072% Na on the kinetics of gas absorption under a steam-satd. atm. was investigated on silumins differing only in the initial Na content.

Silurian, *a.* and *sb.*[1] Add: **2. a.** (Further examples.) In mod. use, the name is restricted to a system of Lower Palæozoic rocks underlying the Devonian and overlying the Ordovician, so corresponding to the Upper Silurian as originally defined.
As orig. defined by Murchison the Silurian included what was subsequently called the Ordovician, and this use continued for a time after the introduction of the Ordovician in 1879.
1879 C. Lapworth in *Geol. Mag.* VI. 3 The Lyell-Hicks division of *Cambrian* and *Lower Silurian* are as rightly entitled to the rank of separate systems as the true or *Upper Silurian* itself. *Ibid.* 9 The general restriction of the title Silurian to the strata that are comprehended between the line marking the base of the Lower Llandovery, and that denoting the commencement of the brackish or fresh-water conditions of the typical Old Red Sandstone, appears..inevitable. *Ibid.* 15 The ideas of the extreme party which claims all the Lower Palæozoics for the Silurian are fated soon to become wholly extinct. 1902 A. J. Jukes-Browne *Student's Handbk. Stratigr. Geol.* vii. 64 Murchison supposed that Sedgwick's Cambrian lay entirely below his Silurian, but when the fossils were collected and described, it was found that the Upper Cambrian was equivalent to the Lower Silurian... Group after group of Sedgwick's Cambrian was gradually absorbed into it [*sc.* the Lower Silurian], till the Lower Silurian came to include the whole of the rocks (below the Upper Silurian) in which any fossils had been found. 1903 A. Geikie *Text-Bk. Geol.* (ed. 4) II. 934 Murchison's 'Lower Silurian' has by many writers been replaced by 'Ordovician', and his 'Upper Silurian' is in a similar manner being ousted by some other term... I shall continue to employ Murchison's terminology. 1912 *Q. Jrnl. Geol. Soc.* LXVIII. 332 In fixing the boundary between the Ordovician and the Silurian the peculiar characters and mode of weathering of the Lower Birkhill rocks have been found useful. 1931 Gregory & Barrett *Gen. Stratigr.* v. 77 The Silurian is now usually restricted to the strata between the Ordovician and Devonian. 1955 [see *Gotlandian, Gothlandian a.]. 1964 *Rep. Internat. Geol. Congr. XXI Sess., 1960* XXVIII. 254 The Commission transmits to the Congress the following proposals on the terminology of the Silurian and Ordovician... 1) Two systems are to be recognized between the Cambrian and Devonian systems. 2) The name of the lower shall be Ordovician. 3) The name of the upper shall be Silurian. 1971 *Jrnl. Geol. Soc.* CXXVII. 106 When the standard classification of the correlation charts is considered in relationship to local successions the Silurian of the British Isles is found to be something of a monument to stratigraphical chaos.
c. *transf.*, loosely designating a primitive age or period in the remote past.
1875 'Mark Twain' in *Atlantic Monthly* Aug. 193/2 In the Old Oölitic Silurian Period, just a million years ago next November. 1962 E. Snow *Red China Today* (1963) xvi. 116 When I last saw Mao..China was weak, disunited and bankrupt. Since then China's Silurian age had ended. 'China has stood up,' as Mao proclaimed.
3. Also with small initial. **a.** Of, pertaining to, or designating a paper showing two or more contrasting colours on its surface; usu. applied to stationery of a blue-grey appearance. Also, of the colour itself.
1892 J. Heywood *Wholesale Catal. Stationery & Stationers' Sundries* 22 Scotch Tinted Writings... Silurian—5 8. 1930 W. de la Mare *On the Edge* 28 The drawer beneath contained only envelopes and letter paper—*Montrésor*, in large pale-blue letters on a 'Silurian' background. 1937 E. J. Labarre *Dict. Paper* 153/1 *Granite* paper, also termed French grey, Ingres, Silurian grey, Mottled, Ingrain, is paper which clearly shows two or more contrasting colours of pulp on its surface. 1964 M. Clive *Day of Reckoning* viii. 73 Their correspondents wrote on double sheets of grey 'silurian' paper which looked hairy but was slippery.
b. As *sb.*, paper or stationery of this type.
1942 H. A. Maddox *Dict. Stationery* (ed. 2) 100 *Silurian*, a tinted writing paper formerly much in favour for note and envelopes. Characterised by a blue-grey mottled colour which gave rise to the occasional term, French Granite. 1954 *Paper Terminol.* (Spalding & Hodge) 54 *Silurian*, coloured paper, usually a writing or cover, produced by introducing into the coloured pulp, fibres dyed a deeper shade. 1960 D. Holman-Hunt *My Grandmothers & I* i. 24, I..sucked the pen and began to scratch at the grey silurian.

Silva (si·lva). The name of José *Silva* (b. 1914), American electrician, used *attrib.* to denote the theory or methods devised by him to improve the functioning of one's mind.
1971 *Nat. Observer* (U.S.) 23 Aug. 16/1 Dr. Green faults the Silva method. 1976 *New Yorker* 15 Mar. 140/3 This book traces the frantic pilgrimage he undertook to pull himself together, running through EST, gestalt therapy,..Esalen, hypnotism, modern dance, meditation, Silva Mind Control, Arica, [etc.]. 1978 Silva & Miele *Silva Mind Control Method* 12 The city planner had been trained in Silva Mind Control.

silva, see Sylva in Dict. and Suppl.

silver, *sb.* and *a.* Add: **I. 1. d.** *ellipt.* for *silver medal* (see sense 21a below).
1960 [see *gold[1] 1 b]. 1968 *Guardian* 22 Oct. 1/1 Major Albusen, aged 55, won the silver in the individual event, and was only two points off taking the gold. 1979 'D. Grant' *Moscow 5000* i. 19 Notes that would help him to win an Olympic medal. Because he would have the Silver, he told himself.
7. a. (Examples of the silver salmon.)
1934 *Nat. Geogr. Mag.* Feb. 211 There are four distinct species of salmon which run up the Columbia: the chinook, silver, sockeye, and chum. 1955 [see *Chinook b].
II. 10. d. (Earlier example.)
1879 *Bradstreet's* 22 Oct. 5/1 The silver men are as violent and rampant as ever.
III. 16. a. *silver-miner* (earlier example), *-plater* (earlier example).
1815 *Niles' Register* VIII. 141/2 There are..2 silver platers; 3 trunk makers [etc.]. 1869 'Mark Twain' *Innoc. Abroad* vi. 57 To speak after the fashion of the silver-miners.
c. *silver-mining* (earlier example).
1872 'Mark Twain' *Roughing It* p. iv, The silver-mining fever in Nevada.
17. a. *silver-bowed, -buckled, -laced* (later example), *-mounted* (earlier example), *-plated* (earlier example), *-sandalled, -stringed, -studded, -topped, -veined.*
1894 'Mark Twain' *Those Twins* v. 393 The Judge..laid aside his silver-bowed spectacles. 1922 Joyce *Ulysses* 506 He carries a silverstringed inlaid dulcimer... He wears dark velvet hose and silverbuckled pumps. 1939 D. Cecil *Young Melbourne* vi. 155 She also created scandal by appearing..imperfectly disguised as a page, in a plumed hat, silver-laced jacket and tight scarlet pantaloons. 1748 Smollett *R. Random* II. xliv. 79 A pair of silver mounted pistols. 1843 Dickens *Mart. Chuz.* (1844) xix. 237 To provide silver-plated handles of the very best description. 1881 O. Wilde *Poems* 67 Sweeter far if silver-sandalled foot Of some long-hidden God should ever tread The Nuneham meadows. 1916 Blunden *Harbingers* 63 So silver-sandalled down those golden ways He triumphs. 1922 Silverstringed [see *silver-buckled* above]. 1944 W. Fortescue *Mountain Madness* i. 22 A wide silver-studded black leather belt. 1928 'Brent of Bin Bin' *Up Country* xv. 258 He was filling a bolster with articles of jewellery, silver-backed brushes, hand mirrors, candlesticks, silver-topped bottles, &c. 1976 'D. Halliday' *Dolly & Nanny Bird* xiii. 170 There's a white leather gift box in every cabin, fitted out with..silver-topped crystal bottles. 1916 Joyce *Portrait of Artist* (1969) v. 176 He would think of the cloistral silverveined prose of Newman.
c. *silver-breasted, -flecked, -sanded, -suited, -winged* (later example).
1881 O. Wilde *Poems* 24 Lure the silver-breasted Helena Back from the lotus meadows of the dead. 1926 *Spectator* 11 Sept. 370/1 Wide silver-breasted rivers flowing to a sunlit sea. 1937 *Burlington Mag.* May 252/2 Two bowls of *Chien yao*... One is of the silver-flecked variety. 1967 *Coast to Coast 1965–6* 32 All manner of fabulous creatures of the deep surged and surfaced amid the molten silver-flecked arrowing lines of foaming waves. 1851 J. G. Whittier in *National Era* 3 Jan. 106/4 Whose small waves on a silver-sanded shore Whisper of peace. 1957 R. Campbell *Coll. Poems* II. 121 On the silver-sanded shores. 1962 *Daily Tel.* 5 Oct. 22/2 The silver-suited astronaut. 1903 Kipling *Five Nations* 2 The in-rolling walls of the fog and the silver-winged breeze that disperses.
18. *silver-shining* (later examples).
1932 D. Gascoyne *Roman Balcony* 33 And through their long-nailed fingers Glide the silver-shining minnows. 1944 W. de la Mare *Coll. Rhymes & Verses* 99 There silver-shining Hesper Smiles at Mars.
19. *silver-blue, -bright* (later example), *-green, -pink, -silent.*
1959 W. Thesiger *Arabian Sands* xii. 242 We came to a succession of dune-chains, each of which..showed up in turn as a wavy silver-blue wall. 1973 J. Seabrook *Loneliness* 117 A quiet, rather subdued woman; smart, with silver-blue hair. 1959 R. Graves *Coll. Poems* 315 And next the silver-bright Hyperborean Queendom. 1914 L. Woolf *Wise Virgins* iv. 110 The silver-green water glided by him. 1923 D. H. Lawrence *Birds, Beasts & Flowers* 19 Silver-pink peach, venetian green glass of medlars and sorb-apples. 1976 'D. Halliday' *Dolly & Nanny Bird* xiii. 165 The pearly capped teeth and silver-pink mouth. 1922 Joyce *Ulysses* 537 Through silver-silent summer air the dummy of Bloom, rolled in a mummy, rolls rotatingly.
IV. 21. a. silver band, a brass band with silver-coloured instruments; **silver beggar** (earlier example); **silver blond(e)** *a.*, of hair: of a very light, silvery colour, esp. as the result of bleaching (cf. *platinum blond(e)* s.v. *platinum 2 c); **silver collection,** a collection of 'silver' coins (or of money of no denomination lower than these) made at a meeting, etc.; **silver cord,** (*a*) used in phr. *the silver cord is loosed* and varr. (in allusion to Eccl. xii. 6) to signify the dissolution of life at death; (*b*) a symbol of excessive devotion between mother and child; **silver doctor,** an artificial fishing fly having a body of tinsel; **silver-fizz,** an effervescing drink based on gin and egg-white (cf. *fizz, fiz *sb.* 3); **silver-fork** (earlier example); also applied to later novelists displaying similar characteristics; **silver handshake,** a gratuity given on retirement or as compensation for dismissal from one's occupation (of less value than a golden handshake); † **silver hell** *slang*, a low-class gambling saloon (cf. Hell *sb.* 8) (*obs.*); **silver jubilee:** see Jubilee *sb.* 3 a; **Silver Lady,** an epithet applied to Miss Elizabeth Baxter (d. 1972), philanthropist, from her custom of giving silver coin to the down-and-outs of the Embankment in London, used *attrib.* to describe a charitable organization (and its appurtenances) which distributes food and hot drinks to vagrants; **silver medal,** a medal made of or resembling silver, awarded as the second prize in a contest, esp. in the Olympic Games; hence **silver medallist; silver point,** (*b*) the freezing point of silver under normal atmospheric pressure (about 962°C), as a thermometric fixed point; **silver-pointed** *a.*, coloured or tinged in the manner of a silver-point drawing; hence, as a back-formation, **silver-point** *v. trans.*, to cause to appear so; **silver polish,** a polish used for cleaning and brightening silver; **silver ring** *Racing* (see quot. 1921) (cf. *Tattersall 1 b); also *attrib.*; **silver screen,** a cinematographic projection screen covered with metallic paint to produce a highly reflective silver-coloured surface; usu. *transf.*, the cinema generically, considered as a medium for such film projection; **silver service** (see quot. 1970); **Silver Shirts** *U.S.*, the name applied to the Silver Legion, an American fascist, anti-Semitic paramilitary group founded in 1933 and disbanded in 1940 (cf. *blackshirt); **silver-side** (earlier example); **Silver Star,** a decoration for gallantry awarded to members of the U.S. Army and Navy (see quot. 1941); also *Silver Star medal*; **silver state** *U.S.*, a state producing silver, or advocating free coinage of silver; *spec.* (with initial capitals) Nevada or, less freq., Colorado; **silver table,** (*a*) a table made of or plated with silver; (*b*) a table used for the display of silverware, freq. with raised edges (and a glass lid); **silver-tail** (later examples); **silver tea** *N. Amer.*, a tea-party at which the guests make contributions (typically, of 'silver' coin) to charity; **silver thaw** (earlier N. Amer. example); **silver wedding** (earlier examples) (see also Wedding *vbl. sb.* 2 b).
1933 *Radio Times* 14 Apr. 126/5 The Tullis Russell Silver Band. 1949 'J. Tey' *Brat Farrar* xxv. 228 'Thump! Thump! Thump!' said the drum of the Bures Silver Band. 1976 *Times* 3 May 12/4 The Eastbourne silver band, in bright red jackets, played *California Here I Come*. *c* 1842 *Exposure of Impositions practised by Vagrants* 4, I shall begin with those vagrants who, generally, obtain the most, and are considered of the first class, and by some

termed 'Silver Beggars', but by travellers they are called 'Lurkers'. 1951 J. C. Fennessy *Sonnet in Bottle* i. 29 Silver-blond hair, silver-grey eyes. 1959 M. Summerton *Small Wilderness* i. 8 The silver-blonde hair that curved..to her shoulders. 1974 D. Francis *Knock Down* iii. 37 She had silver shoes and silver-blonde hair. 1957 B. & S. G. Hulme Beaman *Ernest the Brave* 8 'I was referring to the pence expected as a result of this disgraceful exhibition!' 'Oh, Mr. Growser, sir,' Larry interrupted. 'This is supposed to be a silver collection.' 1972 H. Kemelman *Monday Rabbi took Off* xxii. 145 Imagine, Katz, no charge. Not even a silver collection. 1911 J. A. Thomson *Introd. Sci.* vi. 177 If we can use such a word, the silver cord of the bundle of life is loosed, and earth returns to earth. The microbes of decay break down the dead, and there is a return to air and water and salts. 1934 F. S. Fitzgerald *Tender is Night* i. xiii. 76 'The silver cord is cut and the golden bowl is broken and all that, but an old romantic like me can't do anything about it.' 'I'm romantic too.' They came out of the neat restored trench, and faced a memorial to the Newfoundland dead. 1942 P. Wylie *Generation of Vipers* xi. 185 Our land, subjectively mapped, would have more silver cords and apron strings crisscrossing it than railroads and telephone wires. 1959 J. Braine *Vodi* xxi. 232 Her mother, as usual, had won. And, what was hardest to forgive, had won fairly; she wasn't the Silver Cord type, she'd never been possessive. 1973 G. Mackay Brown *Magnus* vii. 156 Magnus Erlendson would live out his life, until such time as the silver cord was loosed, and the golden bowl broken, and the pitcher broken at the fountain. 1875 *Encycl. Brit.* II. 40/2 *The silver doctor*, also a very great favourite. Tag, silver tinsel; tail, a topping; but, a turn of red crewel; body of silver tinsel entirely; [etc.]. 1931 *Hardy's Anglers' Guide* 31 Your fairy shrimp, just as pretty..as any Jock Scott or Silver Doctor. 1901 O. Wister *Philos. 4* in *Stories of Colleges* 68 It must have been that silver-fizz you took before dinner. 1977 E. Ambler *Send no More Roses* vi. 121 He was drinking a silver-fizz, a long drink made of gin and egg-whites. [1827 *Examiner* 18 Nov. 722/2 A writer of this accomplished stamp..also informs you that the quality eat fish with silver forks.] 1831 *Times* 15 Dec. 5/3 A single chapter of any one of them is worth more than the whole bundle of those contemptible productions of the silver-fork school, which are called 'fashionable novels'. 1974 *Times Lit. Suppl.* 4 Oct. 1092/5 The suspicion grows that this is a new-style 'silver fork' novel, with merchant bankers taking the place of noble dukes. 1958 M. Pugh *Wilderness of Monkeys* 84 Flash Willy is just about to go back to London, pick up his silver handshake, his sacking money, and get himself a corner in the 'I will photograph your child in your home' lark. 1979 C. Dexter *Service of All Dead* ii. 17 A little silver handshake, a little farewell party. 1835 T. Power *Jrnl.* 13 Jan. in *Impressions of Amer.* (1836) II. 196 With here and there a couple of the same sort of gemman to be met with about the silver hells of London. 1843 'W. I. Moncrieff' *Scamps of London* i. i. 5 in *Sel. Dramatic Wks.* (1851) I, He's the principal partner in all the silver hells at the west end. 1961 *Ann. Charities Reg. & Digest* 208/2 *Silver Lady Fund.* Mobile café out on Embankment serving hot tea and food free to the needy. Miss Betty Baxter.., E.C.4. 1978 C. A. Berry *Gentleman of Road* xv. 171 The Silver Lady van arrived and mugs of tea and meat pies were distributed. 1908 Silver medal [see *gold[1] 1 b]. 1958 [see *baton *sb.* 2 b]. 1976 *All about the Games* (Com. Org. des Jeux Olympiques) 24 Canada's 74 athletes won one silver medal. 1911 Silver medallist [listed in Dict. at sense 8 a]. 1976 *Daily Tel.* 20 July 1/7 The cheating by Boris Onischenko, silver medallist at the Mexico City and Munich Olympics. 1928 *Bureau of Standards Jrnl. Res.* (U.S.) I. 637 The constants *a*, *b*, and *c* are to be determined by calibration at the freezing point of antimony, and at the silver and gold points. 1967 Condon & Odishaw *Handbk. Physics* (ed. 2) v. iii. 41/1 In the neighbourhood of 1000°C new determinants of the silver point and gold point have been made in recent years. 1976 I. Murdoch *Henry & Cato* i. 46 A bright half moon was..silverpointing the slates and making pendant shadows beneath the..eaves. 1913 C. Mackenzie *Sinister St.* I. ii. xx. 483 They moved to Geneva, whose silverpointed beauty for a while deceived them. 1930 R. Campbell *Poems* 1 Two sisters... Whose fingers glint with silver-pointed nails. 1895 *Montgomery Ward Catal.* Spring & Summer 193 Thorn's Silver Polish, Liquid Form, 15c. per Bottle. 1974 'D. Fletcher' *Lovable Man* i. 37 He memorised the exact position of the silver polish and..began to polish the lighter. 1921 E. Wallace *Law Four Just Men* ix. 261, I found a poor little bookmaker in the silver ring—the silver ring is the enclosure where smaller bets are made in Tattersall's reservation. 1926 J. Masefield *Odtaa* xv. 257 A vile, taunting, silver-ring tick. 1939 Wodehouse *Uncle Fred in Springtime* xx. 306 She is the daughter of a retired Silver Ring bookie. 1973 'I. Drummond' *Jaws of Watchdog* xvii. 227 Sandro was in Tatt's... He could also go down the social and financial scale into the Silver Ring and the cheapest enclosures. [1921 'M. Pickford' *Let.* in V. Burnett *Romantick Lady* (1927) xxxii. 398 It is not always easy to take a classic like 'Little Lord Fauntleroy' and place it on the cold, silver screen.] 1924 *Amer. Hebrew* 22 Feb. 439 (*heading*) 'Shooting' news for the silver screen; Pathe film editor who brings home to millions timely pictures of world events. 1931 B. Brown *Talking Pictures* i. 19 Somehow there had crept into this new field of endeavour the romance of the silver screen. 1959 *Times Lit. Suppl.* 6 Nov. 636/4 Not a night passes without one aspect or another of the far western frontier holding children from play and old men from the chimney corner on the silver screen or on 'the Telly'. 1979 A. Hailey *Overload* iii. xii. 256 Cameron Clarke objected to Tunipah and the god of the silver screen had spoken. 1970 *Drive* Spring 43/1 Silver service means that your plate is put before you empty and the various parts of your dish are served separately from silver. 1976 *Evening Standard* 14 June 25/3 (Advt.), Commis de rang for our high class Prince's Room Restaurant—must have silver service experience. 1934 *Sun* (Baltimore) 6 Aug. 6/2 A California newspaper has published the exciting news that the Silver Shirts of America plan to put down the Communists and then take over control of the American Government. 1959 W. Faulkner *Mansion* 303 When the Silver Shirts appeared, Clarence was one of the first in Mississippi to join it. [1845 E. Acton *Mod. Cookery* viii. 206 The natural division of the meat will show where the silver-side of the round is to be separated from the upper, or tongue side.] 1861 Mrs. Beeton *Bk. Househ. Managem.* 283 As a whole round of beef..is too large for small families..we here give the recipes for dressing a portion of the silver side of the round. 1932 *U.S. Army Regulations* 8 Aug. No. 600-45 p. 1 The authorized decorations awarded by the United States are: *a. Awarded by the War Department*... (4) Silver Star. 1932 *N.Y. Times* 18 Dec. ii. 2/5 Captain Herbert G. Rosboro..received the Silver Star medal today from the War Department for gallantry in action in October, 1918. 1941 J. McDowell Morgan *Military Medals & Insignia of U.S.* 76 The Silver Star was established..on July 9, 1918. This originally was a unique badge of honor, being a small silver star, 3/16-inch in diameter, designed to be worn on the ribbon of a campaign medal to indicate..'a citation for gallantry in action'... On August 8, 1932, a distinct medal, known as the Silver Star Decoration, was established as a reward to those persons previously cited in orders for gallantry in action. 1948 E. E. Cummings *Let.* 27 Aug. (1969) 185 The hyperscientific climax of this hero (a prominent killer, holder of Silver Stars & Clusters & Purple Hearts galore)'s experience. 1969 I. Kemp *Brit. G.I. in Vietnam* vii. 153 You've been awarded the Silver Star for your action at Dak To. 1982 H. Lieberman *Late Call* lxiii. 308 I'm a veteran with a silver star and a purple heart. 1866 *Eastern Slope* (Washoe, Nevada) 15 Sept. 4/1 The Silver State struck it rich when they elected H. G. Blasdel to the Gubernatorial chair. 1871 *Harper's Mag.* Oct. 799/1 In our early days in the Silver State females were rarely to be seen in the frontier mining camps. 1885 *Weekly New Mexican Rev.* 8 Jan. 4/2 All the silver states and territories [should] organize to resist the effort which the single standard advocates are making in congress to suspend the coinage of silver. 1946 *Trail & Timberline* May 74/1 Colorado miners had been looking for gold but silver became of such importance that when the Territory became a state in 1876, it was known as the Silver State and Georgetown was called the Silver Queen. 1976 *Billings* (Montana) *Gaz.* 20 June 10–c/2 It was a bluish-gray ore—silver—and Nevada is now appropriately nicknamed 'The Silver State'. *c* 1792 C. Fiennes *Journeys* (1947) iii. xii. 279 Here's a silver table and stands and glass frame. 1897 [in Dict., sense 10 c]. 1926 A. Christie *Murder of R. Ackroyd* iv. 33 What..is called a silver table, the lid of which lifts, and through the glass of which you can see the contents. 1975 *Country Life* 10 Apr. Suppl. 48 j/1 (*caption*) A really fine Chippendale period silver table of superb quality. 1908 E. G. Murphy *Jarrahland Jingles* 116 And when they're playing billiards in their flannel tennis suits, We feel like heaving something at these silvertail galoots. 1947 G. Casey *Wits are Out* ix. 125 'Mr Fleming doesn't build for basic-wage earners,' Bill said nastily. 'He hangs around waiting his chance to build for the silvertails.' 1978 *Listener* 9 Feb. 163/3 Mr Whitlam's enemies in his own Labor Party have called him a 'silvertail', meaning a social climber. 1979 *Sunday Mail Mag.* (Brisbane) 11 Feb. 16/1 The Governor-General was, in the Premier's opinion, a super silvertail. 1921 *Daily Colonist* (Victoria) 18 Mar. 9/4 A silver tea will be held at the home of Mrs. H. Lloyd-Young..on Friday afternoon. 1770 G. Cartwright *Jrnl.* 22 Dec. (1792) I. 73 There was a silver thaw in the morning, and it rained freely: very mild weather all the rest of the day. 1845 A. H. Clough in *Ambarvalia* (1849) 28 The Silver Wedding! on some pensive ear..A silvery faint memorial music swells. 1861 Queen Victoria *Let.* 13 Feb. in R. Fulford *Dearest Child* (1964) 307 You must promise to be with us for our silver wedding D. V. which will be in four years.
b. silverback, a mature male mountain gorilla, *Gorilla gorilla beringei*, distinguished by one or more patches of white or silvery hair just below the back of the neck; **silver fox,** (*a*) (earlier example); also *transf.*, the fur of this animal, esp. as a fashion item; (*b*) a fennec, *Vulpes chama*, found in southern Africa; **silver-tip** (earlier example).
1963 G. B. Schaller *Year of Gorilla* viii. 221 He was a silverback in the prime of life. 1975 J. Goulet *Human Ape* (1977) i. 5 The old silverback was having trouble breathing. 1770 G. Cartwright *Jrnl.* 30 Dec. (1792) I. 76 On Niger Sound we saw a good silver fox. 1892 *T. Eaton & Co. Catal.* Fall & Winter 11/2 Three-quarter capes, quilted lined, trimmed silver fox, $4. 1912 J. Stevenson-Hamilton *Anim. Life Afr.* xv. 231 The Silver Fox (*Vulpes chama*)..is silvery-grey, the underneath parts being tawny in hue. 1936 A. Christie *ABC Murders* xix. 143 [She] wears very lovely clothes. That crêpe marocain and the silver fox collar—*dernier cri!* 1940 [see *battle bowler* s.v. *battle *sb.* 14]. 1972 *Stand. Encycl. S.Afr.* VI. 170/2 The Cape or silver fox is found throughout the drier regions of South Africa. 1886 *Turf, Field & Farm* 26 Mar. 238/1 A silver tip is bad enough when he's wounded, and about as active a bear as there is.
c. silver-eye, substitute for def.: one of several birds of the genus *Zosterops*, distinguished by white rings round the eyes; (earlier and later examples).
1875 [see *makomako[1]]. 1911 A. E. Mack *Bush Days* 2 You will hear a whole chorus of bird notes..calling all together—thrushes, thickheads, silvereyes and peewees. 1965 [see *blight-bird* s.v. *blight *sb.*]. 1977 *Kuwait Times* 23 Nov. 6/8 Three fell to Man, including a pigeon and a parrakeet. The rats have been responsible for at least five more (a thrush, a warbler, a fantail, a silvereye and a starling).
d. silver eel, also, a young eel before the adult coloration is developed; (earlier and later examples); **silver salmon** (earlier example); **silver-sides:** also **silver-side** and as one word; (earlier and later examples); **silver trout** *N. Amer.*, one of several silvery trout, esp. *Salmo gairdneri kamloops*; also, = *kokanee.
1735 Swift & Sheridan *Let.* 28 Nov. in *Wks. J. Swift* (1768) XIII. 143 For the rest, we are forced to take up with..silver eels, and such trash. 1952 *New Biol.* XIII. 76 At the silver-eel stage..it is ready to descend the river again. 1878 J. G. Brady *Let.* May in S. Jackson *Alaska* (1880) vii. 209 A silver salmon, weighing thirty-eight to forty pounds, is sold for fifteen or twenty cents. 1820 C. S. Rafinesque in *Western Rev.* II. 240 Silverside Fallfish... Vulgar names, Silverside, Shiner, [etc.]. 1851 R. Glisan *Jrnl. Army Life* (1874) viii. 88 The purer streams from the hills abound in..silver-sides. 1911 *Rep. Fisheries 1908* (U.S.) 316/2 Some of the silversides (*Atherinidæ*) are wrongly called 'smelts'. 1962 K. F. Lagler et al. *Ichthyology* x. 284 In the brook silverside..there is a single elongate filament that serves first for temporary flotation. *Ibid.* xi. 373 Refractive errors change by several diopters during such measurements on schooling fishes, such as the silver-side (*Menidia*). 1873 C. Hallock *Fishing Tourist* i. v. 30 To the above should be added the..brook-trout, the silver-trout, and the..salmon-trout. 1907 T. W. Lambert *Fishing in Brit. Columbia* 43 Every local fisherman speaks of having caught a red side or a silver trout, and firmly believes they are distinct species. 1937, 1970 Silver trout [see *kokanee].
e. silver beech, an evergreen tree, *Nothofagus menziesii*, native to New Zealand (cf. *nothofagus); also, the timber of this tree; **silver beet** *Austral.* and *N.Z.*, the seakale beet, *Beta vulgaris*; = Chard[2]; (later examples); **silver bell (tree)** (earlier example); **silver poplar** *U.S.* (earlier example); = *white poplar* s.v. Poplar 1 b; **silversword,** a perennial herb, *Argyroxiphium sandwicense*, of the family Compositæ, native to Hawaii and bearing linear leaves with silvery hairs and clusters of purplish flowers; **silver wattle** (earlier example); **silver willow,** a variety of the white willow, *Salix alba* var. *sericea*, distinguished by silvery foliage.
1889 T. Kirk *Forest Flora N.Z.* 175 The silver-beech..is known as 'tawhai' or 'tawai' by the Natives. 1950 *N.Z. Jrnl. Agric.* July 8/3 Durability of..less than five years..Silver beech. 1966 *Encycl. N.Z.* I. 177/2 Silver beech..a tree with small, thick, double-toothed leaves and a cherry-like bark on the branches and young trees, reaches heights of about 100 ft. 1915 *N.Z. Jrnl. Agric.* 20 Jan. 75 Early in February is a good time to sow silver-beet. 1951 J. Frame *Lagoon* 98 For dinner I had semolina and silver beet. 1973 *Islander* (Victoria, B.C.) 18 Feb. 2/4 Their hulls loaded down with taro, yams, chinese cabbage (rather like silver beet) and bananas. 1977 *N.Z. Herald* 5 Jan. 2-2/1 The novelty value of spaghetti bolognaise can often get over the hurdle of the silverbeet hidden in the sauce. 1785 H. Marshall *Arbustrum Americanum* 57 Silver-Bell Tree... The Corolla is of one petal, bell'd and bellied. 1847 W. Darlington *Agric. Bot.* 332 Silver Poplar... Some of the grass-plats in the public squares of New York have been quite over-run by the wide-spreading suckers of this tree. 1888 W. Hillebrand *Flora of Hawaiian Islands* 219 The 'Ahinahina' of the natives and 'Silversword' of the foreigners. 1937 *Discovery* Mar. 83 Not the least puzzling of the specialised animal and plant species of the Hawaiian Islands is the six-foot Silver-sword flower. 1965 P. Wylie *They Both were Naked* ii. vi. 302 A silver-sword plant..grows only on a few high places on two islands [of Hawaii]. 1859 D. Bunce *Trav. Dr. Leichhardt* iii. 19 We camped among the butts of the *Acacia affinis*, or silver wattle. 1914 W. J. Bean *Trees & Shrubs Hardy in Brit. Isles* II. 475 Silver Willow.—This is the most striking of all the forms of *S[alix] alba* in the intense silvery hue of its leaves. 1976 *Country Life* 18 Mar. 682/3 A group of silver willows..are annually pollarded.

Silver age. Add: **1. c.** A period of Russian literature and art at the beginning of the twentieth century, considered in comparison with the golden era of the mid-nineteenth century.
1965 P. Benno in Hayward & Crawley *Soviet Lit. in Sixties* 179 The works of the generation of Russia's 'Silver Age' in the first three decades of the present century. 1974 T. P. Whitney tr. *Solzhenitsyn's Gulag Archipelago* I. i. ix. 336 Even though the Silver Age of art, four State Dumas, three wars, and three revolutions had come and gone, all Moscow drank Oldenburger's water. 1976 *Times Lit. Suppl.* 16 Apr. 450/3 The poet's life in pre-war St Petersburg... She grew up in the capital at a time when its artistic life was at its most febrile and brilliant, the height of the 'Silver Age', not only in poetry but in painting, ballet, music.

Silverblu (si·lvəɹblū). Also **Silver Blu(e.** [f. Silver *a.* + Blu(e *a.*)] A mink belonging to a mutated form distinguished by silvery fur; also, the fur of an animal of this kind. Also *attrib.*
1941 *Amer. Fur Breeder* June 8/3 Miss Esther Wyman of Harper's Bazaar magazine was shown a couple of these Platinum pelts... She..suggested that a better fur name be given them such as Silver Blue. 1942 *Ibid.* Oct. 8/3 Of the mutations the platinum or Silver Blu mink are the most prominent. 1944 *Fur Trade Jrnl. Canada* XXI. 10/1 The first offering of 'Silverblu' mink..brought a top price. 1945, 1956 [see *mutation mink* s.v. *mutation 7]. 1966 A. Leonard *Mod. Mink Management* 202 (*caption*) The late William Whittingham developed the Silver Blu mink, the grandaddy of all mutations. 1968 J. Ironside *Fashion Alphabet* vii. 159 Silverblu (also called Platinum): The original mutation mink, introduced in 1942.

silver-foil. Add: **2.** = *SILVER PAPER 2. Cf. *FOIL *sb.*[1] 4 d.
1944 N. MAILER in E. Seaver *Cross-Section* 338 The captain took out a chocolate bar... He separated a piece of silver foil from his teeth. 1974 N. BENTLEY *Inside Information* i. 8 He slid the outside wrapping off his bar of Whole Nut... Hidden between the wrapping and the silver foil underneath was a small piece of paper.

silver-grey, *sb.* **b.** (Earlier example.)
1850 *N.Y. Tribune* 18 Oct. 5/2, I shall gladly fight on in this cause so long as I shall live, and ask no higher post than the proud one of a private in the Silver Grays.

silver-headed, *a.* **2.** (Later example.)
1981 M. McMULLEN *Other Shoe* (1982) ii. 14 She got about slowly..with the help of a silver-headed ebony cane.

silverily (si·lvərili), *adv. rare.* [f. SILVERY *a.* + -LY[2].] = SILVERLY *adv.* 1, 2.
1929 D. H. LAWRENCE *Pansies* 44 This wet white gleam Twitches, and ebbs hitting, washing inwardly, silverily against his ribs. *a* 1930 —— *Phoenix* (1936) 40 You hear the nightingale silverily shouting.

silver-leaf. 3. Delete *local* and substitute for def.: A disease of *Prunus* and other woody plants caused by the fungus *Stereum purpureum*, which is frequently associated with a silvery sheen of the leaves and often fatal to affected branches. (Further examples.)
1902 *Jrnl. Linn. Soc. Bot.* XXXV. 390 The disease known as 'Silver-leaf' is, so far as I am aware, confined to the Pruneæ, and has been the subject of observation and investigation for more than a quarter of a century. 1929 *Trans. Brit. Mycol. Soc.* XIV. 163 Silver leaf, *Stereum purpureum* Pers. 1946 H. WORMALD *Diseases of Fruit & Hops* iii. 57 The Silver Leaf Order of 1923 requires growers to cut off and burn all dead wood of plum and apple trees before 15th of July each year. 1969 P. THROWER *Every Day Gardening* xiii. 292/2 (*caption*) Branches and even complete trees can be killed by Silver Leaf disease, and fungal outgrowths form on the dead wood. First, however, the leaves take on a silvery sheen. 1977 *Field* 13 Jan. 66/1 Pruning [of plum trees] should be carried out in late spring..and preferably in dry weather. This is to avoid infection by the silver leaf fungus.

silver paper. Add: **1.** (Earlier example.)
1800 M. EDGEWORTH *Birth-Day Present* in *Parent's Assistant* (ed. 3) II. 14 She was obliged to go down with her basket but half wrapped up in silver paper.
2. Also, thin metal foil, used chiefly as a damp-proof wrapping for tobacco and confectionery.
1905 *Strand Mag.* XXIX. 274/1 He has been sorting out the pieces of 'silver paper', as he calls them, in which packets of tobacco are wrapped. 1929 *B.B.C. Year-bk.* 1930 404 The balance of the subscriptions..is paid into the local Radio Circle Funds, which are further increased in various ways such as by the sale of 'silver paper'. 1976 W. TREVOR *Children of Dynmouth* iii. 65 The one he'd taken had green silver paper on it, a chocolate-covered toffee.

silver plate. Add: **3.** Used as a jocular representation of Fr. *s'il vous plaît* please. *slang.*
1919 *Yank Talk* 4/1 (*caption*) Silver plate! Loan me a coupla francs! 1920 *Dialect Notes* V. 79 *Silver plate*, s'il vous plaît. 'More of the mutton, Mr. Brown, silver plate.'

silversmith. Add: Hence **si·lversmi:thing.**
1931 E. WENHAM *Domestic Silver* ii. 8 No period in the history of British silversmithing manifests more varying foreign influences than that of the sixteenth century. 1969 T. LLOYD in R. Blythe *Akenfield* xiv. 222 There is something else I do—silversmithing. I learnt it at evening classes. 1981 *Times Lit. Suppl.* 20 Feb. 194/5 Ashbee'n emerges as a many-sided creativity, embracing architecture, silversmithing and printing.

silvery, *a.* Add: **2. a. silvery pout,** a small marine fish, *Gadiculus argenteus thori*, belonging to the cod family and found in north-western Europe and the Mediterranean.
1925 J. T. JENKINS *Fishes Brit. Isles* 155 The Silvery Pout is not often met with close inshore. 1959 A. C. HARDY *Open Sea* II. ii. 229 The little silvery pout..is an even more deep-water species.
5. b. *silvery-tongued* (example).
1885 E. W. HAMILTON *Diary* 29 Jan. (1972) II. 783 The German Ambassador whom Bismarck suspects of being too silvery-tongued over here.

silvex (si·lveks). [f. L. *silv-a* wood, woodland + Eng. *-ex.*] A hormone weedkiller that is also effective against some woody plants; 2-(2,4,5-trichlorophenoxy)propionic acid.
1954 *Proc. Southern Weed Conf.* VII. 268 Silver was active in causing cell elongation and adventitious root formation but not active in causing formative effects. 1973 ASHTON & CRAFTS *Mode of Action of Herbicides* ix. 131 These growth regulators enhance the herbicidal effect of silvex on poison ivy. 1976 *Columbus* (Montana) *News* 3 June (Joliet Suppl.) 4/5 Dandelions are more difficult to control after they flower. Spraying with 2, 4-D, silvex or dicamba (Banvel) is recommended... Silvex also gives excellent control of chickweed.

silvi- (si·lvi), comb. form of L. *silva* wood, woodland. Cf. also SYLVICULTURE, SILVI-.

silvical (si·lvikăl), *a.* [f. as SYLVICS + -ICAL.] Of or pertaining to silvics.
1909 in WEBSTER. 1919 *Jrnl. Forestry* XVII. 276 The Commission of Conservation of Canada has during several seasons conducted silvical investigations... Studies of natural regeneration, with special reference to the effects of repeated fires, have been carried out. 1931 *Ecology* XII. 568 Müller was probably the first to look upon the humus layer in the forest as a natural biological unit, and..he was able to characterize two main types of humus layer and their biological and silvical properties. 1977 *Jrnl. Arnold Arboretum Harvard Univ.* LVIII. 307 (*heading*) Silvical characteristics of sugar maple..in northern Cape Breton Island.

silvichemical (silvike·mikăl). [f. *SILVI- + CHEMICAL *a.*] Any chemical obtained from part of a tree.
1963 L. C. BRATT in *Abstr. Papers 144th Meeting Amer. Chem. Soc.* 11D The name 'silvichemicals' is used in this paper to define chemicals and special products made from tree components, primarily wood, bark, and oleoresins as well as from pulp mill by-products. 1965 *Jrnl. Forestry* LXIII. 163/1 The silvichemicals may be divided into the two broad classifications of complex polymers or mixtures and pure organic chemicals. 1974 *Finnish Chem. Lett.* VII. 262 There are clear opportunities for the creation of a profitable enterprise producing silvichemicals from technical foliage.

silvicide (si·lvisəid). [f. *SILVI- + -CIDE[1].] A substance that kills trees.
1950 in *Forestry Terminol.* (Soc. Amer. Foresters) (ed. 2) 75/1. 1960 *Jrnl. Forestry* LVIII. 403/2 Thinning white pine stands with silvicides might be expected to result in serious backflash damage. 1976 *Amer. Industr. Hygiene Assoc. Jrnl.* XXXVII. 418/1 The increasing use of organic arsenicals..as silvicides in forestry has raised questions concerning the health and safety of exposed workers.

silvics: see SYLVICS in Dict. and Suppl.

silyl (sɔi·ləil, -lil). *Chem.* [f. *SIL(ANE + -YL.] The univalent group or radical $—SiH_3$; any substituted derivative of this, esp. one in which alkyl groups replace the hydrogen atoms. Usu. *attrib.*
1916 *Jrnl. Chem. Soc.* CX. ii. 319 The radicles $·SiH_3$, $·Si_2H_5$, and $·Si_2H_3$ would be respectively designated silyl, disilyl, and disilenyl. 1939 *Ibid.* 1030 Some of the other possible reactions of the silyl radicals are more favoured than that with ethylene. 1970 *Nature* 25 July 335/2 The preparation of silylated ylides of phosphorus, arsenic and sulphur, in which the silyl group is both a stabilizing and an efficient leaving group.
Hence **silyla·tion,** a reaction or process in which a substance is converted into a form having silyl substituents; also **si·lylate** *v. trans.*, to subject to silylation; **si·lylated** *ppl. a.*; **si·lylating** *vbl. sb.*
1938 *Nature* 3 Dec. 997/1 Trimethylsilylammonium chloride is a convenient silylating agent. 1949 *Chem. Abstr. Subject Index 1937–1946* 8856/2 Silylation. 1966 *Jrnl. Amer. Chem. Soc.* LXXXVIII. 3390/1 Bis(trimethylsilyl)acetamide..is a silylating agent superior, in many respects, to the presently used methods. *Ibid.* 3390/2 Silylations of 'good' acceptors—alcohols, amines, carboxylic acids—can be carried out with monosilylamides since the rapidly established equilibria lie far on the product side. *Ibid.* 3391/1 The vapor pressure of the silylated product. 1969 *Kirk-Othmer Encycl. Chem. Technol.* (ed. 2) XVIII. 262 A method of silylating biologically active compounds. 1972 *Science* 12 May 683/3 Silylation was effected in pyridine with a mixture of hexamethyldisilazane and trimethylchlorosilane. 1978 *Experientia* XXXIV. 1380/1 The material..was silylated with 200µl of trimethylsilylimidazole at 65°C for 12–15h.

sima (sɔi·mă). *Geol.* Also **Sima.** [a. G. *sima* (E. Suess *Das Antlitz der Erde* (1909) III. ii. xxiv. 626), f. L. *si-licium* SILICON + *ma-gnesium* MAGNESIUM.] The continuous basal layer of the earth's crust, composed of relatively heavy, basic rocks rich in silica and magnesia, that underlies the sialic continental masses and forms the crust under the oceans; the material of which it is composed.
The lower limit of the sima is generally taken to be the Mohorovičić discontinuity.
1909 [see *NIFE]. 1925 *Glasgow Herald* 29 Sept. 9 In continental regions it [sc. sial] both rises higher to form the land surface, and extends downwards into or displaces somewhat the sima, just as a ship floats on and displaces the water. 1944 A. HOLMES *Princ. Physical Geol.* iv. 41 Certain rocks of the kinds grouped together as sima (e.g. basalt) contain calcic plagioclase, but others are free from felspar. All of the sima rocks, however, are characterized by the abundance or predominance of heavy, greenish silicate minerals. 1950 P. H. KUENEN *Marine Geol.* ii. 127 If America were plowing through the sima westwards.. one would expect a raising of the sima in front of the continent. 1970 L. KNOPOFF in Johnson & Smith *Megatectonics of Continents & Oceans* vi. 120 It seemed plausible to assert that the top of the mantle was a chemically homogeneous material. In many of the older geology texts, this material was simply called sima to describe a dense, basic rock from which the less dense, more acidic sialic crust could be derived by some process of differentiation.

Hence **sima·tic** *a.*, of or pertaining to the basal crust or the material of which it is made.
1942 R. A. DALY *Floor of Ocean* ii. 59 The oceanic sectors of the earth are characteristically simatic... Also simatic is any vitreous basalt which may form 'pockets' in the crust. 1955 [see *SIALIC *a.*[1]]. 1971 *Proc. 2nd Symposium Upper Mantle Project* (Council of Sci. & Industr. Res., New Delhi) p. xix, Resting on the simatic crust, the continental blocks move, as if riding on the conveyor belt.

simandro, simantron, varr. *SEMANTRON.

simazine (si·m-, sɔi·măzīn). Formerly also -in (-in). Also with capital initial. [f. *sim-* (ad. SYM(METRIC *a.*) + TRI)AZINE.] A colourless crystalline compound, 2-chloro-4,6-bis-(ethylamino)-1,3,5-triazine, $C_7H_{12}N_5Cl$, which is a selective weedkiller applied as an emulsion or wettable powder.
1956 *Proc. 13th Meeting N. Central Weed Control Conf.* 57/1 During 1956, an experimental herbicide..designated as Simazin, was released to experiment stations for test purposes. 1957 *Chem. Abstr.* LI. 9995 The herbicides CMU..and Simazine..are used..to prevent weed growth in grafting vineyards after the spring weeding. 1958 *Times* 10 Nov. 19/2 Another chemical on which interest is focused is Simazin..which kills a wide range of weeds. 1971 *Ideal Home* Apr. 120 Keep it weed free with an annual application of a simazine weedkiller like Weedex before weeds appear. 1978 *Financial Times* 2 Dec. 13/3 Rose beds kept clear of weeds by small annual applications of simazine.

simba (si·mbă). Also **Simba.** [Swahili.] A lion. Also *fig.*, a warrior; a leader.
1918 E. R. BURROUGHS *Tarzan & Jewels* iii. 30 The strange white man must certainly succumb to terrible Simba. 1935 E. HEMINGWAY *Green Hills of Africa* ii. 42 A stream of..words in Wakamba ending in the word 'Simba'. 1966 *Transition* (Uganda) XXVI. 39/1 In battle, the simbas had to walk straight forward, and keep their eyes to the front to avoid loss of invulnerability. 1975 T. DINESEN *My Sister, Isak Dinesen* v. 56 A large simba had been lying at the river-crossing and hadn't moved as he went by, only stared menacingly at him. 1976 *Drum* (E. Afr. ed.) Sept. 3/1 Has Amin now recognised his weakness? What does he now think and say? Does he still claim to be the *simba* of Africa?

simchah (si·mχă). Also **simcha, Simchah.** [a. Heb. *śimḥā* rejoicing.] A Jewish private party or celebration.
1932 L. GOLDING *Magnolia Street* ii. ii. 295 Perhaps, after all, it'll be a good match. There will be a *simchah*. 1959 H. PINTER *Birthday Party* ii. 35 Mazoltov! And may we only meet at Simchahs! 1973 *Jewish Chron.* 19 Jan. 42/2 (Advt.), Arkay caterers. Specialists in home, hall and marquee catering for all simchas.

Simchat Torah (si·mχăt tōᵊ·ra). Also **Simchas, Simchath, Simhat** and hyphened. [a. Heb. *śimḥat tōrā*, f. *śimḥat*, construct case of *śimḥā* *SIMCHAH, + TORAH.] The final day of the festival of Succoth, on which the annual cycle of the reading of the Torah reaches its completion.
1891 M. FRIEDLÄNDER *Jewish Relig.* ii. 480 Twice a year we have special occasion for fulfilment of this duty, viz., on Simchath-torah and on the Seder-evening. 1905 *Jewish Encycl.* XI. 365/1 The name 'Simhat Torah' came into use after the introduction of the one-year cycle for the reading of the Law. 1907 OESTERLEY & BOX *Relig. & Worship Synagogue* iii. xx. 374 In some places it has..been customary for the children to tear down the 'booths' (*sukkoth*), and burn them on *Simchath Torah*. 1927 A. FELDMAN *Sabbath Spice & Festival Fare* 30 On Simchas Torah..there is a 'procession' of the Scrolls. 1960 *Jewish Chron.* 8 Apr. 35/3 To dance on *Simchat Torah*, Is there a greater pleasure? 1973 *Synagogue Light* Sept. 56 (*caption*) Israeli Hassidim celebrating Simhat Torah at the Western Wall. 1975 C. POTOK *In Beginning* iv. 213, I thought of the way my father..had danced with the Torah on Simchat Torah. 1978 I. B. SINGER *Shosha* v. 98 Only on Simchas Torah were girls allowed inside a house of worship.

Simeonite. Add: Hence **Si·meonism,** adherence to the doctrines of Simeon.
a 1902 S. BUTLER *Way of All Flesh* (1903) xlvii. 213 These poor fellows formed a class apart..and it was among them that Simeonism chiefly flourished.

simetite (si·m-, sī·mĕtəit). *Min.* Also **Simetite.** [a. It. *simetite* (O. Helm 1887, in *Malpighia* I. 54), f. *Simeto*, name of a river in Sicily: see -ITE[1].] A variety of amber, usu. reddish, found in Sicily.
1892 E. S. DANA *Dana's Syst. Mineral.* (ed. 6) 1005 Simetite... A resin near amber from near Mt. Etna, Sicily. Remarkable for its deep red color and often showing a beautiful fluorescence. 1915 R. LANKESTER *Diversions of Naturalist* ix. 71 The Sicilian amber (called 'Simetite') was not known to the ancients. 1932 G. C. WILLIAMSON *Bk. of Amber* 209 Some of the very clearest pieces of golden Simetite possess bloom and flashes of blue. 1969 *Beaver* Summer 29/2 Reddish amber from Sicily, called simetite, is also a succinite.

Simhat Torah, var. *SIMCHAT TORAH.

simi (si·mi). [ad. Swahili *sime.*] In East Africa: a large knife; a short two-edged sword (see also quot. 1980).
1955 *Times* 7 June 6/6 Small boys in feathers, with cowbells at their ankles and brandishing tiny *simis* (swords), stamp and dance in circles. 1961 *Encounter* Jan. 24/2 He had killed two of the Masai morans..cutting the throat of the other with his simi. 1977 D. BEATY *Excellency* ix. 108 An African was..holding up a thin-bladed simi knife. 1980 *Times* 3 Apr. 8/1 Mrs Joy Adamson, the naturalist and authoress of *Born Free*, was murdered with a simi (a two-edged farming implement like a sword), and an iron bar, a police witness said.

similar, *a.* Add:
3. c. *Math.* Of two square matrices: such that one of them is equal to the other premultiplied by some matrix whose determinant is not zero and postmultiplied by the inverse of the same matrix.
1907 M. BÔCHER *Introd. Higher Algebra* xxi. 283 Two matrices connected by a relation of the form (13) are sometimes called similar matrices. 1937 A. A. ALBERT *Mod. Higher Algebra* iv. 78 Every square matrix is similar to its transpose. 1979 S. H. FRIEDBERG et al. *Linear Algebra* v. 232 Prove that similar matrices have the same trace.

similarity. Add: **3.** *attrib.*, as *similarity continuum, set*; (cf. *SIMILAR *a.* 3 c) *similarity class, group, transformation.*
1952 R. R. STOLL *Linear Algebra & Matrix Theory* vii. 176 The similarity classes of transformations are in one-one correspondence with the similarity classes of $n \times n$ matrices. 1975 S. KOH et al. tr. *Satake's Linear Algebra* iii. 141 The set of all matrices which are similar to a given matrix is called the 'similarity class' of this matrix. 1960 M. ROKEACH *Open & Closed Mind* ii. 46 Recall again that we conceive of the disbelief system as a similarity continuum. 1937 A. A. ALBERT *Mod. Higher Algebra* iv. 76 We are studying the invariants of matrices A under a group of transformations $A \leftrightarrow PAP^{-1}$ called the similarity group. 1977 R. HOLLAND *Self & Social Context* viii. 232 Objects normally grouped together in similarity sets may change their relationships following a revolution in science. 1961 G. HADLEY *Linear Algebra* vii. 239 Either PAP^{-1} or $P^{-1}AP$ represents a similarity transformation on A. 1968 FOX & MAYERS *Computing Methods for Scientists & Engineers* v. 113 A final important iterative method..succeeds in reducing the matrix by similarity transformation to triangular form. 1976 G. STRANG *Linear Algebra & its Applications* v. 222 Similarity transformations leave the eigenvalues unchanged.

similative (si·milātiv), *a.* and *sb.* *Gram.* [f. L. *similis* like + -ATIVE.] **A.** *adj.* Denoting or expressing similarity or likeness. **B.** *sb.* A similative word, case, verbal element, or compound. Cf. SIMILITIVE *a.*
1884 in N.E.D. s.v. AIR *sb.* B. 3. 1903 *Amer. Anthropologist* Jan.–Mar. 13 Besides these, comitatives, similatives, partitives, and suffixes expressing similar ideas, are found. 1911 H. BRADLEY in *Encycl. Brit.* XXV. 209/1 The many jocularly similative uses of ordinary words, such as 'tin' for money. 1930 F. R. BLAKE in J. T. Hatfield et al. *Curme Vol. Ling. Stud.* 37 The immaterial adnominal cases are..similative—an animal *like a pig*. 1954 PEI & GAYNOR *Dict. Linguistics* 197 *Similative*, a declensional case in certain non-Indo-European languages, denoting resemblance. (Also termed *conformative*.)

simile, *v.* (Later example.)
1972 G. JONES *Kings, Beasts, & Heroes* II. i. 75 We are told the colour of her hair and hands, her flesh and bosom, but she stays cool to view as..a wax doll. A clean doll, admittedly... And one most nobly similied.

similize, *v.* **3. b.** (Later examples.)
1925 V. WOOLF *Common Reader* 106 She similised..eternally; the sea became a meadow, the sailors shepherds, the mast a maypole. 1976 *N.Y. Times Mag.* 10 Oct. 111/3 Have a story or anecdote for every point you wish to make. Similize. Exaggerate, euphemize, elide.

simillimum (simi·limɒm). *Homœopathy.* [L., neut. of *simillimus*, superl. of *similis*, like, similar.] The remedy indicated in a particular case, as producing in a healthy person symptoms most like those of the person to be treated.
1849 R. E. DUDGEON tr. *Hahnemann's Organon Med.* 157 But even granting this could be done..the cure is effected only by opposing a *simillimum* to a *simillimum*. 1891 J. C. BURNETT *Greater Dis. Liver* 42 With me it is an axiom to relieve uncomfortable or dangerous organ-states with simple organ-remedies, leaving the more remote and deeper-going to be..treated, if possible, with its pathological simillimum. 1938 [see *POTENTIZATION]. 1972 D. V. TANSLEY *Radionics* viii. 84 This may be dispersed radionically or with the appropriate homoeopathic similimum [*sic*].

simlin. Add: Also **cimbeline, cymblin(g), cymlin. 1.** (Earlier and further examples.)
1775 N. CRESSWELL *Jrnl.* 5 July (1925) 95 The rest plundered about the plantation and got some young cabbages, squashes and Cimbelines. 1785 T. JEFFERSON *Notes Virginia* vi. 68 Cymlings. *Cucurbita verrucosa.* 1796 B. HAWKINS *Let.* 2 Dec. (1916) 21 They made beans, ground peas, cymblins. 1832 J. P. KENNEDY *Swallow Barn* IV. vi. 100 Little garden-patches..where cymblings ..flourished. *a* 1883 G. W. BAGBY *Sel. Misc. Writings* (1885) II. 17 A true Virginian..must have..old hare, butter-beans, new potatoes, squirrel, cymlings, snaps. 1981 *Farmstead Mag.* Winter 41/1 Common pumpkins are actually a form of the same plant from which has also been developed vegetable marrows, cymlings, or cymlins (also spelled simlins), summer crookneck squashes, and yellow-flowered gourds.

Simmental (si·məntāl). Also **Simmenthal, Zimmenthal.** [a. Ger. name of the Simme valley in the canton of Berne, Switzerland.] A bull or cow of the breed of cattle so called, first developed in Switzerland, distinguished by their large size and red and white coats, and used for both milk and meat production. Also *attrib.*
1959 R. B. KELLEY *Native & Adapted Cattle* iv. 62 Red Danes, Simmental, and Friesians..were seen..outside Rome. 1970 *Times* 6 Apr. 10/6 The shorthorn men are also interested in the Simmental, a big dual-purpose animal which is to be found over much of central and eastern Europe. 1973 [see *LIMOUSIN 2]. 1973 *Country Life* 12 July 80/2 Of the exotic breeds, the Simmenthal are second in popularity only to the Charolais. 1980 H. M. & D. M. BRIGGS *Mod. Breeds Livestock* (ed. 4) vii. 156 The first purebred Simmental bull came to the United States in 1971. *Ibid.* 157 Simmentals are large, long, and very muscular cattle.

simmer, *v.*[1] Add: **1. d.** *to simmer down*: to calm down from an angry or excited state. orig. *U.S.*
1871 'MARK TWAIN' *Lett. to Publishers* (1967) 58, I must and will keep shady and quiet till Bret Harte simmers down a little. 1897 W. BEATTY *Secretar* xiii. 102 In a while..he simmered down. 1902 C. J. C. HYNE *Mr. Horrocks, Purser* 42 First Class passengers..don't handicap matters by interference—once they have simmered down. 1972 *Times Lit. Suppl.* 22 Dec. 1561/5 There they simmer down for a space, forget their mundane cares.

Simmerstat (si·məstæt). Also **simmerstat.** [f. SIMMER *v.*[1] + THERMO)STAT.] The proprietary name of a thermostatic control which regulates the temperature of the hotplates or grill of an electric cooker or similar heating appliance.
1938 *Trade Marks Jrnl.* 7 Sept. 1097/1 *Simmerstat*... All goods included in Class 8. Sun-Vic Controls Limited.., London W.C. 2; manufacturers. 1951 *Good Housek. Home Encycl.* 76/1 A radiant hot-plate operated by a 'simmerstat'. 1954 *Archit. Rev.* CXVI. 270 (*caption*) Polished stainless steel and plate glass; fitted with three 60-watt lamps between the heaters and simmerstat control. 1962 *Listener* 13 Sept. 411/2 The middle-priced model is distinguished by simmerstat-controlled radiant plates. 1979 *Nature* 19 Apr. p. xiv/3 The temperature of the hotplate is controlled by a Simmerstat which regulates it between 50° and 325°C.

'simmon (si·mɒn), *sb.*[3] *U.S.* Colloq. abbrev. of PERSIMMON. Freq. *attrib.*, esp. as *'simmon beer.*
a 1775 J. BOUCHER *Gloss. Archaic & Provincial Words* (1832) p. l, Brown linen shirts, and cotton jackets wear, Or only wring-jaw drink, and 'simmon beer. 1839 *Southern Lit. Messenger* V. 378/2, I ask you no odds—the longest pole, you know, takes the simmon. 1883 P. M. HALE *Woods & Timbers N. Carolina* 117 The basis of a beverage, by no means despicable, called 'Simmon Beer. 1909 'O. HENRY' *Roads of Destiny* xxi. 350 That's why you see me cake-walking with the ex-rebs to the illegitimate tune about 'simmon-seeds and cotton. 1945 B. A. BOTKIN *Lay my Burden Down* 66 'Simmon beer was good in the cold freezing weather too. 1949 [see *possum hunt* s.v. *POSSUM *sb.*[1] 1 d].

Simmonds (si·məndz). *Path.* [The name of Morris Simmonds (1855–1925), German pathologist, who described the disease in 1914 (*Deutsch. med. Wochenschr.* 12 Feb. 322).] *Simmonds'* (also *Simmonds's*, ¶ *Simmond's*) *disease*: pituitary insufficiency caused by destruction of the gland, characterized by weakness, loss of body hair, and progressive gonadal, adrenocortical, and thyroidal failure, and occurring chiefly in women; esp. the chronic form (cf. **Sheehan's syndrome*).
1928 *Q. Cumulative Index Medicus* II. 993/2 Pituitary cachexia (Simmonds' disease). 1937 *Jrnl. Path. & Bacteriol.* XLV. 189 The significance of necrosis of the anterior pituitary chiefly concerns its relationship to the series of disorders of pituitary function which culminate in Simmonds's disease. 1970 PASSMORE & ROBSON *Compan. Med. Stud.* II. xxv. 6/2 The necrosis of anterior pituitary cells is followed by pituitary hypofunction (Simmond's disease).

simo (sɔi·mo). [Abbrev. of *simultaneous motion-cycle.*] *simo chart*: a chart in which the bodily movements of a worker are represented in relation to a time scale. Hence *simo-charting* vbl. sb.
1928 L. M. GILBRETH in *Bull. Taylor Soc.* June 127/2 The simo charts and the synthesis work demand both time and effort. 1937 R. M. BARNES *Motion & Time Study* ix. 77 The time for each therblig recorded on the analysis sheet may be shown to scale by means of a simultaneous motion-cycle chart, commonly called a 'simo chart'. 1961 *Engineering* 15 Sept. 353/1 Fig. 2 shows the simo-charts produced by a member of a recent course; they promise particularly high savings and provide a very good example of simo-charting. 1969 P. E. RANDALL *Introd. Work Study & Organization & Methods* iv. 27 A simo chart breaks hand movements down into eighteen categories.

simoleon (simōᵘ·liɒn). *U.S. slang.* Also **samoleon.** [Origin obscure: perh. modelled on NAPOLEON 1.] A dollar.
1896 G. ADE *Artie* vii. 63 He said I could have it for four hundred samoleons. 1913 C. E. MULFORD *Coming of Cassidy* vii. 112 Sixty-two bucks, three score an' two simoleons; all I've got, every cent. 1952 A. LOMAX *Mister Jelly Roll* 194 By the late twenties the golden simoleons began rolling in. 1977 D. ANTHONY *Stud Game* i. 8, I bet the limit, five thousand simoleons.

Simonite (sɔi·mŏnəit). [f. the name of Sir John Allsebrook Simon (1873–1954), Liberal politician + -ITE[1].] A supporter of Sir John Simon; used *spec.* to designate a member of the Liberal National Party which seceded in 1931 from the official Liberal Party led by Sir Herbert Samuel. Freq. *attrib.* Cf. *SAMUELITE.
1931 [see *SAMUELITE]. 1932 [see *MOSLEYITE]. 1957 J. BOWLE *Viscount Samuel* xvii. 287 The Liberals had to accept a measure of Protection... But the rift with the Simonites continued. 1966 T. WILSON *Downfall of Liberal Party 1914–1935* xix. 352 The dismissed employee was to appear in Parliament in 1931 as a Simonite (anti-Lloyd George) M.P. 1977 D. MARQUAND *Ramsay Macdonald* xxvii. 677 On the fiscal question..he sided with the Conservatives and Simonite Liberals against Snowden and the Samuelite Liberals. 1979 G. POTTINGER *Secretaries of State for Scotland 1926–76* vi. 54 It was natural for him to take the Simonites' side when the split with the Samuel faction came in 1931.

simonize (sɔi·mŏnəiz), *v.* Also **Simonize.** [f. *Simoniz* proprietary name of a type of car polish.] *trans.* To polish by the application of Simoniz. Also *transf.* and *fig.* So **si·monizing** *vbl. sb.*; hence **si·monized** *ppl. a.*
1934 *Amer. Speech* IX. 114/1 The work on the car may include..vulcanizing tire cuts, and simonizing. 1942 BERREY & VAN DEN BARK *Amer. Thes. Slang* § 125/6 *Simonize*, to brilliantine the hair. *Ibid.* § 291/4 Curry favour; toady... *simonize the apple* or *orange*. 1949 A. MILLER *Death of Salesman* I. 17 Remember those days? The way Biff used to simonize that car? 1953 *Economist* 25 July 258/1, I have heard it said that Harvard University humanises the scientist and simonises the humanist. 1968 R. H. R. SMITHIES *Shoplifter* (1969) ix. 189 The lovingly simonized fenders, hubcaps, and bumpers. 1975 *New Yorker* 19 May 116/2 He writes of someone's being 'Nixonized', which is apparently like having a car Simonized. 1977 *Chicago Tribune* 2 Oct. XII. 73/3 (Advt.), Will do Blue Coral waxing, Simonizing & interiors.

Simon Pure. Add: Also **Simon-Pure. a.** (Earlier example.)
[1785 'P. PINDAR' *Lyric Odes* ix. 28 Flattery's a Mountebank so spruce—gets riches; Truth, a plain Simon Pure, a Quaker Preacher.] 1795 T. WILKINSON *Wandering Patentee* III. 34 She in her rage denounced vengeance heartily, and said that she would advertise my production as a flimsy disgraceful imposture: But she was mistaken—for mine was the true *Simon Pure*.
b. (Earlier and later examples.) Also, pure, unadulterated; honest, upright.
1869 'MARK TWAIN' *Innoc. Abr.* xli. 436 Soon the bell—a genuine, simon-pure bell—rang. 1913 *Jrnl. Industr. & Engin. Chem.* June 504/1 Willard Gibbs..was simon pure—in his best days he did not soil his hands even with the grime of a laboratory. 1945 R. HARGREAVES *Enemy at Gate* 282 The unsullied, Simon-Pure principles of proletarian ideology. 1951 *Manch. Guardian Weekly* 28 June 2/3 The [drug] peddlers have gone underground and the specified retailers have turned Simon-pure. 1974 *Jrnl. Ecumenical Stud.* Winter 119 The simon-pure neutrality of church leaders, paralyzed by their own myth of 'even-handedness', has embarrassed in a new way the relations between individual Christians and Jews.

simool. Add: Also **semal, semul.** For *malabaricum* substitute *ceiba.* (Earlier and later examples.)
1835 *Penny Cycl.* IV. 2/2 The Semul, or cotton-tree. 1889 *Dict. Econ. Prod. India* I. 489 The semul cotton supplied them was better known as kapok. 1902 T. W. WEBBER *Forests Upper India* xviii. 232 The semal or cotton tree. 1932 PEARSON & BROWN *Comm. Timbers India* 138 The primary use of semul is for tea and rubber boxes.

simoom, *sb.* Add: Also **samoon, samum, samun, semoun. a.** (Further examples.)
α. 1832 Semoum [see *SAMIEL]. 1947 M. A. GARBELL *Tropical & Equatorial Meteorol.* xiii. 200/2 Algerian and Syrian samum.
β. 1926 W. N. SHAW *Man. Metereol.* I. ii. 28 The principal of these are..the bora of the Adriatic, the scirocco of Southern Italy, the samun of Algeria which is also called scirocco. 1931 A. A. MILLER *Climatol.* xiv. 253 Similar winds to the foehn occur in all mountain districts, where cyclonic storms occur... The Chinook..is exactly similar, so are the Samun of Persia, descending from the mountains of Kurdistan,..and many others. 1968 G. R. RUMNEY *Climatol.* xiii. 254/1 The sirocco is known as the khamsin in Egypt, leveche in southeastern Spain, where it is usually quite dry, garbi in the Aegean, samoon in Algeria, sahat in Morocco, and ghibli in Libya.

simp (simp). Colloq. abbrev. of SIMPLE *sb.* 2 b or SIMPLETON 1: a fool, a simpleton.
1903 W. C. THOMPSON *On Road with Circus* i. 23 In circus dialect 'yap' and 'simp' indicate a credulous rustic who is easy prey for sharpers. 1924 WODEHOUSE *Bill the Conqueror* vi. 133 You poor simp, you've got about as much chance of havin' me sneak those books for you as—well, I don't know what. 1937 N. MARSH *Vintage Murder* xv. 165 You looked a big simp, Cass. 1973 *Sat. Rev. Society* (U.S.) Mar. 72/3 Wonder Woman almost (but never quite) loses her head and heart to a weak simp, a U.S. Army Intelligence pilot. 1976 *Publishers Weekly* 19 Apr. 82/3 The book's assumption is that single men are simps who don't know the difference between a pepper mill and a can opener.

simpatico (simpæ·tiko), *a.* Also (fem.) **simpatica.** [It. or Sp.: see SYMPATHIC *a.*] Pleasing, likeable; congenial, understanding; sensitive, sympathetic.
1864 H. SIDGWICK *Let.* 21 Oct. in *Memoir* (1906) ii. 119 The Frau Professorin was less 'simpatica'. 1905 E. M. FORSTER *Where Angels fear to Tread* iii. 86 The person who understands us at first sight, who never irritates us, who never bores..that is what I mean by simpatico. 1908 W. JAMES *Let.* 4 Oct. (1920) II. 314, I find him [sc. Boutroux] very simpatico. 1936 E. AMBLER *Dark Frontier* 80 She was infinitely—as those appreciative Italians put it—infinitely *simpatica*. 1952 A. HUXLEY *Let.* 20 May (1969) 644 There is something *simpatico* about Pascal—he is a kind of Central European Baron Munchausen. 1964 MRS. L. B. JOHNSON *White House Diary* 19 Mar. (1970) 96 Although he is very attached to the Kennedys, I thought we had established a certain simpatico relationship with him. 1976 *Observer* 22 Feb. 26/7 Sylvia reluctantly committed Ralph into the care of a blunt but simpatico medico who didn't talk down either to her or to us.

simple, *a.* and *sb.* Add: **A.** *adj.* **II. 6. b.** *simple life* (further examples); also *attrib.* Hence *simple-lifer*, a follower or proponent of the simple life. Also *simple-liver; simple-living* vbl. sb. and ppl. adj.
1736 J. THOMSON *Liberty* iv. 30 That simple Life, the quiet-whispering Grove. 1862 J. A. SYMONDS *Let.* 7 Aug. (1967) I. 355, I read through..Scotch Bothie, lured on by its intense savour of nature & love of simple life. 1909 H. G. WELLS *Ann Veronica* vii. 138 The Goopes were..following a fruitarian career..and they had reduced simple living to the finest of fine arts. *Ibid.* viii. 165 The chatter of the studios and the..discussions of the simple-life homes. 1927 W. E. COLLINSON *Contemp. Eng.* 38 During my school-days..I remember first hearing the term simple-lifers. 1933 M. ALLINGHAM *Sweet Danger* xvii. 211 If it weren't for the simple-livers on the heath..the affair would be almost plain sailing. 1956 'J. WYNDHAM' *Seeds of Time* 231 It's mostly Janet's economies and simple-living that's built up the savings. 1978 *Listener* 14 Sept. 324/1 Members may be 'arty-crafty' and 'simple-lifers'—rather like the sort of people who nowadays promote *Vole* magazine. 1979 *Ibid.* 11 Jan. 63/3 In Cornwall..simple-living James and earth-mother Anna are variously threatened by authority, rural squalor and true terror.
10. b. (*b*) *U.S. Rhyming slang.* A diamond.
1928 M. C. SHARPE *Chicago May* 287/2 *Simple simon*, diamond. 1929 D. RUNYON in *Cosmopolitan* July 58/1, I do not see any Simple Simon on your lean and linger.
III. 14. b. (*d*) † *simple tone* = *pure tone* s.v. *PURE *a.* 1 e. *Obs.*
1875 A. J. ELLIS tr. *H. L. F. von Helmholtz's Sensations of Tone* vii. 235 Whenever the vibrations of the air or of other elastic bodies which are set in motion at the same time by two generating simple tones, are so powerful that they can no longer be considered infinitely small, mathematical theory shows that vibrations of the air must arise which have the same vibrational numbers as the combination tones. 1878 *Proc. R. Soc. Edin.* IX. 602 According to a usage which has been adopted from the German of Helmholtz by the best English scientific writers on sound, a sound is called a 'simple tone', or without qualification a 'tone', when the variation of pressure of the air..is according to a simple harmonic function of the time. 1910 H. LAMB *Dynamical Theory of Sound* 3 The sensation corresponding to a simple-harmonic vibration is called a 'simple tone',..or merely a 'tone'.
d. *Math.* (*a*) Applied to a group that has no proper normal subgroup; and hence to an algebra or ring that has no proper ideal.
1888 G. G. MORRICE tr. *Klein's Lectures on Ikosahedron* I. i. 7 If a group contains, apart from these improper cases, no self-conjugate sub-groups, it is called *simple*, otherwise it is called *composite*. 1900 *Ann. Math.* I. 151 We have here..an example of two simple abstract groups of the same order which are not identical. 1939 H. WEYL *Classical Groups* iii. 85 A simple algebra..is one capable of a faithful irreducible representation. 1965 PATTERSON & RUTHERFORD *Elem. Abstract Algebra* iii. 100 The above theorem shows that a field is simple. There exist simple rings which are not fields. 1971 E. C. DADE in Powell & Higman *Finite Simple Groups* viii. 255 An algebra A is simple if $A \neq (0)$ and if A and (0) are the only two-sided ideals of A. 1980 *Sci. Amer.* May 68/1 The building blocks of group theory, analogous to the elementary particles of matter or the prime factors of integers, are called simple groups.
(*b*) Used variously (see quots.). (See also sense 15 b (*b*) in Dict.)
1889 *Proc. London Math. Soc.* XX. 70 It may happen that G_p consists of powers of one of its elements *a*, and has no other elements... In this case G_p is called a simple group. 1965 PATTERSON & RUTHERFORD *Elem. Abstract Algebra* iv. 135 Let $p(t) \in P(F)$ be an irreducible polynomial. Then the field $P_p(F)$ contains a sub-field isomorphic with *F*... We call the field $P_p(F)$ a simple algebraic extension of the field *F*. 1972 R. J. WILSON *Introd. Graph Theory* ii. 9 There can never be more than one edge joining a given pair of vertices of a simple graph. 1975 *Sci. Amer.* May 102/2 A simple polyhedron is one that is topologically equivalent to a sphere and whose faces are all simple polygons: polygons topologically equivalent to a disk.
e. *simple structure* (Statistics): a model in which numerous variables, showing various degrees of correlation, have their variances assigned to a smaller number of factors in such a way that no factor affects all of the variables.
1935 L. L. THURSTONE *Vectors of Mind* p. viii, One of the principal problems of factor analysis is to find a unique set of co-ordinate axes, either orthogonal or oblique, which shall represent scientifically meaningful categories in terms of which the tests may be comprehended. This problem has been solved in terms of what I have called 'simple structure' of a trait configuration. *Ibid.* vi. 154 If a set of *r* hyperplanes of dimensionality $(r-1)$ exists such that each trait vector is in one or more of the hyperplanes, then the combined configuration of the trait vectors and the reference vectors will be called a simple structure or an oblique simple structure. 1972 [see *ROTATION 1 d]. 1972 *Jrnl. Social Psychol.* LXXXVI. 188 There is no good scientific reason to expect agreement from analyses on..orthogonal pseudo-simple structure and oblique maximized simple structure.
15. b. (*e*) Applied to those vows which are taken by members of a religious order in the early stage of their profession and from which they may be dispensed; opp. to *solemn* (SOLEMN *a.* 5 a).
1759 A. BUTLER *Lives Saints* IV. 86/2 In some houses these Gray Sisters make solemn vows, but in most they content themselves with simple vows of poverty, obedience, and chastity. 1823 C. BUTLER *Continuation A. Butler's Lives Saints* 191 He entered into the society of Jesus, and made his simple vows. 1884 ADDIS & ARNOLD *Cath. Dict.* 848/1 Whereas a simple vow makes marriage unlawful and deprives the person who has made it of the right to use his property, a solemn vow makes marriage invalid and takes away all dominion over property. 1957 *Oxf. Dict. Christian Ch.* 1451/1 Since c. the 13th cent. canon law has also distinguished between 'simple' and 'solemn' vows. The exact scope of the distinction is disputed, but acc. to a common view the solemnity of vows is determined by their irrevocable acceptance.
(*f*) *Anthropol.*
1929 *N. & Q. Anthropol.* (ed. 5) 63 We might expect the simple patrilineal family to resemble our own, but this is not necessarily so. 1951 *Ibid.* (ed. 6) 70 The elementary or simple family is a group consisting of a father and a mother and their children, whether they are living together or not. 1977 P. LASLETT *Family Life* i. 13 The shape and membership of the familial group. In the West this has been confined for the most part to the parents and children themselves, what is called the nuclear family form or simple family household.
c. In adj. phrs., as *simple-to-follow, simple-to-operate*, etc.
1960 *Times* 18 Jan. 15/7 A simple-to-fit filter removes dust. 1960 *Farmer & Stockbreeder* 15 Mar. (Suppl.) 13 A robust motor with a completely reliable and simple-to-operate starting unit. 1971 *Woman's Own* 27 Mar. 52/1 We have combined simple-to-build shelves with..sliding transparent drawers. 1976 *BSI News* May 11/2 The BS 4264 range is set out in a simple-to-follow table giving 26 sizes.
IV. 18. *simple-natured, -sounding* adjs.
1875 'MARK TWAIN' in *Atlantic Monthly* Aug. 195/1 Good-hearted, simple-natured young Yates. 1930 E. BLUNDEN *De Bello Germanico* 16 The simple-sounding matter of pushing a truck along a French tramway is rather complex on a dark..night.

simple-minded, *a.* Add: Hence **simple-mi·ndedly** *adv.*
1934 A. HUXLEY *Beyond Mexique Bay* 255 An Indian who has given up his Fiestas would not be the simple-mindedly happy peasant beloved of Mr. Chase. 1981 *Christian Order* XXII. 264 Abandoned simple-mindedly and irresponsibly.

simplex, *a.* and *sb.* Add: **A.** *adj.* **2.** *Telegr.* and *Teleph.* Designating a system in which signals can be sent along a line in only one direction at a time. Also, in *Computers*, applied to a circuit along which commands can flow in only one direction, usu. from the central processor to a peripheral. Also *absol.*
1891 C. LANGDON-DAVIES *Explanation of Phonopore* vii. 28/1 The ordinary simplex telegraph is the one most in general use (except, possibly, in Great Britain). 1929 *Amer. Speech* IV. 290 One type of printer, which records the message on tape after the manner of a stock ticker, is known as the 'simplex'. 1967 DAVIDSON & KOENIG *Computers* xv. 525 These data communication lines are generally simplex, that is, they carry data in one direction only. 1975 *Sci. Amer.* Jan. 55 (Advt.), We have a range of modems, from 50 to 2400 bits per second, both for simplex and duplex traffic. 1977 J. R. L. ANDERSON *Death in City* xi. 170 The Radio Officer handed me a telephone... 'It's simplex, remember..you have to take it in turns to speak and listen.'
3. *Biol.* Of an eye: having pigment on the posterior surface of the iris only, not on the anterior surface, and so appearing blue.
1908 C. C. HURST in *Proc. R. Soc.* B. LXXX. 86 The eyes in which the posterior pigment alone is present in the iris, the anterior pigment being absent. Such eyes may be called simplex. 1946 [see *DUPLEX *a.* 1 d].
4. *Genetics.* Of a polyploid individual: having the dominant allele at any particular locus represented once.
1921 [see *NULLIPLEX *a.*]. 1931 *Genetics* XVI. 178 When *R* is simplex (*Rrr*) the expected gametic ratio is 1*R*:2*Rr*:1*rr*:2*r*. *Ibid.* 183 A simplex plant (*Rrr*). 1932, 1963 [see *QUADRUPLEX *a.*].
B. *sb.* **1. b.** *Linguistics.* In transformational grammar, a sentence analysed as having a single kernel structure. Cf. *KERNEL *sb.*[1] 8 b (ii).
1960 R. B. LEES *Gram. Eng. Nominalizations* iii. 101 Within the *matrix*-sentence, or any other simplex, the subject governs *-self*. 1963 *Language* XXXIX. 20 Since the second occurrence of the noun *John* repeats a noun within the same simplex, it is pronominalized to the corresponding *-self* pronoun. 1965 *Ibid.* XLI. 269 These differences in the treatment of repeated and nonrepeated material are evident in both simplexes (single-kernel sentences) and complexes (multikernel sentences).
2. *Geom.* The figure, in any given number of dimensions, that is bounded by the least possible number of hyperplanes: the two-dimensional simplex is the triangle, the three-dimensional simplex is the tetrahedron, and the four-dimensional simplex is bounded by five tetrahedra.
1914 H. P. MANNING *Geom. Four Dimensions* viii. 317 In the space of five dimensions there are only three possible types of regular (convex) figures: the simplex, corresponding to the tetrahedron and pentahedroid, the orthogonal, corresponding to the cube and hypercube, and the figure reciprocal to the latter, constructed on a set of mutually perpendicular diagonals and corresponding to the octahedron and the 16-hedroid. 1929 D. M. Y. SOMMERVILLE *Introd. Geom. Four Dimensions* vii. 96 The simplest polytope in S_n is the simplex $S(n+1)$, which is bounded by $n+1$ hyperplanes. 1975 *Sci. Amer.* May 102/2 The graph is isomorphic with the skeleton of a six-dimensional simplex, the 6-space analogue of the tetrahedron.
3. *Comb.*, as **simplex method,** a method of maximizing a linear function of several variables under several constraints on other linear functions; **simplex tableau,** a table displaying the constraints in problems of the type soluble by the simplex method.
[1951 G. B. DANTZIG in T. C. Koopmans *Activity Analysis of Production & Allocation* xxi. 339 The general nature of the 'simplex' approach (as the method discussed here is known).] 1951 R. DORFMAN in *Ibid.* xxii. 351 The simplex method makes use of the fact that any point in an *n*-dimensional space can be expressed as a sum of *n* linearly independent points. 1966 S. BEER *Decision & Control* viii. 149 There are also variants of the original set of rules for finding the answer (to which the name algorithm is applied); Dantzig's own algorithm is called the Simplex Method. 1980 A. J. JONES *Game Theory* iii. 155 The relationship with our earlier notation is simply that we have replaced A^T by *A* because this is the natural and universally accepted thing to do in setting up the simplex method. 1953 A. CHARNES in W. W. Cooper et al. *Introd. Linear Programming* II. vi. 66 As far as computations are concerned it is most convenient to arrange the data at each stage in a 'simplex tableau' as shown in Table I. 1966 A. BATTERSBY *Math. in Management* v. 125 A more usual form of presentation is the Simplex tableau in which all the variables have columns allocated to them.

simplex munditiis (si·mpleks mundi·ti,is). *Lat. phr.* [L., lit. 'simple in your adornments' (Horace *Odes* I. v. 5).] Unostentatiously beautiful; elegantly simple. Also used substantively.
1766 H. BROOKE *Fool of Quality* II. xii. 274 Even the *simplex Mundities* [sic], that ornament of a clean simplicity, recommended by Horace, can operate only by intimation of deeper purity. 1803 *Edin. Rev.* III. 8 Indeed, the *simplex munditiis* stamped every thing that he did. 1874 A. J. MUNBY *Diary* 4 May (1972) 367 Wearing a rich silk dress.., and simplex munditiis as to her beautiful hair. 1933 E. BLUNDEN *Charles Lamb* 42 The same grace, the *simplex munditiis*, haunted all through his sonnets. 1949 E. POUND *Pisan Cantos* lxxx. 84 To go far and come to an end Simplex munditiis, as the hair of Circe; perhaps without the munditiis.

simplicial (simpli·ʃəl), *a.* [f. L. *simplic-*, *simplex* single (see SIMPLEX *a.* and *sb.*) + -IAL.] Of, being, or pertaining to simplexes (sense *B. 2).
1926 [see *HOMŒOMORPH c]. 1959 E. M. PATTERSON *Topology* (ed. 2) v. 95 A geometrical simplicial complex *K* consists of a finite set of simplexes. 1976 *European Econ. Rev.* VIII. 305 The simplicial sub-division methods modelled by Scarf (1973) can be replaced by quicker Newton methods.

simplicist, *sb.* and *a.* Restrict † *Obs. rare* to sense in Dict. and add: **A.** *sb.* **2.** One who simplifies. *rare.*
1924 *Glasgow Herald* 24 Mar. 8 Can we ever simplify things again? Can we ever produce that statement of scientific method which will..give us an idea of how to conduct our conduct? I believe the day of the great simplicist is beginning to dawn again.
B. *adj.* That simplifies; characterized by simplicity, uncomplicated.
1934 in WEBSTER. 1949 *Sun* (Baltimore) 19 July 12/2 In the best manner of the simplicist versions of Keynes-

ism, Mr. Nathan insists that the recession through which we are moving is due to the heavy capital formation of the last two years. 1951 S. SPENDER *World within World* 63 He had a simplicist view of things which did me good. 1979 *Church Times* 5 Oct. 13/3 Even the somewhat simplicist approach is justified by the need and desire to make the argument intelligible.
Hence **simplici·stic** *a.*, characterized by over-simplicity.
1950 *Archivum Linguisticum* II. 142 Streitberg's and Buck's explanations..are both of this simplicistic..type. 1970 *Nature* 7 Nov. 589/2 Attempts to relate simplicistic finite set theory and automata to faculties of mind and the mechanisms of the brain inevitably leads [*sic*] to such difficulties.

‖ **simpliciter,** *adv.* Delete 'Chiefly in *Sc. Law.*' and add later examples.
1929 A. N. WHITEHEAD *Process & Reality* 69 The Aristotelian phrase suggests the crude notion that one actual entity is added to another *simpliciter*. 1936 C. S. LEWIS *Allegory of Love* v. 222 He [sc. Gower] says too much, not at this point or that, but too much *simpliciter*. 1963 *Times* 28 May 9/2 Substantial damages were not awarded..for physical injury *simpliciter*, but only for the pain and suffering and general loss of happiness which it occasioned. 1970 *N. & Q.* Dec. 453/1, I know of no evidence that 'shambles' *simpliciter* was ever used for 'brothel'. 1977 G. W. H. LAMPE *God as Spirit* vii. 178 It may suggest that the community which the Spirit creates is to be identified *simpliciter* with the institutional Church.

simplicity. Add: **1. c.** *spec.* in *Linguistics*, used *attrib.* with reference to the use of simplicity or economy as a criterion for evaluating a grammatical theory or description, as *simplicity criterion, metric*, etc.
1953 F. J. WHITFIELD tr. *Hjelmslev's Prolegom. Theory Lang.* 11 If..linguistic theory ends by constructing several possible methods of procedure.., that one shall be chosen that results in the simplest possible description. ..This principle, which is deduced from our so-called empirical principle, we call the *simplicity principle*. 1962 M. HALLE in *Word* XVIII. 55, I shall..exhibit the manner in which, by mechanical application of the proposed simplicity measure, certain formulations are chosen from among several alternatives. The plausibility and intrinsic appeal of the descriptions so selected will provide the primary justification not only for the proposed simplicity criterion, but also for the theory of generative grammar, of which the criterion is an integral part. 1968 *Glossa* II. 128 (*title*) Two proposals concerning the simplicity metric in phonology. 1976 J. S. GRUBER *Lexical Structures in Syntax & Semantics* II. ii. 330 The simplicity criterion will be useful in determining which of the two alternate forms of a definition is to be chosen for the lexical entry. 1977 *Canad. Jrnl. Linguistics* Spring 2 Such constraints..entail a fundamental reassessment of current formulations of the simplicity metric.

simplification. Add: **2.** *Logic.* One of the principles of inference used esp. in the calculus of propositions (see quot. 1903).
1903 B. RUSSELL *Princ. Math.* ii. 16 We can now state the six main principles of inference, to each of which ..a name is to be given... If *p* implies *p* and *q* implies *q*, then *pq* implies *p*. This is called *simplification*, and asserts merely that the joint assertion of two propositions implies the assertion of the first of the two. 1934 COHEN & NAGEL *Introd. Logic & Scientific Method* vi. 124 In order to get the calculus [of classes] started, we must state a number of fundamental principles... The following..are those which are usually assumed..Principle of simplification, [etc.]. 1954 I. M. COPI *Symbolic Logic* iii. 43 Simplification (Simp.): *p.q.·.p.* 1969 F. I. DRETSKE *Seeing & Knowing* ii. 57 The logical equivalence which goes under the name of simplification (P is logically equivalent to P and P).

si:mplifica·tory, *a.* [f. SIMPLIFICAT(ION + -ORY[2].] That simplifies.
1936 W. F. R. HARDIE *Study in Plato* xi. 141 Aristotle.. makes use of the misleading simplificatory language according to which the rational soul commands the irrational soul. 1972 *N. & Q.* Dec. 446/2 Simplificatory changes like the loss of unstressed syllables.

simplified, *ppl. a.* Add: *simplified spelling*: a system of writing English with greater phonetic consistency than in conventional spelling; any of various schemes for such a spelling reform. Hence *simplified speller* (nonce-word): an advocate or practitioner of simplified spelling.
1879 J. H. GLADSTONE *Spelling Reform* (ed. 2) p. v, Other advantages of a simplified spelling. 1899 'MARK TWAIN' *What is Man?* (1917) 256, I have had a kindly feeling toward Simplified Spelling from the beginning of the movement three years ago. *Ibid.* 262, I myself am a Simplified Speller. 1907 W. JAMES *Pragmatism* vii. 260 Now the idea of this loose universe affects your typical rationalists in much the same way..as 'simplified spelling' might affect an elderly schoolmistress. 1908 W. W. SKEAT *Presidential Address delivered at first Meeting of Simplified Spelling Society* 3 The object of the Simplified Spelling Society is to consider carefully the whole subject of our modern English spelling, with a view to the initiation of..a moderate system of reforms. 1955 AUDEN in *Encounter* Feb. 11 In my Eden our only source of political news is gossip: In his New Jerusalem there will be a special daily in simplified spelling for non-verbal types. 1977 K. M. E. MURRAY *Caught in Web of Words* vi. 102 In 1905..he [sc. James Murray] joined the American Simplified Spelling Board. 1979 *Guardian* 7 Aug. 3/1 September 30 has been nominated by the Simplified Spelling Society as an international day of Speling [*sic*] Watching.

simplify, *v.* Add: **2. b.** *intr.* To become (more) simple; to admit of simplification.
1955 G. GREENE *Quiet American* IV. ii. 230 When we are young we are a jungle of complications. We simplify as we get older. 1964 K. G. LOCKYER *Introd. Critical Path Analysis* ii. 21 Thus the diagram will simplify to the following [etc.].

simplism. Restrict *nonce-word* to sense in Dict. and add: Also ‖ **simplisme. 2.** [Cf. F. *simplisme.*] A tendency to over-simplify; an over-simplification.
1955 M. REIFER *Dict. New Words* 189/1 *Simplism, n.*, oversimplification of any matter and elimination of all complicating aspects; usually aimed at promoting uncritical conformity. 1969 C. LEECH *Tragedy* iii. 42 Wilbur Sanders..has declared that 'Necessity neither requires nor invites cooperation'.., but this appears to be simplism. 1974 *New Society* 13 June 623/1 A fair number of marxist academics have been alarmed at this anti-fascist *simplisme* coming home to roost. 1976 P. ALEXANDER *Death Thin-Skinned Animal* x. 104 It was impossible to argue against that sort of simplism. 1977 *Times Lit. Suppl.* 11 Feb. 150/2 Gladstone's complaint in 1874 that the opposition fomented by the *Daily News* had been 'one main cause' of the weakness of his late government was, of course, a simplism.

‖ **simpliste** (sænplist, si·mplist), *sb.* and *a.* Also **simplist** [Fr.] **A.** *sb.* One who adopts an over-simplified or one-sided view of something.
1918 W. O'BRIEN *Downfall of Parliamentarianism* i. 5 The simplest of simplists will now own..that the choice of Ireland in 1890 was not that..choice between vice and virtue..which would make human judgments in great affairs an enviably easy process. 1924 *Amer. Mercury* Apr. 465/2 But Nature and History, alas! are not *simplistes*. 1967 A. COMFORT *Anxiety Makers* vi. 197 These are *simplistes*, anxious about ritual obligations, unanxious about phenomena which genuinely threaten us with death and racial extinction. 1970 *Guardian* 12 Nov. 9/1 Robert Ardrey..is..a natural and disabling *simpliste*.
B. *adj.* That over-simplifies, one-sided; plain or uncomplicated in style.
1926 L. A. CLARE tr. *Lévy-Bruhl's How Natives Think* 15 Mental processes are infinitely more elastic, complex, and subtle, and they comprise more elements of the psychic life than a too 'simplist' intellectualism would allow. 1930 *N. & Q.* CLIX. 272/2 Rather a *simpliste*, facile doctrine perhaps. 1950 T. H. MARSHALL *Citizenship & Social Class* 84 The policy, in fact, may not be *simpliste* at all. 1960 *Encounter* Jan. 86/1 He accepts a wholly *simpliste* leftist view. 1966 N. FREELING *Dresden Green* I. 38 The *Golden Age of Landscape Gardening*, simplist, giving you the idea you could do it yourself. 1973 *Times* 25 Apr. 14/6 The familiar story of the parting of the seas is told directly in Williamson's most *simplist* vein. 1977 *Church Times* 13 May 10/2 Shouldn't some kind of theology, even though it's a bit simpliste, come in here.

simplistic, *a.* Add: **2.** (Later examples.) Now usu. with the connotation of excessive or misleading simplification.
1934 *Eng. Studies* XVI. 77 Some of the attempts at elucidating the origin of slang phrases do not seem very successful... It hardly admits of such a simplistic explanation. 1954 *Encounter* July 21/1 The Pluralists oppose the simplistic interpretations which we..inherit.. from Walter Millis' accounts of war as the product of war-mongering. 1967 *Guardian* 3 July 6/1 To see the issue as an attack on middle-class values..is so much simplistic and dangerous nonsense. 1976 T. STOPPARD *Dirty Linen* 40 She's quite right... It is simplistic to speak of malice. 1980 *Jrnl. R. Soc. Arts* Mar. 215/2 It was quite evident that these rather simplistic models were inadequate.
Hence **simpli·stically** *adv.*
1963 F. W. FREY in L. W. Pye *Communications & Political Devel.* xvii. 307 Changes which we simplistically call Westernization. 1974 *Times Lit. Suppl.* 25 Jan. 65/1 The German Democratic Republic from the first made Heine one of its classics, though sometimes reading his politics simplistically in the process. 1978 *Nature* 23 Mar. 306/1 Their basic chain mechanisms can be written simplistically as [etc.].

simply, *adv.* Add: **3. d.** *Logic.* In the simple mode of conversion applicable to propositions; opp. *per accidens* s.v. PER *prep.* 1 b. Cf. CONVERSION 4.
1599 T. BLUNDEVILLE *Art of Logike* III. 69 They say that the disiunct being like to an absolute or simple Proposition, may be converted both simplie and per accidens. 1677 T. GOOD *Brief Eng. Tract of Logick* 31 That is *S*. denotes the Proposition designed by the preceeding vowel, to be converted simply. 1864 [see sense 3 c in Dict.]. 1884 J. N. KEYNES *Formal Logic* II. iii. 74 This, being an E proposition, may be converted simply, giving, No one deserving of the fair is not brave. 1955 A. N. PRIOR *Formal Logic* II. i. 109 *Oba* and *Oab* do not convert either *per accidens* or simply.
6. b. Freq. in phr. *simply and solely.*
1872 C. S. CALVERLEY *Fly Leaves* 111 All least furlable things got 'furled'..simply and solely to rhyme with 'world'. 1920 *Act* 10 & 11 *Geo. V* c. 48 Sched. II. 314 Any loss or damage due simply and solely to the existence of a state of war. 1940 *Daily Tel.* 15 Feb. 6/1 What has been done is simply and solely to adjust our domestic legislation to the pressing requirements of a particular occasion.
7. *Math. simply connected:* (of a surface or other continuous set of points) connected in such a way that every closed curve lying within it forms the boundary of some surface lying within it.
1893 A. R. FORSYTH *Theory of Functions* xiv. 316 A simply connected surface is resolved by *n* cross-cuts into $n+1$ distinct pieces, each simply connected [etc.]. 1939 M. H. A. NEWMAN *Elem. Topology Plane Sets of Points* vi. 135 The complement of a simply connected domain in the open plane may have any number of components. 1976 *Physics Bull.* Sept. 388/2 The torus is not simply connected but is otherwise a uniformly structured 2-manifold.

simpsonite (si·mpsənəit). *Min.* [f. the name of E. S. *Simpson* (1875–1939), Australian mineralogist + -ITE[1].] An oxide essentially of aluminium and tantalum, approximately $Al_4Ta_3O_{13}(F,OH)$, found as colourless hexagonal crystals externally altered to a dull cream.
1939 H. BOWLEY in *Jrnl. R. Soc. W. Australia* XXV. 89 This was sufficient to indicate that it was a mineral not previously recorded so it was decided to give it the name Simpsonite in honour of Dr. E. S. Simpson.., who..has made many outstanding contributions to our knowledge of Western Australian minerals, particularly..tantalum-bearing minerals. 1959 *Mineral. Abstr.* XIV. 274/1 Simpsonite is found in a pegmatite vein near Leshai in the northern part of Kola peninsula. 1964 *Doklady Acad. Sci. U.S.S.R.: Earth Sci. Sect.* CXLVII. 147 The crystal structure of simpsonite is close-packed and has seven octahedra as its repeat distance.

Simpson's rule (si·mpsən). *Math.* [Named after Thomas *Simpson* (1710–61), English mathematician, who proposed the rule in 1743 (*Math. Dissertations* 109).] An arithmetical rule for estimating the area under a curve where the values of an odd number of ordinates, including those at the limits, are known: the approximate area is given by the sum of the first and last ordinates, double all the other odd ordinates, and quadruple all the even ordinates, multiplied by one third of the distance between adjacent ordinates. Also applied to other analogous rules (see quot. 1909).
1875 B. WILLIAMSON *Integral Calculus* vii. 196 This and the preceding are commonly called 'Simpson's rules' for calculating areas; they were however previously noticed by Newton. 1909 *Cent. Dict.* Suppl. 1158/2 *Simpson's rules*... In *Simpson's first rule* the number of ordinates is odd... *Simpson's second rule.* In this rule the area is divided into groups of three intervals... *Simpson's 5–8 rule* is used for obtaining the area of a curve between the first pair of three equally-spaced ordinates. 1930 [see *RUNGE–KUTTA]. 1933 L. M. MILNE-THOMSON *Calculus of Finite Differences* vii. 197 Show that Simpson's rule is tantamount to considering the curve between two consecutive odd ordinates as parabolic. 1980 [see *RUNGE–KUTTA].

simpy (si·mpi), *a. U.S. colloq.* [Perh. f. *SIMP + -Y[1].] Foolish, dull-witted, simple-minded.
1942 BERREY & VAN DEN BARK *Amer. Thes. Slang* § 151/9 Foolish; silly;..*simpy*. 1946 MEZZROW & WOLFE *Really Blues* ix. 139 Some hammy clowning up on the bandstand to tickle the simpy customers. 1976 *Washington Post* 24 Oct. B 7/1 The usual questions—Who is your favorite author? Who in the world would you most like to dine with?—seem out of place at the presidential level. They are, in a word, too simpy.

sim-sim (si·msim). Also **simsin.** [Arab.: see SESAME.] = SESAME. Also *attrib.*
1917 *Chambers's Jrnl.* May 294/1 Sesamum-seed, also known as..sim-sim,..is the product of an annual plant. 1930 C. G. SELIGMAN *Races Afr.* vii. 170 He smears the rain-stone with simsin oil. 1960 *Guardian* 25 Apr. 4/4 Meat and vegetables are fried in groundnut or sim-sim oil. 1972 Y. LOVELOCK *Veg. Bk.* III. 352 Sesame, also known variously as simsim, gingelly, beniseed and til, is grown in Africa, Asia, South and Central America, for the sake of its seeds, which are used to express an oil used on salads and in cooking.

simul (si·mŏl). *Chess.* [f. *SIMUL(TANEOUS *a.* 1 c.] A display in which one player plays simultaneously against a number of opponents at a number of games of chess.
1969 A. GLYN *Dragon Variation* iv. 89 He's playing twenty-four of us at once. It's a simul, simultaneous display. 1973 *Daily Tel.* 27 Jan. 11/4 Up to quite recently 'simuls' were considered very small beer by some players. 1974 SAIDY & LESSING *World of Chess* i. 26/1 (*caption*) Senior Master Edman Mednis conducts a 'simul' exhibition.

simulacral, *a.* For *rare*[−1] read *rare* and add later example.
1957 *Psychol. Rev.* LXIV. 126/1 We have long since given up simulacral theories of representation.

simuland (si·miŭlænd). [f. L. *simuland-um*, gerundive of *simulāre* SIMULATE *v.*: cf. *-AND².] That which is simulated by a (mathematical or computer) model.

1968 R. D. BRENNAN in J. McLeod *Simulation* i. 6/2 There should be a one-to-one correspondence between the mathematical equations and the functions of the elements of the simuland on the one hand, and between the equations and the components or algorithms of the computer on the other. **1972** G. A. MIHRAM *Simulation* v. 222 The goal is to structure entities and events that are as nearly isomorphic as possible to the elements and transformations which exist in the simuland.

simulant, *sb.* (Later examples.)

1979 MILLS & MANSFIELD *Genuine Article* vii. 110 The studio audience were challenged to tell the difference between a genuine diamond and an imitation. A tray of simulants containing a genuine diamond..was produced. **1979** *Nature* 6 Dec. 655/3 Yttrium aluminium garnet (YAG) and cubic zirconia are both used as diamond simulants.

simulate, *v.* Add: **1. d.** To imitate the conditions or behaviour of (a situation or process) by means of a model, esp. for the purpose of study or of training; *spec.* to produce a computer model of (a process).

1947 WILLIAMS & RITSON in *Jrnl. Inst. Electr. Engineers* XCIV. IIA. 123/2 If the control system contains discontinuous devices such as relays, these must be simulated by purely electronic means. **1958** GOTLIEB & HUME *High-Speed Data Processing* xiii. 258 A computer can simulate a warehouse, a factory, an oil refinery, or a river system, and if due regard is paid to detail the imitation can be very exact. **1966** *Guardian* 16 May 3/7 Games in which the situation before the 1914–18 war or the American Mexican war is simulated repeat the original situation with up to 70 per cent reliability. **1972** *Nature* 28 Apr. 462/1 Future population changes were simulated by computer.

simulated, *ppl. a.* Add: **1. b.** Imitative of particular conditions or circumstances, usu. for purposes of experiment or training.

1966 *Word Study* Dec. 3/1 Three-year training in simulated space flight. **1971** *Sci. Amer.* Oct. 44/2 For the purposes of the test four specially trained subjects..spent 25 consecutive days in one of the Institute's high-pressure chambers being exposed to simulated extreme depths. **1978** *Times* 4 Feb. 12/6, I was strapped into a rocket for a simulated flight through space..a ride..not so exhilarating as the old rollercoaster.

2. Of materials, artefacts, etc.: manufactured in imitation of other (usu. more expensive) materials or goods.

1942 *Amer. Speech* XVII. 120 In the trade it is practically impossible to find plain words for *small, artificial,* and *second grade*... *Artificial* and *imitation* appear as *simulated*. **1948** H. LAWRENCE *Death of Doll* ii. 40 'A double strand of pink pearls—' 'Simulated,' said Poke. 'Phony,' agreed Moke. **1960** *Harper's Bazaar* July 67/1 Eyelash curlers in simulated gold. **1973** *Country Life* 17 May (Suppl.) 80b, A set of 8 Regency Period Simulated Rosewood Dining Chairs.

simulation. Add: **3.** The technique of imitating the behaviour of some situation or process (whether economic, military, mechanical, etc.) by means of a suitably analogous situation or apparatus, esp. for the purpose of study or personnel training. Freq. *attrib.*

1947 *Jrnl. Inst. Electr. Engineers* XCIV. IIA. 117/1 The ensuing sections will..describe the simulations of the separate [servo] units. **1958** *Business Week* 29 Nov. 76/3 Men began to raise questions..about their models of the real world. They did this by inventing games such as chess and checkers to simulate battle, games like backgammon and Parcheesi to simulate racing. H. J. R. Murray, in his History of Board Games (Oxford, 1952), finds that such simulation games go back to the beginning of recorded history and are found in every culture. **1966** A. BATTERSBY *Math. in Managem.* vii. 159 Simulation enables a manager to study the system which he controls by imitating or 'simulating' its behaviour. **1972** *Computers & Humanities* VII. 38 The application of computer simulation techniques to the modeling of archaeological situations is one of the newest developments in computer use in archaeology. **1978** *Nature* 28 Sept. 305/1 Simulation studies on the towing of unprotected icebergs to southern continents suggest that the towing distance, ocean currents and the iceberg deterioration rate are of major importance.

simulator. Add: **2. b.** An apparatus designed to simulate the behaviour of a more complicated system; *esp.* one for training purposes that simulates the response of a vehicle, craft, or the like, having a similar set of controls and giving the illusion to the operator of responding like the real thing.

1947 WILLIAMS & RITSON in *Jrnl. Inst. Electr. Engineers* XCIV. IIA. 112 The paper presents an outline of a method which will allow automatic control systems to be studied experimentally by means of an electronic apparatus called a 'simulator', which is constructed so as to have the same characteristic equation as the control system. **1950**, etc. [see *flight simulator* s.v. *FLIGHT *sb.*¹ 15]. **1958** *Engineering* 21 Mar. 374/1 A colour print simulator will be demonstrated... The basis of the new equipment is a closed-circuit colour-television system, whose characteristics can be adjusted to match those of a desired printing system. **1967** *Guardian* 6 Mar. 12/2 The world's first locomotive simulator for driver training at the Willesden electric depot. **1970** *Daily Tel.* 16 Feb. 3/3 For the make-believe flight, Prince Philip and Young had climbed into a lunar-landing craft simulator. **1972** *Sci. Amer.* Apr. 106/3 Both mathematical and experimental demonstrations have established that the behavior of ethyl alcohol in the simulator closely approximates the behavior of propellants such as liquid hydrogen in full-scale, three-dimensional tanks that are subjected to comparable gravitational fields. **1980** P. ABLEMAN *Shoestring's Finest Hour* i. 13 Motorways have never appealed to me. It's like driving in a simulator.

c. *Computers.* Also *simulator program.* A program enabling a computer to execute programs written for a different computer.

1960 GREGORY & VAN HORN *Automatic Data-Processing Systems* viii. 272 A simulator program is essentially a group of subroutines. *Ibid.*, A simulator is useful when changing from one computer to another. **1977** *Sci. Amer.* Sept. 153/1 Thus users of large computer systems and time-sharing services have access to cross-software assemblers, compilers and simulators (programs that enable a computer of one make or model to duplicate the actions of another). **1978** J. C. CLULEY *Programming for Minicomputers* ix. 216 In order to help tracing errors, the simulator program has many of the features of a minicomputer debugging program.

simulcast (si·mɒlkɑst), *v.* orig. *U.S.* [f. SIMUL(TANEOUS *a.* + *BROADCAST *v.* 3.] *trans.* To broadcast (a programme) simultaneously on radio and television. Also, to transmit (a television programme) on two or more channels or networks at the same time. Also *absol.*

1948 *Amer. N. & Q.* May 26/2 *To simulcast*, to broadcast by radio and television simultaneously. **1948** *N.Y. Herald Tribune* 15 June 16/6 A press agent at WCAU-TV in Philadelphia has rather timorously launched the verb 'simulcast' into the uneasy seas of the English language. **1951** *Time* 16 July 58/2 Allen alone of the top announcers 'simulcasts'—broadcasts games simultaneously for both radio and TV. **1977** *Globe & Mail* (Toronto) 8 Jan. 34/3 It used to be they'd play at different times than on the U.S. stations, but not any more. Today the Canadian stations simulcast them as much as possible.

Hence as *sb.*, a programme transmitted simultaneously by radio and television; **si·mulcasting** *vbl. sb.*

1949 *Richmond* (Va.) *News Leader* 30 Aug. 12/1 NBC has announced that it will go in for simulcasting in a big way starting this Fall. **1964** M. McLUHAN *Understanding Media* xxxi. 311 In a group of simulcasts of several media done in Toronto a few years back, TV did a strange flip. **1976** *Broadcast* 29 Nov. 6/2 A performance of Benjamin Britten's Cantata for St Nicholas..will be networked..by both the ITV companies and ILR, making it the first nationwide stereo simulcast link-up. **1977** *Ibid.* 28 Mar. 15/3 No simulcasting of PPBs (except during elections).

simulfix (si·mɒlfiks). *Gram.* [f. L. *simul* at the same time + *-fix* as in AFFIX *sb.*, PREFIX *sb.*, etc.] A formative element occurring as a modification of an element in the basic word or root (i.e. an intonation sequence or a stress pattern). So **si:mulfixa·tion**, the action of affixing a simulfix; **si·mulfixed** *a.*, of an element, employed as a simulfix.

1954 *Word* X. 212 The term 'superfix' is no more apt than 'subfix', and (the bad Latin) 'simulfix' might be even better. **1956** *Language* XXXII. 454 The process of simulfixation may be represented by a pseudo-fractional formula..+X/Y− (=simulfixed aspect marker/underlying root-initial). **1964** [see *INCOMPLETIVE *a.* (*sb.*)]. **1965** *Language* XLI. 74 Chord structures may occur—tagmemes occurring simultaneously (tone replacives, simulfixes).

simulium (simiū·li,ɒm). [mod.L. (P. A. Latreille *Hist. Nat. Crustacés & Insectes* (1802) III. 426), f. L. *simul-āre* to imitate + *-ium*.] A small dark-coloured blood-sucking fly of the genus *Simulium*, which may be the vector of certain diseases. Also *collect.* Cf. *buffalo fly* s.v. *BUFFALO *sb.* 5.

1902 L. O. HOWARD *Insect Bk.* 120 Simulium larvae frequent well aerated and frequently swiftly running streams. **1914** G. D. H. CARPENTER *Jrnl.* 26 July in *Naturalist on Lake Victoria* (1920) iv. 80, I went down to the rocky shore before breakfast, and was set upon by a swarm of viciously biting *Simulium*. **1932** RILEY & JOHANNSEN *Med. Entom.* xviii. 292 The Simulium flies abound in hilly regions of swift-flowing, well-aerated water. **1955** *Times* 8 July 9/7 The disease [*sc.* river blindness]..is caused by the bite of the simulium fly. **1977** *Observer* (Colour Suppl.) 6 Mar. 47/1 The agile simulium can fly twice as far as was previously thought possible.

simultanagnosia (si:mɒltānægnōᵘ·siā). *Psychol.* [ad. G. *simultanagnosie* (I. Wolpert 1924, in *Zentralbl. f. d. Gesamte Neurol. u. Psychiatrie* XXXV. 445), f. G. *simultan* simultaneous + Gr. ἀγνωσία ignorance (cf. *AGNOSIA).] The loss or absence of the ability to experience perceived elements, such as the details of a picture, as components of a whole.

The form *simultagnosia* appears to have arisen as a misprint (see quot. 1961); and afterwards to have been taken, on the basis of its supposed etymology, to be a term for the ability that is absent in simultanagnosia (see quot. 1970).

1936 J. M. NIELSEN *Agnosia, Apraxia, Aphasia* vii. 84 Another term for miscellaneous classification is Simultanagnosia of Wolpert. This is not an agnosia but a psychological loss on a high plane. **1959** *Brain* LXXXII. 437 There may be gross incapacity to combine the elements of the perceptual display into a coherent and integrated whole. To this type of deficit the term 'simultanagnosia' is commonly applied. **1961** W. R. BRAIN *Speech Disorders* 173 The term simultagnosia [*ed.* 2, 1965: simultanagnosia] was coined by Wolpert to describe a condition in which the patient, looking at pictures which exhibited action, failed to recognize the meaning of the whole, while the details were correctly appreciated. **1964** M. CRITCHLEY *Developmental Dyslexia* ix. 58 He observed a veritable simultagnosia, that is, an inability to grasp the meaning of a picture as a whole. **1970** HINSIE & CAMPBELL *Psychiatric Dict.* (ed. 4) 701/2 *Simultagnosia*, inability to describe the action represented in a picture. *Ibid.* 702/1 Simultanagnosia is the lack of, or any disability in, such simultaneous form perception, and is suggestive of a lesion in the anterior part of the left occipital lobe.

simultane (simɒltē¹·n), *v.* orig. *U.S.* [f. SIMULTANE(OUS *a.*] **a.** *trans.* and *intr.* To do (something) simultaneously with something else. **b.** *intr.* To occur or take place at the same time.

Used only by or with reference to 'Mark Twain'.

1880 'MARK TWAIN' *Lett. to Publishers* (1967) 127, I mean to have the 'Atlantic' people delay my articles hereafter, so that I can 'simultane' with you. **1881** *Ibid.* 142 If they don't want it, *then* we'll go to the Century... I want to 'simultane' it with some grave enough London magazine. **1897** —— *Following Equator* lxvii. 668 They..did do them all, but only in turn, not simultaneously. In the nature of things they could not be made to simultane. **1979** *UCT Studies in English* (Univ. Cape Town) Sept. 77 The failure of Bliss to simultane and the pirating of the book by the Canadian, Belford, put a great deal of stress on the young relationship, which nevertheless withstood it.

Simultaneism (simɒltē¹·nī,iz'm, -ē¹·niz'm). Also ‖ **Simultanéisme, Simultanism.** [ad. F. *Simultanéisme*.] **1.** A name given to *ORPHISM 2, because of its use of the principle of simultaneous contrast (*SIMULTANEOUS *a.* 1 b) (see also quot. 1959).

1915 [see *ORPHISM 2]. **1940** R. H. WILENSKI *Mod. French Painters* 238 Delaunay produced a series of Flat-pattern pictures with brightly coloured fragments of revolving discs—a type of Futurist-Dynamism which he called *Simultanéisme* and which Apollinaire christened 'Orphism'. **1959** H. READ *Conc. Hist. Mod. Painting* iii. 94 [Delaunay] strove in particular to combine different aspects of figures and objects in the same painting. He himself gave the name of *Simultanéisme* to this kind of painting and was later to characterize it in these words: 'Nothing horizontal or vertical—light deforms everything, breaks everything up.' **1969** R. MAYER *Dict. Art Terms & Techniques* 273/1 The chief exponent of the style [*sc.* Orphism], Robert Delaunay (1885–1941), preferred the name Simultaneism.

2. A movement in modern French poetry, led by Henri-Martin Barzun (b. 1881), which aimed at the effect of simultaneity of both images and sounds.

1959 *Oxf. Compan. French Lit.* 676/2 *Simultanéisme*, one of the more ephemeral movements in modern poetry, an exaggerated mixture of *Cubisme* and *Unanimisme*. **1964** *Listener* 27 Aug. 315/3 Simultanism..has so far defeated its exponents..because we are not willing to appraise words other than intellectually, which means in logical succession. **1970** C. CAMPOS in J. Cruickshank *French Lit. & its Background* xi. 153 'Simultanéisme' was to be more influential as a technique for gathering impressions about reality than as a means of expressing a broader experience.

Hence **simulta·neist**, a practitioner of Simultaneism; also as *adj.*, in the manner of Simultaneism. Also ‖ **simultanéiste, simultanist** *adjs.*

1923 J. GORDON *Mod. French Painters* viii. 87 It is safer to throw most recent 'isms' out of the window... The..crop of Orphists, Futurists, Synchromists, Purists, Simultaneists, Dada-ists and so on, are as a rule irritating adjectives with no real meaning. **1930** *Times Educ. Suppl.* 3 May 197/2 A short introductory chapter on 'French Poetry after 1870' briefly describes the schools or coteries of the last few decades—the symbolists,.. simultaneists, cubists, unanimists, and so forth. **1959** *Oxf. Compan. French Lit.* 219/2 His [*sc.* Divoire's] collections of *simultanéiste* verse include: La Malediction des enfants (1910). **1975** T. STOPPARD *Travesties* I. 59 Herr Tristan Tzara was the initiator of a performance..of simultanist verse.

simultaneity. Add: **1.** (Earlier examples.) Also *spec.* in *Art*, the simultaneous representation of several views of the same object.

1652 N. CULVERWEL *Light of Nature* 118 There's no succession in God,..there's a compleat simultaneity in all his knowledge. **1798** A. F. M. WILLICH *Elem. Crit.*

Philos. 46 All the predicates of time, simultaneity, succession, &c...belong to it [*sc.* a sensible object]. **1957** *Encycl. Brit.* XVII. 64/2 Both the Cubists and the Italian Futurists used a device known as simultaneity—the practice of combining various parts of an object, or profiles of face and figure within a design concept. **1980** *Illustr. London News* Mar. 58/3 The Italian Futurists took up the notion of simultaneity in a relatively simple-minded way.

simultaneous, *a.* Add: **1. b.** *simultaneous contrast*: the effect of mutual modification of two contiguous areas of colour.

1848 M. CHEVREUL in T. Graham *Chem. Rep. & Mem.* v. 187, I will designate by the term *simultaneous contrast* the modification of colour and height of tone experienced by two differently coloured objects when seen simultaneously. **1890** [see *BORDER-LINE 1]. **1961** G. MILLERSON *Technique Television Production* iii. 45 The final values of any surface will vary with..simultaneous contrast. **1977** *Jrnl. R. Soc. Arts* CXXV. 616/2 Equally clear from the painting in the Fitzwilliam is Titian's awareness of what has since become known as the law of simultaneous contrast: the interaction of forms at the edges, so that the tone and colour of each is intensified by reaction with the other.

c. In *Chess*, denoting a number of games played against a number of opponents simultaneously by one player. Also *absol.*

1883 G. A. MACDONNELL *Chess Life-Pictures* II. 109 One of his strongest opponents in a simultaneous *sans voir* performance lost his game. **1938** P. W. SERGEANT *Championship Chess* ii. 34 Steinitz..gave a simultaneous display against twenty-two opponents at the New Vienna Chess Club. **1964** NABOKOV & SCAMMELL tr. *Nabokov's Defence* v. 77 An onlooker knowing nothing about simultaneous chess would be utterly baffled at the sight of these elderly men in black sitting gloomily behind boards that bristle thickly with curiously cut manikins, while a nimble..lad..walks lightly from table to table. **1974** SAIDY & LESSING *World of Chess* i. 24/1 The simultaneous exhibition places a premium on quick recognition of tactical threats, strong legs, and sheer physical endurance.

d. *Broadcasting.* (See quots.) Cf. S.B. s.v. *S 4 a.

1923 *Radio Times* 28 Sept. 2/3 Simultaneous broadcasting is a combination of ordinary and wireless telephony, whereby it becomes possible to broadcast at one or more stations a performance given at any other station in the country. **1971** *Gloss. Electrotechnical, Power Terms* (*B.S.I.*) III. iv. 6 *Simultaneous broadcast*, broadcast by a number of transmitters of the same programme at the same time.

e. Denoting a running oral translation of the spoken word or one skilled in this art, as *simultaneous interpreter, translation, translator*, etc.

1958 R. GLÉMET in A. H. Smith *Aspects of Translation* 120 With simultaneous interpretation..your 'intellection' of the speech need not be so thorough, but your response to words must be still quicker than before—the speaker speaks, and you are speaking too. **1965** M. SPARK *Mandelbaum Gate* vi. 187 Barbara turned the switch of her earphones to other simultaneous translations—French, Italian, then back to English. **1968** 'D. TORR' *Treason Line* 40 The simultaneous interpreters in their sound-proof boxes adjusting their earphones. **1971** *Guardian* 24 Mar. 3/5 The theatre, which only performs in Yiddish, has to have earphones with simultaneous translations for the audience. **1974** *Spartanburg* (S. Carolina) *Herald* 25 Apr. B 8/9 Simultaneous interpreters returned to their glass booths at the United Nations on Wednesday. **1977** M. T. BLOOM *13th Man* (1978) ii. 18 She's a simultaneous translator... Does a lot of work for the UN and international conferences.

sin, *sb.* Add: **1. a.** *for my sins*: (earlier and further examples).

1808 LADY LYTTELTON *Let.* 9 May (1912) 11 Now, would not you have thought he was a partisan of boxing? I did for my sins. **1906** R. BROOKE *Let.* 1 Apr. (1968) 47 About a year ago I got, for my sins, into the top form of the school. **1961** I. MURDOCH *Severed Head* v. 44 Rosemary..is for her sins a Mrs Michelis, having got married young.., to a dislikeable stockbroker called Bill Michelis, who subsequently left her. **1973** *Times* 2 Nov. 23/3 Take the BSA case in which, for my sins and much against my will, I was concerned.

2. c. Also *like* (or *worse than*) *sin*: vehemently, intensely, vigorously. Cf. *like the devil* s.v. DEVIL *sb.* 16.

1840 T. C. HALIBURTON *Clockmaker* 3rd Ser. viii. 102 Who the plague can live on sugar-candy? I am sure I couldn't. Nothin' does for me like honey; arter a while I get to hate it like sin. **1868** 'MARK TWAIN' *Let.* 8 Jan. (1917) I. 143, I have been working like sin all night to get a lecture written. **1929** W. SCOTT *Mask* i. 16 By the way, Father, dear—*who* is it that Peter Marlin hates worse than sin?

d. *to live in sin*: to cohabit outside marriage.

1838 *Ann. Rep. Bath City Mission* in G. R. Taylor *Angel-Makers* (1958) 67 Front attic, two aged people living in sin. **1855** C. KINGSLEY *Westw. Ho!* II. vii. 213 Why, not..to know whether..she's married to him or not..and I not to know whether she's living in sin or not, Mr. William. **1925** A. P. HERBERT *Laughing Ann* 92 Don't tell my mother I'm living in sin. **1974** R. B. PARKER *Godwulf Manuscript* vii. 56 A couple of freaky kids living in what my aunt used to call sin.

3. a. (Example of mod. colloquial use.) See also *Sc. Nat. Dict.*

1831 C. DARWIN *Let.* 6 Sept. in F. Darwin *Life* (1887) I. v. 207 He takes out twenty chronometers, and it will be a 'sin' not to settle the longitude.

4. a. *sin-stained.*

1843 J. G. WHITTIER *Hum. Sacrifice* in *Lays of my Home* 475 Oh! Never yet upon the scroll Of the sin-stained, but priceless soul, Hath Heaven inscribed '*Despair!*' **1896** E. DOWSON *Let.* May (1967) 363 Except that I want to see your classically sin-stained countenance, I should not even think of a week in Paris.

d. *sin-dark.*

a **1915** JOYCE *Giacomo Joyce* (1968) 10 She stands beside me, pale and chill, clothed with the shadows of the sindark nave.

6. sin-bin *slang* (chiefly *N. Amer.*) = *penalty box* (*a*) s.v. *PENALTY *sb.* 5; also *transf.*; **sin bosun** *Naval slang*, a ship's chaplain; **sin-buster** *U.S. slang*, an evangelist; a clergyman; **sin city** *slang*, a title applied jocularly or otherwise to a city considered to be a place of vice; **sin-shifter** *slang*, a clergyman.

1950 *Amer. Speech* XXV. 104/2 *Sin bin*, the penalty box where hockey players are sent for a few moments for infraction of rules, etc. **1958** *Herald Tribune* (Grand Prairie, Alberta) 11 Mar. 5/3 [The] game saw 37 minutes spent in the sin-bin. **1973** *Times* 10 Dec. 8/2 This game showed that it would be worth while trying the ice-hockey system of on-the-spot discipline with a 'sin-bin' to allow players to cool down. **1982** *Daily Tel.* 25 Feb. 19/5 It often took several months for an infant who has created chaos to be removed to a special school or a 'sin bin'. **1948** PARTRIDGE *Dict. Forces' Slang* 170 *Sin boson* [*sic*], *the*, the Chaplain, R.N. (Lower-deck.) **1964** *Navy News* Dec. (*H.M.S. Royal Arthur* Suppl.) 1/2 Well, at least the Sin Bosun doesn't seem too old, and did you see him get all punchy during deck-hockey yesterday? **1931** L. COCHRAN *Flood Tides* vi. 56 'The Reverend Billy Swinnerton is to conduct a revival here.'..'Not that ole sin-buster?' **1973** *Guardian* 17 Oct. 15/3 Leicester people ..saw Nottingham as a sort of sin city because people there went to the pub at night. **1975** 'A. THACKERAY' *One Way Ticket* II. 95 What's going to happen in Chicago? ..All you want to do is run amok in 'Sin City'. *a* **1912** 'T. COLLINS' *Rigby's Romance* (1946) 187 'Not a proper sin shifter,' objected Dixon. 'You can't chris'n a kid, nor yet say the (adj.) words over people.' **1919** W. H. DOWNING *Digger Dialects* 45 *Sin-shifter*, an army chaplain. **1966** 'L. LANE' *ABZ of Scouse* II. 98 *Sin-shifter*, a parson, priest, or rabbi.

‖ **sinagot** (sinagǫ). [Fr., f. *Séné*, a fishing village on the Gulf of Morbihan, on the west coast of France.] A two-masted Breton fishing-boat.

1927 L. RICHARDSON *Brittany & Loire* 213 Very red are the sails of the *sinagots*. **1975** *Mariner's Mirror* LXI. 93 The name for this type, which, like a Chinese junk, sets its two yards at almost right angles to their respective masts, is *sinagot*.

Sinanthropus (sinæ·nþropɒs). [mod.L. (D. Black 1927, in *Palæontologica Sinica* D.VII. I. 21), f. SIN(O- + Gr. ἄνθρωπος man.] = *Peking man* s.v. *PEKIN 3. Hence **Sinanthro·pic** *a.*, of or pertaining to a fossil hominid of this kind; **Sina·nthropoid** *a.* [-OID], resembling Sinanthropus.

1928 *Daily Tel.* 17 Apr. 9/3 A human tooth as old as the Java ape man..has been named Sinanthropus, or 'the Peking man'. **1931** A. KEITH *New Discoveries Antiquity of Man* xvii. 260 The Sinanthropic mandibular fragment is broken short at the socket for the second incisor. **1937** *Ann. Reg. 1936* 52 Parts of three skulls of Palaeoanthropus showing Sinanthropoid features..were found in the Upper Pleistocene in Tanganyika. **1965** B. E. FREEMAN tr. *Vandel's Biospeleology* iii. 22 *Sinanthropus* used caves for shelter. **1978** *Nagel's Encycl.-Guide: China* 298 Great pride was taken in the discovery of remains of 'sinanthropus' which were older still than those found at Zhou Kou dian by Teilhard de Chardin.

‖ **Sinarquista** (sinaɪkī·sta). Also (anglicized) **Sinarquist** and with small initial. [Amer.-Sp., after Sp. *anarquista* ANARCHIST; cf. SYNARCHY.] A member or adherent of the right-wing authoritarian Unión Nacional Sinarquista in Mexico, which was active between 1938 and 1960 (esp. in rural areas) and sought to restore the old order of the Catholic church and Spanish tradition, and opposed Communism, liberalism, and the policies imposed after the Revolution of 1911. Also *attrib.* or as *adj.*

1941 *N.Y. Times* 27 Dec. 8/8 An organized campaign to combat the Sinarquista Movement in Mexico will be launched next month... The Sinarquistas, often charged with favoring totalitarian nations, have begun large-scale agrarian colonization of Lower California. **1943** *Free World* May 413/1 These are reactionary leaders among the clergy who would go much farther with the Sinarquists. **1946** M. LOWRY *Let.* 2 Jan. (1967) 88 You can even see the German submarine officers taking revenge on the Consul in the form of the *sinarquistas*. **1970** G. HUIZER in I. L. Horowitz *Masses in Lat. Amer.* xiii. 474 The danger from extremely conservative and fascist forces, such as the Sinarquist movement, was alarmingly strong. **1973** P. CALVERT *Mexico* xxii. 284 The Sinarquistas were only the tip of the iceberg.

Hence ‖ **Sinarqui·smo** (anglicized **-ism**), the political doctrine of the Sinarquistas.

1943 *Free World* May 410/2 The enigma of Sinarquism is created by the character of Mexican politics. **1953** H. F. CLINE *U.S. & Mexico* xiv. 293 *Sinarquismo*—Sinarchism—is the opposite of anarchism; the word means 'with order'. Sinarquistas are believers in 'order'. **1963** D. JAMES *Mexico & Americans* xiii. 346 Sinarquismo was essentially a rural movement arising out of peasant discontent. **1971** R. MARETT *Mexico* x. 162 The fascist phenomenon of Sinarquismo, which came to a head in the late 1930's in opposition to the radical policies of Cárdenas.

since, *conj.* Add: **3. c.** In various, chiefly jocular, comparisons, as *since Christ was a corporal*, etc. *colloq.* (chiefly *U.S.* and *Mil.*).

1601 SHAKES. *Twel. N.* III. ii. 18 And they haue beene grande Iurie men, since before Noah was a Saylor. **1816** KEATS *Endymion* II. 443 Never, I aver, Since Ariadne was a vintager, So cool a purple. **1900** I. L. REEVES *Bamboo Tales* 20 Private McCoy..had been in the service since George Washington was a 'lance jack'. **1921** J. DOS PASSOS *Three Soldiers* II. ii. 75 Ain't had any pay since Christ was a corporal. I've forgotten what it looks like. **1961** PARTRIDGE *Dict. Slang* Suppl. 1274/2 *Since Pontius was a pilot*, as in 'He's been in that mob since..' R.A.F. c[atch] p[hrase], testifying to long service: since ca. 1944. **1970** *N.Y. Post* 7 Apr. 5 Dana Stone had been in Vietnam since Christ was a corporal, as the grunts in the field would say.

d. *since when?*: used *ellipt.* as an inquiry into the duration of a state of affairs mentioned in a previous statement (freq. expressing doubt or incredulity); also, with full interrogative clause. *colloq.*

1907 G. B. SHAW *John Bull's Other Island* IV. 84 *Broadbent* (*very solemnly*): No: I am a teetotaller. *Aunt Judy* (*incredulously*): Arra since when? **1966** 'G. BLACK' *You want to die, Johnny?* v. 105 'Lil's disappeared.'..'Since when?' **1977** C. WATSON *One Man's Meat* vii. 64 'Since when has Digger's father been a V.C.?' 'Since when has Digger had a father?'

‖ **sindaco** (si·ndako). [It.: see SYNDIC *sb.*] In Italy: a mayor.

[**1881** *Encycl. Brit.* XIII. 464/1 The syndic (*sindaco*) or chief magistrate of the commune is appointed by the king for three years.] **1902** H. BELLOC *Path to Rome* 342 We came to the house of the Sindaco or Mayor. **1969** M. GILBERT *Etruscan Net* I. v. 64 Broke had been looking at the Sindaco. **1975** S. JOHNSON *Urbane Guerilla* II. 81 He would seek the permission of the Sindaco and Commune of Siena for the staging of a special or *straordinario* Palio.

Sindebele: see *NDEBELE.

Sindhi (si·ndi), *sb.* and *a.* Also † **Sindee; Sindi.** [a. Hind. *Sindhi*, ult. f. Skr. *sindhu* river, *spec.* the Indus or the surrounding area.] **A.** *sb.* **1.** A native or inhabitant of Sind, now a province in the south-east of Pakistan, through which the Indus passes to the Arabian Sea.

1815 M. ELPHINSTONE *Acct. Kingdom of Caubul* IV. v. 500 The Sindees with whom I have conversed. **1836** in *Corresp. relative to Sinde 1836–1838* 20 in *Parl. Papers 1843* XXXIX. 9 The notions of the most enlightened (if I may apply the term) Sindees are..at utter variance with our customs. **1887** *Encycl. Brit.* XXII. 91/2 The Mohammedans [in Sind] may be divided into two great bodies—the Sindis proper and the naturalized Sindis. The Sindi proper is a descendant of the original Hindu. **1927** *Chambers's Jrnl.* Jan. 11/1 The little Sindi could distinguish every hoofprint and point out the goat which had made it. **1978** F. OLBRICH *Desouza pays Price* xxi. 131 Most of the Sindhis..had gathered their wealth together.. by Independence Day and descended on Bombay.

2. The name of a language consisting of several dialects spoken principally in Sind, but also in adjacent districts of north-west India.

1838 *Penny Cycl.* XII. 227/2 *Sindhi*, spoken in Sinde as far as the mouths of the Indus. **1908** [see *NEPALESE *a.* and *sb.*]. **1948** D. DIRINGER *Alphabet* II. vi. 376 Sindhi, spoken by three and a half million people in Sind. **1968** *Guardian* 14 Nov. 3/3 'Jiye Sind' (Long live Sind) has become among many Sindhis a form of salutation and there are demands for the elevation of Sindhi to the status of national language. **1981** V. POWELL *Flora Annie Steel* vii. 52 The strictly local dialect, a variety of Sindhi, made the setting up of female schools impracticable.

B. *adj.* Of or pertaining to this people or their language.

1836 W. H. WATHEN *Gram. Sindhi Lang.* 1 There are several different alphabets used in writing the Sindhi language. **1899** *Folk-Lore* X. 413 The far renowned.. Sindhi story of Sassi and Punnun. **1946** *Civil & Mil. Gaz.* 31 Aug. 8/3 Three Sindhi merchants. **1979** V. S. NAIPAUL *Bend in River* xiv. 244 There was this Sindhi girl who had studied in England.

Also **Si·nd(h)ian** *a.* and *sb.*

1849 J. D. CUNNINGHAM *Hist. Sikhs* i. 19 The occupation, by the Sindhian Daoodpotras of the Lower Sutlej, took place within the last hundred years. **1911** D. S. MARGOLIOUTH *Mohammedanism* i. 26 An outrage committed by a Sindian on a noble Moslem. **1964** in *Panjab Past & Present* (1978) XII. 412 The Sindhian Amirs were of course alarmed to learn that Ranjit Singh positively entertained the intention of attacking Shikarpur.

‖ **sindicato** (sindikā·to). [Sp., Pg.; cf. SYNDICATE *sb.*] In Spain, Portugal, Latin America, etc.: a trade union. *spec.* (usu. in *pl.*) a

Spanish trade union of a type originally established during the regime of General Franco, and subject to close government control. Cf. SYNDICAL *a.* in Dict. and Suppl.

1936 *Times* 21 Aug. 9/1 The newly created *sindicatos*, or labour unions. **1957** LD. HAILEY *Afr. Survey 1956* xx. 1451 The *sindicatos* have powers of investigating grievances or disputes, and provision is made for arbitration; they also try to find employment for members who are out of work. **1964** *Ann. Reg. 1963* 285 During the year it became evident that the Catholic Workers' Brotherhoods had become trade unions in all but name..but legally the government-controlled *sindicatos* maintained their monopoly. **1965** *New Statesman* 16 Apr. 590/1 Before his death Rodolfo Romero was the secretary-general of the peasant *sindicato* at Tapatapa. **1976** *National Observer* (U.S.) 28 Feb. 16/3 Underground leaders of the workers' commissions were widely elected to the above-ground negotiating committees of sindicatos.

sindon. Add: **2. a.** (Later examples in *spec.* use.) Also in It. form ‖ **sindone.**

1902 tr. *P. Vignon's Shroud of Christ* ii. 50 If 'the napkin' of St. John *were* the face-kerchief, where would have been the Shroud (*sindon*)? **1912** H. THURSTON in *Month* Nov. 539 One could well imagine that when the body had been laid out and covered back and front with the long impregnated sindon, strips of linen were used to secure the feet..keeping the sindon in its place. **1933** *Dublin Rev.* Jan. 36 St. Nino, a Georgian princess..was told that the Sindon was formerly in the possession of St. Peter. **1963** *Guardian* 31 May 12/3 The famous Holy Shroud, or Sindone, of Turin.

sindonology (sindǫnǫ·lodʒi). [f. SINDON + -OLOGY.] The study of the Holy Shroud of Turin, in which the body of Christ was reputedly wrapped. So **si:ndonolo·gical** *a.*

1950 *Catholic Digest* May 77/1 Dr Wuenschel..will give an address on 'The Shroud of Turin and the Burial of Christ' this May in Rome at the International Syndonological convention. **1953** E. WUENSCHEL *Holy Shroud of Turin* (rev. ed.) 4 In 1950, he was invited to read a paper at the First Sindonological Congress at Rome—a meeting of internationally recognized authorities on the Holy Shroud of Turin. **1964** J. WALSH *Shroud* xii. 117 By the start of the Second World War the Shroud had been studied more closely than in all its previous history... The investigation..assumed the stature of a separate discipline and was given a name, *sindonology*. **1978** *Church Times* 6 Oct. 6/4 This book..alarms by an account of a visit to a Sindonological Library in New York. **1979** *Radio Times* 7–13 Apr. 86/2 Don Pietro Borga is the general secretary of the International Centre for Sindonology (study of the shroud).

Hence **si:ndono·logist**, a student of the Shroud.

1953 E. WUENSCHEL *Holy Shroud of Turin* (rev. ed.) 4 He collaborated with Dr. Paul Vignon, the greatest of all sindonologists. **1980** *Observer* 28 Dec. 3/1 It has been a disappointment to sindologists (the name for shroud students) that the Roman Catholic authorities have refused to allow this [*sc.* carbon dating].

sine². Add: **4. sine bar** *Mech.*, a device used to set out or measure angles accurately, in which one end of a bar of known length is raised on gauge blocks; **sine tone** = *pure tone* s.v. *PURE *a.* 1 e; **sine wave,** a periodic oscillation of pure and simple form in which the displacement at any point is proportional to the sine of the phase angle at that point; a wave or curve resembling (a segment of) this in form.

1915 *Engineering* 8 Jan. 42/3 A sine-bar is a flat strip of steel, planed true all over, upon which are fixed two hardened and ground plugs, 1 in. in diameter, 10 in. apart. **1975** BRAM & DOWNS *Manuf. Technol.* i. 19 The sine bar is commonly used for marking off and checking the angle of a workpiece. **1962** Sine tone [see *LINE *v.*² 8 a]. **1976** *Times Lit. Suppl.* 3 Dec. 1522/5 Stockhausen's recent memoirs of his early years, which are sometimes strangely at odds with his correspondence of the period (particularly vis-à-vis the whole question of sine-tone synthesis). **1893** D. E. JONES tr. *H. Hertz's Electric Waves* 17 The vibration of the primary conductor is, at any rate to a first approximation, a uniformly damped sine-wave of determinate period. **1916** *Electrician* LXXVI. 800/1 The alternating current produced in the telephones is almost a pure sine wave. **1965** *Wireless World* Sept. 455/1 To verify the rated output of an amplifier a 1 kc/s sinewave is fed through it at a level sufficient to produce the rated output power. **1972** *Islander* (Victoria, B.C.) 21 May 15/2 Some of the flat fishes, such as flounders and rays, move by undulating their bodies in a flattened sine wave. Visualize this movement as a flag held horizontally in a stiff wind.

sine³ (səi·ni). *Eton College slang.* [a. fanciful mod.L. *sine* (*coloribus*), without (colours).] At Eton College: a House team, which excludes those awarded colours; the members of this considered *collect.*

1922 S. LESLIE *Oppidan* xxii. 276 The *sine* was made to forego the sweetness of a long lie in order to run..before breakfast. *Ibid.* 277 As match after match was lost by the *sine*, Mouler put up a notice to say that in future *sine* were expected to do a training walk as well as their morning run. **1940** M. MARPLES *Public School Slang* 112 Idioms such as these are probably not of great age... [e.g.] *sine* (Eton): used to describe a House team excluding colours, from the dog Latin *sine coloribus*, without colours: hence *2nd sine* wrongly used = 2nd XI.

Sinemurian (sinĕmiū·riăn), *a. Geol.* [ad. F. *Sinémurien* (A. d'Orbigny *Paléont. Française. Mollusques et Rayonnés Fossiles. Terrains Jurassiques* (1842–9) I. 604), f. L. *Sinemurum*, ancient name of Sémur-en-Auxois, a town in Côte d'Or department, France + *-ien* -IAN.] Of, pertaining to, or designating a stage of the Lower Jurassic in Europe which in the modern division is next below the Pliensbachian. Freq. *absol.*

1863 J. D. DANA *Man. Geol.* III. iii. 449 The Sinemurian (Lower Lias, named from the locality at Sémur). **1888** J. PRESTWICH *Geol.* II. xvi. 242 M. Gosselet includes the Lower- and Infra-Lias in one division—the Sinemurian, which he divides into two subdivisions—the Upper Sinemurian and the Lower Sinemurian or Hettangian. **1928** E. NEAVERSON *Stratigraphical Palaeont.* xv. 331 The most familiar Sinemurian brachiopod is the broad-plaited *Spiriferina walcotti*. **1969** BENNISON & WRIGHT *Geol. Hist. Brit. Isles* xiii. 289 A cyclic sequence of thin limestones and clays of Blue Lias type comprise the Hettangian followed by clays chiefly, although locally limestones may occur as high as the top of the Sinemurian.

Sineque, var. *SENECA.

‖ **sinfonia** (sinfonī·ā, sinfōᵘ·ni,ā). *Mus.* Pl. **sinfonie, -ias.** [It., = SYMPHONY.] **a.** In early Italian opera: the overture. **b.** A symphony. **c.** (Used in the title of) a small symphony orchestra or chamber orchestra.

1773 C. BURNEY *Present State of Music in France & Italy* 382, I heard a *sinfonia* or overture and a chorus.. which were excellent. **1818** W. GARDNER tr. *Bombet's Lives of Haydn & Mozart* 256 The sinfonia in *The Creation*, which represents the rising of the sun, is an exemplification of this theory. **1828** E. HOLMES *Ramble among Musicians of Germany* 180 The composition given this afternoon was Mozart's Jupiter sinfonia. **1884** *Encyl. Brit.* XVII. 87/1 The sinfonia or overture which is often associated with his [*sc.* Scarlatti's] name. **1924** A. HAM *Outl. Mus. Form* xix. 82 An Overture..is an expansion of the Sinfonia or Symphony. **1946** DAVID & MENDEL *Bach Reader* I. 39 Neither the Inventions and the Sinfonias nor the two books of the *Well-Tempered Clavier*..were printed during Bach's lifetime. **1967** I. SPINK *Hist. Approach to Mus. Form* iii. 98 These operas were not in recitative throughout. They also included choruses, dances and instrumental pieces' or *sinfonie*. **1976** *Leicester Mercury* 16 July, He had recently contacted about 60 top class instrumentalists.. with the intention of forming an orchestra which..would be called the Rutland Sinfonia.

Also **sinfonia concertante** (kontʃerta·nte) [CONCERTANTE], a symphonic work exhibiting characteristics of the concerto.

1903 R. HUGHES *Mus. Guide* I. 266/2 *S. concertan'te, concerta'ta, concertate*... concerto for many instrs., a concerto symphony. **1928** *Musical Times* 1 Feb. 165/1 M. Ansermet..conducted William Walton's new 'Sinfonia Concertante' for orchestra and pianoforte. **1946** MENDEL & BRODER tr. *Einstein's Mozart* xvi. 274 Mozart cultivated this form less and less as the years went on. He abandoned the *sinfonia concertante*, and separated its ingredients. **1950** R. HUGHES *Haydn* vii. 80 At the fourth concert on 9th March [1792] the endearing *Sinfonia concertante* for violin, cello, oboe and bassoon was performed for the first time. **1959** [see *ARGUABLY *adv.*]. **1979** *Early Music* July 415/2 Clarinettists will find in the E flat Symphony by the Earl of Kelly..what amounts to a *sinfonia concertante*.

‖ **sinfonietta** (sinfoni,e·tă, sinfonye·tă). *Mus.* [It., dim. of prec.] **a.** A short, simple form of symphony. **b.** = *SINFONIA c.

1907 T. S. WOTTON *Dict. Mus. Terms* 180 *Sinfonietta*, a little Symphony, e.g. Raff's sinfonietta for 10 wind instruments. **1923** *Daily Mail* 20 Feb. 7 Mr. Eugene Goossens..introduced..his first symphony (shyly entitled a 'Sinfonietta'). **1947** A. EINSTEIN *Music in Romantic Era* xi. 131 But does not the lack of a slow movement make even a sinfonietta into a suite, a more or less disconnected succession of movements? **1970** *Music & Musicians* June 53 Playing at the Elizabeth Hall on March 25 before a miserably small audience, the Bournemouth Sinfonietta, ably directed by Nicholas Braithwaite. **1973** D. EWEN *Orchestral Music* xii. 207 His [*sc.* Hindemith's] major works..included..two compositions with humorous overtones: the Concerto for Woodwinds, Harp and Orchestra (1949) and Sinfonietta in E (1950). **1982** *Times* 29 Nov. 11/1 The London Sinfonietta settled down last night to become a string orchestra in a stabler sort of programme.

sinful, *a.* and *sb.* Add: **2. b.** Also, as a strong intensive: excessive in manner or extent; 'dreadful', 'wicked'. *colloq.*

1880 J. C. HARRIS *Uncle Remus* xxviii. 122 De way he stir up dem bees wuz sinful. **1920** WODEHOUSE *Damsel in Distress* ii. 35 The money that boy makes is sinful.

sinfully, *adv.* Add: **2.** (Earlier and later examples.) Also in weakened use: excessively.

1869 'MARK TWAIN' *Innoc. Abroad* xlv. 475 So sinfully ugly that she couldn't smile after ten o'clock Saturday night without breaking the Sabbath. **1912** [see *JUMP *sb.*¹ 7]. **1976** *New Yorker* 24 May 113/3 It's also guaranteed to be *sinfully* comfortable with its built-in headrest and body-conforming mattress.

sing, *sb.* Add: **1. a.** For 'bullet' read 'bullet or other projectile' and add later examples.

1917 E. C. MIDDLETON *Way of Air* 70 The familiar 'sing' of an approaching shell. **1930** *Carmina* Oct. 45 The sing Of a stone from the sling.

2. a. (Earlier and later examples.) Also, a hearty sing-song or round of collective singing (chiefly *U.S.*).

1850 N. KINGSLEY *Diary* 1 Sept. (1914) 140 We had a fine sing in the Evening which put me in mind of home. **1875** I. L. BIRD *Six Months in Sandwich Islands* xii. 175 There have been pleasant little gatherings for sewing.. and on Sunday evenings what is colloquially termed, 'a sing'. **1932** A. HUXLEY *Brave New World* iv. 73 In the Ealing stadium a Delta gymnastic display and community sing was in progress. **1964** 'J. H. ROBERTS' '*Q' Document* (1965) ix. 211 Skiers were gathered around in an alcoholic community sing. **1972** *Village Voice* (N.Y.) 1 June 96/4 Open sing, Verdi's Requiem. **1981** *Libr. Congr. Inf. Bull.* 16 Jan. (Staff News), Staff members and their families gather in the Great Hall for the annual carol sing.

b. *on the sing*: (of a kettle) singing. Cf. SING *v.*¹ 6 a.

1927 W. DEEPING *Kitty* xxx. 384 'All the kettles—'.. 'Two are boiling, miss; the other's on the sing.'

sing, *v.*¹ Add: **I. 1. d.** *to sing for one's supper* (also † *dinner*): for lack of money. Usu. *fig.*, to provide entertainment or a service in return for a benefit received (often, a meal).

c **1744** *Little Tommy Tucker* in *Tommy Thumb's Pretty Little Song Bk.* 10 Little Tom Tucker Sings for his Supper What shall he Eat White bread and Butter. **1803** J. KENNEY *Raising the Wind* I. i. 4 As you sometimes sing for your dinner, now you may whistle for your breakfast. **1949** N. MITFORD *Love in Cold Climate* I. iii. 30 [At] the various house parties..I had been to..I knew that I was expected..to sing for my supper by being, if possible, amusing. **1972** T. P. McMAHON *Issue of Bishop's Blood* (1973) ix. 134 Thanks for the dinner... Is it too bad a pun to say I'm ready to sing for it?

e. *to sing along*: to sing in accompaniment to a song or piece of music. Also const. *with* the performer. Cf. *SING-ALONG *sb.* and *a.*

1959 *Time* 17 Aug. 60/3 Whether anyone actually sings along with the sing-along albums probably does not bother..Miller. **1973** *Observer* (Colour Suppl.) 29 Apr. 41/3 They all stand on the tables and sing along and stamp their feet. **1977** R. L. DUNCAN *Temple Dogs* (1978) I. ii. 46 He was singing along with the piano player, his voice shrill.

4. d. *Criminals' slang* (now chiefly *U.S.*). = *sing out* (sense 5 c in Dict.). Also *to sing like a canary*. Orig. in proverbial phr. † *he that sings once, weeps all his life after* and varr.

1612 T. SHELTON tr. *Cervantes' Don-Quixote* I. III. viii. 193 Here it is quite contrary, quoth the slave, for He that sings once, weepes all his life after. **1710** S. PALMER *Moral Essays on Proverbs* lxxii. 197 He that Sings in Disaster, shall Weep all his Life-time After. 'Tis generally suppos'd, that this Proverb was born in a Jail. *Sing*..is, when one of the Gang Tattles, Confesses, and Accuses the Rest. **1929** HOSTETTER & BEESLEY *It's a Racket!* 238 *Sing*, to confess. **1937** [implied at *SINGING *vbl. sb.* 1 d]. **1946** *Sun* (Baltimore) 10 Dec. 1/3 A former army colonel 'sang' about the operation of military government there. **1950** R. HIMMEL *I'll find You* (1958) xvii. 117 She's singing like a canary. She turned up at headquarters..and said she had some information on a killing. **1964** L. NKOSI *Rhythm of Violence* 65 Who knows, maybe he's even goin' to sing to the police! **1981** P. NIESEWAND *Word of Gentleman* xix. 126 You don't think they'd sing like canaries?..They'll sing, Claud... If they thought it would help them, they'd tell on their mothers.

II. 13. e. Of the Aboriginal inhabitants of Australia: to endow (an object) with magical properties by singing; to bring a magical influence to bear on (a person or thing) by singing.

1899 SPENCER & GILLEN *Native Tribes Central Austral.* xvi. 537 The wound was not serious..but he persisted in saying that the spear had been sung, and that..he was going to die, which accordingly he did. **1914** B. SPENCER *Native Tribes N. Territory* iii. 140 As soon as the ground was cleared..all the men retired to one side and, to the accompaniment of trumpets..and clapping of hands, it was 'sung'. This 'singing' was supposed to make the ground..in good order so that the performers could dance well. **1959** A. UPFIELD *Bony & Black Virgin* xvii. 158 The aborigines..dug up their rainstones and rubbed them with their magic stones, and 'sang' them in a secret camp. And then it rained. **1975** *Times* 8 Nov. 5/1 The Premier of Queensland was put under a death spell by Aborigines last night... An Aboriginal..said 'These people have sung him and he should start dying from now on.'

III. 14. sing-in [*IN³], a musical performance in which the audience participates in the singing.

1968 *Lebende Sprachen* XIII. 67/1 Neologismen mit *in* im Englischen und Deutschen...sing-in, sit-in, [etc.]. **1970** *New Yorker* 19 Dec. 16 Sing-in at Philharmonic Hall —Handel's 'Messiah' directed (seriatim fortunately) by nineteen directors. **1976** *Flintshire Leader* 10 Dec. 13/2 (*heading*) Penyffordd 'sing-in'.

sing, var. *SHENG.

si·ng-along, *sb.* and *a.* Also **sing-a-long** and as one word. [f. the vbl. phr. *to sing along*: see *SING *v.*¹ 1 e.] **A.** *sb.* **1.** A song or recording to which one can sing along in accom-

paniment (esp. a light popular song with an easy rhythm).

1959 *Time* 17 Aug. 60/3 The nation's mature citizens are merely striking back at rock'n'roll, buying the sing-alongs. **1968** *Globe & Mail* (Toronto) 3 Feb. 23/1 A Gay Nineties room with sing-alongs, familiar tunes of that era. **1971** *Ink* 31 July 16/2 Those ringing certainties which made 'Woodstock' and 'Big Yellow Taxi' into such cosy sing-alongs. **1981** J. WAINWRIGHT *Urge for Justice* I. xii. 84 He could tickle the old ivories..could hammer out a singalong with the best.

2. A sing-song to the accompaniment of a song-leader or tune.

1973 B. BROADFOOT *Ten Lost Years* xxii. 256 There would be a sing-along, or the manager maybe would just pull a lucky number from a hat. **1975** *Daily Mail* 9 June 18/1 Someone in the next room's having a sing-along! **1979** *Guardian* 27 June 11/8 The insulting..notion that working-class audiences want only a beery community sing-along on their night out.

B. *adj.* Of a song, recording, etc.: to which one can sing along in accompaniment. Of or characterized by this unsophisticated but cheerful style.

1959 [see *SING *v.*¹ 1 e]. **1967** *Melody Maker* 1 Apr. 9 How can the Beatles' best..single yet be ousted by so many sing-along melodies and slush-ridden lyrics? **1974** *Financ. Times* 24 Apr. 2/3 Happy music in singalong style. **1977** P. HILL *Liars* ii. 9 A group of relatives were following the words of a sing-along record.

Singapore (siŋāpōᵊ·r, siŋg-). The name of a city and island-republic (formerly, British Crown Colony) in South-East Asia, used *attrib.*, usu. as *Singapore* (*gin*) *sling*, to designate a cocktail with a base of gin and cherry brandy.

1930 *Savoy Cocktail Book* I. 190 Singapore Sling. The Juice of ¼ Lemon. ¼ Dry Gin. ½ Cherry Brandy. Shake well and strain into medium size glass, and fill with soda water. Add 1 lump of ice. **1948** D. S. EMBURY *Fine Art of mixing Drinks* xi. 299 Singapore Gin Sling. Of all the recipes published for this drink I have never seen any two that were alike. Essentially it is simply a Gin Sling with the addition of cherry brandy. **1960** J. J. ROWLANDS *Spindrift from House by Sea* i. 28 Building your own house, he told us after his third Singapore gimlet, is an experience akin to a spiritual awakening. **1969** R. THOMAS *Singapore Wink* xi. 118 I'm going to have a Singapore Sling in the bar of the Raffles Hotel. **1976** *Times* (Singapore Suppl.) 19 July p. iv/4 In 1915 a barman, Mr Ngian Tong Dron, tried mixing two measures of gin with one of cherry brandy and one of orange, pineapple and lime juice..the Singapore gin sling was born.

Singaporean (siŋāpōᵊ·riăn, siŋg-), *a.* and *sb.* Also (occas.) **-ian.** [f. prec. + -AN.] **A.** *adj.* Of or pertaining to Singapore.

1880 [see *moth orchid* s.v. MOTH *sb.* 3]. **1927** *Malaya Tribune* 27 Dec. 7/6 Feckless Singaporean Scots. **1972** *Times* (Singapore Suppl.) 4 July p. ii/3 The Government has been alleged..to have sacrificed important aspects of Chinese culture and language in order to create its image of the Singaporian nation. **1977** *Hongkong Standard* 12 Apr. 16/5 He spoke to 360 striking Singaporean metal workers earlier in the day.

B. *sb.* A native or inhabitant of Singapore.

1927 *Malaya Tribune* 5 Jan. 9/2 (*heading*) World's motor record by Singaporeans. **1956** D. DAVIES *More Old Singapore* 40 A working day in the life of a Singaporean. **1965** *Times* 12 Aug. 8/1 Speaking to Malay journalists, Mr. Lee begged the Malays not to worry. He ..insisted that 'Singapore is not a Chinese country nor a Malay country nor an Indian country... It belongs to the Singaporeans.' **1975** P. THEROUX *Great Railway Bazaar* xxiii. 238 Singaporeans are great assemblers of appliances.

Singatoka, var. *SIGATOKA.

singed, *ppl. a.*¹ Add: **b.** (Earlier examples.) (Parallel Sc. allusive uses incorporate the related form SINGED *ppl. a.*² (cf. quot. 1737).)

1836 *Spirit of Times* 9 Apr. 61/1 Without our Jersey friends bring on a 'singed cat', or some nag, now outside the fence, turns up a trump, the above comprise the entries for the 4 mile day. **1837** J. C. NEAL *Charcoal Sketches* 48 His new friend, however, proved..to be like a singed cat, much better than he looked.

singer¹. Add: **1. c.** An informer. Cf. *SING *v.*¹ 4 d. *Criminals' slang.*

1935 *Amer. Speech* X. 20/2 *Singer*, a stool pigeon or trusty who carries tales to the administration. (Obs.) **1961** *John o' London's* 30 Nov. 610/3 An informer, then a *squealer*, is now more often referred to..as a *singer*.

3. Special Combinations with *singer's*: **singer's node, nodule** *Path.*, a small pale swelling on a vocal cord; **singer's seat** *U.S.*, a choir-seat or bench (cf. *singing-seat* s.v. SINGING *vbl. sb.* 4 b).

1953 C. WAKELY *Faber Med. Dict.* 389/2 *Singers' nodes, or nodules*... Syn. *chorditis tuberosa*. **1961** R. D. BAKER *Essent. Path.* xv. 360 Overuse of the voice can cause traumatic laryngitis and 'singer's nodes', and a biopsy will demonstrate minute hematomas in various stages of scarring. **1967** *Punch* 29 Mar. 458 Singer's Nodule, the name for a minute warty excrescence on overworked vocal apparatus. **1974** PASSMORE & ROBSON *Compan. Med. Stud.* III. xxxii. 22/2 Vocal nodules. These are called singer's nodes because they are seen in singers, particularly sopranos and tenors, but they also occur in others who use their voices excessively, with faulty voice production. **1777** Singer's seat [see *CHORISTER 1 a]. **1861** MRS. STOWE *Pearl of Orr's Island* (1862) ix. 84 Aunt Ruey.. had in her youth been one of the foremost leaders in the 'singers' seats'.

‖ **singerie** (sæ̃ʒəri). [Fr., apish behaviour or trick, a collection of monkeys; cf. CHINOISERIE.] A piece of porcelain, painting, etc., in which monkeys are represented in anthropomorphic (often quasi-Chinese) attitude; work done in this style (esp. popular in the eighteenth cent.). Also *transf.* Cf. *monkey band, orchestra* s.v. *MONKEY *sb.* 17 a.

[**1820** M. EDGEWORTH *Let.* 4 June (1979) 142 The white wainscot..is painted with grey imitation of Indian ink pictures of monkeys in mens and womens clothes... I have some notion of having somewhere read of this cabinet of monkeys.] **1920** A. STRATTON *Eng. Interior* 61 So many influences were tending to shape the arts in that century, that it is not surprising to find reflections of the French 'Chinoiserie' and 'Singerie' styles in English houses. The French painters Jean François Clermont.. and Jean Pillement..both worked in this country, and their fanciful 'Singeries', in which monkeys play the rôle of horsemen and sportsmen, have a certain charm. **1920** E. SITWELL *Wooden Pegasus* 13 (*title*) Singerie. **1957** *Economist* 9 Nov. (Suppl.) 15/2 This style..gives excessive weight to the rest of their lives, which is that of the characters in Miss Mitford's novels: one-tenth genuine emotion (a bit more for Voltaire) and nine-tenths *singerie*. **1963** N. PEVSNER *Wiltshire* 520 Ceiling with *singeries* by *Andieu de Clermont*,..far too finicky for Inigo's architecture. **1977** FLEMING & HONOUR *Penguin Dict. Decorative Arts* 742/1 The vogue for *singeries* did not begin much before the end of the c17 and reached its height of popularity in the c18 when it became associated with *chinoiserie*... *Singeries* were painted on walls, in porcelain and faience, worked in piqué and in marquetry, embroidered and printed in textiles.

Singh (siŋ). Also † **Sing(e.** [a. Hind. *singh* lion, f. Skr. *simha* lion, 'the powerful one'.]

1. A great warrior: a cognomen or title of respect borne by several of the warrior castes of northern India, or a surname adopted by male Sikhs.

1623 N. BANGHAM et al. *Let.* 5 Apr. in W. Foster *Eng. Factories in India* (1908) 218 Beinge soe hotly persued by Abdala Chan and Rajae Sursinge. **1797** *Encycl. Brit.* IX. 213/2 In 1770 the rajah died, and was succeeded by his son Cheit Sing. **1841** *Penny Cycl.* XIX. 276/2 The bravery and talents of the regent Zalim Singh. **1888** KIPLING *Departmental Ditties* (1890) 18 Chimbu Singh from Bikaneer..Jowar Singh the Sikh. **1955** *Times* 5 July 13/3 Maharana Sir Fateh Singh Bahadur, the head of the Sessodia Rajputs..had personal as well as dynastic claims to veneration.

2. As a simple noun.

1851 J. THACKWELL *Narr. Second Sikh War* 227 The British guns were so overwhelming..and their fire so rapid and precise, that the enemy declared there was a ball for every Sing. **1914** J. J. H. GORDON *Sikhs* iv. 39 They were then hailed as 'Singhs' or lions of their race, and declared to be the Khalsa—the select... All the rest of the disciples present were similarly baptised and declared Singhs. **1930** G. B. SCOTT *Relig. & Short Hist. Sikhs* iii. 30 After receiving the Pahal, the novice is no longer a Sikh or scholar only, he is *Singh*, a lion, and is entitled to affix that word to his name. **1973** 'S. HARVESTER' *Corner of Playground* II. viii. 150 He thought about his grandfather Swaran,..first Singh of the family in Africa, a coolie who came..to build the railway from Mombasa to Victoria Nyanza.

singing, *vbl. sb.* Add: **1. b.** Now *N. Amer.* (chiefly *Southern*). A gathering joined for collective singing, esp. at a church; a hearty sing-song.

1860 O. L. JACKSON *Colonel's Diary* (1922) 17, I was at a singing at Woodward Church. **1934** C. CARMER *Stars fell on Alabama* II. ii. 49 Ain't seen him since the singin' down at Samanthy. **1949** B. A. BOTKIN *Treas. S. Folklore* I. iv. 93 All through the South, of course, the church is an important social and cultural force, its sociability running the gamut of church-going..bush-arbor revivals, all-day singings with dinner on the grounds, church suppers, singing schools, [etc.]. **1962** E. LUCIA *Klondike Kate* viii. 172 Families got together for 'singings' around the parlor piano and to play games. **1975** *Budget* (Sugarcreek, Ohio) 20 Mar. 3/5 They all had supper at the Lehman home and a singing was held later in the evening.

d. The action of turning informer or laying information against someone. Cf. sense *4 d of the vb. *Criminals' slang.*

1937 *Sat. Even. Post* 18 Dec. 85/1 One actually preferred a three-year penitentiary term to singing. **1940** *Daily Progress* (Charlottesville, Va.) 21 Mar. 3 (*heading*) 'Singing' at murder syndicate's hunting ground. **1973** *Times* 12 Apr. 7/4 The terrified 'singing' of the Sicilian 'Valachi' to delighted magistrates in Palermo has landed 36 Mafia suspects in jail.

2. b. *Teleph.* A continuous self-excited oscillation of audible frequency in a telephone circuit, normally resulting from excessive positive feedback.

1923 T. E. HERBERT *Telephony* xxvi. 829 If two repeaters are in circuit, spaced so closely that the line loss between them is less than the gain given by each 'singing' or 'howling' will persist continuously. **1962** C. F. BOYCE *Open-Wire Carrier Telephone Transmission* xi. 231 Singing affects not only the channel which is unstable but may also cause crosstalk into another system or overloaning of line amplifiers. **1975** R. L. FREEMAN *Telecommunication Transmission Handbk.* vi. 48 To control singing all four-wire paths must have some loss.

4. a. *singing commercial, -master* (later examples); **singing book** (later U.S. examples); **singing game,** a traditional children's game in which singing accompanies associated actions; **singing point** *Teleph.*, the maximum gain that a telephone repeater can have without being liable to self-oscillation in the circuit.

1793 in *Essex Inst. Hist. Coll.* (1885) XXII. 148 Voted to obtain 6 Psalm Books and 6 Singing Books for the use of the Parish. **1872** MRS. STOWE *Oldtown Fireside Stories* 130 They tore out all the leaves of the hymn-books, and the singin'-books besides. **1948** B. ROSE *Wine, Women & Words* 11, I wrote the first singing commercial. **1955** *New Yorker* 23 Apr. 74/1, I approached Mr. Chayefsky's film with no great hope that the thing would prove to be much more interesting than a singing commercial. **1881** *Folk-Lore Rec.* III. II. 169 The following 'Singing Games' are still played and sung by the children of Bocking, in Essex..I. Mary's gone a-milking..II. Thread the Tailor's needle..III. Nuts in May [etc.]. **1905** G. ADE *Let.* 7 Nov. (1973) 33 One or two of the old-fashioned singing games which went as well in the *Sho-Gun*. **1975** B. MEYRICK *Behind Light* xiv. 183 Playing the singing games 'Jenny is a-weeping', 'In and out the stalky bluebells'. **1754** *N.Y. Mercury* 11 Mar. 3/2 William Tuckey, Singing-Master, Desires to inform all lovers of Psalmody, that.. all persons may be taught by him on very reasonable terms. **1891** *Harper's Mag.* Oct. 813/1 The precentor, or singing-master, as he was called, was a tall young man in a black suit with white ruffles. **1928** W. B. YEATS *Tower* 2 And to be the singing masters of my soul. **1976** J. DRUMMOND *Funeral Urn* xvii. 89 He was planning to make a concert singer of her. He'd engaged a singing-master—ostensibly to train the Amber choir, but in fact for Bess. **1924** K. S. JOHNSON *Transmission Circuits for Telephonic Communications* xiv. 166 The singing point or the limiting condition beyond which satisfactory operation of the repeater cannot be maintained. **1934** *Post Office Electr. Engineers Jrnl.* XXVII. 231/2 The vertical scale gives the singing point of a repeater in decibels.

b. *singing-gallery* (earlier example), *-room*, *-seat* (earlier and later examples).

1774 in *Essex Inst. Hist. Coll.* (1884) XXI. 271 Voted Liberty to beuld a singing Seat in the front of the Gallearry Pues. **1842** F. WITTS *Diary* 22 Oct. (1978) 167 The remains were to be deposited at the west end of Upton St. Leonards church, under the singing gallery, near the font. **1851** J. W. HUDSON *Hist. Adult Educ.* 157 Singing-rooms are numerous, prosperous and constantly well-attended. **1902** A. BENNETT *Anna of Five Towns* ii. 34 Mynors.. should have been in his place in the 'orchestra' (or, as some term it, the 'singing-seat') of the [Methodist] chapel. **1976** *S. Wales Echo* 27 Nov. 12/6 (Advt.), Sing along with Mike and Charles at the newly decorated singing-room upstairs.

singing, *ppl. a.* Add: **4. b.** *singing arc*, a direct current arc across which is connected a tuned circuit, causing the arc to oscillate and emit a sound at the frequency of the tuned circuit; *singing sand*, desert or beach sand that emits a singing, whistling, humming, or other continuous sound when disturbed.

1903 *Sci. Abstr.* VI. 30 The author suggests replacing the ordinary high-capacity condensers necessary to produce Duddell's 'singing arc' by the much less costly aluminium condenser. **1906** *Electrician* 21 Dec. 375/1 Limitations as to frequency..beset the use of the singing arc as a transformer of the direct mains current into uninterrupted high-frequency alternating current. **1950** STARLING & WOODALL *Physics* xxxvi. 874 The singing arc..where electrical and thermal factors are involved in the maintenance of oscillations. [**1884** *Proc. Amer. Assoc. Adv. Sci. 1883* 251 (*heading*) The singing beach of Manchester, Mass.] **1897** Singing sand [in Dict., sense 4 a]. **1941** R. A. BAGNOLD *Physics of Blown Sand & Desert Dunes* xvii. 251, I have found singing sand on the slip-faces of both seif and barchan dunes and of drifts formed under the shelter of cliffs. **1970** R. JOHNSTON *Black Camels of Qashran* viii. 133 The night-long background music of the dunes was silenced. They were through the singing sands.

single, *sb.* Add: **3. h.** (Earlier example.) Also in *Baseball*, *spec.* = *one-base hit* s.v. *ONE *numeral a.* 33.

1851 J. PYCROFT *Cricket Field* ii. 24 Ever and anon a single or a double are safely played away. **1867** *N.Y. Mercury* 2 Aug. 6/5 The sharp fielding of the Athletics caused the retirement of their opponents for a single. **1880** *Chicago Inter-Ocean* 29 June 8/3 Force's winning run came off a wild throw by Ward, a sacrifice and single. **1948** *Herald-Press* (St. Joseph, Mich.) 14 Aug. 7/2 Green also bashed out a triple and single during the game. **1974** *Anderson* (S. Carolina) *Independent* 23 Apr. 6A/2 The Astros broke a scoreless tie in the fourth on Bob Watson's single, a wild pitch, and Doug Rader's double.

j. A locomotive engine having a single pair of driving-wheels. Now only *Hist.*

1901 *Railway Mag.* Jan. 31/2 The engine hauling the 9.45 a.m. was No. 22, one of the rebuilt 8 ft. singles. **1931** *Times Educ. Suppl.* 27 June p. iii/3 A famous locomotive, ..one of the original 9 ft. singles built for the Bristol and Exeter Railway.

k. *pl.* Single-screened coal.

1921 *Glasgow Herald* 7 Dec. 9 Coals used in smithwork say 'pearls' and 'singles', varied from 15s. to 18s. per ton f.o.b. **1931** [see *DOUBLE *sb.* 3 r].

1. *U.S. Theatr.* (See quot. 1923.) Cf. *single act* s.v. *SINGLE *a.* 17 a.

1923 *N.Y. Times* 7 Oct. IX. 2/1 *Single*, an artist working alone as an act. 1955 L. Feather *Encycl. Jazz* (1956) 118/2 After Keaton broke up temporarily in 1949, she worked as a single, but rejoined Keaton for several tours. 1962 J. McCabe *Mr. Laurel & Mr. Hardy* i. 26 Following the *Sleeping Beauty* season, he went on as a single again.. for a few odd engagements. 1976 *National Observer* (U.S.) 24 Jan. 18/5 For the past 18 years George Burns has practiced his profession as a single. He has worked..in night clubs and concert halls; he has appeared..on television talk shows.

m. A one-dollar bill (*U.S.*). Also *occas.*, a one-pound note. Cf. *ONCER 2. *slang.*

1936 J. Weidman in *Amer. Mercury* May 86/2, I took out my wallet... I pulled out two singles. 1961 'J. le Carré' *Call for Dead* vii. 78 There he was..showering old singles on me like used tote tickets. 1964 L. Deighton *Funeral in Berlin* xlii. 258 'Do you have a pistol or a knife or a persuader?' 'I have a persuader... Two hundred dollars in singles.' 1977 H. Fast *Immigrants* I. 35 He.. took out a wad of bills, peeling off two fives and two singles.

n. A gramophone record having only one item (typically, of popular music) on each side; an item of music on such a record.

1949 [see *pop single* s.v. *POP *a.* 1 c]. 1958 *Gramophone* Dec. 328/1 There is a single by Nino Rico and his Orchestra. 1965 G. Melly *Owning Up* xi. 135 His version of 'Rock Island Line', originally part of a Chris Barber in Concert LP, was requested so often on the radio that it was put out as a single and rose to be top of the Hit Parade. 1981 *Listener* 1 Jan. 31/2 A track released as a single..topped the singles chart.

o. An engine with only one cylinder; a motor-cycle or car having such an engine.

1951 B. Osborne *Mod. Motorcycles* iii. 19 The designer of hot-stuff singles will scornfully mention Italy's Monza, ..where riders of high-revving Italian 'fours' have been.. completely licked by one of the finest single-cylinder racers ever turned out of a Birmingham factory. 1955 D. Scott-Moncrieff *Veteran & Edwardian Motor-Cars* vi. 111 The old long-stroke singles and twins were no longer allowed [in 1911]; only four- and six-cylinder cars. 1963 Bird & Hutton-Stott *Veteran Motor Car Pocketbk.* 189 It was soon apparent that the small 4-cylinder engine was destined to supplant the big singles and twins which had served so long for light car work. 1976 *New Motorcycling Monthly* Oct. 24/3 It is every inch a purpose-built motorcycle, and must have been a welcome addition to a scene that still reveres our own BSA 'Gold Star' and Matchless singles.

5. *ellipt.* in general application. **a.** = *single ticket* s.v. *SINGLE *a.* 17 a.

1889 E. Dowson *Let.* 1–2 Apr. (1967) 59 If I could see things..as he does I would take a first class single for La Trappe to-morrow. 1903 L. Merrick *Quaint Companions* iv. 49 She congratulated herself on having taken only a 'third single' at Brighton. 1936 *Punch* 5 Feb. 141/1 'Single to Liverpool Street,' I said with easy hauteur. 1972 'R. Crawford' *Whip Hand* I. ix. 55 He..booked a single on the next flight to London.

b. A single bedroom, esp. in a hotel. Cf. Single *a.* 11 d.

1963 [see *DOUBLE *sb.* 3 n]. 1967 A. Hunter *Gently Continental* ii. 12 He goes up to Clooney's room... Number 7 is a small single at the end of the landing. 1973 E. Pace *Any War* I. 5 Yes, sir, the hotel could provide two singles with bath. 1977 B. Aldiss in *Winter's Tales* 23 12 She opened a door to a narrow room...'It's a bit noisy, but it's the only single I've got.'

c. An unmarried or unaccompanied man or woman; a person living alone. Freq. *pl.*

1964 W. & J. Breedlove *Swap Clubs* ii. 57 A *single* is a man or woman who swings alone, without someone to swap. 1967 D. Francis *Blood Sport* viii. 97 Family groups, mostly, and three married couples. No singles except me. 1972 P. A. Whitney *Snowfire* (1973) iv. 52, I had met most of the guests... Some were married, but there were a few young singles too. 1980 R. L. Duncan *Brimstone* iii. 59 We have a club rule against singles.

6. Special combination, in *pl.*: **singles bar** *U.S.*, a bar which caters esp. for young unmarried people in search of social companions.

1969 S. M. Coy *Single Girl's Bk.* vii. 34 Singles bars.. are generally frequented by those under thirty... The good singles bars are crowded, which provides protective covering for the girl who is timid. 1971 D. E. Westlake *I gave at Office* 141, I looked around to find myself in a sort of New York singles' bar without people. 1974 R. M. Strozier in *Atlantic Monthly* Mar. 44 When I visit the East Side singles bars, some of these upper-class snooty girls look down on you. 1979 *United States 1980/81* (Penguin Travel Guides) 228 Like the Rangoon, the Saloon draws a healthy singles-bar crowd.

single, *a.* Add: **I. 6. a.** Also used for emphasis with a superlative.

1969 [see *LOOP *v.*[1] 7]. 1972 *New Yorker* 8 July 1 (Advt.), The single biggest travel buy to anywhere, ever. 1978 *Church Times* 15 Dec. 11/2 Sir Ronald has also, since last September, been chairman of the Central Board of Finance in Church House, and is thus the single most powerful figure in Church finance.

II. 8. c. Designating a person who is bringing up a child or children without the assistance of a marital partner. Chiefly in phr. *single parent* (*family*).

1969 J. Sprey in B. Schlesinger *One-Parent Family* 16 Stigmatization of the single-parent family, and especially of single parents, does occur. 1976 *Women's Report* Sept./Oct. 7/1 This, coupled with the fact that more women are voluntarily becoming single mothers by refusing to have their babies adopted has caused the government to set up a Cabinet Committee on Family Affairs. 1977 C. Fremlin *Spider-Orchid* vii. 55 Peggy Summers was having teenage troubles at last, and wasn't managing so marvellously as a single parent after all. 1980 *Times* 24 Jan. 9/8 Gingerbread caters for all categories of single parents; the divorced, separated, widowed, unmarried, or those whose partners may be in hospital or prison.

13. b. Of whisky: pure, not mixed or blended.

1920 *Glasgow Herald* 22 July 4 No man who knows whisky when he tastes it would prefer them [*sc.* advertised blends] to a 'single' whisky. 1958 *Spectator* 27 June 838/1 It would have been before the days of proprietary whiskies; it would be interesting to know which 'single' or 'self' whisky he used and its strength. 1968 I. C. Taylor *Highland Whisky* (An Comunn Gaidhealach) 5 It is an excellent dry single malt. 1977 C. McCullough *Thorn Birds* vii. 136 Twelve-year-old single-malt Scotch.

III. 17. a. *single act* (*Theatr.*), a performance (orig. in vaudeville) by one entertainer; also, an entertainer who performs unaccompanied; *single anchor* (earlier *fig.* example); *single bond* (*Chem.*), a chemical bond in which the two atoms share one pair of electrons only; *single cream*, cream with a low fat content; *single crown* (*Naut.*), a single crowning given to a knot (see Crown *v.*[1] 14, Crowning *vbl. sb.* 4); *single end* (*Sc.*, chiefly Glasgow), a one-roomed flat; *single entry*, (*b*) listing of a title at only one place in a catalogue, bibliography, or index, without cross-references; an entry so created; *single fare*, the charge for conveyance on an outward journey (but not back); *single-jack* (*N. Amer.*), a short hammer used in percussive hand-drilling by one person alone; *single premium*, a sum which covers the entire cost of insurance in a single payment; *single reed* (*Mus.*), a reed or blade of other material that serves as the sounding apparatus of certain wind instruments; *single shot*, used, usu. *attrib.*, with reference to a facility for producing a single event where repeated action is normal, as on an automatic weapon; *single side band* (*Telecommunication*), either of the two side bands normally associated with the carrier of a broadcast transmission; freq. *attrib.* with reference to a method of transmission in which only one side band is transmitted, the other being suppressed along with the carrier; also short for *single side band transmission*; *single ticket*, a ticket entitling a traveller to a single journey outward, as opp. to a return ticket (cf. Return *sb.* 1 e in Dict. and Suppl.).

1952 Granville *Dict. Theatrical Terms* 163 *Single act*, a solo performance in vaudeville, e.g. a ballad singer, a juggler, an acrobat, an impersonator, or a *raconteur*. 1960 B. Keaton *My Wonderful World of Slapstick* v. 89, I went to New York to see if I could get work there on my own. As a single act. That's what vaudeville people called it. 1822 C. Arbuthnot *Let.* 2 Sept. (1941) 31, I wish you to keep yrself at single anchor, for shd. the Govt. be broken up, you must hurry to me. [1889 G. M'Gowan tr. *Bernthsen's Text-bk Org. Chem.* i. 50 A double bond between two carbon atoms is looser, and therefore more easily broken than a single one.] 1903 Walker & Mott tr. *Holleman's Text-bk. Org. Chem.* I. 150 When a single bond between two carbon atoms is converted into a double one, the directions of the affinities of each of the two carbon atoms must undergo an appreciable alteration. 1966 Williams & Fleming *Spectrosc. Methods in Org. Chem.* iii. 45 The stretching vibrations of single bonds to hydrogen give rise to the absorption at the high frequency end of the spectrum. 1955 J. G. Davis *Dict. Dairying* 320 The manufacture of cream was prohibited during the war but in 1951 sale was permitted for a few weeks. The standards laid down were: single cream 18 per cent. (usually homogenised), double cream 48 per cent. 1962 *Listener* 26 July 155/1, ½ pint of real single cream. 1979 A. Parker *Country Recipe Notebk.* viii. 108 Single cream is thin cream for pouring. It will not whip. 1808 Single crown [see *DOUBLE WALL]. 1883 *Man. Seamanship Boys' Training Ships* (Admiralty) (1886) 121 Form a double-wall, single-crowned, then lay the strands by the sides of those in the single-crown. 1897 J. Wright *Scenes Sc. Life* 27 'A single en', or one apartment. 1935 McArthur & Long *No Mean City* i. 1 Cavity beds are..a feature of the Glasgow slums... The ordinary 'room-and-kitchen' apartment, and even the one-roomed 'single-end', always include a cavity bed or beds. 1981 P. Turnbull *Deep & Crisp & Even* iv. 60 He took a single end in Maryhill and sent for his wife. 1963 *Guide to Universal Decimal Classification* (B.S.I.) i. 7 Generally speaking, each document gets only one entry in the classified file, and this method of 'single entry' is assumed for the time being. Many UDC users, however, favour a method of 'multiple entry', whereby a document on Harvesting of cereals, for instance, would get an entry under both Cereals and Harvesting. 1976 B. Buchanan *Gloss. Indexing Terms* 123 Single entry systems fail to correct completely the separations caused by the application of a citation order..: for this, multiple entry systems are necessary. 1777 P. Thicknesse *Year's Journey* I. vii. 52, I could not refrain from giving her a double fee, for a single fare. 1841 C. Dickens *Let.* 14 Sept. (1969) II. 383, I wish you'd take an opportunity..of asking all about the Fares—what a single fare is—what a double fare—what a cabin with child-stowage. 1972 C. Fremlin *Appointment with Yesterday* i. 11 A small oblong of cardboard... 'Single fare, £1.40' is what it said. 1961 *Press* (Vancouver) 1 Sept. 11 The mechanization of mines in 1890—the replacing of hand-steel, single-jack and double-jack, by drilling machines—had created new problems. 1877 *Cassell's Family Mag.* Jan. 83/2 Net Single and Annual Premiums. 1880 *Encycl. Brit.* XIII. 171/2 We conclude..that the single premium at age 20 for a whole-term assurance of £1 according to the H[m] mortality table, reckoning interest at 3 per cent., is £·32886 or 6*s.* 7*d.* 1975 R. L. Carter *Handbk. Insurance* II. § 1.2.8 The single premium bond is essentially an investment contract in which a lump sum is paid to the insurance company at the inception of the policy, and..invested in units. 1883 Grove *Dict. Mus.* III. 90/1 It is possible to replace it [*sc.* the Double reed] in both these instruments by a single reed of clarinet shape, beating against a small wooden mouthpiece. 1920 U. Daubeny *Orchestral Wind Instruments* vi. 55 There are no instances of the single reed in Egyptian sculpture, but cylindrical pipes of great age, fitted with single-beating reeds have been found in Greece. 1964 S. Marcuse *Musical Instr.* 478/1 Single reeds are idioglott or heteroglott... The beating reed is the most common among Western single reeds (the clarinet reed, for example). 1942 *R.A.F. Jrnl.* 3 Oct. 29 We fired three rounds with the Tommy gun from the hip (with the single-shot mechanism). 1971 J. M. Smith *Digital Logic* iv. 67 The one-shot or single-shot generator is a device for producing a pulse output from a trigger signal input. 1977 'J. McVean' *Bloodspoor* xx. 263 Hanson switched the Schmeisser regulator to single-shot and started to fire back. 1923 *Proc. IRE* XI. 40 It is of particular importance for long wave radio telephone transmission where the width of a single side-band is so large a fraction of the total frequency range available that the number of independent channels is at best very limited. *Ibid.* 41 The use of single side-band transmission has probably progressed farthest in connection with carrier telephony over wires. 1959 K. Henney *Radio Engin. Handbk.* (ed. 5) xviii. 8 In most commercial single-side-band transmitters the SSB signal is generated in an SSB generator at a frequency in the range of 100 to 500 kc. 1976 *Electronics Today Internat.* July 10/3 In areas around cities, these channels are congested and so many stations have changed over to Single Sideband, which gives higher communications efficiency and an extra 48 channels. 1859 *A.B.C. or Alphabetical Railway Guide* Aug. 121 London, Paris, and the Continent... Fares throughout (Single Tickets, available for four days), First Class, 28/0; Second Class, 20/0. 1979 'J. le Carré' *Smiley's People* (1980) xvii. 212 He bought a second-class single ticket to Hamburg.

b. *single-boater* (see quot. 1933); *single-decker*, † (*a*) *U.S.* (see quot. 1896); (*b*) an aircraft, tramcar, etc., having only one deck; now usu. a single-decked bus; freq. *attrib.*; *single-hander*, (*a*) an action performed single-handedly or without assistance (in quot. 1877, a chase) (*rare*); (*b*) one who sails a boat single-handed; a single-handed yachtsman; *single-seater* (attrib. uses at sense *18).

1933 S. Bradford *Shell-backs* 70 A single boater is a trawler not fishing with a fleet but on its own. 1934 W. Wood *Fleeters* iv. 59 The single-boaters..far outnumbered the fleeters. 1896 C. H. Haswell *Reminisc. Octogenarian in N.Y.* xv. 332 James P. Allaire had constructed..a four-story house designed for many tenants... It is what is now termed a 'single-decker', that is, but one suite of rooms on a floor. 1910 *Sphere* 20 Aug. 176/1 The first aeroplane illustrated is the Santos Dumont aeroplane or single-decker. 1930 *Aberdeen Press & Jrnl.* 12 Feb. 6/3, I see that the L.C.C. has been selling off old single-decker tramcars at £5 apiece. 1935 S. Beckett *Echo's Bones*, The little single-decker. 1935 *Discovery* Feb. 58/2 In these vehicles the main entrance is in front, beside the driver as in many single-decker coaches already on the road. 1954 M. Procter *Hell is City* I. i. 11 A red-and-white bus ..was approaching... It was a..single decker. 1962 L. Deighton *Ipcress File* xxx. 196 At the bottom of the street was a single-decker bus. 1976 P. R. White *Planning for Public Transport* iii. 56 The rear-engine layout was also adopted for single-deckers. 1877 *Coursing Calendar Autumn 1876* 217 War Note never seeing the hare, Adventurer fell in for a single-hander of great length, the first-named being drawn ultimately. 1893 [in Dict.]. 1954 H. Innes *Strange Land* I. 18 'There should be two men on board her.' 'Well, this bloke was single-handed.'..'He was a single-hander all right.' 1976 P. Heaton *Single-handers* i. 26 Alain Colas..is a racing singlehander, a competitor. 1916 H. Barber *Aeroplane Speaks* Pl. xxvi, A 50 h.p. Gnome single-seater. 1972 'M. Yorke' *Silent Witness* II. ii. 19 Twin chairs were not so bad... But this chair-lift was a single-seater and you rode alone.

18. Used attributively. Also, *single-bar*, *-cause*, *-channel*, *-class*, *-coil*, *-colour*, *-column*, *-crystal*, *-deck*, *-electron*, *-engine*, *-family*, *-issue*, *-language*, *-layer*, *-lens*, *-letter*, *-manual*, *-member* (earlier example), *-morpheme*, *-note*, *-pane*, *-particle*, *-party*, *-pass*, *-person*, *-phase*, *-ply*, *-point*, *-pole*, *-purpose*, *-reed*, *-seat*, *-seater*, *-set*, *-sex*, *-speed*, *-stage*, *-storey*, *-stress*, *-syllable*, *-tier*, *-tube*, *-turn*, *-unit*, *-use*, *-word*; *single-cell protein*, protein derived from a culture of single-celled organisms; *single-electrode* (*Chem.*), with reference to a half-cell considered in isolation; *single-lens reflex* (*camera*) (*Photogr.*), a reflex camera in which the lens that forms the image on the film is also used to provide the image in the viewfinder (by means of a mirror behind the lens that is

automatically moved out of the way when the shutter release is operated) (cf. *SLR* s.v. *S 4 a); *single-plate clutch* (see quot. 1940); *single-start* (*Engin.*), designating a screw-thread or worm gear that has one continuous thread along its entire length; *single-vision* (*Ophthalm.*), (of spectacles) of which each lens is a single optical element; not bifocal, etc.; *single-wire*, designating an electrical wiring system in which current is carried by one wire, the return being provided by the chassis or frame of the apparatus or installation or the earth.

1964 *Amer. Speech* XXXIX. 104 There is a break—in speech, a single-bar juncture; in writing, a comma or dash —between the noun and its juncture. 1966 'M. Halliday' *Wicked as Devil* iv. 36 Helen switched on a single-bar electric fire. 1977 P. Johnson *Enemies of Society* ix. 127 Marx, by contrast, has a single-cause theory: all the evils of society arise from private property; abolish that, and they will disappear. 1968 Mateles & Tannenbaum *Single-Cell Protein* i. 7 Some may wonder where the name 'Single-Cell Protein' came from and why it was adopted. It was invented at M.I.T. in May 1966, as a result of the insistent prodding of Professor Carroll Wilson. 1970 *Daily Tel.* (Colour Suppl.) 19 June 14 Single cell protein is one of the most 'exotic' of various new food sources. 1977 Single cell protein [see *single-celled*, sense 19 below]. 1962 A. Nisbett *Technique Sound Studio* ii. 33 Single-channel 'monophonic' recordings. 1977 *Proc. R. Soc. Med.* LXX. 382/1 Single-channel extracochlear stimulation will only provide low frequency information. 1967 *N.Y. Herald Tribune* (Internat. ed.) 11–12 Feb. 3/7 (Advt.), Walk on or drive on to one of these fine ultra-modern single-class ships for an overnight crossing in absolute comfort. 1977 *Listener* 17 Mar. 347/3 They built..for the propagation of middle-class values..a vast process of specialised, single-class development. 1962 Simpson & Richards *Physical Princ. Junction Transistors* iii. 38 One pass of the specimen is then equivalent to several passes in a single-coil apparatus and the removal of impurities is accelerated. 1935 B. Rackham in *Chinese Art* (Burlington Mag. Monographs) 20 The beauty of celadon, turquoise, crackled white and other single-colour glazes. 1964 *Gloss. Letterpress Rotary Printing Terms* (*B.S.I.*) 19 *Single colour unit*, a section of the press embodying one printing couple to print one side of the web in one colour. 1960 *Guardian* 9 Nov. 8/3 The 'News Chronicle'..charged for display advertising at the rate of £13 per single column inch. 1976 J. Bingham *God's Defector* iv. 43 A single-column picture of himself at the church door. 1956 *Nature* 14 Jan. 77/2 In fundamental research, much work is being done on single-crystal specimens, whereby grain-boundary effects and some other variables are eliminated. 1955 *Notes on Science in USA 1954* (Brit. Commonwealth Scientific Office, N. Amer.) 29 The large number of experiments that are now being performed on single crystal specimens. 1929 *Times* 2 Nov. 4/7 The chassis, with rather lighter driving axles and springs is sold for single-deck and coach operation. 1967 M. Chandler *Ceramics in Mod. World* ii. 84 Single-deck tunnel kilns. 1913 *Jrnl. Amer. Chem. Soc.* XXXV. 24 No satisfactory method has been found for determining the absolute magnitude of any single electrode potential. 1965 Phillips & Williams *Inorg. Chem.* I. ix. 312 It has proved very difficult, some believe impossible, to measure absolute single-electrode potentials. 1944 Stewart & Wilson *Rec. Adv. Physical & Inorg. Chem.* (ed. 7) 384 It is probable that single electron bonds..enter into the *average* final structure for diborane. 1968 M. S. Livingston *Particle Physics* iii. 39 The spectra of single-electron atoms such as He[+] and Li[++]. 1942 *R.A.F. Jrnl.* 27 June 1 A service between London and Paris with single-engine modified Service aircraft. 1978 R. Ludlum *Holcroft Covenant* xxxvii. 423 A small single-engine plane circled in the night sky above the flat pasture in Chambéry. 1967 R. Ardrey *Territorial Imperative* iii. 93 Unlike any other ape and like few monkeys, the gibbon lives in a single-family group, paired on a territory usually for life. 1980 *Washington Post* 4 Dec. DC1, 49 buildings containing about 600 units are being transferred from rental status or private single-family ownership to condominiums. 1977 *Time* 19 Dec. 29/3 The right-to-lifers are single-issue individuals... They vote on what he or she says about abortion. 1946 H. Jacob *On Choice of Common Lang.* 39 Four single-language frequency lists. 1978 *Language* LIV. 8 Table 2 shows the number of entries which fulfill the structural conditions for apical displacement, after we eliminate single-language entries and the disqualified ones. 1940 *Chambers's Techn. Dict.* 772/1 *Single-layer winding*, a type of armature winding in which there is only one coil-side per slot. 1946 *Nature* 21 Sept. 422/1 The reflectors consist of single-layer and multiple-layer films. 1936 *Discovery* Aug. 237/1 The camera..has 4½ times the stereoscopic parallax of the single-lens and five-lens cameras. 1940 A. L. M. Sowerby *Wall's Dict. Photogr.* (ed. 15) 547 In a single-lens reflex, pressure on the release first lifts the mirror.., and then releases the focal-plane shutter. 1955 T. A. Longmore *Med. Photogr.* (ed. 5) III. 374 Being a single lens reflex camera there is a complete absence of parallax, so that the image seen on the focusing screen is exactly the same as that which will be recorded on the film. 1957 *Encycl. Brit.* XVII. 825/2 A single-lens camera may be provided with a beam splitter..,giving two adjacent photographs on the normal picture area. 1962 L. S. Sasieni *Optical Dispensing* xiii. 334 Apart from the single-lens magnifiers..there are a number of telescopic units. 1977 L. Gaunt *Olympus Bk.* 8 The final part of the 35 mm single-lens reflex viewing system is the eye-piece lens, focused on the viewfinder screen via the reflecting surfaces of the prism. 1878 H. MacColl in *Proc. Lond. Math. Soc.* X. 26 The monomial (or single-letter) statements. 1964 D. Ward in D. Abercrombie et al. *Daniel Jones* 393 For the convenience of readers not familiar with Cyrillic a single-letter transliteration system is given. 1880 Grove *Dict. Mus.* II. 591/2 Soon after the Restoration, Ralph Dallam built an organ for St. George's Chapel, Windsor... It was a single-manual organ only. 1978 *Early Music* Oct. 585/1 An Italian single-manual harpsichord bearing a spurious inscription dated 1740 but actually of earlier date. 1884 E. W. Hamilton *Diary* 2 Dec. (1972) II. 746 Lord Hampden..likes the system of single-member Districts, for which he has long been an advocate. 1956 J. Lotz in L. White *Frontiers of Knowl.* xiv. 221 The single-word, or better, single-morpheme sentences of the young child..cannot be analyzed into phonemes nor combined into sentences. 1949 L. Feather *Inside Be-bop* i. 6 The single-note solo style was a complete departure from the pattern of solos in chords established by..conventional jazz guitarists. 1922 Joyce *Ulysses* 698 Water closet..with opaque singlepane oblong window. 1892 A. M. Worthington *Dynamics of Rotation* ii. 20 Any rigid body may be regarded as made up of such ideal single-particle systems. 1970 G. K. Woodgate *Elem. Atomic Struct.* v. 87 For *N* non-interacting electrons with no spin-orbit interaction we find it convenient to go back to the single-particle representation. 1941 *Pacific Affairs* XIV. 76 The emergence of a fully totalitarian 'single-party State'. 1979 E. Norman *Christianity & World Order* v. 66 Tanzania—a country with a rigidly enforced socialist collectivization and a single-party constitution. 1964 S. Crawford *Basic Engin. Processes* iii. 93 The vertical plate is tack-welded in position..and then finally welded by either the single-pass or the multiple-pass techniques. 1969 *Computers & Humanities* IV. 43 Clearly, there should be much less occasion for careless errors and omissions in a multiple-pass system like Regener's than in a single-pass system like IML. 1957 *Loneliness* (Women's Group on Public Welfare) iii. 41 The number of single-person households..increased by 104 per cent between 1931 and 1951. 1900 *Jrnl. Brit. Inst. Electr. Engin.* XXIX. 246 The motor is being supplied with single-phase currents. 1946 *Nature* 31 Aug. 307/2 The electrolytic polishing of multi-phase metals is usually more difficult than that of single-phase metals owing to differing properties..of the different phases. 1979 *Dictionaries* I. 31 Editors are not restricted to single-phrase descriptions. 1926 *Motor Man.* (ed. 26) iii. 49 (*caption*) Single-plate clutch, showing internal details. 1940 *Chambers's Techn. Dict.* 772/2 *Single-plate clutch*, a friction clutch..in which the disc-shaped or annular driven member, fabric-faced, is pressed against a similar face on the driving member by springs. 1970 K. Ball *Fiat 600, 600D Autobk.* v. 45/1 The clutch, common to both the 600 and 600D series engines, is a conventional dry, single plate type with a spring cushioned hub as part of the driven plate. 1957 Simpson & Weir *Weaver's Craft* iii. 25 A tightly packed weft of thick hand-spun or single-ply rug wool. 1967 *Jane's Surface Skimmer Systems 1967–68* 13/2 A segmented skirt of single-ply neoprene-nylon fabric. 1935, 1959 Single-point [see *INSTANTANEOUS *a.* 1 d]. 1979 *North Sea Progress* (Shell Internat. Petroleum Co.) 6 The field will be served by tankers at single-point mooring buoys to take the oil production. 1931 *Illustr. London News* 29 Aug. 342/3 The electrical system is 12 volts, with single-pole wiring. 1975 G. J. King *Audio Handbk.* x. 223 The slope of a single pole RC network always ultimately assumes a rate close to 6 dB/octave. 1919 *Daily Mail Year Bk.* 112/2 Many thousands of these are in service, to which the term 'single-purpose machines' is applied. 1943 J. S. Huxley *TVA* v. 27 The single-purpose costs which can be directly allocated to one or other of the functions. 1971 *Fremdsprachen* XV. 46 Each single-purpose use may be justified on its own, but the complete effect of piecemeal development can be chaos. 1931 G. Jacob *Orchestral Technique* iii. 23 *The Clarinet* (single-reed instrument). 1976 D. Munrow *Instr. Middle Ages & Renaissance* vi. 39/4 The French word *chalumeau* has been applied to the shawm as well as a detached double-reed bagpipe chanter..and the single-reed precursor of the clarinet. 1946 *Nature* 5 Oct. 469/1 The aircraft rocket enabling a single-seat fighter to deliver a salvo equal in hitting power to the broadside from a small cruiser. 1967 *Jane's Surface Skimmer Systems 1967–68* 8/1 It has built a single-seat vehicle designated Naviplane N 101. 1910 R. Loraine *Diary* 10 Apr. in W. Loraine *Robert Loraine* (1938) vi. 104 A small single-seater monoplane. 1930 *Engineering* 7 Mar. 316/1 Following upon a recent order for five Bristol Bulldog all-steel single-seater fighter aeroplanes, the Latvian Government has placed a further order..for seven additional Bulldog machines. 1973 J. Leasor *Host of Extras* i. 21 In 1911..with a single-seater body..a Rolls-Royce covered a quarter-mile stretch at Brooklands at 101 miles an hour. 1961 *Guardian* 11 Dec. 7/7 Higher salaries..have made a single-set drama..cost more nearly $100,000. 1971 J. Elsom *Theatre outside London* x. 181 Low-cast, single-set comedies, with the occasional mystery play or classic revival. 1939 A. H. Whipple *Educ. up to Fifteen Years* 36 Experiments should be conducted in single-sex schools to ascertain..whether or not the education of girls should differ in important respects from that of boys. 1980 *Times* 19 Feb. 2 Applications to Oriel, the only remaining men's single-sex college, are down again. 1958 M. L. Hall *Newnes Compl. Amat. Photogr.* ii. 32 Single-speed rotary or single-leaf shutters, placed either in front of or behind the lens. 1975 *Language for Life* (Dept. Educ. & Sci.) vi. 95 Unfortunately, if most of their reading is of the single-speed kind, children will be habituated to becoming single-speed readers. 1922 Glazebrook *Dict. Appl. Physics* II. 891/2 Single-stage amplifiers. 1945 H. D. Smyth *Gen. Acct. Devel. Atomic Energy Mil. Purposes* iv. 40 Single-stage separators had effected the enrichment of the U-235 on a laboratory scale to about the degree predicted theoretically. 1956 *Spaceflight* I. 24/1 The first stage resembles the Viking rocket which attained a record altitude (for single-stage rockets) of 158·4 miles in 1954. 1964 S. Crawford *Basic Engin. Processes* xiv. 301 On a single-start thread the lead and the pitch are identical. 1975 Bram & Downs *Manuf. Technol.* iv. 120 The depth of a single-start thread stands in a definite relationship to its pitch. 1947 A. Waring *Approach to Better Housing* ii. 33 While not advocating the provision of single-storey dwellings for large families..nevertheless these can be planned for small families and aged people..quite as successfully as the two-storey dwelling. 1970 D. Goldrich et al. in I. L. Horowitz *Masses in Lat. Amer.* v. 183 They are housed predominantly in single-story dwellings on individual plots. 1964 C. Barber *Linguistic Change in Present-Day Eng.* iv. 86 The word *greatcoat* is a modern example of a transitional state between a double-stress and a single-stress form. 1959 Wimsatt & Beardsley in *Publ. Mod. Lang. Assoc. Amer.* LXXIV. 595 The single-syllable foot occurs in lines that sound like this: 'Weave, weave, the sunlight in your hair'. 1959 *Daily Tel.* 6 Mar. 21/3 The Ministry has suggested that single-tier highway authorities with less than 200,000 people in their areas should be set up in Greater London. 1977 *Whitaker's Almanack 1978* 624 For the purpose of local government Northern Ireland has a system of 26 single-tier district councils. 1904 A. B. F. Young *Compl. Motorist* xi. 247 There are some single-tube pneumatic tyres in which the whole of the fabric is constructed in one piece. 1935 *Discovery* Feb. 44/1 (*caption*) Guide-cradle for launching single-tube rockets. 1963 *Gloss. Mining Terms* (*B.S.I.*) III. 13 *Single-tube core barrel*, the simplest core barrel, having only a single cylindrical tube. 1940 *Chambers's Techn. Dict.* 772/2 *Single-turn coil*, an armature coil consisting of a single turn of copper bar. 1962 Corson & Lorrain *Introd. Electromagn. Fields* vi. 236 We consider single-turn coils for simplicity. 1936 *Discovery* Aug. 237/1 The world's largest single unit multi-lens aerial mapping camera. 1973 *Tucson* (Ariz.) *Daily Citizen* 22 Aug. 27/2 We have passed the time when we can afford the luxury of building single-unit homes. 1959 *Gloss. Terms Packaging* (*B.S.I.*) 43 *Single-use tube*, a tube with sufficient contents for one use only. 1969 *Computers & Humanities* III. 138 The heavy-duty Selectric typewriter also requires frequent adjustments for uniform impressions, plus a single-use ribbon for publication output. 1962 L. S. Sasieni *Optical Dispensing* vii. 174 Logically a single-vision lens prescribed for reading would be the most comfortable. 1971 *Optometry Today* 24 This team is capable of examining, prescribing, fabricating and dispensing single-vision lenses, mounted in frames, on the spot. 1902 *Encycl. Brit.* XXXIII. 228/2 In 1892..he established communication between Lavernock Point and an island called Flat Holme..by placing at these positions insulated single-wire circuits, earthed at both ends. 1913 V. B. Lewes *Oil Fuel* iv. 108 The single-wire system must not be adopted for any part of the electric lighting installation in vessels carrying petroleum. 1907 W. James *Pragmatism* vii. 239 All the great single-word answers to the world's riddle, such as God, the One, Reason, Law, [etc.]. 1964 R. H. Robins *Gen. Linguistics* 284 Compound verbal expressions..fulfil some of the semantic functions of the single-word tense forms of other languages.

19. *single-barrelled* (earlier and *fig.* examples), *-bedded*, *-celled*, *-coloured*, *-decked*, *-ended*, *-engined*, *-purposed*, *-reeded*, *-roomed*, *-sexed*, *-sided*, *-spaced*, *-storeyed* (*-storied*), *-syllabled*, *-worded*; *single-stranded* (*Biochem.*), (of a nucleic acid) consisting of only one sequence of nucleotides; hence *single-strandedness*.

1821 E. Baker *Remarks on Rifle Guns* (ed. 8) 114 The average weight of a single-barrelled gun should be from 6lbs to 6lbs 4oz. 1880 'Mark Twain' *Speeches* (1923) 83 He possessed a single-barrelled fame before; he will possess a double-barrelled fame now. 1788 J. Woodforde *Diary* 20 May (1927) III. 27, I had a very good single bedded Room to night. 1892 I. Zangwill *Childr. Ghetto* III. 188 Here single-bedded cabins could be had as low as fourpence a night. 1972 'G. North' *Sergeant Cluff rings True* xix. 146 The single-bedded ward had its memories... A constable..watched the bed in which the Sergeant had once lain wounded. 1899 W. James *Talks to Teachers* xiv. 163 In biology, we used to have interminable discussion as to whether certain single-celled organisms were animals or vegetables, until Haeckel introduced the new apperceptive name of Protista, which ended the disputes. 1977 G. Scott *Hot Pursuit* xii. 105 Single cell protein is produced by single-celled animals: bacteria, yeasts, fungi, that sort of thing. 1703 tr. H. van Oosten *Dutch Gardener* II. iv. 60, I think the single colour'd to be the best..because the tulip that is already changed and striped, doth easily mix her colours together; and this is the reason why the single colours that come from them, have not so strong a colour as those from the single coloured ones. 1940 W. Stevens in *Accent* Autumn 12 The single-colored, colorless, primitive. 1869 Single-decked [see *DOUBLE-DECK]. 1972 'G. North' *Sergeant Cluff rings True* xiv. 112 The single-decked bus..laboured up the hill. 1952 *Proc. IRE* XL. 11/1 Since the output is single ended, the feedback can be made directly from the midpoint of the output stage to a preceding single-ended stage. 1975 *Official Transcript Techn. Papers Ann. Nat. Cable Television Assoc. Convention* (New Orleans) 24 Investigations were conducted on an operating cable television system to explore methods of increasing the channel capacity of broadband single-ended amplifiers. 1964 *Oceanogr. & Marine Biol.* II. 47 Transportation was by means of single-engined aircraft. 1978 R. V. Jones *Most Secret War* xxxiv. 301 It happened that a unit of single-engined fighters had been formed in the preceding weeks by Major Hajo Herrmann. 1905 J. London *Let.* 4 Apr. (1966) 169 You and I are both fighters, and single-purposed fighters too. 1933 'R. Crompton' *William—the Rebel* viii. 164 They were all large, single-purposed, unsmiling men. 1920 U. Daubeny *Orchestral Wind Instruments* vi. 55 Somewhat similar rude single-reeded pipes are still used by Italian shepherds and Roman pifferari. 1911 G. B. Shaw *Getting Married* 129 Very few couples can live in a single-roomed tenement without exchanging blows quite frequently. 1903 —— *Man & Superman* III. 112 She [*sc.* Nature] created him in order to produce something better than the single-sexed process can produce. 1934 L. B. Pekin *Progressive Schools* iv. 62, I cannot imagine any teacher who has given co-education a fair trial..ever returning to a single-sexed school. 1937 *Discovery* Sept. 284/1 Single-sided, long-playing, unbreakable durium-type records so popular a few years ago. 1977 *Gramophone* Aug. 262/3 Frank Andrews (London, NW10) points out that Zonophones were single-sided until June 1911. 1956 *Nature* 18 Feb. 334/2 A number of strains [of lucerne] planted as single-spaced plants in the field. 1959 J. Thurber *Years with Ross* xi. 191 Ross sat down at his

typewriter..and wrote..a remarkable five-page single-spaced letter. 1975 T. Allbeury *Special Collection* xv. 99 A foolscap sheet in single-spaced typing. 1835 J. E. Alexander *Sk. in Portugal* v. 113 By the side of the road to Santarem was the quarter of Colonel Shaw, a long single-storied peasant's house in a vineyard. 1967 *Antiquaries Jrnl.* XLVII. 275 The hall probably rose clear above this east room, which may have been roofed as a single-storeyed lean-to. 1954 *Proc. R. Soc.* A. CCXXIII. 94 Most of these earlier formulations..have involved single stranded structures and must be rejected. 1964 G. H. Haggis et al. *Introd. Molecular Biol.* iv. 79 The RNA component of the virus is a long single-stranded nucleic acid chain. 1978 *BioSystems* X. 102/1 Double stranded DNA would be unsuitable because, unlike single-stranded RNA it cannot fold up on its own to form specific and complex 3D structures. 1974 *Nature* 5 Apr. 507/1 They..exhibit greater buoyant densities due to the presence of RNA sequences and/or some degree of single strandedness. 1890 W. James *Princ. Psychol.* I. xiv. 558 The 'cue' was given by single-syllabled words called out by an assistant. 1948 E. Sitwell *Notebk. on Shakespeare* vi. 72 In such lines [from *King Lear*]..the single-syllabled words take on the hugeness of those new-made stones that Deucalion and Pyrrha, the Deluge being over, found and cast behind their backs. *a* 1832 J. Bentham *Logic* vii, in *Wks.* (1843) VIII. 252/1 Finding a more appropriate single-worded denomination for the species. 1893 W. Minto *Logic* I. ii. 68 Whether this is single-worded or many-worded is..a grammatical question.

20. a. *single-tuned* adj. (*Electronics*), having a single tuned circuit between two active devices.

1947 F. E. Terman *Radio Engin.* (ed. 3) vii. 346 The band width in the case of an amplifier system employing double-tuned circuits is defined in the same manner as for a system employing single-tuned circuits. 1975 D. G. Fink *Electronics Engineers' Handbk.* XIII. 43 The two common types [of tuned interstage] are the single- and double-tuned interstage.

b. Also with vbl. sbs., as *single-boating*, *-manning*, *-spacing*.

1934 W. Wood *Fleeters* iv. 59 Men who are only used to Iceland and single-boatin' aren't any good at this fleetin' job. 1965 *Times* 22 Oct. 8/7 The proposal that single manning of locomotives should be introduced within a year or two. 1973 *Guardian* 31 Dec. 13/8 London Transport..has launched..single-manning... But the single-manned buses..take up to five times as long at stops. 1956 F. C. Avis *Bookman's Conc. Dict.* 273/1 *Single spacing*, the style of typewriting in which the lines of characters follow immediately after each other without any interlinear spacing, equivalent to 'Solid' typesetting. 1958 E. Newby *Short Walk in Hindu Kush* ii. 21 A great spate of letters..neatly typed in single spacing. 1978 *Church Times* 21 July 5/4 A letter of more than two pages in single-spacing.

c. With vb. *single-space*.

1961 *Guardian* 6 Feb. 9/5 Press releases would be single-spaced to save paper. 1963 D. Heyes *12th of Never* (1964) i. 8 He..continued typing..and..single-spaced the final line to squeeze it in.

single, *v.*[1] Add: **8. c.** With *up*: to cast off all turns of rope except one. Also *intr. Naut.*

1900 J. C. Cantwell *Diary* 17 May in *Rep. Operations U.S. Revenue Steamer Nunivak* (1904) iv. 57 The lines by which the *Nunivak* was held to the shore were singled up. 1925 R. Clements *Gipsy of Horn* vi. 103 Our moorings had been singled up. 1959 C. S. Forester *Hunting Bismarck* 8 Already sailors at the lines were singling up and then casting off. 1966 T. Pynchon *Crying of Lot 49* ii. 31 The little submarine..was at the quai, singling up all lines.

9. d. *Baseball.* Of a batter: to hit a single (sense *3 h); to make a one-base hit. Also *trans.*, by singling to enable (another player) to reach home base.

1916 *Chicago Tribune* 7 Oct. 13/1 In the ninth, the first man up singled. 1949 *Clarke County Democrat* (Grove Hill, Alabama) 22 Sept. 2/5 The first St. Michael batter singled. 1966 *N.Y. Times* (Internat. ed.) 22 Apr. 12/1 Felipe Alou singled in the first inning off Ray Culp. 1970 *Globe & Mail* (Toronto) 28 Sept. 19/3 Adolfo Phillips singled home Ron Brand in the 11th inning. 1978 *N.Y. Times* 30 Mar. D 19/4 In the third, Ken Henderson doubled, John Stearns singled and it was 4–0.

single-blind, *a.* [f. Single *a.*, after *DOUBLE(-)BLIND *a.*] Applied to a test or experiment conducted by one person on another in which information about the test that may lead to bias in the results is concealed from one of the parties.

1963 *Amer. Jrnl. Psychiatry* CXX. 67/1 The single-blind study showed relief of symptoms with mephenoxalone in 3 of 9 patients. 1976 *Sci. Amer.* Jan. 8/3 The test was not even single-blind. 1978 *Nature* 20 Apr. 729/1 The studies were performed open or single-blind, in most cases with randomised placebo controls. 1981 *Brit. Med. Jrnl.* 11 July 122/2 If the identity of the treatment is concealed only from the assessor then the trial is single-masked or single-blind... The term 'single-blind' has also been applied to trials in which only the patient is in the dark, although this is not nearly so important as the assessor being in the dark.

single grave. *Archæol.* [f. Single *a.* 17, 18: tr. Du. *enkelgraf* single grave.] A barrow-grave containing the remains of only one person. Freq. *attrib.*, *spec.* (esp. with initial capitals) designating a culture characterized by individual burial which first flourished in northern Germany and Scandinavia during the later Neolithic Age, or representatives of this.

[1898 S. Müller in *Aarb. f. nord Oldk.* 157 (*heading*) De jydske Enkeltgraver fra Stenalderen.] 1936 J. G. D. Clark *Mesolithic Settlement of Northern Europe* ii. 73 The triangular-sectioned flake arrowhead..is abundant in passage-graves and occurs in single-graves. 1937 E. V. Gordon tr. *Shetelig & Falk's Scand. Arch.* v. 66 It can be shown that the single-grave people, with their characteristic battle-axes and their own pottery, came from.. central Europe. 1954 S. Piggott *Neolithic Cultures* xii. 378 In Denmark the Single-grave invasion takes place after the LG IV marine transgression. 1955 *Univ. Lond. Inst. Archaeol. Ann. Rep.* XI. 39 The discovery at Ohlenburg, ..of a devolved Single-Grave beaker. 1957 V. G. Childe *Dawn Europ. Civilization* (ed. 6) ix. 160 Archaeologically these graziers are known only by little cemeteries of barrows, and so they are termed the Single Grave folk. 1963 S. Piggott in Foster & Alcock *Culture & Environment* iv. 60 Glasbergen and Van der Waals have shown the Single Grave element to be represented by their Footed Beakers, with an absolute date of *c.* 2200 B.C. 1970 Bray & Trump *Dict. Archaeol.* 212/2 This [burial] rite.. links the Single-Grave cultures with the great Corded Ware-Battleaxe complex.

single-handed, *a.* Add: **2. a.** Also passing into *adv.* (Later examples.)

1900 J. Slocum *Sailing Alone around World* xxi. 272 On the 4th of June, 1898, the *Spray* cleared from the United States consulate, and her license to sail single-handed, even round the world, was returned to her for the last time. 1939 G. B. Shaw *Geneva* I. 4, I have to do it singlehanded too: I havnt even an office boy to help me. 1970 G. M. Fraser *General danced at Dawn* 144 Lance-Corporal Michael O'Leary, who took on crowds of Germans singlehanded.

singlehandedly *adv.* (later examples).

1964 *Duckett's Reg.* Mar. 41/2 America's most famous Franciscan, who almost singlehandedly Christianised America's Pacific Coast. 1979 *N.Y. Rev. Bks.* 25 Oct. 10/1 Barbara Underwood..boasts that she single-handedly made *one quarter of a million dollars* for the cult.

single-line, *a.* Add: **1.** Also, of or pertaining to things ranged in a single line; spec. *single-line traffic* (see quot. 1954).

1924 *Beaver* (Winnipeg) Aug. 410 In the extreme north, single-line hitching [of dogs to sleds], and fan hitching is in general use. 1937 *Burlington Mag.* Feb. 93/2 The National Gallery led the way..towards the open, single- or double-line hanging which is now the rule. 1954 *Gloss. Highway Engin. Terms* (*B.S.I.*) 54 *Single-line traffic*, traffic constrained to movement in one direction in a single traffic lane. 1967 G. F. Fiennes *I tried to run Railway* i. 3, I had learned by heart the Block Regulations for double and single lines, the Guard's Rules and the Rules of Single Line Working. 1967 E. Lemarchand *Death of Old Girl* xx. 226 There was a bad smash and a big pile-up... It was an hour before they got even single-line traffic going again. 1970 O. John *Diamond Dress* iii. 39 Part of the autostrada is down to single line traffic... Road repairs.

single-o, *sb.*, *a.*, and *adv.* *U.S. slang* (chiefly *Criminals'*). [f. Single *a.* 1; cf. O *int.* 3.]

A. *sb.* **a.** In gaming: ? the number one. **b.** A crime perpetrated without an assistant. Also, a solitary or single person, a loner; *spec.* a criminal who works alone.

1916 H. L. Wilson *Somewhere in Red Gap* vi. 262 She exposed some very distressing facts about his [*sc.* her husband's] nature the time she put five apiece on the three numbers and the single-o come up. 1930 R. Chadwick in *Liberty* 5 July 20/2, I have my first experience in single-o jobs... The first single-o is a street heist. 1931 G. Irwin *Amer. Tramp & Underworld Slang* 170 *Single O*, one working a lone 'game' or 'racket'. One travelling alone for preference. 1942 Berrey & Van den Bark *Amer. Thes. Slang* §461/3 *Single O*, who who works without a confederate.

B. *adj.* Solitary, lone; unaccompanied; *spec.* of (one who engages in) criminal activity without an accomplice.

1930 [see the sb. above]. 1950 *Harper's Mag.* Feb. 71/2 There are 'single-o' heist-men, such as the one known in the papers as Slick Willie, who has robbed large and well-protected banks single-handed, but the vast majority of the brotherhood work in mobs. 1955 D. W. Maurer in *Publ. Amer. Dial. Soc.* XXIV. 83 He [*sc.* a lone pickpocket] is usually referred to as a *single o tool*, a *single handed tool*, or a *single o cannon*.

C. *adv.* Alone; independently; without an accomplice.

1948 *Even. Bull.* (Philadelphia) 7 Apr. 39/7 Instead of working single-o as was his custom, Ernie used an accomplice to drive the getaway car. 1955 D. W. Maurer in *Publ. Amer. Dial. Soc.* XXIV. 100 He's a guy that will muzzle around single o. 1962 'K. Orvis' *Damned & Destroyed* xii. 83 Little Faysy wants to go dream-streeting single-o.

singlet. Add: **3. a.** *Physics* and *Chem.* A single line in a spectrum, not part of a multiplet; an atomic or molecular energy level or state possessing (in the case of fine structure) zero electronic spin and orbital angular momenta giving only one value of the quantum number *J*, or (in the case of hyperfine structure) zero electronic and nuclear angular momenta, giving only one value of the quantum number *F*; a molecular state in which all electron spins are paired. Freq. *attrib.*

1920 *Astrophysical Jrnl.* LII. 2 An investigation of the Doppler effect led Stark to the conclusion that the parhelium (singlet) series are due to electronic disturbances in helium atoms which have lost two electrons. 1922 *Phil. Trans. R. Soc.* A. CCXXIII. 137 In the arc spectra of the alkaline-earth elements, in addition to triplet series, there are series of singlets and also certain 'intercombination' lines. 1928 *Proc. R. Soc.* A. CXVII. 147 The general equations of type (18) apply to orthohelium (considered as a singlet system) as well as parhelium. 1934, 1937 [see *INTERCOMBINATION]. 1972 DePuy & Chapman *Molec. Reactions & Photochem.* iii. 33 Molecular states with all electrons paired are called singlet states. 1977 *Nature* 3 Nov. 15/3 A number of papers dealt with experimental techniques, especially interesting were those on the use of dye lasers for excitation of singlet oxygen.

b. *Particle Physics.* A multiplet (sense *b) of one sub-atomic particle.

1937 *Physical Rev.* LI. 119/2 For an element of the mass [*formula follows*] certain states in which the neutrons are in the doublet, the protons in the triplet state, exactly coincides [*sic*] in approximation (1) with a state in which the neutrons are in the quartet, the protons in the singlet state. 1961, 1962 [see *OCTET, OCTETTE 3 c]. 1977 *Nature* 21 July 202/1 Before the advent of charm in 1976 all known mesons were found to come in multiplets of 9 (nonets) and all known baryons in groups of 1, 8 or 10 (singlets, octets or decuplets).

single tax. *Econ.* [f. Single *a.*: tr. Fr. *impôt unique*.] **a.** A tax on that part of land value known as unearned profit (*produit net*), proposed by François Quesnay (1694–1774) and favoured by the Physiocrats. (See esp. quot. 1931.)

1853 J. R. McCulloch *Sketch Life & Writings F. Quesnay* in *Treat. & Ess. Econ. Policy* 436 It is needless to make any remarks on the exploded notion of Quesnay with respect to agriculture being the only source of wealth, or on his project for consolidating all taxes into a single tax (*l'impôt unique*), to be laid direct on land! 1880 *Nation* 12 Aug. 118/2 All this is accomplished by adopting the *impôt unique* of the physiocrats—the single tax on land—provided the tax be high enough to leave the cultivator nothing but wages [etc.]. 1931 *Amer. Econ. Rev.* XXI. 607 The peculiar characteristic of the Physiocrats' single tax is to be found in the idea of *produit net*. It was not essentially a single tax on land at all. It was a single tax on surplus agricultural products. 1966 A. Gilpin *Dict. Econ. Terms* 185 The Physiocrats..favoured a single tax—an 'impôt unique'—on land, arguing that the cultivation of land was the only work really productive of wealth.

b. A tax on land value as the sole source of public revenue, proposed by Henry George (1839–97). Freq. *attrib.*

1879 H. George *Progress & Poverty* VIII. iv. 383 The effect of substituting for the manifold taxes now imposed a single tax on the value of land would hardly lessen the number of conscious taxpayers, for the division of land now held on speculation would much increase the number of tax payers. 1891 *Century Mag.* Sept. 795/1 Here we have the core and essence of the single-tax philosophy—confiscation. 1917 *Amer. City* Apr. 384/1 May I..call serious attention to land value taxation—the single tax— which I believe should be regarded as..a satisfactory and just system of revenue. 1926 *Daily Colonist* (Victoria, B.C.) 14 July 2/5 The single tax method of taxation was praised enthusiastically. 1933 H. G. Wells *Shape of Things to Come* II. 244 Utah had become a practically autonomous Single-Tax State. 1944 G. B. Shaw *Everybody's Political What's What?* xxxvii. 322 If he proves that he knows as much about economic rent as Thomas De Quincey and Henry George did..he must not be asked whether his conclusions are the Conservative ones of De Quincey, the Single Tax of George, or the revolutionary ones of Marx. 1950 *World-Herald Mag.* (Omaha) 26 Mar. 9/2 The once prominent single tax movement, no longer a red hot political issue, is one which has continued on unobtrusively. 1979 *Vole* 8 Nov. 25/1 His [*sc.* George's] Single Tax would confiscate, not the land, but the rents presently paid to owners.

c. *Single Tax Party*, the English name for a Danish political party inspired by the economic principles of Henry George (see sense b above).

1950 *Manch. Guardian* 15 Sept. 6/2 The most remarkable feature, however, was the rise of the Danish single-tax party (disciples of Henry George who call themselves the Union of Justice). 1961 *Denmark* 91/2 Of new parties the Communist Party and the Single-Tax Party proved to be durable. 1970 W. G. Jones *Denmark* vi. 134 A new party called Danmarks Retsforbund—normally known in English as the Single Tax Party..has had parliamentary representation and is still in existence.

Hence **single-taxer** (also with initial capitals), a believer in the advantages of a single tax; a member of the Single Tax Party.

1889 *20th Cent.* (N.Y.) 6 Apr. 102/1 He says that is a fair question which no Single-taxer ever answers, but that if it is evaded the whole single-tax theory vanishes. 1934 B. Russell *Freedom & Organization 1814–1914* xi. 124 Both Socialists and Single-Taxers derived their proposals from him [*sc.* Ricardo]. 1941 [see *JEHOVAH 2]. 1968 *Times Lit. Suppl.* 25 Apr. 436/5 A front-parlour discussion group that included communists, spiritualists, psychic researchers and single taxers.

singleton[2]. Add: **2. b.** *Bibliogr.* (See quot. 1952.)

1952 J. Carter *ABC for Book-Collectors* 166 *Singleton*, a jargon word (of recent origin in this sense), meaning a

single leaf, where a conjugate pair would be expected... A singleton will either be the surviving leaf where the other has been severed for insertion elsewhere, or the severed half in its inset position, or an extra leaf. 1957 [see *BIFOLIUM]. 1975 *Anglo-Saxon Eng.* IV. 116 Leaves 5 and 8 in quire 43 are singletons.

4. a. A child resulting from a single rather than a multiple birth.

1931 A. Gesell in C. Murchison *Handbk. Child Psychol.* vi. 158 Twins have always captured much attention from singletons! 1942 E. B. Hurlock *Child Devel.* vii. 186 In the size of vocabulary, mean length of sentence, and articulation, twins were retarded as contrasted with singletons of the same age. 1980 *Daily Tel.* 5 Nov. 3/2 Identical twins tend to marry less often than singletons.

b. One who is alone or unaccompanied, as an only child or unmarried person. Also *spec.* an undercover agent who operates alone.

1937 E. M. Channon *Son of his Parents* iii. 63 I'm a singleton. But we had an Anglo-Indian kid here for a couple of years, and he and I did our lessons together. 1969 *Daily Tel.* 9 Apr. 14/7 Two wealthy singletons with £3,000 a year apiece would each pay £2,400 10s od and their combined net income would be £5,199; married they would pay £6,083 10s od in tax. 1977 C. McCarry *Secret Lovers* iii. 32 He was alone, a singleton in the jargon, living under deep cover, with an ordinary passport and no protection from his government.

c. The only one of its kind or class; a set having only one member. Also *attrib.*

1966 [see *INJECTIVE *a.*]. 1975 *Language* LI. 648 A singleton like *perdition* (or *conflagration*)..has no relatives like **perdite* **perditive*. 1977 *Canad. Jrnl. Linguistics* 1976 XXI. 144 The speaker uses the definite description as a characterization of a (singleton) set, whose members he wants to say something about.

single track, *sb.* and *a.* [Single *a.*] **A.** *sb.* **1.** A single pair of railway lines (occas. of tramlines). Also (with hyphen) *attrib.*

1832 *Amer. Rail Road Jrnl.* I. 245/1 The entire length of single track [is] yet to be laid. 1837 H. Martineau *Society in Amer.* II. II. ii. 192 On the 26th of November, 1833, the first car traversed the whole length of the single track. 1869 *Bradshaw's Railway Man.* XXI. 422 There are..190 miles of double track, and 130 miles of sidings.., making the entire length of track equal to 1,137 miles of single track. 1898 *McClure's Mag.* Mar. 390/1 Running a first-class train on a single-track branch. 1942 'N. Shute' *Pied Piper* i. 18 The little engine puffed along its single track, pulling its two old coaches through a country dripping with thawing snow. 1955 A. Ross *Australia 55* ix. 108 A river runs through it, also a single-track railway line. 1976 P. R. White *Planning for Public Transport* viii. 164 A number of single-track control systems exist, in which the overriding principle is that only one train has authority to occupy a section at any one time.

2. A recorded strip on magnetic tape that does not have another strip alongside it, usu. occupying almost the full width of the tape. Also *attrib.* and as *adv.*

1959 W. S. Sharps *Dict. Cinemat.* 129/1 *Single-track recorder*, a magnetic tape recorder using a single track, usually the full width of the tape. 1962 A. Nisbett *Technique Sound Studio* vi. 116 It is best to do all original recording single track, and to use other tracks only when copying. 1975 G. N. Patchett *F. M. Reception* III. xvi. 162 The whole tape is normally used for professional recording. It is known as single track.

B. *adj.* Concentrated on or capable of only one line of thought or action, obsessional, esp. in phr. *single-track mind* (chiefly *U.S.*) (cf. *one-track* adj. s.v. *ONE *numeral a.* 33); affording no choice or opportunity of divergence.

1919 *Ladies' Home Jrnl.* Feb. 35/1 The average girl no longer has a single-track mind. 1924 A. J. Small *Frozen Gold* iv. 108 Sitka Charley's was a single-track mind; dour and grim and devilishly dogged, but still, single-track. 1933 *Times Educ. Suppl.* 21 Oct. 349/3 In every school, no matter whether it be 'single-track' or 'multi-bias', there will always be a certain number who fail to come up to the standard set by the school. *c* 1942 L. Mumford *City Devel.* (1946) 152 To make up for a single-track concentration, there must be range of vision and comprehensiveness of understanding. 1964 *Listener* 26 Mar. 508/1 Under the Criminal Justice Act 1948, the double-track system of preventive detention was replaced by a single-track system. 1978 S. Sheldon *Bloodline* xxxi. 302 Max Hornung had a single-track mind.

single-valued, *a.* *Math.* [Single *a.*]. Having a unique value for each value of its argument(s); that maps to one and only one point, number, etc. Hence **single-valuedness**, the property of being single-valued.

1879 [in Dict., s.v. Single *a.* 19]. 1882 [see Reverse *sb.* I a]. 1946 *Nature* 27 July 128/2 In Indian longitudes.. the base of the stratosphere is always sharp and clear-cut and single-valued in summer, but is just one of several inversions in winter. 1968 C. G. Kuper *Introd. Theory Superconductivity* iii. 52 To guarantee single-valuedness, we must impose the flux quantization condition. 1970 G. K. Woodgate *Elem. Atomic Struct.* ii. 17 Thus Φ, like Θ, is a single-valued function of its argument. This single-valuedness is not a necessary postulate of wave mechanics, but a result derived from the condition that we are working in a central field in (r, θ, ϕ) space.

singling, *vbl. sb.* **1.** (Later examples with *out*.)

1911 M. Beerbohm *Let.* 3 Nov. (1964) 210 Your singling-out of that phrase..is a proof that Rothenstein and Ruskin are right about reverence for nature. 1964 F. Bowers *Bibliogr. & Textual Crit.* VI. iv. 200 Sir Walter Greg's singling-out of this as the clinching evidence.

singly, *adv.* Add: **4.** *singly-charged.*

1924 *Phil. Mag.* XLVII. 282 The number of particles due to doubly-charged, singly-charged, and neutral helium atoms..varied in number over..a wide range. 1965 Phillips & Williams *Inorg. Chem.* I. v. 176, *hν* may be taken as the ionization potential of an ion (i.e. for a singly-charged positive ion, the second ionization potential of the element, and for a singly-charged negative ion, the electron affinity of the element).

sing-sing[3]. [Reduplicative pidgin formation f. Sing *v.*[1]] In Papua New Guinea: an occasion of feasting and musical celebration.

1899 C. M. E. David *Funafuti* v. 58 Opataia..came to say that there was to be a sing-sing that night in the schoolroom, in honour of the expedition. 1924 E. Raff *Let.* 10 Jan. in F. E. Williams *Orokaiva Magic* (1928) I. 100 At one of the 'sing sings' (dances) a follower who fell down 'dead' was promptly revived. 1943 S. W. Reed *Making of Mod. New Guinea* vii. 211 In the native villages the initiation ceremonies, tribal dances, and *singsing* will continue to express a mode of life utterly foreign to the European. 1968 *Telegraph* (Brisbane) 3 Sept. 18/6 Cattle given to New Guinean farmers for breeding are being killed and eaten at 'sing sings' (festivals).

sing-song, *sb.* Add: **4. a.** Now more usu. a gathering for, or session of, community singing.

1869 J. Greenwood *Seven Curses of London* ii. 19 The London factory-bred girl..has her 'young man', and accompanies him of evenings to 'sing-songs' and raffles. 1899 Kipling *Stalky & Co.* II. 252 'Had some rippin' sing-songs in camp, too,' said Tertius. 1914 D. O. Barnett *Let.* 18 Dec. (1915) 28 We've got some sing-songs and smokers on for Christmas. 1933 *Sun* (Baltimore) 21 Oct. 14/6 Harvard students will..learn old German drinking songs at the German Sing-Song conducted by James W. Hawkes. 1968 'J. le Carré' *Small Town in Germany* ix. 147 Wednesday was welfare. Ping pong night. Sing song night. 1974 J. Aiken *Midnight is Place* i. 30 Ey, David! Coom to t'sing-song at t'Mason's Arms tonight?

6. Special Comb.: **sing-song girl**, a Chinese girl who entertains men by singing and dancing (*euphem.* one of easy virtue); **sing-song theory**, the theory (propounded by Jespersen) that language evolved from primitive singing.

1934 'A. Bridge' *Ginger Griffin* 326 So there will be more war in China because of a foreigner's idle love-letter to a sing-song girl. 1939 Auden & Isherwood *Journey to War* 157 Even the singsong girls have changed their style. 1965 J. Von Sternberg *Fun in Chinese Laundry* iv. 82 On the first floor were gambling tables, singsong girls. 1978 *China Now* Mar./Apr. 19/2 The shocking thing was that in the cities—where officials danced or played with sing song girls—there were grain and food. 1939 L. H. Gray *Foundations of Lang.* 40 Language has been traced by some to primitive rhythmic chants and to singing (the sing-song theory). 1973 *Current Anthropol.* XIV. 27/3 He makes no reference to Jespersen's classic 'sing-song' theory of language origin (1922). Though this label was applied to Jespersen's hypothesis..by his critics, there seems little doubt that he believed language to have been chanted before it was spoken.

sing-songy *a.* (earlier example).

1892 E. Lytton *Let.* 20 July in E. Lutyens *Blessed Girl* (1953) viii. 155 Vic is made to read poetry aloud to us. He has a very sing-songy voice.

sing-song, *v.* Add: **1. b.** Also, with direct speech as obj.

1931 [see *CALLER *sb.* 1 e]. 1963 R. Wolff *I, Keturah* II. iv. 171 'You can't catch me. You can't catch me,' he singsonged. 1976 P. A. Lake *Leffert's Disease* 173 'I'm sorry, he's not here,' she sing-songed with her nasal twang.

‖ **singspiel** (zi·ŋʃpīl). *Mus.* Also **Singspiel.** [Ger., f. *singen* to sing + *spiel* play.] A semi-dramatic performance in which song and dialogue alternate, popular in Germany in the latter part of the eighteenth century.

1876 F. L. Ritter *Hist. Mus.* 266 The 'Singspiel' or 'operette' as constructed by Hiller, makes use of the spoken dialogue, as does the French comic opera. 1880 *Grove's Dict. Mus.* II. 519/1 That best and truest form of German Opera, the 'Singspiel'. 1911 E. J. Dent *Mozart's Opera 'The Magic Flute'* 4 [Schikaneder] gave a season of German 'Singspiel' (comic opera) at Vienna in the winter of 1784–5. 1930 *Observer* 23 Mar. 25 Many [operettas] of the German 'singspiel' type could be done. 1942 E. Blom *Music in Eng.* ii. 24 English ballad opera crept into Germany by the back door and served as the direct model for the German *Singspiel*. 1962 *Guardian* 7 Feb. 9/1 Cherubini..evolved a style of *opéra comique*.. not unlike the *singspiel* of Mozart ('Seraglio'). 1977 *Times Lit. Suppl.* 10 June 710/5 A more general discussion of *opera seria*, *Singspiel*, etc.

singular, *a.* Add: **I. 3. c.** *singular matrix* (see quots. 1964, 1972); *singular solution*, a solution of a differential equation that cannot be obtained directly from the complete primitive; *esp.* a solution whose graph is the envelope of the graphs of the complete primitive.

1836 A. de Morgan *Differential & Integral Calculus* xi. 191 If there be a singular solution it is $y = a$... We have only found the singular solution from the primitive itself. 1873 A. Cayley in *Messenger of Math.* II. 12, I consider the singular solution to be that given by the equation which belongs to the envelope-locus (viz. I do not recognise any singular solution which is not of the envelope species). 1957 L. Fox *Two-Point Boundary Probl.* i. 2 The most general solution of an ordinary differential equation of order *n* contains *n* arbitrary constants. This general solution is called the Complete Primitive, and a Particular Integral is obtained by giving specific values to these arbitrary constants. Non-linear equations may also have singular solutions, not obtainable from the complete primitive. 1964 N. N. Hancock *Matrix Anal. Electr. Machinery* ii. 18 A 'singular' matrix is one for which the determinant formed by the same array is of zero value. 1972 A. G. Howson *Handbk. Terms Algebra & Anal.* viii. 43 If there is no matrix B..such that $AB = BA = I$, then A is said to be singular.

‖ **singulare tantum** (siŋgiŭlā·re tæ·ntŏm). *Gram.* Also **singularis tantum.** Pl. **singularia tantum.** [L. neut. phr., 'singular only'.] A word which has only a singular form: usu. applied to mass (or uncountable) nouns. Cf. *PLURALE TANTUM.

1940 A. H. Gardiner *Theory of Proper Names* 27 A *singulare tantum* has developed a plural by cutting the designated entity, like a worm, into two parts. 1962 H. M. Hoenigswald in Householder & Saporta *Probl. Lexicogr.* 109 The ordinary coverage of singularia tantum, is mostly limited to mass nouns. 1979 *Trans. Philol. Soc.* 160 This is why the plural of *dahyu-* in the sense of 'nations' would have been *translatable* into Elamite only by *taššup*, a word which at xwāning was doomed to come out as the singularis tantum *kāra-*.

singularism (si·ŋgiŭlăriz'm). [See -ism.] A philosophy which explains the phenomena of the universe from a single principle: opp. to Pluralism. Cf. Monism.

1897 Pillsbury & Titchener tr. *Külpe's Introd. Philos.* iii. 107 It is customary to distinguish the various views that can be held upon this question by the terms *monism*, *dualism* and *pluralism*. But as the difference expressed by the first two is..predominantly qualitative, it seems better to make a..quantitative antithesis, and to speak only of *singularism* and *pluralism*. The former explains or deduces all the phenomena of the universe from one single principle. 1911 J. Ward *Realm of Ends* 24 If the difficulties of Pluralism point the way to Singularism they will at least serve to make the character of the One clearer than any 'cheap and easy monism'.. can ever do. 1931 A. Wolf in W. Rose *Outl. Mod. Knowl.* xiii. 576 The philosophy of James is perhaps best understood as a reaction against the excessive intellectualism, and the monism or singularism of absolute idealism.

singularity. Add: **9. d.** *Math.* A point at which a function takes an infinite value.

1893 A. R. Forsyth *Theory of Functions* xxi. 606 All the essential singularities of a discontinuous group lie on the axis of *a* when the group is real. 1939 [see *Schwarzschild 2]. 1959 *Listener* 27 Aug. 320/1 At the start of the expansion certain quantities in our differential equations become infinite. This frequently happens with differential equations, and when it does the equation is said to contain a mathematical singularity. 1977 *Sci. Amer.* July 126/3 Gauss did much more with complex numbers. In 1811 he discovered what is now called Cauchy's theorem: The integral of a complex analytical function around a closed curve that encloses no singularities is zero.

e. *Astr.* A region in space-time at which matter is infinitely dense.

1965 *Physical Rev. Lett.* XIV. 58/1 An exterior observer will always see matter outside $r = 2m$, the collapse through $r = 2m$ to the singularity at $r = 0$ being invisible to him. 1972 *Nature* 21 Apr. 378/2 In the physical world..we may take the phrase 'space-time singularity' to mean a region in which space and time have become so locally distorted that the present laws of physics are no longer applicable. 1979 *Jrnl. R. Soc. Arts* CXXVII. 579/1 A central 'singularity' where tidal forces (the difference between the gravitational acceleration of his head and his feet) would become infinite, and he would be crushed out of existence.

singulary (si·ŋgiŭlări), *a.* *Logic.* [f. L. *singulāris* Singular *a.*: see -ary[2].] Involving just one element.

1940 W. V. O. Quine *Math. Logic* i. 13 Conjunction and alternation are binary, in that they combine statements two at a time. But denial..is singulary. [*Note*] The series of adjectives 'binary', 'ternary', 'quaternary', 'quinary',..leaves mathematicians in a quandary when $n = 1$. It is customary to stammer out some such makeshift as 'unary' or 'uninary' or 'unitary'. But the proper word is apparent if we reflect that the series of Latin distributives 'bini', 'terni', 'quaterni', 'quini',..begins with 'singuli'. 1954 *Word* X. 227 There are singulary operations, but no singulary relations... An example of a singulary operation in mathematics is 'reciprocal of'. 1965 N. Chomsky *Aspects of Theory of Syntax* iii. 132 Many of the optional singulary transformations of Chomsky..must be reformulated as obligatory transformations.

singulative (si·ŋgiŭlătiv). *Gram.* [ad. F. *singulatif*: cf. Singular *a.* and -ative.] (See quot. 1966); also, a singular form.

[1952 Meillet & Cohen *Langues du Monde* (ed. 2) 1279 *Singulatif*, morphème ayant pour fonction de donner à un mot une valeur de singulier, généralement par opposition à un collectif.] 1966 M. Pei *Gloss. Linguistic Ter-*

minol. 250 *Singulative*, a morpheme having for its function to give a word the force of a singular, usually by way of opposition to a collective (*rice*, *rice-grain*). 1970 J. McN. Dodgson *Pl.-Names Cheshire* I. 151 The *-inn*, *-enn* suffix is not diminutive but a singulative... The singulative effect would indicate some particularised aspect of a location—e.g. a particular piece of moorland in a general area of moors. 1977 *Word* 1972 XXVIII. 194 *Clocs*: plural *clocsiau*..., doubly characterized plural, with a new 'singulative' *clocsen*.

sinh (ʃain, sintʃ, sain,ēi·tʃ). *Math.* Abbrev. of *hyperbolic sine* s.v. Hyperbolic *a.* 2 b.
1873 *Messenger of Math.* II. 190 Sinh x and cosh x are of course the hyperbolic sine and cosine of x, viz. $\frac{1}{2}(e^x - e^{-x})$ and $\frac{1}{2}(e^x + e^{-x})$. 1891 [see *cosh]. 1972 A. G. Howson *Handbk. Terms Algebra & Anal.* xxxiii. 164 Sinh and cosh both have domain R; the image of sinh is R.

Sinhala (sinhā·la), *sb.* and *a.* [a. Skr. (see Sinhalese).] **A.** *sb.* = Sinhalese *sb.* and *a.* 2.
1954 Pei & Gaynor *Dict. Linguistics* 198 *Sinhala*, modern vernacular Sinhalese, mixed with foreign words. 1961 *Times* 27 June 13 The Official Language Act of 1958 came into effect at the beginning of this year, and without reference to Tamil prescribes Sinhala as the sole official language of Ceylon. 1977 *Economist* 3 Sept. 62/1 Mrs Bandaranaike fought an election on a platform of 'Sinhala only' as Sri Lanka's state language.
B. *adj.* = Sinhalese *sb.* and *a.* 3.
1926 M. C. Rasanayagam *Anc. Jaffna* vi. 231 The Pallava king..says in one of his inscriptions that he vanquished 'the Sinhala king who was proud of the strength of his army'. 1962 *Housewife* (Ceylon) Feb. 9 Local hand-made clay tiles based on traditional Sinhala designs. 1963 *Guardian* 2 May 10/3 An Act of 1956 prescribed the Sinhala language as the one official language of Ceylon.

Sinhalese, *sb.* and *a.* Add: The Sinhalese are properly members of an Aryan people deriving from N. India and now forming the majority of the population of Sri Lanka.
1. (Earlier example.) Also *sing.*, a Sinhalese native of Sri Lanka.
1801 *Asiatick Researches* VII. 32 The *Singhalais* assert, from record, the total destruction and regeneration of the universe, many other times. 1913 L. Woolf *Village in Jungle* iii. 45 He was tall for a Sinhalese, broad-shouldered, and big-boned. 1948 [see *Ceylonese *a.* and *sb.*].
2. (Earlier example.)
1801 *Asiatick Researches* VII. 401 The end of the soul is called, in Singalese, *Nivani*.

sinhalite (si·nhălait). *Min.* [f. as Sinhal(ese *sb.* and *a.* + -ite¹.] A borate of aluminium and magnesium, $MgAlBO_4$ (also containing iron), which forms pale yellow to deep brown orthorhombic crystals resembling olivine and frequently of gem quality.
1952 Claringbull & Hey in *Mineral. Mag.* XXIX. 843 For this new mineral the name *sinhalite* is proposed, from Sinhala the Sanskrit name for Ceylon. 1952 *Times* 4 Dec. 3/3 Sinhalite..was previously thought to be an olivine rich in iron. 1965 *Amer. Mineralogist* L. 1979 The sinhalite from Ceylon had a density of 3.494, corresponding to a composition of [illegible]... Judging from the absorption data, most of the iron in sinhalite is divalent. 1971 *Jrnl. Gemmology* XII. 124 One of the few sinhalites we have officially tested..was an attractive golden brown specimen.

Sinico- (si·niko), combining form of med.L. *Sinicus* Sinic *a.*, as in *Sinico-Annamitic*; **Sinico-Japanese** *sb.* and *a.* = *Sino-Japanese* s.v. *Sino-¹ 2, 3. Now *rare*.
1838 P. S. Du Ponceau *Diss. Nature & Char. Chinese Syst. Writing* p. xxix, Those Sinico-annamitic words, if they are really in use, do not belong to the original language. 1866 F. V. Dickins tr. *Teika's Hyak Nin Is'shiu* p. vii, Such ill-sounding Sinico-Japanese syllables as *zeu*, *hasu*, *tatu*,..and the like. 1884 tr. *J. J. Rein's Japan* 396 The Japanese Language... Yamato- and Sinico-Japanese. 1902 *Encycl. Brit.* XXIX. 729/1 Like all fine specimens of the Sinico-Japanese school, the prices are too high to attract wide custom.

sinification. (Later examples.)
1966 D. Wilson *Quarter of Mankind* xxi. 206 There is no separate Hongkong nationality. Sinification of public life proceeds very slowly. 1977 *Daily Tel.* 12 Feb. 16 The..'sinification' of Tibet.

sinify, *v.* (Later examples.)
1942 A. J. Grajdanzev *Formosa Today* iii. 35 Only 95,400 of them are classified as 'savages'..the remaining 60,500 being 'civilized', that is, they are 'sinified'. 1966 *Economist* 1 Oct. 58/1 But he [sc. Mao] did not sinify the basic philosophical principles of Marxism. 1977 *Times* 12 Oct. (China Suppl.) p. vii/7 The central Government has often been accused of 'sinifying' the minority nationality areas of China.

sinister, *a.* Add: **4. b.** *sinister interest* (earlier examples in the works of the Utilitarian philosophers).
1817 J. Bentham *Plan Parl. Reform* iii. p. xi, Here we have one partial, one separate, one sinister interest..with which the universal..interest has to antagonise. 1824 J. S. Mill in *Westm. Rev.* II. 347 When romance assumes the garb of history,..it infallibly allies itself with the sinister interests of the few. 1827 J. Bentham *Rationale Evid.* V. ix. i. 6 Interest when acting in such a direction with such effect as to give birth to falsehood may be termed sinister interest. 1861 J. S. Mill *Repr. Govt.* vi. 118 The evils arising from the prevalence of modes of action in the representative body dictated by sinister interests (to employ the useful phrase introduced by Bentham), that is, interests conflicting more or less with the general good of the community.

sinisterly, *adv.* **1.** Delete *rare* and add later examples.
1930 D. H. Lawrence *A Propos of Lady Chatterley's Lover* 6 The effect is peculiarly depressing, sinisterly high-brow. 1946 G. Millar *Horned Pigeon* xii. 346 The ducks looked happy and healthy, perhaps sinisterly so. 1969 [see *rentier]. 1974 E. Jones *Barlow comes to Judgement* 48 There is nothing sinisterly secret about the WHY Club.

sinistral, *a.* Add: **8.** Also as *sb.*, a left-handed person.
1927 [see *dextral *a.* 1 c]. 1964 M. Critchley *Developmental Dyslexia* xiv. 84 The patient declared herself to be right-handed, but she wore her wrist-watch on the right arm like many sinistrals do.
9. *Geol.* Being or pertaining to a strike-slip fault in which the motion of the block on the farther side of the fault from an observer is towards the left. Opp. *dextral*.
1942 E. M. Anderson *Dynamics of Faulting* v. 55 E. L. Dixon has distinguished the types of displacement as right-hand and left-hand heaves... The two classes of fault-planes may well be termed 'dextral' and 'sinistral'. 1961 ... The Bala fault-zone..is undoubtedly sinistral. 1964 A. Holmes *Princ. Physical Geol.* (ed. 2) xi. 219 In strike-slip faults the relative displacement of the block on the far side of the fault, as viewed from the ground, may have been either to the right (giving a right-lateral or dextral fault) or to the left (giving a left-lateral or sinistral fault). 1978 *Nature* 5 Jan. 50/2 The Bathurst and McDonald faults..., which form the boundaries of the Slave wedge are major sinistral and dextral strike-slip fault systems that extend for hundreds of kilometres.

sink, *sb.*¹ Add: **I. 2. f.** In semi-proverbial phr. *a mind like a sink*, an imagination that tends to put an indecent or lewd construction on events. *slang*.
1932 A. Christie *Thirteen Problems* x. 170 And if one tries to warn them..they tell one that one has a Victorian mind—and that, they say, is like a *sink*. 1949 Wodehouse *Uncle Dynamite* viii. 129 He concluded by saying that it was a pity that some people, whose identity he did not specify, had minds like sinks. 1970 S. Taylor *Murder grows Roots* ii. 16 [She] said he'd probably gone off with some woman. Her mind's like a sink!
g. Used *attrib.* of a (school, estate, etc., in a) socially deprived area.
1972 *Daily Mail* 4 Oct. 25/3 The downward spiral of decline in the 'sink' areas could be broken if the school led the way. 1972 *Guardian* 17 Oct. 17/4 It is a pity..that there is not a 'sink' schools conference, like the Headmasters' Conference of the public schools, to act as a general champion of the rights of urban schools. 1976 *New Society* 18 Nov. 365/2 Somewhere, in every town that has council houses at all, there's a 'sink' estate—the roughest and shabbiest on the books, disproportionately tenanted by families with problems, and despised both by those who live there and the town at large. 1981 *Observer* 8 Feb. 29/4 None of its problems has reduced Callow to a 'sink' school: it has great achievements, including children in its first sixth form about to depart bright-eyed and bushy-tailed to university.
II. 8. Delete *Kinematics* and add: The opposite of *source* in any scientific sense; a place where or a process by which energy (esp. heat) is removed from a system, or some specific component of a system is removed from circulation and either stored or destroyed; a device whose function is to act as a sink. (Earlier and later examples.)
1855 J. C. Maxwell in *Trans. Cambr. Philos. Soc.* (1864) X. 32 If the origin of the tube or filament is within the space under consideration, then we must conceive the fluid to be supplied by a source within that space, capable of creating and emitting unity of fluid in unity of time, and to be afterwards swallowed up by a sink capable of receiving and destroying the same amount continually. 1882 *Electrician* 3 July 174/1 There will..be transfer of energy through the medium from sources to sinks of energy. 1902 *Encycl. Brit.* XXVIII. 167/2 In the case of current flow in plane sheets, we have to consider certain points called sources at which the current flows into the sheet, and certain points called sinks at which it leaves. 1951 *Jrnl. Brit. Interplanetary Soc.* X. 256 The generation of electrical power by means of a heat engine requires that the heat produced at a temperature T_1 be conveyed to a 'sink' at a temperature T_2. 1966 *Economist* 8 Oct. 180/1 They [sc. power stations] could be used as a 'sink' for the gas while the distribution system is geared up to take it elsewhere. 1977 J. M. Campbell *Energy & Atmosphere* viii. 263 The main sink for hydroperoxy radicals in the troposphere appears, at present, to be identified as reaction with nitrogen oxides.
III. 11. c. *gen.* An instance or act of sinking. *rare*.
1818 Keats *Let.* 13 Mar. (1958) I. 240 When a poor devil is drowning, it is said he comes thrice to the surface, ere he makes his final sink.
d. *Aeronautics.* Loss of altitude, esp. in gliding flight; the rate of this.
1943 [see *rate *sb.*¹ 7 b]. 1955 A. Welch et al. *Soaring Pilot* iii. 33 Minimum sink will occur at some lower lift coefficient (i.e. a higher speed). 1962 R. C. S. Allen *Theory of Flight for Glider Pilots* iv. 28 When the power is a minimum, the sink is a minimum. 1973 *Sci. Amer.* Dec. 103/3 The effect of the vulture's lower wing loading is that it can turn in much smaller circles at a similar rate of sink.
12. c. (Earlier example.)
1840 A. Bunn *Stage both before & behind Curtain* III. viii. 280 The scenery..described, in the glowing language of the stage, under the head of flats, wings, side-pieces, borders, sinks, flies, &c.,..has been painted..by Mr. Stanfield and Grieve.
IV. 14. **sink garden**, a miniature garden, comprising a group of small plants (often alpine varieties) grown in an old stone sink or similar container; **sink rate** *Aeronaut.* = *sinking speed* s.v. *Sinking *vbl. sb.* 4; **sink tidy**, a perforated receptacle for kitchen waste, placed on a sink unit; **sink unit**, a kitchen unit comprising a sink and draining-board, usu. with cupboards below.
[1923 *Gardeners' Chron.* 2 June 306/2 Quite a novel feature of the rock garden [at Chelsea] were the miniature gardens..in stone sinks.] 1935 C. Elliott *Rock Garden Plants* 10, I at first intended to devote chapters to the building of rock gardens, the making of screes, to sink gardens, etc. 1954 R. Pearson *Town Gardening* xii. 109 (*heading*) Trough and sink gardens. 1966 *National Observer* (U.S.) 21 Feb. 9/3 Attention so far has focused on the 727's 'sink rate', or rate of descent as it comes down from its 25,000-foot cruising altitude on an approach to landing. 1978 A. Welch *Bk. Airsports* vi. 92/2 Most Para-Commander 'chutes..have a sink rate of about 13 feet per second. 1951 *Catal. of Exhibits, South Bank Exhib., Festival of Britain* 131/1 Deep sink tidy. 1958 *New Scientist* 9 Jan. 13/1 Polyethylene...well known in recent years for its use in the manufacture of..sink-tidies, buckets and washing-up bowls. 1981 R. Barnard *Mother's Boys* v. 52 She took out the sink-tidy, with the rubbish from breakfast, and slapped the contents into the dust-bin. 1939 Martin & Seymour *Flat Bk.* 66 Wringer unit which can be fixed permanently to an 'Easiwork' sink unit in an ideal position between the sink and copper. 1971 R. Rendell *One across, Two Down* v. 45, I would have it painted throughout for you and a sink unit put in.

sink, *v.* Add: **I. 1. a.** Also in fig. phr. *to sink without trace*; also, usu. *pass.* [tr. Ger.: see *spurlos versenkt.]
1925 Fraser & Gibbons *Soldier & Sailor Words* 267 *Spurlos versenkt*, gone entirely. Disappeared. Specifically—sunk without trace, with all on board. 1936 'D. Yates' *And Berry came Too* viii. 313 'That has gone, sir.'..'Sunk without trace,' said Berry. 'What a very beautiful thought.' 1946 W. S. Churchill *Compl. Speeches* (1974) VII. 7337 He has departed 'spurlos versenkt' as the German expression says—sunk without leaving a trace behind. 1965 A. Fairfax-Lucy in Battiscombe & Laski *Chaplet for Charlotte Yonge* 92 Kenneth has sunk without a trace, but *The Little Duke* lives. 1974 'J. le Carré' *Tinker, Tailor* xii. 204 'And the third? Viktorov?' 'Sunk without trace.' ... 'Trained and disappeared.'
b. Also in phr. *to sink through the floor*, used to express deep embarrassment.
1908 L. M. Montgomery *Anne of Green Gables* xii. 118 She thought she would sink through the floor when she saw you come in all rigged out like that. 1956 'C. Blackstock' *Dewey Death* iv. 79 The pause was long enough to make Barbara wish she could sink through the floor. 1969 E. Grierson *Shall I eat You Now?* 96 The fear that he might suddenly say anything—well something really that would make you sink through the floor.
6. c. Of an oil painting: to develop dull spots on the surface where the pigments have sunk into the ground. Also const. *in*. Cf. *sinking *vbl. sb.* 1 e.
1939 H. Hubbard *Materia Pictoria* 231 During the process of painting, and after completion, Oil Paintings are liable to sink-in and become dull in parts. 1968 M. Noakes *Prof. Approach to Oil Painting* ii. 12 Linseed oil can be used for 'oiling out' when a picture shows signs of sinking... Varnish seals and protects the surface of a painting, as well as reviving any areas that have sunk.
III. 17. d. *Golf.* To hole a ball from (a putt); to hole (a ball) by putting.
1926 Travers & White *Winning Shot* 23 After coming up in three and then sinking a ten or a fifteen putt for a four, the situation had suddenly changed. 1940 F. Ouimet *Game of Golf* xiv. 203, I murmured a few mild prayers before taking my stance, and the time I succeeded in sinking the ball. 1955 Keeler & Rice *Bobby Jones Story* xxi. 119 Mrs. Vanderbeck did sink that putt of 25 feet..and..Alexa did sink hers for a win. 1971 'D. Halliday' *Dolly & Doctor Bird* xv. 215 Arrived on the green, he approached the ball, turned the pin..and finally sank it at nine.
e. To consume (an alcoholic drink); to drink down (esp. rapidly); = *down *v.*² I c. *colloq.*
1932 G. Holt *Drums beat at Night* ii. 30 Let's go out and sink a few beers. We can talk at the pub. 1947 L. MacNeice *Dark Tower* 157 I'll sink a pint in The Dog Returns. 1953 A. Neave *They have their Exits* vii. 144 Each man spoke of what he would do first on arrival in England. 'I shall sink three pints of mild and bitter,' said one. 1962 L. Deighton *Ipcress File* ii. 19, I..sank a quick grappa. 1977 M. Kenyon *Rapist* xiv. 182 Get a couple of cups, Sergeant, we'll sink a fast one.
f. *Basketball.* To score a goal or basket from (a shot). Also *absol.*
1935 *N.Y. Times* 24 Feb. III. 4/1 O'Donnell sank a long field goal and Kozloff threw a foul to give Penn a 4–1 lead. 1950 N. Holman *Holman on Basketball* 50 With Norm Mager sinking five and Floyd Layne making four set shots..the City College five won. 1962 *Sports Illustr. Bk. Basketball* iv. 83 Cousy..leaped into the air and sank a left-hander that won the game. 1972 *Sports Illustr.* 3 Jan. 51/1 The Rainbows' John Pennebacker sank from free throws.

20. d. *Hunting.* *to sink the wind*: to move downwind of another; *spec.* of a fox: to pass below the line of scent. Cf. Wind *sb.*¹ 4.
1778 G. Cartwright *Jrnl.* 26 Sept. (1792) II. 374, I saw a large stag upon the south hill..and I let him pass; crossing his route and sinking the wind, I made all possible speed to the foot of Gravel Hills, where I headed him. 1847 R. S. Surtees *Hawbuck Grange* v. 96 We..found a hare by Clipstone Clump, who went as straight as an arrow to Gatley Coppice, from whence, sinking the wind all the way, she ran to Silverspring. 1896 T. Smith *Life of Fox* 299 When men go down wind to hear the cry, it is called sinking the wind of the hounds. 1948 F. Pitt *Hounds, Horses & Hunting* 271 *Sink the Wind*, to go downwind.

sink, var. *sync.

‖ **sinkeh** (si·ŋke). [Malay *singke(h*, a. Hokkien *sinkheh* (also used), f. *sin* new + *kheh* visitor.] In Malaysia, a newcomer (esp. a labourer) recently arrived from China.
1879 J. D. Vaughan *Manners & Customs of Chinese of Straits Settlements* 6 The Chinaman on first landing in the Straits is called a Sinkeh. 1927 R. J. H. Sidney *In Brit. Malaya Today* 145 The Secret Societies were really Friendly Societies which each sinkeh (new-comer) joined. 1948 V. Purcell *Chinese in Malaya* iii. 58 The staple article of local commerce was the *sinkheh* (Hokkien), *sankah* (Cantonese), the new recruit from China. 1972 C. M. Turnbull *Straits Settlements 1826–67* i. 44 The employer obtained full right to their labour for a period usually of one year, during which the sinkhehs were fed, clothed, housed and given a small allowance.

sinker, *sb.*¹ Add: **II. 5. d.** *slang* (orig. *U.S.*). A doughy cake, esp. a doughnut; a dumpling. Now *rare*.
1870 J. H. Beadle *Utah* 223 Our favorite dinner, when we could get the meat, was of fried ham and 'sinkers'. 1903 F. B. Smith *How Paris amuses Itself* 48 The New York Dairy Lunch, with..its elevating Bible texts, and depressing 'sinkers',..would never make a success with Parisians. 1906 *N.Y. Even. Post* 10 Dec. 14 Without 'sinkers', corn cakes, cream puffs, 'cookies', and other standard foodstuffs at reasonable prices to appease the appetite between lectures, it is simply impossible to go on studying. 1926 E. Ferber *Show Boat* xiii. 268 The coffee was hot, strong, revivifying; the sinkers crisp and fresh. 1946 J. Irving *Royal Navalese* 157 *Sinkers*, dumplings. 1975 *Amer. Speech* 1971 XLVI. 172 Round fried cake with hole in the centre..sinker.
III. A person or thing that sinks. Chiefly *N. Amer.* **6*. a.** One who sinks. *rare*.
1851 H. Melville *Moby Dick* II. lxxviii. 78 No sign of either the sinker or the diver.
b. A sunken or partly submerged log.
1884 *Redwood & Lumbering in Calif. Forests* (Edgar Cherry & Co.) 95 The well matured heartwood of the base of these trees is so solid as to sink in water—hence designated as 'sinkers'. 1905 *Terms Forestry & Logging* 34 *Deadhead*, a sunken or partly sunken log... Syn[onym]: sinker. 1915 P. B. Kyne *Cappy Ricks* 28 A sinker is a heavy, close-grained clear redwood butt-log, which, if cut in the spring,..is so heavy it will not float in the mill-pond. 1969 *Marine Digest* 4 Jan. 6/2 Ferry manager..blamed the accident on a sinker.
c. *Baseball.* A ball which drops markedly after being pitched or hit.
1932 *Baseball Mag.* Oct. 496/1 Outfield skill depends a lot on the player's quickness in detecting whether it's a 'sailer' or a 'sinker'. 1943 [see *outcurve 1]. 1952 *Sun* (Baltimore) 19 Apr. 13/1 A baffling repertoire of sliders and lazy sinkers. 1967 *Boston Traveler* 1 June 31/3 I've developed a good sinker and my fastball and curve are moving. 1975 *Cleveland* (Ohio) *Plain Dealer* 29 Mar. 2-c/5 He's missing bad with his sinker.

sinker (si·ŋkəɹ), *sb.*² *Bot.* Also 9 senker. [a. G. *senker* process, shoot, now assimilated to Sinker *sb.*¹] A process of the root system of a mistletoe that grows radially into the tissues of the stem of the host.
1863 J. Harley in *Trans. Linn. Soc.* XXIV. 176 The young plant [sc. mistletoe] first sends into the bark of the nourishing plant a single root, sucker, or senker. 1894 Somerville & Ward tr. *Hartig's Dis. Trees* I. 27 Once a year, very seldom twice, often only each alternate year, a 'sinker' originates on the inner side of the cortex-root near the apex. 1938 J. S. Boyce *Forest Path.* xv. 347 From the cortical haustoria are developed the sinkers which grow radially through the inner bark to the cambium, later becoming embedded in the wood by the formation of new annual rings. 1970 W. H. Smith *Tree Path.* xxi. 220 Generally, sinkers are located within the rays of host xylem tissue, where they appear to grow coincidentally with the host.

sinkful (si·ŋkful). Also sink-full. [f. Sink *sb.*¹ + -ful.] As much or as many as will fill a kitchen sink.
1961 'A. A. Fair' *Shills can't cash Chips* viii. 148, I hate to come home to a sinkful of dirty dishes. 1976 W. J. Burley *Wycliffe & Schoolgirls* iii. 64 The kitchen..had..a sink-full of dirty dishes. 1982 J. B. Hilton *Sunset Law* i. 7 Mock exasperation about the third sinkful of crockery that day.

sink-hole. Add: **1.** (Further examples.)
1949 *Reader's Digest* June 45/1 The French industrialist behaves as if he believed his country were headed toward the sink-hole. 1976 N. Thornburg *Cutter & Bone* v. 124 'In this world!' Cutter's grin did not believe... 'This jailhouse. This sinkhole of piss and misery.' 1978 H. Wouk *War & Remembrance* xlviii. 488 Hollywood's such a sinkhole.
2. (Earlier example.)
1780 W. Fleming *Diary* 20 Mar. in N. D. Mereness *Travels in Amer. Colonies* (1916) 639 Springs..appear again either in Sink holes immediately vanishing or bursting out.

sinking, *vbl. sb.* Add: **1. b.** Also with *in*.
1937 W. B. Yeats *Vision* 178 A sinking-in of the body upon its supersensual life.
e. *Painting.* A dull matt spot on the surface of an oil painting caused by the absorption of the pigments by the ground; the process by which the pigments sink into or become absorbed by the ground. Also *sinking-in*.
1915 P. Young tr. *Vibert's Science Painting* ix. 118 Where, in the execution of a picture on account of repeated re-touching, embus or sinkings appear, it is possible..to make them disappear with a light scumbling of re-touching varnish. 1939 H. Hubbard *Materia Pictoria* 231 The chief causes of sinking-in are: (A) The porousness of the Painting-ground or Priming [etc.]. 1951 R. Mayer *Artist's Handbk. Materials & Techniques* xii. 433 *Embu*, (French), in an oil painting, a dull spot in an otherwise glossy surface, caused by a sinking-in of the oil color. 1971 B. Dorf *Beginner's Guide to Painting in Oils* xiii. 158 *Sinking*, dull patches in oil paint, caused by too absorbent ground, wrong medium, or too much dilutant.
f. Also *attrib.*, as *sinking feeling*.
1890 H. G. Hutchinson *Golf* (Badm. Libr.) ix. 246 The nerves and muscles must be fed for the work before them; otherwise there will ensue a dreadful sinking feeling before the end of the round. 1920 *Poster*, Bovril.. Prevents that sinking feeling. 1920 C. A. W. Monckton *Some Experiences of New Guinea Resident Magistrate* xxv. 302 'Do you feel devilish hungry half an hour before meals?'..'Yes,..sometimes so hungry that I have a sinking feeling.' 1937 *Discovery* Oct. 295/1 The slight sinking feeling experienced by pedestrian members.., when faced by the considerable uphill trudge. 1961 W. Buchan *Helen All Alone* 196, I just have a sinking feeling. 1979 L. Meyer *Fake Front* xiii. 108, I got that nasty sinking feeling again. We..couldn't get the story into the paper.
4. Special comb.: **sinking speed**, the vertical downward component of the velocity of a gliding body.
1930 V. W. Pagé *Henley's ABC of Gliding & Sailflying* (1931) ii. 40 The falling or sinking speed depends on the weight of the glider loaded and the skill of the pilot in manipulating the controls. 1953 *New Biol.* XIV. 72 One requirement of a bird which is to soar in upcurrents is that it should have a low 'sinking speed'; that is to say, it must lose height slowly when gliding in still air. 1973 *Sci. Amer.* Dec. 102/2 If the air through which the bird is flying happens to be rising at a speed greater than the sinking speed, the bird is carried up with it and acquires potential energy it can use later to glide through air that is not rising.

sinneress. (Later example.)
1929 S. Leslie *Anglo-Catholic* xii. 165 Veronica's card was the tell-tale finger pointing to her as a sinneress.

sinnerite (si·nərait). *Min.* [See quot. 1964 and -ite¹.] A sulpharsenite of copper, $Cu_6As_4S_9$, found as brittle, grey, triclinic crystals.
1964 Marumo & Nowacki in *Schweiz. Min. u. Petrogr. Mitt.* XLIV. 440 In the course of the systematical study of the sulfosalt minerals from Lengenbach, Binnatal, a new mineral, a copper arsensulfide was found, which has some similarity to binnite and lautite... The mineral was named sinnerite in honour of the late Rudolf von Sinner, president of the Commission of the Naturhistorisches Museum Bern. 1975 *Amer. Mineralogist* LX. 998/2 Both natural and synthetic crystals of sinnerite are complexly twinned.

Sinn Fein (ʃin fēi·n). [f. Ir. *sinn féin* we ourselves.] The name of an Irish movement founded in 1905 by Arthur Griffith (1872–1922), Irish journalist and politician, orig. aiming at the independence of Ireland and a revival of Irish culture and language and now dedicated to the political unification of Northern Ireland and the Republic of Ireland. Freq. *attrib.*
1905 *United Irishman* 18 Mar. 4/4 The Sinn Fein policy which we have propounded for Ireland, will henceforth be the policy of the National Council. 1906 A. Griffith *Sinn Fein Policy* 4 The policy for which the National Council stands is summarised in its title—'Sinn Fein'. *Ibid.* 32 Increase of employment, industrial effort and advancement should be everywhere the principal means towards the national aims of Sinn Fein Policy. 1907 *Westm. Gaz.* 31 Aug. 7/1 Sinn Fein is not as yet a movement of much political moment. 1920 *Public Opinion* 16 July 58/3 The intellectual leaders of Sinn Fein can by no means bind the extremists. 1930 W. K. Hancock *Australia* x. 213 Labour politicians preached Australia for the Australians and a sort of Sinn Fein exclusiveness. 1936 E. Curtis *Hist. Ireland* xx. 386 Sinn Fein came out as a political force by winning an election in Roscommon in February 1917. 1944 M. J. MacManus *Eamon de Valera* iv. 69 De Valera..had been co-opted on the National Council of Sinn Fein after his release. 1955 *Times* 16 May 3/5 The fact that Sinn Fein has put forward candidates for West Belfast, Mid-Ulster, [etc.]..may mean that a Unionist gain will be recorded. 1962 A. Lurie *Love & Friendship* ii. 29 He looked over-excited, as if he were about to..lead a small raid for the Sinn Fein. 1971 *Eire Nua: Soc. & Econ. Programme of Sinn Fein* 3 The Constitution of Sinn Fein advocates not merely the complete overthrow of English rule in Ireland but also the setting up of a Democratic Socialist Republic. 1972 R. Kee *Green Flag* ix. 452 From May 1905 Griffith's new policy generally began to be called the 'Sinn Fein' rather than the 'Hungarian' policy. The suggestion for the new name had been made to him by a young woman named Mary Butler, late in 1904, though the words had long been fairly commonly used as a motto for Irish self-reliance and had in fact been the early motto of the Gaelic League. 1978 P. Boardman *Worlds of Patrick Geddes* vii. 248 The never-ending nemesis of Sinn Fein Rebellion.
Hence **Sinn Fei·ner**, a member or adherent of Sinn Fein; **Sinn Fei·nism**, the methods, aims, or policies of Sinn Fein.
1907 *Daily Chron.* 13 Aug. 6/2 Sir Thomas Esmonde's action in adopting Sinn Feinism as against Parliamentary agitation. 1907 *Westm. Gaz.* 31 Aug. 7/1 Sinn Feiners further suggest that the Irish representatives, having withdrawn from Westminster, should assemble in Dublin. 1917 A. Huxley *Let.* May (1969) 124 The best part of political life after the war will be an unofficial Sinn Feinism. 1928 *Daily Express* 17 Nov. 3/5 The Sinn Feiners desired to strengthen this protest by obtaining the withdrawal from Parliament of the whole O'Brien party. 1945 R. Chandler *Let.* 1 Jan. (1981) 41, I have a great many Irish relatives..some of them Sinn Feiners. 1979 W. Nelson *Minstrel Code* ix. 75 The spot where Sinn Feiners gunned down a British Field-Marshal, Sir Henry Wilson, in 1922, on the doorstep of his home.

sinningia (siniŋgi,ā). [mod.L. (C. G. Nees von Esenbeck 1825, in *Ann. Sci. Nat.* VI. 296), f. the name of Wilhelm Sinning (1794–1874), German botanist + -ia¹.] A hairy herbaceous plant of the genus so called, belonging to the family Gesneriaceæ, native to Brazil, and bearing bell-shaped flowers. *Sinningia speciosa* is the parent of many varieties commonly known as gloxinias.
1826 *Bot. Reg.* XII. 997 (*heading*) Green Brazilian Sinningia. 1902 L. H. Bailey *Cycl. Amer. Hort.* IV. 1670/2 The Sinningias are first cousins of Gloxinias. 1926 E. Sitwell *Victoria of England* xix. 226 Flowers from the Queen's hothouses at Osborne—the lycaste with its waxlike bells.., the velvety sinningia with dark leaves. 1979 *Homes & Gardens* Feb. 25/1 Gloxinias, or sinningias as they are now called, are old favourites as house and cottage plants.

Sino-¹. Add: Also with pronunc. (sai·no).
1. **Sinological** *a.* (earlier example); **Sinologue** (earlier example); **Si·nophile** [-phile], a lover of China or things Chinese; also as *adj.* and **Sinophi·lia**, love of China or that which is Chinese; contrasted with **Sinopho·bia**, dread or hatred of these; also **Sinophobe** *adj.*; **Sino-xenic** *a.* [f. xen- (see Xeno-) + -ic], of a language: unrelated to a Chinese language but containing some Chinese linguistic elements.
1877 *Trübner's Amer. & Oriental Lit. Record* XI. 2/1 It is significant of the preponderance assigned in sinological studies to the English language, that Mr. von Möllendorff has thought it desirable to publish his work in English. 1970 *Guardian* 26 Nov. 15/6 This very readable and also scholarly collection..has all kinds of Sinological goodies too. 1816 P. du Ponceau *Let.* 31 July in *Trans. Hist. & Lit. Comm. Amer. Philos. Soc.* 1816 I. 400 As I am no Sinologist, I will not undertake to say that the description which I have attempted to give of this language..is very accurate. 1853 *North-China Herald* 21 May 167/2 Prince of Sinologues. 1900 E. R. Scidmore *China* i. 7 One agrees and disagrees, too, with the sinologues, who are usually sinophiles, that the Chinese are the one great race and flower of all Asia. 1977 R. Ludlum *Chancellor MS.* xxxiii. 350 He's a Sinophile..He has one of the most extensive Chinese art collections in the world. 1974 *Daily Colonist* (Victoria, B.C.) 17 July 17/7 Ireland attributed the herbal pill fad to the public's 'sinophilia' or ardent interest in Chinese culture including medicine. 1920 W. J. Locke *House of Baltazar* iii. 31 Water-End became divided into two camps—Sinophile and Sinophobe. 1966 *New Statesman* 22 Apr. 601/2 Siam..exhibits considerable Sinophobia. 1977 *New Yorker* 24 Oct. 177/1 The Korean war and the Sinophobic diplomacy of John Foster Dulles. 1972 *Computers & Humanities* VI. 259 The term 'dialect' in this article refers loosely to all the sources of information in DOC: Middle Chinese,..18 modern Chinese dialects, and 3 Sino-Xenic sources.
2. Combined with adjectives of nationality to mean 'Chinese and..' or 'between China or the Chinese and (the country or people designated)', as *Sino-Albanian*, *-American*, *-Australian*, *-British*, *-Indian*, *-Japanese*, *-Malay*, *-Mongolian*, *-Russian*, *-Soviet*, *-Tibetan*.
1976 W. H. Canaway *Willow-Pattern War* v. 55 The Sino-Albanian axis. 1931 H. B. Morse *Far Eastern International Relations* 750 (*caption*) Significance of the Sino-American treaty of 1928. 1978 D. Bloodworth *Crosstalk* iii. 28 Sino-American exchanges..had yielded a private understanding. 1904 *Amer. Naturalist* Sept. 676 The restriction of the Sino-Australian continent to a certain part of the Cretaceous times consequently would meet the postulates of geography and zoögeography. 1977 *South China Morning Post* (Hong Kong) 15 Apr. 2/1 An insight into how Sino-British relations, already good, are likely to develop. 1959 *Listener* 25 June 1093/2 The Sino-Indian agreement of April 1954. 1903 *Burlington Mag.* Oct. 13/1 We must place to the front the fact that Sino-Japanese design is almost exclusively an art of contours. 1978 *Jrnl. R. Soc. Arts* CXXVI. 652/1 The terms of the Sino-Japanese Trade Agreement are, I would judge, likely to prove more beneficial to both parties than the EEC/China Agreement. 1975 'G. Black' *Big Wind for Summer* ii. 22 The girl..was..Sino-Malay. 1976 *Times Lit. Suppl.* 20 Feb. 206/2 In Central Asia..the Sino-Soviet frontier is straddled by a homogeneous Muslim population, while the Sino-Mongolian frontier is similarly straddled by a Mongolian population. 1926 *Glasgow Herald* 23 Jan. 9 The rift in the Sino-Russian lute. 1929 A. J. Toynbee *Survey Internat. Affairs 1928* 434 The zone..had reverted to Chinese administration in virtue of the Sino-Soviet Russian agreement of the 31st May 1924. 1959 *Listener* 2 Apr. 598/3 The Sino-Soviet zone of nations. 1971 H. Trevelyan *Worlds Apart* x. 125 Sino-Soviet companies were formed to exploit minerals and oil, to develop Sinkiang in which the Soviet Union had a close interest, and to manage civil aviation. 1973 *Times* 14 Nov. 18/3 The Nagas are a group of 20 tribes of Sino-Tibetan origin.

3. Used similarly with sbs. to form sbs. (freq. *attrib.*) with the meaning 'a language (family) or subsection of this, characterized by a relationship between Chinese and the language (family) specified', as *Sino-Japanese*, *-Korean*, *-Siamese*; **Sino-Tibetan**, a family of languages comprising the Chinese, Tibeto-Burman, and (according to some scholars) the Tai languages.
1923 B. Karlgren *Analytic Dict. Chinese & Sino-Japanese* 7 After the Sino-Japanese readings I often add in parenthesis the Kana spelling. 1954 M. Pei *Dict. Linguistics* 198 *Sino-Japanese*, a term applied to Chinese loan-words in the Japanese language, the spoken form of which is different from the form or forms occurring in any of the spoken Chinese vernaculars. 1975 *Amer. Speech* 1973 XLVIII. 122 If Japanese is indeed the source of the term, a more plausible model would be *nemaki*, the colloquial equivalent of *shin-i*, which is the learned or Sino-Japanese term. 1953 Sino-Korean [see *hangul²]. 1975 *Language* LI. 257 Each representing a lexical item with information regarding..its pronunciation in the 21 major dialects of China and in the Sino-xenic languages, i.e. Sino-Japanese and Sino-Korean. 1948 D. Diringer *Alphabet* 402 The Sino-Siamese sub-family of languages. 1954 M. Pei *Dict. Linguistics* 198 Many linguists classify Chinese and Tai into one Tai-Chinese or Sino-Siamese sub-family. 1933 L. Bloomfield *Language* iv. 69 The great *Indo-Chinese* (or *Sino-Tibetan*) family consists of three branches. 1948 R. A. D. Forrest *Chinese Lang.* i. 21 Chinese is reckoned as an independent member of the Sino-Tibetan, Indo-Chinese, or Sinitic family of languages. 1977 C. F. & F. M. Voegelin *Classification & Index of World's Langs.* 307 It is the older, more liberal, classifications of Sino-Tibetan that have now come under critical scrutiny.

sino-² (sai·no), comb. form of Sinus.

sino-atrial (saino,ēi·triăl), *a.* *Anat.* Also sinoatrial. [f. *sino-² + Atrial *a.*] Of, pertaining to, or designating a small body of tissue (the *sino-atrial node*) in the wall of the right atrium of the heart that acts as a pacemaker by producing a contractile signal at regular intervals; (so called because it arises in the embryo at the junction of the sinus venosus and the atrium). Cf. *sinuatrial *a.*
1913 *Gray's Anat.* (ed. 18) 552 The sino-atrial node is situated on the anterior border of the opening of the superior vena cava. 1962 *Listener* 10 May 810/1 A small mass of specialized muscle cells within the heart, the sino-atrial node, discharges a brief electric 'spark' regularly seventy-two times a minute or thereabouts. 1969 [see *dysrhythmia* s.v. *dys-].

sino-auricular (saino,ǭri·kiŭlăɹ), *a.* *Anat.* [f. *sino-² + Auricular *a.*] = prec.
1907 Keith & Flack in *Jrnl. Anat. & Physiol.* XLI. 181 We use the term 'sino-auricular' in preference to 'sino-canalar' because, although a true sino-canalar junction exists on the dorsal side in the most primitive hearts,..yet in all but these the part of the canal between the sinus and the auricle disappears, and the dorsal junction becomes really a sino-auricular junction. 1942 Bramwell & King *Princ. & Pract. Cardiol.* v. 87 Closely related to sinus arrhythmia is the disorder known as sino-auricular heart-block. 1976 *Archives Internationales de Physiologie et de Biochimie* LXXXIV. 81 (*heading*) Chronotropic responses to experimental ischemia of the canine sino auricular node.

sinogram (sai·nogræm). *Med.* [f. *sino-² + -gram.] An X-ray photograph of a sinus into which a contrast medium has been introduced.
1961 *Lancet* 5 Aug. 296/1 A sinogram showed narrowing of the sagittal sinus anterior to the fontanelle. 1974 J. D. Maynard in R. M. Kirk et al. *Surgery* x. 220/1 Such sinograms..can be repeated regularly until the cavity has decreased until it is the size of the tube it contains.
So **sino·graphy**, the radiographic examination of sinuses.
1957 in Dorland *Med. Dict.* (ed. 23) 1254/2. 1974 A. Henry in R. M. Kirk et al. *Surgery* xv. 295/2 When a sinus is present it is worth while performing sinography..in an effort to delineate a communication with an intra-osseous abscess cavity possibly containing a sequestrum.

sinoite (sai·noait). *Min.* [f. the chemical symbols for silicon (*Si*), nitrogen (*N*) and oxygen (*O*) + -ite¹.] Silicon oxynitride, Si_2N_2O, found as colourless orthorhombic crystals in some chondritic meteorites.
1964 C. A. Andersen et al. in *Science* 9 Oct. 257/3 We propose the name *sinoite*, which is derived from the chemical symbols for the elements of this new mineral. 1966 *Geochimica & Cosmochimica Acta* XXX. 367 Sinoite has only been found in the enstatite chondrites. 1972 *Nature* 22 Dec. 461/2 The relatively high nitrogen content in some enstatite chondrites is partly due to the presence of the mineral sinoite, Si_2N_2O, which has been found only in two meteorites.

sinopia (sinō·piä). [It.; cf. Sinoper, Sinople.] **1.** = Sinoper 2 a.
1844 Mrs. Merrifield tr. *Cennini's Treatise on Painting* xxxvi. 25 There is a natural red pigment, which is called sinopia or porphyry. 1930 A. P. Laurie *Materials of Painter's Craft* x. 200 Sinopia..is one of the many names under which red ochres are mentioned. 1978 *Times Lit. Suppl.* 20 Oct. 1208/3 Fresco painters normally drew direct on the wall in sinopia without preliminary drawings.
2. *transf.* The preliminary rough sketch for a fresco, covered by the final work. Pl. *sinopie*.
1958 *Times* 10 Dec. 3/4 During the process of detaching a fresco from the wall it is often possible to separate it also from the *sinopia*—that is, the preliminary rough sketch. 1969 *Daily Tel.* (Colour Suppl.) 28 Mar. 28/3 Their long hidden *sinopie*, as the preparatory full-size wall-drawings are called, are generally our only witness to the evolution of their ideas. 1975 E. H. Gombrich *Let.* 14 Nov. in *Ideas & Idols* (1979) 183 Are the *sinopie* really better than the frescoes? 1981 M. Delahaye *Sale of Lot 236* xxiii. 197 The sinopia..had also to be right technically... He was safe in the materials. Charcoal, ochre, and sinopite were all natural substances.

sinsemilla (sinsəmi·lă). [a. Amer. Sp., lit. 'without seed'.] A plant belonging to a strain of *Cannabis sativa* having a particularly high narcotic content; also, the narcotic produced from a plant of this kind. Also *attrib.*
1975 *High Times* Dec. 68/2 Last year a guy bit the dust in Arizona in a PBY—a big-ass World War II Navy amphibian—full of prime Mexican sinsemilla. 1978 *Time* 12 June 22 Studies show that *sinsemillas* weed contains five times more tetrahydrocannabinol (pot's narcotic ingredient) than the common Mexican variety. 1980 *Daily Tel.* 19 Sept. 11/2 This year sinsemilla buds are fetching from $2,200 (£917) to $3,000 (£1,250) a pound. 1982 *Newsweek* 25 Oct. 60 Sinsemilla retails for up to $250 an ounce these days, and every plant produces up to two pounds of marketable buds.

sinter, *sb.* Add: **3.** Material which has been subjected to sintering; *spec.* iron ore prepared for smelting by sintering the powdered material, usu. together with coke and other materials; (see also quot. 1958).
1909 *Chem. Abstr.* III. 167 Process of treating metal-bearing ore in a uniformly formed mass of fine particles containing combustible elements to form relatively large agglomerated bodies of sinter by internal combustion. 1926 *Jrnl. Iron & Steel Inst.* CXIV. 61 Both the briquettes and the sinter are crushed to a somewhat coarser size than the lump ore before charging in the furnace. 1956 *Planning* 9 Apr. 64 Sinter has other advantages—it may rid the furnace of unwanted sulphur and volatile constituents and..increase furnace output while reducing coke consumption. 1958 A. D. Merriman *Dict. Metall.* 323/2 *Sinter*, a term used in reference to the solid waste from smelting or refining operations. It is also used to denote a product of a sintering operation. 1973 *Times* 12 Feb. (Suppl.) p. ii/4 The mixture of foreign and Frodingham ores will produce a sinter containing about 54 per cent iron.
4. Special Comb.: **sinter plant**, a furnace for sintering iron ore.
1938 R. H. Sweetser *Blast Furnace Practice* I. 57 Three types of sinter plant were erected in this country..all based on the principle of down-draft suction, ignition on the top of a mixture of flue dust, and fine iron pyrites cinder. 1980 *Times* 29 Feb. 2 The coke ovens at Redcar..are part of the steel complex, along with a sinter plant.

sinter (si·ntəɹ), *v.* [f. the sb.] **a.** *intr.* Of particles or particulate material: to coalesce into a solid mass under the influence of heat without liquefaction. Also with *together*. **b.** *trans.* To cause to coalesce in this way.
1871 [implied at Sintering *vbl. sb.*]. 1903 *Amer. Chem. Jrnl.* XXIX. 487. On heating, it [sc. methylmercapto-thymine] sintered at about 225°. 1907 *Trans. Inst. Mining & Metall.* XVI. 321 The formation of ferrous silicate aids..in sintering the material. 1938 R. H. Sweetser *Blast Furnace Practice* I. 58 This plant sinters red ore, fines and flue dust. 1948 *Electronic Engin.* XX. 68 To the surface thus formed a nickel powder is sintered to provide a base for brazing materials. 1953 *Sci. News* XXIX. 43 Many small metal components are now made by sintering metal powders, which have previously been pressed into the desired shape. 1973 *Sci. Amer.* Oct. 129/2 At high temperatures the effect is even stronger: metal powders sinter together under pressure. 1976 *Ceramurgia Internat.* II. 90/2 Attempts were made to sinter material which was dried at 125°C and not calcined.

sintered (si·ntəɹd), *ppl. a.* Substitute for def.: That has been subjected to or formed by sintering; **sintered carbide**, a very hard material manufactured by sintering a pulverized mixture of cobalt or nickel and carbides of metals such as tungsten and tantalum, and used in the cutting parts of tools; **sintered glass**, a porous form of glass made by sintering glass powder and used esp. in chemical filtration apparatus.
1877 [in Dict.]. 1907 *Trans. Inst. Mining & Metall.* XVI. 313 The product obtained consists of a porous sintered mass of ferrous silicate. 1937 *Machinery* L. 777 (*heading*) The production of sintered carbide. 1940 *Jrnl. Sci. Instruments* XVII. 155 (*heading*) Sintered-glass filters. 1951 O. W. Boston *Metal Processing* (ed. 2) vi. 103 Sintered-carbide tools were introduced commercially in this country in 1928. 1959 *Economist* 21 Feb. 719/2 The production of small pellets of sintered fuel..follows the normal practice of the ceramics industry. 1963 J. T. Marsh *Self-Smoothing Fabrics* xv. 281 The liquid is then decanted through a weighed sintered-glass crucible. 1964 S. Crawford *Basic Engin. Processes* iv. 104 Sintered carbide is fully hard when cooled to room temperature. 1982 D. Clark *Down Wales* vii. 147 Front and back wheels have disc brakes with sintered pads to cure the problem of grip on wet roads.

sintering (si·ntərɪŋ), *vbl. sb.* Substitute for def.: The process or action of the vb.; *spec.* as applied to iron ore (see *sinter *sb.* 3). Freq. *attrib.*
1871 [in Dict.]. 1907 *Trans. Inst. Mining & Metall.* XVI. 314 These..difficulties are overcome by the use of the sintering process, as the loss is small owing to the ore not being wet. 1914 *Iron Trade Rev.* LV. 292 (*heading*) Sintering plant at Pottstown. 1930 *Engineering* 16 May 650/2 Sintering occurs when charges of partially reduced ores are maintained for considerable periods at temperatures above 750 deg. C. 1958 [see *sinter *sb.* 3]. 1958 N. Levine *Canada made Me* ii. 78 Above us a steel cable carried large buckets of iron ore from the Mine to the Sintering Plant. 1977 *Western Mail* (Cardiff) 5 Mar. 6/1 South Wales exports about one million tons a year, much of it anthracite duff used in briquetting and sintering in steel production.

sintoc. Add: Also **sintok**.
1900 W. W. Skeat *Malay Magic* v. 278 In Penang a root called *sintok* is usually preferred to limes. 1972 A. Amin tr. *Ahmad's No Harvest but Thorn* iv. 32 The *sintok*-wood *tajak*-handle.

sinuatrial (sainiu,ēi·triăl), *a.* *Anat.* Also sinu-atrial. [f. L. *sinu-*, stem of Sinus + Atrial *a.*] = *sino-atrial *a.*
1935 *Gray's Anat.* (ed. 26) 67 The sinu-atrial and atrio-ventricular nodes, the atrioventricular bundle and its right fasciculus, are supplied by the right coronary artery. 1962 *Ibid.* (ed. 33) 747 The sinuatrial node is a narrow, horse-shoe shaped structure situated in the upper part of the sulcus terminalis of the right atrium. 1968 Passmore & Robson *Compan. Med. Stud.* I. xxviii. 9/2 The part of the heart with the highest spontaneous rate..provides the source of excitation of the whole heart, and this pacemaker is normally the sinuatrial (SA) node.

sinuous, *a.* Add: **3.** Also of people.
1906 B. Vaughan *Sins of Society* 129 The lithesome, sinuous girl trips with it across the stage to her mother.
Hence **si·nuousness** (*b*) in this sense.
1924 R. Macaulay *Orphan Island* xiii. § 4 Like a sturdy little boy without feminine elegances, or any of Flora's wild-animal sinuousness. 1980 *Early Music* July 308/1 Any actor reciting them..would certainly emphasize the contrast between the biting *sk* sound at the beginning of these two words and the powerful sinuousness of 'l'onde'.

sinus. Add: **6.** *sinus infection*; **sinus gland** *Zool.* [tr. G. *sinusdrüse* (B. Hanström 1937, in *K. Svenska Vetenskapsakad. Handl.* XVI. iii. 3)], a structure in the eye stalk or head of crustaceans orig. thought to be a gland but now recognized as a neurohæmal organ in which are stored various hormones concerned with growth, reproduction, and metabolism; **sinus rhythm**, the normal rhythm of the heart, proceeding from the sino-atrial node; **sinus venosus** [mod.L., f. *venōsus* venous], a part of some vertebrate hearts into which the veins lead and which empties into the atrium.
1938 *Arkiv för Zoologi* XXX B. viii. 1 When investigating the nervous system and the organs of sense of the crustaceans, Hanström found (1931–1935) two organs which he called the blood gland (the sinus gland) and the X-organ. 1972 M. S. Gardiner *Biol. Invertebrates* xvii. 714/1 Molting in decapod crustaceans is also influenced by hormones discharged from the sinus glands. 1936 *Discovery* Dec. 380/1, I was suffering from widespread sinus infection. 1911 T. Lewis *Mechanism of Heart Beat* xii. 132 The compensatory pause fails; that is to say, disturbance of sinus rhythm may be demonstrated. 1980 *Brit. Med. Jrnl.* 29 Mar. 922/2 Ventricular fibrillation was converted to sinus rhythm with a 'thump' on the chest. 1836–9 R. B. Todd *Cycl. Anat. & Physiol.* II. 579/2 This division of the auricle into proper auricle and sinus venosus is more distinct in the left than in the right auricle. 1926 J. S. Huxley *Ess. Pop. Sci.* 199 The ventricle of the frog's heart has its own independent rate of beat when isolated. But in the intact animal this independent rate is all the time being speeded up by the faster-beating sinus venosus. 1970 *Encycl. Biol. Sci.* (ed. 2) 400/2 In the fish, amphibians, and reptiles the contraction starts in the thin muscle wall of the sinus venosus... In birds and mammals the sinus venosus is absent.

sinusitis (sainŭsai·tis). [f. Sinus + -itis, prob. ad. F. *sinusite*.] Inflammation of a sinus, esp. a nasal sinus.
1896 *Jrnl. Laryngol.* X. 37 (*heading*) Treatment of sinusitis. 1935 Imperatori & Burman *Dis. Nose & Throat* xi. 146 Morning frontal headache increasing in intensity towards afternoon usually indicates maxillary sinusitis. 1951 M. Lowry *Let.* 25 Aug. (1961) 252 Water..nearly makes him die of dysentery, and milk..gives him sinusitis. 1977 W. Marshall *Thin Air* vii. 87 The smell..came up the rickety stairs..like a poisonous cloud of pollen at the height of the sinusitis season.

sinusoid. Add: **2.** A blood vessel similar in size to a capillary but irregular in shape and without the continuous endothelial lining of capillaries. (Later examples.)
1920 *Nature* 27 May 411/1 A system of wide sinusoids (renal venous meshwork), which has no connection with the intertubular plexus. 1974 D. & M. Webster *Compar. Vertebr. Morphol.* xiii. 315 In the mammalian adrenal cortex the steroidogenic tissues can be divided into zones on the basis of the arrangement of cells and their relationships to the numerous blood sinusoids.

sinusoidal, *a.* Add: Having the form of a sinusoid; varying periodically (with time, distance, etc.) as a sine varies with an angle; (see also quot. 1910). (Further examples.)
1910 N. Hawkins *Electr. Dict.* 400/2 *Sinusoidal alternator*, an alternating current dynamo which generates simple harmonic or sinusoidal currents. 1948 *Nature* 28 Feb. 295/1 Along with the detailed solutions of the differential equations for step velocity input..is given a brief development of the solutions for sinusoidal input. 1957 G. E. Hutchinson *Treat. Limnol.* I. v. 357 The height of the wave at first decreases slightly, then increases as the wave approaches breaking. During the final phase of the rise in height, the sinusoidal or trochoidal form is lost. 1975 *Nature* 23 Oct. 674/1 Patterns of alternating light and dark bars with a sinusoidal luminance profile across the bars—sinusoidal gratings—are commonly used to study spatial interactions in the visual system.
sinusoidally (later examples).
1929 J. A. Ratcliffe *Physical Princ. Wireless* i. 3 The simplest form of alternating current is one in which the current varies sinusoidally with the time. 1947 *Jrnl. Inst. Electr. Engineers* CXIV. iii. 279/1 When the hemispherical end of a reproducing stylus is moved in contact with a sinusoidal surface of comparable curvature, the stylus tip does not itself move sinusoidally. 1974 *Sci. Amer.* Nov. 34/3 Voltages and current vary sinusoidally at a frequency of 60 hertz (cycles per second) in the U.S. and Canada and 50 hertz in most other countries of the world.

‖ **siot** (ʃǫt). [Welsh, evidently a loanword from an unrecorded regional sense of Shot *sb.*¹] In North Wales, a cereal mash of buttermilk and crushed oat-bread.
1936 *Farmhouse Fare* 148 *Siot.* ½ cupful oat bread. 1 pint buttermilk. Crush the bread..and put into a basin. Pour in the buttermilk, and let the bread soak for 1 hour, when it will be ready to serve. 1949 'M. Innes' *Journeying Boy* xiv. 172 For breakfast it was impossible that they should have anything but siot or skirlie-mirlie. 1974 *Country Life* 12 Dec. 1845/2 Siot..was a pint of buttermilk poured on to a half cup of crushed oat bread... The 'mash' was served with bread and butter.

Siouan (sū·ăn), *a.* and *sb.* [f. *Siou(x *a.* and *sb.* + -an.] **A.** *adj.* = *Sioux *a.* b. **B.** *sb.* **a.** = *Sioux *sb.* 2 a. Also *Comb.* **b.** = *Sioux *sb.* 1 b.
1885 J. O. Dorsey in *Smithsonian Inst. Rep. 1883* 919 The term 'Siouan' has been applied to that family of Indians which has been known heretofore as the 'Dakotan Family'. It is unfortunate that we are obliged to use this adjective, which is derived from 'Sioux', as the latter is not a genuine Indian word... In honor of Albert Gallatin, who was the first to classify the Indians of this family as the 'Sioux', the Bureau of Ethnology of the Smithsonian Institution has adopted the new term, 'Siouan', as the name of this family. 1889 *Amer. Naturalist* Jan. 75 The Siouan group had its habitat on the prairies between the Mississippi and Missouri. 1900, 1907 [see *Crow *sb.*⁴ and *a.*]. 1929 E. Sapir in *Encycl. Brit.* V. 139/2 Hokan-Siouan... Eastern group (1) Siouan-Yuchi..(2) Natchez-Muskogian. 1937 R. H. Lowie *Hist. Ethnol. Theory* (1938) vi. 63 [Morgan] recognized the criteria of..the 'Omaha' system and indicated its occurrence among the Algonkian as well as the Siouan family. 1949 B. A. Botkin *Treas. S. Folklore* iii. ii. 497 Enough Biloxi were left..to enable investigators to identify them..as members of the Siouan linguistic family. 1977 H. Landar in T. A. Sebeok *Native Langs. Americas* II. iii. 352 Gallatin (1836) assigned the Shyenne language to his Sioux group on the basis of several names of 'Shyennes' who signed a treaty in July, 1825. The signers were possessed of Siouan, not Algonquian, names. 1978 C. Callender in B. G. Trigger *Handbk. N. Amer. Indians* XV. 610/1 The societies that at the time of European contact were established in the upper Great Lakes area..were mostly Algonquian but included the Siouan-speaking Winnebago.

Sioux (sū), *a.* and *sb.* Also 8 **Sous**, 9 **Suouex**. [a. N. Amer. Fr., earlier *Nadouessioux*, etc., ad. Ojibwa (Ottawa dial.) *nātowēssiwak*: Fr. pl. termination *-x* replaced the equivalent Ojibwa feature *-ak*.] **A.** *adj.* **a.** Of or pertaining to the Sioux people or their language (see the sb. below). **b.** Formerly, of or pertaining to the Siouan languages or language grouping; = *Siouan *a.*
1761 D. Claus *Let.* 19 Mar. in J. Sullivan *Papers Sir W. Johnson* (1921) III. 363, I picked up a pair of shoes made by the Sioux Indⁿˢ. to the Westward. 1805 Z. N. Pike *Jrnl.* 8 Sept. in *Sources of Mississippi* (1810) i. 14 His design was to winter with some of the Sioux bands. 1824 W. H. Keating *Narr. Exped. St. Peter's River* I. viii. 376 Account of the Dacotas or Sioux Indians. 1836 A. Gallatin in *Trans. & Coll. Amer. Antiquarian Soc.* II. 120 The nations which speak the Sioux language may be considered..as consisting of four subdivisions, viz. the Winnebagoes; the Sioux proper and Assiniboins; [etc.]. 1893 L. Wagner *Significance of Names* 36 The Sioux State [is] the territory of the Sioux tribe of Indians. 1919 S. Lewis *Free Air* ix. 101 She fancied that on it the Sioux scout still sat sentinel. 1949 *Amer. Photogr.* Jan. 40/1 Following a speech in Sioux language, in which God was asked for rain, the real Sun Dance now started. 1973 *Black Panther* 1 Sept. 17/1 Chief Fools Crow..damned the Executors in Sioux language.
B. *sb.* **1. a.** = *Dakota *sb.* 1. **b.** Formerly also, more generally, a (member of a) linguistic grouping of North American Indian peoples that includes the Sioux, Crow, Omaha, and others.
1762 D. Claus *Let.* 2 June in J. Sullivan *Papers Sir W. Johnson* (1921) III. 754 Missisages, Ottawawas, Renards & Sioux, were the Nations assembled at Cataracqui. 1768 [see *hot war* s.v. *hot *a.* 12]. 1785 T. Jefferson *Notes on Virginia* xi. 185 Sioux. On the heads of the Mississippi and westward of that river. 1827 J. F. Cooper *Prairie* III. vii. 227 The keen weapon..meeting the naked breast of the impetuous Sioux, the blade was buried to the buck-horn haft. 1836 A. Gallatin in *Trans. & Coll. Amer. Antiquarian Soc.* II. 120 The Indians..east of the mountains are the Sioux; the Pawnees; the Fall, Rapid, or Paunch Indians; [etc.]. 1908 *Rep. Brit. Assoc.* 851 They [sc. the mound-builders] seem to have been followed by the Sioux (Dakotas), Iroquois, who are probably of Aztec origin. 1937 R. H. Lowie *Hist. Ethnol. Theory* (1938) xiv. 262 The signs of Queenslanders and Sioux lend little support to psychic unity. 1957 P. Worsley *Trumpet shall Sound* 222 To the Sioux..the coming of the Dance helped to set off a train of events which culminated in the bloody massacre at Wounded Knee. 1964 Mrs. L. B. Johnson *White House Diary* 11 Jan. (1970) 40 George Catlin..painted the Sioux... His paintings line the second-floor hall of the White House. 1975 D. Pitts *This City is Ours* xiv. 48 His name was Mick Dull Knife; he was a full-blooded Sioux and a graduate of Princeton.
2. a. Formerly, the language family to which the Sioux and related peoples belong; = *Siouan *sb.* **b.** The language of the Sioux or Dakota Indians.
1783 J. O. Justamond tr. *Raynal's Philos. & Polit. Hist. Settlements & Trade of Europeans in E. & W. Indies* VI. xv. 439 Three original languages were spoken in Canada, the Algonquin, the Sioux, and the Huron. 1915 J. Buchan *Salute to Adventurers* ix. 141 He said something in Sioux to one of the warriors. 1971 *Guardian* 18 Sept. 10/4 She spoke only Cheyenne and Sioux.

Sip (sip), *sb.*² Also **sip**. Black English abbrev. of *Mississippi* (cf. *Mississippian *sb.* and *a.*); *the Sip*: the State of Mississippi.
1969 P. Cross in *Folklore Forum* II. vi. 141 There's this nigga who went to the 'Sip', you know, uh—Mississippi, that is. 1971 *Black Scholar* Jan. 41/1 He loaded up the trunk with all his fine clothes and lit' out for the 'sip'.

Sipapu (sī·papŭ). Also **Shipap(u**. [ad. Hopi *sipâ·pɨ*.] In the beliefs of Pueblo Indians, an opening in the earth, variously located by different tribes, through which their mythical ancestors emerged into the present world; a symbolic representation of this opening, as a hole in the floor of a kiva.
1891 A. F. Bandelier in *Jrnl. Amer. Ethnol. & Archaeol.* (1892) III. 111 Cibobe is the same as Shipapu, the lagune where the deceased go to rest. 1896 J. W. Fewkes in *16th Ann. Rep. Bur. Amer. Ethnol.* (1897) 279 The Snake chief at Cipaulovi has no *tiponi*, and consequently no altar. The only objects at the end of the altar..was a row of twenty snake whips leaning against the ledge of the rear wall, behind the *sipapû*. 1931 R. Benedict in *U.S. Bur. Amer. Ethnol. Bull.* xcviii. 3 When the people who came up out of Shipap found these people who had been saved they called them Tsauwan yahana. 1939 E. C. Parsons *Pueblo Indian Religion* I. iii. 216 On the road to the *sipapu* in the west, the place of emergence where the Hopi dead return..an actual spot in the wall of the Grand Canyon.., the breath body is met by..an Agave spirit sentinel. 1955 Priestley & Hawkes *Journey down Rainbow* vii. 115 Although the Sun Father was worshipped as the great cosmic power, Pueblo religion was mainly directed downwards towards the realm of the Earth Mother, the realm that was reached by way of the *Sipapu*, the home of the spirits whence the newborn came and the dead returned. 1977 M. Jenkinson *Land of Clear Light* i. 69 Man came up through the opening, the Sipapu, and flourished upon the surface of the earth... The image of the Sipapu has been likened by non-Indians to the birth process.

Siphnian (si·fniăn), *sb.* and *a.* [f. Gr. Σίφνιος, L. *Siphnius* + -an.] **A.** *sb.* A native or inhabitant of the Greek Cycladic island of Siphnos. **B.** *adj.* Of, pertaining to, or characteristic of Siphnos.
1709 I. Littlebury tr. *Herodotus' Hist.* I. iii. 282 The Siphnians drawing all their forces together, fought a battle, and were defeated by the Samians, who took many prisoners in the pursuit. 1845 *Encycl. Metropol.* XVII. 494/2 The soft stone, whence the ancient Siphnians used to cut pots and cauldrons. 1886 J. H. Frere *Aristophanes* 18 It appears by what Herodotus says of the oracle addressed to the Siphnians, that the 'red cheeks' must have gone out of fashion in his time. 1895 *Jrnl. Hellenic Studies* XV. 208 The school to which the sculpture of the Siphnian treasury must be assigned has already caused some discussion. 1932 R. Fry *Let.* 1 May (1972) II. 669 The early Ionian sculptures—above all the Siphnian treasure. 1956 Parke & Wormell *Delphic Oracle* I. ii. vi. 156 Siphnos had a..rich source of income in gold and silver mines. From the proceeds of these mines the Siphnians became after the mid-sixth century the richest of the islanders. 1977 *Antiquaries Jrnl.* LVII. 348 Chemical tests have shed no further light on the origin of the wares conventionally known as 'Melian' and 'Siphnian'.

siphon, *v.* Add: **1. b.** *fig.* To draw off or from, as if by means of a siphon; to divert. Const. advbs. (chiefly *off*: *spec.* illicitly, of money) and preps.
1940 E. Wilson *To Finland Station* ii. i. 75 All the fervor of which they were still capable was siphoned off into the revolutionary army. 1952 *N.Y. Times* 8 Sept. (late City ed.) 45/7 The police said order was gradually restored as the screaming crowd was slowly siphoned out of the stadium. 1955 H. Roth *Sleeper* ix. 69 We are positive he wasn't siphoning out information. 1957 *Economist* 7 Dec. 842/1 If he were to siphon off the more than 20 per cent of the poll which Liberals won at Gloucester and Ipswich..the Tories' majority..would be reduced. 1965 H. I. Ansoff *Corporate Strategy* (1968) iv. 61 This personal objective [of maximum current earnings] can have a shattering effect on a firm when control is taken over by a person or a group with the explicit aim of siphoning out of the firm most of its liquid..assets. 1976 F. Warner *Killing Time* i. ii. 12 Society depends on the integration of those functions that prostitution siphons off. 1979 'A. Hailey' *Overload* iii. xiii. 261 So how about the remainder [of the income of an organization]? The best guess was that Birdsong, who controlled p & lfp totally, was siphoning it off.

siphonaceous (saifŏnēi·ʃəs), *a.* *Bot.* [f. Siphon *sb.* + -aceous.] Characterized by or being an algal thallus that is tubular and largely without septa; = Siphoneous *a.* in Dict. and Suppl.
1916 G. S. West *Algæ* I. 223 There are a number of undoubted fossil siphonaceous Algæ, the calcified thallus of various forms having lent itself to preservation. 1933 G. M. Smith *Fresh-Water Algae U.S.* 299 Continued growth of such a coenocyte..would lead to a simple siphonaceous form such as *Protosiphon*. 1969 F. E. Round *Introd. Lower Plants* ii. 30 One of the simplest completely siphonaceous genera is *Bryopsis*.

siphonapteran (saifŏnæ·ptərăn), *a.* (*sb.*) [f. mod.L. order name *Siphonaptera* (P. A. Latreille *Familles Naturelles du Règne Animal* (1825) 334), (f. combining form of Gr. σίφων Siphon- + ἄπτερος wingless) + -an: cf. Apteran *a.* and *sb.*] Of or pertaining to a flea of the order Siphonaptera. Also as *sb.*
1842 [in Dict., s.v. Siphon-]. 1941 I. Fox in *Proc. Entomol. Soc. Washington* XLIII. 6 (*title*) The siphonapteran thorax. 1962 Gordon & Lavoipierre *Entomol. for Students of Med.* xxxv. 218 The egg is similar to that of other fleas and the larvae which emerge after 3 to 4 days incubation are typical siphonapteran larvae.

siphonein (sai·fŏnīn). *Biochem.* [f. Siphon(o- + -ein, perh. after Lutein.] An ester of siphonoxanthin present in certain green algæ.
1949 H. H. Strain in Franck & Loomis *Photosynthesis in Plants* vi. 162 Siphonein. In most algae containing siphonaxanthin one finds about equivalent amounts of a spectroscopically similar carotenoid which is converted into siphonaxanthin by saponification with alcoholic potassium hydroxide. 1964 [see *siphonoxanthin* s.v. *siphono-]. 1969 *Tetrahedron Lett.* No. 59. 5141 Siphonein is an ester of siphonaxanthin characterized by the esterification of the primary hydroxyl group with a fatty acid.

siphoneous, *a.* Add: = *siphonaceous *a.* (Later examples.)
1967 I. Morris *Introd. Algae* ii. 21 Multicellular thalli may be regarded as being of five main types: 1. Colonial.. 2. Aggregatious..3. Filamentous..4. Siphoneous..5. Parenchymatous. 1971 Kumar & Singh *Textbk. Algae* ii. 24 A siphoneous thallus is multinucleate lacking septation except during the formation of reproductive organs.

siphoner (sai·fŏnəɹ). [f. Siphon *v.* + -er¹.] One who draws off (liquid, etc.) by siphoning; *spec.* a petrol-thief.
1961 E. S. Turner *Phoney War* vii. 44 The scarcity of petrol bred a new type of sneak-thief, the siphoner, who went round with a length of rubber tubing and a can, helping himself to the contents of car tanks. 1979 *Tucson Mag.* Sept. 35/1 Mass cruising [in cars] is dying rapidly. About the only people who can do it anymore are the filthy rich and the skillful siphoners.

SIPHONET

siphonet. Add: Now *rare* or † *Obs.*, *siphunculus* (q.v., sense *2) or more commonly *cornicle* being used instead.

siphono-. Add: **si·phonostele** *Bot.*, a stele consisting of a core of pith surrounded by concentric layers of xylem and phloem; so **si·phonoste·lic** *a.*; also (*rare*) **si·phonostely**, the state of being or having such a stele; **si·phonoxa·nthin** (also **siphonax-**), a xanthophyll pigment, $C_{40}H_{56}O_4$, present in certain green algae.
1899 E. C. Jeffrey in *Mem. Boston Soc. Nat. Hist.* V. 160 The two primitive types of stele described above may be appropriately designated, protostelic and siphonostelic respectively. *Ibid.*, The primary vascular axes of living Lycopodiales rarely present the phenomena of siphonostely. **1902** Siphonostele [see *mesarch 1]. **1969** F. E. Round *Introd. Lower Plants* x. 123 A simple stele is formed. This consists either of a stellate mass of xylem surrounded by a few rows of simple phloem cells or, in the aerial parts of *Psilotum* and *Tmesipteris*, becoming siphonostelic. *Ibid.*, The central tissue in the siphonosteles is often sclerified. **1949** H. H. Strain in Franck & Loomis *Photosynthesis in Plants* vi. 162 Siphonaxanthin. Green algae of the order Siphonales yield significant quantities of this ketonic pigment that closely resembles fucoxanthin with respect to spectral absorption properties. **1964** *Oceanogr. & Marine Biol.* II. 217 Within the Chlorophyta, the Siphonales differ from the other members in possessing siphonein and siphonoxanthin. **1973** V. J. & D. J. Chapman *Algae* (ed. 2) vi. 137 The discoid chloroplasts..lack siphonoxanthin so characteristic of the siphonaceous Chlorophyceae.

siphuncle. Add: **2.** *spec.* = *siphunculus 2.
1899 *Cambr. Nat. Hist.* VI. 589 Another highly peculiar structure [in aphids] is the siphons, frequently called nectaries, honey-tubes, or siphuncles. **1962** G. A. Tulloch *J. R. de la Torre-Bueno's Gloss. Entomol.* (rev. ed.) 268 *Siphuncle*, the cornicle of the aphids.

siphuncular, *a.* Add: Also, of or pertaining to a siphunculus. (Later example.)
1975 *Jrnl. Zool.* CLXXV. 318 Prodding the abdomen and pinching the legs of a sycamore aphid induces it to produce siphuncular exudate.

siphunculus. Add: **2.** *Ent.* A tubular appendage on the abdomen of aphids, which lets out a waxy substance when the animal is attacked that acts as an alarm pheromone; (formerly believed to be the tube from which honeydew comes). Usu. called a *cornicle*.
1939 V. B. Wigglesworth *Princ. Insect Physiol.* x. 233 At the apex of the abdominal tubes or siphunculi of Aphids are ostioles..which allow wax-laden blood cells to escape. **1975** *Jrnl. Zool.* CLXXV. 280 When prodded or attacked by parasites or predators aphids often exude these cells bathed in fluid from their siphunculi.

Siporex (si·pəreks). A proprietary name for a type of cement or concrete (see quots.).
1938 *Trade Marks Jrnl.* 27 July 913/1 Siporex 583,757. Slates, bricks, blocks, beams, piles, pillars, posts, tiles, pipes, drains and shaped pieces, all made of cement or concrete for use in building or construction. Internationella Siporex Aktiebolaget..Sweden. 24th February 1938. **1943** *Archit. Rev.* XCIV. 68/1 The Mässhallen in Gothenburg—a large covered sports hall in steel, concrete and siporex—is also by Eriksson. **1965** *Economist* 5 June 1176/2 Siporex is an autoclaved lightweight aerated concrete product which was first developed in Sweden some thirty years ago, and is now used all over the world.

sippers (si·pəɹz). *Naut. slang.* [f. Sip *sb.*: see *-er.] A sip (of rum), esp. taken from another's tot, as a reward for some service or in celebration; *spec.* (see quot. 1944).
1944 J. P. W. Mallalieu *Very Ordinary Seaman* 99 For his service in measuring the rum the Leading Hand of the mess was entitled to 'sippers' from every man, a 'sipper' being a taste of each tot. **1945** *Penguin New Writing* XXIII. 49 Old three-badge A.B.'s offer him 'sippers' from their tots of rum and protect him from over-zealous killicks. **1956** H. Tunstall-Behrens *Pamir* 25 A bottle appeared with enough in it to give us all 'sippers'. **1977** R. Baker *Dry Ginger* viii. 190 Two brothers, one old enough to be allowed his tot of rum, and the other too young..known respectively as 'Gulpers' and 'Sippers' Young.

sipunculan (səipʌ·ŋkiŭlăn), *sb.* and *a.* [f. mod.L. name of phylum *Sipuncula*: see *sipunculid.] **A.** *sb.* A burrowing unsegmented worm of the phylum Sipuncula, found in sandy littoral regions. **B.** *adj.* Of or pertaining to a worm of this kind or the group as a whole.
1975 *Nature* 30 Oct. 818/2 There is convincing evidence that myohaemerythrin, a monomeric protein found in the retractor muscles of the sipunculan worm *Themiste pyroides*, and the protomers of haemerythrin have quite similar tertiary structures. **1977** P. E. Gibbs *Brit. Sipunculans* 7 Whilst sipunculans show clear embryological affinities with the Annelida, it is now generally agreed that..they are best regarded as a distinct and separate phylum.

sipunculid. For 'as prec.' in etym. substitute: mod.L. (Linnæus *Systema Naturæ* (ed. 12, 1766) I. II. 1078). (Later examples.) Also as *adj.*
1928 Russell & Yonge *Seas* i. 20 The Sipunculids..have a protrusible proboscis and a tough leathery body. **1941**, **1967** [see *echiuroid *a.* and *sb.*]. **1974** *Sci. Amer.* Apr. 88/2 Probably a fourth type consisted of unsegmented burrowers that fed on surface detritus and gave rise to the modern sipunculid worms.

sir, *sb.* Add: **I. 1. b. Sir Berkeley** *coarse slang* [after *Berkeley* Hunt: see *berk], the female genital organs; hence *transf.*, sexual intercourse, 'sex'; **Sir Garnet**: see *Garnet.
1937 J. Curtis *There ain't no Justice* xvii. 175 She gives me plenty of the old Sir Berkeley, but she knows how to look after herself, that girl.
II. 6. sir knight (later examples).
1939 *Sun* (Baltimore) 30 Sept. 7/8 We stand here today to watch you, sir knights, just in friendly tournament. **1977** *Belfast Tel.* 19 Jan. 2/4 The Officers and Sir Knights of the United Sons of Ulster R.B.P. 1041 regret the death of the Sister of their esteemed Sir Knight Robert Scott, P.M.
7. a. Also, used by schoolchildren in addressing a master.
1838 Dickens *Nickleby* (1839) viii. 69 'Third boy, what's a horse?' 'A beast, Sir,' replied the boy. **1899** Kipling *Stalky & Co.* 108 Please, sir, what am I to do about prep.? *a* **1930** D. H. Lawrence *Phoenix II* (1968) 233 Please Sir, do footmen bite? **1955** E. Blishen *Roaring Boys* i. 31 'The cane,' said Sims vaguely. 'Sir can't,' said Pottell... 'Is it because you're too young, sir?' **1974** 'J. le Carré' *Tinker, Tailor* xiv. 118 'Sir, please sir,' I think he's to do with the church, sir,' said Cole Slaw. 'I saw him talking to Wells Fargo, sir, after the service.'
b. (Earlier and later examples.)
1768 Sterne *Sent. Journey* I. 38 Figure to yourself, my dear Sir, that in giving you a chaise which would fall to pieces before you had got half way to Paris..how much I should suffer. **1776** *Critical Rev.* XLII. 69 It is usual, we are told, with the Scots..to address the person with whom they converse by the appellation My dear Sir. **1893** G. B. Shaw *Widowers' Houses* II. 45 *Sartorius*: Will you excuse me for ten minutes? *Cokane*: My dear sir: certainly. **1983** A. Veysey *Blood on Rocks* xiii. 122 'You must have something to eat, my dear sir.'
c. *yes, sir*: an emphatic assertion; *no, sir*: see *nossir. Chiefly *U.S. colloq.*
1799 *Aurora* (Philadelphia) 8 Aug. (Th.), Yes Sir! and [France] has been successful beyond any former experience. **1889** 'C. E. Craddock' *Despot of Broomsedge Cove* 40 Yes, sir... None like 'em now. **1929** W. Faulkner *Sartoris* II. vi. 114 'Yes sir,' he repeated, 'he's sure some joker.' **1941** J. B. Priestley in *R.A.F. Jrnl.* 1 Oct. 2/1 I could take it and I could dish it out. Yes, Sir!
10. a. Also more recently, a knight or baronet.
1922 W. J. Locke *Tale of Triona* i. 9 A proud old Anglo-Indian family, all Generals and Colonels and Sirs and Ladies. **1932** 'W. Cooper' *Struggles of Albert Woods* iv. I..and Albert thought..there must be a connection between Jameson's appointments and his becoming a Sir. **1974** P. Gore-Booth *With Great Truth & Respect* 374, I argued hard and explicitly on behalf of my diplomatic colleagues because becoming a 'Sir' is one of the tools of the trade.
b. *spec.* a schoolmaster. *colloq.* or *humorous*.
1955 [see sense 7 a above]. **1969** *Guardian* 1 Dec. 7/2 [The] users will be grateful to Sir for providing..a smashing set of answers. **1968**, **1973** [see *Miss *sb.* 3 f]. **1980** *Daily Tel.* 31 Mar. 10/3 Sir never repeated any part of a question.

sirab, var. *serab.

siratro (sirǽ·trəʊ). [f. initial letters (as indicated) of *C*ommonwealth *S*cientific and *I*ndustrial *R*esearch Organization + *atro*(*purpureus* the specific epithet of the parent plants).] A tropical legume of the variety so called, developed at the C.S.I.R.O. Pasture Research Station, Samford, near Brisbane, by E. M. Hutton in the early 1960s from Mexican strains of *Macroptilium* (*Phaseolus*) *atropurpureum*.
1962 E. M. Hutton in *Austral. Jrnl. Exper. Agric.* II. 117/1 Two Mexican strains of *Phaseolus atropurpureus*..were crossed, and this resulted in the development of a new bred strain which has been named Siratro. **1975** *Nature* 31 July 400/1 An isolate from nodules of *Trema cannabina*..is a strain able to nodulate siratro. **1978** *Jrnl. R. Soc. Arts* CXXVI. 628/2 The Legume Siratro is probably the most outstanding example of pasture plant breeding... It is now the most widely grown pasture legume in southern and central Queensland.

sirdar. Add: **I. β.** (Later examples.)
1969 P. S. Allen *Let.* 25 Feb. (1979) 19 A Sardar of this neighbourhood whom we met in Khewra station sent us..a present. **1969** [illegible] *Windward & Other Stories* 36 An arsenal of life pulsed and throbbed between the harsh rasping orders of the sardars echoing through the native compound.

siree, var. Sirree in Dict. and Suppl.

sireen (sairī·n). Also **syreen.** Repr. colloq. (now chiefly *U.S.*) pronunc. of Siren *sb.* (esp sense 7 b in Dict. and Suppl.).
1915 Kipling *Fringes of Fleet* 11 Five damned trawlers with their syreens blowing. **1940** *Economist* 28 Sept. 398/2 The air raids have produced some more new war words... A quite inexplicable new word is the 'sireen', which has widely ousted the siren. Its slight resemblance to Eileen and Doreen suggests that it may have arisen from a desire to give the noise a feminine personality. **1943** G. Greene *Ministry of Fear* I. i. 16 This time of night... It's the sireens. **1957** W. Faulkner *Town* xxiv. 363 Mr. Connors went to his [car] that had the red light and the sireen on it. **1977** J. Cleary *Vortex* i. 20 The siren sounds..'I wish he'd blow up that goddam si-reen.'

sireland (sai·əɹlænd). *nonce-wd.* [f. Sire *sb.*, punningly after *Ireland*.] The land of one's birth; one's native country or fatherland.
1922 Joyce *Ulysses* 182 Cranly's eleven true Wicklowmen to free their sireland. **1939** —— *Finnegans Wake* 448 Sireland calls you.

siren, *sb.* Add: **I. 7. b.** Also, more generally, a device which produces a piercing note (freq. of varying tone), used as an air-raid warning, or to signify the approach of a police car, etc.; the noise itself. Formerly, a motor-horn.
1907 [see *cut-out *sb.* 1 b]. **1917** *Flying* 25 July 2/2 Tests with various sirens were made in Central London in order to ascertain whether they would be audible. **1940** G. O'Casey *Let.* 20 Aug. (1975) I. 866 We have a kind of a cellar that we are to go to when the siren sounds. **1943** *Times* (Weekly ed.) 18 Aug. 11/3 The Luftwaffe helps the Church Army. How?—every night there isn't a siren, a Church Army friend puts sixpence in her box and when there is she puts in 6d. in as soon as the 'All Clear' goes. **1963** Mrs. L. B. Johnson *White House Diary* 22 Nov. (1970) 3 We got in [the car]. Lyndon told the agents to stop the sirens. **1964** [illegible] 43 Then the sirens went. **1971** *Daily Tel.* (Colour Suppl.) 22 Oct. 19/1 Pursued by cops, lights flashing, siren wailing, up Brighton Road.
II. 8. b. *siren-bird.*
1923 E. Sitwell *Bucolic Comedies* 90 Where siren-birds sip Bohea.
10. *siren alarm*; **siren suit,** a one-piece costume resembling overalls or a boiler-suit, orig. designed for wear by women in air-raid shelters; later, worn by either sex, and as a fashion garment.
1939 G. B. Shaw *Farfetched Fables* ii. 107 He is interrupted by a siren alarm, followed by an artillery salvo. **1976** L. Dow? *How Wind Blows* iv. 72 Some time in the early part of the morning there was a siren alarm, and we all trooped down to the basement of No. 10. **1939** *English Autumn* 34 Ladies' dress-shops ambiguously advertise 'siren suits' for the Air Raid Shelter. **1942** [illegible]. **1959** R. Collier *City that wouldn't Die* xii. 210 In a minute he [*sc.* Winston Churchill] came—black silk dressing gown embroidered with gold pheasants over the baby-blue siren suit he called 'my rompers'. **1977** *Belfast Tel.* 19 Jan. 18/7 (Advt.), Good reductions in children's coats and fur siren suits.

siren, *v.* Restrict *rare* to sense 1 and add: **1.** (Later examples.)
1935 L. MacNeice *Poems* 14 Two [women] there are, as I drive in the city,.. The one sirening me to draw up by the kerb. **1967** T. Stacey *Brothers* xxx. 353 They heard brilliant rippling music like some huge wooden xylophone. With this intricate sound still sirening them they found themselves on a hill top.
2. Also (of a police car, etc.), to proceed with siren blaring; to make one's way thus. Also **si·rening** *ppl. a.*
1949 'M. Innes' *Secret Vanguard* x. 106 A sirening ambulance or fire-engine. **1951** R. Bradbury *Silver Locusts* 50 They slammed the police-wagon door and drove off into the morning sun, fast pressed to the rear window, and just before they stirred over a hill, he saw the red fire..on an ordinary morning on the ordinary planet Earth. **1960** *Guardian* 21 Oct. 13/4 He was sirening up the Henry Hudson parkway. **1965** D. S. Davis *False Betrayer* xviii. 206 Fitzgerald sirened his way through the crowd, not leaving the car. **1978** J. I. M. Stewart *Full Term* viii. 86 An ambulance went wildly sirening past.

Sirenian, var. *Siryenian *sb.* and *a.*

Sirex (sai·reks). *Ent.* Also **Sirex.** [mod.L. (Linnæus *Fauna Suecica* (ed. 2, 1761) 396), f. Gr. σειρήν siren, a solitary bee or wasp.] A wood-wasp or horntail of the genus *Sirex*, whose larvæ burrow into the trunks of trees. Also *attrib.*
1895 *Cambr. Nat. Hist.* V. 509 The Sirex will..attack a perfectly healthy tree immediately after it has been felled. **1908** A. T. Gillanders *Forest Entomol.* v. 190 The female Sirex..is armed with a long ovipositor. **1928** *Bull. Entomol. Res.* XIX. 219 (*heading*) The Sirex wood-wasps and their importance in forestry. **1958** *N.Z. Timber Jrnl.* Sept. 57/1 The sirex woodwasp..attacks live and freshly felled trees. **1969** *Sun* (Melbourne) 26 June 16/2 The dreaded sirex wasp has been found in pine plantations.

sirih. Add: Also **sireh.** Substitute for def.: A trailing tropical shrub, *Piper betle*, of the family Piperaceæ, native to Indonesia, where the pungent leaves are chewed with areca nuts; also, the leaves of the shrub; = Betel 1. (Earlier examples.)

1779 T. Forrest *Voyage to New Guinea* vi. 75 The Sulaus call the betel leaf, Cary. **1783** W. Marsden *Hist. Sumatra* 74 The *Seereé*, a creeping plant, whole leaf, of a strong aromatic flavor, they eat with the betel nut. **1795** tr. C. P. Thunberg's *Trav.* (ed. 2) II. 268 The betel leaves, called Siri (*Piper betel*), are therefore brought fresh every day for sale. **1839** T. J. Newbold *Straits of Malacca* I. ii. 87 The parents..offer a small present of plantains, sirih, tobacco, etc.
b. *sirih-box* (earlier example), *juice, leaf* (earlier example), *vine*.
1839 *Chinese Repository* VII. 130 His betel and siri box. **1939** A. Keith *Land below Wind* xiii. 211, I looked back at the nanyang women..; their lips, red with sireh juice, like cerise flowers. **1846** H. Keppel *Exped. to Borneo of H.M.S. Dido* II. i. 13 His majesty chewed his sirih-leaf and betel-nut. **1893** F. A. Swettenham *About Perak* 38 Here the Malay lives under his sirih-vine and durian tree.

sirki. (Earlier example.)
1801 *Asiatick Researches* VII. 463 Hoogla or Sirkee mats.

sirocco. Add: **1. a.** (Earlier *transf.* examples.)
1848 J. S. Robinson *Sk. Gt. West* 17 The dreaded Sirocco..burns us even through our clothes. **1870** *Weekly Standard* (Buenos Aires) 21 Dec. (Suppl.) col. 6 The Sirocco on Wednesday was so terrible that in the effort to keep cool, the mind reverted to icebergs and Polar travels but all in vain.
b. (Earlier example.)
1700 J. Jackson *Let.* 2 Feb. in *Private Corr. S. Pepys* (1926) I. 278 But the weather being changed and the Siroccos now blowing into the place of the Tramontains, this design is become impracticable.

siro·cco, *v. rare.* Also **scirocco.** [f. the sb.] *intr.* and *trans.* To blow (about) like the sirocco.
1922 D. H. Lawrence *Let.* 16 Nov. (1962) II. 677 It has blown, and the wind, and snowed on Calabria, and siroccoed till we are all of us in fragments. **1937** J. Squire *Honeymoon & Ye Bee* vii. 170 The monotonous mandible refrain of a song about the Isle of Capri... As it faded away, I remembered where, when it had already siroccoed the world for six months, I had last heard it.

‖ **sirop** (siro, si·rəp). [Fr.: see Syrup *sb.*] (A drink made from) a sweetened fruit-juice concentrate.
1817 *Monthly Packet* Oct. 369 The sirops and Savoy cakes had been disposed of. **1889** E. Simcox in K. A. McKenzie *E. Simcox & G. Eliot* (1961) ii. 55 Ending up..with sirop at some café in the small hours. **1933** 'G. Orwell' *Down & Out* viii. 126 Two children..sharing a glass of sirop. **1966** H. Yoxall *Fashion of Life* xxiii. 206 We consumed sweetish Continental..sirops, grenadine, framboise. **1978** *Times* 22 July 8/7 The continental 'sirops', in mint, grenadine and blackcurrant flavours, are available in good delicatessens..for under £1.50.

sirree. See also *no siree, *yes siree.

Siryenian (siryī·niăn), *sb.* and *a.* Also **Sirenian, Syrianian, -ienian, Syryenian, Zyrianian, Zyrenian.** [f. mod.L. *Syriænus* (ad. Russ. *zyryanin*, -*dne* *Zyrian *sb.* and *a.*) + -ian.] = *Zyrian *sb.* and *a.*
1851 *Illustr. Catal. Gt. Exhib.* III. 552/1 Sirenian St. Matthew. **1878** *Encycl. Brit.* VIII. 700/1 Finnic or Ugrian represented by.. Siryenian. **1879** *Ibid.* IX. 291/1 The Permian Finns comprise the Siryenians,..the Permian proper,..and the Votyak. **1908** [see *Permian *a.* (*sb.*)]. **1910** *Encycl. Brit.* X. 389/1 The Syryenian headquarters are in the town of Izhma on the Pechora. **1931** *Ibid.* XXVI. 317/2 Syryenians (also Sirianian, Syrjenian, Zyrienian, Zirianian, Zyrian and Zirian), a tribe belonging to the Permian division of the eastern Finns. **1939** Liddell & Scott *Gr.-Eng. Lex.*, Κανναβις..borrowed perh. fr. Ugro-Finnish, cf... Syrianian *pil* 'hemp'.

sis (sis), *sb.* Also **siss.** Colloq. abbrev. of Sister *sb.* (in Dict. and Suppl.)
1656 Dr. Denton *Let.* 10 Nov. in M. M. Verney *Memoirs* (1904) III. ix. 215 We had need call a councell for marryinge and givinge in marriage, you for your sis, shee for hers, and I for mine. **1808** Lady Littleton *Let.* 28 Dec. (1912) II. 53 But oh, Bob, pity your poor Mam and Sis, when they will have to set out on a bleak morning... never such rough, splashy, squashy, jolting and jumbling as ours was. **1835** *Knickerbocker* VI. 293 All the friends would gather around her,...'Sis' was banded about than the whole,..soon dwindled into 'sis'. [see Siss *sb.*]. **1872** C. M. Yonge *P's & Q's* iii. 19 'I knew you were a jolly old sis,' said Horace with a hug. **1881** M. E. Ryan *Pagan of Alleghanies* 133 Folks call boys 'bud' sometimes, just like they call girls 'sis'. **1924** Lawrence & Skinner *Boy in Bush* vi. 86 Skippin' up an' down like Sis. **1935** Z. N. Hurston *Mules & Men* (1970) I. x. 220 Sis Cat, we both got a li'l money. **1948** M. Allingham *More Work for Undertaker* ii. 26 'Who was there? Your love?' 'Gawd, no! My sis.' **1970** M. Walker *Prophets for New Day* 16 in S. Henderson *Understanding New Black Poetry* (1973) ii. 161, I run down to Sis Avery's. **1974** D. Gray *Dead Give Away* i. 14 You'll be wearing clothes in the Private View, won't you, Sis?

‖ **sis** (sis, sɪs), *int. S. Afr.* Also †**cess.** [ad. Afrikaans *sies* (also used), perh. ad. Hottentot *si* or *tsi*.] An ejaculation expressing disgust or disappointment.
1862 A. W. Drayson *Tales at Outspan* 67, I have lost more cattle from the attacks of hyænas than I have from lions, or leopards, and as to sheep, *cess*, I've had nearly a whole flock worried by them. **1909** *Cape* 30 Apr. 6 Sis for her. She gave me nothing to eat but smalina and koffee. **1926** E. Lewis *Mantis* I. iv. 79 'Sis, man!' Hugo had scolded him, 'you go to see one of these private fellows.' **1949** *Forum* 10 Mar. 13 The elegant word 'sies!' is in evidence when persons to whom objection is taken are present or are mentioned. **1972** *Star* (Johannesburg) 15 Nov. 18 The majority of young Afrikaans people.. cannot speak English and..have no intention of doing so. At all? Yuk! English.

-sis, *suffix*, repr. Gr. -σις in nouns of action, as in Analysis, Arsis, *Meiosis, Peristalsis, etc. Also in some nouns denoting a specified diseased state, as *Filariasis, Phthisis, *Psittacosis, Sepsis.

sisal. Add: With pronunc. (sai·s'l).
3. Special combs., as **sisalcraft, -kraft,** a waterproof material with a core of sisal fibres.
1940 *Chambers's Techn. Dict.* 774/1 Sisalcraft. **1945** *Jrnl. R. Army Med. Corps* LXXXIV. 98 The sisalcraft and blankets already sewn up. **1948** *Spectator* 9 Apr. 436/1 The wooden huts in which the occupants of Antarctic bases live are specially designed with layers of tinfoil and sisalkraft between the inner and outer walls. **1961** [see *sark *v.* 2]. **1962** *Economist* 8 Sept. 956/1 The conventional materials..such as bituminised sisalcraft and burlap.

sis-boom-bah (sis,bʊm,bā·), *int.* and *sb.* *U.S.* Also **-ah.** [Echoic, repr. the sound of a skyrocket: a hissing flight (*sis*), an explosion (*boom*), and an exclamation of delight from the spectators (*bah, ah*): see *skyrocket 2.] A shout expressive of support or encouragement to a college team. Hence as *sb.*, enthusiastic or partisan support of spectator sports, esp. football.
[**1867**: see *sky-rocket *sb.* 2.] **1924** *Dialect Notes* V. 275 Sis—boom—ah. **1961** M. Beadle *These Ruins are Inhabited* (1963) iv. 48 Fresh from the land of sis-boom-bah,..the Americans had a hard time at first learning to applaud good play by either team. **1970** *Time* 17 Aug. 64 For the next 2½ years it was girls, flasks and sis-boom-bah. But the public image concealed an all-night reader.

sisertskite, var. *sysertskite.

sisham, var. Shisham.
1890 Kipling *Soldiers Three* 31 We three were comfortably settled under the big *sisham* [ed. 1889 *shisham*].

sisi, var. *seesee. **sisith,** var. *zizith.

siss, *sb.* Substitute for entry:
siss, var. *sis *sb.*

sisserskite, obs. var. *sysertskite.

sissonne. Delete *rare*, substitute for def. '(See quot. 1957.)', and add later examples.
1892, **1923** [see *cisean]. **1930** Cranke & Beaumont *Theory & Pract. Allegro in Class. Ballet* 47 *Sissonne dessus*... Other *sissonnes* are *sissonne dessous*, i. *en avant*, ..*en arrière*, etc. **1947** N. Nicolaeva-Legat *Ballet Education* vi. 109 *Sissonne* is another step which has many variations. *Sissonne simple on the cou-de-pied* is taken from V pos. (right leg in front), pushing with both feet from the ground equally with both feet and with straight knees, and so on, ending on the left leg. **1957** G. B. L. Wilson *Dict. Ballet* 245 *Sissonne*, or *pas de sissonne*, probably from *pas de ciseaux*, a scissor-like movement... With a slight plié, the dancer springs into the air to the fifth position, alighting on one foot with a demi-plié, with the other leg extended to the back, front, or side; the back foot is then closed to the supporting foot (a *sissonne fermée*)... If the dancer lands on one foot with the other extended, the step is a *sissonne simple* (or *ordinaire*). **1968** J. Winearls *Mod. Dance* (ed. 2) iii. 87 This *sissonne* is a combination of the three elements, light—quick—and peripheral.

sissy (si·si). *colloq.* [f. *sis *sb.* + -y; cf. *cissy *sb.* and *a.*] **1.** A sister.
1846 *Dollar Newspaper* (Philadelphia) 21 Apr. 1/7 'Sissy Jane' sauntered back her hair, and smiled at me. **1854** Dickens *Hard Times* I. vi. 41 When Sissy got into the school here..her father was as pleased as Punch. **1859** [see Siss *sb.*]. **1865** K. H. Digby *Short Poems* 39 The little one grasping, with such a tight hold, The frock of sweet sissy, herself not too bold. **1901** M. Franklin *My Brilliant Career* xxii. 207 Don't be frightened, sissy, I never kiss girls. **1939** Joyce *Finnegans Wake* 28 It made ma make merry and sissy so shy.
2. An effeminate person; a coward.
1887 *Lantern* (New Orleans) 27 Aug. 3/2 Look and walk like you didn't do no much fightin'. **1899** T. Hall *Tales* 121 'Well, you were a sissy,' said Blinks contemptuously. **1926** *British Weekly* 2 Sept. 452/3 A religious 'sissy' was anathema to me. **1932** S. Gibbons *Cold Comfort Farm* xvii. 237, I want red blood. I don't want no sissies, no more. **1938** L. MacNeice *Zoo* iv. 72 The Sealyham, say the older breeders, is becoming a sissy. **1969** C. Himes *Blind Man with Pistol* iii. 25 The sissies..were colored and mostly young. They all had straightened hair..; long false eyelashes. **1977** *Time* 21 Feb. 20/2 Smokers proved to be sissies when deprived of cigarettes.
3. *attrib.* or as *adj.* **a.** Effeminate; cowardly.
1891 *Harper's Mag.* Aug. 485/2 He approached and sat near me, deep in conversation with a young gentleman with sissy whiskers. **1893** *Sunday Mercury* (N.Y.) 14 May 15/5 (*heading*) Sissy men in Society.—Powdered, Painted and Laced. They swarm at Afternoon Teas. **1899** T. Hall *Tales* 121 Scotty was, in the newspaper vernacular, 'a sissy boy', or, in other words, a bit effeminate. **1926** *British Weekly* 2 Sept. 452/3 There was nothing 'sissy' about him. He was a born fighter. **1932** S. Gibbons *Cold Comfort Farm* xvii. 241 Thassa sissy sort of a name, but it'll do. **1941** 'R. West' *Black Lamb & Grey Falcon* II. 152 The monuments..had apparently been produced by a pastry-cook under the influence of Persian art. Such sugary little scrolls and swaps, such sissy little flowers in pots, such coy little etchings of swords on the soldiers' tombs. **1959** *Spectator* 25 Sept. 408/2 All the kudos goes to the campaign-scarred, ink-stained veteran: none to the new bug in his sissy clean blazer. **1970** P. Dickinson tr. *Aristophanes' Wasps* in *Plays* I. 192 That sissy son of Chaireas prances In with his mincing walk. **1977** C. McCullough *Thorn Birds* xi. 260 No cutter ever wore gloves. They slowed a man down... Besides, gloves were sissy.
b. sissy bar, a metal loop rising from behind the seat of a bicycle or motor-cycle.
1969 *Daily Colonist* (Victoria, B.C.) 19 June 40/1 (Advt.), The 'Super Cycle' breed with hi-rise handlebars on a cantilever frame..the 36" Sissy bar, [etc.]. **1974** R. B. Parker *God save Child* vi. 49 Another motorcycle... A big one,..small front wheel, sissy bar behind.
Hence **sissifica·tion,** effeminacy; **si·ssified** *a.*, effeminate; **si·ssiness,** effeminacy; **si·ssyish** *a.*, somewhat effeminate.
1889 W. D. Howells *Hazard of New Fortunes* II. 64 The New York fellows carried canes..; and they were both sissyish and fast. **1905** J. C. Lincoln *Partners of Tide* iv. 78 To be seen with girls was not so 'sissified' in his mind as it used to be. **1926** *Harper's Mag.* Feb. 350/2 In spite of his funny sissiness there was not a dog in town that did not love him. **1938** I. Goldberg *Wonder of Words* xv. 305 Mr Sokolsky establishes a correlation between high blood-pressure and masculinity, and between low blood-pressure and femininity or sissification. **1959** E. Pound *Thrones* xcix. 57 In statement, answer; in conversation Not with sissified fussiness (chiao') Always want your own way. **1973** *Guardian* 1 June 10/5 The much-publicised Warhol Factory mystique..thinly veils a highly reactionary, bigoted and sissified neo-Nazi boutique. **1975** *N.Y. Times* 12 Sept. 38/6 Mr. Mahan said that the other cowboys on the rodeo circuit had generally accepted his fashion business, and that none of them considered it 'sissyish or effeminate'.

sister, *sb.* Add: **I. 2.** (Further examples.) *spec.* (*a*) a (fellow) prostitute; (*b*) a (fellow) feminist; (*c*) among Blacks, a Black woman. *sisters under the skin*: see *skin *sb.* 5 j.
1847 A. Harris *Settlers & Convicts* vi. 94 When 'her sister' (so they usually speak in the *sisterhood of sorrow*) came here, she came too. **1870** *Free Lance* 16 Apr. 123/1 The working sisters of this great city are waking up to a sense of what they ought to do in making future provision for themselves. **1889** *Girl's Own Paper* 28 Sept. 824/3 A better day..has dawned..on our sisters, the working girls of this country. **1899** 'J. Flynt' *Tramping* I. iv. 94 If he can only have some outcast woman, or 'sister', as he calls her..he is a comparatively happy fellow. **1912** H. Ellis *Task of Social Hygiene* iii. 104 'La femme libre'.. must be a woman of reflection and intellect who, having meditated on the fate of her 'sisters'..shall give forth the confession of her sex..in such a manner as to furnish the indispensable elements for formulating the rights and duties of woman. **1912** in C. McKay *Songs of Jamaica* 16 Me watch de vine dem grow, S'er t'row dung a de root. **1926** L. Hughes *Weary Blues* 37, I got a railroad ticket, Pack my trunk and ride. Sing 'em sister! Got a railroad ticket, Pack my trunk and ride. **1935** Z. N. Hurston *Mules & Men* (1970) I. vii. 164 De cow went bustin' on down de back-road wid de ole man till they met a sister he knowed. **1940** J. Crad *Traders in Women* v. 130 Then she left her 'selling position' and it was immediately occupied by one of her 'sisters'. **1944** *Publ. Amer. Dial. Soc.* II. 36 *Sister*, *n.*, a woman. W. N[orth] C[arolina]. **1968** *Ramparts* May 12 Our sisters in Vietnam have taught us many lessons. **1973** *Black World* June 90/2 Sister Williams breaks her book down into three major parts. **1976** R. B. Parker *Promised Land* xix. 110 When the sisters call you... Talk to them of obligation and sororal affiliation. **1977** C. McFadden *Serial* (1978) iii. 13/1 The sisters weren't invariably as supportive as she'd hoped they'd be. **1979** *Guardian* 5 May 12/2 Becoming Britain's first woman Prime Minister is one [achievement], whatever the sisters may say, that can only change perceptions of what women can aspire to.
3. d. (Earlier and later examples.) Also, prefixed as a title to the name of a nurse. *Sister Dora* [f. the name of the celebrated nurse Dorothy ('Dora') Pattison, 1832–78], a type of nurse's cap (see quot. 1971).
1860 [see *nursing *vbl. sb.* 1 b]. **1924** 'R. Hall' *Unlit Lamp* xlvi. 314, I made swabs at the Town Hall at Seabourne... I had a Sister Dora arrangement on my head; we all had, it made us look important. Some of the women wore aprons with large red crosses on their bibs. **1949** N. Mitford *Love in Cold Climate* II. viii. 276 Doesn't it seem funny to have talcum powder and..boring old Sister waiting..for somebody who doesn't exist? **1971** J. Manton *Sister Dora* xvi. 266 [Dora Pattison] made herself up a new cap, still tied with a butterfly bow and streamers under the chin but folded smoothly back over her dark hair. It was to be known to generations of nurses as 'a Sister Dora'. **1976** C. Storr *Unnatural Fathers* iii. 34 Kind Sister Tucker..faithful in her professional code of discretion, bustled upstairs..to attend to her interesting patients. **1978** *Church Times* 29 Dec. 11/3 The little white cap worn by nurses everywhere became known within the profession as 'a Sister Dora'. **1979** 'C. Aird' *Some die Eloquent* i. 11 When Sister Casualty..had trouble-makers in her patch she would ring down to the police station.

Ibid. iii. 37 Sister Stork's on the other phone to the delivery ward.
5. (Further examples.) Also *colloq.* as a mode of address to an unrelated woman, esp. one whose name is not known.
1906 H. Green *At Actors' Boarding House* 56 He got up and walked over to her bench. 'You up agin it too, sister?' he said, gently. **1926** E. O'Neill *Great God Brown* I. iii. 40 Blessed are the pitiful, Sister! **1929** W. Faulkner *Sartoris* II. iii. 97 A voice in the other room boomed in rich rolling waves 'Mawnin', sister,' it said. **1934** 'E. M. Delafield' *Provincial Lady in Amer.* 60 Shouted at by a policeman who tells her: Put your lights on, sister! **1943** *Amer. Speech* XVIII. 88 [New Zealand English] Sister is also used for a girl as a term of address. It is not..a recent adoption from American films, but a relic of the whaling slang of a century ago. **1944** M. Laski *Love on Supertax* xi. 103 Just waiting for your boy friend, duckie..? You want the best, we got 'em, eh, sister? **1953** H. Miller *Plexus* (1963) iii. 127 'He doesn't need advice,' she replied. 'He knows what he's doing.' 'O.K. sister, have it your way then!' With this he turned abruptly to me again. **1976** 'R. Boyle' *Cry Rape* i. 6 Come on, sister... Why won't you stay and talk to me? I'm a nice guy.
II. 9. For † read '*Obs.* exc. *arch.* in *sister-son*.' (Later examples.)
1955 J. R. R. Tolkien *Return of King* vi. 255 Fréalaf, Helm's sister-son. *a* **1973** —— *Silmarillion* (1977) xvi. 136 The King..looked with liking upon Maeglin his sister-son.
10. a. *sister woman.*
a **1786** Burns *Address to Unco Guid* in *Poems* (1968) I. 53 Then gently scan your brother Man, Still gentler sister Woman. **1856** Dickens *Little Dorrit* (1857) xxvii. 243 A woman, who..has a perverted delight in making a sister-woman as wretched as she is. **1939** N. Marsh *Overture to Death* vii. 76, I have judged my sister-woman in my heart and condemned her.
b. *sister state* (earlier example).
1777 *Rec. Early Hist. Boston* (1887) XVIII. 285 We are sure, that very large, & much wanted Supplies, the Property of this State & expected here, are now ordered into some of the Sister States.
c. *sister island* (further example), *isle, science, ship* (further examples), *soul.*
1936 *Discovery* June 187/1 The *waganga*, who specialise in *pepo*-exorcism, go through a special training. The art is at its highest in Pemba, the sister-island of Zanzibar. **1838** T. Langton in H. H. Langton *Gentlewoman in Upper Canada* (1950) 58 William Jones is from the sister isle. **1939** Joyce *Finnegans Wake* 51 A native of the sisterisle..by his brogue. **1901** W. James *Mem. & Stud.* (1911) vii. 169 Whether his name [*sc.* F. Myers'] will have in psychology as honorable a place as their names [*sc.* Cuvier and Agassiz's] have gained in the sister-science, will depend on whether future inquirers shall adopt or reject his theories. **1966** N. Nicolson in *Diaries & Lett. H. Nicolson* (1966) 56 The airship, R.101, was contracted by the Government in 1925 as a sister-ship to R.100. **1974** E. R. H. Ivamy *Marine Insurance* (ed. 2) xv. 216 This clause is known as the 'sister ship' clause and was introduced to state the legal position where two ships belonging to the same owner come into collision. **1897** J. Waring tr. *Balzac's Lily of Valley* 74 Feeling now that we were twins of the same nurture, she could not conceive of semi-confidences between sister souls that had drunk of the same spring. **1933** *Times Lit. Suppl.* 16 Nov. 792/2 The influence of the visiting Cousin Nellie, who finds a sister-soul in the comfortable person of an Aug [*sc.* August visitor] of her own age.
d. sister chromatid *Biol.*, each of a pair of chromatids derived from a common parent chromosome.
1942 *Jrnl. Genetics* XLIII. 195 A single chromatid may be broken and fail to rejoin, the chromatid fragment almost always remaining paired to its sister chromatid. **1975** *Nature* 13 Nov. 122/1 Techniques have been developed which distinguish between sister chromatids without using radioisotopes and autoradiography.
11. sister act (see quot. 1952); **sister tutor,** a nursing sister who teaches trainee nurses.
1908 G. V. Hobart *Go to It* 56 Their names were Millie and Tillie, and they..did a sister act. **1908** *Variety* 18 Apr. 4/4 There may be a number of new 'sister' acts in vaudeville next season. **1952** W. Granville *Dict. Theatrical Terms* 163 *Sister act*, a variety act performed by two or more sisters; e.g. the famous Dolly Sisters, the singing Green Sisters of the stage and radio. **1968** R. Rendell *Secret House of Death* vi. 62 An even temperature, that's one thing my sister tutor always impressed on me. **1971** P. D. James *Shroud for Nightingale* iv. 119 I'm not a qualified Sister Tutor. I was only deputizing.

sisterhood. Add: **1.** (Later examples in feminist use.)
1968 *Notes from First Year* June 18 Sisterhood is powerful! **1974** *Time* 6 May 80 Sisterhood across class lines is a myth. **1980** J. R. Richards *Sceptical Feminist* i. 28 Basic to much recent feminist practice has been the idea of *sisterhood*, which..involves a determination among members of the women's movement to work together as equals.
2. b. (Later examples, of feminists.)
1972 *Newsweek* 9 Oct. 104 That book's stinging attack on the women's lib movement achieved for its author a place in the sisterhood's demonology right next to Hugh Hefner. **1981** *Times* 9 Sept. 9/3 At the age of 24 Sarah Daniels must count as a second generation feminist, and if this first play is a portent of what the sisterhood is now brewing up then male chauvinism can breathe again.

Sistine (si·stīn), *a.* and *sb.* [ad. It. *sistino* of Sixtus.] **A.** *adj.* Pertaining to Pope Sixtus IV (1471–84); *spec.* as epithet of the chapel built by him; hence, of or belonging to the Sistine Chapel, as *Sistine Madonna*, a picture by Raphael originally hung there. **B.** *sb.* The Sistine Chapel.
[**1769** J. Reynolds *Discourse delivered at Opening of Royal Academy* 5 On the sight of the Capella Sistina, he [*sc.* Raphael] immediately..assumed that grand style of painting.] **1771** C. Burney *Pres. State Music in France & Italy* p. vi, The Pope's chapel is sometimes called the Sistine chapel, from Sextus Quintus, who built it. **1863** J. A. Symonds *Let.* 19 Dec. (1967) I. 435 The Sistine Chapel too gains everything by being seen in the original. **1869** 'Mark Twain' *Innoc. Abroad* xxvii. 288 He has shown us the great picture in the Sistine Chapel..by Michael Angelo. **1885** *Encycl. Brit.* XIX. 64/1 San Sisto.. lost its chief attraction when Raphael's Sistine Madonna (now in Dresden) was sold by the monks. **1887** 'J. Oldcastle' *Leo XIII* vi. 36 The conclave assembled to-day in the Sistine. **1889** Grove *Dict. Mus.* IV. 122/1 The traditions of the Sistine Choir. **1920** W. B. Yeats *Michael Robartes & Dancer* 2 While Michael Angelo's Sistine roof His 'Morning' and his 'Night' disclose How sinew that has been pulled tight,..Can rule by supernatural right. **1950** *New Yorker* 25 Feb. 82/3 In the vesper Mass, the Sistine's famous boy sopranos let loose their voices. **1968** A. Diment *Bang Bang Birds* x. 182 All I could hear was a choir singing in the Sistine Chapel. **1975** 'R. Player' *Let's talk of Graves* iii. 81 The Czar Nicholas the First.. was kneeling in the Sistine. *Ibid.* 82 The clouds of Sistine incense around the yellow tapers. **1977** R. L. Wolff *Gains & Losses* ii. 131 Guy [Morville]..looks like one of the angels in the Sistine Madonna.

sisyrinchium (siziri·nkiŏm). [mod.L. (L. Plukenet *Almagestum Botanicum* (1696) 348), f. Gr. σισυρίγχιον a plant name used by Theophrastus.] An annual or perennial herb of the genus so called, belonging to the family Iridaceæ, native to North or South America, and bearing linear leaves and clusters of small blue, yellow, red, or white flowers; = *blue-eyed grass* s.v. Blue-eyed *a.* in Dict. and Suppl. and *satin-flower* (*e*) s.v. *Satin *sb.* (and *a.*) 8 b.
1772 R. Weston *Universal Botanist* III. 664 (*heading*) Narrow-leaved Virginian Sisyrinchium. **1919** R. Farrer *Eng. Rock-Garden* II. 367 An unknown Sisyrinchium may often prove to be a *Marica* lurking in ambush for the unwary. **1955** L. D. Hills *Alpine Gardening* v. 137 The Sisyrinchiums are good imitation rushes by the stream side. **1971** B. Miles *Bluebells & Bittersweet* 120/2 Sisyrinchiums will never be accused of flamboyancy [*sic*], but they are darlings.

sit, *sb.*[1] Add: **1. c.** (Earlier example.)
1820 M. Edgeworth *Let.* 10 Aug. (1979) 208 Their bodies look as if..they had taken an eternal *sit* from the stiff square stays of former day.
2. a. (Further example.)
1971 N. Freeling *Over High Side* I. 8 He wanted a nice cup of tea and a sit.

sit, *sb.*[2] For *Printers' slang* read 'orig. *Printers' slang*' and add earlier and later examples. Now esp. in *sit(s) vac*, situation(s) vacant (see *situation 6 b).
1853 'Mark Twain' in *Hannibal* (Missouri) *Jrnl.* 8 Sept. 2/1, I shall look out for a sit; for they say there is plenty of work to be had for *sober* compositors. **1878** W. Whitman *Daybks. & Notebks.* (1978) I. 110 Applied to Bart Bonsall, for a sit. for Harry. **1901** [see *bovrilize *v.*]. **1914** Joyce *Dubliners* 91 But Hogan has a good sit, hasn't he? **1969** *Guardian* 29 July 4/8 Asians.., with the sort of skills that the 'Sits vac' columns of British newspapers are clamouring for—nurses, motor mechanics, turners, secretaries. **1970** *Ibid.* 14 Nov. 9/2 (*heading*) 'I felt I could do the job.'..Peter Terson on the Sit. Vac. he didn't fill. **1973** *Ibid.* 12 Feb. 11/4 June Moelzer wants.. a job... She sifts through the 'sit vac' columns, becoming less and less choosy as the months go by. **1975** *Listener* 8 May 609/3 Raking through the sits. vac., John Timpson ..reported an advertisement in a Devon local paper for a person 'to move Earth—about two days' work'. **1980** D. Francis *Reflex* xiii. 161 No rides, no income. You start looking at 'sits vac'.

sit, *sb.*[3] Also **sitt.** Abbrev. of Sitting-room.
1937 A. Christie *Dumb Witness* vi. 58 We've a nice bungalow at Hemel End, two bed., one sitt. **1961** [see *bed *sb.* 1 e].

sit, *v.* Add: **I. 1. f.** To sit down in a public place as a form of protest; to take part in a sit-in.
1961 *Daily Tel.* 21 Oct. 7/1 A tailor..was yesterday preparing to 'sit' for nine years, if necessary, in the path of a £1,500,000 redevelopment scheme. **1963** G. Butler *Coffin for Baby* iii. 49 My boy said to me this morning: Shall I sit or shan't I? You must do as your conscience tells you, I said. **1966** Wodehouse *Plum Pie* v. 121 Every now and then we march from Aldermaston, protesting like a ton of bricks... And then we sit a good deal. **1970** P. Laurie *Scotland Yard* x. 259 The demonstrators.. could 'sit' as long as they liked.
g. To baby-sit.
1966 J. Gloag *Sentence of Life* xxiii. 192 He wondered if Willy would be able to get Mrs Hillman in to sit. Friday was a bad night. **1968** J. Updike *Couples* iii. 247 It was easier for the Saltzes to leave Bernard, who stays up forever reading anyway, to sit for his brother. **1975** M. Bradbury *History Man* vii. 121 I'll have to get a sitter... I shouldn't have any trouble finding someone to sit. One of the students.
2. c. With *through*.
1889 J. L. Toole *Reminisc.* I. viii. 263 Quite different in its acceptance of fun or pathos from the audience that sits through the same piece the next night. **1932** D. L. Sayers *Have his Carcase* xix. 255 She sat on right through the programme, but when it came to God Save the King, she chucked it. **1981** L. Deighton *XPD* xxix. 236 Sir Sydney..gallantly sat through another half-hour of finer points of script editing.
3. b. *to sit pretty*: see *pretty *adv.* 2 a; *to sit tight*: see Tight *adv.* 2 b in Dict. and Suppl.
c. *to sit at the feet of* (a person), to be the disciple or pupil of (a teacher); *to sit on one's hands*, (*a*) *U.S. colloq.*, to withhold or be sparing of applause; (*b*) to be inactive.
1535 Bible (Coverdale) *Luke* x., Mary..sat hir downe at Iesus fete, and herkened vnto his worde. **1611** *Ibid.* (A.V.) *Luke* x. 39 Mary..also sate at Iesus feet, and heard his word. **1633** [in Dict.]. **1907** Kipling *Sons of Martha*, They sit at the Feet, and they hear The Word—they see how truly the Promise runs. **1926** G. M. Trevelyan *Hist. England* v. iii. 557 They [*sc.* the Whigs] had sat at the feet of Edmund Burke. **1952** G. Sarton *Hist. Sci.* I. xxi. 547 He came to Athens to sit at Plato's feet. **1971** *Nature* 5 Mar. 2/2 It is clear that there are far more universities per square mile in Britain than are necessary to enable students to sit at the feet of some teacher or other.
1926 E. Ferber *Show Boat* vi. 106 Well, they were sitting on their hands to-night, all right. Seemed they never would warm up. **1948** *Newsweek* 16 July 19/1 Listlessly, the convention sat on its hands at all mentions of Mr. Truman in the opening speeches. **1959** *Listener* 1 Jan. 4/1 This helped to reduce the number of constituency associations, who had to be persuaded by MacDonald or Gladstone to sit on their hands or, worse still, toil for their ally. **1961** *IUD Digest* Fall 74/1 Organized labour can ill afford to sit on its hands. **1972** *Guardian* 19 May 12/3 Opposition MPs who only yesterday sat on their hands to let Herr Brandt's Ostpolitik treaties through the Bundestag were back on their feet again this morning. **1976** *New Yorker* 8 Mar. 57/1 'Don't clap too hard—it's a very old building,' he cautions the audience, which is sitting on its hands. **1979** M. A. Sharp *Sunflower* xvi. 148, I should have learned *something* by now. I haven't exactly been sitting on my hands.
6. c. Also, to present oneself *for* an examination, etc.
1929 R. Graves *Goodbye to All That* xxvii. 362 My tutor..warned me that I must on no account disparage the eighteenth century when I sat for my final examination. **1955** *Times* 30 June 6/5 Pupils sitting for the examination for entry to secondary schools. **1963** R. Pedley *Comprehensive School* i. 14 In some of the 3900 'modern' schools in England and Wales it is possible for the cleverer pupils to sit for GCE at ordinary level. **1968** G. Maxwell *Raven seek thy Brother* viii. 102, I appealed to my guardian to be allowed to retire..from the scene on the grounds of ill-health... The refusal was absolute..; I was to sit for my degree, no matter what the outcome. **1980** *Radio Times* 1–7 Mar. 16/4 It is possible to take an A-level without having sat for the O-level.
II. 13. a. (Later examples.)
1976 M. Machlin *Pipeline* liii. 526 Occasionally a guard was left when equipment sat in the field, but now it was lunch hour. **1977** *Oxford Jrnl.* 10 June 1/5 There were a dozen eggs still sitting on the front porch and the dustbin sat at the back of the house where the binmen had left it. **1978** *Sci. Amer.* May 65/2 The idea of an exploding clump of matter sitting somewhere in space offers no natural way to account for the existence of the cosmic background radiation.
c. (Later examples.)
1878 D. Kemp *Man. Yacht & Boat Sailing* 368/2 Sails are said to 'sit' well when they do not girt, pucker, belly, or shake. **1958** *Listener* 28 Aug. 309/3 Slots must be cut in the bottom half of them [*sc.* the horizontal pieces] where they meet the posts, so that they will sit nicely. **1971** *Good Motoring* Sept. 18/1 Where the 33 tended to hang out its back end on fast corners the 44 sat steadily on the road.
16. d. *fig.* Without compl.
1964 G. C. Kunzle *Parallel Bars* ix. 410 Make certain that you can do an individual movement with perfect technique before you include it in the exercise. Then try it out in minor competitions until it 'sits'. **1971** B. Graham *Spy Trap* ii. 19 It was too pat, too smug, like a well-turned-out radio script. It didn't sit.
17. c. Also *poet.*
1878 J. J. Aubertin tr. *Camoens' Lusiads* II. viii. lxiv. 129 With a proud confidence, which sat him well.
IV. *With adverbs.* **20*. sit around,** orig. *N. Amer.* To be idle; to lounge.
1915 N. L. McClung *In Times like These* iv. 42 Personally I sympathize with the young man and believe it would be a happier home if she were as interested in the paper as he and were reading the other half of it instead of sitting around feeling hurt. **1935** *Time* 29 July 42/2 *Accent on Youth* suffers less than most pieces on translation to the screen, for, although its people sit around and talk a lot, they at least talk with wit. **1939** I. Baird *Waste Heritage* xix. 267 A lot of them sat around in shirtsleeves. **1959** J. Thurber *Years with Ross* iv. 61 Ross had asked me..to sit around and talk with him and H. L. Mencken. **1971** C. Whitman *Death Suspended* v. 95 All they did was to sit around looking decorative. **1979** *Times* 13 Nov. 2/8, I could not sit around twiddling my thumbs and doing nothing.
20. sit back.** To be inactive or passive.
1943 D. Powell *Time to be Born* xi. 276 One can't sit back and see one's brother..made a monkey of that way! **1953** E. Simon *Past Masters* III. 156 Harriet is the ideal executive... I can just sit back and let her get on with it. **1970** J. Sangster *Touchfeather, Too* ii. 38 Here then was my contact... I sat back and waited for him to make his play. **1982** *Times* 22 Feb. 1/3 We cannot sit back and let them walk all over us.
21. sit down. e. *spec.* To sit down on strike in one's place of work; to sit down in a public place as a form of protest.
1936 *Sun* (Baltimore) 2 Nov. 14/1 Various of the men reporting for picket duty yesterday came from ships where cooks had been the first to 'sit down' in sympathy with the West Coast strikers and hadn't had any food for forty-eight hours. **1942** Berrey & Van den Bark *Amer. Thes. Slang* § 528/7 *Sit down*, to go on a 'sit-down strike'. **1962** P. Mortimer *Pumpkin Eater* xxi. 179 'She paints Ban the Bomb on everything.'..'I suppose she sits down all over the place?' **1963** D. V. Baker *Door is always Open* ix. 190 Parents sitting down on a dangerous highway, children sitting down to protect a play-street, workers sitting down to achieve some change in conditions. **1965** J. Porter *Dover Three* iii. 34 If she's been sitting down again, she'll have to go... Having members of the staff arrested and flung into jail is bad for the school's reputation.
22. sit in. d. To attend or be present at an event. (*a*) orig. and chiefly *U.S.* To take part in a game or other event. Also const. *on, to, with.* Cf. sense 22 a in Dict.
1868 S. Hale *Let.* 5 Jan. (1919) ii. 44 Before we got to lunch two Englishmen *sat in*. **1916** C. Sandburg *Chicago Poems* 63 He didn't sit in with the big thieves. *a* **1922** T. S. Eliot *Waste Land Drafts* (1971) 5 Sopped up some gin, sat in to the cork game. **1929** Wodehouse *Gentleman of Leisure* x. 79 You'll be able to let me sit in on de game, won't you? **1962** D. Francis *Dead Cert* ii. 19, I took ten of Henry's chips and sat in with them. Joan dealt. **1973** 'H. Howard' *Highway to Murder* viii. 102 You weren't invited to sit in on this deal, but you elected to take a hand.
(*b*) To attend an event or occasion as a spectator or observer. Also const. *at, on, with.*
1919 C. S. Parker *Amer. Idyll* x. 103, I sat in on a meeting of the Building Trades Board. **1931** *Oil & Gas Jrnl.* 5 Mar. 197/4 Any oil man or state or federal official may 'sit in' on the proceedings of the state wide committee. **1945** A. Huxley *Time must have Stop* viii. 91 Paul De Vries had already sat in at a number of the old lady's séances. **1949** 'J. Tey' *Brat Farrar* viii. 66 Kevin Macdermott had 'sat in' at one of these office conferences. **1959** *Times Lit. Suppl.* 10 July 411/1 Mrs. Bennett sat in, in 1928, on the experiment which produced Dr. I. A. Richards's *Practical Criticism*. **1962** 'S. Ransome' *Without Trace* iii. 31 'He has something to talk over with me.'..'Would he mind if I sat in?' **1965** *Listener* 13 May 703/2, I sat in with a class and saw the *Antigone* and *The Caucasian Chalk Circle* cut up into weekly twenty-minute doses. **1967** *Daily Tel.* 15 May 9/4 To sit in at a play of this sort is to realise quite soon that you are being asked a riddle. **1970** *Morning Star* 11 May 4/4 The headmaster will therefore sit in on some lessons and try to pinpoint the weaknesses so that the teacher is helped. **1977** *Spare Rib* July 17/1 If you sat in on some of the interviews I've been through you'd know.
e. To co-operate, to collaborate. Also const. *with, on.* (Only in P. G. Wodehouse.)
1925 Wodehouse *Sam the Sudden* xiii. 96 Do you mean to say..that if Soapy was sitting in with the Archbishop of Canterbury on a plan for skinning a sucker, the archbish wouldn't split Even Stephen? **1937** —— *Lord Emsworth & Others* ii. 96 Can I count on your co-operation?..Sit in, and I shall be able to marry the girl I adore. Refuse to do your bit, and I drift through the remainder of my life a soured, blighted bachelor. *a* **1975** —— *Sunset at Blandings* (1977) xi. 77 Jeff refused to sit in on your chuckleheaded idea of eloping for a very good reason.
f. orig. *U.S.* To join in playing or singing with a (*spec.* jazz) band or orchestra of which one is not a regular member. Also const. *with.*
1936 *Delineator* Nov. 102/ Those on the drawn-up chairs are *sitting in*; they have dropped in with their instruments to jam. **1937** *New Republic* 24 Nov. 69/2 Jess saw Bix Beiderbecke and sat in with him later. **1943** H. L. Mencken *Heathen Days* vii. 91 Once, when a baron sat in for a few sessions, we called him Count. **1949** L. Feather *Inside Be-bop* i. 8 On these occasions Kansas Fields or Jack Parker might sit in on drums. **1956** E. Delaney in S. Traill *Play that Music* v. 57 Another thing which influences the playing of today is the fact that no longer can one 'sit-in'. Before the war any player could walk into a club and 'sit-in' with the band. **1965** G. Melly *Owning Up* vii. 75 Buying a barrel of cider for the musicians who came along to sit in. **1971** *Melody Maker* 4 Sept. 20 Mungo Jerry, with Joe Rush, of the Country Jug band sitting in on washboard, sounded like a five man Jesse Fuller. **1982** *New Yorker* 30 Aug. 63/2 His unique playing..alienated club owners and other musicians, and he found little work. He even had trouble sitting in.
g. orig. *U.S.* (Often written with a hyphen.) To occupy a building as a demonstration of protest.
1941 *Sun* (Baltimore) 15 Apr. 3/3 More than 700 of the city's brighter-than-average students spurned their Easter Monday holiday today and marched to their city college preparatory school to 'sit in and work' as a protest against Mayor LaGuardia's proposal to close the school. **1961** *Look* 25 Apr. 46/2 Negroes who picket, sit-in, crowd our jails, advance on white schools and otherwise approach prevailing privilege. **1967** *Economist* 8 Apr. 142/2 Young people who..defend the virtues of marijuana and LSD and march, demonstrate, sit-in or lie-in to prove their faith in assorted causes. **1974** K. Millett *Flying* (1975) v. 517 Rosset had my friends arrested when they sat in at his publishing house. **1976** C. Hollis *Oxford in Twenties* 128 A few [undergraduates]..sit-in, demonstrate, occupy buildings, demand higher grants and more adequate lodgings.
24. sit out. b. (Earlier example.)
1805 E. Cavanagh *Let.* 20 Aug. in Londonderry & Hyde *Russian Jrnls.* (1934) II. 179, I sat out & made a Gown while we were going along.

25. sit up. b. (Examples with *for*.) Also, *S. Afr.* and *dial.*, to stay up for part of the night (*with* a person) as a sign of or during courtship, to keep company *with*. Cf. *opsit *v.*
1786 J. Woodforde *Diary* 18 Feb. (1926) II. 228 The Captain..did not return till 12 at Night, just as I was going to bed after sitting up for him till that time. **1869** Geo. Eliot *Let.* 15 Nov. (1956) V. 67, I admire your courage and endurance in sitting up for the meteors. **1878** H. A. Roche *On Trek in Transvaal* 136 The question of questions is, whether she will 'sit up and keep company with him!' If she has consented to do this she has virtually consented to 'sit up' with him as long as they both shall live. **1892** O. Schreiner *Thoughts on S. Afr.* iv. 181 Having made up his mind which daughter he desires to pay his attention to, it is now necessary he should request the parents' permission to sit up with her. **1893** *West Cumberland Times* (Holiday No.) 6/2 (E.D.D.), The custom being for the lad to sit up with the lass. **1896** *Dial. Notes* I. 424 *Sit up with*, to receive courtship from. **1951** L. Craig *Singing Hills* 99 Then he asked if he could sit up with me that night. He was merely asking for a date so I said yes. **1961** F. G. Cassidy *Jamaica Talk* x. 221 Courting..is still described by the terms to *sit up* or to *walk out*. **1961** D. Rooke *Lover for Estelle* 36 He had solemnly shown a candle which he carried in his pocket as a hint to Estelle that they should sit up together that night. **1974** *Daily Dispatch* (S. Afr.) 29 Mar. 12 Clinton, you've been sitting up with Nellie..an car riding and nothings come of it.
c. Also *spec.*, to remain in a sitting posture during an overnight train journey, in contrast to taking a sleeper. *to sit up and take nourishment*, to be convalescent.
1909 Wodehouse *Mike* lv. 311 'How's Adair?'..'Sitting up and taking nourishment once more.' **1918** E. V. Lucas *'Twixt Eagle & Dove* 152 The well-worn phrase 'to sit up and take nourishment'. **1947** L. P. Hartley *Eustace & Hilda* I. ix. 162 'But could you cancel your wagon-lit ticket?' 'I don't need to. I'm going to sit up.' **1953** K. Tennant *Joyful Condemned* xxviii. 277 Now there are no sleepers, you'd have to sit up in the train. **1967** O. Wynd *Walk Softly* vii. 107 'You've got a sleeper reservation?' 'No, I was going to sit up.' **1982** N. Painting *Reluctant Archer* v. 82 'Book a sleeper,' said Reggie airily. There were no sleepers. I sat up all night.
f. (Earlier example.) Also *to sit up and take notice*, to become suddenly interested, to pay attention.
1886 H. Baumann *Londinismen* 179/2 *To make a person sit up*, jemand in Erstaunen setzen. [**1898** Kipling *Stalky & Co.* (1899) 84 If they make such a row now, what will they do when she really begins to look up an' take notice?] **1909** *N.Y. Even. Post* 6 Mar. 1/3 The crowd that fell upon Washington was of such a size that the District authorities sat up and took serious notice. **1929** *Burton Evening Gaz.* 1 Jan. 5/1 By the time that the Oxford English Dictionary compilers reach the 'E' section of additions to that great work..they will have to sit up and take some notice of a new and curious borrowing from the French. **1954** A. Huxley *Let.* 18 Jan. (1969) 694 If you want them to sit up and take notice, prepare your way with a barrage of heavy guns from respectable institutions. **1968** H. C. Rae *Few Small Bones* II. viii. 141 Look at him..stuffed full of pet theories, praying I'll sit up and take notice of him. **1977** M. Allen *Spence in Petal Park* xvii. 77 When I see him I sit up and take notice.
g. *to sit up and beg* (cf. *beg *v.* 2 i): used *fig.*; also (with hyphens) as *adj. phr.*
1917 R. Barnes *Let.* 20 July in M. Gilbert *W. S. Churchill* (1977) IV. Compan. I. 105 Pile up the guns & shells & we will make the Hun sit up & beg, but we haven't got enough yet. **1919** W. H. Berry *New Traffic* (*Aircraft*) viii. 46 Our gallant youth is quite prepared..to make his seventy miles per hour motor-bike sit-up-and-beg as he would put it. **1958** *Listener* 16 Oct. 603/2 In the 'sit-up-and-beg' attitude in which some modern fighter aircraft come in to land. **1961** Partridge *Dict. Slang* Suppl. 1275/1 'He can make it sit up and beg' indicates that a man has become extremely proficient in working some material, e.g. a metal. **1963** Bird & Hutton-Stott *Veteran Motor Car* 215 They were usually endowed with handsome 'sporty' bucket seat bodies in marked contrast to the usual sit-up-and-beg auntification of the time. **1968** *New Scientist* 3 Oct. 8/2 The pilot is thus in a sit-up-and-beg attitude. **1973** G. Talbot *Ten Seconds from Now* xvi. 200 He rode, incongruously, in a vintage, sit-up-and-beg, hearse-like black limousine. **1978** *Lancashire Life* Apr. 65/1 Old Luke's sit-up-and-beg bike was propped against one of the sandstone gateposts. **1978** *Times* 4 July 19/4 Ramirez tucked away a net cord by his opponent that sat up and simply begged. **1980** *Radio Times* 4–10 Oct. 13/1 A midwife in London's dockland, travelling around on a sit-up-and-beg bike.
h. *to sit up like Jacky* (Austral.), to sit up straight, to comport oneself in a prim and proper manner.
1941 Baker *Dict. Austral. Slang* 38 *Jacky, sit up like*, to behave, sit up straight. **1969** P. A. Smith *Folklore Austral. Railwaymen* 180 As we were rattling along north to Darwin I happened to look back out of the guard's van and there they [*sc.* the hoboes] were—sitting up like Jacky in the commissioner's car behind us. **1975** H. Porter *Extra* 139 He's telling Edinburgh, and those writers sitting up like jacky in tiers behind him, about the construction of his next book.
26. *With prepositions, in special senses.* **sit on** or **upon. d.** (Examples with *on*.)
1892 I. Zangwill *Childr. Ghetto* I. xi. 238 'Odious prig!' thought Hannah. 'He actually doesn't see I'm sitting on him.' **1894** Somerville & 'Ross' *Real Charlotte* I. vi. 76 If you're going to sit on me every time I open my mouth, I'd better shut up. **1936** N. Streatfeild *Ballet Shoes* xiv. 225 In the tube going home, Pauline and Petrova pestered Posy for criticism of the production; but the moment she made any, they sat on her, asking her what she thought she knew about it. **1969** *New Yorker* 14 June 46/2 Someone should have sat on him when he was young. **1975** *Guardian* 22 Jan. 1/2 The TUC general secretary..proceeded to sit heavily on the CBI's suggestion.
e. To hold back, to keep to oneself without acting upon.
1906 Kipling *Actions & Reactions* (1909) 203 A three-million pound insurrection caused by a deputy Under-Secretary sitting upon a mass of green-labelled correspondence instead of reading it. **1967** F. Clifford *All Men are Lonely Now* I. iv. 64 Where have all the yellows gone?..Seriously, who's sitting on them? **1983** M. Hinxman *Corpse Now Arriving* vii. 45 She'd 'sat' on the article..until..a deadline had galvanized her into putting words on paper.
f. To suppress, to silence.
1915 A. Huxley *Let.* Nov. (1969) 85 What an odd business it was about the suppression of Lawrence's book, *The Rainbow*. It is always the serious books that get sat on. **1925** 'R. Hall' *Saturday Life* iv. 51 But she sat on her conscience. **1972** D. McLachlan *No Case for Crown* iii. 39, I want this story sat on till midnight. **1976** M. Machlin *Pipeline* ii. 32 There had to be a scout on that plane—this one's going to be hard to sit on. **1977** *Undercurrents* June–July 18/1 These huge public charivari were sat upon heavily by the police.
g. To wait for (something or someone) to change or develop; to observe or trail.
1958 *Spectator* 11 July 62/2 Two children..were brought to my hospital..suffering from appendicitis, which had been treated conservatively (or in hospital slang 'sat on') for several days. **1966** I. Jefferies *House-Surgeon* vi. 118, I couldn't make up my mind either, so we decided to sit on her and see what happened.
h. *to sit on the splice* (Cricket): see *splice *sb.* 1 c.
27. sit over. Also, to linger over (a meal, etc.) while sitting. (Further examples.)
1952 M. Laski *Village* x. 156 The dinner was cleared away and they sat over their coffee. **1971** 'E. Ferrars' *Stranger & Afraid* vi. 112 They..had lunch, sitting over it until about two o'clock.
28. sit under. Also, to listen to (a teacher), to be the pupil of.
1899 J. London *Let.* 20 Sept. (1966) 56 Stopped over at Stanford, where I..sat under the various profs. **1952** 'J. Tey' *Singing Sands* iii. 39 He 'sat under' a bank clerk in Glasgow, a chap from Uist, and swotted up some Gaelic.
29. sit with. c. To be received in a specified manner by; to be consonant with.
1961 in Webster s.v. [1]*sit*, Setting an example that may not sit well with the more obedient Communist leaders. **1972** *Listener* 1 June 705/2 In the meanwhile sanctions would be continued, and this did not sit well with right-wing opinion inside the Conservative Party.
V. 31. a. Also *transf.*
1814 Jane Austen *Mansfield Park* II. ii. 33 Poor old coachman would attend us..though he was hardly able to sit the box on account of the rheumatism. **1977** *New Yorker* 11 July 19/1 She sits a bicycle with the feckless insouciance of an eleven-year-old gliding down a country lane.
c. Add to def.: using one's body-weight to adjust its balance. (Earlier example.)
1865 *Etoniana* xi. 170 The time-honoured custom of 'sitting a boat'.
36. d. To act as a baby-sitter for (a child). Also *transf.*
1950 *Here & Now* (N.Z.) Nov. 28/2 He is a nice domestic chap: speaks on international affairs; helps old ladies across the street; can sit a baby. **1971** E. Fenwick *Impeccable People* xx. 110 He can help sit Granny, too. **1976** *Billings* (Montana) *Gaz.* 16 June 9-c/6 (Advt.), 'Grandma' needed to sit 3 pre-school boys, in my home ½ days beginning late August.
e. To take (an examination).
1957 A. Wilson *Bit off Map* 40 With the degree behind me, I shall sit the Administrative in June. **1966** *Rep. Comm. Inquiry Univ. Oxf.* II. 152 Collections are college examinations, usually sat at the beginning of a term. **1980** *Radio Times* 1–7 Mar. 16/4 A child can..enter for and sit an examination without being put forward by the school.
37. sit out. a. (Earlier example of a dance.) Also *transf.*
c **1869** Taylor & Dubourg in M. R. Booth *Eng. Plays of 19th Cent.* (1973) III. 251, I didn't sit out one dance. **1957** *Economist* 19 Oct. 208/2 Only two countries —Holland and Australia—gave a majority in favour of getting involved [in a war with Russia]. In the rest, all but two of which are allies of America, most people said they would rather sit this one out. **1978** G. A. Sheehan *Running & Being* xv. 211, I hadn't realized this..until the 1976 Boston Marathon... Any thinking adult would have sat this one out.
c. (Earlier examples.)
1751 Fielding *Amelia* II. v. v. 127 She resolved to come to an eclaircissement, and having sat out some company that came in, when they were alone together, [etc.]. **1808** Jane Austen *Let.* 9 Dec. (1952) 237 We found Mrs. Lance at home & alone, & sat out three other Ladies who soon came in.

sit-. Add: **sit-me-down (-upon)** *colloq.*, the buttocks, the posterior.
1926 D. L. Sayers *Clouds of Witness* ii. 55 He's left the impression of his sit-me-down-upon on the cushion. **1935** 'G. Orwell' *Clergyman's Daughter* iii. 187 Shift yourself..and make room for my little sit-me-down. **1942** Partridge *Usage & Abusage* 351/2 Euphemism..is often employed, sometimes in such childish form as *sit-me-down*.

sitar. Add: Also with pronunc. (sitā·ɪ). Also **setar.** Delete *Anglo-Ind.* and substitute for def.: A long-necked, guitar-like, Indian musical instrument, having from three to seven strings which the player plucks. Also *attrib.* and *Comb.* (Further examples.)

1954 *Grove's Dict. Mus.* (ed. 5) IV. 459/1 The *Setar* is the most popular instrument in northern India. **1957** *New Oxf. Hist. Music* I. iv. 224 Another favourite plucked instrument is the *setar*, of Persian origin. **1959** D. Cooke *Lang. Music* ii. 55 A recent improvisation on a *raga*, by one of India's leading sitar-players. **1966** *Melody Maker* 7 May 10 Since George Harrison introduced the sitar on 'Norwegian Wood' on the Beatles' 'Rubber Soul' album there has been an intense interest in this Indian instrument. **1975** D. Lodge *Changing Places* ii. 85 Someone dimmed the lights and turned up the sitar music. **1978** P. Griffiths *Conc. Hist. Mod. Music* ix. 139 The use of sitars in rock bands.

sitarist (sitā·rist). [f. Sitar + -ist.] One who plays the sitar.

1966 *Guardian* 16 Aug. 6/5 Jayasri..gave her first public recital at the age of 15 accompanying her father, who is himself a well-known sitarist. **1977** *Times* 23 July 9/3 It provides opportunities for the sitarist to engage in musical small-talk with Menuhin's violin and Jean-Pierre Rampal's flute. **1982** *Listener* 18 Nov. 21/2 They were Indian sitarists, and they went at their sitars with a will.

sitatunga, var. *situtunga.

sitcom (si·tkǫm). orig. *U.S.* Also **sit-com, sit.-com.** Abbreviation of *situation comedy* s.v. *situation 11.

1964 *Life* 18 Sept. 24/2 Even Bing Crosby has succumbed to series TV and will appear in a sitcom as an electrical engineer who happens to break into song once a week. **1970** *Globe & Mail* (Toronto) 25 Sept. 14/4 A domestic sitcom about a pair of newlyweds. **1972** P. Black *Biggest Aspidistra in World* iii. v. 195 The only sitcom shows to see the possibility of success..were *Dear Dotty* and *Friends and Neighbours*. **1973** *Times* 2 June 8/4 *My Good Woman*, the new peak time sit-com spot on Monday evenings. **1978** *Encounter* Feb. 32/2 The nearest thing is the comedian Les Dawson, when he relies on words and not on sit.-com. conventions. **1980** *Times Lit. Suppl.* 2 May 492/5 His [*sc.* N. Simon's] stage comedy is hardly distinguishable from television sitcom.

sit-down, *a.* and *sb.* Add: **A.** *adj.* **3.** Of a strike, demonstration, etc.: in which persons sit down in a work-place, public building, etc.; also *fig.* Of a person: participating in such a strike or demonstration.

1936 *N.Y. Times* 30 Jan. 7/6, 1000 workers of the Firestone Tire and Rubber Company remained idle in a 'sit down' protest. **1936** *Sun* (Baltimore) 2 Nov. 14/2 Claiming 2,300 men already were affected by a 'sit down' strike on eighteen ships, the insurgent seamen's defense committee tonight set out to extend its work embargo on all United States vessels in this port. **1937** *Times* 25 Jan. 19/4 The General Motors Corporation has decided to remain in possession of such of its plants as are occupied by sit-down strikers. *Ibid.* 22 Nov. 12/5 'Business' says it is hampered by Government interference and unwise taxation; the Government seem convinced that political enmity has provoked capital investors to a sit-down strike. **1940** H. Brighouse *Man who ignored War* 23 You're on a sit-down strike, sitting there and saying to Hitler, 'Go away, little man, I'm busy creating beauty.' **1948** *Sun* (Baltimore) 8 Apr. 1/6 A small bloc of Republicans is conducting a deliberate 'sit-down strike' on the ECA and other legislation in an attempt to block as much of the Administration's program as it can. **1958** *San Francisco Examiner* 24 Aug. 1. 25/2 (*heading*) New sit-down battle in Jim Crow cafes. *Ibid.* 25/6 The 'sit-down' endeavor at Brown's [luncheonette] was the latest in a series which began Tuesday night. **1959** *Listener* 15 Jan. 118/1 London Transport appeals for an end to sit-down strikes on the underground railways. **1960** *Washington Post* 24 Feb. 33/7 White college students joined Negroes today in 'sitdown' lunch counter protests in Winston-Salem. **1960** *Sunday Express* 18 Dec. 12/8 Bertrand Russell proposes to stage a 'sit-down' demonstration outside the Ministry of Defence. **1961** *Daily Tel.* 18 Oct. 1/1 Four members of the anti-nuclear Committee of 100 staged a sit-down protest for over four hours last night in a sitting-room at the Russian Embassy in London. **1962** *Listener* 15 Mar. 458/2 The reasonable sit-down demonstrator. **1972** G. Durrell *Catch me Colobus* x. 221 Remember that animals and plants have no M.P. they can write to; they can't perform sit-down strikes..they have nobody to speak for them except us.

B. *sb.* **1. a.** (Earlier and later examples.) Also, an opportunity to sit down and relax.

1861 Mrs. Stowe *Pearl of Orr's Island* I. xii. 104, I am come here for a good sit-down by your kitchen-fire. **1932** D. L. Sayers *Have his Carcase* iv. 56 'A nice sit-down in the lounge,' said Wimsey, sitting down. **1937** 'J. Bell' *Murder in Hospital* i. 6 By the nurse's desk a constant stream came in with..heads done up in dirty, blood-stained rags..some with nerves on edge from witnessing so many..gruesome sights, some all the better for a good sit down. **1967** N. Freeling *Strike out where not Applicable* 69 The sit-down had done his leg..some good. **1982** J. Sherwood *Shot in Arm* iii. 32 If Verney wanted to go back..it would mean a sit-down and a cup of tea.

b. *N. Amer. Tramps' slang.* A free sit-down meal.

1919 *Dialect Notes* V. 42 *Sit-down*, a meal sitting down. 'A sit-down, with hot Java.' **1926** J. Black *You can't Win* vi. 67 She'll give you a sit-down for yourself, chances are, but bring back a 'lump' for us. **1927** F. Niven *Queer Fellows* iv. 45 There must be houses where we could get a hand-out for sure... We might even get a sit-down. **1936** *New Republic* 15 July 289/1 Upon the occasions referred to, 'sit downs', or invitations to eat at the family table, are more apt to be given.

c. A sit-down strike or demonstration.

1936 *N.Y. Times* 2 Feb. 26/1 The Akron rubber industry's second 'sit-down' of the week was in progress today. **1938** *Times* 23 May 13/4 About 1200 single unemployed men entered three Vancouver buildings..on Friday..and began a 'sit down' in order to obtain immediate work and wages. **1958** *Economist* 1 Nov. 421/1 The Japanese government has long been irked by the failure of the police to deal effectively with demonstrators who stage sit-downs at American air bases and also inside government offices. **1960** *N.Y. Times* 3 Apr. iv. 7/1 (*caption*) Negro students at Southern University in Louisiana apply at registrar's office to withdraw from the university to protest expulsion of classmates for lunch-counter sitdowns. **1961** *Guardian* 3 Feb. 5/6, 1,200 volunteers..have so far agreed to take part in a sitdown outside the Ministry of Defence... The sitdown is in protest against the Polaris agreement. **1967** *Times* 28 Feb. (Canada Suppl.) 34 Young Canadians have joined sit-downs against nuclear arms. **1972** R. Thomas *Porkchoppers* (1974) xxvi. 228, I spent forty-one days in that place... It was a sit-down and the old man sent me in to sit with them. **1978** *Peace News* 25 Aug. 3/1 The next day, 2,100 workers turned up for a 'sit-down', and another 300 were dismissed. **1982** M. Wallace *Brit. Govt. N. Ireland* ii. 29 Measures to deal with street sit-downs and the occupation of public buildings.

Hence **si·t-downer,** a participant in a sit-down strike or demonstration.

1936 *Time* 30 Nov. 15/1 Most of the 4,300 workers obeyed, but 1,100 sit-downers sat pat. **1949** *Sun* (Baltimore) 6 July 9/4 The local Communist association aided the sitdowners. **1960** *N.Y. Times* 15 May 12E/1 (*heading*) Sit-downers score a quiet victory. **1961** *Guardian* 27 Sept. 20/6 Lord Russell..has won the sit-downers' enthusiasm. **1963** D. V. Baker *Door is always Open* ix. 187 We had finally reached Hyde Park, where the Committee had arranged for the sit-downers to meet before marching to Whitehall.

site, *sb.*² Add: **2. b.** In scientific use, a position or location in or on something, esp. one where some activity happens or is done.

1950 *Sci. News* XV. 70 Even a perfect crystal will contain a certain number of vacant lattice positions. The proportion of such sites depends only on the temperature. **1954** A. White et al. *Princ. Biochem.* xii. 259 This inhibition may be a result..of combination of the inhibitor with the same site on the enzyme at which the substrate would combine. **1956** M. Demerec in *Publ. Carnegie Inst. Washington* No. 612. 2 The specific properties of an allele are determined by changes at a specific part of the gene locus. Thus it is now evident that a gene locus is composed of a number of units, separable by crossing over, which we call 'sites'. **1966** T. S. & C. R. Leeson *Histology* viii. 140/1 In the fetus..blood cells are formed in different sites at different ages. **1968** Passmore & Robson *Compan. Med. Stud.* I. vii. 2/2 The high specificity of most enzymes..suggests that on the surface of the enzyme molecule there is one or perhaps a few sites specifically adapted for binding the correct substrate and bringing about the reaction. **1971** Levitan & Montagu *Textbk. Human Genetics* xv. 576 On this view the several factors belonging to the same polypeptide..would be determined by different mutable sites on one of the genes. **1982** K. H. Muench in T. M. Devlin *Textbk. Biochem.* xix. 943 (*caption*) Human proinsulin. After cleavage at the two sites indicated..the arginine residues..and the lysine residue..are removed to give insulin and C-peptide.

3. a. (Further examples.) Also, in wider use, a piece of ground or an area which has been appropriated for some purpose; the scene of a specified activity. Freq. in comb. with the first element indicating the (intended) use of the area as *building, caravan, landing, launching, picnic* (etc.) *site*: see these words in Dict. and Suppl.

1888 [see *nesting-site* s.v. Nesting *vbl. sb.* b]. **1930** [see *nest-site* s.v. *nest *sb.* 8]. **1953** N. Tinbergen *Herring Gull's World* xvii. 152 The bird usually developed a clear preference in favour of one of the two nests,..a site-preference. **1963** *Camping* ('Know the Game' Ser.) 43/2 (*heading*) Choosing a camp site. *Ibid.* 46/2 Have consideration for other campers. Do not stroll into someone else's site just as they are in the middle of morning ablutions. **1965** A. J. P. Taylor *Eng. Hist. 1914–1945* xvi. 578 The use of the pilotless aeroplanes and of the rockets was delayed..by bombing their launching sites in France. **1973** E. F. Schumacher *Small is Beautiful* ii. iv. 129 There will be a continuous traffic of radioactive substances..from the stations to waste-processing plants; and from there to disposal sites. **1980** J. McNeil *Spy Game* xix. 189 It was shoe-horned between the radar assembly sheds, a piece of open ground which had miraculously escaped the rash of building covering the rest of the site.

c. *Archæol.* A place containing the remains of former human habitation; an excavation.

1911 T. E. Lawrence *Let.* Apr. (1954) 149 The dig has proved a failure to the present (tho' there is still hope of one part of the site). **1963** E. S. Wood *Collins Field Guide Archaeol.* ii. ii. 200 Bronze Age sacred sites, such as the circles at Knowlton (Dorset), which has a church inside it,..indicate continuity of sacred sites. **1977** *Times* 13 Aug. 14/4 More than four fifths of the villa's walls have been destroyed by ploughing and erosion, and Mr Sumpter feels that the site would not have survived another year's ploughing. **1980** *Rescue News* Dec. 7/5 A cliff site near Northskaill. The site is a great kitchen-midden at least 3·5m thick.

5. *U.S. Naut. slang.* A job, a situation.

1930 *Amer. Speech* V. 393 *Site*, a place as fisherman aboard a fishing vessel. **1957** *Maine Coast Fisherman* July 21/1 Skipper Farrell won't offer a site to a cook who will only cook. **1967** *National Fisherman* June 19-c/2 Palmer was to have a steady job ('site') aboard until he quit or was discharged for cause. **1977** *New Yorker* 15 Aug. 46/3 Joe, who generally keeps his own counsel, tells me that he is hoping to get a site—a job—on the Sniktaw.

6. *attrib.* and *Comb.*, as (sense 3) *site clerk, foreman, manager*; **site assembly,** assembly of building components on the site; **site value,** the amount for which a site may be sold (see also quot. 1893).

1958 *Listener* 6 Nov. 726/2 The mechanization of building and rapid site-assembly of light-weight factory-produced components is a priority item in the current five-year plan. **1961** *Evening Standard* 20 July 18/3 (Advt.), Site Clerk required by..contractors. **1969** T. Parker *Twisting Lane* 57 I'd got a fairly decent job as a site-clerk with a small firm of builders. **1964** K. G. Lockyer *Introd. Critical Path Anal.* iii. 27 Departmental managers, site foremen. **1981** J. B. Hilton *Playground of Death* viii. 97 Three site foremen in succession suffered accidents. **1961** *Technology* May 121/3 There was a need..for a new sort of foreman for the larger projects, a site manager. **1976** R. Lewis *Distant Banner* iii. 85 The bosses had obviously been on the site manager's back. **1893** Site value [in Dict., sense 3 a]. **1904** G. B. Shaw *Common Sense of Municipal Trading* x. 92 The popular remedy is to tax site values directly. **1941** H. Nicolson *Diary* 17 Mar. (1967) 152 It gives a magnificent vista of St Paul's... To get that permanently cleared is worth 40 million pounds in site-value. **1973** E. Page *Fortnight by Sea* vi. 67 Hunston's had been losing trade..had been glad to sell out in the end for the very considerable site value.

site, *v.*² 1, **siting,** *vbl. sb.* (Later examples in general use.) See also *sited *ppl. a.*

1918 *Cornhill Mag.* June 621 The short-sighted policy adopted in the siting and construction of schools. **1920** *Discovery* Apr. 116/1 It is advisable to avoid siting a wireless station close to higher ground. **1931** *Times Lit. Suppl.* 16 Apr. 310/2 A practical handbook on the siting, construction and upkeep of garden pools. **1946** *Nature* 2 Nov. 600/2 Areas of exceptional natural beauty or great historic interest should be avoided if their character would be impaired by the siting of a town. **1955** *Times* 29 June 7/1 New proposals for siting Rodin's sculptured group of the Burghers of Calais. **1980** *New Age* (U.S.) Oct. 37/2 This arrangement..has resulted today in more than sixty so-called 'energy wars' going on in this country between utilities or siting authorities and the politically weaker rural people.

sited, *ppl. a.* Restrict † *Obs.* to sense 2 in Dict. and add: **1.** (Later examples.)

1975 *No Through Road* (Automobile Assoc.), This unusually sited lake, halfway up a steep hillside, was created when a glacial overflow channel became blocked. **1979** *Weekend Bargain Breaks* (Trusthouse Forte) 17/1 The hotel..is well sited in this peaceful town.

sitha, sithee (si·ðā, si·ði), repr. dial. pronunciations of *see thou* (see See *v.* 5 f), used esp. as an interjection to draw attention or as a conversation filler.

For further material see *Eng. Dial. Dict.*

1885 [see *plonker 1 a]. **1887** Kipling *Plain Tales from Hills* (1888) 62 Now, *sitha*, tak' a *tat* an' a *lookri*, an' ride tha domdest. **1920** D. H. Lawrence *Touch & Go* iii. ii. 86 See's motor?—comin' up—sithee? **1932** Kipling *Limits & Renewals* 309 She's proud of hersen!—Sitha! She's tryin' to admire of her own belly! **1969** *Listener* 15 May 700/3 My first awareness of Castleford was as a young sports writer looking round for the press box when an official said: 'Sitha, lad, it's up there, on't top deck.' **1974** J. Aiken *Midnight is Place* iv. 135 They're needing a place to lodge too, sithee.

sit-in, *a.* and *sb.* orig. *U.S.* [The phrase *sit in* (see Sit *v.* 22 in Dict. and Suppl. and -in³) used attrib. and as sb.] **A.** *adj.* Of a strike, demonstration, etc.: in which persons occupy a work place, public building, etc., esp. in protest against alleged activities there. Of a person: participating in such a strike or dmonstration. Also, of or pertaining to such a strike or demonstration.

1937 *Sun* (Baltimore) 3 Apr. 7/2 (*caption*) The Synthetic Yarn Federation Local 2214..is staging a 'sit-in' strike at Covington. **1938** *Topeka* (Kansas) *Capital* 26 Nov. 12/5 [Daladier] used..police to eject sit-in strikers from factories. **1941** *Sun* (Baltimore) 15 Apr. 3/3 (*heading*) New York students stage 'sit-in' protest. Spurn holiday to make demonstration against proposed school closing. **1948** *Ibid.* 25 Nov. 1/4 Passengers aboard the strike-bound luxury liner Queen Elizabeth today were ordered to get off by Friday, but approximately 900 said no and voted for a 'sit-in' strike. **1959** *Times* 10 Jan. 6/7 A warning to passengers not to repeat the recent 'sit-in' strikes on the London Underground was given yesterday. **1960** *Time* 14 Mar. 21/1 Negro 'sit-in' demonstrations at segregated lunch counters. **1960** *Guardian* 25 July 7/3 The two men had agreed..to express 'support for the objectives of the sit-in demonstrators' in the South. **1973** *Black World* Mar. 37 Pressure for opening the main-stream of American life to Blacks mounted in the 1950's—the sit-in kids, the Supreme Court School Desegregation Decision, [etc.]. **1973** *Times* 17 Nov. 8/2 During May [1936] the French

Treasury almost ran out of funds... There was..a great wave of 'sit-in' strikes. **1980** *Washington Post* 1 Feb. A2/1 Their refusal to budge officially launched the sit-in phase of the civil rights movement.

B. *sb.* **1.** A sit-in strike or demonstration.

1937 *N.Y. Times* 29 May 1/7 Fifty members of the Workers Alliance who tried to stage a sit-in at City Hall yesterday were removed..by a dozen policemen. **1941** *Sun* (Baltimore) 15 Apr. 3/3 He approved the 'sit in' as an 'orderly and dignified protest'. **1960** *Newsweek* 22 Feb. 27 What some Negroes were calling the 'sit-down' and some the 'sit-in'. **1960** *Commentary* June 525/2 The spread of similar picket lines to other cities..seems to have been as spontaneous as the sit-ins themselves. **1965** Mrs. L. B. Johnson *White House Diary* 11 Mar. (1970) 250 Some of the Civil Rights marchers had walked into the White House..and refused to budge. A sit-in in the White House! **1973** *Law Reports: Appeal Cases* Nov. 858 A sit-in per se is not threatening, abusive or insulting behaviour. **1976** *Times* 10 May 20/5 Sit-ins and work-ins are used by employees..increasingly..as a tactic in collective bargaining. **1978** *Cornish Guardian* 27 Apr. 1/1 Mothers who last year threatened to stage a sit-in on St. Austell's Truro Road are again worried that a child is going to be knocked down and killed there.

2. A participant in a sit-in strike or demonstration. *U.S.*

1963 R. I. McDavid *Mencken's Amer. Lang.* 557 Most lay newspapers would simply describe such persons [*sc.* sitters-in] as sit-ins. **1970** *Daily Progress* (Charlottesville, Va.) 19 Mar. 1/8 A group of people willing to commit civil disobedience will sit down on the pavement in front of the building, obstructing the entrance. If these sit-ins are arrested, another group will take their place.

Hence **si·t-inner** = *sit-in *sb.* 2.

1946 *Sun* (Baltimore) 26 Oct. 3/1 Sixty American War Department civilian employés..held to their rooms in the luxury Hotel Excelsior tonight as the midnight deadline approached for them to obey army orders to move out. Officially an army spokesman said no action was contemplated against the defiant 'sit-inners' until after midnight. **1960** *New Left Rev.* Sept.–Oct. 39/2 The police hosed and clubbed the sit-inners.

sit-in-'ems, sitinems (si·tinĕmz), *sb. pl. slang* (now *rare* or *Obs.*). [Repr. a colloq. pronunciation of *sit in them* + pl. *-s.*] Trousers.

1886 in H. Baumann *Londinismen* 179/2. **1922** Joyce *Ulysses* 418 Don't stain my brandnew sitinems.

siting, *vbl. sb.* Add: Also in general use: see quots. s.v. *site *v.*²

Sitka (si·tka). [a. Tlingit *sheet'ká* (town of) Sitka, lit. 'outer side of Baranof Island'.] **1.** (A member of) a local group of Tlingit Indians formerly living principally in this North American Indian town. Also formerly, the variety of the Tlingit language spoken there. Also *attrib.* or as *adj.*

1829 J. S. Green in *Missionary Herald* (1830) XXVI. 343 The Sitka Indians have built their village under the guns of the fort, so the Russians can easily defend themselves from their depredations. *Ibid.* 344 The Sitka I think is peculiarly soft and musical. **1836** [see *Chilkat]. **1873** *Alaska Herald* 9 July 4/2 The Hydahs, Chilcats, Tarkous, Ouchanons, and Sitkas participated. **1879** W. G. Morris *Public Service of Alaska* 14 In the fall of 1877, a potlatch was given at Sitka by Jack, chief of the Sitkas, and it is estimated correctly he gave away on that occasion 500 blankets. **1910** F. W. Hodge *Handbk. Amer. Indians* II. 582/2 *Sitka*,..a Tlingit tribe..on the w. coast of Baranof id., Alaska.

2. The name of the town, used *attrib.* or *absol.* to designate trees native to the region, as **Sitka cedar, cypress,** the Alaska cedar, *Chamæcyparis nootkatensis*, of the family Cupressaceæ; **Sitka pine, spruce,** a large conifer, *Picea sitchensis*, of the family Pinaceæ, or its light softwood timber.

1884 C. S. Sargent *Rep. Forests N. Amer.* 580 The most valuable tree of this region [*sc.* Alaska] is the Sitka cedar. *Ibid.* 178 *Chamæcyparis Nutkaensis*... Yellow Cypress. Sitka Cypress. **1884** *N.Y. Times* 5 Oct. 5/2 The white spruce, or Sitka pine..grows to a height of 150 and 175 feet. **1895** Funston & Coville in *Contrib. U.S. Nat. Herbarium* III. 328 The great bulk of this forest is composed of Sitka spruce. **1920** *Nature* 29 July 692/1 The Sitka or silver spruce..might be called the aluminium of timbers. **1928** *Daily Mail* 9 Aug. 13/4 The Duchy estates on Dartmoor, where it is hoped to plant 5,000 acres with sitka, Norway spruce, and Douglas fir. **1948** *Antioch Rev.* Winter 48 This would be the first step toward cutting off the magnificent Douglas fir and Sitka pine from 300,000 acres. **1965** G. Maxwell *House of Elrig* vi. 98 The needles of unthinned sitka. **1975** W. Condry *Pathway to Wild* ix. 151 If that other western American, the Sitka spruce, now planted so multitudinously in Britain, were also going to be given the chance of making forests of giant trees..then conservationists might be less unhappy with it. **1977** *Chicago Tribune* 2 Oct. i. 39/2 (Advt.), This fine piano was made specially for us. Its quality features include..solid Sitka spruce soundboard.

‖ **sitkamer** (si·tkaməɪ). *S. Afr.* Also **sitkamer, zitkamer, zit-kamer.** [Afrikaans, f. Du. *sit* sitting + *kamer* room.] A sitting-room, a lounge.

1902 'Intelligence Officer' *On Heels of De Wet* 88 Those cushions you have on your front seat came out of the Nieuwjaarsfontein *sitkomer* [sic]. **1904** *Argus* (Cape Town) *Christmas Ann.* 12 He saw his mother standing at the door of the *zitkamer*. **1912** F. Bancroft *Veldt Dwellers* 31 It was smoke-room, bar-room, and general zit-kamer combined. **1929** J. G. van Alphen *Jan Venter, S.A.P.* 249 The *sitkamer* was packed with visitors. **1935** P. Smith *Platkops Children* 84 The sit-kamer is so beautiful you know, that Ou-ma Carel never lets the sun shine in it excep' for a little on Sundays. **1955** W. Robertson *Blue Wagon* i. 3 Van Zyl and John..entered the *sit-kamer*, as the general living-room of the place was called. **1964** J. Neintjes *Manor House* 7 We entered a lounge on the right, the *sitkamer*—a large room with a high ceiling on beams. **1971** J. A. Brown *Return* 79 The door of the cuckoo-clock in the *sitkamer* banged in and out.

sitosterol (səitǫ·stĕrǫl). *Biochem.* [ad. G. *sitosterin* (R. Burián 1897, in *Sitzungsber. der K. Akad. der Wissensch.* (*Math.-Nat. Classe*) CVI. IIb. 549), f. Gr. σῖτο-ς grain, bread + G. *-sterin* after *phytosterin* (see Phyto-); cf. *phytosterol, *-sterol.] Any of a number of closely similar crystalline sterols, most of them isomers of formula $C_{29}H_{50}O$, first isolated from corn oil and widely distributed in plants; *spec.* the most common such substance, *β-sitosterol*, also called *cinchol.

Before 1926 considered to be a single substance.

1898 *Jrnl. Chem. Soc.* LXXIV. i. 72 A substance..is obtained which resembles the cholesterol of bile in external appearance and in composition..; it is named sitosterol. **1926** *Jrnl. Amer. Chem. Soc.* XLVIII. 2986 The substance corresponding in composition to sitosterol..is not homogeneous. It is a mixture containing at least three isomeric sterols... It is proposed to name these isomers α-, β- and γ-sitosterol. **1943** *Ann. Reg. 1942* 358 Work on algal chemistry showed that sitosterol, the characteristic sterol of the phanerogams, is common to the green algæ. **1962** H. Burn *Drugs, Med. & Man* v. 57 The substance sitosterol can be given which prevents the cholesterol which is in the diet from being absorbed through the wall of the intestine. **1974** M. C. Gerald *Pharmacol.* xxii. 401 Sitosterols enhance the elimination of cholesterol in the feces.

sitrep (si·t,rep). *Mil.* Also **Sit. Rep., Sitrep, Sit-rep.** Abbrev. of *situation report* s.v. *situation 11.

1943 J. H. Fullarton *Troop Target* xxiii. 173 The daily Sit. Rep. had now identified seven Italian and at least three German divisions in the line. **1947** D. M. Davin *For Rest of our Lives* 307 He thumbed through his log with its patrol reports, phone messages, sitreps, all the raw material from which he must first form in his mind a clear picture of what the enemy was up to. **1955** E. Waugh *Officers & Gentlemen* 264 The B.G.S. said: 'We got a sitrep from the Halberdiers three hours ago.' **1961** I. Fleming *Thunderball* xxiii. 240 I'm going to..get a signal off to Navy Department, give them a Sitrep. **1968** P. Kinsley *Pimpernel 60* ii. 37 Look over a couple of Sit-reps which I received yesterday. **1975** D. W. S. Hunt *On Spot* i. 11 In fact what he [*sc.* Wellington] wrote was a succinct report on the fighting of the past four days. It was in the first place the equivalent of the modern 'Sitrep', which a general, or his staff in his name, telegraphs every night.

sitringee. Add: Also **satrangi, -ji, shatranji;** 7 **citterengee,** 8 **sittringe,** 9 **satrin-, sattran-, satrun-, sut(t)rin-, -gee, -jee.** (Further examples.)

1688 J. Peachey in W. Hedges *Diary* (1888) II. cclxv, 2 Citterengees, charged before. **1785** *Calcutta Gaz.* 3 Mar. 7/2 To be sold by public auction..The valuable effects of Warren Hastings, Esq... Carpets and Sittringes. **1851** *Illustr. Catal. Gt. Exhib.* iv. 917/2 Cotton carpets (*Satrunjees*) of different sizes—from Bengal. **1858** P. L. Simmonds *Dict. Trade Products*, *Sattrangee, Satringee*, a kind of fibrous striped mat or carpet made in India. **1859** M. Thomson *Story of Cawnpore* xii. 189 They provided us with straw to lie upon, and gave us a sutringee each (a piece of carpet) to cover our bodies. **1876** *Encycl. Brit.* V. 129/2 Cotton carpets or *Suttringees* are a cheap substitute for woollen fabrics in almost universal use throughout India. **1881** *Ibid.* XII. 762/1 Carpets and rugs may be classified into those made of cotton and those made of wool. The former, called *satranjis* and *daris*, are made chiefly in Bengal and northern India. **1904** G. Watt *Indian Art at Delhi 1903* 273 If to this list be added..the rug the *dari* or *satranji*, the series of chief artistic textile articles of Native dress and household use may be regarded as complete. *Ibid.* 446 A larger market might be found in India for *shatranjis* than has as yet been attained. **1969** E. Bharnani *Decorative Designs & Craftsmanship of India* iii. 33 In North India..cotton rugs (*Daris* and *Satrangis*) have been produced in several areas since a long time.

sits vac: see *sit *sb.*²

sitter¹. Add: **1. g.** One who has a sitting with a medium.

1909 in Webster. **1928** *Daily Mail* 25 July 6/2 If media were unable to get into a trance the sitting was cancelled and the money returned to the sitter. **1961** W. H. Salter *Zoar* vi. 73 The sitter brings with him marked plates which he gives the medium. **1977** 'L. Egan' *Blind Search* ii. 32 One of their perennial sitters, Claire Ewing..a researcher herself, not an emotion-harried sentimentalist seeking reassurance.

h. *U.S. slang.* (See quots. 1938, 1948².)

1938 S. Hart *New Yorkers* 183 Bowery barkeeps employed homeless men and women as 'sitters' to shiver near the fire on wintry nights and thus evoke the sympathy of cash customers who would treat them to drinks to the great profit of the house. **1948** *Sun* (Baltimore) 7 Aug. 5/4 Violations, particularly of sitters' rules, would mean suspension or barring women from the licensed premises. **1948** H. L. Mencken *Amer. Lang.* Suppl. II. 682 Women who frequent taverns or night-clubs, getting a percentage on the drinks they induce male patrons to buy, are..*sitters*.

i. A baby-sitter.

1943 *Life* 8 Nov. 100/2 (*caption*) Matt Thomson stays home with baby... Once they would have hired capable 'sitter'. Now only inexperienced girls are available. **1951** H. MacInnes *Neither Five nor Three* i. i. 11 She and Jon couldn't come to the party because the baby was sick or they couldn't get a sitter. **1960** *Sunday Express* 26 June 5/3 He has a first-class nurse and a 'sitter' while I am away. **1975** [see *sit *v.* 1 g].

j. A participant in a sit-in or sit-down.

1961 *Britannica Bk. of Year* 537/2 *Sit in*,..and *sitter*, terms used in connection with the attempts by Negroes to eat in cafés and restaurants from which they were normally excluded. **1961** *Guardian* 22 Sept. 12/4 There are still people who think that marchers and sitters can be dismissed because some are oddly dressed.

5. *sitter-in*: (*a*) a baby-sitter; (*b*) one who takes part in a sit-in; (*c*) one who sits in with a band (*sit *v.* 22 f); *sitter-out*: (*b*) one who sits out at cards.

1947 J. L. Burn *Recent Advances in Public Health* ix. 132 To enable husband and wife to go out together, a 'sitters in' service to look after the baby has been established in some areas. **1951** M. Kennedy *Lucy Carmichael* vii. iv. 330 'We shan't even be able to go to the club.' 'We'll get a sitter-in for them.' **1960** *Guardian* 30 Dec. 10/5 The husband..[acted] as the sitter-in when his wife was on evening shift. **1962** *Maroon* (Univ. of Chicago) 20 July 1 (*heading*) Convict Cairo sitters-in; fine each $300. **1963** *Economist* 8 June 1013/1 Negro 'sitters-in'..gather daily to decide who shall court arrest. **1968** *Blues Unlimited* Dec. 12 Jake recorded with his current group..plus assorted sitters-in. **1976** *Southern Even. Echo* (Southampton) 6 Nov. 7/2 Members of the Jess Roden Band are some of our favourite 'sitters-in'. **1976** J. I. M. Stewart *Memorial Service* vi. 92 There would be eruptions. 'Demos' would be held, buildings sat in by sitters-in. **1853** Mrs. Gaskell *Cranford* i. 14 Miss Jessie could not play cards: but she talked to the sitters-out.

6. Anything easy or (apparently) certain of performance; an easy catch, stroke, or shot; a sitting target; a certain winner; a certainty.

1898 *Tit-Bits* 25 June 252/3 A 'sitter' is a catch which falls absolutely into the hands. **1903** G. L. Jessop in H. G. Hutchinson *Cricket* v. 117 The missing of a 'sitter' by some lazy fieldsman whose thoughts were anywhere but on the game. **1908** A. S. M. Hutchinson *Once aboard Lugger* i. iv. 50 'You know I got ploughed?'..'Bad luck, I suppose? I thought it was a sitter for you this time.' **1917** H. A. Vachell *Fishpingle* ix. 183 'Down ours,' enjoined Lionel to his [golfing] partner. 'You'll do it, Joyce. It's a sitter.' **1918** *Chambers's Jrnl.* Apr. 239/1 'A sitter, by the Great Hook Block!' cried Carstairs. 'A transport full of Boches!' **1923** Wodehouse *Inimitable Jeeves* v. 54 An absolute sitter came unstitched in the second race at Haydock Park. **1927** *Observer* 3 July 18/1 A series of very bad shots, including a double fault by Borotra, the missing of absolute 'sitters' by both players and the driving of many easy balls into the net well over the baseline. **1946** *Sunday Dispatch* 8 Sept. 6/2 Midway through the second half Bradley missed a sitter when Roper centred across an open goal. **1951** N. M. Gunn *Well at World's End* xix. 156 A pheasant showed, an old cock. The bird stood. We stood. 'Granville, my boy,' he said to me, 'I always want you to remember this: Confucius never shot at a sitter.' **1973** A. Hunter *Gently French* ii. 16 They use two [routes]... The trouble is they just alternate them... So they were sitters for a villain like Quarles. **1977** *Times of Swaziland* 11 Feb. 14/2 When Wire Kunene was given a sitter on a plate by the evergreen Sugar Ray Zulu, Kunene let slip the chance. **1980** *Amer. Speech 1976* LI. 294 *Sitter*, ball that is soft and easy to return.

sitter² (si·təɪ). *Oxford University slang.* [*-er.⁶: cf. *bed-sitter.] A sitting-room.

1904 [see *-er⁶]. **1925** *Glasgow Herald* 24 Oct. 6 If lectures are to be broadcast, the temptation to listen to them in the quiet and comfort of one's own sitter will be irresistible.

sitting, *vbl. sb.* Add: **1. d.** For *rare*⁻¹ read *rare* and add further example.

1816 Jane Austen *Emma* II. iv. 69 His air as he walked by the house—the very sitting of his hat.

5. c. One in a series of (esp. two) servings of a meal, *spec.* in the restaurant-car of a train.

1959 P. Moyes *Dead Men don't Ski* i. 13 The lights of the train came on..and..the bell sounded..for First Dinner. Jimmy..was..due to dine at the first sitting. **1962** N. Streatfeild *Apple Bough* vii. 101 An attendant from the restaurant-car looked in. 'Luncheon tickets?' he asked. The man said: 'Second sitting, please.' **1965** E. Brown *Big Man* ix. 74 I'll clean up the cabin while you are at lunch. You are second sitting. **1977** C. Allen *Raj* i. 25/2 Gongs were sounded to mark the arrival of each course... Older travellers preferred the second sitting.

6. a. With *in, up* (further examples).

1817 Jane Austen *Let.* 22 May (1932) 493 Words..fail ..to describe what a Nurse she has been to me... There was never any sitting-up necessary. **1900** M. Beerbohm in *Sat. Rev.* 6 Jan. 12/2 Was not a pantomime..the most brilliant occasion for 'sitting-up'? **1946** R. Blesh *Shining Trumpets* x. 237 The band instrumentation was kept normal, and men alternated with one another in the process known as *sitting in.* **1977** 'C. Aird' *Parting Breath*

iv. 54 The students..squatted on the floor... 'Sitting-in means sitting down, I reckon,' remarked one student.

b. (Earlier and further examples.)

1874 W. Lennox *My Recollections* II. 29 A sitting-down supper was announced. **1895** E. F. Benson *Dodo* II. xi. 232 The music..was quite loud enough to be heard distinctly in a small, rather unfrequented sitting-out room. **1900** *Traveller* 4 Aug. 106/2 On the *Caledonia* there was a minimum of space... There were no sitting-out places. **1939** *Sun* (Baltimore) 5 Apr. 26/2 An abundance of unoccupied land for playgrounds, sitting-out areas, walks and service streets. **1964** V. J. Chapman *Coastal Veg.* ix. 212 Maritime cliffs are often the haunt of sea birds, either as nesting areas or as 'sitting out' places. **1976** *National Observer* (U.S.) 19 June 17 (Advt.), It gives full pillow support to head and back for firm sitting-up comfort.

sitting, *ppl. a.* Add: **4. d.** Of a huntsman's target: stationary, and so easily hit. Freq. *fig.* (orig. *Mil.*) in *sitting bird, duck*, etc.

1867 Trollope *Claverings* I. x. 126 The man who fires at a sitting bird is known to be no sportsman. **1944** *R.A.F. Jrnl.* Aug. 270 Carriers are sitting birds for enemy surface craft. **1944** *Reader's Digest* May 53 (*heading*) Why tankers are no longer sitting ducks. **1948** *Harper's Mag.* Apr. 290/2 Ever since Sinclair Lewis gave the first lessons in marksmanship, men of the Senator's type have been sitting ducks for the opposition. **1949** *Sat. Even. Post* 16 July 23/3 (*caption*) Rescuing downed aviators sometimes got a sitting-duck submarine fired on by our own quick-triggered forces. **1954** J. Blish in *If; Worlds Sci. Fict.* No. 4. 38/2 You're a sitting duck for a real infection if you abuse your time during convalescence. **1958** 'A. Gilbert' *Death against Clock* vi. 89 'It could be he was financin' another establishment and didn't want Mrs M. to know. And Wife No. 2 might be makin' trouble.' 'Simpler to put *her* light out in that case,'... 'Ah, but then he'd be a sitting duck for the police. Someone always comes forward in cases like these to say he..saw the dear departed with a gent.' **1961** B. Fergusson *Watery Maze* vii. 183 We had learned that for tanks to land before tank obstacles had been breached was lethal; they would inevitably be halted, and would become sitting ducks. **1977** 'J. D. White' *Salzburg Affair* xiii. 109 Narrow streets, he'd be a sitting target. **1978** J. Wainwright *Thief of Time* 195, I have no 'gentlemanly' distaste at shooting sitting targets; wood pigeons, rabbits, hares..let them be motionless..and I have meat for supper.

b. *fig.* That can hardly be bungled.

1932 *Evening Standard* 28 Jan. 11/2 They had a hand which contained a 'sitting' game in Spades—one of two had five to the four top honours. **1960** I. Peebles *Bowler's Turn* 190 Against these two Dexter with a bit of luck (dropped at square leg at 30 and off a sitting return by Worrell at 32) played confidently.

sitting-room. Add: **1.** (Earlier examples.)

1771 H. Pelham *Let.* 25 Aug. in *Mass. Hist. Soc. Coll.* (1914) LXXI. 147 The Arches at the sides of the Chimnie in the Sitting Room, I like. **1797** J. Woodforde *Diary* 31 Jan. (1931) V. 8 Our sitting Room smoked very much all the Morning.

2. (Earlier example.)

1881 'Mark Twain' *Prince & Pauper* xxxii. 365 Trying to find sitting-room in the galleries.

situate, *v.* For 'Now *rare*' read 'Now *rare* in lit. sense' and add: **1.** (Further examples.) Now usu. *fig.*, to establish or indicate the place of, to put in a context, to bring into defined relations.

1953 *Sunday Times* 20 Sept. 5/7 From Baudelaire to Balchin, Swinburne to Sansom, Virginia Woolf to Angus Wilson, he situates us all, and hands out a ticket to the celestial garden-party of modern literature which assigns us to our table. **1961** *Encounter* May 49/2 The key to 'situating' her..is to grasp..that she was not an 'Edwardian natural' at all. **1964** E. Palmer tr. *Martinet's Elem. General Linguistics* ii. 44 Their linguistic behaviour, whereby *maison* appears in exactly the same contexts in which I would situate it myself. **1973** *Mod. Eucharistic Agreement* 74 We have sought..to situate the role of the presiding minister in relation to the sacerdotal ministry of the Church. **1977** *Proc. Roy. Soc. Med.* LXX. 425/2 Since I know less about medicine than about Molière, I have woven into my paper a brief review of his career with a view to situating the seven comedies in which medicine is parodied or satirized. **1982** *Times Lit. Suppl.* 13 Aug. 872/1 The title of John Allett's book indicates that this is where he wishes to situate Hobson.

situation. Add: **II. 6. b.** (Earlier example.) *situations vacant*: jobs to be filled, *spec.* as advertised in a column or page of a newspaper; a newspaper column or page advertising jobs; also *attrib.* Also *situations wanted.*

1803 G. Colman *John Bull* iii. ii. 36 Service? Nonsense ..I'll put you into a situation in town. *a* **1911** D. G. Phillips *Susan Lenox* (1917) II. iv. 73 Want ads.. closely printed columns of advertisements of help wanted and situations wanted. **1931** M. Allingham *Look to Lady* xx. 207 I'll be readin' the Situations Vacant before I know where I am. 'E aint even left me a reference. **1944** M. Laski *Love on Supertax* ii. 14 'Is it Situations Wanted day?'..If so, give me the outside sheet.'..The Duke pulled his newspaper to pieces. **1949** E. Coxhead *Wind in West* i. 19 She began to look through the situations-vacant columns in the farming papers. **1967** R. Rendell *New Lease of Death* ix. 88 Elizabeth Crilling sat..reading the Situations Vacant in last week's local paper. **1971** D. Lees *Rainbow Company* ii. 29 An obvious commercial traveller reading the situations vacant column of the *Telegraph*.

9. a. Also in mod. usage, preceded by an attributive word or phrase, and designating: (*a*) the state or general circumstances of something at a particular time, as *coal situation*, etc. (and which is acknowledged to change from time to time); (*b*) a particular state of affairs or occasion existing independently, as *standing credit situation, crisis situation*, etc.

Objections have been raised to both usages, but the latter is often especially decried as an ugly and sometimes tautological formation.

1934 *Times* 22 Aug. 11/3 A popular dodge at present is to add the word 'situation' or 'position' to a noun; by this means apparently it has been discovered that the most pregnant meanings can be expressed with the least effort. The 'coal situation' remains unchanged; the 'herring position' is grave. **1935** *Jrnl. Pediatrics* VI. 115 The group that is treated at the Children's Hospital,..may be received after an attempt has been made in an outlying home or hospital to make the best of a premature baby situation. **1941** J. S. Huxley *Uniqueness of Man* ix. 193 A crab can react to various situations—a food-situation, a hunger-situation, a fear-situation, a sex-situation. **1952** *Amer. Speech* XXVII. 13 Corn-shortage situation, draft-evasion situation. **1966** [see *least *a.* 1 e]. **1972** *Where* Jan. 18/1 A school hall or playground, festooned with bunting, with colourful stalls and the sound of the tombola rolling, assumes less of an institutional atmosphere and more of a face-to-face primary group situation. **1973** *Art Internat.* Mar. 26/1 What comes to such galleries is filtered through all kinds of art situations. **1975** *N.Y. Times* 17 Sept. 47/3 We can run innovative or traditional programs here, whereas in an old-fashioned building the architecture limited us to traditional teaching situations. **1976** *Local Council Rev.* Summer 3 (Advt.), These chairs are ideally suitable for multiple seating situations for use in Town Halls, Village Halls, Community Centres, etc. **1977** H. Fast *Immigrants* i. 74 This is not a loan but a standing credit situation. **1978** *Oxford Times* 13 Jan. 4/3 Unless catchment areas are re-drawn Lord Williams's school will go through a crisis situation for at least five years.

11. Special combs.: **situation comedy,** a comedy (serial) in which the humour derives largely from the particular conjunction of characters and circumstances; **situation ethics, morality,** the belief that individual circumstances or particular situations may call for flexibility in the application of moral laws; **situation report** *Mil.* (see quot. 1918); also *transf.*; **situation(s) room,** a room set aside by a military or governmental agency for giving reports on the current state of any action, operation, etc.

1953 *TV Guide* (N.Y. Metro ed.) 23 Oct. 19/1 Ever since *I love Lucy* zoomed to the top rung on the rating ladder, it seems the networks have been filling every available half-hour with another situation comedy. **1967** *Listener* 10 Aug. 161/1 The two best situation-comedy shows this country has ever produced are..*Steptoe and Son* and.. *Till Death Us Do Part*. **1972** *Time* 17 Apr. 39/1 NBC plans *The Little People*, about a Hawaiian pediatrician and his pediatrician daughter, and ABC has *Temperature's Rising*, about the chief surgeon in a big city hospital. Both shows will combine the medical genre with the situation-comedy formula. [**1950** K. Rahner in *Stimmen der Zeit* CXLV. 330 (*heading*) Situationsethik und Sündenmystik.] **1955** *Cross Currents* Winter 79/2 Father Rahner wished to demonstrate..that..situation-ethics was the most modern and fashionable form of laxity; that it was..an attempt of half-Christians to dodge the effort demanded by the notion of morality and the rigor of the law, by appealing to 'the situation'. **1966** J. Fletcher *Situation Ethics* i. 26 Situation ethics goes part of the way with natural law, by accepting reason as the instrument of moral judgment while rejecting the notion that the good is 'given' in the nature of things, objectively. **1979** B. G. Skinner *Robert Exon* vi. 56 The so-called Situation Ethics, where the Christian in a spirit of prayer and love decides for himself what is the right course of action to take. **1962** *Dict. Moral Theol.* 801/1 Situation morality, carried to extremes, must not be understood as an escape from the heavy burden of moral integrity. **1918** E. S. Farrow *Dict. Mil. Terms* 561 *Situation reports*, reports designed to keep superior officers and neighboring units informed of the progress of events and any important changes in the situation or movement of their own or enemy troops. **1960** *Mag. Fantasy & SF* (N.Y.) Nov. 58/1, I assume that I will be attacked, and decide to file a situation report. **1970** C. Whitman *Death out of Focus* vii. 104 No doubt her precious father had telephoned her and given her the latest situation report. **1977** W. H. Saumarez Smith *Young Man's Country* ii. 33 One regular chore was the 'fortnightly confidential report' to the District Magistrate, which was a general 'situation report' about anything interesting happening in the subdivision. **1967** *Sunday Times* 21 May 7 The intelligence 'situation room' will keep the 15 permanent ambassadors of the NATO countries and their key military commands more fully up-to-date on Russia's political and military build-up than they have ever been before. **1970** M. Kelly *Spinifex* ii. 28 Hopkins was sitting on the edge of a table in the Situations Room. **1976** *National Observer* (U.S.) 28 Aug. 2/5 In the Situation Room of the White House, Secretary of State Kissinger convened the Washington Special Action Group, a subcommittee of the National Security Council, to plot possible U.S. responses to the deaths.

situational (sitiu,ē¹·ʃənăl), *a.* [f. Situation + -al.] Of or pertaining to a situation or situations; dependent on, determined by, or in relation to position, situation, or circumstances. *situational analysis, logic* (see quot. 1977²); *situational ethics, morality* = *situation ethics, morality* s.v. *situation 11.

1903 *Academy* 27 June 632/1 As situational drama (if we may coin the term) always is rhetorical. **1927** *Observer* 24 Apr. 14/5 The main defect of this book seems to lie in the way in which literary or dramatic or situational clues are allowed to dictate musical judgments. **1935** *Jrnl. Philos.* XXXII. 650 Psychology has an incontestable claim if it will but stake it out and work it properly. By working at a meaningful level—not of physical stimulus and meaningless sensation—its products will be observable meaningful properties of situational things. **1945** K. R. Popper *Open Soc.* II. xiv. 90 The method of applying a situational logic to the social sciences is not based on any psychological assumption concerning the rationality ..of 'human nature'. **1949** M. Mead *Male & Female* xiv. 286 It [*sc.* the dating pattern] defines the relationship between a male and a female as situational. **1952** *Essays in Criticism* II. 95 The situational analogies are clear enough by the end of the first episode. **1959** J. L. M. Trim in R. Quirk et al. *Teaching of English* iii. 87 Nouns, principal verbs, adjectives and adverbs are indefinite in number and therefore subject to primarily situational constraints. **1968** *Meta* XIII. 16 Situational meaning reflects the influence of context on utterances. Asking someone about his troubles is likely to produce different responses in a bank and in a hospital regardless of the speaker's intention. **1969** *Observer* 21 Dec. (Colour Suppl.) 38/3 As for sin—situational ethics could take care of that. **1972** K. R. Popper *Objective Knowl.* iv. 179 By a situational analysis I mean a certain kind of tentative or conjectural explanation of some human action which appeals to the situation in which the agent finds himself. **1975** *Language for Life* (Dept. Educ. & Sci.) x. 157 It is this 'situational context', as a linguist would term it, that calls for improvisation. **1977** J. D. Douglas in Douglas & Johnson *Existential Sociol.* i. 14 Man is fundamentally grounded, situational—existential. **1977** in Bullock & Stallybrass *Fontana Dict. Mod. Thought* 575/1 *Situational analysis; situational logic*.., an approach to the explanation of social action in which a detailed reconstruction of the circumstances of action (incuding both objective conditions and the participant's aims, knowledge, beliefs, values, and subjective 'definitions' of the situation) is taken as a basis for hypothesizing rational courses of action for the individual involved, through which their observed behaviour may be rendered intelligible. **1978** J. M. Gustafson *Protestant & Roman Catholic Ethics* ii. 48 Rahner's criticisms warned against the radical extension of situational morality. **1980** *English World-Wide* I. i. 3 Relating linguistic to situational factors.

Hence **situa·tionalism** = *situationism 2; **situa·tionalist** *a.*, of or pertaining to situation ethics; **situa·tionally** *adv.*, with respect to situation; in a situational manner.

1935 *Word Study* Feb. 1/2 *Situationally* is a regular adverbial formation from the adjective situational. **1939** P. Christophersen *Articles* 38 The article is of course situationally determined here. **1964** R. H. Robins *Gen. Linguistics* 191 A sentence is by definition grammatically complete (the alleged 'incomplete' or 'elliptical' situationally tied sentences are complete in those situations). **1970** W. K. Frankena in Pahel & Schiller *Readings in Contemp. Ethical Theory* 542 Views variously referred to as antinomian,..existentialist, situationalist, or contextualist. **1971** N. H. G. Robinson *Groundwk. Christian Ethics* ix. 242 Nor indeed may we expect any other outcome, unless..the logical successor to Bonhoeffer is to be found either in secularization or in situationalism. **1977** J. D. Douglas in Douglas & Johnson *Existential Sociol.* i. 60 There are other reasons why we see experience as necessarily problematic, as necessarily free and situationally contingent. **1977** J. M. Johnson in *Ibid.* v. 161 This extreme situationalism is perhaps best illustrated by Zimmerman and Pollner's (1970) discussion of the *occasioned corpus*. **1979** *Guardian* 9 June 10/3 Since the Bible was written in a very different age from ours, its commands must be interpreted situationally.

situationer (sitiu,ē¹·ʃənəɪ). [f. Situation + -er¹.] In *Journalism*, an article or report constituting a general essay on a situation.

1959 *Observer* 15 Nov. 20/5 I.T.V.'s *This Week*..was struggling to condense an African situationer—ranging from Oxford undergraduates boycotting South African sherry to forecasts of the future of Federation and the White Highlands—and a profile of Mao Tse-tung into a few minutes each. **1972** D. Bloodworth *Any Number can Play* xi. 93 Ivansong had written and cabled a one-thousand-word situationer on Mekong. **1977** *Radio Times* 2–8 July 13/2 Newspapers..had far more word-space than radio..so newspapers found it easier to accommodate 'situationers'.

situationism (sitiu,ē¹·ʃəniz'm). [f. Situation + -ism.] **1.** The revolutionary ideas relating to culture associated with the Situationist International (see quot. 1971 s.v. *situationist *sb.* 1).

1964 *Times Lit. Suppl.* 3 Sept. 781/4 Our International ..coming after the development both of our philosophy and of our art, at once refuses to proclaim any sort of doctrine and rejects the term 'situationism' as used only by enemies of the situationist programme. **1973** *Listener* 2 Aug. 152/1 The Angry Brigade['s]..communiqués suggest..a combination of situationism and Syndicalism. **1978** *Radio Times* 28 Jan.–3 Feb. 15/3 One philosophical strain, peculiar to developed countries, was 'Situationism'. At its simplest, its followers believe the working class has been bamboozled out of its legitimate rights by a capitalist conspiracy that 'appropriated' trade unionism and socialism, then 'laundered' and returned them as harmless institutions.

2. Adherence to situation ethics.

1966 J. Fletcher *Situation Ethics* i. 29 There are various names for this approach: situationism, contextualism, occasionalism, circumstantialism, even actualism. **1977** A. Kolnai *Ethics, Value, & Reality* vii. 145 The wide variety of reductionist and constructivist types of Ethics

ranging, say, from utilitarianism to prescriptivism or from metaphysical perfectionism to situationism.

situationist (sitiu,ē¹·ʃənist), *a.* and *sb.* [f. Situation + -ist; in sense A. 1 ad. Fr. *situationniste.*] **A.** *adj.* **1.** Of or pertaining to certain revolutionary views about the situation of man in modern culture (see quot. 1971, sense B. 1 below); *Situationist International*, a movement started in Paris in the 1950s to promote these views.

1958 *Archit. Rev.* CXXIV. 1/2 Snap judgments on the publications of the Situationist International had best be restrained until the documents have been frisked for hidden persuaders. **1963** *Listener* 31 Jan. 202/2 What she has to say about the uses of diversity seems to derive as uniquely from this particular urban scene as does the Situationist vision from the psychogeography of Paris. **1975** *Observer* (Colour Suppl.) 13 July 26/1 The sergeant discovered that the word 'spectacles' was a concept, an emblem, almost, of a group subscribing to the views of the so-called Situationist International. **1980** *Times Lit. Suppl.* 4 Apr. 387 The walls and statues of the Sorbonne were plastered with posters of Marx, Lenin, Che and Mao, with situationist slogans and Red Flags.

2. Dependent on or determined by circumstances; situationalist.

1970 G. Greer *Female Eunuch* 328 Women's revolution is necessarily situationist. **1972** *Times Lit. Suppl.* 19 May 580/4 His own view involves contextual considerations without being situationist. The relevant context is that of a particular social system.

B. *sb.* **1.** An adherent of the Situationist International or of situationism.

1963 *Listener* 31 Jan. 201/1 The Situationists are best-known as one of the most subversive anti-art groups of the post-war epoch. **1964** *Times Lit. Suppl.* 3 Sept. 781/5 True situationists are much more strongly opposed to all the prevailing mechanisms of culture and information. **1971** R. Gombin in Apter & Joll *Anarchism Today* 19 For the situationists, the bureaucratic system of industrial society has considerably increased the sum total of the exploitation and repression of man... The tremendous development of science and technology has led to the individual being completely taken over by the system; the individual is no more than a commodity..manipulated by the specialists in cultural repression: artists, psychiatrists,..sociologists and 'experts' of all kinds. To fight against a 'spectacular' society, in which everything is treated as a commodity and in which creative energy spends itself in the fabrication of pseudo-needs, one must attack on all fronts simultaneously. **1977** *It* May 5/1 Debord was (is?) a Situationist—a member of perhaps the most radical group to emerge in France in the years approaching the 1968 eruption: they were radical in the sense that they explored most deeply the critique of modern industrial society, which formed the ideological basis for the French upheaval.

2. An upholder of situation ethics.

1966 J. Fletcher *Situation Ethics* i. 26 The situationist follows a moral law or violates it according to love's need.

situla (si·tiulă). *Archæol.* Pl. **situlae, -las.** [L., = bucket.] Any of various bucket-shaped vessels. Also *attrib.*

1897 *Knowledge* 1 Oct. 229/1 Situla (Bucket), of Apulian fabric, with scene representing Dionysos espousing Ariadne. **1905** *Brit. Mus. Guide Early Iron Age* 14 The succeeding (iron-sword) period..is richly represented by articles decorated in the *situla* style. Such is the name given to a method of ornamenting vessels of the bucket-type..by means of horizontal bands. **1928** D. Randall-MacIver *Italy before Romans* 61 An early example of the situla or bronze bucket for which Bologna no less than Este was to become famous in future years. **1942** *Oxoniensia* VII. 45 (*caption*) Everted rim; general form approaching carinated situlas of Long Wittenham type. **1970** *Ashmolean Mus.: Rep. Visitors 1969* 14 Luristan bronze situla decorated with a feast scene. **1972** *Times* 18 May (Egypt Suppl.) p. iv/6 (*caption*) A *situla*, or temple ritual vessel.

Hence **si·tulate, situ·liform** *adjs.*, having the form of a situla.

1937 *Oxoniensia* II. 26 The Iron Age A2 wares also call for some remark. The situliform jar passes through the usual stages of degradation, and finger-tip ornament goes out of use. **1945** *Proc. Prehistoric Soc.* XI. 32 Shouldered and situlate vessels, either plain or with simple finger-tip ornament. **1946** *Ibid.* XII. 125 A tall situliform urn which may owe some features to Iron Age 'A' forms. **1967** *Antiquaries Jrnl.* XLVII. 181 The rim of a coarse, situlate jar with short upright rim and weak shoulder of Early Iron Age affinities.

sit-up, *sb.* and *a.* Add: **A.** *sb.* **2. a.** An act of sitting up.

1843 R. Carlton *New Purchase* I. ix. 64 This sit-up we instantly performed—as well, at least, as we could.

b. A physical exercise in which the upper half of the body is raised to a sitting from a supine posture. Also *attrib.*

1955 V. Fallon *Figure Correction & Beauty for You* xx. 112 Sit-up—2 sets of twenty repetitions. **1960** J. Hewitt *Yoga* iii. 56 The well-known 'sit-up' exercise... Paschimatanasana differs from an ordinary sit-up in that the movement is continued until the face comes close to the knees. **1971** A. A. Michele *You don't have to Ache* 93 Do five to ten sit-ups in a curling fashion bringing your body straight forward. Then do five sit-ups twisting your body to the left, and five twisting your body to the right. **1977** J. F. Fixx *Compl. Bk. Running* xv. 179 To strengthen your stomach muscles,..do twenty or so sit-ups with your knees bent.

B. *adj.* Also, at or against which one sits up.

1960 *Times* 16 Mar. (Canberra Suppl.) p. v/4 A tavern, with a sit-up bar. **1980** L. Birnbach et al. *Official Preppy Handbk.* 93 To Buy (or otherwise obtain as soon as possible:..Sit-up pillow.

sit-upon. Add: **2.** The buttocks, the posterior.

1920 C. A. W. Monckton *Some Experiences of New Guinea Resident Magistrate* xxv. 302 You have a big boil on your sit-upon. **1955** E. Waugh *Officers & Gentlemen* ii. i. 169 Mrs Stitch immediately sat in the place he vacated. 'Hot-sit-upon,' she remarked. **1976** 'J. Welcome' *Grand National* viii. 130 I've got something to eat off and to put me sit-upon on in the evenings.

situs. Restrict *rare* to gen. sense in Dict. and add: **2.** *Law.* **a.** Chiefly *U.S.* The place to which for purposes of legal jurisdiction or taxation a property is deemed to belong.

1834 J. Story *Commentaries on Conflict of Laws* xiv. 462 Moveables are, for many purposes, to be deemed to have no *situs*, except that of the domicil of the owner. **1884** R. Desty *Amer. Law of Taxation* I. v. 97 The legislature has the power to fix the *situs* of property for purposes of taxation. **1926** *Pacific Reporter* CCXLVIII. 341/1 Actual situs of personal property is necessary for taxation. ..In order to tax the cars in controversy as personal property, they must have actual situs at common law in this state. **1956** *All England Law Reports* 19 Jan. 134 If, however, the situs of the debt be German, he submits that the moratorium law is confiscatory. **1970** *Southwestern Reporter* CCCCXLV. Ser. ii. 57/1 Bulldozers, which were regularly moved from one temporary location to another, did not have actual situs of their own, and were not subject to ad valorem taxation. **1977** Johns & Greenfield *Dymond's Capital Transfer Tax* xxiii. 491 Seven of the agreements provide *situs* codes (for determining the locality of assets).

b. *U.S.* A work-site, esp. (*common situs*) one occupied by two or more employers. Freq. *attrib.*

1950 *Fed. Suppl.* XCI. 698/2 The Act..was intended to keep the situs of a labor dispute confined to actual functions of the parties involved... It is 'stranger picketing', i.e., picketing in aid of a secondary boycott. **1952** *Cornell Law Q.* Winter 247 In such 'common-situs cases' additional criteria necessarily must be invoked. **1959** *Missouri Law Rev.* Jan. 89 Common-premises or common-situs picketing occurs where a labor organization pickets premises where the employees of two or more employers are working and the labor organization has a dispute with only one of the employers. **1977** *Time* 4 July 47/1 With smoothly coordinated pressure, business lobbyists have managed..to defeat organized labor's bid to pass a common situs picketing bill that would have allowed a single union to shut down a construction site.

situs inversus (səi·tǒs, sī·tǒs invɔ̄·ɪsǒs). *Med.* [L., in full *situs inversus viscerum* inverted disposition of the internal organs.] The condition in which the organs of the body are transposed through the sagittal plane (so that the heart lies on the right side, etc.).

1896 J. T. Whittaker in T. L. Stedman *Twentieth Century Pract. Med.* IV. 59 Malformations of individual organs are very rare in typical situs inversus. **1966** Wright & Symmers *Systemic Path.* I. xv. 484/2 Situs inversus due to malrotation of the foetal gut..is very rare. **1976** *Sci. Amer.* Sept. 68/2 Patients exhibiting situs inversus frequently suffer from chronic sinusitis and bronchitis.

situtunga (situtǒ·ŋgă). Also **sitatunga.** [Swahili.] A medium-sized brown or greyish antelope, *Tragelaphus spekei*, found in east and central Africa, and distinguished by elongated, splayed hooves that enable the animal to walk on marshy ground, and spiral horns in the male. Also *attrib.*

1881 F. C. Selous *Hunter's Wanderings Afr.* x. 158 These men told me that in some thick beds of reeds near their town were some situtunga antelope. **1899** [see *harnessed *ppl. a.* 4]. **1920** G. D. H. Carpenter *Naturalist on Lake Victoria* iv. 80 The thunderous snortings of hippos, the muffled bark of the Situtunga, break in. **1947** L. Hastings *Dragons are Extra* viii. 191 The highly specialised sitatunga, lord of the swamps. **1949** *Cape Argus* 11 Aug. 7/6 Lions in the swamp appeared to live on Situtunga bush buck. **1955** P. A. Buxton *Nat. Hist. Tsetse Flies* vi. 144 On some uninhabited islands in Lake Victoria, Uganda, the sitatunga antelope (*Limnotragus s. spekii*) multiplies and with it *G. palpalis.* **1973** *Stand. Encycl. S. Afr.* IX. 650/1 The sitatunga has coarse, shaggy hair.

sit vac: see *sit *sb.*²

Sitwellian (sitwe·liăn), *a.* and *sb.* [f. the name *Sitwell* (see below) + -ian.] **A.** *adj.* Of, pertaining to, or characteristic of the writers Edith Sitwell (1887–1964), and her brothers Osbert (1892–1969) and Sacheverell (b. 1897). **B.** *sb.* An admirer of the Sitwells.

1923 A. Bennett in *Adelphi* Aug. 237 This book..is a characteristically Sitwellian beauty. **1927** *Observer* 5 June 4 The Sitwells are known to everyone who has even a casual acquaintance with modern literature, though many who talk of them seem to have read about their doings rather than studied them in their own works. This is a necessary consequence of the Sitwellian methods of publicity. **1937** *Times Lit. Suppl.* 1 May 322/1 Mr. Courtenay's extremely modern simile with its oozy vowels —Sitwellian, surely. **1952** *Scrutiny* Oct. 6 Rajan's examination of Milton's verse keeps turning into Sitwellian clap-trap. **1960** V. Sackville-West *Let.* 23 Mar. in H. Nicolson *Diaries* (1968) 382 Edith has built up her personality in many fortuitous ways—her strange appearance,..and all the Sitwellian legend. **1978** J. Pearson *Façades* xi. 196 The society was conducted with true Sitwellian panache. *Ibid.* 212 Thomas Balston at Duckworth's—himself an enthusiastic Sitwellian—had taken over from Grant Richards as the trio's publisher.

So **Si·twellism,** the style or behaviour of the Sitwells.

1927 R. L. Mégroz *Three Sitwells* vi. 105 The phenomenon of Sitwellism. **1932** F. R. Leavis *New Bearings in Eng. Poetry* ii. 73 The opposition to the Georgians was already..(just after the war) Sitwellism. **1981** V. Glendinning *E. Sitwell* v. 80 The Sitwells defence was attack... Sitwellism at its silliest and most inflated.

‖ **Sitzfleisch** (zi·tsflaiʃ). Also **sitzfleisch.** [Ger., f. *sitz-en* to sit + *fleisch* flesh.] The ability to endure or persist in some activity.

a **1930** D. H. Lawrence *Lovely Lady* (1932) 165 They simply hadn't enough *Sitzfleisch* to squat under a bho-tree. **1971** *Atlantic Monthly* May 106 It takes not only special training but a liberal endowment of *Sitzfleisch* to hear one of his [*sc.* Messiaen's] pieces out from one end to the other. **1975** *Harpers & Queen* May 127/3 Lenny hadn't got the patience, the concentration, the sitzfleisch. **1977** *Time* 26 Dec. 31/1 Some of the games now filtering into the general consciousness are distance runs indeed, taking anything from several hours to several months to play, and requiring formidable *Sitzfleisch* (German for sitting flesh).

‖ **Sitz im Leben** (zi·ts im lēbən). *Theol.* Also hyphened. [Ger., lit. 'place in life'.] In Biblical criticism, the circumstances in which a tradition developed, considered as determining the form of that tradition.

1934 M. Dibelius *From Tradition to Gospel* i. 7 The categories enable us to draw a conclusion as to what is called the 'Sitz im Leben', i.e. the historical and social stratum in which precisely these literary forms were developed. **1955** D. E. Nineham *Stud. Gospels* 230 They [*sc.* the Form-critics] have shown the importance of the factor they call *Sitz-im-Leben* in the preservation of the material included in our gospels. **1956** *Scottish Jrnl. Theol.* IX. 403 Events in Jesus' life are seen by Mark from the *Sitz im Leben* which they have in the early Church. **1976** *Expository Times* LXXXVII. v. 139/2 We may therefore accept without hesitation the doctrine of *Sitz im Leben*, provided that we clearly understand what it is and what are its limitations.

sitzkrieg (si·tskrīg). Also **Sitzkrieg.** [Formed on the analogy of *Blitzkrieg, as if f. G. *sitz-en* to sit + *krieg* war.] A war, or part of a war, marked by a (relative) absence of active hostilities; *spec.* that phase of the war of 1939–45 lasting from September 1939 to May 1940; a 'phoney war'. Also *fig.*

1940 *N.Y. Times* 21 Feb. 4/7 The R.A.F. referred to the war as a 'Sitzkrieg', which it translated as 'sit-down war'. **1940** *Newsweek* 11 Mar. 28/2 The European 'sitzkrieg' has definitely taught the usually turbulent Balkan states that their safety and independence can best be attained through at least a modicum of unity. **1943** *Sun* (Baltimore) 12 Nov. 1/6 (*heading*) Temporary 'sitzkrieg' on Garigliano sector. **1954** W. K. Hancock *Country & Calling* vii. 187 The *sitzkreig* had at least given me a golden chance to visit West Africa and had thereby enabled me to finish my book. **1967** B. B. Gilbert *Britain Since 1918* iv. 118 As the German conquest of Poland had been termed a '*blitzkrieg*', a lightning war, the new phase of the war that lasted from the end of September [1939] until the beginning of April was frequently designated a '*sitzkrieg*'. **1970** *Courier-Mail* (Brisbane) 22 Dec. 3/1 (*heading*) Old men prepare for sit-in battle... Thirty-six old men..yesterday settled down to a 'sitzkrieg' in the Battle of Rosebank—the Glebe, Sydney, aged men's home. **1980** *Christian Sci. Monitor* (Midwestern ed.) 4 Dec. 6/1 Most observers think the two armies will become involved in a 'sitzkrieg', meaning a long military waiting game.

sitzmark (si·tsmāɪk). *Skiing.* [App. f. G. *sitz-en* to sit + Mark *sb.*¹] The impression in the snow made by a skier falling backwards on his posterior; an act of so falling. Hence **si·tzmark** *v. intr.*, to fall in this manner.

1935 *Punch* 13 Feb. 176/2 One seems to see as in an inspired flash..the insidious vileness of a sitzmark. [*Note*] Deliberately sitting down in the snow when no other method of stopping seems possible. Very degrading. **1935** *Sierra Club Bull.* Feb. 7 One should be ashamed to make a long descent by 'sitzmarking' at every turn. **1947** F. S. Smythe *Again Switzerland* ii. 22 While to fall backwards and *sitzmark* is the hallmark of the craven. **1964** *Harper's Bazaar* Nov. 140/2 The wintersporter who prefers..a bar stool to sitting repeatedly in a *sitzmark*. **1973** P. A. Whitney *Snowfire* xvi. 306, I sat down and let my skis go out from under me, sliding a little way on the seat of my pants, leaving sitzmarks behind me. **1977** *Globe & Mail* (Toronto) 16 Nov. 37/6 My rump has had much practice from last winter's skiing—sitzmarks were almost second nature to it.

Siva (ʃī·va, sī·vă). Also **Shiva.** [a. Skr. *śiva*, lit. 'the auspicious one'.] **1.** The third deity of the Hindu triad, to whom are attributed the

powers of reproduction and dissolution. Also *transf.*, a representation of this deity.
1788 *Asiatick Res.* I. 248 Siva is believed to have *three* eyes. **1862** J. B. Speid *Our Last Years in India* iv. 73 The horrid Churruckpuja..festival..in honour of Kali, the consort of Siva, the God of Destruction. **1903** J. C. Oman *Mystics, Ascetics, & Saints of India* vii. 110 Siva, regarded by his special followers as the Supreme Being, commands their adoration in many different and even seemingly contradictory characters. **1931** *Times Lit. Suppl.* 28 May 429/2 Soon he had brought a haunting fear to the minds of several people, turning upon the possession of an old bronze dancing Siva. **1963** *Times* 23 May 14/6 They [sc. deformed cattle] are taken as avatars of..the riding bull of the Lord Shiva. **1977** *Jrnl. R. Soc. Arts* CXXV. 583/2 The famous dancing Siva from Ujjain in the Gwālior Museum.
2. *attrib.* and *Comb.*, as *Siva-complex*, *-worship*.
1876 *Encycl. Brit.* V. 243/1 The Mahávinayaka Peak..has been consecrated for ages to Siva-worship by ascetics and pilgrims. **1937** H. Nicolson *Helen's Tower* ix. 191 Was it (as the Viennese might further contend) some Siva-complex, some dread of the destruction-principle? **1967** Spate & Learmonth *India & Pakistan* (ed. 3) vi. 174 Artistically, the best things are a handful of..seals. These..carry an undeciphered script, and their symbolism suggests that some elements in Hinduism, notably Siva-worship, may stem from Harappan culture.

Sivaism. Add: Also **Shivaism.** Cf. *S(h)aivism* s.v. *Saiva.
1931 *Times Lit. Suppl.* 19 Mar. 228/2 Vishnuism and Shivaism. **1962** A. Huxley *Island* xi. 178 The local brand of Mahayana Buddhism, with a bit of Shivaism, probably, on the side.

Sivaist (ʃi·vă,ist, sī·vă,ist), *a.* [f. as prec. + -ist.] Of or pertaining to the worship of Siva.
1937 M. Covarrubias *Island of Bali* i. vii. 173 It was within this period, from the seventh to the ninth centuries,..that the finest monuments of Java were built, the Buddhist *Borobudur* and the Sivaist *Lora Djongrang* in Prambanan.

Sivaite. Add: Also **Shivaite, shivaite.** Cf. *S(h)aivite* s.v. *Saiva.
1958 A. Huxley *Let.* 22 June (1969) 850 The process of turning old Shivaite-cum-Mahayana-Buddhist society into something combining the best features of East and West was inaugurated in the eighteen-forties by a Scottish surgeon. **1962** —— *Island* vi. 82 We're still Buddhists or Shivaites. **1970** *Guardian* 10 Aug. 9/4 A 1,000-year-old shivaite temple..on the border between Cambodia and Thailand.

Sivan (sī·vān). Also 4–5 **Ciban, Siban, Siwan.** [a. Heb. *sīwān*.] The ninth month of the Jewish year, though named third in the traditional month-list, corresponding to the latter part of May and the earlier part of June.
c **1382** Bible (Wycliffe) *Esther* viii. 9 The thridde moneth, that is clepid Ciban [*later text* Siban], that is June. **1535** *Ibid.* (Coverdale), In the thirde moneth, that is the moneth Siuan. **1737** W. Whiston tr. *Josephus' Jewish War* iii. vii. § 29 This fight happened upon the twentieth day of the month Desius (Sivan). **1816** J. Allen *Mod. Judaism* xxi. 386 The feast of Pentecost is on the *sixth* day of the month of *Sivan*, the *fiftieth* of the *Omer*. **1891** M. Friedländer *Jewish Relig.* ii. 393 The Feast of Weeks, the 6th and 7th of *Sivan*, commemorates..an historical event. **1952** *Jewish Q. Rev.* XLIII. ii. 181 On Siwan 7..Moses ascended the mountain. **1977** *Jewish Chron.* 20 May 17/1 Friday, May 20 (Sivan 3), Sabbath begins in London at 8.38.

Siwalik (siwā·lik). Also †**Sewalik; Sivalik.** [Hind.]
a. The name of the southern outlying foothills of the Himalayas, extending from Sikkim through Nepal and India to Pakistan, used *attrib.* with reference to the thick sequence of fluviatile and lacustrine sediments, rich in fossil vertebrates, of which they are composed and the time in the Pliocene (or late Miocene) to early Pleistocene when they were deposited.
1836 Cautley & Falconer in *Asiatic Researches* XIX. xiii. 200 There existed along with the Mastodon, Sivatherium, Fossil Camel, &c. of the Sivalik deposits, a large distinct species of Bear, equalling if not exceeding the largest known of the genus. **1864** *Q. Jrnl. Geol. Soc.* XX. 383 He refers the rocks observed [in the north-western Himalayas] to the following formations:—1. The fluvio-lacustrine series. 2. The Siwalik series... 6. The Palæozoic series. **1902** *Encycl. Brit.* XXV. 466/1 Dr. Dubois..excavated from a bed, considered by him to be of Sivalik formation (Pliocene), a thighbone which competent anatomists decide to be human. **1908** H. B. C. Sollas tr. *Suess's Face of Earth* III. iv. vi. 218 Sediments possibly of Siwalik age. **1955** Brown & Dey *India's Mineral Wealth* (ed. 3) xix. 697 The outer margin of the Himalayas is formed by a continuous fringe of foot-hills built of the rocks of the Siwalik Series. **1975** A. M. Davies et al. *Tertiary Faunas* (ed. 2) II. vi. 359 From the Indian subcontinent the Middle Siwalik succession is of Late Miocene equivalent.
b. *absol.* The Siwalik series or period.
1877 *Rec. Geol. Survey India* X. 121 Upper Siwalik. This division includes the great conglomerates and associated beds which terminate the tertiary series of the country. **1938** *Q. Jrnl. Geol. Soc.* XCIV. 407 The alluvial deposits of the Bannu and Derajat plains probably conceal..a geological record more or less continuous from the latest Siwalik. **1974** *Encycl. Brit. Micropædia* IX. 246/2 Remains of the first clearly hominid forms..are known from the Siwalik. **1978** Mitchell & Reading in H. G. Reading *Sedimentary Environments & Facies* xiv. 464/2 Stratabound deposits of uranium with minor vanadium occur in the late Miocene to early Pliocene Middle Siwaliks.

Siwash (səi·wɒʃ), *sb. N. Amer.* Also **Si-wash, siwash.** [Chinook Jargon, a N. Amer. Fr. dial. form of Fr. *sauvage* (Savage *a.*) in same sense.] **1. a.** An Indian, *spec.* of the North Pacific Coast. Freq. *attrib.* (Now considered perjorative.)
1847 J. Palmer *Jrnl. Trav. Rocky Mts.* 150 Si-wash Indians. *a* **1861** T. Winthrop *Canoe & Saddle* (1883) ii. 18 The three unsavory..mat-haired, truculent siwashes. **1869** [see *Haida *a.* and *sb.*]. **1870** [see *Aleutian *a.* and *sb.*]. **1897** *Outing* XXX. 541/1 As we neared the Narrows other Siwashes in other queer-looking canoes paddled out. **1904** E. Robins *Magnetic North* 293 You soon learn it is the Siwash custom. **1949** *Boston Globe* 15 May (Fiction Mag.) 3/2 The Siwash showed him a poke of coarse gold. **1967** C. L. Evans *Newel Post* 6 He was looking portly in a heavy Siwash sweater, and unselfconsciously wearing the knitted hat to match.
b. *transf.* A name of opprobrium; occas. joc.
1882 *Edmonton Bull.* 3 June 4/3 Does this great chieftain think new settlers are a community of Siwashes or cringing dependants. **1924** C. E. Mulford *Rustlers' Valley* xiii. 158 So-long, you Siwash! **1964** P. Berton *Golden Trail* 23, I wouldn't go across the river on that old Siwash's word.
2. Chinook Jargon, the lingua franca of the North Pacific Coast Indians.
1902 *Skagway Daily Alaskan* 23 Aug. 3/1 The governor was forced back upon his ability to talk siwash, hoping thereby to control the Indian vote. **1908** R. E. Beach *Barrier* 56 Address me in Siwash or in English unless we are alone. **1936** W. B. Mowery *Paradise Trail* 14 That's what Saghelia means in Siwash—the purty land..paradise.
3. *Comb.*, as **siwash camp,** an open camp with no tent; **Siwash duck,** a scoter of the genus *Melonitta*.
1922 *19th Cent.* Feb. 267 At night they would build a 'siwash' camp, digging a big hole in the snow, lining it with green spruce boughs and building up a three-foot wall of green spruce trees for a windbreak on back and sides. **1962** M. E. Murie *Two in Far North* ii. x. 197 We had only about twelve miles to travel from our siwash camp to Tramway Bar. **1911** *Daily Colonist* (Victoria, B.C.) 30 Apr. 10/1, I finally caught Mr. Indian just as he was coming ashore with his ducks, he had about 60 or 70 in the canoe, but they were mostly scoter or what is more commonly called Siwash ducks. **1927** *Blackw. Mag.* Aug. 207/2 He could see when any siwash ducks were on a shallow part of the lagoon. **1966** *Daily Colonist* (Victoria, B.C.) 20 Mar. 11/4 It is a rare occurrence for a Siwash duck, as the species [sc. surf scoter] is commonly called, to be found on such a shoreline.

siwash, *v. N. Amer.* [f. prec.] **1.** *intr.* To camp without a tent, like an Indian.
1938 T. C. Stanwell-Fletcher *Driftwood Valley* (1946) v. 94 Since we can't carry the additional weight of a tent, we'll have to siwash under trees. **1977** *New Yorker* 20 June 64/3 In discrete valleys were a few cabins, and they stayed in them or siwashed (camped on the trail).
2. *trans.* To bar (a person) from purchasing alcoholic drink. *colloq.*
1948 C. W. Holliday *Valley of Youth* 144 It [sc. Painkiller] was in great demand by the old inebriates during the periods when they had been 'Siwashed'—which meant that it was illegal to serve them with a drink over the bar or sell them liquor. **1957** A. R. Barratt *Coronets & Buckskins* 9 Wen a wite man gets so's no one will sell him drinks—well folks say e's been siwashed.
So (sense 1) **si·washing** *vbl. sb.*
1904 *Churchman* 21 May 626, I have a lame shoulder, the result of continuous 'siwashing' and sleeping in the snow. **1938** T. C. Stanwell-Fletcher *Driftwood Valley* (1946) vi. 110 The day after our siwashing trip, we lounged about the cabin, luxuriating in a paradise of warmth and rest. **1962** M. E. Murie *Two in Far North* ii. ix. 194 South Fork Henry no doubt thought we were young fools to be looking forward to a night of siwashing.

six, *a.* and *sb.* Add: **A.** *adj.* **1. d.** *Six Acts* (earlier example); *Six Counties*, the Ulster counties of Antrim, Down, Armagh, Londonderry, Tyrone, and Fermanagh, which have since 1920 comprised the province of Northern Ireland; (cf. *twenty-six counties* s.v. *Twenty-six *a.*); *Six Dynasties*, a collective term for the Chinese dynasties of Ch'en, Eastern Chin, Liang, Liu-Sung, Southern Ch'i and Wu, belonging to the period AD 220–589; freq. used *attrib.* to denote this period of history in China; *Six Nations* (earlier examples).
1834 *Times* 22 Apr. 5/6 Every man that dared to open his mouth against the Castlereagh and Sidmouth despotism must have set the Six Acts at defiance. **1921** *Notes from Ireland* 40/1 The Unionists of the 'Six Counties'. **1922** C. J. C. Street *Ireland in 1921* ix. 226 The House of Commons contained only Unionist members, who were obviously deeply concerned at the position of affairs while anxious to say nothing which might compromise the position of the Six Counties. **1935** *Frontier Sentinel* (Newry) 22 June 4/4 The Six-County Premier. **1949** [see *Englishize *v.*]. **1960** J. Stroud *Shorn Lamb* iv. 49 Suppose..she turns out to be some one-eyed horror in the Six Counties looney-bin. **1974** *Irish Democrat* Nov. 5/2 The significance of the six county election results has been widely debated. **1934** K. S. Latourette *Chinese* I. iv. 155 The fall of the Eastern Chin..is usually said to mark the beginning of the era known to the Chinese as the..Southern and Northern Dynasties, which lasted until 589. Another classification—inclusive of a longer period—employed by Chinese historians is the Six Dynasties, by which are meant the six kingdoms and dynasties between the downfall of the Han and the reunification of China in 589. **1966** F. Schurmann *Ideology & Organization in Communist China* vii. 407 The most ambitious attempt of this sort occurred during the Six Dynasties Period (third to sixth century A.D.). This was a period of serious political and social breakdown. **1973** T. R. Tregear *Chinese* i. 26 Monks from India..so impressed the barbarians of the Six Dynasties..that they adopted Buddhism. **1980** E. Behr *Getting Even* v. 59 A huge 'Six Dynasties' celadon jar and a horseman of the Northern Wei period. **1710** in J. W. Lydekker *Faithful Mohawks* (1938) ii. 28 And as a sure Token of the sincerity of the six Nations, We do..present Our Great Queen with these Belts of Wampum. **1785** T. Jefferson *Notes Virginia* 390 The Mingo or Six-nation Indians. *Ibid.*, The Mingos are a war colony from the six nations.
e. Colloq. phr. *six feet under* and varr.: dead and buried; in or into the grave.
1942 Berrey & Van den Bark *Amer. Thes. Slang* § 117/19 *Dead and buried*..six feet under. **1968** J. Sangster *Touchfeather* xv. 184 Bill didn't realise it, but he was as dead now as he would be when they lowered him six feet under. **1976** A. Price *War Game* v. 107 He never cared for nobody born... He never did, and he never will. Not till he's six foot under. **1979** J. Gerson *Omega Factor* 78 In Islay..we make sure the dead are stiff and cold and six feet under.
2. g. Naut. slang phr. *six upon four*: (on) short rations (i.e. four men's food shared between six men). ? *Obs.*
1829 D. Jerrold *Black-ey'd Susan* ii. ii. 31 May you live a life of ban-yan days, and be put six upon four for't. **1843** J. F. Cooper *Ned Myers* 78 As to food, we were kept 'six upon four' the whole time I was prisoner. *Ibid.* 86 Put at 'six upon four' again.
h. *Cricket.* With omission of *runs.* (i) Six runs scored by striking the ball clear over the boundary. Cf. sense 6 of the sb., below.
1857 T. Hughes *Tom Brown's School Days* ii. viii. 392 When you or Raggles hit a ball hard away for six, I am delighted. **1951** G. Brodribb *All Round Wicket* vi. 39 A lusty hit for six gives many people intense pleasure... There were..many outstanding feats of six-hitting. **1979** *Daily Tel.* 29 May 17/3 The ball was hit for six..into the vicarage grounds.
(ii) Also *transf.* and *fig.* in various colloq. phrases: *to knock* (someone) *for six*, to wrench from a state of composure; to defeat (soundly); to astonish; also in extended uses, and with inanimate obj. Similarly, *to go for six* (see also quot. 1943); *to hit for six*: see *Hit *v.* 8 f.
1902 J. Milne *Epistles of Atkins* vi. 107 'It knocked me for six', is the statement we have about a bullet in the knee. **1934** A. Berkeley *Panic Party* iv. 69 It's a crashing bore..to think of those dim cads knocking us for six like this, but..it's no use getting strenuous about it. **1941** L. A. G. Strong *John McCormack* viii. 129 John..hurled his Santuzza from him with such vigour that she went for six, landing in the wings on the far side of the stage. **1943** C. H. W. Jackson *It's a Piece of Cake* 32 *Gone for six*, killed, missing. **1949** R. Gow *Ann Veronica* i, in *Plays of Year 1949* I. 240 You're just like an angel yourself sitting there. You knock me for six, if I may borrow a sporting metaphor. **1955** *Times* 6 Aug. 7/4 It is a song that knocks for six the illusion..that little creatures probably know by instinct how to do their stuffs. **1973** 'B. Graeme' *Two & Two make Five* vi. 58 A glass of his home-made cider..knocks you for six.
i. Chiefly as Fr. phr. *Les Six* (le sis) a Parisian group of six composers, Louis Durey (1888–1979), Arthur Honegger (1892–1955), Darius Milhaud (1892–1974), Germaine Tailleferre (b. 1892), Georges Auric (b. 1899), and Francis Poulenc (1899–1963), formed after the war of 1914–18, whose music represents a reaction against romanticism and impressionism.
[**1920** H. Collet in *Comœdia* 16 Jan. 2/6 Les artistes conscients reconnaissent en Satie un maître. Voyons maintenant ce que valent les 'Six' qu'il précéda sur la voie nouvelle.] **1927** *Grove's Dict. Mus.* (ed. 3) II. 662/1 Although a member of the group of French musicians known as the 'Six', he [sc. Honegger] is of Swiss parentage. **1934** C. Lambert *Music Ho!* iii. 194 Chabrier..may be considered..the father of the post-war movement associated with the names of Les Six. **1952** B. Ulanov *Hist. Jazz in Amer.* (1958) x. 113 The lessons the composer had learned from..the music of Ravel and *les six* were poorly applied. **1978** P. Griffiths *Conc. Hist. Mod. Music* vi. 72 Les Six did not exist as a group for more than a few years.
j. *the Six*, the group of countries (Belgium, France, the German Federal Republic, Holland, Italy, and Luxembourg) which were the original members of the European Economic Community from 1958 until the admission of others in 1973.
1957 [see *European *a.* 1 c]. **1958** *Economist* 18 Oct. 207/1 The Six are the European Community, forerunner of the continental political union..that is to be when General de Gaulle is gone. **1981** *Times* 30 June

7/1 In 1977 Britain..threatened to bring down the whole European edifice built up painstakingly by the original Six.
B. *sb.* **2. b.** (*b*) *spec.* a group of six Brownie Guides or Cub Scouts.
1916 R. Baden-Powell *Wolf Cub's Handbk.* ii. 59 Each Six is called after a Wolf by its colour. **1920** —— *Brownies or Blue Birds* ii. 11 The Brownies are divided into parties of six; each 'Six' is under the leader who is called a 'Sixer'. **1972** Trinky & Peters *Cub Scout Games* 26 Cubs enjoy Six relays and will compete fiercely for their Six.
3. i. Substitute for def.: Large flower-pots, six of which are formed from a cast of clay. (Earlier example.)
1824 J. C. Loudon *Encycl. Gardening* ii. 327 The Flower Pot, is a cylindrical tapering vessel of burnt clay, with a perforated bottom, and of which there are ten sorts, distinguished by their sizes thus: the..third [size has] 6 [to the cast, and are called] sixes [, being] 9 [inches diameter] 8 [inches deep].
k. *U.S. slang.* A prison sentence of six months.
1844 J. H. Ingraham *La Bonita Cigarera* vi. 27/1 I've served two sixes in her, and that's enough for me to give to the service o' my country! **1928** J. O'Connor *Broadway Racketeers* xvii. 182 Even if its only a sixer in the pen, too many sixes are bad for the health.
l. A six-cylinder motor car or engine.
1920 *Motor Man.* (ed. 23) 19 The chief constructional difference between the six and the four is in the crankshaft. **1977** *Chicago Tribune* 2 Oct. xiii. 18/2 The use of fours, sixes, and small V-8's is more prevalent in 1978.
4. a. Also *old six* (see quot. 1890).
1890 Barrère & Leland *Dict. Slang* II. 98 *Old six* (common), old ale at sixpence a quart. *Spoken*—Look what I've got to do tonight! There's fourteen 'pubs' on my beat... That means that I've got fourteen pints of old six to get down me. **1898** J. D. Brayshaw *Slum Silhouettes* 155 Tell 'er ter send for an hexty pint of old six. *Ibid.* 156 The foaming jug of 'old six' was placed on the table.
6. *Cricket.* A score of six runs made by striking the ball clear over the boundary; a shot which achieves this.
1920 D. J. Knight in P. F. Warner *Cricket* i. 34 If accurately timed—remember that flick of the wrists at the psychological moment—there goes the easiest six in cricket to your credit! **1933** H. Larwood *Body-Line?* 8 Mr. Wyatt again bats well... He ends a fiery rubber with a crashing six! **1949** J. Symons *Bland Beginning* 217 The ball sailed high into the air... The umpire signalled a six. **1957** G. Lyttelton *Let.* 27 June in *Lyttelton-Hart-Davis Lett.* (1979) II. 120 His [sc. F. Trueman's] three sixes off consecutive balls were worth seeing. **1976** J. Snow *Cricket Rebel* 16 Shortly after my six-hitting effort the school was sold and the ground used for building.
7. *the deep six*: used in various slang phrs. to denote death or the grave (perh. from the custom of burial at sea, at a depth of six fathoms); also *fig.* Hence as *v. trans.*, to submerge in water; also *fig.*, to reject, abandon, conceal. orig. and chiefly *U.S.*
1929 M. A. Gill *Underworld Slang* 4/2 *Deep six*, grave. **1947** S. Palmer *Miss Withers Regrets* (1948) xii. 135 My old lady went over the hill with my bank account before I was out of boot camp. I'd have given her the deep-six if I coulda got a furlough. **1966** T. Pynchon *Crying of Lot 49* iii. 50 Attack, retaliation, both projectiles deep-sixed forever and the Pacific rolls on. **1973** *Times* 26 July 8/3 Mr Dean has testified that Mr Ehrlichman told him to 'deep six' the documents... He said he threw the documents into the Potomac river. **1975** *Publishers Weekly* 28 July 116/1 They discovered that Americans..have been hung up on the wisdom of Franklin's 'Poor Richard's Almanack'..: 'work hard, be thrifty, don't borrow.'.. They tell their readers to 'deep six' Poor Richard and put his advice into reverse. **1976** *Listener* 28 Oct. 524/3 The more serious charge from Dean that he [sc. President Ford] tried to 'deep six' the Watergate investigation. **1977** *Islander* (Victoria, B.C.) 21 Aug. 2/3, I heard later that Bruce had taken them [sc. three guns] out into the saltchuck and deep-sixed them. **1978** *Sunday Mail Color Mag.* (Brisbane) 7 May 13 'Fraid the rest of the treasure and your gear have gone for the deep six!..It's over three hundred feet to the bottom of that trench!
C. 1. a. *six-ball* (over), *six-bit* (code, row), *six-cylinder* (engine, motor vehicle) (also *absol.*); *six-piece* (band), *six-water* (grog) (earlier example).
1910 *Blackw. Mag.* Jan. 97/2 He covered something between 250 and 300 yards in the course of each six-ball over. **1979** *Times* 29 Nov. 19/1 England..made 211 for eight in their 50 six-ball overs. **1964** T. W. McRae *Impact of Computers on Accounting* i. 9 A computer using a 'six-bit' binary code. **1964** C. Dent *Quantity Surveying by Computer* vi. 72 These word trains, forming blocks of data, are directed to the tape via the tape control unit, which breaks them into six-bit rows. **1905** G. B. Shaw in *Grand Mag.* Feb. 116 An old crock of a 1904 six-cylinder car. **1938** *New Statesman* 22 Jan. 142/2 At one period the Morris interests were building a couple of dozen six-cylinders, all different. **1977** 'D. Rutherford' *Return Load* iv. 83 Under the shiny cellulose lid..lurked a six-cylinder engine. **1948** A. Baron *From City from Plough* 39 On the dais at the end of the ballroom a six-piece band thumped and brayed. **1959** *Encounter* Oct. 49/1 In the dining-room a six-piece dance-band plays. **1829** D. Jerrold *Black-Ey'd Susan* ii. i. 27 May I be put on six-water grog for a lubber.
2. *six-holed, -membered, -pointed* (earlier example), *-toothed* (earlier example), *-wheeled* (later example); *six-belted, -striped*.
1908 R. South *Moths Brit. Isles* II. 358 Six-belted Clearwing... The body of the male has seven yellow belts, and that of the female one less. **1958** W. J. Stokoe *Caterpillars Brit. Moths* II. 244 The Six-belted Clearwing..*Dipsosphecia scopigera*. The haunts of this moth are on chalk downs. **1955** E. Pound *Classic Anthol.* ii. 135 With six-holed flutes That were bamboo shoots. **1956** I. L. Finar *Org. Chem.* II. viii. 252 Bicyclic monoterpenes contain a six-membered ring and a three-, four-, or five-membered ring. **1978** A. J. Birch in *Further Perspectives Organic Chem.* (Ciba Symposium) 6 A similar cyclization to a six-membered ring from the C_{20}-precursor geranylgeranyl pyrophosphate would not yield this group. **1764** T. H. Croker et al. *Compl. Dict. Arts & Sciences* I. s.v. *Diamond*, These [diamonds] the jewellers call six-pointed stones. **1907** R. South *Moths Brit. Isles* I. 227 The Six-striped Rustic..is also generally distributed over our islands. **1742** W. Ellis *Mod. Husbandman* Sept. xxvi. 124 The Sheep..are generally six-toothed Wethers. *a* **1976** A. Christie *Autobiogr.* (1977) viii. i. 372 It was at this time of day when the big six-wheeled cars most often went off the track.
3. *six-bitter* [*Bit *sb.*[2] 8 b], *-roomer* (earlier example), *-seater*.
1928 *Coast Guard* Aug. 6/1 (*caption*) A fleet of 'six-bitters', or 75-footers, at maneuvers. **1964** M. F. Willoughby *Rum War at Sea* vii. 88 Many six-bitters were used offshore to picket larger rum vessels. **1853** Dickens *Bleak Ho.* lxiv. 612 'It's a six roomer, exclusive of kitchens,' said Mr. Guppy, 'and in the opinion of my friends, a commodious tenement.' **1932** T. S. Eliot *Sweeney Agonistes* 24 There's no motor cars No two-seaters, no six-seaters. **1977** *Herald* (Melbourne) 18 Jan. 2/2 A six-seater antique dining table, and two antique chairs were among the $4000 haul.
5. **six-ale** (earlier example); **six by six** *U.S. Mil. slang* (see quot. 1966); also written *6 × 6*, *6 by 6* and *ellipt.* as **six-by; six chamber:** also in full **six chamber revolver; Six Day(s) War,** an Arab–Israeli war that lasted from 5 to 10 June 1967; **six-eight tempo, time** *Mus.*, time or rhythm having a bar length of six quavers' duration divided into two equal beats; also *ellipt.* and as $\frac{6}{8}$; **six-figure** *a.*, (*a*) evaluated to or containing six significant figures or six decimal places; (*b*) containing or represented by six digits; *spec.* worth hundreds of thousands (of pounds, dollars, etc.); also in phr. *in six figures*; **six-four** (later example); **six-four measure, meter, time** *Mus.*, time or rhythm having a bar length of six crotchets' duration divided into two equal beats; also *ellipt.* and as $\frac{6}{4}$; **six-gun** *N. Amer.* = Six-Shooter; cf. *six-chamber* (*revolver*) in Dict. and Suppl.; **six o'clock:** see sense 2 c of the adj.; also denoting any position resembling that of the hands of a clock at six o'clock; **six-pack** orig. and chiefly *U.S.*, a package containing six cans or bottles of a drink; **six-two time** *Mus.*, time or rhythm having a bar length of six minims' duration divided into two equal beats; also written $\frac{6}{2}$.
1871 *N.Y. Almanac* 40/1 And a glass of 'six ale', punctually every morning at eleven o'clock, was absolutely necessary to his existence. **1942** *Infantry Jrnl.* Sept. 41/1 A group of upturned faces stands out to an aerial observer like a 6 × 6 in a flock of jeeps. **1943** *Yank* 3 Sept. 7/2 I'm herding a 6 by 6. **1966** *Sunday Times* (Colour Suppl.) 4 Dec. 73/4 GI Jargon. *Six by six*, six wheel truck with six-wheel drive. **1973** D. Fairbairn *Shoot* xiv. 109, I want you to load everything onto a six-by, and I want you to have the six-by all gassed up and ready to go. **1922** Joyce *Ulysses* 642 The sixchamber revolver anecdotes verging on the tropical. **1967** *Times* 14 June 1/3 The purpose of this move is clearly to have the Assembly condemn Israel and demand that she withdraw her armed forces to the armistice demarcation lines as they existed before the six-day war that started on June 5. **1967** *Listener* 17 Aug. 196/3 In the euphoria that followed the Six Days War, the Israelis appear to be convinced that they can work miracles. **1977** P. Johnson *Enemies of Society* xviii. 241 The impetus which created the political terrorism of the 1970s was undoubtedly provided by the Arab defeat in the 1967 Six Day war. **1873** *Illustr. London News* 2 Aug. 114/2 'Sleep, baby darling,' a lullaby..is in the six-eight tempo conventionally associated with slumber-songs. **1884** Grove *Dict. Mus.* IV. 119/1 Six-eight time..with two Beats in the Bar, each represented by a dotted Crotchet—or its equivalent, three Quavers. **1936** F. G. Hawkes *Stud. in Time & Tempo* xii. 74 $\frac{6}{4}$, $\frac{6}{8}$, and $\frac{6}{16}$ are much alike so far as the movements of the baton are concerned. **1965** *New Yorker* 8 May 173/1 McFarland's rhythms are full of stop-times, double time-passages, six-eight time, and shuffle rhythms. **1978** G. Read *Mod. Rhythmic Notation* v. 158 In essence, the $\frac{6}{8}$ violin part constitutes triplets in duple time. **1840** R. Farley (*title*) Tables of six-figure logarithms. **1873** 'Mark Twain' & Warner *Gilded Age* xiii. 123 He always talked in six figures. It was as natural for the dear boy to be rich as it is for most of us to be poor. **1963** P. Drackett *Motor Rallying* iii. 40 To illustrate, let's take a six-figure reference, the type normally employed. It may be, say 386 443. **1970** J. McN. Dodgson *Pl.-Names Cheshire* I. p. xliv, A four- or six-figure National Grid reference to the location of the principal hamlet. **1971** *Daily Tel.* 7 July 14, 64 pictures in 75 minutes, three in the six-figure class, and Monet's great painting..not far behind. **1981** R. Adams *Girl in Swing* (rev. ed.) xxi. 291 If you were to decide to put it into auction it would be almost bound to go for a very large sum—in the six-figure range. **1884** Grove *Dict. Mus.* IV. 119/1 Six-four Time..with two Beats in the bar, each represented by a dotted Minim—or its equivalent, three Crotchets. **1936** Six-four [see *six-eight* above]. **1938** *Oxf. Compan. Mus.* 409/1 If the fifth [is in the bass] it is a Second Inversion (also spoken of as 'six-four' chord). **1968** *Listener* 6 June 748/3 An unmetrical vocal line in six-four against an accompaniment in four-four. **1978** G. Read *Mod. Rhythmic Notation* v. 159 Combining two $\frac{3}{4}$ measures into one $\frac{6}{4}$ measure does not simplify the issue, nor would altering the $\frac{6}{4}$ meter to $\frac{6}{8}$. **1912** W. M. Raine *Brand Blotters* 336 My carbine was gone. It was too far for a six-gun. **1968** E. McCourt *Saskatchewan* v. 61 The Canadian cowboys rode unarmed, the Americans carried six-guns. **1979** G. Swarthout *Skeletons* 30 I'll..order a shot of red-eye and lay my six-gun on the bar. **1684** *Phil. Trans. R. Soc.* XVII. 672 The Courses [i.e. veins of ore] usually lying from East to West, or at Six a Clock as their Term is. **1915** 'I. Hay' *First Hundred Thousand* vii. 82 He..touched 'six o' clock' on the distant bull..and took the second pull for the last time. **1927** W. E. Collinson *Contemp. Eng.* 92 Expressions I learnt..when doing target-practice..viz. the six o'clock aim (from position on target focussed). **1961** *Wall St. Jrnl.* 19 Oct. 1/5 One leading brand which retailed for $2·09 a six-pack last January now is sold for $1·79. **1972** M. J. Bosse *Incident at Naha* iii. 138, I took her a pound of bacon and a six-pack of diet cola. **1981** *TV Picture Life* Mar. 46/2 'I went out and borrowed a pickup truck and wore my hat out in the rain for a couple of days, got a six-pack of beer and didn't shower,' John remembers. **1884** Grove *Dict. Mus.* IV. 119/1 Six-two time, $\frac{6}{2}$; with two beats in each Bar; each represented by a dotted Semibreve—or its equivalent, three Minims. **1978** G. Read *Mod. Rhythmic Notation* ii. 19 In more traditional symbology, $\frac{6}{2}$ or 6/𝅗𝅥.

sixer. Add: **2. a.** Also, six months' imprisonment.
1849 *Session Papers* 1 Feb. 324 The prisoner said he should not mind if he got off with a *sixer*—that means six months' imprisonment. **1887** J. W. Horsley *Jottings from Jail* i. 23 Neddie, from City Road, smugged for attempt up the Grove, expects a sixer..and is reconciling himself to an absence from his oriental home for half a year. **1903** [see *Carpet *sb.* 7]. **1926** J. Black *You can't Win* xii. 161 We'll both get a 'sixer' in the morning if we go in front of the judge. **1955** D. W. Maurer in *Publ. Amer. Dial. Soc.* xxiv. 151 Maybe he will get off with a *bit*..or a *sixer*, which is six months in jail.
d. Six strokes of the cane administered as punishment in school. Cf. *six of the best* s.v. *Best *a.* 10 d.
1927 *Chambers's Jrnl.* 10 Sept. 645/2 About a dozen boys..ranged up in front of the Doctor, who, stepping off his pedestal, administered a 'sixer' to each of the culprits with a long and stout cane. **1977** C. McCullough *Thorn Birds* ii. 29 They all got sixers, but Meggie was terribly upset because she thought she ought to have been the only one punished.
3. A Brownie Guide or Cub Scout in charge of a six.
1916 R. Baden-Powell *Wolf Cub's Handbk.* iii. 216 The average Cub when promoted to Sixer does not carry so much authority. **1920** [see *six *sb.* 2 b (*b*)]. **1965** G. McInnes *Road to Gundagai* x. 161, I was promoted to be Sixer of the Whites. **1978** *Lochaber News* 31 Mar. 7/5 On Thursday, March 23 the sixers and seconds went to Raigmore Hospital to present a cheque to Dr John Burton for the kidney unit at the hospital.

sixpence. Add: **1.** (Earlier *Comb.* example.)
1780 J. Woodforde *Diary* 20 June (1924) I. 286 Gave on going—O.I.O. For which you have 6d worth of anything at the Bar.
2. b. (Further examples.)
1818 H. B. Fearon *Sk. Amer.* 13 A beggar came in, and was relieved with a Spanish silver piece called a sixpence. **1891** S. M. Welch *Home Hist.* 169 It was common, particularly in New England, to call a sixpence or a half dime, a *fip*.

sixpenny, *a.* and *sb.* Add: **2.** (Later examples.) Also of things.
1878 H. H. Jackson *Travel at Home* 11 She didn't never want to see any o' them sixpenny towns agin. **1911** G. B. Shaw *Doctor's Dilemma* p. xxvi, The sixpenny doctor, with his low prices and quick turnover of patients, visibly makes much more than you. **1927** Kipling *Limits & Renewals* (1932) 164 When I was a sixpenny doctor at Lambeth.
3. c. (Earlier example.)
1840 *Knickerbocker* XV. 138 The larger newspaper establishments, satirically termed by their Lilliputian rivals, 'the respectable sixpennies'.
d. As *sb.* A cinema seat that costs sixpence.
1958 *Listener* 4 Dec. 927/2 A small boy sitting in the sixpennies at the Bijou Cinema.
4. *sixpenny piece* (earlier example).
1897 M. H. Kingsley *Travels in W. Afr.* xxvi. 589 A piece of ground the size of a sixpenny piece.

sixpennyworth. Add: Also **sixpennorth.**
1933 [see *Bacco, Baccy].

six-shooter. Add: (Earlier example.)
1844 *Nauvoo* (Illinois) *Neighbor* 24 July 3/1 Joseph..opening the door..discharged one barrel of a six shooter (Pistol) in the entry.
six-shooting *ppl. a.* (earlier example). Also **si·x-shoo:ter** *v.*; **si·x shoo:tering** *vbl. sb.*
1858 T. Viele *Following Drum* 224 A belt full of pistols,..and a six-shooting rifle. **1904** P. Fountain *Great North-West* xx. 242, I never was so near six-shootering myself as I was that night. **1909** 'O. Henry' *Roads of Destiny* xxii. 371 We heard a yelling and a six-shootering.

sixteen, *a.* and *sb.* Add: **B.** *sb.* **5.** A medium-sized flower-pot, sixteen of which are formed from a cast of clay.

1802 W. Forsyth *Treat. Culture & Managem. Fruit-Trees* viii. 114 The 1st size of 8 in the Cast is called Eights. 2[nd size of] 12 [in the Cast is called] Twelves...Sixteens...Twenty-fours. **1852** G. W. Johnson *Cottage Gardener's Dict.* 392/2 Nine-inch pot..16s. [= sixteens]. **1895** *Culture of Veg. & Flowers* (Sutton & Sons) (ed. 6) 323 Small 60..2¼ [inches]. Mid. 60..3. Large 60..3½... 16..8½...6..12½. **1955** W. E. Shewell-Cooper *Pot Plants* ii. 19 The tendency is to use smaller pots and, whereas years ago we grew all our late Chrysanthemums in 12's, we try to do them today in 16's or even 24's. **1962** [see *sixty *sb.* 4 c].
C. 1. **sixteen millimetre,** (more usually) **16mm,** a cine film which is sixteen millimetres wide; in full **sixteen millimetre film;** also *attrib.*
1926–7 *Army & Navy Stores Catal.* 971/1 The Bell Howell 'Filmo' automatic camera..Accommodates newly standardized 16 m/m films. **1951** R. Spottiswoode *Film & its Techniques* i. 9 The producer must also make up his mind whether to shoot in color or black and white, in standard theater 35 mm. film, or nontheatrical 16 mm. film. **1969** G. Greene *Travels with my Aunt* i. viii. 82 The films, of course, had all been shot on sixteen millimetre, and..they were enlarged practically to cinerama size. **1977** *New Yorker* 29 Aug. 66/2 Travelling, then, at almost six hundred miles an hour, we watch this innovative sixteen-millimetre movie, projected with infinite difficulty at thirty-six feet a minute.

sixteener. 2. (Later example.)
1966 *Daily Tel.* 5 Nov. 7/6 The world's biggest organization of young motor-cyclists..launched a petition against the sixteener ban.

sixteensome (sikstī·nsəm). [f. Sixteen *a.* + -some[2].] A group of sixteen persons. Usu. *attrib.* in *sixteensome reel*, a Scottish dance performed in sets of sixteen persons. Also *absol.*
1926 [see *eightsome *a.*]. **1938** *St. Andrews Citizen* 13 Aug. 9 He is much in request to teach..eightsomes, sixteensomes, and thirty-twosomes. **1954** H. A. Thurston *Scotland's Dances* 51 The same process can be applied to the eightsome reel: the result is the *double eightsome* or *sixteensome reel*. **1964** J. & T. Flett *Trad. Dancing* i. 18 Mr Reid taught..the Eightsome Reel, and the Sixteensome Reel. **1979** *Harper's & Queen* July 157/3 They led the sixteensome of Atholl Highlanders down the stairs two by two.

sixth, *a.* and *sb.* Add: **A.** *adj.* **2.** With omission of *form* (later examples), *former*.
c **1898** W. Lewis *Let.* (1963) 6 First a fellow got a 'sixth licking' (stripes from every sixth in the house). **1906** R. Brooke *Let.* 3 Feb. (1968) 39 [He] has been discovered..showing up proses done for him by a wee & terrified Sixth. **1914** 'I. Hay' *Lighter Side School Life* i. 5 The Head..probably takes the Sixth for an hour or two a day. **1963** *Sunday Times* 8 Sept. 29/3 Cool Shakespeare thrives in the sixth and phrases like 'Pox on't'..are in present usage. **1977** R. Rendell *Judgement in Stone* vi. 50 You're no longer the naughtiest girl in the sixth.
C. 2. *sixth day*, the name given to Friday by members of the Society of Friends; *sixth form*: see Form *sb.* 6 b (later examples); hence *sixth-former*: see *-former; *sixth-form college*, a college for pupils over the age of sixteen, chiefly providing A-level courses.
1655 G. Fox *Jrnl.* (1694) 152 On the Sixth day of that Week I had a meeting near Colchester. **1858** M. Tuckett *Diary* 26 Sept. in H. Fox *Mariana's Diary* (*c* 1975) 8 Sixth day morning was bright and fine. **1976** *Minutes Ohio Yearly Meeting of Friends* 30 (*heading*) Sixth day afternoon session, eighth month 27th. **1938** C. Morgan *Flashing Stream* 31 He who wrote the Sonnets, or Hamlet's bidding to Ophelia..had no moderation, no smell of the sixth form, no sense of humour. **1965** H. L. Elvin *Educ. & Contemp. Soc.* ii. vii. 134 There is little doubt that the sixth form college would be welcomed by most of the young people who would go to it. **1967** *Listener* 18 May 645/1 The eighteen-year-old emerging from our sixth forms has a level of knowledge as good as a second year student in a North American university. **1976** *Times* 18 Aug. 3/2 Tameside council has decided not to introduce two proposed sixth form colleges.

sixth sense (siksþ sens). [f. Sixth *a.* + Sense *sb.*] A supposed intuitive faculty by which a person or animal perceives facts and regulates action without the direct use of any of the five senses. Hence **sixth-sense** *v. trans.*, to discover by means of a sixth sense; **sixth-sensed** *a.*, possessing a sixth sense.
[**1687** W. Domvile tr. *B. de Fontenelle's Plurality of Worlds* iii. 50 It has been thought that we want a sixth natural Sense, by which we might know many things more than we do.] **1761** L. Sterne *Tristram Shandy* IV. i. 75 There seems in some passages to want a sixth sense to do it rightly. **1807** R. Southey *Lett. from England* II. xl. 176 It was surprising to see them [*sc.* the blind] move about the room,..as if they had possessed that sixth sense, which experimental naturalists..are said to have discovered in bats, when they have put out their eyes. **1841** Dickens *Barnaby Rudge* xxxiii. 122 People..doing exactly the same things for a great many years, acquire a sixth sense, or some unknown power of influencing each other. **1903** *Science Siftings* 31 Oct. 46/1 The 'sixth sense' by which blind persons perceive certain objects. **1958** R. Godden *Greengage Summer* xvi. 199 Did she have some sixth-sense warning? **1967** *Punch* 25 Oct. 609/2 This, I sixth-sensed, could well be it. **1976** J. Crosby *Nightfall* xxx. 176 About Elf I am second-sighted, sixth-sensed, magicked. **1979** R. Jaffe *Class Reunion* (1980) ii. iii. 199 There was something about living together and being very close that gave people a sort of sixth sense.

sixty, *a.* and *sb.* Add: **A.** *adj.* **2. b.** *sixty-six* (earlier example); *sixty-nine, 69* = *Soixante-neuf.
1857 T. Frere *Hoyle's Games* 4 The German game of 'Sechs und Sechszig', or Sixty-six, has never before, that we are aware of, been dressed in an English garb. **1888** [see *soixante-neuf]. **1973** D. Lang *Freaks* 90 We spent many hours lying on her bed, more or less in the classical 69 position, but motionless. **1978** *Guardian Weekly* 23 Apr. 21/5 When I first met him, I thought 69 was a bottle of Scotch.
B. *sb.* **1. b.** (Later examples.)
1910 *Dialect Notes* III. 445 That child cuts up like sixty. **1975** J. D. Fitzgerald *Great Brain does it Again* ii. 20 We ran like sixty to the front porch.
3. (Example of years in a person's life.) Now *spec.* the period 1960–9.
1964 M. McLuhan *Understanding Media* ii. xxxi. 320 TV in the Fifties and Sixties spread to the entire population. **1978** *Listener* 3 Aug. 145/1, I was, alas, one of those who spent the Sixties sneering at the notion of parish-pump broadcasting. **1981** 'D. Shannon' *Murder most Strange* ii. 34 They were both in the sixties, middle-sized, sandy coloring. **1983** D. Gethin *Wyatt* xiv. 99 An ageing sixties swinger with the elegant mannerisms of a professional hotelier.
4. c. A small flower-pot, sixty of which are formed from a cast of clay.
1802 W. Forsyth *Treat. Culture & Managem. Fruit-Trees* viii. 114 There are some [pots] smaller than sixtys, for seedlings and heaths. **1895** [see *sixteen *sb.* 5]. **1962** A. J. Huxley *Garden Terms Simplified* 69 Above are shown, to scale top row from left to right, an 8½in. pot (16),..and a 3½ in. pot (large 60).
C. a. *sixty-miler* (Austral.), a small cargo vessel which transports coal along the coast from Newcastle to Sydney; *sixty-pounder* (later example).
1933 I. Hamilton *Nights Ashore* 29 The Five Stars was a few tons larger than the average ancient sixty miler. *Ibid.* 210 During the slack 'sixty-miler' season. **1940** Blunden *Poems 1930–40* 202 And as the stream's last murmer stilled, Our sixty-pounders started talking. **1948** *Sydney Morning Herald* 28 Jan. 1/7 Sydney's gas supply now depends on the '60-milers'.
c. Also called *sixty-fours*.
1805 in E. Howe *London Compositor* (1947) ii. 92 Forty-eights to be paid two shillings per sheet extra, and sixty-fours two shillings and sixpence per sheet extra.
d. *sixty-four dollar question, $64 question*, orig. the question posed at the climax of a U.S. radio quiz for a prize of sixty-four dollars, used *transf.* to denote a difficult or crucial question; also *sixty-four dollar answer*, *sixty-four thousand dollar question*, and varr.
1942 J. R. Tunis *All American* vii. 240 Here's the sixty-four dollar question. Will the team go to Miami? **1942** *Time* 18 May 22/3 The Jap..could still sweat over the $64 question. **1955** M. Gilbert *Sky High* xii. 176 'What have these receivers got to do with us?..'That's the sixty-four dollar question.' **1957** R. Hoggart *Uses of Literacy* vi. 150 All the time he had the sixty-four dollar answer but did not know it. **1957** *Observer* 21 July 1/3 Mr. Macmillan said..there was only one answer to the 64,000-dollar question—to increase production. **1958** *Listener* 4 Dec. 930/1, I come now to what you probably feel is the sixty-four-dollar question. How is all this to be paid for? **1963** *N.Y. Times* 2 Dec. 37/1 Mr. Baker..left the air, to return in 1942 as master of ceremonies on 'Take it or Leave it'... He posed 'the $64 question', a term that became part of everyday language. **1967** *N.Y. Rev. Bks.* 7 Dec. 27/1 On June 1, 1955, 'The $64,000 Question' was born and commercial television was never the same again. **1979** *Jrnl. R. Soc. Arts* CXXVII. 143/2 Like his predecessor on this rostrum he left it to Mr. Tyrrell Burgess, our lecturer tonight, to tackle the sixty-four dollar question—What now? **1981** B. Healey *Last Ferry from Lido* vi. 101 It still leaves the sixty-four thousand dollar question. Where do we go from here?

size, *sb.*[1] **II. 10. d.** For *rare* read: *rare* exc. in phr. *to cut* (*chop*, etc.) *down to size*: see *Cut *v.* 53 h. *colloq.*
1953 *Time* 20 July 40/3 He kept Stalin down to size. **1962** *Listener* 17 May 883/1 The complexity and psychological depth abandoned in hacking the novel down to size. **1972** *N.Y. Times* 3 Nov. 39/6 Once the warlord armies supporting him were chopped down to size.., he was content to fade away.
f. In colloq. phr. *that's* (*about*) *the size of it*, etc.: that is what it amounts to, that is the situation.
1860 [in Dict., sense 12 a]. **1880** 'Mark Twain' *Tramp Abroad* viii. 71 'Bloodshed!' 'That's about the size of it,' I said. **1914** G. Atherton *Perch of Devil* i. 89 That's the size of it, only I couldn't ever say it like that. **1922** Joyce *Ulysses* 305 Talking about the Gaelic league and the antitreating league and drink, the curse of Ireland. Antitreating is about the size of it. **1966** D. Francis *Flying Finish* ii. 21 'He just went to Italy and didn't come back?' 'That's about the size of it,' Simon agreed. **1973** 'M. Innes' *Appleby's Answer* xv. 134 It's money that's really in his head... That's about the size of it, wouldn't you say?
11. c. *to try* (something) (*on*) *for size*: to consider (an idea, theory, etc.) to see whether it fits the facts. Also *loosely*, to try out or sample. *colloq.* (orig. *U.S.*).
1956 'E. McBain' *Cop Hater* (1958) viii. 70 'Try this for size,' Bush said. 'I'm listening,' Carella said. **1967** 'E. Queen' *Face to Face* xxix. 127 'All right,' said the Inspector. 'Let's try this on for size: You knew what Spotty had to sell, didn't you?' **1969** 'J. Fraser' *Cock-pit of Roses* x. 81 'I know some bugger's been pinching 'em, if that's what you're getting at.' Try that on for size, you devil, his look seemed to say. **1979** A. Boyle *Climate of Treason* (1980) viii. 258 Trying his boss's desk for size, Philby noticed the untidy array of memoranda and pending files in the in-tray. **1980** J. McClure *Blood of an Englishman* xxiv. 221 She was shaking the sardines into a saucer. 'There, kitty! Try those for size.'
13. **size distribution,** the way in which size varies among members of a population of particles; **size effect,** an effect due to size; **size-group,** those constituents of a population whose sizes fall within a specific range; **size-range,** a range of sizes; a size group.
1925 *Trans. Faraday Soc.* XXI. 381 (*heading*) A simple method of obtaining the size distribution of particles in soils and precipitates. **1966** D. G. Brandon *Mod. Techniques Metallogr.* v. 250 In the past, size-distribution analysis has usually involved time-consuming measurements on individual grains or particles. **1943** *Ann. Appl. Biol.* XXX. 216/2 The loss in weight increases progressively from *I* to *IV* indicating the existence of a size effect. **1968** C. G. Kuper *Introd. Theory Superconductivity* v. 92 The discrepancy between the experiments and the London theory has been interpreted as a size effect, arising from the scattering of normal electrons by the n-s boundary. **1944** J. S. Huxley *On Living in Revolution* 110 The total population can be separated into four size-groups, corresponding to the produce of the four successive years that each grub lives in the soil before it turns into a beetle. **1971** I. G. Gass et al. *Understanding Earth* ix. 132/2 They contain both filamentous and globular structures, and the latter occur in more than one size-group. **1924** *Industr. & Engin. Chem.* XVI. 930/2 The figures given..for the size range were calculated in this way. **1955** *New Biol.* XIX. 95 Some particles in the mitochondrial size-range appear to be proplastids. **1962** *Science Survey* III. 296 Shoals of large salmon were observed to remain inactive at the tail of the pool while a smaller size-range of salmon and trout ascended successfully.

size, *v.*[1] Add: **4. a.** (Later U.S. examples.)
1862 *Trans. Ill. Agric. Soc.* (1865) V. 233 There is no provision made for 'sizing the gavels' [in reaping]. **1897** F. C. Moore *How to build Home* 90 The first-story beams are to be sized and leveled upon the sill and upon the foundation wall. **1981** *Sci. Amer.* Apr. 30/1 Second, says Dr. Derry, the book was sized to fit into a briefcase.
fig. **1899** [see *panhandler 1].
6. For *rare*$^{-1}$ read † *Obs.* and add later examples.
1853 J. G. Baldwin *Flush Times Alabama* 113 The jury shortly after returned into court with a verdict which 'sized their pile'. **1873** J. H. Beadle *Undevel. West* xii. 198 They are satisfied to 'size your pile' and take quarter of it. **1889** 'Mark Twain' *Conn. Yankee* 300, I was resolved he should have at least one [bath]..if it sized up my whole influence and bankrupted the pile.
7. a. (Later examples.) Hence **si·ze-up** *sb.*, an estimate. *U.S.*
1924 E. O'Neill *All God's Chillun got Wings* 103 John scrutinizes their faces keenly, sizing up the situation. **1945** E. S. Gardner *Case of Golddigger's Purse* v. 35, I always like to plan my campaign after I've sized up my man. **1949** *Security* (Charlottesville, Va.) May 1/2 Reminded him how wrong a bright man can be in his size-up of other folks. **1952** *Sun* (Baltimore) 22 Oct. 19/7 Casey's size-up of Woodling as a pinch-hitter was verified in the eighth inning. **1978** *N.Y. Times* 30 Mar. B3/1 A teen-ager in sneakers, sizing up the drinkers in the darkened bar as easy marks, whispered to a friend, 'I wonder if they've got any change.'
b. (Later example.)
1912 A. Conan Doyle *Lost World* vi. 87, I want a man I can bank on. So I sized you down, and I'm bound to say that you came well out of it.
8. b. (Earlier and later examples.)
a **1631** Donne *Poems* (1633) 63 So As they waxe lesser, fall, as they sise, grow. *c* **1866** G. M. Hopkins *Poems* (1967) 36 That a quince I pore upon? O no it is the sizing moon. **1926** *Daily Colonist* (Victoria, B.C.) 13 July 4/5 Pears and apples are sizing well and developing nicely. **1940** *Sun* (Baltimore) 5 Sept. 7/5 Sweet potato tubers are developing and sizing in Western Maryland.
c. With *up*: to develop or take shape; to amount (*to* something); to reach the necessary standard. Cf. *to measure up to* s.v. *Measure *v.* 4 c. *U.S.*
1884 E. W. Nye *Baled Hay* 126 Time, at last, makes all things size up in proper shape. **1905** *N.Y. Even. Post* 21 Nov. 1 If the President does not think that Mr. Halpin sizes up as chairman of the County Committee, [etc.]. **1910** J. Hart *Vigilante Girl* ii. 28 Burke isn't a bad sort—he sizes up about as well as most of them. **1917** H. James *Ivory Tower* ii. i. 89 The question of what Gray's 'interest'..might size up to.

sizeless, *a.* (Later examples.)
1896 L. T. Hobhouse *Theory Knowl.* 47 Can we now stigmatise the sizeless, figureless, positionless extension in the same way? **1935** W. de la Mare *Early One Morning* xv. 195 A universe that was nothing but a sizeless point.

sizing, *vbl. sb.*[1] Add: **4.** *sizing up*, the process of assessing or evaluating.
1967 *Coast to Coast 1965–6* 136 No doubt he guessed I'd been doing a bit of sizing up and decided he had better help me get my ideas into order.

sizzle, *sb.* Add: **1.** Also *fig.*
1964 *Economist* 1 Feb. 400/1 The Jesuit sanctuary at Loyola..was a sizzle of sex-appeal. **1976** *National Observer* (U.S.) 17 July 6/1 You've reached middle age, and your marriage has lost its sizzle.
2. *Comb.*, as **sizzle cymbal,** a cymbal, used chiefly in jazz and dance bands, with several small rivets set loosely through it to make a sizzling sound when the cymbal is struck.
1944 W. Apel *Harvard Dict. Mus.* 198/1 Various modifications are used in jazz bands, e.g., the Choke cymbal, the Sizzle cymbal. **1964** R. Burns *Selection, Care, & Use of Cymbals* 7 Sizzle cymbals are immensely popular with drummers playing with jazz groups. **1967** *New Yorker* 21 Jan. 52/2 Marsala told me just to play with wood-blocks and a sizzle cymbal.

sizzle, *v.* Add: **2.** (Later *fig.* examples.)
1928 'Brent of Bin Bin' *Up Country* xv. 266 As Little River was in the same direction he sizzled away to old Healey at the point of the spur. **1966** *Daily Tel.* 15 Nov. 13/3 Some colours vibrate when they are used together and I often use them to make a dull corner sizzle. **1979** *Fisherman's Weekly* 21 June 27/1 When a run comes, line can sizzle off the reel, so be sure to use an open bale-arm plus indicator.

sizzler (si·zləɹ). *colloq.* [f. Sizzle *v.* + -er[1].] **1.** *U.S.* A sizzling heat, day, etc.; a 'scorcher'.
1901 *Emporia* (Kansas) *Gaz.* 1 July 9 The drought which is a sizzler and frier and boiler is a good thing for Kansas. **1904** G. H. Lorimer *Old Gorgon Graham* ii. 37 Satan may be down in Arizona cooking up a sizzler for the corn belt.
2. a. Something salacious or *risqué*. Cf. *Scorcher 3 d.
1957 'N. Blake' *End of Chapter* xi. 169 A lot of morons who only wanted to curl up with a nice sizzler by Elinor Glyn. **1977** *News of World* 17 Apr. 9/9 A blue movie being shown to a judge in court was a real sizzler. In fact, it was so hot that it began to melt.
b. A very fast shot or hit. Cf. *Scorcher 3 c.
1960 E. W. Swanton *West Indies* iii. 51 The one chance he gave..was a sizzler to the gully. **1976** *Sunday Times* (Lagos) 26 Sept. 1/3 Buffaloes goalkeeper Chilongwe positioned himself to record an amazing save off a sizzler.

sizzling, *ppl. a.* Add: Also *fig.*
1923 [see *picture *sb.* 2 j]. **1947** *Sporting Mirror* 7 Nov. 8/1 His intended tap back to goalkeeper Hesford, from a long way out, became a sizzling shot which won the game for Charlton. **1977** *Time* 30 May 51/2 The sizzling increases [in personal income] of February and March.

sizzlingly (si·zliŋli), *adv.* [f. prec. + -ly[2].] So as to sizzle. Used to give emphasis to expressions of warmth, intensity, etc.
1956 *Essays in Crit.* VI. 204 You, too, were once not allegoric But blazed with passions sizzlingly phosphoric. **1963** *Listener* 21 Feb. 327/2 Unless one has some sizzlingly hot contributions to make, one somehow feels that a mere dialogue is perhaps not enough.

sizzly (si·zli), *a.* [f. Sizzle *v.* + -y[1].] Sizzling, effervescent, exciting.
1936 J. Dos Passos *Big Money* 17 The sizzly smell of champagne and welshrabbit. **1948** D. Ballantyne *Cunninghams* xii. 70 Carole Plowman..was the most beautiful girl in the hall... 'A few good-looking girls here,' he said. 'About three,' Phil said. 'That Plowman tart's sizzly.'

sjambok, *sb.* Add: Also 8 **chanboc(k),** 9 **samboc, sambok, schambok, shambo.** (Earlier and further examples of var. forms.)
α. **1790** E. Helme tr. *Le Vaillant's Trav. Interior Parts Afr.* I. xxi. 412 The next day my men employed themselves in cutting the skin off the Hippopotamus, to make what, in this country, the[y] call *Chanboc*, which are whips, used to drive the oxen. **1804** J. Barrow *Trav. Interior S. Afr.* II. ii. 96 One of those infernal whips, made from the hide of a rhinoceros or sea-cow, known by the name of *sambocs*. **1808** J. Read in G. E. Cory *Rise of S. Afr.* (1921) I. vii. 203 Terribly flogged with a sambok or whip made of the skin of a rhinoceros. **1850** T. Shone *Diary* III. 19 Mar. in *Voorloper* (1976) 717 Lost my Samboc, the Horse Broke his bridle. **1852** M. B. Hudson *S. African Frontier Life* I. 40 No coaxing nor threats, after sambok persuasion, Could cure him of sulks on the present occasion.
β. **1791** tr. *Le Vaillant's Trav. Interior Parts Afr.* I. 195 My people were busied in cutting to pieces the hide of the hippopotamus, to make what the country folks call *chanboc*. **1801** J. Barrow *Trav. S. Afr.* iii. 145 These sort of whips which they call *shambos* are most horrid instruments. **1812** A. Plumptre tr. *Lichtenstein's Trav. S. Afr.* I. i. vii. 98 The skin is the only thing valuable to the colonists, to cut into strips for making the driving whips known here by the Malay name of Schamboks. **1822** W. J. Burchell *Trav. Interior S. Afr.* I. 86 Sometimes encouraged by good words, at other times terrified into exertion by ten blows of the *shambok*. **1911** L. Cohen *Reminisc. Kimberley* xviii. 320 'All right,' he replied... 'Bring me a shambok.'

sjambok, *v.* Add: Also **sjambook.** (Earlier examples.)
1881 *Blackw. Mag.* Dec. 756/1 To associate or have anything to do with blacks, except to make them work, or *sjambook* them if they don't work hard is an unpardonable offence in a Boer's eyes. **1894** E. Glanville *Fair Colonist* xv. 116, I would cheerfully sjambok a stock-lifter until he dropped.
Hence **sja·mbokker,** one who uses a sjambok.
1953 *Cape Times* 30 Mar. 1/1 The sjambokker came outside and hit..Mr. Eddy..on the legs.

sjamboking, *vbl. sb.* Add: Also spelt *sjambokking*. Also *fig.*
1908 D. Blackburn *I came & Saw* 208 Your sjambokking of Sixpence gave me the idea for the Humanitarian Company. **1953** *Cape Times* 30 Mar. 1/1 A police investigation into the alleged sjambokking of two United Party canvassers. **1980** *Listener* 17 Apr. 487/3 Lilford, landowner and power behind Smith and the Rhodesian Front, gave me a verbal sjambokking over the telephone.

Sjögren (ʃɜ·grĕn). *Path.* [The name of H. S. C. *Sjögren*, Swedish ophthalmologist, who described the condition in 1933 (*Acta Ophthalm.* Suppl. No. 2. 1–151); used after G. *Sjögrens-syndrom* (Weber & Schlüter 1937, in *Deutsch. Arch. für klin. Med.* CLXXX. 333).] *Sjögren's disease* or *syndrome*: a condition characterized by chronic inflammatory swelling of the salivary and lachrymal glands and by auto-immune antibodies in the blood.
1938 *Proc. R. Soc. Med.* XXXII. 255 (*heading*) Sjögren's syndrome associated with pigmentation and sclerodermia of the legs. **1954** [see *keratoconjunctivitis* s.v. *Kerato-]. **1961** *Lancet* 26 Aug. 456/2 The diagnosis of Sjögren's disease was made from the case-history and clinical condition, supported by positive Schirmer and rose-bengal tests. **1974** Passmore & Robson *Compan. Med. Stud.* III. xxv. 18/2 Sjögren's syndrome rivals systemic lupus erythematosus in the multiplicity of auto-antibodies present in the serum.

sjögrenite (ʃɜ·grĕnəit). *Min.* [f. the name of S. A. Hjalmar *Sjögren* (1856–1922), Swedish mineralogist + -ite[1].] A hydrated basic carbonate of iron and magnesium found as yellowish or brownish thin transparent plates formed by hydrothermal action.
1941 C. Frondel in *Amer. Mineralogist* XXVI. 303 The hexagonal mineral commonly admixed with pyroaurite is described on a following page under the name sjögrenite. **1968** *Acta Crystallographica* B. XXIV. 972/1 The carbonate-hydroxides $Mg_6M_2^{III}(OH)_{16}CO_3.4H_2O$ (M^{III} = Fe, Al, or Cr) are known to occur in two dimorphic forms: the hexagonal sjögrenite group has $a \sim 3{\cdot}1$ and $c \sim 15{\cdot}5$ Å; the rhombohedral pyroaurite group has the same *a* value but *c* is about 23·2 Å. **1973** *Mineral. Mag.* XXXIX. 378 Two sub-groups could be distinguished, which would today be described as polytypes differing only in layer stacking; these are represented by pyroaurite and sjögrenite respectively.

ska (skā). [Origin unknown, perh. echoic.] A kind of popular music of Jamaican origin, characterized by a fast tempo and emphasis of the off-beat. Also, a dance to such music. Cf. *Reggae, *Rocksteady.
1964 [see *game *sb.* 3 b]. **1969**, etc. [see *rocksteady]. **1971** *Guardian* 25 Feb. 10/4 West Indian ska or blue beat music, latterly taken up by skinheads. **1980** *Rolling Stone* (Austral.) 26 June 16/5 'We don't see it as a ska revival,' Davies says, 'because we don't play pure ska.'..The group is also quick to point out the differences between the various 'ska' bands.

skaapsteker (skā·pstēˑkəɹ). Also **scarpsticker, scha(a)psteeker, -steker, -sticker.** [Afrikaans, f. Du. *skaap* sheep + *steker* stinger.] A venomous but usually harmless snake of the genus *Psammophylax*, esp. the spotted skaapsteker, *P. rhombeatus*, or the striped skaapsteker, *P. tritæniatus*, which are both greyish-brown with darker markings.
1818 C. I. Latrobe *Jrnl. Visit S. Afr.* xxii. 353 Our good-natured Hottentots, perceiving that I had begun to collect serpents, brought me several kinds, among which were the..schaapsteker, (sheep-stinger); and copra di capella. **1834** T. Pringle *Afr. Sketches* 280 There are several species of snakes..such as the nacht-slang (night adder), the schaap-steeker (sheep-stinger), [etc.]. **1856** F. Fleming *S. Afr.* 406 The 'Scarpsticker' of the Dutch, or Night-Adder; a small dingy-brown Adder, spotted with black, about eighteen inches long. **1887** *Encycl. Brit.* XXII. 197/1 The second African snake of this family is the 'schapsticker' (Sheep Stinger). **1915** *Chambers's Jrnl.* June 437/2 The schaap-sticker..is a short, silvery snake about the thickness of a man's finger. **1931** *Discovery* Mar. 73/2 We have droves of skaapstekers, night-adders, and house snakes. **1952** *Cape Argus* 7 June (Mag. Sect.) 2/4, I..'bagged' a good collection of the local snakes; mostly of a harmless variety, such as mole snakes, water snakes, herald snakes, schaapstekers. **1973** *Stand. Encycl. S. Afr.* IX. 651/2 The name skaapsteker is quite misleading and unfortunate, as there is certainly no truth in the widespread belief that these snakes are in the habit of biting and killing sheep.

skad, var. *Scad[7]. **skaffie,** var. *Scaf. **skag,** var. *Scag. **skål,** var. Skoal *sb.* in Dict. and Suppl.

‖ **skandalon** (skæ·ndălɒn). *Theol.* Also **scandalon.** [Gr. σκάνδαλον stumbling-block; cf. I *Cor.* i. 23 Χριστὸν ἐσταυρωμένον, Ἰουδαίοις μὲν σκάνδαλον Christ crucified, to the Jews a stumbling-block.] A stumbling-block, cause of offence, scandal (sense 1 b).
1945 *Theology* XLVIII. 104 Any particular 'Thou' may become a *skandalon* at any time. **1948** *Scott. Jrnl. Theol.* I. 113 It is not a truth which can be arrived at by scientific investigation, and on that account it is a *scandalon* to the scientist, and to this scientifically minded generation. **1957** *Ibid.* X. 86 The scandalon of Christian faith is accepted, and the essay aims at being the account which biblical thinking must give of itself, taking itself seriously and simply. **1972** *Times Lit. Suppl.* 28 Jan. 105/4 Belief in God is *the* skandalon barring men and women in their thousands from a secular faith in Jesus Christ.

skans, var. Schanse in Dict. and Suppl.

skarn (skāɹn). *Geol.* Also **scarn.** [a. Sw. *skarn* lit. 'dung, filth', in same sense, f. ON. *skarn* (cf. north. dial. *scarn* dung (E.D.D.)).] Orig. applied to the silicate gangue of certain Archæan iron-ore or other mineral deposits, esp. where these occur in limestone or dolomite; now extended to any lime-bearing siliceous rock produced by metamorphism (esp. of limestone or dolomite) and the introduction of new elements. Freq. *attrib.*
1901 H. Louis in *Trans. Inst. Mining & Metallurgy* X. 49, I have of late years adopted a very convenient Scandinavian word, *skarn*, for the zone of altered rock, usually calcareous, that contains such minerals as epidote, schorl, idocrase, at times augite, hornblende, mica, etc., and I speak of these altered rocks as epidote skarn, etc. **1911** *Econ. Geol.* VI. 708 Very characteristic of this zone, especially in connection with ore deposits, are the 'scarn' rocks—iron-rich silicate rocks produced by the addition of iron and silica to limestones. **1932** A. Harker *Metamorphism* ix. 129 The skarn type of metasomatism is found at numerous British localities, but usually as a narrow belt and with little of the impregnation with sulphides. **1954** *Econ. Geol.* XLIX. 633 The skarn is a hard dense green rock characterized by the presence of quartz, green ferro-magnesian silicates..and garnet. **1966** [see *sakhaite]. **1978** [see *serendibite].

skate, *sb.*[1] **3. skate-barrow** (earlier example).
1851 Thoreau *Jrnl.* 27 July in *Writings* (1906) VIII. 354 Skates' eggs, called in England skate-barrows from their form, on the sand.

skate, *sb.*[2] Add: **1. a.** (Examples = Roller-skate.) In slang (orig. *Mil.*) phr. *to get* (or *put*) *one's skates on*, to hurry up (see also quot. 1925).
(*a*) **1876** J. A. Harwood *Rinks & Rollers* iii. 39 The skates used had four wheels of iron placed in one line from the foot to the heel. **1925** *Sears Roebuck & Co. Catal.* 751 Children's Extension Skates With Steel Self Contained Ball Bearing Rolls. **1959** *Ice & Roller Skating* 20 Wooden wheels are essential for rink skating and the skate should be screwed onto the boot by an expert. **1975** *Man. Artistic Roller Skating* 24 The change-of-edge..should be as short as possible, not materially longer than the length of the skate.
(*b*) **1895** W. C. Gore in *Inlander* Dec. 113 *Get your skates on*, hurry up. **1919** *War Slang* in *Athenæum* 8 Aug. 727/2 To evade duty or get clear, you 'put your skates on'. **1925** Fraser & Gibbons *Soldier & Sailor Words* 260 *Skates, to put on*, to hurry up. Also to evade duty. To desert. **1938** F. D. Sharpe *Sharpe of Flying Squad* xxvi. 263 'Jack, guy for your b— life. The Squad are here.'..Said Jack: 'I very soon put my skates on.' **1969** G. Lyall *Venus with Pistol* ix. 54 It was Carlos telling me to get my skates on and down to the Doelen plenty chop-chop. **1976** W. J. Burley *Wycliffe & Schoolgirls* i. 33 I'd better be getting my skates on, I'm catching the night train and I haven't done a thing about getting ready.
c. *U.S.* A sledge runner.
1781 S. Peters *Gen. Hist. Connecticut* 320 In the winter, the sleigh is used; a vehicle..carrying six persons in its box, which hangs on four posts standing on two steel sliders, or large scates. **1907** *St. Nicholas* July 781/1 You make a framework of timbers..and stick a skate or runner at each corner.
d. *transf.* A device with a set of rollers or wheels on which something moves; a device which can be placed under a heavy object to facilitate its movement.
1905 *Engineering Rev.* XIII. 103/1 The Dolter system.. consists of a skate suspended from the under part of the car; this makes contact with a small iron block embedded in the road. **1940** *Chambers's Techn. Dict.* 774/2 *Skate, sidetracking.* (1) A device to move an aeroplane sideways on the ground, for manoeuvring in confined spaces, as when packing into sheds. (2) A shoe for slipping beneath the wheels for handling an aeroplane on soft snow. **1961** *Daily Tel.* 4 Oct. 15/2 An engineering firm has offered to supply skates for the Arch. *Ibid.*, The skates, or tracked skids, are made of steel plates. **1972** *Police Rev.* 10 Nov. 1477/1 It appears that the 'skates' shown in..last week's *Police Review* (page 1405) would combat the above criticisms. **1976** *Southern Even. Echo* (Southampton) 18 Nov. 17/7 Terry Maine..invented a 'skate' to go under the damaged wheel—which enables the aircraft to be moved quickly into a hangar for repairs.
e. N. Amer. A set of tackle for halibut-fishing, etc., used chiefly on the Pacific Coast of N. Amer.
1882 J. W. Collins in *Fishermen's Own Book* 96 We set twenty skates of trawl—the whole string. **1897** Kipling *Captains Courageous* iii. 75 'How many skates you reckon we'll need?' ''Baout three. Hurry!' 'There's three-hundred fathom to each tub,' Dan explained. **1960** M. Sharcott *Place of Many Winds* vii. 127 Either the night before or in the morning before the skates

of gear are set they must be baited. *Ibid.* 129 Trolling fishermen often curse the skates of halibut gear. **1972** F. Ford *Atush Inlet* ix. 85 The marker, then the anchor, then two-three hundred yards of halibut line with a baited hook every ten feet, then another anchor and a marker. That's a skate.

3. b. *skate-iron* (earlier example); **skate key,** a key for tightening roller-skates; **skatepark,** a park or rink for skateboarding.

1838 J. H. Ingraham *Burton* I. x. 143 It was placed on runners sixteen inches high, shaped like skate-irons. **1962** 'E. McBain' *Like Love* xiv. 193 A little girl..was sitting on the steps tightening her skates with a skate key. **1977** *Montgomery Ward Catal.* Spring–Summer 509/1 Clamp-on sidewalk skates... Skate key included. **1976** *N.Y. Times Mag.* 12 Sept. 85/2 A $60,000 15 thousand-square-feet-of-concrete skatepark. **1977** *Sunday Times* 27 Nov. (Colour Suppl.) 27/4 Use purpose-built skate-parks as they have a variety of bowls and slaloms which allow you freedom to develop tricks away from other skaters and spectators.

skate (skē^ɪ^t), *sb.*³ *slang* (chiefly *U.S.*). [Origin uncertain.] **1.** A poor, worn-out, decrepit horse.

1894 Kipling in *Cent. Mag.* Dec. 295/2 This yaller-backed skate comes to our pastur'. **1923** E. Hemingway *Three Stories* 29 They'd kill that bunch of skates for their hides and hoofs up at Paris. **1935** H. Davis *Honey in Horn* vi. 61 Joel Hardcastle's horses were underfed, badly shod, and skates. **1978** E. Tidyman *Table Stakes* I. iv. 68 The man was a gambler... A pony player. Used to bet thousands on the worst-looking skates you've ever seen.

2. a. A mean or contemptible person. Esp. in *cheap skate* (also *attrib.* or as *adj.*).

1896 Cheap skate [see *horse *sb.* 17]. **1898** F. P. Dunne *Mr. Dooley in Peace & War* 198 If th' skate fr'm Oklahoma is allowed f'r to belch anny in this here assimblage, th' diligates fr'm th' imperyal Territ'ry iv New Mexico'll lave th' hall. **1904** J. C. Lincoln *Cap'n Eri* xxi. 383 Offered me a hundred dollars a week, the skate! **1935** D. L. Sayers *Gaudy Night* xix. 399 'It would suit them very well,' thought Harriet, 'the cheap skates!' **1947** *Partisan Rev.* XIV. 259 Samuel lost his temper and told the boss what he thought of him, what a cheap skate he was. **1958** *New Statesman* 4 Oct. 444/2 A cheapskate doctor he employed to save a few dollars gave his wife, Mary, morphine to ease her pains after delivering her youngest son, Edmund, and she has become an addict. **1960** H. Pinter *Caretaker* I. 9 *Aston:* I saw him have a go at you. *Davies:*..The filthy skate, an old man like me. **1973** J. Porter *It's Murder with Dover* xii. 119 They were hardened women of the world and knew a cheap skate when they saw one.

b. *labour skate* (U.S.), a trade-union official.

1930 *Amer. Mercury* Dec. 456/2 *Labor-skate*, an official of a labor union. **1978** *Washington Post* 27 Jan. D7/3 Most of the crowd consisted of labor skates, members of Jewish groups, and friends of Jackson and Moynihan.

skate, *v.* Add: **1. a.** *fig.* Esp. in phr. (*a*) *to skate over* (or *on*) *thin ice*; (*b*) *to skate over* or *round* (a fact, subject, etc.), to pass by or over hurriedly, to avoid mentioning.

(*a*) **1897** A. Beardsley *Let.* 15 Sept. (1970) 368, I hardly like to think now of all the thin ice I must have skated over since March 31st—a miraculous patinage! **1897** [in Dict.]. **1926** P. Guedalla *Palmerston* V. iii. 356 Even *Punch* regaled its readers with a princely figure of slightly sinister aspect skating perilously on the thin ice of foreign affairs. **1945** E. Waugh *Brideshead Revisited* I. v. 98 He..could talk at length of..how this or that Jesuit or Dominican had skated on thin ice or sailed near the wind in his Lenten discourses. **1978** H. Carpenter *Inklings* IV. i. 216 He skated on thin ice in the opening chapter of *The Problem of Pain*, where he offered his readers a 'proof' of the existence of God which..tackled this immense issue 'on the scale of a pamphlet in a church porch'.

(*b*) **1928** *Manch. Guardian Weekly* 30 Mar. 243/1 The Premier did not do more than skate round the problem. **1948** 'N. Shute' *No Highway* v. 123 We both skated over the implications of that. **1957** *Economist* 7 Dec. 860/1 The reason for the outbreak of the second Balkan war in 1913..is gracefully skated over. **1965** *New Statesman* 16 Apr. 622/3 Mr Brown's latest paper on prices and incomes skates carefully around this point. **1971** *Where* Sept. 266/1 It also skates over the fact that it is an offence to be in possession of the drugs listed if they have not been legally prescribed. **1979** C. Moule in M. Goulder *Incarnation & Myth* v. 135 It has been claimed that Mark's christology is authoritative and as much part of the New Testament as Paul's... But this is to skate over the question, What was Mark's intention?

c. *colloq.* To depart speedily.

1915 in C. Johnson *Battleground Adventures* liv. 418 Holt met the ol' man comin' from the barn as hard as he could run. Oh! he was comin' from thar skatin'. *c* **1926** 'Mixer' *Transport Workers' Song Bk.* 31 Well, I'm skating. Coming, 'Slasher'? **1937** G. Frankau *More of Us* v. 63 When one's happy—well, time simply flies. Me for the hay. Let's get our bill, and skate.

d. *U.S. slang.* (See quots.)

1945 L. Shelly *Jive Talk Dict.* 17/1 *Skate*, to get away with something. **1977** *Amer. Speech* 1975 L. 66 *Skate vi*, shirk duties. 'The new pledges are really skating this week.' **1979** *Observer* 18 Mar. (Colour Suppl.) 56 I'm not a woman's libber but I don't want to skate (shirk).

2. c. To slide or glide over. Also *fig.*

1900 [see *fenner]. **1970** G. F. Newman *Sir, you Bastard* i. 22 Sneed skated the passing out examination with the highest marks on record. **1971** B. Patten *Irrelevant Song* 27 Quick as the autumn marigold Skates the borders of whitening grass.

skateboard (skē^ɪ^·tbō^ə^ɹd), *sb.* orig. *U.S.* Also **skate board** [f. Skate *sb.*², after *surfboard*.] A narrow platform mounted on roller-skate wheels, on which the rider coasts along, usu. in a standing position (orig. developed from surf-riding, chiefly as a pastime). Also *attrib.*

1964 *Life* 5 June 89 Skateboards appeared last fall in southern California. **1964** *Surfer* Sept. 72 No one, except possibly the skateboard manufacturers, took them seriously until recently. *Ibid.* 76 When they can't go out in the water and surf, they can do it on land with a skateboard. **1965** *Globe & Mail* (Toronto) 21 Apr. 5/6 Hundreds of vacationing teen-agers..wheeled to City Hall yesterday on their skate boards to demand an off-the-street area to carry on their latest fad. **1974** *Daily Colonist* (Victoria, B.C.) 21 Aug. 18/6 He was..a legless cripple whose only means of locomotion was a skate board. **1976** A. Cassorla *Skateboarder's Bible* i. 9 Already in the works are extensive skateboard parks. **1976** C. Weston *Rouse Demon* (1977) xxi. 96 A boy on a skateboard whirled by on the sidewalk. **1978** K. Amis *Jake's Thing* xx. 204 The local authorities wanted..a skateboard park built on the site. **1978** *Cornish Guardian* 27 Apr. 23/1 (Advt.), Used Car Parts (fronting the quay, adjoining skateboard rink). **1980** P. Harcourt *Tomorrow's Treason* I. i. 28 An even smaller boy..seemingly unable to get off his skateboard..was careering straight into the path of my car.

Hence **ska·teboard** *v. intr.*, to ride on a skateboard; **ska·teboarder**; **ska·teboarding** *vbl. sb.*

1964 *Life* 5 June 89 Skateboarding requires only a tapered piece of wood flexibly mounted on roller-skate wheels and a stretch of pavement. *Ibid.*, A good skateboarder can do all a surfer's tricks and more. **1964** *Surfer* Sept. 74 Some of the skateboarders have set up slalom courses, timing each other from stand-still starts. **1968** W. Warwick *Surfriding in N.Z.* 19/1 To skateboard properly the rider should have a reasonable sense of balance. **1976** *National Observer* (U.S.) 3 July 12/2 They're..'taking it to the limit': skateboarding up the sides of empty swimming pools and pipelines. **1977** *Times* 15 Oct. 24/2 The odd skateboarder en route for the adventure playground. **1978** *Morecambe Guardian* 14 Mar. 4/5 Coun. Mrs Taylor said that if any children still felt strongly about having no skateboarding park they should contact her so that a united effort could be made. **1979** V. S. Naipaul *Bend in River* xv. 249 A wide, sloping avenue..with boys skateboarding.

skater. Add: Occas. = *skateboarder.

1977 *Times* 19 Nov. 26/6 Many a skate park, rightly, refuses admittance unless skaters wear protective helmets and pads. **1977** [see *skatepark* s.v. *skate *sb.*³ 3 b].

skating, *vbl. sb.* Add: **a.** (Examples = Roller-skating.)

1876 J. A. Harwood *Rinks & Rollers* iv. 63 There are few more lively..sights than..the young engaged in skating and flirting, while the elder ones sip tea between the trees. **1948** C. Beastall *Int. Roller Skating Annual* 15 The RSROA, who banned metal wheels, introduced skating and proficiency tests..in their rinks. **1975** *Man. Artistic Roller Skating* 38 The candidate is required to compose..an original dance..suitable for skating in public skating sessions.

b. *skating pond, -rink* (further examples; also *fig.*), *skirt.*

1903 Skating pond [see *live wire* s.v. *live *a.* 8]. **1981** *Northeast Woods & Waters* Jan. 24/3 Osborndale in Derby even offers a large skating pond, lighted for night skating. **1875** *Building News* 19 Nov. 579/1 Skating Rinks. The [roller-]skating rink is an institution that has rapidly grown in..popularity. **1926** Maines & Grant *Wise-Crack Dict.* 14/2 *Skating rink for flies*, baldheaded man. **1958** *Roller Dance Skating Man.* 32 (Advt.), Skating Rinks at: Skating Rink & Cafeteria, Granby Halls, Leicester. **1976** *Flintshire Leader* 10 Dec. 2/8 Freezing conditions this week turned a road in Carmel near Holywell into a skating rink. **1968** J. Ironside *Fashion Alphabet* 62 *Skating skirt.* Traditionally this is cut circular and very short, worn with tights. **1971** *Petticoat* 24 July 3/4 They'll be taking hemlines just about as high as they can go, even to skating-skirt length.

‖ **skaz** (skaz). [Russ.] First-person narrative in which the author assumes a persona. Also *attrib.*

1926 D. S. Mirsky *Contemp. Russ. Lit.* vii. 311 Michael Zoshchenko..is a more narrative writer: he is also an ornamentalist, but his ornamentalism is a purely colloquial skaz, which proceeds from Leskov. **1957** W. E. Harkins *Dict. Russ. Lit.* 360 *Skaz*, a Russian word designating a narrative told by a fictitious narrator, rather than by the author directly. **1967** *Russ. Review* XXVI. 177 In succeeding works Solzhenitsyn resorts to *skaz* only occasionally. **1974** Moore & Parry *Twentieth-Cent. Russ. Lit.* i. 10 Remizov was a sharp stylist who often wrote in the vein of folklore; he carried on Nikolai Leskov's tradition of *skaz*, an attempt to reproduce the exact idiom of each speaker in a story. **1980** *Times Lit. Suppl.* 7 Nov. 1264/4 The narrator [is] a typically Russian busybody in the 'skaz' tradition.

sked (sked), colloq. abbrev. (orig. *U.S.*) of (*a*) Schedule *sb.*; (*b*) Schedule *v.*; (*c*) Scheduled *ppl. a.*; also *ellipt.*, a scheduled flight.

1929 *Amer. Speech* IV. 289 A 'sked' (schedule) is a message to the central office briefly sketching the contents of a story offered for transmission. **1942** Berrey & Van den Bark *Amer. Thes. Slang* § 597/5 *Bill* [verb]; *schedule*, sked. **1949** *Daily Progress* (Charlottesville, Va.) 26 Aug. 12/3 (*heading*) W[illiam] & M[ary] skeds tour for court team. **1953** Scott & Fisher *Thousand Geese* ix. 103 If we were to get back for our radio sked..we thought we should start home. **1972** *Sat. Rev.* (U.S.) 4 Mar. 33/2 The ultimate answer..is for the skeds to become competitive with the charters. **1977** *New Day* Summer 6/1 All mission stations [in Papua New Guinea] are connected by radio, and different times (skeds) are assigned for their use. The Post Office has radio skeds for Government traffic. **1981** *Beautiful Brit. Columbia* Fall 22 Near the Dean, at Eliguk or Gatcho lakes you can be flown out to Nimpo Lake at Highway 20; thence by car, 'sked-flight' (a Cariboo phrase for a scheduled flight). **1981** L. Leamer *Assignment* i. 8, I don't mind when it's skedded right..but..I don't ever want to work with Henderson again.

skedaddle, *v.* Add: Also **skiddaddle. 1.** (Earlier example.)

1861 *N.Y. Tribune* 10 Aug. 5/5 No sooner did the traitors discover their approach than they 'skiddaddled', (a phrase the Union boys up here apply to the good use the seceshers make of their legs in time of danger).

Skee-Ball (skī·bǭl). Also **skeeball.** [f. Ski *sb.* or *v.* + Ball *sb.*¹] An indoor game in which balls set rolling down an alley are projected over a hump or otherwise into targets. Also *attrib.*

Skee-Ball is a proprietary term in the U.S.

1923 *Daily Mail* 11 June 3 The game of Skeeball is an American invention which has been in increasing operation in the United States for some six years. It is the latest development of the game of bowls, sharply distinguished from all similar games by a hump in the centre of the alley which causes the ball to leap high in the air and enter a target. **1930** A. P. Herbert *Water Gipsies* xviii. 258 Ernest played skeeball because it was the nearest thing to skittles, and he won a china dog. **1974** J. Wainwright *Evidence I shall Give* xxiv. 121 A Skee-Ball set-up with six lanes.

skeeler (skī·ləɹ). [Perh. f. Ski *sb.* + Rol)ler *sb.*¹ (See quots.) Hence **skee·ling** *vbl. sb.*, skating on skeelers.

1968 *Daily Tel.* 28 Dec. 23/4 (*caption*) 'Skeelers', a cross between ice and roller skates. *Ibid.* 23/6 The skates, called 'Skeelers'.., are just part of the fun for the children at the *Daily Mail* New Year Show. **1969** *Ibid.* 31 Jan. 17/5 Skeeling, launched last autumn, gives some of the exhilarating feeling of skating—but without ice. Skeelers are like ice skates but three wheels replace the blade; they will slide on any hard surface. **1972** *Kingston* (Ontario) *Whig-Standard* 13 June 15/6 Skeeling, according to the publicity releases, is iceless skating; a summer hockey equivalent. Skeelers are a cross between roller skates and ice skates and were originally developed by the Russians to help condition speed skaters during the off-season.

skeesicks, var. *skeezicks.

skeet (skīt), *sb.*² orig. *U.S.* [Proposed for the name of the sport (see quot. 1926) as an 'old' form of Shoot *v.*] A form of clay-pigeon shooting in which targets are projected at a variety of shooting angles in a semicircular range. Also, in some *attrib.* uses (see below), a clay pigeon; so *to shoot skeet.*

1926 *National Sportsman* (U.S.) May 18 (*heading*) Skeet, the new sport. *Ibid.*, Since the prize of $100 was offered for the most suitable name for the new shooting sport.., nearly 10,000 suggestions have been received... After careful consideration, the name that seemed to apply itself the best was '*skeet*', a very old form of our present word 'shoot'... Mrs. Gertrude Hurlbutt, Dayton, Montana, sent in the suggestion. **1931** *Daily Progress* (Charlottesville, Va.) 26 Oct. 3/2 Skeets [*sic*] is unlike regular trap-shooting in that the birds come from oppositely located houses and that, in one position two birds come at once. **1931** L. B. Smith (*title*) Better trapshooting. With a section on skeet. **1939** *Country Life* 11 Feb. p. xxi/3 (Advt.), Clay bird shooting.—Practice and Coaching; every flight imitated; skeet; automatic traps. **1955** R. Churchill *Game Shooting* III. i. 173 In 1927, I myself introduced the game of skeet to England. **1976** *Webster's Sports Dict.* 395/1 A round of skeet consists of 25 shots. **1979** R. Jaffe *Class Reunion* II. ii. 139 He liked to dance, play golf, drink, shoot skeet, and laugh.

b. *attrib.* and *Comb.*, as *skeet championship, contest, ground, gun, match, range, shoot, shooter, shooting.*

1942 *Tee Emm* (Air Ministry) II. 75 He has won the Skeet championship several times. **1975** *Oxf. Compan. Sports & Games* 927/1 The first U.S. national skeet championships were held in 1935. **1952** *Times* 14 July 5/5 Colonel C. T. Edwinson..won the skeet (clay pigeon) contest in the world shooting championships here yesterday. **1926** *National Sportsman* (U.S.) Sept. 22 (*heading*) News from the skeet grounds. **1975** *Oxf. Compan. Sports & Games* 925/2 On a skeet ground layout there are two spring-release traps. **1976** *Shooting Times & Country Mag.* 9–15 Dec. 5/1 (Advt.), David Price..can testify to the suitability of his Miroku 800SW Skeet gun for game shooting. **1970** *Cape Times* 28 Oct. 24/4 Evgeny Petrov, of Russia, set an unofficial world record here, breaking 200 straight targets in the International skeet match of the world shooting championships here. **1942** *Tee Emm* (Air Ministry) II. 75 We had an article on Training with Clay Targets and the Skeet Range in our April, 1942, issue. **1970** G. Jackson *Let.* 21 May in *Soledad Brother* (1971) 255 Quietly..I would have my fronts open as many skeet, trap, rifle, and pistol ranges as I could rent space for in and around the black community. **1926** *National Sportsman* (U.S.) June 18 A group of about twenty good sportsmen in the town of Maynard, Mass. held their first skeet shoot. **1959** *Times* 1 Aug. 9/5 In addition to tests which simulate the different types of game-shooting, 'skeet' and 'down the line' shoots are held. **1926** *National Sportsman* (U.S.) June 18 If the skeet shooter were in hot competition..he would probably choose small shot. **1967** *Boston Sunday Herald* 26 Mar. vi. 6/5 A rifle range for the skeet shooter. **1926** *National Sportsman* (U.S.) May 19/2 (Advt.), The Western Practice Trap is ideal for 'Skeet' shooting, or for any kind of shotgun practice. **1971** *Courier-Mail* (Brisbane) 11 June 7/3 Winner..was Graham Bailey,..who took the final of the Winchester Australia skeet shooting championship at the Belmont range.

Hence **skee·ting** *vbl. sb.*², participating in the sport of skeet, skeet shooting.

1926 *National Sportsman* (U.S.) May 18/2 It is as easy to say *skeet, skeeting, skeeter* as it is to say shoot, shooting, shooter. *Ibid.* Sept. 22/3 The game of *skeeting* is just like a ball game. *Ibid.*, Hunt's Club and *skeeting* field is a public place. **1968** *Daily Mail* 23 Oct 16/7 Skeeting is a precise form of shooting which is in the Olympic Games for the first time.

skeet (skīt), *v.*² *dial.* [Alteration of (esp. *U.S.*) Scoot *v.* or (esp. *Sc.*) Skite *v.*² or Skate *v.* in Dict. and Suppl.] **1.** *intr.* To move swiftly; to hurry; to run. Usu. with advbs.

1838 J. C. Neal *Charcoal Sk.* 97 You must skeete, even if you have to cut high-dutchers with your irons loose. *a* **1855** J. F. Kelly *Humors of Falconbridge* (1856) 251, I skeeted down them steps into the Common to let off my corked up risibilities. **1861** in L. C. Baker *Hist. U.S. Secret Service* (1867) v. 101 Burn the letter..and then get in your hole and skeet for Dixie. **1877** G. Stewart *Fireside Tales* 89 Skeet howe hame, guid folk! **1922** Joyce *Ulysses* 748 That icy wind skeeting across from those mountains. **1924** C. Greer-Petrie *Angeline of Hill Country* 18 Here comes a mighty impudent lookin' darky a-skeetin' towards us. **1929** L. Robinson in *Lett. Sean O'Casey* (1975) I. 358 Johnny Perrin..got married yesterday and skeeted off to Wales for a few days.

2. *trans.* To squirt, to eject (fluid). Also *absol.*

1880 Courtney & Couch *Gloss. Words in Use in Cornwall* 51/2 *Skeet*, *v.*, to eject saliva through the teeth. **1886** J. J. H. Burgess *Shetland Sk. & Poems* 114 Every platch 'at he med skeetit it up and doon ower every ane 'at cam' near. **1908** *Dial. Notes* III. 370 *Skeet, v. tr.*, to scoot (water), spew out of the mouth, especially between the teeth. **1912** J. Nicolson *Hame-Spun* 45, I was skeetit frae head ta foot, sae 'at I'm not onlookin'. **1935** Z. N. Hurston *Mules & Men* I. iii. 64 Julius spat out into the yard, trying to give the impression that he was skeeting tobacco juice like a man. **1946** C. McCullers *Member of Wedding* I. 48 She loved to..lightly meddle with their things—with Mrs Marlowe's atomizer which skeeted perfume, the grey-pink powder puff, [etc.].

skeeter, *sb.*¹ (Earlier example.)

1839 *Spirit of Times* 21 Dec. 495/2, I was fas asleep, and dreaming dat a big skeeter was a biten me.

skeeter (skī·təɹ), *sb.*² [f. *skeet *sb.*² + -er¹.] One who participates in the sport of skeet, a skeet shooter.

1926 *National Sportsman* (U.S.) May 18/1 From this station Skeeter faces each trap in turn and shoots at the target coming directly over his head. **1968** *Daily Mail* 23 Oct. 16/8 The skeeter loses a point if he fails to destroy the pigeon as it passes invisibly over his head.

skeeter (skī·təɹ), *v.* Var. Skitter *v.*² 1.

1964 J. Hillaby *Journey to Jade Sea* 101 Three geese promptly took off, skeetering up into a gust of wind. **1971** C. McCullers in *Redbook* Oct. 196/1 She had one of those grotesque little imaginings... She saw herself and Marshall... Skeetering angrily up and down the cold blank glass like minute monkeys. **1972** *Observer* (Colour Suppl.) 23 Apr. 35/4 Fiennes and Stanley Cribbett skeeter through the savage waters of the Bridge River rapids.

skeeze (skīz), *v. rare.* [Origin uncertain.] *intr.* ? To peer, to glance obliquely.

1922 Joyce *Ulysses* 298 Old Garryowen started growling again at Bloom that was skeezing round the door.—Come in, come on, he won't eat you, says the citizen. *Ibid.* 731 Hes mad on the subject of drawers..always skeezing at those brazenfaced things on the bicycles with their skirts blowing up to their navels.

skeezicks (skī·ziks). *U.S. slang.* Also **skeesicks, -zacks, -zecks.** [? Fanciful.] A good-for-nothing, a rascal, a rogue. (Now usu. playfully of children.)

1850 *Frontier Guardian* (Kanesville, Iowa) 2 Oct. 3/5 Though Kister that skeezecks with Hall at his back, Should come again thieving [etc.]. **1869** B. Harte in *Overland Monthly* June 572/1 Thar aint nobody but him within ten mile of the shanty, and that 'ar d—d old skeesicks knows it. **1875** J. G. Holland *Sevenoaks* iii. 40 If there's anything awful bad..in [the word] Skeezacks—I should say that Tom Buffum was an old Skeezacks. **1908** *Everybody's* Dec. 796/2 This is a poor skeezicks that's got nothing to eat but an onion. **1939** P. A. Rollins *Gone Haywire* v. 117 Eb Hawkins, that ol' skeesicks you met on th' railway train an' liked, is th' feller that's acted as th' owners' agent in sellin' rights to your uncle.

skeg, *sb.*¹ Add: **2. b.** *Surfboarding.* The fin of a surfboard.

1962 T. Masters *Surfing Made Easy* 65 Skeg, the rudder or fin of a surfboard. **1964** *Sunday Mail Mag.* (Brisbane) 17 May 1 These were the now famous Malibu boards. 10 feet long, and made of balsa, with a fin or skeg for greater control. **1968** *Surfer Mag.* Jan. 73/2 He showed them skeg-first take offs.

skein, *sb.*¹ Add: **1. b.** (Further examples.)

1932 W. B. Yeats *Words for Music* 26 For love is but a skein unwound Between the dark and dawn. **1935** T. S. Eliot *Murder in Cathedral* I. 37 You hold the skein: wind, Thomas, wind The thread of eternal life and death. **1939** Dylan Thomas *Map of Love* 16, I with a living skein, Tongue and ear in the thread, angle the temple-bound Curl-locked and animal cavepools of spells and bone.

† **2. c.** *Cytology.* The chromosomal strands in a cell undergoing mitosis; used *attrib.* to denote the stage of mitosis now known as *prophase; = *spireme. *Obs.*

1889 *Q. Jrnl. Microsc. Sci.* XXX. 164 The first stage of karyokinesis, the so-called 'dense skein' ('dichter Knäuel'). *Ibid.* 173 Rable says definitely that he has always found the longitudinal splitting of the chromatic threads to be completed at the end of the skein phase. **1904** *Science* 4 Mar. 393/1 No sign of chromatin thread (linin or skein) is apparent.

3. *skein-winder.*

1920 L. Hooper *Weaving for Beginner* x. 76 A skein winder..is only required if the weft is supplied to the weaver in skeins. **1964** O. G. Tod *Joy of Hand Weaving* (ed. 2) xviii. 87 If winding from a skein, place the skein around an adjustable *skein-winder*.

skein, *v.* Add: Also *fig.* Hence **skeined** *ppl. a.*

c **1885** G. M. Hopkins *Poems* (1967) 98 Let life, waned, ah let life wind Off her once skeined stained veined variety upon, all on two spools. **1955** E. Bowen *World of Love* xi. 219 Water skeined the landscape. **1971** 'D. Halliday' *Dolly & Doctor Bird* v. 62 The Florida coast. Flat land skeined with sheets of flat water. **1977** P. Scupham *Hinterland* 8 When Vulcan beat new armour out for Rome..Skeined cupids hooded their toy bacchanals.

Hence **skei·ner**, one who or that which makes yarn into skeins.

1921 *Dict. Occup. Terms* (1927) 169/1 *Skeiner* (twine); minds skein or rand machine, which winds finished twine into skeins. *Ibid.* 367/2 *Bundler* (flax and hemp); *skeiner*; puts together necessary number of hanks of yarn to form a bundle. **1931** M. L. Davies *Life as we have known It* 74 As a 'skeiner' her work was to separate and twist up the skeins from the 'bond' (on a silk mill). **1969** E. H. Pinto *Treen* 318/2 The niddy-noddy was a combined measure and skeiner.

skeletal, *a.* Add: The pronunciation (skelī·tăl) is occasionally heard. **b.** *skeletal muscle*, add to def.: or transmitting force to connective tissue sheets, and in most cases under voluntary control; striated muscle other than cardiac muscle. (Further examples.)

1936 L. B. Arey *Developmental Anat.* (ed. 3) xiii. 361 With the exception of those muscles of the head and neck which differentiate out of the branchial arches, the skeletal muscles originate from that portion of the mesodermal segment designated a myotome, or muscle plate. **1978** D. R. Lamb *Physiol. of Exercise* ii. 15 The skeletal muscles consume most of the oxygen and require most of the body's blood during heavy exercise. **1982** *Sci. Amer.* June 48/2 It was found that in skeletal and cardiac muscle (which is called striated muscle because of its striped appearance in a micrograph) calcium binds to a protein called troponin *c*.

c. *skeletal soil* = *lithosol* s.v. *litho-.

[**1928** *Proc. 1st Internat. Congr. Soil Sci.* IV. 31 [Soils] developing in a normal way..whose profile is imperfect wholly because of lack of time to complete their development have never been given any designation covering them as a whole. They have been designated as *skeleton soils* but this term is not applicable to the group as a whole.] **1932** G. W. Robinson *Soils* xvi. 320 Immature skeletal soils are found in the south [of Germany]. **1939** [see *lithosol* s.v. *litho-]. **1977** J. C. F. Tedrow *Soils of Polar Landscapes* xxii. 568 Ohsumi reviewed the characteristics of the alpine soils in Japan and established four varieties: (1) Alpine grassland soils, (2) Alpine podzols, (3) Alpine wet meadow soils, and (4) Skeletal soils.

d. Having or consisting of only a framework or outline; bare, meagre.

1961 W. Brown *Bedeviled* 106 Once Dr. Hazel had pieced together this skeletal tale, he notified Captain Brill. **1967** T. Keneally *Bring Larks* ii. 17 She stood business-like against the skeletal tracery of her master's sick vines. **1967** A. N. Sherwin-White *Racial Prejudice in Imperial Rome* Pref. p. vii, They have been printed much as delivered, with the addition only of source references, a skeletal bibliography, and translations of most quotations. **1971** *Physics Bull.* Aug. 462/1 Some body can be given to these skeletal facts by an order-of-magnitude calculation.

Hence **ske·letally** *adv.*, as regards the skeleton.

1956 *Nature* 18 Feb. 342/2 The history of the Amphibia Salientia can readily be traced back to the Jurassic when it appears that, skeletally at least, they were already typical, modern Anura. **1974** *Ibid.* 13 Sept. 137/2 In Detroit..serum antirachitic activity..was significantly less in symptomatic osteoporotic subjects than in the skeletally normal.

skeleton, *sb.* Add: **1. d.** *Hist.* A member of a 'skeleton army' (see sense 7 d below).

1882 *Eastern Post* 4 Nov. 3/3 There was nothing to fear from the latest born army; there would be 'skeletons' enough in it. **1950** R. Sandall *Hist. Salvation Army* xxxiii. 196 The police..dispersed the 'skeletons'. **1981** C. Scott *Heavenly Witch* viii. 120 The Judge of Assize..condemned the Skeletons as the aggressive party.

3. b. *Chem.* The basic atomic framework of a molecule, disregarding substituents (and sometimes also side chains or bond type).

1907 J. B. Cohen *Org. Chem. Adv. Students* I. xiii. 472 It will be at once perceived how very large a number of possible menthadienes can be derived from these two skeleton structures. **1910** *Jrnl. Physiol.* XLI. 29 The carbon-skeleton of β-phenylethylamine is..identical with that of adrenine. **1926** [see **isoprene unit*]. **1956** I. L. Finar *Org. Chem.* II. viii. 310 The nature of the sesquiterpene skeleton is also characterised by the number of double bonds present in the molecule. **1975** *Jrnl. Chem. Soc. Dalton Trans.* 31/1 Internal vibrations of nitrate, the imidazole ligands.., and the ML$_4$ skeleton.

6. d. A skeleton forme.

1938 F. T. Bowers in *Library* XIX. 315 When the term skeleton is used it will indicate the imposed cross-bars, furniture, and running-titles of a forme. **1950** *Studies in Bibliography* III. 246 The first five sheets of the play were printed with three skeletons used in a pattern somewhat different from that in *Lear.* **1978** *Studies in Eng. Lit.: Eng. Number* (Tokyo) 22 Skeleton II was used for the outer forme of sheets B, C, D.

7. c. (Further examples.) Also, in wider use, applied to any staff, company, etc., of the minimum size for carrying on the work to be done; so *skeleton service*, a service reduced to a bare minimum.

1914 in W. S. Churchill *World Crisis* (1923) xix. 445 There is only a skeleton force of patrol vessels available on the East Coast. **1925** *Strand Mag.* Sept. 255/2 A skeleton staff were working nervously under the direction of a chartered accountant. **1926** *Times* 6 May 3/1 On the railways skeleton services were run on main and suburban lines, and more trains are promised to-day. **1928** *Daily Mail* 7 Aug. 2/5 The establishment of skeleton air defence formations on the northern coast. **1937** W. H. Saumarez Smith *Let.* 23 Jan. in *Young Man's Country* (1977) ii. 53 All my officers have gone out to the mofussil..and all the clerks except 6, a mere skeleton staff to keep the work going. **1938** *Sun* (Baltimore) 20 Jan. 6/3 A skeleton crew of 125 men will take the ship to Scotland. **1957** P. Kemp *Mine were of Trouble* ii. 29 Skeleton crews whose task was to train Spaniards in the use of their weapons. **1973** 'I. Drummond' *Jaws of Watchdog* xviii. 255 A skeleton staff was still on duty, juniors, message-takers. **1976** *S. Wales Echo* 25 Nov. 4/2 Buses will not run in Cardiff for three consecutive days over Christmas despite a last-ditch effort to have a skeleton service on one day.

d. *skeleton frame, framework*; also applied to a vehicle or other conveyance of basic or light construction, as *skeleton bob* [*bob *sb.*¹ 2 e], *brake, break* [Break *sb.*²], *car, gig, sleigh, wagon*; **skeleton army** *Hist.*, a group of people attempting to disrupt the activities of the Salvation Army or Church Army; **skeleton brass,** a memorial brass representing a skeleton; **skeleton construction** (see quot.); **skeleton drill** *Mil.*, infantry drill for the instruction of officers, in which a small number of men represents a battalion; also *fig.*; **skeleton forme** (see quot. 1972); **skeleton weed,** a perennial herb, *Chondrilla juncea*, of the family Compositæ, native to the Mediterranean region and naturalized in Australia, where it is a troublesome weed of cereal crops.

1881 *War Cry* Christmas No. 6/2 The chief officers of 'The Skeleton Army', raised to oppose us at Exeter, were converted. **1888** C. M. Yonge *Beechcroft at Rockstone* I. iii. 64 The Salvation Army was marching that way, and ..yells and cat-calls behind showed that the Skeleton Army was on its way to meet them. **1920** H. Begbie *Life W. Booth* I. xxix. 482 'Skeleton Armies'..set themselves up to break up the processions of the Salvation Army. **1950** R. Sandall *Hist. Salvation Army* xxxiii. 196 These skeleton armies carried flags usually bearing a skull and crossbones device. **1980** F. K. Prochaska *Women & Philanthropy in 19th-Cent. England* vi. 193 A 'Skeleton Army' of rowdies often shadowed volunteers of the Church Army and..pelted them. **1954** R. Martin *Your Ski Holiday* xiii. 91 Racing on the 'skeleton bob' is a very different affair... The 'skeleton bob' is a steel chassis with two steel runners, about 3 feet long and about 13 inches apart. On this chassis is a sliding seat. **1963** I. Fleming *On H.M. Secret Service* xii. 133 A little 'garage' that housed the bob-sleighs and one-man skeleton-bobs. **1898** *Carriage Builders' Jrnl.* Nov. p. viii/2 (Advt.), Wanted, Pair-horse Skeleton Brake. **1935** *Automobile & Carriage Builders' Jrnl.* Apr. 68/2 The skeleton brake has a high driving seat with the fore and hind carriages connected by a perch only. **1890** H. W. Macklin *Monumental Brasses* i. 17 Shroud and skeleton brasses came into general use. **1956** A. C. Bouquet *Church Brasses* vii. 147 There is a skeleton brass at Weybridge, Surrey, with three effigies. **1972** R. le Strange *Compl. Descriptive Guide Brit. Monumental Brasses* 9 Between the two, the shroud and skeleton brasses, lay the cadavers. **1974** S. Walrond *Encycl. Driving* 237 Skeleton Break... This vehicle..was used for breaking and training when a youngster would be put in alongside an older schoolmaster. **1936** '*ABC*' *Brit. Columbia Lumber Trade Directory* 73 Elco Logging Co. Ltd... Three High Leads;..45 Skeleton Cars. **1942** R. L. Haig-Brown *Timber* 253 *Skeleton car*, a railroad car made up of two sets of four wheels joined by a heavy timber across which the steel bunks are set to carry the logs. **1891** *Archit. Record* Oct.–Dec. 228 Within the past three or four years a new method of constructing very high buildings in New York has come into vogue. It is known as the skeleton construction and consists in the use of iron or steel columns, with thin curtain walls between, in place of solid thick brick walls. **1876** Voyle & Stevenson *Military Dict.* 387/2 *Skeleton drill*, which is a method of instructing officers and non-commissioned officers in drill, when a sufficient number of men cannot be collected to form a battalion in single rank. **1897** Hardy *Well-Beloved* III. v. 278 Pierston..could consider, and practise thoroughly a species of skeleton-drill in receiving visitors when the pair should announce themselves as married. **1888** Skeleton forme [in Dict.]. **1964** F. Bowers *Bibliogr. & Textual Crit.* I. i. 10 An examination of the running-titles..discloses that the text of the play in sheet B was imposed in two different skeleton-formes. **1972** P. Gaskell *New Introd. Bibliogr.* 109 All these re-usable parts, the typographical parts which left their mark upon the paper, and the chase, quoins, and furniture which did not, are known collectively today as the 'skeleton forme'. **1951** Koestler *Age of Longing* iv. 84 Towers of wood and towers of metal, towers which had merely a skeleton-frame and towers that were panelled in from all sides. **1897** *Building Construction* (new ed.) I. 17 The combination of columns and girders which form the 'skeleton' framework. **1867** 'T. Lackland' *Homespun* II. 181 At all hours of the day..a fly, a sulky, or a skeleton gig could be seen somewhere about the yard. **1974** S. Walrond *Encycl. Driving* 238 *Skeleton gig*, a light gig with a curved open stick-back seat which is suspended by iron stays on two side and one cross-spring. The shafts run outside the bootless body. **1902** *Hub* Aug. 172/1 The skeleton sleigh as seen last winter had a black body, cream gear and black irons. **1955** E. A. Collard *Canadian Yesterdays* 232 Next comes a stunner—a skeleton sleigh, red as fire, drawn by a trotter black as coal. **1868** H. Woodruff *Trotting Horse Amer.* x. 112 If the race is to be run in harness, it will be advisable to change the sulky for a skeleton wagon occasionally. **1974** S. Walrond *Encycl. Driving* 238 *Skeleton waggon*, an American four-wheeled single-seat vehicle which was built for racing. **1935** Ross & Taylor in *Agric. Gaz. N.S.W.* XLVI. 16/1 Skeleton weed is well liked by sheep, especially when it is in the young stages. **1965** *Austral. Encycl.* IX. 225/2 Skeleton weed..is a close relative of the dandelion and chicory, having a spindly habit of growth.

8. *skeleton-gaunt* adj.

1929 W. B. Yeats *Winding Stair* 3 When withered old and skeleton-gaunt.

skeletonized, *ppl. a.* Add: Also, possessing or having developed a skeleton. (Further examples.)

1976 *Nature* 29 Jan. 271/1 Some 500 million years ago.. all but two of the living phyla that are well skeletonised had already appeared. **1978** *Sci. Amer.* Sept. 108/1 These durable skeletonized invertebrates seem to have one thing in common: they all originally lived on the sea floor rather than burrowing in it.

skellum, *sb.* and *a.* Add: The usual spelling is now **skelm.** Also **schelm** (for examples attributed to German speakers see Schelm).

A. *sb.* **1.** (Further examples.)

1827 *Scenes & Occurrences in Albany & Caffer-Land* ii. 38 The Caffer flew into a violent passion, and said that he was no *schelm*:— that a *schelm* was a man that ought to be strangled. **1858** A. W. Drayson *Sporting Scenes* xviii. 314 A thorough Cape 'schelm' would..beat the best English swindler living. **1910** D. Fairbridge *That which hath Been* xxiii. 281 That poor nervous woman called out in Dutch: 'I will open the door, you schelms'. **1916** J. Buchan *Greenmantle* iii. 35, I got into German territory all right, and then a *skellum* of an officer came along, and commandeered all my mules. **1939** S. Cloete *Watch for Dawn* i. 7 A thieving skelm of a Hottentot. **1950** *Cape Argus* 26 July 2/5 There are very few fools in Johannesburg, but a lot of skelms. **1956** E. U. T. Huddleston *Naught for your Comfort* ii. 29 As for Absalom—he is a 'skellum', a 'tsotsi'—the kind of Kaffir who ought to be sjambokked every day: it would teach him sense. **1961** L. van der Post *Heart of Hunter* I. v. 95 'Moren!' he said hoarsely, 'you're more of a skelm than any spirit.' **1976** J. McClure *Rogue Eagle* v. 90 Vorster pardoned that *skelm* who was always breaking out of jail and shooting policemen.

2. (Earlier and later examples.)

1827 G. Thompson *Travels in S. Afr.* 467 Both the lion and the saddle had disappeared, and nothing could be found but the horse's clean picked bones. Lucas said he could excuse the schelm for killing the horse. **1909** P. Fitzpatrick *Jock of Bushveld* 260 The natives told us it was quite useless to follow it up as it was a real 'schelm'. **1939** S. Cloete *Watch for Dawn* v. 67 'I am not dead,' Kaspar said, having quieted his horse, 'and I do not think anything is broken,..but your horse is a skelm.'

B. *attrib.* or as *adj.* Rascally, villainous; sly; untrustworthy; of an animal: vicious, bad-tempered. Chiefly and now only *S. Afr.*

1673 [in Dict., sense 1]. **1801** G. M. Theal *Records Cape Colony* (1899) IV. 442 Owing to the present hostile disposition of the Skellam Hottentots.., I sent the Euphrosyne with the Dispatch. **1827** *Scenes & Occurrences in Albany & Caffer-Land* vii. 151 Diederik..determined on shooting it, declaring, that no *schelm* beast should kill his horse. **1828** T. Pringle *Ephemerides* 114 'Tis his lair—'tis his voice! from your saddles alight, For the bold skelm-beast is preparing to fight. **1829** C. Rose *4 Yrs. in S. Africa* 115, I joined a party of Schelm (robber) Hottentots and Kaffers, and we had horses, and arms, and we would attack the boors' houses. **1852** M. B. Hudson *S. Afr. Frontier Life* I. 16 We had in our drove a most skellum young mare. *c* **1902** I. Vaughan *Diary* (1958) 3 Joseph it is my feet that are skelm. They go by themselves, before I know. **1911** L. Cohen *Reminisc. Kimberley* xxiii. 397 'Hi, skilum Pontac!' (crack) 'Ah! you verdompt England!' (crack, whack, bang) and poor England would plunge into the yoke mad with pain and terror. **1972** *Sunday Tribune* (Natal) 25 June 23 You've got to be 'skelm' (sly) when you're working on a project like this.

skelp, *sb.*² **2.** *skelp-forger* (earlier example).

1804 *Aris's Gaz.* (Birmingham) 23 Apr. 3/3 (Advt.), Wanted a good skelp-forger, who has a perfect knowledge of drawing skelps for all kinds of binding, military, and African gun barrels.

skelp, *v.*² Add: Hence **ske·lping** *vbl. sb.*² (in quot. *attrib.*).

1803 *Aris's Gaz.* (Birmingham) 26 Dec. 2/4 (Advt), Lot 1. A forge and mill..recently used as a plating or skelping forge.

Skeltonic, *a.* and *sb.* Add: **a.** *adj.* (Later examples.)

1938 L. MacNeice *Mod. Poetry* 190 Witness his [*sc.* Auden's] Skeltonic polemic. **1954** C. S. Lewis *Eng. Lit. in Sixteenth Cent.* I. i. 72 *Pars Prima* is in rough trimeters of Skeltonic type. *Ibid.* ii. 137 There would be no problem if all Skelton's Skeltonic poems had been on this level.

b. *sb.* (Later examples.) Also *sing.*

1923 A. Huxley in *Athenæum* 12 Nov. 655/2 Skelton, whose..variations on the decasyllable are mostly..rough skeltonics. **1936** *N. & Q.* 21 Nov. 362/1 The Skeltonic consists of short verses of two, three or four accents.. varying in syllabic content..and rhyming in groups of anything from two to five or more lines at a time. **1954** C. S. Lewis *Eng. Lit. in Sixteenth Cent.* I. ii. 136 The problem about the source of Skeltonics sinks into insignificance beside the critical problem. **1976** *Times Lit. Suppl.* 12 Mar. 295/3 [Hood's] 'A Public Dinner' catches exactly in its breathless Skeltonics the noise and hurry of the occasion.

Skene (skīn). *Anat.* The name of Alexander Johnston Chalmers *Skene* (1838–1900), Scottish-born U.S. gynæcologist, used in the possessive to designate two small, blind ducts which open into the female urethra and the glands which they drain, homologous to the ducts of the male prostate gland; (described by Skene in 1880).

1890 Billings *Med. Dict.* II. 543/2 *Skene's tubules*, small blind canals, 3 to 6 mm. in length, lying along the urethra of the female and opening near the meatus. **1910** *Practitioner* Mar. 393 Two of these spots..probably represent the opening of Skene's ducts. **1932** C. J. Miller *Clin. Gynecol.* ii. 35 The cure of a chronic urethritis cannot be expected while there remains an active infection in Skene's glands. **1963** *Lancet* 5 Jan. 23/2 The urethra is often tender on palpation, and thick yellow mucus can be expressed from Skene's glands.

skeow-ways (skiau·wē^ɪ^z), *adv. Ir. dial.* [f. Skew *a.* and *adv.* + -ways.] = Askew *adv.*

1869 P. Kennedy *Evenings in Duffrey* xii. 108 Down I flew *Skeow ways* across the river. **1922** Joyce *Ulysses* 99 Horse looking round at it with his plume skeowways. **1947** P. C. O'Neill *North-County Dublin Gloss.* xvii. 279 He went skeow-ways down the field.

‖ **skepsel** (ske·psəl). *S. Afr. colloq.* Also **schepsel.** [Afrikaans *skepsel*, Du. *schepsel*, f. *scheppen* to create.] A creature; freq. used as a derogatory designation for a Black or Coloured person. Also *attrib.* or as *adj.*

1844 J. Backhouse *Narr. Visit to Mauritius & S. Afr.* xxxiv. 620 The coloured, who are generally styled Heathens, Schepsels, *Creatures*. **1899** B. Mitford *Weird of Deadly Hollow* 125 Your baas? Confound you, it's my buck. Leave it alone, you schepsel. **1920** R. Y. Stormberg *Mrs. Pieter de Bruyn* 96 Swartz was called a Hottentot, a Shangaan, a skepsel and a few other elegant compliments, and being a full-blooded and very proud Basuto his eyes rolled at the insult. **1943** J. Burger *Black Man's Burden* 67 The Boer farmer treats his workers in a kindly and tolerant fashion; he does not refer to them as 'bloody niggers', but as *skepsels*. **1951** P. Abraham *Wild Conquest* I. i. 48 All they are going to do is to stop working for us. So that *skepsel* Johannes says. **1953** M. Murray *Fire Raisers* xiv. 131 How can we bring up our children decently when there are skepsels like that about? **1968** K. McMagh *Dinner of Herbs* 32 You know how the volk love meat and don't mind where it comes from. Poor skepsels! **1975** W. M. MacMillan *My S. Afr. Years* 143 There was I remember one terrible case of a horrible old rascal who sold his step-daughter to a Hottentot. I remember that the *schepsel* was soundly punished.

skerm, var. Scherm in Dict. and Suppl.

skerrick (ske·rik). Now chiefly *Austral. colloq.* (orig. *dial.*). Also **skerrik,** *Sc.* **skourick;** 9 **scurrick, skirrack, skirrick, skurrick.** [Origin uncertain: cf. Scuddick.] † **1.** (See quot.) *Obs.*

1823 [see Scuddick].

2. A small amount; a small fragment; the slightest bit. Usu. in neg. contexts.

1825 Jamieson *Etym. Dict. Sc. Lang.* Suppl. 407/2, I care nae a *skourick*. **1841** R. W. Hamilton *Nugae Literariae* 359 *Skerrick*, the smallest thing or fraction. 'Not a skerrick remaining.' 'Not worth a skerrick.' **1859** W. Dickinson *Gloss. Words & Phr. Cumberland* 104 Nay, aal nut give a *skurrick* mair. **1863** *Bairnsla Foak's Annual* 14 A son ov hiz woddant gie a skirrick a nowt ta noabdy. **1873** *Halifax Orig. Illuminated Clock Almanack* 11 He coarted a lass 'at didn't care a skirrack fur him. **1890** J. D. Robertson *Gloss. Dial. & Arch. Words used in County of Gloucester* 135 You shan't use a *scurrick* of anything that belongs to me. **1916** *Bulletin* (Sydney, N.S.W.) 16 Mar. 47/1 Nothing found at all of them? Not a skerrick. **1936** F. Clune *Roaming round Darling* xxv. 269 These wadless blokes of the Never-Never have to pay road, car, petrol, State, Federal and Unemployment Relief taxes, and never get a skerrick in exchange. **1947** H. Drake-Brockman *Fatal Days* 116 Eddie had rushed off without leaving a skerrick of kindling; he often did. **1962** A. Upfield *Will of Tribe* i. 13 'And no tracks..you said.' 'Not a skerrick of a track.' **1969** D. Clark *Nobody's Perfect* v. 139 'Any luck?' 'Not a skerrick,' said Green. **1972** *South China Morning Post* 18 Aug. 5/2 The felon was made to pick up every skerrick of refuse. **1977** C. McCullough *Thorn Birds* vii. 143 If I had paid you a skerrick of attention it would have been all over Gilly in record time.

sketch, *sb.* Add: **4. a.** (Earlier examples.)

1789 W. Dunlap *Darby's Return* (title-page), A comic sketch. As performed..for the benefit of Mr. Wignell. **1829** H. Foote *Compan. to Theatres* 74 Satirical sketches, or slight comic pieces on the follies of the day, have likewise been produced here with good effect.

5*. A ridiculous sight, a very amusing person; so *hot sketch*, a comical or colourful person. *slang.*

1917 S. Lewis *Job* xx. 299 You women cer'nly are a sketch! **1921** H. C. Witwer *Leather Pushers* x. 269 This Roberts is a hot sketch for a fighter, anyways! **1925** E. Hemingway *In Our Time* (1926) 84 You're a hot sketch. Who the hell asked you to butt in here? **1926** Maines & Grant *Wise-Crack Dict.* 9/2 *He's a sketch*, he's comical. **1930** J. Dos Passos *42nd Parallel* v. 399 'He's a hot sketch,' said one of the girls to the other. **1930** J. B. Priestley *Angel Pavement* xi. 604 You do look a sight, Dad... I never saw such a sketch.

6. *sketch-pad* [Pad *sb.*³].

1961 M. Spark *Prime of Miss Jean Brodie* iii. 64, I went to get a new sketch pad. **1981** *Listener* 5 Nov. 546/2 The drawings..offering imaginative ideas to any child with a sketch-pad.

sketching, *vbl. sb.* Add: **2.** *sketching-basket, -block* (earlier example), *-club* (earlier example), *-tour, -umbrella.*

1843 D. G. Rossetti *Let.* 7 July (1965) I. 16 There have been two meetings of the Sketching Club since your departure. **1852** C. M. Yonge *Two Guardians* i. 9 They set off..Marian carrying her little sketching-basket. **1865** G. M. Hopkins *Jrnls. & Papers* (1959) 86 Shewing a sketching-block, he asked if there would be any objection to his sketching there. **1890** C. M. Yonge *More Bywords* viii. 269 There's..our sketching tour in August. **1902** A. Bennett *Anna of Five Towns* x. 247 Beatrice, with easel and sketching-umbrella. **1939–40** *Army & Navy Stores Catal.* 375/1 *Sketching umbrellas.* Cream cover, lined green, wind valve, jointed stick, and spike, closes up to 34½ in.—42/-.

sketchy, *a.* Add: **3.** Also *fig.*

1943 E. B. White *Let.* 13 Mar. (1976) 238, I am hoping that my health (which has been rather sketchy lately) will improve. **1977** *Horse & Hound* 14 Jan. 7/3 He survived some decidedly sketchy jumps in the early stages.

skeuomorph (skiū·omǭɹf). [f. Gr. σκεῦος vessel, implement + μορφή form.] **1.** An ornament or ornamental design on an artefact resulting from the nature of the material used or the method of working it.

1889 H. Colley March in *Trans. Lancs. & Cheshire Antiq. Soc.* VII. 166 The forms of ornament demonstrably due to structure require a name. If those taken from animals are called zoomorphs, and those from plants phyllomorphs, it will be convenient to call those derived from structure, skeuomorphs. **1979** *Nature* 6 Dec. 632/2 So-called 'skeuomorphs' in architecture that involve conversion of originally necessary features into purely decorative patterns.

2. An object or feature copying the design of a similar artefact in another material.

1938 *Proc. Prehistoric Soc.* IV. 82 This necklace type is best known in jet from northern Britain, where it has.. provided the type of which the gold lunula is a skeuomorph. **1943** *Antiquity* XVII. 7 Stone skeuomorphs of wooden fences. **1977** [see skeuomorphic *a.* below]. **1981** *Chartered Mechanical Engineer* Sept. 20/3 'Skeuomorphs' seem to be very common in motor car design, particularly where changes are made from ferrous metals to non-ferrous metals or plastics.

Hence **skeuomo·rphic** *a.*, of, pertaining to, decorated with, or having the character of a skeuomorph or skeuomorphs.

1889 H. Colley March in *Trans. Lancs. & Cheshire Antiq. Soc.* VII. 168 The transfer of thong-work from the flint axe, where it was functional, to the bronze celt, where it was skeuomorphic. **1895** A. C. Haddon *Evol. Art* 6 The reader is referred to the section on skeuomorphic pottery. **1905** [see *anthropomorphic *a.* 2]. **1928** R. A. S. Macalister *Archaeol. Ireland* vi. 277 Celtic interlacement..is *skeuomorphic*; that is, it is derived from the patterns produced by a technical process—in this case, by weaving. **1930** J. L. Myres *Who were Greeks?* viii. 464 When a potter, working in red clay..fashions clay vessels so that..they resemble metal work or leather work or basketry..his style is 'skeuomorphic'. **1951** B. Z. Seligman *N. & Q. Anthropol.* (ed. 6) III. 311 Note whether the design is..skeuomorphic. **1959** J. D. Evans *Malta* iii. 127 The skeuomorphic carving of some of the inner halls of the rock-cut monument of Hal Saflieni. **1977** T. Shaw *Unearthing Igbo-Ukwu* 15 When something is originally made in one material and is then translated into another, but by its form and decoration reveals the original model which it imitates, this is called a 'skeuomorph', and the object in the new material is said to be 'skeuomorphic'. Thus the bronze pot described is skeuomorphic of an ordinary pottery vessel.

skew, *sb.*³ **1.** For *Obs.*⁻¹ read *Obs. rare* and add further example.

1884 G. Forbes in W. Thomson *Molecular Dynamics* 289 So the coefficients sighed and gave a last tangential skew And *a* shook hands with *b* & *c* and S and T and U, And with a tear they parted.

4. *Statistics.* Skewness.

1974 *Listener* 7 Nov. 595/2 The skew in the graph is at both ends. **1978** *Nature* 2 Mar. 39/1 The distribution is not symmetrical but displays positive skew, a feature held in common with observations at lower frequencies.

skew, *a.* Add: **1. c.** *Statistics.* Of a statistical distribution: not symmetrical about its mean. Cf. *skewed *a.*² 2.

A distribution is said to be *skew* (or *skewed*) *positively* or *to the right* if its third moment about its mean is positive, so that its larger tail lies to the right; and conversely.

1894 *Phil. Trans. R. Soc.* CLXXXV. 107, I have succeeded in resolving this mortality-curve into components which are not..all of the normal type, but become, as we approach infinite mortality, of the skew form. **1905** *Drapers' Co. Res. Mem.* (Biometric Ser.) II. 22 The theory of skew variation will give regression curves..containing product terms in *x* and *y*. **1929** *Jrnl. du Conseil* IV. 219 The area of the curve has been reduced to about half its original dimensions, but it has not been rendered very skew. **1936** *Bot. Rev.* II. 229 The distributions of the less common grasses are markedly skew. **1968** *Brit. Med. Bull.* XXIV. 210/2 The first is fairly symmetrical but discloses one outlying value; the second is notably skew to the right.

2. a. *skew gearing*, gearing consisting of two cog-wheels having non-parallel, non-intersecting axes; so *skew gear*; *skew nail, nailing* (cf. *skew-nail* vb., sense 3 in Dict.).

1908 J. Richardson *Mod. Steam Engine* ix. 159 Fig. 150 shows the usual bevel gear, and Fig. 151 the skew gear now used in preference. **1929** *Times* 2 Nov. 4/7 The oil pump, driven by skew gear from the camshaft, is in the sump. **1975** Ryder & Bennett *Mechanics of Machines* iv. 112 Skew or spiral gears (which are helical gears of differing helix angles forming a mating pair) are used to transmit motion between non-intersecting shafts. *a* **1877** Knight *Dict. Mech.* III. 2194/2 *Skew-gearing*, cog-wheels with teeth placed obliquely, so as to slide into each other and avoid clashing. **1902** A. C. Harmsworth et al. *Motors & Motor-Driving* x. 191 Another plan..substitutes for the bevel gearing what is known as skew or screw gearing. **1954** W. E. Kelsey *Carpentry, Joinery & Wood-cutting Machinery* xiv. 394 In practice, skew nails are driven in various places..to prevent any movement. **1958** *Times* 27 Mar. 5/2 In one case there was a double skew nail. **1929** T. Corkhill in R. Greenhalgh *Joinery & Carpentry* VI. 1561 *Skew nailing*, nails driven with an inclination to the surface to give greater security. **1958** *Times* 27 Mar. 5/2 Double skew nailing was an old traditional practice. **1973** P. Hutchinson *Home Carpenter* ii. 13 (*caption*) Skew nailing locks timber framing firmly in place.

b. *skew field*, a ring whose non-zero elements form a group with respect to multiplication; a set which satisfies the axioms for a field except that multiplication is not commutative.

1965 Patterson & Rutherford *Elem. Abstract Algebra* iii. 75 In certain cases we encounter systems which satisfy all the required properties for a field with one exception, the commutative law of multiplication. Such systems are known as..skew fields. **1969** F. M. Hall *Introd. Abstr. Algebra* II. iv. 114 The only skew field of any importance is the set of quaternions.

3. a. *skew-eyed* adj. (later example; also as *adv.*).

1922 Joyce *Ulysses* 30 And skeweyed Walter sirring his father. **1976** I. Levin *Boys from Brazil* ii. 57 He.. smiled skew-eyed at him.

b. **skew-symmetric** *a. Math.*, (of a matrix or other square array of elements) having all the elements of the principal diagonal equal to zero, and each of the remaining elements equal to the negative of the element in the corresponding position on the other side of the diagonal; more generally, applied to an array of any dimension in which every element having a repeated subscript is zero and every other element is equal to the negative of elements having an odd permutation of the same subscripts; also **skew-symmetrical** *a.*; hence **skew-symmetry.**

[**1849** A. Cayley in *Jrnl. für die reine und angewandte Math.* XXXVIII. 93 On a $\lambda_{r,s} = -\lambda_{s,r}$ $(r \neq o)$; $\lambda_{r,r} = o$. Ces déterminants peuvent être nommés 'gauches et symmétriques'.] **1911** T. Muir *Hist. Determinants* II. ix. 255 Any skew determinant is expressible in terms of skew symmetric determinants and those of the original determinant which are not included in the latter. *Ibid.* 269 The identity..is the twin theorem to one given in his previous paper regarding a bordered skew symmetrical determinant of even order. **1955** W. Pauli *Niels Bohr* 45 The vectors and skew-symmetric tensors transform just like the analogous electromagnetic quantities. **1967** [see *bilinear *a.* 2]. **1980** A. J. Jones *Game Theory* i. 44 Thus a matrix game is symmetric if its matrix is skew-symmetric. **1927** *Proc. R. Soc.* A. CXVI. 249 In any example the quickest way of showing the skew-symmetry is to write T symbolically as a determinant. **1980** A. J. Jones *Game Theory* i. 46 Skew-symmetry of the matrix is preserved by this operation.

skew, *v.*² Add: **6.** Also, to distort, bias.

1975 *Amer. Speech* 1972 XLVII. 284 If he sings the song, he pronounces the name and possibly skews the results. **1979** *Sci. Amer.* Feb. 105/2 Whatever was skewing the eye-color ratio had its effect only in the course of sperm formation, not in egg formation. **1981** *Amer. Speech* LVI. 45 If we count those informants using both *big daddy* and *big mamma* only once, to avoid falsely skewing the data, fully 39 percent of the 38 different informants..are black.

8. *Statistics.* To make skew (*skew *a.* 1 c).

1929 *Jrnl. du Conseil* IV. 219 The frequency curve has been reduced to a very small proportion of the original, the mode has been shifted 1½ cms., and the group has been distinctly 'skewed'. **1931** *Brit. Jrnl. Psychol.* XXII. 85 The raising of the level of difficulty of test *A* will tend to skew the score-scatter positively.

skewed, *a.*² Add: **1. b.** *fig.*

1960 *Economist* 25 June 1330/1 This triple structure [of a book] is a magnificent achievement,..but its magnificence is..skewed at its foundation... The skewness derives from Professor Hayek's perfectly legitimate definition of freedom. **1977** *New Yorker* 17 Oct. 78/2 She quickly came to share her husband's feelings about the skewed state of the world.

2. a. *Statistics.* = *skew *a.* 1 c.

1940 *Brit. Jrnl. Psychol.* XXX. 259 A difficult test tends to produce a positively skewed score-scatter. **1953** E. Mayr et al. *Methods & Princ. Systematic Zool.* vii. 134 A skewed curve is a curve in which the mode..is above or below the mean. **1977** *Lancet* 5 Feb. 311/1 In our hands, the crude breath-test results are highly skewed, but logarithmic transformation does produce a distribution which is indistinguishable from normal. **1977** R. E. Megill *Introd. Risk Analysis* iii. 22 The distribution shown in Fig. 3.1 is skewed to the right; i.e. it has more values to the right of its highest frequency (mode) than to the left.

b. Of a sample or data: biased, not representative.

1975 *Amer. Speech* 1973 XLVIII. 6 Since the data used for restructuring would have been external and skewed, one would expect attempts at restructuring to be only partially successful. **1977** *Times* 12 May 23/2 Quotas.. are at present heavily skewed in favour of those countries which were economically powerful when the [International Monetary] Fund was set up.

3. Distorted; shifted in emphasis or character.

1935 G. K. Zipf *Psycho-Biol. of Lang.* iii. 105 A range of frequency where the skewed phoneme *t* in *ts* would be most stable. **1965** E. Haugen *Norwegian Eng. Dict.* 40/1 The typical East Norwegian (Oslo) [vowel] system is markedly skewed (in the same direction as Swedish)... The whole system has undergone a counterclockwise movement. **1980** *Nature* 17 Jan. 234/1 Critics..argued that the military control of research meant such research would inevitably be skewed towards the production of weapons of mass destruction. **1981** *Time* 7 Dec. 79/2 The wildness of the cat, its..skewed version of reality.

skewer, *sb.* Add: **1. d.** (Further examples.)

1848 *Sinks of London Laid Open* 124/2 *Skewer*, sword. **1934** A. Russell *Tramp-Royal in Wild Australia* xxxviii. 254 There'd be 'skewers' flying in all directions. Rotten wounds they'd make—barbed, you know.

Skewes (skiū·ėz, skiūz). *Math.* The name of S. *Skewes* (b. 1899), South African born mathematician, used *attrib.* and in the possessive to designate an extremely large number, relevant in the theory of the distribution of prime numbers, having the value exp(exp (exp 79)).

The first pronunc. is that used by the Skewes family.

In 1933 (*Jrnl. London Math. Soc.* VIII. 277) Skewes claimed that, for some value of *a* less than this number, the number of primes less than *a* exceeds $\int_0^a \frac{dx}{\ln x}$.

1949 Kasner & Newman *Math. & Imagination* ii. 32 A veritable giant is Skewes' number, even bigger than a gogolplex. **1955** *Math. Rev.* XVI. 676/1 In a previous paper..the author has obtained a larger value of X_1 (the so-called Skewes number) on the assumption of the Riemann hypothesis.

skewgy-mewgy, var. *soogee-moogee.

skewing, *vbl. sb.*¹ (in Dict. s.v. Skew *v.*²). Add: **2.** Bias, distortion.

1969 *Language* XLV. 487 In reality the transfer at the kernel level can generally be made with far less danger of skewing than if one follows the highly involved processes. **1975** *Nature* 13 Mar. 139/2 The degree of skewing would depend on factors such as the length of time between the act of volunteering and actual participation in the experiment.

skewness. (Further examples.)

1905 K. Pearson in *Biometrika* IV. 173 The chief physical differences between actual frequency distributions and the Gaussian theoretical distribution are:... (ii) The ratio of this separation between mean and mode to the variability of the character—a quantity I have termed the skewness. **1935** G. K. Zipf *Psycho-Biol. of Lang.* iii. 99 The varying degrees of difficulty in the articulation of a phoneme resulting from the different combinations in which it occurs, together with the various relative frequencies of occurrences of the phoneme in its different combinations, may introduce a modification in the normal distribution of speech-sounds about the phonemic norm which, it seems, may well be termed *skewness*. **1936** *Hereditas* XXI. 330 The difference is probably not significant (the standard error could not be calculated owing to a marked skewness in both series). **1965** *Language* XLI. 189 A 'sound system'..of a complexity and skewness out of line with anything attested by languages known to us. **1968** *Economist* 30 Nov. 63/1 The latest crisis..has been about the skewness of certain European currencies and not about the dollar or gold at all. **1970** *Watsonia* VIII. 124 A similar distribution of this character was found in most of the plants included in Table 1; in only four of them was the skewness more markedly positive.

skew-whiff (skiū,wi·f, -hwi·f), *a.* and *adv. dial.* and *colloq.* Also **skew-wiff, -wift,** etc.; 8 **scew-.** [f. Skew *a.* and *adv.* + Whiff *sb.*¹ or *v.*¹] Askew, awry (*lit.* and *fig.*).

1754 *Scots Mag.* July 337/2 Behind, with a coach-horse short dock, cut your hair; Stick a flower before, scew-whiff, with an air. **1839** W. Holloway *General Dict. Provincialisms* 154/1 *Skew-whift, adj.* (Askew, from Skef, Belg. oblique; and perhaps Whiffed, blown.) Awry. **1854** A. E. Baker *Gloss. Northamptonshire Words & Phrases* II. 239 *Skew-whiff*, awry, aslant. 'It's all *skew-whiff*.' Probably blown on one side by a whiff or puff. **1879** G. F. Jackson *Shropshire Word-Bk.* 386 *Skew-wiff*,.. *adv.* awry; irregular; zigzag. **1895** J. T. Clegg *Stories, Sketches, & Rhymes in Rockdale Dialect* 228 Her judgment's getten thrut skew-wift. **1899** *Shetland News* 20 May 7/2, I hed ta geng skewquieff. **1935** A. P. Herbert *What a Word!* iv. 101 Go on cackling..until the orator has to stop and ask you why you cackle. Then tell him. He won't get *Frankenstein* skew-whiff again. **1946** D. L. Sayers *Unpopular Opinions* 59 When Neptune shouldered Britain out of the sea, he did not make a neat engineering job of it. Characteristically, Britain came up skew-wiff, with one edge thick and hard and the other soft and thin, like a slice of wedding-cake. **1959** I. & P. Opie *Lore & Lang. Schoolch.* iii. 47 If a boy's cap is on skew-whiff: 'Are you wearing that cap or just walking underneath it?' **1959** *N.Z. Listener* 12 June 20/3 A breaker turned the bow skew-wiff. **1974** J. Cleary *Peter's Pence* iii. 82 Our plans seem to have gone a bit skew-wiff, don't they? That's the trouble with the Irish. **1977** *Lancashire Life* Feb. 53/4 Thi tie's put on skew-wiff.

skewy (skiū·i), *a. colloq.* [f. Skew *a.* and *adv.* + -y¹.] Somewhat askew, awry, or twisted. So **skew·iness,** the quality of being skewy.

1862 'G. Hamilton' *Country Living* 62 Though freedom from foreign growth discovered an intention of straightness, the most casual observer could not but see that skewiness had usurped its place. **1898** A. T. Slosson *Dumb Foxglove* 71 Narrer an' p'inted like a pear, or skewy an' knobby like a quince. **1940** *Amer. Speech* XV. 131/2 They have the same meaning as awry or 'skewy'. **1960** E. H. Gombrich *Art & Illusion* viii. 248 The right-hand one is really a distorted, skewy object which only assumes the appearance of a chair.

‖ **skhod** (sχǫd). [Russ.] In the U.S.S.R. (and pre-Revolutionary Russia), an assembly of villagers. Also *selskii skhod.*

1877 D. M. Wallace *Russia* viii. 120 All important communal affairs are regulated by the *Selski Skhod*, or Village Assembly. **1914** H. W. Williams *Russia of Russians* xi. 348 The affairs of the community are managed by a *skhod*, or mote of which all the adult males are members. **1948** J. Towster *Political Power in U.S.S.R.* x. 201 In small rural settlements the place of a village soviet was to be taken by a *skhod*—a general meeting of electors. **1959** E. H. Carr *Socialism in One Country* II. 307 The *skhod*, which dated from Tsarist times, was the village meeting... It was not, strictly speaking, a public body. It had no constitutional status, no officially recognized duties..but it sometimes performed primitive functions of local government. **1972** T. Shanin *Awkward Class* ix. 164 A 'rural gathering' (*sel'skii skhod*) was to be established in parallel with the 'land gathering'. The 'rural gathering' would consist of all the inhabitants with Soviet electoral rights within the area of a Rural Soviet.

ski, *sb.* Delete ‖ and add: Now only with pronunc. (skī) and usu. with pl. **skis. 1. a.** (Earlier and later examples.) Also *fig.*

Quot. 1755 is an isolated early use.

1755 *Monthly Rev.* XII. 451 He says they have *skies*, or long and thin pieces of board, so smooth, that the peasants wade through the snow with them. **1923** D. H. Lawrence *Birds, Beasts & Flowers* 177 So she..goes off in slow sad leaps On the long flat skis of her legs. **1933** *Illustr. London News* 9 Dec. 942/1 A German soldier named Schuhmacher, said to have belonged to an infantry battalion training on ski in the Bavarian Alps, was shot dead by Austrian frontier guards. **1960** A. S. Neill in *Id* III. 4 The snow was deep and we all had to go out on skis.

b. Each of two or three runners forming part of the landing gear of an aeroplane designed to land on snow or ice. Cf. *ski-plane*, sense 2 b below.

1912 *Flight* 17 Feb. 137 (*caption*) Note the special skis attached for landing purposes. **1931** F. D. Bradbrooke *Light Aeroplane Man.* vi. 96 For work on ice and snow skis are fitted to the undercarriage instead of wheels. **1948** 'N. Shute' *No Highway* ii. 41 They flew up in a Norseman fitted with skis and landed in deep snow. **1959** Green & Pollinger *World's Fighting Planes* (ed. 3) 21 The Otter may be fitted with wheels, floats or skis.

c. *transf.* = *water ski. Usu. *pl.*

1930 *Literary Digest* 11 Oct. 48/3 Many games have been introduced this summer for those who walk on water with skis. **1964** G. McDonald *Running Scared* xii. 160 Tom suggested water skiing... Tom's skis surfaced easily and beautifully. **1974** *Encycl. Brit. Micropædia* X. 574/3 Typical all-purpose skis are of wood 6½ inches wide and 5 feet 9 inches..long, with a stabilizing fin on the bottom near the heel.

d. A launching site for flying bombs. Freq. *attrib.*

1953 P. C. Berg *Dict. New Words* (ed. 2) 146/1 *Ski*, the launching apparatus for the flying bomb, from its fancied resemblance to a gigantic ski. (1944.) **1957** P. J. de la Ferte *Rocket* vi. 89 The Allied Air Forces launched a massive assault on the ramps, ski constructions and preparation huts. **1958** C. B. Smith *Evidence in Camera* ix. 224 A launching site..which also matched up with the foundations for ramps at the ski sites... The ski buildings provided storage space for twenty flying bombs on each site. **1978** R. A. Young *Flying Bomb* ii. 27 (*caption*) The left-hand 'ski' has not yet been roofed. *Ibid.* 30 When components arrived at the sites they could be most easily stored in one of the 'ski' buildings.

2. a. *ski-runner* (earlier example); *ski-racing, -running* (later examples); (= *ski-ing* vbl. sb.), as *ski boot, -cap, centre, chalet, clothes, club, goggles, -hut, instructor, -jacket -lodge, pants, parka, -race, resort, -room, school, shop, slope, suit, track, trail, troops, trousers, -wear*.

1907 E. C. Richardson *Ski-Running* 48 In the Black Forest ski boots are often made of dog or calf skin. **1972** *Guardian* 31 Oct. 11/3 Ski boots are now injection-moulded plastic shells lined with foam padding. **1937** *Sierra Club Bull.* Feb. 46 There follows, however, our usual clothing list..ski-boots and ski-caps. **1975** E. Hillary *Nothing Venture, Nothing Win* ii. 39, I had an icicle about five inches long hanging from the strap of my ski-cap. **1942** Ski center [see *ski trail* below]. **1948** [see *orienteering]. **1960** *Sunday Express* 27 Nov. 15/5 On the same railway line..there are nine other ski centres. **1971** *Country Life* 23 Dec. 1814/4 (Advt.), Ski Chalets. Top resorts in France, Switzerland and Austria. **1975** *Times* 2 Jan. 4/7 President Ford continued to study a report he received at his ski chalet. **1965** 'J. le Carré' *Looking-Glass War* i. 3 A group of children... Some wore ski clothes. **1913** F. H. Harris *Dartmouth out o' Doors* 104 In our own country ski clubs are flourishing. **1963** *Ski-ing* ('Know the Game' Series) 12 There are a number of ski clubs which are affiliated to the Ski Club of Great Britain. **1971** C. Bonington *Annapurna South Face* 244 Ski-goggles..proved ideal in bad weather conditions. **1958** E. Dundy *Dud Avocado* i. vi. 93 Dressed for the Select as for a ski-hut..in a checkered wool-shirt, G.I. pants and ski-boots. **1973** *Times* 8 Jan. 5/1 The party.. had arrived back at a skihut at Abisko. **1959** P. Moyes *Dead Men don't Ski* iv. 45 It is..traditional for ski instructors to be handsome. **1978** S. Sheldon *Bloodline* ii. 28 On her thirty-fifth birthday Anna had gone to Kitzbühel, in Austria, and there she had met Walther Gassner, a ski instructor thirteen years younger than she. **1968** M. Woodhouse *Rock Baby* xxiv. 234 He was wearing a blue ski-jacket and denim trousers. **1966** —— *Tree Frog* xii. 86 A hunting-lodge modernised to make a ski-lodge. **1978** *Chicago* June 36/1 Once a ski lodge, then a theatre, this rural listening room now provides one of the most relaxed settings for good pop, jazz, and folk. **1937** *Sierra Club Bull.* Feb. 46 There follows..our usual clothing list... Ski-pants. **1977** C. Forbes *Avalanche Express* xxv. 267 Six men clad in ski-masks and ski-pants. **1974** *Amer. Speech 1970* XLV. 180 A variety of wearing apparel, such as children's clothes, men's slacks, ski parkas and sweaters. **1936** C. M. Dole *Amer. Ski Ann.* 53 There should be a competent course patrol for all ski races made up from the personnel of the club ski patrols. **1973** D. Francis *Slay-Ride* ii. 23 He used to win across-country ski races. **1898** *Encycl. Sport* II. 379/1 About a hundred competitors come from various parts of the country to strive for these blue ribbons of the ski-racing world. **1975** *Times* 19 Dec. 9/5 In ski racing..one's position in the starting order is based on one's previous results. **1965** 'J. le Carré' *Looking-Glass War* iii. 33 The town..feeds the main ski resorts. **1972** D. Haston *In High Places* vi. 72 We..tried to be as inconspicuous as four laden climbers can be in a ski resort. **1963** I. Fleming *On H.M. Secret Service* xii. 129 There was a ski-room and workshop to the left of the exit. **1887** *Appleton's Ann. Cycl. 1886* XI. 805/1 The distance covered by an expert skee-runner is from fifty to sixty miles a day. **1911** A. Bennett *Card* xi. 273 No sport was true sport save the sport of ski-running. **1969** H. MacInnes *Salzburg Connection* ii. 37 We have no ski lift here..no special slopes. But there is good ski-running. That is the best sport anyway... Let me show you on this map... You can ski for thirty kilometres. **1934** *Leisure* Jan. 23 This run..will probably be used by the official Ski School of the U.S. Eastern Amateur Ski Association. **1954** R. Martin *Your Ski Holiday* vi. 44 Some people nowadays go to one of the 'dry ski schools' which are being set up in several of the large towns of England. **1980** J. Cartwright *Horse of Darius* iii. 42 She..made her way along to the office of the ski school. **1969** H. MacInnes *Salzburg Connection* viii. 114 Just beyond Bad Aussee..where Johann Kronsteiner has his ski-shop. **1934** *Discovery* Oct. 297/2 (*caption*) The new 'tower-hotels' at Colle di Sestrières in the Italian Alps, with a background of ski-slopes. **1976** A. White *Long Silence* xi. 104 A fall of snow at the wrong time can start an avalanche on the ski slopes. **1956** R. Braddon *Nancy Wake* vi. 55 She accordingly bought herself a new ski suit. **1978** J. A. Michener *Chesapeake* 809 They walked together, bundled in ski suits, to all corners of their estate. **1948** H. Innes *Blue Ice* vii. 194 Three ski tracks ran off at an angle, crossing the tracks we were following. **1975** D. Bagley *Snow Tiger* xxxii. 281 Here is an enlargement of the breakaway point of the avalanche. There is a ski track going into it. **1942** *Economic Geogr.* July 318/2 Lack of overnight accommodations, or their location at inconvenient distances from the ski trails and slopes, has, and will, retard the popularity of many otherwise excellent ski centers. **1973** P. A. Whitney *Snowfire* vi. 108 The immediate problem of getting myself down a ski trail without breaking my neck. **1934** S. Spender *Vienna* iii. 32 Like diving monoplanes..curled down on them the ski-troops. **1974** *Encycl. Brit. Macropædia* XVI. 835/1 Ski troops were..used in Sweden as early as 1452. **1946** P. Bottome *Lifeline* vii. 71 Ida, dressed in long ski trousers and a black pullover. **1969** N. Freeling *Tsing-Boum* xxii. 158 Tight black ski trousers and excessively brilliant orange sweater. **1961** *Sports & Camping Goods Dealer* Nov. 9/1 Big demand for Ski-wear... Sales of skiwear continue to rise. **1980** L. Birnbach et al. *Official Preppy Handbk.* 152/2 The latest and most expensive skiwear and equipment.

b. Special Combs. **ski-boat,** (*a*) *S. Afr.*, a raftlike boat with two outboard motors used esp. for offshore fishing; (*b*) a small powerboat used for towing water-skiers; **ski-bob** [*Bob *sb.*[1] 2 d, e], a vehicle resembling a bicycle with skis instead of wheels, which slides quickly over snow; hence as *v. intr.*, to ride a ski-bob; **ski-bobber,** one who ski-bobs; **ski-bobbing** *vbl. sb.*, the action of riding a ski-bob, esp. as a sport; **ski bum** *N. Amer. slang*, a skiing enthusiast who works casually at a resort in order to ski; hence **ski bumming; ski carrier** = *ski rack* below; **ski flying** (see quot. 1974); **skijamas** *N. Amer.*, a pair of pyjamas in the style of a ski suit; **ski-jump,** (*a*) the artificial structure built on a natural slope, from which a ski-jumper takes off; also *transf.* and *fig.*; (*b*) a leap made by a ski-jumper; **ski-jumper,** one who takes part in ski-jumping; **ski-jumping,** a winter sport in which skiers 'jump' from the end of a snow-covered chute built high on a slope, marks being usu. awarded for style and distance covered by the leap; also, this action; **ski-lift:** see *lift *sb.*[2] 10 b; **ski-mask,** a protective covering for the face, of the type worn by skiers (and adopted by others to conceal identity); hence **ski-masked** *a.*; **ski pack,** an arrangement whereby a tour company offers holiday facilities and the hire of skiing equipment at one inclusive price; **ski patrol** *N. Amer.*, a group of expert skiers who patrol ski slopes to check on conditions and assist skiers in difficulties; hence **ski patrolman,** a member of a ski patrol; **ski-plane,** an aeroplane having its undercarriage fitted with skis (sense 1 c above) for landing on snow or ice; **ski pole** *U.S.* = *ski stick* below; **ski rack,** a frame (usu. fixed to the roof of a car) on which skis are placed for transportation; **ski ramp,** a ramp constructed for skiing practice; also *transf.*; **ski run,** (*a*) a spell of travelling on skis; (*b*) a skiing piste; **ski stick,** one of two long sticks held by a skier to assist in propulsion or braking and in balancing; **ski tour,** a tour made by cross-country skiing; hence **ski-tourer, ski-touring; ski tow,** (*a*) a mechanical device for conveying skiers up a slope, in the form of an endless moving rope or of bars or seats suspended from an overhead cable; (*b*) a tow-rope for water-skiers; hence **ski-tow** *v. trans.*, to pull with a ski tow; **ski-walking,** cross-country skiing (cf. *Nordic *a.* b); so **ski-walk** *v. trans.*, to travel over by ski-walking; **ski-wax,** wax applied to the undersides of skis to improve performance; **ski-wheel,** on the undercarriage of an aircraft: a combination of ski and wheel (see quots.); hence **ski-wheeled** *a.*, having a ski-wheel landing gear.

1964 A. Trew *Smoke Island* ii. 43 José brought the ski-boat round and Andy recovered some of the lost line as he scrambled into the fighting-chair. **1971** 'D. Halliday' *Dolly & Doctor Bird* viii. 103 The holidaymakers..hissing past..in the ski boat. **1974** *Argus* (Cape Town) 2 Aug. 11/4 Many of the ski-boats operating in that area had 60, 80 or 100 snoek. **1966** *Skier* Sept. 7/1, 1962: I became German and Bavarian Women's ski-bob Champion. **1968** *Guardian* 21 Sept. 10/3 The ski-bob is a small bicycle on skis with short skis for the feet with braking claws on them. **1969** *Winter Sports Ann.* 20/1 One can learn to ski-bob very quickly. **1976** *Daily Tel.* (Colour Suppl.) 6 Aug. 6/2 A ski-bob can exceed speeds of 80mph on a steep, straight run. *Ibid.* 6/3 Captain John Beckett..who once taught a lady journalist to ski-bob in one-and-a-half hours. **1967** Ski bobber [see *mini-ski* s.v. *mini-]. **1976** *Daily Tel.* (Colour Suppl.) 6 Aug. 6/1 Ski-bobbers are now eligible for grants from the National Council for Physical Recreation. **1966** *Skier* Sept. 6/2 Ski-bobbing has come a long way since those first laughing days. **1971** *Daily Tel.* 30 Jan. 7/5 Ski-bobbing (not yet so popular in France as elsewhere, but rapidly catching on). **1976** *Ibid.* (Colour Suppl.) 6 Aug. 6/1 In 1971, the Army included ski-bobbing in their own skiing championships. **1960** *Washington Post* 4 Mar. c5/3 John Kerr..is firmly settled in the picturesque ex-mining town of Aspen, Colo., and happily pursuing one of the world's newest professions—ski bumming... By ski bum standards, John Kerr has struck it rich. **1978** *N.Y. Times* 16 Jan. c8/1 (*headline*) Ski-bum shortage shakes the resorts. **1978** S. Sheldon *Bloodline* ii. 34 'What can a ski bum contribute to Roffe and Sons?' he asked. **1965** 'D. Shannon' *Death-Bringers* xiv. 187 It was the right Anglia. It had ski carriers on its roof. **1970** *Globe & Mail* (Toronto) 25 Sept. 32/7 (Advt.), 100 Accessories..from ski carriers to repair bases, we have them all at spectacular savings. **1952** *Sun* (Baltimore) 3 Mar. (B ed.) 15/7 Toivo Lauren..won the international 'ski-flying' contest today with a jump of approximately 429 feet 9 inches. **1974** R. Scharff *Ski Magazine's Encycl. Skiing* 420/2 *Ski flying*, a form of jumping on hills where distances of 100 meters or more can be reached. **1958** L. Whishaw *As Far as you'll take Me* vii. 104, I..then, dressed in my skijamas, ate my dinner in peace. **1964** *N.Y. Times* 29 Nov. 132 Reis 'ski-jamas' are set for winter slumber. **1907** E. C. Richardson *Ski-Running* 89 The outlook from the top of a ski-jump of any magnitude is indeed alarming. **1922** E. E. Cummings *Enormous Room* i.18 A face all ski-jumps and toboggan slides. **1948** H. Innes *Blue Ice* x. 250 He was going to do a ski jump..on to the top of the moving train. **1953** X. Fielding *Stronghold* 42 At last I reached a scree, as long and steep as a ski-jump. **1960** C. H. Gibbs-Smith *Aeroplane* 3/5 A later illustration shows a long 'ski-jump' ramp. **1971** L. Koppett *N.Y. Times Guide Spectator Sports* xvi. 222 One of the most spectacular sights in any sport is the ski jump. **1978** *Navy News* Aug. 40/4 The invention of the Ski Jump take-off ramp to be fitted in Royal Navy ships carrying Harrier aircraft has won an award of £25,000. **1894** *Engineering News* 1 Mar. 169/3 Ski-jumpers..have for some time been holding tournaments in Minnesota. **1981** 'E. Lathen' *Going for Gold* i. 15 People..were taking up stations from which they could watch the last of the ski jumpers. **1904** *Sandow's Mag.* Mar. 200 (*caption*) A ski-jumping competition. **1927** A. Huxley *Let.* 14 Feb. (1969) 283 We had an international ski jumping competition here last week. **1960** *Guardian* 8 Dec. 10/6 In 1950 and 1951 there was ski jumping..on Hampstead Heath. **1973** *Country Life* 29 Nov. 1807 Ski-jumping and cross-country (*langlauf*) racing. **1973** 'D. Shannon' *No Holiday for Crime* xiv. 208 It had been a professional job: ski-masks, a look-out at the door. **1980** J. Ball *Then came Violence* xiii. 103 Since ski masks had been used..there were no descriptions. **1976** *Time* 20 Dec. 7/2 Graziella Quartuccio, 43, was snatched away in her nightgown..by a machine-gun-toting gang of ski-masked Mafiosi. **1969** *Guardian* 11 Oct. 10/6 See which holidays give the best value in terms of ski packs. **1936** C. M. Dole *Amer. Ski Ann.* 52 A definite organization is necessary. The Ski Patrol has been suggested..and..has been instituted in one locality. **1963** *Amer. Speech* XXXVIII. 204 The warning signs erected in ski areas by the ski patrol. **1978** *Globe & Mail* (Toronto) 8 Feb. 31/1 The Canadian Ski Patrol system is again this season doing on-the-slopes testing of bindings for skiers. **1957** *Today's Health* Jan. 54/3 Two ski patrolmen.. lashed their skis together to make an emergency toboggan. **1930** O. H. Kneen *Everyman's Bk. Flying* xiii. 231 The Fairchild Airplane Company and others use the term *ski-plane* for Canadian machines, convertible to seaplanes in the summer. **1936** J. Grierson *High Failure* xiii. 281 For Canadian conditions the most useful machines are seaplanes in summer and ski-planes in winter. **1964** G. Lyall *Most Dangerous Game* xxi. 172, I came in from Spitzbergen, in an old Noorduyn Norseman ski-plane. **1975** E. Hillary *Nothing Venture, Nothing Win* vi. 82 Access to the mountains is often long and difficult—or used to be before the development of ski-planes and helicopters. **1920** *Literary Digest* 14 Feb. 115, I need not describe these poles to you, as any dealer will know what you mean by ski-poles. **1978** W. F. Buckley *Stained Glass* xv. 155 The accordion player, without ski poles, and making music all the way, began the three-mile ski down the mountain. **1968** *Globe & Mail* (Toronto) 5 Feb. 26/8 (Advt.), 67 Austin Cooper, tachometer ski rack. **1980** L. Birnbach et al. *Official Preppy Handbk.* 204/2 In the winter, the car is heading north, topped with a ski rack. **1973** *Houston Post* (Spotlight Suppl.) 14 Oct. 9/5 The fair is free, including hourly ski demonstrations by Ed Williams on a ski ramp. **1976** *Southern Even. Echo* (Southampton) 11 Nov. 17/2 Ships with 'ski-ramps' at the bow for the launching jump-jet aircraft were forecast by the chief designer of the Harrier at Southampton University last night. **1924** O. Poulsen *Skiing* 72 It is on one of these long ski-runs alone that a man can think out his problems best. **1951** M. Kennedy *Lucy Carmichael* V. iii. 222 The next few seconds had been as good as a first class ski run. **1953** Dylan Thomas *Let.* 27 Feb. (1966) 395 We could go anywhere on that, except Laughlin's heart and ski-run. **1977** N. Freeling *Gadget* ii. 92 The climb..was steep as a downhill ski-run. **1907** E. C. Richardson *Ski-Running* 46 (*caption*) Disc for bottom of ski stick. **1924** E. Hemingway in *Transatlantic Rev.* Dec. 635 Nick knocked his clamps loose with one of his ski sticks. **1970** N. Fleming *Czech Point* (1971) i. 7, I hooked my skisticks on to the T-bar. **1949** E. Coxhead *Wind in West* vii. 180 Ilse knew..nothing of nature except what she could see from a ski-tour. **1972** *Guardian* 11 Mar. 13/1 The standard yardstick for these ski-tourers and ski-mountaineers has been..a mountain obstacle race between..Chamonix, and Zermatt. **1960** *Ibid.* 24 Oct. 11/4 In Scotland..after April only high-level ski-touring is possible. **1972** Dean & Smith *Wisconsin* 161/1 Call it ski touring, Nordic skiing, or simply X-C, but it all boils down to the same thing—the art of walking on skis. **1942** *Economic Geogr.* July 307/2 The invention of the ski tow at about this time.. gave great impetus to the growing interest in skiing. **1967** *Spectator* 15 Sept. 301/1 One laird explained his resistance ..to..building a ski tow in the snowy heights he owned—by saying simply, 'Look what happened to Switzerland.' **1971** J. Yardley *Kiss a Day* vii. 121 He..trailed the ski-tow over the side for her to catch. Five minutes later they were cutting a white wake across the lagoon. **1976** P. Cave *High Flying Birds* iii. 28 Ski-towing a hang-glider is reckoned on being a pretty dangerous sport at the best of times. **1970** R. Lowell *Notebk.* 148 We ski-walked the eggshell at the Mittersill. **1974** *Observer* 3 Nov. 34/2 The age-old Nordic ski-ing that is both easier to learn and cheaper to enjoy. Essentially it is walking, or running, on skis over undulating country-side... Ski-walking, Nordic ski-ing, *Skiwandern, ski de randonnée, Ski du fond, Langlauf*—the variety of names given to the sport perhaps causes confusion. **1910** W. R. Rickmers *Ski-ing* 35 Ski-wax can be obtained hard or in tubes. **1979** R. Fiennes *Hell on Ice* v. 68, I needed feeling in my fingers to..feel for my map, change the ski-wax and many other things. **1938** C. Winchester *Wonders of World Aviation* 979/2 It sometimes happens that flights to the northern parts of Canada necessitate a take-off fit only for wheels and a destination with ground fit only for skis. This indicates that there is a need for a ski-wheel combination. **1958** *Edmonton* (Alberta) *Jrnl.* 24 June III. 17/8 A pilot on ski-wheels who runs into trouble will pick a lake or river, set down parallel to the shore, and ski neatly up onto the beach before his aircraft loses its forward momentum and sinks. **1976** *Jrnl. R. Soc. Arts* CXXIV. 634/2 Two ski-wheeled de Havilland Twin Otter aircraft are operational in the Antarctic between November and March.

ski, *v.* Add: **1. b.** To water-ski.

1947 *Life* 17 Feb. 7/3 Lumber dealers..have seen pictures of her skiing among the cypresses and want to know how they can purchase the trees. **1956** Peterson & Fisher *Wild Amer.* ix. 101 A girl came by at breakneck speed on one water ski,..then three daredevil young men

..skied up and jumped over an inclined platform. **1970** 'D. Halliday' *Dolly & Cookie Bird* viii. 130 'Does Louie ski?' He swooped away, the spray flying.

2. *trans.* To travel over (a slope, etc.) on skis; to ski at (a place).

1973 P. A. Whitney *Snowfire* xii. 236 It's easier to ski a steep slope than a gentle one. **1980** *Sunday Times* 21 Sept. 8 Ski the top resorts in Europe..and now the U.S.A.

skiable (skī·ăb'l), *a.* [f. Ski *v.* + -able.] Of a slope, snow, etc.: capable of being skied on; fit for skiing.

1961 in Webster. **1963** *Amer. Speech* XXXVIII. 205 Inferior..heavy snow that is hardly 'skiable' for the average skier. **1969** *Observer* 14 Sept. 35/1 In every skiable side-valley, new resorts have sprouted. **1979** *United States 1980/81* (Penguin Travel Guides) 486 This..blown-out volcano is skiable from late October.

skiagraph, *sb.* Add: (Later examples.)

1925 [see *roetgenological* adj. s.v. *roentgen-, roentgeno-]. **1940** E. F. Benson *Final Edition* x. 208 When.. one of them suggested that an X-ray skiagraph should be taken, it showed osteo-arthritis in an advanced stage.

skiagrapher, (*b*) = Sciagrapher.

1957 V. Nabokov *Pnin* iv. 98 Lads..would spend years grinding colors in the workshop of some great Italian skiagrapher.

skiascopy (skəiæ·skŏpi). *Med.* [ad. F. *skiascopie* (Chibret 1886, in *Arch. d'Ophthalmol.* VI. 147), f. Gr. σκιά shadow: see *-scopy.] Retinoscopy, esp. by means of a skiascope.

1886 *Amer. Jrnl. Med. Sci.* XCII. 248 Chibret proposes to employ the term 'skiascopy', in place of the terms keratoscopy, retinoscopy, pupilloscopy, and phantoscopy ..to designate this..method of determining the ocular refraction based upon the examination of the shadows which are formed in the eye when light is thrown into it. **1933** *Arch. Ophthalmol.* X. 689 Cycloskiascopy is more difficult than the classic method of cylinder skiascopy. **1958** *Ophthalmic Lit.* XII. 557 The term 'skiascopy' was proposed by Chibret instead of retinoscopy, as the retina is not normally in focus. **1976** *Ophthalmic Res.* VIII. 115 Data of skiascopy are not very exact because of the difficult procedure of measurement.

Hence **ski·ascope,** an instrument that directs light into a patient's eye along the line of sight of the examiner, so that the latter can judge the refraction of the eye from the movement of the illuminated area and the shadows as the light source is moved; **skiasco·pic** *a.*, of or pertaining to skiascopy; **skiasco·pically** *adv.*

1892 *Arch. Ophthalmol.* XXI. 422 A skiascope is now used at the Utrecht Clinique with glasses on each side which by combinations make a series of plus and minus D from 1 to 10. **1897** *Ibid.* XXVI. 598 A new hand at the sciascopic test. **1903** E. Clarke *Errors Accommodation & Refraction of Eye* vi. 77 Marple's skiascopes, made by Meyrowitz of New York, are very useful, and obviate the necessity for keeping a separate test case in the dark room. **1930** *Amer. Jrnl. Ophthalmol.* XIII. 102/1 Jackson measured skiascopically the symmetrical aberration in one hundred eyes. **1958** *Ophthalmic Lit.* XII. 557 Four important skiascopic fields are to be considered. **1971** *Biol. Abstr.* LII. 5247/1 (*heading*) Focal ophthalmoscopy with the help of a modified electric skiascope.

Skiatron (skəi·ătrŏn). *Electronics.* Also **skiatron.** [f. Gr. σκιά shadow + *-tron.] A proprietary name for a type of cathode-ray tube in which the electron beam produces a dark trace. Also *attrib.*, esp. as *Skiatron tube*.

1940 A. H. Rosenthal in *Electronics & Television & Short-Wave World* XIII. 52 (*heading*) The Skiatron—a new scophony development towards large-screen television projection. **1946** *Trade Marks Jrnl.* 26 June 330/2 *Skiatron*... Scientific apparatus and instruments, electrical apparatus and instruments included in Class 9. Scophony Limited... 17th Sept. 1945. **1947** L. J. Haworth in L. N. Ridenour *Radar Systems Engin.* xiii. 483 The alkali halides..have the property of darkening for a time at a point where they have been struck by an electron beam of sufficient energy. This phenomenon has been made use of in one form of cathode-ray tube, known as the 'skiatron'..used for projection purposes. **1948** *Electronic Engin.* XX. 21/2 (*caption*) Large screen plan position indicator with 'Skiatron' tube, as used by the Admiralty. **1953** *Official Gaz.* (U.S. Patent Office) 3 Nov. 21/2 *Skiatron*..for cathode ray tubes... Claims use since Aug. 1, 1942. **1966** H. H. Poole *Fund. Display Systems* ii. 32 The dark trace tube, also referred to as the Skiatron tube, is currently manufactured by National Union. It is available in screen sizes of 5, 7, 10, and 12 inches. **1973** C. J. Richards *Electronic Display & Data Systems* iii. 76 During the Second World War..the skiatron projection tube was used... Today a family of 'deformographic' tubes exists which can project a picture several metres across.

skid, *sb.* Add: **1. c.** *Oil Industry.* A skid beam (see sense 5 below).

1975 *Offshore Engin.* Oct. 66/1 The order for the skids went to Hopson, Co Durham. **1976** *Offshore Platforms & Pipelining* 20/2 Major items of drilling equipment represent skid sizes of available components.

2. c. (Earlier example.)

1782 'J. H. St. John de Crèvecoeur' *Lett. from Amer. Farmer* iii. 110 The logs were placed with skids, and the usual contrivances: thus the rude house was raised.

f. A runner attached to the underside of an aircraft. Cf. *tail skid, *wing skid.

1909 *Aëronaut. Jrnl.* XIII. 119/2 The tail of the machine rests on a small skid fixed to the frame. *Ibid.* 120/1 The wing tips are provided with light wheels and skids. **1909** A. Berget *Conquest of Air* ii. iii. 194 The planes rest upon two skids which form a kind of sleigh. **1947** A. C. Douglas *Gliding & Advanced Soaring* x. 259 Very close to the usual position of the release, will be the nose fitting of the skid. **1981** 'A. Hall' *Pekin Target* i. 11 We put the chopper down..on one skid and a rotor tip.

g. *pl.* In *fig.* phrases describing impending downfall or defeat; the way to failure or ruin: *to put the skids under* (a person or thing), to arrange the downfall of; to remove or oust (esp. from office), to get rid of; to cause to fail; *to hit the skids*, to enter a rapid decline or deterioration; *on the skids*, on the way to defeat or ruin; in a steadily worsening state. *colloq.* (orig. *U.S.*).

1918 H. C. Witwer in *Collier's* 9 Mar. 16/3 Me and Jeanne is gonna have a flat over in Brooklyn as soon as we put the skids under the Kaiser. **1920** —— in *Ibid.* 5 June 36/3 Kane Halliday, as the butlers was wonted to announce him previous to the time he hit the skids, was merely engaged to this gold mine. **1921** A. G. Empey *Madonna of Hills* lxi. 363 As yet it would be too dangerous to flaunt Davis openly, even if he was on the 'skids'. **1929** D. Hammett *Red Harvest* xii. 119 'What did Noonan put the skids under you for?' 'Skids? What skids? I quit.' **1938** A. J. Liebling *Back where I came From* 48 The men on the Bowery..had taken to hanging around saloons, drinking on empty stomachs. Eventually they had hit the skids and been forced to abscond from the genteel communities where their wives still lingered. **1943** M. McCarthy *Company she Keeps* v. 128 Capitalism was on the skids, and everybody ought to know about it. **1954** A. Melville in *Plays of Year 1954* (1955) XI. 26 You're too pigheaded to realise that the skids are well and truly under you. **1962** 'K. Orvis' *Damned & Destroyed* ix. 61 I'm on the skids. For good, it looks like. **1975** J. Symons *Three Pipe Problem* xvi. 165 A plan by one gang to put the skids under another. **1976** *Daily Mirror* 12 Mar. 23/4 They were only 378p when the £ hit the skids a week ago. **1977** *Irish Press* 29 Sept. 12/6 The Irish shoe industry, after being on the skids for six years, may be finding its feet again.

4. a. (Examples.)

1907 A. Bennett *Grim Smile of Five Towns* 192 At the bottom they had a severe skid. **1933** Kipling in *Strand Mag.* Feb. 131 The Cars put up an average bag of twenty dead per diem... And so began, in skid and stink, the real blood-sport of Britain. **1972** 'H. Calvin' *Take Two Popes* xi. 111 The cars touched gently at 100 kph, went into a skid together and slid on to the roadside verge.

b. *Aeronaut.* A movement of an aircraft that includes a sideways component, esp. away from the centre of curvature of a turn. Cf. *slip *sb.*[3] 9 j.

1916 H. Barber *Aeroplane Speaks* 83 The velocity of the 'skid', or sideways movement. **1942** *Tee Emm* (Air Ministry) II. 65 Do a quick barrel half roll with plenty of skid. **1952** A. Y. Bramble *Air-plane Flight* xiii. 199 Some pilots use the term 'slip' loosely to mean a skid. It is important to note the difference. A side-slip may occur whilst flying straight or *inwards* during a turn, but 'skidding' of the machine can occur only during yawing motion, and strictly, is always in the direction outwards, away from the turn.

5. skid beam: also *transf.*, a horizontal beam supporting a deck on an oil or natural gas platform; **skid-lid** *slang*, a motor-cyclist's crash-helmet; **skid mark,** the mark made on the road by the tyre of a skidding vehicle (usu. *pl.*); **skid-mounted** *a.*, mounted on runners; **skid-pan,** (*b*) a slippery road surface prepared to enable drivers to practise skid-correction; also *transf.*; **skid-proof** *a.*, of a road, etc.: on which one cannot skid or slip (esp. in wet conditions); designed to prevent skidding; hence as *v. trans.*, to render skid-proof; **skid road,** (*b*) N. Amer., orig. a down-town area frequented by loggers; now *gen.* = *skid row.

1976 *Offshore Platforms & Pipelining* 23 The distance between skid beams is the first dimension to consider when designing a platform rig. **1958** *Oxf. Mail* 8 Feb. 8/7 Skid-lids' importance stressed. **1968** R. V. Beste *Repeat Instructions* vii. 71 He wore the leather jacket and round skid-lid..usual to ton-up tearaways. **1977** C. Watson *One Man's Meat* xiii. 120 This bird in motor-cycle get-up ..with that great skid-lid hiding half her face. **1937** M. Allingham *Dancers in Mourning* iv. 59 The skid marks were easily discernible on the flint road. **1978** M. Gilbert *Empty House* v. 47 There were no skid marks, which would..indicate that he made no attempt to brake. **1960** *Farmer & Stockbreeder* 16 Feb. Suppl. 24/2 Being skid-mounted, it can readily be moved to a new site. **1975** *Petroleum Rev.* XXIX. 135/1 AOT will provide complete prefabricated skid mounted systems into which are built flowmeters, [etc.]. **1958** *Times* 17 June 9/6 On the skid pan (which is equivalent to a road covered with wet ice). **1959** *News Chron.* 12 Oct. 1/4 Rain turned many roads into skid-pans and caused hundreds of accidents. **1972** C. Mudie *Motor Boats & Boating* 108 The control of a car is exact and direct whereas a boat is on the skid pan of the sea. **1976** *Norwich Mercury* 17 Dec. 10/3 This thrilling game played at a terrific pace..on a greasy treacherous skidpan at Skinners Lane on Saturday. **1980** *Sunday Times* (Colour Suppl.) 21 Sept. 58/2 The first days of the course are instruction in evasive driving, done on a nearby skidpan. **1937** *Daily Express* 20 Feb. 10/2 Some roads are practically skid-proof, others not. **1958** *Washington Post* 25 Oct. B10/1 To skidproof concrete porch steps, it has long been the custom to sprinkle sand on freshly painted surfaces. **1906** *Log of 'Columbia'* 1. 8/1 'We'll likely see him in town.'.. 'He'll be in the Skid road somewhere.' **1925** *Amer. Speech* I. 135 When the logger of to-day speaks of the 'skid-road' he means the place where loggers gather when they are in town. **1940** *Amer. Mercury* Dec. 412 Most of the skid-road bars provide either a floor show..or a hill-billy band. **1962** E. Lucia *Klondike Kate* iii. 98 Arriving in Dawson like some little tramp from Seattle's Skid Road. **1971** *Daily Colonist* (Victoria, B.C.) 15 May 5/5 Young heroin users are replacing the Skid Road addicts in British Columbia. **1980** *Washington Post* 1 Mar. A13/4 In Seattle, where the 'Skid Row' term originated 'from flophouses built along the 'Skid Road' of lumber being 'skidded' to the water front', reminders of that age remain.

Hence **ski·dded** *a.*, provided with a skid or skids.

1935 *Charlottesville* (Va.) *Daily Progress* 22 Feb. 1/1 He streaked off into the stratosphere at dawn today on a 2,447 mile flight to New York City where he planned to land on the wood skidded belly of the ship. **1961** *Flight* LXXX. 471/2 The lift platform was variously adjusted to handle simple cargo, general 'skidded' cargo and fully loaded vehicles.

skid, *v.*[1] Add: **3. b.** Also, of the vehicle itself.

1907 G. B. Shaw in *Neolith* Nov. 3 'The bus skidded.' 'So would any bus skid in this mud, going at that rate.' **1926** E. F. Spanner *Naviators* i. 13 A car that looks as though it is going straight can be dodged, but one that skids about like a demented dog chasing its tail is not to be faced with assurance. **1927** Kipling *Limits & Renewals* (1932) 159 It was raining hard, and the car skidded badly. **1966** G. Greene *Comedians* III. iii. 280 The car because of my momentary inattention skidded sideways.

c. *Aeronaut.* Of an aircraft: to move sideways, esp. away from the centre of curvature while turning. Cf. *slip *v.*[1] 9 c.

1911 *Daily Colonist* (Victoria, B.C.) (Mag. Section) 9 Apr. 9/5 If the rag points up, the operator knows his machine is going down..but if it points to the side, the operator knows he is 'skidding'. **1916** H. Barber *Aeroplane Speaks* 22 When an aeroplane is turned to the left or the right the centrifugal force of its momentum causes it to skid sideways and outwards away from the centre of the turn. **1942** *R.A.F. Jrnl.* 3 Oct. 35 He lost speed suddenly and skidded to the starboard. **1965** C. N. Van Deventer *Introd. Gen. Aeronaut.* x. 233/2 If the ball moves in the direction of the turn, it indicates that the airplane is slipping toward the inside of the turn... If the ball moves in the opposite direction, the airplane is skidding toward the outside of the turn.

d. *fig.* To follow an incorrect or unprofitable course; to make a mistake, to err or fail. Also of prices, etc.: to fall or decline rapidly. *colloq.* (chiefly *U.S.*).

1920 *Literary Digest* 22 May 120 Time was when the unsuccessful man merely failed, but these days, in a world scurrying about in motor-cars and breathing gasoline, he is said to 'skid'. **1937** G. Frankau *More of Us* vi. 67 Next morning woke a damsel heavy-lidded To wonder had she not, or had she, skidded. **1962** K. Orvis *Damned & Destroyed* iv. 29 He had skidded until he had become merely a brilliant young pianist. **1970** *Computers & Humanities* V. 11 Alabama Populists..tended mainly to be unsuccessful yeoman farmers who were skidding toward tenancy. **1976** *National Observer* (U.S.) 21 Aug. 8/6 A good investment..if some unforeseen misfortune causes currencies to skid.

4. *trans.* To cause (a vehicle, its wheels, etc.) to skid; to turn (a corner, etc.) by means of a skidding movement.

1924 'J. Sutherland' *Circle of Stars* xxiii. 238 Carter skidded the Ford to a standstill. **1928** *Sunday Express* 19 Aug. 1/1 Viscount Curzon..was cheered as he skilfully skidded the corners. **1931** G. Le Q. Martel *In Wake of Tank* 84 A turn through to a right angle absorbed so much power in skidding the track round that it was usually necessary to change down to first gear on the machine.

skidder, (*a*) (earlier example); (*b*) a tractor or other machine for skidding logs.

1870 *Overland Monthly* V. 56/1 Another, called the 'skidder', skids the road. **1905** [see *bummer 4]. **1945** B. Macdonald *Egg & I* (1946) 184 These toots were the signals given by the 'whistle punk' to direct the operations of the skidder bringing in the logs. **1965** *Weekend Mag.* 2 Oct. 3/3 Then it lays the denuded trunk on the ground and another machine, called 'a skidder', takes it to a landing area where it is cut into pulpwood lengths. **1976** *New Yorker* 3 May 52/2 The skidder, which vaguely resembles a pair of tractors coupled together..replaced the horse not long ago as the means of 'twitching' a tree ..from the cutter to the truck.

skid, *v.*[2] Add: **1.** Also, = Scud *v.*[1] 2.

1940 L. MacNeice *Last Ditch* 32 Toy sail skidding on Whitestone Pond at the peak of London.

2. *trans.* = Scud *v.*[1] 5 a.

1891 in *Eng. Dial. Dict.* **1897** G. B. Shaw *Let.* 1 Mar. in *Ellen Terry & Shaw* (1931) 159 Get ten sovereigns and skid them out from the beach into the sea.

skiddaddle, var. Skedaddle *v.* in Dict. and Suppl.

Skiddavian (skidē̆·viăn), *a.* *Geol.* [f. *Skiddav-*, stem of latinized form of next + -ian.] = *Skiddaw *b.* Also *absol.*

1905 J. E. Marr in *Q. Jrnl. Geol. Soc.* LXI. p. lxxxi, The time-divisions which I would..propose to adopt for the rocks of this [Ordovician] system are as follows:—Ashgillian. Caradocian. Llandeilian. Skiddavian. *Ibid.* p. lxxxii, The beds tabulated above may..be taken as

defining the upper and lower limits of the Skiddavian Series. *Ibid.* p. lxxxiii, *Didymograptus bifidus*..belongs to the top of the Skiddavian. **1929** W. W. Watts in Evans & Stubblefield *Handbk. Geol. Brit. Isles* 63 The Skiddavian fauna includes many species of *Didymograptus*. *Ibid.*, The Manx Slates of the Isle of Man are probably in part of Skiddavian age. **1955** G. G. Woodford tr. *Gignoux's Stratigr. Geol.* iii. 83 (*caption*) Upper Skiddavian: shales and sandstones of the upper Arenig. Lower Skiddavian: basal sandstone of the Arenig.

Skiddaw (ski·dǭ). *Geol.* The name of a mountain in the English Lake District, used *attrib.*: (*a*) in **Skiddaw slate(s)** (or **Slate(s)**), a thick group of slates, flags, and mudstones that outcrops in the northern part of the Lake District and out of which Skiddaw and neighbouring mountains have been eroded; (*b*) to designate the lowest division of the Ordovician in Britain, esp. in the Lake District.

1832 *Proc. Geol. Soc.* I. 401 Skiddaw slate.—The author briefly describes the range and extent of this group. **1855** J. Phillips *Man. Geol.* vi. 108 These..occupy a long range of mountains parallel to the Skiddaw slates. **1897** *Index Vols. I.–L. Q. Jrnl. Geol. Soc.* 357/1 Skiddaw age of Easdale Slates. **1900** *Proc. & Trans. R. Soc. Canada* VI. iv. 200 There is a remarkable similarity between the Ordovician of..Quebec and New Brunswick and the Ordovician of western Europe as developed in Great Britain: The Skiddaw and Arenig, the Hartfell and Llandeilo formations, being easily recognized in Canada. **1933** H. H. Symonds *Walking in Lake District* x. 255 Thus came the 'Skiddaw slate', that earliest rock of Cumberland, found *par excellence* in Skiddaw himself. **1969** Bennison & Wright *Geol. Hist. Brit. Isles* v. 102 The Skiddaw Slates have been slightly metamorphosed.. by later intrusions... However, the early Ordovician seems to have been devoid of igneous activity—unlike Wales—and only late in Skiddaw times did it commence. **1977** R. Prosser *Geol. explained in Lake District* i. 13 The oldest rocks [of the Lake District massif], the Skiddaw Slates.., have their main exposures north of a line from Ennerdale Water, past Derwentwater to Troutbeck and include the hill masses of the Skiddaw range.

skiddy (ski·di), *a.* [f. Skid *v.*[1] + -y[1].] Of surfaces, etc.: on which one is liable to skid; treacherously slippery. Also, characterized by skidding.

1902 *Car* 4 June 72/2 The state of a road which constitutes a 'skiddy' and a 'non-skiddy' surface. **1928** *Daily Express* 10 Dec. 19 Motorists drove with caution on skiddy roads. **1935** *Punch* 27 Mar. 358/3 Drive frightfully cautiously and go slowly round the corners.., because you *know* how skiddy it is these days. **1947** A. C. Douglas *Gliding & Soaring* 56 Another range of difficulties..caused by..dislike of being out of normal position in the air, resulting in flat skiddy turns, and so on. **1977** D. Murphy *Where Indus is Young* xi. 233 Skiddy mud making it impossible to keep upright.

skidoo (skidū·), *v.* *N. Amer. slang.* Also **skiddoo.** [Orig. uncertain, perh. f. Skedaddle *v.*] **1.** *intr.* To go away, leave, or depart hurriedly. Freq. *imp.*

1905 'H. McHugh' *You can search Me* i. 13 Skidoo, skidoo, and quit me, Mr. Josheimer! **1911** G. B. Shaw *Shewing-up of B. Posnet* 404 Outside, Nestor. Out you go... Skiddoo, Nestor. **1949** *New Yorker* 2 Apr. 26/3, I skiddoo and take a trip. **1963** B. Malamud *Idiots First* 135 'If you skidoo now..you'll get spit.' 'Who's skidooing?'

2. In catch-phrases. **a.** Used as an exclamation of disrespect (*for* a person). Esp. in nonsense association with *twenty-three*. (*temporary*.)

1906 J. F. Kelly *Man with Grip* (ed. 2) 99 As for Belmont and Ryan and the rest of that bunch, Skidoo for that crowd when we pass. *Ibid.* 118 'I can see a reason for 'skidoo',' said one, 'and for '23' also. Skidoo from skids and '23' from 23rd Street that has ferries and depots for 80 per cent. of the railroads leaving New York.' **1911** *Maclean's Mag.* Oct. 348/1 Surrounded by this conglomerate procession as I went on my way, the urchins would yell 'Skidoo,' '23 for you!'

b. *spec.* as *twenty-three skidoo*: formerly, a exclamation of uncertain meaning; later used *imp.*, go away, 'scram'.

1926 C. T. Ryan in *Amer. Speech* II. 92/1, I really do not recall which appeared first in my vocabulary, the use of 'some' for emphasis or that effective but horrible '23-Skiddoo'—perhaps they were simultaneous. **1929** *Amer. Speech* IV. 430 Among the terms which the daily press credits Mr. Dorgan with inventing are:..twenty-three skiddoo (go away). **1957** W. Faulkner *Town* iii. 56 Almost any time now Father would walk in rubbing his hands and saying 'oh you kid' or 'twenty-three skidoo'. **1978** D. Bagley *Flyaway* xi. 80 This elderly, profane woman..used an antique American slang... I expected her to come out with 'twenty-three, skidoo'.

Ski-doo, Skidoo (skidū·, skī·dū). orig. *N. Amer.* Also **skidoo.** The proprietary name of a motorized toboggan; hence *gen.* (with small initial), any motorized toboggan.

1961 *Time* (Canada ed.) 29 Dec. 10/1 To hear the Eskimos tell it, the Ski-doo is the greatest thing to hit the north since stripped blubber. **1963** *Globe & Mail* (Toronto) 2 Mar. 30/2 Fishermen..were getting to the choice spots the easy way, using skidoos. **1964** *Star Weekly* (Toronto) 19 Dec. 13/1 The first of these open-air snowmobiles was the Skidoo, originated by the late Armand Bombardier of Valcourt, Que., seven years ago. **1966** *Beaver* Winter 29/2 A trapper with a fast skidoo is able to cover his lines in a third of the time that it took him to do it by dog-team. **1969** *Trade Marks Jrnl.* (Canada) 23 July 961/2 Ski-doo... Filing date: Oct. 31 1968. Bombardier Limited, Valcourt, Quebec. **1969** *Guardian* 30 Aug. 4/5 It was this valley which one had originally proposed to travel along with the skidoos. **1973** C. Bonington *Next Horizon* xvii. 238 Do you want to go by Skidoo, or dog-team? **1977** *Courier-Mail* (Brisbane) 8 Mar. 5/5 One man..gave me a thrilling but hair-raising lift on his skidoo.

Hence as *v. intr.*, to travel by Skidoo; **skidoo·ing** *vbl. sb.*, **skidoo·er.**

1966 *North* (Ottawa) Nov.–Dec. 38 She now substitutes skidooing under the Midnight Sun for surfing on subtropical seas. **1968** J. G. Vermandel *So Long at Fair* xvii. 129 It was near one of those small hills that a skidooer came upon the body. **1975** W. S. Avis in *Occasional Papers Dept. English R. Military Coll. Canada* (1978) No. 2. 25 To skidoo. **1979** R. Fiennes *Hell on Ice* viii. 120 Charlie and I skidooed back to our abandoned tents. **1980** *Beautiful British Columbia* Winter 9 (*caption*) Skidooing near Fort St. John.

skid row (skid rōu). Chiefly *N. Amer.* Also with capital initials. [Altered f. *skid road* (*b*) s.v. *skid *sb.* 5.] **a.** Any run-down area of a town where the unemployed, vagrants, alcoholics, etc., tend to congregate. Also *fig.*

1931 G. Irwin *Amer. Tramp & Underworld Slang* 170 *Skid row*, the district where workers congregate when in town or away from their job. **1935** A. J. Pollock *Underworld Speaks* 107/1 *Skid row*, district in a city where tramps (bums) congregate. **1942** *Crisis* Oct. 314/1 Here, a short walk up from 'Skid Row',..is haven for men of all races. **1944** *N. & Q.* Nov. 120/2 A skidrow..is a district (mostly in western cities) where unskilled workers.. gather to look for jobs—a district of employment agencies, cheap flop-houses, etc. **1953** W. Burroughs *Junkie* viii. 80 When the time came for my sendoff shot, I was assigned to Ward B—'Skid Row', it was called. **1959** *New Statesman* 26 Dec. 899/2 Described in a report by a church mission as the nearest thing to 'Skid Row' that we have, it is a festering slum hidden in a narrow valley between the backs of lush shops in Firth Street, and Caxton Hill. **1963** *Economist* 13 July 125/1 The grisly inhabitants of the numerous urban 'skid rows'. **1977** D. M. Smith *Human Geogr.* xi. 338 The concentration of social deviants in the local 'Skid Row' produces freak figures.

b. *attrib.*

1948 *Sun* (Baltimore) 18 Sept. 3/2 Salisbury was a skid-row alcoholic when he was committed to Eloise. **1962** *Times* 1 Feb. 5/3 The 'skid-row' pictures painted in Seattle. **1973** E. B. Ritson in Howe & Loraine *Environmental Med.* xvi. 216 The grim environment of the skid-row districts of cities tends to attract social misfits. **1980** N. Marsh *Photo-Finish* i. 18 He disguises himself..like a Skid Row drop-out.

skier. (In Dict. s.v. Ski *sb.*) Add: (*a*) (later examples); (*b*) a water-skier. See also *ski-ing *vbl. sb.*

1924 K. Furse *Ski-Running* 56 By his tracks shall a Ski-er be judged! **1941** *Life* 4 Aug. 54/1 In water-skiing the skiers hang on to a 75-ft. rope behind a speeding motorboat and skim along the water as fast as 45 m.p.h. **1959** P. Moyes *Dead Men don't Ski* iv. 48 Not a bad skier... Could have used him in the Team in the old days. **1973** V. Canning *Finger of Saturn* v. 97 The boat was moving into the small quay now. Its speed slackened and the skier dropped low into the water.

skiff, *sb.*[2] For *Sc.* read Chiefly *Sc.* and add: **1.** Also, a light flurry or cover *of* snow (N. Amer. examples).

1928 *Dialect Notes* VI. 88 *Skiff or skift* as applied to snow..means a thin coating. **1959** E. Collier *Three against Wilderness* xix. 193, I saw the track in a skiff of snow, half a mile from the cabin. **1966** M. & O. Murie *Wapiti Wilderness* iv. 44 We were glad enough to have a skiff of snow, it made study and counting of tracks so much easier. **1975** *Budget* (Sugarcreek, Ohio) 20 Mar. 10/5 The weather..has been colder with skiffs of snow here and there.

skiffle (ski·f'l). orig. *U.S.* [Origin unknown.] **1.** Formerly (*U.S.*), a style of jazz music popular at rent parties, deriving from blues, ragtime, and folk music, and played on standard and improvised instruments. Later, a form of popular music developed from this in the 1950s (esp. in the United Kingdom), in which the vocal part is supported by a rhythmic accompaniment of guitars or banjos and other more or less conventional instruments; a song written in this style. Cf. *washboard.

1926 (*jazz-music title*) Chicago skiffle. **1930** *Paramount Dealers' List* Apr. 2/1, 12886—Home Town Skiffle—Part I. Descriptive Novelty—All Star. **1946** in Carey & McCarthy *Jazz Directory* (1949) 167 Dan Burley and his Skiffle Boys..Skiffle Blues. **1948** *N.Y. Age* 9 Oct. 2/7 Fletcher insisted on looking over the Skiffler's shoulder trying to dig the riffle that make the skiffle. **1957** *Times* 3 May 13/4 Earnest young women will not consent to hear even skiffle, unless they are sure of the reverberation factor of the sitting room. **1959** *Times* 27 June 7/3 Skiffle is a form of Do it Yourself that depends entirely on song and has the supreme merit of persuading its devotees to make music for themselves. **1966** P. J. Kavanagh *Perfect Stranger* xiv. 199 It was the age of skiffle but the ethos had reached Djakarta before the technique. **1973** 'J. Marks' *Mick Jagger* (1974) 55 Chris Barber and Ken Colyer were the leaders of the skiffle movement. I was with the very first Barber-Colyer Skiffle Band, and when they split up, I left because I didn't actually enjoy skiffle. **1976** *Jrnl. R. Soc. Arts* CXXIV. 603/2 One of the most cheering things that came out of skiffle was the fact that it got children singing, because they sang in groups.

2. *U.S. Blacks.* = *rent party* s.v. *rent *sb.*[1] 4 c. (Perhaps the original sense.)

1946 [see *percolator c]. **1956** S. Longstreet *Real Jazz Old & New* 126 You could always get together and charge a few coins and have a skiffle... The money paid the rent. **1974** [see *percolator c].

3. a. *attrib.*, as *skiffle band, group, music*.

1957 *Sing Out!* Spring 30 In the first decade of the 20th Century, these New Orleans boys called themselves a 'Skiffle' band. **1981** *Washington Post* 8 Jan. DC7 The Sunshine Skiffle Band..has become a favorite at area folk festivals. **1953** *Melody Maker* 10 Oct. 15/1 London Jazz Club..Ken Colyer's Sensational Jazzmen and Skiffle Group. **1957** C. Brooke-Rose *Languages of Love* 215 A skiffle group—consisting of two guitarists, a thimble-fingered drummer with a wooden washboard, and a man sweeping a carpet-brush rhythmically over three metal strings drawn taut across a saucepan. **1976** *Dumfries & Galloway Standard* 25 Dec. 7/8 The 'Vipers' comprise five well-known artistes in their own right who get together to form a skiffle group. **1948** *Record Changer* June 5/2 The Jazz Room..will feature only the best of jazz and opening night featured..Freddy Moore, Kansas Fields, Dan Burley and his skiffle music. **1958** J. Asman in P. Gammond *Decca Bk. Jazz* xiv. 173 The phenomenon of skiffle music, peculiar only to Britain as yet, is well under way. **1978** *New York* 3 Apr. 73/2 Lonnie Donnegan was already a veteran performer in England when, in 1961, his recording of 'Does Your Chewing Gum Lose Its Flavor on the Bedpost Overnight?' introduced skiffle music to America.

b. General *attrib.* uses.

1946 [see sense 1 above]. **1948** *Record Changer* Aug. 5 Dan Burley, the skiffle man, is now a disc jockey. **1953** *Melody Maker* 19 Sept. 15/2 Club Calendar... London Jazz Club... Ken Colyer's Sensational Jazzmen, every Monday, with skiffle party. **1957** *Universe* 30 Aug. 8/3 There were lots of children's playclothes..'skiffle outfits' of jersey and jeans. **1965** G. Melly *Owning Up* xi. 135 He [*sc.* Ken Colyer] called these interludes 'Skiffle Sessions', to differentiate them from the more serious activity of playing blues, rags, [etc.].

Hence as *v. intr.*, to play skiffle music; **ski·ffler,** one who plays skiffle; a devotee of skiffle; **ski·ffling** *vbl. sb.*[2]

1948 *N.Y. Age* 9 Oct. 3/2 The skiffler's Club is one in which members control the titles of records selected for such programs. **1956** *Observer* 21 Oct. 13/3 The clientele whom the skifflers serve are almost wholly under-thirty, non-drinkers.., mainly middle-class. **1957** *Auckland* (N.Z.) *Weekly News* 2 Oct. 5 'Skiffling' is the latest craze to come to New Zealand. **1957** [see *rock and roll *v.*]. **1959** H. Hobson *Mission House Murder* iii. 21 Anybody with floppy hair, a pair of tight jeans and a cheap guitar can be a skiffler in three easy lessons. **1974** *Times* 9 Jan. 12/7 That was the time when..those who did not rock skiffled away merrily.

skiffling (ski·fliŋ), *vbl. sb.*[1] [Cf. Scabbling, Scaffling, Scappling *vbl. sbs.*] (See quot. *a* 1877.)

a **1877** Knight *Dict. Mech.* III. 2195/1 *Skiffling*, knobbing. Knocking off the rough corners of ashlar in the preliminary dressing. **1901** J. Black *Masonry* 23 For ragstone ashlar work, the stone, when quarried, has its rough projections knocked off with a heavy double-pointed hammer... This operation is locally [*sc.* in Kent] called 'skiffling', and is the same as that known in the neighbourhood of London and other parts of the country by the term of 'knobbling'.

skift, *sb.*[3] (Earlier and later N. Amer. examples.)

1808 B. Hunt *Diary* 25 Dec. in *Chester Co.* (Pa.) *Hist. Soc. Bull.* (1898) 17 May be call'd green Christmas: a small skift of snow. **1927** *Amer. Speech* II. 364 *Skift of snow*, a small amount of snow. **1947** *Canad. Cattlemen* (Winnipeg) Dec. 148 Farmers regard 'a skift o' rain' as an adjunct to the fermentation of the natural juices in the semi-green corn blown into the [silo]. **1970** I. Petite *Meander to Alaska* iv. 34 We had seen practically every rock, headland, light, skift of sea birds rising, tree, and deserted beach cabin.

skift (skift), *sb.*[4] Var. Skiff *sb.*[1] *U.S. dial.*

1656 *Suffolk Co.* (Mass.) *Deeds* (1880) I. 2 [We are] desired by Jno. Blackman to App[rize] a smale skifte taken vp adrift. **1807** J. R. Bedford in *Tenn. Hist. Mag.* (1919) V. 118 They would board us in their skift without the inconvenience to us of going to shore. **1816** U. Brown *Jrnl.* 12 Sept. in *Maryland Hist. Mag.* (1916) XI. 222 Wm. Wells..prevails with me to go with him down the River 1½ Miles to what he called a skift. **1885** *Century Mag.* Aug. 505/2 Visitors call it a skift, natives a skift. **1935** G. Santayana *Last Puritan* III. x. 395 You were perfectly happy here, sculling in your skift.

ski-ing, *vbl. sb.* (in Dict. s.v. Ski *sb.*). Add: Also **skiing. 1.** (Later examples.)

1911 A. Bennett *Card* xi. 273 Ski-ing became the rage... The Captain said 'skee', but he did not object to 'shee'... People with no shame..said brazenly 'sky'. **1927** E. Hemingway in V. W. Brooks *Amer. Caravan* 46 In the Silvretta the ski-ing had been all right, but it was spring ski-ing. **1975** *New Yorker* 21 Apr. 40/3 It has been raining a lot, ruining the skiing.

2. Water-skiing.

1971 'D. Halliday' *Dolly & Doctor Bird* xii. 161, I have never yet met..a man who with greater clarity

could teach me to water-ski... I made a reasonable success, for a beginner, at skiing.

3. *attrib.*

1921 A. Lunn *Alpine Ski-ing* 16 (*caption*) Ski-ing slopes at Scheidegg. **1932** Auden in *Rev. Eng. Stud.* (1978) Aug. 283 A waxen sandboy in skiing kit. **1946** G. Millar *Horned Pigeon* xx. 306 The first passengers to arrive.. wore ski-ing clothes. **1951** 'J. Wyndham' *Day of Triffids* vi. 107 She had chosen a dark-blue skiing suit. **1961** 'J. le Carré' *Call for Dead* xviii. 191 She met Dieter on a skiing holiday in Germany. **1978** S. Sheldon *Bloodline* ii. 30 He skipped skiing lessons in order to go into the village with Anna.

ski-joring (skī·dʒōᵊ·riŋ). Also **skijoring.** [Semi-naturalized alteration of Norw. *skikjøring*, f. *ski* Ski *sb.* + *kjøring* driving (f. *kjøre* to drive).] A winter sport in which a skier is pulled over the snow by a horse or horses (or by a motorized vehicle).

1920 *Punch* 17 Mar. 204/1 Skating, sliding, curling and yodelling in the intervals of ski-ing, skijoring, skilacking and skihandlung. **1927** *Sunday Times* 13 Feb. 17/2 In the skijoring races on the Lake,..the well-known Swiss hockey player fell. **1946** G. Stimpson *Bk. about Thousand Things* 473 The winter sport in which a person on skis is drawn over the snow or ice by a horse is called *skijoring*. **1963** *Guardian* 13 Feb. 13/3 Every type of winter sport.. skating, skijöring, bob-sledding. **1980** G. M. Fraser *Mr. American* xvi. 296 Ski-joring..is when you have horses to pull you along on skis.

Also **ski-jo·rer,** a skier who engages in ski-joring.

1936 'F. Breding' *Nine Waxed Faces* 193 Then..I led t back to where Granby was bending over the skijörer.

skil (skil). [a. Haida *sqil.*] = *sablefish. Also **ski·lfish.**

1886 *Encycl. Brit.* XX. 170/1 Halibut, herring, salmon, cod, and coal-fish or 'skil' (this last also rich in oil and a valuable food-fish) are likewise abundant [off the coast of British Columbia]. **1897** C. R. Tuttle *Golden North* 124 One of the most delicious of deep water fish is the skil, or black cod, as it is sometimes called. **1910** F. W. Hodge *Handbk. Amer. Indians* II. 591/1 Skil. A local name of the black candle-fish. **1923** D. K. Tressler *Marine Prod. Commerce* 736/2 Skilfish (*Anoploma fimbria*). A common food fish from Unalaska to Monterey. **1964** G. C. Carl *Some Common Marine Fishes Brit. Columbia* 48 The giant skilfish..may attain a weight of 200 pounds.

skilful, *a.* and *adv.* For 6 skillful read 6, 9—(chiefly *U.S.*) **skillful** and add examples of this form.

1891 J. W. Stevens *Leather Manuf.* vii. 75 There are several machines for this work each of which will perform satisfactory work when managed by skillful hands. **1952** G. Sarton *Hist. Sci.* I. ix. 225 The power of Athens was based..also..upon the skillful use of all the resources that such places as Delos and Delphi offered. **1969** R. Buckminster Fuller *Operating Man. Spaceship Earth* i. 19 The myriad of physical, muscle, and craft-skill specializations which their intellect and their skillful swordplay commanded. **1972** P. H. Kocher *Master of Middle-Earth* ii. 25 Their skillful blending as achieved by Tolkien requires some sophistication of understanding.

skill, *sb.*[1] Add: **6. a.** Also, an ability to perform a function, acquired or learnt with practice (usu. *pl.*).

1932 H. G. Wells *Work, Wealth & Happiness of Mankind* xvi. 808 Unforgettable memories, obstinate prepossessions, life-worn traditions, obsolete skills and responses. **1938** *English* II. 20 It is the function of the educator..to enable the pupils to appropriate and use all that preceding generations have learnt, the useful skills, the practical knowledge, the social organizations, the moral principles. **1945** *Times* 29 Sept. 4/6 There is a sizeable body in Congress which believes..that this country should secure the greatest possible political advantage from its present monopoly of the actual manufacturing skills. **1958** *Listener* 12 June 976/2 There are ages of maturation at which it is appropriate to teach children skills like reading. **1964** P. Strevens *Papers in Lang.* (1965) ii. 25 The national needs for foreign language skills in the nineteen-sixties are of a different order. **1975** *Language for Life* (Dept. Educ. & Sci.) xiii. 198 The advocates of this form of organisation say that these conditions lead to an assured attention to the 'basic skills'. **1980** *Times* 29 Feb. 19 For the advertising agencies a restricted market means that their skills will be needed more than ever.

9. a. skill centre orig. *U.S.*, a local training institution providing instruction in practical and technical skills, *spec.* in U.K. (**Skillcentre, skillcentre**), one sponsored by the government (cf. *job centre* s.v. *job *sb.*[2] 7).

1963 *Amer. Vocational Jrnl.* Dec. 33/2 The industrial situation assures the future of the area schools, but two problems involved are: (a) a common agreement on the type of regional education programs (i.e. state vocational schools,..vocational departments in comprehensive high schools, state skill centers), [etc.]. **1975** *Manpower Services Commission Ann. Rep.* 1974–75 8/3 The government training centres under their new name of 'skillcentres' were to be expanded. **1976** *Ibid.* 1975–76 16/3 These services include sponsored training at Skillcentres designed to enable firms to send employees to be trained to meet the firms' own precisely defined needs. **1977** *Daily Tel.* 12 Sept. 11 Technicians from the Government Skillcentres, who are of much higher standard. **1978** *Church Times* 27 Jan. 2/2 Skill centres and other training provisions for school-leavers and the young unemployed.

b. In *pl.*

1967 Coulthard & Smith in Wills & Yearsley *Handbk. Management Technol.* 196 Techniques of management by objectives, performance planning, and skills analysis are being more widely applied as they become increasingly effective in contributing to success. *Ibid.* 212 *Skills analysis*, the setting down of the underlying knowledge and dexterity which an operative will require in order to perform a given industrial operation. **1971** R. N. Evans *Foundations of Vocational Education* iii. xiv. 231 Unlike earlier manpower and anti-poverty training programs, Skills Centers could accept trainees whenever a training slot was open. **1976** *National Observer* (U.S.) 17 Apr. 14/5 The English teaching profession..has progressed..well beyond thinking of writing instruction solely or principally in terms of basic skills instruction. **1977** P. Strevens *New Orientations Teaching of English* vi. 78 Shortcomings in demonstration and practice facilities affect the skills component.

skillet[1]. Add: **1.** Now chiefly *N. Amer.*, a frying-pan, a (heavy) cooking-pan.

1917 C. Mathewson *Second Base Sloan* vi. 73 [Nearby lay] an iron skillet with the handle broken off. **1932** W. Faulkner *Light in August* xvii. 382 He stands, tall, misshapen, lonely in his lonely and illkept kitchen, holding in his hand an iron skillet in which yesterday's old grease is bleakly caked. **1959** A. Sexton in *Audience* Autumn 31, I have found the warm caves in the woods, filled them with skillets, carvings, shelves, closets. **1968** *Globe & Mail Mag.* (Toronto) 13 Jan. 16/3 Cook, stirring constantly, over medium heat, until slightly thickened. Add frankfurters. Cover skillet, simmer 8 to 10 minutes. **1979** *Arizona Daily Star* 5 Aug. (Parade Suppl.) 12/4 Heat oil in paella pan or wide skillet.

skillet[2]. Add: **1.** (Later examples.) Also in more general applications.

1959 *Gloss. Terms Packaging* (*B.S.I.*) 19 *Skillet*, a piece of scored timber rotarily cut into veneers which forms part of the finished punnet. **1968** *Guardian* 18 Nov. 6/4 Mr Moreland keeps a selection of 'skillets' (flattened out matchboxes) in his wallet to pass round at social gatherings.

skillful, chiefly U.S. var. *skilful *a.* and *adv.*

skilling, *sb.*[1] Add: Also 9 **skillen. 1. b.** (Earlier examples.)

1799 R. Johnson *Let.* 26 Aug. in *Evangelical Mag.* (1800) July 299 Blood was discovered in different parts, particularly in a small skilling, where, as afterwards appeared, my friend was dragged. **1826** J. Atkinson *Acct. State of Agric. & Grazing in N.S.W.* 100 The barn may be built with lean-to's or skillings all round.

attrib. **1852** *Colonial Church Chron.* VI. 294 As is usually the case with bush houses, two skillen-rooms were to be added behind.

skillinger (ski·liŋgəɹ). *East Anglian local.* [f. corruption of *Terschelling*, the name of one of the West Frisian Islands in north-western Netherlands + -er[1].] An East Anglian oyster smack operating off the north-western coast of Holland (see quots.).

1933 *Yachting Monthly* Aug. 265 Used for oyster dredging off the Terschelling light on the Dutch coast they [*sc.* large Brightlingsea smacks] were known locally as 'Skillingers'. **1959** P. Norton *End of Voyage* xv. 107 Many of these larger smacks were ketch rigged... They were called 'Skillingers'. **1970** E. March *Inshore Craft Gt. Brit.* I. vi. 192 Another ground lay off the Terschilling Light on the Dutch coast, which gave its name to the vessels working there—'Skillingers'.

skillion. For *Austr.* read *Austral.* and *N.Z.*, and add: Also 9 **skilion. 1.** (Earlier examples.)

1843 C. Rowcroft *Tales of Colonies* I. vii. 120 At the back of the long room of twenty feet, a skillion, to serve as a kitchen, &c.

attrib. **1866** Mrs. N. Chevalier *Reminisc. Journey* 23 The house consisted of a front room..a skillion kitchen with big fire, and one end of this skillion had a kind of shed.

2. (See quot. 1933.) Now *Hist.*

1846 C. P. Hodgson *Reminisc. Austral.* 39 Skillions formed by a sloping verandah to receive the sheep in from the fold as required. **1863** S. Butler *First Year Canterbury Settlement* x. 158 The wool-shed..[has] a large central space, and an aisle-like partition on each side.. for holding the sheep at night... In a wool-shed, the aisles would be called skilions. **1871** M. A. Broome *Christmas Cake in Four Quarters* iv. iii. 279 There was the skillions standing empty, and the shearers lounging about idling when Christmas Day came. **1933** L. Acland in *Press* (Christchurch, N.Z.) 2 Dec. 15/7 *Skillion*. In old Australian and New Zealand books (up to about 1880) it is common and means the sheep-holding part of a wool-shed.

3. Special Comb. **skillion roof,** a roof sloping from the side of a building; hence **skillion-roofed** *a.*

1911 J. Nangle *Australian Building Practice* 171 *Lean-to Roof*. This kind (sometimes called a Skillion Roof) ..is generally used only in rear buildings, or verandahs where no ceilings are required. **1977** *N.Z. Jrnl. Science* XIX. 312 A skillion roof and other thin roof forms. **1967** J. Morrison in *Coast to Coast 1965–66* 136 It was little more than a skillion-roofed shack.

skilly, *sb.* Add: **2.** *transf.* An insipid beverage; tea or coffee. Also *attrib.* and *Comb.* Chiefly *Naut. slang.*

1927 [see *madam *sb.* 3 c (*e*)]. **1933** J. Masefield *Bird of Dawning* 214 The skilly-can, which may have contained a hot drink of some sort, was rolling with the whack-pots. **1935** —— *Victorious Troy* 37 There was a big old battered tin coffee pot containing skilly, or a brown, hot liquid, which the crew called 'tea' at night, and 'coffee' in the morning. **1936** B. M. Adams *Ships & Women* iii. 59 They said it was tea. Skilly, they called it. **1953** J. Masefield *Conway* (ed. 2) III. 250 A cup of skilly completed the repast.

skim, *sb.* Add: **1. c.** (Earlier examples.)

1807 J. R. Bedford *Jrnl.* 22 Jan. in *Tennessee Hist. Mag.* (1919) V. 50 Nothing worth noting..but the intense severity of the cold..occasioning a very thin skim of ice on the river. **1869** 'Mark Twain' *Innoc. Abr.* xx. 206 It never has even a skim of ice upon its surface.

d. The fraction of latex which is poor in globules of rubber and is separated from the cream by centrifugation in the manufacture of rubber.

1928 *Brit. Pat.* 319,410 2/1 The concentrate contained approximately 0·07 and the skim about 0·14 per cent of ammonia. **1937** H. Barron *Mod. Rubber Chem.* v. 57 Latex is separated into two portions, one containing about 60 per cent of rubber and very little serum constituents, while the 'skim' contains about 6 per cent rubber and nearly all the serum constituents. **1952** *Ann. Rep. Progress of Rubber Technol.* xvi. 21 A producer of latex concentrate has described a procedure for recovering the residual rubber from centrifuge skim or cream underlayer. **1972** *Materials & Technol.* V. xiv. 471 The whey or skim obtained after the concentration of latex still contains small amounts of rubber, which can be recovered as skim rubber.

5. *skim ice*; **skim-board,** a type of surf-board used for riding shallow water; **skim money** *slang*, a portion of the takings at a casino illicitly diverted in order to evade taxes.

1965 P. L. Dixon *Compl. Bk. Surfing* 143 Riding a skim board is simple to define, difficult to accomplish. **1972** *National Geographic* Nov. 688 Riding a skimboard, a youth glides over a comber-laved beach. **1938** W. Faulkner *Unvanquished* 211 It lay with its body on the land and its head fixed in the skim ice like it was set into a mirror. **1974** J. Keats *Of Time & Island* v. 74 You can see them [*sc.* fish] lying under skim ice in the shallows in the spring. **1973** W. McCarthy *Detail* iii. 144 They used her as a courier for skim money.

skim, *v.* Add: **2. b.** (Later *fig.* example.)

1926 W. R. Inge *Lay Thoughts of Dean* ii. x. 157 Civilisation tends to sterilise the ablest part of a nation. In each generation it skims off the cream and leaves the milk thinner.

d. To conceal or divert (some of one's earnings or takings, freq. from gambling) to avoid paying tax on them; also *absol.* Also with *off*. *U.S. slang.*

1966 *Nat. Observer* (U.S.) 5 Sept. 7/3 Certain Las Vegas gamblers have been 'skimming' millions of dollars in casino winnings—taking a cut of the receipts before the tax collector had a chance to get his share. **1973** *Sun* (Baltimore) 24 July A7/8 Noting the frequent discrepancies in amounts of cash transferred between people and the huge sums involved, he asked: 'Do you have any information, Mr Strachan, of anyone skimming?' **1978** M. Puzo *Fools Die* xviii. 194 Gronevelt felt that hotel owners who skimmed money in the casino counting room were jerks, that the FBI would catch up with them sooner or later.

skimble-skamble, *sb.* Add: Also, writing of this nature.

1818 Byron *Let.* 1 June in *Works* (1900) IV. xvii. 238 Did you read his skimble-skamble about Wordsworth being at the head of his own profession, in the eyes of those who followed it?

skimi, var. *shikimi.

skimish (ski·miʃ). *slang.* [ad. Shelta *škimis* to drink, *škimišk* drunk.] Alcoholic drink; liquor. Also **ski·mished** *a.*, drunk.

1908 W. H. Davies *Autobiogr. Super-Tramp* xxiv. 211, I seldom lie down at night but what I am half skimished (half drunk), for I assure you I never go short of my skimish. **1936** J. Curtis *Gilt Kid* iv. 40 He had been drinking all that skimish without having had a bite to eat.

skimmer, *sb.* Add: **1. c.** Esp. the black clam, *Cyprina islandica* (examples.)

1881 E. Ingersoll *Oyster Industry* 248 *Skimmer*, the *Cyprina islandica*, or big beach clam. (South shore of Long Island.) **1949** R. J. Sim *Pages from Past* 65 The big surf clam, or skimmer (Mactra solidissima Chemn.), lies bedded down in great colonies off shore.

d. A device or craft designed to collect oil spilled on water.

1971 *Petroleum Rev.* May 203/2 (*caption*) The skimmer straddles the boom and the suction box is about to be immersed to suck up the oil floating on the surface of the water. **1976** M. Machlin *Pipeline* li. 516 Heavy duty floating skimmers will be deployed to recover as much oil as is feasible. **1977** *Times* 25 Apr. 1/4 A fleet of skimmers is steaming from Stavanger to suck up the oil and transfer it to waiting tankers.

3. b. One who conceals or diverts some of his earnings or takings in order to avoid paying tax on them. *U.S. slang.* Cf. *skim *v.* 2 d.

1970 *Wall St. Jrnl.* 23 Mar. 13/1 Some skimmers..give themselves away by keeping track of their true earnings. A New York dentist, for instance, devised a dot-dash code for his office records. 1978 S. Brill *Teamsters* vi. 241 The cash was being split, some to be counted for taxes and the rest to go to the skimmers.

6. c. A metal hook for trundling a child's iron hoop. Now chiefly *Hist.*

1891 R. Pearse Chope *Dial. Hartland, Devonshire* 70 *Skimmer*, a hooked iron rod used by children for trundling iron hoops without striking them. 1953 L. Daiken *Children's Toys* iii. 38 In Victorian times the old-fashioned metal hoop was controlled by a 'skimmer', the vernacular name for the hook-and-handle apparatus held in the hand. 1961 *Listener* 12 Oct. 549/2 For boys they [sc. hoops] were of iron, driven along and steered by an iron hook or called a skimmer. 1979 *This England* Winter 66/3 The tool used to drive and check the hoop had a hook at the end of a short length of steel with a wooden handle, and was called the 'skimmer'.

7. b. (Later examples.) Also, a hydroplane, hydrofoil, hovercraft, or other vessel that has little or no displacement at speed.

1909 J. I. Thornycroft in *Engineering* 12 Mar. 365/1 Vessels which greatly reduce their displacement when travelling at high speeds are generally called 'hydroplanes', but this name is not altogether satisfactory, as the surfaces on which they glide are not always planes. To call such vessels 'gliders' or 'skimmers' has been suggested as more appropriate... The latter word will be used to describe boats which at high speeds are heavier than the water they displace. 1930 *Yachting Monthly* XXIX. 20 Owing to the fact that she was an unballasted skimmer she had an unfortunate habit of capsizing at moorings. 1945 J. J. Fahey *Pacific War Diary* 198 The General, Admiral, Captain and a few other officers left the ship in a skimmer for a picnic. 1967 (*title*) Jane's surface skimmer systems 1967–68. 1971 *Morning Star* 30 Mar. 9/1 Soviet sea-going skimmers type Kometa-M are furnished with log, radio, radar and other navigation safety instruments. 1975 *Sunday Mail Mag.* (Brisbane) 22 June 5/1 They are not hovercraft or hydrofoils but 'skimmers'—ships which can take off from the water and 'thunder along a few metres above it, supported on huge airliner-type wings.

c. A hat; a broad-brimmed boater, esp. of straw. Formerly, **skimmer hat**. *slang* (chiefly *U.S.*).

1830 J. F. Watson *Annals of Philadelphia* 176 Other articles of female wear..[include] a 'skimmer hat',..of a very small flat crown and big brim, not unlike the present Leghorn flats. 1929 *Amer. Speech* IV. 430 Among the terms which the daily press credits Mr. Dongan with inventing are:..skimmer (hat). 1939 M. B. Picken *Lang. Fashion* 73/1 *Skimmer*, flat-crowned sailor, usually of straw, having wide, straight brim. Worn and so-called by students at Eton College. 1946 *Sun* (Baltimore) 14 Jan. 12/1 New Yorkers who patronize such places pay several times over the original cost of their skimmers, in tips alone, during the course of a year. 1974 H. De Vries *Glory of Hummingbird* ii. 13 The thoroughly incompatible straw hat..the brightly banded boater, or 'skimmer' or 'katy'.

d. A sheath-like dress that fits closely to the lines of the body. Chiefly *U.S.*

1964 *N.Y. Times* 9 Dec. 5 Irish linen skimmer with a flirty scalloped hemline! 1968 *Tel.* (Brisbane) 2 Feb. 14/7 Cotton crepe skimmers.. finished with scrolls on sleeves and a self bow trim. 1974 *News & Press* (Darlington, S. Carolina) 25 Apr. 3 (Advt.), Our large collection includes wraps, skimmers, pleated coat dresses, fit 'n flares [etc.].

8. b. *Cricket*, etc. A ball that travels with a low trajectory.

c 1868 in H. Chadwick *Scrapbk.* XI. 3/1 An over-throw of Hatfield allowed Wilkins to seize second; he then stole to third, and ran in on Fisler's 'skimmer' to left field. 1897 K. S. Ranjitsinhji *Jubilee Bk. Cricket* iii. 119 Strawyard promptly drives it just as expected—a real 'skimmer' 6 feet over extra-cover's head. 1908 *N.Y. Even. Post* 11 June 11/1 Twice Honus dug up skimming grounders that Mike shot past Leach. 1911 P. F. Warner *Bk. of Cricket* v. 114 Haigh was bowling, and.. Palairet batting, when a 'skimmer' came towards the pavilion straight for me. 1930 A. P. F. Chapman in Lonsdale & Parker *Game of Cricket* vi. 114 The catches sent him are a varied assortment—'skiers and skimmers'—but he rarely gets an easy one. 1980 *Amer. Speech* 1976 LI. 294 Tennis slang..skimmer, ball gliding lightly and rapidly over the net.

9. skimmer-cake (earlier example); † **skimmer hat**: see sense 7 c above; **skimmer shell** *U.S.*, the shell of a clam or scallop (cf. sense 1 c in Dict. and Suppl.).

1795 J. Woodforde *Diary* 9 Feb. (1929) IV. 172 Dinner to day, boiled Beef & a Skimmer-Cake. 1880 *Golden Hours* XII. 520/1 Two pretty shells of the kind that children call 'skimmer shells'.

skimmi, var. *SHIKIMI.

skimmia. In etym. for (Thunberg, 1784) substitute (C. P. Thunberg *Nova Genera Plantarum* (1783) III. 57) and for def. read: An evergreen shrub of the genus so called, belonging to the family Rutaceæ, native to Japan, China, or the Himalayas, and bearing panicles of small white flowers followed by red berries. Cf. *SHIKIMI. (Earlier and later examples.)

1853 *Curtis's Bot. Mag.* LXXIX. 4719 (*heading*) Japan Skimmia. 1908 G. Jekyll *Colour in Flower Garden* xi. 104 Here are green Aucubas and Skimmias. 1925 A. J. Macself *Flowering Trees & Shrubs* xi. 179 Cuttings of Skimmias root very well in sandy peat under bell-glasses. 1960 *Times* 24 Sept. 9/3 There are..the skimmias, but here we have to be..more careful, because some of them are monosexual. 1980 *Plantsman* I. 237 Skimmias prefer deep heavy fertile acid moist soils.

skimming, *vbl. sb.* Add: **2. b.** The practice of concealing or diverting some of one's earnings or takings to avoid paying tax on them. *U.S. slang.* Cf. *SKIM *v.* 2 d.

1966 *Economist* 10 Sept. 1023/3 The report ascribed the allegations of skimming and unofficial connections to excessive zeal displayed by the department. 1966 *Wall St. Jrnl.* (Eastern ed.) 23 Nov. 32 The term 'skimming' refers to the alleged practice of some casino operators of failing to report the full amount of their gambling revenues to state and Federal tax authorities, and often distributing this unreported income to alleged secret interests in their casinos. 1970 *Ibid.* 23 Mar. 13/1 A few years ago in Philadelphia, the owner of a large pizza parlour was suspected of 'skimming'. 1976 *National Observer* (U.S.) 23 Oct. 4/4 To correct the common practice of 'skimming' (underreporting the income from bingo games), the commission suggests that all states require operators to report the percentage of the take that actually goes to the intended charity. 1984 *Daily Tel.* 24 Aug. 11/2 Charged in connection with a 'skimming' operation—siphoning off money to avoid tax—at a Las Vegas casino.

4. skimming net, a fishing-net with a handle, a dip-net (earlier example).

1806 Lewis & Clark *Orig. Jrnls. Lewis & Clark Exped.* (1905) IV. 249, 250 Three people have a number of buffalow robes. They have great number of skimming nets.

skimming, *ppl. a.* Add: **4.** *Cricket.* Of a bowler: that bowls the ball with a low trajectory (*now rare*). Also, of a shot which carries low and fast. Cf. *SKIMMER 8 b.

1851 W. Clark in W. Bolland *Cricket Notes* 132 Suppose you have what I call a skimming low-bowler. 1888 A. G. Steel in Steel & Lyttelton *Cricket* iii. 165 The low skimming fast bowler is generally an easy man to play. 1930 *Morning Post* 9 Aug. 14/1 Woolley..made a low skimming drive over the ring at long-on for 6 off S. Staples.

skimobile (skī·mōbīl). *N. Amer.* [f. SKI *sb.* + *-MOBILE.] **1.** A car or chain of cars used to carry skiers up a mountain; a ski-lift.

1946 *Richmond* (Va.) *News Leader* 9 Jan. 4/2 A skimobile ascends Mount Cranmore, near North Conway, N.H., taking skiers to the top of a 2,052 foot rundown. 1970 *United States 1980/81* (Penguin Travel Guides) 483 The oddball skimobile is one of the oldest lifts in New England.

2. A small vehicle for travelling over snow, with caterpillar tracks at the back and steerable skis in front. Cf. *SNOWMOBILE. (Now the usual sense.)

1955 *Kingston* (Ont.) *Whig-Standard* 6 Apr. 25/3 With the advent of snowmobiles and skimobiles, most dog teams are now used for dog races. 1969 'R. Stark' *Blackbird* (1970) xx. 127 There were a couple of skimobiles down there, little open scooters with skis in front and treads in back. 1974 R. B. Parker *God save Child* iii. 8 Shopping centers, a fish market, a skimobile shop.

skimp, *a.* (Later examples.)

1928 [see *long-eat* s.v. *LOVE *sb.*[1] 16]. 1970 G. F. Newman *Sir, You Bastard* viii. 244 Sneed turned his attention to considering the skimp possibilities; there were no probables for the job.

skimp (skimp), *sb.* Chiefly *dial.* and *colloq.* [f. the adj.] A small or insignificant piece of something; a small or scanty article, esp. a fashionably skimpy garment.

1943 C. Robinson *Dial. Leeds* 118 He thowt 'at t' moin wa necessary til his existence, and.. he'd one miaeke o' white plaster, an' oiled here an' there, so as to resemble skimps o' cloud, as seen to the physical eye. 1925 *Bulletin* (Glasgow) 11 Apr. 10/2 She surreptitiously clutched a little skimp of a handkerchief in her eyes. 1966 *Seventeen* June 92 A skimp.. of striped green French cotton. 1979 B. Malamud *Dubin's Lives* vi. 211 She wore a yellow skimp, her bosom snug in the fabric, her legs.. good to see in short dresses.

ski·mping, *vbl. sb.* [f. SKIMP *v.* + -ING[1].] The action of the verb.

1898 A. Beardsley *Let.* Jan. (1970) 425, I must try and boil the book down but it's so rich and full of chances that skimping would be a sin. 1977 *Hot Car* Oct. 125/3 (Advt.), No skimping to cut cost.

skimpy, *a.* (Earlier example.)

1842 C. Ridley *Let.* 9 Oct. in *Cecilia* (1938) viii. 101 She had on her wedding dress which was very skimpy.

skin, *sb.* Add: **I. 1. c.** (Earlier and further examples in sense: a purse.) Also, a wallet, or a pocket-book.

a 1790 H. Potter *New Dict. Cant & Flash* (1795) 53 *Skin*, a purse. 1902 S. Clapin *New Dict. Americanisms* 365 *Skin*... A purse; a pocket-book. 1935 A. J. Pollock *Underworld Speaks* 107/1 *Skin*, a pocketbook or wallet. 1936 J. Curtis *Gilt Kid* xii. 137 Proper jobs I mean. Not nicking skins from blokes what are lit up. 1955 D. W. Maurer in *Publ. Amer. Dialect Soc.* xxiv. 114 Synonymous terms [of billfold] are *hide*, *skin*, or *poke*.

d. The integument of a bird or mammal, which is preserved but not mounted.

1840 W. Swainson *Taxidermy* i. iii. 84 The preservation of birds in skins, or, more properly, in an unmounted state, is above all others, the best for scientific purposes. 1888 *Encycl. Brit.* XXIII. 90/1 Powders consisting of tannin, pepper, camphor, and burnt alum are sometimes used for 'making skins', but they dry them too rapidly for the purposes of 'mounting'... When 'skins' only are to be made for the cabinet, it is sufficient to fill the head and neck with chopped tow, [etc.]. 1964 G. Corbet in H. N. Southern *Handbk. Brit. Mammals* i. 117 There are two current methods of preparing study skins, resulting in 'round' and 'flat' skins respectively. The round skin, in which the skin is filled to simulate the shape of the body, is the traditional method.. but flat skins are now usually preferred.

e. *spec.* A piece of sealskin or the like attached to the running surface of a ski to prevent slipping backwards during climbing. Also called *climbing skin*. Usu. *pl.*

1904 E. C. Richardson '*Ski-ing*' *Ski-Runner* (ed. 3) 11 Sealskins.. are useful where long, unbroken ascents are to be made. By far the best kind are those which are stuck temporarily to the bottom of the ski... They are called, after their inventor, 'Sohm' skins. 1924 K. Furse *Ski-Running* 39 Skins are used for climbing uphill on tour. They consist of strips of sealskin, which are attached to the running surface of the skis. 1946 H. Innes *Blue Ice* vii. 192 The Norwegians use different waxes, not skins, for climbing through snow. 1966 M. Woodhouse *Tree Frog* xvi. 123 The back room, damp and musty with stored rugs and skis and skins. 1980 J. Cartwright *Horse of Darius* xii. 175 He fastened his skin, attached some skins and made the climb.

2. b. *U.S. slang.* A dollar.

1930 [see *few* pp. 33 c]. 1950 [see *LAY *sb.* 3 d]. 1976 R. B. Parker *Promised Land* xx. 121, I got a buyer with about a hundred thousand dollars..a hundred thousand skins.

3*. Chiefly *Jazz.* (*a*) A drum-head; (*b*) *slang*, a drum. Usu. *pl.*

1927 *Melody Maker* Aug. 756/1 Moisture from the breathing of the dancers will also condense on your side drum and the side absorb this immediately. 1938 *Manch. Guardian Weekly* 2 Sept. 188/3 The swing musicians' cats talk' play.. 'skins' (drums) and 'woodpiles' (xylophones). 1945 L. Shelly *Jive Talk Dict.* 21 *Beatin' the skins*, striking the drums. 1980 *Musicians Only* 26 Apr. 15/3 Two years on skins, the drummer...

II. 4. d. The bare (human) skin.

1933 Joyce *Ulysses* 748, I in my skin hopping around. 1956 H. Gold *Man who was not with It* (1965) xxv. 142, I asked her to bring her swimming clothes...because we were not going to swim in our skins today. 1976 *Western Mail* (Cardiff) 27 Nov. The great day dawned, Wales v Africa, Wales in skins and Jack Sharkey's and S. Africa in white (skins meant no jerseys).

c. *U.S. Blacks' slang.* The skin of the palm of the hand, as making contact in shaking or slapping hands in friendship or solidarity. Freq. in phr. *to give (some) skin*, imp. *gimme some skin* (also as *sb.*).

1942 Z. N. Hurston in A. Dundes *Mother Wit* (1973) 223/2 'Gimme some skin!' 'Lay de skin on me, pal!' Sweet Back grabbed Jelly's outstretched hand and gave it a good shake. 1944 D. Burley *Handbk. Jive* 85 The act of 'Gimme-some-skin' involves some theatrics, an intricate sense of timing, plenty of gestures. 1967 *Harper's Mag.* Nov. 62/2 Once—when I came in on the break behind him at precisely the right point—Pops gave me some skin. He reached out his dark old hand.. and I turned my hand, palm up... Pops lightly brushed my open palm with his, a rare sign of approval. 1972 B. G. Cooke in T. Kochman *Rappin' & Stylin' Out* 33 The gesture of 'giving skin' and 'getting skin' are very common in the black community. 1974 H. L. Foster *Ribbin', Jivin', & Playin' Dozens* iv. 119 The viewer of TV sporting events will often observe black athletes, and whites too now, giving skin after a home run, a touchdown, or at the start of a basketball game.

5. b. skin and bone(s) (further examples); hence **skin-and-bony** adj. Also **skin and grief**.

1886 H. Baumann *Londinismen* 180/2 *Skin-and-bones*...*skin-and-grief*.. haut und knochen, dürre Person. 1906 [see *CHAMBERPOT]. 1912 D. H. Lawrence *Let.* 24 Dec. (1962) I. 172 They want me to have very fat. You're nothing but a laugh and a daft skin-and-bony man. 1955 G. Greene *Loser takes All* i. vii. 43 The horse was all skin and bone and I had forgotten that the road was uphill. 1958 B. Granger *Schone* ii. 9 The old man..was just skin and bones. Maybe they could fatten him up.

c. (Later examples.)

c 1885 A. W. Pinero in M. R. Booth *Eng. Plays of 19th Cent.* (1973) IV. 338, I'm wet to the skin and frightfully hungry! 1938 R. D. Finlayson *Brown Man's Burden* 60 It was useless to try and find shelter, and the two runaways were soaked to the skin in a minute. 1974 S. Milligan *Rommel* 128 The rain had temporarily stopped... We were all soaked to the skin and bloody miserable.

i. (Further examples: cf. quot. 1896, sense 4 b, and sense *5 k (*b*).)

1886 G. B. Shaw *How to become Musical Critic* (1960) 115 The one [*sc.* actor] gets into the skin of one character: the other only puts on the clothes of twelve. 1916 J. R. Towse *Sixty Years of Theater* xxiii. 361 In the church scene, Miss Rehan won her audience by a fine display of honest womanly indignation, but she never really 'got into the skin' of Beatrice. 1959 M. Summerton *Small Wilderness* i. 11 He got under the skin of the rôle and lived it... He was given a small part in the spring production. 1963 *Listener* 28 Mar. 564/2 Those who enjoy the fun of getting inside someone else's skin.

j. *under the skin*, in reality, as opposed to superficial appearances. Esp. in phr. *sisters under the skin* (after quot. 1896).

1896 Kipling *Seven Seas* 193 For the Colonel's lady an' Judy O'Grady Are sisters under their skins! 1946 A. Christie *Hollow* iii. 31 They were the same, sisters under the skin, Mrs. Pearstock from Tottenham and Mrs. Forrester of Park Lane. 1959 [see *IMPULSE *sb.* 5 c]. 1960 P. Gallico *Mrs Harris goes to New York* ii. 24 Mrs Schreiber poured it all forth to her sympathetic sister-under-the-skin, Mrs Harris. 1966 *Economist* 8 Oct. 129/1 The old ladies who booed him...are sisters under the skin to the dockers who met him with a placard clearly advising him to 'drop dead, you bum'. 1975 D. Francis *High Stakes* xi. 161 Merchant bankers are pirates under the skin. 1977 *Times* 19 Apr. 14/2 *Sub specie aeternitatis*, you might say, the Richmond dustmen and Jimmy Edwards are brothers under the skin.

k. *to get under* (a person's) *skin*, (*a*) to affect the deep feelings of; to irritate, to annoy; (*b*) to come to an understanding of, to empathize with.

1896 Ade *Artie* vi. 54 Say, Miller, if I was to beat his whole face off I couldn't kelch even. He got way under the skin on me. 1927 H. T. Lowe-Porter tr. *T. Mann's Magic Mountain* I. v. 290 What's the matter? Has anything got under your skin? 1927 H. Crane *Let.* 12 Sept. (1965) 307, I think I really succeeded in getting under the skin of this glorious and dying animal [sc. the Indian]. 1933 F. Baldwin *Innocent Bystander* (1935) vii. 122 That pleased her, she had got under his skin, he had at least admitted something. 1938 E. Bowen *Death of Heart* I. vii. 143 'That is why she always gets under my skin.' 'So she'd been very kind.' 'Indeed she has—that's her way of getting under my skin.' 1944 A. Christie *Five Little Pigs* I. iv. 32, I think you are interested in—character, shall we say? to get under the skin, as it were, of your criminal. 1948 L. A. G. Strong *Trevannion* xxi. 297 'Aren't you perhaps afraid that inadequacy may be on your side?' 'Damn you, Walter. You do get under a man's skin.' 1972 D. Delman *Sudden Death* iii. 58 Do I bug you, Mr Mathews? Do I get under your skin? 1977 C. McCullough *Thorn Birds* xvii. 455, I can keep you, because I'll never let you get under my skin.

l. *no skin off one's nose* and vars. (*colloq.*), a matter of indifference to one.

1920 S. Lewis *Main Street* xxv. 312 Go to it. No skin off my nose, Nat.. Think I want to be fifth wheel in the coach? 1928 —— *Mantrap* viii. 99 If you think.. that it's any skin off my nose to lose the pleasures of your company, you got another think coming. 1930 *Amer. Mercury* Dec. 420/1 It ain't no skin off of Hymie's bugle. 1934 J. O'Hara *Appointment in Samarra* vi. 136 Okay, No skin off my nose. 1938 D. Baker *Young Man with Horn* i. iv. 30 It was no skin off Jeff what color his old lady painted the piano. 1955 A. Miller *View from Bridge* 100 Don't thank me... It's no skin off me. 1962 D. Lytton *Goddam White Man* v. 113 But it was no skin off my nose that she was dead. 1963 *Australasian Post* 14 Mar. 51/2 If you want to go up like a drongo in the DTs it's no skin off my bugle. Go ahead: be a gig! See if I care! 1966 L. Porter *Four Fevers of Clea* 6 Our arrival was no skin off her nose and she didn't pay all that much attention to us. 1971 B. Malamud *Tenants* 35 Make it like eight [o'clock] or around that if it's no skin off you. If I miss a day don't fret on it. 1973 R. Milner in W. King *Black Short Story Anthol.* 376 Then Clyde said it was no skin off his ass. 1978 J. Meynell *Papeesmake* xiv. 188 It was no skin off my nose... My heart wasn't hurt, even if my pride was.

m. *(here's) to the skin off your nose* and vars.: used as a toast.

1925 Fraser & Gibbons *Soldier & Sailor Words* 260 *Here's to the skin off your nose*: Your good health! 1936 Wodehouse *Young Men in Spats* ii. 42 'Well, skin off your nose,' said Pongo. 'Fluff in your latchkey,' said Barmy. 1949 [see *MUD *sb.*[1] 3 e]. 1970 D. Elder *Sleeping Bride* ix. 85 Philip handed her a drink and she lifted it, 'Here's the skin off your nose.'

n. *skin and blister*, sister. *Rhyming slang.*

1925 Fraser & Gibbons *Soldier & Sailor Words* 260 *Skin and blister*, sister. [Rhyming slang.] 1961 Ingram *Cockney Cavalcade* xiv. 170, I saw your skin and blister last night. 1972 G. F. Newman *You Nice Bastard* 348 *Skin and blister*, sister.

7. f. The outermost layer of a pearl.

1886 *Encycl. Brit.* XVIII. 446/2 A pearl of the first water should possess, in jewellers' language, a perfect 'skin' and a fine 'orient'. 1908 L. Kornitzer *Pearls & Men* xix. 165 Keep on inspecting your pearl... When the blemish has been removed and a clean bright skin shows up, the worst is over. Smooth the skin with the finest emery paper you can obtain. 1976 B. W. Anderson *Gemstones for Everyman* xxiii. 300 The finest cultured pearls have come from waters off the north Australian coast... The oyster used here is the large *Pinctada maxima*. The resultant cultured pearls are also large and have very thick skins.

g. The outer or surface layers of a conductor, in which alternating current tends to be concentrated at high frequencies.

1891 [see *skin effect*, sense 13 below]. 1893 J. J. Thomson *Recent Res. Electr. & Magn.* iv. 266 When the vibrations are very rapid the currents are practically confined to a thin skin on the outside of the conductor. 1943 C. L. Boltz *Basic Radio* vii. 121 As the frequency is increased, the current is concentrated more and more in the outer layer—the 'skin'—of a conductor. 1958 J. Shepherd et al. *Higher Electr. Engin.* vii. 182 The effect increases with frequency, until at high frequencies the current is almost entirely in the 'outer skin' of the conductor.

h. *slang.* A tyre.

1954 *Rocky Mountain News* (Denver) 2 Sept. in *Amer. Speech* (1956) XXXI. 305 *Skin*, a tire. 1977 *Hot Car* Oct. 62/1 The answer is to run at the *same pressure* as the standard tyres, as by dropping the pressure any more than two pounds, you could cause sidewall failure, even in the big American skins.

i. A duplicating stencil; *spec.* the part that actually goes on the duplicator.

1966 G. M. Beck *Machines for Office Workers* iv. 73 When the [correcting] fluid is applied [to the stencil] it will..seep through the incisions and make the carbon..adhere to the wax sheet; subsequently, at the duplicator, the carbon and backing sheets are removed, and in doing this it is..possible that the re-formed skin will also be detached with the torn off the incorrect letter, and the correction cover it, appear on the duplicated sheet. 1972 T. Lilley *'K' Section* xi. 203 She had typed the 'skin', duplicated it, and then ran off about four hundred copies. 1973 *Daily Tel.* 25 Apr. 23/8 It was then discovered that one foolscap duplicating skin could produce only 10,000 copies. Four skins had to be typed and 'run off'.

8. c. *transf.* The outer covering of any craft or vehicle (or a constituent layer of this); esp. of an aircraft or spacecraft.

1921 *Flight* XIII. 247/2 The skin when the chines is formed of two thicknesses of mahogany planking... One ply of varnished cotton fabric is laid between the two. 1937 *Jrnl. R. Aeronaut. Soc.* XLI. 846 It is proposed to form the skin of the wing from two separated sheets of plywood. 1948 'N. Shute' *No Highway* iii. 64 Here in the aircraft everything was firm and steady and secure; the even tremor of the engines, the faintly heard rush of air over the outer skin, these bred confidence. 1962 G. Cooper in *Into Orbit* 29 The crews are surprised to find a lot of cutting through the capsule's skin. 1973 *Times* 19 May 4/1 They found that the whole of the skin of the caravan was full of blocks of cannabis. 1973 Terry & Baker *Racing Car Design & Development* xv. 135 Increasing safety-consciousness caused the FIA to stipulate that, for 1972, the outer skins of all Formula I monocoques had to have a maximum thickness of 16 swg. 1977 D. Beaty *Excellency* 6 The company to which it [sc. an aircraft] belonged had been painted out... What remained against the silver duralumin skin was AN—.

9. b. *U.S. slang.* = SKINFLINT.

1900 Ade *More Fables* 30 Some of the Folks... used to say that Henry was a Skin, and was too Stingy to give his Family enough to eat.

c. Without contemptuous implications: a person (of a specified kind). Chiefly *Anglo-Ir.*

1914 Joyce *Dubliners* 152 Ah, poor Joe is a decent skin. 1939 F. O'Brien *At Swim-Two-Birds* 166 A decent skin if ever there was one, said Slug with warmth, a man that didn't stint the porter. 1958 B. Behan *Borstal Boy* iii. 236 These were fine.. that Craggs was muttering about the Colonel, who wasn't a bad old skin. 1966 *Listener* 23 June 906/1 And he seemed a decent old skin. 1966 F. Shaw et al. *Lern Yerself Scouse* 22 *Ee's a good skin*, he is an agreeable fellow.

d. *slang.* A horse or mule.

1923 E. Hemingway *Three Stories & Ten Poems* 31 They take the first batch of skins out to gallop. 1925 Fraser & Gibbons *Soldier & Sailor Words* 260 A skin, a horse; mule. 1941 Baker *Dict. Austral. Slang* 67 *Skin*, a horse, 'generally the property of a professional wayfarer'.

e. *slang.* = *SKINHEAD 2 (*b*).

1970 *Daily Progress* (Charlottesville, Va.) 15 Apr. 7/1 You gotta decide what family you are going to join, the hairies or the skins. 1978 R. Westall *Devil on Road* iv. 26 Those Midland sods must be crazy... I shouted the loudest thing you can shout at skins. 1981 *Times* 22 July 11/3 'There's good and bad skinheads,' as he said of the skins.. And there are non-violent skins.. Certainly, many of the skins are things.

9*. *U.S.* A card game in which each player has one card which he bets will not be the first to be matched by a card dealt from the pack.

1925 *Messenger* Dec. 386/1 Playing 'skin' for matches. 1935 Z. N. Hurston *Mules & Men* i. iii. 72 Ah played skin wid de Devil for his life. 1973 J. Scarne *Scarne's Encycl. Games* xvi. 409/1 The game of Skin is dead even; that is, dealer and player have exactly equal chances of winning. 1978 Moore & Levine *Big Paddle* (1979) i. 15 Lovern loses skin. He'll go all over looking for a skin game.

9.** *the Skins*, the nickname of the 5th Royal Inniskilling Dragoon Guards or, formerly, the Royal Inniskilling Fusiliers [properly a corruption of *Inniskilling*, assimilated to the sb.].

1938 H. Hayward *In Praise of Ulster* 235 The Indian Mutiny, South Africa and the Great War brought fresh glories to the valiant 'Skins'. 1949 St. J. Ervine *Craigavon* iv. xlvi. 213 The history of the 'Skins', the nickname of the Inniskilling Fusiliers, is rich with the jewels of courage. 1954 L. MacNeice *Autumn Sequel* 67 The Skins have gone to Kenya with their trousers smartly creased. 1981 J. Johnston *Christmas Tree* 25 Did you have a brother in the Skins?

III. 10. a. *skin care, colour, -flake, -tissue.*

1954 V. Dengel *All about You* vi. 141 There are four points to proper, daily skin care. 1969 V. J. Kehoe *Technique Film & Television Make-Up* (ed. 2) iii. 40 (*heading*) Skin care products. 1972 *Sat. Rev.* (U.S.) 27 May 18/2 The idea of discrimination based on skin color is beyond their comprehension. 1930 E. Pound *XXX Cantos* xv. 66 Skin-flakes, repetitions, erosions. 1944 *Horizon* Mar. 172 The grey matter of the brain-rind was originally skin-tissue. 1949 M. Mead *Male & Female* i. 19 The sensitivity of our skin-tissues.

b. *skin-bag, -boat* (earlier and later examples), *-cover.*

1860 Skin-bag [see *ATTA]. 1910 W. de la Mare *Three Mulla-Mulgars* xxviii. 237 Having cut one of their skin-bags to pieces. 1804 W. Clark in *Orig. Jrnls. Lewis & Clark Expedition* (1904) I. 87 The Indians pass this river in Skin Boats which is flat and will not turn over. 1968 G. Jones *Hist. Vikings* i. i. 17 These hunters, fishermen, and food-gatherers from the south...developed the skin-boat. 1954 J. R. R. Tolkien *Fellowship of Ring* ix. 408 They drew the skin-covers over their boats.

11. a. *skin-fitting.*

1915 D. H. Lawrence *Rainbow* iv. 91 She wore an elegant, skin-fitting coat. 1947 *Science News* IV. 11 The men who went into enemy ports during the war wore skinfitting dresses.

b. *skin-grafting* (earlier examples).

1870 *Lancet* 27 Aug. 306/2 (*heading*) Skin-grafting. *Ibid.* 22 Oct. 566/2 Mr. Francis Mason has performed the operation of skin grafting on granulating surfaces in nine instances.

12. a. *skin-covered.*

1897 Yeats *Secret Rose* 1 A large house with skin-covered wattles for the assembly. 1977 *Proc. R. Soc. Med.* LXX. 234/1 Any open lesion is more dangerous than a skin-covered one.

b. *skin-thin* adj.

1946 W. de la Mare *Traveller* 12 Their skin-thin gills. 1966 P. Scott *Jewel in Crown* iv. 171 The tough little shell of skin-thin masculinity that used to harden the outward appearance of the British military wives.

13. skin-beater *slang* (now *Obs.* or *rare*), in a jazz- or dance-band: a drummer; **skin beetle** *U.S.*, a greyish-brown beetle of the genus *Trox* or a brown, hairy beetle of the family Dermestidæ, feeding on carrion or other organic material; **skin-changer**, one supposedly able to metamorphose himself or herself; **skin cream**, an oleaginous cosmetic preparation for care of the skin; **skin depth** *Electr.* [cf. sense *7 g], the distance from the surface of a conductor at which an electromagnetic wave of a given frequency is attenuated by a factor of $1/e$ ($e = 2{\cdot}718\ldots$); also *fig.*; **skin-drying** *vbl. sb.* (*Founding*), drying of the surface of a greensand mould before casting; so **skin-dried** *a.*; **skin effect** *Electr.*, the tendency of an alternating current of high frequency to flow through the outer layers only of a conductor, resulting in an increase in effective resistance; **skin flap** *Surg.*, a portion of living skin attached to the body by one edge so that it remains alive while it is used to close a wound after amputation, or in plastic surgery; **skin-flick** *slang*, a film of a pornographic type; **skin-food** (examples); **skin friction**, add: *esp.* the friction between the surface of an aircraft or the like and the air; (further examples); **skin game**, (*a*) (earlier example); also *transf.* and *fig.*; (*b*) the pornographic trade; (*c*) = sense 9* above; a game of this; **skin graft**, a piece of living skin which has been surgically transferred to a new site or to a different individual; also, the process of making such a transfer; **skin house** *slang*, (*a*) a gambling establishment; (*b*) an establishment providing pornographic entertainment; **skin magazine** *colloq.*, a magazine containing nude photographs, a pornographic magazine; **skinman** (further examples); **skin packaging**, a method of packaging in which the article, placed on a backing plate which is to form part of the package, has a plastic film cover thermoformed on to it; **skin pass** *Metallurgy*, a final cold-rolling, effecting a small reduction in thickness, given to heat-treated strip steel in order to improve surface and mechanical properties; **skin-pop** *v. intr.* (*slang*, orig. *U.S.*), to inject a drug subcutaneously (cf. *MAIN-LINE *v.*); so as *sb.*, the action of skin-popping; also *fig.*; **skin-popper**; **skin-popping** *vbl. sb.*; **skin potential**, the electrical potential between different points on the skin, esp. as exhibited in the galvanic skin response; **skin resistance**, † (*a*) = *skin friction* above; † (*b*) the resistance of the skin (sense *7 g) of an electrical conductor; (*c*) the electrical resistance of the skin of an organism; **skin-search** *sb.* and *v.* (*slang*) = *strip-search sb.* and *vb.* s.v. *STRIP *v.*[1] 24 a; **skin test** *sb.*, a test to see whether an immune reaction is elicited when a substance is applied to or injected into the skin; so **skin-test** *v. trans.*, **skin testing** *vbl. sb.*; **skin tonic**, a cosmetic astringent for the skin; **skin trade** (orig. *U.S.*), commerce in animals' skins; also *fig.*: also = *skin game* (*b*) above.

1936 *Amer. Mercury* XXXVIII. p. x/2 *Skin beater*, the drummer man. 1953 *N.Y. Times Book Rev.* 13 Sept. 33/3 Red, the reefer-smitten skin beater. 1842 T. W. Harris *Treat. Insects New Eng. Injurious to Vegetation* 11 Skin-beetles.., bone-beetles..act the useful part of scavengers. 1895 J. H. & A. B. Comstock *Man. Study Insects* 559 The skin-beetles..are small or of medium size. 1942 [see

larder beetle s.v. *LARDER[1] 3]. 1972 Swan & Papp *Common Insects N. Amer.* xx. 436 Skin beetles feed on carrion, skin, feathers, and dung. 1927 E. V. Gordon *Introd. Old Norse* 224 Berserks were probably named 'bear-shirts' from a superstition that they were 'skin-changers'. 1937 J. R. R. Tolkien *Hobbit* vi. 121 He is a skin-changer. He changes his skin: sometimes he is a huge black bear, sometimes he is a great strong black-haired man with huge arms and a great beard. 1907 *Yesterday's Shopping* (1969) 537/1 Violet oatmeal skin cream. 1979 P. Ferris *Talk to me about England* III. 133, I appear to be missing a pot of special vitamin skin-cream. 1941 J. A. Stratton *Electromagn. Theory* ix. 536 One may assume for conductors of arbitrary cross section that the field and current distributions near the surface differ negligibly from those near the surface of an infinite plane provided the radius of curvature is very much greater than the skin depth. 1962 Corson & Lorrain *Introd. Electromagn. Fields* x. 338 The skin depth decreases if either the conductivity σ, the permeability K_m, or the frequency *f* increases. 1966 *Listener* 5 May 653/3 Myshkin's apparent niceness and gentleness, his ridiculousness, are the surface, the skin depth of his assumed role. 1954 Skin-dried [see *skin-drying* vbl. sb. below]. 1970 E. Parkes et al. in K. Strauss *Appl. Sci. in Casting Metals* ix. 321 (*heading*) Skin dried and dry sand moulding. 1888 *Lockwood's Dict. Mech. Engin.* 324 Skin drying effects the removal of a portion of the moisture and diminishes the risk of a blown or a scabbed casting. 1954 J. E. Garside in A. J. Murphy *Non-Ferrous Foundry Metall.* v. 171 Pouring should be completed as soon as possible after skin-drying owing to the fact that the moisture from the backing sand slowly penetrates towards the skin-dried mould face. 1891 *Electrician* 29 May 91/1 Sir William Thomson recalled attention to the tendency of alternating currents to avoid the central portions of metallic conductors, thereby giving rise to an increase of resistance which has been occasionally alluded to under the name of the 'skin effect'. 1965 *Wireless World* Aug. 401/1 The h.f. resistance is increased partly by skin effect, and more significantly by eddy currents induced in the lossy magnet system. 1873 *Brit. Med. Jrnl.* 15 Mar. 286/2 He took a large skin-flap from the front below the knee, a smaller flap behind, and left just enough of the bones to fit an apparatus. 1974 R. M. Kirk et al. *Surgery* v. 73/2 Skin flaps are used to close large defects in situations where sound healing is essential, when good quality of skin is desirable and when the local blood supply would not sustain free grafts. 1968–70 *Current Slang* (Univ. S. Dakota) III–IV. 110 *Skin flick*, n. A pornographic movie. 1969 *Daily Colonist* (Victoria, B.C.) 7 June 13/3 We ran family movies for nine years and almost went broke. For the last three years, we've been showing skinflicks and doing much better financially. 1975 P. Webb *Erotic Arts* viii. 280 In the '60s..film-makers became aware of the commercial possibilities of the voyeur film, or 'skin-flick'. 1898, etc. Skin-food [see *FOOD *sb.* 2 b]. 1977 B. Pym *Quartet in Autumn* ix. 81 Turning her attention to the wash basin she noted..a jar of skinfood and a tube of Steradent tablets. 1907 F. W. Lanchester *Aerodynamics* vi. 220 In actual planes it is impossible to do away with thickness, so that in addition to skin friction there must be the possibility of a longitudinal pressure component due to the shape of the plane. 1919 R. H. Goddard *Method of Reaching Extreme Altitudes* 9 The resistance, R, may be taken as independent of the length of the rocket by neglecting 'skin friction'. 1948 *Sci. News* VII. 24 In the same way a body moving through air loses energy by skin friction (analogous to conduction). 1978 *Jrnl. R. Soc. Arts* CXXVI. 683/1 During acceleration to supersonic speeds the external surface of the structure becomes hotter due to skin friction from the air flow. 1868 M. H. Smith *Sunshine & Shadow in New York* 405 The square game..is played only by gentlemen, and in first-class houses;..the skin game..is played in all the dens and chambers, and in the thousand low hells of New York. 1904 W. H. Smith *Promoters* 98 We built the bridges finally,..for we weren't really working a skin game. 1920 Galsworthy *Skin Game* I. 19 She wants to sell, an' she'll get her price, whatever it is. *Hillcrist.* (*With deep anger*) If that isn't a skin game..I don't know what is. 1958 *Economist* 1 Feb. 398/2 The..ironies of German political life: the strange mixture of elements.. that mingle in the Bonn skin game. 1970 *Times Educ. Suppl.* 18 Dec. 1/1 The censor and the skin game. 1973 J. Scarne *Scarne's Encycl. Games* xvi. 308 (*heading*) The skin game. 1973 E. McGirr *Bardel's Murder* i. 10 As a very small [antiques] dealer, I was no opposition... His business is rather a skin game. 1976 *Globe & Mail* (Toronto) 7 Jan. 10/3 The long-respected publication had been sold and new publishers had changed big game to the skin game. 1978 Skin game [see sense 9* above]. 1871 *Lancet* 22 Apr. 535/1 On taking off the plaster the skin-grafts were found adhering. 1930 A. H. Davis *Burns* xxi. 195 Most surgeons..find that heteroplastic skin grafts are universally unsuccessful. 1935 P. H. Mitchiner *Mod. Treatment of Burns & Scalds* v. 54 Riverdin's or Thiersch's skin grafts give excellent results. 1977 D. Bagley *Enemy* xxxi. 251 Gillian..had just had the operation for the first of the skin grafts. 1871 *Galaxy* XII. 61 A 'skin' house, as the dens where cheating games are played are called. 1902 Farmer & Henley *Slang* VI. 227/1 *Skin-house*, a gambling den. 1970 *Harper's Mag.* July 34 The skin houses were mostly playing short subjects—a girl taking a bath in a sylvan stream, a volleyball game in a nudist camp. 1972 *Dict. Contemp. & Colloq. Usage* (Eng. Lang. Inst. Amer.) 27/1 *Skin house*, a theater featuring nude women or films of nude women. 1972 J. Wambaugh *Blue Knight* (1973) i. 29 Some gunsel I'd heard was hanging out in the skin houses and taxi-dance joints. 1968 *Rat* 13–16 May 11/1 Two prophylactics and a skin magazine was found in President Kirk's drawer. 1980 *Cosmopolitan* May 319/1 Men often use pictures as stimulation when they masturbate (hence the popularity of so-called skin magazines), but women do so much less often. 1829 P. Egan *Boxiana* 2nd Ser. II. 220 At a proper age, JEM turned out to earn an honest penny, and was apprenticed to a *skyver*, or skinman, in Newcastle-upon-Tyne. 1970 M. Tarmey *Skinman* vi. 128 He sat hunched and helpless in the chair..., a skinman without any skins. 1962 A. L. Griff *Plastics Extrusion Technol.* vi. 122 In skin-packaging, the coated board can now be the base, while coated flexible film can be the skin. 1971 *Engineering* Apr. 63/2 Two commercial systems of skin packaging are in general use. The Soag-Stanley process uses an uncoated board, the other a coated and perforated board. 1939 J. Dearden *Iron & Steel Today* x. 149 A single pass through a skin pass mill then brings it to its final thickness. 1977 R. B. Ross *Handbk. Metal Treatments & Testing* 351 On the production side, the Skin pass will be used to produce the final surface finish and simultaneously achieve slightly improved mechanical properties. 1953 Kramer & Karr *Teen-Age Gangs* i. 35 You get a big fat mouth every time you give that leg of yours a skin-pop. *Ibid.* 243 *Skin-pop*, to inject drugs, usually heroin, under skin into body. 1959, 1964 [see *MAIN-LINE *v.*]. 1971 'D. Halliday' *Dolly & Doctor Bird* xvi. 242 You can't deny we gave your social habits a skin-pop. 1970 *Daily Tel.* 30 Jan. 19/1 She had also 'skin-popped' (injected drugs just below the surface of the skin) and taken a vast assortment of pills. 1953 Kramer & Karr *Teen-age Gangs* i. 35 A very expert skin-popper, Hoppy is. 1970 H. Waugh *Finish me Off* 48 No marks. She must be a skin-popper. 1952 *Sunday Times* 3 Feb. 5/4 'Skin popping'..consists of scratching open a place in the skin and injecting heroin or morphine there. 1970 *Observer* 3 May 3/1 When the addicts run out of veins to inject, because of scars and ulcers, they try skin-popping—injecting just under the skin or into a muscle. 1936 *Amer. Jrnl. Physiol.* CXVII. 189 (*heading*) Skin potential and impedance responses with recurring shock stimulation. 1967 Venables & Martin *Man. Psychophysiol. Methods* ii. 58 The permeability of the cell membrane is a physiological phenomenon, and measurements of skin resistance and skin potential must be made within physiological limits. 1875 *English Mechanic* 3 Sept. 634/3 We have sufficient data from which the skin-resistance [of a ship's hull] can be determined. 1891 *Jrnl. Inst. Electr. Engineers* XX. 479 When we deal with conductors of about a centimetre in diameter there is no apparent effect of this skin resistance. 1895 H. Lamb *Hydrodynamics* xi. 575 The frictional or 'skin-resistance' experienced by a solid of 'easy' shape moving through a liquid. 1904 *Jrnl. Franklin Inst.* CLVII. 248 The skin resistance of copper bonds increases with time. 1927 *Brain* L. 231 We have learned that the skin resistance is invariably decreased both by pain and by elevation of body temperature. 1942 S. R. Hathaway *Physiol. Psychol.* xi. 236 The level of skin resistance has a low inverse correlation with neuroticism. 1971 *Jrnl. Gen. Psychol.* LXXXV. 88 Each record was analyzed by recording the averaged skin resistances for the last two minutes of the resting period and for each minute of the stimulus film period. 1935 A. J. Pollock *Underworld Speaks* 107/2 *Skin search*, an arrested person who is stripped naked and his body thoroughly searched for narcotics. 1970 G. Jackson *Let.* 4 Apr. in *Soledad Brother* (1971) 212 Our cells were being invaded by the goon squad: you wake up, take your licks, get skin-searched. 1973 *Time* 26 Mar. 64 So far, none of the three new guards in California's state prison system for men have been assigned to conduct 'skin searches' of nude prisoners for contraband. 1979 F. Forsyth *Devil's Alternative* xvii. 386 If you are thinking of giving me a weapon, don't bother. On my return I am to be skin-searched. 1925 W. W. Duke *Allergy* xv. 207 She gave positive skin tests to a number of extracts including wheat. 1943 W. C. Boyd *Fund. Immunol.* xi. 424 Patients should not be skin tested without previous adequate history and physical examination. 1961 *New Scientist* 16 Mar. 696/1 All the components cause a positive reaction in the skin tests commonly used for allergic responses. 1971 R. Scott *Wedding Man* ii. 65 Every Asian child was skin-tested [for tuberculosis] as soon as possible after arrival. 1925 W. W. Duke *Allergy* xv. 206 One's first impression of skin testing is likely to be one of disappointment. 1963 L. V. Crawford in F. Speer *Allergic Child* xxvii. 420 Although the mechanics of skin testing are simple, considerable experience is required for proper interpretation. 1906 *Daily Colonist* (Victoria, B.C.) 27 Jan. 5/1 (Advt.), Special Sales. Toilet Goods..Skin Tonics. Perfumes. 1971 M. Lee *Dying for Fun* xxxiv. 167 The fragrance of Lapsang Souchong, mingled with the tang of skin tonic. 1710 W. Byrd *Secret Diary* (1941) 186 About 5 o'clock Robin Hix and Robin Mumford came to discourse about the skin trade. 1885 *List of Subscribers, Classified* (United Telephone Co.) (ed. 6) 101 (*heading*) Fur and skin trade. a 1953 Dylan Thomas (*title*) Adventures in the skin trade. 1977 *Time* 19 Sept. 41/1 The city [*sc.* Boston] set aside this seedy downtown area three years ago for X-rated movies, porn shops and other facets of the skin trade—in hopes of being able to contain them.

skin, *v.* Add: **II. 4. d.** *to skin the cat* (U.S.), to perform a gymnastic exercise involving passing the feet and legs between the arms while hanging by the hands from a horizontal bar and so drawing the body up and over the bar; also *transf.* and *fig.*

1845 S. Judd *Margaret* II. i. 199 Their several diversions, snapping-the-whip, skinning-the-cat, racing round the Meeting-house, or what not. 1888 'C. E. Craddock' *Story of Keedon Bluffs* v. 88 He did not wait a second but 'skinned the cat' among the rafters. 1905 *N.Y. Even. Post* 14 Oct., We have learned how to hide behind the back log of 'environment' or to 'skin the cat' in morality on the score of 'heredity'. 1907 C. E. Mulford *Bar-20* viii. 80, I used to shinny up this here wall an' skin th' cat getting through that hole up there. 1931 *Sun* (Baltimore) 29 May 12/7 You saw them skin the cat On the high trapeze. 1946 B. Treadwell *Big Bk. Swing* 125/2 Skin he cat; ride, brother, ride.

e. To keep (one's eyes) open. *U.S. colloq.*

1865 *N.Y. Herald* in Farmer & Henley *Slang* (1891) II. 361, Keep a padlock on yer mouth and skin yer weather eye. 1875 J. G. Holland *Sevenoaks* x. 133 Skin yer eyes, now, Mr. Balfour, we're comin' to a lick.

f. *fig.* To beat or overcome completely. *U.S. slang.*

1862 *Charleston* (S. Carolina) *Mercury* 9 Aug. 1/5 They were 'skinning' the soldiers of other regiments the 'tallest kind'. 1911 H. Quick *Yellowstone Nights* iv. 110 'Purty good little places,' said he, 'but the home place skins 'em all.' 1981 *Verbatim* VII. III. 7/2 Puns ('Eagles *skin* Washington')..offer limitless possibilities to the enterprising sports journalist.

7. b. (Earlier examples.)

1819 *Massachusetts Spy* 24 Mar. 3/1 They will not be able to skin the people as deep as they did during their former reign. 1839 C. F. Briggs *Harry Franco* II. vi. 76, I wish I may be blown into a gin shop if I warnt skinned clean O! The young woman had..picked my pockets of every cent.

9. a. (Earlier examples.) *Obs.*

1835 J. Todd *Student's Manual* (ed. 3) 115 Should you allow yourself to think of going into the recitation-room, and there trust to 'skinning', as it is called in some colleges. 1837 *Yale Lit. Mag.* Feb. 138 A student is said to skin a problem, when he places the most implicit faith in the correctness of his neighbor's solution of it, or at least sufficient to warrant bestowing upon it the rites of adoption.

b. For *trans.* read *intr.* and add: Also with *through*, to slip through, to pass by a narrow margin. *U.S.*

1902 G. H. Lorimer *Lett. Self-Made Merchant* xi. 141 If you would make a downright failure or a clean-cut success once in a while, instead of always just skinning through this way. 1920 W. Camp *Football without Coach* 57 The best a runner can hope for is a chance to skin through that opening before it ceases to exist.

c. (Earlier example.)

1873 J. Miller *Life amongst Modocs* iv. 44 Four aces! and what else? Skin 'em out, skin 'em out!

III. 11. *trans.* and *intr.* To inject (a drug) subcutaneously. Cf. *skin-pop* v. s.v. *SKIN *sb.* 13. *slang.*

1953 W. Burroughs *Junkie* vi. 57 He had to shoot in the skin about half the time. But he only gave up and 'skinned' a shot after an agonizing half-hour of probing and poking and cleaning out the needle, which would clog up with blood. 1970, 1972 [see *MAIN *v.* 2].

skinch (skintʃ), *v.* and *int.* *north* and *Midland dial.* [Origin unknown.] **A.** *v. intr.* To encroach, to cheat. **B.** *int.* A formula used by children in a game to demand a truce.

1891 S. O. Addy *Gloss. Words Sheffield* Suppl. 52 *Skinch*, *v.* to encroach, to shorten distance. When a boy playing at marbles moves his taw nearer to the ring than he ought to do he is said to *skinch*, *i.e.*, to encroach unfairly. 1893–4 O. Heslop *Northumberland Words* II. 649 *Skinch!* the cry for parley in a boys' game. 1914 D. H. Lawrence *Prussian Officer* 262 Willy could hear the endless calling and shouting of men's voices. 'Tha'rt skinchin!' 1959 I. & P. Opie *Lore & Lang. Schoolch.* viii. 150 'Croggie' is..general in West Hartlepool although the usual term in County Durham is 'skinch'.

skin-diver. [f. SKIN *sb.* + DIVER.] One who dives or swims underwater without a full diving suit or a fixed line to the surface. Hence **skin-diving** *vbl. sb.* and *ppl. a.*; [as back-formation] **skin-dive** *v. intr.*, to dive or swim underwater as a skin-diver; also as *sb.*, an act or instance of skin-diving.

1932 *Blackw. Mag.* Jan. 20/1 They relied solely upon their skin divers—the divers' technical term for a naked man. 1938 D. Long *Sailing all Seas in Idle Hour* ix. 121 The finest 'skin' divers in the world come from lonely atolls such as Penrhyn, where only 'skin' diving is allowed. 1950 Schenk & Kendall *Shallow Water Diving* iv. 74 This equipment is used in the sport commonly known as 'Goggling' or 'Skindiving'. 1951 *Skin Diver* Dec. 2 The name of *The Skin Diver* was picked because it includes everyone interested and participating in underwater fishing and hunting. *Ibid.*, We want to publish any and all items of interest to our skin diving readers. 1952 *Time* 17 Nov. 50/3 Bucher, poised on the rail of the small ship bobbing in the rough water, was aiming to become the first man ever to 'skin-dive' (*i.e.*, without the aid of artificial breathing apparatus) deeper than 115 ft. 1953 J. Y. Cousteau *Silent World* i. 9 Vanity coloured our early skin dives. We plumed ourselves at the thought that we latecomers could attain the working depths of pearl and sponge divers who had made their first plunges as infants. *Ibid.*, Dumas's skin-diving technique consisted of floating face under water and breathing through a schnorkel tube. 1959 L. Smith *One Hour* v. 69 For hours, we'd engage in this philosophical skin-diving. 1964 M. McLuhan *Understanding Media* xxiii. 233 The smarter advertisers have made free with fur and fuzz, and blur and buzz. They have, in a word, taken a skin-dive. For that is what the TV viewer is. He is a skin-diver, and he no longer likes garish daylight on hard, shiny surfaces. 1966 T. Pynchon *Crying of Lot 49* iii. 57 There stood Di Presso, in a skin-diving suit and wraparound shades. 1970 *Daily Colonist* (Victoria, B.C.) 14 Oct. 7/6 At Taveuni, the prince took time off to swim and skindive. 1973 J. Leasor *Host of Extras* vii. 127 He adjusted his mask and went over backwards in the way of the trained skin-diver. 1975 *N.Y. Times* 6 Nov. 20/2 Skindiving Israeli scientists are exploring the sunken fortifications of the city of Acre. 1977 B. Pym *Quartet in Autumn* iv. 37 'It's the swimming that would attract me.' 'You mean skin-diving and that sort of thing?'

skinflinty, *a.* (Earlier example.)

1886 F. R. Stockton *Casting away of Mrs. Lecks & Mrs. Aleshine* II. 122 If he undertook to be skinflinty he'd better try it on somebody else besides us.

skin fold. *Med.* Also **skin-fold, skinfold.** [f. SKIN *sb.* + FOLD *sb.*[2]] A fold of skin and underlying fat formed by pinching, as a measure of nutritional status; freq. *attrib.*, as

skinfold thickness; **skinfold cal(l)iper(s)**, a pair of callipers for measuring the thickness of such a fold.

1921 *Amer. Jrnl. Physical Anthrop.* IV. 224 For the estimate of the quantity of the skin and of the subcutaneous fat, the writer uses the thickness of the skin fold on the upper arm, above the biceps; on the forearm ..and on the abdomen. 1950 *Nutrition Abstr. & Rev.* XX. 250 Pinching a skinfold to obtain a rough estimate of the thickness of the subcutaneous adipose tissue and, by inference, of the fatness or leanness of a subject, is an old clinical procedure. 1954 *Jrnl. Lab. & Clin. Med.* XLIII. 969 The skinfold thickness is read to the nearest half millimeter on the slide scale. *Ibid.* 970 A skinfold caliper..which is easily manipulated with one hand. 1961 L. Martin *Clin. Endocrinol.* (ed. 3) ii. 54 Measurements by skin-fold calipers have shown that from birth until three months..there is a rapid increase in fat deposition. 1973 *Times* 17 May 20/1 Using skinfold calipers, designed by and for doctors working on obesity, the thickness of the fat layer at four sites on the body is measured. 1977 *Lancet* 1 Jan. 17/2 Skinfold thickness measurements provide a non-invasive and reproducible means of measuring subcutaneous fat in newborn babies.

skinful. **3. a.** (Earlier example.)

1788 *Columbian Mag.* Oct. 557/2 Determined, as they said, once more to get *a skin-full of liquor*.

ski·nhead. *colloq.* [f. SKIN *sb.* + HEAD *sb.*] **1.** (A person with) a bald head.

1953 Berrey & Van den Bark *Amer. Thes. Slang* (1954) §430/5 Bald-headed man,..cue ball, skinhead, turret top. 1957 M. Shulman *Rally round Flag, Boys!* vi. 66 Oscar was a bow-legged, barrel-chested man with a skin head. 1976 'O. Jacks' *Assassination Day* v. 85 'Ow long you gonna be, skinhead?.. Your wig's slipping.'

2. A person with a shaven head or closely cropped hair; *spec.* (*a*) a recruit to the U.S. Marine Corps; (*b*) in the U.K., a youth (often one of a gang), also typically characterized by wearing workman-like clothing and heavy boots, and by a tendency to aggressive behaviour.

1953 Berrey & Van den Bark *Amer. Thes. Slang* (1954) §825/2 Cue ball, skinhead, a fellow with a crew haircut. 1956 *Amer. Speech* XXXI. 190 He will administer a royal chewing out (tongue lashing) to the hapless skinheads (recruits; so-called because of their completely shaven heads). 1969 *Daily Mirror* 3 Sept. 12/1 A group of teenagers..wear tight and rather short jeans, collarless T-shirts, exposed braces, big steel-capped boots and hair erased almost to their scalps. The lack of hair is what gives them their generic names..crop-heads, skin-heads or peanuts. The boots are good for kicking. 1971 *Daily Tel.* 13 Apr. 1 Gangs of Hell's Angels and skinheads marred Easter Monday seaside outings. 1973 C. Mullard *Black Britain* iv. xi. 131 According to an eye-witness a gang of white skinheads savagely attacked a black youth who today is nearly blind as a result. 1975 I. Shaw *Nightwork* vii. 84 The skinheads are preparing the ground... One morning we'll wake up and the tanks will be rolling down Pennsylvania Avenue and the machine guns will be on every roof. 1978 *Lancashire Life* Apr. 27/1 The index can be visualised: Beats, Jesus Freaks, Groupies, Skinheads, Punks..the procession seems endless, and to many is evoked by one word—Hippies. 1980 *Herald* (Melbourne) 9 Apr. 5/3 Thousands of skinheads shouting Nazi slogans invaded the resort in special trains.

skinless, *a.* Add: *spec.* of sausages and similar meats.

1954 *Food Manufacture* 1 Nov. p. xi (Advt.), Visking food and sausage casings are..available through Viskase Limited.... Visking Nojax for skinless sausages. 1959 E. H. Clements *High Tension* vi. 119 'He'd bring.. *cevabcici*—'.. 'Sausages, you mean.' 'Oh yes... Special skinless ones..from a shop in Soho.' 1968 R. Clapperton *No News on Monday* v. 52, I helped myself to a couple of skinless sausages from the refrigerator. 1972 D. Bloodworth *Any Number can Play* xvii. 170 He..had..eaten a tin of skinless frankfurters with a purée of mashed yam.

skinned, *ppl. a.* Add: **II. 4. b.** (Earlier example.)

1833 *Political Examiner* (Shelbyville, Kentucky) 22 June 4/1, I wish I may be shot if I dont think you had better keep your eyes skinned so that you can look powerful sharp, lest we get rowed up the river this heat.

c. Beaten, bested, overcome completely; esp. in *to have* (*got*) (a person or object) *skinned*. *colloq.* (orig. *U.S.*).

1904 'O. Henry' *Cabbages & Kings* iii. 56, I guess you've got us skinned on the animal and vegetation question. 1908 'Yeslah' *Tenderfoot S. Calif.* ii. 22 When it rains in California, it's got all the rest of the country skinned to death. 1913 R. Brooke *Let.* c 23 July in *Coll. Poems* (1918) Mem. p. lxxxiv, 'Sir, I may tell you that in my opinion you have Mr. Noyes skinned.' That means I'm better than him. 1927 E. Wallace *Feathered Serpent* iv. 47, I came down here to make a few inquiries... I've got these reporter guys skinned to death!

d. = *SKINT *a.* Also with *out*. *colloq.*

1935 A. J. Pollock *Underworld Speaks* 107/1 *Skinned out*, broke; without funds. 1957 C. MacInnes *City of Spades* I. xii. 93 Why's he left me skinned in hopeless destitution? 1958 *Observer* 14 Dec. 7/8 I'm skinned, I know I can always count on someone helpin' me.

skinner[1]. Add: **2. b.** An implement used for skinning animals.

1872 *Amer. Naturalist* VI. 223 The specimen could have been used as a knife, or 'skinner', although now its edge is too irregular and dull for skinning.

4. b. (Later examples.)

Esp. common in Australia to mean (*a*) a horse that wins a race at very long odds; (*b*) any betting coup.

1903 A. M. Binstead *Pitcher in Paradise* xii. 292 A skinner!—great-balls-of-fire! a skinner! 1907 A. Wright *Keane of Kalgoorlie* 66 Although he had gone up in the weights considerably, his owner decreed that he should win the Rosehill handicap, and give the 'shop' another 'skinner'. 1930 *Technique of Betting* 7 Frequently a race is won by a horse against which the bookmaker has not laid any bet, and the book then shows a profit of 100%—the bookmaker has what he calls a 'skinner'. 1934 T. Wood *Cobbers* viii. 96 Charles..would lay two to one port-wine jelly, five to apple-pie... Tonight we had college pudding and jam tart. Charles..said it was a skinner for the books. 1974 *Sydney Morning Herald* 8 Oct. 17 Skinner for bookmakers. 1977 A. C. H. Smith *Jericho Gun* v. 60 At twelve to one, which is the forecast SP here, it's a skinner.

7. A driver of a team of horses or mules; occas. also, a lorry driver. *N. Amer.*

1870 [see *mule skinner* s.v. *MULE[1] 5 c]. 1910 E. Ferguson *Janey Canuck in West* 91 The teamsters are called 'skinners'. I met them all on the log road. 1924 *Scribner's Mag.* Dec. 645/1 The skinner with the longest words travels the fastest. 1929 *Amer. Speech* V. 147 Since the driver of the old time orecar was called a *mule-skinner* or *mule-whacker*, the driver of the modern motor-propelled car is a *motor-skinner*, sometimes just a *trammer*. 1939 J. Steinbeck *Grapes of Wrath* ii. 14 A guy that never been a truck skinner don't know nothin' what it's like. 1954 E. F. Hagell *When Grass was Free* 3 A single line attached to the next leader's bit and passed back along the teams to the teamster or 'skinner'.

8. *a skinner* semi-adj.: 'skint', broke; empty. *N.Z. colloq.*

1943 *New Writing* XVIII. 68 So I paid for the pair of us, which left me practically a skinner. 1967 *Landfall* XXI. 241 Sure you're a skinner? Not a drop in the place, I mean? 1981 *Macquarie Dict.* (s.v. *skinner*), The beer's a skinner.

Skinner[2] (ski·nəɹ). The name of the American psychologist, B. F. *Skinner* (b. 1904), used *attrib.* to indicate the theories or methods concerned with conditioning human or animal behaviour associated with him; esp. as *Skinner box*, a box in which an animal is isolated, equipped essentially with a bar or other device that it learns to use either to obtain a reward or to escape punishment.

1938 *Jrnl. Exper. Psychol.* XXIII. 507 A modified form of the Skinner apparatus. 1940 *Ibid.* XXVI. 614 (*heading*) The variability of extinction scores in 'Skinner-box' experiments. 1951 E. R. Hilgard in S. S. Stevens *Handbk. Exper. Psychol.* 530/1 It is possible to train animals in the Skinner box to do what the experimenter wishes. 1962 *Listener* 13 Sept. 390/1 The technique for measuring the pressure applied to a lever by a rat in a Skinner-box. 1973 *Nature* 27 July 241/3 The method of 'operant conditioning' in a Skinner box can be used to investigate physiological changes that accompany habit formation. 1980 Brown & Wallace *Physiol. Psychol.* xv. 457 Most of the principles of learning were derived from studies on a single animal (the white rat) in a single learning situation (instrumental conditioning in the Skinner box).

Skinnerian (skinɪə·riăn), *a.* and *sb.* [f. prec. + -IAN.] **A.** *adj.* Of or pertaining to B. F. Skinner's behaviourist theories or methods. **B.** *sb.* A follower or adherent of B. F. Skinner.

1958 *Psychol. Bull.* LV. 148/2 The studies to be reviewed in this paper follow this Skinnerian paradigm. 1964 H. J. Eysenck *Exper. Behav. Therapy* II. 187 One possible reason why Skinnerians have restricted themselves in this fashion. 1965 *Language* XLI. 98 The extreme Skinnerian view that the correct inductive generalization can be accomplished with no need for anything more than positive instances. 1977 H. J. Eysenck *You & Neurosis* v. 168 Skinnerian behaviourism has become a 'school', in the same way as Freudianism did many years earlier. Skinnerians concentrate exclusively on a very small area of psychology. 1979 *Nature* 29 Nov. 440/1 The sales techniques used in Iran, and no doubt they are now being used in other Third World countries, were a balanced compound of Skinnerian Psychology and gangsterism.

So **Ski·nnerism**, Skinnerian behaviourism.

1969 *Times Educ. Suppl.* 16 May 1640/2, I fear the growth of Skinnerism and its rats and pigeons. 1979 *Nature* 24 May 355/1 They bring us a metatheoretical commitment to a hard-line Skinnerism, according to which even the editorial policy of the *Journal of the Experimental Analysis of Behavior* is dangerously revisionist.

skinning, *vbl. sb.* Add: **2. c.** (Earlier example.)

1856 *Trans. Michigan State Agric. Soc.* VII. 171 The old plan of constant cropping without manure, or 'skinning', will ruin the land.

d. A physical or verbal beating; a hammering. *U.S. slang.*

1929 *Chicagoan* 17 Aug. 22/2 In this period he [*sc.* Carl Sandburg] wrote the poetic denunciation of the Rev. Billy Sunday that..remains as the most thorough skinning that the evangelist ever received. 1972 J. W. Thompson in W. King *Black Short Story Anthol.* 260 Daddy..has taught me several different ways to skin a cat, and that redhead doesn't know it yet, but he's got a skinning coming.

2*. (See quot.) *slang.* Cf. *SKIN *v.* 11.

1973 *Daily Mail* 3 Apr. 19/4 *Skinning*, injecting drugs under the skin.

3. *skinning knife* (earlier example).

1859 G. A. Jackson *Diary* 8 Jan. in F. Hall *Hist. Colorado* (1890) II. 521 Dug and panned to-day until my belt knife was worn out; so will have to quit or use my skinning knife.

skinny, *a.* Add: **A.** *adj.* **5.** Of clothing: tight-fitting.

1970 'D. Halliday' *Dolly & Cookie Bird* vii. 96 Janey's friends..in skinny sweaters and bell-bottomed corduroy trousers. 1972 *Vogue* Feb. 63 Long skinny jacket over beautifully cut pants. 1982 *Times* 2 Apr. 10/3 Teddy bear fur over skinny suede skirts.

6. Special collocations, as **Skinny Liz**, a thin girl or woman (see also quot. 1940); **skinny-malink, -links, -linky** (chiefly *Sc.*), a thin or emaciated person or animal; also *attrib.* or as *adj.*; **skinny-rib** *a.*, of a sweater, etc., fitting tightly across the ribs [the apparent connection with *rib stitch* is accidental]; also *ellipt.* as *sb.*

1940 M. Marples *Public School Slang* 190 St. Bees.. used *wimp* (a corruption of *women*) and the Arabic *bint* of women in general, while *skinny liz* was applied, almost as a nickname, to any elderly woman. 1959 I. & P. Opie *Lore & Lang. Schoolch.* ix. 169 Thin people..skin and bones, skinny,..skinny guts, Skinny Liz, skinny-malink. 1961 N. Fitzgerald *Black Welcome* iv. 95 She takes no interest in..eatin'. That's why she's such a Skinny Liz. 1892 *Brechin Advertiser* 6 Sept. 3 Twa skinamalinks o' the genus horse. 1904 'H. Foulis' *Erchie* iii. 15 Wee skinamalink craturs dottin' up the passages in U.F. kirks carryin' the books. 1916 *Dialect Notes* IV. 280 *Skinny-malink*, a very thin person. 'O, she's a regular skinny-malink.' Usage jocular. 1935 S. Beckett *Echo's Bones*, The chagrin of the old skinnymalinks. 1956 *Sunday Times* 22 Jan. 2/5 There used to be a children's song in Aberdeen relating the adventures of a thin man called 'Skinamalinky Lang Legs', which is still sung as a skipping song, etc.: Skinamalinky lang legs Umbrella feet. 1979 L. Derwent *Border Bairn* vi. 71 A skinnymalink of a laddie with holes in his stockings. 1973 *Tucson* (Arizona) *Daily Citizen* 22 Aug. 48/2 (Advt.), Men's fashionable, skinny-rib..acrylic pullover. 1976 *Milton Keynes Express* 25 June 11/1 (Advt.), Skirts, blouses, and skinny ribs.

B. *sb.* **1. a.** *Austral.* A girl or woman. ? *Obs.* **b.** A thin person.

1941 Baker *Dict. Austral. Slang* 67 *Skinny*, a girl or young woman. 1959 [see *Skinny Liz*, sense A. 6 above]. 1977 *Time* 6 June 48/2 The skinnies of the world have, in effect, righteously established fitness standards that reward their own strengths and forgive their weaknesses.

2. Information; rumour (see also quot. 1959). *slang* (chiefly *U.S.*).

1959 *Amer. Speech* XXXIV. 156 *What's the skinny* means 'What's up?' 1974 E. Brawley *Rap* (1975) II. xxiii. 363 Come to lay some skinny on you that I picked up off the vine. 1979 D. Anthony *Long Hard Cure* xxi. 162 Who killed her, Butler? Let's have the skinny. 1980 L. Cody *Dupe* xxiii. 165 Give them the skinny but keep the kudos.

ski·nny, *v.* [f. the adj.] *intr.* To lose flesh; to become skinny or skinnier. Usu. with *down*.

1939 J. Steinbeck *Grapes of Wrath* xxii. 423 She thinned out and she skinnied out, an'—she dropped that baby, dead. 1976 *Billings* (Montana) *Gaz.* 30 June 6-B/3 She joined TOPS after a neighbor started skinnying down to the system. 1981 *TV Picture Life* Mar. 61/1 (Advt.), All the difficult 'skinnying-down' has been done for you while you slept.

ski·nny-dip, *v. slang* (orig. *U.S.*). [f. SKINNY *a.* 3 + DIP *v.* 8.] *intr.* To swim naked. So as *sb.*, a naked swim; **ski·nny-dipper**, a person who swims naked; **ski·nny-dipping** *vbl. sb.*

1966 *Punch* 12 Oct. 557/2 Nearly a year has passed since three members of the San Francisco Sexual Freedom League went skinny-dipping in the San Francisco bay. 1967 Wentworth & Flexner *Dict. Amer. Slang* Suppl. 704/2 *Skinny-dip*, to swim in the nude. 1970 J. Howard *Please Touch* 14 Except for a couple of furtive midnight skinny-dips I had never taken off my clothes in public. 1971 *Daily Colonist* (Victoria, B.C.) 26 May 19/4 It was never resolved if the chickens got over the light problem or the skinny dippers were apprehended by the constable. 1971 *Daily Progress* (Charlottesville, Va.) 17 July 1/7 Two young men decided to skinny-dip near Rexburg, Idaho and left their clothes in their sail boat, but the boat sailed out of reach. 1975 *Sunday Mail* (Brisbane) 9 Mar. 30/6 *Perth.*—Skinny-dippers breathed a collective sigh of relief last week when Police Minister Ray O'Connor said ..they could carry on stripping. 1977 *Times* 11 Apr. 5/7 His guests took skinny-dips in baths of champagne. 1980 L. Birnbach et al. *Official Preppy Handbk.* 190/1 Once every summer, teenagers are caught skinny-dipping after dark. 1981 *Times* 20 Apr. 4/8 One weekend no less than 36 people were arrested for 'skinny-dipping'.

skint (skint), *a.* *colloq.* [Var. *SKINNED *ppl. a.* (see sense 4 d).] Penniless, broke.

1925 Fraser & Gibbons *Soldier & Sailor Words* 260 *Skint, to be*, hard up. 1935 G. Ingram *Cockney Cavalcade* vii. 97 Edina [*sic*] offered him a shilling. 'That's all right... I ain't "skint" yet.' 1955 G. Freeman *Liberty Man* III. iv. 158 If he had enough to pay, it would just about leave him skint. 1962 *New Statesman* 18 May 708/3 All I want is a bike and ten pounds a week in my pocket—there's one thing I can't stand and that's being skint. 1977 S. Milligan in *Observer* (Colour Suppl.) 6 Nov. 32/2 McGonagall..journeyed on foot (he was skint) from Dundee to Balmoral. 1981 *Times* 27 Aug. 17/3 Are the British really as skint as we tend to make out?

SKIN-TIGHT

skin-tight, *a.* Add: Also *fig.*
1916 JOYCE *Portrait of Artist* v. 280 They..gave orders to jarvies in highpitched provincial voices which pierced through their skintight accents. 1977 *Sounds* 9 July 31/1 'Cathedral', 'Dark Star', 'Cold Rain' and 'In My Dreams' are all skintight songs, questioning, but personal and without pretension.

skip, *sb.*[1] Add: **2. d.** *Poker.* = *skip straight* s.v. *SKIP *v.*[1] 8.
1880 J. BLACKBRIDGE *Compl. Poker-Player* vii. 48 'Skips' consisting of alternate cards in sequence for instance, 3, 5, 7, 9, Jack. 1905 R. F. FOSTER *Practical Poker* 75 A skip is almost twice as difficult to get as any other straight, the exact odds against it being 423 to 1.
e. *Radio.* The phenomenon of the poor or non-existent reception of signals from a particular station which occurs between points where signals propagated directly from the station become undetectable and points where signals begin again to be received owing to reflection in the upper atmosphere. Also applied to the silent region itself, and to radio signals received from beyond it.
1925 [see *skip region*, sense 5 below]. 1927 O. F. BROWN *Elem. Radio-Communication* xvi. 203 The existence of the skip is explained by there being insufficient electrons to bring the wave down again until the angle of incidence becomes that corresponding to the 500 range. 1931 *Observer* 8 Nov. 18/5 Because of 'skip' it will hardly ever be audible in this country. 1965 B. SWEET-ESCOTT *Baker St. Irregular* iv. 114 The 'skip' was explained..as being the heavenward arc made by the path of the waves emitted by the short-wave transmitters. 1976 PERKOWSKI & STRAL *Joy of CB* vii. 68 The FCC purposely limited CB operations to distances under 150 miles to preclude the use of skip. 1976 *S9* (N.Y.) Feb. 88/2 When CB skip starts rolling in, he says that's the time to start tuning 25 to 50 MHz.
f. In automatic data processing, the action of a machine (e.g. a punch) in passing over material not requiring the functioning of the machine; a computer instruction or routine specifying such action.
1946 [see *skip bar*, sense 5 below]. 1962 *Gloss. Terms Automatic Data Processing* (*B.S.I.*) 91 Machines in current use can perform the function called skip wherein a field in which no punching is required is rapidly passed under the punch knives, which are not active at the time. 1966 H. P. HARTKEMEIER *Data Processing* iv. 199/1 All functions of the machine are stopped while a skip is taking place. 1969 P. B. JOURDAIN *Condensed Computer Encycl.* 468 An unconditional skip is a computer instruction demanding that the next *n* instructions be ignored. 1976 KERNIGAN & PLAUGER *Software Tools* iii. 80 Skip produces n blank lines.
4. *N. Amer. colloq.* One who absconds, *spec.* to avoid paying debts; one who defaults in payment.
1915 J. R. FOOTE *Mod. Collection Methods* 32 In some lines of business, much, and in some, most of the collection department work is the tracing of skips. A skip is a handy term used to describe a debtor who finds it easy to forget to leave any tracks when he moves his earthly possessions. 1939 *Amer. Speech* XIV. 240 *Skip*, guest who leaves without paying his bill. 1949 *Collier's* 8 Jan. 27/1 Kleinman's book of procedures lists exactly 110 ways to trace a skip. 1978 *Detroit Free Press* 14 Apr. 2c/2 Jean Phelan traces all kinds of hard-to-locate 'skips'—the defaulters who have 'skipped' out.
5. *attrib.* and *Comb.*, as (sense *2 e) *skip distance, region, zone*; (sense *2 f) *skip bar*; (sense *4) *skip-trace, -tracer, -tracing* vbl. sb.
1946 *Ann. Harvard Computation Laboratory* I. 274 Cards may be punched containing a function in the first columns of the cards and a serial number in the last columns of the cards. After the function is punched, a duplicating card and skip bar control the punch. 1926 *Physical Rev.* XXVII. 189 Larmor's theory of refraction due to the electrons of the Kennelly-Heaviside layer does not explain the 'skip distances' for short radio waves. 1977 T. ALLBEURY *Man with President's Mind* vii. 75 The radiated strength was fifty kilowatts..a power of about seventy-five kilometres due east. It would be the skip distance that carried it to Washington, or a relay from London. 1925 *Proc. IRE* XIII. 680 An uncertain region not far from the transmitter has been introduced between 100 and 350 miles during the summer night range and a skip, or entirely-missed, region, occurs in the winter night ranges between 100–350 miles. 1970 K. CONWAY *Naked Nemesis* ii. 18 The last one hadn't paid me... There wasn't enough involved for me to start a skip-trace on him. 1980 J. GARDNER *Garden of Weapons* II. i. 119 The Yanks think we need it [*sc.* a safe house] for a skip-trace outfit. They think we've lost somebody. 1953 BERREY & VAN DEN BARK *Amer. Thes. Slang* (1954) § 460/18 *Skip tracer*, a tracer of defaulting debtors. 1960 P. S. BEAGLE *Fine & Private Place* i. 12 You ran away from it [*sc.* life] nineteen years ago, and it follows you like a skip-tracer. 1978 *Globe & Mail* (Toronto) 14 Sept. 3/2 Mr. Lillie testified that he is a skip-tracer who tracks down persons who default on their debts then change addresses. 1960 J. BLISH *Galactic Cluster* 124 If he has rebuilt..the Universe to accommodate a private skip-tracing firm..I..see no reason why we can't counter-check him. 1977 B. GARFIELD *Recoil* xi. 134 This is..better than repossessing cars and skip-tracing. 1926 *Physical Rev.* XXVII. 192 The skip zone was not very sharply defined. 1946 *Richmond* (Va.) *Times-Dispatch* 27 Jan. I. 16/6 The skip zone is one of the knottiest problems of present-day radio communications.

skip, *sb.*[2] Add: Also *gen.*, a large container for the reception and conveyance of materials or rubbish.
1940 *Chambers's Techn. Dict.* 775/2 *Skip*,..a bucket used for the transport of spoil or materials and hung for this purpose from a crane or cableway. 1950 *Landfall* (N.Z.) IV. 125 We start loading seasoned timber into one of the skips. 1972 *Daily Tel.* 17 Jan. 3/3 Householders who leave builders' skips—large containers which can be hired to take away rubble—at the kerbside for collection by a special vehicle face fines of up to £100 under a law coming into force today. 1978 *Cornish Guardian* 27 Apr. 8/8 There will be a skip placed at the Town Hall, St. Columb and at the entrance to Halloon Avenue, St. Columb Road, on Friday, 28th. April, and at the Town Hall and Public Conveniences, Indian Queens on Friday, 12th. May, 1978 for Bulk Household refuse collections.
attrib. (Further examples.)
1951 J. CLEMO in D. V. Baker *One & All* 260 He had worked as a loader in a clay-pit near Pengarth, and one winter's day he had been crushed by a skip-waggon. 1972 CONYUS in A. Chapman *New Black Voices* 219 Shoveling straw Into the mouth of the skip loader. 1976 *Star* (Sheffield) 29 Nov. 12/5 (Advt.), Sale, TK skip lorry. 12 months' test.

skip, *sb.*[3] For *Sc.* read 'orig. *Sc.*' and add further examples. Also *gen.*, a captain, a commanding officer, a manager, a boss.
1921 *Amer. Legion Weekly* 28 Jan. 7 The skip wanted to investigate. 1930 T. FREDENBURGH *Soldiers March!* xxv. 201 Better get into a wagon somewhere..in case the Skip starts prowling. 1948 M. ALLINGHAM *More Work for Undertaker* xiii. 163 I've been chinning with the old Skip and he says Bang on, jolly good show. 1955 *Times* 15 Aug. 8/5 In rink games the 'skip', or captain, of each side stands near the jack to direct his men by voice or signal where their next shot should arrive. 1968 *Globe & Mail* (Toronto) 3 Feb. 35/5 In addition to winning several minor bonspiels, the Thornhill skip is in the last 16 for the Ontario Curling Association Championship. 1970 *Wall St. Jrnl.* 8 July 18/6 If you're ever called up to play baseball in the big leagues, be sure to call the manager 'Skip'. Managers like to be called Skip. 1973 D. KYLE *Raft of Swords* (1974) viii. 81 On the flight deck..the young navigator said, 'I don't really understand what we're looking *for*, skip.' 'Just Russian warships.' 1977 *N.Z. Herald* 8 Jan. 1-10/3 Who are Arthur Connew's great heroes in all those many years and thousands of ends of bowling? J.S. Martin in the singles and Jimmy Mingins and Mort Squire as skips. 1977 *S. Wales Guardian* 27 Oct. 4/5 Skip Mr. Cliff Davies invested new members to the scout troop.

skip (skip), *sb.*[7] orig. *Sc.* [Origin obscure.] The peak of a cap.
1888 A. G. MURDOCH *Scotch Readings* (Ser. 2) 29 Ye're surely no ettlin' to put on that ugly twa-faced kep..wi' the skip baith back an' fore? 1969 M. PUGH *Last Place Left* ii. 11 He adjusted his American fatigue cap so that the skip almost covered his eyes. 1974 H. MACINNES *Climb to Lost World* xi. 207 'Hiya, Jo. Did you make it?' asked Don, peering up from beneath his cap skip.

skip, *v.*[1] Add: **I. 2. b.** For 'Now *U.S. colloq.*' read 'Now *colloq.*' and add: Also with *out* and as *to skip it*.
1865 M. GRIGSBY *Diary* 3 Jan. in *Smoked Yank* (1888) xxi. 179 Thirteen [paroled men]..skipped out to-day. 1902 'MARK TWAIN' in *Harper's Mag.* Jan. 265/2 Skip out for the coast some night. 1959 I. & P. OPIE *Lore & Lang. Schoolch.* x. 193 Juvenile language is well stocked..with expressions inviting a person's departure, for instance:.. skip it, sling your hook, [etc.]. a 1966 'M. NA GOPALEEN' *Best of Myles* (1977) 308 The son turned out to be a very bad bit of work, sold all the furniture to buy drink and then skipped it to America. 1969 G. LYALL *Venus with Pistol* xxxv. 231 He tells Dona Margarita we seem to have skipped out together. 1977 J. THOMSON *Case Closed* ii. 21 Bibby hadn't turned up. He wondered if he had skipped out.
II. 5. e. Phr. *skip it, let's skip it*: an exhortation or command to drop a subject or forget something. orig. *U.S.*
1934 M. H. WESEEN *Dict. Amer. Slang* 395 *Skip it*, drop the matter. 1939 R. CHANDLER *Big Sleep* xiii. 97, I started to say: 'What the hell—!' 'Oh, skip it,' Eddie Mars sighed. 1943 M. MCCARTHY *Company she Keeps* vi. 195 'Oh, Dr James,' she sighed. 'Let's skip it this time.' 1945 E. WAUGH *Brideshead Revisited* 17 Oh, very conscientious, I'm sure. Skip it and get a move on. 1955 E. CADELL *Lark shall Sing* x. 116 'I hate to seem to butt in on your—' 'Skip it. Go ahead and help me.' 1971 R. DENTRY *Encounter at Kharmel* ii. 31 At home..we cope and never give it a second thought. Out here we—oh, skip it! 1977 *New Yorker* 3 Oct. 40/3 Forgive me... Let's skip it, then, she says.
f. To forgo, to abstain from; to omit to take part in or to do.
1961 in WEBSTER s.v. [1]*skip*, The president skipped his regular Thursday press conference. 1970 K. H. COOPER *New Aerobics* ix. 137 Women suffering from cramps find exercise extremely uncomfortable. Common sense alone tells them to skip exercise during those days. 1979 R. JAFFE *Class Reunion* II. ix. 209 They picked at their dinner, unable to eat the roast pigeon..or the salad, and skipping dessert.
6. b. (Earlier and later examples.)
c 1810 W. HICKEY *Memoirs* (1960) ii. 28, I had intended to skip school, and take the usual march with the Guards to Kensington. 1951 J. D. SALINGER *Catcher in Rye* xxv. 270 If I let you skip school this afternoon and go for a little walk, will you cut out the crazy stuff? 1976 *National Observer* (U.S.) 17 Jan. 1/2 School phobia is a fairly common reason why some kids skip school.
c. For '(the country)' read '(a place)' and add further examples.
1885 *Santa Fé Weekly New Mexican* 10 Sept. 4/7 George Handley, a laundryman at Albuquerque, has skipped the town. 1906 U. SINCLAIR *Jungle* xxv. 307 The offending gambler had got wind of what was coming to him, and had skipped the town. 1977 *Detroit Free Press* 11 Dec. 11-B/1 Cliff won't go along with Molly's scheme to take Olive's $10,000 and skip town.
d. *to skip* (*one's*) *bail* = *to jump* (*one's*) *bail* (see JUMP *v.* 10 a in Dict. and Suppl.).
1900 *Congr. Rec.* 5 Feb. 1521/2, I should like the gentleman to know that one lot of those ballot-box stuffers are in jail and every one of the others has skipped his bail. 1930 P. W. SLOSSON *Great Crusade* (1931) 88 The I.W.W. leader who had 'skipped bail' and fled abroad. 1973 *Black Panther* 16 June 3/3 Eldridge Cleaver..skipped bail to avoid prosecution.
III. 8. skip-bombing (see quot. 1973); also *attrib.*; **skip-read** *v. trans.* and *intr.*, to read (a book) while skipping the passages of less importance; so **skip-reader; skip straight** *Poker*, a straight (STRAIGHT *sb.* 5) consisting of cards of alternate values.
1943 *Time* 18 Jan. 68/3 A U.S. Flying Fortress thundered into the Jap Harbor at Rabaul..to make the first test in the South Pacific of a new technique—'skip-bombing'. 1944 W. W. ELTON et al. *Guide Naval Aviation* ix. 172 A skip-bombing airplane must be fast and maneuverable... Tanks and ships are often attacked with skip bombing. 1964 D. MACARTHUR *Reminisc.* VI. 171 Special preparations were made to carry out a new technique of skip-bombing in the event of unfavorable weather and low cloud formations. 1973 J. QUICK *Dict. Weapons & Mil. Terms* 401/2 *Skip bombing*, a method of aerial bombing in which the bomb is released from such a low altitude that it slides or glances along the surface of the water or ground and strikes the target at or above water level or ground level. 1977 M. T. BLOOM *13th Man* (1978) p. ii, Skip-read all you want through the book, but go through the last page word by word. 1977 *Modern Railways* Dec. 490/1 Once taken up it is not a book which can be skip-read, for every page is packed solid with information. 1973 *Howard Jrnl.* XIII. 342 A very clear and easy to read book which should present no difficulties to the skip-reader. 1887 J. W. KELLER *Game of Draw Poker* 17 Efforts have been made to introduce into the game of Draw Poker what is known as the 'skip' straight—a sequence of alternate cards. 1944 A. H. MOREHEAD *Mod. Hoyle* 31 *Skip straight*, a sequence of cards once separated in rank. Examples: A—Q—10—8—6, or J—9—7—5—3.

skipjack, *sb.* Add: **4.** Also *attrib.*, esp. in **skipjack tuna**, a tropical pelagic food fish, *Katsuwonus pelamis*, of the family Scombridæ, distinguished by its large size and striped body. (Later examples.)
1920 BLUNDEN *Waggoner* 2 Where flock and shine the skip-jack dace. 1936 P. S. BARNHART *Marine Fishes S. Calif.* 36 Skipjack..a pelagic fish of a wide range. 1937 L. A. WALFORD *Marine Game Fishes Pacific Coast* 17 The skipjack is the smallest of the tunas, rarely exceeding 25 inches in length. 1949 THOMAS & LOVETT in Vesey-Fitzgerald & Lamonte *Game Fish of World* II. 132 There are the oceanic bonito, known locally as the skipjack, and the California bonito. 1961 E. S. HERALD *Living Fishes of World* 229 (*caption*) Skipjack Tuna..world-wide in tropical waters. 1973 *Sunday Times* 10 June (Colour Suppl.) 44/3 The most valuable fish in Papua New Guinea waters are skipjack tuna.
6. *U.S.* A kind of sailing-boat (see quot. 1976). Also *attrib.*
1887 *Forest & Stream* IX. 75 The 'skip-jack' is a connecting link between the skiff and the round-bottom boat. 1941 H. I. CHAPELLE *Boatbuilding* i. 36 The well-known Chesapeake Bay Skipjacks may be taken to represent the next type, having a good deal of beam and more dead rise than the modified sharpies. 1968 *Washington Star* 27 May B-1/2 They were watching last October's Chesapeake Appreciation Day skipjack races, off Annapolis. 1976 *Oxf. Compan. Ships & Sea* 807/2 *Skipjack*, a work-boat of the east coast of the U.S.A., sloop-rigged with a jib-headed mainsail and a foresail set on a bowsprit. They were hard-chined boats with a large wooden centreboard. 1978 J. A. MICHENER *Chesapeake* 725 In winter months he labored aboard a white man's skipjack dredging oysters... In spring he helped the skipjack captain haul timber to Baltimore.

skipper, *sb.*[2] Add: **4.** *Services' slang.* A commanding officer in the army; the captain of an aircraft or squadron.
1906 *Soldier Slang* in C. McGovern *Sarjint Larry an' Frinds*, *Skipper*, the commanding officer. 1926 *Sat. Even. Post* 6 Mar. 154/3 The skipper told us to look after yuh. 1929 *Papers Mich. Acad. Sci., Arts & Lett.* X. 323 *Skipper*, the squadron commander in the Royal Naval Air Service. 1958 P. KEMP *No Colours or Crest* v. 86 The Skipper wondered if you'd like to go forward to the flight deck for a look around. 1977 *R.A.F. News* 11–24 May 9/1 The headmaster..will join his wartime Whitley skipper, Gp Capt Leonard Cheshire.
5. *slang* (orig. *U.S.*). A police captain or sergeant; a police chief.
1929 HOSTETTER & BEESLEY *It's a Racket!* 238 *Skipper*, a police captain in command of a police station, bureau, or district. 1930 *Amer. Mercury* Dec. 457/2, I goes to the skipper and fronts for the mutt. 1962 *John o' London's* 25 Jan. 82/2 A police sergeant is called *skipper*. 1976 D. BARNES *Yesterday is Dead* (1977) II. 262 Good piece of police work... I'll fill the skipper in. I'm sure he'll be pleased.

skipper, *sb.*[3] Restrict † *Obs.* to sense in Dict. and add: **1. b.** Any sleeping-place for a vagrant.

1925 *Flynn's* 3 Jan. 661/2 *Skipper*,..a lodging house; a tramp. 1935 'G. ORWELL' *Clergyman's Daughter* ii. 101 We ain't got a brown between us, and we..got to tap for our tommy and skipper at nights. 1939 J. WORBY *Spiv's Progress* i. 9 I'm going north. Do you know any good skippers up the road? 1978 *Country Life* 20 July 189/2 He had painfully to learn the rudiments of vagrant survival: to make sure of his 'skipper' or kip before dark.
2. A vagrant; one who sleeps rough.
1925 [see sense 1 b above]. 1965 *Guardian* 9 Dec. 9/1 It was the night of the big Government census of the 'skippers'—the people who sleep rough. 1977 *Listener* 28 July 103/3 On the rubble-strewn redevelopment sites of central Glasgow, you find the groups of 'skippers', the men who live rough... These are the ones who admit that alcohol has won.
3. An act of sleeping rough; esp. in *to do a skipper*.
1935 H. NEVILLE *Sneak Thief on Road* 347 *Skipper*, a liedown in a spinney or anywhere where no rent is paid. 1937 J. CURTIS *You're in Racket Too* i. 9 It would be no fun doing a skipper on a November night. 1962 *Observer* 11 Mar. 35/1 There are not enough beds. Many will be turned away and have to do a 'skipper' in station, park or ruin.

skipper, *v.*[1] (Further examples.)
1950 *Sport* 7–11 Apr. 14/1 Dick skippered the City side who, in 1938, wrote a chapter of Cup history. 1951 N. M. GUNN *Well at World's End* xxvi. 239 The idea..was to get the old man to skipper her for a couple of seasons. 1977 *R.A.F. News* 27 Apr.–10 May 5/1 Brian, then a flight lieutenant, had skippered the Sunderland on three flights.

skipping, *vbl. sb.*[1] **1.** (Earlier example of the pastime using a skipping-rope.)
1800 *Infant's Library* IX. 7 Skipping. This is a very healthful play in winter; it will make you nice and warm in frosty weather.
attrib. (Further examples.)
1959 I. & P. OPIE *Lore & Lang. Schoolch.* ii. 38 Norman Douglas gives it as a skipping rhyme in 'London Street Games'. 1977 N. FREELING *Gadget* II. 70 How many skipping rhymes do I know? Sixty? A hundred?
2. (Earlier example in sense 5a of SKIP *v.*[1])
1824 SCOTT *Redgauntlet* II. i. 15 Such as are addicted to the laudable practice of *skipping*.

skipping-rope. (Earlier example.)
1802 F. BURNEY *Jrnl.* (1975) V. 388 He interests himself warmly about them, since he has seen the Cuttings, especially of the skipping ropes.

skirlie (skɜ̄·ɹli). *Sc.* Also **skirley.** [Shortened dim. form of *skirl-in-the-pan* s.v. SKIRL *sb.* 2 b: see -Y[6], -IE.] A dish of oatmeal and onions, etc., fried together.
1914 *Trans. Banffshire Field Club* 26/27 June 26 Crackens, bayheads, and skirlie are formed of fish livers and oatmeal cooked together. 1929 F. M. MCNEILL *Scots Kitchen* 205 *Skirlie*... Chop two ounces of suet finely. Have a pan very hot and put in the suet. When..melted, add one or two finely chopped onions and brown them well. Now add enough oatmeal to absorb the fat. 1947 *Sc. Women's Rural Inst. Cookery Bk.* x. 184 *Skirley*..3 handfuls of oatmeal.., ½ lb. suet, 2 onions.., pepper, salt. 1969 *Observer* 12 Jan. 33/4 *Skirley* (oatmeal, onions, suet, seasoned and fried in a cake).

skirmish, *v.* Add: **1. c.** *colloq.* (orig. *U.S.*). To make excursions in order to see what one can find; to scout *round* in search of something.
1864 'MARK TWAIN' in Harte & 'Twain' *Sk. Sixties* (1926) 129 His first cousin..is a skirmisher and is with the parson—he goes through the camp-meetings and skirmishes for raw converts. 1869 —— *Innocents Abroad* ix. 86 When the commissary department fails they 'skirmish', as Jack terms it in his sinful, slangy way. 1893 M. HOLLEY *Samantha at World's Fair* xix. 608 The males, from creation down, have been left free to skirmish round and git a livin' for themselves. 1894 'R. ANDOM' *We Three & Troddles* xxiii. 220 He left the room to skirmish after a clean handkerchief. *Ibid.* xxiv. 231 We had them downstairs and into the cab before they could skirmish after more substantial fare. 1907 S. E. WHITE *Arizona Nights* 17 We skirmished around and found a condemned army pack saddle with aparejos.

skirrack, obs. var. *SKERRICK.

skirret[1]. **1.** (Earlier *attrib.* example.)
1728 E. SMITH *Compleat Housewife* (ed. 2) 113 (*heading*) To make a skirret-pye.

skirret[2]. (Earlier example.)
1825 *Republican* 29 July 123 The skirret is an implement which acts on a centre pin, from whence a centre line is drawn, chalked and struck, to mark out the ground for the foundation of the intended structure.

skirrick, obs. var. *SKERRICK.

skirt, *sb.* Add: **I. 1. a.** *divided skirt* (earlier example).
1885 'V. LEE' *Let.* in P. Gunn *V. Lee* (1964) x. 127 A very bright blue paper dress suggestive of divided skirts and ulster to match.
b. (Further examples.) *a bit of skirt*: a woman; esp. an attractive one.
1914 S. LEWIS *Our Mr. Wrenn* iv. 55 Pete was..singing hoarsely, 'Dey was a skoit and 'er name was Goity.' 1916 C. J. DENNIS *Doreen & Sentimental Bloke* 89 *Skirt, or bit of skirt*, a female. 1928 D. H. LAWRENCE *Woman who rode Away* 283 And what about your American skirt?—I told him, there was nothing to say about her. 1934 J. BROPHY *Waterfront* ii. 42 A nice juicy bit of skirt, eh? 1958 'N. CULOTTA' *They're a Weird Mob* 190 'Reckon we better stick ter beer?' 'Until them other two skirts turn up.' 1974 K. MILLETT *Flying* (1975) v. 469 The two patriarchs, never tired of chasing twenty-year-old skirts in their old age. 1977 J. I. M. STEWART *Madonna of Astrolabe* xx. 280 They mustn't quarrel over a bit of skirt.
c. An underskirt or petticoat.
1862 *Catal. Internat. Exhib., Industr. Dept., Brit. Div.* II. No. 3674 Counterpanes, toilette-covers, skirts. *Ibid.* No. 4935 Spiral Crinoline Steel and Bronze for Ladies' Skirts. 1908 M. E. MORGAN *How to dress Doll* v. 51 The flannel skirt is cut from a straight piece of fine white flannel.
3. *to hide behind the skirts of*, to take refuge behind, to use for protection.
1938 G. GRAHAM *Swiss Sonata* 356 Is *she* hiding behind your skirts too? 1975 *Current Hist.* Dec. 230/2 In terms of foreign dominance, Thailand asserted her independence from China only as she was able to hide behind the skirts of the Western giants, Great Britain, France and, most important, the United States.
II. 4. c. Also in other *techn.* senses (see quots.).
1951 *Gloss. Aeronaut. Terms* (*B.S.I.*) III. 15 *Skirt*, the lower portion of the canopy [of a parachute]. 1962 J. GLENN et al. in *Into Orbit* 245 Parachutes used on Mercury capsules are reefed by means of ropes tied around the skirt of the parachute. 1964 J. L. NAYLER *Dict. Astronautics* 252 *Skirt*, the lower outer part of a rocket vehicle. It acts as a fairing to the rocket motor or booster. 1969 *Times* 22 July (Moon Rep. Suppl.) p. iii/7 There's one picture I'm taking now of the right rear of the spacecraft looking at the skirts of the descent stage. 1970 *Gloss. Aeronaut. & Astronaut. Terms* (*B.S.I.*) VI. 2 *Skirt*, an aerodynamic fairing to influence the airflow in the vicinity of the propelling nozzles.
e. A surface that conceals or protects the wheels or underneath of a vehicle or aircraft; *spec.* (*a*) a detachable panel concealing part of the wheel of a car and fitted flush with the bodywork; (*b*) a surface designed to deflect the air so as to produce a downward force on the car aerodynamically.
1912 C. B. HAYWARD *Practical Aeronautics* 286 There are six landing wheels forward, three on each side of the center and enclosed in what is termed a 'skirt'. 1953 FRAZEE & SPICER *Automotive Collision Work* i. 55 Sometimes fender skirts are used on the rear fenders of cars. These skirts cover the wheel opening and are attached to the fender by clamps. 1965 M. C. OAKS *Fell's Guide to Mobile Home Living* vi. 97 Many mobile home owners enclose the space beneath their mobile homes with skirts or siding.. These skirts..provide enclosed storage space, protect your tires from the sun, and..provide extra insulation. 1974 *Country Life* 21 Mar. 659/1 The Triumph Dolomite Sprint..is recognized by its..discreet spoiler beneath the front skirt. 1981 *Times* 5 Feb. 13/1 The South African Grand Prix will take place at Kyalmi... The cars will be equipped with skirts, almost certainly for the last time. 1981 *Sci. Amer.* Aug. 25/3 Most new main battle tanks have lightly armored 'skirts' to cover vulnerable treads and wheels.
f. *Mech.* The lower part of the curved surface of a piston in a piston engine, below the grooves for piston rings. Also *piston skirt*.
1913 W. E. DOMMETT *Motor Car Mech.* 12 For the sake of lightness and more particularly for use on racing cars, holes are drilled around the lower part or skirt of the piston and two rings only may be used. 1929 NEWTON & STEEDS *Motor Vehicle* vi. 57 In order that finer clearances may be used without risk of seizure, many different designs of semi-flexible skirt have been introduced. 1970 K. BALL *Fiat 600, 600 D Autobook* i. 14/2 In each case, the number is on the opposite side to the slot in the piston skirt.
g. A flexible surface that projects downwards underneath a hovercraft to contain or divide the air-cushion.
1962 *Daily Tel.* 12 Apr. 15/7 The cushion of pressurised air can take any proportion of the weight off the wheels. It is in an adjustable synthetic rubber 'skirt' below the waist-line of the vehicle. 1968 *Economist* 7 Sept. 81/1 The skirt of the hovercraft is one of its most sensitive parts. If the design is not right, the ride is uncomfortable and skirt edges flap up and down on the surface of the sea causing excessive wear. 1977 *Hovering Craft & Hydrofoil* XVII. 18/2 SEDAM, the French manufacturer, claims that their skirt is better than ours, but these claims will be put to the acid test when the N 500 and the SR.N4 Mk 3 run alongside each other on the Channel next year.
6. c. Chiefly *pl.* = SKIRTING *vbl. sb.* 5.
1851 F. A. WELD *Hints to Intending Sheep-Farmers N.Z.* 8 The Merino has the more valuable wool, being finer, and particularly superior in the 'skirts', which are remarkably deficient in the crossed sheep. 1886 [see sense 6 b in Dict.]. 1965 J. S. GUNN *Terminol. Shearing Industry* II. 23 *Skirt, skirting.* This word is generally used in the plural and refers to the wool round the edge of the fleece which is pulled off by the 'skirter'..or woolroller... In original English practice the skirts were handled by the 'wool sorter'..and not processed quite so carefully as in Australia.
III. 8. c. The lower sloping portions of a peak or rise on a graph, esp. of one representing electrical resonance.
1940 *Chambers's Techn. Dict.* 775/2 *Skirt*, the lower side portions of a resonance curve. 1962 SIMPSON & RICHARDS *Physical Princ. Junction Transistors* xiv. 341 This would be partly overcome if the top of the amplitude response curve were made flatter and the 'skirts' made steeper. 1965 *Wireless World* Sept. 33 (Advt.), Bandwidth skirts are better than 80-dB down. 1970 J. EARL *Tuners & Amplifiers* ii. 43 In such sets..the element is arranged in the form of a bandpass coupling or filter, giving sharp response skirts while handling signals in the required bandwidth.
11. *skirt-guard, -length*; *skirt-like* adj.; **skirt-board,** (*b*) a board to iron skirts on; **skirt-chaser** *slang*, one who pursues women with amorous attentions; hence **skirt-chase** *v. intr.*; **skirt-chasing** *vbl. sb.*; **skirt-dance** *sb.* (examples); **skirt-dancer** (examples); **skirt duty** *slang*, (*a*) acting in a way designed to attract men; (*b*) keeping company with women, regarded as a military duty; **skirt-knicker(s)** (see quot. 1913); **skirt-land,** land having skirt soil (see below); **skirt patrol** *slang* (orig. *U.S.*) (see quot. 1941); **skirt soil,** a loam composed of a mixture of peat and clay or sand or silt (cf. SKIRTY *a.*).
1861 MRS. BEETON *Bk. Househ. Managem.* 1013 The skirts of muslin dresses should be ironed on a skirt-board covered with flannel. 1932 D. C. MINTER *Mod. Needlecraft* 107/2 Almost indispensable to successful dressmaking are..sleeve and skirt board for pressing, [etc.]. 1943 J. B. PRIESTLEY *Daylight on Saturday* xi. 70 Don't be a dam' fool, Percy. I'm not skirt-chasing. 1942 BERREY & VAN DEN BARK *Amer. Thes. Slang* § 438/2 Lascivious man... *Skirt* or *woman chaser*. 1962 L. PETERS *Snatch of Music* iii. 45 He had always despised..the indiscriminate skirt-chaser. 1974 L. LAMB *Man in Mist* xvi. 106, I don't suppose that Settle is a skirt chaser. He probably wanted to frighten the girl away. 1950 'S. RANSOME' *Deadly Miss Ashley* xiv. 167, I always told you you'd regret your skirt-chasing... A man should stick with his wife and family. 1981 D. BOGGIS *Time to Betray* vii. 40 Chevalier went skirt-chasing at a disco. 1895 G. B. SHAW *Let.* 1 Nov. in *E. Terry & B. Shaw* (1931) 17 Mrs Pat Campbell entrances all London as Juliet, with a skirt dance. 1961 WODEHOUSE *Ice in Bedroom* vi. 47, I feel like dancing a skirt dance. 1974 D. SMITH *Look back with Love* xii. 113 There was usually one skirt-dance, during which the boys lolled..looking tolerant and slightly cynical. 1895 G. B. SHAW in *Sat. Rev.* 6 Apr. 445/1 Our skirt dancers are all petticoats. 1922 Skirt-dancer [see *high-kicker* s.v. *HIGH *a.* 21]. 1922 JOYCE *Ulysses* 758 He was throwing his sheeps eyes at those two doing skirt duty up and down. 1925 in *Amer. Speech* 1972 (1975) XLVII. 102 That evening, Jim detailed himself to some more 'skirt duty'. 1932 C. MORGAN *Fountain* II. ii. 100 A woman's bicycle with the broken strings of its skirt-guard dangling in a melancholy fringe over its spokes. 1982 J. HONE *Valley of Fox* vii. 105 A big, black old-fashioned woman's bicycle, with cord skirt-guards forming a fan over the back wheel. 1908 in C. W. Cunnington *Eng. Women's Clothing in Present Cent.* (1952) ii. 84 The skirt-knickers which the up-to-date maiden delights in. 1913 *Queen* 13 Dec. 1091/2 The tango and peg-top fashion between them are responsible for an entirely new form of skirt-knicker... The characteristic of the new garment..is that it is formed entirely of one length of material falling from the waist in front to the knees and up again to the waist at the back, slits or openings occurring at the sides through which the legs are passed. 1946 J. W. DAY *Harvest Adventure* x. 145 Those cows are fed for more than nine months of the year on by-products of the farm—such as beet-tops, beet-pulp, kale—and on skirt-land, and marsh grazings. 1981 P. SALWAY *Roman Britain* 268 The skirtlands of the southern Fens were the worst hit by these troubles. 1920 *T. Eaton & Co. Catal.* Spring & Summer 1/1 Skirt Lengths 35 ins. 37 ins. 38 ins. 1980 L. LEWIS *Private Life Country House* xii. 166 Skirt lengths remained what you had been wearing for some time. 1862 W. C. BRYANT *Tale of Cloudland* in *Poet. Wks.* (1883) II. 315, I plainly saw a chariot cushioned deep With sides that seemed of down, and skirt-like wings On which they nestled. 1980 *Motor* 16 Feb. 31/1 Deep, 'skirt-like' door sills. 1941 *Amer. Speech* XVI. 168/2 *Skirt patrol*, search for feminine companionship. 1967 Skirt patrol [see OAO s.v. *O 5 d]. 1960 *Times* 5 July (Suppl. on Agric.) p. vi/3 Lying between the areas of silt and peat there are indeterminate areas of what are now called 'skirt' soils. The soil physicist has classified them as organic silty clay loams. 1968 *Economist* 27 Apr. 52/2 Only two-thirds of the original acreage of peat in the fens—over 300,000 acres—are now covered with more than a 'skirt soil'.

skirted, *ppl. a.* Add: **1. c.** Of a hovercraft, having a skirt. Cf. *SKIRT *sb.* 4 g.
1967 *Jane's Surface Skimmer Systems 1967–68* 1/2 The relative sophistication of the 80 knot skirted hovercraft.

skirting, *vbl. sb.* Add: **4.** *skirting radiator*, a radiator running along a wall at the level of the skirting.
1970 *Home & Garden* Mar. 94/4 Skirting radiators have a great deal to recommend them. 1978 *Cornish Guardian* 27 Apr. 16/6 (Advt.), Superbly renovated cottage with full central heating, unobtrusive skirting radiators [etc.].

skirty (skɜ·ɹti), *sb. colloq.* [f. SKIRT *sb.* + -Y[6].] A skirt or underskirt.
1922 JOYCE *Ulysses* 47 A woman and a man. I see her skirties. Pinned up, I bet. 1977 *Sounds* 9 July 18/2 I'd like to dirty up Their little skirties up.

skish (skiʃ). *U.S.* [perh. f. *SK(EET *sb.*[2] or SK(ILL *sb.*[1] + F)ISH *sb.*[1]] A game in which participants use fishing tackle to cast a plug or fly at a target on dry land. Also *attrib.*

1940 *Outdoor America* Jan. 11/3 *Skish*—the new name chosen for the casting game originally known as Fish-O was selected on January 6 by the judges of the Fish-O change-of-name contest. 1942 *Sun* (Baltimore) 8 July 13/3 He said the best score he ever made in 'skish'—dry-land 'fishing' in which participants cast at thirty-inch rings at distances from forty to eighty feet—was in the eighties out of a possible 100 points. 1965 *Richmond* (Va.) *Times-Dispatch* 5 Feb. 25/2 Skish..is a game devised for indoor anglers who want to improve their casting techniques by casting at targets rather than for fish. 1976 *Webster's Sports Dict.* 399/1 There are 3 different skish events: skish bait, skish spinning, and skish fly casting.

skit (skit), *sb.*[4] *colloq.* [Origin obscure.] A large number, a crowd; *pl.* 'lots'.
1913 C. MACKENZIE *Sinister St.* I. II. ix. 287, I met an odd sort of chap..who told me a skit of things—you—know—about a bad life. 1925 A. S. M. HUTCHINSON *One Increasing Purpose* III. ix. 268 'What was that little red rosette he had on his left arm? I see skits of people with it.' 'Been vaccinated, of course.' 1927 *Blackw. Mag.* Nov. 594/1 The Kachins were in the jungle, a skit of them, trying to stop us at the ford.

skit (skit), *a.*[2] *Anglo-Ir.* [App. colloq. adjectival use of Ir. *sciót* cut, bit, laugh.] Amusing; *to be right skit*, to be a great laugh (cf. *LAUGH *sb.* 4 b).
1914 JOYCE *Dubliners* 26 Mahoney said it would be right skit to run away to sea on one of those big ships.

skite, *sb.* Add: Also *Austral.* and *N.Z. colloq.*
3. a. (Earlier example.)
1790 W. MACLAY *Jrnl.* 28 June (1890) x. 310 Hamilton has a very boyish, giddy manner, and Scotch-Irish people could well call him a 'skite'.
4. *Austral.* and *N.Z. colloq.* **a.** Boasting, boastfulness; ostentation, show; conceit.
1860 C. THATCHER *Victoria Songster* v. 160 You don't often see a chap given to 'skite, Can do very much when it comes to a fight. 1910 E. W. HORNUNG *Boss of Taroomba* 180 'Then none o' your skite, mate,' said Bill, knocking out a clay pipe against his heel. 1918 G. WALL *Lett. Airman* 85 This notepaper is a part of it, quite unnecessary skite. I thought you might like a sample of it, though. 1933 N. LINDSAY *Saturdee* 115 Ponk's the bloke to take the skite outer him. 1958 I. CROSS *God Boy* vi. 44 You started us off with your skite about not caring about fifty bangs and now you say shut up. 1965 S. T. OLLIVIER *Petticoat Farm* ii. 31 'Alister Bridgeman says it's mostly skite,' Sarah said breezily. 1972 P. NEWTON *Sheep Thief* xviii. 149, I thought I had a good district run but you've taken the skite out of me.
b. A braggart, a boaster; a conceited person.
1906 'T. COLLINS' in *Barrier Truth* (Broken Hill, N.S.W.) 1 June, In spite of Rigby's very complimentary insinuation that I'm a skite and a liar, the wagon was gone. 1928 *Bulletin* (Sydney) 29 Feb. 21/1 To the mug, the skite and the liar, to the nark and the hypocrite A fellow can always jerry. 1941 S. J. BAKER *N.Z. Slang* vii. 61 Our borrowings [from Australia] of what might be described as 'social' slang..[include] *skite*. 1952 D. NILAND in *Coast to Coast 1951–1952* 198 And what a skite! You should have seen him. 1958 I. CROSS *God Boy* i. 8 I'm not a skite but if I was fighting a man it would be the same. 1965 G. MCINNES *Road to Gundagai* ii. 30 He had no time for 'skites', that is, boastful know-it-alls. 1969 *Australian* 23 Sept. 2 Australians should not see themselves as boastful, arrogant skites, the Governor-General, Sir Paul Hasluck, said yesterday.
5. Orig. and chiefly *Sc.* A jollification, a spree, a binge. Freq. in phr. *on the skite*.
1869 *St. Andrews Gaz.* 27 Nov. 3/6 A correspondent.. sends the following..catalogue of synonyms of whisky and whisky drinking in the West of Scotland:—..on the skyte. 1895 W. STEWART *Lilts* 65 When ye went on the 'skite' an' sent railway things glee'd. 1909 J. J. BELL *Oh! Christina* xiv. 112 You an' me's gaun to ha'e an' awfu' skite, eh, auntie? 1946 J. IRVING *Royal Navalese* 158 *Skite* (*skyte*), *on the*, indulging in an orgy; on 'the tiles'. 1954 *Times* 16 Nov. 4/1 The Bejant skite (the party of the first-year students). 1972 N. SMYTHE in E. Berman *Ten of Best* (1979) 113, I was a bit too fond of the old jar, Went on the skite once too often.

skite, *v.*[2] Restrict '*Sc.* and *dial.*' to senses in Dict. and add:
For further material see also *Sc. Nat. Dict.*
3. *Austral.* and *N.Z. colloq.* To brag, to boast.
1857 C. R. THATCHER *Colonial Songster* 18 If ever you get into a fight, Of course you'll not forget to skite. 1896 E. TURNER *Little Larrikin* xxiv. 295 It used to make me a bit sick sometimes to hear him skite, knowing how much chance I had. 1902 *N.Z. Illustr. Mag.* V. 486 They had him skiting about his moonlight stroll with someone. 1940 F. SARGESON *Man & his Wife* (1944) 79, I suppose he went back to his ship and skited about the time he'd had. 1956 G. CASEY in *Coast to Coast 1955–56* 82 Spent most of the forty-eight hours skiting to his wife and young Les that he had got old Spend's goat thoroughly. 1968 A. HOLDEN *Death after School* xxv. 181 They did save my life... I don't mind *how* much they skite. 1978 P. GRACE *Mutuwhenua* xiv. 102 Everyone laughing, hugging Nanny who was skiting about her hat.
Hence (sense *3) **ski·ter; ski·ting** *vbl. sb.*
1898 *Bulletin* (Sydney) 17 Dec. (Red Page) An incessant talker is a *skiter* or a *fluter*. 1916 *Anzac Bk.* 99 If there's one thing I hate, it is skiting... So you won't think I boast when I tell you. 1936 F. CLUNE *Roaming round Darling* ix. 77 A fellow, fed up with city skiters, came out west to the edge of beyond, and began this village, where he could do his own skiting. 1957 R. LAWLER *Summer of Seventeenth Doll* III. 113 Lyin' comes as natural to him as skiting.

skither (ski·ðəɹ), var. SKITTER *v.*[2]
1904 *Eng. Dial. Dict.* V. 483/1 Leeak at mah scoperil, hoo it skithers across teeable. 1922 H. QUICK *Vandemark's Folly* viii. 143, I remembered..how she had skithered back to the carriage. 1929 S. LEACOCK *Iron Man* v. 205 When we drive the ball..skithers off sideways. 1977 *Meanjin* (Austral.) XXXVI. I. 68, I skither down barefeet first.

skitter, *sb.*[1] Add: Also *colloq.* **1.** (Further examples.) Now freq. in *pl.*
1939 J. STEINBECK *Grapes of Wrath* xxii. 431 They et green grapes. They all five got the howlin' skitters. Run out ever' ten minutes. 1940 M. MARPLES *Public School Slang* 159 *Skitters, squitters*, diarrhœa. 1948 PARTRIDGE *Dict. Forces' Slang* 180 *Squitters, skitters*, a symptom of dysentery and other stomach troubles.

skitter, *sb.*[2] Delete *U.S.* and for def. read: A light scampering or skipping movement or the sound caused by this. (Further examples.)
1959 E. ALLEN *Man who chose Death* xiii. 130 A quick skitter of footsteps like mice in the rafters. 1961 S. BUNCE *No Sainted City* xxiii. 170 A confusion of sounds. A skitter of light footfalls.

skitter, *v.*[2] Add: **1. a.** (Further examples.) Freq. with advbs.
1903 KIPLING in *Windsor Mag.* Sept. 363/2 She skittered about in the bracken, being a 'citable child. 1935 M. EBERHART *Cases of S. Dare* 64 The monkey darted out from under the sofa and was suddenly skittering across the room again. 1946 C. MCCULLERS *Member of Wedding* III. 178 Frances watched the Portuguese who..played a mock piano on the counter to the music-box tune. He swayed as he played and his fingers skittered up and down the counter. 1949 B. MARSHALL *To Every Man a Penny* xlix. 151 The limousines, the taxicabs, the lorries and the buses roared round the church, skittering away to Neuilly, Auteuil and Montmartre. 1968 B. HINES *Kestrel for Knave* 57 The boys began to swallow their Adam's apples, their eyes skittering about in still heads. 1976 *Church Times* 26 Nov. (Bk. Suppl.) p. iv/2 He skittered through a fantastic mass of scientific evidence for the scarcely believable and downright unbelievable. 1977 *Listener* 17 Feb. 215/3 Not only is the tenor-saxophonist playing his usual devious game but..Basie is joining in... Both men skittering around the melody line in high good humour. 1977 *New Yorker* 24 Oct. 33/1, I found, skittering in nervous computer printout across the bottom of my bill, the words 'Thank you very much for your prompt payment'.
b. For *U.S.* read 'orig. *dial.*' and add: Also with other advbs. and advb. phrases. (Earlier and later examples.)
1847 J. O. HALLIWELL *Dict. Archaic & Provincial Words* II. 750 A countryman who was leading me up a steep hill, when we came to a place which was inaccessible, said 'We had better *skitter* under here, and it won't be so steep.' (Kent). 1931 W. FAULKNER *Sanctuary* xxv. 298 A second man flew out and skittered along the floor on his back. 1951 J. C. FENNESSY *Sonnet in Bottle* II. i. 40 The little whitish-silver flying fish skittering over the ship's bow wave. 1956 C. EVANS *Kanchenjunga* xii. 121 Fragments of snow kept skittering down the slope and bombarding the tent. 1969 *New Yorker* 12 Apr. 127/1 The astronauts will start back to the LM—first tossing the universal handling tool..across the black *mare*, where it may skitter to rest inside a small crater. 1970 *Globe & Mail* (Toronto) 28 Sept. 21/7 His long blast was deflected by defenseman Jim McKenny's glove and skittered past a surprised Bruce Gamble in goal who was moving the opposite way to cover. 1978 M. PUZO *Fools Die* ii. 15 He was tired of the glittering red dice skittering across green felt.
2. b. In various senses, with reference to the impartation of a rapid or sliding motion (see quots.).
1902 KIPLING *Just So Stories* 61 Let's say things to the bunnies, and watch 'em skitter their tails! 1907 *Harper's Mag.* Feb. 460 The younger boy skittered rocks at a chicken-hawk. 1919 J. MASEFIELD *Reynard* 112 The great hooves skittered The Blood Brook's shallows to sheets that glittered. 1968 A. DIMENT *Bang Bang Birds* ii. 22 She produced a 6 × 4 glossy..and skittered it across the desk to me. 1972 M. J. BOSSE *Incident at Naha* i. 54 Edgar Gear blinked and skittered his hand through his hair.

skittery, *a.* Add: **2.** Skittish, restless.
1941 *Sun* (Baltimore) 29 Aug. 17/2 The little fellers—the skittery moths, and the 11½-foot penguin dinghies, complete the list. 1974 D. SEARS *Lark in Clear Air* x. 123 You can see that kind of eye in a skittery horse. 1976 *Gramophone* Oct. 607/1 The only two points at which I was aware of any lack of finish were in the finale of Op. 13 (sometimes a bit skittery) and the 'Dance' of Op. 14.
3. *Textiles.* Producing or taking on an undesired speckly appearance in dyeing.
1955 *Jrnl. Soc. Dyers & Colourists* LXXI. 707/1 A dye which is selective, i.e. skittery, in dyeing behaviour, however, will show marked differences in the rate of exhaustion. 1970 E. R. TROTMAN *Dyeing & Chem. Technol. Textile Fibres* (ed. 4) xvii. 442 It [*sc.* water solubility] increases the tendency to give 'skittery' shades, caused by the emphasis of variations in affinity from one fibre to another.
Hence **ski·tteriness.**
1952 *Jrnl. Soc. Dyers & Colourists* LXVIII. 306/1 *Skitteriness*, an undesired speckled effect in a yarn or fabric arising from differences in colour or depth of dyeing between adjacent fibres or portions of the same fibre. 1955 *Ibid.* LXXI. 707/1 Skitteriness is a specific property of the dye... Skitteriness comes about through selective dyeing of different wool fibres. 1964 *Dyeing of Polyester Fibres* (I.C.I.) (ed. 3) viii. 237 Although the..dyes recommended..are those with superior levelling properties, it is still essential to ensure initial level application of the wool dye if excessive 'skitteriness' is to be avoided.

skittle, *sb.* Add: **1. b.** (Further examples.)
1857 T. HUGHES *Tom Brown's School Days* I. ii. 46 Life isn't all beer and skittles. 1931 A. CHRISTIE *Sittaford Mystery* xxvi. 211 'It's an experience, isn't it?' 'Teach him life can't be all beer and skittles,' said Robert Gardner maliciously. 1963 D. OGILVY *Confessions Advert. Man* (1964) i. 12 Managing an advertising agency isn't all beer and skittles.
c. (Earlier examples.)
1864 *Orchestra* 12 Nov. 104/1 *Se faire applaudir* is not 'to make oneself applauded', and 'joyous comedian' is simply skittles. 1886 KIPLING *Departmental Ditties* (ed. 2) 43 'Where is your heat?' says he, 'Coming,' says I to Pagett. 'Skittles!' says Pagett, M.P.
d. *colloq.* Chess played without serious application.
1856 C. TOMLINSON *Chess Player's Ann.* 61 Nor will our royal Game less royal sound, If shallow men play skittles on the ground, Where first-rate Chess sedately sits in state, And spends long hours accomplishing a mate. 1894 *Daily News* 30 May 3/6 There is, as every experienced chessist knows, all the difference in the world between what is known as off-hand play or 'skittles' and chess. 1940 PRINS & WOOD tr. *Euwe's Meet Masters* i. 14 Every game of chess, serious or 'skittles',
3. a. *skittle-ground* (earlier and later examples).
1737 *London Mag.* Sept. 477/2 Such days would still be much better employed in that Way, than in sotting at an Ale-House, or loitering in a Skettle or Nine-Pin Ground. 1971 *Country Life* 9 Dec. 1673/3 In 1773, the spring was covered over, and the site reverted to a simple public house with a skittle-ground attached.

skittle, *v.* Add: **2. b.** To knock down (skittles, etc.); *Cricket*, to bowl out (batsmen) in rapid succession. Also *fig.*, to kill, defeat easily.
1880 *Wisden's Cricketers' Almanack* 18 Mr. Chatterton 'skittled' the wickets down so rapidly. 1919 W. H. DOWNING *Digger Dial.* 45 *Skittled*, killed. 1928 *Daily Express* 31 Mar. 3/4 Mine host and Mr. Herbert swung their arms, flung the cheeses, and skittled the pins. 1977 *World of Cricket Monthly* June 92/3 The Warwickshire bowling attack..skittled the students for a mere 59 in just 2½ hours.
c. *Cricket.* Similarly with *out*. Also, to dismiss (a team) cheaply.
1906 A. E. KNIGHT *Compl. Cricketer* v. 172 Jim Jones thinks Sir Arthur Squire a rotten captain, who never gives him a chance to 'skittle the rabbits out'. 1949 J. SYMONS *Bland Beginning* 216 Now that Anthony had found a length, he began to skittle out the batsmen. 1979 *Daily Tel.* 9 Aug. 1/3 Somerset's West Indian fast bowler, Joel Garner, took five wickets for 11 runs, helping to skittle out Kent for 60.

skittler. Add: **1.** (Later example.)
1958 G. USHER *Death in Bag* vii. 64 He'd go and see how the skittlers was doin'.
2. One who plays chess without serious application.
1868 *Westminster Chess Club Papers* I. 87 We consider it quite possible to diffuse the game [of chess] without affording encouragement to the mere 'skittler'. 1911 *Daily News* 24 Apr. 4 The spread of chess literature, which has made every 'skittler' a book player.

skive (skaiv), *sb.*[3] *slang.* [f. *SKIVE *v.*[3]] An act of shirking; an opportunity for avoiding a difficult or unpleasant task, an easy option.
1958 F. NORMAN *Bang to Rights* III. 90 Not many of us wanted to learn english and only went on the class for a skive. 1960 [see *CHUFFED *a.* a]. 1968 *Guardian* 1 Oct. 6/1 'Isn't just a skive,' he told them. 'At any time, any time, you may be parachuted behind the Russian lines.' 1976 *Times Higher Educ. Suppl.* 26 Mar. 7/1 I'd always thought that science degrees with a non-scientific element would be attractive but perhaps students associate them with the general studies they do in the sixth form and think of them either as a skive or a nuisance. 1980 J. DITTON *Copley's Hunch* I. ii. 68 He thought the sentry was on the skive. Thought he'd come down..for a cup of coffee.

skive (skaiv), *v.*[3] *slang* (orig. *Mil.*). Also **scive, skyve.** [perh. ad. Fr. *esquiver*, to dodge, slink away, but cf. SKIVE *v.*[2]] *intr.* To evade a duty, to shirk; to avoid work by absenting oneself, to play truant. Also with *off*.
1919 *Athenæum* 1 Aug. 695/1 'To skive,' to dodge a fatigue. 1925 FRASER & GIBBONS *Soldier & Sailor Words & Phrases* 260 *To skive*, to dodge a duty or fatigue. 1960 *Twentieth Cent.* Nov. 390 Who hasn't bought black-market, possibly stolen goods, who hasn't skived off work? 1960 J. BIRLEY *Time of Cuckoo* 139 'But a second is what you aimed for, isn't it? I mean one has to be pretty brilliant for the other, anyway, and if not, why work all that hard?' 'Oh, yes, yes, I dare say... I've been skyving all my life, certainly, if that's what you mean.' 1961 *New Statesman* 21 July 82/2 If one of the other cleaners offered advice, it was usually on how to scive off better. 1962 *Listener* 27 Dec. 1104/3 Anyone who thinks of the Fire Service as a soft option that literary types skived into during the war ought to have watched 'Fire Rescue'. 1965 [see *free period* s.v. *FREE *a.* D. 2]. 1971 *Times Educ. Suppl.* 25 June 69 People work—and skyve—openly at any time. 1973 J. MANN *Only Security* vi. 58 The girls who dig are always glad of an excuse to skive off and have a rest. 1976 *Sunday Post* (Glasgow) 26 Dec., A Dundee bus conductor was chatting to three young boys as he took their fares last

Monday. He asked if they were on holiday. They replied they were 'just skiving'. 1982 *Sunday Times* 31 Jan. 3/6 These people work, skive and fiddle in packs.
Hence **ski·ving** *vbl. sb.*[2]
1958 *Daily Mail* 18 July 3/3 You do two hours' work a day and spend the rest of the time dodging. In the Army we called it swinging the lead, but on the railway it's called skiving. 1974 *New Society* 19 Sept. 727/1 It is among managerial and professional workers that sponging, skiving and malingering is epidemic. 1977 *Ibid.* 25 Aug. 381/3 Some of the overseers connive at, even join in, this skyving. 1978 P. MARSH et al. *Rules of Disorder* ii. 33 Skiving was not infrequent—the pupils slipping out once the register had been called.

skiver, *sb.*[2] **2.** (Earlier examples.)
1829 P. EGAN *Boxiana* 2nd Ser. II. 220 At a proper age, Jem turned out to earn an honest penny, and was apprenticed to a *skyver*, or skinman, in Newcastle-upon-Tyne. 1850 *Rep. Comm. Patents 1849* (U.S.) I. 313, I claim..the application of a gauge or gauges to a skiver.

skiver (skai·vəɹ), *sb.*[3] *slang.* Also **sciver, skyver.** [f. *SKIVE *v.*[3] + -ER[1].] One who avoids work; a shirker; a truant.
1941 G. KERSH *They die with their Boots Clean* 219 Well, mud in your eye, old skivers! 1942 —— *Nine Lives Bill Nelson* xi. 71 You end up like them old skivers that mooch about the Naffy Library. 1959 I. & P. OPIE *Lore & Lang. Schoolch.* xvii. 372 A few children in Kirkcaldy give 'sciver' as a name for a truant. 1961 *New Statesman* 26 May 830/1 As it was day-time, everyone in the coffee bar was a sciver, on the dole or on the fiddle or just plain hopeful. 1971 [see *fruit gum* s.v. *FRUIT *sb.* 9]. 1977 *Daily Tel.* 24 Nov. 8/6 A Labour-controlled council is to crack down on 'skivers' following a report which alleges large scale absenteeism and sick leave among its manual workers. 1982 *Times* 20 Apr. 4/8, I frequently came across cases at depots where there were skivers galore.

skiving, *vbl. sb.*[2]: see *SKIVE *v.*[3]

skivvy (ski·vi), *sb.*[1] *colloq.* (usu. *derogatory*). Also **scivey, skivey.** [Of obscure origin.] A female domestic servant, esp. a maid-of-all-work. Also *transf.*
1902 H. BAUMANN *Londinismen* (ed. 2) 211/2 *Skivey*,.. slav(e)y. 1913 C. MACKENZIE *Sinister St.* I. II. iv. 195 The ball had landed twice..on the same balcony to the great annoyance of the 'skivvy', who was..invited to bung it down. 1915 W. OWEN *Let.* 23 Nov. (1967) 367, I never thought myself capable of such strenuosities as to do skivvy's drudgery. 1938 L. DURRELL *Black Book* 240 Gwen, that dirty little skivvy, smelling of..grease from the sink! 1947 E. WILSON *Europe without Baedeker* viii. 206 You were waited on by slovenly skivvies so pallidly unappetizing that they made the meager food seem more tasteless. 1961 'F. O'BRIEN' *Hard Life* vii. 52 The young people of today think the daddy is a tramp and the mammy a poor skivvy. 1974 *Times* 1 May 4/5 This represents a change in the nurses' attitude. No longer will you be the skivvies of the health service. 1977 *Punch* 31 Aug.–6 Sept. 331/1 Blossom, the strikingly handsome new scivvy, is mixing her a posset.
Hence **ski·vvy** *v. intr.*, to work as a skivvy; **ski·vvying** *vbl. sb.*
1931 R. CAMPBELL *Georgiad* ii. 44 Every Nellie, Gertie Or Daisy that..ever left her own unlettered stews To skivvy in the kitchen of the Muse. 1968 *Punch* 21 Feb. 287/1 There are now better jobs than skivvying to look for, and wages sometimes comparable with what the job-seeking housewife herself might be earning. 1973 J. THOMSON *Death Cap* ii. 33 It wasn't no skivvying job... Mrs King treated me like a friend. 1974 W. FOLEY *Child in Forest* II. 156 She come up to skivvy in the same place as you, then?

skivvy (ski·vi), *sb.*[2] Also **scivvy, skivie, skivvie.** [Of unknown origin.] **1.** *N. Amer. slang* (orig. *Naut.*). **a.** An undershirt, a vest. Also *skivvy shirt*. **b.** *pl.* Underclothes.
1932 J. L. D'ESQUE *Count in Fo'c'sle* 216 He cut away the unconscious engineer's shirt and skivie with his jack-knife. 1942 BERREY & VAN DEN BARK *Amer. Thes. Slang* § 900/6 *Skivvie shirt*, an undervest. 1944 C. MACKENZIE *Sailors of Fortune* 95 The Red Cross woman brought me khaki trousers, scivvy shirt and a left shoe. 1945 *Seafarers' Log* 9 Nov. 6/3 Chips appeared, also gassed up, in his skivvies. 1946 T. HEGGEN *Mister Roberts* 7 It now contained a soiled scivvy shirt. 1953 S. BELLOW *Adventures Augie March* 28 We had to brush our teeth with salt..and sleeping in skivvies was outlawed; we had to wear pajamas. 1967 'J. CROSS' *To Hell for Half-a-Crown* ii. 36 The..biceps straining at the white skivvy shirt. 1972 R. WHITE *Be not Afraid* xvii. 206 We jumped out of our sleeping bags.., only realizing after we had recovered the gear that we were barefoot and in skivvies in subzero weather. 1978 W. HJORTSBERG *Falling Angel* xxxviii. 184 Ethan Krusemark, wearing boxer shorts and a skivvy, lay on his back..doing leg presses.
2. A high-necked lightweight pullover or jumper.
1967 *Telegraph* (Brisbane) 27 Feb. 14/4 (*caption*) This waistcoat should be a winner on the winter scene teamed with white long-sleeved skivvy. 1972 *New York* 8 May 20/1 (Advt.), Sleeveless skivvy (shown under the cardigans) in peach or pale blue. 1972 D. SALE *Love Bite* xix. 239 She was wearing a navy-blue jump suit over a black skivvy, so that she wouldn't be seen in the dark.

sklodowskite (sklǭdǭ·vskait). *Min.* [a. F. *sklodowskite* (A. Schoep 1924, in *Compt. Rend.* CLXXIX. 415), f. *Skłodowska*, maiden name of Marie Curie (1867–1934), Polish-born chemist and co-discoverer of radium: see -ITE[1].] A hydrated uranyl magnesium silicate, $Mg(UO_2)_2Si_2O_7.6H_2O$, which occurs as a secondary mineral, forming lemon-yellow monoclinic crystals.
1924 *Chem. Abstr.* XVIII. 3577 (*heading*) Sklodowskite, a new radioactive mineral. 1957 *Amer. Mineralogist* XLII. 617 Sklodowskite, which is considered iso-structural with uranophane, and beta-uranophane may have structures based on uranophane. 1972 *Canad. Mineralogist* XI. 562 A mixture of uranophane and sklodowskite forms a coating on a specimen of quartz with magnetite.

skoal, *sb.* Add: Now usu. with pronunc. (skǫl) and in form **skol.** Also **skal, sköl.** (Further examples of use as a toast.)
1924 *Vogue* late Jan. 64/2 If only I could make out why nobody dared touch his glass before someone else had said 'Skål', and why you yourself were at it the whole time. 1935 J. D. CARR *Death-Watch* xix. 187 'Skoal,' wheezed Dr. Fell absently. He pushed his glass away. 1948 F. BROWN *Murder can be Fun* (1951) ii. 24 'Skoal!' he said. They drank. 1961 *Guardian* 8 Feb. 12/5 'Skol' which we imagine to be a jolly kind of 'Cheerio' on lifting a glass, is an essential part of Swedish etiquette. 1961 M. BEADLE *These Ruins are Inhabited* (1963) ix. 116 We.. learned the etiquette of *sköl*. The Swedes are enthusiastic drinkers. 1973 D. FRANCIS *Slay-Ride* i. 17 '*Skol*' they said. '*Skol*' I repeated. They watched interestedly while I drank.

skoal, *v.* Delete † *Obs.*, for *Sc.* read 'orig. *Sc.*', and add: Also **skol. a.** (Later examples.) **b.** *trans.* To drink the health of.
1909 E. POUND *Personae* 17, I skoal to the eyes as grey-blown mere..Wineing the ghosts of yester-year. 1928 H. CRANE *Let.* 27 Mar. (1965) 320 Many bottles of dubious gin and whiskey—with much 'skoling'. 1935 G. GREENE *England made Me* I. 1 She swallowed it at a draught.. skoal, skoal, but there was no one to skoal. 1963 *Times* 23 Jan. 12/7 There was ample time to *skol* one another. 1980 P. HARCOURT *Tomorrow's Treason* I. iv. 67 We had skolled in champagne the bonfires being lit around the fjord.

Skoda (skǭ·dā). The name of the Czech engineer and industrialist Emil von *Skoda* (1839–1900), used *attrib.* and *ellipt.* to designate guns manufactured in the factories established by him.
1902 *Encycl. Brit.* XXIX. 166/1 In the Hotchkiss and also in the Skoda systems the mechanism is of the vertical breech block type. 1933 J. BUCHAN *Prince of Captivity* I. 78 Skoda mountain howitzers which had once been destined for the Hedjaz. 1973 J. QUICK *Dict. Weapons & Mil. Terms* 402/1 The first Skoda machine gun appeared in 1888 and was a delayed-blowback weapon.

‖ **skoff** (skǫf). *S. Afr.* Also **schoft, skof, skoft.** Pl. **skoffs, skofte.** [Afrikaans *skof*, f. Du. *schoft* (see SCOFF *sb.*[2]).] A stage of a journey, a period of travel between outspans.
1785 G. FORSTER tr. *Sparrman's Voyage to Cape of Good Hope* I. 132 Four such hours with a horse, or with eight oxen, are reckoned to make one *skoft*. 1801 J. BARROW *Acct. Trav. S. Afr.* I. ii. 55 Each day's journey is called a *skoff*; and the length of these is generally regulated by local circumstances, being from five to fifteen hours. 1835 H. I. VENABLE in D. J. Kotzé *Lett. Amer. Missionaries* (1950) 70 The Dutch call a day's journey a *schoft*. For oxen in good condition a schoft is from twenty to twenty-four miles. 1892 R. CHURCHILL *Men, Mines & Animals in S. Afr.* ix. 134 We have done twenty-five miles from Silika in three 'skoffs', which is excellent trekking. 1932 L. FOUCHE in C. Fuller *Louis Trigardt's Trek* p. xv, It became imperative to ascertain the value of Trigardt's unit, the length of a 'skof'. 1969 A. FUGARD *Boesman & Lena* I. 2 That last skof was hard. Against the wind... Heavier and heavier. Every step. 1972 L. G. GREEN *When Journey's Over* (1973) v. 50 The normal day on the road was made up of two or three stages known as *skofte*.

skoff, var. SCOFF *sb.*[2] and *v.*[2] in Dict. and Suppl.

skokiaan (skǫ·kiān). *S. Afr.* Also **skokian.** [Poss. of Nguni origin.] An intoxicating home-brewed liquor fermented with yeast. Also *attrib.*
1926 S. G. MILLIN *S. Africans* VII. ii. 223 Sometimes a group of Kaffirs are found drinking a rapidly fermenting preparation called *skokiaan* in the outbuilding of a European's home. 1936 WILLIAMS & MAY *I am Black* IV. xvii. 175 The white men call me a skokiaan queen. 1946 P. ABRAHAMS *Mine Boy* xi. 135 He knew she was one of the foremost Skokiaan Queens—for that is what they call the women who deal in illicit liquor. 1953 D. LESSING *Five* 286 Skokian is a wicked and dangerous drink, and it is illegal. It is made quickly, in one day, and may contain many different substances. On this night it has mealie-meal, sugar, tobacco, methylated spirits, boot polish and yeast. Some skokian queens use magic, such as the limb of a dead person. 1970 G. LORD *Marshmallow Pie* iv. 33 He didn't hate the blacks... All they wanted was a quiet life and some *skokiaan* on Saturday nights. 1976 *Globe & Mail* (Toronto) 13 Nov. 16/5 The rougher shebeens lack refinements and sell often-questionable products, ranging from commercial and homemade beers to a variety of moonshine liquors, including the legendary skokiaan.

skol, var. SKOAL *sb.*, *v.*, in Dict. and Suppl.
sköl, var. SKOAL *sb.* in Dict. and Suppl.

skolly (skǫ·li). *S. Afr.* Also **scolly, -ie.** [a. Afrikaans, prob. ad. Du. *schoelje* scoundrel, rascal.] A Black or Coloured African hooligan, gangster, hoodlum; a vagrant. Also *attrib.*, as *skolly boy*. Hence **sko·llydom,** the condition or activity of a skolly; **sko·llyism,** the way of life of a skolly.
1934 *Cape Argus* 8 Jan. 10/6 The accused..were actually several degrees lower than the average 'scolly-boy' who commits most of the crimes of violence and theft in the Peninsula. 1934 *Cape Times* 12 Jan. 9/7 The 'scollie' boys could be rounded up and sent to labour colonies. 1949 *Ibid.* 10 Sept. 8/8 The cure for skollydom is not mollycoddling. 1954 R. ST. JOHN *Through Malan's Africa* v. 32 Experts say that dagga is one of the least harmful of the narcotics, and yet in District 6 it leads to the shebeens and skollyism. 1961 *Cape Times* 28 Jan. 11/1 Don't you realize that your son is becoming a White skolly?.. He is going to land in gaol unless you do something about him. 1972 *East Cape Post* 14 Mar., Tsotsis and skollies who lurk in street corners, waiting to assault and rob innocent people. 1980 J. MCCLURE *Blood of Englishman* ix. 82 Look at the clientele... All the top-notch... You won't find skolly boys in here.

skoob (skūb). [Reversal of *books*: not in general use (see quots.).] A pile of books assembled in order to be destroyed as a gesture against the proliferation and undue veneration of the printed word (see quot. 1967); the ceremonial burning of a book or books.
1963 *Guardian* 14 Feb. 6/5 The Skoob Image—the mangled books. 1966 *Archit. Rev.* Dec. 441/2 A skoob tower by John Latham—a construction of art books destined to be detonated. 1967 *Listener* 18 May 654/3 Latham's earlier works were assemblages of torn and paint-covered books... His 'skoob' (books spelt backwards) are a cathartic topsy-turvying of the natural reaction of horror at the destruction of the printed word. 1968 *Guardian* 14 Sept. 3/3 Mr. Stone said it was the first time that a 'skoob', the burning of a book, had taken place in Kensington Church Walk.

skookum (skū·kǫ̆m), *sb.* and *a.* *N. Amer.* Also † **scocum, skokum,** etc. [a. Chinook Jargon.] **A.** *sb.* An evil spirit; a disease. *Obs.* exc. *Hist.*
1838 S. PARKER *Jrnl. of Exploring Tour beyond Rocky Mts.* 336 Evil spirit, skookoom. Hell, skookoom. 1844 LEE & FROST *Ten Yrs. in Oregon* xvi. 180 He [*sc.* the medicine man] transfers the 'sko-koms', or 'tam-an-a-was', or disease, wholly or in part from the patient to himself. 1846 JOHNSON & WINTER *Route Across Rocky Mts.* iii. 54 Several loud shouts are uttered in as frightful a manner as they are able. They then open their fingers gradually, to allow the terrified Scocum, (evil spirit,) to make his escape. 1900 *Oregon Hist. Soc. Q.* I. 185 The benefits of his fishery had gone, not to the people, but to the wicked skookum.
B. *adj.* **a.** Strong, stout, brave; fine, splendid.
1847 J. PALMER *Jrnl. Trav. Rocky Mts.* 151/1 *Skokum*, strong, stout. 1891 H. W. SETON-KARR *Bear Hunting in White Mts.* viii. 83 He believed that a bear would hold out its paw towards a man at a distance and feel whether he was *skookum*—brave. 1901 *Daily Colonist* (Victoria, B.C.) 29 Oct. 6/3, I would have gone up myself but could not stand the five days' tramp with 40 or 50 pounds on my back. It takes a 'skookum' man to stand a trip like that. 1913 [see *ROUGH LOCK, ROUGH-LOCK]. 1941 J. SMILEY *Hash House Lingo* 50 *Skookum*, good; all right. 1949 *Sierra Club. Bull.* (San Francisco) June 105 Billy and Pete were skookum, and I was pretty good myself in those days. 1962 E. LUCIA *Klondike Kate* 12 As Klondike Kate, she was a mighty skookum gal. 1975 *Islander* (Victoria, B.C.) 6 July 4/2 Ted, by this time a skookum young fellow of 20, then turned his eyes further west.
b. Special collocations. **skookum chuck** [*CHUCK *sb.*[6]], a fast-moving body of water, a torrent, rapids; an ocean; **skookum house,** a gaol.
1888 LEES & CLUTTERBUCK *B.C. 1887: Ramble in British Columbia* xvii. 184 We arranged to meet at the Skookumchuck Creek ('the stream of the rapid torrent'). 1899 *Bull. U.S. Fish Comm. 1898* XVIII. 73 The passage is a 'skookum chuck', through which the water runs in whirls and rapids almost constantly and with great velocity. 1911 E. P. JOHNSON *Legends of Vancouver* 47 You have listened to the call of the Skookum Chuck, as the Chinook speakers call the rollicking, tumbling streams that sing their way through the canyons. 1940 K. S. PINKERTON *Three's Crew* x. 102 A tidal rapids is a 'skookum chuck'. 1959 *Times* (Queen in Canada Suppl.) 18 June p. xii/1 This is the salt chuck, the skookum chuck..in fact the Pacific. 1873 R. C. L. BROWN *Klatsassan* 165 It was only after much *waw-waw* (parley) and sundry threats of the skookum-house (gaol)..that one of them was got to undertake to carry him. 1901 *Daily Colonist* (Victoria, B.C.) 5 Oct. 3/2 There were no less than fourteen inmates of the 'skookum house' on Cormorant Street last night. 1965 *Islander* (Victoria, B.C.) 9 May 6/2 In 1872, Frederick Brent was appointed Justice of the Peace and a skookum-house (jail) was built on his land.

skoptophilia, var. *SCOPOPHILIA.

Skoptsi (skǫ·ptsi), *sb. pl.* Also occas. **Skoptzi, Skopzy,** etc. [Russ., pl. of *skopéts*, eunuch, member of Skoptsi.] An ascetic Russian Christian sect, known since the eighteenth century and now forbidden, given to self-mutilation (see quots.). Also *rarely* as *sing*. Hence **Sko·ptsism**, the faith and practice of the Skoptsi.

1856 [see *KHLIST]. 1874 J. H. BLUNT *Dict. Sects* 564/1 *Skoptsi*, a name signifying 'eunuchs', given to a Russian sect of the Bezpopoftschin Dissenters, and derived from their practice of self-mutilation, which they supposed to be warranted by Scripture (Matt. xix. 12). 1887 A. F. HEARD *Russ. Ch. & Russ. Dissent* xi. 270 Notwithstanding their precautions, the Skoptsi are betrayed by their pale, sallow complexion, their scanty beard, shrill voice, effeminate, peculiar gait, and hesitating, wavering look. 1888 'STEPNIAK' *Russ. Peasantry* II. iii. 439 The *Skopzy* or *Castrati*, founded by Selivanov at the close of the eighteenth century. 1911 *Encycl. Brit.* XXV. 194/1 Skoptsism was, however, not exterminated, and grave scandals constantly arose. 1957 *Ibid.* XXIII. 873/1 The Skoptsi..settled in Yakut in the 1860s and introduced agriculture... The clean, well-built Skoptsi villages were a striking contrast to the dirty Yakutsh settlements. 1960 O. MANNING *Great Fortune* I. 25 A *Skopit*. One of the sights of the city. The *Skopits* belong to a Russian sect. 1970 B. WALKER *Sex & Supernatural* ix. 84 The best known of the modern castrant cults called the Skoptsi, or 'eunuchs', a mystical Russian sect which first came into prominence in the middle of the 18th century but which was said to have been in existence for at least three centuries before that.

‖ **skothending** (skǫ·tˌhendiŋ). *Pros.* [ON.] Chiefly in Scaldic verse: rhyme formed with the same consonant or consonant cluster preceded by differing vowels; half-rhyme.

1838 G. P. MARSH *Compendious Gram. Old-Northern or Icel. Lang.* IV. 144 The same consonants with different vowels (skothending, *half-rhyme*, or *assonance*). 1860, 1873–4 [see *half-rhyme* s.v. *HALF- II. n]. 1945 L. M. HOLLANDER *Skalds* 10 The *skothending* in the odd, and the *aðalhending* in the even, half-line always involve the second but last syllable. 1977 J. MILROY *Lang. G. M. Hopkins* v. 144 We are not accustomed to notice *skothending* (end-consonant rhyme) when it occurs in English. Hopkins uses it very liberally.

skotophil(e, var. *SCOTOPHIL *a.* **skourick**, Sc. var. *SKERRICK.

Skraeling (skrǣ·liŋ). *Scandinavian Hist.* Also † **Schrelling, Skrelling, Skrælling**, etc.; **Skræling**. [ad. ON. *Skræling(j)ar* pl., the Norse name for the inhabitants of Greenland at the time of the Norse settlement.] A member of a savage people encountered by the early Norse settlers on Greenland, of uncertain origin but often considered to be of Eskimo descent. Also applied similarly to the inhabitants of Vinland (sometimes identified with the NE. coast of N. America). Usu. *pl.*

1767 tr. *Crantz's Hist. Greenland* I. III. i. 132 The Greenlanders call themselves..*Innuit*... The Icelanders, who many hundred years ago discovered..this country and the neighbouring coasts of America, called them in form *Skrællings*, because they are little of stature. 1797 *Encycl. Brit.* VIII. 129/2 This nation, called *Schrellings*, at length prevailed against the Iceland settlers who inhabited the western district [of Greenland]. 1875 *Ibid.* I. 706/2 They had some intercourse..with a people who came in leathern boats, and were called *Skrælings*, from their dwarfish size... The Skrælings were of course Esquimaux... The hostilities of the Skrælings was no doubt the principal cause of the abandonment of the colony [of Vinland]. 1891 KIPLING in *Contemp. Rev.* July 21 He..said:—'When they heard *our* bulls bellow the Skrælings ran away!' 1921 G. M. GATHORNE-HARDY *Norse Discoverers of Amer.* II. iv. 172 In so far..as the descriptions of the Skrælings of Wineland are realistic, and differ materially from anything which can have been derived from Eskimo sources, these descriptions form probably the most convincing proof of the historical accuracy of these stories. 1979 N. DAVIES *Voyagers to New World* 227 When spring came the skraelings once more appeared.

Skraup (skraup). *Chem.* The name of Zdenko Hans *Skraup* (1850–1910), Czech chemist, used *attrib.* and in the possessive to denote a reaction which he discovered (*Sitzungsber. der K. Akad. der Wissensch. (Math.-Nat. Classe)* (1880) LXXXI. II. 593), in which a quinoline is synthesized by heating a primary aromatic amine with glycerol, sulphuric acid, and an oxidizing agent.

1886 *Jrnl. Chem. Soc.* L. 79 Chloromethylquinoline..is prepared by Skraup's method from parachlorometatoluidine. 1935 L. F. MAREK in P. H. Groggins *Unit Processes in Org. Synthesis* vii. 350 Quinoline is prepared from aniline and glycerol by the Skraup reaction. 1954 I. L. FINAR *Org. Chem.* (ed. 2) I. xxxi. 691 Alizarin Blue..may be prepared by first reducing Alizarin Orange to the corresponding amino-compound and then heating this with glycerol, sulphuric acid and nitrobenzene (Skraup's synthesis). 1975 R. F. BROWN *Org. Chem.* xxviii. 908 The Skraup synthesis involves a dehydration, a Michael addition, an electrophilic substitution (the ring-closing reaction), and an oxidation, all occurring in one flask during a relatively short period of refluxing.

skreef, obs. var. *SCREEF *sb.*

‖ **skrik** (skrik). *S. Afr. colloq.* Also **schreik, schrick, schrijk, schrik, scrick.** [Afrikaans, f. Du. *schrik* fright.] A sudden fright, start; a shock or *frisson*. Freq. in phr. *to get, have* or *give* (someone) *a skrik*.

1887 A. A. ANDERSON *Twenty-Five Years in Waggon* I. 21 They heard the rattling of wheels in a manner which made them think that the oxen must have had a 'scrick' (scare) from a lion. 1896 H. A. BRYDEN *Tales of S. Afr.* 68 It gave me a very nasty *schrijk* at the time. 1897 E. GLANVILLE *Tales from Veld* xxiii. 173 Lor' bless yer, the *schreik* he gave me. 1899 D. BLACKBURN *Prinsloo of Prinsloosdorp* 30 Piet had a bad schrick, but he thought out a plan. 1913 J. J. DOKE *Secret City* xxx. 255 How you do frighten me. You gave me quite a schrik. 1942 S. CLOETE *Hill of Doves* xxxi. 440 'It's Reuter,' someone said. 'What are you doing to give us a skrik like this? It is a joke. It is Reuter.' 1969 A. FUGARD *Boesman & Lena* I. 7 Remember that night the water came up so high?.. You got such a skrik you ran the wrong way. 1975 *Darling* 12 Feb. 119 'Who, me?' I tune him, meantime, feeling the skriks running up and down my spine.

skrike, *v.* Now the more usual form of SCRIKE *v.* Add: **2.** To weep, cry.

1905 *Eng. Dial. Dict.* V. s.v., Hoo skrite't so when hur mother deed I thow't hoo'd ne'er ha done. *Ibid.*, I can tell by yur een as yo'n bin skrikin'. 1977 P. CARTER *Under Goliath* xxvi. 142, I stood there..skriking my eyes out like a mammy's boy... I really cried my eyes out in the loft. 1978 *Lancashire Life* Apr. 42/3 Second un poor little soul Did nuthin' else but skrike.

skriking *vbl. sb.* (later example); **skriker** (later example).

1937 J. R. R. TOLKIEN *Hobbit* iv. 76 The yells and yammering, croaking, jibbering and jabbering; howls, growls and curses; shrieking and skriking, that followed were beyond description. 1959 I. & P. OPIE *Lore & Lang. Schoolch.* x. 186 In the area of Blackburn, Bolton, Manchester, Stockport, and Halifax the term 'skriking' [*sc.* for 'crying'] is common, the noun being 'skriker'.

skulduggery. (In Dict. s.v. SCULDUDDERY 3.) For *U.S.* read orig. *U.S.* and add: 9 **scull-duggery; skullduggery.** Substitute for def.: underhand dealing, roguish intrigue or machination, trickery. (Earlier and later examples.)

1867 A. D. RICHARDSON *Beyond Mississippi* xi. 134 From Minnesota had been imported the mysterious term 'scull-duggery', used to signify political or other trickery. 1911 H. QUICK *Yellowstone Nights* ix. 239 It began to look to me like Hen was up to some skulduggery. 1929 M. A. GILL *Underworld Slang* 11/1 *Skull duggery*, dirty work. 1936 H. HAGEDORN *Brookings* iii. 49 America at ..its worst in financiering, political machination and the skullduggery of the stock market. 1949 J. STEINBECK *Russian Jrnl.* ix. 215 The political skulduggery of the Kremlin. 1957 A. GRIMBLE *Return to Islands* vii. 130 Disgraceful stories of all the skulduggeries he had got away with by suborning government officials. 1962 D. FRANCIS *Dead Cert* ix. 108 The skulduggery that goes on in respectable little old Brighton. 1980 *Times* 3 Jan. 10/2 Watergate was such a sensational piece of skulduggery.

Hence [as back-formation] **sku·ldug** *v. trans.*, to extract by trickery. *nonce-wd.*

1936 W. FAULKNER *Absalom, Absalom!* vi. 178 This Faustus who..skulduggled a hundred miles of land out of a poor ignorant Indian.

skulk, *sb.* **1.** (Earlier modern U.S. example.)

1838 *Knickerbocker* XI. 448 Spotswood had told the middie that Tudor was a great 'skulk', and would probably be reluctant to turn out.

skull, *sb.*[1] Add: **1. b.** Also in slang phr. *out of one's skull*, out of one's mind, crazy. Also succeeding pa. pple., as *bored out of one's skull*, beside oneself with boredom, bored stiff.

1967 *Listener* 7 Dec. 740/2, 12 good men and true, glumly spruce, resigned to a long haul and bored, bored out of their skulls. 1968 T. WOLFE *Electric Kool-Aid Acid Test* xv. 205 They [*sc.* the Beatles] have brought this whole mass of human beings to the point where they are.. out of their skulls. 1973 W. SHEED *People will always be Kind* II. v. 301 You'd have had to be out of your skull not to in those days. 1978 G. VIDAL *Kalki* iii. 83, I thought that Kalki was out of his skull.

d. Also *skull and crossbones*, a representation of a bare skull with two thigh-bones crossed beneath it as an emblem of death, esp. as depicted on a pirate's flag (later examples). Cf. *The Jolly Roger* s.v. ROGER[2] 4. Also *attrib.* and *fig.* Hence *skull-and-crossboned* adj.

1911 D. H. LAWRENCE *Let.* c11 May (1979) I. 268 I've got a grinning skull-and-crossbones headache. 1924 WODEHOUSE *Bill the Conqueror* xvii. 254 This was open rebellion. This was hoisting the skull and cross-bones. 1928 J. M. BARRIE *Peter Pan* in *Plays* v. 73 We see what is happening on the deck of the *Jolly Roger*, which is flying the skull and crossbones. 1930 *Times Lit. Suppl.* 5 June 481/4 The pirates on the Spanish Main in the old skull-and-crossbones days were pleasant and picturesque fellows. 1931 A. RANSOME *Swallowdale* iii. 50 A small varnished dinghy..was sailing in between the headlands. At the masthead was a black flag with the skull and crossbones on it in white. 1955 J. KENWARD *Suburban Child* xxxii. 94 Further down the street where I lived there lived a pirate five years old, the very thing in appearance as in temperament, with a cutlass (silver painted) and a black triangular hat (skull-and-crossboned) both home-made by his understanding parents. 1982 *Times* 5 July 4/3 The nuclear submarine..[was] flying the Jolly Roger to denote their success in sinking the Argentine cruiser... The Skull-and-Crossbones denotes a 'kill'.

e. *slang.* (So much) *a skull*, per person. Cf. HEAD *sb.* 7 b.

1922 JOYCE *Ulysses* 299 They chop up the rope after and sell the bits for a few bob a skull. 1950 *Chambers's Jrnl.* Apr. 213/2 'What difference would the five of clubs make? Sure he had a cast-iron hand.' The Sergeant drew slow caressing fingers along his jaw. That'll be two bob a skull, boys,' he reminded them pleasantly.

5. a. *skull-eye, -neck.*

1928 BLUNDEN *Retreat* 32 The stone skull-eyes look down most drearily. 1922 JOYCE *Ulysses* 509 His eye agonising in his flat skullneck.

6. b. *skull-hunter* (earlier example).

1866 'MARK TWAIN' *Lett. from Hawaii* (1967) 62 In spite of the depredations of 'skull hunters', we rode a considerable distance over ground..thickly strewn with human bones.

7. skull-buster *U.S. slang*, something that taxes the mind; a complicated problem; **skull session** *U.S. slang*, a discussion, conference.

1926 *University Mag.* (Univ. Va.) Oct. 17 *Skull-buster*, a particularly hard course. 1946 MEZZROW & WOLFE *Really Blues* I. 18 Most of my skullbusters got solved at The School. 1959 J. BLISH *Clash of Cymbals* iv. 97 Web and Estelle..had become accepted silent partners at such skull-sessions. 1973 'D. JORDAN' *Nile Green* xi. 49 Joe was ready for the skull session.

skull (skʌl), *v.* [f. SKULL *sb.*[1]] **1.** *trans.* (with *up*) and *intr.* *Metallurgy.* Of molten metal: to freeze and form a skull (in). Cf. SKULL *sb.*[1] 4.

1941 *Engineers' Digest* II. 409/2 Very low sulphur iron, or slow-running iron, would skull up the ladles if much scrap were used. 1953 D. J. O. BRANDT *Manuf. Iron & Steel* xxiii. 174 Neither may the ladle be emptied too slowly, for if it is the steel will get too cold and will 'scull' [*ed.* 2: skull], i.e., freeze.

2. *trans.* To strike (someone) on the head. *slang* (chiefly *U.S.*).

1945 BAKER *Austral. Lang.* viii. 157 *Skull*, to strike (someone). 1952 B. MALAMUD *Natural* 32 My father? Well, maybe I did want to skull him sometimes. 1956 F. CASTLE *Violent Hours* vii. 58 'You didn't get skulled backing away from him,' Webb said dryly. 1975 A. BERGMAN *Hollywood & Le Vine* (1976) viii. 97 My waking came in drugged stages... I had been skulled.

skull-cap. Add: **1. b.** *fig.*

1960 S. BECKER tr. *A. Schwarz-Bart's Last of Just* (1961) VI. 305 Everything under the skullcap of the heavens that called itself a democracy. 1978 *Amer. Poetry Rev.* July/Aug. 35/2, I remember the moon is a skullcap not placed properly on the head.

skullduggery, var. *SKULDUGGERY.

skun, occas. pa. t. or pa. pple. of SKIN *v. dial.* and *colloq.* (chiefly *U.S.*). (See also Wentworth *Amer. Dial. Dict.*)

1917 R. FROST *Let.* 3 Dec. (1972) 20 They might have skun him alive if he had been a mere pupil in their classes. 1927 *Bulletin* (Glasgow) 26 Sept. 12/3 When it comes to breakfast foods America has got Great Britain skun a mile. 1936 J. G. BRANDON *Pawnshop Murder* xxvi. 260 The toff Wibley is working for might be connected with the old dame Widgett skun for her 'ice'? 1942 W. FAULKNER *Go down, Moses* 23 You run a hard race and you run a good one, but you skun the hen-house one time too many.

skunge, var. *SCUNGE *sb.*

skunk, *sb.* Add: **2. b.** Something unpleasant or rotten; rubbish, nonsense.

1929 D. H. LAWRENCE *Pansies* 148 Once and for all, have done with it, all the silly bunk of upper-class superiority; that superior stuff is just holy skunk. 1976 *Daily Times* (Lagos) 8 Oct. 7/6 However, for throwing away the skunk of a national anthem that was unashamedly saddled to this federation for 16 long years, a thousand cheers to the father of this nation, late Murtala Muhammed.

c. *U.S. Mil. slang.* An unidentified surface craft. Cf. *BOGY[1], BOGEY[1] 6.

1945 J. BRYAN *Diary* 24 Mar. in *Aircraft Carrier* (1954) 112 'Skunk' is code for a surface contact, a companion term to 'bogey' in the air. 1952 *N.Y. Times Mag.* 19 Oct. 14/4 The cruiser is..useful at times for coastal bombardment or to seek out and destroy enemy 'skunks' (surface craft). 1957 *Ibid.* 19 May 22/3 A Skunk is an unidentified surface ship, as opposed to a Bogie, which is an unidentified aircraft.

3. skunk bear = WOLVERENE, -INE, 1; **skunk spruce**, one of several aromatic North American spruces, esp. the eastern *Picea glauca* or the western *Picea engelmannii*.

1876 G. B. GRINNELL in W. Ludlow *U.S. Army Corps of Engineers Rep. Reconn. to Yellowstone Nat. Park* i. 65 *Gulo luscus*... In this region, they were spoken of as the 'Skunk-bear'. 1911 J. E. ROGERS *Wild Animals* 112 The wolverine, largest of all the weasels, looks more like a bear and a skunk combined. 'Skunk-bear' is one of his many nicknames. 1961 *Tamarack Rev.* Spring 9 He slouched.. but never preventing the fear from settling in him, never preventing it from turning his eye wary and cruel as any skunkbear's. 1894 *Amer. Folk-Lore* VII. 99 *Picea alba*,.. skunk-spruce. 1921 P. B. KYNE *Go-getter* iii. 32 Have you ever had any experience selling skunk spruce?.. It's coarse and stringy and wet and heavy and smells just

like a skunk. 1948 *Sun* (Baltimore) 21 Dec. 14/2 They are not, it seems, 'real Christmas trees: only skunk spruce'.

skunky *a.*: also, evil-smelling.

1946 D. C. PEATTIE *Road of Naturalist* I. 21 Nicolletia with off-shade yellow-pinks and mustardy green-yellows and a skunky odour that simply would not wash off the fingers. 1960 *Tamarack Rev.* Winter 127 He called.. 'Skunky small one?' Tommy Moore turned his soapy face upward. He was used to being summoned this way. 1973 *Globe & Mail* (Toronto) 3 July 28/1 The thousands of blue-jeaned teenagers..drink beer in the sun, drinking it fast so it doesn't go skunky. 1981 P. THEROUX *Mosquito Coast* xxi. 277 In very hot weather..the jungle odour is skunky and as strong as garbage.

skunk, *v.* Substitute for entry:

skunk (skʌŋk), *v. slang* (orig. and chiefly *N. Amer.*). [f. the sb.] **1. a.** *intr.* To fail. *rare.*

1831 *Constellation* 1 Jan. 54/1 It is a common expression in New-England, to say of a person, who does not get a king in the game of chequers, he *skunked*.

b. *trans.* To defeat or get the better of; to inflict defeat upon.

In some quots. in passive = 'defeated without making any score'.

1843 *Quincy* (Illinois) *Herald* 24 Nov. 2/1 The Legislature will be Democratic by an overwhelming majority; it is more than probable that the Whigs have been skunked. 1845 *Spirit of Times* 9 Aug. 273/2 In the second hand of the third game, I made high, low, game, and 'skunked' him, outright. 1848 BARTLETT *Dict. Amer.* 409 In games of chance, if one of the players fails to make a point, he is said to be *skunked*. A presidential candidate who fails to secure one electoral vote is also skunked. 1876 W. WRIGHT *Big Bonanza* lxxi. 541 'Skunked, by the holy spoons', cried he. 1898 N. BROOKS *Boys of Fairport* ii. 37 Their only hope now was to 'skunk' the White Bears, who were coming to bat. 1904 F. CRISSEY *Tattlings* xvii. 365 A certain trio of choice scamps from the city hall gang would make a strong committee that could skunk the enemy. 1921 *Daily Colonist* (Victoria, B.C.) 18 Oct. 10/1 Very few hunters who went after the pheasants..did not get some at any rate, and the man who was 'skunked' probably would be very hard to find. 1939 *Sun* (Baltimore) 4 Dec. 18/1 The outcome of that battle..was Navy 10, Army 0... 'Which means,' one of them explained to his girl friend, 'We got 10 and they got skunked.' 1944 DUNCAN & NICKOLS *Mentor Graham* xvi. 170 Lincoln, with a short, logical speech in which no words were wasted, 'skunked' his adversary. 1948 *Field & Stream* June 86/2, I have fished on opening day in the snow.., only to get skunked. 1971 D. CONOVER *One Man's Island* 33 When the Colonel comes home from fishing, by his vociferous oaths we know that he was skunked. 1972 D. DELMAN *Sudden Death* (1973) vi. 64 She'll skunk Nell Duncan today, and win.

2. a. *trans.* To fail to pay (a bill or a creditor).

1851 B. H. HALL *College Words* 284 *Skunk*, at Princeton College, to fail to pay a debt; used actively; e.g. to skunk a tailor, i.e. not to pay him. 1859 [in Dict.]. 1961 WEBSTER s.v., Made a practice of skunking hotels.

b. To cheat; in *pass.*, to be cheated *out of*.

1890 C. W. HASKINS *Argonauts of Calif.* xvii. 250, I got skunked once out of a good claim. 1971 E. FENWICK *Impeccable People* iii. 21 I'm beginning to think we skunked you over the price.

skunk-cabbage. Add: Also **skunk's cabbage.** Also used, esp. on the Pacific Coast, to designate *Lysichiton camtschatcense*, another member of the family Araceæ, or a false hellebore of the genus *Veratrum*, of the family Liliaceæ. (Earlier and later examples.)

1751 J. ELIOT *Ess. Field-Husbandry* iii. 66 Take the Roots of Swamp Hellebore, sometimes called Skunk Cabbage, Tickle Weed. 1849 N. KINGSLEY *Jrnl.* 3 May (1914) 15 The fruit grows on the extreme top with a blow or flower resembling our Skunks Cabbage. 1868 H. W. BEECHER *Norwood* 91 The great, succulent leaves of the skunk's cabbage were fully expanded. 1906 *Atlantic Monthly* Oct. 495 The first flower to bloom in this latitude when the winter frost loosens its grip upon the sod is the gross, uncouth, and noisome skunk cabbage. 1950 *Chicago Tribune* 28 Mar. 14/3 Some watch for skunk cabbages poking mottled brown snouts thru the swamp muck. 1968 PETERSON & MCKENNY *Field Guide to Wildflowers* 368 Skunk Cabbage... The sheathing, shell-like spathe, mottled and varying from green to purple-brown, envelops the heavy rounded spadix. 1976 *Hortus Third* (L. H. Bailey Hortorium) 1148/1 *Veratrum..californicum* E. Durand. Corn lily, skunk cabbage.

‖ **Skupština** (sku·pʃtinā). Also 9 **Scubsch'tina, Skoupschina**, etc.; **Skupshtina.** [Serbo-Croatian *skupština*, f. *skupa* together, *skupiti* to assemble.] The national assembly of Yugoslavia; formerly, of Serbia or Montenegro.

1847 A. KERR tr. *von Ranke's Hist. Servia* x. 190 Soon after New Year's Day, all the Woiwodes, with their suites, assembled at a Diet called Skupschtina. 1862 W. DENTON *Servia & Servians* xii. 241 Immediately after the Servians had succeeded in liberating themselves from the Turkish rule they set about forming political institutions for themselves. Among the earliest is the assembly called the Skoupschina. 1866 *Chambers's Encycl.* VIII. 629/2 Each circle..sends a deputy to the *Scubsch'tina.* 1883 *Encycl. Brit.* XVI. 781/2 In 1851, Danilo..prevailed on the 'skuptchina' to declare Montenegro a secular state with the hereditary government of a prince. 1902 *Ibid.* XXX. 821/1 A *Skupshtina*, or popular assembly, is summoned on rare occasions of national importance. 1911 R. W. SETON-WATSON *Southern Slav Question* ix. 194 The Skupština had sufficient self-restraint and sanity to decide against war. 1923 [see *Serbo-Bulgarian* adj. s.v. *SERBO-]. 1940 C. SFORZA *Fifty Years of War & Diplomacy in Balkans* iv. 22 In the elections..Pashich won a complete victory. In the Skupshtina, an imposing majority was ready to follow him. 1968 F. W. HONDIUS *Yugoslav Community of Nations* ii. 74 This Constitution.. created a precedent by recognizing the concept of popular sovereignty, reflected in a powerful one-chamber Skupština. 1976 F. SINGLETON *Twentieth-Cent. Yugoslavia* I. iv. 48 The regency introduced constitutional reforms which provided for a Skupština of 120 members.

skurrick, obs. var. *SKERRICK. **skut**, var. *SCUT *sb.*[4]

skutterudite (sku·tərŭdəit). *Min.* [ad. G. *skutterudit* (W. von Haidinger *Handb. der bestimm. Mineral.*, etc. (1845) IV. 560), f. *Skutterud*, name of a village in SE. Norway (now called *Skotterud*): see -ITE[1].] An arsenide of cobalt, ideally $CoAs_3$, that commonly contains other elements, esp. nickel, iron, bismuth, and sulphur, and is found as grey cubic crystals with a metallic lustre.

1850 J. D. DANA *Syst. Mineral.* (ed. 3) 474 Skutterudite... Lustre bright metallic. Color between tin-white and pale lead-grey, sometimes iridescent. 1892 [see *nickel-skutterudite* s.v. *NICKEL *sb.* 3 b]. 1902 H. A. MIERS *Mineralogy* 332 It is remarkable that there is another arsenide of cobalt of quite different composition, $CoAs_3$, known as skutterudite, which is also cubic and pyritohedral. 1947 *Bull. Geol. Soc. Amer.* LVIII. 317 In spite of repeated attempts, varying the conditions of temperature and using different fluxes, no higher cobalt arsenide other than skutterudite $CoAs_3$ was obtained. 1968 I. KOSTOV *Mineral.* 135 Skutterudite, smaltite, and chloanthite are typical intermediate- or high-temperature hydrothermal minerals.

sky, *sb.*[1] Add: **3. d.** (*a*) Also, *out of a clear* (or *blue*) *sky* and varr. = *out of the blue* s.v. *BLUE *sb.* 5 a; (*b*) *to the skies* (later examples); (*d*) *the sky's the limit*, there is no apparent limit.

(*a*) 1875 TENNYSON *Q. Mary* v. iii. 264 So from a clear sky falls the thunderbolt! 1897 [in Dict.]. 1903 WODEHOUSE *Tales of St. Austin's* 2 To spring an examination on you in the middle of the term out of a blue sky, as it were, was underhand and unsportsmanlike. 1924 E. O'NEILL *Welded* I, in *All God's Chillun got Wings* I. 98 It was revelation, then—a miracle out of the sky! 1958 G. GREENE *Our Man in Havana* III. ii. 115 She's had two unhappy *coups de foudre* herself. They came quite suddenly, out of a clear sky.

(*b*) 1915 W. S. MAUGHAM *Of Human Bondage* xlii. 198 Red-nosed comedians were lauded to the skies for their sense of character. 1955 D. GARNETT *Golden Echo* II. i. 16 If he had praised it to the skies or damned it, or even told the truth, not much harm would have been done. 1973 P. J. SEYBOLD *Revolutionary Educ. in China* xiv. 156 At one time they shouted 'Long live the teachers'.., praising them to the skies.

(*d*) 1920 *Current History* (U.S.) Oct. 142/2 (*caption*) The sky is now her limit. 1933 *Daily Mirror* 26 Oct. 12/4 To say 'the sky was his limit' definitely adds something to the usual 'he succeeded' or 'he rose in the world'. 1934 WEBSTER s.v. *sky*, the sky is the limit. 1936 C. SANDBURG *People, Yes* 160 'Did you say the sky is the limit?' 'Yes, we won't go any higher than the sky.' 1942 E. PAUL *Narrow St.* xxiv. 211 Every municipality, excepting small villages, had its official Mont-de-Piété, and the sky was the limit. 1952 W. R. BURNETT *Vanity Row* vii. 68 If there's ever anything we can do for you... You know. Sky's the limit, as people say. 1961 L. MUMFORD *City in History* ii. 52 The cult of power exulted in its own boundless display... The sky was the limit. 1977 H. FAST *Immigrants* II. 97 As far as the Pacific passage is concerned, rates are going up and the sky's the limit.

4. c. In joc. phr. *the* (or *that*) *great —— in the sky*: with personal subj., God considered as the omniscient exponent of an earthly art or profession; of a place or structure, the type of a paradise especially suited to the deceased.

1977 MCKNIGHT & TOBLER *Bob Marley* v. 62 Chuck Willis, the 'Sheik of the Stroll' became one of the first members of the great rock group in the sky. 1979 *Times* 24 Nov. 15/7 It is up to that Great Film Critic in the sky to deal with *Life of Brian* in His own way. 1980 D. BLOODWORTH *Trapdoor* xvii. 107 There's a Director of Central Intelligence up there in that great Langley in the sky. 1982 *Times* 26 Jan. 10/3 Daphne, the pelican, has gone to that great aviary in the sky after 25 years residence..in St James's Park.

5. a. (Examples illustrating use as a fashion shade.)

1851 *Illustr. Catal. Gt. Exhib.* III. 506/2 Pink, white, sky, and maize gros de Naples for ladies' bonnets. 1894 [see *HELIO[2]]. 1923 *Weekly Dispatch* 11 Feb. 14 (Advt.), Will not Fade... Silky finish. Ivory, Biscuit, Sky, Coral [etc.]. 1949 *Radio Times* 15 July 44/3 Lovely pastel shades—Peach, Apple, Sky. 1976 *National Observer* (U.S.) 22 May 17/4 (Advt.), The plain shades... Rust, Beige, Tan, Sky, White, Black.

7*. *Rhyming slang. ellipt.* for *SKY-ROCKET *sb.* 3: pocket.

1890 in Barrère & Leland *Dict. Slang* II. 248/2 The Oof Bird's scarce and the landlady's fly, And there isn't a mash with a mag in his sky. 1898 A. M. BINSTEAD *Pink 'Un & Pelican* xi. 237 After thirty-six 'ands 'ad bin all over him, tore his trowseys an' left 'im as naked as Barth-Sheber—why, even *then* we never found his sky! 1928 E. WALLACE *Gunner* xviii. 140 'Put that in your sky... In your pocket,' she said impatiently. 1979 P. HILL *Washermen* lx. 132 Said 'ee found it [*sc.* a gun] on the rattler. Put it in 'is sky when 'ee got off at Leicester Square.

8. Now chiefly *Lit.* and *poet.* **a.** Simple *attrib.* (Later examples.)

1904 W. B. YEATS *Pot of Broth* in *Hour-Glass* 78 Give me some vessel till I give this sky-woman a taste of it. 1916 BLUNDEN *Harbingers* 64 He stells the meadows in similitude Of stars in black sky-spaces. 1920 D. H. LAWRENCE *Lost Girl* xiv. 320 White clouds, in the sort of hollow sky-dome. 1930 W. H. AUDEN *Poems* 9 Though heart fears all heart cries for, rebuffs with mortal beat Skyfall, the legs sucked under, adder's bite. 1946 L. B. LYON *Rough Walk Home* 11 Lift arm or lift an eyebrow, He'll weave his sky-brow Spell round the offender. 1959 D. DAVIE *Forests of Lithuania* vi. 59 Fires cluster and dart Cross over, light over light Overarches the sky-round. 1979 D. WILLIAMS *Genesis & Exodus* vii. 127 He enjoyed the wide sky-sweep of the fens.

b. *sky-gazer.*

1930 V. WOOLF *On being Ill* 19 Pedestrians would be impeded and disconcerted by a public sky-gazer.

c. With pa. pples. (Earlier and further examples.)

1589 R. GREENE *Menaphon* Sig. F1v, A Skie borne forme. 1807 J. BARLOW *Columbiad* III. 110 Far beneath, the sky-borne waters ride, Veil the dark deep and sheet the mountain's side. 1923 D. H. LAWRENCE *Birds, Beasts & Flowers* 17 And slipped down, perhaps, with a sip of Marsala So that the rambling, sky-dropped grape can add its music to yours. 1934 L. B. LYON *White Hare* 14 Windscoured and sky-burned The fell was. 1946 DYLAN THOMAS *Deaths & Entrances* 27 May his hunger go howling on bare white bones Past the statues of the stables and the sky roofed sties. 1977 *Hongkong Standard* 12 Apr. 3/3 Aides say Mr Peres originated and pushed the idea of last July's skyborne rescue from Uganda of 100 hostages.

d. With pres. pples. (Further examples.)

1788 P. FRENEAU *Hermit of Saba* in *Misc. Works* 31 When thou, sky-pointing Saba, Shall tremble on thy base most fearfully! 1819 Sky-aspiring [see *BALMILY *adv.*]. 1844 J. R. LOWELL *Poems* 274 They tell us that our land was made for song, With its huge rivers and sky-piercing peaks. 1922 JOYCE *Ulysses* 563 Stephen with hat ashplant frogsplits in middle highkicks with skykicking mouth shut hand clasp part under thigh. 1933 C. DAY LEWIS *Magnetic Mountain* 9 Void are the valleys.. And dumb the sky-dividing hills. 1957 L. MACNEICE *Visitations* 42 Felt suddenly harassed, a sky-splitting headache with nothing to cause it. 1977 *New Scientist* 24 Feb. 478/1 A spiky, sky-piercing crenellation of buildings running down the Royal Mile from the hunched bulk of the castle to the rounded towers of the Palace of Holyroodhouse.

9. sky bear *N. Amer. slang*, (an officer in) a police helicopter (cf. *SMOKEY BEAR 2); **sky border** *Theatr.*, a border of painted cloth, used both to represent sky and to conceal the top of the stage from the audience; **skycap** *N. Amer.* [after *REDCAP 5], a porter at an airport; **sky-clad** *a. slang*, nude, unclothed (esp. in *Witchcraft*); **sky cloth** *Theatr.*, a backcloth painted or coloured to represent the sky (cf. *sky border* above); **sky-clothed** *a.* = *sky-clad* adj. above; **sky-diving**, the sport of parachuting from an aeroplane with a long period of (freq. acrobatic) free fall before the parachute is opened; also as adj., **sky-diver** and (as back-formation) **sky-dive** *v. intr.*; **sky-drop** *Theatr.* = *sky cloth* above; **sky fighter**, an aeroplane or airman that engages in aerial combat; hence **sky-fight**; **sky filter**, a filter (usu. yellow and denser at the top than at the bottom) for improving the rendering of a bright sky in black and white photography; **sky-fire** (later *poet.* example); **sky-flower**, a shrub or small tree, *Duranta repens*, of the family Verbenaceæ, native to Central and South America and bearing clusters of pale blue flowers followed by yellow berries (cf. *PIGEON-BERRY 1); **sky-god** *Religion* and *Mythol.*, a god of or in the sky; also **sky-goddess**; **skyman** *Journalists' slang*, a paratrooper; **sky-marker** *Mil.*, a parachute flare used by raiding aircraft to mark a target (cf. *PARACHUTE *sb.* 5); also Comb., as *sky-marker bomb*; also *sky-marking*; **sky marshal** *U.S.* a plain-clothes armed guard on an aeroplane employed to counter hi-jacking; **sky-path**, a route taken through the sky, a skyway; **sky pilot** *slang* (earlier and later examples); also, a chaplain in any of the armed forces, prison service, etc., and *gen.*, a priest or parson, a clergyman; † **sky race**, in British India, an amateur steeplechase; **sky-ride** *U.S.*, a device for conveying passengers at a considerable height above ground, spec. one at the World Fair at Chicago in 1933–4; **sky screen**, an array of photocells used to record or detect the travel of an aircraft, projectile, etc.; **sky shade** *Photogr.* (see quots.); **sky-ship**, a very large craft for air or space travel; **sky-shouting**, the sending of advertisements or messages from an aircraft to the ground by means of a loudspeaker; also as adj. and **sky-shouter**; **sky-surfing** *U.S.* = *hang-gliding* s.v. *HANG-; hence **sky-surfer**; **skywatch** orig. *U.S.*, the process or activity of watching the sky for aircraft or other phenomena; hence **skywatcher**; **sky wave**, a radio wave reflected back towards the earth's surface by the ionosphere (cf. *ground wave* s.v. *GROUND *sb.* 18 a).

1975 L. DILLS *CB Slanguage Dict.* 54 *Sky bear*, police helicopter. 1977 *Daily Colonist* (Victoria, B.C.) 3 July 1/2 (*heading*) Sky bear keeps eye on Island's drivers. 1846 G. A. A'BECKETT *Quizziology Brit. Drama* 16 Pointing with a property sword to the sky borders. 1896 W. ARCHER *Theatr. World of 1895* iv. 28 Above it hang mathematically horizontal 'sky-borders', apparently representing a flat layer of fog in the upper air. 1918 G. B. SHAW in *Nation* 22 June 310/1 The scenery made Old Drury feel young again. Wings, sky-borders, set pieces: nothing was missing. 1950 *Official Gaz.* (U.S. Patent Office) 26 Dec. 1066/2 Sky Cap. For General Porter Service. 1966 *National Observer* (U.S.) 7 Nov. 6/5 They would reduce the number of Negro 'sky caps' employed at the airport. 1972 T. KENRICK *Tough One to Lose* ii. 33 He took a job as a skycap at the International Airport. 1977 J. WAMBAUGH *Black Marble* (1978) xv. 342 He spotted a skycap carrying some bags towards the front. 1909 WEBSTER, Sky-clad. 1970 R. BUCKLAND in K. Singer *Tales from Unknown* 296 Witches always work naked or, as they call it, *skyclad*. 1933 Sky-cloth [see *BATTEN *sb.*[1] 1 b]. 1981 *Times Lit. Suppl.* 4 Sept. 998/3 Instead of a naturalistically painted backcloth he used a plain sheet of colour (what we should call a sky cloth). 1924 EARL OF RONALDSHAY *India* xxiv. 305 The Digambara, or sky-clothed ascetic, must live stark naked. 1965 *N.Y. Times* 24 Apr. 21 Mary Cushing..snorkles, surfs, skis and sky-dives. 1961 *Times* 3 July 6/4 The screech of the skydiver was heard above Hereford this weekend as.. people saw a demonstration of the latest 'official' British Army sport. 1970 *Daily Tel.* 5 Oct. 1/1 Two women sky-divers and a man were injured when they parachuted at 2,500ft from a De Havilland aircraft..last night. 1979 P. NIESEWAND *Member of Club* xii. 84 A flypast might be nice... How about some sky-divers? Everyone likes them. 1959 *News Chron.* 8 July 4/7 The sport..of 'sky-diving' in which certain adventurous types turn somersaults in the air before opening their parachutes. 1962 *Daily Tel.* 7 Aug. 11/7 A sky-diving team will leave London Airport today..to compete in the sixth world sport parachuting championships. 1979 R. JAFFE *Class Reunion* (1980) II. ix. 275 Apparently she's taken up skydiving... She's going to kill herself. 1901 C. MORRIS *Life on Stage* xii. 84 In this tableau the circular opening in the flat, backed by a sky-drop and with blue clouds hanging about the opening, represented heaven. 1969 G. MACBETH *War Quartet* 36 He was dead to this Antarctic sky-fight. 1937 *Sun* (Baltimore) 2 Mar. 1/3 The first of the army's super sky-fighters, a four-engined Boeing bomber, dropped to a perfect landing on snow-covered Langley Field at 2.09 P.M. today. 1943 R. WHELAN *Flying Tigers* ix. 97 George Paxton..moved to break up this shocking attack on a helpless airman, which violated the code of sky fighters. 1930 G. E. BROWN *Clerc's Photogr.* xi. 83 Sky filters are made in the shape of a long rectangle, which is carried in a mount, allowing it to be raised or lowered. 1970 M. J. SETHNA *Photogr.* v. 102 Where the sky is light and bright, but the landscape is less bright or is dark, the balance of tones can well be secured through the use of what is known as 'the graduated sky filter'. 1906 Sky-fire [see *night-web* s.v. *NIGHT *sb.* 13 a]. 1938 D. WYMAN *Hedges, Screens & Windbreaks* II. 77 Tall Broad-leaved Evergreens... *Duranta plumieri*, Skyflower. 1971 Sky-flower [see *PIGEON-BERRY 1]. 1907 H. M. CHADWICK *Orig. Eng. Nation* x. 245 On the strength of this passage [in *Gylfaginning*] it has been supposed that Frey was originally a sky-god or sun-god. 1938 E. BEVAN *Symbolism & Belief* ii. 30 The belief in the Sky-God may have, of course, two forms according as the sky itself is personified, is identified with the Person up there, or as the Person is conceived more anthropomorphically. 1948 B. G. M. SUNDKLER *Bantu Prophets in S. Afr.* i. 24 The lightning-magician, the priest of the sky-god, arrives much sooner..at proficiency in his particular speciality. 1979 *N.Y. Rev. Bks.* 25 Oct. 19/3 Now the Heavenly Father, or Aryan sky-god, is found to be..simply irrelevant. 1959 *New Larousse Encycl. Mythol.* 23/2 *Hathor*... A sky-goddess, she was originally described as the daughter of Ra and the wife of Horus. 1982 N. FRYE *Great Code* iii. 70 Zeus..third in a line of sky-gods. The earth-mother..tends to take on the characteristics of a sky-goddess. 1952 *John o' London's Weekly* 18 Jan. 54/1 The failure of Fleet Street to make paratroopers into skymen. 1958 *Daily Mail* 18 July 1/2 Skymen saved Hussein's life... But for the arrival of our paratroops yesterday,..Hussein..would almost certainly have been assassinated. 1964 *Sunday Tel.* 14 June 3/4 (*heading*) Skymen hit the target. 1943 *Times* 31 Dec. 4/6 The Pathfinder force used parachute flares known as 'sky-markers' which drift downwards very slowly, to mark the target area. 1944 *Times* 17 Feb. 4/4 Flak was so violent when the first sky-marker bombs were dropped that it was evident that the main night fighter force was late. 1946 *R.A.F. Jrnl.* May 169 The red, yellow and green T.I.s and the skymarker flares, remained the principal weapons of P.F.F. throughout the war... The 'wanganni' skymarkers went down over Germany and load after load of destruction followed. 1944 R. DIMBLEBY in *War Report* (B.B.C.) (1946) 281 Our job was to replenish the flares already dropped by the Pathfinders ahead... This was 'sky-marking'. 1968 *Sunday Mail* (Brisbane) 23 June 32/6 He is a member of a new elite breed of law enforcement officers in America—the sky marshals. His job is to prevent airliner hi-jacking... All the sky marshals are volunteers from the ranks of the F.A.A.'s regular inspectors. 1971 *Daily Colonist* (Victoria, B.C.) 13 Mar. 25/6 A number of women trained in marksmanship and hand-to-hand combat will join the men assigned to the skymarshal force recently created to protect U.S. airliners from hijackers. 1931 F. SIMPICH in *Nat. Geogr. Mag.* Jan. 1 (*heading*) Skypaths through Latin America. 1958 *Times* 27 Oct. 10/1 Nearly 400 Glacier dry bearings assist each Comet 4 on its smooth skypath. 1883 G. W. PECK *Peck's Bad Boy* 177 Look-a-here you sky-pilot, this thing has gone far enough. 1910 *Busy Man's Mag.* Mar. 71/1, I was hailed as a 'sky pilot' by the trio and invited to be sociable over a whisky bottle. 1922 JOYCE *Ulysses* 305 One or two sky pilots having an eye around that there was no goings on with the females. 1935 AUDEN & ISHERWOOD *Dog beneath Skin* II. v. 114 Ort ter 'ave bin a sky pilot, you ort! 1973 B. BROADFOOT *Ten Lost Years* xii. 140 At the missions you would get a sermon, say 15 minutes of religion from a sky pilot. 1982 *New Scientist* 18 Mar. 740/3 The first issue includes an attack on the accuracy of radiocarbon dating by an English skypilot called Charles Foley. 1858 G. F. ATKINSON *Curry & Rice* (ed. 2) xviii, The Sky Races, which to the uninitiated may be explained as a meeting for horses that have enjoyed no specific training beyond what could be accomplished during the interval of the 'get-up' and the 'come-off'. 1885 LADY DUFFERIN *Jrnl.* 11 June in *Our Viceregal Life in India* (1889) I. iv. 157 The Simla sky races began today. 1933 *Sun* (Baltimore) 22 July 10/3 Two Concordia women, who attended the World's Fair at Chicago recently ventured on the 'sky ride', the device which carries passengers across the grounds at a height of 200 feet. 1966 E. MCCULLOCH *World's Fair Midways* viii. 93 One of the features of the fair [*sc.* Century of Progress Exposition] was the Sky Ride, a monorail structure whose two giant towers stretched up sixty-four stories... They were connected at the twenty-fourth story by cables, from which so-called rocket cars were suspended. 1945 L. E. SIMON *German Sci. Establishments* (PB Rep. No. 19849) II. ii. 48 One thing that was particularly notable was a large number of photoelectric 'sky screens'. 1969 *New Scientist* 2 Oct. 25/3 The accuracy of the sky screens is said to be two metres in azimuth, and 1·5 metres in elevation. 1889 E. J. WALL *Dict. Photogr.* 177 *Sky shade*, a piece of wood or card used to shade the lens during exposure, to prevent reflections from the sky or sun. 1909 G. L. JOHNSON *Photographic Optics & Colour Photogr.* ii. 152 Skyshades are of great value in colour photography..as without some such screen the skies are invariably spoilt through overexposure. 1930 G. E. BROWN *Clerc's Photogr.* xi. 83 The most usual form of commercial sky shade..consists of a uniformly graduated filter of gelatine or glass. 1973 D. A. SPENCER *Focal Dict. Photogr. Technol.* 568 *Sky shade*, any form of shield attached to the lens mount for preventing direct rays of the sun reaching the camera lens. In USA, the term is sometimes used as another name for lens hood. 1923 L. PAUER *Day of Judgment* 16 If possible ..we are going to board that sky-ship. 1960 *Analog Science Fact/Fiction* Dec. 10/1 When we first heard of the Sky Ship, we were on an island whose name..was Yarzik. 1975 *Times* 18 Apr. 6/3 The Sky Ship, with a diameter of 30ft and in the shape of a flying saucer..is a scaled-down prototype of a planned vehicle..which will be 700 ft in diameter..able to carry a payload of up to 400 tons. 1932 *Children's Newspaper* 23 Jan. 6/1 The inventor..can now quote terms for Sky Shouting or Sky Advertising. Concerning the sky-shouters a really alarming invention has been successfully tried. 1932 *Flight* 8 July 638/1 They recommend that sky-shouting (by means of a loud speaker) should be prohibited by law for all private purposes. 1955 *Times* 9 June 8/2 The withdrawal of the surrender offer was being conveyed to the terrorists by all possible means, including radio and sky-shouting aircraft. 1962 *Engineering* 26 Oct. 564 The practicability of long range speech transmission was seen during British army operations in Malaya when a 'sky-shouting' installation in an aircraft was used for propaganda purposes. 1973 *Times* 15 May 8/4 Last year 2,285 hours were logged by aircraft flying 'sky-shouting' patrols in which recorded propaganda messages were relayed to Africans living in the bush. 1972 *Popular Mechanics* June 102/2 How long a sky surfer can stay in the air depends on wind strength and skill. 1972 *Popular Sci.* June 94/2 Today's hang-glider pilots like to call their sport 'sky-surfing'. 1974 *Sci. Amer.* Dec. 138/1 The rapidly evolving sport, which is known as sky surfing or hang gliding, makes about equal demands on the enthusiast's skills as a pilot and as an aerodynamicist. 1952 *Sun* (Baltimore) 17 July B7/5 (*heading*) 18,000 in Britain serve on skywatch. 1958 A. BUDRYS in Aldiss & Harrison *Decade 1950s* (1976) 68, I made it. Got to this Navy skywatch station. 1972 *Oxford Times* 21 Jan. 3/7 A full house is expected for next week's UFO convention in Banbury... 'Many of those attending are the people who made sightings over North Oxon last year which ended with us having a skywatch, which was unfortunately rained off.' 1973 *Daily Tel.* 30 July 1/5 Skylab, at present, is not visible to British skywatchers since its orbital track does not take it over Britain. 1928 STERLING & KRUSE *Radio Manual* xiv. 524 The day signal which reappears at 850 miles may be considered the sky wave. 1944 *Proc. IRE* XXXII. 668/1 Design of directional antennas for broadcast stations to prevent skywave interference to another station. 1971 K. KENT in C. Bonington *Annapurna South Face* 277 The near-link radios had to be h/f sets capable of voice/cw and able to utilize a sky-wave and surface-wave mode of operation.

sky, *v.*[1] Add: **1. a.** Also used of other objects, as a hat, etc.; spec. *to sky the wipe* Austral. Boxing slang = *to throw up the sponge* s.v. SPONGE *sb.*[1] 1 c.

1898 A. M. BINSTEAD *Pink 'Un & Pelican* x. 215 He skied his tile in the most approved fashion..literally beaming with good-nature as he shook his jockey by the hand. 1916 C. J. DENNIS *Songs of Sentimental Bloke* xi. 87 Fer 'arf a mo' I 'as a fight; Then conscience skies the wipe... Sez I 'Orright'. 1933 *Bulletin* (Sydney) 14 June 27/2 It is generally understood that a boxer must consider himself beaten when his seconds 'sky the wipe'.

b. (Earlier example.) Also in *Golf*, etc.

1868 *J. Lillywhite's Cricketers' Compan.* 102 He..sometimes gets deceived by a short one, and 'skies' it. 1909 [see *FLUFF *v.*[1] 5 e]. 1922 E. F. BENSON *Miss Mapp* iii. 73 Major Flint drove, skying the ball to a prodigious height. 1976 *Morecambe Guardian* 7 Dec. 8/1 Worksop came back into the attack when Wall started a good move but Joe Johnson skied the ball high over.

c. At an auction: to raise the price of (an item) by high bidding; to raise (the bidding) by a considerable amount.

1892 STEVENSON & OSBOURNE *Wrecker* ix. 146 All of a sudden he appeared as a third competitor, skied the *Flying Scud* with four fat bids of a thousand dollars each, and then as suddenly fled the field. 1928 D. L. SAYERS *Ld. Peter views Body* x. 236 Wimsey, entering into the spirit of the thing, skied the bidding with enthusiasm. The dealers,..fancying that there must be some special excellence about the book.., joined in.

sky-blue, *sb.* Add: **1. b.** Comb., as *sky-blue-pink*, a fantasy colour. Also as adj. *joc.*

1942 G. KERSH *Nine Lives Bill Nelson* xi. 66 Bill could swear that black was white, or green was sky-blue-pink. 1967 L. J. BRAUN *Cat who ate Danish Modern* xv. 131 What colours do you mix to get sky-blue-pink? 1982 S. T. HAYMON *Ritual Murder* xvii. 116 Caught him out red-handed... Not to say green and yellow an' sky-blue-pink.

3. b. Barley broth (*Naut.*); any vegetable soup.

1887 S. SAMUELS *Forecastle to Cabin* iv. 53 Sky-blue (boiled barley), hard tack, and tea sweetened with treacle was Jack's fare for the morning meal. 1908 J. M. SULLIVAN *Criminal Slang* 23 *Sky-blue*, vegetable soup. 1910 H. Y. MOFFAT *Ship's Boy to Skipper* xiii. 204 Once a week we had for dinner what we called 'Sky Blue'. It was made by putting a small quantity of barley into a large quantity of water.

Skybus (skəi·bʌs). *U.S.* Also **skybus, sky bus.** [SKY *sb.*[1]] **1.** The proprietary name of a regular air service for which passengers need not book in advance. Cf. *SKYTRAIN 2.

1945 *Aviation* Feb. 132 Douglas Skybus. Designed for feeder line operations, seating 24 passengers. 1966 *Official Gaz.* (U.S. Patent Office) 8 Feb. TM 65/1 New York Airways, Inc... *Skybus* for air service for passengers, mail, and freight. First use Oct. 15, 1952. 1972 *Times* 28 Sept. 1/1 Trans International Airlines today proposed to the Civil Aeronautics Board a daily New York to London 'sky bus' service at a fare of $75. 1977 *Time* 10 Oct. 5/1 Trans International Airlines..borrowed a leaf from Laker's book and last month proposed a 'Skybus' service from the U.S. to Tokyo and Hong Kong.

2. (See quots.)

1966 *Daily Tel.* 21 Apr. 16/4 (*caption*) In Pittsburgh.. tests are being carried out with a Skybus... The bus itself is a lightweight aluminium car with light rubber wheels, which run on two small concrete rails. It picks up current from a centre beam. 1967 *Guardian* 2 Feb. 3/3 Westinghouse Electric Corporation, of Pittsburgh..have developed a driverless 'skybus' for the US Government... The skybus is a cross between a bus and a train and can run in tunnels, on the ground, or elevated.

Skye. Add: **a.** *Skye terrier* (earlier and later examples).

1847 H. D. RICHARDSON *Dogs* vii. 71 The Skye Terrier, so called from its being found in greatest perfection in the Western Isles of Scotland, and the Isle of Skye in particular. 1971 F. HAMILTON *World Encycl. Dogs* 478 The show Skye Terrier's profuse, well groomed coat.., its veil of thick hair shielding fore-face, and the silky feathering of ears and tail, are not attained and maintained without considerable trouble.

skyer. (Earlier examples.)

1853 *Bell's Life in London* 24 July 6/2 Mr Walker gave a few 'skyers' in scoring 20, for which number he brought out his bat. 1867 *Australasian* 26 Jan. 107/4 Fowler hit a skyer to leg, no one in the road.

sky-ful. Add: Now usu. **skyful.** (Later examples.)

1910 KIPLING *Rewards & Fairies* 152 Presently I heard guns... I stopped fiddling to listen, and I heard a whole skyful o' French up in the fog. 1966 D. VARADAY *Gara-Yaka's Domain* xiii. 146 They slept under the trees, with a skyful of stars for their blankets.

sky-high, *adv.* and *a.* Add: **A.** *adv.* Also, freq. in *fig.* phr. *to blow sky-high*, to refute utterly; = EXPLODE *v.* 3.

1845 A. JACKSON *Let.* 6 June in M. James *A. Jackson* (1937) xxiii. 498 Put your veto upon them both, or you and your Secretary will be blown sky high. 1948 *Daily Tel.* 29 May 2/6 [He] blew sky-high the complacent superstition of a war-time 'renaissance'. 1951 N. MITFORD *Blessing* I. viii. 83 At this dinner Grace's preconceived ideas about the French..were blown sky high. 1955 *Times* 26 May 3/3 Matters which had been put forward in the course of that case had either fallen by the wayside or had been blown 'sky-high'.

B. *adj.* Also *fig.*, very high.

1959 *Times* 7 Sept. 11/6 The current miracle of labour-saving flats, sky-high wages, and welfare state. 1975 A. BERGMAN *Hollywood & Le Vine* (1976) v. 67 We don't pay writers whatever sky-high figures their agents talk them into demanding.

sky-hook. Also **sky hook, skyhook.** [SKY *sb.*[1]] **1. a.** orig. *Aeronaut.* An imaginary contrivance for attachment to the sky; an imaginary means of suspension in the sky.

1915 *Aeroplane* 10 Mar. 222 The battery signaller sent a message: 'Battery out of action for an hour, remain aloft awaiting orders.' Back came the reply with remarkable promptitude: 'Submitted; that this machine is not fitted with sky-hooks.' 1933 *Ibid.* 11 Jan. 47/1 They [*sc.* slots] have in fact been used as..sky-hooks on the wing tips. 1948 *Register* Sept. 262 Mr. Shaw's plea..is, at bottom, a plea for a skyhook. For that common agreement which it assumes hung from a skyhook and was broken when we severed it from its support. 1960 R. W. MARKS *Dymaxion World of B. Fuller* 28/1 With each thrust of his wings, the duck generates a momentary vacuum sky-hook above each wing. 1970 D. WATERFIELD *Continental Waterboy* I. 6, I..was prepared to..fetch him a left-handed wrench, a skyhook, anything.

b. Applied to various devices or craft capable of lifting something into the air, as: a hook on an aircraft; an aerial cableway; a balloon, helicopter, etc., designed for lifting. orig. *U.S.*

1935 MEIER & LINDBERGH in *Sci. Monthly* (N.Y.) Jan. 5 An instrument new to transatlantic airplanes..which, being untried, was noncommittally called the 'sky hook'. 1945 *Collier's Mag.* 27 Jan. 38/4 He tried building a model of one of these seed pods, blowing it up to the size of a man and loading its hollow container with cargo. It spiraled down vertically and could safely bring 100 pounds to earth from any height... The device, now nicknamed the 'skyhook', is fluttering down on many Allied battlefields. 1949 *Sun* (Baltimore) 13 Aug. 12/3 That long renowned but hitherto incredible gadget, the sky-hook, is now a reality... It is known as 'The Aerial Track Airport: a cable runway for landing and launching airplanes'. 1951 *N.Y. Times* 13 Feb. 1/7 The balloons, called skyhooks by the Navy researchers, were released at many points in the country. 1966 *Science* 11 Feb. 682 (*heading*) Satellite elongation into a true 'sky-hook'. 1970 [see *dogman* s.v. *DOG *sb.* 18 a]. 1982 *Monitor* (McAllen, Texas) 25 Mar. 1-A/1 The shuttle's robot arm lifted a space environment monitor out of the payload bay today in an important 'first' that proved the remotely controlled skyhook will be able to launch satellites.

2. *Mountaineering.* A hook fixed into a rock-face, from which ropes, etc., may depend.

1957 *Appalachia* 15 Dec. 452 On the one occasion when I carried a 'skyhook' among my pitons I was no less than delighted to find it really could be used to circumvent the placing of a bolt. 1968 P. CREW *Encycl. Dict. Mountaineering* 108/2 *Sky-hook*..consists of a flattened hook of high quality steel... Sky-hooks have a small eye, to which a sling and étrier can be attached... They are particularly useful on loose flakes..which would probably break off if a piton was used. 1976 [see *RURP].

skyhoot, *v.* ? Fanciful perversion of SCOOT *v.* Freq. as pres. pple.

1888 *Boy's Own Paper* Summer 38/1 Something's skyhooted in my shoulder... That brute threw me on my head. 1900 H. F. DAY *Up in Maine* 56 That air pessle ..come sky-hootin' like a ten-inch bomb. 1911 [see *LICKETY *adv.*]. 1928 E. WALLACE *Double* i. 7 Why I'm sky-hooting down to Brighton for two days, heaven knows!

skyjack (skəi·dʒæk), *v.* Also **sky-jack.** [f. SKY *sb.*[1] + *HI)JACK *v.*] *trans.* To hijack (an aeroplane). Also const. *to* (a destination).

1961 *N.Y. Mirror* 10 Aug. 1 (*heading*) Pan Am Jet skyjacked to Havana. 1968 *Tel.* (Brisbane) 13 Mar. 24/7 Over the years scores of light planes, in addition to larger passenger airliners, have been skyjacked and flown to Cuba. 1970 *Time* 12 Oct. 38/3 Aboard the plane was Leila Khaled, 24, the Palestinian guerrilla who attempted to skyjack an El Al airliner over Britain last month. 1971 *Daily Tel.* 1 Feb. 20/3 Sentence of death on a Lithuanian ..for trying to skyjack a Soviet airliner to Sweden, has been commuted. 1977 *Whitaker's Almanack 1978* 604/2 A Russian engineer who skyjacked a Soviet airliner surrendered to Swedish police when the plane landed at Stockholm.

Hence **sky-jacked** *ppl. a.*, **sky·jacking** *vbl. sb.*

1961 *N.Y. Mirror* 5 Aug. 3 (*heading*) Rush stiff skyjacking law. 1969 *Daily Tel.* 1 Nov. 1/1 All air traffic was warned..to avoid the sky-jacked airliner as special preparations were made at Shannon Airport to receive the Trans-World airliner. 1977 D. E. WESTLAKE *Nobody's Perfect* (1978) 227 Listen to this. We fake a skyjacking, but what we *really* do—.

skyjack (skəi·dʒæk), *sb.* [f. the vb.] An instance of skyjacking.

1968 *Tel.* (Brisbane) 13 Mar. 24/7 In the third skyjack in 16 days, a U.S. National Airlines DC8 with 59 people aboard landed at Havana, Cuba, today. 1969 *Cavalier Daily* (Univ. Va.) 5 Feb., United Airlines Tuesday reported an unsuccessful skyjack attempt. 1975 A. OSMOND *Saladin!* I. v. 39 The wars and the riots, the skyjacks and terrorists' bombs. 1978 R. JANSSON *News Caper* iv. 46 This is a skyjack. We have a hostage at either end of the plane.

skyjacker (skəi·dʒækəɹ). Also **sky-jacker.** [f. SKY *sb.*[1] + *HI)JACKER.] One who hijacks an aeroplane.

1961 *N.Y. Mirror* 4 Aug. 1 (*caption*) JFK directed capture of skyjackers. 1970 *Pacifist* Sept. 5/2 The death of hostages will not necessarily lead to the release of..the woman sky-jacker, Leila Khaled. 1976 P. HENISSART *Winter Quarry* ix. 91 You're probably a skyjacker or a bigamist on the run. 1982 *Daily Tel.* 24 July 4/6 The skyjackers..were said to have threatened crew members.

Skylab (skəi·læb). [f. SKY *sb.*[1] + *LAB *sb.*[2]] The name of a space laboratory launched into earth orbit by the U.S. in 1973. Freq. *attrib.* Occas. (with small initial) in *gen.* use.

The laboratory itself was more precisely designated *Skylab 1*; *Skylabs 2, 3*, and *4* were the spacecraft which successively ferried crews to it. The laboratory was manned until 1974, and disintegrated in the atmosphere in 1979.

1970 *Nature* 13 June 1001/2 There are now only four journeys to the Moon before autumn 1972 when the engineers at the Cape and in Houston turn their attention to the Skylab programme—the renamed Apollo applications programme. *Ibid.* 22 Aug. 774/1 Cancelling two of the [Apollo] flights so as to conserve the Saturn V launcher vehicles for ambitions nearer home, such as Skylab and the space station project. 1972 *Daily Progress* (Charlottesville, Va.) 19 Jan. B9/8 The first three-man crew is scheduled to spend 28 days aboard the Skylab mission starting next May. 1973 *Guardian* 11 Apr. 4/3 Preparations are precisely on schedule at Cape Kennedy for the launching on May 14 of Skylab-1, the first US experimental space station. 1975 *McGraw-Hill Yearbk. Sci. & Technol.* 43 With the recovery of the *Skylab 4* command module and crew on Feb. 8, 1974, the active operational phase of the Skylab Program was brought to a successful conclusion. 1979 *Guardian* 12 July 1/8 Skylab fell to earth last night... Its debris scattered into the sea and on to a sparsely inhabited Australian desert. 1982 M. DUKE *Flashpoint* xiii. 98 Scientifically equipped satellites and sky labs.

skylark, *sb.* Add: **3.** With initial capital in catch-phr. *any more for the Skylark?* (from a boatman's cry at seaside resorts), used to offer an open invitation, usu. for a ride or lift.

1931 M. ALLINGHAM *Look to Lady* x. 107 What a good job there's no more for the Skylark... I love riding in other people's motor-cars. 1977 E. DEWHURST *Curtain Fall* v. 46 You hoping for a lift, Miss Rhoda?.. Come along, then, ducky. Any more for the *Skylark*?

sky-larker. Add: † **2.** *Cricket.* = SKYER. *Obs.*

1839 *Bell's Life* 30 June 4/1 Wells..soon retired, giving a 'skylarker', which Button caught.

skylight, *sb.* Add: **4.** Special Comb.: **skylight filter**, a pale filter for use when taking colour photographs to counter excessive blueness from skylight.

1955 G. R. SHARP tr. *Lorelle's Colour Bk. Photogr.* 77 A pale yellow haze or skylight filter may be desirable for shots of people or groups in the shade under a blue sky. 1965 *Focal Encycl. Photogr.* I. 615/2 General correction filters for colour include haze and skylight filters, which mainly absorb ultraviolet and a certain amount of blue light. 1978 *SLR Camera* Sept. 45/1 In addition to performing this very useful warming task, the Skylight filter also gets rid of a lot of UV and gives a mild degree of haze penetration.

skylighted *ppl. a.* (earlier example).

1849 DICKENS *Dav. Copp.* (1850) xxiii. 246 The skylighted offices of Spenlow and Jorkins.

sky-line. Add: **1. a.** (Earlier examples.) Also, the representation of this in painting or another art.

1824 SCOTT *St. Ronan's* I. iv. 84 Some boy's daubing, I suppose... Eh! What..is this?.. Who can this be?.. Do but see the sky-line—why, this is..an exquisite little bit. 1849 LYTTON *Caxtons* III. xiv. ii. 10 Seeing only the roof of that palace boldly breaking the sky-line, how serene your contemplations!

b. The outline or silhouette of a building or number of buildings or other objects seen against the sky.

1896 G. B. SHAW in *Sat. Rev.* 10 Oct. 386/2 A tall and beautiful figure, rising like a delicate spire above a skyline of city chimney-pots. 1928 *Daily Mail Year Bk.* 48/1 A traveller returning to the metropolis after some years' absence has difficulty in recognising some of our famous streets; the sky-line is different, salients have disappeared. 1932 *News Chron.* 5 July 9/5 The city's skyline of roofs was silhouetted against a blaze of gold. 1962 *Economist* 27 Oct. 347/1 The decision to free London from the sixty-year-old 100-foot ceiling which cramped ground space and deadened the skyline. 1971 P. GRESSWELL *Environment* 230 Consideration of skylines should be one of the prerequisites of planning.

2. *Forestry.* An overhead cable for the transport of logs.

1925 A. PHILIP *Crimson West* 144 Preparing the spar-tree for 'high-lead' or 'sky-line' rigging, is the most spectacular and thrilling performance in the logging industry. 1942 R. L. HAIG-BROWN *Timber* 254 In the case of a yarder the sky line is the same as the main-line, going from the spar tree through a block attached to a tree stump at the back of the setting. 1958 W. F. MCCULLOCH *Woods Words* 168 Perhaps the first skyline in the Northwest was rigged by Bob Barr at the Bridal Veil Lumber Company in Oregon in 1899. 1963 *Press* (Vancouver) Apr. 9 There's where the donkeys puff and strain As they pull the logs on the road And the skyline moans, as if in pain As it bears its heavy load.

Hence **sky-lined** *ppl. a.*, visible or silhouetted on the skyline.

1946 G. MILLAR *Horned Pigeon* ii. 33 We heard trucks behind us. When we dropped flat we saw them sky-lined. 1969 'J. FRASER' *Cock-pit of Roses* xvi. 126 Out past the hedge you'll be sky-lined, flashing like neon. 1978 L. HEREN *Growing up on The Times* v. 170 The bear was briefly skylined on another false crest, and I fired..and finally killed it.

skylit (skəi·lit), *ppl. a.* [Pa. pple. of SKYLIGHT *v.*] = SKYLIGHTED *ppl. a.*

1923 A. E. HOUSMAN *Shropshire Lad* xli. 62 And like a skylit water stood The bluebells in the azured wood. 1978 *Jrnl. R. Soc. Arts* CXXVI. 237/2 It occupies eight of the Royal Academy skylit galleries. 1979 *United States 1980/81* (Penguin Travel Guides) 191 Rooms full of medieval and Renaissance art,..ranged around a skylit, plant-filled courtyard.

Skylon, skylon (skəi·lǫn). *Archit.* [f. SKY *sb.*[1], prob. after PYLON.] The name of a spindle-shaped filigree spire, illuminated at night, forming a prominent feature of the South Bank exhibition at the Festival of Britain in 1951. Also used *transf.* of other similar structures.

1950 *Times* 11 Nov. 4/4 Mrs. A. G. S. Fidler, the wife of the chief architect to the Crawley Development Corporation, was informed yesterday that her suggestion of the name of 'Skylon' for the vertical feature of the Festival of Britain exhibition had been accepted. 1951 *John o' London's Weekly* 11 May 280/2 Skylon..denoting something streamlined and precariously poised. 1959 I. & P. OPIE *Lore & Lang. Schoolch.* ix. 169 The term Skylon..was found two years later [*sc.* after the Festival of Britain] to be a name for the lanky not only in London but as far away as Ruthin. 1961 *Daily Tel.* 9 Nov. 28/5 The centrepiece of decoration is, perhaps, a gold skylon on the road from the airport [in Accra]. 1966 *Guardian* 5 Nov. 7/5 There is an elegant skylon with a revolving restaurant on top. 1977 *Times* 9 Feb. 16/3 Dominating the exhibition will be a Skylon-type structure recalling the Festival of Britain.

skyphos (skəi·fǫs). *Gr. Antiq.* [Gr. σκύφος SCYPHUS.] A large drinking-cup or bowl, having two handles not extending above the rim, and no foot.

1858 S. BIRCH *Anc. Pottery* II. 103 The Heracleotan *scyphos* had its handle ornamented with the Heraclean knot. 1921 *Brit. Museum Return* 61 in *Parl. Papers 1921* XXVII. 555 Attic black-figure skyphos. 1942 J. D. BEAZLEY *Attic Red-Figure Vase-Painters* xii. 301 The vases..are by Makron, whose name appears on his masterpiece, the Boston skyphos. 1960 *Oxf. Univ. Gaz.* 4 Mar. 805/2 From Cyprus:..a painted skyphos imitating a class of pottery made at al Mina under East Greek influence (8th century B.C.). 1979 *Nature* 29 Feb. 643/2 Samples were taken from the base, sides and handles of an Attic skyphos which had been fired between 440 and 425 BC.

sky-rocket, *sb.* Add: **2.** *transf.* An enthusiastic cheer, raised esp. by college students; = *SIS-BOOM-BAH. *U.S. slang.*

1867 *Ball Players' Chron.* (N.Y.) 25 July 2/2 After cheers had been interchanged, and the Nationals had let off a 'sky rocket'—namely a sort of finish to three cheers, with a 'hiss—boom—ah!'—an adjournment was had to the clubhouse. 1894 [in Dict.] 1947 G. S. PERRY *Cities of Amer.* 222 He is such a stimulating lecturer that many of his classes are preceded by a 'skyrocket': a Wisconsin yell reserved for its heroes.

3. *Rhyming slang* for 'pocket'. Cf. *SKY *sb.*[1] 7*.

1879 *Macmillan's Mag.* Oct. 502/1 A slavey piped the spoons sticking out of my skyrocket (pocket), so I got smugged. 1898 [see *OUTER colloq. var. OUT OF *prep. phr.*]. 1936 J. CURTIS *Gilt Kid* ix. 98 There's no sense in hanging around in the West End with a cane and Christ knows what else in my sky-rocket. 1962 F. NORMAN *Guntz* i. 8 He put the letter in an envelope and..handed it to me. I took it and stuffed it in my sky rocket. 1973 'B. MATHER' *Snowline* xv. 180 Ten trouble-free runs.. and you're back in England with five thousand quid in your skyrocket.

sky-rocket, *v.* [f. the sb.] **1.** *trans.* In *Cricket*, = SKY *v.*[1] 1 b. *rare*[−1].

1851 W. CLARKE *Pract. Hints Cricket* in W. Bolland *Cricket Notes* 134 It's enough to make you bite your thumbs to see your best balls pulled and sky-rocketed about.

2. a. *intr.* To rise abruptly and rapidly; to increase dramatically (in number, amount, etc.). Cf. *ROCKET *v.* 2 c. Chiefly *U.S.*

1895 in *Funk's Stand. Dict.* 1923 *Nation* (N.Y.) 22 Aug. 181 The supply runs short and prices go skyrocketing. 1935 *Motion Picture* Nov. 40/2 Frances Dee..skyrockets to new importance with an amazingly fine performance. 1936 M. MITCHELL *Gone with Wind* xii. 217 Even the cheapest cotton goods had skyrocketed in price. 1943 J. STEINBECK in *N.Y. Herald Tribune* 29 Sept. 21/8 The incidence of GI dysentery skyrocketed. 1951 E. PAUL *Springtime in Paris* iv. 80 After World War I,.. the mark skyrocketed from 100 to the dollar to 3,000,000 or more. 1968 MRS. L. B. JOHNSON *White House Diary* 14 Mar. (1970) 638 The headlines ran:—'Mad Rush in Europe'..'Gold Buying Skyrockets'. 1974 'E. LATHEN' *Sweet & Low* xviii. 173 It had taken only twenty-four hours in the hinterland for his opinion of Milan to skyrocket.

b. To jump or fly up suddenly, in the manner of a sky-rocket. *rare.*

1907 G. B. SHAW *John Bull's Other Island* II. 30 Here! where are you jumpin to? Wheres your manners to go skyrocketin like that out o the box in the middle o your confession? 1946 J. W. DAY *Harvest Adventure* iv. 48 Partridges sky-rocketed and screwballed overhead and fled to safety.

2. a. *trans.* **b.** To destroy utterly; = EXPLODE *v.* 3. **b.** To cause to rise abruptly and rapidly: to propel rapidly forward or upward; to increase sharply.

1928 *Daily Express* 24 Nov. 3/5 A careful 'once-over' of some of London's most prominent public men has.. sky-rocketed the popular American idea that all Englishmen are snappy dressers! 1950 BLESH & JANIS *They all played Ragtime* v. 102 His 1896 success at Keith's and Tony Pastor's in New York skyrocketed him into the public eye. 1976 *National Observer* (U.S.) 31 Jan. 6/1 If I had a disease that skyrocketed my chances of dying early.., would I take the drugs that would control my disease.

So **sky-rocketing** *vbl. sb.* and *ppl. a.* Cf. *ROCKETING *ppl. a.*

1849 Poe *Marginalia* in *Compl. Wks.* (1902) XVI. 166 The German 'Schwärmerei'—not exactly 'hum-bug', but 'sky-rocketing'—..that peculiar style of criticism which has lately come into fashion, through the influence of certain..people who live..about Boston. **1933** *Sun* (Baltimore) 18 Aug. 16/5 Skyrocketing of all grain values immediately ensued. **1962** [see *EXPLOSION 4 b]. **1979** *Vole* 8 Nov. 26/1 Land-prices..are a key element in the sky-rocketing costs of both housing and food.

sky-scraper. Add: **2. a.** [A horse named Skyscraper, sired by Highflyer, won the Epsom Derby in 1789:

1788 *Racing Calendar* 269 Mr. Dutton named the D. of Bedford's c. Skyscraper, by Highflyer. **1810** T. H. Morland *Geneal. English Race Horse* 147 Skyscraper mare produced Brainworm by Buzzard. *Ibid.* 160 [Death] Skyscraper, 1807.]

† **d.** A tall hat or bonnet. *Obs.*

1800 W. Scott *Let.* 5 Apr. (1937) XII. 159 The trumpets call me to swagger in a cockd skyscraper and sword. **1847** J. A. Eames *Budget of Lett.* 397 She gave me a black silk bonnet..which stuck right up in the air after the fashion of the old 'sky scrapers'.

e. In *Baseball, Cricket*, etc., a ball propelled high in the air; a towering hit, a skyer.

1866 *N.Y. Herald* 27 June 5/5 Goodspeed made three handsome fly catches; Mehl, Sweet and Dupignac each paying their share of attention to the 'skyscrapers'. **1907** *St. Nicholas* (N.Y.) Sept. 996 A 'skyscraper' throw to first. **1943** *Amer. Speech* XVIII. 104 Fly balls include the *skyscraper*, the *cloud-buster*, [etc.]. **1963** *Times* 28 Feb. 3/6 Alabaster's skyscraper to Titmus at midwicket demonstrated only the extraordinary sureness of Titmus in the field.

4. (Earlier and later examples.)

[**1883** J. Moser in *Amer. Architect & Building News* 30 June 305 The capitol building should always have a dome. I should raise thereon a gigantic 'sky-scraper', contrary to all precedent in practice.] **1888** *Inter-Ocean* 30 Dec. 10/5 The 'sky-scrapers' of Chicago outrival anything of their kind in the world. **1903** O. Kildare *My Mamie Rose* xix. 288 We reach our stoop in the yawning dark cañon of the skyscrapers. **1928** W. A. Starrett *Skyscrapers & Men who build Them* i. 1 The skyscraper is the most distinctively American thing in the world. **1942** *Short Guide Gt. Brit.* (U.S. War Dept.) 7 London has no skyscrapers. **1951** *Manch. Guardian Weekly* 19 Apr. 5 Theatres will have skyscrapers superimposed on them. **1976** *Sunday Mail* (Glasgow) 28 Nov. 20/2 Babs Marchant ..lives 18 storeys up in an Ibrox, Glasgow, skyscraper.

Hence **sky-scrapered** *a.*, characterized by the presence of or full of sky-scrapers; surrounded by sky-scrapers; built very tall.

1947 *Ann. Reg. 1946* 212 The new home [for the U.N.] would be sky-scrapered, congested and expensive. **1963** *Harper's Bazaar* Jan. 21/2 Cagliari..is now a busy sky-scrapered seaport. **1963** C. L. Cooper *Black!* x. 151 The skyscrapered trillion-bricked dwellings. **1965** *Guardian* 4 Oct. 9/4 Salisbury is brittle, skyscrapered, centralised, and carefully zoned into European and African residential areas.

sky-scraping, *a.* (Further examples.)

1884 *Chicago Daily Tribune* 9 Mar. 19/1 The sky-scraping buildings that have been and will be erected down-town will never endanger human life, because they will only be occupied by people wide awake. **1904** *N.Y. Even. Post* 12 Mar. 10 (*heading*) Sky-scraping prices at last Brandus picture sale. **1905** Kipling *Actions & Reactions* (1909) 123 What under the stars are you doing here, you sky-scraping chimney-sweep? **1961** P. Fleming *Bayonets to Lhasa* xviii. 230 The force camped..less than a mile from the sky-scraping Potala. **1973** *Washington Post* 13 Jan. C1/1 Man's most wondrous, skyscraping community, the 110-story Brobdingnagian shafts stand ..at the western edge of Manhattan island.

sky-sign. Add: **1.** (Later example.)

1940 C. Day Lewis tr. *Virgil's Georgics* I. 23 Well for us that we watch the rise and fall of the sky-signs And the four different seasons that divide the year equally.

2. a. (Later examples.) Also *spec.* an electrically illuminated sign or message similarly placed.

1903 *Building News* 10 Apr. 510/2 Mr Caudwell..contended..that a sky sign was any sign which could be seen against the sky by a person standing beneath. **1916** A. Bennett *Lion's Share* ix. 64 They crossed a thoroughfare that twinkled and glittered from end to end with moving sky-signs. **1926** *Socialist Rev.* June 13 In Piccadilly a large crowd—mostly strikers—watched a *Daily Dispatch* news bulletin thrown up on a huge sky-sign. **1938** F. Brett Young *Dr. Bradley Remembers* iii. 121 He was to see the dim, discreet streets of the city centre..garish with neon lights and sky-signs. **1954** *Times* 17 Dec. 9/4 London County Council gives notice of a Bill..'To repeal the provisions of the London Building Acts, 1930 to 1939, prohibiting the erection of sky signs.' **1966** *Punch* 16 Mar. 368/2 There's to be one of those electric sky-sign newscasters in Piccadilly Circus at last. But remember how easy it is to get hold of the wrong end of those revolving messages.

b. An advertisement or other message in sky-writing.

1922 *Flight* XIV. 330/2 This week has witnessed the first practical application of the invention of Major J. C. Savage, by means of which chemical smoke trailing from an aeroplane can be used for writing words or tracing figures in the sky... His first 'sky sign' took the form of the word 'Castrol'. **1931** C. Day Lewis *From Feathers to Iron* 53 Now shall the airman vertically banking Out of the blue write a new sky-sign.

skytrain (skəi·trēⁱn). Also with hyphen and as two words. [Sky *sb.*[1]] **1.** *U.S.* A convoy consisting of a number of gliders towed in line by an aeroplane, and used for the transport of freight, etc. and in *Mil.* applications. (*temporary.*)

1934 *Sun* (Baltimore) 4 Aug. 4/2 Glider lands from airmail 'train' in pioneer flight... Lustig sky trains. **1935** *Evening Sun* (Baltimore) 14 May 1/7 The sky-train..—two gliders towed by an airplane—took off at Miami on its one-stop flight under perfect weather conditions. **1944** C. Milburn *Diary* 18 June (1979) 221 Wounded men..are evacuated from Normandy by 'sky trains', which carry a flight nurse and two medical officers. **1946** A. Lee *German Air Force* xii. 165 The fighter units in Western Germany..carried out only one or two thrusts against the long sky train of tempting Dakotas and their gliders.

2. Skytrain. Also **skytrain.** The name of a privately-owned passenger air service seeking to provide regular, low-cost, transatlantic travel facilities. (No longer in operation.)

A proprietary name in the U.S.

1971 *Daily Tel.* 20 Oct. 2/6 Laker Airways applied yesterday..for permission to operate a 'sky train' service to New York for a winter single fare of £32.50 per passenger. **1972** *Ibid.* 27 Sept. 1 Mr Freddy Laker..said last night: 'Skytrain is a walk on, walk off, no frills, no reservation service similar to a train service.' **1977** *Arab Times* 14 Dec. 6/2 Freddie Laker, creator of the cut-rate London-New York Skytrain air service, said Monday his airline has applied to operate a similar Skytrain service between London and Los Angeles. **1978** *Official Gaz.* (U.S. Patent Office) 23 May 345/2 Laker Airways Limited ..*Skytrain*..For air passenger services—namely, the transportation of passengers by air.

skyve, -er, -ing, varr. *SKIVE *v.*[3], *SKIVER *sb.*[3], *SKIVING *vbl. sb.*[2]

skyward, *a.* (Earlier and later examples.)

1831 W. Howitt *Bk. of Seasons* 325 The sky-ward and inaccessible pinnacles. **1935** C. Day Lewis *Time to Dance* 22 Speak up, speak up, you skyward man, Speak up and tell us true.

skyway (skəi·wēⁱ). Chiefly *U.S.* [Sky *sb.*[1]]

1. The sky as a medium of transport or a route used by aircraft; = *AIRWAY 2.

1919 *Pearson's Mag.* Dec. 547/1 The letters will tell.. the nationality of the ship passing along the world's skyways. **1928** L. Thomas (*title*) European skyways. The story of a tour of Europe by aeroplane. **1934** *Sun* (Baltimore) 15 May 1/3 Theirs is the great-circle route over the North Atlantic—the skyway emblazoned by Byrd, Acosta, [etc.]. **1952** *N.Y. Times* 7 Aug. 39/8 The team of precision flyers will fly the skyway route on the way to their home base at Corpus Christi, Tex. **1967** M. Craven *I heard Owl call my Name* xi. 68 High in the skyways wild geese called exultantly on their first early passages back from the south.

2. An overhead motorway. Also *transf.* (see quot. 1960).

1940 *Sun* (Baltimore) 6 Sept. 13/4 It has been suggested that a skyway be built along Pratt street from Monroe to the Fallsway, the western end to connect with the route to Washington. **1958** *Wall St. Jrnl.* 10 Nov. 18/3 A reduction in toll rates charged on the Sunshine Skyway ..is expected to be approved by the State Road Department Board. **1960** *Daily Tel.* 10 Feb. 20/6 The system, known as Skyway, runs on a concrete way 15 ft above ground. It has a speed of 50 m.p.h. and it is estimated operates for as little as a penny a mile. **1972** J. Gores *Dead Skip* (1973) i. 7 Heslip unlocked the chain-link storage lot under the concrete abutments of the skyway adjacent to the DICA office.

3. Chiefly in Minneapolis: (see quot. 1975[1]); an aerial walkway between buildings. *U.S.*

[**1968** *Archit. Forum* Jan.–Feb. 83/2 The first pedestrian bridge of an elaborate second-story network has been completed in downtown St. Paul. Part of the 12-block urban renewal project, the 'Skyway' will eventually comprise an elevated system of bridges and concourses.] **1969** *Ibid.* Jan.–Feb. 75/3 The overpass idea, now called a 'Skyway', is being carried out several blocks to the east. **1970** *Business Week* 26 Dec. 48/2 Shelter from summer heat and winter weather is one advantage of skyways—especially in Minneapolis, where it is below or near freezing for five months of the year. **1975** *Country Life* 2 Jan. 36/3 There has been a great incentive to construct the Skyway system... Skyways are broad, carpeted, heated, air-conditioned, glass-covered corridors, linking the main city buildings at first-floor level. **1975** *New Yorker* 24 Mar. 106/2 Walking there is a pleasure—down the carless mall, or through a glazed 'skyway' system if snow starts to fall. **1979** *United States 1980/81* (Penguin Travel Guides) 266 The most extraordinary feature of downtown Minneapolis is its internal skyway, an interconnected belt of pedestrian malls and escalators.

sky-writing, *vbl. sb.* [Sky *sb.*[1]] The tracing of legible signs in the sky, esp. for advertising purposes, by means of smoke trails made by aircraft or (occas.) by letters and devices projected by searchlight. Also, the writing so produced.

1923 *Western Gaz.* 16 May 8/4 Sky-writing..did not commend itself to the general body of advertisers. **1926** H. Barr in *Mod. Advertising* I. 9/1 Invented in 1910 by John Clifford Savage, it was not until twelve years later that sky-writing became an accomplished fact. *Ibid.* 9/2 The corporation..is stated to have spent £320,000 on sky-writing publicity alone. **1932** *Flight* 8 July 638/1 Night sky-writing, by projection of light on to clouds at night, was considered at no little length. **1935** H. G. Wells *Things to Come* ix. 91 Sky writing by the new planes. **1959** *Belfast Tel.* 9 Oct. 7/1 It is believed to be the first time skywriting has been used in a general election. **1966** *Punch* 13 July 54/1 Long ago You fancied many wonders that we know, The 'Telephonoscope', for one, 'Sky Writing', and the Long Range Gun. **1978** F. Ross *Sleeping Dogs* 157 He couldn't have made his intentions universally clearer if he'd spelled them out in sky-writing.

So as *ppl. a.* and hence [as back-formation] **sky-write** *v.* (*a*) *intr.*, to practise sky-writing; (*b*) *trans.*, to trace in the sky by means of smoke trails; **sky-writer**; **sky-written** *ppl. a.*

1922 *Daily Mail* 8 Aug. 5/3 The Daily Mail sky-writing aeroplane was over Margate and Folkestone yesterday. **1926** *Mod. Advertising* I. 9/2, 1,450 sky-written demonstrations were carried out by eight pilots. **1927** *Pictorial Weekly* 17 Sept. 198 A window sign which reproduces in miniature the work of the sky-writer. **1932** *Times Educ. Suppl.* 22 Oct. p. iv/3 A battery of sky-writing projectors will be installed on the roofs of buildings. **1934** Webster, Skywrite. **1941** B. Schulberg *What makes Sammy Run?* xii. 219 Feeling his words fade off into the air like a sky-writer's. **1959** J. Cary *Captive & Free* xlvii. 203 She was always calling him on the telephone with new ideas—aeroplanes sky-writing the message..; pavement-writing; ..and loudspeaker vans. **1960** *Guardian* 29 Dec. 3/2 Sky-written slogans. **1973** J. Drummond *Bang! Bang!* i. 2 The plane swooped... The smoke spurted... People wondered if it was possible for a sky-writer to dot his 'i's. *Ibid.* 3 They've been sky-writing to advertise their shows.

slab, *sb.*[1] Add: **1. a.** (Earlier and later *transf.* and further *techn.* examples.)

1881 *Encycl. Brit.* XII. 839/1 The rubber is glossy, of a bright pink colour and mottled appearance, and occurs in the form either of small balls pressed together or of irregular masses called 'slabs or 'loaf' rubber. **1903** *Imperial Inst. Techn. Rep.* 153 The 'slabs' of blackish rubber alone being worth 1*s.* 11*d.* per pound. **1964** *Amer. Speech* XXXIX. 274 *Slab*, a solid piece of rubber used as an ingredient to be melted and mixed with solvents to form rubber cements or to be milled and stripped off. *transf.* **1882** *Harper's Mag.* July 32/2 From one of our exchanges..we chip off the following slab of scientific knowledge. **1951** Partridge *Dict. Slang* (ed. 4) Add. 1172/2 *Slab*..a long paragraph. **1958** *Listener* 14 Aug. 249/3 The conventional slabs of Brahms, Beethoven, and Tchaikovsky. *a* **1974** R. Crossman *Diaries* (1976) II. 623 When I started, I used to dictate slabs without any real preparation.

b. *spec.* in *Metallurgy*, such a piece of metal produced from an ingot for subsequent rolling into sheet or plate.

1863 [in Dict.]. **1910** [see *BILLET *sb.*[2] 9]. **1931** *Economist* 21 Mar. 608/1 The biggest decline occurring in billets, blooms and slabs. **1968** D. R. Cliffe *Technical Metallurgy* iv. 70 Ingots are broken down into blooms or slabs, as a hot working process, in a cogging mill. **1972** *Times* 18 Sept. 21/6 Production of slab zinc was 35,000 tons.

c. = *slab-cake*, sense 6 below.

1908 J. Kirkland *Mod. Baker* III. 462 Rice Slab at 6d. per lb... Lemon Madeira at 3d. per lb... Fruit Slab at 3d. per lb. **1948** *Good Housek. Cookery Bk.* 580 Cut the Genoese slab into 2-inch squares. **1974** W. Foley *Child in Forest* I. iii. 39 Plain slab was a delicate luxury and this was no plain slab!

d. *Archit.* A rectangular block of pre-cast, reinforced concrete used in building, esp. in multi-storey constructions.

1927 *Archit. Rec.* Dec. 452/2 A number of experimental houses have been built with 'textile-block slab construction'... The system consists of concrete block slabs about two or three inches thick of unit sizes which can be handled, laid on end with interlocking grooves, reinforced horizontally and vertically by means of steel rods. **1930** *Amer. Architect* Apr. 32/2 A type of floor and ceiling construction light in weight, quickly erected..is being used... The system consists of two types of slabs, one for floors and another for ceilings, used in conjunction with the ordinary supporting members of steel construction. **1938** *Archit. Rev.* LXXXIII. 223 (*caption*) One of the two 'porte-cochères', in reinforced concrete column and slab construction. **1951** *Ibid.* CX. 92 The café terrace, which disappears beneath its lightly supported slab roof to become a two-level café-bar. **1973** D. Francis *Slay-Ride* x. 122 A modern square-built glass and slab affair a mile out of the city centre.

e. *Archit.* A high-rise block of impersonal aspect.

[**1933** L. Mumford in *New Yorker* 23 Dec. 29/2 What does one find? First, a gigantic slab of a building.] **1952** *Archit. Rev.* CXI. 119/1 As is well known, the term 'slab' was coined in the 1930's in connection with publicity on the RCA Building at Rockefeller Center. **1958** *Listener* 20 Nov. 827/1 A point block of government offices is now going up at Wellington..and other high-rise slabs for offices and flats. **1969** Burchard & Bush-Brown *Archit. of Amer.* iv. 353 The early skyscrapers were massive blocks... The new characteristic form became the slab, a term applied to the buildings erected at the Rockefeller Center beginning about 1930. The slab form had appeared briefly in the early history of the skyscraper, notably in the Monadnock Building... It remained for the architects of Rockefeller Center..to modernize the slab, to make it thinner in relation to its height, to simplify it and to treat it with characteristic but underemphasized setbacks.

2. b. (Earlier and later examples.) Also *N.Z.*

1829 H. Widowson *Present State Van Diemen's Land* 86 Logs, or as they are more commonly called, slabs, for erecting barns or small buildings are erected in the same manner. **1905** W.B. *Where White Man Treads* 259 It is a low whare of split slabs, adzed over, and sunk into the earth as closely as the inequalities of adze-jointing will

permit. **1950** *N.Z. Jrnl. Agric.* Apr. 375/2 Floors [in very early milking sheds] were of wooden slabs, bricks, stones, or even clay. **1957** P. White *Voss* vi. 154 She was standing in front of a house, or hut, of bleached slabs, that melted into the live trunks of the surrounding trees.

3. d. A flat piece of stone, etc., immediately in front of a fire-place; a stone hearth.

1876 *Encycl. Brit.* IV. 466/2 The slab is that part of the floor of a room which is immediately before the fire-place, and along the extent of its front. In basement rooms, this slab is supported by a brick wall brought up from the ground; but in upper rooms the slab is supported by a flat half brick arch called a brick trimmer. **1883** R. L. Stevenson *Treasure Is.* xix. 153 Little had been left beside the framework of the house; but in one corner there was a stone slab laid down by way of hearth, and an old rusty iron basket to contain the fire. **1963** B. Goodson *Pract. Guide to House Repairs* iv. 50 The tiled hearth slab is bedded in about ½ in. of mortar on the existing hearth. **1977** J. S. Curl *Eng. Archit.* 154/1 *Slab*..the hearth of a fireplace.

e. The stone on which a corpse is laid in a mortuary. Also *transf.* and *fig.*

1903 A. H. Lewis *Boss* viii. 101 I've seen a bloke take a slab in th' morgue for less. It was Benny the Bite; he gets a knife between his slats. **1924** G. C. Henderson *Keys to Crookdom* 417 Slab... Undertaker's table. **1930** H. C. Bailey *Mr. Fortune Explains* 111 On a slab in the mortuary the woman's body lay and the divisional surgeon turned from it to nod at Reggie. **1932** E. Wallace *When Gangs came to London* ii. 26 My best friend is a forty-five..and the day he puts you on the slab I'm going to put diamonds all round his muzzle. **1977** 'C. Aird' *Parting Breath* x. 127 Pathologists had hobby-horses, too, and obesity was..Dr Dabbe's. He was always having a go at Sergeant Gelven..about his weight. 'See you soon,' was his favourite form of greeting to the portly detective, 'on my slab.'

4. a. (sense 2 b) *slab-and-bark house, hut*; *slab-and-shingle hut*; *slab house* (earlier and later examples), *hut* (later examples), *whare* [Whare].

1862 R. Henning *Let.* 18 Oct. (1966) 111 It is not much to move a slab house; all the woodwork takes down and puts up again. **1901** M. Franklin *My Brilliant Career* i. 5 Our comfortable, wide-veranda'ed..slab house..was ever full to overflowing. **1905** W.B. *Where White Man Treads* 293 He..who lives in a slab whare, and on a fare which his dainty collie sniffs at and rejects. **1908** E. J. Banfield *Confessions of Beachcomber* I. i. 12 According to the formula neatly printed in official journals, the building of a slab hut is absurdly easy. **1933** *Bulletin* (Sydney) 25 Jan. 20 A good three-roomed slab-and-shingle hut that had been vacated by a white family. **1945** *Salt* 12 Feb. 4/1 His earliest years were spent in a slab-and-bark hut. **1949** F. Sargeson *I saw it in my Dream* 100 A slab *whare* in a narrow valley. **1959** J. Wright *Generations of Men* 53 A slab-and-bark house in lonely fever-ridden country. **1969** F. Sargeson *Joy of Worm* i. 7 It then became pleasant to look forward to hot food, the shelter of the farmer's slab hut, and talk with the man himself.

b. (sense 2 a) *slabwood.*

1877 [in Dict.]. **1921** *Daily Colonist* (Victoria, B.C.) 2 Oct. 10/1 (Advt.), Cordwood, Slabwood, Blockwood, Dry Kindling. **1962** M. E. Murie *Two in Far North* I. ii. 20 Off at one side, a lean-to bedroom built of slab wood. **1976** *Newmarket Jrnl.* 16 Dec., (Advt.), Hardwood and softwood, slabwood and off-cuts for sale.

5. a. *slab-bridged* (earlier *fig.* example), *-walled*. Cf. Slab-sided *a.* in Dict. and Suppl.

1859 *Atlantic Monthly* Nov. 642/1 Anyone who has driven over a mountain-stream by one of those bridges made of *slabs* will feel the force of a term we once heard applied to a parson so shaky in character that no dependence could be placed on him,—'A slab-bridged kind o' feller!' **1930** L. G. D. Acland *Early Canterbury Runs* 1st Ser. viii. 213 The slab-walled, earthen-floored hut.

c. *slab-like* adj.

1899 W. James *Talks to Teachers* 214 The even forehead, the slab-like cheek, the codfish eye, may be less interesting for the moment. **1970** R. J. Small *Study of Landforms* iv. 122 On the high valley slopes above Glen Rosa great slab-like outcrops of granite, tilted at between 30° and 60°, are developed where glacial erosion and frost weathering have exposed dilatation joints.

6. Special combinations: **slab avalanche,** an avalanche in which a sheet of snow breaks cleanly away along a fracture line; **slab bacon,** unsliced bacon; **slab-cake,** (a) cake baked in a large rectangular tin (cf. sense 1 c above).

[**1920** A. Lunn in G. W. Young *Mountain Craft* ix. 431 The wind-slab is the most treacherous of all avalanches. **1936** G. Seligman *Snow Structure & Ski Fields* vii. 160 If this 'slab' of snow has formed on a steep slope it will shatter into countless blocks of hardened snow, and these, sliding downhill, will precipitate the most insidious of all avalanches—the wind-slab avalanche.] **1953** *Avalanche Handbk.* (U.S. Dept. Agric. Forest Service) iv. 34 (*caption*) Major slab-avalanche. Depth of fracture: 10 feet. **1978** C. Fraser *Avalanches & Snow Safety* v. 79 Criterion 1 is the form of break which started the avalanche and this leads to the broad division of all avalanches into two types: 'loose-snow avalanches'..and 'slab avalanches'. **1932** *Even. Sun* (Baltimore) 7 Nov. 12/7 (Advt.), Boneless Slab Bacon lb 13½c. **1975** L. & S. Lobel *All about Meat* vii. 115 Take the trouble to hunt for unsliced slab bacon. **1902** J. T. Law *Grocer's Man.* (ed. 2) 854/1 Slab Cakes, or Cut Cakes (sold by the pound).—The introduction of these ready-made cakes, as an extension of the birthday, wedding, and Christmas cake system, appears to be displacing or supplanting much of the old-fashioned retail business in currants, raisins,..spices, etc. **1935** *Economist* 22 June 1439/1 Scribbans and Company, the well known makers of slab cake, find their activities in fields where steady progress..is the natural order of things. **1974** W. Foley *Child in Forest* I. v. 57 A slice of bright slab cake.

slab, *v.*[3] Add: **5.** *trans.* Of a path, climber, etc.: to traverse (the side of a slope) horizontally or at a gentle angle. *U.S.*

1889 Farmer *Americanisms* 492/2 *To slab*, to make roads round the sides of mountains. **1892** *Outing* Jan. 268/1 So we started blindly up the bank and into the forest, continuing for an hour and a half to 'slab' the mountain, as the backwoodsmen say. **1907** *Guide Paths & Camps White Mountains* (Appalachian Mountain Club) 62 The path now slabs the east side. **1916** *Ibid.* (ed. 2) 265 The path..rises by easy zigzags slabbing the S.W. flank of Eagle Cliff. **1963** *Appalachian Trailway News* Sept. 43/2 We zigzagged and slabbed mountains, finally coming..down a beautiful grassy glade where stood Big Stamp Shelter. **1968** *Ibid.* Sept. 43/2 Route slabs northwestern slope of ridge.

slabber, *sb.*[2] Add: **c.** A workman who cuts or forms materials into slabs.

An early form of the word may occur in the proper name *Ric. Sclaber* (1327), *Ric. le Sclabber* (1333): see *N. & Q.* (1963) July 256/1.

1921 *Dict. Occup. Terms* (1927) § 118 *Slabber*, cements together tiles and other shaped articles for hearths, kerbs, etc. *Ibid.* § 159 *Slabber*,..cuts blocks of gelatine..into slabs. **1977** *New Society* 25 Aug. 387/1 His heavy manual work as a fireplace 'tile slabber'.

slabby (slæ·bi), *sb. N.Z. colloq.* [f. Slab *sb.*[1] 2 a.] A timber worker dealing with slabs of timber.

1907 'G. B. Lancaster' *Tracks We Tread* vi. 87 The.. clumsiest slabby that lumped in the mill. **1916** G. Thornton *Wowser* v. 72 Barnabas the slabby (the man who wheels away the unused portions of timber).

slab-sided, *a.* For *U.S.* read orig. *U.S.* and add earlier and later examples. Also in *Archit.*: see *SLAB *sb.*[1] 1 d.

1817 J. K. Paulding *Lett. from South* II. 122 He was what is usually called a tall slab-sided Virginian. **1927** A. Conan Doyle *Case-Bk. Sherlock Holmes* iii. 89 The prize-fighter, a heavily built young man with a stupid, obstinate, slab-sided face. **1933** *Flight* 23 Mar. 268 The fuselage is not quite 'slab-sided'. **1964** L. Deighton *Funeral in Berlin* v. 32 The large slab-sided department store. **1981** E. Corlett *Revolution in Merchant Shipping* 35/1 High slab-sided deckhouses.

slack, *sb.*[1] Add: **2.** (Later examples.) Also, a depression among sand-dunes.

1929 [see *LOW *sb.* 3 a]. **1934** *Geogr. Jrnl.* LXXXIII. 498 The question that always comes to my mind in looking at dune formations is what is the primary cause of the rhythmical or ripple effect, the succession of ridges and slacks. **1963** *Times* 27 Feb. 11/6 In the lee of the high dunes lie wet slacks and attractive freshwater pools out of which grow strands of reed and reed-mace. **1964** V. J. Chapman *Coastal Vegetation* vi. 153 The damp soil of the slacks is colonized by a carpet of the Creeping Willow.

slack, *sb.*[3] Add: **3. c.** In critical path analysis, the length of time by which a particular event can be delayed without delaying the completion of the overall objective.

1962 *NASA PERT & Compan. Cost System Handbk.* (U.S. Nat. Aeronaut. & Space Admin.) B-3 The accomplishment of event #3 could be delayed by three weeks without jeopardizing meeting the expected date for the end objective. This difference or cushion is called slack. **1964** K. G. Lockyer *Introd. Critical Path Anal.* v. 46 A different expression of the ability of activities to move is given by considering the head and tail events. These have 'earliest' and 'latest' times, and slack is the difference between these times. **1970** O. Dopping *Computers & Data Processing* xxii. 346 After the critical path has been found, it may be possible to transfer resources from activities with a big slack to certain critical activities.

4. a. Also *fig.*, esp. in phr. *to take up the slack*, to use up a surplus or make up a deficiency, thereby maintaining or returning to a stable condition.

1915 J. London *Jacket* viii. 63 Jones was forcing his foot into my back in order to cinch me tighter, while I was trying with my muscle to steal slack. **1930** *Economist* 21 June 1391/2 There is general agreement that the termination of the Stevenson Scheme left the industry with much more 'slack' to take up than was realised two years ago. **1933** *Sun* (Baltimore) 12 Sept. 1/6 The American Federation of Labor..was framing demands for further..wage-boosting to take up employment slack. **1957** *Economist* 16 Nov. 610/2 Sir Alexander Fleck's investigations may show just how much slack has crept into an organization that should be, if anything, over-cautious. **1967** *Times* 28 Feb. (Canada Suppl.) 33 There is the feeling that slack will develop as the year progresses. **1972** A. MacVicar *Golden Venus Affair* vi. 61 It was a comfort having somebody like Mary Jo to take up the slack of decision-making. **1980** B. Paul *First Gravedigger* v. 60 We'd no longer be handling his speciality... Our new rare books department in London would take up the slack.

c. Phr. *to give* (or *cut*) (a person) *some slack*, to show (a person) understanding or restraint, to give (one) a chance. *U.S. slang* (chiefly *Blacks'*).

1968 M. F. Jackmon in Jones & Neal *Black Fire* (1969) 555 Say, baby, light'n up on me—gimme some slack. **1969** H. Rap Brown *Die Nigger Die* ii. 29 Now, if the brother couldn't come back behind that, I usually cut him some slack. **1971** *Current Slang* (Univ. S. Dakota) VI. 3 *Cut me some slack!*, to give one a chance (imperative). **1973** *Black World* May 39/1 Tradesmen give them no slack in the unfamiliar bargaining processes.

5. *pl.* Now *spec.* loosely-cut trousers for informal wear, esp. those worn by women.

1932 D. L. Sayers *Have his Carcase* ii. 31 He wore a pair of old flannel slacks, and a khaki shirt. **1937** *Night & Day* 29 July 22/2 Deeply to be deplored are such things as sandals..slacks and sun-top dresses. **1942** A. Christie *Body in Library* xii. 139 She was wearing grey slacks and an emerald jumper. **1947** W. S. Maugham *Creatures of Circumstance* 303 He changed from his business clothes into slacks and an old coat. **1956** A. H. Compton *Atomic Quest* i. 32 Dressed casually in slacks and a sweater, he invited me cordially into his study. **1960** J. Betjeman *High & Low* 56 The debs may turn disdainful backs On Pearl's uncouth mechanic slacks. **1968** *Listener* 10 Dec. 790/3 In Jordan, girls at Amman University have been instructed not to wear..slacks either and, moreover, to keep off heavy make-up. **1979** R. Jaffe *Class Reunion* (1980) I. i. 27 Nor could you wear slacks or any other sort of pants to class, even in the snow.

6. (Earlier example.)

1825 J. Neal *Bro. Jonathan* I. 156 'None o' your slack,' says I..'none o' your pokin' fun at me.'

7. *Pros.* A syllable or part of a foot which does not receive stress.

c **1883** G. M. Hopkins *Poems* (1967) 45 Every foot has one principal stress or accent, and this or the syllable it falls on may be called the Stress of the foot and the other part, the one or two unaccented syllables, the Slack. **1970** J. Malof *Man. Eng. Meters* i. 2 In the freer varieties of accentual verse, meter is determined simply by counting the number of stresses in the line.., ignoring the relatively unemphatic or unstressed syllables, which we call slacks. **1973** *Word 1970* XXVI. 56 None or as many as six slacks may appear between such isochronous accents though one, two, or three slacks are more normal.

8. A street-walker or prostitute. *slang.*

1959 *Encounter* May 24 *Slack*, which is the call-girls' word for a street-girl. **1963** *Observer* 29 Sept. 31/4 A young master was asked by a boy..'Can a slack (prostitute) work hard enough to earn a living?' **1965** W. Young *Eros Denied* xiv. 141 The slack is afraid of disease, and afraid of the sex maniac who thinks it'd be fun to strangle her.

9. *Comb.* **slack suit,** a pair of slacks with a matching jacket, as a fashionable garment for women; **slack variable** *Math.*, a variable which expresses the difference between the two sides of an inequality.

1940 R. Chandler *Farewell, my Lovely* xxvii. 203 Miss Anne Riordan stood there, in a pale green slack suit. **1973** H. Nielsen *Severed Key* viii. 91 She had..changed into a lime-green knitted slack-suit. **1953** Cooper & Henderson *Introd. Linear Programming* I. ii. 6 These values λ_i (i = 10,11,.. 16) so introduced may be referred to as slack variables... The requirement that the slack variables be non-negative merely extends the range of the subscript. **1974** Adby & Dempster *Introd. Optimization Methods* v. 156 At least some evidence exists which suggests that the use of slack variables is an effective method for handling inequality constraints, both linear and non-linear.

slack, *a.* and *adv.* Add: **A.** *adj.* **III. 7. e.** *Phonetics.* Of a vowel: = *LAX *a.* 5 c.

1909 H. C. Wyld *Elem. Lessons Eng. Gram.* ii. 28 Vowels formed with the tongue tense we call Tense Vowels, those with the tongue soft we call Slack Vowels. **1934** C. Davies *Eng. Pronunc. from 15th to 18th Cent.* 8 It was probably a slack, round mid-back, vowel. **1970** B. M. H. Strang *Hist. Eng.* 285 The letter *æ* represents a long, low, slack front vowel /æ:/ in *dælan*.

10. *slack key* (Mus.) [tr. Hawaiian *kī hōʻalu* f. *kī* key + *hōʻalu* slack], used *absol.* and *attrib.*, esp. as *slack-key guitar*, with reference to a style of guitar-playing originating in Hawaii, in which the strings are slightly relaxed to produce strong bass resonances; *slack party* (Naut slang) (see quots.); *slack wire* (earlier and later examples).

1975 G. S. Kanahele in *Haʻilono Mele* Jan. 2/2 Our first concert [in 1972] was memorable. It featured the slack key guitar, the first time that an entire concert was devoted to this unique style of playing. **1976** *Guitar Player* Apr. 14/2 The original style is kept alive solely by those guitarists who insist on playing only slack key. **1977** *Zigzag* Mar. 20/1 Could you explain about slack-key guitar? **1933** J. Masefield *Conway* iv. 145 For official punishments there was an institution known as 'slack party', which meant employment upon every available job..from morning till night. **1945** 'Tackline' *Holiday Sailor* 133 Jimmy's pet form of punishment was his 'slack party'. Hands tardy in going on watch or performing some allotted task with lack of zeal were enrolled in the slack party. And the slack party did not lead an especially restful existence. **1753** *N.-Y. Mercury* 20 Aug. 3/3 The Surprizing Performances of the celebrated Anthony Joseph Dugee..On a Slack Wire scarcely perceptible and without a Balance. **1866** M. Mackintosh *Stage Reminisc.* xi. 138 Andrew was at once a good tight-rope dancer and slack-wire vaulter. **1977** E. Ambler *Send no more Roses* viii. 183 A slack-wire baggy-pants act out of a third-rate circus.

IV. 11. a. *slack-jawed, -mouthed.*

1901 Kipling *Kim* iii. 76 Our colonel used to send for slack-jawed down-country men who talked too much. **1936** W. Faulkner *Absalom, Absalom!* 44 Wild-eyed and considerably slack-mouthed. **1942** —— *Go down, Moses* 32 He stood for perhaps ten seconds, slackjawed with amazed and incredulous comprehension. **1976** J. Carroll *Mortal Friends* II. iii. 170 The people were Catholics nearly to a person, and they stared slack-jawed at the line of nuns. **1976** P. Cave *High Flying Birds* iii. 42, I continued to gaze at Sonya with slack-mouthed adoration.

B. *adv.* **b.** *slack-tethered.*

1922 Joyce *Ulysses* 98 On the towpath by the lock a slacktethered horse.

slackening, *vbl. sb.* Add: Also with *off.*

1903 Conrad *Romance* v. i. 402, I wanted rest, woman's love, slackening off. **1951** J. M. Fraser *Psychol.* xiv. 157 It is rather more usual for a slackening-off process to take place.

slacker. Add: **2.** (Later examples.) Also, *spec.* used in *Mil.* contexts in the war of 1914–18.

1914 'Bartimeus' *Naval Occasions* v. 37 What about a song, you slacker! **1917** 'Taffrail' *Sub* ii. 62 An habitual slacker at Osborne, however, was soon sized up, and if repeated warnings did not cause him to mend his ways, he generally got the Order of the Boot at the end of his third term. **1921** H. Williamson *Beautiful Years* 217 He was a little slacker, the kind that used cribs later on in school life. **1933** H. C. Bailey *Mr. Fortune Wonders* 109 Eston strode past him with a laugh, 'Had enough already?' and Cicely pattered after crying, 'Slacker!' **1964** A. Whitehouse *Epics & Legends of First World War* vi. 134 The..League..rendered important service in..handling the 'slacker' and conscientious-objector groups. **1969** R. Maugham *Link* II. 16 'You're a slacker and you're a shirker,' he said. 'You're a little runt in many ways. But you're the best of the lot of them.'

Slack-ma-girdle (slæ·kmăgə·ɪd'l). Also **Slack-my-girdle.** [f. phr. *slack my girdle*: cf. Slack *v.* 6.] A variety of cider apple (see quots.).

1885 Hogg & Bull *Herefordshire Pomona, List of other Cider Apples*, Slack-My-Girdle, or Slack-my-girl.—A striped Somersetshire apple of large size. It is also much grown in Devonshire. **1967** [see *HANG-DOWN *sb.* and *a.*]. **1981** *Countryman* LXXXVI. III. 42 Others..use only true cider apples, from trees with names as evocative as..Slack-ma-Girdle.

slackness. Add: **5.** *Naut.* (See quot. 1877.)

1877 W. H. White *Man. Naval Archit.* 484 The contrary condition, where the resultant resistance acts abaft the resultant wind pressure, and makes the head of the ship fall off from the wind, is termed 'slackness', and can only be counteracted by keeping the helm a-lee. **1922** E. L. Attwood *Theoret. Naval Archit.* (ed. 8) xi. 380 If the centre of effort is forward of the C.L.R. the bow of the ship tends to fall off from the wind (which is termed 'slackness').

slack-water. Add: **3.** (Earlier example.)

1837 J. M. Peck *Gazetteer Illinois* (ed. 2) III. 264 Fox river is susceptible of improvement by slack water at small expense.

4. *slack-water basin.*

1836 J. Hall *Statistics of West* 38 At low stages the [Ohio] river becomes resolved into a succession of ripples, with extensive slack water basins between them.

sladang, var. *SELADANG.

Slade (slēⁱd). The name of Felix Slade (1790–1868) used: **a.** *attrib.* to designate the School of Fine Art (founded 1871) at University College London and its members, and scholarships and professorships in fine art endowed by him at Oxford, Cambridge, and London.

1869 *Proc. A.G.M.* (U.C.L.) 13 At their Session on May and the Council received notice of the bequest of £45,000 made by the late Mr Felix Slade, for the purpose of founding 'three or more Professorships for promoting the study of Fine Arts, to be termed the Slade Professorships of Fine Arts..one..in..Oxford, another..in.. Cambridge, and one more in the University College of London'. **1872** *U.C.L. Calendar, 1871–1872* 44 Slade Scholarships. Under the will of the late Mr Felix Slade, six Scholarships of £50 per annum each..have been founded in the College. **1885** Kipling *Let.* 18 Dec. in C. Carrington *Rudyard Kipling* (1955) iv. 71 Do you ever come to know anything about the Slade Art School and the students there—the female ones. **1925** E. Budge *Mummy* (ed. 2) p. vii, In the year 1892, on the recommendation of J. H. Middleton, Slade Professor of Fine Art in the University of Cambridge, the Syndics..commissioned me to make a Catalogue. **1958** *Observer* 15 June 15/2 Joan Mitchell's rather beautiful painting..which has a sensitive, tight-lipped, almost Slade School quality. **1964** K. Clark *Ruskin Today* I. 11 In 1869 he [*sc.* John Ruskin] was appointed Slade Professor at Oxford. **1978** *Ann. Rep. 1977–78* (U.C.L.) 19 The College acknowledges ..the co-operation of the Fine Art Society Gallery.., which put on an exhibition of work by Slade artists.

b. *absol.* with *the*: the School of Fine Art itself.

1890 C. M. Yonge *More Bywords* 249 There are the art classes at the Slade, and the lectures I am down for. **1904** R. Fry *Let.* 22 June (1972) I. 222 You will..be as much disappointed, almost, as I am about the Slade. It is a very serious blow to my hopes... They have long ago realized that Waldstein was a failure as Slade Professor. **1928** R. Campbell *Wayzgoose* I. 10 And surely from the stir that this one made He might have been a student at the Slade. **1961** G. Spencer *Stanley Spencer* v. 102 His entry into the Slade proved to be a most far-reaching and valuable decision. **1980** I. Murdoch *Nuns & Soldiers* 201 Tim described the Slade and his early experiments in painting.

slag, *sb.*[1] Add: **2. a.** (Examples referring to the use of slag in the construction of roads.)

1951 *Sources of Road Aggregate in Gt. Brit.* (Dept. Scientific & Industrial Research) (ed. 2) 4 Slags from smelting operations form a valuable source of roadstone in England and Wales... With one exception the slags are derived from iron smelters or steel works. **1955** I. D. Margary *Roman Roads in Britain* I. i. 15 In districts where iron was being worked the hard slag provided an almost ideal metalling [for Roman roads]. **1958** *Optima* Mar. 14/1 At Maresfield Hempstead, near Benenden, and at Slinfold, Roman roads are partly constructed of slag, and large quantities of slag have since been used.

b. (Further examples.)

1888 [see *BASIC *a.* 2 d]. **1937** *Nature* 20 Feb. 318/2 The Industrial Research Council of the British Iron and Steel Federation will show..the applications of foamed slag as an aggregate for lightweight concrete. *c* **1957** *Story of Slag* (Brit. Slag Federation) 16 (*caption*) Airfield, showing main runway coated with three-quarter inch graded bituminous slag. *Ibid.* 17 Dry slag makes excellent rail track ballast. *Ibid.* 18 Dry slag is used by many local authorities as filter-bed media. *Ibid.* 21 Foamed slag building blocks used in construction of houses at Gateshead and Middlesbrough. **1971** *Arable Farmer* Feb. 69/2 Dressings of basic slag, potassic basic slag or PK compound fertiliser at the beginning and near the end of the ley break should prove sufficient to carry three successive corn crops.

4*. *slang.* **a.** A worthless or insignificant person (freq. used as a term of contempt): *spec.* (*a*) a coward; (*b*) a rough or brutal person; (*c*) any objectionable or contemptible person; (*d*) a vagrant or a petty criminal; also, such persons collectively; (*e*) (the most usual sense) a prostitute or promiscuous woman; a slattern.

(*a*) **1788** Grose *Dict. Vulgar Tongue* (ed. 2), *Slag*, a slack-mettled fellow, one not ready to resent an affront. **1958** F. Norman *Bang to Rights* II. 62 You'v got the guts of a slag.

(*b*) **1934** P. Allingham *Cheapjack* xix. 237 The Newcastle 'slag' is the sort of man who makes up the personnel of the race-gangs, and..he will pick a fight with anyone. **1961** *New Statesman* 14 Apr. 576/2 As the underworld put it, 'he steamed in like a slag and roughed them up as he topped them.'

(*c*) **1943** W. Buchanan-Taylor *Shake It Again* xxi. 199 It seemed slaggy to me. (A *slag* is a person who is not much *bottle*—not much good;..for whom you have no respect or time.) **1958** M. Pugh *Wilderness of Monkeys* 89 Sit down..you slag. **1962** Parker & Allerton *Courage of his Convictions* iv. 159 When I got out there was some slag on the door, all gold braid and nose in the air, wouldn't let me in! **1981** *Daily Tel.* 8 July 3/1 As sentence was announced, the dead boy's father..shouted: 'I hope you rot in it, you slag.'

(*d*) **1955** P. Wildeblood *Against Law* III. 120 Several different kinds of burglars..the rank amateurs or 'slags' who had stolen paltry sums. **1962** R. Cook *Crust on its Uppers* iii. 43 Marchmare lent it to this hatful of slag. **1962** D. Warner *Death of Bogey* I. ii. 14 In my day, the strong-arm boys were slags. Nobody in the big-time would look at 'em. **1963** T. & P. Morris *Pentonville* xi. 227 Pentonville also contains at any given time a number of vagrants, drunks and similar social derelicts... Prison argot classifies them all as *slags*. **1968** J. Lock *Lady Policeman* xii. 108 Only prostitutes, their friends, layabouts, tom watchers, petty criminals and the like are left—'the slag' we call them. **1970** P. Laurie *Scotland Yard* vi. 141, I could get them up the nick and take their prints with ink, but that's really for slag.

(*e*) **1958** *N.Z. Listener* 10 Oct. 6/3 A 'slag' is a white girl who lives with or is friendly with coloured people of either sex. **1959** Anon. *Streetwalker* iv. 87 It's my pride he hurts, with his little slags. **1966** *New Statesman* 23 Dec. 934/2 You have to wear glasses these days if you don't want to be called a slag. **1970** 'D. Craig' *Young Men may Die* x. 72 Does anyone care what happens to a slag? **1973** J. Seabrook *Loneliness* 185, I went out with a girl called Angie, who was really a bit of a slag.

b. Worthless matter, rubbish; nonsense.

1948 V. Palmer *Golconda* v. 35 'Listen,' he said... 'There's some men in every camp will get a kick out of throwing dirt. Who's been filling you up with slag about me? That old crank up on the mountain, was it?' **1970** *New Yorker* 12 Sept. 32/3 It is very depressing to think about the wonderful..letters people used to get..and then look at the slag on one's desk.

5. a. *slag-dump, -heap* (also *fig.*), *inclusion, -mound, -tip, wool* (earlier and later examples).

1923 Kipling *Irish Guards in Gt. War* I. 109 It was a jagged, scarred and mutilated sweep of mining-villages, factories, quarries, slag-dumps, pit-heads, chalk-pits and railway embankments. **1974** *Times* 11 Jan. 16/6 Various stages of iron-working are represented by ore-roasting areas, three slag dumps, 36 smelting furnaces [etc.]. **1917** A. G. Empey *Over Top* 308 *Slag heap*, a pile of rubbish, tin cans, etc. **1931** C. Day Lewis *From Feathers to Iron* xxix. 54 Wherever radiance from ashes arises—Willow-herb glowing on abandoned slagheaps. **1963** *Times* 24 Jan. 6/2 It was not enough to put one or two council houses near the slag heaps. **1974** F. Warner *Meeting Ends* II. ii. 39 What a bore the slagheap of matrimony! **1913** Slag inclusion [see *INCLUSION 2]. **1934** *Jrnl. R. Aeronaut. Soc.* XXXVIII. 249 Thirteen micro-photographs show forms of slag inclusion and distortion of films in butt welding. **1945** Greaves & Wrighton *Practical Microsc. Metallogr.* vi. 96 Of commercial wrought irons, Swedish Bar Iron is most free from slag inclusions. **1955** J. R. R. Tolkien *Return of King* 374 Frodo reaches the slag-mounds. **1960** C. Day Lewis *Buried Day* vii. 130 The prevailing wind brought the acrid smell of slag-tips from the Mansfield collieries. **1878** J. Deby *Rep. Iron & Steel Industries Foreign Countries* II. i. 13 The manufacture of blast furnace slag wool has recently been established commercially for the first time in America. **1979** *Nature* 19 July 183/2 The fibres are produced from glass, rock, slag and metallic oxides and include fibreglass and rock and slag wools which are widely used for home insulation as a substitute for asbestos.

b. slag notch, a hole in a furnace, above the level of the molten metal, which can be unstopped to let out slag.

1895 E. L. Rhead *Metallurgy* 96 In the top of the dam is a groove, the 'slag-notch', through which the slags flow continuously, after reaching that level, and through which the blast blows to keep it clear. **1929** W. Lister *Practical Steelmaking* xxxviii. 392 A 'Dewhurst' ladle can stand under the middle door in order to catch the slag running over the slag notch.

† **slag** (slæg), *sb.*[2] *Obs. Criminals' slang.* [Prob. f. Slang *sb.*[4], under influence of Slag *sb.*[1] in Dict. and Suppl.] = Slang *sb.*[4] 1.

1857 'Ducange Anglicus' *Vulgar Tongue* 19 *Slag, n.* Chain, a gold or silver one. **1926** *Clues* Nov. 159/1 Then we'll take the hot hoops and slags up to the block dealers. **1929** *Detective Fiction Weekly* 2 Mar. 694/2 One 'nips the slag' when one cuts the watch chain, a practice practically defunct now.

slag, *v.* Add: **3.** *trans.* To abuse or denigrate (a person); to criticize, insult. Also with *off*. *slang.*

1971 J. Mandelkau *Buttons* v. 75 He was doing a good job of bad mouthing and slagging me to a number of the Angels. **1972** *Guardian* 17 Aug. 1/1 Mr Jack Jones, general secretary of the Transport and General Workers' Union, was 'slagged off'—in dockland jargon—several times during the day. A gang of furious dockers invaded his press conference. **1974** G. F. Newman *Price* v. 156 She always put out warnings, invariably slagging Terry. **1976** E. Dunphy *Only a Game?* v. 150 When the game starts, if things start going wrong, everyone blames them. Everyone slags them off. **1978** *Broadcast* 29 May 2/2 There's been a growing tendency for some sales organisations to slag the [television] company which used to be.. the brand leader in British sales. **1981** *Daily Tel.* 25 Feb. 17/6 He followed me down the street, slagging me off.

slagging *vbl. sb.* (examples in sense *3 of the vb.).

1971 *News-Advocate* (Barbados) 20 Mar. 9/6 You get so much slagging of bands nowadays. **1977** *Zigzag* Aug. 20/3 The gig met with quite a bit of slagging in the rock press.

slaggy, *a.* Add: **2.** *slang.* **a.** Of a person or thing: objectionable, unpleasant, offensive.

1943 [see *SLAG *sb.*[1] 4* a (*c*)]. **1962** R. Cook *Crust on its Uppers* x. 82 That slaggy basement of yours. **1972** R. Quilty *Tenth Session* I. 20 Some sort of hippie, maybe? No roadie, he could swear to that, having tangled with some real slaggy ones.

b. Of a woman: promiscuous, 'cheap'; slatternly.

1973 H. Miller *Open City* xiv. 160 The writer and his slaggy girl friend. **1980** R. Connolly *Sunday Kind of Woman* II. 22 He thought about some of the slaggy models he had known.

slagless (slæ·glės), *a.* [f. Slag *sb.*[1] + -less.] Of iron, steel, etc.: free from slag. Hence **sla·glessness.**

1902 *Encycl. Brit.* XXIX. 571/1 (*in table*) Slagless or 'Ingot-metal' Series. *Ibid.* 571/2 But the former lack the essential quality—slaglessness—which makes the latter steel.

‖ **slainte** (slā·ntʃə), *int.* Also **slainté.** [a. Gael. *sláinte*, lit. 'health'.] A Gaelic toast: good health!

1824 Scott *Redgauntlet* II. vii. 159 He then took up the tankard, and saying aloud in Gaelic, '*Slaint an Rey*', just tasted the liquor. **1880** A. P. Graves *Irish Songs & Ballads* 71 Here's a health to you, Father O'Flynn, Slainté, and slainté, and slainté agin. **1922** Joyce *Ulysses* 44 Well: *slainte!* **1949** E. Coxhead *Wind in West* iv. 114 '*Slainte mhor*, Rory!' cried Kurt, who knew what was necessary in every language. **1952** 'J. Tey' *Singing Sands* vii. 105 '*Slainte!*' he said, and took a swig of it. **1966** S. Forbes *Terror touches Me* iv. 43 'Here, Brendan. Slainté.' Brendan's hand trembled as he took the full glass. **1980** M. McMullen *My Cousin Death* (1981) vi. 78 Upshaw gave him his drink..'*Slainte.*'

slalom (slā·lǫm). [a. Norw. *slalåm*, f. *sla* sloping + *låm* track.] **1.** A downhill race in which skiers, descending singly, describe a zigzag course between artificial obstacles, usu. flags. Freq. *attrib.*

1921 *British Ski Year Bk.* 274 Slalom race on Inner-Arosa practice slopes. **1927** A. Lunn *Hist. Skiing* xviii. 227 However, the Slalom was worth a trial, and in 1922 the Alpine Ski Challenge Cup became a Slalom race. **1950** *Times* 13 Feb. 7/5 The Kandahar Ski Club, which originated the modern downhill racing movement (the slalom is a British invention worked out at Mürren), still insists that its candidates shall pass a test in soft snow. **1966** L. Deighton *Billion-Dollar Brain* I. vii. 58 Three sets of skis..one set of which were slalom skis. **1972** C. Short *Naked Skier* I. 2 She did a slalom turn... I stood for a while looking at the ski tracks she had left. **1980** *Daily Tel.* 26 Jan. 32 The shadow of Switzerland's Marie-Therese Nadig, who beat her in both the downhill and giant slaloms in the 1972 Winter Games.

2. *Water-skiing.* A run along a zigzag course defined by buoys. Also *attrib.*

1949 *Sun* (Baltimore) 25 July 14/1 Mary Lois Thornhill ..yesterday added the slalom to the trick riding and jumping titles she won on Martin Lagoon. **1963** *Newsweek* 23 Sept. 66/3 Billy Spencer, the youngest of the 99 competitors in the world water-skiing championship at Vichy, France, finished his first slalom run. **1978** G.

Wright *Illustr. Handbk. Sporting Terms* 154/1 In Slalom —a timed run through two lines of buoys—half a point is awarded for every buoy successfully rounded and also for returning to within the boat's wake before the next buoy.

3. A race in which canoeists weave between obstacles, esp. along a course of rapid water. Freq. *attrib.*

1956 N. McNaught *Canoeing Man.* vii. 86 Most local slaloms take place on weir-type courses. *Ibid.* 88 Slalom organizers must ensure that a rescue boat is always manned and ready. **1964, 1969** [see *KAYAKING *vbl. sb.*] **1969** *Publ. Amer. Dial. Soc.* LI. 7 *Slalom kayak*,..a highly maneuverable kayak constructed with more curve rocker in the keel than a downriver or combination kayak. **1973** R. Fiennes *Headless Valley* vii. 111 Then in 1969 two Russians, expert slalom canoeists, set out to navigate the Liard. **1977** *Herald* (Melbourne) 17 Jan. 2/5 Kaine had a junior kayak for a year, but found it did not perform as well as the slalom boat his father built.

4. An exercise or contest in which a motor vehicle is driven along a zigzag course defined by markers. Also *attrib.*

1965 *Listener* 15 Apr. 578/1 There is a keen following of motor racing among the Swiss: entries are not lacking in their speed hill climbs, rallies, and slaloms (or 'wiggle-woggles' between pylons). **1972** *National Observer* (N.Y.) 27 May 19/2 Next came a *slalom* exercise through a long row of traffic cones. You must swerve the car to the right of the first cone, to the left of the second, back right around the third... It teaches braking and shows how much more your car can take and 'do' than you thought. **1974** *Rules of Game* 297/1 Slalom, or autotest, competitions are a test of maneuverability. Cars attempt the course singly. Each car starts with 0 points and receives 1 point for each second taken and 10 points for each marker touched. The winner is the driver with the fewest points at the finish.

5. A similar race or activity in *Skateboarding*. (See also quot. 1976[2].) Also, a track suitable for this.

1976 A. Cassorla *Skateboarder's Bible* 11 Slaloms could not be held on a steep grade or be set up with widely spread cones because of the poor turning capacity of the boards. **1976** *National Observer* (U.S.) 3 July 12/3 Slalom, as in skiing, going downhill and weaving around markers. **1978** *Skatcat's Quiz Bk.* (R. Soc. Prevention of Accidents) 2/1 You need a 6" wide deck to start with. Flexi for slalom. Stiffer for free-style. **1978** *Cornish Guardian* 27 Apr. 13/4 The playgrounds were flat and the youngsters wanted the excitement of bowls and slaloms which the Polkyth park would have.

Hence as *v. intr.*, to perform or compete in a slalom, to make frequent rapid (slalom) turns; **sla·lomer, sla·lomist,** one who slaloms; **sla·loming** *vbl. sb.*

1956 N. McNaught *Canoeing Man.* vii. 90 An individual slalomist who overturns is disqualified on that particular run. *Ibid.*, Strength and skill are needed for successful slaloming. **1973** *Times* 28 Sept. 36/5 Forty miles of pistes where you can schuss, trek, slalom, langlauf. **1976** *Daily Tel.* (Colour Suppl.) 30 July 9/1 What they are all doing is skateboarding—zooming and slaloming and 'hanging ten' and catamaraning on four-wheeled boards. **1977** *Skateboard Special* Sept. 5/1 The best slalomists still use wide wheels to prevent the board wobbling at high speed. **1978** *Guardian Weekly* 12 Feb. 24/3 Whether he is the world's greatest ever slalomer no-one can say.

slam, *sb.*[1] Add: **1. b.** A violent blow administered to a ball. *slang* (chiefly *U.S.*).

1931 *Lit. Digest* 18 Apr. 40, I remember when a hit was a..clout,..slam..but never..just a hit. **1978** *Chicago* June 274/1 Engrossed as each team was in setting up the ball for a slam, the players' concentration was marred by the explosion of tear-gas canisters some distance away.

2. a. (Earlier example.)

a **1817** Jane Austen *Persuasion* (1818) IV. viii. 147 The various noises of the room, the almost ceaseless slam of the door.

3. An insult or 'put-down'. *U.S. slang.*

1884 I. M. Rittenhouse *Jrnl.* in *Maud* (1939) 296 Oh! did I tell you that Mr. Hough to atone for his 'slams', said, 'I did want to make one gallant speech, but I hardly dared, about how remarkably well you looked Tuesday night.' **1944** B. A. Botkin *Treas. Amer. Folklore* III. 410 Certain formulae are identified with disparaging or insulting wisecracks, or 'slams'. **1980** R. L. Duncan *Brimstone* I. 22, I don't take that description as a slam. I was a great piece of ass.

4. = *SLAMMER 3. Usu. with *the*. Chiefly *U.S. slang.*

1960 R. G. Reisner *Jazz Titans* 164 *Slam*, jail. **1965** A. Lurie *Nowhere City* (1966) xi. 118 That was really thinking fast. I guess you saved me a night in slam. **1972** J. Wainwright *Requiem for Loser* vi. 132 Reginald Drover. Escapee from one of Her Majesty's slams. **1972** S. Greenlee in W. King *Black Short Story Anthol.* 95 Uncle Benny told him that getting a bad teacher for a year was like being in the slam, and you just did your time and didn't let it bug you. **1978** J. Gores *Gone, no Forwarding* vii. 40 You're going to the slam for fifteen.

5. Special *Comb.* **slamdunk** *U.S. Basketball* [cf. *DUNK *v.*], a forceful shot in which a player jumps and slams the ball down into the basket.

1976 *N.Y. Times* 25 May 35 The only one-eyed candidate who would know how to put in a slamdunk on a New York playground has new financial life. **1981** *Washington Post* 25 Feb. E7/3 Robinson had 32 points and Jones put on a slam-dunk show to finish with 17.

slam, *sb.*[2] Add: **2. b.** *grand slam* (earlier examples, from the game of Boston). Also *small slam* = *little slam*, in Bridge.

1814 C. Jones *Hoyle's Games Improved* 188 These declarations will supersede that of Boston simply... The highest, called Grand Slam, is undertaking to get 13 tricks. **1921** F. Irwin *Compl. Auction Player* I. 25 To take all, or all but one, of the tricks is to make a slam. The former is called a grand slam, and is worth 100 above the line. The latter is called a small slam, and is worth 50. **1937** N. de V. Hart *Slams à la Culbertson* I. vi. 53 If he has his maximum count.., he will bid Small Slam even without a five-card suit. **1959** *Listener* 15 Jan. 146/2 West might then content himself with the small slam. **1977** *Times* 3 Sept. 7/2 South made an overtrick in the small slam.

c. *grand slam* (transf.): (*a*) a complete success; *spec.* victory in all of a series of matches or competitions; (*b*) an attack in force; forceful or decisive behaviour; (*c*) in Baseball, *grand slam* (*homer, home run*): (see quot. 1974).

(*a*) **1920** D. H. Lawrence *Let.* 5 Feb. (1962) I. 619, I feel that this is the time to make our grand-slam. **1966** D. F. Galouye *Lost Perception* viii. 86 We're going to try for a grand slam—knock them all out at once with nuclear stuff the next time we pin down their locations. **1967** *Boston Sunday Herald* 26 Mar. II. 7/1 He was the world's No. 1 amateur in 1962 and only the second player in history to accomplish the 'Grand Slam'. **1976** *Scottish Daily Express* 24 Dec. 14/1 We are the Home International champions after a Grand Slam of victories against England, Wales and Northern Ireland.

(*b*) **1933** F. Richards *Old Soldiers never Die* ix. 123 Dawn broke..and we were anxiously waiting for the time when the Grand Slam commenced. **1959** *Times Lit. Suppl.* 29 May 322/5 Churchillian impetuosity and grand-slam rashness. **1963** *Times* 7 Mar. 8/3 The most persuasive argument for retaining *Checkmate* in the Royal Ballet's repertory..is the opportunity for grand slam acting it offers to its protagonist, the malevolent Black Queen.

(*c*) **1953** *Sport* June 58/1 Bill Carr knocked his pitch over the fence for a grand-slam homer. **1961** *Sports Illustr.* 2 Oct. 10/2 Baltimore's Jim Gentile hit a grand slam homer off Don Larsen in Chicago one rainy night last week. **1967** *Boston Sunday Herald* 30 Apr. I. 8/2 (Advt.), It's as exciting as a ninth inning grand-slam! **1974** *Rules of Game* 168/3 A grand slam home run is a home run hit when the bases are loaded, i.e. when three men are on base. It scores four runs, the maximum possible from one hit. **1978** *Detroit Free Press* 16 Apr. E 3/1 The Pirates finally put it together and broke a five-game skid with a little help from a pair of homers from Bill Robinson, one a grand slam.

3. *attrib.* in *Bridge*, as *slam bid*(*ding*), *contract, hand.*

1929 M. C. Work *Compl. Contract Bridge* p. xi, The partner..must jump if his hand warrant it—either one step..or a vault toward a slam bid. **1947** E. Klein *Enjoy your Bridge* II. xiii. 114 Be content to win a perfectly safe contract and leave your slam bids for a later stage. **1927** *Work-Whitehead Auction Bridge Bull.* Feb. 141 Contract in its original form, minus the recently introduced 'vulnerable' feature and slam bidding, made its first appearance abroad some fifteen years ago. **1974** *Times* 16 Feb. 13/2 Slam bidding is treated in most text books as if it..cannot be covered by ordinary approach bidding. **1938** Slam contract [see *CONTROL *sb.* 3 e]. **1959** *Listener* 8 Jan. 84/2 There is every reason to hope for a slam contract in some suit. **1977** *Times* 16 Apr. 11/8 He.. bid Three Diamonds which..in conjunction with the cue-bids, produced a slam contract in the wrong suit. **1937** J. Crane *Crane Syst. Contract Bidding* 78 (*heading*) Examples of a bidding game and slam hands from matches. **1979** Reese & Flint *Trick 13* 17, I didn't know anything was wrong till that slam hand near the finish.

slam, *v.*[1] Add: **2.** Also with *on*, as in *to slam on the brakes*, to apply the brakes of a motor vehicle, etc. suddenly. Also *fig.*

1958 L. Uris *Exodus* IV. iii. 501 Zev slammed on the brakes and pulled over to the side of the road. **1975** *Business Week* 1 Sept. 23 The rule is designed to prevent a truck from jackknifing or jumping a lane when drivers slam on the brakes at 20 mph to 60 mph. **1975** *Economist* 4 Oct. 11 Can the driver [*sc. contextually* General Franco] be persuaded to look forward instead of back, or can someone else intervene to slam on the brakes? **1976** *Business Week* 11 Oct. 96 A radar unit in the nose..that warns the driver of road hazards ahead—and slams on the brakes if he fails to do so. **1982** *Chr. Sci. Monitor* 13 Apr. 7 Inflation has dropped dramatically as the quasi-independent Federal Reserve Board slams on the brakes of high interest rates.

c. (Earlier example.)

1870 'Mark Twain' *Lett. to Publishers* (1967) 49, I can slam you into the lecture field for life and secure you ten thousand dollars a year as long as you live.

4. (Later examples.)

1914 G. B. Shaw *Misalliance* 42 Theyre coming slam into the greenhouse. **1930** E. Pound *XXX Cantos* xviii. 82 An' he run damn slam on the breakwater.

5. a. To be severely critical, to utter insults. *U.S. slang. rare.*

1884 I. M. Rittenhouse *Jrnl.* in *Maud* (1939) 291 When I and Mr Hough arrived late Dr Benson and Mr Parsons slammed right and left at the tardiness.

b. *trans.* To criticize severely. *colloq.* (orig *U.S.*).

[**1914** 'High Jinks, Jr.' *Choice Slang* 18 *Slamming contest*, a condition where two or more individuals are engaged in criticism. 'A Knockfest.'] **1916** H. L. Wilson *Somewhere in Red Gap* ii. 57 Couldn't even agree on the same kind of cocktail. Both slamming the waiter. **1932** G. Atherton *Adventures of Novelist* vi. xiii. 380 She took care I should constantly be slammed. **1958** 'N. Shute' *Rainbow & Rose* 252 They come with bright and tinkling vivacity until I slam them down. **1962** J. Symons *Killing of Francie Lake* ii. 19 You go on the air and slam negro landlords and they'll be saying you're anti-negro. **1978** J. Irving *World according to Garp* v. 89 A long, cocky letter, quoting Marcus Aurelius and slamming Franz Grillparzer.

6. Const. *prep.* To move violently, to crash.

1973 *Times* 2 Nov. 13/6 Rosa..savagely slamming around the kitchen. **1976** M. Machlin *Pipeline* lvi. 566 The lifeboat was now slamming through the choppy two and three foot high waves at over twenty knots. **1979** R. Jaffe *Class Reunion* (1980) II. iv. 209 She didn't even see the small stone wall until she had slammed into it.

slamming *vbl. sb.* (examples *spec.* of boats: see quot. 1948).

1935 *Engineering* 18 Jan. 55/1 'Pounding' or 'slamming' damage is looked upon as no more than a normal circumstance..of cargo vessels trading across the North Atlantic. **1948** R. de Kerchove *Internat. Maritime Dict.* 685/1 Slamming almost always takes place forward... Slamming damage is usually ascribed to dynamic pressures arising from impact of the ship's hull upon the surface of the water, and from the actual impulsive displacement of water caused by the downward movement of the ship when pitching. **1972** C. Mudie *Motor Boats & Boating* 17 If such a craft were to be taken to sea she would rapidly be found to be..apt to break her back from slamming when pitching.

slam-bang, *adv., a.,* and *v.* Add to etym.: cf. Slap-bang *adv., a.,* and *sb.* **A.** *adv.* **1.** (Earlier example.)

1840 R. M. Bird *Robin Day* 25 Five or six hundred field pieces blazing away slambang.

B. *adj.* **1.** (Earlier and later examples.)

1823 'J. Bee' *Dict. Slang* 158 *Slam-bang shops*, places where gourmands of the fourth rate regale;..probably, from..the 'slam-banging' of the doors, plates, and tools. **1957** *Economist* 28 Sept. 999/1 With all the diplomatic finesse of a runaway bulldozer, the governments of the major powers are conducting a slambang exchange of public accusations. **1981** H. R. F. Keating *Go West, Inspector Ghote* ii. 25 Fred Hoskin's slam-bang voice broke in on his thoughts.

2. In weakened use: exciting, impressive, first-rate. Also, vigorous, energetic. *colloq.*

1939 *Sun* (Baltimore) 4 Dec. 13/4 The balance of the card will be made up of some real slam-bang preliminaries. **1942** *Ibid.* 17 June 1/8 American heavy bombers have entered the Mediterranean sea war in slambang fashion. **1952** B. Malamud *Natural* 20 A slambang young pitcher who'd soon be laying them low in the big leagues. **1965** *Times Lit. Suppl.* 25 Nov. 1061/2 We have this plot for a slam-bang topical novel about the Johnson administration. **1972** D. Delman *Sudden Death* (1973) v. 126 You were good today. I watched you. Slam-bang. **1975** *Publishers Weekly* 2 June 49/1 A very cerebral English mystery, with..a finale that is full of slambang action. **1979** *Radio Times* 5–11 May 23/2 It was described by Judith Crist as a 'slam-bang top-quality grown-up adventure which thumbs its nose at authority and morality'.

slam-banging *vbl. sb.* (earlier example).

1823 [see sense 1 of the adj., above]. **1843** *Knickerbocker* XXII. 41 The creaking on its rusty hinges and slam-banging of the sign of the Devil-Tavern.

slammable (slæ·măb'l), *a. rare.* [f. Slam *v.*[1] + -able.] Of a door, etc.: capable of being slammed (shut).

1976 T. Stoppard *Dirty Linen* 9 Separate table with good slammable drawers for *Maddie*.

slammakin, slammerkin, *a.* Add: (Examples of the spelling *slummocking*.)

1960 R. Collier *House called Memory* iv. 53 A big slummocking girl with cropped hair and braying laugh. **1978** S. Radley *Death & Maiden* iv. 37 They're a slummocking family..always have been.

slammer. Add: **3.** Prison, gaol. Usu. with *the*: occas. *the slammers*. Cf. *SLAM *sb.*[1] 4. *slang* (orig. *U.S.*).

1952 G. Mandel *Flee Angry Strangers* 358 I'm hip what you was doin wit Ange while I was in the slammer. **1961** Rigney & Smith *Real Bohemia* p. xvii, *Slammer*, jail. **1970** E. Bullins *Theme is Blackness* (1973) 177 I'm into a heavy petty criminal thing, man. The Man is always ready to vamp on you and take you off 'round here or put you in the slammers. **1975** B. Garfield *Death Sentence* (1976) v. 31 Less than one per cent of Chicago's crimes are solved, in the sense that some joker gets tried and convicted and sent to the slammer. **1977** D. Bagley *Enemy* xv. 131 This one's not for the slammer. He'll go to Broadmoor for sure.

slan (slæn). [Invented word, from the novel by A. E. van Vogt: see quot. 1940.] In works of Science Fiction: a being of superior intelligence, physique, etc.; a superman. Hence used *gen.* among fans of this type of literature.

1940 A. E. van Vogt *Slan* in *Astounding Sci. Fiction* Oct. 27/1 They..accuse Samuel Lann, the human being and biological scientist who first created slans, and after whom slans are named—Samuel Lann: S. Lann: Slan—of fostering in his children the belief that they must rule the world. **1955** Koestler *Trail of Dinosaur* 143 Fen gather in clubhouses called slanshacks, 'slan' meaning a biologically mutated superman. **1969** H. Warner *All Our Yesterdays* ii. 42 'Fans are slans!' became the rallying cry of the Cosmic Circle. **1975** E. Weinstein *Fillostrated Fan Dict.* 119 *Slan shack*, a place where more than two fans live. **1980** *Verbatim* Autumn 10/1 'He's a slan' is the fannish equivalent of 'He's a helluva guy'.

slanchways, var. *SLAUNCHWAYS, -WISE *adv.* and *a.*

slang, *sb.*³ Add: **1. b.** (Earlier example.)
1801 *Encycl. Brit.* Suppl. I. 723/1 A studied harangue, filled with that sentimental slang of philanthropy, which costs so little, promises so much, and has now corrupted all the languages of Europe.
c. (Later examples.)
1914 J. M. Barrie *Admirable Crichton* IV. 227 In the regrettable slang of the servants' hall, my lady, the master is usually referred to as the Gov. **1925** T. Dreiser *Amer. Tragedy* (1926) I. II. i. 156 Don't say 'swell'. And don't say 'huh'. Can't you learn to cut out the slang? **1937** Partridge *Dict. Slang* p. ix, *A Dictionary of Slang and Unconventional English*, i.e. of linguistically unconventional English, should be of interest to word-lovers. **1976** *Times Lit. Suppl.* 30 Apr. 520/4 In Australia, slang simply has a quite different status from slang in England. It is a *part* of 'Standard English' there, not outside 'Standard English'. Slang words are used informally, casually and naturally by all Australians regardless of class or education.
attrib. and *Comb.* **1856** G. Meredith *Let.* 15 Dec. (1970) I. 28 Have you..a book of Hampshire Dialect?.. Also a slang Dictionary, or book of the same with Gloss. **1926** *Variety* 29 Dec. 5/3, I was hep that the slang slingers were not crowding each other. **1977** K. F. Kister *Dictionary Buying Guide* II. 240 The more substantial slang dictionaries provide detailed word histories and thus complement the etymological dictionaries.
d. (Earlier example.)
1805 T. Campbell *Let.* 9 Feb. in W. Partington *Private Letter-bks. W. Scott* (1930) 100 In five weeks, however, her slang broke out, and within the seventh she discovered the whole catalogue of Vices of which a very ugly woman can be guilty.

slanging, *vbl. sb.* (In Dict. s.v. Slang *v.*) Add: *slanging match*: an exchange of abuse; a vituperative argument.
1896 T. E. Taylor *Running Blockade* vi. 74 A slanging match went on between us, like that sometimes to be heard between two penny steamboat captains on the Thames. **1936** *Sun* (Baltimore) 14 Oct. 12/3 They [*sc.* the speeches] were made by the man who commands the highest authority and the greatest power in all the Reich, and they provoked an international slanging match. **1938** 'G. Orwell' *Homage to Catalonia* xii. 243 The slanging-match in the newspapers. **1978** J. Porter *Dead Easy for Dover* xiv. 140 Mrs Vincent very sensibly decided not to get involved in a slanging match with Dover. Their views on unmarried mothers were poles apart.

slangish, *a.* (Earlier example.)
1813 J. B. S. Morritt *Let.* 3 Apr. in W. Partington *Private Letter-bks. W. Scott* (1930) 110 Poems of their own which are dull and slangish.

slangster. (In Dict. s.v. Slangism.) (Later examples.)
1926 *Variety* 29 Dec. 5/4 Most slangsters use the exaggerated simile when breaking into print. **1933** *Times Lit. Suppl.* 16 Nov. 781/4 Rhoda Broughton..would probably have thought Galsworthy far too much of a 'slangster'. **1945** *Gen* 5 May 24/1 His [*sc.* Walter Winchell's] slangster column..in the New York Daily Mirror. **1965** *English Studies* XLVI. 465 A *slangster* [is] a user of slang.

slanguage. (In Dict. s.v. Slangism.) (Earlier and later examples.)
1879 *Harvard Lampoon* 21 Nov. 88/1 (*title*) Slanguage on Angele. **1911** *Daily Colonist* (Victoria, B.C.) 7 Apr. 4/2 The 'slanguage' of a sporting reporter is a fearful and wonderful thing. **1926** *Irish Statesman* 18 Dec. 355/1 (*heading*) The American slanguage. **1926** *Variety* 29 Dec. 7/4 Every phase of our complex civilization, and every class have contributed something to what is fast becoming a national slanguage. **1927** *Vanity Fair* XXIX. 67/2 Jack Conway..is conceded to be the ace 'slanguage' hurler in the world. **1935** [see *cat *sb.*¹ 2 c]. **1958** *Inside the ACD* (Amer. College Dict.) Nov. 2/1 Max Shulman shows in *Rally Round the Flag, Boys*, a new novel, that he can capture and record living speech including the 'slanguage' of the current cool crop of hipsters. **1963** [see *cube, *sb.*¹ 1 c]. **1974** *Trailer Life* Nov. 92 Our slanguage is so off-beat that during World War II American military men were able to foil the enemy by resorting to American vernacular.

slangwhang, *v.* (Earlier example.)
1880 *Punch* 11 Sept. 117/1 Over the Census Bill Honourable Members got senselessly incensed.., Orthodoxy and the other Doxies slang-whanging each other just as if they were really in earnest.

† **slangwhang** (slæ·ŋhwæŋ), *sb. U.S. Obs.* Also **slang-wang.** [f. as prec. + Whang *sb.*²] Violent or abusive language.
1834 H. M. Brackenridge *Recollections* xvi. 183 The young lawyer..who has acquired nothing of the ordinary slangwhang. **1859** *Harper's Mag.* July 164/1 Don't allow their vulgar slang-wang to have the slightest effect upon you.

slangwhanging, *vbl. sb.* (Earlier example.)
1809 *Essex Register* (Salem, Mass.) 20 May 2/2 (*heading*) Federal Slang-whanging..or, a new comical, heroic, quizzical, serio-farcical melo drama.

slangwhanging, *ppl. a.* Add: Also **slang-wanging.** (Later example.)
1959 *Listener* 19 Feb. 325/2 A slang-wanging, stump speaker.

slangy, *a.* Add: **2. b.** (Earlier example.)
1842 R. H. Dana *Jrnl.* 14 Nov. (1968) I. 103 His letters have been careless, pretentious, & with a kind of off-hand, slang-ey, defying tone.

slant, *sb.*¹ Add: **1. e.** *Typogr.* = *oblique *sb.* 5, Solidus¹ 2 in Dict. and Suppl. Used esp. of either of a pair of lines enclosing the representation of a linguistic (esp. phonemic) element.
1962 *Gen. Systems* VII. 299/2 Its mate is suffixed with a slant (virgule), thus: 4006 How to Silence. 4006/ How to Sound. **1964** E. Palmer tr. *Martinet's Elem. Gen. Linguistics* i. 24 This [*sc.* a significans] we represent between slants (/ʒ e mal a la tet/, /ʒ e mal/, /mal/). **1972** Hartmann & Stork *Dict. Lang. & Linguistics* 172/1 Phonemic transcription is usually written between slants, e.g. /haus/.
3. a. Also *on a slant*.
1951 E. Paul *Springtime in Paris* xv. 286 Busse.. leaped quickly, hit the lower level of the street pavement on a slant, and almost turned his ankle. **1957** D. Lessing *Going Home* ii. 35 The night was magnificent; the Southern Cross on a slant overhead.
b. *Microbiol.* A sloping surface of culture medium, usu. prepared by letting it solidify in a sloping test-tube, and used for the culture of micro-organisms. Cf. *slope *sb.*¹ 3 a.
1899 T. Bowhill *Man. Bacteriol. Technique* ii. 60 Take three freshly prepared tubes of oblique surface agar-agar—usually called 'agar-slants'—with plenty of water of condensation in the bottom. **1924** *Jrnl. Bacteriol.* IX. 398 Loops were transferred, at intervals up to four hours, to agar slants, and these were incubated overnight. **1949** *Amer. Jrnl. Path.* XXV. 7 Growth on plated media, while not unlike that on slants, was somewhat slower. **1972** *Sci. Amer.* Sept. 187/1 Dried yeast is typically sealed in an airtight envelope filled with nitrogen. Cultures can be perpetuated by inoculating slants of fresh nutrient agar under sterile conditions every 90 days.
6. (Later example.)
1868 H. Woodruff *Trotting Horse Amer.* iii. 58, I have known many that will be always watching slants to get an extra quart of oats for their colts.
8. A way of regarding something, a point of view or 'angle'; an interpretation; a bias. orig. *U.S.*
1905 *N.Y. Even. Post* 28 Jan. 5 The titles of articles on this subject bear an extremely pessimistic slant. **1927** C. Connolly *Let.* 26 Jan. in *Romantic Friendship* (1975) 230 The slant at which I write betrays an unbearable optimism. **1935** M. M. Atwater *Murder in Midsummer* xv. 138 Mentally he was going over his 'story'..to change the slant of some of the phrases. **1948** *Sunday Pictorial* 18 July 12/3 A new and intriguing slant on the Borgias by Nigel Balchin. **1965** *Amer. N. & Q.* Mar. 99/2 The book has a pro-Galvão slant showing the man as a romantic hero. **1973** J. Wood *North Beat* ii. 19 New slant—timing the lunch-hour, eh? When did we have that one before?
9. *U.S. colloq.* A glance, look.
1911 E. Ferber *Dawn O'Hara* viii. 109 You're supposed t'take a slant at th'things an' make up your mind w'at you want. **1934** [see *petting *vbl. sb.* 3].
10. *U.S. slang.* A slant-eyed person, *spec.* used as a term of contempt for one of Oriental stock. Cf. *slant-eye(s)* s.v. *slant *a.* 3.
1942 Berrey & Van den Bark *Amer. Thes. Slang* §385/19 Oriental..*slant*. **1969** *Time* 5 Dec. 26/1 To the G.I. the Vietnamese..is a 'gook', 'dink', 'slope' or 'slant'. **1976** M. Machlin *Pipeline* vii. 79 And the fuckin' Eskimo slants are tryin' to get the rest of it. **1978** J. Gores *Gone, no Forwarding* (1979) 191 He took me back to the slant broad... A slant or a Buddha-head.

slant, *a.* Add: **3. slant-drill** *v. intr. Oil Industry*, to drill a bore hole at an angle to the vertical; also *trans.*; so **slant-drilling** *vbl. sb.*; **slant-eye(s)** *slang* (orig. *U.S.*), a slant-eyed person, *spec.* an Asian (cf. *slant *sb.*¹ 10); **slant-line** = *slant *sb.*¹ 1 e; **slant-rhyme** = *half-rhyme* s.v. *half- II. n.
1969 *New Scientist* 24 Apr. 169/1 They suggest the search for oil..should be restricted to slant-drilling from the shore. **1975** *Offshore* Sept. 244/2 Much of the area covered by the sale can be slant-drilled from the shoreline or the barrier islands. **1976** L. St. Clair *Fortune in Death* i. 8 We've wasted enough time fishing drill pipe out of this hole. Let's plug back and slant-drill. **1977** *Time* 28 Feb. 17/2 Two weeks ago the Israelis began sinking another hole on the shore at El Tur, slant-drilling into the waters whose ownership it disputes. **1929** *Amer. Speech* IV. 344 *Slant eye*, an oriental. **1962** E. Snow *Red China Today* (1963) xii. 85 One might assume that contempt for American imperialism would by now have produced Chinese equivalents of insulting American epithets such as slopeys, slant-eyes and chinks. **1966** *Publ. Amer. Dial. Soc.* 1964 XLII. 31 A few terms [for Orientals] reflect stereotype racial characteristics, i.e., *yellow-belly, yellow-man, slant eyes*. **1972** *Times* 20 May 3/4, I have engaged in campaigns against blacks, yellows and slant-eyes. Why should we have one rule for the whites and one for coloureds? **1974** *Times Lit. Suppl.* 26 July 795/4 And those Jap Ph.D.'s, their questionnaires! (Replying 'Sod off, Slant-Eyes' led to friction.) **1977** 'J. le Carré' *Hon. Schoolboy* i. 36 Renting a cottage in the New Territories, he..proposed to expire under a slanteye heaven. **1954** F. G. Cassidy *Robertson's Devel. Mod. Eng.* (ed. 2) iv. 61 Phonemic symbols are placed between virgules (or 'slant-lines', or 'diagonals'). **1966** *Publ. Amer. Dial. Soc.* XLVI. 16 In line with Haugen's procedure..*slant lines*..used in this article perform double duty for phonemic and diaphonic representations. **1944** *Mod. Lang. Q.* V. 324 Traditional prosodists have discussed rhyme as a degree of likeness in word sounds and have catalogued its approximations, alliteration, assonance, slant rhyme, eye rhyme, [etc.]. **1976** *Times Lit. Suppl.* 16 Jan. 50/2 Wilfrid Owen and Yeats opted for slant-rhyme because it suited their poetic purposes; Fry seems to chime because sometimes he cannot rhyme.

slant, *v.* Add: **5. b.** *fig.* To give a slant (*slant *sb.*¹ 8) or bias to (something). orig. *U.S.*
1939 *Writer's Digest* Sept. 26/2 These types of articles are exceptionally valuable in slanting the writing for certain magazines and trade journals. **1951** H. MacInnes *Neither Five nor Three* II. xi. 166 Did that fool Weidler see that Blackworth was 'slanting' his use of material? **1960** *New Left Rev.* May–June 66/1 There is no suggestion ..that Mr. Bullock is deliberately slanting the picture he paints. **1980** M. Babson *Dangerous to Know* vi. 41 'I suppose it *could* be slanted that way.' It was obvious that May had been considering a different slant.

slanted, *ppl. a.* Add: **2.** Biased, tendentious.
1959 *Listener* 2 Apr. 580/2 A consistently slanted picture of Soviet policy. **1967** 'R. Simons' *Taxed to Death* v. 80 Why are you asking me all these slanted questions about Nathin? **1978** *Detroit Free Press* 16 Apr. E8/1 'The material coming our way is so slanted on gun control and hunting's role in wildlife management, I can't, in good conscience, pass it on to our youngsters,' he had said.

slanter, var. *schlenter *sb.* and *a.*

slantindicular, *a.* (*sb.*) and *adv.* Add: Also -diclar, -dickelar, slantendikular. **A.** *adj.* **a.** (Earlier examples.)
β. **1832** *Mem. of Nullifier* iv. 37 This is sorter a slantindickelar road, stranger. **1835** P. H. Nicklin *Lett. Descr. Virginia Springs* 30 [He] makes his bivouac among the trees..under a slantindicular shed.
B. *adv.* (Earlier and later examples.)
1831 *Daily Louisville Public Advertiser* 17 Oct. 2/3 He looked up at me slantindicler [*sic*], and I looked down on him slantindiclar. **1873** C. H. Smith *Bill Arp's Peace Papers* xxx. 202 If we could have slid into it quietly and slantendikular, if slavery could have sorter tapered out and freedom sorter tapered in, everybody could have got used to it.

slanting, *vbl. sb.* (Later examples.)
1959 *Times* 15 May 6/7 It is still an even chance that any University production of [Shakespeare's plays] one samples will not be heavily flavoured with gimmickry or slanting. **1980** M. McMullen *My Cousin Death* (1981) 7 Not a gentle..rain..but an ill-natured heavy slanting from the northeast.

slantingways (slɑ·ntiŋwēiz), *adv. rare.* [f. Slanting *ppl. a.* + -ways.] Slantwise. Cf. Slantways *adv.*
1899 H. G. Wells *When Sleeper Wakes* vi. 59 He walked slantingways across the room. **1916** —— *Mr. Britling* II. iv. 332 We were busy..pushing our trench out from an angle slantingways forward.

slap, *sb.*¹ Add: **1. a.** *slap on the back* (or *shoulder*): as a hearty gesture of friendship or congratulation. Also *fig.* Cf. *back-slapping *ppl. a.* and *vbl. sb.*
1820 W. Irving *Sketch Bk.* vi. 94 His hospitable attentions were brief, but expressive, being confined to a shake of the hand, a slap on the shoulder..and a pressing invitation to 'reach to, and help themselves'. **1863** G. Meredith *Let.* 19 Feb. (1970) I. 193 Did you say in it you are sorry for your virulent offensive letter that I received? ..If so, a slap on the back and we're friends again. **1883** Stevenson *Treasure Isl.* viii. 62 He seemed in the most cheerful spirits..with a merry word or a slap on the shoulder for the most favoured of his guests. **1929** L. MacNeice *Blind Fireworks* 5 There are lines which may, by the incautious, be (wrongly) read in a merry slap-on-the-back fashion.
e. *slap and tickle*: (a bout of) light amorous play.
1928 E. Robertson *Cullum* ix. 178 She gave me a playful push... She was one of the dreadful type that Cullum called 'slap-and-tickle' girls. **1936** N. Coward *To-Night at 8.30* 77 She won't [come back]—she's out having a bit of slap and tickle with our Albert. **1958** 'N. Shute' *Rainbow & Rose* vi. 243 When I want a bit of slap and tickle I'll arrange it for myself, thank you. **1977** C. McCullough *Thorn Birds* x. 236 He'd woo her the way she obviously wanted, flowers and attention and not too much slap-and-tickle.
f. *Mech.* = *piston slap* s.v. *piston *sb.* 4.
1930 *Engineering* 7 Mar. 304/1 Large engines..used,.. until recently,..cast-iron pistons. This is because of required durability..and because of expansion troubles—slap, leakage, &c.
2. a. Esp. in phrs. *a slap in* (or † *on*) *the face, in the eye, on the wrist*.
1861 [in Dict.]. **1895** Lloyd George *Let.* 3 June (1973) 85 So there's another slap in the eye for the Bryn party. **1898** G. B. Shaw *Philanderer* I. 96 I'll have to apologize for her... Her going away is a downright slap in the face for these people. **1914** *Dialect Notes* IV. 112 *Slap on the wrist*, mild rebuke or criticism. **1920** D. H. Lawrence *Touch & Go* 5 How much will you give me for my syllogism? Not a slap in the eye, I hope. **1932** L. Golding *Magnolia Street* II. iii. 308 The Great War was..a slap in the face, quite simply. **1966** *Economist* 26 Feb. 801/1 The Administration has been trying to choke off North Vietnam's supplies for some time... The latest move—a 'slap on the wrist', according to the dockers—has been to blacklist ships which visit Haiphong. **1970** P. Carlon *Death by Demonstration* vi. 73 All she said was, 'I don't want to talk about it.' That was as good as a slap in the eye. **1977** *Rolling Stone* 7 Apr. 17/3 We think we can get Anita off with a fine and a slap on the wrist, but the thing with Keith is much more serious. **1979** *Guardian* 17 Jan. 1/4 Industry will regard action to tighten price control.. as a slap in the face.
b. (Earlier example.)
1840 A. Bunn *Stage before & behind Curtain* III. 38 Enabled me to have a slap at the pretenders.

slap, *v.*¹ Add: **1. a.** Also *to slap* (someone) *on the back*: to clap (someone) on the back as a gesture of goodwill or congratulation; to treat in a hearty or jovial manner.
1908 E. M. Forster *Room with View* x. 173, I said, 'Hooray, old boy!' and slapped him on the back. **1914** A. C. Benson *Jrnl.* in D. Newsome *On Edge of Paradise* (1980) x. 320 M.F. was always the sort of man who slapped everyone on the back. **1931** R. Campbell *Georgiad* iii. 54 Nicolson who in his weekly crack Will slap the meanest scribbler on the back. **1941** 'R. West' *Black Lamb & Grey Falcon* I. 496 The stocky little men were..lifting their glasses to him and slapping him on the back.
d. *U.S. slang.* To play (a double-bass) without a bow in jazz style, *spec.* to pull the strings so as to let them snap back on to the fingerboard.
1933 [see *dog-house 2 b]. **1935** *Swing Music* June 83/1 The lyric, which was a masterpiece of fatuity.., had to do with the vogue of 'picking and slapping' the double bass. **1958** *Times Lit. Suppl.* 11 Apr. (Children's Literature Suppl.) p. iii/3 He takes up the violin, viola and cello, but happily stays just the right size for the double-bass, which he 'slaps' with such proficiency that he ends up in a famous jazz orchestra.
2. Delete † *Obs.* and add later example.
1884 'Mark Twain' *Huck. Finn* xvii. 158 She could rattle off poetry like nothing... She would slap down a line, and if she couldn't find anything to rhyme with it she would just scratch it out and slap down another one.
3. b. (Later *fig.* examples.)
1922 *Collier's* 1 July 26/1 Judge Tuckerman..slapped on the fines and costs with a lavish hand. **1924** E. M. Forster *Passage to India* vii. 61 The College itself had been slapped down by the Public Works Department, but its grounds included an ancient garden. **1968** J. Wainwright *Web of Silence* 12 A 'D Notice' gets slapped on the inquiry... The newspapers are gagged from the word 'go'. **1976** *Milton Keynes Express* 23 July 38/5 The Berks and Bucks FA have slapped a severe six-week ban on Hurrell following his sending off in a charity match.
7. (Later *poet.* example.)
1966 S. Heaney *Death of Naturalist* 57 And one Was scaresome for there, out of ferns and tall Foxgloves, a rat slapped across my reflection.
8. Also *trans.*, to throw. (In quot. *fig.*)
1957 [see *prisoner's dilemma* s.v. *prisoner² 1 b].
9. *trans.* To punish (someone) *with* a penalty, sentence, etc. *N. Amer.*
1968 *Globe & Mail* (Toronto) 13 Jan. 37/2 Late in the contest coach John Petrushchak and centre Bruno Marcocchio were slapped with technical fouls for disagreeing with the referees. **1972** *Newsweek* 10 Jan. 17/3 For his indiscretion, he is slapped with a stiff $17,000 fine. **1973** *Tucson* (Arizona) *Daily Citizen* 22 Aug. 4/1 Two young stepbrothers involved in a drug-crime..were slapped with five years partial probation.
10. In *fig.* phrs. *to slap* (a person or thing) *down*: to snub, suppress, or rebuke; *to slap* (a person's) *wrist*: to scold or reprimand; *to slap* (a person's) *face*: to administer a sharp reproof or rebuff.
1938 'E. Queen' *Four of Hearts* xi. 153 She's been.. leading me on just so she could turn around and slap me down. **1949** L. A. G. Strong *Maud Cherrill* 40 Any hint of affectation or pretentiousness she would have slapped down hard. **1960** 'E. McBain' *See them Die* xvi. 209 You're God, and there isn't anyone who's going to slap your wrist, no matter how you do it. **1973** *Times* 13 Aug. 4/2 Sales of this have gone up in recent years and we cannot think that the Government is going to slap the face of a very large number of users to save a little money. *Ibid.* 17 Oct. 20/3 The police sergeant who conducted the prosecutions was often slapped down by the clerk of the court for leading his witnesses. **1977** E. Ambler *Send no More Roses* iii. 53 Thinking that he was about to deliver the admonition, I let him go ahead. He didn't even slap her wrist. **1978** *Lancashire Life* July 63/2 His seniors might well have felt he was a publicity-seeker who needed slapping-down.
11. *Comb.* **slap-bass,** a double-bass played in jazz style (see quot. 1956); **slap-you-on-the-back** *attrib. phr.*, hearty, jovial.
1949 *Sun* (Baltimore) 22 Jan. 6/3 The slap-bass virtuoso who accidentally kicks a hole in his instrument in the middle of a jam session. **1956** S. Traill *Play that Music* iv. 46 This was the era of the slap bass: so called because the strings were pulled away from the fingerboard of the bass and..let smartly back to the fingerboard—thereby making a loud clicking sound. **1932** B. Worsley-Gough *Public Affairs* x. 182 'Lord,' said Venetia, 'I had forgotten the Bishop. What is he like?' 'Jovial. Jolly. Slap-you-on-the-back-for-tuppence.' **1957** *Times Lit. Suppl.* 22 Nov. 708/4 Mr Matthews has a jolly, slap-you-on-the-back approach. **1962** *Listener* 28 June 1114/2 Newbolt himself was no hearty, bluff, slap-you-on-the-back sort of man.

slap, *a.* (Earlier example.)
1840 H. Cockton *Valentine Vox* xiv. 108 But it's a werry nice place; werry private and genteel. None o' your public 'uns!—everything slap and respectable!

sla·p-back. *U.S.* Also **slapback.** [f. Slap *v.*¹ + Back *adv.*] A counter-attack, retaliation.
1931 *Sun* (Baltimore) 31 Mar. 10/7 The working of Canada's tariff slap-back against our beloved Smoot-Hawley law. **1941** *New Yorker* 27 Dec. 37/1 A resounding Allied slapback at the enemy in the Pacific.

slap-bang, *adv.* Add: Also of position: directly or precisely (*in* the centre); completely, absolutely. Cf. *bang *adv.* a.
1963 A. Smith *Throw out Two Hands* xiv. 143 That gas was contentedly holding over three-quarters of a ton 1,500 feet above a lake and slap bang in the middle of the sky.

slap-dab, *adv. N. Amer. dial.* and *colloq.* [f. Slap *adv.* + Dab *adv.*] = Slap-bang *adv.* in Dict. and Suppl. (see also quot. 1896). Cf. *smack-dab *adv.*
1886 *Turf, Field & Farm* XLII. 174/3 He was goin' that fas' he run slap-dab agin me afo' he seed me. **1896** *Dialect Notes* I. 399 *Slap-dab*.., violently or awkwardly. 'He rushed in slap-dab and broke things.' **1949** *Publ. Amer. Dial. Soc.* XI. 11 It jumped slap-dab in the middle. **1973** *Victorian* (Victoria, B.C.) 1 Aug. 1/1 It occurred slap dab in the middle of what is now Centennial Square.

slap-dash, *adv.*, *a.* and *sb.* Add: Hence **slapda·shness.**
1929 *Daily Tel.* 15 Jan. 7 If he has the defects of his virtues—a certain slap-dashness visible enough in one or two of these stories—he has also the virtues of his defects. **1965** *Punch* 27 Jan. 146/2 By halfway I was finding the slapdashness of the overall pattern rather self-indulgent and the separate delights were suffering from this.

slapdashery. Delete '(in *nonce use*)' and add later examples. Also (*rare*) **slapdasherie.**
1908 Kipling *Lett. of Travel* (1920) 144 Here and there the people are infected with the unworthy superstition of 'hustle', which means half-doing your appointed job and applauding your own slapdasherie for as long a time as would enable you to finish off two clean pieces of work. *a* **1913** F. Rolfe *Desire & Pursuit of Whole* (1934) vii. 60 That really huge romance..which his friend..wrote with such reprehensible slapdashery. **1966** *New Statesman* 14 Oct. 547/2 Sensibility and earthiness, obsession with detail and romantic slapdashery. **1982** *Times* 28 July 11/3 What it loses in slapdashery it gains in exuberance.

sla·p-happy, *a. colloq.* (orig. *U.S.*). Also **slaphappy.** [f. Slap *sb.*¹ + *-happy.]
1. Dazed, punch-drunk; dizzy (with happiness).
1936 J. Tully *Bruiser* x. 89 A slap-happy bum. **1938** *Newsweek* 23 May 22/1 A sample [of talk] designed to knock philologists slap-happy. **1940** *Detective Tales* Apr. 8/1 He was a little slap-happy from a decade of slug-festing. **1947** [see *punch-drunk *a.*]. **1973** *Maclean's Mag.* (Toronto) Feb. 32/2 It was so exhilaratingly ludicrous..that I felt quite slaphappy.
2. Carefree, casual; careless, thoughtless, irresponsible.
1937 *N.Y. Herald Tribune* 28 Aug. 14/1 After the dust had settled he and Ernest Hemingway, the slaphappy litterateur, toured Spain together. **1940** *Nation* 6 Apr. 448/1 Unless production [of television programmes] is slap-happy, the costs promise to compare with those of Broadway shows. **1951** R. Hoggart *Auden* iii. 78 This is Auden's characteristic 'Love' again, allied to his leftism—a lyrical but vague coda to a slap-happy knocking-down of many old guys. **1958** E. H. Clements *Uncommon Cold* i. 26 The irresponsible slap-happy manners of extreme modern youth. **1977** *Meanjin* XXXVI. i. 131 The real point is that Sydney—slap-happy and extroverted—is the favoured haunt of the cultural bureaucrats.
Hence **slap-ha·ppily** *adv.*, **slap-ha·ppiness.**
1958 S. Hyland *Who goes Hang?* xiv. 63 The impression of boisterous slap-happiness. **1968** M. Bragg *Without City Wall* II. xxvii. 249 Spray the window with froth of slap-happiness. **1969** *Guardian* 20 Oct. 22/2 He is unlikely to display himself so slap-happily before those surging crowds.

slapjack. 1. For *U.S.* read *N. Amer.* and add earlier and further examples.
1805 'An American Lady' *New Amer. Cookery* 60 *Indian Slapjack.* One quart milk, 1 pint of Indian meal, 4 eggs, 4 spoons of flour. **1867** J. K. Lord *At Home in Wilderness* viii. 132 Then I can bake bread in my frying-pan, make and fry pancakes, or 'slap-jacks', as trappers call them. **1895** W. Elkington *Five Years in Canada* xiii. 111 Another favourite dish..is also made of flour and water, mixed into a batter and fried in fat; it is eaten with syrup or sugar, and is called 'slap-jack'.

slapper¹. Add: **2. b.** In jazz, one who plays the double-bass (see *slap *v.*¹ 1 d).
1934 S. R. Nelson *All about Jazz* vi. 126 So a race of pickers and slappers..sprang into being. **1936** *Swing Music* Mar. 9/2 Steve Brown, that tremendous string-bass slapper.

slapping, *vbl. sb.*¹ Add: **c.** In jazz, the action of playing a double-bass (see *slap *v.*¹ 1 d).
1931 *Melody Maker* Dec. 1029/3 Slapping, too, becomes next to impossible with a high bridge. **1959** 'F. Newton' *Jazz Scene* 289 *Slapping* for pizzicato playing.

slappy (slæ·pi), *a. U.S. slang.* [f. *slap(-happy *a.* + -y¹.] = *slap-happy *a.* Also as *sb.*
1937 [see *punchy *a.*⁴]. **1942** Berrey & Van den Bark *Amer. Thes. Slang* § 151/9 *Foolish; silly;*..slappy. *Ibid.* § 702/32 '*Punch-drunk*'; *dazed*..slappy.

slap shot. *Ice Hockey.* Also **slapshot.** [f. Slap *v.*¹ + Shot *sb.*¹] A shot made with a sharp slapping movement of the stick, which usu. lifts the puck off the ice. Hence **sla·p-shoot** *v. trans.*, to hit (a puck) in this manner.
1942 E. Jeremiah *Ice Hockey* iii. 12 A slap shot..is very effective because its suddenness has a valuable surprise element. **1956** *Roy. Canad. Air Force Coach's Manual Hockey* 40 The slap shot can be made by stopping the puck first, then teeing off on it. **1968** *Globe & Mail* (Toronto) 3 Feb. 35/1 Athlete-columnists are more common than writers who can also skate and slapshoot a hockey puck. **1972** 'E. Lathen' *Murder without Icing* (1973) ii. 18 Billy Siragusa, the Huskies' new center, was practicing slap shots. **1974** *Los Angeles Times* 13 Oct. III. 2/3 Rene Robert got Buffalo's only goal at 14:06 of the opening period with a 30-foot slapshot.

sla·pstick. orig. *U.S.* Also **slap-stick.** [f. Slap *v.*¹ + Stick *sb.*¹] **1.** Two flat pieces of wood joined together at one end, used to produce a loud slapping noise; *spec.* such a device used in pantomime and low comedy to make a great noise with the pretence of dealing a heavy blow (see also quot. 1950).
1896 *N.Y. Dramatic News* 4 July 9/3 What a relief, truly, from the slap-sticks, rough-and-tumble comedy couples abounding in the variety ranks. **1907** *Weekly Budget* 19 Oct. 1/2 The special officer in the gallery, armed with a 'slap-stick', the customary weapon in American theatre galleries, made himself very officious amongst the small boys. **1925** M. W. Disher *Clowns & Pantomimes* 13 What has caused the playgoers' sudden callousness? The slapstick. Towards the end of the seventeenth century Arlequin had introduced into England the double-lath of castigation, which made the maximum amount of noise with the minimum of injury. **1937** M. Covarrubias *Island of Bali* iv. 77 Life-size scarecrows are erected, but soon the birds become familiar with them... Then watchmen circulate among the fields beating bamboo drums and cracking loud bamboo slap-sticks. **1950** *Sun* (Baltimore) 10 Apr. 3/1 The 50-year-old clown..said that when he bent over another funnyman accidentally hit him with the wrong side of a slap-stick. He explained that a slap-stick contains a blank ·38-caliber cartridge on one side to make a bang.
2. a. *attrib.* passing into *adj.* Of or pertaining to a slapstick; of or reminiscent of knockabout comedy.
1906 *N.Y. Even. Post* 25 Oct. 10 It required all the untiring efforts of an industrious 'slap-stick' coterie..to keep the enthusiasm up to a respectable degree. **1914** *Photoplay* Sept. 91 (*heading*) Making slap-stick comedy. **1923** *Weekly Dispatch* 4 Mar. 9 He likes good comedies.. but thinks the slapstick ones ridiculous. **1928** *Daily Sketch* 7 Aug. 4/3 The jokes..are rapier-like in their keenness, not the usual rolling-pin or slapstick form of humour. **1936** W. Holtby *South Riding* IV. v. 258 She took a one-and-threepenny ticket, sat in comfort, and watched a Mickey Mouse film, a slapstick comedy, and the tragedy of Greta Garbo acting Mata Hari. **1944** [see *pocho]. **1962** A. Nisbett *Technique Sound Studio* x. 173 Decidedly unobvious effects, such as the cork-and-resin 'creak' or the hinged slapstick 'whip'. **1977** R. L. Wolff *Gains & Losses* II. iv. 296 The prevailing tone of the book is highly satirical, with strong overtones of slapstick farce.
b. *absol.* Knockabout comedy or humour, farce, horseplay.
1926 *Amer. Speech* I. 437/2 *Slap-stick*, low comedy in its simplest form. Named from the double paddles formerly used by circus clowns to beat each other. **1930** *Publishers' Weekly* 25 Jan. 420/2 The slapstick of 1929 was often exciting. The Joan Lowell episode was regarded as exposing the gullibility of the critics... The popularity of 'The Specialist' made the whole book business look cockeyed. **1955** *Times* 6 June 9/1 A comic parson (Mr. Noel Howlett) is added for good measure, mainly to play on the piano while other people crawl under it. Even on the level of slapstick the farce seemed to keep in motion with some difficulty and raised but moderate laughter. **1967** M. Kenyon *Whole Hog* xxv. 253 A contest which had promised..to be short and cruel, had become slapstick. **1976** *Oxf. Compan. Film* 640/1 As it developed in the decade 1910–20..slapstick depended on frenzied, often disorganized, motion that increased in tempo as visual gags proliferated.

sla·p-tongue, *v.* Also **slaptongue, slap tongue.** [f. Slap *v.*¹ + Tongue *sb.*] *intr.* To produce a staccato effect in playing the saxophone by striking the tongue against the reed. So **sla·p-tonguing** *vbl. sb.*; hence **sla·p-tongue** *a.*, playing or played with this technique.
1926 Whiteman & McBride *Jazz* ix. 203 Slap tonguing is accomplished by sucking on the reed, thus creating a vacuum, then hitting the vacuum with the tongue, causing a pop. **1927** *Melody Maker* May 503/1 How to slap-tongue is not the easiest thing to explain in words. **1954** *Grove's Dict. Mus.* (ed. 5) VII. 433/2 Tongued staccato playing..was at one period developed by dance musicians into a curious explosive effect termed 'slap-tonguing'. **1963** *Down Beat* 3 Jan. 20 His first solo with Henderson, a clownlike, slap-tongue effort. **1969** *Listener* 12 June 838/1 Experimentation was limited to exploiting its [*sc.* the saxophone's] agility..or producing comic effects by slap tonguing. **1971** *Daily Tel.* 18 Sept. 7/8 Henderson's Columbia sessions..despite occasional farmyard noises and slaptongue reeds offer good Redman and Charlie Green.

slap-up, *a.* Add: **a.** (Earlier and later examples.) Now used esp. of meals.
1823 'J. Bee' *Slang* 161 *Slap-up*, used for 'bang-up'. 'Tis *northern*. **1931** W. S. Maugham in *Nash's Pall Mall Mag.* Dec. 24/1 A bottle of pop tonight, my pet, and a slap-up dinner. **1977** *Lancashire Life* Nov. 74/2 There was a slap-up tea at the institute.

slart (slɑːt). *dial.* [Origin unknown: for related senses see *Eng. Dial. Dict.*] *pl.* Left-overs, scraps.
1913, 1917 [see *ort]. **1977** Scollins & Titford *Ey up, mi Duck!* III. 50 *Slarts*, left-overs, scraps of food.

slash, *sb.*¹ Add: **1. b.** In *Cricket*, any unorthodox attacking stroke played with a great swing of the bat.
1906 A. E. Knight *Complete Cricket* ii. 78 A slash at the ball, the bat slicing the ball instead of meeting it with the full face. **1948** *Sporting Mirror* 21 May 7/1 The first shot he made after arriving at the wickets was a glorious 'slash' to the boundary. **1977** *Daily Express* 29 Jan. 35/1 Yajuvendra, never looking the part in his first Test innings, took a full-blooded slash outside off stump at a short one from Willis.
c. *fig.* A reduction; a (swingeing) cut. Cf. *slash *v.*¹ *1 d.
1950 *N.Y. Times* 20 Apr. 2/1 (*heading*) House group bars overall 50% slash in wartime excises. **1951** [see *consumer 2 c]. **1973** *Tucson* (Arizona) *Daily Citizen* 22 Aug. 1 This would be the second wave of base slashes in about a year. **1983** *Guardian Weekly* 6 Mar. 14/1 A 50 per cent slash in the army's budget.
3. b. (Earlier *absol.* and later *attrib.* examples.)
1839 H. Brandon *Poverty, Mendicity & Crime* 165/1 *Slash*, outside coat pocket. **1969** *Sears Catal.* Spring/Summer 47 No-iron reversible jacket... Plaid side has two slash pockets. **1973** W. Hallahan *Ross Forgery* iii. 3 The watchman pushed his hands into the slash pockets of his jacket.
4. a. An open tract or clearing in a forest, esp. one strewn with debris resulting from felling or logging, high wind, or fire. Cf. *slashing *vbl. sb.* 4 b. *N. Amer.*
1825 A. Anderson *Diary* 30 Aug. in G. Sellar *Narr.* (1916) vii. 102 We have been here scarce three months and there is a great slash. **1849** J. E. Alexander *L'Acadie* I. 272 After various difficulties..getting with our horses into 'slashes' or parts of the forest cut down.. we at last reached the small wooden hostel. **1923** H. E. Williams *Spinning Wheels & Homespun* 154 Raspberries are found oftenest in what are called 'slashes' in the woods, where the older timber has been cut down, and the new has not yet grown up to replace it. **1963** *Sun* (Vancouver) 23 Nov. 21/1 The rolling hills along the.. rivers are parklike with their copses of fir, tamarack, poplar and willow..left standing in old log slashes or burns.
b. Felled trees and other debris left in a forest after logging or the clearing of a tract, or resulting from high wind or fire. Cf. *slashing *vbl. sb.* 4 c. orig. and chiefly *N. Amer.*
In mod. use, *slash* denotes the branches and other trimmings cut from trees preparatory to removing the logs from a forest.
1841 *Bytown* (Ottawa) *Gaz.* 17 Feb. 1/3 To end of month clearing up old 'slash', which term has previously been defined. **1917** F. D. Adams in J. O. Miller *New Era in Canada* 85 In Quebec and British Columbia, settlers who desire to burn their slash must now obtain permits from the Government forest ranger, who supervises the burning. **1928** *Indian Forest Rec.* XIII. VII. 3 Comprehensively defined, *chir* slash includes all débris resulting from operations involving the felling and utilization of *chir* trees, and also from the destruction of trees of this species by such agencies as wind, snow, fire, lightning, floods, landslips, insects and fungi. **1952** P. W. Richards *Tropical Rain Forest* xvii. 379 Soil impoverishment will in turn depend on..the quantity of debris and 'slash' left on the ground after clearing. **1965** *Wildlife Rev.* Mar. 19/2 Cougars travel over long ranges and are found in various ecological types of terrain such as slash, mature forest and second growth. **1980** *Search* XI. 71/1 Planting of tubed stock on corridors cleared of slash may be another means of establishing eucalypt seedlings.
c. *attrib.* and *Comb.*, as *slash area, fire*.
1971 *Islander* (Victoria, B.C.) 30 May 12/3 After about a third of a mile you break out into a slash area where logging operations have been carried out. **1949** *Pacific Discovery* Jan.–Feb. 4/1 The river knew well the flashing draft of lightning fires in the grass but not the consuming roar of a slash fire. **1980** *Search* XI. 69/1 Slash fires result.. in mobilisation of large amounts of nutrients both during the fire..and subsequently as a result of stimulated biological mineralisation in the soil.
5. A thin sloping line, thus /; = *oblique *sb.* 5, Solidus¹ 2 in Dict. and Suppl. *U.S.* Also *slash-mark*.
1961 in Webster. **1964** *Amer. Speech* XXXIX. 103 The number to the right of the slash is the total number of occurrences of that type of clause. **1976** T. Allbeury *Only Good German* x. 76 Reference SC49 slash two. **1979** C. E. Schorske *Fin-de-Siècle Vienna* vii. 331 Breaking the phrase with slash marks at unsuspected nodes. **1980** *Maledicta* III. II. 206 Although it is true that . , : ; () - [] – ! ? / and * have names—in the case of /, several names: solidus, virgule, slash-mark, diagonal—there is a gap in the naming of *.

slash, *sb.*² Restrict † *Obs. rare* to sense in Dict. and add: **2.** *slang.* An act of urination.
1950 P. Tempest *Lag's Lexicon* 192 *Slash*, *to go for a*, to visit the urinal. **1953** *Chambers's Jrnl.* June 325/1 'I'm leaving my turret for a moment. I want a slash.' 'Okay, kid, you know where to find it?' 'I should do. I've had to empty them often enough!' **1977** N. J. Crisp *Odd Job Man* i. 5 He decided to risk a quick slash, which..he needed.

slash, *sb.*³ **a.** Substitute for def.: Swampy ground; a swamp. (Earlier and later examples.)
1652 in N. M. Nugent *Cavaliers & Pioneers* (1934) I. 239/2 Neer a wett slash, running N.N.W. to an Easternmost branch of Richard Cr. **1717** *Prince George County* (Virginia) *Deed Bk.* 202 in *Amer. Speech* (1940) XV. 393/1 A white Oake Standing in a round Slash. **1897** *Geogr. Jrnl.* IX. 538 There are many successive ridges of shingle running in varying directions, and often with narrow strips of marsh enclosed between successive ridges. Such bands of marsh have been given the very appropriate name of 'slashes' in New Jersey. **1903** *Dialect Notes* II. 330 [S. E. Missouri] *Slash*, wet bottom land. A slash differs from a slough in having no perceptible channel. **1966** *Publ. Amer. Dial. Soc.* XLVI. 29 *Slash*, a swamp.— 'It was in that slash down on the river.'
b. slash-pine, for '(see quots.)' read: a pine growing in a slash or low-lying coastal region, esp. *Pinus caribæa*, the principal native pine of south-eastern North America; also, the wood of this tree. (Later examples.)
1934 *Sun* (Baltimore) 7 Dec. 3/2 Dr. Herty recently made newsprint from slash and other southern pine. **1949** *Clarke County Democrat* (Grove Hill, Alabama) 28 July 1/3 Both have plots on which they have set out slash pine seedlings. **1974** *Calhoun Times* (St. Matthews, S. Carolina) 18 Apr. 2/1 Loblolly and slash pines are most susceptible [to rust galls].

slash, *sb.*⁴ (Later example.)
1916 [see *slatch 1].

slash, *v.*¹ Add: **1. c.** To clear (land) of vegetation, to cut (trees or undergrowth) *down*, esp. preparatory to burning off the resulting slash. Chiefly *N. Amer.*
1821 T. McCulloch *Stepsure Letters* (1960) 20 He had slashed down a large piece of wood; and now he determined to raise a crop. **1849** C. Hursthouse *Acct. Settlement New Plymouth* vii. 93 The cane-like fern stalks.. should be cut at once,..and the 'Tutu' slashed down with a bill-hook. **1889** W. H. Withrow *Our Own Country* 362 The native forest had been 'slashed' in that particular locality. **1931** *Beaver* Sept. 276 Five acres of virgin land were slashed. **1962** A. Fry *Ranch on Cariboo* 66 Sometimes we built fence or slashed brush to extend the yard.
d. *fig.* To reduce (something) severely in size or quantity. Freq. used with reference to prices, payments, etc.
1906 *Washington Post* 29 Apr. 6 A disposition was manifested in the Senate Committee to slash the salaries of members of the commission. **1910** *Springfield* (Mass.) *Weekly Republican* 8 Dec. 8 It is not a pleasant thing to slash a presidential message to this extent. **1931** *Evening Standard* 4 Aug. 10/1 The big department stores have not merely reduced their prices; they have 'slashed' them. **1958** *Listener* 13 Nov. 777/2 After that I stuck to one garage and slashed expenditure by 50 per cent. at a single stroke. **1976** *Daily Mirror* 16 July 1/1 Labour held their seat in yesterday's vital Thurrock by-election. But their majority was slashed.
2. b. (Later U.S. example.) Also in cricket, to play a vigorous attacking stroke. Occas. *trans.* (in quot. with bowler as object).
1955 [see *gully *sb.*¹ 2 d]. **1974** *Plain Dealer* (Cleveland, Ohio) 26 Oct. 7-D/7 Tailback Mike Newman slashed across from two yards out to cut Westlake's margin to 7-6 with 8:42 left. **1977** C. Martin-Jenkins *MCC in India* iii. 51 Viswanath slashed, snicked and was caught by Knott. **1977** *World of Cricket Monthly* June 87/1 The self-appointed England exile slashed Sarfraz for two boundaries in the first over.
9. *Comb.* **slash-hook** = Slasher 2 b.
1891 R. Wallace *Rural Econ. Austral. & N.Z.* xv. 231 Vines, creepers, supplejacks, and small saplings..require to be carefully cut by slash-hooks. **1930** Blunden *Poems* 188 Some harsh slash-hook Slit my skull and poured out all the fountains of my senses. **1942** [see *fagging *vbl. sb.*²].

slash (slæʃ), *v.*² *slang.* [f. *slash *sb.*² 2.] *intr.* To urinate.
1973 M. Amis *Rachel Papers* 189 If you can slash in my bed (I thought) don't tell me you can't suck my cock.

slash-and-burn, *attrib. phr.* Also **slash and burn, slash-burn, slash, burn.** [f. phr. *to slash and burn*: see Slash *v.*, Burn *v.*¹] Designating a method of shifting cultivation in which vegetation is cut down in an area of virgin or rejuvenated forest, allowed to dry, and then burned off before seeds are planted. Also occas. in non-*attrib.* use.
1942 Chapple & Coon *Princ. Anthropol.* viii. 189 Linton has a theory that the development of confederacies by the Maya and the Indians of the southeastern States was a consequence of this aspect of the slash-and-burn system. **1951** [see *swidden 2]. **1969** G. Clark *World Prehistory* (ed. 2) vi. 126 The rapidity of their spread.. was due in part to their use of slash and burn agriculture,..and in part to the lack of opposition. **1973** J. J. McKelvey *Man against Tsetse* iii. 190 The Kibori River eradication program..obviated the need to continue with a slash-burn program. **1976** *Conservation News* Sept./Oct. 5/1 The slash, burn, short-term cultivation and migration habit of tribal populations has been forced for centuries by the intractable fact that when an area of forest is cleared crops can be grown in it for only two or three seasons before the soil nutrients are exhausted. **1977** G. Clark *World Prehistory* (ed. 3) III. 126 The regime of slash and burn, involving the shift of cultivation every few years and the clearance of new patches of forest, inevitably implied the temporary abandonment of settlements. **1978** *Guardian Weekly* 2 Apr. 13/1 Eking out a living by fishing, picking fruit and berries and by slash-and-burn farming.
Hence **slash and burn** *v. trans.*; **slashed and burnt** *ppl. a.* Also **slash-bur·ning** *vbl. sb.*, the process or practice of slash-and-burn agriculture.
1919 *Jrnl. Forestry* XVII. 277 It is proposed..to conduct slash-burning operations under control in selected localities [in Canada] and observe the establishment of natural regeneration through a period of years. **1949** *Malayan Forester* XII. 83 The plot was established in order to ascertain the extent of soil erosion..and to study natural plant succession of the slashed and burnt..areas. **1954** *Brit. Columbia Lumberman* Mar. 78 Slash burning in a good seed year should precede seedfall, if possible. **1955** J. D. Freeman *Iban Agric.* vi. 128 Deleterious results.. ensue when healthy, young vegetation..is slashed and burnt before proper regeneration..has taken place. *Ibid.* 129 Three sample plots were established to study 'natural plant succession', on (*a*) the slashed and burnt area [etc.]. **1980** *Search* XI. 71/2 Since slash-burning has been widely used [in Australia] only for the last 20–25 years, there has been no opportunity in Australian native forests to observe its effect on the productivity of successive forest rotations. **1981** P. Theroux *Mosquito Coast* xii. 137 Slash and burn the whole area and we've got four or five acres of good growing land.

slashed, *ppl. a.* Add: **4.** Of timber: felled, esp. in an unplanned or destructive manner. *N. Amer.*
1843 *Yale Lit. Mag.* VIII. 332 His eye wandered far away over acres of slashed timber.
5. *Cricket.* Played with or resulting from a slash (*slash *sb.*¹ 1 b).
1974 *Observer* 9 June 24/8 A slashed catch to first slip accounted for Johnson. **1976** *S. Wales Echo* 27 Nov. 18/6 Reaching his half century in 138 minutes, he gave one difficult chance off a slashed stroke when he was 55, but Glenn Turner could not hold the ball at slip.

slasher. Add: **2. b.** (Earlier U.S. and later Austral. and N.Z. examples.)
1858 J. A. Warder *Hedges & Evergreens* 98 The slasher with a wooden handle set at an angle with the edge of the blade. **1916** J. B. Cooper *Coo-oo-ee* xi. 147 Cathead ferns and bracken..soon sprang up if he neglected to use the 'slasher', a large broad half-moon knife fixed to a stout ash handle. **1947** J. Bertram *Shadow of War* 70 Native convicts kept the lawns trimmed with formidable slashers.
d. A form of circular saw used to cut logs into predetermined lengths, usu. having several blades mounted on the same shaft.
1892 P. Benjamin *Mod. Mechanism* 777 A power-feed slab slasher, which differs greatly from the ordinary type of slabbing machines..has but one saw. **1915** *Saw in Hist.* III. 38 Then there are Slashers—Circular Saws used in a gang, and averaging four or more to each set. **1947** N. C. Brown *Lumber* ii. 108 The slasher consists of a set of circular cut-off saws, generally arranged 49" apart on a single shaft, to cut slabs, edgings, and other sawmill refuse into suitable lengths for lath stock, broomhandle stock, fuelwood, or other purposes. **1963** R. R. A. Higham *Handbk. Papermaking* v. 111 It will be assumed that they [*sc.* the logs] have already arrived at the slasher

slashing, *vbl. sb.* Add: **2. b.** (Earlier example.)
1842 *Illustr. London News* 14 May 9/1 The slashing being fully studded..with diamonds, rubies, emeralds.
4. For *U.S.* read *N. Amer.* Delete def., transfer quot. to sense *4 b, and add: **a.** The action of felling trees, esp. as a preliminary to clearing a tract of forest by burning.
1822 *Port Folio* XIII. 68 The act of hewing down the timber is called *slashing*. **1833** A. Fergusson *Pract. Notes* 42 A mode of chopping is in use hereabouts, termed slashing. It consists in merely prostrating the trees, without any further operation for a season, and then at leisure consuming the whole by fire. **1899** W. A. MacKay *Pioneer Life in Zorra* 167 There were three ways by which the first settlers cleared the land. The first was called 'slashing'. **1963** E. C. Guillet *Pioneer Farmer & Backwoodsman* I. 309 Another easy and cheap method was 'slashing'... felling the trees and allowing them to stay where they fell for a season or two.
b. = *slash *sb.*¹ 4 a.
1840 *Jamestown* (N.Y.) *Jrnl.* 1 July 2/5 On Monday, the body of Mr. Brown was found in a slashing. **1894** [in Dict., sense 4 a]. **1912** 'R. Connor' *Corporal Cameron of N.W. Mounted Police* 269 At the 'slashing' the wagon ruts faded out and the road narrowed to a single cow path.
c. = *slash *sb.*¹ 4 b. Freq. *pl.*
1864 T. Weed in T. L. Nichols *40 Yrs. Amer. Life* II. 215 Cattle..were turned out to 'browse' in the 'slashings'. **1928** P. A. Taverner *Ornithol. Invest. near Belvedere, Alberta* 104 *Mountain Bluebird*, not uncommon in the burnt spruce and slashings, but scarce elsewhere. **1964** *Islander* (Victoria, B.C.) 20 Sept. 5/2 Another small fire was burning in slashing on the west side of Reef Point. **1980** *Northeast Woods & Waters* Dec. 7/3 It headed into a small swamp made up of very thick and very low alders, criss-crossed with blowdowns and slashings.
5. The sizing of yarn by means of a slasher (sense 4) (see also quot. 1960).
1895 R. Marsden *Cotton Weaving* 514 Blackburn prices for Tape-sizeing or Slashing. **1921** *Dict. Occup. Terms* (1927) § 369 *Dresser*,..one who prepares delicate and fine yarns by passing them through sizing or slashing frame. **1960** *Textile Terms & Definitions* (Textile Inst.) (ed. 4) 136 *Slashing*,..(1) A synonym for slasher sizing. (2) This term has also been adopted to indicate the process which is used to reduce the extensibility of rayon yarns... The process consists in stretching the yarn in the wet state and then drying it while maintaining the stretched length.

slashing, *ppl. a.* Add: **2.** Also of weapons with cutting edges.
1950 H. L. Lorimer *Homer & Monuments* i. 34 To give better protection against the slashing sword. **1964** C. Willock *Enormous Zoo* ix. 162 Their wrists adorned with semi-circular slashing knives and their fingers with the slashing blades they mount on rings.
3. a. Now used esp. of horses. (Later examples.)
1951 *Sport* 16–22 Mar. 20/2 A big slashing powerful chestnut with powerful limbs to support his considerable weight, he is equally good in front and behind the saddle. **1976** *Horse & Hound* 10 Dec. 8/1 Fred Winter was able to give Bula the run he needed..in Wednesday's Sundew 'Chase, named after the fine, big, slashing chestnut whom he rode to victory in..1957.
c. *spec.* in *Cricket*, playing or played in a vigorous or unrestrained manner. Cf. *slash *v.*¹ 2 b.
1832 P. Egan *Bk. Sports* 346/1 A free slashing hitter, who holds it a crime To get any less than six runs a time. **1849** *Boy's Own Bk.* 69 Such implements as these [*sc.* the old bats] were but ill adapted even for what is termed 'slashing hitting'. **1885** [see *free *a.* 8 f].
4. (Later examples.) Now chiefly *Austral.*
1861 *Harper's Mag.* Mar. 470/1 Capin Clapp..relaxed his rigid features..as he thought of the 'slashin' cargo we had aboard. **1969** *Telegraph* (Brisbane) 16 May 15/4 Elke Sommer in some slashing fashions takes up crime in this study of computerised skullduggery. **1973** *Ibid.* 16 Aug. 26/3 Who'll win a slashing $700 wardrobe just for looking her smart, bright self?

slat, *sb.*¹ Add: **4. c.** *pl.* The ribs. *slang* (orig. and chiefly *U.S.*).
1898 H. E. Hamblen *General Manager's Story* 33 There's nothing much the matter with him; few of his slats stove in, that's all. **1911** J. Masefield *Everlasting Mercy* 11 Billy bats Some stinging short-arms in my slats. **1916** C. J. Dennis *Ginger Mick* 28 Why don't ole England belt 'em in the slats? **1928** *New Yorker* 3 Nov. 44/2 When Mr. Kaplan pokes M. de Vos in the slats he (or it) [*sc.* the crowd] halloos rapturously for Mr. Kaplan. **1944** W. Stevens *Let.* 12 Sept. (1967) 473, I want to give the office a kick in the slats. **1976** *Observer* (Colour Suppl.) 29 Feb. 33/1 The crunch probably came with the V & G report where, to my mind unfairly, certain civil servants got a real kick in the slat.
d. *Aeronaut.* The part of an aeroplane wing that is forward of a slot near the leading edge, or that can be moved forward to create such a slot and so provide additional lift. Cf. *slot *sb.*² 2 d.
1931 *Man. Rigging for Aircraft* (H.M.S.O.) (ed. 3) i. 12 Slots are a device for varying the air flow over the surface of an aerofoil, by the use of an auxiliary aerofoil, or slat, set parallel to and in front of the leading edge of the main aerofoil. **1935** C. G. Burge *Compl. Bk. Aviation* 451/2 The slat behind which the slot itself lies is mounted so that it swings forward automatically when an angle of incidence some few degrees below stalling point is reached. **1960** C. H. Gibbs-Smith *Aeroplane* I. xiii. 104 The slotted wing, matured in 1919,..was a device consisting of a curved slat (at first manually operated and then automatic) which was made to project from the leading edge of the wing and thus force air through the resulting slot and over the upper surface of the wing: the effect was to.. postpone stalling.
6. b. slat-back *a.* orig. and chiefly *U.S.*, of a chair: having a back constructed of several horizontal ribs (cf. *ladder-back* (*chair*) s.v. *ladder *sb.* 6); also *absol.* as *sb.*; **slat conveyor** (see quot. 1957); **slat fence** *U.S.*, a fence made of slats.
1891 I. W. Lyon *Colonial Furnit. New England* 165 They were called in their day 'bannister back', 'split back', 'slit back', and sometimes 'slat back' chairs. **1904** W. B. Ware *Seats of Colonists* 12 *Slat-back Chair*: Now often known as *Shaker Chair*, is the simplest expression of the *Turned Chair*. **1952** J. Gloag *Short Dict. Furnit.* 434 *Slat back*, a name sometimes used for a primitive form of ladder back chair, with four or five slats between the seat and the top rail: a type made in the countryside. **1976** *Billings* (Montana) *Gaz.* 30 June 4-C/1 (Advt.), Slat-back rocker is constructed of selected hardwoods with an antique pine finish. **1916** G. F. Zimmer *Mech. Handling & Storing* (ed. 2) vii. 101 Slat conveyors are used largely to carry substance in bags, also general merchandise packed in boxes and crates. **1957** J. A. W. Huggill in H. W. Cremer *Chem. Engin. Practice* III. 413 The slat conveyor, for packages, sacks and similar unit loads, has its carrying surface made of wooden or metal slats.., each attached to the chain links. **1790** W. Bentley *Diary* 22 June (1905) I. 180 The Principal Garden is in three parts divided by an open slat fence painted white. **1938** M. K. Rawlings *Yearling* xxxiii. 424 He came to the slat fence. He felt his way along it.

slatch. 1. (Later example.)
1916 T. C. Cantrill et al. *Geol. S. Wales Coalfield* XII. xii. 116 It is probable that the circular or elliptical pockets of coal known as 'slatches' or 'slashes' are the remains or short closed synclinal masses of coal abnormally swollen out by the squeezing-down of the two sides of the syncline.

slate, *sb.*¹ Add: **1. c.** (Earlier example.)
1854 J. E. Millais *Let.* 25 May in M. Lutyens *Millais & Ruskins* (1967) 216 Ruskin..is certainly *mad* or has a slate loose.
2. b. Also in phrs. *to wipe* (*off*) *the slate*, *to wipe the slate clean*: to obliterate or cancel a record, usu. of a debt, misdemeanour, etc.; hence *loosely*, to make a fresh start.
1899 [see *active *a.* 4]. **1921** G. B. Shaw *Back to Methuselah* p. lxix, We are helpless before a slate scrawled with figures of National Debts... the sensible thing to do is to wipe the slate and let the wrangling States distribute what they can spare. **1937** A. Huxley *Ends & Means* iv. 27 Where violence is pushed to its limits and the victims are totally exterminated, the slate is wiped clean and the perpetrators of violence are free to begin afresh on their own account. **1960** *Times* 2 Mar. 14/1 Tactically, Wolves must bank on all-out attack to wipe the slate clean. **1973** *Times* 28 Apr. 11/4 What I try to do each year is to wipe the slate clean. 'Now what can I do this year?'
c. For *U.S.* read 'orig. and chiefly *N. Amer.*' and substitute for def.: a list of candidates proposed for election or appointment to an official (esp. political) post; also *transf.*, the group of candidates so nominated; a group of candidates (occas. also of electors) with a set of shared political views. (Earlier and later examples.)
1842 *N.Y. Tribune* 24 Jan. 3/1 The Regency are obliged to put them on the *slate* to be rid of them, and then rub names out at leisure. **1913** R. M. LaFollette *Autobiogr.* 12 Well, the fraternities made their slate and put it through. **1931** W. G. McAdoo *Crowded Years* xii. 182 The Governor had Brandeis on the slate as Secretary of Commerce. **1952** *Manch. Guardian Weekly* 1 May 2/2 There were..nine contests between slates of delegates pledged to Taft and slates pledged to Eisenhower. **1963** *Economist* 2 Nov. 18/1 Electors were originally independent agents, not bound to any party. However, 'slates' of electors soon appeared, usually, though not always, pledged..to one or other of the parties. **1968** *Globe & Mail* (Toronto) 13 Feb. 3/2 The Eglinton Federal Liberal Association last night..selected a complete slate of delegates pledged to vote for Finance Minister Mitchell Sharp at the national Liberal leadership convention. **1970** *New Yorker* 15 Aug. 78/3 Only three slates, or thirty candidates, can be elected. **1972** R. Thomas *Porkchoppers* (1974) xxvi. 230 Cubbin voted without hesitation for himself and his slate. **1976** *National Observer* (U.S.) 12 June 2/2 Uncommitted slates led the voting in the Democratic Presidential primary. **1977** *N.Y. Rev. Bks.* 23 June 19/3 Eliav..who abandoned Labor in 1975, led a socialist and dovish slate ('Sheli') in this election. **1979** *Observer* 27 May 9/2 It was possible to see lists—Labour backbenchers are great ones for lists—giving the 'slates' of the Tribune and Manifesto Groups for the Shadow Cabinet elections.
d. A written record of a debt made when purchase of goods is allowed on credit. Also *fig.*, esp. in phr. *on the slate*, on account. (See also quot. 1909.)
1909 J. R. Ware *Passing Eng.* 188/1 *On the slate* (*Lower Peoples'*), written up against you—from the credit-slate kept in chandlers' shops. **1922** Joyce *Ulysses* 369 Lose your customers that way. Pubs do. Fellows run up a bill on the slate and then slinking around the back streets into somewhere else. **1954** *Sun* (Baltimore) 30 Oct. 1/5 [London] Many food stores are putting the bills 'on the slate' until the men go back to work. **1966** 'J. Hackston' *Father clears Out* 114 The Site Committee..made history by going on the slate and ticking up a few rounds of drinks. **1973** J. Marks *Mick Jagger* (1974) 39 He let them run a slate because they seemed like good sorts. **1980** *Observer* 7 Dec. 3/3 He knew of pharmacists who had been asked to put the bill 'on the slate' by families needing four or five prescriptions.
5. (Earlier example.)
1813 Jane Austen *Let.* 16 Sept. (1952) 327 There was but 2 yᵈ and a qʳ of the dark slate in the Shop, but the Man promised to match it.
6. b. *slate-floored, -hung, -pointed, -roofed*.
1978 J. L. Hensley *Killing in Gold* (1979) xi. 151 The slate-floored entrance hall. **1948** J. Betjeman *Sel. Poems* 116 The slate-hung, goodly-builded house. **1930** J. Dos Passos *42nd Parallel* iv. 292 The shining gray slate-pointed roofs of Quebec. **1960** *Times* 26 Mar. 9/5 Indeed, everything here recalls France—the squares and cobbled streets, the whitewashed walls and dormered slate-pointed houses. **1981** V. Glendinning *Edith Sitwell* xi. 151 A whitewashed slate-roofed village.
d. *slate-grey* (later examples).
1937 *Discovery* Dec. 384/2 Its black or slate-grey body. **1976** *National Observer* (U.S.) 6 Nov. 21 (Advt.), Generously cut in quality wool gabardine—Mid-Fawn, Slate Grey or Lovat.
7. slate-writer, a person who practises slate-writing; **slate-writing,** in spiritualism: writing performed on a slate, attributed to the agency of a medium, but without physical contact of the medium and the writing instrument.
1902 F. Podmore *Mod. Spiritualism* II. IV. ii. 221 Professional slate-writers. **1949** G. B. Shaw *Buoyant Billions* 7 They have a cohort of Slate Writers and Writing Mediums. **1885** *Century Mag.* July 382/2 She can do the trance business, and knocks, and slate-writing, and all that sort of thing. **1898** *Sci. Amer.* 8 Oct. 229/2 There has probably been nothing that has made more converts to spiritualism than the much talked of 'Slate Writing Test'. **1930** H. Carrington *Story of Psychic Science* VI. 147 The majority of messages..have been upon *slates*—hence the former popularity of 'slate-writing' mediums. **1977** B. Inglis *Natural & Supernatural* xxviii. 277 Slade was one of the practitioners of the new technique: slate-writing.

slate, *v.*¹ Add: **1.** (Later *absol.* example.)
1941 *Cross & Plough* Ladyday 9/2 To shelter him, man had to fell timber,..to burn bricks and tiles, to thatch and to slate.
2. a. (Later examples.) Also const. *to*. Also, to plan, propose, or schedule (an event). Chiefly *U.S.*
1904 F. Lynde *Grafters* xxvii. 343 Griggs was on for the night run eastward with the express; and 'Dutch' Tischer had found himself slated to take the fast mail west. **1936** Wodehouse *Laughing Gas* ix. 94 You ought to be thanking me on your knees for warning you. Yes, sir, unless you pull up mighty quick, you're slated to get yours. **1944** *College Topics* (Univ. of Virginia) 30 Mar. 3 No one has been slated for the 220, but Wenger may run in that event. **1960** *Times* 14 Sept. 12/6, I was intrigued to see this heading in a Charleston paper 'Church Tour slated'... It turned out to be nothing more than the announcement of an annual plantation tour..to raise funds for the local Protestant Episcopal Church. **1966** [see *locomotive *sb.* 4]. **1971** *Wall St. Jrnl.* 22 July 1/2 The Treasury is offering new 7%, 10-year bonds... Other cash-raising moves are also slated. **1973** *Oxf. Mag.* 4 May 10/1 When Americans mean to do something they *slate* it, rather than timetable or table it. When they do *table* it, they don't mean to do it. **1979** *Farmington* (New Mexico) *Daily Times* 27 May 3A/3 Gov. Bruce King and.. Navajo Tribal Chairman Peter MacDonald are slated to attend the ceremony.
b. *spec.* to propose or nominate a candidate for political office; to form a slate (*slate *sb.*¹ 2 c) of candidates. *U.S.*
1804 J. Pearson *Let.* 26 Nov. in J. Steele *Papers* (1924) I. 441 The Federalists have not, nor do they intend slating a candidate. **1912** T. Dreiser *Financier* xxvii. 297 Stener, although he had served two terms, was slated for re-election. **1961** T. H. White *Making of President 1960* iv. 100 On one huge ballot the Charlestonian was offered *fifty-three* individual choices of candidates if he wished to ponder his selections. Such a mystifying ballot requires simplification..supplied by 'slating'. The local bosses, the union chiefs, the statewide candidates, the education-board candidates, even the veterans organizations, all make cross-alliances to settle on, then print, a 'slate' of approved candidates among the multitude of names.
3. To scrape (a skin or hide) with a slater to remove loosened hairs.
1885 C. T. Davis *Manuf. Leather* xxxii. 527 Upon removal from the bate the skins are 'slated', which is the removal of the fine hair remaining upon the skins after the unhairing operation.

slate-colour. (Earlier and later examples.)
1799 in M. Edgeworth *Parent's Assistant* (1800) VI. 119 Mr. Davis, slate-colour and straw. **1919** E. O'Neill *Rope* in *Moon of Caribbees* 180 The sea is a dark slate color.

slater¹. Add: **2.** (Later Austral. and N.Z. examples.)
1951 J. Frame *Lagoon* 45 She collected things, slaters and earwigs and spiders. **1965** *Austral. Encycl.* IX. 349/2 The best-known isopods are the terrestrial forms commonly called wood-lice, slaters, carpenters, or sow-bugs. **1979** *Sunday Mail Color Mag.* (Brisbane) 23 Sept. 21/3 (Advt.), Snails slugs slaters millipedes. Now Baysol kills them all.
3. A blade of slate or the like used for slating skins and hides.
1885 C. T. Davis *Manuf. Leather* xxxii. 527 The 'slater' is a tool closely resembling a 'slicker'; but the edge of the 'slater' is ground sharp.

Slater³ (slēi·təɹ). *Physics.* [The name of John C. *Slater* (1900–76), U.S. physicist.] *Slater determinant*: a determinant which expresses the total wave function of an atom and is totally anti-symmetric with respect to an interchange of electrons, the elements being single-electron wave functions.
Described by Slater in *Physical Rev.* (1929) XXXIV. 1293.
1952 *Jrnl. Chem. Physics* XX. 769/1 The wave function of the ³Eᵤ state is $|\phi_0\phi_0\phi_1\phi_1\phi_{-1}\phi_{-2}|$ or $|\phi_0\phi_0\phi_{-1}\phi_{-1}\phi_1\phi_2|$, where |...| means a Slater determinant, and the spin is assumed to be suitable for each state. **1964** L. Wilets *Theories Nucl. Fission* iv. 56 A Slater determinant of plane waves replaces the highly correlated true wave function. **1976** *Nature* 28 Oct. 804/2 Such determinants have ever since been known as Slater determinants and have found applications in many fields.

slath (slaþ). = Slat *sb.*¹ 5 a.
1875 [see Slat *sb.*¹ 5 a]. **1906** M. Corelli *Treasure of Heaven* 288 He..was now patiently mastering the technical business of forming a 'slath'. **1949** K. S. Woods *Rural Crafts of England* III. x. 162 A base is woven first, called in the Midlands a 'slath'. **1968** J. Arnold *Shell Bk. Country Crafts* 257 A round or oval one [*sc.* base] is started with the 'slath', an interlaced cross-work of rods of at least four each way, around which the weaving begins.

slather, *v.* For 'Chiefly *dial.*' read 'Chiefly *dial.* and *U.S.*' and add: **2. a.** *trans.* To spill or slop; to scatter. Also, to use in large quantities; to squander; to paste, spread, or smear liberally. Usu. with advbs.

1866 'Mark Twain' *Sk. Sixties* (1926) 210 You have slathered too many frivolous sentimental tales into your paper. **1875** —— *Let.* 26 Jan. (1917) I. 247 The partialities of Providence do seem to be slathered around (as one may say). **1876** C. C. Robinson *Gloss. Words Dial. Mid-Yorks.* 126/1 *Slather*.., to spill. **1877** F. Ross et al. *Gloss. Words used in Holderness* 128/2 Leeak at him! he's *slatherin* pig-meeat all across hoose fleear. **1895** J. T. Clegg *Stories, Sk., & Rhymes* 43 Some carless chilt had bin a buyin peawdher blue an' slattherr it. **1904** G. Ade *True Bills* 131 A very Rich Man who wishes to be Respected must fill his Clothes with Currency and go out and slather it around. **1928** Pease & Fairfax-Blakeborough *Dict. Dial. N. Riding Yorks.* 118/2 'Eh bud things is slather'd aboot i' that hoos!' 'T'hay an' strae war blawn clean oot o' t'stagarth an' slather'd aboot t'field ayont.' **1937** M. Hillis *Orchids on your Budget* v. 82 You can get a good make-shift facial by slathering cream on your face after a shampoo. **1954** *Columbia* (S. Carolina) *Record* (Sat. Comic Suppl.) 27 Nov. 6, I got some American toast here, an' I'll gladly slather some jam on to sweeten it up. **1967** E. B. Nickerson *Kayaks to Arctic* iii. 24 Slather it [*sc.* mosquito repellent] on where common sense indicates, and have confidence. **1978** J. Irving *World according to Garp* xvii. 363 Perfume, which he remembered Roberta slathering over him.

b. To besmear; to spread or splash liberally on. Usu. with *with*.

1941 W. A. Percy *Lanterns on Levee* xvi. 184 Another rest-camp, half completed, leaking the ample French rain, slathered with mud, awaited us. **1961** Webster s.v. *slather, Slathering the cars with paint. **1977** *Time* 25 Apr. 17/1 His top chef offers such specialties as veal tongue slathered with *foie gras*. **1979** R. Gillespie *Crossword Mystery* iii. 53 She burned painfully. Rocky slathered her externally with Solarcaine.

3. *slang.* To thrash, defeat thoroughly, castigate.

1910 O. Johnson *Varmint* viii. 112 He turned on the Coffee-colored Angel and slathered him, drove him hither and thither with terrific blows. *c* **1926** 'Mixer' *Transport Workers' Song Bk.* 36 And all their Simon Puritans Are infused with great delight When an officer is slathered As chair-warmer, parasite. **1968** *Globe & Mail* (Toronto) 13 Feb. 27/1 Canadians can get slathered in Olympic hockey.

slather (slæ·ðəɹ), *sb.* [f. prec.] **1.** *U.S. colloq.* Usu. *pl.* A large amount, lots, lashings.

1857 M. D. Sanford *Jrnl.* 30 Mar. in D. F. Danker *Mollie* (1959) 6, I believe there are 500 passengers, lots and 'slathers' of young men. **1876** 'Mark Twain' *Tom Sawyer* vii. 68 They get slathers of money—most a dollar a day. **1906** *N.Y. Globe* 20 Aug. 6 There is the same slather of indefinite charges. **1966** 'E. Lindall' *Time too Soon* (1967) vii. 69 What he's giving her I don't want anyway. A whole slather of animal passion. **1970** —— *Gathering of Eagles* vi. 63 They..ate the damper with slathers of sweet, brown treacle.

2. *north.* and *Sc. dial.* Thin mud; a sloppy mass.

1876 C. C. Robinson *Gloss. Words Dial. Mid-Yorks.* 126/1 *Slather*, puddle, in a thin state. **1928** Pease & Fairfax-Blakeborough *Dict. Dial. N. Riding Yorks.* 118/2 *Slather*.., mud and mess. **1939** A. Borthwick *Always Little Further* vi. 126 Two big slabs o' breed wi' a slather o' jam in atween. **1980** J. Gardam *Sidmouth Lett.* 134 Wear yer wellies or you'll get in a slather int yard.

3. *open slather* (*Austral.* and *N.Z. colloq.*), freedom to operate without interference, a free-for-all.

1919 J. V. Marshall *World of Living Dead* 71 They say she's an open slather up there. Not a demon [*sc.* a policeman] in the burg. **1949** J. Morrison *Creeping City* xxviii. 227 You're asking to be allowed an open slather at an essential public service without being challenged. **1959** G. Slatter *Gun in Hand* xiii. 180 It's worth a go. Come round, she said, it'll be open slather. **1974** *Sunday Sun* (Brisbane) 1 Sept. 5/1 The beef was marked and butcher shop customers had the opportunity to check the grading... Now it's open slather. **1979** *Financial Rev.* (Austral.) 9 May 5 A problem..was how to prevent an 'open slather' in the sale of tickets.

slating, *vbl. sb.*[1] Add: **4.** The process of removing hairs from skins or hides with a slater. Freq. *attrib.*

1885 C. T. Davis *Manuf. Leather* xvi. 313 In many..tanneries fleshing machines have been tried..in..other tanneries experiments were made to convert them into slating..machines. **1903** L. A. Flemming *Pract. Tanning* 12 In some cases it is necessary to work the skins through the slating machine, or upon the beams.

slating, *vbl. sb.*[2] **1.** (Earlier example.)

1860 P. H. Rathbone in *Trades' Societies & Strikes* 368 The society defended all men prosecuted for trade assaults or 'slatings', as the term was.

‖ **slatko** (slæ·tko). [Serbo-Croatian, lit. sugared fruit.] (See quots.)

1941 'R. West' *Black Lamb & Grey Falcon* II. 111 The gallery, here walled in though it is open in most monasteries, where the visitors are given slatko, the ceremonial offering of sugar or jam and glasses of cold water. **1961** *Times* 9 Sept. 11/3 In the hostelry guests..are offered *slatko*, the ceremonial offering of sugar or jam.

slatted (slæ·tėd), *a.* [f. Slat *sb.*[1] or *v.*[1] + -ed[2], -ed[1].] Made or furnished with slats.

1886 [see Slat *v.*[1] 2]. **1948** W. Faulkner *Intruder in Dust* ix. 180 The truck (it was another pickup; they..had commandeered it, with a slatted cattle frame on the bed). **1960** *Farmer & Stockbreeder* 2 Feb. 82/1 Calves..were housed in individual slatted-floor boxes until seven or eight weeks old. **1968** C. Brooke-Rose *Between* 30 The light pours through the slatted shutters making a slatted pattern on the left pale green wall. **1972** *Daily Tel.* 11 July 3/2 A vertically slatted radiator grille..and an XJ12 motif at the back distinguish the new car from the XJ6. **1980** *Amat. Gardening* 4 Oct. 31/3 An advantage of slatted benching against solid trays is that the slats can be opened out to permit better air circulation between the plants.

slaty, *a.* Add: **2.** *slaty cleavage*: see Cleavage 1 c.

4. (Later example.)

1981 *Woman's Jrnl.* Mar. 135/3 Duck is..on the fatty side, so we need a dry, even slaty, wine to accompany it.

5. *slaty-blue* (earlier and later examples).

1854 [see *jacinth 1 e]. **1975** H. R. F. Keating *Remarkable Case of Burglary* i. 1 The slaty-blue eyes in his thin pale face.

slaughter, *sb.* Add: **3. a.** (Later *transf.* example.)

1971 *Rand Daily Mail* 27 Mar. 5/3 The slaughter on our roads and damage to property are apparently accepted with equanimity.

4. (Later examples.)

1911 M. Beerbohm *Zuleika Dobson* viii. 137, I..am going to die for the love I bear this woman. And let no man think I go unwilling. I am no lamb led to the slaughter. **1926** W. R. Inge *Lay Thoughts of Dean* ii. ii. 98 The Russians..were driven like sheep to the slaughter, in some cases unarmed, and always insufficiently protected by artillery. **1955** J. Masters *Coromandel!* iii. 203 They are on their way now... They will be goats for the slaughter. **1982** *Daily Tel.* 10 Feb. 16/5 The rank-and-file membership of the union are meekly following their so-called leaders like lambs to the slaughter.

9. a. (Further examples.)

1899 C. J. Cutcliffe Hyne *Further Adventures of Captain Kettle* v. 123 The foreign crew of the lifeboat, limp with scare, would have been mere slaughter-pigs on board, even if they could have been lured there. **1958** *Johannesburg Sunday Times* 14 Dec. 7/1 The highest price for slaughter stock at the Ladysmith Farmers' Association stock sale last week was £52 10s. **1968** *Globe & Mail* (Toronto) 13 Feb. B10/3 Slaughter cattle of mixed quality. **1977** *West Briton* 25 Aug. 6/1 (Advt.), We have received Ministry approval under this Order for the sale of slaughter sheep and store and breeding sheep on the same day. **1978** *Morecambe Guardian* 14 Mar. 22/1 (Advt.), Usual Sale of Livestock including..Fat Cattle and Slaughter Cows.

b. *slaughter-pen, -yard* (further examples).

1796 *Deb. Congress U.S.* 28 Dec. (1849) 1720 Georgia was a slaughter-pen during the war. **1856** W. G. Simms *Charlemont* ii. 27 These lads..raise hogs for the slaughter-pen. **1878** *Rep. Indian Affairs* (U.S.) 151 For the first time in the history of this agency Indians have been induced..to perform the labor of the slaughter-pen. **1928** Blunden *Undertones of War* iv. 37 The casualties caused by the mine were sixty or more. Cuinchy..was a slaughter-yard. **1968** T. Kinsella *Nightwalker* 28 Pigs in a slaughteryard that turn and savage each other.

slaughter, *v.* Add: **2. b.** To destroy by excessive felling.

1896 *Vermont Agric. Rep.* XV. 85 Our lumber forests are being slaughtered. **1903** S. E. White *Blazed Trail Stories* 27 Fitzpatrick would not have the pine 'slaughtered'.

c. To defeat or demolish completely. *colloq.*

1903 *N.Y. Even. Post* 5 Oct. 3 McLaughlin's lieutenants are openly declaring that they will 'slaughter' the McClellan-Grout-Fornes ticket. **1929** C. E. Merriam *Chicago* 280 He was hopelessly beaten..in the primaries of 1907; and again slaughtered..in the primaries of 1915.

slaughterable (slǭ·tərăb'l), *a.* [f. Slaughter *v.* + -able.] That may be slaughtered; fit for slaughter.

1911 *Daily News* 25 Sept. 4 There will simply be a dearth of slaughterable cattle. **1966** *Punch* 17 Aug. 261/3 Even the angriest demon drivers are reduced to the status of slaughterable black sheep.

slaughter-house. Add: **2. a.** (Later *fig.* example.)

1918 [see *cutting-room* s.v. *cutting *vbl. sb.* 10].

c. *slang.* A cheap brothel.

1928 E. Sutton tr. *Londres' Road to Buenos Ayres* vii. 55 She had got into a slaughter-house at two dollars instead of five. **1962** W. Faulkner *Reivers* viii. 164 Both of you get to hell back to that slaughterhouse.

slaughtery (slǭ·təri). Delete † *Obs.* and add: **1.** (Later example.)

1904 Hardy *Dynasts* I. vi. iii. 196 If it indeed must be That this day Austria smoke with slaughtery, Quicken the issue as Thou knowest how.

2. (Later examples.)

1917 G. J. Nicholls *Bacon & Hams* 26 The exporting slaughteries or factories. *Ibid.*, There is appointed to each slaughtery at least one veterinary officer who acts as inspector.

slaum (slǭm), *v. dial.* Also **slorm.** [perh. related to Slime *sb.*, *v.*[1]; for other uses see *Eng. Dial. Dict.*] *intr.* To slobber, to blubber; also, to flatter obsequiously. Hence **slau·ming** *ppl. a.*, (*a*) muddy, sticky; (*b*) slobbering, obsequiously flattering. Also *vbl. sb.*

1787 W. Taylor *Scots Poems* 99 He has a dreadfu' drouth, Whilk *slawmin* canna put awa'. **1904** in *Eng. Dial. Dict.* V. 506/2 The wet maks the road a bit slaumin'. *Ibid.*, Yer needn't come slawmin 'ere, for a don't believe yer. (Notts.) **1911** A. Warrack *Scots Dict.* 527/1 *Slaum*, *v.*, to slobber; to blubber; to smear. **1920** D. H. Lawrence *Lost Girl* xi. 278 I'd rather have him than your smarmy slormin sort.

slaunchways, -wise (slǭ·ntʃwēˈz, -waiz), *adv.* and *a. U.S. colloq.* Also **slanch-, slawnch-.** [Alteration of Slantways *adv.*, Slantwise *adv.* and *a.*] Slanting(ly), oblique(ly); out of true. Also *fig.*

1913 H. Kephart *Our Southern Highlanders* xiii. 294 Slaunchways denotes slanting. **1923** *Dialect Notes* V. 236 *Slawnch-wise*, adj., slantwise, slanting, out of true. **1933** *Amer. Speech* VIII. iii. 82 A Texas colleague of mine..said he wanted a full-width bed so that he could lie *slaunchwise* part of the time. **1941** *Sat. Even. Post* 5 Apr. 23/2 The race tide would slap him slanchways pronto. **1944** *Publ. Amer. Dial. Soc.* ii. 61 'The road runs *slaunchways* across the field.' 'Don't hold your cup *slaunchways*; you'll spill your coffee.' **1981** *N.Y. Times Mag.* 19 Apr. 10 If you think Professor Cassidy's ideas are slaunchwise or skewhiffy, harass him directly.

Slav, *sb.* Add: **2.** = Slavonic *sb.* (Cf. Fr. *slave.*) Also *Comb.*

1924 G. G. Walsh *Emperor Charles IV* iii. 34 The right of the monks, in his presence, to recite the Offices in Slav. **1935** Huxley & Haddon *We Europeans* vii. 203 The Slav-speaking population of central Europe. **1972** D. Dakin *Unification of Greece* 265 The Greek Church and the Greek communities had maintained schools where even the Slav-speaking Orthodox could acquire a knowledge of Greek.

‖ **slava** (slā·vā). [Serbo-Croatian, lit. honour, renown.] A festival of a family saint in Yugoslavia, a name-day.

1900 'Odysseus' *Turkey in Europe* viii. 372 The Slava, or festival of the family saint. **1920** *Glasgow Herald* 6 July 9 They told us of that country beyond, with its mountains and rivers, its peasant homes, its Slavas and songs of heroes—an Arcadia in truth. **1970** J. Brown *Un-melting Pot* v. 76 The household gods have had a central place in Yugoslav life since pre-Christian times and families today still have their household saints and keep their slava—their annual family days. **1976** *New Yorker* 22 Mar. 68/3 He remembers the priest blessing the house on his father's *slava*, or name day.

slave, *sb.*[1] (and *a.*) Add: **I. 1. d.** *Slave of the Lamp*, in the story of Aladdin in the *Arabian Nights*: a genie summoned by rubbing a magic lamp and bound to perform the wishes of the lamp's possessor; hence, allusively, one who performs swift miracles, or one who is under an inescapable obligation.

c **1840** Lady Wilton *Art of Needlework* xv. 238 The accommodations provided for the king..on this occasion [*sc.* The Field of Cloth of Gold] were more than magnificent; a vast and splendid edifice that seemed..to rise almost with the celerity of that prepared by the slaves of the lamp. **1841** Dickens *Let.* 1 July (1969) II. 319, I am bound to be..constant to my plans. I am a poor Slave of the Lamp. **1853** C. Brontë *Villette* II. xxi. 90, I almost looked to see if a huge, dark, cloudy hand—that of the Slave of the Lamp—were not..guarding its wondrous treasure. **1897** Kipling *Stalky & Co* (1899) 38 (*title*) Slaves of the Lamp. **1953** E. Coxhead *Midlanders* v. 120 Their working life was a deadening one. They were as near as possible machines themselves, slaves of the implacable lamp. **1959** *Encounter* Aug. 67/2 The physical scientist..is the magician. He is the contemporary equivalent of that old friend of our children, the Slave of the Lamp. Which means that he is very much an underling: he makes his magic at the command of his masters.

4*. Applied to a thing. **a.** *Naut.* = *slave jib*, sense 9 below.

1934 *Yachting Monthly* LVII. 119/2 These craft [*sc.* Bristol Channel pilot cutters], when in the pilot service, carried a heavy mainsail roped up the leech, a heavy staysail with two sets of reef points, and a working jib, generally known as 'the slave'. **1970** E. March *Inshore Craft* II. vii. 263 A 'slave' slightly larger [than the storm jib] and so called because it was almost permanently set.

b. A slave device (see sense 5 c below).

1940 F. Hope-Jones *Electr. Timekeeping* (1942) xvi. 156 Using a remontoire impulse as a synchronizing signal to control a Graham dead-beat escapement clock employed as a slave. **1965** *Jrnl. Scientific Instruments* XLII. 444/2 The whole seconds of the slave and chronometer can be matched regardless of the position of synchronization. **1969** J. J. Sparkes *Transistor Switching* v. 129 Which flip-flop is called the master and which the slave is quite arbitrary. **1975** D. Pitts *Target Manhattan* (1976) xxix. 126 'The first move is to get hold of that master computer.'..'That would stop the explosion?' 'It will if they haven't given final instructions to the slave.'

II. 5. a. *slave-boy, -girl* (later examples), *-labourer, woman.*

1848 Mill *Pol. Econ.* I. ii. v. 294 No slave-labourers are worse fed, clothed, or lodged, than the free peasantry of Ireland. **1897** M. Kingsley *Trav. W. Afr.* iii. 70, I have myself seen..slave women who had suffered for theft. **1920** J. Masefield *Enslaved* 13 They took my lady with them as a slave-girl to be sold. **1957** H. Roosenburg *Walls came tumbling Down* iv. 79 The Nazis had used Russian and Polish POWs as slave labourers.

1962 H. R. Loyn *Anglo-Saxon England* i. 34 From this district [of Yorkshire] came the slave-boys seen and questioned by Pope Gregory in the Roman slave-market. **1977** P. Johnson *Enemies of Society* ii. 15 We find the great fourth-century senator, Symmachus.., asking a Danubian official to buy him twenty slave-boys, 'because on the frontier it is easy to find slaves and the price is usually tolerable'. **1980** F. Warner *Light Shadows* ii. 10, I gave the Emperor that slave-girl, Acte.

c. *techn.* Used to denote a subsidiary device, *esp.* one which is controlled by, or which follows accurately the movements of, another device.

1904 D. Gill in *Rep. H.M. Astronomer at Cape of Good Hope for 1903* 7 The Clock consists of two separate instruments:—(a) A pendulum... (b) The 'slave-clock' with a wheel train and dead-beat escapement, the pendulum of which has a period of vibration slightly shorter than one second. **1930** *Engineering* 11 Apr. 466/2 A micrometer..bearing against a 'slave' micrometer introduced to allow of the main one being set to a zero reading anywhere over a considerable range. **1938** *Jrnl. R. Aeronaut. Soc.* XLII. 907 This is a double acting liquid pressure remote control system having a number of slave cylinder units fed from a common source of pressure... The slave cylinders are arranged to effect a number of operations in a predetermined sequence when pressure is fed through one pipe line. **1945** *Electronics* Nov. 94/1 (*caption*) Master and slave stations transmit synchronized pulses, and the difference in their times of arrival determines the position of the ship or aircraft. **1963** *Wall St. Jrnl.* 13 Feb. 20/1 The 'slave' locomotive..is being tested... The 'slave' would ride in the middle of a train perhaps 200 cars long; automatic sensing devices would keep it pulling the back end of such a train in rhythm with the manned master locomotive up front. **1972** *Amat. Photographer* 12 Jan. 41/1 (*caption*) A slave flash... This gun can be clipped anywhere and will trigger its own bulb, being actuated by the flash from another bulb gun. **1976** *Pract. Householder* Nov. (Heating Suppl.) 2/1 A houseful of remote slave units (electronic, of course) and a central control system is no longer eccentric gadgetry. **1977** *Grimsby Even. Tel.* 27 May 3/9 (Advt.), 100 watt slave amplifier, £40. **1980** Keene & Haynes *Spyship* xiii. 152 With this information, beamed off a slave satellite.., set and drift can be measured.

6. a. *slave labour* (earlier examples), *-master, song, work.*

1820 *Deb. Congr. U.S.* 9 Feb. (1855) 1213 Free labor and slave labor cannot be employed together. **1842** Dickens *Amer. Notes* II. i. 16 The system of employing a great amount of slave labour in forcing crops. **1822** *Sunday Times* 20 Oct. 1/4 The Continental Monarchs were but so many slave-masters. **1869** Mill *Subj. Women* iii. 142 Servitude, except when it actually brutalizes, though corrupting to both, is less so to the slaves than to the slave-masters. **1924** W. Deeping *Three Rooms* ii. 16 His manner towards women was Oriental, playfully and tolerantly casual, but behind it was the thick hand of the slave-master. **1971** *Black Scholar* June 11/1 We were so abhorrent to our slavemasters that legal barriers were instituted to prevent the natural process of assimilation. **1881** *Harper's Mag.* May 818/2 The plaintive slave songs..have won popularity wherever the English language is spoken. **1939** *Jrnl. Negro Educ.* VIII. 641/2 The slave song was an awesome prophecy, rooted in the knowledge of what was going on and of human nature, and not in mystical lore. **1973** *Advocate-News* (Barbados) 19 Feb. 4/4 He sang an old slave song his grandmother had taught him. **1916** D. H. Lawrence *Twilight in Italy* 303 Life is now a matter of selling oneself to slave-work. **1926** J. Pedersen *Israel, its Life & Culture* I. ii. 381 They..set him to the ignominious slave-work of grinding in the prison house.

b. *slave block, camp, house, pen* (further examples), *pit, quarter.*

1907 Kipling *Actions & Reactions* (1909) 188 The Hajji had often gloatingly appraised his skill..at five thousand rupees upon any slave block. **1966** *Keystone Folklore Q.* XI. 74 She was only eight when she was sold on the slave block. **1953** K. Tennant *Joyful Condemned* xxii. 208 Men all over the world making weary marches to prison camps and slave camps. **1973** R. Dougall *In & out of Box* xi. 127 The great bulk of the work was carried out manually by wretched, scarecrow figures dressed in rags... They were gangs from the notorious Stalin slave camps in the Arctic. **1939** J. Masefield *Live & Kicking Ned* 145 Inside the slave-house. **1943** H. T. Kane *Bayous of Louisiana* iii. 256 A barn and a few slave houses are all that can be found today of the former grandeur of the Durands, among the trees. **1901** W. Churchill *Crisis* i. iv. 35 A score of miserable human beings waiting to be sold at auction. Mr. Lynch's slave pen had been disgorged that morning. **1951** E. M. Graham *My Window looks down East* x. 86 In the cellar they have a slavepen. **1931** *Times Lit. Suppl.* 4 June 440/3 Characteristic..are the winding entrance gangways, a feature..in the..'slave pits'. **1959** J. D. Clark *Prehist. Southern Africa* xi. 302 These pits have been popularly referred to as 'Slave pits' suggesting that they were the places where slaves were confined on their journey from the interior to the coast. This explanation is, however, more fanciful than factual. **1982** *Evening Post* (Bristol) 19 Jan. 12 (Advt.), 2 further rooms and basement, with slave pit. **1837** H. Martineau *Society in America* II. i. 49 The slave-quarter is large. **1911** G. B. Shaw *Getting Married* Pref. 160 The people whose conception of marriage is a farm-yard or slave-quarter conception are always more or less in a panic lest the slightest relaxation of the marriage laws should utterly demoralize society. **1956** G. P. Kurath in A. F. C. Wallace *Men & Cultures* (1960) 152 The hot rhythms of jazz..have emerged from slave quarters..to respectable society. **1981** P. Mallory *Killing Matter* vii. 82, I bought some groceries..and, once inside my undistinguished slave-quarter, made myself a drink.

c. *slave-class.*

1935 Huxley & Haddon *We Europeans* ix. 279 A slave-class..of markedly different ethnic type from their masters. **1977** W. M. Spackman *Armful of Warm Girl* 108 No birthright Philadelphia Quaker ever bothered his head over their slave-class preoccupation with the safety of their unappetizing souls.

7. a. (Later examples.)

1939 J. Masefield *Live & Kicking Ned* 71 A white slave who knew medicine might be worth double that to a slave-owner. **1946** *Nature* 2 Nov. 607/2 Slave-raiders were exhausting a wasting asset, the chief export of tropical Africa. **1957** V. W. Turner *Schism & Continuity in Afr. Soc.* vi. 193 The mechanisms which formerly maintained the norms governing the relations of slave-owners and slaves could no longer operate. **1973** *Black World* Oct. 68/2 Unable to live in the slave-holder's kingdom, Walker fled to Boston. **1982** *English World-Wide* III. i. 19 A majority of the planters attempted to surmount the moral conflict inherent in being a slave-holder.

b. *slave-holding* (earlier and later examples), *-making* (later examples), *-owning* (earlier and later examples).

1798 *Deb. Congr. U.S.* 29 June (1851) 2058 At present the slaveholding parts of the State are burdened with the heaviest part of the State taxes. **1959** D. K. Wilgus *Anglo-Amer. Folksong Scholarship* 353 White songs in the slave-holding areas. **1928** Slave-making [see *dulosis]. **1944** J. S. Huxley *On Living in Revolution* 61 The raids of the slave-making ants are..a curious combination of predation and parasitism. **1977** Richards & Davies *Imms' Gen. Textbk. Ent.* (ed. 10) II. iii. 1243 Slave-making ants are confined to the northern hemisphere and are members of four genera only. **1828** J. F. Cooper *Notions of Americans* II. xiii. 296 The confederation is nearly equally divided into slave-owning, and what are called free states. **1934** E. O'Neill *Days without End* i. 34 They..had supplanted Him [*sc.* Almighty God] with the slave-owning State—the most grotesque god that ever came out of Asia! **1971** *Black Scholar* June 3/1 The human capital of the slave-owning class.

c. *slave-dealing* (earlier example), *-holding* (earlier example), *-raiding.*

1835 J. E. Alexander *Sk. in Portugal* ix. 212 Many of the governors have held office solely for the purpose of enriching themselves by slave-dealing. **1841** J. Sturge *Let.* 30 June in *Visit to U.S. in 1841* (1842) 33 If slave-holding were to be justified at all, the slave-trade must be also. **1933** A. N. Whitehead *Adv. of Ideas* iii. 34 Mediaeval wars were dissociated from slave-raiding expeditions. **1957** V. W. Turner *Schism & Continuity in Afr. Soc.* p. xx, The slave-trading and -raiding of the nineteenth century.

9. slave ant = sense 4 in Dict.; **slave bangle, bracelet** orig. *U.S.*: formerly, a slave's identity bracelet worn on the wrist or ankle; now, a bandle of metal, glass, bone, etc., worn for ornament, freq. above the elbow; **slave jib** *Naut.* (see quot.); **slave king** *Indian Hist.*, one of a dynasty founded by a former slave, Qutb uddin Aibak, which ruled the Delhi Sultanate from 1206 to 1290; usu. *pl.*; **slave-maker,** an ant belonging to a species that uses ants of a different species as slaves; **slave market,** a market at which slaves are bought and sold; also *fig.*, esp. (*N. Amer. slang*) an employment exchange; **slave morality** [tr. G. *sklaven-moral* (Nietzsche, *Jenseits von Gut und Böse* (1886) 231)], a morality characteristic of the weak, and rooted in resentment of the powerful, that exalts the lowly virtues of meekness, obedience, etc.; **slave worker,** in the war of 1939–45: a person put to enforced labour by the German Nazi regime, esp. a foreigner deported to Germany for this purpose.

1862 *Chambers's Jrnl.* 15 Feb. 97/2 Reaumur discovered how the ants of South America sally forth to kidnap hundreds of the black 'slave ants'. **1895** J. H. & A. Comstock *Man. Study Insects* 641 The Slave-ant, *Formica subsericea*..is usually a dark-brown or ash-colored ant with reddish legs. **1923** U. L. Silberrad *Lett. Jean Armiter* ii. 33 A green-glass slave bangle. **1931** N. Cunard *Black Man & White Ladyship* 4 The thick old Congo ivories she thinks, are slave bangles. **1975** D. Gray *Ride on Tiger* ii. 20 She wore..a silver slave bangle on her right arm. **1934** Webster, Slave bracelet. **1940** R. Chandler *Farewell, My Lovely* xxi. 164 An emerald..that..managed to look as phony as a dime-store slave bracelet. **1976** Botham & Donnelly *Valentino* xxii. 169 Her special gift for her husband—a platinum slave bracelet. **1948** R. de Kerchove *Internat. Maritime Dict.* 685/2 *Slave jib*, a term used by yachtsmen to denote a working jib, almost permanently set. **1841** M. Elphinstone *Hist. India* II. vi. i. 1 (*heading*) Slave kings. **1882** W. W. Hunter *Indian Empire* ix. 223 Kutab-ud-din had started life as a Túrkí slave, and several of his successors rose by valour or intrigue from the same low condition to the throne. His dynasty is accordingly known as that of the Slave Kings. **1958** O. Caroe *Pathans* i. 17 The slave-kings who followed them in Delhi were, every one, a Turk. **1971** *Illustr. Weekly India* 11 Apr. 55/1 Following Mohammed Ghori's establishment in Delhi, we had several dynasties of Muslim rulers like the so-called Slave Kings. **1859** Slave-maker [in Dict., sense 7 a]. **1915** H. St. J. K. Donisthorpe *Brit. Ants* 282 *Formica sanguinea*, the blood-red Robber Ant, our only slave-maker, is one of the most interesting species, showing great intelligence in adapting its habits to varying circumstances. **1978** *Nature* 16 Mar. 209/1 Wilson surveys a range of behaviour including that of 'slavemaker' ants who have become dependent on workers of other species. **1835** W. E. Channing *Slavery* iv. 87 Slave-markets..turn to mockery the language of freedom in the halls of Congress. **1838, 1871** [in Dict., sense 6 b]. **1911** G. B. Shaw *Getting Married* Pref. 179 We are in the slave-market, where the conception of our relations to the persons sold is..simply commercial. **1931** 'D. Stiff' *Milk & Honey Route* 214 *Slave market*, that part of the main stem where jobs are sold. **1960** *Voice of Idle Worker* (Vancouver) 8 Feb. 2/2 The chances are that he will be a regular customer at the slave market for a few months. **1907** H. Zimmern tr. *Nietzsche's Beyond Good & Evil* ix. 227 In a tour through the many finer and coarser moralities which have hitherto prevailed or still prevail on the earth, I found certain traits recurring regularly together and connected with one another, until finally two primary types revealed themselves to me, and a radical distinction was brought to light. There is *master-morality* and *slave-morality*. **1907** G. B. Shaw *Major Barbara* Pref. 153 Nietzsche..regarded the slave-morality as having been..imposed on the world by slaves making a virtue of necessity and a religion of their servitude. Mr Stuart-Glennie regards the slave-morality as an invention of the superior white race to subjugate the minds of the inferior races whom they wished to exploit. **1960** J. O. Urmson *Conc. Encycl. Western Philos.* 282/2 It deals at length with slave morality which contrasts good and *evil*. **1946** *Ann. Reg.* 1945 185 The all-overshadowing event after the collapse of Nazi Germany was the liberation of the concentration camps and the huge army of foreign slave-workers in Germany. **1956** I. Serraillier *Silver Sword* vii. 47 Everyone over twelve had to register, and he would almost certainly have been carried off to Germany as a slave worker. **1971** P. D. James *Shroud for Nightingale* viii. 264 [They] were Jewish slave workers in Germany..they were given lethal injections.

Slave (slēˈv), *sb.*[2] Also **Slavey;** 9 **Slavé, Slavi.** [tr. Cree *awahkān* captive, slave; the disyllabic Eng. forms reflect a local jargon var. with Fr. suffix *-ais*.] (A member of) a grouping of Athapascan-speaking North American Indians living in the boreal forest region of northwestern Canada; the language of this people. Also *attrib.* or as *adj.*

1789 A. MacKenzie *Let.* 22 May in L. R. Masson *Les Bourgeois de la Compagnie du Nord-Ouest* (1889) 30 (*Récits* section), Mr. Leroux arrived on the 22nd March from the other side of Slave Lake where he had seen a great number of Red Knives and Slave Indians. **1801** —— *Voyages from Montreal* (1903) I. viii. 340 When this country was formerly invaded by the Knisteneaux, they found the Beaver Indians inhabiting the land about Portage la Roche; and the adjoining tribe were those whome they called slaves. They drove both these tribes before them; when the latter proceeded down the river from the Lake of the Hills, in consequence of which that part of it obtained the name of the Slave River. **1851** J. Richardson *Arctic Searching Exped.* I. viii. 242 The comfort, and not unfrequently the lives, of parties of the timid Slave or Hare Indians are sacrificed. **1862** R. G. Latham *Elements Comparative Philol.* lv. 391 The *Beaver* Indian is transitional to the Slave and the Chepewyan proper. **1875** H. H. Bancroft *Native Races Pacific States* III. 587 A greater divergence from the stock language is observable in the dialect of the Tutchone Kutchin, which, with those of..the Slavé of Francis Lake..might almost be called a dialectic division of the Tinneh language. **1890** W. C. Bompas (*title*) Hymns in the Tenni or Slavi language. **1907** F. W. Hodge *Handbk. Amer. Indians N. of Mexico* I. 440 Petitot restricted the term [*sc.* Etchareottine] to the Etcheridiegottine, whom he distinguished from the Slaves proper. **1932** D. Jenness *Indians of Canada* xxiii. 390 In summer the Slave lived in conical lodges covered with brush or spruce bark. **1938** E. M. North *Bk. Thousand Tongues* 870/1 *Slave*... Spoken by Indians living along the Mackenzie River, northwestern Canada. **1946** J. J. Honigmann *Ethnogr. & Acculturation of Fort Nelson Slave* 16 He is married to a Slave woman and in his cultural affiliations and background is more Slave than Cree. **1959** E. Tunis *Indians* x. 132 There was a group, the Etchaottine (Slaves), who were kind to old people. **1974** *Sunday Tel.* 18 Aug. 5/5 In the Territorial Capital of Yellowknife barmen have noticed a substantial reduction in the number of Dogribs seeking drink—and a corresponding increase in the number of Indians claiming to be members of the Chepeweyan and Slavey tribes. **1979** M. E. Krauss in Campbell & Mithun *Languages of Native Amer.* 862 These are all to a significant degree mutually intelligible, with Dogrib being the most divergent (not counting Slavey). **1981** *Handbk. N. Amer. Indians* VI. 79 No convenient name for this language exists, although Slave or Slavey was in 1980 commonly used as a self-designation by most speakers of Mountain, Bearlake, and Hare, as well as of Slavey proper.

slave, *v.*[1] Add: **1. d.** *techn.* To subject (a device) to control or regulation by another device. Const. *to* the device.

1952 [see *manipulator 2 f]. **1958** C. C. Adams *Space Flight* v. 132 The camera is synchronized with the National Bureau of Standards radio transmitter WWV, whose chief function is to broadcast time signals of incredible accuracy... It does this by means of a crystal clock..which is 'slaved' to WWV. **1978** *Broadcast* 21 Aug. 5/3 (Advt.), Picture stabilization provided by an oscillating mirror slaved to the film perforations.

2. a. (Later example.)

1925 E. O'Neill *Compl. Wks.* II. 154 Didn't he slave Maw t' death?

slaved, *ppl. a.* Add: **3.** With prep., as *slaved-for. rare.*

1952 Dylan Thomas *Coll. Poems* 132 My paid-for slaved-for own.

sla·ve-drive, *v.* [Back-formation from next.] *intr.* To exploit slave labour; to demand hard or servile labour. Also *trans.*, to demand an excessive amount of work from (a person). So **sla·ve-driven** *ppl. a.*, **sla·ve-driving** *ppl. a.* and *vbl. sb.*

1830 *Reg. Deb. Congr. U.S.* 10 May 939/1 Here they may live and flourish, until some slave-driving politician and planter of South Carolina..again chains them to a miserable dependence on South Carolina cotton and British looms. **1836** *Blackw. Mag.* Oct. 562/1 Ye slave driving hosts of factory and poor law commissioners. **1859** J. B. Jones *Wild Southern Scenes* vi. 46 The reason alleged was his alliance with the 'slave-driving' Blounts of the South. **1878** G. Meredith *Lett.* (1970) II. 565, I hope to propose myself to you for a night in January. At present I have the Devil behind me slave-driving. **1889** Slave-driving [in Dict. s.v. Slave *sb.* (and *a.*) 7 c]. **1889** G. B. Shaw in *Star* 1 June 4/1 Bully him; slave-drive him. **1933** Slave-driven [see *medium close-up* s.v. *medium B. 3 d]. *a* **1935** T. E. Lawrence *Mint* (1955) ii. xiv. 136 Corporal Hemmings again supervised the gymnasium-work today. He slave-drove us as usual. **1952** E. O'Neill *Moon for Misbegotten* i. 13 He's gone like Thomas and John before him to escape your slave-driving. **1957** W. Camp *Prospects of Love* ii. v. 61 You're not to let her slave-drive you. **1982** P. Fitzgerald *At Freddie's* iv. 32 There was no need for her to go back... Indeed it was probably a mistake, and might give Freddie the notion that slave-driving encourages slavery.

sla·ve-driver. orig. *U.S.* [Slave *sb.* + Driver.] An overseer of slaves; *transf.* an exacting taskmaster.

1807 *Salmagundi* 13 Feb. 42 Beautiful, Oh most puissant slave-driver, as are my wives, they are exceeded by the women of this country. **1828** *Athenæum* 29 Feb. 162/2 The scourge shall no longer sound among the Antilles, nor the image of God be trampled by the slave-driver into the likeness of the beasts that perish. **1830**, etc. [in Dict. s.v. Slave *sb.* (and *a.*) 7 a]. **1854** Thoreau *Walden* 10 It is..worst of all when you are the slave-driver of yourself. **1857** T. B. Gunn *Physiol. N.Y. Boarding-Houses* xv. 127 He'd been an overseer—or, as they termed it, *slave-driver*—down South. **1901** Merwin & Webster *Calumet 'K'* x. 189 Do you think it would be worth something to the men who hire you for a dirty slave-driver to be protected from a strike? **1922** E. O'Neill *Hairy Ape* iii. 32 Bloody slave-driver! **1948** H. L. Mencken *Amer. Lang.* Suppl. II. 674 The [prison] guards are *shields, screws,..slave-drivers* or *herders.* **1975** *Investors Chron.* 28 Feb. 589/1 He is not a slave driver but somehow generates an extremely hard-working atmosphere.

slavey[1]. **1.** Delete † *Obs.* and add later examples.

1901 M. Franklin *My Brilliant Career* xvii. 141 Harold Beecham kept a snivelling little Queensland black boy as a sort of black-your-boots, odd-jobs slavey or factotum. **1967** *Atlantic Monthly* Apr. 103/2 All his years a loyal slavey he had worked his heart out for peanuts.

Slavey, var. *Slave *sb.*[2] **Slavi,** obs. var. *Slave *sb.*[2]

Slavic, *sb.* Add: **a.** (Earlier examples.)

1812 A. Murray *Let.* 8 Aug. in T. Constable *A. Constable* (1873) I. 333, I wish, however, to have about 100 or 150 printed pages additional on the Latin, Slavic, Persic, and Celtic. **1850** [see *Church Slavic* s.v. *church *sb.* 18].

b. *Comb.*, as *Slavic-speaking* adj.

1942 *Amer. Council of Learned Societies Bull.* No. 34. 58 (*heading*) The Slavic-speaking groups of the United States and Canada. **1980** *Word* 1979 XXX. 19 Albanians in Yugoslavia are classified as a nationality (*narodnost*) within a population consisting predominantly of Slavic-speaking peoples.

Slavicist (slā·v-, slæ·visist). [f. Slavic *sb.* + -ist.] = Slavist[2].

1930 K. Malone in *Stud. in Honor of H. Collitz* 328 *Slavist*..actually has two meanings:..(2) an authority on Slavic, and for the second meaning *Slavicist* would be a more appropriate term. **1964** *Slavic Rev.* XXIII. 707 There does not seem to be an organized correlation between trained Slavicists and potentially available posts. **1976** *Language* LII. 108 Most Slavicists agree on two cardinal classes of denominal qualitative adjectives.

slavikite (slæ·vīkəit). *Min.* [ad. Czech. *slavíkit* (Jirkovský & Ulrich 1926, in *Věstník Státního Geol. Ústavu Česk. Repub.* II. 345), f. the name of František *Slavík* (1876–1957), Czech mineralogist: see -ite[1].] A trigonal greenish-yellow hydrated basic sulphate of ferric iron found as an oxidation product of pyrite and usu. containing additional magnesium and sodium.

1927 *Mineral. Abstr.* III. 365 This mineral, named slavikite (after the abstractor), forms minute, almost microscopic, crystals with the rhombohedron (10$\bar{1}$1) and the base (0001) equally developed, or tabular parallel to (0001). **1957** *Ibid.* XIII. 369 Slavikite..associated with gypsum, pickeringite, and jarosite, occurs in Belgium at Stavelot on Revinian phyllites and at Val-Dieu on Famennian sandstone. **1968** I. Kostov *Mineral.* 500 Slavikite is trigonal, found in minute uniaxial negative crystals tabular on {0001}.

slaving, *vbl. sb.* Add: **2.** The automatic control or regulation of one device by another.

1960 *How TV Works* ii. 17/2 This technique—known by the jargon of 'slaving'—ensures that the programmes sent out from the transmitters are no less excellent in quality when they originate from very distant studios. **1974** *Some Technical Terms & Slang* (Granada Television), *Slaving*, interlocking electronic signals between disparate sources.

Slavistic. Add: In *pl.*, Slav linguistic studies.

1956 *Archivum Linguisticum* VIII. ii. 176 Slavic synchronic comparative grammar, a branch of Slavistics which does not yet exist. **1965** [see *jat' *sb.*[2]].

Slavo-. a. *Slavo-Lithuanian* (earlier example).

1874 H. Bendall tr. *Schleicher's Compar. Gram.* i. 7 Teutonic and Sclavo-Lithuanian.

slavocracy. (Earlier example.)

1840 *Illinois State Reg.* (Springfield) 22 Jan. 2/2 The reign of the slaveocracy is hastening to a close.

slavocrat. (Earlier example.)

1842 S. M. Gates *Let.* 24 Jan. in J. G. Birney *Lett.* (1938) II. 666 Some Slaveocrats in Georgia have lately attempted to cast odium upon Mr. Adams.

Slavonian, *a.* Add: **2.** *Slavonian oak*, the silvery timber of a European oak, esp. *Quercus robur* or *Q. petræa*, grown in the Slavonian region of Yugoslavia.

1938 E. H. P. Boulton *Dict. Wood* 127 Oak, Slavonian... Similar to English Oak but softer. **1956** *Handbk. of Hardwoods* (Forest Prod. Res. Lab.) 167 Slavonian oak, from Yugoslavia, is typically of slow, even growth, has a uniform colour and straight grain, and is mild and easy to work. **1966** A. W. Lewis *Gloss. Woodworking Terms* 106 Slavonian oak (throughout Europe).

slaw. For *U.S.* read *N. Amer.* and add: (Earlier and later examples.) Also, any dish the main ingredient of which is sliced cabbage.

1794, etc. [see *cold-slaw]. **1861** T. Winthrop *Cecil Dreeme* 157 Pad of butter. Plate of slaw, ready vinegared. **1905** *N.Y. Even. Post* 23 Sept. 2 Mince pie, hokey-pokey ice cream, over-ripe watermelon, frankfurters with hot slaw—all the less expensive and less desirable articles of diet go to stunt the gamin's growth. **1916** *Chambers's Jrnl.* Feb. 143/1 In Canada it [*sc.* celery cabbage] is used for cold slaw. **1944** *Sun* (Baltimore) 1 Nov. 10/7 It was customary in his family in his boyhood to serve a 'hot slaw' with turkey, the slaw consisting of cabbage cooked with vinegar and sugar. **1977** *National Observer* (U.S.) 22 Jan. 9/1 If she craves tossed salad when the price of lettuce is high, she resolutely buys cabbage and makes slaw instead.

slawnchwise, var. *slaunchways, -wise *adv.* and *a.*

slay, sley, *sb.*[1] Add: **2. slay sword**, each of the supports upon which the slay of a loom oscillates during the process of weaving.

1895 R. Marsden *Cotton Weaving* v. 166 The shaft is cranked, and by means of arms from these cranks is attached to the 'slay' or lathe..which oscillates upon the 'slay-swords'. **1963** A. J. Hall *Textile Sci.* iii. 142 This reed is fastened to the sley sword *S*, which is pivoted..so that as required it can swing to and from position *X* after the insertion of each weft thread.

slay, *v.*[1] Add: **II. 5.** *fig.* (Further examples.) *esp.* To overwhelm with delight, to convulse (someone) with laughter. Cf. *kill *v.* 6 a.

1863 G. Meredith *Let.* May (1970) I. 203, I have lately been slain by a pretty face. **1927** L. Mayer *Just between Us Girls* i. 2 Well, anyways, my dear, it simply slayed me. **1937** *Amer. Speech* XII. 181 (*heading*) Satchelmouth slays 'em. **1943** H. A. Smith *Life in Putty Knife Factory* xiii. 225 The boys who slay me..are the ones who have set pieces to recite when they answer the phone. **1953** R. Chandler *Long Good-Bye* xi. 68 A hoodlum with sentiment... That slays me. **1958** *Spectator* 21 Nov. 728/1 Frost,..reading naturally and roughly but with a high degree of contrivance, slaying them into calls for encores and favourite poems. **1965** D. Francis *Odds Against* xi. 150 'Oh God, Dolly, you slay me,' said Chico, laughing warmly. **1975** D. O'Sullivan in D. Marcus *Best Irish Short Stories* (1977) II. 98 They're fun... They'll slay you! **1977** *Guardian Weekly* 23 Oct. 4/3 The earliest comment on these lines that I can find comes from Denis Thatcher in October, 1970. 'Who could meet Margaret..without being completely slain by her personality and intellectual brilliance?'

slean (slīn), var. Slane.

1951 *Engineering* 6 Apr. 389/2 'Sleans' are used to cut the peat. **1976** J. Hayes *Missing* (1977) v. 203 The man's had his slean out for me since we cut the turf together... Had me unfrocked, he did. **1977** J. Hodgins *Invention of World* iii. 74 When some turf-cutter drives his slean into the peat in that desolate valley [etc.].

sleaze (slīz), *sb. slang.* [Back-formation from Sleazy, sleezy *a.*] **1.** Squalor; sordidness, sleaziness; dilapidation; (something of) inferior quality or low moral standards. Also *attrib.*

1967 *Listener* 14 Sept. 326/2 For all its brazen sleaze, Soho is a pretty fair working model of what a city neighbourhood should be. **1975** *Publishers Weekly* 29 Dec. 68/2 Obviously written to cash in on 'Mandingo', this isn't even readable sleaze: the plot's sloppy, Gilchrist hasn't the knack for writing commercial sex, and the hero is too despicable to be seductive. **1976** *National Observer* (U.S.) 17 July 16 (*heading*) At home with the sleaze king. **1981** *New Yorker* 9 Mar. 104/1 These stores are vast, computerized sleaze centers, where you can buy almost anything—pills, toys, candy, liquor, stockings, pillows, and gadgetry.

2. A person of low moral standards.

1976 *Telegraph* (Brisbane) 3 Aug. 10/3 When I made the mistake of calling them 'sleazy' to their faces, their reaction was outrage. 'Don't call me a sleaze,' said Miss Currie. **1977** *Time* 28 Feb. 48/1 Oh God, red nail polish—I look like a sleaze.

sleaze, sleeze (slīz), *v.* [f. as prec.] **1.** *dial.* (See quots.) ? *Obs.*

1777 in *Eng. Dial. Dict.* **1825** J. Jennings *Observations on Some of Dialects in W. of England* 69 *To sleeze*,..to separate; to come apart: applied to cloth, when the warp and woof readily separate from each other. **1904** *Eng. Dial. Dict.* V. 513/1 *Sleeze*,..*v.* Of loosely or badly woven cloth: to separate, part asunder; to wear away; also with *away*.

2. *slang.* To move in a sleazy fashion.

1964 *Punch* 30 Dec. 986/2 Other plays, sleazing across the West End boards. **1978** W. F. Buckley *Stained Glass* xxii. 211 The depressing, unseasonal föhn that sleazes over Europe with dumpy barometric pressures that enervate and depress. **1978** *Whig-Standard* (Kingston, Ontario) 31 Mar. A-6 When not thumping.., they [*sc.* a rock group] just kind of sleaze along, a little like Lou Reed at his best.

slea·zily, *adv.* [f. Sleazy, sleezy *a.* + -ly[2].] In a sleazy manner.

1959 *Listener* 5 Nov. 793/2 A bloated society staggering sleazily towards decay. **1974** M. Fido *Kipling* 119/2 Huneefa and her sleazily exotic magic. **1976** *Listener* 14 Oct. 486/4 Despising and envying the order and bourgeois contentment of the new ménage, sleazily growling in her basement flat.

sleaziness. Delete *rare* and add later examples.

1959 *Guardian* 3 Dec. 7/2 The pop-eyed sleaziness of a Master Elliot's night-boy. **1978** *Fortune* 18 Dec. 107 The classifieds represent a sleaziness of practice that I hope will be corrected by the more ethical newspapers and publishers.

sleazo (slī·zo), *a.* and *sb. U.S. slang.* [f. Sleazy, sleezy *a.* + *-o[2].] (Something) sleazy, (something) pornographic.

1972 E. Sanders *Family* iii. 68 Manson used to hang out on the Sunset Strip using the name Chuck Summers. There were a bunch of sleazo bars and cafés on or near the Sunset Strip. *Ibid.* 69 What was it that caused Manson's death-trip? The factors that seem to have fed the violent freak-out shall be termed here sleazo inputs. **1975** *Guardian Weekly* 2 Aug. 20 Her sleazo routines seem half-rehearsed. Her dirty jokes sound almost like last-gasp fill-ins for the cleaner lines she was supposed to deliver but forgot. **1978** *Courier-Mail* (Brisbane) 4 Jan. 16/2 Norman Mailer said he liked sex movies, especially 'love pornies' and 'the sleazos'.

sleazy, sleezy, *a.* Add: **2. c.** Dilapidated, filthy, slatternly, squalid; sordid, depraved, disreputable, worthless.

1941 J. Faulkner *Men Working* i. 31 Gwendolin..had been hanging on to her dress and peering around her wide sleazy hips. **1941** W. A. Percy *Lanterns on Levee* x. 111, I was always happening on a Hermaphrodite, in some discreet alcove, and I would examine the sleazy mock-modest little monster. **1946** 'P. Quentin' *Puzzle for Fiends* 248 Her glamour dissolved... Suddenly Selena seemed sleazy. **1951** S. Kaye-Smith *Mrs Gailey* iv. xvi. 211 In her now was a real distaste for the sleazy comforts of Mrs Turner's kitchen. **1956** L. McIntosh *Oxford Folly* viii. 128 Beyond it was the cemetery and some sleazy suburb whose name Adrian did not know. **1958** *Punch* 27 Aug. 286/2 A kind of sleazy, leering sex for its own sake. **1959** F. Usher *Death in Error* x. 145 I've seen all the sleazy joints of Paris. **1966** *Listener* 30 June 947/2 He scratches a middle-aged living in Paris as a gigolo in a sleazy night club. **1971** B. W. Aldiss *Soldier Erect* 46 Some sleezy and probably malevolent god. **1976** *National Observer* (U.S.) 1 May 19/1 The methods used to slur the innocent grew sleazier. **1977** *Listener* 3 Mar. 282/1 The entrance of Salomé is greeted by the world's sleeziest tune, 'La Paloma'. **1979** R. Jaffe *Class Reunion* i. iv. 42 Scollay Square, an area that was sleazy, neon lit, and disreputable.

sled, *sb.*[1] Add: **1. c.** Any of various devices made to be towed along the sea bed.

1939 *Sun* (Baltimore) 25 Jan. 3/4 As the sled passes over a buried cable both coils develop electric current which is wired to the mother ship above. **1967** *Petroleum* XXX. 158/2 The jets and suction dredge are mounted on a sled lowered from a frame at the stern of the vessel and straddle the pipe, along which they move as the barge proceeds. **1978** *Nature* 9 Mar. 156/2 This hypothesis is consistent with..the detection of a large *He excess in a 'thermal plume' (thermal anomaly ~0.1°C, sampled using a deep-tow sled) over the Galapagos Spreading Centre.

2. b. Also *rocket sled*. A rocket-propelled vehicle running on rails for subjecting things to controlled high accelerations and decelerations.

1948 *Richmond* (Va.) *Times-Dispatch* 13 Jan. 9 The rocket-powered sleds moved over a standard-gauge railroad track.., covering the 2,000 feet in less than two seconds. **1956** L. Mallan *Men, Rockets & Space* vi. 84 When he decelerated from 421 m.p.h. on the new sled, he reached a peak of only 22 G's. **1967** *Technology Week* 20 Feb. 18/1 Feasibility of a new segmented solid rocket sled motor designed for multiple re-use in sled testing

has been shown in recent tests. **1974** *Encycl. Brit. Macropædia* XV. 942/1 Braking is accomplished by a parachute or, more often, by extending a scoop beneath the sled into a trough of water beneath the track rails.

5. *sled-dog* (N. Amer.).

1692 H. Kelsey *Indian Belief* in *Kelsey Papers* (1929) 21 Now as for a woman they do not so much mind her for they reckon she is like a Slead dog or Bitch when she is living & when she dies they think she dyes to Eternity. **1777** G. Cartwright *Jrnl.* 18 Dec. in *Trans. Labrador* (1792) II. 277 Finding my sled-dog lame, I defered my journey. **1914, 1966** [see *mush *v.*[2]]. **1980** *Beautiful British Columbia* Winter 9 (*caption*) Sled dogs at Atlin's Long Distance Dog Sled Race.

sledding, *vbl. sb.* For *U.S.* read *N. Amer.* and add: **b.** With qualifying adj. (Further examples.)

1908 R. E. Beach *Barrier* 127 Now them kind of places is all right for married men but they're tough sleddin' for single ones. **1939** L. M. Montgomery *Anne of Ingleside* xxvii. 185 Ye *can* keep a house without a woman, but it's hard sledding. **1954** T. P. Kelley *Black Donnellys* 57 Damnation—had he somehow, unknowingly, revealed himself? If so, he could surely expect some rough sledding ahead. **1968** M. Woodhouse *Rock Baby* iii. 20 It was hard sledding, but after a while I got him to say something definite. **1979** *Arizona Daily Sun* 19 Apr. 4/2 This part of the administration's plan likewise faces tough sledding on Capitol Hill.

sledge, *sb.*[2] Add: **1. b.** = *sled *sb.*[1] 2 b.

1957 *New Scientist* 20 June 16/3 The isolation of such a fault may lead to simpler performance studies on restricted imitative devices... They may be built into the man-carrying centrifuge or rocket sledge to observe changes due to acceleration.

4. *sledge-driver* (earlier example), *-head*; **sledge-meter,** a wheel and counting device towed behind a sledge to measure the distance travelled.

1819 *Theatrical Inquisitor* Apr. 314 Of the literary talent of the stage manager, we have never thought highly, and his 'Land Storm; or the Sledge Driver and his Dogs' seems only a little alteration from a piece, called 'Lowina of Tobolski'..by the same author. **1924** A. J. Small *Frozen Gold* i. 15 A..human thunderbolt which hurled at him... A sledge-head knocked up from nowhere and connected with his chin. **1966** S. Heaney *Death of Naturalist* 41 The cap juts like a gantry's crossbeam, Cowling plated forehead and sledgehead jaw. **1902** R. F. Scott *Jrnl.* 18 Nov. in *Voyage of 'Discovery'* (1905) II. xiii. 24 A dull day..., but we plodded on... Starting at 11 A.M., we pushed on for two and a half miles by our sledge-meter. **1929** J. G. Hayes *R. E. Peary* x. 166 Peary never used a sledgemeter on the Arctic pack, saying it would have been smashed by the rough surface. **1958** *Times* 2 Jan. 6/5 During last night's painful run the tractors were biting down 2 ft... The 12-hour haul wound a laborious 22 miles through the sledgemeter.

sledge-hammer, *sb.* (Earlier *attrib.* and *fig.* examples.)

1799 T. Holcroft *Jrnl.* 13 Jan. in *Memoirs* (1816) III. 123 Yet having read mine, you come with a sledge hammer of criticism, describe it as absolutely contemptible. **1843** Dickens *Let.* 24 Sept. (1974) III. 572, I sent Miss Coutts a sledge-hammer account of the Ragged schools.

sledge-hammer *v.* (later examples).

1963 A. Smith *Throw out Two Hands* xvi. 167 Sledge-hammering a steel spike into the ground. **1976** *CRC Jrnl.* July 19/1 It is perfectly possible to understand what is going on on stage without having the point sledge-hammered home.

sle·dging, *vbl. sb.*[2] *Austral. Cricket slang.* [f. Sledge *sb.*[1] + -ing[1].] (See quot. 1982.)

1977 *World of Cricket Monthly* June 5/1 Lillee had his views on intimidating batsmen on Melbourne television. And on *sledging*—the term comes from *subtle as a sledge-hammer*. **1979** *Age* (Melbourne) 2 July 9/6 A year or so earlier the Australian team coined 'sledging' for needling or gamesmanship. **1982** *London Portrait Mag.* May 68/3 *Sledging*, trying to disrupt batsmen's concentration by abusing, teasing them. **1983** *Guardian* 8 Feb. 23/1 Geoff Howarth says he intends to complain about the amount of swearing, sledging and unchecked short-pitched bowling New Zealand have faced.

sleek, *v.* Add: **2. d.** To draw (a comb) through hair with a smoothing effect.

1959 *Listener* 21 May 904/1 The boys sleeking combs through their hair. **1967** G. B. Mair *Girl from Peking* vii. 73 The Admiral..was sleeking a comb through his thinning hair.

sleep, *sb.* Add: **1. d.** (Later examples.) Also *spec.*, the solid substance found in the corners of the eyes and along the edges of the eyelids after sleep.

1905 in *Eng. Dial. Dict.* **1922** 'R. West' *Judge* i. iv. 195 Richard was sitting in front of the fire, rubbing the sleep out of his eyes. **1951** L. MacNeice tr. *Goethe's Faust* 241 Children, you have scarcely scrubbed your eyes of sleep—and bored already? **1955** J. D. Salinger in *New Yorker* 29 Jan. 27/1 He began to massage the side of his face.., removing..a bit of sleep from one eye. **1973** P. White *Eye of Storm* vii. 300 The girl stood..washing the sleep out of her eyes.

e. *to lose sleep over*, etc.: see *lose *v.*[1] 3 b.

2. Also, in hyperbolic phrase *could do something in one's sleep* and varr.

1953 E. Coxhead *Midlanders* viii. 187 There's no difficulty. We could make them in our sleep. **1970** J. Braine *Stay with me till Morning* i. 9 His job didn't claim much of his energy. He could, as they say, do it in his sleep.

3. a. Also, in phr. *to have* or *get one's sleep out*, to sleep until one wakes naturally.

1685 C. Gardiner *Let.* in M. M. Verney *Mem.* (1899) IV. ix. 341 Your grandsons shall have their sleep out beefore they goe. **1911** F. H. Burnett *Secret Garden* xvii. 183 'You must go back and get your sleep out,' she said. **1930** A. Bennett *Imperial Palace* lvii. 434 She had told him to call her. He had refused; she must have her sleep out.

b. (Earlier *N. Amer.* and later examples.)

1670 N. Carteret in *S. Carolina Hist. Soc. Coll.* (1897) V. 166 The Caseeka..was within one sleep of us. **1896** C. Whitney *On Snow-Shoes to Barren Grounds* 182 The one 'sleep' did not bring us up to the caribou, but it took us north to the lodge of another Indian. **1919** Cody & Cooper *Memories of Buffalo Bill* 312 It was many sleeps away. **1953** D. Cushman *Stay away, Joe* 53 From three-four sleeps came riders to the tepee of our father.

c. *fig.* A prison term, usu. comparatively short. *slang* (orig. *U.S.*).

1911 D. Lowrie *My Life in Prison* vi. 63 A year sentence is known as a 'sleep'. **1931** 'D. Stiff' *Milk & Honey Route* vi. 65 Any time you want to retreat to some such place for a short or long rest or 'sleep', just go to the social worker displaying the proper symptoms. **1938** J. Phelan *Lifer* xix. 202, I wasn't interested myself [in escaping]. Three years was nothing—just a sleep, as you chaps put it. **1971** D. Bagley *Freedom Trap* iii. 59 In prison jargon, a 'sleep' is a sentence from six months to two years; a 'cut' is from two to four years, and a 'stretch' is anything over four years.

4. a. *to put to sleep*, to kill, esp. painlessly; also *fig.*

1942 R. Godden *Breakfast with Nikolides* v. 118, I want you simply to give him an injection and put him to sleep. I will muzzle him. **1966** K. A. Saddler *Gilt Edge* ii. 37, I had to have the Allard [*sc.* a car] put to sleep... She started coughing up oil... I couldn't bear to see her in agony. **1967** A. Lewin *Unaltered Cat* ii. vi. 134 Her cousin's Siamese cat..had..a litter of four adorable seal-point kittens. Ethel's husband was for putting them to sleep, but Ethel wouldn't hear of it. **1970** *Women's Household* July 10/1 She had started to suffer, so the humane thing to do was to put her to sleep. **1975** tr. *Melchior's Sleeper Agent* (1976) iii. vii. 154 The Führer's Alsatian dog, Blondi..had been ordered put to sleep. Dr. Haase had given her poison.

5. b. (Earlier example.)

1859 Princess Royal *Let.* 12 Dec. in R. Fulford *Dearest Child* (1964) 212 Wegner..pinches his arm to see whether he feels it... He feels just a little but not much, like a part that is gone to sleep.

6. a. *sleep deprivation*.

1966 I. Jefferies *House-Surgeon* xii. 230 Those who are still not weakened sufficiently are eroded remorselessly by sleep-deprivation. **1980** E. Behr *Getting Even* i. 18 We tried another tactic... Truth drugs... Even some subtle sleep deprivation. Inconclusive.

b. *sleep-producing*.

1844 E. A. Poe in *Godey's Lady's Bk.* Apr. 177/2 This *rapport* extended beyond the limits of the simple sleep-producing power... At the first attempt..the mesmerist entirely failed. **1907** W. James in *Amer. Mag.* Nov. 64/2 The best sleep-producing agent which his practice had revealed to him was *prayer*.

c. *sleep-bound, -dazed, -twisted*; also with ppl. adjs., as *sleep-drunk*.

1841 J. G. Whittier in *Knickerbocker* May 369 Bend o'er us now, as over them, And set our sleep-bound spirits free. **1889** *Cent. Dict.*, Sleep-drunk. **1938** E. Bowen *Death of Heart* iii. v. 417 He..saw in a sleep-bound way how specious wisdom was. **1951** Koestler *Age of Longing* i. iii. 41 Sleep-drunk and frightened, Hydie begins to cry. **1954** —— *Invisible Writing* iv. xxxvii. 394 Sleep-dazed, he is unable to decide which of the two hostile dictators is reaching out for him this time. **1960** T. Hughes *Lupercal* 21 Our lantern's little orange flare Made a round mask of our each sleep-dazed face. **1960** S. Plath *Colossus* 39 Sleep-twisted sheets.

7. sleep apnœa, apnœa occurring in sleep; **sleep-coat**, a knee-length front-fastening night-shirt or dressing-gown; **sleep-learn** *a.*, pertaining to sleep-learning; **sleep-learning, -teaching** *vbl. sbs.*, learning, teaching, during sleep, esp. by exposure to radio, tape-recordings, etc. (cf. *hypnopædia); **sleep movement** *Bot.*, a movement of a part of a plant, esp. a leaf, that occurs each nightfall and, in reverse, each daybreak; **sleep-shorts,** shorts as an item of nightwear; **sleep sofa** *U.S.*, a sofa which may be used as a bed; **sleep-talk** *v. intr.*, to speak (as) during sleep; **sleep-talker** (later examples); **sleep-talking** (later examples); **sleep-teaching** *vbl. sb.*: see *sleep-learning* vbl. sb. above; **sleep-waker** (earlier example); **sleep-wear,** night wear.

1976 *National Observer* (U.S.) 18 Dec. 16/4 Sleep apnea has three forms: central apnea [etc.]. **1980** *Brit. Med. Jrnl.* 29 Mar. 895/2 Sleep apnoea was defined as cessation of airflow at the nose and mouth lasting for at least 10 seconds. **1948** *Sun* (Baltimore) 11 Feb. 3/5 (Advt.), Rayon knit sleep-coats... What young, gay, pretty and practical sleepers! **1966** *Punch* 23 Feb. 290/2 The nearest Hardy Amies equivalent is a..short-sleeved garment called a 'sleep-coat', designed for warm climates and centrally heated homes... It is a loose negligé, fastening with a sash. **1976** *National Observer* (U.S.) 6 Nov. 21 (Advt.), Sleepcoats for cool knights. **1968** *Punch* 4 Dec. 804/2 Two new 'sleep-learn' devices. **1972** D. Lees *Zodiac* 151 By using an adaptation of the sleep-learn technique we can turn you into anything we want. **1953** M. L. Coyne in *Jrnl. Exper. Psychol.* (1956) LI. 97/1 (*heading*) Some problems and parameters of sleep learning. **1966** *Listener* 8 Dec. 852/3 Equally dangerous might be so-called 'sleep-learning' courses which try to teach you something special, like a new language, while you sleep. **1972** J. Gores *Dead Skip* (1973) iii. 18 Through his mind, like a sleep-learning tape, reeled Bart's words. **1880** Sleep-movement [in Dict., sense 6 a]. **1906**, etc. [see *nyctinastic *a.*]. **1965** Bell & Coombe tr. *Strasburger's Textbk. Bot.* 391 *Mimosa* also shows sleep movements, and at nightfall appears almost as if stimulated by mechanical shock. **1964** *Women's Wear Daily* 30 Nov. 50 A pair of sleepshorts in all sleepwear colors to be sold separately, worn with anything the consumer wishes. **1976** *National Observer* (U.S.) 29 May 2/6 (Advt.), Classic travel shave coat & matching sleep shorts. **1973** *Washington Post* 13 Jan. A13/3 (Advt.), Elegant traditional sleeper. Refined traditional sleep sofa from Waynline. **1960** S. Plath *Colossus* 48 The sleep-talking virgin. **1980** *Times Lit. Suppl.* 23 May 578/5 She has a tendency to over-point,..and I can feel death around too, very little hope or heart—'although I admit I desire', she intones, sleep-talking. **1972** R. Adams *Watership Down* x. 44 His voice sank and became that of a sleep-talker. **1981** *Maledicta* V. 287 Each sleeptalker was asked..a series of some 25 questions and sub-questions about his/her verbalizations. **1899** Sleep-talking [see *somniloquy* s.v. Somniloquacious *a.*]. **1939** Joyce *Finnegans Wake* 459 She's a fright, poor old dutch, in her sleeptalking. **1981** *Maledicta* V. 286 There has always existed the phenomenon of somniloquy, commonly known as sleeptalking. **1932** Sleep-teaching [see *hypnopædia]. **1957** A. Huxley *Let.* 12 Dec. (1969) 837 A dictator..could, by the use of drugs, sleep-teaching, hypnosis, subliminal projection..establish a high degree of control over his subjects. **1970** *Times* 19 Sept. 13 In Russia more than 180 educational institutes..are now fully equipped for sleepteaching. **1844** E. A. Poe in *Columbian Mag.* Aug. 70/2 As the sleep-waker pronounced these latter words..I observed upon his countenance a singular expression. **1935** A. P. Herbert *What a Word!* iv. 115, I have been implored by many to attack..'sleep-wear', and 'swim-wear'. **1964** [see *sleep-shorts* above]. **1979** N. Hynd *False Flags* xxiv. 213 You travel light... Just..some sleepwear.

sleep, *v.* Add: **I. 1. a.** *to sleep rough*: see Rough *adv.* 1 b; *to sleep tight*: see *tight *adv.* 1.

b. (Further examples.) Also, with *around*: to engage in sexual intercourse casually with a variety of partners; to be sexually promiscuous (*colloq.*).

1898 *Sessions Paper of Central Criminal Court* Feb. 266 He has been sleeping with my wife. How would you like it? **1928** A. Huxley *Point Counter Point* xxvii. 445 'Sleeping around'—that was how he had heard a young American girl describe the amorous side of the ideal life, as lived in Hollywood. **1936** R. Lehmann *Weather in Streets* ii. 185 A child's out of the question now, they don't sleep together any more. **1940** W. Faulkner *Hamlet* ii. 92 All we want anyway is to keep her out of trouble until she gets old enough to sleep with a man without getting me and him both arrested. **1952** M. Laski *Village* xvi. 218, I don't think for a minute she's been sleeping around..but you know what gossip is. **1967** J. Potter *Foul Play* xiii. 161 He's only interested in George and Freda and whether Johnnie and Freda slept together. **1975** P. Loraine *Wrong Man in Mirror* 78 Rose Maddox was not a loose girl; she did not sleep around with just anybody.

c. (Further examples.)

a **1907** F. Thompson *St Ignatius Loyola* (1909) x. 192 He discussed all measures with his brethren; and ever enjoined them to sleep on the matter, and pray the next morning before decision. **1926** V. McNabb *Church & Land* 83 My friend rose from his seat. 'I see—we must do things ourselves. I must sleep on this.' **1959** S. Salton-Vane *Black Whippet* ii. 32 Sleep on it. Think it over, and come and see me early tomorrow morning. **1962** P. Gregory *Like Tigress at Bay* ix. 99 Let me think about it, though. I'd like to sleep on it. **1983** 'W. Haggard' *Heirloom* viii. 90 He simply looked at a problem hard and then slept on it.

g. For *Sc.* read 'orig. *Sc.*' at sense 'to oversleep' and add earlier and later examples. Also, to lie in (*lie *v.*[1] 23 d), to sleep late.

1827 C. I. Johnstone *Eliz. de Bruce* I. iii. 56 Ye whiles sleep in on a morning. **1883** W. Aitken *Lays of Line* 58 A'e mornin' last March, when Rab Black sleepit in. **1931** *Amer. Speech* VII. 20 *Sleep in*, to sleep late. 'I'm going to sleep in tomorrow.' **1931** D. L. Sayers *Five Red Herrings* i. 16 Shall I tell Mrs. McLeod to let you sleep in, as they say? And call you with a couple of aspirins on toast? **1935** *Beaver* Dec. 66/2 On Sundays the chief guide usually allows his voyageurs to sleep in, which means that instead of getting up at four o'clock they get up about six-thirty. **1967** E. Taylor *Second Thursday* ii. 31 Susan dear, you *must* have slept in this morning. **1975** J. Grady *Shadow of Condor* v. 90 Because he slept in and had an appointment, Malcolm excused himself from the exercises.

h. Also as *ppl. adj.*

1939 W. Fortescue *There's Rosemary, There's Rue* xi. 79 We crawled up to our bedroom..and got under our bed-quilts, *not* between those slept-in sheets. **1966** N. Freeling *King of Rainy Country* 140 He..changed his slept-in suit. **1976** J. Crosby *Snake* (1977) xi. 53 She pulled the crinkly blue-and-white striped blouse as taut as she could..trying to make it look a little less slept-in.

i. With *on*: to continue sleeping, to sleep late.

1739–40 Richardson *Pamela* (1741) I. 267 I'll wake her,—said I.—No, don't,—said she,—let her sleep on; we shall lie better without her. **1939** 'J. Bell' *Death at Half-Term* v. 92, I let our lot sleep on, but the other four came over from the San, and woke them up. **1958** P.

Scott *Mark of Warrior* I. ii. 44 In four hours Hussein would wake him... Esther would sleep on. **1969** J. Fraser *Clap Hands if you believe in Fairies* iv. 48 Don't worry too much if he's not sleepy. I'll let him sleep on in the morning.

j. With *over*. † (*a*) to sleep late (*obs.*); (*b*) to spend the night at a place other than one's own residence (chiefly *U.S.*).

1827 *Harvard Reg.* Sept. 202 They have indulged in the luxury of 'sleeping over'. **1871** L. H. Bagg *Four Years at Yale* 570 On Sunday mornings, too, there is an unusual amount of 'sleeping over',—breakfast being often cut as well as chapel by the votaries of Morpheus. **1975** *Sunday Advocate-News* (Barbados) 15 June 7/2 The sleep-overs will be the night of the last Thursday of each session. On that night campers sleep-over if they choose to do so. **1977** D. Anthony *Stud Game* iii. 23, I begged him to sleep over... But he had an early appointment the next day. **1978** M. Puzo *Fools Die* xxviii. 331 Those were nights I'd hop a plane to Vegas for the evening, sleep over and come back in the early morning.

k. With *out*: to spend the night in the open air; also, to sleep away from the premises on which one is employed.

1852 [implied at *sleeping *vbl. sb.* 1 c]. **1890** Kipling *Departmental Ditties, Barrack-Room Ballads* 56 'E's sleepin' out an' far to-night,' the Colour-Sergeant said. **1908** R. Brooke *Let.* 18 Aug. (1968) 139, I should love to sleep out with nothing but a few extra socks on. **1912** [used in Dict., sense 1 g]. **1936** 'J. Tey' *Shilling for Candles* xiv. 159 He might have been sleeping out, the first three nights. But you know what last night was like. Torrents... He must have found shelter. **1974** *Whig-Standard* (Kingston, Ontario) 11 Jan. 7/1 There are more dossers sleeping out in London today than there were at the turn of the century.

l. Of a bed or mattress: to afford sleep of a specified quality.

1942 W. Faulkner *Go down, Moses* 83 This here pallet sleeps all right to me. **1977** *Austral. House & Garden* Jan. 115/1 A foam mattress is generally lighter, is non-allergenic and resists mildew... Foam sleeps cooler in warm temperatures, it is claimed, and warmer in cool temperatures.

m. With *up*: to catch up on one's sleep, to stop for a sleep.

1951 *Manch. Guardian Weekly* 1 Feb. 3/3 The General.. would say nothing more than that he was off for a rest. He headed..for the..hotel to sleep-up until Wednesday. **1968** *Listener* 11 July 50/2 It was..so full and exhausting ..that, at the end of term..we used to sleep up for several days before venturing out onto the streets. **1968** in P. G. Hollowell *Lorry Driver* vii. 183 You want to give your mate a bit of a shaking up. What you do is to spot him 'sleeping up' and go quietly by him and turn your wagon round... Then hold your hands on the horn.

n. With *through*: esp. of a baby, to sleep uninterruptedly through a period of time, usually the night.

1967 'L. Egan' *Nameless Ones* i. 3 She's the most beautiful baby ever... Sleeping right through. God, when I think—. **1971** O. Norton *Corpse-Bird Cries* ii. 33, I..slept through until nearly ten o'clock on Sunday morning. **1976** 'J. Charlton' *Remington Set* xx. 104 'Baby going to be all right in the guest room?' 'She'll sleep right through,' Fran said.

4. a. *spec.*, to act as a sleeping partner (see Sleeping *ppl. a.* 5) or as a sleeper (sense *2 d).

1949 D. Leon *Ruskin* I. i. 8 Telford supplied adequate capital and otherwise 'slept' most gracefully. **1975** J. Hone *Sixth Directorate* II. 48 Once they were sure the man was with the KGB..they had watched him... And no one had come near him... They assumed the man was sleeping.

II. 6. a. (Further examples of *to sleep the sleep of the just.*)

1927 R. Lehmann *Dusty Answer* III. i. 139 Jennifer's peaceful flushed countenance and regular breathing greeted her astonished senses. She was sleeping the sleep of the slightly intoxicated just. **1944** W. S. Maugham *Razor's Edge* vii. 276 Gray's conversation was composed of clichés... He never went to bed, but hit the hay, where he slept the sleep of the just. **1977** *Lancashire Life* Dec. 57/3 Tired and ready to go home, we went back to the river where Edith's father was still sleeping the sleep of the just.

11. (Further examples.)

1861 [see *mother's blessing* s.v. *mother *sb.*¹ 15 a]. **1923** *Daily Mail* 24 Feb. 7 A gasworks foreman..said that at present the gas included 13 per cent of carbon monoxide... 'One per cent is sufficient to kill. It sleeps you to death.'

12. (Earlier and further examples.)

1848 Bartlett *Dict. Amer.* 306 She could eat fifty people in her house, but could not sleep half the number. **1883** *Gringo & Greaser* (Manzano, New Mexico) 1 Sept. 2/1 If we can find some other philanthropist who will kindly hash, beer and sleep us, we'll be there. **1919** Mencken *Amer. Lang.* i. 24 A sleeping-car sleeps thirty passengers. **1941** W. A. Percy *Lanterns on Levee* i. 12 Welcome, messieurs, I can eat you but I cannot sleep you. **1949** *Spectator* 11 Nov. 631/2 During August some hotels were sleeping four or five guests to a room. **1965** E. O'Brien *August is Wicked Month* xi. 113 Oh we got beds, we can sleep..eighty. **1977** *Western Mail* (Cardiff) 5 Mar. 10/7 (Advt.), Farmhouse holiday flat. Sleeps six.

C. *Comb.*: **sleep-away** *a.* (*U.S.*), at which one sleeps away from home. See also *sleep-in, -out, -over *sbs.* and *adjs.*

1976 *Woman's Day* (U.S.) Nov. 244/2 Diana is told she must go to a sleep-away camp because the doctor says she needs more exercise. **1978** *Chicago* June 157/2 All the sleep-away camps have separate living quarters for boys and girls, but the tennis itself is mixed.

sleeper. Add: **I. 2. c.** A sleeping partner.

1901 *Edin. Rev.* Apr. 385 If..a director can be treated as a purely sleeping partner, it can do that company no harm that..the sleeper, on becoming a Minister, should cease to be a director. **1983** 'W. Haggard' *Heirloom* xi. 122, I might fix that. At a price... No consortium takes in a sleeper for nothing.

d. A spy, saboteur, or the like, who remains inactive for a long period before engaging in spying or sabotage or otherwise acting to achieve his ends; *loosely*, any undercover agent.

1955 H. Roth *Sleeper* ix. 66 Hollister..was a sleeper—a member of the Communist Party whose whole life was dedicated to the one big moment. **1963** J. Joesten *They call it Intelligence* I. iv. 45 A 'sleeper' is an agent planted in a strategic place for a specific purpose only. **1966** M. Woodhouse *Tree Frog* xxv. 187 Bought, or brainwashed? Or had he been a sleeper for years, waiting for just this job? **1975** *Daily Mail* 16 Aug. 2/1 They had been responsible for a year-long campaign of bombings in the city... When police cleaned up the cell, the IRA activated a reserve unit of 'sleepers'. **1976** *Times* 7 June 12/5 There almost certainly exists within our political establishment, what is known as a 'sleeper'—a high level political figure who is in fact a Soviet agent, infiltrated into the system many years ago. **1981** *Observer* 29 Mar. 15 Key members of the..Committee concluded that only the existence of a 'mole' or 'sleeper' (the preferred 'trade' word) could explain the many leaks and failures of the 1950s and 1960s.

4. b. *Gambling.* (See quots. 1864, 1897.)

1856 *San Francisco Bulletin* 4 Dec. 2/2 Some were waiting for 'sleepers', others were telling some other betters a certain card was going to win, dead 'sure'. **1864** W. B. Dick *Amer. Hoyle* 208 A bet [in faro] is said to be a sleeper, when the owner has forgotten it, when it becomes public property, any one having a right to take it. **1897** [in Dict., sense 4 a]. **1939** P. A. Rollins *Gone Haywire* 16 A Dakota miner had been detected attempting to steal 'sleepers' from the faro table. **1944** [see *heave ho b].

5. a. (Earlier and *attrib.* examples.) Also, a train made up of or including sleeping-cars.

1875 *Chicago Tribune* 11 Sept. 3/2 Every item of wood, iron, or upholstery which enters into the make-up..of a Pullman sleeper is Selected with Skilled Care. **1880** W. Whitman *Daybks. & Notebks.* (1978) I. 188 On a first-class sleeper. **1881** 'Mark Twain' *Speeches* (1910) 258, I..must change cars there and take the sleeper train. **1950** O. S. Nock *Brit. Locomotives from Footplate* vi. 118 The run with No. 46235 on the Inverness 'sleeper'..shows how great sometimes is the gulf between maximum locomotive capacity under ideal conditions and practical application on the road. **1952** 'J. Tey' *Singing Sands* iv. 62 A friend who came across him on the train. Saw his name on the sleeper list, or noticed him in passing. **1967** O. Wynd *Walk Softly, Men Praying* vii. 106 'You've got a sleeper reservation?' 'No, I was going to sit up.' 'All right... How about the eleven o'clock express.' **1969** M. Pugh *Last Place Left* xxi. 164 He made out sleeper tickets for us..and we shared a two-berth compartment. **1970** R. Adam *Stalk to Kill* xi. 158 A restless night in the sleeper train from London. **1979** P. Theroux *Old Patagonian Express* iv. 53, I was glad to be on this sleeper to the coast.

b. Used *attrib.* and *absol.* to designate a vehicle with sleeping facilities.

1939 *Nat. Geogr. Mag.* Feb. 133/2 Their covered wagons are now shiny streamliners (40 hours from Chicago), or mammoth sleeper buses (which they still call 'stages'). **1951** *Amer. Speech* XXVI. 308/2 *Sleeper*, any truck or tractor that is equipped with a sleeping berth. **1969** *Sydney Morning Herald* 24 May 9/7 Mr Ferrier plans to buy a sleeper-van and, with his wife, spend the next 18 months travelling around Australia. **1971** M. Tak *Truck Talk* 146 *Sleeper cab*, a tractor in which an adjoining bunk area is located behind the driver's seat. **1976** *Eastern Daily Press* (Norwich) 19 Nov. 14/2 (Advt.), Scania 32-ton tractor, sleeper cab.

5*. Something whose quality or value proves to be greater than was generally expected; a 'dark horse'. orig. *U.S.*

1892 *Outing* (U.S.) Mar. 454/2 Williams won the high and low hurdles in record time,..and Harmar a second in the mile, being beaten by Wells, a 'sleeper' from Amherst. **1903** J. P. Paret *Lawn Tennis* III. vi. 350 Sleeper, a slang expression meaning a player who is much better than was thought. **1926** *Clues* Nov. 162/2 *Sleeper*, something of value that has been overlooked. **1945** *Richmond* (Va.) *Times-Dispatch* 31 July 7/4 In film parlance, this is a *sleeper*—a picture made with the thought that it would be just another light Summer item, but which has turned out to be a surprisingly popular box-office success. **1962** W. & M. Morris *Dict. Word & Phrase Origins* I. 267 A sleeper is a stamp more rare—and thus more valuable—than the catalogue listings indicate. **1968** J. D. Watson *Double Helix* xx. 141 Bill's appearance was the sleeper of the three-day gathering... As soon as he had finished his unassuming report..everyone in the audience knew that a bombshell had exploded. **1978** *Detroit Free Press* 5 Mar. B. 12/2 In any given week of new book arrivals, there occasionally is a sleeper, a book which comes in virtually unannounced with 'best seller' written all over it. **1981** J. D. MacDonald *Free Fall in Crimson* v. 49 He made a couple of motion pictures... on a very small budget, and they were what is called sleepers. They made a lot of money, considering what they cost.

5.** Miscellaneous uses. **a.** An unbranded calf which has had a notch cut in its ear. *U.S.*

1893 O. Wister *Jrnl.* 31 Dec. (1958) 198 *Sleeper*, a cow with earmark and no brand. **1933** J. V. Allen *Cowboy Lore* I. 12 *A sleeper* is a calf ear-marked by a cattle thief who intends to come back later and steal the animal. **1949** *Boston Sunday Globe* 1 May (Fiction Mag.) 2/1 He..gave a tally of the sleepers and mavericks he had branded.

b. An earring, esp. one in the form of a simple hoop, worn not primarily as ornament but to keep the hole in a pierced ear-lobe open.

1896 G. F. Northall *Warwickshire Word-bk.* 215 *Sleepers, sb. pl.*, fine, small rings of gold, first put into the ears after boring, and afterwards worn whenever the larger ear-rings, or 'droppers', are inconvenient. Their use is to prevent the closing of the perforations of the lobes. **1959** *News Chron.* 6 Oct. 6/2 That ghastly business of turning sleepers in a fresh and often painful wound. **1971** R. Scott *Wedding Man is Nicer than Cats, Miss* i. 38 Even quite small girls had had their ears pierced... No one wore sleepers, but the ear-rings could be taken out and the holes plugged with tiny pieces of wood. **1978** J. Updike *Coup* (1979) vii. 293 Substituting for her great hoop earrings little sleepers of agate.

c. A sleeping-suit for a baby or a small child. Also *pl.* orig. *N. Amer.*

1921 *Daily Colonist* (Victoria, B.C.) 12 Oct. 7/1 (Advt.), Children's sleepers at $1.95 a suit. Made of strong quality flanelette in neat colored stripe. **1944** C. Himes *Black on Black* (1973) 247 Norma sat on the side of the bed and helped Lucy into her sleepers. **1970** *Toronto Daily Star* 24 Sept. 23/1 (Advt.), Minute imperfections should not affect wear, appearance or comfort of these cosy sleepers. **1975** *Daily Tel.* 14 Feb. 15/3 (*caption*) Lucy is in her Mothercare sleeper with vinyl feet.

d. A particle of sleep (*sleep *sb.* 1 d).

1942 Berrey & Van den Bark *Amer. Thes. Slang* § 251/1 *Sleepers*, particles in the eyes after a sound sleep. **1944** H. Croome *You've gone Astray* i. 9 He had sleepers in his eyes, ugh!

e. A sleeping-pill. *slang.*

1961 Rigney & Smith *Real Bohemia* p. xvii, *Sleepers*, barbiturates (sedatives), usually seconal, nembutal, amytal, etc. **1967** M. M. Glatt et al. *Drug Scene* vii. 91 A lot of addicts are taking liquid Methedrine with 'sleepers' now—it is getting worse. **1979** C. Dale *Helping with Inquiries* I. 11 Take a sleeper, I would, put yourself right out.

II. 10. *sleeper-beam.*

1937 *Discovery* Dec. 377/2 The house had been divided into rooms by lath and plaster walls, the sleeper-beams for which were let into shallow trenches in the chalk or gravel floors. **1970** Bray & Trump *Dict. Archæol.* 213/2 In early timber-framed buildings, Roman, Saxon and medieval, the framing was often erected not on a wall foundation but directly on a horizontal beam resting on or slightly recessed into the ground. From its recumbent position this is known as a sleeper beam.

III. 11. Special combs., as **sleeper agent** = sense 2 d above; **sleeper pass** *N. Amer. Football*, a pass unexpectedly involving a player hitherto ignored; **sleeper seat**, a reclining seat on which one can sleep during a journey; **sleeper wall** *Building*, a low wall built under a ground floor to support joists where there is no basement; so **sleeper walling.**

1973 *TV Times* (Austral.) 3 Feb. 11/1 A sleeper agent is someone who, over the years, has worked himself up into a position of trust. **1977** H. Kaplan *Damascus Cover* (1978) v. 54 Operative Sixty-six is a member of the Syrian Parliament. He was a sleeper agent for twelve years. **1954** *Sun* (Baltimore) 4 Dec. (B ed.) 11/3 The Rams pulled the old corner lot 'sleeper' pass on the first running play of the new season for a touchdown. **1966** *Globe Mag.* (Toronto) 20 Aug. 7/3 Part of Canadian football folklore is the sleeper pass Keith Spaith threw..in 1948. **1960** *Times* 11 Feb. 9/5 Whether sleeper-seats and bunks should be provided. **1980** *Sunday Times* 21 Sept. 11 (Advt.), TWA's First Class Sleeper-Seats make it easy to lie back and relax peacefully. **1836, 1893** Sleeper wall [in Dict., sense 10]. **1972** S. Smith *Brickwork* viii. 37 The sleeper walls supporting the floor are built 'honey-comb', that is, with holes left through them to permit through ventilation. **1971** *Power Farming* Mar. 9/1 The latest aid to producing a 12ft-high stack at Kexby—a 50ft square of sleeper walling—is illustrated.

sleeper, *v. rare.* [f. the sb.] **1.** *trans.* To mark (a calf) with a notch in its ear.

1910 C. E. Mulford *Hopalong Cassidy* xii. 79 Either the H2 was sleepering Bar-20 calves for their irons later on, or rustlers were at work.

2. *intr.* To travel in a railway sleeping-car.

1978 A. Fraser *Wild Island* xvii. 155 Beauregard was off..on the overnight sleeper to London. 'Flying visit... Back in the morning. Sleepering both ways.'

Hence **slee·pering** *vbl. sb.*

1910 C. E. Mulford *Hopalong Cassidy* xii. 80 'I saw a H2 sleeper, up just above th' Bend.'.. 'Lazy trick, that sleepering.'

Sleeperette (slīːpəre·t). Also **sleeperette.** [f. Sleeper + -ette.] The proprietary name of a kind of reclining seat; = *sleeper seat* s.v. *sleeper 11.

1950 *Official Gaz.* (U.S. Patent Office) 11 July 443/2 Pan American Airways Inc., New York... Sleeperette. For transportation of passengers by aircraft. **1959** *Radio Times* 2 Jan. 28/2 (Advt.), Travel by sleeperettes on the 'Austropa Express'. **1966** *Punch* 31 Aug. 341/1 We could put Emma, who is three.., beside a VIP in his long-lie sleeperette. **1968** *Guardian* 27 Apr. 10/2 (Advt.), Fred Olsen car ferry from Harwich; single sleeperette fare £8 5s. **1982** *Daily Tel.* 24 Dec. 11/5 British Airways' new first class 'sleeperette' service.

slee·p-in, *sb.* and *a.* **A.** *sb.* **1.** [*-in³.] A form of protest in which the participants sleep overnight in premises which they have occupied.

1965, 1971 [see *-in³].

2. [f. vbl. phr. *to sleep in*: see sense 1 g of the vb. in Dict. and Suppl.] An act of sleeping later than usual, a lie-in.

1977 P. G. Winslow *Witch Hill Murder* II. 67 Those bells do make the damnedest racket. Nobody..can have a sleep-in on Sunday.

B. *attrib.* or as *adj.* [Sleep *v.* 1 g.] Of a person: that sleeps on the premises, resident. Of a place: at which one stays overnight, residential.

1961 in *Webster* s.v., Five sleep-in servants. **1970** L. Sanders *Anderson Tapes* xxi. 56 Apartment Four A... A sleep-in maid. **1974** H. L. Foster *Ribbin', Jivin', & Playin' Dozens* vii. 329 The first was a three-week sleep-in camp experience for 62 children and their teachers in June, 1947, at Life Camps. The second was a three-day sleep-over school camping experience..at Hudson Guild Farm, Nokong, New Jersey. **1981** J. D. MacDonald *Free Fall in Crimson* x. 106 He managed to hustle me into bed... I told him I had to quit. I wasn't going to be a sleep-in secretary.

sleeping, *vbl. sb.* Add: **1. c.** With advbs. as *around* (see *sleep *v.* 1 b), *out* (see *sleep *v.* 1 k).

1852 *Rep. Committee on Criminal & Destitute Juveniles* App. III. 427 in *Parl. Papers* VII. 389 It is his fourth committal; his offence being, 'sleeping out'. **1945** S. Lewis *C. Timberlane* xxiv. 155 Going to be none of this 'modern civilized, urbane' sleeping around and getting complicated in *our* house. **1957** J. Braine *Room at Top* xxv. 204 I'm glad you've decided to settle down. You're too old for sleeping around. **1973** E. J. Bahr *Nice Neighbourhood* xiii. 137 Her mother..did some sleeping around to help make ends meet. **1974** 'M. Innes' *Appleby's Other Story* ix. 73 One very large bush has been curiously hollowed out... The badgers use it as a sleeping-out place. **1976** V. Canning *Doomsday Carrier* iv. 68 A sleeping-out pass until six tomorrow. **1976** *Howard Jrnl.* XV. i. 43 It seems wrong to assume..that non-indictable assaults, malicious damage, begging and sleeping out [etc.],..are all associated with social dereliction, homelessness, or disturbed behaviour.

2. a. *sleeping-berth, -car* (further examples), *-place* (further examples), *-platform, -porch, -quarters, -room* (further examples).

1834 *Chambers's Edin. Jrnl.* III. 316/2 The Calais boats being small, and mounting few sleeping berths. **1939** Auden & Isherwood *Journey to War* v. 123 The bugs must have been nesting in the upholstery of the shabby old Belgian sleeping-berths. **1979** O. Sela *Petrograd Consignment* 259 The..relative comfort of a first-class sleeping berth..all the way to Stockholm. **1872** Sleeping-car [see *Pullman a]. **1903** Mrs. H. Ward *Lady Rose's Daughter* xviii. 313, I will go and get a sleeping car for you to Calais. **1954** T. S. Eliot *Confidential Clerk* I. 32, I shall go and rest now. In a sleeping-car it is quite impossible To get one's quiet hour. **1978** J. Simmons *Railway in England & Wales 1830–1914* I. viii. 195 The first sleeping cars in Britain appeared in 1873, on trains running between London and Glasgow. **1910** W. de la Mare *Three Mulla-Mulgars* 52 Let us hobble on, Mulla-mulgars, until we find a quieter sleeping-place. **1957** P. Worsley *Trumpet shall Sound* viii. 150 Existing huts were to be..replaced by communal houses..one as sleeping-place for the men and one for the women. **1935** *Discovery* Dec. 361/2 Mr E. W. Savory did not actually see any of these apes in the district, but the presence of their sleeping-platforms is proof of their residence there. **1940** R. Finnie *Lure of North* 199 His original excavation, from which most of the snow blocks needed for the house had been taken, now constituted the floor, and the rest of that space—more than half—was the sleeping platform, a foot or so higher. **1915** J. Webster *Dear Enemy* 44, I want two hundred feet of sleeping-porch running along the outside of our dormitories. **1971** *Sunday Express* (Johannesburg) (Homefinder) 28 Mar. 2/2 (Advt.), Flats 6 × 2 plus sleeping porch. **1919** E. O'Neill *Moon of Caribbees* 117 A door leading to the captain's sleeping quarters. **1944** J. S. Huxley *On Living in Revolution* 111 Many people have to share crowded sleeping-quarters. **1982** V. Mehta *Vedi* iii. 52 Mrs. Ras Mohun marched me up to her sleeping quarters. **1833** *Chambers's Edin. Jrnl.* I. 386/1, I..was shown to my sleeping-room by the waiter. **1903** W. B. Yeats in *Fortn. Rev.* Apr. 752 Maeve walked through that great hall, and with a sigh Lifted the curtain of her sleeping-room. **1978** *Chicago* June 135/2 He..leads me into a small sleeping room, which has a cot, a small desk, and a chair.

b. *sleeping-bag* (earlier and later examples), *-chair, -mask, -mat* (later examples); **sleeping dictionary** *slang*, a foreign woman with whom a man has a sexual relationship and from whom he learns her language.

1850 S. Osborn *Jrnl.* 11 Oct. in *Stray Leaves from Arctic Jrnl.* (1852) 147 Friday morning, at seven o'clock, we rolled up our beds, or rather sleeping-bags. **1933** *Discovery* Sept. 284/1 We retired to sleep in our sleeping-bags in the tiny room allotted us. **1978** *Times* 22 Nov. 5/4 She had put him..in his carrycot wrapped in a sleeping bag, romper suit and cardigan. **1675** in J. Gloag *Short Dict. Furnit.* (1969) 619 For a sleeping chaire to fall in the back of Iron Worke. *a* **1877** Knight *Dict. Mech.* I. 481/2 Car-seats..made reclining, for night travel..are termed 'sleeping-chairs'. **1924** Macquoid & Edwards *Dict. English Furnit.* I. 215 In the Queen's Closet at Ham House are two winged 'sleeping chairs'..with ratchets to let down the backs. **1928** J. B. Wharton *Squad* 21 We picked up two beauties... Oo-la-la—I've learned French—out uv a sleepin' dictionary—dat's what dey're called. **1965** *Listener* 25 Mar. 461/3 He paints the old China of bound feet,..the endless dinners, the mistress (sleeping dictionary) as fragile as a butterfly. **1979** M. Lindsay in C. Allen *Tales from Dark Continent* i. 14 In East Africa.. 'East African officers as a whole maintained a..stricter code in the matter of sleeping with African women'—sometimes referred to as 'sleeping dictionaries', from their obvious advantages as language instructors—than did their fellow-officers in West Africa. **1908** S. Ford *Side-Stepping with Shorty McCabe* ix. 139 They'd kept his face in a steam box by the hour..made him wear a sleepin' mask, and done everything but peel him alive. **1944** S. Bellow *Dangling Man* 79 A black cotton sleeping mask hung around her neck. **1965** A. Nicol *Truly Married Woman* 3 He then severely flogged his eldest son..for wetting his sleeping-mat last night. **1979** J. Melville *Wages of Zen* ii. 21 Hanae always folded the sleeping mats up and put them away during the day.

c. *sleeping-draught* (further examples), *pill, powder* (further examples), *tablet*.

1664 Lady Hobart *Let.* 23 Mar. in M. M. Verney *Mem.* (1899) IV. ii. 53 Thay had no way but to give hur a sleping pell, & she slep all night. **1790** J. Woodforde *Diary* 17 Sept. (1927) III. 214 Her Child is dead..owing it is supposed to her [having] given him a Sleeping Pill. *c* **1900** H. A. Jones in M. R. Booth *Eng. Plays of 19th Cent.* (1969) II. 414 You'll find a sleeping powder in the second drawer... We must manage to give Lionel a little sleep tonight. **1934** G. B. Shaw *Too True to be Good* I. 42 It must be that new sleeping draught the doctor gave me. **1936** *Time* 19 Oct. 66/2 It was then that she attempted to find independence in an overdose of sleeping powder. **1938** W. S. Maugham *Writer's Notebk.* (1949) 300 He feels the moment can never be excelled and so takes an overdose of sleeping-pills. **1941** C. Milburn *Diary* 4 Oct. (1979) vii. 110 A sleeping tablet last night gave me a good rest. **1973** M. Amis *Rachel Papers* 30 Dozy afternoons slugging on opiate cough mixtures, sleeping-draughts dropped at noon, stolen handfuls of Valium, a sheet of aspirins before breakfast. **1976** H. Wilson *Governance of Britain* iv. 105, I have never had a sleeping pill in No. 10. I have never needed one. **1977** K. O'Hara *Ghost of T. Penry* xv. 148 She'd taken a quadruple dose of sleeping tablets last night.

d. *sleeping sickness*: see as main entry.

e. *sleeping partner, time* (later examples; also *fig.*).

1707 E. Settle *Siege of Troy* III. 18 'Tis high sleeping Time, and so let's all home to Bed. **1833** E. B. Browning tr. *Aeschylus' Prometheus Bound* 96 The close and subtle clasping of a chain..Whose links are furnished from the common mine..From work-times, diet-times, and sleeping-times. **1903** Farmer & Henley *Slang* VI. 247/1 *Sleeping-partner*... 2. (common).—A bed-fellow. **1926** Maines & Grant *Wise-Crack Dict.* 14/1 *Sleeping time*, one year in Jail. **1959** T. S. Eliot *Elder Statesman* 5 The rhythm that governs the repose of our sleepingtime. **1967** P. D. James *Unnatural Causes* I. vi. 42 Your sleeping partner would provide you with an alibi. **1979** G. Mitchell *Mudflats of Dead* xiv. 144 Camilla wasn't the only predator... People are always changing their sleeping partners.

sleeping, *ppl. a.* Add: **1. a.** *Sleeping Beauty* (occas. *Princess*), the heroine of a fairy tale (Charles Perrault's *La belle au bois dormant*) who slept for a hundred years, until woken by the kiss of her prince; also (sometimes with small initial) applied allusively and joc. to any sleeping or unconscious person; also *attrib.* and *transf.*

1729 R. Samber *Perrault's Tales* iv. 32 (*heading*) The sleeping beauty in the wood. **1830** Tennyson (*title*) The sleeping beauty. **1893** S. J. Weyman *Gentleman of France* III. xxviii. 91 The Castle before us..might have been that of the Sleeping Princess, so fairylike it looked. **1907** E. Glyn *Three Weeks* iv. 64 The Austrians..are naturally awake, whereas you English are naturally asleep, and you yourself are the Sleeping Beauty, Paul. **1909** Mrs. H. Ward *Daphne* ii. 40 It had been a Sleeping Beauty story so far. Treasure for the winning—a thorn hedge—and slain lovers! **1936** C. Day Lewis *Friendly Tree* vii. 97 Who could wake the Sleeping Beauty with a kiss of friendship? **1955** E. Bowen *World of Love* vii. 126 Sleeping-beauty briars..swung at her. **1965** J. Porter *Dover Two* ii. 26 A rather smudgy photograph of Curdley's Sleeping Beauty lying motionless in her hospital bed. **1967** V. Nabokov *Speak, Memory* vi. 136 The terra-incognita blanks map makers of old used to call 'sleeping beauties'. **1977** D. Bagley *Enemy* xxiii. 179 You can go in and wake the sleeping beauties. **1979** A. Price *Tomorrow's Ghost* ii. 28 You can be our Sleeping Princess in the Library, and I shall come and wake you with a kiss.

d. *sleeping lizard* (earlier example).

1859 D. Bunce *Trav. with Dr Leichhardt* ix. 94 We disturbed many of the short, knobby-tailed sleeping lizard (*Agama*).

f. *sleeping policeman*: see *policeman 1 e.

sleeping sickness. (In Dict. s.v. Sleeping *vbl. sb.* 2 d.) Substitute for def.: **1.** In general and *fig.* senses.

1551 R. Robinson tr. *More's Utopia* II. M.v.r, Is there annye man so possessed wyth stonyshe insensibilitie, or with the sleping sicknes, that he wyll not graunt health to be acceptable to hym and delectable. **1647** [in Dict. s.v. Sleeping *vbl. sb.* 2 d]. **1904** *Jrnl. R. Microsc. Soc.* Apr. 179 Sleeping Sickness of Silk-worms..is in no wise due to the micro-organisms of the mulberry leaves.

2. *Path.* Any of several similar diseases caused by protozoans of the genus *Trypanosoma* and transmitted by flies of the genus *Glossina*, prevalent in tropical Africa, and characterized by the proliferation of the trypanosomes in the blood and changes in the central nervous system leading to apathy, coma, and death. Also *attrib.*

1875, 1897 [in Dict. s.v. Sleeping *vbl. sb.* 2 d]. **1905** [see *Gambia]. **1908** W. S. Churchill *My African Journey* v. 96 On April 28th, 1903, Colonel Bruce, whose services had been obtained for the investigation of 'sleeping sickness'.., announced that he considered the disease to be due to a kind of trypanosome, conveyed from one person to another by the bite of a species of tsetse-fly called *Glossina palpalis*. **1926** *Encycl. Brit.* III. 558/1 Sleeping sickness is now treated by compounds of arsenic..; by compounds of antimony..; and by a drug of undisclosed composition called Bayer 205, or Germanin. **1958** L. van der Post *Lost World Kalahari* (1961) vi. 109 A small African outpost on the edge of the sleeping sickness country of Northern Bechuanaland. **1970** Passmore & Robson *Compan. Med. Stud.* II. xix. 7/2 Sleeping sickness in West Africa differs clinically and epidemiologically from the condition in East Africa. In the West, the disease generally runs a chronic course, in the East it is acute.

3. = *sleepy sickness 2. Now *rare*.

1918 *Proc. R. Soc. Med.* XII. (Med. Section) p. xvii, The term 'sleeping sickness'..would not be an inappropriate name for this epidemic [*sc.* encephalitis]. **1920** *Lancet* 13 Mar. 620/2 Some popular term for encephalitis lethargica less cumbrous than 'lethargic encephalitis' and free from the objection to 'sleeping sickness'. **1921** *Times* 3 Feb. 7/2 The Registrar-General's returns for the week.. show that there were 21 cases of sleeping sickness (*encephalitis lethargica*) notified..in London alone. **1961** L. E. Bollo *Introd. Med. & Med. Terminol.* xiv. 148 The von Economo type of encephalitis (encephalitis lethargica, or sleeping sickness) is said to be of unknown etiology... African sleeping sickness, caused by protozoa of the genus *Trypanosoma*, is discussed later.

slee·p-out, *sb.* and *a.* [f. vbl. phr. *to sleep out*: see *sleep *v.* 1 k.] **A.** *sb.* A veranda, porch, or outbuilding providing sleeping accommodation; a sleeping area not in the main building. *Austral.* and *N.Z.*

1941 *Coast to Coast* 84 'A nice scone and a nice cup of tea and a lay-down in the sleep-out,' Mrs Smith was saying in her warm, motherly voice. 'You'll be fine then.' **1962** A. Seymour *One Day of Year* 7 A multiple set, main areas being the kitchen; the 'lounge'; and Hughie's study, which is a glassed-in sleepout at the side of the house. **1977** *N.Z. Herald* 5 Jan. 2-15/8 (Advt.), Well-established country tearooms with 3-brm attached acom... There is almost an acre of grounds plus a sleepout to go with this bargain. **1979** *Sunday Mail Mag.* (Brisbane) 29 Apr. 26/3 The sleep-out..is now back to veranda play space.

B. *attrib.* or as *adj.* Of a person: that sleeps away from the premises, non-resident.

1958 V. P. Johns *Servant's Problem* i. 11 It concerned the household in which she was the sleep-out, full-time maid. **1961** in *Webster* s.v., The sleep-out cooks and maids were coming to work.

slee·p-over, *sb.* (and *a.*) Chiefly *U.S.* [f. vbl. phr. *to sleep over*: see *sleep *v.* 1 j.] **A.** *sb.* **a.** (See quot. 1935.) **b.** An occasion of spending the night at a place other than one's residence. **B.** *attrib.* or as *adj.* Involving spending the night away from one's residence. Of a person: that stays the night.

1935 *Amer. Speech* X. 236/1 A contributor testifies that in part of Pennsylvania, in college use, a sleep-over is a permission to stay away from church and remain in bed on Sunday morning. **1974** [see *sleep-in *a.*]. **1975** [see *sleep *v.* 1 j]. **1979** *Sunset* (Desert ed.) Apr. 156/2 (*caption*) At night wall hanging unhooks to become a feather-soft mat for sleep-over guests.

slee·p-walk, *v.* [Back-formation f. Sleep-walking *vbl. sb.* and *ppl. a.*] *intr.* To walk while asleep; to be in a state resembling that of a sleep-walker. Also *fig.*

1923 in *Englische Studien* (1935) LXX. 119 The heroine sleep-walks. **1954** *Gramophone* XXXI. 445 She sleep-walks noisily, but with dramatic vigour. **1976** E. O'Brien in *New Yorker* 16 Aug. 30/1 Every night Mrs. Reinhardt sleepwalked. **1981** I. McEwan *Comfort of Strangers* i. 18 She sleepwalked from moment to moment, and whole months slipped by without memory, without bearing the faintest imprint of her conscious will. **1982** *Daily Tel.* 21 Dec. 4/6 Mr Reagan..accused Congress of 'sleep-walking' into the future.

sleepy, *a.* Add: **1. d.** Substitute for def.: One of several Australian lizards of the family Scincidæ, esp. the shingleback, *Trachylosaurus rugosus*, found in the southern part of the country. (Earlier example.)

1883 [see *blue tongue, blue-tongue 2].

2. b. (Further examples.) See also *sleepy sickness.

1831 W. Youatt *Horse* 103 Some say that there is a yellowness of the eye..in the early stage of sleepy or stomach-staggers. **1913** Dorland *Med. Dict.* (ed. 7) 887/1 *Sleepy staggers, stomach staggers*, a disease of horses, of unknown causation but usually associated with the eating of moldy hay and grain. **1922** *Times Lit. Suppl.* 3 Aug. 511/2 The tomato suffers from the so-called 'Sleepy Disease' manifested in a wilting of the plant.

4. sleepy-bye(s), *Sc.* **-baw** *sb.*, a nursery name for sleep; also as *v. intr.*, to go to sleep; **Sleepy Hollow, sleepy hollow,** (*a*) a name given to a place with a soporific atmosphere or characterized by torpidity (in quot. 1820, to which some quots. allude, the name of a valley near Tarrytown (Irving's home) in Westchester county, N.Y. State); (*b*) a type of comfortable deep-upholstered armchair; also called *sleepy-hollow chair*; **sleepy-time** *U.S.*, bedtime.

1907 N. Munro *Daft Days* x. 85 Just you lie down there pet, and sleepy-baw. **1925** M. Beerbohm *Observations* 37 Before you go to sleepy-bye I'll read it to you. **1968** A. Diment *Bang Bang Birds* x. 177 'Sleepy byes time, Lex,' and I just felt the prick in my arm before I was blotted out again. **1820** W. Irving *Sk. Bk.* vi. 51 (*heading*) The legend of Sleepy Hollow. *Ibid.* 53 This sequestered glen has long been known by the name of Sleepy Hollow, and its rustic lads are called the Sleepy Hollow boys. **1834** M. Edgeworth *Helen* I. xv. 321 Beauclerc, who had not yet tried the chair, sank into its luxurious depth, and leaning back, asked if it might not be appropriately called the 'Sleepy-hollow'. **1836** *Madrid in 1835: Sk. by Resident Officer* I. v. 94 No friendly arm-chair; none of that somniferous form, not unaptly termed 'sleepy hollow'. **1868** L. M. Alcott *Little Women* v. 79 There were..Sleepy-Hollow chairs, and queer tables. **1897** [in Dict., sense 2 c]. **1922** Joyce *Ulysses* 371 Rip van Winkle we played... Then I did Rip van Winkle coming back. She leaned on the sideboard watching. Moorish eyes. Twenty years asleep in Sleepy Hollow. **1924** Galsworthy *White Monkey* I. vii. 53 'First time I remember anything of the sort on that Board.' 'Sleepy hollow,' said Soames. **1955** 'A. Gilbert' *Is she Dead Too?* v. 91 There's plenty of work to be had, you don't have to stop in Sleepy Hollow. **1957** M. Swan *Brit. Guiana* 98 It is a charming little sleepy hollow of a town with long strait streets. **1966** M. M. Pegler *Dict. Interior Design* (1967) 412 *Sleepy hollow chair*, a mid-19th-century American chair which is upholstered and has a curved back and low, comfortable arms. The seat is usually scooped out. **1976** N. Freeling *Lake Isle* viii. 43 This bastard in Soulay is merely wanting to make a fuss. Sleepy hollow. If he were any good he wouldn't be there. **1981** *London Rev. Bks.* 2–15 July 6/4 The restless forces were never as strong in Britain as the laws of inertia and the politics of Sleepy Hollow. **1862** K. Stone *Jrnl.* 19 Aug. in *Brokenburn* (1955) 137 He was satisfied I would not sleep a wink, but at sleepy time.. we all went to bed and slept soundly. **1918** S. C. Bryant *Stories to tell Little Ones* p. ix, I have been in the habit of singing them rhymes..a while before sleepy-time. **1950** O. Nash *Family Reunion* 81 At sleepy-time he beats a path Straight to the bedroom or the bath.

sleepy sickness. [Sleepy *a.*] †**1.** = *sleeping sickness 2. *Obs.*

1803, 1903 [in Dict. s.v. Sleepy *a.* 2b].

2. Encephalitis lethargica, an often fatal disease widespread between 1916 and 1928, characterized in many of those who survived it by extreme somnolence due to physiological brain damage.

1922 Ashby & Wright *Dis. Children* (ed. 6) 389 Encephalitis lethargica. Sleepy sickness. Epidemic stupor. This disease is more common in adults. **1923** *Daily Mail* 26 Feb. 9 Sleepy sickness (encephalitis lethargica) is attacking prominent people in Winnipeg. **1930** A. Christie *Murder at Vicarage* xiv. 112 There's nothing radically wrong with him... He's had Encephalitis Lethargica, sleepy sickness, as it's commonly called. **1962** 'J. Bell' *Crime in our Time* iv. 99 Encephalitis lethargica, or sleepy sickness (not to be confused with trypanosomiasis, sleeping sickness, carried by the tsetse fly). **1973** *Daily Tel.* 27 Jan. 11/1 Survivors of the great sleepy-sickness epidemic, which claimed five million victims between 1916 and 1927, lived on in a kind of somnolence for 40 years till this drug awoke them.

3. *Vet.* A disease of pregnant ewes, provoked by imbalance between the degree of nourishment and the stage of pregnancy, and characterized by somnolence and neuromuscular disturbances; pregnancy toxæmia. Also *Comb.*

1937 *Vet. Jrnl.* XCIII. 213 In the early stages of sleepy sickness they lie in a perfectly normal recumbent position, but before long..on their side with legs stretched out. **1950** D. Gascoyne *Vagrant* 8 Baa, Baa, O sleepy-sickness-rotted sheep, in your nice fold Are none but marketable fleeces. **1964** *N.Z. Jrnl. Agric.* Mar. 546/2 More than 80 per cent of deaths in the ewe flock occur over lambing, when losses due to blood poisoning, bearing trouble, sleepy sickness..and lambing difficulties occur.

sleet, *sb.*¹ Add: **2.** *sleet-gust*; *sleet-bound* adj.

1782 J. Trumbull *M'Fingal* III. 65 He glitter'd to the Western ray Like Sleet-bound trees in wintry skies. **1929** C. Day Lewis *Transitional Poem* II. 34 When bullying April bruised mine eyes With sleet-bound appetites and crude Experiments of green. **1928** Blunden *Retreat* 48 Where the lashed sleet-gust foams, buffeting and blinding.

sleet, *v.* **1. b.** (Later example.)

1955 C. M. Kornbluth *Mindworm* 39 Her manicured hand gripped his arm in excitement and terror. Unfelt radiation sleeted through their loins.

sleety, *a.* **2.** (Earlier example.)

1816 Jane Austen *Emma* II. xvii. 329 The evening of a cold sleety April day.

sleeve, *sb.* Add: **1. a.** (Further examples denoting a separate article of dress.)

1805 Jane Austen *Let.* 21 Apr. (1952) 154, I wore my crape sleeves to the Concert, I had them put in on the occasion. **1897** *Montgomery Ward Catal.* 297/1 Ladies' Gossamer Rubber Sleeves, 16 inches long. **1967** G. Bellairs *Single Ticket to Death* v. 61 He was without jacket and wore black calico detachable sleeves reaching to the elbows of his white shirt.

2. f. *another pair of sleeves* (later examples); *to put the sleeve on* (someone): (*a*) to beg or borrow money from (someone); (*b*) to arrest (someone): to cause (someone) to be arrested; *a sleeve across the windpipe*, an assault or severe blow (usu. *fig.*).

1904 H. James *Golden Bowl* I. III. xxiv. 395 'Decide to live—ah yes!—for her child.' 'Oh, bother her child!.. To live..for her father—which is another pair of sleeves!' **1930** D. Runyon in *Sat. Even. Post* 5 Apr. 72/2 These coppers..know who he is very well indeed and will take great pleasure in putting the old sleeve on him if they only have a few charges against him, which they do not. **1931** *Amer. Speech* VI. 440 *Put the sleeve on*, to borrow; to make a touch from a fellow convict. **1934** H. N. Rose *Thes. Slang* iii. 29/1 Wait'll I put the sleeve on Joe fer some chewin'. **1937** *Nature* 23 Jan. 130/1 Prof. Furnas's exasperating,..naive volume is altogether another pair of sleeves. **1952** Wodehouse *Barmy in Wonderland* i. 13 My wardrobe perished in the holocaust, of course. When you're being given the sleeve across the windpipe by Acts of God, you don't waste time fumbling around for socks and trousers. **1960** Wentworth & Flexner *Dict. Amer. Slang* 486/1 *Put the sleeve on* (*someone*), 1. To arrest someone; to identify someone to the police for arrest. 2. To stop a friend on the street in order to ask for a loan of money; to ask for a contribution or for money owed. **1972** Wodehouse *Pearls, Girls, & Monty Bodkin* ii. 17 Just as if it looked as though all they had to do was collect the bridesmaids, order the cake and sign up the Bishop and assistant clergy, along came the sleeve across the windpipe. Her father refused to give his consent to their union.

7. b. *spec.* part of a celt or prehistoric axe.

1929 V. G. Childe *Danube in Prehistory* 78 Possibly they were shafted with the aid of horn sleeves. *Ibid.* 107 Axes and adzes hafted in deer-horn sleeves. **1970** Bray & Trump *Dict. Archaeol.* 20/1 *Antler sleeve*, a section of deer antler carved into a mortice at one end to hold a stone axe head.

c. *Electr.* A metal cylinder fitted round the full length of the core of an electromagnetic relay to modify the speeds of opening and closing. Cf. *slug *sb.*² 3 c.

1921 W. Aitken *Autom. Telephone Systems* I. 45 The copper sleeve and heavy ring on the core of F gives it a greater range of adjustment. **1969** S. F. Smith *Teleph. & Telegr. A* ii. 45 The skin effect, due mainly to the iron core, tends to confine alternating magnetic fluxes at speech frequencies to the nickel-iron sleeves to give the required impedance.

d. *Aeronaut.* = *drogue 3 (*b*) and (*c*).

1933 C. K. Stewart *Speech Amer. Airman* (thesis, Univ. Akron) 90 *Sleeve*, a towed target for anti-aircraft guns to practice shooting at. **1933** S. Spender *Poems* 45 The air-liner..Glides over suburbs and the sleeves set trailing tall To point the wind. **1937** *Times* 12 June 16/4 The target was the usual sleeve, towed behind a Fairey Gordon. **1942** *Tee Emm* (Air Ministry) II. 67 There's the old Henley and there's the sleeve coming up—. They're off!—..A grand salvo after a week's weary waiting.

e. A close-fitting protective cover or case, esp. one for a gramophone record; a slip-case. Cf. *record sleeve* s.v. *record *sb.* 14.

1953 *N.Y. Times* 22 Mar. II. 40/6 Another group of buyers is swayed more by the art on the 'sleeve' or jacket than by the quality, or even by the title, of the music. **1954** *Melody Maker* 11 Dec. 15/2 His first LP to be released in this country..reached me without sleeve. **1976** W. Goldman *Magic* III. ii. 119 She lifted the tone arm off, and..put the record back in the sleeve. **1981** *Verbatim* Spring 20/2 This is a man's pocket wallet with some plastic credit card sleeves.

8. a. (sense *7 d) *sleeve target*; (sense *7 e) *sleeve artist, design, information, -picture.*

1977 *Times* 18 Apr. (Gram. Suppl.) p. iv/6 Individual sleeve artists such as Roger Dean or Patrick Woodroffe. *Ibid.* iv/7 Jazz musicians..insisted on good sleeve design. **1966** *Melody Maker* 23 July 16/5 Bob Houston's review of the John Coltrane album 'Ascension' blindly copies the sleeve information that Freddie Hubbard plays the first trumpet solo. **1959** *Times* 10 Jan. 9/5 A disc with an imaginative if punning sleeve-picture showing wind blowing through barley. **1932** *Aeroplane* 11 May 839 (*caption*), A Fairey IIIF seaplane towing a sleeve target for gunnery practice. **1955** 'N. Shute' *Requiem for Wren* iii. 68 Firing the Oerlikon at a sleeve target towed by an aeroplane. **1979** A. Fox *Threat Warning Red* iii. 27 Sleeve target this afternoon, an AA shoot.

8. b. sleeve bearing, a form of bearing in which an axle or shaft turns in a lubricated sleeve; **sleeve-board** (later examples); **sleeve-cap** *U.S.*, the topmost part of a sleeve; **sleeve dog**, for '(see quot.)' read: a very small Pekinese dog, usually under six pounds in weight; (later examples); **sleeve gun** *U.S.*, a miniature gun which can be concealed in the clothing; **sleeve-note**, an informative or critical note about a gramophone record, printed on the sleeve; **sleeve Pekinese** = *sleeve dog* above; **sleeve-valve**, a kind of valve, employed in certain types of internal-combustion engine, consisting of a hollow cylindrical sleeve fitting closely inside the engine cylinder and moving with the piston in such a way that inlet and exhaust ports are opened and closed at appropriate times; freq. *attrib.*; hence **sleeve-valved** *a.*

1907 W. S. Boulton *Pract. Coal-Mining* III. 9 The 'sleeve' bearing..is intended to obviate this waste [of oil], and to secure continuously good lubrication. **1967** *Times Rev. Industry* Aug. 22/2 Exhaust silencers, anti-vibration mounts, the substitution of sleeve bearings for ball or roller bearings, [etc.]..are only a few of the attempts made to reduce the noise at source. **1975** *Sci. Amer.* July 50/1 Since many motors, engines and other machines incorporate journal bearings (sometimes designated plain bearings, sleeve bearings, fluid-film bearings or bushings), the annual production of journal bearings is in the billions. **1916** *Daily Colonist* (Victoria, B.C.) 2 July 7/6 (Advt.), Sleeve Boards, regular 75c. Our price 32c. **1969** D. Clark *Death after Evensong* ii. 31 Trousers, with no fore-and-aft creases..., ironed on a sleeve board. **1964** *McCall's Sewing* xi. 158/1 All set-in sleeves are cut with a sleeve cap that is larger than the armhole section into which it must fit. **1978** *Detroit Free Press* 2 Apr. 50/1 All too often a knitted sweater is ruined by puckers where the sleeve cap joins the shoulder section. **1931** A. C. Dixey *Lion Dog of Peking* iv. 12 They were petted and pampered, the smallest—the highly-prized 'sleeve' dogs—being carried in the voluminous sleeves of the long robes worn at Court by both sexes. **1970** P. Tamony *Americanisms* (typescript) No. 27. 7 The toy or small breeds such as the Pekinese and Shih Tzu had been the sleeve dogs of women at the Imperial Court of China. **1944** R. F. Adams *Western Words* 146/1 *Sleeve gun*, a derringer such as a gambler carried up his sleeve. **1971** K. Wheeler *Epitaph for Mister Wynn* xxxii. 396 He..took out a snub-barreled Sharps derringer, a sleeve gun. **1974** E. McGirr *Murderous Journey* 153 Have a look at Pout ... Sleeve gun and I'd guess an envelope in his breast pocket. **1956** *Gramophone* Oct. 184/2 Each soloist is given his fair share of the spotlight (the sleeve notes helpfully identify which plays when). **1980** *Early Music* Jan. 85/2 A pity, though, that the programme book's translation of the libretti kept so close to the impenetrable one in the sleeve-notes to the Electrola/Reflexe recording. **1949** I. Harman *Pekingese* iv. 37 A Miniature or Sleeve Pekingese is, officially, any Pekingese which is not more than six pounds in weight. **1978** 'J. Melville' *Axwater* i. 12 A tiny, bright-eyed, button-faced creature appeared.. and behind it another even smaller. Two sleeve Pekingese. **1910** *Engineering* 18 Nov. 688/3 Both Messrs. Panhard and Levassor..and Messrs. Milnes Daimler, Limited.. showed examples of the Knight sleeve-valve engine. **1911** *Ibid.* 3 Nov. 590/2 The usual tappet-valves are replaced by a single sleeve-valve. **1958** Gibson & Tuteur *Control System Components* xi. 424 A Vickers two-land sleeve valve. **1982** P. Dickinson *Last House-Party* iv. 49 'This car makes a remarkable amount of smoke.' 'That's the trouble with these sleeve-valve engines.' **1932** *World Today* Feb. 261/2 The news that Daimlers had taken over the Lanchester Company suggested that it might be a sleeve-valved job.

sleeved, *ppl. a.* Add: **b.** Fitted or covered with a sleeve or sleeves (in sense 7 of the sb.).

1905 *Engineering Rev.* XIII. 272/1 The Hyatt bearing has..been successfully applied to sleeved axles, both in small cars and heavy 'buses and lorries. **1970** 'S. Harvester' *Moscow Road* i. 13 A stereophonic phonograph and two racks of sleeved discs. **1976** *Gramophone* June 51/3 The pressing is beautifully smooth, and the disc is attractively sleeved; the cover is a water colour of the composer by his son.

sleeveen (slīvīn, slīvī·n). *Ir.* and *Newfoundland*. Also **sleiveen, slieveen.** [ad. Ir. *slighbhín, slíbhín* sly person, trickster.] An untrustworthy or cunning person.

1834 S. Lover *Legends & Stories of Ireland* (Ser. 2) 295 How the man was chated by a *sleeveen* vagabone. **1888** W. B. Yeats *Fairy & Folk Tales Irish Peasantry* 220 In trust took he John's lands,—*Sleiveens* were all his race. **1892** J. Barlow *Irish Idylls* viii. 215 He isn't the *slieveen* to be playin' fast and loose wid your dacint little slip of a girl. **1955** L. E. F. English *Historic Newfoundland* 36 *Slieveen*, a deceitful person. *a* **1966** 'M. na Gopaleen' *Best of Myles* (1968) 104 A thief, a fly-be-night, a sleeveen and a baucagh-shool. **1973** G. Pinsent *Rowdyman* 42 Mr. Lowe told Will all about his friendship with our father and about how decent a fellow he was, but that Will had 'the looks of a sleeveen about him'. *Ibid.* 53 Well, I took my eyes off him for a half a second and that sleeveen jabbed me in the gut with two hard fingers. **1975** D. O'Sullivan in D. Marcus *Best Irish Short Stories* (1977) II. 95 'O, the crabbed, conniving little sleeveen.' 'She's up in London now, dancing in the Harp.'

sleeveless, *a.* **2. b.** Delete 'Now *rare*' and add later examples.

1931 L. Storm *Dragon* xvii. 293 I'd never have the courage like you..to venture forth on what might be a sleeveless errand. **1948** *Chambers's Jrnl.* 320/2 And, as soon as they were settled in, he had McGilchrist ride openly away, putting it about he was satisfied all the talk of whisky-'stilling was a pack of lies by ill-doing ones willing to give the King's officer a sleeveless errand. **1959** I. & P. Opie *Lore & Lang. Schoolch.* iv. 58 In some [*sc.* schoolchildren's tricks] he [*sc.* the dupe] is sent on sleeveless errands.

sleever. Add: (Later example.) Also *Austral.* and *N.Z.* Also, a measure of drink (usu. of beer) contained in a sleever. Cf. *long-sleever* s.v. Long *a.*¹ 18.

1936 'R. Hyde' *Passport to Hell* v. 89 Places where the police weren't so quick off the mark if the landlord passed a few sleevers over the counter after six o'clock. **1941** Baker *Dict. Austral. Slang* 67 *Sleever*, a drink, esp. a large drink. **1970** 'H. Carmichael' *Remote Control* i. 8 'I haven't got a glass with a handle so you'll have to make do with a sleever.'..She brought his beer. **1975** B. Meyrick *Behind Light* viii. 99 Herby used to nip up to the New Inn, buy a 'sleever' of beer and bring it back.

sleeving, *vbl. sb.* Add: **3.** A tubular covering for a cylindrical object, esp. of insulating material for an electric cable, etc.; material used for this purpose.

1923 *Wireless World* 12 May 168/2, ½ lb. No. 20 tinned copper wire and a quantity of insulating sleeving of various colours. **1933** *Electrician* 10 Feb. 185/2 Woven sleevings, stockingettes or circular tapes of cotton or silk are varnished and used extensively for insulating wire connections of windings, radio sets,..etc. **1978** *SLR Camera* Nov. 31/3 Should it prove to be undersize, the best policy will be to use suitable piece of thin-walled aluminium tube (available from model shops) as sleeving.

sleigh, *sb.* Add: **4. sleigh bed** *N. Amer.*, a type of bed resembling a sleigh, having head- and footboards curving outwards; a French bed.

1902 F. C. Morse *Furnit. of Olden Time* iii. 77 Plainer bedsteads in this [French bed] style were made, veneered with mahogany, and they are sometimes called sleigh beds, on account of their shape. **1950** W. Bird *Nova Scotia* iii. 87 Those who spend the night in the ancient bedrooms, perhaps sleeping on the great 'sleigh bed' that remains in one. **1976** *National Observer* (U.S.) 14 Feb. 4/4 (Advt.), Antique Marketry Furniture, Ca. 1790, Dresser w/mirror, highboy, desk, two sleigh beds.

sleigh-bell. (Earlier example.)

c **1780** in *Amer. Poems* (1793) I. 208 Mind and have the sleigh-bells sent.

slei·gh-ride, *sb.* Also **sleighride, sleigh ride.** **1.** A ride in a sleigh. Also *fig.*

1770 J. Hiltzheimer *Diary* (1893) 2 Apr. 20 Took a sleigh ride, the 'five mile round', with wife, sister, and son Tommy. **1828** H. J. Finn et al. *Whimwhams* 22 Such worthy gentlemen happen to remember..a winter's breakfast at a country inn, after a sleigh-ride of ten miles for an appetite. **1849** [in Dict. s.v. Sleigh *sb.* 4]. **1902** W. D. Hulbert *Forest Neighbors* (1903) 181 Not even a sleigh-ride on a winter's night can set the live blood dancing as it will dance and tingle up there above the clouds. **1956** E. B. White *Let.* 14 Jan. (1976) 412, I am cheered up when I see our political giants discovering that the lil ole writing game isn't quite the sleighride they like to think it is.

2. *U.S. slang.* The action of taking a narcotic drug, usu. cocaine; the euphoria resulting from taking a narcotic drug. Usu. in phr. *to take* (*go on*) *a sleigh ride* and varr. Cf. *snow *sb.*¹ 4 d.

1925 *Flynn's* 4 Apr. 818/2 *Sleigh-ride*, a jab of morphine from a hypodermic syringe, or the resulting state of intoxication. **1928** M. C. Sharpe *Chicago May* xxxi. 286 *Taking a sleigh ride*, getting morphine. **1938** D. Castle *Do your Own Time* xxix. 251 'He took to going on sleigh rides.' 'No! Where the hell did he get the snow?' **1942** *Detective Fiction* Apr. 56/2 Julio is very fond of his hop. Anything from the weed to a sniff of snow. Suppose he gets on a big sleigh ride and talks out of turn. **1963** 'D. Shannon' *Death of Busybody* ix. 52 It was just some dope out on a sleigh-ride.

3. *U.S. slang.* An implausible or false story; a hoax, a deliberate deception. Freq. in phr. *to take* (someone) *for a sleigh ride*, to mislead (someone). Cf. *ride *sb.*¹ 1 f (*a*).

1931 G. Irwin *Amer. Tramp & Underworld Slang* 172 An absolutely impossible or unlikely idea or action, or.. the cheating or fleecing of a victim... 'We gave him a sleigh ride'—we cheated him by a false story or by sharp practice. **1942** Berrey & Van den Bark *Amer. Thes. Slang* § 202/2 Incredible story, *sleigh ride.* **1950** *Sun* (Baltimore) 13 Mar. 1/1 House Republicans, charging that the taxpayers are being taken for a 'bureaucratic sleighride'. **1960** Wentworth & Flexner *Dict. Amer. Slang* 486/2 *Sleighride*, an instance of being cheated, believing a lie, or being taken advantage of. Almost always in the expression 'taken for a sleighride'.

Hence as *v. intr.*, (*a*) to ride in a sleigh; (*b*) to take a narcotic drug. Also **slei·gh-rider; slei·gh-riding** *vbl. sb.*

1807–8 Sleigh-riding [in Dict. s.v. Sleigh *sb.* 4]. **1833** *Knickerbocker* I. 207 Arrived at the Plains, the sleigh riders stopped at a tavern. **1845** Sleigh-ride [in Dict. s.v. Sleigh *sb.* 4]. **1883** *Wheelman* (Boston, Mass.) I. 434, I was making my first trial of it [*sc.* a bicycle] in the snow, among the sleigh-riders. **1915** G. Bronson-Howard *God's Man* vi. iii. 376 Whadda you been doing?—sleigh-riding? Stick to the long bamboo, Charley—that snow's awful bad for the imagination. *Ibid.* vii. i. 409 Petty's kind had been profitable 'sleigh-riders' when he provided 'snow' on Seventh Avenue. **1929** *Detective Fiction Weekly* 13 Apr. 599/1 He's a sleigh rider. You know, sniffs coke. Made a fortune writing papers for booze hustlers and has spent every dime of it on snow. **1934** C. de Lenoir *Hundredth Man* i. 13 Sniffing heroin or cocaine is 'sleigh-riding'. **1949** *Summit Valley Times* (Argo, Illinois) 1 Dec. 4/3 Santa, who now reigns the Christmas card realm, in 1919 managed to sleighride onto only a handful of cards for children. **1977** H. Waugh *Secret Room of Morgate House* (1978) xxxiv. 164 Between times they sleighrode, and even walked. **1982** J. Adair *Founding Fathers* xii. 267 The New Englanders adopted..skating and sleigh-riding from their Dutch neighbours.

sleiveen, var. *sleeveen.

‖ **slendang** (sle·ndæŋ). Also **selendang.** [a. Javanese *slèndang*, Indonesian *seléndang*.] In Indonesia, a long scarf or stole worn by women.

1885 H. O. Forbes *Naturalist's Wanderings Eastern Archipelago* III. ii. 147 Above this the body is girt with a silk *slendang*, half concealing the breasts. **1911** B. Miall tr. *Cabaton's Java* v. 125 An indispensable article of the feminine toilet is the *slendang*: a scarf in *batik*, often ornamented with fringes. **1937** M. Covarrubias *Island of Bali* iii. 48 It is necessary to be properly dressed to pay or to receive a visit. The breasts of men and women should be covered by a special breast-cloth, a *saput* for men and a *selendang* for women. **1941** *Lincoln* (Nebraska) *Sunday Jrnl. & Star* 12 Oct. D-6/1 Maria Montez, in 'White Savage', will wear a scant bit of cloth called a *slendang*, something like a sarong, only more so, or less so! **1963** J. Kirkup *Tropic Temper* 168 The men approach the women and claim their selendangs or long gauzy stoles which are used most effectively in the dance movements. **1964** Langewis & Wagner *Decorative Art in Indonesian Textiles* 27 In a certain *batik* cloth, the *slendang*, a centrally placed undecorated area always appears in the form of an elongated rhomboid.

slender, *v.* Delete † *Obs. rare*⁻¹ and add: **a.** *trans.* (Later example.)

1965 H. Porter *Cats of Venice* 81 These shape-gripping *cheong sams* had slendered them to elegance.

b. *intr.* To become narrower, to narrow. Also with *down*.

1871 G. M. Hopkins *Poems* (1967) 13 And slendering to his burning rim Into the flat blue mist the sun Drops out and all our day is done. **1955** E. Bowen *World of Love* iii. 55 Her strong forearms, which slendered down..to the wrists.

slender-bodied, *a.* (Later example.)

1963 *Times* 4 Feb. 13/2 Dresses..with a slender-bodied look.

slenderize (sle·ndəraiz), *v.* Also **slenderise.** [f. Slender *a.* + -ize.] **a.** *intr.* To make oneself slender, to slim. **b.** *trans.* To make (something) slender, to make (the figure) appear slender. Also *absol.*

1923 *Weekly Dispatch* 4 Feb. 15/5 A slight figure will be more essential than ever. 'You must slenderise,' said one, coining a useful word. **1923** *Daily Mail* 21 Mar. 6/1 (Advt.), Corsets for slenderising full figures. **1946** *Sun* (Baltimore) 10 May 12/1 He brought it down to within a few inches of the ground, slenderized the column on which it hangs in merry-go-round fashion. **1964** *McCall's Sewing* i. 11/1 Lines for the 'Viking' type are a problem. You'll want those which slenderise without adding height. **1976** *National Observer* (U.S.) 13 Mar. 11/5 (Advt.), Starting age 6 yrs. Slenderize. Charm... Clover Lodge [establishment for overweight girls].

Hence **sle·nderizer**; **sle·nderizing** *ppl. a.* and *vbl. sb.*

1927 *Daily Express* 9 Sept. 5/5 There is the high straight line that is suited to the office frock, and the graceful slenderising V-cut accentuated by removable front of deep ivory crêpe-de-chine. **1928** *Sunday Express* 29 Apr. 15/4 Chefs are searching their brains for the slenderising sweet that will tempt both men and women diners. **1932** *Woman & Beauty* Apr. 87/1 (Advt.), Slenderizing. Free book tells how you can become slim. **1935** *Amer. Speech* X. 192/2 Black velvet, we find, is the best *slenderizer*. **1958** *People* 4 May 7/1 (Advt.), A most flattering and slenderising style. **1969** *Daily Tel.* 18 Apr. 17 The skirt does a slenderising job where most women want to look slim. **1970** *Globe & Mail* (Toronto) 25 Sept. 10/2 (Advt.), The jacket is a little longer. It continues the slenderizing effect of the waist. **1978** *Detroit Free Press* 5 Mar. B12/1, I made a slenderizing lunch of cottage cheese and grapefruit sections.

slenderness. Add: **3.** *attrib.* **slenderness ratio** *Engin.*, the ratio of the effective length of a column or pillar to its least radius of gyration (formerly, to its least diameter).

1905 M. Merriman *Mechanics of Materials* (ed. 10) ix. 191 The ratio *l/r* is called the 'slenderness ratio' of the column. **1931** *Engineering* 9 Jan. 61/3 Experiments on eccentrically-loaded steel columns... The slenderness ratios ranged from 50 to 150. **1976** R. F. Warner et al. *Reinforced Concrete* xiii. 308 The effective strength of the column section..is divided by a reduction factor R... R is given as R = 1·2 − 0·01(*l/r*) in which *l/r* is the maximum slenderness ratio for the column, for bending in any plane.

slenter, var. *schlenter, *sb.* and *a.* **sleughi,** var. *saluki.

sleuth, *sb.*² **2. b.** For *U.S.* read 'orig. *U.S.*' and add earlier and further examples. Also *transf.*

1872 *N.Y. Fireside Compan.* 13 May 4/3 The name of the story is *Sleuth, the Detective* and a more remarkable and thrilling story has seldom ever been written. **1904** 'O. Henry' *Cabbages & Kings* iv. 73 Goodwin followed at increased speed, but without any of the artful tactics that are so dear to the heart of the sleuth. **1949** *Manch. Guardian Weekly* 22 Dec. 2/3 A school of newspaper sleuths who attributed every declaration of American foreign policy to the hidden hand of George Kennan. **1958** 'J. Byrom' *Or be he Dead* v. 69, I gather you have Miss Canning as your assistant sleuth. **1979** *Oxf. Jrnl.* 16 Nov. 1 (*caption*) Amateur sleuths Gordon Murray and Jane Lawton... Their investigations launched a top-level probe into an Oxford business.

sleuth, *v.*² Add: **a.** Also, to investigate (something or someone). Also with *out* (in quot. 1939: to detect or expose).

1939 [see *record *sb.* 5 g]. **1949** *Sun* (Baltimore) 16 Nov. 14/3 Men who qualify for the tremendous job of sleuthing a single big industry like steel or coal—and determining the facts to make wage, hour and pension recommendations. **1956** A. Christie *Dead Man's Folly* xviii. 240 'Who hired you to sleuth me?'.. 'You are in error,' replied Poirot. 'I have not been sleuthing you.' **1968** P. Dickinson *Skin Deep* v. 108 It had been something private he'd sleuthed out, something secret. **1979** *Amer. Speech* 1978 LIII. 285 Ten years ago, sleuthing a clue from Lenneberg, I wrote..'The use of tools may be much older than language.'

b. *intr.* To act as a detective; to conduct an investigation. Also with *around*.

1912 L. J. Vance *Destroying Angel* xx. 276 So I went sleuthing; traced you through the canal to Peconic. **1930** 'Sapper' *Finger of Fate* 99 My poor friend..labours under the delusion that he is a detective. He goes about with magnifying glasses, and sleuths. **1975** *High Times* Dec. 31/2 If you sleuth around—beginning at the roach-infested gringo palace, the Hotel Astorial—you can get directions to the mushroom fields overlooking San José. **1980** E. Dewhurst *Drink This* ii. 28 He had been sleuthing, unconsciously..all the time he had thought he was relaxing.

Hence **sleu·thing** *vbl. sb.*

1900 Ade *More Fables* 193 He called himself a Reformer, and he did all his Sleuthing in the line of Duty. **1924** *Weekly Westm. Gaz.* 13 Sept. 580/2 One always knew all about his theories and his sleuthing. **1946** *Reader's Digest* Sept. 76/1 Izzy knew nothing of sleuthing procedure; he simply knocked on the door. **1958** T. F. T. Plucknett *Early Eng. Legal Lit.* v. 83 Teasing as these references are, they seem too obscure and divergent to permit any plausible conclusion as the authorship of *Brevia Placitata*, in spite of the very clever sleuthing of Mr Turner. **1979** *Dædalus* Summer 111 It is possible, through conscientious sleuthing, to decode the secondary associations of symbols.

sleuth-hound. **2.** (Examples in sense 'detective'.)

[**1849** A. B. Reach *Clement Lorimer* xiii. 130 There is an awful mystery which the sleuth hounds of the law may trace—a mystery of suspicion, perhaps a mystery of crime.] **1902** Wodehouse *Pothunters* iv. 66 Jim's respect for the abilities of our national sleuth-hounds was greater than Tony's, and a good deal greater than that of most people. **1929** *Bookman* Nov. 264/1 'What is it, Fra Diavolo?' he asked... 'A peeler, fellow, a sleuth-hound.' **1948** *Amer. Speech* XXIII. 306/2 The hunt for it would be engrossing to a literary sleuth-hound.

slew, *sb.*¹ Add: **1. b.** More generally, an expanse or mass of water. *rare.*

1915 D. H. Lawrence *Rainbow* ix. 227 Tilly, an old woman now, came in saying that the labourers who had been suppering up said the yard and everywhere was just a slew of water. **1941** *Penguin New Writing* II. 20 Great slews of water flushed along the deck.

slew, *sb.*² Add: **2.** = *slewing *vbl. sb.* 2. Usu. *attrib.*

1958 J. G. Truxal *Control Engineers' Handbk.* III. 6 (*table*) Slew rate. **1981** *Popular Hi-Fi* Mar. 78/2 Measurements of slew rate do appear to be relevant to the performance of an amplifier.

slew (slū), *sb.*³ *colloq.* (orig. *U.S.*). Also **slue.** [ad. Ir. *slua*(*gh*), crowd, multitude.] A very large number *of*, a great amount *of*. Also in pl.

1839 D. P. Thompson *Green Mountain Boys* II. x. 145 He has cut out a road, and drawn up a whole slew of cannon clean to the top of Mount Defiance. **1858** *Harper's Mag.* May 767/2 By gracious! three thousand dollars is a 'tarnal slue of money. **1897** R. E. Robinson *Uncle Lisha's Outing* i. 2 I've seen slews on 'em [*sc.* ducks] on the ma'shes. **1937** *Sun* (Baltimore) 13 Nov. 8/1 This fable furnishes an excuse for a whole slue of low-comedy gags and wheezes. **1958** *Listener* 19 June 1015/2, I got up and checked with another inspector. There seemed to be slews of them lounging around. **1970** *Guardian* 9 Apr. 3/2 The offer has brought in 'piles of letters'... In addition.. he has received a 'slew of calls' from other bankers asking about the offer. **1978** J. Carroll *Mortal Friends* IV. vi. 458 Should I ask a slew of questions just to draw his gaze my way? **1982** *Radio Times* 11–17 Sept. 86/2 Roger Dennhardt had served three years of a 13-year sentence for armed robbery when..he offered to give evidence for the Crown against a slew of former associates.

slew (slū), *sb.*⁴ *Basketry.* [Orig. uncertain: perh. a new sense of Slew *sb.*²] A filling made of two or more strands worked together. Hence **slew** *v.*³; **sle·wing** *vbl. sb.*

1902 P. N. Hasluck *Basket Work* 50 Next fill in by working two rods together; this process is known by basket-makers as slewing. *Ibid.* 53 Start slewing with one rod, add another a few stakes farther on. **1907** [see *fitch *sb.*⁴]. **1912** T. Okey *Art of Basket-Making* vi. 27 The slath being now finished he slews up the bottom to its required size. *Ibid.* vii. 59 Any small modification..may be made, during the slewing up of the bottom. **1953** A. G. Knock *Willow Basket-Work* 47 The upsetting, which is begun with tops, consists of four rounds of three-rod waling, and the siding is a three-rod slew. **1960** E. Legg *Country Baskets* 79 It was made of coarse brown willows in the familiar slew beloved of the worker anxious to turn out as many baskets as he could. **1964** H. Hodges *Artifacts* x. 146 Both slewing and randing require an odd number of stakes.

slew, *v.* Add: **1. d.** Also *Austral.* and *N.Z.*, to outwit, to trick. Also in phr. *to get slewed*, to lose one's bearings in the bush, to be 'bushed'.

1813 V. Pyke *Wild Will Enderby* (ed. 2) I. xi. 62 The general impression seemed to be that Jack Ketch had been 'slued' (*anglice*, robbed of his dues) by the trio. **1929** K. S. Prichard *Coonardoo* xvii. 167 We separated, followin' tracks, and I managed to get slewed. **1944** *Living off Land* iv. 65 Many a bushman has become bushed before now, while even a good bushman may get slewed for a few hours in strange and difficult country.

2. Also with *over*. More recently, of motor vehicles, to skid uncontrollably (*across* a surface); to slide and turn out of the proper course, to 'career'.

1914 KIPLING *Diversities of Creatures* (1917) 389 'We overtake on the right as a rule in England.' 'Thanks!' Mr Lingnam slued over. 1943 *Sun* (Baltimore) 8 Sept. 3/2 The..luxury train..slewed crazily over four tracks when its locomotive boiler blew up. 1965 M. BRADBURY *Stepping Westward* viii. 380 On one sharp bend the car slewed across the road and angled round again just short of the edge of a deep ravine. 1982 B. CHATWIN *On Black Hill* xx. 97 The car slewed off down the yard.

3. Of a control mechanism or electronic device: to undergo slewing (*SLEWING *vbl. sb.* 2).

1958 GIBSON & TUTEUR *Control System Components* v. 237 A servo using this circuit tends to have relatively poor synchronizing characteristics when slewing, i.e., when large and rapid changes of the input are made. 1962 L. A. STOCKDALE *Servomechanisms* vii. 112 The slewing time may form part of the servo specification, i.e. the servo to slew through 90° in the minimum time.

slewed, *ppl. a.* (Earlier and later examples.)

1801 A. ELLICOTT in C. V. Mathews *Andrew Ellicott* (1908) 201 He was two thirds slewed (as the Rahway people call being in liquor). 1886 [see *BOILED *ppl. a.* c]. 1935 H. H. FINLAYSON *Red Centre* xiii. 129 When questioned closely he admitted rather sheepishly that he was 'sleued'. 1975 D. LODGE *Changing Places* iii. 106, I was somewhat slewed by this time and kept calling him Sparrow.

slew-foot. *U.S. slang.* Also **slough-, slue-foot,** etc. [f. SLEW *v.*] A person who walks with his feet turned out; a clumsy person. Also *transf.* Hence **slew-foot** *v.*; **slew-footed** *ppl. a.*

1896 BOGERT & O'BRIEN *Slew Foot Sal* (song) 3 I'll tell you of a lady, her name is Slew Foot Sal,..She's heavy and she weighs five-thirty-three. 1922 F. SCOTT FITZGERALD *Beautiful & Damned* I. iii. 122 A man in a striped blue suit, walking slue-footed in white-spatted feet. 1945 L. SAXON et al. *Gumbo Ya-Ya* xxiii. 496 She is hoping that her galloping, slue-foot, light-brown, lazy husband ..will soon find a job. 1950 R. STARNES *Another Mug for Bier* xx. 130 Haggis [*sc.* an Airedale] shrugged and slough-footed away. 1961 J. B. PRIESTLEY *Saturn over Water* iv. 53 Leaving your work to go slewfooting in South America.

slewing, *vbl. sb.* Add: **I. 2.** The response of a control mechanism or electronic device to a sudden large increase in input, esp. one that causes the device to respond at its maximum rate (the *slewing rate* or *speed*). Usu. *attrib.*

1958 O. J. M. SMITH *Feedback Control Systems* vii. 201 The infrequent large-magnitude changes of the average input to a servo produce slewing of the output at maximum velocity. 1962 R. N. CLARK *Introd. Automatic Control Systems* vii. 270 The large signal (or slewing speed) response characteristics of a system. 1975 G. J. KING *Audio Handbk.* ii. 47 The maximum frequency at which full power can be obtained is a function of the amplifier's slewing rate, which is different from rise time.

slice, *sb.*[1] Add: **2. b.** *Geol.* A relatively thin, broad mass of rock situated between two approximately parallel thrust faults, esp. when these make a small angle with the horizontal. Also *thrust slice.*

1914 PEACH & HORNE *Guide Geol. Model Assynt Mts.* 18 The slices of strata thus repeated have been driven westwards by major thrusts along planes which truncate the overlying reversed faults. 1942 M. P. BILLINGS *Structural Geol.* xvii. 327 Surrounding the basin is a zone..of outwardly-driven thrust slices. In still another zone.. rootless slices and isolated blocks of various slices are common. 1957 *Q. Jrnl. Geol. Soc.* CXIII. 59 They occur.. as infolds, and slices brought up along the Strathconon tear-fault. 1969 BENNISON & WRIGHT *Geol. Hist. Brit. Isles* iii. 46 In it [*sc.* the Laxfordian orogeny] are found fragments of earlier orogenic belts brought up as thrust slices.

c. *Electronics.* A small, thin slab of semiconducting material on which circuit elements have been formed.

1964 *Proc. IEEE* LII. 1713 (*heading*) Evolution of the concept of a computer on a slice. 1975 *Sci. Amer.* May 36/2 National Semiconductor..introduced a four-bit PMOS slice that could be used as a modular unit in the design of machines ranging from four to 32 bits.

3. b. *slice of life* [tr. F. *tranche de la vie*, a term orig. applied to French Naturalist literature: see quot. 1890], a realistic and detailed portrayal in drama, narrative, painting, etc., of incidents typical of everyday life. Freq. (usu. with hyphens) *attrib.*

[1890 J. JULLIEN in *Art et Critique* 9 Aug. 500/2 Ce n'est donc qu'une tranche de la vie que nous pouvons mettre à la scène.] 1895 G. B. SHAW in *Sat. Rev.* 19 Oct. 503/1 The substitution of a homogeneous slice of life for the old theatrical sandwich of sentiment and comic relief. 1914 H. JAMES in *Times Lit. Suppl.* 19 Mar. 134/4 The Orgreaves..come..as near squaring aesthetically with the famous formula of the 'slice of life' as any example that could be adduced. 1938 R. G. COLLINGWOOD *Princ. Art* p. v, We have..a new drama, taking the place of the old 'slice of life' entertainment, in which the author's chief business was to represent everyday doings of ordinary people as the audience believed them to behave. 1954 M. EWER *Heart Untouched* ix. 154 This is a costume picture, not a slice-of-life drama. 1962 *Listener* 14 June 1028/2 The pure landscape, the still life, the 'slice of life', the painting for painting's sake, is a late development. 1976 *National Observer* (U.S.) 4 Dec. 20/3 Mrs McFarland is one of the just-folks who appeared in one of those slice-of-life commercials. 1981 *Daily Tel.* 19 Feb. 13/1 Yet another indigestible slice of life about 'a warm, winning, and wise and wonderful Jewish family'.

slice, *sb.*[2] Add: **2.** Similarly in *Tennis.* Cf. *SLICE *v.*[1] 5.

1969 *New Yorker* 14 June 47/3 He hits a slice so hard and with such sharp placement, close to the sideline, that the ball jumps cleanly past Graebner's racquet for a service ace. 1971 LAVER & COLLINS *Educ. Tennis Player* xi. 144 My slice (a left-hander's) will move to a right-hander's backhand, and that's convenient.

slice, *v.*[1] Add: **1. c.** In colloq. phr. *no matter how* (or *whichever way*, etc.) *you slice it*: however you look at it. orig. and chiefly *U.S.*

1936 C. SANDBURG *People, Yes* 160 No matter how thick or how thin you slice it it's still baloney. 1941 WODEHOUSE *Berlin Broadcasts* in *Performing Flea* (1961) I. 261 Slice it where you like, it is still a German prison camp. 1968 J. SANGSTER *Touchfeather* xvii. 198 Whichever way you sliced it, I had absolutely nothing on Roger Gerastan except what I had guessed. 1979 'A. HAILEY' *Overload* III. xii. 257 Whichever way you slice it,.. Cameron Clarke has done our cause a lot of harm.

5. Also in other sports, to make a sharp stroke across a ball rather than straight on it, causing it to be propelled forward at an angle (on purpose or unintentionally); in *Lawn Tennis*, etc., to impart spin or swing in this manner. Cf. *CHOP *v.*[1] 7 d, e.

1905 H. A. VACHELL *Hill* xii. 255 Scaife has been transformed into a tremendous human machine, inexorably cutting and slicing, pulling and drawing. 1954 J. B. G. THOMAS *On Tour* 68 Birt, normally the safest of place kickers, made his mark, only for the ball to be sliced towards the corner flag. 1969 *New Yorker* 14 June 61/2 He'll slice. He'll lob. 1976 J. SNOW *Cricket Rebel* 63 Alan Smith..started to hit out boldly, slicing the ball repeatedly over and through the covers.

sliceable (slai·săb'l), *a.* [f. SLICE *v.*[1] + -ABLE.] That can be sliced or divided.

1976 *Evening Post* (Nottingham) 16 Dec. 11/2 Other useful buys here for Christmas catering are..1 litre packs Walls sliceable vanilla ice cream. 1978 *Aslib Proc.* XXX. 33 On-line data bases..are tools which are infinitely sliceable into subsets.

sliced, *ppl. a.* Add: **1. b.** Of food: sold already cut into slices, esp. *sliced bread.* Also in colloq. phr. *the best* (or *greatest*, etc.) *thing since sliced bread*: an expression of enthusiastic appreciation, esp. of a new invention or discovery.

1958 J. MORTIMER *What shall we tell Caroline?* ii. 109 The trouble with living here, the butter gets as hard as the rock of Gibraltar. It blasts great holes in your sliced bread. 1963 L. DEIGHTON *Horse under Water* xxxi. 124 They prodded..their product... Sliced, sterilized and Cellophane-wrapped; a loaf. 1969 R. JAFFE *Fame Game* (1970) v. 106 You're the greatest thing since sliced bread. 1970 *Guardian* 28 Feb. 10/1 The biggest thing since sliced bread is now in the shops. It is called part-baked bread. 1972 WODEHOUSE *Pearls, Girls, & Monty Bodkin* xii. 183 Bodkin regards you as the best thing that's happened since sliced bread. 1973 J. THOMSON *Death Cap* ii. 34 Mrs King asked me to..buy half a pound of sliced ham and I cut the sandwiches. 1976 *National Observer* (U.S.) 21 Aug. 6/3 They're the best thing since sliced bread and they're selling like wildfire here,' says Dan Wagner, proprietor of the Georgetown Cycle Shop. 1981 *Austral. Forest Industries Jrnl.* Suppl. (Plywood & Veneer) Oct. 35 (*heading*) Sheathing—the greatest thing since sliced bread!

3. Also in tennis (see *SLICE *v.*[1] 5).

1971 LAVER & COLLINS *Educ. Tennis Player* xi. 144 Most of the time I used a sliced serve, and that seems to be the common delivery.

slicer. 1. (Further examples.)

1907 *Yesterday's Shopping* (1969) 215/2 The Sterling Slicer. Slices any kind of vegetable or fruit evenly. 1956 F. S. ATKINSON in D. L. Linton *Sheffield* xiv. 269 Intensive efforts are being made to replace the hand-loading of coal on to conveyors by mechanical power-loaders and, in seams over 3 ft. 3 in. considerable success is being achieved with Meco Moore Loaders..and Huwood Slicers. 1971 [see *PLOUGH, *PLOW *sb.*[1] 5 i].

slick, *sb.*[1] **3. a.** For *U.S.* read 'orig. *U.S.*', and add: Also, a floating mass of oil. Also *transf.*

[1889, etc.: see *oil slick* s.v. *OIL *sb.*[1] 6 e.] 1938 *Daily Progress* (Charlottesville, Va.) 30 July 1/8 The slick.. caused by oil from the 'Hawaii Clipper'. 1950 *Jrnl. Marine Res.* IX. 69 Artificial slicks form in harbor waters contaminated with refuse and oil. 1973 *Daily Tel.* (Colour Suppl.) 9 Mar. 12/2 You can see the mouse-run quite clearly because of the slick of oil which all rodents leave behind on walls and floors if they regularly move along a particular route. 1982 H. INNES *Black Tide* II. i. 28 The slick now stretched in a great smooth, brown, greasy layer right across the bay.

5. *U.S.* A wild, unbranded horse, cow, or other range animal; a maverick.

1890 *Stock Grower & Farmer* 12 July 6/3 Seven of them were branded, the remainder were 'slicks', or horses which had run wild from birth. 1934 in J. A. & A. Lomax *Amer. Ballads* xvi. 411 No maverick or slick will be tallied In that great book of life in His home. 1965 G. SHEPHERD *West of Yesterday* xiv. 127 By picking up slicks or unbranded cattle on the way—gathered a nice little herd.

6. *U.S.* An expensive or 'glossy' magazine (opp. *pulp magazine* s.v. *PULP *sb.* 5 c).

1934 *Writer* Mar. 73/2 Perhaps he [*sc.* the author] gets an offer for two hundred dollars from one of the 'slicks'. 1952 S. KAUFFMANN *Philanderer* (1953) iv. 65 We're going to change one of our present magazines—from a confession to a woman's slick. 1958 *Manch. Guardian* 26 Sept. 4/3 Jack Finney's stories, which have been popular in the better American slicks, point the trend all the more for not all being science fiction. 1977 *Transatlantic Rev.* LX. 57 One will keep on about 'the slicks' he wants to write for.

7. a. A smooth tyre used on various kinds of racing vehicle.

1959 *Wall St. Jrnl.* (Eastern ed.) 11 Aug. 1/4 Both Goodyear Tire & Rubber Co. and General Tire & Rubber Co. have jumped into the business..developing and building 'slicks'—smooth racing tires—for the little vehicles. 1965 *Daily Mail* 2 Oct. 5/7 The slicks (smooth, £20-a-piece racing tyres) burst into smoke and then into flame. 1978 'D. RUTHERFORD' *Collision Course* 56 Everybody had fitted 'slicks', the smooth treadless tyres used on dry roads.

b. (See quots.)

1969 I. KEMP *Brit. G.I. in Vietnam* iii. 45 The 'slicks'—small Huey helicopters—would fly the troops to the battle zone, while the larger Chinooks brought in their heavy equipment. 1974 N. MEYER *Target Practice* (1975) v. 61 We were lifted by choppers called 'Hueys' (or 'slicks', because they land on runners instead of wheels).

8. *U.S. slang.* A clever or smart person; a cheat or swindler. Cf. *SLICKER 3; *SLICKSTER.

1959 N. MAILER *Advts. for Myself* (1961) I. 28 To try a major novel about the last war in Europe without a sense of the past is to fail in the worst way—as an over-ambitious and opportunistic slick. 1970 R. D. ABRAHAMS *Positively Black* iv. 88 These stories commonly turn on some way in which the 'slick' manages to trick the white storekeeper 'Mr. Charlie' into giving him respect and service. 1971 E. BULLINS in W. King *Black Short Story Anthol.* (1972) 76 Dandy's mother had a civil-service job in the city, and the city slick Dandy was from Philly. 1973 [see *RUN *v.* 52 h].

9. Special Comb.: **slick-licker** *Canad. colloq.*, an apparatus for removing oil floating as a slick.

1970 *Globe & Mail* (Toronto) 7 May 4 (*caption*) Slick-lickers—barges in Nova Scotia's Chedabucto Bay designed to suck in oil off the water's surface from the wrecked tanker Arrow and transfer it to 45-gallon drums. 1972 *Daily Colonist* (Victoria, B.C.) 12 Feb. 9/5 The federal transport department has placed a $150,000 order for 22 slicklicker machines... The machine..picks up oil from water surfaces by using an endless belt that has been treated chemically. 1975 *Lamp* (Exxon Corporation) Winter 24/1 Imperial Oil Limited, Exxon Corporation's Canadian affiliate, is employing a 'slicklicker' to counter the menace of oil spills on water.

slick, *a.* Add: **1.** Also, of a surface: slippery (chiefly *U.S.*). (Later examples.)

1901 A. H. RICE *Mrs Wiggs of Cabbage Patch* xi. 143 When the floor was dry and the candle sprinkled over it, Australia and Europena were detailed to slide upon it until it became slick. 1936 M. MITCHELL *Gone with Wind* xxxv. 590 The horse..plodded off, picking its way carefully down the slick road. 1979 C. FREEMAN *Portraits* (1980) vi. 31 The streets were covered with a white blanket of snow and ice so slick it was almost impossible for him to walk. 1981 *Railway Mag.* Mar. 115/3 No. 765 [*sc.* a steam locomotive] was true to her breed, losing her footing temporarily on the slick rails as she fought for adhesion.

2. b. Of range animals: unbranded, wild. *U.S.*

1955 R. HOBSON *Nothing too Good* xvii. 181 The pounding of slick horses hitting across the range. 1973 R. SYMONS *Where Wagon Led* I. iii. 39 Then brand everything that's 'slick'—provided you know they're off your own mares.

3. (Of persons or things personified.) (Further example.) See also sense 4 below.

1936 AUDEN & ISHERWOOD *Ascent of F6* (1937) I. i. 17 Evening. A slick and unctuous Time Has sold us yet another shop-soiled day.

4. (Of persons.) (Earlier and further examples.) Also (merging with sense 3), glibly clever, having easy assurance.

1807 *Lancaster* (Pa.) *Jrnl.* 16 Oct. 3/1 You are getting too slick. What a charming thing it is to see men under good discipline. 1921 E. O'NEILL *Diff'rent* I, in *Emperor Jones* 213 Jim Benson's one o' them slick jokers, same's Jack; can't keep their mouths shet or mind their own business. 1953 R. LEHMANN *Echoing Grove* 155 Give her a pen and she cannot be trusted not to express herself in clichés, like a schoolgirl with a smear of the popular slick journalist. 1951 [see *REGISTER *v.* 3 c]. 1966 'J. HACKSTON' *Father clears Out* 98 The rest of our local colour was made up of a community of slick, quick-off-the-mark jumpers.

5. (Of things, actions, etc.) (Earlier and further examples.) Also, neat, in good order; smart, efficient, that operates smoothly; superficially attractive, glibly clever.

1833 *Jamestown* (N.Y.) *Jrnl.* 25 Sept. 2/1 Of all the inventions I've hearn on of Mr. Van Buren's, this is about the slickest. 1837 *Baltimore Commercial Transcript* 4

Sept. 2/3 Prudence guessed strawberries and cream were slick. Jonathan thought they wa'n't so slick as Pru's lips. 1860 J. G. HOLLAND *Miss Gilbert's Career* viii. 131, I love to see a young man that keeps things slick around him. 1891 *Fur, Fin & Feather* Mar. 169 They reckons to make mighty slick work in cleaning everything up on the way back. 1901 W. CHURCHILL *Crisis* I. xii. 104 'You'd die laughing, Lige, to hear how he did it.' 'Some *slickness*, I'll gamble,' grunted Captain Lige. 'Well, I reckon 'twas slick.' 1904 W. H. SMITH *Promoters* i. 19 I've seen the thing done a hundred times, with a slick word every time. 1920 E. O'NEILL *Beyond Horizon* I. ii. 35 'He'll make this one of the slickest, best-payin' farms in the state.'.. 'Seems to me it's a pretty slick place right now.' 1921 GALSWORTHY *To Let* 286 He could not go on staying here, walled in and sheltered, with everything so slick and comfortable. 1927 *New Republic* 12 Oct. 218/2 His dialogue is of that slick and well oiled kind that you may meet in good vaudeville or in the Saturday Evening Post. 1931 E. F. BENSON *Mapp & Lucia* i. 18 Let us practise that scene where I knight you. We must get it very slick. 1933 E. O'NEILL *Ah, Wilderness!* III. i. 95 How was that for a slick way of getting rid of him? 1940 M. V. HUGHES *London Family between Wars* xii. 157 America.., with its slick, fervid haste and its terrifying efficiency. 1958 *Woman* 22 Feb. 23/3 That's what 'Six-Five Special' does to you!.. We must admit that it's about the slickest light show on TV today. 1972 E. H. GOMBRICH *Story of Art* (ed. 12) xxvi. 439 He had more and more become convinced that art was in danger of becoming slick and superficial. 1978 *Electronics & Power* Nov./Dec. 824/1 When that [*sc.* the robbing of banks] became difficult we went for the cash in transit, until the professionals got their drill so slick that the game was not worth the candle. 1978 *Lancashire Life* Oct. 155/2 The Accord four-door I drove had a slick, finger-light, five-speed manual gearbox with a well chosen set of ratios. 1979 *Church Times* 6 July 16/5, I will scrap Series Three And invent something slicker. How grand to be free—A real alive Vicar!

6. a. **slick ear** *U.S.* = *SLICK *sb.*[1] 5; also *fig.*

1914 *World's Work* XXVII. 447/1 Any 'slick-ear' (steer not marked on the ears or branded) found on the range about which inquiry was made was promptly assigned to his ownership, and 'slick-ears' eventually became known as 'mavericks'. 1958 'W. HENRY' *Seven Men at Mimbres Springs* vii. 74 I'd clean forgot the slick-ear son of a bitch! 1966 H. MARRIOTT *Cariboo Cowboy* iv. 48 Sometimes the cow would die for some reason or another and the calf would be left without a mother, in which case it grew up an orphan calf or a 'slick-ear'.

b. **slick-paper** *U.S.*, a kind of glossy paper used esp. for printing popular magazines (cf. *SLICK *sb.*[1] 6); usu. *attrib.*; hence **slick magazine, story.**

1930 D. WILHELM *Writing for Profit* iv. 102 There are between 100 and 125 pulp-paper magazines alone beside all the illustrated and 'slick paper' magazines! 1936 *Amer. Mercury* XXXVII. 286/2 Occasionally I have put aside two weeks..in which to attempt a slick magazine story. 1949 H. E. NEAL *Writing & Selling Fact & Fiction* ii. 17 The average slick story is divisible into six parts. 1976 *National Observer* (U.S.) 23 Oct. 22/5 They will come again next year with..their suitcases stuffed with slick-paper brochures full of self-praise. 1980 *TWA Ambassador* Oct. 82/3 Two distinct genres of regional business publications are trying to serve this market: the tabloid and the slick magazine.

slick, *adv.* **2.** (Earlier example.)

1818 H. B. FEARON *Sk. Amer.* 123 Did she die slick right away?

slick, *v.* Add: **1. b.** (Further examples with *up*.)

Sometimes with derogatory overtones.

1831 [see *BETTER *a.* 4 b (*b*)]. 1863 *Harper's Mag.* 55/1 He got into Peter's way by attempting..to 'slick up' the barn. 1953 R. LEHMANN *Echoing Grove* 187 Stop slicking it up into cheap melodrama. 1973 E.-J. BAHR *Nice Neighbourhood* iii. 25 I'm going to get her all slicked up in her new outfit from Aunt Joan and show her off.

d. *intr.* To smarten or tidy *up*. *U.S.*

1841 *Knickerbocker* XVII. 41 In a little while he recovered his self-possession, or, to make use of one of his own expressions 'he slicked up'. 1887 M. E. WILKINS *Humble Romance* 395 I'm going to slick up here a little for you while I stay... He watched her..as she flew about putting things to rights. 1948 *Family Circle* June 96/2 It's always serious when they slick up for a girl!

slicked, *ppl. a.* Add: **a.** (Later examples in sense 3 of the vb.) Also with *back, down.*

1921 E. O'NEILL *Diff'rent* I, in *Emperor Jones* 215 His hair [is] wet and slicked in a part. 1937 A. CHRISTIE *Murder in Mews* 163 It was the good-looking young man with the slicked-back hair who had spoken. 1949 E. DE MAUNY *Huntsman in his Career* II. 128 There was dandruff on the collar from his slicked-back, fair hair. 1964 *Economist* 20 June 1365/2 Mr McNamara's slicked-down patent leather hair. 1975 B. GARFIELD *Death Sentence* (1976) xi. 57 Latins, with slicked hair. 1976 N. THORNBURG *Cutter & Bone* iv. 86 He had a distinctly Arabic look about him, slicked-down black hair. 1977 *Listener* 17 Nov. 653/2 People with slicked-back hair..jitter-bugging round the radio.

b. With *up*: dressed up, smart, elegant; also *fig.*, made more sophisticated. orig. *U.S.*

1836 [in Dict. s.v. SLICK *v.* 1 b]. 1867 *Atlantic Monthly* Jan. 109/2 Is this my farm?... It looks more slicked up than ever it used to. 1928 J. GALSWORTHY *Swan Song* I. xi. 81 Montpellier Square..was all slicked up since he was last there... Builders and decorators must have done well lately. 1957 R. HOGGART *Uses of Literacy* xi. 276 The new-style popular publications..are pallid but slicked-up extensions even of nineteenth-century sensationalism. 1979 *Tucson Mag.* Apr. 34/1 In the same wave came..the new City Hall and the slicked-up El Presidio Park.

slicker. 2. a. For *U.S.* read 'orig. and chiefly *U.S.*' and add: *spec.* a loose-fitting oilskin outer garment, usu. of a bright yellow or orange colour.

1910 C. E. MULFORD *Hopalong Cassidy* xx. 126 After throwing his saddle on his horse he went back to the house to get his 'slicker', a yellow water-proof coat. 1953 M. PEAKE *Mr Pye* xxii. 176 Tintagieu..now wore a black oilskin slicker. 1971 *N.Z. Listener* 19 Apr. 56/5 One was a nuggety bloke in a sou'-wester, oilskin slicker, and bowyangs. 1978 R. LUDLUM *Holcroft Covenant* iii. 43 Police and maintenance crews were everywhere, distinguished from one another by the contrasting black and orange of their slickers.

3. = *city slicker* s.v. *CITY 9. orig. and chiefly *U.S.*

1900 'J. FLYNT' *Notes Itinerant Policeman* iii. 62 Pick-pockets!.. You just bring the slickers in. 1932 J. T. FARRELL *Young Lonigan* vi. 230 Swan, the slicker, who wore a tout's gray checked suit with narrow-cuffed trousers [etc.]. 1936 *Sat. Morning Advertiser* (Durant, Okla.) 14 Mar. 1/3 (*heading*) 'Slicker' insurance agents better be a bit wary now. 1946 WODEHOUSE *Money in Bank* xii. 106 I'm going to put it across that slicker..if it's the last thing I do. 1978 *Morecambe Guardian* 14 Mar. 15/4 He becomes a sort of Midnight Cowboy, lost and confused by the slickers around him.

4. *U.S.* = SILVER-FISH 2.

1902 L. O. HOWARD *Insect Bk.* 380 (Order *Thysanura.*) The insects of this order are usually of very small size... They comprise the little insects known as springtails, bristletails, fishmoths or slickers. 1962 METCALF & FLINT *Destructive & Useful Insects* (ed. 4) xix. 905 The silverfish or slicker..is uniform, silvery or greenish gray.

Hence as *vb. trans.*, to cheat; to defeat by being 'slick'.

1935 H. DAVIS *Honey in Horn* xxii. 376 His entertainment had mostly been swindling and slickering them. 1971 LAVER & COLLINS *Educ. Tennis Player* xiii. 157 It happens all the time, an older, less powerful team who understand the principles of doubles slickering a couple of youths who might be considerably superior in singles. 1974 *Tel.* (Brisbane) 4 Sept. 30/4, I thrive on guys who try to slicker me. 1979 *Globe & Mail* (Toronto) 16 May 41/8 What was it Charlie had said about being slickered?

slickered (sli·kəɹd), *a.* *U.S.* [f. SLICKER *sb.*[2] + -ED[2].] Wearing a slicker.

1972 C. L. COOPER *Black* 151, I could see..slickered cops smoking on duty. 1975 *New Yorker* 10 Mar. 29/1 About twelve horses on the track, a few running hard, their slickered riders standing tall in the stirrups.

slicking, *vbl. sb.* Add: **1*.** With *up*: the action of making (oneself, a place, etc.) neat and tidy. *U.S. colloq.*

1855 *Trans. Mich. Agric. Soc.* VI. 495 The farm needs a good deal of slicking up to make the general appearance equal to what nature has done for the land. 1907 *Springfield* (Mass.) *Weekly Republ.* 9 May 1 Denver has been having her period of spring slicking up.

slickly, *adv.* Add: **1.** (Later example.)

1973 *Daily Tel.* 11 Jan. 17/1 Walter Albini likes to see his male models with their hair oiled slickly down like George Raft's.

2. (Later examples.)

1927 *Daily Tel.* 16 Aug. 12/5 The play..needs to be more slickly produced and better acted in order to be made convincing. 1978 K. ROYCE *Satan Touch* vi. 99 Slickly he turned his hand down to indicate the chair.

slickness. Add: **1.** (Later U.S. example.)

1976 *Billings* (Montana) *Gaz.* 28 June 3-A/1 It's a three-quarter ton pickup with skid-trailer, painted the bright orange of the Montana Highway Department. It tests the slickness of the road surface.

3. (Later examples.) Now freq. *derog.*

1927 *Music & Letters* July 321 Pure artistry may decline to a mere slickness and facility. 1967 E. SHORT *Embroidery & Fabric Collage* iv. 112 Anything cheap, shoddy, shallow or clever to a degree of slickness, is out of place.

slickster (sli·kstəɹ). *U.S. slang.* [f. SLICK *sb.*[1] + -STER.] A swindler.

1965 C. BROWN *Manchild in Promised Land* xiv. 332 All the Muslims now felt as though 125th Street was theirs. It used to belong to the hustlers and the slicksters. 1973 [see *RUN *v.* 52 h].

slide, *sb.* Add: **I. 1. c.** (Examples of the sense 'portamento'.)

1908 *Grove's Dict. Mus.* (ed. 2) IV. 482/1 To violinists the 'slide' is one of the principal vehicles of expression, at the same time a means of passing from one note to another at a distance. 1913 F. THISTLETON *Mod. Violin Technique* xv. 74 The slide is one of the principal mediums of expression on the violin. 1938 L. TERTIS *Beauty of Tone in String Playing* 14 The celerity with which this is done is the secret of discreet natural portamento... There must be no drawling, languishing, or lingering in the action of the slide.

d. *fig.* A rapid decline; a downturn. Also in phr. *on the slide.*

1884 [in Dict., sense 1 b]. 1931 *Economist* 14 Mar. 569/2 Unsatisfactory traffics, and the passing of the B.A. Great Western dividend, accentuated the 'slide' of prices. 1969 N. COHN *Pop from Beginning* ix. 86 He began to flag. By early 1964, he was definitely on the slide. 1981 *Times* 5 May 18/1 A 20 per cent slide in profits at the half-way stage.

e. *Baseball.* A plunging or sliding approach made to a base along the ground.

1886 H. CHADWICK *Art of Batting* 68 A slide in time saves an out. 1934 *Baseball Mag.* Apr. 497/2 Chapman has a natural talent as a base stealer... He knows how.. to make a perfect slide. 1944 E. ALLEN *Major League Baseball* xv. 204 There are four types of slides: the hook slide, the bent leg slide, the feet first slide and the head first slide. 1972 J. MOSEDALE *Football* iii. 32 He stretched it into a triple with a daring slide [contextually in Baseball].

f. *Surfing.* A ride across the face of a wave (see quot. 1963); a wave suitable for this.

1935 D. KAHANOMOKU in T. Blake *Hawaiian Surfboard* i. 32 (*caption*) A fine illustration of the slide. The wave is coming on while the rider is sliding left across the face of the swell. 1946 J. A. BALL *Scrapbk. of Surfriding* 57 (*caption*) Tom Blake streaking along on a long belly slide. 1963 J. POLLARD *Austral. Surfrider* ii. 20/1 A 'slide' can be either 'left' or 'right', angling down the wave to one side or the other. 1968 *Surfer Mag.* Jan. 47/3 Ten-foot waves that peel off in good right and left slides.

g. *Curling.* A delivery in which a curler slides some distance forward in launching his stone.

1950 K. WATSON *Curling* i. 42 By delivering the stone at the end of a long slide, a player could be more accurate in delivery. 1962, 1969 [see *long slide* s.v. *LONG *a.*[1] A. 18].

h. *Jazz.* = *GLISSANDO.

1959 'F. NEWTON' *Jazz Scene* 289 Technical terms either duplicate existing, but unfamiliar, ones—e.g. *slide, smear*, for glissando..or they describe things for which no proper academic equivalent exists. 1973 *Black World* Nov. 48/1 Performance practices require a 'slur' and/or 'slide' when moving from tone to tone.

3. c. *Geol.* A fault formed at and associated with a fold.

1910 E. B. BAILEY in *Q. Jrnl. Geol. Soc.* LXVI. 593 'Fold-fault' itself is too cumbrous for constant repetition, and accordingly 'slide' has been introduced to take its place. 1934 —— in *Ibid.* XC. 467 'Fold-fault' is an old word for a fault formed in close causal connexion with folding... I employed the word 'slip' in this sense, but on Lapworth's suggestion exchanged it for 'slide'. Now that reversed and unreversed limbs are often distinguishable through the help of current-bedding, the word 'slide' is less necessary, since it can often conveniently be replaced by 'thrust' or 'lag'. The following are helpful though incomplete definitions:— A thrust is a slide replacing an inverted limb, actual or ideal. A lag is a slide replacing a normal limb. 1969 BENNISON & WRIGHT *Geol. Hist. Brit. Isles* iii. 55 An important structural break, the Iltay Boundary Slide, separates two contrasted sedimentary successions both in Scotland and northern Ireland.

II. 5. j. *N.Z.* A serving-hatch.

1949 J. R. COLE *It was so Late* 92, I was standing by the slide in the lounge one night. 1955 *Numbers* I. iv. 14 Charlie ordered for them both and Kate dragged back to the slide.

k. *transf.* A musical instrument equipped with a slide.

1976 A. BAINES *Brass Instruments* iv. 94 (*heading*) Renaissance slides. 1977 J. WAINWRIGHT *Do Nothin' till you hear from Me* viii. 129 Walter Green—trombonist... As a slide-player he is average.

6. Now *spec.* a clasp for fastening in the hair. (Later examples.)

1932 L. GOLDING *Magnolia Street* III. ix. 593 That slide which has just slid out of her hair on to the parquet floor. 1952 M. LASKI *Village* ii. 33 Her soft brown hair caught back with a slide. 1981 J. B. HILTON *Playground of Death* ii. 10 There wasn't a grip, clip or slide on the market that would keep my mam's hair up.

7. b. (Earlier example.) Now also, a photographic transparency for use in a slide projector.

1819 M. EDGEWORTH *Let.* 17 Apr. (1971) 199 You know him and his magic lantern of good things. Some new figures on the slides. 1940 P. E. BOUCHER *Fundamentals of Photogr.* (1941) xiii. 200 Valuable slides..which are to be subjected to considerable use should be mounted in glass. 1978 M. J. LANGFORD *Step by Step Guide to Photogr.* 176 Before presentation, your slides must be inserted in holders ready for projection.

III. 10. c. (Earlier example.)

1855 W. HOWITT *Land, Labour & Gold* I. 206 We take the fine gravel out of the slide of the cradle.

d. A structure with a smooth sloping surface used as a toy or piece of playground equipment down which children slide, or as an entertainment at a fairground. Cf. *CHUTE *sb.* 3 c.

1890 *Century Mag. Advertising Suppl.* Dec. 70 Wood's parlor toboggan slide. The most satisfactory toy yet invented for children. 1924 *Ladies' Home Jrnl.* Nov. 126/1 This kiddies' slide is more fun than the old cellar door! 1954 R. DAHL *Someone like You* 210 All week the swings and the see-saws and the high slide with steps going up to it stood deserted. 1975 I. STARSMORE *English Fairs* iv. 92 (*caption*) The Portable Slide. Height 29 ft... The three 100 ft. lanes guide 'passengers' down the slide. 1979 'J. LE CARRÉ' *Smiley's People* (1980) xxi. 257 There was a children's slide in the garden.

11. b. *U.S. slang.* (See quot. 1932.)

1932 *Even. Sun* (Baltimore) 9 Dec. 31/5 *Slide*, a trouser's [*sic*] pocket. 1967 'I. SLIM' *Pimp: Story of my Life* (1969) iii. 68 How would you like a half a 'G' in your 'slide'?

12. (Earlier example.)

1842 J. E. DEKAY *Zool. N.Y.* I. 40 The steel trap is placed..at the bottom of one of their slides.

slide, *v.* Add: **B. I. 1. d.** *Baseball.* To perform a slide (sense 1 e).

1891 *Harper's Weekly* 23 May 391/4 His base running, in spite of his care about sliding, is of the old-time quality that has already won two championships for Yale. 1904 J. J. MCGRAW *Science of Baseball* 67 He shouldn't slide unless his pants are properly padded. 1932 *Baseball Mag.* Oct. 501/2 George Watkins, quick to grasp Dazzy's slight slip, turned on a full burst of speed and slid across home plate with the only run of the game. 1977 *Rolling Stone* 30 June 76/2 Do you always think about baseball players when you're making love?.. I couldn't figure out why you kept yelling, 'Slide!'

e. *Surfing.* To ride across the face of a wave.

1931 *Country Life in Amer.* Jan. 57 If the wave proves exceptionally steep, keep to the stern of the board and then, after you 'catch' the wave, head the board at an angle to it. This will enable you to 'slide' with the wave. 1959 J. BLOOMFIELD *Know-How in Surf* iii. 27 The gradually breaking crest enables the body to slide down its front at an angle of approximately 45 degrees.

f. *Curling.* To move forwards while delivering the stone.

1936 F. B. TALBOT *Mr Besom starts Curling* xiii. 34 Many good players slide out of the hack as they deliver the stone. 1950 K. WATSON *Curling* i. 1 Whether you slide or do not slide, that follow-through is essentially the smooth delivery of the stone.

4. b. (Later examples.)

1904 [see *BUSY *sb.*[2]]. 1932 E. WALLACE *When Gangs came to London* xxvii. 269 There's only one word that any sensible man can read in this situation, and that word is —slide!

slide-. Add: **a.** *slide cornet, trombone, trumpet* (earlier example), *whistle*; **slide fastener** chiefly *U.S.*, a zip-fastener; **slide-wire** *Electr.*, a resistance wire along which a contact slides in a Wheatstone bridge or similar device.

1926 WHITEMAN & MCBRIDE *Jazz* ix. 206 The jazz band has introduced some little known instruments such as..the slide cornet and the czimbalon. 1946 MEZZROW & WOLFE *Really Blues* (1957) i. 12 He showed up in the band room with a slide cornet. 1934 *Newsweek* 21 July 29/2 The Prince of Wales uses a slide fastener on his trousers. 1944 *Sun* (Baltimore) 16 Sept. 7/4 The B. F. Goodrich Company announced today development of a 'zipped lip' construction that makes a metal slide fastener watertight and airtight. 1971 N. MARSH *Tied up in Tinsel* vi. 135 'Let's have a look at the robe.'.. A slide fastener ran right down the back. 1891 C. R. DAY *Descr. Catal. Musical Instruments, R. Military Exhib.* x. 180 Slide trombone... In this instrument, as should always be the case, the taper of the bell is carried right through the tuning slides. 1934 *Hound & Horn* July–Sept. 595 The harpsichord seems a very complicated instrument to compare alongside the single-noted valve trumpet, or a slide trombone. 1977 'E. CRISPIN' *Glimpses of Moon* xiii. 269 How we could brighten our Church Fêtes up, short of breaking all the Ten Commandments simultaneously to a fanfare of slide-trombones. 1885 G. B. SHAW in *Mag. of Music* II. 112/1 These slide trumpets are not the instruments Bach wrote for. 1939 *Sears, Roebuck Catal.* Fall–Winter 914/3 Slide whistle. Professional model... Has full chromatic scale of two octaves. 1976 *Gramophone* Feb. 1355/2 The flight grows slower to reveal gentle tones of slide-whistles, zither and harp. 1885 J. DREDGE *Electric Illumination* II. I. 53 The Slide-Wire, or Metre Bridge..is a modification of the bridge due to Kirchoff, and is especially useful for the measurement of low resistances. 1922 GLAZEBROOK *Dict. Appl. Physics* II. 714/2 The first bridge to employ a slide wire was devised by Fleeming Jenkin in 1862 and was used to intercompare the standard coils made for the British Association Committee on Electrical Standards. 1964 *Oceanogr. & Marine Biol.* II. 359 The depth element is a Bourdon tube coupled to a slide-wire potentiometer. 1969 A. BRODGESELL in B. G. Lipták *Instrument Engineers' Handbk.* I. ix. 942 Potentiometric displacement sensors consist of a slide wire and wiper. The slide wire is powered by a constant voltage representing full scale travel.

c. *slide carrier, changer, projector, show, viewer*; *slide-in* adj.; **slide area** *U.S.*, an area in which landslips or avalanches are likely to happen; **slide-back** *Electronics*, the alteration of the grid bias of a thermionic valve which is necessary to restore the anode current to zero after the application of a signal voltage to the grid; apparatus to measure this alteration, freq. as an indirect measure of the signal voltage; also **slide-back voltmeter; slide guitar,** a style of guitar-playing characterized by a glissando effect produced by moving an object along the strings; usu. *attrib.*; **slide-rock,** talus rock; **slide-tape** *attrib. phr.*, involving photographic slides shown in a predetermined sequence to the accompaniment of a synchronized commentary recorded on magnetic tape.

1959 *Sunday Times* 7 June 16/6 The 'slide area' itself is that part of the Californian coast which is physically slipping, dropping and sliding towards the sea. 1970 *Wall St. Jrnl.* (Eastern ed.) 19 May 1/4 The Kildares live in what is euphemistically called here a 'slide area'. 1925 *Year-bk. Wireless Telegr.* 847 When a control room is some distance from the transmitter it is usual to install a valve voltmeter with a slideback which either measures the voltage across the output of the main amplifier or indicates when a certain voltage is exceeded. 1931 *B.B.C. Year-bk.* 1932 356 The 'slide-back'..consisted of a valve or similar device so biassed that no indication occurred until there was present and superimposed upon the bias a voltage greater than, and opposing in phase, the biassing voltage. 1938 H. A. BROWN *Radio-Frequency Electr. Measurements* (ed. 2) vi. 279 Peak, or slide-back, voltmeters are coming more and more into use. 1948 A. L. ALBERT *Radio Fundamentals* ix. 354 There is an error involved with this slide-back voltmeter, but with large signals..the error is small. 1965 *Wireless World* July 19 (Advt.), Slide-back measurement of time and amplitude by means of directly-calibrated shift controls. 1953 A. PEARLMAN *Rollei Manual* xxiii. 357 Sticky exudations may foul the slide carrier of the projector. 1971 *Sci. Amer.* Sept. 224/2 An adequate beam can be formed by making a pinhole aperture in the slide carrier of a 35-millimeter projector. 1959 *IRE Trans. Mil. Electronics* III. 97/1 Like a projection slide changer, we can observe one slide while discarding the slide already observed and replacing it with a new one. 1962 *Which?* Mar. 69/1 We did not test fully automatic projectors, but 6 had semi-automatic slide changers built in. 1968 P. OLIVER *Screening Blues* i. 35 Another..version was recorded in 1937 by Black Ace (B. K. Turner) who accompanied himself with brilliant slide guitar playing. 1976 *Morecambe Guardian* 7 Dec. 23/2 And inevitably those ubiquitous sessioners, Klaus Voorman (bass)..and Jesse Ed Davies (slide guitar), have played a major part. 1977 MCKNIGHT & TOBLER *Bob Marley* ix. 111 The Wailers version is decorated by an ethereal slide guitar solo. 1973 G. DAVEY *Fun with Hi-Fi* v. 35 The BSR MacDonald playing deck which I use has slide-in facilities for fitting the cartridge of one's choice. 1977 *Gramophone* Apr. 1629/2 The head-shell has a slide-in cartridge carrier. 1956 E. S. BOMBACK *Retina Manual* xxi. 222 (*caption*) The Leitz Prado 150-watt slide projector. 1979 P. NIESEWAND *Member of Club* ii. 17 Two slide projectors were being positioned... 'Remember, when the lights go out, we'll be showing some slides.' 1901 *Yearbk. U.S. Dept. Agric. 1900* 195 In the mountains we often find the hillside slopes covered with broken rock of various sizes. This we call slide rock. 1974 FLINT & SKINNER *Physical Geol.* vii. 121/2 Weathering converts the sliderock into fine-grained regolith which, with its pores of extremely small diameter, can hold much more moisture than sliderock and thus acquire both vegetation and soil. 1956 E. S. BOMBACK *Retina Manual* xxi. 222 The color slide show has quite a lot in common with the motion picture film. 1978 *Peace News* 25 Aug. 18/3 On their last visit to Britain four years ago they did a slide show and a question and answer session. 1971 *Publishers' Weekly* 22 Mar. 20/2 The final part of the program..consisted of a slide-tape commentary. 1977 J. HEDGECOE *Photographer's Handbk.* 305 Sound for slide-tape presentations can be prepared from studio recordings using a microphone direct. 1960 *Which?* Oct. 228/2 A slide viewer should give a good *optical performance.*

slideable, *a.* Restrict *rare*[−1] to sense in Dict. and add: Also **slidable. b.** That may be slid.

1888 *Engineer* LXV. 538/4 A screw mounted in a bearing slidable in a right line. 1925 A. W. JUDGE *Carburettors & Carburation* iv. 63 M is a slidable clip. 1969 *Jane's Freight Containers 1968–69* 408/2 The bogie will be slidable along the full length of the chassis.

Hence **sli·dably** *adv.*

1907 F. W. LANCHESTER *Aerodynamics* 348 A square plane of thin brass, mounted 'slidably' on anti-friction rollers. 1954 *Patents for Inventions. Abridgments of Specifications* Group xxxiv. 23/1 The spiders 8 of which are also slidably mounted on the shafts.

slider. Add: **1. c.** (Earlier example.)

1877 *Scribner's Monthly* Nov. 11/1 'Sliders', the common river turtles of almost all the rivers of the region, grow to a much larger size.

d. *Baseball.* A fast pitch that breaks or slides away from its original path.

1936 *Sun* (Baltimore) 14 Aug. 12/6 It looks like what some of the modern pitchers call 'a slider'. 1980 *Washington Post* 1 Aug. D3/3 The human body isn't meant to throw the slider.

2. (Further example.)

1805 R. SUTCLIFF *Jrnl.* 6 Jan. in *Trav. N. Amer.* (1811) iv. 67 They make use of a boat that has two sliders, one on each side the keel.

4. e. *Bell-ringing.* (See quot. 1901.)

1871 [see STAY *sb.*[2] 2 h]. 1901 H. E. BULWER *Gloss. Techn. Terms Bells & Ringing* 4 *Slider*, usually a bar of wood pivoted at one end on one of the lower members of the 'frame', and extending across the bottom of the 'bell-pit' so that its free end may move to and fro on a bed provided for it on the opposite side of the 'pit'. 1931 E. MORRIS *Hist. & Art Change Ringing* i. 15 For many years ..bells were rung without stay or slider as we now know them. *Ibid.*, Stedman..mentions what would be the forerunner of the stay and slider adjustment. 1974 J. CAMP *Bell Ringing* ii. 30 (*caption*) The stay has pushed the slider to the limit of its movement and the bell cannot turn any further.

f. A sliding electrical contact, forming part of a variable resistance or the like, or serving as a control on electrical equipment.

1872 *Jrnl. Soc. Electr. Engineers* I. 202 The slider *n* is moved on the compensating wire so as to destroy the deflection of the galvonometer. 1923 *Popular Wireless* 13 Oct. Suppl. 1 Suppose we have a coil consisting of 500 turns of No. 22 wire,..fitted with a slider, and we wish to know approximately where to put the slider to receive the Dutch concerts. 1965 *Wireless World* Sept. 432/1 The base..is taken to the slider of a potentiometer connected across the output terminals of the power supply. 1975 *Hi-Fi Answers* Feb. 76/3 Set the input level sliders to about three-quarters of full travel and route the signal back through the amplifier by means of the tape monitor button. 1978 *Gramophone* Aug. 392/3 The latest in Bang and Olufsen's range of Beomaster tuner-amplifiers is.. distinguished by its..absence of switches and knobs—all functions being handled by sliders, press-keys, and wheels.

7. b. Ice-cream served in a sandwich form between two wafers. *colloq.*

1915 J. J. BELL *Wee Macgregor Enlists* ii, Come on oot wi' me an' I'll stan' ye a dizzen sliders. 1935 L. MACNEICE *Poems* 63 Ice cream in sliders Bought in dusty streets. 1967 R. MACKAY *House & Day* 75 'I'll have a slider too.'.. The woman took a wafer..and covered it with the thin yellow ice-cream... She put a second wafer on the top.

8. *slider bridge, clutch, control, potentiometer, switch.*

1919 S. F. WALKER *Electr. Mining Machinery* xxiii. 186 Use of the slider bridge in connection with the loop test for finding a fault to earth in a cable. 1972 *World of Wild Wheels* (Custom Car) 57/2 One of the latest innovations being tried in the States..is the slider clutch. 1978 *Detroit Free Press* 16 Apr. F14/1 (Advt.), Enderle fuel injection, 2 spd Lenco, new slider clutch, Airheart disc brks. 1973 *Wireless World* Oct. 72 (Advt.), Top quality slider controls. 1972 *Ibid.* Jan. 88/2 (Advt.), New slider potentiometers. As used on only the most exclusive of Audio Amplifiers and Mixers. 1970 *Ibid.* July 87/2 (Advt.), Slider switches. Double pole, double throw.

slide-rule. Also **slide rule.** Add: *spec.* A device whereby multiplication and division, and sometimes other mathematical operations, may be performed with speed but limited accuracy, consisting essentially of two rules marked with logarithmic scales and capable of being slid along one another, and usually also a transparent cursor marked with a line crossing the scales, so that a required result may be obtained by inspection after proper movement of these. Also *fig.* (Further examples.)

This, not *sliding rule*, is now the usual name.

1890 *Engineering News* 20 Dec. 543 In France the slide rule is seen on the desk of almost every manufacturer, miller or engineer. 1930 *Engineering* 31 Jan. 131/3 A special form of slide rule..for the purpose of simplifying calculations relating to structures of reinforced concrete. 1958 *Times Lit. Suppl.* 10 Jan. 19/1 They bring out their slide-rule at the sight of a signature, and then, by nice calculation, decide whether there has been any sense in the prose equations which precede it. 1958 *Listener* 11 Sept. 375/1 Towards the end of the war many of the operations of war, especially those concerned with aircraft, were kept under close scientific scrutiny and control. This was the era of the so-called slide-rule strategy. 1974 *Encycl. Brit. Macropædia* XI. 653/2 The first known slide rule in which the slide worked between parts of a fixed stock was made by Robert Bissaker of Great Britain in 1654. *Ibid.*, Amédée Mannheim, an officer of the French artillery, invented in 1859 what may be considered the first of the modern slide rules... This rule.. brought into general use a cursor. 1976 NICHOLS & ARMSTRONG *Workers Divided* 68 Workers were incensed by the efforts of a new manager to run their plant flat out all the time: 'You have to have a real feel for the plant not just be all slide-rule like him.'

sliding, *vbl. sb.* Add: **2. b.** *sliding motion* (earlier example).

1690 LOCKE *Essay Hum. Und.* II. xxiii. 144 All parts of Bodies must be easily separable by such a lateral sliding motion.

sliding, *ppl. a.* Add: **3. b.** (Later examples.)

1948 *Assoc. Football* ('Know the Game' Series) 31/2 A sliding tackle done fairly is not dangerous.., especially when clear contact is made with the ball, and should therefore not be penalised. 1974 *Liverpool Echo* (Football ed.) 12 Oct. 1/2 Foggon went racing through again, but Boersma took the ball off him with a splendid sliding tackle.

6. a. *sliding contact*, a connection in an electric circuit that can be slid along a length of resistance wire; see also s.v. SLIDING *vbl. sb.* 2 b; *sliding valve* = SLIDE-VALVE.

1872 *Jrnl. Soc. Telegr. Engineers* I. 208 The wire with sliding contact was apt to wear if much used. 1926 R. W. HUTCHINSON *Wireless* 77 Sliding contacts can be moved to and fro along two brass sliding rails. 1971 B. SCHARF *Engin. & its Language* xxi. 307 In order to vary the value of one of the known resistances a rheostat may be used, or two of the known resistances may be replaced by a single wire of known resistance with a sliding contact. 1909 *Westm. Gaz.* 11 Nov. 5/1 The new Daimler engine may be said to have brought us to the end of the first stage of the sliding-valve principle.

b. *sliding panel* (earlier example), *roof, shutter* (earlier example), *window.*

a 1817 JANE AUSTEN *Northanger Abbey* (1818) II. v. 76 Have you a stout heart?—Nerves fit for sliding pannels and tapestry? 1929 *Motor World* 29 Mar. 199/1 One or other of the various types of sun-saloons (folding or sliding roof) may be offered at an extra charge. 1959 *Observer* 1 Mar. 21/6 Although many will welcome the sliding roof, the handle is rather prominent. 1766 T. H. CROKER et al. *Compl. Dict. Arts & Sciences* II., s.v. *Madder* col. 4, The sliding-shutters are pulled down. 1724 in *Maryland Hist. Mag.* (1911) VI. 1 Two sliding windows ..with good frame shutters. 1880 *Dict. Leading Techn. & Trade Terms Archit. Design & Building Construction* 207/1 Another form of opening and closing window is one used in domestic structures of a humble character, and termed a 'sliding', sometimes a 'rolling window'. 1976 H. MACINNES *Agent in Place* xxv. 260 A stretch of sliding windows opening onto a balcony.

9. b. (Earlier example.)

1842 C. GUEST *Jrnl.* 14 Feb. (1950) vii. 129 His opinion that in times of scarcity the fixed duty he proposes would have to give way, which is exactly the argument the Tories use when advocating the sliding scale.

10. a. *sliding hernia* (Med.) (see quot. 1958).

1910 SPENCER & GASK *Pract. Surg.* xvii. 995 Retro-

peritoneal hernia... The cæcum or sigmoid flexure may slide up and down behind the peritoneum, 'Sliding hernia', 'Hernie par glissement'. 1936 COLE & ELMAN *Textbk. Gen. Surg.* xxv. 753 Sliding hernias descend so readily that a truss is rarely satisfactory in maintaining reduction. 1958 D. L. B. FARLEY in L. Oliver *Basic Surg.* xiii. 198 Hernia 'en glissade' ('sliding' hernia) refers to herniation of a viscus such as the cæcum or bladder which has an extraperitoneal surface. 1974 R. M. KIRK et al. *Surgery* vi. 78 Sliding hernia. If the lower oesophagus and cardia straighten out and slide into the chest, the competence of the cardiac sphincter may be impaired so that gastric contents reflux into the oesophagus.

b. *sliding filament* (Physiol.), used *attrib.* to designate the model of the action of striated muscle in which contraction results from filaments of actin and of myosin sliding past one another.

1957 *Jrnl. Biophys. & Biochem. Cytol.* III. 640 The results which have been described above give full support to the 'sliding filament model' of striated muscle. 1973 *Times* 14 Aug. 14/7 This..provided the main basis for the 'sliding filament' theory of muscle contraction, now universally accepted.

11. *sliding parity* (Econ.) = *crawling peg* s.v. *CRAWLING *ppl. a.* b.

1966 *Economist* 25 June 1440/2 A tiny minority advocates completely free exchange markets... Possibly a majority favours continued official intervention to set limits to market fluctuations... But an increasing minority favours a compromise system, variously called the crawling peg..or..the sliding parity. 1970 *Times* 9 Feb. 20/1 It sets out to demolish the arguments of those who are..downright hostile to the introduction of the so called sliding parity or crawling peg.

slieveen, var. *SLEEVEEN.

slight, *v.* **2.** For † *Obs.* read *Obs.* exc. *Hist.*, and add later examples.

1974 *Country Life* 28 Mar. 747/3 In March 1645–46, Parliament gave orders that Corfe [Castle] should be slighted. 1976 E. N. LUTTWAK *Grand Strategy of Roman Empire* ii. 57 It was standard practice to slight the defenses once the site was left. 1977 *Brit. Med. Jrnl.* 24 Dec. 1619/1 Mrs Barbara Castle shattered the political confidence of consultants as effectively as Henry II slighted his opponents' strongholds.

slighten, *v.* For † *Obs.* read *rare* and add: **2.** To make smaller or more slight.

1954 P. TOYNBEE *Friends Apart* vii. 90 Dysentery.. had thinned his face and slightened his broad body.

Hence **sli·ghtening** *ppl. a.*

1916 A. QUILLER-COUCH *Art of Reading* (1920) i. 14 God forbid that anyone should hint a slightening word of what our sons and brothers are doing just now.

slighting, *vbl. sb.* **2.** For † *Obs.* read *Obs.* exc. *Hist.*, and add later examples.

1936 *Times Lit. Suppl.* 6 June 479/4 In spite of Cromwellian 'slighting' and the quarrying of local builders and road-makers, so much..still remains. 1977 H. R. LOYN *Vikings in Britain* v. 95 Evidence of possible slighting of fortifications at Cricklade and Cadbury, may well indicate the confidence of the new régime at least in Wessex under Cnut and Earl Godwin.

† slightually (slai·tiuăli), *adv.* *Obs.* *U.S. joc.* [f. SLIGHT(LY *adv.* + ACT)UALLY *adv.*] Actually slightly.

1859 E. H. N. PATTERSON *Jrnl.* 29 Mar. in L. Hafen *Overland Routes to Gold Fields* (1942) 79 The weather has been beautiful, although last night was 'slightually' frosty. 1873 'MARK TWAIN' *Gilded Age* xxix. 266 The Hooverville *Patriot and Clarion* had this 'item':—Slightually Overboard.

slim, *sb.* Restrict † *Obs.* to sense in Dict. and add: **2.** A course of slimming, a diet; usu. in phr. *sponsored slim.*

1977 *Gay News* 7–20 Apr. 8/1 Barrie announced his intention to go on a sponsored 'slim'. People were asked to sign pledge forms to give a certain amount of money for every pound Barrie would lose between Cambridge and Oxford. 1977 *Navy News* Aug. 30/6 Bill Skilliter went on a sponsored slim and lost 3st.

slim, *a.* Add: **1. f.** Of clothes: cut on slender lines; designed to give an appearance of slimness.

1884 [see *scoop-shovel bonnet* s.v. *SCOOP *sb.*[1] 7]. 1970/71 *Kay's Catal.* Autumn–Winter 145/1 Crimplene Skirt elegantly slim with a raised stripe effect. 1979 *Daily Tel.* 4 June 17/3 (Advt.), One of the most handsome and utterly wearable of this spring's suits is the linen-y slim one in a soft lilac-grey.

4. a. *slim-faced, -footed, -legged, -muzzled, -pointed, -sandalled, -shanked, -tailed, -textured, -waisted* (later example).

1838 POE *Narr. A. G. Pym* xix. 160 Slim-legged hogs. 1862 G. M. HOPKINS *Poems* (1967) 10 And spread Slim-pointed seagull plumes. 1866 —— *Jrnls. & Papers* (1959) 142 After six a very slim-textured and pale causeway of mare's tail cloud running N.E. and S.W. 1872 HARDY *Under Greenwood Tree* I. I. ii. 22 The slim-faced martel had knocked 'em down to me because I nodded to en in my friendly way. 1880 'MARK TWAIN' *Tramp Abr.* x. 95 A long, slim-legged boy, he was, encased in quite a short shirt. *Ibid.* 96 Into the midst of this peaceful scene burst that slim-shanked boy in the brief shirt. 1912 W. DE LA MARE *Child's Day* 14 But there, Ann dear, You'd rather be A slim-tailed mermaid In the sea. 1914 D. H. LAWRENCE *Prussian Officer & Other Stories* 20 His slim-legged, beautiful horse. 1922 JOYCE *Ulysses* 530 Milly Bloom, fairhaired, greenvested, slimsandalled.., breaks from the arms of her lover. 1923 D. H. LAWRENCE *Birds, Beasts & Flowers* 61 Cyclamens putting their ears back. Long, pensive, slim-muzzled greyhound buds. 1927 —— *Mornings in Mexico* 83 Donkeys, mules, on they come..great bundles bouncing against the sides of the slim-footed animals. 1978 W. F. BUCKLEY *Stained Glass* 225 Himmelfarb..grinned at his long-legged, slim-faced, lightly freckled assistant. 1981 A. SEWART *Close your Eyes & Sleep* vi. 57 He was slim-waisted and muscular looking.

b. **slim hole** *Oil Industry*, a drill hole of smaller diameter than normal; usu. *attrib.* (with hyphen); **slim volume,** a book of verse by a little-known poet (freq. mildly *derog.*); hence **slim vol** *colloq. abbrev.*

1953 *World Oil* June 112/3 Analysis of areas favorable and unfavorable to slim hole drilling may have been influenced by the performance of available bits in the various types of formations encountered. 1959 *Wall St. Jrnl.* (Eastern ed.) 20 July 15/3 'Slim-hole' drilling..employs conventional equipment—but less of it. The idea is to substitute small pipe and other tinier tools to do a job historically done by larger, more costly units. 1972 L. M. HARRIS *Introd. Deepwater Floating Drilling Operations* iii. 16 Some companies use slim-hole designs and special-clearance couplings on casings. 1975 A. P. SZILAS *Production & Transport of Oil & Gas* iv. 345 (*caption*) Bottom-hole arrangements for the sucker-rod pumping of slim holes. 1920 E. WALLACE *Daffodil Mystery* iii. 23 Thornton Lyne was a store-keeper, a Bachelor of Arts,.. and the author of a slim volume. 1953 R. LEHMANN *Echoing Grove* 242 The accent, I gather, is on culture—lots of slim vols in the house now. 1979 *Church Times* 1 June 5/3 Friends will..welcome a slim volume of his poems that has come out.

slim, *v.* Restrict 'Chiefly *dial.*' to senses in Dict. and add: **3. b.** *fig.* To reduce in size or extent. Freq. const. *down.*

1963 *Richmond* (Va.) *Times-Dispatch* 16 Dec. 19/1 He set out to slim the budget. 1971 *Daily Tel.* 16 Aug. 11/6, I have slimmed down my holding of Westland Aircraft.. by selling 1,250 at 46½p. 1976 A. WHITE *Long Silence* xix. 169 We'd been able to slim our plan down considerably, to make do with the minimum of men. 1980 *Daily Tel.* 18 Jan. 1/8 Sir Charles Villiers, British Steel's chairman, ended weeks of speculation over the future of two plants..when he said both works would be slimmed.

4. *intr.* To try to reduce one's weight by dieting; to become slim. Also with *down.* Also *fig.*

1930 *Punch* 2 Apr. 366/2 The hostess ate hardly any. She is slimming. 1937 L. C. DOUGLAS *Forgive us our Trespasses* xiii. 254 As the minutes slimmed down to four, three, two, Dinny found his heart beating rapidly. 1963 'E. MCBAIN' *Ten Plus One* xii. 158, I was too fat... But the funny part was, once I slimmed down, I didn't want to be an actor any more. 1975 G. HOWELL *In Vogue* 1/2 'Dressing on a war income' was a regular feature [in British *Vogue*],..recommending that women should slim in order to use less fabric.

Hence **sli·mmed-down** *ppl. a.*; also **sli·m-down** (usu. *attrib.*).

1978 *Detroit Free Press* 5 Mar. B14/5 Other corporations are impressed by their strong cash positions, their slimmed-down lease structures and their good earnings. 1978 *N.Y. Times Mag.* 23 July 23/2 Frye-boot chic was swept aside as big-top looks became slimdown looks. 1980 *Daily Tel.* 4 Nov. 1/2 There may be a repeat of the late-1979 confrontation over the British Leyland slimdown and efficiency programme. 1981 *Times* 7 Aug. 20/3 Mercantile House continued to advance in its slimmed-down form.

slime, *sb.* Add: **4. b.** Also *anode slime.* The deposit of insoluble material formed at the anode in the electrolytic refining of copper and some other metals; = *anode mud* s.v. *ANODE c.

1902 J. MCCRAE tr. *Arrhenius' Text-bk. Electrochem.* xvi. 276 The other impurities, such as gold, silver,..and lead, remain undissolved, or form insoluble compounds ..and falling from the anode, collect in the so-called anode slime. 1935 W. A. KOEHLER *Princ. & Applic. Electrochem.* II. xxiii. 170 A large part of the silver produced is obtained from the slimes which are a by-product from the electrorefining of baser metals, especially from the refining of copper, lead, nickel, and zinc. 1954 M. C. SNEED et al. *Comprehensive Inorg. Chem.* II. ii. 128 Copper refinery slime is a dirty-black mixture of very finely divided copper and metallic and nonmetallic anode impurities. 1969 H. T. EVANS tr. *Hägg's Gen. & Inorg. Chem.* xxxvi. 749 Silver and gold by-products of the production of copper are collected in the anode slime during copper electrolysis.

6. c. *slimes dam* (S. Afr.); *slime-silvered* adj.

1927 JOYCE *On Beach at Fontana* in *Pomes Penyeach*, A senile sea numbers each single Slimesilvered stone. 1956 *Archit. Rev.* CXX. 48/3 There are three main varieties of dump, the sand dumps.., the rock dumps.., and the slimes dams, 50 to 100 feet high, covering wide areas, flat-topped. 1971 *Sunday Times* (Johannesburg) (Mag. Section) 28 Mar. 11/5 That square outline you see at the corner of what looks like a Witwatersrand slimes dam is, in fact, the remains of a Roman army camp.

slime, *v.*[1] **3.** (Earlier examples.)

1723 J. NOTT *Cook's & Confectioner's Dict.* sig. R5[v], *To fry Lampreys*, Bleed them, preserve their Blood, slime them, and cut them in pieces. 1747 H. GLASSE *Art of Cookery* ix. 86 Slime your Tenches.

slimed, *ppl. a.* (Later examples.)

1940 W. FAULKNER *Hamlet* ii. 201 He..stooped and began to drag away the slimed and rotten branches. 1952 *Chambers's Jrnl.* Feb. 110/1 A dark, dank, and sepulchral sphere of silent stone, where passage succeeded passage in an unending monotony of slimed and moss-grown solitude. 1972 F. WARNER *Lying Figures* IV. 43 Two bodies..still slimed from the womb.

sli·m-in. Also **slim in.** [f. SLIM *v.* + *-IN[3].] A course of (usu. sponsored) slimming undertaken by several people in competition with or in support of one another.

1973 *Inverness Courier* 31 July 5/4 A sponsored 'slim-in'..recently raised £15. 1977 *Cornish Times* 19 Aug. 10/1 The sponsored 'slim in' was won by Mrs V. Humphries.

sliming, *vbl. sb.* Add: **2.** *Mining.* The reduction of ore to slime (SLIME *sb.* 4).

1920, etc. [see *all-sliming vbl. sb.* and *ppl. a.* s.v. *ALL 13]. 1965 G. J. WILLIAMS *Econ. Geol. N.Z.* v. 56/1 The fine grinding which was necessary to liberate the gold caused losses in scheelite through sliming.

slim jim (slim dʒim). Also **slim Jim, Slim Jim.** [Rhyming combination of SLIM *a.* and the proper name *Jim.*] A very slim or thin person.

1889 'MARK TWAIN' *Conn. Yankee* xxxix. 500 Go it, slim Jim!

2. *transf.* (See quot. 1925.)

1916 JOYCE *Portrait of Artist* ii. 104 He had..eaten slim jim out of his cricket cap. 1925 —— *Let.* 31 Oct. (1966) III. 129 This is a kind of sweet meat made of a soft marshmellow jelly which is coated first with pink sugar and then powdered, so far as I remember with cocoanut chips. It is called 'Slim Jim' because it is sold in strips about a foot or a foot and a half in length and an inch in breadth.

3. *attrib.* Designating something long and thin or narrow, as *slim-jim pants, tie,* etc. Also *ellipt.* in pl.

1916 R. FROST *Mountain Interval* 60 That ought to make you An ideal one-girl farm, And give you a chance to put some strength On your slim-jim arm. 1956 *Amer. Speech* XXXI. 307 *Frontier pants*..*slim jims*, long pants, tapered so closely that they must be zipped at the ankles, or must have slits. 1957 [see *JEAN 2 b]. 1960 *Guardian* 9 Mar. 6/1 Warm flannelette lined corduroy 'slim jims'. 1962 *Times* 13 Apr. 6/4 Worthing Museum..is asking for a Teddy boy costume, 'with narrow trousers, fancy waistcoat, three-quarter jacket, Slim Jim tie and thick-soled crepe sneakers'. 1962 *Spectator* 27 Apr. 536 The spade ties everyone wore before the slim jims came in. 1973 O. SELA *Portuguese Fragment* (1974) xi. 61 He wore a long sleeved white shirt and slim-jim tie.

slimline, slim-line (sli·mlain), *a.* [f. SLIM *a.* + *LINE *sb.*[2] 14 b.] **a.** Slim, narrow, gracefully thin in style or appearance. Occas. *absol.* as *sb.*

1949 PENDER & DEL MAR *Electr. Engineers Handbk.* (ed. 4) xv. 33 (*headings*) Technical data on fluorescent lamps... Standard line lamps... Slimline lamps. 1961 *Economist* 28 Oct. 328/3 Stereophonic equipment, slimline television, the second wireless set. 1964 *N.Y. Post* 21 Oct. 18/2 (Advt.), Elegant long gowns..slinky slimlines and ultra-feminine bouffants. 1971 *New Scientist* 11 Feb. 323/1 Whenever I switched on the box I seemed to find pontificating doctors: podgy ones telling me to smoke cigars, slimline ones telling me to eat less. 1976 C. EGLETON *State Visit* viii. 82 'What if you wanted to speak to someone outside the airport?' Mostyn pointed to a red slim-line. 'Then I would use that one [*sc.* telephone],' he said. 1978 *Times* 10 Aug. 11/1, I put on my white crepe slimline dress.

b. *fig.* Exiguous, economical, stripped of unnecessary elaboration.

1973 *Listener* 26 July 125/2 This slim-line plot is made the occasion for many a worthy animadversion on man's rape of the surface of the earth. 1975 B. J. ENRIGHT in Barr & Line *Ess. on Information & Libraries* v. 66 A hesitant and doubtful response greeted Maurice Line's forecast of 'lean muscular libraries' to replace 'fat bloated ones of the past'—the 'new slimline library' seemed a somewhat dubious proposition. 1980 *Daily Tel.* 18 Jan. 1/8 The board has decided to concentrate further consultations and discussions with the unions and the workforce on a 'slimline' operation.

slimly, *adv.* Add: **2. b.** So as to give an effect or appearance of slimness.

1970–1 *Kay's Catal.* Autumn–Winter 60/1 A casual classic that's never out of fashion; slimly styled in knitted Courtelle. 1979 *Homes & Gardens* June 103/1 The front [of the skirt] double wraps, so that it fits sleekly and slimly.

slimmer (sli·məɹ). [f. *SLIM *v.* 4 + -ER[1].] One who practises slimming.

1974 *Country Life* 12 Dec. 1845/1 The horrors of a slimmer can be as awful as those of an alcoholic, at least until his stomach shrinks. 1978 J. PUDNEY *Thank Goodness for Cake* 124 It was not designed as a slimmer's diet but I lost weight. 1980 *West Lancs. Evening Gazette* 11 Aug. 10 (Advt.), Slimmers, have you heard that the.. new slimming method..is now available?

slimming (sli·miŋ), *ppl. a.* [f. SLIM *v.* + -ING[2].] Producing an appearance of slimness; conducive to slimness.

1925 *Daily Express* 18 Nov. 6/3 The Lord Chamberlain took out the words 'slimming over the hips', which is a phrase used every day by fashionable costumiers. **1927** *Daily Chron.* 29 Mar. 15/4 Orange juice with a dash of gin in it..is said to be slimming! **1952** *Observer* 14 Sept. 8/6 Youthlines '1980'. The so-slimming girdle in marvellous American Leno. **1980** *Daily Tel.* 13 Oct. 19/3 Slimming Nehru jackets, buttoned close to the throat.

slimming (sli·miŋ), *vbl. sb.* [f. SLIM *v.* + -ING^{1}.] **a.** The practice of using special means, such as dieting and exercises, to produce slimness of body. Also *fig.*

1931 GALSWORTHY *Maid-in-Waiting* xi. 101 Perhaps the young of today will nevah grow fat. They do slimming—ah-ha! **1958** *Spectator* 7 Feb. 165/1 If such a drastic slimming is to be enforced on agriculture, there is little doubt that [etc.]. **1974** *Times* 17 Apr. 10/5 A medical view of slimming. **1982** *Daily Tel.* 3 Aug. 13/2 The slimming of prime rates brought investors back in force late in the session.

b. *attrib.* (passing into *adj.*)

1932 *Times* 1 Feb. 9/3 She was a bit exercised about getting too stout and might have been going in for 'slimming' exercises as sometimes ladies did. **1951** M. McLUHAN *Mech. Bride* (1967) 154/1 The plump wife who went off for a prolonged slimming course. **1979** A. MORICE *Murder in Outline* x. 85 She was taking slimming pills... She was..worried about her weight.

slimnastics (slimnæ·stiks), *sb. pl.* (const. as *sing.*). *U.S.* [Blend of *SLIM(MING *vbl. sb.* and GYM)NASTICS.] (See quots. 1970.)

1967 *New Yorker* 22 Apr. 138/2 The calendar on the Community Center bulletin board lists citizenship classes, pinochle and bridge games,..'slimnastics', ceramic classes, and dances of the Teen Club. **1970** M. PEI *Words in Sheep's Clothing* ii. 12 'Slimnastics' (gymnastics that slim you down). **1970** NOTTIDGE & LAMPLUGH *Slimnastics* i. 9 'Slimnastics' is a combination of slimming and gymnastics in a group. **1979** *Honolulu Advertiser* 8 Jan. A-4/3 Women's Slimnastics Classes. **1983** *N.Y. Times* 12 Jan. B 2/5 In a gymnasium, 40 women were exercising in a 'slimnastics' class.

sling, *sb.*^{1} **1. a.** (Earlier example in sense 'boy's catapult'.)

1828 W. CLARKE *Boy's Own Bk.* (ed. 2) 25 The sling... Whirl it round several times, let go the shorter thong, and the stone will be shot to a great distance.

sling, *sb.*^{2} Add: **1. b.** In mountaineering, rock-climbing, etc., a short length of rope used to provide additional support for the body in abseiling or belaying.

1920 G. W. YOUNG *Mountain Craft* iv. 194 Not only is the single sling more likely to snap under the rub of the hard ring. **1946** J. E. Q. BARFORD *Climbing in Britain* ii. 24 *Slings.* Most parties doing exposed or difficult rock climbs nowadays carry one or more slings. **1965** A. BLACKSHAW *Mountaineering* II. vii. 204 Most British climbers carry nylon slings... The use of slings has been very highly developed by British mountaineers; mainly, no doubt, because natural running belays have come to be used much more here than elsewhere. **1976** G. MOFFAT *Over Sea to Death* v. 53 She placed her slings, clipped in her rope and, watching it fall, caught her second's eye.

3. d. *to have* (one's) *ass in a sling*, etc.: (see quot. 1960); to be in trouble. Cf. *ASS, arse. *U.S. slang.*

1960 WENTWORTH & FLEXNER *Dict. Amer. Slang* 10/1 *To have one's ass in a sling*, to be or to appear to be sad, rejected, tired, or defeated. **1976** 'B. SHELBY' *Great Pebble Affair* (1977) 157, I figure there's no money in it for me, but I sure as hell want Rosale's ass in the sling. **1982** S. F. X. DEAN *Such Pretty Toys* (1983) vi. 94 Gonna get my ass in some sling if I miss that plane.

7. sling-back, used *attrib.* and *absol.* to designate (*a*) a woman's shoe which has an open back and is held on by a strap across the heel; so **sling-backed** *a.*; (*b*) a type of chair characterized by a fabric seat suspended from a rigid frame; **sling-bag,** a bag with a long strap which may be hung from the shoulder; **sling chair** *U.S.*, a sling-back chair (see *sling-back* (*b*) above); **sling-jacket** (see quot. 1900); **sling-load, sling load,** a load which is lifted in a sling; also (with hyphen) as *v. trans.*; **sling pump** *N. Amer.*, a sling-back shoe (see *sling-back* (*a*) above).

1949 *10 Eventful Years* (Encycl. Brit.) II. 312/2 They were soft suede slippers, little leather sling-backs, ankle-high boots, and ballet slippers of all colours and materials. **1950** 'S. RANSOME' *Deadly Miss Ashley* ii. 25 Neat black sling-back wedgies on small feet. **1973** 'D. JORDAN' *Nile Green* xxvii. 119, I was sitting on a sling-back chair looking out on the Nile. **1974** *Country Life* 21 Mar. 688/1 A canvas sling-back with a rope wedge sole for £5. **1976** B. BOVA *Multiple Man* iii. 34, I walked across to the Scandinavian sling back that I usually sat in... I eased myself into the slingback chair. **1978** *Vogue* 1 Mar. 131 (*caption*) Sling-back high heels..£37. **1948** 'J. BELL' *Wonderful Mrs Marriott* vii. 86 A pair of toeless sling-backed wedge-heeled shoes. **1965** M. SPARK *Mandelbaum Gate* vii. 218 Sturdily clutching with one thumb the shoulder strap of her sling-bag. **1976** *Woman's Weekly* 6 Nov. 11/1 The letter still lay in the bottom of her sling bag. **1957** *Holiday* Nov. 141/1 The sling, or Hardoy, chair, a leather suspension from a rigid metal cradle, adapted from a wooden folding chair used by Italian officers in North Africa. **1978** L. BLOCK *Burglar in Closet* ix. 75 Jillian..sat in a sling chair. **1900** HARDY in *Sphere* 21 Apr. 419/2 In those days the Hussar regiments still wore over the left shoulder that attractive attachment, or frilled half-coat, hanging loosely behind like the wounded wing of a bird, which was called the pelisse, though it was known among the troopers themselves as a 'sling-jacket'. **1908** —— *Dynasts* III. II. 1 Will the gay sling-jacket glow again beside the muslin gown? **1933** M. LOWRY *Ultramarine* v. 213 The cargo, chests of tea, was hoisted in slingloads of ten from the piles. **1968** *Globe & Mail* (Toronto) 3 Feb. B1/2 The union was asking for a 22-man basic work gang; extensive sling-load limitations, [etc.]. **1969** *Jane's Freight Containers 1968–69* 8/1 The line of action of the sling load is assumed to be parallel to and not more than 38 mm..from the outer face of the corner fitting. **1969** I. KEMP *Brit. G.I. in Vietnam* x. 176 A Chinook to sling-load our chopper back to Phuoc Vinh. **1941** *Women's Wear Daily* 31 Oct. I. 13/1 The shoe which so many retailers claimed they could not sell, the sling pump, is due to make another trip, very definitely an evidence that women want them and like them. **1968** *Globe & Mail* (Toronto) 17 Feb. 12 (Advt.), Cool sandal sling pump with adjustable T-strap, low heel.

sling, *sb.*^{3} Add: **3.** *Austral.* A gratuity; a bribe. Also *sling back.* Cf. *SLING *v.*^{1} 9.

1948 K. S. PRICHARD *Golden Miles* viii. 92 'There's some hungry bastards,' the men said, 'making big money on their ore, never give the poor bugger boggin' for 'em a sling back.' The sling back might be ten bob on pay-day, or no more than a few pots of beer, but was always appreciated. *Ibid.* ix. 102 Sling backs to the shift boss got some men their jobs. **1953** K. TENNANT *Joyful Condemned* xxiv. 232 Say I take twenty per cent of the cop for myself..all the rest goes in slings. **1969** *People* (Austral.) 15 Jan. 21/2 It is not uncommon for a [poker] machine to go into a club with what is known in the trade as 'a sling'..to someone or other of the 'power men'... These 'slings' can range up to $300 a machine sold. **1973** *Nation Rev.* (Melbourne) 31 Aug. 1450/1 The hospital.. must have been quite notorious in police circles. As far as I knew, we were exceptional in refusing to pay the customary sling.

sling, *sb.*^{5} **1.** (Earlier U.S. example.)

1792 P. FRENEAU in *National Gazette* (Philadelphia) 28 June 280/1 Rum ne'er shall meet my lips..In shape of toddy, punch, grog, sling or dram.

sling, *v.*^{1} Add: **2. a.** (Later *fig.* examples.)

1959 E. H. CLEMENTS *High Tension* vii. 128 He'd been slung out of the test because he'd hurt his foot. **1977** W. MARSHALL *Thin Air* vi. 69 He was so bloody stupid we slung him out.

3. b. Also in phr. *to sling hash*, to wait at tables. *U.S.* Cf. *hash-slinger* s.v. *HASH *sb.*^{1} 6.

1876 [see *Calamity Jane* s.v. *CALAMITY 3]. **1906** 'O. HENRY' *Four Million* 106 I'm going back there and ask her to marry me. I guess she won't want to sling hash any more when she sees the pile of dust I've got. **1949** *Life* 24 Oct. 20/2 She..slung hash for a couple of weeks.

d. See also *HOOK *sb.*^{1} 16 b.

e. (Earlier example.) Also, to speak or utter (language, etc.) well or fluently. Also, *to sling the bull*, to talk glib nonsense (*U.S.*). Cf. sense a in Dict.

1874 E. EGGLESTON *Circuit Rider* vii. 72 He was beginning to sling his rude metaphors to the right and left. **1892** KIPLING *Barrack-Room Ballads* 67 An' 'ow they would admire for to hear us sling the *bat* [= speak the language]. **1904** G. B. SHAW *Lett. to Granville Barker* (1956) 27 One of them, the stage Irishman,..might be done by, say, Neville Doone, if he can sling the dialect. **1934** T. E. LAWRENCE *Let.* 8 June (1938) 806 In such an eyewash job as this of mine, the power to sling the gab would be very helpful. **1940** A. H. MARCKWARD *Scribner Handbk. Eng.* vii. 212 Undoubtedly the chief reason for the conversational effectiveness of many individuals is their inherent ability to sling it. **1982** *Verbatim* Autumn 14/2 Watch out for..the low-down curs and dirty dogs, who sling the bull and then send you on a wild goose chase.

g. (Later example.) Also *to sling in* or *up.*

1911 G. B. SHAW *Blanco Posnet* 384 Stow it, Boozy. Sling it. Cut it. Cheese it. Shut up. **1910** G. B. LANCASTER *Jim of Ranges* ii. 48 I've slung her [*sc.* Queensland] up. Guv her the go, the ole jade. **1953** K. TENNANT *Joyful Condemned* xxxii. 309 We both slung in our jobs.. and went off after him.

8. *to sling off* (*at*), to jeer (at). *Austral.* and *N.Z. colloq.*

1911 S. RUDD *Dashwoods* 24, I heard yer both slingin' off. **1916** *Anzac Bk.* 31, I could not understand them slinging off at 'im and 'im thinking they were treatin' 'im like as 'e was one of themselves. **1921** K. S. PRICHARD *Black Opal* xiii. 112 The rest of the men continued nevertheless to 'sling off', as they said, at Bully and Roy O'Mara as they saw fit. **1941** *Coast to Coast* 232 'Why was he so wild?' 'Aw, it was just some chaps'd been slinging off at him,' I said. **1960** N. HILLIARD *Maori Girl* III. vii. 221 The *pakehas* think you're slinging off about them or saying something rude. **1963** J. CANTWELL *No Stranger to Flame* v. 86 'Stop it,' Barry said, flushing. 'Stop slinging off.' **1975** M. R. LIVERANI *Winter Sparrows* II. xv. 232 She glowered at the driver suspiciously. Was he slinging off at her?

9. To pay a bribe or gratuity. Occas. with *it.* Cf. *SLING *sb.*^{3} 3. *Austral.*

1939 K. TENNANT *Foveaux* II. 172 'I'm slinging it to Hamp,' Bardy said sullenly. **1949** L. GLOSSOP *Lucky Palmer* 5 Clarrie, he ain't gone off in six months. Must sling to the cops. Wonder how much he pays 'em. **1953** T. A. G. HUNGERFORD *Riverslake* vi. 130 'Sling, Stefan!' When the Pole looked at him uncomprehendingly Murdoch whipped a ten-pound note out of the bundle and handed it to the ring-keeper. 'He don't know,' he explained. 'It's the first time he's played.' **1971** F. HARDY *Outcasts of Foolgarah* 56 On first name terms with every shire President so long as they didn't forget to sling when backhanders came in.

sling, *v.*^{2} Add: **4. a.** Also in phr. *to sling one's hammock*, to have a period of time off-duty to get used to a new ship. *Naut.*

1913 T. T. JEANS *John Graham Sub-Lieutenant R.N.* iii. 58 There was no 'school' till morning, the Padré had a day off 'to sling his hammock'. **1917** 'TAFFRAIL' *Sub* iii. 92 'You'll have to-morrow to sling your hammock and to get used to the ship, youngster,' he went on. **1946** G. HACKWORTH JONES *Sixteen Bells* I. iv. 67 Reggie was hardly given a day to 'sling his hammock' before he was instructed to take over the afternoon watch.

sling-. Add: **sling-net** (later example); **sling-trot** (earlier example).

1866 MRS. GASKELL *Wives & Daughters* I. x. 112 He had been out dredging in ponds and ditches, and has his wet sling-net, with its imprisoned treasures..over his shoulder. **1853** J. PALLISER *Solitary Rambles & Adventures Hunter in Prairies* vi. 144, I saw my stag begin to fall in the rear of the band, and his pace slacken to a sling trot.

slinger, *sb.*^{1} Add: **2.** (Later examples.)

1920 D. J. KNIGHT in P. F. Warner *Cricket* i. 45 The man [*sc.* fast bowler] who bowls with his arm more or less horizontal—a slinger. **1926** *Variety* 29 Dec. 5/3 The erudite word slingers. *Ibid.*, The slang *slingers.* **1944** E. BLUNDEN *Cricket Country* iv. 52 T. was soon displaced for a less expensive slinger. **1979** *Vole* June 45/2 The slingers of the mid-nineteenth century terrified lesser batsmen.

3. Chiefly *Services' slang.* Bread soaked in tea. Usu. *pl.* Also applied *loosely* to other food.

1882 F. W. P. JAGO *Anc. Lang. & Dial. Cornwall* 267 Slingers. Kettle broth made of boiling water, bread, salt, and pepper, with sometimes a little butter. **1890** BARRÈRE & LELAND *Dict. Slang* II. 256/2 *Slingers*,..bits of bread floating in tea. **1895** KIPLING in *Pall Mall Gaz.* 30 May 2/2 You won't have no mind for slingers, not to-morrow —..bein' sick! **1925** FRASER & GIBBONS *Soldier & Sailor Words* 261 *Slingers*, tea or coffee with bread soaked in it. Dumplings. **1962** W. GRANVILLE *Dict. Sailors' Slang* 108 *Slingers*, ship's biscuits or bread soaked in cocoa. A snack eaten before *slinging* one's hammock. Supper is very early in the Navy and *slingers* help to fortify the men until breakfast the next day. **1965** 'J. LE CARRÉ' *Looking Glass War* ii. 31 We had the canteen in the old days. Slinger and wadge.

slinger, *sb.*^{2} (Later examples.)

1969 *Daily Tel.* 2 Apr. 1/1 The slip was dropped from an overall pocket by a slinger—a man who guides car bodies and other parts into position on the assembly track—and it showed that he was earning as much as the assembly workers themselves. **1974** P. WRIGHT *Lang. Brit. Industry* viii. 69 In heavy electrical engineering, where the crane-driver can see the slinger directing his load, they communicate by shouts. **1982** *Sunday Times* 27 June 3/1 I've been in the ropery 12 years and my husband's been in the yard, as a slinger, for 17 years.

sli·ngful. [f. SLING *sb.*^{2} 1 + -FUL.] As much as a sling will hold.

1913 *Cassell's Mag. Fiction & Popular Lit.* June 183/1 A stevedore was killed by a falling slingful of railway sleepers. **1941** E. P. O'DONNELL *Great Big Doorstep* xxii. 313 She heard the donkey engines clanking aboard an approaching ship, and slingfuls of fruit splashing into the water.

slinging, *vbl. sb.*^{2} (Later example.)

1901 *Business Terms, Phrases & Abbrev.* (ed. 2) 109 The buyer must attend to their being put on board, and pay the dues or the charges for slinging, should any be incurred.

sli·ngshot. orig. *U.S.* Also **sling-shot.** (In Dict. s.v. SLING-.) Add: **1. a.** (Earlier and later examples.)

1849 N. KINGSLEY *Diary* 23 Oct. (1914) 77 Many are getting up sling-shots,..but I hope we shall never have occasion to use them, but we must all do something to pass away the time. **1895** [in Dict. s.v. SLING-]. **1901** *Daily Colonist* (Victoria, B.C.) 15 Oct. 5/2 The police have started a crusade against the use of slingshots and air and pea guns by boys. **1966** *Economist* 2 July 28/2 In peasant style, they placed women and children to the fore, and used slingshots with shepherds' accuracy to defend themselves against the detachment of police cavalry. **1977** C. McCULLOUGH *Thorn Birds* i. 3 She played happily with the whistles and slingshots..her brothers discarded.

b. In various *fig.* and *transf.* uses implying propulsion from or as from a catapult (see quots.) Also as quasi-*adv.*

1951 *Sun* (Baltimore) 10 July 15/3 [In a game of softball] I've never seen a pitcher get away with an illegal pitch in the national or league play... Most of them throw just like I do—straight windmill—and out on the Coast, they throw slingshot. **1964** W. GOLDING *Spire* ix. 159 The weather..began to squeeze bursts of rain out of the air like slingshot, so that although the men were wet, they were warm with the stinging. **1975** *New Yorker* 30 June 81/1 That's the slingshot there—where the rubber is stretched between bumpers [in a pinball machine]. **1976** J. JAMES *Bridge of Sand* iii. 62 There came immediately on us a great storm of hail. The slingshots of the Gods rattled on our helmets, bounced off our shields, almost cut our faces. **1976** *6,000 Words* 184 *Slingshot*, n, 1: a maneuver in auto racing in which a

drafting car accelerates past the car in front by taking advantage of reserve power 2: a dragster in which the driver sits behind the rear wheels. **1979** 'A. HAILEY' *Overload* III. vii. 222 Harry London's feet hit the floor like slingshots.

c. *Astronaut.* A space flight which uses the gravitational pull of a celestial body in order to accelerate sharply and, usu., to change course. Freq. *attrib.* (see sense 3 below.)

1970 N. ARMSTRONG et al. *First on Moon* viii. 174 The three men on Apollo II had to decide whether to allow themselves to be 'captured' by lunar gravity—or take the slingshot and come home. **1971** *Encyclopedia Science Suppl.* 350 The next flight now scheduled is for 1973, a 'slingshot' that will pass close to Venus on its way to Mercury.

2. A weapon consisting of a heavy weight wrapped in a cloth or equivalent and used as a cosh; a blackjack. ? *Obs.*

1891 H. HERMAN *His Angel* ix. 149 He made a ghastly horrible sling-shot by filling a heavy tumbler with the iron tops screwed off from the fire-irons, and tying the lot in a handkerchief. **1904** *N.Y. Evening Post* 24 June 2/6 The guards..are authorized to carry slingshots... [They] are heavily loaded with lead, and are securely attached to the wrist.

3. *attrib.*, as (sense 1 a) *slingshot crotch*; (sense 1 b) *slingshot dragster*; (sense 1 c) *slingshot action, attitude, flight, manœuvre, trajectory.*

1979 *Guardian* 3 Sept. 11/3 Accelerated and turned by the slingshot action of Saturn's gravitational field, the spacecraft is heading out past Titan on a new trajectory. **1970** N. ARMSTRONG et al. *First on Moon* iv. 83 We've completed our maneuvers to observe the slingshot attitude. **1923** W. S. CATHER *Lost Lady* ii. 16 They had behaved like wild creatures all morning;..cutting sling-shot crotches. **1962** *Engineering* 5 Jan. 5/1 Sydney Allard's immaculately prepared 'slingshot' dragster, with its 480 hp. Roots-blown Chrysler V8 engine. **1971** *Guardian* 16 Dec. 13/5 Designed..to use the gravitational fields of the planets for 'sling-shot' flights which—using Jupiter as the primary springboard—visit Saturn and Pluto, then Uranus and Neptune. **1970** N. ARMSTRONG et al. *First on Moon* iv. 82 A 'slingshot' maneuver, a trajectory which would take it behind the trailing edge of the moon. **1976** *Sci. News* 18 June 391 The 'slingshot' trajectory between the worlds has carried Pioneer 11 about 16° above the plane of the ecliptic.

Hence **sli·ngshot** *v. intr.*; **sli·ng-shooting** *ppl. a.*

1969 *Daily Progress* (Charlottesville, Va.) 5 July 6/4 'I could stay with him in a draft (the two cars running one behind the other).'..Yarborough said he purposely gave Baker two chances to slingshot past to learn if he was fast enough. **1975** B. GARFIELD *Death Sentence* xxi. 106 The lights were all gone: stone-throwing and sling-shooting kids routinely used them for target practice. **1980** *Dirt Bike* Oct. 68/1 Ward's body inched forward with increasing speed until he slingshotted at blurring velocity to and beyond the Yamaha.

slink, *sb.* **2. d.** (Earlier example.)

1858 THOREAU *Jrnl.* 29 Aug. in *Writings* (1906) XVII. 134 F. says they call the cardinal-flower 'slink-weed', and say that the eating it will cause cows to miscarry.

slink, *v.* Add: **2. d.** To withdraw from. *rare.*

1853 J. G. BALDWIN *Flush Times Alabama* 26 Many a witness..'slunk his pitch mightily' when old Kasm put him through on the cross-examination.

e. To turn (the eyes) *round* in a stealthy or slinking manner. *rare.*

1923 GALSWORTHY *Captures* 162 Leaning down to our scoundrel and slinking her eyes round at the Countess, she murmured something malicious.

sli·nker, *sb.*^{2} [f. SLINK *v.* + -ER^{1}.] One who slinks about; a shirker. So as *v. intr.*, to shirk.

1880 G. SMITH *Gipsy Life* ii. 48 When the task-master perceived the 'gang' had begun to 'slinker' he would shout out. **1919** G. W. DEEPING *Second Youth* xxviii. 238 It makes a man so mean, so sly, such a slinker round corners. **1923** —— *Secret Sanctuary* x. 97 He had seen the most inveterate slinker change into a creature of crude and bounding energy when a piece of leather was to be kicked about a field. **1954** J. R. R. TOLKIEN *Two Towers* II. iii. 247 Sam's guess was that the Sméagol and Gollum halves (or what in his own mind he called Slinker and Stinker) had made a truce and a temporary alliance.

slinky (sli·ŋki), *a.* [f. SLINK *v.* + -Y^{1}.] Of a woman, esp. from the manner of her dress: sinuous, slender, gliding; of a garment: close-fitting, as if moulded to the figure. In extended use (with varying degrees of approval): stealthy, dextrous, furtive.

1921 *Ladies' Home Jrnl.* Jan. 8/1 Even now I seem to see in memory a slinky, slant-eyed person with long, slender finger nails, who wears green. **1923** *Glasgow Herald* 21 July 6/5 Jessica was swathed in a slinky gown of flat crepe in a deep blue shade. **1932** D. L. SAYERS *Have his Carcase* xviii. 235 She now selected a slinky garment, composed of what male writers call 'some soft, clinging material'. **1944** R. CHANDLER *Lady in Lake* 116 One of those slinky glittering females. **1951** N. C. HUNTER *Waters of Moon* III. i. 69 Some really classy young man with heaps of money. A Guards officer, for instance, or something rather slinky on the Stock Exchange. **1962** *Times* 28 Dec. 9/5 Miss Ross sings a variety of songs—slinky, torchy, witty. **1973** *Country Life* 8 Mar. 633/2 Slinky dresses that have the finest of straps or are completely strapless. **1980** J. WAINWRIGHT *Eye of Beholder* 18 His missus. The slinky, brittle bint.

Hence **sli·nkily** *adv.*

1935 *Amer. Speech* X. 192/2 A long flowing gown trails slinkily on the floor.

Slinky (sli·ŋki), *sb.* Also **slinky.** [f. SLINK *v.* + -Y^{6}.] The proprietary name of a toy consisting of a flexible helical spring which can be made to 'somersault' down steps as an inverted U. Also *Slinky toy.* Also *fig.*

1948 *Trade Marks Jrnl.* 11 Feb. 114/2 Slinky. 654,972. Games (other than ordinary playing cards) toys and sporting articles (except clothing). James Industries Inc.., 4932, Portico St., Germantown, Philadelphia..— 24th Dec. 1946. **1968** *Habitat Xmas Catal.*, Slinky toy— fun for childish adults and grown up children! **1975** *Publishers Weekly* 7 Apr. 83/1 The tangled slinky-toy lives of these three turned-on folk heroes. **1979** *Early Learning 1979–80* (Early Learning Centre) 32/2 Plastic Slinky... Walks down stairs!

slinter, var. *SCHLENTER *sb.* and *a.*

slip, *sb.*^{1} Add: **4. b.** *slip house* (later examples), *kiln* (earlier example); *slip decoration, glaze*; **slip casting,** the manufacture of ceramic articles by allowing slip to solidify in a porous mould.

1901 W. P. RIX tr. *E. Bourry's Treat. Ceramic Industries* iv. 231 (*heading*) Moulding by slip casting. **1959** *Jrnl. Iron & Steel Inst.* CXCI. 208/1 A description is given of the technique of slip casting, and its application to the casting of high-temperature materials including ceramics and metallo-ceramics. **1974** F. H. NORTON *Elem. Ceramics* viii. 95 The slip-casting method is much used in ceramic production as it is possible by this means to reproduce very complicated shapes in plaster molds. **1960** R. G. HAGGAR *Conc. Encycl. Cont. Pottery & Porc.* 458/1 Slip decoration consists of applying to the unfired clay surface of the ware, before it has been dried and fired, contrasted coloured slips, either by trailing from a quill, [etc.]. **1973** R. FOURNIER *Illustr. Dict. Pract. Pottery* 211/2 (*caption*) Slip decoration. An English slip-trailed dish—possibly Tickenhall. **1960** *Times* 6 Aug. 9/7 Without the slip-glaze upper decoration finish. **1902** A. BENNETT *Anna of Five Towns* viii. 167 The clay travelled naturally in a circle from the slip-house by the canal to the packing-house by the canal. **1961** M. JONES *Potbank* viii. 30 The maker breaks a lump of clay off the hunk brought from the sliphouse. **1769** J. WEDGWOOD *Let.* 9 Apr. (1965) 73 The Slip Kiln is nearly finished.

c. *slip-decorated, -decorator, -glazing, -painting, -trailer, -trailing.*

1883 L. M. SOLON *Art Eng. Potter* 27 (*heading*) Slip-decorated ware. **1900** F. LITCHFIELD *Pottery & Porcelain* ii. 25 At Wrotham..were produced..quaint, slip-decorated posset-pots, tygs, and dishes. **1979** *Essex Jrnl.* XIV. 20 There were also a number of small sherds of slip-decorated lead-glazed jugs and plates produced at Harlow. **1921** *Dict. Occup. Terms* (1927) § 105 *Slip decorator*, applies a pattern to pottery in the green state by blowing on coloured clay slips. **1960** *Times* 6 Aug. 9/6 The dual colouring..was obtained by slip-glazing before firing. **1902** *Encycl. Brit.* XXXI. 874/2 Turning to the decorative side of pottery work, we have in *slip-painting* a method as old as primitive pottery itself. **1964** H. HODGES *Artifacts* i. 33 One particularly elaborate form of slip painting, *feather combing*, involves the use of a brush with multiple points. **1940** B. LEACH *Potter's Bk.* ii. 33 There are at least half a dozen potteries in Japan where the slip-trailer is employed. **1960** H. POWELL *Beginner's Bk. Pottery* II. 64 *Slip trailer*.., a small rubber bag with a narrow neck into which is fitted a thin glass tube. **1940** B. LEACH *Potter's Bk.* vi. 147 One glaze can be trailed over another with the same instrument that is used in slip trailing. **1964** H. HODGES *Artifacts* i. 33 The clay may be applied in a fairly fluid form using for the purpose a container with a nozzle, much as a cake is iced by bakers. This method is called *slip-trailing.* **a 1977** *Harrison Mayer Ltd. Catal.* 27/1 A range of coloured slips prepared for slip trailing.

slip, *sb.*^{2} Add: **I. 3. a.** (Further examples.) Also *N.Z.*

1950 *N.Z. Jrnl. Agric.* Dec. 559/1 The usual practice is to buy the pigs as slips. **1977** *Cornish Times* 19 Aug. 1/1 (Advt.), Strong quality Slips and Pigs for slaughter accepted.

II. 4. Restrict † *Obs. rare* to sense in Dict. and add: **b.** A light under-waistcoat with the edge showing to form a border to a waistcoat worn with morning dress.

1933 C. ST. J. SPRIGG *Fatality in Fleet St.* viii. 98 Oakley looked like..a monkey which had surprisingly been trained to wear a morning-coat and grey slip. **1941** H. G. WELLS *You can't be too Careful* III. x. 158 And you looking *lovely* in a silk hat and light grey trousers. You'll have, you know, white slips to your waistcoat.

6. b. (Later examples.)

1903 G. JACK *Wood Carving* iii. 43 For sharpening the insides of tools, 'slips' are made with rounded edges of different sizes. One slip of 'Washita' stone and one of 'Arcansas'. **1960** C. H. HAYWARD *Cabinet Making for Beginners* (ed. 2) iv. 104 Drawer Making... Grooved slips are glued to the sides to hold the bottom.

10. a. Also *betting slip*: see *BETTING *vbl. sb.*^{1} 2.

III. 13. b. *slip proof.*

1892 A. OLDFIELD *Pract. Man. Typogr.* iii. 37 Proofs are required in various stages, and have a distinct name in each stage, as follows:—'slip', or galley proofs [etc.]. **1908** W. S. CHURCHILL *Let.* 8 Sept. in R. S. Churchill *Winston S. Churchill* (1969) II. Compan. II. 839 Messrs Hodder & Stoughton should let me have all the *Strand* articles up to date in slip proof as soon as possible. **1973** S. JENNETT *Making of Books* (ed. 5) I. vi. 99 Paging is a manual operation carried out by compositors... Each man has his share of slip proofs.

slip, *sb.*^{3} Add: **II. 4. c.** (Further examples.) Now, an underskirt or petticoat dependent from the waist or the shoulders and having no sleeves. Colloq. phr. *your slip is showing*: see *SHOW *v.* 28 g.

1825 H. WILSON *Mem.* II. 103 What do you call a slip? do you mean a petticoat, or an intrigue? **1903** M. M. *How to dress & what to Wear* 185 Slips. This term is applicable either to a skirt or a bodice. A skirt slip is made of silk, satin, or even batiste, and is employed for wearing under a thin upper dress... Slips may, or may not, be provided with sleeves. **1920** M. S. WOOLMAN *Clothing* ix. 135 Slips or underfrocks with detachable sleeves have also been designed... Many of the slips are made without sewed-in lining. **1944** H. CROOME *You've gone Astray* xv. 158 He glowered at Linda, sitting on the edge of the bed in her slip with one stocking off. **1957** J. BRAINE *Room at Top* xi. 109 She came over in her slip... She was already a different person in the blue silk garment. **1979** R. JAFFE *Class Reunion* (1980) I. vi. 85 In her slip and pants and garter belt and stockings she would lie down.

d. (Later example.)

1977 *New Yorker* 27 June 72/3 What I want is my pillow... The slip is homemade.

f. *pl.* In full *bathing slips*: bathing-drawers. (No longer in use.)

1904 *Times* 11 Aug. 10/3 He wore a pair of bathing slips and a broad-brimmed white linen cap. **1927** W. E. COLLINSON *Contemp. Eng.* 62 Bathing togs consisted of a bathing suit and slips, a reduced type of bathing-drawers.

5. (Earlier example.)

1771 C. BURNEY *Present State of Mus. France & Italy* 244 Printed sonnets, in praise of singers and dancers, were thrown from the slips.

III. 9. a. (Further *techn.* examples.)

1888 *Lockwood's Dict. Mech. Engin.* 329 *Slip*, the sliding of riveted joints one over the other to such an extent as to be visible. **1950** *Sci. News* XV. 143 The copper-rich oxide layer..acted as a lubricant between billet and container... This particular type of oxide layer seems to favour slip or slide of the metal under it.

c. Substitute for def.: The difference between the pitch of a propeller (on a ship or aircraft) and the distance it moves through the ambient medium in one revolution. (Examples in *Aeronaut.*)

1910 R. W. A. BREWER *Art of Aviation* viii. 110 A certain amount of slip is necessary in order to obtain thrust. **1919** [see *PITCH *sb.*^{2} 25 e]. **1946** H. ROUSE *Elem. Mech. Fluids* ix. 293 At peak efficiency the effective pitch of the propeller is somewhat below the geometric pitch, which results in a so-called 'slip' of the blades. **1965** C. N. VAN DEVENTER *Introd. Gen. Aeronautics* viii. 194/1 (*caption*) A comparison of geometric pitch with working pitch and slip.

e. The sudden descent of material within a blast furnace. Cf. *SLIPPING *vbl. sb.*^{1} 1 c.

1881 *Encycl. Brit.* XIII. 305/2 A 'slip' (or sudden jerky motion downwards of a mass of material that had previously more or less 'scaffolded'). **1911** *Jrnl. Iron & Steel Inst.* LXXXIII. 587 The causes of accidents peculiar to blast-furnaces, especially explosions, slips, and break-outs. **1948** G. R. BASHFORTH *Manuf. Iron & Steel* I. x. 166 The sudden slip of cold solid material into a hotter zone.. may result in serious explosions. **1969** K. R. HALEY in J. H. Strassburger *Blast Furnace* II. xii. 592 A slip.. causes wear on the lining.

f. Movement relative to a solid surface of the fluid immediately adjacent to it.

1887 *Encycl. Brit.* XXII. 771/1 While greater surface than is offered by [a swimmer's] hands and feet was always given, with the evident intention of reducing 'slip', much resistance took place at the neutral or negative part of the stroke. **1891** *Phil. Trans. R. Soc.* A. CLXXXI. 560 From equation (i.) it follows that the effect of slip varies inversely as the radius of the tube. **1937** DODGE & THOMPSON *Fluid Mech.* xii. 308 The principal reason..is the fact that the hypothesis of zero slip at the boundary of the solid has been abandoned. **1967** R. S. BRODKEY *Phenomena of Fluid Motions* vii. 91 If slip at the boundary were allowed, the flow rate would become Q = [etc.]. **1979** *Nature* 22 Mar. 350/2 Circumferential slip is essential if a helical object is to develop thrust in a true liquid, that is, if it is to propel itself using viscous forces.

g. *Electr. Engin.* The proportion by which the speed of an electric motor falls short of the speed of rotation of the magnetic flux inside it.

1893 *Jrnl. Inst. Electr. Engineers* XXII. 328 The machine has a frequency of 50 and a slip of 6 per cent. **1936** SAY & PINK *Performance & Design of Alternating Current Machines* xii. 211 On no load the slip is generally less than..1 per cent. **1976** A. R. DANIELS *Introd. Electr. Machines* vii. 122 The rotor is driven at a small slip with respect to the armature rotating m.m.f.

h. *Cryst.* The movement of one layer of ions over another in a stressed crystal.

1899 *Proc. R. Soc.* LXV. 86 The real character of the lines is apparent when the crystalline constitution of each grain is considered. They are not cracks, but *slips* along planes of cleavage or gliding planes. **1932** *Jrnl. Iron & Steel Inst.* CXXVI. 600 These results are in accord with the theory of deformation by slip. **1966** C. R. TOTTLE *Sci. Engin. Materials* iii. 64 In slip, a restricted number of planes are involved, and a restricted number of directions, so that whole areas of the crystal are affected. **1976** M. C. NUTT *Metallurgy & Plastics for Engineers* v. 70 Since metals deform by slip only on certain planes of

atoms, it follows that anything that interferes with the slip process hardens the metal.

i. The turning of one plate of a clutch relative to the other when they are in contact.

1902 A. C. HARMSWORTH *Motors & Motor-Driving* vi. 95 On the road also, if a clutch does not act, due to slip, a small dose of water puts matters right at once if the mechanical portions are in order. **1925** *Morris Owner's Man.* 22 The more pressure there is on the foot-board the less pressure is available in the clutch, and consequently there is a danger of slip starting. **1976** C. WEBB *Be your own Car Mechanic* vii. 96 When a clutch is worn it begins to slip. The slip generates heat and can cause the clutch spring or springs to lose their strength.

j. *Aeronaut.* A movement of an aircraft that includes a sideways component, esp. downwards towards the centre of curvature of a turn. Cf. *SKID *sb.* 4 b.

1916 GRAHAME-WHITE & HARPER *Learning to Fly* v. 50 The machine being near the ground, it came into contact with the surface of the aerodrome before the 'slip' had time to develop any high rate of speed. **1929** HALL & NILES *One Man's War* 114 Our slip was a slow one. It would be impossible to come out of a fast slip because that was done by putting on the rudder nearest to the direction of the slip. **1930** R. DUNCAN *Stunt Flying* ix. 79 A slip sideways into a landing is invaluable if it is necessary to land in a small area. **1952** [see *SKID *sb.* 4 b]. **1965** C. N. VAN DEVENTER *Introd. Gen. Aeronaut.* x. 233/2 Information on slip and turn is nearly always wanted at the same time. **1983** D. STINTON *Design of Aeroplane* xiii. 465 A spiral dive..is marked by increasing airspeed and, usually, no slip or skid.

10. d. (Later examples of *a slip of the tongue.*)

1906 H. C. WYLD *Hist. Study of Mother Tongue* iv. 72 He at once perceives the difference [in his pronunciation], and 'corrects' the result as a 'mistake' or a 'slip of the tongue'. **1928** E. O'NEILL *Strange Interlude* VIII. 314 'You said "Navy".' 'Slip of the tongue! I meant Gordon.' **1939** G. B. SHAW *In Good King Charles's Golden Days* I. 13 'What did you call the gentleman, Mr Fox?'.. 'A slip of the tongue, Mistress Basham.' **1958** J. WAIN *Contenders* vii. 151 The Canon was still beating down Robert's attempt to explain away that slip of the tongue. **1975** *Economist* 21 June 31/2 Transcripts of the call, accurate to the last slip of the tongue, have been sent to the magazine Stern.

14. b. (Earlier example.)

1816 W. LAMBERT *Cricketer's Guide* (ed. 6) iii. 41 In *backing up*, he [*sc.* point] should take care to give the man at the slip sufficient room.

c. With qualifying words indicating the various positions in the slips, and the fieldsmen stationed there, as † *extra slip*, a man who stands outside second slip; third slip; *first slip*, the slip fielder who stands immediately to the right of the wicket-keeper (for a right-handed batsman); his position; similarly, *second* (*third*, etc.) *slip*, ranged in a line out from the wicket-keeper; *leg slip*: see *LEG *sb.* 17b; † *long slip*, a fieldsman placed as first slip, but deeper (or wider) in the field; † *middle slip*, short third man; † *short slip* = *first slip* above. Also COVER-SLIP 1 (*Obs.*).

1816 W. LAMBERT *Cricketer's Guide* (ed. 6) iii. 41 The Fieldsman that can best be spared is placed between the first Slip and Point. *Ibid.* 43 The Long Slip to cover the Short Slip. This man must stand..about the same distance from the Wicket as the Long Stop. *Ibid.* 44 This man should stand the same distance, playing between the Point and second slip. **1851** J. PYCROFT *Cricket Field* v. 75 A third man on, and a forward point,..with slow bowling, or an extra slip with fast, made a very strong field. *Ibid.* x. 193 A third man up, or a middle slip, is at times very killing. *Ibid.* xi. 222 A third slip can hardly be spared. **1892** W. G. GRACE in G. A. Hutchison *Outdoor Games* i. 26 Third man, who is, perhaps, rather a middle-slip, being long-slip placed in close enough to save the run. **1900** P. F. WARNER *Cricket in Many Climes* iii. 45, I was missed at extra slip..when I had only made a few runs. **1921** —— *My Cricketing Life* xii. 227 Jack was..a short slip of the same class as Tunnicliffe [, etc.]. **1955** *Times* 4 July 3/3 He swung the ball both ways, supported by a hostile, close-set field, Holliday taking two sharp low catches at first slip. **1976** *Times* 23 July 9/4 After adding 43 with Murray, Rowe was out to a tumbling catch at first slip; when Snow took over from Ward, Murray was well caught at second slip; when Willis came on, Holder gave third slip a catch.

IV. 15. a. (sense 4 c) *slip-dress* (U.S.); (sense 14) *slip-catch, -catcher, -catching, -fielder, -fielding, fieldsman.*

1903 S. L. JESSOP in H. G. Hutchinson *Cricket* v. 119 This range [of hits for practising catches] will include different kinds of chances, from 'slip' catches to catches in the long field. **1977** *World of Cricket Monthly* June 29/3 Raja took his third wicket through a slip-catch. **1920** LYTTELTON & WILSON in P. F. Warner *Cricket* vii. 268 As a slip catcher he was worthy to rank with R.E. and G.N. Foster. **1963** *Times* 17 Apr. 3/2 Downside are looking for proficient slip-catchers to give the required support to R. F. Thompson, a fast bowler, for whom they have high regard. **1950** W. HAMMOND *Cricketers' School* vi. 64 His slip-catching is first-rate. **1964** *Glamour* May 149 Andrea wears a bare blue linen slip-dress. **1912** P. F. WARNER *Eng. v. Australia* i. 2 The Committee..invited George Gunn, Woolley, and Mead, slip-fielders all of them. **1963** *Times* 13 June 3/5 The slip fielder is Titmus. **1906** A. E. KNIGHT *Complete Cricketer* iv. 153 The possibilities of slip fielding are so very great. **1976** *Times Lit. Suppl.* 16 July 896/5 There are a number of mistakes of fact (the editors' slip-fielding is not infallible). **1906** A. E. KNIGHT *Complete Cricketer* iv. 153 In degree it is true of all fieldsmen, yet it is more true of slip fieldsmen, that a position in the field is largely what the individual fieldsman cares to make it. **1920** G. L. JESSOP in P. F. Warner *Cricket* iv. 167 The importance of a slip fieldsman is only second to that of a wicket-keeper.

b. Special Combinations. **slip angle,** (*a*) a parameter of a screw propeller (see quots. 1878, 1902); (*b*) *Motoring*, (see quot. 1959^{2}); **slip band,** a slip line, or a cluster of such lines; **slip face,** the steepest face of a sand dune, down which sand slips; **slip flow,** in fluid dynamics, a mode of flow of a gas over a surface, the gas in contact with the surface having a definite velocity relative to it; **slip line,** a fine line visible on a polished crystalline surface where it is cut by a slip plane; (*b*) a line in a solid whose tangent at any point is one of the shear directions at that point; **slip plane,** a plane along which slip occurs in a crystal; **slip ratio,** the ratio of the slip of a propeller to its pitch.

1878 W. FROUDE in *Trans. Inst. Naval Architects* XIX. 50 The difference between the direction of the plane itself ..and the direction of its motion through the water..may be called the slip angle. **1902** *Encycl. Brit.* XXXII. 587/1 The slip angle (obliquity of surface to the line of its motion) ought always to have the same value (proportional to the square root of the coefficient of friction). **1936**, **1959** [see *OVERSTEER *v.*]. **1959** *Manch. Guardian* 27 July 2/3 The slip angle is the difference between the direction in which the wheels are pointing and the actual direction in which the car travels. **1899** EWING & ROSENHAIN in *Proc. R. Soc.* LXV. 87 Rotation of the stage to which the strained specimen is fixed makes the bands on one or another of the grains flash out successively, with kaleidoscopic effect. In what follows we shall speak of these lines as slip bands. **1976** C. BRADSHAW *Metallurgy for Schools* vi. 60 (*caption*) Photomicrograph showing slip bands formed on the surface of brass strip that has been stretched. **1941** R. A. BAGNOLD *Physics of Blown Sand & Desert Dunes* xiv. 224 The seif dune differs more markedly from the barchan in that its slip-face, instead of running mainly transverse to the prevailing wind, runs parallel with it. **1976** *Nature* 22 July 284/2 The beetles either burrowed into dune slipfaces when returned or remained active. **1946** HSUE-SHEN TSIEN in *Jrnl. Aeronaut. Sci.* XIII. 654/2 It was found that gas no longer sticks to the surface but slips over the surface with a definite velocity... This type of flow can be called the slip flow. **1978** *Jrnl. Fluid Mech.* LXXXV. 731 (*heading*) Slip flow past a tangential flat plate at low Reynolds numbers. **1900** *Phil. Trans. R. Soc.* A. CXCIII. 369 In gold or copper, it is very usual to find, on examining a strained specimen that one portion of a grain is covered with simple slip lines. **1931** NÁDAI & WAHL *Plasticity* xvii. 110 (*caption*) Helical slip lines on polished marble cylinder after compression. **1950** *Sci. News* XV. Plate 16 (*caption*) Slip-lines in pure zinc after exposure to 50 thermal cycles between 30°C and 150°C (× 500). **1973** JOHNSON & MELLOR *Engin. Plasticity* xii. 383 In order to determine the load necessary for a particular plastic forming operation, we must first of all obtain the slip-line field pattern. **1976** M. C. NUTT *Metallurgy & Plastics for Engineers* v. 66 In Fig. 5–2 the intersection of individual slip planes with the polished surface is observed, thus forming slip lines. **1925** *Jrnl. Iron & Steel Inst.* CXII. 87 The authors consider that the lowest value for tensile strength in an iron crystal will be obtained when two slip planes of the crystal make angles of 45° to the axis of stress. **1975** *Nature* 10 Apr. 489/1 Granular xenoliths..in which olivine and pyroxene show various strain effects, including undulose extinction.., slip-planes, and subgrain development. **1878** *Trans. Inst. Naval Architects* XIX. 50 The area which will drive the ship with a given slip ratio, is directly as the ship's resistance and is inversely as the square of her speed. **1902** *Encycl. Brit.* XXXVIII. 587/2 In combining the results from the four propellers great assistance was derived from the discovery that the curves expressing the variation of efficiency with slip-ratio had a close similarity. **1920** A. FAGE *Airscrews* vi. 70 In the writer's opinion the notion of slip is superfluous; and the introduction of a slip-ratio as a performance parameter quite unnecessary.

slip, *v.*^{1} Add: **I. 2. a.** (Earlier example with *off.*)

c **1810** W. HICKEY *Mem.* (1960) xix. 309, I..might slip off *sans cérémonie* and proceed to join the Oxfordshire party.

5. b. (Examples of accidental disclosure in conversation.)

1942 T. BAILEY *Pink Camellia* xxiv. 180, I didn't mean it, darling. It just slipped out. **1979** *Homes & Gardens* June 77/2, I always know if he's worried but he never tells me the details straight out. It sometimes slips out in conversation when the crisis is over and I think, Oh, that's what it was about.

II. 8. d. For *U.S.* read orig. *U.S.* and add earlier and later examples. Freq. const. *on.*

1855 *Jrnl. Discourses* II. 67/2 Some men think the way they are going to be saviors is to get as many wives as they can, and save them; now, they may slip up on that. **1866** *Weekly New Mexican* 14 July 2/1 The knowledge that he has 'slipped up' and been exposed is more than sufficient punishment for the offense. **1923** C. J. DUTTON *Shadow on Glass* xviii. 247 All of us slipped up. **1940** J. REITH *Diary* 31 Jan. (1975) v. 240, I wish I had been City member instead of Southampton. I slipped up on that. **1959** J. VERNEY *Friday's Tunnel* viii. 80, I couldn't help feeling that Daddy had slipped up pretty badly this time. **1981** A. MORICE *Men in her Death* x. 108 Somewhere along the line I had slipped up.

e. Of a person: to fall away from a standard (in behaviour or achievement); to deteriorate; to lose one's command of things. Chiefly as *pres. pple. colloq.*

[**1907** G. B. SHAW *Major Barbara* III. 286 You are fencing, Euripides. You are weakening: your grip is slipping. **1914** 'HIGH JINKS, JR.' *Choice Slang* 18 *Slipping*, failing, 'losing out', 'going under'.] **1930** *Publishers' Weekly* 22 Feb. 933/2, I must be slipping for I turned in a measly 78 on No. 4 in the Lenz-Rendel book. **1949** 'J. TEY' *Brat Farrar* xxvii. 242 I'm behaving very badly to-night, aren't I? I seem to be slipping. **1962** 'E. FERRARS' *Busy Body* ix. 104 He'd been slipping lately, drinking too much and boasting. **1976** H. MACINNES *Agent in Place* xiii. 137 The journalist was the first to know he was slipping; next his editors; and then the public. End of a career.

9. c. *intr.* and *trans. Aeronaut.* = *SIDE-SLIP *v.* b.

1911 *Aero* July (Suppl.) 2/2 The extra weight caused the machine to slide down sideways when steeply banked round the end corner. The Blériot slipped downward. **1930** R. DUNCAN *Stunt Flying* iii. 15 Side-slipping.. enables the machine to be put down in a far shorter space than would be possible through a normal glide, forward speed being reduced to a minimum by slipping the air-plane sideways down to within a few feet of the ground. **1941** POPE & OTIS *Elem. Aeronaut.* iii. 19 If the banking is insufficient for such a turn, the plane will skid, and if the banking is too great, the plane will slip toward the inside of the curve. **1952** A. Y. BRAMBLE *Air-plane Flight* xiii. 199 Slipping may be used deliberately with useful effect, providing the air-plane is of the type that may be 'slipped'. **1965** [see *SKID *v.*^{1} 3 c].

10. a. (Later examples of *to slip through one's fingers.*)

1915 W. S. MAUGHAM *Of Human Bondage* xlvii. 236 He was mad to have let such an adventure slip through his fingers. **1970** J. A. T. ROBINSON *Christian Freedom in Permissive Society* p. ix, Try to net it [*sc.* the concept of freedom] in the categories of discursive knowledge,..it slips through your fingers, and you end up..by concluding that it does not exist.

c. *to slip through the net*: to evade detection or apprehension; to escape someone's vigilance; to be overlooked.

1902 G. B. SHAW *Mrs. Warren's Profession* p. xviii, Nothing can really shake the confidence of the public in the Lord Chamberlain's department except a remorseless and unbowdlerized narration of the licentious fictions which slip through its net. **1970** *Times* 21 Feb. 7/5 All those in the 'know' in the underworld..maintain that it was a man who was never on trial but who slipped through the net. **1977** M. DRABBLE *Ice Age* I. 67 The real poor..were better off than they would have been in the thirties, for Britain is, after all, a welfare state, and not many slip through its net.

III. 15. a. *spec.* (*Motoring*) in phr. *to slip* (*in*) *the clutch*, to let in, release the clutch (*CLUTCH *sb.*^{1} 6 a), slightly or momentarily.

1904 A. B. F. YOUNG *Compl. Motorist* 214 When the brake lever is in the 'on' position, it is impossible to start the car by slipping in the clutch until it has been released. **1912** *Motor Man.* 73 The metal clutch..can be 'slipped' to any extent without affecting the surface of the discs. **1965** PRIESTLEY & WISDOM *Good Driving* ix. 63 It is permissible [in reversing] to slip (feather) the clutch a little so as to maintain an even rate of travel. **1972** HILLIER & PITTUCK *Fundamentals of Motor Vehicle Technol.* 20 Most modern engines have a speed range from about 400 revolutions per minute..unless the clutch is partly disconnected or slipped.

18. c. (Later examples.)

1922 E. O'NEILL *Hairy Ape* vii. 78 Man in de Moon, yuh look so wise, gimme de answer, huh? Slip me de inside dope. **1935** [see *DUCAT 2 b]. **1936** WODEHOUSE *Laughing Gas* xix. 210 You tell me..and I'll slip you that money you wanted. **1952** 'N. SHUTE' *Far Country* ix. 257 Jim must have known the man was a boozer, and he might have thought some of his mates would try to slip him something. **1968** P. H. NEWBY *Something to answer For* iii. 88 If it's money you want, give me a little time, I can slip you a few hundred. **1978** S. BRILL *Teamsters* iv. 133 At one lunch, he testified, he slipped Provenzano $1,500.

d. In *slang* phr. *to slip* (something) *over* (*on*) (someone), to take advantage of someone by trickery, to hoodwink; *to slip a fast one over on* (someone) = to pull or put over a fast one (see *FAST *a.* 11).

1912 C. MATHEWSON *Pitching in Pinch* iii. 63, I attempted to slip a fast one over on Cooley and got the ball a little too high. **1927** *Daily Tel.* 29 Mar. 10/7 If one only had the nerve and audacity one could 'slip it over' the German every time. **1936** WODEHOUSE *Laughing Gas* v. 63 Can you imagine my lawyer letting them slip that over! **1960** 'B. McCORQUODALE' *Price is Love* iii. 53 It was something he really wanted to know and was trying to slip it over on her unexpectedly.

19. b. *to slip a disc*: to sustain a 'slipped disc' (*SLIPPED *ppl. a.*^{1} 2).

1958 'J. BYROM' *Or be he Dead* v. 68 An unfortunate tramp who had slipped a disk. **1974** G. MITCHELL *Javelin for Jonah* ii. 33 He told Margot to rake the long-jump pit ..and she slipped a disc.

VI. 26. c. (Earlier and later examples.)

1840 J. GAUGAIN *Lady's Assistant* 13 Slip a stitch having wool in front, then pass the wool to the back under the left pin. **1926** E. K. MIDDLETON *New Knitting* 15 To decrease two at a time. Slip one. Knit two together. Draw the slipped stitch over. **1951** E. CLOSE *Knitting* ii. 29 Slip one stitch from the left hand to the right hand needle as if you were about to knit it. **1973** M. STRADAL *Knitting, Crochet & Looping* i. 26 Slip one stitch purlwise, thread over needle and knit together the slipped stitch and the thread-over-needle of previous row.

VII. 30. *Shipbuilding.* To place (a boat) on

a slip (SLIP *sb.*^{3} 1 b) for inspection, repair, etc.

1950 H. M. DENHAM in *Jrnl. R. Cruising Club 1949* 122, I got Korby slipped (only £4) and put on a coat of anti-fouling. **1964** *Roving Commissions 1903* 207 We crossed to Hermione in the hopes of finding a caique yard which would slip us for a reasonable fee to check up on the bump we received at Finike. **1975** R. BUTLER *Where all Girls are Sweeter* iii. 23 The boat looked new. Short of slipping her she was in prime condition.

slip-. Add: **1. a. slip-case,** a close-fitting box with one side open into which a book or books are placed for protection, while allowing the spine to remain visible; also, a similar case for gramophone records or photographic equipment; hence **slipcased** *a.*, contained in a box of this kind; **slip-cover** *U.S.* = *loose cover* s.v. *LOOSE *a.* 9; hence as *v. trans.*, to cover with a slip-cover; **slip crew** *Aviation*, an aircrew stationed at an intermediate point or carried to take over the operation of an aeroplane on a long-distance flight; **slip edition,** a special (usu. local) edition of a newspaper, carrying news items not included in the main issue; **slip gauge** *Engin.*, a Johannson block (see *JOHANNSON); **slip-gear,** a gear designed to slip if loaded above a predetermined limit; **slip-horn,** a slide-trombone; **slip-jig** (earlier and later examples); **slip joint,** (*b*) a joint in a pipe, one section of which can move telescopically within another, to allow longitudinal expansion and contraction and so prevent damage by temperature changes or jolts; **slip-noose** (earlier example); **slip ring** *Electr. Engin.*, a ring of conducting material which is attached to and rotates with a shaft, so that electric current may be transferred to a stationary circuit through a fixed brush pressing against it; also *attrib.*; **slip road,** a short (usu. one-way) road giving access to or exit from a main highway, esp. a motorway; an approach road; **slip rope** (later examples); **slip scraper** *U.S.*, a horse-drawn earth-moving device; **slip sheet** *Printing*, a sheet of paper interleaving newly printed sheets to prevent set-off or smudging; hence **slip-sheet** *v. trans.*, **slip-sheeting** *vbl. sb.*; **slip-stitch** *sb.* (earlier and later examples); also in sewing, dressmaking, etc. (see also quot. 1964); **slip winder** (see quot. 1921); also *slip winding.*

1930 A. E. NEWTON *This Bk.-Collecting Game* ii. 20 Many collectors, in binding their books or in having slip cases made for them,..have their novels..in one colour, their poetry in another. **1942** W. STEVENS *Let.* 17 Sept. (1967) 420, I should like the general effect of the binding to be light... There should, of course, be a slip case. **1966** P. J. KAVANAGH *Perfect Stranger* xiii. 187 Why not put all three [volumes] into a slip-case..and sell them as a set? **1977** *Gramophone* May 1738/2 Decca have also made again available this month an integral recording of the five Beethoven Piano Concertos, previously issued in a slip-case but now in a box. **1979** *Amat. Photographer* Feb. 54/3 Both versions are supplied in a gift outfit with flash unit, slip case and wrist chain at a price of about £85. **1969** *Times* 15 Nov. p. iv/6 (Advt.), *Magellan's Voyage*.. 2 volumes slipcased. **1978** *Amer. N. & Q.* Nov. 44/2 A handsome slip-cased volume..seems to be under-priced. **1886** *Home Decoration* 3 Apr. 79/1 The slip covers for the furniture are of..toile. **1911** [see *KLAXON *sb.*]. **1920** *T. Eaton & Co. Catal.* Spring & Summer 395/2 Slip Covers for Ford Cars... Set consists of cover for each seat and back of seats, doors, kicking pad for front seat and complete cover for hood. **1952** S. KAUFFMANN *Philanderer* (1953) iii. 36 Cora had, of course, made all the curtains, slip covers and bedspreads herself. **1965** T. CAPOTE *In Cold Blood* ii. 78 The couch..that Nancy had slip-covered. **1978** T. GIFFORD *Glendower Legacy* 75 The slipcovers were wearing out at the arm. **1947** *Shell Aviation News* No. 112. 8/3 One of the most important problems is that of aircrew fatigue, and research on this question includes investigation into..provision of 'slip' crews at strategic points, facilities for crew rest in aircraft. **1973** C. EGLETON *Seven Days to Killing* xix. 196 The RAF are not carrying a slip crew on this trip. They would need to rest before the return flight. **1961** 'B. WELLS' *Day Earth caught Fire* iii. 39 This is terrific stuff. We'll have a slip edition. **1975** T. ALLBEURY *Palomino Blonde* iv. 12 Issues of the *Northumberland Gazette* with slip editions for Morpeth and Berwick. **1919** *Engineering* 11 July 33/3 The minimeter for comparing slip gauges to an accuracy of one-millionth of an inch is another of the new precision instruments. **1971** B. SCHARF *Engin. & its Language* vii. 47 In order to combine two slip gauges, they are slid together with slight pressure. **1897** E. K. SCOTT *Local Distribution of Electric Power in Workshops* 53 The current at starting a heavy lift can be materially reduced by having a slip gear or belt, so as to enable the motor to get up speed. **1930** *Engineering* 4 Apr. 431/2 An electric motor fitted with a centrifugal clutch and slip-gear. **1923** G. McKNIGHT *Eng. Words & their Background* iv. 45 *Sliphorn*, trombone. **1938** D. BAKER *Young Man with Horn* III. i. 120 He doesn't play a valve trombone either, just a regular slip-horn. **1957** *Melody Maker* 4 May 6/2 Wilbur himself was somewhat subdued, using both slip-horn and valve, but what he did was pleasant trombone. **1829** G. GRIFFIN *Collegian* I. ii. 19 Eily was dancing with a strange young gentleman..and..he would not let her go until she had finished the slip jig. **a 1966** 'M. NA GOPALEEN' *Best of Myles* (1968) 284 The lads who believe that in slip-jigs we have a national prophylaxis make life less stark. **1930** WALKER & CROCKER *Piping Handbk.* vii. 504 Slip joints are used very extensively in water and saturated-steam lines. **1972** L. M. HARRIS *Introd. Deep-water Floating Drilling Operations* ii. 6 Slip joints compensate for vertical motion in the lower section of the drill string. **1837** Slip noose [see *LASSO *sb.* 1]. **1896** S. P. THOMPSON *Dynamo-Electric Machinery* (ed. 4) iii. 35 There must..be sliding contacts to maintain the coils of the revolving field-magnet part in continuous metallic connexion with the auxiliary exciting circuit. In either case the appropriate device consists of a pair of slip rings, against each of which a brush presses. **1958** *Times Rev. Industry* Feb. 46/1 Automatic starting of slip-ring electric motors has always presented many problems. **1974** *Physics Bull.* May 204/2 Special emphasis is placed on the study of aluminium as a contact material in place of copper for sliprings and commutators. **1953** *Times* 11 Feb. 3/3 A 'slip-road' a mile and a half long..would draw away from the narrow streets of High Barnet the great number of heavy lorries now passing through this congested centre. **1968** W. GARNER *Deep, Deep Freeze* iii. 36 He took the slip road on to the *Autobahn*. **1979** A. PRICE *Tomorrow's Ghost* ii. 17 He waited to leave the slip-road for the motorway proper. **1909** *Man. Seamanship* (Admiralty) II. III. ix. 177 If there is much strain on the slip rope, it should be eased before letting it go. **1964** Slip rope [see *HONDA]. **1934** *Sun* (Baltimore) 9 Nov. 15/3 Ringle, while clearing loose dirt with the aid of a horse and slip scraper, lost his footing and fell. **1942** W. FAULKNER *Go down, Moses* 113 Throwing dirt..faster than a slip scraper could have done it. **1917** F. S. HENRY *Printing for School & Shop* xiv. 237 Never use enameled paper for slipsheets, or the sheets will stick together. **1924** —— *Essent. Printing* x. 149 While an expert feeder can sometimes do his own slipsheeting, it is customary to have an assistant place the slipsheets as the printed sheets are stacked on the feedboard. **1949** MELCHER & LARRICK *Printing & Promotion Handbk.* 279/2 This slip sheeting assures a clean job of mimeographing where it is necessary to print on non-absorbent paper. **1957** JACKSON & CLEVERDON *Printing* III. 139/2 Clean unprinted newspaper ..makes good slip sheets. **1967** V. STRAUSS *Printing Industry* vii. 516/2 In two-sided printing..it may be necessary to 'slip-sheet' the job. (Slip-sheeting means that a sheet of waste paper is inserted after each printed sheet. When the ink is dry the slip sheets are removed.) **1872** *Young Englishwoman* Oct. 558/1 Work 2 slip-stitches on the first 2 chain. **1932** D. C. MINTER *Mod. Needlecraft* 92/2 *Single Crochet* (or *Slip-stitch*)..is used for making a very narrow row. **1951** *Good Housek. Home Encycl.* 58/1 Tack, and slip-stitch by hand. **1964** *McCall's Sewing* ii. 32/1 *Slip-stitch*, tiny hand-stitches taken through and under a fold of fabric where the stitching must be invisible. **1921** *Dict. Occup. Terms* (1927) § 371 *Slip winder*; winds silk threads, for use in lace making, from hanks or cops on to spools or bobbins. **1976** *Evening Post* (Nottingham) 16 Dec. 17/5 (Advt.), Wanted male or female experienced slip winder (cone to spool). **1940** *Chambers's Techn. Dict.* 778/2 *Slip winding* (Textiles), the process of transferring yarn from a hank to flanged bobbins in lace manufacture.

b. *slip-lid.*

1938 *Shelf Appeal* July 26/2 The slip-lid tin was evolved rather over fifty years ago. **1979** *Gloss. Packaging Terms* (*B.S.I.*) II. 8 *Slip lid*, a lid that fits over the mouth of the container body.

c. *slip coach* (later examples); also in connection with other vehicles and craft, as **slip-tank,** a fuel tank that may be jettisoned from an aircraft when empty.

1932 G. GREENE *Stamboul Train* I. i. 8 The party.. belonged to the slip-coach for Athens. **1978** *Lancs. Life* Apr. 52/2 The L & Y..brought in slip coaches, notably at Rochdale where two trains from Bradford to Manchester unhitched the back two coaches without stopping. **1920** *Flight* XII. 957/1 Seventeen tanks may be readily slipped overboard to act as ballast. These slip-tanks have no bottom connections, and petrol is drawn from them by means of a semi-rotary pump.

2. slip-on (later examples); also (chiefly *U.S.*) a glove and now usu. a shoe; *attrib.*: now also as *adj.*; **slip-over** *a.*, of a garment: made without an opening at the front, and to be slipped on over the head; hence as *sb.* (usu. one word), a sweater or pullover, usu. with a V-neck and no sleeves; **slip-up,** the act of slipping up (see SLIP *v.*^{1} 8 d), a failure, mistake, blunder.

1920 *T. Eaton & Co. Catal.* Spring & Summer 395/3 Running Board Mat. Fastened with patent slip-on fasteners. **1923** *Ibid.* 214/3 One of the smartest of the new models [of coat]..a double-breasted 'slip-on', with yoke. **1938** [see *SCUFF *sb.*^{1} 4]. **1949** *Sun* (Baltimore) 8 Sept. 5/1 (Advt.), Wear Right gloves... Shorties and slipons all hand-sewn. **1956** *People* 13 May 3/7 (Advt.), Slip-on Casual. Brown Willow uppers, leather sole. **1959** *Wall St. Jrnl.* 13 Dec. (Eastern ed.) 17/4 'Slim slip-ons', or dressy shoes without laces, will be promoted for men. **1965** *N.Y. Times* 9 Dec. 5 (Advt.), Hand-in-Glove with fashion:..Elbow-deep slipons. **1972** J. BALL *Five Pieces of Jade* xiv. 178 Tibbs..noted the slip-on shoes which.. could be shed within a second or two. **1978** L. CHARTERIS *Saint in Trouble* (1979) I. iii. 33 His eyes..started with the suede slip-ons, journeyed up..the light grey suit. **1982** BARR & YORK *Official Sloane Ranger Handbk.* 33/2 These are smart tough slip-ons, in black or dark blue patent or leather, with a chunky heel and gilt snaffle or chain across the top. **1919–20** *T. Eaton & Co. Catal.* Fall & Winter 153/3 Slip-over Nightgown. **1923** *Daily Mail* 6 Mar. 1 (Advt.), Elastic bust bodice of special value, slip-over shape. **1936** *Times* 10 Jan. 7/4 Bargains for men include poplin shirts at 9s. 6d. and all wool slip-overs at 10s. **1941** *Picture Post* 3 May 32/1 Pullovers, Slipovers, and Beach Shirts for men and boys. **1945** *Richmond* (Va.) *Times-Dispatch* 9 Jan. 16 (Advt.), All-wool slipover sweater. **1962** *Punch* 26 Sept. p. xvii, Burberrys have.. reversible alpaca slip-overs. **1981** *Times* 6 Jan. 12/7 According to numerous shop window displays observed during the current sales, the garment I have always called a pullover is now known as a slipover. **1909** *N.Y. Even. Post* (semi-weekly ed.) 30 Sept. 1 Should there be any slip-up in the present plans. **1948** 'G. ORWELL' *Let.* 21 Dec. in *Coll. Essays* (1968) IV. 459, I suppose there *may* be some slip-up, but if not my address..will be The Cotswold Sanatorium. **1978** H. JOBSON *To die a Little* ii. 26, I..had a slight feeling of apprehension... Had there been some legal slip-up?

slip form. *Engin.* Also **slip-form, slipform.** [f. SLIP *v.*^{1}] A mould open at both ends in which a structure of uniform cross-section is cast by filling it with concrete and continually moving and refilling it as the concrete sets. Freq. *attrib.* with reference to this technique of construction, esp. as *slipform* (*concrete*) *paver*, a machine which continually forms the concrete surface of a road or the like by this technique.

1958 *Construction Methods & Equipment* Aug. 72/2 Slip-form placing of concrete reduced material and man-power requirements to a fraction of what they might have been with timber forms and scaffolding. *Ibid.* 73/2 The contractor supported slip forms with a number of steel yokes. **1963** *Contractors & Engineers* Jan. 40/4 Slip-form pavers are shorter this year... The early slip-form machines, with 40 to 60 feet of sliding form, amazed many people. **1964** [see *FORM *sb.* 18 b]. **1974** *Sci. Amer.* Dec. 145/1 The road itself, that great ribbon of concrete, is being placed by the slip-form paver, which spreads, smooths, levels, cuts the edges squarely and even finishes the surface. **1975** *Offshore* Sept. 147/1 The steel slipform is guided on rails cast into the interior surface of the cylinder wall.

Hence **sli·pform** *v. trans.*, to cast by this technique; **sli·pformed** *ppl. a.* **sli·pforming** *vbl. sb.*

1968 J. J. WADDELL *Concrete Construction Handbk.* xxxiv. 1 The ultimate approach in incorporating a slip-form into building construction is to slipform all vertical concrete. *Ibid.*, Slipforming is a faster method of construction than conventional forming. *Ibid.* 2 Slipformed-concrete. **1975** *Offshore* Sept. 147/1 Tests are complete on the slip-forming of inclined legs for a deepwater concrete platform. **1981** *Times* 12 Mar. 3/6 The contractors are slip forming 2,000 metres of double track. This track-laying method was first used in 1967.

slip-knot. (Earlier *transf.* example.)

1842 T. MOORE *Jrnl.* Sept. 14 in *Mem. Thomas Moore* (1856) VII. 330 Had already formed a sort of slip-knot with Easthope to dine at his country house, but he had luckily put me off till to-morrow.

slippage. Add: **2.** *Mech.* The difference between the expected and the actual output of a system.

1905 W. ROGERS *Pumps & Hydraulics* II. 384 Pump slip or slippage represents the difference between the calculated and the actual discharge of a pump. **1936** *Kent's Mech. Engineers' Handbk.* (ed. 11) II. 1. 41 Actual Volumetric Efficiency/Indicated Volumetric Efficiency = Slippage Efficiency [of an air compressor]. This is the ratio of the volume of measured air delivered to the apparent volume shown by the indicator diagram.

3. *transf.* and *fig.* Falling away from a standard; the measure of this. *spec.* with reference to (*a*) failure to meet a deadline or fulfil a promise, delay; (*b*) loss of public esteem, of a candidate for office in popularity ratings; (*c*) *Econ.*, decline in value.

1920 in WEBSTER. **1960** *Washington Post* 1 Jan. A16 It is almost as if a deliberate decision had been taken to accept second-place status. This continued slippage also affects starkly the challenges that lie ahead. **1960** *Guardian* 13 May 10/5 The failure (or 'slippage' as delays are now called) of their 400-mile anti-aircraft missile. **1968** *Ibid.* 22 Aug. 9/7 The latest Gallup poll..shows that Nixon would get 45 per cent of the votes compared with Humphrey's 29 per cent. That is an astonishing slippage for the Vice-President. **1970** *Nature* 13 June 1011/2 ESRO's biggest project..is the half-ton TD1 astronomy satellite which was to have been launched in spring 1972, although some slippage now seems likely. **1972** *Guardian* 1 Nov. 12/2 British living standards could be eroded by a continued slippage of sterling. **1976** *Times* 20 Apr. 13/3 There is widespread concern among parents that standards of achievement and behaviour in schools have been allowed, if not to collapse, at least to slip... This apparent slippage has taken place at the time when the number of comprehensive schools has increased rapidly. **1980** M. LEE *Govt. by Pen* 211 His health had been giving way, and there were signs of mental slippage. **1982** *Sunday Times* 10 Oct. 54/3 Given the traditional slippage that occurs in the timetable on such projects, the company will not know at the outset just when it will have to draw down the cash for each stage payment.

slipped, *ppl. a.*^{1} Add: **1.** (Examples in *Knitting.*)

1840 J. GAUGAIN *Lady's Assistant* 11 Take in back stitch of three, by slipping off backwards without working the first,..lift over the first slipped one over the taken-in loop. **1872** *Young Englishwoman* Nov. 607/1 In decreasing..a stitch may be slipped, the following stitch knitted and the slipped stitch drawn over it. **1926**, etc. [see *SLIP *v.*^{1} 26 e]. **1958** J. NORBURY *Knitting Adventure* i. 15 Purl slipped stitch from cable needle.

2. *slipped disc*, an intervertebral disc that is ruptured or injured, causing pain in the back or (if nerve roots are compressed) in other parts of the body. *colloq.*

1953 E. SIMON *Past Masters* III. 142 The slipped disk everybody nowadays is suffering from. 1959 *Listener* 10 Sept. 397/3 Achondroplasic breeds [of dogs] are also more apt to develop so-called 'slipped discs'. 1972 J. MINIFIE *Homesteader* x. 79 God help the farmer who 'pulled his back', as they used to say before 'slipped disc' became the fashionable term. 1974 PASSMORE & ROBSON *Compan. Med. Stud.* III. xxv. 56/2 If a load exceeds the ultimate tolerance of a disc..the nucleus may be forced through the annulus or cause it to bulge locally, a condition known as ruptured or prolapsed or slipped disc.

slipped (slipt), *ppl. a.*[3] [f. SLIP *v.*[3]] Painted or ornamented with slip.

1914 *Oxf. Univ. Gaz.* 11 Mar. 574 A boar's head in slipped painted ware of Late Hittite date. 1976 *Nature* 15 Apr. 581/2 This pottery includes sophisticated slipped wares.

slipper, *sb.* Add: **I. 1. e.** As an instrument of punishment with which a child (etc.) is disciplined by beating. In phr. *to take a slipper to* (someone).

[1682, etc.: implied at SLIPPER *v.*[2] 1]. 1876 [see *REAR *sb.*[2] 2 b]. 1924 GALSWORTHY *White Monkey* III. viii. 266 Teach him a sense of other people, as young as possible, with a slipper if necessary. 1932 A. J. WORRALL *Eng. Idioms* ii. 18 He is very impudent in his manner, and I should dearly like to take my slipper to him. 1978 R. MILLS *Comprehensive Educ.* 21 'I haven't done my homework, my History.' 'Ha, ha, it's the slipper for you then.' 1982 *Daily Tel.* 1 Mar. 10/7 The cane and slipper never did the likes of us any harm.

f. A temporary shoe for a horse.

1903 SOMERVILLE & 'ROSS' *All on Irish Shore* iii. 82 He [*sc.* the smith] examined each hoof in succession..and then, turning to Mr Fennessy, remarked:—'Ye'd laugh if ye were here the day I put a slipper on this one.' 1953 G. BROOKE *Introd. Riding & Stablecraft* vi. 52 You should pull up at a shoeing-smith's, and..he will tack on what they call a slipper; that is, a shoe that fits well enough, and will see you through the day.

4. d. *Mech.* Also *slipper block*. A guide block attached to a reciprocating rod, esp. a piston rod or its cross-head, so as to slide with the motion of the rod against a fixed plate and prevent any tendency of the rod to bend.

1881 N. P. BURGH *Mod. Marine Engin.* (rev. ed.) 72/1 The guide channels..are the ordinary kind, arranged to receive slipper blocks. 1883 A. E. SEATON *Man. Marine Engin.* viii. 144 To preserve the piston-rod in its true course, a guide is provided, and the piston-rod end fitted with blocks or slippers to work in it. 1952 FOX & MCBIRNIE *Marine Steam Engines & Turbines* ix. 155 A single slipper crosshead is generally made as shown. 1971 *Naval Marine Engin. Practice* (Min. of Defence) (ed. 2) I. vi. 137 A slipper is fitted to the crosshead, and this slipper slides in a vertical guide to maintain straight reciprocating motion against the thrust of the connecting rod. The slipper may be circular..or in the form of a flat plate.

e. *Mech.* A part that is capable of sliding in the direction of its length.

1903 *Electr. World & Engin.* 21 Nov. 845/2 The buckets of the Riedler turbine resemble those of the well-known Pelton type. The steam jet is divided by a central 'slipper' in two parts. 1930 *Engineering* 25 Apr. 539/3 About half-way along this arm is attached a slipper, D, which is kept in contact with the straight-edge, E, by a weight, F. 1966 *McGraw-Hill Encycl. Sci. & Technol.* XI. 611/2 Sleds are carried on shoes or slippers that grip the railhead in order to prevent derailing.

5. (Example.)

1818 L. D. CLARK *Diary* 11 July in *Firelands Pioneer* (1920) XXI. 2314 Made a slipper for Mrs. Caufield's baby.

II. 7. b. *Cricket.* One who fields in the slips. *colloq.* Somewhat *rare*.

1903 D. L. G. JEPHSON in H. G. Hutchinson *Cricket* iv. 102 There were good slips, bad slips, fast-asleep slips, and since his time every variety of 'slipper' has passed across the stage. 1973 *Daily Tel.* 23 July 32/1 If Hayes be sufficiently rated as a slipper, Roope could be spared.

III. 10. slipper-bath: now usu. one of a number of single baths of the modern domestic style installed for hire at public baths; **slipper chair** *U.S.*, a low-seated, freq. upholstered chair with a high back; **slipper satin**, a strong, closely woven fabric with a semi-glossy appearance, used for making slippers, dresses, furnishings, etc.; **slipperslapper** *nonce-wd.*, a loose, sloppily-fitting slipper (cf. *SLIP-SLOP *sb.* 4); **slipper socks, sox** *pl.*, a pair of slippers with socks combined.

1960 L. WRIGHT *Clean & Decent* xii. 172 Confusion may arise from the fact that single baths of the ordinary modern kind, on hire at Public Baths, are there still called 'Slipper Baths'. Such, no doubt, they originally were; the term has survived the change. 1981 *Times* 25 Feb. 8/5 The council closed down a slipper bath which had been used by pensioners living in bed-sitters. 1938 *House Beautiful* Jan. 41 Slipper chair... Typical of the French Regency. Broad proportions, beech frame delicately carved. Circa 1715. 1957 M. MILLAR *Soft Talkers* x. 94 Her mother sat in a slipper chair. 1979 M. MCMULLEN *But Nellie was so Nice* (1981) I. v. 42 He..made himself pleasant to Ursula in her slipper chair. 1937 M. SHARP *Nutmeg Tree* xxiii. 297 They saw the Disgusted Lady.. marvellous in ice-blue slipper satin. 1970 D. CLARK *Deadly Pattern* v. 104 A gilt-legged Chesterfield upholstered in cream slipper satin. 1922 JOYCE *Ulysses* 86 Slop about in slipperslappers for fear he'd wake. 1950 *Sears, Roebuck Catal.* Fall/Winter 1324/6 Slipper socks. 1951 *Ibid.* Fall/Winter 371 (*heading*) Footease slipper socks for every member of the family. 1970 *Guardian* 15 Dec. 9/3 Just tee-shirts with dungarees or jeans—plus slippersox. 1973 *Sunday Advocate-News* (Barbados) 9 Dec. 25/1 Slipper socks and bow ties can be great ways to gift him.

slippered, *ppl. a.* Add: **1. b.** *fig. Lit.* and *poet.*

1851 H. MELVILLE *Moby Dick* II. xvii. 136 The slippered waves whispered together. 1912 R. BROOKE *Coll. Poems* (1970) 68 A vague unpunctual star, A slippered Hesper.

slippering, *vbl. sb.* [f. SLIPPER *v.*[2] + -ING[1].] Beating with a slipper.

1851 H. MELVILLE *Moby Dick* I. iv. 41 Beseeching her as a particular favour to give me a good slippering for my misbehaviour. 1919 H. WALPOLE *Jeremy* ii. 33 A slippering from his father or idiotic punishments from the Jampot. 1934 M. V. HUGHES *London Child of Seventies* iv. 35 A threatened 'slippering' of the boys by my father if they are too noisy. 1953 J. CARY *Except the Lord* ix. 32 Beating was a punishment that could be graded from a random cuff.., through a hard smack, to a formal slippering. 1972 *Times* 9 Oct. 13/5 Children..respected striking or slippering as common..in their schools.

slipper-slopper, *a. dial.* and *colloq. rare.* [Redupl. f. SLIPPER *sb.*] Wearing loose slippers; down-at-heel, sloppy.

1825 J. JENNINGS *Observations Dial. W. Eng.* 70 *Slipper-slopper*,..having shoes or slippers down at the heel; loose. 1888 F. T. ELWORTHY *West Somerset Wd.-bk.* 680 Father, be sure you baint gwain out all slipper-slopper like that. 1904 in *Eng. Dial. Dict.* V. 525/1 Slippers trodden down at the heels are said to be slipper-slopper. 1951 J. FRAME *Lagoon* 29 Their shoes were slipper-slopper.

Hence as *v. intr.*, to walk about in loose slippers; also **slipper-slopping** *vbl. sb.*

1929 R. HUGHES *High Wind in Jamaica* viii. 192 Jonsen slipper-sloppered up and down his side of the deck. 1933 L. A. G. STRONG *Sea Wall* I. i. 13 He was roused from his meditations by a..slipper-slopping on the stairs.

slippery, *a.* Add: **1. c.** *slippery slope* fig., a course leading to disaster or destruction.

1951 J. FLEMING *Man who looked Back* x. 132 You go off down the slippery slope; it'll do you good. 1964 *Daily Tel.* 6 Jan. 12/2 While Western feet thus approach what some fear may be a 'slippery slope' towards recognition of the East, Ulbricht's ground seems as firm as ever it was. 1979 N. LASH *Theology on Dover Beach* iv. 74 It could be argued..that to give priority to love, to trust, to action, to commitment, is to start down the slippery slope along which rationality, objectivity and—eventually—truth are abandoned.

2. d. *slippery elm* (earlier examples). Cf. *red elm* s.v. RED *a.* 17 d in Dict. and Suppl.

1748 D. DRAKE *Let.* 1 Jan. in *Pioneer Life Kentucky* (1870) iv. 73 Of the whole forest the red or slippery elm was the best. 1780 W. FLEMING *Jrnl.* 20 Mar. in N. D. Mereness *Trav. Amer. Colonies* (1916) 640 Bear fat is preserved sweet and pure by putting in a bunch of the Slippery Elem [*sic*] bark into it when rendering. 1810 [see *moose elm* s.v. *MOOSE b].

4. c. Prov. phr. *as slippery as an eel*. Cf. EEL 1 c.

[*c* 1412 HOCCLEVE *Reg. Princes* (1897) 1985 Mi wit is also slipir as an eel. 1562 [see SLIPPER *a.* 1b].] 1601 HOLLAND tr. *Pliny's Nat. Hist.* I. IX. xx. 247 All that be long and slipperie as Yeels and Congres. 1622 S. ROWLANDS *Good Newes & Bad Newes* sig. B1[v], Fie vpon giddie Fortune, and her wheele, Vnconstant, and as slipperie as an Eele. 1739–40 RICHARDSON *Pamela* (1740) I. 245 You'll find her as slippery as an eel, I'll assure you. 1855 MRS. GASKELL *North & South* I. xvii. 209 He's as slippery as an eel, he is. He's like a cat,—as sleek, and cunning. 1914 T. DREISER *Titan* xxvi. 220, I am morally certain he uses money to get what he is after as freely as a fireman uses water. He's as slippery as an eel. 1980 J. GARDNER *Garden of Weapons* II. i. 128 He's big, but slippery as an eel.

7*. *look slippery* = *look slippy* s.v. SLIPPY *a.*[1] 2. *dial.* or *colloq. rare.*

1922 JOYCE *Ulysses* 418 Two Ardilauns. Same here. Look slippery.

9. *slippery-footed*; **slippery hitch** *Naut.*, a knot made fast by catching part of the rope beneath the bight, released at a pull on the free end; also *fig.*; **slippery pole** = *greasy pole* s.v. GREASY *a.* 9 (in quots. *fig.*); **Slippery Sam,** a card-game (cf. *blind-hookey* s.v. BLIND *a.* 16).

1903 Slippery-footed [see *HOT-FOOT *v.*]. 1832 Slippery hitch [see HITCH *sb.* 6 b]. 1903 'T. COLLINS' *Such is Life* vii. 273 Alf, it appeared, had left the station six or eight weeks before, bound for no-one knew where. Jack's opinion was that in so doing he had made a slippery-hitch. 1944 C. W. ASHLEY *Bk. Knots* i. 19 The *Slippery Hitch* is often found in the sheets and halyards of small boats. 1972 *Village Voice* (N.Y.) 1 June 20/1 The Voice had this wonderfully sobering policy of rejecting your stories right when you thought you had a firm grasp on the slippery pole. 1977 H. GREENE *FSO-1* xviii. 167 All of these years of clambering up the slippery pole only to find there was no top to it. 1923 L. H. DAWSON *Hoyle's Games Modernized* I. 162 Slippery Sam is a variation of and by many considered an improvement on Blind Hookey. 1954 A. S. C. ROSS in *Neuphilologische Mitteilungen* LV. 22 Solo whist..is non-U, though much 'lower' games (e.g. pontoon, nap and even slippery sam) are not necessarily so. 1978 C. STORR *Winter's End* iv. 58 The others were exclaiming and shouting at Racing Demon... They'd changed to Slippery Sam.

slippiness. Add: **b.** Speed, promptitude, alacrity. *colloq.*

1974 WODEHOUSE *Aunts aren't Gentlemen* xii. 98, I think I outlined the position..rather well, making it abundantly clear that..the cat..must be restored to its proprietor with all possible slippiness.

slipping, *vbl. sb.*[1] Add: **1. c.** = *SLIP *sb.*[3] 9 e.

1912 *Q. Rev.* Jan. 182 The danger attendant on the 'slipping' of a charge in a blast furnace—the descent of hunks of limestone and ore that are hurled into the air when such a 'slip' occurs. 1948 G. R. BASHFORTH *Manuf. Iron & Steel* I. x. 166 Slipping is the aftermath of scaffolding or hanging. *Ibid.*, Slipping or irregular descent of the stock may..occur due to badly designed bosh walls. 1969 K. R. HALEY in J. H. Strassburger *Blast Furnace* II. xii. 592 (*heading*) Hanging and slipping.

2. (*Lit.* and *fig.* examples in sense *15 a of the vb.)

1925 *Morris Owner's Man.* 22 Persistent slipping of the clutch must not be resorted to. 1959 *Listener* 17 Dec. 1085/3 Such statements as that on page 169..may be regarded as a momentary slipping of the clutch.

slippy, *a.*[1] **2.** (Later examples.)

1915 [see *buck-rabbit* s.v. *BUCK *sb.*[1] 3]. 1924 D. MOORE *Fen's First Term* i. 4 There is a hurry... Go and change and look slippy. 1972 WODEHOUSE *Pearls, Girls, & Monty Bodkin* xi. 178 Both of you get out of the car... And make it slippy, because I haven't got all day.

slip-slap, *v.* Restrict *rare*[−1] to sense in Dict. and add: **2.** = SLIP-SLOP *v.* 3. *rare.*

1926 T. E. LAWRENCE *Seven Pillars* v. lvii. 302 At Cairo my sandalled feet slip-slapped up the quiet Savoy corridors. 1965 'LAUCHMONEN' *Old Thom's Harvest* v. 58 She slip-slapped in her cut-back slippers to the door.

slip-slop, *sb.* **4.** Restrict *U.S.* to sense in Dict. and add: **b.** A kind of beach sandal; = *FLIP-FLOP *sb.* f. Chiefly *S. Afr.*

1971 *Studies in English* (Univ. Cape Town) Feb. 29 Beach-thongs, sandles [*sic*] made of rubber..have a great many names here—sloppies, slip-slops, plakkies, etc. 1974 'G. BLACK' *Golden Cockatrice* iv. 66, I couldn't believe they had my shoe size too, almost relieved to find only a pair of slip-slops. 1976 J. MCCLURE *Rogue Eagle* ii. 31 Knotted blouse, blue jeans and slip-slop sandals.

slip-sloppism. (Earlier example.)

1803 *Lett. Miss Riversdale* III. 228 Slip-sopism [*sic*] is not confined to females, now-a-days, I perceive.

slip stone, slipstone (sli·pˌstōᵘn). [f. SLIP *sb.*[6] + STONE *sb.*] A shaped oilstone used for sharpening gouges.

1927 H. HUBBARD *Colour Block Print Making* 209 *Slip stone*, a thin sharpening stone so shaped as to remove the burr or ragged edge from the inside of a gouge. 1947 J. C. RICH *Materials & Methods of Sculpture* x. 300 With gouges, the sharpening method is generally reversed. Specially shaped slip stones..are employed and the stone is rubbed on the gouge. 1958 J. R. BIGGS *Woodcuts* 36 Remember to have one slip-stone with its smallest curve small enough to sharpen the smallest gouge. 1976 *Billings* (Montana) *Gaz.* 27 June 4-G/5 With a slipstone, the tool is held firm while the sharpener is moved along the edges. With a whetstone, the tool is moved across it to get sharp edges.

slip-stream (sli·pstrīm). Also **slip stream, slipstream.** [f. SLIP *sb.*[3] + STREAM *sb.*] **1. a.** The current of air or water driven backward by a propeller or downward by a rotor.

1913 A. E. BERRIMAN *Aviation* viii. 79 Each blade deflects air backwards as it moves; the combined effect of both blades operating always in the same region when the machine is standing still produces a concentrated flow of air, which becomes a very pronounced draught. Technically, this draught is called the slip stream. 1919, 1920 [see *air scoop* s.v. *AIR *sb.*[1] B. II]. 1935 *Sun* (Baltimore) 19 Jan. 9/2 The slip stream from the propeller cast bits of mud and ice into the air during the take-off. 1963 J. ROWLAND *North to Adventure* ix. 132 Her rudder was one of those absurd little metal plates which are utterly ineffective unless the slipstream from the propeller is impinging directly upon them. 1973 C. BONINGTON *Next Horizon* xiii. 190 The snow leapt up at me, and I was there, in one piece, with the slip-stream of the helicopter hammering at me.

b. Any localized current associated with an object, esp. a moving one.

1947 in J. A. Carruth *Loch Ness & its Monster* (1950) 31 'We unmistakably sighted,' said Mr. Cottier, 'on the placid surface of the loch a fairly long slip-stream which quickly developed in length.' 1963 *Times* 8 June 5/2 He [*sc.* a cyclist] sat at the back, carried along in the slipstream of his adversaries, and so was the least tired rider when it came to the final sprint. 1973 'A. HALL' *Tango Briefing* v. 62 The slipstream didn't cool anything: it just circulated the heat. There were gnats already sticking to the windscreen. 1974 M. BABSON *Stalking Lamb* xxv. 183 The candles flared and died..leaving little slipstreams of smoke.

2. *fig.* An assisting force considered to draw

something along with or in the wake of the principal.

1957 *Universe* 16 Aug. 4/2 The kaleidoscope of life which has moved with such tremendous upheavals..for the people caught up in its slipstream since 1933. 1961 *Times* 30 June 4/1 Sangster it was..who proved himself the leader, drawing in his slipstream Wilson. 1970 D. MATHEW *Courtiers of Henry VIII* III. vi. 203 They stood in contrast to all the slipstream of the modern State. 1980 *Times* 29 Feb. 1/2 Some of those who were trying to get into the slipstream behind the Labour Party.

Hence as *v. trans.*, to follow closely behind (another vehicle) so that the resistance of the air to one's progress is less; also, to pass after travelling in another's slipstream; also **sli·p-streaming** *vbl. sb.*

1957 S. MOSS *In Track of Speed* ii. 18, I tried for the first time the art of slip-streaming... What you do is to tuck in behind a faster car, so that you are more or less sucked along in the 'partial vacuum'. 1960 *News Chron.* 18 July 5/4, I planned to slip-stream him on the last lap. 1969 *Man* (Austral.) Mar. 43/1 If you are in near-equal machinery, it becomes a case..of slip-streaming an inch behind him and diving out for a pass at the vital moment. 1979 L. PRYOR *Viper* viii. 156, I caught them on the pit straight as they slipstreamed each other on the left side of the road.

slipware, slip-ware (sli·pwɛ̄ᵊɹ). [f. SLIP *sb.*[1] + WARE *sb.*[2]]. Pottery coated with slip. Also *fig.* and *attrib.* passing into *adj.*

1883 L. M. SOLON *Art Eng. Potter* 41 We shall conclude by mentioning another sort of Slip-ware also made now-a-days, the sham 'old Slip', of which we have to confess the possession of several pieces, bought..for genuine specimens. 1929 H. READ *Staffordshire Pottery Figures* 2 The Staffordshire potters, who had already acquired considerable skill in the manipulation of clays in the making of the type of pottery known as slipware. 1935 *Burlington Mag.* Oct. 180/1 A Wrotham slipware dish. 1957 *Listener* 17 Oct. 608/1 My second preference among English pots..would be slip ware, which seems to be the earliest which can definitely be assigned to Staffordshire. 1968 J. ARNOLD *Shell Bk. Country Crafts* 232 Slip-ware is a porous earthenware on which a creamy clay has been applied on the still damp body, before firing, either by immersion or brush trailing. 1976 'D. HALLIDAY' *Dolly & Nanny Bird* vii. 92 Two girls in curled hats..and pink slip-ware faces.

slip-way. Add: **1.** (Examples of a ramp used by a seaplane.) Also *fig.*

1912 *Flight* 7 Dec. 1133 (*caption*) The machine being launched from the slipway that leads down to the water from the hangar. 1922 *Encycl. Brit.* XXX. 48/1 The management of these [seaplane] stations is very similar to that of an aerodrome, with the exception of slipways up and down which aircraft are moved on leaving and entering the water. 1936 *Sun* (Baltimore) 6 July 9/1 The giant four-motored plane, prototype of the airliner with which Britain is to face the Atlantic voyage, made its first appearance on the slipway at Rochester this week. 1938 G. GREENE *Brighton Rock* I. iii. 46 The coffin slid smoothly down into the fiery sea... The clergyman smiled gently from behind the slipway. 1982 P. CONNON *In Shadow of Eagle's Wing* 28 On the evening of Wednesday, August 3, the hydromonoplane was launched down the slipway into Bowness Bay.

slit, *sb.* Add: **1. d.** The vulva. *coarse slang.*

Its currency is restricted in the manner of other coarse terms: see small-type note s.v. *FUCK *v.*

1648 R. HERRICK *Hesperides* 47 Scobble for Whoredome whips his wife; and cryes, He'll slit her nose; but blubb'ring, she replyes, Good Sir, make no more cuts i' th'outward skin, One slit's enough to let Adultry in. 1714 *Cabinet of Love* 18 His tarse, as soon as to my slit applied Up to the hilt into my cunt did slide. 1970 G. GREER *Female Eunuch* 265 The vagina..belittled by terms like..*slit*. 1977 *Rolling Stone* 24 Mar. 41/4 What am I going to call it? Snatch, Twat? Pussy? Puss puss, nice kitty, nice little animal that's so goddam patronizing it's almost as bad as saying 'slit'.

e. A narrow, usu. straight aperture in an optical instrument through which a beam of light can be received.

[1832 *Nat. Philos.* (Libr. Useful Knowl.) II. II. iii. 25/1 Instead of a row of holes, he formed one narrow slit in the shutter... By this means a spectrum of any required breadth may be formed.] 1863 E. ATKINSON tr. *Ganot's Elem. Treat. Physics* VII. iv. 410 A telescope, the eyepiece of which can be regulated by a micrometric screw... The slit is in the focus of the object-glass of the telescope. 1888 *Proc. R. Soc.* XLIII. 130 Huggins's photograph of the spectrum of Comet Wells, taken with a wide slit. 1905 E. C. C. BALY *Spectroscopy* iii. 48 As generally used at the present time the slit is formed between two metal jaws, one of which is fixed while the other is moved by a fine-pitched screw. 1926 *Jrnl. Optical Soc. Amer.* X. 186 If the spectrophotometer has a second collimator, the continuous spectrum may be formed from an incandescent light placed in front of its slit. 1969 D. W. TENQUIST et al. *University Optics* I. xi. 293 (*caption*) Fraunhofer diffraction at a double slit.

5. slit drum, a primitive percussion instrument made out of a hollowed log with a longitudinal slit; **slit fricative** *Phonetics*, a fricative or spirant sound made by expelling the breath through a narrow aperture; **slit-gong** = *slit drum* above; **slit lamp** *Ophthalm.*, a lamp which emits a narrow but intense beam of light, used for examining the interior of the eye; freq. *attrib.*; **slit pocket,** a side pocket in a garment, with a vertical opening; **slit sampler,** a device for studying the bacterial content of the air, having a slit through which it is drawn; **slit sound, spirant** *Phonetics* = *slit fricative* above; **slit-trench,** a narrow trench made to accommodate and protect a soldier or weapon in battle.

1933 *Africa* VI. 155 The deep-toned slit-drum..is assigned by ethnologists..to the matriarchal 'two-class' culture circle. 1957 *New Oxf. Hist. Music* I. ii. 185 The wooden fish, still in use among Taoists and Buddhists, is a slit-drum. 1974 *Encycl. Brit. Macropædia* XIV. 61/2 In Vietnam the slit-drum is both a temple and a watchmen's instrument. On Java slit-drums can be traced to the Hindu-Javanese period (1st–9th century AD). 1955 H. A. GLEASON *Introd. Descr. Linguistics* ii. 22 Because of the slit-like shape of the opening, these sounds are called slit fricatives. 1973 J. C. WELLS *Jamaican Pronunc. in London* 127 This feature is an extension of the..distinction between 'groove' and 'slit' fricatives. 1938 *Jrnl. R. Anthrop. Inst.* LXVIII. 241 Samara returned to the village and after beating the slit-gong gave vent to his grievance. 1970 *Times* 23 Feb. 10/4 (*caption*) A 'slit gong' which has been accepted by the Queen as a gift... The gong is used in the New Hebrides to convey messages, summon people and sound the alarm. 1922 *Arch. Ophthalm.* LI. 271 Many new phases of examination of the living eye are made possible by the use of the slit lamp. 1961 [see *GONIOSCOPE]. 1978 *Jrnl. R. Soc. Med.* LXXI. 100 Slit-lamp examination revealed a moderate number of cells in the anterior and posterior vitreous. 1933 J. E. LIBERTY *Practical Tailoring* ix. 169 Slit pockets are somewhat similar to trouser pockets and are made with jettings, or welts, but the pocket is not sewn twice. The mouth of the pocket is almost upright and at least 7 in. long for a normal size coat. 1978 M. SICHEL *Costume Reference 8: 1918–1939* 36 Many skirts worn with blouses had pockets at the sides or slit pockets and belts of the same material. 1941 R. B. BOURDILLON et al. in *Jrnl. Hygiene* XLI. 220 The range of concentrations which can be measured accurately with the slit sampler is from about 1 to 10,000 per cu. ft. 1963 WALTER & ISRAEL *Gen. Path.* xix. 295 The slit-sampler consists of a narrow slit through which air is sucked on to a rotating culture plate beneath it. 1912 Slit sound [see *RILL *sb.*[1] 4]. 1958 Slit spirant [see *RILL *sb.*[1] 4]. 1970 *Publ. Amer. Dial. Soc.* 1968 L. 21 The phoneme /θ/ of *thirty, Martha, hearth* is a voiceless dental slit spirant [θ]. 1942 Slit trench [see *FOX-HOLE]. 1944 *Times* 12 May 3/2 At Manus Island the prisoners were made to dig slit trenches for the Japanese. 1971 B. W. ALDISS *Soldier Erect* 162 My slit-trench is the first on the right, next to the cookhouse. Cheerio, Ali, you old robber!

slit, *ppl. a.* Add: **3. c.** *slit iron* (earlier Amer. example); **slit skirt,** a tight skirt slit upward from the hem for ease of movement or sexual allurement.

1789 *Deb. Congress U.S.* 17 Apr. (1834) I. 167 To lay an impost of seven and a half per cent..upon..slit or rolled iron. 1913 *Punch* 30 July 101/2 Four young women who last week promenaded Fifth Avenue, New York, in slit skirts..were surrounded by an enraged mob. 1954 C. G. BRADLEY *Western World Costume* xxi. 342 The hobble skirt of 1914 was worn even on long walking excursions. The slit skirt of the same year brought protests from bishops and ministers. 1976 'M. DELVING' *China Expert* i. 7 The slit skirt of the *ch'i pao* she always wore.

slither, *sb.* Add: **4. a.** Also *transf.* and *fig.*

1915 E. WALLACE *Man who bought London* ii. 23 So many people were following closely in that hurried slither to the platform. 1970 *Guardian* 12 Nov. 12/2 If the whole slither into inflation is not to accelerate..some private employers will have to stand firm.

b. Something smooth and slippery; a smoothly sliding mass; = SLIVER *sb.*[1] 1.

1919 E. POUND *Quia Pauper Amavi* 40 If she goes in a gleam of Cos, in a slither of dyed stuff, There is a volume in the matter. 1955 N. NICHOLSON *Lakers* xi. 188 Only after rain, when..the rocks are hung with slithers of water like lace curtains against the black slate. 1966 G. GREENE *Comedians* I. v. 153 Little fenced saucers of earth where a few palm-trees grew and slithers of water gleamed between. 1981 *Daily Tel.* 27 May 15/1 Calvin Klein's newest dress is a slither of silk shaped simply like an overgrown T-shirt.

slithering, *ppl. a.* (Earlier mod. example.)

1840 M. EDGEWORTH *Let.* 30 Dec. (1971) 573 Not one name when introduced had I been able to make out from Mrs. Hollands slithering pronunciation.

slithery, *a.* For Chiefly *dial.* read orig. *dial.* and add: **a.** (Later examples).

1922 E. R. EDDISON *Worm* xxvi. 332 With rocks and pits hidden in the heather, and slithery slabs of granite. 1946 M. PEAKE *Titus Groan* 298 His lordship..was gazing at his daughter with a slithery smile upon his mouth that had once been so finely drawn. 1952 L. A. G. STRONG *Darling Tom* i. 12 Against the wall stood four slithery horse-hair chairs. 1975 E. HUXLEY *Gallipot Eyes* (1976) 168 Oaksey Wood is a morass with yellowish water lying deep in slithery tractor ruts.

b. *Comb.*, as *slithery-eyed, -slobbery* adjs.

1921 D. H. LAWRENCE *Sea & Sardinia* iii. 123 Give me the old, salty way of love. How I am nauseated with sentiment and nobility, the macaroni slithery-slobbery mess of modern adorations. 1977 *New Yorker* 8 Aug. 11/1 He picks out the very lad:..a twin..identical to the slithery-eyed boy destined to grow up into Marty Feldman.

slithy (slai·ði), *a.* Also † **slythy.** [Presumably a blend of SLIMY *a.* and LITHE *a.*] A word invented by 'Lewis Carroll': 'smooth and active' ('Carroll', 1855, 140) and popularized esp. in phr. *slithy toves* from *Through the Looking-Glass* (1871). Also in subsequent allusive uses.

1855 'L. CARROLL' *Rectory Umbrella & Mischmasch* (1932) 139 Twas bryllyg and the slythy [1871: slithy] toves Did gyre and gymble in the wabe. 1920 'K. MANSFIELD' *Let.* 27 Sept. (1928) II. 48, I watched him [*sc.* a lizard] come forth to-day—*very* slithy—and eat an ant. 1928 A. S. EDDINGTON *Nature of Physical World* xiii. 291 Eight slithy toves gyre and gimble in the oxygen wabe; seven in nitrogen. 1937 G. FRANKAU *More of Us* 2 While the free-versifier gyres and gimbles The slithy tove—with his own 'private symbols'. 1960 H. MARCHAND *Categories* x. 368 Lewis Carroll's *slithy..., chortle*..have become common property. Shakespeare's *glaze* (f. *glare* and *gaze*) has not. 1981 *Time Out* 20–26 Mar. 54/1 Pity the slithy toves of academe.

sli·tted, *a. rare.* [f. SLIT *sb.* + -ED[2].] Having a slit or slits; shaped like a slit.

1936 B. BROOKER *Think of Earth* I. i. 20 Stale fumes of beer..were belched at the Canon as he passed the slitted swing doors of the bar. 1947 W. STEVENS *Transport to Summer* 139 A face of stone in an unending red, Red-emerald, red-slitted-blue, a face of slate. 1974 R. ADAMS *Shardik* xxiv. 184 The window-openings, rounded and slitted like key-holes, which lit the spiral stairways.

slitty (sli·ti), *a.* [f. SLIT *sb.* + -Y[1].] Of the eyes: long and narrow. Also in *Comb.*, as *slitty-eyed*, slit-eyed.

1908 R. BROUGHTON *Mamma* x. 104 Her slitty eyes, opened so wide as almost to look large. *Ibid.* xxiii. 218 Her slitty eyes dancing with mirth and benevolence. 1926 G. FRANKAU *My Unsentimental Journey* 276 Where the slitty-eyed Chinese are. 1947 M. MORRIS *Township* 9 She was a sturdy young woman.., with..bright, slitty black eyes. 1976 *Listener* 18 Mar. 324/1 The English couple who turned down the Thai baby that was procured for them 'because its eyes are too slitty'.

sliver, *sb.*[1] Add: **1. a.** (Later *fig.* examples.)

1967 T. KINSELLA *Nightwalker* I. 5 Bone-splinters, silvery slivers of screams. 1978 J. CARROLL *Mortal Friends* III. vi. 327 People on buses and on the streets of Boston traded slivers of information as if they were coins.

c. (Earlier example.)

1869 *Maine Acts & Resolves* 24 Any pumice, scraps or other offal arising from the making of oil or slivers for bait.

sliving, *ppl. a.* (Later examples.)

c 1909 D. H. LAWRENCE *Collier's Friday Night* (1934) iii. 75 I'm not a fool, if you think so. I can pay you yet, you sliving bitch! 1913 —— *Sons & Lovers* iv. 66 What should go runnin' up my arm but a mouse... They'll get in your pocket an' eat your snap, if you'll let 'em..the slivin', nibblin' little nuisances.

slivovitz (sli·vǒvits). Also **sliwowitz;** ‖ **slivovic(e), slivovitza, sljivovica.** [ad. G. *slibowitz*, f. Bulg. *slivovitza*, Serbo-Croat *šljivovica* plum brandy, f. OSl. *sliva* plum; cf. SLOE.] A central and eastern European plum brandy.

α. 1885 *Encycl. Brit.* XVIII. 692/2 Other important articles of commerce are wine, wool, cattle, timber, hides, honey, wax and 'slivovitza', an inferior spirit made from plums. 1938 C. MORGAN *Flashing Stream* I. i. 72 They have a drink called Šljivovica. It's a kind of plum-brandy. 1950 [see *plum brandy* s.v. *PLUM *sb.* 6]. 1973 S. JACKMAN *Guns covered with Flowers* vi. 98 They drank two small glasses of sljivovica.

β. 1897 B. STOKER *Dracula* i. 11 There is a flask of slivovitz..underneath the seat, if you should require it. 1900 'ODYSSEUS' *Turkey in Europe* viii. 374 The plum-trees [of Northern Servia] that produce the national drink of slivovitz, or plum-brandy. 1940 M. HEALY *Stay me with Flagons* 273 Kirsch and Kirschwasser..never have appealed to me; and Slivovitz I have never drunk, I think. 1951 F. BROWN *Murder can be Fun* viii. 104 What are you drinking? Me, I like slivovitz. 1960 A. E. BENDER *Dict. Nutrition* 114/2 *Sliwowitz*, plum brandy originating in Yugoslavia. 1960 R. ST. JOHN *Foreign Correspondent* ix. 192 There was a young German in his compartment who had drunk too deeply of Bulgarian slivovitz. 1976 *New Yorker* 22 Mar. 47/1 He sits by the kitchen window of his little flat,..drinking the slivovitz he smuggles into Sweden each September in carefully emptied beer bottles.

γ. 1958 P. KEMP *No Colours or Crest* viii. 172 A flask of excellent Prizren slivovic. 1961 E. WAUGH *Unconditional Surrender* III. ii. 226 Once, after a jack-pot, he was offered a glass of Slivovic. 1971 *Southerly* XXXI. 17 Here's to scotch, to bourbon, to slivovic, to uyzo.

Sloane Ranger (slōᵘn rēⁱ·ndʒəɹ), *sb.* and *adj. phr.* [Blend of *Sloane* Square, London, and *Lone Ranger*, a well-known hero of western stories and films.] (Of, pertaining to, or characteristic of) a fashionable but upper class and conventional young woman in London. Also occas. extended to any member of the class to which such young women belong, and *ellipt.* as *Sloane*. Hence **Sloa·neness.**

1975 P. YORK in *Harpers & Queen* Oct. 190/3 The Sloane Rangers..are the nicest British Girl. *Ibid.* 191/2 The Sloane Rangers always *add tone*. They never put on prole accents, like self-conscious Oxford boys in the sixties. *Ibid.* 191/3 Once a Sloane marries and moves to Kennington and starts learning sociology through the Open University, she is off the rails. *Ibid.*, Sloaneness, some people would say, is a track to be liberated from

Ibid. 192/3 Sloane Ranger pet hates..incense, Norman Mailer. 1978 *Evening Standard* 21 Aug. 13/2 A way of life neither Mayfair, nor West End nor Sloane Ranger, but which is summed up in the words Cafe Society. 1978 D. MACKENZIE *Deep, Dark & Dangerous* i. 55 Emma.. was a hell of a lot different to the succession of Sloane Rangers who had been her predecessors, harpies bent on getting far more than they gave. 1980 S. ALLAN *Dead Giveaway* xv. 154 She wore a cashmere sweater..a Sloane ranger type. 1981 J. MANN *Funeral Sites* xviii. 111 The all-English Phoebe with her Sloane Ranger voice and manners. 1982 BARR & YORK *Official Sloane Ranger Handbk.* 10/1 Sloane Rangers hesitate to use the term 'breeding' now (of people, not animals) but that's what background means. 1983 *Times* 16 Apr. 3/7 (*headline*) Bogus Sloane Ranger lived like a lord. *Ibid.* 3/8 He even emulated the voice of those known as Sloane Rangers and men about town.

slob, *sb.*[1] Add: **1. d.** *Canad.* = *slob-ice*, sense 4 below.

1878 *North Star* (St. John's, Newfoundland) 30 Mar. 3/1 The bay here was caught over last week, and a string of 'slob' made its appearance across the mouth, but the heavy sea of Thursday broke it all up. 1907 J. G. MILLAIS *Newfoundland* ii. 44 They themselves had hooked seventeen white coats out o' the slob (shore ice). 1920 W. T. GRENFELL *Labrador Doctor* ix. 174 This ice is of very different qualities. Now it is 'slob' mixed with snow born on the Newfoundland coast. 1951 *Beaver* Sept. 20 Wind half a gale, temperature away down and slob in harbour and around schooner turning to ice.

3. (Further examples.) Also, a lout, a fat person; one who is gullible or excessively soft-hearted, a fool; a person of little account. *slang.*

1904 'No. 1500' *Life in Sing Sing* 252/2 *Slob*, a person easy to impose upon; an untidy person. 1904 G. V. HOBART *Jim Hickey* i. 16 You're a warm young guy When you start to buy—You're a slob when you lose the price! 1927 H. V. MORTON *In Search of England* x. 185 He was no beauty to look at, but then women seem to like the ugly slobs, don't they? 1938 R. FLANNAGAN *County Court* 36 That praying old slob, Jones, has three boys and every one of 'em has run away. 1950 WODEHOUSE *Nothing Serious* 29 'The poor old slob,' she murmured. 1953 *If; Worlds of Sci. Fiction* Sept. 40/1 Speaking..as an ordinary slob that doesn't follow rarefied reasoning very well. 1958 S. ELLIN *Eighth Circle* II. xv. 123 A big, fat, gutless slob. 1959 I. & P. OPIE *Lore & Lang. Schoolch.* ix. 168 The unfortunate fat boy..is known as:..slob, [etc.]. 1960 'E. MCBAIN' *Heckler* v. 60 There are people..who always look like slobs... The tendency toward sloppiness first exhibits itself when the subject is still a child. 1966 A. CAVANAUGH *Children are Gone* II. vi. 49 'I'm a slob,' Shirley said. 'I'm not an intellectual.' 1970 R. PRICE *Gt. Roob Revolution* 7 The hucksters who control the..mass media think they are manipulating what they refer to as 'the slobs'. 1972 T. P. MCMAHON *Issue of Bishop's Blood* (1973) ii. 17 He's a real slob for his employees. He buys them houses, goes to their bar mitzvahs. 1978 J. IRVING *World according to Garp* ix. 184, I think you're an irresponsible slob.

4. slob ice orig. *Canad.*, densely-packed, sludgy ice, esp. sea ice; **slob trout,** a brown trout, *Salmo trutta*, which stays in a river estuary instead of going further out to sea.

1835 E. WIX *Six Months Newfoundland Missionary's Jrnl.* (1836) 16, I crossed through the 'slob ice', which was very thick in Conception Bay, to Port de Grave. 1920 W. T. GRENFELL *Labrador Doctor* vi. 132 The slob ice had already made ballicaters and the biting cold of winter so far north had set in with all its vigour. 1955 *Sci. Amer.* Apr. 52/3 On the way to Little America, its first Antarctic port of call, the Atka saw very little of the drifting ice pack that surrounds the continent. It passed through a few 'bergy bits' and pieces of 'slob ice'—melting remnants of the pack. 1965 F. RUSSELL *Secret Islands* vii. 88 The island was isolated because it was surrounded by impassable slob ice. 1907 W. L. CALDERWOOD *Life of Salmon* i. 6 In the West of Ireland we have.. the so-called slob trout. 1930 G. H. NALL *Life Sea Trout* vi. 75 These brown Trout, feeding in brackish and salt water, are numerous, and special names have been given to them, such as 'Slob Trout', and 'Estuarine Trout'. 1960 C. WILLOCK *Angler's Encycl.* 190/1 Slob-trout: sometimes called bull trout, are brown trout that migrate only as far as the estuary.

slob (slǫb), *v.* [f. SLOB *sb.*[1]] *trans.* To slop (*out*); to express by slobbering. So **slo·bbed** *ppl. a.*

1887 PARISH & SHAW *Dict. Kentish Dial.* 152 *Slobbed*, slopped, spilt. *a* 1918 W. OWEN *Poems* (1963) 69 Drooping tongues from jaws that slob their relish. 1946 B. MARSHALL *George Brown's Schooldays* 56 The master began to slob out the tapioca.

slobber, *v.* Add: **1. d.** fig. *to slobber over*, to be over-attentive or over-affectionate towards (someone); to be exaggeratedly enthusiastic about (something).

1825 [in Dict., sense 1 b]. 1892 'MARK TWAIN' *Amer. Claim.* xiii. 139 They treat you as a tramp until they find out you're a congressman, and then they slobber all over you. 1914 W. OWEN *Let.* 28 Aug. (1967) 282 He received me like a lover. To use an expression of the Rev. H. Wigan's, he quite slobbered over me. 1927 D. L. SAYERS *Unnatural Death* III. xxii. 257 Miss Climpson had little difficulty in reconstructing one of those hateful and passionate 'scenes' of slighted jealousy... 'I do everything for you—you don't care a bit for me..!' And 'Don't be.. ridiculous... Oh, stop it, Vera! I hate being slobbered over.' 1978 P. THEROUX *Picture Palace* viii. 50 Even if they had slobbered over every blessed picture in the place they would not have understood.

slobberhannes (slǫ·bəɹhænès). [Origin unknown. Perh. ad. Du.: cf. Du. dial. *slabberjan* the name of a game, and Du. *Hannes* Jack.] A card-game for four persons played with only high-ranking cards, in which the object is to lose tricks. Also, a point scored at this game.

1877 W. B. DICK *Mod. Pocket Hoyle* (ed. 8) 211 *Slobberhannes*... The object of the game is to *avoid* making points, as the player who first gets ten points loses the game. *Ibid.* 212 If a player scores *all* of the three foregoing points, he receives one point extra, which is called 'Slobberhannes'. 1952 J. B. PICK *Phoenix Dict. Games* 276 *Slobberhannes* (4 players). Played with a pack from which all cards below seven have been withdrawn. 1964 A. WYKES *Gambling* vii. 165 *Greek hearts*..and *slobberhannes* are slightly simpler variations on the trick-losing theme.

slobby, *a.* Add: **2. a.** Sloppy, sentimental. **b.** Of or pertaining to a slob (SLOB *sb.*[1] 3 in Dict. and Suppl.).

1913 R. BROOKE *Let.* 3 July (1968) 479, I had a bad fit of home-sickness this morning... I threw up quite a lot of slobby old memories. 1967 *Spectator* 4 Aug. 131/1 To be honest, backbone isn't, as it were, at the forefront of my character. I am pretty slobby within. 1970 W. BURROUGHS JR. *Speed* (1971) v. 108 Vicki told me that I looked like a slobby bum. 1976 W. GOLDMAN *Magic* III. ix. 181 She'd end up stranded some place maybe with..some slobby dummy. 1976 *New Yorker* 9 Feb. 84/3 Peter Boyle's role is small, but..he does slobby wonders with his scenes as the gently thick Wizard.

slobgollion (slǫbgǫ·lyən). *Whaling slang.* [Origin unknown: cf. SLUMGULLION.] A substance found in sperm-oil (see quot.).

1851 H. MELVILLE *Moby Dick* III. viii. 65 It is called slobgollion; an appellation original with the whalemen... It is an ineffably oozy, stringy affair, most frequently found in the tubs of sperm, after a profound squeezing, and subsequent decanting.

sloe. Add: **3.** *sloe eye.*

1957 V. J. KEHOE *Technique Film & T.V. Make-Up* ix. 107 (*caption*) Effect of sloe eye after applying latex to outer ends of lashes and eyelid and pressing them together. 1977 N. MARSH *Last Ditch* ii. 37 His sloe eyes looked out of a pale face.

slog, *sb.* **2.** (Earlier examples.)

1846 *Swell's Night Guide* 76 And she felt inclined to mug her rival, only she thought it would be no bottle, cos her rival could go in a buster at a slog. 1865 *Lillywhite's Cricketers' Compan.* 139 Too fond of losing his wicket for a 'slog'.

slog, *v.* Add: **1. a.** (Earlier example.)

1824 *Session Papers Central Criminal Court* 21 Sept. 535/1 One of them said, 'Go back and slog him.'

3. a. (Earlier example.)

1846 *Swell's Night Guide* 37 Most of them can slog, that is to say,..fight.

b. *Cricket.* To hit, or attempt to hit, the ball hard and with abandon.

1869 *Baily's Mag.* July 21 Not only did he 'slog', in the true sense of the word, which we take to be hitting blindly and high in the air, but [etc.]. 1904 F. C. HOLLAND *Cricket* 36 You should go to the nets, not to slog, but to play. 1935 J. C. MASTERMAN *Fate cannot harm Me* viii. 167 At the fifth ball the Admiral slogged with even crookeder bat and even more mighty effort; he missed it, and all three stumps were spreadeagled. 1980 *Cricketer International* Feb. 11/1 The incredible thing is that he never had to slog once to make his runs.

slogan. Add: **1. b.** (Further examples.)

1922 *Times* 20 June 7/4 'Post early.' New P.O. slogan on letters. 1928 *Publishers' Weekly* 9 June 2386 As an advertising man, Mr. Calkins believes the slogan 'a cent a copy to sell the art of reading', a great and revolutionary one. 1951 H. ARENDT *Burden of Our Time* I. ii. 38 Antisemitic slogans were highly effective in mobilizing large strata of the population. 1958 P. H. GIBBS *Curtains of Yesterday* xix. 157 On the other side [of an ancient gateway] with big letters deeply carved was the new slogan of Lenin's Russia. 'Religion is the Opium of the People.' 1968 V. S. PRITCHETT *Cab at Door* ix. 163 All sects have their jargon and Father, eager as an advertising man is for slogans, had picked them all up and lived by them. 1971 H. MACMILLAN *Riding the Storm* xv. 478 The somewhat disingenuous slogan of 'ban the bomb'. 1972 F. FITZGERALD *Fire in Lake* viii. 277 Thousands of soldiers and civil servants marched with the dock workers shouting anti-government and occasionally anti-American slogans. 1980 R. SCRUTON *Meaning of Conservatism* iii. 59 One particular slogan will later occupy our attention—'equality of opportunity'.

2. Also *Comb.*, as *slogan-shouter, -shouting* vbl. sb. and ppl. adj.; *slogan-like* adj.

1936 WIRTH & SHILS tr. *Mannheim's Ideology & Utopia* I. 36 Two slogan-like concepts 'ideology and utopia'. 1975 *Language for Life* (Dept. Educ. & Sci.) xiii. 199 Public debate..has often been conducted through a series of slogan-like headings: progressive, formal, integration, basics, and several more. 1940 G. CUNNINGHAM *Jrnl.* 14 Dec. in N. Mitchell *Sir G. Cunningham* (1968) iv. 83 Heard from Peshawar that slogan shouters had been told to go home. 1968 N. MITCHELL *Ibid.*, On the 12th, the slogan shouters sent notices to the Deputy Commissioner where and when they would shout their slogans on 14th December. 1940 G. CUNNINGHAM *Jrnl.* 24 Dec. in *Ibid.* 84 Slogan shouting by Satyagrahis has stopped as Gandhi has declared a holiday for Christmas. 1972 J. BIGGS-DAVISON *Africa—Hope Deferred* x. 98 These are but a few of the problems facing independent Africa that can no longer be concealed by an excess of slogan-shouting. 1976 M. ZIA-UD-DIN *Memoirs* 63 Wherever it stopped there were large crowds of slogan-shouting Muslims. 1978 *Jrnl. R. Soc. Arts* CXXVI. 696/2 *Medical Inspection* destroys the 'nobility' of war more readily than 1,000 slogan-shouting demonstrators.

Hence **slo·ganed** *a.*, marked with a slogan.

1966 'G. DOUGLAS' *Odd Woman Out* ix. 60 This tall girl ..in jeans and a sloganed sweater. 1978 *Church Times* 1 Sept. 15/2 With..leather waistcoats or sloganed T-shirts. 1979 *Sci. Amer.* Apr. 30/1 A graffiti remover which was developed by the organic chemistry section in response to a government campaign to clean-up the much-sloganed areas of Belfast and Londonderry.

sloganeer (slōᵘgănīᵊ·ɹ). orig. *U.S.* [f. SLOGAN + -EER.] One who devises or who uses slogans.

1922 R. CONNELL in *Sat. Even. Post* (U.S.) 29 Apr. 100/2 (*heading*) Once a sloganeer. 1935 *Sat. Rev. Lit.* 11 May 30/3 The day may come when a West Coast sloganeer will proudly proclaim, 'If it isn't at San Marino, it isn't a book.' 1963 D. OGILVY *Confessions Advertising Man* vii. 127 Posters are for sloganeers. 1971 *N.Z. Listener* 31 May 5 'Sloganeers'—young, sometimes older people, who do not analyse a problem but pick up a current catchcry. 1978 *Times* 7 Aug. 12/4 Questions are a favourite device of envelope sloganeers. 'Is he to be our next President?' asked one.

Hence as *v. intr.*, to express oneself in slogans (now usu. in a political context); **sloganee·ring** *vbl. sb.* and *ppl. a.*

1941 H. S. JOHNSON *Hell-Bent for War* ii. 37 In this modern sloganeering day,..the constant repetition of a lie has become the..weapon of the totalitarian propagandist. *Ibid.* iv. 85 We are..getting all ready to do it all over again with hardly a variation in timing sequence or superficial sloganeering. 1944 *Sun* (Baltimore) 18 Mar. 6/6 To speak of the German dead in terms of carpets is not to exaggerate or 'sloganeer'. 1949 *Ibid.* 13 Oct. 18/3 Eastern Germany's tireless Communists, still a bit breathless from the ten-day marathon of sloganeering over the new 'East German Republic', [etc.]. 1967 *Philos. Rev.* LXXVI. 105 An area where superficiality and sloganeering too often hold sway. 1970 K. MILLETT *Sexual Politics* III. v. 265 What she does 'become' is only a nonentity, utterly incorporated into Birkin, his single follower, proselytizing and sloganeering. 1978 *New Statesman* 27 Oct. 556/3 The islanders have learnt to deploy the bullying sloganeering and empty hard sell of 'Westminster' politics. 1981 *Encounter* Apr. 48/2 To distinguish truth from sloganeering licence and exaggeration.

sloganize (slōᵘ·gănaiz), *v.* [f. SLOGAN + -IZE.] **1.** *trans.* **a.** To make (something) the subject of a slogan; to express in a slogan.

1929 *Bull. Amer. Library Assoc.* Apr. 70 Infected with the economy virus sloganized by recent high national executives. 1941 AUDEN *New Year Let.* II. 39 Round a provincial régime that sloganized the Rights of Man. 1949 *Archit. Rev.* CV. 96/2 To sloganize the present trend in Sweden: 'Personal, organic creation in a harmonious democracy'—which should be vague enough to please everybody. 1981 *R. United Services Inst. Jrnl.* June 7/1 If you try to sloganise it I think it is quite dangerous.

b. To influence by means of slogans.

1954 *ADA World* Feb. 1M/1 Of all the economic prescriptions that have been written in Washington in the past 25 years, none is stranger or more dangerous than the current attempt to sloganize the country out of an economic decline.

2. *intr.* To compose slogans, to utter slogans.

1960 *Encounter* June 38/1 Mao's ability to sloganise and reduce complex matters to simple formulas has been one of his key assets. 1975 *Daily Tel.* 24 June 11/2 The conspirators met in a pub and sloganised about tyrants.

Hence **slo·ganized** *ppl. a.*, expressed in the form of a slogan or slogans; **slo·ganizer,** one who uses slogans; **slo·ganizing** *vbl. sb.* and *ppl. a.*

1940 *New Statesman* 2 Mar. 274/1 Snappy presentation and slick sloganising. 1965 *Ibid.* 29 Oct. 660/3 This could become mere sententiousness, the jingling or sloganising final flourish in a poem like 'Waiting'. 1970 *Times* 31 Mar. (Austral. Suppl.) p. ii/7 Throughout the 1950s and early 1960s Mr. (later Sir) Robert Menzies encapsulated this *weltanschauung* in the sloganized search for 'great and powerful friends'. 1974 *Daily Tel.* 27 June 18 The cliché-ridden sloganisers of the extreme Left. 1975 R. LEWIS *M. Thatcher* v. 42 The manifesto gave an easy sloganised summary of the 'action' to which the Party was committing itself. 1981 *Times Lit. Suppl.* 15 May 548/1 The success of 'All You Need is Love'..encouraged the Beatles to try their hand at sloganizing.

slogger, *sb.* Add: **2. a.** (Earlier example.) Also, a heavy blow.

1829 P. EGAN *Boxiana* 2nd Ser. II. 19 He got away from a *slogger*, but immediately commenced an exchange of blows. 1846 *Swell's Night Guide* 75 No one dares to dispute the ability of the boshman; 'cos he's..a numming slogger.

b. (Earlier example.)

1850 *County Herald* 31 Aug. 7/3 Who went in a slogger, scoring two's and three's, till the scorers called game.

c. A person or machine that works hard or with effort, often with a suggestion of ponderousness or lack of sparkle.

1928 G. B. SHAW *Intelligent Woman's Guide* xlix. 208 The employers, to find out how much work can be got out of a man, pick out an exceptionally quick and indefatigable man called a slogger. 1968 J. SANGSTER *Touchfeather* xiii. 138 Bill was a bloody genius... Harvey's a

slogger. It takes him a year to arrive where Bill could in twenty-four hours. 1977 *Drive* Mar.–Apr. 56/2 The Manta's 1.9 engine is a solid, reliable slogger in the General Motors tradition.

sloggering, *ppl. a.* Add: **b.** (See quot. 1977.) *rare*[−1].

1876 G. M. HOPKINS *Wr. Deutschland* in *Poems* (1967) 57 The inboard seas run swirling and hawling; The rash smart sloggering brine Blinds her. 1977 J. MILROY *Lang. G. M. Hopkins* 245 *Sloggering*..: Certainly imit., belonging to phonaesthetic series (*slither*, etc.), a derivative (*slog*, etc.), and a blend... Complex associations suggest the meaning: 'dashing (against the ship) repeatedly and drawing back with a sucking gurgling noise'.

slogging, *vbl. sb.* Add: (Earlier and later examples.) Also *attrib.*

1857 *Bell's Life* 31 May 6/4 Nor ought we to omit notice of Reynolds's 15 by a kind of paralytic slogging, which *sometimes* tells. 1934 [see *RUN *sb.*[1] 2 a]. 1958 F. C. AVIS *Boxing Ref. Dict.* 103 *Slogging match*, a contest in which both men hit each other heavily and very often.

sloke[2]. (Earlier example.)

1788 *Asiatick Researches* I. 127 Thus speak the following *Slokes* from the *Dhormo Onoosaason*.

sloop, *sb.*[1] **3.** *sloop-man* (earlier examples).

1676 T. GLOVER *Acct. of Virginia* in *Phil. Trans. R. Soc.* XI. 625 The sloop man dropped his grap-line. 1737 *Calendar Virginia State Papers* (1875) I. 229 Paid a Sloop man for 2 gals of rum.

sloosh (slūʃ), *sb. dial.* and *colloq.* [Echoic: cf. SLOSH *sb.*, SLUSH *sb.*[2] But perhaps partly a variant of SLUICE *sb.*] A pouring of water; a wash; a noise of, or as of, heavily splashing or rushing water.

1919 *Athenæum* 11 July 582/2 Among the brand-new slang one may discern some that had an onomatopœic or at any rate an imitative origin; for instance 'sloosh', a wash. 1920 *Blackw. Mag.* Apr. 500/2 There was the sloosh of bilge-water. 1926 W. DE LA MARE *Connoisseur* 187 Mr. Thripp..not only tidied up his own and Tilda's bedroom..but even gave a sloosh to the bath. 1973 C. BONINGTON *Next Horizon* v. 88 He got about half-way up, and suddenly there was a sloosh, and he came shooting down the steep snow. 1981 P. THEROUX *Mosquito Coast* xiv. 173 The only sounds were the flap and splash..and the sloosh of water in the culverts.

sloosh (slūʃ), *v. dial.* and *colloq.* [Echoic: cf. SLOSH *v.*[1], SLUSH *v.* But perhaps partly a variant of SLUICE *v.*] **1.** *trans.* **a.** To wash with a copious supply of water; to pour water or other liquid copiously over.

1912 W. DE LA MARE *Child's Day* 10 Elizabeth Ann.. stands slooshing herself With that 'normous sponge. 1933 G. MURRAY tr. *Acharnians* in *Aristophanes* i. 27 Niagara'd me and slooshed me, till—almost—In so much sewage I gave up the ghost.

b. To pour with a rush, to dash (water).

1952 L. A. G. STRONG *Darling Tom* 129 The butcher's boy in his blue Sunday suit had got the gardener's wheeled tank from the park, and was slooshing water down the area.

2. *intr.* To make a heavy splashing or rushing noise; to flow or pour with a rush.

1914 [see *CLICKY *a.*]. 1920 *Blackw. Mag.* Apr. 502/2 The slooshing bilge-water. 1971 C. BONINGTON *Annapurna South Face* xii. 153 Tom knocked it [*sc.* the primus stove] over, sending the liquid slooshing over the tent floor. *Ibid.* xiii. 167 Instant porridge...a hot, palatable food that simply slooshed down the throat. 1981 P. THEROUX *Mosquito Coast* xxvi. 345 She gazed at the torrent of water slooshing downstream.

So **sloo·shy** *v. trans.* = *SLOOSH *v.* 1 a.

1907 W. DE MORGAN *Alice-for-Short* xlix. 531 But Cook was turning cataracts of water into her sink, to slooshy it well out after a real good wash-up.

sloothering (slū·ðəriŋ), *vbl. sb. Anglo-Ir.* Also **sluthering.** [Perh. ad. Ir. *lúdar* fawning, flattery with prosthetic *s-* (A. J. Bliss).] Cajoling, wheedling. Hence as *ppl. a.*

1892 J. BARLOW *Irish Idylls* viii. 215 There do be girls will get round a man wid their slootherin'. 1896 G. B. SHAW in *Sat. Rev.* 1 Feb. 123/1 Boucicault had a charming brogue: not even the speech of the eminent journalist.. is more musical in sound or irresistible in insinuation—'sloothering' would be the right word, were it current here—than his. 1901 J. BARLOW *From Land of Shamrock* 79 Coaxing and sluthering, reinforced by a couple of.. florins. 1919 G. B. SHAW *O'Flaherty V.C.* 188 You foul-mouthed, dirty minded, lying, sloothering old sow. 1922 JOYCE *Ulysses* 756 He used to amuse me the things he said with the half sloothering smile on him.

slop, *sb.*[1] Add: **7. b.** *slop-clothing* (earlier example).

1802 D. COLLINS *Acct. Eng. Colony in New S. Wales* II. xiii. 131 For want of slop clothing and bedding, indeed, they were much distressed.

c. slop-chit *Naut.*, a note offered at a ship's stores in exchange for clothing; also *transf.*, an expense sheet.

1946 'TACKLINE' *You met such Nice Girls in Wrens* i. 11 Sailors come in at different times with what we call Slop-Chits, but which are really nothing but shopping lists. So the slop-chits are the things they want, such as shoes and socks and caps and things. 1969 S. HYLAND *Top Bloody Secret* II. 123 Superb autobahn all the way, and a reasonably fast car on his slopchit, made Karlsruhe almost one of the nearer suburbs.

slop, *sb.*[2] Add: **2. b.** *Naut.* A choppy sea, chop.

1956 *Sun* (Baltimore) 21 Nov. 23/6 Mostly southerlies, they pick up to twenty knots and kick up a good slop. 1974 F. MOWAT *Boat who wouldn't Float* xi. 120 A slop was banging against her bows, giving her life and motion. 1977 *Austral. Sailing* Jan. 65/1 The Tornados and Hobie 16s revelled in the slop.

3. b. *colloq.* (orig. *U.S.*). Sentimentality, affected sensibility.

1866 'MARK TWAIN' *Lett. from Hawaii* (1967) 33 You can go on writing that slop about balmy breezes and fragrant flowers, and all that sort of truck. 1917 E. POUND *Let.* Mar. (1971) 108, I would suggest that a series of this sort [*sc.* essays on French poets] by me, Eliot, and De Bosschère would at least keep out a certain amount of *slop* from the prose section. 1924 GALSWORTHY *White Monkey* I. xiii. 106 Sentiment being 'slop', and championship mere condescension. 1927 *Sunday Express* 24 July 4 'Seventh Heaven', the swamp of sentiment into which the critics were invited to plunge a few days ago. Personally I should describe it as the sublimity of slop. 1942 BERREY & VAN DEN BARK *Amer. Thes. Slang* § 265/1 *Sentimentality*,..slop.

c. *U.S.* and *Austral. slang.* Beer. Usu. *pl.*

1904 'No. 1500' *Life in Sing Sing* 252/2 *Slop*, beer. 1919 *Dial. Notes* V. 42 *Slops*, beer. 1949 L. GLASSOP *Lucky Palmer* i. 5 Keep your shirt on. There's no harm in having a jug of slops, now is there? 1953 T. A. G. HUNGERFORD *Riverslake* x. 197 His wife and both of his kids got burned to death when his house went up... They reckon that's what sent him onto the slops in the first place. 1963 *Australasian Post* 14 Mar. 51/2 Bung me and me mate over a droppa slops, will yer love?

4. b. *U.S. dial.* and *colloq.* Kitchen refuse or swill fed to cattle or pigs. Usu. *pl.*

1805 R. PARKINSON *Tour in America* I. 39 It was natural for me to inquire, what they kept their cows and horses on during the winter. They told me—their horses on blades, and their cows on slops. 1912 T. DREISER *Financier* xii. 127 A slop-man,..who could come with a great wagon filled with barrels and haul away the slops from your back door, was absolutely essential. 1961 *Publ. Amer. Dial. Soc.* XXXVI. 7 *Slop*, food for pigs.

c. *fig.* Nonsense, rubbish; insolence.

1952 B. MALAMUD *Natural* 214 Roy tore it up and told the usher to take no more slop from him. 1978 J. IRVING *World according to Garp* vi. 120 'Sometimes I feel it is my responsibility to say no,' the editor was quoted as saying, 'even if I know people *do* want to read this slop.'

5*. A dance (see quot. 1962).

1962 *Punch* 16 May 761/1 In the Slop the partners face each other and perform rhythmic movements with feet stationary, the arms swinging pendulum-like in front of the body. 1966 T. PYNCHON *Crying of Lot 49* v. 131 Each couple on the floor danced whatever was in the fellow's head: tango, two-step, bossa nova, slop. 1969 N. COHN *Pop from Beginning* ix. 84 There was the Hully Gully,..the Slop and..the Frug.

6. *slop-barrel* (earlier example), *-bowl* (earlier examples), *-can, -pail* (earlier examples), *-tank, -tub*; **slop-stone** *dial.*, a stone slab used as a surface for washing.

1831 J. M. PECK *Guide for Emigrants* 172 With..a dairy and slop barrel..pork may be raised from the sow. 1810 *Columbian Centinel* 25 Aug. 4/2 For sale at Davis & Brown's Silver Ware and Jewelry Store..Sugar Basons, ..Slop-Bowls. 1861 TROLLOPE *Orley Farm* (1862) I. xi. 85 A small pile of buttered toast on the slop-bowl, kept warm by hot water below. 1926 *Scribner's Mag.* Aug. 204/1 A strange black dog..supporting himself by raiding the slop-cans of Nigger Town. 1854 B. P. SHILLABER *Life & Sayings Mrs. Partington* 212 My boy knows very well how to manage it when the slop-pail is within reach. 1860 F. NIGHTINGALE *Notes on Nursing* 14 A slop-pail should never be brought into a sick room. 1882 NODAL & MILNER *Gloss. Lancs. Dial.* 245 *Slopstone*, a place for washing. 1911 A. BENNETT *Card* viii. 197 A gas range and a marble slopstone with two taps. 1978 *Lancashire Life* Oct. 83/2 Ah want thad slop-stooan scrubbin'. 1968 *New Scientist* 25 Jan. 196/2 One of the ship's tanks is selected to serve as a slop tank... The tanks to be cleaned are washed in turn and washings are pumped continuously into the top of the slop tank. 1979 F. FORSYTH *Devil's Alternative* ix. 223 She had sixty giant tanks, or holds... One of these was the slop tank, to be used for..gathering the slops from her fifty crude-carrying cargo tanks. 1867 'MARK TWAIN' *Celebrated Jumping Frog* 169 You will proceed toward the window and sit down in that slop-tub.

slop, *v.*[2] Add: **1. c.** *Prison slang.* To empty the contents of (esp. a chamber-pot). Usu. *absol.*

1955 [implied at *SLOPPING *vbl. sb.* b]. 1957 *Listener* 28 Nov. 893/3 He watched the prisoners 'slopping out', working, and attending classes. 1963 T. & P. MORRIS *Pentonville* v. 104 The next two hours are given over to getting washed, shaved, breakfasted and 'slopped out'. 1967 *Guardian* 2 June 6/1 Along the landing from his cell is a single recess for 'slopping out' chamber pots and wash-basins. 1973 *Times* 20 Dec. 2/3 At Brixton..they queue at communal lavatories to slop out their pots. 1978 'A. GARVE' *Counterstroke* I. xvi. 67 Prisoners rise at 6.30 a.m. Slop out. Clean their cells.

4. b. With *up*: to become intoxicated. *U.S. slang.*

1899 [implied at *slopping-up* s.v. *SLOPPING *vbl. sb.* b]. 1916 'W. SCOTT' *Seventeen Years in Underworld* xi. 64 The illgotten gains are spent 'slopping up' (getting drunk). 1919 *Bookman* (N.Y.) Apr. 208/2 Discuss the effect of the Prohibition Amendment on a white liner. What would be his chances after its passage of procuring sufficient powders to enable him to get slopped up (a) in the State of Maine, (b) in New York City? 1926 J. BLACK *You can't Win* ix. 108 No use takin' a bunch of thirsty bums along and stealin' money for them to slop up in some saloon the next day.

5. b. *colloq.* With *about, around*, etc.: to wander aimlessly, to move in a slovenly manner; to mess *about*.

1907 W. DE MORGAN *Alice-for-Short* xlv. 471 Old pictures do slop about the world in a vague way, till some aesthomenous person detects quality in them. 1922 M. ARLEN *'Piracy'* III. xi. 236 When I said that you were too fine and I was too old to slop about Europe in a hole-and-corner way, I meant that this disorderly kind of life is unworthy of you. 1958 X. FIELDING *Corsair Country* i. 22 So many of them slop about in cast-off men's shoes several sizes too big. 1973 *Times* 28 Nov. 13/6 Jimmy always says I mustn't slop about while I'm learning a piece, and I think that's good advice. 1982 *TV Times Mag.* 9–15 Oct. 47/1 At home..there's nothing she likes better than to 'slop around in anything that's comfy to wear'.

7. *U.S. dial.* and *colloq.* To feed (pigs or cattle) with slops.

1848 D. DRAKE *Pioneer Life in Kentucky* (1870) 92 To slop the cows..was another [labour]. 1920 C. RUSSELL *Story of Nonpartisan League* 63 An angry representative told them to 'go home and slop the hogs'. 1947 *Time* 27 Jan. 21/2 Did you ever slop a hawg? 1966 R. G. TOEPFER *Witness* iii. 19 First off, he'd better feed the chickens and slop the pigs. 1976 *New Yorker* 17 May 34/1 Your hosts.. will be up at dawn to slop their pig.

8. The vbl. stem in comb., as **slop-over** *U.S.*, an act or instance of slopping over; also *fig.*

1908 Z. GALE *Friendship Village* 275, I see 'em all sprinkled along comin' from the funeral—neighbours an' friends an' just folks—an' most of 'em livin' in Friendship peaceful an'—barrin' slopovers—doin' the level best they could. 1952 *Richmond* (Va.) *Times-Dispatch* 5 Jan. 17/4 The ordinary foams used in fire-fighting do the rest of the job. The job is easy because there is no slop-over, frothing or expansion of the hot oil layer. 1977 *Time* 9 May 26/2 My motive in everything I was saying or certainly thinking at the time was not to try to cover up a criminal action but—to be sure that as far as any slip-over—or should I say slop-over, I think, would be a better word—any slop-over in a way that would damage innocent people.

slope, *sb.*[1] Add: **1. a.** (Examples in *pl.* in *spec.* sense 'ski-slopes'.)

1924, etc. [see *nursery slope* s.v. *NURSERY 8 c]. 1972 P. A. WHITNEY *Snowfire* (1973) vi. 100 Snow bunny..was a term applied to beginners, usually female, who haunted the slopes. 1976 J. FARRIS *Fury* (1977) xviii. 306 He bought..clothing for the slopes and for après-ski.

c. (Earlier example.)

1863 *Harper's Mag.* Sept. 459/2 There is an entrance to the mine by means of an inclined plane, called a slope.

2. c. The tangent of the angle between a line and the horizontal; the ratio of the projection on the y-axis of an infinitesimal segment of a graph to its projection on the x-axis; the value of the first differential of some quantity, esp. with respect to distance.

1889 J. A. FLEMING *Alternate Current Transformer* I. iii. 92 We shall call the trigonometrical tangent of the angle PTN, the slope of the tangent at the point P. *Ibid.* 93 The firm line curve is a curve of sines... The dotted line is a curve of sines, whose ordinate QN at any point represents the slope of the tangent at P on the original curve. 1898 *Proc. R. Soc.* LX. 478 If the slope of RR is positive we may say that large values of x are on the whole associated with large values of y, if it is negative large values of x are associated with small values of y. 1905 *Physical Rev.* XX. 174 The difference of temperature slope at different parts of the two bars was measured by means of thermoelectric couples. 1933 G. VAN PRAAGH *Introd. Calculus* i. 9 If y is a function of x, the differential coefficient or derivative of y with respect to x measures the rate of change of y with x for some particular value of x, x_1, and is the slope of the graph of $f(x)$ at the point x_1. 1959 *Listener* 26 Feb. 371/2 The ball will slow up because of the gravitational 'slope'. 1971 *Physics Bull.* Feb. 86/1 A ln σ against $1/T$ plot should, at the temperature of conversion, exhibit a change of slope.

3. a. (Further examples, in sense of *SLANT *sb.*[1] 3 b.)

1928 L. F. H. WHITBY *Med. Bacteriol.* iii. 46 Secondary cultures, or subcultures, are made by picking colonies from the plate and planting them on to slopes. 1951 —— & HYNES *Ibid.* (ed. 5) iii. 24 Slopes are made by allowing the medium to set in test-tubes or bottles tilted about 10° from the horizontal. 1974 R. K. PAWSEY *Techniques with Bacteria* iv. 51 The loop is introduced to the base of the slope and a wavy line made on the slope with the loop gradually rising to the top.

5. *Electronics.* The mutual conductance of a valve (so called because it is numerically equal to the slope of one of the characteristic curves of the valve).

[1918 *Wireless World* Nov. 458 (*heading*) A thermionic valve slopemeter. *Ibid.*, The effectiveness of a valve as a relay and amplifier depends primarily on the slope of the grid voltage-plate current characteristic.] 1932 *B.B.C. Year-bk.* 395 It is now the common practice of valve manufacturers to give a figure for the mutual conductance (or slope) of each of their products. 1948 C. A. QUARRINGTON *Mod. Practical Radio & Television* (ed. 2) I. x. 78 The measurement of slope may be carried out under any conditions of grid voltage. 1953 A. H. W. BECK *Thermionic Valves* ix. 246 The mutual conductance or slope = $(\partial I_a/\partial V_g)$, V_a const.

6. *U.S. slang.* An oriental; more recently,

spec. a Vietnamese. (Abusive.) Cf. *slopehead* s.v. *SLOPE- a, *SLOPY *sb.*, and *SLANT *sb.*¹ 10.
1948 G. H. Johnston *Death takes small Bites* v. 121 He seemed a hell of a lot more concerned with his bunch of flea-bitten slopes and his pots of medicine. 1966 *Publ. Amer. Dial. Soc.* 1964 XLII. 45 *Slope* and *slopehead* were the most popular terms applied to all 'indigenous personnel' [in Korea in 1950–3]. 1966 *New Statesman* 25 Mar. 436/3 He confirms the soldiers' contempt for the Vietnamese ('slopes' and 'gooks'). 1969 [see *DINK *sb.*²]. 1978 R. Thomas *Chinaman's Chance* iii. 35 All the Chinaman's gotta do is get into Saigon.. Once he's in nobody's gonna notice him, because all those slopes look alike.

7. Used *attrib.* to designate a quantity defined as a rate of change or derivative instead of as a ratio; chiefly in *slope resistance*.
1931 L. B. Turner *Wireless* vii. 203 In the metallic parts of the circuits..the slope resistance δe/δi and Ohm's resistance e/i are equal. *Ibid.* viii. 235 It is necessary to allow..for the very small slope or differential permeability dB/dH of the core. 1971 *Gloss. Electrotechnical Power Terms (B.S.I.)* II. ii. 18 *Slope resistance*, value of forward resistance calculated from the slope of the straight line used when determining the threshold voltage from the forward current/voltage drop characteristics of a diode or thyristor in the on-state.

slope, *v.*² Add: **1. a.** (Earlier example.)
1830 *Palladium of Brit. N. Amer.* (Toronto) 29 Aug. 224/1 Bad climate indeed, wonder people dont all slope.
b. (Further examples.) Also, to move (*off*, *in*, etc.) in a leisurely manner; to amble (*in*, etc.); to depart surreptitiously, sneak off.
1876 Trollope *Prime Minister* II. xvi. 265 You should have seen the policeman sloping over and putting himself in the way. 1922 Joyce *Ulysses* 298 Come in, come on, he won't eat you... So Bloom slopes in with his cod's eye on the dog. 1980 *Private Eye* 26 Sept. 13/1 Anyway, he sloped in for a chinwag with the Boss.

slope-. Add: **slope circuit** *Electronics* = *slope filter* below; **slope current,** (*a*) an air current produced when wind is deflected upwards by a hill; (*b*) an ocean current that arises when the surface of the sea slopes as a result of wind action; **slope detection** *Electronics*, the detection of a frequency-modulated signal by means of a slope filter followed by a detector for amplitude-modulated signals; so **slope detector; slope filter** *Electronics*, a filter whose response increases or decreases more or less uniformly over the frequency range to which it is used; **slopehead** *U.S. slang* = *SLOPE *sb.*³ 6 (abusive; cf. also *SLOPY *sb.*); **slope wash, slopewash** *Geomorphol.*, the downhill movement of soil or rock under the action of gravity assisted by running water not confined to a channel.
1966 M. Schwartz et al. *Communication Systems & Techniques* v. 239 One side of a resonance curve is used as a 'slope circuit' in a frequency detector. 1978 S. Haykin *Communication Systems* iv. 341 Basically, a frequency discriminator consists of a slope circuit followed by an envelope detector. 1931 V. W. Pagé *Henley's ABC Gliding & Sailflying* vii. 150 This ascending air current, which is defined as a slope current, forms the source of energy for sailing flight. 1959 Revelle & Shepard in P. D. Trask *Recent Marine Sediments* 277 In the Southern California region..the chief role in the transportation of débris must be played by tidal currents and non-permanent eddying 'slope currents' resulting from wind action. 1968 G. Neumann *Ocean Currents* iv. 193 The resulting horizontal pressure gradients in meridional direction, which, in turn, cause slope currents to develop. 1949 B. Grob *Basic Television* xxi. 433 While slope detection is seldom used in television receivers for reception of the associated sound signal, the principle is important because it illustrates how an FM signal can be received with an AM system. 1978 D. Cameron *Audio Technol. Systems* iv. 129 Economy-type FM tuners occasionally use some form of slope detection. 1958 *Wireless World Frequency Modulation* iii. 83 This circuit, with the resonant frequency adjusted above or below the highest value of the f.m. signal frequency, or below the lowest, was the first to be used..as an f.m. detector and is known as a 'slope' detector. 1977 F. G. Stremler *Introd. Communication Systems* vi. 293 Although the slope detector is economical, it has a very limited range and its use is restricted to input signals with small frequency variations. 1937 *Proc. IRE* XXV. 474 The intermediate-frequency output of a superheterodyne receiver is fed through a limiter to a slope filter or conversion circuit which converts the frequency modulation into amplitude modulation. 1942 A. Hund *Frequency Modulation* i. 84 For the upper cutoff frequency..of the slope filter a voltage 2E is obtained at the output of this filter. 1966 Slopehead [see *slope *sb.*³ 6]. 1968 *Listener* 23 May 656/2 As Cao Do, two years ago, I heard American Air Force men sing a ballad about the Vietnamese, whom they then called 'slopeheads' or 'slopes'. 1938 C. F. S. Sharpe *Landslides & Related Phenomena* i. 8 The series can be represented as follows: Stream-flow (much water, small load, low angle), Slopewash, Sheetflood, Mudflow, Earthflow, Debris-avalanche, Landslide (little water, large load, moderate to high angle). 1966 J. Wyckoff *Rock, Time, & Landscape* iv. 71 As the ravine is cut deeper, its walls are worn back by weathering, mass wasting, and slope wash. 1978 *Nature* 23 Feb. 740/1 Slopewash seems important in sediment transport and fallen sandstone blocks and trees accumulate sediment on their upslope sides.

slopey: see *SLOPY *a.*

sloppage. For *rare*⁻¹ read *rare* and add: Also, the action of slopping.
1962 J. Onslow *Bowler-Hatted Cowboy* viii. 79, I filled the barrel among the boulders of the creek, and put a few sticks to float in it. These served as baffles and minimized the loss from sloppage.

sloppery. (Earlier example.)
1832 *Chambers's Edin. Jrnl.* I. 130/1 Their pails, and buckets, and brushes, and all their slopery, are just as rife..as they were a month earlier.

sloppily, *adv.* (Further examples.)
1909 *Chambers's Jrnl.* Sept. 572/2 The swish of the water licking sloppily against the yacht's side had a very lonesome sound. *a* 1911 D. G. Phillips *Susan Lenox* (1917) II. xv. 354 She regarded his play as mediocre claptrap..fit only for the unthinking, sloppily sentimental crowd. 1942 E. Paul *Narrow St.* xi. 83 In first class, one was slighted by the poverty-stricken employees, and served sloppily in the dining car, if any. 1952 S. Kauffmann *Philanderer* (1953) iii. 39 I've done it six summers now and its all a lot of crudney. Rush around and push shows together sloppily. 1979 *Country Life* 18 Jan. 134/3 The subject..is perilously near the sloppily mawkish.

sloppiness. (Further examples.)
1928 [see *animal-lover* s.v. *ANIMAL *sb.* and *a.* C. 1]. 1939 A. Thirkell *Before Lunch* vii. 187 It had also crossed her mind that Daphne liked young Mr Bond, but..Daphne was all against what she called sloppiness. 1960 [see *SLOB *sb.*¹ 3]. 1980 *Maledicta* Summer 157 Because of the sloppiness of most original references..I have spent many years tracking down complete bibliographical data.

slopping, *vbl. sb.* Add: **b.** With advbs., as *-out* (see *SLOP *v.*² 1 c), *-over*; *slopping-up* (N. Amer. slang), a drinking-bout.
1899 'J. Flynt' *Tramping* II. iv. 271 The bums intended to have a great 'sloppin'-up' (drinking-bout). 1922 H. Kemp *Tramping on Life* 133 'Slopping up' is what the tramps call a drinking jamboree. 1945 H. Read *Coat of Many Colours* lxii. 304 But there is no slopping-over of irrelevant emotion. 1948 *Richmond* (Va.) *Times-Dispatch* 9 Jan. 16/1 This [*sc.* a fulsome speech] is what is known among the hamlets of his native State as 'slopping over', and is considered as serious a breach of etiquette as..beating one's grandmother in public. 1955 *Times* 25 May 6/6 A hospital officer at the prison, said that on May 7 he was unlocking doors for 'slopping out' to be done. 1968 L. Deighton *Only when I Larf* xviii. 232 You can tell a new warder; they just can't stand the smell of the slopping-out each morning. 1976 O. Jacks *Assassination Day* v. 80 A long term in prison..reviled him, from slopping out to terrible food.

sloppy, *a.* Add: **3. b.** *colloq.* Of the sea: choppy.
1970 *Studies in English* (Univ. of Cape Town) I. 26 Mushy, or sloppy surf, indicates a troubled, choppy water surface which would be difficult to ride. 1977 *Austral. Sailing* Jan. 69/1 However the sloppy Botany Bay conditions plus a series of freakish thunder storms made the series wide open.
4. b. *colloq.* Weakly sentimental.
1883 'Mark Twain' *Life on Miss.* xlviii. 482 The sloppy twaddle in the way of answers, furnished by Manchester. 1919 J. C. Snaith *Love Lane* xxxi. 163 The Corporal stopped suddenly, took Melia in his arms and kissed her. It was a sloppy thing to do, unworthy of old married people. 1936 R. Lehmann *Weather in Streets* ii. 53 Kate said with a funny look, as if she were saying something a tiny bit embarrassing, on the sloppy side. 1959 [see *DRIP *sb.* 3 c].
c. *Comb.*, as *sloppy-minded, -mindedness*.
1903 G. B. Shaw *Let.* 15 Sept. (1946) 18 Sloppyminded lunatics. 1965 B. Sweet-Escott *Baker Street Irregular* ii. 43 He had a sharp tongue for the sloppy-minded and the half-baked. 1976 *Listener* 22 July 89/2 This final hymn to sloppy-mindedness.
6. Special collocations: **Sloppy Joe, sloppy joe** *colloq.*, (*a*) used *attrib.* and *absol.* to designate a loose-fitting sweater; (*b*) *U.S.*, a kind of hamburger in which the minced-beef filling is made into a kind of meat sauce; (*c*) a slovenly person.
1942 Berrey & Van den Bark *Amer. Thes. Slang* § 87/32 *Sloppy Joe*, a loose cardigan sweater. 1943 *Knitted Outerwear Times* 15 May 1/1 You can't look like Lana Turner in a Sloppy Joe. Well—maybe Lana could. 1944 *Life* 15 May 67/1 (*caption*) Traditional garb of all school girls is 'Sloppy Joe' sweater, single string of pearls, pleated skirt, socks and shoes. 1958 E. Hyams *Taking it Easy* II. 147 The young men were dressed in fashionable jeans and sloppy joes. 1961 R. E. Church *Burger Cook Bk.* I. 42 *Sloppy Joes*...ground beef..onions..celery..sweet pickle relish..brown sugar..Worcestershire sauce..chili sauce..vinegar..green pepper..hamburger buns. 1961 Webster, *Sloppy joe*.., a man who is negligent of his clothes or personal appearance. 1966 'L. Lane' *ABZ of Scouse* 99 *Sloppy Joe*, a careless, shiftless person. 1974 *Washington Post* 1 Aug. D-1/2 Bill Myer..sits in the cafeteria..eating sloppy joes. 1980 *Ibid.* 17 Jan. B-5/1 Teen-agers wore a baggier variety [of sweater], often pairing 'sloppy joe' sweater..with pleated skirts.

slop-worker. Add: (Earlier example.) So **slo·p-working** *vbl. sb.*
1850 C. Kingsley *Cheap Clothes & Nasty* 22 It served as a blanket to the fever-stricken slopworker. *Ibid.* 26 Fresh victims are being driven by penury into the slop-working trade.

slopy, *a.* Add: Also **slopey. B.** *sb. U.S. slang.* An oriental, a Chinese. (Abusive.) Cf. *SLOPE *sb.*³ 6, *slopehead* s.v. *SLOPE- a.
1948 G. H. Johnston *Death takes small Bites* v. 107 'And you'll find it mighty hard, son, to convince Petroleum Developments that the fields aren't as important as a bunch of flea-bitten slopeys!' 'Well, go on.' 'Okay—as long as you understand that oil's a pretty important commodity—and Chinese ain't.' 1962 E. Snow *Red China Today* (1963) xli. 85 One might assume that contempt for American imperialism would by now have produced Chinese equivalents of insulting American epithets such as slopeys, slant-eyes and chinks.

slorm, var. *SLAUM *v.*

slosh, *sb.* Add: **2. a.** Also, watery, sodden, or unappetizing food.
1923 Blunden *Christ's Hospital* 201 *Slosh*, boiled rice. 1959 I. & P. Opie *Lore & Lang. Schoolch.* ix. 163 Any kind of milk pudding is 'slosh' or 'baby pudding'. 1980 *Telegraph* (Brisbane) 16 Jan. 2/3 Honesty in advertising. A sign outside a Noosa Heads fast food shop 'American Slosh'.
b. (Earlier and later examples.)
1894 G. B. Shaw *Let.* 2 June (1965) I. 440 The assumption that society takes the sort of loyal, constitutional, jingo, pietistic slosh it has to pretend to like. 1915 E. M. Forster *Let.* 2 Aug. in P. N. Furbank *E. M. Forster* (1977) II. 13 He [*sc.* Rupert Brooke] was essentially hard: he hated all slosh.
4. A blow, an act of striking.
1936 Wodehouse *Laughing Gas* x. 107, I recalled that I had noticed her hand quiver once or twice, as if itching for the slosh. 1977 *Listener* 21 Apr. 20/5 (*caption*) I'll give you such a slosh when I get up from here.
5. A game played on a billiard table with six coloured balls and one white, with which each player tries to pot the coloured balls in a certain order.
1928 [see *SNOOKER *sb.*]. 1951 G. Frankau *Oliver Trenton* xxiii. 180 His brother-in-law was..good at slosh. 1961 E. Waugh *Unconditional Surrender* i. iv. 60 Guy spent the remaining hours of his fortieth birthday at Bellamy's playing 'slosh'. 1976 *Daily Tel.* 29 Jan. 17/8 The equipment was suitable not only for billiards but also for 'slosh'.

slosh, *v.*¹ Add: **3. b.** Of liquid: to splash; to flow in streams.
1945 E. O'Neill *Touch of Poet* (1957) i. 35 When he attempts to raise the glass to his lips the water sloshes over his hand. 1969 L. Michaels *Going Places* 59, I might, as I toppled, blood sloshing through my lips, beg forgiveness. 1977 'J. Le Carré' *Honourable Schoolboy* i. 29 The rain poured off them..sloshing in red rivulets round their ankles.
4. *trans.* **a.** To pour or dash (liquid); to splash, throw, pour, or swallow carelessly. Also *fig.* Usu. with advbs. *colloq.*
1875 *Chicago Tribune* 5 Sept. 2/5 The Ring-paid scribblers and papers will slosh on the usual amount of whitewash. 1885 *Century Mag.* Nov. 63/2 If mining records was ever kep' as they'd ought to be, and not sloshed round so public like. 1949 G. B. Shaw *Let.* 26 Apr. (1972) II. 85, I dipped into the book..and sloshed down a heap of words... But it is a scandalously poor job of a review. 1952 E. Ferber *Show Boat* vi. 221 Often he sloshed down whole gallons of river water before the rounds were over. 1936 M. Mitchell *Gone with Wind* iv. 318 picked up the decanter and sloshed a glassful, untidily. 1945 *Everybody's Digest* Aug. 86 He sloshed on his shoes, and..went outta there, heatin' his axles. *a* 1953 E. O'Neill *Touch of Poet* (1957) III. 100 He sloshes whiskey from the decanter into both their glasses. 1963 J. Stroud *Sharon Lamb* xxiii. 197 Ted..attic bedrooms and Harry used to go up there and slosh paint about. 1964 L. Deighton *Funeral in Berlin* xv. 91 He laughed a deep, manly laugh. 1978 'J. Lymington' *Waking of Stone* vi. 149 She sloshed out gin for Nick.
b. To pour or dash liquid upon, to douse. *colloq.*
1911 G. W. Deeping *Sincerity* ii. 18, I can't stand these counter-bouncing little beasts like Threadgold. He's only fit to slosh people with treacle and water. 1917 H. Garland *Son of Border* xxxiii. 371, I generally managed to slosh myself with cold water from the well. 1979 *Amer. Poetry Rev.* Mar./Apr. 16/2 Rain began to pelt the car and slosh the yard and spatter down the flivvers.
5. To hit, to strike; to crush, to defeat. Also *fig.* Cf. Slosh *v.* 2 b.
1890 Kipling *Barrack-Room Ballads* (1892) 11 We sloshed you with Martinis, an' it wasn't 'ardly fair; But for all the odds agin' you, Fuzzy-Wuz, you broke the square. 1904 E. Nesbit *Phoenix & Carpet* v. 94, I say, slosh 'em..and get clear off with the swag. 1914 C. Mackenzie *Sinister Street* II. iv. iii. 881, I wouldn't half slosh his jaw in, if I was a man. 1958 R. P. Fleming *Let.* in D. Hart-Davis *Peter Fleming* (1974) ii. 33, I saw one [adder] coiled up asleep in some bushes, and picked it up by the tail..and we took it into the open and sloshed it. 1921 A. S. M. Hutchinson *If Winter Comes* II. vi. 148 The Balkan chap set to, to slosh Turkey. 1933 *Punch* 18 Oct. 421/1 'I wish to contradict the rumour that I wish to slosh Sir Stafford Cripps,' says Mr. Ernest Bevin. 1967 N. Freeling *Strike out where not Applicable* 75 Somebody sloshed him, if I may be allowed the word. *a* 1974 R. Crossman *Diaries* (1975) I. 399 Characteristically enough, at the end the committee chairman who had been wildest in his wrath moved the vote of thanks and said, 'Well, we have to have a good go sometimes at sloshing our Labour Minister.' 1977 'J. Gash' *Judas Pair* x. 118 I've sloshed her..sometimes when she'd got me mad.

sloshed (slǫʃt), *ppl. a. slang.* [f. Slosh *v.*¹ + -ed¹.] Drunk, tipsy. Also *absol.*
1946 *Word Study* May 3/1 Synonyms for *drunk* now current in England..*tiddley*, *oiled* or *well oiled*, *sloshed*. 1952 'R. Gordon' *Doctor in House* xi. 120 Tony, you're sloshed already. 1966 L. Southworth *Felon in Disguise* viii. 118 The D.S.O. was due to an overdose of Italian vino. I was too sloshed to care. 1978 R. Ludlum *Holcroft Covenant* iii. 41 They drank a *great* deal... They appeared quite sloshed. 1979 *Logophile* III. ii. 17/1 It started as a pub game, played for the amusement of the slightly sloshed.

sloshing, *vbl. sb.* (In Dict. s.v. Slosh *v.*¹ 3.) Add: **2.** = Hiding *vbl. sb.*²
1931 T. R. G. Lyell *Slang, Phrase & Idiom* 693 He ran up against a gang of roughs in the street last night, and they gave him an awful sloshing!

sloshy, *a.* **1.** (Earlier example.)
1797 B. Hawkins *Let.* 22 Feb. in *Coll. Georgia Hist. Soc.* (1916) IX. 88 Flat piney sloshy land.

sloshy, *a.* Add: (Further examples.) Also, sloppy, sentimental.
1913 A. Lunn *Harrovians* ii. 35 Because..if they had form games some sloshy old chaw [*Note*, i.e. Old Harrovian, a term of affection] would write to the papers. 1920 *Glasgow Herald* 30 Apr. 9 They are living not merely upon vegetables, but sloshy vegetables. 1924 E. F. Benson *David of King's* vii. 136 'Positively his last appearance,' said David. 'Rather theatrical, but not sloshy.' 1933 *Daily Mirror* 26 Oct. 12/4 'Sloshy talk' is simply the perpetration of such phrases as 'too, too marvellous' and 'utterly divine'. 1949 [see *BOOB *sb.* 4]. 1957 D. Piper *English Face* x. 244 Broader, indeed sloshy, effects were needed by an emulator of Reynolds, Daniel Gardner. 1978 C. Storr *Winter's End* ii. 37 You thought it was like in those rotten stories... I've read enough sloshy stuff in my time.

slot, *sb.*² Add: **2. b.** Also (*slang*), a slot-machine.
1950 R. Bissell *Stretch on River* xiii. 135 The slots are going night and day. 1978 M. Puzo *Fools Die* xviii. 197 The slots usually brought in a profit of about a hundred thousand dollars a week.
c. The middle of the semi-circular or horse-shoe-shaped desk at which a newspaper's sub-editors work, occupied by the chief sub-editor. *U.S. slang.*
1917 H. Grant *Two Sides of Atlantic* iii. 44 The man who 'sits in the slot' (the chief-sub.), will know for a certainty that the decision of 'Bill' to invade Windy City will automatically entail the departure of all who 'hunt' with Bill. 1923 [see *RIM *sb.*¹ 3 c]. 1970 R. K. Kent *Lang. Journalism* 123 *Slot*, the middle of the horseshoe-shaped copy desk where the news editor or copy editor (sometimes called *slotman*) sits. To be *in the slot* is to be in charge of the copy desk.
d. *Aeronaut.* A linear gap in an aerofoil, running parallel to its leading edge, which allows the passage of air from the lower to the upper surface and so increases the lift. Cf. *SLAT *sb.*¹ 4 d.
1920 *Flight* XII. 1124/1 It has already been mentioned that the slot separating the false from the main leading edge is contracted towards the upper surface. 1936 *Discovery* Mar. 73/2 The Weick and Hammond have..a control which is a combination of slot and aileron. This is intended to obviate the need of a rudder. 1960 C. H. Gibbs-Smith *Aeroplane* ii. 221 Ingenious as the machine undoubtedly was,..there is no visible trace of any slots, or of any wires or other gear attached to, or passing anywhere near, the wings that could be associated with slots.
e. The vulva. *coarse slang.* Cf. *SLIT *sb.* 1 d.
1942 Berrey & Van den Bark *Amer. Thes. Slang* § 121/38 *Female pudendum*...shape, slot, snatch, tail. 1977 Miller & Swift *Words & Women* vii. 117 No such positive connotations attach to prick, but even this word does not convey the absolute scorn of slit, slot, snatch, and gash.
f. A marked-out parking space. Chiefly *U.S.*
1944 R. Chandler *Lady in Lake* xiii. 74, I..parked in one of the diagonal slots at the side of the Prescott Hotel. 1968 A. Diment *Great Spy Race* iii. 35, I shunted my car into a small slot near the fire station. 1978 R. Ludlum *Holcroft Covenant* xvii. 195 Holcroft backed the car out of its slot, then drove through the entrance posts onto the country road.
g. A prison cell; also, = *CELL *sb.*¹ 4 b. *Austral. slang.*
1947 *Pix* 20 Sept. 15 *Peter* or *slot*, cell. 1969 *Sydney Sunday Tel.* 21 Dec. 14/4 'I'd hate to try and tot up the number of hooks and badges (rank and good conduct insignia) that little lot's whipped away since the ship came out.'.. 'And what about the slot (cells) they've dished out. It must run into years!' 1976 *Cleo* Aug. 33 Some of the old heads are in the slot, he says. The slot is jail.
5. b. *Austral.* and *N.Z.* A crevasse.
1959 *Tararua* XIII. 46 *Slot*, for a crevasse, sometimes used by climbers, is not necessary and is merely slang. 1968 K. Weatherly *Roo Shooter* 58 This was the roughest bit of country yet—short, miserable, scrubby trees and stringy bushes; broken country, high slots and hollows full of water.
5*. *fig.* **a.** A position in a list, hierarchy, system, or scheme; a position to be filled; a category; a place or division in a timetable, esp. in broadcasting.
1942 Berrey & Van den Bark *Amer. Thes. Slang* § 672/10 Rank or rating in league, division, percentage ladder, slot. 1947 Auden *Nones* (1952) 64 Among those staring blemishes that mark War's havocking slot. 1956 *Sat. Rev.* (U.S.) 2 June 50/2 No management slot is harder to fill today than the research director's post. 1956 W. H. Whyte *Organization Man* ii. viii. 104 Sales work..is about the only slot they would qualify for if they took English or history. 1964 *Economist* 25 Jan. 327/2 There is a 'slot' in the market for a medium-range supersonic airliner. *Ibid.* 20 June 1369/3 The 'slots' once gained, were never given up [by regional TV stations]. 1966 *Listener* 6 Oct. 518/1 It seems perverse that when the Monday evening 'slot' has been extended to as much as two hours for lesser fry, it should be made to stand at ninety minutes for *The Merchant of Venice* (Home Service, September 26). 1967 *Ibid.* 6 Apr. 467/3 'Theology, during the great controversies of the mid-nineteenth century, was anti-scientific.'.. (Try fitting Newman or even Kingsley into that slot!) 1969 *Times* 21 Nov. 23/5 How serious is the situation at Heathrow?.. There are no spare 'slots' into which landing and taking-off airliners can be fitted. 1970 *Daily Tel.* 3 Sept. 13/2 The first Radio London slot each day will be from 6.45 a.m. to 9 a.m., a blend of news, music, personalities and information called 'Rush Hour'. 1972 *Business Week* 18 Mar. 81/1 Although he held a top slot at Snia, he was lured away for the even bigger job at Alitalia. 1973 C. Bonnington *Next Horizon* xxii. 298 An Italian millionaire..had permission for an autumn reconnaissance in 1972 to be followed by a spring attempt in 1973. Owing to sickness he gave up his autumn slot. 1974 *Guardian* 26 Mar. 14/1 Welland's script was accepted by the BBC for its 'Play For Today' slot. 1976 *National Observer* (U.S.) 24 Apr. 19A/3 Stanford has 10,009 applications for 1,450 freshman slots next fall. 1976 A. Davis *Television: First Forty Years* viii. 86 Suitable slots are normally of 90 to 120 minutes, with time for commercials to be taken out of this, but films are rarely obliging enough to run to exactly the length required. 1977 *Film & Television Technician* Mar. 6/4 The British programme-makers are actually pushing the Americans out of the number one slot in key Western Europe and Scandinavian countries. 1977 *Times* 25 Aug. 1/8 The importance of taking a flight 'slot' when it comes up. 1978 M. Puzo *Fools Die* xii. 131 After six months of free-lance work he offered me a magazine editor slot. 1980 *Jrnl. R. Soc. Arts* July 529/2 Many place Kokoschka in the slot 'Expressionist'.
b. *spec.* in *Linguistics* (see quot. 1960). Also *attrib.* and *Comb.*
1957 K. L. Pike in *General Linguistics* Spring 36 The characteristics of a grammeme which in many instances are perhaps most readily recognized in current descriptions of a grammemic system are the functional slot with its class filler. 1959 [see *DIAPHONEME]. 1959 W. H. Mittins in Quirk & Smith *Teaching of English* iv. 116 Some teachers..seek to achieve a kind of concentration and continuity by methodically working through batteries of vocabulary exercises in slot-filling..and the like. 1960 Elson & Pickett *Beginning Morphology-Syntax* ii. 16 A *slot* is a grammatical position or function (e.g. subject) which is *filled* by a list of mutually substitutable items (e.g. nouns). The tagmeme is the unit of grammatical arrangement involved in or resulting from this slot-class correlation. *Ibid.* 37 Make a chart with the slot names.., listing fillers below each slot name. *Ibid.* iii. 40 As a filler class they can only be united by some such term as 'subject slot fillers'. 1962 W. A. Stewart in F. A. Rice *Study of Role of Second Languages in Asia, Africa, & Latin Amer.* 21 Under various conditions, the same language may occupy more than one functional slot. 1964 E. Bach *Introd. Transformational Gram.* iii. 44 But the basic points of the slot-symbol and class-symbol description and the lack of context-sensitive rules remain untouched. 1964 *Language* XL. 314 American slot-and-filler grammatical description. 1965 *Word Study* Feb. 3/1 Word groups filling noun slots and verb slots comprise the chief building blocks of utterances. 1970 B. M. H. Strang *Hist. English* 25 When, through cross-cultural experience, speakers of one language are conscious of an 'empty slot' in their language which is filled in another language.., they may..fill the gap by borrowing the filler. 1972 *Archivum Linguisticum* III. 22 The lexical co-occurrence restrictions which hold between the fillers of predicate and subject slots are different from the restrictions which hold between the fillers of predicate and instrument slots. 1972 *Computers & Humanities* VII. 14 Some recent publications include..'Computerized Japanese Haiku', which describes how the poems are created by slot-filling. 1972 M. L. Samuels *Linguistic Evolution* 65 A new slot-filler may arise from borrowing or creation, or the 'pull' of the empty slot may hasten a new process of extension in another existing word. 1981 *Word* 1980 XXXI. 230 He makes use of the slot-and-filler infrastructure, characteristic of tagmemics.
7. b. Special combs., as **slot aerial, antenna,** an aerial in the form of one or more slots in a metal surface; **slot-back,** *N. Amer. Football*, (the position of) a back who stands behind a gap in the forward line; **slot car,** a miniature racing car, powered by electricity, which travels in a slot in a track; **slot-machine:** see sense 6 b in Dict.; also *fig.* and *attrib.*; **slot man** *U.S. slang*, a newspaper's chief sub-editor, a news editor; **slot racer** = **slot car*; so **slot-race** *v. intr.*, **slot-racing** *vbl. sb.*; **slot radiator** = **slot aerial*; **slot seam,** a clothing seam reinforced underneath; also, = *channel seam* s.v. *CHANNEL *sb.*¹ 12; **slot television,** a coin-operated television; **slot winding** *Electr. Engin.*, an armature winding in which the conductors are laid in slots or grooves in the core; so **slot-wound** *a.*
1946 *Jrnl. Inst. Electr. Engineers* XCIII. IIIA. 626/1 It would appear that slot aerials are capable of making a contribution to the problem of designing a radiating element that produces circular polarization in all directions of radiation. 1956 *B.B.C. Handbk.* 1957 56 To keep staff up to date, training supplements on such items as slot aerials, television lighting, frequency modulation, and other developments are issued. 1946 *Jrnl. Inst. Electr. Engineers* XCIII. IIIA. 749/2 In spite of the length of the slot antenna, this load can be treated as lumped at the position of the centre of the slot. 1975 D. G. Fink *Electronics Engineers' Handbk.* xviii. 43 Very low profile slot antennas utilizing shallow cavities fed by coaxial cables have been designed for aircraft use. 1959 *Washington Post* 21 Nov. A14/4 He helped develop Elroy (Crazy-legs) Hirsch into a slotback with the Los Angeles Rams. 1970 *Globe & Mail* (Toronto) 28 Sept. 18/2 Slotback Dick Smith took a 12-yard pass from Sonny Wade in the second quarter for one Montreal touchdown. 1974 *Anderson* (S. Carolina) *Independent* 24 Apr. 5B/1 At slot back..Rut Livingston..has the makings of a great player. 1966 *Maclean's Mag.* 22 Jan. 9A The track on which slot cars race is a tabletop affair. 1971 *Publishers' Weekly* 27 Sept. 129/3 (Advt.), *The New Zealand Boys' Book of Crafts, Pets, Sports and Hobbies* by Anthony Harvey and Peter Snell provides information on..model-making, rugby, sailing, slot cars, and wood-carving. 1891 *Brooklyn Daily Eagle* Index July–Dec. 155/2 Slot machine. 1892 [in Dict., sense 6 b]. *a* 1910 'O. Henry' *Rolling Stones* (1912) 196 Mac McGowan was to..drop his silver talent into the slot of the slot-machine of fame and fortune that gives up reputation and dough. 1929 *Sun* (Baltimore) 15 Nov. 1/6 'Spike' O'Donnell,..beer baron and bootlegger,..is acting as his own lawyer in the 'slot machine' trial. 1933 C. Day Lewis *Magnetic Mountain* 12 Eating chocolate creams from the slot-machines. 1957 *Observer* 1 Sept. 13/4 At Las Vegas the plane empties of passengers who, with cold passion, play the slot-machines in the concourse until ten minutes later, when it is time to go. 1978 J. Wainwright *Jury People* lxiv. 216 A slot-machine arcade. One of these pin-table places. 1928 *Amer. Speech* IV. 134 Presiding over the copy readers is the 'head of the desk' or 'slot man'. His chief duty is to judge the amount of space to be given any 'story' or news article and to designate the size of the 'headline' or 'head'. 1972 Slot man [see *REVISE *sb.* 3]. 1966 *Daily Progress* (Charlottesville, Va.) 4 Feb. 13 (*caption*) J. P. Evans..and Ed Johnson get ready to put their slot racers through a gruelling five-lap race. 1965 *Wall St. Jrnl.* 27 Aug. 22 A rapidly growing number of Americans..have caught the slot-racing bug. 1967 J. Symons *Man who killed Himself* i. vi. 69, I haven't joined a slot racing club... I like slot racing on my own. 1946 *Jrnl. Inst. Electr. Engineers* XCIII. IIIA. 748/2 When slot radiators which were very loosely coupled to the guide had to be measured, standing-wave measurements of single slots became unreliable. 1967 *IEEE Trans. Antennas & Propagation* XV. 826/1 If the boundaries of the dielectric cover are kept within the local reactive fields of the slot, the primary effect upon the radiation is a change in the impedance of the slot radiator. 1918 E. & M. Wallbank *Dress Cutting & Making* x. 69 *Slot Seam*, in which both edges are overlapped on to a wrap piece or 'slot'. 1968 J. Ironside *Fashion Alphabet* 99 *Channel Seam* (*slot seam*): to make a channel or slot seam, the seam turnings should be basted together. An additional strip of fabric should be laid under the basted seam and should then be machined approximately half-inch or so away from the basted seam line. 1958 *Kinematograph Weekly* (Studio Rev.) 29 May p. iv/3 If slot television gets a real hold on the public, commercial cinema..is doomed. 1977 *Grimsby Even. Tel.* 26 May 3/8 (Advt.), Slot television, bargains galore. 1900 *Jrnl. Brit. Inst. Electr. Engineers* XXIX. 802 A hole-winding produces a somewhat smoother pole than a slot-winding. 1968 Fink & Carroll *Standard Handbk. Electr. Engineers* (ed. 10) vi. 10 Fractional slot windings, where the number of slots per phase per pole is not an integer, have unequal coil groups. 1931 L. B. Turner *Wireless* xiv. 471 The calculation of E.M.F. in a slot-wound dynamo.

slot, *v.*² Add: **4.** To thread (material, etc.) *with* (ribbon).
1922 Joyce *Ulysses* 344 She had four dinky sets,..each set slotted with different coloured ribbons. 1975 G. Howell *In Vogue* 243 The flowerpot hat..in coffee cream satin slotted with a brown ribbon.
5. *intr.* **a.** To admit of being threaded through a hole or slot.
1928 *Daily Express* 9 Jan. 13 The unique collar slots through the buckle.
b. *fig.* To fit *in* or *into*; to take up a position *in* a space or slot (*SLOT *sb.*² 5*).
1940 H. G. Wells *Babes in Darkling Wood* i. i. 40 We are not the people we were yesterday. We slot into the new order. 1965 *New Statesman* 7 May 715/1 Personally I never knew yet quite what I was nor where I slotted in; I suppose I was ready for total identification somewhere, but never where I happened to be. 1966 'A. Hall' *9th Directive* xx. 187, I..watched the police-car slot in between us and the car ahead; then it pulled out and one lost it. 1971 *Daily Tel.* 9 June 2/4 Initially the trains will operate at a maximum 125 mph to 'slot in' with new-type conventional diesels. 1976 *Times* 15 Apr. 27/1 The French company augments its range in Britain with the GTL, which slots in between the 956cc TL and the high-performance TS. 1976 *Ilkeston Advertiser* 10 Dec. 19/1 They produced a great team display with new boy Henshaw slotting in well. 1980 S. Brett *Dead Side of Mike* xiii. 149 There are quite a few details which haven't slotted into position yet, but..the outline's right.
6. *trans.* **a.** To fit (something) *in* or *into* a position, space, or slot (*SLOT *sb.*² 5*).
1966 A. Battersby *Math. in Management* viii. 211 Certain complex calculations..are available in a form which can be readily 'slotted in' to bigger programs. 1968 *Listener* 4 Jan. 27/1 The television slotting system..separates programmes into categories... But slotting also creates a climate in which surprise is unwelcome... Slotted into one of the arts programme times the Beatles' film would hardly have raised a whisper. 1970 O. Norton *Dead on Prediction* i. 14, I..managed to slot the Mini into the corner of the temporary car park. 1971 *Country Life* 8 Apr. 801/1 The richly Italianate façade of the Finsbury Bank for Savings..was slotted into the terrace in 1840. 1972 M. Williams *Inside Number 10* xiii. 339 The National Agent went to great pains to slot this function into the election tour. 1973 *Scotsman* 13 Feb. 8/4 Steady progress up the scale leads to a salary of £3638. It is inconceivable that the Bishop of Bath and Wells..would be slotted in at the minimum. 1977 C. Dexter *Silent World N. Quinn* viii. 71 He slotted the book back into its shelf. 1977 *Irish Times* 8 June 8/4 Why, for instance, wasn't Síle de Valera slotted into this constituency once Vivion de Valera stood down?

b. *spec.* in *Football*. To kick (the ball, a pass) accurately through a narrow space, esp. *in* or *into* the goal; to score (a goal) in this way. Also *absol.*
1970 F. C. Avis *Soccer Dict.* (ed. 3) 86 *Slot in*, to pass, or to score a goal, by the very skilful placing of the ball through a narrow gap between players. 1974 *Observer* 1 Sept. 18/4 Boersma hardly needed to leave the ground to slot his..header into the net. 1975 *Liverpool Echo* (Football ed.) 1 Feb. 1/4 United took the lead through Jones who slotted home. 1975 *Evening News* (Edinburgh) (Sports Final ed.) 15 Mar. 10/2 McDowell slotted the ball into the net. 1977 *Wandsworth Borough News* 7 Oct. 10/1 He beat Newton, centre-half Robinson and goalkeeper Stevenson before slotting the ball in. 1978 *Cornish Guardian* 27 Apr. 5/3 The home team were showing good touches and Hargreaves slotted in a third goal after Barker had sent a good shot screaming goalwards

slote, var. *SLUIT.

slotted, *ppl. a.* Add: **1.** (Further examples in *Aeronaut.*: cf. *SLOT *sb.*² 2 d.) Also, *slotted armature* (Electr. Engin.), an armature having slots or grooves to contain the conductors; a slot-wound armature; *slotted line*, a length of coaxial cable or wave guide having a slot running lengthwise in its outer conductor to receive a probe for investigating standing waves.
1902 *Encycl. Brit.* XXVII. 582/2 These wires lie side by side in the smooth-core armature with one layer, or one on the top of the other if there are two layers, as is usually the case in slotted armatures. 1921 *Jrnl. R. Aeronaut. Soc.* XXV. 274 (*heading*) Flap experiments with slotted wings. *Ibid.* 275 (*caption*) Slotted aerofoil with flap. 1926 *Ibid.* XXX. 357 With a combination of slotted aileron and forward aerofoil, complete lateral control at the stall has been obtained. 1935 C. G. Burge *Compl. Bk. Aviation* 240/1 The Handley-Page Slotted Wing is one of the best-known devices for preventing this stalling, at least within the normal range of angles which an aeroplane is likely to reach. 1947 E. A. Yunker et al. in H. J. Reich *Very High-Frequency Techniques* II. xxiv. 592 A notched coaxial cable is a cheap and easily made substitute for a slotted line and is satisfactory for measurements not requiring great accuracy. 1966 *McGraw-Hill Encycl. Sci. & Technol.* VII. 44/2 A commercial slotted line for measuring impedances at the lower microwave frequencies is shown. 1973 J. D. Edwards *Electr. Machines* ii. 53 The basic machine equations are the same for a slotted armature as for a smooth cylindrical armature, provided that the total pole flux..is unchanged. 1977 *R.A.F. Yearbk.* 31/2 On rolling or overshooting, the lifting power of the wing with full slotted flaps is an experience to behold.
2. Threaded through a hole or slot.
1932 *Woman & Beauty* Apr. 27 A slotted scarf gives a great many opportunities for freshening up an everyday frock.

slotting, *vbl. sb.* Add: **4. a.** The action of threading through a hole or slot. Also, ornamentation with threading.
1923 *Daily Mail* 18 June 1 Filet lace and hem stitching, finished ribbon slotting at low waist line.
b. The action or condition of fitting in a slot (*SLOT *sb.*² 5*). Also *attrib.*
1959 *Washington Post* 26 Dec. A19/1 Ratings have climbed despite the show's unhappy time slotting between 'Laramie' and 'Bronco', without a big show preceding it on the network. 1968 [see *SLOT *v.*² 6].

slouch, *sb.* Add: **1. b.** For *U.S. slang* read 'orig. *U.S. slang*' and add earlier and later examples.
1796 A. Barton *Disappointment* III. i. 73 He's no slouch of a fellow. 1840 C. F. Hoffman *Greyslaer* II. ii. x. 23 You are no slouch of a woodsman to carry a yearling of such a heft as that. 1924 Galsworthy *White Monkey* III. x. 281 'No slouch of a miracle!' he thought, 'modern town life!' 1956 *People* 13 May 4/5 But Tony, no slouch when it comes to showmanship, helped it along by wearing..a rose brocade dinner jacket. 1961 *Coast to Coast* 1959–60 42 Speaking of Bradman, Mr Stulpnagel, they say you were no mean slouch with the bat yourself. 1967 G. F. Fiennes *I tried to run a Railway* ii. 7 He, Happy, himself no slouch at basic English, was revolted by George's language. 1978 R. Holles *Spawn* v. 42 He was making his pile... He's certainly no slouch in the business world.

slouch, *a.* **3.** (Earlier example.)
1812 E. Weeton *Let.* 15 June (1969) II. 34, I had on a small slouch straw hat, a grey stuff jacket, and petticoat.

slouch-eared, *a.* (Later example.)
1855 *Trans. Mich. Agric. Soc.* VI. 511 The original English breed [of hogs]—those long-legged,..slouch-eared, big-headed..animals.

slough, *sb.*¹ **4.** For *U.S.* read *N. Amer.* and add: (Earlier and *Canad.* examples.) Also, a side channel of a river, or a natural channel that is only sporadically filled with water.
1714 *Rep. Record Commissioners* (Boston Registry Dept.) (1877) III. 217 Between his old house & the Slough or Small Bridge. 1858 W. P. Blake *Rep. Geol. Reconnaissance in California* i. 10 There lay outstretched the broad and green Tulares—great swamps or lowlands overgrown with rushes and threaded by the sinuous channels and sloughs of the river. 1859 *Brit. Colonist* (Victoria, B.C.) 17 Dec. 3/2 At Old Langley, the slough is entirely frozen up. 1888 D. M. Gordon *Mountain & Prairie* 143 At the same time there are many sloughs, or 'slews' so-called, where part of the river flows by some devious and half-hidden course. 1913 Thomas & Watt *Improvement of Rivers* (ed. 2) I. i. 30 In valleys with narrow bottom lands the result is a slough or drain close to the hills which returns the water to the main channel further down, one slough succeeding another along the valley. 1924 M. H. Mason *Arctic Forests* 225 There was an Indian toboggan trail on the long slough, past Jenny Island up to the eight-mile point. 1932 C. R. Longwell et al. *Textbk. Geol.* I. iii. 60 Most [short cuts]..are abandoned as the flood subsides and are left as sloughs, which are slowly undercut as the meander shifts downstream. 1939 W. Häntzschel in P. D. Trask *Recent Marine Sediments* III. 202 The sloughs (Priele) on the tidal flats are comparable to rivers and brooks. *Ibid.*, Where the range in tides in Jade Bay is as high as 3·6 meters, the sloughs are deeply incised. 1962 W. Stegner *Wolf Willow* I. i. 8 In deep sloughs tules have rooted, and every such pond is dignified with mating mallards. 1970 Leopold & Wolman in G. H. Dury *Rivers & River Terraces* vii. 199 Opposite the gravel island is a slough aligned with a grassed depression. Both features undoubtedly carry water during flood flow. 1974 P. Gzowski *Bk. about this Country* 20, I remember seeing a bunch of geese sitting in a little slough. 1976 *Prof. Paper U.S. Geol. Survey* No. 929. 150/2 The ecological model is designed to relate the wildlife in the Shark River Slough to the availability of food and water.
5. slough bass, for def. read: a black bass of the genus *Micropterus* (earlier example); **slough grass,** for def. read: one of several coarse grasses of swampy ground, esp. a species of the genus *Muhlenbergia* (earlier and later examples); **slough hay** *Canad.*, (hay made from) slough grass.
1877 C. Hallock *Sportsman's Gaz.* 276 Locally they are termed perch..slough bass, etc. 1860 *Trans. Ill. Agric. Soc.* IV. 488 Then [I] make a band of whatever material I have at hand, (slough grass is preferable). 1907 L. H. Bailey *Cycl. Amer. Agric.* II. 454/1 In wet and swampy places, slough-grass (*Spartina*) furnishes a supply of coarse hay. 1980 *Country Life* 13 Nov. 1819/3 The hay is made of wild slough grass. 1934 G. Bettany *Valley of Lost Gold* 264 In the tall slough hay beside them orange lilies raised their heads waist high. 1948 T. Onraet *Sixty Below* 135, I have often seen them kneeling on their forelegs to feed in comfort on short willows and slough hay. 1955 *Sentinel-Courier* (Pilot Mound, Manitoba) 31 Mar. 4/1 (Advt.), For sale—Baled slough hay, wire tied. 1968 S. E. Roberts *Of Us & Oxen* ii. 14 This 'slough hay' is said to be less nutritious than the 'upland hay' cut from buffalo grass.

slough, *v.*¹ Add: **a.** Also without *up*. (Earlier example.)
1861 in *Daily Colonist* (Victoria, B.C.) (1911) 16 Apr. 1/6 Several of the wagons while conveying passengers and freight from the steamer on Sunday night became sloughed and the passengers were compelled to 'foot it' to town.
b. *slang.* To imprison, to lock (*up*). Usu. in passive.
1848 *Ladies' Repository* Oct. 317/1 *Slough*, to lock. 1894 'J. Flynt' in *Century Mag.* Feb. 518/2 I've boozed around this town..for seven years, and I've not been sloughed up yet. 1926 J. Black *You can't Win* vii. 87 They'll..haul us over to Martinez..an' slough us in the county jail. 1935 A. J. Pollock *Underworld Speaks* 108/2 *Sloughed*, arrested.

slough, *v.*² Add: **5.** *intr.* Of soil, rock, etc.: to fall *away* or slide *down* into an adjoining hole or depression.
1897 W. Starling *Floods of Mississippi* i. 14/1 Water leaking through the old bank infiltrates the new earth and it sloughs away bodily. 1942 W. Faulkner *Go down, Moses* 30 As though the whole mound had stooped roaring down at him—the entire overhang sloughed. 1955 Hennes & Ekse *Fund. Transportation Engin.* ii. 30 The processes of weathering tend to loosen surface material and cause it to slough and drift down any slope greater than the angle of repose of the dry loose material. 1957 A. C. Clarke *Deep Range* v. 54 Sometimes, in deep ocean waters far from the eternal rain of silt which sloughs down from the edges of the continents, it was possible to see as much as two hundred feet. 1974 P. L. Moore et al. *Drilling Practices Manual* iii. 46 Shale sloughs into the hole.
sloughed *ppl. a.* (further example).
1897 W. Starling *Floods of Mississippi* i. 14/1 A good thick dressing of brush is laid on the sloughed mass.

sloughi, var. *SALUKI.

sloughing, *vbl. sb.* Add: **3.** The collapse of soil or rock into a hole or down a bank.
1897 W. Starling *Floods of Mississippi* i. 14/1 There is no incident..more alarming..than the sloughing or slipping of the inside slope of a levee. 1948 Terzaghi & Peck *Soil Mech. in Engin. Pract.* viii. 336 To prevent sloughing of the toes of the slopes, the small quantity of water that flows through the gaps between the wells is removed. 1957 *Nature* 13 July 100 (*heading*) Vernal sloughing of sludge deposits in a sewage effluent channel. 1972 L. M. Harris *Introd. Deepwater Floating Drilling Operations* vi. 90 On completion of drilling, the hole is normally filled with gel-water mud to prevent sloughing and fill.

sloughing, *ppl. a.* Add: **2.** Of soil or rock: collapsing, sliding.
1974 P. L. Moore et al. *Drilling Practices Manual* iii. 46 It has been common practice to reduce water-loss when sloughing shale becomes a problem.

slougi, var. *SALUKI.

Slovak, *sb.* and *a.* Add: Now usu. with pronunc. (slǝᵘ·væk). **A. 1.** For 'dwelling.. Hungary' read 'dwelling in Slovakia, formerly part of Hungary, now the Slovak Socialist Republic and part of Czechoslovakia'.
Slovakian *a.* and *sb.*: now usu. with pronunc. (slovæ·kiăn); **Slovakish** *a.* and *sb.* (earlier examples).
1850 'Talvi' *Hist. View Lang. Lit. Slavic Nations* III. i. 217 A voluminous Slovakish dictionary. *Ibid.* 219 Books written in Slovakish.

sloven, *sb.* Add: **5.** *Canad.* (See quots. 1895, 1941.) Also called *sloven-wagon*.
1895 *Dialect Notes* I. 381 *Sloven*, a low truck wagon. 1907 *Canad. Mag.* XXIX. 442/1 It is called a 'sloven-waggon' (doubtless for some good reason). 1941 H. MacLennan *Barometer Rising* 11 Grinding on the cobblestones behind a pair of plunging Clydesdales came one of Halifax's most typical vehicles, a low-slung dray with a high driver's box, known as a sloven. 1964 *Atlantic Advocate* Aug. 79 As evening approached the horses were hitched to a long, low wagon, known in our country as a 'sloven', and the apples were hauled to the house.
6. *Forestry.* (See quot. 1957.)
1946 F. Sargeson *That Summer* 175 The stumps still had the sloven sticking up. 1953 H. L. Edlin *Forester's Handbk.* xiii. 201 Only when the crown of the tree strikes the ground will the last link be broken; the 'hinge' will then break, the tree pulling an irregular splinter of wood out of the stump below it. This splinter, or sloven, is then sawn off. 1957 *Brit. Commonwealth Forest Terminol.* II. 178 *Sloven*, the torn splintered portion of timber left on a stump or the end of a butt, where the key finally broke when the tree fell.

sloven, *v.* Add: Hence **slo·vened** *ppl. a.*, done in a slovenly manner.
1937 'G. Orwell' *Road to Wigan Pier* I. i. 17 It was not only the dirt.., but the feeling..of having got down into some subterranean place where people go creeping round and round..in an endless muddle of slovened jobs and mean grievances.

Slovene, *sb.* Add: **2.** The language of the Slovenes.
1911 *Encycl. Brit.* XXV. 245/2 Except for a few 15th-century prayers and formulæ we do not find any more specimens of Slovene until the Reformation. 1960 O. Manning *Great Fortune* III. 174 David smiled down modestly. 'My Slovene is a little rusty,' he said. 1972 W. B. Lockwood *Panorama Indo-Europ. Lang.* ix. 161 Slovene is the official language of the Constituent Republic of Slovenia. 1980 *English World-Wide* I. 256 Of the remaining essays not involving English, most are on minority languages, such as..the individual cases of Slovene in Southern Austria.

Slovenian, *a.* and *sb.* Add: **a.** *adj.* (Further examples.)
1902 *Encycl. Brit.* XXVI. 16 Of late years attempts have been made to turn the Slovenian national movement into this direction, and to attract the Slovenians also towards the Orthodox non-Austrian Slavs. 1922 M. S. Stanoyevich *Early Jugoslav Lit.* viii. 69 The works of Croatian and Slovenian authors..hardly contributed anything towards progress in linguistics science and literature. 1960 O. Manning *Great Fortune* I. 5 When they woke the next morning they were on the Slovenian plain.
b. *sb.* (Further examples.) Also, a Slovene.
1885 [see *CROATIAN *sb.* and *a.*]. 1902 [see sense a above]. 1955 [see *MACEDONIAN *sb.*¹ c]. 1978 *Language* LIV. 451 The beginner thus has to look in the glossary if he wishes to ascertain whether the forms cited in these examples are younger or older Old Slavonic, Old Slovenian, or 15th-century Russian spellings.

Slovincian (slovi·nsiăn). Also **Slovintian, Slovinzian.** [f. F. *Slovince*, ad. G. *Slowinze*, f. Kashubian *Słowińci* + -IAN.] An extinct dialect of Kashubian.
1883 *Trans. Philol. Soc.* 1880–1 373 With regard to Baltic Slavonic, represented by its only surviving dialects, Slovintian and Cassubian, I shall treat it as an independent language. 1934, 1935 [see *KASHUBE]. 1939 L. H. Gray *Foundations of Lang.* 355 Polish (closely connected with Kashubian and Slovincian). 1965 G. Y. Shevelov *Prehist. Slavic* i. 1 Slovincian was still extant in the early twentieth century in the district of Słupsk, now in north-western Poland. 1972 W. B. Lockwood *Panorama Indo-Europ. Lang.* 158 Due west of the present Kashubian area lies the district of Słupsk (Stolp), where autochthonous Slavonic survived until the end of the last century, when it was entirely supplanted by German; here the dialects were locally termed Slovinzian. 1974 [see *POMERANIAN *a.*].

slow, *sb.* Add: **2. b.** A slow train.
1956 *Railway Mag.* Mar. 163/2 There is a daily slow, stopping at all stations between Damascus and Deraa. 1976 P. Lovesey *Swing, swing Together* xiii. 55 We can take a train... We can catch a slow to Oxford.
c. A slow tune in popular music.
1956 B. Burns in S. Traill *Play that Music* ii. 34 His style is hot and aggressive—pushing the beat in fast numbers and rhapsodic in slows. 1977 J. Wainwright *Do Nothin' till you hear from Me* xi. 184 In that set of standard slows, this ten-piece [band] of mine was like Ellington's piano.
3. a. (Earlier example.)
1854 F. Lillywhite *Guide to Cricketers* 84 [He] is a good bat, and can bowl 'slows' well.
4. b. *the slows* (colloq.), an imaginary disease or ailment accounting for slowness.
1843 *Ainsworth's Mag.* IV. 124 If somewhat troubled with 'the slows', not a hound but was as true as the sun. 1843 J. C. Shairp *Let.* 25 Sept. in W. A. Knight *Principal Shairp* (1888) vi. 71 This..makes good my summer's work. 'The Slows' are my bane, but I must be courageous and face what remains... If I could but secure a *second*, I should be happy. 1927 *Daily Express* 13 Dec. 16/2 Rimell's mare, How Nice, had a fit of the slows, for she was always in the next division from start to finish. 1970 D. Francis *Rat Race* viii. 102 They might as well send him [*sc.* a racehorse] to the knackers. Got the slows right and proper, that one has.

slow, *a.* Add: **I. 4.** (Later examples.) *slow learner*: spec. in *Educ.* (see quot. 1981).
1938 *High Points* Apr. 33/1 The problem of the slow learner continues to grow involved as increased numbers flow into high school. 1945 'O. Malet' *My Bird Sings* ii. 121 Your sister does not benefit so swiftly by her education... She is a slow learner. 1946 P. Bottome *Lifeline* xvi. 144 She wanted everything... I'm exactly the opposite in love. I'm a slow starter. I want little. 1963 *Times* 25 May 9/4 Here is comfort for the slow-starter who begins at the bottom. 1975 *Times* 26 Apr. 7/4, I was 24 and I'd run away from home... I was kind of a slow starter. 1977 *Wandsworth Borough News* 7 Oct. 28/2 (Advt.), Tutorials: 'A' and 'O' level, most subjects; public examinations; slow learners welcome. 1981 D. Rowntree *Dict. Education* 286 *Slow learner*. A term often used rather loosely of any child whose attainments have always fallen noticeably behind those of other children of the same age, without any implication as to what might be thought to be the cause..or whether the child might be enabled to speed up or catch up. Sometimes, however, the term is used to indicate children who are not only expected to remain slow learners but also to be unable ever to learn as much as others. Some people would even restrict the term to pupils who are educationally subnormal.
6. (Further examples, of solemn or tragic music.) *slow handclap*: see *HANDCLAP c.
1826 F. Reynolds *Life & Times* II. 144 The curtain to the new piece having risen, the heroine entered to slow music. 1895 *World* 4 Dec. 27/1, I could see the conductor of the orchestra waiting eagerly for the word 'mother'—the cue for the slow music—and I was, oh! so thankful when it came. 1926 G. B. Shaw *Translations & Tomfooleries* 78 You were not found..with the limelight streaming on your white face, and the band playing slow music.
7. b. Of an oven: of such a temperature that it cooks slowly.
1747 H. Glasse *Art of Cookery* ix. 113 Bake it in a slow oven, with crust as above. 1846 A. Soyer *Gastronomic Regenerator* 571 Place them in a slow oven to bake. 1917 F. Klickmann *Between Larch-Woods & Weir* xiii. 242 She had told Dick to put the patties into a slow oven for ten or twelve minutes before eating. 1973 *Times* 1 Dec. 11 Place the casserole in a slow oven.
c. *slow burn* fig.: (*a*) (see *BURN *sb.*³ 1 d); (*b*) *Theatr.*, delayed response or slow reaction to a joke; also *attrib.*
1975 *Daily Tel.* 10 Feb. 12 For some of her jokes in public Margaret Thatcher relies..on what professional comedians call the 'slow burn'. 1975 D. Lodge *Changing Places* i. 22 The realization..strikes him, like a slow-burn gag in a movie-comedy.
8. d. *slow poke, slowpoke* (colloq., chiefly *U.S.*) = Slow-coach; also *attrib.* or as *adj.*, slow, idle. Cf. Poke *sb.*³ 3.
1848 Bartlett *Dict. Americanisms* 255 'What a slow poke you are!' A woman's word. 1877 F. Ross et al. *Gloss Words used in Holderness* 128/2 *Slaw-pooak*.., a dunce; a driveller. 1920 G. Ade *Hand-Made Fables* 226 He placed the Experiences of an ordinary slow-poke year into one Week. 1935 J. T. Farrell *Judgment Day* xvi. 367 Slackers, slow-pokes, easy-going, unambitious fellows, I neither want, nor can tolerate. 1959 *Guardian* 31 Aug. 2/3 'Slowpokes'—unnecessarily slow drivers—cause other road users to take additional risks. 1971 W. H. McNeill in A. Bullock *Twentieth Century* 54/1 What could a slow-poke airplane do against intercontinental missiles? 1981 S. Rushdie *Midnight's Children* I. 112 Come on, slowpoke, you don't want to be late.
II. 9. a. *slow pass*: spec. in *Bridge* (see quot. 1934); *slow-scan* (Television), scanning at a much slower rate than in ordinary television, so that the resulting signal has a much smaller bandwidth and can be transmitted more cheaply; usu. *attrib.*
1934 *Amer. Speech* IX. 11/1 A *slow pass* is a pass preceded by a long period of thought, and since it conveys the impression that some strength is held without paying the regular price of a bid, is regarded as unethical. 1955 *Sun* (Baltimore) (B ed.) 7 Dec. 3/4 The so-called slow-scan system paints a picture every two seconds and requires only 8,000 cycles. *Ibid.* 3/5 Slow scan cannot handle objects in motion. 1960 *Guardian* 20 Apr. 4/1 The National Broadcasting Company will show short excerpts to American audiences at breakfast-time by using the 'slowscan' process and transmitting film by cable to Montreal. 1970 N. Armstrong et al. *First on Moon* v. 110 TV was tried on Gordon Cooper's Mercury flight, using a slow-scan black and white camera. 1973 *Times* 2 June 10/6 It was agreed when North South protested the score that East's pass over Three Spades was a 'slow pass'. 1975 D. G. Fink *Electronics Engineers' Handbk.* xi. 62 Slow-scan tubes are useful for remotely located cameras requiring a data link to the monitor wherein the band-width reduction significantly lowers the data-link cost.
b. *slow bell* (Naut., N. Amer.), a bell signalling that a ship should proceed slowly; chiefly in prepositional phrases, at a reduced speed; also *fig.*; *slow puncture*, a puncture from which the air escapes gradually.
1901 *Daily Colonist* (Victoria, B.C.) 5 Nov. 3/2 Early in the evening she [*sc.* S.S. City of Seattle] had run among a number of small icebergs and she was coming down the channel under a slow bell. 1944 *Amer. Speech* XIX. 108 Another of the best [phrases], used especially in declining a drink or an extra job of work, is 'Not me, thanks; *I'm taking it on the slow bell*.' 1946 *Seafarers' Log* 19 Apr. 3/4 There will be no slow bell on the organizing drive. 1958 *Wall St. Jrnl.* 17 Dec. 26/2 We were at slow-bell for much of '58 because of the recession. 1958 E. Newby *Short Walk in Hindu Kush* xviii. 217 Our airbeds had slow punctures and the ground was hard. 1968 C. P. Bracken *Roman Ring* xv. 147 He had left the Fiat at a garage outside Rome to have a slow puncture repaired. 1974 A. Morice *Death of Heavenly Twin* xii. 124 I've got a flat tyre. It's probably a slow puncture
10. c. (Further examples.) Also of other photographic items: necessitating longer exposures (e.g., in the case of a lens, because its aperture is small).
1902 *Encycl. Brit.* XXXI. 695/1 Owing to the small working aperture it [*sc.* a lens] seems slow, but it is not so for the definition and flatness of field obtained. 1915 D. Grant *Manual of Photogr.* 74 'Slow' papers give plucky results from flat negatives. 1957 E. S. Bomback *Photogr. in Colour* x. 107 The use of slow contact printing paper..may result in negative or partially negative images in the print. 1973 *Sci. Amer.* Dec. 39/3 Telescopes such as the 100-inch reflector on Mount Wilson, which has a focal ratio of f/5, are quite 'slow', that is, they require long exposure times.
d. Med. *slow* (*virus*) *disease* or *infection*, any of various progressive diseases caused by a virus or virus-like organism that multiplies slowly in the host organism and having incubation periods of months or years; hence (by a false analysis) *slow virus*.
1954 B. Sigurdsson in *Brit. Vet. Jrnl.* CX. 350 If the word chronic is taken to mean not only protracted, but also something which lingers on, has an irregular and unpredictable course and may end in any one of several different ways, then the expression should not be used about the diseases I have discussed here; these infections should perhaps rather be called *slow* infections. *Ibid.* 352 It seems as if an important group of slow virus infections is gradually coming to light. 1967 *New England Jrnl. Med.* CCLXXVI. 392/1 (*heading*) Slow-virus infections of the nervous system. 1971 *Jrnl. Virol.* VII. 301 Progressive pneumonia virus, the causative agent of slow, pulmonary disease of Montana sheep, was shown to be antigenically related to two other slow viruses of sheep, visna and maedi. 1976 R. H. Kimberlin *Slow Virus Diseases of Animals & Man* i. 5 One major distinguishing feature of slow diseases..was this: once clinical signs of disease have appeared the disease then follows a regular progressive course which always ends in serious illness and usually death. 1977 *Sci. Amer.* May 140/2 A dozen fatal diseases of the human central nervous system stand suspect of slow-virus origin.
III. 13. a. *slow-pitch* (*softball*) (U.S.), a type of softball in which each pitch must travel in an arc of a specified minimum height; *slow wheel* spec., a type of potter's wheel turned at a slow speed.
1946 W. B. Honey *Art of Potter* II. ii. 12 A device was introduced for rotating the pot while it was being built, on a horizontal pivoted table or disc..; this 'slow wheel' presumably led to the invention of the fully developed potter's wheel. 1964 H. Hodges *Artifacts* i. 28 Many archaeologists are in the habit of distinguishing between a *slow wheel*, that is to say a device in which the movement of the wheel-head is either intermittent or relatively slow, and a fast wheel. 1971 *Canadian Antiques Collector* Apr. 16/1 Most frequently Ipswich ware is formed on a 'slow wheel' which is in principle a freely-revolving turntable. Both pot and wheel were revolved by hand. 1971 *N.Y. Times* 6 June 95/5 With the opening of the annual Long Beach Slow Pitch Softball League..policemen and addicts met for the first time in friendly competition. 1976 *Billings* (Montana) *Gaz.* 2 July 1-c/1 Softball fans will have a chance to glimpse some of the top men's slowpitch teams in the northwest during the Town & Country-Corner Pocket Men's Slowpitch Softball Tournament.
c. Nuclear Physics. *slow neutron*, a neutron with little kinetic energy, esp. as a result of being slowed down by a moderator; freq. = **thermal neutron*; also *attrib.*; so *slow reactor*, a reactor in which fission is produced primarily by moderated neutrons; a thermal reactor.
1934 *Chem. Abstr.* XXVIII. 2263 The slow neutrons are probably emitted from Be when an excitation of a Be nucleus by α-particles occurs without capture. 1938 R. W. Lawson tr. *Hevesy & Paneth's Manual of Radio-activity* (ed. 2) x. 112 These slow neutrons are produced by allowing fast neutrons, such as those emitted by a radon-beryllium source, to pass through water, paraffin wax, or other substances containing hydrogen. 1945 H. D. Smyth *Gen. Acct. Devel. Atomic Energy Mil. Purposes* ii 23 For a slow-neutron chain reaction using a moderator and unseparated uranium it was almost certain that tons of metal and of moderator would be required. 1949 *Atomics* Sept. 45/2 The so-called 'slow neutron reactor'. These reactors take advantage of the fact that neutrons produce fission in uranium more easily as they go slower. 1958 *Chambers's Techn. Dict.* Add. 1014/1 Slow reactor. 1964 M. Gowing *Britain & Atomic Energy 1939–1945* iv. 115 Bombs were discussed in terms of slow neutron chain reactions. 1973 *Daily Tel.* 12 Oct. 8/5 Fast reactors, cooled by sodium instead of carbon dioxide, as in slow reactors now used, can provide more heat more rapidly.
14. a. *spec.* (i) *slow motion*, (*a*) motion of slower speed than normal; also *attrib.*; (*b*) *Cinemat.*, the technique of shooting a film at a faster speed than normal so that when it is projected the action will appear to be slowed down; also *transf.*, and *attrib.*
1801 J. Strutt *Sports & Pastimes* III. ii. 130 At a show in the country, about forty years ago, which was contrived in such a manner, that the whole group descended and ascended with a slow motion to the sound of music. 1834 *Mechanics' Mag.* 4 Oct. 16/2 The gear was changed from the quick to the slow motion. 1860 [in Dict.]. 1878 [in Dict., sense 16 d]. 1903 *Work* 4 July 341/2 Next make the ball and vertical slow-motion screw. 1924 *Spectator* 3 May 720/1 Its deliberation becomes a separate quality, akin to the slow motion of the cinema. 1929 E. Wilson *I thought of Daisy* ii. 102 A slow-motion diving picture of champion woman swimmers. 1938 W. S. Churchill *Let.* 9 June in M. Gilbert *W. S. Churchill* (1976) V. xlvi. 946 In all my experience of public offices..I have never seen anything like the slow-motion picture which the work of this Committee has presented. 1943 J. B. Priestley *Daylight on Saturday* xi. 70 The sound of their slow-motion patter..just gets your goat. 1956 J. Baldwin *Giovanni's Room* I. ii. 57 Like those figures in slow motion one sometimes sees on the screen. 1962 F. I. Ordway et al. *Basic Astronautics* vii. 316 This test is accomplished by firing the model engine in a wind tunnel and observing the flow of the exhaust gases with the aid of slow motion cameras. 1973 *Listener* 23 Aug. 244/1 Time on Friday moved very slowly. Everything seemed to be as if in a slow motion film.
(ii) Of a ballroom dance: with steps at walking-pace.
1928 '*H.M.V.' Catal.* 76/2 Slow F[ox]-T[rot]. 1949 A. Chujoy *Dance Encycl.* 76/2 *Baston*, a form of slow waltz..in which the couples turn in circles in several directions. 1949 A. Wilson *Wrong Set* 56 She was almost lying in Bruce's arms as he carried her through the slow foxtrot. 1962 J. Braine *Life at Top* ii. 54 They were taking their partners for a waltz now, a slow waltz, an old waltz from the days of Carroll Gibbons and the Orpheans. 1966 'M. Halliday' *Wicked as Devil* ix. 79 The floor was crowded for a slow foxtrot. 1976 *Listener* 29 July 105/2 Going to the dancing class to practise the slow waltz and the tango.
c. (Earlier example.)
1868 *J. Lillywhite's Cricketers' Compan.* 61 The wickets were in excellent order, though somewhat 'slow' for Gravesend.
d. Also of a lane of a dual carriageway or motorway: intended for vehicles which are not overtaking; also *fig.*
1967 'M. Carrel' *Dark Edge of Violence* ix. 80 The police car..dropped back and sulked along in the slow-lane. 1969 R. Petrie *Despatch of Dove* III. 146 He'd put himself in a silly position, in the slow lane without the chance of a turn-off. 1972 B. Everitt *Cold Front* v. 38 'Are we conversing or making love?'..'Let's go into the slow lane for a minute. What are you doing between leaving Medicom and joining the Grand Old Man?' 1977 R. Rendell *Judgement in Stone* xix. 151 Joan's driving had become erratic, and her jerky zigzagging from slow lane to fast was a frightening experience.
V. 16. a. Parasynthetic, as *slow-minded, -motioned*. Also *slow-mindedness*.
1930 D. H. Lawrence *A Propos of Lady Chatterley's Lover* 9 The evocative power of the so-called obscene words..perhaps are [*sic*] still too strong for slow-minded, half-evoked lower natures to-day. 1935 Kipling *Two Forewords* 16 The seller..berated, for their slow-mindedness, men who, but for being too much urged to buy, would have bought. 1856 'Mark Twain' *Let.* 10 June (1917) I. 33 They are either excessively slow motioned or very lazy. 1951 N. G. Annan *Leslie Stephen* x. 285 Progress for him was an incalculable and slow-motioned operation.
b. With verbs, as *slow-clap, -foot* (later example), *-handclap* (see *HANDCLAP c), *-march* (later examples), *-steam, -waltz*.
1960 Slow-clap [see *OUT-SCORE *v.*]. 1979 *Guardian* 26 Oct. 2/1 Mrs Thatcher was also slow-clapped and heckled..during her speech. 1885 W. B. Yeats in *Dublin Univ. Rev.* May 82/2, I see the night, Deep-eyed, slow-footing down the empty glade. 1927 T. E. Lawrence *Let.* 19 May (1938) 516 These long dreary slow-marching books are invaluable friends in Drigh Road. 1977 *Times* 30 July 2/3 The RAF colour squadron slow-marched the old colour..off the parade ground. 1975 *Petroleum Economist* Sept. 356/3, 22 per cent of the active fleet..was estimated to be slow-steaming in July. *Ibid.* 341/2 Due to slow-steaming, fuel consumption..fell. 1977 *Living with Tanker Surplus* (Shell Internat. Petroleum Co.) 5 The financial rewards for slow-steaming small vessels are greater than for large vessels, because a small ship consumes relatively more bunkers per tonne of cargo. 1976 'P. B. Yuill' *Hazell & Menacing Jester* viii. 81 He recovered a bit as I was slow-waltzing him to the stairs.

c. With adjs., as *slow-syruppy*.
1922 JOYCE *Ulysses* 261 Neatly she poured slowsyruppy sole.

d. With sbs., as *slow-beat*, *-burn* (see sense *7 c), *-gait*, *-growth*, *-neutron* (see sense *13 c), *-pitch* (see sense *13 a), *-poke* (see sense *8 d), *-scan* (see sense *9 a), *-speed*, *-tempo*; *slow-release* adj., (*a*) *Electr.*, applied to a relay in which the delay between a de-energizing signal and the opening of the contacts is intentionally increased; (*b*) characterized by the slow release of an active substance (as by a pharmaceutical preparation); *slow-wave* adj., *spec.* applied to the commonest form of non-REM sleep, in which brain waves having a frequency of ½ to 3 hertz, lower than that of alpha or beta waves, are detectable.
1965 *Economist* 4 Sept. 857/2 Egypt's slow-beat socialism is not so agreeable to Moscow as was Algeria's hot rhythm under Mr Ben Bella. **1977** J. WAINWRIGHT *Do Nothin' till you hear from Me* viii. 125 He can blow a beautiful slow-beat chorus. **1940** C. DAY LEWIS tr. *Virgil's Georgics* II. 41 Waggonloads drawn home by the slow-gait oxen. **1965** H. I. ANSOFF *Corporate Strategy* vi. 109 The electronics industry ranges from high growth in technologically sophisticated areas, such as optical electronics, to slow-growth consumer oriented product-markets, such as radio and television. **1970** *Globe & Mail* (Toronto) 26 Sept. B2/2 Ontario Development Corp. has granted loans totalling more than $1-million to 10 companies planning to set up plants in slow-growth areas of the province. **1928** *Jrnl. Inst. Electr. Engineers* LXVI. 342/2 The No. 2 armature is especially suitable for slow-release relays. **1961** *Lancet* 29 July 230/1 Possibly the diminished efficacy of penicillin in the treatment of uncomplicated gonorrhœa may be due to the wide use of slow-release preparations giving a prolonged, though low, blood concentration. **1969** S. F. SMITH *Teleph. & Telegr. A* ii. 43 One particularly useful way of obtaining a slow release feature without making the relay slow to operate is to connect a permanent shunt across the operating coil of the relay. **1974** *Nature* 25 Jan. 199/2 The efficiency of slow-release nitrogen was investigated as an initial and maintenance fertiliser treatment. **1958** *Newnes Compl. Amat. Photogr.* 33 An older box camera with a slow-speed shutter can be speeded up very simply. **1971** *Engineering* Apr. 20/1 In the case of slow-speed engines, machining can often be confined to pin and journal diameters and web faces. **1980** *Redbook* Oct. G12/2 Slow-speed film lets you get good pictures even on a sun-drenched sandy beach or glistening ski slope. **1962** D. FRANCIS *Dead Cert* vi. 55 We swayed lazily round the floor to some dreamy slow-tempo music. **1978** T. WILLIAMSON *Technicians of Death* viii. 60 The invisible disc jockey began to play a slow-tempo number. **1963** G. TROUP *Masers & Lasers* (ed. 2) iv. 71 There are three kinds of structures suitable for slow-wave propagation in the paramagnetic maser. **1967** *Physiol. Rev.* XLVII. 118 Slow wave sleep. **1968** *Brit. Med. Bull.* XXIV. 257/1 There is a reduction in the amplitude of the evoked response during rapid eye-movement sleep as compared with slow-wave sleep. **1974** W. P. KOELLA in van Praag & Meinardi *Brain & Sleep* 9 Stages 1 to 4 are referred to as NREM-sleep; the term 'slow-wave-sleep' should be reserved for stages 3 and 4 which are characterized by the 'slow' delta-waves.

slow, *adv.* Add: **1. b.** *slow back*: a direction to a golfer when the club is swung back from the ball in making a stroke.
1886 H. G. HUTCHINSON *Hints on Golf* 17 Golfers have gone so far as to instruct their caddies to say to them, 'Slow back,' so as to keep them in mind of this precept each time they addressed themselves to drive the ball. **1909** P. A. VAILE *Mod. Golf* viii. 136 Gather speed as you go up, but do not get up much speed on the upward swing, for if you do you have to waste energy fighting it at the top of the swing. This is the reason for 'slow back'. **1922** WODEHOUSE *Clicking of Cuthbert* iii. 73 Slow back—keep the head.

2. a. (Further examples.) spec. *slow-reacting substance* (Physiol.), any of various substances that are produced in the body in response to various stimuli and cause slower and longer lasting contraction of smooth muscle than does histamine; *spec.* one released in anaphylaxis; abbrev. SRS (see *S 4 a); *slow-releasing* = *slow-release* adj. (*a*) s.v. *SLOW *a.* 16 d.
1876 'MARK TWAIN' *Tom Sawyer* xxxi. 237 The slow-dragging ages. **1894** KIPLING *Jungle Bk.* 95 Asleep in the arms of the slow-swinging seas. **1904** —— *Traffics & Discoveries* 368 Warn them of seas that slip our yoke Of slow conspiring stars. **1904** H. G. WELLS *Food of Gods* II. i. 142 They were children—slow-growing children of a new race. **1910** J. POOLE *Pract. Telephone Handbk.* (ed. 4) xi. 169 A slow-acting relay. **1912** W. DE LA MARE *Listeners* 24 Whose feathers..Gleam with slow-gathering drops of dew. **1915** D. H. LAWRENCE *Rainbow* i. 1 They were fresh, blond, slow-speaking people. **1916** A. HUXLEY *Burning Wheel* 22 The chime Of bells slow-dying. **1920** E. SITWELL *Wooden Pegasus* 101 A slow-leaking tap. **1921** D. H. LAWRENCE *Tortoises* 10 You..set forward, slow-dragging, on your four-pinned toes. **1921** R. GRAVES *Pier glass* 44 Slow-rising smoke and nothing wrong! **1923** T. E. HERBERT *Telephony* xiii. 334 The special feature of the slow releasing relay..is the extension of the core on which is placed a solid copper collar. **1924** R. CAMPBELL *Flaming Terrapin* iii. 43 The death-cry and the agony supreme Of the slow-drowning world. **1925** Slow-running [see *idling passage]. **1925** F. SCOTT FITZGERALD *Great Gatsby* iii. 59 But I am slow-thinking. **1927** T. WILDER *Bridge of San Luis Rey* II. 38 That city of large girdled Women..slow-moving and slow-smiling, a city of crystal air. **1930** E. POUND *XXX Cantos* v. 20 A Knight with slow-lifting eyelids. **1930** BLUNDEN *Poems* 315 At last slow-mending From the hacked wounds of our proud error's field. **1931** A. HUXLEY *Cicadas* 33 Death in the Scorpion hunts him up the sky..round the vault of time, round the slow-curving year. **1932** W. FAULKNER *Light in August* i. 23 The slowspitting and squatting men. **1932** AUDEN *Orators* II. 75 O charged-to-the-full-in-secret slow-beating heart. **1939** KELLAWAY & TRETHEWIE in *Austral. Jrnl. Exper. Biol.* XVII. 227 The perfusate was also tested for the presence of the substance described by Feldberg and Kellaway (1938) which, after a latent interval, causes contraction of the gut followed by slow relaxation and by characteristic after-changes in the excitability of the muscle. This substance we have called S.R.S. (slow reacting substance). **1947** CROWTHER & WHIDDINGTON *Science at War* 160 The slow-sinking depth charges may be evaded by the submarine. **1948** R. GRAVES *Coll. Poems, 1914–1947* 240 Under your Milky Way And slow-revolving Bear. **1954** J. R. R. TOLKIEN *Fellowship of Ring* 212 A long slow-climbing slope. **1954** A. G. L. HELLYER *Encycl. Garden Work* 18/2 Basic slag is a relatively slow-acting fertilizer. *a* **1957** R. CAMPBELL tr. *Rimbaud's Drunken Boat* in *Coll. Poems* (1960) III. 18 Where with slow-pulsing and delirious fires,..ferments the crimson bitterness of love. **1958** J. W. FREEBODY *Telegraphy* iv. 85/1 For slow operating or slow releasing relays three lengths of slug are employed. **1958** *Pharmacol. Rev.* X. 419 Two entirely different groups of slow reacting substances have been established; one..consists of polypeptides, the other one of lipid-soluble acids. **1964** L. MACNEICE *Astrol.* ii. 66 With slow-running planets the *transits* are the things to look for. **1970** PASSMORE & ROBSON *Compan. Med. Stud.* II. xxiv. 3/2 Histamine, the slow reacting substance (SRS-A) and bradykinin may be the most important agents involved in the pathogenesis of human bronchial asthma. **1975** *New Phytologist* LXXIV. 367 In alpine regions of Africa, there are trends from erect pachycaul plants towards stout slow-growing plant forms.

b. (Further examples.)
1896 KIPLING *Seven Seas* 53 To the grist of the slow-ground ages. *Ibid.* 65 Each man drew his watchful breath slow taken 'tween the teeth. **1910** W. DE LA MARE *Three Mulla-Mulgars* xi. 153 Between his slow-drawled, shakety notes of deep and shrill Nod listened for the least stir in the forest. **1914** KIPLING *For All We have & Are* 3 Comfort, content, delight, The ages' slow-bought gain. **1916** A. HUXLEY *Burning Wheel* 39 While on the windy down-land..The slow-driven sun beheld us. **1944** BLUNDEN *Shells by Stream* 43 Your eyes..Enfold the slow-bloomed scenes. **1949** —— *After Bombing* 34 All other lightning Might be as honey or kind balms slow-melted.

slow, *v.* Add: **3.** Also with *down*.
1971 *Engineering* Apr. 34/1 Most neutron radiography has been carried out using slowed-down thermal neutrons. **1976** 'J. ROSS' *I know what it's like to Die* xxiv. 151 A creeping slowed-down vision with a dreamlike clarity.

slow-down. Also **slowdown.** [f. SLOW *v.* + DOWN *adv.*] The action or process of slowing down; an instance of slowing down; *spec.* (*a*) *U.S.*, a form of industrial protest in which employees work at a deliberately slow pace; (*b*) an economic recession, a decline in productivity or demand. Also *attrib.*
1897 [in Dict. s.v. SLOW *sb.* 5]. **1937** *Daily Progress* (Charlottesville, Va.) 16 Apr. 1/5 An official of the U.A.W.A. said the objective of workers in a slow-down was to 'try to see how little they can do without actually stopping work entirely'. **1939** *Sun* (Baltimore) 10 Jan. 16/3 The slow-down strike was called off late today. **1942** J. STEINBECK *Moon is Down* v. 70 'I think I fixed the mine for a while.' 'What's your trouble?'..'Oh, the usual thing with me—the slow-down and a wrecked dump car.' **1944** *Sun* (Baltimore) 28 Oct. 12/1 Low-quoted stocks..came out in blocks running to 33,000 shares and propped volume, although slowdowns were plentiful. **1950** *Ibid.* 7 Feb. 7/4 Helmstedt, Germany, checkpoint for the Russians' off-again, on-again 'slowdown' of Berlin-bound highway traffic. **1953** *Encounter* Nov. 70/1 In dozens of factories there were slow-downs..or sit-down strikes. **1957** *Economist* 28 Dec. 1132/1 At least 5,000 employees in lead-zinc mines and smelters..have lost their jobs this year... The slow-down extends even to the largest and most diversified producers. **1969** *Nature* 19 Apr. 218/2 The search will now be on for further eccentric behaviour in..the Crab Nebula pulsar which has the shortest period and the fastest slowdown rate of all. **1969** *Guardian* 2 Oct. 12/1 Has the Government any plans to protect Britain from a slowdown? **1973** *Nature* 3 Aug. 260/1 The observed evolutionary slowdown in higher primates may simply be the consequence of progressive prolongation of their generation time. **1977** *New Yorker* 24 Oct. 79/2 The mess-hall workers staged a work slowdown. **1977** J. W. KENDRICK *Understanding Productivity* iv. 35 The factors explaining the 1966–69 slowdown also help to explain why productivity advance recovered..during the final complete subperiod. **1978** *Sci. Amer.* July 28/2 Energy is required, for example, to accelerate the vehicle from traffic stops and slowdowns, to climb hills and to overcome the rolling resistance of the tires. **1980** A. COPPEL *Hastings Conspiracy* xxviii. 184 The switchboard operators are starting a slow-down over wages. **1981** *Washington Star* 14 Jan. F4/2 The University of the District of Columbia barely overcame the visitors' slow-down tactics and escaped with a bizarre 28–27 victory last night at D.C. Armory.

slow drag. Also **slow-drag.** [f. SLOW *a.* + DRAG *sb.*] A slow blues rhythm; a piece of music in this rhythm; a dance characterized by dragging of the feet, performed to this rhythm. Also *attrib.* Hence **slow-drag** *v. intr.*, to dance the slow drag.
1911 S. JOPLIN *Treemonisha* (music score) XXVII. 215 Directions for the Slow Drag. 1. The Slow Drag must begin on the first beat of each measure. 2. When moving forward, drag the left foot; when moving backward, drag the right foot, [etc.]. **1924** *Etude* Aug. 527/1 An exceedingly popular jazz is the slow drag. **1926** A. NILES in W. C. Handy *Blues* ii. 9 The songs were woven of the same stuff as the other overlapping items..the over-and-overs, slow-drags, pats, and stomps. **1935** Z. N. HURSTON *Mules & Men* I. ix. 185 The piano was throbbing like a stringed drum and the couples slow-dragging about the floor were urging the player on to new lows. **1941** W. C. HANDY *Father of Blues* (1957) vi. 76 It was not easy for me to concede that a simple slow-drag and repeat could be rhythm itself. **1949** B. A. BOTKIN *Treas. S. Folklore* v. i. 745 This is one of the best-known of the slow-drag work songs sung by Negro prisoners in South Texas. **1968** P. OLIVER *Screening Blues* iii. 94 The shuffle of the 'slow drag' would have been effective at the medium tempo of the tune. **1970** —— *Savannah Syncopators* 57 On occasions..'slow-dragging' couples may be seen to be shuffling in a manner reminiscent, perhaps, of the 'shout'. **1976** A. MURRAY *Stomping Blues* vi. 83 There is Bessie Smith's slow-drag version..recorded in 1925. **1980** *Amer. Speech* LV. 210 They slow-dragged to low-down blues.

slower, *sb.* For *rare*⁻¹ read *rare* and add later examples. Also with *down*.
1947 CROWTHER & WHIDDINGTON *Science at War* 144 The most suitable substance for use as a slower-down was heavy water. **1961** P. FRANKAU *Pen to Paper* 64, I know also that one fatal slower of the pace can be found in the adjectives.

slowing, *vbl. sb.* (Examples with *up*.)
1900 [see *CONTROL *sb.* 3 c]. **1960** E. DELAVENAY *Introd. to Machine Translation* 82 This means a slowing-up in the matching of input words with words stored in the dictionary. **1965** J. POLLITT *Depression & its Treatment* vi. 82 In the elderly, particularly those in the senium, the effects of retardation are added to the existing slowing-up process associated with aging.

slowly, *adv.* **3. a.** (Further examples.)
1921 D. H. LAWRENCE *Tortoises* 33 That mud-hovel of his slowly-rambling spouse. **1926** J. S. HUXLEY *Essays in Pop. Science* 195 They [*sc.* frogs in Arctic zones] would be so slowly-working that their growth would be prolonged over an uneconomic number of seasons. **1931** R. CAMPBELL *Georgiad* ii. 47 A blush began to dawn As rosy as a slowly-cooking prawn. **1937** *Discovery* July 196/1 Long strings of slowly-moving camels. **1953** S. SPENDER tr. *Rilke's Life of Virgin Mary* 15 And if the gleam from distant flambeaux plays On slowly-nearing ceremonial dresses. **1965** PHILLIPS & WILLIAMS *Inorg. Chem.* I. viii. 292 Now in many of these inter-combinations *dA/dx* is a slowly-varying function, sometimes over relatively small ranges but often over very wide ranges of composition. **1967** *Oceanography & Marine Biol.* V. 468 On soft substrata two biocoenoses may be recognized, namely, one on beaches..with quickly-drying wracks and a second where slowly-drying wracks are present.

sloyd, *sb.* Add: Also **slöijd.** (Earlier example.)
1884 [see FAD *sb.*[1] 1].

slub, *sb.*[2] Add: **2.** A yarn containing thickened parts, or slubs, at intervals; a fabric woven from such a yarn. Also *attrib.*, having an irregular effect given by a warp of uneven thickness.
1928 *Daily Express* 14 Aug. 4 Slub reps, artificial silk velour, chenille combined with artificial silk..are shown. **1940** *Chambers's Techn. Dict.* 779/1 *Slub yarns*, fancy yarns with thickened parts at frequent intervals. **1946** *Daily Progress* (Charlottesville, Va.) 30 Apr. 12/7 Solid color blue slub-weave and white striped blue chambray. **1957** *Textile Terms & Defs.* (Textile Inst.) (ed. 3) 42 Spun slubs may be produced by an intermittent acceleration of one pair of rollers during spinning. **1958** *Observer* 24 Aug. 7/7 Structured skirt fabrics come from Manchester as well as from Switzerland, but there they go by the general, somewhat unfeminine designation of 'slubs'. **1974** *News & Reporter* (Chester, S. Carolina) 24 Apr. 6-A (Advt.), Textured rayon with interesting slub highlights. **1979** *Radio Times* 5–11 May 14/1 In his choice of clothes—slub silk tie, silk pocket handkerchief, hound's-tooth jacket with turn-back cuffs—as in his choice of claret he declares himself a man of some style.

Hence **slubbed** *a.*, of fabrics: containing slubs.
1961 *Harper's Bazaar* June 55 Three piece in white slubbed rayon. **1980** M. DRABBLE *Middle Ground* 171 Her vast off-white, gold-tasselled, slubbed-silk settee.

slubber, *v.* Add: **4.** Also with *through*.
1941 AUDEN *New Year Let.* I. 24 Time and again have slubbered through With slip and slapdash what I do.

sludge, *sb.* Add: **1. a.** (Further examples: cf. *SLUDGE *v.* 6.)
1946 L. D. STAMP *Britain's Struct. & Scenery* iii. 20 In tundra lands the sub-soil remains permanently frozen whilst the surface thaws in summer and, where there are steep slopes, masses of sludge slide downhill. **1959** G. H. DURY *Face of Earth* xv. 181 The loess records cold, very dry climate. The underlying sludge was formed in a preceding episode of moister conditions when the topsoil thawed annually.

c. The colour of sludge. Also *attrib.*
1962 *Sunday Express* 4 Feb. 19/2 Pyjama stripes—navy, 'sludge' or 'smog' on white. **1977** *Vogue* Feb. 90/1 From a commanding frilly and deep sludge dress, Merle Park changes into her working clothes.

2. *spec.* Such material formed as waste in

various industrial and mechanical processes. (Further examples.)
1920 CROSS & BEVAN *Paper-Making* 144 It constitutes a 'sludge', practically devoid of useful felting properties. **1933** *Charlottesville* (Va.) *Daily Progress* 22 May 6/5 The death of five men in a drainage vat at the Hess and Drucker Tannery was being investigated... The fifth was a would-be rescuer, who plunged into the tank of sludge. **1965** *New Statesman* 19 Nov. 809/4 (Advt.), Industrial Chemist required by Company dealing in sludges & effluents of all kinds. **1974** J. DYSON *Prime Minister's Boat is Missing* xxx. 180 The sludge of crude oil in the bottom of tanks.

d. A loose sediment that forms in boilers and other vessels in which water is habitually heated.
1839 [in Dict., sense 2 a]. **1912** R. B. DOLE in Rogers & Aubert *Industr. Chem.* iii. 47 If magnesium and sulphates are comparatively low or if suspended matter is comparatively high the scale is soft and bulky and may be in the form of sludge that can be blown or washed from the boiler. **1937** E. PULL *Boiler-House Practice* xv. 147 Every opportunity should be taken to remove accumulations of grease, scale, sludge and soot. **1955** KIRK & OTHMER *Encycl. Chem. Technol.* XIV. 940 This led to the system..of maintaining at all times in the boiler water a small but sufficient excess of phosphate ion so that all calcium ion entering with the feed water would be precipitated as a loose sludge of calcium phosphate rather than as a hard scale.

e. The mixture of water or mud fluid with cuttings that is produced in rock drilling.
1871 [see SLUDGER]. **1911** DANA & SAUNDERS *Rock Drilling* ii. 21 The shales will often form a sludge containing such proportions of large and small particles as to cake on the bit. **1933** R. S. LEWIS *Elements Mining* xiv. 422 Where much sludge is made it is very important to have the water return with sufficient velocity to lift the heaviest particles in the sludge. **1963** *Gloss. Mining Terms (B.S.I.)* III. 13 *Sludge*, rock cuttings produced by the drill bit.

f. A dark viscous liquid or semi-solid mass deposited when a petroleum distillate is mixed with strong sulphuric acid during refining. Also called *sludge acid* (see sense 4 below), *acid sludge*.
1885 *Encycl. Brit.* XVIII. 719/1 The acid 'sludge', consisting of the oil of vitriol combined with the impurities of the oil and forming a black tarry liquid, settles to the bottom..and is drawn off. **1938** OLIVER & SPANGLER in A. E. Dunstan et al. *Sci. Petroleum* IV. 2765/1 At refineries on the sea coast acid sludge was frequently discharged into the ocean. **1954** KIRK & OTHMER *Encycl. Chem. Technol.* XIII. 493 Most refineries hydrolyze the light nonlube sludges, effecting an incomplete separation of acid tars from the weak impure acid. **1965** O. T. FASULLO *Sulfuric Acid* v. 127 The tendency of sludges to evolve sulfur dioxide.., in addition to the enormous quantity in which sludges are necessarily produced by the petroleum industry, makes these materials worthy subjects for the application of pollution control measures.

g. = *SLIME *sb.* 4 b.
1900 *Jrnl. Brit. Inst. Electr. Engin.* XXIX. 274 In copper refining with high current densities less anode sludge is formed. **1948** T. C. ELLIOTT *Electric Accumulator Man.* iii. 26 The piling up of sludge..and the creation of possible short circuits through the growths clinging to the tubes..must be avoided. **1977** BRODD & KORDESCH tr. *Bode's Lead-Acid Batteries* iii. 222 A Pb content of more than 5% is supposed to cause difficulties during the formation of positive plates (swelling or warping or extensive sludge formation).

h. A thick, semi-solid deposit that tends to form in oil when it is heated, exposed to the air, or mixed with another kind of oil.
1920 *Whittaker's Electr. Engineer's Pocket-bk.* (ed. 4) 245 When the transformer is examined, it is found that the windings and core are covered with a reddish-brown flocculent deposit or sludge. **1927** *Jrnl. Soc. Chem. Industry* 8 Apr. 135T/2 The sludge deposits which are sometimes found in hollow crankpins..are due evidently to decomposition products of the oil being thrown out by centrifugal action. **1941** D. F. MINER *Insulation Electr. Apparatus* iv. 81 The purification of oil used in circuit breakers and transformers consists principally in the removal of water, carbon, and sludge. **1973** G. ZWICK *Everyman's Guide Car Maintenance* (1974) vi. 153 If the motorist has consistently been using a high grade heavy duty oil, the chances are that the interior of the engine is clean, with a bare minimum of sludge and other residues.

i. *Med.* (A quantity of) sludged blood. Cf. *SLUDGING *vbl. sb.* 4.
1947 *Science* 7 Nov. 436/2 Many human patients have various degrees of unexplainable edema... These can now be examined for sludges. **1972** H. I. BICHER *Blood Cell Aggregation* ii. 27 All of this provides strong additional support for considering sludge as..a possible antecedent to thrombosis.

4. *sludge brown, green, grey* sbs. and adjs.; *sludge-coloured* adj.; **sludge acid** = sense 2 f above.
1885 Sludge acid [in Dict.]. **1938** OLIVER & SPANGLER in A. E. Dunstan et al. *Sci. Petroleum* IV. 2766/2 Formerly it was possible to use some of this weak separated sludge acid for the manufacture of superphosphate fertilizer, but the objections..to the presence of evil-smelling hydrocarbon derivatives in superphosphate has practically stopped this. **1965** O. T. FASULLO *Sulfuric Acid* v. 127 Heavy sludges are generally mixed with lighter sludge acids to produce a blend with a middle-range viscosity. **1977** D. CLARK *Gimmel Flask* v. 82, I thought this sort of sludge brown varnish paint went out with Queen Victoria. **1962** A. LEJEUNE *Duel in Shadows* vii. 91 An Englishman in a sludge-coloured raincoat. **1979** S. GAINHAM in G. Hardinge *Winter's Crimes* XI. 69 The woman in sludge-coloured tweed. **1971** *Vogue* 1 Oct. 127 Sludge green knit tights. **1972** *Guardian* 8 Feb. 11/2 Cheesecloth smock shirt..natural, yellow, sludge green, red or blue. **1976** *Times* 3 Feb. 9/7 Single breasted coat..sludge green with maroon overcheck. **1977** *Listener* 22–29 Dec. 855/4 He's a moray eel, a sludge-grey reptilian lurker.

sludge, *v.* **I. 1.** Delete (2 b) and add later examples.
1950 *Brit. Birds* XLIII. 383 The bird was on a pool used for sludging boiler ash, which has been constructed within the last two years. **1978** *Sci. Amer.* Jan. 85 (Advt.), In industry, chromic acid and oil-water emulsions usually live only once. They do their job; then they get discharged, sludged, trucked away, and buried.

4. *intr.* To form or deposit sludge.
1941 D. F. MINER *Insulation Electr. Apparatus* iv. 80 Transformer oil that has begun to sludge will continue to do so after it has been purified by means of the centrifuge or filter press. **1977** BRODD & KORDESCH tr. *Bode's Lead-Acid Batteries* iii. 271 Those plates that have been formed at 40°C with 1·06 kg/liter (acid concentration) and 74 A/m² (current density) are the last to sludge.

II. 5. *intr.* To trudge, to tramp; to labour. Cf. SLUTCH *v.* 3.
1908 M. & J. FINDLATER *Crossriggs* xxiii. 170 She had got to sludge back to the station in the rain, and then go home and give a cheerful account of her day. **1913** D. H. LAWRENCE *Love Poems & Others* 44 A widow o' forty-five As has sludged like a horse all her life. **1954** *New Statesman* 3 Apr. 432/3 'Well, goodnight,' he said, sludging away.

III. 6. *intr.* To move slowly by solifluction.
1938 *Geol. Mag.* LXXV. 254 Only the upper 800 feet is precipitous, the lower 600 feet..consisting of vegetated scree down which recent waste is sludging. **1940** C. F. C. HAWKES *Prehistoric Foundations of Europe* ii. 7 Seasonal thawing will cause surface deposits to sludge over more deeply frozen subsoil: this phenomenon is called solifluxion. **1959** G. H. DURY *Face of Earth* xv. 177 Sludging downslope, rock-waste tends to mask the break between high and low ground. **1964** *New Scientist* 8 Oct. 105/2 Half frozen material sludges rapidly downhill in the spring thaw.

Hence **sludged** *ppl. a.*
1941 D. F. MINER *Insulation Electr. Apparatus* iv. 80 No method is yet available that will..bring sludged oil back to its original condition. **1947** *Science* 7 Nov. 435/2 The resistance of sludged blood to its own passage through the bottlenecks of the circulatory system forcibly reduces the rates of blood flow through all the open vessels of the body. **1972** H. I. BICHER *Blood Cell Aggregation* ii. 35 Ischemic changes in the myocardium as a result of sludged blood.

sludging (slʌ·dʒiŋ), *vbl. sb.* [f. SLUDGE *sb.* or *v.* + -ING[1].] **1.** The action of filling up crevices in dried clay by means of free-flowing mud; also, the mud so used.
1852 J. WIGGINS *Embanking* 19 As these spits contract in drying, the crevices outside are therefore filled with mud, which is called 'sludging'. *a* **1877** KNIGHT *Dict. Mech.* III. 2217/1 *Sludging*, stopping the crevices incident to the contraction of clay piled in embankments, by mud sufficiently fluid to run freely. **1940** *Chambers's Techn. Dict.* 342/1 The crevices left when the clay has dried out are filled with sludging.

2. The formation or deposition of sludge, esp. in oil.
1922 GLAZEBROOK *Dict. Appl. Physics* II. 928/1 Various schemes and devices are used to prevent sludging, such as..by using a float on the oil surface. **1931** HOFFERT & CLAXTON *Motor Benzole* vi. 143 The difference in the solubilities of the resinous substances in different types of wash oils, also accounts for the sludging that usually occurs when such oils are mixed. **1969** *Practical Motorist* Jan. 557/1 Most modern engine oils incorporate additives which are designed to reduce sludging. **1973** *Times* 1 June 11/5 No black sludge will be able to form in your central heating system. How horrid. I never knew sludging was going on behind those pipes.

3. = *SOLIFLUCTION.
1946 *Amer. Jrnl. Sci.* CCXLIV. 626 The down-slope movement of the fine-grained components is presumed to be largely a flow as mud and is called by English writers 'sludging'. **1959** G. H. DURY *Face of Earth* xv. 177 Wide-spread sludging is mainly responsible for the featureless aspect of many Arctic landscapes. **1964** *New Scientist* 8 Oct. 105/3 He also showed..that the much-discussed movement by freeze-thaw sludging (solifluction) is quantitatively of very little significance [in a Lappland valley]. **1975** J. G. EVANS *Environment Early Man Brit. Isles* i. 7 The Upper Gravel..was most likely formed by downhill sludging under conditions of impeded drainage.

4. *Med.* The aggregation of blood cells into jelly-like masses in such a way as to impede circulation.
1950 *Amer. Jrnl. Med. Sci.* CCXIX. 538/2 We have..found a close correlation between the blood sedimentation rate and the degree of sludging in the capillaries. **1976** *Path. Ann.* XI. 3 This hemorrhage is sustained for up to 24 hours until the deeper intact capillaries adopt a ballooned appearance due to stasis and sludging of red cells.

sludgy, *a.* Add: **3.** Of the colour of sludge.
1975 *Country Life* 18 Dec. 1770/2 'Workwear', made in sludgy colours and materials. **1978** *Daily Tel.* 6 Dec. 17/4 All colours are sludgy at the fashionable end of the market at present... Pinks, yellows and purples are just unsaleable.

sluff (slʌf), *v.* *U.S.* var. SLOUGH *v.*[2] 3.
For other uses see *Engl. Dial. Dict.* s.v. *Slough* sb.[1] and vb.[1].
1934 *Amer. Ballads & Folk Songs* i. 24 And sluffed their coin for 'dago red'. **1959** T. REESE *Bridge Player's Dict.* 206 *Sluff*, to discard; to throw a card, other than a trump, of a suit different from the one led. **1964** *N.Y. Times Mag.* 6 Dec. 20 Its water-repellent finish sluffs off snow. **1966** J. DOS PASSOS *Best Times* (1968) ii. 56, I had sluffed off Harvard indifference, but Harvard snobbery still hung on. **1972** *New York* 8 May 43/3 His [*sc.* a dog's] shedding mechanism, which now goes about building up and sluffing off the coat. **1972** *Village Voice* (N.Y.) 1 June 50/4 When I consulted a urologist he complained that he was sick of other doctors sluffing the problem off on him all the time. **1976** *National Skat & Sheepshead Q.* Mar. 5 The picker sluffed off the club king. **1978** *Detroit Free Press* 2 Apr. 19C/3 East is now squeezed in the red suits—he must either give up a trick to the jack of hearts or sluff two diamonds, which sets up declarer's third diamond. **1980** *Amer. Speech* LV. 210 Black jazzmen returned to the linguistic roots of their art which had been sidetracked and sluffed off in the *bebop/bop* movement of the 1940s.

slug, *sb.*[1] Add: **1. b.** A contemptible person; a fat person.
In some instances the influence of sense 4a is probable.
1931 A. HUXLEY *Let.* 25 Sept. (1969) 355, I am making notes for a short study..and tho' this cannot be specifically a retort to Murry it will in effect try to undo some of the mischief that that slug has undoubtedly done. **1940** G. & S. LORIMER *First Love Farewell* iv. 133 'He didn't love me and I felt pretty bad about it!' 'The complete and utter slug!' **1959** I. & P. OPIE *Lore & Lang. Schoolch.* ix. 168 The unfortunate fat boy..is known as..slug. **1966** 'J. HACKSTON' *Father clears Out* 104 No more big slugs turning up, the rush lost its enthusiasm.

6. **slug pellet,** a pellet of bait containing a poison to kill slugs.
1960 *Do it Yourself Gardening Ann.* 100/3 The most effective way of controlling slugs and snails is to use a metaldehyde-bait mixture... The easiest way to do this is to use a proprietary brand of slug pellets. **1976** L. THOMAS *Dangerous Davies* xvi. 188 A large tin of garden slug pellets.

slug, *sb.*[2] Add: **1. b.** For 'Now *U.S.*' read 'Now chiefly *U.S.*' and add further examples.
1916 H. L. WILSON *Somewhere in Red Gap* vi. 245 Even the new Episcopalian minister..took a slug of rye and said it was undeniably delightful. **1940** R. CHANDLER *Farewell, my Lovely* v. 34, I poured her a slug that would have made me float over a wall. **1958** C. WILLIAMS *Man in Motion* (1959) i. 6 Pouring another cup of coffee, I dropped a slug of bourbon in it. **1973** C. BONINGTON *Next Horizon* xv. 216 The scene was Hogarthian—with a soldier lying flat in the gutter,..a mother giving her eighteen-month babe a slug of the fire-water, to stop it crying. **1978** L. HEREN *Growing up on The Times* v. 182 Their simple niceness was almost as good as a slug of scotch and a cigarette which I..could not enjoy in their company.

c. A compact mass of liquid regarded as retaining its identity as it travels.
1947 I. THOMAS *Injection Molding of Plastics* vi. 353 If the reservoir were omitted the cold slug of material would enter the cavity and possibly cause smudge or flow marks in the molded article. **1967** *Guardian* 13 Feb. 14/6 In each pipe will be methane gas plus liquid oil called the 'slug'... Once ashore the gas and slug have to be separated and cleaned. **1971** *Nature* 21 May 181/1 A rapid intravenous injection leads to a 'slug' of relatively undispersed drug traversing the arteries on the first circuit. **1975** *Petroleum Rev.* XXIX. 315/3 A collection of liquid, known as a 'slug', can amount to several hundred thousand gallons and will travel along the pipeline at a speed of up to 10 miles per hour. **1979** *Nature* 8 Feb. 441/1 The velocity of the ejected slug of [volcanic] debris.

3. a. (Earlier examples.)
1849 *Picayune* (New Orleans) 6 June 1/6 The gold from that stream is generally in large pieces, more generally termed slugs or coarse, but very fine gold, if you please. **1855** *Golden Era* (San Francisco) 21 Jan. 2/7 We took out one slug weighing 60 ounces of pure gold, in the shape of an ox's tongue.

c. *Electr.* A metal cylinder fitted round the end of the core of an electromagnetic relay to modify the speeds of opening and closing. Cf. *SLEEVE *sb.* 7 c.
1928 *Jrnl. Inst. Electr. Engineers* LXVI. 346/1 A thick copper cylinder—or 'slug'—placed over part of the core, or a thin cylinder—or 'sleeve'—placed over the whole of the core, will provide a closed path of very low resistance in which currents of considerable magnitude may flow. **1943** A. L. ALBERT *Fund. Telephony* ix. 199 The amount of time delay can be regulated by the size and location of the copper slug. **1969** M. L. GAYFORD *Mod. Relay Techniques* iii. 60 A slug at the armature end slows down both operation and release. A slug at the rear or heel end has little effect on the operate time but produces suitable delay on release.

d. *Nucl. Engin.* A rod or bar of nuclear fuel.
1945 H. D. SMYTH *Gen. Acct. Devel. Atomic Energy Mil. Purposes* vii. 69 The uranium would react chemically with the water,..probably to the point of disintegrating the uranium slugs. **1952** [see *PLUTONIUM 2 b]. *a* **1958** K. EDWARDS in 'E. Crispin' *Best SF 3* (1958) 34 The uranium slugs were short and the aluminium cans that held them in the centre of the pile..were long. **1967** J. T. LONG *Engin. Nuclear Fuel Reprocessing* xi. 838 The slugs were cooled by natural convection of water in a finned tank. **1973** *Trans. Amer. Nuclear Soc.* XVII. 508/2 A charge of fuel particles, a preformed matrix slug, and a top punch are inserted.

e. *Electr.* An adjustable magnetic core used to vary the inductance of a coil containing it. Chiefly in *slug tuning* (so *slug-tuned* adj.).
1957 *Practical Wireless* XXXIII. 566/1 Now adjust trimmers or slugs of discriminator transformer to obtain a symmetrical pattern. **1959** R. L. SHRADER *Electronic Communication* xiv. 390 A few receiver RF amplifiers..use slug tuning, having a mechanical means of pulling the slug into the desired position in the coil by a dial on the front of the equipment. **1960** *Practical Wireless* XXXVI. 416/2 Coil L2 is also heavily damped and variable tuning is hardly worthwhile; it is accordingly slug-tuned to the centre of the three transmissions to be received. **1979** A. A. LIFF *Color & Black & White Television Theory & Servicing* iv. 108 The individually tuned circuits in the oscillator section are all slug tuned.

4. c. (Earlier examples.)
1851 *Oregon Statesman* 23 Sept. 2/6 He accordingly 'pungled down' two of Moffat's $50 slugs, and of course, cut the black, there being no red spots in the pack. **1872** S. POWERS *Afoot & Alone* 303 A shining 'slug', fresh from the San Francisco mint, [was] laid scrupulously in the place.

e. *U.S. slang.* A dollar; a counterfeit coin; a token.
1887 *Lantern* (New Orleans) 4 June 5/2 She'd sooner put up her ten slugs and go back to the pipe. **1913** *Dial. Notes* IV. 28 *Slug*,..round piece of metal for slot machines. **1934** J. T. FARRELL *Young Manhood of Studs Lonigan* xvii. 259 He bought a slug from the cashier in the Chain drugstore at Prairie and walked back to the telephone booths.

f. *Journalism.* = *slug-line*, sense 6 below.
1925 G. M. HYDE *Newspaper Editing* (ed. 2) ii. 89 Use expressions that will not offend readers..if the slugs slip into print..'Kill widow'..may be misunderstood. **1927** *Amer. Speech* II. 240/2 'Slug'..is a brief title..placed above a story for the guidance of the copy-reader and the printer. **1949** T. F. BARNHART *Weekly Newspaper Writing & Editing* xxii. 224/2 In many newspaper plants some other term is substituted for guidelines, such as catchlines, slugs and slug-lines. **1979** 'A. HAILEY' *Overload* IV. vii. 324 The newspaper put a copyright slug over her story.

5*. *Engin.* A unit of mass equal to 32·1740 lb., being the mass of a body which accelerates at one foot per second per second when acted on by one pound force.
1902 A. M. WORTHINGTON *Dynamics of Rotation* (ed. 4) p. viii, I have ventured to give the name of a 'slug' to the British Engineer's Unit of Mass, *i.e.*, to the mass in which an acceleration of one foot-per-sec.-per-sec. is produced by a force of one pound. **1923** A. R. LOW in W. L. Marsh *Rep. Internat. Air Congr.* 62 The 'slug' of 32·2 pounds avoirdupois mass, which has actually been imposed on British aeronautics by the Advisory Committee. **1936** F. W. LANCHESTER *Theory of Dimensions & its Application for Engineers* v. 37 Even amongst the advocates of Perry's system.., the slug has never taken shape except on paper; it has, and has had no real material existence. **1944** N. A. V. PIERCY *Compl. Course Elem. Aerodynamics* iv. 86 At 15° C. and 760 mm. pressure, ρ [*sc.* the density of air]..= 0·00238 slug per cubic foot. **1973** *Nature* 20 July 184/3 The statement that the unit of mass in the British system is the slug is several years out of date.

6. *slug-gun*; **slug-line** *Journalism*, an identifying title, usu. occupying one slug, accompanying a news story in draft and galleys; **slug-setting** *Printing*, the method of setting an entire line of type on a single slug; so *slug-set a.*
1940 *Illustr. London News* CXCVI. 53 (*caption*) Their training includes the use of tear-gas, while they are armed with 'slug-guns', ·303 rifles and staves. **1973** M. AMIS *Rachel Papers* 218 The youth, handsomely reading Tennyson on summer evenings, or trying to kill birds with feeble, rusted slug-guns. **1930** K. E. OLSON *Typogr. & Mechanics of Newspaper* xi. 360 Sometimes the printer forgets to remove the slugline when he places a story in its column, and this would appear above the story unless caught in the final check. **1976** J. MCCLURE *Rogue Eagle* ii. 27 He tapped out the name of his freelance agency and the slug-line. **1963** Slug-setting [see *FILMSET *v.*]. **1970** *Times Lit. Suppl.* 9 Apr. 387/2 It is an instance of their attention to quality that even novels are 'Monotype' set, whereas most publishers use slug-setting. **1975** J. BUTCHER *Copy-Editing* ii. 7 Slugsetting is unsuitable for complex tables, though Monotype may be used for tables in a slugset book.

slug, *sb.*[3] Add: **2.** *attrib.* and *Comb.*, as **slug-fest** [*FEST] *U.S. slang*, a hard-hitting contest, *spec.* in boxing and baseball; **slug-nutty** *a.* (*U.S. slang*), punch-drunk; hence **slug-nuttiness.**
1916 *Nebraska State Jrnl.* 27 July 3/1 (*heading*) Denver wins in slugfest. **1933** G. TUNNEY *Man must Fight* 14 If Dempsey would gamble with me in a slug-fest I would beat him to the punch every time. **1943** *Amer. Speech* XVIII. 105 A good inning at bat..is a..*slug-fest*. **1969** *Daily Colonist* (Victoria, B.C.) 21 Mar. 1/4 A meeting between Canadian MPs and a high French government official turned into a verbal slugfest as the Canadians raked the Gaullist government for its policies towards Quebec and Ottawa. **1976** M. MACHLIN *Pipeline* x. 116 For a while it looked as though there was going to be a real slugfest. **1979** *Arizona Daily Star* 1 Apr. B5/4 'Back to Basics' is today's biggest educational debate topic. But so far, it has generated a mostly muddled, emotional slugfest. *Ibid.* 8 Apr. C1/2 Powers gave up four runs on seven hits, a contrast from the 33-hit slugfest of Friday night. **1943** *Gen* 16 Jan. 30/1 Though no medical man, I know enough about slug-nuttiness to tell you..how it comes about. **1933** 'P. CAIN' *Fast One* vi. 196 He shook his head sadly without looking up. 'Slug-nutty.' **1936** J. STEINBECK *In Dubious Battle* ii. 16 Don't mind Joy. He's slug-nutty. He's been smacked over the head too much. **1950** E. HEMINGWAY *Across River & into Trees* iv. 26 He's been beat up so much he's slug-nutty.

slug, *v.*[1] Add: **3.** *Journalism. trans.* To mark with a slug-line; with compl.: to give as a slug-line or other heading.
1925 G. M. HYDE *Newspaper Editing* (ed. 2) ii. 27 What the copywriter does..may be summarized... 'Slugs' story for record and make-up. **1928** *Amer. Speech* IV. 135 The hardened deskman merely grins, squeezes the gist of a column story into ten words and 'slugs' the article to designate the 'head' that is to accompany it. **1940** N. MACNEIL *Without Fear or Favor* v. 74 The foreign editor..may slug his stories by correspondent. **1953** B. WESTLEY *News Editing* xii. 197 When they moved a story about the possible use of bacteriological warfare.., the United Press slugged it: 'Germs'. **1974** J. BANNING *How I fooled World* xix. 77 The [news] story, slugged 'Television', ran something like this. **1976** *Daily Tel.* 13 Oct. 6/7 Members of the National Union of Journalists..called..for all relevant copy to be marked, or 'slugged', 'Not for Preston'. *Ibid.*, There was a previous precedent for slugging copy in an outside dispute.

slug, *v.*[3] Add: **1. b.** *fig.* To treat roughly; to drug; to exploit; to overcharge; to force *out* of; to churn *out*.
1925 A. S. M. HUTCHINSON *One Increasing Purpose* I. xviii. 112, I know..that really you were court-martialled and hoofed out... But what the facts were, *why* you were slugged, *how* they found out your hideous goings-on, I never could discover. **1938** J. STEINBECK *Long Valley* 116, I guess the doctor slugged me pretty hard... I feel all right now, only a little dopey. **1941** S. J. BAKER *Dict. Austral. Slang* 68 *Get slugged*, to be charged excessively. **1946** K. S. PRICHARD *Roaring Nineties* 326 Alf knew the mine-owners were slugging the prospectors and alluvial diggers. **1962** 'K. ORVIS' *Damned & Destroyed* vi. 46 She had..slugged herself insensible with a terrific overdose of heroin. **1974** P. DE VRIES *Glory of Hummingbird* ii. 14 The thankless task of slugging contributions out of the congregation. **1976** *Australasian Express* 11 June 26/2 Canberra: Big cars will be slugged for extra insurance if a report to the Government is adopted. **1977** *New Yorker* 19 Sept. 96/2 I've been slugging out so many notes of American music I'm turning into a piece of apple pie.

c. *to slug it out*: to fight it out; to stick it out.
1943 *Time* 10 May 98/3 Twice it screens exciting action: once when the sub slugs it out with a disguised German raider. **1952** M. LOWRY *Let.* 11 Jan. (1967) 286 Somehow we slugged it out, without having to abandon the house. **1970** *Listener* 23 July 127/3 It decided not to slug it out directly with *News at Ten*. **1973** *Black World* Mar. 58, I saw the two shadows boxing on the side of the brick building... It was Bernie and Bennie Speakes, twins about 10, slugging it out in the alley. **1978** *Detroit Free Press* 16 Apr. 14C/1 They'll slug it out, week by week, blow by blow, for all the world to see.

2. *intr.* To slog. Also with *along*, *away*.
1943 *Fortune* Feb. 122/1 Guns slugging at close range. **1943** *Newsweek* 9 Aug. 27/1 This found them..still slugging along at a point between the coastal villages of Tetere and Zovi. **1944** *Reader's Digest* Nov. 62/1 But always he was slugging away at novel writing on the side. **1959** *Times* 9 Mar. 3/1 In the second [half], largely an affair of forwards slugging away with barely diminished vigour, the Navy's packing became untidy.

sluggard, *sb.* **a.** *Comb.* (Later examples.)
1892 J. LUMSDEN *Sheep-head & Trotters* 301 Harvest, however, came at length, lagging on a-pace, sweet and sluggard-wise. **1910** W. DE MORGAN *Affair of Dishonour* vii. 87 That he should, simply from an idle indulgence of laziness, lie sluggard-wise till near mid-day.

sluggardry. For † *Obs.*⁻¹ read *rare* and add later examples.
1925 V. WOOLF *Common Reader* 79 The extremes of passion are not for the novelist;..he must tame his swiftness to sluggardry. **1935** W. DE LA MARE *Early One Morning* 145 Unusual discomfort of mind or body may slow down the wheels of existence to a sluggardry almost beyond endurance.

slugged (slʌgd), *a.*[2] [f. SLUG *sb.*[2] + -ED[2].] Having a slug or slugs.
1906 *Daily Colonist* (Victoria, B.C.) 5 Jan. 9/1 (Advt.), Boys' Calf Boots, double soles, slugged bottoms, sizes 1 to 5. **1922** *Daily Mail* 10 Nov. 4 Boys' unlined slugged soles Derbys. **1928** *Jrnl. Inst. Electr. Engineers* LXVI. 341/1 The release of slugged relays is complicated by saturation. **1958** J. W. FREEBODY *Telegraphy* iv. 84/1 The actual release lag of a slugged relay depends also on the spring tension.

slugger[2]. For *U.S.* read 'orig. and chiefly *U.S.*' and add: **1.** (Earlier and further examples.) Also in *Baseball*, a hard-hitting batter.
1877 [see *COON *sb.* 4 b]. **1883** *Chicago Tribune* 3 July 6/5 Poor Burns fell an easy victim to the Cleveland sluggers. **1894** G. MEREDITH *Let.* 19 Feb. (1970) III. 1155 Harrison is a controversial slugger. I did not expect he would descend to the use of street missiles. Your answer hit the right tone. **1946** *Sun* (Baltimore) 27 May 15 (*caption*) Charlie Keller, the Yankee's slugger, is shown being caught in a rundown. **1952** *Manch. Guardian Weekly* 16 Oct. 2/1 Their team includes Mr. Gromyko and the notorious slugger Vyshinsky. **1967** *Boston Sunday Herald* 26 Mar. II. 3/1 Tony has the build of a slugger... Maybe some club will make an offer for him, giving a dependable pitcher in return. **1970** *Daily Tel.* (Colour Suppl.) 9 Oct. 27/3 No amateur karate exponent stands a halfway chance against an old-fashioned slugger and kicker who has real experience of street brawls. **1972** J. MOSEDALE *Football* vii. 100 One of the National League's most powerful sluggers as a Cincinnati Red. **1977** *Times of Zambia* 7 Sept. 10/7 The new holder of baseball's all-time Home Run record, Japanese slugger Sadaharu Oh.

3. Usu. *pl.* Ear-to-chin whiskers. Also called *slugger whiskers*. *U.S. slang.*
1898 F. P. DUNNE *Mr. Dooley in Peace & War* 211 Ganderbilk he was there, too, standin' out on th' steps in th' cold, combin' his whiskers... He wears a pair iv sluggers..with his fingers. **1900** G. ADE *More Fables* 118 The mild old Gentleman with the straw-colored Sluggers ..came near. **1960** WENTWORTH & FLEXNER *Dict. Amer. Slang* 489/2 *Slugger*,..an ear-to-ear chin beard, as worn by a stage Irishman. **1960** B. KEATON *My Wonderful World of Slapstick* i. 12 A fright wig, slugger whiskers, fancy vest.

slugger[3]. Add: **2.** One who attaches the top piece of the heel of a shoe to the seat.
1911 *Rep. Labour & Soc. Cond. Germany* (Tariff Reform League) III. vi–vii. 30 The average wages paid in the district were:.. Sole layer 20s. to 32s. Slugger 30s. to 35s. **1921** *Dict. Occup. Terms* (1927) § 414 *Slugger*,..operates, by means of a lever, slugging machine, which attaches top piece of heel to seat.

slugging, *vbl. sb.*[2] Add: (Further examples.) Also *attrib.* and *fig.*
1942 BERREY & VAN DEN BARK *Amer. Thes. Slang* § 675/3 *Game with many hits*..slugging match. **1956** A. H. COMPTON *Atomic Quest* 315 The war would become a heavy slugging match. **1959** *Economist* 30 May 819/1 The abortive Shawwaf revolt in Mosul set off a slugging match between Cairo and Baghdad. **1974** *Sumter* (S. Carolina) *Daily Item* 23 Apr. 7A/5 His slugging percentage? A whopping .860. **1978** J. CARROLL *Mortal Friends* III. vi. 339 A month's worth of slugging along a coastline through a series of towns.

sluggish, *a.* Add: **3. b.** [Rendering of Russ. *stértaya* (*shizofreniya*), worn, hackneyed (schizophrenia).] Applied to an alleged type of schizophrenia ascribed to political or religious dissidents confined in state psychiatric hospitals in the U.S.S.R.
1977 *Science News* 10 Sept. 165 'Even if one should accept the diagnosis of sluggish schizophrenia in these and similar cases,' Chodoff said, 'one must wonder why a disease without delusions, hallucinations or agitated behavior should require injections of chlorpromazine (an antischizophrenic drug) for its treatment.' **1978** *Nature* 4 May 6/2 Yet, when committed, Shikhanovich was diagnosed as a 'psychopath with the possibility of the onset of sluggish schizophrenia' (the latter 'disease', of course, being unknown to non-Soviet diagnosis). **1980** *Prisoners of Conscience in USSR* (Amnesty Internat.) (ed. 2) 184 Schizophrenia, often in its 'sluggish' form, has been the diagnosis most commonly made of dissenters.

sluice, *sb.* Add: **5.** (Earlier example.)
1851 *San Francisco Picayune* 14 Oct. 2/4 In the neighbourhood of Rough and Ready, a sluice of fourteen miles in length has been constructed.

6. a. **sluice-fork,** a fork used to break up lumps of gravel in a gold-miner's sluice-box; **sluice-head** orig. *U.S.*, a supply or head of water sufficient for flushing out a sluice; also *fig.*
1856 *San Francisco Call* 16 Dec. 4/2 As he went—took it puss'nal—it commenced raining 'sluice-forks'. **1874** A. BATHGATE *Colonial Experiences* viii. 92 The large stones..lifted out by hand..while the smaller ones are sometimes taken out with a long handled long pronged sluice-fork. **1909** H. THOMPSON in A. E. Currie *Centennial Treasury of Otago Verse* (1949) 59 Slinging stones out with his sluice-fork—what a pleasant little game. **1855** *Golden Era* (San Francisco) 4 Mar. 1/6 At Eureka there are only twelve sluiceheads of water running. **1863** V. PYKE in *App. Jrnls. House Reps. N.Z.* D. vi. 15 Head races..represent about 200 sluice heads. **1901** E. DYSON in *Austral. Short Stories* (1951) 58 Mrs. Mooney..wept sluice-heads... She had been replenishing the fountain of tears with whisky. **1906** *Daily Colonist* (Victoria, B.C.) 6 Jan. 12/3 Although little opened, the springs now have a flow of two sluice-heads. **1935** G. L. MEREDITH *Adventuring in Maoriland in Seventies* xiii. 145 The one we went to is just a boiling spring, running about five 'sluice-heads' of boiling water.

sluice-box. (Earlier examples.)
1857 *Hutching's Mag.* July 7/1 A continuous line of these troughs or 'sluice boxes', the smaller and lower end of each, inserted for three or four inches into the larger end of the next one below, form the 'sluice'. **1864** *Richmond-Atkinson Papers* (1960) II. ii. 110 You will not be quite out of the rattle of the shovel, long tom, and sluice box, for we are very golden just now.

sluice-way. (Earlier example.)
1779 W. MCKENDRY *Jrnl.* 8 Aug. in *Proc. Mass. Hist. Soc.* (1886) II. 461 The sluce way was broke up and the water filld. the river immediately.

sluit. Add: Also **slote.** (Earlier examples.)
1818 C. I. LATROBE *Jrnl. Visit to S. Afr.* x. 187 It has..water in abundance, brought by a *slote*, or canal, from a considerable distance. **1852** C. BARTER *Dorp & Veld* iv. 33 Going one dark night to a friend's house, and keeping in the middle of the road to avoid the 'sloots', I stumbled over..a large black ox. **1862** L. DUFF GORDON *Let.* 29 Dec. in F. Galton *Vacation Tourists* (1864) 157 There is no water but what runs down the streets in the sloot, a paved channel, which brings the water from the mountain and supplies the houses and gardens.

slum, *sb.*[1] Add: **I. 2. a.** Also *rarely*, a house materially unfit for human habitation.
1955 *Times* 25 Aug. 5/5 Nowadays people who live in so-called slum houses (a 'slum', as officially defined

means a house materially unfit for habitation), set a good standard of cleanliness. **1972** *Observer* 31 Dec. 8/2 He had inherited nearly two million slums.

4. a. *slum area*, *-dweller*, *property*, *street*, etc. (further examples); *slum-bred* adj.
1891 *Contemp. Rev.* Oct. 548 Tens of millions will be exposed to the physical and mental blight of the 'submerged' slum-dweller. **1898** E. HOWARD *To-morrow* xiv. 147 What will become of this slum property?.. These wretched slums will be pulled down. **1924** *Glasgow Herald* 8 Mar. 7 The slum problem is fundamentally not one of stone and lime or cubic space, but of mental and social outlook. **1928** GALSWORTHY *Swan Song* I. vii. 57 Slum-dwellers were such good sorts! **1932** L. GOLDING *Magnolia St.* III. vi. 537 A tribe of inconceivable people who lived in a slum street in a dark English town. **1935** C. S. FORESTER *Afr. Queen* iv. 86 His slum-bred father and mother. **1939** C. DAY LEWIS *Child of Misfortune* II. ii. 151 The Church..held slum-property, helped to exploit innocent native tribes and ruined their morals and physique. **1940** 'G. ORWELL' *Inside Whale* 20 Low wages and the growth and shift of population had brought into existence a huge, dangerous slum-proletariat. **1959** I. & P. OPIE *Lore & Lang. Schoolch.* xviii. 389 A row of crowded slum houses with front doors cheek by jowl. **1959** J. CARY *Captive & Free* ix. 41 The mid-town terraces which can and have so easily become slum tenements. **1960** 'F. NEWTON' *Jazz Scene* vi. 100 The penalties of the isolated, community-less life of the slum-bred entertainer. **1968** *Globe & Mail* (Toronto) 13 Feb. 27/1 Maple Leafs can muddle into the slum area of fifth place and there are few hoots of derision. *a* **1974** R. CROSSMAN *Diaries* (1975) I. 182 The town itself has some 30,000 or 40,000 inhabitants..who don't want Birmingham slum dwellers dumped on them as overspill. **1979** G. ST. AUBYN *Edward VII* viii. 379 She insisted on being shown his slum property in the East End of London. **1980** J. MELVILLE *Chrysanthemum Chain* 135 Walker made his way out and into the shabby slum street.

b. Special Combinations: **slum clearance,** the evacuation and demolition of slums, usu. accompanied by the rehousing of the inhabitants; freq. *attrib.*; also **slum clearer, clearing** *ppl. a.*; **slumland** (later examples); **slum landlord,** one who lets slum property to tenants, esp. one who allows his property to fall into disrepair; hence **slum landlordism,** the practice of letting slum property; **slumlord** *U.S.* = *slum landlord* above; hence **slumlordship.**
[**1900** A. SMITH *Housing Question* iii. 60 The clearance of slums should not be taken to make room for housing schemes.] **1907** E. R. DEWSNUP *Housing Problem in Eng.* xi. 227 Local Authorities..have hesitated to shoulder the financial burden that would result from any general application of the powers of slum clearance placed at their disposal by statute law. **1930** T. E. LAWRENCE *Let.* 19 Jan. (1938) 678 The area it occupied turned into a public garden, in pursuance of the slum-clearance scheme. **1936** T. S. ELIOT *Essays Ancient & Modern* 132 We recognize that possibility in every work of slum-clearance and housing reform. **1953** E. SMITH *Guide to English Traditions & Public Life* 133 The necessity of slum-clearance had to be faced. **1961** L. MUMFORD *City in History* viii. 220 Such systematic slum clearance projects as Nero's great fire naturally increased the housing shortage. **1979** *Punch* 28 Nov. 1032/3 Ms Greer..does not establish much that is positive; but she has performed a monumental work of intellectual slum-clearance. **1934** 'R. CROMPTON' *William—the Gangster* viii. 177 They were all lofty and spacious enough to satisfy the most determined and particular of slum clearers. **1977** *Listener* 28 Apr. 531/1 If the developer had to pay over a slice to the slum-clearing authorities, this would allow them to rehouse the slum-dwellers. **1929** S. LESLIE *Anglo-Catholic* i. 10 He felt at home in the East End and refreshed..when he lay back at night and sniffed the indefinable steam of slumland. **1978** D. MURPHY *Place Apart* vi. 110, I..cycled back to slumland to spend the rest of the day with Catholic families. **1893** G. B. SHAW *Widowers' Houses* III. vii. 84 The worst slum landlord in London. **1931** W. HOLTBY *Poor Caroline* iii. 97 There was so much to be done..slum landlords to be confronted. **1972** C. DRUMMOND *Death at Bar* vii. 179 A slum landlord who augmented his meagre rents in strange and unlawful ways. **1892** *Black & White* 17 Dec. 698/1 As a discussion, with open doors, of the pros and cons of slum-landlordism ..Mr Shaw's *Widowers' Houses* is..a very considerable piece of work. **1967** *Sunday Times* 30 Apr. 11/1 Slum landlordism..has not been seriously curtailed. **1953** *Chicago Daily News* 12 Sept. 3/7 Reporters..found that slumlords frequently twist Illinois' trust laws into blinds for escaping detection. **1957** *N.Y. Times Mag.* 12 May 36/3 The landlord had bitterly protested..that he was not a 'slumlord' and avowed that he was ready to put the building in condition *if* he could get a guarantee that it would stay that way. **1978** S. WILSON *Dealer's Move* vii. 121 A big place in Surrey—it belonged to one of the king slum-lords. **1966** *Atlantic Monthly* Nov. 128 Within the chivalric order of slumlordship he is a very minor vassal.

II. 5. a. (Earlier example.)
1812 P. EGAN *Boxiana* I. 122 The flowing harangue of some dusty cove..lavish with his slum on the beauties possessed by some distinguished pugilist.

8. *N. Amer.* **a.** Cheap or imitation jewellery. Also as *adj. Criminals' slang.*
1914 JACKSON & HELLYER *Vocab. Criminal Slang* 77 *Slum*, jewelry of any description, but lately reduced in scope of meaning to include only the less valuable kinds of jewelry... 'He's got a bale of slum for sloughings.' **1924** G. C. HENDERSON *Keys to Crookdom* 418 *Slum*, plated jewelry. **1931** *Amer. Speech* VII. 102 Nail the stones but blow the slum. **1946** S. S. JACOBS in *Mag. Digest* Aug. 89/2 A guy buys a slum ring for ten cents. **1955** *Publ. Amer. Dialect Soc.* XXIV. 122 Any kind of jewelry, usually exclusive of watches, was—and still is—referred to as slum. **1962** 'K. ORVIS' *Damned & Destroyed* vii. 53 Jewellery... Top stuff. No slum.

b. Cheap prizes at a fair, carnival, etc.
1929 *Sat. Even. Post* 19 Oct. 26/2 Business Opportunities hammer at every door in the advertising columns of this trade paper. Slum, 1008 pieces for tie pins, collar pins, brooches, cigarette holders, rings, $695 the lot. **1956** H. GOLD *Man who was not with It* (1965) i. 5 The slum prizes dripped from their hands, taffy, teddy-bears, streamers of paper.

Hence **slu·mism** [-ISM], the existence of slums; the deprivations and other ills associated with or characteristic of life in the slums.
1967 *Britannica Bk. of Year 1966* 804/2 *Slumism*, the existence of highly congested urban residential areas characterized by deteriorated unsanitary buildings, poverty, and social disorganization. **1967** *Harper's Mag.* Feb. 83 We must show the same unhesitating commitment to fighting slumism, poverty, ignorance, prejudice, and unemployment that we show to fighting Communism. **1971** L. CHESTER *Martin Luther King* xi. 262 We are victims of slumism!

slum, *sb.*[3] *slang.* [App. abbrev. of SLUMGULLION.] **1.** = SLUMGULLION 2 c.
1847 J. MITCHELL *Reminisc. College* 117 Though the son of Vulcan found the pork and cabbage harmless, I am sure that slum would have been a match for him. **1865** 'MARK TWAIN' *Notebook* 28 Jan. (1935) i. 6 Chili-beans and dish-water three times today as usual and some kind of 'slum' which the Frenchman called 'hash'. **1898** E. H. BLATCHFORD *Let.* 17 July (1920) 37 Beef stew, commonly known as slum. **1918** *Stars & Stripes* 5 Apr. 4 Everyone knows that there are at least three different kinds of slum —the watered kind, the more solid variety and the occasional special sort that wears a pie-crust. The Marines describe these three types in sea-lingo: 'slum with the tide in', 'slum with the tide out', and 'slum with an overcoat'. **1972** J. M. MINIFIE *Homesteader* xx. 182 There would be white table-cloths and sparkling glass and silver, instead of a mess-tin of slum on a dirty table in barracks.

2. Special Combinations. **slum burner,** an army cook; **slum gun,** a field-kitchen.
1930 *Our Army* Aug. 33 The..cook..is a 'slum-burner'. **1943** M. HARGROVE *See here, Private Hargrove* xlii. 118 Oscar of the Waldorf, in the Army, would still be..a *slum-burner*. **1917** R. BATCHELDER *Watching & Waiting on Border* vii. 90 The regiment owned a field-kitchen, or 'slum-gun', a bulky vehicle in which food might be prepared on the march. **1947** D. RUNYON *Poems for Men* 213 Our slum-gun busted down.

slum, *v.* Add: **2. b.** *intr.* (See quot. 1965.) Also *trans.*, to shear (a sheep, etc.) in this manner. *Austral. Sheep-shearing.*
1965 J. S. GUNN *Terminol. Shearing Industry* II. 24 A shearer slums if he works as fast as he can, and perhaps carelessly, while the pen is full, and thus takes a large proportion of the easy sheep. **1966** J. CARTER *People of Inland* (1967) xvii. 165 Then at shearing time, these same 'guns' can slum pen after pen of fine, clean sheep, because the opportunity to set a new record has presented itself.

4. a. Also with *it*.
1899 W. JAMES *Let.* 8 Feb. (1920) II. 88 Kipling knows perfectly well that our camps in the tropics are not college settlements or our armies bands of philanthropists, slumming it.

b. To accept, temporarily and voluntarily, a standard (of living, travel, etc.) lower than that to which one is accustomed; to mix with one's inferiors. Freq. as *pres. pple.* and with *it*. Also *fig.*
1928 E. WALLACE *Gunner* xxiii. 192 'What are you doing down here?' 'Slumming,' said Gunner Hayes coolly. 'I like now and again to establish contact with the underworld.' **1944** N. COWARD *Middle East Diary* 95 We quite enjoyed slumming it in the ordinary pullman. **1946** R. G. COLLINGWOOD *Idea of Hist.* IV. 145 It is necessary to go slumming among the most unsavoury relics of third-rate historical work. **1951** E. COXHEAD *One Green Bottle* ii. 57 He isn't quite a professor yet. She's just slumming till he becomes one. **1959** *Ann. Reg. 1958* 192 Mr Rockefeller, as he slummed it in New York in the battle with his fellow millionaire the Democratic Mr Harriman. **1978** P. PORTER *Cost of Seriousness* 35 On its dorsal, a monster is drumming Messages for the new world—each wraith Is a spirit of old Europe slumming. **1981** *Birds* Autumn 68/1 It [*sc.* a brambling] was quite unabashed by the proximity of the feeding area to the back door and was happily 'slumming it' with the resident sparrows, chaffinches and greenfinches.

slumber, *sb.* Add: **4.** **slumber cap,** a light, close-fitting cap of lace, ribbon, etc., worn in bed to keep the hair tidy; **slumbercoach** *U.S.*, a railway car which provides economical private sleeping accommodation; **slumber net,** a slumber cap made of net; **slumber party** *U.S.*, a party for youngsters (esp. girls) who stay on to sleep overnight; **slumber room** *U.S.*, a room in which a corpse is laid out by an undertaker until the funeral takes place; **slumberwear,** night-clothes.
1928 *Sunday Dispatch* 8 July 16 Shingle caps or slumber caps for the seaside..in Nottingham lace, bound with pink, blue or any coloured satin ribbon. **1971** 'A. GILBERT' *Tenant for Tomb* iii. 47 She rolled up the plaits under..a slumber cap, an affair of bright blue silk and lace and a ribbon bow. **1958** *Washington Post* 26 June C19/3 B&O charges regular coach fare plus a $6 service charge for a single room..for its slumbercoaches, which are operated on only one train, the Baltimore–Washington–Chicago Columbian. **1979** *United States 1980/81* (Penguin Travel Guides) 19 Long-distance trains offer sleeping accommodations..slumbercoaches, private rooms,..roomettes. **1930** J. RHYS *After leaving Mr Mackenzie* II. xiii. 217 Out of the second door emerged a lady in a pink dressing-gown, with her hair hidden by a slumber-net. **1950** A. WILSON *Such Darling Dodos* 79 The artifice of the black waved hair..beneath the neat mesh of the slumber net. **1966** *Olney Amsden & Sons Ltd. Price List* 23 Hair and slumber nets. [**1942** BERREY & VAN DEN BARK *Amer. Thes. Slang* § 251/1 *Sleep*,..slumber party.] **1949** *Senior Prom* Nov. 22/2 For a girls' party you might have a brunch, lunch, dinner,..or slumber party. **1954** *Life* 26 Apr. 186/2 Because it was to be a slumber party, the 19 girls..came carrying pillows, blankets and floppy animals—but no one really expecting to get much sleep. **1974** A. LURIE *War between Tates* ix. 181 'I'm invited to Elsie's slumber party.' 'Oh? And what is a slumber party?' 'Don't you even know that? You have a party, and then you sleep overnight.' **1936** Slumber-room [see *funeral-home* s.v. *FUNERAL *sb.* 6]. **1963** J. MITFORD *Amer. Way of Death* iv. 61 The slumber rooms are elusively reminiscent of some other feature of American life... 'So then you've got a slumber room tied up for three days or more,' he said... 'How much would it cost you to stay in a good motel for three days?'.. Motels for the dead! That's it, of course. **1979** *Sun-Times* (Chicago) 28 Sept. 4 Would it be considered improper to take a photograph of a deceased friend or relative in the slumber room during viewing hours? **1909** *Punch* 24 Mar. 206/3 The famous house so long consecrated to the habiliments of Morpheus, or 'slumberwear', as of late we have been taught to call them. **1961** L. P. HARTLEY *Two for River* iv. 74 If he called her now she would probably be in bed, and come down in her nightgown or her pyjamas, or whatever slumberwear she favoured, and that would never do.

slumdom. Add: Also, the condition or character of slums or slum-dwellers.
1927 G. B. SHAW in *Yorks. Even. News* 30 Nov. 9/2 Civilisation means 'Respect my life and property and I will respect yours.' Slumdom means 'Disregard my life and property and I will disregard yours.' **1962** *Economist* 18 Aug. 593/1 Another 60,000 old houses slip into slumdom. **1973** *Daily Tel.* 7 Nov. 13/1 A pre-war, cottage-type housing estate that is slipping into slumdom as fast as the downward slope will take it.

slumgullion. Add: **1.** (U.S. example.)
1872 'MARK TWAIN' *Roughing It* iv. 44 He poured for us a beverage which he called '*Slumgullion*'.

2. b. (Earlier and later examples.)
1887 B. HARTE *Millionaire & Devil's Ford* 146 We preach at them for playing in the slumgullion, and getting themselves splashed. **1906** C. DE L. CANFIELD *Diary of Forty-Niner* 27 The mud we were sending down the stream buried them under slumgullion. **1948** O. WESTON *Mother Lode Album* 82 The miners..insisted on calling it 'Slumgullion', because when it rained the knee-deep adobe mud was no small problem.

c. (Earlier and later examples.) Occas. used outside U.S.
1902 J. LONDON *Daughter of Snows* 45 'What do you happen to call it?' 'Slumgullion,' she responded curtly, and thereafter the meal went on in silence. **1932** J. DOS PASSOS *1919* 17 Bedbugs in the bunks in the stinking focastle, slumgullion for grub. **1959** A. SILLITOE *Loneliness* 7 The first thing a long-distance cross-country runner would do..would be to run as far away from the place as he could get on a bellyful of Borstal slumgullion. **1976** T. WALKER *Spatsizi* x. 115 For want of a better word we called it slumgullion.

slumgum (slʌ·mgʌm). *U.S.* Also **slum gum.** [f. SLUM *sb.*[2] + GUM *sb.*[2]] The residual wax, propolis, and other impurities that remain when the honey and most of the wax are extracted from honeycombs by warming them.
1890 *Gleanings Bee Culture* XVIII. 704/2 The cappings are laid on this perforated tin, and, when they melt, the wax and honey run through into the chamber below, leaving what Californians call the 'slumgum' on the tin above. **1917** *Rep. Iowa State Apiarist* 36 A third of the weight of this slumgum is wax and should be saved. **1946** R. A. GROUT *Hive & Honeybee* xxii. 544 Slumgum is the material remaining after some rendering treatment has been performed on comb material. It may be more or less rich in beeswax and is usually dark brown or almost black in color. **1980** *Bee Craft* LXII. 154/2 The water..had to heat to the extractor temperature and this allowed the dross or slum gum to coagulate on the under surface of the straining cloth.

slumless (slʌ·mlės), *a.* [f. SLUM *sb.*[1] + -LESS.] Containing no slums.
1924 *Glasgow Herald* 8 Mar. 7 The difference between almost slumless Düsseldorf and slummy Glasgow is not altogether in municipal policy or school education. **1946** P. BOTTOME *Lifeline* XXX. 235 Berlin..bustling, self-determined, ordered, slumless. **1966** *Guardian* 16 May 15/3 Any big-city mayor..can successfully achieve a relatively slumless city.

slumminess. (Later examples.)
1926 A. HUXLEY *Two or Three Graces* 176 It was a slummy street... It was not hard to know where respectable slumminess ended and gay Bohemianism began. **1961** *Guardian* 29 Mar. 9/3 Being overcrowded does not necessarily imply slumminess.

slumming, *vbl. sb.* Add: **2. c.** *fig.* with defining adj. Cf. sense *4 b of the vb.
1933 DYLAN THOMAS *Let.* (1966) 70 Few understand the works of Cummings, And few James Joyce's mental

slummings. **1958** [see *DEMOTIC *a.* 2]. **1977** M. DRABBLE *Ice Age* I. 35 She accused Anthony of hypocrisy, of intellectual slumming, of *folie de grandeur*, of brain fever.

slu·mmock, *sb. dial.* and *colloq.* [Var. of dial. *slammock*: see *Eng. Dial. Dict.* and SLUMMOCK *v.*] A dirty, untidy, or slovenly person; a slut. Freq. as a disrespectful term of address. Cf. SLAMMAKIN *sb.* 2.

19th.-cent. dial. examples in *Eng. Dial. Dict.* s.v. *Slammock* sb.

1932 'L. G. GIBBON' *Sunset Song* 186 Chris found herself dancing with Mistress Mutch, the great, easy-going slummock. **1953** L. HILL tr. *Anouilh's Waltz of Toreadors* in J. C. Trewin *Plays of Year* VIII. 444 A slummock, a girl who hasn't even washed! **1966** M. KELLY *Dead Corse* i. 10 'You are the greatest slummock,' she said. 'How can you bear to lie on an unmade bed?' **1974** P. FLOWER *Odd Job* ix. 59 He wiped Norah's table-top... Norah was a slummock.

slummocker (slɒ·mŏkəɪ). *dial.* Also **slummicker.** [Of obscure ulterior etym.: see SLAMMAKIN *sb.* and *a.*; SLUMMOCK *v.* This form is not recorded in dialect dicts.] = SLAMMAKIN *sb.* 2; an awkward or careless person.

1905 G. B. SHAW *Let.* 13 Aug. in A. T. Schwab *James Gibbons Huneker* (1963) xiii. 167 You will never be anything but a clever slummocker in America. **1905** —— *Let.* 16 Sept. in J. G. Huneker *Steeplejack* (1920) II. 258 The reason I call you a slummocker and heap insults on you, is that you are very useful to me in America, and quite friendly; consequently, you must be educated or you will compromise me. **1940** C. STEAD *Man who loved Children* (1941) ix. 376, I have to let that great big slummicker wash the dishes and smash every glass and plate in the house.

slummocking: see SLAMMAKIN *a.* in Dict. and Suppl.

slummocky, *a.* Add: *dial.* and *colloq.* Also (rarely) **slammocky, slommachy, slummucky.** (Further examples.) Hence **slu·mmockiness.**

? **1861** MRS. GASKELL *Let.* 28 Feb. (1966) 643 A tall, gentlemanly, slammocky-as-to-figure man. **1914** KIPLING in *Nash's Mag.* June 278/1 The rough-ironed table-linen, ..the slummocky set-out of victuals at meals. **1926** W. DE LA MARE *Connoisseur* 65 A help from the village—precious little good *she* was. Slummocky—and *stupid*! **1947** M. PENN *Manchester Fourteen Miles* iii. 34 Grandma Winstanley was..a slattern... Lizzie couldn't abide her slummockiness. **1953** J. CARY *Except the Lord* xxxviii. 169 Girls after a few months service would return on holiday not only in smart clothes but with quite new scorn for what they called our slummucky ways. **1962** J. CANNAN *All is Discovered* ii. 29 'An attractive woman?' 'No, sir. A slummacky sort. More like a gyppo.' **1973** P. WHITE *Eye of Storm* viii. 376 Her hands had been coarsened by menial grind, her body made slommacky by childbearing. **1974** H. R. F. KEATING *Underside* xix. 187 The doom-laden slummockiness of his bohemian days.

slummy (slɒ·mi), *sb. colloq.* Also **slummie.** [f. SLUM *sb.*^{1} + -Y^6.] A slum-dweller.

1934 P. O'MARA (*title*) Autobiography of a Liverpool Irish slummy. **1964** A. PRIOR *Z Cars Again* xvi. 158 The remains of many meals stood on..a newspaper... It was a typical slummie's house. **1973** 'J. PATRICK' *Glasgow Gang Observed* xii. 111 Big Fry..tauntingly called out: 'We're the slummies!'

slump, *sb.*^{2} Add: **2. a.** Also *spec.* in *Econ.*, a sharp or sudden decline in trade or business, usu. accompanied by widespread unemployment. Freq. with reference to a particular instance, esp. the Great Depression of 1929 and subsequent years.

1922 H. A. SILVERMAN *Substance of Economics* xv. 231 Industries grew to depend increasingly on one another... It became inevitable..that a 'boom' or a 'slump' in one branch should synchronize with similar conditions elsewhere. **1930** *Engineering* 10 Jan. 42/2 To discover opportunities for employment on such jobs during industrial slumps. **1936** J. M. KEYNES *Gen. Theory Employment, Interest & Money* IV. xvi. 218 In the succeeding 'slump' the stock of capital may fall for a time below the level which will yield a marginal efficiency of zero. **1936** N. STREATFEILD *Ballet Shoes* vi. 89 'Well, I can't go back to Kuala Lumpur.' 'Why?' 'A thing called a slump.' **1952** *Granta* 15 Nov. 12/1 We wanted to fight Fascism, War and the Slump. **1953** M. SCOTT *Breakfast at Six* iii. 24 Bought all this land—got it cheap in slump time. **1957** I. CROSS *God Boy* (1958) iii. 27 Then there was the slump ..and then I never did get a chance with that hotel in Wellington. **1976** *Economist* 16 Oct. 13/2 A record rise in mortgage charges during a building slump.

b. *Geomorphol.* A landslide in which soil, sediment, or the like slides a short distance with some degree of cohesion and usu. a slight backward rotation owing to the concavity of the surface of separation from the parent mass; movement of this kind; also, a mass of material that has so fallen.

1905 CHAMBERLIN & SALISBURY *Geology* I. iv. 218 (*heading*) Creep, slumps, and landslides. **1949** F. J. PETTIJOHN *Sedimentary Rocks* iv. 145 A structure of similar appearance..is reported from some limestones and dolomites. The cause of the folding may be due to subaqueous slump. **1954** W. D. THORNBURY *Princ. Geomorphol.* v. 104 Mantle rock..is moved downslope by creep, slump, other types of mass-wasting, and by sheet-wash. **1963** D. W. & E. E. HUMPHRIES tr. *Termier's Erosion & Sedimentation* vii. 166 Water-laid phenomena (slumps, low-angle cross bedding) are observed, and suggest that eolian sands have been blown into a shallow sea. **1964** V. J. CHAPMAN *Coastal Veget.* i. 2 A large scale change induced by a major cliff-fall or slump. **1970** W. H. MATTHEWS *Geol. made Simple* (rev. ed.) viii. 117 Slump is a common occurrence along the banks of streams. **1978** A. L. BLOOM *Geomorphology* viii. 178 An elaborate engineering technology has been developed to predict the surface of rupture beneath a slump in order to drill into it and drain the water from the vicinity.

c. *Engin.* The height through which the top of a mass of fresh concrete sinks when the mould containing it is removed, as in the slump test (see sense 4 below).

1920 D. A. ABRAMS *Design of Concrete Mixes* (Bull. No. 1, Structural Materials Res. Lab., Lewis Inst., Chicago) 13 Normal consistency..requires the use of such a quantity of mixing water as will cause a slump of ½ to 1 in. in a freshly molded 6 × 12-in. cylinder of about 1:4 mix. **1934** S. C. HOLLISTER in L. C. Urquhart *Civil Engin. Handbk.* VII. 562 Concrete for buildings ranges from 4 to 6 in. slump. **1977** D. E. BRANSON *Deformation of Concrete Structures* i. 48 Creep correction factors... May be marginal but normally can be neglected for slumps up to 4 in.

3. *gen.* A slumping movement or fall.

1850 S. JUDD *Richard Edney* i. 12 Move carefully! It is a slip, or a slump, all the way through. **1867** 'T. LACKLAND' *Homespun* I. 90 A..black snake..slid down with a slump ..into the water. **1900** S. HALE *Let.* 29 Apr. (1919) 361, I let my huge bulk down with a slump.

4. *attrib.*, as (sense *2 b) *slump bed, bedding, block, series, sheet, structure*; **slump test** *Engin.*, a test of the consistency of fresh concrete in which the slump is measured following the removal of a mould of specified size and shape (usu. the frustum of a cone).

1974 *Sedimentology* XXI. 2 Exposures of banks and slump beds extend along the whole of the coast. **1949** F. J. PETTIJOHN *Sedimentary Rocks* iv. 145 The disturbance is restricted to layers a mere inch or two thick. Such deformation is usually due to subaqueous slump or gliding and has been termed 'slump' or 'glide bedding'. **1964** *Gloss. Mining Terms* (*B.S.I.*) v. 13 *Slump bedding*, disturbed strata interbedded between undisturbed strata, caused by flow of newly deposited sediment. **1969** D. J. EASTERBROOK *Princ. Geomorphol.* xi. 228 During movement of a slump block, secondary slumps may develop and produce a stair-step-like series of parallel slump blocks. **1978** A. L. BLOOM *Geomorphology* viii. 178 Vegetation or even houses may be carried intact on the surface of a large slump block. **1937** O. T. JONES in *Q. Jrnl. Geol. Soc.* XCIII. 272 In view of the fact that a thick mass.. may have been formed by successive sliding or slumping of sediments, it is proposed to speak of it as a slump series, and where it is reasonably certain that a mass is the result of a single episode, that mass is referred to as a slump sheet. A slump series is or may be..made up of several slump sheets separated by a greater or lesser thickness of normal mudstones. **1976** *Jrnl. Geol. Soc.* CXXXII. 125 Sequence 4, in the upper part of the slump sheet, is most complexly deformed, showing closed and contorted folds. **1963** *Geol. Mag.* C. 205 The slump structures which characterize the Torridonian red sandstones of North-West Scotland. **1975** J. L. WILSON *Carbonate Facies Geol. Hist.* viii. 238 The limestone..has graded beds, lamination, microbreccias and slump structure. **1920** F. L. ROMAN in *Engin. & Contracting* 3 Mar. 241/1 Cone No. 1..was far better than a cylinder for determining the consistency of concrete by means of a 'slump' test. **1975** *Concrete Inspection Procedures* (Portland Cement Assoc.) iv. 41 A slump test is made at the start of the operation each day and whenever the appearance of concrete indicates a change in consistency.

slump, *v.*^{2} Add: **2. b.** Also, to fall or collapse clumsily or heavily. *spec.* in *Geomorphol.* of soil, sediment, etc.: to fall in a slump (sense *2 b).

1905 CHAMBERLIN & SALISBURY *Geology* I. iv. 220 Where a stream's banks are high..considerable masses sometimes slump from the bank. **1920** *Engin. & Contracting* 3 Mar. 241/1 Large voids or stone pockets tend to cause the concrete specimen to slump on one side rather than vertically. **1937** *Q. Jrnl. Geol. Soc.* XCIII. 276 Sediments accumulating on a sub-aqueous slope would slide or slump if the weight increased beyond a certain amount. **1978** FRIEDMAN & SANDERS *Princ. Sedimentol.* xii. 400/1 Strata that slumped and were deformed may be..overlain by turbidites.

c. (Earlier example with † *off.*)

1888 in Farmer *Americanisms* (1889) 495/2 'How's North-western this morning, Uncle Zeke?' asked Dick... 'Slumped off six points, hang it!' scowling viciously over his paper.

d. *transf.* and *fig.*

1925 *Sunday Times* 20 Sept. 12/6 Where one's sympathy slumps and all one's optimism fails is in face of two depressing facts. **1970** *Daily Tel.* 16 May 12 Better pay and conditions are essential if police morale is not to go on slumping. **1977** *Cork Examiner* 6 June 7/1 The overnight leader..slumped to an 80 in his second round for 150.

Hence **slumped** *ppl. a.*

1937 *Q. Jrnl. Geol. Soc.* XCIII. 277 Local after-slides.. added low ridges on the surface of the major slumped mass. **1965** G. J. WILLIAMS *Econ. Geol. N.Z.* iii. 31/2 There is a good deal of glacial debris and slumped ground under the thick forest. **1976** J. E. SANDERS et al. *Physical Geol.* vii. 244 A slumped mass usually does not travel very far nor spectacularly fast.

slumpflation (slɒmpflēɪ·ʃən). *Econ.* [Portmanteau blend of SLUMP *sb.*^{2} + *IN)FLATION 6: cf. *STAGFLATION.] A state of economic depression in which decreasing output and employment in industry are accompanied by increasing inflation.

1974 W. REES-MOGG *Reigning Error* iv. 75 So-called stagflation and slumpflation are the inevitable reflection of the progressive divergence between a rising nominal and a falling real supply of money. **1976** *Economic Jrnl.* LXXXVI. 171 Chronic slumpflation has given rise to much agonising reappraisal of doctrines that were hardening into orthodoxies. **1980** *Economist* 23 Feb. 13/1 The government can get less slumpflation in British industry only by making life easier for the employers' wage negotiators. **1981** J. SUTHERLAND *Bestsellers* xix. 201 Portugal wallows in the slumpflation that will eventually lead to fascism.

slumping (slɒ·mpiŋ), *vbl. sb.*^{2} *Geomorphol.* [f. SLUMP *v.*^{2} + -ING^1.] The fall of soil, sediment, or the like in a slump (*SLUMP *sb.*^{2} 3 b).

1907 R. D. SALISBURY *Physiography* vi. 106 Slumping is very common on slopes composed of unconsolidated material, such as clay or accumulations of loose rock. **1944** A. HOLMES *Princ. Physical Geol.* x. 148 Similar conditions favour landslides on a bigger scale, wherever slumping (Fig. 63) or sliding (Fig. 64) can occur on the sides of undercut slopes, precipices, and cliffs. **1979** *Geogr. Mag.* July 668/3 Many sub-circular pans on the Essex marshes may be formed by the blocking-off by slumping and vegetation overgrowth of the large number of creek heads which appear rounded in outline.

slumscape (slɒ·mskēɪp). [f. SLUM *sb.*^{1}, after LANDSCAPE, etc.: cf. SCAPE *sb.*^{3}] Slum scenery, or a picture of this.

1947 WYNDHAM LEWIS *Let.* Apr. (1963) 405 Down another [road] moved a great slumscape painter. **1967** *N.Y. Times* 4 May 41/5 They walked slowly through the scarred and dreary slumscape.

slup, *v.* Delete † *Obs.*$^{-1}$ and add later examples. [Modern examples may represent an echoic form. Cf. SLURP *v.* in Dict. and Suppl.] Hence **slup** *sb.*, the noise of slupping; **slu·pping** *ppl. a.* and *vbl. sb.*

1947 *Time* 7 Apr. 74/2 A julep-slupping burlesque of a Southern politico. **1949** *Daily Progress* (Charlottesville, Va.) 2 Mar. 10 (*caption*) Put a muffler on that soup sluppin'. **1949** H. HORNSBY *Lonesome Valley* 11 The mule slupped the clear water. **1952** J. STEINBECK *East of Eden* 23 There was no talk at supper. The quiet was disturbed only by the slup of soup and gnash of chewing. **1971** G. EWART *Gavin Ewart Show* I. 28, I am a bottle of wine... slup me rough and homely and I'll taste fine.

slur, *sb.*^{2} **3.** (Examples of *slur-cock.*)

1927 T. WOODHOUSE *Artificial Silk* ix. 98 The jack sinkers..are operated directly or indirectly by means of the cam of a 'slur-cock'. **1962** *Engineering* 15 June 771/1 Straight bar knitting machines..depend largely for their successful operation upon the motion given by a linear cam known as the 'slurcock'.

slurb (slɔ̄ɹb). orig. *U.S.* [App. f. *sl-* (as in *sloppy, sleazy*, etc.: see quot. 1962) + URB(AN *a.*, though later re-analysed as if from SL(UM *sb.*^{1} + SUB)URB.] An area of unplanned suburban development.

1962 WOOD & HELLER *California going, Going...* 10 The character and quality of such urban sprawl is readily recognizable... These are the qualities of most of our new urban areas—of our *slurbs*—our sloppy, sleazy, slovenly, slipshod semi-cities. **1966** *Guardian* 22 Apr. 12/3 About 35 miles west of London there is a new town that no one knows about. It is in Berkshire, between Reading and Sandhurst, and it includes Wokingham and Crowthorne. It is what Californians, who have plenty of them, call a 'slurb'—an amorphous and intermittent spread of houses. **1967** *Economist* 8 July 120/2 The pattern which has turned Los Angeles into an 'un-community', into 'twenty suburbs in search of a city', into the archetypal 'slurb'. **1979** B. WARD *Progress for Small Planet* xxi. 235 The basic concept of 'urban villages' (the Chelseas, the Trasteveres, the Greenwich villages)..often allow a vitality and an attractiveness which the world's concrete suburban deserts, slurbs, and sprawls so demonstrably lack.

slurp, *v.* **1. a.** *trans.* Delete *rare* and add: Also, to eat in this manner. (See also quot. 1976). Also with *down, up.*

1917 *Dialect Notes* IV. 329 *Slurp, v.t.* and *i.*..to eat liquid food with audible inhalation of air. **1947** *Richmond* (Va.) *Times-Dispatch* 30 Dec. 15/1 The stars just whirl in.., slurp a cup of coffee and zoom out again. **1952** C. ARMSTRONG *Black-Eyed Stranger* ii. 10 'You know—' Baby slurped food, 'If you want to live you got to eat.' **1962** R. LOWELL *Imitations* 39 They are slurping their dinners quite happily. **1974** P. CAVE *Mama* (new ed.) xiii. 107 The Angels obediently slurped down the remainder of their teas and rose to their feet noisily. **1976** *Daily Tel.* 21 Jan. 15/5 After about seven years of 'slurping' (the correct professional word) tea in the tasting room he [*sc.* a tea-taster]..has prospects of becoming a buyer or blender.

b. *transf.* and *fig.*

1968 *Punch* 11 Dec. 858/1 The idiocies of British holiday habits, which tend to waver between 'slurping up the kilometres' and getting away from it all amid insanitary *souks*. **1973** M. AMIS *Rachel Papers* 15 Mrs Bladderby had an even wreckier mother, who..had, moreover, during a recent outing, got her left leg slurped into a dreadful piece of agricultural machinery.

2. a. *intr.* To make a sucking noise in drinking or eating.

1917 [see sense 1a above]. **1961** B. CRUMP *Hang on a Minute, Mate* 108, I had my head inside the can slurping happily away. **1975** A. A. THOMPSON *Message from Absolom* iv. 24 The Americans ate hungrily. At the other table, the Elberts slurped audibly.

b. *transf.* and *fig.*

1958 J. KEROUAC *On Road* vii. 155 He..stuck his finger in Marylou's dress, slurped up her knee. **1963** [see *MATIÈRE]. **1971** B. W. ALDISS *Soldier Erect* 24 On bed immediately... Fanny swimming with juice, slurps when touched. Marvellous tits, delicious underarms. **1976** P. CAVE *High Flying Birds* iv. 51 With a couple of pints of champagne slurping around her insides, she found it increasingly difficult to wrap her tongue round the hard Anglo-Saxon consonants.

Hence **slu·rping** *ppl. a.* and *vbl. sb.*

1960 W. SHEED *Middle Class Educ.* (1961) 31 The slurping roar of the undergraduates. **1976** *Daily Tel.* 21 Jan. 15/5 Mr Ronald Calvert, chief taster for Ridgways,.. wonders if women are daunted by a job entailing the 'slurping' of 200 to 400 mouthfuls of tea on an average working day. **1980** *Sunday Express* 27 July 16 No one would want them to endanger their striped pants by sitting next to spotty Coca-Cola slurping children.

slurp (slɔ̄ɹp), *sb.* and *int.* [f. the vb.] **A.** *sb.* A slurping sip or lick; the noise of slurping.

1949 H. HORNSBY *Lonesome Valley* xxi. 270 The second cow..slammed her nose against her side and swiped at flies with her tongue. Almost before the slobbery slurp was over with the cow trotted after the other. **1959** A. BAILEY *Making Progress* 142 The slurps and gulps of the Danes moodily drinking their beer. **1960** I. CROSS *Backward Sex* 140 He..took a huge slurp of tea. **1977** D. CLARK *Gimmel Flask* iv. 79 Green took a slurp of his coffee, grimaced at the taste.

B. *int.* An interjection imitating the sound of a slurp.

1966 L. COHEN *Beautiful Losers* I. 16 It is recorded that she prayed incessantly. Glog, glog, dear God..slurp, flark, glamph, hiccup, jerk. **1967** W. H. CANAWAY *Mules of Borgo San Marco* iv. 51 She dipped her spoon in the soup, sipped, and said, 'Perfect!' '*Slurp*,' said Major Widdicombe. **1970** *Private Eye* 22 May 16 Leetle Germaine eez at votre service!!! (slurp!).

slurry, *sb.* Add: **1. a.** Also in extended use, any fluid mixture of a pulverized solid with a liquid (usu. water), freq. used as a convenient form in which to handle solids in bulk. (Later examples.)

1935 *Discovery* Apr. 119/1 The strata are pounded into a 'slurry' by the constant rising and falling of the heavy drilling tools. **1948** *Chambers's Jrnl.* July 388/1 The idea here is to break up the turf bank face by means of high-pressure jets, like firemen's hoses, into slurry. **1955** *Sci. News Let.* 20 Aug. 115/2 The U.S.-designed power plant.. uses..thorium oxide slurry in heavy water in another part of the device, called a homogenous reactor. **1961** *Aeroplane* CI. 342/3 Whatever form of contaminant is selected, it is first mixed with kerosene to form a 'slurry', so representing the normal state in which it would pass through a filter in service. **1975** *Nature* 30 Oct. 818/1 In a second experiment, another batch of about 3 ml of the packed cell slurry was resuspended in the culture medium.

b. *spec.* A mixture of water and fine particles of coal, produced esp. as a by-product of the washing of coal; such material in dried form, used as fuel.

1913 *Trans. Inst. Mining Engineers* XLV. 429 Where the moisture percentage is not over 5, the small dust measuring less than 1 cubic millimetre (which is the cause of all the slurry trouble) can be just as efficiently treated by dry-percussive screening as by any other method. **1930** *Ibid.* LXXVIII. 27 It cannot be claimed that slurry is a suitable fuel for pulverizing. **1955** *Times* 4 July 15/5 They [*sc.* the N.C.B.] have been using a new type of mechanical stoker, which is fully automatic, to burn slurry—fine particles of coal and soluble shale which is rejected from washeries. **1976** 'R. LEWIS' *Witness my Death* v. 180 That tip up there, it's full of water. The slurry is drifting down.

c. *spec.* A mixture of manure or farmyard waste and water; manure in fluid form.

1965 *Punch* 22 Dec. 932/2 In a modern fattening house ..the manure from several hundred swine falls through slatted floors into tanks beneath the building where a daily dose of water soon turns it into a forbidding quantity of evil-smelling slurry. **1971** *Farmers Weekly* 19 Mar. 48/4 It takes one man about six minutes a day to clear away the slurry and a bit longer to put out the hay. **1970** R. JEFFRIES *Dead Man's Bluff* i. 5 He went through from the dairy into the herringbone parlour and stared..at the two days' accumulation of slurry.

3. *attrib.*, as *slurry disposal, pipeline, pit, pump, refiner, tank, tanker*; **slurry seal** (see quot. 1967).

1970 R. JEFFRIES *Dead Man's Bluff* i. 5 He stared angrily at..the slurry disposal unit. **1969** *Daily Colonist* (Victoria, B.C.) 13 Dec. 9/3 Kaiser..was contemplating a slurry pipeline to the coast as an alternative to rail. **1976** *Cumberland News* 3 Dec. 13/4 There was a slurry pit also under a byre, made of slate. **1940** KRISTAL & ANNETT *Pumps* 338/2 (Index), Slurry pumps. **1976** *Cumberland News* 3 Dec. 34/3 (Advt.), Alfa Laval slurry pumps. **1916** *Trans. Inst. Mining Engineers* LI. 272 With the combination of this slurry-refiner and the elevated settling-tank.., it has been found possible to work a washery, year in and year out, without any outlet whatever. **1967** *Gloss. Highway Engin. Terms* (*B.S.I.*) 30 *Slurry seal*, a mixture of binder, fine aggregate and mineral filler with water added to produce a material of slurry consistency. **1974** *Globe & Mail* (Toronto) 7 Feb. 5/6 Slurry seal is a tar-like chemical substance that is spread on city streets to preserve them. It is cheaper than asphalt. **1936** *Economist* 25 Apr. 213/2 Property account has been increased to £12,557 by the addition of the slurry tanks and quartz deposit. **1971** *Farmers Weekly* (*Extra*) 19 Mar. 5/1 (Advt.), We know we have the best slurry tanker.

slurry (slɔ̄·ri), *a.* [f. SLUR *v.*^{1} + -Y^1.] Blurred, indistinct: now usu. of speech. Also **slu·rrily** *adv.*

1937 *Daily Express* 12 Feb. 5/3 This is the way to detect a forgery—look for notes that are rather blurred or 'slurry'. **1969** 'H. CALVIN' *Chosen Instrument* ii. 22 'We own nothing, we need nothing,' the fat boy said slurrily. **1977** J. MCCLURE *Sunday Hangman* xv. 174 'Why leave the bodies everywhere?' Willie demanded, driven..to speak his mind, if a little slurrily.

slush, *sb.*^{1} Add: **2. c.** Food, esp. of a watery consistency. *slang.*

1941 J. SMILEY *Hash House Lingo* 51 *Slush*, hash. **1955** J. THOMAS *No Banners* ix. 79 It was years since he had tasted anything but jail slush. **1962** W. GRANVILLE *Dict. Sailors' Slang* 108/2 *Slush*, any 'sloppy' food: e.g., soup or stew.

4. a. (Earlier and later examples.) Also *gen.*, nonsense, drivel; sentimental rubbish. Also as *int.*

1869 'MARK TWAIN' *Innoc. Abr.* x. 91 He'll..grind out about four reams of the awfullest slush. **1906** *Dialect Notes* III. 156 O, slush! What nonsense. **1919** C. E. VAN LOAN *Score by Innings* 332 A woman reporter..took one look at Conley..and tore off a whole page of slush. **1937** PARTRIDGE *Dict. Slang* 786/1 As = sickly sentiment, *slush* is familiar S.E. **1944** [see *blush-making* adj. s.v. *BLUSH *sb.* C.]. **1949** [see *KITSCH]. **1953** F. SCOTT FITZGERALD *Tender is Night* I. ix. 51, I mean, would I have been the sort of girl you might have—oh, slush, you know what I mean. **1961** *Observer* 26 Nov. 27/4 The ending is purest slush, and there are some cheap dramatics in the camera work. **1970** R. K. KENT *Lang. Journalism* 124 *Slush*,.. cheaply sentimental copy; trash; drivel.

b. Counterfeit paper money. *slang.*

1924 E. WALLACE *Room 13* i. 11 Young Legge's..the biggest printer of slush in the world! And it's not ord'nary slush. Experts..can't tell 'em from real Bank of England stuff. **1933** D. HUME *Crime Unlimited* vii. 64 We've been handling slush lately—ten bobs and quids. Where they were printed doesn't matter to you.

6. *slush-lamp* (earlier example), *-light*; (sense 4) *slush melodrama, novel*; **slush casting,** a method of making hollow castings in which molten metal is poured into a mould and then poured out again after a layer of metal has solidified on the inner surface of the mould; a casting produced by this method; also **slush-cast** *v.*; **slush-money** orig. *U.S.*, money paid out from a *SLUSH FUND b; **slush moulding,** a process identical to the slush casting of metal but carried out with plastic or latex; so **slush mould,** a mould for use in slush moulding; **slush-moulded** *ppl. a.*; **slush oil** *U.S.*, crude oil found in association with certain shales or sandstones (? *obs.*); **slush pit** (see quots.); **slush pump,** (*a*) a pump used to circulate mud through a rotary drilling column; (*b*) *U.S. slang*, a trombone.

1934 WEBSTER, Slush-cast, *v. t.* & *i.* **1965** E. TUNIS *Colonial Craftsmen* iv. 75/2 Feet and knobs were cast solid, but spouts and handles were slush cast, a system used also for such things as sand shakers and nursing bottles whose inner surfaces wouldn't be visible. **1930** M. STERN *Die-Casting Pract.* i. 16 The thickness of a slush casting depends upon the length of time that the metal is left in the mold. **1934** CHARNOCK & PARTINGTON *Mech. Technol.* (ed. 2) xxxv. 485 Slush casting is a method of producing light hollow castings without the use of cores. **1936** H. L. CAMPBELL *Metal Castings* iii. 53 Slush castings are produced by pouring the low-melting alloys of lead, antimony, and zinc into metal molds and, after a short interval, slushing out the metal which remains in a liquid state. **1963** JONES & SCHUBERT *Engin. Encycl.* (ed. 3) 1161 The process known as slush-casting is employed extensively in the production of ornamental objects made of spelter or zinc. **1871** C. L. MONEY *Knocking about N.Z.* vi. 77 An old volume of 'Household Words' to spell over at my little fire in the evenings by the light of my slush-lamp. **1887** S. SAMUELS *Forecastle to Cabin* vi. 76 This thump, we found out afterwards, was caused by a handspike; the jar from it put the slush light out. **1972** *People* (Austral.) 13 Feb. 12/2 They..lit their earth-floored bunkhouses at night with slushlights made from treacle-tins filled with fat. **1916** 'B. M. BOWER' *Phantom Herd* vii. 112 You want those stories worked up in a lot of darned, sickly slush melodramas. **1842** J. F. COOPER *Wing-and-Wing* II. 20 They were only put there yesterday..a little slush-money did it all. **1976** *National Observer* (U.S.) 24 Jan. 3/3 There have been a series of well-founded reports..that the multinational corporations have been shoveling slush money into Christian Democrat coffers. **1957** V. J. KEHOE *Technique Film & Television Make-up* xii. 148 It [*sc.* dental stone] has low absorption qualities so is not suitable for slush molds. **1965** E. TUNIS *Colonial Craftsmen* iv. 75/2 The slush mold's two halves shaped only the outside of the article. The caster poured hot metal into it, sluiced it around carefully, and then poured it out again. **1954** N. J. RAKAS *Plastics Engin. Handbk.* x. 314 The physical qualities of slush-molded plastisol are such that it is possible to do exceptionally accurate work. **1943** SIMONDS & ELLIS *Handbk. Plastics* 971 *Slush molding*, a process for molding hollow castings with accelerated thermoplastic phenolic resins. **1957** V. J. KEHOE *Technique Film & Television Make-up* xii. 154 Slush molding requires the use of a fillered latex as the unfillered variety will not build-up on itself in a plaster mold. **1963** H. R. CLAUSER *Encycl. Engin. Materials & Processes* 491/2 Vinyl foam products such as armrests..are manufactured by first forming a tough vinyl skin by spraying, slush molding or rotational molding. The interior then is formed by casting..a vinyl plastisol foam within the pregelled skin. **1977** *Listener* 3 Mar. 284/4 Communications—meaning journalism, detective fiction, slush novels, and film-making. **1880** J. F. CARLL *Geol. Oil Regions of Warren* [etc.] *Counties* 254 The measures above the Warren and Bradford 'Third sands' have produced considerable 'shale or slush oil', which may perhaps be attributed to a fissured condition of these rocks. **1884** *U.S. Tenth Census* X. 13 The first well sunk to the Bradford sands was drilled..2 miles northeast of Bradford. 'Slush oil' was found at a depth of 751 feet, and in November, 1871, producing sand was struck at 1,110 feet. **1931** *Sun* (Baltimore) 29 Apr. 1/4 Four lay in the slush pit, an earthen depression intended to catch drilling refuse. **1975** L. CROOK *Oil Terms* 105 *Slush pit*, pit used for storing drilling mud. **1913** *Oil & Gas Man's Mag.* VIII. 822/2 Two slush pumps are usually installed with each drilling outfit—one pump to operate and the other one to act as a relay. **1921**, etc. [see *mud-laden fluid* s.v. *MUD *sb.*^{1} 5]. **1937** *Amer. Speech* XII. 48/2 *Slushpump*, a trombone. **1938** [see *gob-stick* s.v. *GOB *sb.*^{3} b]. **1943** *N.Y. Times* 9 May II. 5/4 That man with the Slush Pump was a fine sender. **1949** *Our Industry* (Anglo-Iranian Oil Co. Ltd.) (ed. 2) ii. 37 The circuit of this mud-laden fluid..commences at the slush pumps. **1962** J. WAIN *Strike Father Dead* 141, I could see at once why he preferred the valve trombone to the ordinary slush-pump. **1974** *BP Shield Internat.* Oct. 19/3 There's always a lot of work we can do in the sheltered places—like maintaining the slush-pump. **1977** J. WAINWRIGHT *Do Nothin' till you hear from Me* x. 176 Get Walt to help on the slushpump try-outs. Walt stays first trombone.

slush, *v.* **II. 5.** (Earlier example.)

1853 MOSSMAN & BANISTER *Australia visited & Revisited* iv. 45 The dirty work, mud, and slushing in water.. are so contrary to the habits of the many, that few can stand the training.

slusher. Add: **2.** *Mining.* A mechanical device for loading or packing broken material in which a bucket is drawn to and fro through a pile of the material by ropes wound round a drum at each end of its length of travel.

1923 *Mine & Quarry* July 1267/1 The contractor..has what is practically the double drum slusher in the ordinary drag line excavator and also in his back-filling machine. **1946** *Trans. Inst. Mining Engineers* CVI. 115 Four scours, totalling 1,200 yds., have been driven with the slusher. **1976** *Times Lit. Suppl.* 8 Oct. 1281/2 Diamonds... To mine them you need, among other things,..a double drum slusher.

slush fund. orig. and chiefly *U.S.* [SLUSH *sb.*^{1} 2.] **a.** In the Navy: money collected from the sale of slush, etc., and used to buy luxuries for the crew. Also, a similar fund in the Army.

1839 W. MCNALLY *Evils & Abuses in Naval & Merchant Service* xvii. 162 The sailors in the navy are allowed salt beef... From this provision, when cooked..nearly all the fat boils off; this is carefully skimmed..and put into empty beef or pork barrels, and sold, and the money so received is called the *slush fund*. **1884** *Naval Encycl.* 759/2 *Slush-fund*, money obtained from the sale of slush. It is to be used for premiums for target firing, etc., and not for ship's purposes. **1963** T. PYNCHON *V.* i. 14 Pappy ended up borrowing 500 for 700 from Mac the cook's slush fund.

b. *transf.* A fund used to supplement the salaries of government employees; a fund used to bribe, or influence the action of, a person or group of persons, *spec.* for political ends; a fund used to support a favoured political candidate.

1874 *Congr. Record* 17 Apr. 3166/1 We have had this 'slush-fund' since 1866... It was divided among these officers to increase their salaries. **1894** *Ibid.* 16 Jan. 904/1 [Cleveland] was not elected in 1888..because of pious John Wanamaker and his $400,000 of campaign slush funds. **1924** *Glasgow Herald* 16 Feb. 10 A huge fund alleged to have been deposited in a Washington bank to the credit of a widely-known citizen very intimate with men prominent in public life... The name given to the mysterious fund is the 'slush fund'... 'Slush', in the American acceptance of the word, means illicit commission, bribery, corruption, and graft. **1931** *Economist* 10 Oct. 658/1 How candidate-deputies will react against this suppression of what has usually, in the past, been regarded as an electoral 'slush' fund remains to be seen. **1962** *Guardian* 15 Mar. 9/6 Eisenhower's running mate was accused of being the beneficiary of a 'slush fund' subscribed by wealthy backers. **1977** *Whitaker's Almanack 1978* 567/2 The Prime Minister on May 23 commissioned Mr. Dell, the Trade Secretary, to review Government policy and commercial practices in the light of the 'slush fund' allegations. **1980** *Washington Post* 1 Feb. A1/1 The company had a secret $600,000 slush fund for entertaining Pentagon officials.

slushing, *vbl. sb.* (In Dict. s.v. SLUSH *v.*) Add: *spec.* in *Mining*, the action or process of moving or scraping broken ore into a dump or on to a wagon or chute.

1923 *Mine & Quarry* July 1267/1 The scraping or slushing of one into chutes or raises or into mine cars is only one of the many uses for the portable double drum hoists. **1946** *Trans. Inst. Mining Engineers* CVI. 116 Both for hand-filling and slushing one haulage-hand was engaged in getting the tubs away from the contractor.

1966 S. D. WOODRUFF *Methods of Working Coal & Metal Mines* III. B. ii. 287 As soon as a cut was completed the tramming, or slushing, floor was filled with waste before another cut was started.

slushing, *ppl. a.* (In Dict. s.v. SLUSH *v.*) Add: *spec.* pertaining to or designating a viscous oil or grease used to protect bright metal surfaces, when paint or other fixed coatings cannot be used.

1920 *Technologic Papers Bureau of Standards* (U.S.) No. 176. 4 Slushing oils or slushing compounds are of a varied nature, sometimes being straight mineral greases ..sometimes mixtures of mineral and animal greases.. finely divided mineral matter, and volatile solvents. **1939** BURNS & SCHUH *Protective Coatings for Metals* xvi. 381 The successful use of slushing compounds depends to a considerable extent upon proper cleaning of the metal surface beforehand. **1975** *Chem. Abstr.* LXXXIII. 13232 A slushing oil was prepd. from a mixt. of a metal mahogany or naphthalene sulphonate, a C_{11-40} fatty acid, a cosolvent, and mineral oil.

slushy, *sb.* Also **slushey, slushie.** Add: **a.** (Earlier example as a nickname.) Also in more gen. application: a cook; any unskilled kitchen or domestic help.

1876 F. W. H. SYMONDSON *Two Years abaft Mast* xiii. 261 An unexpected roll of the ship sent both pig and cook sliding... There was 'Slushy' sometimes over, sometimes under the pig. **1919** W. H. DOWNING *Digger Dial.* 46 *Slushey*, a mess orderly. **1936** [see *POISONER b]. **1953** 'CADDIE' *Sydney Barmaid* 25 Nellie, a wisp of a girl who was slushie at Mrs Murphy's boarding-house. Slushie was the name given to anyone who worked at a camp boarding-house. **1967** C. DRUMMOND *Death at Furlong Post* xii. 153 'You could get a job as a slushy in any restaurant,' he pronounced, 'good short-order cookery.' **1970** K. GILES *Death in Church* iv. 85 A grey-headed woman was crying in a corner—'The part-time slushy,' said Porterman.

b. (Later examples.)

1904 *Bulletin* (Sydney) 15 Sept. 39/1 Terms as usual, 4s. per man per week, the cook to find his own slushy. **1936** A. RUSSELL *Gone Nomad* iii. 14, I had to take my turn at butchering the ration sheep and as 'slushy' to 'Doughboy' Terry, the cook. **1959** H. P. TRITTON *Time means Tucker* 25/2 Having over 100 men to cook for, he selected two off-siders, sometimes called 'slushies', and started his duties immediately.

slut, *sb.* Add: **3.** (Earlier example.)

1821 J. FOWLER *Jrnl.* 13 Nov. (1898) 42 A large Slut Which belongs to the Party atacted the Bare.

5. slut's wool (earlier example).

1862 *Edin. Rev.* Apr. 410 Upstairs there is 'slut's wool' under the beds.

slut, *v.* Add: **2.** (Later examples.) Also, to behave like a slovenly woman or a woman of loose morals. Also with *about.*

1913 A. LUNN *Harrovians* ii. 37 They groise their horrid eyes off and get out of fagging in a term or two, while we poor devils have to slut about 'on boy' for three years. **1948** G. GREENE *Heart of Matter* II. III. i. 200 Perhaps I'd have slutted with Bagster or killed myself. **1955** D. BARTON *Glorious Life* 199 In winter she had less of the Cinderella look that slutting about in summer cottons imposed on her.

sluthering, var. *SLOOTHERING *vbl. sb.*

sluttishly, *adv.* (Later example.)

1976 J. COOPER *Harriet* xiii. 106 They had a nice, relaxed evening..sluttishly eating curry off their knees.

sly, *a.*, *adv.*, and *sb.* Add: **A.** *adj.* **5. b.** (Earlier and later Austral. examples.)

1829 H. WIDOWSON *Present State of Van Diemen's Land* iv. 24 To these [inns] also, I believe, I may add a like number of 'sly grog shops', as they are called. **1840** T. P. MACQUEEN *Australia* 23 Increased powers ought to be given the magistrates and police to prevent the nuisances usually termed sly grog shops. **1936** F. CLUNE *Roaming round Darling* xxiv. 265 What a promotion—a sly-grogger to king of the Toko blacks! **1941** BAKER *Dict. Austral. Slang* 68 *Sly-groggery*, a sly-grog shop. **1959** M. SCOTT *White Elephant* xiii. 137 It's a sly-grogging hole and..I didn't go there. **1969** W. DICK *Naked Prodigal* 64 We were on our way to the sly grog joint to buy a dozen bottles.

7. slypuss [PUSS 3 b], a cunning or deceitful girl, a minx; so **slypussness.**

1908 W. DE MORGAN *Somehow Good* ix. 79 Laetitia, whose speech..appeared to impute insight, or penetration, or sly-pussness..to her young friend. **1942** [see *NITWITTED *a.*].

8. *sly-eyed, -looking.*

1967 G. KELLY in *Coast to Coast 1965–6* 97 He hated blonde women—sly-eyed, breasts..undulating above her ..tunic. **1795** P. FRENEAU *Poems* (1902) II. 341 A youngster was order'd to hold himself ready, A sly looking lad that was 'prentice to Snip. **1945** 'G. ORWELL' *Animal Farm* vi. 47 He was a sly-looking little man.

B. *adv. Comb.* (Later *poet.* example.)

1931 H. BELLOC *Sonnets & Verse* 80 Wine, bright avenger of sly-dealing wrong.

C. *absol.* or as *sb.* **2.** (*b*) (Earlier example.)

1818 KEATS *Let.* 18 Dec. (1958) II. 13 It might have been a good joke to pour on the sly bottle after bottle into a washing tub and roar for more.

slype. Substitute for def.: A covered way or passage, esp. one lying between the transept of a cathedral or monastic church and the chapter-house, and commonly leading out from the cloister. (Earlier example.)

1860 M. E. C. WALCOTT *Cathedrals in U.K.* (ed. 2) 261 The slype is the passage on the south-west side of the nave.

slype (sləɪp), *v. Basket-making.* Also **slipe.** [Prob. var. SLIPE *v.*^{2}] *trans.* To cut away one side of (a rod or cane) with a long slanting cut, so that it comes to a point. Hence **slype** *sb.*; **slyped** *ppl. a.*; **sly·ping** *vbl. sb.*

1910 *Encycl. Brit.* III. 482/2 If the bottom is made on a hoop the butts of the stakes are 'sliped', *i.e.* cut away with a long cut of the shop-knife, and turned tightly round the hoop; they are then said to be 'scalloped' on. **1912** T. OKEY *Introd. Art of Basket-making* vii. 68 Having prepared the stuff, slype six bottom-sticks. *Ibid.* 154 *Slype*, a long cut. **1953** A. G. KNOCK *Willow Basket-Work* 27 The blackberry basket,..requires a bow of stout rod fitted across the basket..by inserting its slyped ends down into the siding. **1959** D. WRIGHT *Baskets & Basketry* vi. 136 *Slype*, a slanting or flat cut. **1960** E. LEGG *Country Baskets* 52 The liners are duly slyped before insertion... Slyping is done whenever you need to shape a rod or cane to fit it snugly and securely against another.

smack, *sb.*^{3} Add: **3. a.** *a smack in the face* (examples); also *a smack in the eye.*

1895 A. W. PINERO *Second Mrs. Tanqueray* III. 105 One gets so many smacks in the face through interfering in matrimonial squabbles. **1930** A. P. HERBERT *Water Gipsies* viii. 84 'I'm leaving here.' 'Leaving us, Mr. Bryan?' Jane thought 'Oh, Lord, another smack in the face!' **1941** BAKER *Dict. Austral. Slang* 68 *Smack in the eye, a*, a disappointment, a refusal, a rebuff. **1949** E. COXHEAD *Wind in West* vii. 191 Well, but it was a nice smack in the face! To keep him, with all his experience.. so completely in the dark. **1958** I. CROSS *God Boy* xii. 98, I could tell that what he said was supposed to be a smack in the eye for me.

5. *Comb.*, as **smackwarm** *nonce-wd.* (see quot.).

1922 JOYCE *Ulysses* 262 She let free..her nipped elastic garter smackwarm against her smackable woman's warmhosed thigh.

smack (smæk), *sb.*^{4} *slang* (orig. *U.S.*). [Prob. alteration of *SCHMECK.] A drug, *spec.* heroin.

1942 BERREY & VAN DEN BARK *Amer. Thes. Slang* § 509/8 *Smack*,..a small packet of drugs. **1960** R. G. REISNER *Jazz Titans* 164 *Smack*, heroin. **1964** *N.Y. Times Mag.* 23 Aug. 64/2 Cocaine..referred to as..*smack*. **1968** *Sunday Truth* (Brisbane) 6 Oct. 36/3 When I first came here you were a big swinger if you turned on with pot. Now they're going straight on to smack (another slang word for heroin). **1969** *Oz* May 36/1 In the paper today it said that Jimmy Hendrix got busted for smack. **1976** R. CONDON *Whisper of Axe* II. vii. 208 She began by having the smack pushers recruit from 53 street gangs. **1980** P. KINSLEY *Vatchman Switch* xii. 91 You're dealing and.. I'm going to prove it. You're into opium and smack.

smack, *v.*^{2} Add: **5. a.** (Earlier and *fig.* examples.) Also *spec.* to chastise (a child) in this manner.

1835 DICKENS *Seven Dials* in *Bell's Life* 27 Sept. 1/1 Mrs. A. smacks Mrs. B.'s child for 'making faces'. **1892** G. B. SHAW *Let.* 12 Aug. (1965) I. 359 Smacking Bebel & Singer in the eye for their *dénigrement* of our programme. **1918** E. POUND *Let.* 1 Jan. (1971) 127, I liked your comment p. 89, Nov. no. Naturally pleased to see the folk song idea smacked again. **1976** *Evening Post* (Nottingham) 14 Dec. 18/9 It appeared to have been put there by her two-year-old son, who had been smacked for moving things about the house.

c. *to smack it about* (see quot. 1962). *Naut. slang.*

1914 'BARTIMEUS' *Naval Occasions* i. 5 Better man your boat from the boom and shove straight off. Smack it about! **1915** —— *Tall Ship* iv. 71 It's three o'clock now, so I advise you to smack it about and clean if you're going ashore. **1962** W. GRANVILLE *Dict. Sailors' Slang* 108/2 *Smack it about!*, naval exhortation to the hands to 'get a move on';..from the smacking of paint brushes about the ship's hull by the side party.

8. In *Comb.*, as **smack-bottom,** a childish expression for a smack on the bottom given in chastisement.

1970 P. LAURIE *Scotland Yard* iii. 89 'Put them down, Pop, or you'll get a smack-bottom.'.. The old man has gone back to babyhood. **1978** K. AMIS *Jake's Thing* xvii. 182 What he needs is a damn good smack-bottom and being told not to be so boring.

smackable (smæ·kăb'l), *a. rare.* [f. SMACK *v.*^{2} + -ABLE.] That may be smacked.

1922 [see *smackwarm* s.v. *SMACK *sb.*^{3} 5].

smack-dab, *adv. U.S. dial.* and *colloq.* Also **smack dab.** [f. SMACK *v.*^{2} 7 + DAB *adv.*] Exactly, precisely; with a smack. Cf. *SLAP-DAB *adv.*

1892 *Dialect Notes* I. 232 He hit him smack dab in the mouth. **1893** H. A. SHANDS *Some Peculiarities of Speech in Mississippi* 75 *Smack-dab*, a term used by all classes, but more especially by the uneducated, to mean *exactly, precisely*; as, 'I hit him *smack-dab* in the face'. **1934** D. RUNYON in *Collier's* 3 Mar. 8/2 The old King tumbles smack-dab into the street. **1949** H. HORNSBY *Lonesome Valley* 16 He gave a little hop and landed smack-dab in the water. **1953** *Sun* (Baltimore) 30 Apr. (ed. B) 19/6 An April rain fell today where no man..would have dared tread last night—smack dab between the bristling New York Yankees and St. Louis Browns. **1967** *Boston Globe* 18 May 27/2 The university trustees apparently want to put it [*sc.* a university] smack-dab into Copley sq. **1970** N. ARMSTRONG et al. *First on Moon* xiii. 321 Here was the LM..right smack dab where it should have been. **1979** *United States 1980/81* (Penguin Travel Guides) 395 Whether you want to be..smack-dab downtown,..or conveniently near Lambert Field Airport, quality hotels are available.

smacker, *sb.*^{2} Add: **3.** *slang* (orig. *U.S.*). A coin or note of money; *spec.* a dollar; a pound.

1920 *Chicago Herald & Examiner* 2 Jan. 14/2 Along comes Earl Gray and knocks off the U.S. treasury for 13,000,000 smackers. **1924** WODEHOUSE *Bill the Conqueror* xvi. 261, I asked him to lend me a hundred smackers. **1928** [see *GRAFT *sb.*^{4} a]. **1935** C. W. T. CRAIG *Paraguayan Interlude* xxvi. 302 'I will give you a thousand beautiful smackers for your church.'.. Mac took out a thousand peso bill and handed it to me. **1937** C. W. PARMENTER *Kings of Beacon Hill* I. x. 70 Easy to touch, too, whether for a cigarette or a hundred smackers. **1946** F. SARGESON *That Summer* 86, I gave him a couple of smackers. **1953** R. LEHMANN *Echoing Grove* 205 Could I touch you for a smacker? I'm stony broke. **1968** *Landfall* XXII. 42, I wouldn't mind a few smackers. Enough for a case of whisky. **1979** 'L. BLACK' *Penny Murder* i. 5 'Gone at twelve thousand pounds.'.. Twelve thousand smackers for a tray of old coins. Whew!

smackeroo (smæ:kĕrū·). *slang* (orig. and chiefly *U.S.*). [f. SMACKER *sb.*^{2}: see *-EROO.] Used in senses of SMACKER *sb.*^{2} in Dict. and Suppl.: a coin or note of money; a kiss; a blow. Also as *int.*

Amer. Speech (1942) XVII. 14/1 gives citations of *smackeroo* 'dollar' used on U.S. radio programmes in 1940 and 1941.

1942 BERREY & VAN DEN BARK *Amer. Thes. Slang* § 29/2 *Something excellent*...smackeroo. *Ibid.* § 559/16 *Silver dollar*...smacker,..smackeroo. *Ibid.* § 702/1 *Blow*. ..smackeroo. **1951** P. BRANCH *Lion in Cellar* ix. 102 She grimps up the ladder... And what happens?..*Smackeroo!* **1961** S. PRICE *Just for Record* viii. 71, I got out the crisp crackling smackeroos and counted out two hundred of them. **1964** C. CHAPLIN *Autobiogr.* xvii. 300 You're getting the Legion of Honour, kid... That's the wrong colour—that's what they give to school-teachers; you don't get the smackeroos on the cheek for that one. **1977** 'E. V. CUNNINGHAM' *One-Penny Orange* (1978) vii. 90 The price is eight thousand pounds, and the pound was five dollars then, so that makes it forty thousand smackeroos.

smahan (smæ·hăn, sma·han). *Anglo-Ir.* [ad. Ir. *smeathán.*] A drop (to drink); a taste or nip.

1914 JOYCE *Dubliners* 117 Pony up, boys. We'll have just one little smahan more. **1961** 'F. O'BRIEN' *Hard Life* iii. 21 That reminds me—I think I deserve a smahan. Where's my crock?

small, *a.* and *sb.*^{2} Add: **A.** *adj.* **II. 3. a.** Phr. *small is beautiful*: expressing a belief that small-scale institutions, systems, etc., are more desirable than large-scale ones. Also as *adj. phr.*

1973 E. F. SCHUMACHER (*title*) Small is beautiful. **1975** *Country Life* 25 Dec. 1784/1 Adapting Schumacher's phrase, we decide that not only small but piecemeal is beautiful. **1976** *Seed* V. v. 6/3 Included are articles on self-sufficiency, 'small-is-beautiful' politics and agriculture and nutrition. **1977** D. JAMES *Spy at Evening* xxiv. 193 Small Is Beautiful—but big pays more. **1978** *Times* 23 Mar. 16/3 Mr. St John-Stevas..has proclaimed that 'small is beautiful'. There will be 'no more of those monster schools'. **1979** *Jrnl. R. Soc. Arts* July 468/1 It is worth mentioning another and different pressure upon the nature and shape of the hospital: and that is the vague but pervasive notion that 'small is beautiful'.

b. (*it's a*) *small world* and varr.: a comment on an unexpected meeting with an acquaintance or other similar coincidence.

1895 A. W. PINERO *Second Mrs. Tanqueray* III. 147 Mr. Ardale and I have met in London... They say the world's very small, don't they. **1959** M. GILBERT *Blood & Judgement* i. 17 He was in the pub tonight... It's a small world, isn't it? **1967** R. RENDELL *Wolf to Slaughter* iii. 21 'Small world,' he said... 'That bloke was in here yesterday.' **1973** 'H. CARMICHAEL' *Too Late for Tears* v. 69, I might've guessed even if I hadn't seen..you at the inquest in Aylesbury. Small world, isn't it? **1979** S. BARLAY *Crash Course* I. 13 'How did you know?'..'It's a small world.'

c. *small boy* (earlier examples); also *small girl*. Also, of a sibling: younger.

1786 G. WASHINGTON *Diary* 5 July (1925) III. 86 That Cowpers Jack and Day, with some small boys and girls,.. were assisting the farmer. **1821** J. F. COOPER *Spy* I. i. 14 A small boy was directed to guide him to his room. **1876** C. M. YONGE *Three Brides* II. xv. 292 He still looked on the tall, young man as the small brother to be patronized, and protected. **1891** L. T. MEADE *Sweet Girl Graduate* xxii. 182 Three small girls were making themselves busy with holly and ivy. **1923** E. E. CUMMINGS *Let.* 28 July (1969) 99 What happens to my 'small' 'sister'..is not among the interests of my own completely erratic..existence. **1936** N. STREATFEILD *Ballet Shoes* vi. 58 About twenty small girls..were learning tap-dancing. **1949** 'J. TEY' *Brat Farrar* xii. 100 If he had ever had a small sister he would have liked her to be just like Jane. **1973**

M. MACKINTOSH *King & Two Queens* i. 14 'Go away, small girl,' Frances commanded. **1977** A. WILSON *Strange Ride R. Kipling* i. 23 The strength of Rudyard's love for his small sister.

4. a. *small print*: freq. applied to the detailed information or conditions qualifying the principal text of a document, and printed in a smaller type. Also *attrib.* and *fig.* Cf. *fine print* (*FINE *a.* 7 i).

1698 [in Dict.]. **1785** T. JEFFERSON *Notes on State of Virginia* xxii. 323 A large octavo volume of small print. **1856** [see PRINT *sb.* 8]. **1944** BLUNDEN *Cricket Country* xvi. 161 What all these curious titles [of games said by Rabelais to have been played by Gargantua] meant in practice may be left to the small print of the commentaries. **1970** 'W. HAGGARD' *Hardliners* xi. 126 His contract had been terminated under some small-print clause he hadn't much studied. **1971** *Daily Tel.* 2 Nov. 14 Some interest attaches therefore to the 'small print' of the Queen's speech and how far it avoids firm undertakings on some of the more controversial measures. **1972** A. PRICE *Col. Butler's Wolf* xii. 136 There was nothing in the small print about having to like the men one served with. **1974** *Times* 17 Aug. 12/5 The collapse will.. strengthen Government moves to reform the 'small print' [holiday] booking conditions. **1979** *Sunday Times* (Colour Suppl.) 18 Nov. 35/2 The Prince of Wales reckoned he got the better of the PM in one or two exchanges on the small print of recent Cabinet memoranda.

5. c. (Later examples used of playing-cards.)

1863 *Hoyle's Games* ii. 23 Having only a few small trumps, make them when you can. **1910** W. DALTON *'Saturday' Bridge* iii. 63 Suppose that he holds ace, king, and three small diamonds, and ace, knave, and two small hearts. **1973** *Country Life* 10 May 1331/2 The declarer.. led a small Spade to dummy.

9. c. Introducing a repetition of the initial letter of a word to show that it has general rather than specific reference or that it is a less serious variety of the thing denoted. Cf. *CAPITAL *a.* 5 b, *BIG *a.* 3 h.

1952 *Observer* 18 May 7/6 Back to liberalism-with-a-small-l they trail. **1960** *Times* 22 Oct. 8/1 A newspaper that is serious, lively, and radical with a small r. **1968** *Globe & Mail Mag.* (Toronto) 13 Jan. 2/4 A general mood of small-c conservatism in the country. **1971** 'A. GARVE' *Late Bill Smith* v. 139 They were both ardently liberal with a small 'l'. **1974** W. GARNER *Big enough Wreath* xii. 165 You're not out of trouble but I'm just beginning to spell it with a small t. **1979** L. MEYER *False Front* ii. 14 They came from different sides of the track, but that wouldn't have been enough to stop a small 'd' democrat like Winston.

IV. 16. b. (Further examples.) Spec., *the small man*, the typical small businessman.

1835 C. F. HOFFMAN *Winter in West* I. 79 They were chiefly plain people, small farmers and graziers. **1850** C. KINGSLEY *Alton Locke* II. xi. 154 In helping to pass the Reform Bill, [they had] only helped to give power to the two very classes who crushed them—the great labour kings, and the small shopkeepers. **1926** CHESTERTON *Outl. Sanity* IV. iii. 170 If the small man found his small mechanical plant helped to the preservation of his small property, its claim would be very considerable. **1931** V. WOOLF *Waves* 255 How comforting it is to watch the lights coming out in the bedrooms of small shopkeepers on the other side of the river. **1935** M. EGAN *Dominant Sex* I. 27 In these days of trusts and combines the small man hardly stands an earthly. **1947** MCCALLUM & READMAN *Brit. Gen. Election 1945* iii. 63 The Conservatives..professed that they were not in favour of the growth of monopoly, and that they were anxious to curb it for the sake of the 'small man'. **1948** KOESTLER in *Partisan Rev.* XV. 38 The petite bourgeoisie—the store-keeper, artisan, small businessman, white-collar worker. **1960** A. CLARKE *Horse-Eaters* 9 Thousands Bred yearly will fatten Small farmers. **1973** *Sat. Rev. Soc.* (U.S.) Mar. 58/1, I was a small farmer..and there's no way you can do it today. You need technology and you need efficiency..and you can't afford..it if you're down there on a small farm. **1974** *Times* 12 Nov. 14/2 The Smaller Businesses Association..set up to champion the interests of the small business man. **1976** R. BARNARD *Little Local Murder* ix. 106 'Aven't you 'eard 'ow difficult things are for the small shopkeeper?

20. c. Also with *look.* Cf. sense 16 in Dict.

1784 E. SHERIDAN *Jrnl.* 6 Oct. (1960) 31 Linley came to see my Father, he received him very kindly but poor L. look'd very small. **1863** T. TAYLOR *Ticket-of-Leave Man* IV. i. 74 You've no right to be..coming after a chap, to make him look small this way.

V. 21. a. *small ad, advertisement*, a small advertisement in a newspaper, usu. in a separate section devoted to such and printed with lack of display; *small-bourgeois* adj. = *PETIT BOURGEOIS; so *small bourgeoisie*; *small caps* = *small capitals* in Dict.; *small chop* [*CHOP *sb.*^{6}] (W. Afr. colloq.), small items of food; *small end*, in a piston engine, the end of a connecting rod nearer to the piston; also *attrib.*; *smallest room* (colloq.), the lavatory of a particular building (cf. *small room* below); *small-for-dates*, (of a new-born baby) smaller than would be expected in view of the time since its conception; *small fortune*: see FORTUNE *sb.* 6 in Dict. and Suppl.; *small fruit* (N. Amer.) (see quot. 1892); *smallgoods* (Austral.), sausages, bacon, etc.; *small-hand* (earlier examples); *small-holder*, one who owns or works a small-holding; *small-holding*, (*b*) the practice or occupation of working a small-holding; *small paper* (later example); *small-pipe(s)*, a Northumbrian bellows-filled bagpipe; *small room* (colloq.), a lavatory (cf. *smallest room* above); *small screen*, television; also *attrib.*; *small seed* (see quot. 1950); *small slam*: see *SLAM *sb.*^{2} 2 b; *small stores* (Naut.), (*a*) items for personal use or consumption on a sea-voyage; (*b*) *U.S.* articles of regulation issue clothing; a shop selling these; also *attrib.* (in *sing.* or *pl.*).

1922 Small ad [see *AD]. **1961** 'F. O'BRIEN' *Hard Life* vi. 45, I put a small ad. in one of the papers. **1969** *Sunday Times* 2 Mar. 8/3 To recruit models, they ran a string of small-ads asking: 'Are you really ugly?' **1978** J. WAINWRIGHT *Ripple of Murders* 11 A small ad. in the Personal Column..will read, 'J.D. Message received.' **1919** *Times* 6 Nov. 2 Small Advertisement order form. **1937** M. ALLINGHAM *Dancers in Mourning* xxiii. 285 Uncle William put down *The Times*. He had been looking at the small advertisements. **1930** A. BENNETT *Imperial Palace* lv. 414 Customers of the small bourgeois class. **1974** N. FREELING *Dressing of Diamond* 134 A very small-bourgeois existence, with a canary. **1970** F. C. WEFFORT in I. L. Horowitz *Masses in Lat. Amer.* xi. 398 They did not feel so much like members of a decadent small bourgeoisie, but rather like operators with a stable position moving toward the better. **1856** GEO. ELIOT in *Westm. Rev.* Oct. 454 She informs us, with all the lucidity of italics and small caps, that '*function*, not *form*,..weakly engrossed her'. **1967** *Style Man.* (U.S. Govt. Printing Off.) (rev. ed.) iii. 31 In matter set in caps and small caps..capitalize all principal words. **1963** M. LAURENCE in R. Weaver *Canad. Short Stories* (1968) 2nd Ser. 140, I use the shallow ones to put groundnuts in..for small-chop with drinks. **1971** J. SPENCER *Eng. Lang. W. Afr.* 11 Pidgin words, known and used by almost everyone,.. who has lived in the coastal areas of West Africa..*chop* n and v, 'food' and 'eat', and its recent extensions in phrases such as *small chop*, 'cocktail eats', *chop box*,..etc. **1850** T. TREDGOLD *Steam Engine* (ed. 3) I. ix. 7 Into these guide-blocks are fixed the cross-heads, forming the bearings for the small ends of the connecting rods. **1908** *Autocar Handbk.* (ed. 2) ii. 38 The gudgeon pin end of the connecting rod is called the small end, and the other end the big end. **1922** *Encycl. Brit.* XXX. 36/2 The ordinary small-end bronze bush system with gudgeon pins fixed in the piston was used. **1948** A. W. JUDGE *Mod. Motor Engineer* (ed. 4) I. ix. 208 After a considerable period of running..the small-end bush may wear oval, due to the more or less vertical thrust of the connecting rod. **1971** B. SCHARF *Engin. & its Language* xii. 122 Connecting rods... They comprise a big end..and a small end through which the gudgeon pin..passes. **1930** Smallest room [see *GEOGRAPHY 1 d]. **1933** P. GODFREY *Back-Stage* xvi. 202 The smallest room in the house invariably has prohibitory notices of a stern and intimate character. **1960** *Guardian* 29 Apr. 12/6 Soon she had become unable..to take herself into her 'smallest room' in the midget backyard. **1973** 'H. CARMICHAEL' *Candles for Dead* xii. 150 At a guess, I'd say he's probably in the smallest room. **1965** *Clinics in Developmental Med.* XIX. 1 Not all 'small for dates' babies should necessarily be regarded as suffering from pathological growth retardation. **1978** *Nature* 30 Mar. 404/1 During the past twenty years it has become accepted that some babies are born small not because they are premature but because their growth has been retarded in the uterus. These so-called small-for-dates babies..are at greater than normal risk at birth. **1822** J. C. LOUDON *Encycl. Gardening* III. 537 This operation [*sc.* gathering] in the case of small fruits, as the gooseberry, strawberry, &c. is generally performed by the under gardeners. **1892** A. A. CROZIER *Dict. Bot. Terms* 164/2 Small fruits, a horticultural term for certain low-growing perennial, fruit-bearing plants and their product, including the strawberry, raspberry, blackberry, gooseberry, currant, huckleberry, and cranberry. **1926** *Daily Colonist* (Victoria, B.C.) 11 July 12/3 Small fruit continue to show prominence, but some varieties are becoming scarce. **1950** *Sci. Monthly* Apr. 212 The story of the evolution of the groups from which our small fruits have been derived is shrouded in the mists of the geological past. **1969** *Northern Territory News* (Focus '69) 22 (Advt.), It sells the best cuts of meat—beef, lamb, veal and pork—as well as smallgoods. **1973** *Bulletin* (Sydney) 25 Aug. 20/3 What are the smallgoods manufacturers putting in their sausages now that mutton, the backbone of their industry, has ceased to be cheap and plentiful? **1821** Small hand [see *NARY *a.*]. **1847** DICKENS *Dombey* (1848) xiv. 134 He would be expected to inform Doctor and Mrs. Blimber, in superfine small-hand, that Mr. P. Dombey would be happy to have the honour of waiting on them. **1837** Small-holder [in Dict., sense 16 b]. **1915** H. R. HAGGARD in *Times* 15 Mar. 3/2 The wives and daughters of smallholders will help their menfolk because they are labouring for themselves. **1977** *Times* 18 Aug. 14/6 A smallholder living in a hut on the duneland plantation. **1927** S. L. BENSUSAN *Latter-Day Rural Engl.* viii. 73 Smallholding thrives in Lincolnshire. **1936** W. M. SALE *S. Richardson* 31 It was not until the autumn of 1742 that he began to see the need for publishing his second small paper edition. **1855** in Wright *Eng. Dial. Dict.* (1904) V. 548/1 The torch was lit on point of spear—And small pipes they did sound. **1927** *Observer* 30 Oct. 8 A humble performer on the Northumbrian small-pipes. **1967** A. L. LLOYD *Folk Song in Eng.* vi. 333 The silvery-toned Northumbrian small-pipes struggling to be heard above the full shrill singing. **1975** S. MARCUSE *Mus. Instruments* 482/1 The mid-18th c., when the characteristic feature of the small-pipe was developed: its chanter was stopped. **1858** QUEEN VICTORIA *Let.* 7 Feb. in R. Fulford *Dearest Child* (1964) 35 Has the railway carriage got a small room to it? **1979** D. SANDERS *Queen sends for Mrs. Chadwick* 134 Where..was the nearest small room to the Queen's drawing-room, where the President would be received? **1956** *B.B.C. Handbk.* 1957 79 This unit.. provides the means for young writers to become acquainted with the requirements of the small screen. **1963** 'E. CRISPIN' *Best SF Five* 9 The success of the *Quatermass* series has given science fiction a limited yet tolerably regular share in the hotchpotch purveyed by the small screen. **1966** *Guardian* 29 Mar. 10/3 Both sides have studiously kept their small-screen liabilities out of the studios. **1971** *Oxford Times* 26 Nov. 31/4 The stars and the characters they portray are as in the small-screen version. **1980** *Times Lit. Suppl.* 31 Oct. 1229/4 Jonathan Miller had little success in finding the play's natural rhythms and adapting them to the small screen. **1840** W. DEANS *Let.* 30 Oct. in J. Deans *Pioneers Canterbury* (1939) 29, I..have got in about two acres of potatoes and ..numerous small seeds. **1950** *N.Z. Jrnl. Agric.* Apr. 359/1 Small seeds [grown in a Canterbury district] include perennial and Italian rye-grass, cocksfoot, crested dogstail, and clover. **1814** G. COGGESHALL *Jrnl.* 24 Mar. in *Voyages* (1851) 55 We took on board six casks of fresh water, some fresh provisions, and sundry small stores. **1877** *Nautical Mag.* XLVI. 195 Small stores as coffee, tea, or sugar was called by the 'geordies'. **1927** *U.S. Navy Bluejackets' Man.* 379 The storekeeper who issues this clothing will make out a small-store slip in duplicate. **1938** *Ibid.* 20 Besides the small stores where you can get all your clothing at very cheap prices, your station will have the following shops. **1950** *Ibid.* 124 The storekeeper who issues this clothing makes out a small-stores slip, in duplicate. **1966** NOEL & BUSH *Naval Terms Dict.* 93 *Clothing and small stores*, standard articles of uniform for officers and enlisted men with such related articles as buttons, brushes, etc.

22. a. *small-angle, -band, -bore* (also *fig.*), *-boy, -budget, -city, -claims, -end* (see sense *21 a), *-farm, -fry* (earlier example), *-gauge, -girl, -group, -letter* (later examples), *-plane, -power, -print* (see sense *4 a), *-sample, -screen* (see sense *21 a), *-shop, -signal, -step, -store(s)* (see sense *21 a), *-type*; **small-cell** *Path.*, used *attrib.* to designate various tumours of uncertain origin composed of small cells, esp. an oat-cell carcinoma of the bronchus; **small-scale** *a.*, operating or executed on a small scale; drawn to a small scale; of small size or extent; **small-yield** *a.* = *low-yield* adj. s.v. *LOW *a.* 23.

1960 *Proc. R. Soc.* A. CCLIV. 242 (*heading*) The small-angle scattering of photons. **1979** *Jrnl. R. Soc. Arts* Jan. 106/2 *Small-angle x-ray scattering*, which is another versatile structural tool for phase transformations, defects and voids, supported catalysts, polymers and various biological materials. **1941** *Jazz Information* Nov. 25/2 They are the best jazz recordings of the New Orleans small-band type. **1977** *New Yorker* 12 Sept. 122/2 Hampton was enough of a sensation to be asked by Victor in 1937 to lead a series of small-band recordings. **1898** W. S. CHURCHILL in *Morning Post* 7 Oct. 5/7 The Dervish gunboat *Bordaine*..returned from its quest with nearly a hundred men wounded by the small-bore bullets of a civilised force which was in occupation. **1900** *Congress. Rec.* 14 Feb. 1804/2 No small-bore, two-by-four, radical politicians can hurt that great court. **1932** *Sun* (Baltimore) 15 Nov. 10/2 The thing to do is to sweep this ghastly mess out of doors, once and for all. It will not be easy to do—if small-bore statesmanship is to continue to rule us. **1976** *R.A.F. News* 11–24 May 19 (*caption*) Brawdy's small bore rifle team pictured with..the Nobel Challenge Cup. **1861** *Harper's Mag.* June 133/1 We sometimes sell skates: and buying a lot at auction last fall, we thought to monopolize all the small-boy trade by posting a flaming placard announcing that we sold skates at twenty cents a pair. **1937** M. ALLINGHAM *Dancers in Mourning* xiii. 178 He was..handsome in the downy, small boy fashion of his kind. **1973** J. STRANGER *Walk Lonely Road* xiii. 92 He grinned at her, and she grinned her small-boy grin back at him. **1961** *Times* 31 July 14/7 Hardly anyone in Hollywood makes small-budget pictures any more. **1979** 'P. O'CONNOR' *Into Strong City* xxxiv. 122 Someone in London..would rush out a small-budget film into which they would cram every Scottish actor and actress. **1929** CLUTE & SMITH in *Arch. Surg.* XVIII. 11 In the fourth and last group..we have another debatable form of tumor. For the purposes of classification, we have designated it as the small cell carcinoma. This is the type of tumor which is sometimes called lymphosarcoma, as it is made up of small round cells, poorly differentiated..with a prominent nucleus and an almost negligible amount of cytoplasm. **1966** WRIGHT & SYMMERS *Systemic Path.* I. v. 265/1 In the earlier days of histopathology, although lymphocytic and lymphoblastic types of lymphosarcoma were recognized, various tumours were frequently confused with them, including..small-cell anaplastic carcinomas (for example, of the thyroid and bronchus). **1976** *Path. Ann.* XI. 319 In the differential diagnosis of acute leukemia the following should be considered: small cell epithelial tumors (eg, oat cell carcinoma, carcinoid, islet cell tumor), neuroblastoma, [etc.]. **1977** V. COLEMAN *Paper Doctors* xiii. 139 Surgery for small-cell cancer of the bronchus. **1964** S. M. MILLER in I. L. Horowitz *New Sociology* 292 The small-town and small-city poor suffer from the demise of local industry. **1977** *Sat. Rev.* (U.S.) 5 Mar. 12/2 This array of small-city labor disputes. **1961** WEBSTER, Small-claims court. **1972** M. KAYE *Lively Game of Death* vii. 38, I simply threatened him with Small Claims Court... The maximum claim there is five hundred dollars. **1978** *Listener* 3 Aug. 159/3 The London Small Claims Court..is being run on an experimental basis and only handles claims up to £350. **1960** *Farmer & Stockbreeder* 16 Feb. 60/3 This pinpoints the need for a cheaper small-farm tank of from 80 to 125 gallons capacity. **1970** S. L. BARRACLOUGH in I. L. Horowitz *Masses in Lat. Amer.* iv. 112 If any group has been systematically discriminated against, it is the small-farm producers and landless workers. **1874** 'MARK TWAIN' in *Atlantic Monthly* Nov. 592/2 Dey wa'n't no small-fry officers, mine you; dey was de biggest dey is. **1957** R. FRANKENBERG *Village on Border* 10 The small-gauge railway. *a* **1976** A. CHRISTIE *Autobiogr.* (1977) VI. i. 296 An expedition in a small-gauge train. **1980** 'A. SKINNER' *Mind's Eye* v. 73 'Thank you,' she said with careful, small-girl politeness. **1951** in Rohrer & Sherif *Soc. Psychol. at Crossroads* 333 The very few small-group studies made in recent years. **1955** Keepnews & Grauer *Pictorial Hist. Jazz* xi. 117 Alternating big-band work with prolific small-group recording activity. **1964** I. L. Horowitz *New Sociology* 25 Even at the level..of small-group research, time must be recorded. **1972** G. Little in G. W. Turner *Good Austral. Eng.* vii. 135 The pupils sit in groups face to face, and pursue a variety of small-group activities. **1974** *Melody Maker* 13 Apr. 50/7 A lovely example of small-group jazz by players who have worked together. **1979** in W. G. Lawrence *Exploring Indiv. & Organiz. Boundaries* vi. 91 The students were told that they would participate in a small group experience. **1945** E. Johnston *Writing & Illuminat. & Lettering* xv. 263 More time and material than a Small-letter MS. entails. **1950** *Language* XXVI. 13 Small-letter vowels. **1964** G. Lyall *Most Dangerous Game* vi. 42 He was the only other pilot doing small-plane charter work in Lapland that year. **1956** *Nature* 18 Feb. 322/2 Specifications for two light-weight small-power sprayers. *Ibid.* 28 Jan. 160/1 Techniques of small-sample loading employing micro-pipettes. **1968** *Brit. Med. Bull.* XXIV. 220/2 The calculations in the program have been written for the particular needs of the medical research worker; thus 'small sample' statistical theory is well represented. **1852** Small-scale [in Dict.]. **1887** [see Scale *sb.*³ 11 a]. **1935** *Discovery* Mar. 78/1 The small-scale structure of the ice will be examined. **1951** R. Firth *Elem. Social Organization* ii. 43 These are small-scale units..the members of which are in close personal contact in daily life. **1954** M. Rickert *Painting in Britain: Middle Ages* v. 131 Softer, finer vellum, more suitable for the small-scale miniatures. **1960** *Farmer & Stockbreeder* 16 Feb. 75/3 The disadvantages of small-scale farming heavily outweighed the advantages. **1964** W. L. Goodman *Hist. Woodworking Tools* 56 The difficulties of showing detail in small-scale carvings. **1979** *N. & Q.* Feb. 16/2 Orrery was very much the literary dilettante and small-scale Mæcenas. **1937** *Discovery* Feb. 45/1 Flicker, 'rain' and a rather dim appearance make the demonstration noticeably inferior in quality to that obtained with the small screen receivers now on the market. **1868** G. Meredith *Let.* Oct. (1970) I. 376 Eleven a.m. plates of small-shop ham, thick cut, grisly with brine: four smashed eggs on it. **1967** Karch & Buber *Offset Processes* ii. 9 The graphic arts is a small-shop industry. Only about 1,000 plants employ more than 100 employees, and the average is 17. **1949** *Bell Syst. Techn. Jrnl.* XXVIII. 401 An explicit calculation of the transient phenomena outside the range of small-signal theory. **1962** Simpson & Richards *Physical Princ. Junction Transistors* xv. 372 The equivalent circuits..may be used in the determination of the transient response of video amplifiers by estimating their time of response to a small-step input of current or voltage. **1962** R. Williams *Communications* iii. 58 Regular columns of close print, with small-type headlines. **1976** *Amer. Speech 1974* XLIX. 267 Good lexicography keeps open a variety of approaches to defining, which include..exploiting small-type notes for added comment. **1959** *N.Y. Times* 19 Mar. 16/1 The three Project Argus detonations involved relatively small yield nuclear devices. **1963** *Listener* 7 Feb. 251/2 The introduction of compact, mobile 'small yield' weapons into service with American forces in Europe.

b. *small-boyish* (earlier example).
1882 W. James *Let.* 9 Nov. (1920) I. 214 Baginsky's torrent of words was even more overwhelming than Munk's. I never felt quite so helpless and small-boyish before.

23. *small-faced* (later examples), *-headed, -printed, -propertied, -spored, -typed, -windowed.*
1932 D. H. Lawrence *Etruscan Places* 143 A small-faced, weedy sort of youth. **1967** Karch & Buber *Offset Processes* iv. 78 Sizes range from a small-faced six point to a 13-point full point sized capital letter font. **1851** H. Melville *Moby Dick* I. xli. 303 The White Steed of the Prairies; a magnificent milk-white charger, large-eyed, small-headed, bluff-chested. **1931** W. Faulkner *Sanctuary* xvii. 159 The thick small-headed shape of him would be clinging to the bars, gorilla-like. **1956** P. A. Larkin *Less Deceived* 40 Nor the shelves stuffed with small-printed books for the Sabbath. **1951** C. W. Mills *White Collar* i. iii. 34 The ideology suitable for a nation of small capitalists persists, as if that small-propertied world were still a going concern. **1899** *Allbutt's Syst. Med.* VIII. 854 The large-spored fungus first attacks the root of the hair and grows upwards. **1964** M. Hynes *Med. Bacteriol.* (ed. 8) xxvii. 418 (*caption*) Ringworm of hair—Small-spored ectothrix. **1891** T. Hardy *Let.* 4 Mar. (1978) I. 230 My own occupation at present is that of correcting a bundle of miserably small-typed proofs. **1902** M. Beerbohm in *Sat. Rev.* 15 Nov. 612/2 The authentic 'star'..is a no greater person than small-typed, smuggled-in 'J. M. Barrie'. **1930** W. de la Mare *Poems for Children* 35 The small-windowed moonlit house. **1976** *Times* 21 Aug. 12/3 One-storeyed, small-windowed cottages.

B. *absol.* or as *sb.*² **1. b.** *sing.* A child, a little one.
1907 W. De Morgan *Alice-for-Short* xxx. 300 How much can you remember of all that time, Alice? You were only a small, you know. *Ibid.*, I wasn't such a small as all that. **1947** *Forum* (Johannesburg) 5 Apr. 37/1 In a dozen other ways the prefects are the right-hand men of their Housemasters and me, and many a small owes a very great debt indeed to them. **1968** *Guardian* 1 Apr. 7/3 Leave two smalls to the tender mercies of a baby sitter? **1981** P. Dickinson *Seventh Raven* vi. 75 After each performance there's always a dozen smalls wandering miserably around.

9. a. Also, underclothes.
1943 N. Coward *Middle East Diary* (1944) 80 Their mothers stood nearby washing out a few 'smalls' in the shallows. **1951** *People* 3 June 4/6 Most of those who do send out to the laundry still wash smalls and personal linen themselves. **1973** *Guardian* 12 Mar. 10/2 Not many Americans..can have a clear idea of what to use the bidet for, apart from soaking the smalls.

d. (Further examples.)
1919 H. Etheridge *Dict. Typewriting* 125 May be in either capitals and smalls or all capitals. **1947** M. Morris in 'B. James' *Austral. Short Stories* (1963) 344 Make it another small... All round. **1976** *Wymondham & Attleborough Express* 10 Dec. 21/2 Mediums [*sc.* eggs] from 46p to 43p per dozen with only smalls below this rate.

e. Small advertisements.
1942 *New Statesman* 11 July 25/3 The members of staff usually responsible for selecting 'smalls' and rejecting those that are undesirable. **1959** *Times* 2 Dec. 9/4 The Press figure does not take into account the booming Classified advertising revenue, or 'Smalls'.

small, *adv.* Add: **5.** *small-drawn* adj.
a **1918** W. Owen *Poems* (1920) 13 And terror's first constriction over, Their hearts remain small drawn.

6. *Naut.* Close to the wind.
1848 J. F. Cooper *Oak-Openings* II. xiv. 203 All the difficulty was reduced to steering so 'small', as seamen term it, as to prevent one or the other of the lugs from jibing. **1911** J. Barten *Compl. Naut. Pocket Dict.* 192 Steer small.

small, *v.* **1.** Delete † *Obs.* and add later example. (Still *rare.*)
1962 'K. Orvis' *Damned & Destroyed* xiv. 95 Welch smalled his hands against his desk.

small-leaved, *a.* Also small-leafed. [Small *a.*] Having small leaves.
1699 M. Lister *Journey to Paris* 209 The Tree most in use here, was the small-leaved Horne-Beam. **1731** P. Miller *Gardeners Dict.* s.v. Tilia, The small-leav'd Lime-tree. **1827** [see Small *a.* 23 b]. **1884** [see Small *a.* 23 a]. **1889** [see Small *a.* 23 b]. **1962** R. Page *Educ. Gardener* v. 159 The small-leaved elm *Ulmus pumila*..seems resistant to elm-disease. **1967** *Coast to Coast 1965–6* 209 The dunes gave way to the..hardy small-leafed bushes. **1979** P. Theroux *Old Patagonian Express* xxi. 322 These sparse, small-leaved thorn bushes create the illusion of green.

small-pox, *sb.* Add: **2. c.** *Comb.*, as *smallpox-pitted* adj.; also *fig.*
1897 'Mark Twain' *Following Equator* xlix. 465 Saharas of sand, smallpox-pitted with footprints. **1926** D. H. Lawrence *Plumed Serpent* v. 92 The pug-faced Mexican in charge, and his small-pox-pitted assistant.

small-poxed, *a.* Add: Also *fig.*
1901 'Mark Twain' *Let.* 28 July (1917) II. 711 Water, small-poxed with rain-splashes. **1952** *Landfall* Mar. 28 The smallpoxed plaster walls of the railway station that the English had shelled from the sea.

small-talk, *sb.* Add: Also *attrib.*
1905 A. Bennett *Tales of Five Towns* I. 22 'Will Harry be late at the works to-night again?' she asked in her colder, small-talk manner, which committed her to nothing. **1977** *Gay News* 7–20 Apr. 16/3 He will begin in a very small talk kind of way about anything and before long he builds up a kind of empathy between him and the other person.

small time, small-time, *sb.* and *a.* (*phr.*) orig. and chiefly *U.S.* Also (as adj.) small-time. [Small *a.*] **A.** *sb. Theatr.* (See quot. 1926.) Also *transf.*
1910 *Variety* 30 Apr. 9/4 Acts on the 'small time' of any merit or quality will not play in the houses calling for 'five shows daily'. **1917** Wodehouse *Man with Two Left Feet* 38 He's booked me in the small time at thirty-five dollars a week. **1926** *Amer. Speech* I. 436/2 *Small time*, the lower salaried circuits or where acts must work three or more times a day. **1960** B. Keaton *My Wonderful World of Slapstick* v. 88 Rather than play the small time, other big-time acts preferred not to work. **1977** B. Langley *Death Stalk* ix. 103 Small-time hoodlums looked very much alike. Perhaps it was that streak..of uniformity which kept them in the small-time.

B. *adj.* Of or pertaining to second-rate vaudeville; *gen.* operating on a small scale, second-rate, unimportant, insignificant.
1910 *Variety* 30 Apr. 9/4 The Hartford Opera House has been taken under a five-year lease by S. Z. Poli who will operate it as a 'small time' vaudeville house. **1921** A. G. Empey *Madonna of Hills* iii. 24 She had been given a 'tryout' before a booking agency, and had made good to the extent of working in 'small time' vaudeville. **1934** *Punch* 7 Mar. 280/2 He lumbered, mined and starved in Canada; he became an itinerant 'small-time' wrestler and pugilist all over the United States. **1938** Wodehouse *Code of Woosters* xiii. 278 Sidney Carton..was small-time stuff compared with you, Bertie. **1940** R. Chandler *Farewell, my Lovely* xxxii. 244 You think I'm a small time private dick trying to push ten times his own weight. **1949** 'J. Tey' *Brat Farrar* xv. 139 Timber..was a deliberate and intelligent rogue... There was nothing small-time about Timber. **1961** *John o' London's* 5 Jan. 22/2 A gang of big-time smugglers want to eliminate the small-time hero. **1968** P. Oliver *Screening Blues* iv. 130 The words [of a blues song] were gauche but revealing of the anxieties of the small-time player. **1976** P. R. White *Planning for Public Transport* x. 204 Many economists expert in the finer points of judging small-time savings, or the effect of taxation in determining net resource costs, have only the haziest idea of costs. **1977** I. Shaw *Beggarman, Thief* III. viii. 307 Do you intend to be a small-time tennis pro.. all your life?

Hence **small-timer,** (*a*) a small-time theatre (*rare*); (*b*) a small-time operator, an insignificant person.
1910 *Variety* 30 Apr. 9/4 S. Z. Poli..will operate it as a small time' vaudeville house, under the booking direction of J. J. Clancy, and in conjunction with the other Poli 'small timers'. **1935** A. J. Pollock *Underworld Speaks* 109/1 *Small timer*, a person who doesn't amount to much; a piker. **1950** *Sport* 24–30 Mar. 21/4 The small-timer is now not only demanding protection from the Guild but a rightful place in boxing affairs. **1959** R. Simons *Houseboat Killings* viii. 90 She was a small timer when I met 'er... Then she got so 'igh and mighty she wouldn't speak to me. **1976** 'Trevanian' *Main* (1977) x. 189 A couple of small-timers..who make their money by 'laundering' men for the American organized-crime market.

small-town, *a.* Also small town. [Small *a.*] Of, pertaining to, or characteristic of a small town; unsophisticated, provincial.
1824 *Blackw. Mag.* June 659/2 Nothing can be better than Miss Austin's [*sic*] sketches of that sober, orderly, small-town, parsonage, sort of society in which she herself had spent her life. **1859** *Bentley's Q. Rev.* July 437 If 'George Eliot' proves to be..a very young man, son of a small town tradesman [etc.]. **1881** *Harper's Mag.* Jan. 223/2 Cosmopolitans, they do not sink into the ruts of small-town life. **1930** R. Macaulay *Staying with Relations* xvii. 254 It looked even a little more decayed and small-town than the Miramar. **1937** J. M. Murry *Necessity of Pacifism* iii. 47 Their natural mode of feeling is still small-town and parochial. **1949** Koestler *Promise & Fulfilment* I. ii. 14 The Jews had..only a homogeneous lower-middle-class of Eastern European small-town origin. **1959** *Times* 30 Sept. 13/4 A lawyer who is by no means so simple and small-town as he makes himself out to be. **1964** R. Miliband in I. L. Horowitz *New Sociology* 80 A rural, small-town, one-man-one-gun America. **1969** A. Lurie *Real People* 105 Not a slick professional show like Saratoga; but very amateurish, small-town. **1975** *Verbatim* May 12/1 Can a small-town girl from Kansas and Indiana make good as a reporter in the wilds of Greater Boston? **1980** *Times Lit. Suppl.* 19 Sept. 1047/4 Pinning her characters..in small-town Oregon.., she uncovers the hidden principles and ambitions of little men and women as they battle against the smothering ordinariness of suburban life.

So **small-towner,** an inhabitant of a small town, one who comes from a small town; **small-townish, -towny** *adjs.*, characteristic or suggestive of a small town.
1920 S. Lewis *Main Street* xxix. 353 It's dreadfully tabby and small-towny. **1931** H. Concannon *St. Patrick* xiv. 189 The smooth paths of a smug small-townish officialdom. **1945** *Sun* (Baltimore) 17 Apr. 1/2 A friendly, small-townish man of the Middle West, called suddenly to the presidency..by the death of Franklin D. Roosevelt, [etc.]. **1969** *Listener* 5 July 3/2 The small-towners gratefully seized upon this idealised portrait, using it as psychological support and insulation from the cold reality of their dependence on the mass society. **1974** Small-townish [see *Oregonian *a.*].

smally, *adv.* For † *Obs.* read 'Now *rare*' and add: **4.** (Later example.)
1946 B. Marshall *George Brown's Schooldays* xxx. 128 Abinger moved smally in between them.

5. Of a sound: with low volume.
1958 T. H. White *Once & Future King* II. ii. 224 All these noises came up to the two on the tower smally, as though they were listening through the wrong end of a megaphone. **1964** F. O'Rourke *Mule for Marquesa* 55 Hoover said, 'Good luck,' and his words died smally in the thud of seats hitting saddles.

smalm, *v.*: see *smarm *v.* **smalmy,** var. *smarmy *a.*

smalt, *sb.* (and *a.*). Add: **4. a.** *smalt-maker.*
1921 *Dict. Occup. Terms* (1927) § 143 *Smalt maker*, places powdered mixture of cobalt oxide, silica (pure sand) and potassium carbonate into a wheeled fireclay crucible [etc.]. **1923** *Chambers's Jrnl.* 15 Dec. 36/1 The supreme obstacle confronting the smalt-makers.

Hence **sma·lter,** one who prepares smalt.
1923 *Chambers's Jrnl.* 15 Dec. 36/1 The Venetian glass-makers..were content to purchase their requirements from the 'smalters' of Germany.

smarm, *v.* Now the usual form of Smalm *v.* For *dial.* read '*colloq.* orig. *dial.*' and add:
1. b. To make smooth with an oily or greasy substance; to smooth or slick *down.* Also *fig.*
1937 Kipling *Something of Myself* viii. 211 The Provincial Press has been syndicated, standardized, and smarmed down out of individuality. **1953** *Chambers's Jrnl.* June 356/1, I remembered in time the need to attach a canvas, oil-filled bag to the sea-anchor, so that the thin trickle of oil would smalm the crippling seas. **1953** *Landfall* June 126 He twitted me about my orange brogues, my smarmed-down hair, my nervous attempts at wit. **1977** M. Kenyon *Rapist* i. 5 The rapist looked in the looking-glass at the smarmed hair.

c. To treat in a wheedling, flattering way.
1902 *Little Folks* Mar. 221/2 You can go and smarm him over if you want to.

2. a. *intr.* To behave in a fulsomely flattering or toadying manner, to suck *up to* a person. Also const. *about, over.*
1911 O. Onions *Widdershins* ix. 265 It had been the usual thing..twenty years ago—smarming about Art and the Arts. **1920** 'O. Douglas' *Penny Plain* xx. 218 The people you try to help will smarm to your face and blackguard you behind your back. **1924** H. de Selincourt *Cricket Match* v. 184 Loving her to tease him..and not sort of smalm over him like some chaps' maters did. **1928** A. P. Herbert *Trials of Topsy* xxiii. 138, I will *not* spend

week-ends *smarming* about in your *ulcerated* town. **1942** G. R. Gilbert *Free to Laugh and Dance* vi. 29 The..children of his two previous marriages were indignant but they couldn't do anything about it, they all hanging round and smarming up to the old man, and his marrying again in spite of them. **1950** J. Cannan *Murder Included* vi. 108 Murder was the last thing that a toady like that doctor would look for in the family he smarmed over. **1960** F. Raphael *Limits of Love* III. vi. 343 Dulles and Eden were smarming up to Nasser.

b. *to smarm one's way*, to make one's way by flattery or toadying.
1940 G. D. H. & M. Cole *Counterpoint Murder* i. 15, I am going to report you to Mr. Marston. So it's no use you thinking you can smarm your way out of it. **1975** *Times Lit. Suppl.* 20 June 692/4 He is as convincing an uncaught villain as any in crime fiction; cautious and false, smarming his way into the upper reaches of such a society as only such a man would seek to crawl into.

Hence **smarmed** *ppl. a.*; **sma·rming** *vbl. sb.* and *ppl. a.*
1950 J. Cannan *Murder Included* ii. 22 Bunny strove to use tact and courtesy: Elizabeth accused her of 'smarming'. **1953** Smarmed [see *smarm *v.* 1 b]. **1957** L. Durrell *Justine* II. 143 He is a complete *parvenu* of course and rose on..judicious smarming of powerful people. **1970** E. Berckman *She asked for It* ix. 108 It's wonderful how greedy people are for these smarming little compliments. **1974** D. Gray *Dead Give Away* iii. 31 He hated her smarming ways. **1977** Smarmed [see *smarm *v.* 1 b].

smarm (smɑːm), *sb. colloq.* Also **smalm.** [f. the vb.] An unctuous bearing; fulsome flattery; flattering or toadying behaviour.
1937 H. C. Bailey *Clunk's Claimant* xiii. 84 That smarm of holiness..was pretty near the ruddy limit. **1962** M. Urquhart *Frail on North Circular* xiii. 72 'That's a nice new hair-do, Lil.'.. 'Don't come your smarm with me.' **1966** C. Fenn *Pyramid of Night* iii. 60 The manager escorted her right to the door. There was no doubt that this kind of smarm was cute. **1969** B. Cobb *Scandal at Scotland Yard* xvi. 154 He was ready to turn on the smarm with the 'Kiss and be friends again' talk. **1978** *Guardian Weekly* 19 Feb. 22/1 'George' did this, 'George' did that, all the way through. 'George' is the victim of bonhomie and smarm.

smarmy (smā·ɹmi), *a.* (and *sb.*). *colloq.* Also **smalmy.** [f. *smarm *v.* or *sb.* + -y¹.] **A.** *adj.* **a.** Smooth and sleek. **b.** Ingratiating, obsequious; smug, unctuous.
1909 C. Hamilton *Plain Brown* i. 4 A tall, slight, smarmy-headed man. **1924** 'L. Brock' *Deductions Col. Gore* iv. 51 Don't you be taken in by that smarmy swine. **1928** K. Clark *Gothic Revival* ix. 236 If he [*sc.* Gilbert Scott] saved a few cathedrals from ruins, we must, I suppose, look more kindly on their smalmy surfaces. **1929** St. J. Ervine *First Mrs. Fraser* II. 49 *Janet.* But, Ninian dear, you shouldn't have said anything. *Ninian.* He was looking so smarmy and self-satisfied that I had to. **1930** *London Mercury* Aug. 380 Our forefathers rejected 'smarmy' overtures with 'Fair words butter no parsnips'. **1955** H. Spring *These Lovers fled Away* 133 Uriah Heep's sort of smarmy humbleness. **1962** S. Raven *Close of Play* I. vi. 77 He's a smarmy, ingratiating swine. **1976** J. B. Hilton *Gamekeeper's Gallows* viii. 66 She used to..tell her how grateful she was. Not smarmy, like; genuine. **1979** R. Jaffe *Class Reunion* I. ix. 98 He always treated her as if she were a little bit better than he was; an attitude that was smarmy in other men.

B. *absol.* as *sb.* A smarmy person. *rare.*
1957 T. Gunn *Sense of Movement* 22 You understand..the speculative man or passionate You know the smarmies, but side-step the grease.

Hence **sma·rmily** *adv.*; **sma·rminess.**
1934 H. Nicolson *Let.* 17 Nov. (1966) 189 There is something about the smarminess of Americans which makes me see red. **1958** 'A. Bridge' *Portuguese Escape* 184 The art of making himself agreeable in his current surroundings, whatever they were; but..he rather over-did it..'smarminess' described his conversation with painful accuracy. **1968** J. Porter *Dover goes to Pott* i. 15 Dover, sycophant and snob that he was, smarmily agreed. **1974** 'D. Kyle' *Raft of Swords* II. xii. 129 With deliberate smarminess, Gawthorpe thanked him. **1982** *Sunday Express* 30 May 6/8 It is smarmily hypocritical.

smart, *sb.*² Add: **3.** Usu. *pl.* Intelligence, cleverness, acumen; wits. *U.S. slang.*
1970 *It* 27 Feb.–13 Mar. 14/4 Now Abbie's a very brilliant cat. He's very far out, very freaked out, but he's still got some smarts. **1970** *New Yorker* 28 Nov. 126/3, I knew I had the smarts—the business smarts—even then. **1972** H. Kemelman *Monday Rabbi took Off* xlv. 263 The whole story is a little weak... I mean, this kid of yours has the normal amount of smarts. **1977** *N.Y. Times* 9 Oct. 26 Mrs. Maynard said that Mr. Miller ..'doesn't have enough smart to run a union as big as the United Mine Workers'. **1979** *Sunday Star* (Toronto) 6 May B7/4 You know—streetwise, lots of smart, lots of quick, playing pinochle with vice cops. **1981** *Guardian Weekly* 26 July 15/4 They complain that the level of intelligence is low and that the soldiers have neither the smarts nor the education to work the complicated weapons of modern warfare.

smart, *a.* Add: **I. 6. b.** (Further example.)
1870 G. M. Hopkins *Jrnls. & Papers* (1959) 201 The day had been very bright and clear, distances smart.

7. a. (Earlier example.)
1778 S. Foote *Devil upon Two Sticks* II. 44 in *Wks.* IV, [Scot loq.] Ah! for the mater of that, it is a praty smart little income.

b. *right smart* (earlier and later examples).
1842 J. S. Buckingham *Slave States of America* II. 327, I asked here, whether the people made much maple-sugar in this neighbourhood; when the gentleman..answered, 'Yes, they do, I reckon, right smart.' **1932** W. Faulkner *Light in August* i. 25 There is a right smart of folks in Jefferson I don't know. **1938** M. K. Rawlings *Yearling* vi. 54 'Howdy, Mr. Forrester. Proud to see you. How's your health?' 'Howdy, sir. I'm right smart tol'able, seein' as how I be near about done for.' **1949** L. Nordyke *Cattle Empire* 81 Heard a right smart about you, Pincham.

II. 8. For † *Obs. rare* read '*Obs. rare* exc. *predic.* in *to be* or *get smart* (U.S.)'.
1933 E. O'Neill *Ah, Wilderness* II. 60 *Tommy*... Uncle Sid's soused again. *Mrs. Miller*... You be quiet! Did I ever! You're getting too smart! **1955** W. C. Gault *Ring around Rosa* v. 59 Don't get smart, Callahan. **1956** B. Holiday *Lady sings Blues* (1973) i. 4 This time Cousin Ida beat me for being smart with her.

9. c. Healthy, well. *U.S.*
1788 J. May *Jrnl.* 31 Aug. (1873) 116 Didn't feel smart enough to go to meeting. **1832** J. J. Strang *Diary* 23 Aug. in M. M. Quaife *Kingdom of St. James* (1930) 205 This commenced the sickest day I ever suffered since my remembrance but now (evening) I am again smart for a sick person. **1956** B. Holiday *Lady sings Blues* (1973) i. 1 By the time she worked her way out of hock in the hospital and took me home to her folks, I was so big and smart I could sit up in a carriage.

10. c. Of a device: capable of some independent and seemingly intelligent action. Cf. *smart bomb* in sense 15 below.
1972 *Proc. IEEE* LX. 1282/1 The term 'smart terminal' is used here to identify an interactive terminal in which part of the processing is accomplished by a small computer or processor contained within the terminal itself. **1977** *Sci. Amer.* Sept. 188/1 When smart traffic signals become ubiquitous and are linked to a control center, the traffic cop at the intersection will become obsolete. **1980** *Economist* 15 Mar. 84/3 Tomorrow's vehicles are likely to have a series of 'smart' transducers attached to the engine, gear-box, brakes, etc, all sending coded messages via a common wire to the dashboard. **1982** *Times* 1 June 15/5 Smart sensors ensure a direct hit on a target.

13. *the smart set* (examples); also *transf.*
1885 E. W. Hamilton *Diary* 20 May (1972) II. 867 Dined at Brook House... This is a house at which one meets the 'grand set' as distinct from the 'smart set'—two totally different sections of the best London Society. **1900** *Smart Set* Apr. 137 The Smart Set of London has for the last ten or fifteen years..been the chief influence of our English playwrights, plays and players. **1937** K. Blixen *Out of Africa* I. i. 12 Our Quasi Smart Set of the Colony. **1949** P. Hastings *Cases in Court* v. 265 Mrs Barney's family were well-known in Mayfair and both Mrs Barney and the dead man were notorious members of the so-called young 'smart set'. **1974** P. Dickinson *Poison Oracle* ii. 66 Dinah [*sc.* an ape]..had indeed left the slums to join the evolutionary smart set, Man. **1981** V. Glendinning *E. Sitwell* ix. 131 Edith's interest in the intrigues of the smart set was minimal.

14. *smart-looking* (earlier example), *-suited.*
1770 'P. Pennyless' *Sentimental Lucubrations* ii. 31 A smart-looking waiter came up to me. **1922** Joyce *Ulysses* 249 James's wax smartsuited freshcheeked models.

15. Special collocations: **smart bomb,** a powered missile which is guided to its target by an optical system; **smart money** *U.S.*, money bet or invested by persons with expert knowledge; *transf.* knowledgeable persons; **smart mouth** *U.S. slang*, one who is good at repartee, one who gives cheek; so **smart-mouth** *v. trans.* to be cheeky to, to be witty at the expense of; **smart-mouthed** *a.*
1972 *Guardian* 29 June 4/2 Three out of four [missions] have been using 'smart' bombs. **1975** *N.Y. Times* 8 Sept. 2/4 Since 1973 the Israeli Air Force has been furnished with a variety of 'smart bombs' guided by laser beams of television. **1982** *Sunday Tel.* 9 May 17/5 The American 'smart' bomb, which homes on a laser beam shone on to the target by a spotter aircraft. **1926** *Amer. Mercury* Dec. 464/2 In referring to money wagered by persons with good tips or information, the term used is *smart* money. **1930** W. R. Burnett *Iron Man* I. 5 'Well,' said Regan, 'all the smart money's on the black boy.' **1947** *Sun* (Baltimore) 6 Aug. 10/5 Bookmakers and layoff men are gamblers and many times they add personal wages to 'smart money'. **1977** H. Fast *Immigrants* II. 87 Germany has declared war on Russia, and the smart money says that this is only the beginning. **1981** *Times* 7 Nov. 6/8 Mr Weinberger..is close to the President..much closer than Mr Haig has ever been. In a battle for Mr Reagan's ear, all the smart money would be on Mr Weinberger. **1968** *Sun Mag.* (Baltimore) 13 Oct. 19/1, I was a smart mouth, a troublemaker in school. **1976** R. B. Parker *Promised Land* (1977) xii. 65 Don't smart-mouth me, man. You wising off at me? **1978** J. L. Hensley *Killing in Gold* (1979) viii. 97 He..beat up three kids..when one of them smart-mouthed him. **1976** *Publishers Weekly* 19 Apr. 82/3 Smart-mouthed film critic for *Gotham* magazine. **1978** J. Irving *World according to Garp* xii. 229 Some smart-mouthed motorist..will..ask.., 'What are you in training for?'

smart alec (smɑːt æ·lĕk). *colloq.* (orig. *U.S.*). Also **aleck, alick,** hyphenated, and with capital initial(s). [f. Smart *a.* + *Alec*, dim. of personal name *Alexander.*] A would-be clever person; a 'know-all'; occas., a man who is ostentatiously smart in dress or manner. Also *attrib.* or as *adj.*
1865 *Carson* (Nevada) *Appeal* 17 Oct. 2/3 Halloa, old smart Aleck—how is the complimentary vote for Ashley? **1873** J. H. Beadle *Undevel. West* vii. 140, I had the pleasure of seeing at least a score of 'smart Alecks' relieved of their surplus cash. **1887** F. Francis *Saddle & Mocassin* 312 You may talk about your Smart Alicks, and your Joe-dandies and daisies. **1902** [see *run *v.* [illegible]]. **1904** W. N. Harben *Georgians* ix. 92 That was a smart Alec of a feller from Little Dogtrot over in the mountains. **1934** C. Stead *Seven Poor Men of Sydney* vi. 184 Rawson, from the Trades Hall, ready, assured, blatant, a political opportunist, a pushing young man, 'a real smart-alec'. **1934** H. G. Wells *Exper. Autobiog.* I. v. 276 Nowhere was there protection from those Smart Alecs, the primary poisons of the whole process, who piled up the rents. **1941** B. Schulberg *What makes Sammy Run?* xii. 221 He's a smart-aleck. I can see already he thinks he knows [illegible]. **1952** C. Barrett *On Wallaby* x. 202 One smart Alec [illegible] for a large tin of pineapple slices. **1956** C. V. Wedgwood *Lit. & Hist.* [illegible] managed by their smart-aleck antics to [illegible] decade what [illegible] biography. **1961** L. Nkosi *Rhythm of Violence* ii. 13 I'm bored with their smart-aleck talk! **1976** T. Stoppard *Dirty Linen* 40 Smart alec-paragraphs about innocent tripe-and-onions with titillating voluptuaries? **1979** L. Meynell *Hooky & Villainous Chauffeur* viii. 99 Smart alec, aren't you? Smart and smug like all you intellectual lot.

Hence **smart-alec(k)ism,** smart-aleckry, an instance of this; **smart-aleckry,** behaviour characteristic of a smart alec, ostentatious or smug cleverness; **smart-alecky** *a.*, characteristic of a smart alec, ostentatiously or smugly clever.
1905 *Dialect Notes* III. ii. 157 *Smart Elecky*, impertinent, impudent. 'He's too Smart Elecky for me.' **1926** G. J. Nathan *House of Satan* 6 Ripples of smart-aleckry. **1929** J. B. Priestley *Good Companions* III. v. 579, I would have you know that he hadn't been so Smart Alecky about it. **1934** *Sun* (Baltimore) 28 Aug. 8/7 The futility of those moderns who have substituted smart aleckism for sophistication and vulgarity for art. **1938** *Scrutiny* VI. 400 [illegible] supreme disregard of public opinion is difficult to parallel in an age in which composers seek consolation for their lack of popular appreciation in the formation of cliques and the cultivation of a pert smart-alecism. **1958** *Spectator* 5 Sept. 306/2 *Brouhaha* remains a hodgepodge of smartaleckry aimed..at the intellectual teddy-boy set. **1962** *Listener* 21 June 1091/3 A slight tendency towards smart-alecry like the 'nine-minute operas' *Introductions and Goodbyes.* **1965** *New Statesman* 30 Apr. 690/1 The last thing the Hansons go in for is the shrug of the shoulders or the smart-alecism—unless it is at the expense of psychoanalysing critics who..suggest that Tchaikovsky was..rather odd. **1976** *National Observer* (U.S.) 10 July 8/3 She now writes regularly for New York magazine, though her typically measured manner of speaking doesn't seem to fit that magazine's usually brash, smart-alecky tone. **1981** *Times Lit. Suppl.* 6 Mar. 247/3 The various lists in the book have an air of Christmas competition-setting smart-aleckry.

smart-arse, -ass (smɑːt-ɑːs, -æs), *sb. slang.* Also as one word without hyphen and as two words. [f. Smart *a.* + Arse *sb.*, *ass.] = *Smart alec.
Only the form with *-ass* is current in the U.S.
1962 J. D. MacDonald *Girl, Gold Watch & Everything* ix. 118 Some smart-ass crowd set up all the confusion so as they could clean out them stores. **1962** W. Dick *Bunch of Ratbags* 245 Anyhow, where else could we go, you know, smart-arse? **1970** R. Oakley *Salute to Great McCarthy* 28 That smart-arse from the East. **1972** D. Delman *Week to Kill* 115 But you've got a smart-ass answer for that, haven't you? **1972** P. Marks *Collector's Choice* ii. 102 You're too pretty to be such a smart-ass. Just watch your step. **1973** *Guardian* 24 Feb. 8/5 A picture of a rose..needn't be just beautiful, or decorative, or a matter of smart-arse graphics. **1973** *Time Out* 4 May 13/2 When I'm fighting a bit, as I'm doing now, I try not to get too smartass and articulate. **1977** *Private Eye* 1 Apr. 23/4 (Advt.), Get it together to put down pseuds, poseurs and general smartasses. **1981** J. Burnett *Flying Squad* x. 92 He had indulged in reckless speculation.. He was just as much a smart-arse as the [illegible].

So **smart-ass** *v. trans.*, to say in a smart-alecky manner; *intr.* and *trans.*, to behave in a smart-alecky manner (towards). **smart-arsed, -assed** *adj.*, smart-alecky.
1960 P. Blondal *Candle to Light Sun* 191 A lot of smart-assed novelists would have you believe [illegible]. **1962** A. Seymour [illegible] smart-arsed [illegible] wrong, son. **1971** [illegible] supposed to [illegible]. **1974** [illegible] *Shaft* (1971) [illegible] had smart-arsed [illegible] my rate for [illegible]. **1976** R. B. Parker *Promised Land* xix. 115 Don't smart-ass with me, Johnny, or you'll be looking very close at the floor. **1977** J. Wainwright *Day of Peppercorn Kill* 68 No smart-arsed lawyer's going to shift my evidence. **1978** F. Ross *Sleeping Dogs* 152 'I guess it's something to do with the generation gap, sir.' 'Don't smart-ass me!' **1979** *Globe & Mail* (Toronto) 15 Sept. 42/1 It is tempting to be smart-assed when reviewing a Richard Rohmer novel.

smartie, var. Smarty in Dict. and Suppl.

smartish, *a.* Add: **B.** *adv.* Somewhat smartly.
1877 *Coursing Calendar Autumn 1876* 5 Vicar-of-Bray..won a smartish-run trial with his extra pace. **1896** G. F. Northall *Warwickshire Word-book* 218 *Smartish*, adj. and *adv.*, fairly well. 'How are you?' 'Smartish, thank you.' 'I'm getting on smartish now.' **1973** W. H. Canaway *Harry doing Good* II. iii. 155 You nipped off..a bit smartish, didn't you? **1979** N. Freeling *Widow* xiii. 79 Whip out smartish to the Italian grocer.

smartness. 6. (Earlier examples.)
1800 M. Edgeworth *Little Merchants* in *Parent's Assistant* (ed. 3) III. 111 His son's *smartness* was no longer useful in making bargains. **1819** M. Wilmot *Let.* 8 Dec. (1935) 34 We have got ourselves settled in..with fewer plagues than almost any family of strangers could boast. This we owe..to Willys smartness in his quiet way.

smarty, *sb.* For *U.S.* read 'orig. *U.S.*' and add: Also **smartie.** (Earlier and later examples.) Also, a smartly-dressed person; a member of a smart set.
1861 *Calif. Mag.* Aug. 39/2 'Juvenile smartys' are interesting, even to a vagabond. **1876** 'Mark Twain' *Tom Sawyer* i. 22 Smarty! You think you're *some*, now. *Ibid.* xviii. 156 That Saint Louis smarty that thinks he dresses so fine. **1929** D. H. Lawrence *Pansies* 89 But it is hard to be tolerant with the smarties. **1932** Auden *Orators* II. 44 Poops and smarties, Who pilfer always but are never whipped. **1933** R. Strachey *Many Happy Returns* II. 123 A gala night at the 'Shadwell Palace', dancing those Limehouse blues with her smartie. **1956** L. McIntosh *Oxford Folly* viii. 118 It's amazing how easy it is for anyone like me, with no background, to pose as an Oxford smartie. **1957** M. Millar *Soft Talkers* xxi. 204 'Do you happen to know how much it was?' 'No, and neither do you, smartie.' **1962** A. Bourne *Doctor's Creed* ii. 46 The worst payers were what we used to call the 'West End Smarties', flimsy young women who would appear from nowhere with no doctor's letter or tangible recommendations.

B. *adj.* **a.** Smart-alecky; ostentatiously smart.
1883 'Mark Twain' *Life on Mississippi* xxxiii. 370 The barkeeper..was gay and smarty and talky. **1940** *Horizon* Feb. 68 Another line of attack is to concede that the first number is interesting, but to add that it is middlebrow and 'smarty'. **1948** M. Allingham *More Work for Undertaker* vi. 83 He was full of smarty ideas and had no manners. **1960** D. Potter *Glittering Coffin* vi. 96 A smarty gossip column. **1967** G. Kelly in *Coast to Coast 1965–6* 104 The local smarty boys, the privileged class.

b. Special Combs. (also written *smarti-*), as **smarty-boots,** (orig. *U.S.*) **-pants** *colloq.*, an overly clever person, a know-all, a smart alec; also *attrib.* or as *adj.*
1962 *Times* 7 June 16/3 The phoney-ness of a smartiboots Ivy League undergraduate. **1962** *John o' London's* 22 Nov. 467/1 A cold, well-bred English smarty-boots. **1965** J. Porter *Dover Two* xiii. 166 He was grateful that smartie-boots MacGregor had overlooked the obvious, too. **1970** *Sunday Times* (Colour Suppl.) 6 Dec. 31/3 His self-confidence and satisfaction in his own life got under people's skin...phrases like 'Smartie Boots' were attached to him. **1979** *Guardian* 12 Nov. 13/7, I am not trying to be wilfully iconoclastic or smarty-boots when I say that.. Picasso simply is not the greatest painter of the 20th century. **1941** B. Schulberg *What makes Sammy Run?* iv. 57 One of those Vassar smarty-pants. **1953** M. Dickens *No More Meadows* vi. 266 [Amer. loq.] He jumped right in with his slick talk... That smarty pants. **1967** N. Marsh *Death at Dolphin* viii. 199 Hawkins, Mr. Smartypants, has a little chat. **1969** *Punch* 26 Mar. 465/3 *Cage Me a Leacock* (BBC-2) owed everything—though not, I hope, the smarty-pants title—to Braden's enormous enthusiasm for his subject. **1976** *Listener* 24 June 815/2 Technologically outclassed and outsold by pinstriped smartipants from foreign business schools. **1981** *Times Lit. Suppl.* 13 Feb. 158/3 The smarty-pants youthfulness is very period.

smash, *sb.*¹ Add: **1. a.** (Earlier example.)
1725 E. Wigan *Let.* 25 Sept. in N. D. Mereness *Trav. Amer. Colonies* (1916) 156 They design to keep out look-outs every way and be ready to give them a Smash in their Towns.

b. Also in Badminton, Table Tennis, etc.
1950 *Badminton* ('Know the Game' Ser.) 30/1 The smash is perhaps the most vulnerable stroke in the game. **1981** G. MacBeth *Kind of Treason* xvi. 156 'Fourteen-fifteen,' said Yoshida, serving again. He'd lost the point ..by a wasted smash.

3. a. (Earlier and later examples in sense of 'collision'.)
1841 [see *railroad line* s.v. *railroad *sb.* 3 a]. **1909** *Westm. Gaz.* 27 Apr. 4/2 If the brakes fail to hold it is impossible to avoid a bad smash. **1957** M. Spark *Comforters* viii. 166 If I hadn't had the smash I'd have got you last autumn.

5. a. (Earlier and later examples.)
1850 [see *brandy-smash* s.v. *brandy *sb.* 2]. **1871** [see *corpse reviver* s.v. *corpse *sb.* 6]. **1909** [see *brandy-smash* s.v. *brandy *sb.* 2]. **1958** A. L. Simon *Dict. Wines, Spirits & Liqueurs* 147/2 *Smashes*, mixed iced drinks always with a spirit foundation and some mint flavouring. **1973** Wodehouse *Bachelors Anonymous* xiii. 170 What more likely than that he should have fetched up in Hollywood, made a packet, perished of a surfeit of brandy smashes, and left that packet to that nephew.

b. An alcoholic drink, esp. wine. *N. Amer. slang.*
1959 *Maclean's Mag.* 15 Aug. 28/2 So I had a couple of smashes and marched in. **1966** *Globe & Mail* (Toronto) 15 Mar. 35/8 Every time you wanted a smash the check girl would hand the coat to you over the counter, so you could get your mickey without actually taking the coat out. **1975** *Amer. Speech 1972* XLVII. 153 Let's get in the wind and belt some smash.

6. A great success; a film, person, play, song, etc., which enjoys popular success; a hit (*hit *sb.* 4). Also *attrib.*, esp. in **smash hit.**
1923 *Variety* 11 Oct. 16 (*heading*) 'Rosie O'Reilly' and 'The Fool', Loop's Two Smash Hits. **1930** *Times Lit. Suppl.* 16 Oct. 841/1 An entirely strange girl; whom anyone would have admitted to be a 'smash'. **1931** *Daily Express* 21 Sept. 9/3 The magnates who had contracted to buy the picture indulged in fits of doubt concerning its prospects as a box-office 'smash'. **1935** *Amer. Speech* X. 193/2 Terminology from other fields aids the fashion editor... The sports writer is also responsible for the *smash hit* dinner dress. **1935** Wodehouse *Blandings Castle* xii. 305 Our whole programme is built around it. We are relying on it to be our big smash. **1948** W. S. Maugham *Colonel's Lady* in *Quartet* 201 The English publisher said to him: 'We've not had a success like this with a book of verse for twenty years.'.. The American publisher said to him: 'It's swell. It'll be a smash hit in America.' **1949** R. Chandler *Let.* 23 Apr. (1981) 174 You can't make me into a smash best seller. **1956** B. Holiday *Lady sings Blues* (1973) xix. 158 *Holiday on Broadway* was a sellout, and the first performance made us think we had a smash. **1961** *Amer. Speech* XXXVI. 110 It was a smash commercial success. **1969** R. Lowell *Notebk. 1967–68* 71 Eliot dead, you [*sc.* Ezra Pound] saying, 'And who is left to understand my jokes? My old Brother in the acts..and besides, he was a smash of a poet.' **1973** *Black World* Apr. 18/2 All smash Broadway musical hits. **1975** D. Francis *High Stakes* 232 The oddly mixed party proved a smash-hit success. **1978** *Times* 1 Nov. 13/1 [His] aim..has been to expand a truthful little ethnic comedy into a popular smash.

smash, *sb.*² Add: **1. b.** Loose change.
1821 D. Haggart *Life* 13 M'Guire got L.7 of smash; I got a L.10 banknote. **1953** W. Burroughs *Junkie* (1972) ii. 22 Soon I was buying his drinks and meals, and he was hitting me for 'smash' (change) at regular intervals. **1953** K. Tennant *Joyful Condemned* iii. 21 Giving her his smash on pay-night so's she can blow it. **1965** *Australasian Post* 4 Mar. 47 Russell goes on to point out that all loose change is sometimes known as 'smash'.

smash, *v.*¹ Add: **I. 3. b.** Also in Badminton, Squash Rackets, etc.
1965 *Badminton* ('Know the Game' Ser.) (ed. 2) 31/1 The man should take the shuttle as early as possible, playing drives when the shuttle is too low to smash. **1968** *Squash Rackets* ('Know the Game' Ser.) 43/1 You cannot smash a good high lob as you can at lawn tennis. *Ibid.* 44/2 A lob that was too high above the player's head has been smashed on to the tin.

III. 8. a. (Earlier example.) Also with *run.*
1818 [see Go *v.* B. 10]. **1849** T. T. Johnson *Sights Gold Region* xxii. 211 The afternoon of our 'first day out' was signalized by running smash into a big sycamore tree.

b. *to play smash*: to come to grief; to wreak havoc *with. dial.* and *U.S. colloq.*
1841 *Spirit of Times* 2 Jan. 523/2 Bill Spence got drunk and played smash with all the arrangements. **1842** D. Vedder *Poems* 84 Slates an' tiles, frae aff the houses, On the causey crown played smash. **1887** *Courier-Journal* (Louisville, Kentucky) 17 Jan. 1/7 (*caption*) Plays Smash With a Passenger Train on the Fitchburg Railroad. **1903** W. N. Harben *Abner Daniel* ii. 11 Yore pa's as bull-headed as a young steer, an' he's already played smash anyway. **1912** *Dialect Notes* III. viii. 585 *Play smash*,..a euphemism for *play hell* or *play the devil.* **1915** *Ibid.* IV. iii. 187 *Play smash*,..to make a great blunder; do a thing wholly wrong.

smash, *v.*² **1.** (Earlier and later examples.)
1801 *Sporting Mag.* XIX. 88/1 He had never seen any [forged notes] that were better done; he had *smashed* several. **1898** A. M. Binstead *Pink 'Un & Pelican* x. 229 The small tradesman, afraid to smash his notes at a bureau, had them still intact when the police called upon him. **1905** —— *Mop Fair* ii. 28 The imaginary 'bailiff' who spoke about the handcuffs is well known in the neighbourhood..while the counterfeit 'tipstaff' who smashed the cheque is a dog-fighting publican.

smash-and-grab. Also smash and grab, smash'n-grab. [f. Smash *v.*¹, Grab *v.*] Used *attrib.* to designate a type of robbery in which the thief smashes a shop-window and grabs the goods there displayed. Also *transf.* and *fig.*, and *absol.* Hence **smash-and-grabber; smash and grabbing** *vbl. sb.*
1927 J. C. Goodwin *Crook Pie* ii. 52 'Smash and Grab' raids seem to be the order of the day. **1928** *Daily Tel.* 9 Oct. 12/2 Three men in a motor-car were concerned in a smash-and-grab raid... One of them threw a stone through the window. They then seized all the cameras available and returned to the car, which was driven away before a chase could be started. **1932** [see *screwer 2]. **1933** Blunden *Charles Lamb* vii. 206 A literature of the smash-and-grab type..seems to have some chance of superseding the thorough, persuasive, modulated and interwoven style. **1937** A. L. Rowse *Sir Richard Grenville* v. 106 A smash-and-grab run upon the Isthmus of Panama. **1938** [see *pick-up *sb.* a (vi)]. **1939** T. S. Eliot *Old Possum's Practical Cats* 23 They were..remarkably smart at a smash-and-grab. **1944** G. B. Shaw *Everybody's Political What's What?* xxvi. 232 Monstrous world wars and smash-and-grab revolutions. **1951** M. McLuhan *Mech. Bride* (1967) 145/2 A commercial society dedicated to the smash and grab and one-man fury of enterprise. **1960** *Observer* 24 Jan. 5/1 A sausage team always had to work three-handed... One to do the smash and grabbing. **1965** H. I. Ansoff *Corporate Strategy* (1968) iii. 39 In reaction to the public outrages at the 'smash'n-grab imperialism' of the nineteenth century, business has acquired a sense of social responsibility. **1970** *Oxf. Times* 23 Oct. 1/6 A smash and grab raid was carried out on the shop of Horns..shortly before midnight. **1973** W. M. Duncan *Big Timer* iv. 29 They tell me there was a smash-and-grab at Shader's, miss. **1978** *Daily Mail* 25 Jan. 12/2 Robbery with violence..used to be a 'snatch' or a 'smash-and-grab'.

smashed, *ppl. a.* Add: **1.** Also *fig.* Also *smashed-down, -up.*
1909 J. R. Ware *Passing Eng.* 227/1 *Smashed* (Navy), reduced in rank. **1915** J. Webster *Dear Enemy* 325 Our poor smashed-up doctor. **1918** W. S. Churchill in M. Gilbert *Winston S. Churchill* (1977) IV. Compan. I. 365 Ought we to build our lives & policy & the future arrangement of the world on the unreal basis of a smashed-up Russia & an invincible Germany. **1935** A. J. Pollock *Underworld Speaks* 109/1 *Smashed*, to have lost all material possessions. **1938** E. Blunden in *Times Lit. Suppl.* 8 Oct. 633/2 Nor the dead in smashed-down den. **1982** J. Hansen *Gravedigger* iii. 24 No abandoned or smashed-up Rollses.

2. Intoxicated, drunk; under the influence of drugs; 'stoned'. *slang* (orig. *U.S.*).
1962 J. D. MacDonald *Key to Suite* (1968) viii. 139 Are you figuring on getting smashed? **1968** *New Scientist* 26 Sept. 679/2 The males rapidly acquired a taste for the stuff [*sc.* alcohol], bent their elbows with great application, and soon became smashed. **1968** A. Young in A. Chapman *New Black Voices* (1972) 147 Turns out he was half-smashed and half-drunk because he'd smoked some dope when he got up that morning, then on the way to school he'd met up with Wine, so the two of them did up a fifth of Nature Boy, a brand of sweet wine. **1973** D. Laing *Freaks* 20 He would get smashed on two and a half pints of Worthington E from the wood, and fall about misquoting the poetry of the beat generation. **1977** *New Society* 27 Jan. 185/3 If you're smashed out of your skull all the time on peyote, then even the bizarre patronage of Marlon Brando must seem tolerable.

smasher¹. Add: **1. b.** A very pretty or attractive woman; an attractive man.
1948 Partridge *Dict. Forces' Slang* 173 To a Scotsman an attractive girl was 'a wee smasher'. **1949** J. R. Cole *It was so Late* 61 'Yes. No kidding,' Don said. 'But she was easier to look at than anything around here. She was a smasher—straight she was!' **1957** A. Wilson *Bit off Map & Other Stories* 74 When the jeunes filles met Rodney, Jackie..put her head on one side and said, 'I say, isn't he a smasher!' **1963** *Security Gazette* V. I. 13/2 The applicant for a shorthand typist's job—she was a smasher. **1977** C. McCullough *Thorn Birds* xviii. 458 In a long black wig, tan body paint and my few scraps of metal I look a smasher.

2. b. (Earlier example.)
1829 P. Egan *Boxiana* 2nd Ser. II. 706 Tom, by the effect of this *smasher*, lost his equilibrium.

4. (Later examples.)
1921 [see *double-faulter* s.v. *double *a.* C. 3]. **1928** B. Nuthall *Learning Lawn Tennis* vii. 114, I am not an expert smasher myself.

5. (Earlier example.) Also *ellipt.*
1891 E. Glanville *Fossicker* xviii. 156 The Dutchmen stared at him from under the brims of their felt 'smashers'. **1892** J. R. Couper *Mixed Humanity* i. 4 A wide-awake, called in South Africa a 'smasher'. **1894** C. H. W. Donaldson *With Wilson in Matabeleland* ix. 189 Brown cord jackets and 'smasher' hats, bandoliers and rifles.

smasher². Add: **3.** A receiver of stolen property. *rare.*
1929 C. Humphreys *Gt. Pearl Robbery* i. 60 The goods might be disposed of to a 'smasher', that is, a receiver of stolen property.

smasheroo (smæʃərū·). *slang* (orig. and chiefly *U.S.*). [f. Smash *sb.*¹ + *-eroo.] A great success.
1948 *Life* 26 Jan. 47/2 The press snickeringly reported Virginia's coming-out party, a smasheroo, right down to the 500 bottles of ginger beer, the spots left on the furniture. **1960** S. Kauffmann *If it be Love* III. iv. 211 A smasheroo. All seven reviews were great. **1962** *John o' London's* 11 Jan. 43/3 In the historical hokum-pokum bracket there's been one real smasheroo, *El Cid.* **1967** *Punch* 29 Nov. 822/2 A smasheroo musical will bring its creators vastly more. The most spectacular example of this in recent years is *Hello, Dolly!* **1975** *New Yorker* 17 Mar. 92/1 Is one going to make the burning a big Broadway smasheroo of a scene?

smashing, *ppl. a.*¹ Add: **2.** *colloq.* Very good; greatly pleasing; excellent; sensational.
a **1911** D. G. Phillips *Susan Lenox* (1917) II. vi. 164 When you get dressed up a bit..you'll do a smashing business. **1914** W. Owen *Let.* 27 Dec. (1967) 310, I come in hungry to find a 'smashin' dinner. **1922** [see *crack *sb.* 1 d]. **1944** M. Paneth *Branch Street* 8 When the children came..to play in the house they thought it 'smashing'. **1948** *Mind* LVII. 418 The fact is, the verification principle is a metaphysical proposition—a 'smashing' one if I may be permitted the expression. **1959** *Times Lit. Suppl.* 2 Oct. 564/2 It is not her fault that the publishers, in big letters on the jacket, promise 'as smashing a last sentence as we can recall!' That promise is not fulfilled. The final twist is surprisingly unsurprising. **1977** *Chem. in Brit.* XIII. 118/2 This is a smashing book for anyone interested in surface chemistry and physics to have available on his bookshelf.

smashingly *adv.* (later examples).
1923 *Daily Mail* 8 Sept. 6/6 Her volleying is splendid, And smashingly she serves. **1956** 'C. Blackstock' *Dewey Death* iv. 75 He was so smashingly handsome. **1970** *Daily Tel.* 21 July 13/4 The Férand evening midis of white organdie, smashingly printed with swirls of grey and cocoa.

smash-up. Add: Also (*U.S.*) **smashup.** **a.** (Later examples.)
1940 W. Empson *Gathering Storm* 67 Politicians, etc., living now, who made a smash-up of international affairs. **1974** [see *military police* s.v. *military *a.* 3 b]. **1978** H. Wouk *War & Remembrance* xxxix. 399 Historians tend to miss the awful simultaneity of the fourfold smashup.

b. *spec.* A collision, esp. of road or rail vehicles; a crash. Chiefly *U.S.*
1856 M. J. Holmes *'Lena Rivers* 36 The old lady, sure of a *smash-up* this time, had attempted to rise. **1875** H. W. Shaw *Josh Billings' Farmer's Allminax* 13 Got the orfull smash up on the rale rode. **1923** M. B. Watts *Luther Nichols* 354 There had been..a smash up; a delivery-wagon..had run head-on into that there stone. **1931** *Kansas City Times* 3 Oct. E/6 What could run more typically true to form than a smashup when that bee got up a motorist's pants leg a short time ago? **1978** J. Irving *World according to Garp* xii. 236 They all drive so fast... If it weren't for you, I sometimes think they'd be having their smashups right in my living room.

S-matrix: see *S II. 11.

smatter, *v.* **1.** Restrict † *Obs.* to sense in Dict. and add: **b.** *U.S.* To splash, splatter. Also *intr.* Hence **sma·ttered** *ppl. a.*
1893 *N. & Q.* 15 July 45 In the daily reports of the interesting Lizzie Borden murder trial, recently held in Massachusetts, I notice the peculiar use of the words *smatter, smattering,* and *smattered* in reference to splashes of blood. **1958** S. A. Grau *Hard Blue Sky* III. 125 The first heavy drops fell and smattered in the dust. **1974** D. Richards *Coming of Winter* v. 144 The man had on a long grey coat, smattered with mud.

smattering, *vbl. sb.* Add: **1. c.** A small amount or number.
1973 *Nation Rev.* (Melbourne) 31 Aug. 1442/6 The news that *does* appear—other than a smattering of inconsequential rubbish—is..thinly disguised opinion. *Ibid.* 1464/2 There were 10,000 men (and a smattering of women) of letters in the UK. **1975** D. Nobbs *Death of Reginald Perrin* 154 There was a surprised pause, then a smattering of applause, which grew slowly into a tolerable ovation.

'smatter of fact. Also smatterer fact, smatter fact, etc. Repr. colloq. pronunc. of phr. *as a matter of fact.* Cf. Fact 6 a.
1922 S. Lewis *Babbitt* iv. 44 Besides, smatter fact, I'll tell you confidentially. *Ibid.* vi. 85 Smatter of fact, there's a whole lot of valuable time lost even at the U., studying poetry and French and subjects that never brought in anybody a cent. **1957** V. S. Naipaul *Mystic Masseur* iii. 39 But, smatterer fact, I don't like the idea. **1968** A. Clarke *Darkened Room* v. 59 Oh, I don't mind... 'Smatter o' fact..I like it. **1972** 'J. & E. Bonett' *No Time to Kill* vi. 67 'Smatter of fact, my real name's Gladys, but..I took to Carmen... 'Sides, now I've picked up the lingo, a lot of 'em tell me what good English I speak. **1978** D. Bloodworth *Crosstalk* vi. 45 S'matter of a fact, that one's a cinch.

smaze (smēⁱz). [Blend of Smoke *sb.* and Haze *sb.*] A mixture of smoke and haze.
1953 *Daily Progress* (Charlottesville, Va.) 21 Nov. 1/6 (*caption*) Manhattan skyscrapers look like misty dream-castles as a combination of smoke and haze drifts around them. Called 'smog' by most people, the smoke-haze combination might more aptly be termed 'smaze'. **1953** *N.Y. Times* 22 Nov. 1/1 From smaze and smog, the city got down to an old-fashioned ocean fog yesterday. **1958** *Manch. Guardian* 22 Nov. 4/4 Over seven million domestic chimneys emit at low level smoky particles and tarry substances which cause the urban 'haze' (or 'smaze') in industrial areas. **1960** *Daily Tel.* 16 Nov. 1/8 A Weather Bureau official described the condition as a kind of smog-like haze. 'Call it smaze,' he said. **1968** *Courier-Mail* (Brisbane) 3 July 3/7 That smoky, cloudy, dirty stuff that's been hanging around blurring the buildings of Brisbane these last few mornings is not *smog*, it's *smaze.*

smear, *sb.* Add: **3. b.** (Further examples.) *esp.* a sample of human or other cells obtained without surgery; *vaginal smear*, a smear of cells obtained from the vagina, studied to detect cervical cancer of the womb.
1917 Stockard & Papanicolaou in *Amer. Jrnl. Anat.* XXII. 227 In order to examine the vagina [of a guinea-pig] thoroughly we have introduced a small nasal speculum which facilitates clear view of the interior and a smear is made of any fluid that may be present. *Ibid.*, A study of the vaginal smears from guinea-pigs. **1920** *Proc. Nat. Conf. Social Work 1919* 58 Dr. Knight's plan of requiring a smear [for the detection of venereal disease] from every female child coming under their care would seem a wise precaution. **1925** *Jrnl. Amer. Med. Assoc.* 8 May 1422/2 The guinea-pig is a particularly suitable animal for such an investigation, on account of the regularity of its estrual cycle. The use of vaginal smear examinations makes it possible to detect the return of estrus in a very exact way. **1928** G. N. Papanicolaou in *3rd Race Betterment Conf.* 530 In a case of benign tumor everything you find in a vaginal smear is more or less normal... In contrast to this, in..cases..of malignant tumors, there are some definite characteristic changes. **1943** —— *Diagnosis Uterine Cancer* vi. 34 Vaginal smears made after the operation continued to show the carcinoma cells in considerable numbers. **1958** Cervical smear [see *Papanicolaou]. **1966** *Listener* 4 Aug. 151/1 Cervical smear centres for the early diagnosis of womb cancer..have..been outstandingly successful. **1969,** etc. [see *Pap *sb.*⁴]. **1975** *Nature* 9 Oct. 480/1 The presence of sperm cells in vaginal smears taken [from rats] the following morning was taken as positive indication of pregnancy.

c. A slanderous or defamatory remark; an attempt to defame by slander. *colloq.* (orig. *U.S.*).
1943 *Sun* (Baltimore) 22 Oct. 8/3 'This is an outright smear,' Stromberg asserted. **1953** E. Simon *Past Masters* IV. v. 256 Our only hope is to get some sort of official enquiry..to scotch all the smears. **1958** *Spectator* 15 Aug. 225/2, I would have expected from Mr. Lehmann not that implied smear but approving pats on both our heads. **1959** *Listener* 25 June 1115/1, I became aware of a gentle campaign of smear. **1977** E. Ambler *Send no More Roses* x. 246 There is the smear, and I'm the subject of it.

5*. In *Jazz*, a short glissando; a slurring or sliding effect produced by a brass instrument, esp. a trombone.
1926 [see *portamento]. **1944** *New Yorker* 1 July 29/2 Someone may advocate extending a note or cutting it off. The sax section may want to put an additional smear on it. **1959** M. T. Williams *Art of Jazz* (1960) iv. 36 Those devices that gave..the illusion of smear and roughness to his tone.

6. (sense *3 c) *smear document, interview, job, journalism, merchant, -monger, process, sheet, story*; **smear campaign,** a plan to discredit someone or something, or to destroy a reputation, by means of smears; **smear-glaze** (examples); hence *smear-glazed* adj.; **smear tactics,** the tactics used in a smear campaign; **smear test,** a test for cancer of the womb made by microscopic examination of a smear (see sense 3 b in Dict. and Suppl.); **smear-word,** a word which in spite of its literal meaning is used to imply something derogatory.
1938 *Sun* (Baltimore) 7 May 1/5 He called the Lobby Committee 'a snooping committee' which was engaged in 'a smear campaign', a campaign of 'terror and intimidation' against newspapers and magazines which dare to criticize activities of the New Deal. **1978** N. Freeling *Night Lords* vii. 31 We'll get accused of a smear campaign against Rolls-Royce cars. **1940** *Sun* (Baltimore) 18 Oct. 22/6 This was the pamphlet attacked by Republicans as a 'smear document'. **1977** M. Walker *National Front* vii. 183 Tyndall denied responsibility for the smear document. **1893** E. A. Barber *Pottery & Porcelain of United States* vi. 82 Glaze, which in the kiln would vaporize and form a slight deposit on the ware, technically known as '*smear*' glaze. **1971** L. A. Boger *Dict. World Pottery & Porcelain* 320/1 Smear glaze was a development following salt glaze and is frequently mistaken for it. **1963** *Times* 26 Jan. 11/7 The delicately smear-glazed porcelain more usually associated with small Victorian statuary and so suggestive of marble that it was known as parian ware. **1960** *New Statesman* 23 Jan. 96/1 But the elaborate smear interviews and paragraphs in Sydney were nobody's mistake, but just the boys obeying orders. **1970** E. Ambler *Intercom Conspiracy* iii. 73 It was a smear job hashed up to discredit one or another of his clients' competitors. **1967** *Punch* 8 Feb. 190/2 This touched the muddiest depths of smear journalism, full of cheap sneers and nasty innuendo. **1963** *Times* 15 May 9/2 The finding on this point was a bitter disappointment to the smear merchants. **1967** *Punch* 8 Feb. 190/2, I read with great satisfaction the editorial.. on the smearmongers of the Press and other vehicles of opinion. **1958** *Times* 26 Feb. 9/5 Mr Gaitskell's intention was, I imagine, to minimize the value of the report and rob it of its influence.., another application of the now well-known 'smear' process. **1951** *Observer* 16 Dec. 7/4 He is instructed to dismiss five people..accused by a smear-sheet of Communist sympathies. **1947** *New Statesman* 22 Nov. 404/3 The Garry Allighan affair has done great harm in confirming in the minds of thoughtless people the smear stories about politicians that are put about by people who are not thoughtless. **1955** 'E. C. R. Lorac' *Ask Policeman* v. 56 They didn't want the Sunday papers to write up Rosetta Towers as a smear story. **1945** *West Va. Rev.* Nov. 40/1 In recent years there has developed in his country a group of press agents who have adopted 'smear' tactics. **1974** *Times* 14 Feb. 22/3 In the old days they used to be called smear tactics but this year..mini-Watergates. **1950** *Consumer Rep.* XV. 367/1 The smear test for uterine cancer is done by scraping tissue..from the rear of the vagina. **1977** *Spare Rib* May 19/1 Yearly smear tests are important as they give early warning of a disease which takes 15 years to develop. **1938** I. Goldberg *Wonder of Words* xv. 298 The term *Bolshevik* ..becomes so encrusted with non-political significance that it loses any sharpness of outline..and grows into what has been called 'a smear-word'—a word that takes on whatever emotional color the speaker may..desire. **1961** *Twentieth Cent.* Jan. 87 'Philology', that smear-word among students everywhere.

smear, *v.* Add: **4. c.** Also (without const.), to attempt to discredit (a reputation, etc.). *colloq.* (orig. *U.S.*).
1936 W. Irwin *Propaganda & News* xxii. 292 When the Republicans began calling this line of attack the 'smear Hoover' campaign, Michelson..faced the microphone with a masterpiece of ingenious invective. **1945** G. Endore *Methinks the Lady* xi. 268 You had plenty of time..to hold up the jury while you smeared the science of psychoanalysis. **1948** *Manch. Guardian Weekly* 23 Sept. 9 He is afraid of being smeared. **1951** *Here & Now* (N.Z.) May 23/2 Mary Jane Keeny has been a target for the red-baiters who from time to time try to smear the UN as a Communist-dominated organisation. **1966** *Listener* 30 June 934/2 What the successors are trying to do seems to me not so much to prove particular corruption as to smear the previous regime. **1978** G. McDonald *Fletch's Fortune* (1979) xix. 130 Is the campaign against me going to continue? Are the March newspapers going to continue to smear me?

e. In Southern Africa, to coat over (the floor of a hut, etc.) with a mixture of cow-dung and water.
1839 W. C. Harris *Wild Sports S. Afr.* xvii. 143 The space was smeared with a mixture of mud and cow-dung, resembling that used in all parts of India for similar purposes. **1878, 1880** [see *dagga¹]. **1893** Blennerhassett & Sleeman *Adv. Mashonaland* ii. 32 We were unsuited physically for such work as 'daghering' huts or 'smearing floors'.

6. a. (Further examples.)
1854 W. Collins *Hide & Seek* III. viii. 231 They seemed to get smeared out of my head—like we used to smear old sums off our slates at school. **1920** 'K. Mansfield' *Bliss & Other Stories* 73 And then there is the waiter... When he is not smearing over the table or flicking at a dead fly or two, he stands with one hand on the back of a chair. **1964** W. Golding *Spire* ii. 37 One delver relaxed, and smeared a hand over his sweaty face. *Ibid.* viii. 155 He peered in [a metal sheet] closer and closer until his breath dimmed his own image and he had to smear it off with his sleeve.

b. To thrash or kill; to wipe out or destroy by bombing. *slang.*
1935 A. J. Pollock *Underworld Speaks* 109/1 *Smear*, to kill. **1941** Baker *Dict. Austral. Slang* 68 *To smear someone*, to thrash a person in a bout of fisticuffs. **1944** *Amer. Speech* XIX. 187 He [*sc.* S. J. Baker] gives examples of Australian argot, of which several follow:..*smear*, to murder, [etc.]. **1957** P. Frank *Seven Days to Never* ix. 245 We can smear every base, every industrial complex, once and for all. **1958** P. Bryant *Two Hours to Doom* 43 The report on the..Russian I.C.B.M. site had removed his..doubt..whether his bombers could smear it before the missiles were fired off.

smear-case. Also smearcase. (Earlier and later examples.)
1829 A. Royall *Pennsylvania* I. 471 A dish, common amongst the Germans,..is curds and cream. It is very palatable, and called by the Germans *smearcase.* **1934** *Sun* (Baltimore) 21 July 4/7 Under the seat in the old wagon two jars of smearcase. **1978** *Amer. Speech* LIII. 201 The 'miscellaneous' category includes..*clabber cheese* (but more commonly *cottage cheese, Dutch cheese, smearcase*).

smeared, *ppl. a.* Add: **3. b.** *smeared out*: spread out, distributed; averaged over a volume of space or a period of time.
1931 *Nature* 8 Aug. 211/1 It can be represented..by supposing each electron to move in the field of the nucleus and the other electrons, representing it by the 'smeared out' continuous distribution which the solution of the wave equation gives, and then taking the electron atmosphere to be the sum of these distributions. **1977** *New Scientist* 21 Apr. 120/2 By measuring their extinction properties at a single wavelength we could show that the smeared-out density of dust grains is a few per cent of the density of all matter in interstellar space. **1979** *Nature* 18 Jan. 188/1 This behaviour..will be completely smeared out at experimental temperatures. **1979** *Sci. Amer.* Nov. 128/1 A wave function..often describes the electron as if it were smeared out over a large region of space.

smearer. (Later examples.)
1955 E. Pound *Classic Anthol.* II. 117 Take therefore, I say, these smearers And fellow travellers, chuck 'em To wolves and tigers. **1960** *Guardian* 7 Oct. 9/2 A common trick is for a smearer to..tell..drinkers that a candidate is a teetotaller.

smearing, *vbl. sb.* Add: **1.** (Further examples.)
1893 [see *dagga¹]. **1940** G. Seldes *Witch Hunt* p. xiii, But surely there ought to be some resort to the American spirit of fair play in combating underhanded smearing campaigns. **1948** E. Rosenthal *African Switzerland* iii. 34 Smearing..is a Basuto custom. You may have noticed that all the native huts are covered on the outside with a layer of mud, and this again is worked into all kinds of pretty and complicated patterns. That is smearing.

2. In *Jazz*, the production of a smear or glissando.
1934 S. R. Nelson *All about Jazz* ii. 61 The subtle slurring and smearing of to-day would have astonished some of the early players. **1952** B. Ulanov *Hist. Jazz in Amer.* (1958) vi. 66 Filhe..contrasted his low-register looping and smearing with Bab Frank.

smearing (smiᵊ·riŋ), *ppl. a.* [f. Smear *v.* + -ing².] **a.** In *Jazz*, pertaining to a smear or slurring effect. **b.** Slanderous.
1958 P. Gammond *Decca Bk. of Jazz* iv. 57 The smearing, broad glissandi of Christian's trombone. *a* **1974** R. Crossman *Diaries* (1976) II. 627 We also have to overcome something else—the stream of anti-government propaganda, smearing, snarky, derisive, which comes out of Fleet Street.

smeck, var. *schmeck.

smectic, *a.* Restrict *rare* to sense in Dict. and add: **2.** *Physical Chem.* Applied to (the state of) a mesophase (a liquid crystal) in which the molecules all have the same orientation and are arranged in well-defined planes. Also as *sb.*, a smectic substance. Cf. *nematic *a.* [ad. F. *smectique* (G. Friedel 1922, in *Ann. de Physique* XVIII. 276).]
1923 [see *nematic *a.*]. **1936** *Mineral. Abstr.* VI. 237 In addition to the fuller's earths and smectic clays the series includes montmorillonite, confolensite, [etc.]. **1940** Glasstone *Text-bk. Physical Chem.* vii. 505 In the smectic state normal liquid flow does not occur: the movement is of a gliding nature, in one plane. **1971** *New Scientist* 14 Jan. 63/2 Once the smectic mesophase had formed, the continuous or 'closed' bimolecular sheets would isolate the aqueous compartments from their neighbours. **1974** *Nature* 25 Jan. 178/3 As yet uncategorised smectics probably exist, for example, the smectic phase of 4′-n-octyl-4-cyanobiphenyl described by Gray.

smectite. Add: **a.** (Later examples.) Now *Obs.*
1932 *Amer. Mineralogist* XVII. 198 It seems hardly worth while to retain the two names smectite and montmorillonite for what appears to be the same mineral... It seems in the best interests of science to continue the name montmorillonite, and to drop that of smectite. **1942** *Ibid.* XXVII. 810 The differential thermal curve for the smectite sample..is like that of montmorillonite... This is in agreement with Kerr's..conclusion [*prec. quot.*] that smectite is not a valid species because of its similarity to montmorillonite.
b. = *montmorillonoid. So *smectite group* = *montmorillonite group* s.v. *montmorillonite b.
1955 G. Brown in *Clay Minerals Bull.* II. 296 *Smectites* is the name proposed for the minerals at present variously known as montmorillonoids, montmorins, minerals of the montmorillonite group and frequently even montmorillonites. Smectites are defined as minerals composed of 2:1 or triphormic layers, which, when the readily exchangeable cations are replaced by Na^+ and the material is saturated with glycerol, give a basal spacing of 18Å approximately. **1957** R. Greene-Kelly in R. C. Mackenzie *Differential Thermal Investigation of Clays* v. 140 By far the most abundant dioctahedral smectite is montmorillonite. **1966** [see *montmorillonite a]. **1975** *Amer. Mineralogist* LX. 66/1 The significant chemical and physical properties of smectites often depend on the nature of the interlayer exchange ions. **1979** *Sci. Amer.* Apr. 78/1 (*caption*) Smectite group of clay minerals, formerly called the montmorillonite group, has extremely fine-grained, irregular and thin-layered crystals.

Smee[2] (smī). The name of Alfred *Smee* (1818–1877), English surgeon and experimenter, used *attrib.* and in the possessive to designate an obsolete type of primary cell or battery he invented (*Phil. Mag.* (1840) XVI. 315), consisting of zinc and platinum (or platinized silver) electrodes in dilute sulphuric acid.
1852 F. S. Williams *Iron Roads* 314 Great inconvenience arose from the spilling of the acid solution used in Smee's batteries. **1873** F. Jenkin *Electr. & Magnetism* xv. 215 The Smee battery is better than the copper zinc battery. **1950** G. W. Vinal *Primary Batteries* i. 16 Smee's cell of 1840 avoided..the difficulty experienced with polarization in other single-fluid batteries. *Ibid.*, The electromotive force of the Smee cell was low, about 0·5 volt, which was its principal disadvantage.

smell, *sb.* Add: **3. a.** Also, the special, indefinable, or subtle character of something.
1948 'N. Shute' *No Highway* ii. 38 Fifteen years in the aircraft industry... One gets to know the smell of things like this. **1974** J. Thomson *Long Revenge* iii. 40 The smell of the case had come back to him..and he had the feeling that there was a great deal more to it.
b. (Later examples.)
1973 *Times* 19 Dec. 14/7 Things are looking up: there is a smell of success in the air. **1981** *Listener* 2 July 3/1 There's a smell of success: people really think they can shift governments.
5. smell fox, the wood anemone, *Anemone nemorosa*.
1892 C. M. Yonge *Old Woman's Outlook in Hampshire Village* 49 The beloved *Anemone nemorosa*—the windflower—or, as the village children unpoetically call it, 'smell foxes'. **1898** —— *John Keble's Parishes* xv. 172 Smellfox, anemone. **1931** M. Grieve *Mod. Herbal* I. 34 Anemone (Wood)..Synonyms. Crowfoot, Windflower, Smell Fox.

smell, *v.* Add: **I. 2. b.** *to smell a rat*: see Rat *sb.*[1] 2 a.
II. 6. (*c*) (Later U.S. examples.)
1912 F. J. Haskin *Amer. Govt.* 276 He took out the cork, smelled of it, and then replaced it. **1919** E. O'Neill *Moon of Caribbees* 30 His foot hits a bottle. He stoops down and picks it up and smells of it.
III. 8. c. (Later examples.) Also in phr. *not to smell right*, to have an air of being not quite in order.
1939 'N. Blake' *Smiler with Knife* x. 154 It doesn't sound like Fascism. It doesn't smell like Fascism. **1950** 'J. Tey' *To love & be Wise* xvii. 219 It's..the whole set-up... It doesn't smell right. **1969** *Sunday Times* (Colour Suppl.) 21 Dec. 11/1 Jock could not have been nicer... As a matter of fact he has been so nice that it smells bad. **1974** J. Thomson *Long Revenge* iii. 33 Finch was inclined towards accepting the case... And yet..he hesitated... It still did not smell right to him.
d. To give rise to suspicion; to have an air of dishonesty or fraud.
1939 *Sun* (Baltimore) 12 Dec. 3/3 What 'smelled' about the..case appeared to have been saved by committee counsel for later inquiry. **1950** *Austral. Police Jrnl.* Apr. 118 *It smells*, it is something to be wary about; highly suspicious. **1970** G. F. Newman *Sir, You Bastard* ii. 78 Things..wouldn't always get past the sharp-eyed QC. If a case smelt, he would smell it. **1973** 'H. Howard' *Highway to Murder* viii. 103 There's a wrong slant to this affair. I can't put my finger on it—but it smells.
9. c. (Later examples of *to smell of the lamp*.)
1887 [see Lamp *sb.*[1] 1 b]. **1927** Galsworthy *Castles in Spain* 154 At times he wrote stories unworthy of him. At times his work smelled of the lamp. **1953** G. S. Fraser *Modern Writer & his World* iii. iv. 254 This desire of his..to be 'complex' and to bring in a wide range of cultural references at all costs does make his work sometimes smell a little of the lamp.
11. Also with *out*.
1978 *Lancashire Life* Oct. 83/3 Ah must 'a' smelt the class-room a'et When a' them odours mingled. **1979** 'J. Ross' *Rattling of Old Bones* ii. 17 How..can you have a dead body smelling out the house and not know it?

smellage. Substitute for etym.: [Alteration of Smallage.] and add to def.: *Levisticum officinale*, of the family Umbelliferæ. (Earlier and later examples.)
1836 A. H. Lincoln *Familiar Lect. Bot.* (ed. 5) 110 [*Ligusticum*] *levisticum* (smellage,) leaves many..Medicinal. **1889** R. T. Cooke *Steadfast* iii. 43 A nosegay of lavender, damask roses, smellage, old man, clove pinks [etc.].

smeller. Add: **5.** *fig.* Anything remarkable for exceptional violence, severity, strength, etc. **a.** = Snorter[1] 2 b. **b.** *slang.* A heavy fall; usu. in phr. *to come a smeller*. Cf. *stinker 6 c.
1898 Kipling *Fleet in Being* v. 55 Good old gales—regular smellers. **1923** J. Manchon *Le Slang* 278 *Smeller*, (2) to come a smeller, ramasser une pelle. **1934** Wodehouse *Right Ho, Jeeves* ix. 92 A man's brain whizzes along for years exceeding the speed limit, and then something suddenly goes wrong with the steering gear and it skids and comes a smeller in the ditch.

smellie (sme·li). [f. Smell *v.* + -ie, after *talkie.] A (hypothetical) cinema or television film in which smell is synchronized with the picture. Usu. pl. Cf. *feely.
1929 A. P. Herbert in *Punch* 8 May 508/3 These early smellies made a great sensation, particularly *Fish*, a strong story written 'around the life of a San Francisco fishwife of homicidal tendencies'. **1949** *Sun* (Baltimore) 12 Apr. 6/6 (*cartoon caption*) Soviet Smellies present 'Uncle Joe's Pipe'. **1958** *Spectator* 20 June 801/3 We had an Esther Williams picture and I wanted to advertise it as a smellie, with an ozonair machine in the foyer... The circuit wouldn't wear it, though. **1977** *Time* 11 Apr. 33/2 Another treat in the works: smellies—a futuristic device attached to the [T.V.] set will emit aromas into the living room.

smelling, *vbl. sb.* Add: **3. c.** (Earlier and later examples.)
1722 Defoe *Memoirs of Plague* 239 In a Word, the whole Church was like a smelling Bottle. **1936** W. Faulkner *Absalom, Absalom!* vi. 195 Clytie..stood impassive beside the wagon that last day, following the second ceremonial to the grave with the silk cushion and the parasol and the smelling-bottle.

smellsip (sme·lsip), *v. nonce-wd.* [f. Smell *v.* + Sip *v.*] *trans.* To smell and sip almost simultaneously.
1922 Joyce *Ulysses* 170 He smellsipped the cordial juice and..set his wineglass delicately down.

smelly, *a.* Add: **a.** Also *fig.* **b.** Suspicious. *rare.*
1923 J. Manchon *Le Slang* 278 *Smelly*,..louche, véreux. **1944** [see *ride *v.* 9 c]. **1970** G. F. Newman *Sir, You Bastard* viii. 227 Seems a bit smelly, Terry. I should blow him out.

SMERSH, Smersh (smɜːʃ). [Russ. abbrev. of *smert' shpionam*, lit. 'death to spies'.] The popular name of the Russian counter-espionage organization, originating during the war of 1939–45, which is responsible for maintaining security within the Soviet armed and intelligence services.
1953 I. Fleming *Casino Royale* xxvii. 217 He would take on smersh and hunt it down. Without smersh, without this cold weapon of death and revenge, the M.W.D. would be just another bunch of civil servant spies. **1955** H. Hodgkinson *Doubletalk* 1 Still to be met with are..SMERSH (*smert shpionam*: death to spies), the war-time name for the Soviet Army Counter-espionage organisation. **1961** D. Moore *Highway of Fear* iv. 27 Faster!.. The Smersh zombies..are right behind you. **1967** E. Grierson *Crime of One's Own* ix. 73 All this nonsense of Calvert being some type from SMERSH—and Mason's the Man from Uncle, I suppose? **1977** *Times Lit. Suppl.* 29 Apr. 534/3 Missing..from..the *Great Soviet Encyclopedia*..are two Abakumovs, Andrei Ivanovich..and Viktor Semenovich, head of Smersh during the Second World War and Minister for State Security after it.

smetana (sme·tānā). Also with Fr. spelling smitane (smitan). [a. Russ. *smetána* sour cream, f. *smetat'* to sweep together, collect.] Sour cream. Freq. *attrib.* as *smetana* (or *smitane*) *sauce*, a sauce made with sour cream and seasonings, usu. served with meat.
1909 M. Ronald *Century Cook Bk.* (rev. ed.) 606 Smetana sauce... Pour a cup of thick sour cream into the pan, let it just brown, and then pour over steaks. **1938** *Zionist Rev.* 11 Aug. 13/1 For supper there is bread and *smetana*. **1939** A. L. Simon *Conc. Encycl. Gastron.* I. 48/1 *Smitane*, Sauce. Sour cream and onions. **1963** I. Fleming *On H.M. Secret Service* (1964) xxiii. 242 Rehrücken mit Sahne. That's saddle of roebuck with a smitane sauce. **1968** A. Tack *Spy who wasn't Exchanged* xii. 91 He ordered a sour milk and cream mixture called *Smetana*. **1978** *Chicago* June 242/2 Our favorites are the eggy cheese blintzes..Lithuanian schnitzel with smetana (sour cream) sauce, veal cordon bleu, etc. **1979** N. Freeling *Widow* xiv. 83 Any sort of sauce you like except tomato. Smitane maybe.

S mèter: see *S II. 10.

smi·ckering, *ppl. a. rare*[−1]. That smirks or smickers.
1930 W. de la Mare *Poems for Children* 17 All turned their heads with a smickering smile.

smidge (smidʒ). [Origin unknown, perh. f. Smitch *sb.*[2]: cf. next.] A tiny amount. Orig. and chiefly *U.S.*
1905 *Dialect Notes* III. 65 *Smidge, smitch*, n., smallest piece, tiniest particle. **1934** *Sun* (Baltimore) 21 Sept. 1/6 Every last smidge of his record will be investigated. **1965** Auden *About House* (1966) 17 Surrender my smidge Of nitrogen to the World Fund. **1967** 'E. Queen' *Face to Face* xli. 184 Do you suppose I might have a smidge of that, Inspector? It looks so good, and I haven't had any breakfast. **1973** *Observer* 28 Oct. 48/4 Inviting us to buy their mail-order course for a smidge under £13. **1976** *Washington Post* 15 June A7/2 A Democratic Party that can't even afford..a smidge of debate.

smidgen (smi·dʒěn). orig. and chiefly *U.S.* Also **smidgeon, smidgin, smitchin,** etc. [Origin unknown, perh. f. Smitch *sb.*[2] + *-en, -in*, repr. dial. pronunc. of -ing[1]: cf. prec.] A tiny amount, a trace; a very small person or thing.
1845 C. M. Kirkland *Western Clearings* 71 They wouldn't have left a smitchin o' honey. **1878** J. H. Beadle *Western Wilds* 611 Not a smidgeon left—just bodaciously chawed up and spit out. **1886** *Trans. Amer. Philol. Assoc.* XVII. 43 *Smidgen*, 'a small bit, a grain', as 'a *smidgen* of meal', is common in East Tennessee. **1913** [see *clip *sb.*[2] 4]. **1930** *Va. Quarterly Rev.* 6 Apr. 249 He can testify perhaps..that he has had a bait, a snack, or a mere smidgen of them. **1952** J. Steinbeck *East of Eden* xxiii. 289 You little, silly, half-pint, smidgin of a wife. **1954** R. Millar *Waiting for Gillian* in *Plays of the Year* X. 346 There's a smidgin of Gordon's in the whisky decanter. **1960** Wodehouse *Jeeves in Offing* iv. 45 'No will of her own?' 'Not a smidgeon.' **1968** *Globe & Mail* (Toronto) 17 Feb. 37 (Advt.), Whether you're nine months or ninety years old, plump or twiggy, tall as a tree or small as a smidgeon. **1971** *N.Z. Listener* 18 Oct. 11/5 It's an unknown quantity often combined with just a smidgen of skill. **1973** *People's Jrnl.* (Inverness & Northern Counties ed.) 15 Dec. 4/5 My family would eat mince pies to a band playing so long as there's at least a smidgeon of rum butter to wipe over the top crust. **1982** R. Conquest in *Times Lit. Suppl.* 17 Dec. 1385/4 Any writer allowing the merest smidgin of Soviet reality into his work was headed straight for Magadan.

smilacina (smailăsai·nă). [mod.L. (R. L. Desfontaines 1807, in *Ann. Mus. Hist. Nat.* IX. 51), f. *smilac-*, Smilax + -ina[2].] A perennial herb of the genus so called, belonging to the family Liliaceæ, native to North America or temperate parts of Asia, and bearing terminal clusters of small white flowers; also called false Solomon's seal.
1808 *Curtis's Bot. Mag.* XXIX. 1155 (*heading*) Oval-Leaved Smilacina. **1890** *Harper's Mag.* Apr. 709/1 The little smilacina lifts its spike of tiny, fragrant blossoms. **1970** B. Miles *Bluebells & Bittersweet* viii. 141/3 Though perhaps not as well known [as Solomon's seal], smilacina wants the same situation.

smile, *sb.*[1] Add: **1. a.** (Later example of phr. *all smiles*.)
1916 G. B. Shaw *Pygmalion* iii. 154 Higgins:..Dont be nervous about it. Pitch it in strong. Clara (all smiles): I will. Good-bye.
2. (Earlier example.)
1839 *Spirit of Times* 24 Aug. 294/3 We all agreed to take another smile.
3. b. *attrib.*, as *smile-line, -wrinkle*.
1921 W. de la Mare *Memoirs of Midget* xxiv. 158, I looked at his long, fair eyelashes and the smile-line on his cheek. **1977** *New Yorker* 19 Sept. 58/3 Firkusny is a tall, lithe, trim man with gray hair, blue eyes, and smile lines in his face which soften an almost austere handsomeness. *a* **1930** D. H. Lawrence *Phoenix II* (1968) iii. 254 The smile-wrinkles on the fresh, pleased face, they give odd quivers.

smile, *v.* Add: **I. 1. c.** (Earlier U.S. examples.) Also in phr. *to come up smiling*: see *come *v.* 69 b.
1883 C. H. Hoyt *Bunch of Keys* iii, in *Five Plays* (1944) 48 'Single room?' 'Well, I should smile.' **1889** 'Mark Twain' *Conn. Yank.* ix. 113 They actually wanted *me* to put in! Well, I should smile.
d. Of eyes: to express pleasure, amusement, etc.
1759 C. Wesley in J. & C. Wesley *Funeral Hymns* 2nd Ser. 37 Those laughing eyes shall smile no more. *a* **1889** G. M. Hopkins *Poems* (1967) 37 And May has come, hair-bound in flowers, With eyes that smile thro' the tears of the hours. **1938** Dylan Thomas in *Seven* Winter 17 She wept in her pain and made mouths, Talked and tore though her eyes smiled. **1940** W. Faulkner *Hamlet* iv. i. 289 His face changed—something fleeting, quizzical, but not smiling, his eyes did not smile.
2. a. Also with *advbs.*, as *back, down, over, up*.
1845 [see *rank *sb.*[1] 3 b]. **1860** A. J. Evans *Beulah* xx. 177 He smiled down into her tearful face. **1866** Mrs. C. J. Newby *Common Sense* II. 88 The cheerful rallying

tone awoke something of the old pride in him, and he smiled up. **1889** R. L. Stevenson *Master of Ballantrae* iv. 111 How was he to smile back on the deceiver? **1905** E. M. Albanesi *Brown Eyes of Mary* iii. 38 She smiled up at him from under the white sunshade. **1908** *Smart Set* June 89/1 The girl pushed aside the screen and smiled over at her. **1949** A. Miller *Death of Salesman* 132 He's a man way out there in the blue, riding on a smile and a shoeshine. And when they start not smiling back—that's an earthquake. **1952** E. O'Neill *Moon for Misbegotten* i. 40 She smiles down at Jim, her face softening.
II. 6. b. Also with *down, out*.
1860 G. Meredith *Evan Harrington* xi. 111 'Another!' the hostess instantly smiled down the inhospitable outcry. **1936** R. Campbell *Mithraic Emblems* 52 It is the blossom in our blood With folded petals smiling out the sere, Brown, shuffled slippers of the limping year.
8. b. Also with direct speech as obj., to say with a smile.
1860 [in Dict.]. **1886** 'M. Gray' *Silence of Dean Maitland* III. ii. 35 'This is alarming,' smiled the dean. **1902** H. James *Wings of Dove* xv. 231 'He won't..make up his mind about me.' 'Well,' Milly smiled, 'give him time.' **1936** W. Holtby *South Riding* iii. iii. 175 'Well, Dolly, I hope you're looking after these young men,' smiled Carne shyly. **1976** H. MacInnes *Agent in Place* xix. 208 'Not for me,' Georges smiled.

Smilesian (smai·lziăn), *a.* Also **Smilesean.** [f. the name of Samuel *Smiles* + -ian.] Of, pertaining to, or characteristic of Samuel Smiles (1812–1904), author of *Self-Help* (1859) and other works for those who wish to 'improve' themselves by personal effort and initiative, or his thought and writings. Also occas. **Samuel-Smi·lesian.**
1889 G. B. Shaw *Fabian Essays in Socialism* 23 With all its [*sc.* private property's] energy, its Smilesian 'self-help', its merchant-princely enterprise,..what has it heaped up, over and above the pittance of its slaves? **1928** A. Huxley *Point Counter Point* xvii. 298 Most Smilesian souls must smell rather nasty, I should think. **1929** —— *Do what you Will* 151 There are..occasions—and this is what..the Samuel-Smilesian morality refuses to admit—when a man ought to permit himself to be subdued to things. **1968** *Punch* 29 May 792/2 It was up to the individual in a Samuel Smilesian way to restore his own fortunes through personal effort. **1969** G. M. Brown *Orkney Tapestry* ii. 53 Religion was Smilesian—heaven looked with favour on those who strove to improve themselves—unto him that hath shall be given. **1981** J. S. Bratton *Impact of Victorian Children's Fiction* ii. 45 In the year after the Great Exhibition, a thrusting optimistic appeal to Smilesean self-confidence..was certainly the shape of things to come.

smilesmirk (smai·lsmɜːk), *v. nonce-wd.* [f. Smile *v.* + Smirk *v.*] *intr.* To smile in a smirking manner.
1922 Joyce *Ulysses* 262 She smilesmirked supercilious.

smilet. (Further example.)
1845 J. J. Hooper *Taking Census* in *Some Adventures Capt. Simon Suggs* 175 The wrinkles on Mr. Kuncker's face formed themselves into fifty little smilets.

smiley, var. Smily *a.* in Dict. and Suppl.

smiling, *ppl. a.* Add: **5.** *smiling-lipped*.
1936 C. S. Lewis *Oxford Mag.* 14 May 575/2 The smiling-lipped Assyrian, cruel-bearded king.

smily, *a.* Add: Now usu. with spelling smiley. **1. a.** (Later examples.)
1969 D. Francis *Enquiry* vii. 97 It's your eyes... Dark and sort of smiley and sad. **1976** H. Kemelman *Wednesday Rabbi got Wet* iii. 21 A short man..with a round head and smiley face.
b. Smiling; cheerful.
1970 *Sunday Times* 8 Feb. 15/2 If the instruction wasn't nice and smiley, it'd upset me. **1972** *Times* 1 Dec. 20/2 *China*—new from France with the smiley face in black on yellow. **1976** R. B. Parker *Promised Land* (1977) vii. 35, I..said goodby with smily pleasant overtones in my voice.
2. Caused by or causing a smile.
1974 *Times* 22 Jan. 11/5 For smiley lines which are too deep, just paint out the start of them with a highlight. **1977** *Rolling Stone* 13 Jan. 52/2 To the extent that he includes these fast, cheerful numbers and smiley oddities such as Cole Porter's 'True Love' and the impenetrable fable 'Crackerbox Palace', Harrison seems hoping to achieve fresh popularity.

S-mine. Also **S mine.** [Abbrev. and anglicization of G. *schützenmine*, lit. 'infantryman mine'.] Used, esp. in the war of 1939–45, to designate a variety of enemy anti-personnel mine.
1944 J. H. Fullarton *Troop Target* 206 The enemy had sown elaborate mine fields—heavy Teller mines and deadly anti-personnel S mines. **1945** [see *anti-personnel* s.v. *anti-[1] 4 (iii)]. **1972** *Daily Tel.* (Colour Suppl.) 1 Sept. 16/2 In the last war the Germans devised a series of anti-personnel devices, including the S-mine & the 'butterfly-bomb'. **1977** C. McCullough *Thorn Birds* xv. 351 Sometimes a man would step on an unexploded S-mine.

smi·rkily, *adv.* [f. Smirky *a.* + -ly[2].] = Smirkingly *adv.*
a **1974** R. Crossman *Diaries* (1975) I. 135 There they were today looking at Jennie rather smugly and smirkily. **1978** *Guardian Weekly* 5 Feb. 18/3 Cryptic allusions..to padlocks and fetters, which most recent biographers have smirkily interpreted as meaning that he was a masochist.

Smirnoff (smɜ·ɹnof). The proprietary name of a brand of vodka.
1948 *Official Gaz.* (U.S. Patent Office) 26 Apr. 1008/1 *Smirnoff*..for vodka and kummel. Claims use since 1914. **1959** *Trade Marks Jrnl.* 15 Apr. 411/2 Smirnoff... Vodka. W. & A. Gilbey Limited. **1961** *Twining Bros.* (Oxford) *Wine List* Autumn 15 *Vodka etc.* Smirnoff 65.5° 39/4. **1965** 'M. Fallon' *Keys of Hell* v. 50 One of the bottles contained Smirnoff, his favourite vodka. **1977** *New Yorker* 3 Oct. 46/2, I got out the Smirnoff,..and finished it off.

smish. (Earlier example.)
1807 H. Tufts in E. Pearson *Autobiogr. of Criminal* (1930) ii. iv. 292 *Smish*, a shirt.

smitane: see *smetana.

smitch, *sb.*[2] Add: Also used adverbially. *rare.*
1963 M. McCarthy *Group* viii. 177 Neat but not gaudy, in a plain skirt and shirtwaist,..an old cameo brooch..—general effect a smitch Victorian.

smitchy (smi·tʃi), *a. colloq., rare.* [f. Smitch *sb.*[2] + -y.[1]] Tiny.
1888 Kipling *Plain Tales from Hills* 245 The Copper that takes you up is an old friend that tuk you up before, when you was a little, smitchy boy.

smite, *sb.*[1] Add: **1. b.** *Cricket.* A hard hit made by a batsman.
1888 R. H. Lyttelton in Steel & Lyttelton *Cricket* ii. 39 There is one hit in particular that in these days is very seldom seen—that is, the smite to long-leg with a horizontal bat. **1898** G. Giffen *With Bat & Ball* ii. 19 Since Trott hit one over [the boundary] at square leg, Jack Lyons and Tom Garrett have effected a similar smite.
2. b. A small or least amount; a particle. *U.S.* and *dial.*
1843 'R. Carlton' *New Purchase* I. xix. 175 Not a smite of noise, only my breathing and a sort of pittin-pattin sound of my heart. *a* **1852** F. M. Whitcher *Widow Bedott P.* (1856) v. 50 But it didn't do a smite o' good. **1913** [see *ream *sb.*[3] d].

smite, *v.* Add: **B. II. 3. a.** Also in phr. *to smite* (someone) *hip and thigh*: see Hip *sb.*[1] 2 d.
c. *Cricket.* To hit with great force; to defeat by hard hitting.
1891 W. G. Grace *Cricket* iv. 127 Mr. I. D. Walker..smote them to the tune of 90. **1904** F. C. Holland *Cricket* 28 After you have smitten him [*sc.* the bowler of yorkers] full-pitch two or three times, he will soon stop bothering you in this way. **1982** P. Tinniswood *More Tales from Long Room* vii. 87 My next sermon will take as its text: 'And, lo, Harry Halliday was a plump man, yet many a six did he smite for Yorkshire.'

smiter. Add: **1. e.** *Cricket.* A hard-hitting batsman.
1897 W. J. Ford in K. S. Ranjitsinhji *Jubilee Bk. Cricket* vii. 267 A little mercy should be shown to the muscular if unscientific 'smiter'. **1897** D. Moffat *Crickety Cricket* 25 Stoddart the smiter has carried his bat For a grandly made three-figure score. **1944** E. Blunden *Cricket Country* iv. 53 Poor Charles could not solve the problem of getting this smiter caught in the deep field.

smith, *sb.* Add: **3.** *smith-shop* (earlier examples); also *smith('s) shop*. chiefly *U.S.*
1651 *Early Rec. Dedham, Mass.* (1892) III. 179 Whensoever the said shopp shall be no longer vsed for a smithes shopp.., then it shall be remoued out of the high way. **1710** *Rec. Early Hist. Boston* (1884) XI. 105 Ordered that complaint be made..against Enoch Greenliefe for making a Smith Shop in his buildings. **1743** W. Ellis *Mod. Husbandman* Oct. xxii. 236 The Ploughman here has seldom Occasion to go to a Smith's Shop. **1755** *New Hampsh. Probate Rec.* (1916) III. 705 The Corner where Geo. Warrens Smith Shop Stands. *a* **1816** B. Hawkins *Sk. Creek Country* (1848) 30 At the public establishment there is a smith's shop. **1882** *Econ. Geol. Illinois* III. 150 The coal..is..used in an adjoining smith-shop.

Smith & Wesson (we·sǫn). [The names of Horace *Smith* (1808–93) and Daniel B. *Wesson* (1825–1906), founders of a firm of gunsmiths in Springfield, Mass.] The proprietary name of a make of firearm, esp. a type of cartridge revolver.
1860 *Charleston* (S. Carolina) *Mercury* 6 Nov. 3/3 (Advt.), Smith & Wesson's seven shooters. **1865** [see *Remington]. **1881** G. W. Romspert *Western Echo* 115 The second ball from my Smith & Wesson stretched him struggling upon the earth with a bullet through his lungs. **1893** *Official Gaz.* (U.S. Patent Office) 16 May 1058/1 Revolving firearms. Smith & Wesson, Springfield, Mass... Essential feature, the word and character *Smith & Wesson*. Used since 1857. **1928** *Trade Marks Jrnl.* 14 Nov. 1822/1 Smith and Wesson. 482,778. Revolvers and Pistols. Smith and Wesson Inc. **1957** J. E. Parsons *Smith & Wesson Revolvers* i. 13 The Smith & Wesson revolver had arrived as a de luxe presentation item. **1964** E. S. Gardner *Case of Phantom Fortune* (1970) xix. 192, I show you a Smith and Wesson revolver. **1981** *Daily Tel.* 18 June 18/5 They prefer to throw truncheons rather than draw a 0.38 Smith & Wesson.

smither, *sb.* (Later example.)
1976 *Star* (Sheffield) 3 Dec. 19/4 (Advt.), Retired Smither required for part-time light Machine Knife Smithing work.

smithereen (smiðərī·n), *v.* [f. Smithereens *sb. pl.*] *trans.* To smash or blow up into tiny fragments. Hence **smitheree·ning** *ppl. a.*
1927 H. Crane in *Transition* Dec. 136 Lo, Lord, Thou ridest! Lord, Lord, Thy swifting heart Naught stayeth, naught now bideth But's smithereened apart! **1959** *Listener* 5 Mar. 429/2 A bomb in a suit-case timed shortly to do its smithereening work. **1964** *Economist* 16 May 699/2 Dum-dum bullets or smithereening explosives. **1973** *N.Y. News* 21 Aug. 53/1 He'd like to smithereen the crystal ball.

smithereens, *sb. pl.* Add: **a.** (Earlier and later examples.)
1829 G. Griffin *Collegians* II. xxii. 157 A body would tink it hardly safe to stand here under 'em, in dread dey'd come tumblin' down, may be, an' make *smiddereens* of him, bless de mark! **1922** Joyce *Ulysses* 372 Crew and cargo in smithereens. **1927** D. H. Lawrence *Mornings in Mexico* 16 The sun went bang, with smithereens of birds bursting in all directions. **1933** *Sun* (Baltimore) 22 Dec. 22/6 A substantial charge of dynamite—enough, in fact, to blast the bridge to smithereens. **1961** J. I. Packer *Evangelism & Sovereignty of God* ii. 31 Books like Deuteronomy and Isaiah and John's Gospel and Romans smash it [*sc.* a man-centred outlook] to smithereens. **1976** *Time* 27 Dec. 36/3 The result is another kind of supernova, a fantastic explosion that blows the star to smithereens, dispersing into space most of the remaining elements that it had manufactured during its lifetime.

Smithfield[2] (smi·þfīld). *U.S.* The name of a town in Virginia, used *attrib.* to designate a type of ham cured by a special process which originated there.
Properly applied only to hams cured within the corporate limits of Smithfield.
1908 *Sat. Even. Post* 31 Oct. 25/2 Next to singing a hymn, nothing gives him so much pleasure as a Smithfield ham. **1947** R. Berolzheimer *U.S. Regional Cookbk.* 188 *Smithfield ham*... The hogs fatten rapidly by foraging in the peanut fields after the crop is harvested, special care is taken in curing and smoking. **1973** M. R. Crowell *Greener Pastures* 113 Smithfield ham is truly worth it all. **1977** *Times* 15 Oct. 13/5 Seek out a Virginian friend..and sip mint juleps..until the Smithfield ham and spoonbread are ready.

smithiantha (smiþi,æ·nþă). [mod.L. (P. Magnus in C. E. O. Kuntze *Revisio Generum Plantarum* (1891) ii. 977), f. the name of Matilda *Smith* (1854–1926), botanical artist + Gr. ἄνθος flower (with fem. ending to conform with the gender of names it superseded).] A small, perennial, rhizomatous herb of the genus so called, belonging to the family Gesneriaceæ, native to Mexico, and bearing hairy, cordate, variegated leaves and clusters of red, yellow, or orange bell-shaped flowers; = Gesnera in Dict. and Suppl.
1961 *Times* 27 Sept. 6/6 Colourful smithianthas, better known under their familiar name of gesneria. **1979** A. Huxley *Success with House Plants* 364/3 Smithianthas grow best in medium light.

smithite (smi·þəit). *Min.* [f. the name of G. F. Herbert *Smith* (1872–1953), English mineralogist + -ite[1].] A sulpharsenite of silver, $AgAsS_2$, found as red tabular monoclinic crystals.
1905 *Nature* 13 Apr. 574/2 Further crystallographic and chemical details were given of the three new red minerals from the Binnenthal originally described by R. H. Solly, and named by him Smithite (after G. F. Herbert Smith), Hutchinsonite (after A. Hutchinson), and Trechmannite after C. O. Trechmann). **1905** R. H. Solly in *Mineral. Mag.* XIV. 74 Smithite is associated with hutchinsonite, sartorite, and rathite in the white dolomite of the Lengenbach. **1938** *Econ. Geol.* XXXIII. 155 Proustite occurs..in magnificent eutectic relationships with smithite. **1968** I. Kostov *Mineral.* 173 Pyrostilpnite, aramayoite, and smithite have perfect pinacoidal cleavage... Smithite is very soft, the other minerals have a hardness between 2 and 3.

Smith-Trager (smiþ,trē[i]·gəɹ). *Linguistics.* = *Trager-Smith.
1959 *Canadian Jrnl. Linguistics* V. 1. 8 The Smith-Trager schema for plotting the English phonemes and the diaphonemic relations between idiolects and dialects is a case in point. **1964** W. S. Allen in D. Abercrombie et al. *Daniel Jones* 9 This should not be taken to imply an acceptance of the 4-term (Smith-Trager) stress system.

smithy, *v.* Add: **1. a.** Also *fig.*
1910 J. Masefield *Ballads & Poems* 66 Until this case, this clogging mould, Be smithied all to kingly gold. **1929** A. Clarke *Pilgrimage* 19 Smithied in gloom the low day Had glowed upon the axle.
2. (Earlier example.)
1733 L. Theobald in *Works of Shakespeare* VII. 96 To smithy, is, to perform the Work and Office of a Smith.

smithying, *vbl. sb.* (Later *fig.* example.)
1934 E. Blunden *Challenge to Death* 336 History's smithying should not disappear Without reverberation.

smock, *sb.* Add: **2. b.** A loose garment worn by artists over their other clothes to keep them clean; a woman's or child's loose dress or blouse resembling a smock-frock in shape.
1907 *Yesterday's Shopping* (1969) 790/1 Girls' cashmere smock. In cream, sky, cardinal. **1938** N. Marsh *Artists in Crime* (1941) xiii. 192 He found..an evening dress in close proximity to a painting-smock. *Ibid.* 193 He was amused to find that even the Seacliff painting-bags and smock smelt of Worth. **1969** R. T. Wilcox *Dict. Costume* 326/2 The smock is now much worn as a coverall by professional people at work, especially artists. **1971** [see *maternity 3 b].
c. In full *camouflage smock*, a loose outer tunic of coarse material dyed brown and green and worn by troops as camouflage.
1964 L. Deighton *Funeral in Berlin* xxxii. 171 They wore camouflage smocks and steel helmets... They were front-line troops, not Waffen S.S. **1974** C. Ryan *Bridge too Far* iv. vii. 295 The only thing I could do for most of them was to take off their smocks and cover their faces. **1978** M. Walker *Infiltrator* xxi. 224 He tossed me an assault rifle and..a camouflage smock.
3. a. *smock dress, jacket, linen* (earlier example), *shirt* (later example), *sleeve* (later example); **smock-ravelled** *dial.*, perplexed.
1980 J. Hone *Flowers of Forest* i. 28 The woman in the pale smock dress. **1976** *Bridgwater Mercury* 21 Dec. 3/2 (Advt.), Half price smock jackets. **1880** L. Higgin *Handbk. Embroidery* ii. 12 *Smock Linen* is a strong even green cloth..an excellent ground for working screens. **1904** in *Eng. Dial. Dict.*, Smock-ravelled. **1913** D. H. Lawrence *Let.* 15 Sept. (1962), I feel a bit smock-ravelled—don't know where the east is, nor the north and west. **1972** *Guardian* 8 Feb. 11/6 A smock shirt is less voluminous than a smock, but has..smock details: the neck is round or square, there is usually a yoke, and often pleated or pin-tucked fronts. **1870** A. J. Munby *Diary* 25 June in D. Hudson *Munby* (1972) 288 A stout fair girl..who looked well in her cotton hoodbonnet and red neckerchief, and smocksleeves.

smock, *v.* **4.** (Later examples.)
1963 N. Streatfeild *Vicarage Family* iii. 27 Louise..was still small enough for smocks and her mother smocked beautifully. **1980** *Daily Tel.* 24 Apr. 14/5 His mother brought him up alone on a war pension plus what she could make by smocking children's clothes.

smocked, *ppl. a.* Add: **2.** (Later examples.)
1918 E. A. Archer *Needlecraft* x. 112, I am sure you will not be contented till Sally Ann has a smocked dress. **1934** A. M. Miall *Compl. Needlecraft* 92 Smocked garments readily stretch as a child grows.

smock-faced, *a.* (Later example.)
1923 E. Sitwell *Bucolic Comedies* 16 Forlorn the smock-faced sheep sit.

smock-frock, *sb.* Add: **2.** (Earlier example.)
1858 Thackeray *Virginians* I. xv. 112 The smock-frocks did not seem to heed, and clamped out of church quite unconcerned.

smocking, *vbl. sb.* Add: (Later examples.)
1916 *T. Eaton & Co. Catal.* Spring & Summer 23/1 Shoulder yoke with neat smocking trims both sides of the front. **1934** A. M. Miall *Compl. Needlecraft* 92 Smocking is a special favourite for children's clothes. **1961** A. Liley *Craft of Embroidery* iv. 174 Smocking generally takes up three times as much fabric as the final width required. **1964** *McCall's Sewing* xiii. 242/1 The same smocking stitches can be used in both types of smocking. **1976** P. Clabburn *Needleworker's Dict.* 245/2 (*caption*) Smocking: detail of a child's dress, English, 1930.

smock-racing, *vbl. sb.* (Earlier example.)
1790 J. Woodforde *Diary* 24 May (1927) III. 192 Smock-racing at the Heart this Aft. being Whit-Monday.

smog (smǫg). [Blend of Smoke *sb.* and Fog *sb.*[2]] **1. a.** Fog intensified by smoke. Cf. *photochemical smog* s.v. *photochemical *a.*
1905 *Daily Graphic* 26 July 10/2 In the engineering section of the Congress Dr. H. A. des Vœux, hon. treasurer of the Coal Smoke Abatement Society, read a paper on 'Fog and Smoke'. He said it required no science to see that there was something produced in great cities which was not found in the country, and that was smoky fog, or what was known as 'smog'. **1905** *Globe* 27 July 3/5 The other day at a meeting of the Public Health Congress Dr. Des Vœux did a public service in coining a new word for the London fog, which was referred to as 'smog', a compound of 'smoke' and 'fog'. **1938** *Daily Progress* (Charlottesville, Va.) 12 May 7 In the opinion of many medical authorities, 'smog' is the principal reason why Pittsburgh has the highest pneumonia death rate in the United States. **1950** *Economist* 25 Feb. 432/2 Smog is a problem, far from completely understood as yet, of air contamination not by smoke, but by the fumes and gases—sulphur compounds, chlorine and so on—given off by modern industrial processes such as oil refining, chemical manufacturing and metallurgy. **1955** *Sci. Amer.* May 63/3 At first it was thought that smoke, dust, sulfur dioxide and hydrofluoric acid were responsible for the smog [in Los Angeles], but soon it became clear that these known pollutants, in the concentrations measured on smoggy days, could not cause the physiological effects observed... It was then that A. J. Haagen-Smit..suggested that peroxides and ozonides of hydrocarbons were responsible for smog. **1961** L. Mumford *City in History* xv. 479 Nor have they eliminated the unburned hydrocarbons which help produce the smog that blankets such a motor-ridden conurbation as Los Angeles. **1975** D. Lodge *Changing Places* ii. 71 It was difficult to tell whether the sediment thickening the atmosphere was rain or sleet or smog.
b. *fig.* A state or condition of obscurity or confusion; something designed to confuse or obscure.
1954 *Ann. Reg. 1953* i. 54 Lord Reading..described it [*sc.* the Russian Note] in the House of Lords as 18 pages of 'somewhat dismal and turgid "smog"'. **1976** *Billings* (Montana) *Gaz.* 30 June 1-a/1 When the political smog clears, Billings city government somehow continues to function. **1978** D. Bloodworth *Crosstalk* xxiv. 191 He hoped..Zoe's gift might pierce the gathering smog? Because things were getting tough, and the Russians were..accusing the Maoists of trying to flood Moscow with narcotics.
2. *attrib.* and *Comb.*, as *smog-bank, -burner, mask, producer*; *smog-bound, -free, -producing* adjs.
1975 *Country Life* 16 Jan. 130/2 Take a commuter jet from Los Angeles to San Francisco... You rise above the smog-bank. **1970** *New Scientist* 1 Jan. 8/3 Smogbound, noise-deafened, misanthropic Londoners..might be taking their high blood pressure with them. **1961** *Engineering* 27 Jan. 175/3 The smog-burner is a mechanical rather than a chemical or catalytic device. **1959** *News Chron.* 19 June 4/3 The six-bedroom houses hardly get dirty in California's smog-free climate. **1981** *Times* 6 Aug. 7/7 Smog-free sunsets over the Indian Ocean. **1954** *Ann. Reg. 1953* iv. 391 The year 1953 might well be remembered as the one in which 'smog' masks first appeared. **1979** *Listener* 5 July 6/1 Visiting journalists [to Tokyo]..were amazed to find they didn't have to wear smog-masks. **1951** *Sun* (Baltimore) (B ed.) 31 Dec. 14/2 More than a dozen Baltimore firms have been definitely albeit informally tagged as smog producers. **1970** *New Scientist* 13 Aug. 324/1 Efforts to curb auto-pollution concern the directly poisonous or smog-producing colourless emissions of carbon monoxide, unburnt hydrocarbons and nitrogen oxides.
Hence as *v. trans.* N. Amer. colloq., (*a*) with *out, up*: to cover or envelop in smog; (*b*) with *in*: to confine or imprison because of smog; (freq. *pass.*); **smogged** *ppl. a.*
1966 P. Tamony *Americanisms* (typescript) No. 14. 2 The era of the motor-car smogged up greenery. **1970** *Globe & Mail* (Toronto) 28 Sept. 4/1 Mr. Lewis was 'smogged in' at Sudbury..and was unable to arrive in time for the Ottawa meeting. **1974** *Science News* 24–31 Aug. 136 Conventional geodesy depends on clear lines of sight, and in the Los Angeles basin these are often smogged out. **1982** *Chr. Sci. Monitor* (Mid-Western ed.) 8 Dec. 12 Yet you can't do it because they have to meet the same pollution standards they do in heavily smogged areas.

smogger (smǫ·gəɹ). [Origin unknown.] (See quots.).
1958 F. Jennings *Men of Lanes* 217 Veteran tramps have their road-sign language, by which they communicate with each other... Vagabonds call it 'smogger', and it is said to have been introduced into England by gipsies in the time of Henry VIII. **1975** *Indexer* IX. 131/2 Supposing, for example, you see a chalk mark (a 'smogger') made by a tramp on your gate-post, consisting of two large and slightly overlapping circles, you could identify this as meaning 'tell a pathetic story', but if the sign consisted of three small circles in a line, it would mean 'money usually given here'.

smoggy (smǫ·gi), *a.* [f. *smog + -y[1].] Characterized by the presence of smog. Hence **smo·ggily** *adv.*
1905 *Daily News* 26 July 9 Observations had proved that, even in the most 'smoggy' periods, there was far less of the evil element in the early hours of the morning. **1948** *Smog Problem in Los Angeles County* 17/1 Smoggy days are usually days of high temperature. **1965** P. Wylie *They Both were Naked* ii. vi. 291 When the fourth day was smoggily born in the black-smudged, yellow opacity that was air I ordered coffee. **1978** P. Niesewand *Underground Connection* 25 A smoggy wind blew across the crowded platform as a train thundered in.

smogless (smǫ·glės), *a.* [f. *smog + -less.] Free from smog; characterized by the absence of smog.
1948 G. Marx *Let.* 27 July (1967) 192 On a smogless day our gleaming skulls can be seen as far east as Cedar Rapids. **1960** *Spectator* 30 Sept. 501/1 In an attempt to bring the smogless society a little closer, the Minister asked..the official black areas to submit their plans for smoke control. **1971** *Nature* 19 Feb. 549/1, I measured the atmospheric extinction in three wavelengths on 31 clear, smogless nights at the Mount Wilson Observatory.

smokable, *a.* and *sb.* Add: **A.** *adj.* **2.** Able to be ridiculed. Cf. Smoke *v.* 9. *rare.*
1818 Keats *Let.* 31 Dec. (1958) II. 19 The Dress Maker, the blue Stocking and the most charming sentimentalist..are equally smokeable.
B. *sb.* Also in *sing.*
1897 'Mark Twain' *Autobiogr.* (1924) I. 98 There are people who strictly deprive themselves of each and every eatable, drinkable and smokable which has in any way acquired a shady reputation.

smoke, *sb.* Add: **I. 1. d.** Also, any large city or town. Chiefly *Austral.*
1848 H. W. Haygarth *Recoll. Bush Life in Australia* 6 As he gradually leaves behind him the 'big smoke' (as the aborigines picturesquely call the town), the accommodations become more and more scanty. **1893** J. A. Barry *Steve Brown's Bunyip* 21 You want to get away amongst the spielers and forties of the big smoke? **1903** Farmer & Henley *Slang* VI. 270 *The Smoke* = any large city: spec. London: also *The Great Smoke*. **1971** *Sunday Australian* 8 Aug. 5/3 The unhappy pilgrimage from bush to big smoke.
f. A shade of grey.
1882 *Cassell's Family Mag.* Apr. 314/2 Charming colouring.., smoke, bright blues and drabs. **1923** *Daily Mail* 13 Feb. 13 (Advt.), Wool hose..in..lt. grey, shoe grey, smoke, mole. **1971** [see *kasha[1] 2]. **1978** *Hot Car* June 98/1 (Advt.), All Portholes are supplied *domed* either in clear, blue, green, smoke, bronze or black perspex.
2. a. (Further examples.) Also, a particular kind of smoke.
1919 *Jrnl. Amer. Chem. Soc.* XLI. 312 The rate of disappearance of a finely divided smoke of a given concentration was greater than for a coarser smoke. **1950** *Thorpe's Dict. Appl. Chem.* (ed. 4) X. 787/1 Determining the particle size of a smoke. **1972** *Combustion Sci. & Technol.* VI. 55/1 Carbon smokes are generated by combustion.
c. † (*a*) *N. Amer.* = Smudge *sb.*[2] 2. *Obs.* (*b*) (See quot. 1961.)
1689 H. Kelsey *Jrnl.* 29 June (1929) 26 Abundance of Musketers & at night could not gett wood Enough for to make a smoke to Clear y^m. **1765** R. Rogers *Conc. Account N. Amer.* 140 It is difficult to sleep without a smoak in your bed-chamber, to expell [mosquitoes]. **1860** H. Y. Hind *Assiniboine* in *Canad. Red River Exped.* I. xiii. 286 At each camping place we were obliged to make 'smokes' to drive away these tormentors [*sc.* mosquitoes]. **1961** *Amateur Gardening* 4 Nov. Suppl. 47/2 Many of the modern insecticides and fungicides are sold in the form of small containers which when ignited give off clouds of vapour carrying fine chemical into all parts of the greenhouse. These devices are known simply as 'smokes'.
4. e. *spec.* in *Espionage*, false information to distract opponents.
[**1859** G. W. Matsell *Vocabulum* 82 *Smoke*, humbug; any thing said to conceal the true sentiment of the talker; to cover the intent.] **1966** 'A. Hall' *9th Directive* xxi. 200 'No go. I got myself cornered. One dead.'.. 'Do you need any smoke out?' **1977** 'J. le Carré' *Honourable Schoolboy* iv. 91 For smoke.. Molly chose a dozen other R's.
i. *to watch someone's smoke* (slang, orig. *U.S.*), to watch someone go, to observe someone's actions; chiefly imp. in phr. *watch my smoke.*
1905 G. W. Peck *Peck's Bad Boy with Circus* ix. 114 The elephant..winked at the other elephants, as much as to say: 'Watch my smoke.' **1921** R. D. Paine *Comrades Rolling Ocean* i. 10 Suspend judgement and watch my smoke. That's all I ask. **1927** Wodehouse *Meet Mr Mulliner* iii. 82 'You are a curate, eh?' 'At present. But,' said Augustine, tapping his companion on the chest, 'just watch my smoke.' **1928** C. Sandburg *Good Morning, America* 18 Let's go. Watch our smoke. Excuse our dust. **1947** Wodehouse *Full Moon* ii. 27 Look at Henry the Eighth... And Solomon. Once they started marrying, there was no holding them—you just sat back and watched their smoke.
j. *in(to) smoke* (slang, chiefly *Austral.*), in(to) hiding.
1924 C. J. Dennis *Rose of Spadgers* 72 'Jist now,' says Brannigan, 'Spike Wegg's in smoke. Oh, jist concerns a cove 'e tried to croak.' **1938** P. J. Smith *Con Man* ix. 179 Denman advised Naysmith to remain 'in smoke'—an expression meaning to hide himself—and play golf until Denman had stood his trial alone for the offence in Glasgow. **1943** K. Tennant *Ride on Stranger* xvii. 203 The New Zealand delegate returned anonymously, slipped ashore and 'went into smoke' like some famous criminal. **1967** K. S. Prichard *Subtle Flame* 252 Meanwhile Tony's got to be kept in smoke?
k. *to go up in smoke*, to be consumed by fire; to be destroyed completely; also *fig.*, to lose one's temper.
1933 [see *go *v.* 94 h]. **1939** 'N. Blake' *Smiler with Knife* 94 Oh, glory no! He'd go up in smoke. **1946** [see *interimsethik]. **1955** [see *highbrow, high-brow *sb.*].
5. a. (*b*) Opium; (*c*) marijuana.
1884 Kipling *Plain Tales from Hills* (1888) 238 The coffin is gone—gone to China again—with the old man and a couple of *tolahs* of Smoke inside, in case he should want 'em on the way. **1946** C. McCullers *Member of Wedding* iii. 192 Made crazy one night by a marihuana cigarette, by something called smoke or snow. **1956** S. Longstreet *Real Jazz Old & New* xiii. 104 He mixed..with studs shying a toy of opium. But there isn't much record that he went for tea-sticks or the smoke himself. **1963** H. Parkhurst *Undertow* (1964) v. 84 To her 'smoke' and the 'kicks' were the things that seemed to count. **1977** *Rolling Stone* 16 June 76/2 He wondered aloud if there were 'smoke' in the house, prompting people in the front rows to toss lit joints upon the stage.
b. Also, a marijuana cigarette.
1967 [see *joint *sb.* 14 c]. **1980** 'D. Kavanagh' *Duffy* iii. 52 He'd known who handled smokes, who handled snort and who handled smack.
6. (Earlier and further examples.) *to have* (or *take*) *a smoke*, to smoke a pipe, cigarette, or cigar.
1835 A. B. Longstreet *Georgia Scenes* 213 *Mrs. B.* [to *Mrs. S.*]. Well, let's light our pipes, and take a short smoke, and go to bed. **1860** W. H. Russell *My Diary in India* II. iii. 53 Here..were to be seen a few soldiers,..lounging about, taking an early morning smoke. **1887** *Lantern* (New Orleans) May, in *Amer. Speech* (1948) XXIII. 247/2 A book-keeper for a large firm here begged an acquaintance for twenty-five cents to go and have a smoke with. **1922** Joyce *Ulysses* 112 Ideal spot to have

a quiet smoke and read the *Church Times*. **1926** J. Black *You can't Win* xi. 141, I..found my way into the hop joints. Curiosity was my only excuse for my first 'smoke'. **1978** O. White *Silent Reach* xix. 206 Can't say I blame you. Have another smoke?
7. b. Cheap whisky; a concoction based on raw alcohol, etc., used as a substitute for whisky. *N. Amer.*
1904 'O. Henry' *Cabbages & Kings* iii. 52 Brandy, anisada, Scotch 'smoke' and various wines. **1928** *Daily Tel.* 9 Oct. 11/3 Twelve additional deaths today are attributed to week-end 'jags', which have been traced to 'speak-easies' in the New York east-end, where the liquor is known as 'smoke'. **1940** *Sun* (Baltimore) 14 Nov. 8/2 Judge Eugene O'Dunne yesterday ruled that the sale of denatured alcohol diluted with water and known as 'smoke'—comes within the effect of the liquor laws. **1950** [see *rubby-dub* s.v. *rubby]. **1959** *Washington Post* 18 Aug. A3/4 It was the smoke that made Heaton a loner and junk peddler in the demolition jungles of the Southwest area. **1980** *Amer. Speech* 1977 LII. 117 Blends of anti-freeze and water, sometimes including methyl alcohol, solvent or paint remover, cleaning fluid, canned heat, or other alcohol mixtures: *smoke*.
8. Also, a short-haired cat with similar blue-grey or black colouring. (Later examples.)
1933 E. Buckworth-Herne-Soame *Cats* xviii. 99 A smoke is one of the most handsome cats living. **1958** *Listener* 28 Aug. 298/2 With two exceptions, the Chinchilla and smoke, short-hairs have the same variety of colour as do long-hairs. **1972** Ing & Pond *Champion Cats of World* 79/1 The first Smokes were bred by chance. *Ibid.* 99/2 Black and Blue Smoke. Except that the fur is short rather than long, the standard is the same as..the long-haired varieties. **1976** *Southern Even. Echo* (Southampton) 11 Nov. (Advts. Suppl.) 14/3 Pedigree long-haired male kittens, black and blue smokes.
b. An abusive and offensive term for a Black. *U.S. slang.*
1913 J. T. Foote *Blister Jones* viii. 242 'Who you callin' a smoke?' says Snowball, startin' fur Micky. **1932** J. T. Farrell *Young Lonigan* i. 21 He had bashed the living moses out of that smoke who pulled a razor on him over in Carter playground. **1940** R. Chandler *Farewell, my Lovely* iii. 24 There was five smokes carved Harlem sunsets on each other. **1970** L. Sanders *Anderson Tapes* xxxiii. 109 Five men. One's a smoke.
II. 9. a. *smoke-barrage, -column, -devil, -dust, -fog, -pall, -plume, -puff, -ring* (further examples), *-trail, -wreath* (later examples).
1923 Kipling *Irish Guards in Great War* I. 317 A Guards Battalion..came up..under cover of what looked like a smoke-barrage. **1932** Blunden *Fall in Ghosts* 10 The youth scurried away to the problem of preventing that smoke-column from the cookers. **1919** J. Masefield *Battle of Somme* 5 The No Man's Land, into which our men advanced, was a strip of earth without life, made smoky, dusty, and dim by explosions which came out of the air upon it, and left black, curling, slowly fading, dust and smoke-devils behind them. **1970** R. Lowell *Notebook* 247 Smoke-dust the Chinese draftsman made eternal. **1933** *Gloss. Aeronaut. Terms* (*B.S.I.*) xiv. 88 *Smoke fog*, fog due to particles of smoke in the atmosphere. A thick haze. **1918** G. Frankau *Poetical Wks.* xx. 153 Southward, gray skies with smoke-pall overcast. **1940** W. Faulkner *Hamlet* iii. 159 He actually heard the cow's voice..from beyond the smokepall on the other hill. **1920** *Glasgow Herald* 3 July 6 We may even be deprived of some of these interesting smoke-plumes that float proudly and unafraid over public works. **1978** *Sci. Amer.* May 162/1 There are two basic types of smoke plume: the momentum jet..and the buoyant plume. **1897** W. S. Churchill in *Daily Tel.* 9 Nov. 7/6 The mountain battery..came into action and began shelling the summits, from which the smoke-puffs were most frequent and continuous. **1909** F. Barclay *Rosary* xxiv. 257 'And this pleases you?' inquired the doctor, blowing smoke rings into the air. **1959** I. & P. Opie *Lore & Lang. Schoolch.* xi. 220 A 14-year-old girl, Newbridge, Monmouthshire, writes: 'If you see a smoke ring coming from an engine you can wish once, and if you see two smoke rings you can wish twice.' **1971** *Wall St. Jrnl.* 22 July W.1/5 The industry tries all sorts of promotions from a cigar smoke ring blowing contest at Palisades Amusement Park..to cigar and cognac tasting sessions. **1933** *O.E.D.* (Suppl.) s.v. Sky *sb.*[1] 9, Smoke-trails made by aircraft. **1979** 'P. O'Connor' *Into Strong City* xxxviii. 142 A puffer making its way up the Clyde. A smoke trail. The sea calm. **1913** D. H. Lawrence *Love Poems & Others* 36 To-day a thicket of sunshine with blue smoke-wreaths. **1939** R. Campbell *Flowering Rifle* vi. 138 Through rolling smoke-wreaths, there, like ant-hills rise The kopjes in the nitre-breathing skies.
b. *smoke-hood, -vent.*
1969 'M. Renault' *Fire from Heaven* (1972) v. 199 King Archelaos had hung a smoke-hood over the hearth. **1912** J. L. M. Myres *Dawn of Hist.* viii. 185 In Crete the climate is mild enough..for portable braziers to suffice, and this release from anxiety for smoke-vents encouraged the architects to daring experiments. **1936** Smoke-vent [see *smoke-blackening*, sense 9 c below].
c. *smoke-blackening, -burn, -mark.*
1936 *Discovery* Feb. 55/2 The semi-conical apartment at the east end was evidently a fire-chamber, as traces of smoke-blackening were found on stones that had fallen from the roof and had once surrounded a smoke-vent. **1971** S. Hill *Strange Meeting* ii. 141 He had very pale, almost white eyelashes, and a curious mark, like a smoke burn, across his forehead. **1876** 'Mark Twain' *Tom Sawyer* xxxi. 23 Holding their candles aloft and reading the tangled web-work of names..with which the rocky walls [of the cave] had been frescoed (in candle smoke)... They made a smoke-mark for future guidance.
d. *smoke-grey* (later examples).
1903 [see sense 9 e in Dict.]. **1924** [see *Harris]. **1976** *Billings* (Montana) *Gaz.* 30 June 9-D/3 (Advt.), For sale 1976 Corvette... Red w/ smoke gray interior.
10. a. *smoke-belching, -chaser, -control, -detecting, -detection, -detector, -discharger, -generator.*
1963 Bird & Hutton-Stott *Veteran Motor Car* 81 They were extremely refined,..smoke-belching, costly,..and slow. **1974** Smoke-belching [see *samlor]. **1935** L. Luard *Conquering Seas* xii. 153 Shot thirty-five miles east of Cape. Towed for three hours. Double bag. Fish hard and golden. Worked edge of Strunda four days. Good living. Shifted to avoid smoke-chasers.—Trial shoot in 45 fathoms. Nowt. **1942** *Sun* (Baltimore) 14 May 4/2 Lookouts, smoke-chasers, firemen and organized stand-by crews to prevent, detect and fight forest fires. **1956** *Ann. Reg. 1955* 4 Densely populated areas designated for smoke-control. **1967** *Economist* 30 Dec. 1277/1 The real trouble is still the old one: too few smoke control orders in the Black Areas. **1962** *Flight International* LXXXI. 190/2 The Pyrene Co Ltd have contributed smoke-detecting..equipment to the D.H.121 Trident. **1936** *Discovery* May 146/2 Other uses [of light-sensitive cells] were for..smoke detection in factory chimneys. **1957** *Practical Wireless* XXXIII. 683/2 Such devices as..a smoke detector and fire indicator. **1978** *N.Y. Times* 30 Mar. B16/5 (Advt.), 117 West 58th St... Completely remodeled Prewar Bldg—featuring—..smoke detectors—ceramic tile kitch & bath. *a* **1944** K. Douglas *Alamein to Zem Zem* (1946) 16 There were also two smoke dischargers to be operated by me. Stacked round the sides of the turret were the six-pounder shells. **1979** P. Alexander *Show me Hero* xx. 210 Kemp threw a canister of tear-gas... Quinn threw a smoke-discharger. **1933** *Gloss. Aeronaut. Terms* (*B.S.I.*) xiii. 3 *Smoke generator*, a pyrotechnic device, placed on the ground, emitting smoke for indicating wind direction. **1961** W. Vaughan-Thomas *Anzio* ix. 211 The smoke generators, manned by cheerful coloured troops, tried to blot out observation and protect the shipping from sneak raiders.
b. *smoke-bemuffled, -blackened, -defiled, -dyed* (earlier example), *-filled, -laden, -logged, -palled, -reddened, -scorched, -sodden* (later example), *-stained, -warmed.*
1912 W. de la Mare *Listeners* 81 Vainly 'gainst that thin wall The trumpets call, Or with loud hum The smoke-bemuffled drum. **1890** A. Conan Doyle *Firm of Girdlestone* xiv. 111 Puffing at his weed and staring up at the smoke-blackened ceiling. **1976** M. Machlin *Pipeline* lvi. 567 The small smoke-blackened figure walked slowly toward the forepeak of the Globtik Alamo. **1919** Kipling *Years Between* 42 Witness thy portrait, smoke-defiled. **1817** *Jrnl. Salem Mechanic* 19 Oct. in *Essex Inst. Hist. Coll.* (1866) VIII. 234, I have not seen a handsome woman since I left Salem; they are here [*sc.* Pittsburgh] all smoke-dyed. **1920** *Evening Star* (Washington) 14 June 1/2 Harry Daugherty..predicted that about 2.11 a.m., 'in a smoke-filled room', on a certain night during the republican national convention, the next nominee would be chosen. **1965** G. McInnes *Road to Gundagai* v. 77 These damp and smokefilled holes. **1979** *Now!* 21–27 Sept. 74/3 Presidential candidates are not selected by political pros in smoke-filled rooms these days. **1903** *Work* 21 Mar. 105/1 Incrustations due to the smoke-laden atmosphere. **1975** *Economist* 6 Sept. 20/1 If you've ever tried to make clearheaded decisions in a stuffy, smoke-laden conference room, you'll appreciate what we mean. **1963** *Times* 25 May 8/4 A fire officer said: 'When we arrived ammunition was exploding everywhere. Our job was made even more difficult because the building was completely smoke-logged.' **1976** *Southern Even. Echo* (Southampton) 13 Nov. 16/2 Firemen had to wear breathing equipment to get into the smoke-logged electrical input room on the first floor. **1922** Joyce *Ulysses* 583 Two shafts of light fall on the smokepalled altarstone. **1893** Kipling *Seven Seas* (1896) 97 To the smoke-reddened eyes of Loben. **1888** —— *Departmental Ditties* (1890) 81 Anger and pain and terror Stamped on the smoke-scorched skin. **1959** C. Devlin *Sermons & Devotional Writings of G. M. Hopkins* 5 In this smoke-sodden little town he [*sc.* Hopkins] came up against people who needed him desperately. **1849** *Chambers's Edin. Jrnl.* 24 Nov. 327/2 Smoke-stained walls. **1965** J. A. Michener *Source* (1966) 71 Along the smoke-stained walls hung spears and clubs, animal skins drying for later use and baskets containing grain. **1951** L. MacNeice tr. *Goethe's Faust* ii. 193 Yon smoke-warmed garment.
c. *smoke-dim, -foul.*
1937 'G. Orwell' *Road to Wigan Pier* iv. 71 You walk through the smoke-dim slums of Manchester. **1888** G. M. Hopkins *Let.* 1 May (1938) 145, I..dislike any town..for its bad and smokefoul air.
11. smoke alarm, a device that automatically gives a warning of the presence of smoke; **smoke-boat** *Naut. slang*, a steamship; **smoke-bomb,** a bomb which produces a smoke-screen; **smoke bush** = *smoke plant*; **smoke candle** (see quot. 1962); **smoke canister,** a canister whose contents can be ignited to produce smoke; **smoke concert** *N.Z.*, a concert at which smoking is allowed; **smokefall** [after Nightfall] *rare*[−1], 'the moment when the wind drops and smoke that had ascended descends' (Dame Helen Gardner); **smoke goggles,** goggles that protect the eyes against smoke; **smoke grenade,** a grenade that emits a cloud of smoke on impact; **smoke-head,** (*a*) the head of a column of smoke; (*b*) *Naut.*, a funnel; **smoke helmet,** (*a*) also, a similar helmet used by others; (*b*) a form of respirator used for counteracting poison gas, etc., in the war of 1914–18; **smoke-hound** *U.S. slang*, an alcoholic who drinks smoke (see sense 7 b above); **smoke joint** [*joint *sb.* 14] *U.S. slang*, a bar selling inferior liquor; **smoke-meter,** an instrument for measuring the density or the composition of smoke; **smoke Persian,** a long-haired smoke-coloured cat (cf. sense 8 in Dict. and Suppl.); **smoke plant:** for *Rhus cotinus* substitute *Cotinus coggygria* and add further examples; **smoke point,** (*a*) the lowest temperature at which an oil or fat gives off smoke; (*b*) the height of the tallest flame with which a particular sample of kerosene will burn; **smoke-pole** *slang*, a firearm; **smoke pot,** a tin containing substances that produce smoke or a similar opaque vapour; **smoke rocket,** a rocket that emits smoke; **smoke shell** *Mil.*, a projectile that generates a dense cloud of smoke after it is fired; **smoke-shop,** restrict † to sense in Dict. and add: now *U.S.*; (*b*) a tobacconist's shop; a place where people gather to smoke and talk; (*c*) a bar, esp. one selling inferior or cheap liquor; **smoke-signal,** a column of smoke used as a signal (cf. sense 2 a in Dict.); also *transf.* and *fig.*; **smoke-stick** *slang*, = *smoke-pole* above; **smoke tunnel,** a wind tunnel into which smoke can be introduced to make the airflow visible; **smoke-wagon** *U.S. slang*, a firearm; **smoke-writing** = *sky-writing *vbl. sb.*
1936 *Discovery* Nov. 359/2 A smoke alarm apparatus for the small industrial chimney is also provided. **1977** *Chicago Tribune* 2 Oct. xii. 17/9 (Advt.), Full basement, low assoc. fee, central humidifier and smoke alarm. **1867** G. E. Clark *Seven Years of Sailor's Life* xii. 116 Capen and de missis go munyana in de big smoke boat. **1901** *Rudder* Jan. 9/2 The magnificent steam yacht Mayflower passed us close aboard. We had a fine contempt for any and all 'smoke boats', but the sweetness of her lines..compelled admiration. **1929** F. Bowen *Sea Slang* 127 *Smoke boat*, the old sailing ship man's term of contempt for a steamer. **1917** A. G. Empey *Over Top* 308 *Smoke bomb*, a shell which, in exploding, emits a dense white smoke, hiding the operations of troops. **1931** V. Bruce *Bluebird's Flight* iii. 29 Throwing over a smoke bomb, I descended on a nice hard piece of ground. **1973** 'I. Drummond' *Jaws of Watchdog* xv. 205 A little smoke-bomb. I put it through a window... Then I went in through another window. **1902** L. H. Bailey *Cycl. Amer. Hort.* IV. 1529/1 *Rhus..Cotinus*, Linn. Smoke Bush. Venice Sumach..fl[ower]s purple, in ample loose panicles. **1940** [see *partridge vine* s.v. *partridge 5]. **1977** *Weekly Times* (Melbourne) 19 Jan. 29/4 The other sample is from the smoke bush. **1932** C. Gilson *Wild Metal* iii. i. 248 We had been provided with smoke-candles; and when we had cleared the Boche out of his trenches, the wind changed and the smoke masked our own fire. **1950** *Times* 13 May 4/5 The pilot started to descend in order to round the turning point at not more than 300 ft. for recognition purposes. The point was near a golf course..where white strips had been laid out and smoke candles were fired as the Meteor approached. **1962** *Ordnance Technical Terminol.* (U.S. Army Ordnance School, AD 660 112) 282/1 *Smoke candle*, munition which produces smoke by vaporizing a smoke producing oil. **1973** 'I. Drummond' *Jaws of Watchdog* x. 135 The smoke-canisters were not of a recognized pattern..used by any NATO army. **1974** H. MacInnes *Climb to Lost World* ix. 148 We had smoke canisters all ready, in case we heard a plane. **1888** J. D. Wickham *Casual Ramblings* 42 They had a smoke concert with a Salvation Army accompaniment till a clock was 'ayont the twal'. **1935** A. Mulgan *Pilgrim's Way in N.Z.* xiv. 90 'A man should go on to the football field,' declared a representative forward at a 'smoke' concert, 'prepared to meet his God.' **1936** T. S. Eliot *Coll. Poems 1909–1935* 188 But only in time can..The moment in the draughty church at smokefall Be remembered. **1962** *Flight International* LXXXII. 487/2 Stowage provision is made for smoke goggles at each duty station. **1976** B. Jackson *Flameout* iv. 59 Fumes and smoke..surged forward into the flight deck. The crew put on smoke goggles. *a* **1944** K. Douglas *Alamein to Zem Zem* (1946) 16 Stacked round the sides of the turret were..hand-grenades, smoke grenades and machine-gun ammunition. **1980** *Globe & Laurel* July/Aug. 229/2 Twice during the night we were attacked by a small enemy force who ran through our position throwing Chinese crackers and smoke grenades. **1915** Kipling *France at War* 11 He pointed to the large deliberate smoke-heads. **1942** H. Bloomfield *Sailing to Sun* xvi. 164 There was smoke coming from the smoke-head of the *Owl*. **1906** *Royal Mag.* Feb. 338/1 A safety smoke helmet. **1915** D. O. Barnett *Let.* 10 June 171 We've got a wonderful new respirator issued, a 'smoke helmet' made of cloth..which is soaked in a solution. **1972** *Police Rev.* 17 Nov. 1491/2 Constables equipped and wearing diving or smoke helmets. **1932** *Sun* (Baltimore) 23 Nov. 20/4 If a downpour has just started, the jungles are literally emptied into the stations... They come in by the dozen.., ancient smoke hounds and middle-aged rovers. **1931** 'D. Stiff' *Milk & Honey Route* xii. 128 Mark my word, out of the muck hole that was a 'smoke joint' will rise a lily that will outdo the old-time saloon in all those old virtues. **1941** *Jrnl. Soc. Automotive Engineers* XLVIII. 188/2 The smokemeter itself is our only means of measuring smoke density precisely. **1961** *Guardian* 24 Mar. 6/3 In..a letter..regarding..the control of diesel exhaust smoke..Mr. Marples has indicated that further detailed investigation into the possibilities of using a smokemeter would be worth while. **1904** 'Saki' *Reginald* 3 You want one of her smoke Persian kittens. **1973** *Country Life* 25 Jan. 226/3 My smoke persian was an individualist like Mr. Fearon's cat. **1888** *Garden* 18 Aug. 159/1 The Venetian Sumach, Wig Tree, or Smoke Plant..is one of the most remarkable of late-flowering shrubs. **1948** N. Catchpole *Flowering Shrubs & Small Trees* 177 Smoke Plant or Burning Bush... The

SMOKE

common names relate to the fine, feathery inflorescence. **1933** *Petroleum Handbk.* x. 181 The smoke point only gives an indication of the burning quality of a kerosine immediately the lamp is lit. **1951** McMichael & Bailey in M. B. Jacobs *Chem. & Technol. of Food & Food Products* II. xxv. 1150 If a fat is to be used for frying..its smoke point, or smoking temperature, is of some importance. **1958** *Jrnl. Home Econ.* L. 778/1 An emulsifier lowers the smoke point of the fat to which it is added. **1975** E. M. Goodger *Hydrocarbon Fuels* vii. 134 Two kerosine types of burner fuel are classified as C1 and C2, with minimum smoke-points of 35 and 25 mm, respectively. **1929** M. A. Gill *Underworld Slang* 11/1 *Smoke pole*, gun. **1970** *N.Z. Listener* 21 Sept. 14/5 A long time since he'd fired the old smoke-pole, anyway. **1980** *Hunting Ann. 1981* 55/1 This requires the hunter to decide in advance whether he wants to hunt with an antique or modern... There is no going out later using a scoped rifle after getting zilched with a smokepole. **1950** *N.Z. Jrnl. Agric.* June 534/3 By taking one or two smoke pots into a poultry house and seeing where the smoke goes and what happens to it, improvements in ventilation are often suggested. **1965** 'Lauchmonen' *Old Thom's Harvest* x. 135 They all sit down near..their mosquito smoke-pots. **1978** J. Gardner *Dancing Dodo* xiii. 92 She would.. begin her let down west of Brussels, and start up the smoke pots housed in the starboard nacelles. **1891** A. Conan Doyle in *Strand Mag.* July 70/1 'It is nothing very formidable,' he said, taking a long cigar-shaped roll from his pocket. 'It is an ordinary plumber's smoke rocket, fitted with a cap at either end to make it self-lighting.' **1954** [see *ripple-fired* adj. s.v. *ripple *sb.*[3] 7]. **1964** J. S. Scott *Dict. Building* 265 *Smoke Rocket*.., a rocket which gives off a dense, lasting smoke which is directed into a drain under test. **1919** P. R. Worrall *Smoke Tactics* 27 Smoke shells may be used as a visible sign to Infantry and Tanks to mark the barrage. **1937** *Times* 16 Apr. 8/6 The howitzers used smoke shells mixed with their high explosive to give a screen effect. **1966** *McGraw-Hill Encycl. Sci. & Technol.* I. 539/1 The 81-mm and 4·2-in. mortars, capable of lobbing high-explosive or smoke shells onto enemy positions, round out the category of infantry weapons. **1937** C. Himes *Nigger* in *Black on Black* (1973) 125 Harold Price..was just leaving the house for his afternoon tonk session down at the smoke shop at 100th Street and Cedar. **1959** R. M. Dorson *Amer. Folklore* vii. 267 The enterprising folklorist need not journey into the back hills to scoop up tradition. He can set up his recording machine in the smokeshop or the union grill. **1972** J. Wambaugh *Blue Knight* (1973) i. 19, I walked down to the smoke shop. I picked up half a dozen fifty-cent cigars. **1977** *New Yorker* 27 June 31/1 Send out to the smoke shop for three cartons of straw-tipped Melachrinos. **1873** S. W. Cozzens *Marvellous Country* iv. 65 After leaving the Organos Mountains we had noticed Indian smoke-signals. **1923** *Beaver* Dec. 108 Another smoke signal was seen curling upward away to the north. **1944** *Living off Land* iv. 84 A very useful mode of attracting attention is by means of smoke signals. **1962** *Amer. Speech* XXXVII. 135 *Smoke signals, n.* Sometimes trains were taken on a logging railroad without authority. The men had to keep a sharp lookout for smoke from other trains in order to get onto a side track or to back up quickly: 'We had to watch for smoke signals.' **1978** *Times* 20 Jan. 4/8 Mr Enoch Powell['s].. delphic remarks certainly got Mrs Thatcher asking herself what smoke signals he intended. **1927** *Flynn's* 22 Jan. 376/2, I ups and prods him and says, 'Hand it over, er this smokestick'll do the talking.' **1940** in S. J. Baker *Austral. Lang.* (1945) viii. 153 A rifle is a smoke-stick, a machine-gun is a death-adder. **1931** *Flight* 18 Dec. 1243/2 The type of smoke tunnel used by Mr. Farren for his demonstrations had cost approximately £65. **1964** P. Bradshaw *Exper. Fluid Mech.* vi. 151 Smoke tunnels are usually of open-circuit design to prevent the accumulation of smoke in the airstream. **1975** L. J. Clancy *Aerodynamics* xiii. 365 The principal requirement of a smoke tunnel is for uniform flow with low turbulence. **1891** J. Maitland *Amer. Slang Dict.* 255 *Smoke-wagon* (Am.), a revolver. The word is used by the negroes of the Chicago levee. **1926** J. Black *You can't Win* x. 152 I'll have her buy me a pair of 'smoke wagons'. **1950** *Western Folklore* IX. 158 Familiar epithets for the revolver were equalizer, shootin' iron,..smoke wagon. **1932** *Flight* 8 July 638 The committee recommend that smoke-writing should not be prohibited or controlled.

smoke, *v.* Add: **I. 2. d.** Also const. *off.*

1961 P. White *Riders in Chariot* 415 Dubbo had gone all right. Had taken his tin box, it seemed, and smoked off.

4. (Further U.S. examples.)

1818 A. Royall *Let.* 19 Feb. (1830) 104 It's as fair chestin, says I, as I ever seed in my life; and you can make him smoke for it. **1878** J. H. Beadle *Western Wilds* xxviii. 447 The residents will make him 'smoke' with high taxes on his land.

II. 6. b. Also *fig.*, to force *out* into the open (a conspirator); to bring *out* publicly. *orig. U.S.*

1914 *Dialect Notes* IV. 164 *Smoke one out*, *v. phr.*, to find and bring from concealment. 'I'll try and smoke him out again.' **1948** *Times* 25 Feb. 5/7 Speculators were 'smoked out' by a Congressional inquiry. **1959** *Listener* 25 June 1105/2 We were using a food guide, compiled by some daring spy who was determined to smoke out tasty food if it cost him his citizenship. **1977** G. V. Higgins *Dreamland* viii. 83, I had done it to smoke them out, and had succeeded.

III. 11. (Further examples with reference to marijuana, opium, or other illegal drugs.)

1895 Conrad *Almayer's Folly* xi. 287 'And they both smoke,' added Ali. 'Phew! Opium, you mean?' Ali nodded. **1957** *Sun* (Baltimore) 12 Jan. 12/2 Asked how he took dope, Harrod replied that he 'smoked, snorted and skin-popped'. **1972** *Guardian* 29 Jan. 9/2 Mr Williams had three previous convictions for possession of cannabis.. 'I've said I smoke sometimes.' **1977** J. le Carré *Honourable Schoolboy* xv. 382 For a large Kuan secrecy was vital... The safest place to smoke would undoubtedly be upstairs.

12. a. (Further examples with marijuana as object.) Also *fig.*

1946 E. O'Neill *Iceman Cometh* i. 34 Bejees, Jimmy's started them off smoking the same hop. **1952** *Life* 23 June 21/1, I heard and saw guys who skin pop, ..smoke pot, banging and shoot up the main vein in your arm and leg. **1976** *New Yorker* 8 Mar. 48/2 Suzanne, at one time, everybody in the place had smoked pot except the lieutenant.

13. Also, to 'rag' by smoking. *U.S.*

1850 in B. H. Hall *College Words* (1851) 285, I would not have you sacrifice all these advantages for the sake of smoking future Freshmen. **1880** *Harper's Mag.* Nov. 950/2 They hazed and smoked Freshmen.

15. *trans.* To furnish with tobacco. *rare.*

1897 'Mark Twain' *Following Equator* xi. 129 He will.. feed you and take you and smoke you with the best that money can buy.

IV. 16. To shoot (a person) with a firearm. *U.S. slang.*

1926 J. Black *You can't Win* xi. 144 Git inside an' stay there or I'll smoke both of youse off. **1942** *Detective Fiction* May 13/1 You chiseling rat. You didn't figure Tommy and these here heels could hold me, did you? I smoked them just like I'm gonna smoke you, Bugs.

smoked, *ppl. a.* Add: **1. b.** *smoked sheet*, a form of raw rubber that is preserved for transportation by drying the coagulated latex in a smoky atmosphere.

1909 *Westm. Gaz.* 26 Oct. 9/2 Buyers..appeared willing to give higher prices for smoked sheet than for crêpe. **1950** *Thorpe's Dict. Appl. Chem.* (ed. 4) X. 552/1 For the preparation of smoked sheet the strained, diluted latex is poured into rectangular tanks carrying vertical slots at 1¼-in. intervals. **1972** P. W. Allen *Natural Rubber & Synthetics* iii. 69 The new grading method would free producers from the need to make rubber in those forms such as ribbed smoked sheet which had evolved around the need to fit the traditional grading procedures.

4. (Further examples: cf. quots. 1755, 1819, sense 2 in Dict.)

1910 H. James *Let.* 29 July (1981) III. 391 You all melt away in this hard Swiss light. But I have just bought a tinted (I believe they call it a 'smoked') pince nez, and I am attempting to focus you again. **1947** [see *bathy-thermograph* s.v. *bathy-]. **1964** *Lancashire Life* Apr. 14/1 The sun visors..are made of a dark red smoked plastic and slide away completely out of sight. **1979** G. Watson *Black Jack* xii. 82 A limousine with smoked-glass windows.

smoke-dried, *ppl. a.* Add: Also *fig.*

1942 V. Woolf *Death of Moth* (1942) 108 The dead weight of smoke-dried culture.

smoke-ho, -oh. Add: Chiefly *Austral., N.Z.,* and *Naut.* Also **smoke-o.** [See *-o².] (Earlier and further examples.) More generally, a tea-break, a rest period. Also, a cup of tea or a snack taken at work.

1865 A. L. Kennaway *Crusts* 114 Sawyers, and stockmen, carpenters, packers, shinglers and loafers, Smoke as they work to assist them, and then knock off for a 'smoke oh!' **1882** *Adelaide Observer* 31 Dec. 46, I must go to 'Smoke O'. **1930** V. Palmer *Passage* 147 At smoko, when they took a spell in the paddock of loading the boat, the talk was of Lew. **1938** F. Worsley *First Voyage* iii. 56 'Five minutes' smoke-oh,' was the mate's reply. **1942** S. Campion *Bonanza* i. 21 C'm on, now, cobbers, we'll go make ourselves smoke-o, eh? Nuthin' like a cuppa. **1953** A. Upfield *Murder must Wait* iii. 26 A billy of tea and a slice of brownie..a smoko..suitable for a half caste. **1954** T. Ronan *Vision Splendid* 179 If you blokes aren't coming down for your smoko I'll throw it away. **1961** G. Foulser *Seaman's Voice* ix. 136 It was then 'smokeho' (smoke time). **1961** P. White *Riders in Chariot* 436 They had come together at the smoke-oh. **1969** H. C. de Mierre *Long Voyage* i. This stevedores broke off for their mid-morning 'smoke-oh'. **1970** D. M. Davin *Not Here, Not Now* vi. i. 274 An hour off to eat and then on again from one till three. Another smoko then, and on again till five. **1972** P. Newton *Sheep Thief* ii. 27 By 'smoko' time..the three drovers had arrived in from the back. **1978** O. White *Silent Reach* viii. 85 Margan..let the big truck roll to a standstill. 'Smoke-oh,' he announced.

2. *a smoke concert* s.v. *smoke *sb.* 11.

1908 C. G. D. Roberts Three Heroes ... The State Governor was present, and it was a rare incident for that distinguished party to grace an Art Society 'Smoke'. **1957** D. Niland *Call me when Cross turns Over* vii. 170 In chopped wood for hotels and boarding-houses, and sometimes was roped in as an entertainer at smokos and concerts. **1976** *Australian* 24 Apr. 18 The Leader of the Opposition, Mr Whitlam, and ..in the Sydney office and attended a 'smoko' at Wentworthville RSL club last night.

smoke-house. Add: **3.** Chiefly *N. Amer.* (Earlier examples.)

1746 in *Lower Norfolk Co. Virginia Antiquary* (1896) I. iv. 111 ..bequeath to my wife Mary the free use & occupation of my dwelling house..with the Kitching, Store house Smoke house, Hen house [etc.]. **1841** A. Langton *Let.* 29 May (1904) 282 We intend putting up a smoke-house immediately, as this is the best way of keeping hung meat.

smoke-in (smōᵘ·k,in). [*-in².] A gathering for the purpose of smoking or otherwise inhaling cannabis.

1968 *Courier-Mail* (Brisbane) 14 Sept. 7/9 Hashish burnt in a teaspoon held over a gas stove had been inhaled through a metal tube during a 'smoke-in' at a Chelsea flat. **1972** *Guardian* 1 July 1/5 The yippies and the Street People staged a 'smoke-in'.

smo·ke-jump, *v. N. Amer.* [f. Smoke *sb.* + Jump *v.*] To jump by parachute from an aircraft, in order to extinguish a forest fire. Chiefly as *vbl. sb.*

1942 *Fire Control Notes* VI. 95 Parachute smoke jumping..has proved, according to all reports, that such a method of attack on small fires is practical. **1949** *Amer. Forests* Oct. 18/3 Smoke jumping is a hazardous occupation. **1958** *Amer. Speech* XXXIII. 26 *Smokejumping and the verb *to smokejump* appear frequently in newspapers. **1976** *Billings* (Montana) *Gaz.* 21 July 1-6/4 As the blaze spread Friday, the state forestry agency supplied forces, and Missoula sent in two smoke-jumping crews by bus totaling about 40 men.

So **smo·ke-jumper** a forest-fire fighter who arrives by parachute.

1940 *Sci. Amer.* Feb. 97/3 These experiments have proved entirely successful and the 'smoke jumpers' will be of inestimable value in preserving our forests. **1956** Peterson & Fisher *Wild America* xxi. 237 Fires are spotted and fought while they are still small. 'Smoke jumpers' parachute from planes to fires in the roadless back country. **1979** *Arizona Daily Star* 5 Aug. A 10/3 The lone survivor..was pulled from the wreckage by smoke jumpers called in to fight a small forest fire touched off by the crash.

smokeless, *a.* Add: **1.** (Further examples.)

1904 *Sci. Amer.* 4 June 446/3 *Smokeless fuel*...Mr. Weeple employs a simple method of treating such carbonaceous substances as bituminous coal, coal-dust, oil residue, and the like as will produce a fuel that will burn free from 'black smoke' during combustion. **1932** *Economist* 19 June 1479/1 The main product of low-temperature carbonization is smokeless fuel. **1977** M. Russell *Mr T* xvii. 145 A fireplace..in which smokeless fuel glowered.

2. *smokeless zone*, a district in which the creation of smoke is forbidden by law.

1953 *Interim Rep. Comm. Air Pollution* 27 in *Parl. Papers* (Cmd. 9011) VIII. 655 The 'smokeless zone' provision of some local Acts..apply to domestic as well as industrial smoke. **1969** A. H. Lindop *Sight Unseen* iv. 34 In spite of the fact that we were a smokeless zone I had a comfortable coal fire. **1976** *Daily Record* (Glasgow) 4 Dec. 12/3 It came at a public meeting after Stirling District Council had tried to get a smokeless zone order on the village quashed.

smoker. Add: **2. b.** For *Obs.*⁻¹ read *Obs. rare* and add further example.

1840 H. A. Wise *Los Gringos* xiv. 340, I..took passage in one of them smokers, bigger than a three-decker.

e. A motor vehicle or engine that emits excessive exhaust fumes (see also quot. 1951). *colloq.*

1951 *Amer. Speech* XXVI. 309/1 *Smoker, n.*, a Diesel-motored truck. **1963** *Daily Tel.* 18 Aug. 13/4 Roadside checks..have resulted in about one diesel lorry in eight being termed a 'smoker' because it is making too much exhaust. **1976** *Globe & Mail* (Toronto) 30 Jan. 1/6 The Ontario Environment Ministry has laid its first pollution charge against the driver of a 'smoker'—a car emitting dense smoke.

3. *smoker's cough*, a cough caused by excessive smoking.

[**1907** B. M. Croker *Company's Servant* i. 7 Dye hear the cough of him? That's the real Ganja smoker's cough.] **1927** F. Harris *My Life & Loves* III. xii. 178 He smoked incessantly though the cigarettes plagued him with smoker's cough. **1942** R.A.F. *Jrnl.* 13 June 17 There was one case of a heavy smoker being sent back to his unit. He had a smoker's cough which betrayed him on night exercises. **1962** *Guardian* 14 Apr. 3/1 His smoker's cough, his overdraft anxiety, his impending divorce. **1967** E. Taylor *Second Thoughts* 14 ... the old truck burst into life drowning out..the hacking smoker's cough of its owner.

4. b. (Earlier and later examples.)

1882 *Referee* 9 Jan. 6/3 (Advt.), East Hill Smoking Concert Club.—The first smoker for the season will be given at the East Hill Hotel, Wandsworth, on Thursday, at 8 o'clock. **1894** W. T. Vincent *Recoll. Fred Leslie* I. xvi. 25 Come down to our concert, A Smoker 'tis called. **1939** Joyce *Finnegans Wake* 433 Spring trip..rachet to ..the Pontiff's Smoker ... **1955** D. M. Davin ..—smoker.. **1961** E. Williams *George* xx. 319 He was..in the Oods and last term leading lady in the 'smoker', Oxford for smoking-concert. **1976** W. Goldman *Magic* ii. 79 Merlin..brought him along to an Elks' smoker.

c. *U.S.* A social gathering of men, sometimes with organized entertainment.

1890 *N.Y. Jrnl.* 7 Sept. 11/8 The smoker at the Waldorf-Astoria for the sailors of the Olympic. **1921** H. S. Harrison *Queed* xvi. 206 After the bouts or the 'smoker' of a Saturday, there was always a smoker. **1956** E. N. Rogers *Queenie's Brood* 42 A smoker was scheduled frequently at which boxing bouts were featured, or a pie race, a wrestling match, etc. **1969** A. R. Bosworth *My Love Affair with Navy* xii. 168 Both the tin cans and the subs have long been famed for the smokers they hold ashore.

smoke-room. Add: Also *attrib.*

1906 *Nature* 17 May 53/2 The book is full of smoke-room gossip and snatches of sailors' songs. **1927** [see *lounge bar* s.v. *lounge *sb.* 4]. **1945** *Auden Coll. Poetry* 134 That caged rocketed question the Occasionally let out at clambakes or College reunions, and which the smoke-room story Alone, ironically enough, stands up for.

smokery. 3. b. = Smoke-house 3. *rare.*

1794 T. Cooper *America* 132 His smokery for bacon, hams, &c. is a room about twelve feet square. **1961** N. Froud et al. tr. *Montagné's Larousse Gastronomique*

494/1 In modern smokeries, the racks are fitted into the chimney, thus avoiding a great deal of handling.

smoke-screen (smōᵘ·k,skrīn). Also smoke screen, smokescreen. [f. Smoke *sb.* + Screen *sb.*¹] **1.** A screen of smoke, *spec.* one produced to conceal military or naval forces or operations, or a stretch of land or sea.

1915 F. A. Talbot *Aeroplanes* 172 The 'smoke screen', ..accepted and extensively practised ruse in naval strategy, and..now adopted by its mosquito colleagues of the air. **1937** [see *flash *a.* 4]. **1946** K. Douglas *Alamein to Zem Zem* (1946) xiii. 107 A straight path would take me behind the smoke-screen rising and slanting from the carrier. **1977** O. Jacks *Autumn Heroes* xiv. 200 The smoke screen was breaking up patchily to reveal..charred bodies.

2. *fig.* Something designed to conceal or mislead; a deliberate distraction or diversion. Also *attrib.*

1926 R. Macaulay *Crewe Train* II. iv. 107 The winds, doubtless, were a smoke-screen put up to conceal an Anglo-French naval accord. **1928** *Manch. Guardian Weekly* 7 Sept. 184/3 The diplomatic correspondents'..are putting up a smoke-screen of excuses for the Anglo-French naval accord. **1935** A. Kennedy *Current Eng.* xiii. 567 'Orientological sesquipedalianism'..has been much employed of late as a form of smoke-screen writing intended to assist advertisers. **1943** H. Read *Politics of Unpolitical* ii. 13 The incursions of democracy..are always accompanied by a smoke-screen of righteousness which hides their real nature and dimensions. **1954** *Encounter* Mar. 73/2 Behind the 'scientific' smoke-screen of statistical tables, graphs, codes, and rebarbative language there is a continuous propaganda for more, and more varied, sexual 'outlets' as physiologically good in themselves. **1973** 'H. Carmichael' *Too Late for Tears* viii. 109 Telling you about his other women was just a smokescreen.

Hence **smo·ke-screen** *v. trans.* (*a*) to deceive by a smoke-screen; (*b*) to conceal or divert attention from by a smoke-screen; **smo·ke-screened** *ppl. a.*, hidden by smoke; **smo·ke-screening** *vbl. sb.*, concealment by smoke-screen, the use of smoke-screens.

1922 O. Parkes *Ships of R. Navy* 12 The sphere of usefulness could be extended to include..smoke screening. **1928** *Nature* 7 Feb. 194/2 Advisory duties at H.Q. Bomber Command..were undertaken. **1954** ... **1950** J. D. MacDonald *Brass Cupcake* ix. 82 Don't let her smoke-screen you, Chief. **1958** *This Week Mag.* 18 May 35/1 Earlier that day, John Foster Dulles had fully smoke-screened the operation by issuing a statement about summit talks with Russia. **1963** M. Allingham *China Governess* iii. 47 Some silly little bit..had got him into a scandal which had to be smoke-screened. **1971** N. Freeling *Over High Side* ii. 106 He lit it [*sc.* a cigar]. Smoke-screened, he looked back. **1979** J. S. Black *Journey to Safe Place* xix. 252 You made a balls of it.. Nothing's going to smoke-screen that.

smoke-stack. Add: **1. a.** For *U.S.* read 'Chiefly *U.S.*' and add earlier and later examples.

1859 *Harper's Mag.* Apr. 606/1 The hoarse breath of the smoke stacks..came from the rosin-fed furnaces. **1903** [see *salt-caked* adj. s.v. *salt *sb.*¹ 8 b]. **1942** E. Paul *Narrow St.* ix. 79 A miniature steamboat that could be tilted flat when the craft passed under a bridge. **1976** *National Observer* (U.S.) 7 Aug. 7/6 Her smokestacks are lowered..so she can get under some bridges.

2. (Earlier example.)

1859 *Cairo City* (Illinois) *Gaz.* 8 July 3/1 A number of mischievous boys,..lighting a bunch of fire crackers,..threw them into the smoke stack.

smo·ke-up. *U.S. slang.* [f. Smoke *v.* + Up *adv.*] An official notice that a student's work is not up to the required standard.

1937 *Amer. Speech* II. 278/1 [Stanford Univ.] *Smoke up*, official notice of unsatisfactory work. **1960** *Indiana Daily Student* 23 Nov., Sikes says 56 p.c. of Frosh probably had one Smoke-up.

smokey, var. Smoky *a.* and *sb.* in Dict. and Suppl.

Smokey Bear (smōᵘ·ki bēᵊɹ). *U.S. slang.* Also **Smoky Bear**; in sense 2 **the Smokey Bear** and *ellipt.* as **Smok(e)y.** [f. *smokey* var. Smoky *a.* and *sb.* + Bear *sb.*¹: the name of an animal character used in U.S. fire-prevention advertising.] **1.** Used *attrib.* and *absol.* to designate a type of wide-brimmed hat.

1969 I. Kemp *Brit. G.I. in Vietnam* ii. 23 Sergeants Sullivant, McKane and Rothweiler..wore the round, soft-brimmed hats known to Americans as 'Smokey Bear' —similar to those of the Royal Canadian Mounted Police. **1974** R. M. Pirsig *Zen & Art of Motorcycle Maintenance* xiii. 142 He..puts on his Smokey Bear hat and is a Smokey Bear hat.

2. A state policeman; *collect.* state police.

1974 *Rolling Stone* 26 Sept. 86 Truckers..keep Don advised for the location of 'Smokies'..so he'll know when to gear the tour bus down from its maximum speed of 82 mph. **1975** *Courier-Mail* (Brisbane) 26 Oct. 36/2 They [*sc.* CB radios] are used by lorry drivers to warn 'Smokey down the line.' **1975** *High Times* Dec. 67/2 The cab is better outfitted..with..CB radio (for trackin' those Smokeys in the unwrapped package). **1976** Perkowski & Strål *Joy of CB* ii. 16 Truckers warned each other of the location of 'Smokey Bears' (the name state troopers were given because their hats resembled that worn by the fabled firefighter). **1976** Liebermann & Rhodes *Compl. CB Handbk* vii. 137 *Smokey the bear*, State Police. **1978** *Weekend* 20–26 Dec. 22 Long distance lorry drivers in America try to avoid smokey bears and tend to drink coffee. Or, in English, avoid state police and drink coffee. **1979** O. McNab *Horror Story* (1980) xviii. 72 That Smoky looking at us? *Ibid.* xix. 79 We've got a Smoky Bear on the side.

smokie, obs. var. Smoky *a.*, *Sc.* var. Smoky *sb.* in Dict and Suppl.

smoking, *vbl. sb.* Add: **2. b.** *ellipt.* A (railway) smoking-carriage or compartment.

1880–90 Kipling Let. in C. Carrington *Rudyard Kipling* (1955) vi. 140 Went home with him as far as Charing Cross in a 3rd smoking.

c. *ellipt.* A smoking-jacket; also (chiefly as a gallicism), a dinner jacket (see *dinner *sb.* 2).

1922 M. Arlen *Piracy* II. ix. 127 He put on a dress-suit.. It suited Argentines very well, *le smoking*. But Englishmen were made of sterner stuff. **1934** S. Beckett *More Pricks than Kicks* 77 Cinched beyond reproach in the double-breasted smoking. **1960** R. St. John *Foreign Correspondent* viii. 142, I will make you a 'smoking' of fine English material for twenty thousand lei. **1977** T. Sharpe *Just Desserts* vii. 171 Guests wore tuxedos if they were American males and black dinner jackets if..European..the odd Italian in tobacco brown 'smoking'.

5. *smoking box* [Box *sb.*² III], *-cap* (earlier example), *car*, *carriage* (earlier example), *compartment*, *-lamp*, *lounge*, *-suit*; **smoking machine**, a device which draws air through a cigarette, etc., so that the smoke may be used for scientific study; **smoking weed** (*a*) = Bearberry a; cf. Kinnikinic 2; (*b*) = *cannabis 1; cf. *marijuana, *marihuana.

1841 Dickens *Old Curiosity Shop* lxxiii. 220 A little summer-house in its garden a smoking-box. **1841** J. Romilly *Diary* 9 Mar. (1967) 211 Dined with Bayne: he..wore an embroidered smoking cap. **1846** in Wallis *Railroad Lat.* 380/3 Smoking Cars—..we mean cars expressly provided for the lovers of the 'weed'. **1931** W. Faulkner *Sanctuary* xix. 202 Horace..went forward into the smoking car. It was full too. **1958** 'E. McBain' *Killer's Payoff* (1960) vi. 59 She boarded the train and went into the smoking car. **1888** *Amer. Humorist* (London) 5 May 1/1 Came over from New York..in the smoking compartment of a parlor car. **1862** J. Simmons *Railway Traveller's Handy Book* 83 Some lines have smoking carriages provided. **1889** *Cent. Dict.*, ..a lamp hung up on board of a man-of-war..where smoking is permitted, for the men to light their pipes by. **1940** D. Potter *Sailing Sulu Sea* 58 Smoking was prohibited except when the smoking-lamp—a name now almost forgotten—was authorized to be lighted. **1966** I. V. Noel *Naval Terms Dict.* 304 *Smoking lamp*: If 'lighted', it means that smoking is permitted; if out, it means smoking prohibited. **1951** E. Paul *Springtime in Paris* iii. 45 He could relax like a tomcat, in an easy chair in the smoking lounge. **1968** *Globe & Mail* (Toronto) 17 Feb. 6/2 Why should the teachers have a smoking lounge and not the students? **1953** *Life* 21 Dec. 20/1 [caption] Smoking machine puffs on 60 cigarets in front of Dr. Evarts Graham, who, with his collaborator in the mouse cancer research, Dr. Ernest Wynder, perfected the robot. **1963** *Times* 22 Apr. 19/3 A 'smoking machine' is used to demonstrate the amount of chemicals inhaled, including those which cause cancer. **1971** *Nature* 26 Nov. 227/1 Lung explants..were..exposed in a Filtrona (Type SM 2) smoking machine to puffs of fresh cigarette smoke. **1898** M. Beerbohm *Let.* 12 Jan. (1964) 127, I have..a smoking-suit of purple silk, with dark red facings. **1968** *Listener* 11 Aug. 261/2 Did you know..that women wore smoking suits in the 'twenties? **1857** J. Hector *Jrnl.* 31 Oct. in *Capt. Palliser's Exploration in Brit. N. Amer.* 63 in *Parl. Papers* 1863 XXXIX. 441 A gravelly soil supports a low growth of grass, but in some parts covered with a dense matting of the smoking weed.., the bright red berries of which afford food for large coveys of the prairie hen. **1957** C. MacInnes *City of Spades* I. v. 45 Mr. Peter Hazlitt placed it was in flower-pots inside there... 'It's smoking weed,' I said.

c. *smoking-concert* (later examples).

1934 T. S. Eliot *Rock* ii. 40 Dance 'alls, picture palaces, swimmin' baths, smokin' concerts, restaurants. **1945** *Penguin New Writing* XXIV. 64 Anyone who could perform at a private party or a smoking concert was sought. **1973** *Sunday Nation* (Nairobi) 11 Apr. 42/5 The smoking concert will be held on Sunday evening and not on the Saturday as I stated.

d. In other senses, as *smoking-party* [Party *sb.* 8]; **smoking point, temperature** = *smoke point* s.v. *smoke *sb.* 11.

1898 'Mark Twain' in *Century Mag.* Nov. 100/1 This smoking-party had been gathered together partly for business. **1903** Kipling *Irish Guards in Great War* I. 206 Leave was possible; smoking-parties made themselves in this period. **1915** *Jrnl. Home Econ.* VII. 538 For each case the addition of acid resulted in the lowering of the smoking point. **1931** *Industr. & Engin. Chem.* (Analyt. Ed.) 15 Oct. 348/1 Take the temperature at which the first wisp of smoke is seen rising from the top of the flask as the smoking point. **1915** *Jrnl. Home Econ.* VII. 535 The smoking temperature of a fat may be defined as the temperature at which the fat gives off visible fumes. **1945** *ABC of Cookery* (Ministry of Food) xii. 46 A good frying fat is one that can be heated to a high temperature.. before it smokes and burns, that is, it has a high smoking temperature. **1951** Smoking temperature [see *smoke point* s.v. *smoke *sb.* 11].

smoking, *ppl. a.* Add: **1.** fig. *smoking gun, pistol* (U.S.), a piece of incontrovertible incriminating evidence.

1974 *New Yorker* 21 Oct. 135/1 Some are still searching for what has come to be termed 'the murder weapon'—or 'the smoking gun'—the definitive piece of evidence that the President committed a crime. **1975** *Collier's Year Bk.* 10/2 After the new transcripts were disclosed..members of Congress abandoned Nixon in droves. 'I guess we have found the smoking pistol, haven't we?' asked Representative Barber Conable. **1976** Woodward & Bernstein *Final Days* 269 Buzhardt felt that here was a potential smoking gun. *Ibid.* 271 He had heard the President approve the plan, he had heard him suggest the exact wording. Buzhardt had found the 'smoking pistol'. He had heard the President load it, aim and fire. **1977** *Time* 19 Sept. 24/2 In fact, there may well be no 'smoking gun' —no incontrovertible, black-and-white evidence of wrongdoing by Lance. **1979** *N.Y. Times* 12 Jan. D14 We haven't got a smoking pistol. Unfortunately, everyone is zeroing in on this as a cause, but the case isn't that strong.

smoky, *a.* and *sb.* Add: **smokie** (also *Sc.* as *sb.*), **smokey** (common in *U.S.*). **A.** *adj.* **3. c.** Foggy, misty. Now *rare* exc. in proper names. *U.S.*

1769 in *Essex Inst. Hist. Coll.* (1877) XIV. 262 This week much smoky. **1824** J. Doddridge *Notes on Virginia* xxxi. 266 The smokey time commenced, and lasted for a considerable number of days. **1825** J. Neal *Bro. Jonathan* I. 105 See'd him jess now, comin' over the smoky mountain there. **1971** *N.Y. Times Encycl. Almanac 1971* 243/1 In East Tennessee are the Great Smoky and Cumberland Mountains of the Appalachian range.

5. (Further examples.)

1951 E. David *French Country Cooking* 26 The smoky wines of Pouilly-sur-Loire. **1978** *Sunday Times* (Colour Suppl.) 19 Feb. 17/3 *Smoky*, a delicate aroma that is found in several white wines, often originating from the volcanic soil, and also on the bouquet of Madeira as a result of the latter's special *estufado* heat treatment.

b. *fig.* Of the sound of a musical instrument or voice.

1958 G. Boatfield in P. Gammond *Decca Bk. Jazz* xxiv. 312 Noone's deceptively easy clarinet and Kelly's smoky trumpet are noteworthy. **1966** *Cavalier Daily* (Univ. of Virginia) 11 Nov. 1 The smooth, smoky sound of the Platters combined with their expressive hand jive will entertain students from 9 to 1 Friday night.

7. (Further examples.)

1869 [see *Kentish *a.* 3 b]. **1934** Webster, *Smoky-blue*. **1934** *Discovery* June 166/2 Plain burnished red ware and smoky grey pottery. **1974** *Men's Wear* 29 Aug. 17/2 Smokey-navy motifs on beige. **1976** H. Tracy *Death in Reserve* xi. 87 A smoky-blue spring evening. **1980** *New Age* (U.S.) Oct. 58 (Advt.), Danish Souperbag..In wilderness rust brown, battleship grey, parrot green, smokey black.

12. *smoky-tasting, -voiced, -winged.*

1925 H. Crane *Let.* 28 Feb. (1965) 199 Delicious smoky tasting sardines. **1973** J. J. McKelvey *Man against Tsetse* ii. 67 He did, however, add entomology to his accomplishments by studying the life cycle of the dark-eyed, smoky-winged tsetse that was causing nagana in Zululand. **1976** *New Yorker* 29 Mar. 6/3 A promising, smoky-voiced jazz and rhythm-and-blues singer.

B. *sb.* **2.** (Further examples.)

1948 R. de Kerchove *International Maritime Dict.* 690/2 *Scotch haddie* (*U.S.*)... In Great Britain called smokie. **1965** *Arbroath Guide* 3 Apr. 4 The older generation argue that the old time 'smokie' is a haddock freshly caught by line and smoked in the Arbroath way. **1974** *Sunday Tel.* 23 June 15/5 Let us make the distinction between the genuine Arbroath smokie, which gets its colour from the actual smoke when the fish is cured, and the common kipper. **1976** *Daily Record* (Glasgow) 29 Nov., Arbroath's famous 'smokie' industry will be hit by the ban on catching haddock in the North Sea.

4. See *Smokey Bear.

smole (smōᵘl), joc. var. of Smile *sb.*¹ or *v.* or *smiled* (pa. t. of Smile *v.*). Now *rare* or *Obs.*

1858 J. C. Thomson *Almae Matres* i. 5 Tick, Esquire, rose at our entrance, smole blandly, and mumbled something to the effect that he was glad..to make our acquaintance. **1894** 'Mark Twain' in *St. Nicholas* Mar. 400/2 Then he smole a smile that spread around and covered the whole Sahara. **1909** J. R. Ware *Passing Eng.* 227/1 *Smole*.., a grotesque variation of smile. **1937** Partridge *Dict. Slang* 789/1 (He) smoled a smile (or smole).

smolyaninovite (smǫlyănī·nǒvait). *Min.* Also **smolia-**. [ad. Russ. *smolyaninovit* (L. K. Yakhontova 1956, in *Doklady Akad. Nauk SSSR* CIX. 849), f. the name of N. A. *Smolyaninov*: see -ite¹.] A hydrated arsenate of iron, cobalt, nickel, and other metals found as a yellow oxidation product of cobalt and nickel ores.

1957 *Chem. Abstr.* LI. 4885 (*heading*) Smolyaninovite, a new mineral. **1977** *Mineral. Mag.* XLI. 388 A specimen purchased by the [British] Museum in 1927..from Schneeberg, Saxony, has been found to carry small amounts of smolyaninovite, constituting a third occurrence of the mineral. **1981** K. Frye *Encyl. Mineral.* 715/2 *Smolianinovite*, orth.

smon (smǫn). *Path.* Also **SMON.** [Acronym f. the initial letters of *s*ubacute *m*yelo-*o*ptico-

*n*europathy.] A disease of the nervous system characterized by recurrent motor, sensory, and visual symptoms, freq. including numbness of the legs.

1971 *Lancet* 3 Apr. 697/1 The Japanese S.M.O.N. syndrome occurs in patients who have not taken clioquinol or other hydroxyquinolines. **1977** *Arab Times* 31 Oct. 2/3 In June last year, the two firms admitted that the drug quinoform, also called chinoform, had 'a causal relationship' with a disease of the nervous system called smon (subacute myelo-optico-neuropathy), which affects limb movement and can cause blindness. **1979** *Guardian Weekly* 1 Apr. 13/4 Smon (subacute myelo-optico-neuropathy) outbreaks in several communities [in Japan] occurred as early as the 1950s. **1982** *Lancet* 28 Sept. 716/2 They stated that rise and fall in SMON was synchronous with consumption of clioquinol... This is a continuing controversy.

smooch, *sb.*¹ (Earlier example.)

1823 J. Neal *Bro. Jonathan* II. 424 Cowhide shoes—newly greased..which left a 'smooch' upon whatever they came near.

smooch (smūtʃ), *sb.*² orig. and chiefly *U.S.* [f. *Smooch *v.*² or *v.* Smouch *sb.*¹] **a.** A kiss; a fondling embrace or caress, a cuddle. Also, slow, close dancing; (music suitable for) a dance of this nature. Freq. *attrib.*

1942 Berrey & Van den Bark *Amer. Thes. Slang* §147/4 *Smooch*, a kiss. *Ibid.* §830/5 *Smoochbuggy*,..an automobile used for 'necking'. **1945** *Tacoma* (Washington) *News Tribune* 27 Oct. 4/3 I'd rather have hootch And a bit of a smooch—The air cops will always do me. **1957** *Times* 29 Apr. 28/3 Ethel Merman and Fernando Lamas..found that their nightly onstage smooch grated too harshly on their star-crossed sensibilities. **1971** *New Scientist* 24 June 730/1 Two Antipodean couples joined their lips in a long-term kiss aimed at beating the existing smooch record. **1973** J. Wainwright *Pride of Pigs* 43 The smooch classics—*Mood Indigo, Lazy River, Georgia On My Mind*—slow and draggy. **1977** *Record Mirror* 16 Apr. 27/5 Perfect even-tempoed bland MOR until it slows up for a smooch.

smooch (smūtʃ), *v.*¹ *dial.* and *colloq.* [App. an altered form of Mooch, Mouch *v.*] **1.** *intr.* To sneak, creep; to wander or prowl *round* (somewhere).

1904 in *Eng. Dial. Dict.* **1950** R. Moore *Candlemas Bay* 213 Then he realized his mother would probably send him back for the dish, so he smooched glumly in to retrieve it. **1960** J. Jefferies *Dignity & Purity* iv. 76 'What are your plans?' 'I'm going to smooch round here, if that's all right.'

2. *trans.* To steal.

1941 J. M. Cain *Mildred Pierce* xi. 229 Then she..went over to the cash box, and smooched four $10 bills.

smooch (smūtʃ), *v.*² orig. *U.S.* [Var. Smouch *v.*²] *intr.* To kiss; to neck or pet; *spec.* while dancing to a lazy, romantic melody.

1932 *Amer. Speech* VII. 336 *Smooch*,..to kiss. **1937** *Sat. Even. Post* 20 Feb. 69/2 Once upon a time you 'spooned',..but now you may 'smooch' or 'perch'. **1952** K. Vonnegut *Player Piano*... **1959** ... **1964** L. Nkosi *Rhythm of Violence* 4 Mary and Gama are..smooching and sharing a cigarette... **1972** 'M. Yorke' *Silent Witness* ii. 39 'I prefer to smooch to subtle melodies,' said Patrick, clasping her closely.

Hence **smoo·ching** *vbl. sb.* and *ppl. a.*

1941 J. Smiley *Hash House Lingo* 51 *Smooching*, employee making love to one of the opposite sex while on duty. **1951** S. J. Perelman in *New Yorker* 20 Oct. 29/2 No parenthetical smooching is going to upset her applecart. **1964** John o' London's 16 Aug. 163/3 When Miss Baxter pokes her head through the window of a smooching couple's car..it's merely funny. **1977** *Time* 8 Aug. 30/2 Stapleton..kissing..and a lot of smooching..affectionate smooching from the guest of honor. **1978** H. Jobson *To Tie a Little* iii. 52 He moved on to the smooching.

smoocher (smū·tʃəɹ). orig. *U.S.* [f. *Smooch *v.*² + -er¹.] **a.** One who smooches. **b.** A song or piece of music suitable for accompanying slow, close dancing.

1946 *Sun* (Baltimore) 18 May 2/6 Seeking out 'necking spots' to catch 'smoochers'. **1976** *Record Mirror* 3 Apr. 19/5 Ella's dreamy reading of the 'How strange' the change from major to minor' tune is of course an ace smoocher at any time.

smoochy (smū·tʃi), *a.* [f. *Smooch *v.*² or *sb.*² + -y¹.] Amorous, sexy; *spec.* of music: suitable for accompanying slow, close dancing.

1966 A. E. Lindop *I start Counting* xvi. 190 Cooings and mewings and smoochy murmurings. **1971** *Women's Own* 27 Mar. 16/2 His heart was broken. **1976** The next smoochy number was played. **1976** G. Sims *End of Web* xii. 120 Dated, smoochy music was being played. **1980** I. Watson *Gardens of Delight* xvii. 111 Partnering the wench in a body-rubbing, smoochy glide.

Hence **smoo·chily** *adv.*, **smoo·chiness.**

1970 'D. Halliday' *Dolly & Nanny Bird* xi. 138 Three or four couples were dancing smoochily at the end of the room. **1977** E. W. Hildick *Loop* xi. 69 I'm still not sure whether today's callous outspokenness isn't preferable to ..sly smoochiness.

smoodge, smooge (smūdʒ), *v. Austral.* and *N.Z. colloq.* [Prob. var. Smudge *v.*²: see *Smudge v.*¹ in *Eng. Dial. Dict.*] *intr.* To act in an ingratiating or fawning manner; to display affection, to behave amorously.

1910 C. Dennis *Fairy Tale* 'And' v. 34 He would smooge to the when the boss wasn't about, 'n he said we could have a good little time on the lonesome. **1916** C. J. Dennis *Songs Sentimental Bloke* 19 An' there they smooge a treat, wiv pretty words Like two love-birds. **1936** M. Franklin *All that Swagger* xiii. 395 They smoodge around Roger, but they order me about like a rouseabout. **1947** ... **1977** N. Kirkwood ... all were ...

So **smoo·dger**, a flatterer, a sycophant; **smoo·dging** *ppl. a.* and *vbl. sb.*

1898 *Bulletin* (Sydney) 30 July 32/3 Another undesirable specimen is the 'crawler'. Always carrying yarns to the boss about the other men... He is first cousin to the 'smoodger', who is only superficially a white man. **1899** *Ibid.* 7 Jan. 15/1 Lawson..feels for the wretch who is not faking, and is not kept on by servile, 'smooging' station lifers. **1918** T. Skinhill *Soldier Boy*... **1946** F. D. Davison *Woman at Mill* 128 He was a mean customer,...a petty bureaucrat, and a smooger. **1959** D. M. Davin *In Landfall* VII. 20 He would be putting his arms around her, 'Smooging'—tell her to cut it off,' she said. **1958** R. Stow *To Islands* i. 19 'Smooging,' she sighed lovingly, hiding her face against Helen's neck. 'You old smooger,' Helen said. **1963** B. Pearson *Coal Flat* xvii. 304 You do your smooging somewhere else.

smooth, *a.* Add: **I. d.** In tennis, squash, etc., of one of the two sides of the racket (see quot. 1901): used as a call when the racket is spun to decide the right to serve first or to choose ends. Opp. *rough *a.* 1 d.

1890 J. Marshall in *Tennis, Rackets, Fives* 16 *Smooth*, the front of the racket, which shows no knots. *Spin*, the decision by a racket, thrown spinning up into the air by one player, while the other calls 'rough' or 'smooth'. **1901** *Encycl. Sport* II. 521/2 *Smooth side of racket*, the side from which the twisted gut does not project. **1911** [see *rough *a.* 1 d]. **1961** *Times* 1 July 11/5 The 'rough or smooth' confusion was cleared up. **1969** C. M. Jones *Your Book of Tennis* 11..the racket's..**1973** ...'Rough.' 'Rough.'

I*. Specialized uses in the sciences. **a.** *Anat.* Applied to those muscles of vertebrates that are neither skeletal (sense *b) nor cardiac, such as those forming the gut wall, being capable of sustained but not rapid contraction and generally not under voluntary control; also to the non-striated muscle of invertebrates.

1866 Busk & Huxley tr. *Kölliker's Man. Human Microsc. Anat.* I. xxxiv. 112 In the areola of the nipple, the smooth muscles are especially well developed in the female, are disposed circularly in a delicate layer. **1866** [see *Striated ppl. a.* 1 a]. **1947** Haldane & Huxley *Animal Biol.* ii. 117 The nervous system controls striped muscles, heart muscle, smooth muscle, and glands. **1960** W. Andrew *Textbk. Compar. Histol.* viii. 335 Alternating with these elastic tissue laminae are the masses of smooth muscle fibers with some collagenous fibers. **1962** *Lancet* 8 Dec. 1192/2 There was swelling of the vessel walls with separation of the smooth-muscle fibres. **1971** N. Gratzer... **1982** ...

b. *Bacteriol.* Applied to a bacterial phenotype characterized by smooth-looking colonies of regular outline, and by cells having polysaccharide capsules.

[**1920:** see *S 4 a]. **1921** [see *rough *a.* 1 e]. **1947** *Ann. Rev. Microbiol.* I. 15 The sharply distinct antigenic pattern observed in the smooth colony of encapsulated organisms is not preserved in the rough colony of unencapsulated variants. **1973** Klinger & Geis *Agents of Bacterial Dis.* I. 9 Smooth (S) colonies are convex, round, and slimy and are usually regarded as the 'normal' form.

c. Of a graph, function, or distribution: having no breaks, discontinuities, or irregularities.

1929 *Jrnl. du Conseil* IV. 221 The result was a smooth unimodal curve but very skew. **1933** *Econometrica* I. 243 If the values of a variable extend over a wide range, there will be little likelihood that the distribution will be smooth and unimodal. **1946** M. G. Kendall *Adv. Theory Statistics* II. xxix. 386 The conception of a trend as a 'smooth' or 'regular' movement is equivalent to the supposition that the trend can be represented, at least locally, by a smooth mathematical function. **1959** *Listener* 2 July 14/1 The light-curve is not entirely smooth, as the increase to maximum is steeper than the subsequent drop. **1962** A. Nisbett *Technique Sound Studio* iv. 82 Microphones with a smooth response in the upper middle frequency range have come into general use. **1966** *Rep. Comm. Inquiry Univ. Oxf.* II. 400 Scales A and B bring the total college and university stipend to the same level as far as possible with a smooth college scale.

6. Superior, excellent, 'classy'; clever, 'neat'. *colloq.* (orig. *U.S.*).

1893 W. K. Post *Harvard Stories* 210 'Well, you'll have a rattling good time down there.' 'A smooth time, you mean,' corrected Eastleton. **1909** *Ade Fables for Today* ... The Benevolent Lady..derived much Joy from the Knowledge that People..were remarking..'Say, ain't we the Smooth Article?' **1924** Wodehouse *Bill the Conqueror* iii. 33 How did you come to think of this stunt?'..It was the smoothest idea. **1928** R. W. ... **1942** ...

b. Of manners, dress, etc.: stylish, suave, chic. *colloq.*

Sometimes indistinguishable from senses 6 b and *c.

1922 Wodehouse *Jill the Reckless* xix. 285 'What charming manners Major Selby has.' 'Smooth.' **1925** 'Smooth, said Mr Pazzi..**1955** ...

12. a. *smooth breathing*: see Breathing *vbl. sb.*; *smooth mouth*, the worn teeth without cusps found in horses more than seven or eight years old; so *smooth-mouthed* adj.

1746 *Smooth breathing* [see Breathing *vbl. sb.* 9]. **1888** Kiss & Cushion *Sound & Dullen Gr. of Horse* 51 The stage of 'smooth mouth'... **1974** H. S. Thomas *Horses* ix. 185 At age nine the cups are gone from the corner incisors. The horse is said to be smooth-mouthed.

b. *smooth-head*, a deep-sea fish belonging to the family Alepocephalidæ, resembling a herring with a larger body and dark-coloured skin.

1931 J. R. Norman *Hist. Fishes* viii. 150 A species of Smooth-head (*Leptoderma*) captured in the Bay of Bengal has been described as having the skin covered all over with a thick, opalescent, and uniformly luminous epidermis. **1969** A. Wheeler *Fishes Brit. Isles & N.-W. Europe* 115 Smooth-heads..are deep-water fishes, relatives of the herring family. **1975** *Times* 5 Dec. 12/3 Smooth-head is abundant, but..its flesh has the consistency of custard.

14. a. *smooth-cheeked* (later example), *-fleshed, -grained* (later example), *-lined, -lipped, -paced* (later example), *-perfumed, -plumed, -surfaced.*

1927 V. Woolf in *Nation & Athenæum* Aug. 661/2 One of those smooth-cheeked, steady-eyed men. **1923** D. H. Lawrence *Birds, Beasts & Flowers* 41 Fig-trees, weird fig-trees Made of thick, smooth silver. Thick, smooth-fleshed silver. **1982** *Times* 10 May 16/1 Viennese singers are apparently smooth-grained in comparison with their Italian colleagues. **1930** W. B. Yeats *Wild Apples* 23 And lined and sand-grained And all are fair. As that smooth-lined up-tilted boat From which the Foam-Born Queen ... **1868** Marston... **1929** ... **1932** ...

smooth, *adv.* Add: **2. a.** *smooth-oiled, -rounded, -trodden, -worn.*

1935 E. Pound *Classic Anthol.* i. 49 Double trams matched, smooth-oiled reins. **1916** D. H. Lawrence *Amores* 102 A new night pouring down shall swill Us away in an utter sleep, until We are one, smooth-rounded. **1958** R. Graves *Steps* 249 There are some words carry a curse with them... Smooth-trodden, the Pentagram... **1940** A. Huxley *Leda* 14 Smooth-worn silver, polished through the years. **1922** Joyce *Ulysses* 325 Lifting your billowy flounces on the smooth-worn throne.

b. *smooth-rolling, -running, -weeping.*

1885 Hazlitt *Lib. Amorist* ... 156 Thousands of years of smooth-rolling eternity and balmy, sainted repose. **1927** 'Contact' *Airman's Outings* v. 117 The smooth-running ambulances bring broken soldiers. **1941** J. Masefield *Gautama the Enlightened* 14 The black-bright, smooth-running, typewriting machine. **1971** J. P. Anderson in Douglas & Johnson *Existential Social.* vi. 186 To make the screening interview a smooth-running interaction, the patient has to be able to talk about the topics the screening worker thinks are important. **1964** E. Sitwell *Green Song* 10 The amber blood of the smooth-weeping tree.

smooth, *v.* Add: **I. 1. c.** To transform or modify (a graph, distribution, or function) so as to make it smooth; to lessen irregularities or fluctuations in (something that can be represented by a graph). Cf. sense 11 c below.

1889 F. Galton *Natural Inheritance* vii. 100 These relations came out distinctly after I had 'smoothed' the entries. **1898** *Knowledge* 1 Oct. 233/1 Then the thirteen year series of these numbers is smoothed with averages of some length. **1934** *Brit. Jrnl. Psychol.* Oct. 249 The theoretical periodogram was..made to conform... It was therefore smoothed out by the..numbers. **1962** D. F. Shaw *Introd. Electronics* x. 203 The performance of the diode rectifier is improved by the use of a filter circuit to smooth the output. **1979** *Sci. Amer.* May 52/3 This generalized chart contains a number of stars..smoothed out in the chart representing the distribution... It does not rule out the existence of isolated patches of vigorous star formation.

II. 11. b. Also, = sense 1 a in Dict.

1962 *Listener* ... **1977** ...

c. = sense 1 c above; also, to lessen (irregularities or fluctuations) in something which can be described by a graph, esp. a time series.

1933 *Econometrica* I. 238 An elaborately weighted moving function..prevents the smoothing curve from smoothing out fluctuations. **1963** L. A. Maverick *Time Series Analysis* p. vii, In smoothing out the moving cycle ..Wardwell's moving cyclical average of changing lengths is used. **1967** *Encycl. Brit.* XXIII. 432/1 By the provision of storage facilities at the source and of the main aqueduct service reservoirs, these various fluctuations [in demand] can be smoothed out. **1962** A. Nisbett *Technique Sound Studio* i. 30 At the 'back' [of the microphone] the response is fairly flat—the effect of the pad being to smooth out the peak. **1973** *Sci. Amer.* Oct. 69/1 Tests of nuclear weapons have shown that atmospheric mixing is rapid and that irregularities in composition are smoothed out after a few years. **1978** *Daily Tel.* 6 Jan. 17 The Americans will be very reluctant to do more than smooth out fluctuations in the exchange rate.

smooth-bore. Add: **1.** (Earlier examples.)

1812 *Niles' Weekly Reg.* II. 308/1 It was the best smooth bore he ever shot with in his life. **1834** W. A. Carruthers *Kentuckian in N.Y.* I. 21 Your smooth bores waste a deal of powder and lead.

2. a. (Earlier example.)

1799 in *Deb. Congress U.S.* (1851) 7th Congress 2 Sess., App. 1402 One had a rifle, and the other a smooth-bore piece.

smoothed, *ppl. a.* Add: **2. b.** Of graphs, statistical fluctuations, etc. Cf. *smooth *v.* 1 c, 11 c.

1888 *Proc. R. Soc.* XLV. 140 These smoothed values were obtained by plotting the observed values, after transmuting them as...described into their respective Q... **1939** *Science* 17 July 91/2 Smoothed rainfall curve for the British Isles. **1933** *Econometrica* I. 242 The points of inflection should be marked in the smoothed curve. **1962** D. F. Shaw *Introd. Electronics* x. 204 The advantage of this circuit is that the smoothed output voltage has a high value and the residual ripple voltage may be reduced to a fraction of 1%. **1964** K. G. Lockyer *Introd. Critical Path Anal.* vii. 78 Clearly it is desirable to try to lift the..smoothed curve. **1969** ... if this could be completely done, then the load would be said to be 'smoothed'.

smoothen, *v.* Add: **3. b.** Also const. *back, out.*

1913 R. Kane *Good Friday to Easter Sunday* iii. 126 She may smoothen back his hair. **1936** ... smoothen out... **1945** N. Collins *London belongs to Me* ... 189 We thought that it made so me smoothened it out for him.

smoothie (smū·ði), *sb.* and *a.* *colloq.* (orig. *U.S.*). Also **smoothy.** [f. Smooth *a.* + -y⁴, -ie.] **A.** *sb.* A person who is 'smooth' (sense 6 in Dict. and Suppl.); one who is suave or stylish in conduct or appearance; *usu.* a man. Occas. with unfavourable sense: a slick but shallow or insinuating fellow, a fop.

1929 *Princeton Alumni Weekly* 24 May 981/3 *Smoothie* indicates *savoir faire*, a certain *je ne sais quoi*... **1932** ... **1936** R. Chandler *Big Sleep* xxv. 213 It might be a smoothie in the detective business trying to get a snootful of somebody else's case. **1943** Hunt & Pringle *Service Slang* 60 *Smoothie*, a lady's man... **1954** F. Frankau *Wreath for Enemy* I. i. 5 Laurent is a smoothy, and I do not see how anybody could be in love with him. **1957** *Listener* 14 Nov. 801/2 A television smoothie. (This last is now gossip-writer's English for a slick commentator or smiling interviewer.) **1958** F. Rice *April Robin Murders* v. 54 This poetic-looking smoothie makes a thing out of marrying women with money. **1960** *Spectator* 8 July 66 The usual smoothie with a pseudo-American accent, 'Swanky'. **1969** *Listener* 14 Aug. 212/1 'They think where's the money,' said the man..who was '..' .. **1971** *Cambr. Cuttings*...'Smoothies'... **1972** ... **1976** ... **1977** ...

B. *adj.* = Smooth *a.* 6 in Dict. and Suppl.

1959 J. Jefferies *Thirteen Days* iv. 54 Stern..was a real smoothie boy..like what I intended to be when I grew up. **1972** *Times* 1 Sept. 1/3 ..his ..a 'smoothie' character. **1976** N. Freeling *Dressing of Diamond* 101 Bernard.. is good at the smoothie commercial stuff.

smoothing, *vbl. sb.* Add: **1. a.** (Further examples.) Also with *out.*

1931 *Jrnl. du Conseil* IV. 228 The purpose of the smoothing is to eliminate minor fluctuations of the character of hebdomadal (monthly) cycles which are known to exist in the sardine. **1933** *Econometrica* I. 238 (*heading*) Time series of regressions by successive smoothings. **1939** *Proc. R. Soc.* A. CLXXI. 81 The smoothing out of the stress distribution becomes less and less effective, and the maximum stress at the stress peaks rises. **1955** L. D. Landau in W. Pauli *Niels Bohr* 68 The great difficulties which arise in a physical 'smoothing-out' of particles, as opposed to a purely formal 'smoothing-out'..are well known. **1971** *Nature* 13 Aug. 824/1 The output signal was..displayed..after smoothing with a 1 s time constant.

b. (Later examples.)

1953 *Harvard Univ. Summ. Ph.D. Theses* 288 Under smoothing are included all monophthongizations and monophthong-retaining effects which are due to WG [original] velar consonants. **1977** *Archivum Linguisticum* VIII. 71 It seems most probable that S[econd] [smoothing] was a rather late change, taking place after i-umlaut had occurred although almost certainly before the beginning of the.. smoothing.

2. Also designating devices for reducing ripple in electrical signals, as *smoothing capacitor, choke, circuit, filter.*

1959 *Engineering* 20 Feb. 230/2 A small smoothing capacitor in the leading amplifier prevents the output from the transmitter from..the transport of the wiper from ... **1941** *P.O. Electr. Engineers Jrnl.* XXXIV. 118/1 With the development of floating battery power supply systems in telephone exchanges and repeater stations, there is a considerable demand for smoothing chokes. **1940** *Chambers's Techn. Dict.* 780/1 Smoothing circuit. **1979** G. J. King *Audio Handbk.* iv. 96 Separate smoothing and filtering circuits are used for the supplies of the preceding stages and for the preamplifier stages of the control section. **1941** *P.O. Electr. Engineers Jrnl.* XXXIV. 118/1 The cost of the complete smoothing equipment compares favourably with that of the associated batteries.

smooth-leaved, *a.* Add: *smooth-leaved elm*, a large tree, *Ulmus carpinifolia*, native to Europe, North Africa, and western Asia, and distinguished by smooth leaves with shiny upper surfaces.

1731 P. Miller *Gardeners Dict.* s.v. Ulmus; folio glabro. The Smooth-leav'd or Witch-Elm. **1811** J. E. Smith *Eng. Bot.* XXXII. 2248 (*heading*) Smooth-leaved, or Wych Elm. **1913** Elwes & Henry *Trees Gt. Brit. & Ireland* VII. 1887 Smooth-leaved Elm... A tree, with a straight bole, and wide-spreading branches. **1971** *Country Life* 23 Dec. 1772/2 We can be pretty certain that these trees..are forms of the smooth-leaved elm.

smoo·th-talk, *v.* *colloq.* (orig. *U.S.*). [f. Smooth *a.*] *trans.* To address or persuade with bland, persuasive language. Also, to win (one's way) by smooth-talking.

1950 A. Lomax *Mister Jelly Roll* (1952) 205 If a cop stopped him, he'd smooth-talk his way right out of it just like they were relatives. **1956** *Sun* (Baltimore) 2 Mar. 32/6 A Michigan man was accused today of smooth-talking bankers..into sending him cash by posing over the telephone as a depositor. **1958** J. Wain *Contenders* 171 She was transferring herself..as a result of being smooth-talked. **1960** *Washington Post* 2 Nov. 3/2 He and his wife had been 'smooth-talked' and later threatened. **1962** 'E. Lathen' *Murder against Grain* ii. 18, I let him smooth-talk me into doing the Sloan's errands. **1979** F. Olbrich *Sweet & Deadly* viii. 98 The practised politician smooth-talked his way round an awkward question.

|| smørbrød (smǿ·rbrǿd). Also **smorbrodt.** [Norw.: cf. *smørrebrød.] A Norwegian open sandwich; also *collect.*

1933 M. Lowry *Ultramarine* ii. 96 Nevermore sit in a park in the rain..holding each other's hands, or eating smørbrød. **1958** R. Barnard *Death in Cold Climate* i. 8 He..collected on a plate a ham smørbrød and a cheese roll.

smorgasbord (smǿᵊ·ɹgăsbǿᵊɹd). Also **|| smörgåsbord,** *erron.* **smörgosbrod,** etc. [a. Sw., f. *smörgås* (slice of) bread and butter (f. *smör* butter, cogn. w. Smear *sb.* + *gås* goose, lump of butter) + *bord* Board *sb.*, table: cf. *smørrebrød.] **1.** The Swedish hors d'œuvres, typically comprising a cold table of open sandwiches served with an assortment of delicacies; also provided as a separate meal or buffet.

1893 *Figaro* (Chicago) 26 Jan. 345/1 In every household..before sitting down to dinner an appetizer, or smörgåsbrod, is partaken of. **1895** *Baedeker's Norway, Sweden & Denmark* p. xxiii, The Smörgåsbord or Brännvinsbord, with its..dishes, bread-and-butter, and cheese, the special form of stimulant to the appetite, is peculiar to Sweden. **1926** *Ladies' Home Jrnl.* Nov. 150/2 The 'Smorgasbord', a bountifully supplied relish table, is a Scandinavian institution, though it is said the custom originated in ancient Russia. **1936** *Discovery* Apr. 109/2 Meals, which begin with the smörgåsbord, as in Sweden, are excellent and cheap. **1948** 'J. Tey' *Franchise Affair* i. 7 'Sandwiches without tops' the girl called them. 'Smorgasbord.' **1959** *Good Food Guide* 1959–60 314 The first level offers a Swedish—Swedish, not Danish, because of the smorgasbord which is included—smorgasbord at 8/6. **1975** D. Ramsay *Descent into Dark* ii. 66 The smorgasbord restaurant Joyce had been fond of. **1978** *New York* 3 Apr. 19 (Advt.), Wine & Dine! Gourmet menu. Smorgasbord..2 bars. Duty-free prices.

2. *fig.* A medley, miscellany; a rich variety or selection.

1948 Menjou & Musselman *It took Nine Tailors* xxii. 176 Instead the studio offered me the lead in a piece of *smörgåsbord* called *The Sorrows of Satan*, a novel by Marie Corelli. **1961** *Encounter* Apr. 56/2 The messy smörgasbord of his hysterical whimsical ideas. **1969** T. E. B. Howarth *Culture, Anarchy & Public Schools* iii. 55 The smorgasbord system is a form of shorthand to describe the multiplicity of optional 'subjects' pupils in American high schools elect to study. **1978** M. Puzo *Fools Die* xii. 129 Everyday Magazines..was a group of publications that drowned the American public with information, pseudoinformation, sex and pseudosex... A real smorgasbord.

'smorning (smǭᵊ·ɹnɪŋ), colloq. or dial. abbrev. of 'this morning'.

1932 S. Gibbons *Cold Comfort Farm* xx. 267 She took on something awful about Miss Judith going off 'smorning. **1967** 'G. Douglas' *Death went Hunting* xviii. 155 Well, 'smorning Mr Bolton comes over to me. **1973** J. Mann *Only Security* xiv. 184, I found her, I did. 'Smorning, when I come rabbiting.

|| smørrebrød (smǿ·rəbrǿð). Also **smörrebröd.** [Da., f. *smør* butter + *brød* Bread *sb.*¹; cf. *smørbrod.] A Danish open sandwich; also *collect.*

1902 M. Thomas *Denmark, Past & Present* II. xviii. 133 Your true Dane takes a piece of bread, plasters it with butter, then searches..for the slice of meat or fish he prefers and puts it on the bread and butter... This is called 'smörrebröd' and is the national dish. **1932** G. Bröchner *Wayfarer in Denmark* vi. 76 All Danish cafés ..have..a 'Smörrebröds seddel': a long narrow card with a list of the different kinds of Smörrebröd served. **1959** *Good Food Guide* 1959–60 242 It has opened a Scandinavian smörrebröd cold table in its basement. **1975** tr. *Melchior's Sleeper Agent* (1976) III. 247, I will make you some *Smørrebrød*, Rudi, the kind you like.

smother, *sb.* Add: **1. a.** (Later *fig.* example in literary use.)

1975 N. Nicholson *Wednesday Early Closing* ix. 176 A dull smother of hopelessness hung over the town like the smutch from a smoking rubbish dump.

2. e. *N.Z.* An incident in which sheep are lost by suffocation caused by others falling on top of them, as during a round-up.

1930 L. G. D. Acland *Early Canterbury Runs* 1st Ser. vi. 128 They once had a bad smother there. *Ibid.* Mt. Peel was unlucky with smothers. **1933** —— in *Press* (Christchurch, N.Z.) 2 Dec. 15/7, I believe there was a still worse smother [of sheep] on a station called Roxburgh in Otago. **1949** S. S. Crawford *Sheep & Sheepmen of Canterbury* v. 42 Mt. Peel [station] was unlucky with smothers [of sheep].

3*. *Rugby Football.* A high tackle in which the player 'smothers' (sense *3 d) his opponent. In full *smother-tackle.*

1927 Wakefield & Marshall *Rugger* iv. 248 Hoping ..that by the swiftness of your advance you may get him in a smother-tackle, taking both man and ball. **1929** *Illustr. Sporting & Dramatic News* 19 Oct. 183 (*caption*) Getting the ball away from a smother. **1960** E. S. & W. J. Higham *High Speed Rugby* v. 48 *The Smother Tackle.* This tackle is a high tackle and is used when you want to prevent a player from passing or touching down.

smother, *v.* Add: **I. 1. a.** (Later *fig.* example.)

1944 [see *blanket. *sb.* 2 c].

b. Also *spec.* of sheep, to suffocate others by falling on top of them, as during a round-up; to cause (sheep) to die in this manner. *N.Z.*

1871 M. A. Barker *Christmas Cake in Four Quarters* iv. iii. 290, I had to bring 'em [*sc.* the mob of sheep] down uncommon easy, for it was a nasty place, and I didn't want half of 'em to be smothered in the creek. **1930** L. G. D. Acland *Early Canterbury Runs* 1st Ser. vi. 128 They once smothered 5000 in the gully. *a* **1948** —— *Ibid.* (1951) 397 Run sheep..are very easy to s[mother] on broken hill ground... They s[mothere]d 1,200 once..at Mount Peel.

3. c. (Earlier example.) Also in *Association Football* (see quot. 1954).

1845 N. Wanostrocht *Felix on Bat* I. iv. 18 Should it be pitched an inch too far, be sure to get well out at it, and smother it. **1954** F. C. Avis *Soccer Dict.* 112 *Smother*, to put oneself in the way of an opponent's shot, especially by the goalkeeper advancing from his goal towards the opponent. **1976** *Northumberland Gaz.* 26 Nov., His shot was smothered as the final whistle went.

d. *Rugby Football.* To tackle with a bear-like hug embracing the body and arms, preventing one's opponent from releasing the ball or touching it down.

1920 W. Camp *Football without a Coach* vii. 132 Unless experience shows that there is a certain definite play to watch or a certain player to smother. **1928** *Sunday Times* 5 Feb. 24/7 He kicked well ahead on the slippery turf, and after Hunt had smothered the full-back, scored.

II. 11. *Boxing.* (See quot. 1954.)

1916 [see *infight *v.* 2]. **1954** F. C. Avis *Boxing Dict.* 103 *Smother*, to prevent, by clever positioning of the arms, the development of an opponent's attack.

smother crop. [f. Smother *v.*] A crop which is grown to suppress weeds.

1920 W. E. Brenchley *Weeds on Farm Land* iii. 51

When a smother crop is grown it is of course necessary that the crop seed should be free from weed seed. 1937 A. F. Hill *Econ. Bot.* xv. 336 Barley is also used for hay and pasturage and as a smother crop to kill out weeds. 1973 P. A. Colinvaux *Introd. Ecol.* xxi. 303 Barley is known as a 'smother crop' because it keeps down weeds in the field. This it does with root secretions, weed killers of its own.

smothered, *ppl. a.* Add: **2. b.** (Earlier example.) See also Mate *sb.*^{1} b.

1804 *Introd. Hist. & Study of Chess* v. 82 *Smothered mate* is when the king is so surrounded by his own friends that he cannot move out of check for them; and this mate is generally given by the knight.

3. (Later examples.)

1877 E. S. Dallas *Kettner's Book of Table* 369 Smothered Rabbit. This is the name given in England to boiled rabbit. It is smothered with a white onion sauce. 1923 *Mrs. Beeton's Bk. Househ. Managem.* 582 (*heading*) Chicken, Smothered (Poulet étuvé). 1948 *Good Househ. Cookery Bk.* ii. 237 *Smothered tongues*... Cook preferably in a double saucepan or in a basin in a steamer. 1975 tr. *Melchior's Sleeper Agent* (1976) iii. 290 A juicy steak heaped with smothered onions.

smothering, *vbl. sb.* (Examples in sense *1 b of the vb.)

1950 *N.Z. Jrnl. Agric.* July 5/2 It is undesirable to have yards on a very steep slope, as the danger of smothering, particularly with lambs in a large yard, is greatly increased. 1956 G. Bowen *Wool Away!* (ed. 2) iv. 47 The 'sheepo' must keep an eye on them for smothering, the first sign of which is sheep jumping up in the pen.

smoulder, *v.* Add: **2. d.** To show suppressed anger, hatred, resentment, etc.

1934 in Webster. 1957 L. Durrell *Justine* i. 69 She seemed to smoulder like a tar-barrel on the point of explosion. 1983 'J. Gash' *Sleepers of Erin* iv. 42 Kurak smouldered his way to the Rolls, vibing pure hate in my direction.

smudge, *sb.*^{1} Add: **4.** *smudge pan*; **smudge cell** *Med.*, a degenerate leucocyte in a blood film.

[1935 Whitby & Britton *Disorders of Blood* iv. 89 Degenerate lymphocytes usually appear as smudges and are known as 'smear cells'.] 1937 Kracke & Garver *Dis. Blood & Atlas Hematol.* vi. 84 It has been stated that smudge forms are degenerating lymphocytes and that basket cells..are degenerating granulocytes... It seems more probable that the smudge cell is an early stage and the basket cell a later stage of the same process. 1971 W. M. Dougherty *Introd. Hematol.* iii. 70/1 Most often the bare nuclei that we call smudge cells or basket cells are in fact the bare nuclei of the lymphocytes. 1798 J. Constable *Let.* 2 Dec. (1964) II. 16, I should be glad of the smudge pan as soon as convenient.

smudge, *sb.*^{2} **1.** For Now *U.S.* read Now *N. Amer.* and add: esp. a smoke made to repel mosquitoes, etc. (Further examples.)

1887 E. Custer *Tenting on Plains* ii. 77 Eliza.. brought old kettles with raw cotton into our room, from which proceeded such smudges and such odors as would soon have wilted a Northern mosquito. 1939 F. P. Grove *Two Generations* v. 32 If there had been a moon, a person coming over the hills would have seen that smudge as a perfectly level sheet closing the bowl like a lid. 1971 [see *repellent *a.* 2 d].

2. a. (Earlier and later examples.)

1806 A. Henry in E. Coues *New Light on Early Hist. Greater Northwest* (1897) I. 287 The women closed the openings of the cabins, and made a smudge inside. 1936 B. Brooker *Think of Earth* iii. vii. 278 'We'll make a smudge,' said Bundy, and..began gathering twigs and handfuls of scorched grass. 1952 *Chambers's Jrnl.* Aug. 503/2 Laddash greeted her, squatting in the smoke of a smudge against the mosquitoes. 1959 [see *punky *a.* a].

b. *attrib.* *smudge bonfire, fire* (earlier and later examples), *-smoke*; designating containers for the smouldering fire, as *smudge box, can, kettle, pot*, etc.

1846 *Knickerbocker* XXVIII. 241 You make a large 'smudge' fire outside that the smoke may drive these [insects] away. 1860 *Harper's Mag.* Oct. 584/1 Through the smudge-smoke issuing from the half-breeds' quarters we could catch glimpses of dark eyes. 1882 G. C. Eggleston *Wreck of Red Bird* 55 'What is a "smudge box", Ned?' 'Simply a shallow box of earth set upon a post, to build a smudge upon.' 1902 S. E. White *Blazed Trail* xx. 148 Thorpe's old tin pail was pressed into service as a smudge-kettle. 1903 *Outing* XLIII. 166/1 Other settlers keep the smudge-pot going and live in smoke. 1909 H. Bindloss *Lorimer of Northwest* 3 The dun smoke of a smudge-fire shows that Harry is in prairie fashion protecting our stock. 1923 F. Waldo *Down Mackenzie* 116 One sees the horses after a trip..released for rest, huddling to windward of smudge bonfires, or in default of these standing in a forlorn group together to get in one another's shade. 1944 *Living off Land* ii. 31 Professional beekeepers use a small smoke bellows, but the best substitute is to light a smudge fire and let the smoke drift past the entrance to the hive. 1954 A. M. Bezanson *Sodbusters invade Peace* 134 A smudge can was my constant companion in or in front of the house. 1965 H. Johnson *Bay of Pigs* iii. i. 106 While one of the men put up the signs, another lighted the smudge pots. 1978 J. A. Michener *Chesapeake* 18 If he kept a smudge-fire going..he could survive. *Ibid.* 36 The mosquitoes were terrible..and people stayed close to smudge pots when the sun went down.

smudge, *v.*^{1} Add: Hence **smu·dging** *vbl. sb.*^{3} (in quot. *fig.*).

1873 J. Brown *Let.* 27 Dec. (1912) 288, I always feel insulted by these smudgings and besmearings.

smudge, *v.*^{2} **1. b.** For *U.S.* read *N. Amer.* and add examples. Also, to cause (a fire) to smoke; to drive (mosquitoes, etc.) *away* by smoke. Now *rare.*

1860 *Harper's Mag.* Aug. 296/2 The blankets were spread in the tents, the tents smudged or mosquito nets hung. 1866 *Ibid.* Jan. 265/2 The others sat by the fire and 'smudged' it. 1880 D. Currie *Lett. of Rusticus* 56/1 Before going to bed we smudged the tent, which made the mosquitoes so drunk that they did not molest us again before morning. 1921 *Daily Colonist* (Victoria, B.C.) 30 Oct. 21/1, I piled on some brush and tried to smudge 'em away.

c. Among North American Indians, to smoke (pottery) in order to give it a black shiny finish. See also *smudging *vbl. sb.*^{2}

1936 K. M. Chapman *Pottery of Santo Domingo Pueblo* 7 The ware turns light red in firing, though this is often purposely smudged to a more or less dense black after firing is complete.

Hence **smu·dging** *vbl. sb.*^{2} (*spec.* in sense *1 c of the vb.).

1846 E. W. Farnham *Life in Prairie Land* II. x. 314 This process is more briefly designated by its technical name of '*smudging*'. 1955 Bushnell & Digby *Anc. Amer. Pottery* iv. 32 In these examples the colour is due rather to smudging with carbon in the fire than to chemical reaction. 1973 A. H. Whiteford *North Amer. Indian Arts* 15 Smudging is achieved by smothering the fire with fine damp manure.

smudge, *v.*^{4} Add: So **smu·dging** *vbl. sb.*^{1}

1789 W. Maclay *Jrnl.* 11 May (1890) 30 He will.. dimple his visage with the most silly kind of half smile which I can not well express in English. The Scotch-Irish have a word that hits it exactly—*smudging*.

smudgeless (smʌ·dʒlės), *a.* [f. Smudge *v.*^{1} or *sb.*^{1} + -less.] **a.** That will not smudge or smear. **b.** Without a smudge, clean.

1913 *Chambers's Jrnl.* Mar. 270/1 So-called indelible and smudgeless inks have been placed upon the market. 1924 W. Deeping *Three Rooms* xxxvi. 320 She had dealt with the silver, and it lay bright and smudgeless on a sheet of green baize. 1976 *Sci. Amer.* Jan. 15/2 (Advt.), Its built-in silent recorder prepares a permanent cardiotocogram on smudgeless thermal paper.

smug, *v.*^{1} Add: **3.** [Perh. a different word: cf. Smuggle *v.*^{2}] *intr.* To caress, fondle. *dial. rare.*

1813 E. Picken *Poems* I. 176 We'll cuddle baith amang the fug An' while we hug, an' kiss, an' smug, I'll haud thee firm by ilka lug. 1922 Joyce *Ulysses* 308 Blind to the world up in a shebeen in Bride street after closing time..and hugging and smugging.

smug, *v.*^{2} **4.** (Earlier example.)

1887 J. W. Horsley *Jottings from Jail* i. 6 We used to go and smug snowy (steal linen) that was hung out to dry.

smuggery (smʌ·gəri). *nonce-wd.* [f. Smug *a.* + -ery.] The quality or condition of being smug, or an instance of this; smugness.

1928 A. Huxley *Point Counter Point* xi. 170 Enlargements,..by contrast with our bourgeois and Pecksniffian smuggeries. 1961 G. Frankau in P. Frankau *Pen to Paper* 221, I must pray to be redeemed from the sin of Smuggery.

smugging, *vbl. sb.*^{1} (In Dict. s.v. Smug *v.*^{1}) (Later *poet.* example.)

1932 Auden *Orators* iii. 104 Only hard On smugging, smartness, and self-regard.

smush (smʌʃ), *sb.*^{1} [Alteration of Mush *sb.*^{1}: cf. Smash *sb.*^{1}] **1.** = Mush *sb.*^{1} 3 a; a messy pulp. *dial.*

1825 Jamieson *Sc. Dict.* (Suppl.) II. 429/1 *Gane to smush*, reduced to a friable or crumbled state, like potatoes too much boiled, &c. 1929 D. H. Lawrence *Pansies* 130 Then suddenly the mastodon rose with the wonderful lady And trampled all the listeners to a smush.

2. The mouth; = *mush *sb.*^{1} 3 d. *U.S. slang. rare.*

1930 [see *haul *v.* 3 c]. 1935 D. Runyon in *Hearst's* Jan. 160/2 He grabs Miss Amelia Bodkin in his arms and kisses her kerplump on the smush.

smut, *sb.* Add: **7.** (sense 1) *smut machine* (earlier example), *mill*; (sense 5) *smut book, shop*, etc.; *smut-hunting* ppl. adj.; **smut-hound** [cf. *hound *sb.*^{1} 4 e] *colloq.*, one who seeks to censor or suppress smut (sense 5), esp. in literature.

1818 *Niles' Reg.* XV. 80/1 *A smut mill*, for cleaning wheat of smut, is in operation at Plattsburg. 1850 *Mary Wedlake's Priced List Farming Implements* 25 A *Smut Machine*, to clean damaged grain. 1927 H. L. Mencken *Let.* 2 Dec. (1961) 305 Of my inventions I am vainest of Bible Belt, booboisie, smut-hound and Boobus americanus. 1928 D. H. Lawrence *Let.* 9 Mar. (1962) II. 1042 Mason wrote me rather scared about the censor and smut-hunting authorities. 1930 Auden *Poems* 69 Lawrence was brought down by smut-hounds, Blake went dotty as he sang. 1930 *Publishers' Weekly* 31 May 2737/2 The confiscation of dirty picture postals and smut books. 1961 *John o' London's* 28 Sept. 357/3 The bulk of *The High Price of Pornography* is devoted to a survey of the rancid avalanche of smut magazines..which are pulped out in the States. 1965 E. L. Myles *Emperor of Peace* i. xiii. 135 He bought..a two-inch stone burr mill complete with smut mill, cleaner and water wheel. 1967 *Spectator* 1 Dec. 683/1 Eminent men of letters would not be dismissed as fools or smuthounds. 1977 *Zigzag* Apr. 28/3 He said we were turning lunchtime into a 42nd street smut shop.

smuttily, *adv.* (Later examples.)

1737 *London Mag.* May 261/2, I suppose you do not mean an old Woman, seeing that to talk smuttily to such, would be no great Insult. 1974 P. Cave *Mama* (new ed.) xiii. 109 'Got some..business, down at the docks, have you?' asked one of the drivers and leered smuttily.

smuttiness. 2. (Later example.)

1973 *Times* 16 Nov. 5/5 Viewers are also asked.. whether..smuttiness in comedy programmes..is found offensive.

Smyrna. Add: **a.** *Smyrna carpet, fig, rug.*

1881 C. C. Harrison *Woman's Handiwork* iii. 165 Curtains of French make, stamped with patterns taken from Turkish or Smyrna rugs. 1897 *Sears, Roebuck Catal.* 12/2 Imported Smyrna Figs, very choice. 1904 W. D. Ellwanger *Oriental Rug* 153 Smyrna carpets, 97, 98. 1956 S. Bedford *Legacy* iii. vi. 198 Smyrna figs, grapes in cotton-wool, Turkish delight. 1966 N. Freeling *Dresden Green* i. 16 Two smyrna rugs on the polished wooden floor, that he had made on winter evenings. 1977 'R. Player' *Month of Mangled Models* vii. 125 They were treading silently on the best Aubusson in Paris although ..the little Smyrna carpet in the boudoir was worth three times as much.

b. Also, a Smyrna carpet.

1904 W. D. Ellwanger *Oriental Rug* 97 Most other carpets are of Turkish weaving..and come under the general title of Smyrnas.

Smyrniote, *a.* (Earlier example, with spelling *Smyrniot.*)

1867 C. M. Yonge *Pupils of St. John* xii. 191 The strong spirit of contending for the purity of the faith had descended from St. John upon the great Smyrniot bishop.

smythite (smai·ðait). *Min.* [f. the name of Charles H. *Smyth* (1866–1937), U.S. geologist + -ite^1.] A sulphide of iron and probably nickel found as opaque bronze-coloured crystals that are strongly magnetic and have a metallic lustre.

1956 Erd & Evans in *Jrnl. Amer. Chem. Soc.* LXXVIII. 2017/1 We have found minute, plate-like crystal inclusions in calcite crystals..to be a new iron sulfide... The mineral is named smythite (pronounced smīth'ite) in honor of Professor C. H. Smyth, Jr. 1972 *Amer. Mineralogist* LVII. 1571 Smythite was originally described by Erd *et al.*.. as having a rhombohedral structure..and Fe_3S_4 composition. These data are in error and smythite is hereby redefined... Nickel is present in all smythites present to date..and it is suggested that smythite is *not* a phase in the Fe–S system but possibly in the Fe–Ni–S system. 1976 *Minerals Sci. & Engin.* VIII. 119/2 There is still considerable doubt about the composition range and thermal stability of natural and synthetic smythite, and even whether it should be included as a phase in the iron-sulphur system.

snack, *sb.*^{2} Add: **4. c.** Also, designating a place at which snacks are sold, as *snack booth, counter, shop*; in appositive use, as *snack lunch, meal*; *snack-sized* adj. Cf. *snack bar.

1976 D. Heffron *Crusty Crossed* xxii. 147, I sat alone on the sand, watching my sisters parade over to the snack booth with their boyfriends. 1977 W. J. Weatherby *Home in Dark* viii. 44 A large woman who served behind the snack counter. 1964 N. Marsh *Dead Water* iii. 85, I..had a snack lunch in the new bar. 1977 W. Hildick *Loop* ix. 47 After a snack lunch, I walked round to the School House. 1962 *Punch* 28 Nov. 773/1 We are becoming 'increasingly a nation of tea and soft-drink consumers and snack-meal eaters'. 1976 'K. Royce' *Bustillo* xii. 157 Bustillo was eating a snack meal. 1977 *Chicago Tribune* 2 Oct. xii. 18/2 (Advt.), Partial bldg. standing due to fire, selling as is, formerly snack shop, restaurant. 1974 E. Ambler *Doctor Frigo* iii. 153 A snack-sized gobbet of raw flesh.

5. *Austral. slang.* Something easy to accomplish, a 'snip'.

1941 S. J. Baker *Austral. Slang* 68 *Snack*, a certainty. 1952 T. A. G. Hungerford *Ridge & River* 138 There was nothing to it... It was a snack. 1961 M. Calthorpe *Dyehouse* 150 In Hughie's day he'd made this a snack. 1970 R. Beilby *No Medals for Aphrodite* 274 'How could I do that, Harry?' 'Easy. It'll be a snack.'

snack, *v.* Add: **1. a.** Also *fig.*, to utter or exchange sharp, snapping words or remarks. (Not in general use.) Cf. Snap *v.* 2 a.

1956 C. P. Snow *Homecomings* l. 357 They quarrelled and snacked. *Ibid.* 358 The prickles and self-assertiveness which made them snack. 1959 P. H. Johnson *Humbler Creation* xii. 85 The usual strung-up celebrations at home, with..Libby and her mother gently snacking at each other in tones of excessive goodwill. 1960 C. P. Snow *Affair* viii. 91 Irene and I glanced at each other with discomfort, a discomfort different from just looking on at her husband and my wife snacking.

Hence **sna·cking** *vbl. sb.*

1959 P. H. Johnson *Humbler Creation* v. 30 The meeting petered away as it usually did, into desultory snackings and exchanges of fellowship. 1969 W. Cahn *Out of Cracker Barrel* xxiii. 318 Premium Saltines and Ritz crackers were used as snacks long before snacking came into vogue. 1978 *Radio Times* 18–24 Feb. 67 There are three meals a day, 'more or less regular, with no snacking in between'. 1980 P. S. Powers *Obesity* ix. 219 Snacking while watching television is notorious for increasing the total daily calories ingested.

snack bar. [f. Snack *sb.*^{2} + Bar *sb.*^{1} 28.] **1.** A bar or counter at which snacks are served to customers; (part of) a restaurant containing such a bar.

1930 *Punch* 16 Apr. 433 A vegetarian snack-bar. 1937 *Archit. Rev.* Nov. p. lxii (*caption*) A restaurant and snack bar in Regent Street, London. 1943 *Sun* (Baltimore) 14 Oct. 5/1 A 'snack bar', offering sandwiches and soft drinks, is maintained at the Gold street center. 1958 S. Hyland *Who goes Hang?* xliii, 210 Arthur's highly polished snack-bar. 1965 D. Henderson *Heart of Newfoundland* 54 Marty's run a chain of snack bars in different parts of the city. 1978 S. Brill *Teamsters* vii. 284 Barkett pulled the truck into a gas station snack bar for a coffee break.

2. *attrib.* Also *fig.*

1940 'G. Orwell' *Inside Whale* 138 It is a flowing, swelling prose..with rhythms in it, something quite different from the flat cautious statements and snackbar dialects that are now in fashion. 1957 *Observer* 1 Sept. 11/7 Amongst the things most desired are seats bookable by telephone, good coffee and snack-bar service. 1982 J. Scott *Uprush of Mayhem* vi. 59 The lunch-time rush for snackbar sandwiches.

sna·ckery. Also **snackerie.** [f. Snack *sb.*^{2} + -ery.] A snack bar or other public eating-place serving snacks.

1936 *Amer. Speech* XI. 374/1 *Steve's Snackerie* is the sign over a small café in Lincoln, Nebraska. 1967 *Punch* 30 Aug. 295/2 If we want better pubs, pubs catering for the family, we must..convert them all into decent snackeries, and let them stay open for at least twelve hours a day. 1969 'R. Crawford' *Cockleburr* i. vi. 55, I..ate a nondescript pizza in a nondescript snackery. 1981 *Nordic Skiing* Jan. 48/3 The ski shop sports rentals, needed accessories, waxing area and snackery.

snacke·tte. *W. Indies.* [f. Snack *sb.*^{2} + -ette, after *launderette.] A snack bar.

1973 *Advocate-News* (Barbados) 20 Jan. 3/1 The snackette also offers a take-away service, and customers can make their orders by telephone, and then collect them. 1973 *Trinidad & Tobago Overseas Express* 28 May 11/3 Her husband was returning from a snackette in Marabella. 1974 *Sunday Advocate-News* (Barbados) 3 Mar. 16 (Advt.), Experienced person required to manage and control Snackette and Restaurant.

snaffle, *sb.*^{1} Add: **2. snaffle-mouth,** the mouth of a horse which can be managed with a snaffle alone.

1910 *Chambers's Jrnl.* 1 Oct. 703/1 A jungle-fowl which ..causes my horse to dance a gavotte.., a feat for which his snaffle mouth and indolent disposition eminently unfit him. 1932 J. E. Hance *School for Horse & Rider* x. 85 From time to time one hears of such and such a horse possessing a 'snaffle mouth'. To be entitled to such a designation the animal would have to be capable of flexing and bending to this form of bit at all paces, and such animals are extremely rare. 1977 *Horse & Hound* 14 Jan. 36/3 (Advt.), Chestnut gelding... Snaffle mouth, quiet in every way.

snaffle, *v.*^{4} Add: **2.** To appropriate, seize, catch, snatch. Also with *up*.

1895 Kipling in *Century Mag.* Dec. 273/2 A year's leave was among the things he had 'snaffled out of the campaign', to use his own words. 1915 D. O. Barnett *Let.* 7 Aug. 218, I see they've snaffled Warsaw. 1916 'Peter' *Trench Yarns* 10 A certain airman had engine trouble up aloft and had to come down behind the German lines. Of course they snaffled him. 1928 *Sunday Express* 15 Apr. 11/6, I soon snaffled a double role in a big spectacle. 1959 *Times* 28 May 4/7 Slade threw down Cook's wicket when Meyer tried to snaffle the strike. 1964 M. McLuhan *Understanding Media* xxxi. 331 Jack discovered how to extend the TV mosaic image..seemingly snaffling up just anybody from anywhere. *a* 1974 R. Crossman *Diaries* (1975) I. 149 There was a good deal of bleating, but I got my way and was able to snaffle the Statement for myself in the process.

snafu (snæ·fu), *phr.*, *a.*, and *sb.* Also **SNAFU.** *slang* (chiefly *U.S.*, orig. *U.S. Mil.*). [Acronym f. the initial letters of *s*ituation *n*ormal: *a*ll *f*ouled (or *f*ucked) *u*p.] **A.** Used acronymically (often with an explanation) as an expression conveying the common soldier's laconic acceptance of the disorder of war and the ineptitude of his superiors.

1941 *Amer. N. & Q.* Sept. 94/2 *Snafu*, situation normal. 1943 *Amer. Mercury* Nov. 555/2 *Snafu*—politely translated as 'situation normal; all fouled up', to indicate that things are not going too well. 1946 *Amer. Jrnl. Sociol.* Mar. 419 Interestingly, the expression 'snafu', derived from this, 'Situation normal, all f—ed up', is coming into general civilian use. 1966 *Sunday Times* (Colour Suppl.) 4 Dec. 73/4 GI Jargon... *Snafu*, Situation normal, all fouled up. 1975 *Listener* 13 Mar. 349/1 There was a barrack-room mnemonic which fits the ill-starred Dieppe raid: SNAFU, or Situation Normal, All Fouled Up.

B. *adj.* Confused, chaotic.

1942 *Time* 15 June 11/1 Last week U.S. citizens knew that gasoline rationing and rubber requisitioning were snafu. 1950 'D. Divine' *King of Fassarai* (1951) xxviii. 245 Situation Snafu... Send for the Seabees.

C. *sb.* Now usu. with *a* and *pl.* A confusion or mix-up; a hitch, mishap; muddle, confused state.

1943 *Yank* 10 Sept. 9 They worked hard and steadily, with a minimum of snafu. 1945 *Richmond* (Va.) *Times-Dispatch* 11 Dec. 10/7 Corporations struggling with the problems of reconversion, strikes, shortages and snafu in general. 1956 C. W. Mills *Power Elite* viii. 182 The key to the bureaucratic snafu that has often characterized the navy is that as the ships and the guns and the logistics became more technically complicated, the men who ran them acquired rank less by technical specialty than by seniority. 1958 'Castle' & 'Hailey' *Flight into Danger* i. 17 It would have to be a big show in Vancouver to justify this snafu. 1963 Mrs. L. B. Johnson *White House Diary* 28 Dec. (1970) 23 Pretty soon the German plane rolled in, overshooting the red carpet by a few feet, so there was a slight snafu and they had to hop around to get onto it. 1965 *Times Lit. Suppl.* 25 Nov. 1039/1 He must have seen enough 'snafus' to make him sceptical of 'the Brass'. 1976 *Guardian* 20 Oct. 4/8 As that monumental snafu at Domadedovo attests, Aeroflot's shortcomings are also big ones. 1978 W. F. Buckley *Stained Glass* xxi. 206 Singer then rehearsed Blackford in emergency instructions to be followed in the hideous event of a snafu. 1980 B. Mason *Solo* 117 And Holy Moses, *what* a snafu! Why foul up poor, harmless, gormless Glad?

Hence as vb. *U.S. slang*, (*a*) *trans.* to mess up, to play havoc with; (*b*) *intr.* to go wrong; also **sna·fued** *ppl. a.*

1943 *Yank* 19 Nov. 9 Then the Army snafued the romance by transferring Kinser to this post. 1944 *Life* 16 Oct. 20/2 It is a symbol of SNAFU.., and a star is rated for each snafued campaign after Guadalcanal. 1953 *Sun* (Baltimore) 1 Sept. (B ed.) 17/2 Eddie had twice this season snafued a batting order and caused men to be called out for swinging out of turn. 1955 'J. Christopher' *Year of Comet* i. 15 Of course I didn't bring you up here simply to tell you P & M snafued your psychoplan sixteen years ago. 1975 J. Grady *Shadow of Condor* i. 31 Every now and then something snafus and there is one hell of a mess. 1981 G. Markstein *Ultimate Issue* 38 My arrangements seemed snafued. I guess the lines got crossed.

snag, *sb.*^{1} Add: **1. c.** (Later examples.) Also, a disadvantage, a hitch; a defect.

1903 *N.Y. Times* 20 Oct. 1 A conference lasting three hours took place which was plain sailing until the last moment, when a snag was struck. 1923 Wodehouse *Inimitable Jeeves* vi. 66 At this point the scenario struck another snag. 1936 W. H. Saumarez Smith *Let.* 26 Dec. in *Young Man's Country* (1977) ii. 48 The only snag..was that Grindlay's had failed to send my tickets to the U.S. Club. 1940 *Economist* 27 Jan. 142/2 Mr Gandhi referred to the 'undoubted snags' in Lord Linlithgow's statement, but a pre-requisite to tackling the 'snags' is a better spirit and some measure of confidence. 1945 C. H. Ward-Jackson *Piece of Cake* 56 *Snag*, aircraft defect. 1950 J. Cannan *Murder Included* iii. 39 The house would be all right; the snag would be the skivvies. 1962 *Daily Tel.* 19 Nov. 22/5 (*heading*) Bonn Air Force finds snags in U.S. plane. 1977 B. Pym *Quartet in Autumn* v. 44 There had been a good deal of discussion..as to whether he should go by coach or by train and the advantages and snags of each method were endlessly weighed up.

d. *N. Amer.* A standing dead tree.

1904 *Dialect Notes* II. 421 There was a big snag with a woodpecker's nest in it south of our house. 1936 *Sun* (Baltimore) 1 Aug. 11/1 Flames..have turned more than 30,000 acres of once-green forest into charred and smoldering snags. 1946 B. MacDonald *Egg & I* 94 Incredibly tall, spindly snags leaned threateningly towards me. 1960 M. Sharcott *Place of Many Winds* i. 14 The sun touched the hills behind us, lighting the dead white snags that so liberally sprinkled the live spruce. 1975 *Islander* (Victoria, B.C.) 8 June 16/1 Osprey will choose a nest site atop a dead snag from which he can command a view of the habitat around him.

4. snag-boat (earlier example).

1832 *Reg. Deb. Congress U.S.* (1833) 3 May 2722 The snag boat had been employed in improving the navigation of the Mississippi.

snag (snæg), *sb.*^{4} *Austral. colloq.* [Origin unknown: cf. *Snag* vb.^{2} and sb.^{3} 6 in E.D.D.; *Snag* sb.^{1} 2 in S.N.D.] A sausage.

1941 Baker *Dict. Austral. Slang* 68 *Snags*, sausages. 1949 R. Park *Poor Man's Orange* 33 'Let's have sausages.' .. Good old snags. They were always there to be fallen back on. 1972 *Sunday Mail Mag.* (Brisbane) 26 Mar. 13/1 The dog had an uncanny capacity for nicking in to the butcher's shop, snatching a snag and getting out again. 1980 *Bulletin* (Sydney) 6 May 112/3, I make my own snags, my own pies and pasties. The Yanks love them after you've twisted their arms to try them.

snag, *v.*^{2} Add: **1. b.** (Earlier and later examples.) Also, to impede, to inconvenience. Also with *up*.

1833 *Polit. Examiner* (Shelbyville, Kentucky) 22 June 4/1, I will agree to be shot with a paper wadding if there ar' room enough in the whole clearing for a man of ordinary parts to stand on five minutes at a time, without getting snagged by some tape and cotton yarn dealer in the street. 1929 Hall & Niles *One Man's War* 131, I decided to try clipping the German's tail with my propeller or snag him in some way. 1962 *Guardian* 19 Dec. 8/3 No place to work and insufficient funds..had snagged her. 1968 C. Burke *Elephant across Border* v. 193 He was going to do whatever he could to snag things up.

4. *N. Amer.* **a.** To catch, get hold of, grab, steal, pick up. *colloq.*

1895 *Dialect Notes* I. 399 *Snag*, to steal. 1927 *Amer. Speech* II. 278/1 *Snag a pick up*, get a free ride. *Ibid.*, *Snag the current*, get the drift. 1930 D. Runyon in *Collier's* 1 Feb. 44/2 There is plenty of trouble over Lillian snagging her Peke. 1941 B. Appel in C. Grayson *New Stories for Men* 32 Red would snag a dollar or two out of me with a promise that I should take it out of his wages. 1946 Mezzrow & Wolfe *Really Blues* xii. 226 You didn't come of age on the welfare, snagging butts out of the gutter. 1962 J. Glenn in *Into Orbit* 221 Two sailors reached over with a shepherd's hook to snag the capsule. 1966 L. J. Braun *Cat who could read Backwards* (1967) xiv. 160 I'll get to the club early and snag a quiet table. 1978 J. Carroll *Mortal Friends* v. i. 503 Colman and Janet did not snag each other with their eyes at the mention of the Ritz.

b. *Sport.* To catch or field (a ball); to receive (a pass).

1942 Berrey & Van den Bark *Amer. Thes. Slang* § 679/5 *Field a ball*..snag the oval. 1968 *Washington Post* 4 July C1/3 (*caption*) Mantle was safe as Ron Hansen's throw, after snagging a line drive by Andy Kosco, was a trifle tardy. 1977 *Time* 30 May 40/2 In their place came players tailored to Big Bill's skills: quick, sure-handed guards to snag his crisp outlet passes and start the fast break rolling.

c. *Angling.* To catch (fish), *spec.* with a bare hook; to catch illicitly or improperly.

1946 *Richmond* (Va.) *Times-Dispatch* 17 Mar. B-11/2 When the herring are in in numbers, you snag a herring about every third or fourth attempt. 1960 *Washington Post* 11 Mar. D5 Unbelievable as it may sound, the herring are so thick at the height of the run that simply jerking a bare hook through the water will snag fish. 1974 *Evening Herald* (Rock Hill, S. Carolina) 18 Apr. 7/1 The largest striper of the first week of the 10 week-long derby went to Sumter's John Benenhaly who also used cut bait to snag a 26 pound, 8 ounce fish. 1979 *Globe & Mail* (Toronto) 7 Feb. 5/6 His six-man patrol is fed up with those who snag fish illegally.

5. *intr.* **a.** To strike a snag, to get caught *on* a projection or obstacle.

1866 *Harper's Mag.* Nov. 810/1 A Mississippi steamer, that snagged and went down on 'Yazoo Bend'. 1929 W. Faulkner *Sound & Fury* 3 You snagged on that nail again. Cant you never crawl through here without snagging on that nail. 1970 G. F. Newman *Sir, You Bastard* 262 He could neither see the thorn he had snagged on, nor the path by which to pull clear. 1976 M. Machlin *Pipeline* lv. 559 The rock itself..was some seventy-five yards from the ship, which apparently had snagged on an underwater rock projection some hundred feet down.

b. Of a fabric: to be rendered imperfect by a pulled thread.

1970 *Which?* Oct. 300/2 A few brands suffered slightly from pilling..and several of the Crimplene ones snagged.

6. The verb-stem in combinations, as (sense *4 c) *snag-fishing, -hook, -line.*

1936 *Sun* (Baltimore) 8 July 8/6 They..charged all three with using snag lines and snag hooks with intent to do bodily harm to the aforesaid sturgeon. 1952 B. Harwin *Home is Upriver* viii. 86 He could get an old gasboat somewhere, cheap, some fishhooks and lines enough for snag-fishing. *Ibid.* x. 99 Kip and Lenny fished: snaglines and bushlines. 1960 *Washington Post* 11 Mar. D5 A snag hook is a huge treble hook which can be fished from either a rod or hand. Usually two or three are tied to a strong line (chalk line is a favorite), a singer is attached and the whole shebang is cast into the drink and retrieved in a series of short jerks.

snagged, *ppl. a.* Add: **2.** (Earlier and later examples.) Also *fig.*

1851 E. C. E. Stuart-Wortley *Travels in U.S.* 112 In the papers will you often see whole columns, headed, 'Snagged', containing a melancholy list of boats. 1867 A. D. Richardson *Beyond Mississippi* i. 21 A snagged steamer. 1930 D. Runyon in *Collier's* 22 Mar. 53/1 Basil is snagged if ever I see a guy snagged, and personally I do not blame him, because Miss Harriet Mackyle may not look like a million, but she has a couple. 1977 *Time* 3 Jan. 34/1 The talks have been snagged for months on how to deal with two new weapons.

snagger. Add: **2.** *Austral.* A slow, inexpert, or poor sheep-shearer.

1887 *Tibb's Popular Songbk.* 11, I found a lot of snaggers Not a shearer in the mob. *a* 1914 *Click go Shears* in R. Ward *Austral. Ballads* (1964) 120 The ringer looks round and is beaten by a blow, And curses the snagger with the bare-bellied yeo. 1945 Baker *Austral. Lang.* 63 *Snagger*, a shearer who is learning the trade and handling less than fifty sheep a day. 1969 B. Hardy *West of Darling* 106 Since they were slow, inexpert, and rough in their performances, the poorest shearers in the shed were nicknamed 'snaggers'. 1975 *Sunday Mail* (Brisbane) 14 Sept. 6/6 The younger men who have taken the old 'snaggers'' places, stand up on 'the board'..and shear along with the best of them.

3. *N. Amer.* [*snag *v.*^{2} 4 c.] One who snags fish; one who catches fish illicitly.

1946 *Richmond* (Va.) *Times-Dispatch* 17 Mar. B-11/2 Alewives which have been snagged by the many 'snaggers' on the dam. 1976 *Globe & Mail* (Toronto) 20 Oct. 36/3 Among this group were many snaggers, fish hogs using any means to trap the big fish. One man, armed only with a landing net, scooped away at fish in one pool.

snagging, *vbl. sb.*^{2} (In Dict. s.v. Snag *v.*^{2}) (Earlier and later examples.)

1775 in J. J. Henry *Campaign against Quebec* (1812) 53 The paths and carrying places we had sufficiently developed..by strong blazing and snagging of bushes. 1851 A. O. Hall *Manhattaner in N. Orleans* 179 There may sometimes occur a snagging, or a fire, with perhaps a collision. 1874 *Rep. Vermont Board Agric.* II. 550 The stumps are rotting, and it is nearly ready for the..work of snagging and bogging. 1960 *Washington Post* 11 Mar. D5/1 They're [*sc.* herrings] caught mostly by dipnets and the gentle method known as snagging. 1980 *Outdoor Life* (U.S.) (Northeast ed.) Oct. 56/2 Because Pacific Coast salmon die after spawning, snagging was introduced as a sporting way to harvest huge numbers of fish in a short time.

snaggle (snæ·g'l), *sb.* Chiefly *dial.* and *colloq.* [app. f. Snag *sb.*^{1}: cf. Snaggle-tooth in Dict. and Suppl.] **1.** A snaggle-tooth; one who has snaggle-teeth. *rare.*

1823 M. Wilmot *Let.* 1 Oct. (1935) 197 Blanche [has] become *alas* a snaggle! Those dear little pearls of teeth are going. 1880 Courtney & Couch *Gloss. Words Cornwall* 52/2 What snaggles the cheeld has.

2. A tangle; a knotted or projecting mass.

1904 *Eng. Dial. Dict.* V. 567/1 *Snaggle*,..a knotted, entangled condition. 1968 C. Helmericks *Down Wild River North* ii. xxii. 336 The girls pitched our tent in the sparse, pristine plant population between rock snaggles. 1978 T. Hughes in *Times Lit. Suppl.* 14 Apr. 409/1 All eyes watch The weathered, rooty, bushy pile of faces, A snaggle of faces.

3. *attrib.*, as *snaggle-tusk.*

1922 Joyce *Ulysses* 424 The famished snaggletusks of an elderly bawd protrude from a doorway.

snaggled, *a.* Add: **1.** Also *transf.*

1938 J. Steinbeck *Long Valley* 58 The sharp snaggled edge of the ridge stood out above them. 1942 W. Faulkner *Go down, Moses* 69 A few snaggled trees of what had been an orchard.

snaggle-tooth. Add: (Earlier example.) Also, one with snaggle-teeth.

1820 M. Wilmot *Let.* 12 Jan. (1935) 51 Catherine has actually lost one of her teeth!.. The poor Cat will be a rare frightful snaggle tooth. 1821 —— *Let.* 17 Mar. 99 Instead of being hideous in the snaggle tooth age..she is..improved. 1906 *Dialect Notes* III. 157 You'll be a snaggle-tooth before you're twenty, if you don't quit eating so much candy. 1909 J. R. Ware *Passing Eng.* 227/2 *Snaggle-tooth*, woman of lower order..who, lifting her upper lip when scolding, shows an irregular row of teeth.

snaggle-toothed, *a.* Delete *rare* and add later examples. Also *fig.*

1945 B. MacDonald *Egg & I* (1946) 85 On grey winter days its snaggle-toothed horizon could be seen plainly. 1954 *Caribbean Quarterly* III. iv. 231 Albert is a bright-eyed, snaggle-toothed little man. 1971 B. W. Aldiss *Soldier Erect* 32 That snaggle-toothed chap in the comic button-up white suit,..—put him in a proper pinstripe and he'd pass for an Eastbourne estate agent! 1977 *Time* 14 Feb. 21/3 Entertainment is provided by..a Hollywood drop-cloth view of snaggle-toothed Mount Kenya.

snaggly (snæ·gli), *a.* Chiefly *dial.* and *colloq.* [f. as *snaggle *sb.*: see -y^1.] Irregular; tangled; ragged.

1794 W. Clark *Jrnl.* 4 Aug. in *Mississippi Valley Hist. Rev.* (1914) I. 422 The army was conducted..through intolerable thick woods & the earth covered with Snagley underwoods. 1882 F. W. P. Jago *Anc. Lang. & Dial. Cornwall* 269 'Snaggly teeth', i.e., very irregular or ill-shaped teeth. 1968 P. S. Beagle *Last Unicorn* xiv. 206 Squat, snaggly trees that had never yet bloomed were putting forth flowers in the wary way an army sends out scouts. 1978 R. Jansson *News Caper* vi. 55 The bullet.. bounced off something metal before it hit her, because the wound is all snaggly.

snaggy, *a.*^{1} **3.** (Earlier examples.)

1806 W. Clark in Lewis & Clark *Orig. Jrnls. Lewis & Clark Expedition* (1905) V. 380 The Sand bars..confined the [river] to a narrow Snagey Chanel. 1843 'R. Carlton' *New Purchase* ix. 58 To learn the nature of 'mash land'—'rooty and snaggy land' [etc.].

snail, *sb.*^{1} Add: **2. c.** (Later examples.)

1915 *Dialect Notes* IV. 198 We'll have to wait for Edith. She's such a snail. 1959 I. & P. Opie *Lore & Lang. Schoolch.* xvii. 366 He [*sc.* a latecomer] is a..Snail.

6. a. *snail-trace, -track*; *snail-green, -nacreous, -nosed* adjs.

1931 V. Woolf *Waves* 25 Louis regards the wall opposite with snail-green eyes. 1923 D. H. Lawrence *Birds, Beasts & Flowers* 60 Cyclamen leaves..Spurned with mud Snail-nacreous Low down. 1960 S. Plath *Colossus* 10 In their jars the snail-nosed babies moon and glow. 1966 J. Merrill *Nights & Days* 42 The brief snail-trace Of her withdrawal dries upon our faces. 1930 D. H. Lawrence *Nettles* 20 All those nasty police-eyes like snail-tracks smearing the gentle souls that figure in the paint.

7. snail darter [cf. Darter 5], a small fresh-water fish, *Percina tanasi*, belonging to the family Percidæ, and found in certain rivers of the U.S.; **snail trail** *Needlework* (see quots.); **snail-wheel** (earlier example).

1975 *U.S. Federal Reg.* 17 June 25597/2 The Fish and Wildlife Service has evidence on hand that the snail darter *Percina* (*Imostoma*) sp. is an endangered species. 1977 *Time* 11 Apr. 17/2 The Mississippi sand-hill crane and the three-inch snail darter of the Little Tennessee River have already halted state and federal bulldozers. 1981 *Science* 15 May 761/3 Populations of snail darters have been found in three new places... If they turn out to be established populations the fish may no longer be an endangered species... The snail darter may be reclassified as a 'threatened'..species. 1899 W. G. P. Townsend *Embroidery* vi. 94 *Snail-trail*,..the same principle as *single coral*, only worked more on the slope. 1948 C. Christopher *Compl. Bk. Embroidery Stitches* iii. 64 Snail Trail, or Knot Stitch, makes a series of simple knots connected with each other on the surface of the fabric. 1973 E. Wilson *Embroidery Bk.* (1975) vi. 322 Snail trail worked very close becomes Broad Rope stitch. When worked with the needle at right angles to the thread, instead of slanting, it becomes Coral. 1831 M. Edgeworth *Let.* 6 May (1971) 535 Inkstands that shut impervious to ink—insured by the snail wheel tightener.

snail, *v.* Add: **2. c.** To make (one's way) very slowly.

1936 M. Franklin *All that Swagger* v. 56 Two bullock drays were snailing their way from the Port.

snailish, *a.* Add: Hence **snai·lishness.**

1905 M. Beerbohm in *Sat. Rev.* 24 June 835/1 Usually, this pace in elocution does not madden me. But..I am moved to cry out against this idiotic tradition of snailishness.

snail-slow, *a.* Add: Hence as *adv.*

1901 [in Dict., sense 1]. 1951 W. de la Mare *Winged Chariot* 44 Snail-slow moves *everything* for which we wait. 1974 V. Canning *Painted Tent* x. 203 The time of waiting..had passed snail-slow for Smiler.

snaily, *a.* (and *sb.*). Add: **1.** (Later examples.) Also *fig.*

1928 D. H. Lawrence *Let.* 17 Mar. (1932) 710 We *must* put salt on the hypocritical and snaily tails, the good public. 1979 *Sci. Amer.* Mar. 27/2 Periwinkles, snaily bivalves and the plants and animals (such as barnacles and algae) that live on other organisms each get a detailed chapter.

snake, *sb.* Add: **I. 1. c.** (Earlier and further examples.)

1839 *Spirit of Times* 17 Aug. 283/3 Snakes! such a row! 1888 'R. Boldrewood' *Robbery under Arms* II. xi. 190 So the muchacha went back on yer—snakes alive! I kinder expected it. 1922 E. Raymond *Tell England* ix. 122, I thought we'd be last for the Swimming Cup. But snakes alive! we'll get in the semi-final. 1927 G. D. H. & M. Cole *Murder at Crome House* xxii. 271 But, snakes, Flint—this is Exeter! 1930 G. B. Shaw *Apple Cart* i. 15 Holy snakes! look at Bill.

2. b. (Later examples.)

1907 E. Gosse *Father & Son* xi. 281 He did not scruple to remind the Deity of various objections to a life of pleasure and of the snakes that lie hidden in the grass of evening parties. 1978 J. Irving *World according to Garp* xiv. 271 We were playing in Dallas, when that snake in the grass..came up on my blind side.

d. *to wake snakes*: (*b*) to rouse oneself, to look lively; (*c*) see Wake *v.* 8 c.

1835 A. B. Longstreet *Georgia Scenes* 6 Oh, wake snakes, and walk your chalks! *c* 1859 in Bartlett *Dict. Amer.* (1860) 498 Well, here I be; wake snakes, the day's a-breaking.

e. *snakes in Iceland*: used allusively (see quot. 1758) of something posited only to be dismissed as non-existent.

[1758 tr. *N. Horrebow's Natural Hist. Iceland* lxxii. 91 No snakes of any kind are to be met with throughout the whole island.] 1791 Boswell *Life of Johnson* II. 220 Johnson had said that he could repeat a complete chapter of 'The Natural History of Iceland', from the Danish of *Horrebow*, the whole of which was exactly thus:—'Chap. lxxii. *Concerning Snakes*. There are no snakes to be met with throughout the whole island'. 1906 *Spectator* 5 May 716/1 'The Value of a Public School Education' reminds one of the chapter on the snakes in Iceland... 'So far as the school at large is concerned every Greek and Latin book should be destroyed.' 1978 C. Sykes in R. Buckle *U & Non-U Revisited* 60 And what about hats? Of them it may be said as was said of snakes in *The Natural History of Iceland*.

f. *lower than a snake's belly*: despicable, very low indeed. *Austral. slang.*

1932 L. Mann *Flesh in Armour* 191 'It was a dirty trick. He knew about me and her.' 'Dirty! Lower than a snake's belly.' 1951 D. Cusack *Say no to Death* 20 He'd only have to take one look at Jan to be convinced in his honest old heart that his son was lower than a snake's belly. 1965 J. Beede *They hosed them Out* 175, I thought, 'if I have to crawl to this illegitimate I'll get lower than a snake's belly.'

3. b. *U.S.* and *Austral. slang.* (See quots.) Cf. *snake charmer*, sense 11 a below.

1929 *Bookman* (U.S.) July 526/1 A Snake has many jobs. If he's a Hump-brakey he handles the cars rolled onto a series of tracks placed on a slight incline. The engine shoves them 'over the hump' and it is his job to handle the brakes. 1934 *Amer. Speech* IX. 73/2 *Snake*, switchman. His work requires him to crawl around and over cars, and he has a reputation for never hurrying. 1945 Baker *Austral. Lang.* xiv. 249 There are terms like ..*snake-charmers, snakes* or *lizards*, railway platelayers.

c. *Austral. Mil. slang.* (See quot. 1945.)

1945 Baker *Austral. Lang.* viii. 160 *Snakes*, a sergeant. 1948 [see *snake-pit 2]. 1951 E. Lambert *Twenty Thousand Thieves* 314 Baxter reckoned the officers and snakes are pinching our beer.

4. e. In various technical uses.

1947 *Britannica Bk. of Year* 841/1 *Snake*, nickname of a device used during an advance to destroy wires and detonate mines. 1957 *Daily Progress* (Charlottesville, Va.) 8 Jan. 5/3 A plumber's 'snake' has succeeded where a mixed pack of rats and mongooses failed. *Ibid.*, The snakes are thin flexible cables used to clean or carry wires inside pipes. 1961 W. Vaughan-Thomas *Anzio* ix. 207 The Snake was a 300-foot tube of steel packed with TNT up to about fifty feet from the tank, which first towed the tube into battle and then swung around and pushed it out over a minefield. The crew..exploded the TNT by fire from their machine-guns. 1964 'E. McBain' *Ax* v. 88 The plumber's snake had caught on one of the cross supports... Hawes reached up and shoved at the snake, coiling it back into the drawer.

f. *Econ.* A narrow range of fluctuation in rates of exchange, agreed to by certain member countries of the EEC (see quot. 1973). Hence *snake in the tunnel*: this range in relation to a wider range of fluctuation agreed in the foreign exchange markets.

1972 *Economist* 11 Mar. 87/1 Europe's currencies will try to be held inside the celebrated 'snake' wriggling within the overall 4.5 per cent dollar 'tunnel'. 1972 *Accountant* 12 Oct. 451/2 It would take over the day-to-day running of the so-called 'snake in the tunnel' system of exchange rate margins which Britain opted out of when the £ was floated on June 23rd. 1973 *Business Week* 10 Mar. 37/3 In March 1972, the six charter members of the EEC and the three nations then awaiting membership agreed to keep their currencies trading within a narrower band against one another than they do in trading against the dollar. When set down on graph paper, the snake is the narrow EEC band and the tunnel the wider dollar band. *Ibid.*, A year-old technique that is dubbed, whimsically enough, the 'snake in the tunnel'. 1975 *Sunday Tel.* 11 May 24/4 There may be an agreement on the amount the pound should be devalued..followed by a return to the European currency arrangement (the 'snake in the tunnel'). 1976 *Times* 14 Aug. 15/1 Finance ministers from the 'snake countries' (Belgium, Denmark, Germany, Luxembourg, The Netherlands, Norway and Sweden). 1979 *Dædalus* Winter 63 But the idea offers an opportunity..of avoiding the pitfalls of previous efforts that had aimed prematurely at stabilizing exchange rates in a European 'snake'. 1980 T. Barling *Goodbye Piccadilly* viii. 155 An illuminated wallchart showed the present float of the European Money Snake.

7*. With capital initial. Applied to American Indians of various Shoshone groups, esp. those of Oregon. Freq. *attrib.*, esp. as *Snake Indian.*

1791 in *Mass. Hist. Soc. Coll.* (1794) III. 24 The tribes of Indians..were called..the Blackfeet tribe, the Snake Indians [, etc.]. 1805 P. Gass *Jrnl.* 22 Oct. (1807) xiv. 154 This..is the same river whose head waters we saw at the Snake nation. 1831 *Niles' Weekly Reg.* IV. 265/2 They happily fell in with a small party of Snake Indians. 1821 J. Fowler *Jrnl.* 24 Nov. (1898) 55 Last night on Counting them over find now four Hundred of the following nations—Ietans—Arrapohoes—Kiawa Padduce—Cheans—Snakes. 1843 T. Talbot *Jrnl.* 7 Sept. (1931) 45 The trappers prefer Snake Indians and Snake horses before any race of men or horses in the world. 1890 N. P. Langford *Vigilante Days* xiii. 161 [With] a band of Snakes.., we can run off two thousand of the best of those animals. 1920 S. M. Drumm in J. C. Luttig *Jrnl. Expedition Upper Missouri* 166 Snake Indians. This tribe was so generally known by this term as to almost obscure the family name of Shoshoni. 1938 M. Thompson *High Trails Glacier Nat. Park* 53 The Snake warriors got ready for an attack as soon as the moon should come up. 1940 *Places to see in Wyoming* p. xxiv/2 Shoshones were also referred to as Snakes or the Snake People. 1977 [see *Paiute *sb.* a].

II. 8. a. *snake farm, meat, -poisoning* (later example), *-venom.*

1934 *Discovery* July 207/2 The Pasteur Institute in India, the Snake Institute at Port Elizabeth, South Africa, and the Butantan 'Snake-farm' near São Paulo, Brazil, are the headquarters of snake research and cure. 1979 *United States 1980/81* (Penguin Travel Guides) 34 Traveling by car you can be flexible—making any number of stops at souvenir shops or snake farms. 1976 H. Kemelman *Wednesday Rabbi got Wet* xxxix. 226 As alien and outlandish as snails or snakemeat or fried termites. 1965 R. & D. Morris *Men & Snakes* v. 106 Normally death by snake-poisoning is a prolonged and unpleasant business. 1951 Whitby & Hynes *Med. Bacteriol.* (ed. 5) xx. 317 It is possible that other toxins of these soil bacteria are (like the snake-venoms) primarily digestive ferments rather than aggressive mechanisms.

b. *snake ceremony, cult, cultist, -priest, -stick.*

1959 E. Tunis *Indians* ix. 128/1 Nearly all of the rituals had the same purpose: to cajole rain from the gods. The famous Snake Ceremony had that object. 1958 C. Achebe *Things fall Apart* iii. xxi. 159 His father was the priest of the snake cult. 1965 R. & D. Morris *Men & Snakes* iii. 67 It has been argued that the Egyptian contingent of the Jews in the Exodus may have been snake cultists and Moses himself a kind of snake shaman. 1900 *Outing* June 305/2 Then, like a flash, the Snake priests dart upon them grabbing in their hands all they can pick up. 1958 C. Achebe *Things fall Apart* iii. xxii. 165 One of them was Enoch, the son of the snake-priest who was believed to have killed and eaten the sacred python. 1902 *Chambers's Jrnl.* Feb. 81/1 Readers of that delightful novel, *The World went very well Then*, will remember Mr Brinjes of the fiery eye and the snake-stick, who made every negro do his bidding. 1974 H. MacInnes *Climb to Lost World* vi. 97 We were all hypersensitive about the possibility of being stung or bitten, and kept our snake sticks handy.

c. *snake-god* (later examples), *goddess, monster.*

1863 W. K. Kelly *Curiosities Indo-Europ. Trad. & Folk-Lore* i. 9 The bird, beast, and snake-gods. 1965 R. & D. Morris *Men & Snakes* ii. 28 The snake god Danh-gbi of Whydah, Dahomey. 1925 A. Evans *Ring of Nestor* 15 Besides the well-known Snake Goddess of the Temple Repository at Knossos, a series of other figures have now come to light showing this attribute. 1979 *Jrnl. R. Soc. Arts* July 511/2 But in this region they also make paper caskets on bamboo frames which are used in festivals, especially for that of the snake-goddess, Bishahari. 1965 R. & D. Morris *Men & Snakes* ii. 49 A similar snake monster, Typhon, who in Greek mythology merges with Typhoeus, was said to be the cause of earthquakes as well as many springs.

d. *snake bracelet, -buckle, -hook, -mark, -mask, -spiral.*

1968 *New Larousse Encycl. Mythol.* (ed. 2) 484/1 (*caption*) Snake bracelet from Dahomey. 1979 F. Morton *Nervous Splendour* (1980) ix. 89 He had long wanted to give Martha a gold snake bracelet, a status symbol. *a* 1882 H. Kendall in *Penguin Bk. Austral. Ballads* (1964) 92 A hero..With a jumper and snake-buckle belt on. 1971 P. D. James *Shroud for Nightingale* iv. 123 A schoolboy's belt..clasped with a snake buckle. 1978 M. Dickens *An Open Book* i. 6 Dining room lunch meant putting on a dress instead of the boy's shirt and flannel shorts and snake-buckle belt we wore at Chilworthy. 1944 Blunden *Cricket Country* xi. 122 Wearing a revolver holster on a snake-hook belt. 1929 D. H. Lawrence *Pansies* 39 In the odd pattern, like snake-marks on the sand It leaves its trail. 1968 K. Weatherly *Roo Shooter* 16 Two large [kangaroo] does..came in,.. their tails dragging long snake marks in the dust. 1965 R. & D. Morris *Men & Snakes* ii. 41 A snake mask set with turquoises, the emblems of the god [*sc.* Quetzalcoatl]. 1922 Joyce *Ulysses* 715 The snakespiral springs of the mattress being old.

9. *snake-bearing* adj.; *snake-charming* vbl. sb.; *snake-handling* vbl. sb. and adj.

1927 D. H. Lawrence *Mornings in Mexico* 162 The shoulders of the young, snake-bearing men. 1897 'Mark Twain' *Foll. Equat.* xliii. 388 The girls went through a performance which represented snake-charming. 1978 *Amer. Poetry Rev.* Nov./Dec. 25 Adam and eve because they had a snakecharming act. 1940 *Sci. News Let.* 17 Aug. 103/2 Snake-handling religious cultists of Georgia are 'all of a piece' with followers of other cults who go to unusual lengths to show their faith or their access to supernatural powers. The same thing, with or without snake-handling, has been seen in various cultures and various times. 1973 R. L. Fox *Alexander the Great* iii. 45 Snake-handling is a known practice in the wilder sorts of Greek religion.

10. *snake-green, -haired* (later example), *-hipped, -locked, -tailed, -tressed* (later example).

1948 C. S. Lewis in *Punch* 23 June 543/2 Sea-chances brought To her forest-silent And crimson-fruited And snake-green island Her guests unsought. 1921 W. de la Mare *Veil* 59 Snake-haired, snow-shouldered, pure as flame and dew,..Rises the Goddess. 1976 'G. Black' *Moon for Killers* i. 7 He looked like a Hollywood top actor of the fifties..still almost snake-hipped, with long, thrust-out legs. 1954 G. Barker *Vision of Beasts & Gods* 39 The snake-locked image of dream Hanging ahead. 1946 R. Graves *Poems 1938–45* 32 By noting that the snake-tailed chthonian winds Were answerable to fate alone, not Zeus. 1894 O. Wilde *Sphinx* 28 What snake-tressed fury fresh from Hell.

11. a. snake-bit(ten) *a.*, (*a*) bitten by a snake; (*b*) *U.S.* irremediably doomed to misfortune; **snake boot** *N. Amer.*, a boot with a high ankle worn for protection against snake-bites, or a fashion boot resembling this; **snake charmer** *Austral. slang* (see quots.); **snake eyes,** (*a*) *U.S. slang*, tapioca; (*b*) *N. Amer. slang*, a throw of two ones with a pair of dice; also *fig.*, bad luck; **snake-headed** *a. slang* (see quot. 1941); **snake hips,** (*a*) very narrow hips; (*b*) the name of a popular dance (see quot. 1970); so **snake-hip** *attrib.*; **snake juice**: for *Austr. slang* read '*slang* (chiefly *Austral.*)' and add: also *loosely*, any alcoholic drink; **snake oil,** a quack remedy or panacea; also *fig.*; freq. *attrib.*, esp. as *snake-oil salesman*; **snake poison**: for *U.S.* read *U.S.* and *Austral. slang* and add later example; **snake rail fence** *N. Amer.* = Snake-fence; **snake room** *Canad.* (see quot. 1912); **snake story** (earlier U.S. examples).

1807 Snake-bitten [in Dict., sense 10]. 1938 M. K. Rawlings *Yearling* xiv. 149 He sobbed, 'Pa—He's snake-bit.' 1942 W. Faulkner *Go down, Moses* 111 Ah'm snakebit and de pizen cant hawm me. 1957 *Daily Progress* (Charlottesville, Va.) 18 Nov. 14/1 It was another long afternoon Saturday at Scott Stadium for Coach Ben Martin, his assistants and his 'snake-bitten' football players as they fell before South Carolina, 13-0. *Ibid.*, Commenting on the game last Saturday afternoon Martin said: 'We're just snake-bit that's all there is to it.' Snake-bit is a term used by coaches when referring to a team which never seems to have a break in its favor. 1965 Mrs. L. B. Johnson *White House Diary* 10 June (1970) 283 From the first moment of the day we were 'snake-bit' —everything went wrong. 1976 *Columbus* (Montana) *News* (Joliet Suppl.) 17 June 2/3 We managed to get back to the house, not snake-bitten and not smelling *too* much like a skunk. 1965 Snakeboot [see *kinky *sb.*]. 1972 R. Reid *Canadian Style* (1973) iv. 144 'Say, what is, or are, galoshes?' 'Like rubber snake boots, but they buckle or zip up the front.' 1937 A. W. Upfield *Mr Jelly's Business* 16 'And what are the Snake Charmers?' 'They are the permanent-way men.' 1969 P. A. Smith *Folklore Austral. Railwaymen* 279 Fettlers are invariably referred to as 'snake charmers'. 1918 L. E. Ruggles *Navy Explained* 20 Tapioca is 'snake eyes'. 1929 M. A. Gill *Underworld Slang* 11/2 *Snake eyes*, aces up on the dice. 1935 *Jrnl. Abnormal Psychol.* XXX. 364 *Snake eyes*, tapioca. 1964 A. Wykes *Gambling* vi. 134 Modern craps players use..slang for various combinations of two dice: 'snake eyes' for Two, [etc.]. 1972 *Islander* (Victoria, B.C.) 30 July 12/3 But this time Baychimo's annual throw of the dice came up 'snake-eyes', and the ice closed about trapping her forever. 1978 R. Moore *Big Paddle* iv. 88 Cliff..let the dice go... He didn't have to look to know they'd come up snake eyes. 1978 G. Vidal *Kalki* vi. 138 It's like throwing dice. Let's just hope it won't be snake eyes for Jim Kelly. 1920 B. Cronin *Timber Wolves* viii. 137 Anyhow, they's no need to get snake-headed about it. 1941 S. J. Baker *Dict. Austral. Slang* 68 *Snake-headed*, angry, vindictive. 1932 *Daily Progress* (Charlottesville, Va.) 20 Apr. 4/3 There is a distinct class clash between the Harlem intelligentsia and snake-hip dancers and chanters of hot-cha-cha and skiddle-de-scow in the black and tan auberges. 1977 *Melody Maker* 26 Mar. 43/2 The biggest sensation of all..was the 'snake-hip' dancer, Bessie Dudley, waggling her bottom, clad in black satin knickers. 1933 *Fortune* Aug. 48/1 Dancers like the gelatinous 'Snake Hips' Tucker. 1956 G. P. Kurath in A. F. C. Wallace *Men & Cultures* (1960) 153 Restraints were shaken off..in an epidemic of angular, foot-twisting gyrations—the Charleston, Snake Hips, Susie-Q, and Truckin'. 1970 C. Major *Dict. Afro-Amer. Slang* 106 *Snake hips*, a Baltimore- and New York-oriented jazz dance. 1977 N. Slater *Crossfire* iii. 62 The fellows..all seem to have snake-hips, painfully tight trousers and platform shoes. 1904 E. S. Emerson *Shanty Entertainment* 70 Then he started them on snake-juice, known as Boot and Blacking Rum. 1965 M. McIntyre *Place of Quiet Waters* xii. 224, I wonder if that snake juice is fit to drink. 1973 R. Robinson *Drift of Things* 290 Broke into Eric's hut, threw the 'pickled' specimens out of the jars, and drank the methylated spirits. That must have been the real 'Snake-Juice'. 1927 S. V. Benét *John Brown's Body* 294 Crooked creatures of a thousand dubious trades, ..sellers of snake-oil balm and lucky rings. 1946 E. O'Neill *Iceman Cometh* i. 90 I'll bet he's standing on a street corner in hell right now, making suckers of the damned, telling them there's nothing like snake oil for a bad burn. 1961 *Washington Post* 10 May A4/2 Advertisers who try to 'lubricate the wheels of our economy with snake oil'. 1976 *Listener* 25 Mar. 382/1 Jimmy Savile has always had more *chutzpah* than a wagonload of snake-oil salesmen. 1977 *Rolling Stone* 21 Apr. 66/2 It was, after all, the Jew who was the perennial doubter, the archetypal outsider, longing for redemption while dismissing the claims of would-be redeemers as so much snake oil. 1978 *Times* 21 Jan. 12/7 The pseudo-graphic industry.. are snake-oil salesmen deceiving the public. 1947 K. Tennant *Lost Haven* iv. 66 If Bee-Bonnet ever again wants me to sample his snake poison, I'll pour it on him and set it alight. 1889 B. Harte *Cressy* ii. 38 Mr. McKinstry's 'snake rail' fence was already discernible in the lighter opening of the woods. 1958 H. Symons *Fences* 48 One of the early Canadian fences most popular in the east was the snake rail fence. 1912 J. Sandilands *Western Canad. Dict.* 42/1 *Snake-room*, a side room of a basement where saloon-keepers accommodate doped or drunken people until they recover their senses, presumably a place where they 'see snakes'. 1921 *Daily Colonist* (Victoria, B.C.) 29 Oct. 15/2 'Tommy' was one of a bunch who were swapping stories recently in the snake room. 1975 F. Kennedy *Alberta was my Beat* vi. 73 All adjourned to the 'snake room' in the basement. 1826 *Virginia Herald* (Fredericksburg) 6 Sept. 3/2 The New-York Spectator will probably class this with the Snake stories of the day. 1867 *Harper's Mag.* Aug. 281/3 We told snake and fish stories.

b. snake doctor *U.S.* = Dragon-fly or Hellgrammite; **snake feeder** *U.S.* = prec.

a 1883 G. W. Bagby *Old Virginia Gentleman* (1910) 92 [The water is] full of all manner of nasty and confounded 'mud-kittens', 'snap'n turtles', and snake doctors. 1948 *Field & Stream* July 42/2 Various stages of the dobson are known as..flip-flaps, snake doctors. 1978 *Amer. Speech* LIII. 201 The flora and fauna terms include.. *snake feeder* (listed as the common name for the dragon-fly, *snake doctor* being listed as 'slightly known'). 1861 *Trans. Illinois Agric. Soc.* IV. 341 A particular species of dragon-fly, or snake-feeder, as it is absurdly called in this country. 1904 G. Stratton-Porter *Freckles* xiv. 289 He shifted restlessly, and the movement sent the snake-feeders skimming. 1949 H. Kurath *Word Geogr. Eastern U.S.* 14/1 The line of demarkation over against the Midland snake feeder is remarkably clear and sharp.

c. snake-locked anemone = Opelet; **snake plant,** (*b*) = *mother-in-law's tongue* s.v. *mother-in-law.

1853 P. H. Gosse *Naturalist's Rambles Devon. Coast* iv. 96 The Snake-locked Anemone..is by no means common. 1928 Russell & Yonge *Seas* 37 Especially common in the pools is the 'Snake-locked anemone'. 1979 J. D. & J. J. George *Marine Life* 32/1 *Anemonia sulcata* ..(snakelocks anemone). A species with many sinuous tentacles. 1946 M. Free *All about House Plants* xviii. 271 The common Snake-plant..is one of the most inelegant of all plants, with its stiff, 30-inch, upright leaves. 1973 *Daily Colonist* (Victoria, B.C.) 21 Nov. 24/1 He had poured his heart out to a hardy sansevieria, otherwise known as snake plant or mother-in-law's tongues.

snake, *v.*^{1} Add: **II. 5. a.** (Earlier example.)

1829 T. Flint *George Mason* ii. 21 It was so contrived that..logs..could be drawn, or, as it is technically phrased, *snaked* into church.

b. (Earlier example.)

1856 M. Thomson *Plu-ri-bus-tah* xii. 135 First he pulled the pillow-case off. Then he snaked the stars and stripes off.

snake, *v.*^{2} For '*dial.* (chiefly *north.* and *Sc.*)' read '*dial.* and *U.S.*' and add: **2.** (Further examples.) Also, to cheat (someone) at cards.

a 1861 T. Winthrop *John Brent* (1862) xvi. 183 They snaked me to the figure of a slug at their cheatin' game. 1886 Kipling *Departmental Ditties* (ed. 2) 36 You will find excuse to snake Three days' 'casual' on the bust. 1921 T. Dreiser *Let.* 2 Jan. (1959) I. 333 Start the ball and if I snake the forty thousand..you get five thousand. 1959 [see *shaft *sb.*^{1} 9* c]. 1977 *Amer. Speech* 1975 L. 66 *Snake*,..steal (one's date) 'Carol tried to snake my date last night'.

snake-bark (snēi·kbāɹk). [f. Snake *sb.* + Bark *sb.*^{1} 1.] In full, *snake-bark(ed) maple.* A maple, esp. *Acer pennsylvanicum* from eastern North America or *Acer davidii* from eastern Asia, belonging to a group distinguished by bark streaked with white.

1838 J. C. Loudon *Arboretum & Fruticetum Britannicum* I. 407 The striped-barked Maple..Snake-barked Maple, Moose Wood. 1914 W. J. Bean *Trees & Shrubs Hardy in Brit. Isles* I. 153 *A. pennsylvanicum*..Snake-bark Maple.... This maple is remarkable chiefly for the exceedingly handsome striping of its younger branches and stem. 1974 A. Mitchell *Field Guide Trees Brit. Isles* 341 The Snake-barked Maples are a difficult and confused group. *Ibid.* 342 Similar species. *A. distylum*..although not truly a Snake-bark. 1977 *Harpers & Queen* Nov. 274/4 Suburban gardens..are now enriched with azaleas, ..snakebark maple, winter-sweet and all sorts of curiosities. 1980 *Amat. Gardening* 25 Oct. 29/1 With the exception of *Acer pennsylvanicum*..the Snakebarks are native to China or Japan.

sna·ke dance. Also **snake-dance.** [Snake *sb.*] **1.** Among the Hopi Indians, a religious dance involving the handling of live rattlesnakes. Also, among other American Indian groups, various dances so called from the motion of the dancers or the function of the dance.

1772 D. Taitt *Jrnl.* 6 Mar. in N. D. Mereness *Trav. Amer. Colonies* (1916) 517 The women danced the Snake dance, the leader haveing her legs Covered with Turpin shells which is filled with small stones on purpose to make a noise. 1883 [in Dict. s.v. Snake *sb.* 8 b]. 1891 *Rep. Bureau Amer. Ethnol.* VIII. 136 Among the Hopi, particularly at Walpi, the snake-dance is renowned. 1927 D. H. Lawrence *Mornings in Mexico* 136 The snake dance (I am told) is held once a year. *Ibid.* 138 Three thousand people came to see the little snake dance this year. 1940 R. Chandler *Farewell, my Lovely* xx. 158 I'm no school-marm at the snake dances. 1965 R. & D. Morris *Men & Snakes* ii. 38 The snakes used in the famous Hopi Snake Dance are not worshipped, but sent as messengers to the raingods of the underworld. 1970 K. Platt *Pushbutton Butterfly* ix. 100 A Shoshone snake dance.

2. a. A dance performed as a stage entertainment in imitation of the movement of a snake or involving the handling of a snake.

1895 *N.Y. Dramatic News* 23 Nov. 4 Ida Siddons in her snake dance, two Italian pantomimists [etc.]. 1971 R. Russell tr. *Ahmad's Shore & Wave* vii. 66 'What cabaret girl?' 'The one who does the snake dance.'

b. A dance performed by a group of people linked together in a long line and moving about in a zig-zag fashion, as at parties, celebrations, etc. orig. and chiefly *U.S.*

1911 G. Burgess *Find Woman* x. 244 So he..went, reminding them of the [football] score and the snake-dance every time he opened a bottle. 1946 E. B. Thompson *Amer. Daughter* 234 A few minutes later [I] was a link in a howling, writhing snake dance that weaved itself in and out of the business section. 1960 *Daily Tel.* 15 July 19/2, 23,000 in Tokyo anti-treaty snake-dance. 1965 G. B. Schaller *Year of Gorilla* viii. 206 One grabbed the rump hairs of the first one with both hands, the third animal did the same to the second one, and then all three careened wildly down a slope in snake-dance fashion. 1976 *National Observer* (U.S.) 10 July 11/1 During the rebellious days of the 1960s, college students spurned many traditional campus pleasures, from spring proms and snake dances to the social reassurance offered by membership in fraternities and sororities.

Also (with hyphen) as *v. intr.*; so **snake-dancing** *ppl. a.*; hence **snake-dancer.**

1922 *Chicago Daily Maroon* 3 Oct. 2/1 The public.. picture..howling, snakedancing crowds whenever colleges and universities are mentioned. 1931 F. L. Allen *Only Yesterday* 17 Eight hundred Barnard College girls snake-danced on Morningside Heights in New York. 1960 *Daily Tel.* 15 July 19/2 They snake-danced through the city with their paper lanterns and red banners and dispersed quietly without incident, police said. 1977 C. McCullough *Thorn Birds* v. 93 Princess Houri the Snake Dancer (See Her Fan the Flames of a Cobra's Rage!).

snake-fence. For *U.S.* read *N. Amer.* and add earlier and further examples.

1805 R. Parkinson *Tour Amer.* I. i. 48 Snake-fences; which are rails laid with the ends of one upon the other, from eight to sixteen in number in one length. 1844 F. Marryat *Settlers in Canada* 53 A herd of cattle were grazing on a portion of the cleared land; the other was divided off by a snake-fence..and was under cultivation. 1904 C. G. D. Roberts *Watchers of Trails* 239 The snake fence of split rails which bounded the pasture. 1973 L. Russell *Everyday Life Colonial Canada* ii. 33 In constructing a snake fence the rails of adjacent bays were overlapped at a wide angle... Such a fence was a zig-zag of bays.

snake-head. Add: **2.** (Earlier example.) Now *Hist.*

1845 *Yankee* (Boston) 9 Aug. 3/4 Mr. John F. Wall ..was near being killed..by what is technically called a *snakehead*.

4. Substitute for def.: A tropical marine or fresh-water carnivorous fish of the family Channidæ, esp. one of the genus *Ophiocephalus*, found in Africa or Asia, usually mottled grey, brown, or black in colour. (Examples.)

1905 D. S. JORDAN *Guide Study of Fishes* II. xxi. 370 Snake-head mullets..seem to us nearer the labyrinthine fishes. 1961 E. S. HERALD *Living Fishes of World* 244/2 Snakeheads will live for many hours and sometimes days out of water.

sna·ke-pit. Also **snakepit.** [SNAKE *sb.*]

1. Among primitive peoples, a large pit containing poisonous snakes into which victims are thrown for execution or as a test of endurance.

1883 VIGFUSSON & YORK-POWELL *Corpus Poeticum Boreale* II. IX. 346 Anslaug..gives him a charmed coat, which preserves him even in the snake-pit into which he is cast by Ælla, king of the Northumbrians. 1909 *Saga-Book of Viking Club* VI. 1. 73 Ragnarr was captured by King Ella, and cast into a snake pit. 1940–41 *Scandinavian Stud.* XVI. 32 We may say that the theme of the Snake Tower or Snake Pit is foreign to Europe but is a typical Oriental importation. 1961 H. TREECE *Jason* III. xxvii. 224 Medea halted at the edge of the snake-pit... Then she lowered herself to the marble rim of the pit and gently eased down among the little snakes. 1977 A. P. SMYTH *Scandinavian Kings in Brit. Isles 850–880* iii. 36 The account of Ragnarr's invasion of Northumbria and his death in the snake-pit at the hands of King Ælla is found in its most elaborate form in *Ragnars saga*.

2. *transf.* and *fig.*

1941 *Argus* (Melbourne) *Week-End Mag.* 15 Nov. 1/4 *Snake pit*, sergeants' mess. 1948 S. L. ELLIOTT in E. Hanger *Khaki, Bush & Bigotry* (1968) 91 Andy Edwards has been promoted and moved up to the snake pit with you and the other snakes. 1956 A. L. ROWSE *Early Churchills* v. 82 He was a man..simple and rigid, in that snake-pit of a Court with its twisting..creatures,..deceitful and insincere. 1966 B. GLEMSER *Dear Hungarian Friend* ii. 36 He is at the United Nations... That is the only way to survive in the political snakepit. 1969 *N.Y. Review Bks.* 21 Aug. 8/1 The venal and compromising snakepit of American politics. 1976 *Publishers Weekly* 20 Sept. 74/3 All depicted as guilty as hell, conniving, scheming, fighting and feuding. It's a snakepit of a scene. 1977 *Listener* 25 Aug. 245/1 A snake-pit of imperfect chromatic scales.

3. *spec.* A mental hospital (after the title of the novel by M. J. Ward: see quot. 1947).

1947 M. J. WARD (*title*) The snake pit. 1960 *Sunday Express* 15 May 17/4 The snake-pit women's ward. 1968 A. LASKI *Keeper* ii. 22 They had visited him in the snake-pit. 1976 *Courier-Mail* (Brisbane) 30 Apr. 5/1 It's like going back to the days when psychiatric hospitals everywhere were called snakepits.

snake-root. Add: **2. b.** = *RAUWOLFIA.

1955 *Sci. Amer.* Oct. 81/1 Reserpine is an alkaloid extract from the snakeroot plant. 1976 W. A. R. THOMSON *Herbs that Heal* ix. 147 The root, popularly known as 'snake-root' because of its long, tapering, crooked nature, contains most of the medicinal properties of the plant.

Snakes and Ladders. Also with hyphens and small initials. [See below.]

1. The name of a board-game for children in which the hazards and advantages are provided by snakes and ladders depicted on the board.

A counter that chances to arrive on a square at the head of a snake must be withdrawn to the snake's tail, while one that arrives at the foot of a ladder can be advanced to its top.

1907 *Yesterday's Shopping* (1969) 1031/1 *Snakes and Ladders.* An interesting and most exciting game of chance. 1933 N. STREATFEILD *Tops & Bottoms* xii. 146 Felicity thought that bringing up Beaty was rather like playing Snakes and Ladders, through no fault of your own stepping on the head of a snake and sliding to the bottom again; in this case, with no ladder in view up which to shoot to regain lost ground. 1946 E. LINKLATER *Private Angelo* viii. 89 Promotion in war-time was like..a game of snakes-and-ladders. 1964 A. WYKES *Gambling* vi. 128 Backgammon..is the precursor of practically every modern board-and-pieces game in which the moves are decided by dice—even parlor games like *snakes-and-ladders*. 1980 *Daily Tel.* 25 Jan. 15/3 New and more sophisticated versions of snakes and ladders and noughts and crosses will also be provided [on trains].

2. *fig.* A series of unpredictable successes and set-backs. Hence *snake-and-ladder* adj.

1930 M. ALLINGHAM *Mystery Mile* xxiii. 213 The [fire] engines were still drawn up outside Number Thirty Four. 'Still playing Snakes and Ladders, I see,' said Campion. 1961 *Listener* 31 Aug. 319/2 The artists whose life-work does not find a place in the historical snakes and ladders of *avant-garde* development. 1978 G. GREENE *Human Factor* v. iii. 267 After so many years of concealment he was beginning to enjoy this snake-and-ladder game. 1978 G. SIMS *Rex Mundi* xix. 117 Snakes and ladders progress, old man—that's all we can hope for. A lucky throw that puts us on the ladder. 1982 *Church Times* 23 July 6/1 Nothing could illustrate more graphically the snake-and-ladder aspect of a political career than the last year of Tony Crosland's life.

snakesman. For *Obs.* read 'Now *Hist.*' and add later examples.

1973 G. BUTLER *Coffin for Pandora* vii. 153 Perhaps I did train up one or two snakesmen..and perhaps we did work together. 1975 M. CRICHTON *Great Train Robbery* v. 31 A snakesman was a child adept at wriggling through small spaces.

snaking, *vbl. sb.* Add: **4.** A rapid oscillation of an aircraft about a vertical axis; a similar motion of a caravan or trailer.

1945 *Jrnl. R. Aeronaut. Soc.* XLIX. 463/2 There is a possibility of the aeroplane developing an undamped short period oscillation in which rapid movement of the rudder from side to side plays an essential part—the tail wagging the dog. Such an oscillation is known as 'snaking'. 1949 *Aircraft Engin.* Oct. 311/1 Snaking can be cured by sticking on strips to the trailing edge of the rudder. 1966 *Caravanning* ('Know the Game' Ser.) 11 Stabilisers have been devised to overcome any side movement or 'snaking' of the caravan in relation to the direction of the car. 1972 *Nature* 18 Aug. 377/2 Detailed studies of gust effects, and of 'snaking'.

snakishly (snēⁱ·kɪʃlɪ), *adv.* [f. SNAKISH *a.* + -LY[2].] In the manner of a snake; treacherously, venomously.

1935 E. R. EDDISON *Mistress* xi. 204 The Vicar, regarding him snakishly, drew back his thin lips in a smile. 1963 D. HUGHES in Sissons & French *Age of Austerity* iv. 87 Heath..was accepted..as a quite normal manifestation of that time..which made his vicious sallies into madness..all the more snakishly repellent.

snaky, *a.* Add: **3. a.** (Later N. Amer. examples.)

1933 J. V. ALLEN *Cowboy Lore* iv. 101 If you reckon your mounts are some snakey and raw Just try ridin' herd on a stove that won't draw. 1966 M. & O. MURIE *Wapiti Wilderness* iv. 71 Oh, I believe he's a pretty good horse. He may be a little bit snaky. 1980 [see *shavetail* s.v. *SHAVE *v.* 13].

b. *Austral.* and *N.Z. slang.* Angry, annoyed.

1919 W. H. DOWNING *Digger Dialects* 46 *Snaky*,..(1) angry (e.g., to turn snaky); (2) irritable. 1941 K. TENNANT *Battlers* 86 Now lay off, sport... Don't go snaky on the kid. 1943 *Amer. Speech* XVIII. 90 [In New Zealand] *To go crook* is to show anger or annoyance... The verbs *to go snaky, to go maggoty*..have the same implications. 1945 N. MARSH *Died in Wool* vii. 155 There was a hold up... Everyone was snakey. Young Doug says the sheep are dry and I say they're not. 1974 D. WILLIAMSON *Three Plays* 34 What are you snaky about this time? 1981 *Courier-Mail* (Brisbane) 28 Nov. 23/1 They remain very snaky indeed about allegedly non-impartial treatment from players and umpires in Perth.

snallygaster (snæ·ligɑstəɪ). *U.S. dial.* [ad. G. *schnelle geister*, lit. 'quick spirits'.] A mythical monster supposedly found in Maryland. Cf. *SNOLLYGOSTER.

1940 *Maryland* (Writers' Program) 348 Residents of a Negro settlement near the distillery are firm in their belief that the neighborhood has a 'snallygaster'—a fabulous reptilian bird of vast size that preys on poultry and Negro children after nightfall. 1949 *Sun* (Baltimore) 28 July 14/1 (*heading*) Could it have been a snallygaster? 1954 *Sunday Sun* (Baltimore) *Mag.* 31 Oct. (*recto front cover*), Is this, at last, the snallygaster that has been said to terrorize Western Maryland but that most people have considered legendary?

snam (snæm), *v. slang.* ? *Obs.* [Origin unknown.] *intr.* To snatch; to steal.

App. recorded only in Dicts.

1824 J. MACTAGGART *Scottish Gallovidian Encycl.* 429 *Snam*, to snap at any thing greedily. 1874 HOTTEN *Slang Dict.* 298 *Snam*, to snatch, or rob from the person. Mostly used to describe that kind of theft which consists in picking up anything lying about, and making off with it rapidly.

Hence as *sb.*; also **sna·mmer**, one who snams; a thief.

1839 H. BRANDON in W. A. Miles *Poverty, Mendicity & Crime* 164/2 *Pudding Snammer*, one who steals from a cook shop. 1887 Snam [see *DUB *sb.*[4]]. 1950 PARTRIDGE *Dict. Underworld* 649/1 *On the snam*, engaged in stealing.

snap, *sb.* Add: **I. 2. b.** (Earlier examples.)

1882 *Adventures Billy Shakespoke* v. 89, I dropped in to see my old partner..and he proposed that we should try another 'snap' in Lynn. 1885 *Santa Fé Weekly New Mexican* 24 Sept. 4/6 It is the custom, during the summer months, for 'snap' companies to travel through the country and gather shekels.

c. = *soft snap* s.v. SOFT *a.* 27. chiefly *N. Amer.*

1877 H. RUEDE *Sod-House Days* 120 It is no snap, for the straw rolls out fast enough to keep them very busy. 1901 *Daily Colonist* (Victoria, B.C.) 27 Oct. 3/4 Formerly porters received as low as $15 a month and this wage in a buffet car was at one time considered a snap, as tips were wont to bring a man's income up to all the way from $100 to $200 a month. 1924 P. MARKS *Plastic Age* xxiv. 287 He had three classes in literature, one in music —partly because it was a 'snap' and partly because he really wanted to know more about music—and his composition course. 1936 V. SHEEAN *Personal Hist.* i. 3 The football players, the social lights, the pretty co-eds, and all the other students who regarded study as an inconvenient detail in college life, rushed to inscribe themselves for 'snap' courses. 1962 A. LURIE *Love & Friendship* viii. 180 The new semester has started, and I have a whole new selection of little 'creative writers' on my hands...the course is rumored to be a Snap (one thinks of those paper crackers at children's parties). 1967 *Technology Week* XX. 95/2 (Advt.), Blazing a path to the moon is no snap. Neither is charting a career.

4. (Later examples.) Also in *Comb.*

1913 D. H. LAWRENCE *Sons & Lovers* i. 25 She..put him out a clean scarf and snap-bag. *Ibid.* iv. 65, I went to put my coat on at snap-time. 1935 A. J. CRONIN *Stars look Down* I. ix. 67 'Come on, ye old beggor, and have yer snap,' Tom called out with his mouth full of bread and cheese. 1960 C. DAY LEWIS *Buried Day* vii. 131 The black-faced miners cycling home from work with their snap-tins bumping at their sides. 1980 *Guardian* 11 Nov. 8/3 At 10 o'clock the regular farm hands disappeared to the dutch barn for their 'snap'.

5. d. Add to def.: The call of 'snap' is made when two matching cards are exposed. (Earlier examples.) Also *attrib.*, as *snap-card*.

1881 *Cassell's Bk. Indoor Amusements* 144 The game of Snap may either be played with the ordinary Whist cards or with special cards prepared for the purpose. 1890 CHAMPLIN & BOSTWICK *Young Folks' Cyclopaedia Games & Sports* 659/1 *Snap*, a game played by any number of persons with 36 cards. 1916 *N. & Q.* 9 Sept. 210/1 Who designed the illustrations that appear on snap cards, and when did they first appear? 1966 J. DERRICK *Teaching English to Immigrants* v. 188 Much pre-reading apparatus can be used for this purpose, such as word-matching cards, snapcards, word lotto, and other sets of apparatus where identifying and matching single words is involved.

e. A temporary faro game.

1845 J. J. HOOPER *Some Adventures Capt. S. Suggs* x. 133 I'll never bet on two pair agin! They're peart at the snap game, theyselves; but they're badly lewed this hitch! 1864 W. B. DICK *Amer. Hoyle* 208 *Snap*, [in Faro] a temporary bank, not a regular or established game. 1938 H. ASBURY *Sucker's Progress* 280 A few of the river gamesters ran Faro snaps when ashore in St. Louis, but most of them concentrated on Poker.

f. A U.S. party game in which one of the players chases another round a ring formed by the rest.

1865 B. L. RIDLEY *Battles & Sk. Army of Tennessee* (1906) 481 Games [in Georgia] soon began—'Thimble', 'Snap', and kissing songs. 1930 *Virginia Jrnl. Educ.* Oct. 73 Social intercourse [in the mountains of Virginia] was very limited. Monthly religious meetings at widely scattered churches, occasional parties at which 'Boston', 'Snap' and 'Shaker's Dance' were played all night long. 1944 G. WILSON *Passing Institutions* 93 Our liveliest game was Snap, a game that used to seem very exciting but now somewhat resembles Drop the Handkerchief.

g. *U.S.* and *Canad. Football.* = *SNAP-BACK 1 b.

1922 P. D. HAUGHTON *Football & how to watch It* 30 Watch the offensive ends begin their mad rush downfield at the snap of the ball. 1947 *Richmond* (Va.) *Times-Dispatch* 9 Nov. B7/7 From a single-wing formation to the right, Deuber, the tailback, took the snap and set sail wide around Virginia's left end. 1958 *Edmonton Jrnl.* 7 Aug. 7/2 London kicker Legg fumbled a snap on the third down. 1974 *Plain Dealer* (Cleveland, Ohio) 27 Oct. 2-C/3 Penn State..converted a fumbled snap into the game's first score only three minutes into the first period.

9. b. (Later examples.)

1899 *St. George's Hosp. Gaz.* VII. 91 An extremely pretty set of Kodak 'snaps' are contributed by Mr. Peck. 1950 *Nat. Geogr. Mag.* Apr. 514/1 We..eventually secured a few satisfactory snaps of the ordinary garden variety of jump. 1977 *Time* 26 Sept. 31/2 They even had a prospectus put together for publishers and included some sample snaps.

11. a. (Earlier example.)

1865 *Harper's Mag.* Jan. 145/2 [They were] good enough people in their way, but had no snap about them. She liked people with snap.

b. (Earlier example.)

1870 'MARK TWAIN' *Lett. to Publishers* (1967) 49, I should write the book as if *I* went through all these adventures myself—this in order to give it snap and freshness.

II. 12. e. *U.S.* A trick, deception, trap; also in phr. *to give the snap away*.

1844 *Lexington* (Kentucky) *Observer* 18 Sept. 3/1 Mr Van Buren..with his characteristic politeness *declined to be caught in any such snap*. 1885 *Weekly New Mexican Rev.* 2 July 4/3 He was roped into this snap by Chicago sharpers. 1900 *Congress. Rec.* 15 Feb. 1850/2 Ex-Senator Vilas gave the snap away when he said [etc.]. 1919 E. HOUGH *Sagebrusher* 501 If that girl's not blind she'll get out and give this snap away.

13. e. (Later examples.)

1976 *Eastern Daily Press* (Norwich) 16 Dec. 13/4 The friction strips which make the bang are known as 'snaps' in the industry. 1980 *Daily Tel.* 25 Nov. 15/5 Those who felt inspired to make their own crackers..may have encountered some difficulty in finding the vital bangers, called snaps in the trade.

III. 14. c. *pl.* Hand-cuffs. *slang.*

1895 J. CAMINADA *Twenty-Five Years Detective Life* 49, I put the 'snaps' on 'Pudding', and conveyed him..to Livesey Street police station. 1910 [see *NIPPER *sb.*[1] 4 c]. 1958 M. PROCTER *Man in Ambush* x. 119 We got the snaps on him and locked him up. 1967 —— *Exercise Hoodwink* xxv. 178 Sergeant, we'd better have the snaps on these three.

d. *U.S.* A press-stud or snap-fastener. *Usu. pl.*

1964 *McCall's Sewing* xii. 221/1 Snaps are used to hold fabrics together where there is little strain on the garment. They give a neat flat closure. 1968 J. UPDIKE *Couples* (1970) ii. 160 Frank's delicate hand uncoupled her bra snaps. 1977 *New Yorker* 27 June 72/3 The [pillow]slip is homemade, with snaps at one end.

15. a. *snap, crackle,* (*and*) *pop*, an advertiser's catchphrase representing the lively sound produced by a brand of breakfast cereal when milk is added; used allusively and in *transf.* senses for breakfast cereal or for vigour or energetic behaviour.

1954 *Daily Mail* 15 Dec. 2/1 (Advt.), With their fascinating 'Snap! Crackle! Pop!' as the milk's poured on, *Rice Krispies* really do seem to be talking. 1959 *Times Lit. Suppl.* 27 Mar. 179/4 In the 1950s the whole of

America, Canada, Australia, England and parts of Europe are eating cereals for breakfast—snap, crackle, pop. 1960 *Guardian* 17 Mar. 9/2 Marples..always acting with that zestful snap, crackle, pop that entertains as much as it nourishes. 1962 'R. GORDON' *Doctor in Swim* xi. 66 Now the poor fellow was as jumpy as a plate of snap-crackle-pop when you pour the milk on. 1962 F. WILLIAMS *Amer. Invasion* ii. 20 With every snap, crackle and pop on the breakfast table the American accent carries farther. 1963 *Trade Marks Jrnl.* 1 June 730/2 Snap Crackle Pop. 851,181. Cereal preparations made of rice for food for human consumption. Kellogg Company of Great Britain Limited..Manchester, 3rd July 1963. 1965 *Times Lit. Suppl.* 22 Apr. 315/5 When due allowance has been made for the chapter's snap crackle pop style. 1977 *N.Y. Rev. Bks.* 28 Apr. 11/4 But the few paragraphs of real information are hard to find in the snap, crackle, and pop of gossip and insult. 1979 P. LEVI *Head in Soup* vii. 128 Snap, crackle, pop. The telephone went dead.

b. (Earlier U.S. examples.)

1833 S. SMITH *Life & Writings J. Downing* 43 As long as I have President Jackson to look to for paymaster, I don't care a snap about sending in any bills. *a* 1852 F. M. WHITCHER *Widow Bedott Papers* (1856) xxii. 232 If you don't care a snap for him, what makes you go with him to lecters, and concerts, and sleigh rides?

18. (Earlier examples.)

c 1770 J. RANDOLPH *Treat. Gardening* in Gardiner & Hepburn *Amer. Gardener* (1818) 275 French beans and snaps are the same. 1842 C. M. KIRKLAND *Forest Life* II. xli. 165 'Snaps' are green beans.

snap, *v.* Add: **II. 5. a.** (Earlier example in Cricket.)

1872 *Wisden* 23 John Smith stayed with Mr Grace until 63 runs were made, when Pooley snapped him.

f. *U.S.* and *Canad. Football.* To put (the ball) in play by passing it quickly backwards to begin a scrimmage; to make a snap (sense *5 g). Also with *back*.

1887 *Outing* Oct. 70/1 In a scrimmage he places it on the ground, and at a signal from his quarter, snaps the ball back by a downward and backward pressure with his foot. 1920 W. CAMP *Football without Coach* iii. 48 Now let us say the quarter calls the signal..the play would get under way and the center would snap him the ball. 1968 *Globe & Mail* (Toronto) 3 Feb. 37/6 During such periods, when a pass is incomplete or a ball goes into touch, time will not resume until the ball has been snapped on the next play. 1973 *Philadelphia Inquirer* (Today Suppl.) 7 Oct. 42/1 Moss lights up and hunches over his desk, like a linebacker waiting for signals. The ball is snapped; he's off.

g. To match (an exposed card in a game of snap); to call out 'snap!' to (an opponent).

1935 *Encycl. Sports* 568/2 In case a player calls snap when there is nothing to snap on the table, the cards in front of him go to a pool. *Ibid.*, Grimace snap is extremely simple. Instead of snapping each other, the two players are under contract to make each other laugh, to which they may do anything except speak.

7. b. (Later examples.)

1950 T. S. ELIOT *Cocktail Party* I. ii. 60 Have you looked in your bag?.. Well, don't snap my head off. 1976 J. I. M. STEWART *Memorial Service* i. 12 He adores the place... That's why he snaps your head off if you venture to say a good word for it.

III. 12. a. Also, to switch *off* or *on*, or to shut *to*, with a snapping sound; to cause (fabric, elastic, etc.) to make such a sound.

1911 H. S. HARRISON *Queed* 68 Queed cleverly bethought him to snap on an electric light. 1922 JOYCE *Ulysses* 21 Haines helped himself [to a cigarette] and snapped the case to. 1925 F. SCOTT FITZGERALD *Great Gatsby* viii. 191 About five o'clock it was blue enough outside to snap off the light. 1926 J. MASEFIELD *Odtaa* xvi. 277 He snapped-to the breech of his rifle. 1949 B. A. BOTKIN *Treas. S. Folklore* II. iii. 252 Both Bilbo and Gene Talmadge were famous for their red suspenders, which Talmadge loved to snap. 1962 J. UPDIKE *Centaur* (1963) vi. 178 He laughed and behind me I could hear all the Caucasus laughing and snapping their towels and flipping their silvery genitals. 1979 R. JAFFE *Class Reunion* (1980) II. vii. 248 Ken was dressed, snapping on his wrist-watch. *Ibid.* 254 Ken snapped off the TV with his remote control.

13. b. Also, to fit *in*, to come *off*, with a snap.

1967 *Boston Sunday Herald Mag.* 26 Mar. 26/2 (Advt.), Quality absorbent reusable cotton pad snaps in—removes easily for laundering. 1976 *Columbus* (Montana) *News* 1 July 3/3 (Advt.), Safety grilles snap off to clean. 3-speed 20" Fan.

IV. 15. b. (Later examples.)

1927 WODEHOUSE *Meet Mr Mulliner* ix. 310 Something seemed to snap in James. The scales seemed to fall from James's eyes. 1933 E. O'NEILL *Days without End* (1934) I. 49 He knew..she was going to die... He..saw that no miracle would happen... Something snapped in him then. 1970 A. FRY *How a People Die* xxiv. 212 Something snapped. I lost my temper and I chewed that poor guy out from hell to breakfast.

d. *colloq.* To change one's behaviour or position quickly, esp. *to snap back*: to recover; *to snap* (*in*)*to*: to throw oneself smartly into (an action); *to snap out of*: to desist from (an attitude, etc.), to change a mood, pattern of behaviour, etc., by sudden effort. Freq. as imp. *snap out of it*.

1918 in F. A. Pottle *Stretchers* (1929) ix. 239 Oh, snap into it! We want to get this done. 1918 [see *HIT *v.* 22* e]. 1928 *Sat. Even. Post* 7 Jan. 9/3 Oh, for heaven's sake, Lucia, snap out of it and act like a human being. 1941 N. MARSH *Death & Dancing Footman* (1942) vi. 114 Do snap out of being all Freudian. 1943 K. TENNANT *Ride on Stranger* xviii. 205 Time we were getting a move on... Snap into it, Joe. 1944 *Sun* (Baltimore) 13 Jan. 5/1 If the Government acts quickly.., the aircraft industry will snap back quickly. 1962 J. GLENN in *Into Orbit* 18 We had to demonstrate how well we could undergo all kinds of stress and discomfort and then snap back again. 1967 [see *OWNSOME]. 1967 *Boston Sunday Herald* 14 May (This Week Mag.) 15/3 The Senator..spent half an hour persuading a very reluctant repairman to come. 'Why,' asked a guest, 'didn't you just tell him to snap to it?' 1981 M. SPARK *Loitering with Intent* x. 158 We mustn't get morbid. Let's snap out of it.

16. b. Also *spec.* in sport, to break a tie or a pattern of performance. *U.S.*

1951 *Amer. Speech* XXVI. 230 Michigan snaps Gopher streak. 1967 *Boston Herald* 8 May 16/6 His run-scoring single in the fifth inning climaxed a two-run rally that snapped a 3-3 tie. 1973 *Internat. Herald Tribune* 15 June 15/6 California held on to score a 7-5 home victory over Boston, snapping a four-game losing streak. 1976 *Washington Post* 19 Apr. D3/1 Danny Lawson's goal at 14:53 of the third period snapped a 4-4 tie.

d. To get (a person) *out of* a certain frame of mind. Cf. sense 15 d above.

1957 A. GRIMBLE *Return to Islands* iv. 78 Once they had struck their noble attitude officially..nothing but the crack of doom would ever snap them out of it. 1964 M. MCLUHAN *Understanding Media* I. v. 55 The parallel between two media holds us on the frontiers between forms that snap us out of the Narcissus-narcosis. 1968 *Globe & Mail* (Toronto) 13 Jan. 28/5 Brisk way to snap yourself out of the post holiday lethargy is to get out your little or not so little lists and decide to do some entertaining.

snap (snæp), *int.* The call in the card-game snap (SNAP *sb.* 5 d in Dict. and Suppl.); hence as an exclamation used when two similar objects turn up or two similar events take place.

1890 CHAMPLIN & BOSTWICK *Young Folks' Cycl. Games & Sports* 659/2 When a player turns a card having the same design as one on the top of another player's exposed pile, both must say 'Snap'. 1958 N. F. SIMPSON *Hole* 7 *Cerebro.* He seems to be biding his time at the moment. *Soma.* Snap! 1962 J. BRAINE *Life at Top* v. 88 He passed me his cigarette-case. The cigarettes..bore his initials. I reached for my cigarette-lighter then took out instead one of the books of matches I'd taken away from the Savoy. He looked at the matches and grinned. 'Snap,' he said. 1971 M. RUSSELL *Deadline* viii. 95 'I've read your stuff.' 'Snap.' 1980 J. WAINWRIGHT *Venus Fly-Trap* 39 Daphne, too, was wearing dark glasses... Harry ..murmured, 'Snap.'

snap-. Add: **A. a. snap action**: also used *attrib.* to designate switches and relays that make and break contact rapidly, independently of the speed of the actuating mechanism; **snap-bean** (earlier example); **snap-brim**, used *attrib.* to designate a type of hat for men with a brim which may be arranged in different ways; also *absol.*; hence **snap-brimmed** *a.*; **snap gauge** *Mech.*, a form of caliper gauge that can be used to check that a component is neither too large nor too small within stated tolerances.

1951 *Chambers's Jrnl.* Oct. 639/2 A snap-action switch ..cuts out the power supply at short and timed intervals. 1962 *Newnes Conc. Encycl. Electr. Engin.* 726/1 Many rotary switches are manufactured for d.c. operation with snap-action mechanisms. 1977 R. W. SMEATON *Switchgear & Control Handbk.* III. 4 Snap-action contacts reduce the arcing time. 1770 M. AMBLER *Jrnl.* Sept. in *Virginia Mag. Hist. & Biogr.* (1937) XLV. 156 A Breast of Veal for Dinner Snap Beans & gooseberry tart. 1928 *Daily Express* 5 July 9/4 The snap-brim (or turn-down) soft felt hat. 1941 [see *BRETON *sb.* and *a.*]. 1969 V. C. CLINTON-BADDELEY *Only Matter of Time* 45 Davie..put on his soft green hat... Why a 'snap brim', by the way? He always wondered. 1972 B. F. CONNERS *Don't embarrass Bureau* (1973) I. 3 The man was wearing a nondescript raincoat and a gray snap-brim hat. 1949 B. A. BOTKIN *Treas. S. Folklore* II. iii. 252 Loud checked suit, flaming necktie, diamond stickpin, and rakish snap-brimmed felt hat. 1976 *Maclean's Mag.* (Toronto) 15 Nov. 28/1 Al Capone wasn't talking through his snap-brimmed hat. 1918 D. T. HAMILTON *Gages, Gaging & Inspection* iii. 67 Snap gages used in general manufacturing are made in three types; namely, solid, adjustable, and built-up gages. 1964 S. CRAWFORD *Basic Engin. Processes* iv. 298 When using snap gauges for checking external diameters the component should be gauged in a number of positions along and around its surface.

b. *snap button, closing, closure, fastener, fastening, -joint, -lid, -lock, shackle*; **snap-link** (later examples); *spec.* = *KARABINER; **snap-ring**, (*a*) (later examples); (*b*) = *KARABINER; **snap switch**, a snap-action switch.

1897 *Sears, Roebuck Catal.* (1968) 228/3 Gloves... Set in thumbs, and patent snap buttons. 1976 *National Observer* (U.S.) 23 Oct. 20/2 His pearl snap-button shirt is open, his belt loosened a notch, his head propped up with a bulky, weathered arm. 1964 *McCall's Sewing* 171/2 (*heading*), Sleeve with snap closing. 1969 *Sears Catal.* Spring/Summer 16 Band waist, snap closure. 1976 *National Observer* (U.S.) 10 Apr. 16/3 (Advt.), The waist band and two patch pockets are fitted with permanent and attractive metal snap closures. No buttons to fall off. 1895 *Montgomery Ward Catal.* Spring & Summer 289/3 Men's Genuine Oil Tanned Calfskin Gloves,..one button, patent snap fastener. 1976 M. MAGUIRE *Scratchproof* xi. 165 My eyes instinctively traced the zipper, buttons and snap fasteners on her suit. 1898 *T. Eaton & Co. Catal.* Spring & Summer 124/3 Men's Klondike sleeping bags,..envelope top, snap fastenings, riveted corners. 1981 *Sunday Express Mag.* 11 Oct. 17/2 (Advt.), And each of the two key fobs has its own snap fastening. 1956 *Archit. Rev.* CXIX. 213/2 In this case the positive fixing and covering are achieved by a simple snap-joint incorporated in the edge corrugations. 1968 *Gloss. Terms Mechanized & Hand Sheet Metal Work* (*B.S.I.*) 21 *Snap joint*, the junction between two pieces of sheet or strip, the edges of which are formed so that they can be clipped together to form a rigid joint. 1932 *N. & Q.* 13 Feb. 123/2 Portable Ink-Bottles: these were in quite common use until recent times... I frequently had in a waistcoat pocket, the small square, or oblong leathered covered box, with a snap-lid. 1946 D. R. BROWER *Man. Ski Mountaineering* (ed. 2) iii. 28 The army aluminium carabiner (snaplink) has equal strength, weighs much less. 1981 L. DEIGHTON *XPD* xxxvii. 294 Never mind all these modern contraptions—pitons, snap links and stirrups. 1913 *Strand Mag.* Nov. 103 (Advt.), Smart, neat, snap-lock links. Of all Hosiers, Outfitters, etc... *Snap-lock Link Co.* 1971 B. MALAMUD *Tenants* 12 His apartment, stoutly protected by two patent locks plus a strong snap-lock enclosing heavy circular bolts. 1941 T. A. H. PEACOCKE *Mountaineering* ii. 25 Each climber carries a loop of line... A 'karabiner' (snap-ring) is hung on the loop. 1941 *Engineers' Digest* II. 321/1 Snap rings provide an economical and effective means of facilitating machine assemblies in applications where the loading is not excessive. 1957 *Listener* 28 Nov. 882/2 The less fortunate..of us have to..join the modern school of artificial rock climbers..with their pitons, snap rings, miniature ladders, [etc.]. 1970 K. BALL *Fiat 600, 600D Autobook* i. 12/2 With a special tool..depress the valve springs, remove the split cotters and snap rings. 1974 *Islander* (Victoria, B.C.) 21 July 11/1 Another day ..a snap shackle on the mizzen staysail broke. 1926 *Gloss. Terms Electr. Engin.* (Brit. Engin. Stand. Assoc.) 159 Snap switch. 1977 R. W. SMEATON *Switchgear & Control Handbk.* III. 6 (*heading*) Resistance of control-circuit duty contacts and snap switches.

d. (Further examples.) Also, **snap freezing**, freezing done by reducing the temperature suddenly to well below freezing point; hence **snap-freeze** *v. trans.*

1841 *Congress. Globe* X. App. 42/3 This extra session of Congress, called in time of peace to take snap judgments on the American people. 1867 *J. Lillywhite's Cricketers' Companion* 13 Get some one to give you difficult catches... You will soon be good at snap-catching. 1888 A. G. STEEL in Steel & Lyttelton *Cricket* iii. 179 The object of short-slip is to pick up snicks which just miss the wicket-keeper, and although he may hold a larger proportion of these quick snap catches when a long way from the wicket, he will get an infinitely greater number when closer in. 1896 *Congress. Rec.* 27 Feb. 2214/2 When the snap tally was taken, he..went to the clerk. 1898 A. P. ATTERBURY tr. *Sombart's Socialism & Social Movement in 19th Cent.* vi. 137 A snap resolution of a working-men's congress. 1932 'N. SHUTE' *Lonely Road* xi. 229 This isn't any snap decision on my part. I've been thinking of it for some time. 1933 *Mod. Lang. Notes* XLVIII. 393 In most cases they reflect, not conclusions drawn from research, but snap judgments based on chance observation and personal likes and dislikes. 1940 *Sun* (Baltimore) 23 Sept. 12/2 (*caption*) Charley Gehringer takes a snap throw from Shortstop Dick Bartell. 1954 X. FIELDING *Hide & Seek* v. 67 Unless we were unlucky enough to be held up by a snap Gestapo..check. 1955 *New Biol.* XVIII. 93 Slow cooling is supposed to cause more damage than rapid cooling since it results in the formation of larger ice crystals, and the use of 'snap' or ultra-rapid freezing has been advocated on the grounds that ice crystal formation may be avoided altogether, the system cooling in an amorphous state. 1959 'A. GILBERT' *Death takes Wife* v. 54, I had to keep this snap appointment. 1959 *Punch* 27 May 704/3 You have to be constantly on the *qui vive* if you are not going to risk being behindhand with the timing of the first drop [of hail], or even missing a snap shower altogether. 1960 Snap-test [see *BREATHALYSER]. 1960 V. JENKINS *Lions down Under* xi. 161 There was no doubt about his speed and capacity for 'snap-thinking'. 1965 *Listener* 21 Oct. 637/2 The snap opinions in this programme which I have always respected are those of Patrick Campbell. 1973 *Times* 8 Jan. 3/3 The investigation..will involve snap checks at garages. 1974 HAWKEY & BINGHAM *Wild Card* ix. 95 If he builds a cryogenic tank into the spacecraft, it'll rupture on explosion, spray the tissue with liquid oxygen, snap-freeze it, and provide perfect histological specimens for the investigators. 1974 B. A. NEWTON *Trypanosomiasis & Leishmaniasis* 277 Annear (1956) described a technique for 'snap freezing' *Crithidia oncopelti* on a 'peptone plug'. 1974 'M. INNES' *Appleby's other Story* viii. 62 One must make no snap judgement. 1974 *Socialist Worker* 26 Oct. 15/1, 2800 teachers in 90 schools staging snap strikes in support of a £15-a-week interim increase. 1976 *Alyn & Deeside Observer* 10 Dec. 16/3 The information arrives at a time which can allow only for a 'snap decision'. 1977 *Times of Zambia* 7 Sept. 2/2 A snap survey of clinics found that medical assistants in the health institutions had not been able to administer prescribed drugs for some time now.

e. *snap bill, division* (earlier example), *election.*

1883 E. W. HAMILTON *Diary* 19 July (1972) II. 461 The Agricultural Holdings Bill..is on the whole being well received, though the Government were on Tuesday defeated by a snap division on a landlords' amendment. 1940 *Sun* (Baltimore) 28 Mar. 12/3 The outcome of the voting in Canada is no surprise. It was a 'snap' election engineered with the greatest skill by Prime Minister W. L. Mackenzie King and the results have fully justified his astute political strategy. 1973 *Times* 8 Dec. 3/8 The strong majority..who do not think the Government should hold a snap general election on this issue are mainly Conservative voters. 1975 J. P. MORGAN *House of Lords & Labour Govt.* viii. 217 The reformers themselves therefore argued that the announcement should be made immediately, and the Lords' reform, including any mention of a snap Bill, indefinitely deferred.

B. Forming combinations, esp. with prepositions, used *attrib.* to designate things or appliances operating with a snap (cf. sense A. b in Dict. and Suppl.); as *snap-in, -off, -on,* etc.

1905–6 *T. Eaton & Co. Catal.* Fall & Winter 158/4 Snap-on Hose Supporters, snaps on to edge of corset. 1939–40 *Army & Navy Stores Catal.* 590 Sports seats... Pigskin covered handle, 'snap-on' ground disc. 1963 *Punch* 17 Apr. 560/3 This is a hygienic, snap-top canister-ette. 1967 *Electronics* 6 Mar. 81/1 (Advt.), Printed Circuit Pins (A) may be attached at rates up to 4,000 an hour; snap-in design holds leads in position for easy solder dipping. 1972 *N.Y. Times* 3 Nov. 3/4 (Advt.), 4 snap-close shirts. 1977 *National Observer* (U.S.) 15 Jan. 21/4 Old wine is also served up in snaptop cans in an entertainment called *The Club*, now at the downtown Circle in the Square. 1978 *Detroit Free Press* 16 Apr. (Parade Suppl.) 14/2 (Advt.), Our very special features, such as a new one-piece, snap-in drapery rod. Works like a traverse. No tabs to tear! 1979 *Nature* 29 Mar. p. xxix/2 A unique one-handed operation, using inexpensive snap-on snap-off plastic, disposable tips.

sna·p-back. Also **snapback.** [f. SNAP-.]

1. *U.S.* and *Canad. Football.* **a.** A centre player; the centre-rusher. ? *Obs.*

1887 *Outing* Oct. 69/2 He it is who, receiving the ball from the 'center-rusher', or 'snap-back', as he is more commonly called, passes it to some other player for a kick or run. 1893 W. C. CAMP *College Sports* 99 This name [*sc.* center rusher] has since given place almost entirely to 'snap-back'. 1901 *Encycl. Sport* II. 426/1 The influence of the snap-back will at once be realised from the consideration that the moment he puts the ball into play through his legs behind him, he also makes six men besides himself immediately off-side.

b. A backward pass from the centre which puts the ball in play to begin a scrimmage.

1910 [see *SCRIMMAGE, SCRUMMAGE *sb.* 4 c (*a*)]. 1947 *Sun* (Baltimore) 8 Nov. 10/3 The first [attempt] fizzled when Wally Wilson fumbled the snapback, while Jones booted wide of the goal posts following the final six pointer.

2. a. A recovery of an earlier position or circumstances. **b.** A reaction or retaliation.

1949 *Sun* (Baltimore) 15 Oct. 13/1 A mild recovery movement in mid-morning cut losses... In the final hour [of trading] there was a snap-back from the lows that gave the whole market a lift. 1961 *Economist* 28 Oct. 370/3 There is little prospect of a swift snapback until the economy shows more signs of strength. 1972 P. TAMONY *Americanisms* (typescript) No. 32. 2 The snap-back of the childish and immature took the form of re-mailings of obscene scrawls. 1979 *Financial Rev.* 11 June 7/1 Behind the upward move of the dollar a pressure cooker effect is building—the kind of pressure that could cause a snap-back and make last year's weakness in the currency look like strength.

3. In Boxing, a swift backward movement of the body to evade an opponent's blow.

1950 J. DEMPSEY *Championship Fighting* xxii. 169 The last and worst type of evasion is the pull-away. Some fighters call it the 'snap-back'. 1952 *Amateur Boxing* ('Know the Game' ser.) 23/1 *Snap back*, a sway backwards —a quick 'snap' back from the hips, sufficient to be out of range of the blow.

snapdragon. **5.** (Earlier example in Glass-making.)

1869 *Our Young Folks* V. 85 This was taken up by a second boy on a 'snap-dragon',—a rod something like a ponty, but with a socket at the end for holding articles of glass,—and carried to a glory-hole.

snapped, *ppl. a.* Add: **2.** Designating matching exposed cards which have prompted the call 'snap!' in the game of snap. Cf. *SNAP *v.* 5 g.

1935 *Encycl. Sports, Games & Pastimes* 568/2 The one who calls first obtains all the cards which lie beneath the two snapped ones.

snapper, *sb.*[1] Add: **2. d.** (Later example.)

1980 *Times* 22 Dec. 12/8 People write to me of 'snappers' which are available at posh, probably preppy, parties in Boston, and which go pop like crackers.

e. (Later examples.) Also *attrib.*, as *snapper ending*.

1895 'MARK TWAIN' in *Youth's Companion* 3 Oct. 464/1 The..humorous story finishes with a nub, point, snapper, or whatever you like to call it. 1949 *Newsweek* 19 Dec. 13/3 Then came the snapper: 'No matter by what method we achieve security, we'll not achieve it in a bankrupt economy.' 1962 E. LACY *Freeloaders* ix. 186 This is the end of the story. I hardly think I've been steering you towards a twist, or snapper ending. 1973 *Publishesr Weekly* 10 Dec. 31/1 The second story is written as a correspondence between a clerk and an alien, and has an O. Henry snapper at the end. 1976 *New Yorker* 24 May 143/1 The first, a male ensemble with some very good martial-arts-style acrobatics, has a snapper ending that doesn't snap.

f. (Earlier and later examples.)

1817 J. SANSOM *Sk. Lower Canada* 15 One had proposed to put *a snapper* on the driver's whip. 1841 *Knickerbocker* XVII. 277 All the whips were provided with red snappers. 1903 *N.Y. Even. Post* 29 Sept. 8/2 Senator Carmack..is simply adding a snapper to the lash of his vigorous denunciation of the Philippine policy. 1949 B. A. BOTKIN *Treas. S. Folklore* I. v. 117 Showing off his prowess..he first split a horsefly into pieces, and then tore a bumblebee into shreds with the snapper on the end of his whip.

g. *pl.* Teeth; a set of false teeth. *slang.*

1924 WODEHOUSE *Leave it to Psmith* i. 36 You see, this fellow understands my snappers. 1958 *Listener* 31 July 154/2 Do your snappers fit snugly?

h. A sea-bed sampler that operates by enclosing material between two or more jaws that come together on contact with the bottom. Also *snapper grab, sampler*.

1925 *Proc. & Trans. R. Soc. Canada* (*Math., etc. Sciences*) 3rd Ser. XIX. IV. 51 The 'snapper' is a simple and inexpensive instrument which has long been used in connection with submarine cable laying. 1942 H. U. SVERDRUP et al. *Oceans* x. 344 Bottom samplers used for oceanographic work fall into three general categories: dredges (drag buckets), snappers, and coring tubes. *Ibid.* 345 Snapper samplers of the clamshell type have been widely used for obtaining samples of the superficial layers of the sediments. 1968 R. V. TAIT *Elements Marine Ecol.* iii. 49 For larger samples, various small spring-loaded, snapper grabs have been devised which take a shallow bite out of the sea-floor.

3. b. *U.S. Football.* Also *snapper-back* = *SNAP-BACK 1 a.

1887 in P. H. Davis *Football* (1911) 475 Rule 12 altered so as to prohibit interference with the snapper-back until the ball is in motion. 1920 W. CAMP *Football without Coach* 30 That involves a great deal harder work from the center rush or snapper back in getting the ball back to him. 1961 J. S. SALAK *Dict. Amer. Sports* 409 The snapper is the player who snaps the ball. 1974 *Rules of Game* 148/1 The snapper may not slide his hands along the ball before grasping it, nor move his feet or lift a hand until after a snap.

3*. a. A taker of snapshots; a casual photographer.

1910 *Chambers's Jrnl.* 13 Aug. 589/1 There is no relief in a protest, for the rampant 'snapper' knows that the law is on his side. 1921 *Ibid.* 30 July 546/1 He who was but a snapper, a presser of a button, and next became a photographic enthusiast. 1977 *Ripped & Torn* VI. 7/2 And thanks a lot to all you budding photographers for the offers of photos, just send 'em in you snappers.

b. *slang.* A ticket inspector.

1938 F. D. SHARPE *Sharpe of Flying Squad* 333 *A snapper*, ticket inspector. 1957 'N. CULOTTA' *They're a Weird Mob* (1958) x. 142 'E doesn't want yer ticket. The snapper's got yer ticket.

6. a. (Further examples.) Cf. SCHNAPPER in Dict. and Suppl.

1842 W. R. WADE *Journey in Northern Island N.Z.* vii. 180 Some snappers which the lads had caught furnished us with a hearty supper. 1896 [see *NANNYGAI]. 1959 A. MCLINTOCK *Descr. Atlas N.Z.* 48 Snapper..is the most important species in the commercial catch. 1977 *Best of Austral. Angler* 9/1 The floating gar system used for tailor is also one of the very best ways to catch snapper from the rocks.

b. *red snapper*: see RED *a.* 17 c in Dict. and Suppl.

c. (Earlier example.)

1796 [see *mud-turtle* s.v. *MUD *sb.*[1] 5 b].

snappily, *adv.* Add: **2.** Smartly, nattily; crisply, deftly.

1936 J. T. FARRELL *$1,000 a Week* (1942) 141 He looked unobtrusively at two snappily dressed young fellows on his left. 1947 *People* 22 June 5/1 Another snappily-togged Ascot-bound party of bright young things noisily piling into a glittering £5,000 limousine. 1977 'A. STUART' *Snap Judgement* 16 A strong, blue-eyed, snappily dressed young man. 1981 E. AGRY *Assault Force* v. 53 Mac reversed the Audi snappily into the nearest driveway.

snapping, *vbl. sb.* Add: **3. snapping-point,** the point at which something will snap, or someone's strength or endurance will fail.

1933 G. ARTHUR *Septuagenarian's Scrap Bk.* 272 And like all good artists, like Sarah herself, she is a '*traqueuse*' whose head feels hot and hands are cold on a first night, and who, with fever in the veins and nerves strained to snapping-point, will yet perhaps give the most inspired performance of the whole run. 1946 K. TENNANT *Lost Haven* (1947) i. 20 To have not only mud but sticky honeycomb all over her shiny, clean linoleum was the snapping-point. 1982 *India Today* 15 Feb. 125/2 Relations between the Government and the judiciary are stretched to snapping point.

snapping, *ppl. a.* Add: **2. b.** That makes a sharp cracking or snapping noise.

1891 *Outlook* Dec. 238/1 In the tender light of the rising sun he creeps downstairs, avoiding that squeaking board and that snapping step. 1942 W. FAULKNER *Go down, Moses* 170 They emerged from the narrow, roofless tunnel of snapping and hissing cane, still galloping, onto the open ridge below. 1968 B. HINES *Kestrel for Knave* 132 Every time he tried to escape [from the shower] the three boys bounced him back, stinging him with their snapping towels as he retreated.

3. b. (Earlier examples.) Also *snapping tortoise*.

1784 J. F. D. SMYTH *Tour U.S.A.* I. 338 One kind of them bites very fiercely when incensed..; these are called Snapping Turtles. 1808 T. ASHE *Trav. Amer.* II. 234 The Indians call this by a name which implies the snapping tortoise. 1828 [see *SALT RIVER 2 a]. 1840 *Knickerbocker* XVI. 54 The..snapping-tortoises, frogs, squirrels, and such small deer, are their flocks and herds.

c. *snapping shrimp*, a shrimp of the family Alpheidæ, which uses its large chelæ to make a snapping noise; also called the pistol shrimp.

1941 STEINBECK & RICKETTS *Sea of Cortez* 194 Sponges and tunicates under which small crabs and snapping shrimps hid themselves. 1964 *Oceanogr. & Marine Biol.* II. 431 The clicking of snapping shrimp..is a form of ambient sound when one is concentrating on the sounds of fish.

4. Violent, severe, extreme; usu. as quasi-*adv.*

1845 *Knickerbocker* XXV. 87 I've got a snapping headache. 1876 *Wide Awake* (Boston, Mass.) July 19/1 The night was snapping cold. 1905 K. D. WIGGIN *Rose o' the River* 93 The snapping cold weather and the depth to which the water was frozen were aiding it.

snappingly, *adv.* Add: Also, briskly, smartly. Cf. *SNAPPILY *adv.* 2.

1976 *Gramophone* Dec. 1016/1 The second subject [is] held in a more or less strict tempo but flecked with fine, subliminally caught colours, the mordents now fluttering like spread acciaccaturas, now expressive, now snappingly exact. 1978 *Ibid.* Apr. 1756/2 Bernstein, as one would expect, is ideally alive and snappingly rhythmic at the opening of the last movement's allegro section.

snappy, *a.* Add: **6. a.** (Earlier and later examples.)

1871 'MARK TWAIN' in *Galaxy* Apr. 615/2, I compressed it into a snappy foot-note at the bottom. 1955 *Times* 27 Aug. 8/4 Her clean texture, snappy rhythm, and general strength of tone and purpose all betokened a true grasp of the composer's style. 1977 *N.Z. Herald* 8 Jan. 2-12/4 (Advt.), Painting roofs, for free quotes phone the expert. Snappy service.

b. *snappy dresser*, someone who dresses in a stylish or natty manner.

1925 *New Yorker* 9 May 27/1, I always used to be a snappy dresser. 1958 [see *DUDE *v.*]. 1977 F. THEROUX *Consul's File* 174 A woman waiting for her lover..whom she would describe as a snappy dresser, a riot, a real card.

d. *U.S.* Designating weather characteristic of a cold snap (SNAP *sb.* 7 a, b).

1928 J. C. LINCOLN *Silas Bradford's Boy* 149 It was a clear, snappy early winter day. 1951 *Publ. Amer. Dial. Soc.* xv. 60 *Snappy*,..said of crispy cold weather.

7. b. Phr. *to make it snappy*: to make haste, to get a move on.

1926 G. FRANKAU *My Unsentimental Journey* ii. 31 After that we 'made it snappy' (Anglicé—got a move on). 1945 A. HUXLEY *Let.* 10 Apr. (1969) 520, I wish there had been space in my review to quote you at length on these subjects, But, alas, I had to 'make it snappy'. 1976 J. I. M. STEWART *Young Patullo* ix. 195 Make it snappy. Taxi's waiting.

snap-shooter. Add: **2.** Also **snapshooter.**

1904 *Car* X. 240 (*caption*) Mr. W. K. Vanderbilt, Jr.,..is reluctant to pose before a camera, but occasionally falls a victim to wily 'snapshooters'. 1973 C. BONINGTON *Next Horizon* iv. 72, I had always taken a camera with me on my climbs, but had been little more than a holiday snapshooter.

snap-shooting, *vbl. sb.* (Example in *Photogr.*)

1979 G. MACDONALD *Camera* iv. 57 Snapshooting was..a..haphazard affair... Most snaps were still portraits of family and friends.

snap-shot, *sb.* Add: **1. b.** (Earlier example.)

1845 F. TOLFREY *Sportsman in Canada* II. v. 131 It is capital practice is this snipe-shooting for a youngster; at least it makes a man a good snap-shot.

2. a. (Later examples.) Also *fig.*

1903 'O. HENRY' in *Everybody's Mag.* Aug. 194/1 You see a man doing nothing but loafing around making snapshots. 1928 *Observer* 17 June 10/2, I asked President Masaryk..if he could give me a snapshot of the difference between what he found when he came to Prague in 1918, and what he has the satisfaction of seeing now. 1930 [see *HUSTLE *v.* 5]. 1950 G. B. SHAW *Farfetched Fables* iii. 109 What are you doing here?..Only hiking round the island. May I take a snapshot? 1962 M. MCLUHAN *Gutenberg Galaxy* 241 He [*sc.* Montaigne] bred up a great race of self-portrayers by means of the mental snapshot. 1975 P. FUSSELL *Gt. War & Mod. Memory* i. 10 British and German soldiers..meeting in No Man's Land to exchange cigarets and to take snapshots. 1978 P. O'DONNELL *Dragon's Claw* iii. 47 Snapshots of sight and sound, of touch, taste, and smell.

b. *Computers.* A record of the contents of some or all of the storage locations in a computer at a particular stage in the execution of a program (see quot. 1963). Freq. *attrib.*

1963 GREGORY & VAN HORN *Automatic Data-Processing Systems* (ed. 2) xii. 473 Some simplified forms of post-mortem routines give only a storage snapshot, which is a complete copy of all storage locations at the time the processor stopped. A snapshot routine may also list the instruction that caused the program to stop, the current contents of arithmetic units and indexes, and perhaps, several of the most-recently executed jumps thus indicating the path of program control. A differential snapshot lists the contents of storage locations that have changed from their initial value or from their value in a prior snapshot. 1966 *IFIP-ICC Vocab. Information Processing* 85 When a trace program gives output only on selected instructions, or for selected conditions, it is called a snapshot program. 1973 C. W. GEAR *Introd. Computer Sci.* vi. 244 An alternative is to take a series of snapshots at points in the program section.

2*. In various sports, a quick shot (of the ball, etc.) at goal.

1961 *Times* 29 May 4/3 [In Polo.] After Hanut had

scored with a lovely snapshot to make it 3–2. 1963 *Globe & Mail* (Toronto) 21 Jan. 16/3 [In Hockey.] Hull responded by taking a quick pass from Balfour and scoring on a quick snap-shot. 1976 *Oadby & Wigston* (Leics.) *Advertiser* 26 Nov. 15/4 [In Football.] Saints hit back and a snapshot by Jim White hit the crossbar.

3. (Further examples.)

1894 [see *ENLARGER 1 b]. 1901 MERWIN & WEBSTER *Calumet 'K'* xv. 288 Young men with snap-shot cameras waylaid Bannon. 1967 J. PHILIP et al. *Best of Granta* I. 17 The winning photo in *The Granta* Holiday Snapshot Competition shows a couple kissing on a beach. 1977 R. E. HARRINGTON *Quintain* iii. 24 He searched the terrain, storing quick snapshot impressions. 1977 *N.Y. Rev. Bks.* 23 June 25/3 The crudely chronological order of snapshot-sequences pasted in family albums.

snap-shot, *v.* Add: **b.** Also *fig.*

1907 *Outlook* 17 Aug. 206/2 All the peculiar attitude of our race toward dancing was suddenly snapshotted in that absurdity. 1932 *Essays & Stud.* XVII. 84 Thackeray found them [*sc.* the railways] vulgar, but amusing for the opportunities they gave of snapshotting people. 1980 *Daily Tel.* 21 Nov. 15/1 The play snapshots pretty sharply Jimmy's furtive park meetings with his waif.

c. To shoot (something) quickly without taking deliberate aim.

1928 *Daily Express* 6 Dec. 19/3 Mr Blyth...was a fairly deliberate shot, and liked to take his high birds neatly and quietly, but could nevertheless snapshot a woodcock in thick covert with an effortless ease.

Hence **sna·pshotting** *ppl. a.*

1978 *Nature* 7 Dec. 647/2 Mr Sankhala also remarks that the snap-shotting tourist is so preoccupied with shutter speeds, lens apertures and focussing that he fails to see anything around him.

snare, *sb.* Add: **2. a.** Now often made of wire.

b. *ellipt.* for *snare-drum*, sense 3 b.

1938 D. BAKER *Young Man with Horn* I. iv. 28 He could, of course, play his snare and..sooner or later he'd have money enough to buy a piano. 1950 A. LOMAX *Mister Jelly Roll* 64, I had a drummer that hit his snares so loud that one night I gave him a couple of fly swatters for a gag. 1960 'E. MCBAIN' *Give Boys Great Big Hand* xii. 137 The big one is the bass drum, and that round black case is what they call the snare. 1973 J. WAINWRIGHT *Pride of Pigs* 31 The drummer giving his snare a series of flicks with the wire brushes.

3. a. *snare wire*.

1804 LEWIS & CLARK *Orig. Jrnls. Lewis & Clark Exped.* (1905) VI. 274 Baling Invoice of Sundries for Indian Presents..3 Rolls Ear Wire, 3 do Snare Wire. 1953 P. PROVENCHER *I live in Woods* iv. 36, I am in no danger because I have my axe, matches, fishing lines and snare-wire. 1964 C. WILLOCK *Enormous Zoo* iv. 56 Snare wire began to make its appearance most frequently... Catching animals with wire snares attached to large logs is understandably popular.

b. *snare-drum* (earlier example), *drummer, drumming, -head* (later examples).

1873 T. B. ALDRICH *Marjorie Daw* 130 Morning and evening we heard the spiteful roll of their snare-drums. 1926 E. FERBER *Show Boat* v. 96 A snare drummer who was always called a 'sticks', and the bass drum, known as the bull. 1941 W. C. HANDY *Father of Blues* (1957) i. 5 The youngster would..beat on the strings in the manner of a snare drummer. 1961 A. BAINES *Mus. Instruments* xiv. 335 'Snares'..consist of a number of gut or wire strings stretched across the lower skin or 'snare head'. 1966 *Crescendo* Apr. 30/3 Complete with snare drumming that would make the Dagenham Girl Pipers turn green with envy. 1976 *New Yorker* 8 Mar. 108/3 He would hit the snare directly, or hit the snarehead and the rim.

snare, *v.* Add: **1. c.** *U.S.* To catch, to win by a small margin.

1942 BERREY & VAN DEN BARK *Amer. Thes. Slang* § 650/6 *Win*...snare a win. *Ibid.* § 679/5 *Field a ball; catch*...snare a hit. 1948 *Sun* (Baltimore) 26 Nov. 17/1 Double Brandy..came from next to last to snare the second money about a half length in front of Brookmeade Stable's Gnu.

snareless, *a.* Add: **b.** Without a snare.

1978 *Early Music* Jan. 29/1 Drums are often snareless, though the pictorial evidence is that the cylindrical drum almost always had a snare.

snaring, *vbl. sb.* (In Dict. s.v. SNARE *v.*) Add: Also *attrib.*

1923 *Beaver* Mar. 236/1 Having some snaring twine she killed sufficient rabbits to keep herself and child alive. 1971 A. FRY *Long Journey* iv. 18 The snaring wing, a long brush fence with a few 'escapes'.., each set with a snare. *Ibid.* ix. 52 Three or four men could work together, driving animals..toward a snaring fence.

snarky (snā·ɪki), *a. colloq.* [f. SNARK *v.* + -Y[1].] Irritable, short-tempered, 'narky'.

1906 E. NESBIT *Railway Children* ii. 49 Don't be snarky, Peter. It isn't our fault. 1913 J. VAIZEY *College Girl* xxiv. 326 'Why should you think I am "snarky"?' 'Because—you *are*! You're not a bit sociable and friendly.' 1953 E. COXHEAD *Midlanders* x. 247 I've known you were the soul of kindness, under that snarky way. *a* 1974 R. CROSSMAN *Diaries* (1976) II. 627 We also have to overcome something else—the stream of anti-government propaganda, smearing, snarky, derisive, which comes out of Fleet Street.

Hence **sna·rkily** *adv.*; **sna·rkiness**; **sna·rkish** *a.*

1912 R. FRY *Let.* 16 Mar. (1972) I. 355 So sorry I seem so snarkish just now. 1960 *Economist* 28 May 859/2 In some of his comments on bureaucracy there is a relapse into snarkiness. 1967 *Listener* 20 July 91/3 Viewers' letters are not just read out. They are commented upon by Kenneth Robinson (usually rather snarkily).

snarl, *sb.*[1] Add: **2. a.** (Later examples of sense 'a knot in the hair'.)

1966 J. S. COX *Illustr. Dict. Hairdressing & Wigmaking* 139/2 *Snarl*, a tangle in the hair. 1976 'TREVANIAN' *Main* (1977) xiii. 243 A young slattern who tugs a snarl out of her hair with her fingers.

b. (Further examples.) *traffic snarl* (U.S. colloq.), a traffic jam.

1933 E. B. WHITE *Let.* Mar. (1976) 113 At noon I happened to be driving north on Fourth Avenue, and got held up in a traffic snarl. 1950 J. D. MACDONALD *Brass Cupcake* ii. 18, I dove slowly back into the traffic snarl. 1968 S. CHALLIS *Death on Quiet Beach* iii. 30 The traffic was a slow snarl that cost him forty minutes. 1975 *New Yorker* 19 May 99/1 The traffic snarls were impenetrable. 1979 *Arizona Daily Star* 5 Aug. D 3/3 Production snarls kept cars out of the showroom. 1980 R. L. DUNCAN *Brimstone* iv. 65 The congressional reorganization studies ..were resulting in a snarl of immense proportions.

3. (Earlier and later examples.)

1775 in O. E. Winslow *Amer. Broadside Verse* (1930) 141/2, I see another snarl of men. 1825 J. NEAL *Bro. Jonathan* I. 76 There being 'a pootty consid'r'ble snarl o' gals, I guess' the supper was bravely furnished. 1836 W. DUNLAP *Mem. Water Drinker* (1837) II. iii. 24 They swarm like a snarl of bees before hiving. 1904 *N.Y. Tribune* 10 Apr. (Suppl.) 7/3 A veritable snarl of street urchins took possession of several benches in Lincoln Park.

snarl, *v.*[1] Add: **2. c.** *to snarl up*: to throw into confusion, to mess up; to entangle, to impede the smooth running of (something). *colloq.*

1937 C. DAY LEWIS *Starting Point* I. iii. 49 He short-punted ahead, snarling up the defence. 1957 J. F. HORNER *Summary of Scientology* vi. 67 Self-processing tends only to snarl-up the person attempting it. 1960 *Economist* 22 Oct. 317/2 A..wish to snarl up the relations between the western governments. 1962 *Listener* 5 July 36/2 Private cars are increasing at such a rate.. that the roads will be snarled up. 1976 *Daily Tel.* 22 June 1/3 The Conservatives snarled up Government business by ceasing the pairing of MPs and other co-operation. 1981 *Sunday Express* 25 Oct. (*heading*) 150,000 marchers snarl up London.

3. Also *fig.* and with *up*.

1951 *Manch. Guardian Weekly* 19 Apr. 5/3 The traffic clears quickly at times, when it might otherwise snarl—which is the American way of saying it might 'cause an inextricable jam'. 1963 *Listener* 14 Feb. 300/2 The action ..snarls up into an obtrusive expressionism. 1970 G. F. NEWMAN *Sir, You Bastard* viii. 201 Traffic snarled eastwards along Brompton Road at a snail's pace.

snarled, *ppl. a.* **2.** (Further *fig.* examples.) Of road traffic: congested (orig. *U.S.*). Also with *up*.

1967 N. MARSH *Death at Dolphin* viii. 230 We've caught a snarled-up little job this time. 1973 R. HILL *Ruling Passion* I. vii. 73 Another diversionary tactic. What a snarled-up lot of people they were! 1976 R. MOORE *Dubai* i. 8 Fitz pushed his way through the..streets, walking through the snarled traffic. 1980 *Times* 8 July 6/2 One of the worst spots for snarled-up traffic.

sna·rl-up. *colloq.* [f. vbl. phr. *to snarl up*: see *SNARL *v.*[1] 2 c, 3.] A muddle, state of confusion; a mistake; a traffic jam; a blockage. Also *attrib.*

1960 M. PHILLIPS in *Analog Sci. Fact/Fiction* Nov. 24/1 Both courses..resulted in more snarl-ups. Reports that should have been sent in weeks before arrived too late; reports meant for the eyes of only one man were turned out in triplicate. 1962 J. BRAINE *Life at Top* xi. 152 There was going to be a huge snarl-up very soon. 1963 *Daily Tel.* 3 June 1/1 On what the AA described as 'snarl-up Sunday' there were queues of up to 19 miles on several major arteries. 1966 *Musical Opinion* Aug. 691/1 The main cause of the brouhaha about electronic instruments is a snarl-up in terminology. 1969 *Daily Tel.* 10 Jan. 1/2 Sixty-three people were injured and more than 100 vehicles smashed up as freezing fog gripped the M1 and M10 yesterday... The AA described it as 'the worst snarl-up since the M1 opened'. 1974 *Financial Times* 15 Mar. 23/7 Small organisation snarl-ups, such as failing to get out the Speakers' Handbook in time. 1977 'E. CRISPIN' *Glimpses of Moon* xi. 220 A helicopter..dipped to examine the snarl-up in the lane below.

snatch, *sb.* Add: **3. d.** (*a*) An unexpected and quick robbery; an act of forcibly robbing someone; (*b*) *slang* (orig. and chiefly *U.S.*), a kidnapping; also *attrib.*

(*a*) 1866 [in Dict., sense 3 a]. 1885 M. DAVITT *Leaves from Prison Diary* I. I. xvi. 152, I did a snatch near St. Paul's. 1939 *Forum* Dec. 275/2 A piece of paper covering the slit was rolled aside in the course of a snatch. 1976 *Southern Even. Echo* (Southampton) 17 Nov. 17/2 Basingstoke police warned women to hang on to their handbags after a sixth attempted snatch in recent weeks. 1980 *West Lancs. Even. Gaz.* 6 June 1 An engineering labourer who was stabbed..during a wage snatch on Merseyside.

(*b*) 1931 D. RUNYON in *Collier's* 26 Sept. 7/2 Harry the Horse and Spanish John and Little Isadore pay no attention whatever to local sentiment and go on the snatch. 1932 E. D. SULLIVAN *Snatch Racket* p. x, Bootleg millions ..have provided the sound support for two hundred standard rackets..in the United States and among them is kidnapping—the 'snatch racket'. 1934 'D. HUME' *Too Dangerous to Live* xix. 200 Where did this snatch take place, Inspector? 1945 —— *Come back for Body* ii. 21 Their only child..has vanished. It looks like a straightforward snatch. 1950 J. D. MACDONALD *Brass Cupcake* ii. 21, I handle it just like a snatch payment. The ransom for Junior. 1980 C. MOOREHEAD *Fortune's Hostages* ii. 26 By 1932 America was in the middle of.. the 'snatch racket'. Dozens of children had been seized... Kidnapping was happening everywhere.

e. *Weight-lifting.* A lift in which the weight is raised in a single motion from the floor to a position overhead with the arms straight.

1928, etc. [see *PRESS *sb.*[1] 6 c]. 1950 *Sun* (Baltimore) 1 May 15/4 Sheppard lifted 240 in the snatch. 1968 *Globe & Mail* (Toronto) 17 Feb. 11/3 Modern lifters do not perform one-arm lifts. Once, two of them were on the Olympic agenda: the one-arm press and the one-arm snatch. 1976 *All about Games* (Com. Org. des Jeux Olympiques) 81 There are two lifts in modern weight-lifting—the snatch and the clean and jerk.

f. Jerkiness in the working of the transmission of a motor vehicle.

1932 *Motoring Encyl.* 137/1 A little thin oil..will soften the [clutch] surfaces sufficiently to avoid snatch. 1955 *Times* 2 Aug. 10/5 Upward and downward changes were made without any trace of snatch. 1962 *Which? Car Suppl.* Oct. 139/1 The car was in excellent condition, apart from..a little 'snatch' in the transmission.

14. *dial.* and *slang.* The female pudenda. Also *attrib.* Cf. sense 6 b in Dict.

1904 in *Eng. Dial. Dict.* 1955 W. GADDIS *Recognitions* III. iv. 851 She said, See? and pulled up her dress to show me her...to show there weren't any marks on her...anywhere else on her body.—You mean on her snatch. 1961 J. HELLER *Catch-22* xxvii. 303 She..twisted away, fleeing far enough..for Yossarian to lunge forward and grab her by the snatch again. 1969 P. ROTH *Portnoy's Complaint* 193 Know what I did when I was fifteen? Sent a lock of my snatch-hair off in an envelope to Marlon Brando. 1971 B. W. ALDISS *Soldier Erect* 128, I was vexed and disappointed that the contact with the *bibi* had been so commercial, so perfunctory—why, I had not even seen or touched her snatch. 1978 J. UPDIKE *Coup* (1979) v. 191 Sooner a black man mate with a lazy shit-smeared sow..than entrust his ebony penis to the snatch of a white devil mare.

snatch, *v.* Add: **2. d.** *spec.* (*a*) To steal, esp. by snatching; (*b*) *slang* (orig. and chiefly *U.S.*), to kidnap.

1765 [in Dict., sense 2 a]. 1887 G. W. WALLING *Recollections N.Y. Chief of Police* xviii. 254 His most brilliant exploit was his 'snatching' of $100,000 from the Royal Insurance Company's office in Broadway in broad daylight. 1919 WODEHOUSE *Coming of Bill* (1920) I. i. 12 As if she had caught him in the act of endeavouring to snatch her purse. 1932 *Detective Fiction Weekly* 17 Dec. 23/2 It's dollars to doughnuts the kid was snatched up in the park. 1934 *Sun* (Baltimore) 10 Mar. 1/7 Banghart had introduced him to the Touhy mob just before the market speculator was 'snatched'. 1936 *Detective Fiction Weekly* 6 June 12/1 'It's one more sweet-running crate. Just about the sweetest I ever snatched. 1973 'I. DRUMMOND' *Jaws of Watchdog* xii. 156 Why didn't we snatch him in the street and take him away someplace?

e. To partake hurriedly of (food, sleep, etc.).

1803 M. WILMOT *Let.* 6 Aug. in *Russian Jrnls.* (1934) I. 34 We rose with one accord, dress'd, snatch'd a cup of Coffee and got into Mdm R—'s Carriage. 1942 BERREY & VAN DEN BARK *Amer. Thes. Slang* § 94/13 *Eat a small or hurried meal*,..snatch a bite. *Ibid.* § 251/6 *Take a nap*,..snatch a wink. 1952 M. STEEN *Phoenix Rising* i. 27 I'm snatching a sandwich at the club. 1977 M. KENYON *Rapist* x. 121 He might snatch two hours' sleep..if he swallowed a couple of sleepers.

f. *to snatch it* or *one's time*: to resign, to leave a job and take the wages due. *Austral. slang.*

1941 *Argus* (Melbourne) *Week-End Mag.* 15 Nov. 1/4 *Snatch your time*, resign from the Army, or threaten to leave. 1944 A. MARSHALL *These are my People* 158 'I suppose you struck some bad bosses in your time?' 'If they're bad, I snatch it.' 1962 T. RONAN *Deep of Sky* 55 What's more, when we pass Silverton I'm snatching my time. 1973 F. HUELIN *Keep Moving* 83 What are yous goin' to do? Snatch it or stay?

10. *intr.* Of a mechanism or its control in a motor vehicle, aircraft, etc.: to operate in a jerky or rough manner.

1932 *Motoring Encyl.* 137/1 When..fabric disks have settled down and worn smooth, they are sometimes prone to snatch and engage fiercely. 1942 B. J. ELLAN *Spitfire* xii. 65 His ailerons were obviously snatching too, as first one wing and then the other would dip violently. 1955 *Times* 12 July 12/6 The car tested was inclined to 'snatch' in the transmission if the speed was allowed to drop too low in top gear.

snatch-. Add: **c.** (Earlier example.)

1884 E. W. HAMILTON *Diary* 15 Mar. (1972) II. 577 A motion..which..the Opposition supported in the hope of taking the Government by surprise and putting them in a minority by a snatch division.

d. *snatch-thief* (earlier example); **snatch-back**, the action of taking back; also *attrib.*; (see also quot. 1905); **snatch crop**, a crop grown for quick returns without regard to the future productivity of the soil; also *attrib.* and *fig.*; **snatch squad** *Mil.*, a group of soldiers detailed to seize troublemakers in a crowd; also *transf.*

1905 *Dialect Notes* III. 94 *Snatchback*, change for the worse in circumstances... 'That's a snatchback for him.' 1949 *New Statesman* 24 Dec. 750/3 The distress caused by the snatch-back is no less tragic than would be suffered by natural parents who were forcibly deprived of their children. 1962 A. Sampson *Anat. of Britain* xxiii. 377 Hire-purchase..companies..could be ruthless in enforcing 'snatch-backs' if payments had lapsed. 1965 E. Gundrey *Foot in Door* xvii. 123 'Snatchback' machines, that is ones which had to be returned to dealers by people who failed to keep up their H.P. payments. 1979 H. S. Kent *In on Act* ix. 101 The main objects of the Bill were, first, to make sure that the hire-purchaser knew what he was paying..secondly, to restrict the seller's rights to 'snatch back' the goods on default... The most dear to Ellen's heart was the ban on the snatch-back. 1937 H. G. Wells *Brynhild* v. 58 Fellows like Blatch can reap a harvest..at ten per cent... There are too many authors. Blatch is able to live by snatch crops. 1959 *Listener* 30 July 179/2 The heart of the soil..had been weakened by the greed of the snatch-crop farmers. 1970 *Financial Times* 23 Mar. 1/1 About 150 youths moved out of Bogside.., smashing windows..and stoning the Army 'snatch squad'. 1976 *Western Mail* (Cardiff) 22 Nov. 1/2 A snatch squad of animal lovers seized 11 beagle pups in a night-time commando-style raid on a top-security breeding centre in West Wales. 1982 *Times* 1 Sept. 3/1 Snatch squads tried ineffectively to combat roaming gangs of pickpockets. 1887 *Courier-Jrnl.* (Louisville, Kentucky) 1 May 13/2 Where the bonnet-buyer is there is the pickpocket and snatch-thief also.

snatcher. Add: **1. d.** *slang* (orig. *U.S.*). A kidnapper.
1932 *Tulsa* (Okla.) *Daily World* 7 Mar. 10/5 'Snatchers' or kidnapers have not been as busy in Tulsa as they have in other cities. 1940 'D. Hume' *Invitation to Grave* xviii. 235 Mick was reflecting upon the fact that his father's 'snatchers' had entered through this door.

snatching, *vbl. sb.* **1.** (Further examples.)
1931 D. Runyon in *Collier's* 26 Sept. 7 (*heading*) The snatching of Bookie Bob. 1955 *Times* 26 July 6/6 As soon as the speed drops to 20 m.p.h., a change to third gear is essential to avoid 'snatching'. 1972 J. Philips *Vanishing Senator* (1973) iii. ii. 127 They're all in a state over the snatching of Mrs. Lloyd.

snavel (snæ·v'l), *v.*[2] *slang and dial.* (now chiefly *Austral.*). Also **snavvle.** [Perh. var. Snabble *v.* or Snaffle *v.*[4]] *trans.* To steal; to appropriate, to grab.
For further material see *Eng. Dial. Dict.* s.v. *Snavel* vb.[2] and sb.[2]
a 1790 H. T. Potter *New Dict. Cant & Flash* (ed. 2, 1795) 54 *Snavel*, to steal when running. 1823 'J. Bee' *Slang* 162 *Snavel*, to steal, by snatching, probably, or concealing any small property by piece-meal. 1903 'T. Collins' *Such is Life* 18 Well, we had a bunch o' keys at the camp. I had snavelled 'em at the railway station. 1919 W. Downing *Digger Dialects* 46 *Snavvle*, take by stealth; steal; capture. 1933 *Bulletin* (Sydney) 4 Oct. 10/1 Could we but snavel *that* We'd incontestably be home and dried In this keen race. 1948 V. Palmer *Golconda* xiii. 100 They're booming the notion o' a new township and snavelling all the land within a mile o' it.
So † **sna·veller,** a thief. *Obs. rare.*
1781 G. Parker *View of Society* II. 168 The Snaveller.. coaxes the child up some by-alley,..and grabbles the whole.

snax (snæks), commercial var. *snacks*, pl. of Snack *sb.*[2]
1947 I. Brown *Say the Word* 17 Why does such shop-window spelling, Sox and Snax, irritate me so? 1965 I. Fleming *Man with Golden Gun* v. 70 A hand-painted sign said '*Snax*'. 1980 'M. Yorke' *Scent of Fear* vii. 62 She bought sandwiches at Takeaway Snax.

snazzy (snæ·zi), *a. slang* (orig. *U.S.*). [Origin unknown.] Excellent; attractive; classy, stylish, flashy.
1932 *Amer. Speech* VII. 336 *Snazzy*, agreeable; attractive. 1935 N. Ersine *Underworld & Prison Slang* 68 That's a snazzy dressup you've got. 1938 W. Chambers *Once too Often* i. 17 It was indeed a very snazzy setup and I wondered how many months he was in arrears with his rent. 1944 R. Chandler *Let.* 16 Dec. (1966) 43, I had a very snazzy beginning which they cut out, because it didn't really have anything to do with detective stories. 1946 'P. Quentin' *Puzzle for Friends* iii. 27 Think what a snazzy life you've got. All the money in the world. No worries. No work. 1956 'J. Wyndham' *Seeds of Time* 214 You come here with your ritzy ways and your snazzy talk. 1960 *News Chron.* 30 Sept. 4/5 Rod..takes off his snazzy smoking jacket to rescue his..girl friend. 1968 J. Lock *Lady Policeman* ix. 82 They've made the plain uniforms look as snazzy as possible with whiter-than-white hat-bands, belts and gaiters. 1971 B. W. Aldiss *Soldier Erect* 115, I see you're all togged up... It's really snazzy you look. 1974 *Early Music* Jan. 24/2, I am convinced that the snazzy rhythms and syncopations that one often hears are wrong. 1978 C. Leopold *Casablack* 1 A snazzy, loose-fitting pinstripe that was emphatically fashion in..1942.

sneak, *sb.* Add: **2. a.** Also more generally *on the sneak*, on the sly, by stealth, under concealment.
c 1863 T. Taylor *Ticket-of-Leave Man* i. 9 Pottering about on the sneak, flimping or smashing a little when I get the chance. 1930 *Amer. Mercury* XXI. 458/1 You got to work strictly on the sneak. All the spots are hot. 1935 *Sun* (Baltimore) 13 July 9/6 A few of them [*sc.* betting spots] were 'sneaking' with just as many customers as ever... These spots 'on the sneak' usually are located in the upper floors of Loop skyscrapers. 1955 *Publ. Amer. Dial. Soc.* xxiv. 86 If the *road mob* decides to work *on the sneak*, that is, without advance arrangements in any locality, [etc.]. 1982 *Chicago Sun-Times* 6 Aug. 71/1 He does so with all the glee of a schoolkid reading Playboy magazine on the sneak.

3. (Earlier example.)
1851 J. Pycroft *Cricket Field* vii. 105 Cowley..put on one Tailor Humphreys to bowl twisting underhand sneaks.
4. (Earlier example.)
1862 *Female Life in Prison* I. xvii. 211 The night-officer is generally accustomed to wear a species of India-rubber shoes or goloshes on her feet. These are termed 'sneaks' by the women [of Brixton Prison].
5. *U.S. colloq.* = *sneak preview* s.v. *Sneak-* a.
1941 B. Schulberg *What makes Sammy Run?* iv. 60 We'll know better after the sneak... And..when we see Chicago and L.A. 1976 *Boston Globe* 5 Apr. 57/1 (heading) Sneak attack. 1978 E. Tidyman *Table Stakes* ii. vi. 265 The studio agreed to give the production three previews... The first 'sneak'..took place at a small theater in Redlands.

sneak, *v.* Add: Pa. t. and pple. also (orig. and chiefly *U.S.*) **snuck.**
1887 *Lantern* (New Orleans) 17 Dec. 3/1 He grabbed ten dollars from de bums an den snuck home. 1932 J. T. Farrell *Young Lonigan* ii. 55 They had all snuck in and were having a good time, making trouble. 1940 R. Chandler *Farewell, my Lovely* vi. 36, I snuck in there. 1958 J. Kerouac *On Road* ii. viii. 159 Four sullen fieldworkers, snuck from their chores to brawl in drinking fields. 1960 *Oz* May 3/1 It was sticking out of a dustbin..so I snuck off to the park and had a good old read. 1976 S. Brett *So Much Blood* xvi. 191 At the interval Charles and Frances snuck out to the pub. 1979 *Vague Q.* Summer 73/1, I have come around the back way and snuck up, as we say in Nebraska, on my subject.
I. 1. d. (Earlier example.)
1866 G. Ade *Artie* 7 I'd a' sneaked early in the game.
II. 4. (Examples with a person as obj.)
1968 J. M. Ullman *Lady on Fire* (1969) xii. 100 You've got a new lead. Maybe something the sister told you after you snuck her out of that hotel. 1973 D. E. Westlake *I gave at Office* (1972) 12 There was some cousin of mine, a couple of guys had snuck friends in. 1979 R. Jaffe *Class Reunion* i. vii. 69 He wanted to sneak her into..
6. b. (Further examples.) Also, to take or partake of surreptitiously.
1900 *Dialect Notes* II. 58 *Sneak*, to appropriate. 1921 E. O'Neill *Emperor Jones* i. 160 When I sleeps, dey sneaks a sleep, too, and I pretends I never suspicions it. *a* 1953 — *More Stately Mansions* (1964) ii. 136 Each sneaks a..glance. 1955 J. O'Hara *Ten North Frederick* iv. I can sneak us another drink. 1956 R. Duggan *Immanuel's Land* 107 The conductor stood on the bucking platform, sneaking a cigarette. 1968 *Globe & Mail* (Toronto) 17 Feb. 6/2 If they did have these smoking areas..the students wouldn't have to sneak a smoke in the washroom. 1978 J. Irving *World according to Garp* ii. 30 He was happy to run errands for the patients, deliver messages, sneak them food.

sneak-. Also: **a.** **sneak-boat** (earlier example); **sneak-current** (examples); **sneak-guest,** one who makes public the events of private social gatherings at which he is a guest; **sneak-hunting,** hunting from an unobserved approach; **sneak preview** orig. *U.S.*, a showing of a (usually unnamed) cinematic film prior to regular release, to test audience reaction; also *transf.* and *fig.*; hence **sneak-preview** *v. trans.*, (*a*) to show (a film) in a sneak preview; (*b*) to have a sneak preview of (something); **sneak-thief** (earlier and later examples); also, a pickpocket, a snatch-thief; also *attrib.*; hence **sneak-thief** vb. *trans.* (*nonce-wd.*); **sneak-thievery, -thievery.**
1853 *Laws General Assembly of Maryland* 220 Any person or persons..shall not use any sink boats, sneak boats or floats. 1899 K. B. Miller *Amer. Telephone Practice* xxiii. 275 It frequently happens..that a very small current..will not be sufficient to blow the fuse... These currents are very appropriately termed 'sneak currents'. 1934 A. L. Albert *Electr. Commun.* xii. 325 Currents slightly in excess of the normal operating values..are often called 'sneak' currents. 1930 *Times Lit. Suppl.* 9 Jan. 18/1 [He] was in fact (if a very modern term may be permitted to be used of him) a 'sneak guest'. 1958 *Listener* 18 Dec. 1045/1 He [*sc.* Boswell] was regarded in society as something of a 'sneak guest'. 1878 E. B. Tuttle *Border Tales* 45 By sneak-hunting, one man can kill a whole band of elk. 1960 *Outdoor Life* (U.S.) Oct. 84/3 Sneak hunting is a difficult and time-consuming sport. 1938 *Daily Progress* (Charlottesville, Va.) 28 Nov. 1/6 A double-barreled, two-blizzard 'sneak' pre-view of the 1938 edition of winter. 1939 *Chambers's Jrnl.* Nov. 858/1 In America, pre-views, frequently called 'sneak pre-views', have always been allowed. 1949 *Sun* (Baltimore) 28 Jan. 13/4 (Advt.), Sneak preview—tonight at 11.40 p.m... Even though the producers say we mustn't tell—we can hint it's..one of the funniest comedies you've ever seen! 1950 *Ibid.* 14 Sept. 16/1 The film was sneak-previewed in Hollywood. 1952 *Art Digest* 15 Sept. 5/1 Sneak Preview. On the theory that our readers like to know in advance about important art events, we summarize..the 1952–53 season. 1960 *Sunday Express* 18 Dec. 9/3 Paris-bound passengers were given a sneak preview of Britain's 'pennyfarthing' airliner, the Vickers Vanguard. 1972 *Guardian* 24 May 13/7 The old-established Oregon primary ..served as a sneak preview of the multimillion dollar Californian entertainment. 1975 *New Yorker* 22 Dec. 31/2 Our pal..delights in opportunities to see things in advance, so he was easily persuaded last week to accept our invitation to sneak-preview the new open-air observation platform twelve feet above the roof. 1980 *Times Lit. Suppl.* 12 Sept. 990/2 This selection brings together poems from all five of her [*sc.* P. Beer's] published collections, plus a satisfying sneak preview of what one hopes will be her sixth. 1859 G. W. Matsell *Vocabulum* 82 *Sneak-thief*, a fellow who sneaks into areas, basement-doors or windows, or through front doors by means of latch-keys, and entering the various apartments, steals any thing he can carry off. 1866 *Harper's Mag.* Nov. 690/1 A female 'sneak thief' and a 'longshoreman' now appear. *a* 1930 D. H. Lawrence *Last Poems* (1932) 100 That is why business seems to me despicable, and most lives pedantic, sneak-thief and pick-pocket picking of dressed-up people. *Ibid.* 242 The jizzy perhaps never picked a man's pocket But my god, how sneak-thieved his very genitals away from him. 1959 M. Cumberland *Murmurs in Rue Morgue* xix. 117 He is the sneak-thief type and the petty blackmailer. 1976 *Liverpool Echo* 6 Dec. 7/9 Wrexham Police to-day warned shoppers to be on the lookout for sneak thieves after a woman shopping in a chemist shop had her bag. 1923 Sneak-thievery [see *gangsterdom]. 1963 V. Gielgud *Goggle-Box Affair* xvii. 177 Nothing else was taken, so it wasn't just sneak-thievery. 1973 E. Beckman *Victorian Album* 82 So there I was, practising deceit on Christabel and sneak-thievery on Mrs Rendell.
b. In misc. other uses, passing into adj.: that acts or is effected by stealth, deceit, or surprise; unexpected.
1938 *Sun* (Baltimore) 19 July 8/3 His 'sneak hop' from New York to Ireland terminated successfully. 1943 [see *prince sb. 2 b]. 1943 *Sun* (Baltimore) 27 Dec. 5/2 A sneak air attack might be attempted by the enemy on Christmas Day. 1944 *Ann. Reg.* 1943 1. 21 The soundness of the air defence..had compelled them to confine themselves largely to 'sneak' raids on coastal towns. 1952 *Sun* (Baltimore) 17 Jan. 4/1 The small-spread sneak disease, bilharzia. 1959 *Publ. Amer. Dial. Soc.* xxxi. 59 The act of theft from the person by stealth is..referred to as a *sneak job*. 1970 [see *lead sb.[1] 5 c]. 1971 'L. Black' *Death has Green Fingers* iii. 58 Anyway, a sneak photographer. 1976 *Evening Post* (Nottingham) 14 Dec. 11/6 A sneak raider stole £740 takings from the Triangle toy shop.

sneaker. 3. For *U.S. colloq.* read '*colloq.* (orig. and chiefly *U.S.*)' and add further examples.
1900 G. Ade *More Fables* 193 His Job on the Earth was to put on a pair of Pneumatic Sneakers with Gum Soles and go out and investigate Other People's Affairs. 1914 S. Lewis *Our Mr. Wrenn* iv. 56 First but fearful in his rubber sneakers. 1929 'S. S. Van Dine' *Scarab Murder Case* iv. 61 He got relief by wearing white canvas sneakers with rubber soles. 1936 Wodehouse *Laughing Gas* xii. 126 You could scarcely expect to turn up in sneakers and a sweater, my good fellow. 1948 J. Steinbeck *Russ. Jrnl.* (1949) 13 She wore canvas sneakers. 1959 *Manch. Guardian* 24 June 7/2 The international uniform of jeans and sneakers over brown ankles. 1974 A. Lurie *War between Tates* v. 95 It was Jeffrey who started it; he could not find his left sneaker. 1981 *Sunday Express Mag.* 26 July 16/3 [caption] Shades of throwaway chic for pop singer, Graham Bonnet. Old sneakers and a borrowed suit?
4. (Earlier example.)
1851 J. Pycroft *Cricket Field* vi. 63 With the primitive fashion of ground bowling, called sneakers, forward play could have no place.
Hence **snea·kered** *a.*, clad in sneakers.
1962 'E. Fenwick' *Friend of Mary Rose* (1962) iv. 39 He heard..a soft jump—as of sneakered feet. 1976 McBain *Guns* (1977) i. 34 He floats on sneakered feet to the back door of the shop. 1980 *Listener* 3 May 613/2 Their crew-cut, pony-tailed, sneakered sons and daughters.

sneakishly, *adv.* For *rare*⁻¹ read *rare* and add. Also, in a sneaking or stealthy manner.
1867 'T. Lackland' *Homespun* i. 55 He begins with throwing a glance at her sneakishly. 1922 *Daily News* 11 May 6 When men come together to profess a creed they come courageously... When they come together in a clique they come sneakishly.

sneaky, *a.* Add: **3.** *Sneaky Pete, sneaky pete*: a name given to any of various illicit or cheap intoxicating beverages. Also *attrib. slang* (orig. and chiefly *U.S.*).
1946 *Collier's* 1 Sept. 49/1 A group which was..discussing the effects of 'sneaky-pete', a generic term for fortified wines. 1951 [see *rot-gut b.]. 1955 *Amer. Speech* XXX. 85 *Sneaky pete*, marijuana mixed in wine. 1962 J. S. Gunn *Terminal Shearing Industry* i. 32 Sneaky Pete, one of the many violent sneaks, which shearers and others give to cheap wine. 1971 J. H. Jones in J. H. Clarke *Harlem* 310 He walked around an unconscious Sneaky Pete drinker. 1978 J. Gores *Gone, no Forwarding* (1979) vi. 40 The stranger bought a pint of sneaky pete and he and Sammy went to sit in the bright end of the dance.
B. as *sb.* A small concealed microphone or other device for surveillance or espionage.
1974 M. Copeland *Real Spy World* 317 A 'sneaky' differs from the other technical devices in that it is planted inside an intelligence target. 1977 P. Wyke *Go for Gold* v. 28 His car had been fitted with a sneaky ..searching for a microscopic radio transmitter was pointless. 1978 D. Bloodworth *Crosstalk* xxi. 168, I never actually found the mikes or the cameras, but..I just thought I'd get away from all the sneakies.
Hence **snea·kily** *adv.*
1966 D. Varaday *Gara-Yaka's Domain* vii. 75 As fast as he drove off one crowd of fluttering birds, another swooped in, nipping sneakily with slashing beaks. 1974 P. Cave *Mama* (new ed.) ix. 77 Peter the publican was hovering sneakily around with his ears wide open.

sned, *v.* Restrict 'In later use *Sc.* and *north. dial.*' to sense 2 in Dict. and add: **1. b.** (Later examples.)
1953 H. L. Edlin *Forester's Handbk.* xiii. 205 As soon as the tree is down it should be lopped or snedded, by cutting the branches away from the trunk. 1971 *Timber Trades Jrnl.* 3 Apr. 58/2 Although the chainsaw has long been used for limbing hardwoods the technique of snedding softwoods with a power saw is relatively new.

sneery, *a.* (Later examples.)
1919 D. Ashford *Young Visiters* viii. 53 Ethel patted her hair and looked very sneery. 1949 D. Smith *I capture Castle* iii. xv. 289 'Does he believe in it?' 'No, he's always very sneery.' 1967 *Punch* 22 Nov. 776/1 It has taken twenty years to get rid of the amateurs, and the professionals..are having a hard time breaking down the sneery reputation gained. 1977 *N.Y. Rev. Bks.* 9 June 10/4 The sneery attitude toward surgeons that many physicians..were at one time wont to affect.

sneeze, *sb.* Add: **3.** *attrib.*, as **sneeze gas,** a substance used to incapacitate people by causing them to sneeze when it is inhaled or absorbed through the skin.
1918 E. S. Farrow *Dict. Mil. Terms* 567 *Sneeze-gas*, a gas which produces paroxysms of sneezing, so that it is difficult to keep on a mask if any of the gas is inhaled. 1966 *McGraw-Hill Encycl. Sci. & Technol.* III. 45/1 Sternutators, sometimes called sneeze gases or vomiting gases, cause physical discomfort.., and general malaise to such an extent that a casualty results.

sneeze, *v.* **3.** (Later examples.)
1930 R. Campbell *Adamastor* 76 Their horses..Vast phantom shapes with eyeballs rolling white That sneeze a fiery steam about their knees. 1961 G. Durrell *Whispering Land* viii. 194 Anyway, when I had sneezed some of the dust out of my nose, I clapped dutifully outside the gate.

sneezing, *vbl. sb.* Add: **3. a.** *sneezing-powder* (later examples); **sneezing gas** = *sneeze gas* s.v. *Sneeze *sb.* 3.
1918 H. H. Tudor *Let.* 13 Nov. in M. Gilbert *Winston S. Churchill* (1977) IV. Companion 1. 415 The shell you speak of may be sneezing gas, which is not deadly. 1939 H. F. Thuillier *Gas in Next War* xvi. 145 The General Disarmament Conference of the League of Nations sitting in 1932 sought to divide the known chemical war agents into..two categories [lethal and non-lethal], and to obtain agreement for the use of the non-lethal kind, *i.e.* the lachrimatory (tear gases) and the sternutatory (sneezing gases) in war. 1979 *Guardian* 25 Aug. 12/6 Soup ..always tastes the same when you have accidentally emptied the pot of fine grey sneezing powder into it. 1982 *London Mag.* June 31 Once he bought sneezing-powder from the joke-shop in New Oxford Street.

‖ **snekkja** (sne·kyā). Pl. **snekkjur,** (*erron.*) **snekkar.** [Icel.: see Snack *sb.*[1]] An ancient Icelandic or Scandinavian longship.
1847 N. H. Nicolas *Hist. R. Navy* I. i. 10 The Scandinavians are said to have possessed small boats with.. twelve seamen, and a longer kind of vessel called 'snekkar' or serpents, chiefly used for war, with twenty rowers. 1889 P. B. du Chaillu *Viking Age* II. ix. 137 The *snekkja* was a somewhat smaller long-ship, of which frequent mention is made; but sometimes it must have been as large as a dragon-ship. 1911 *Encycl. Brit.* XXIV. 865/1 The famous *snekkjur* or serpents, said to be represented on the Bayeux tapestry. 1970 Foote & Wilson *Viking Achievement* vii. 236 The longship was the real warship, with at least twenty benches. One common sort was called *snekkja*.

snell, *sb.* (Earlier example.)
1846 *Spirit of Times* 9 May 126/2 [The bass] was taken with a jointed rod, with a single gut snell, after half an hour's play.

snell, *a.* and *adv.* Add: Hence **sne·llness,** sharpness, keenness.
1915 J. Buchan *Salute to Adventurers* i. 15 That bold girl singing a martial ballad to the storm and taking pleasure in the snellness of the air.

snell, *v.* Add: Hence **snelled** *ppl. a.*
1893 [in Dict.]. 1960 *Washington Post* 29 Apr. D9 The somewhat cumbersome assembly of one or more spinner blades ahead of a snelled or long-shanked hook. 1976 *Billings* (Montana) *Gaz.* 28 June 2-D (Advt), Packages of 6 snelled hooks.

Snellen (sne·lən). *Ophthalm.* The name of Hermann *Snellen* (1834–1908), Dutch ophthalmologist, used *attrib.* and in the possessive to designate: (*a*) a scale of similar square-serifed type-faces of different sizes, all subtending the same angle at different rated distances, proposed by him in 1862 (in his *Échelle Typographique*) and used to print test cards which are presented at known distances to ophthalmic patients who are asked to read out as many lines as they can; also, the letters, test cards, etc., associated with this scale; (*b*) a fraction which expresses a patient's visual acuity as the actual reading distance over the rated distance of the smallest Snellen letters read.
1864 T. Longmore *Man. Instructions Defective Vision in Soldiers* ii. 10 The emmetropic eye can read Snellen's types at any of the indicated distances. 1866 H. W. Williams *Rec. Adv. Ophthalm. Sci.* 29 Two other series have been added as reading tests... Both of these are almost perfectly accurate in their gradations of sizes,— and correspond, the first with the same numbers of Snellen's scale, the second with those of Jaeger's test. 1912 L. Laurance *Visual Optics & Sight Testing* iv. 73 The visual acuity, as expressed by a Snellen fraction, varies..with the health of the person. 1934 C. S. Price *Improvement of Sight by Natural Methods* ii. 21 The large cards or charts bearing a series of sizes in the types are conveniently known as 'Snellen Charts'. 1960 N. Bier *Correction Subnormal Vision* i. 4 A person with visual acuity of 6/60 Snellen or better should not ordinarily be regarded as blind. 1971 *Jrnl. Gen. Psychol.* LXXXIV. 85 He measured visual acuity, using Snellen letters at 10 meters.

Snell's law (snelz). *Optics.* [Named after Willebrord van Roijen *Snell* (1591–1626), Dutch astronomer and mathematician, who formulated the law in 1621.] The law which states that for a ray of light passing from one uniform medium to another the sines of the angles of incidence and refraction are always in the same ratio.
1873 J. Tyndall *On Light* i. 24 Snell's law of refraction is one of the corner-stones of optical science, and its applications to-day are millionfold. 1935 Dawson & Porritt *Rubber* 404/1 Stretched rubber shows double refraction, i.e. a ray of light entering the rubber from air or a vacuum is split up into two rays, one of which.. obeys Snell's law of refraction, whilst the other..does not unless it travels in a certain critical direction. 1974 *Nature* 18 Jan. 156/2 Snell's law and a value of 1·67 for the refractive index of the lens were used to determine the angles of refraction at the surfaces of the lens.

‖ **snelskrif** (sne·lskrif). *S. Afr.* [Afrikaans, f. *snel* rapid + *skrif* writing.] A system of shorthand for the Afrikaans language. Also *attrib.*
1949 *Cape Argus* 16 Apr. 11/2 (Advt.), Take a rapid course..in book-keeping, Afrikaans, snelskrif, shorthand, [etc.]. 1952 *Cape Times* 2 Aug. 9/7 Typists who qualify for shorthand and *snelskrif* tests. 1972 *Grocott's Mail* (Grahamstown) 1 Sept. 2 Bilingualism, shorthand, snelskrif, typing..are all essential.

snib, *v.*[2] For *Sc.* read orig. *Sc.* and add: **1.** (Further examples.)
1889 A. Conan Doyle *Sign of Four* vi. 92 Your ally would..shut the window, snib it on the inside. 1934 R. Knox *Still Dead* xxii. 269 Trying to shut the door quietly, she left it not quite snibbed properly, so that it came a bit ajar. 1953 A. Upfield *Murder must Wait* i. 4 The Yale-type lock was snibbed. 1962 W. H. Murray *Maelstrom* v. 73 He snibbed all the ground-floor windows. 1967 I. Hamilton *Man with Brown Paper Face* ix. 132, I went softly to the main entrance and snibbed it from the inside and put the bolt across. 1971 *Islander* (Victoria, B.C.) 21 Nov. 2/3 The windows were not only unbroken but snibbed shut.

snick, *sb.*[2] **2.** (Earlier example.)
1857 *Bell's Life* 19 July 7/5 The last jump from 135 to 158..included many 'snicks', not hits.

Snick (snik), *sb.*[5] *U.S.* Also **SNICK.** [Alt. of *SNCC* (see S 4 a and below).] The Student Non-violent Co-ordinating Committee, an organization of Black Americans campaigning for civil rights and Black power. Also *attrib.*
1962 *Time* 12 Jan. 15/1 To fight segregation in their own way, young Negroes have organized themselves into a federation called the Student Nonviolent Coordinating Committee ('Snick' for short). 1967 *National Observer* (U.S.) 27 Nov. 1/4 Snick had its origins in the sit-in movements of 1960. *Ibid.*, Snick leaders consider elections a white man's device to deceive Negroes with false promises. 1967 *Telegraph* (Austral.) 5 Aug. 2/1 The white man has been violent towards the Negro for 400 years... If we are violent to him he deserves every bit. SNICK is respected because if we say burn, baby, burn, we'll be the first to strike a match. 1978 L. Heren *Growing up on The Times* ix. 292 In 1964..white students..met members of the student non-violent coordinating committee. Snick, as it was usually called, was still very much a genuine student movement with religious roots.

snick, *v.*[2] **2. b.** (Earlier example.)
1871 'Thomsonby' *Cricketers in Council* 3 The new trundler then put down a tice..which the Surrey colt snicked cleverly through the slips.

snick, *v.*[3] Add: **1. b.** To turn *on, off, out, up*, to push *open*, with a clicking noise.
1927 *Daily Express* 30 Aug. 3/4 As she snicks open the trellised door. 1927 *Observer* 4 Dec. 12 A Foreman with a lantern..walks down a dark platform snicking on lights. 1959 I. Jefferies *Thirteen Days* vi. 75 This corner was all wrong for..firing a pistol round..but I snicked the heat out and exposed an eye. 1973 R. Hayes *Hungarian Game* xxxiv. 207 Hagopian crested the hill and snicked off the ignition. 1977 *Detroit Free Press* 11 Dec. 15-C/1 The electric vacuum system that snicks lamps up with the tap of a toggle on modern cars.
2. b. To move *back*, to come *open*, with a click.
1963 C. D. Simak *They walked like Men* xiv. 76 The lock snicked back and the door came open. 1972 J. Potts *Trouble-Maker* (1973) xviii. 146 The back door snicked open.

snicker, *sb.*[1] (Earlier example.)
1836 *Knickerbocker* VI. 562, I was partially 'ware of a general *snicker* through the room.

snicker-snack (sni:kəɹsnæ·k, sni·kəɹsnæ:k), *adv.* and *sb.* Also **snickasnack.** [Imit.: cf. *Snick-snack *adv.* and *sb.*] (With) a snipping or clicking sound.
1871 'L. Carroll' *Through Looking Glass* i. 22 The vorpal blade went snicker-snack! 1913 C. Mackenzie *Sinister Street* I. ii. xv. 403 Mrs Carthew snipped away, talking in sentences that matched the quick snickasnack of her weapon. 1979 P. Way *Sunrise* viii. 79 The little man next door was chopping his hedge. The shears were going snicker-snack. *Ibid.* 85 He glanced back at them fiddling with the dominoes... The pattern fell snicker-snack over.

snicket (sni·kėt). *north. dial.* [Origin obscure.] A narrow passage between houses, an alley-way.
For further senses of the word see *Eng. Dial. Dict.*
1898 B. Kirkby *Lakeland Words* 136 *Snicket*, a narrow passage between buildings. 1947 I. Brown *Say the Word* 65 We have vennels, gunnels, and snickets in our northern towns. 1957 R. Hoggart *Uses of Literacy* i. ii. 52 Street after regular street of shoddily uniform houses intersected by a dark pattern of ginnels and snickets (alley-ways) and courts. 1968 B. Hines *Kestrel for Knave* 31 He cut down a snicket between two houses, out into the fields. 1981 J. Stubbs *Ironmaster* xx. 276 We are cramming poor people into ginnels and snickets and foetid courts.

snickety (sni·kėti), *a. rare.* [Origin obscure: cf. *persnickety *a.* (*adv.*).] Fussy, pernickety.
a 1960 E. M. Forster *Maurice* (1971) iv. xxxix. 186 Maurice hated cricket. It demanded a snickety neatness he could not supply.

snick-snack (sni·ksnæk, sniksnæ·k), *adv.* and *sb.* Also **snic-snac.** [Imit.: redupl. from Snick *sb.*[2]] = *snicker-snack *adv.* and *sb.*
1925 C. Day Lewis *Beechen Vigil* 11 Meanders around the rose-beds, gnarled, clay-brown, Old Tom the pruner, snic-snac up and down. 1970 *New Yorker* 28 Nov. 151/1 Big, straight scissors, from Finland..have a nice feel in the hand and a reassuring snick-snack.

sniddy, var. *snidey *a.*

snide, *a.* and *sb.* For *Cant* read '*colloq.* (orig. *Cant*)' and add: **A.** *adj.* **1.** (Earlier and further examples.) Also more widely, inferior, worthless.
1859 G. W. Matsell *Vocabulum* 83 *Snide stuff*, bad money. 1861 J. Clay *Prison Chaplain* viii. 537 The observant and experienced E.R. says 'The utterers of "*Snide pewter*" (base silver) are almost all Irish.' 1887 F. Francis *Saddle & Mocassin* i. 3 These here men don't want none of your..snide outfits, but jest good *bronchos* and a waggon, and strong harness. 1887 *Lantern* (New Orleans) 9 Apr. 2/3 Who runs dat snide hash house. 1899 'J. Flynt' *Tramping with Tramps* ii. 277 Utica..is sort of a snide place, this time of the year. 1906 E. Dyson *Fact'ry 'Ands* xiv. 180 'Tain't ther liquor wot's snide, it's ther dead hookity hides what it gets chuted into. 1926 [see *jungle *sb.* 2 c]. 1973 'J. Patrick' *Glasgow Gang Observed* 235 Snide 'boggin': used in phrase 'snide gear', i.e. clothes that are out of fashion, contemptible, inferior.
2. Of a person: cunning, sharp.
1883 E. J. Milliken *Childe Chappie's Pilgrimage* ii. 15 They self-deemed astute and 'snide', Of *nous* bereft, low chaff the bar-queen golden dyed. 1889 *Cent. Dict.* 5730/3 *Snide*,..sharp; characterized by low cunning and sharp practice. 1950 P. Tempest *Lag's Lexicon* 193 'He's a "snide" so-and-so' = he's a slippery customer.
3. Insinuating, sneering, slyly derogatory.
1933 *N. & Q.* 14 Oct. 261/2 Our snide way of saying it was cheap, false, and counterfeit. 1939 *Sun* (Baltimore) 15 Apr. 8/1 Snide trick. Any reprehensible bills..enacted in secrecy at a session of the Legislature are bound to come to the surface after the State's lawmakers have left Annapolis. 1943 *Ibid.* 22 Apr. 30/6 It was a horrifying thing..to hear the President..making a snide attack against a group of Americans instead of defending America. 1954 M. Davenport *My Brother's Keeper* 182 She lived in a tenement..and she would fill the whole neighbourhood with snide gossip. 1961 J. Heller *Catch-22* (1962) x. 103 Ex-P.F.C. Wintergreen was a snide little punk who enjoyed working at cross-purposes. 1978 G. Greene *Human Factor* vi. ii. 318 Next day when Ivan made his snide references to 'gratitude' he broke furiously out: 'You call this gratitude.' 1981 *Maledicta* V. 123 He ..curses a great deal and writes snide letters to careless authors.
B. *sb.* **1. b.** A base, contemptible person; a swindler, cheat, liar.
1874 Hotten *Slang Dict.* 299 'He's a snide,' though this seems but a contraction of *snide 'un*. 1883 J. Hay *Bread-Winners* xix. 297 'I am right glad I got here to save you from that—' he paused, searching for a word which would be descriptive and yet not improper in the presence of a lady,..'that snide.' 1919 *Dial. Notes* V. 67 That fellow is a snide, do not trust him. 1935 Auden & Isherwood *Dog beneath Skin* ii. iii. 99 Young Waters is playing too. He's no snyde at the game. 1972 L. Henderson *Cage until Tame* xii. 103 Tolly's not a snide, he's better than most, and he's been bloody unlucky.
c. Hypocrisy, pretence; malicious gossip.

1902 G. H. Lorimer *Lett. Self-made Merchant* vii. 90 Courtesy without condescension,..simplicity without snide. 1966 *New Statesman* 8 Apr. 499/2 She analysed.. the nasty state of affairs on the gossip beat. The result was spectacular—some of the popular papers changed the titles of their columns, keyhole snide was banned, [etc.].
Hence **sni·dely** *adv.*; **sni·deness; sni·dery** = *Snide *sb.* 1 c; also, an instance of this.
1942 Berrey & Van den Bark *Amer. Thes. Slang* § 317/1 *Treachery*...snideness. 1953 *Britannica Bk. of Year* 639/2 *Snidery*,..hypocrisy, pretence. 1956 D. Karp *All Honorable Men* 39, I drew aside the people from *Time* and asked them if they were going to treat the Institute snidely or soberly. 1961 'B. Wells' *Day Earth caught Fire* viii. 119 'If you're right this means no private water at all.' 'Correct,' said Pete snidely. 'Just turn on the taps and hear the rude noise of progress.' 1965 P. Wylie *They both were Naked* i. ii. 92, I was 'in'...by great good luck and the use of some small snideness. 1967 *Punch* 8 Nov. 697/2 Those sardonic snideries which come too readily to one's lips. 1969 *Daily Tel.* 24 Apr. 21/5 The snidery of the humour may escape those playgoers who cannot look at it with fairly Irish eyes. 1975 *Country Life* 30 Oct. 1160/1, I have often snidely remarked, that the flowers on each spike are not half of them opened before the first are already brown. 1978 A. Noakes *William Frith* iv. 68 Frith's success with *Derby Day*.. sparked off some ill-tempered snideries.

snidey (snai·di), *a. slang.* Also **sniddy, snidy.** [f. Snide *a.* + -y[1].] **a.** Bad, contemptible. **b.** Insinuating, cutting.
1890 in Barrère & Leland *Dict. Slang* II. 267/2 Since Bill George was nabbed for liftin' them sax things is been very sniddy, so you'll be glad to learn as I have got on a new hook. 1903 Farmer & Henley *Slang* VI. 281/1 *Snide*... As *adj.* (also *sniddy* or *snidey*) = bad, wretched, contemptible. 1928 F. Hurst *President is Born* xxii. 232 'Fraid! Snidey! Poof! 'Fraid. Poof! Poof! Poof! 1972 *Guardian* 20 Jan. 13/2 Miss Duncan will not allow snidy little one-liners to upset her. 1977 *Sounds* 9 July 33/3 The journalists thought he was being 'Hip' when he was snidey about the Dolls on TV.
Hence **sni·diness.**
1976 E. Dunphy *Only a Game?* iii. 98 Because there is glory and money and your career at stake. And that entails backbiting, snidiness, scapegoating and a whole host of other things.

snie, var. *snye.

sniff, *sb.* Add: **1. d.** *fig.* A hint, intimation.
1936 C. Day Lewis *Friendly Tree* ii. ix. 124, I have been..wondering if I shall ever get a job... I have just got a sniff of one—experimental work.
5. *U.S.* A domino game in which the first double played has special significance; the first double played.
1917 J. Hergesheimer *Three Black Pennys* iii. xxiv. 289 After dinner, when they were playing sniff. 1930 J. H. Appel *Business Biogr. J. Wanamaker* xxii. 336 His own favourite game was 'sniff', played with dominoes. 1961 D. C. Armanino *Pop. Domino Games* 37 A *singles* may be played off the end of singles, the sides of doubles, and the ends of Sniff. Sniff is the only double on which plays can be made on the ends. 1974 F. Berndt *Domino Bk.* 33 Sniff is yet another variation of Muggins. *Ibid.*, The first double played is called the Sniff.

sniff, *v.* Add: **1. b.** *spec.* To inhale cocaine, the fumes of glue, etc., through the nose. *slang.*
1925 *Flynn's* 4 Apr. 819/2 *Sniff*,..to use powdered cocaine as snuff. 1931 E. Wallace *On Spot* ii. 24 Red, you're..a hop-head... We got no room in this outfit for guys who sniff. 1967 C. Drummond *Death at Furlong Post* v. 62 So they send us a dipso who sniffs! 1970 *New Scientist* 13 Aug. 352/1 These young people generally 'sniffed' from a plastic bag into which they first squirted aeroplane glue, cleaning fluid or whatever. 1975 *Weekend Mag.* (Montreal) 8 Feb. 21 The Whitebear sisters began sniffing almost two years ago, Janice says. 'A friend of ours used to sniff. At first, we didn't know what he was doing, so we asked and then we tried it too.' 1977 J. van de Wetering *Death of Hawker* vii. 73 He's sniffing too... Cocaine powder.
2. a. (Later *fig.* examples.)
1973 A. Mann *Tiara* ix. 76, I want to..sniff around the Vatican again. 1977 R. Player *Month of Mangled Models* vii. 133 Sniffing around Chelsea and Kelmscott.
5. b. Also with *out*.
1946 *Sun* (Baltimore) 12 Aug. 1/2 A pilotless aircraft that is sent into the air to 'sniff out' its own enemy target. 1979 J. Barnett *Backfire is Hostile!* xi. 111 You should concentrate more on sniffing out the sex fiends than speculating on spies.
c. Phr. *to sniff the wind*: see *wind *sb.*[1]

sniffable (sni·fǎb'l), *a.* [f. Sniff *v.* + -able.] That can be sniffed.
1975 *Weekend Mag.* (Montreal) 8 Feb. 23/1 Stocks of the old sniffable product were being bootlegged by merchants all over Regina. 1977 J. Wambaugh *Black Marble* (1978) iv. 36 Lopez boasted that he could..have enough sniffable paint left to get three of his pals loaded.

sniffer (In Dict. s.v. Sniff *v.*). Add: **1. b.** *spec.* One who sniffs a drug or toxic substance. Cf. *glue-sniffer* s.v. *glue *sb.* 6. orig. *U.S. slang.*
1920 E. S. Bishop *Narcotic Drug Problem* iii. 23 The heroin 'sniffer' of idle and curious adolescence. 1928 *Amer. Mercury* Aug. 485/2 The Baron was..a 'sniffer' himself. 1942 J. Henry *Henry's Famous Cases* iv. 40 Cocaine addicts are known as 'sniffers'. 1968 *Guardian* 22 Mar. 11/1 Doreen was also a 'sniffer'. This is the name given to people who inhale a mixture of ether and methylated spirits and become 'blocked'. 1981 *Daily Tel.* 24 Apr. 3/1 A glue sniffer is under the influence of a drug for the purposes of the 1972 Road Traffic Act, magistrates decided yesterday when a self-confessed 'sniffer' denied being unfit to drive through drink or drugs while in charge of a motorcycle.
2. *slang.* The nose.
1858 [see *pile-driver 2]. 1962 R. Cook *Crust on its Uppers* ii. 34 They'll..look down their sniffers at you.
3. a. Any device for detecting gas, radiation, etc. *colloq.*
1945 *Richmond* (Virginia) *Times-Dispatch* 10 Oct. 2/5 The hydrogen content in copper wire annealing furnaces.. is now continuously indicated by a new sensitive apparatus called a sniffer nose. 1946 *Sun* (Baltimore) 21 June 10/3 Louis E. De La Fleur..demonstrated a small hand-borne radio fixer, known as a 'sniffer'. He said that it was so accurate that he had been able..to locate an outlaw transmitter in a New York apartment house where hundreds of legal radios and electrical devices were putting out potential inteference. 1950 *Listener* 5 Jan. 12/1 These tiny Geiger counters first came to public attention last spring in New York... Uranium can turn up anywhere so there is no reason why, if you had a 'sniffer', as they are called, you should not start prospecting here in Great Britain. 1968 *Guardian* 5 Sept. 2/7 Perch a radar sensor on the tail..: insert a diesel fume 'sniffer'. 1972 'J. Lange' *Binary* 170 The sniffer..had been developed for use in Vietnam and had been adapted for customs operations... If the sniffer said plastic explosive was behind the door, he had to believe it. 1979 F. Pohl *Jem* iii. 26 The car was..an indispensable necessity in what he did for the agency; twice a day, other employees of the agency went over it with electronic sniffers and radio probes to make sure it had been neither bombed nor bugged.
b. Usu. *sniffer dog.* A dog trained to detect specific odours, esp. those of drugs or explosives. *colloq.*
1964 *N.Y. Times Mag.* 23 Aug. 62/3 *Sniffer*, police dog. 1975 A. Beevor *Violent Brink* iii. 66 We are using.. sniffer dogs at ports and airports so as to increase our chances of catching the explosive coming in. 1977 *Air Mail* Spring 7/1 In the first two months 'sniffer' dogs and handlers trained by the RAF Police Dog Training Flight had helped British Customs and Excise officers detect £125,000 worth of smuggled drugs. 1979 *Daily Tel.* 17 Apr. 1/6 Forty-five 'sniffer' dogs were flown into Yugoslavia from Switzerland and set to work to smell out casualties from debris in towns around Kotor Bay. 1982 *Times* 3 Sept. 10/5 Sniffer-dogs for drugs.

sniffing, *vbl. sb.* (Examples in sense 1 b of the vb.) Cf. *glue-sniffing* s.v. *glue *sb.* 6.
1968 *Guardian* 22 Mar. 11/2, I asked her what attraction there was in 'sniffing'. 1975 *Weekend Mag.* (Montreal) 8 Feb. 23/1 In any case, legislation doesn't eliminate the sniffing problem which is nation-wide. 1977 *Lancet* 8 Jan. 84/1 Investigation of the 42 patients..showed that 'sniffing' was a group activity involving mainly adolescents aged 12–19 years, all of whom had a previous history of solvent abuse.

sniffly (sni·fli), *a.* [f. Sniffle *v.* + -y[1].] Sniffling; characterized by sniffling. Also *fig.*
1927 W. E. Collinson *Contemp. Eng.* 59 They vary in intensity from the sniffling or sniffly cold to the church-yard cough! 1929 G. Ade *Let.* 8 Feb. (1973) 139 The warm weather will be welcome. Most of us have sniffly colds. Otherwise we are all right. 1960 *Guardian* 9 Apr. 6/6 We sat there in the bare little room, wet and sniffly with sentiment. 1966 R. H. Rimmer *Harrad Experiment* (1967) 79 Get out your handkerchiefs, wipe your sniffly nose. 1974 [see *luxe 1].

sniffy, *a.* Add: (Further examples.)
1915 W. S. Maugham *Of Human Bondage* lxxiv. 382 You were rather sniffy about meeting him. 1925 S. Baring-Gould *Further Reminisc.* iv. 45 Their wives were especially sniffy towards Mrs Jervis. 1965 *Listener* 3 June 834/3 He was catty about Balzac, sniffy about Stendhal, stuffy about Flaubert, and cagey about Baudelaire. 1979 *Jrnl. R. Soc. Arts* July 511/1 Sniffy comments of a patronizing nature about Victorian buildings so regrettably sprinkled throughout earlier books in *The Buildings of England* are carefully avoided.
Hence **sni·ffily** *adv.*, **sni·ffiness** (examples); **sni·ffish** *a.* (*rare*), somewhat sniffy.
1900 Kipling *Just So Stories* (1902) 108 'What will happen it I do?' said the Jaguar, most sniffily and most cautious. 1927 *Blackw. Mag.* Dec. 834/1, I didn't think sniffiness was usual under the circumstances. 1928 'M. Neville' *Kiss Proof* xviii. 163 'Oh, if that's the way you feel about it,' Toddles said sniffily. 1933 'G. Orwell' *Let.* June (1968) I. 121 There is also a certain sniffish 'I told you so' implication. 1968 M. Collis *Somerville & Ross* iii. 45 In her *Irish Memories* (published in 1917) Edith, recalling that time, writes of the sniffyness of her brothers and uncles. 1973 *Guardian* 30 June 11/3 'He wasn't up to much as a sub-editor,' said one of the older hacks, sniffily. 1981 *Economist* 24 Jan. 22/2 Although for the past two years the United States has been Algeria's largest trading partner, the relationship between the two countries has been marked by a suspicious sniffiness.

snifter, *sb.* Add: **4.** A (small) quantity of intoxicating liquor, a drink, a 'nip'. *colloq.* (orig. *U.S.*).
See note, sense 5 below.
1844 *Spirit of Times* 20 Apr. 86/2 He swallowed a cool 'snifter' at the nearest cabaret. 1910 G. B. McCutcheon *Rose in Ring* v. 90 You need a snifter of brandy...Joey handed her a drink from his flask. 1924 Wodehouse *Ukridge* iii. 56 And now, old horse, you may lead me across the street to the Coal Hole for a short snifter. 1934 *Bulletin* (Sydney) 26 Dec. 41/1 The postboy brought George a telegram, and..on opening it George smiled and shouted snifters all round. 1942 E. Paul *Narrow St.* xxvii. 246, I..was on the point of suggesting that he step across to the Café St. Michel for a snifter. 1963 B. Pearson *Coal Flat* i. 14 Do you want a drink—or would you rather have a bit of a snifter with the boys? 1978 R. V. Jones *Most Secret War* vii. 59 What happened was that he had taken it from his own station to another for a lunch which was preceded, and doubtless followed, by a surfeit of what he termed 'lightning snifters'.
5. A glass with a wide body narrowing towards the top, used for brandy, etc. orig. and chiefly *U.S.*
The sense 'the contents of a snifter' is usu. indistinguishable from sense 4 above and may be represented in some examples there.
1937 G. Frankau *More of Us* xvi. 170 And sought that other room to drain a snifter With Herr Staatsschauspiel-hausmeister Kohn-Goering. 1943 D. Baker *Trio* 155 She was sitting beside me holding a brandy snifter. 1970 J. Hansen *Fadeout* i. 4 A bottle of brandy warmed on the hearth... She poured splashes from it into two small snifters. 1978 G. Vidal *Kalki* x. iii. 241, I drank brandy from a huge Baccarat snifter.
6. a. *U.S. slang.* A cocaine addict. Cf. *sniffer 1 b.
1925 *Flynn's* 4 Apr. 819/2 *Snifter*, a cocaine fiend. 1929 *Detective Fiction Weekly* 27 Apr. 31/2 A certain cocaine addict, known as Snifter Selton. 1955 *Amer. Speech* XXX. 85 *Snifter*, an addict who inhales cocaine.
b. *slang* (orig. *U.S.*). A small quantity of cocaine inhaled through the nose.
1930 *Detective Fiction Weekly* 5 July 357/1 Well, boys, take me down [to the police station]. Just one snifter of snow and I'm with you. 1934 'D. Hume' *Too Dangerous to Live* viii. 85 He's been doping for a few months— cocaine. When he was picked up he hadn't had a snifter for nearly twenty-four hours. 1974 J. Wainwright *Evidence I shall Give* xxi. 99 A snifter when the pain's bad... It ain't for kicks. You're no junkie.
7. *U.S. slang.* A portable radio direction-finder. Cf. *sniffer 3 a.
1944 *Sci. News Let.* 12 Aug. 103 'The snifter'..is a portable, one-man direction finder that 'smells out' by radio the very room in which an illegal radio transmitter is hidden. 1949 *Life* 5 Dec. 166/2 At the start hunters with radio direction finders, called 'snifters', collect at Brookfield Zoo.

snifty, *a.* Add: **2.** *slang* (orig. and chiefly *U.S.*). Haughty, disdainful.
1889 K. Munroe *Golden Days* xvii. 188 If you notice me getting anyways snifty..you just bump me down hard. 1902 G. H. Lorimer *Lett. Self-made Merchant* xviii. 268 Clytie said..that spirits were mighty snifty and high-toned. 1909 H. G. Wells *Tono-Bungay* i. i. 40 'Snifty beast!'..That governess made things impossible. 1942 Berrey & Van den Bark *Amer. Thes. Slang* § 301/6 *Arrogant*,..snifty.

snig (snig), *v.*[2] *north. dial., Austral., N.Z.,* and *Canad. local.* [Origin obscure.] *trans.* To drag (a heavy load, esp. timber) by means of ropes and chains. Hence **sni·gging** *vbl. sb.* (also *attrib.*).
For further material see *Eng. Dial. Dict.*
1790 F. Grose *Provinc. Gloss.* (ed. 2), *Snig*, to drag wood without a cart. 1866 J. T. Staton *Rays fro th' Loominary* 127 He wur one ut wur brought up to sniggin timber. 1933 L. Acland in *Press* (Christchurch, N.Z.) 2 Dec. 15/7 *Snig*, to drag along the ground by horse or bullocks, especially to drag logs or other timber. The stout chain which goes round the log has a ring at one end, and a hook to which the horses' chains are attached. It is called a *snigging chain*. 1946 B. James in Murdoch & Drake-Brockman *Austral. Short Stories* (1951) 251 Peter cut timber on the hills, and snigged it down with the plough horses. 1961 B. Crump *Hang on a Minute Mate* 44 They dug their axes into a handy stump and trudged off down the snigging-track. 1968 E. R. Buckler *Ox Bells & Fireflies* xv. 221 A group of men have gathered to help another lay a new sill under his barn. 'Hadn't I better hitch up the team and snig her closer the foundation there?' 1969 *Parade* (Austral.) Dec. 17/2 He would get his horse and snig Trompson's body off the claim. 1975 *Sunday Mail* (Brisbane) 1 June 6/2 Bullock teams would snig the logs to the winder.

snig (snig), *v.*[3] *dial.* and *slang.* [Origin obscure.] *intr.* and *trans.* To steal.
1862 C. C. Robinson *Dial. Leeds* 415 *Snig*, to steal after a mean fashion, as a man who undertakes any business, or interests himself any way in the property of a person, and is 'snigging' away at it all the time. 1864 J. Ramsbottam *Phases of Distress* 37 They'll pitch an' toss an' swear, An' snig an' snatch owt wheer they con. 1892 Kipling *Barrack-Room Ballads* 31 If you've ever snigged the washin' from the line.

snip, *sb.* Add: **II. 8.** (Later examples.) Cf. *tinsnips.
1940 I. L. Idriess *Lightning Ridge* xiii. 88 As the miner's hand gently closed on the snips the jaws came together and bit a chip from the edge of the nobby. 1966 D. F. Galouye *Lost Perception* xviii. 188 He fished his snips out of the kit. 'Now we have only to cut the cables.' 1979 *Sunset* Apr. 170/2 The home owner made a pattern first, then used tin snips to cut long pieces of copper trimmings into strips.
10. a. (Earlier and later examples.)
a 1890 *Sporting Life* in Barrère & Leland *Dict. Slang* (1890) II. 268 D. is in glorious form with his wires, and is certain to keep it up next week at the above meetings, for which he knows of several snips. 1913 C. Mackenzie *Sinister St.* I. ii. xi. 323 You'll get your Third Fifteen cap for a snip. 1923 E. P. Oppenheim *Inevitable Millionaires* xxviii. 285 'You think it will be a good speculation, then?' Stephen observed, a little sadly. 'A dead snip,' Sir Philip assured them. 1945 'N. Shute' *Most Secret* viii. 187 It is a snip; we will get both of them. 1954 Wodehouse *Jeeves & Feudal Spirit* ii. 19 Wooster..is the deadest of snips. He throws a beautiful dart.
b. A bargain, a good buy. *slang.*
1926 H. V. Morton *Spell of London* 94 She sees a tea-gown with the authentic plainness..about it that tells her it began life in higher circles. 'Now, that's a snip, miss. Just your style!' 1933 *Camera* Aug. 7 (Advt.), Exchange your present camera for one of these guaranteed 'snips'. 1935 L. A. G. Strong *Tuesday Afternoon* 20 The smart man comes along, looks in the window, spots the real snip. 1956 'N. Shute' *Beyond Black Stump* ix. 254 Got them for only a couple of quid each, a snip. 1963 *Punch* 30 Jan. 162/1 A snip at forty bucks. 1977 *Times* 29 Oct. 10/6 At a time when Beaujolais prices are soaring it is a snip at £1·90.
c. A piece of good fortune. In phr. *a snip of a* (thing) to designate something simpler, more excellent, or more pleasing than one could have expected, a 'gift'.
1932 W. S. Maugham *For Services Rendered* ii. 47 It's been a snip for me having this house to come to. Except for all of you I should have had a pretty thin time. 1952 M. Tripp *Faith is Windsock* xiv. 210 A snip of an op. Cloud over the target thwarted any searchlights, fog kept the fighters down and there was no flak. 1953 Dylan Thomas *Under Milk Wood* (1954) 47 Llaregyb this snip of a morning is wildfruit and warm, the streets, fields, sands and waters springing in the young sun.

snipe, *sb.* Add: **4. c.** *U.S.* The discarded stub of a cigar or cigarette.
1891 H. Campbell *Darkness & Daylight* iv. 124 The 'Snipe-Shooter' was guilty of smoking cigar-stubs picked out of the gutter a habit known among the boys as 'snipe-shooting'. 1899 'J. Flynt' *Tramping* ii. iv. 274 This 'snipe' chewing and smoking is the most popular use of tobacco in trampdom. 1914 'High Jinks, Jr.' *Choice Slang* 18 *Snipe*, a cigar or a cigarette stub. 1939 J. Steinbeck *Grapes of Wrath* x. 129 Winfield was..an inveterate collector and smoker of snipes.
d. One of a group of workers, esp. on board ship (see quots.). *U.S.*
1918 L. E. Ruggles *Navy Explained* 139 *Snipe*— Firemen in the 'black gang' always refer to each other as 'snipes'. In a gang of snipes below there is generally one dude who is known as the 'king snipe'. He is considered the leading snipe of the watch. 1932 *Santa Fé Mag.* XXVI. ii. 34/1 A foreman of a section gang is a *Jerry* or a *king*; a section laborer is a *snipe*. 1951 H. Wouk *Caine Mutiny* xxvi. 289 A big sloppy chowhound named Wagner, a snipe, had made himself a wax impression of the cook's key. 1953 M. Dibner *Deep Six* xv. 169 A snipe chief wearing a blue shirt and an oil-soiled khaki cap stood legs apart, drinking coffee.
e. *Logging.* A sloping surface or bevel cut on the fore end of a log to facilitate dragging.
1958 W. F. McCulloch *Woods Words* 172 *Snipe*, a bevel hewed on the ride side of the end of a log, making it easier to pull over the skids. 1975 *Islander* (Victoria, B.C.) 2 Mar. 13/1 He took out the long, beautiful 155-foot timbers, 18 inches at the butt and tapered with a four-foot snipe to a four-inch square point.
5*. Also **Snipe.** A type of sloop-rigged sailing boat approximately 15½ ft. long and used for racing; also, the name of this class of boat.
1931 *Rudder* (U.S.) July 46 Snipe... Designed especially for the *Rudder* by William F. Crosby. *Ibid.* 47/1 Snipe is a design for a small racing sloop. 1941 *Sun* (Baltimore) 20 Aug. 13/6 For the small-boat sailors races have been arranged in the following classes: Snipe, moth, penguin, winabout, Hampton, 20-foot roundbottom, 20-foot (and under) chine built knockabout. 1942 *E. African Ann.* 1941–2 57/1 There is always some fine sailing to be had.. some craft beating out into the bay or later in the evening some homeward bound 'Snipe' coming in before the wind. 1969 H. Horwood *Newfoundland* xix. 149 Holyrood has a small harbour at the mouth of a brook, and a junior sailing club with instructors and racing snipes.
5.** A long-range shot or attack from a sharp-shooter; the sound of a sniper's bullet. Also *fig.*
1969 G. Macbeth *War Quartet* 72 The return snipe struck His mouth below the helmet. 1973 E. Bullins *Theme is Blackness* 6 For paeans of Blackness were videoed throughout Black America, between the stoccado snipe of the assassin's slug. 1977 *Rolling Stone* 13 Jan. 10/1 Rod Stewart..has transcended two years of snipes for his romance with actress Britt Ekland. 1977 *Sunday Times* 3 July 17/5 The difficulty of organising a 'snipe' (assassination by a single gunman) in the tight security of Belfast.
6. b. *snipe-nosed* (later examples referring to pliers).
1968 M. Woodhouse *Rock Baby* xvi. 157, I took the Allen Keys and a small pair of snipe-nosed pliers. 1969 *Gloss. Terms Dentistry* (*B.S.I.*) 48 *Snipe-nosed pliers*, pliers with square nosed flat beaks... Used for bending wire.

snipe, *v.* Add: **2. b.** *fig.* To assault with harsh sly criticism; to rebuke or censure sharply; to make a carping attack *at* (someone).
1892 [implied at *sniping *vbl. sb.* 1 b]. 1959 I. & P. Opie *Lore & Lang. Schoolch.* xvi. 343 Although adult factions may have made peace with each other, their children on the way to school may continue sniping at each other for generations. 1979 'A. Hailey' *Overload* i. xiv. 79 The press representatives had eaten and imbibed with gusto, then in published reports, some had sniped at GSP & L for extravagant entertaining at a time of rising utility bills.
3. *trans. Logging.* To cut a snipe or bevel on (a log) to ease dragging.
1870 *Overland Monthly* 5 July 56/1 The fourth man is the 'hook-tender', whose duties are to wait on the team and 'snipe the logs'. 1902 *N.Z. Illustr. Mag.* V. 375 If the weather is favourable, the log is 'sniped' or rounded at one end, an iron grip driven into it, and to this the team is fastened. 1958 W. F. McCulloch *Woods Words* 173 *Snipe*,..to hew a snipe on the end of a log. *Snipe for the ride*, to put the snipe on the side of the log which would ride on the bottom, saving the work of sniping a bevel around the entire end of the log.
4. *trans.* and *intr.* To pilfer, steal; to pick up or obtain (from the roadside, etc.); *spec.* to prospect for gold, as in old diggings. Cf. Sniper 3. *slang* (chiefly *N. Amer.*).
1909 R. Service *Ballads of Cheechako* 122, I panned and I panned in the shiny sand, and I sniped on the river bar; But I know, I know, that it's down below that the golden treasures are. 1923 J. Manchon *Le Slang* 280 *Snipe*, escamoter [*sc.* to steal, filch]. 1932 J. T. Farrell *Young Lonigan* iv. 169 He walked down to Fifty-seventh St, furtively looked round to see if anyone saw him, and when the coast was clear, he sniped a butt from the street. 1974 F. W. Luddit *Campfire Sketches of Cariboo* vi. 27 They..made small amounts of money sniping for gold. 1977 *New Yorker* 20 June 81/1 He 'sniped' a lot of his gold—just took it from likely spots without settling down to the formalities of a claim.

sniper. Add: **4.** *Logging.* One who cuts a snipe on a log.
1905 *Terms Forestry & Logging* (U.S. Dept. Agric. Bureau Forestry) 48 *Sniper*, one who noses logs before they are skidded. 1906 *Log of 'Columbia'* June 8/1 'He ought to be chased out of the woods,' said Jim, the sniper. 1914 *Chambers's Jrnl.* 3 Oct. 696/1 Summoning the sniper for an occasional undersnipe. 1956 R. W. Andrews *Glory Days of Logging* 14 These were..snipers who shaped the butt ends, so they [*sc.* logs] wouldn't hang up on the skids.
5. *Austral.* (See quot. 1945.)
1945 Baker *Austral. Lang.* xiv. 248 A waterfront term of fairly recent origin is *sniper*, a non-union labourer. 1955 J. Morrison *Black Cargo* 14 It will need only one shout of 'Sniper!' and Lamond will be lucky to get out without being knocked down. 1957 T. Nelson *Hungry Mile* 72 The W.W.F. had preference of work, wharf by wharf. The outsiders (snipers) would stand back at the gate until the W.W.F. men were all used.

sniperscope (snai·pəɹskōᵘp). [f. Sniper + -scope.] † **1.** A device incorporating a periscope, whereby a rifle may be fired by a soldier who remains concealed. *Obs.*
1918 R. H. Knyvett *Over There* 135 Many of the inventions are forgotten, but some are in use in France today, notably the 'periscope rifle' or 'sniperscope'.
2. A small device which converts infra-red radiation to a visible image and may be fixed to a gun so that it can be aimed in the dark.
1941 *Sun* (Baltimore) 1 Jan. 6/3 The selected few American infantrymen who tried out the 'sniperscope' at Okinawa have already had a glimpse of future warfare. 1954 W. Tucker *Wild Talent* (1955) xi. 418 'He must have used a rifle. And had good eyesight.' 'He probably had a sniperscope on it.' 1971 F. Forsyth *Day of Jackal* i. i. 14 A rifle with sniperscope was found at Poinard's flat.

sni·piness. [f. Snipy *a.* + -ness.] Undue length and pointedness of the muzzle of an animal, suggestive of a snipe's bill.
1938 J. W. Day *Dog in Sport* xv. 204 The jaws should be long and powerful, and quite free from snipiness or exaggeration in length. 1963 B. S. Vesey-Fitzgerald *Cat Owner's Encycl.* 121 The nose is longish, but the cheeks being very prominent do away with any snipiness, which is a bad fault.

sniping, *vbl. sb.* Add: **1. b.** *fig.* The making of sly critical assaults; sharp fault-finding or carping. Freq. with *at*.
1892 Mrs. H. Ward *David Grieve* I. vi. 128 Hannah's appetite for snipin' returned. 1935 E. Pound *Let.* 7 Feb. (1971) 267 Eng. print so smeared with personal sniping and clique politics that any definition of limitations..is likely to be taken as 'anti-'. 1945 *Sun* (Baltimore) 25 June 1/3 There has been a lot of unfortunate sniping at the project. 1955 *Essays in Criticism* V. 64 Mr. Liddell's 'criticisms' of his essay..are mere sniping and peripheral eroding which say next to nothing about their subject. 1969 H. Perkin *Key Profession* vi. 229 Hence that perpetual sniping..at the universities for their supposed idleness and inefficiency. 1977 *Time* 4 July 2/3 Your sniping at the U.N. and its jobs-for-the-boys was not premature. 1980 M. Fonteyn *Magic of Dance* 214 Taglioni, tired of petty sniping and endless comparisons favouring her rival, journeyed to St. Petersburg.
3. Prospecting for gold, esp. in old diggings. *N. Amer. slang.*
1897 M. H. E. Hayne *Pioneers of Klondyke* 93 There is little 'snipping' [*sic*]—i.e. working old bars—on Forty Mile Creek, but it does not pay much. 1963 *Placer Mining B.C.* 7/1 Chinese miners were particularly adept at sniping... Sniping and bar-combing are..carried out by individuals who would rather do this than work for wages.
4. *Austral.* Working as a non-union labourer.
1951 *Meanjin* (Melbourne) X. 334 You never scabbed, or anything like that? You never did any sniping?

snippy, *a.* Add: **2.** (Later examples.) Also, putting on airs, supercilious.
1894 P. L. Ford *Hon. Peter Stirling* xxx. 171 Before I could possibly have said or done anything to offend her, she treated me in the snippiest way. 1934 E. Carr *Jrnl.* 12 Feb. in *Hundreds & Thousands* (1966) 95, I don't want to be mean and snippy but I don't think they know. 1952 H. Garner *Yellow Sweater* 93 There followed an explanation of why her son hadn't been attending school as regular as he should, and how snippy the teacher was getting to be. 1961 *Insurance Salesman* Jan. 47/1 One of our clerks was a snippy, opinionated girl who kept everything unsettled and rubbed everyone the wrong way. 1970 J. Potts *Affair of Heart* viii. 65, I must say, she was very snippy. Downright rude. 1974 *Times* 28 Feb. 10/6 This irritates Mr Heath... Privately, he is quite snippy about it. 1977 D. Ramsay *You can't call it Murder* i. 52 Daughter Sarah described as uppity and snippy.

snit[2]. *slang* (orig. and chiefly *U.S.*). [Of uncertain origin (see quot. 1939[2]).] A state of agitation; a fit of rage or bad temper; a tantrum, sulk. Freq. in phr. *in a snit*.
1939 C. Boothe *Kiss Boys Good-bye* ii. i. 105 'I declare, Mrs. Rand, I cried myself into a snit.' 'A snit?' 'I do deplore it, but when I'm in a snit I'm prone to bull the object of my wrath plumb in the tummy.' 1939 *Sat. Rev. Lit.* 23 Dec. 12/1 The membership could hardly be said to be in a snit,..as nobody in Georgia seems ever to have heard of either the word or the state of being until Miss Clare Boothe isolated and defined it. 1962 J. Potts *Evil Wish* x. 136 If you hadn't been in such a snit when I came upstairs I'd have told you so. 1971 *Daily Progress* (Charlottesville, Va.) 21 Jan. 4A/3 If New York solves its problems through gambling, every state in the union is going to follow suit except Nevada, which will probably secede from the nation in a snit. 1975 J. Goulet *Oh's Profit* xxxvii. 208 The President of the United States had bawled him out and left Cambridge College in a fierce snit. 1980 *N.Y. Times* 8 Jan. D 16, I was recently..put in charge of six other copywriters, two of them men. The men are in a quiet snit.

snitch, *sb.* Add: **3.** (Later examples, chiefly U.S.)
1906 *Atlantic Monthly* Nov. 589 He employs that phenomenon of despicability..in Western parlance called a snitch..to work up the lawsuit. 1930 *Forum* Dec. 375/1 A police informer in New York, for instance, is a stool or snitch. 1959 I. & P. Opie *Lore & Lang. Schoolch.* x. 189 The tell tale is..a sly, a snitch or snitcher (common, especially in the Midlands). 1965 J. Wainwright *Death in Sleeping City* 142 The 'snitches' and the 'grassers' and the 'stoolpigeons' whispered out of the corner of their mouths, and money changed hands. 1979 S. Rifkin *McQuaid in August* ix. 97 Lopez was an informant..a paragon among snitches.
4. Phr. *to have* (or *get*) *a snitch on* (someone): to have a grudge against or 'down' on; to dislike. *N.Z. slang.*
1943 J. A. W. Bennett in *Amer. Speech* XVIII. 90 A person [in New Zealand] complaining of another's ill-will might also say, 'He's got a proper snitch on me'— obviously a variant of 'to snitch upon' (to inform against). 1948 *Landfall* II. 109 These jokers didn't understand the snitch Myers had on you, seemed to think it was right that Myers should always be tormenting you. 1953 O. E. Middleton *Short Stories* 28 He wasn't a man to get a snitch on his neighbours because of a bit of bad luck and it wasn't long before he was his own self again. 1959 G. Slatter *Gun in my Hand* viii. 91 Got a snitch on me and put me in crook with the boss.
5. *Comb.*, as **snitch-rag** *slang*, a handkerchief.
1940 H. G. Wells *Babes in Darkling Wood* i. i. 25 Can I borrow your snitch-rag, Gemini?

snitch, *v.* Add: **1.** (Later examples.) Now usu. with *on*. Also, to reveal or give information *to* (someone). *slang.*
1910 'O. Henry' *Whirligigs* xiii. 157 Say, don't snitch to the tenants about this, will yer? 1926 J. Black *You can't Win* xix. 279 If I get a job some copper will snitch on me to my boss. 1933 *Daily Progress* (Charlottesville, Va.) 26 Jan. 1/8 He did it, he said, because she 'snitched' on him when he played truant from school. 1941 B. Schulberg *What makes Sammy Run?* v. 83, I felt a little guilty about snitching on my neighbor. 1957 A. Miller *View from Bridge* i. 33 The family had an uncle that they were hidin' in the house, and he snitched to the Immigration. 1966 P. Moloney *Plea for Mersey* 14 The Captain.. had snitched to the police that his cargo was being pilfered.
3. To take surreptitiously, purloin; to steal or 'pinch'. *slang.*
1904 *N.Y. Times* 6 June 9 They reached Coney Island by snitching rides. 1933 D. L. Sayers *Murder must Advertise* iii. 46 He first of all snitched people's ideas without telling them, and then didn't give them the credit for it. 1948 L. A. G. Strong *Trevannion* xvii. 323 You love a girl faithfully for years, and some glib sod comes along at the heel of the hunt and snitches her from you. 1958 [see *booksy *a.*]. 1976 M. Machlin *Pipeline* xxx. 348 How about that guy who snitched a whole D-9 tractor, brand-new?
Hence **sni·tching** *vbl. sb.*
1923 W. S. Maugham *Our Betters* iii. 172 You really might have left Tony alone. This habit you have of snitching has got you into trouble before. 1933 *Sun* (Baltimore) 24 Aug. 6/7 Not long ago we had the fine stirring story by Neil Swanson 'The Judas Tree', and now comes a snitching of that title by Leslie Ford, who calls his new detective thriller 'The Clue of the Judas Tree'. 1961 B. Malamud *New Life* (1962) 298 He had been thinking of discussing with him Bullock's concern with athletes but it was too much like snitching. 1972 J. Wambaugh *Blue Knight* (1973) ii. 39 'Okay,' I said, giving

him a chance to rationalize his snitching, which all informants have to do when they start out.

sni·tter, *v.*² *Sc.* and *north. dial.* [Cf. SNICKER *v.*, SNIGGER *v.*¹, SNIRT *v.*, and TITTER *v.*¹] To laugh in a suppressed, nervous manner (*at* something). Also as *sb.*

1825 JAMIESON Suppl., *To snuister*, or *snuitter*,..to laugh in a suppressed or clandestine way through the nostrils. *Snuister, snuitter*,..a laugh of this kind. **1892** M. C. F. MORRIS *Yorkshire Folk-Talk* 374 What's ta stannin' theer snitterin' an' laffin' at? **1896** 'G. UMBER' *Ayrshire Idylls* 71 Hoo her words should provoke sae muckle snitterin' an' lauchin'. **1975** *New Society* 31 July 235/2 A prevailing snitter (cross between snigger and titter) greeted the preview of..a new play... There was plenty to snitter at. **1975** W. MCILVANNEY *Docherty* III. v. 270 'Ye micht never be heard o'.' 'Sen' in David Livingstone,' Conn said. Tom snittered. **1977** —— *Laidlaw* xxxviii. 177 Harkness began to laugh. Laidlaw stared at him, then..snittered at himself.

snivelization (sniv'ləizē¹·ʃən). *nonce-wd.* [Melville's factitious blend of SNIVEL *sb.* and CIVILIZATION.] Civilization considered derisively as a cause of anxiety or plaintiveness. Also **sni·velize** *v. trans.*, to reduce (someone) to a state of whimpering civilization; **sni·velized** *ppl. a.*

1849 H. MELVILLE *Redburn* I. xxi. 200 Ye wouldn't have been to sea here, leadin' this dog's life, if you hadn't been snivelized... Snivelization has been the ruin on ye. *Ibid.*, Snivelized chaps only learns the way to take on 'bout life, and snivel. **1892** 'MARK TWAIN' *Satires & Burlesques* (1967) 169 He was working this character into an elaborate satire on civilization to be called 'Affeland (Snivelization)'. *c* **1938** L. MUMFORD *Report on Honolulu* in *City Development* (1946) x. 106 The restrictions and burdens imposed by what one of Herman Melville's characters derisively called 'snivelization'.

snivellingly (sni·v'liŋli), *adv.* [f. SNIVELLING *ppl. a.* + -LY².] In a snivelling or whimpering manner; abjectly.

1959 *Times* 31 July 9/1 Mary [Queen of Scots] is presented as a creature of radiant perfection surrounded by snivellingly devoted waiting women and surly guards. **1970** N. FLEMING *Czech Point* (1971) i. 9 Beside him I would have appeared a snivellingly puny specimen.

snob, *sb.*¹ Add: **1. b.** The last sheep to be sheared; hence, the roughest or most difficult sheep to shear; = *COBBLER 1 b. *Austral.* and *N.Z. slang.*

1945 C. E. W. BEAN *On Wool Track* (new ed.) 135 The sheep most difficult to shear, which naturally is left last in the pen, is also called the 'snob'. **1955** G. BOWEN *Wool Away!* 157 *Snob*, the last sheep in the pen. **1971** J. S. GUNN *Distrib. Shearing Terms N.S.W.* 9 As it is the practice to leave rough sheep until last it is only to be expected that *snob* and *cobbler* for both 'rough' and 'last' will occur... *Snob* and *cobbler* meant 'last' before specialising to 'rough'. **1975** L. RYAN *Shearers* i. 49 'Get on to this wrinkled bludger!' he said. It was the last sheep in the pen... 'Real snob, ain't it?'

3. d. One who despises those who are considered inferior in rank, attainment, or taste. Freq. in extended sense with defining word that limits its reference to a particular sphere.

Overlaps with sense 3 c in Dict.

1911 G. B. SHAW *Getting Married* 228 All her childish affectations of conscientious scruple and religious impulse have been applauded and deferred to until she has become an ethical snob of the first water. **1925** F. SCOTT FITZGERALD *Great Gatsby* vii. 146 Listen, Tom. If you're such a snob, why did you invite him to lunch? **1931** A. HUXLEY *Music at Night* 121, I have met several adolescent consumption-snobs...these ingenuous young tubercle-snobs. **1935** C. ISHERWOOD *Mr Norris changes Trains* iv. 58, I rather enjoyed playing with the idea that he was, in fact, a dangerous criminal... Nearly every member of my generation is a crime-snob. **1939** [see *INTELLECTUAL *a.* 1 b]. **1959** G. FREEMAN *Jack would be Gentleman* iii. 54 God knows, Moyra, I'm not a snob but that sort of person just wouldn't understand. **1960** J. O'HARA *Sermons & Soda-Water* I. 26 He doesn't want to know her any better and neither would my mother. That isn't snobbishness... You're the snob of us two. **1977** T. HEALD *Just Desserts* i. 16 He does..that frightful column in the Chronicle... The wine snob's guide to an early cirrhosis.

e. *inverted snob*: see *INVERTED *ppl. a.* 9.

5. a. *snob jargon, school, word*, etc.; *snob-free* adj.; **snob appeal,** attractiveness to snobs; **snob value,** value as a commodity prized by snobs or as an indication of superiority.

1933 LEAVIS & THOMPSON *Culture & Environment* 15 (*heading*) The snob appeal. **1943** *Scrutiny* XI. 289 There is, of course, the same snob-appeal, and just as Mr. Richards is always introducing a Shakespearean phrase.., so Jeeves is always quoting Pope. **1958** M. DICKENS *Man Overboard* xii. 192 There's a snob appeal about having a retired officer as bursar. **1978** J. PEARSON *Façades* vii. 127 Osbert and Edith [Sitwell]..had inherited..style; their snob appeal was undeniable. **1961** D. L. MUNBY *God & Rich Society* iv. 68 Americans and Scandinavians have a lot to teach us about real social equality and snob-free education. **1952** E. PARTRIDGE *From Sanskrit to Brazil* 59 The most dangerous snob jargon of all is that used by ordinarily well-educated..men and women. **1953** R. CHANDLER *Let.* 16 Sept. (1981) 351 If your boy won't behave himself..you can send him to one of the New England snob schools like Groton. **1978** M. BIRMINGHAM *Sleep in Ditch* 113 She'd been married, very young, almost the moment she'd left her snob school. **1936** *Proc. Inst. Automobile Engineers* XXX. 762 Generally, if the big luxury car leads with any new refinement sooner or later the lower and lowest-priced cars follow, the new feature acquires from its aristocratic origin what has been aptly termed 'snob-value'. **1955** T. H. PEAR *Eng. Social Differences* 131 The terms of normal psychology have never achieved snob-value. **1969** M. FISH in A. S. C. ROSS *What are U?* 78 It was an example of faulty handcraft giving a snob value to a product that could have been made more efficiently by machine. **1935** A. P. HERBERT *What a Word!* iv. 92 'Beginning' is musical and 'commencement' is not. Also, it is a Snob-word.

b. Used *predicatively* as *adj.*, fashionable, snobbish, pretentious.

1958 *Spectator* 14 Feb. 209/3 A little slower than Buchan, a little less naively snob than Dornford Yates. **1970** *Daily Tel.* 9 Apr. 17/2 Champagne we consider too snob, and we're all off hard liquor. We drink wine now as an aperitif.

snob, *sb.*² Add: In full, *snob-cricket.* ? Now *obs.*

1888 A. LANG in Steel & Lyttelton *Cricket* i. 1 There is a sport known at some schools as 'stump-cricket', 'snob-cricket', or ..'Dex'. **1893** J. W. BAINES in A. G. Bradley et al. *Hist. Marlborough Coll.* xxii. 220 The great thing was 'Snob' cricket, which speedily became a most popular and fashionable pursuit. **1901** *Blackw. Mag.* Oct. 490/2 The game known as 'snob-cricket', little cricket, 'stump-and-ball', and so forth, might be introduced.

snobbery. Add: **2. c.** With defining word: pretension to superior knowledge, taste, etc., in a particular sphere.

1903 [see *INTELLECTUAL *a.* 1 b]. **1937** LD. SAMUEL *Belief & Action* iii. 30 It is a kind of cosmic snobbery to expect us to feel 'humble' in the presence of astronomical dimensions merely because they are big. **1977** *Times* 15 Nov. (Italian Wine Suppl.) p. i/3 It is perhaps another instance of declining wine snobbery when people want to offer a wine that is good but cheap.

snobbism. Add: Also **snobism** [cf. next]. **1.** (Earlier and later examples.) Also, an instance or manifestation of this.

1845 *Punch* June 254/1 We never saw any living creature in such a high state of snobbism. **1895** [see *EGOMANIA]. **1923** J. M. MURRY *Pencillings* 36 Dickens is safe, so safe indeed that within the next twelve months he may become a snobism in his turn... Although I have floundered into most of the artistic snobisms of my time.. I have never deserted Mr. Micawber. **1932** P. BALFOUR *Society Racket* i. 48 'Belle's Letters', in which the great social figures..liked to be mentioned and to have their dresses described, is nauseous in its unabashed and luscious snobism. **1940** H. G. WELLS *All aboard for Ararat* iii. 84 Some mysterious process of snobism. **1966** *Punch* 20 July 124/2 It deflates a lot of irritating snobbisms and pomposities. **1972** *Science* 12 May 620/1 A union representative must cope with the inevitable snobbism of the better educated engineer or chemist.

‖ **snobisme** (snobīz'm). Also *erron.* **snobbisme.** [Fr.] = SNOBBISM in Dict. and Suppl.

1913 E. MARSH in S. Hynes *Edwardian Turn of Mind* (1968) ix. 343, I went with Denis to the Ballet... It's *delicious*, I went thoroughly meaning to dislike it, so it isn't *snobisme* on my part. **1920** A. HUXLEY *Let.* 4 Mar. (1969) 182 What you must go and see in Paris is the Cirque Medrano... It's become rather a snobisme to go. **1931** *Punch* 23 Sept. 334 For a womanly yet talented woman of good family, with enough English *snobisme* to relish Society and enough Irish irresponsibility to take its awful procedure lightly. **1958** *Spectator* 18 July 86/2 The one-eyed monster..? Oh dear, must that particular *snobbisme* be perpetuated? **1968** J. M. WHITE *Nightclimber* vii. 46 Always marvellous, that English nation, compounded equally of the Bible and *Snobisme*. **1977** T. HEALD *Just Desserts* vii. 145 It was undrinkable... No ludicrous snobisme about it being an English wine, could possibly persuade any normal palate of anything else.

snobocracy. Add: (Earlier example.) Also **snobocra·tic** *a.*

1853 J. M. RICHMOND *Let.* 17 July in *Richmond-Atkinson Papers* (1960) I. 129 His impartiality..is not satisfactory to the 'snobocracy', as Jas calls the genteel of this place. **1960** *Times Lit. Suppl.* 16 Sept. 589/1 These days of tax-evader farmers and snobocratic huntsmen.

snobographer. (Later example.)

1966 *Punch* 29 June 965/3 Auchincloss has surpassed himself in this magnificent comedy of manners... He is *the* post-Freud snobographer.

SNOBOL (snō͞u·bǫl). *Computers.* [Acronym f. the letters of 'string-oriented symbolic language', after *Cobol*, etc.: cf. *STRING *sb.* 15.] A high-level programming language used chiefly in literary research and symbolic computation.

1964 D. J. FARBER et al. in *Jrnl. Assoc. Computing Machinery* XI. 21 Interest in language translation, program compilation and combinatorial problems has increased... The string-orientated symbolic language SNOBOL has been developed with these problems in mind. **1969** *Computers & Humanities* IV. 74, I..began first to implement a year-long research project on rhythm in the Spanish language, using SNOBOL for text-manipulation and FORTRAN for statistical operations. **1971** *Ibid.* V. 156 SNOBOL IV..is a string manipulation and pattern-matching language and in this area makes both ALGOL and FORTRAN look clumsy... SNOBOL programs are typically shorter in character and line count than a FORTRAN or ALGOL counterpart. **1971** R. A. WISBEY *Computer in Lit. & Ling. Research* 165 The computer used in this study is the IBM 360/91. The language is SNOBOL 4, Version 3, a string manipulation language developed by Bell Telephone Laboratories, Inc.

Sno-cat (snō͞u·kæt). orig. *U.S.* Also **sno-cat, Snocat, snocat.** [f. *sno*, an arbitrary respelling of SNOW *sb.*¹ + *CAT(ERPILLAR 1 b.] A proprietary name in the U.S. for a type of snowcat (see *SNOWCAT).

1946 *Official Gaz.* (U.S. Patent Office) 10 Sept. 212/1 *Sno-cat.* No claim is made to the exclusive use of the word 'Sno' apart from the mark. For automotive vehicles for traveling over snow. Claims use since Sept. 1, 1941. **1957** *Times* 4 Dec. 11/7 The Snocat..came to the rescue and easily hauled both the Weasels and sledges on to better surfaces. **1958** *Listener* 13 Nov. 793/1 The popular impression that Antarctica is always and only white is matched by the suspicion that a modern expedition in those regions is all sno-cats and telecommunications. **1968** MRS. L. B. JOHNSON *White House Diary* 26 June (1970) 691 We got into a strange vehicle called a 'Sno-Cat', a long cab on caterpillar treads. **1970** *Observer* 20 Dec. 25/1 A wealthy industrialist seriously proposed running a Snocat service along the entire Haute Route. **1977** *New Scientist* 20 Jan. 123 The unloading operation was carried out with heavy cargo sledges towed in relays by snocats on the sea ice and caterpillar tractors on the ice shelf.

snockered (snǫ·kəɹd), *ppl. a. slang.* Also **snookered.** [Perh. arbitrary alteration of *SNOOKERED *ppl. a.*] Drunk, intoxicated. Cf. *SCHNOCKERED *ppl. a.*

1961 S. PRICE *Just for Record* x. 105 You rolled along half-snockered after Sunday lunch. **1969** 'R. STARK' *Dame* xx. 121 'I may be a little high,' she said, 'but I'm not snockered.' **1977** *Amer. Speech* 1975 L. 66 *Snockered adj*, drunk. 'She was snockered; she didn't mean it.' **1980** *Globe & Mail* (Toronto) 4 Oct. 6/6 I'll get a bottle of Jack Daniel's for cocktails. Get them snookered on bourbon and they won't know the difference.

snodger (snǫ·dʒəɹ), *a.*, (*adv.*, and *sb.*). *Austral.* and *N.Z. slang.* ? *Obs.* [Of uncertain origin: cf. SNOD *a.* and SNOG *a.*] Excellent, very good, first-rate. Also as *adv.* and *sb.*

1919 W. H. DOWNING *Digger Dial.* 46 *Snodger* (adj.), excellent. **1924** C. J. DENNIS *Rose of Spadgers* 40 It was a snodger day!.. The apple trees was white with bloom. All things seemed good to me. **1941** BAKER *N.Z. Slang* vi. 51 Expressions..in constant use by our youngsters.. stunner, snorter, snodger, ripsnorter. **1946** *Sunday Sun* (Austral.) 11 Aug. (Suppl.) 15 There they find the conships fitted up snodger with bulkheads studded with nails. **1950** *Austral. Police Jrnl.* Apr. 119 If something is a snodger it is 'mighty' in Queensland, it is 'colossal' in N.S.W., and just 'very nice' everywhere else.

snoek. Add: The S. Afr. pronunc. is (snuk). Substitute for def.: A snake mackerel, *Thyrsites atun*, of the family Gempylidæ, a large marine food fish found in large shoals in colder parts of Southern Hemisphere oceans. Cf. *BARRACUDA. (Earlier and further examples.)

1797 A. BARNARD *Jrnl.* in *Lives of Lindsays* (1849) 388 The fish called *snook*..when salted and dried, was one of the best fish at the Cape. *a* **1823** J. EWART *Jrnl. Stay Cape Good Hope* (1970) ii. 13 Snoek, a long oily fish which being caught in great quantities and consequently cheap, forms the principal food of the slaves. **1833** *Graham's Town* (Cape Province) *Jrnl.* 14 Feb. 3 Phosphorescent glimmerings of a decayed Snoek. **1880** A. C. L. G. GÜNTHER *Introd. Study of Fishes* 436 In New Zealand it is called 'barracuda' or 'snoek'. **1913** D. FAIRBRIDGE *That which hath Been* 73 An old Malay fisherman, carrying his baskets of snoek. **1931** *Times Lit. Suppl.* 16 Apr. 301/2 The snoek..is not a pike..but a distant cousin of the mackerel. **1946** L. G. GREEN *So Few are Free* iv. 57 Snoek are caught by each boat's crew at the rate of a thousand to three thousand a day. **1963** S. COOPER in Sissons & French *Age of Austerity* 51 In October 1947.. the hungry British first heard the word 'snoek'. Ten million tins of it from South Africa were to replace Portuguese sardines. **1974** *Stand. Encyl. S. Afr.* X. 28/1 The snoek is also an important food fish in Australia.

snoek (snuk), *v. S. Afr.* [f. the sb.] *intr.* To fish for snoek. Hence **snoe·ker; snoe·king** *vbl. sb.*

1913 W. W. THOMPSON *Sea Fisheries Cape Colony* ii. 50 It is a pretty sight to watch a fleet of fishing boats snoeking under sail. **1937** L. G. GREEN *Great Afr. Mysteries* xii. 137 The total catch by all the snoeking vessels often amounts to a million fish. **1950** *Cape Argus* 28 Oct. (Mag. Section) 3 Fishermen declare that China snoek are caught after the ordinary snoeking season is over. **1952** L. G. GREEN *Lords of Last Frontier* xxi. 299 Snoeking, a trade that has prospered here for forty years, keeps a grand fleet of small craft in commission. **1959** *Cape Times* 5 May 2/7 (*heading*) Snoeker found ringed bird. *Ibid.*, While snoeking at St. Helena Bay, John Mentor..found a dead black sea-duiker.

snogging (snǫ·giŋ), *vbl. sb. slang.* [Origin unknown: cf. SNUG *v.*] Engagement in light, amorous play, esp. kissing and cuddling.

1945 C. H. WARD-JACKSON *It's a Piece of Cake* (ed. 2)

56 *Snogging*, courting, running around with the opposite sex. Comes from India. Thus, 'On my leave I'm going up to the hills for a bit of snogging.' Also used as a verb. **1951** *Sunday Pictorial* 28 Oct. 10/6 Few hounds can get in more than half an hour of 'snogging'—their elegant term for not-too-serious courtship. **1960** N. EPTON *Love & English* vi. 341 It is all right..to cuddle. (The current term among teen-agers is 'snogging'.) **1966** P. WILLMOTT *Adolescent Boys* iii. 40, I went upstairs with Jill and we did a bit of snogging on the bed. **1975** *Weekend* 4 Feb. 19/1 If a cinema manager tolerates snogging among his audience he is liable to lose his licence.

Also **snog** *v. intr.*, to engage in snogging; **snog** *sb.*, an instance of this; **sno·gger, sno·gging** *ppl. a.*

1945 [see *SNOGGING *vbl. sb.*]. **1958** 'J. BROGAN' *Cummings Report* xv. 156 He is a..girl-snogging.. bounder. **1959** W. CAMP *Ruling Passion* xii. 82 Let's pretend we're teenagers and stop for a nice snog. **1962** A. SAMPSON *Anat. Britain* xxxvi. 574 The cinema has lost its hold—except among unmarried teenagers, two-thirds of whom go at least once a week, perhaps to snog in the doubles. **1965** J. GASKELL *Fabulous Heroine* 94 A most experienced snogger. **1973** M. AMIS *Rachel Papers* 21 They were enjoying a kiss—well, more of a snog really. **1981** R. BARNARD *Mother's Boys* ii. 20 They had..taken the side way through the little cutting known popularly as 'Snoggers Alley'.

Snohomish (snō͞uhō͞u·miʃ), *a.* and *sb.* Also **Snow-.** [19th-cent. Puget Salish *snuhumš.*] **A.** *adj.* Of or pertaining to the Snohomish or their language. **B.** *sb.* (A member of) a Salish Indian people of western Washington; also, their language.

1856 N. D. HILL *Let.* 30 Sept. in *U.S. Congr. House Exec. Doc.* (1857) XXXVII. 77, I received a letter from you appointing me the local agent for the Snohomish, the Snoqualmi, and the Skiquamish tribes of Indians. **1874** *Field & Stream* 20 Aug. 18/3 This assertion was verified afterwards by a Snohomish Indian. **1910** F. W. HODGE *Handbk. Amer. Indians* II. 606/2 Snohomish. A Salish tribe formerly on the s. end of Whidbey id., Puget sd., and on the mainland opposite at the mouth of Snohomish r., Wash. Pop. 350 in 1850. The remnant is now on Tulalip res., Wash., mixed with other broken tribes. **1940** M. W. SMITH *Puyallup-Nisqually* 17 In the Puyallup-Nisqually dialect the word means butte or rump but informants thought it might mean something else in the dialects of either the Twana or Snohomish. **1966** *Internat. Jrnl. Amer. Linguistics* XXXII. 350/2 Snowhomish..was spoken in the region around Port Gardner Bay..and along the Snowhomish River. **1977** C. F. & F. M. VOEGELIN *Classification & Index World's Lang.* 301 Puget Sound Salish = Puget = Toughnowawmish. D[ialect]s Northern Puget Sound (Skagit, Snohomish), Southern Puget Sound (Duwamish, Muckleshoot, Nisqualli, Puyallup, Snoqualmie, Suquamish). 10–20. Washington.

snollygoster (snǫ·ligǫstəɹ). *U.S. dial.* and *slang.* [Perh. connected with *SNALLYGASTER, which is, however, of more recent appearance.] A shrewd, unprincipled person, esp. a politician. Also in other more or less fanciful uses (see quots.).

1846 *Commonwealth* (Frankfort, Kentucky) 7 Apr. 2/6 Now here I am a rale propelling, double revolving locomotive Snolly Goster, ready to attack anything. **1863** D. EMMETT *Black Brigade* 5 We am de snolly-gosters, An' lubs Jim Ribber oysters. **1895** *Columbus* (Ohio) *Dispatch* 28 Oct. 4/3 A Georgia editor kindly explains that 'a snollygoster is a fellow who wants office, regardless of party, platform or principles, and who, whenever he wins, gets there by the sheer force of monumental talknophical assumnacy'. **1912** *Dialect Notes* III. 590 *Snolly-goster*, a shyster. **1915** *Nebraska State Jrnl.* 7 Sept. 6/3 We once knew a miserly old *snollygoster* who used to look in a mirror to see the reflection of a saint. **1952** *N.Y. Herald Tribune* 3 Sept. 17/2 President Truman..said some people like to pray in public so that others will view them as honorable and religious men... 'I wish some of these snollygosters would read the New Testament and perform accordingly.' **1953** *Cavalier Daily* (Univ. of Virginia) 12 Nov. 1/2 Former President Truman may have been making a talknophical assumnacy when he said a snollygoster is what Southerners call a man born out of wedlock. **1972** A. ROUDYBUSH *Sybaritic Death* xx. 168 The deaths of a middle-aged tart and an elderly snollygoster are of little moment.

snomobile, var. *SNOWMOBILE.

snood, *sb.* Add: **1. a.** More recently, a fashionable bag-like or closed woman's hairnet, usu. worn at the back of the head. (Later examples.)

1938 *Sun* (Baltimore) 22 Oct. 5/6 (*caption*) New hats in vivid colors... Shakos, pill boxes, turbans, brims, pie plates and snoods. **1939** in C. W. Cunnington *English Women's Clothing* (1952) vii. 262 A spate of hoods and snoods. **1944** M. LASKI *Love on Superiax* x. 92 She carefully placed on the top of her head a little forward-tilting black hat whose draped jersey snood just failed to conceal the mass of yellow wrinkles. **1947** E. JENKINS *Young Enthusiasts* 47 They..wore ribbon snoods secured under their buns. **1968** J. IRONSIDE *Fashion Alphabet* 148 A knitted or open-work 'bag' over the back of the hair. Sometimes a snood is attached to a hat.

snoodle (snū·d'l), *v. dial.* (chiefly *north.*) or *nursery.* Now *rare.* [See *Eng. Dial. Dict.*: prob. rel. to SNUDGE *v.*², SNUGGLE *v.*, etc.] *intr.* To snuggle, nestle. Also *trans.*

1887 in T. DARLINGTON *Folk-Speech S. Cheshire* 355. **1898** R. DOTTIE *Rambles & Recoll.* ' "R" Dick' 115 Eaur snug, white hostelry snoodlin' i' th' valley. **1904** J. VAIZEY *More about Pixie* (1910) i. 9 She snoodled her head along the pillow so as to lean it against the nurse's shoulder. **1908** E. J. BANFIELD *Confessions of Beachcomber* I. i. 32 Snoodling beside lumps of coral or beneath weather-beaten drift-wood, they [*sc.* young birds] afford startling proof of the effect of sympathetic coloration.

snook: now the commoner form of SNOOKS derisive gesture (q.v. in Dict. and Suppl.); var. SNOEK in Dict. and Suppl.

snooker, *sb.*² Add:

It is commonly held that the word represents an allusive use of SNOOKER *sb.*¹, a newly joined cadet, first applied to the game by Col. Sir Neville Chamberlain (1856–1944), a subaltern in the Devonshire Regiment stationed at Jubbulpore in central India in 1875, with reference to the rawness of the play of a fellow officer. The story is often repeated, e.g. in *The Times* (1980) 29 Dec. 9.

snooker *v.*: also *fig.* (chiefly *pass.*), to place in an impossible position; to balk, 'stymie'; **snoo·kered** *ppl. a.*, **snoo·kering** *vbl. sb.*

1915 *Morning Post* 8 Apr. 5/1 If we had fired the Germans might have sent up a light and then we should have been snookered all right. **1927** C. MACKENZIE *Vestal Fire* I. i. 5 One of the recognized amusements of a Sirene dinner-party was to try to snooker Joseph R. Neave over Dante. **1935** *Times* 5 Oct. 5/6 The snookering all through was clever. **1970** K. GILES *Death in Church* iii. 65 In France they might have had a chance..but here they were snookered.

snooks. Add: The form **snook** is now usual. Chiefly in phr. *to cock a snook* (*at*). (Earlier and later examples.)

1791 E. WYNNE *Diary* 7 Dec. (1935) I. 90 They *cock snooks* at one on every occasion. **1929** H. S. WALPOLE *Hans Frost* I. vii. 78 He was like a dirty street boy cocking a snook at Sappho. **1938** E. AMBLER *Cause for Alarm* viii. 128 The Rome–Berlin axis..cocked the biggest snook yet at the League of Nations idea. **1959** M. CUMBERLAND *Murmurs in Rue Morgue* v. 38 With his right hand he made the somewhat coarse gesture known as 'cocking a snook'. The thumb and extended fingers, spread in front of the face, made a baffling disguise. **1961** B. FERGUSSON *Watery Maze* ii. 48 It would be idle to pretend that it was of much importance; it was really only cocking a vulgar snook. **1965** *Listener* 9 Sept. 374/2, I walked past the Thatched House..where I and other young journalists used to cock snooks at our superiors. **1980** *Times* 29 Feb. 10 East German craft last spring embarked upon a new ploy..to net a Danish torpedo,..cocking a snook at Nato's Baltic muscle.

Hence **snoo·k-cocking** *vbl. sb.*, **snook-cockingly** *adv.*; **snook-cocker.**

1950 D. GASCOYNE *Vagrant* 57 And not think them impudent snook-cocking. **1958** *Economist* 6 Dec. 880/2 The seven Liberal councillors of Finchley..in snook-cocking protest against the local shortcomings of London Transport's bus service have launched a free private service of their own. **1962** *Spectator* 13 Apr. 478 A snook-cockingly 'blasphemous' film. **1965** E. GOWERS *Fowler's Mod. Eng. Usage* (ed. 2) 535/2 Mere snook-cockers of whom it has been said..that their only concern is to 'find someone who is doing something..and fling a few insults at him'. **1978** CADOGAN & CRAIG *Women & Children First* viii. 167 Spike Milligan's snook-cocking record of his war-time experiences.

Snooks² (snūks). A proper name or familiar appellation applied to a hypothetical person in a particular case (see quots.); also, any individual person. Cf. *Joe Bloggs* s.v. *JOE *sb.*² 5 c.

1860 HOTTEN *Dict. Slang* (ed. 2) 221 *Snooks*, an imaginary personage often brought forward as the answer to an idle question, or as the perpetrator of a senseless joke. **1919** G. B. SHAW *Inca of Perusalem* in *Heartbreak House* 214 Well, what about Snooks? **1922** LD. RIDDELL *Some Things that Matter* ix. 108 '50 per cent of the inhabitants of Bunkumville who use this valuable adjunct to health and personal beauty wear Snooks's Expanders, which are undoubtedly the best.' That may be true, but I omit to mention that only two persons in Bunkumville wear chest expanders, one of whom is Snooks himself. **1959** *Times* 8 Dec. 13/4 The recommended formula goes something like this: 'This is Flaxway 5768. Mr. Snooks is out. If you wish to leave a message, go ahead.'.. Snooks, returning eventually to base, presses a button, and the machine reels off all the messages.

snookums (snū·kŏmz). [Nonsense formation: cf. *DIDDUMS and prec.] A trivial term of endearment, usu. applied to children or lapdogs.

1919 *Ladies' Home Jrnl.* May 153/1 Even 'Snookums' knows and appreciates the soothing qualities of Johnson's Toilet and Baby Powder. **1928** *Chambers's Jrnl.* 21 Jan. 128/2 She is now a 'city-lady', with a couple of dear little 'snookums'.

snoop, *v.* Restrict *U.S.* to sense 1 and add: **2. a.** (Earlier and later examples.) Also with other advbs. orig. *U.S.*

In quot. 1832 unusually without following *adv.* or *prep.* **1832** R. C. SANDS *Writings* (1834) II. 291 The world has realms wherein to *snoop*, And I am not a noddy. **1840** C. F. HOFFMAN *Greyslaer* II. III. i. 105 Our scouts would make us believe that both he and Bradshawe are snooping about the country among the Tories. **1855** *Knickerbocker* XLVI. 317 The level which the..engineers 'snooped' round and found out, hasn't 'a *parallel*' in all the adjacent region. **1902** H. L. WILSON *Spenders* iii. 26 Work..is something you want to get done; play is something you jest like to be doin'. Snoopin' up these gulches is both of 'em to me. **1931** D. L. SAYERS *Five Red Herrings* xviii. 185 It is hardly possible for a local policeman in a country place to snoop about, wheedling information out of the inhabitants. **1943** J. B. PRIESTLEY *Daylight on Saturday* xxviii. 220 What are you doing here? Snooping around and then sending in a report in triplicate—eh? **1951** J. FLEMING *Man who looked Back* xv. 195 You..have all the fun snooping round and I've got to wait.

b. To pry into matters one need not be concerned with. Often const. *on* (a person). *colloq.*

1921 *Daily Colonist* (Victoria, B.C.) 3 Apr. 9/3 There is the landlady who 'snoops' too much to suit her boarders. **1946** [see *FERRET *sb.*¹ 1 b]. **1950** *Chicago Daily News* 14 Apr. 18/3 Another thing is that snoopers often get snooped on, in retribution. **1965** M. SPARK *Mandelbaum Gate* vii. 254 Has he had any opportunity to snoop? **1975** R. STOUT *Family Affair* vi. 55, I wouldn't ask you to snoop on a friend.

3. *trans.* To steal, to misappropriate. Also *absol. rare.*

1924 GALSWORTHY *White Monkey* I. viii. 57 If we let you snoop copies, all the packers will snoop copies. *Ibid.* III. viii. 270 Yes, and look at that little snooper himself; he snooped to keep her alive after pneumonia.

Hence **snoo·ping** *vbl. sb.* and *ppl. a.*

1936 J. STEINBECK *In Dubious Battle* vii. 109 The health authorities are going to do plenty of snooping. If they can catch us off base, they'll bounce us. **1946** K. TENNANT *Lost Haven* (1947) xvii. 288 He was just a snooping tourist. **1952** *Manch. Guardian Weekly* 14 Feb. 13 Whose job it is to engage in political snooping. **1965** D. FRANCIS *Odds Against* iv. 49 All very normal... It was my snooping which seemed unreal. **1974** 'M. INNES' *Appleby's Other Story* xi. 87 'If you want my help—' 'A snooping copper's *help*?' **1977** *Rolling Stone* 13 Jan. 30/2 She collected the results of her snooping in a manila folder.

snoop (snūp), *sb. colloq.* (orig. *U.S.*). [f. the vb.] **1.** = SNOOPER 1 in Dict. and Suppl.; *spec.* one who makes official or other investigation, a detective.

1891 *Amer. Folk-lore* IV. 160 *Snoop.*—This word I have frequently heard in New England, used both as a verb and as a noun. It implies sneaking, spying, prying around. **1929** *Amer. Speech* V. 152 *Snoop*, one who noses something out. 'That woman is a snoop.' **1942** *New Statesman* 19 Sept. 186/3 *Snoops* are the Service Police, corresponding to the Army's Military Police. **1944** DYLAN THOMAS *Let.* 21 Sept. (1966) 267 There stinks a snoop in black. I'm thinking it Is Mr. Jones the Cake. **1948** *Time* 3 May 19/3 Every cop,..stool pigeon and neighborhood snoop in Detroit was working overtime. **1970** A. SILLITOE *Start in Life* VI. 318 His snoops already know I left Beirut. **1978** R. THOMAS *Chinaman's Chance* xxxvii. 360 The Congressman seems to have been an awfully fine snoop. But then, he used to be a cop.

2. An act of snooping, prying, or investigation; a surreptitious inspection. Freq. with (*a*)*round.*

1908 G. H. LORIMER *Jack Spurlock—Prodigal* xi. 274 She couldn't keep her servants, for she was torn with dark doubts of their honesty... Life for her was one long snoop about the house. **1939** 'N. BLAKE' *Smiler with Knife* xii. 172 Why not have a snoop round in Chilton's study? **1969** M. PUGH *Last Place Left* xxii. 167 'You're going to take this to Brunner's house?' 'Not straight. Once I've had a good snoop round it.' **1972** G. LYALL *Blame Dead* xv. 107, I did a little unpacking and then went for a general snoop.

snooper. (In Dict. s.v. SNOOP *v.*) Add: **1.** (Earlier and later examples.) *spec.* one who makes an intrusive official investigation. orig. *U.S.*

1889 in *Cent. Dict.* **1928** *Chicago Tribune* 11 July 10/4 Prohibition Commissioner Doran has warned dry snoopers to stop gunplay against innocent citizens. **1939** 'N. BLAKE' *Smiler with Knife* i. 19 What a snooper you are! **1948** *Jrnl. R. Aeronaut. Soc.* LII. 719 The difficulty with this is that the potential user is unlikely to come into the picture in the detail design stage, and the designing firm would probably not, in any case, welcome yet another 'snooper'. **1959** E. H. CLEMENTS *High Tension* vi. 103 We should consider ourselves lucky to have a professional snooper as a neighbour. **1965** M. SPARK *Mandelbaum Gate* iii. 80 We know Ramdez. He's a snooper for his government. **1978** D. GRYLLS *Guardians & Angels* iii. 89 The parents..are tip-toeingly attentive... Of course, the adults are not depicted as snoopers.

2. A sneak-thief, a misappropriator. *rare.*

1924 [see *SNOOP *v.* 3]. **1927** 'J. BARBICAN' *Confessions Rum-Runner* II. xxiii. 257 You rotten little cross-eyed snooper.

snooperscope (snū·pəɹskō͞up). [f. SNOOPER + -SCOPE; cf. *SNIPERSCOPE.] A device which converts infra-red radiation to a visible image; esp. a pair of such devices fitted together and worn on the head to provide binocular vision in the dark.

1946 *Times-Dispatch* (Richmond, Va.) 16 Apr. 2/1 The snooperscope can be used over a special helmet. It weighs from six to seven pounds, and looks like something out of this world. *Ibid.*, The snooperscope had another use. With one on his noggin, a jeep or truck driver could go barrelling down the road to the front without lights. **1955** *Sci. News Let.* 7 May 295/3 A modified snooperscope is being used to 'see' through silicon crystals, spotting im-

perfections produced in manufacturing transistors, rectifiers and other semi-conducting devices. **1962** *Appl. Physics Lett.* I. 91/2 We looked at the diode output through a 'snooperscope' and above the threshold observed a very intense and narrow beam [of infrared radiation] radiating from the junction region. **1972** J. MILLS *Report to Commissioner* 227 Hanson says why not darken the floor and watch the elevator with snooperscopes.

snoo·pery. orig. *U.S.* [f. SNOOP *v.* + -ERY.] The activity of snooping or prying; surreptitious investigation, *spec.* into another's private affairs.

1935 *Sun* (Baltimore) 13 Feb. 2/6 C. Jasper Bell (Dem., Mo.) turned the Capitol Hill fight against 'innovations into snoopery' upon another law, the NIRA. Thus, the number of enactments now known to contain provisions making public the private financial affairs of citizens was brought to five. **1964** *Spectator* 13 Mar. 337/1 In time private enterprise snoopery could become a growth industry and major job-supplier for our unemployed. **1972** G. LYALL *Blame Dead* xiv. 104 'He sounds shifty as hell.'.. That..might help justify David's snoopery. **1981** A. PATON *Towards Mountain* xxii. 187 The rules were simple—no sharing of blankets, the doors to stand open, no boy to sleep in any other dormitory except the one to which he had been assigned. These rules could be evaded, but their evasion was preferable to a reign of snoopery and an encouragement of informers.

snoo·py, *a.* [f. SNOOP *v.* + -Y¹.] Inquisitive, excessively curious or prying.

1895 in *Funk's Stand. Dict.* **1921** S. FORD *Inez & Trilby May* xii. 212 With the cops so snoopy, we can't afford a scene. **1930** P. MACDONALD *Link* xi. 216 I'm not snoopy, but I opened my door and listened. **1952** *Chambers's Jrnl.* Feb. 81/2 This was a depot where the Canadian rum-runners met the American bootleggers to exchange Canadian and Scotch whisky for American dollars, and I knew enough about these operations to realise that life did not amount to much when snoopy guys butted in. **1978** J. WAINWRIGHT *Jury People* xii. 41 Don't think I was being snoopy—but..I saw you arrive, this morning.

Hence **snoo·piness.**

1969 L. HELLMAN *Unfinished Woman* iv. 39 The vicarious, excited snoopiness I knew was mixed with the kindness.

snoose (snūs, snūz). *Western N. Amer.* Also **schnoose, snooze.** [ad. Da., Norw., and Sw. *snus* snuff, shortening of Da., Sw. *snustobak*, Norw. *snustobakk* snuff tobacco; cf. SNUSH *sb.*] Chewing snuff, esp. taken by loggers. Also *fig.*

1912 H. FOOTNER *New Rivers of North* 21 Loud were the lamentations of his foreigners when his 'snooze' gave out, 'snooze' being the local familiarity for snuff. **1925** *Amer. Speech* I. 138/1 He 'fogs-up' on his pipe, or takes a 'rear of snoose'. 'Snoose' is a certain brand of Swedish snuff; it is moist and hot with pepper, and the man who is not used to it will find his gums burning and his head swimming when he tries his first 'rear'; but nearly every logger in this neck of the woods [*sc.* Northwest] has abandoned the old-time American plug for this terrific Nordic concoction. **1942** *Ibid.* XXVII. 221/2 *Give her snoose*, an order to increase power. **1955** R. HOBSON *Nothing too Good* xiii. 136 Larkie was a snoose chewer. **1965** *Sun* (Vancouver) 22 Apr. 51/7 (*caption*) Just before ya face the old lady ya shove a wad of schnoose in yer mouth.. she'll never smell yer breath. **1977** J. HODGINS *Invention of World* i. 5 He spat snoose out the broken window onto the pavement.

snoot (snūt), *sb.* [dial. var. SNOUT *sb.*¹] **1.** = SNOUT *sb.*¹ 2. *dial.* and *slang.*

1861 J. BARR *Poems* 33 Like harrow teeth they're stickin' out, To catch the dirt below their snoot. **1866** *Galaxy* 1 Oct. 277, I had supposed that such phrases as 'I'll mash your head!' 'I'll bash you on the snoot!' 'I'll mawl yer jaws,' and similar expressive threats, were invented in the New World. **1884** E. W. NYE *Baled Hay* 209 Read our..'Ode to the Busted Snoot of a Shattered Venus de Milo'. **1905** [see SNOUT *sb.*¹ 2]. **1924** WODEHOUSE *Bill the Conqueror* v. 101 He seethed with generous indignation and even went so far as to state his intention ..of busting the fellow one on the snoot. **1938** D. RUNYON *Furthermore* v. 86 A bust in the snoot. **1956** D. M. DAVIN *Sullen Bell* II. iv. 136 At first I was all for poking the bloke in the snoot. **1971** J. AIKEN *Nightly Deadshade* iii. 33 Snell is sticking his long snoot into the middle of things.

2. The nose of an aircraft, esp. of adjustable construction (cf. *droop-snoot* s.v. *DROOP *sb.* 3). Also, the nose of a car, etc.

1945, etc. [see *droop-snoot* s.v. *DROOP *sb.* 3]. **1962** *New Scientist* 18 Jan. 135/1 As the flaps are depressed, so the snoot is tilted downwards until at full flaps it is depressed at an angle of 35°. **1977** *Drive* Mar.–Apr. 52/3 Drivers are in a poor position to judge the droop-snoot of the car. **1980** A. COPPEL *Hastings Conspiracy* iv. 32 Through the open door of the flight-deck Brede could see that the snoot had been lowered for better visibility.

3. A tubular or conical attachment used to produce a narrow beam from a spotlight.

1952 *Cinema* 7 Jan. 108/1 (Advt.), Viking Films Ltd... Lighting equipment... Spots... Cans, bashers, overhead banks, snoots, barndoors, diffusers, niggers, etc. **1972** QUICK & LA BAU *Handbk. Film Production* xi. 73 Snoots consist of metal tubes that are mounted on the front of spotlights to control the spread of their beams. **1977** J. HEDGECOE *Photographer's Handbk.* 34 Spotlight accessories include folding barn-doors..and conical snoots.. both of which restrict the beam.

snoot, *v. U.S.* [f. prec.] **1.** *intr.* = NOSE *v.* 8 b; = SNOUT *v.* 2. (In quot. *fig.*) *U.S. dial. rare.*

1890 *Dialect Notes* I. 75 *Snoot* (snût), of the human face or nose, apparently the same word as *snout*. A vulgar word in New England. 'I'll bu'st your snoot'; 'hit him on the snoot'. As a verb in 'to snoot round', *i.e.* to nose around, it is reported from *Poughkeepsie*, N.Y.

2. *trans.* To snub; to treat scornfully or with disdain. *U.S.*

1928 E. HATCH *Couple of Quick Ones* IV. 198, I followed him..up the street to where the Wright limousine was snooting the world in general at the kerb. **1939** J. P. MARQUAND *Wickford Point* xi. 124 Don't try to snoot Sue Jaeckel. **1959** V. PACKARD *Status Seekers* iii. 44 Many intellectuals..develop their own ways of snooting. **1977** *Time* 17 Jan. 28/3 Cinderella (Gemma Craven) gets snooted by her Stepsisters and gazes sorrowfully into the flames of the scullery fire.

snooter (snū·təɹ), *v.* [f. as prec. + -ER⁵.] *trans.* To harass, to bedevil; to snub. (Only in P. G. Wodehouse.)

1923 WODEHOUSE *Inimit. Jeeves* iii. 30 My Aunt Agatha ..wouldn't be on hand to snooter me for at least another six weeks. **1929** —— *Mr. Mulliner Speaking* viii. 286 'As far', replied Mr. Finch, frigidly, 'as a bloke can be said to be all right..who has been..chivvied and snootered and shot in the fleshy part of the leg—.' **1932** —— *Let.* 13 Aug. in *Performing Flea* (1953) 66 Downtrodden young peer, much snootered by aunts, etc., has become engaged to two girls at once.

snootful (snū·tful). [f. *SNOOT *sb.* + -FUL.] As much (alcohol, etc.) as one can take; a quantity of alcohol, esp. one sufficient to induce drunkenness. Cf. SKINFUL 3 a.

1918 R. LARDNER *Real Dope* 43 When somebodys else husband pulls something its O.K. but if their own husband does it he must of had a snoot full. **1935** WODEHOUSE *Luck of Bodkins* xvii. 205 His whole mind was manifestly intent on reaching the smoking-room and getting a snootful. **1953** W. R. BURNETT *Vanity Row* xvi. 117 He was drunk... 'A snootful, eh?' **1969** K. VONNEGUT *Slaughterhouse-Five* ii. 40 Billy didn't usually drink much..but he certainly had a snootful now. **1977** H. GREENE *FSO-1* xi. 103 Kim..had gotten a snootful of the tear gas.

snootily (snū·tili), *adv.* [f. *SNOOTY *a.* + -LY².] In a snooty manner, superciliously.

1940 'G. ORWELL' *Let.* 16 Apr. in *Coll. Essays* (1968) II. 22, I get quite a lot of letters..from people snootily pointing out some mistake I've made. **1954** KOESTLER *Invisible Writing* III. xix. 220 And now, when I am down, you snootily refuse to help me with my business. **1961** *Guardian* 29 Mar. 14/3 An old lady..remarked..rather snootily..'It's where you were born that counts.' **1980** I. HUNTER *Malcolm Muggeridge* ii. 36 Vidler was unimpressed by all this and replied somewhat snootily, rejecting all his arguments.

snoo·tiness. [f. as prec. + -NESS.] The character or quality of being snooty; conceitedness, superciliousness.

1932 J. T. FARRELL *Young Lonigan* iii. 124 She said it served Helen right that she had gotten a crush on a guy like Weary, because Weary would take some of the snootiness out of her. **1942** *R.A.F. Jrnl.* 2 May 22 They in turn mistook our English reserve for 'snootiness'. **1956** S. HOPE *Diggers' Paradise* 155 One or two tourists I met were annoyed about what they described as this 'snootiness'. **1977** *Sunday Times* 15 May 40/1 All these books steer with tact between the contrasting risks of sycophancy and snootiness.

snooty (snū·ti), *a.* [f. *SNOOT *sb.* + -Y¹.] Supercilious, haughty, conceited; affecting superiority, snobbish; 'highbrow', 'stuck-up'. Occas., irritable, short-tempered.

1919 A. HUXLEY *Let.* 12 Aug. (1969) 180 A very snooty cousin and a sporty one. **1922** S. LEWIS *Babbitt* xx. 252, I didn't like..the snooty way you talked. **1931** E. LINKLATER *Juan in Amer.* II. xvi. 172 She says you were kinda snooty with her. Tried to high-hat her. **1938** E. BOWEN *Death of Heart* II. vii. 303 Reproaches and rather snooty laughs were exchanged. **1940** in Harrison & Madge *War begins at Home* xiv. 379 They're quite snooty, because you don't buy anything else. **1947** 'A. P. GASKELL' in D. M. Davin *N.Z. Short Stories* (1953) 282, I was lucky to have a girl like Betty who was keen on football. Some of the girls used to go very snooty when the blokes couldn't take them to the Friday-night hops. **1955** E. CADELL *Lark shall Sing* v. 67 One of those snooty little cafés..run by bony gentlewomen. **1959** I. & P. OPIE *Lore & Lang. Schoolch.* x. 178 A short-tempered person is spoken of as being..snappy, snooty (meaning easily irritated), and sharp-edged. **1960** O. MANNING *Great Fortune* II. 142 The English wives were a bit snooty with me. **1980** R. BARNARD *Death in Cold Climate* vi. 60 You know how the English can say 'Really?'—all cold and snooty.

snooze, *sb.* Add: **3.** *Comb.* **snooze alarm,** an alarm on a bedside clock which may be preset or reset to repeat after a short interval, allowing the sleeper a further nap; **snooze button,** a button on a clock which sets the snooze alarm.

1973 *Electrical Wholesaler* Sept. 76/1 (Advt.), Snooze alarm clock housed in finely tooled aluminium case. **1976** *Washington Post* 19 Apr. A15/6 (Advt.), Multiband clock radio. Digital Numbers. Wake to Music. Extra Snooze Alarm. **1974** *Sci. Amer.* Oct. 63/2 For years, alarm clocks were dull..even those with snooze buttons and fancy dials.

snoozer. Add: **1. b.** As a vague appellation: a fellow, a chap. *colloq.* (orig. *U.S.*).

1884 [see *BANK *v.*³ 4 b]. **1891** 'E. PERKINS' *Thirty Years of Wit* 296 I'm the snoozer from the upper trail; I'm the reveler in murder and in gore. **1903** 'O. HENRY' in *Ainslee's* Sept. 116/2 She knows what a wild kind of a snoozer I've been. **1916** *Anzac Book* 99 The chaps of the 16th Battalion Are not easy snoozers to beat. **1923** R. D. PAINE *Comrades of Rolling Ocean* iv. 65 Do you mean to say that the wonderful old snoozer had the grit to cruise out to your country at his age? **1939** JOYCE *Finnegans Wake* 174 They had cornered him about until there was not a snoozer among them but was utterly undeceived. **1946** *Sunday Sun* (Austral.) 20 Oct. (Suppl.) 15 They'd have lamped a snoozer rigged up as an army skipper clop-clopping along on a nag just behind them. **1966** H. MARRIOTT *Cariboo Cowboy* v. 52 Zim was a tough old snoozer. I know that he cut his knee open with an axe and sewed it up with some worsted yarn and his wife's darning needle.

† **2.** A thief who steals from the hotel or house in which he is staying. Cf. SNOOZE *sb.* 2. *slang. Obs.*

1862 H. MAYHEW *London Labour* Extra vol. 242/1 Some two years ago a robbery was committed by a 'snoozer' or one of those thieves who take up their quarters at hotels for the purpose of robbery. **1882** *Sydney Slang Dict.* 8/1 *Snoozers*, men and women who sleep at hotels and boarding-houses and decamp with other people's effects in the morning. **1889** FARMER *Americanisms* 501/2 *Snooser*, an hotel thief who lives in the place, and thus seeks for opportunities to carry out his depredations.

Snopes (snō͞u·ps). Also **snopes**; *erron. pl.* **Snopes.** The family name of a series of vicious characters in the fiction of William Faulkner (first described in *Sartoris*, 1929), used as a type of an unscrupulous or heartless person.

1962 *Guardian* 3 Oct. 1/6 There are plenty of snopes waiting to take over and act as though Mr Meredith has never been. **1970** *Times* 10 Apr. 10 In the heated circumstances of the present Mr. Nixon would be joining the yahoos and the snopes. **1977** *Time* 17 Oct. 48/2 His brother, the Snopes in the woodpile, satirizes the theme by assuming the very worst of the American people and braying at them.

snore, *v.* Add: **1. b.** Also *U.S.*

1935 W. FAULKNER *As I lay Dying* 40 Beyond the porch Cash's saw snores steadily into the board.

snore-. Add: **b.** With advb. **snore-off** *colloq.* (chiefly *Austral.* and *N.Z.*), a sleep or nap, esp. after drinking.

1950 *Landfall* June 127, I notice Little Spike's legs sticking out from an empty tallow cask where he is having a snore-off. **1967** K. GILES *Death & Mr Prettyman* vi. 120 He always vowed to cut out these afternoon snore-offs. **1968** D. O'GRADY *Bottle of Sandwiches* 49 He surfaced from his plonk-induced snore-off.

snorer. Add: **1. b.** *slang.* The nose.

1891 FARMER *Slang* II. 168/1 *Conk*,..the nose... English synonyms..snorer; [etc.]. **1925** O. JESPERSEN *Mankind, Nation & Individual* viii. 156 Lastly we have Slang-words for..the Nose. Danish, *snude*... Engl., a number of expressions:..snorter, snorer. **1959** I. & P. OPIE *Lore & Lang. Schoolch.* ix. 155 Children go in for short sharp words, as in their more usual names for parts of the body:..'snorer', 'snozzle', and 'boko' for nose.

snoring, *vbl. sb.* Add: **a.** Also *transf.*

1935 A. J. CRONIN *Stars look Down* I. ix. 69 There was a silence, broken only by the snoring of air through the wind-bore cast of the pump. *Ibid.* 70 The snoring of the pump had stopped. **1951** R. HARGREAVES *This Happy Breed* ix. 102 The obscene snorings of the saxophone.

snork, *sb.* Add: **3.** *Austral.* and *N.Z. slang.* A baby.

1941 BAKER *Dict. Austral. Slang* 68 *Snork*, a baby. **1941** —— *N.Z. Slang* vi. 57 Other twentieth century New Zealand expressions of varied use include..*snork*, a baby. **1944** L. GLASSOP *We were Rats* 273 Got a scar on his hand, but probably he's had it since he was a little snork. **1956** D. M. DAVIN *Sullen Bell* II. v. 136 What I wasn't expecting was to find her living with the same bloke again and well on the way to having another snork. **1963** B. PEARSON *Coal Flat* x. 194 It's better to knock it on the head at birth, isn't it? Like a snork you don't want. **1970** D. M. DAVIN *Not Here, Not Now* II. vii. 108 Have to give up being on the bum once there's a snork or two to be looked after.

snorkel, schnorkel (snǭ·ɹkĕl, ʃnǭ·ɹkĕl). Also ‖ **Schnorchel** (ʃnǭ·rχəl), **Schnorkel.** [ad. G. *schnorchel.*] **1. a.** Usu. in forms **schnorkel, Schnorkel.** An airshaft, invented in the Netherlands and developed in Germany, which was fitted to diesel-engined submarines so that air could reach the engines, allowing them to function, and exhaust gases to be expelled, while the vessel was submerged; also a submarine fitted with such an airshaft. Also *Schnorkel Spirall.*

1944 *News Chron.* 11 Dec. 4/2 They are the new submarines fitted with what the Germans call the Schnorkel Spirall, the purpose of which is to extend under-water endurance. **1945** *Engineer* 19 Jan. 52/3 We hear that the

Germans are fitting their U-boats with what is called the Schnorkel. **1945** *News-Leader* (Richmond, Va.) 12 Mar. 13/3 The 'schnorkel', or stovepipe breather, and the folding kite are chief among the new German gadgets. **1946** [see *breathing-tube* s.v. *BREATHING *vbl. sb.* 10]. **1946** *Collier's* 11 May 69/2 The other the Germans called the 'Schnorchel'. That was a pipe or tube of about periscope height, that extended from the ventilating system of the engines to the surface. **1950** *Sat. Even. Post* 11 Nov. 79 Chief credit for this went to the snorkel, a device which enables subs to breathe under water. Invented by the Dutch, stolen by the Nazis and perfected by the U.S. Navy, the snorkel has revolutionized naval warfare. **1959** *Sunday Times* 8 Feb. 13/4 In the spring of 1944 operational U-boats of the older types began to be equipped with the 'Schnorkel'. **1969** *New Scientist* 28 Aug. 418/1 The invention of the schnorkel reduced the area exposed during recharging to a single pipe extending a few feet above the surface. **1974** L. DEIGHTON *Spy Story* xviii. 192 We came up to periscope depth and let a blow of fresh air through the schnorkel.

b. Usu. in form **snorkel.** A short breathing-tube used by underwater swimmers.

1953 J. Y. COUSTEAU *Silent World* 6 They claimed we drove away fish, damaged nets, looted their seines, and caused mistrals with our schnorkels. **1958** *Oxf. Mail* 17 Apr. 6/7 The American film television series, *Sea Hunt*, claims to be boosting the sport of skin-diving. If that is true there is soon going to be a big demand for snorkels and spear-guns in the Midlands. **1962** *Underwater Swimming* ('Know the Game' ser.) 9/1 By lying on the surface with the face in the water and breathing through the snorkel, the diver can watch the underwater scene continuously. **1968** *T.V. World* 10 Feb. 18/1 It is only when she sees a swimmer's snorkel in her stepfather's room that Mandy realises how the crime could have been committed. **1977** G. DURRELL *Golden Bats & Pink Pigeons* v. 110 We had only masks and no snorkels, and my mask let in water.

2. Usu. in form **Snorkel.** A proprietary name for a piece of apparatus used in fighting fires in tall buildings, consisting of a platform which may be elevated and extended.

1959 *Official Gaz.* (U.S. Patent Office) 27 Oct. TM 140/1 Pitman Manufacturing Company, Grandview, Mo. Filed July 6, 1959. *Snorkel.* For Aerial Platform Apparatus, Particularly Such Apparatus Adapted for Use in Fire Fighting. First use June 11, 1959. **1960** *Amer City* Jan. 83/2 After re-design and further testing of pilot models, the Pitman Aerial Platform, now known as the Snorkel, was offered in July 1958. **1963** R. I. MCDAVID *Mencken's Amer. Lang.* 258 *Snorkel*... The Chicago Fire Department uses it to designate a piece with an elevated pumping platform, for fighting fires in tall buildings. **1969** *Trade Marks Jrnl.* 26 Nov. 1955/1 *Snorkel* 940,580. Mobile hydraulically operated rotatable and elevatable platforms for use in fire fighting. Simon Engineering Dudley Limited,..Dudley, Worcestershire; Manufacturers. **1973** *Lebende Sprachen* XVIII. 69/2 At Newcastle upon Tyne the Chief Fire Officer has installed a closed circuit television camera on an aerial platform known as a *snorkel.* **1977** *Monitor* (McAllen, Texas) 28 June 1A/9 The fire was declared out at 8 a.m. The Edinburg snorkel was used to wet down all parts under the collapsed roof.

3. *attrib.*

1944 [see sense 1 a above]. **1945** *Illustr. London News* 3 Mar. 229/2 The most recent move in this never-ceasing battle was the introduction..of the 'Schnorkel' apparatus. **1949** *Sun* (Baltimore) 2 Apr. 7/2 (*caption*) First photo of damaged sub... The periscope and snorkel equipment are bent. **1953** J. Y. COUSTEAU *Silent World* 1 My wife, Simone, would swim out on the surface with a schnorkel breathing-tube and watch me through her submerged mask. **1954** E. CLARK *Lady with Spear* xix. 187 He.. wasn't a strong swimmer, but he wanted to try a face mask and snorkel tube. **1958** *Times Lit. Suppl.* 9 May 257/4 'There is no tactical requirement for such a fitting,' was the crushing reply to his early suggestion for the designing of a *schnorkel* apparatus. **1962** F. I. ORDWAY et al. *Basic Astronautics* xiii. 510 These devices had snorkel attachments to prevent the entry of water into the systems. **1967** *New Scientist* 9 Mar. 457/1 Modern conventional submarines can proceed at speed for many days with nothing except their 'schnorkel' air-breathing tubes breaking the surface. **1967** O. WYND *Walk Softly, Men Praying* xii. 187 The men..might have got out in snorkel-suits and been picked up by a deep-sea fishing fleet. **1973** *People's Jrnl.* (Inverness) 28 July 10/3 (*caption*) Firemen use the 85ft. snorkel escape ladder to rescue a 'casualty' from the training tower. **1980** P. MOYES *Angel Death* iv. 47 They climbed ashore, with their snorkel masks and fins slung in a string shopping-bag.

Hence as *v. intr.* (also *erron.* **snorkle**), to use a snorkel; to swim underwater using a snorkel; **sno·rkeller, sno·rkelling** *vbl. sb.*

1959 *New Scientist* 26 Mar. 695/2 Pressure variations due to this 'snorkeling' might disturb sensitive instrumentation systems. Mark I could snorkel. **1959** *Elizabethan* June 21/2 But archaeology is really work for trained specialists. You will most likely want to know where to 'snorkel' and fish. **1960** F. M. ROBERTS *Basic Scuba* ii. 27 Snorkeling through weeds..might pull the mouthpiece from the diver's lips because the crook gets caught. **1963** *Harper's Bazaar* Jan. 30/2 Many skin divers are content to remain snorkelers, but some want to go deeper and deeper. **1968** J. UPDIKE *Couples* ii. 171 Ken liked to snorkel. **1974** *Country Life* 24 Jan. (Suppl.) 32 Vast golden bays... Each..an adventure for intrepid snorklers. **1975** D. MARLOWE *Nightshade* x. 116 Water sports, shuffleboard, scuba and snorkling. **1977** G. DURRELL *Golden Bats & Pink Pigeons* iv. 81 We went snorkling on the reef. **1980** P. MOYES *Angel Death* viii. 104 Henry and Emma swam and snorkelled and sunbathed.

snort, *sb.*¹ Add: **3. a.** *slang* (orig. *U.S.*). An alcoholic drink; a measure of spirits; a 'snifter'.

1889 FARMER *Americanisms* 501/2 A *snort* of whiskey is a dram; a nip; a small quantity. **1912** J. SANDILANDS *Western Canad. Dict. & Phrase-Book* 42/2 The sporting Canadian asks his friends, 'Will you have a snort?' **1925** WODEHOUSE *Carry on, Jeeves* iv. 80 We were taking a quiet snort in a corner. **1945** J. STEINBECK *Cannery Row* xxix. 189 She..took out a bottle and a glass and poured herself a snort. **1962** 'R. GORDON' *Doctor in Swim* xii. 75 'How about an—ah—quick snort?' I stared at him. 'But you never drink except at Christmas.' **1966** M. LAURENCE *Jest of God* xii. 199 Ladies often feel it wouldn't be very nice to drink rye at such a time [as bereavement], but a snort of sherry is usually acceptable. **1981** M. E. ATKINS *Palimpsest* viii. 83 We'll have another snort... C'mon, drink up, I'll fill your glass.

b. *slang* (orig. *U.S.*). A dose or measure of cocaine or heroin which is taken by inhalation.

1951 [see *joy-pop* s.v. *JOY *sb.* 10]. **1959** [see *MAIN LINE 1 c]. **1962** [see *HORSE *sb.* 14*]. **1972** H. C. RAE *Shooting Gallery* II. 73 How did McDowell pick up a big enough snort to do for himself? **1978** G. VIDAL *Kalki* iv. 88 'Want a snort?' Bruce produced a cocaine snifter.

snort (snǭɹt), *sb.*² *Naut. slang* (now only *Hist.*). [Anglicized corruption of G. *schnorchel*, after SNORT *sb.*¹] = *SNORKEL, SCHNORKEL 1 a. Freq. *attrib.*

1944 *News Chron.* 11 Dec. 4/2 The first 'snort' U-boats are probably already at sea... 'Snort' is the Navy's nickname for them. **1944** *N.Y. Herald Tribune* 12 Dec. 1/7 (*heading*) 'Snorts' said to enable vessels to stay under 20 days. **1950** *Times* 26 Apr. 6/6 Under the programme 10 existing submarines are to be equipped with the 'Snort' breathing apparatus. **1954** H. M. BURTON tr. *Diolé's Under-Water Exploration* v. 69 The chief improvements to the standard type submarine which were introduced during the last war were the work of the Germans. They were responsible, in particular, for the *Schnorchel*, or 'snort'. **1976** *Oxf. Compan. Ships & Sea* 759/1 In the British Navy the schnorkel tube was given the name snort.

Hence as *v.*² *intr.*: of a submarine, to travel underwater by means of a snort; **sno·rter**³, a submarine fitted with a snort; **sno·rting** *vbl. sb.*²

1953 *John o' London's Weekly* 3 July 602/2 Since the *Andrew* crossed the Atlantic in total submergence, the word snort has acquired a different significance. Said her captain after she had achieved her object: 'All we were told was: "You are going to snort back"—so we snorted.' **1957** *Jane's Fighting Ships 1957–8* 51 On 15 June 1953 *Andrew* completed a 2500 sea miles voyage under water from Bermuda to the English Channel in 15 days, a record for 'snorting' in the Royal Navy. **1962** W. GRANVILLE *Dict. Sailor's Slang* 109/1 *Snorter*, submarine fitted with the snorkel device which enables her to keep at sea for a considerable period. **1974** 'M. HEBDEN' *Pride of Dolphins* III. ii. 230 'Open Three Main vents. Periscope depth. Stand by to snort.'.. They were snorting slowly back up the Solent. **1979** *Daily Tel.* 3 May 3/3 Since Olympus could reach safety from snorting depth in about a minute, the order to dive was given 45 seconds too late.

snort, *v.* Add: **7.** *slang* (orig. *U.S.*). To inhale (a narcotic drug in powder form, esp. cocaine or heroin). Also *absol.*

1935 A. J. POLLOCK *Underworld Speaks* 110/1 *Snort*, to sniff cocaine or heroin. **1958** H. BRADDY in *Southern Folklore Q.* Sept. 134 Since ma was a viper And daddy would snort, There wasn't much more I had to be taught. **1967** M. M. GLATT et al. *Drug Scene* iii. 32, I started snorting cocaine through the nose. **1972** M. J. BOSSE *Incident at Naha* i. 38 She snorted Methedrine. I saw her do it many times. **1974** M. C. GERALD *Pharmacol.* xv. 291 Cocaine is usually administered intravenously, although some prefer to 'sniff' or 'snort' it. **1980** M. BOOTH *Bad Track* ii. 46 'Are you snorting?'.. He nodded... He inhaled the cocaine. **1982** *Daily Tel.* 4 Oct. 3/3 Mrs Pulitzer's lawyers claim that she started snorting cocaine after being sucked into the vortex of the 'Palm Beach lifestyle'.

snorter¹. Add: **2. c.** (Examples in *Cricket.*)

1888 R. H. MITCHELL in Steel & Lyttelton *Cricket* xiii. 380 How different this..from being compelled to play a real 'snorter' before the breath is fairly recovered after the effort of running several fourers in succession! **1898** G. GIFFEN *With Bat & Ball* xi. 189, I know of no bowler whom one has to watch so closely [as T. R. McKibben], for you never know when you are going to get a 'snorter' of a break, from one side or the other. **1929** *Morning Post* 11 Mar. 16/4 When in the next Test, at Lord's, McDonald bowled him [*sc.* Hendren] for 0 with a 'snorter'. **1954** J. H. FINGLETON *Ashes crown Year* xxiv. 257 May..now hit another 'snorter' through the covers.

d. (Earlier example.)

1829 P. EGAN *Boxiana* 2nd Ser. II. 119 The latter got a severe snorter, which not only uncorked the claret, but left a stupifying quality behind it.

e. The nose itself.

1829 P. EGAN *Boxiana* 2nd Ser. II. 353 The snorter of Raines looked red! For why? Jones's mauley had given it a rum tap! **1846** *Swell's Night Guide* 132/2 *Snorter*, the nose. **1925** [see *SNORER 1 b].

snorter². Add: **2.** = *SNOTTER *sb.*² 2.

1950 BOWN & DOVE *Port Operation & Admin.* iv. 138 The snotter, or snorter, is a length of cordage or S.W.R. with an eye spliced in each end. **1965** R. B. ORAM *Cargo Handling* v. 93 Rope snorters are used at Sydney and Brisbane to discharge the pallets and these are left on the cargo at loading.

snorter³: see *SNORT *sb.*² **snorting,** *vbl. sb.*²: see *SNORT *sb.*²

snorting, *ppl. a.* Add: **4.** Exceptionally remarkable for excellence, size, strength, etc. (In quot. as *advb.*) *colloq. rare.*

1924 GALSWORTHY *White Monkey* II. ix. 195 I've played bridge with him,..—snorting good player.

snot, *sb.* Add: **3.** (Later examples.)

1939 JOYCE *Finnegans Wake* 494, I would misdemean to rebuke to the libels of snots from the fleshambles, the canalles. **1952** B. HARWIN *Home is Upriver* xvii. 172 You want that damn' little snot now, hah? A damn' little snot ain't even dry behind the ears. **1974** N. FREELING *Dressing of Diamond* 159 She wasn't going to cry in front of that rotten-toothed snot. **1981** J. MELVILLE *Sort of Samurai* iii. 25 We've let the boy go home on bail... Miserable little snot, but no real harm in him.

5. *snot-green, -smeared* adjs.; **snotnose** *slang*, a term of contempt applied to a childish, despicable, or conceited person; = SNOTTY-NOSE; **snot-nosed** *a. slang*, foul with nasal mucus; conceited; inexperienced and contemptible; = SNOTTY-NOSED *a.*; **snot-rag** *slang*, a pocket-handkerchief; also *transf.* as a term of opprobrium.

1922 Snotgreen [see *SCROTUM b]. **1975** T. STOPPARD *Travesties* I. 23 The swiftly-gliding snot-green (mucus mutandis) Limmat River. **1941** T. WOLFE *Hills Beyond* ix. 338 How do *you* know whether it's round or flat—a little two-by-fo' snotnose like you that ain't *been* nowhere. **1949** A. MILLER *Death of Salesman* II. 97 That snotnose. Imagine that? **1963** 'E. MCBAIN' *Ten plus One* xi. 143 He was not enjoying this little snotnose..and the college girl talk. **1977** H. FAST *Immigrants* I. 72 So don't be young snotnose with me. I like serious boys. **1941** B. SCHULBERG *What makes Sammy Run?* i. 13 A snot-nosed little office boy. **1960** H. LEE *To kill Mockingbird* iii. 34 Ain't no snot-nosed slut of a schoolteacher ever born c'n make me do nothin'! You ain't makin' me go nowhere, missus. **1972** M. WOODHOUSE *Mama Doll* viii. 100 A persuasive manner you picked up at some snot-nosed advertising agency. **1886** F. T. ELWORTHY *W. Somerset Word-Bk.* 690 *Snot-rag*.., a pocket-handkerchief. **1916** 'TAFFRAIL' *Pincher Martin* vi. 95 Any schoolboy will tell you what a 'snot rag' is. **1929** T. WOLFE *Look homeward, Angel* xiv. 170, I don't give a good goddam..if you're the President's snotrag. **1959** N. MAILER *Advts. for Myself* (1961) 84 One of them said he was going to take my shirt and use it for a snotrag, and they all laughed. **1973** J. WAINWRIGHT *High-Class Kill* 241 You are a self-opinionated idiot. You, and every snivelling little snot-rag like you. **1939** AUDEN & ISHERWOOD *Journey to War* i. 48 The averted, snot-smeared, animal faces of the very humble.

snotter, *sb.*² Add: **2.** A length of rope with an eye spliced in each end.

1950 [see *SNORTER² 2]. **1956** C. L. SAUERBIER *Marine Cargo Operations* vii. 416 The sling is constructed in the same manner as the fiber rope snotter. **1961** COURSE & ORAM *Gloss. Cargo-Handling Terms* 72 The snotter is stretched out to its full length and the package placed on it centrally. The ends of the snotter are brought over it and one eye rove through the other and placed on the lifting hook.

sno·ttily, *adv.* (Examples.)

1927 *Blackw. Mag.* Dec. 816/2 'Of course I did,' he replied, rather snottily I thought. **1937** J. T. FARRELL *Fellow Countrymen* 180 A man in a hurry bumped into him, and hastening on, snottily suggested that he quit taking up the whole sidewalk. **1973** M. AMIS *Rachel Papers* 59, I had a face looking over my shoulder, no matter how snottily equivocal its expression.

snottiness. (Later examples.)

1973 *Guardian* 18 June 4/6 The snottiness of the elitist eastern [US] establishment. **1976** *New Society* 19 Aug. 407/3 The snottiness of the French..Surly waiter..snapping concierge.

snotty, *sb.* (Later examples.)

1916 'TAFFRAIL' *Pincher Martin* vi. 95 No boat ever left the ship under steam or sail without a 'snotty' in charge. **1943** HUNT & PRINGLE *Service Slang* 61 *Snotty*, midshipman. (So called after the buttons on his sleeve, which are said to be there for a purpose not unconnected with the nickname.) **1974** P. DICKINSON *Poison Oracle* ii. 47 A British Naval Party under the command of a snappily saluting little snotty.

snotty, *a.* Add: **1. b.** (Later examples.)

1958 J. C. HEROLD *Mistress to Age* (1959) III. xiii. 263 Albertine had slapped the Crown Prince and called him a snotty brat. **1967** P. WELLES *Babyhip* ii. 36 My brother tried to date her, but she rejected him. She told him she didn't admire Catholics. I think that's pretty snotty. We did go to the same Sunday School. **1974** S. ELLIN *Stronghold* 60 'Did it strike you,' Coco asks at his snotty meanest, 'that if we came properly prepared, we could have stopped him from taking off?'

c. (Later examples.) Now *esp.* supercilious, aloof, 'snooty'.

1905 JOYCE *Let.* 7 Feb. (1966) II. 80 Are the 'girls' 'snotty' about Nora? **1916** W. OWEN *Let.* 9 Dec. (1967) 417 A snotty, acid, scot, impatient, irritated wretch. **1926** E. HEMINGWAY *Sun also Rises* xviii. 218, I won't eat down-stairs with that German head waiter. He was damned snotty. **1936** J. REITH *Diary* 13 May (1975) ii. 170 This is an insult... I was very snotty and reserved with the prig. **1968** *Globe & Mail Mag.* (Toronto) 13 Jan. 12/3 Francois is not always snotty, thank heaven. **1978** T. GIFFORD *Glendower Legacy* (1979) 39 He..thought for a moment of taking up the possibility of an exchange program with the snotty bastards in Cultural Affairs.

snotty-nose. For ? *Obs.* read 'Now *rare*' and add later example.
1932 L. GOLDING *Magnolia St.* III. i. 495 A little snotty-nose like that..and he's the [boxing] champion from all the world!

snotty-nosed, *a.* Delete Now *dial.* and add later examples.
1948 *Sun* (Baltimore) 20 Aug. 15/8 It was Walker who once told the Babe on a memorable occasion never to let down 'those snotty-nosed kids' who always loved him. 1971 P. AUDEMARS *Stolen like Magic Away* v. 66 All that love and..passion—thrown away every day on a bunch of snotty-nosed kids who take it..for granted. 1978 N. J. CRISP *London Deal* vii. 109 There's a snotty nosed young DC from the Yard sitting in his car outside.

snous (snaus). Also **snouse.** [ad. Da. or Sw. *snus* snuff.] Powdered tobacco.
1962 J. ONSLOW *Bowler-Hatted Cowboy* xxi. 204 His lower lip bulged with a wad of 'snouse', or Copenhagen snuff. 1979 *Guardian* 31 Mar. 13/5 The narcotic to which most [Swedish] young people are addicted is..'snous', a concoction of powdered tobacco stuffed under the upper lip.

snout, *sb.*[1] Add: **2. c.** Phr. *to have a snout on* (someone), to bear ill-will towards someone. *Austral.*
1941 BAKER *Dict. Austral. Slang* 69 *Snout on, have a,* to bear a grudge against a person. 1949 L. GLASSOP *Lucky Palmer* 212 He's got a snout on the Kid for something. 1966 T. RONAN *Once there was Bagman* 39 The reason you blokes have such a snout on him..is that he's forgotten more Law than you've ever learned.
5*. *slang.* A police informer.
1910 C. E. B. RUSSELL *Young Gaol-Birds* xii. 176 He was in reality a 'snout' or 'nark',..and from time to time had 'given away' many of his comrades. 1938 F. D. SHARPE *Sharpe of Flying Squad* xvii. 189 A 'sneak' or 'snout' is looked upon more or less as a leper in the Underworld. 1954 [see *GRASS *sb.*[1] 11*]. 1964 *Sunday Mail Mag.* (Brisbane) 5 Apr. 5/5 Then a 'snout' (or informant) called Big Ears made a long trip just to tell me: 'You're in trouble, Monty.' 1977 'E. CRISPIN' *Glimpses of Moon* xii. 235 His previous arrests had all been..the work probably of some anonymous snout. 1982 *Observer* 15 Aug. 22/6 You may have been 'grassed'..by a 'snout'.
7. snout-face: used as a personal insult.
1923 D. H. LAWRENCE *Birds, Beasts & Flowers* 184 But you, you snout-face, you reject nothing. 1979 *Amer. Poetry Rev.* Mar./Apr. 6/2 And the people In the streets, speechless, saw them passing: The scrawny guy, the barefoot one, the fellow with The bicycle, The black, Snout-face, that gal in yellow, [etc.].

snout, *sb.*[2] Add: **1. a.** (Earlier example.)
1885 A. GRIFFITHS *Fast & Loose* III. xii. 202 He knows Joe; worked for him, with regard to snout (tobacco); and he's straight—as a rod.
b. A cigarette.
1950 P. TEMPEST *Lag's Lexicon* 193 *Snout.* Word used collectively to cover all tobacco, hand-rolled and factory-made cigarettes, cigarette ends, and pipe dottles. 1954 *Evening News* 7 Jan. 2/2 Savage was seen and said: 'You will not find any export snouts here.' 1959 H. HOBSON *Mission House Murder* xxix. 187, I would smoke it slowly and..save the butt—*snouts,* the old lags called them. 1961 R. LONGRIGG *Daughters of Mulberry* 94 'Snout?' said her Ronnie, offering the Rothman's Kingsize. 1966 P. MOLONEY *Plea for Mersey* 54 Goin down the city fer a booze an a snout. 1976 J. O'CONNOR *Eleventh Commandment* vii. 91 If you were wise you chose non-smokers as your friends because they wouldn't shop you to an unscrupulous warder for a couple of snouts.
2. *attrib.*, as *snout ash, baron* [*BARON 2 c], *case, gaff* [*GAFF *sb.*[4] 3], *paper.*
1962 R. COOK *Crust on its Uppers* i. 21 Ever had someone put some snout ash in your rosie? 1950 P. TEMPEST *Lag's Lexicon* 194 Snout-baron. 1964 *Economist* 25 Jan. 317/1 The 'snout barons'—prisoners who make a profit from the shortage of tobacco within prisons. 1962 R. COOK *Crust on its Uppers* iv. 48 'I'm going to give our Brian a fag,' an' he..brings out this heavy old snout case. 1936 Snout gaff [see *GAFF *sb.*[4] 3]. 1958 *Encounter* Apr. 18/1 He hardly ever spoke to me unless he wanted something, like a smoke or a snout paper.

snout, *v.* Add: **3.** *trans.* To bear ill-will towards; to treat with disfavour, to rebuff. Freq. as pa. pple. and ppl. adj. *Austral. slang.*
1916 C. J. DENNIS *Moods of Ginger Mick* 11 An' snouted them that snouted 'im, an' never give a dam. 1916 —— *Songs of Sentimental Bloke* 13 The world 'as got me snouted jist a treat. 1944 A. MARSHALL *These are my People* 155, I was sore as a snouted sheila for weeks. 1970 R. BEILBY *No Medals for Aphrodite* 149 That officer happened to have me snouted because I got you across the river, against his orders.
4. *intr.* To act as a police informer. *slang.*
1923 E. WALLACE *Missing Million* xx. 161 The gang found he was snouting. 1930 —— *White Face* xiii. 206 Dr. Marford knows, but he's not the feller that goes snouting on his patients. 1962 D. WARNER *Death of Bogey* II. iii. 71 No one wanted to be seen talking to him in case they were afterwards accused of snouting. Nevertheless, a great many did snout. 1973 'B. MATHER' *Snowline* x. 116 I've got to live in London when I go back. How long do you think I'd last if word got round that I'd been snouting?
Hence **snou·ting** *vbl. sb.* (also *attrib.*).
1937 PARTRIDGE *Dict. Slang* 795/2 *Snouting,* vbl. n., giving information to the police. 1962 [see sense 4 above]. 1973 J. WAINWRIGHT *Pride of Pigs* 55 Arranging a 'snouting service' with those villains; the lesser hooks being pulled in for the piffling crimes, while the big boys work the blinders without..being pushed too hard. 1978 F. BRANSTON *Sergeant Ritchie's Conscience* iv. 56 He started on his snouting expedition.

Snovian (snǭ·viăn), *a.* [f. the name of the English writer Charles Percy *Snow* (1905–80), on the model of *SHAVIAN *a.* and *sb.*, etc.] Of or pertaining to the writings or ideas of C. P. Snow. Hence **Sno·vianism,** the beliefs or theories of C. P. Snow.
1966 *Listener* 19 May 733/1 This twinkling dancing life and soul of these almost edible pages is also one of the Snovian Olympians. *Ibid.*, Nothing could be more depressing than this total acceptance of the doctrines of Snovianism, even down to the use of the term *mana* and the reluctance to take seriously anyone below the rank of knight bachelor. 1969 *Observer* 17 Aug. 21/2 The attack on scientific rationalism..first appears in a paper on the Snovian conception of the Two Cultures. 1977 P. JOHNSON *Enemies of Society* xii. 163 The United Kingdom, compared with other..countries..had (in 1964) 'the greatest concentration on science and technology in higher education and the biggest proportion of qualified scientists and technologists..in relation to population and labour-force'—the exact opposite of the conventional Snovian thesis.

snow, *sb.*[1] Add: **I. 2. b.** (Earlier example.)
1778 J. CARVER *Trav. N.-Amer.* 250 Those [Indians] in the interior parts..count their years by winters; or, as they express themselves, by snows.
3*. Ellipt. for *snow tyre*, sense 7 b below. *N. Amer.*
1968 *Globe & Mail* (Toronto) 13 Jan. 26/2 (Advt.), 67 Fiat,..special exhaust, snows. 1977 *Detroit Free Press* 11 Dec. 22-D/8 (Advt.), '73 F-350 V8 4spd, dual tanks, PsPb, Ranger, snows.
II. 4. b. (Further examples.) *spec.* Solid carbon dioxide.
1913 J. HALL-EDWARDS *Carbon Dioxide Snow* 28 Having prepared our cone, or stick of snow..the first step is to place the patient in a comfortable and easy position. 1931 DOUGHERTY & KEARNEY *Fire* 243 The 'snow' does not freeze the fire as is sometimes erroneously believed, but blankets or smothers it. 1951 WHITBY & HYNES *Med. Bacteriol.* (ed. 5) ii. 20 Many bacteria and viruses..may be preserved by rapid freezing to −70°C, with CO_2-snow. 1974 L. E. LONG *Geology* i. 19 A frozen 'snow' of methane and ammonia glued the dust particles into globs that eventually grew to about the size of basketballs. 1979 *Nature* 30 Aug. 738/1 Much of the distributed SO_2 snow would be expected to fall within a few tens of kilometres of the scarps [on Jupiter's satellite Io].
d. *slang* (orig. *U.S.*). Cocaine; occas. heroin or morphine.
1914 JACKSON & HELLYER *Vocab. Criminal Slang* 78 *Snow,*..derived from the extremely flocculent nature of cocaine when pulverized. 1915 *Policeman's Monthly* Dec. 17/3 One day, his pal found him depressed and told him to take a little sniff of 'snow', as heroin is known to the vernacular of the criminal. 1925 A. P. HERBERT *Laughing Ann* 92 Don't let her know about whisky and 'snow'. 1933 N. DOUGLAS *Looking Back* II. 364 He..walked up and down the room..taking, every now and then, a pinch of cocaine... 'I didn't know you took snow.' 1956 [see *JAB *v.* e]. 1966 'A. HALL' *9th Directive* iii. 25 Pangsapa was a narcotics contrabandist and would therefore know people ..prepared to kill for a fix of snow. 1967 N. LUCAS *C.I.D.* x. 135 Luckier still not to have graduated from pep pills to..'Snow'..—morphine. 1979 P. DRISCOLL *Pangolin* xx. 151 'Tell me how much this roll will get me.' 'I guess around a hundred twenty grams. That's..the purest snow you'll ever see.'
e. *slang.* (Silver) money.
1925 FRASER & GIBBONS *Soldier & Sailor Words & Phrases* 263 *Snow,* money. Silver. 1936 J. CURTIS *Gilt Kid* 173 Count up that snow while I go through the other drawers. 1970 F. MCKENNA *Gloss. Railwaymen's Talk* 38 *Snow,* small silver i.e. sixpences.
f. Spots that appear as a flickering mass filling a television or radar screen, caused by interference or a low signal-to-noise ratio.
1946 *Proc. IRE* XXXIV. 428/2 These [current] fluctuations give rise to a masking effect, often referred to as 'snow', in the transmitted picture. 1950 HELLER & SHULMAN *Television Servicing* vi. 121 Low signal input may be recognized by the characteristic presence of 'snow' in the received picture. 1977 J. CHEEVER *Falconer* 209, I took my TV... I had a little snow and asked the repairman to come in. 1978 *Sci. Amer.* Apr. 18/1 The most commonly encountered white noise is the thermal noise produced by the random motions of electrons through an electrical resistance. It causes most of the static in a radio or amplifier and the 'snow' on radar and television screens when there is no input.
III. 7. a. *snow-bank* (earlier example), *block, bridge, cave, -cloud* (earlier example), *cover, -crust, -flurry, -glare, -hut* (earlier and later examples), *-light* (earlier example), *-patch, -squall* (earlier example).
1779 E. PARKMAN *Diary* (1899) 194 Snow-Banks very high one nigh my saddle-house 6 feet high. 1893 'MARK TWAIN' in *Cosmopolitan* Nov. 54/1 My father..built this great mansion of frozen snow-blocks. 1973 W. S. AVIS in *Occasional Papers Dept. English R. Military Coll. Canada* (1978) No. 2. 152 A knife..used primarily in cutting snow blocks for igloo-building. 1982 S. B. FLEXNER *Listening to America* 22 Alaskan Eskimos often built their igloos out of animal skins, driftwood, etc., using snow-block ones only for temporary or emergency shelters. 1890 *Moose Jaw* (Saskatchewan) *Times* 20 June 1/4 Every observant passenger on the Canadian Pacific Railroad had noticed the snow bridge on the Illecillewaet, but there are records of ice bridges also. 1921 A. LUNN *Alpine Skiing* vii. 83 On the Grenz glacier a snow-bridge fourteen feet thick, and in the recent Oberaarfoch accident a snow-bridge six feet thick, collapsed beneath men on skis. 1939 [see *SCHRUND]. 1979 C. KILIAN *Icequake* xiii. 228 The snow bridges seem good and thick, but the quake probably weakened them. 1972 D. HASTON *In High Places* ix. 103 On descending they found Mick at the col installed in a snow-cave that he had dug out. 1981 *Nordic Skiing* Jan. 21/2 You can imagine me huddled in my own hastily dug snow cave waiting out the blizzard. 1879 I. BIRD *Lady's Life in Rocky Mountains* x. 168 Looming vaguely through a heavy snow-cloud. 1919 *Sci. Monthly* IX. 397 A winter snow-cover prevents deep freezing of the ground. 1956 A. GARNETT in D. L. Linton *Sheffield* 48, 1947..was phenomenal for the prolonged and severe cold weather experienced and for the long duration of a snow cover. 1824 S. BLACK *Jrnl.* 24 May (1955) 14 They left the Fort in March on the snow crust. 1957 G. E. HUTCHINSON *Treat. Limnol.* I. iii. 214 Teis (1946) examined various snow crusts and firn samples. 1979 R. FIENNES *Hell on Ice* iv. 63 The wind-firm snowcrust. 1879 I. BIRD *Lady's Life in Rocky Mountains* ix. 124 The wild flowers are gorgeous..though..the recent snow-flurries have finished them. 1936 *Geogr. Jrnl.* LXXXVII. 133 On September 1 came the first snow-flurries of the season. 1860 M. REID *Odd People* 394 More likely it is the snow-glare to which the Laplander, as well as the Esquimaux, is much exposed, that brings about the copious *watering* of the eyes. 1962 L. S. SASIENI *Optical Dispensing* xiii. 326 In snow glare protection is required against the ultra-violet. 1970 R. D. TARING *Daughter of Tibet* xix. 246 Between the smoke and the snow-glare of the day our eyes were red and watering and very sore. 1823 *Lit. Gaz.* 25 Oct. 673/3 A tribe of about fifty Esquimaux who were erecting their snow-huts. 1930 V. SACKVILLE-WEST *Edwardians* i. 28 He had been marooned..somewhere near the South Pole in a snow-hut. 1830 M. O'BRIEN *Jrnl.* (1968) I. ix. 87 It was dark—as dark as it can be with snowlight. 1909 Snow-patch [see *FLORA 3 b]. 1979 B. JOHN *World of Ice* 26 (*caption*) The peaks and mountain-sides at this time of year are almost free of snow and ice, and only a few perennial snowpatches remain. 1775 E. WILD *Jrnl.* 6 Dec. in *Mass. Hist. Soc. Proc.* (1886) II. 287 The weather is attended with Snow Squalls.
b. *snow-anchor, -board, -boot* (later examples), *buggy, chain, -coat, -fence* (earlier example), *-fencing, gallery, gauge* (further examples), *-glasses, -pants, -scoop, scooter, -shed* (earlier and later examples), *-stake, -suit, tractor, tyre, vehicle.*
1971 C. BONINGTON *Annapurna South Face* 248 The 'dead men' were an outstanding success and..gave by far the most reliable..snow anchor we were able to use on the expedition. 1972 D. HASTON *In High Places* xi. 120 Using devious combinations of snow-stakes, 'dead men' (or snow-anchors).., they took two days to come out of those overhangs. 1881 W. P. BUCHAN *Plumbing* (ed. 3) xi. 70 A style which serves both as a snow-board and as a preventive of broken chimney cans, loose slates, &c., falling over the roof. 1971 *Country Life* 14 Oct. 964/1 Notices warning of snow-board avalanches had been posted..that very morning. 1856 S. OSBORN *M'Clure's Discovery North-West Passage* xii. 160 The heavy falls the men experienced in their thick winter clothing and cloth snow-boots. 1962 A. LURIE *Love & Friendship* I. viii. 142 She came..to ask if she could borrow my snowboots to walk in the snow with. 1970 *Toronto Daily Star* 24 Sept. 16/3 (Advt.), Tamarack snow boots. The new style. 1949 *Sun* (Baltimore) 8 Feb. 15/3 Second Army headquarters.. is sending 48 'snow buggy' operators..to the aid of snow-bound Nebraskans... Their main job will be to drive weasels, the Army's special vehicle for snow-covered terrain. 1965 *Kingston* (Ontario) *Whig-Standard* 27 Dec. 17 (*caption*) Roaring through the snow at speeds.. approaching 35 miles-an-hour on the..new snow buggy. 1975 *Islander* (Victoria, B.C.) 9 Feb. 12/1 If you don't have snow chains, don't even try to get up the steep logging road. 1981 P. TURNBULL *Deep & Crisp & Even* i. 8 An ambulance with snow chains drove along the street. 1963 *N.Y. Times* 15 Dec. 18/7 (Advt.), This jaunty..pile-lined 'snowcoat' gets you ready for Winter's worst! 1965 *Harper's Bazaar* Nov. 95 Fir green quilted snowcoat. 1873 G. M. GRANT *Ocean to Ocean* ix. 261 The high mountains..act as natural snow fences. 1953 *Canad. Geogr. Jrnl.* XLVI. 68/2 Others made cribs out of snow fencing and piled the grain in the open fields. 1972 L. HANCOCK *Sleeping Bag* viii. 181 We dug an extensive salt-water pool and walk-in aviary..then snow-fencing enclosures for the raptorial birds. 1874 Snow gallery [see *round timber* s.v. *ROUND *a.* 15 a]. 1975 D. BAGLEY *Snow Tiger* xix. 157 They build snow galleries over roads..in Switzerland. The snow goes straight over the top. 1939 *Meteorol. Gloss.* (Met. Office) (ed. 3) 172 In the Hellmann-Fuess snow-gauge the snow is caught in a receiver supported on a balance, the displacement of which is continuously recorded. 1952 E. F. DAVIES *Illyrian Venture* ii. 32 The snow gauges on the mountain passes, dead tree trunks with marks nailed to them to show the depth of the winter drifts. 1927 E. HEMINGWAY *Men without Women* 162 Around the major's eyes were two white circles where his snow-glasses had protected his face from the sun on the snow. 1975 E. HILLARY *Nothing Venture, Nothing Win* xi. 175 Wilkins..seemed comparatively unhurt, although his snowglasses had cut his forehead. 1948 T. ONRAET *Sixty Below* 100 The ordinary snow pants and parka are made with the least possible openings. 1962 *N.Y. Post* 9 Oct. 22 (Advt.), Infants' pile snowsuits... Matching, contrasting snowpants. 1978 *Detroit Free Press* 5 Mar. D-1/1 It was still cold and your mother made you put on your coat, hat and mittens, but you could never-mind the 'snow pants' by now. 1961 J. W. ANDERSON *Fur Trader's Story* x. 80, I struck the tent, loaded the toboggan with tent, stove,..snow scoop..and so forth, and set off. 1963 *Engineering* 18 Jan. 79 The manufacturers are now considering adding the snow-scoop to their range of standard attachments. 1964 *Star Weekly* (Toronto) 19 Dec. 13/1 The odd little snow scooters you see cavorting about ..represent the newest phenomenon to revolutionize Canadian sport, family living—and business. 1969 *Daily*

Colonist (Victoria, B.C.) 5 Sept. 27/3 Reindeer-tending Lapps of northern Norway use snow scooters to round up strays and transport supplies. 1981 *Times* 14 Dec. 22/8 Four policemen..have been..to North Cape, in Norway, for charity. They reached there on snow scooters. 1868 *Oregon State Jrnl.* 22 Aug. 2/3 The Pacific Railroad advertises for a thousand men to build snow sheds on the summit. 1965 E. MCCOURT *Road across Canada* 177 In Glacier [B.C.] more than half a mile of snowsheds, solidly built of steel and concrete.., guard the most vulnerable spots. 1971 *Daily Tel.* 9 Jan. 9/2 The railway line runs through numerous long snow-sheds in these high lands [in Norway]. These are built over the line to keep it free of snow in winter. 1971 C. BONINGTON *Annapurna South Face* viii. 95, I pushed in a snow-stake, but it went in too easily and would almost certainly be pulled out if I fell on it. 1972 Snow-stake [see *snow-anchor* above]. 1942 D. POWELL *Time to be Born* i. 37 The red snow suit her mother had promised. 1962 A. LURIE *Love & Friendship* I. iii. 53 Emmy put Freddy into his snow-suit. 1980 *Daily Tel.* 9 Jan. 1/8 There was no sign of the guerrillas in the rugged terrain, but Russians and their armour, including tanks, were everywhere. Some were in white snow suits. 1936 *Canad. Geogr. Jrnl.* XII. 34/2 Somebody began to work on the idea of snowmobiles and snow tractors. 1971 *Country Life* 14 Oct. 964/1 Hardly had the two children been freed when they [*sc.* a rescue team] were on the spot, having covered the ground in a snow-tractor. 1954 *Sun* (Baltimore) 23 Jan. 8/1 Now it's chains vs. snow tires, the treachery of the steep hill by the lake and stern telephone calls to warn the little woman off the roads. 1968 'E. MCBAIN' *Fuzz* xii. 197 The snow.. presented no major traffic problems as yet, especially if.. one had snow tyres on one's automobile. 1978 *Times* 23 Jan. 12/7 Avis..had only one car they could rent me and it had no snow tyres or chains. 1968 *Globe & Mail* (Toronto) 3 Feb. 46/3 (*heading*) Snow vehicles.
c. *snow-pinion.*
1879 G. M. HOPKINS *Poems* (1967) 80 If a wuthering of his palmy snow-pinions scatter a colossal smile Off him.
d. *snow cake* (earlier example), *eggs* (earlier example).
1861 MRS. BEETON *Bk. Househ. Managem.* 747 *Snow eggs,*..4 eggs, ½ pint of milk,..sugar..vanilla, lemon-rind. *Ibid.* 864 *Snow cake*...½ lb. of *tous-les-mois,* ½ lb. of ..sugar, ¼ lb. of..butter, 1 egg,..1 lemon.
8. a. *snow-backed, -blanched, -blanketed, -born, -bowered, -cooled, -dazed, -dimmed, -drowned* (later example), *-fed* (later examples), *-hooded, -packed, -shouldered, -suited.*
1897 KIPLING *Five Nations* (1903) 18 While thick around the homestead Our snow-backed leaders graze. 1945 W. DE LA MARE *Burning-Glass* 23 The snow-blanched sunshine. 1971 R. DENTRY *Encounter at Kharmel* ix. 151 The snow-blanketed hills. 1879 I. BIRD *Lady's Life in Rocky Mountains* vii. 97 From this side rise, snow-born, the bright St. Vrain, and the Big and Little Thompson. 1930 R. CAMPBELL *Adamastor* 62 Fair siren of the snow-born lake. 1919 W. DE LA MARE *Flora* 42 Still from the snow-bowered, link-lit street The muffled hooves of horses beat. 1920 R. GRAVES *Country Sentiment* 63 Or toys or meat or snow-cooled drink. a 1918 W. OWEN *Poems* (1963) 48 We cringe in holes, back on forgotten dreams, and stare, snow-dazed, deep into grassier ditches. 1957 BLUNDEN *Poems of Many Years* 295 In snow-dimmed moonlight. 1978 G. GREENE *Human Factor* VI. ii. 322 Outside the silence of the snow-drowned street was so extreme that Castle hesitated to break it. 1936 R. CAMPBELL *Mithraic Emblems* 31 The lily-scented blood, the snow-fed wine of scarlet stain. 1963 *Times* 6 Feb. (New Zealand Suppl.) p. vii/3 The Rangitata itself —snow-fed and treacherous. 1880 'MARK TWAIN' *Tramp Abroad* xxv. 245 The stately border of snow-hooded mountain peaks. 1945 W. DE LA MARE *Burning-Glass* 44 A moth, snow-hooded, delicate past belief. 1973 J. M. WHITE *Garden Game* 188 Teague drove his Mercedes..on to the snow-packed verge. 1921 W. DE LA MARE *Veil* 59 Snake-haired, snow-shouldered, pure as flame and dew. 1936 R. CAMPBELL *Mithraic Emblems* 17 Each great snow-shouldered beast. 1961 'E. LATHEN' *Banking on Death* (1962) ix. 71 Snowsuited toddlers frolicking merrily in the snow. 1971 A. BAILEY *In Village* (1972) xix. 189 Snow-suited small children.
b. *snow-blower, -clearer, -loader, -scraper* (earlier example), *-shifter, -thrower.*
1955 *Hamilton* (Ontario) *Spectator* 25 Jan. 24/3 Street sweepers, snow blowers, and other city equipment stored outdoors at the Elgin Street yard. 1964 S. FORBES *Long Hate* (1966) x. 92 'We'll have to shovel, I guess.'.. 'Can't you use the snow blower?' 1978 *Daily Tel.* 1 Feb. 1/7 Extra snow-clearing equipment was being sent to the area and the RAF was bringing in a large snowblower from Switzerland. 1963 *Times* 18 Feb. 4/1 The efforts of dedicated Kingsholm snow-clearers were rewarded, and the surface was unbelievably good in the circumstances. *Ibid.* 28 Jan. 9/6 Clearing is done by a continuous moving belt operation with a plough in front followed by a specially built snow loader which digs into drifts with rotating blades and funnels it into a line of waiting lorries. 1974 *Globe & Mail* (Toronto) 12 Feb. 5/3 The combined snow loader and melter was designed by Metro roads department and consultants after testing a small 75-ton snow melter during the past three winters. 1851 in H. Greeley *Recoll. Busy Life* (1868) 559 We met with a bad accident ..45 miles from Baltimore, our snow-scraper catching against some part of the track. 1962 Snow-shifter [see *MACK *sb.**]. 1966 *Wall St. Jrnl.* 28 Dec. 1/4 The power-driven snowblower (or snowthrower, if you prefer), a gadget with reel-type blades that chew through the snow and push it into a chute, from whence it's blown aside. 1978 *Detroit Free Press* 16 Apr. (Gardening Guide) 6 (Advt.), Attachments include 60-inch rotary mower, 48-inch snow thrower, [etc.].
c. *snow-clear, -cool, -deep* (later example), *-proof, -soft* (later examples).
1925 E. SITWELL et al. *Poor Young People* 15 Or peck Anne's snow-clear cheek. 1919 R. GRAVES *Treasure Box* 11 Where Sweetheart, my brown mare,..May loll her leathern tongue In snow-cool water. 1964 J. MICHIE tr. *Horace's Odes* I. xii. 41 The snow-cool shoulder Of Haemus. 1920 T. S. ELIOT *Ara Vos Prec* 25 Buried beneath some snow-deep Alps. 1972 'M. YORKE' *Silent Witness* ii. 26 A small figure lightly encased in snow-proof garments. 1978 J. COWLEY in *Islands* (N.Z.) Aug. 25 Padded nylon windbreakers and snow-proof pants. 1924 E. SITWELL *Sleeping Beauty* xvi. 54 Far from snow-soft sleep. 1959 E. POUND *Thrones* civ. 92 The small breasts snow-soft over tripod.
9. a. snow-belt *U.S.* [*BELT *sb.* 5 a], a region subject to heavy snowfalls; also *attrib.*; **snow-break** (later examples); **snow bunny** *N. Amer. slang*, an inexperienced (usu. female) skier; a pretty girl who frequents ski slopes; **snow-cone** *U.S.* (see quot. 1969); also *attrib.*; **snow course,** a line along which the depth of snow is periodically sampled at fixed points; **snow cruiser** *N. Amer.*, a motor vehicle designed to travel over snow; *spec.* (with capital initials) a Canadian proprietary term for a type of motorized toboggan; also *attrib.*; hence **snow-cruising** *vbl. sb.*; also *attrib.*; **snow devil,** a column of snow whirled round by the wind (cf. DEVIL *sb.* 11); **snow-dropper** (examples); **snow-dropping** (further examples); also as gerund; **snow-eater** *Meteorol.* [tr. G. *schneefresser*], a warm wind, esp. a föhn, that causes rapid melting of snow; **snow grain** *Meteorol.*, a small, opaque, precipitated ice particle, usu. flattened and less than 1 mm. in diameter, that does not bounce on a hard surface; cf. *snow pellet* below; **snow gun** *U.S.* = **snow-maker*; **snow-hole,** (*b*) a hole in snow used as a temporary shelter; **snow job** *slang* (orig. *U.S.*), a concerted attempt at flattery, deception, or persuasion; also *attrib.*; hence **snow-job** *v. trans.*, to do a snow job on (someone); **snow-jobbing** *vbl. sb.*, the performing of a snow job; **snow machine** *N. Amer.*, a motor vehicle designed to travel over snow; also *attrib.*; **snow-maker** (orig. *U.S.*), a device used for the artificial production of a snow-like precipitate for ski-slopes and the like; also, one who makes snow by the use of such a device; so **snow-making** *vbl. sb.* and *ppl. a.*; **snow-melt,** the melting of fallen snow; also, the water that results; **snowpack** *U.S.*, lying snow that is compressed and hardened by its own weight; **snow pellet** *Meteorol.*, an opaque precipitated ice particle, usu. a few millimetres in diameter, that will bounce on a hard surface; a soft hailstone; cf. *snow grain* above; **snow plane** *N. Amer.*, a type of snowmobile that is mounted on skis and propelled by an engine-driven propeller; **Snow Queen,** the chief character in a fairy-tale of this name by Hans Christian Andersen, used allusively to designate a cold-hearted woman; also *attrib.*; **snow-raking** *N.Z.* (see quots.); **snow roller,** a cylinder of snow formed by the action of the wind rolling it along; **snow scene,** a landscape covered with snow; **snow-skiing** *vbl. sb.* = SKI-ING *vbl. sb.* (*SKI-ING *vbl. sb.* 1), opp. to water-skiing; so **snow-ski** *v. intr.*; **snow-skier; snow-snake(s)** *N. Amer.*, 'an Indian game played with a straight wooden rod having a weighted head resembling that of a snake, this rod being slid over a smooth field of snow or down specially constructed runways; the rod used in this game' (Dict. Canad.); hence **snow-snaking; snow-sports,** sports that take place on snow, *spec.* skiing; also *attrib.*
1874 *Los Angeles County Ten Thousand Questions Answered* 11/1 There are two great continental railroad routes within the snowbelt. 1933 *Amer. City* Sept. 53/1 Old-fashioned winters have not been as prevalent in the snow belt in the last few years as they were ten or twenty years ago. 1967 *Wall St. Jrnl.* 1 Feb. 1/4 Some makers predict snowmobile sales soon will surpass boat sales in snowbelt states. 1981 *Nordic Skiing* Jan. 39/1 Thanks to a 120–140 inch snowbelt location, Temple Mountain offers skiing from early December to mid-April. 1895 W. R. FISHER tr. *Hess's Forest Protection* 482 The term *snow-break* is used to denote the breakage of stems or branches. 1905 *Terms Forestry* (U.S. Dept. Agric. Bureau Forestry) 21 *Snowbreak.* 1. The breaking of trees by snow. 2. An area on which trees have been broken by snow. 3. Shelterbelt. 1928 R. S. TROUP *Silvicultural Systems* v. 70 Its uneven-aged condition up to the pole stage is considered to be a protection against both snowbreak and sliding snow on steep hill-sides. 1933 *Forestry* VII. 146 In spite of the relatively high elevation there was no indication of snowbreak. 1953 P. C. BERG *Dict. New Words in Eng.* 147/2 *Snow bunny,*..n. Skiing. A beginner, esp. a girl. 1964 *Star Weekly* (Toronto) 19 Dec. 39/1 December used to be a dull month, but that was before our pretty Canadian snow bunnies..started brightening up the Canadian snow scene. 1968 *Globe & Mail* (Toronto) 13 Jan. 49/6 'Watching you for only two runs, I can see you're not just a 'snow bunny', Coral!' 'No, I was on the women's ski-team at college.' 1972 P. A. WHITNEY *Snowfire* (1973) vi. 100 Snow bunny..was a term applied to beginners, usually female, who haunted the slopes. 1969 *Daily Tel.* 6 June 18 A snowcone is a paper cup of flavoured shaved ice, highly popular among children. 1976 *Billings* (Montana) *Gaz.* 4 July 2-B/4 The Jolly Wagons had competition in those days from a snow-cone vendor driving an identical Cushman which contained only ice and flavored syrups. 1933 *Geogr. Rev.* XXIII. 540 It was only necessary to maintain a series of measurements carefully taken in the same spot each year. These measurements, laid out at definite intervals.., were named 'snow courses'. 1965 R. G. KAZMANN *Mod. Hydrol.* ii. 36 This type of measurement, made at frequent intervals over very elaborately organized snow courses.. is the accepted practical method of measuring solid-state precipitation. 1939 *Sun* (Baltimore) 14 Nov. 11/3 A twenty-seven-ton snow cruiser..designed to serve as an igloo on wheels to help the forces inspect vast areas of unexplored ice and snow. 1956 *Canad. Trade Mark* 102,409 13 Jan., Wares: Small engine driven snow remover. Trade Mark: Snow-Cruiser. 1966 *Canad. Geogr. Jrnl.* Sept. 79/3 Outboard Marine makes..Snow Cruiser.. a small motorized toboggan on rubber tires and skis, a variation of the original snowmobile invented by Armand Bombardier of Quebec ten years ago. 1969 *Sears Catal.* Spring/Summer 14 A tent of this type would be ideal for sportsmen, hunters and Snowcruiser enthusiasts. 1966 *British Columbia Digest* Dec. 10 (Advt.), '67 is the big year for snow cruising..and you have 3 fabulous OMC Snow Cruisers to choose from! 1968 *Globe & Mail* (Toronto) 15 Jan. 24/3 (Advt.), Wonderful snow-cruising parklands. 1932 F. S. SMYTHE *Kamet Conquered* xii. 169 From the serene skyline of Meade's Col little 'snow devils' were rising against the deepening green of the evening sky. 1962 W. H. MURRAY *Maelstrom* xiv. 183 Whirling snow-devils came charging across the plateau, driving spiculae in their faces. 1847 G. W. M. REYNOLDS *Mysteries of London* III. xxix. 85/1 A stranger looked like a snow-dropper. 1963 T. & P. MORRIS *Pentonville* viii. 190 The larcenist who steals feminine underwear from clothes-lines (the 'snowdropper') is often a pathetic object of derision and contempt. 1977 *Western Mail* (Cardiff) 5 Mar. 8/1 A 'snowdropper' is a man who steals women's underwear. 1882 *Sydney Slang Dict.* 9/2 Dick's a broker and has gone out *snow-dropping.* 1930 [see *cattle-duffing* s.v. *CATTLE 9]. 1967 *Telegraph* (Brisbane) 1 Mar. 26/4 Patfield had set out last November to steal sheets, but in the most systematic manner of 'snow dropping' (clothes-line thefts) he had stolen everything he could find. 1972 *Observer* 31 Dec. 3/4 He couldn't resist the temptation to go 'snow dropping' (stealing clothes from lines). 1886 *Science* 12 Mar. 242/2 Warm west winds answering to the 'Chinook' winds occur as far south as southern Colorado, though I have seldom heard the name 'Chinook' applied to them in this region. They are here [*sc.* in Colorado Springs] often called Pacific winds, also 'snow-eaters' and 'zephyrs'. 1933 F. H. CHELEY *Camping Out* 197 It was the Chinook wind... The Indians call it the 'snow eater'. 1967 R. W. FAIRBRIDGE *Encycl. Atmospheric Sci. & Astrogeol.* 1151/2 The rapid melting of the snow caused by the chinook ('Snow-eater') is welcomed because it frees the higher pastures. 1944 H. R. BYERS *Gen. Meteorol.* vi. 125 Granular snow, snow grains... White, opaque, snow-like grains, similar to soft hail but more or less flattened or oblong. 1967 R. W. FAIRBRIDGE *Encycl. Atmospheric Sci. & Astrogeol.* 772/1 Snow grains..neither bounce nor break when hitting the ground. 1971 *Industr. & Engin. Chem.* (*Process Design & Devel.*) Jan. 75/1 To cover a bare ski slope, 10 to 15 commercial snow guns (nozzles in which water and air are combined, usually at 100 psig) are used. 1974 *Compressed Air* Apr. 9/1 The snow-guns are 'very efficient, inexpensive and can be moved easily'. 1953 P. PROVENCHER *I live in Woods* vii. 64 To make a snow hole, dig to a depth of five feet at the foot of a steep incline or cliff. 1965 B. E. FREEMAN tr. *Vandel's Biospeleology* xiii. 195 Nivicoles, the inhabitants of snow-holes. 1978 *Daily Colonist* (Victoria, B.C.) 7 May 7/8 The six men and three women spent..three nights in snow-holes—man-made snow caves—before reaching..the summit. 1943 *Amer. Mercury* Nov. 555 There he tries a *snow job* on her (hands her a line) and if she falls for it she's been *snowed under.* 1953 K. TENNANT *Joyful Condemned* xx. 192 He..made a bee-line for the red-head. 'Now for the snow job,' Geechi murmured. 1962 'K. ORVIS' *Damned & Destroyed* xxi. 155 Are you going to snow-job me about finding substitutes? 1966 S. MORROW *Moonlighters* (1967) v. 53 Possibly her scepticism accounted for her success with the teenagers..kids were most apt to trust the adults who were immune to a snow job. 1969 C. BURKE *God is Beautiful, Man* (1970) 52 It's better to say yes or no and mean it—than to give a lot of snow job promises anyway. 1979 D. ROBINSON *Eldorado Network* xliii. 291, I just saw you do another snow job. You were in North Wales..which is why it sounds so convincing. Nice try, Luis. 1966 *National Observer* (U.S.) 19 Dec. 12/2 Democratic county chairmen hereabouts have, of necessity, worked out a terrific combination of railroading, arm twisting, and snow jobbing, not necessarily involving consent or persuasion. 1973 *Whig-Standard* (Kingston, Ontario) 14 Jan. 15/7 Roads are not for snowmobiles—the snow machines and other vehicles using the highways simply do not mix. 1976 *News Miner* (Fairbanks, Alaska) 6 Nov. B17/2 Snow machine driving, in which participants may cross miles of wintry terrain on a weekend outing. 1977 *New Yorker* 4 July 42/1 Their snow machine—Ski-Doo Alpine—rests on the floor below the furs. It goes ten miles an hour on the trail, and the two of them ride it. 1955 *N.Y. Times* 30 Jan. II. 31/4 The snow makers provided a long-needed answer on how to cope with the snowless situation..in the Southern Catskills. 1963 *Engineering* 13 Sept. 321/3 Snow-makers mix air and water under pressure and blow the resulting mixture in dense 50 ft arcs. 1965 *Economist* 25 Dec. 1416/1 While the rainmakers have been failing, for a decade or more the snowmakers have been succeeding beyond their wildest dreams and as a result..more American skiers than ever are assured of at least enough snow to try out the new skis which they have been given for Christmas. 1980 J. KRANTZ *Princess Daisy* xxvi. 461

The snow-making machines had started... The snow-makers continued to cover the path. 1954 *U.S. Pat.* 2,676,471 7 At an ambient temperature of 31°F and less, snow has been made at any pressure from 25 to 200 lbs. per square inch by varying the water pressure to give a snow making mixture. 1956 *Compressed Air Mag.* LXI. 101/3 Snow-making at Fahnestock consists..of bringing compressed air and water together at a nozzle that acts in the same manner as a paint spray gun. 1960 *N.Y. Herald-Tribune* 13 Nov. VII. 8/1 Across the country.. dozens of snow-making machines are poised, ready to transform bare hillsides into Alpine paradises. 1976 'A. CROSS' *Question of Max* i. 8 There is a damn snow-making machine on some blasted ski slope. 1927 *Q. Jrnl. Geol. Soc.* LXXXIII. 167 We arrived just as the spring snow-melt was finishing. 1941 *Yearbk. Agric. 1941* (U.S.) 560 In cleared areas snow depths are intermediate..and snow melt is rapid. 1971 W. HILLEN *Blackwater River* ii. 16 Snowmelt starting to run from exposed mountain slopes. 1979 *Field* 17 Oct. 1048/3 So far as rainfall is concerned,..the total amount of this element..in meterological records includes snowmelt. 1952 *Trans. Amer. Geophysical Union* XXXIII. 874 The water equivalent of the seasonal snow pack was observed after individual falls. 1955 *Sci. News Let.* 1 Oct. 214/3 Winter snowpack is the source of 40% of California's streamflow. 1973 R. HAYES *Hungarian Game* xxxvi. 215 Beneath the thin, brittle crust there was an inch of powder before the snowpack. 1935 *Jrnl. Faculty Sci. Hokkaido Imp. Univ.* 2nd Ser. I. 215 The snow pellet or the graupel..is one of the modified forms of snow crystal. 1967 R. W. FAIRBRIDGE *Encycl. Atmospheric Sci. & Astrogeol.* 442/2 Small hail, under 5 mm, is officially classified as ice pellets or snow pellets. 1953 R. MOON *This is Saskatchewan* ii. 9 Bob Fudge's manufacturing is not confined to snow planes. 1967 E. B. NICKERSON *Kayaks to Arctic* xix. 186 He had a snow plane—an enclosed cabin on ski runners shoved along by an aeroplane propeller in the fashion of an Everglades swamp buggy. 1972 T. MCHUGH *Time of Buffalo* xii. 145 We rented two snowplanes for a trip into the snow-bound heartland of Yellowstone Park. 1935 MARSH & JELLETT *Nursing-Home Murder* vi. 75 A very cold fishy sort of lady... A Snow Queen, in fact. 1974 L. DEIGHTON *Spy Story* xi. 111 She gave me the inscrutable Snow-queen smile. 1977 *N.Y. Rev. Bks.* 27 Oct. 14/. Charlotte was a Snow Queen who flirted coldly and shamelessly with her son. 1919 *N.Z. Jrnl. Agric.* 20 Feb. 90 After a heavy snowfall..send out as many men as can be got together..to get the sheep on to the sunny faces, where a certain amount of thaw may have taken place... This is what is generally known as 'snow-raking'. 1958 J. PASCOE *N.Z. Sheep-Station* in *People of World* 1st Ser. 19 Then the men must stamp out a trail through the snow—a job called 'snow-raking'—and lead the sheep down to the valley flats. 1866 G. J. SYMONS *British Rainfall, 1865* p. vii, Snow Rollers... The snow ripples up.., and the ripples breaking into sections, the wind rolls each..until, just like a..snow-ball, they rapidly increase in size. 1876 *Meteorol. Mag.* XI. 52 This is the first instance recorded of the formation of 'Snow Rollers' in England. 1959 *Weatherwise* XII. 63/2 The area cleared of snow during the formation of snow rollers is usually V-shaped, accounting for their peculiar shape, which is cylindrical with concave ends. 1836 H. C. ROBINSON *Diary* 15 Jan. (1967) 152, I found a snow scene quite pleasant in this mountainous country. 1921 R. FRY *Let.* 14 Dec. (1972) II. 518 A stupendous Courbet snow scene. 1978 'L. BLACK' *Foursome* i. 6 It was incongruous against the background of..correspondence files..stacks of catalogues, the snow-scene on the calendar. 1975 *New Yorker* 1 Sept. 28/1 You don't play tennis, you don't snow-ski, you don't water-ski, you don't ride a bicycle... Albert, we have nothing in common. 1941 *Life* 4 Aug. 55/2 (*caption*) Bending her knees like a snow skier, Hallie rides over the wake. *Ibid.* 54 Combining aquaplaning and snow skiing, water-skiing was imported from the Riviera several years ago. 1977 *Chicago Tribune* 2 Oct. XII. 33/3 (Advt.), We're looking for a bright, enthusiastic gal, who knows the retail clothing business, especially snow skiing attire. 1844 *Chambers's Edin. Jrnl.* I. 327 They [*sc.* Cherokee Indians]..in winter amuse themselves with their snow-snakes, which are long smooth sticks of hard wood..which they send to an extraordinary distance over the smooth surface of the snow. 1888 *Trans. R. Soc. Canada* VI. II. 44 If this is the game spoken of by other writers as 'Snow-snakes', there is nothing in the [Abenaki] name to so indicate. 1959 E. TUNIS *Indians* 56/2 Snow snake was played by all the northern tribes on a level track made by dragging a log or a boy through the snow. 1973 M. CROWELL *Greener Pastures* 81 The wall photograph..of Indians playing the venerable game of snow-snake. 1978 *Whig-Standard* (Kingston, Ontario) 11 Feb. A8/1 The snow snake is a smooth, thin stick about 2m long. It is thrown along a crust of smooth, hard snow. The player whose snake slides the farthest is the winner. 1979 *Ibid.* 1 Feb. 9/1 It is called snow-snaking and the Mohawk Indians have played it for centuries. It is not recognized at the Canada Winter Games but maybe it will some day. [1905 *Country Life* Dec. 181 (*heading*) Practical side of snow and ice sports.] 1966 *Guardian* 15 Oct. 5/2 (Advt.), Snowsports. 2 weeks including full-board £29.15.0! 1974 *Country Life* 3/10 Jan. 52/1 (Advt.), *Off-season winter rates*..for skiers and snow-sports enthusiasts.
b. snow bear, a buff or brown bear, *Ursus arctos isabellinus*, found in the Himalayan region; **snow-camel,** the Bactrian camel, *Camelus bactrianus*; **snow flea,** esp. one of the genus *Achorutes*; (earlier and later examples); **snow-wolf,** a wolf that lives in snowy regions; the (imitation) fur of this animal; **snow-worm:** esp. = *ICE-WORM a; (later examples).
1869 A. A. KINLOCH *Large Game Shooting Thibet & N. West* I. xv. 46 The Snow Bear varies a good deal in size. 1884 R. A. STERNDALE *Nat. Hist. Mammalia India & Ceylon* 111 The bear of which we have the oldest record is almost the same as our Indian or Snow Bear. 1910 *Blackw. Mag.* Oct. 433/2 One of them..got three really good heads, and two snow-bears, in one day. 1901 KIPLING *Kim* viii. 204 Nor is even a Balkh stallion..of any account in the great Northern deserts beside the snow-camels I have seen. 1850 THOREAU *Jrnl.* 16 Dec. in *Writings* (1906) VIII. 125 The snow everywhere was covered with snow-fleas like pepper. 1868 *Amer. Naturalist* II. 53 The little insects called snow-fleas..are found in winter at the foot of trees. 1943 B. DAMON *Sense of Humus* 106 Snow fleas..have a disagreeable habit of putting an end to their brief existence by drowning themselves in sap buckets. 1910 W. DE LA MARE *Three Mulla-Mulgars* 192 So brave are these snow-wolves. 1976 *Sunday Mail* (Glasgow) 28 Nov. 46/1 (Advt.), De luxe heavy pile Silver Mink, Ocelot, Tiger, Snow Wolf, they are beautiful. 1899 H. G. BRYANT in *Proc. Acad. Nat. Sci. Philadelphia* 134 The snow-worms were first observed a few hundred yards from our first camp. 1916 *Trans. Amer. Microsc. Soc.* XXXV. 102 Nothing definite is known concerning the food of these snow-worms.
d. snow bush, esp. the small silvery shrub, *Calocephalus brownii*, of the family Compositæ, native to Australia; (examples); hence *snow-bushed* adj.; **snow-grass,** substitute for def.: one of several coarse grasses of upland regions, esp., in New Zealand, a tussock grass of the genus *Danthonia*; cf. *DANTHONIA; also *attrib.*; (further examples); **snow gum,** a shrub or small tree, *Eucalyptus niphophila*, with white bark and glaucous leaves, native to high regions of New South Wales; **snow lily,** a perennial herb, *Erythronium grandiflorum*, belonging to the family Liliaceæ, native to alpine regions of western North America, and bearing white or yellow flowers; **snow pea** = *MANGE-TOUT; **snow plant,** (*b*) *Sarcodes sanguinea*; (earlier and later examples).
1909 A. E. MACK *Bush Calendar* 12 Where the trees were fewer, 'snow bushes' grew white. 1965 *Austral. Encycl.* III. 158/1 Snow-bush, a dense and intricately branched shrub..forms large and rounded, white-woolly growths. 1946 DYLAN THOMAS *Deaths & Entrances* 28 And the dancers move On the departed snowbushed green. 1906 T. F. CHEESEMAN *Man. N.Z. Flora* 887 Snow-grass. 1918 F. W. HILGENDORF *Pasture Plants & Pastures N.Z.* ii. 42 Snow Grass (*Danthonia raoulii*).—This is another Tussock, growing 4 to 6 feet high. It has broad leaves shining below, and feathery oat-like heads... Its presence in quantity frequently marks the limit above which it is not safe to carry sheep in winter..as indeed its popular name of Snow Grass would indicate. 1930 L. G. D. ACLAND *Early Canterbury Runs* 1st Ser. vi. 131 When he was first thatching the cob house..he put the top of each bundle of snow-grass outside the bottom of the one above so that all the rain ran inwards. 1968 *N.Z. Listener* 10 May 10/4 The beast, a young stag, had its antlers hopelessly entangled in the tough-rooted snowgrass. 1972 P. NEWTON *Sheep Thief* ii. 18 The roof consisting of bare birch rafters with a thick layer of snow grass thatch. 1928 'BRENT OF BIN BIN' *Up Country* xiv. 237 The snow-gums stood like brides in veils of perfumed lace. 1964 D. STEWART in R. Ward *Penguin Bk. Austral. Ballads* 278 Hard to say where he came from—..out of a hollow snowgum Or out of a granite boulder. 1981 *Garden* CVI. 275/1 There are..very large trees of the Tasmanian snow gum at Inverewe in Ross-shire. 1907 S. BROWN *Alpine Flora of Canadian Rocky Mountains* 44 (*heading*) *Erythronium grandiflorum* Pursh. *Snow Lily.* 1936 D. MCCOWAN *Animals Canad. Rockies* xxix. 250 Great quantities of the bulbs of Snow Lilies. 1972 *Islander* (Victoria, B.C.) 2 Apr. 13/3 The snow lily..pops its bright yellow head out as soon as the snow has left the hill-sides. 1949 *Nature Mag.* XLII. 35/2 The snow pea.. is commonly listed by all large seed-firms as an edible-podded pea. 1956 'E. MCBAIN' *Cop Hater* (1958) xx. 172 Chinese vegetables; luscious snow peas, and water chestnuts. 1978 *Times* 17 July 14/3 We had a prolific crop of sugar peas, which the Americans call snow peas. 1870 *Old & New* Mar. 349/2 The strange snow-plant.. must be passed as a railroad traveller passes all things. 1940 *Oregon: End of Trail* 20 Deeper in the forest grow the waxy Indian pipe, the blood-red snow plant, and the rare moccasin flower. 1959 MUNZ & KECK *California Flora* 436 Snow Plant. Red fleshy usually pubescent saprophyte.

snow, *v.* Add: **4. b.** *fig.* To deceive or win over with plausible words; to kid, to dupe. Also with *under. slang* (orig. and chiefly *U.S.*).
1943 [see *snow job* s.v. *SNOW *sb.*[1] 9 a]. 1945 D. DEMPSEY in M. Mayorga *Best One-Act Plays of '44* 18 Give me the lid, Greenberg..who you tryin' to snow, Lou-i-siana? 1956 'E.S. AARONS' *Assignment Treason* (1967) v. 43 Were you snowing me about Hackett doing the clobber job on you? 1963 N. FREELING *Because of Cats* xi. 175, I won't get mad. Just don't snow me with any sob-sister business. 1966 H. WAUGH *Pure Poison* (1967) xiv. 87 Roger'd be alone in a corner with some girl and..looked like he was really snowing them. 1980 *Australian* 9 Dec. 6/5 Mr J. C. Moore (the new minister in charge of the Customs Bureau) has taken the most immediate and active interest in the workings of the bureau. Unfortunately, it is most likely that he also will be snowed by the bureaucrats as has been the case with previous ministers.
6. d. With *in*. To block, imprison with snow. Chiefly *N. Amer.*
1857 G. F. MCDOUGALL *Eventful Voy. 'Resolute'* xiii. 331 The fore and after parts of the upper deck were now snowed in, to the depth of nine inches on the starboard side. 1887 C. B. GEORGE *40 Yrs. on Rail* ix. 188 My train was snowed in during one of the terrible storms. 1970 *Daily Colonist* (Victoria, B.C.) 1 Jan. 1/3 Picture above taken a year ago as worst blizzard in years blanketed area shows cars snowed-in on King's Road.
7. *U.S. slang.* To drug, to dope. Also with advbs. Usu. in pa. pple.
1927 *Amer. Speech* Dec. 167/2 *Snowed in,* dopey, as if full of cocaine. 1934 R. CHANDLER in *Black Mask* July 70/2 She looked snowed, weaved around funny. 1942 BERREY & VAN DEN BARK *Amer. Thes. Slang* § 509/30 *Snowed, snowed in, up* or *under,*..under the influence of cocaine. 1956 H. GOLD *Man who was not with It* xxiii. 222 But I figured on how to get snowed.

snowball, *sb.* Add: **1. b.** Phr. *a snowball's chance in hell*: see *HELL *sb.* 10 b. Also ellipt. as *a snowball's chance.*
1934 *Esquire* Sept. 27 He wouldn't have a snowball's chance with you. 1977 *Amer. Machinist* 1 June 27 There is not a snowball's chance in Haiti of making the deadline on an across-the-board basis. 1979 'A. HAILEY' *Overload* I. i. 4 'Told 'em there wasn't a snowball's chance,' a woman assistant dispatcher called over.
e. *transf.* A scheme or project that relies for its growth on a snowball effect (see quots.).
1892 *Whitehall Rev.* 17 Sept. 7/1 The system of 'Snowballs' is multiplication at a very rapid rate, each giver being obliged to bind himself to find a certain number of others who will not only give, but bind themselves each to find an equal number of contributors on the same terms. 1923 H. C. BAILEY *Mr Fortune's Practice* v. 141 It's just like a snowball... When you want subscriptions and have a snowball where every one has to get some one else to subscribe. 1927 E. F. BENSON *Lucia in London* iii. 70 Will she just pick up acquaintances, and pick up more from them, like one of those charity snowballs?
f. In bingo, etc.: a cash prize which accumulates through successive games until it is won.
1949 S. P. LLEWELLYN *Troopships* 5 Last house... May I remind you, gentlemen, that the snowball is now worth over fourteen pounds? 1960 *Guardian* 2 Dec. 23/5 The British Legion..club..was more or less built on Bingo... The crowds, drawn by a 'snowball' on a lucky number which had reached £16, had been growing..too large. 1971 A. ROSS *Huddersfield Job* 129 The snowball—a sort of continuing competition in which the cash prizes, if not won, are carried forward to swell next week's total. 1976 *Evening Post* (Nottingham) 15 Dec. 13/2 Tote Baseball Nos. 20 & 13 & 6 Three winners. Snowball not won.
2. b. One of various cocktails (see quots.).
1930 *Savoy Cocktail Bk.* 150 Snowball Cocktail. ⅙ Crème de Violette. ⅙ White Crème de Menthe. ⅙ Anisette. ⅙ Sweet Cream. ½ Dry Gin. 1963 D. A. EMBURY *Fine Art of Mixing Drinks* (ed. 2) 289 Snow Ball. A Silver Fizz with whisky in place of the gin and ginger ale in place of the charged water. 1966 J. DOXAT *Booth's Handbk. Cocktails & Mixed Drinks* xiv. 145 Snowball. Ice cube in tall glass. Generous measure of Advocaat; top with Fizzy Lemonade; decorate with slice of Lemon. 1972 A. DRAPER *Death Penalty* ii. 16 Ben ordered the drinks—a snowball for Jeannie and whisky mac for himself. 1979 R. BARNARD *Posthumous Papers* xvii. 158 She ordered a snowball... 'I'm not used to coming into a pub on my own.'
c. *U.S.* and *W. Indies.* An ice-cream; a confection made of shaved or chipped ice covered in syrup, etc.
1941 J. SMILEY *Hash House Lingo* 51 *Snowball,* dip of vanilla ice cream. 1946 K. DUNHAM *Journey to Accompong* 93 My Maroon neighbors..were lolling around..drinking the penny 'snowball' made from chipped ice with a sweet purple syrup poured over it. 1953 H. P. MORRISON in *Caribbean Anthol. Short Stories* 137 Customers of every age milled round to buy 'snow-ball'—cool crushed ice in cheap glass tumblers with red, yellow or even green syrup oozing slowly through the crystalline mass. 1962 [see *MAUBY].
4. a. Also used for other species of *Viburnum.* (Later example.)
1948 W. ARNOLD-FORSTER *Shrubs for Milder Counties* iv. 184 *V. Opulus sterile,* the familiar 'Snowball', is..quite good as a hedge.
5. a. *snowball cocktail, fight, vendor.*
1930 Snowball cocktail [see sense 2 b above]. 1890 CHAMPLIN & BOSTWICK *Young Folks' Cycl. Games & Sports* 660/1 *Snowball fights,* contests between two parties armed with snowballs. 1948 *Sun* (Baltimore) 27 Aug. 24/3 Snowball vendors did a rush business.
b. snowball bush, tree, also = sense 4 a in Dict. and Suppl.; (later examples).
1931 W. N. CLUTE *Common Names Plants* 48 Guelder rose, a common name of the snow-ball bush.., is said to be properly elder rose. 1979 *Seymour* (Indiana) *Daily Tribune* 19 May 1/3 The 'snowball bush' in his side yard is in full bloom. 1902 E. T. COOK *Trees & Shrubs for Eng. Gardens* 443 Snowball tree..is too well known to need description. 1973 A. BONAR *Shrubs & Decorative Trees* III. 86 The snowball tree..is more attractive florally.
c. *snowball effect, prize.*
1941 I. L. IDRIESS *Great Boomerang* xxxii. 251 It will not be the amount to be spent that will be considered, but the snowball effect of the resulting benefits. 1963 *Daily Tel.* 23 Jan. 20/8 A 'snowball' prize played for evening after evening at a bingo club is legal, provided the management gives the prize money. 1979 P. NIESEWAND *Member of Club* xviii. 142 Hundreds of families emigrate [from South Africa] every month... Each one has a snowball effect. Other families start thinking: should we leave also?

snowball, *v.* Add: **1. b.** *fig.* To increase or grow like a snowball rolled across snow; to accumulate or gather momentum at an ever-increasing rate.
1929 E. N. NICHOLSON *Study of Birds* 39 Some flocks are freshly formed each day, and recruits can be watched joining the original members at intervals until it snowballs up to its full size. 1934 *Sun* (Baltimore) 9 Nov. 26/7 The [housing] program in Maryland is 'snowballing'. 1967 R. LEHMANN *Swan in Evening* III. 104 The success of those classes delighted and amused her. How polyglot they became and how they snowballed. 1969 *New Yorker* 19 Apr. 94/2 When a man knows what to look for, his value snowballs. 1973 *Lebende Sprachen* XVIII. 69/2 Management must appreciate the extra profit that snowballs from making use of advanced techniques. 1976 *Ilkeston Advertiser* 10 Dec. 15/4 Anyone is welcome to join in at any time during the day. A coach will ferry people around the circuit and singers usually 'snowball' throughout the day.
2. (Earlier example.)
1850 L. SAWYER *Way Sk.* (1926) iii. 46 Our men amused themselves with snowballing each other.
3. (Earlier example.)
1852 F. A. BUCK *Let.* 18 Dec. in *Yankee Trader* (1930) 112 At first we snow-balled, the whole town engaging in the sport like school boys.
Hence **snowballing** *vbl. sb.* (earlier and later examples); **snow-balling** *ppl. a.*
1861 F. A. BUCK *Let.* 20 Jan. in *Yankee Trader* (1930) 186 Christmas we had a nice lot of egg nog and cake and snow balling. 1941 *Sun* (Baltimore) 3 Nov. 14/1 The constantly snow-balling defense effort may cut into the everyday things we use in normal civilian existence. 1966 *Word Study* Dec. 4/1 Dubious meaning..starts a snowballing that soon places the intended meaning beyond retrieval. 1971 *Daily Tel.* 29 Dec. 10 The snowballing success of Alan Ayckbourn's plays abroad..is a constant surprise to him. 1973 *Times* 14 Aug. 3/3 The 'snowballing' technique by which researchers were introduced to one drug taker; who introduced a second and so on. 1977 *N.Y. Times* 16 Jan. IV. 19/3 Mr. Kissinger's pet theory of 'linkage', a kind of snowballing of détente, had to be given up.

snowberry. 2. (Earlier example.)
1813 T. JEFFERSON *Let.* 8 Dec. in *Orig. Jrnls. Lewis & Clark Exped.* (1905) VII. 393 We call it the snow-berry bush, no botanical name being yet given to it.

snow-bird. Add: **3.** *U.S. slang.* One who sniffs cocaine (cf. *SNOW *sb.*[1] 4 d); *gen.* a drug addict.
1914 JACKSON & HELLYER *Vocab. Criminal Slang* 78 A 'snowbird' is the customary designation of the cocaine habitue. 1923 [see *LOADED *ppl. a.* 3 b]. 1952 *Sunday Times* 3 Feb. 5/4 Present-day New York is not..a city overrun by 'snowbirds' jabbing needles into their arms. 1963 'M. CORRIGAN' *Why do Women—?* xxiii. 175 Don't tell me you never heard that name for a dope addict—a snowbird.
4. *U.S. slang.* **a.** (See quots.)
1905 *N.Y. Even. Post* 20 Nov. 6, 28 per cent. deserted after three months, and were presumably 'snow-birds', that is, men who enlist to get food and clothing during the winter months. 1918 *Sat. Even. Post* 23 Nov. 11/1 They belonged to a shiftless class, the members of which often enlist in the army late in the fall because they want a job for the winter—the boys call them snowbirds. 1930 W. H. WALDRON *Old Sergeant's Conferences* vii. 123 A 'Snow bird' is a deserter who surrenders in the fall to get a place to stay through the winter.
b. (See quot. 1924.)
1923 *Nation* 31 Oct. 487 In winter, when building is at a standstill in the North, northern workmen, 'snow birds' or 'white doves' in Negro parlance, flock south. 1924 'DIGIT' *Confessions 20th Century Hobo* 12 *Snowbird,* in the Southern States a Northerner who migrates south to avoid the winter. 1962 *Economist* 22 Dec. 1206/1 The Negro, who regularly loses his job to the 'snowbirds' from New York in the winter holiday season. 1979 *United States 1980/81* (Penguin Travel Guides) 243 This figure swells..during the winter months when 'snowbirds' arrive. ('Snowbird' is a tricky term as used in Miami, it refers primarily to tourists escaping the Northeastern freeze.)
5. *colloq.* A person who likes snow; a snow-sports enthusiast.
1928 D. H. LAWRENCE *Let.* in F. Lawrence *Not I* (1934) 269, I am no snow-bird, I hate the stark and shroudy whitemen, white and black. 1973 *Globe & Mail* (Toronto) 8 Dec. 43/8 No joy yet for snowbirds. Snow enthusiasts will have to wait at least one more week before they can start up their snow-mobile engines or put on their skis.

snow-blind, *a.* Add: Also *fig.*
1946 DYLAN THOMAS *Deaths & Entrances* 26 It is a winter's tale That the snow blind twilight ferries over the lakes.

snowcat (snǭ·kæt). orig. *N. Amer.* [Respelling of *SNO-CAT.] A tracked vehicle designed for travelling over snow (heavier and more rugged than a snowmobile). Cf. *SNO-CAT.
1955 *Sun* (Baltimore) 10 Dec. 2/3 Two tractor-treaded snowcats will set out for a point in Byrd Land about 600 miles away. 1960 *Maclean's Mag.* 2 Jan. 15/2 The snow is brushed and groomed between one day's skiing and the next by tread-driven machines called snow cats. 1971 *Country Life* 25 Feb. 436/2 The final requiem to Isaac's services came with the Snowcat, a sort of caterpillar contrivance that could do most things bar swim. 1973 *Observer* (Colour Suppl.) 21 Jan. 35/2 You can..make the magical and distinctly tricky trip by snowcat across Iceland's vast Vatnajokull Glacier.

Snowcem (snǭ·sem). Also **snowcem.** [f. SNOW *sb.*[1] + CEM(ENT *sb.*] The proprietary name of a cement-based (typically white) paint, used for covering external walls. Hence **Snow-cemmed** *a.*, painted with Snowcem.
1939 *Trade Marks Jrnl.* 20 Sept. 1301 Snowcem. 608,464. Paints having a base of cement. The Cement Marketing Co. Ltd...—1st Aug. 1939. 1947 *E. African Ann.* 1946–7 120 (Advt.), Building Materials..Snowcrete, Colorcrete and Snowcem. 1966 D. FRANCIS *Flying Finish* i. 13 Its drab walls..badly needed a coat of 'Snowcem'. 1969 R. BLYTHE *Akenfield* 19 The old farmsteads, snowcemmed and trim, ride high on the hills. 1972 J. BLACKBURN *Devil Daddy* i. 15 Batterday shamed his neighbours with Snowcemmed walls. 1979 B. HINES *Price of Coal* 25 I've heard they've ordered ten thousand gallons of Snowcem to whitewash the shaft with.

snowdrop. Add: **4.** *slang.* An American military policeman; hence, any military policeman.
1944 *N.Y. Times* 9 Apr. IV. 1/7 'Snowdrops'—the London nickname for white-helmeted American military police—were patroling the sidewalks. 1946 H. NICOLSON *Diary* 30 Apr. (1968) 59 Schacht sits in the witness box ..flanked by two young Americans in white helmets. Every hour, two other snowdrops..take over from their comrades the white batons of office. 1967 'A. CORDELL' *Bright Cantonese* ix. 100 Where are you goin'?..Running to the first Snowdrop you can find? 1978 D. KYLE *Black Camelot* xvii. 270 Special detachments of military police 'snowdrops' went from house to house.

snowdrop tree. 2. For *Halesia tetraptera* substitute 'of the genus *Halesia*, bearing clusters of drooping white flowers'. (Earlier example.)
1813 H. MUHLENBERG *Catal. Plant. Amer.* 46 Silverbell tree, or Four-winged Snow-drop tree.

snowed, *ppl. a.* Add: **3.** Also *snowed-in* (see *SNOW *v.* 6 d).
1904 *N.Y. Even. Post* 5 Feb. 3 The Wabash is devoting all its energies to clearing the line of delayed and snowed-in trains. 1982 N. FREELING *Wolfnight* 128 A winter sun on a snowed-in landscape.

snowfall. Add: **1.** Also *fig.*
1964 M. A. JOHNSON in *Oceanogr. & Marine Biol.* II. 31 In recent years there has been a fundamental change in our picture of the deep-sea floor; previously conditions were thought to be essentially 'static', with negligible currents and the only variation with time being the steady 'snowfall' of sediment to the sea bed. 1969 MRS. L. B. JOHNSON *White House Diary* 10 Jan. (1970) 762 Dinner became a very snowfall of menu cards being passed around the tables for autographs.

snowflake. Add: **6.** = *hair-line crack* s.v. *HAIR-LINE 7. *U.S.*
1919 *Bull. Amer. Inst. Mining Engineers* Feb. 183 The appearance of 'snow-flakes' is unmistakable... The white silvery area, which always has the appearance of being of a very coarsely crystalline structure, in the specimen stands out in bold contrast to the darker background, and readily justifies the use of the term 'snow-flakes'. 1925 [see *hair crack* s.v. *HAIR *sb.* 10]. 1942 [see *fish-eye* s.v. *FISH *sb.*[1] 7].
7. *attrib.*, as **snowflake curve** *Math.*, a mathematically conceived curve (see quot. 1975) whose sixfold symmetry is reminiscent of that of a snowflake, of interest because its infinite length bounds a finite area.
1956 W. G. WALTER *Further Outlook* III. iv. 100 Jim Bursley had explained the snowflake curve to me and we had discussed the projection of such a curve into three dimensions. 1975 *Sci. Amer.* Nov. 144/2 Take the analyst's 'snowflake' curve.., which is made in an elementary way from an equilateral triangle of unit side by replacing the middle third of each side with a 'cape', itself the two jutting equal sides of a triangle a third as large as the original, and so on, repeating indefinitely. 1978 *Ibid.* Apr. 21/2 Among the fractals that exhibit strong regularity the best-known are the Peano curves that completely fill the finite region and the beautiful snowflake curve discovered by the Swedish mathematician Helge von Koch in 1904.

snow-man. Add: **1. b.** *Archæol.* Used *attrib.* and *absol.* to designate a technique of clay-modelling (see quot. 1955) or the figurines so produced.
1908 [in Dict., sense 1]. 1955 L. WOOLLEY *Alalakh* viii. 244 The vast majority [of figurines]..were hand-modelled more or less in the round in what is called the 'snow-man' technique, i.e. the clay is pinched into shape with the fingers and details are added by sticking on small pellets of clay, as well as by incision with a blunt stick or..by dotted lines made with a roulette. 1962 D. HARDEN *Phoenicians* ii. 42 The site..has produced remains of the seventh century, including some surprisingly primitive snow-man figurines of clay. 1974 J. CHESTERMAN *Classical Terracotta Figures* ii. 29 A group from Cyprus.., known as 'snowmen', was being produced in relative abundance... I think they are jolly little fellows with their peaked caps, pointed beards and stumpy arms.
c. Used *attrib.* and *absol.* to designate a type of pottery figure (see quot. 1957).
1931 D. MACALISTER in *William Duesbury's London Account Bk. 1751–53* p. xxv, I have illustrated several figures in white porcelain... Certain figures of the 'snow-man' type belong to this class. 1933 *Trans. Eng. Ceramic Circle* I. 46 The 'snow man' bag-piper... Another 'snow man' in white porcelain... He looks like a *poilu*, but may be meant for a Chinaman. 1957 MANKOWICZ & HAGGAR *Encycl. Eng. Pottery & Porcelain* 205/1 '*Snowman' figures,* porcelain figures heavily glazed with a thick, opaque, glassy glaze, obscuring the modelling,..now known..to be the production of William Littler of Longton Hall, or of Jenkinson at the same works. 1974 *Encycl. Brit. Micropædia* IX. 303/2 *Snowman porcelain...* Called snowmen because of their thick white enveloping glaze, they include figures of human beings and animals.
3. *Abominable Snowman*: see *ABOMINABLE *a.* 1 c. Also simply *snowman.*
1931 J. CANNAN *Ithuriel's Hour* iv. 131 His gods, to say nothing of hairless Snow Men, and the shades of his ancestors. 1937 *Times* 31 Dec. 8/3 Mr. Smythe says that the snowman superstition is known only to the Tibetan or semi-Tibetan peoples. 1959 *Times* 8 Jan. 13/5 In the case of the famous Himalayan 'snowman' the evidence.. seems to point to the existence of perhaps two unknown species..of anthropoid ape.
4. *U.S. slang.* One who snows (*SNOW *v.* 4 b) someone.
1967 P. MCGIRR *Murder is Absurd* iii. 44 You're a great snow man, Warren. But I'm not in dreamland yet. 1977 *Amer. Speech 1975* L. 66 *Snowman,* male who easily wins the affections of females.

snowmobile (snǭ·mŏbīl). orig. *N. Amer.* Also **snow-mobile, snow mobile, snomobile, Snomobile.** [f. SNOW *sb.* + *AUTO)MOBILE *a.* and *sb.*] Any motor vehicle designed for travelling over snow; *spec.* a small, light passenger vehicle supported on runners at the front and a traction chain at the rear. Also *attrib.*
1931 *Times Lit. Suppl.* 5 Feb. 89/1 The American expedition to the South Pole under..Admiral Byrd was.. carried out in the grand manner... It had a 'snowmobile'; but this did not travel very far before meeting disaster on the rough surface. 1934 *Canad. Patent Office Rec.* 30 Oct. 2464/1 A snowmobile comprising an automobile having a pair of runners in place of the automobile front wheels, beams extending longitudinally of the automobile and rearwardly of the rear axle, a third axle mounted adjacent the rear ends of said beams, wheels forming the sole supporting means for the rear of the snowmobile journalled on said third axle, driving gears mounted on the automobile rear axle in place of the ordinary rear wheels, and traction chains connecting said wheels and driving gears. 1947 *Times* 8 Mar. 5/6 They [*sc.* the Eskimos] have seen..drums of petrol parachuted to the snowmobiles of the Canadian Army exercise 'Musk-Ox'. 1961 *Times* 24 Apr. 16/6, I can..watch the teenagers race their stripped down jalopies... There are snow-mobiles and snocats. 1966 *Popular Science* Jan. 117 Snow really flies as this snowmobile owner takes his Larson Eagle on a long, thrilling jump. 1968 *Globe & Mail* (Toronto) 3 Feb. 46/4 (Advt.), Snow mobiles, demonstrator sale. 1969 *Daily Tel.* 11 Mar. 24/8 Organisers of a fox-hunt, mounted on snowmobiles, on a frozen lake in Minnesota have been threatened with police action. 1970 *Toronto Daily Star* 24 Sept. 10/4 (Advt.), Snowmobile suits on sale. 1970 *Nature* 29 Aug. 880/1 It is powered by..air-cooled engines originally developed for the 'sno-mobiles' now so popular during the Canadian winter. 1972 C. MUDIE *Motor Boats & Boating* 82 The question of how to use your Snomobile in high summer has been met by one manufacturer who markets a float attachment and a modified propeller drive so that the basic unit quickly becomes a lakeside holiday flier. 1977 *New Yorker* 2 May 60/2 Dog teams have largely been replaced by snowmobiles. 1980 *Harpers & Queen* Jan. 79/4 Snowmobiles alone use over 117 million gallons of fuel each year in the United States.
Hence **snow-mobiler,** one who drives or rides upon a snowmobile; **snow-mobiling** *vbl. sb.*, the action or sport of using a snowmobile; **snow-mobilist** = *SNOWMOBILER.
1964 *Star Weekly* (Toronto) 19 Dec. 14/1 We'd like to see a broader approach to snowmobiling as a family-style sport. 1967 *Daily Tel.* 14 Mar. 21/1 This weekend, some 11,000 'snowmobilists' were skimming across the snow-covered Adirondacks at speeds up to 30 mph, twice those on downhill slopes. 1968 *Globe & Mail* (Toronto) 13 Jan. 41/5 Snowmobilers would be well advised to obtain from the Ontario Safety League a copy of the Ten Commandments of Snowmobile Safety. 1977 *Time* 7 Feb. 60/2 The arduous route..led the snowmobilers along busy, narrow roads, through woods and across ditches. 1981 *Northeast Woods & Waters* Jan. 4/1 Years of driving and snowmobiling have prepared me and I always carry spare parts and tools and we fixed it in a jiffy.

snow-on-the-mountain. (In Dict. s.v. SNOW *sb.*[1] 5 d.) Substitute for def.:
1. *U.S.* An annual spurge, *Euphorbia marginata*, of the family Euphorbiaceæ, native to the central southern United States, and bearing white bracts.
1873 *Kansas Mag.* June 502/2 Their miller's-plant, or snow-on-the-mountain, is nothing but our spurge. 1878–80 [in Dict. s.v. SNOW *sb.*[1] 5 d]. 1918 W. CATHER *My Antonia* IV. iv. 364 Every sunflower stalk and clump of snow-on-the-mountain drew itself up high and pointed. 1966 Mrs. L. B. JOHNSON *White House Diary* 27 Aug. (1970) 416 And there is snow-on-the-mountains—everywhere—especially in the meadows.
2. One of several low-growing, white-flowered, cruciferous plants, esp. white alyssum, *Arabis alpina*, or *Alyssum maritimum.*
1882 H. FRIEND *Gloss. Devon. Plant Names* 51 Snow-on-the-Mountain. *Alyssum maritimum,* L. 1886 [in Dict. s.v. SNOW *sb.*[1] 5 d]. 1890 J. D. ROBERTSON *Gloss. Dial. & Archaic Words Gloucestershire* 145 Snow-on-the-mountain. *Arabis alpina,* L. 1933 *Downside Rev.* LI. 523 The 'snow-on-the-mountains' was in full bloom, groups of crocuses were holding out their golden fingers just behind it, and the yellow and white looked like strips of spring sunshine.

snow-plough, *sb.* Add: **2.** *Skiing.* = *double stem* s.v. *DOUBLE *a.* A. 6. Also *attrib.*

1905 RICKMERS & RICHARDSON in D. M. M. C. Somerville et al. *Ski-Running* (ed. 2) 69 Stemming is akin to snow-ploughing, and by some German writers the stemming position is termed the half-snow-plough position. 1922 V. CAULFEILD *Ski-ing Turns* vii. 134 Although it is unsafe for any one but an expert to take the snow-plough position when travelling at all fast, this Snow-plough Christiania can be done at a good deal higher speed than the Pure Snow-plough turn. 1936 [see *double stem* s.v. *DOUBLE *a.* A 6]. 1948 P. LUNN *Ski-ing Primer* xi. 56 The snow-plough..is not only a useful manœuvre in itself, but is also the basis of the stem turn. 1953 A. WOODBURN tr. *Jacques's Downhill Skiing* I. iii. 79 The snow plough position..is one of the few that the beginner finds quite natural. 1966 A. N. GOODING *Basic Ski-ing* vi. 37 You first learn how to stop by using the Snowplough Brake. 1978 *Observer* 29 Jan. 35/1 Our instructor starts us on snowplough turns (with the tips of the skis pointing inwards). 1981 *Northeast Woods & Waters* Jan. 27/2 The snowplow, the 'dinking on the ski tips inward to make turns or to stop' is definitely a beginner maneuver.

So **snow-plough** *v. intr.*, to execute a snow-plough in skiing; **snow-ploughing** *vbl. sb.*

1904 D. M. M. C. SOMERVILLE et al. *Ski-Running* 39 The Norwegians call it 'snow-ploughing', but 'stemming' is shorter, and, we think, more expressive. 1928 E. JESSUP *Skis & Ski-ing* ix. 124 Very often it suffices to brake with the skis fairly flat on snow. In such cases, bend your knees inward only slightly... When you 'snowplough' in the foregoing fashion, both of your skis are serving as brakes. 1959 P. MOYES *Dead Men don't Ski* iv. 50 They had learnt, now, to snow-plough—putting the tips of their skis together to slow down or stop. 1961 *Times* 14 Feb. 14/7 The usual portly Frenchman..who charges along, skis wide apart and braking hard by stemming or 'snowploughing'. 1979 N. SLATER *Falcon* ix. 161 He snow-ploughed down towards her..snapped off his ski-bindings.

snowshoe. Add: **3. snowshoe hare,** the North American varying hare, *Lepus americanus*; **snowshoe rabbit** = *snowshoe hare* above (further examples).

1921 *Frontier* May 11 In a zig-zag pattern in the snow were the tracks of the snow-shoe hares. 1977 *New Yorker* 9 May 96/2 We kicked at some wolf scat, old as winter. It was woolly and white and filled with the hair of a snowshoe hare. 1903 J. LONDON *Call of Wild* 90 Leap by leap, like some pale frost wraith, the snowshoe rabbit flashed on ahead. 1971 W. HILLEN *Blackwater River* iii. 21 Snowshoe rabbits, so called because their large hairy feet serve as snowshoes.

snow-shoed *a.*: also *fig.*; **snow-shoeing** *vbl. sb.* (earlier example); **snow-shoer** (earlier example).

1946 DYLAN THOMAS *Deaths & Entrances* 30 The singing breaks in the snow shoed villages of wishes. 1867 *Territorial Enterprise* (Virginia City, Nevada) 12 Mar. 3/2 A race for a gold buckle, free to all lady snow-shoers, was also announced. 1884 H. CHADWICK *Sports & Pastimes Amer. Boys* 205 One of the favorite winter sports of the Canadians is snowshoeing, which is enjoyed to a great extent by the clubs of Montreal, who engage in races and long tramps over the hills on snowshoes.

snow-storm. Add: **1.** (Earlier *lit.* and *fig.* examples.)

1771 A. G. WINSLOW *Diary* 6 Dec. (1895) 8, I was prevented dining at unkle Joshua's by a snow storm. 1869 'MARK TWAIN' *Innocents Abroad* xiii. 125 A snow-storm of waving handkerchiefs.

2. A paperweight or toy in the form of a transparent dome or globe containing a representation of a scene and loose snow-like particles, which, when shaken, creates the appearance of a snow-storm. Also *attrib.*

1926 'O. DOUGLAS' *Proper Place* xvii. 149 A round glass globe containing a miniature cottage, which, when shaken, became surrounded with whirling snow-flakes. 'It's a snow-storm,' she declared triumphantly. 1931 E. SACKVILLE-WEST *Simpson* II. 144 Salathiel held up a glass globe, inside which was a minute Scotch-baronial castle..He shook the globe and a whirlwind of white flakes swirled up.... The Snowstorm jerked downwards in his hand. 1939 C. MORLEY *Kitty Foyle* (1940) xxxii. 332 It's good to have a person call your attention to something you're so used to you almost forgot thinking about it. I mean the glass snowstorm ball. Molly's back in Chicago and I take the glass ball and give it a whirl. 1947 'D. YATES' *Berry Scene* x. 273 My eye was caught by a snowstorm—one of those little glass balls, with a baby cottage inside. And when you shake it, snow-flakes begin to fall. 1967 M. DRABBLE *Jerusalem the Golden* v. 101 Toys..a tower of bricks, a weather house, a huge pendant snowstorm globe containing a small palace and a small forest. 1975 S. LAUDER *Killing Time on Corvo* ix. 85, I recalled, as a child, staring entranced into Modrinka's snowstorm paper-weight.

3. *fig.* An appearance of dense snow on a television or radar screen. Cf. *SNOW *sb.*[1] 4 f.

1948 *Nature* 31 Jan. 167/1 The visual effect was that of a violent snowstorm of the type well known to televiewers due to motor-car ignition interference, but at a very much more intense level. 1974 L. DEIGHTON *Spy Story* xviii. 195 The radar screen was a snowstorm that dashed..in a mad rhythm. 1980 J. B. HILTON *Anathema Stone* i. 16 The television set..produced a snow-storm on every channel.

snowy, *a.* Add: **3. b.** Of the picture on a television screen: affected with snow (*SNOW *sb.*[1] 4 f).

1959 LEVY & FRANKEL *Television Servicing* xiv. 442 The picture may become weak and snowy. 1976 H. KEMELMAN *Wednesday Rabbi got Wet* xxiii. 73 She turned the set on... There was a lot of static, and the picture wavered and became snowy.

snozzle (snǫ·z'l), var. *SCHNOZZLE. Also *transf.*

1930 D. RUNYON in *Collier's* 20 Dec. 13/2 He no sooner pokes his snozzle into the joint than a guy by the name of Louie the Lug..jumps up. 1931 *Ibid.* 26 Sept. 9/2, I put the old convincer on him by letting him peer down the snozzle of my John Roscoe. 1959 I. & P. OPIE *Lore & Lang. Schoolch.* ix. 155 Children go in for short sharp words, as..'snozzle'..for nose. 1968 D. O'GRADY *Bottle of Sandwiches* (1969) iv. 60 The poor Old Girl [*sc.* a truck] was mud from anus to snozzle.

snub, *sb.*[1] Add: **I. 1. b.** *pl.* As *int.*, expressing total indifference or contempt. *slang.*

1934 *Neuphilologische Mitteilungen* XXXV. 130 Prep-school slang..*snubs* interj. accompanied by making a long nose. 1945 E. WAUGH *Brideshead Revisited* I. vi. 135 Now I shall tell her I have had it straight from a real artist, and snubs to her.

II. 4. For † *Obs.* read *rare* and add later example.

1925 W. DE LA MARE *Broomsticks* 220 Not so much as an ole scrubbin'-brush or a snub of soap.

5. Also *fig.*

1973 R. D. SYMONS *Where Wagon Led* p. xiii, A man who cannot put a snub on his temper had better leave horses alone.

snub, *sb.*[3] and *a.* Add: **A.** *sb.* **2.** *Geom.* A snub polyhedron or polytope.

1948 H. S. M. COXETER *Regular Polytopes* viii. 151 (*heading*) The snub {3, 4, 3}. 1952 CUNDY & ROLLETT *Math. Models* iii. 94 It can be proved..that..there are only thirteen Archimedean solids, two of which occur in two forms. Those two are the two 'snubs'. 1971 M. J. WENNINGER *Polyhedron Models* III. 179 Great inverted snub icosidodecahedron. This polyhedron is another snub that is simpler in construction than most of the others in this set.

B. *adj.* **3.** *Geom.* Used to designate certain symmetrical polyhedra and polytopes; in general, they have no mirror symmetry and occur in enantiomorphic pairs. [tr. L. *sīmus* squashed (Kepler *Harmonices Mundi* (1619) V. II. xxviii. 62).]

A snub cube has as its faces 6 squares and 32 equilateral triangles; a snub dodecahedron has 12 pentagons and 80 equilateral triangles: they are more nearly spherical than the cube and dodecahedron to which Kepler related them, hence the name.

1934 *Proc. London Math. Soc.* XXXVIII. 338 We might symbolize such a snub polytope by ringing all the dots in the graph. [*Note*] The word 'snub' is a free translation of Kepler's *simus*. 1952 CUNDY & ROLLETT *Math. Models* iii. 101 (*caption*) Snub cube. *Ibid.* 108 (*caption*) Snub dodecahedron. 1971 M. J. WENNINGER *Polyhedron Models* I. 32 The snub dodecahedron. This polyhedron has the same relation to the regular dodecahedron that the snub cube has to the regular hexahedron.

snubbed, *ppl. a.* Add: **2. b.** Shortened, stumpy.

1835 J. H. INGRAHAM *South-West* I. iii. 27 With swallow-tailed sterns, snubbed bows, and black hulls. 1903 *Trans. Inst. Naval Archit.* XLV. 26 Conditions in which you may have the 'snubbed' finish of the curve of areas.

snubber. Add: **3. a.** A simple form of shock-absorber used esp. in motor vehicles. Freq. *attrib.*

1921 *Daily Colonist* (Victoria, B.C.) 22 Oct. 10/7 (Advt.), Gabriel snubbers increase the riding comfort of any car. 1928 *Sunday Dispatch* 19 Aug. 8 New snubber plates are fitted on the front springs. 1956 TOBOLDT & PURVIS *Automotive Encycl.* 30d/1 The first so called 'shock absorbers' were simple rebound absorbers or snubbers. 1961 *Aeroplane* C. 429/1 The second-stage rotor blades have been fitted with 'snubbers' at approximately their mean height. 1970 *Telegraph* (Brisbane) 18 Dec. 8/1 The trouble with our washing machine..was that it needed a new snubber spring. 1973 R. L. ESHLEMAN in Snowdon & Ungar *Isolation of Mech. Vibration, Impact, & Noise* 228 The friction snubber utilizes coulomb friction between a liner material and the snubber wall. 1980 *Truck & Bus Transportation* (Austral.) Jan. 32/3 The snubber consists of an endless circle of chain encased in a solid block of rubber.

b. *Electronics.* A circuit intended to suppress voltage spikes.

1968 *IEEE Trans. Industry & General Applications* IV. 666/2 With properly damped snubber circuits the commutation transient waveshapes are such that the initial *dv/dt* is highest. 1977 *Design Engin.* July 27/1 All thyristor installations normally require a snubber, or dv/dt suppression network consisting of a capacitor and resistor in series across the device.

c. A device used to damp pulsations in, or check the flow of, a fluid.

1972 L. J. LORTIE in Blake & Mitchell *Vibration & Acoustic Measurement Handbk.* xxvii. 542 Using the simple snubber (expansion chamber)..which is made of standard piping components, an original pulse amplitude of 100 pressure units may usually be reduced to the order of ten or five units. 1977 *Sci. Amer.* June 46/2 The rate of descent of the piston is predetermined by the setting of the flow snubber in the dividing plate, which admits the water into the lower half of the cylinder.

snubbing, *vbl. sb.* Add: **3.** The action of reducing or suppressing oscillation; damping.

1951 C. E. CREDE *Vibration & Shock Isolation* iii. 112 (*heading*) Effect of snubbing. 1961 HARRIS & —— *Shock & Vibration Handbk.* III. xlv. 24 Because car loading can change from trip to trip, practical considerations dictate a compromise both in the selection of load springs and in the design of a snubbing means.

snub-nosed, *a.* Add: **2.** *fig.* Stumpy; short and broad at the front; abbreviated.

1925 F. SCOTT FITZGERALD *Great Gatsby* i. 9 A snub-nosed motor-boat..bumped the tide offshore. 1961 E. S. GARDNER in *Webster* s.v., A snub-nosed revolver. 1963 *Times* 5 Feb. 7/5 These snub-nosed,..two-stroke machines. 1966 P. O'DONNELL *Sabre-Tooth* x. 137 A Smith & Wesson Centennial, a snub-nosed hammerless revolver. 1978 S. SHELDON *Bloodline* l. 405 Others carried snub-nosed tear gas rifles.

snuck, chiefly U.S. pa. t. and pple. of SNEAK *v.* in Dict. and Suppl.

snuff, *sb.*[1] Add: **I. 1.** *Comb.* (Earlier and later examples.)

1709 in J. S. Moore *Goods & Chattels of our Forefathers* (1976) 196 Two brasse Snuffers and Snuff panns, one latten hoop for Cakes, a greater Chayr. 1963 *Times* 11 May 11/1 Candle-snuffers fitted with snuff pans were being used in Britain by the mid-fifteenth-century.

II. 7. Used *attrib.* to designate pornographic photographs or films involving the actual killing of a woman. Cf. *SNUFF *v.*[1] 1 d.

1975 *Whig-Standard* (Kingston, Ontario) 2 Oct. 3/6 New York City police detective Joseph Horman said..that the 8-millimetre, eight-reel films called 'snuff' or 'slasher' movies had been in tightly controlled distribution for a month. 1975 *Globe & Mail* (Toronto) 20 Nov. 7/4 There are reports of 'snuff films' in the United States, pornographic movies that contain all the usual perversions but culminate with women being mutilated and killed—for real. 1976 *New Musical Express* 31 Jan. 11/3 The 'snuff movie', a kind of ultimate pornography that has at its climax the supposedly unfaked murder of a young woman. 1977 *Daily Colonist* (Victoria, B.C.) 23 July 3/2 Charged with attempted murder in the making of 'snuff' photographic stills. 1978 S. SHELDON *Bloodline* xlii. 360 For the last several years we have been hearing increasing rumors of snuff films, pornographic films in which at the end of the sexual act the victim is murdered on camera. 1981 *Observer* 12 July 39/4 The merchants and devotees of 'snuff porn' require that the masochistic models who pose for it actually die while receiving some hideous sexual punishment. At the actual moment of death the models are photographed in full colour, with very good lighting.

snuff, *sb.*[3] Add: **1. a.** Add to def.: In the Southern United States, usually taken orally. (Further examples.)

1849 [see *DIP *v.* 5]. 1891 M. E. RYAN *Pagan of Alleghanies* 105 [Does] your deity of the lower world..chew snuff? 1907 *Dialect Notes* III. 230 *Dip* (*snuff*), *v.t.*, to smear snuff on the gums with a brush made by chewing the end of a small stick. 1913 [see *DIP *v.* 5]. 1951 W. FAULKNER *Knight's Gambit* 87 We watched him take..a tin of snuff and tilt a measure of it into the lid and then into his lower lip, tapping the final grain from the lid with..deliberation.

c. The colour of snuff. Also *attrib.* or as *adj.*

1951 [see *MUSTARD *sb.* 1 f]. 1974 *Times* 26 Nov. 19/6 Colour combinations..snuff/ice, blue/white.

3. a. *up to snuff*, also, up to the required or usual standard, up to scratch.

1906 J. LONDON *Let.* 31 May (1966) 204 As usual, your criticisms are right up to snuff. 1931 *Punch* 4 Nov. 495/2 Now Romney painted well enough, And Reynolds too, they say, And Gainsborough's things are up to snuff, And Lawrence had his day. 1943 E. B. WHITE *Let.* 20 Mar. (1976) 239 The Central Park piece..is up to snuff or better. 1944 R. LEHMANN *Ballad & Source* 204 Madame Jardine says you're to go and see her for a few minutes. Only a *few* to-day. She's not quite up to snuff. 1974 S. ELLIN *Stronghold* 33 He did not..go as far as some Quakers by convincement and suggest that birthright Quakers are not quite up to snuff. 1982 *N. & Q.* Feb. 83/1 The publisher's rejection of such received proceedings reflects, I suppose, their commitment to a databank, annual slices of which will suffice to bring future editions of their guides up to snuff.

4. *snuff-bottle* (earlier example), *-stain.*

1850 *Spirit of Times* 16 Mar. 41/3 Did you see..that old snuff-bottle? 1914 JOYCE *Dubliners* 13 The red handkerchief,..blackened..with the snuff-stains of a week.

5. snuff-dipper (earlier example); **snuff-dipping** (earlier example); **snuff-gourd,** a bottle gourd, the dried shell of the fruit of *Lagenaria siceraria*, a white-flowered annual vine; = *snuff-box gourd* s.v. SNUFF-BOX 3.

1845 T. J. GREEN *Texian Exped.* x. 137 We believe the most filthy of all practices is that of your..'snuff-dippers'. 1860 E. M. COWELL *Jrnl.* 22 Apr. in M. W. Disher *Cowells in Amer.* (1934) 65 The ladies have a habit.., 'snuff dipping' which is openly practised in the South, and privately indulged in, in the North. 1901 KIPLING *Kim* iv. 99 The lama dipped deep into his snuff-gourd. 1921 *United Free Church Missionary Rec.* June 190/2 Her dress consists simply of a ragged apron of goatskin, and a snuff-gourd hung round her neck.

snuff, *v.*[1] Add: **1. d.** *slang.* = sense 2 d below.

1973 C. ALVERSON *Fighting Back* xxv. 129 Wait'll you see what you've got when Speranza finds out that you put Gino in a position to get snuffed. 1976 F. WARNER *Killing Time* I. i. 7 They had to sneak over and come

back with a prisoner, and most got snuffed themselves. 1978 T. GIFFORD *Glendower Legacy* (1979) 158 We should have snuffed this little shit when we had the chance.

2. c. Also, to terminate (life).

1929 HALL & NILES *One Man's War* iii. 25 We..knew ..that many lives would be snuffed out ere long. 1981 *Telegraph* (Brisbane) 23 Jan. 4/5 A prosecution witness today admitted she was content that the life of the man known as 'Mr Asia' should be snuffed out.

d. *slang.* To kill, to murder.

1932 E. WALLACE *When Gangs came to London* xxviii. 285 Eddie would have snuffed out Cora. 1973 *Philadelphia Inquirer* 7 Oct. (Today Suppl.) 12/1 'You're saying you're going to snuff that guy out before you know?'.. 'The people who are murdered didn't get an equal chance.' 1980 E. BEHR *Getting Even* xv. 174 If I cause too much embarrassment, they'll just snuff me out.

3. Also without *out.*

1916 C. J. DENNIS *Songs of Sentimental Bloke* v. 43 They think she's snuffed, an' plant 'er in 'er tomb.

snuff-box. 2. (Earlier example.)

1829 P. EGAN *Boxiana* 2nd Ser. II. 251 He came up with a frown..and, without the slightest ceremony, opened with a fillip on the Gipsy's snuff-box.

snuffer[1]. Add: **3.** *snuffer(s)-box, -handle, -stand, -tray* (later examples).

1843 *Ainsworth's Mag.* III. 180 Spectacle cases and snuffer-stands. 1898 G. B. SHAW *Man of Destiny* 203 With a couple of candles alight, and a broad snuffers tray in the other. 1952 B. & T. HUGHES *Three Centuries of Eng. Domestic Silver* v. 82 The upright snuffer stand, with a vertical socket to receive the snuffer box, was a late Charles II innovation. *Ibid.* 84 An immense amount of ingenuity now began to be lavished upon the ornamentation of snuffer handles. 1960 H. HAYWARD *Handbk. Antique Coll.* 260/1 *Snuffer-tray*, oblong or oval tray with or without small feet and scroll and ring handle at side for holding snuffers. 1971 *Country Life* 10 June 1434/1 A snuffers tray inscribed four years after it was made 'In Memory of Mrs. Jane Parsons, Oct. 11th, 1750'.

snuffer[2]. **3.** (Earlier example.)

1882 J. SNODGRASS tr. *Heine's Relig. & Philos. in Germany* II. 89 You know that he [*sc.* Frederick the Great] composed French verses,..was a prodigious snuffer, and believed in nothing but cannon.

4. (Earlier example.)

1829 T. C. HALIBURTON *Hist. & Statist. Acct. Nova-Scotia* II. ix. 404 Fish—Whale Species... Snuffer.

snuffing, *vbl. sb.*[1] Add: **3.** With *out*: dying.

1922 P. A. ROLLINS *Cowboy* iii. 55 His demise was sometimes referred to as his 'snuffing out'..or 'passing in his checks'.

snuffliness. (Earlier example.)

1862 J. A. SYMONDS *Let.* c 28 Feb. (1967) I. 336 Nothing cd have exceeded the snuffliness of my journey yesterday. I..did my cold no good.

snuff-mull. (Earlier example.)

1808 *Monthly Pantheon* I. 598/2 He was..famous for making Highland dirks and snuff mulls.

snuffy, *a.*[1] Add: **2.** Of cattle or horses: excitable, spirited, wild.

1955 R. HOBSON *Nothing too Good* xviii. 186 Any ball-up or milling around business up in front of the mile-and-a-half line of snuffy range beef could easily cause the critters to split into the spruce. 1964 *Penguin Bk. Austral. Ballads* 131 I'll yard them snuffy cattle in a way that's safe to swear. 1973 R. SYMONS *Where Wagon Led* vi. xix. 290 When he worked, he worked. Otherwise he played, mostly at breaking-in the snuffy ones.

snuffy, *a.*[2] Add: **2. b.** (Earlier and *fig.* examples.)

1765 STERNE *Tr. Shandy* VIII. 51 A plan..upon the lower corner of which..there is still remaining the marks of a snuffy finger and thumb. *a* 1846 B. R. HAYDON *Autobiogr.* (1927) III. xiii. 229 Brighton gay, gambling, dissipated..Dieppe dark, old, snuffy and picturesque. 1925 E. SITWELL *Troy Park* 67 Trees periwigged and snuffy.

3. (Earlier example.)

1823 'J. BEE' *Slang* 162 *Snuffy*—drunk, with a nasal delivery. *Snuffy*—drunk in the feminine application, and applied but seldom to puling fellows.

snug, *sb.*[2] Add: **1.** (Earlier example.)

1757 S. FOOTE *Author* I. i. 8 You love the snug, the Chimney-Corner of Life; and retire to this obscure Nook.

2. a. (Earlier and later examples.) Also *snug bar.*

1838 *Actors by Daylight* I. 84 Act-drop..the signal for the stage-manager to run in from the snug, and bully everybody. 1903 SOMERVILLE & 'ROSS' *All on Irish Shore* 226 'Don't be afraid,' said our hostess reassuringly, 'he'll never see ye—sure I have him safe back in the snug!' 1956 J. M. MOGEY *Family & Neighbourhood* 105 Solitary women drinkers prefer the snug in a larger public-house. 1967 *Punch* 17 May 736/3 Old women's gossip in the snug-bar..at 'The Garibaldi'. 1977 *Ibid.* 31 Aug.–6 Sept. 331/2 Angus Beakley's cart-horse went berserk and crashed into the snug of *The Flat Pig And Hat.*

b. Delete *Sc.* and add earlier example.

1860 DICKENS *Uncommercial Traveller* (1861) v. 63 Across the room, a series of open pews for Jack..at the other end, a larger pew..entitled *snug*, and reserved for mates.

snug, *a.*[1] **2. b.** (Further examples.)

1886 F. T. ELWORTHY *W. Somerset Word-bk.* 691 'So snug's a bug in a rug' is the common superlative expression. 1934 J. BUCHAN *Free Fishers* xiii. 211 Jem hung up his hat and ever since has been as snug as a flea in a blanket. 1936 D. POWELL *Turn, Magic Wheel* I. 93 The fetish of permanency, the snug-as-a-bug-in-a-rug fetish. 1974 P. DICKINSON *Poison Oracle* ii. 42 You just wait here, snug as a bug in a rug, learning it all second hand.

snug, *v.* Add: **1.** *transf.* and *fig.* (Later example.)

1978 T. L. SMITH *Money War* (1979) I. 74 Folding bipod which snugs under the barrel when not in use; large winter trigger.

c. *U.S.* With *up.*

1868 M. M. POMEROY *Nonsense* xxvi. 248 She 'snugged up' toward us as gently as a juvenile dove. 1873 J. H. BEADLE *Undeveloped West* v. 106 He used to complain that I 'snugged up' altogether too much.

6. b. Also *transf.*

1918 KIPLING *Land & Sea Tales* (1923) 116 They snugged her down. I don't know how one snugs down an aeroplane.

snuggery. Add: **1. a.** (Earlier example.)

1812 M. EDGEWORTH *Tales of Fashionable Life* V. 268 Let me establish you comfortably in this, which I call my sanctuary—my *snuggery.*

b. (Earlier example.)

1829 P. EGAN *Boxiana* 2nd Ser. II. 206 Shelton, on Friday evening after the fight, made his *bow* to the *Daffy Club*, at the Castle Tavern, Holborn: and, in the *snuggery*, Tom received the £100 stakes, as the reward of his victory.

2. b. (Further example.)

1953 DYLAN THOMAS *Under Milk Wood* (1954) 2 It is night neddying among the snuggeries of babies.

3. For *rare*[−1] read *rare* and add later example.

1958 [see *DO-GOOD].

snuggle, *sb.* Add: Also, a group of persons or things which are snuggled together.

a 1910 'MARK TWAIN' *Autobiogr.* (1924) I. 103 In the early cold mornings a snuggle of children..occupying the hearthstone. 1935 E. BOWEN *House in Paris* II. i. 87 A snuggle of gothic villas. 1966 J. S. COX *Illustr. Dict. Hairdressing & Wigmaking* p. ix, Words which, when used in conjunction with the word *curls*, are indicative of either the quantity,..or arrangement of a *group of curls*: ..Ruffle, Snuggle, Soufflé.

snu·ggle-pup. *U.S. slang.* ? *Obs.* Also **-pupper, -puppy.** [f. SNUGGLE *v.* + *PUP *sb.*[1] 2 b.] An attractive young girl.

1922 [see *jazz queen* s.v. *JAZZ *sb.* 5]. 1925 LINDSEY & EVANS *Revolt Mod. Youth* v. 58 We go to parties with these young crumpet munchers and snuggle pups. 1933 *Forum & Century* (N.Y.) Dec. 367/2, I glimmed him with a snuggle-puppy.

snuggler (snʌ·gləɹ). [f. SNUGGLE *v.* + -ER[1].] One who snuggles.

1887 E. B. CUSTER *Tenting on Plains* xii. 379 It finally dawned upon us that the little horse was a constitutional snuggler. 1939 JOYCE *Finnegans Wake* 548, I chained her chastemate to grippe fuming snugglers.

snuggly (snʌ·gli), *a. colloq.* [f. SNUGGLE *v.* + -Y[1].] Characterized by or inviting snuggling; snug, close-fitting. Also redupl. as *snuggly-wuggly.*

1928 A. HUXLEY *Point Counter Point* xvi. 287 Such a dear snuggly-wuggly, lovey-dovey little chap. 1966 R. H. RIMMER *Harrad Experiment* 36 It's [*sc.* sexual intercourse is] going to be nice and snuggly, and lots of fun. 1976 *National Observer* (U.S.) 28 Aug. 15/4 (Advt.), Our own grown-up version of the classic favorite for children. The snuggly fashion that will make any 'Lig kid' a beautiful baby all over again. Deliciously tantalizing on *her*, and..ruggedly good-looking on *him.*

snum (snʌm), *v. U.S. colloq.* Alteration of *swear*, esp. in *I snum* as exclamation. Also as *sb.* in *by snum!* (*obs.*). Cf. VUM *v.*, *sb.*

1825 J. NEAL *Brother Jonathan* II. 315 By snum; but you're a precious fellow! 1839 *Yale Lit. Mag.* IV. 357 I snum, 'tain't the thing for me. 1904 J. C. LINCOLN *Cap'n Eri* I. 4, I ain't quite a fool yit, Eri Hedge. I guess I know—well, I snum! I forgot that upper vest pocket! 1916 H. L. WILSON *Somewhere in Red Gap* viii. 333 Now, I snum! Here she's two-thirty! 1951 *Publ. Amer. Dial. Soc.* xv. 67 *I snum!*, exclamation of amazement.

snurge (snɜ̄ɹdʒ), *sb. slang.* [Cf. SNEAK *sb.*] **a.** In schools, an informer, a tell-tale. **b.** One who curries favour, a toady; generally, an obnoxious person.

1933 M. HODGE *Wind & Rain* (1934) I. i. 19 He's probably only a kid. He may be a perfect little snurge...for all you know. 1955 M. GILBERT *Sky High* ii. 29 He's such a little snurge... He's so bogus. 1955 *People* (Austral.) 13 July 47/4 She was going to cut off my allowance. But do you know, it was that out-and-out snurge that pleaded for me? 1956 C. P. SNOW *Homecomings* xliv. 308 He had got on a good deal better as a snurge than he would have done as a malcontent.

snye (snai). *Canad.* and *local U.S.* Also **snie, sny.** [ad. Canad. Fr. *chenail*, Fr. *chenal* CHANNEL *sb.*[1]] A side-channel, esp. one creating an island.

1819 W. KEYES *Diary* 17 Apr. in *Wisconsin Mag. Hist.* (1920) III. 457 Evening, anchor a little above the upper snie (or channel) that leads to the Mississippi. 1826 *Kingston* (Ontario) *Chron.* 3 Nov. 2/5 We are also busy forming a channel through the rapids, for the sake of the raftsmen—this is done by building two strong dams, and deepening what is called a *dry snie.* 1829 J. MACTAGGART *Three Years in Canada* I. 136 At this place, there are numbers of islands formed by snies winding round the Falls. 1886 in *Alberta Hist. Rev.* (1971) Summer 16/2 And from there to the snye which is a short cut into Fort Resolution. 1893 'MARK TWAIN' in *St. Nicholas* Nov. 24/2 Ef we..slips acrost de river to-night arter de moon's gone down, en kills dat sick fam'ly dat's over on the Sny. 1908 C. MAIR *Through Mackenzie Basin* 40 Much of [the tracking]..is in the water, wading up 'snies', or tortuous shallow channels..floundering in gumbo slides. 1921 *Beaver* Aug.–Sept. 15/1 The Imperial Oil Company narrowly escaped the loss of their machines, which were lying on the snye at the back of the Fort awaiting favorable weather. 1948 *Canad. Geogr. Jrnl.* Mar. 150/2 The word *snye*, *sny* or *snie* has been used for many years to describe a channel behind an island, with slack current or partly dried, or some such similar feature. 1967 E. B. NICKERSON *Kayaks to Arctic* ii. 17 There is a snye for float planes. 1969 E. W. MORSE *Fur Trade Canoe Routes* II. v. 57 The brigades shot the Allumette Rapids in their main (north) channel, the 'Timber Snye', where a safe canoe course passes.

so, *adv.* and *conj.* Add: **B. I. 5. g.** Ellipt. for *is that so?* expressing (*a*) recognition or realization of a fact or (*b*) questioning or dismissal of a statement (cf. *so what*, sense 10 c below).

1803 G. COLMAN *John Bull* I. 7 *Peregrine.* Is your house far from the sea-shore? *Mrs. Brulgruddery.* About three miles, Sir. *Peregrine.* So! I have been wandering about since day-break. 1886 *Liverpool Even. Express* 9 Jan. 3/4 'Oh, Mr. Blobbs, you can form no idea of the terrible dream I had last night.'.. 'So?' remarked Mr. Blobbs, continuing the perusal of the morning paper. 1903 FARMER & HENLEY *Slang* VI. 289/1 'The King returns to town to-day' 'So?' 1973 H. NIELSEN *Severed Key* I. 6 'Small craft warnings are out.' 'So?' Simon queried. 1977 W. TUTE *Cairo Sleeper* vii. 122 'You will see whoever Major Masri decides you should see,' the officer said curtly... 'So!' she said to herself. 1978 A. MORICE *Murder by Proxy* i. 13 'He's an estate agent.'.. 'So?' 'So nothing.'

h. Used to add emphasis to a statement contradicting a negative assertion made by the previous speaker. *dial.* or *colloq.* (chiefly *U.S.*).

1913 *Dialect Notes* IV. 55 *So, adj.*, used sometimes as 'too' and 'just the same' are used to intensify an assertion in reply to an expression of scepticism. 'You don't know anything about it!' 'I do so!' 1931 *Amer. Speech* VII. 20 *So*, emphatic in absolute use. 'I was—so!' 1937 L. B. MURPHY *Social Behavior & Child Personality* ii. 62 Eunice, 'I don't.' Anne, 'You do so.' 1951 N. M. GUNN *Well at World's End* xiv. 101 'You don't like butter!' she cried. 'I do so like butter!' 'You don't! You don't!'.. 'I do so!' he yelled. 1953 K. TENNANT *Joyful Condemned* xii. 103 'How old are you?' 'Eighteen.' 'Eighteen, my fat aunt.'.. 'I am so eighteen.' 1979 G. SWARTHOUT *Skeletons* 28 'I've published nineteen!' 'You haven't.' 'I have so.'

7. a. so-fashion *adv.*, in this or that manner. *U.S. dial.*

1890 *Dialect Notes* I. 23 *So fashion*, meaning *so, in that way.* Is this known all over New England? 1903 G. S. WASSON *Cap'n Simeon's Store* v. 86 It don't look right for nobody..to take and hang on to them tormented ole witch-bridles so-fashion! 1913 R. FROST in *Poetry & Drama* Dec. 415 I'll knock so-fashion and peep round the door When I come back, so you'll know who it is.

† b. *slang.* Homosexual. *Obs.*

1937 in PARTRIDGE *Dict. Slang.* 1963 C. MACKENZIE *Life & Times* II. 254 'I've come to the conclusion,' he told me, 'that I'm not really "so" at all. I much prefer girls.' At this date [*sc.* 1899] the cant word among homosexuals for their proclivities was 'so'. That seems to have vanished completely from current cant. 1968 J. R. ACKERLEY *My Father & Myself* xvi. 192 A young 'so' man, picked up by Arthur in a Hyde Park urinal. 1973 *Daily Tel.* (Colour Suppl.) 23 Feb. 51/4 Wilde used to call him 'the architect of the moon'. Rothenstein, Beerbohm,..and Epstein were his more predictable friends, as he was not..at all 'gay', as it is now called, or, as it was then called, 'so'.

II. 10. b. (*b*) [Reflecting Yiddish idioms.] Without implication of a preceding statement, or with concessive force: = well then, in that case, very well; also (introducing interrogative clauses) with adversative force: = but then, anyway.

1950 B. MALAMUD in *Partisan Rev.* XVII. 666 Miriam returned after 11.30... 'So where did you go?' Feld asked pleasantly. 1952 M. PEI *Story of English* 182 The adverb *so* at the beginning of a sentence ('So I'll pay for it!'), probably of Yiddish origin, occurs frequently in conversation. 1960 'E. MCBAIN' *Give Boys Great Big Hand* i. 4 'I warn you..I ain't got no wine.' 'So who wants wine?' 1977 F. BRANSTON *Up & Coming Man* v. 49 'How much profit..?' 'Impossible to do more than make a wild guess.' 'So make a wild guess.'

c. *so what?*: a retort made to an assertion, implying that the problem expressed has no immediate interest or obvious solution. Also as *attrib. phr.* orig. *U.S.*

1934 M. H. WESEEN *Dict. Amer. Slang* 399 So what?—What of it? What does it matter?.. What does that have to do with the matter? Your remark has no bearing or significance. 1935 F. BALDWIN *Innocent Bystander* v. 83

He has a wife,' said the girl gloomily. 'So what?' asked Angela carelessly. 1938 C. LANDERY (*title*) So what? a young man's odyssey. 1949 *Hansard Commons* 21 Nov. 104 That is unfortunate and disappointing but, to use an American expression, 'So what?' 1953 in *Shorter Oxf. Eng. Dict.* (1955). Add., The tragedy of the 'So what?' generation. 1960 M. A. SINDALL *Matey* xiii. 177 She suddenly yawned and flung the magazine on to the seat. 'So what!' she murmured. 1968 C. WATSON *Charity ends at Home* x. 126 No, the fact is that Henny and I got along as well as most. Not around each other's necks all the time, but so what? 1970 T. HILTON *Pre-Raphaelites* viii. 201 Burne-Jones pushed art so far away from this world that our reactions to some of his paintings are of a merely so-what kind.

III. 13. *not so* preceding an adj., in the sense 'not very, none too—': see *NOT *adv.* and *sb.* 15 d.

14. g. With an adj. of size or quantity, with the implication of an accompanying gesture: = as—as this. Esp. in phr. *when I* (*he*, etc.) *was so high*, when I (etc.) was a small child. Hence *so-high* adj. Cf. THAT *dem. adv.* b, quot. 1870.

1876 GEO. ELIOT *Dan. Der.* IV. VII. liii. 89 'You would have me love what I have from the time I was so high'—here she held her left hand a yard from the floor. 1899 KIPLING *Stalky & Co.* 226 'Do 'ee lov' me, Mary?' 'Iss—fai! Talled 'ee zo since yeou was zo high!' 1916 A. HUXLEY *Let.* c 12 July (1969) 105 Vassall..seems..to have known me when I was 'so high'. 1963 'B. GRAEME' *Almost without Murder* xiv. 157 As a so-high kid I had 'liked' ice cream.

V. 23. For '† rarely' read 'also', and add later, chiefly *colloq.*, examples of *so* alone.

In the revived use, orig. *U.S.*

1851 H. MELVILLE *Moby Dick* III. 564 Take your leg off from the crown of the anchor here, though, so I can pass the rope. 1902 E. L. BANKS *Autobiogr. Newspaper Girl* xii. 143 One of the books in front of mine was six shillings. I bought it so mine would show. 1913 [see sense 7 a above]. 1949 W. ROGERS *Autobiogr.* 44 The reason they leave some of our boys over there..is so they can get mail that was sent to them during the war. 1951 C. P. SNOW *Masters* i. 3 Shovelling coal up the back of the chimney, throwing it on so it would burn for hours. 1968 *Los Angeles Times* 3 Mar. E6/3 The main reason Gender is back in the classroom is so he can converse in the many languages he knows. 1977 A. THWAITE *Portion for Foxes* 28, I shall make it simple so you understand.

VII. 35. f. (Earlier examples.)

1721 J. KELLY *Scottish Proverbs* 300 *So far, so good.* So much is done to good purpose. 1754 RICHARDSON *Sir Charles Grandison* V. x. 56 'So far, so good,' said aunt Eleanor.

soak, *v.* Add. **I. 1. d.** Of metal: to become heated uniformly throughout its mass.

1843 [in Dict., sense 1 b]. 1939 J. DEARDEN *Iron & Steel To-day* x. 134 Here the ingots are allowed to 'soak' until they are the same heat all through, and then they are rolled.

2. e. *transf.* Of heat: to penetrate *through* the mass of an ingot until it is at a uniform temperature.

1902 *Encycl. Brit.* XXIX. 587/1 Bringing such an ingot..to the rolling temperature is not really an operation of heating,..but one of equalizing the temperature, by allowing the internal excess of heat to 'soak' through the mass. 1970 F. SIMONS *Dict. Ferrous Metals* 191 *Soaked steel*, steel heated in a furnace and held at the chosen temperature sufficiently long for the heat to have 'soaked' right through to the centre of the mass, which is only then uniformly heated.

II. 6. b. To maintain (metal or ceramics) at a constant temperature for a period to ensure that they are uniformly heated.

1925 *Jrnl. Iron & Steel Inst.* CXII. 491 The ingots should be stripped, soaked, and forged before reaching the temperature of the critical range. 1956 A. K. OSBORNE *Encycl. Iron & Steel Industry* 392/1 The pit was for soaking the ingots or permitting the heat contained in the still molten steel in its interior to penetrate to the outer portions until the temperature of the entire ingot was reasonably uniform. 1966 *McGraw-Hill Encycl. Sci. & Technol.* VII. 346/1 In this type of kiln the cycle of setting ware in the kiln, heating up, 'soaking' or holding at peak temperature for some time, cooling and removing or 'drawing' the ware is repeated for each batch.

7. e. (Earlier and further examples.) Also, to criticize harshly, to 'knock'; *to soak it to* (one) = *to sock it to* (one) (see SOCK *v.*[2] 1 c in Dict. and Suppl.).

1892 *Columbus* (Ohio) *Even. Dispatch* 29 July 1/4 To-day's Washington Post 'soaks' it to the Southern Democrats in the House who were so rallied in 1885 in their support of the bill making an appropriation to the New Orleans Exposition, but are now opposed to a similar appropriation for the World's Fair. 1896 S. CRANE *George's Mother* xiii. 152 At the gang's corner, they asked: 'Who soaked yeh, Fidsey?' 1908 G. H. LORIMER *Jack Spurlock* ii. 44 Yes, he done it! Soak it to him good! *Ibid.* vi. 107 My troubles came at me from all sides, and soaked it to me till my conscience fairly ached. 1915 H. L. WILSON *Ruggles of Red Gap* (1917) xii. 210 If he gets fancy with you, soak him again. You done it once. 1920 WODEHOUSE *Coming of Bill* II. xiv. 239 Soak it to him, kid. 1925 H. L. FOSTER *Trop. Tramp Tourists* iii. 21, I found that we had on board..the man whose newspaper soaked my last book. 1936 [see *BUTTON *sb.* 5 g].

f. To impose upon (a person, etc.) by an extortionate charge or price; to charge or tax heavily; to borrow or extort money from; to cost a high price. Freq. const. *for* or with indirect object expressing a sum of money. *slang* (orig. *U.S.*).

1895 *N.Y. Dramatic News* 23 Nov. 2/2 This little scheme sometimes..enables the photographer to 'soak' them. 1904 *Newspaperdom* 21 Apr. 8 When a local merchant asks you to give his business a friendly notice, soak him 10 or 15 cents a line. 1915 WODEHOUSE *Something Fresh* ii. 37 Especially after poor old Percy had just got soaked for such a pile of money. 1932 D. L. SAYERS *Have his Carcase* xiii. 164 Poor, but not mercenary or dishonest, since he refused to soak Mrs W. 1936 N. COWARD *To-night at 8.30* II. 60 She soaked her old man plenty, I'm sure—before he took to soaking himself! 1949 [see *BRASS *v.*[1] 2]. 1958 *Times* 17 Mar. 12/6, I hope the Court-Leet soaked the Air Ministry, and I expect it did, for these Berkshire men knew their own value. 1966 'L. LANE' *ABZ of Scouse* 101 Can I soak yer for a coupler bob? 1971 *Farmers Weekly* 19 Mar. 42/3 If you think this is a soak-the-housewives review, nobody knows more about soaking the housewives than you. 1977 *Time* 21 Nov. 59/2 Then add the investment in sophisticated equipment: a single stainless-steel 1,000-gal. vat can soak the vintner for some $6,000.

Hence **soak-the-rich,** *attrib. phr.* applied to a policy of progressive taxation (*PROGRESSIVE *a.* 3 f); also in similar phrases, as *soak-the-poor*, etc. orig. *U.S.*

1935 J. WARBURG *Hell Bent for Election* 72 He [*sc.* F. D. Roosevelt] thought he was being 'clever' when he tried to steal Huey Long's thunder by suddenly coming out with his 'soak the rich' tax message. 1935 H. L. ICKES in *Lit. Digest* 14 Dec. 6/3 Soak the Rich (Antonym, Soak the Poor)—Newspaperese for a system of taxation founded upon the absurd and revolutionary theory that a man should be assessed taxes in proportion to his ability to pay. 1949 A. CHRISTIE *Crooked House* i. 8 No Soak-the-rich taxes would have any effect on him. He'd just soak the soakers. 1959 *Economist* 7 Feb. 498/1 The Democratic cry that this is a 'soak-the-poor' Budget. 1970 *Wall St. Jrnl.* 29 Apr. 1/5 Soak-the-sinner tax policy remains a stand-by... Taxes on alcoholic beverages and on cigarets have been the most frequent targets for increases. 1972 *Listener* 28 Dec. 898/3 Advocates of populist soak-the-rich policies.

III. 9. d. *pass.* with *in*: to be imbued with, to be profoundly acquainted with (a subject of study).

1937 *Ann. Reg. 1936* 56 It was generally agreed that the best speeches were made more or less extempore by speakers who were 'soaked' in their subject. *a* 1960 E. M. FORSTER *Maurice* (1971) I. i. 4 Mr Ducie would smile, for he was soaked in evolution.

IV. 11. *Comb.*, as *soak-dike, -ditch, -drain*; † **soakpit** = *SOAKAWAY (*obs.*); **soakway** [WAY *adv.*] = *SOAKAWAY.

1970 S. J. HALLAM in C. W. Phillips *Fenland in Roman Times* 23 Settlers clung tenaciously to these rich soils, and we can read from the air the story of their constant efforts to cope with deteriorating drainage: silting, the digging of soak-dykes, renewed silting, re-digging. *Ibid.* 33 The modern soak ditches take the drainage from the field ditches and discharge it into the main channel at regulated points; the Roman soak ditches must have had a similar function. 1963 *Times* 1 Feb. 13/7 The southern fringes of the planned settlement went first; water courses were provided with parallel soak-drains, which in their turn proved inadequate. 1970 P. SALWAY in C. W. Phillips *Fenland in Roman Times* 18 On the silts, continued occupation must have depended on keeping the system of soak-drains in operation. These drains imply sluices to let the water drain out of field and settlement ditches into the main watercourses at low tide and to prevent or control river water entering the ditches at high tide. 1898 E. C. S. MOORE *Sanitary Engin.* i. 5 If made in porous soils so that the liquid soaks away they are called soak-pits; they are dangerous to neighbouring wells. 1956 C. D. PIGOTT in D. L. Linton *Sheffield* 83 Now only *Sphagnum recurvum* is at all frequent and this is no doubt due to its occupation of the wettest soakways which are avoided by sheep. 1978 *Jrnl. R. Soc. Arts* CXXVI. 438/2 These latter [*sc.* rain water channels]..empty into soakways at the base of the building.

soakage. Add: **3.** Also *attrib.*, in *Austral.* and *N.Z.* use.

1921 H. GUTHRIE-SMITH *Tutira* xx. 196 These surface swellings are the result of a blocked soakage system. 1936 I. L. IDRIESS *Cattle King* iv. 30 Often you can dig in a dry creek-bed and obtain soakage water if you dig in the right place. 1937 E. HILL *Great Austral. Loneliness* vi. 53 At the crude soakage wells provided, he [*sc.* a white man] camps in the evenings.

soakaway (sōᵘ·kăwēⁱ). Also **soak-away.** [f. SOAK *v.* + AWAY *adv.*] A pit, usu. filled with hard-core, into which water or other liquids may flow and from which they may percolate slowly into the surrounding subsoil.

1916 H. G. WELLS *Mr. Britling* II. iv. 331 Every now and then someone stumbles into a soakaway for rain-water. 1928 *Daily Express* 31 May 5/3 Be sure to find out if your kitchen sink drains to a 'soak-away'. If it does, you must not allow much water go down it, but throw out washing-up water and suchlike on the garden. 1951 *Archit. Rev.* CIX. 291 The drainage system..consists of a series of septic tanks for soil drains and soakaways for stormwater. 1976 *Sunday Times* (Lagos) 26 Sept. 4/3 He says he can as well repair blocked soak-aways or any job a qualified plumber can do.

soaked, *ppl. a.* Add: **3.** Intoxicated. Freq. as second element of a *Comb.*

Some of these quotations may be regarded as further examples of SOAK *v.* 7 b.

1737 *Pennsylvania Gaz.* 6–13 Jan. 2/2 He carries too much Sail, Stew'd, Stubb'd, Soak'd, Soft. 1899, 1908 [see *gin-soaked* adj. s.v. *GIN *sb.*[2] 2 b]. 1939 JOYCE *Finnegans Wake* I. 85 The prisoner, soaked in methylated, appeared in dry dock. *a* 1953 E. O'NEILL *Touch of Poet* (1957) IV. 158 Like a rum-soaked trooper, brawling before a brothel on a Saturday night.

soaker. Add: **4. b.** (Earlier example.)

1789 J. BYNG *Jrnl.* 29 May in *Torrington Diaries* (1938) IV. 95 An approaching Storm made me pull up near the Grey Hound Inn; and well I did, for it came down a Soaker.

c. A soaking pit.

1928 *Jrnl. Iron & Steel Inst.* CXVII. 201 The heated soakers are fired by blast-furnace gas. 1959 *Ibid.* CXCIII. 368/2 The soaking pit used was an electric soaker with a single trough coke resistor. 1976 *Steel USSR* VI. 194/2 The holding period from the end of casting to charging into the soakers was maintained constant as far as possible.

soaking, *vbl. sb.* Add: **2.** Also, a similar process in which ingots of other metals or ceramic objects are brought to a uniform temperature in a furnace or kiln. (Further examples.)

1926 *Jrnl. Iron & Steel Inst.* CXIII. 648 A cooling curve taken of the 0·48 carbon steel after heating up to 1000°C., and cooled immediately without soaking, gave the ferrite point at the normal temperature. 1964 H. HODGES *Artifacts* i. 39 The early stages of firing [of pottery] are thus slow..; the temperature is allowed to rise slowly by stages, each rise being followed by a period at which a steady temperature is maintained for a time, a process known as soaking. 1966 *McGraw-Hill Encycl. Sci. & Technol.* VI. 379/1 If the alloy is heated to a temperature not far below its freezing temperature and held at that temperature for a long time, interdiffusion of the alloy constituents will tend to eliminate segregation. Such homogenization treatment is frequently called soaking.

3. *soaking pit* (further examples).

1913 *Jrnl. Iron & Steel Inst.* LXXXVII. I. 67 With a view to having a regular sequence of hot ingots delivered to Gjers soaking-pits, whilst the centre of the ingot was still liquid,..a central casting-pit was substituted, designed on the Bessemer principle. 1962 *Gloss. Terms Glass Ind.* (*B.S.I.*) 12 *Soaking pit*, a conditioning furnace used to bring glass in open pots to a uniform temperature for castings. 1976 *Steel USSR* VI. 196/2 The duration of holding in the soaking pits for 13–18 t ingots from the end of casting to the start of stripping must not exceed 1 h 10 min.

Soamin (sōᵘ·ămin). *Pharm.* Also **soamin.** [f. So(DIUM + *AMIN(O-.] A proprietary term for, sodium *p*-aminophenylarsonate (= *ATOXYL), formerly used to treat skin diseases.

1908 *Trade Marks Jrnl.* 13 May 763/2 *Soamin*... Chemical substances prepared for use in Medicine and Pharmacy. Henry Solomon Wellcome, trading as Burroughs, Wellcome and Co.,..London EC.; Manufacturing Chemist. 1909 *Official Gaz.* (U.S. Patent Office) 19 Jan. 759/2 Henry Solomon Wellcome, London, England. Filed Nov. 2, 1908. *Soamin*... Sodium Para-amino-phenylarsonates. 1918 J. H. PARSONS *Dis. Eye* (ed. 3) xvii. 354 Arsenic is specially liable to cause optic atrophy, usually total, when administered in the form of trivalent benzol-ring compounds such as atoxyl or soamin. 1920 J. M. H. MCLEOD *Dis. Skin* vi. 111 Arylarsonate Group, which consists of atoxyl or soamin, arsacetin, and orsudan, was at one time much in favour but has fallen into disuse recently.

so-and-so, *sb.*, *a.*, and *adv.* Add: Also **soandso.** **A.** *sb.* **2.** Used *euphem.* as a term of abuse for a person (occas. a thing). Also, with weakened force, as a term of affection.

1897 W. S. MAUGHAM *Liza of Lambeth* iii. 42 'You little so-and-so!' said Liza, somewhat inelegantly, making a dash at him. 1931 D. L. SAYERS *Five Red Herrings* xii. 132 Some rigmarole about always finding the so-and-so hanging round his place and he wanted to have it out with him. 1943 *Lafayette Alumnus* (Lafayette College, Easton, Pa.) Nov. 5/1 Hiya, Joe, you old so-and-so, haven't seen you since that time, etc., etc. 1945 *Penguin New Writing* XXVI. 55, I told 'em all that but they wouldn't listen, the ignorant soandsos. 1956 B. GOOLDEN *At Foot of Hills* vi. 124 'He felt he oughtn't to leave his work.' 'Poor old so-and-so.' 1958 'A. BRIDGE' *Portuguese Escape* i. 13 The Countess is a hard-baked, publicity-minded old So-and-so, with about as much consideration for other people as a sack of dried beans! 1968 K. WEATHERLY *Roo Shooter* 107 It's not much good you staying out if some other so-and-so is going to work it, is there? 1973 *Times* 28 Nov. 13/5 The set [of an opera] is an absolute so-and-so to walk about on. 1977 B. PYM *Quartet in Autumn* i. 9 'Hoping to get off early, lazy little so-and-so,' said Norman.

B. *adj.* **2.** *euphem.* as a term of abuse.

1929 E. WALLACE *Kennedy the Con Man* iv. in *Red Aces* 173 'That's what we pay rates and taxes for, and no so-and-so policemen in sight!' He did not say 'so-and-so', but Mr. Reeder thought his profanity was excusable. 1942 B. HIMES *Lunching at Ritzmore* in *Black on Black* (1973) 177 You would..resume your discussions..on defense.. or the F.B.I., or the 'so and so' owners of Lockheed, or that (unprintable) Aimee Semple McPherson. 1959 *Listener* 30 July 186/2 Some [clients] are good, some are indifferent, some are a so-and-so nuisance.

C. *adv.* **4.** As a mere intensive.

1959 'A. FRASER' *High Tension* v. 60 'Why can't Hugh help then? Or won't he?' 'Not so-and-so likely.'

Soanean (sōᵘ·nian), *a.* [f. the name of Sir John *Soane* (see below) + -AN.] Of, pertaining to, or characteristic of Sir John Soane (1753–1837), British architect, or the buildings designed by him. Also **Soane·sque** [-ESQUE], **Soa·nic** [-IC], *adjs.*

1842 *Penny Cycl.* XXII. 168/2 He thought proper to limit the time of the 'Soanean Museum' being opened to the public to two days in each week for three months in the year. 1945 E. WAUGH *Brideshead Revisited* I. iv. 72 It was an aesthetic education to live within those walls, to wander from room to room, from the Soanesque library to the Chinese drawing-room. 1948 *Archit. Rev.* CIV. 64/1 An extravagant blend of Soanic abstraction and Italian *grotesquerie.* 1974 SHERWOOD & PEVSNER *Oxfordshire* 815 A doorway with Soanean incised decoration.

soap, *sb.* Add: **I. 1. a.** Now usu. distinguished from DETERGENT *sb.* in Dict. and Suppl.: soap is prepared from natural oils and fats and is precipitated by the ions (notably calcium) present in hard water. (Further examples.)

1940 J. H. WIGNER *Soap Manufacture* i. 20 In the textile trades soap is largely used for removing the natural impurities from the fibre and detergent properties are the main consideration. 1966 *McGraw-Hill Encycl. Sci. & Technol.* XII. 393/1 In ordinary usage the term soap specifies an alkali metal or substituted ammonium salt of a straight-chain carboxylic acid 10–18 carbon atoms in length, and the name detergent is given to synthetic materials of similar structure. 1972 *Materials & Technol.* V. x. 295 In synthetic anionic detergents, the main weaknesses associated with the traditional carboxylate soaps, namely, precipitation in hard water and decomposition in acidic solutions, are avoided by the use of other hydrophilic groups in place of the carboxylate group.

d. (Earlier and later examples.)

1854 D. G. ROSSETTI *Let.* 11 May (1965) I. 193, I heard from MacCrae who offers £50 for the water-colour, with all manner of soap and sawder into the bargain. 1957 W. FAULKNER *Town* (1958) x. 149 'The pattern,' Uncle Gavin said. 'First the soap, then the threat, then the bribe.'

f. *no soap*: an announcement of refusal of a request or offer, failure in an attempt, etc.; 'nothing doing'. *slang* (orig. and chiefly *U.S.*).

1926 MAINES & GRANT *Wise-Crack Dict.* 11/2 *No soap*, can't talk business. 1929 E. WILSON *I thought of Daisy* iii. 153 If he tries to cut in on you, don't letum—I'll just tellum, no soap! 1932 J. T. FARRELL *Young Lonigan* vi. 216 Studs said he'd take a dozen or two when Nate brought them around. Nate tried to collect in advance; but Studs was no soap for that. 1939 W. FAULKNER *Wild Palms* 42, I told him. Not that I was to meet you at a hotel. I just said, suppose I did. And he still said no soap. 1948 A. N. KEITH *Three came Home* iii. 72 We would..call across.. 'No-soap!' or 'Not to-night!' 1957 J. KEROUAC *On Road* (1958) I. xiii. 86 Terry and I tried to find work at the drive-ins. It was no soap anywhere. 1977 'E. CRISPIN' *Glimpses of Moon* vi. 93 'The police tried to trace the handkerchief, I take it?' 'They did, but no soap.'

g. *not to know* (someone) *from a bar of soap*: not to have the slightest acquaintance with. *Austral. colloq.*

1938 *Smith's Weekly* 26 Nov. 23 (*caption*), I don't know you from a bar of soap. 1943 K. TENNANT *Ride on Stranger* xxv. 319 'Why doesn't she marry the child's father?'..'It's my belief she doesn't know him from a bar of soap.' 1970 J. CLEARY *Helga's Web* vii. 130 I've never met any of his—interests. Certainly not this girl. I dunno her from a bar of soap.

h. = *SOAP OPERA 1 a.

1943 *N.Y. Times Mag.* 28 Mar. 19 Within these specifications, there is a deal of shrewd craftsmanship in the preparation of the 'soaps'. 1958 *New Statesman* 12 Apr. 455/3 Pay-TV will lure whatever is good in television now and leave those who cannot afford to pay for programmes stuck with an unvaried diet of soap and corn. 1969 A. ARENT *Laying on of Hands* vi. 46, I was one of five writers doing a daytime soap. One script a week. 1974 *Anderson* (S. Carolina) *Independent* 20 Apr. 5A/2 Agnes..had landed a job dialoging soaps for well-known television writer Irna Phillips. 1978 *Amer. Poetry Rev.* July/Aug. 19/3 If you turn on day-time T.V. you will see most of his actors playing rather similar roles in the soaps.

II. 4. a. *soap-film, -lather* (earlier example) *-pad, -pipe, -tablet.*

1924 R. M. OGDEN tr. *Koffka's Growth of Mind* iii. 105 A soap-film is produced upon a wire-frame..and upon it a little noose of thread is cast in whatever form it may take. 1976 *Sci. Amer.* July 93/3 The area-minimizing principle alone is sufficient to account for the overall geometry of soap films and soap bubbles. 1771 SMOLLETT *Humph. Cl.* I. 238 His face frothed up to the eyes with soap lather. 1958 *Listener* 16 Oct. 627/1 Scour round the inside with a steel wool soap-pad. 1956 S. BECKETT *Malone Dies* (1958) 21, I remember the soap-pipe with which, as a child, I used to blow bubbles. 1920 D. H. LAWRENCE *Lost Girl* iv. 52 Happiness is a sort of soap-tablet—he won't be happy till he gets it.

c. *soap basket.*

1926–7 *Army & Navy Stores Catal.* 122 Wire sponge and soap basket. For hanging on bath, etc.—1/-. 1975 *New Yorker* 17 Nov. 145/1 We found an assortment of brass soap baskets to hook over the side of the tub.

d. (in sense *1 h), *soap fan, land, star, watcher.*

1976 *National Observer* (U.S.) 10 July 16/2 Real soap fans have a dozen or so fan magazines, newspapers, and newsletters. 1948 Soapland: see *SOAP OPERA 1 a. 1977 *Guardian Weekly* 17 Apr. 18/1 People who resent the behaviour of a character she plays don't walk up and slap her, as has happened to more than one American soap star. 1978 *Times* 29 Aug. 1/8 The dedicated soap watcher..can switch channels for a solid five hours.. until, at 4.30 p.m...soap-land is closed for another day.

5. a. *soap-saver.*

1919 *T. Eaton & Co. Catal.* Spring & Summer 366/4 Wire soap saver 7c. 1973 *Listener* 25 Jan. 117/2 The soap-saver..was made like a diffuser-spoon, but bigger, with a basket of open wire-mesh.

c. *soap-smooth* adj.

1949 E. POUND *Pisan Cantos* lxxvi. 45 By the soap-smooth stone posts.

d. Instrumental, as *soap-filled* adj.

1970 *Which?* May 149/2 The cheapest soap-filled pads cost nearly 2d each.

6. a. soap flakes *pl.*, soap in the form of thin flakes for washing clothes, etc.; **soap leaf,** a leaf of soap (see *LEAF *sb.*[1] 10 b); **soap-lock** (earlier examples); **soap powder** (earlier example); also *loosely*, detergent in the form of a powder; **soap-stock, soapstock,** a crude, partially saponified mixture of fatty acids formed as a by-product in the refining of natural fats.

1926–7 *Army & Navy Stores Catal.* 38/2 Soap Flakes—lb., –/8. 1933 'G. ORWELL' *Down & Out in Paris & London* xii. 91 There are no soap-flakes, only the treacly soft soap. 1967 N. FREELING *Strike out where not Applicable* 137 It was a little like a copywriter presenting an advertising campaign to a soapflakes manufacturer. 1909 *Cent. Dict. Suppl.*, Soap-leaf. 1925 [see *LEAF *sb.* 10 b]. 1978 *Times* 4 Nov. 24/5 Good presents for adults: a book of soap leaves. 1840 *Picayune* (New Orleans) 30 Aug. 2/2 Howard..is described as..wearing moustaches and soap-locks. 1842 'UNCLE SAM' *Peculiarities* I. 119 You are an incendiary, a robber by profession, a soap-lock and a loafer. 1865 H. MAYHEW *Shops & Companies of London* 199/1, I can always make quick work of my washing by using 'Harper Twelvetrees' Glycerine Soap-Powder', and it makes the clothes beautifully clean and white. 1964 M. DRABBLE *Garrick Year* xii. 193 Putting in my second instalment of soap-powder. 1970 G. GREER *Female Eunuch* 325 Some of the mark-up on soap powders.. could be avoided. 1895 J. LEWKOWITSCH tr. *Benedikt's Chem. Analysis of Oils, Fats, Waxes* xii. 632 (*heading*) Examination of the fatty matter ('soap stock'). 1924 MYDDLETON & BARRY *Fats* iii. 35 The recovery of oil from the soap-stock depends for its commercial success upon the ruling prices of the edible oil. 1972 *Materials & Technol.* V. x. 279 [In soap-making] use is also made of by-product fatty materials such as soapstocks and curd oils.

soap and water. [SOAP *sb.*] The commonest method of washing, used in phrases referring to standards of personal cleanliness. Also *attrib.*

1837 H. MARTINEAU *Society in Amer.* II. III. ii. 151 The demand of society for fresh air and soap and water has considerably increased. 1861 GEO. ELIOT *Silas Marner* xiv. 243 A great ceremony with soap and water, from which baby came out in new beauty. 1907 G. B. SHAW *Major Barbara* in *John Bull's Other Island* 168 Transfigured men and women carry their gospel through a transfigured world..practising what the world will let them practise, including soap and water, color and music. 1922 E. O'NEILL *Hairy Ape* iv. 37 Their faces and bodies shine from a soap and water scrubbing. 1961 L. MUMFORD *City in History* xv. 469 The spread of the soap-and-water habit might well account for the lowering of infant mortality rates. 1973 A. MACVICAR *Painted Doll Affair* iii. 39 She was an enthusiast for soap and water, as her schoolgirl complexion showed.

Hence **soap-and-water** *v. trans.*

1848, 1883 [in Dict. s.v. SOAP *v.* 1 *Comb.*].

soapberry. Add: **2. c.** *N. Amer.* A deciduous shrub, *Shepherdia canadensis*, of the family Elæagnaceæ, native to North America, and bearing small yellow flowers followed by edible red berries; also, the berries of this shrub; = *buffalo-berry* s.v. BUFFALO 5.

1904 A. G. MORICE *Hist. N. Interior Brit. Columbia* 61 The soap-berries were ripening. 1923 *Beaver* Dec. 104 In Central British Columbia it [*sc.* Indian ice cream] was made by working to a lather the dried soap berry. 1957 J. R. & I. M. CHRISTIE *Story Okanagan Falls* 42 Soap-berry..bears its gay red-currant-like berries now only for the birds to enjoy. 1963 *Beaver* Autumn 40/1 The interior fresh-water Indians..readily gave soapberries, kinninnick leaves and bark for smoking.

soa·p-box. [f. SOAP *sb.* + BOX *sb.*[2]] **a.** A box for holding soap; orig. and still occas., a small receptacle for a ball or bar of soap; later *esp.* a wooden case in which soap is or may be packed, traditionally used as a makeshift stand for a speaker; hence used *fig.* and allusively.

1660 [in Dict. s.v. SOAP *sb.* 4 c]. 1834 *Chambers's Edin. Jrnl.* III. 143/3 A soap-box! A thing with a lid which is found on almost every wash-stand in Great Britain. 1862 [in Dict. s.v. SOAP *sb.* 4 c]. 1907 J. LONDON *Road* 211, I get up on a soap-box to trot out the particular economic bees that buzz in my bonnet. 1912 *Town Topics* 16 Nov. 3/4 The days when a couple of..clerks on inverted soap boxes..were his staff. 1926–7 *Army & Navy Stores Catal.* 104/3 Soap box. Useful for travellers—each 1/6. 1928 *Observer* 1 Apr. 21/4 To use the language of Australian politics, 'Soapbox must be met by soapbox.' 1933 E. O'NEILL *Days without End* I. 32 If you knew what a burden he made my life for years with his preaching. Letter upon letter—each with a soap box inclosed, so to speak. 1943 K. TENNANT *Ride on Stranger* xxiii. 295 She made no answer to this outburst which, she felt, was only old Shanno blowing off steam... 'Back to your soap box,' she said briefly. 1945 N. MITFORD *Pursuit of Love* xiii. 97 She became an out-and-out Communist..preaching her new-found doctrine..from a soap-box in Hyde Park. 1948 M. LASKI *Tory Heaven* xi. 153 At his feet, a wax-faced baby moaned incessantly in a soap-box. 1960 H. HAYWARD *Antique Coll.* 260/1 *Soap-box*, spherical box of silver, pewter or brass for soap-ball, standing on moulded base with screw-on or hinged pierced or plain cover. 1968 *Daily Tel.* (Colour Suppl.) 13 Dec. 19/2, I look upon my wealth, and now the House of Lords, as useful soap boxes. 1977 *Time* 14 Nov. 59/3 The primary U.S. condition for rejoining is that the I.L.O. get off its political soapbox.

b. *attrib.* (chiefly with reference to public speaking from a soap-box); **soap-box cart,** a child's cart made from a soap-box; so **soap-box derby** [*DERBY 1 d].

1918 *National Geographic* July 8/1 (*caption*) The soap-box orator and his auditors. 1924 *Telephone Topics* XVIII. 262 (*heading*) Public address system supersedes soap box oratory. 1927 T. C. PEASE *United States* 546 The choice of party candidates by manipulation of party conventions and soap-box primaries. 1933 DYLAN THOMAS *Let.* Sept. (1966) 21 You must excuse my slight soap-box attitude. 1942 E. WAUGH *Put out More Flags* I. 72 Soap-box orators screaming their envy of the rich. 1950 *Manch. Guardian Weekly* 4 May 3/2 The 'Soap Box Derby' is an American festival rather more important to some..youngsters than the Fourth of July. 1960 *Times* 1 Mar. 13/1 The theatrical appeal of her soap-box oratory. 1977 J. VAN DE WETERING *Death of Hawker* xiii. 119, I like inventing. I was always making soap box carts when I was a child.

Hence (U.S.) **soap-box** *v. intr.*, to speak from or as from a soap-box; **soap-boxer,** one who speaks from a soap-box; **soap-boxing** *vbl. sb.*

1913 *Industrial Worker* (Spokane, Washington) 10 Apr. 4/1 They do want all the publicity that can be given them by the press, by the locals, soap-boxers, and by individual conversation. 1919 U. SINCLAIR *Jimmie Higgins* iv. 42 If he could have an assistant..the soap-boxing could go on every night. 1926 E. O'NEILL *Great God Brown* I. iii. 41 When you got to love to live it's hard to love living. I better join the A.F. of L. and soap-box for the eight-hour night! 1972 *Village Voice* (N.Y.) 1 June 78/3 He's been a soap boxer for the IWW (actually he sang and his brother spoke).

soap-bubble. (Earlier example.)

1800 M. EDGEWORTH *Parent's Assistant* (ed. 3) V. 100 Two other little children..came to him to beg, that he would blow some soap bubbles for them.

soaper. Add: **1. d.** A manufacturer of soap.

1965 *Economist* 16 Oct. 303/2 Denied any real difference to exploit, the soapers have not even got an expanding market to sell in. 1979 *Jrnl. R. Soc. Arts* Dec. 60/1 The glassmakers and soapers responded to the growing shortage of domestic potash in several ways. 1982 *Shell Technol.* No. 3. 6/2 Manufacturing the surfactant molecules known as detergent active matter falls, normally, into the domain of the chemicals industry. Combining this active matter with the other constituents of a modern synthetic detergent and marketing the finished product is the concern of 'soapers'.

3. = *SOAP OPERA 1 a. *N. Amer.*

1946 *Time* 26 Aug. 56/3 The result: *Pepper Young's Family*, one of radio's most popular soapers. 1972 *Daily Colonist* (Victoria, B.C.) 4 Feb. 2/1 The CBC soaper Whiteoaks of Jalna rates only slightly higher than a documentary on the mating habits of the tsetse fly. 1981 *TV Picture Life* Mar. 6/1 Daytime soapers were dealing with sex and violence far more explicitly than their night-time brothers for quite a while.

soapie (sōᵘ·pi). *colloq.* [f. SOAP *sb.* + -IE.] = *SOAP OPERA 1 a.

1964 F. POHL in *Galaxy* Oct. 190/2 You had a nervous breakdown..space cafard, as they call it on the soapies. 1978 *N.Y. Times* 30 Mar. C21/6 Movie: 'Daughters Courageous'... A soapie, granted. And dated. But pleasantly cheerful.

soapily, *adv.* Add: (Later examples.) Also in sense 4 of SOAPY *adj.*

1976 A. E. LINDOP in H. Watson *Winter's Crimes 8* 200 Soapily he would say, 'Darling, don't worry.' 1979 *Sci. Amer.* Jan. 26/1 The breakdown of the insulator may have many causes from the death of the maintaining cells to the appearance of a detergentlike metabolite capable of soapily cleaning off the invaluable lipid layers.

soapless, *a.* Add: **a.** (Later *lit.* example.)

1906 [see *CANDLELESS *a.*].

b. Of shampoo, detergent, etc.: not containing soap.

1936 *Chemist & Druggist* CXXIV. 56/2 The alternative to sulphonated lorol [*sic*] for soapless shampoos is saponin. 1959 *Which?* Nov. 152/2 All the liquid and cream shampoos, since they are all based on very similar soapless detergents, would clean the hair effectively in hard or soft water. 1966 J. S. COX *Illustr. Dict. Hairdressing & Wigmaking* 140/1 Sulphonated vegetable oils are also used in another type of soapless detergent.

soapolallie (sōᵘ·polæli). *N. Amer.* Also **soapol(l)ali(e), soopolallie, sopelalee,** etc. [f. SOAP *sb.* + Chinook Jargon *olallie* berry.] **1.** A thick drink made from crushed soap-berries. Cf. *SOAPBERRY 2 c.

1895 *Canad. Mag.* Aug. 344/1 We were fortunate enough to see some Indians eating 'soapolali'. 1944 C. BARBEAU *Mountain Cloud* 199 Here are the stems of blackberries and the wild fruit of the hills that gives sopelalee. 1966 *Islander* (Victoria, B.C.) 27 Feb. 6/2 There was [at the potlatch] also a great deal of oolachan-grease and soapollalie.

2. = *SOAPBERRY 2 c. Also *attrib.*

1937 T. STANWELL-FLETCHER *Jrnl.* 23 Sept. in *Driftwood Valley* (1946) 33 On drier, more open ridges..are..dense thickets of small Shepherdia, or soopolallie, bushes. 1953 A. F. FLUKE *Kwakiutl* 21 The berries of the 'soop-olally' bush..were dried and stored whole. 1957 J. R. & I. M. CHRISTIE *Story of Okanagan Falls* 42 One little shrub, the Indians' 'soopolallie'..bears its gay red currant-like berries now only for the birds to enjoy. 1976 T. WALKER *Spatsizi* xii. 139 The soap olallie leaves were a darker green, and the fruit larger.

soap opera. *colloq.* (orig. *U.S.*). [f. SOAP *sb.* + OPERA.

So called because some of the early sponsors of the programmes were soap manufacturers. For the use of *opera*, cf. *horse opera* s.v. *HORSE *sb.* 27 a.]

1. a. A radio or television serial dealing esp. with domestic situations and freq. characterized by melodrama and sentimentality; this type of serial considered as a genre.

[1938 *Christian Cent.* 24 Aug. 1011/1 These fifteen-minute tragedies..I call the 'soap tragedies'..because it is by the grace of soap I am allowed to shed tears for these characters who suffer so much from life.] 1939 *Newsweek* 13 Nov. 44/2 Transcontinental Network bubbled up out of the 'soap operas'. 1948 *Time* 11 Oct. 40/3 *The Beast in Me* also includes such matter as Humorist Thurber's grimly unhumorous 'Soapland' (studies in contemporary soap opera). 1953 M. DICKENS *No More Meadows* iv. 180 More and more soap operas had hit the air to sell detergents and deodorants and headache pills. 1978 J. IRVING *World according to Garp* xvi. 321 Hoping that the visceral reality of Garp's language..somehow rescued the book from sheer soap opera. 1980 *Times Lit. Suppl.* 24 Oct. 1210/5 Some advertising campaigns [on ITV] have become mini soap-operas.

b. *transf.* and *fig.*

1944 R. CHANDLER *Lady in Lake* v. 36, I haven't heard a word from Muriel in the whole month... I don't have any idea at all where's she's at. With some other guy maybe. I hope he treats her better than I did... Thanks for listening to the soap opera. 1958 *Spectator* 19 Sept. 369/2 Eugene O'Neill's wordy autobiographical play is an endlessly tragic soap-opera, a sort of Mrs. Dale's Diarrhœa. 1962 [see *HUFF *sb.* 1]. 1971 'A. BURGESS' *MF* ii. 25 The act of robbery..near 39th Street... This was daily soap-opera of the streets.

2. *attrib.*

1942 W. STEGNER *Mormon Country* 347 They deal with impressionable virgins caught in the net of polygamy and agonizing worse than any soap-opera heroine through endless difficulties. 1951 M. MCLUHAN *Mech. Bride* (1967) 157/1 Soap-opera serials are short on action, long on situations. 1958 *Punch* 9 July 59/2 The revival of 'The Royalty' (BBC) is the latest development in the soap-opera world, on a Channel that has no soap to sell. 1978 S. BRILL *Teamsters* ix. 349 Most of the soap-opera intrigue of innuendo and in-fighting was not terribly subtle.

Hence **soap-opera·tic, -operatical** *adjs.*, of or characteristic of a soap opera.

1963 *New Yorker* 1 June 66 'The L-Shaped Room'... A sentimental piece of work, but so justly and successfully sentimental that it nearly always avoids seeming soap-operatic. 1975 *Country Life* 20 Mar. 742/3 A few weeks ago the BBC concluded a soap-operatical version of the loves of Georges Sand. 1979 *Boston Globe* 18 May 39 From her soap-operatic point of view, Watergate was not a national tragedy but rather was the personal pathos of a woman with nothing to give a husband in need.

soapy, *a.* Add: **5. a.** (Earlier example.)

1854 E. TWISLETON *Let.* 22 June (1928) xi. 202 The Bishop of Oxford I never do like..his manner, when Lords are in presence, richly merits his popular sobriquet of 'Soapy Sam'.

5*. Of style, tone, etc.: smooth, bland, sickly, sentimental.

1889 G. B. SHAW in *Star* 12 Aug. 3/4 Miss Nettie Carpenter played Svendsen's Romance for Violin, and played it very well, though her tone is just a little soapy—if I may be permitted to use such an expression. 1926 C. CONNOLLY *Let.* 1 June in *Romantic Friendship* (1975) 139 Benson's style is pretty soapy. 1973 *Publishers Weekly* 17 Sept. 59/3 Romance, which gets a bit soapy at times.

soar, *v.* Add: **I. 1. f.** *Aeronaut.* Of an aircraft or its pilot: to fly without the aid of an engine, esp. for an extended period without significant loss of altitude.

1893 O. CHANUTE in *Amer. Engineer & Railroad Jrnl.* Feb. 85/2 M. de Sanderval..is to be commended for having made an earnest if unsuccessful effort to learn how to soar in a wind like a bird. 1903 W. WRIGHT in *Jrnl. Western Soc. Engineers* VIII. 402 On trial we found that the machine would soar on the side of a hill having a slope of about 7 degrees. *Ibid.* 407 It would be easy to soar in front of any hill of suitable slope, whenever the wind blew with sufficient force to furnish support. 1931 V. W. PAGÉ *ABC of Gliding* vii. 159 An expert in Germany recently soared for a distance of 42 miles. 1940 L. B. BARRINGER *Flight without Power* xii. 218 After being checked out in two-seaters, they are allowed to soar in single-seaters. 1976 D. PIGGOTT *Gliding* (ed. 4) viii. 49 In general, it is not wise to attempt to soar by circling if you are below 500 feet.

3. c. Of an amount, price, etc.: to rise or increase rapidly. Hence, of a commodity: to increase rapidly in price.

1929 T. WOLFE *Look Homeward, Angel* xv. 196 She realized that in a very short time land values would soar beyond her present means. 1965 *New Statesman* 30 Apr. 672/3 The improvement..cannot be more than a stopgap whilst numbers continue to soar. 1978 I. B. SINGER *Shosha* i. 12 The price of meat soared. 1979 *Tucson* (Arizona) *Citizen* 20 Sept. 1A/4 Gold soared to another record of $380 at London's five major bullion firms.

7. To cause to soar.

1661 J. HEATH in J. W. Draper *Cent. Broadside Elegies* (1928) No. 43 A Cherubs wing hath soar'd him to this Hight. 1930 R. CAMPBELL *Adamastor* 88 Partaking the strain of the heavenward pride That soars me away from the earth I deride. 1978 A. WELCH *Book of Airports* ii. 28/1 Soaring the glider all the way back to where you started from is both exciting and satisfying. 1982 *Sci. Amer.* July 60/1 With the engine off the craft can be soared like a hang glider.

soarable (sōᵊ·răb'l), *a.* [f. SOAR *v.* + -ABLE.] Suitable for soaring flight. Hence **soarabi·lity**, a soarable condition.

1922 *Nature* 17 June 799/1 When the air at the level of the fin-ray was 'soarable', as shown by the behavior of dragon-flies. 1922 *Flight* XIV. 620/2 How machines will fare..remains to be seen. The southern slopes are not nearly so steep, and the extent to which they give soarability is at present a matter for speculation. 1961 *Aeroplane & Astronautics* CI. 163/2 The second day..suffered from clamp in the middle but produced soarable periods at either end. *Ibid.*, Nobody went away in the morning soarability, because the post-frontal sky, when it came in late afternoon, should theoretically have been worth waiting for.

soaraway (sōᵊ·răwēⁱ), *a.* [f. SOAR *v.* + AWAY *adv.*] Soaring, making rapid or impressive progress.

1977 *Zigzag* Aug. 6/1 All the great American pop styles rolled into one but fueled with the energy of the super soaraway seventies. 1978 *Oxford Jrnl.* 6 Jan. 1/1 The team which has made the Journal a soaraway success. 1982 *Observer* 26 Sept. 9/4 He'll soon be writing for Britain's best and liveliest soaraway Sunday newspaper.

soarer. Add: **1.** (Further examples.)

1900 W. WRIGHT in M. W. McFarland *Papers W. & O. Wright* (1953) I. 34 Hawks are better soarers than buzzards. 1910 *Times* 18 May 14/5 The albatross is pre-eminently the gliding soarer of the bird kingdom. 1978 *Sci. Amer.* July 102/1 Those master soarers, the great albatrosses.

2. An aircraft designed for soaring.

1909 *Flight* 20 Feb. 110/1 For a machine heavier-than-air, the true distinctive expression should decidedly be, 'flying machine', comprising 'flyers', 'gliders', 'soarers', &c. 1931 V. W. PAGÉ *ABC of Gliding* viii. 164 The primary training or school machines..are gliders rather than soarers. 1941 S. P. JOHNSTON *Horizons Unlimited* 30 Sailplanes or soarers are simply light and efficient gliders that may be made to take advantage of up currents of air to attain altitudes far above their launching points.

soaring, *vbl. sb.* Add: **1. b.** *Aeronaut.* Gliding; now *esp.* gliding for extended periods without significant loss of altitude. Freq. *attrib.*

[1864 *Leisure Hour* 21 May 328/1 The sciences of aerostation and meteorology must progress together as wedded sciences... The effect of a mutual reaction upon each other we are unable to conjecture, further than to anticipate..more than probable extension of the properties and simple soaring power of the balloon.] 1893 *Amer. Engineer & Railroad Jrnl.* LXVII. 396/1 It seems now reasonably possible for designers of soaring machines..to experiment with their apparatus without further search for some hidden secret. 1894 O. CHANUTE *Progr. Flying Machines* p. iv, Aeroplanes for soaring flight. 1896 [see *GLIDING *vbl. sb.* 2]. 1903 W. WRIGHT in *Jrnl. Western Soc. Engineers* VIII. 401 In principle Soaring is exactly equivalent to gliding, the practical difference being that in one case the wind moves with an upward trend against a motionless surface, while in the other the surface moves with a downward trend against motionless air. 1931 V. W. PAGÉ *ABC of Gliding* p. vii, Soaring machines or sailplanes are usually monoplanes with a higher aspect ratio than found in the training planes. 1931 P. & M. WHITE *Gliding & Soaring* xviii. 145 Soaring differs from gliding in that the ship, instead of losing altitude, either pursues a level course or gains height. 1952 F. GREEN *ABC of Gliding* 90 Ridge soaring depends basically on the wind. 1958 D. PIGGOTT *Gliding* xviii. 118 Many glider pilots become anxious to start cross-country flying as soon as they have made one or two soaring flights. 1974 *Sci. Amer.* Aug. 14/2 His memberships..reflect several of his outside interests, which he lists as 'camping, canoeing, gliding and soaring and gardening'. 1979 *Yale Alumni Mag.* Apr. (Suppl.) cn 20/1 He is fully recovered..and still believes that soaring is a great sport.

soave (soā·ve), *adv. Mus.* [It.] As a direction to the performer: softly, gently, with delicacy and tenderness. Also **soaveme·nte.**

1740 J. GRASSINEAU *Mus. Dict.* 228 *Soave*, or *Soavement*, sweetly or agreeably. 1876 STAINER & BARRETT *Dict. Mus. Terms* 399/2 *Soave, Soavemente* (It.), agreeably, delicately, gently, softly, sweetly. 1959 *Collins Mus. Encycl.* 609/2 *Soave*,..in a smooth and gentle manner.

Soave (soā·ve). The name of a town in northern Italy, used *attrib.* and *absol.* to designate a dry white wine made there.

1935 SCHOONMAKER & MARVEL *Compl. Wine Bk.* v. 130 The most widely sold and the best white wine of Veneto is the dry Soave. 1960 *Spectator* 15 July 114 Soave in Italian means 'suave' in English, but the wine gets its name..after the battlemented little town of Soave... It is a white wine, very dry indeed as Italian wines go, with a refreshing acidity. 1969 R. AIRTH *Snatch!* ix. 90 We had a bottle of Soave Bolla, chilled, with the lobster. 1975 *Observer* (Colour Suppl.) 3 Aug. 12/2 Soave, a light dry white from Italy, is getting into more and more shops. It comes from around Verona.

Soay (sōᵘ·ə). Also **Soa.** The name of an island in the Western Isles, used *absol.* or *attrib.* to designate a small, brownish, short-tailed sheep, *Ovis aries*, belonging to a variety once restricted to the island.

1906 J. G. MILLAIS *Mammals Gt. Brit.* III. 210 The history of the Soay sheep is unknown. 1912 R. LYDEKKER *Sheep & its Cousins* iv. 59 These small and half-wild Soa sheep belong to a group of breeds, or sub-breeds, which are widely distributed over Northern Europe. 1922 *Nature* 6 May 595/1 It will be gathered that the primitive sheep of Europe was of the Soay type. 1949 E. COXHEAD *Wind in West* vi. 161 A chap..on a wee island north of Skye who's experimenting with the Soay sheep. 1970 *Observer* 26 Apr. (Colour Suppl.) 35/2 The unusually leggy sheep..is a Soay. 1974 R. N. CAMPBELL in P. A. Jewell *Island Survivors* ii. 28 The Soay sheep..is the most primitive domestic form in Europe. *Ibid.* 31 The flocks..of Soays on Soay were left behind. 1979 *Vole* Feb. 45/2 The sheep that are used are Soays, small, brown, attractive animals. A few years ago this very ancient breed faced extinction. Today its future is assured.

sob, *sb.*¹ Add: **3. b.** *colloq.* (orig. *U.S.*) with reference to sentimental appeals to the emotions, as *sob act, -raiser, -reporter, -singer, -song, specialist, squad, -talk, tune*; **sob brother,** *U.S. colloq.*, a sentimental man; **sob sister,** a female journalist who writes sentimental reports or articles; a writer of sob stories; hence in various *transf.* uses, *esp.*: an actress who plays pathetic roles; a sentimental, impractical person, a do-gooder; a journalist who gives advice on readers' problems; **sob story,** a report or article designed to make a sentimental appeal to the emotions; *transf.* a narrative of one's misfortunes, a 'hard luck story'; **sob-stuff,** speech or writing which makes a sentimental appeal to the emotions; also *attrib.*

a1953 E. O'NEILL *Long Day's Journey into Night* (1956) iv. 157 He's been putting on the old sob act for you, eh? 1914 J. LONDON *Let.* 23 Sept. (1966) 430 All I can say is that he is a weak-brother, a sob-brother. 1917 S. GRAHAM *Priest of Ideal* xxix. 278 Our great sob-raiser who persistently pleads in the *Primer* for all causes which obviously evoke pity and rage. 1929 *McGraw-Hill Book Notes* 11 Feb., The story in that announcement..looked too much like the efforts of a newspaper sob-reporter. 1955 *Star* (Johannesburg) 10 Oct. 8/2 Should a squad of police be seconded..to guard the American 'sob singer' Johnnie Ray? 1912 *Sat. Even. Post* (N.Y.) 7 Dec. 9/3 Of the Daily Blatt's seven sob sisters six had husbands; and of the six it was more or less pure coincidence that five were supported by their wives. 1922 *Opportunities in Motion Picture Ind.* (Photoplay Research Soc.) 5 Some sob-sisters have gratified their ambition to play comedy, and have played it well. 1927 *Sat. Even. Post* (N.Y.) 24 Dec. 62/3 The sob sisters and the sob brothers..who didn't raise their boys to be soldiers. 1936 WODEHOUSE *Laughing Gas* xviii. 196 It's one of the things the sob-sisters are sure to write up. 1939 *Sun* (Baltimore) 21 Feb. 9/8 Forecasting opposition to his plan by 'sob-sisters' Goodwin said 'it wouldn't do any harm to give these sob-sisters a couple of wallops too'. 1963 J. MITFORD *Amer. Way Death* x. 153 Mrs. St. Johns is best known as one of the original sob sisters, a Hearst reporter in her youth. 1967 *Boston Herald* 8 May 19/5 Now that Svetlana has become America's newest millionaire glamor girl sob-sister, American interest in peeking or looking through the iron curtain is at a new all-time high. 1972 *Listener* 20 July 72/3 Sob sisters, those ladies who advise the unhappy about their problems. 1927 *New Republic* 12 Oct. 211/1 He has possibly scored some moderate hits: in 'Manhattan Mary', 'Broadway', 'The Five Step'..,a curiously constructed sob-song called 'Memories', and the title-piece. 1964 J. P. CLARK *Three Plays* 114 So you turn your broad back Upon me and will continue with your sob-songs? 1931 *Kansas City* (Missouri) *Star* 3 Nov. 22/5 It is gratifying..that the sob specialists can find practically nothing..to be sorry about. 1912 G. M. HYDE *Newspaper Reporting* 236 The search for human interest material is a modification of the 'sob squad' work of the sensational papers, on more delicate lines. 1913 *Writer's Mag.* Nov. 174/2, I wrote the 'sob' story of 'the City that Turned Down Santa'. 1923 C. E. MONTAGUE *Fiery Particles* 177 Thomas Curtayne, the greatest of Irishmen, was to be buried in homely state... Here was a sob-story, manifestly. 1949 *Los Angeles Times* 15 June II. 4/4 How anyone could heed such a sob story is beyond me. 1979 N. HYND *False Flags* xxi. 188 'Sometimes a man tries to do too much.'..'I'm familiar with the old sob story.' 1982 A. MATHER *Impetuous Masquerade* xi. 170 And give him some sob-story? 1918 H. C. WITWER in *Collier's* 11 May 15/2 Well, Joe, we gotta lot of new songs over here now, besides 'Where Do We Go From Here?' which same is our favourite and a lot more of the old stand-bys, which runs more to the sob stuff. 1922 C. SIDGWICK *Victorian* xxvi. 193 When the girls talked sobstuff at school I always told them I meant to marry a millionaire. 1929 D. H. LAWRENCE *Pansies* 128 A sickly people will slay us If we touch the sob-stuff crown of such martyrs. 1937 A. CHRISTIE *Murder in Mews* ix. 223 Of course I'm sorry. I don't indulge in sob-stuff. But I shall miss him. 1978 N. MARSH *Grave Mistake* iii. 90 He puts on a bit of an act like a guide doing his sob-stuff over Mary Queen of Scots in Edinburgh Castle. 1946 KOESTLER *Thieves in Night* 219 'That's so much sob-talk,' said Matthews. 1926 E. O'NEILL *Great God Brown* II. i. 46 I love those rotten old sob tunes.

sob (sǫb), *sb.*² *slang.* [prob. altered form of Sov.] A pound.

1970 G. F. NEWMAN *Sir, You Bastard* iii. 113 Two hundred sobs was a small piece of fifty grand. 1973 'K. ROYCE' *Spider Underground* v. 79 Norman could have back his fifty sobs; when I failed I didn't want compensation.

sob, *v.*¹ Add: **3. b.** (Earlier example with direct speech.)

1861 C. M. YONGE *Young Step-mother* iv. 42 'Things didn't use to be stupid when Ned was there!' sobbed Gilbert.

‖**soba** (sōᵘ·bā). [Jap.] A type of noodle that is made from buckwheat and is a popular Japanese food. Also *attrib.*

1896 *Far East* 20 Dec. 33/1 A strange custom of eating *Soba* (a kind of vermicelli made of buckwheat) on the last day of December prevails among a large class of people. 1928 K. YAMATO *Shoji* vi. 90 We were presented with bowls of vermicelli, or *soba*. 1936 K. TEZUKA *Jap. Food* 25 Sometimes *udon*..or *soba* (buckwheat noodles) are used in place of boiled rice. 1965 W. SWAAN *Jap. Lantern* i. 6 Bowls of steaming *soba*, a type of Chinese noodle dearly beloved by the Japanese. 1971 *Ashmolean Mus. Rep. of Visitors 1970* 54 Soba cup, blue and white decoration of bamboos, probably Arita ware.

Hence **sobaya** (soba·ya), in Japan, a shop or restaurant which serves *soba*.

1958 *Japan* (Unesco) (1964) 727/1 The term *sobaya* is used in eastern Japan for vendors and shops that sell *udon* and *soba*. 1960 B. LEACH *Potter in Japan* vi. 132 We made our way to a Sobaya (buckwheat macaroni restaurant) and ate 'Zaru Soba'.

sober, *a.* Add: **II. 4. b.** *sober sadness* (later *arch.* example).

1819 J. KEATS *Let.* 21 Sept. (1931) II. 426 Isabella is what I should call..'A weak-sided Poem' with an amusing sober-sadness about it.

IV. 14. *sober-coloured, -living, -looking, -spoken* adjs.

1851 BORROW *Lavengro* I. xxv. 317 They were dressed in sober-coloured habiliments. 1892 'MARK TWAIN' *Amer. Claimant* xvi. 168 He drops into the stoodio as sober-colored as anything you ever see. 1960 *Times* 4 Mar. 13/7 There is a hard-working, sober-living, self-respecting section among them. a1817 JANE AUSTEN *Northanger Abbey* (1818) I. xi. 183 The morrow brought a very sober looking morning. 1934 W. S. CHURCHILL *Marlborough* II. xiv. 304 These were very unusual expressions for the sober-spoken and matter-of-fact Marlborough.

sober, *v.* Add: **II. 5.** Also with *up*. Hence as *attrib. phr.*

1884 [see SOBERING *vbl. sb.*]. 1901 *Daily Colonist* (Victoria, B.C.) 2 Nov. 5/2 The police yesterday gathered in an Indian woman who was rolling along in the street in a drunken condition with a baby in her arms. She was released as soon as she had sobered up. 1938 E. WAUGH *Scoop* III. ii. 284 'Aunt Agnes and I very much fear that he has taken too much.'..'Oh, he'll sober up,' said Uncle Theodore, from deep experience. 1963 AUDEN *Dyer's Hand* 261 When he [sc. Cassio] sobers up, his regret is..that he has lost his reputation. 1967 *Listener* 23 Nov. 669/3 The National Federation of Licensed Victuallers announced that they're to back the search for a sober-up pill—an alcohol antidote.

sobering *ppl. a.* (earlier examples); also with *down*.

1816 JANE AUSTEN *Emma* II. ii. 24 The sobering suggestions of her own good understanding. a1817 —— *Persuasion* (1818) III. vii. 141 These were words which could not but dwell with her... They were of sobering tendency. 1975 B. MEYRICK *Behind Light* xiv. 183 The sobering-down item of community hymn singing.

Soberano (soberā·no). [Sp., lit. 'sovereign'.] A Spanish brandy; also, a drink of this.

1963 'D. CORY' *Hammerhead* iv. 59 He called the barman over and ordered a Soberano. *Ibid.* vii. 109 Downing three fingers of cognac, more exactly Soberano. 1969 R. V. BESTE *Next Time I'll pay my own Fare* vii. 88 Gage opted for..a 'Soberano' which he found to be the least syrupy of the Spanish brandies. 1974 R. JEFFRIES *Mistakenly in Mallorca* xxii. 199 He..went through to the larder for a bottle of Soberano and three glasses... Back on the patio, he poured out three brandies.

sobersides. Add: (Earlier *transf.* example.)

1779 J. WEDGWOOD *Let.* 25 Feb. (1965) 229, I am sorry you have been again out of luck with a horse: but do not despair. I have got a *sober sides* on trial for a week past.

sober-sided *a.* (later examples).

1880 G. W. CABLE *Grandissimes* i. 4 Honoré in mask? he is too sober-sided to do such a thing. 1892 [see *CHORAL *a.*¹ 2 c]. 1950 *Psychiatry* Feb. 8/2 A sober-sided, meticulous investigator. 1970 N. ARMSTRONG et al. *First on Moon* ii. 41 The Apollo news center at Cape Kennedy issued a sobersided 'status report'.

sobful (sǫ·bfŭl), *a. rare.* [f. SOB *sb.* or *v.*¹ + -FUL.] Full of sobs, given to sobbing; provocative of sobs.

1921 W. J. TURNER *Music & Life* 8 The composer of the most sobful ballad that ever made a drunkard weep. 1924 *Blackw. Mag.* Nov. 692/2 He was not really in a very sobful mood.

‖**sobornost** (sobǭ·rnǫst). *Theol.* [a. Russ. *sobórnost'* conciliarism, catholicity.] A unity of persons in a loving fellowship in which each member retains freedom and integrity without excessive individualism.

1935 O. F. CLARKE tr. *Berdyaev's Freedom & Spirit* iii. 91 The revelation of the Trinity is, however, not that of a heavenly monarchy..but that of heavenly love, the divine *sobornost*. 1962 *Listener* 30 Aug. 317/1 How are we to achieve what Berdyaeff would call a valid *sobornost*—a really felt community? 1976 N. V. RIASANOVSKY *Parting of Ways* iv. 177 Khomiakov's concept of *sobornost*, an association in love, freedom, and truth of believers. 1977 *Church Times* 21 Jan. 13/3 *Sobornost* furthermore provides a further incentive to Roman Catholic officialdom not to regard Church unity too exclusively from a juridical point of view.

Sobranie (sǫbrā·ni). [Proprietary name: cf. next.] A kind of tobacco or a cigarette made from it. Also *Balkan Sobranie*, and *attrib.*

1899 *Tobacco* 1 Mar. p. lxvi/2 Register of specialities... cigarettes... Balkan Sobranie—Robt. Lewis, 20, St. James St., Lon., S.W. 1923 *Trade Marks Jrnl.* 19 Sept. 1969 *Sobraine* [sic]... Cigarettes. Isaiah Redstone..cigarette manufacturer. *Ibid.* 19 Dec. 2707 The Balkan Sobranie... Cigarettes. 1927 D. L. SAYERS *Unnatural Death* iii. 31 Lord Peter wriggled into the window seat, lit a Sobranie and clasped his hands about his knees. 1955 N. FITZGERALD *House is Falling* xi. 188 When three Sobranies had been lighted, Lake continued. 1966 L. SOUTHWORTH *Felon in Disguise* vii. 116, I will leave it at the tobacconists... All you have to do is to..ask for the pound tin of Balkan Sobranie you ordered. 1966 'A. YORK' *Eliminator* iii. 49 She put the pistol away in her bedside drawer and lit a Balkan Sobranie. 1977 *Punch* 31 Aug.–6 Sept. 355/1 Insouciant Senior Lecturers smoked Sobranie.

Sobranye (sǝbrā·n(i)ye). Also **Sobraniye, -je, Subranie.** [ad. Bulg. *săbrănie* assembly; cf. Russ. *sobrănie* and quot. 1902.] The parliament or national assembly of Bulgaria.

1894 E. DICEY *Peasant State* xv. 142 There are..three Estates in Bulgaria: the Crown, the Ministers, and the Sobranje. 1902 *Encycl. Brit.* XXVI. 448/2 The national representation is embodied in the Sobranye, or ordinary assembly (Bulgarian, *Săbranie*, the Russian form *Sobranye* being usually employed by foreign writers), and the Grand Sobranye, which is convoked in extraordinary circumstances. 1923 G. BUCHANAN *My Mission to Russia* I. ii. 22 In spite, however, of his declaring the elections invalid, the Grand Sobranje met and occupied itself with the difficult task of finding a prince willing to accept the thorny crown which Prince Alexander had laid down. 1957 *Times* 21 Dec. 5/4 Some 4,500,000 to five million Bulgarians will go through the motions of 'electing' a new Sobranye. 1974 J. ROTHSCHILD *East Central Europe between Two World Wars* vii. 334 The Peasantists won 85 seats in the Sŭbranie, the unicameral national legislature.

soc. (sǫk), abbrev. of SOCIETY 8 and 10.

1890 BARRÈRE & LELAND *Dict. Slang* II. 274 *Soc* (printers), this is an abbreviation of the word 'Society'. To be a member of the *Soc.* (compositors'), hence not a 'rat'. 1903 FARMER & HENLEY *Slang* VI. 291/2 *Non-Soc-man*, a *rat*.., a blackleg, a non-Union-man. 1980 'J. MARCUS' *Marsh Blood* v. 73 The overwhelming number of the Art Soc.'s members were amateurs.

so-called, *ppl. a.* Add: Also as one word. **2.** (*a*) Also *loosely* or *catachr.* as a term of abuse. (*b*) Now freq. used without implication of incorrectness.

(*a*) 1888 O. WILDE in *Woman's World* I. 134/2 'This so-called nineteenth century'—as an impassioned young orator once termed it, after a contemptuous diatribe against the evils of modern civilisation. 1960 C. S. LEWIS *Studies in Words* ix. 226 Rose Macaulay noticed a tendency to prefix 'so called' to almost any adjective when it was used of those the speaker hated; the final absurdity being reached when people referred to the Germans as 'these so-called Germans'. 1980 W. SAFIRE in *N.Y. Times Mag.* 13 Jan. 6/1 Examples of sneer words are 'self-proclaimed', 'would-be', 'purported' and that Soviet favorite, 'so-called'.

(*b*) a1961 in WEBSTER, s.v., His heavy working schedule did not keep the student out of so-called campus politics. 1962 R. CARSON *Silent Spring* viii. 86 The so-called Dutch elm disease entered the United States from Europe about 1930. 1966 G. GREENE *Comedians* I. ii. 46 New buildings..built for an international exhibition in so-called modern style. 1968 *Physics Bull.* Nov. 373/1 The socalled Schrödinger representation. 1977 C. SAGAN *Dragons of Eden* ii. 41 Many spinal-cord neurons seem to have about 10,000 synapses, and the so-called Purkinje cells of the cerebellum may have still more. 1979 P. NIHALANI et al. *Indian & Brit. English* I. 164 A number of so-called transformational grammarians are to attend the teachers' conference at Krishnapur next week.

soccer. Now the usual form of SOCKER in Dict. and Suppl.

sociability. Add: **2.** *Ecol.* The extent to which the plants of a species are found in proximity to one another. [The sense is due to Braun-Blanquet and Pavillard, who used F. *sociabilité* (*Vocabulaire de Sociologie végétale* (1922) 3).]

1922 *Jrnl. Ecol.* X. 246 Where the French terms are practically identical with the English equivalents..they are simply translated... La Sociabilité (Soziabilität, Geselligkeit): disposition of individuals in the interior of an association. Five grades of sociability are expressed as follows. 1932 FULLER & CONARD tr. *Braun-Blanquet's Plant Sociol.* iii. 36 Gregariousness or 'sociability' expresses a space relationship of individual plants, answering the question, how are the individuals or shoots of a species grouped? 1961 HANSON & CHURCHILL *Plant Community* iii. 97 Species that spread only by seed may also show a high degree of sociability, especially in the early stages of succession, as in abandoned fields where certain annual weeds may become very dense. 1973 P. A. COLINVAUX *Introd. Ecol.* v. 65 *Zea mays*..had a cover abundance rating of only '3' since, although the commonest plant, it by no means covered nearly all the ground as it must to rate a '5', and a sociability of '1', earned because it was evenly spaced and thus the extreme loner.

sociable, *a.* and *sb.* Add: **A.** *adj.* **5.** *Math.* Designating a cycle of three or more integers such that each is the sum of the factors of the previous one; cf. PERFECT *a.* B. 8. [The sense is due to P. Poulet, who used F. *sociable* (*L'Intermédiaire des Mathématiciens* (1918) XXV. 101).]

1970 *Math. Computation* XXIV. 428 Until now only two groups of sociable numbers were known, respectively of order 5 and 28... I have made an exhaustive search for sociable groups of order 4 of which the lesser number is smaller than 6.10^7. This search has yielded 9 new groups,..all of order 4. 1972 C. S. OGILVY *Tomorrow's Math* (ed. 2) v. 113 The numbers 12496→14288→15472→14536→14264→12496 form what has been called a sociable chain of 5 links... The sum of factors of 12496 is 14288, the sum of factors of 14288 is 15472, and so on around the chain.

B. *sb.* **1.** Restrict † *Obs.* to sense in Dict. and add: **b.** A sociable person. *rare.*

1927 A. HUXLEY *Proper Studies* 190 The ratio of solitaries to sociables will remain much as it is.

2. a. (Earlier example.)

1780 *Pennsylvania Jrnl. & Weekly Advertiser* 15 Mar. 4/1 Wanted to exchange, a neat sulkey, almost new, for a sociable or handy one horse chair, equally good.

c. (Examples.)

1851 C. CIST *Sketches & Statistics of Cincinnati in 1851* 202 Dressing bureaus, sociables, and *vis-à-vis* are sure to catch the visitor's eye, and to open the visitor's purse. 1872 *Atlantic Monthly* May 544 She was lying on a little sociable or sofa, as he entered. 1930 V. SACKVILLE-WEST *Edwardians* iii. 133 In the centre of the room stood a sociable..on which two persons might sit, facing one another, but properly divided by the arm and wriggle of the S. 1959 *Times* 8 Aug. 9/4, I would venture to claim for something like the double sofa of the illustration the honour of representing the ideals of this whole decade... The seats revolve so that the two occupants—there is ample space for broad backs and expansive crinolines—have room for manoeuvre. It was called a 'sociable'. 1961 L. G. G. RAMSEY *Connoisseur New Guide Antique Eng. Furnit.* 120 The variant known as the 'sociable', 'conversation sofa', or 'tête-à-tête', with the two ends facing each other on the lines of the French 'causeuse'..was popular for a short time during the 1840's, but seems to have already gone out of favour by the mid-1850's.

social, *a.* and *sb.* Add: **4. a.** (Further examples.) *social evening* (later examples); similarly *social tea.*

1785 BOSWELL *Jrnl. Tour Hebr.* 142 His benevolent, gay, social intercourse. 1857 DICKENS *Little Dorrit* II. xiv. 441 He took pains, on all social occasions, to draw Mr Sparkler out. 1877 *Independent* 8 Feb. 4/3 The social event of the season! 1887 [see VISIT *sb.* 1 a]. 1896 W. JAMES *Mem. & Stud.* (1911) i. 3 On this social occasion it has seemed that what Agassiz stood for in the way of character and influence is the more fitting to commemorate. 1899 J. LONDON *Let.* 17 Apr. (1966) 28 So you grow a-weary of the social whirl. 1911 G. STRATTON-PORTER *Harvester* xvi. 342 Wait until afternoon, and pretend you are making a social call. 1915 F. M. HUEFFER *Good Soldier* III. iv. 178 He would dine and pass the evening..at social functions of one kind or another. 1926 Social tea [see *bridge roll* s.v. *BRIDGE *sb.*¹ c]. 1943 G. GREENE *Ministry of Fear* III. ii. 193 Haven't we met—?..On one of the doctor's social evenings. 1946 L. P. HARTLEY *Sixth Heaven* iii. 67 We saw him chattering away... He loves the social round. 1958 A. HUXLEY *Let.* 20 Oct. (1969) 855, I have been revolving in the social whirl—seeing everybody. 1959 B. BERNARDI *Mugwe, Failing Prophet* i. 6 Social intercourse between the main section of the Tharaka and the Thagichu has never been broken off. 1976 *Eastern Even. News* (Norwich) 9 Dec. 12/7 Patients of St. Andrews Hospital, Thorpe, enjoyed a social evening in the Octagon Centre at the hospital. 1977 P. SCOTT *Staying On* (1978) xiii. 204 In all the years they'd known one another they had never exchanged social visits. 1980 *Jewish Chron.* 4 Jan. 9/3 Brighton and Hove Emunah held a social tea at the Talmud Torah hall of Hove Hebrew Congregation.

c. Of, relating to, or connected with fashionable or leisured society (cf. SOCIETY 3 c). See also *social column, social register,* sense 12 below.

1873 TROLLOPE *Eustace Diamonds* III. lxxviii. 331 The police..had..succeeded in sending two scoundrels out of the social world, probably for life. 1894 *Harper's New Monthly Mag.* Oct. 697/2 But who looked so far from their faces, so certain to reveal the types of all styles of the beauty of our theatrical and social queens? 1896 M. CORELLI *Thelma* II. II. ii. 137 In the social world, Fashion, the capricious deity, must be followed. 1903 A. BENNETT *Truth about Author* vi. 80 The editor was enchanted with my social paragraphs. 1911 M. CORELLI *Life Everlasting* xi. 237 It was supposed then..that as I found myself the possessor of an income of between five and six thousand a year, I would naturally..enter upon what is called a social career. 1925 *Ladies' Home Jrnl.* Apr. 163/3 After you married Jack Hollsworth you went into a sort of social eclipse and almost kept out of things entirely. 1930 G. B. SHAW *Apple Cart* I. 39 The King's

displeasure is still a sentence of social death within range of St James's Palace. 1938 L. BEMELMANS *Life Class* II. v. 164 Their committee selected their dates..at the beginning of the season, but late enough to give them some knowledge of the social calendar. 1977 G. SCOTT *Hot Pursuit* v. 51 Little country towns where the social calendar revolved gently around race meetings and the seasons.

5. c. (Earlier and further examples.)

1792 N. WEBSTER in E. E. Ford *Notes on Life of N. Webster* (1912) I. 363 A number of Gentlemen meet at my house for the purpose of forming a social Club. 1817 COLERIDGE *Biographia Literaria* II. xxii. 136 In the social circles of private life we often find a striking use of the latter put a stop to the general flow of conversation. 1843 MILL *Logic* II. IV. v. 264 The accident that one of the words was used and not the other on a particular occasion or in a particular social circle. 1872 B. JERROLD *London* xix. 155 The [Covent Garden] piazzas..where a few noteworthy social clubs still linger. 1935 *Burlington Mag.* Apr. 161/2 All social circles allied to the Court. 1966 J. CLEARY *High Commissioner* ix. 185 He..belonged to none of the social clubs. He played golf..at a public course. 1977 *Evening Post* (Nottingham) 24 Jan. 7/6 But the couple who lived there escaped with their lives—because just two hours earlier a neighbour had persuaded them to go with him to a local social club.

d. Of a room, a building, etc.: used for friendly intercourse or association. See also *social centre*, sense 12 below.

1889 KIPLING *From Sea to Sea* (1899) I. xxii. 426 The ladies' saloon..according to American custom, was labelled 'Social Hall'. 1975 C. POTOK *In Beginning* (1976) iv. 234 After the service we all went to the social hall downstairs and there was wine and whiskey and cake.

6. a. (Further examples.)

1842 *Boston Quarterly Rev.* 184 Man is a social Being. 1966 G. N. LEECH *Eng. in Advertising* i. 3 Yet the study of language can be regarded as central to man's study of himself, whether as an individual or as a social being.

c. *social whale* (earlier example); = *pilot whale* s.v. PILOT *sb.* 6.

1865 H. D. THOREAU *Cape Cod* vii. 130 In the summer and fall sometimes, hundreds of blackfish (the Social Whale..)..are driven ashore.

7. a, b. (Earlier and further examples.)

1695 LOCKE *Some Thoughts concerning Educ.* (ed. 3) 191 Careful guard ought to be kept over them [sc. children]; and every least slip in this great social vertue taken notice of and rectified. a1704 —— *Conduct of Understanding* (1754) 164 We should love our neighbour as ourselves, is such a fundamental truth for the regulating human society, that, I think, by that alone, one might without difficulty, determine all the cases and doubts in social morality. 1801 M. EDGEWORTH *Belinda* (1833) I. xvi. 135 His social prejudices were such as..to supply the place of the power and habit of reasoning. 1814 M. BIRKBECK *Journey through France* 22 The labouring class here is certainly much higher, on the social scale, than with us. 1830 J. S. MILL *Let.* 9 Feb. in *Wks.* (1963) XII. 48 Those parts of our social institutions and policy which at present oppose improvement. 1833 —— in *Monthly Repos.* VII. 801 The St. Simonians are, just now, the only association of public writers existing in the world who systematically stir up from the foundation all the great social questions. 1843 Social phenomenon [see HISTORICAL *a.* (*sb.*) 2 c]. a1854 MILL *Draft Autobiogr.* (1961) 173 The social problem of the future we considered to be, how to unite the greatest individual liberty of action with an equal ownership of all in the raw material of the globe & an equal participation of all in the benefits of combined labour. 1856 GEO. ELIOT in *Westm. Rev.* X. 70 The study of at least one social group—namely, the factory operatives. 1857 *Edin. Rev.* CVI. 223 Goethe's early experiences at first led him to view the whole social fabric with contempt. 1858 TROLLOPE *Dr. Thorne* I. i. 2 Its social graces, and the general air of clanship which pervades it [sc. Barsetshire]. 1859 THACKERAY *Virginians* II. xxxiii. 266 To marry without a competence is..a crime against our social codes. 1861 J. S. MILL in *Fraser's Mag.* Dec. 672/1 This is the highest abstract standard of social and distributive justice. 1863 *Home & Foreign Rev.* Oct. 546 The multiplicity of her characters,..the richness of her social backgrounds. 1864 Social hierarchy [see HIERARCHY 4]. 1869 MILL *Subj. Women* iv. 163 Self-respect, self-help, and self-control..are the essential conditions both of individual prosperity and of social virtue. 1871 A. C. FRASER *Life of Berkeley* ii. 88 He was shocked by the tone of social morality, which so appallingly greeted him on his return. 1876 H. SPENCER *Princ. Sociol.* I. III. ii. 629 After welfare of the social group and welfare of progeny, comes welfare of parents. 1887 J. BASCOM *Sociol.* i. 9 But, with fitting modifications, they shape also the social contact of diverse ranks. 1887 J. PAYN *Holiday Tasks* 123 If people would only say what they really think concerning this and that..social life would be much more interesting. 1892 Social welfare [see WELFARE *sb.* 1]. 1897 *Amer. Jrnl. Sociol.* Nov. 343 As we say in sociological language, there was a very low degree of social consciousness. 1901 W. JAMES *Mem. & Stud.* (1911) vii. 150 There are social prejudices which scientific men themselves obey. 1901 *Amer. Jrnl. Sociol.* Nov. 399 He makes a better home and moves upward in the social scale, perhaps, faster than the immigrant from any other country. 1902 Social justice [see *ENVIRONMENTAL *a.*]. 1911 M. CORELLI *Life Everlasting* ix. 210 He doesn't fit into any accepted social code at all. 1927 B. RUSSELL *Outl. Philos.* ii. 27 Knowledge..as a social phenomenon..is something displayed in bodily movements. 1935 H. EDIB *Clown & his Daughter* xxv. 137 The shadow of a social barrier was added to the damnable shadow of separation! 1935 T. S. ELIOT *Murder in Cathedral* ii. 80 There are times when violence is the only way in which social justice can be secured. 1938 L. MACNEICE *I crossed Minch* ix. 130 A mind must be conditioned by education and social context. 1940 *Economist* 5 Oct. 428/1 More than two decades of extremely divergent developments in contrasting social climates had to be undone. 1947 *Mind* LVI. 327 If we speak of 'social problems', that is something different. 1949 M. FORTES *Social Structure* 55 The British House of Commons is a familiar instance of growth in social institutions and organization. 1955 T. WILLIAMS in S. J. Kunitz *20th Cent. Authors* Suppl. I. 1088/1 In St. Louis we suddenly discovered there were two kinds of people, the rich and the poor, and that we belonged more to the latter... It was the beginning of the social-consciousness which I think has marked most of my writing. 1957 *Practical Wireless* XXXIII. 727/2 (Advt.), You get a welcome break from the usual routine, with sports, games and a great social life. 1964 M. ARGYLE *Psychol. & Social Probl.* xvi. 199 In limited spheres advice is also given by social scientists,..on the social welfare of the old, young, and poor. 1966 G. N. LEECH *Eng. in Advertising* v. 49 Slang and familiar forms of language..help..to fix the identity and social background of the speaker. 1967 Social mix [see *MIX *sb.*¹ 1 a]. 1970 *Guardian* 3 June 8/6 The Arts Council and its affiliated agencies..are seen as a vital part of the social fabric for which society must be responsible. 1970 F. C. WEFFORT in I. L. Horowitz *Masses in Lat. Amer.* xi. 391 It legalized the 'social question'; that is, it formally recognised that the masses have a right to express their aspirations. 1974 in *Wertheim's Evolution & Revolution* 91 In looking for the structural features of social life we look first for the existence of social groups of all kinds. 1978 *Bookseller* 17 June 3186/3 A school with only 170 children, a high percentage of whom have severe social problems. 1981 G. PRIESTLAND *Priestland's Progress* ii. 34 The compilers of the gospels had other things on their minds than..Jewish social problems. 1982 *Times* 14 June 9/5 The bribery, abuse of privilege, and indifference to social welfare on his own [sc. the Labour Party] side.

c. (Earlier and later examples.)

1835 H. REEVE tr. *de Tocqueville's Democracy in Amer.* II. ix. 256 The Anglo-Americans settled in the New World in a state of social equality. 1840 *Ibid.* IV. III. xii. 106 They have allowed the social inferiority of woman to subsist. 1840 J. S. MILL in *Westm. Rev.* Mar. 262 The demoralizing effect of great inequalities in wealth and social rank. a1876 H. MARTINEAU *Autobiogr.* (1877) (ed. 3) I. 297 Norwich..has now no claims to social superiority. 1885 W. HARRIS *Hist. Radical Party* xvii. 429 Whigs and Conservatives alike desired..conditions and limitations which should preserve power to the same social class which had now the control of so many of the constituencies. 1888 E. BELLAMY *Looking Backward* xi. 164 Who are willing to be domestic servants..where all are social equals? Our ladies found it hard enough to find such even when there was little pretense of social equality. 1917 N. DOUGLAS *South Wind* xxxix. 453 Her home broken up; her child a bastard; herself and Meadows—social outcasts. 1925 *New Yorker* 11 July 6/1 The fact that Davis is a social outcast because of his want of faith..wouldn't cut any ice. 1928 MRS. BELLOC LOWNDES *Diary* 20 Feb. (1971) 113 One of Curzon's most unfortunate peculiarities was his rudeness to those whom he considered his social inferiors. 1934 M. V. HUGHES *London Child of Seventies* xiii. 159 One must never suppose that any other people whatsoever are one's social superiors. 1944 L. P. HARTLEY *Shrimp & Anemone* xiv. 200 The social superiority of the South over all parts of England. 1948 'G. ORWELL' in *Observer* 28 Nov. 4/4 The social misfit..should learn to be contented in his own station. 1958 J. K. GALBRAITH *Affluent Society* vi. 55 In the central tradition of economic theory, the existence of social classes—of capitalists, middle class, and proletarians—was only surreptitiously conceded. 1964 T. B. BOTTOMORE *Elites & Society* i. 23 S. F. Nadel..emphasizes 'social superiority' as the distinguishing feature of an elite. 1967 A. L. LLOYD *Folk Song in Eng.* ii. 86 The medieval peasant..his illiteracy, his social inferiority. 1970 N. A. VICTORIA in I. L. Horowitz *Masses in Lat. Amer.* xv. 549 Among the remaining social classes, the split was reflected in general skepticism. 1973 *Listener* 28 June 863/2 The Industrial Revolution..became in time a social revolution and established that social equality on which we all depend.

e. *Social Sciences.* Pertaining or due to the interrelations resulting from an individual's association with others or connected with the functions and structures necessary to membership of a group or society. Also *transf.* in *Zool.*

There is no rigid demarcation between this sense and the primary meaning of 7 a and b; the examples given illustrate some of the uses commonly found among writers on the social sciences.

1840 J. S. MILL in *Edin. Rev.* Oct. 5 By Democracy M. de Tocqueville understands equality of conditions; the absence of all aristocracy, whether constituted by political privileges, or by superiority in individual importance and social power. 1843 Social organism [see DYNAMICS 1 b]. 1852 J. S. MILL in *Westm. Rev.* II. 380 Attention is due to those opinions and feelings,..not as matter of history, but as social forces in present being. 1861 —— *Repr. Govt.* iii. 68 Where this school of public spirit does not exist, scarcely any sense is entertained that private persons, in no eminent social situation, owe any duties to society, except to obey the laws and submit to the government. 1876 H. SPENCER *Princ. Sociol.* I. II. iii. 487 That social integration that results from the clustering of clusters, is joined with augmentation of the number contained by each cluster. 1877 G. H. LEWES *Problems* III. 5 A new factor, namely, the social factor. 1878 W. JAMES *Let.* 25 Nov. in R. B. Perry *Tht. & Char. of W. James* (1935) II. 35 Their only weakness would lie in the fact of their social environment not recognizing this as the ultimate interest. 1890 —— *Princ. Psychol.* I. x. 293 A man's Social Self is the recognition which he gets from his mates. 1904 —— *Mem. & Stud.* (1911) vi. 138 It would never occur to a reader of his [sc. Spencer's] pages that a social force proper might be anything that acted as a stimulus of social change. 1905 A. W. SMALL *Gen. Sociol.* xxvii. 381 It is the factor which is essential in the end, to economize and co-ordinate all the details of social adjustment. 1920 THOMAS & ZNANIECKI *Polish Peasant* IV. p. xii, Many individuals..consider the social isolation and relatively low cultural level of the peasant communities an undesirable phenomenon. 1934 C. W. MORRIS in G. H. Mead *Mind, Self & Soc.* p. xvi, Though not used by Mead, the term 'social behaviorism' may serve to characterize the relation of Mead's position to that of John B. Watson. 1936 M. SHERIF *Psychol. of Social Norms* 3 We shall consider customs, traditions, standards, rules, values, fashions, and all other criteria of conduct which are standardized as a consequence of the contact of individuals, as specific cases of 'social norms'. 1936 *Amer. Jrnl. Orthopsychiatry* VI. 416 (*title*) Trends in social therapy. 1941 MILLER & DOLLARD (*title*) Social learning and imitation. 1941 *Mind* L. 396 Boodin's reflections on society and the social behaviour of men have, obviously, been deeply influenced by these two special sorts of experiences. 1944 *Mind* LIII. 351 It is said to be evident 'on evolutionary grounds' that the individual is 'higher than the state or the social organism'. 1945 E. MAYO *Social Problems of Industrial Civilization* i. 13 Social skill shows itself as a capacity to receive communications from others, and to respond to the attitudes and ideas of others in such fashion as to promote congenial participation in a common task. 1949 M. MEAD *Male & Female* i. 10 None of these powers—to kill individuals, to destroy the social integration of groups..are new. 1951 GERTH & BRAMSTEDT tr. *Mannheim's Freedom, Power & Democratic Planning* i. 6 By social techniques I refer to all methods of influencing human behavior so that it fits into the prevailing patterns of social interaction and organization. 1959 W. F. LEOPOLD in J. A. Fishman *Readings Sociol. of Lang.* (1968) 349 The colloquial standard [speech] of an individual has several layers suitable for a variety of social situations. 1964 M. ARGYLE *Psychol. & Social Probl.* viii. 108 Social factors are of considerable importance in job satisfaction, according to a number of early studies. 1975 E. O. WILSON *Sociobiol.* II. vii. 160/1 Although the development of 'social behavior' has not been analyzed in these animals, the visible responses are..elementary and stereotyped. 1978 D. GRYLLS *Guardians & Angels* i. 31 Traherne..blames the social environment instead of original sin.

9. a. (Earlier and further examples.)

1833 J. S. MILL in *Monthly Repos.* VII. 269 An error which many..of our social reformers, habitually fall into. 1859 G. A. SALA *Twice Round Clock* 36, I am glad to observe, for the edification of social economists. 1885 W. HARRIS *Hist. Radical Party* vii. 120 Some immediate remedy such as the Spencean and other social theorists had to offer. 1919 BRANFORD & GEDDES *Coming Polity* (ed. 2) III. i. 223 Compound these three insurgent types of social critic..and you have the disorders of Revolution. 1926 B. WEBB *My Apprenticeship* v. 217 In comparison with the preceding generation of social researchers, I suggest that his [sc. Charles Booth's] method of analysis constitutes..the first sign-post directing the student on one of the main ways to discovery. 1931 *Times Lit. Suppl.* 22 Oct. 809/2 Like the work of so many of the 'social' novelists of his period, it is to a large extent a sort of narrative journalism of contemporary events. 1941 J. S. HUXLEY *Uniqueness of Man* xi. 251 Our social planners would undoubtedly benefit from a study of the evolution of individuality in animals. 1949 M. FORTES *Social Structure* p. ix, Their theme was the comparative study of human society by the methods of the natural sciences, and the difference between such studies and those of social philosophers. 1970 R. A. H. ROBINSON *Origins of Franco's Spain* v. 224 In the terminology of European Catholic thought, the Spanish Monarchists of the 1930s were social-romantics, the *Cedistas* social-reformists. 1970 *Guardian* 7 Aug. 10/5 [Notting Hill's] over-exposure is the result of the number of social agencies and sheer do-gooders that have moved into the area.

b. (Earlier and further examples.)

1828 J. S. MILL in *Westm. Rev.* IX. 257 In political and social philosophy his [sc. Sir Walter Scott's] principles are all summed up in the orthodox one, that whatever is English is best. 1835 H. REEVE tr. *de Tocqueville's Democracy in America* I. ii. 23 The..main ideas which constitute the basis of the social theory of the United States. 1836 J. S. MILL in *Westm. Rev.* XXVI. 11 Laws of society, or laws of human nature in the social state..form the subject of a branch of science which may be aptly designated from the title of *social economy*; somewhat less happily by that of *speculative politics*, or the *science* of politics, as contra-distinguished from the art. 1837 —— in *Ibid.* XXVIII. 100 These men raised the cry of social reform. 1887 B. WEBB *My Apprenticeship* (1926) 418 Seeking justification in social research. 1899 *Amer. Jrnl. Sociol.* May 765 But they have all come by experience to discover that the social ethics of Christianity can indeed supply a moral basis of a general kind for social work and social politics. 1901 *Amer. Jrnl. Sociol.* Jan. 472 We have a social technology—a system of conscious and purposeful organization of persons in which every actual, natural social organization finds its true place, and all factors..cooperate to realize an increasing aggregate and better proportions of the 'health, wealth, beauty, knowledge, sociability, and rightness' desires. 1914 G. B. SHAW *Dark Lady of Sonnets* 129 Our plays of poverty and squalor..will then be..read only by historical students of social pathology. 1914 *New Republic* 14 Nov. 28/2 The author believes that one way to write best sellers is to write filth. This is not as it should be. These two propositions, taken together, are social criticism. 1920 W. R. SORLEY *Hist. Eng. Philos.* xii. 265 Maurice's..work, both in social reform and in religion, derived stimulus and direction from philosophical ideas. 1931 *Proc. Nat. Conf. Social Work 1930* 448 Social planning suggests activity on the part of groups in making planning effective in action. 1937 L. C. KNIGHTS in *Scrutiny* VI. 137 It is just possible to claim that Restoration comedy contains 'social criticism' in its handling of 'the vulgar'. 1949 *Mind* LVIII. 383 'Social theory' is the general study of the whole field of social phenomena. 1957 P. COVENEY *Poor Monkey* iv. 54 The social novel of Disraeli, Mrs. Gaskell, and Kingsley..represents something essential to the literary consciousness of the age. 1964 M. ARGYLE *Psychol. & Social Probl.* xvi. 202 Another objection to social planning is that it is felt to increase the power of the state and restrict individual freedom. 1976 *Listener* 3 June 705/3 A different style of thought in social philosophy..post-Marxist critique. 1979 A.

EASSON *Elizabeth Gaskell* ii. 61 *Mary Barton* and *North and South* are often spoken of in the context of social fiction. 1980 *Times* 9 Jan. 10/3 The British tradition of politically committed social research.

c. Of activities, etc., carried out (esp. by government agencies) to improve the condition of society or for the benefit of society as a whole.

1964 *Times Rev. Industry & Technol.* Jan. 65/3 The large backlog in 'social' investments—schools, hospitals, public and private housing. 1965 B. PEARCE tr. *Preobrazhensky's New Economics* 222 Where the tractor is acquired by society as a whole it will facilitate the transition to the social cultivation of the land throughout the countryside. 1973 *Listener* 1 Mar. 287/2 Love of the poor needs social legislation to stiffen it up. 1977 *Spare Rib* Jan. 25/1 Legislation is ineffectual while cuts in social spending continue.

10. *social chauvinism*, a communist term for the attitude or action of a socialist party which supports the non-socialist government of its country in the prosecution of a war; so *social chauvinist* sb. and adj.; *social democracy*, (the advocacy of) a socialist system achieved by democratic means (formerly also a general term for socialism and communism); *social democrat* (further examples); also applied *spec.* to a member of the communist party in Russia and elsewhere; now applied chiefly to one who advocates the achievement of socialism by democratic means; *esp.* in the U.K., a member of the Social Democratic Party; hence *social democratic* adj., now *esp.* in the U.K. designating a party founded in March 1981 by a group of former Labour MPs; also in extended use; *social fascist*, term used by communists for a member of any other left-wing party (implying the identity of non-communist socialism with fascism); hence *social fascism*; *social-imperialism*, a term used at one time in China (and occas. elsewhere) for policies held to conceal imperialist aims beneath a socialist veneer; hence *social imperialist*; *social revolutionary* adj., advocating or supporting social revolution (applied *spec.* to a former political party in Russia); also as *sb.*

1976 H. T. WILLETTS tr. *A. Solzhenitsyn's Lenin in Zurich* 256 'The development of the international socialist movement', he [sc. Lenin] wrote, 'is moving slowly..but definitely in the direction of "a break" with opportunism and social chauvinism.' 1957 R. N. CAREW HUNT *Guide to Communist Jargon* xliv. 147 The social chauvinists are the socialist leaders who were supporting their bourgeois governments in prosecuting the war as one of national defence. 1974 J. WHITE tr. *Poulantzas's Fascism & Dictatorship* IV. ii. 185 It was during this same period, mainly after 1930, that the *social-chauvinist* side of KPD policy grew decisively. 1888 G. B. SHAW in *Fabian Essays in Socialism* (1889) 183 What then does a gradual transition to Social Democracy mean specifically? It means the gradual extension of the franchise; and the transfer of rent and interest to the State. 1928 [see *CENTRIST b]. 1947 *Vogue* May 104/3 Sweden represents the best way of existence a flourishing social-democracy has yet found. 1974 J. WHITE tr. *Poulantzas's Fascism & Dictatorship* IV. i. 151 Social democracy, except sometimes in revolutionary periods, has in principle a permanent mass basis in a capitalist formation. 1981 *Times* 10 Mar. 1/6 Mr Jenkins said that 'well before Easter' the Council for Social Democracy would have been turned into the Social Democrat Party. 1881 Social democrat [see *PROLETARIATE, -AT 2 a]. 1918 [see *MINIMALIST 1]. 1947 *Partisan Rev.* XIV. 312 Barea was one of those gray murky middle-class Social Democrats who made of his adherence a means of avoiding rather than of engaging in political thought. 1965 *New Statesman* 14 May 753/2 A preference for the Social Democrat stronghold [sc. Berlin] over the Christian Democrat capital [sc. Bonn]. 1974 tr. *Sniečkus's Soviet Lithuania* 13 During the revolution, increasing numbers of Lithuanian Social-Democrats urged that their party unite with the Russian Social-Democratic Labour Party. 1981 *Times* 26 Mar. 14/8 We have gained the sympathy of a quarter of Britain's voters, and between a third and two-fifths if Social Democrats fight together with Liberals. 1870 *Times* 10 Oct. 3/2 General von Falkenstein has issued the following order:— 'The prohibition to hold social Democratic meetings is rescinded.' 1887 [see INTERNATIONAL *a.* b]. 1893 [in Dict., sense 11]. 1966 I. DEUTSCHER in *Marxism in Our Time* (1972) 36 Trotsky..represents the Marxist school of thought in its purity, as it existed before its debasement by the social-democratic and Stalinist orthodoxies. 1975 *Times Lit. Suppl.* 23 May 563/4 Recommendations which ranged from orthodox socialist models to moderately social-democratic solutions. 1981 *Guardian* 26 Mar. 4/8 The Social Democratic Party will be launched today. 1981 *Times* 12 Dec. 7/1 It is essential..to divine in good time which of the two mutually exclusive positions on the subject [of snow] will be taken by each person one meets. As far as snow is concerned, there is no middle ground, no Social Democratic stance. 1941 'G. ORWELL' in V. Gollancz *Betrayal of Left* 210 All orthodox Communists were committed to the belief that 'Social-fascism' (i.e. Socialism) was the real enemy of the workers. 1974 J. WHITE tr. *Poulantzas's Fascism & Dictatorship* IV. i. 149 It was no accident that the theory of social fascism was unfailingly accompanied by the identification of fascism with the other forms of bourgeois State. a1937 J. BELL in *Essays, Poems & Lett.* (1938) II. 294 'The adversary.' He takes the form of an enthusiastic member of the Young Communist League and he bellows incessantly. That I am a social-democratic, social-fascist, weak-kneed traitor. That I am a bourgeois intellectual. 1961 [see *FASCIST *sb.* and *a.*]. 1975 *Guardian* 19 Mar. 2/3 Their [sc. the Portuguese Maoists'] stated philosophy is that the Armed Forces movement, the Communist Party, and other left-wing groups are 'Social Fascists'. 1965 tr. *Lenin's Tasks of Third International* in *Coll. Works* XXIX. 502 'Fabian imperialism' and 'social-imperialism' are one and the same thing: socialism in words, imperialism in deeds. 1971 Social imperialism [see *REVISIONIST *a.*]. 1978 HUA KUO-FENG in *Peking Rev.* 10 Mar. 35/2 The people's struggle against them, and in particular against Soviet social-imperialism, is on a higher upsurge than ever. 1974 *China Reconstructs* July 4/1 The Soviet revisionist social-imperialists bill Confucius as 'the most holy sage and foremost teacher of China.' 1907 Social revolutionary [see *INTEGRALIST *sb.*]. 1918 [see *MINIMALIST 1]. 1931 E. WILSON *Axel's Castle* viii. 270 He [sc. Rimbaud] had flamed up, at the fall of the Second Empire, with social-revolutionary idealism. 1978 *Listener* 17 Aug. 206/2 Lenin proposed that the land should be given to the peasants. The indignant social revolutionaries shouted out, 'But that's our programme..and you have opposed it.'

11. *social-conscious, -cultural, -economic, -emotional, -ethical, -minded, -philosophical, -political* (earlier example), *-relational, -situational.*

1856 GEO. ELIOT in *Westm. Rev.* X. 68 The views at which he has arrived.., he sums up in the term—*social-political-conservatism*. 1919 M. BEER *Hist. Brit. Socialism* I. I. v. 71 A serious contribution to social-economic speculation. 1932 *Addresses & Proc. Nat. Educ. Assoc. of U.S.* LXX. 231 It has been thought that social-economic planning in the United States would break down our democratic form of government. 1939 A. HUXLEY *After Many a Summer* II. v. 229 For these..'normality' in sexual behaviour would be quite different from what it was for the more social-minded. 1940 K. MANNHEIM *Ideology & Utopia* 35 The discovery of the social-situational roots of thought at first, therefore, took the form of unmasking. 1942 C. HIMES *Black on Black* (1973) 183 His social-conscious protestations of hurt had leapt the bounds of amateur sincerity. 1951 PARSONS & SHILS *Toward General Theory of Action* I. i. 18 Fundamentals of behavior psychology..primary viscerogenic and possibly social-relational needs, cognition and learning. 1956 J. KLEIN *Study of Groups* viii. 118 Social-emotional behaviour. 1960 C. S. LEWIS *Studies in Words* i. 22 Thus from the very first the social-ethical meaning, merely by existing, is bound to separate itself from the status-meaning. 1970 A. G. FRANK in I. L. Horowitz *Masses in Lat. Amer.* vi. 220 There are undoubtedly differences in..various social-cultural indices between the self-built and the other two types of low-income urban settlements. 1977 A. GIDDENS *Stud. in Social & Polit. Theory* viii. 291 Durkheim..was very critical of some features of Comte's social-philosophical writings.

12. Special collocations. **social action,** deliberate action that results in the restructuring of institutions or a change in the conditions of life in a society; **social anthropology,** the study of (esp. primitive) peoples comparatively through their kinship systems, associations, institutions, culture, etc., and the forces that affect their social systems; hence **social anthropological** *a.*, **social anthropologist; social benefit,** (*a*) a benefit to society resulting from technological innovation and the like; (*b*) a benefit (*BENEFIT *sb.* 4 d) payable under a system of social security; **social butterfly,** a person who flits from one social entertainment to another or is chiefly occupied with social activities, a socialite; **social case-work, case-worker** = *CASE-WORK, *CASE-WORKER; **social causation,** the causation of human actions by social factors, or their analysis in these terms; **social centre,** any place in which people gather for communal activities, recreations, etc., *esp.* a building designed for this purpose; **social change,** change in the customs, institutions, or culture of a society brought about by some new, esp. technological or ideological, element; **social character** (see quots.); **social climber,** one who seeks to advance himself socially, esp. by gaining acceptance in fashionable society; hence (as back-formations) **social climb** *sb.* and *v. intr.*, **social climbing** *vbl. sb.* and *ppl. a.*; **social column,** a column in a newspaper or magazine that reports the activities of members of fashionable or leisured society; hence **social columnist; social compact** = *social contract*, below; **social conscience,** a conscience that is sensitive to or preoccupied with the problems and injustices of society; **social contract,** (*a*) (see sense 7 b in Dict.) (*b*) *transf.*, a mutual agreement between specific groups or elements within a society; **social control,** control of the individual by the social group to which he belongs; control by government, on behalf of society as a whole, of particular sectors of society; **social cost,** the cost to society in terms of effort, ill-health, inconvenience, etc., of some enterprise or innovation; **social cycle** (see quot. 1963); **social Darwinism** *Sociol.*, the Darwinian theory of evolution extended and applied to various aspects of the concept of social progress; hence **social Darwinist** *sb.* and *a.*; **social deprivation,** deprivation of social interaction or of the ordinary benefits of social life; **social dialectology,** the study of the dialects spoken by particular social groups; hence **social dialectologist; social differentiation,** the process whereby a group or community becomes separate or distinct; the process whereby the different roles and functions of individuals become institutionalized; **social disease,** any social evil such as poverty, starvation, etc.; *spec. U.S.*, venereal disease (orig. a euphemism); **social disorganization** (see quot. 1920); **social distance** *Social Psychol.*, (*a*) the degree of remoteness that a member of one social group would like to exist or feels to exist between himself and the members of another, expressed (for example) in terms of the relationships to which he would admit them; (*b*) the physical distance between individuals that they find acceptable in social contexts; **social document,** a literary work embodying an authentic and informative description of the social conditions of its time; also *transf.*; **social drinking** *vbl. sb.*, the drinking of alcoholic liquor as a stimulus to, or an accompaniment of, social intercourse; hence as *ppl. a.* and (as back-formations) **social drink** *sb.*, **social drinker; social dynamics,** (the branch of sociology treating of) the forces at work in social change; **social engineering** orig. *U.S.*, the application of sociological principles to specific social problems; hence **social engineer,** a specialist in this field; **social evolution:** see *EVOLUTION 9; **social fact,** something originating in the institutions or culture of a society which affects the behaviour or attitudes of the individual member of that society; **social geography** (see quot. 1929); hence **social geographer; social gospel,** an understanding of the gospel as having especially a social application; used esp. with reference to many U.S. churchmen of the late nineteenth and early twentieth centuries who advocated social reform through the Christian gospel; also *gen.*, a message of salvation for society; **social history,** (*a*) the history of social behaviour or of society (the usual sense); (*b*) the background and circumstances of a social worker's client; hence (in sense (*a*)) **social historian, social-historical** *a.*; **social inquiry report** (see quot. 1967); **social insurance,** the insurance of the citizen against loss of income through sickness, unemployment, etc., with the participation of the government and the employer; also in extended use and *attrib.*; **social lie,** an untrue statement designed to facilitate social relations; hence **social liar; social medicine** (see quot. 1925); **social mobility** = *MOBILITY¹ 1 c; **social mobilization** *Sociol.*, (see quot. 1961); **social morphology,** (the study of) the various forms of social structure and the changes that take place within them or govern them; **social order,** (*a*) orderliness within society, absence of disorder and unrest; (*b*) the way in which society is organized at a given time, the constituted social system; **social organization:** see *ORGANIZATION 2 c; **social position:** see *POSITION *sb.* 9 b; **social process,** a pattern that can be discerned in the way a society coheres and adapts to change over a period of years; **social psychiatry,** the branch of mental health concerned with the social causes and social consequences of mental illness and with the various social methods which may be used to treat such illness; hence **social psychiatric** *a.*, **social psychiatrist; social psychology,** the study of human behaviour as it is affected by social factors; hence **social-psychological** *a.*, **social psychologist;** cf. *collective psychology* s.v. *COLLECTIVE *a.* 2 d; **social realism,** realism in art and literature that has a specifically social or political content or message; sometimes applied *spec.* to a movement in U.S. art in the 1930s; also *attrib.*;

hence **social realist** sb. and a., **social realistic** a.; **social reality**, a conception of what exists that is affected by the customs and beliefs of the group; **social register** orig. *U.S.*, a register or directory of those who are socially prominent; *transf.*, a union black list; also *attrib.* or as *adj.*; **social releaser** = *RELEASER c; **social revolution**, a revolution in the structure and nature of society; *spec.* that anticipated or fostered by socialists and communists (cf. *social revolutionary* a. and *sb.*, sense 10 above); **social role** = *RÔLE 2; **social secretary**, a secretary whose function it is to make arrangements for the social activities of a person or society; **social space** *Sociol.*, the 'space', in terms of the difference in social position or individual freedom of action, that is felt to exist between one person and another; **social statics** *Sociol.*, the study of the organization and structure of a stable society or social group; **social status**: see *STATUS 3; **social stratification**, the division of society into strata based on social position or class; **social strata, stratum**: see *STRATUM 6; **social structure**, the established set of customs, relationships, institutions, etc., of which a social system is composed; hence **social structural** a.; **social studies**, an inclusive term for various aspects or branches of the study of human society; **social survey**, a comprehensive and detailed examination of some aspect of the social life, history, problems, etc., of a particular locality; **social system**, a set of interdependent relationships, customs, institutions, etc., that constitute a society; **social table** (see quot. 1952); **social unit**, an individual considered as one of the separate parts of which a society or group is composed; a community or group considered as having a separate identity within a larger whole; **social wage** (see quot. 1975); also *transf.*; **social will**, a term for the desires regarding the affairs of a society or group expressed by its members in general.

1853 H. Martineau tr. *Comte's Positive Philos.* II. viii. 246 Its distinctive social action..was well represented by the noble Fabricius. 1873 H. Spencer *Study of Sociol.* i. 2 Minds in which the conceptions of social actions are thus rudimentary, are also minds ready to harbour wild hopes of benefits to be achieved by administrative agencies. 1937 T. Parsons (*title*) The structure of social action. 1951 R. Firth *Elements of Social Organization* i. 33 No social action, no element of culture, can be adequately studied or defined in isolation. 1953 in S. Tax et al. *Appraisal of Anthropology Today* xiii. 220/2 The outline does not vary much from subjects..considered to be in the cultural/social-anthropological field. 1927 Ogburn & Goldenweiser *Social Sciences* ii. 11 Then there are the social anthropologists who make economic activity the basis of social anthropology. 1896 *Academy* 18 Jan. 49/1 Dr. Steinmetz..found processes of moral, political, and religious development so intricately entwined, that his researches had to spread far over the field of social anthropology. 1908 J. G. Frazer *Psyche's Task* (1913) 159 (*heading*) The scope of social anthropology, an Inaugural Lecture. 1975 M. Bradbury *History Man* iv. 147, I worked in social anthropology with him..he's certainly not a racist. 1872 Geo. Eliot *Middlemarch* III. iv. 241 Your neatly-carved argument for a social benefit which they [sc. rustics] do not feel. 1963 *Listener* 23 May 855/1 Dr Beesley..and I have attempted a so-called social-benefit study or, as it is sometimes called, a cost-benefit analysis, of the London Victoria line. 1963 J. R. Sargent in M. Shanks *Lessons of Public Enterprise* xv. 250 A balancing of social benefits against social costs. 1972 *Guardian* 30 Dec. 4/1 From Monday every citizen of the nine nations..is entitled to the same pay and social benefits..as fellow Europeans. 1910 A. E. Housman *Let.* 4 Mar. (1971) 108 People are asking me out a great deal too often; I am not a social butterfly. 1938 D. du Maurier *Rebecca* ix. 117 A social butterfly, very modern and plastered with paint. 1975 D. Ramsay *Descent into Dark* i. 15 To be a social butterfly and make my wife happy. 1917 M. E. Richmond *Social Diagnosis* i. The methods and aims of social case work were or should be the same in every type of service. *Ibid.*, The ground which all social case workers could occupy in common. 1896 F. H. Giddings *Princ. Sociol.* i. 20 Thus the cycle of social causation begins and ends in the physical process. 1937 R. M. MacIver *Society* xxvi. 476 In social causation there is a logical order of relationship between the factors that we do not find in physical causation. 1964 Gould & Kolb *Dict. Social Sciences* 647/1 There has been much discussion about the relationship between ideas of social causation and the problem of the freedom of the individual will. 1901 *Amer. Jrnl. Sociol.* Sept. 206 Is there not room for the school..in providing accessible and agreeable social centers? 1922 L. Mumford in H. Stearns *Civilization in U.S.* 6 The social centre and the community centre, which in a singularly hard and consciously beatific way have sought to organize fellowship and mutual aid.., are products of the last decade. 1937 *Discovery* Feb. 47/1 Social amelioration through a special milk scheme, a special housing scheme, and the encouragement of camps and social centres. 1978 J. Anderson *Angel of Death* v. 42 The main saloon..was the social centre of the yacht. 1836 J. S. Mill in *London & Westm. Rev.* Jan. 28 The main thing which social changes can do for the..higher classes..is gradually to put an end to every kind of unearned distinction. 1869 A. C. Fraser *Ess. Philos.* i. 11 The most memorable religious and social change which the world has witnessed since the introduction of Christianity. 1952 Gerth & Martindale tr. *Weber's Ancient Judaism* p. xviii, A second sociological issue of concern to Weber is the examination of social changes due to territorial organization and urbanization. 1942 E. Fromm *Fear of Freedom* 239 The social character comprises only a selection of traits, the essential nucleus of the character structure of most members of a group which has developed as the result of the basic experiences and mode of life common to that group. 1961 R. H. Williams *Long Revol.* i. iii. 79 The 'social character' is a selective response to experience, a learned system of feeling and action, in a majority of the community into which the child is born. 1975 G. Howell *In Vogue* 131/2 Sweaters appear..Their social climb touches its peak..over a grandly outsize evening skirt. 1962 *Sunday Express* 21 Jan. 1/3 Allegations that I was social-climbing among royalty. 1926 S. Lewis *Mantrap* viii. 95 You sniveling little social climber! 1941 A. Christie *Evil under Sun* ix. 178 There are many of your English idioms that describe him. The rough diamond! The self-made man! The social climber! 1973 *Black World* Mar. 21 Whatever the shortcomings of our early literary efforts, they ought not be condemned in the mass as merely representing the fumblings of middle-class social-climbers. 1924 W. Holtby *Crowded Street* xxxviii. 294 The careful tact of years of social climbing. 1927 P. Sorokin *Social Mobility* vii. 133 There are two types of vertical mobility:..social climbing and social sinking. 1938 Auden & Isherwood *On Frontier* iii. ii. 107 A social-climbing ex-public-school boy. 1960 *Twentieth Century* Dec. 588 A political thriller about a middle-aged reporter who hears of a major scandal involving a Minister of the Crown..Good pictures of the social-climbing Minister,..and various Fleet Street characters. 1978 *Trans. Yorks. Dial. Soc.* LXXVIII. 12 He takes the opportunity to castigate the creeping hypocrisy and social climbing which had always called forth his most bitter satire. 1936 'K. West' *Thinking Reed* vii. 223 A crowded paragraph in the social column of the Paris *New York Herald*. 1952 M. Allingham *Tiger in Smoke* i. 3 Every social column in the country had announced that she was about to marry him. 1976 'R. Macdonald' *Blue Hammer* ix. 43, I don't intend to write a social column for the rest of my life. 1979 M. Green *Children of Sun* vi. 209 Hannen Swaffer was the first social columnist who knew from inside the world he wrote about. 1793 etc. Social compact [see Compact sb.² 1 b]. 1979 *Times* 26 Feb. 12/3 Mr Wilson has sought to defend this rubbish by arguing that a great new 'social compact' with the unions will be achieved, by the terms of which they will agree to forgo wage-increases in return for all the splendidly socialist things Mr Wilson's Government will be doing. 1883 B. Potter *Let.* July in *Let. Sidney & Beatrice Webb* (1978) I. 16 So ends the London Season! and I shall retire from this social conscience to my dowdy dress. 1888 G. B. Shaw in *Fabian Essays in Socialism* (1889) 185 The value of Trade Unionism in awakening the social conscience of the skilled workers. 1925 A. Huxley *Those Barren Leaves* I. iii. 35 When our dividends came rolling in..we did, it is true, feel something a twinge of social conscience. 1978 *Architectural Design* 8/1 Public libraries and art galleries were built by..millionaires with a social conscience. 1967 B. R. Williams *New Social Contract* i. 2 In a modern industrial nation..the individual and the community enter in effect into a Social Contract, by which the group as a whole agrees to dedicate a certain part of its total assets..to the provision of benefits for its members. 1972 *Times* 3 Oct. 1/1 'We say that what Britain needs is a new social contract,' Mr Callaghan said. 1974 *Socialist Worker* 23 Nov. 3/4 The last week has seen two massive holes punched through the Social Contract. First by the Rolls-Royce workers in getting their £8 a week rise, then [etc.]. 1977 *Guardian Weekly* 11 Sept. 5/5 The greatest strength of the social contract policy lies..in the social contract with the public. 1859 Mill *Liberty* i. 14 The practical question, where to place the limit—how to make the fitting adjustment between individual independence and social control—is a subject on which nearly everything remains to be done. [illegible] *Social Control* i. 1 [illegible] Social control, on the other hand, means that ascendancy over the aims and acts of the individual which is exercised in behalf of the group. 1898 F. H. Giddings *Elem. Sociol.* xvi. 217 Social control, manifesting itself in the authoritative organization of society..., is the state, and acting through the organs of government, is sovereignty. 1913 L. T. Hobhouse *Devel. & Purpose* p. xviii, A new demand for the extension of collective responsibility and the social control of industrial life. 1951 N. Annan *Leslie Stephen* vii. 218 The field of social science over which the..ideas..on law, custom, religion and moral codes govern men's actions? 1972 *Guardian* 19 Jan. 14/3 By closing its doors relatively early, the public transport system is an effective measure of social control over non-car owners. 1901 *Amer. Jrnl. Sociol.* July 137 For marginal social cost always equals marginal social utility. 1927 G. D. H. Cole *Econ. System* vi. 62 Social cost simply cannot be measured in terms of money, but only in the last resort in terms of human effort and destruction of natural resources. 1977 *N.Y. Rev. Bks.* 31 Mar. 21/2 Rehabilitation [of housing]..has proved to be cheaper in social costs. 1961 *B.S.I. News* Nov. 11/2 The new regulations..also lay down requirements about the lights to be carried on four-wheeled 'social cycles', now common in holiday areas. 1963 *Daily Tel.* 8 July 11/8 A young woman holidaymaker riding a 'social-cycle' was killed... The cycles are often seen at seaside resorts. They are like tricycles with two people sitting beside each other on a bench. 1887 *Mind* XII. 627 There can be no 'social Darwinism'. Social progress is not essentially the result of a struggle, but of intelligence. 1972 P. B. Medawar *Hope of Progress* 71 Social Darwinism in the form expounded by Haeckel provided a theoretical justification for the great biological crimes of Fascism. 1907 *Amer. Jrnl. Sociol.* XII. 709 The great writers on race-struggles never use the term 'social Darwinism' but a number of sociologists have called them 'social Darwinists'. 1945 R. Hofstadter *Social Darwinism in Amer. Thought* ii. 25 In applying evolution to society, Spencer, and after him the Social Darwinists, were simply doing poetic justice to its origins. 1981 J. Sutherland *Bestsellers* iv. 37 There is no room for..cosiness in Hailey's social-Darwinist universe. 1958 *Jrnl. Abnormal & Social Psychol.* LVI. 49 (*title*) The effect of brief social deprivation on behaviors for a social reinforcer. 1970 W. J. Fishman *Streets of E. London* 135/2 The social deprivation inherent in East End life. 1977 *Publ. Amer. Dial. Soc.* 1974 LXI–LXII. 4 Social dialectologists in recent years have made numerous attempts to describe the speech of black Americans. 1980 *Amer. Speech* LVI. 104 Social dialectologists..have also neglected important work of the area linguists. 1979 *Jrnl. Eng. Linguistics* IV. 46 (*title*) Social dialectology in America: a critical survey. *Ibid.*, Although..Mencken..and others had discussed American social dialects, the systematic study of the sociology of American English really began in the late 1950s. [illegible] XVI. 32 Rustic is an example of social dialectology at its thoroughly honest best. 1872 H. Spencer in *Contemp. Rev.* XX. 317 The primary social differentiation which we have noted between the regulative part and the operative part, is presently followed by a distinction..between the internal arrangements of the two parts. 1903 L. F. Ward *Pure Sociol.* ix. 221, I propose to use..the sufficiently vague..term race..for all the different kinds of social groups that have resulted from the process of social differentiation. 1926 C. C. North *Social Differentiation* i. 3 It is essential to any proper understanding of social differentiation that some effort be made to distinguish between the biological and the social in the sources or causes of social distinctions. 1971 R. F. Allen *Socio-Cultural Dynamics* iii. 92 Social differentiation as a major view of change. 1891 T. H. Huxley (*title*) Social diseases and worse remedies. 1907 *Amer. Jrnl. Sociol.* July 20 (*title*) Prophylaxis of social diseases. 1945 G. Endore *Methinks the Lady* vii. 158 What rights? Overtime pay, maybe? Union hours? Sure, social security, maybe? Or social diseases? 1970 *Guardian* 28 Apr. 10/1 Hard drugs addiction is a comparatively social disease. 1978 R. Ludlum *Holcroft Covenant* xxvii. 314 She was probably an Odessa agent and you've come down with a social disease, as planned. 1920 Thomas & Znaniecki *Polish Peasant* IV. i. 2 The question of social disorganization. We can define the latter briefly as a decrease of the influence of existing social rules of behavior upon individual members of the group. 1958 M. Argyle *Relig. Behaviour* xi. 158 The evidence showing how the level of mental disorder increases with social disorganization. 1924 R. E. Park in *Jrnl. Appl. Sociol.* VIII. 344 Prejudice is..a sort of spontaneous conservation which tends to preserve the social order and the social distances upon which that order rests. 1948 M. Sherif *Outl. Social Psychol.* xiv. 341 The average..member of a group exhibits the degree of prejudice toward the member of another group prescribed by the social distance scale of his group. [illegible] According to its species, each individual keeps at a certain social distance [illegible]. 1950 L. & P. Opie *Lore & Lang. Schoolch.* p. v, This pioneer work and social document of first importance is..something of a curiosity. 1974 *Country Life* 30 May 1319/1 This painting of the Lawn at Goodwood 1886..has been purchased by the Goodwood Trust for 7,000 gn. I find it a decided eye-opener in the world of art, but a social document of real importance. 1976 *New Yorker* 12 Jan. 47/2 He's been boning. And I don't mean he's just had a 'social' drink or two. 1969 in Halpert & Story *Christmas Mumming in Newfoundland* 82 Deep Harbour fishermen have traditionally celebrated Christmas by group visiting, whether as mummers or as 'social drinkers'. 1977 E. Leonard *Unknown Man* i. 10 I wasn't an alcoholic. I was a heavy social drinker. 1901 B. S. Rowntree *Poverty* iii The [public] house is evidently one where 'social' drinking is carried on. 1958 Keller & Seeley *Alcohol Lang.* 32 Social drinking..1. Moderate drinking on social occasions. 2. Drinking to comply with the expectation of companions. 3. Drinking in a way and within the limits accepted by a cultural group. 1969 in Halpert & Story *Christmas Mumming in Newfoundland* 82 At night throngs of 'social-drinking' men threading their way along the narrow footpaths sing between visits. 1843 Social dynamics [see Dynamics 1 b]. 1938 B. Russell *Power* i. 11 The laws of social dynamics are laws which can only be stated in terms of power, not in terms of this or that form of power. 1974 *Howard Jrnl.* XIV. 97 Overcrowding and..the introduction of strangers into the resident group, or other alterations in the social dynamics. 1900 W. H. Tolman *Industrial Betterment* ii. 81 'Of course you are too busy..and..need someone on your staff whose sole business will be the planning and direction of movements to improve industrial conditions; in other words you need a social engineer.' Social engineering is a new profession. 1980 *Gazette* (Montreal) 22 Mar. 109/2 Woodward and Armstrong..revile the efforts of President and Congress to appoint Supreme Court judges who will not act as left-of-center social engineers. 1899 *Social Engineering* Sept. 18 The following subjects were adopted for A Course of Lectures on Modern Social Problems... Social Engineering a New Profession. 1919 M. Beer *Hist. Brit. Socialism* II. xiv. 287 The Fabian Society appears to form an institute for social engineering. 1945

K. R. Popper *Open Society* I. ix. 138 The Platonic approach..can be called Utopian engineering, as opposed to that kind of social engineering..which may be described by the name of piecemeal engineering. 1980 *Jrnl. R. Soc. Arts* May 351/2 A sort of extension of Architecture..as a vast subject in its own right with a powerful social-engineering content. 1843 J. S. Mill *Logic* II. iv. v. 273 There is hardly a single name, expressive of any moral or social fact calculated to call forth strong affections.., which does not carry with it..a connotation of those strong affections. 1887 Moore & Aveling tr. *Marx's Capital* I. i. 44 The mutual exchangeability of all kinds of useful private labour is an established social fact [illegible]. 1938 Solovay & Mueller tr. *Durkheim's Rules Sociol. Method* p. iii, We give a definition of social facts as ways of acting or thinking with the peculiar characteristic of exercising a coercive influence on individual consciousnesses. 1977 P. Laslett *Family Life* i. 8 History stands as much in need of a theory of itself as of any other form of generalization about social facts. 1929 P. Geddes in *Sociol. Rev.* XXI. 7 [illegible] 1980 *Verbatim* Autumn 1/1 Social geographers are aware of the [illegible] 1907 G. W. Howe in *Geogr. Jrnl.* XXIX. 67 In addition to the physiographical group of factors which are by common consent held to be fundamental, the sociological factors are no less fundamental [illegible]. 1933 Huntington & Carlson *Environmental Basis of Social Geogr.* i. 5 Social geography [illegible] the relationship between man and his environment] from the point of view of man and his activities, that is, the social aspect. 1886 C. O. Brown *Talks on Labor Troubles* i. 9 Three views are being read as a new social gospel by hundreds of thousands of people. 1896 *Dawn* II. Suppl. 1 In man's relation to man, Jesus Christ preached a social gospel; accordingly, in those relations, his disciples must be socialists. 1917 W. Rauschenbusch *Theol. for Social Gospel* i. 1 We have a social gospel. We need a systematic theology large enough to match it. *Ibid.* 2 The social gospel has become orthodox. *Ibid.* xix. 279 The social gospel is the voice of prophecy in modern life. 1958 M. Argyle *Relig. Behaviour* v. 45 A small college at which it seems that a modernistic and social gospel was widely held among the staff. 1969 A. Richardson *Dict. Christian Theol.* 313/2 The christology of the social gospel focuses on the way the divine life of Christ can get control of human society. *Ibid.*, There were other conspicuous leaders of the social gospel. 1921 A. Conan Doyle in *Strand Mag.* Dec. 603/1 There are few social historians of those days who have not told of the long and close friendship..between Sir Charles Tregellis and Lord Barrymore. 1973 *Listener* 25 Oct. 571/1 Social historians must do the best they can with such evidence as they have. 1897 *Library Jrnl.* Mar. 239 'Colonial Days in Old New York'..is really of the social historical order rather than a book of travel. 1937 *Burlington Mag.* June 291/2 The historical section..is sketchy and its lack of concentration results in an apparent insufficiency of social-historical facts to explain stylistic changes. 1977 A. Wilson *Strange Ride of R. Kipling* vii. 342, I prefer..a social-historical description of long generations of Evangelical belief ending in post-Darwinian doubt. 1856 G. Roberts *Social Hist. of People of Southern Counties of England in Past Centuries* p. v, Researches..disclosed many particulars of the former condition of our countrymen..These have been made available for the express correction of a very general ignorance of our Social History that prevails. 1907 H. Morgan *Ancient Society* i. 5 [illegible] 1950 McDougall & Cormack in C. Morris *Social Case-work in Great Britain* ii. 40 The social history [illegible] against his father's heirs. 1977 *Times* 10 Sept. 2/6 Research in this field is..a piece of social insurance. 1976 R. Harris *Three Candles for Dark* iv. 27 I'm not a liar, or not a real one. A social liar, maybe, like everyone else. 1941 Auden *New Year Letter* iii. 64 And yet although the social lie Looks double to the dreamer's eye. 1969 M. Drabble *Waterfall* 94 She might have invented the information as a social lie (a felicitous duplicity). 1971 W. Tute *Tarnham Connection* vii. 135 [He] said he was expected back home, which Mado knew to be a social lie. 1919 *Lancet* 24 May 921/1 (*heading*) Social medicine in Vienna. 1925 F. L. Dunham *Approach to Social Medicine* i. 14 A need arises in welfare work for a field of preventive science to which social science, psychology, psychiatry and various other departments shall contribute... It may be called Social Medicine. 1977 *Lancet* 24 & 31 Dec. 1336/1 What has been the contribution of social medicine to social policy in general and to health-services policy in particular. 1925 P. Sorokin in *Social Forces* May 635/2 We used to think that in the United States 'social mobility' was greatest. 1954 D. V. Glass (*title*) Social mobility in Britain. 1978 *Listener* 26 Jan. 107/2 The patterns of social mobility over the past generation. 1953 K. W. Deutsch *Nationalism & Social Communication* vi. 114 If there is economic growth, social communication will probably spread and social mobilization will progress. 1961 *Amer. Pol. Sci. Rev.* Sept. 493/1 Social mobilization is a name given to an overall process of change, which happens to substantial parts of the population in countries which are moving from traditional to modern ways of life. 1905 E. A. Ross *Foundations Sociol.* viii. 182 The term..social morphology..will describe, not only human relations and groupings, but also their [illegible] *Sociol. Rev.* XXV. 191/1 Durkheim proposed the examination and comparison of languages..in order to find the manner in which social representations adhere to and repel one another; and how 'our social morphology' is to be explained. 1797 *Encycl. Brit.* XVII. 570/2 Though social order is no longer unknown nor undervalued, yet the form of government is still extremely simple. 1817 [in Dict., sense 7 b]. 1853 H. Martineau tr. *Comte's Positive Philos.* II. vi. i. 6 A polity that could not hold its ground against the natural progress of intelligence and of society can only stand in the way of social order. 1909 H. G. Wells *Ann Veronica* vi. 129, I know that our social order is dreadful enough [illegible] 1928 B. Russell *Pract. & Theory Bolshevism* ii. 28 He [sc. the Communist] is..aiming at the creation of a new social order. 1955 M. Gluckman *Custom & Conflict in Africa* i. 17 What emerges, I think, is that if there are sufficient conflicts of loyalties at work, settlement will be achieved..and social order maintained. 1977 P. Laslett *Family Life* i. 46 The traditional social order on our continent. 1835 H. Reeve tr. *de Tocqueville's Democracy in Amer.* II. x. 429 If republican principles are to perish in America, they can only yield after a laborious social process. 1887 [see *process sb. 5 b]. 1947 Henderson & Parsons tr. *Weber's Theory of Social & Econ. Organization* i. 96 Charisma..is thus the bearer of many dynamic tendencies of social processes. 1974 tr. *Wertheim's Evolution & Revolution* 76 The aim of this study is to deal with revolutions as social processes. 1966 G. Tannenbaum in S. Arieti *Amer. Handbk. Psychiatry* xxv. 577/1 The social psychiatric model is based on public health principles rather than on the traditional clinical prototype. 1964 *Observer* 23 Aug. 1/1 The social psychiatrists..believe that the answers to mental illness can be found by..studying the patient in relation to the groups he moves in. 1924 *Amer. Jrnl. Psychiatry* LXXXI. 149 The Round Table Conferences..were well attended (76 formed the group which discussed problems of social psychiatry. [illegible] *Preventive & Social Psychiatry* [illegible] sessions..are [illegible] other clinical [illegible] years. 1976 [illegible] social [illegible] A collection [illegible] *Language* LIV. [illegible] refer to the [illegible] attitudes. [illegible] social psychologists [illegible] are phenomena [illegible] process of [illegible] *of Work* i. 3 [illegible] problems in [illegible] of the social [illegible] in *Amer.* [illegible] an *Origin of* [illegible] Psychology? [illegible] study the [illegible] This is social [illegible] *Soc. Sciences* [illegible] include social [illegible] matter. 1977 [illegible] On the other [illegible] with psychology [illegible] are socio- [illegible] trained social [illegible] Nov. 186/2 [illegible] mass movements. The artists [illegible] not organic [illegible] in *Horizon* [illegible] is also a [illegible] an aesthetic [illegible] xxix/4 The [illegible] during the [illegible] is aloof and [illegible] Even Ben [illegible] years ago, [illegible] through [illegible] political satire [illegible] The dismal [illegible] art in the [illegible] -inspired s.v. [illegible] Perhaps [illegible] one called [illegible] [see *social* [illegible] Twenty years [illegible] called the [illegible] 'kitchen-sink school'... John Berger wrote social realist reviews of their exhibitions. 1960 *Guardian* 18 Oct. 5/5 The group has..a distinct social realistic bias. 1859 D. Masson *Brit. Novelists* iv. 308 It may be that the representation of social reality is..the proper business of the Novel. 1887 Moore & Aveling tr. *Marx's Capital* I. i. 15 The value of commodities has a purely social reality. 1949 W. L. Warner in M. Fortes *Social Structure* 4 Agreement among the informants assures the status analyst that the social class sytem derived from their statements is..an ever present..social reality. 1978 *Language* LIV. 449 It may be noted that O's social realities of the West German job market refer more to the over-employment of the 1960's than to the under-employment of the 1970's. 1889 (*title*) Social Register, New York. 1945 *Seafarers' Log* 6 July 6/4 The crew recommended that the 'advantages' of the social register be extended to William Chance and J. D. Bell, both trip carders. 1949 *Sat. Even. Post* 15 Oct. 142/3 The student body..has a heavy sprinkling of millionaires' sons and Social Register families. 1981 *Newsweek* 20 July 24 Martha von Bulow was pure Social Register, born into wealth, educated in the best private schools, and married for a time to an Austrian prince. 1953 N. Tinbergen *Herring Gull's World* ii. 13 Colour can act as a 'social releaser' by releasing a response in another individual just as a call often does. 1962 *Listener* 9 Aug. 207/2 In such tribes we see something very like social releasers. 1831 J. S. Mill *Ess.* (1962) 20 There must be a moral and social revolution,..which shall leave to no man one fraction of unearned distinction or unearned importance. 1890 W. Booth *In Darkest England* i. ix. 80 The Socialist tells me that the great Social Revolution is looming large on the horizon. 1910 H. G. Wells *New Machiavelli* (1911) i. iv. 121 The social revolution and the triumph of the Proletariat after the class war. 1941 *Time* 19 May 98/1 World War II is a social revolution, but not the kind of social revolution almost everybody thinks it is. 1974 tr. *Wertheim's Evolution & Revolution* 173 However, true social revolutions in earlier times showed a religious component as well. 1928 *Psychol. Abstr.* 889 Social role of language. 1949 [see *rôle 2]. 1977 Social role [see *role set* s.v. *rôle 4]. 1905 E. Wharton *House of Mirth* ii. viii. 417 Carry promised to find somebody who wants a kind of social secretary—you know she makes a speciality of the helpless rich. 1949 'J. Tey' *Brat Farrar* xxii. 195 She moved him on from one group to another as expertly as a social secretary. 1978 'J. Horbury' *Diplomatic Affair* i. iv. 43 'Never neglect *placement* if you wish to rise.'..'By the time you rise to the point where it matters, some charming social secretary will worry about it for you.' 1925 P. Sorokin *Sociol. of Revol.* xii. 250 Some of their members in two or three years cover an enormous distance in 'social space': from ordinary lawyers,..workers, peasants,..etc., they become persons..occupying high posts. 1927 —— *Social Mobility* i. 3 Persons near each other in geometrical space—*e.g.*, a king and his servant..—are often separated by the greatest distance in social space. 1961 J. N. Findlay *Values & Intentions* ix. 398 The first encounter with social space is normally that of a being without formed views, abilities or values, with a range of beings surpassingly mature,..decided, powerful and in general benign. 1977 T. M. Kando *Social Interaction* xi. 260/1 The study of space becomes truly interesting when it is conceived of as social space. 1843 Social statics [see Dynamics 1 b]. 1851 H. Spencer (*title*) Social statics. 1958 A. R. Radcliffe-Brown *Meth. in Social Anthropol.* i. v. 128 For social anthropology the task is to formulate and validate statements about the conditions of existence of social systems (laws of social statics) and the regularities that are observable in social change (laws of social dynamics). 1927 P. Sorokin *Social Mobility* ii. 11 Social stratification means the differentiation of a given population into hierarchically superposed classes. 1979 G. Ritzer et al. *Sociol.* ix. 238 Their contention that because social stratification is universal, it must be a functional necessity. 1972 P. Laslett *Household & Family* 58 In the relations of children to servants we may..find..important and revealing social structural differences. 1835 H. Reeve tr. *de Tocqueville's Democracy in Amer.* I. v. 69 The Constitution of the United States..consists of two distinct social structures, connected, and..encased one within the other. 1872 H. Spencer in *Contemp. Rev.* XX. 311 Social influences which..facilitate further aggregation with consequent further complexity of social structure. 1949 M. Fortes *Social Structure* p. xiii, The study of kinship systems and..the concept of social structure. 1968 Jacobson & Schoepf tr. *Levi-Strauss's Structural Anthropol.* I. xv. 277 Studies in social structure have to do with the formal aspects of social phenomena. a 1854 Mill *Early Draft Autobiogr.* (1961) 108 The social studies of myself and several of my companions assumed a shape which contributed very much to my mental development. 1926 B. Webb *My Apprenticeship* v. 217 A subtle combination of quantitative and qualitative analysis is a necessary factor in social studies. 1938 S. Chase *Tyranny of Words* vii. 78 The social studies are..backward compared to the physical sciences. 1977 *Lancashire Life* Nov. 136/3 In some schools R.E. is lumped together with history and geography and called Humanities or Social Studies. 1927 *Amer. Jrnl. Sociol.* XXXIII. 424 (*title*) The social survey of Tyneside: an English regional social survey. 1948 [see *participant sb. 1]. 1956 *B.B.C. Handbk.* 1957 106 Two major enquiries by this section, both making use of social-survey methods. 1979 J. MacKenzie *Victorian Courtship* iii. 40 Mary Booth..welcomed the help Beatrice [Potter] proposed to give her husband in his ambitious social survey. 1782 'J. H. St. John de Crèvecœur' *Lett. from Amer. Farmer* iii. 50 New laws, a new mode of living, a new social system. 1853 H. Martineau tr. *Comte's Positive Philos.* II. vi. i. 11 The passage from one social system to another can never be continuous and direct. 1917 Kipling *Diversity of Creatures* 335, I cannot think it right that any human being should exercise mastery over others in the merciless fashion our tom-fool social system permits. 1951 E. E. Evans-Pritchard *Social Anthropol.* i. 11 The social anthropologist studies societies as wholes. He studies their oecologies, their economics.., their technologies, their arts, *etc.* as parts of general social systems. 1971 P. Worsthorne *Socialist Myth* vii. 143 Socialism has a conception of the role of government which can only be realized in a social system that exalts authority. 1797 in J. Gloag *Short Dict. Furniture* (1969) 626 A Gentleman's social table. 1952 *Ibid.* 436 *Social table*, a small, kidney-shaped table with four legs, and a revolving, cylindrical receptacle for wine bottles, supported on a pillar-and-claw stand, which fitted into the concave curve of the table. . It allowed two or three people to sit with their wine near a fire. 1962 'M. Innes' *Connoisseur's Case* xvi. 190 A gentleman's social table by Hepplewhite. 1873 H. Spencer *Study of Sociol.* xv. 372 No one doubts that the spendthrift or the gambler..is inferior as a social unit. a 1899 [see *super-organism* s.v. Super- 6 b]. 1907 W. James *Pragmatism* vi. 232 Must my thoughts dwell night and day on my personal sins and blemishes..or sink may I and ignore them in order to be a decent social unit? 1939 Auden in *I Believe* (1940) 26 Recent technical advances, such as cheap electrical power, are making smaller social units more of a practical possibility than they seemed fifty years ago. 1978 *Country Life* 17 Aug. 467/1 The family as a social unit. 1969 *Daily Tel.* (Colour Suppl.) 14 Mar. 5/1 Wage rises might have been slightly checked but 'the social wage' has gone up steadily. ..'The social wage', in plain English, means Government hand-outs, the exact opposite of a wage. 1971 *Guardian* 7 July 12/4 The real social wage of many low-paid workers has been adversely affected. 1975 M. Thatcher in *Let Children Grow Tall* (1977) 18 People..complain that government takes too much of their incomes for what is now called the 'Social Wage'—the estimated annual value of the services provided out of public funds for each individual. 1977 *Daily News* (Perth, Austral.) 19 Jan. 8/1 If, on becoming a mother, every woman became entitled to a social wage of $100 a week, tax free, then this might achieve a great deal in terms of righting the lot of some women. 1892 L. F. Ward *Let.* 17 Mar. in *Amer. Sociol. Rev.* (1938) June 371 (*heading*) The social will. 1911 J. Ward *Realm of Ends* vi. 118 Is there in any exact sense a social spirit, a social will, a social end, a social conscience? 1942 R. G. Collingwood *New Leviathan* xxi. 152 There is always a discrepancy between the social will and its products

B. *sb.* **2.** (Earlier example.)

1870 *Mainland Guardian* (New Westminster, Brit. Columbia) 8 Jan. 3/4 A very pleasant Social was given by the ladies and friends in our new mission church on Christmas day.

3. *ellipt* for *SOCIAL SECURITY 1 a or b. *colloq.*

1981 *Times* 20 May 3/8 I'm getting two wages, one from prison, and one from the social. 1983 J. Wainwright *Their Evil Ways* 17 They were both 'on the social'. *Ibid.* 26 She applied for extra 'social'. She was..sure she was *entitled* to some extra.

Social Credit. [f. Social *a.* + Credit *sb.*] **1.** A political theory advocated by C. H. Douglas (see *Douglas³), according to which the supposed chronic deficiency in the purchasing power of consumers was to be remedied through a reduction of prices by means of subsidies to producers or through the giving of additional money to consumers; occas. also, a subsidy under this system. Also short for *Social Credit Party* or *League*.

1920 A. R. Orage in C. H. Douglas *Credit-Power & Democracy* 166 The effect is inherent in the separation of Real Credit from Financial Credit—Social Credit, that is to say, from Financial Credit privately controlled. 1922 C. H. Douglas *Labour Party & Social Credit* 17 It is an important Report, not..as advancing any valid..criticism of the principles or details of Social Credit. 1935 *Calgary Typo News* 15 Mar. 1/1 We wonder..what is the policy of The Albertan? Liberal, Social Credit or Independent? 1936 S. E. Thomas *This Social Credit Business* 16 If the A+B Theorem is sound, what is the necessity for 'National Dividends' *as well as* Social Credits to retailers? 1944 G. B. Shaw *Everybody's Political What's What?* xi. 84 The apostles of Social Credit once actually persuaded a Canadian legislature to budget on its imaginary riches. 1966 R. S. Milne *Political Parties in New Zealand* iii. 78 *Potentially*, therefore, the Democrats, Democratic Labour, and Social Credit were 'national' parties. 1974 E. McGirr *Murderous Journey* 127 Ricardo's political grin..calculated to warm the cockles of any voter way back to Douglas Social Credit.

2. *attrib.* and *Comb.*, as *Social Credit analysis*, *group*, *leader*, etc.; **Social Credit League** (*a*) *Canad.*, an organization outside party politics corresponding to the Social Credit Party; (*b*) *N.Z.* (also *Social Credit Political League*), a political party advocating Social Credit; **Social Credit Party** *Canad.*, a political party whose policies are based on the theory of Social Credit.

1922 C. H. Douglas *Labour Party & Social Credit* 26 This article successfully avoided any mention or indication of the Social Credit Proposals. *Ibid.* 29 The Social Credit Movement..is in sharp opposition to the official Labour Party and High Finance jointly. 1931 V. A. Demant *This Unemployment* vi. 121 The Social Credit analysis of C. H. Douglas, costing expert for the Government during the War. 1934 T. S. Eliot *Rock* i. 12 Ain't you ever 'eard me speak o' the principles of Social Credit Reform? 1935 *Economist* 29 June 1474/1 A Social Credit scheme for Alberta under which he promised to guarantee every adult inhabitant of the Province $25 per week. 1958 *Maclean's Mag.* 10 May 97/4 When the two factions separated Wicks and Savage took the name and formed the Social Credit League. 1958 *Time* (Canadian ed.) 15 Sept. 16/3 The leader of the badly mauled Social Credit party. 1961 *Canada Month* Nov. 11/3 The B.C. Social Credit League offered alternative solutions to his problem. 1965 *Oxf. N.Z. Encycl.* 277/2 Two splinter parties, the Social Credit League and the Communist Party, have contested seats in elections, but neither has ever secured a seat in the House of Representatives. 1966 *Encycl. N.Z.* II. 812/1 In May 1953, the New Zealand Social Credit Political League was formed. 1966 'H. MacDiarmid' *Company I've Kept* iv. 106 Douglas's Social Credit system (or a partial form of it) was tried out by Alberta, Canada, in the inter-war years. 1967 *Canad. Ann. Rev.* 1966 5 Both the NDP and the small Social Credit group indicated that they were willing to co-operate with the government. 1968 *Globe & Mail* (Toronto) 17 Feb. 8/1 The national Social Credit Party could drive forward to counteract its 'inaccurate image'. 1968 *Landfall* XXII. 365 The men armed with Social Credit conversation for the interval. 1969 *N.Z. News* 5 Nov. 8/5 The Social Credit Leader, Mr. Cracknell, may have all too often been dismissed as something of a political lightweight.

Hence **Social Crediter, Social Creditor**, an advocate of Social Credit; also occas. **Social Creditist** (cf. *Creditiste).

1938 *Social Crediter* 17 Sept. 2/1 What I (a Social Crediter) should say to that, is that I don't know whether Sir Oswald agrees with me or not. 1953 M. Lowry *Sel. Lett.* (1967) 337 The Social Crediters fell, after having made some demonstrations of power. 1963 *Daily Tel.* 31 Oct. 14/2 The Social Creditists, appealing mainly to Quebec nationalism, could muster only 27 votes against 210. 1965 *Economist* 13 Nov. 713/1 In the new [Canadian] house there will still be nine Creditistes attached to Mr Caouette, and five Social Crediters of the Thompson variety. 1976 H. Wilson *Governance of Britain* viii. 161 In past times, not recently, they were told, the number of committed Henry Georgeites, or Douglas Social Creditors, might have put forward similar claims.

socialism. Add: **1.** (Earlier example.) *Christian socialism* (example).

1837 *Leeds Times* 12 Aug. 5/1 *Socialism.*—Messrs. Fleming and Rigby.—On Monday evening..these two gentlemen attended [*sic*] an audience..on the topics of the real nature of man. 1850 *Daily News* 13 Mar. 5/2 The infection of..'Christian Socialism' is spreading to Whitehall.

socialist. Add: **1. a.** (Earlier example.)

1827 *Co-operative Magazine* Nov. 509 The chief question..between the modern,..Political Economists, and the Communionists or Socialists, is whether it is more beneficial that this capital should be individual or in common.

b. *Comb.*, as *socialist-controlled*, *-dominated*, *-ridden*.

1929 *Times* 16 Aug. 11/3 Mr. Moore, the Leader of the Opposition, appealed to the people of this Socialist-ridden State on a promise to substitute for Socialism 'private effort and enterprise'. 1976 N. O'Sullivan *Conservatism* v. 124 Whenever important bills were opposed by the House of Lords, he suggested, there might be an appeal, in the form of a mass referendum, over the heads of a socialist-dominated House of Commons to the people. 1976 *Southern Even. Echo* (Southampton) 1 Nov. 3/3 There are many Socialist-controlled local authorities on which the dominant group doesn't give the opposition any places on any committees.

2. c. In *Combs.* used *attrib.* or as *adj.*, as *socialist-chauvinist*, *-feminist*, *-rebel*, *-revolutionary*, *-workman*.

1919 A. Sirnis tr. *Lenin's Collapse of Second International* x. 65 This group, the only one which had performed systematic work amongst the masses..turned Socialist-Chauvinist. 1921 D. H. Lawrence *Sea & Sardinia* ii. 76 He immediately put on the socialist-workman indignation. 1952 E. Hobsbawm in *Granta* 15 Nov. 10/2 The Radical age of the 1820s and 1830s, and the mixed socialist-rebel age which overlapped the first world war. 1976 H. T. Willetts tr. *Solzhenitsyn's Lenin in Zurich* 253 The Socialist-Revolutionary Party was born at the end of 1901 out of the merging of the Populist groups. 1976 *Women's Report* Sept./Oct. 17/1 They would like feedback from the last issue, articles from individuals and groups giving a socialist-feminist analysis of activities.

socialist realism. [tr. Russ. *sotsialisticheskiĭ realizm*.] The official theory of art and literature of the Soviet Communist party, according to which the artist's or writer's work should reflect and commend the life and ideals of socialist society. Also *attrib.*

1934 M. Eastman *Artists in Uniform* (rev. ed.) i. ii. 16 The present creative method in proletarian art and literature..is 'Socialist Realism'. *Ibid.* 17 Socialist Realism means seeing all reality as a development toward socialism. 1940 Graves & Hodge *Long Week-End* xxiii. 402 The hope that fact-finding would bring to fruition the theory of socialist realism. 1967 G. Steiner *Lang. & Silence* 390 Her recent novels betray the contortions of a genuine artist trying to come to terms with the grey half-truths of 'socialist realism'. 1978 *Jrnl. R. Soc. Arts* Dec. 64/2 In 1934 the term 'socialist realism' came into current use, and architecture suffered from various interpretations of it.

Hence **socialist-realist** *sb.* and *a.*

1935 G. Struve *Soviet Russ. Literature* xiv. 247 It is difficult to say why Sholokhov's *Upturned Soil* should be regarded as a work of a Socialist Realist and not of a realist *tout court*. 1945 H. Read *Coat of Many Colours* vi. 27 The 'socialist-realists' in Russia, who believe that the function of art is primarily to further the cause of socialism. 1958 *Listener* 6 Nov. 727/2 The whole fabric of the so-called 'socialist-realist' architectural philosophy [in the U.S.S.R.], which was actually pseudo-classical, has collapsed. 1964 *Ann. Reg.* 1963 229 He [*sc.* Mr Gomulka]..announced that socialist-realist art would receive special support. 1977 V. S. Pritchett *Gentle Barbarian* xiii. 218 The crude, black and white, schematic works of the Socialist Realists of our time.

socialite (sōᵘ·ʃăləit). *colloq.* (orig. *U.S.*). [f. Social *a.* and *sb.* + -ite¹.] A person who is prominent in fashionable society; one who is fond of social activities and entertainments. Also *attrib.*

1928 *Time* 31 Dec. 30/3 Splendorous as hostess & socialite was Princess Clara in both Germany and England. 1937 D. B. Wyndham Lewis in L. Russell *Press Gang!* 249 *Eric Daintee*, wealthy lissom Mayfair filmamateur, socialite, dressdesigner, surrealist. 1938 E. Lyons *Assignment in Utopia* ii. xiv. 226 The main body of amateur sociologists,..earnest probers, socialite thrill-hunters, and miscellaneous neurotics did not take possession [of Moscow] until the following years. 1944 R. Manvell *Film* ii. vi. 139, I do not think a working girl should take her standards from a socialite. 1956 J. C. Masterman *Case of Four Friends* vii. 125 A rather irresponsible *bon viveur* or socialite (as the young call them). 1960 *Harper's Bazaar* July 19/1 Private citizens in the socialite belt even distribute matches bearing their own monograms. 1978 J. Krantz *Scruples* iii. 79 At *Vogue* there are something like twenty-one editors of varying degrees of importance, including those stationed in Paris, Rome, and Madrid who are socialites first, editors second. 1980 *Times Lit. Suppl.* 12 Sept. 1000/4 Although a professional naturalist.., he [*sc.* Victor Jacquemont] was by nature something of a dilettante and very much a socialite.

sociality. Add: **2.** (*a*) (Further examples.)

1932 S. Zuckerman *Soc. Life Monkeys & Apes* xvii. 291 The monkey's sociality. 1966 R. M. Lockley *Grey Seal, Common Seal* ix. 125 The puzzling feature of the sociality of the Farne grey seals is that they continue to crowd together to nurse their pups on certain of the islets only. 1978 *Sci. Amer.* Sept. 139/1 The order also includes many nonsocial species, and the surprising fact is that sociality has originated on a number of separate occasions among the bees, the ants and the wasps.

socialization. Add: **1.** (Earlier and later examples.)

1884 W. Morris in *Justice* 31 May 2/1 The socialisation of labour which ought to have been a blessing to the community has been turned into a curse by the appropriation of the products of its labour by individuals..; the result of which to the workers has been a dire slavery, of which long hours of labour,..and complete repulsiveness in the work itself have been the greatest evils. 1957 *Ann. Reg.* 1956 311 The Chinese People's Republic remained largely preoccupied with the drive for greater industrialization and for the 'socialization' of agriculture. 1965 B. Pearce tr. *Preobrazhensky's New Economics* 6 The socialization of industry means by its very essence a transference of responsibility in economic leadership to science, to an extent quite unknown in capitalist economics. 1977 D. M. Smith *Human Geogr.* xii. 359 The effective socialization of production under capitalism..stands in supreme contradiction with private appropriation.

2. *Social Sciences*. The process of forming associations or of adapting oneself to them; esp. the process whereby an individual acquires the modifications of behaviour and the values necessary for the stability of the social group of which he is or becomes a member. Also *attrib.*, esp. in *socialization process*.

1841 *London Phalanx* 6 Nov. 505/1 Fourier in his analysis of universal movement in society..forms a scale of seven degrees or periods of general progress between.. *Edenism*, or the primitive state of humanity..and.. *Socialization*, or complete simple association. 1895 G. Simmel in *Ann. Amer. Acad. Pol. Sci.* VI. iii. 417 The investigation of the forces, forms and development of socialization, of co-operation, of association of individuals, should be the single object of sociology as a special science. 1924 F. H. Allport *Soc. Psychol.* xiv. 378 It is..possible that the evils mentioned are not *necessary* accompaniments of contemporary civilization or of the socialization of mankind. 1924 E. S. Bogardus *Fundamentals Soc. Psychol.* xx. 235 The best way to understand the socialization process is to consider the experiences of persons who have..a broad social vision and understanding. 1951 *Mind* LX. 288 This process of 'socialisation'—to use the standard term—produces..the 'personality' of the individual. 1953 Whiting & Child *Child Training & Personality* xi. 247 A relatively unimportant role in the total socialization process. 1964 M. Argyle *Psychol. & Social Probl.* v. 72 The socialization factor [in reducing the crime rate] is already being manipulated to some extent by placing children from very bad homes in foster homes. 1965 G. P. Murdock *Ess.* 44 The processes by which it [*sc.* culture]..is transmitted from one generation to the next (education and socialization). 1980 in N. Warren *Stud. in Cross-Cultural Psychol.* II. v. 215 We know relatively little about an area I shall label 'intellectual socialization'.

socialize, *v.* Add: **1.** esp. in *Social Sciences*. Also, to transmit to an individual the cultural values and standards of behaviour of the social group of which he is regarded as a member.

1932 M. Gabain tr. *Piaget's Moral Judgment of Child* iv. 350 This assumption gains force if..social constraint does not really suffice to 'socialize' the child but accentuates its egocentrism. 1957 O. R. McGregor *Divorce in Eng.* iii. 79 With horrifying overcrowding..the family could not be a stable, ongoing concern. Children were brutally socialised. 1971 *Mod. Law Rev.* XXXIV. vi. 643 New lawyers are..informally 'socialised', taught the norms and behaviour patterns which are the basis of a stable legal profession. 1976 *National Observer* (U.S.) 31 July 1/2 She is convinced..that women traditionally have failed to excel physically because they were socialized to think they could not.

2. (Later examples.) Also in extended uses: to administer or organize with social aims in view (rather than predominantly for profit); to finance with public funds; to bring under public control. Also *absol.*

1920 M. Beer *Hist. Brit. Socialism* II. iii. ix. 181 Maurice's main idea was to socialise the Christian and to Christianise the socialist. 1920 *Westm. Gaz.* 2 Dec. 6/1 The Fehrenbach Cabinet is being increasingly pressed to Socialise, and first of all to Socialise coal. 1926 J. M. Keynes *End of Laissez-Faire* iv. 42 One of the most interesting and unnoticed developments of recent decades has been the tendency of big enterprise to socialise itself.

1938 *Sun* (Baltimore) 15 June 2/3 Discussions as to what the medical association and the Government might do about 'socializing' or financing with public funds the medical treatment of economically submerged individuals. 1948 *News Chron.* 20 Sept. 3/3 In fact, we have to socialise men's minds—without destroying their individuality and enterprise—as well as socialising the physical assets. 1975 J. De Bres tr. *Mandel's Late Capitalism* vii. 233 Within the company labour is directly socialized in the sense that the overall plan of the company..directly determines the output of the various factories, workshops and conveyor belts. 1976 *Ilkeston Advertiser* 10 Dec. 20/3 It could do this by compulsory purchase, saying that it was merely 'socialising' the land by buying it at current use value and preventing a capitalist developer from doing the same, and then re-selling the land at a profit.

3. *intr.* To be sociable, participate in social activities. Freq. const. *with*. orig. *U.S.*

1895 in *Funk's Stand. Dict.* 1900 *Dialect Notes* II. 62 *Socialize*,..to talk with one of the opposite sex. 1934 *Amer. Speech* IX. 76 Did you socialize much in Atlantic City?..I am not going to socialize much this winter. 1939 Joyce *Finnegans Wake* (1964) iii. 498 After plenty of his fresh stout and his good balls of malt,..socializing and communicanting in the deification of his members. 1959 V. Packard *Status Seekers* xii. 167 Trying to socialize across class barriers can be a strain. 1966 J. Pearl *Crucifixion of Pete McCabe* ii. 21 McCabe could socialize with any of his three neighbours. 1970 *Daily Tel.* 25 Feb. 15/3, I prefer to socialise outside the team but wouldn't go out with any man who was shattered by my job. 1978 *Detroit Free Press* 5 Mar. 6/2 They're hard to spot because they don't like to sit around in trees and socialize with other birds.

socialized, *ppl. a.* Add: (Later examples.) *socialized medicine*: a system of medical care that is financed and administered by the state.

1892 G. B. Shaw *Fabian Soc.* (Fabian Tract No. 41) 20 Thousands of thoroughly Socialized Radicals..who would have resisted Socialism fiercely if it had been forced on them. 1935 *Sun* (Baltimore) 16 Feb. 9/1 Traditionally hostile to socialized medical care, the organized medical profession will decide tomorrow whether to temper its stand to the trend of the day toward State control. 1936 J. M. Keynes *Gen. Theory Employment, Interest & Money* v. xix. 267 Except in a socialised community where wage-policy is settled by decree, there is no means of securing uniform wage reductions for every class of labour. 1939 *Time* 30 Jan. 52/2 Although socialized medicine would certainly limit a patient's free choice of a physician, few people today are free to choose their doctors. 1949 Koestler *Insight & Outlook* xiv. 195 The listing together of such different phenomena as, for example, optical projective illusions, hypnosis, and socialized behaviour as manifestations of 'self-transcendence' is arbitrary. 1949 *Sun* (Baltimore) 15 June 12/3 For instance, at this moment the socialized railways of Britain face labor unrest of great intensity, the basic quarrel being over wages. 1964 R. Braddon *Year Angry Rabbit* ix. 82 'Goddam,' muttered the American sourly, 'why can't you Australians have Socialized Medicine like the Limies?' 1973 *Sci. Amer.* Jan. 33/1 One [*sc.* primate] in particular—the chimpanzee—not only cooperates in the work of the chase but also engages in a remarkably socialized distribution of the prey after the kill. 1976 *Howard Jrnl.* XV. i. 51 The juvenile courts dispensed unchallenged what the Americans call socialized justice. (Socialized simply means individualized civil justice which focuses upon social conditions. It has no party political connotations.) 1977 *Lancet* 17 Sept. 596/1 Why socialised medicine should so often be equated with socialist medicine I fail to understand

socializee (sōᵘ·ʃăləizī·). [f. Socialize *v.* + -ee¹.] One who is being socialized (in sense 1 of the vb.)

1952 T. Parsons *Social System* vi. 209 Thus not only the socializing agents *but the socializee* must be conceived as acting in roles. 1975 *Jrnl. Politics* XXXVII. 83 Yet socialization within the family is not simply a two-generation phenomenon, for the parents themselves were once the socializees instead of the socializers.

socializer (sōᵘ·ʃăləizəɹ). [f. Socialize *v.* + -er¹.] **1.** One who socializes an industry, an economy, or the like. Cf. *socialize *v.* 2.

1947 [see *decartel(l)ization].

2. One who or that which makes a person social or sociable; that which induces a sociable atmosphere.

1974 H. L. Foster *Ribbin', Jivin', & Playin' Dozens* iv. 162 The right language at the proper time can serve as a socializer, a relaxing agent, and a positive catalyst to enhance communication. 1975 [see prec.]. 1976 *Word* 1971 XXVII. 476 They are *socializers* which function as the verbal oil to ease friction in communication... A's phrase 'How are you?' is a socializer; A does not usually expect to learn B's physiological or psychological problems. 1981 *Underground Grammarian* Oct. 4/1 Our educationists are socializers with political intentions.

3. One who likes to participate in social activities.

1977 J. Anderson *Appearance of Evil* iv. 21 She was the quietest member of the family, was less of a socializer than Victor. 1978 G. A. Sheehan *Running & Being* ii. 29 His 'good life' is quite different from that of an aggressive football player and the relaxed socializer.

socializing, *vbl. sb.* (In Dict. s.v. Socialize *v.*) (Later examples.)

1944 J. S. Huxley *On Living in Revolution* 21 This humanizing and socializing of sectional groups is one way in which the new social order will differ from the old. 1959 V. Packard *Status Seekers* xii. 168 Status is attached to the act of socializing. 1965 B. Pearce tr. *Preobrazhensky's New Economics* 76 The whole system of regulation of the economy cannot but be affected by the socializing of industry and transport. 1970 *Globe & Mail* (Toronto) 28 Sept. 13/2 Having no brothers or sisters to toughen her, socializing is probably bewildering. 1976 Botham & Donnelly *Valentino* vi. 45 Lunchtime socialising..brought him into contact with members of a motion picture company.

socializing, *ppl. a.* Add: Also, that revels in participating in social activities.

1965 F. L. Utley in Bessinger & Creed *Medieval & Linguistic Stud.* 308 A multiple martyr, to a fretful socializing wife, to the chthonic Mother, to the Judas Wilkes, to and for the Nation. 1980 *London Mag.* July 81/2 Mostly he [*sc.* Somerset Maugham] dealt in types, such as the rich bitch, the honest whore, the socializing snob.

socially, *adv.* Add: **4.** (Later examples.) Freq. regarded as forming a *Comb.* with a following *adj.* or *ppl. adj.*

1890 W. H. Dawson *Unearned Increment* iv. 47 The socially-created value of land. 1909 *Daily Chron.* 3 Nov. 1/1 The Budget taxes one form of 'socially-created wealth'. 1937 L. Cheskin in *Education* Nov. 186/2 They are socially conscious, they study their society and interpret it, each in his own manner. 1951 M. McLuhan *Mech. Bride* (1967) 51/1 Panic enters the socially spotlighted host or hostess. 1955 *Sci. News Let.* 22 Jan. 56/1 Staphylococcus germs that are common even on 'socially clean' hands. 1961 D. Jenkins *Equality & Excellence* vi. 130 The great public schools and socially-privileged private schools. 1964 L. Woolf *Beginning Again* iii. 232 Socially it was the prehistoric era in which one still had servants living in one's house. 1964 *Language* XL. 92 A leading predecessor in socially-oriented thought. 1970 *Daily Tel.* 26 Nov. 9/3 One such smoker said: 'The claustrophobic atmosphere makes pot-smoking socially acceptable.' 1976 *Times* 20 Aug. 4/3 (*caption*) Socially deprived pupils from a West Berlin school. *Ibid.* 4/4 The Government is..deciding to spend £50,000 on job schemes for socially disadvantaged people. 1977 *Daily Tel.* 13 Oct. 10/4 In the past it has been opposed to any 'head counts' of this kind. They were considered socially divisive.

socialry. (Earlier example.)

1896 A. Small *Let.* 22 May in *Social Forces* (1932) Mar. 313/1 'Socialry' has an archaic sight sound & sense which will hardly get itself naturalized in modern society, and as to his division of the factors of life in society with 'economics' & 'socialry'—it is a classification of things that he [*sc.* F. H. Giddings] hasn't begun to correlate in his own thinking.

social science. [Social *a.* 9 b.] The scientific study of the structure and functions of society; any discipline that attempts to study human society, either as a whole or in part, in a systematic way. Also *attrib.*

[1785 J. Adams *Let.* 10 Sept. in *Works* (1854) IX. 540 The social science will never be much improved, until the people unanimously know and consider themselves as the fountain of power. 1791 D.-J. Garat *Let. à M. Condorcet* 82 Ces vérités..qu'il etoit important de découvrir, de rendre incontestables, sont les premières données de la science sociale, mais elles ne sont point la science.] 1811 tr. *Destutt de Tracy's Commentary Montesquieu's Spirit of Laws* 4, I have no other ambition..than to contribute my effort to the progress of social science, the most important of all to the happiness of man, and that which must necessarily be the last to reach perfection, because it is the product and the result of all the other sciences. 1821 J. Bentham *Let.* 21 Apr. in J. H. Burns *J. Bentham & Univ. Coll.* (1962) 8 The minds of the ruling few in their growing state should be turned towards the science so aptly stiled by you *the social science*—that science, in the progress of which the allied powers of tyranny, corruption, and delusion have so long..beheld their final downfall. 1846 [in Dict. s.v. Social *a.* 9 b]. 1849 *Southern Lit. Messenger* XV. 77/2 On the Importance of the Social Sciences in the present day. 1874 [in Dict. s.v. Social *a.* 9 b]. 1908 W. McDougall *Introd. Social Psychol.* p. vii, I hope that the book may be of service to students of all the social sciences. 1949 M. Mead *Male & Female* 435 The relationship between our social-science skills and our world. 1966 G. N. Leech *Eng. in Advertising* i. 3 Linguistics..has so far been the least influential of the major social sciences. 1969 *Times* 7 Jan. 8/6 Students reading social science were more sceptical than either the arts students or the scientists. 1971 *New Scientist* 18 Mar. 591/1 It seems more difficult to present a programme of viable-looking research in the social sciences than in the natural sciences. 1977 A. Giddens *Stud. in Social & Polit. Theory* ix. 306 Such a view is founded upon an erroneous idea of the relation between lay and social-science concepts.

Hence **social scientific** *a.*, **social scientist.**

1875 R. J. Wright *Principia; or, Basis of Social Sci.* p. v, *As to Spencer*; we admit he is *the* King of the Social Scientists. 1882 W. James *Let.* 2 Nov. (1920) I. 211 As for Prague, *veni, vidi, vici*. I went there with much trepidation to do my social-scientific duty. 1920 J. M. Williams *Foundations of Social Sci.* p. xiii, The trend of thought of the psychological social scientists signifies an aim to arrive at truer assumptions, and to keep an open mind toward the psychological, as well as the other aspects of those assumptions. 1971 *Nature* 25 June 538/1 Why do social scientists, particularly American social scientists, murder the English language? 1977 J. M. Johnson in Douglas & Johnson *Existential Sociol.* v. 166 Investigations of..social scientific situations.

social security. [Social *a.* 7.] **1. a.** A system whereby the state provides financial assistance for those citizens whose income is inadequate or non-existent owing to disability, unemployment, old age, etc.

1908 [see sense 2 below]. 1933 *Old Age Security Herald* June 1/1 Transformed in the *American Association for Social Security*, the organization [*sc.* the American Association for Old Age Security] will continue to function. 1936 A. Epstein *Insecurity* (rev. ed.) p. vi, The Act bars the realization of the relief promised by social insurance even though the law is presented under the glittering title of 'Social Security'. 1941 *Atlantic Charter* (Messages of Presidents, U.S.) (1943) 5 They desire to bring about the fullest collaboration between all nations in the economic field, with the object of securing for all improved labour standards, economic advancement and social security. 1942 *Times* (Weekly ed.) 2 Dec. 2/1 Social security as envisaged in this report is a plan to secure to each citizen an income adequate to satisfy a national minimum standard. *Ibid.*, As regards unification, Sir William Beveridge suggests that there should be a Ministry of Social Security. 1959 G. Slatter *Gun in My Hand* 44 What in hell are ya doin in Christchurch anyway? Bludgin on the social security, I spose. 1969 N. W. Pirie *Food Resources* i. 31 The Ministry of Social Security..found that there are now a million children at or below the poverty line [in the U.K.]. a 1974 R. Crossman *Diaries* (1975) I. 23 For years I've been a specialist on social security and I know enough about it.

b. The money paid out under this system.

1959 in J. Reid *Kiwi Laughs* (1961) 217 'Use your brains,' said Father Christmas testily. 'When you get to my age you have to live on social security. There's not much over to buy racing cars for every kid that wants one.' 1971 *Sunday Times* (Colour Suppl.) 5 Dec. 28 He has seven of his children to support on social security and family allowances. 1975 J. Symons *Three Pipe Problem* xv. 140 And you draw social security? 1976 *National Observer* (U.S.) 2 Oct. 7/2 Ethel..takes in tourists to complement her Social Security.

2. *attrib.*, as *social-security benefit*, *legislation*, *policy*, etc.

1908 W. S. Churchill *Let.* 4 Jan. in R. S. Churchill *Winston S. Churchill* (1969) II. Compan. ii. 759 If we were able to underpin the whole existing social security apparatus with a foundation of comparatively low-grade state safeguards, we should in the result obtain something that would combine the greatest merits both of the English & the German systems. 1935 *N.Y. Times* 15 Aug. 1/4 The Social Security Bill, providing a broad program of unemployment insurance and old-age pensions..became law today. 1936 *U.S. Statutes* XLIX. 635 There is hereby established a Social Security Board..to be composed of three members to be appointed by the President. 1940 *Economist* 31 Aug. 282/1 The old social security tax and the new national security tax. 1941 J. S. Huxley *Uniqueness of Man* p. ix, Subsidized housing, free milk, social security legislation, health insurance, free education ..—these are all symptoms of the change. 1942 *Rep. Commissioners Insurance & Allied Services from Organisations* App. G. 35 in *Parl. Papers* 1942–3 (Cmd. 6405) VI. 419 Post-war social policy should aim at establishing a *national Plimsoll line* of goods and services for all citizens. .. Such a social security policy grants the citizens new opportunities and freedoms. 1962 *Listener* 26 July 130/1 Poor people in Mexico City who can attend a social-security clinic. 1964 W. Markfield *To Early Grave* vii. 131 He spread out..his army discharge papers, his social security card and a B'nai B'rith newsletter. 1966 N. Freeling *Dresden Green* i. 16 Chemists' bills and a social security number. 1969 *Times* 13 Jan. 11/2 £250m. available for higher social security benefits in the very low income range. 1972 *Accountant* 21 Sept. 347/1 The social security structure of individual countries depends heavily on their political climate. a 1974 R. Crossman *Diaries* (1976) II. 587 As chairman of the Pensions Committee I just had to get down to thinking what to do about social security payments after devaluation. 1976 W. Trevor *Children of Dynmouth* iii. 80 She was tired of listening to Mrs Slewy complaining about the social security man. 1976 *Ilkeston Advertiser* 10 Dec. 2/2 The Social Security 'scroungers', some of whom deliberately swindle the taxpayers to the extent of £10 million per year. 1978 S. Sheldon *Bloodline* xxxix. 347 People were on record if they had a Social Security number, an insurance policy a driver's license or a bank account.

social service. [Social *a.* 7.] **1.** Service to society or to one's fellow-men, esp. as exhibited in work on behalf of the poor, the underprivileged, etc.

1851 J. S. Mill *Lett.* (1910) I. 169 Scientific inquiry into the production and distribution of wealth, as a branch of social service. 1890 W. James *Princ. Psychol.* I. xiv. 599 The other social affections, *Benevolence, Conscientiousness, Ambition*, etc., arise in like manner by the transfer of the bodily pleasure experienced as a reward for social service. 1921 R. H. Tawney *Acquisitive Society* x. 219 If medical officers of health, directors of education, and directors of the Co-operative Wholesale be assumed to be quite uninfluenced by any consciousness of social service. 1926 E. Hemingway *Sun also Rises* ix. 85 'I rather thought it would be good for him.' 'You might take up social service.' 'Don't be nasty.' 1977 R. L. Wolff *Gains & Losses* vii. 404 Sublimation of doubt in sex, social service among the poor as a substitute for faith.. we shall encounter them again and again in..novels of doubt.

2. With *a* and *pl.* A service supplied for the benefit of the community, esp. any of those provided by the central or local government, such as education, medical treatment, social welfare, etc.

1933 J. Buchan *Prince of Captivity* ii. iii. 230 He is not prepared to go back on our social services... All parties go on sluicing out..new benefits from the public funds. 1941 'R. West' *Black Lamb & Grey Falcon* II. 506 This Cockney taxi-driver would be..able to rely on an amplitude of social services in any emergency. 1945 *Ann. Reg.* 1944 i. 82 The basic principle that compensation of workmen for industrial injury should be made a public social service. **1959** *Times Lit. Suppl.* 10 Apr. 203/2 Not even the welfare state and social services, and certainly not the nationalized industries, are now viewed by the mass of adult Britons as immaculate, brought down from Sinai. **1976** *Times* 22 July 16/1 There is more demand to cut taxes than to expand social services.

3. *attrib.* **a.** *sing.* (senses 1 and 2), as *social-service cut, work*, etc.

1900 J. P. Smyth (*title*) Social service ideals. **1911** J. B. Haldane *Social Workers' Guide* 127/1 Diocesan Social Service Committees. **1921** *Daily Colonist* (Victoria, B.C.) 8 Apr. 9/4 Mrs. Hallam, secretary of the Social Service Committee, requested that Miss Thompson, now acting for the organization as a social worker, be recognized as a woman police officer. **1929** Hughes & Stuenkel (*title*) The Social Service Exchange in Chicago. **1932** *Bombay Chron.* 20 Dec. 5 The work of the Social Service League has been good all round. **1937** M. Hillis *Orchids on your Budget* iii. 57 She also does the marketing by car, does social-service work in it once a week. **1956** F. Lafitte in A. Pryce-Jones *New Outl. Mod. Knowl.* 574 Britain's heavy heritage of obsolescent social-service buildings. *a* **1974** R. Crossman *Diaries* (1976) II. 444 He made an immensely demagogic speech saying that we'd spent more on social services than the Tories in their last three years, denying there was any question of social-service cuts. **1975** *Language for Life* (Dept. Educ. & Sci.) xix. 279 They include the Social Service Departments, Youth Service Departments, probation officers, and officers of the L.E.A. Careers Service.

b. *pl.* (sense 2), as *social services department*, etc.

1973 *Guardian* 30 May 9/2 Croydon's social services department is setting up a special training programme for new foster parents. **1974** *Times* 13 Nov. 16/2 The whole social services structure leads to good fieldworkers being lost to the management side. *Ibid.*, Management training in the social services field is possible. **1976** *Derbyshire Times* (Peak ed.) 3 Sept. 1/1 The party was from a Social Services home in Wartsones Road, Wolverhampton.

social work. [Social *a.* 7.] Work of benefit to those in need of help, esp. professional or voluntary service of a specialized nature concerned with community welfare and family or social problems arising mainly from poverty, mental or physical handicap, maladjustment, delinquency, etc. Also (with hyphen) *attrib.*

1890 *Girl's Own Paper* 27 Dec. 197/3 'Stump oratory' may safely be regarded as quite beyond the limits of a woman's social work. **1892** S. M. Lindsay in *Annals Amer. Acad. Pol. & Social Sci.* Nov. 76 The growth of the Krupp cast steel works and the motives of the firm in their social work. **1908** *Busy Man's Mag.* Jan. 88/1 Few who read Mrs. Humphrey Ward's interesting books know of her social work in London. **1914** *Lancet* 31 Jan. 345/2 (*heading*) The sixth international congress of social work and service. **1931** *Economist* 10 Jan. 57/1 For in spite of the very considerable development which organised social work has attained in the United States.. it is not capable of shouldering the responsibility of caring for the wreckage of a major industrial depression. **1964** *New Statesman* 10 Apr. 581/1 An experienced social work teacher from Canada. **1975** *Listener* 14 Aug. 212/1 By the use of social-work skills, probation officers seek to help the offender cope. **1979** G. Wagner *Barnardo* xvii. 299 No teacher training colleges or social work courses existed. **1980** *Times Lit. Suppl.* 28 Nov. 1347/1 Notwithstanding the vagueness of its aims, social work became sanctified as a discrete discipline under the Social Services Act of 1970.

Hence **social worker,** one who undertakes social work, esp. someone professionally trained.

1904 *Ann. Reg. Univ. Chicago* 245 A training center for social workers. **1912** F. G. D'Aeth in H. Bosanquet *Social Conditions in Provincial Towns* iv. 50 A Local Committee of Social Welfare..consists of clergy, ministers, and social workers. **1931** *Economist* 10 Jan. 57/1 Social workers in the United States are meeting an unusually 'hard winter' with feelings almost of despair. **1955** Earl Winterton *Fifty Tumultuous Years* 117 Few, if any, of the social workers of today are old enough to remember conditions then. **1964** [see *almoner[1] 4]. **1975** *New Yorker* 11 Aug. 27/1 Ma says her social worker will not permit her to move her furniture until the rent is paid for August. **1977** B. Pym *Quartet in Autumn* xxii. 201 A real bossy social-worker type.

sociation. Restrict †*Obs. rare* to sense in Dict. and add: (sōᵘʃiˌēɪ·ʃən). **2.** *Ecol.* = *society 11*.

1930 *Svensk Bot. Tidskrift* XXIV. 492 The most elementary units in the series of phytocoenoses, or the sociations, have until now been studied nearly only by the Scandinavian School of Ecologists (= Phytosociologists). Until 1928 they were called 'associations', but in order to facilitate an international agreement, Scandinavian ecologists have now agreed to accept this term in its Middle-European sense, following Rübel's proposition to apply the new term 'sociation' to the earlier Scandinavian 'associations' (or 'micro-associations'). **1936** *Jrnl. Ecol.* XXIV. 276 It is here proposed to call the aspect society a *sociation* and the layer society a *lamiation*, while the corresponding seral terms would be *socies* and *lamies*. **1973** P. A. Colinvaux *Introd. Ecol.* v. 67 (*caption*) Hypothetical species area curve as used by the Uppsala school for determining both the number of species in the sociation and the minimum area of that sociation.

socies (sōᵘ·ʃiˌīz). *Ecol.* [mod.L., f. Society after *species* (cf. *associes).] The term answering to *society 11* in analyses of immature plant communities.

1916 F. E. Clements *Plant Succession* vii. 138 The socies bears exactly the same relation to consocies and associes that the society does to consociation and association... The term socies comes from the root *seq-* (*sec-, soc-*), *follow*. **1926** [see *associes]. **1929** *Ecology* X. 133 Each group of seasonals plus the constantly present predominants make up a socies, that is a seasonal subdivision or aspect-phase of the presocies. **1964** [see *associes].

‖**sociétaire** (sosyetɛ̄r). [Fr., f. *société* society.] An actor who is a full member of the Comédie Française, Paris, and thereby has a share in its management and profits.

1881 W. H. Rideing *Dramatic Notes, 1880–81* xii. 67 Of the three parts in which London audiences have now seen her [*sc.* Mme Modjeska], this of the gifted and passionate *sociétaire* of the Comédie Française [*sc.* Sarah Bernhardt] illustrates most fully the measure of her genius. **1901** *Scotsman* 16 Apr. 8/7 The vexed question of the distribution of rôles among sociétaires and pensionnaires—full members of the company and salaried aspirants. **1909** Beerbohm in *Sat. Rev.* 30 Oct. 529/1 He, too, has schooled himself in the traditions of the Gaiety, and is a worthy sociétaire. **1923** G. Arthur *Sarah Bernhardt* iv. 66 To the Directors of the Français the *Sociétaires* ranked only by seniority. **1959** *Times* 23 Feb. 12/2 If we had a National Theatre properly based and founded, all of us could be Sociétaires as in the Comédie-Française and be called upon when required.

societal, *a.* Add: (Earlier and later examples.) Hence **soci·etally** *adv.*

1898 *Catal. Yale Univ. 1898–9* 204 A study of the evolution of the institutions of the democratic republic, of the societal organization, and of the history of the money of account. **1907** [see *mores 1]. **1956** *Kenyon Rev.* XVIII. 411 Trager and Smith's results are amazingly coherent, both practically and societally. **1959** *Sunday Times* 22 Nov. 16/4 His [*sc.* Kipling's] passion..for the values of what D. H. Lawrence (borrowing the revolting word from Dr. Trigant Burrow) called 'societal man'. **1964** [see *egocentric *a.*]. **1973** *Black World* Mar. 28 Educational systems are constructed to socialize individuals to perform societally defined and necessitated tasks. **1976** *Times Lit. Suppl.* 19 Nov. 1449/1 Leading educators believe that the schools must keep pace with broad societal changes. **1979** 'A. Hailey' *Overload* iv. vii. 328 The Sequoia Club has been something we all needed—part of our societal system of checks and balances.

soci·etified, *ppl. a. rare.* [f. Society + -fy + -ed[1].] Of or made fitting for cultured or fashionable society.

1934 in Webster. **1936** E. M. Forster *Abinger Harvest* 108 The societified lady and the obscure maniac are in a sense the same person.

societology (sosəiˌetǫ·lŏdʒi). *U.S. rare.* [f. Societ(y + -ology.] The study of human society; sociology.

1895 *Catal. Yale Univ. 1895–6* 62 In connection with this is a course of lectures on systematic sociology (Societology). This course is strictly academical..and does not take up topics popularly classed under 'social science'. **1915** *Nation* (N.Y.) 14 Oct. 467 He has made a real contribution to the study of society, and has demonstrated (what some of us had begun to doubt) that there is such a subject as sociology—or at least societology. **1924** C. M. Case *Outlines Introductory Sociol.* p. xxxvi, It [*sc.* sociology] probably *should* be, and more than probably will *not* be, known as 'societology'.

society. Add: **I. 3. c.** *alternative society*: the aggregate of (predominantly young) persons whose cultural values and habits of association purport to represent a preferable and cogent alternative to those of the established social order. Usu. with definite article.

1969 *It* 13–25 June 21/3 Brother Simon Tugwell is planning a 3-day talk-in on the alternative society. **1971** *Guardian* 16 Mar. 10/4 American cities seem full of young people wanting to 'drop out'—but what do they drop into? It is called 'The Alternative Society' and it is already becoming a vogue term. **1971** *Times Lit. Suppl.* 31 Dec. 1621/5 Sorel, like Nietzsche, preached the need for a new civilization of makers and doers, what is now called a counter-culture or an alternative society. **1975** D. Lodge *Changing Places* v. 164 A middle-aged parasite on the alternative society.

III. 9. c. *Zool.* A group of animals of the same species organized in a co-operative manner.

1902 *Encycl. Brit.* XXIX. 503/2 Perhaps the most remarkable fact as regards the higher societies of insects is that though the individuals composing a community are the offspring of one mother..yet they do not resemble their parents. **1925** A. D. Imms *Gen. Textbk. Entomol.* 522 In certain species of the order [Hymenoptera] the individuals have acquired the habit of living together in great societies, as in the case of the ants. **1964** V. B. Wigglesworth *Life of Insects* xiv. 237 All insect societies are overgrown families. **1971** E. O. Wilson *Insect Societies* ii. 6/2 Bird flocks, wolf packs, locust swarms, and groups of communally nesting bees are good examples of elementary societies.

11*. *Ecol.* A community of plants within a mature consociation characterized by one or more subdominant species.

1899 *Bot. Gaz.* XXVII. 111 A plant society is defined as a group of plants living together in a common habitat and subjected to similar life conditions. The term is taken to be the English equivalent of Warming's *Plantesamfund*, translated into the German as *Pflanzenverein*. **1905** F. E. Clements *Res. Methods in Ecol.* 296 For these areas controlled by principal species, but changing from aspect to aspect, the term *society* is proposed. **1916** —— *Plant Succession* vii. 130 The society is a community characterized by a subdominant or sometimes by two or more subdominants... The society comes next below the consociation in rank, but it is not necessarily a division of it, for the same society may extend through or recur in two or more consociations, *i.e.*, throughout the entire association. **1932** Fuller & Conard tr. *Braun-Blanquet's Plant Sociol.* xiii. 306 The..'societies' of Clements and Weaver are based entirely upon the dominance of certain species; they are, thus, quite incapable of replacing our association in any system of classification. **1932** *Ecology* XIII. 118 A single pair of terms, society and socies (developmental), has been quite generally applied to subordinate assemblages within associes and associations. **1952** P. W. Richards *Tropical Rain Forest* xi. 259 *Shorea curtesii*..dominates small societies on steep slopes in the hill rain forests of the Malay Peninsula.

IV. 12. b. With reference to cultured or fashionable society: further examples.

1875 Mrs. Stowe *We & Our Neighbours* 205 My sisters ..are society girls in the best sense. **1880** J. C. Harris *Uncle Remus* viii. 203 'The old man's mind is wandering,' said the society editor. **1882** J. D. McCabe *N.Y. by Sunlight & Gaslight* 228 An engagement..is promptly announced in one of the 'Society journals'. *c* **1884** (*title*) A society beauty. **1888** *St Louis* (Missouri) *Globe-Democrat* 29 Apr. 22/2 The brainy paragraphs thrown off by one society reporter. **1891** *Girl's Own Paper* 21 Mar. 385/1, I..said I was tired of society life, and..liked nursing better than anything. **1893** 'S. Grand' *Heavenly Twins* I. xv. 109 You would not counsel a son of yours to marry a society woman of the same character as Major Colquehoun. **1895** T. K. Gavon *Fancy Notions by a Yankee Notion Clerk* 28 Already we have cattle kings, coal barons, merchant princes and society queens. **1910** E. M. Forster *Howards End* iii. 19 She did not..pretend that nothing had happened, as a competent society hostess would have done. **1910** Chesterton *G. B. Shaw* 152 A pleasant society lady, Lady Cicely Waynefleet. **1911** G. S. Porter *Harvester* xx. 508 He scanned the society columns of the papers. **1924** Galsworthy *White Monkey* I. ix. 73 A society painter and his wife. **1947** 'N. Blake' *Minute for Murder* ii. 32 He had been a society photographer before the war. **1949** H. MacLennan *Precipice* I. 144 A picture I saw of her in the society page of *The New York Times*. **1950** *New Yorker* 8 Apr. 76/3 Hearst's society columnist, Cholly Knickerbocker. **1950** E. H. Gombrich *Story of Art* xxiii. 349 Vandyke had established a standard of society portraits. **1955** L. Feather *Encycl. of Jazz* x. 347 *Society band*,..band that plays innocuous commercial dance music. **1956** C. Cockburn *In Time of Trouble* xvii. 228 The secretary was away attending some society wedding. **1957** D. Piper *Eng. Face* viii. 199 Behind almost all society portraiture before Reynolds there is a basic, and dead, symmetry. **1959** G. D. Painter *Marcel Proust* I. vii. 85 In the name 'Le Gandare' Proust alludes to the society portraitist La Gandara. *Ibid.* xi. 181 A little bird..informed the society columnist of *Le Gaulois*. **1976** C. Storr *Unnatural Fathers* iii. 36 He had had a long liaison with a society beauty. **1977** *Time* 26 Sept. 36/1 The society columns buzzed regularly for years with accounts of their parties and travels aboard an assortment of yachts.

socio-. Add: Also with pronunc. (sōᵘ·sio). **2.** **socioce·ntric** *a.*, tending to focus one's interest on the community or one's own group; hence **socioce·ntrism**; **sociodyna·mic** *a.*, tending to produce change in a society or group; hence **sociodyna·mics** *sb. pl.*; **socio-eco·logy**, the study of the interactions both among the members of a species and between them and their environment; so **socio-ecolo·gic, -ecolo·gical** *adjs.*; **socio·geny**, the origin and development of society; hence **socioge·nic, socio·genous** *adjs.*, originating in society or social interaction; **so·ciogram** *Sociol.*, a diagrammatic representation of the ratings for popularity, leadership, etc., that members of a small group give each other; a sociometric diagram; **socio·graphy**, an empirical method of sociological analysis that makes use of both quantitative and qualitative data; such an analysis; hence **sociogra·phic** *a.*; **so·ciogroup** *Sociol.*, a group of people who associate for some reason or purpose other than personal preference; **sociome·dical** *a.*, of or pertaining to the relations between medicine and society; **socio-no·mic**, relating to the environmental conditions affecting the formation and development of social groups, esp. as *socionomic sex ratio*, the ratio of females to males in relatively stable social groups; hence **sociono·mics** (*rare*).

1881 L. F. Ward in *Trans. Anthrop. Soc. Washington* (1882) I. 97 Those in consequence of which social progress tends to defeat itself—*anti-sociocentric facts*. **1930** [see *cerebrocentric *a.*]. **1970** J. D. Caute *Fanon* iv. 49 There was the Fanon who condemned such insularity as egocentric and sociocentric. **1947** G. Murphy *Personality* xv. 386 Side by side with egocentrism there was sociocentrism. **1934** J. L. Moreno *Who shall Survive?* vii. 74 This demonstrates what we may call the process of slowing down of interest, the cooling off of emotional expansiveness, the sociodynamic decline of interest. **1977** R. Holland *Self & Social Context* vii. 221 The hypothesis that there are psychodynamic and sociodynamic processes

at work even in this relatively 'self-conscious' area of knowledge production. **1978** *Nature* 18 May 184/2 The number of administrators in an organisation expands irresistibly in response to the first law of sociodynamics. **1972** *Biol. Abstr.* LIV. 6370/2 Geographic vegetational zones are given, as are vegetational stages. Socioecologic groups are described. **1961** *Ibid.* XXXVI. 5947/1 (*heading*) A socio-ecological study of pastured domestic rabbits in Mae-sima. **1973** W. P. J. Dittus in R. H. Tuttle *Socioecol. & Psychol. Primates* 149 Its net reproductive rate is attuned to the availability of food, and is regulated through socioecological mechanisms. **1972** *Biol. Abstr.* LIV. 3830/1 (Index), Socio ecology. **1975** L. L. & D. J. Klein in R. H. Tuttle *Socioecol. & Psychol. Primates* 83 A dietary factor..that is generally overlooked in discussion of primate socioecology, was the degree to which specific primates are able to utilize.. varied substances in any single day. **1979** *Nature* 8 Feb. 433/3 The main aims are to study the socio-ecology of the lesser-known species such as the slow loris and the pig-tailed macaque. **1969** Zigler & Child in Lindzey & Aronson *Handbk. Social Psychol.* III. xxiv. 484 Two main types of interpretations have been employed to explain the cross-cultural findings, the sociogenic and the psychogenic. **1978** *Language* LIV. 228 His separation of cultures, making sociogenic interaction impossible. **1941** W. Dennis in *Genetic Psychol. Monographs* XXIII. 187 Prior to the second year of life sociogenous responses, those which are learned through the intercession of other persons, are few. **1967** C. L. Markmann tr. *Fanon's Black Skin, White Masks* (1968) 13 Besides phylogeny and ontogeny stands sociogeny. **1933** J. L. Moreno in *Proc. Amer. Assoc. Mental Deficiency* 236 To visualize how each individual is affected by the maze of attractions and repulsions coming from any individual or going out from any individual of a group we mapped these relations graphically into a *sociogram*. **1972** M. Argyle *Social Psychol. of Work* vi. 109 The choices may be plotted to form a sociogram, and this can show the affective structure of a group very clearly. **1954** *Encounter* Dec. 55/1 A sociographic study attempts to relate all the data concerning a given locality in a meaningful way. **1881** O. T. Mason in *Ann. Rep. Smithsonian Inst.* (1883) 501 Observing and descriptive stage..sociography. **1940** C. P. Loomis *Tönnies's Fundamental Concepts Sociol.* p. xxiv, (*heading*) Empirical sociology or sociography. **1966** *Listener* 12 May 677/1 We do not yet have a sociography of English Catholicism. **1968** *Internat. Encycl. Social Sci.* XV. 49/1 To this he [*sc.* Tönnies]..opposed his own notion of 'sociography', in which systematic observation, case studies, and other qualitative methods were included, together with statistics. **1950** H. H. Jennings *Leadership & Isolation* (ed. 2) xiii. 276 This collective, more or less formalized setting, where concerns must be shared and obligations held in common..might appropriately be called..the sociogroup. **1956** J. Klein *Study of Groups* 179 In sociometric theory a sociogroup is based on preferences involving work in the group. **1934** Webster, Sociomedical. **1961** *Lancet* 2 Sept. 549/1 Euthanasia..is more than a sociomedical problem. **1977** *Time* 10 Jan. 41/1 The flourishing condom market is only one sign of a growing sociomedical phenomenon in the U.S.: a back-to-basics movement in birth control. **1902** J. M. Baldwin *Social & Ethical Interpretations* (ed. 3) 3 The *Biogenetic* method is valuable mainly in investigating the *socionomic* forces (those which condition or limit social change, but are not themselves social in their character). **1935** *Jrnl. Mammology* XVI. 176 As yet..it is impossible to ascertain the central grouping tendencies and the characteristic socionomic sex ratio (ratio of males to females living within groups). **1976** *Nature* 12 Feb. 459/1 Most body weights..do not take into account interspecies differences in socionomic sex ratio. **1902** J. M. Baldwin *Social & Ethical Interpretations* (ed. 3) xi. 484 Socionomics —the science of the relation of social life to its environment, including other social groups.

3. Also in comb. with advbs., in the sense 'socially and..': *socio-critical, -cultural, -culturally, -demographic, -educational, -emotional, -environmental, -geographic, -geographical, -historical, -industrial, -literary, -politically, -psychological, -psychologically, -regional, -religious* (earlier example), *-scientific, -sexual, -technic* (hence *socio-technics* sb. pl.), *-technical*.

1963 *Cambr. Rev.* 4 May 401/1 The Leitmotiv of the 'socio-critical' movement [in literature] is 'human destiny and national destiny'. **1929** H. Schreuder *Pejorative Sense Devel. in English* II. vi. 65 (*heading*) The socio-cultural group. **1958** *New Biol.* XXVI. 123 In discussing human potentiality it is not useful to dissociate socio-cultural from biological factors. **1978** *Language* LIV. 188 Linguistics consists partly in an investigation of certain aspects of human socio-cultural behavior. **1980** *Times Lit. Suppl.* 19 Sept. 1044/4 For various socio-cultural reasons, the short story has always occupied a more central place in Australian fiction than in European literatures. **1968** C. A. Doxiadis *Betw. Dystopia & Utopia* 66 We are not in a position yet to define how far he [*sc.* man] is adjustable either biologically or socioculturally. **1972** *Bankers Mag.* Winter 68/2 Socio-demographic characteristics of the population such as age, sex, income, occupation, race, [etc.]. **1961** *Times* 16 Sept. 9/3 There is here an old socio-educational dispute in a new context. **1974** *Times Lit. Suppl.* 31 May 580/5 While the mother's strategies correlate with the cognitive development of the child..this correlation is totally unrelated to social class—which effectively deflates some hoary socio-educational myths. **1973** *Word* 1966 XXII. 112 Interaction situations, with men preponderantly performing the task role, and women, the socioemotional. **1969** *Punch* 22 Jan. p. vi/2 Socio-environmental pressures on me had taken an entirely different turn: I had been seduced by advertising. **1962** in H. O. Beecheno *Introd. Business Stud.* p. iii, A condensed survey of the historical background of our present economy..and the more basic socio-geographic factors. **1945** *Archit. Rev.* XCVII. 70/1 The socio-geographical differentiation in our own cities is simple and peaceful compared with that of a newer country like the United States. **1949** *Sci. Amer.* Oct. 53/3 Their philosophical ideas stem from the same socio-historical conditions of our epoch. **1965** *English Studies* XLVI. 390 Melville embodies the concept of revolt in specific socio-historical frames. **1909** W. H. Tolman *Social Engineering* xii. 366 A new profession necessitated by the complexity of socio-industrial relations. **1933** A. Parry *Garrets & Pretenders* p. ix, Since Bohemianism is pre-eminently a socio-literary phenomenon, the periods of its rise and decay coincided fairly well throughout the world. **1980** *Times Lit. Suppl.* 7 Nov. 1249/2 He [*sc.* Edmund Wilson] acquired half a dozen languages that served as socio-literary tools. **1968** *New Left Rev.* Jan.–Feb. 63 Germany did not merely invent the *ersatz* industrially, it produced it socio-politically as well. **1899** *Amer. Jrnl. Sociol.* IV. 661 Such a theory can be developed only along socio-psychological lines. **1970** *Nature* 19 Dec. 1136/1 Too little work has been done on the socio-psychological aspects of spaceflight. **1971** J. J. Shapiro tr. *Habermas's Toward Rational Society* ii. 30 Insensitivity to what in more naive times philosophers called 'the good life' can only be broken through today under the socio-psychologically exceptional conditions of university study. **1964** M. A. K. Halliday et al. *Linguistic Sci.* iv. 86 Our dialects and accents are no longer simply regional: they are regional and social, or 'socio-regional'. **1871** *Aldine* Jan. 18/2 It is a prime merit of such writers as Mr. Keeler and Bret Harte to give a rousing nudge to the saintly self-complacency of our socio-religious 'priggishness'. **1891** *Nation* 3 Sept. 182 The current socio-scientific use of *environment* is first found in Carlyle (1827). **1932** S. Zuckerman *Soc. Life Monkeys & Apes* xiv. 215 Facts concerning sub-human primate socio-sexual responses. **1931** Sociotechnic [see *nuttery 3]. **1937** *Burlington Mag.* Nov. 246/1 The concreteness with which socio-technical problems are realized. **1975** *Times Lit. Suppl.* 28 Feb. 229/3 P. G. Herbst's *Socio-technical Design*.. explores the interaction between the social and technical parts of industrial, educational and scientific organizations. **1976** A. Cherns (*title*) Sociotechnics. *Ibid.* p. ix, Sociotechnics..is..a quest for a methodology of bringing the knowledge and concepts of the social sciences to bear on human and social issues.

sociobiological (sōᵘ:ʃiobəiˌolǫ·dʒikăl, sōᵘ:sio-), *a.* [f. Socio- + Biological *a.*] Of or pertaining to sociobiology. Hence **so:ciobiolo·gically** *adv.*

1921 Hannay & Collingwood tr. *Ruggiero's Mod. Philos.* v. 185 Espinas was one of the first to apply the socio-biological method to the treatment of social questions. According to him the instinct of sociability is found in all grades of being, and is shared by animals and men alike. **1948** *Amer. Scientist* XXXVI. 567 We may regard the mechanism, for sociobiological purposes, as given. **1978** M. S. Gregory et al. *Sociobiology & Human Nature* 3 Pure sociobiological theory, being independent of human biology, does not imply by itself that human social behavior is determed by genes. **1980** *Nature* 24 Apr. 682/1 Sociobiologically sophisticated readers.

sociobiology (sōᵘ:ʃiobəiˌǫ·lodʒi, sōᵘ:sio-). [f. Socio- + Biology.] The study of the biological, esp. the ecological and evolutionary, bases of social behaviour.

1946 J. P. Scott in *Minutes Conf. Genetics & Social Behaviour* 5 The central objective of the scientific method is the development of generalizations which are as nearly true as possible... The zoologists and psychologists who work with animals can do their part to help extend these generalizations by working toward the development of comparative sociology, or perhaps it may be called psychobiology or sociobiology. **1948** C. F. Hockett in *Amer. Scientist* XXXVI. 564 Since there is no generally accepted word for just what we mean, we can, for the purposes of the present discussion, coin one. On the analogy of *biophysics* we shall speak of *sociobiology*. **1950** J. P. Scott in *Ann. N.Y. Acad. Sci.* LI. 1004 An interdisciplinary science which lies between the fields of biology (particularly ecology and physiology) and psychology and sociology. Many names have been given to it, but perhaps the best and most descriptive is 'sociobiology'. **1975** E. O. Wilson *Sociobiology* i. 4/1 Sociobiology is defined as the systematic study of the biological basis of all social behaviour. **1976** *Ann. Rev. Microbiol.* XXX. 236 One of the most exciting aspects of this new field of sociobiology is the chemical control of behavioral responses. **1979** *Nature* 2 Aug. 427/1 The widespread identification of 'sociobiology' with a school of thought is an undeniable fact of current biological sociology.

Hence **so:ciobio·logist,** one who studies sociobiology.

1975 *Times* 29 Dec. 5/1 Interest is growing in the theories of sociobiologists who seek to prove that most human social behaviour has genetic origins. **1980** *Daily Tel.* 10 Mar. 12/8 Sociobiologists hope their theory of behaviour, which seems at present to apply only to individuals, can be expanded into a science which could analyse humanity as a whole.

so·ciodrama. *Sociol.* Also with hyphen. [f. Socio- + Drama.] An improvised play acted by or for those involved in a situation of social tension in order to portray different perceptions of the same situation and represent objectively what each experiences in his or her role; a form of psychiatric treatment based on this type of play. Hence **sociodrama·tic** *a.*, **sociodra·matist**, someone directing a sociodrama.

1943 J. L. Moreno in *Sociometry* VI. 331 Sociodrama which deals with inter-group relations and with collective ideologies. *Ibid.* 438 Let us consider first two broad fields of application of sociodramatic procedures. **1952** W. J. H. Sprott *Social Psychol.* ii. 36 The 'spontaneity training', from which the so-called 'Socio-drama' has been developed. **1958** —— *Human Groups* 188 The elaborate technique of training for foremanship, often employing the socio-drama technique in which foremen take the part of workers and act out a scene. **1964** *Telegraph* (Brisbane) 17 June 8 No one knew what to expect when Jean Jacques Lebel staged his 'sociodramatic event' the other night. **1972** H. J. Eysenck *Psychology is about People* i. 16 The psychiatrist asks, Is psychotherapy better than sociodrama? **1979** Glassner & Freedman *Clinical Sociol.* xiv. 326 In important respects all of the sociodramatic procedures are forms of spontaneity training. *Ibid.*, It is important to distinguish the sociodramatist's view of spontaneity from some others.

so:cio-econo·mic, *a.* Also as one word. [f. Socio- + Economic *a.*] That derives from both social and economic factors; that combines both factors to provide an indication of a person's or a group's effective social situation, esp. as *socio-economic class, status*.

1883 L. F. Ward *Dynamic Sociol.* I. vii. 525 It is not from an anthropological or ethnological stand-point that our treatise proceeds but rather from a strictly sociological or socio-economic one. **1937** L. B. Murphy *Social Behaviour & Child Personality* i. 32 Although the socio-economic status was important..it was by no means directly correlated with the educational stimulus which the children received. **1949** M. Mead in M. Fortes *Social Structure* 18 An account of Arapesh socio-economic life. **1966** Bereiter & Engelmann *Teaching Disadvantaged Children* i. 3 Studies of three- to five-year old children from lower socio-economic backgrounds have shown them to be retarded..in every intellectual ability. **1972** *Jrnl. Social Psychol.* LXXXVI. 207 The subjects were required to be from the same socioeconomic class, in this case the middle class. **1978** P. A. Cowan *Piaget: with Feeling* xii. 310 There are studies which compare infants and..children who differ in socioeconomic status.

Hence **socio-econo·mically** *adv.*

1955 *Social Problems* July 98 The strong influence of abortion law is seen in the refusal of hospitals to grant socio-economically indicated abortions. **1977** *Lancet* 30 Apr. 955/2 In socioeconomically advanced West Germany the vaccine is specifically recommended for infants with chronic diseases of the heart.

sociolect (sōᵘ·siolekt, sōᵘ·ʃio-). [f. Socio- + *-lect.] A variety of a language that is characteristic of the social background or status of its user. Also *attrib.* Hence **sociole·ctal** *a.*

1972 [see *-lect]. **1976** *Archivum Linguisticum* VII. 158 Here the functional shift from dialect to sociolect indicators is of particular interest. In fact, dialectal, sociolectal, and stylistic varieties will perhaps be the first 'partial features' which lend themselves to a more ambitious attempt at establishing a complex network of linguistic and social functional co-variation. **1978** *Language* LIV. 227 The book contains 491 pages, written in a highly academic sociolect of German which makes rich use of embedding devices, some sentences being over a page long. **1980** *English World-Wide* I. I. 129 A collection of twelve articles on a variety of sociolectal and dialectal phenomena in British English.

sociolinguistic (sōᵘ:sioliŋgwi·stik, sōᵘ:ʃio-), *a.* and *sb.* Also socio-linguistic. [f. Socio- + Linguistic *a.* and *sb.*] **A.** *adj.* Of or pertaining to the study of language in its social context.

1949 E. A. Nida *Morphology* (ed. 2) vi. 152 The reactions of language-users to the sociolinguistic environment. **1952** *Word* VIII. III. 261 Therefore we may expect to come across socio-linguistic situations which we may hesitate to class in one or another of our four categories. **1959** *Amer. Speech* XXXIV. 118 Enumerating these sounds without giving clear indications of all the pertinent sociolinguistic facts can be dangerous. **1964** L. Kaiser in D. Abercombie et al. *Daniel Jones* 102 Ladefoged has discerned three kinds of information: linguistic, socio-linguistic, and personal. **1971** J. Spencer *Eng. Lang. W. Afr.* 7 Before we can understand the processes of stabilisation and change in the English of West Africa, we need a great deal more sociolinguistic evidence. **1978** *Verbatim* Feb. 10/1 Dillard organizes his work around 'socio-linguistic domains' and stresses the importance of discourse over sentence as the primary carrier of meaning.

B. *sb. pl.* (usu. const. as *sing.*). The study of language in relation to social factors. Also *attrib.*

1939 T. C. Hodson in *Man in India* XIX. 94 (*title*) Socio-linguistics in India. **1951** E. Haugen in *Language* XXVII. 213 If semantics should be an undesirable term, there is always 'ethno-linguistics' or perhaps 'socio-linguistics'. **1951** *Directory of American Scholars* 1061/2 Currie, Prof. Haver C(ecil)... History of American thought; socio-linguistics. **1952** H. C. Currie in *Southern Speech Jrnl.* XVIII. I. 28 This field is here designated *socio-linguistics*. *Ibid.* 36 The present projection of socio-linguistics proposes a fresh start toward researches into the social significance of language in all respects. **1964** *9th Internat. Congress Linguistics 1962* 1129 Those of us who work in the interdisciplinary area of 'socio-linguistics' may feel that we are here at this Congress on sufferance. **1967** *Language* XLIII. 586, I find that sociolinguistics connotes a branch of linguistics or, at best, a neutralization of the fruitful distinction between sociological linguistics and sociology of language. **1979** *London Rev. Bks.* 25 Oct. 4/3 (Advt.), Shows how social psychological theories and methods can increase the explanatory power of sociolinguistics. **1980** *English World-Wide* I. 179 Sociolinguistics intends to produce a linguistic

description as its end result, although it uses social facts and methods to arrive at this end.

Hence **socio-li·nguist,** a student of or specialist in sociolinguistics; **so:ciolingui·stically** *adv.*

1960 *Amer. Anthropologist* LXVI. 86 Sociolinguists study verbal behavior in terms of the relations between the setting, the participants, the topic, the functions of the interaction, the form, and the values held by the participants about each of these. **1968** W. A. Stewart in J. A. Fishman *Readings in Sociol. of Lang.* 539 [Pidgins] and [Creoles] usually function sociolinguistically as special kinds of dialects of their lexical-source languages. **1972** J. L. Dillard *Black English* v. 193 Men's dialects and languages as well as women's dialects and languages are well known to the sociolinguist. **1973** *Archivum Linguisticum* IV. 70 The following would seem most likely to be sociolinguistically of significance. **1979** *Amer. Speech 1976* LI. 118 Perhaps,..through the combined efforts of dialectologists, sociolinguists, and other observers of language, a more accurate picture will emerge.

sociologese (sōᵘ:siǫlŏdʒī·z, sōᵘ:ʃiǫ-). [f. Sociolog(y + -ese.] A derogatory term used to describe the style of writing supposedly typical of sociologists; a style which is over-complicated or jargonistic and abstruse. Also *attrib.*

1963 *Times Lit. Suppl.* 29 Mar. 211/1 They are written ..not in Sociologese. **1965** E. Gowers *Fowler's Mod. Eng. Usage* (ed. 2) 570/1 Sociologese, like Commercialese and Officialese, deserves an article to itself. **1969** R. Blackburn in Cockburn & Blackburn *Student Power* 184 The notion of structural contradiction emerges in Sociologese as 'lack of fit' between, for example, the economy and the 'core institutional framework'. **1973** *Times* 8 Nov. 16/3 His first chapter or two is marred by heavy sociologese. **1977** *New Statesman* 1 July 16/3 So, it's 'thwokk', is it? What a pity..for Mick Shepherd..that he was unable to find a better acoustic whizz-word than this in his dictionary of fifth-rate sociologese clichés. **1982** *Times Lit. Suppl.* 26 Mar. 356/5 *The Culture of Consent* contains, perhaps, too much sociologese for some tastes.

sociological, *a.* (Earlier example.)

1843 Mill *Logic* II. vi. x. 585 There are two kinds of sociological inquiry.

sociologism (sōᵘsiǫ·lŏdʒiz'm, sōᵘʃiǫ-). [f. Sociolog(y + -ism.] The tendency to ascribe a sociological basis to other disciplines. Hence **sociologi·stic** *a.*

1945 K. R. Popper *Open Society* II. xxiii. 202 This theory of Hegel's..is sometimes called 'historicism'... The sociology of knowledge or 'sociologism' is obviously very closely related to or nearly identical with it. *Ibid.* 205 If scientific objectivity were founded, as the sociologistic theory of knowledge naively assumes, upon the individual scientist's impartiality,..we should have to say good-bye to it. **1958** W. Stark *Sociology of Knowledge* 331 We have to reject 'sociologism' (a pendant to psychologism). **1964** I. L. Horowitz *New Sociology* 17 The recent work in some quarters, ostensibly critical of excessive sociologism, seems to point precisely in the direction of the self-liquidation of sociology. **1965** E. E. Evans-Pritchard *Theories Primitive Relig.* iii. 70 But, masterly though it was, its conclusions are an unconvincing piece of sociologistic metaphysics. **1977** *Language* LIII. 398 The fact that the autonomy principle underlies K's attack on Doroszewski is shown by the following remark on Saussure's alleged sociologism. **1978** E. A. Tiryakian in Bottomore & Nisbet *Hist. Sociol. Anal.* vi. 212 Durkheim's 'sociologism' in this respect amounts to no less than an epistemological revolution.

sociologize, *v.* Add: Also *trans.*, to render sociological in character; to study from the standpoint of sociology. Hence **socio·logizing** *ppl. a.* and *vbl. sb.*

1924 *Amer. Jrnl. Sociol.* XXX. 302 (*title*) The sociologizing movement within political science. **1960** *20th Cent.* May 443 The 'sociologizing' of these disciplines [*sc.* economics and politics] in America..has increased their current vitality. **1971** *New Society* 7 Jan. 24/2 How many sociologists are there in Britain? We do not know, for the profession has not seriously sociologised itself. **1980** *Times Lit. Suppl.* 26 Sept. 1073/3 *The People's Choice* of 1944 is central to the beneficial sociologizing of more than one sector of the political science with which we now live.

sociology. Add: Also with pronunc. (sōᵘsi-). **1. a.** Also, the study of social organization and institutions and of collective behaviour and interaction, including the individual's relationship to the group; also *attrib.*

1897 L. F. Ward *Social Philos.* I. iv. 66 The nature of this being, man, whose associative habits form the chief subject of sociology. **1927** P. Sorokin *Social Mobility* p. ix, Speculative sociology is passing over. An objective, factual, behavioristic, and quantitative sociology is.. superseding it. **1932** S. & B. Webb *Meth. of Social Study* i. 3 Sociology is concerned not with the individual man, regarded as a living organism.., but with relations among men. **1951** Spaulding & Simpson tr. E. Durkheim (*title*) Suicide. A study in sociology. **1965** Howard & Weaver tr. *Aron's Main Currents in Sociolog. Thought* 8 Sociology may be said to be characterized by two specific aims which account for its nature. On the one hand, sociology lays claim to objective and scientific knowledge. On the other, what it claims to know objectively and scientifically is some vaguely defined thing we call society or societies or social phenomena. **1977** R. Holland *Self & Social Context* ix. 260 A college would rather fall below its intake targets and lose revenue than take in sociology students. **1980** *Daily Tel.* 17 Apr. 10/8 He told me that he taught sociology and I told him straight that hogwash was not one of my subjects.

b. The application of sociological concepts and analysis to the social context of other disciplines or fields; a particular sociological system.

1916 E. Ehrlich in *Harvard Law Rev.* XXIX. 582 Mr. Justice Holmes has suggested a criticism of my book on the sociology of law..in that he finds therein no reference to Montesquieu. **1928** P. Sorokin *Contemp. Sociological Theories* I. 40 It will be more convenient to discuss his [*sc.* Weber's] sociology in the chapters on the sociology of religion. **1936** Wirth & Shils tr. *Mannheim's Ideology & Utopia* ii. 69 The simple theory of ideology develops into the sociology of knowledge. **1947** J. Wach (*title*) Sociology of religion. **1965** Howard & Weaver tr. *Aron's Main Currents in Sociological Thought* 7 The specialized sociologies—the sociology of science, the sociology of language, the sociology of art, the sociology of literature—are attempts to explain the evolution of human phenomena in relation to the social milieu. **1978** *Amer. N. & Q.* Feb. 92/1 The first volume.. contains seven essays on the sociology of reading. **1980** *English World-Wide* I. 179 The sociology of language.. uses linguistic information as a means of describing social phenomena.

2. The study of plant or animal communities. Cf. *phytosociology.

1932 Fuller & Conard tr. *Braun-Blanquet's Plant Sociol.* 1 We may divide all biology into (1) idiobiology: the science of individual organisms; and (2) sociology: the science of organic communities. The latter is divided into the social science of man (sociology in the usual sense), zoosociology, and phytosociology or plant sociology. **1932** S. Zuckerman *Social Life Monkeys & Apes* i. 9 Mammalian sociology has developed without any real regard for questions concerning the validity of anecdotal and anthropomorphic evidence. **1953** D. A. Bannerman *Birds Brit. Isles* I. 35 We are learning more and more regarding what may be termed the sociology of birds. **1960** N. Polunin *Introd. Plant Geogr.* iii. 92 Plant sociology, where considerations of life-forms may help in the description of the structure of the communities.

sociometric (sōᵘsiome·trik, sōᵘʃio-). *a. Sociol.* and *Social Psychol.* [f. Socio- + *-metric.] Connected with or pertaining to sociometry or the assessment of relationships in groups. Hence **sociome·trically** *adv.*; **sociome·trics** *sb. pl.* = *sociometry.

1933 J. L. Moreno in *Proc. Amer. Assoc. on Mental Deficiency* 224 An instrument to measure the amount of organization shown by social groups is called the sociometric test. **1943** —— in *Sociometry* VI. 316 How a community can be sociometrically tested. **1952** *Personnel Psychol.* V. 178 (*heading*) Sociometrically selected work teams increase production. **1956** C. W. Valentine *Normal Child* xi. 189 The 'sociometric' technique is flexible. **1964** D. F. Dowd in I. L. Horowitz *New Sociol.* 61 A mindless procession that moves from sociology to sociometrics. **1978** P. Marsh et al. *Rules of Disorder* iv. 86 Simply by looking at who meets whom and in what order we can begin to get some idea of the sociometric structure of the football fan culture.

sociometry (sōᵘsiǫ·mětri, sōᵘʃiǫ-). *Sociol.* and *Social Psychol.* [f. Socio- + -metry.] The qualitative and quantitative analysis of the structure of groups, esp. through charting the relationships that exist between the members of small groups. Hence **socio-metrist.**

1908 W. de Morgan *Somehow Good* xi. 99 'I suppose you'd admit there *are* such things as social distinctions?' Sally wouldn't admit anything whatever. If sociometry was to be a science, it must be worked out without axioms or postulates. **1933** J. L. Moreno in *Proc. Amer. Assoc. on Mental Deficiency* 224 The mathematical study of psychological properties of populations, the experimental technique of and the results obtained by the application of quantitative and qualitative methods is called sociometry. **1937** *Sociometry* I. 219 The sociometrist has the task of breaking down..the misunderstandings and fears..in the group he is facing. **1956** J. Klein *Study of Groups* 180 The insistence of the sociometrist that the two kinds of criterion must be presented simultaneously now begins to sound suspicious. **1962** K. M. Evans (*title*) Sociometry and education. **1976** *Times Lit. Suppl.* 26 Mar. 343/2 Mr Lucas deserves a loud cheer for writing a whole informative chapter on the way members of groups affect each other in communities without once using that dreadful word, beloved of educational theories, sociometry.

sociopath (sōᵘ·siopæþ, sōᵘ·ʃio-). *Psychol.* [f. Socio-, after Psychopath.] Someone with a personality disorder manifesting itself chiefly in anti-social attitudes and behaviour. Hence **sociopa·thic** *a.*; **socio·pathy.**

1930 G. E. Partridge in *Amer. Jrnl. Psychiatry* X. 55 A conspicuous number who..may justly be termed 'sociopathic'. *Ibid.*, We may use the term 'sociopathy' to mean anything deviated or pathological in social relations. *Ibid.* 56 We may exclude from the class of essential sociopaths those whose inadequacy is primarily related to physical weakness, fear, hypersensitiveness, shyness and self-blame. **1940** Hinsie & Schatzky *Psychiatric Dict.* 493/1 *Sociopathy*, this term has generally been used to designate an abnormal or pathological mental attitude toward the environment. **1962** L. Yablonsky *Violent Gang* (1967) xii. 216 The violent-gang structure recruits its participants from the more sociopathic youths living in the disorganized-slum community. **1968** *Listener* 26 Sept. 408/1 In America 'psychopathy' has been replaced by 'sociopathy'. **1976** Smythies & Corbett *Psychiatry* iii. 29 Many sociopaths come from appalling backgrounds or from genetically afflicted families.

socius. Add: **3.** *Philos.* Applied to God, as the 'Great Companion' of man.

1890 W. James *Princ. Psychol.* I. x. 316 The impulse to pray is a necessary consequence of the fact that whilst the innermost of the empirical selves of a man is a *self* of the *social* sort, it yet can find its only adequate *socius* in an ideal world. **1917** A. S. Pringle-Pattison *Idea of God* xv. 297 The idea of a divine socius has been one of the most abiding inspirations of religious experience.

4. The individual person, considered as the unit of human society; the social self.

1895 J. M. Baldwin *Mental Devel. Child & Race* xi. 338 Both *ego* and *alter* are thus essentially social; each is a *socius*, and each is an imitative creation. **1898** F. H. Giddings *Elements of Sociol.* 10 What, now, is the unit of investigation in Sociology?... In its simplest form society exists whenever an individual has a companion or associate. The socius, then, is the unit of any social group or society. **1912** C. A. Ellwood *Sociol. in Psychol. Aspects* ii. 21 The *socius*, or associated individual,..the unit out of which all the simpler social groups are composed. **1963** S. Koch *Psychol.* VI. p.v, (*title*) Investigations of man as socius: their place in psychology and the social sciences.

sock, *sb.*[1] Add: Pl. also **sox** (see as main entry). **2. a.** Slang and colloq. phrases: *in one's socks*, as a condition of measurement of stature; = *in one's stockings* s.v. Stocking *sb.* 5 a (cf. *in one's shoes* s.v. Shoe *sb.* 2 c); *to knock the socks off* (someone), and varr. (*U.S.*): to beat thoroughly, to trounce; similarly *to rot the socks off*; *to pull one's socks up*: to make an effort, to pull oneself together; *to put a sock in it*: to stop speaking or making a noise, to shut up; to 'stop it'; usu. in *imp.*; *old socks* (orig. *N. Amer.*): a familiar form of address.

1835 Dickens *Sk. Boz* (1836) II. 2 He..stood four feet six inches and three-quarters in his socks. **1913** T. H. S. Escott *Anthony Trollope* x. 191 Those who knew Anthony Trollope in the flesh saw in him one who, at his prime, had stood some six feet in his socks. **1927** in C. J. Finger *Frontier Ballads* 69 This Floyd stood six feet in his socks and passed for mighty fly. **1977** *Horse & Hound* 14 Jan. 40/3 (Advt.), 11 hands in his socks... Smart, free-moving pony.

1845 C. Beecher *Let.* in M. Rugoff *Beechers* (1981) xi. 212 'Beecher you must put in your best licks today!' 'You must knock the socks off those Old School folks!' And so they stood by to see me fight. **1964** J. Porter *Dover One* vi. 65 This thick, fruity beverage..was guaranteed by one revolting old boozer as being strong enough to rot the socks off you. **1979** *Arizona Daily Star* 22 July D3/1 'Trucks have been beating our socks off,' said..a spokesman for the Atchison, Topeka & Santa Fe Railway in Chicago. 'But now we have a chance to get some of the business back.'

1893 H. F. McClelland *Jack & Beanstalk* 31 Pull up your socks! I'll see naught goes wrong with you. **1906** *Daily Mail* 14 Feb. 6/6 The 'smart set' have got hold of another neat expression. 'You must pull your socks up' is the latest form of saying 'Never mind', or 'Pull yourself together'. **1914** 'Bartimeus' *Naval Occasions* xi. 78 Pull your socks up, Ah Chee, an' think of something. **1936** M. Kennedy *Together & Apart* iv. 294 There's hope for you if you pull your socks up. **1976** *Southern Even. Echo* (Southampton) 11 Nov. 17/5 The dismissal was unfair because Mr. Collier had not been given adequate warning and a chance 'to pull his socks up' before dismissal.

1919 *Athenæum* 8 Aug. 729/2 The expression 'Put a sock in it', meaning 'Leave off talking, singing or shouting'. **1930** J. B. Priestley *Good Companions* I. vi. 232 Two or three members of his audience laughed, but a young man in a green cloth cap was very annoyed. 'Oh, put a sock in it,' he said to the ripe gentleman, who immediately and very loudly asked him what he meant by it. **1933** M. Lowry *Ultramarine* vi. 243 Aw, put a sock in it. Well, I'm going to sleep, chaps, and if you wake me again, the fellow that does it I'll slip him thirteen inches of saltpetre. **1944** 'N. Shute' *Pastoral* v. 107 'For Christ's sake put a sock in it,' he had said..'and tell them I want an ambulance down here.' **1978** A. Price *'44 Vintage* vi. 69 He..drew his finger across his throat, grinning horribly. 'Put a sock in it, Taf,' said Sergeant Purvis sharply.

1925 T. Dreiser *Amer. Tragedy* I. II. iii. 185 Gee, it's good to have a look at you, old socks! **1934** H. G. Radcliffe in *Passing Show* 27 Jan. 5/4 Hey, Morrison, old socks. How's things! **1950** R. Moore *Candlemas Bay* 19 Ninety..pounds, Jebby, old socks.

c. *colloq.* A sock used as a receptacle for storing one's money; hence, a store of money. Cf. Stocking *sb.* 2 a.

1930 Wodehouse *Very Good, Jeeves!* x. 263 Her name was Maudie and he loved her dearly, but the family would have none of it. They dug down into the sock and paid her off. **1949** H. MacLennan *Precipice* III. 248 Once we've got enough put away in the sock I'm going to..go back to M.I.T. **1951** Cusack & James *Come in Spinner* 54 He just can't adjust himself to not having the best. And that's what marriage without a sock in the bank would mean. **1956** B. Holiday *Lady sings Blues* (1973) x. 95, I opened Café Society as an unknown; I left two years later as a star. But you couldn't tell the difference from what I had in my sock.

d. *N.Z.* (See quot. 1965.)

1955 G. Bowen *Wool Away!* iii. 45 Many sheepowners do not like socks taken off, as it puts hair in the wool. **1965** *N.Z. Listener* 26 Feb. 15/2 *Socks*, wool between the knee and the foot. In some sheds and competitions the instruction 'leave the socks on' means not to shear this wool, which usually contains a proportion of hair.

6. *attrib.* and *Comb.*, as **sock foot** *U.S.* =

Stocking-foot c; hence **sock-footed** *a.*; **sock-suspender** = Suspender 4 b. See also sense 2 in Dict.

1934 *Jrnl. Amer. Folk-Lore* XLVII. 52 No boots could he find. He was about to..go to his wedding in his sock feet, when a Voice told him to crawl out from under the bed. **1935** Z. N. Hurston *Mules & Men* I. viii. 177 Pull off yo' shoes and set in yo' sock feet. **1981** *Nordic Skiing* Jan. 34/2 Chairback is for the making of memories... unforgettable skiing, bone-easing hot shower and sauna, a fine dinner, a tumbler of mulled wine, sock-footed by your woodstove. **1912** E. C. Bentley *Trent's last Case* iii. 36 He had on a complete outfit of underclothing, studs in his shirt, sock-suspenders. **1922** Wodehouse *Jill the Reckless* xviii. 268 Give me your share of the show for three dollars in cash and I'll throw in a pair of sock-suspenders and an Ingersoll. **1978** S. Wilson *Dealer's Move* ix. 185 One of his trousers had been dragged up to his knee... He was wearing sock suspenders.

sock, *sb.*[4] Add: **1.** Also in phr. *a sock in the eye* (also *fig.*).

1972 Wodehouse *Pearls, Girls, & Monty Bodkin* vi. 87 He's asked you to lunch in the hope of talking you into giving me the sock in the eye on which his heart is set. **1974** —— *Aunts aren't Gentlemen* xiv. 119, I knew that her name would be mud. I still wasn't sure she couldn't even be jugged, and what a sock in the eye that would give Uncle Tom's digestion. **1979** *Woman & Home* June 154/2 The return to tradition; a sock in the eye for the mechanisation that was creating unemployment.

2. *U.S. slang.* A strong impact, emphasis, a 'kick'.

1936 *Swing Music* Mar. 10/1, I used to get a terrific sock out of Rappolo riding high on his clarinet. **1937** B. Goodman *This Thing called Swing* 9 *Sock*, emphasis, usually referring to the last chorus. **1950** *Audio Engin.* Sept. 14/3 More low middles increase the *Punch* until the program is *solid*, and has *sock*. **1972** *Publishers' Weekly* 21 Aug. 15/1 (Advt.), Here's solid history with the sock of unforgettable fiction. **1979** *Arizona Daily Star* 22 July C3/2, I figure we have enough speed and sock in our lineup to score runs.

3. *attrib.* and *Comb.*, as **sock chorus** *Jazz* (see quot. 1936); **sock cymbal** *Jazz* = *high hat, high-hat 3; also *attrib.*

1936 *Delineator* Nov. 49/2 *Sock chorus*, last chorus of an arrangement. **1937** *Metronome* Mar. 31/1 The full sock chorus..hits you between the eyes. **1956** E. Hunter *Second Ending* iv. 69 They rode into the sock chorus like a storm cloud of marauders. **1936** Sock cymbal [see *off-beat *a.* 1]. **1949** L. Feather *Inside Be-bop* III. 80 Kenny originally played the old Jo Jones sock cymbal style. **1972** *Jazz & Blues* Sept. 7/3 You wouldn't play your sock cymbal the same as your ride cymbal.

sock, *v.*[2] Add: **1. a.** (*a*) (Further examples.) (*b*) *fig.* To give a hard blow to; *esp.* to take large sums of money from (someone). *U.S.*

(*a*) **1870** R. B. Mansfield *School-Life at Winchester College* 234 *Sock*, to hit hard at Cricket. **1916** [see *gas *sb.*[1]]. **1926** *Variety* 29 Dec. 5/3 No craving for expression motivated me when I hung up the finger glove and sliding pads in favor of socking a typewriter. **1933** G. B. Shaw *Political Madhouse in America* 21 Why do you applaud these screen heroes who, when they are not kissing the heroine, are socking jaws? It is a criminal offence to sock a citizen in the jaw. **1982** B. Chatwin *On Black Hill* xiv. 67 The porter had socked him on the jaw, and he now lay, face down on the paving.

(*b*) **1939** J. Steinbeck *Grapes of Wrath* xvi. 248 Find out how much they gonna sock ya for the lessons. **1941** B. Schulberg *What makes Sammy Run?* x. 191 When a moving picture is right, it socks the eye and the ear and the solar plexus. **1943** *Sun* (Baltimore) 8 Nov. 1/2 Cost rises are so precipitate..that one Congressman..suggested 'we're being socked everywhere in foreign countries.' **1973** J. Cleary *Ransom* ii. 44, I don't know what sort of demands they're making. If they're socking the Mayor.. the price is gonna be high—he's a very rich man. **1978** *Detroit Free Press* 5 Mar. A.8/1 The township socked the company with a building permit violation.

c. Substitute for def.: orig. and chiefly *U.S.* (*a*) To drive or thrust *in* or *in(to)* something. (Earlier examples.) (*b*) Phr. *to sock it to* (one): to strike, deal a blow to (that person); to 'give it' to (one). *fig.* Freq. in imp., as catch-phrase *sock it to me* (*them*, etc.)*!*, used to express encouragement, sexual invitation, etc. Also in sb. phr. *sock-it-to-me*, a loud and violent style of music; a piece of such music. (Earlier and later examples.)

(*a*) **1843** *Spirit of Times* 15 July 234/1 About one hundred yards from home, Spicer pulled Beppo out, and 'socked in' his spurs. **1845** T. J. Green *Texian Exped.* xvii. 321 The corporal 'socked' it [*sc.* a shoe-maker's awl] in the thick of his back. **1878** J. H. Beadle *Western Wilds* ii. 37 The very next day they put me in jail— socked me right in with them two Hodges.

(*b*) **1877** Bartlett *Dict. Americanisms* (ed. 4) 623 Two loafers are fighting; one of the crowd cries out, 'Sock it to him.' **1883** 'Mark Twain' *Life on Mississippi* xliii. 438 A rich man won't have anything but your very best; and you can..pile it on and sock it to him. **1889** —— *Connecticut Yankee* xxxiii. 422 'Well, observe the difference: you pay eight cents and four mills, we pay only eight cents.' I prepared, now, to sock it to him. I said: 'Look here..,what's become of your high wages you were bragging so about, a few minutes ago?'—and I looked around on the company with placid satisfaction. **1927** O. W. Holmes in *Holmes-Laski Lett.* (1953) II. 975, I have heard an English judge sock it to the jury in a murder case. **1963** B. J. Chute *Shift to Right* 153 There was a shriek from the panting Trenton stands: 'Yea, Rusty. Sock it to 'em.' **1968** *Tel.* (Brisbane) 15 June 2/3 'Sock it to me' is a catch-phrase which is sweeping America... It's all due to Judy Carne,..who cheekily used the phrase in a weekly comedy show called 'Laugh In'. **1969** R. Lowell *Notebook* 137 The little girl's bedroom, perfect with posters: 'Do not enter,' and 'Sock it to me, Baby.' **1969** *Times* 19 July 9/6 The black American phrase 'sock it to me' (with an obscene connotation). **1969** *Times* 29 July 1/3 If President Nixon is going to 'sock it' to anyone, the likeliest recipient is the South Vietnamese government. **1970** S. Sheldon *Naked Face* (1971) ii. 16 She reached between his legs and stroked him, whispering, 'Go, baby. Sock it to me.' **1970** *Melody Maker* 11 July 19/7 It's good to hear Pickett getting away from the sock-it-to-me and into gentler songs. **1971** *West Indian World* 12 Nov. 14/3 Back to the sock-it-to-me's with Jesse James's 'Don't Nobody Want to Get Married'..which storms breathlessly along complete with hard-hitting bass and wow-wow guitars. **1977** *New Yorker* 2 May 34/2, I can't afford a second divorce. Daphne would really sock it to me. **1978** *Railway Age* 25 Dec. 25/2 Does all this boil down to some kind of accounting legerdemain that, in the end, will be socking it to the taxpayers?

d. *Jazz.* To perform (music) in a swinging manner. Freq. in phr. *to sock it* (*out*).

1927 *Melody Maker* July 697 Sock out your last chorus on that, my friends. **1933** [see *get *v.* 62 l]. **1935** *Vanity Fair* XLV. 71/3 Hot artists or bands that can put across their licks successfully are 'senders';..they can '*sock it*'. **1955** Shapiro & Hentoff *Hear me talkin' to Ya* vii. 79 'Blow it, kid. Sock it out,' Tig and Jones kept shouting, until I finally loosened up and did tricks with that slide that I probably never did before or since. **1968** *Radio Times* 28 Nov. 47/1 He's spent his evenings singing in pubs..'socking' out the rhythm and blues. **1976** *New Yorker* 12 Jan. 37 (*caption*) From the top—'Watermelon Man'. Let's sock it out and give Mrs. Ritterhouse a chance to really cook.

Hence **so·cking** *vbl. sb.*

1978 *N.Y. Times* 30 Mar. D 17/2 Harrelson played no part in all the socking because the Mets stopped scoring after four innings and the Phillies after five, and he didn't get into the game until the seventh.

sock, *v.*[3] **a.** (Earlier example.)

1842 *Eton Bureau* 162 Sock means prog, but when you sock a boy anything, he eats it, and you pay for it... I was asked by A— to sock him a verse the other day, and I had to sock him a construe of his lesson too.

sock, *v.*[5] Add: **2.** *colloq.* (orig. *U.S.*). To put (money) aside as savings. Also with *away*.

1942 Berrey & Van den Bark *Amer. Thes. Slang* § 376/5 *Save*..sock one's money away. **1951** Cusack & James *Come in Spinner* 297, I bet he's socked a pretty packet away. **1962** E. Lucia *Klondike Kate* iii. 97 Instead of gambling a fortune away at the wheels..[Kitty] was wisely socking it into the bank. **1963** C. D. Simak *They walked like Men* ix. 53 They've been busy for the last week scooping it in. People come in loaded and are socking it away. **1971** *Maclean's Mag.* Sept. 11/2 Now they seem to believe that a buck earned is a buck to be socked away. **1978** R. Doliner *On the Edge* v. 84 He's got to have money... How much you figure he socked away?

3. *N. Amer.* Of fog, cloud, etc.: to close *in*, to enshroud.

1950 Webster Add. Sock in. **1953** Berrey & Van den Bark *Amer. Thes. Slang* (1954) § 761/2 *Socked in*, ceiling zero. **1955** *Sci. News Let.* 26 Feb. 136 Man-made ice-fog that 'socks in' Arctic airfields can now be licked by a new device developed by the Armour Research Foundation, Chicago. **1969** *Daily Tel.* 21 May 1/6 All of Europe, the Soviet Union,..are socked in cloud cover. **1975** *High Times* Dec. 70/3 Pilots..are often completely socked in by fog and haze. **1976** C. Egleton *State Visit* iii. 25 Wednesday is always a busy day... As long as the airfield isn't socked in.

sockdolager. Add: Also **sock dologer, -doliger. 1.** (Earlier example.)

1830 *Virginia Literary Museum* I. 479 *Sock dologer*, a decisive blow.

3. (Earlier examples.)

1838 J. F. Cooper *Home as Found* II. 72 There is but one 'sogdollager' in the universe, and that is in Lake Oswego. **1842** *Knickerbocker* XIX. 223 This seemed to be a 'socdoliger' (which translated into Latin, means a ne plus ultra).

socked, *ppl. a.* (Later examples.)

1918 W. J. Locke *Rough Road* xxiii. 293 Her thoughts winged themselves back to an afternoon, remote almost as her socked and sashed childhood. **1937** L. Durrell *Panic Spring* vii. 116 Her brown socked legs stretched out under the dashboard. **1976** 'J. Ross' *I know what it's like to Die* i. 7 Five feet and eight inches..in his socked feet.

socker. Add: Also **socca.** The form *soccer* is now usual. (Earlier and later examples.) Also in *Comb.* Hence **so·ccerite,** a player of soccer; also *attrib.*

1889 E. C. Dowson *Let.* 21 Feb. (1967) 38, I absolutely decline to see 'socca' matches. **1916** Blancroft & Pulvemacher *Handbk. Athletic Games* (1922) 429 Soccer football, as it is called in America, is the English Association Football. **1924** H. de Selincourt *Cricket Match* iv. 83 However any sane person could prefer soccer to cricket the good little Horace totally failed to comprehend. **1935** *Punch* 24 Apr. 476/2 No one more thoroughly qualified to write the history of 'soccer'..can be imagined. **1945** *Gen* 13 Jan. 30/1 Many Soccerites..took to Rugby. **1951** R. Campbell *Light on Dark Horse* 69 My father had founded the Technical College, a 'soccerite' school. **1951** *Sport* 7–13 Jan. 9/1 We had the F.A. scheme to bring the big professional clubs and the soccer-playing schools into closer contact. **1971** L. Koppett *N.Y. Times Guide Spectator Sports* xii. 193 All you need to play soccer is a ball, a field and players. **1976** *Field* 18 Nov. 989/2 They roar around, fighting and frolicking beneath like soccer hooligans. **1978** P. Marsh et al. *Rules of Disorder* iv. 97 The soccer terraces offer..a chance to escape from the dreariness of the weekday world.

sockeroo (sǫ:kĕrū·). *slang* (orig. *U.S.*). [f. Sock *sb.*[4] + *-eroo.] Something with an overwhelming impact, a 'smash'.

1942 *Time* 9 Nov. 77/2 The act was an old-fashioned Hippodrome sockeroo. **1943** *Sat. Rev. Lit.* 18 Dec. 16/3 This program ..was a boff, a wham, a sockeroo. **1964** *Spectator* 7 Feb. 178 This latest box-office sockeroo also provides a modest example of the industry's throat-cutting activities. **1977** *Daily Tel.* 22 Apr. 13 The Royal Court's new régime opens with a great loud sockeroo of a play, a thumping American drama of a divided family, rich in purple prose and loaded with gutsy symbolism.

socket, *sb.* Add: **2. c.** An object in which the terminals of an electricity supply are inside holes made to receive the pins of a plug; *spec.* one that is fixed to a wall.

1885 C. J. Wharton tr. *Hospitalier's Domestic Electr.* vii. 113 The whole [lamp-holder] is fitted to a wooden socket C, which may be screwed into an ordinary chandelier or in place of a gas burner. **1892** [see *plug *sb.* 1 c]. **1914** S. C. Batstone *Electric-Light Fitting* vi. 127 The wires come into the socket from behind the skirting. **1938** J. W. Sims *Electr. Installations* 155 Apparatus requiring not more than 50 watts may be supplied from..one 15-amp socket. **1955** N. W. Kay *Mod. Building Encycl.* 637/1 The fuse can be renewed only when the plug-head is withdrawn from the socket. **1977** F. Hall *Building Services & Equipment* II. v. 35/2 The sockets will only accept plugs for 110V, single-phase, 50Hz supply.

d. *Golf.* That part of the head of a club into which the shaft is fitted; a shot made off the socket.

1887 W. G. Simpson *Art of Golf* I. iv. 22 Irons and cleeks ..have sockets instead of necks. **1922** C. Leitch *Golf* 98 There is no bad shot in golf which flurries a player so much as a shot off the socket. **1927** *Daily Express* 12 Feb. 3/7 In the down stroke, the left hand and arm get ahead of the right, and that causes either a socket or a 'push out'. **1963** J. Jacobs *Golf* 78 The socket is simply explained—the club head is being brought down further away from the body than it should be.

7. a. For 'sense 1' read 'sense 2'. **socket outlet,** a socket (sense *2 c) fixed to a wall and connected to an electricity supply; cf. *outlet *sb.* 1 d, *point *sb.*[1] A. 19 e; **socket set,** a number of sockets for use with a socket wrench; **socket wrench,** a wrench equipped with a set of detachable sockets of different sizes.

1934 *Two-Pole & Earthing-Pin Plugs & Socket-Outlets* (Brit. Standards Inst.) 6 When the plug and the socket-outlet are in complete engagement no live parts shall be accessible. **1977** *Jrnl. R. Soc. Arts* CXXV. 119/2 There will probably be more storage space [in today's new house] and almost certainly more socket outlets. [**1918** A. L. Dyke *Automobile & Gasoline Engine Encycl.* 613/2 (*caption*) No. 12 combination socket wrench set.] **1935** *Gen. Catal. Tools & Supplies* (Buck & Hickman Ltd.) 270/2 'Ratchet handle' socket sets. **1976** *Star* (Sheffield) 29 Nov. 5/5 Hinds pleaded guilty to stealing an electric drill and a 52-piece socket set. **1905** W. Rogers *Pumps & Hydraulics* II. 344 An interchangeable socket wrench is shown in Fig. 629. **1921** *Car* 31 Mar. 313/2 (Advt.), Your car will be kept in perfect tune if a socket wrench set is in your kit. **1977** *New Yorker* 9 May 34/3 Bicycling accessories for every contingency:. monkey wrenches, socket wrenches, wrench holders.

b. *socket-eyed* adj.

1964 F. Warner *Early Poems* 76 Laws That rule this meaningless and cancered globe In socket-eyed, gigantic merriment. **1975** *New Yorker* 26 May 104/1 Raskolnikov ..is played by Georgi Taratorkin, a socket-eyed figure ransacked by self-inquiry and staring at us out of a very lonely desert.

socket, *v.* Add: **2.** *Golf.* To strike (the ball) inadvertently off the socket or heel of a club; to make (a shot) in this way. Also *absol.* Cf. *shank *v.* 4.

1911 C. Leitch *Golf for Girls* 87 If you socket and don't want to, here's the cure. Keep your left elbow close to your side. **1920** *Isis* 27 Oct. 9/1 He socketed a couple of iron shots into the gorse. **1927** *Daily Express* 31 Jan. 8/2 A mashie that persists in socketing the ball. **1961** F. C. Avis *Sportsman's Gloss.* 210/2 *Socket*, to hit the ball with the shank of the club; also known as Shank.

Hence **so·cketer,** one who sockets the ball.

1912 *New Bk. Golf* 341 Even the most confirmed socketer will find that with such a club socketing is a sheer impossibility. **1952** H. Longhurst *Golf Mixture* 113 Frostick, of St George's Hill, tells the socketer to keep his head down.

socketed, *ppl. a.* Add: **2.** *Golf.* Of a ball or shot: that has been played off the socket.

1911 C. Leitch *Golf for Girls* 86 At the finish of a socketted shot. **1963** P. Campbell *How to become Scratch Golfer* v. 57 Tom's contribution is a socketed recovery shot out of the bushes.

socketing, *vbl. sb.* Add: **2.** *Golf.* The action of hitting the ball off the socket.

1911 C. Leitch *Golf for Girls* 86 Socketting, that is, hitting the ball off the socket, or shank, as some call it, of the club. **1959** D. Rees *Dai Rees on Gol f* xxx. 126

Socketing is something which strikes suddenly, li tear poisonous adder.

sockette (sǫke·t). [f. Sock *sb.*¹ + -ette.] A short sock.
1950 *Landfall* IV. 309 Tanned legs with neat navy sockettes. **1976** *Times* 26 Mar. 10/4 Sales of hosiery (which includes tights and sockettes etc) run at around 620,000,000 pairs a year.

socking (sǫ·kiŋ), *adv.* and *ppl. a. slang.* [? f. Sock *v.*²: see also B below.] **A.** *adv.* As an intensive, esp. qualifying *big* or *great*: very.
1896 *Dialect Notes* I. 425 That was a socking big fish. **1942** *Tee Emm* (Air Ministry) II. 67 A socking great Wellington has just gate-crashed the range. **1951** J. B. Priestley *Festival at Farbridge* iii. iii. 548 A teeny drink before lunch, and it turned out to be a socking great double gin and Dubonnet. **1958** M. Dickens *Man Overboard* viii. 122 A socking great button-hole. **1976** D. Francis *In Frame* iv. 65 A brooch I had..with a socking big diamond in the middle.
B. *ppl. a.* A euphemistic substitute for *fucking *ppl. a.*
1941 *Penguin New Writing* X. 114 That socking kid's playing a game with me. **1945** S. J. Baker *Austral. Lang.* xiv. 257 *Socker* and *socking*, as synonyms for an old English vulgarism widely current in this country, are recent inventions.

sockless, *a.* (Later examples.)
1970 P. Dickinson *Seals* vii. 145 His bruised feet slipping sockless in the unfamiliar boots. **1981** *Daily Tel.* 20 Feb. 17/1 Going sockless is the preferred style, 'to give the beachside look that is so desirable'.

socko (sǫ·ko). *int.*, *a.*, *sb. slang* (orig. and chiefly *U.S.*). [f. Sock *sb.*⁴ + *-o².] **A.** *int.* An interjection imitative of the sound of a violent blow.
1924 *Dialect Notes* V. 258 *Sock-o* (blow). **1931** E. Linklater *Juan in Amer.* II. i. 63 He hung a lullaby on the Frog's chin—socko! **1936** Wodehouse *Laughing Gas* xxi. 226 And then, as she stood there with the love-light shining in her eyes...socko! **1966** L. Cohen *Beautiful Losers* i. 71 We're fat, F.—Smack! Wham! Pow!—Fat.—Socko! Sok! Bash!
B. *adj.* Stunningly effective or successful, 'knock-out'.
1939 J. B. Priestley *Johnson over Jordan* ii. 67 And now, friends, a new novelty act, the first time here, and I know it will be a socko number. **1942** *Photoplay & Movie Mirror* Mar. 6/3 Van Heflin..almost steals the show—and he must be good to rob Taylor of one iota of glory, Bob's that socko. **1960** *Sales Managem.* 6 May 96/1 Automated manufacture, sock-o selling and all-out advertising. **1961** *John o' London's* 14 Dec. 664/2 The religious plays which are at the moment filling our theatres, religion being Socko box office these days. **1972** T. P. McMahon *Issue of Bishop's Blood* vii. 83 The blue of the incense rising to the white gold of the altar..the soaring voices of the seventy or so nuns..provided a socko finish. **1981** *Underground Grammarian* Feb. 3/1 Their latest brochure starts *right off* with this absolutely socko bit of dialog: 'What is cooperative education? In it's simplist [*sic*] definition, it is learning by doing.'
C. *sb.* orig. *U.S.* A success, a 'hit'.
1937 *Amer. Speech* XII. 317/2 *Socko*, a success. **1942** Berrey & Van den Bark *Amer. Thes. Slang* § 591/1 *Successful show hit*... socker,..socko, sock show. **1973** Wodehouse *Bachelors Anonymous* iii. 23 Triumph or disaster, socko or flop, he went on forever like one of those permanent officials at the Foreign Office.

Socratean (sǫkrătī·ăn), *a. rare.* [See -an.] Pertaining to or resembling the celebrated Greek philosopher Socrates or his way of life; Socratic.
1930 Belloc *Richelieu* i. vi. 114 Father Joseph was short, bullet-headed, of a vivacious Socratean ugliness. **1976** *S. Wales Echo* 26 Nov. 5/1 It's hardly the kind of job that Plato would have relished—and there's nothing Socratean about filling in a VAT return.

Socratist. Delete † *Obs.* and add later example.
1866 Mill in *Edin. Rev.* CXXIII. 337 There are..two complete Plato's in Plato—the Sokratist and the Dogmatist.

Socred (sōᵘ·kred). Abbrev. of *Social Credit, *Social Crediter.
1955 *Pictou* (Novia Scotia) *Advocate* 24 Feb. 1/1 The addition of the British Columbia Socreds has given them just the monkey glands they needed to restore them! **1962** *Canada Month* Feb. 21/2 Social Credit has been badly damaged by the highhanded methods of B.C.'s Socred government. **1970** J. Blackburn *Land of Promise* xv. 191 It was a landslide for the Socreds, the name soon applied to members of the Social Credit Party. **1975** *Australasian Express* 24 Oct. 10/2 (*heading*) Socred fields more women.

sod, *sb.*¹ Add: **3. a.** Also *N. Amer.*, more generally, soil which is grass-covered; sward which has never been cultivated; the surface of a lawn.
1968 *Globe & Mail* (Toronto) 17 Feb. 47/1 (Advt.), 1st class sandy land with substantial buildings. This irrigated land would produce excellent sod. **1976** *National Observer* (U.S.) 12 June 5/2 Some Postal Service employes also think that 'a lot of people don't want us to cross their lawns, tear up yards, and stomp holes in the sod.' **1976** *Billings* (Montana) *Gaz.* 5 July 9-C/1 (Advt.), 761 Acres cropland; 600 Acres former cropland, grassed; 800 Acres sod to break.
Phr. *under the sod*: dead and buried; *to put under the sod*: to kill. *colloq.* and *dial.*
1847 Trollope *Macdermots* III. vii. 286 I've heard the boys say that he would be undher the sod that day six months. *Ibid.* 288 A lot of boys swore together..to put him undher the sod. **1894** H. Pease *Mark o' Deil* i. 19 'Fear-nowt Charlie,' who was put under the sod, poor chap, a year come Michaelmas. **1972** K. Bonfiglioli *Don't point that Thing* xviii. 159 Happiness is..being alive and wonderful-for-his-age when old so-and-so is under the sod.
b. Restrict † *Obs.* to sense in Dict. and add: Also, the surface of a cockpit (sense 1); the institution, practice, or action of cock-fighting, the cock-fighting world.
1814 W. Sketchley *Cocker* p. iii, The author having been attached to the sod at a very early period of life.. he flatters himself that..his attempt at writing 'The Cocker' will be found to contain..instruction. **1840** D. P. Blaine *Encycl. Rural Sports* ix. i. 1208 His chief opponent was Potter, who was feeder for that veteran sportsman, the Earl of Derby, whose attachment for the sod continued unwearied. **1912** W. Gilbey *Sport in Olden Time* 41 So closely was the grass-covered pit associated with the sport, that 'the sod' bore to cocking the same significance as 'the turf' bears to racing. **1977** *Verbatim* Feb. 1/1 Although the cockpit is as remote from the lives of most of us as a brontosaurus wallow, our language has been richly endowed by The Sod, and few of us get through a single day without recourse to at least one phrase from the lexicon of cocking.
4. b. *spec.*, Ireland. Also without *old*.
1812 P. Egan *Boxiana* I. 315 O'Donnel..was a native of Ireland, who left the sod at a very early period of his life. **1891** [in Dict.]. **1892** W. G. Lyttle *Life in Ballycuddy* 12 (E.D.D.), A'll niver lee the auld sod again. **1939** Joyce *Finnegans Wake* 19 To say too us to be every tim, nick and larry of us, sons of the sod. *Ibid.* 194 Dry yanks will visit old sod. *a* **1953** E. O'Neill *Long Day's Journey into Night* (1956) ii. ii. 80 Then before his father can react to this insult to the Old Sod, he adds dryly, [etc.]. **1955** J. P. Donleavy *Ginger Man* v. 35 I'll give you the jug to remember me when I'm gone from the ould sod, sacked in with some lovely French doll.
5. a. *sod cabin, -cloth, crop, ground, -house* (further examples), *-hut* (earlier and later examples), *land*; **sod widow** (see quot. 1946).
1932 W. Faulkner *Light in August* xi. 232 He found the sod cabin. **1905** Sod-cloth [see *mudwalling *vbl. sb.*]. **1956** C. Evans *On Climbing* viii. 128 Round the bottom of the tent, on the outside, a foot-wide strip of cloth should run, the 'sod-cloth', on which stones and snow can be put to anchor the tent. **1848** *Rep. U.S. Comm. Patents 1847* 539 This gave a sod crop without tending of thirty to forty bushels per acre. **1950** *Jrnl. Illinois State Hist. Soc.* Spring 37 They learned to plant a 'sod crop' by cutting upturned furrows with an ax, then dropping in a few kernels of corn. **1839** W. Sewall *Diary* 23 Aug. (1930) 207/1 Broke up the sod ground in the prairie up the hollow for a yard in which to make brick. **1932** *Randolph Enterprise* (Elkins, W. Va.) 4 Feb. 4/2 Sod ground is about all ploughed and some stubble ground also. **1937** J. Ise (*title*) Sod-house days. Letters from a Kansas homesteader, 1877–78. **1948** B. Sundkler *Bantu Prophets* vi. 183 On the door of the sod house used as a church there is painted a green and white cross. **1977** *Westworld* (Vancouver, B.C.) May–June 6/2 There certainly was nothing like a sod house for being cool in summer and warm in winter. **1869** *Harper's Monthly* June 25/1 A warmer abode than the sod hut..he will never have. **1930** L. G. D. Acland *Early Canterbury Runs* 1st Ser. ii. 13 Ford built a six-roomed wooden cottage to replace the original sod hut. **1972** *Science* 19 May 747/2 While others were planning fancy facilities, Herzberg's helpers built a sod-hut similar to those used as homes by the early settlers of Saskatchewan. **1856** *Rep. U.S. Comm. Patents 1855: Agric.* 262 They were mostly sown upon sod-land. **1946** *Sun* (Baltimore) 11 Nov. 2/1 Anderson cautioned growers of wheat and flax not to break up sod or grass lands which are not adapted to continued cultivation and which would create erosion hazards in the future. **1927** *Amer. Speech* II. 278 *Sod* (widow), husband dead. **1946** G. Stimpson *Bk. about Thousand Things* 349 A grass widow's husband was alive while a *sod widow's* husband was under the sod. **1973** *Raleigh* (N.C.) *News & Observer* 12 Mar. 34/2 Last but not least is our large number of widows (sod). There are 70.
b. *sod-soaker*; **sod planting** *Agric.*, the sowing of seed in unploughed ground, herbicides being used to kill or control any existing vegetation; so **sod-plant** *v. trans.*, **sod-planted** *ppl. a.*
1965 *Proc. Southern Weed Control Conf.* XVIII. 146 A second screening test was conducted in 1963 in connection with a new sod planting research program. The aim.. was to develop a high producing summer grain or silage crop grown in a chemically controlled perennial sod which would return to productive pasture in the fall, winter and spring, maintaining a protective mulch cover at all times. **1967** *Agronomy Jrnl.* LIX. 549/1 Removal of the rye immediately before sod-planting corn in the stubble lowered yields as compared to mulched crops. *Ibid.* 548/1 Inclusion of the winter legumes..did not increase sod-planted corn yields. *Ibid.* 550/1 The moisture conserving aspect of sod planting is most pronounced for droughts of short duration. **1978** *McGraw-Hill Yearbk. Sci. & Technol.* 78/2 Various reduced tillage systems are referred to as direct drilling, minimum tillage, no-tillage, sod planting,..depending upon the operations used, the crops grown, and the locale. **1903** Sod-soaker [see *gully-washer* s.v. *gully *sb.*¹ 4]. **1953** *Manch. Guardian Weekly* 18 June 3/1 'Oh Lord, send us a sod-soaker and not a gully-drencher.' The prayer of the prairie farmer is specific.
c. *sod-built* (earlier example).
1805–6 Wordsworth *Prelude* (1959) viii. 613 The Woodman languish'd Within his sod-built cabin.

sod, *pa. pple.*, *ppl. a.*, and *sb.*² Add: **2. a.** Also as *sb.*: a damper (Damper 6) that has not risen. *Austral.*
1900–10 in G. A. Wilkes *Dict. Austral. Colloquialisms* (1978) 309/1. **1931** I. L. Idriess *Lasseter's Last Ride* v. 36 He made dampers so light that they were in danger of blowing away [and]..had not baked one 'sod' during the entire trip. **1957** R. S. Porteous *Brigalow* 206 His dampers were leaden sods. **1975** X. Herbert *Poor Fellow my Country* xvi. 838, I want to cook our own damper, too...don't want one of their sods.

sod (sǫd), *sb.*³ [Short for Sodomite.] **1.** One who practises or commits sodomy. *coarse slang.*
c **1855** *Yokel's Preceptor* 6 It is not long since, in the neighbourhood of Charing Cross, they posted bills in the windows of several respectable public houses, cautioning the public to 'Beware of Sods!' **1859** G. W. Matsell *Vocabulum* 83 *Sod*, a worn-out debauchee, whom excess of indulgence has rendered unnatural. **1934** V. Woolf *Let.* 24 Jan. (1979) V. 273, I am writing about sodomy at the moment and wish I could discuss the matter with you; how far can one say openly what is the relation of a woman and a sod? **1949** Wyndham Lewis *Let.* ?8 Mar. (1963) 484 When you come to write your book, its scene our day to day life, I should put in the sods. Sartre has shown what a superb figure of comedy a homo can be. **1968** S. Jameson *White Crow* xxxiv. 291 Homosexuals are always getting themselves assaulted. You read that some respectable middle-aged bachelor has been beaten insensible on the stairway of his Mayfair flat, and invariably it turns out that he was a sod.
2. a. Used as a vulgar term of abuse for (usu.) a male person. Also with weakened force, as the equivalent of 'fellow', 'chap', freq. affectionately or in commiseration; *odds and sods*: see *odds *sb.* 7 b.
1818 *Sessions* 17 June 283/2 As he passed me he said the other was a b—y s—d. **1931** K. O'Brien *Without my Cloak* iii. xi. 380 That auld sod of a husband making her black and blue every night of his filthy life. **1931** W. V. Tilsley *Other Ranks* 12 Lucky sods, getting this far and then going back. **1942** G. Kersh *Nine Lives Bill Nelson* x. 61 There are plenty of sods in this battalion that get their pleasure by exercising their two-penny-ha'penny authority. **1942** T. Rattigan *Flare Path* iii. 164 Johnny, you old sod! Is it really you? **1957** I. Murdoch *Sandcastle* xiii. 210 He thought to himself, what a sod I am, what a poor confused sod. **1958** 'E. O'Conner' *Steak for Breakfast* 28 Good on yer, Martha, yer old sod! **1963** T. Parker *Unknown Citizen* i. 40 Don't you call me a daft sod, you daft sod. **1968** J. Braine *Crying Game* i. 18 It's time he was dead... If you want to destroy the sod, Frank, I'll give you absolutely all the dirt. **1969** D. Wallace *Turtle* xiv. 148 That's a shame, the poor little dawg, but if that was moine I'd hev that put down. That can't help but make no end o' work, the poor little sod. **1974** N. Freeling *Dressing of Diamond* 201 Yes, now I remember him, cheeky young sod. **1978** *Globe & Mail* (Toronto) 16 Aug. 31/7 And when they do, these lucky sods will forget years of fish-fib training and head pell mell for shore, seeking, of all things, the truth.
b. Something difficult; a great nuisance. *slang.*
1936 'G. Orwell' *Keep Aspidistra Flying* i. 11 'Bare' is a sod to 'rhyme'; however, there's always 'air'. **1950** C. MacInnes *To Victor Spoils* i. 84 It'd be a sod if they got through to the Meuse. **1971** V. Canning *Firecrest* i. 3 At least..he'd seen them come back, though it was a sod he'd missed them going off. **1977** [see *satin *sb.* (and *a.*) 6 b].
c. *sod-all*, nothing, no. Cf. *all A. 8 f. *slang.*
1958 K. Amis *I like it Here* i. 12 There's been sod-all since. **1961** I. Jefferies *It wasn't Me!* iii. 39 When I was at that pharmaceutical firm..I did sod-all for months on end. **1972** J. Wainwright *Requiem for Loser* viii. 167 Like the concert hall... A bit of a stage—and sod-all else. **1978** 'K. Blake' *Where Jungle Ends* iii. 37 Here he was in this cold chill room, and two maniacs sitting playing cards at the table and taking sod-all notice of him.
d. *not to give a sod* = *not to give a damn* s.v. *damn *sb.* 2. *slang.*
1961 B. Aldiss *Primal Urge* i. 29 Nobody gave a sod. Euphoria had its high tide. **1973** D. Storey *Temporary Life* v. ii. 224, I don't give a sod for any of them, Phil.
e. *Sod's Law* = *Murphy's law* s.v. *Murphy² 3.
1970 *New Statesman* 9 Oct. 460/1 Sod's Law..is the force in nature which causes it to rain mostly at weekends, which makes you get flu when you are on holiday, and which makes the phone ring just as you've got into the bath. **1978** *New Scientist* 7 Sept. 744/1 The great unshakeable list of interdisciplinary laws—Sod's Law, Newton's Fourth Law of Motion, the Inverse Midas Touch and their kin. **1980** *SLR Camera* July 56/2 Even if you're using a masking frame this can easily overbalance. According to Sod's Law, that's going to happen when you're halfway through exposing a sheet of 20 × 16in colour paper costing the best part of £1·30.

sod, *v.*³ *slang.* [f. prec.] **1.** *trans.* = Damn *v.* 5.
1904 *Eng. Dial. Dict.* V. 605/2 Phr. *sod him*, may mischief befall him. *w. Yks.* Sod him, he can go to —. **1942** G. Kersh *Nine Lives Bill Nelson* i. 3 Well, sod the Drill Pig. **1945** *Penguin New Writing* XXIII. 51 Sod that, chummie. **1953** P. Scott *Alien Sky* i. viii. 131 At seven-fifteen they had to go out to dinner. Sod it. **1958** —— *Mark of Warrior* ii. 131 'Look, you'd better go sick.'

'Sod you, Bob. I wouldn't miss it for the world.' **1967** J. Wain *Smaller Sky* 170 'He'll come out,' said Swarthmore. 'And if he doesn't, we'll sit where we are and you'll get paid for a full day's work, with overtime if necessary, and you won't have to do a stroke.' 'I'd rather be at home,' said the chief cameraman, 'and sod the overtime. I'm definitely sickening for something.' **1971** B. W. Aldiss *Soldier Erect* 209 Quite a road! Sod me! I'll say it is! **1977** *Chainsaw* Sept./Oct. 3/2 Sod it! There goes my banner headline.
2. *intr.* With *off*: to go away, depart. Also = *get away* s.v. *get *v.* 54 b, and *fig.* Usu. in imp.
1960 J. Symons *Progress of Crime* xv. 92 Now sod off and get your identification parade done. **1968** *Listener* 14 Nov. 650/3 It's good to learn what Mr Reed said to the vicar who complained that boys had damaged a valuable rose tree: 'I told him to sod off.' **1971** F. Forsyth *Day of Jackal* xx. 335 The policeman asked for papers. The Jackal giggled seductively... 'Sod off,' said the policeman and withdrew. **1976** P. Cave *High Flying Birds* ii. 16 'Sod off,' I said, 'How can you call a glider a vehicle?' **1977** *Observer* 4 Sept. 14/2, I am simply waiting for the day when I can say 'sod off' to your institution. **1978** I. Murdoch *Sea* 168 That's right, sod off just when ..the light of understanding has dawned... Oh all right, sod off then!

soda¹. Add: **4. a.** (Earlier example.)
1834 J. R. Planché *Olympic Revels* in *Extravaganzas* (1879) I. 47 Make him sit down—give him some hock and soda.
b. A glass or drink of soda-water; also, an ice-cream soda.
1933 E. O'Neill *Ah Wilderness!* ii. 58 Ever drink anything besides sodas? **1962** A. Lurie *Love & Friendship* viii. 145 Vanilla sodas with strawberry ice-cream. **1973** 'E. McBain' *Hail to Chief* vi. 94 Toy..put the straws between her lips, and busied herself with the soda.
4*. *Faro.* In full *soda card.* (See quot. 1975.) Phr. *from soda* (*card*) *to hock*: see *hock *sb.*⁶ b.
1845 J. H. Greene *Exposure of Arts & Miseries Gambling* (ed. 2) 135 The top card, when the deal is first commenced, is called the *deal card*; this card neither wins nor loses, and on that account is sometimes called the *soda card.* **1975** *Way to Play* 206/2 The exposed top card is called the 'soda'. It is ignored for betting.
4.** *Austral. slang.* Something easy to accomplish, a simple task; a 'pushover'.
1930 V. Palmer *Passage* i. i. 22 'Just one more guess.' ..'Umph, that's a soda! Must be the old doctor.' *Ibid.* x. 83 They're getting ready for the long dive now, and it ought to be a soda for you. If I hadn't rheumatics down the back of my leg, I'd give it a fly myself. **1943** G. H. Johnston *New Guinea Diary* iv. 136 'The Middle East was a soda beside this,' one of them told me. **1955** A. Marshall *I can jump Puddles* 108 Swipe him on the knuckles if you can. If he's like his old man he's a soda. **1966** H. Porter *Paper Chase* 74 The job, for which I have no really specialized training, is nevertheless a soda.
5. a. *soda pan* [Pan *sb.*¹ 5 a], *plain*; *soda-chapped* adj.; **soda-acid**, used *attrib.* to designate a fire extinguisher containing sulphuric acid and sodium bicarbonate (or sometimes the carbonate), which are mixed just before use to provide the gas for expelling the water; **soda cellulose**, a form of cellulose heavily impregnated with soda, produced by the action of caustic soda on wood-pulp esp. in the manufacture of paper or rayon; **soda glass**, glass containing a high proportion of soda; sometimes = *soda-lime glass; **soda lake** (further examples); **soda-lime glass**, the standard form of glass in everyday use, manufactured essentially from silica, soda, and lime; **soda process**, a method of pulping wood by boiling with caustic soda; **soda pulp**, wood-pulp made by the soda process.
1928 R. Northwood *Fire Extinguishment & Fire Alarm Syst.* xxi. 185 (*caption*) Method of recharging 'The Conquest' soda-acid extinguisher. **1966** *McGraw-Hill Encycl. Sci. & Technol.* V. 279/1 In the small first-aid water fire extinguishers, a propellant must be provided. Usually this is carbon dioxide, which is either generated when needed (the soda-acid extinguisher) or stored in a cartridge. **1890** *Jrnl. Soc. Chem. Industry* 28 Feb. 225/1 Their production is estimated at 30,000 tons sulphite (wet) and 9,000 soda cellulose (dry). **1948** J. T. Marsh *Textile Sci.* ii. 19 Sheets of pulp are converted into soda-cellulose by steeping in caustic soda solution. **1973** *Materials & Technol.* VI. iv. 306 The swollen, and still further purified 'soda cellulose' so formed, is broken down into 'crumbs', and these are then transferred to reactor vessels. **1922** Joyce *Ulysses* 59 Sodachapped hands. **1897** A. Hartshorne *Old English Glasses* 39 Venetian soda-glass is much lighter than that made in the Low Countries with potash. **1947** J. C. Rich *Materials & Methods of Sculpture* xi. 329 Sodium carbonate, calcium carbonate, and sand yield a soft or 'soda glass'. **1965** Phillips & Williams *Inorg. Chem.* I. xiv. 553 Soda glass is conveniently worked at a lower temperature than borosilicate glass. **1937** *Discovery* Feb. 58/1 A new exhibit presented by the Magadi Soda Lake Co...at the Imperial Institute..illustrates..the exploitation of one of the most remarkable natural soda lakes in the world. **1976** K. Thackeray *Crownbird* ix. 199 The road..ran steeply down beside a soda lake... The soda was firm near the edge. **1917** A. B. Searle in G. Martin *Industr. & Manufacturing Chem.* (*Inorg.*) II. 245 For soda-lime glasses the mean coefficient of expansion lies between 0·000023 and 0·000027 per 1°C. **1955** E. B. Shand *Glass Engin. Handbk.* i. 2/1 Soda-lime glasses are commonly used for bottles, jars, window sheet and plate glass, electric lamp bulbs, and ophthalmic (sight-correcting) lenses. **1971** *Materials & Technol.* II. vi. 340 Soda-lime glasses in commercial use have devitrification temperatures in the region of 900 to 1000°C. **1976** K. Thackeray *Crownbird* v. 82 The plane was flying..over one end of an enormous soda pan. **1946** D. C. Peattie *Road of Naturalist* i. 20 You find mariposas all over the West; they change height, change shape and colour, as you trace them from the mountains of Colorado, over the Utah soda plains. **1885** *Encycl. Brit.* XVIII. 226/1 The pulp produced by all those processes is of excellent quality; and, according to the statements of the patentees, it can be prepared at a cost greatly lower than by the soda process. **1907** *Jrnl. Soc. Chem. Industry* 15 June 561/2 In the period of 1865 to 1875 a large number of mills were erected throughout Canada and the United States, for the cooking of wood by the soda process. **1967** V. Strauss *Printing Industry* viii. 532/2 The soda process..has lower yields than the sulfate process..and the recovery of caustic soda is costly. For these and other reasons the soda process is losing ground fast to the sulfate process. **1893** *Jrnl. Soc. Chem. Industry* 30 Sept. 793/2 The product of 1892 included only 12,500 tons of soda pulp, the remaining 137,500 tons being produced by the sulphite process. **1962** F. T. Day *Introd. Paper* ii. 20 Deciduous or broad-leafed trees such as the poplar are used in the production of soda pulp.
c. For 'minerals' read 'minerals and rocks'. (Further examples.)
1889 Soda-felsite [see *keratophyre]. **1913** Soda-amphibole, -richterite [see *imerinite]. **1926** *Proc. U.S. Nat. Mus.* LXVIII. Art. 17. 4 It may be chemically classed as a soda-rhyolite, but none of the calculated normative minerals of rhyolite are found in its mode. **1931** *Mineral. Mag.* XXII. 453 The brown augite has in places fringes of a green soda-augite. **1935** *Amer. Mineralogist* XX. 58 The high percentage of soda..justifies the classification of the mineral as soda-alunite in the broad sense. **1965** G. J. Williams *Econ. Geol. N.Z.* xi. 167/1 Watters *et al.* (1961) noticed fergusonite in small water-worn grains from a restricted locality in the Canaan area where the Separation Point soda-granite invades Paleozoic marbles. **1966** *McGraw-Hill Encycl. Sci. & Technol.* XII. 407/2 Soda niter is by far the most abundant of the nitrate minerals. **1968** I. Kostov *Mineral.* 494 As 'alums' are denoted the following double sulphates:..Soda alum $NaAl(SO_4)_2.12H_2O$.
6. *soda-biscuit* (earlier example), *-bread* (earlier example), *-cake* (earlier example), *-cocktail*, *-cracker* (earlier example), *-fritter*, *-mint* (examples), *-powder*.
1830 *Albany Jrnl.* 25 Aug. 3/5 Fresh Soda Biscuit, just received from Treadwell's Bakery. **1850** N. Kingsley *Diary* 3 Aug. (1914) 134 They raised some bread with it, which he said was the best soda bread ever tasted. **1846** *Jewish Manual* vii. 155 (*heading*) A soda cake. **1818** *N.Y. Herald* 2 July 4/1 We have the Fourth of July thrown in with..its exhilarating associations so conducive of headaches and soda cocktails. **1863** *Harper's Mag.* Feb. 313/1 This repast, whatever its name might be, consisted of perhaps half a pound of soda crackers, two red herrings, and one red apple. **1837** M. R. Walker *Diary* in C. M. Drury *Elkanah & Mary Walker* (1940) iii. 91 In the morning baked soda biscuit and fried soda fritters. **1895** *Montgomery Ward Catal.* Spring & Summer 261/2 Soda Mint tablets, for sour stomach, colic, flatulency, etc. **1928** D. L. Sayers *Unpleasantness at Bellona Club* xv. 170 Suppose..somebody had dropped a poisoned pill into his usual bottle of soda-mints. **1975** C. Mott-Radclyffe *Foreign Body in Eye* iv. 77 Joyce Britten-Jones asked me one evening whether I had any soda-mints. **1820** *Columbian Centinel* 1 July 3/6 Maynard & Noyes continue to prepare Soda Powders, of superior quality. **1843** Mill *Logic* I. iii. xiii. 575 The old but not undisputed empirical generalization that soda powders weaken the human system.
7. Also, that dispenses soda water. *soda-clerk* (hence *soda-clerking* vbl. sb.), *-siphon*, *-straw*; **soda-counter**, the counter of a soda fountain; any counter or bar where soft drinks, ice cream, etc., are sold; **soda-fountain** (also † **-font, -fount**) orig. *U.S.*, (*a*) (see quot. 1875 in Dict.); (*b*) an apparatus for supplying ice-cream sodas, sundaes, etc.; a counter or an establishment of which this is a feature; **soda-jerk, -jerker** [*jerk *v.*¹ 7], one who mixes and sells soft drinks, etc., at a soda-fountain; **soda-pop**, flavoured soda-water.
1941 N. Coward *Australia Visited* iii. 16 That initial contact with the ordinary people [of New York]—the soda clerks, the cops, the struggling young theatre people. **1925** T. Dreiser *Amer. Tragedy* I. ii. iii. 180 He had wandered on..dishwashing in a restaurant, soda-clerking in a small outlying drug-store. **1846** *Dollar Newspaper* (Philadelphia) 19 Aug. 4/2 He..went up to the soda counter, and 'reckoned they'd take a little whisky'. **1939** A. Huxley *After Many a Summer* i. x. 135 Virginia was at the soda-counter, pensively eating a chocolate-and-banana split. **1976** J. Lee *Ninth Man* 31 A customer took a stool at the soda fountain... Dietrich..forced himself to..move..to the soda counter. **1848** *Knickerbocker* XXXI. 40 They had not a theatre, nor an oyster-saloon, nor a soda-font. **1848** in N. E. Eliason *Tarheel Talk* (1956) 296 My soda fount cost me in ninety dollars. **1908** *Home Herald* (Chicago) 13 May, Here is the popular soda-fount drink known as Coca-Cola. **1824** *Independent Chron.* (Boston) 9 Oct. 3/3 This luxury in a hot and dusty season, together with an ever-flowing Soda Fountain,.. he flatters himself will ensure a continuance of public patronage. **1876** *Napa* (California) *Reg.* 29 July 4/2 A Woodward avenue drug-store hired a new soda-fountain boy the other day. **1918** G. Frankau *One of Them* (1923) II. xv. 108, I loved thy daughters, daintiest as dowdiest; Cadby's tea'd Halls as Fuller's soda-fountain. **1955** R. Blesh *Shining Trumpets* (ed. 3) x. 232 A New Orleans Rhythm Kings' disc, playing in a soda fountain, gave these schoolboys the incentive. **1977** *New Yorker* 6 June 50/2 Afterward, at the soda fountain, they went over the day's movies. **1922** *Collier's* 17 June 4/1 You can tell a big-league head soda jerk by the way he picks up a glass, but the acid test is what kind of chocolate sirup he can make. **1958** *Daily Herald* 24 Mar. 3/7 This bustling little man never forgot his early years when he worked as a fairground barber and soda jerk. **1978** J. Updike *Coup* (1979) iv. 132 The counter-boy, ingloriously dubbed the soda jerk.. These 'soda jerks', I came to understand, were recruited from the adolescent ranks of the 'townies'. **1883** G. W. Peck *Groceryman & Peck's Bad Boy* 137 A sensitive soda jerker..feels that it is worse than three card monte. **1932** Wodehouse *Louder & Funnier* 48 He..is now a soda-jerker in a small town in Kansas. **1959** N. Mailer *Advts. for Myself* (1961) 35 The soda jerker.. from some outside compulsion had been forced to grow a beard. **1863** W. Whitman *Daybks. & Notebks.* (1978) III. 655 The continual soda-pop-like burstings of members calling 'Mr. Speaker! Mr. Speaker!' **1963** *Listener* 14 Feb. 301/3 The man who drove the soda-pop lorry. **1977** *Time* 11 Apr. 5/1 Were a visitor from another planet to read about the saccharin ban, he would conclude that earthlings' basic nutritional needs required large amounts of soda pop, jelly and chewing gum. **1926** *Daily Colonist* (Victoria, B.C.) 4 July 6/1 (Advt.), Warm Weather Supplies. Ice Cream Pails. Soda Straws. Lily Drinking Cups. **1911** *Ibid.* 21 Apr. 6/6 (Advt.), Soda Syphon Holder. This is a handsome silver plated stand into which the syphon fits. **1963** W. Soyinka *Lion & Jewel* 24 The foreman..unpacks the usual box of bush comforts—soda siphon, whisky bottle and geometric sandwiches.

sodar (sōᵘ·dāɹ). [f. So(und *sb.*³ + *ra)dar.] A system for investigating the state of the atmosphere, which works on the principle of radar but uses ultrasonic sound waves instead of microwaves.
1955 M. Reifer *Dict. New Words* 192/2 *Sodar*,..an instrument for detecting weather conditions by recording on an oscilloscope..the reflected sound waves which have been projected directly overhead. **1974** *Globe & Mail* (Toronto) 1 June 5/1 These new instruments, called lidar ..and sodar.., could be used for things like tracing and identifying particles in the atmosphere..and detecting dangerous wind patterns and turbulence around airport runways. **1976** *McGraw-Hill Yearbk. Sci. & Technol.* 402/2 The active network includes not only the standard pilot balloons and radiosondes with which meteorologists customarily measure wind, temperature, and humidity aloft, but also weather radars.., lidars..and sodars.

soda-water. Add: **1. b.** *soda-water-bottle* (earlier example), *manufacturer* (earlier example), *-room.*
1813 E. Gerry Jr. *Diary* 4 July (1927) 168 We had not gone far, before the girls ordered the charioteer to stop at the soda water room. **1822** *Sunday Times* 20 Oct. 2/4 William Clarke..soda-water and ginger-beer manufacturer. **1852** Dickens *Bleak Ho.* (1853) v. 35 Ginger-beer and soda-water bottles.

so·dbuster. *western N. Amer.* [f. Sod *sb.*¹ + Buster.] A term, chiefly opprobrious, for a farmer, farmworker, or homesteader in a cattle-grazing region, esp. one who ploughs virgin grassland. Hence **so·dbusting** *vbl. sb.*, ploughing virgin grassland; farming.
1922 R. Stead *Neighbours* 126 Between your fine words I figger that you pick up a dollar now an' again by tottin' these tenderfoot sod-busters out over the bald-headed. **1927** C. Sandburg *Amer. Songbag* 89 Its tune was familiar to the lonely 'sodbuster'. **1958** H. B. Allen in *Publ. Amer. Dialect Soc.* xxx. 11 The persistence of *sodbuster* and *soddy* as not necessarily opprobrious designations of the new settlers [in North Dakota]. **1965** G. Shepherd *West of Yesterday* x. 72 We owned cows and found that ranching was more attractive than sod-busting. **1970** [see *nester 2]. **1972** G. Beine *Land of Coyote* 85 (*heading*) Sod busting a new field. **1979** *Guardian* 24 Mar. 12/1 The conflicts between sodbusters and cattlemen.

sod corn. *western N. Amer.* [f. Sod *sb.*¹ + Corn *sb.*¹] **a.** Corn or maize planted in ploughed up grassland. **b.** Whisky made from sod corn. In full, *sod-corn whisky*.
1835 P. Shirreff *Tour* xxiv. 248 Indian corn is dropped into every third furrow..and covered with the next cut turf. This crop receives no farther cultivation of any kind, is termed sod corn, and said to yield fifty bushels per acre. **1838** *Bytown* (Ottawa) *Gaz.* 19 Sept. 1/3 The sod corn does not make up more than half a crop, and is..stacked for fodder stock. **1857** E. F. Beadle *Diary* 2 Aug. (1923) 73 Found the family enjoying themselves over their 'Sod corn whiskey'. **1878** [see *bare-footed *a.* c]. **1913** W. Cather *O Pioneers!* 27 John Bergson says to his boys, 'Try to break a little more land each year; sod corn is good for fodder.' **1927** K. Eubank *Horse & Buggy Days* 94 They..ate their dinners,..munching cheese,.. which helped along on its onward course by a tumbler or so of sod corn, made in a moonshine still especially for the occasion. **1940** L. I. Wilder *Long Winter* iii. 29 He cut and shocked the sod corn.

so·ddenly, *adv.* [f. Sodden *ppl. a.* + -ly².] In a sodden manner: heavily and dully; damply.
1901 Kipling *Kim* xv. 390 Kim had reeled to a room with a cot in it, and was dozing soddenly. **1920** *Blackw. Mag.* Apr. 501/2, I slept, fitfully at first, soddenly later. **1939** Joyce *Finnegans Wake* 514 Or (soddenly) Schott, furtivfired by the riots. **1976** *Church Times* 26 Nov. 5/1, I trudged soddenly along unfamiliar terrain.

soddenness. (Earlier example.)
1883 H. James *Let.* 25 Nov. (1980) III. 14 Yes, I have read Trollope's autobiography and regard it as one of the most curious and amazing books in all literature, for its density, blockishness and general thickness and soddenness.

so·dding, *ppl. a.* and *adv. slang.* [f. *sod *v.*³] A vague epithet expressing anger or contempt; freq. as a mere intensive.
1912 D. H. Lawrence *Let.* 3 July (1962) I. 134 The miserable sodding rotters..that make up England today. **1929** 'H. Green' *Living* iii. 25 'It [sc. Australia] am a grand country' 'e said to me, 'this [sc. England] be a poor sodding place for a poor bleeder', 'e said. 'I'm for goin'.' **1933** M. Lowry *Ultramarine* 36, I don't know what you think you're doing idling round this sodding ship. **1950** R. Godden *Breath of Air* xix. 235 'That's my sodding business,' said McGinty. **1954** K. Amis *Lucky Jim* xvii. 168 Cuts his own hair now, you see. Too sodding mean to pay out his one-and-six, that is what it is. My God. **1966** M. Woodhouse *Tree Frog* xxi. 155 Hundred and twenty two semiconductors in there, all radiating heat... What are we supposed to do, sodding blow on them? **1968** J. Braine *Crying Game* vi. 141 The bastard who was giving me dinner stood me up, and I shall sodding well ring him and tell him I'm going out with someone much nicer. **1976** C. Storr *Unnatural Fathers* ii. 27 My sodding brother got into hospital and then dropped out before his finals. **1980** D. Bogarde *Gentle Occupation* i. 24 I'll remember this sodding day until the day I die.

so·ddish, *a.* [f. *sod *sb.*³: see -ish¹.] Awful, 'rotten', terrible.
1959 W. Camp *Ruling Passion* xxvii. 237 What a bloody soddish thing to do.
So **so·ddishness** [-ness], behaviour characteristic of a 'sod' (*sod *sb.*³ 1, 2).
1938 L. MacNeice *I crossed Minch* vi. 76 Charles Edward..sank into chambering and soddishness. **1970** 'D. Craig' *Young Men may Die* xvi. 121 Happily there was no opportunity for soddishness about whom I should go with.

soddite: see *soddyite.

soddy, *a.* and *sb.* **B.** *sb.* For *U.S.* read *western N. Amer.* and add: **1.** (Earlier and later examples.)
1877 H. Ruede *Let.* 24 Apr. in J. Ise *Sod-House Days* (1937) 57 Many of the young bachelors..were building their own 'soddies'. **1970** *Islander* (Victoria, B.C.) 29 Nov. 10/4 It was a sort of soddy, the rear dug into a cut-bank.
2. One who occupies or who has occupied a sod-house.
1958 [see *sodbuster]. **1977** *Westworld* (Vancouver, B.C.) May–June 6/2, I..received a nicely decorated certificate to the effect that I was a Soddy.

soddyite (sǫ·di,əit). *Min.* Orig. † **soddite.** [Coined in Fr. as *soddite* (A. Schoep 1922, in *Compt. Rend.* CLXXIV. 1067), f. the name of Frederick *Soddy* (1877–1956), English chemist and physicist: see -ite¹.] A hydrated uranyl silicate found as yellow orthorhombic crystals.
1922 *Nature* 13 May 631/2 Soddite, a new radioactive mineral. This is a yellow crystalline mineral found associated with curite from Kasolo (Belgian Congo). **1927** *Mineral. Abstr.* III. 233 Soddyite (= soddite), $12UO_3.5SiO_2.14H_2O$. **1937** *Mem. Geol. Survey S. Afr.* No. 31. 108 Soddyite. This mineral occurs as an encrustation on quartz with malachite in the pegmatite at Norrabees. **1965** *Amer. Mineralogist* L. 919 Soddyite..occurs in blocky, fibrous, deep-yellow to yellow-green crystals which vein other uranyl minerals in the matrix.

sodian (sōᵘ·diăn), *a. Min.* [f. Sod(ium + *-ian 2.] Of a mineral: having a proportion of a constituent element replaced by sodium.
1930 W. T. Schaller in *Amer. Mineralogist* XV. 572 The adjectival endings thus formed for the names of all the chemical elements are given below... Sodium—sodian. **1951** C. Palache et al. *Dana's Syst. Min.* (ed. 7) II. 1022 On the dispersion of sodian and manganoan romeite ('atopite') from Brazil see Rose (1919). **1963** W. A. Deer et al. *Rock-Forming Minerals* II. 113 Titanaugites and sodian augites also have more ferric iron than most other pyroxenes.

sodic, *a.* Add: **b.** *Geol.* Of a mineral or rock: containing an appreciable or a greater-than-average quantity of sodium, often as compared with calcium or potassium. Also applied to a metamorphic process in which such minerals are formed.
1902 *Jrnl. Geol.* X. 574 The standard SO_3-bearing feldspathoid is therefore considered to be a purely sodic noselite. **1927** S. J. Shand *Eruptive Rocks* xi. 200 More sodic types are also known, in which both orthoclase and quartz are present. **1952** T. F. W. Barth *Theoret. Petrol.* 96 Plagioclases of gneisses and of crystalline schists, if zoned, usually show inverse order, that is, sodic core, calcic shell. **1967**, **1971** [see *potassic *a.* b]. **1978** *Nature* 7 Sept. 23/1 In addition to the phenocrysts appropriate to magmas of intermediate bulk composition, andesite and dacite contain both anomalously calcic and sodic plagioclase.

sodipotassic (sōᵘ:dipǫtæ·sik), *a. Geol.* Also **sodo-** (sōᵘ:do-). [Blend of Sodic and Potassic *adjs.*] Containing both sodium and potassium in appreciable quantities.
In quot. 1902 used to denote a specified compositional range in the classification scheme of Cross, Iddings, Pirsson, and Washington.
1902 *Jrnl. Geol.* X. 596 The minerals of the sodalite group are only present in the sodic or sodipotassic Subrangs of Classes I, II and III. **1927** S. J. Shand *Eruptive Rocks* ix. 163 A typical example of the Quincy granite gave quartz 33, sodipotassic felspar 55, and ægirine and riebeckite 10 per cent. **1974** T. G. Sahama in H. Sørensen *Alkaline Rocks* 98/1 The rocks are mainly potassic to perpotassic, but sodopotassic to sodic varieties are known among the jumillites and fortunites.

sodium. Add: **2. a.** *sodium vapour* (freq. *attrib.*); **sodium Amytal** *Pharm.*, the sodium salt of *Amytal, used as a sedative and hypnotic; sodium 5-ethyl-5-isopropylbarbiturate, $C_{11}H_{17}N_2O_3Na$; **sodium Pentothal** (see *Pentothal).
1929 *Proc. Soc. Exper. Biol. & Med.* XXVI. 709 Anesthesia has been produced in human beings by the intravenous injection of solutions of the anhydrous sodium amytal. **1937** [see *Seconal]. **1938** [see *Amytal]. **1955** A. Huxley *Genius & Goddess* 9 One escapes into reminiscence as one escapes into gin or sodium amytal. **1933** *Discovery* Feb. 50/2 In the construction of sodium vapour lamps this difficulty is overcome by introducing a rare gas into the tube. **1968** M. S. Livingston *Particle Physics* iii. 39 A well-known example [of a multiplet] is the sodium D-line doublet which gives the yellow color to the light from a sodium-vapor lamp.
b. **sodium-cooled** *a.*, that employs liquid sodium as a coolant; *spec.* of (*a*) an aero-engine exhaust valve, or (*b*) a nuclear reactor; **sodium pump** *Physiol.*, a pump (*pump *sb.*¹ 1 e) which operates on sodium ions.
1934 *Jrnl. R. Aeronaut. Soc.* XXXVIII. 223 The Americans have attained and even exceeded [500 lbs. per sq. in.]..with a poppet valve (sodium cooled) engine. **1954** *Sci. Amer.* Dec. 38/3 A sodium-cooled reactor, which can be operated at very high temperatures, has greater thermal efficiency than a water-cooled system. **1956** E. Molloy *Automobile Engineer's Ref. Bk.* iii. 242 In engines where exhaust-valve cooling is a serious problem, the sodium-cooled valve has been adopted. **1971** *New Scientist* 11 Mar. 529/1 Rudzinski was mainly concerned about the immense complexity of the sodium-cooled technology. **1977** *Time* 15 Aug. 11/3 What worries them in particular is that Super Phénix will produce energy from a sophisticated sodium-cooled reactor eight times more powerful than smaller, water-cooled plants. **1951** *Jrnl. Physiol.* CXIV. 143 An active sodium pump cannot be ruled out on the grounds that it would require more energy than is available from the resting metabolism. **1964** [see *pump *sb.*¹ 1 e]. **1974** D. & M. Webster *Compar. Vertebr. Morphol.* ix. 183 Instead of osmotically equilibrating, this imbalance of cations is maintained by the cell membrane's physical characteristics plus an enzyme system, called the sodium pump, which actively removes sodium from inside the cell.
c. Used *attrib.* and in *Comb.* with reference to (the intense yellow light emitted from) discharge tubes containing sodium vapour, used esp. for street lighting.
1888 Sodium light [in Dict., sense 2 b]. **1912** *Jrnl. Soc. Chem. Industry* 31 Oct. 1010/2 (*heading*) Polarisation; Sodium lamps for. **1933** *Discovery* Oct. 318/2 The electrical impulses corresponding to the television picture signals are converted into the light variations of a specially designed sodium tube. **1956** R. Fuller *Image of Society* viii. 195 The sodium lamps of the bypass. **1959** *New Statesman* 8 Aug. 152/2 Their fight is often associated with the campaign against concrete lamp-posts, sodium lighting and similar 'outrages' on our towns and cities. **1967** A. J. Marshall in L. Deighton *London Dossier* 138 Walking in the sodium-lit Kilburn High Road. **1973** J. Mann *Only Security* x. 132 Thea pressed the whole row of light switches and..the sodium strips flickered into brilliance. **1977** D. Harsent *Dreams of Dead* 35 Midnight, bruised insomniacs, alive to the growing silence and opening to the sodium glare like flowers.
d. Objective, as *sodium-demanding, -retaining* adjs.
1977 J. L. Harper *Population Biol. Plants* xxi. 655 The life cycle strategy is likely in such a case to be influenced by the optimal allocation of sodium between parents and offspring and between the various sodium-demanding activities. **1977** *Proc. R. Soc. Med.* LXX. 692/1 One of the hypotheses..has been that these patients fail to escape normally from the sodium-retaining effect of aldosterone.

sodoku (sǫ·doku). *Path.* [Jap.] The form of rat-bite fever caused by *Spirillum minus.*
1926 *Trans. Soc. Tropical Med. & Hygiene* XIX. 183 Apert and his colleagues suggested the use of stovarsol or tréparsol by the mouth in the case of persons who had been exposed to the infection of sodoku. **1955** W. L. Jellison in T. G. Hull *Dis. transmitted from Animals to Man* (ed. 4) xxvii. 539 The term 'sodoku' is from the Japanese (*so*, a rat, *doku*, poison) and is being resorted to more commonly by American workers to avoid controversy and confusion over the correct application of the term rat-bite fever. Sodoku is primarily an infection of rats, mice, and other rodents. **1970** *Scand. Jrnl. Infectious Dis.* II. 71/1 Two aetiologically different but clinically similar diseases may occur as results of rat-bites: the Japanese Sodoku caused by Spirillum minus, and the bacillary form.

Sodom. Add: **1.** Also coupled with *Gomorrah* (see Gomorr(h)ean *a.* and *sb.*), the name of the other of the two wicked cities of the plain in Gen. xviii–xix.
1862 Queen Victoria *Let.* 7 June in R. Fulford *Dearest Mama* (1968) 67 It was intended he should come home through Paris stopping only a day in order to have got over his visit to that Sodom and Gomorrah. **1864** Trollope *Can you forgive Her?* I. xxiii. 179, I always regarded the States as a Sodom and Gomorrah, prospering in wickedness. **1972** I. Hamilton *Thrill Machine* xv. 63 It wasn't exactly Sodom and Gomorrah—the ladies kept their clothes on. **1974** *Listener* 24 Jan. 121/3 Heliogabalus ..reduced Rome to a kind of post-Christian Sodom and Gomorrah.

sodomist. Add: (Example.) Also as *adj.* = Sodomitical *a.*
1948 *Rep. Native Laws Commission 1946–48* (Dept. Native Affairs, South Africa) 38/1 We may quote from a memorandum submitted by..the Reverend H. P. Junod:—'..He was the son of one of our Evangelists, and refused to submit to the sodomist suggestions and solicitations of old mine workers.' **1950** M. Hay *Foot of Pride* iv. 95 The inhabitants of Cahors..joined in the [money-lending] business, and Dante put them in Hell alongside the sodomists.

sodomize (sǫ·dəmaiz), *v.* [f. Sodom(y + -ize.] *trans.* (occas. *absol.*) To practise sodomy upon (a man or a woman). Also **so·domized** *ppl. a.*, **so·domizing** *vbl. sb.*
1868 tr. *Index Expurgatorius of Martial* 89 You must give up sodomising and womanising. **1888** [see *irrumation]. **1969** P. Barnes *Ruling Class* II. vi. 94 Earl of Gurney: You want two seconds of dripping sin to fertilize sodomized idiots. **1971** K. Millett *Sexual Politics* (1972) I. i. 13 Her gratitude at being sodomized is positively astonishing. **1972** *Times Lit. Suppl.* 2 June 622/4 Everything, animate and inanimate, appears to be either sodomizing, sodomized, ejaculating or bleeding. **1980** *Ibid.* 19 Sept. 1020/3 Cassady roughly sodomizing another man in a public lavatory.

sodopotassic, var. *sodipotassic *a.*

sody, *U.S. dial.* and *colloq.* var. Soda¹ (esp. in sense *4 b).
1900 *Dialect Notes* I. 241 *Soda.* Always *sōdi* in Kansas City. **1907** J. London *White Fang* I. ii. 14 Swallow a spoonful of sody, an' you'll sweeten up wonderful. **1907** *Dialect Notes* III. 200 *Sody, n.*, soda; either bicarbonate of soda or soda water. 'Have a glass of sody with me?' The normal pronunciation seems affected. **1951** W. Faulkner *Requiem for Nun* III. 267 She's usually got a bottle of sody pop in the icebox. **1975** B. Garfield *Hopscotch* viii. 86 There was even a sody-cracker barrel by the fountain.

soebak, var. *subak. **soeju,** obs. var. *shoyu.

soetkoekie (sutku·ki). *S. Afr.* Also **zoete-koekie, soet-koekie.** [Afrikaans, lit. 'a little sweet cake', f. Du. *zoet* sweet + *koek* cake + *-ie* dim. suff.] A traditional South African spiced biscuit.
1910 D. Fairbridge *That which hath Been* ix. 115 Juffvrouw wanted very bad to help hand zoete-koekies. **1949** L. G. Green *In Land of Afternoon* xii. 165, I have heard of a special ginger beer which is brewed during Christmas week and served with soetkoekies. **1939** 'D. Rame' *Wine of Good Hope* III. iii. 368 They ended their tea and the thin bread and butter and soet-koekies of Grim's ceremony. **1973** *Fair Lady* 7 Mar. 23 With visions of my Voortrekker ancestors embarking on hazardous journeys with tinfuls of 'mebos', biltong, and 'soet-koekies', I scratched through my recipe book.

sofa. Add: **3. a.** *sofa-cover, settee, -table* (earlier and further examples).
1805 *Times* 7 Nov. 4/2 Card, sofa, and Pembroke tables. **1807** Jane Austen *Let.* 8 Feb. (1932) I. 49 There will then be the Window-Curtains, sofa-cover, & a carpet to be altered. **1848** Mrs. Gaskell *Mary Barton* I. vii. 115 The dead body..which she was laying out on a board, placed on a sort of sofa-settee. **1861** C. M. Yonge *Young Stepmother* xxiii. 333 She felt the misfortune to the beautiful new sofa-cover as a most serious calamity. **1937** *Burlington Mag.* May 240/1 Attempts to introduce gothic ornament into a sofa-table, a bookcase or a chair. **1968** *Canad. Antiques Collector* Aug. 6/2 (Advt.), Exceptionally fine Rosewood Regency sofa table. **1978–9** *House & Garden* Dec./Jan. 78/2 Striped Welsh flannel chair and sofa covers.
4. sofa-back, (*a*) an antimacassar; (*b*) the back of a sofa; **sofa-bed** (earlier example).
1878 Geo. Eliot *Let.* 27 June (1956) VII. 33 The sorrows of those who can afford..to think of anything better than sofa-backs. **1880** L. Higgin *Handbk. Embroidery* 63 Design for sofa-back cover. **1894** 'Mark Twain' in *Century Mag.* Jan. 338/2 Tom..hoisted a leg over the sofa-back. **1805** Sofa bed [see *chair-bed* s.v. *chair *sb.*¹ 15].
sofaed *ppl. a.*, (*b*) furnished with a sofa or sofas.
1802 T. Campbell *Let.* 28 Aug. (1849) I. xv. 397 A lord's house, fashionable *strangers*, sofa'd saloons, and winding galleries. **1842** Dickens *Let.* 3 Jan. (1974) III. 7 A comfortable room,..well-lighted, sofa'd, mirrored, and so forth. **1860** G. Vandenhoff *Dramatic Reminiscences* vi. 104 A very good-sized room had been fitted up as my dressing-room, cleaned, carpeted, sofa'd, well lit. **1934** 'A. Bridge' *Ginger Griffin* ii. 22 The Grant-Howards sat in the green-sofa'd saloon.

Sofar (sōᵘ·fāɹ). Also **SOFAR.** [See quot. 1948.] A system in which the sound waves from an underwater explosion (either artificial or natural) are detected at a number of listening stations so that its position can be fixed; more generally, detection of deep explosions a great distance away.
1947 *Britannica Bk. of Year* 841/2 An underwater sound system, called 'Sofar'.., made possible the location of air and ship survivors as far as 2,000 mi. from shore. **1948** Ewing & Worzel *Long-Range Sound Transmission* 12 A network of four listening stations is being established in the Pacific by the Navy Department, and the name SOFAR, from the words SOund Fixing And Ranging, has been assigned to the system. **1961** H. H. Koelle *Handbk. Astronaut. Engin.* xxviii. 28 The missile must carry a Sofar bomb when a deep-water detection system is used, in order to provide a deep sound source. **1966** *McGraw-Hill Encycl. Sci. & Technol.* XIV. 429/2 Shock waves used in Sofar propagate long distances under water, being refracted by the isothermal layers in the oceans. **1979** *Nature* 20–27 Dec. 820/2 The deep ocean sound channel is used to obtain very long range (typically > 2,000 km) acoustic transmission via totally refracted propagation paths (SOFAR propagation).

soft, *sb.* Add: **2. b.** (Earlier and further examples.)
1821 [see *fancy-piece* s.v. *fancy *sb.* and *a.* B. 2]. **1955** *Publ. Amer. Dialect Soc.* xxiv. 115 Paper money is known, in general, as *scratch* or *soft*.
d. *pl.* Soft commodities (see *soft *a.* 20 b (*b*)).
1979 *Financial Times* 28 Mar. 37/1 Will 'softs' boom next? **1981** *Times* 5 May 17/2 Softs are less homogeneous in outlook because crop conditions vary so much.
6. *Hist. rare.* A Menshevik.
1950 E. H. Carr *Bolshevik Revol.* I. ii. 30 But the withdrawal of seven delegates who had voted with the 'softs'..had the result of shifting the balance of votes in favour of the 'hards'. **1955** H. Hodgkinson *Doubletalk* 17 Lenin's group..was described as 'Iskraists'..or 'hards'... Its rivals were 'softs', because it approached the membership problem in the spirit of Martov's 'The more people there are called Party members, the better it will be'.

soft, *a.* Add: **I. 1. d.** Of a photographic film or paper: producing an image of low contrast.
1892 [in Dict., sense 1 c]. **1910** W. Wallington *Chats on Photogr.* xiii. 113 The paper may be obtained in a number of speeds..the slower varieties being more suitable for printing soft negatives. **1937** *Amat. Photogr.* ix. 120 The 'soft' grade has the..merit of registering tone in dense high-lights without clogging shadows. **1966** D. G. Brandon *Mod. Techniques Metallogr.* 15 An approximately linear dependence of the blackening is only obtained over a limited range of exposure times, and this range is far greater for 'soft' emulsions than for 'hard' ones. **1979** *SLR Camera* Jan. 59/1 If the photographic image shows a large number of tones between the extremes of light and dark, it is said to have soft gradation.
e. Of a lens: having low resolving power. Cf. *soft-focus *a.*, *sb.*, and *v.*
1958 M. L. Hall *Newnes' Compl. Amat. Photographer* i. 26 Soft lenses. Some highly complex lenses are said to be 'soft' in definition. **1974** *Publishers' Weekly* 24 June 56/2 'Soft lens' photos well suited indeed to the muted inner dignity of these deeply religious black people. **1978** *SLR Camera* Sept. 37/3 How good is the lens on your camera? Is it a bit soft?
2. c. Applied in the Soviet Union and China to a class of railway carriage (esp. a sleeper) having soft, upholstered seats.
1928 *Cook's Continental Timetable* 15 May 102 Sleeping car of direct communication, soft and hard class. **1949** F. Maclean *Eastern Approaches* (1951) i. iii. 39 In the train I found myself in a 'soft' compartment with three senior and somewhat supercilious officers of the Red Army. **1954** Koestler *Invisible Writing* v. 61 It entitled me..to travel in the 'soft class' on trains. **1968** Bethell & Burg tr. *Solzhenitsyn's Cancer Ward* i. xiv. 228 They found it quite unbearable, of course, to travel in ordinary railway carriages... The Rusanovs now travelled only in reserved compartments or 'soft class'. **1974** *Times* 1 Apr. 15/5 All the railways in Europe, and even those in China, have first and second class, although it may be wrapped up as 'hard' and 'soft' in USSR and China. **1978** G. E. Newby *Big Red Train Ride* ii. 22 We now took a closer look at our deluxe, 'soft-class' compartment. **1982** *Brit. Med. Jrnl.* 3 Apr. 1031/1 One end of our 'soft-class' carriage [in China].
III. 13. b. (Earlier example of *to have a soft spot in one's heart.*) Hence phr. *to have a soft spot for*, to have a tender regard for, be fond of.
1857 C. M. Yonge *Dynevor Terrace* I. xi. 160 Jane has a soft spot in her heart, and will not think true love is confined within the rank that keeps a gig. **1902** [see Spot *sb.*¹ 10 b]. **1947** K. Tennant *Lost Haven* i. 24 She always did have a soft spot for him. **1971** *New Scientist* 13 May 400/1 He won a scholarship advertised in New Scientist and has had a soft spot for the magazine ever since.
18. d. *to be* or *go soft on*: to be or become excessively lenient or partial to.
1883 'Mark Twain' *Life on Mississippi* xxxiv. 373 If he was soft on the Arkansas mosquitoes, he was hard enough on the mosquitoes of Lake Providence to make up for it. **1911** G. B. Shaw *Shewing-up of Blanco Posnet* 405 Why did He make me go soft on the child if He was going hard on it Himself? **1959** *New Statesman* 30 May 751 Why ..were we all 'going soft' on Dulles? **1971** *Guardian* 19 Nov. 13/7 In the eyes of the militants, the Guardian is still 'soft on the rebels'.
IV. 19. c. fig. *soft spot*, a weak or vulnerable place.
1933 [see *infiltration 1 e]. **1956** A. L. Rowse *Early Churchills* 239 The French..withdrew behind their fortifications... Marlborough was all for an assault on these; he had proved and found a soft spot opposite Ramillies. **1958** *Engineering* 21 Mar. 361/1 Even if the country as a whole was in the best of economic health, the local soft spots could not be ignored without serious political repercussions. **1965** H. I. Ansoff *Corporate Strategy* vii. 126 If the problem appears as a minor 'soft spot' in an otherwise healthy product-market position, temporary *ad hoc* arrangement..may suffice. **1975** *Times Lit. Suppl.* 17 Oct. 1233/3 There is probably much truth in Mr Levison's analysis.., but his obvious sincerity and dedication mask a few soft spots in the argument.
d. *Mil.* Of a military vehicle: unarmoured. Of a missile base: vulnerable to a direct nuclear explosion because of its construction or location.
1944 A. Jacob *Traveller's War* vii. 129 The tanks crunch forward like a battle fleet: our 'soft' vehicles in the middle of the phalanx, with the armoured cars of the Dragoon Guards moving on both flanks. **1948** Partridge *Dict. Forces' Slang 1939–45* 175 *Soft-skinned vehicles; soft stuff*, unarmoured vehicles. (Army.) Both of these terms started as slang; the latter remained unofficial, although it did become colloquial; the former very rapidly became colloquial and then official. **1958** [see *hard *a.* (*sb.*) 14 f]. **1961** E. Burgess *Long-Range Ballistic Missiles* vi. 182 All forms of long-range missiles now appear to be developing towards the abandonment of the soft base, but there are..difficulties associated with the hardening concept. **1972** *Dict. Military & Associated Terms* (U.S. Department of Defense) 275/2 *Soft missile base*, a launching base that is not protected against a nuclear explosion.
e. *Physics.* Of a mode of vibration in a crystal lattice: such that its frequency decreases to zero as the temperature of the crystal approaches that of a phase transition.
1964 *Physical Rev.* CXXXVI. a. 429/1 Cochran proposed a theory which links the cause of ferroelectricity in the perovskites to the existence of a temperature-dependent 'soft' lattice vibrational mode. **1967** *Ibid.* CLVII. 396/1 The energy of this soft mode can be extracted from reflectivity measurements in the infrared region. **1973** G. R. Wilkinson in A. Anderson *Raman Effect* II. xi. 813 A number of 'soft' modes of vibration whose frequencies depend upon temperature have been found.
20. b. (*a*) (Earlier example.) Also *soft furnishing*(*s*). (*b*) *Comm.*, designating relatively perishable consumer goods such as clothes, foods, and drugs; *spec.* in the commodity market, used of commodities produced from vegetables, such as textiles, rubber, and foodstuffs.
1833 *Chambers's Edin. Jrnl.* I. 385/3, I could occasionally hear a detached sentence on politics..the price of stocks—soft goods. **1925** *Daily Tel.* 13 May 20/6 Soft furnishing department. **1927** *Ibid.* 11 May 18/6 Manageress wanted... Must have thorough experience in the sale and scheme side of soft furnishings. **1946** *Sun* (Baltimore) 5 Oct. 2/6 These points stand out with respect to the production of 'soft' consumer goods (clothing, food, drugs and the like). **1961** *Ann. Reg. 1960* 502 The value of retail sales rose by 4 per cent, a higher demand for 'soft' goods more than offsetting the drop in purchases of durable household goods. **1967** E. Short *Embroidery & Fabric Collage* iii. 78 Probably the most useful and versatile object in soft furnishings is the cushion. **1976** L. Deighton *Twinkle, twinkle, Little Spy* xvi. 169 She did nothing except sink lower in the soft furnishings and continue to drink. **1979** *Daily Tel.* 9 Oct. 21 While the metal markets continue their volatile course the 'soft' commodities, with the notable exception of sugar, have largely been untouched by the urge to get out of the dollar and currencies in general.
22. d. *Astronautics.* Of a landing made by a spacecraft: slow enough for no serious damage to be incurred. Chiefly in *soft landing *vbl. sb.*
1958 *Times* 28 Mar. 10/3 Next (in difficulty) would be a 'soft' (controlled) landing [on the moon] by an unmanned vehicle. **1966** E. Burgess *Assault on Moon* v. 151 A soft lunar landing is the landing of a payload on the Moon with a small shock to that payload. **1975** *Daily Tel.* 11 Aug. 11/4 One vehicle will make a soft touchdown on Mars while the large spacecraft which carried it on its journey will remain in orbit.
e. Of a substance: readily magnetized by an ambient magnetic field but retaining no permanent magnetization in the absence of such a field.
1839, **1873** [see *soft iron*, sense 24 b in Dict.]. **1900** *Jrnl. Iron & Steel Inst.* LVII. 403 The author divides the different varieties of iron and steel into those which are magnetically 'soft' and those which are magnetically 'hard'. **1948** F. Brailsford *Magnetic Materials* iv. 69 The most important source of internal strains in the higher grade soft magnetic materials is that due to the presence of impurities held in solution in the metal. **1976** *Nature* 5 Feb. 381/1 In a few cases large randomly directed magnetically soft components were removed in low alternating fields.
f. Of glass: softening at a relatively low temperature when heated.
1925 Hodkin & Cousen *Textbk. Glass Technol.* vi. 51 A..vessel of soft soda-lime glass in which water is boiled, will liberate so much alkali in 15 minutes as to make impossible correct titrations with decinormal solutions. **1961** G. R. Choppin *Exper. Nuclear Chem.* viii. 121 Since Pyrex glass contains boron..it is better to use a soft glass tube to hold the source. **1965** Phillips & Williams *Inorg. Chem.* I. xiv. 545 Soft glasses are made by adding soda to the silica. About 25 per cent soda reduces the viscosity of a glass by a factor of 10^{10}.
g. Of a contact lens: made of a soft, yielding material.
1964 *Highlights Ophthalm.* VII. 252 (*heading*) The new hydrophilic gel, soft, contact lenses. **1971** *Time* 31 May 46, I intend using the soft lenses on every patient I possibly can. **1978** H. Hamand in M. Ruben *Soft Contact Lenses* viii. 128 The thickness of cornea of a rabbit was found to increase by 20% with 17 h of hard lens wear and about 5% with soft lens.
24. b. *soft hammer, paste* (earlier examples), *porcelain* (earlier examples) (see Paste *sb.* 3 b, Porcelain 1 a note).
1964 S. Crawford *Basic Engin. Processes* i. 16 Hammers with heads made of lead, copper, rubber, or rawhide are known as soft hammers. The head is usually in the form of a cast tube with a recess at each end to locate the soft inserts. **1977** G. Clark *World Prehistory* (ed. 3) v. 212 The introduction of the soft hammer technique resulted in the production of thinner bifaces. **1848** H. R. Forster *Stowe Catal.* 8 Chantilly Porcelain is a fine kind of 'soft paste'. **1879** Soft paste [see Paste *sb.* 3 b]. **1819** A. Rees *Cycl.* XXVIII. s.v. *Porcelain*, The porcelain is made of these substances [*sc.* porcelain clay and felspar]..but other materials are employed to give the required transparency at a lower temperature. This has received the name of soft porcelain. **1839** Ure *Dict. Arts* II. 1025 The manufacture of soft porcelain is longer and more difficult than that of hard.
c. Applied to a coal of low rank, usu. a bituminous or a brown coal.
soft coal in sense 24 b in Dict. is coal that is easily cleft.
1857 J. B. Jukes *Student's Man. Geol.* iv. 133 All these minute varieties are commonly included under four principal heads:—1, Caking coal; 2, Splint or hard coal; 3, Cherry or soft coal, and 4, Cannel or parrot coal. **1903** *Bull. U.S. Geol. Surv.* No. 213. 265 By far the largest part of the coal mined [in the Eastern Interior field] is soft bituminous, making a good steam fuel. **1926** J. Roberts *Mining Educator* I. 677/1 American cities where the use of 'soft' or smoky coal is forbidden by law. **1958** I. C. F. Statham *Coal Mining Practice* (1960) I. ii. 77 Practically all black coals, such as are worked in British coalfields, show banding of soft bright and hard dull coal..parallel to the bedding. **1979** *Sci. Amer.* Jan. 28/3 Hard coal (anthracite and the various grades of bituminous coal) and soft coal (brown coal and lignite).
25. b. For *dial.* and *U.S.* read orig. *dial.* and *U.S.* Now usu. restricted to cold fruit drinks and the like. (Later examples.)
1911 *Chambers's Jrnl.* Feb. 111/2 In the matter of 'soft' drinks the chemist or druggist is not in it [in Canada]. **1919** P. B. Clayton *Tales Talbot House* 29 The House was always what the Canadians called a 'soft drink' establishment, but no one resented this, lapping up tea or cocoa or Bovril with thanksgiving. **1936** G. B. Shaw *Simpleton of Unexpected Isles* Prol. iii. 27 A feast of fruit and bread and soft drinks is spread on the ground. **1944** Auden *For Time Being* (1945) 113 Soft drinks and sandwiches may be had in the inns at reasonable prices. **1960** Koestler *Lotus & Robot* i. i. 36 A soft-drink cocktail party in the house of a leading Parsee politician. **1964** I. Murdoch *Italian Girl* iii. 36 You haven't anything soft, ginger beer? All right, tomato juice. **1966** 'A. Hall' *9th Directive* xii. 116 The sun was hot..soft-drinks men worked their way through [the crowd]. **1978** R. Ludlum *Holcroft Covenant* xviii. 203 They had proceeded to a second hotel on the rue Chevalle, where a soft-drink sign provided him with a name for the registry: N. Fresca.
c. Of a detergent: biodegradable.
1963 *New Scientist* 27 June 716/1 A soft detergent is one that is biologically soft; that is, readily oxidised in a modern sewage plant. *Ibid.* 717/1 In most western countries the consumption of soap, which is a soft detergent, remains fairly steady or is falling. **1966** *Economist* 23 July 385/1 Replacing this branched chain with a straight chain makes the detergent 'soft'—that is, easily munchable by the hard-working bugs. **1971** *Daily Tel.* 16 Oct. 10/6 The detergent industry switched over to 'soft', biodegradable detergents.
26*. a. *Electronics.* Of a thermionic valve or discharge tube: (*a*) having had an inert gas introduced into it at the time of manufacture in order to modify or enhance its performance; (*b*) containing gas at low pressure as a result of a leak or of outgassing by component parts. [tr. G. *weich*, used in sense (*a*) by W. C. Röntgen 1897, in *Sitzungsber. d. K. Preuss. Akad. d. Wissensch. zu Berlin* 584.]
1899 [see *hard *a.* (*sb.*) 16* b]. **1901** *Phil. Trans. R. Soc.* CXCVI. 42 The value of λ obtained..for a much 'softer' bulb was ·001, or about four times the absorption of the bulb employed in these experiments. **1921** *Wireless World* 6 Aug. 288/2 One 'soft' triode will give as much amplification as two 'hard' valves. **1931** [see *hard *a.* (*sb.*) 16* b]. **1932** *Discovery* July 216/1 These hard valves were found to be very much more reliable and uniform in their action than earlier 'soft' valves. **1948** *Electronic Engin.* XX. 384/1 In recent years the soft valve (thyratron) counter has been replaced by the hard valve counter. **1956** G. A. Montgomerie *Digital Calculating Machines* ix. 178 The soft valves are usually beam-switching tubes of the multicathode type.
1919 R. D. Bangay *Oscillation Valve* 203 There are several indications which enable one to tell when a valve is going 'soft'. The first is loss of power in the oscillatory circuit. **1929** Duncan & Drew *Radio Telegr. & Teleph.* xi. 214 The degree of vacuum in the tube would change and some tubes became soft (having less vacuum) while others became hard (having a higher vacuum, with little or no gas present). **1958** W. F. Lovering *Radio Com-*

munication viii. 173 A valve in which the vacuum is poor is said to be soft; the presence of a small number of molecules of gas adversely affects the performance.

b. *Physics.* Of X-rays and gamma rays: of relatively long wavelength and low penetrating power. Of sub-atomic particles: of relatively low energy.

Orig. so called because soft X-rays are emitted by a soft tube (see prec. sense).

1902 *Phil. Trans. R. Soc.* CXCVI. 516 The radiation from the barium compound is enormously reduced by the interposition of so thin a screen as an ordinary piece of tinfoil; these 'soft' rays accordingly form much the greater part of the whole. **1925** *Proc. Cambr. Philos. Soc.* XXII. 834 The soft γ-radiation gave a well-marked spectrum containing two strong lines. **1940** *Nature* 20 July 94/2 The beryllium oxide was tested inside a counter so designed that even very soft particles could be detected. **1950** D. H. Wilkinson *Ionization Chambers & Counters* vi. 160 The use of a proportional counter for counting extremely soft electrons which give very few ion pairs, has been proposed. **1960** [illegible] V. 163/2 The soft X-rays emitted from television screens cause great concern to some radiologists. **1978** *Nature* 30 Mar. 396/3 There are several isotopes..which emit radiation sufficiently soft to be shielded even by the syringe wall but which can be presented in a form such that all the tissues of an animal become labelled a short time after injection.

26.** Miscellaneous *transf.* and *fig.* uses. **a.** Of facts, information, etc.: insubstantial, impressionistic, imprecise (opp. *Hard *a.* 7 b, c). Of a science or its method: not amenable to precise mathematical treatment or to experimental verification or refutation; *esp.* in *soft science.*

1923 *Sci. Amer.* Feb. 77/2 Its functions and its limitations are to get the facts from the top of [illegible]. **1966** *Time* 3 June 43 Project simile Director Hall T. Sprague says these are the 'soft' sciences: what a laboratory is to the hard sciences of physics, chemistry and biology. **1968** *Physics Bull.* Oct. 351/2 One of the striking features of the present times is the penetration of 'hard' methods (quantitative, physical analysis) into subjects which were hitherto 'soft' (descriptive, non-numerical). **1970** *Publishers' Weekly* 8 June 154 Hardscience is science (physics, math, chemistry), softscience is the humanities, sociology or psychology. **1972** *Lancet* 25 Nov. 1138/1 Clinical departments..must report a new respect for the 'soft' data of sociology. **1976** *National Observer* (U.S.) 20 Apr. 20/5 All the *President's Men* is [illegible] a 'soft' story—breezily entertaining but short on hard facts. **1976** *Times Lit. Suppl.* 25 June 766/2 The soft areas in revolutionary theory, which he sorts into a series of Hegelian opposites: [illegible]. **1980** *Dædalus* Spring 94 One might view these various expressions..spanning (from the 'hard' end) science, history and anthropology..to (at the 'soft' end) dreams and personal fantasy. **1982** *Daily Tel.* 23 Apr. 22 Most academic articles in all the sciences (hard and soft) are read by very few people.

b. *Comm.* Of markets, commodities, etc.: depressed, characterized by falling prices or excess supply. (Cf. Firm *a.* 7 a.)

1930 *Morning Post* 19 Aug. 3/4 Oils were generally soft, while Coppers were far from being buoyant. **1935** *Commercial & Financial Chron.* 7 Sept. 1488/2 Gold mining stocks were soft, and international shares neglected. **1968** *Globe & Mail* (Toronto) 13 Feb. B11/2 At the end of last year, the fish industry in Newfoundland was reeling from the effects of the extremely soft U.S. market. **1981** *Times* 8 May 26/5 Disappointing trading news also left.. [illegible] 1p softer at 99p. **1982** *Times* 5 May 17/1 Britain must cease being the soft market for the so-called developed world and action was needed against countries which blocked imports of British goods by crippling duties while having free access to the United Kingdom.

c. *Econ.* Of currency: (see quot. 1949). (Cf. also sense 20 c in Dict. and *hard currency* (b) s.v. *Hard *a.* (*sb.*) 22 b.)

1940 *Economist* 6 Apr. 609/1 There are some currencies..'soft' currencies, notably the lira and the yen—for which no official rates are fixed. **1949** *Times* 20 Sept. 5/7 Soft currency..is a relative rather than an absolute term. It means a currency of which other countries (or some other countries) have earned more than they are willingly spend in the country whose currency it is... A soft currency is by definition, non-convertible—i.e., cannot be converted into gold or dollars... A currency may, however, be 'transferable' (within limits) and yet remain a soft currency in relation to some other currencies. **1960** *Economist* 15 Oct. 241/1 The United States now 'sells' abroad each year surplus farm products worth more than $1 billion, taking in exchange soft currencies—as one bureaucrat calls them, 'clam shells, coloured buttons and other forms of local currency'. **1967** A. Diment *Dolly Dolly Spy* vii. 97 I had changed some of my hard Swiss francs into soft pesetas. **1980** *Times* 23 May 14/6 Some African countries can only make ends meet in Moscow by smuggling in soft roubles bought abroad... This is categorically forbidden by..most..western embassies.

d. Designating a kind of technology that uses renewable resources such as wind or solar power and human or animal exertion and is not harmful to the natural environment. Also, of energy employed in or derived from this technology.

1974 *Harper's Mag.* Apr. 6 The term 'soft technology' was coined amid the British counter-culture in 1970. Technology which is soft is gentle on its surroundings, responds to it, incorporates it, feeds it. A nuclear power-generating station doesn't qualify. A wooden windmill with cloth sails grinding local grain does. **1977** A. Lovins *Soft Energy Paths* ii. 38, I shall call them 'soft' technologies: a textural description, intended to mean not vague, mushy, speculative, or ephemeral, but rather flexible, resilient, sustainable, and benign. Energy paths dependent on soft technologies..will be called 'soft' energy paths. **1978** *Dædalus* Summer 188 The Progressives.. have tended to emphasize economics and technology, yet they have been 'soft' technological determinists in a way that maintains..Rousseau's fascination with the potential..of human autonomy. **1978** *Internat. Relations Dict.* (U.S. Dept. State Library) 2/1 Other terms used synonymously with appropriate technology are alternative technology, intermediate technology, and soft technology.

V. 27. (Special combinations, corresponding in formation to those (undefined) under senses 28–30, are included here.) *softback* attrib., (of books) bound in paper or limp covers; *U.S.*; *soft-board* sb., a relatively soft form of fibreboard (see also sense 30 in Dict.); *soft cancer*, a cancer in which the affected tissue is soft and yielding; now *rare* or *Obs.*; *soft chancre*, a venereal disease caused by local infection with *Hæmophilus ducreyi*; chancroid; also, one of the characteristic lesions of this disease; *soft copy*, a legible but transient presentation of information, as on a VDU screen; information so displayed; *soft-core* attrib. [after *hard-core* (*pornography*): see *hard core* (*b*) s.v. *Hard *a.* 22 b], (of pornography) less obscene than hard-core pornography; also *absol.*; *soft corn* (examples); *soft-cover* attrib., of, pertaining to, or designating a book bound in a limp or paper cover; also (usu. as two words) *ellipt.* for some such phrase as *soft-cover edition*; hence *soft-covered* adj.; *soft drug*, a drug that is held to be comparatively non-addictive and safe to use; also *fig.*; *soft food*, the partly digested food which pigeons regurgitate to feed to their young; = Pigeon's milk 1 in Dict. and Suppl.; *soft-foot* vb. intr. (N. Amer.), to go with quiet footsteps, to tiptoe; *soft fruit* = *small fruit* s.v. *Small *a.* 21; *soft ground*, a sticky covering of wax mixed with grease for an etching plate; also = next; *soft ground etching*, a process of print-making using plates covered with a soft ground, producing prints with softened lines resembling chalk or pencil drawings; also, a print produced by this process; *soft hail*, precipitation of snow pellets (see *Snow *sb.*[1] 9 a); *soft line*, a flexible or conciliatory policy; freq. (with hyphen) *attrib.*; hence *soft-liner*, *soft-lining* adj.; *soft loan* orig. *U.S.*, a loan, esp. one to a developing country, made on especially favourable terms; † *soft meat* = *soft food* above; *soft money*, (*a*) (see sense 20 c in Dict.); (*b*) (see quot. 1976); *soft-nosed* adj., (of a bullet) expanding; *soft option*: see *Option *sb.* 1; *soft palate*: see Palate *sb.* (*a*) 1 b; *soft porn*(*ography*): see *Porn, Porno *sb.* 2, *Pornography 2 a; *soft rock*, a type of rock music which is less strident than hard rock; hence *soft-rocker*; *soft rot*, any of various bacterial or fungal diseases of vegetables, fruit, and herbaceous plants in which the tissue becomes soft and pulpy; also, a condition of timber in which a fungus renders it soft and brittle; freq. *attrib.*; *soft sculpture* (Pop Art, etc.), a form of sculpture in cloth, foam rubber, or other pliable materials; *soft second* (Bowls) (see quot. 1905); *soft sell* orig. *U.S.*, advertising or salesmanship that is subtly persuasive rather than aggressive (opp. *hard sell* s.v. *Hard *a.* 22 b); also *transf.*, *fig.*, and *attrib.*; *soft-sell* vb. trans.; *soft-selling* ppl. adj.; *soft-shoe* attrib. (orig. *U.S.*), designating, of, or pertaining to a kind of tap-dance performed in soft-soled shoes without metal taps; also *fig.*; hence as *sb.*, a dance of this kind, and as *vb.*; *soft-shoulder* orig. *U.S.* [*Shoulder *sb.* 6 j], an unmetalled strip of land at the side of a road; *soft silk*, silk from which the gum has been removed; *soft-skinned* adj., (*a*) having a soft skin (see sense 29 in Dict.); (*b*) *Mil.*, of a military vehicle, unarmoured (cf. sense 19 d above); *soft snap* U.S. (earlier example); *soft solder*: see Solder *sb.*[1] 4 a; *soft sore* = *soft chancre* above; *soft tissue*, body tissue other than bone and cartilage; also *attrib.*; *soft-top*, a roof of a motor vehicle that is made of soft material and can be opened; a car so fitted, a convertible; freq. *attrib.*; *soft touch*: see *Touch *sb.* 20 c; *soft toy*, a toy animal stuffed with a soft material; *soft wart*, a small, soft, pedunculated growth of skin occurring most frequently on the neck.

1958 B. Malamud *Magic Barrel* 141 He had got some of his softback books read. **1966** *Time* 8 July 60/2 Shimkin and three other men in 1939 founded his Pocket Books, Inc., the world's most voluminous soft-back-book producer. *Ibid.* 60/3 Pocket books will be better able to assure authors of bonuses for softback reprint rights. **1966** C. Sweeney *Scurrying Bush* v. 68 The sagging soft-board ceiling. **1976** P. Hill *Hunters* v. 43 The inside wall was partially covered by a large sheet of soft-board pinned to which was a large-scale map. **1804** J. Abernethy *Surg. Observ. containing Classification of Tumours* 51 The sarcoma which is..generally found in the testis, and is distinguished by the name of the soft cancer of that part. **1834** [see Spongoid *a.* 1]. **1871** *Med. Times & Gaz.* 20 May 568/1 A soft cancer of the uterus. **1894** R. Quain *Dict. Med.* (new ed.) I. 269/2 Encephaloid, medullary, or soft cancer, so named from its usually brain-like appearance and consistence, is softer and grows more rapidly..than scirrhus. **1961** A. S. MacNalty *Brit. Med. Dict.* 247/2 *Soft cancer*, medullary carcinoma. [**1965** I. Macdonald in T. F. Nealon *Managem. Patient with Cancer* xvii. 451 The designation of medullary (soft) is less than ideal, but is not of significance unless it refers to bulky, pseudo-encapsulated tumors.] **1859** C. F. Maunder tr. *Ricord's Lect. Chancre* 9 Numerous examples of the soft chancre. **1887** H. Raphael tr. *Zeissl's Path. & Treatm. Syphilis* ii. 116 In the female the soft chancre is most frequently met with upon the labia majora and minora. **1917** *Act* 7 & 8 *Geo. V* c. 21 § 4 In this Act the expression 'venereal disease' means syphilis, gonorrhœa, or soft chancre. **1961** R. D. Baker *Essent. Path.* ix. 157 Chancroid (soft chancre), a venereal sore on the genital organs, resembles the chancre of syphilis in location but differs in being the starting point of a purely regional process and never a systemic disease. **1968** *Internat. Solid-State Circuits Conf. Digest Technical Papers* 76/1 This scheme permits one to share the same phone link for both hard and soft-copy output. **1982** *New Yorker* 17 May 34/2 'Soft copy and hard copy' (words on a television screen and words on a piece of paper in hand). **1966** *N.Y. Times* 25 Sept. D 15/4 The soft-core pornography of advertisements like 'Have you had any lately?' **1971** *Guardian* 9 Mar. 8/2 The market for pure pornography is insatiable, while soft-core porno is waning in appeal. **1973** *Publishers Weekly* 6 Aug. 67/3 Lots of softcore sex scenes as Bertha finds revenge. **1977** *Time Out* 17–23 June 47/2 Borowczyk's slightest and most commercial offering has provoked wildly different responses: great pagan art or ultimate soft-core? **1979** *Listener* 5 July 21/3 The soft-core entertainment end of the television spectrum. **1834** W. A. Carruthers *Kentuckian in N.Y.* I. 98 He's feeding me on soft corn, thought I. **1948** *Antioch Rev.* Autumn 161 He was all soft corn.., but you couldn't be sure, not with a man like Malcolm. **1958** *Times* 12 Feb. 9/4 Earnings from soft-cover rights, movie rights, etc. **1961** *Guardian* 20 Oct. 17/5 Creative fiction will eventually prove itself in soft cover in this mass market. **1965** *Amer. N. & Q.* Sept. 13/1 Olms has also initiated an important new series of 'Paperbacks', soft-cover reprints of important scholarly works. **1975** *Bookseller* 16 Aug. 1306/1 What makes a paperback publisher pay nearly two million dollars for the softcover privilege of a book? **1977** *Time* 17 Jan. 54/2 In 1976 U.S. softcover publishers issued more than 150 historical novels, many of them as paperback originals. **1960** *Times* 15 Oct. 7/4 Now British presses also are producing 'soft-covered' volumes that can cost anything up to a guinea. **1959** *Oxford Mail* 14 Jan. 4/4 Dr. D. C. M. Yardley of Oxford found that of about 50 university users of soft drugs (mostly marihuana) about 20 were regular takers, and that although the latter were convinced they could give it up at any time, in fact they hardly ever did so without professional help. **1968** *Sunday Mail Mag.* (Brisbane) 7 July 6/1 Some 300,000 people in Britain are estimated to be using some form of 'soft drug' such as marijuana, amphetamines or barbiturates. **1969** *Punch* 12 Mar. 376/2 For most of my life I've thought flattery was only a soft drug. It was nice; but I could take it or leave it. **1976** J. Archer *Not Penny More* i. 11 He was a sly, smart little boy, unloved by the school authorities..for his control of the underground school market in soft drugs and liquor. **1876** R. Fulton *Bk. Pigeons* iv. 39 This 'soft food'..is pumped up by the old ones with a sort of vomiting action. **1969** C. R. Hill *Pigeon Guide* vi. 91 At first, the parents will feed the youngsters on soft food (pigeons' milk). **1939** *Ottawa Jrnl.* 22 July 12/8 He softfooted to the window and looked inside. **1972** J. Mosher *Adultery* iii. xiv. 133 As he soft footed it through the kitchen, father began to whistle a tune she liked. **1918** W. P. Seabrook *Mod. Fruit Growing* vii. 60 In the case of..soft (or bottom) fruit cutting back at once may be done. **1956** H. H. Crane *Fruit* i. 9 If the area is very small, it may be possible to grow only soft fruits. **1981** *Observer* 26 Apr. (Colour Suppl.) 47/3 It's obviously much easier to fit soft fruit into a small garden. **1840** S. Fuller *Let.* 31 Jan. in N. N. Solly *Mem. Life David Cox* (1873) iv. 57 We propose to republish... The plates have been proved, and found to be in good condition, particularly the soft ground, which I consider as good as ever. **1925** E. S. Lumsden *Art of Etching* xiii. 113 Soft-ground (*vernis mou*) is nearly allied to etching proper. **1965** Zigrosser & Gaehde *Guide Coll. Orig. Prints* iv. 56 A soft ground is prepared from hard ground by the addition of tallow or Vaseline. *Ibid.*, Both linear and textured or tonal effects can be created by soft ground. **1868** P. G. Hamerton *Etching & Etchers* v. xi. 340 In some books on engraving, ordinary etching is called soft-ground etching, to distinguish it from etching done in a hard ground, by the old masters. This old hard ground now being disused, a modern writer may call common etching hard-ground etching, and reserve the title of soft-ground etching for that to which it is here applied. **1873** N. N. Solly *Mem. Life David Cox* iii. 36 Cox had been employed..to make soft ground etchings on copper from his own drawings. **1914** G. T. Plowman *Etching & Other Graphic Arts* xiii. 111 For soft ground etching melt together lard or tallow and an equal amount of etching ground. **1976** P. Coker *Etching Technique* 70 The principle of soft ground etching is that the ground adheres to anything that is brought into contact with it. **1881** W.

Marriott *Hints Meteorol. Observers* 16 International symbols... Soft hail △. **1894** [see Graupel]. **1945** F. A. Berry et al. *Handbk. Meteorol.* iii. 257 Soft hail usually accompanies the less severe winter or spring storms. **1970** R. M. Longley *Elem. Meteorol.* iv. 91 At the tops of these clouds pellets of soft hail are formed by the collision of the snowflakes found there. **1966** *Sunday Times* 5 June 1 But Canada, Norway, Denmark and Italy prefer a 'soft' line [illegible]. **1975** *New Left Rev.* Nov.–Dec. 70 The ambassador [illegible] American faction. **1977** *Time* 30 May [illegible] collaborators. **1967** *Economist* 18 Feb. 614/2 Life is made even more difficult for the guerrillas, hard-liners by definition, because of their open conflict with the soft-liners in the Venezuelan communist party. **1980** *N.Y. Times* 17 Jan. A23 Soft-liners will say that the Hayner mission prevented a bloodbath, with the Iranian Army battling the mobs. **1977** *Time* 28 Mar. 15/2 They accuse it [sc. the Communist Party] of betraying the revolution and joining the Establishment with only its soft support of Premier Giulio Andreotti's minority government. **1958** *N.Y. Times* 2 Mar. iv. 3/6 The fund is authorized to make some 'soft loans', that is with long maturities and low rates of interest. **1959** *Washington Post* 8 Oct. A12/2 The American proposal for an International Development Association, equipped to make softer loans to supplement the 'bankable' advances of the World Bank, is vitally important. **1965** [see I.D.A. s.v. *I II. 3]. **1979** *Financial Times* 11 Sept. 9/1 'Hidden subsidies' to the paper industry provided by most European governments—in the form of tax incentives, soft loans and regional employment schemes—will continue. **1765** *Treat. Pigeons* 22 Soft meat is then a kind of liquid pap, prepared as it were by instinct by the parents, by a digestion of the hard grains in their craw. **1853** 'B. Moubray' *Pract. Treat. Poultry* (ed. 4) xii. 183 Soft meat is a sort of milky fluid or pap secreted in the craw of pigeons, by the wise providence of nature. **1879** [illegible] *Penny [illegible]. **1976** *Daily Tel.* 13 Dec. 8/5 'Soft money' is the research money which charitable foundations give to research projects... [illegible] the sponsorship [illegible] concerned [illegible]. **1978** *Bull.* [illegible] teaching units that [illegible] grants [illegible]. **1908** [illegible] in R. S. Churchill *Winston S. Churchill* (1969) II. Compan. II. 957 [illegible] lances [illegible] Acts [illegible]. **1922** Joyce *Ulysses* [illegible]. **1977** J. Blackburn [illegible] nosed [illegible]. **1969** *Harper's Mag.* [illegible] invaded [illegible]. *Time* [illegible] soft rock [illegible] sleek [illegible] soft rock [illegible] David [illegible] wonder [illegible] a soft [illegible] *Station* [illegible] bacillus [illegible] *Path.* [illegible] heavy [illegible] bacteria [illegible] toes b[illegible] *Forest* [illegible] soft rot [illegible]. **1969** C[illegible] and on [illegible] always [illegible] Baksh[illegible] show [illegible] may [illegible] G. Bar[illegible] emphasis [illegible] made [illegible]—which [illegible] *No Lady* [illegible] objects [illegible] human [illegible] II. 884 [illegible] of four [illegible] second [illegible] weakest [illegible]. **1955** [illegible] sell'—[illegible] just get [illegible] iii. 24 [illegible] the mo[illegible] 260 Th[illegible] G. F. [illegible] sceptic [illegible] (Victoria) [illegible] have t[illegible]. **1978** [illegible] master [illegible] more f[illegible]. **1982** T[illegible] approach [illegible] *Rev.* J[illegible] the be[illegible] P. Dur[illegible] of soft-[illegible] *Soft-shoe*, generally acknowledge[d] to be dancing of the star type and containing certain ballet steps. Any tap dancing done without metal taps or clogs. **1941** *Life* 25 Aug. 74/2 George Primrose was famed as an exponent of 'soft-shoe' dancing 35 years ago. *Ibid.* 77/1 The misleading phrase 'soft-shoe' was applied to an increasingly popular form of mobile, eccentric step dance in rolling 4–4 time. **1962** J. D. Salinger *Franny & Zooey* 180 Les and Bessie did a lovely soft-shoe on sand swiped by Boo Boo from the urn in the lobby. **1965** P. O'Donnell *Modesty Blaise* xviii. 190 We'll make a soft-shoe job of it if we can... Straight for the diamonds..and away. [illegible] I don't want to use the gun, not while there's any chance of keeping this soft-shoe. **1967** M. Stewart *Gabriel Hounds* x. 139 He..beckoned. I soft-shoed after him. **1975** W. McIlvanney *Docherty* [illegible] Only a few couples still soft-shoed around the floor. **1982** *Daily Tel.* 27 Jan. 12/6 They [sc. Adele and Fred Astaire] regularly soft-shoed their way to acclaim in musical comedies on Broadway. **1939** *Time* 20 Feb. 28/3 Driving toward his home on the outskirts of Indianapolis..he got off the road on a soft shoulder. **1978** J. Irving *World according to Garp* [illegible]. **1862** M. Merryweather *Experiences of Factory Life* (ed. 3) iii. 31 In this town, a [illegible]. **1877** Knight *Dict. Mech.* III. 2180/1 Silken thread..is called 'soft silk'. **1861** C. Salter tr. [illegible] *Dyeing Silk* i. 155 The bath temperature must..be modified..being lukewarm for soft silk. **1942** *Hutchinson's Pictorial Hist. of War* 10 June–1 Sept. 129/1 That is the protection of all this paraphernalia of supply and maintenance, all what may be called the 'soft-skinned stuff'. **1968** *Times* 18 Jan. 14/6 Many of the 'soft-skinned' vehicles [illegible]. **1856** *Spirit of Times* 25 Oct. 378 One of them, however..suddenly lamed herself, and another..'found a softer snap,' so they paid forfeit to The Heiress. **1884** A. Cooper *Syphilis & Pseudo-Syphilis* iv. 33 There are two principal theories with regard to the relationship between the hard and the soft sore. **1940** E. T. Burke *Venereal Dis.* xi. 236 The term 'soft sore' or '*ulcus molle*' and the term 'chancroid' should be used. **1974** Passmore & Robson *Compan. Med. Stud.* III. [illegible] Chancroid or soft sore [illegible] *Hæmophilus ducreyi* [illegible]. **1892** G. M. Gould *Pocket Med. Dict.* 272 [illegible] soft tissues. **1964** L. Martin *Clinical Endocrinol.* (ed. 4) iii. 113 Soft-tissue [illegible]. **1977** [illegible] Advocates of early soft-tissue repair..have reported successful results in a significant number of patients [illegible]. **1966** [illegible] First soft-top model in the so-called compact car size. **1967** *Guardian* [illegible]. **1976** *Milton Keynes Express* 18 June [illegible] Triumph Spitfire [illegible] soft top. [illegible] **1917** E. A. Hickman *Soft Toys & How to Make Them* 7 The object of this book is to give instruction in the art of making stuffed toys for sale. **1952** *Dryad Handicraft Catal.* 93 Soft toy making. **1964** M. Laski in S. Nowell-Smith *Edwardian England* iv. 203 The named soft toy was now starting its long run of popularity..Golliwog..Teddy Bear..Cæsar. **1970** *Guardian* 24 Sept. 11/1 Wendy Boston, pioneers in safe soft toys. [illegible] lessons known by the laity as moles, mothers'-marks, and by the profession as [illegible] soft warts. **1967** Soft wart [see *Molluscum [illegible]].

c. *soft corn*, a variety of maize, *Zea mays* var. *amylacea*, whose seeds are rich in soft starch; also, maize containing a high quantity of moisture, making it unlikely to keep well; *soft maple*, one of several maples with less durable wood, esp. the red maple, *Acer rubrum*, or the silver maple, *A. saccharinum*; also, the timber of these trees; *soft wheat*, one of several varieties of wheat having a soft grain rich in starch.

1751 J. Bartram *Observations* 60 Last of all we served a great bowl full of Indian dumplings, of new soft corn, cut [illegible] off the ear. **1868** *Mich. Agric. Rep.* VII. 160 Early frosts made considerable soft corn. **1902** A. S. Hitchcock in L. H. Bailey *Cycl. Amer. Hort.* IV. 2004/2 Brazilian Flour Corn sold by seedsmen is a type of the Soft Corn. **1947** *Chicago Tribune* 23 July 9/4 The state must prepare for a soft corn crop this year. [illegible] [see *Maple tree]. [illegible] soft maple [illegible] soft wheat [illegible].

c. *soft-back*, a soft-shelled turtle of the genus *Trionyx*; (earlier example); *soft clam* = *Long-neck 2 b; (earlier examples); *soft crab*, a crab that has shed its shell and is awaiting the hardening of the new one; (earlier examples); *soft tick*, a tick of the family Argasidae, lacking a dorsal shield.

1859 P. H. Gosse *Lett. Alabama* 99 Another Tortoise of even greater size and equal ferocity is the Soft-back (*Trionyx ferox*). **1806** D. Roe *Jrnl.* 27 Feb. (1904) 25 Got Sum Soft Clams. **1855** *Knickerbocker* XLVI. 222 Along the strand..these great delicacies, 'soft clams' and sand-crabs may be found. **1772** L. Carter *Jrnl.* 10 Oct. in *William & Mary Coll. Q.* (1906) 1st Ser. XIV. 38 Like the shell of a soft crab, the body of the crab after the shell is off seems by much too large for the shell. **1805** R. Parkinson *Tour Amer.* 315 Soft crabs..are reckoned great dainties. **1932** R. Matheson *Med. Entomol.* iii. 40 The family Argasidae contains those ticks which lack a scutum and hence have been called the soft ticks. **1974** *Nature* 25 Jan. 226/1 This is the first proven example of transmission of a mammalian piroplasm by an Argasid ('soft') tick.

28. a. *soft-foot* (later examples); *-ground*, *-top*.

1916 D. H. Lawrence *Amores* 73 When I carried my mother downstairs..at the beginning of her soft-foot trip. **1959** *Listener* 8 Jan. 60/2 The soft-foot priest. **1916** *Daily Colonist* (Victoria, B.C.) 26 July 5/5 Muckers, sewermen, blacksmiths, softground workers, and timbermen are wanted. **1977** P. Geddes *Hangman* xi. 98 She carried everything she owned in the soft-top suitcase.

29. a. *soft-balled, -bellied, -chaired, -edged, -fleshed* (later example), *-grained, -horned, -minded* (later example; hence *soft-mindedness*), *-palmed, -sandalled, -sized, -soled, -topped, -worded.*

a **1930** D. H. Lawrence *Last Poems* (1932) 42 Kisses of the soft-balled paws. **1923** —— *Birds, Beasts & Flowers* 113 He..trailed his yellow-brown slackness soft-bellied down. a **1918** W. Owen *Poems* (1931) 98 Comforted years will sit soft-chaired in rooms of lamplight. **1970** *Jrnl. General Psychol.* Apr. 183 The other two paintings were non-representational 'soft-edged' geometric abstractions. **1928** D. H. Lawrence *Woman who rode Away* [illegible] looked pleasant and soft-fleshed..they were ardent and generous. **1848** J. R. Lowell *Poems* 2nd Ser. 167 The red-oak, softer-grained yields all for lost. **1962** *Listener* 1 Mar. 392/1 [illegible] soft-grained voice cover a wide range of human experience. **1868** G. M. Hopkins *Jrnl. & Papers* (1959) 175 Beauty of the sycamores here, native to the soil, soft-horned, and falling apart like stars. **1919** E. O'Neill *Moon of Caribbees* 185 It's soft-minded she is, like I've always told you, ye'r stupid. **1925** F. Scott Fitzgerald *Great Gatsby* vi. 120 The transactions in Montana copper that made him many times a millionaire found him physically robust but on the verge of soft-mindedness. **1848** J. R. Lowell *Uncoll. Poems* (1950) 59 The soft palmed tradesman coming home at eve. **1978** *Times* 3 July 11/5 [illegible] The American farm worker and farmer have subsidized the American dinner plate... There will be the usual soft-palmed protesters. **1886** Kipling *Departmental Ditties* (ed. 3) 13 From rockridge to spur fly the soft-sandalled feet. **1923** D. H. Lawrence *Birds, Beasts & Flowers* 101 Now that in England is silence, where before was a moving of soft-sized women. **1931** J. Buchan *Prince of Captivity* III. ii. 287 The soft-soled shoes of the pursuit did not slip. **1976** *Milton Keynes Express* 23 July 39/1 Heavy rain during the day provided a soft-topped, damp wicket which gave a lot of help to the bowlers. **1916** Joyce *Portrait of Artist* iv. 180 It was only amid soft-worded phrases..that he dared to conceive of the soul or body of a woman moving with tender life.

30. *soft-palated* (earlier example), *-talk*. See also *soft-land *v.*

1903 G. B. Shaw *Let.* 21–2 Dec. (1972) II. 384 Mrs Robertson..would have had to get her stories soft-talked and her son knocked down by a flood of soft-palatable words. **1970** H. McLeave *Question of Negligence* (1973) ii. 27 [illegible] the pressure in number-two boiler room is hardly high enough to soft-boil an egg. **1968** *Daily Progress* (Charlottesville, Va.) 15 Apr. 21/2 Young Century used artificial bait and he soft-talked the fish as he pulled him in. **1968** B. Mather *Springers* x. 100 He asked peevishly when he'd be moving me—and where? I soft-talked him and said any minute.

31. With ppl. adjs., as *soft-looking*.

1860 C. M. Yonge *Hopes & Fears* I. ii. 33 Honora thought her the prettiest child she had ever seen..such a soft-looking little creature. **1924** 'R. Crompton' *William—the Fourth* vi. 87 He said he'd come to his party if he'd go an' do dancing class anyway, with that soft-looking kid.

soft, *adv.* Add: **I. 1. c.** *ellipt.* for *soft class* (in travelling by train in China or the U.S.S.R.). Also *transf.*, first-class.

1939 'M. Innes' *Stop Press* i. 11. 44 As a matter of fact, he's on the train now. But of course travelling soft. **1976** *Times* 13 Nov. 11/1 Trains in China are made up of 'soft' classes coaches but you travel hard or soft according to your pocket.

II. 9. a. *soft-falling, -flaming, -going, living.*

1845 F. W. Faber *Lett.* 29 Jan. in J. E. Bowden *Life F. W. Faber* (1869) vi. 167 [illegible] I know how miserably sinful and soft-living I have been, I ought never to have stepped out in the way that I have done. a **1918** W. Owen *Poems* (1963) 103 And through those snows my looks were soft-flaming. **1955** [illegible] Like curd soft-falling. **1944** E. Blunden *Shells by Stream* 49 [illegible] soft-going.

b. *soft-falling, -living, -sailing, -speaking.*

1829 D. Jerrold *Black-Ey'd Susan* i. 13 That pretty soft-speaking [illegible]. **1888** W. Whitman *November Boughs* 33 And who art thou? said I to the soft-falling shower. **1916** D. H. Lawrence *Amores* 24 Soft-sailing waters where fears no longer shake. **1965** F. Sargeson *Memoirs of Peon* iv. 73 The shapely soft-speaking Ma— girl..brought me a plate of sandwiches. **1975** *New Yorker* 5 May 109/1 He hit a great, soft-falling fly that fell for a bloop double. **1977** *Times* 17 Mar. 18/6 A dark-haired, soft-living Mayfair clientele.

10. b. *soft-tinted*; *soft-spun*, loosely twisted in spinning; also *transf.*; opp. *hard-spun* s.v. Hard *adv.* 8 e. See also *soft-landed *ppl. a.*

1869 'Mark Twain' *Innoc. Abr.* xxx. 324 A lace-work of soft-tinted crystals of sulphur. **1902** W. de la Mare *Songs of Childhood* 96 As if it were a perfect jewel in the morning's soft-spun hair. **1906** Soft spun [see *hard spun* s.v. *Hard *adv.* 8 e]. **1940** E. Blunden *Poems 1930–40* 193 Choose this soft-tinted willow tree. **1964** H. Hodges *Artifacts* ix. 129 Excessively twisted, or *hard-spun*, yarns may kink..while *soft-spun* threads with little twist may untwist further.

softball (sǫ·ftbǫl). Also soft-ball, soft ball. [f. Soft *a.* + Ball *sb.*[1]] **1.** *Confectionery.* (As two words or with hyphen.) A soft globule of sugar formed (e.g. by dropping into water) as a means of testing that the mass of sugar being boiled has reached a certain stage; hence used *attrib.* and *absol.* to designate this stage.

1894 E. Skuse *Compl. Confectioner* 70 Add the cocoanut slices and allow the whole to boil, for say two minutes, or until the sugar comes to a soft ball. **1907** J. Kirkland *Mod. Baker, Confectioner & Caterer* IV. iv. 13 The soft-ball or full-feather degree is tested by making a small ball of sugar between the fingers while cooling in cold water. **1921** [see *Fondue 2]. **1961** J. Stobart *Cook's Encycl.* 404/1 Soft ball—115°C (240°F). The sugar clinging to the skimmer will now, when shaken, produce a feathery, downy effect. The syrup is now beginning to thicken and will form a soft ball if a little of it is dropped into cold water.

2. *orig. N. Amer.* **a.** A game resembling baseball but played on a smaller field with a larger ball that is pitched underarm.

1926 *Daily Colonist* (Victoria, B.C.) 1 July 5/3 The remainder of the morning was occupied by the younger members of the party in playing soft ball and other less strenuous games. **1947** J. Steinbeck *Wayward Bus* 54 A big and muscular young woman who taught ice hockey and softball and archery at the university. **1977** J. F. Fixx *Compl. Bk. Running* p. xvi, Someone who all his life had played tennis, touch football and Saturday-afternoon softball shouldn't be thus laid low.

b. A ball of the kind used in the game of softball.

1914 *Vanity Fair* Feb. 49 (caption) Soft ball, soft hands and a soft game. **1958** [illegible] 1932 Suffice it to say that *Playground Baseball*..differs from ordinary baseball in four ways, namely: [1] A big ball is used (softball). **1974** [see *softball throw*, sense c below]. **1975** R. Kostetski *Badlands* 136 Lumps of ice the size of softballs.

c. *attrib.*, as *softball court, field, game, team*; **softball question**, a question that is easy to answer; **softball throw**, an athletic event in which a softball is thrown as far as possible.

1943 J. S. Huxley *TVA* ix. 73 The playground is floodlit to give the adults a soft-ball court after their day's work is over. **1958** J. Kerouac *On Road* iii. i. 180 A softball game was going on under floodlights. **1974** *News & Press* (Darlington, S. Carolina) 25 Apr. 11/1 Beverly Robinson won first-place in the long jump and the softball throw. **1976** *New Society* 28 Oct. 209/3 'Why Switzerland?' may seem the ultimate softball question, [illegible] along Zurich's Bahnhofstrasse. **1977** J. Cheever *Falconer* 4 There was a softball field where the gallows had stood. **1977** *Sci. Amer.* Nov. 15/2 He has also collaborated with Cross..on a number of double plays for the physics department softball team.

3. Tactical slow and gentle play in lawn tennis. Also *attrib.* So **soft-balling** *ppl. a.*

soft-ball *v. trans.* and *intr.*

1961 *Times* 18 May 5/2 Not only did Sangster adapt his game to the slow court [illegible] opponent. **1962** *Times* 25 Apr. 4/7 It was the Chilean's soft ball game that ruffled his opponent's feathers. *Ibid.* 27 June 4/6 Playing soft ball, and apparently resigned to defeat, he was offered a reprieve [illegible]. **1963** *Times* [illegible] Miss Mottram [illegible]. **1976** [illegible] *Amer. Speech* **1976** LI. 294 *Softball*, play slow soft shots.

soft-boiled, *a.* Add: [So **soft-boil** vb. s.v. Soft *a.* 30.] **1.** Of an egg: boiled but not hard-boiled. Also *transf.*

1889 Kipling in *Macmillan's Mag.* Dec. 113/1 You never had a soft-boiled egg. **1896** *Woman's Home Companion* Nov. 5/3, I have plenty of towels and soap and soft-boiled eggs. **1948** W. Stevens *Let.* 2 Apr. (1967) 582 How good grated Parmesan is on soft-boiled eggs. **1954** —— *Let.* 23 July (1967) 841 [illegible] soft-boiled eggs thrown in now and then. **1983** [illegible] *Snake* iii. 37 His big, soft-boiled eyes, with puckers beneath them like black egg-cups.

2. Of a person: mild, easy-going; naïve, impractical; opp. *hard-boiled *a.* 2.

a **1930** D. H. Lawrence *Last Poems* (1932) 258 O you hard-boiled conservatives and you soft-boiled liberals, don't you see how you make bolshevism inevitable? **1942** E. Waugh *Put out More Flags* iii. 191 Father's friends were all hard-boiled and rich... And then I met Cedric who was poor and very, very soft-boiled. **1963** *Times* 28 Feb. 15/1 But, say the T.V.A. enthusiasts, there is a psychological stimulus in receiving a rebate as opposed to merely not paying tax from the start. Is the British businessman really so soft-boiled? **1978** D. Gryll *Guardians & Angels* iv. 142 Dickens's..pantheon is crammed with a soft-boiled array of credulous infantile adults.

soft centre. Also soft-centre. [f. Soft *a.* + Centre *sb.*] **1. a.** A soft filling inside a chocolate. Also *attrib.*

1930 H. W. Bywaters *Mod. Methods of Cocoa & Chocolate Manuf.* xiv. 251 During recent years chocolates containing soft cream centres..have increased in popularity. **1947** 'G. Orwell' in *Tribune* 7 Mar. 11/3 The same kind of charm as belongs to a pink geranium or a soft-centre chocolate. **1959** [see *Centre *sb.* 11 e].

b. *transf.* A chocolate having a soft centre.

1970 H. McLeave *Question of Negligence* (1973) ii. 18 Every time Lord Blye turned nasty..her consumption of petit fours and soft centres soared. **1974** P. Haines *Tea at Gunter's* xiv. 149 Me..lying about on the settee eating soft centres, and Mantovani on the radiogram.

2. A soft heart, esp. in contrast with a tough exterior; a vulnerable or weak core. Also *attrib.*

1955 E. Blishen *Roaring Boys* iv. 94 The fierceness and roughness of the boys was of the surface... The [illegible] of this soft centre. **1973** M. Amis *Rachel Papers* [illegible] I gathered [illegible] that the [illegible] considered [illegible]. **1974** [illegible] They love [illegible] by their authors.

soft-centred, *a.* [f. prec.] **1.** Of a person or his attitudes: soft-hearted; of works of literature, art, music, etc.: having a weak, vulnerable, or sentimental core.

1957 *Times Lit. Suppl.* 8 Nov. 674/4 Like his attack on the Monarchy this attachment to Socialism is essentially emotional and soft-centred. **1962** *Listener* 5 Mar. 425/2 The fact that the catalyst is a woman friend..does not make the play any less soft-centred. **1963** *Times* 7 Mar. 15/3 [illegible] Harryboy [illegible] resulting [illegible] soft-centered. **1977** [illegible] Michael [illegible] soft-centred [illegible].

2. Of a chocolate: having a soft centre.

[illegible] A soft-centred milk chocolate. **1978** *Times Lit. Suppl.* 21 Apr. 438/3 [illegible] a twelve-course dinner..ending with five pounds of soft-centred chocolate.

Hence **soft-ce·ntredness**.

1967 *Guardian* 10 Apr. 6/6 The critics..began to note..a reluctance to scrap and build. **1981** *Economist* 28 Nov. 47/1 A mixture of Saudi soft-centredness, Arab pettiness and Syrian bloody-mindedness has led the Arabs to miss a rare opportunity.

soften, *v.* Add: **I. 4. c.** Usu. with *up*. To reduce the strength of (a defensive position) by bombing or other preliminary attack; also *transf.* Hence *fig.*: to undermine the resistance of (a person). *colloq.* (orig. *U.S.*).

1940 W. L. Shirer *Berlin Diary* (1941) 378 Stuka dive-bombers are softening the Allied defense positions, making them ripe for the tanks. **1941** *Sun* (Baltimore) 3 Feb. [illegible] for a final invasion thrust. **1949** F. Maclean *Eastern Approaches* [illegible] at least [illegible] a system of [illegible] them. **1952** [illegible] 'He's been making me [illegible]' Then in a moment she smiled. 'You said that just to soften me up.' **1962** *Listener* 1 Nov. 706/1 The farming industry is being softened up by the [illegible] coming in the subsidy system, whether or not we go into Europe. **1974** *Daily Tel.* 21 Sept. 15 A man who had told police of goings on in Soho was kidnapped in broad daylight, tied to a chair and 'softened up' before he had told them. **1980** [illegible] 'Mortar? What do you [illegible] soften [illegible].'

II. 7. (Later examples in sense *26** b of Soft *a.*)

1947 *Kiplinger Washington Let.* (Kiplinger Washington Agency) 5 Apr. Consumers are not buying quite enough to keep up with output, so luxury markets began to soften last fall. **1982** *Daily Tel.* 6 July 15 American rates might soften following the publication of reassuring money supply figures on Friday evening. **1982** *Times* 17 Aug. 13 The shares softened 2p to 168p after announcement of [illegible].

softener. **1.** (Further examples.)

1954 A. J. Hall *Stand. Handbk. Textiles* (ed. 4) iv. 265 It is important not to use softeners which reduce the fastness of coloured goods to light. **1973** *Times* 1 June 11/5 [illegible] the little two-inch square brain-box attached to the softener.

softening, *vbl. sb.* Add: **1.** Also, the action or process of becoming soft. (Later examples.)

1942 R. D. Bangay *Oscillation Valves* [illegible] serious 'softening' of the valve will [illegible] characteristics [illegible]. **1945** *Electronic Engin.* XVII. 338/3 A softening of the [illegible]. **1946** *Ann. Reg.* 1945 214 The constant harping on the subject of [illegible] in the Soviet Press, accompanied by attacks on the régime in power in Turkey, conveyed the impression that the familiar process of softening would in due course be followed by a formal demand. **1957** *Economist* 30 Nov. [illegible] softening [illegible]. **1960** N. [illegible] softening [illegible] settled [illegible] (Toronto) [illegible] marked a [illegible] loans [illegible] softening-up' [illegible] consolidation [illegible] and repetition.

3. b. With *up*, in sense *4 c of the vb., as *softening-up period, process, raid, technique, trick*.

1951 *R.A.F. Rev.* Apr.–May 32/3 They helped to cover Allied bombers on 'softening-up' raids on the European fortress. **1953** L. P. Hartley *Go-Between* x. 127 The softening-up process, as we should call it now, which be had put me through had been enough. **1954** J. Stein *Basic Everyday Encycl.* 558/1 This 'softening-up' technique broke communications, disrupted production, disorganized civilian existence, destroyed the German air force. **1971** B. W. Aldiss *Soldier Erect* 229 After a softening up period, another attack was launched. Our attack! **1976** B. Lecomber *Dead Weight* viii. 90 Leaving the suspect to stew..is the oldest softening-up trick in the book.

soft-fo·cus, *a.* and *sb.* Also **soft focus**. [f. Soft *a.* + Focus *sb.*] **A.** *adj.* **a.** *Photogr.* Characterized by or producing a deliberate slight lack of clarity and definition in a photograph.

1917 P. L. Anderson *Pictorial Photogr.* ii. 42 It should be noted that the soft-focus lens..has greater apparent depth, both of field and of focus. **1940** A. L. M. Sowerby *Wall's Dict. Photogr.* (ed. 15) 600 It is generally considered that the most pleasing soft-focus effects are obtained by superposing a diffused image upon a more sharply-defined one, this giving a kind of 'halo' round the lights. **1958** [see sense *B]. **1975** *Publishers Weekly* 13 Jan. 58/1 Aided by lovely soft-focus photographs.., Miss Bailey tells the story of a little seedling blown away from its mother tree too soon. **1977** J. Hedgecoe *Photographer's Handbk.* 31 The design of soft focus lenses leaves one pronounced optical error, 'spherical aberration'. This gives halos to highlights and a general softness of outline. **1978** P. Theroux *Picture Palace* ix. 71 Photographs looked freckled and corpse-like, soft-focus poses that might have been painters' instant fossils.

b. *fig.* Diffuse, blurred, unclear, imprecise.

1961 W. T. Jones *Romantic Syndrome* viii. 227 We might begin..by making a count of soft-focus imagery in the works of various poets. **1965** *Punch* 17 Nov. 712 You will see the advantage of reviewing single instalments. It allows elbow-room to savour the glittering detail. Handle the whole book and you fall back on soft-focus generalities, the tiny individual flavours lost. **1975** *New Yorker* 24 Feb. 127/1 The voice is soft-focus, not keenly projected, but of pleasant quality in the middle ranges. **1979** *Listener* 16 Aug. 214/4 This [play]..was full of winsome Irishness and soft-focus sentimentality.

B. *sb.* A deliberate slight lack of clarity and definition in a photograph. Also *fig.*

1958 P. Pollack *Picture Hist. Photogr.* xx. 261 Dr. P. H. Emerson held that soft focus corresponded to natural vision and that soft-focus photography was an art superior to all other graphic arts. **1961** W. T. Jones *Romantic Syndrome* viii. 235 Once the critic replaces a vague liking for 'romantic qualities' with a preference for 'soft-focus'..he is much more likely to make an adequate assessment of the work of the poets and painters. **1977** *Practical Photogr.* Jan. 23/4 To suggest that this method of obtaining soft-focus costs 1p is ridiculous.

Hence **soft-fo·cus** *v. trans.*, **soft-fo·cused** *ppl. a.* (both *fig.* in the examples).

1957 *Archit. Rev.* CXXI. 319 The whole effect is to blur and 'soft-focus' the precision-made look which has been one of the chief qualities of the curtain wall. **1977** *New Yorker* 27 June 35/1 Rose did not like to look at them, at their soft-focussed meekly smiling gratitude.

soft-footed, *a.* (Later *fig.* example.)

1947 Auden *Age of Anxiety* (1948) ii. 47 In the soft-footed Hours of darkness.

soft-la·nd, *v. Astronautics.* Also **softland.** [Back-formation from next.] *trans.* and *intr.* To land slowly without serious damage, esp. on another planet or a satellite.

trans. **1960** *Aeroplane* XCIX. 540 (*caption*) The Surveyor-type probe..should be capable of soft-landing between 100 and 300 lb. of scientific equipment on the Moon. **1960** *Times Mag.* (Seattle) 29 May 22 The first lunar vehicle may be a small robot to be soft-landed on the moon within the next five years. **1966** *Times* 7 Apr. 9/3 This was to have been the final test flight before the United States attempts to softland a real Surveyor spacecraft on the moon. **1967** *New Scientist* 1 June 549/3 The balloon itself was constructed of fine polythene and helium-filled... It was soft-landed and recovered for further use. **1970** *Guardian* 21 Sept. 3/7 Russia today soft-landed its Luna 16 unmanned spacecraft on the moon's barren Sea of Fertility.

intr. **1964** *Yearbk. Astron.* 1965 135 The LEM will detach from the orbiting parent and soft-land on the Moon. **1967** *New Scientist* 3 Aug. 242/2 *Surveyor III*, which soft-landed on 20 April this year returned over 6000 TV pictures. **1969** *Daily Tel.* 18 July 1 Reports from Moscow..say that Luna 15 is ready 'to soft-land on the moon today' and collect moon dust. **1975** *Times* 7 Apr. 6/1 A manned Soyuz spacecraft..soft-landed southwest of the western Siberian town of Gorno-Altaysk.

So **soft-la·nded** *ppl. a.*; **soft-la·nder**, a vehicle that is capable of making or has made a soft landing.

1958 *Proc. Lunar & Planetary Exploration Colloquium* July 13/1 Would you..talk about the design and weight distribution of this soft-landed payload? **1961** *Aeroplane* C. 510/3 Russia's long-range space programme is said to include sending two spaceships to the Moon by 1967. They would be preceded by unmanned 'soft-landers' depositing supplies and propellants. **1966** *Guardian* 28 Dec. 7/1 Luna-13..is returning more information than any other soft lander, whether Russian or American. **1969** *Nature* 12 July 123/1 Between 1966 and 1968, seven Surveyor softlanders were launched which among other things tested the surface strength [of the moon]. **1971** *Ibid.* 26 Nov. 211/2 Such data will be of value in interpreting the findings of soft-landed spacecraft on other planets. **1977** A. Hallam *Planet Earth* 20/1 The surface panoramas transmitted back to Earth from the Soviet soft-landers Veneras 9 and 10 show a barren landscape that contains both angular and rounded rocks.

soft landing, *vbl. sb. Astronautics.* Also **softlanding.** [f. Soft *a.* + Landing *vbl. sb.*] **1. a.** A landing of a spacecraft that is slow enough for no serious damage to be incurred. Cf. *soft *a.* 22 d.

1958 *Proc. Lunar & Planetary Exploration Colloquium* July 13/1 With a soft landing on the moon one might put down a payload of 225 to 800 pounds, but..only about 10 to 25 percent of this would be usable for instruments. **1959** *Washington Post* 24 Mar. A6/4 He said the first 'soft landing' on the moon and return probably will be made with a multi-stage chemical-fueled vehicle. **1966** *Listener* 24 Mar. 427/1 The main events [of 1966] have been the soft landing on the Moon.., and the progress of the two Venus rockets. **1967** *New Scientist* 25 May 448/2 Each of these craft will consist of..an experimental capsule to enter the Martian atmosphere and..make a soft landing either by parachute or..by means of retro rockets. **1971** *Guardian* 1 July 1/5 The Soyuz made a normal re-entry and soft landing.

b. *fig.*

1969 *Sci. Jrnl.* Jan. 54/3 Should one bombard with excess energy or attempt..a 'soft landing' where the energy of the incoming particle is just enough to allow it to penetrate the barrier, be captured and form the compound nucleus. **1973** *Newsweek* 17 Sept. 65/2 Even if the President succeeds in pulling in the rampaging economy for a soft landing, of course, the arrival will be nonetheless bumpy for many.

2. *attrib.* or as *adj.*

1960 *Aeroplane* XCIX. 541/2 Mr. Stoller said that in 1962 three Ranger vehicles were planned to rough-land payloads on the surface of the Moon. These will be followed by the soft-landing mission. **1962** [see *landing-place 1 c]. **1965** *Guardian* 29 Dec. 3/7 The Apollo moon landing programme..depends on highly sophisticated soft-landing techniques. **1969** *New Scientist* 27 Feb. 439/2 The upshot of the planetary experiments should be to discover more about the suitability of Mars to support life; and to select possible sites for future soft-landing craft.

softly, *adv.* Add: **4. c.** *softly, softly, catchee monkey* (and varr.): proverbial phr. advocating caution or guile as the best way to achieve an end. Also *ellipt.* as *softly softly* and (with hyphen) *attrib.*

1907 G. Benham *Cassell's Bk. of Quotations* 849/1 'Softly, softly' caught the monkey.—(Negro.) **1942** N. Balchin *Darkness falls from Air* x. 176 Softly catch monkey... That's the answer. **1960** *Times* 1 Oct. 7/2 That colloquial adage—'softly, softly, catchee monkey'. **1962** P. Brickhill *Deadline* xiii. 152, I didn't pursue it any further then. Softly, softly catchee monkey—and I hated that phrase. **1967** *Autocar* 28 Dec. 7/1 Softly, softly is our policy too, but not at such expense in time. **1970** *Sunday Times* 5 July 11 (*heading*) Ulster: end of 'softly, softly'. **1971** E. F. Schoeters in B. de Ferranti *Living with Computer* viii. 71 Users are naturally applying a 'softly-softly' approach. **1979** *Now!* 14 Sept. 53/2 Sadat and Begin..are both adopting a softly-softly' approach to the Palestinian problem.

11. a. *softly-featured.*

1922 Joyce *Ulysses* 342 This..lent to her softly-featured face at whiles a look..that imparted a strange yearning tendency to the beautiful eyes.

b. *softly-burning, -gliding* (earlier example), *-smiling, -stirring, -swaying.*

1864 W. C. Bryant *Thirty Poems* 38 The softly gliding, bashful stream. **1907** Joyce *Chamber Music* p. xv, Eastward the gradual dawn prevails Where softly-burning fires appear. **1918** D. H. Lawrence *New Poems* 48 Ah, love, Could I but..remove Its softly-stirring, crimson welling-up Of kisses! **1923** —— *Birds, Beasts & Flowers* 38 Vicious, dark cypresses: Vicious, you supple, brooding, softly-swaying pillars of dark flame. **1924** E. Sitwell *Sleeping Beauty* xvii 68 To catch the softly-smiling wind.

softness. Add: **II. 4. e.** The state or property (of a material or device) of being soft, in extended technical usage.

1900 *Sci. Trans. R. Dublin Soc.* VII. 121 The addition of 2 to 5½ per cent. of silicon to steel..increases the magnetic softness. **1919** R. D. Bangay *Oscillation Valve* 204 [A blue glow] is produced by the energy expended by the electrons as they collide with the atoms, and if noticeable is a certain indication of the softness of a valve. **1945** *Electronic Engin.* XVII. 338 The maximum value [of the grid leak resistance]..is limited by the danger of causing softness to develop in the succeeding valve. **1980** *Sci. Amer.* Apr. 94/3 The magnetic 'softness' and high resistivity of glassy alloys also make them likely candidates for the 'read' and 'write' heads in magnetic tape recorders and magnetic disk memories.

f. *Econ.* With reference to commodities, prices, etc.: a state of or tendency towards depression. Cf. *soft *a.* 26** b.

1927 *Comm. & Financ. Chron.* 20 Aug. 961/1 When sterling is firmer a stronger tone develops in the entire European list. On the other hand when sterling reacts, softness develops in the rest of the list. **1930** *Economist* 27 Sept. 569/1 Apart from the recent weakness in grain and cotton prices, and softness in the copper market, the commodity price structure seems to be strengthening. **1970** *Globe & Mail* (Toronto) 25 Sept. B5/1 William S. Brewster, chairman, attributes the disappointing results to softness in the economy.

soft pedal, *sb.* [f. Soft *a.* + Pedal *sb.* 1 b (*b*).] A foot-lever on a pianoforte which softens the tone. Also *fig.* (in senses corresponding to those of the vb.: see next) and (with hyphen) *attrib.*

1856 M. C. Clarke tr. *Berlioz' Treat. Mod. Instrumentation & Orchestration* 79 A pedal much less used than that which raises the dampers..is the soft pedal (or one-string pedal). **1861** [see Pedal *sb.* 1 b]. **1880** Grove *Dict. Mus.* II. 723/1 He..thus produces something of the shifting soft pedal *timbre*. **1911** A. B. Reeve *Poisoned Pen* 255 But can't it be done with the soft pedal? **1936** *Times* 12 Oct. 8/5 Some people thought that the 'soft pedal' should be put on the entertainment factor of a zoo, but he [*sc.* J. S. Huxley] did not agree. **1958** *Times* 8 Oct. 6/1 What this particular play demands from a film director is a certain application of the soft pedal. **1961** *Sunday Express* 23 Apr. 1/2 President de Gaulle—whose soft-pedal policy ..has sparked off this third rebellion. **1973** Junkin & Ornadel *Piano can be Fun* 69/2 Calling the damper pedal the 'loud' pedal neatly distinguishes it from the left foot pedal which is called the 'soft' pedal.

soft-pedal, *v.* [f. prec.] *trans.* and *intr.* (freq. const. *on*). To reduce the loudness or volume of (a noise); to reduce in force or effect; to tone down, play down, go easy on.

1915 R. Wagner in *Sat. Even. Post* 16 Oct. 15/2 The scene was rehearsed time and again, but always the action looked faked because of the necessity of soft-pedaling such a blow. **1916** G. A. England *Pod, Bender & Co.* i. 6 Can that! Soft pedal on that chatter, Ben! **1926** R. H. Davis *Over my Left Shoulder* xxix. 204 'Captain Sam heap mad!' replied the Peacemaker, with no effort to soft pedal the announcement. **1927** *Daily Express* 19 July 3/4 Both parties are at present 'soft pedalling' on the world-revolution thesis. **1931** F. F. Bond *Mr. Miller of 'The Times'* 170 The leading educational centres tended to stress the utilitarian studies and soft-pedal those courses which sought merely cultural ends. **1932** K. S. Prichard *Kiss on Lips & Other Stories* 20 He soft-pedalled about Rose, and the skinflint of an aunt who threatened to take her away. **1944** Auden *Sea & Mirror* in *For Time Being* i. 8 Be frank about our heathen foe, For Rome will be a goner If you soft-pedal the loud beast. **1953** A. Upfield *Murder must Wait* v. 47 'We'll get something out of her.' 'You will soft-pedal,' Bony said quietly. **1957** W. H. Whyte *Organization Man* 52 Out of respect for the sense of the meeting you tend to soft-pedal that which would go against the grain. **1965** *Listener* 27 May 764/1 He must have the drug or endure the sheer agony which the romanticists soft-pedal: high-fever, shivering, fits of vomiting, [etc.]. **1978** R. Hill *Pinch of Snuff* vi. 55 You're noted for soft-pedalling on these squatters.

Hence **soft-pe·dalling** *vbl. sb.*

1952 G. Raverat *Period Piece* x. 197 Uncle Lenny was far too judicious to need soft-pedalling. **1979** *New Statesman* 6 July 3/3 The concern of homophile organisations to make homosexuality an acceptable alternative has led to a distinct soft-pedalling on ticklish issues like paedophilia and transvestism.

soft sawder, *sb.*: see Sawder *sb.*

soft-shell. Add: **1.** *soft-shell clam, crab* = *soft clam, crab* s.v. *soft *a.* 27 c; (earlier and later examples); **soft-shell turtle**, a freshwater turtle of the genus *Trionyx*; (earlier and later examples).

1805 J. Ordway *Jrnl.* 26 May in *Wisconsin Hist. Coll.* (1916) XXII. 218 Passed 2 creeks..in one of them saw Soft Shell Turtle. **1818** *Amer. Monthly Mag.* II. 296 Soft shell Clam. These animals..are excellent eating. **1844** J. E. DeKay *Zool. N.Y.* vi. 11 During this interval, they are known under the name of Soft-shell Crabs, or Shedders. **1847** *Knickerbocker* XXIX. 494 A battle between a soft-shell turtle..and a terrier puppy. **1887** Soft-shell crab [see *diamond-back *a.*]. **1977** E. Leonard *Unknown Man No. 80* xx. 202 Softshell crabs, very good fish, steak. **1980** *Washington Post* 27 June (Weekend Suppl.) 36/1 You mustn't rush through the throng, hoagie in one hand, soft-shell crab sandwich in the other.

2. (Earlier example.)

1845 *Knickerbocker* XXVI. 285 The 'Hard and Soft Shell Baptists'.

3. a. (Earlier example.)

1845 [see *hardshell *sb.* 2].

b. Also, a soft-shelled crab or turtle.

1830 R. C. Sands *Writings* (1834) II. 230 The soft-shell of the Red River. **1846** T. B. Thorpe *Myst. Backwoods* 156 It is Turtle Lake from its abundance of 'green, amphibious soft-shells'. **1935** Z. N. Hurston *Mules & Men* i. iii. 79 Ah'm gointer prune a gang of soft-shells (turtles). **1941** *Louisiana* (Writers' Program) 227 'Soft-shells' and 'busters' (shedding crabs from which the old shell is pried off) are coated with cracker meal and fried. **1942** [see *hardshell *a.* 1]. **1958** R. Conant *Field Guide Reptiles & Amphibians* 70 The Florida Softshell lives chiefly in lakes; all the others are river turtles to a large degree.

soft-shelled, *a.* Add: **1.** Esp. of the soft-shell crab or turtle. (Further examples.)

1796 *Rec. Smithtown, N.Y.* (1898) 129 Any person not an inhabitant..taking Soft shelled clams within the limits of said Town shall pay six pence for every bushel. **1835** J. J. Audubon *Ornith. Biogr.* III. 239 He knows.. how to watch the soft-shelled turtle's crawl. **1856** *Rep. Comm. Patents 1855: Agric.* (U.S.) p. xviii, The 'soft-shelled' almond..is the variety recently introduced and distributed by this Office. **1948** [see *paper-shelled* adj. s.v. *paper *sb.* 12]. **1953** G. M. Durrell *Overloaded Ark* xiii. 223 It was a species known as the Soft-shelled Turtle: the shell is fairly smooth and domed, and it protruded round the edge in a great soft rim.

soft soap, *sb.* Add: **2.** (Earlier examples.) Also *attrib.* orig. *U.S.*

1830 *Reg. Deb. Congress U.S.* 12 Apr. 774, I will not use the vulgar phrase, and say he has been pouring soft soap down the backs of the New York delegation. **1842** *People's Organ* (St. Louis) 15 Apr. 2/2 The magnificent bombshell, rammed full of pride, aristocracy,..soft-soap, curiosity, folly, display, nonsense, man-worship and small-talk, was touched off. **1934** *Sun* (Baltimore) 6 Nov. 2/2 Assailing Governor Lehman for his 'soft soap' manner of campaign, the park commissioner..renewed his assault on the Lehman banking family. **1961** *Radio Times* 6 Apr. 21/2 'Nobody likes to watch a soft-soap interview. People *want* the facts and they can take them,' says Robin Day. **1977** *Irish Times* 8 June 9/6 The public would not be fooled by this 'crazy parade of soft-soap offers'.

soft-soaper. (Earlier example.)

1852 'Mark Twain' in *Hannibal Jrnl.* Sept. 16 He was narrowly watching this soft-soaper of Democratic rascality.

soft-soapy, *a. rare.* [f. Soft soap *sb.* + -y[1].] Flattering, ingratiating, unctuous.

1904 J. C. Lincoln *Cap'n Eri* iii. 36 The thing to do is to be sort of soft-soapy and high-toned.

soft-solder, *v.* Add: Also soft-sodder. **2.** = Soft-sawder *v.* *U.S. colloq.* Cf. Solder *sb.*[1] 4 b.

1866 C. H. Smith *Bill Arp, so called: Side Show of Southern Side of War* 159 Wouldn't you think that as a matter of policy they would soft sodder us a little and quit their slanderin? **1905** J. C. Lincoln *Partners of Tide* vi. 111 He soft-soddered me till I felt slippery all over.

software (sǫ·ftwɛ̄ᵊɹ). [f. Soft *a.* + Ware *sb.*[3], after *hardware 1 c.] **1.** *Computers.* **a.** The programs and procedures required to enable a computer to perform a specific task, as opposed to the physical components of the system (see also quot. 1961). **b.** *esp.* The body of system programs, including compilers and library routines, required for the operation of a particular computer and often provided by the manufacturer, as opposed to program material provided by a user for a specific task.

In early use, the word was interpreted widely to include program material written by a user, as well as systems programs, and also occas. the cards and tapes by means of which programs and data are read into the system. Popular usage, as represented by sense 2, is freq. wider in meaning than the current more restrictive technical usage (sense b).

1960 *Communications Assoc. Computing Machinery* June 381 Nearly every manufacturer is claiming compatibility with all other equipment via such software as Cobol. **1961** *Computer Bull.* June 42 The programming expertise, or 'software', that is at the disposal of the computer user comprises expert advice on all matters of machine code programming, comprehensive libraries of subroutines for all purposes, and the pegasus/sirius scientific autocode. **1962** D. S. Halacy *Computers* iii. 54 Punched cards, which fall into the category called computer 'software' are cheap, flexible, and compatible with many types of equipment. **1964** *Observer* 13 Dec. 1/1 The toughest problem was the 'software'—particularly the 'supervisory programme', the complex instructions which enable the machine to handle many tasks simultaneously. **1965** Hollingdale & Tootill *Electronic Computers* 192 The cost of developing and making the computer itself (the *hardware*) is matched by the cost of making programming schemes for it (often, regrettably, termed *software*). **1966** *New Scientist* 25 Aug. 433/3 The cost of providing 'software'—the programmes for operating the computer on a wide range of problems—is enormous... The user needs to find the bureau which has the appropriate software for his problems. **1967** Cox & Grose *Organization & Handling Bibl. Rec. by Computer* 1 About three years ago, it became clear..that the computer software which was provided and maintained by the manufacturers was not suited to some of the problems of handling and processing large files of data. **1969** P. Dickinson *Pride of Heroes* 187 A rather wet young man who sells software for computers. **1971** B. de Ferranti *Living with Computer* 89 *Software*, all computer programs, or that part of a computer system that is not hardware. **1972** *Computer Bull.* XVI. 85/1 In those days [*sc.* 1966] the term 'software' was still thought rather disreputable, and the concept was probably thought rather vague... More recently, 'software' has become more particularised and often seems to refer to what we might call 'system software', that is, excluding any programs written for specific applications. ..Thus we have 'software packages' and 'application packages', and people who write software consider themselves superior to mere programmers. **1977** K. Hegstad in P. G. J. van Sterkenburg et al. *Lexicologie* 163 The unit price of hardware is going down... On the other hand software costs are rising equally dramatically. **1978** J. McNeil *Consultant* i. 30 Hardware is what you can touch—the actual computer, all its peripheral devices... Without software all that is quite useless... Software, computer programs—they're the same thing... My software staff are very strictly monitored.

2. *transf.* and *fig.*

1963 *Flight International* LXXXIII. 186/1 To get at the total commitment one has to consider the 'software' aspect very closely: for every controller at the scope there may need to be five in the background. **1966**

National Observer (U.S.) 21 Feb. 8/3 This deal..is the latest..in a series of corporate marriages combining..'the software and the hardware' of education. 1967 *Punch* 24 May 770/3 This documentary was a refreshing change from most space-age reportage, dealing sympathetically with the families of the astronauts living outside the perimeter fence of the Manned Spacecraft Centre in Texas: the software rather than the hardware. 1969 *Guardian* 29 Mar. 4/8 The 'Talking Page'..is..being launched with a mass of matching software—a maths course, a reading course, an English course for immigrants. 1978 *Gramophone* June 136/3 They [*sc.* players for digitally recorded discs] will be usable with normal stereo amplifiers and speakers but, of course, they will be incompatible with existing software (records and cassettes). 1979 *Observer* 11 Nov. 33/2 It was phrased in terms of Israel giving the United States 'software'—a more flexible attitude on the Middle East—in return for 'hardware'—arms and military equipment.

3. Special Combs.: **software engineering,** the professional development, production, and management of system software; so **software engineer; software house,** a company that specializes in producing and testing software; also *fig.*

1969 Naur & Randell *Software Engin.* (NATO) 81 Is it possible to have software engineers in the numbers in which we need them, without formal software engineering education? 1979 Jensen & Tonies *Software Engin.* 14 The software engineer is not a theoretician as is the computer scientist. 1969 (*title*) Software engineering; report of a conference sponsored by the NATO Science Committee, Garmisch, Germany, 7th to 11th October, 1968. 1973 K. W. Morton in F. L. Bauer *Adv. Course Software Engin.* i. A. 4 When we sit down at a console to write an Algol program, it is software engineering which determines how easy it is to achieve this end. 1982 I. Sommerville *Software Engin.* i. 3 Software engineering is now maturing into a fully fledged discipline. 1969 *New Scientist* 6 Nov. 285/1 Today there are just over 2000 software houses throughout the world, mostly in America. 1982 *Listener* 23–30 Dec. 31/1 If the world's wealth is maximised by specialisation, Britain should become its 'software house'.

soft wood. Also as one word. Add: **1.** Esp. coniferous trees or their timber. Also *attrib.* (Later examples.)

1905 *Terms Forestry & Logging* (U.S. Dept. Agric.) 48 Softwood..As applied to trees and logs, needle-leafed, coniferous... Softwood..A needle-leafed, or coniferous, tree. 1914 Moon & Brown *Elem. Forestry* 218 Many of our hardwoods are much softer in their wood structure than certain conifers or so-called softwoods. 1930 *Observer* 26 Jan. 20/4 Every year in Finland, Sweden and Russia millions of pine trees are felled and shipped to London... The trade name for such timber is softwood. 1968 J. Arnold *Shell Bk. Country Crafts* xxxi. 321 Yew, though as hard and heavy as oak, is classified as a softwood. 1977 J. L. Harper *Population Biol. Plants* iv. 94 In some hardwood and softwood forests in Maine the buried seed population diverges remarkably in species composition from that of the vegetation.

softy, *sb.* Add: **b.** A very soft-hearted person.

1886 *19th Cent.* Jan. 80 The sentimental softy..who loses his heart at seventeen, is a father at eighteen, and at nineteen is the husband of a dirty trollop. 1914 *Maclean's Mag.* July 88/3 'It's cruel,' said Steve... 'You're a softy!' he said. 1964 Mrs. L. B. Johnson *White House Diary* 16 Jan. (1970) A trip that I fear will not meet with the approval of all the members of our family, but which I—maybe I am a softie—very much want her to have. 1970 'D. Halliday' *Dolly & Cookie Bird* v. 66 You didn't know Daddy like I did. He was an awful old softie inside.

c. One who is considered cowardly, weak, or unmanly; a weakling; an effeminate man.

1895 *Cent. Mag.* Oct. 943/2 If the well-initiated inmates discover that he is unwilling to enter into all their schemes and customs, they call him a 'sucker' or 'softy', and shun his company. 1912 Beerbohm *Christmas Garland* 16 There was nothing of the softy about Smithers. 1924 J. M. Murry *Voyage* xii. 227 'It's no go,' he said. 'I'm not going to bed to-night'. 'But'..She didn't know what to say. Was he a softy? Or was she his first? 1960 T. McLean *Kings of Rugby* xi. 204 Mr Jenkins declared that the All Blacks of the morrow were 'softies' for wearing such impedimenta [*sc.* shoulder-pads]. 1975 *Liverpool Echo* (Football ed.) 11 Jan. 8/6 He never lost his temper, but he was no softie. 1979 *Beano* 2 June 20/3 (*caption*) Who did that? No-one to be seen except those softies playing soppy games.

so·fty, *a.* [f. Soft *a.* + -y[1].] *rare. N. Amer.* Characterized by softness.

1884 'Mark Twain' *Huck. Finn* xxvii. 272 When the place was packed full, the undertaker he slid around in his black gloves with his softy soothering ways. 1970 *Globe & Mail* (Toronto) 25 Sept. 16/2 (Advt.), Fringed shoulder pouches in softy suede.

sog, *sb.*[3] Delete *rare*$^{-1}$ and substitute ? *Obs.* Add earlier and later examples.

1839 *Knickerbocker* XIII. 379 He was a most extraordinary fish; or, in the vernacular of Nantucket, 'a genuine old sog', of the first water. 1851 H. Melville *Moby Dick* II. xxxix. 261 Such a sog! such a sogger! Don't ye love sperm!

SOGAT, Sogat (sō^u-gæt). [Acronym f. the initial letters of *S*ociety *o*f *G*raphical and *A*llied *T*rades.] A trade union now composed of paper-workers, etc. (see below), in the printing industries

The union was formed in 1966 by the amalgamation of the National Union of Printing, Bookbinding, and Paper Workers and the National Society of Operative Printers and Assistants. In 1972 this union was divided, with the paper workers retaining the acronym SOGAT. In 1982 SOGAT amalgamated with the National Society of Operative Printers Graphical and Media Personnel. The new union was called SOGAT 1982.

1966 *Paperworker* Mar. 3/1 Formation of SOGAT. The Registrar of Friendly Societies informed us today of his approval of the..establishment of the Society of Graphical and Allied Trades (SOGAT). 1967 *SOGAT* Feb. 6/1 Only one year ago the Society of Graphical and Allied Trades, now using the coined name SOGAT, came into being. 1969 *Times* 2 May 1/8 About 500 Sogat members ..stopped work..at the Stationery Office's printing plant. 1971 H. Wilson *Labour Government* xiii. 210 SOGAT rejected a multilateral meeting. 1977 in R. Crossman *Diaries* III. 723 Richard Briginshaw, General Secretary of the National Society of Operative Printers, Graphical and Media Personnel (as SOGAT became) 1951–75. 1982 *Times* 8 Nov. 2/4 Sogat '82 hopes that several hundred trade unionists will demonstrate.

Sogdian (sǫ·gdiăn), *a.* and *sb.* Also **Soghdian, Sughdian.** [ad. L. *Sogdiānus*, a. Gr. *Σογδιανός*, f. O. Persian *Suguda*, later *Sugud*.] **A.** *adj.* Of or belonging to Sogdiana, an ancient Persian province corresponding to the modern Samarkand and Bokhara in the Uzbek S.S.R. **B.** *sb.* **a.** A native of this province. **b.** The Middle Iranian language of this province. Also *attrib.*

1553 J. Brende tr. *Quintus Curtius' Hist.* vii. sig. Ui, When he had ordred all thinges amonges the Sogdians, he..remoued into Bactria. 1700 G. Booth tr. *Diodorus Siculus' Historical Libr.* 785 How the King led his Army against the Sogdians and Scythians. *Ibid.* 787 How the Sogdian Noblemen being led forth to be put to Death, were unexpectedly preserv'd. 1729 J. Rooke tr. *Arrian's Hist. of Alexander's Exped.* I. iv. xvi. 24 He then, with part of his army, march'd straight into the country of the Sogdians. *Ibid.*, Spitamenes, at the head of a band of Sogdian exiles, who had fled into Scythia,.. attack'd a certain castle in Bactria. 1909 *Indogerm. Forsch.* XXV. 182 The Sughdian rendering of the Syriac version of the Greek. 1923 H. G. Wells *Outl. Hist.* (rev. ed.) xxx. 295/2 A very considerable literature..in Sogdian and another Aryan language has been discovered. 1947 C. P. Snow *Light & Dark* i. iv. 47 It was written in an unknown variety of Middle Persian called Early Soghdian. 1954 I. Gershevitch (*title*) Grammar of Manichean Sogdian. 1973 R. L. Fox *Alexander the Great* iii. xxii. 309 Heavy drinking is the corollary of survival for a traveller in a Sogdian summer. *Ibid.* 314 By now, Sogdians and Bactrians were serving in Alexander's army.

sogged (sǫgd), *ppl. a.* [f. Sog *v.* + -ed[1].] Soaked, saturated.

1860 [see Sog *v.* 2]. 1929 H. Williamson *Beautiful Yrs.* xxiii. 237 A weary, misty dawn dispelled the phantasms and presented a reality of sogged ground and wet drippings from the trees. 1947 K. Tennant *Lost Haven* iii. 53 Ground's too cold, everything's too sogged for the feed to grow. 1966 S. Heaney *Death of Naturalist* 23 For days I sadly hung Round the yard, watching the three sogged remains Turn mealy and crisp as old summer dung.

soggily (sǫ·gili), *adv.* [f. next + -ly[2].] In a soggy manner.

1939 Auden & Isherwood *Journey to a War* 222 The bastard, I thought soggily, he's sneaking off. 1960 *Times* 1 Nov. 16/4 The result [*sc.* a film]..rather soggily directed by Mr. Peter Brook. 1975 C. Weston *Susannah Screaming* (1976) xxii. 118 'You can see what happened,' Haynes said soggily, mopping his nose. 1981 M. Kenyon *Zigzag* xx. 131 Three slabs of still soggily magnificent date cake.

soggy, *a.* Add: Now common in standard English in the U.K. **2.** (Earlier example.) Also of persons.

a 1852 F. M. Whitcher *Widow Bedott Papers* (1856) xxix. 375 Oh yes, to be sure it needs sugar, the best o' sugar, too; not this wet, soggy, brown sugar. 1964 [see *plastic-macked* s.v. *plastic *sb.*[3] 5].

4. b. Of things, in various *transf.* and *fig.* uses: dull, lifeless, lacking in vigour, sluggish; (of steering) unresponsive.

1928 [see *dawk *sb.*[2] b]. 1932 [see *blimey *int.*]. 1957 J. Braine *Room at Top* xxv. 205 The steering [of a car] was low-geared and more than a trifle soggy. 1965 G. McInnes *Road to Gundagai* v. 79 The poor fellow found the evening hanging pretty soggy on his hands. 1966 Williams & Fleming *Spectrosc. Methods in Org. Chem.* iii. 44 The soggy vibrations of the molecule as a whole give rise to a series of absorption bands at low energy, below 1500 cm.$^{-1}$ 1977 *Gramophone* Dec. 1080/1 The brass fanfares at the start of the finale are hardly of the brightest in sound—but then tuttis show the general orchestral sound to be pretty soggy.

Soho (sō^u·hō^u, sō^uhō^u·), *sb.*[2] **a.** The name of a district in the West End of London, noted for its foreign population, prostitutes, and restaurants, and latterly for its night clubs, striptease shows, pornography shops, etc. Freq. *attrib.* of things connected with or characteristic of Soho.

1818 Keats *Let.* 14 Mar. (1931) I. 127 Then who would go Into dark Soho, And chatter with dack'd hair'd critics. 1890 E. Dowson *Let.* 10 or 11 June (1967) 153 The two artists—with some other artistic & Bohemian types might meet in the early part of the book in a Soho restaurant based on Poland. 1905 Chesterton *Club of Queer Trades* iv. 165 Some dirty Soho restaurant. 1913 W. Whitten *Londoner's London* vi. 140 He had always a nice Soho taste in wine. 1930 W. S. Maugham *Cakes & Ale* i. 12 You drive away wondering whether when he comes he will think you are swanking if you ask him to Claridge's or mean if you suggest Soho. 1937 L. MacNeice in Auden & MacNeice *Lett. from Iceland* 129 To pore on picture catalogues and Soho menus. 1959 *Good Food Guide* 221 There is now a good Soho restaurant in Birmingham with a wide menu properly cooked. It is also resolutely described in Soho French. 1964 *Times Lit. Suppl.* 12 Nov. 1019/4 [Frank] Harris..wrote about sex in the manner and style of the cheapest Soho trash. 1976 *Listener* 26 Feb. 239/2 Sex was..a mean..sleezy, Soho-type thing.

b. *Hist.* Used *attrib.* to designate a type of tapestry produced in England after 1685, usu. in Soho.

[1914 W. G. Thomson *Tapestry Weaving in England* xvii. 139 At Christmas 1685 the arras-workers and tailors employed in the Great Wardrobe changed the scene of their labours to offices in Great Queen Street, Soho, which lies between the northwest corner of Lincoln's Inn Fields and Drury Lane.] 1930 H.-C. Mariller *Eng. Tapestries of Eighteenth Cent.* p. xvii, In the absence of records it would be impossible to say which Soho tapestries were executed in the late seventeenth and which in the eighteenth century. *Ibid.* p. xx, There is no clear line of demarcation between the later Mortlake and the earlier Soho tapestries. 1963 *Times* 18 Jan. 10/6 An anonymous purchaser secured four early-eighteenth-century walnut chairs covered in Soho tapestry, for 450 gns. 1978 *Country Life* 13 Apr. 973/1 The Soho tapestries which now line the room were originally bought..in 1720.

‖ **soigné** (swanʸe), *a.* Fem. **soignée.** [Fr., pa. pple. of *soigner* to take care of, f. *soin* care.] Dressed, adorned, tended, or prepared with great care and attention to detail; well-groomed.

1821 M. Edgeworth *Let.* 27 Nov. (1971) 281 Which would become me best..to pin or not to pin it. I think rather *not to pin*. It looks less soignée but then I may lose the frill. 1900 G. Arthur *Let.* 25 July in *Letters from Man of No Importance* (1928) 123 The Boers may not be particularly *soigné* in their habits, but the Japanese who are, use very little soap, and swear by vapour baths. 1907 E. Glyn *Three Weeks* xii. 137 This lady was so intensely *soignée*. 1927 A. E. W. Mason *No Other Tiger* xi. 98 As she stood there in that flood of radiance, *soignée*, polished from head to foot. 1936 *Punch* 24 June 710/2 William Powell is of course William Powell—suave and *soigné* and perfectly poised. 1959 *Good Food Guide* 35 The prices remain fairly high but the cooking is genuinely *soigné*. 1978 J. Gardner *Dancing Dodo* xxxv. 276 The soignée women and immaculate men..who could afford places like the Hilton.

soil, *sb.*[1] Add: **II. 7. a.** (Further examples.) Usu., but not always, such material as will support the growth of plants, as contrasted with subsoil.

1906 E. W. Hilgard *Soils* viii. 120 Universal experience has long ago recognized and established the distinction between soil and subsoil: by which are ordinarily meant, respectively, the portion of the soil-material usually subjected to tillage, and what lies beneath. 1932 G. W. Robinson *Soils* i. 2 Soil consists essentially of (*a*) mineral matter,..(*b*) organic matter,..(*c*) soil moisture,..and (*d*) soil air. 1952 L. M. Thompson *Soils & Soil Fertility* i. 3 Soil is the mixture of mineral and organic material at the land surface of the earth that is capable of sustaining plant life. 1976 D. Steila *Geogr. Soils* 2 Soil serves as an anchorage for plants and as their nutrient reservoir.

b. *Engin.* Fragmentary or unconsolidated material occurring naturally at or near the earth's surface, regardless of its suitability for plant life. Cf. *regolith.

1934 L. C. Urquhart *Civil Engin. Handbk.* viii. 632 The earth consists of various rock formations covered with a mantle of unconsolidated products of rock disintegration, called the regolith or, more commonly, the soil, although agriculturists use the term soil in a somewhat different sense. 1967 A. Singh *Soil Engin.* i. 1 Soil is considered to include all naturally occurring loose or soft deposit overlying the solid bedrock crust. 1972 C. B. Hunt *Geol. Soils* i. 5 In engineering, 'soil' refers to the ground that can be excavated by earth-moving equipment without blasting.

c. Friable or powdery material occurring naturally on another planet.

1967 *Sci. Amer.* Nov. 43/1 Some of the objects observed on the lunar surface were clodlike clumps of soil. 1970 *Nature* 28 Nov. 795/2 (*caption*) Lunakhod-1 tracks in the lunar soil. 1976 *Daily Tel.* 4 Aug. 10/6 The mechanical digging arm on the Viking I lander was activated again yesterday, scooping up fresh soil to explore for basic life forms on Mars. 1977 J. M. Pasachoff *Contemporary Astron.* iii. xiii. 336 The Venera landers also made measurements of the soil, determining that its chemical composition and density correspond to that of basalt, in common with the Earth, the Moon, and Mars.

8. b. *Engin.* A particular kind of fragmentary material (sense 7 b above).

1913 Blanchard & Drowne *Text-bk. Highway Engin.* vi. 127 Some of the more common soils encountered in highway work are classified as gravel, sand, clay, loam, marl, peat and muck. 1966 *McGraw-Hill Encycl. Sci. & Technol.* XII. 450/1 Soils range from deep-lying geologic deposits to agricultural soils.

9. *soil aggregate, amelioration, bacterium* (usu. *pl.*), *characteristic, classification, compaction, condition, cover, depletion, development, drainage, fertility, formation, genesis, geography,*

geology, layer, management, material, micro-biology, micro-organism, mineral, moisture, nutrient, organic matter, organism, particle, population, pore, restoration, -restorer, sterilization, structure, study, temperature, test, texture; soil-binding, -borne, -building, -depleting, -dwelling, -forming, -inhabiting, -restorative adjs.; *soil-testing, -warming* sbs. and adjs.

1934 *Discovery* July 198/2 Important chemical properties are indicated by the form of the soil aggregates. 1967 Soil aggregate [see *Krilium]. 1969 *Gloss. for Landscape Work* (B.S.I.) v. 20 Soil amelioration. 1972 Edwards & Lofty *Biol. Earthworms* vii. 171 (*heading*) Soil amelioration by earthworms. 1900 *Knowledge* 2 July 161/2 In removing from the land his annual crop, the farmer carries off the greater part of the year's supply of potential humus whence the soil looks to be provided with Nitrates—by the action of the soil-bacteria—for the coming season. 1973 R. G. Krueger et al. *Introd. Microbiol.* xxx. 743/1 Two groups of aerobic soil bacteria are in large part responsible for the conversion of ammonia to nitrate. Representatives of the *Nitrosomonas* group oxidize ammonia to nitrite; those of the *Nitrobacter* group oxidize nitrite to nitrate. 1913 *Bull. Bureau of Soils U.S. Dept. Agric.* No. 96. 19 This wasteful wash can be checked..by seeding the land to soil-binding grasses. 1943 J. S. Huxley *TVA* vi. 42 Protective, soil-binding crops. 1946 *Nature* 9 Nov. 661/2 *Verticillium Malthousei* is the causal fungus; it may be soil-borne, or carried by flies. 1968 *Times* 16 Dec. 7/2 The soil-borne diseases, take-all and eyespot. 1977 J. L. Harper *Population Biol. Plants* v. 139 It is..not particularly easy to discriminate between direct toxic action of soil chemical conditions and indirect effects due to soil-borne pathogens which are themselves determined in distribution by the chemical conditions. 1920 W. W. Weir *Productive Soils* i. 9 Because of the source of soil building materials, the nature of soil formation, [etc.]..all soil can not be the same. 1938, 1962 Soil-building [see *soil-depleting* adj. below]. 1902 P. McConnell *Elem. Agric. Geol.* iv. 122 (*heading*) Soil characteristics. 1954 W. D. Thornbury *Princ. Geomorphol.* iv. 73 No geomorphologist today is adequately trained who lacks an appreciation of the soil-forming processes and a basic understanding of soil characteristics. 1923 *Soil Sci.* XVI. 95 On the basis of this concept of soils and soil classification, field and laboratory studies of soils in Michigan have been undertaken. 1946 L. D. Stamp *Britain's Struct. & Scenery* xi. 92 The basis of the soil classification used for these maps is a textural one:..broadly the purpose was to separate sands, light, medium and heavy loams, clays and peats. 1972 J. G. Cruickshank *Soil Geogr.* i. 23 Senior members of the national soil surveys such as G. W. Robinson (1932) and B. W. Avery (1956) have been responsible for soil classifications which developed from the work of soil survey in Britain. 1933 *Engin. News-Record* 31 Aug. 245/2 The basic principles of soil compaction..apply to all types of earthfills and to foundation design. 1971 *Power Farming* Mar. 80/1 (Advt.), The Salo [harrow] produces a fine, level, shallow bed with only one or two tractor passes. Soil compaction is reduced to the minimum. 1905 *Jrnl. Agric. Sci.* I. 78 The clover crop feels the effect of the changed soil conditions to a much greater extent. 1966 Soil condition [see *soil survey*, sense *10 below]. 1964 W. C. Putnam *Geology* x. 249/1 Soil cover serves to alleviate the starkness of a rock-dominated landscape. 1938 *Sun* (Baltimore) 14 Sept. 4/8 Payments will be made for keeping within soil-depleting acreage allotments and for attaining soil-building goals. 1962 *Times* 12 Oct. (Stand. Bank Suppl.) p. vii/5 The ability to overcome..problems by applying research findings such as..soil-building rotations as opposed to soil-depleting rotations and harmful practices such as monoculture. 1925 Soil depletion [see *mosaic *a.*[1] 6]. 1921 *Proc. 41st Ann. Meeting Soc. Promotion Agric. Sci., 1920* (U.S.) 118 A great deal of fundamental work has been done in Russia. It has been concerned..with the working out of the principles and the formulation of the laws of soil development. 1972 J. G. Cruickshank *Soil Geogr.* ii. 57 The origin of parent materials is not significant for soil development except as an indication of the soil properties that may be expected. 1946 L. D. Stamp *Britain's Struct. & Scenery* xi. 100 The relationship between vegetation cover and soil drainage is far from being sufficiently realised. 1970 Gay & Calaby in Krishna & Weesner *Biol. Termites* II. ix. 440 Soil-dwelling colonies commonly build radiating gallery systems on the soil surface to adjacent grass tussocks. 1901 *Proc. 22nd Ann. Meeting Soc. Promotion Agric. Sci.* (U.S.) 62 The subject of humus in its relation to soil fertility. 1968 R. W. Fairbridge *Encycl. Geomorphol.* 1235/2 Deposition of fine dust (obvious in the case of thick loess) has occurred in very thin mantles..over broad areas. The latter is a vastly underrated factor in the maintenance of soil fertility in wide regions. 1912 *Bull. Bur. Soils U.S. Dept. Agric.* No. 85. 14 The most important agency of soil formation is moisture. 1963 [see *pedogenesis]. 1967 M. J. Coe *Ecol. Alpine Zone Mt. Kenya* 69 The processes of soil formation are also very evident. 1902 P. McConnell *Elem. Agric. Geol.* iii. 66 (*heading*) List of the principal soil-forming minerals. 1936 [see *pedogenesis]. 1972 J. G. Cruickshank *Soil Geogr.* ii. 34 Climate was regarded as the principal and dominant soil-forming factor for the greater part of the short history of pedological thought. 1927 C. F. Marbut in tr. *Glinka's Great Soil Groups* p. i, The development of the first and only comprehensive theory of soil genesis. 1946 S. A. Wilde *Forest Soils & Forest Growth* iii. 20 An essential factor of soil genesis, the composition of vegetative cover. 1972 Soil genesis [see *pedocal]. 1927 C. F. Marbut tr. *Glinka's Great Soil Groups* 7 The study of the soil geography of North America has..enforced the recognition of the close relationship between the soil and climatic conditions. 1972 J. G. Cruickshank *Soil Geogr.* i. 30 Pedology is, by this definition, very close to soil geography except that the latter is concerned with all kinds of distributions involving soil, from those of natural genesis to limitations for soil cultivation. 1907 J. R. Kilroe (*title*) A description of the soil-geology of Ireland, based upon Geological Survey maps and records. 1972 J. G. Cruickshank *Soil Geogr.* i. 16 A further distinction differentiating soil from weathered rock was subsequently made.., but nevertheless the view of soil science as soil geology prevailed for the rest of the nineteenth century in Western Europe and America. 1939 Melhus & Kent *Elem. Plant Path.* vii. 97 Mercuric chloride in dilute solution has been used in the control of..certain soil-inhabiting pathogens, etc. 1969 W. L. Nutting in Krishna & Weesner *Biol. Termites* I. viii. 274 There is even less information on the wood- and soil-inhabiting Hodotermitidae. 1911 *Encycl. Brit.* XXV. 351/2 The general evidence indicates that the specific bacteria of cholera discharges are capable of a much longer existence in the superficial soil layers than was formerly supposed. 1964 W. C. Putnam *Geology* x. 249/2 Soil layers and particles may be lifted up by the expansion of freezing water. 1968 Soil layer [see *pedosphere]. 1909 *Bull. Bur. Soils U.S. Dept. Agric.* No. 55. 26 This is borne out by the experience of farmers, who testify as to differences in soil management. 1979 W. L. Pritchett *Properties & Managem. Forest Soils* p. v, Significant advances have been made in silviculture, especially in reforestation technology and soil management of short rotation forests for fiber production. 1912 *Bull. Bur. Soils U.S. Dept. Agric.* No. 85. 23 Numerous kinds of rocks or soil material, subjected to the action of many agencies and processes,..have resulted in the formation of many varieties or types of soil. 1971 A. R. Jumikis *Foundation Engin.* vii. 179 The soil materials to use for building earth cofferdams are sandy clay and clayey sand. 1925 *Soil Sci.* XIX. 201 Agricultural practice has hardly been modified as a result of the development of soil microbiology. 1972 J. G. Cruickshank *Soil Geogr.* i. 15 We are not concerned with the foundation or the history of soil chemistry, soil physics, soil microbiology, soil mineralogy, and other member parts of soil science, but rather with the inception and growth of pedology within the last century. 1916 *Soil Sci.* I. 99 The study of soil microörganisms has attracted the attention of many investigators. 1972 J. G. Cruickshank *Soil Geogr.* 170 Easily attacked by a wide variety of soil micro-organisms are substances like protein, sugars, and pectins. 1913 *Phil. Trans. R. Soc.* B. CCIV. 181 The soil solution may not be of constant concentration, because the soil minerals may not be so similar as is supposed, especially after the application of fertilizers. 1980 *Amateur Gardening* 4 Oct. 23/3 Grass needs potassium.., but it is extraordinarily efficient about extracting it from naturally-occuring soil minerals. 1926 *Phytopathology* XVI. 582 Soil temperature and soil moisture were believed by many writers to be responsible in part for the variation in potato mosaic symptoms. 1980 *Amateur Gardening* 4 Oct. 23/1 Nitrogen..applied in the form of a fertiliser dissolves in the soil moisture and is very rapidly lost. 1926 *Phytopathology* XVI. 583 Some attempt has been made to modify the symptoms of mosaic by varying the soil nutrients. 1915 T. L. Lyon et al. *Soils* viii. 126 The source of practically all soil organic matter is plant tissue. 1971 *Gloss. Soil Sci. Terms* (Soil Sci. Soc. Amer.) 17/1 *Soil organic matter*, the organic fraction of the soil; includes plant and animal residues at various stages of decomposition, cells and tissues of soil organisms, and substances synthesized by the soil population. 1979 W. L. Prichett *Properties & Managem. Forest Soils* xvi. 290 While soil organic matter can be increased by use of green manure crops and the additions of composts.., such increases are temporary due to the decomposition of these materials by soil organisms. 1901 H. M. Ward *Dis. Plants* xv. 143 Cuttings..stuck into ordinary soil in dirty boxes covered with equally dirty glass, present every chance for infection by soil organisms. 1967 M. J. Coe *Ecol. Alpine Zone Mt. Kenya* 70 On account of the low temperature..and the consequent puacity of soil organisms, there is a marked inhibition of the chemical breakdown of parent materials. 1900 R. Warington *Lect. Physical Properties Soil* i. 11 Any group of particles obtained by subsidence will not be entirely of the same size in cases where the soil particles consist of substances having different specific gravities. 1914 [see *crumb *sb.* 1 c]. 1964 W. C. Putnam *Geology* x. 245/1 The C-horizon..is a mixture in varying proportions of altered and unaltered rock fragments and soil particles. 1927 Soil population [see *edaphon]. 1971 *Gloss. Soil Sci. Terms* (Soil Sci. Soc. Amer.) 17/1 Soil pores. 1976 *Physics Bull.* Aug. 342/3 Marshall's equation.. implies a certain connectivity of soil pores which may not always be justified. 1946 *Nature* 2 Nov. 605/1 A combination of cereal agriculture and tree-fruit crops, with subsidiary pasturage, hunting, and fishing, as an approximately stable regime,..presumes a cycle of soil-restoration. 1962 E. Snow *Red China Today* (1963) lxvii. 508 Many of these toy dams are already used for local power, irrigation and soil-restorative purposes. 1910 *Chambers's Jrnl.* Aug. 560/1 The new alfalfa..is expected to yield an ideal forage and act as a soil-restorer. 1913 L. C. Corbett *Garden Farming* ii. 23 Soil sterilization has for its direct object the treatment of soil in such a way as to render it free from injurious enemies. 1923 W. F. Bewley *Dis. Glasshouse Plants* viii. 154 The practice of soil sterilization is now an accepted method of increasing the fertility of infertile soils. 1920 W. W. Weir *Productive Soils* ii. 13 Texture should not be confused with soil 'structure' which means the arrangements of the soil grains..or..the relation of the soil particles to each other. 1954 W. D. Thornbury *Princ. Geomorphol.* iv. 87 Not all soil structures are..solifluction features, for downslope movement may be either lacking or of minor importance. 1976 *Physics Bull.* Aug. 343/3 This amount of tillage..can also be harmful to the stability of the soil structure. 1927 C. F. Marbut tr. *Glinka's Great Soil Groups* 10 Natural exposures..can be utilized as aids to soil study but artificial exposures, such as dry wells,..are better. 1972 Soil study [see *soil profile*, sense *10 below]. 1923 W. F. Bewley *Dis. Glasshouse Plants* ii. 37 Investigations upon the *Verticillium* wilt of tomatoes..illustrate the importance of air and soil temperatures in conditioning the progress of disease. 1976 L. F. Curtis et al. *Soils in Brit. Isles* xii. 221 Another effect of good drainage is that it allows the soil temperature to rise more quickly in the spring. 1926 *Public Roads* VII. 153 (*heading*) Simplified soil tests for subgrades and their physical significance. 1978 Friedman & Sanders *Princ. Sedimentol.* xiii. 417/1 It suffices for most engineering purposes to put the samples from a soil-test boring into small jars. 1934 *Proc. Amer. Soc. Testing Materials* XXXIV. ii. 693 (*heading*) Subgrade soil testing methods. 1979 S. Smith *Survivor* vi. 74 A local horticulturist giving a demonstration of soil testing. 1980 *Amateur Gardening* 4 Oct. 23/3 Home soil-testing kits can be purchased at quite reasonable prices at garden centres. 1912 R. L. Watts *Vegetable Gardening* iii. 25 (*heading*) Soil texture. 1971 *Arable Farmer* Feb. 62/1 Soil texture is an important factor in determining the equilibrium of organic matter level. 1938 C. P. Quarrell *Intensive Salad Production* vi. 72 Before undertaking any system of electrical soil warming the grower should consult the cable manufacturers. 1954 A. G. L. Hellyer *Encycl. Garden Work* 97/1 Electric soil-warming cables may be buried in the soil.

b. *attrib.* in *pl.*

1925 P. Emerson *Soil Characteristics* i. 22 The soils student should become acquainted with the common soil-forming minerals and rocks. 1945 P. Work *Vegetable Production & Marketing* x. 164 See soils textbooks for discussions of the principles and practices of land drainage. 1969 *Civil Engin.* June 43/2 The stratum, our soils consultant recommended, could be used for safe bearing pressures of 1,200 and 1,800 psf, for dead load and total load respectively. 1973 [see *pedology].

10. soil air, air present in the soil; **soil amendment,** a substance added to the soil to improve its properties, esp. its physical properties; also, the use of such substances; **soil analysis,** the scientific investigation of the composition and structure of soil or soil samples; **soil association,** a group of soils that are related geographically or topographically, esp. ones derived from a common parent material; **soil auger,** a rotary tool (either powered or operated manually) for boring into or taking samples of soil; **soil bank,** (*a*) land taken out of use for agricultural production (? *temporary*); (*b*) the soil as a continuing store of seeds, pathogens, nutrients, etc.; **soil biology,** the study of soil organisms and their life; **soil catena:** see *catena c; **soil-cement** *a.* and (also without hyphen) *sb.*, (material) composed of soil or a soil substitute that has been strengthened and stabilized by the admixture of cement; **soil chemistry,** the branch of soil science concerned with the chemical properties and reactions of soil; so **soil chemist; soil class,** a group of soils similar to one another in texture or (in mod. use) some other physical property; **soil climate,** the prevailing physical conditions in the soil, esp. as they affect soil organisms and plant life; **soil colloid,** a substance present in the soil as a colloid, i.e. in the form of very small particles; **soil conditioner,** a substance added to the soil to improve its physical characteristics, *esp.* one made synthetically for the purpose; **soil conservation,** the protection and safeguarding of the soil against erosion, loss of fertility, and damage; **soil deficiency,** an insufficiency in the soil of some substance necessary for the proper growth of plants; **soil erosion,** the removal of soil by the action of wind or running water; **soil exhaustion,** the disappearance of fertility from the soil; **soil extract** (see quot. 1971); **soil group,** a group of soils; *spec.* in *Soil Sci.* (also *great soil group*), each of the relatively small number of groups into which the world's soils are divided on the basis of their profiles and the climate in which they exist; **soil horizon** = *horizon *sb.* 5 b; **soil mantle,** the soil as a covering of the underlying rock; **soil map,** a map showing the location and nature of the various kinds of soil in a region; so **soil mapping** *vbl. sb.*; **soil mark** *Archæol.*, a trace of a levelled or buried feature indicated by differences in the colour or texture of the soil, usu. on ploughed land; **soil mechanics,** the science concerned with the mechanical properties and behaviour of soil as they affect its use in civil engineering; **soil phase,** each of a number of soils that belong to the same soil type or soil series but differ in some feature such as stoniness, slope, etc.; **soil physics** (see quot. 1976); hence **soil physicist; soil polygon** = *polygon *sb.* 2 b; **soil profile** = *profile *sb.*[1] 4 d; **soil province:** see *province 6 d; **soil resistivity,** the electrical resistivity of the soil; usu. *attrib.*; **soil sample,** a sample of soil taken for scientific investigation; **soil sampler,** any device for taking soil samples; so **soil sampling** *vbl. sb.*; **soil science** = *pedology; so **soil scientist** = *pedologist; **soil separate,** a separate (sense *6) obtained from soil; **soil series,** a group of soils similar in profile, origin, and other characteristics but varying in the texture of the surface horizon; **soil sickness,** a condition of soil in

which it has become unable to support the healthy growth of a crop; so **soil-sick** *a.* (*rare*); **soil solution,** the water present around and between soil particles as a dilute solution of mineral salts; **soil stabilization,** the treatment of soil to give it increased resistance to movement, esp. under load, and erosion; **soil stripe** *Geomorphol.*, one of the low ridges of stony soil which occur in cold environments and form parallel, evenly spaced lines; **soil survey,** a systematic examination and mapping of the different kinds of soil present in a region or on a site; a report of the results so obtained; a body of people engaged in such work; so **soil surveyor; soil type,** a particular kind of soil; *spec.* in *Soil Sci.*, a subdivision of a soil series made according to the texture of the surface horizon, and representing the lowest unit in the system of classification; (see also quot. 1928); **soil wash,** the movement of soil by ground water; **soil water,** the water present in soil.

1920 *Mem. Cornell Univ. Agric. Exper. Station* No. 32. 326 Before seeding, some preliminary studies were made in order to ascertain the best method of obtaining the sample of soil air for analysis. 1972 J. G. Cruickshank *Soil Geogr.* iii. 81 Differences between the composition of soil air and atmospheric air become greater with depth.. provided organisms remain present. 1915 T. L. Lyon et al. *Soils* xxiv. 542 Gypsum..was a popular soil amendment in this country before the common commercial fertilizers were used to any great extent. 1967 *Boston Sunday Globe* 28 Apr. b. 67/4 Whenever the garden has to be in a new housing development, liming is particularly needed and all the other additions of manure, peat and fertilizer. This is now called 'soil amendment' by the more technical. 1978 R. C. Oelhaf *Organic Agric.* iii. 37 Many 'organic' soil amendments are now on the market which are mainly crushed rock, selling at prices as high as 200 times the price of the ingredients. 1873 *Amer. Jrnl. Sci.* CVI. 289 In soil analysis special importance attaches to these finer sediments. 1891 R. Wallace *Rural Econ. Austral. & N.Z.* x. 169 No analyst, using the ordinary processes for soil analysis, can determine whether or not such infinitesimal amounts [of minerals] as are required by the crop are present or are not present in an available form in a soil. 1946 R. J. C. Atkinson *Field Archaeol.* ii. 62 Another technique which is becoming increasingly valuable to the excavator is that of soil-analysis. 1939 *Yearbk. Agric. 1938* (U.S. Dept. Agric.) 1163 *Soil association*, group of soils, with or without common characteristics, geographically associated in an individual pattern. 1952 L. M. Thompson *Soils & Soil Fertility* vi. 87 The most important grouping of series, in so far as the farmer is concerned, is that of the soil association. 1970 E. M. Bridges *World Soils* v. 34/1 The Scottish soil scientists have grouped topographically related soils developed on one geological parent material into a soil association. 1927 E. L. Worthen *Farm Soils* vi. 224 A soil auger, if available, should be used instead of a spade for sampling both surface soil and subsoil. 1975 *Sci. Amer.* May 93/1 The oak-hickory-tulip stand and the bigtooth aspen stand are on coarse, well-drained soil, which is aerated to as great a depth as I can reach with a two-meter soil auger. 1955 *Sun* (Baltimore) 26 Nov. 8/2 'Soil bank' is the current farm bloc slang for a scheme by which farmers are paid by the Government for taking acreage out of the production of surplus crops. 1958 J. K. Galbraith *Affluent Society* xx. 221 Wherever possible euphemisms were employed—as this is written, instead of taking acres out of production they are being put into a soil bank. 1977 J. L. Harper *Population Biol. Plants* iv. 95 In a sense there is a circular argument here; species which have pioneered the succession are strongly persistent in the soil-bank and so appear as pioneer species in the next succession on the area. Do they persist because they are pioneers or do they become pioneers because they have persisted. 1928 *Proc. & Papers 1st Internat. Congr. Soil Sci.* III. 325 Soil biology is essentially a science of observation and experimentation. 1967 Burges & Raw *Soil Biol.* p. vi, One of the stimulating developments in soil biology in recent years has been the general recognition that the soil cannot be studied solely from a chemical, microbiological, botanical or zoological stand-point. 1936 *Proc. Highway Res. Board* (U.S.) XVI. 324/2 Tests were conducted to determine the moisture-density relations of the raw soil and the soil-cement mixtures. *Ibid.* 348/2 Would the mixtures of soil-cement when compacted at optimum moisture to maximum density, maintain these characteristics under natural weathering conditions. 1950 *N.Z. Jrnl. Agric.* May 481/1 Soil cement (silty material mixed with cement) ..gives a building material which has a definite structural strength, besides being resistant to the action of water and frost. 1966 R. Ashworth *Highway Engin.* ix. 161 By far the greatest proportion of soil-cement construction has been carried out in the U.S.A. 1979 R. J. Salter *Highway Design & Constr.* i. 28 Soil cement may be formed by the addition of cement to a wide range of materials, including natural soils, chalk, pulverised fuel ash..and processed granular material. 1927 N. M. Comber *Introd. Scientific Study Soil* xiii. 130 Soil chemists and agriculturists frequently speak about soil 'types', and yet the definition of the various types is a matter which presents very considerable difficulty. 1959 J. D. Clark *Prehist. Southern Africa* ii. 36 It should be possible..to amass information concerning the main vegetation patterns..and for this we need the help of the..soil chemist. 1971 *Power Farming* Mar. 34/1 That handbook..is a chemical engineer's handbook and a 'natural' for all contractor services. Soil chemists have played only a minor role in assembling it. 1927 C. F. Marbut in tr. *Glinka's Great Soil Groups* p. ii, In the more detailed study of the soil profile, the clearer recognition of the nature of soil horizons, soil structures, soil colors and in the relation of soil chemistry to the processes of soil development in Nature, this book will be of great suggestive value. 1941 J. S. Huxley *Uniqueness of Man* vi. 103 What began as a study of local cattle diseases has turned into a problem of the soil chemistry of grasslands. 1972 J. G. Cruickshank *Soil Geogr.* i. 15 In this discussion we are not concerned with the foundation or the history of soil chemistry, soil physics,..and other member parts of soil science, but rather with the inception and growth of pedology within the last century. 1913 *Bull. Bur. Soils U.S. Dept. Agric.* No. 96. 8 A soil class..includes all soils having the same texture, such as sands, clays, loam, etc. 1951 *Soil Survey Man.* (U.S. Dept. Agric. Handbk. No. 18) 135 Soil class is observed in the field by feeling the soil with the fingers. 1900 R. Warington *Lect. Physical Properties Soil* p. xii, If seeds are to germinate in a soil,.. there must be a suitable soil climate. 1976 A. Young *Tropical Soils & Soil Survey* i. 7 The factor which directly influences soil-forming processes is soil climate rather than air climate. 1915 *Chem. Abstr.* IX. 1084 R. discusses the importance of soil colloids for agriculture. 1935 *Nature* 24 Aug. 307/2 Much attention was directed.. towards the base-exchange properties of soil colloids, particularly from the mineralogical point of view. 1970 J. A. Daji *Textbk. Soil Sci.* xiii. 120 Soil colloids are of two kinds: (1) inorganic and (2) organic... The organic colloid..is more commonly known as humus. 1952 *Sci. News Let.* 5 Jan. 8/2 The new soil conditioner changes the structure of clay making it porous and crumbly. 1976 L. F. Curtis et al. *Soils in Brit. Isles* xv. 285 Soil conditioners may be applied to add stability. 1978 Friedman & Sanders *Princ. Sedimentol.* v. 141/2 Zeolites..are mined from sedimentary deposits for use as fillers in the paper industry; as soil conditioners; [etc.]. 1932 *Yearbk. U.S. Dept. Agric.* 349 The national plan for soil and water conservation calls for the establishment of experiment stations. 1935 *U.S. Laws, Statutes* XLIX. i. 164 The Secretary of Agriculture shall establish an agency to be known as the 'Soil Conservation Service'. 1944 Auden *For Time Being* (1945) 90 The Committees on Fen-Drainage and Soil-Conservation. 1952 W. L. Miner *World of W. Faulkner* ii. 61 Since 1933 the various soil conservation programs..have done much for Lafayette county. 1971 *E. Afr. Standard* (Nairobi) 13 Apr. 2/1 The committee stressed that unlike in the colonial era, farmers in the rural areas should now take great pains in soil conservation. 1925 J. F. Cox *Crop Production & Soil Managem.* vii. 116 (*heading*) The elements of fertility, common soil deficiencies, and fertilizers carrying nitrogen, phosphorus and potassium. 1935 *Discovery* Oct. 294/1 Non-parasitic diseases of plants, due principally to soil deficiencies. 1896 *Nat. Geogr. Mag.* Nov. 368 (*heading*) The economic aspects of soil erosion. 1944 J. S. Huxley *Living in Revolution* iii. 30 It neglected conservation and amenities: the result was deforestation, soil erosion, the dust bowl. 1980 *Sci. Amer.* Sept. 114/2 Major problems related to land use, soil erosion and water pollution are likely to place further limits on the recovery of these nonconventional oil resources. 1920 W. W. Weir *Productive Soils* vii. 84 In some instances soil exhaustion may be attributed largely to the removal, mainly through cropping and leaching, of some one or all of the three named elements. 1934 A. Toynbee *Study of Hist.* I. 126 As regards the possibility of soil-exhaustion, an observation of latter-day native agriculture in the area..seems to show that a repeated clearing and burning-off of the tropical forest..does tend..to exhaust the soil. 1946 J. S. Huxley *Unesco* ii. 28 It is possible to exploit new agricultural methods in a way that is..disastrous to agriculture itself, by causing soil exhaustion or erosion. 1957 G. E. Hutchinson *Treat. Limnol.* I. xvii. 896 The vitamins and accessory growth substances in soils and soil extracts. 1971 *Gloss. Soil Sci. Terms* (Soil Sci. Soc. Amer.) 15/2 *Soil extract*, the solution separated from a soil suspension or from a soil by filtration, centrifugation, suction, or pressure. 1921 Soil group [see *soil surveyor* below]. 1927 C. F. Marbut tr. *Glinka's Great Soil Groups* p. iii, Some of the great soil groups have not been studied by Russian investigators. 1954 W. D. Thornbury *Princ. Geomorphol.* iv. 76 Mature and old soils in areas that are climatically alike are strikingly similar, and it is possible to classify them in soil groups that developed under similar climatic conditions. 1972 J. G. Cruickshank *Soil Geogr.* iv. 110 There is only limited regional association in the world distribution of type profiles or great soil groups. 1976 A. Young *Tropical Soils & Soil Survey* xiii. 241 Although many of the soil groups [of the FAO classification of 1974] are natural soil types, this is structurally an artificial classification. 1923 *Soil Sci.* XVI. 97 The relative amount of silica in the gray horizon appears to be higher than in the soil horizons below. 1964 W. C. Putnam *Geology* x. 245/1 The C-horizon is essentially a transitional zone between the true soil horizons above and the unaltered parent material below. 1972 J. G. Cruickshank *Soil Geogr.* iii. 93 All soil horizons have a three-dimensional form, but those that have a clearly visible colour and texture..are perhaps the most convincing examples. 1961 *Listener* 12 Oct. 559/1 The changes that it [*sc.* soil science] recognizes in soil mantles and geological solids are termed 'weathering'. 1972 J. G. Cruickshank *Soil Geogr.* i. 28 Although the soil body or soil mantle obviously has a three-dimensional form, it has been represented traditionally by a so-called 'two-dimensional' section or slice called the soil profile. 1898 *Yearbk. U.S. Dept. Agric.* 43 One of the first necessities in the development of a new district or in the improvement of an established district is an accurate soil map of the locality. 1927 N. M. Comber *Introd. Scientific Study Soil* xiii. 132 Two important bases of classification..have been invoked in the attempts to prepare soil maps of agricultural significance. 1975 J. G. Evans *Environment Early Man Brit. Isles* vi. 138 W. F. Grimes was one of the first archaeologists to appreciate the importance of detailed soil maps in studying the settlement distribution of early man. 1920 W. W. Weir *Productive Soils* ii. 22 (*heading*) Soil mapping. 1928 *Proc. & Papers 1st Internat. Congr. Soil Sci.* IV. 34 In the soil survey of the United States the term Soil Type has been applied to the unit of soil mapping. 1972 J. G. Cruickshank *Soil Geogr.* i. 24 These men and many others, who have been responsible for selective soil mapping in countries like France..and Yugoslavia, have a history of soil research behind them going back to the nineteenth century. 1939 G. Clark *Archæol. & Society* ii. 38 In chalk regions subjected to heavy ploughing, soil-marks, especially when seen from the air, preserve the sites of ancient monuments. 1950 *Oxoniensia* XV. 7 The best results of an air-survey of Celtic field-systems may be expected from photographs taken during the winter months... Soil-marks..will be more evident. 1963 E. S. Wood *Collins Field Guide to Archæol.* iii. i. 284 Another type of mark is the soil-mark. When earthworks or barrows are levelled, or when grass is stripped, or on bare (ploughed) land, differences in soil-colour become apparent. 1920 *Engin. News-Record* 30 Sept. 630/1 (*heading*) Research in soil mechanics. 1965 A. B. Carson *Foundation Constr.* iii. 70/1 Despite the relative youth of the science of soil mechanics, the literature on the subject is extensive, particularly that relating the foundation structure to the..soil or rock formation upon which it will be built. 1977 A. Hallam *Planet Earth* 104 The engineering geologist works with experts in the related fields of soil mechanics and rock mechanics. [1928 C. F. Marbut in *Proc. & Papers 1st Internat. Congr. Soil Sci.* IV. 51 *Phase*, a subdivision of the soil type covering departures from the typical soil characteristics, insufficient to justify the establishment of a new type, yet worthy of recognition.] 1939 *Yearbk. Agric. 1938* (U.S. Dept. Agric.) 1174 Soil phase. 1946 Lutz & Chandler *Forest Soils* xii. 424 Soil phases as currently recognized appear to have more significance for the agriculturist than for the forester. 1972 J. G. Cruickshank *Soil Geogr.* viii. 235 Soil phase..can only be shown in detail and with precision on maps of 1:10,000 scale or larger. 1937 C. A. Hogentogler et al. *Engin. Properties Soil* p. vii, Publications by the soil scientist, the soil physicist, the agronomist, the pedologist, and the geologist have been drawn upon. 1976 *Physics Bull.* Aug. 341/2 Soil physicists around the world are researching into an incredibly diverse range of phenomena, of which the diffusion of gases to and from plant roots..and the break-up of soil by tillage implements are just a few examples. 1900 R. Warington *Lect. Physical Properties Soil* p. xi, The only early investigation on soil physics is that of Schübler, made more than sixty years ago. 1935 *Nature* 24 Aug. 307/1 In the Soil Physics Section, the main interest centred round problems of soil moisture. 1972 [see *soil chemistry* above]. 1976 *Physics Bull.* Aug. 341/2 Soil physics is the branch of soil science that is concerned with the physical constitution and geometrical structure of soil, with the potentials and movements of water, gases and heat in soil, and with the deformation of soil in response to mechanical stress. 1927 *Q. Jrnl. Geol. Soc.* LXXXIII. 190 Soil-polygons are divided into (*a*) 'mud-polygons', containing few or no stones, and (*b*) 'stone-polygons', in which stones are arranged in curious patterns over the surface of the mud. 1963 D. W. & E. E. Humphries tr. *Termier's Erosion & Sedimentation* iv. 86 The periglacial zones are equally rich in detrital material and display phenomena completely comparable with those of hot deserts: loess, reg, soil polygons, 'dreikanters' and dunes. 1967 M. J. Coe *Ecol. Alpine Zone Mt. Kenya* 71 On ridge tops, which are usually scattered with boulders and small stones, soil polygons are particularly common. 1906, 1923 Soil profile [see *profile *sb.* 4 d]. 1928 *Forestry* II. 15 A natural basis for the classification of soils..resulted from a study of what have come to be known as soil profiles, vertical exposures of sections of soil down to the unaltered parent rock. 1954 W. D. Thornbury *Princ. Geomorphol.* iv. 76 A mature soil profile exhibits well-developed horizons. 1972 J. G. Cruickshank *Soil Geogr.* i. 28 Sometimes the soil profile is accepted as the basic unit of soil study. 1940 C. A. Heiland *Geophysical Explor.* x. 646 As an example of the galvanic application of intermediate frequency methods, the soil resistivity bridge..is illustrated. 1964 R. F. Ficchi *Electrical Interference* viii. 153 Probably the biggest stumbling block in such analytical calculations is the vaguely defined soil-resistivity measurements. 1967 *Gloss. Terms Gas Industry* (B.S.I.) 68 *Soil resistivity survey*, the determination of the electrical resistivity of the soil at intervals along the route of a main to assist in designing a cathodic protection system. 1902 *Bull. U.S. Fish Commission 1901* XXI. 58 For taking soil samples an instrument was made after drawings in Delbecque. 1975 *New Yorker* 28 Apr. 112/2 District-level officials are now collecting soil samples, so that in the future they can advise the co-ops on the most productive way to use their land. 1902 Soil sampler [see *sampler *sb.*[2] 2]. 1950 *N.Z. Jrnl. Agric.* June 553/3 Holes may be dug with a spade, sunk with a post-hole borer, or cored with a soil sampler (in order to examine the structure of the soil). 1927 E. L. Worthern *Farm Soils* 409/1 (Index), Soil sampling. 1958 J. Blish *Case of Conscience* (1959) iv. 41 We will start a soil-sampling programme. 1960 *Farmer & Stockbreeder* 9 Feb. 97/3 Soil-sampling for ectoparasitic eel-worms may be worth a thought. 1915 *Chem. Abstr.* IX. 1084 (*heading*) The development of soil science from the earliest attempts to the beginning of the twentieth century. 1916 (*periodical title*) Soil science. 1935 *Nature* 24 Aug. 308/1 The very large number of papers dealing with the practical side of soil science..emphasised the dominating influence exerted on the development of soil science by modern economic conditions. 1938, 1958 [see *pedology]. 1972 *Nature* 28 Jan. 231/2 The most characteristic Russian contribution to science was soil science. 1979 W. L. Pritchett *Properties & Managem. Forest Soils* p. v, Most of the basic principles of soil science apply to forest soils as well as to agricultural soils. 1921 *Proc. 41st Ann. Meeting Soc. Promotion Agric. Sci., 1920* (U.S.) 117 The soil scientist must be concerned primarily with the accumulation or assimilation of knowledge concerning the soil without reference to the use to be made of that knowledge. 1958 *Times* 1 July (Agric. Suppl.) p. ii/2 The plant breeder and the soil scientist have worked hand in hand; together they are responsible for disproving the gloomy prophecies of increasing world hunger so commonly heard only 10 years ago. 1928 *Proc. & Papers 1st Internat. Congr. Soil Sci.* IV. 34 *Soil separate*, one of the several grain-size groups into which the soil is separated by mechanical analysis. 1951 *Soil Survey Man.* (U.S. Dept. Agric. Handbk. No. 18) 207 (*heading*) Size limits of soil separates from two schemes of analysis. 1972 J. G. Cruickshank *Soil Geogr.* ii. 55 These categories of particle size—sometimes called the soil separates—are mixed in any soil into what is called its texture. 1905 *Field Operations of U.S. Bur. Soils, 1904* 35 Whenever there is a general relationship between these two classes of soils, due either to their geological origin, their method of formation, or their location within an area, a common distinctive locality name is used, and the soils thus grouped together are called a soil series. 1946 L. D. Stamp *Britain's Struct. & Scenery* xi. 95 Within each soil series there may be a considerable range of texture which is important ecologically. 1972 J. G. Cruickshank *Soil Geogr.* i. 29 A soil series is a composite unit, but being the basic unit of soil mapping it is expected to be predominantly composed of one named soil profile type and confined to one parent material. 1962 *Listener* 25 Jan. 174/2 The ground beyond the filter-bed is what we call soil-sick. 1934 Webster, Soil sickness. 1938 *Encycl. Brit. Bk. of Year* 111/1 Among more recent concepts is that of the possibility of beneficial root-excretions, to which the older view ascribed toxic properties and the responsibility for soil-sickness. 1960 *Farmer & Stockbreeder* 9 Feb. 97/1 We have long been familiar with the potato root and sugar-beet eelworm..but other types are now known to cause 'soil sickness'. 1901 *Bull. Div. Soils U.S. Dept. Agric.* No. 17. 5 Soil solutions from which plants draw their food are for the most part aqueous solutions of the mineral components of the soil. 1957 G. E. Hutchinson *Treat. Limnol.* I. viii. 556 The over-all composition of soil solutions is in fact very similar to that of average river water. 1973 *Sci. Amer.* May 48/2 Perhaps 5 percent of a plant's dry weight is minerals. Eight elements account for the bulk of this amount... All are normally present in the 'soil solution', as the water of the soil is called, at very low concentrations. 1934 *Proc. Amer. Soc. Testing Materials* XXXIV. ii. 737 Investigations along the line suggested by Mr. Housel are also considered in our soil stabilization work. 1950 *Engineering* 13 Jan. 44/3 They were used in conjunction with processes of soil stabilisation in which the stability of the soil under traffic load is improved by adding clay, sand or gravel. 1969 Capper & Cassie *Mech. Engin. Soils* (ed. 5) xi. 234 An important method of soil stabilization is by the use of resins. 1910 *12th Rep. Michigan Acad. Sci.* 52 A comparison of the Asulkan and Greenland soil stripes with the great barrancas, suggests that the width of ridge..is in some way a function of the viscosity of the rock paste. 1954 W. D. Thornbury *Princ. Geomorphol.* iv. 89 Earth stripes or soil stripes are similar to stone stripes except that they have finer textures. 1900 *Yearbk. U.S. Dept. Agric. 1899* 26 A detailed soil survey has been undertaken of the soils of Maryland. 1924 [see *separate *sb.* 6]. 1966 R. Ashworth *Highway Engin.* iii. 49 The soil survey involves an exploration of the soil conditions along the proposed road alignment by means of boreholes or trial pits. 1972 J. G. Cruickshank *Soil Geogr.* i. 23 Some of the later private surveys have..introduced soil terms and definitions later adopted by the national soil surveys. 1921 *Proc. 41st Ann. Meeting Soc. Promotion Agric. Sci., 1920* (U.S.) 119 Before the soil surveyor had mapped textural soil units over any considerable area he discovered that these units are not all alike, that they are not ultimate soil units therefore but soil groups. 1902 *Instructions to Field Parties & Descr. Soil Types* (U.S. Bur. of Soils) 15 The selection of a provisional name for each soil type should be made, and in all correspondence and reports this name should be used when speaking of the type. 1928 *Proc. & Papers 1st Internat. Congr. Soil Sci.* IV. 41 The soil type is a subdivision of the soil series based primarily and almost wholly on the texture of the surface soil... The term Soil Type has been used by some writers with a more inclusive meaning, sometimes to indicate the general characteristics of the soils of a region. 1954 W. D. Thornbury *Princ. Geomorphol.* iv. 78 A common soil series in the middle western states is the Miami series... Included in this series are such soil types as the Miami fine sandy loam, Miami loam, Miami silt loam, and Miami silty clay loam. 1967 M. J. Coe *Ecol. Alpine Zone Mt. Kenya* 71 Soil movement and deformation are of great significance in governing the distribution of soil types and in their effect on vegetation. 1972 J. G. Cruickshank *Soil Geogr.* i. 26 In 1904 the soil series was introduced to include all soil types developed on the same parent material. 1926 *Sci. Amer.* Feb. 97/3 Erosion or soil wash is impoverishing our sloping farm lands. 1962 Soil wash [see *sailab]. 1892 J. M. H. Munro *Soils & Manures* i. 25 What this soil water contains we may see by examining the water running from the drain-pipes of any arable field when no crop is growing on it. 1921 *Discovery* Feb. 47/1 Plants require moisture, and in taking this up by the root-hairs, they also take up their food, consisting of salts dissolved in the soil water. 1976 *Physics Bull.* Aug. 343/3 One of the main obstacles to progress in the physics of soil water is the lack of quantitative methods for describing adequately the geometrical structure of soil at its various levels of organization.

soil, *sb.*[3] Add: **II. 3. c.** Dirt or discolouring matter on cloth.

1959 Meredith & Hearle *Physical Methods Investigation Textiles* xiv. 376 Both the soiling of textiles and the removal of dirt from them can be investigated by using soils containing radioactive materials. 1968 E. R. Trotman *Textile Scouring & Bleaching* iii. 74 It [*sc.* the material] is then scoured under controlled conditions with the detergent under investigation, and the amount of soil removed is measured. 1975 J. Labarthe *Elem. Textiles* vii. 325 Soil may be deposited on and be made to cling to some of these fabrics as the result of static electricity.

IV. 7. (Further examples.) In *techn.* use, liquid matter likely to contain excrement. Cf. *waste *sb.* 12 c.

1928 E. T. Swinson *Sanitation of Buildings* xiv. 246 Lead pipes used for soil, ventilating, and waste purposes in London must be of *drawn* lead. 1973 H. King *Do your own Home Plumbing* ix. 87 Many older houses have a two-pipe plumbing system consisting of separate waste and soil services. 1977 E. Hall *Home Plumbing* vii. 57/1 From the point of view of drainage, bidets are regarded as being 'waste', not 'soil' fittings.

9. soil-pipe, *spec.* (see quot. 1928); (further

examples); **soil-release** *a.*, causing the loosening of dirt from cloth during washing; also as *sb.*, such a substance.

1876 W. P. Buchan *Plumbing* xiv. 90 When soil-pipes and waste-pipes are put up *inside* the house, great care should be taken that they are properly fitted up. 1928 H. E. Babbitt *Plumbing* ix. 156 A soil pipe is any drainage pipe which carries human excrement... A waste pipe carries waste water which does not include human excrement. 1962 *New Statesman* 21 Dec. 897/3 Any fit man, given a certain amount of practice, can climb a soil pipe up to the first floor. 1978 T. Pettit *Home Plumbing* x. 53/2 Other waste pipes can be run into the soil pipe by means of a range of solvent-welded bossed fittings. 1969 A. J. Hall *Stand. Handbk. Textiles* (ed. 7) v. 340 To overcome such difficulties..'soil-release' agents have become available. These can be applied to the textile materials..during their production, or immediately before washing. 1969 [see *Oxford cloth* s.v. *Oxford]. 1977 D. S. Lyle *Performance Textiles* v. 219 Soil release finishes permit relatively easy removal of soils (especially oily soils) in laundering.

soil, *v.*[1] Add: **2.** Also *spec.*, of a child or patient: to make foul by defecation (esp. when involuntary). Freq. *absol.*, to defecate involuntarily. Hence **soi·ler.**

1943, etc. [implied in *soiling *vbl. sb.*[1] 1 b]. 1943 *Our Towns* (Women's Group on Public Welfare) iii. 85 The mother of the enuretic and the soiler does not teach her child..control of its natural functions. 1956 *Brit. Med. Jrnl.* 15 Dec. 1390/1 The mother or other adults show no resentment or disgust when the child soils the floor or the body of the person caring for it. *a* 1961 in Webster, s.v. [1]*soil*, Patients also showed infantile reactions..continually wetting and soiling. 1977 *New Society* 17 Feb. 333/1 When she started school she still wet and soiled by day and night.

soilage. Restrict † *Obs.* to senses in Dict. and add: **3.** The act or process of soiling; the condition of being soiled. *U.S. rare.*

1926 *Publishers' Weekly* 22 May 1679/2 One of the practical problems of retail bookselling is the depreciation of stock due to soilage.

soile: the variation also occurs in Newfoundland English (see *Dict. Newfoundland Eng.*). Cf. *swile.

soiled, *ppl. a.*[1] Add: *Comb.* (Earlier and later examples); **soiled dove** *Austral.* and *N. Amer. slang.*, a prostitute.

1882 *Sydney Slang Dict.* 8/1 Soiled doves, the 'midnight meeting' term for prostitutes and 'gay' ladies generally. 1897 'S. Grand' *Beth Book* xvi. 140 A white sheet filched from the soiled-clothes bag. 1907 *Yesterday's Shopping* (1969) 325/3 Soiled Linen Bags..Sack shape. 1929 Soiled dove [see *hustler 2 c]. 1939 M. Allingham *Mr Campion & Others* 181 He's only over here for four days and yet he's brought..a neat little soiled-linen bag embroidered with his monogram. 1962 E. Lucia *Klondike Kate* iii. 95 The line between the dance-hall girls and those of Lousetown was a thin one..because the soiled doves from across the river intermingled in the variety halls to pick up customers.

soiling, *vbl. sb.*[1] Add: **1. b.** Defecation (usu. when caused by incontinence or stress in a patient or child).

1943 *Our Towns* (Women's Group on Public Welfare) iii. 83 Some evacuated children were guilty of deliberate wetting and soiling. 1960 I. Bennett *Delinquent & Neurotic Children* iii. 113 Faecal incontinence, and soiling episodes. 1980 *Jrnl. R. Soc. Med.* LXXIII. 217 The affected children themselves are liable to behavioural problems such as temper tantrums, soiling and school refusal.

soilless, *a.*[1] Add: **a.** (Further example.)

1971 *Daily Tel.* 2 Oct. 8/3 Put each young plant in a pot of its own..using..one of the peat-based soilless mixtures.

b. Applied to methods of growing plants without soil. Cf. *hydroponics.

1938 *Sat. Even. Post* 20 Aug. 14/2 Having caught the public imagination, soilless farming has the trappings of another 'bubble'. 1946 *Soil Sci.* LXII. 71 Artificial or soilless cultures have been used very extensively during the last 30 years in plant nutrition studies. 1974 D. Harris *Hydroponics* ii. 45 Although called variously 'aggregate culture', 'soil-less culture', 'nutriculture', or 'chemiculture', Dr Gericke's term is so universally used that all forms of growing plants without soil are loosely referred to as 'hydroponics'.

soirée, *sb.* Add: (Earlier examples.) *soirée dansante*: see *dansant *a.*

1793 F. Burney *Jrnl.* 8 Apr. (1972) II. 58 He asked how my Mother did? I said if he came any *soiree*, he would probably see. 1802 C. Wilmot *Let.* 3 Jan. in T. U. Sadleir *Irish Peer on Continent* (1920) 22 We have had abundant specimens of Plays, Balls, Soirées, Thé's, &c.

‖ **soit** (swa), *int.* [Fr., third pers. sing. pres. subj. of *etre* to be.] So be it.

1889 E. Dowson *Let.* 16 Nov. (1967) 116 Your letter greatly cheered me—especially by the hope it held out of a meeting at Philippi. Soit! 1912 T. E. Lawrence *Let.* 23 June (1954) 217 It seems the Turks suffered a defeat the other day somewhere: *soit*: it won't hurt Turkey. *a* 1935 —— *Mint* (1955) 141 The R.A.F. claims to order our sitting and standing, our lying down and our going forth. *Soit*: but let its direction be extremely good. 1958 L. Durrell *Mountolive* iv. 90 Well, *soit!*

‖ **soixante-neuf** (swasañt nōf). [Fr., lit. 'sixty-nine'.] Simultaneous cunnilingus and fellatio. Cf. *sixty-nine* s.v. *sixty *a.* 2 b.

1888 P. Perret *Tableaux Vivants* xiii. 109 In familiar language this divine variant of pleasure is called: *faire soixante neuf* (literally, to do '69'). 1970 E. M. Brecher *Sex Researchers* iv. 98 By a delicate turn of phrase, van de Velde awards his post-Victorian *nihil obstat* to the practice of *soixante-neuf*. 1973 M. Amis *Rachel Papers* 53 The other couple were writhing about still, now seemingly poised for a session of fully robed soixante-neuf.

sôk, var. *souk.

Soka Gakkai (sōka gakai). Also **Sōka-gakkai, Sokagakki.** [Jap. f. *so* to create + *ka* value + *Gakkai* (learned) society.] In Japan, a lay religious group whose teachings are based on Buddhism.

1958 *Jap. Christian Quarterly* Apr. 104 (*title*) Sōka Gakkai, strange Buddhist sect. 1964 *Asia Mag.* 18 Oct. 3/2 'Let us propagate Buddhism with high and bright spirit to save the world!' The dynamic society with such a transcendental goal is Sokagakkai—the startling new Japanese society. 1964 *Listener* 24 Dec. 998/2 The Soka Gakkai—a kind of mixture of Moral Rearmament and Goldwater republicanism. 1968 P. S. Buck *People of Japan* xiv. 174 The Soka Gakkai philosophy is an ancient one based on the only Buddhist sect which was, like Christianity, intolerant of all other religions. 1974 *Encycl. Brit. Micropædia* IX. 328/2 The Sōka-gakkai follows an intensive policy of conversion..which increased its membership within a seven-year period (1951–57) from 3,000 families to 765,000 families.

Sokol (so·kol). [Czech, lit. 'falcon'.] A Slav gymnastic society first formed in Prague in 1862 (and disbanded in Czechoslovakia in 1952), bearing the falcon as its ensign, and aiming to promote a communal spirit and physical fitness. Also, (a member of) a club in this society.

1910 W. S. Monroe *Bohemia* x. 189 The organization of the Sokols in 1862 has undoubtedly been the most forceful factor in the social unification of the Bohemian people. *Ibid.* 194 A great gathering of all the Sokol unions of the world was called at Prague in 1887. 1915 *Scotsman* 10 Feb. 10/1 The gymnastic volunteer organizations ('sokols') which are popular among all the Slav nationalities of Austria. 1920 *Public Opinion* 2 July 17/2 Over 100,000 Sokols have responded to the call and of that number 50,000 have been selected—27,000 men and 23,000 women. 1925 E. I. Robson *Wayfarer in Czecho-Slovakia* viii. 126 It is a fine sight to see a really big Sokol exhibition, hundreds of men or girls moving like one. 1941 *Ann. Reg. 1940* 200 The systematic arrest of leading Czechs..teachers, and Sokol workers. 1966 *Inland* (Inland Steel Co., Chicago) Autumn 15/2 In Czech settlements the athletic club might still be a *sokol*. 1978 *Chicago* June 56/2 The program will include folk dancing as well as calisthenics and apparatus work. Sponsored by the Central District of the American Sokol Organization.

sol (sǫl), *sb.*[6] *Physical Chem.* [Orig. a suffix f. the first syllable of *solution* (as in *alcosol, *hydrosol).] **1.** A liquid solution or suspension of a colloid. Cf. *gel *sb.*

1899, etc. [see *gel *sb.*]. 1936 W. Stiles *Introd. Princ. Plant Physiol.* ii. 16 If the ability to flow is our criterion of the sol state, then protoplasm is usually, but by no means always, a sol; but there are other indicators of the colloidal state such as elasticity, rigidity, and inhibition, and these are gel characteristics. 1940 Glasstone *Textbk. Physical Chem.* xiv. 1213 The characteristic colors shown by many sols are related in some degree to the particle size; in the course of coagulation, for example, the color of a gold sol changes from red to violet and then to blue. 1970 Ambrose & Easty *Cell Biol.* xv. 479 When two hydrophilic sols carrying opposite charges are mixed, viscous drops known as coacervates often form instead of a continuous liquid phase.

2. *Comb.*: **sol–gel,** used *attrib.* with reference to the interconversion of sol and gel.

1915 W. W. Taylor *Chem. Colloids* i. 10 (*heading*) Sol-gel transformation. 1922 *Jrnl. Amer. Chem. Soc.* XLIV. 1313 (*heading*) The sol-gel equilibrium in protein systems. 1951 *New. Biol.* X. 14 These interchanges, the so-called sol-gel transformations, are constantly going on in the amoeba. 1967 *Oceanogr. & Marine Biol.* V. 191 Cytoplasmic movement results from sol-gel reactions within cells.

sol (sǫl), *sb.*[7] [f. L. *sōl* sun: cf. Sol *sb.*[1]] A solar day on the planet Mars (24 hours 39 minutes).

1976 *Times* 22 July 1/8 The squat little lander seemed to get through its first sol (as the Martian day..is called) without any problems. 1977 *Sci. Amer.* Nov. 58/3 The release of gas tapered off soon after the first sol. 1979 *New Yorker* 5 Feb. 41 On sol 8..the craft's sampler arm extended straight out and then dropped to the ground.

-sol, an ending [f. L. *solum* floor, ground, soil] used to form the names of different kinds and states of soil, as *lithosol* s.v. *litho-, *pergelisol.

sola, *a.* **2.** (Earlier Amer. examples.)

1737 W. Stephens *Jrnl.* 27 Oct. in *Jrnl. Proc. in Georgia* (1742) I. 5 He brought a small Box with sola Bills for a large Sum. 1750 *Colonial Rec. Georgia* (1906)

VI. 323 The last Issue of Sola Bills was not sufficient to defray the Estimate to Michaelmas.

Solacet (sǫ·lăset). Also **solacet**. [f. Sol(uble *a.* + Acet(ate.] A proprietary name for any of a range of azo-dyestuffs which contain sulphate ester groups and were formerly much used for direct dyeing of artificial fibres.
1938 *Times* 11 Jan. 9/5 Another is the 'solacet' range, produced within the last half of 1937, for giving bright fast colours to acetate rayons. 1939 *Trade Marks Jrnl.* 17 May 647/1 *Solacet*... Water soluble dyes for cellulose acetate silk. British Dyestuffs Corporation Limited. 1952 K. Venkataraman *Chem. Synthetic Dyes* I. xxi. 646 The Solacets undergo no chemical change during the dyeing process. 1955 *Official Gaz.* (U.S. Patent Office) 7 June TM 7 *Solacet*... For water soluble dyes for cellulose acetate silk. 1964 E. R. Trotman *Dyeing & Chem. Technol. of Textile Fibres* (ed. 3) xxiii. 515 The Solacet dyes..show little redistribution of colour on continued boiling. 1971 R. L. M. Allen *Colour Chem.* vi. 71 In consequence of the development of dispersed dyes with improved dyeing and fastness properties the Solacet range has now been superseded.

solapsone (sǫlæ·psōᵘn). *Pharm.* [f. Sol(uble *a.* + d)*apsone*, name of a drug of which solapsone is a more soluble derivative (f. *di*(*p*-*amino*phenyl)sulph*one*).] A white powder given as tablets or by injection of an aqueous solution for the treatment of leprosy; the hydrated tetrasodium salt of di(*p*-3-phenyl-1, 3-disulphopropylamino)phenylsulphone, $C_{30}H_{28}N_2O_{14}S_6Na_4.xH_2O$.
1952 *Brit. Pharm. Codex* Suppl. 66 Solapsone consists mainly of the tetrasodium salt. 1959 R. G. Cochrane *Leprosy in Theory & Practice* xvii. 212 The following sulphone drugs are those most commonly employed in the treatment of leprosy: (*a*) the parent sulphone.. Dapsone (B.P.)..(*b*) Sulphetrone (Solapsone (B.P.)) administered orally or parenterally; [etc.]. 1974 R. M. Kirk et al. *Surgery* ii. 23 Specific treatment [of leprosy] is with dapsone..or solapsone (Sulphetrone) given orally, or by injection,..twice weekly. 1977 *Martindale's Extra Pharmacopoeia* (ed. 27) 1505/1 In general, the results with solapsone have not been impressive and it has been superseded by more active drugs.

solar, *a.* and *sb.*¹ Add: **A.** *adj.* **1. d.** (Further examples.)
1876 J. Ericsson *Centennial Exhibition* xlv. 561 The solar engine..is composed of three distinct parts—the engine, the steam-generator, and the mechanism by means of which the..energy of the sun's rays..is increased. 1880 J. P. Mauzey *U.S. Patent* 227,028 27 Apr. 1, I..have invented a new and Improved Solar Heater. 1914 *Metal Worker, Plumber & Steam Fitter* LXXXII. 758/2 Ordinarily a solar heater is mounted upon the roof of the house. 1929 C. G. Abbot in *Smithsonian Sci. Ser.* II. ix. 222 The solar cooker is a delightful luxury. 1955 E. Burgess *Frontier to Space* viii. 150 The system envisaged is a solar engine which would intercept a relatively minute portion of the 92,000 calories which each square centimetre of the Sun's surface radiates every minute. 1962 A. Shepard in *Into Orbit* 83 There is a solar distiller in the kit which will help you convert salt water into the pint of water you need a day to stay alive. 1967 *Daily Tel.* 30 Jan. 10/6 Cooking food at practically no cost has been achieved by scientists at the Hebrew University by the use of a 'solar cooker'. 1979 *Tucson Mag.* Apr. 62/3 A solar heater costs between \$2000 and \$2600 to buy and install.

e. Concerned with or pertaining to the utilization of the sun's rays as a source of energy.
1972 *Guardian* 17 Oct. 15/6 Ambitious schemes..are under study—including a huge solar farm on earth feeding heat to power stations. 1979 *Washington Star* 8 May A14/2 The crowd wasn't sure how well it liked *him*, impeccably solar though he is when it comes to energy. 1979 *N.Y. Rev. Bks.* 17 May 15/1 It will take a long time for solar energy to become the dominant source of world energy. It will first be necessary to eliminate inefficiencies in solar technologies. 1979 *Guardian* 6 Sept. 4/4 Solar systems were immediately viable.

4. a. (Further examples.)
1884 *Nature* 3 Jan. 217/2 This heater..contains the acting medium, steam or air, employed to transfer solar energy to the motor. 1939 A. Huxley *Many a Summer* I. x. 130 It's a gadget..for making use of solar energy. 1976 *National Observer* (U.S.) 21 Feb. 8/3 He's living in a house heated mostly by solar energy.

7. solar battery, a solar cell, or an assembly of such cells; **solar cell**, a photovoltaic device which converts solar radiation into electrical energy; **solar collector**, a device which absorbs solar radiation as heat or reflects it to a focus; **solar flare** = *FLARE *sb.*¹ 1 b; **solar furnace**, an apparatus in which high temperature reactions are carried out at the focus of a system which concentrates the sun's radiation, usually by reflection; **solar glass**, tinted glass for large windows; **solar house** orig. *U.S.*, a solar-heated house; **solar neutrino unit**, a unit used in expressing the detected flux of neutrinos from the sun, equal to 10^{-36} neutrino captures per target atom per second; abbrev. SNU s.v. *S 4 a; **solar paddle**, a large, flat array of solar cells projecting from a spacecraft like a paddle; **solar panel**, a panel designed to absorb the sun's rays for the purpose of generating electricity (by means of solar cells) or heating; **solar pillar** = *sun-pillar* s.v. Sun *sb.* 13; **solar plasma** = *solar wind* below; **solar pond, pool**, a pool or lake of very salty water in which convection is inhibited, allowing considerable heating of the bottom water by solar radiation; **solar power**, power derived more or less directly from solar radiation; **solar sail**, a surface designed to utilize the pressure of solar radiation to provide the propulsive force for a spacecraft to which it is attached; so **solar sailing** *vbl. sb.*; **solar salt**, salt obtained by allowing sea water to evaporate in sunlight; **solar still**, a still, often portable, in which solar radiation is employed to evaporate salty or impure water and produce fresh water; **solar wind**, the stream of ions and electrons which constantly emanates from the sun and which permeates the solar system.
1954 *N.Y. Times* 26 Apr. 1/2 A solar battery, the first of its kind, which converts useful amounts of the sun's radiation directly and efficiently into electricity, has been constructed. 1962 Simpson & Richards *Physical Princ. Junction Transistors* iv. 73 Most of the radiation from the sun is in the region of 1·0eV and above so that 'solar batteries' can be made from germanium or silicon. The efficiency of the process is greater when silicon is used and commercial solar batteries (sometimes called 'solar cells') are at present made from this material. 1978 W. Palz *Solar Electr.* iii. 179 The direct conversion of sunlight into electric power is achieved by means of solar batteries, made up of solar cells. 1955 G. L. Pearson in *Bell Lab. Rec.* July 241/1 The Bell Solar Battery consists of a number of individual silicon solar cells. 1962 [see *solar battery* above]. 1967 *New Scientist* 25 May 463/2 The *Mariner IV* solar cell surface..provided power for the first close photographs of the planet Mars. 1980 *Solar Energy* XXIV. facing p. 1 (*caption*) The world's largest solar cell electric power generation station... The 60kW system consists of nearly 98,000 individual silicon solar cells. 1955 *Trans. Conf. Use of Solar Energy* (Tucson, Ariz.) II. 1. vi. 75 For optimum performance a solar collector should face true south (north in the southern hemisphere). 1976 *Toronto Star* 24 Jan. E1/4 A well-insulated detached house requires about 35,000 kilowatt hours of energy per year to heat it. A solar collector of the size on the Mississauga house would be able to supply about half of the heat needed. 1980 *Solar Energy* XXV. facing p. 1 (*caption*) Giant parabolic dish solar collectors ..have produced efficiencies of 71 per cent at 750°F operating temperature. 1938 *Nature* 17 Sept. 500/2 It would..scarcely be permissible to deduce from the single observation of Carrington..that the three phenomena, solar flare, radio fade-out and..magnetic disturbance, were associated. 1957 *Practical Wireless* XXXIII. 722/1 Solar flares are shortlived, sudden increases in the intensity of the surface brightness in the neighbourhood of sunspots. 1979 C. Kilian *Icequake* vii. 120 Your plane's electronics go bonkers now and then when another solar flare hits. 1924 M. E. Moreau *U.S. Patent 1,479,923* The primary object of the present invention is to provide a new and improved solar furnace for producing an intense heat to be used for scientific purposes. 1951 *Bull. Amer. Ceramic Soc.* XXX. 163/1 A new and important instrument for high-temperature research is the recently developed solar furnace, which, by using the radiant energy of the sun, is able to produce extremely high temperatures in a small area. 1974 *Encycl. Brit. Micropædia* IX. 330/3 Because of its unique ability to heat materials for long periods without contamination, the solar furnace has become an important tool in high-temperature research. 1979 J. F. Kreider *Medium & High Temp. Solar Processes* vi. 232 (*caption*) White Sands solar furnace showing the single heliostat on the right, the concentrator on the left, and flux control shutters in the center. 1977 *Whitaker's Almanack 1978* 1058/1 The window wall was designed to contain outer sheets of brown solar glass. 1978 J. McNeil *Consultant* iii. 55 The Waterman building soared in dark grey metal and matching tinted solar glass. 1946 *Fortune Mag.* Apr. 166 Solar houses will be erected in forty-eight states. 1957 *Economist* 28 Sept. 1027/1 The Massachusetts Institute of Technology is building a series of 'solar houses'. 1976 *Toronto Star* 24 Jan. E1/2 At Provident House, another experimental solar house now being built in King Township, the storage tank contains 60,300 gallons of water. 1970, 1976 Solar neutrino unit [see SNU s.v. *S 4 a]. 1980 *McGraw-Hill Yearbk. Sci. & Technol.* 397/1 The low rate, if attributed to solar neutrinos, would correspond to a neutrino capture rate in the tank of $0·41 \pm 0·07$ per day. To compare this rate to [*sic*] the theory, one expresses the neutrino capture rate in solar neutrino units... The rate corresponds to $2·2 \pm 0·4$ SNU. 1962 *Listener* 29 Nov. 902/1 The probe was first instructed to 'find the Sun'—by means of sun-sensors on the bottom of the vehicle and on the 'solar paddles'. 1968 *New Scientist* 2 May 230/2 An even greater increase in power can be obtained by adding solar paddles to the stabilised platform. 1964 *IEEE Trans. Aerospace* II. 770/1 The solar panel was set on a surface plate with a piece of frosted glass flush against the end of the panel. 1968 *Times* 16 Oct. 8/8 The spacecraft has large solar panels. 1974 P. Dickinson *Poison Oracle* i. 13 The roof offered the widest possible expanse to the solar panels that provided much of the energy for the palace's gadgetry. 1976 *Pract. Householder* Nov. (Heating Suppl.) 23/2 The past summer..produced astonishingly high air temperature and sunshine hours so that reservoir temperature and storage tank levels rose and solar panels heated the water quickly. 1978 *Sci. Amer.* Apr. 149/1 The solar pillar, a commoner phenomenon, is a vertical shaft of light extending upward from the sun. 1962 *Listener* 29 Nov. 902/1 The solar plasma (commonly called 'solar wind'), which consists of low-energy charged particles which continually stream outward from the Sun. 1972 A. Hewish in C. P. Sonett et al. *Solar Wind* vii. 477 Radio waves traversing the solar plasma are scattered by irregularities of plasma density. 1961 *Sci. News Let.* 12 Aug. 106/2 Instead of using vast expanses of expensive mirrors, a solar pond traps heat in shallow water. 1971 *Sci. Amer.* June 127/1 The hot, salty water is selectively withdrawn from the solar ponds and used to drive a generator. 1979 *Nature* 11 Jan. 91/2 A solar pond is a water-filled pond, 1–2 m deep, with a blackened bottom for greater heat absorption and a gradually increasing salt concentration towards the base to eliminate convection, which is the main cause of heat loss. 1960 *Daily Tel.* 18 Aug. 16/2 (*heading*) Solar pools as rival to nuclear power stations. 1975 *Globe & Mail* (Toronto) 15 Dec. 2/8 Scientists have failed to find a way of releasing the potential energy at the bottom of solar pools where water approaches the boiling point. 1979 *Arizona Daily Star* 5 Aug. (Advt. Section) 20/2 Featuring solar pool, putting green, horse facilities & country atmosphere. [1908 *Sci. Amer.* 8 Feb. 97/1 (*heading*) A new solar power plant.] 1915 *Jrnl. R. Soc. Arts* LXIII. 564/1 Solar power was quite within the range of practical matters. 1956 *Sci. Amer.* July 97/2 Any attempt to produce solar power means collecting the energy falling on a large area. 1976 *Times* 26 Mar. (Energy Suppl.) p. vi/4 If an equal number of existing houses were converted, domestic solar power could then substitute for about 4 per cent of Britain's energy needs. 1960 *Aeroplane* XCIX. 693/1 Another interesting concept which has not yet really undergone feasibility determination is that of the solar sail. With this device, a space 'ship' may some day be able literally to sail through interplanetary space. 1978 *Listener* 6 July 13/3 The 'solar sail' makes space travel in the inner solar system very cheap. 1960 *Aeroplane* XCIX. 744/1 The subject of solar-sailing. 1973 C. Sagan *Cosmic Connection* xxiii. 162 Solar sailing, the use of the pressure of sunlight and of the protons and electrons in the solar wind for tripping through the solar system. 1861 J. S. Muspratt *Chem., Theoret., Pract., & Analytical* II. 906/1 This Onondaga solar or coarse salt is unsurpassed..in the world. 1950 *Thorpe's Dict. Appl. Chem.* (ed. 4) X. 844/2 Some solar salt-factories produce high-quality table-salt by re-dissolving solar salt in water, purifying the resultant brine, and then evaporating it. 1972 *Times* 16 Oct. 19/7 Ventures ranging from solar salt to uranium. 1946 W. R. P. Delano *U.S. Patent 2,413,101* (*heading*) Solar still with nonfogging window. 1970 *Guardian* 15 Jan. 11/2 At the end of 1969..the Clan McIlwraith sailed..for Mombasa, carrying a prefabricated solar still for Aldabra, a lonely coral atoll in the Indian Ocean. 1979 *Solar Energy* XXIII. 271/1 The basin-type pitched roof solar still is the commonest and cheapest. 1958 *Physical Rev.* CX. 1448/1 The geomagnetic field can be penetrated to a considerable depth by tongues of ionized gas from the solar wind. 1969 *Times* 22 July (Moon Rep.) p. iii/6 Buzz is erecting the solar wind experiment now. 1977 D. Harsent *Dreams of Dead* 21 Unimaginable, the solar winds roared through space, putting the earth awry. 1978 Pasachoff & Kutner *University Astron.* viii. 227 (*caption*) The solar wind causes the wavy streaming of the tails of comets.

8. *solar-charged, -generated, -terrestrial* adjs.; **solar-heated** *a.*, heated by means of the sun's rays; equipped with a solar heating system; also **solar-heat** *v. trans.* (also *absol.*); **solar heating**, heating by means of the sun's rays, esp. when utilized for water or space heating; also *attrib.*; **solar-powered** *a.*, using power derived directly from the sun's rays.
1968 G. M. B. Dobson *Explor. Atmos.* (ed. 2) xi. 195 The effect of the earth's magnetic field on the solar-charged particles is to deflect them back, away from the earth. 1978 *N.Y. Times* 30 Mar. A-14/6 The costs of solar-generated electric power currently did not 'stand up' in comparison with other energy sources. 1952 Ayres & Scarlott *Energy Sources—Wealth of World* xv. 208 Houses can be solar-heated completely without fuel when the mean atmospheric transmissivity is above 55 percent. 1977 *National Observer* (U.S.) 8 Jan. 8/4 When solar heating a house, the calculations become more complex. 1979 *Sunset* Apr. 132/1 (Advt.), The *Poolsaver* Automatic Solar Pool Cover solar heats and thermal insulates. By day *Poolsaver* absorbs solar energy. 1950 *Heating & Ventilating Engineer* XXIV. 148/1 The practicability of employing solar-heated air as a source of heat in removing moisture from a dehumidifying agent. 1956 *World Symposium on Applied Solar Energy* 107/2 The first solar-heated house was built in 1939 at M.I.T. as part of the Godfrey L. Cabot Solar Energy Conversion Research Project. 1977 *Time* 24 Jan. 19/1 Carter and Mondale will watch the parade in front of the White House from a 60-ft. by 40-ft. solar-heated reviewing stand. 1903 C. H. Pope *Solar Heat* i. 43 Another patent for solar-heating devices was obtained..from the British Government. 1951 *Archit. Rev.* CIX. 291/1 A limited supply of hot water is provided by solar heating systems on the roof of each ward block. 1958 *Times Rev. Industry* Feb. 94/1 The chief advantage claimed for solar heating, compared with other methods, is that the region of high temperature is localized, and contamination of material, while in the molten state, can thus be avoided. 1968 R. A. Lyttleton *Mysteries Solar Syst.* v. 175 No serious effects of solar heating could be expected at such a distance from the sun, though some intense local heating on [*sic*] the particles through collisions could occur. 1977 'E. Trevor' *Theta Syndrome* iii. 43 David Pryor, a solar heating engineer. 1959 *Time* 26 Oct. 58/2 The satellite is shaped like a gyroscope... It squeals like a bagpipe as it signals from two transmitters—one powered by a chemical battery, the other solar-powered. 1978 *Illustr. London News* Nov. 19/2 The Administration succeeded in gaining agreement for..conservation measures, including tax relief for householders insulating their homes or for installing solar-powered systems. 1946 *Nature* 7 Sept. 329/1 Study of precise solar-terrestrial relationships has been a major Smithsonian activity for many years. 1966 *McGraw-Hill Encycl. Sci. & Technol.* I. 619/2 Solar

physics..overlaps with geophysics in the consideration of solar-terrestrial relationships.

B. *sb.*¹ **2.** A solar lamp.
1853 M. J. McIntosh *Lofty & Lowly* i. 9 The astral lamp—solars were not yet invented—..throws its rays on cases filled with richly gilded volumes. 1976 H. R. F. Keating *Filmi, Filmi, Inspector Ghote* iii. 29 We are using a great number of different lights for different purposes in filming, Five-Ks, Two-Ks, Sunspots, Solars, Babies.

3. *U.S.* Solar radiation as a source of domestic or industrial energy.
1976 *National Observer* (U.S.) 17 July 9/2 What one is trying to do is go out and demonstrate for the building industry that solar is here for heating buildings. 1978 *Tucson* (Ariz.) *Mag.* Dec. 77 Passage of the solar tax credits by Congress will result in the rapid growth of solar in Southern Arizona.

solarimeter (sōᵘlări·mĭtəɹ). [f. Solari- + -meter.] A device for measuring the total intensity of radiation incident upon a surface.
1926 L. Gorczyński in *Monthly Weather Rev.* LIV. 381/2 To these direct-reading instruments, designed for both solar and sky radiation, we propose to give the name of 'Solarimeters' in order to distinguish them from pyrheliometers, which serve generally for radiation intensity of the sun at normal incidence. 1940 *Sci. Abstr.* XLIII. 383 The radiative properties of the snow cover were investigated... Two solarimeters were employed. 1969 McIntosh & Thom *Essent. Meteorol.* vii. 105 A solarimeter..detects both direct and diffuse solar radiation. *Ibid.*, An inverted solarimeter detects the short-wave radiation reflected by the surface.

solarium. Delete ‖ and add: **2. a.** Also, a sun-parlour. Chiefly *N. Amer.*
1911 G. W. James *Grand Canyon* iii. 20 It is called the solarium or sun-parlor. 1968 *Globe & Mail* (Toronto) 17 Feb. 45 (Advt.), Separate dining room with adjoining solarium—both overlook ravine. 1978 *Morecambe Guardian* 14 Mar. 25/1 (Advt.), Large..working kitchen, frontal solarium, porches [etc.].

b. A room equipped with sun-lamps.
1960 *Playboy* Mar. 70/2 There's a poolside soda fountain, as well as bar, massage rooms, steam cabinets, solaria, and Finnish baths. 1972 *Homes & Gardens* Sept. 20/2 The uncertain perils of the British climate..have finally induced one of our hotels to install a solarium. 1978 *Cornish Guardian* 27 Apr. 16/3 (Advt.), Heated swimming pool, sauna, solarium.

solarization. Add: **1. a.** More generally, the progressive reduction in the developable density of an emulsion (corresponding to a progressive darkening of the picture) following initial exposure beyond a certain light intensity. (Further examples.)
1948 James & Higgins *Fund. of Photogr. Theory* iv. 59 The curve representing developable density as a function of exposure passes through a maximum. If the exposure is increased beyond that which produces the maximum density, a decrease in developable density will occur... This effect is known as solarization. 1956 *Focal Encycl. Photogr.* 1079/1 Strictly speaking, solarization is the reversal of the image on a film or plate by an extreme amount of over-exposure... The term has by this time almost lost its original meaning. Nowadays it is applied.. to the technique for producing a partly reversed image by exposing the negative to unsafe light during development —actually, the phenomenon known as the Sabattier effect. 1961 *Jrnl. Photogr. Sci.* IX. 195/1 Solarization.. is usually attributed to a reduction in the number of developable grains. 1973 *SPSE Handbk. Photogr. Sci. & Engin.* vi. 427 The addition of halogen acceptors..to the emulsions prevents solarization.

b. *Photogr.* = *Sabatier effect* s.v. *Sabatier. Also called *pseudo-solarization.
1937 *Photogr. Jrnl.* LXXVII. 21/1 If the original negative consists of a well-exposed object on an unexposed background, the fogging of the background with the second exposure does not come right up to the edge of the image, but leaves a clear white line... This line was formerly only of academic interest. Recently, however, it has been used as the basis of the so-called 'Solarization Process'. 1939 [see *Sabatier effect* s.v. *Sabatier]. 1956 [see sense a above]. 1969 M. J. Langford *Adv. Photogr.* xi. 233 The image exposure relative to solarisation fogging exposure makes decisive changes in tone rendering. 1977 J. Hedgecoe *Photographer's Handbk.* 278 The solarization, below, was produced by the black and white method, but with the solarized positive printed onto color negative film, using a colored light source.

c. *Plant Physiol.* [a. G. *solarisation* (A. Ursprung 1913, in *Ber. d. Deut. bot. Ges.* XXXV. 57).] The inhibition of photosynthesis as a result of prolonged exposure to high light intensities.
1925 W. Stiles *Photosynthesis* vii. 97 Solarization appears to have no permanently injurious effect on the activities of the leaf. 1960 B. S. Meyer et al. *Introd. Plant Physiol.* (1963) xi. 219 Solarization effects appear to result principally..from the phenomenon of photo-oxidation, in which leaves consume oxygen in the light, and use it in the oxidation of certain cell constituents. 1974 R. G. S. Bidwell *Plant Physiol.* vii. 170 Very high light intensity may be damaging to plants—solarization is the photodestruction of chlorophyll by excessive illumination. Shade plants are more susceptible to solarization than are sun plants.

d. The alteration of the light transmission characteristics of glass as a result of prolonged exposure to visible or ultraviolet light.
1928 W. W. Coblentz *Let. Circular U.S. Bureau Standards* No. 235 (3rd revision) 4 A sample [of glass] which had been in a hospital window..for a year was found to have a transmission of 25 per cent at 302 $m\mu$... Further exposure to the quartz mercury arc reduced the transmission but little, showing that solarization was complete. 1955 E. B. Shand *Glass Engin. Handbk.* xvi. 145/2 Mercury-vapor lamps utilize a number of glasses of different properties... Because of their exposure to strong ultra-violet radiations, the glasses must be capable of resisting solarization effects to a large degree. 1972 F. L. Harding in L. D. Pye et al. *Introd. Glass Sci.* 422 Long term exposure to the ultraviolet radiation in sunlight can result in another type of coloring phenomenon known as solarization. If certain multivalent ions..are present in the glass, their valence can be changed by ionizing radiation.

solarize, *v.* Add: **1.** More widely, to affect by solarization of any kind. (Earlier and later examples.)
1853 C. Goodyear *Gum-Elastic* I. vii. 114 Another effect yet more remarkable in the treatment of gum-elastic, is that of the sun's rays upon it. When combined with sulphur, and exposed to the action of the sun..it becomes solarized, or divested of its adhesive quality. 1950 *Jrnl. Amer. Ceramic Soc.* XXXIII. 257/2 From the general appearance of the glasses that were solarized in the mercury arc it became obvious that the depth of color change was slight. 1966 LaCow & Lathrop *Photo Technol.* xix. 229/2 To solarize a negative the developing process is carried on in a normal manner for about 2/3 to 3/4 of the developing time [etc.].

2. (Further examples.)
1955 E. B. Shand *Glass Engin. Handbk.* xvii. 159 Some glasses will discolor perceptibly, or 'solarize' when exposed to ultra-violet radiations. 1977 *Jrnl. Photogr. Sci.* XXV. 103/2 The emulsion investigated..was a chlorobromide, halide-exchange emulsion which had a low surface/internal speed ratio and which solarized readily.

solarized, solarizing *ppl. adjs.* (further examples); also **so·larizing** *vbl. sb.*
1853 C. Goodyear *Gum-Elastic* I. vii. 112 The use of acid gas in connection with the solarizing process. 1923 B. D. W. Luff *Chem. of Rubber* i. 18 Articles made with such a composition, when exposed to the sun's rays, a process termed solarising, were said to lose their adhesive nature. 1969 *Focal Encyl. Photogr.* (rev. ed.) 1415/2 The solarizing exposure itself must also be sufficient. Too little leads to..only a partially solarized image with weak border outlines. 1977 Solarized [see *solarization 1 b]. 1977 R. Hattersley *Photogr. Printing* ii. 20 It involves exposing a print twice: once in the usual way and once with a solarizing light source after development has gone about halfway.

solation² (sǫlē¹·ʃən). *Physical Chem.* [f. *sol *sb.*⁶ + -ation.] The change of a gel into a sol. So **sola·te** *v. intr.*, to undergo solation; *trans.*, to convert into a sol.
1915 [see *gelation²]. 1926 *Jrnl. Morphol. & Physiol.* XLI. 351 Locomotion in Amoeba is associated with gelation and solation. 1951 *New Biol.* X. 16 The plasmagel must presumably contract at the hind end, as well as solating. 1958 *Jrnl. Cellular & Compar. Physiol.* LII. 270 The plasmagel system of the intact *Amoeba* undergoes complete solation under suitably high pressure. 1977 *Jrnl. Cell Biol.* LXXIV. 909/1 The gel solated slowly at room temperature after forming. *Ibid.* 921/1 Elevated KCl concentrations that solate the *Dictyostelium* gel.

sold, *ppl. a.* Add: **3.** *slang.* Tricked, deceived. Cf. Sell *v.* 9.
1876 'Mark Twain' *Tom Sawyer* xvii. 147 As the 'sold' congregation trooped out they said they would almost be willing to be made ridiculous again.

4. *sold-out.* **a.** *colloq.* Bankrupt; exhausted, 'finished'.
1859 Hotten *Dict. Slang* 98 *Sold up, or out*, broken down, bankrupt. 1958 F. C. Avis *Boxing Ref. Dict.* 104 *Sold out*: said of a boxer who is nearly exhausted. 1973 *Observer* 3 June 25/6 At the end of last season he was physically and mentally sold out. 1977 *New Yorker* 4 July 24/1 A pool player who was vaguely associated with the big-money barracudas and sold-out types hanging back in the pool-hall shadows.

b. That has sold all its stock, seats, etc.
1903 Kipling *Five Nations* 191 The sold-out shops and the bank And the wet, wide-open town. 1960 *Farmer & Stockbreeder* 8 Mar. 60/1 (*heading*) Sold-out at Bath. All available stand space at the Bath and West Show..has been sold. 1975 *High Times* Dec. 21/1 They..are a sold-out attraction wherever they perform. 1976 *Early Music* Oct. 447/2 Bodies whose present idea of an 18th-century orchestra is a group playing Haydn and Mozart badly to a sold-out Festival Hall.

solder, *sb.*¹ Add: **4. b.** (Earlier example.)
1845 J. R. Planché *Golden Fleece* II. 23 Begone, I charge you, none of your soft solder: Your downy words don't weigh with me a feather.

6. *solder-pin*; *solder-coated, -sealed* adjs.
1964 R. F. Ficchi *Electr. Interference* x. 193 A third method of connecting bus bars is by bolting two solder-coated bus bars, and applying heat to make a continuous connection. 1965 *Wireless World* Sept. 464/2 This unit.. is available with either solder pins for direct connection.. or with valve base pins. 1964 R. F. Ficchi *Electr. Interference* v. 72 (*caption*) Input and output connections ..are solder-sealed terminals and an A–N connector.

so·lderable, *a.* [f. Solder *v.* + -able.] Able to be joined by means of solder. So **solderabi·lity**, the property of being solderable.
1949 *Iron Age* 8 Dec. 96/2 The solderability of electro-deposited lead-tin alloy remains excellent for at least 9 months under normal operating conditions. 1959 *Trans. Inst. Metal Finishing* XXXVI. 203 The solderability of various coatings of tin with lead, zinc, cadmium, and of cadmium and silver has been compared. 1961 Webster, Solderable. 1967 E. R. Wells in C. R. Martens *Technol. Paints, Varnishes & Lacquers* xiii. 214 The urethane single-package low solids..enamels..had the unique advantage of being solderable without removal from the ends of the wires to be joined. 1976 *Wireless World* Nov. 52/1 The transistor terminal pad was quite easily solderable. 1978 *BSI News* Mar. 7/1 The test methods included in the BS 9760 series cover solderability..of surface conductors.

soldered, *ppl. a.* Add: **1. b.** *soldered dot* (Building), a means of fastening sheet lead to woodwork, consisting of a mass of solder put in a depression in the lead after the latter has been fitted into a corresponding depression in the wood and a screw fixed through the bottom of it.
1893 J. W. Clarke *Lect. Plumbing* 70/1 Soldered dots never last for any great length of time and the lead invariably breaks away from them. 1930 P. Manser *Plumbing & Gasfitting* VII. xxii. 1606 Soldered dots are not quite satisfactory as they hold the lead too rigidly, and where the load on them is heavy the screws work through due to the strain on them. 1966 G. E. Evans *Pattern under Plough* iii. 51 A circle of plumber's black is painted round the outside of the hollow, partly to confine the soldered dot, and partly to give the whole a neat decorative finish.

so·lderless, *a.* [f. Solder *sb.*¹ + -less.] Made without solder; that does not require solder.
1920 C. T. Schaefer *Motor Lorry Design* iii. 43 A solderless copper radiator is obtained,..which..withstands severe vibration without failure. 1957 '*Motor Cycling' Workbench Wisdom* 75/2 A particularly neat type of solderless nipple. 1974 *Sci. Amer.* Apr. 79/1 The Logic Lab utilizes a solderless, plug-in connection technique for components and wires.

soldier, *sb.* Add: **1. d.** A member of the Salvation Army.
1876 W. Booth *Salvation Soldiery* (1882) 70 Get fixed in your mind the ungainsayable truth that every soldier can do something. 1890 —— *In Darkest England* II. v. 168 Emma Y.—Now a Soldier of the Marylebone Slum Post. 1935 *Chambers's Encycl.* IX. 64/2 In some of the jails there is now a regularly organised corps of Salvation Army soldiers. 1978 *Lochaber News* 31 Mar. 3/2 At the evening service four young soldiers..were enrolled by Major Holstead.

e. *to play* (*at*) *soldiers*: said of children; also derisively of volunteers.
1911 in *Conc. Oxford Dict* s.v. *Soldier*. 1969 I. & P. Opie *Children's Games* xii. 338 There is a noteworthy difference between playing at 'Soldiers' and playing at 'War' with two opposing sides. *Ibid.*, If they [*sc.* the boys] were playing soldiers, she took it as a warning that it was time for her to arm. 1977 *Daily Mirror* 16 Mar. 10/2 (Advt.), I can tell you that digging a trench in pouring rain when you've had no kip is hardly playing at soldiers.

f. A rank-and-file member of the Mafia.
1963 *Organized Crime & Illicit Traffic in Narcotics* (Comm. Govt. Operations, U.S. Senate) I. 80 Then we had what we call a caporegima which is a lieutenant, and then we have what we call soldiers. 1970 L. Sanders *Anderson Tapes* lxxii. 218 The organization variably known as Cosa Nostra, Syndicate, Mafia, etc., even has military titles for its members—don for general or colonel, *capo* for major or captain, soldier for men in the ranks, etc. 1974 J. Gardner *Corner Men* xv. 248 Vescari was coming to him. There were several men around him, the don's soldiers. 1977 *Time* 16 May 35/3 Since then scores of new soldiers have signed up [in the Mafia].

2. c. *old soldier*: one practised or experienced in a thing, or one who pretends to be so. (See also quot. 1912.) Also *attrib.* Cf. senses 2 b in Dict. and *2 d.
1722 [see Old *a.* 5 b]. 1858 Geo. Eliot *Lett.* (1954) II. 511 He..will be as much interested as I shall be in knowing about the vicissitudes of Coventry journalism, when any new phase or crisis comes of which you can tell us. He is an old soldier, and cares for battles of that sort. 1912 R. A. Freeman *Singing Bone* ii. 119 Poor Pratt was what you'd call an old soldier—sly, you know, sir—and a bit of a sneak. 1949 [see *head *sb.* 49**]. 1950 N. Cardus *Second Innings* 93 The umpire (an old soldier) confidentially tells you he could see it all coming.

d. *Naut. slang* (orig. and chiefly *U.S.*). A worthless seaman; a loafer, a shirker. Also *old soldier*. Cf. Soldier *v.* 1 d.
1840 R. H. Dana *Two Yrs. before Mast* xvii. 154 The captain called him a 'soger', and promised to 'ride him down'. 1849 [see *soldiering *vbl. sb.* 2]. 1850 H. Melville *White Jacket* II. xxx. 205 Off Cape Horn some 'sogers' of sailors will stand cupping, and bleeding, and blistering, before they will budge. 1898 A. J. Boyd *Shellback* ii. 28 Some are good men, some mere 'sojers' (useless as seamen—loafers). 1933 E. P. Mitchell *Deep Water* xxi. 184, I hear that you have shipped as an A.B. You don't look like one, and if you're a soldier you'll get soldier's jobs and be disrated. 1958 B. Hamilton *Too Much of Water* vi. 140 He's a bit of an old soldier, but a first-rate seaman, and a hundred per cent reliable at sea.

e. *dead soldier* (*U.S. slang*): an empty bottle. Cf. (*dead*) *marine* s.v. Marine *sb.* 4 d.
1917 in *Dialect Notes* IV. 322. 1929 *New Yorker* 9 Feb. 42/3 His aim with a dead soldier was..unerring. 1940 R. Chandler *Farewell, my Lovely* v. 33, I held up the dead soldier and shook it. Then I..reached for the pint of bonded bourbon. 1979 R. B. Gillespie *Crossword Mystery* ii. 50 There weren't any prints on that bottle... That dead soldier was as clean as a whistle.

f. *colloq.* A strip or finger of bread or toast.
1966 N. Freeling *Dresden Green* I. 73 Potato soup with fried onions and 'soldiers' of fried bread. 1971 J. Grigson *Good Things* 120 First dip the asparagus into the butter, then into the runny egg yolk, as if it were a child's bread 'soldier'. 1979 *Woman's Own* 21 Apr. 8/3 Our medical writer..advises: 'Bread, butter and milk is a good idea, but you can't really beat a boiled egg and "soldiers".'

3. f. Substitute for def.: One of several deep-water fishes with reddish skins, esp. one of the genus *Hoplostethus*. (Later examples.)
1935 'R. M.' *Trawler* x. 51 By far the most plentiful animals in all the catch were the 'soldiers'. 1953 [see *dageraad]. 1971 *Grocott's Mail* (Grahamstown, S.Afr.) 28 May 3 Mrs. E. Birch took both the ladies' awards with a soldier of 0·963 kg, another unusual fish and decidedly a deep sea species. 1974 *Nature* 22 Mar. 306/3 The berycoid fishes comprise a mixture of deepwater 'soldiers', *Hoplostethus*, and other genera.

6*. In allusion to the resemblance to a line of soldiers on parade. **a.** *Carpentry.* Each of a series of short vertical pieces of wood to which a skirting-board is fixed.
1927 T. Corkhill in R. Greenhalgh *Building Educator* II. 817/2 The vertical grounds, or soldiers.., are plugged to the wall about every 3 ft. apart. 1950 M. T. Telling *Carpentry & Joinery* v. 200 The skirting is fixed with nails to the horizontal ground and to the short vertical grounds called 'soldiers'.

b. *Building.* (See quot.) Cf. *soldier arch, course*, sense 8 below.
1929 W. C. Huntington *Building Constr.* iv. 130 Belt courses and flat arches may be formed of brick[s] set on end with the narrow side exposed. Such bricks are called soldiers.

c. *Building.* Each of a series of vertical members of timber or metal used to hold formwork in position or support the lining of an excavation.
1932 Dowsett & Bartle *Practical Formwork & Shuttering* ii. 19 The ribs are held in position by uprights made from 3"×6" material; these uprights—frequently referred to as 'soldiers'—are in turn held by 3"×6" horizontal timbers called 'walings'. 1932 T. Corkhill *Conc. Building Encycl.* 197 *Soldiers*,..heavy vertical timbers placed across several walings and strutted. This is done in stages, to remove the lower struts for a deep excavation, as the wall is built. 1961 *Engineering* 8 Dec. 739/1 Aluminium 'soldiers' are being used..to support the shuttering for the concrete shields of the reactors. 1970 W. G. Nash *Brickwork Three* viii. 175 When a sufficient depth has been supported in this way the whole system is held back by soldiers which are secured by the permanent struts.

7. a. *soldier-boy, -poet, -servant* (earlier example).
1861 in *Rebellion Rec.* (1862) I. III. 91 My hungry soger-boys shall soon have meat and drink. 1978 J. Barnett *Head of Force* viii. 72 This was his field. The soldier-boy was out of his depth. 1912 D. H. Lawrence in *Eng. Rev.* Jan. 373 Liliencron is well represented. But this soldier poet is so straight, so free from the modern artist's hyper-sensitive self-consciousness, that we would have more of him. 1958 Blunden *War Poets 1914–18* i. 13 The number and the activity of the soldier-poets of Britain in the First World War were bewildering. 1794 W. B. Stevens *Jrnl.* 13 Feb. (1965) 135 Stables has displayed a boisterousness of temper..to his Soldier-Servant..which I cannot palliate.

b. *soldier-settlement, -suit.*
1921 *Daily Colonist* (Victoria, B.C.) 25 Mar. 1/5 The opening for soldier settlement of about two townships from the Riding Mountain, Manitoba, reserve, will be held at the Dominion land office in Dauphin, Manitoba, in about two weeks. 1930 W. K. Hancock *Australia* vii. 141 It would..be not altogether unfair to separate soldier settlement from closer settlement, and to consider the former as part of the cost of the war. 1977 *Weekly Times* (Melbourne) 19 Jan. 39/2 The property remained in the Bell family until taken up under Soldier Settlement by Mr J. Smedley after World War 2. 1944 S. Bellow *Dangling Man* 182 I d murder him, soldier suit or no soldier suit. 1977 H. Fast *Immigrants* II. 141 If you have to put on that lousy soldier suit to live with yourself, then for Christ's sake become a medic or a clerk or something like that.

8. soldier arch *Building*, a soldier course serving as a lintel; **soldier bean** *N. Amer.*, the mottled kidney-shaped seeds of certain varieties of *Phaseolus vulgaris*; **soldier-bug**, a predacious North American bug of the genus *Podisus* of the family Pentatomidæ, esp. *P. maculiventris*, which is yellowish brown and has a spine on the under-side of its head; **soldier course** *Building*, a course of bricks set on end with their narrower long face exposed; **soldier-fish**: delete *U.S.* and add: = Squirrel 4 and *squirrel-fish* s.v. Squirrel 7 b; (later examples); **soldier fly**: for *U.S.* read orig. *U.S.* and add: [tr. mod.L. *Stratiomys*], an often brightly coloured fly of the family Stratiomyidæ, the larvæ of which damage the roots of certain grasses; (earlier and later examples); **soldier orchid** = *military orchid* s.v. *military *a.* 3 b; **soldier-termite** = sense 3 e in Dict.
1963 Seakins & Smith *Practical Brickwork* xiv. 174 (*caption*) Flexible D.P.C. behind soldier arch. 1972 S. Smith *Brickwork* xiv. 73 A method of supporting a soldier arch by means of wire ties built into a concrete lintel at the rear, is shown. 1931 W. G. McGregor *Field Beans in Canada* 8 In Nova Scotia..four leading varieties ..are Navy Ottawa 711, White Marrowfat, Soldier, and Yellow Eye. 1968 E. R. Buckler *Ox Bells & Fireflies* vi. 101 Yellow-eyed soldier beans to be threshed on the barn floor with the leather-jointed flail. 1868 *Mich. Agric. Rep.* VII. 175 [I] found [them] to be soldier-bugs, with their long harpoon bills thrust into a fine fat slug. 1946 *Richmond* (Va.) *Times-Dispatch* 4 Feb. 4/1 More than two tons of it [*sc.* an insecticide made from sabadilla] was used this year to kill..the soldier bugs in Illinois. 1948 Dalzell & Townsend *Masonry Simplified* I. vii. 268 Soldier courses are used mainly as a water table around a building at the level of the first floor. 1979 *Arizona Daily Star* 1 Apr. (Advt. Section) 22/3 Burnt adobe hacienda with Soldier Course on parapet. 1905 D. S. Jordan *Guide to Study of Fishes* II. xv. 253 The soldier-fishes (*Holocentridæ*) also known as squirrel-fishes ..are shore fishes very characteristic of rocky banks in the tropical seas. 1931 J. R. Norman *Hist. Fishes* iv. 69 Soldier-fishes..of the coral reefs of tropical seas derive their name from the stout and sharply pointed spines with which the fins are provided. 1961 E. S. Herald *Living Fishes of World* 157/1 Squirrelfishes or soldier-fishes..tend to hide in crevices and cracks. 1842 T. W. Harris *Treat. Insects New England Injurious to Vegetation* 408 Most of the soldier-flies..are armed with two thorns or sharp spines on the hinder part of the thorax. 1905 V. L. Kellogg *Amer. Insects* xiii. 329 The soldier-flies, Stratiomyidæ, are unfamiliar insects... Many of the species have bright yellow or green markings, and most of them have the abdomen curiously broad and flattened. 1952 J. Clegg *Freshwater Life Brit. Isles* xiv. 238 The Soldier-flies are the first of the stouter, short-horned flies to be considered. 1975 *N.Z. Jrnl. Agric.* Sept. 7/1 Infestations of grassgrub and soldier fly..have the effect of inducing a clover-strong pasture which increases the danger of bloat. 1934, 1969 Soldier orchid [see *military orchid* s.v. *military *a.* 3 b]. 1963 Soldier-termite [see *nasute *a.* 3].

9. soldier's breeze = *soldier's wind*; **soldier's farewell** *slang*, an abusive farewell (cf. *sailor's farewell* s.v. *sailor 5 c); **soldier's heart** (further examples); = *irritable heart*; **soldiers' home**, a place of stay for soldiers.
1894 Stevenson & Osbourne *Ebb-Tide* II. vii. 125 The *Farallone* made a soldier's breeze of it. 1909 J. R. Ware *Passing Eng.* 229/1 *Soldier's farewell*, 'Go to bed', with noisy additions. 1936 J. Curtis *Gilt Kid* viii. 82 'Good-bye. I hope they'll poke you into the Lock Hospital.' 'Soldier's farewell to you.' 1938 F. D. Sharpe *Sharpe of Flying Squad* xviii. 184 As you pass through the door, you'll sometimes hear a raspberry... No one wants to accept responsibility for that soldier's farewell. 1979 *Guardian* 12 Nov. 2/5 One school of thought within ITN.. is that..the darling [newsreader] of millions then decided to say a soldier's farewell. 1967 *Punch* 29 Mar. 458/3 World War One produced, besides Trench Foot, a syndrome called Soldier's Heart, caused by great anxiety coupled with severe physical strain. 1971 Soldier's heart [see *irritable *a.* 2 b]. 1860 Mrs. Gaskell *Let.* 10 Dec. (1966) 640 This autumn [I]..helped Florence Nightingale ..in establishing a Soldier's [*sic*] Home in Gibraltar where they can have cheap refreshments, can read, play games, write letters, &c. 1866 J. C. Greg *Life in Army* xxvi. 224 The idea of a Soldiers' Home is, I believe, original with the American people... It is said to have been first instituted in the city of Baltimore in 1861. 1900 *Congress. Rec.* 19 Jan. 1001/1 Part of his [*sc.* the veteran's] meager pension [is] confiscated at Soldiers' Homes.

soldier, *v.* Add: **1. f.** *to soldier on*: to persevere, to carry on doggedly.
1954 K. Amis *Lucky Jim* vii. 77 The eeriness..disconcerted him..but he soldiered pluckily on to his objective. 1959 *Times* 20 Aug. 3/6 Lomax soldiered on and at tea had made 68. 1963 *Times* 21 Feb. 13/2 To give the maximum increase to the new recruits..and to offer a much smaller percentage to men and women who have soldiered on into the thirties and forties is a division hard to justify. 1978 *Jrnl. R. Soc. Med.* LXXI. 648 The alternatives are to let the patient soldier on, or to take the radical approach of abdominoperineal resection.

soldiering, *vbl. sb.* **2.** (Earlier examples.)
1840 R. H. Dana *Two Yrs. before Mast* xii. 91 '*Sogering*' was the order of the day. 1849 H. Melville *Redburn* I. xii. 118 Sailors were always bitter against any thing like *sogering*..though this Jackson was a notorious old *soger* the whole voyage.

sole, *sb.*¹ Add: **8. d.** *Geol.* The underlying or lowest thrust plane of a thrust.
1889 H. M. Cadell in *Trans. R. Soc. Edin.* XXXV. 347 This experiment shows that underneath a series of beds, repeated and heaped together by small thrusts, inclined perhaps at considerable angles, there runs..a major thrust or 'sole', inclined at a lower angle, along which the whole mass may have travelled for considerable distances. 1907 J. Horne in B. N. Peach et al. *Geol. Struct. N.W. Highlands Scotland* xxxii. 464 Owing to..the friction along the unyielding lower plane or 'sole' of the thrust, there was a tendency in the materials to fold over and curve under. 1965 A. Holmes *Princ. Physical Geol.* (ed. 2) ix. 225 Along the sole of a major thrust severe crushing and grinding of the rocks is to be expected.

e. *Geol.* The lowest layer of ice in a glacier, containing rock debris.
1930 *Amer. Jrnl. Sci.* CCXIX. 13 A rock fragment subjected to the abrasion processes in action on the sole of a glacier. 1952 *Jrnl. Glaciol.* II. 128 This deposit is then pulled along in a continuous manner by the movements of the glacier, thus forming the sole. 1977 A. Hallam *Planet Earth* 87 In the roofs of these cavities we see the rock-studded glacier sole.

f. *Geol.* The under-side of a sedimentary stratum.
1957, etc. [see *sole marking*, sense 9 below]. 1972 H. Blatt et al. *Origin Sedimentary Rocks* v. 170 Such structures are normally observed in the field on the sandstone sole. 1972 F. J. Pettijohn et al. *Sand & Sandstone* iv. 114 The flute..preserved as a raised structure or flute cast on the underside or sole of the overlying sand bed.

9. a. sole-bar (also **solebar**), *spec.* a longitudinal member forming part of the underframe of a railway carriage or wagon; (further examples); **sole mark, marking** *Geol.*, a feature that is found on the undersurface of sedimentary strata which overlie softer beds, and is the cast of a depression originally formed in the surface of the lower bed; **sole-plate** (also **soleplate**), *spec.* the metal plate forming the base of an electric iron; (further examples).
1930 *Engineering* 24 Jan. 102/2 The main frame.. consists of two longitudinals or solebars. 1977 *Modern Railways* Dec. 486/1 Current new stock..has an all-aluminium underframe with the solebars made from continuous extrusions. 1961 J. Challinor *Dict. Geol.* 185/1 Sole-marks. 1972 J. F. Pettijohn et al. *Sand & Sandstone* I. iv. 113 Although they occur in almost all sands, sole marks are particularly abundant in turbidites where they provide the best means of determining current flow. 1978 Friedman & Sanders *Princ. Sedimentol.* iv. 110/2 (*caption*) Sole mark assemblage dominated by counterparts of flutes..and transverse scour marks. 1957 P. H. Kuenen in *Jrnl. Geol.* LXV. 231/1 In a number of papers..mention is made of the occurrence of various types of markings on the sole of the graywackes. The present paper aims at presenting a coherent record of these sole markings. 1976 R. C. Selley *Introd. Sedimentol.* vii. 211 Flutes, grooves and tool marks are three of the commonest sole markings found as interbed sedimentary structures. 1960 *Housewife* Apr. 86/2 The sole-plate, being extra thin, heats quickly. 1974 *Spartanburg* (S. Carolina) *Herald* 24 Apr. (Sears Advts. Suppl.) 2 Has a 21-vent soleplate. Steams up to 30 minutes at a low setting.

sole, *sb.*² Add: **1. c.** In the names of various dishes, as *sole bonne femme* [*bonne femme* s.v. *bonne C]; *sole* (*à la*) *Colbert* (see quot. 1877); *sole* (*à la*) *meunière* [*meunière *a.* and *adv.*]; *sole Véronique* (see quot. 1960).
1846 A. Soyer *Gastronomic Regenerator* 115 *Sole à la Colbert*,..sole..butter..chopped parsley..chopped tarragon and chervil..lemon juice. *Ibid.*, *Sole à la Meûnière*,.. sole..chopped onions..butter..lemon..cayenne pepper. 1877 E. S. Dallas *Kettner's Bk. of Table* 136 The sole of Colbert..is a fried sole which after being cooked is boned and then filled with maître d'hôtel butter and with lemon-juice. 1928 D. L. Sayers *Unpleasantness at Bellona Club* vii. 71 He..had a sole Colbert very well cooked. 1930 R. Lehmann *Note in Music* v. 205 He..ordered sole *bonne femme*, a mixed grill, salad, trifle, a welsh rarebit. 1960 *Good Housek. Cookery Book* 95/2 *Sole Véronique*,.. sole..mushrooms..wine..cream..grapes..butter. 1966 *Harper's Bazaar* Sept. 87/3 She..does *sole Véronique* with lichees instead of grapes. 1967 G. Greene *May we borrow your Husband?* 185 For a while the *sole meunière* gave them an excuse not to talk. 1978 F. Mullally *Deadly Payoff* xi. 142 The two burly men would..plough through a hearty meal of sea-food, sole *meunière* and Stilton. 1979 'L. Black' *Penny Murders* iv. 38 Kate.. declared she would make it a fish day..a sole. Should it be *Véronique* with white grapes, or *à la Duglére* [*sic*], cooked in white wine with tomatoes and shallots?

sole, *a.* Add: **10. c. sole-charge** *attrib. N.Z.*, (*a*) of a teacher: that has sole charge of a school; (*b*) of a school: having only one teacher; also *absol.*
1941 A. Curnow *Island & Time* 12, I am the sums the sole-charge teachers teach. 1944 H. Wilson *Moonshine* ii. 21 It's [*sc.* the school's] a sole charge. 1955 D. O. W. Hall *Portrait of N.Z.* ix. 171 Small country settlements have their 'sole charge' or single-teacher schools.

Soledon [sǫ·lĕdǫn]. Also **soledon.** [Blend of Soluble *a.* and *Caledon*, proprietary name of an earlier range of dyes.] A proprietary term for a range of water-soluble vat dyes derived mainly from anthraquinones. Freq. *attrib.*
1924 *Trade Marks Jrnl.* 24 Dec. 2898 *Soledon*... Dyes and dyestuffs,..Scottish Dyes Ltd. 1929 *Textile Colorist* XLVII. 568/1 Considerable interest has been expressed.. regarding another water-soluble vat dye, Soledon Jade Green. 1938 *Times* 11 Jan. 9/5 A third important recent advance was represented by the 'soledon' colours for cotton fabrics and linens and for rayons. 1952 K. Venkataraman *Synthetic Dyes* II. xxxiv. 1050 About 35 Indigosols and 25 Soledons have been marketed. 1964 *Official Gaz.* (U.S. Patent Office) 22 Dec. TM153 Imperial Chemical Industries Limited... *Soledon*... For dyes and dyestuffs. 1970 *Times* 30 Apr. 12/7 A limited edition of Shaw's works was bound in his [*sc.* J. E. G. Harris's] Soledon jade green cloth. 1971 R. L. M. Allen *Colour Chem.* x. 179 The water-soluble sulphuric ester salts were marketed as Soledon dyes by Scottish Dyes Ltd, and are now produced under the same name by ICI.

solemn, *a.* (*adv.* and *sb.*). Add: **10. b.** *solemn-eyed.*
1889 W. B. Yeats *Wanderings of Oisin* 59 Away with us he's going, The solemn-eyed. 1930 M. Mead *Growing up in New Guinea* ix. 151 A tiny curly grass skirt is fashioned.., and the solemn-eyed baby arrayed in it for a feast day.

solemncholy, *a.* Add: **A.** *adj.* (Earlier example.)

1772 P. V. Fithian *Jrnl.* 18 Dec. (1900) 27 Being very *Solemncholly* and somewhat tired, I concluded to stay there all night.

B. *sb. pl.* A solemn or serious mood. *rare.*
1834 W. A. Carruthers *Kentuckian in New York* I. 214 It drives away the solemncholies, and makes a fellow feel so good-natured, and so comfortable.

solemnify, *v.* (Earlier example.)
1780 J. Mainwaring *Sermons on Several Occasions* p. xxv, Some divines delight to sadden and solemnify their sermons with Scripture passages.

solemnsides (sǫ·ləmsəi:dz), *a.* and *sb.* [f. Solemn *a.* + *-sides* as in Sobersides.] **A.** *adj.* Excessively solemn or serious. **B.** *sb.* An excessively solemn or serious person.
1922 J. M. Barrie *Courage* 36 Courage. I do not think it is to be got by your becoming solemnsides before your time. 1957 J. Kirkup *Only Child* ii. 50 People who are 'solemn-sides' cannot account for self-mockery. 1959 I. & P. Opie *Lore & Lang. Schoolch.* ix. 155 Sobersides, solemnsides.

solenoglyph (solĭ·noglif). *Zool.* [ad. F. *solénoglyphe*, mod.L. *Solenoglypha* (A. H. A. Duméril 1853, in *Mém. Acad. Sci.* XXIII. 417), f. Soleno- + Gr. γλυφή carving.] A venomous snake belonging to a group characterized by extra venom glands and grooved fangs which can be retracted. Also as *adj.* Hence **solenogly·phous** *a.*, of or pertaining to a snake of this kind.
1913 G. A. Boulenger *Snakes of Europe* v. 56 The Proteroglyphs (Cobras, Coral-snakes, Sea-snakes) and the Solenoglyphs (Vipers, Pit-vipers, Rattlesnakes) may be regarded as the diverging extremes in the development of the poison apparatus. 1965 R. & D. Morris *Men & Snakes* viii. 177 With the vipers, pit vipers and rattlesnakes, we come to..the Solenoglyphous forms... These are the so-called 'folding-fang' snakes. 1968 R. D. Martin tr. *Wickler's Mimicry in Plants & Animals* xii. 112 The vipers have the most specialised tooth-form,..a longitudinal canal that functions as a poison syringe. This condition is called solenoglyph. 1969 A. Bellairs *Life of Reptiles* I. v. 195 If the skull of a poisonous snake is examined, whether it be a proteroglyph or a solenoglyph, it very often appears that there are two fangs in each maxilla, set side by side.

solenoid. Add: Also with pronunc. (sǫ·lĕnoid); the pronunc. (solĭ·noid) is no longer current. **1.** (Earlier example.)
1827 J. Cumming *Man. Electro Dynamics* 240 In the case of a straight solenoid (an electro-dynamic cylinder) θ is the angle between the axis of the solenoid and the extremity of the cylinder.

3. *attrib.* and *Comb.*, as *solenoid-operated* adj.; **solenoid brake**, a brake actuated by the movement of a core into or out of a solenoid when an electric current is passed through the latter; similarly **solenoid lock**.
1914 *Machinery* Dec. 328/1 A new automatic solenoid brake for crane, hoist, lift bridge and similar service has recently been developed. 1963 Jones & Schubert *Engin. Encycl.* (ed. 3) 1168 One type of solenoid brake adapted for mill, crane and hoist motors and similar classes of service, is so arranged that the brake mechanism is held in the off or release position by a coil and plunger. 1976 L. Deighton *Twinkle, twinkle Little Spy* iv. 41 The glass door..had an electric solenoid lock. I had to push the override. 1956 *Nature* 14 Jan. 84/2 Two solenoid-operated valves are incorporated in the flow line. 1971 *Jrnl. Gen. Psychol.* LXXXIV. 325 The four match stimuli were presented by the second projector through a solenoid operated shutter which varied exposure duration.

solenoidal, *a.* Add: *spec.* of a vector field: having no divergence anywhere, and hence expressible as the curl of another vector field. (Further examples.)
1873 J. C. Maxwell *Electr. & Magnetism* I. 23 The whole space can be divided into tubes of this kind provided $dX/dx + dY/dy + dZ/dz = 0$, a distribution of a vector quantity consistent with this equation is called a Solenoidal Distribution. 1909 J. G. Coffin *Vector Anal.* vi. 154 (*heading*) Separation of a vector point-function **W**, which has a vector-potential, into solenoidal or rotational and lamellar or irrotational components. 1933 H. Phillips *Vector Anal.* viii. 195 If two vectors satisfy these conditions their difference, being irrotational, solenoidal, piecewise continuous, and zero at infinity, is zero. 1972 A. G. Howson *Handbk. Terms Algebra & Anal.* xxxv. 175 A vector field satisfying div $\mathbf{f} = 0$, i.e. div $\mathbf{f}(\mathbf{x}) = 0$ for all $\mathbf{x} \in S$, is said to be solenoidal.

solenoidally *adv.* (earlier example).
1851 *Phil. Trans. R. Soc.* CXLI. 270 The distribution of magnetism in it is said to be solenoidal, and the substance is said to be solenoidally magnetized.

solera. Add: **1.** Also *attrib.*, as *solera wine*; **solera system** (see quot. 1965).
1961 *Times* 16 Jan. 13/5 Sherry is made on the solera system. 1965 A. Sichel *Penguin Bk. of Wines* III. 227 The solera system is..a..method of refreshing delicately flavoured old wines still in cask with small quantities of slightly younger wine of the same character in order to keep a continuous stock of mature wine of one type and character always available. 1972 *Times* (Spain Suppl.) 11 May p. iii/2 The sherries are *solera* wines.

‖ **solfège** (solfɛ̄ʒ). *Mus.* [Fr.] **a.** = Solfeggio *sb.* **b.** (See quot. 1954.) Also *attrib.*
1912 E. Ingham in J. W. Harvey *Eurhythmics Jaques-Dalcroze* 52 The solfège lessons are chiefly for ear-training and practical harmony. 1914 M. Gibb *Chassevant Method of Musical Educ.* ii. 19 In the 'First Solfège' there are no dots used, though there are many examples of tied notes. 1921 H. F. Rubinstein tr. *Jaques-Dalcroze's Rhythm, Music & Educ.* v. 78 One of these groups confined itself to studying solfège, the other commenced pianoforte lessons... The solfège students were then initiated into the study of the piano. 1936 *Times Educ. Suppl.* 13 June 218/2 The course [of Eurhythmics] will consist of daily lessons in rhythmic movement and *solfège*. 1954 *Grove's Dict. Mus.* (ed. 5) VII. 877/2 *Solfège*... Although derived from the Italian *solfeggio*, and originally confined to the same meaning, the French word is now much more comprehensive. It stands for the teaching of the rudiments of music, which includes ear-training as an important factor. 1970 W. Apel *Harvard Dict. Mus.* (ed. 2) 786/1 Extensive courses in solfège..were first introduced in France and Belgium.

solicit, *v.* Add: **4. d.** More recently, also with a homosexual (or a pimp) as subj. Also *intr.*
1956 *Act 4 & 5 Eliz. II* c. 69 § 32 It is an offence for a man persistently to solicit or importune in a public place for immoral purposes. *Solicitation by men.* 1962 *Law Rep.* 13 Mar. 666 The appeal of the appellant..against his conviction..that he, being a man, persistently solicited in a public place for immoral purposes..contrary to s. 32 of the Sexual Offences Act, 1956. 1983 J. Gardner *Elephants in Attic* xvii. 153 She was soliciting to cover her air fare.

solicitancy. (Later example.)
1890 W. James *Princ. Psychol.* II. xxvi. 551 Those persons obey a curiously narrow teleological superstition who think themselves bound to interpret them [*sc.* the impulsions] in every instance as effects of the secret solicitancy of pleasure and repugnancy of pain.

solicitor. Add: **3. c. Official Solicitor** (see quot. 1977).
1875 *Minutes of Evidence taken by Commissioners appointed to inquire into Administrative Departments of Courts of Justice* 344/1 in *Parl. Papers* (C. 1245) XXX. 163, I hold the office of official solicitor to the Court of Chancery. 1896 *Law Rep. Chancery Div.* I. 368 We have an officer of this Court who is called the Official Solicitor. 1961 [see *ad litem]. 1977 *Jowitt's Dict. Eng. Law* (ed. 2) II. 1281/2 The Official Solicitor of the Court of Chancery was an officer whose functions consisted of protecting the Suitors Fund... He is now known as the Official Solicitor of the Supreme Court... He acts for persons suffering under a disability; he acts generally as solicitor in cases in which the Chancery Division requires his services as solicitor; he visits persons in custody for contempt.

7. *U.S.* One who solicits business orders, advertising, etc.
1897 *Scribner's Mag.* Oct. 463/2 A small army of solicitors is despatched to a neighborhood to go from house to house telling people about the features of the paper. 1903 E. L. Shuman *Practical Journalism* 200 Have as many good solicitors out as necessary and make your rates low enough to invite this form of advertising. 1916 *John Bull* 13 May 10/2 He called himself a grocer's solicitor, meaning a canvasser for orders. 1926 *Publishers' Weekly* 15 May 1589 Why can't he leave it to the judgment of the printers? Or to the advertising solicitor? 1952 S. Eisenberg *How to earn Income selling Products & Services by Phone* i. 4/2 If you can handle the English language..you can be a telephone solicitor. 1976 D. Barnes *Yesterday is Dead* (1977) II. 207 'No Solicitors', a sign on the glass double doors announced.

solicitrix. Restrict ? *Obs.* to sense 2 and add: **1.** (Later example.) *rare.*
1961 L. P. Hartley *Two for River* 48 When accosted I had not, as some men have, a polite formula of refusal ready: I swerved or..walked straight on. But one evening I couldn't, for my solicitrix..planted herself in front of me and blocked my way.

solid, *sb.*¹ Add: **4. a.** (Earlier and later examples.)
1786 J. Woodforde *Diary* 18 July (1926) II. 258, I.. could eat no solids all day long. 1973 *Jrnl. Genetic Psychol.* CXXIII. 103 It can be seen also in Table 4 that infants did not eat solids at two weeks, but did at five weeks. 1977 W. H. Manville *Good-bye* iv. 40 Junior just began eating solids.

f. Special Comb.: **solids-not-fat** *Dairying*, the components of milk other than water and fats and other lipids (being largely lactose and proteins); the proportion of such components in a sample of milk; **solids pump**, a machine for forcing lumpy or granular material, or liquid containing it, through a pipe or chamber against the force of gravity.
1874 J. A. Wanklyn *Milk-Analysis* ix. 38 The effect of skimming is to diminish the proportion of fat, and to leave the proportion of 'solids not fat' unaltered. 1930 *Analyst* LV. 543 The resulting curves..do not indicate that the highest proportion of solids-not-fat is contained in the bottom part of the milk. 1960 *Farmer & Stockbreeder* 5 Jan. 99/1 Changes in solids-not-fat were small, and nearly 19 per cent of the tests in Friesian herds and 7 per cent of those in Ayrshire herds were below the legal minimum standard of 8·5 per cent. 1957 T. G. Hicks *Pump Selection & Application* xiv. 313 Solids pumps.. are designed to handle solutions containing large percentages of suspended abrasive materials. 1966 *McGraw-Hill Encycl. Sci. & Technol.* XII. 474/2 The solids pump has found its principal application in the operation of oil-shale retorts. Here it is used to feed crushed shale into the bottom of a conical vessel.

5. A solid rubber tyre. (No longer current.)
1919 *Brit. Manufacturer* Nov. 38/1, 50,000 pneumatic tyres a week, in addition to solids. 1924 A. W. Judge *Mod. Motor Cars* II. 178 Both ordinary and Giant Solids.

solid, *a.* Add: **I. 3. e.** *Astronautics.* Using solid fuel.
1949 G. P. Sutton *Rocket Propulsion Elem.* i. 10 Long duration solid rocket units require an excessively heavy and large combustion chamber. 1961 *Flight* LXXX. 650/2 A study of the requirements associated with the transport, handling, checkout, assembly, and launch of extremely large solid boosters. 1967 *Technology Week* 20 Feb. 13/1 The program will cover the design, development and demonstration of a controllable solid propulsion system using integral propulsion. 1979 J. W. Cornelisse et al. *Rocket Propulsion* ix. 169 Solid rockets find widespread military and civil applications.

II. 7. e. *Cards.* (See quot. 1927.)
1927 M. C. Work *Contract Bridge* 145 *Solid suit*, one of such length and strength as to be practically sure of winning every trick in that suit. 1959 *Listener* 12 Feb. 309/1 The jump after a forcing opening bid shows a solid suit. 1976 *Country Life* 1 Apr. 846/2 Even when your trump suit is solid, it may still be fatal to touch it too early.

8. b, c. (Earlier and later examples.) *solid South*, the politically united southern States of America; the unanimous vote of the white electorate in these States for the Democratic party.
1855 in P. S. Foner *Business & Slavery* (1941) 114 We are now beaten by the solid vote of the City of New York. 1858 S. Colfax *Let.* in O. J. Hollister *Life Schuyler Colfax* (1887) 137 We have fallen on strange times when the solid South in the House and a score of Northern Democrats dare to vote '*No*' on a resolution approving existing laws against the African slave trade. 1872 *Chicago Tribune* 14 Oct. 1/3 The Democrats are solid for Greeley in this county. 1876 *Harper's Weekly* 26 Aug. 691/2 We must recognize the solid South as the core of the Democratic party... The solid South is the Southern Confederacy seeking domination of the United States through the machinery of the Democratic party. 1880 *Sen. Rep. 46th U.S. Congress 2 Sess.* No. 693. 326 Q. These gentlemen..are both good Democrats? A. Yes, sir; they are solid Democrats. 1884, 1890 Solid South [in Dict.]. 1974 *Socialist Worker* 23 Nov. 16/5 With the exception of a handful of white scabs on one shift all four shifts are solid. 1977 *Chicago Tribune* 2 Oct. II. 2/5 It is possible for the GOP to revitalize itself by becoming the necessary counterweight to the newly reconstituted Solid South.

d. For *U.S.* read orig. *U.S.* and add earlier and later examples.
1882 G. W. Peck *Peck's Sunshine* 161, I was pretty solid with him. 1951 E. Paul *Springtime in Paris* iv. 80 He..went back to his native village in the Ain, where there are plenty of unregenerate Pétainists to this day, quite solid with the bishop and the anti-Jewish provincial administration.

10. a. (The 'material' is not necessarily pure: the implication is of homogeneity rather than purity, so that, e.g., articles made of plate are excluded but not those made of an alloy.) (Further examples.)
1910 *Jrnl. R. Soc. Arts* LVIII. 260/2 There remained a mere film..like silver foil... That is, I believe, a fair example of the so-called 'solid silver' sold in our swell shops, with the aid of much electric light and many bowing salesmen. 1926 J. P. De Castro *Law & Practice of hall-marking Gold & Silver Wares* I. 138 Though often legitimately used to differentiate between sterling silver and electro-plated silver, the expression 'solid silver' is frequently a much-abused term. 1962 L. S. Sasieni *Optical Dispensing* i. 3 Solid gold is not necessarily pure gold, but is an alloy of pure gold with other metals. The unalloyed pure metal is known as fine gold. *Ibid.*, The colour of solid gold can be varied by altering the proportions and the metals in the alloy. 1970 Choate & De May *Creative Gold- & Silversmithing* ii. 10 Any karat gold is called solid gold to distinguish it from gold-filled metals. 1981 *Daily Tel.* 24 Sept. 17 (Advt.), An absolutely perfect solid gold neckchain... These are not seconds or plated gold—but sound 9 carat gold!

III. 18*. *Austral.* and *N.Z. slang.* Severe, difficult; unfair.
1916 C. J. Dennis *Moods of Ginger Mick* 155 *Solid*, severe; severely. 1943 N. Marsh *Colour Scheme* ii. 35 You'd think it was royalty. They've been making it pretty solid for everybody down there. Hauling everything out and shifting us all round. 1948 R. Park *Harp in South* v. 62 After all, Auntie Josie's got all them kids to look after. It must be pretty solid for her with Grandma as well. 1959 E. Lambert *Glory thrown In* 66 They'll be solid on him for that, won't they?

18.** *U.S. slang.* In the language of jazz: excellent, first-rate, 'great'; (see also quot. 1937). Also as *int.* *solid sender*: see *sender d.
1935 *Vanity Fair* (N.Y.) Nov. 38/1 He puts a solid man like Joe on suitcase. 1937 *Amer. Speech* XII. 182/2 *Solid*, describes a player whose improvisation indicates that he is en rapport with the rhythm of the band... A band that is solid has a psychic unanimity of feeling. 1943 *N.Y. Times* 9 May II. 5/4 There has [*sic*] been some solid trumpet players who can really send. 1959 'F. Newton' *Jazz Scene* xii. 220 The hipster classifies what other people would call good as 'solid' or 'in there'. 1978 W. Hjortsberg *Falling Angel* (1979) xii. 54 'Park your axe and have a drink.' 'Solid.' He placed his saxophone case carefully on the table.

IV. Quasi-*adv.* **19. a.** (Later U.S. dial. and colloq. examples.) Also, certainly, surely.
1937 *Amer. Speech* XII. 232/2 'Are you taking Amelia

to the Charcoal Dance?' 'I solid am.' **1944** *Richmond* (Va.) *Times-Dispatch* 5 Oct. 6/3 Dowdy [said] he was going to leave, whereupon the Bayer woman said she'd kill him if he did... Dowdy told her that 'You'll solid have to kill me.' **1946** Mezzrow & Wolf *Really Blues* xii. 226 Not looking for trouble but solid ready for it. **1950** L. Hughes *Simple speaks his Mind* xx. 108 Man, if I had a rocket plane, I would rock off into space and be solid gone. Gone. Real gone! I mean gone!

c. *to book solid*: to sell all the tickets of (a theatre, cinema, etc.). Usu. in *pass.* Also *absol.*

1916 *Variety* 27 Oct. 12/1 The Boston opera house is booked solid until March. **1921** *Kinematograph Monthly Rec.* Feb. 4 So many individual exhibitors are refusing to book 'solid'. **1955** M. Allingham *Beckoning Lady* vii. 105, I told him the show was booked solid. **1967** N. Marsh *Death at Dolphin* ix. 234 We're booked out solid for another four months.

d. Of time: consecutively, without a break.

1938 M. Allingham *Fashion in Shrouds* xx. 379 I've had forty-eight hours solid and I'm no longer intelligent. **1964** L. Deighton *Funeral in Berlin* xliii. 270 He'll be out for eight hours solid.

20. *solid angle* (Math.), † (*a*) a vertex of a three-dimensional body; (*b*) a quantity associated with a vertex or the like in three dimensions, being proportional to the fraction of a sphere centred on it which would subtend it, and conventionally measured in steradians, of which 4π make up the whole sphere; *solid circuit* (Electronics) = *integrated circuit* s.v. *integrated *ppl. a.* b; *solid diffusion*, migration of atoms within the crystal lattice of a solid; *spec.* in *Geol.*, considered as a possible mechanism for a metasomatizing process in rock masses; *solid fuel*, fuel that is solid, rather than liquid or gaseous; *spec.* (*a*) coal, coke, etc., as opposed to oil, gas, or electricity for domestic heating; (*b*) as used in rocketry; freq. *attrib.*; hence *solid-fuelled* adj. (esp. of rockets); *solid geology*, the geological features of a given region specifically excluding superficial deposits such as clay, sand, etc.; opp. *drift*; *solid injection*, in diesel engines, the use of a mechanical pump to spray fuel into the cylinder at high pressure, without the use of compressed air; = **airless injection*; *solid solution*, a solid phase consisting of two or more substances uniformly mixed in proportions that can be varied; also, the state of being a constituent of such a phase; *solid stowing* (Coal Mining), the process of filling abandoned workings with solid material, esp. spoil; *solid system* (Electr. Engin.), a system of cable-laying in which insulated cables are laid in a trough which is then filled with bitumen; *solid tyre*, a tyre made of solid rubber, with no pneumatic cavity; so *solid-tyred* adj., fitted with such tyres.

1704 Solid angle [in Dict., sense 2 a]. **1798** C. Hutton *Course in Math.* I. 327 Similar Solids, contained by plane figures, are such as have all their solid angles equal..and are bounded by the same number of similar planes. **1814** P. Barlow *New Math. & Philos. Dict.* s.v., Solid angles may be computed and compared with each other, as to quantity, by considering the angular point as the centre of a sphere, and the portion of its surface intercepted between the bounding planes as the measure of the angles. **1820** N. J. Larkin *Introd. Solid Geom.* 5 [The tetrahedron] differs from every other solid, whose faces are all equal, by having a solid angle opposite to each face. **1928** *Bureau Standards Jrnl. Res.* (U.S.) I. 34 The integrals are to be evaluated..over the surface of a hemisphere (solid angle 2π). **1948** *Research* I. 394/2 Substances..similar to a black body radiating in a solid angle of 2π. **1958** *Proc. Internat. Symp. Electronic Components, Malvern, 1957* 4 The increasing tempo of work on solid state physics may result in solid circuits of another form. **1961** *Solid-State Electronics* II. 20/1 Work on solid circuits was begun in the United Kingdom in 1956... The objective is..the fabrication of working circuits using doping, shaping and other techniques in single crystals of silicon. **1966** *New Scientist* 30 June 846/3 (*caption*) Solid-circuit amplifier. The chip measures 0·075 × 0·075 in and contains 30 resistors and 30 transistors. **1913** *Rep. Brit. Assoc. Adv. Sci. 1912* 367 The question whether true solid diffusion ever occurs in minerals is very difficult to answer. **1947** *Geol. Mag.* LXXXIV. 218 More recent work has shown that such transformations are most easily explained as a result of solid diffusion. **1965** P. C. Badgley *Structural & Tectonic Princ.* ix. 346 The advocates of solid diffusion propose..large-scale granitization of country rocks without the presence of contemporary magmatic granites in the district. **1891** H. J. Phillips *Fuels* 1 Determining the value of a solid fuel such as coal, coke, or patent fuel. **1936** *Archit. Rev.* LXXX. 45/1 In electric fires, Tudor and Adam surrounds flourished, and some characteristics of the traditional solid-fuel fireplace were introduced into the designs. **1952** E. Burgess *Rocket Propulsion* ii. 38 Most British war rockets employed cordite, whereas America used ballistite for their solid-fuel rockets. **1960** *Which?* Jan. 7/2 In general..the most economic fuels are solid fuel and oil. **1971** P. J. McMahon *Aircraft Propulsion* x. 295 The most important factors which have led to the increase in interest in solid fuel rockets in recent years..have been concerned with the convenience and ease of storage of the solid propellants. **1979** H. McLeave *Borderline Case* xiv. 145 Who had cached these ..boots, solid-fuel heater, and two sleeping bags? **1958** *Economist* 13 Dec. 985/2 Work is likely to continue on the Titan as a reinforcement for the Atlas..until the solid-fuelled Minuteman is ready some years hence. **1972** *Nature* 21 Apr. 368/3 The decision to use a solid fuelled disposable booster for the proposed space shuttle effectively limits the number of possible launch sites to Cape Kennedy and Vandenberg Air Force Base. **1937** J. S. Flett *First 100 Yrs. Geol. Survey Gt. Brit.* v. 113 Many maps of north Yorkshire were issued only with 'Solid' geology. **1946** L. D. Stamp *Britain's Structure & Scenery* iv. 30 Following the practice of the Geological Survey in some of their detailed maps, there are to be two maps—one to show the 'solid' geology as it would appear if superficial deposits such as boulder clay, glacial sands and gravels and clay-with-flints..were removed and the other to show the 'drift' geology with all those surface deposits indicated. **1970** *Watsonia* VIII. 171 A map of the 'drift' as well as the solid geology would have been valuable. **1915** A. P. Chalkley *Diesel Engines for Land & Marine Work* (ed. 4) iii. 122 Solid injection is now being employed with a large number of engines installed in British submarines. **1936** [see *cold starting* s.v. *cold *a.* 19]. **1969** J. Flack et al. *Marine Combustion Practice* ii. v. 154 Airless or solid injection superseded the blast system... Mechanically operated injectors have given way in their turn to automatic injectors. **1890** *Jrnl. Chem. Soc.* LVIII. 1044 As instances of solid solutions, we have isomorphous mixtures and mixed crystals, amorphous solutions, as in the case of the glasses and certain minerals; and then such cases as the solution of hydrogen by palladium and other metals. **1900** *Proc. R. Soc.* LXVII. 109 Silver and copper are each capable of holding a small percentage of the other in solid solution, but..if both metals are present in considerable amounts, the two solidified solutions exist side by side. **1964** H. Hodges *Artifacts* xix. 215 In brasses containing less than 36% zinc a solid solution, the α-phase, is formed in which the zinc atoms enter the space lattice of the copper. **1977** A. Hallam *Planet Earth* 119 Such a series is called a solid solution series, and all members have the same crystalline structure. **1929** *Trans. Inst. Mining Engineers* LXXVI. 258 Solid stowing of the goaf is universal, and in most cases consists of dry stowing done by hand. **1964** *Times Rev. Industry* Feb. 48/3 The industry in this area has pioneered a method of dealing with subsidence known as 'solid stowing' in which colliery spoil is dampened and blown under pressure into abandoned seams. **1977** Down & Stocks *Environmental Impact of Mining* xii. 313 Solid stowing reduces *a* [*sc.* the 'subsidence factor'] to about 0·4–0·5. **1891** *Electr. Engineer* 30 Jan. 121/2 In the Callender Solid system the insulated cables are laid in an iron trough and the whole filled in solid with melted bitumen. **1898** [see *built *ppl. a.* 1 b]. **1920** *Whittaker's Electr. Engineer's Pocket-bk.* (ed. 4) 352 The B.O.T. raise no objection to the omission of the copper tape on unarmoured cables laid on the solid system; a great deal of such cable is in use without any leakage trouble. **1953** C. C. Barnes *Power Cables* xv. 154 The solid system is more expensive than burying cables direct in the ground and requires a greater measure of skilled supervision and favourable weather conditions... For the above reasons the solid system is seldom used today. **1891** *Bicycling News* 31 Jan. 77/1 Given a solid tyre..it will be found that about one half of its diameter is available for tractive and cushioning purposes. **1895** G. B. Shaw *Let.* 6 Aug. (1965) I. 540 It would be a bad machine even of its own kind, the art of building for solid tyres being a decaying one. **1946** W. H. Crouse *Automobile Engin.* xxvii. 572 Solid tires have very limited usage, their use being confined largely to specialized industrial applications. **1891** Solid-tyred [see Pneumatic *sb.* 3 a]. **1963** Bird & Hutton-Stott *Veteran Motor Car* 17 A horizontal-engined, twin-cylinder, chain-driven, solid-tyred 'dog-cart' was in production by the end of the year.

21. a. *solid-fuelled* (see sense *20), *-tyred* (see sense *20).

b. *solid-looking* (earlier example); **solid-drawn** *a.*, made or shaped by deep drawing (see *deep *a.* IV. c).

1888 *Lockwood's Dict. Mech. Engin. Terms* 335 The copper piping for feed, bilge, blow-off, and similar purposes in connection with marine engines..are all solid drawn. **1909** F. W. Raynes *Domestic Sanitary Engin. & Plumbing* vii. 167 Solid drawn lead pipes have many advantages as soil pipes. **1966** A. W. Lewis *Gloss. Woodworking Terms* 47 Best-quality brass hinges are 'solid drawn' because they are made by the leaf and the tube for the knuckle being drawn out in a long solid strip which is then cut off into lengths. **1840** Poe in *Graham's Mag.* Dec. 268/1 These were known by their coats and pantaloons of black or brown..with white cravats and waistcoats, broad solid-looking shoes, and thick hose.

c. *solid-propellant*, *-stem*; **solid-shot** *U.S.* (see quot. 1949).

1946 *Jrnl. Brit. Interplanetary Soc.* VI. 45 The missile is driven by a solid-propellant rocket unit. **1961** Solid-propellant [see *apogee 4]. **1982** *Navy News* Mar. 18/3 The missiles would be accelerated from rest by two solid-propellant boosters. **1935** *Daily Progress* (Charlottesville, Va.) 25 July 1/5 'Solid shot' votes are within the law in primaries as well as in general elections. **1949** *Richmond* (Va.) *Times-Dispatch* 25 Nov. 4/5 Under the 'solid-shot' method a voter casts his ballot for a single candidate when two or more persons are to be elected to the same office. **1906** E. Johnston *Writing and Illuminating* xii. 213 A solid-stem pattern cuts up the ground into small pieces. **1961** *B.S.I. News* July 23/2 Solid-stem calorimeter thermometers.

‖ **solidaire** (sǫlidēᵊ·ɹ), *a.* [Fr.] = Solidary *a.* 2.

1845 Macaulay *Let.* 1 Mar. (1977) IV. 244, I certainly did suppose that you considered yourself as solidaire for doctrines which the cabinet has repeatedly and emphatically proclaimed. **1877** W. R. Alger *Life of Edwin Forrest* I. i. 25 When volition put rigidity into his muscles the centre was solidaire with the periphery. **1942** Wyndham Lewis *Let.* 27 Jan. (1963) 316 But one cannot help feeling *solidaire* with the nation to which one belongs. **1962** *Times Lit. Suppl.* 3 Aug. 556/2 They must be made not 'solidaire' (to use M. Neveux's words) in a shared, superior understanding of the plight of the figure on the stage, but made 'solitaire' as they realize that they each, alone, share his fate.

solidarist. (In Dict. s.v. Solidaric *a.*) Add: Also *attrib.* or as *adj.* Hence **solidari·stic** *a.*

1957 [see *personalism b]. **1968** *Economist* 28 Dec. 27/3 Further evidence of an 'instrumental collectivism' as opposed to a traditional 'solidaristic collectivism'. **1969** P. Worsley in Ionescu & Gellner *Populism* 224 These independent commodity-producers..were not simply 'petty-bourgeois' individualists, as..their solidarist political associations demonstrate. **1974** B. Jessop *Traditionalism, Conservatism & Brit. Pol. Culture* ii. 32 The distinctive attribute of secular voters is an absence of solidaristic class consciousness rather than commitment to deferential norms.

solidarity. Add: **1.** (Earlier and further examples.) Also, *spec.* with reference to the aspirations or actions of trade-union members. Also *attrib.* and *Comb.*

Also the English rendering of Polish *Solidarność*, the name of an independent trade-union movement in Poland, registered in September 1980 and officially banned in October 1982.

1841 H. Doherty *False Assoc. & its Remedy* 24 Solidarity, Solidary. Collective responsibility. Collectively responsible. **1885** *To-day* III. 83 [Strike manifesto] But if, on the contrary, you design this strike as a step toward a final and definite solution of the great labour question, if you would make it the means of teaching the worker the absolute necessity of combination and of unity, if having secured the adoption of Solidarity you will build upon this a superstructure of Education, if you will learn why you are poor, [etc.]. **1962** *Listener* 31 May 935/1 These gangs have group-cohesiveness (in our present jargon) or solidarity (in socialist jargon), but they are against society. **1963** *Daily Tel.* 5 Feb. 10/2 Twice as many countries are attending this conference as were at the Afro-Asian States conference in Bandung in 1955; but the great difference is that those now meeting are merely 'solidarity organisations'. **1968** *Listener* 6 June 713/1 Well before the last election, sociologists were telling us that an increasing number of working-class people were beginning to look at politics instrumentally rather than in terms of class solidarity or ideological allegiance. **1969** *Ibid.* 30 Jan. 131/3 Afro-Asian Peoples Solidarity Committees display predictable signs at airports. **1971** I. Deutscher *Marxism in our Time* (1972) v. 109 The perennial conflict between national egoism and international solidarity becomes more and more visible. **1974** *Socialist Worker* 9 Nov. 6/4 The building workers called a solidarity strike. **1977** *Time* 4 July 7/3 In the months since then, Soviet ideologues have opened a campaign to increase 'fidelity to the principles of internationalist solidarity'—party jargon for rallying round Moscow's flag. **1979** *Time* 13 Aug. 12/3 'Solidarity' marchers arrived from Sanandaj, the Kurds' provincial capital. **1980** *Times* 26 Sept. 6/4 The Warsaw daily *Zycie Warszawy* quoted members of the Solidarity free trade union movement as rejecting reforms of the old unions as mere name-changing. **1980** *Economist* 18 Oct. 46/1 Over 20 unions, including Mr Lech Walesa's Gdansk-based Solidarity (an umbrella organisation representing 50 small unions, and claiming a total membership of over 4m), have applied to register with the courts in Warsaw. **1982** *Times* 9 Oct. 1/5 The Polish Parliament..yesterday voted..for a new trade union law that sounds the death knell of Solidarity. *Ibid.*, In broad outline, the bill dissolves all registered trade unions including Solidarity.

solidary, *a.* **2.** (Earlier example.)

1841 [see *solidarity 1].

solidity. Add: **3. c.** The ratio of the area of the blades of a propeller (counting one side only) to the area of the circle they turn in.

1926 H. Glauert *Elements Aerofoil & Airscrew Theory* xvi. 213 This quantity σ represents the ratio of the area of the blade elements to the area of the annulus at the radial distance *r*, and may be termed the solidity of the blade element. **1953** D. O. Dommasch *Elem. Propeller & Helicopter Aerodynamics* ii. 61 As far as propeller operation is concerned, increasing solidity has much the same effect as decreasing the aspect ratio of the wing. **1980** *Sci. Amer.* July 114/2 The twin screws of the 31-knot *Queen Elizabeth 2*, each of which absorbs 55,000 h.p., are appreciably smaller, turn much faster and have an even greater solidity than the screws of the largest tankers.

solid state. [f. Solid *a.* + State *sb.*] **1.** The condition or state of being solid rather than fluid.

1866 E. Atkinson tr. *Ganot's Elem. Treat. Physics* (ed. 2) vii. iv. 425 A metallic salt is introduced either in a solid state or in a state of solution. **1908** *Chem. Abstr.* II. 1654 The electrical conductivity of a series of natural silicates was measured in the solid and liquid states. *Ibid.* 2757 (*heading*) The solid state. **1959** *Cambr. Rev.* 6 June 597/1 The section on nuclear magnetic resonance is almost entirely concerned with wide-line studies on the solid state.

2. *attrib.* (Usu. with hyphen.) **a.** Concerned with the structure and properties of solids, esp. with their explanation in terms of atomic and nuclear physics; as *solid-state physics* (hence *solid-state physicist*).

1953 C. Kittel (*title*) Introduction to solid state physics. **1956** *Electronic Engin.* XXVIII. 63/1 Research in solid-state physics in recent years has yielded much information on the relations between the electrical

magnetic and elastic properties of metals. **1964** *New Scientist* 4 June 595 In his studies, the solid-state physicist must have some means of knowing what compounds are likely to be semi-conducting if he makes them. **1973** *Sci. Amer.* Jan. 98/3 Solid-state physics is currently faced with the problem of constructing an effective theory of noncrystalline, amorphous substances. **1974** *BP Shield Internat.* Oct. p. iii/3 He graduated as an applied physicist and subsequently as a solid state physicist. **1975** *Nature* 20 Nov. 274/2, I think that the solid-state theorist will enjoy this part of the monograph particularly.

b. (Employing devices) utilizing the electronic properties of solids (as in transistors and other semiconductor devices, in contrast to the partial vacuum of valves). Occas. *absol.*, such devices collectively.

1959 *Economist* 23 May 769/2 The transistor, best known of the 'solid-state' devices that employ these materials [*sc.* semiconductors], can do most of the jobs of a valve while taking far less space. **1961** *Times* 30 June 9/2 In this system solid-state electronic circuits..will replace conventional devices for interlocking signals and points. **1965** *Wireless World* July 350/2 Marconi's..showed their new solid-state 4½-inch image orthicon. **1968** *Times* 1 Nov. 23/4 In [computer] hardware, solid-state electronic devices have completely transformed the reliability, cost, speed and size of central processors. **1970** *New Scientist* 15 Oct. (Suppl.) 13/2 Solid state radars are required for airborne applications where size and weight are of prime importance. **1971** *Hi-Fi Sound* Feb. 40/1 (Advt.), All silicon solid state circuitry using 20 transistors and 2 diodes provides a full 20 watts r.m.s. output. **1975** G. J. King *Audio Handbk.* ix. 202 A high-gain solidstate amplifier. **1978** *Broadcast* 5 June 21/2 The system's information is stored in solid state, not (as one might expect) on floppy disc.

solidungular, *a.* For *rare*⁻¹ read *rare* and add later example.

1922 Joyce *Ulysses* 699 Smart phaeton with good working solidungular cob.

solidus[1]. Add: **2.** Also, such a mark used in writing fractions and for other separations of figures and letters. Cf. *oblique *sb.* 5.

1891, etc. [in Dict.]. **1923** N. Shaw *Forecasting Weather* i. 35 A solidus (/) such as occurs in the combination 'bc/r' separates weather at the time of observation from the preceding weather, bc/r thus indicating 'fine or fair after rain or drizzle'. **1947** [see *non-linear *a.* b]. **1971** *Archivum Linguisticum* II. 4 Johnson/Jenkinson's 'oblique dash'.., which is otherwise called a 'solidus' or 'virgule'.

solidus[2]. Add: [Adopted in this sense in Ger. by H. W. B. Roozeboom 1899, in *Zeitschr. f. phys. Chem.* XXX. 387.] Substitute for def.: A line or surface in a binary or ternary phase diagram respectively, or a temperature (corresponding to a point on the line or surface), below which a mixture is entirely solid and above which it consists of solid and liquid in equilibrium. Freq. *attrib.*, as *solidus curve, temperature*, etc. (Earlier and later examples.)

1901, etc. [see *liquidus]. **1933** Liddell & Doan *Princ. Metall.* iii. xvi. 501 The solidus plane of the system..lies at a lower temperature than do any of the binary eutectics concerned. **1935** [see *homogenized *ppl. a.* 1]. **1959** B. Chalmers *Physical Metall.* iii. 85 For a ternary alloy the liquidus and solidus are surfaces. **1965** G. V. Raynor in R. W. Cahn *Physical Metall.* vii. 325 The solidus temperature for a given alloy composition. **1967** A. H. Cottrell *Introd. Metall.* xv. 230 This gives a phase diagram with a retrograde solidus curve,..with the striking property that alloys of certain compositions..can become completely solid and then melt again on cooling.

solificatio (sōᵘ:lifikā·tio). *nonce-wd.* [An invented Latin word, formed on Solific *a.*: see -fication.] A radiating warmth as of sunshine.

1941 Auden *New Year Letter* iii. 74 Who on the lives about you throw A calm solificatio.

soliflual (sǫli·flⁱūǎl), *sb.* and *a.* *Physical Geogr.* [f. *soliflu(ction + -al.] **A.** *sb.* Material that has moved by solifluction. *rare.* **B.** *adj.* = *solifluctional *a.*

1941 *Trans. R. Soc. Edin.* LX. 376 Generally the layers basal to an aggradational series, following production of an erosion scarp, are re-sorted soliflual. **1965** B. T. Bunting *Geogr. of Soil* xii. 144 Sørensen classified high-arctic soliflual soils into those of inhomogeneous material ..and those of stoneless fines.

solifluction (sōᵘ-, sǫliflʌ·kʃən). *Physical Geogr.* Also **-fluxion.** [f. L. *sol-um* ground, earth + -i- + *fluction*, Fluxion.] **1.** The gradual movement of waterlogged soil or other surface material down a slope, esp. where the subsoil is frozen and acts as a barrier to the percolation of surface water.

1906 J. G. Andersson in *Jrnl. Geol.* XIV. 96 This process, the slow flowing from higher to lower ground of masses of waste saturated with water.., I propose to name *solifluction*. **1916** T. G. Taylor *With Scott* 115 These symmetrical polygons are due to a slow movement of half-frozen soil, which has been noted in polar lands, and is called solifluxion or soil-creep. **1936** *Geogr. Jrnl.* LXXXVII. 449 Solifluction also is known from the Alps, as for instance in the flysch region of the Segnas Pass. **1938** C. F. S. Sharpe *Landslides & Related Phenomena* iii. 35 The definition given by Andersson..does not limit solifluction to cold climates. From the cases he mentions, however, and from subsequent usages by Eakin.., Nichols.., and others, the word solifluction has come to be intimately associated with frost action. **1957** G. E. Hutchinson *Treat. Limnol.* I. i. 144 The presence of such ponds can lead to striking solifluction if the pressure of the water on the down-slope wall is great enough. **1965** M. Fieldes et al. in G. J. Williams *Econ. Geol. N.Z.* xx. 364/2 This solifluxion is thought to account for the emplacement of plastic clays occurring at..localities in the Wellington area. **1974** C. Taylor *Fieldwork in Medieval Archaeol.* iv. 85 Deep hollows on the faces of chalk scarps in Wessex are usually the result of periglacial action and solifluction.

2. *attrib.*

1946 F. E. Zeuner *Dating Past* 119 Solifluction deposits..are a conspicuous feature of sections from the periglacial zone. **1968** C. R. Twidale *Geomorphol.* ix. 274 Small solifluxion terraces and lobes are typical of slopes in periglacial regions. **1970** R. J. Small *Study of Landforms* i. 2 Our knowledge of the vital role of periglaciation..is based almost wholly on analysis..of the distribution and character of solifluxion gravels.

Hence **soliflu·ctional** *a.*, pertaining to or produced by solifluction; also (as a back-formation) **so·lifluncted** *a.*, that has moved by solifluction.

1924 *Geogr. Jrnl.* LXIII. 225 On well-developed solifluctional slopes of mixed material, the different-sized stones move downwards at different rates. **1954** *Sci. News* XXXIII. 70 On the continent, the task of dating solifluncted material is aided by interbedded sheets of loess. **1956** *Antiquity* XXX. 99 Weathered carboniferous sandstones (whose upper solifluxional levels are in several places separated by multi-coloured clays). **1971** J. N. Jennings in Jennings & Mabbutt *Landform Studies from Australia & New Guinea* xii. 271 Their hummockiness and the way they spread out distally with a bulging lower margin are features in favour of an end-moraine rather than a solifluctional origin. **1977** *Antiquaries Jrnl.* LVII. 187 Some sarsens..could be derived from chalk or Greensand as could the solifluncted spread in the Vale of Pewsey.

solifuge. Restrict † *Obs.*⁻⁰ to sense in Dict. and add: **2.** = Solpugid in Dict. and Suppl. Also **soli·fugid** (-dʒid) [-id³], in the same sense.

1925 R. W. G. Hingston in E. F. Norton *Fight for Everest, 1924* iii. 286 Solifugids occur up to 15,000 feet. **1935** *Discovery* Sept. 282/2 The dread of a Spider or a Solifuge is due to the speed of its movement. **1964** J. Hillaby *Journey to Jade Sea* 182 The total bag was one grasshopper, one dead beetle..and a spider-like animal called a solifuge. **1968** R. D. Barnes *Invertebrate Zool.* (ed. 2) xiii. 399/2 Solifugids possess voracious appetites and feed on all types of small animals, including vertebrates.

Solignum (sǒli·gnǒm). Also **solignum.** Proprietary name for a preservative for timber. Hence **soli·gnumed** *a.*, treated with this material.

1900 *Trade Marks Jrnl.* 17 Oct. 1105 *Solignum*... Preparations for the destruction of weeds, vermin, and insects. Major and Co. Ltd. Chemical manufacturers. **1909** *Chambers's Jrnl.* 27 Feb. 204/1 The preparation known as 'solignum'..is a fluid composition, applied..with a brush, possesses great covering capacity, and..improves with age both in appearance and resistant qualities. **1925** *Glasgow Herald* 19 June 8/8 One [way of dealing with white ants] is to keep painting the wood with solignum, and breaking down the trails whenever they appear. **1932** 'Dane' & Simpson *Re-enter Sir John* xx. 270 His neat little dressing-room, matchboard, salignumed [*sic*], sham-antique in the film-studio manner. **1969** *Observer* (Colour Suppl.) 12 Jan. 17/3 The walls continue with a cladding of weatherboarding, solignumed black, to the 12-ft level of the eaves.

soliloquacity (sǫ:lilǒkwæ·sɪti). [Blend of Soli(loquy *sb.* and Loquacity: cf. Soliloquacious *a.*] Soliloquizing at great length.

1895 *World* 30 Oct. 25/1 As he is soliloquising to this effect (he out-Hamlets Hamlet in soliloquacity), enter a letter from Mexico. **1967** *Time* 31 Mar. 40/3 Stephen's soliloquy on the beach, Bloom's trip to Paddy Dignam's funeral... Molly Bloom's magnificent end-spurt of soliloquacity.

solion. *Electronics. temporary.* [f. Sol(ution + Ion.] An electrochemical device consisting of two or more electrodes sealed in an electrolyte in which a reversible electrochemical reaction is monitored, versions of which are used as amplifiers, integrators, and as pressure transducers which also sense low-frequency sound and changes in temperature or acceleration. Freq. *attrib.*

1957 *N.Y. Times* 23 June 24/4 It has been nicknamed 'solion', which is short for ions in solution. *Ibid.*, The laboratory [*sc.* the U.S. Naval Ordnance Laboratory] expects that the solion will make possible cheaper, smaller and simpler electronic control systems. **1962** *New Scientist* 2 Aug. 254/1 The solion tetrode..is used as an integrator or as a low-frequency amplifier. **1966** *McGraw-Hill Encycl. Sci. & Technol.* XII. 482/2 The solion diode uses platinum electrodes in an aqueous solution that may be iodine and potassium iodide. *Ibid.* 483/1 The solion pressure transducer..measures fluid flow through an orifice separating two electrolyte chambers.

solipsism. Add: (Earlier and later examples.) Also, = Egoism 1, and in weakened sense.

1874 A. C. Fraser *Sel. from Berkeley* 47 Ueberweg suggests that Berkeley's reasoning implies that we can know only *our own* notions of what we call *other* spirits—thus leading, by a *reductio ad absurdum*, to Egoism or Solipsism. **1978** *Poetry* Aug. 298 The deep underlying motive of Mark Strand's poetry is solipsism or loneliness of the individual imagination.

solipsist. Add: **B.** *adj.* Favouring or characterized by solipsism; also in weakened sense.

1903 A. E. Taylor *Elements of Metaphysics* iii. ii. 202 Why..did Berkeley..accept neither the solipsist nor the sceptical conclusion? **1927** V. McNabb *Cath. Ch. & Philos.* iii. 101 His [*sc.* Kant's] own words are ingenuously solipsist! **1972** *Last Whole Earth Catalog* 16/2 Solipsist tyrants, believing that their will, like their eyeballs, could move mountains, have come to believe that it should trample over these small annoying figures in their visual field.

solipsistic, *a.* Add: (Earlier and later examples.) Also in weakened sense.

1885 W. James in *Mind* X. 37 Men who see each other's bodies sharing the same space..will never practically believe in a pluralism of solipsistic worlds. **1952** A. Wilson *Hemlock & After* ii. iii. 145 His intense, solipsistic world of personal ambition. **1958** I. Murdoch *Bell* xiv. 184 But now, driven by this fit of solipsistic melancholy one degree more desperate, she felt the need of an act. **1968** A. Storr *Human Aggression* xi. 104 Psychopaths share with the schizophrenic the characteristic of living in a world which is predominantly solipsistic; that is, in which people and events are not valued in and for themselves, but only in so far as they affect the subject. **1971** E. Shorris *Death of Great Spirit* iv. 66 The vision of Western man—seeing himself in the central role—might have been considered solipsistic only two hundred years ago. **1977** *Dædalus* Summer 42 It expanded, not by conflicts and deals with equals, but by short spurts of solipsistic exuberance at the expense of much weaker neighbors.

Hence **solipsi·stically** *adv.*

1898 W. James in R. B. Perry *Tht. & Char. of W. James* (1935) II. 370 Take me solipsistically if you will. My talk is merely a description of my present field of experience. **1923** *Times Lit. Suppl.* 23 Aug. 549/2 There are traits of experience which almost all of us are accustomed to explain solipsistically—our dream-worlds. **1952** *Mind* LXI. 10 We each solipsistically confined ourselves to statements which we may properly claim to know to be true directly. **1981** *Times Lit. Suppl.* 8 May 512/3 The real tramps aren't..made miserable by self-consciousness. For the most part, they seem solipsistically content, boozily quarrelling.

solitary, *a.* Add: **1. e.** *solitary wave*, a travelling, non-dissipative wave which is neither preceded nor followed by another such disturbance.

[**1837** J. S. Russell in *Trans. R. Soc. Edin.* (1840) XIV. 61 This accumulated mass..appeared to roll forward alone along the surface of the quiescent fluid, a large, solitary, progressive wave.] **1838** Russell & Robison in *Rep. Brit. Assoc. Adv. Sci. 1837* 418 This wave had been called the great solitary wave of the fluid. **1876** *Phil. Mag.* I. 262 The very different behaviour of solitary waves according as they are positive or negative... In the former case, the wave has a remarkable permanence, being propagated to great distances without much loss. **1899** [see *equivoluminal *a.*]. **1952** Russell & Macmillan *Waves & Tides* i. ii. 44 The velocity of solitary waves of small height is: √[g(depth of water + wave height)]. **1976** *Nature* 8 Apr. 510/2 Figure 2*a* and *b* shows typical streamline patterns for waves corresponding to solitary waves of elevation (E solitons) and of depression (D solitons).

3. c. (Later examples.)

1916 J. Buchan *Greenmantle* v. 62 There was nothing the Boche liked so much as an excuse for sending a poor devil to 'solitary'. **1924** W. M. Raine *Troubled Waters* xxvi. 262 'He's been in solitary for a week,' explained the warden. **1963** M. Duggan in C. K. Stead *N.Z. Short Stories* (1966) 101 Bread and water and solitary and take that writ on his eyeballs. **1978** T. Allbeury *Lantern Net* xi. 164, I visited prisoners in solitary every other day.

soliton (sǫ·litǫn). *Physics.* [f. Solit(ary *a.* + *-on¹.] A solitary wave (see *Solitary *a.* 1 e); a quantum or quasiparticle propagated in the manner of a solitary wave.

1965 Zabusky & Kruskal in *Physical Rev. Lett.* XV. 240/1 Each such 'solitary-wave pulse' or 'soliton' begins to move uniformly at a rate..which is linearly proportional to its amplitude. **1967** *Ibid.* XIX. 1096/1 The solitons exhibit a remarkable stability in that their identity is preserved through nonlinear interactions. This property of solitons..was discovered numerically and justifies the name suggestive of particles. **1968** *Trans. Amer. Geophysical Union* XLIX. 209/2 Steep waves in shallow water have nonlinear properties similar to those exhibited by interacting 'solitons', nonlinear dispersive wave entities that arise in solutions of the Korteweg-de-Vries..equation. **1976** [see *solitary wave* s.v. *solitary *a.* 1 e]. **1979** *Physica Scripta* XX. 306/1 Solitons appear in many fields of our life ranging from classical fluids, solid state and elementary particle physics to biophysics.

solitudinous, *a.* (Earlier example.)

1803 S. Pegge *Anecdotes of Eng. Lang.* 312, *-ous* is a termination which carries weight with it, and might be

admitted, as in *multitudinous*, and other similar words in which it has obtained a situation; as,—magnitudin*ous*, gratitudin*ous*, solitudin*ous*, plenitudin*ous*, &c.

soljanka, var. *solyanka.

sollar, *sb.*¹ Add: The form *solar* is now usual in sense 1 a, esp. when used *Hist.*

sollicker (sǫ·likəɹ). *Austral. slang.* Also **soliker.** [Of unknown origin.] Something very big, a 'whopper'. Hence **so·llicking** *a.*, 'whopping'.

1898 R. Graeme *From England to Back Blocks* 82 Who was it I heard that in cutting-out some cattle on one of the Methvin plains, did come down a soliker and broke his horse's knees? **1899** 'S. Rudd' *On our Selection* 64 He kicked Farmer what he afterwards called 'a sollicker on the tail'. **1939** Franklin & Cusack *Pioneers on Parade* 168 She gave me a sollicker of a dose out of a blue bottle. **1946** K. Tennant *Lost Haven* (1947) x. 155 It was a great big sollicking stitch if ever there was one. **1956** P. White *Tree of Man* i. vii. 91 'You can jump down, can't you? You're quite big, you know.' 'Of course he can... He's a sollicker.'

sol-lunar, *a.* Delete *Med.* and add: Also **solunar.** (Further examples.)

1936 J. A. Knight *Mod. Angler* xviii. 198 It is well to use a descriptive term or word instead of referring to 'inland tides' or 'the conditions which cause tides'. For this purpose the word 'Solunar' has been coined. Solunar ..may be defined as follows: The time at which the conditions which cause ocean tides (*i.e.* the pull of the sun and the moon) pass the longitudinal meridian of any given point is the Solunar period at that point. **1962** E. Bruton *Dict. Clocks & Watches* 162 *'Solunar' dial*, daily tidal times, as shown on some wrist-chronographs and used in conjunction with 'solunar tables' (J. Alden Knight) to forecast feeding times of fish and game for sportsmen. **1971** *Nature* 31 Dec. 537/2 The phase lag in the incidence of large earthquakes with respect to overhead position of the Sun and Moon is coincident with the imposition of horizontal force in such a direction and sense as to increase accumulated shear strain if this is due to the solunar torque.

solo, *sb.*¹ and *a.* Add: **A.** *sb.* **III. 6. a.** (Earlier examples.)

1814 C. Jones *Hoyle's Games Improved* 189 The quotient shews the number of fish to be paid to each of the successful players by the other two; or in event of a Solo to be paid him by each of the three others. **1875** W. B. Dick *Mod. Pocket Hoyle* (ed. 7) 144 *Solo.*—This is an announcement to accomplish the same ends as in bidding to play in 'suit', but without the aid of the Scat cards.

b. *Solo Whist* (earlier example); also *ellipt.*; *Heart Solo* (earlier example).

c **1875** W. B. Dick *Mod. Pocket Hoyle* (ed. 6) 146 Thus a player announcing Heart Solo (worth six counters), and having in his hand four Matadores, can bid Heart Solo *with* four Matadores, equal to ten counters, [etc.]. **1888** Wilks & Pardon (*title*) How to play solo whist. **1907** [in Dict.]. **1972** C. Drummond *Death at Bar* v. 129 They had looked forward to a cosy evening of cocoa and solo with..the other boys.

7. Solo flying; a solo flight.

1911 *Flight* 16 Sept. 805/1 Capt. Watt made a very good solo round Fargo and Stonehenge, landing exceptionally well. **1928** T. E. Lawrence *Let.* 20 Jan. (1938) 569 All decent birds hop it when their infants have done their first solo. **1942** R. Hillary *Last Enemy* iii. 74 The flight immediately following our first solo was an hour's aerobatics. **1976** B. Jackson *Flameout* iii. 44 'How long to get your license?' 'Thirty-five hours to solo, if you're good.'

B. *adj.* **1. a.** *spec.* with reference to flying.

1914 H. Rosher *In R.N.A.S.* (1916) 15 Hope to be flying solo by Thursday or Friday. **1928** *Daily Mail Year Bk.* 24/1 They can obtain the use of a machine in which to fly solo for £1 an hour. **1934** *Sun* (Baltimore) 22 Oct. 2/1 Lieut. M. Hansen,..who is flying solo, left Athens at 9.11 A.M. for Baghdad. **1946** *Happy Landings* July 4/1 The sergeant-pilot..had amassed the considerable sum of two hours solo on Oxfords. **1955** *Times* 24 Aug. 6/4 Eight years later Colonel Lindbergh took 33hr. 30min. in his monoplane to fly solo non-stop from New York to Paris. **1977** 'J. Herriot' *Vet in Spin* x. 113 'I said take her up.' 'You mean, on my own?.. Go solo?'

b. Acting alone or without assistance (*spec.* in *N.Z.* of single parents).

1934 *Sun* (Baltimore) 2 Mar. 19/8 The sportsman pilot ..is variously designated as 'private', 'solo' and 'student' pilot. **1965** A. Blackshaw *Mountaineering* 19 The risks to the inexperienced *solo* rock climber or snow-and-ice climber are very great. **1966** P. O'Donnell *Sabre-Tooth* vii. 100 She wondered..if an army of mercenaries was being assembled... But no—he was very much a solo man. **1977** *N.Z. Herald* 8 Jan. ii. 2/7 So far, the research has shown that few solo mothers are out to skin the welfare state by claiming a domestic purposes benefit and living on boyfriends.

2. (Later example.)

1927 *Glasgow Herald* 18 Mar. 11 London..has only two 'solo' machines. Its members, nevertheless, contrived to put in a total of 84 hours flying during..January.

3. (Earlier example.)

1862 [see *euphonion].

4. That is achieved or performed unaccompanied or unassisted.

1909 *Flight* 18 Sept. 576/2 After making a short 'solo' flight he came down. **1914** *Daily Express* 22 Sept. 7/4 A solo effort by Cantrell, who weaved his way prettily through the defence, brought the winning goal. **1927** *Glasgow Herald* 30 Sept. 11 He has made the longest solo flight so far achieved by an airman. **1940** *Daily Progress* (Charlottesville, Va.) 25 Jan. 9/4 Police claim she admits one solo holdup, made to prove her nerve. **1944** *Ibid.* 12 Sept. 6/6 The trends within medicine which make solo practice no longer..the best..kind of service for the physician or for the patient. **1955** *Times* 22 Aug. 8/5 Lord de L'Isle and Dudley, V.C., Secretary of State for Air,..has logged 13 hours' solo flying. **1974** *Times* 6 Dec. 5/5 (*caption*) The Prince of Wales after making his first solo deck landing of a Royal Navy Wessex helicopter.

Solo (sōᵘ·lo), *sb.*² The name of a river in Java, used *attrib.* with reference to a neanderthaloid fossil hominid, *Homo soloensis*, known from skulls discovered at Ngandong in the valley of the Solo river in 1931.

1932 *Discovery* Aug. 240/2 A new type of early man..is to be called 'Solo Man' from the name of the river in the Pleistocene gravels of which the skull was found. **1951** *Anthrop. Papers Amer. Mus.* XLIII. 205 (*title*) Morphology of Solo Man. **1973** B. J. Williams *Evolution & Human Origins* xi. 185/1 The Solo skulls..appear to be more primitive than Neandertal. *Ibid.*, Solo Man tends to have an occipital torus higher on the skull than did the African Neandertals. **1978** B. G. Campbell *W. E. Le Gross Clark's Fossil Evidence for Human Evolution* (ed. 3) iii. 120 The main problem in this connection lies in the status of the Solo population.

solo (sōᵘ·lo), *v.* [f. Solo *sb.*¹ and *a.*] **1.** *intr.* To perform an action on one's own; *spec.* (*a*) to perform a vocal or instrumental solo (now usu. in jazz); (*b*) to fly solo; *spec.* to make one's first solo flight; (*c*) *Mountaineering*, to climb without a partner.

1886 W. Booth *Orders & Regul. Salvation Army* iii. ii. 96 All cannot solo or speak eloquently. **1917** J. M. Grider *War Birds* (1927) 57, I have been flying for three days and Capt. Harrison says I can solo to-morrow if it's calm. **1931** V. W. Pagé *ABC of Gliding* 164 In training glider pilots the student usually 'solos' from the very start. **1932** *Daily Progress* (Charlottesville, Va.) 25 Feb. 4/3 The most magnificent bodega was recently ordered to remove its enormous bar and is now trying to solo to prosperity as a restaurant only. **1942** R. Hillary *Last Enemy* iii. 72 Here for the first time was a machine in which there was no chance of making a dual circuit as a preliminary. I must solo right off. **1956** B. Holiday *Lady sings Blues* (1973) vi. 60 Whenever Basie had an arranger work out something for me, I'd tell him I wanted Lester to solo behind me. **1958** P. Gammond *Decca Bk. of Jazz* xx. 249 With Reinhardt and Grappelly soloing over the pulsating guitars—bass rhythm section. **1962** E. Snow *Red China Today* (1963) xxiv. 183 All the items I have mentioned are primary sinews of a modern industrial civilization, the development of which enables a nation to 'solo' as a major industrial power. **1964** J. E. B. Wright in Murray & Wright *Craft of Climbing* v. 35, I solo-ed up the Slab Climb. **1971** M. Tak *Truck Talk* 148 *Solo*, 1. to drive a tractor without a trailer... 2. to drive a rig alone when the driver in question is usually part of a two-man operation. **1972** D. Haston *In High Places* iv. 57 But what to do? Thoughts of soloing down and alerting a rescue party, but that would have meant a major operation. **1977** *National Observer* (U.S.) 1 Jan. 13/3 If the river is easy, it's fun to solo. But on a formidable stream you need a partner.

2. *trans.* To perform (a piece of music) as a solo. *rare.*

1858 *Punch* 8 May 184/2 The sweetness of his oratory would be completely wasted on the air of '*Keemo Kimo*' soloed by the ophicleide.

3. *Mountaineering.* To climb (a mountain, etc.) without a partner.

1962 *Listener* 8 Nov. 758/2 Not that you climb alone; only very talented fanatics or complete fools 'solo' hard routes. **1975** G. Moffat *Miss Pink at Edge of World* xiv. 194 Solo-ing steep rock at Clive's age is just not on.

Hence **so·loing** *vbl. sb.*; also *attrib.*

1929 *Papers Mich. Acad. Sci., Arts & Lett.* X. 324/2 *Soloing*, flying alone. **1971** C. Bonington *Annapurna South Face* 324 *Soloing*, climbing without the security of a rope. **1973** —— *Next Horizon* v. 90 Dick..was unaccustomed to fast soloing, and eventually they had turned back before even reaching the foot of the climb proper. **1977** *Rolling Stone* 24 Mar. 66/2 His guitar style combines Hendrix-inspired production technique with virtuoso soloing ability.

Solochrome (sōᵘ·lǒkrōᵘm). Also **solochrome.** A proprietary name for a range of synthetic dyestuffs used esp. in chemical analysis in colour tests for various metals, notably aluminium. Usu. *attrib.* in names of particular dyes.

1924 *Trade Marks Jrnl.* 23 Apr. 910 *Solochrome*... Chemical substances used in dyeing... British Dyestuffs Corporation Limited. **1938** *Analyst* LXIII. 266 A test for aluminium..has been based on the vivid orange-red fluorescence obtained with aqueous alcoholic solutions of Solochrome Red ERS and Solochrome Violet RS. **1960** A. G. E. Pearse *Histochem.* (ed. 2) xxiv. 693 Other Solochrome dyes give coloured, non-fluorescent complexes with aluminium hydroxide. **1976** *Lancet* 4 Dec. 1209/2 Undecalcified 6μm sections were prepared and stained with toluidine blue, pH 2·8, and solochrome cyanin.

solod (sǫ·lət). *Soil Science.* Also **soloth.** Pl. (sometimes const. as *sing.*) **solodi, soloti**; also **solods.** [a. Russ. *sólod'*, f. *sol'* salt.] A type of soil derived from a solonetz by leaching of saline or alkaline constituents, having a pale, leached subsurface horizon, and occurring characteristically under grass or shrub vegetation in semi-arid and desert regions.

1925 S. A. Waksman tr. *K. K. Gedroits's Soil Absorbing Complex & Absorbed Soil Cations* (U.S. Dept. Agric.) 13 The process of forming a new type of soil from an alkali soil as a mother soil leads to secondary soils ('soloti'), which are distinguished from primary podsol soils both in origin and in the properties of the formation. **1933** *Soil Sci.* XXXVI. 181 The solodi profiles of the semiarid short grass plains zone. *Ibid.* 184 (*caption*) A well-developed solodi. **1934** *Ibid.* XXXVIII. 484 After the complete formation of Soloth with the removal of the mobile colloids, the continued growth of the native grasses again changes the soil to that normal for the region. **1953** *Proc. Soil Sci. Soc. Amer.* XVII. 287 Solod soils ordinarily occur in the lowest local position in the landscape, usually a deep depression, while Solonetz and solodized Solonetz soils develop on level areas or in slight depressions. **1963** D. W. & E. E. Humphries tr. *Termier's Erosion & Sedimentation* xv. 324 The soloti are the degraded alkaline soils derived from solonetz soils by solotization, a process analogous to podsolization. *Ibid.* 325 Solotis are known in the U.S.S.R. and in the western United States. **1974** E. A. Fitzpatrick *Introd. Soil Sci.* vii. 119 Solods can be regarded as leached solonetzes in which the upper horizons are strongly bleached becoming pale grey or white.

Hence **solo·dic** *a.*, being, resembling, or characteristic of a solod; **so·lodize** *v. intr.*, to change into a solod; **solodiza·tion** (also **solot-**), the formation of a solod by the leaching of salts from a solonetz; **so·lodized** (**solot-**) *ppl. a.*, altered by this process.

1925 S. A. Waksman tr. *K. K. Gedroits's Soil Absorbing Complex & Absorbed Soil Cations* (U.S. Dept. Agric.) 14 Just as the carbonates of calcium and magnesium protect the soil of forest zone from podsolization, these salts protect alkaline soils from 'solotization'. *Ibid.* 15 The question of the presence of absorbed hydrogen in solotized soils is still not clear. **1932** E. J. Russell *Soil Conditions & Plant Growth* (ed. 6) iv. 269 In this process of solodisation, the percolating water continues to remove sodium and other products of decomposition from the soil complex, replacing the exchangeable sodium by hydrogen, and depositing the products of decomposition lower down. **1934** *Soil Sci.* XXXVIII. 484 As soon as solonetz forms it immediately begins to solotize with the development of a profile approaching the soloth. **1964** *Jrnl. Soil Sci.* XV. 176 The absence of solodized-solonetz from the deep sands is related to the coarser texture of the material of which they are composed. **1968** H. C. T. Stace et al. *Handbk. Austral. Soils* vi. 153 In the solodized solonetz, solodic soils and red-brown earths the topsoil is acid.., but all..contain secondary carbonate in the deeper horizons. **1978** Faniran & Areola *Essent. Soil Study* viii. 183 Salinization, solonization, and solodization..resulting in the formation of solonchaks, solonetz, and solodic soils respectively.

solograph. For *rare*⁻⁰ read *rare* and add *attrib.* example.

1851 C. Cist *Sketches & Statistics of Cincinnati in 1851* 187 Hawkins, in addition to his daguerrotypes, produces what he terms a *solograph* picture.

soloistic (sōᵘlo,i·stik), *a.* [f. Soloist + -ic.] Of, pertaining to, containing, or of the nature of, soloists or solo parts. Also *transf.* Hence **solo,i·stically** *adv.*

1947 A. Einstein *Music in Romantic Era* xvi. 285 Three acts..: each one in two parts, the first part 'soloistic', the second always uniting all the figures of the farce, as the tradition of *opera buffa* demanded. **1952** B. Ulanov *Hist. Jazz in Amer.* (1958) xiv. 165 Edward Inge an alto, and Horace Henderson and Don Kirkpatrick on piano were other soloistic assets of the band. **1961** *Times* 9 June 17/1 The cellos are strung out in front of the basses, which..can cause the cello tone to be diffused when the section is used soloistically. **1974** *Daily Tel.* 11 Feb. 10/4 Every member of the choir seems to command a soloistic projection yet the overall sonority is magnificently blended. **1975** *Gramophone* Aug. 316/3 Above all, the position of the solo viola is too soloistic. **1977** *Country Life* 2 June 1494/3 The commitment, enthusiasm, and general lack of soloistic egotism of almost all involved was delightful to observe. **1980** *Early Music* Jan. 19/1 Soloistic music needs a more generous portion of personal sound.

Solomon Islander (sǫ·lǒmǒn ai·lǽndəɹ). [f. *Solomon Islands* (see below) + -er¹.] A native or inhabitant of the Solomon Islands in the southwest Pacific.

1864 in C. M. Yonge *Life John Coleridge Patteson* (1874) II. ix. 96 Two of the Solomon Islanders distinguished themselves by jumping off the foreyard, and diving under the ship. **1911** F. Coombe *Islands of Enchantment* iii. i. 239 Perchance it is as well that the Solomon Islanders had about 300 years in which to forget the first Christian emissaries before the next visited them! **1951** R. Firth *Elements of Social Organization* iii. 92 The Solomon Islander is no less sensitive to the appellation of country bumpkin than is his European counterpart. **1968** *World Book Encycl. Australasia* II. 323 In 1896, Solomon Islanders killed five Austrian scientists who were attempting to climb their sacred mountain.

Solon. Add: Also (with small initial) in the *U.S.*, in weakened sense: a legislator, congressman.

1903 'O. Henry' *Art & Bronco* in *Ainslee's* Feb. 59/1 The season of activity and profit that the congregation of the solons bestowed. The boarding houses were corralling the easy dollars of the gamesome lawmakers.

1948 E. Pound *Pisan Cantos* (1949) lxxxiv. 128 Thus the Solons, in Washington. **1959** *New Scientist* 29 Jan. 230/3 Congress was not in session at the time of the AAAS meeting... But the solons are back in the Capitol now. **1976** *National Observer* (U.S.) 21 Aug. 11/2 The great national political community of solons, scribes, policemen, [etc.].

solonchak (sǫ·lǫntʃæk). *Soil Science.* Also **solontschak,** etc. [a. Russ. *solonchák* salt marsh, salt lake, f. *sol'* salt.] A type of salty, alkaline soil that has little or no structure, is characteristically pale in colour, and occurs typically under salt-tolerant vegetation in poorly-drained semi-arid or desert regions.

1925 S. A. Waksman tr. *K. K. Gedroits's Soil Absorbing Complex & Absorbed Soil Cations* (U.S. Dept. Agric.) 9 The stage of formation of a saline soil (solontshak) will in this case always take place before the stage of formation of an alkaline soil (solonetz). **1927** C. F. Marbut tr. *Glinka's Great Soil Groups of World* 40 There is no word in English carrying the same meaning conveyed by the expression 'alkali soils with definite structure'. To avoid the use of so long a phrase the Russian word Solonetz will be used in the following pages while the word Solontschak will be used for 'alkali soils without pronounced structure'. **1939** *Agric. in 20th Cent.* 177 Where the soil water tended to move upwards, as in parts of the hot, dry countries, and especially where it contained salts in solution, another set of changes took place and the resulting soils were called 'solontchaks'. **1963** D. W. & E. E. Humphries tr. *Termier's Erosion & Sedimentation* xv. 325 At the research station at Hamadena, the soils are dominantly solonchak, secondarily characterized by being sodic, magnesian and calcic. **1972** J. G. Cruickshank *Soil Geogr.* iv. 145 Where sodium salts exceed 2 per cent of the mineral matter, a salic horizon is produced which may even be a salt crust on the soil surface under extremely dry conditions and high groundwater table. The soil is called a solonchak.

solonetz (sǫ·lǫnets). *Soil Science.* Also **-nez, -nietz.** [ad. Russ. *solonéts* salt marsh, salt lake, f. *sol'* salt.] A type of alkaline soil that is rich in carbonates, consists characteristically of a hard, dark, columnar subsoil overlain by a thin, friable surface layer, and occurs in conditions similar to those associated with solonchaks but having better drainage.

1924 S. A. Waksman tr. *K. K. Gedroits's Ultramechanical Composition of Soils* (U.S. Dept. Agric.) 17 These are soils which have become saline with sodium salts (saline soils) and alkaline soils ('solonez' soil) formed from these as a result of removal of the salt. **1925, 1927** [see *solonchak]. **1932** E. J. Russell *Soil Conditions & Plant Growth* (ed. 6) iv. 268 The last stage, when the solonietz has been exposed to prolonged leaching, is the solod of the Russians. **1938** [see *planosol]. **1968** H. C. T. Stace et al. *Handbk. Austral. Soils* vi. 153 The essential features of solonetz are prominent texture differentiation with an abrupt boundary between loamy A horizons and clay B horizons, neutral to alkaline surface soil and strongly alkaline subsoil [etc.]. **1972** J. G. Cruickshank *Soil Geogr.* iv. 127 (*caption*) A solonetz profile in South Australia showing strong prismatic structure in the B horizon.

Hence **solone·tzic** *a.*, being, resembling, or characteristic of a solonetz.

1935 *Soil Sci.* XL. 465 A strongly solonetzic soil from Chongar. **1974** E. C. Stacey *Peace Country Heritage* ii. 74 If the parental material has a high salt content, a hard-pan solonetzic soil will result.

solonization (sǫ:lǫnaizēⁱ·ʃən). *Soil Science.* [f. *solon(etz + -ization.] The formation of a solonetzic soil by the leaching of salts from a solonchak. So **so·lonized** *ppl. a.*

1934 *Soil Sci.* XXXVIII. 483 This process is called solonization and might be said to consist of desalinization plus alkalinization. **1945** *Ibid.* LIX. 420 The physical and morphological feature is the compactness of the solonized section of the profile. **1948** *Queensland Jrnl. Agric. Sci.* V. 19 These solonized patches are comparatively widespread throughout the habitat of 'sandalwood'. **1963** D. W. & E. E. Humphries tr. *Termier's Erosion & Sedimentation* xv. 324 In a region of saline soils, the areas undergoing solonization become depressions where structural change brings about compaction of the soil.

soloth, var. *solod.

solpugid. (Earlier and later examples.)

1869 A. S. Packard *Guide to Study of Insects* 655 Under the term Pedipalpi we would embrace..the Solpugids and Phalangids. **1912** J. H. Comstock *Spider Bk.* i. 34 Most solpugids spend the day under stones or other rubbish.. and come forth at night. **1954** Borror & DeLong *Introd. Study of Insects* xxx. 787 The body of a solpugid is about ½ inch long and is somewhat constricted in the middle, and the chelicerae are very large.

solubility. Add: **4.** Special Combs.: **solubility curve** *Chem.*, a curve showing how the solubility of a substance varies with temperature; **solubility product** *Chem.*, the product of the concentrations (*spec.* the activities) of each of the component ions present in a saturated solution of a sparingly soluble salt.

1892 *Jrnl. Chem. Soc.* LXII. 1384 (*heading*) Solubility curves of pairs of salts. **1933** A. K. Goard *Physical Chem.* ii. 36 A break in a solubility curve always indicates that the composition of the solid in equilibrium with the saturated solution has changed. **1971** Solubility curve [see *invariant *a.* b]. **1899** J. Walker *Introd. Physical Chem.* xxvi. 307 The solubility product of silver chloride in water is very small, corresponding to the very slight solubility of the salt. **1978** P. W. Atkins *Physical Chem.* xii. 361 The magnitude of K_s, the solubility product, can be predicted from a knowledge of standard electrode potentials.

solubilization (sǫ:liŭbiləizēⁱ·ʃən). [f. next + -ation.] The process of making something (more) soluble.

1930 *Brit. Chem. Abstr.* B. XLIX. 1061/1 (*heading*) Solubilisation of a perylenetetracarboxylic di-imide dye and dyes produced thereby. **1946** *Industr. & Engin. Chem.* June 642 (*heading*) Solubilization of insoluble organic liquids by detergents. **1977** *Jrnl. Protozool.* XXIV. 17/2 The selective solubilization of protein C permitted us to determine its source.

solubilize (sǫ·liŭbɪləiz), *v.* [f. L. *solūbil-is* Soluble *a.* + -ize.] *trans.* To increase the solubility of; to convert into a soluble form. Also *absol.*

1926 [implied at solubilizing *vbl. sb.* below]. **1930** *U.S. Patent 1,776,971* (*title*) Process of solubilizing a perylenetetracarboxylic di-imide dyestuff. **1947** *Jrnl. Soc. Chem. Industry* Jan. 4/2 Free lauryl sulphonic acid solubilises slightly, even in chloroform. **1970** *Jrnl. Neurobiol.* I. 331 The tissue segments were solubilized. **1977** *Nature* 17 Nov. 272/1 After adding hydroxylamine to complex the excess retinal, the rhodopsin was solubilised by adding..octyl glucoside.

So **so·lubilized** *ppl. a.*, **so·lubilizing** *vbl. sb.*; also **so·lubilizable** *a.*, capable of being solubilized.

1926 *Brit. Pat.* 275,267 3/2 The use of products..as emulsifying agents or solubilising agents for organic solvents. *Ibid.* 280,647 1 (*title*) Manufacture of a solubilised vat dye and process of dyeing therewith. **1949** E. Chain in H. W. Florey et al. *Antibiotics* II. xviii. 748 A solubilizing effect of impurities on other salts of benzylpenicillin..was observed. **1955** *Times* 1 July 16/2 The therapeutic advantages of solubilized aspirin are as much appreciated in a compound including phenacetin and codeine as they are in Disprin itself. **1978** *Nature* 6 Apr. p. xxv/3 Bio-Rad introduce a new solubilisable gel for polyacrylamide gel electrophoresis. **1979** *Experientia* XXXV. 280/2 For the various thicknesses of sections, the ratio of radioactivity level of the intact section to that of the solubilized one was expressed in percent.

solubilizer (sǫ·liŭbiləizəɹ). [f. prec. + -er¹.] A solubilizing agent.

1963 *Chem. Abstr.* LIX. 6203 As solubilizers, Tween 20, 60, 80, Carbowax 1500, and Myrj 52 were used. **1979** *Experientia* XXXV. 280/2 The solubilizer..was added to yield the same concentration of scintillator molecules as in the section undergoing solubilization.

soluble, *a.* Add: **2. b.** (Further examples.) *soluble blue* (also *Soluble Blue*), any of a class of water-soluble dyes that are di- and trisulphonic acid derivatives of aniline blue and are now used chiefly in papers and inks. In *Biochem.* applied to those species of RNA now usu. known as **transfer RNA*.

[**1862** E. C. Nicholson *Brit. Patent 1857* 3 A colourless solution is obtained which, when neutralized.., developes the improved soluble blue dye.] **1879** *Jrnl. Chem. Soc.* XXXVI. 418 In preparing the soluble blues, the monosulphonic acids and higher-substituted acids must be of great purity. **1893** T. E. Thorpe *Dict. Appl. Chem.* III. 562/1 Perhaps the readiest way of preparing soluble starch is that recommended by Zulkowski.., who finds that starch dissolves in hot glycerin and is converted into the soluble modification. **1952** K. Venkataraman *Chem. Synthetic Dyes* II. xxiii. 723 Sulphonic acids of phenylated Rosanilines, which are old dyes (Nicholson Blue, Soluble Blue, Water Blue, Alkali Blue; CI 703–707) continue to be extensively used. **1958** M. B. Hoagland et al. in *Jrnl. Biol. Chem.* CCXXXI. 256 Evidence is presented that a soluble ribonucleic acid..binds amino acids in the presence of adenosine triphosphate. **1961** *Nature* 13 May 582/1 The other principal (10–15 per cent) form of RNA in *E. coli* is soluble RNA (now more appropriately called transfer RNA), which functions in the movement of activated amino-acids to the ribosomes. **1964** G. H. Haggis et al. *Introd. Molecular Biol.* xii. 306 Because of its special function this RNA is now generally called transfer-RNA, although it is also sometimes referred to as soluble-RNA, or as acceptor-RNA. **1971** R. L. M. Allen *Colour Chem.* viii. 115 A marked bathochromic effect is obtained by phenylation of the amino groups in rosanilines... These Soluble Blues are now little used for textile coloration, but are applied to leather, paper and.. in printing inks. **1973** *Times* 18 Oct. (Brazil Suppl.) p. v/2 Soluble (instant) coffee exports have grown at a remarkable pace. **1974** E. Ambler *Doctor Frigo* iii. 185, I gave her some soluble aspirin and left. **1977** *Whitaker's Almanack 1978* 914/1 The chief exports are cotton, coffee, beef, gold, sugar, cottonseed, bananas, copper and soluble coffee.

5. b. *Math.* = *solvable *a.* 3 b.

1902 *Encycl. Brit.* XXIX. 140/1 A group defines uniquely the set of factor-groups that occur in its composition series... When the orders of all the factor-groups are primes the group is said to be soluble. **1940** *Trans. Amer. Math. Soc.* XLVII. 393 A group may be termed soluble, if it may be swept out by an ascending (finite or transfinite) chain of normal subgroups such that the quotient groups of its consecutive terms are abelian groups of finite rank. **1972** [see *quotient 1 b].

B. *sb.* A soluble constituent, esp. of a foodstuff.

1952 *Poultry Sci.* XXXI. 937/1 There was no change in the hatchability of eggs from the hens receiving condensed fish solubles. **1962** M. N. Hill *Sea* I. vi. 305 (*heading*) Solubles. **1972** *Brit. Jrnl. Nutrition* XXVIII. 221 The main growth-promoting effect of fish solubles has been shown to be mediated through the intestinal microflora of the chick.

solum. Delete ‖ and add: **2.** *Soil Science.* (Pl. **solums, sola.**) The upper part of a soil profile, in which the soil-forming processes predominantly occur; *spec.* the A and B horizons. [Introduced (as G. *solumhorizont* solum horizon) by B. Frosterus 1922, in *Compt. Rend. de la Conf. Extraordinaire (IIIème Internat.) Agropédologique* (1924) 361.]

1928 *Proc. & Papers 1st Internat. Congr. Soil Sci.* IV. 7 To the soil body has been given definitely the status of a well defined concept by the application to it of a name consisting of a single term, the *Solum*, by Frosterus. **1942** *Technical Bull. U.S. Dept. Agric.* No. 834. 49 All but small quantities or traces [of carbonates] are leached from the solums of all but the Clyde profile. **1956** *Soil Sci.* LXXXII. 451 In the coarser sand fractions (0.5–2mm.) weatherable mineral material such as granite, diorite, and basalt is abundant in all horizons of the sola. **1972** J. G. Cruickshank *Soil Geogr.* ii. 41 The minimum moisture storage in the solum will be equivalent to between 10 and 15 cm water per unit area.

solunar, var. Sol-lunar *a.* in Dict. and Suppl.

solus, *a.* Add: **1. b.** *solus cum sola* [L.]: alone with an unchaperoned woman; *solus cum solo* [lit. 'alone with (oneself) alone']: all on one's own.

1611 Coryat *Crudities* 404 They sing merily together, but especially that sweet & most amorous song of *solus cum sola*. **1700** Dryden *Fables* 226 Stretching his Neck, and warbling in his Throat, *Solus cum Sola*, then was all his *Note*. **1742** R. North *Life of Francis North, Ld. Guilford* 242 But he was in the midst of all the court, *solus cum solo*, alone by himself. **1818** Lady Morgan *Florence Macarthy* II. iv. 202, I shall have the honour to drink your ladyship's health, *solus cum solo*. **1831** [see *chaperonless *a.*]. **1940** H. H. Henson *Jrnl.* 24 Mar. in *Retrospect* (1950) III. ii. 88 In the afternoon I had tea with Mrs. Carnegie *solus cum sola*.

3. Of things. **a.** *Advertising.* Of an advertisement: that stands alone; sometimes, that deals with one item (e.g. one book) only. Also, pertaining to such advertisements. Also *transf.* and *absol.*

1937 Partridge *Dict. Slang* 800/1 *Solus*, an advertisement on a page containing no other advertisement. **1952** *Economist* 5 Apr. 18/2 The advertiser would no doubt be charged varying rates equivalent to 'solus position'. **1958** *Times Lit. Suppl.* 5 Dec. 699/1 Within four days of publication *The Middle Age of Mrs. Eliot* had received solus feature reviews in *The Observer*, *The Sunday Times*, [etc.]. **1974** *Bookseller* 4 May 2154 (Advt.), Massive promotion. 10 giant solus advertisements in the National Press. **1980** *Financial Rev.* (Sydney) 18 Apr. 42/3 Rod Muir's station does have a rate card which concentrates on selling solus spots—60 seconds to two minutes. Mr Muir thinks radio has lost a lot of its effectiveness with the clutter of 30 second ads.

b. *Comm.* Of an outlet for the sale of oil and petrol: that sells the products of one company only. Also, of or pertaining to such an arrangement.

1957 *Economist* 7 Dec. 836/2 Whether a solus site offering three grades of one brand of petrol provides less choice than a mixed site offering three brands of the same grade is a question for chemists and advertising men to answer. **1958** *Spectator* 11 July 62/3 The Government had struck a blow against monopoly..by banning 'solus' petrol stations from the new motorways. **1965** *Economist* 7 Aug. 542/1 The major companies selling petrol in Britain have been sure for many months now that the Monopolies Commission would not find the 'solus' or 'tied garage' system, in itself, to be against the public interest. **1976** *Drive* Jan.–Feb. 10/1 One of the busiest solus sites (where only one brand of petrol is sold) in the country.

solute, *sb.* Add: Now usu. with pronunc. (sǫ·liūt). **2.** (Earlier and later examples.)

1893 F. G. Donnan in *Nature* 27 Dec. 200/2 Corresponding to the words 'solvent' and 'solution', some word is very badly wanted to express 'the dissolved substance'. The analogous word is evidently 'solute', and it is as short and euphonious as the others. **1978** P. W. Atkins *Physical Chem.* viii. 216 When a solute is present there is an extra randomness present in the solution that was not present in the pure solvent.

solution, *sb.* Add: **II. 6. b.** = *rubber solution* s.v. *rubber *sb.*¹ 14.

1897 A. C. Pemberton *Complete Cyclist* vii. 190 The edges of the cut must be well cleaned and coated with solution. **1930** F. Gardner *How to repair your Cycle* 20 Spread a thin layer of solution over the part that has been rubbed. **1974** S. Townroe *How to mend your Bike* 44 Never put on the patch until the solution has dried.

8. b. *Physical Geogr.* Denoting features and phenomena resulting from the solvent action of water, as *solution basin, depression, subsidence*, etc.

1894 Solution lake [in Dict., sense 8 a]. **1931** *Jrnl. Geol.* XXXIX. 641 (*heading*) Solution depressions in sandy sediments of the coastal plain in South Carolina. **1934** C. R. Longwell et al. *Outl. Physical Geol.* iv. 73 In some regions sinks, caverns, and solution valleys are so numerous that they give rise to a peculiar and characteristic topography with many surface depressions, irregular drainage patterns, and disappearing streams. **1939** A. K. Lobeck *Geomorphol.* iv. 145 Very large sinks or areas of depression are known as solution basins. **1954** *Geol. Mag.* XCI. 225 A lowering of surface by solution subsidence has been proceeding more or less continuously since Triassic times on this high moorland terrain. **1957** E. E. Evans *Irish Folk Ways* ii. 18 The lowland of which it is the centre is..diversified by occasional low hills, by intricate solution lakes and by glacial eskers. **1977** *Antiquaries Jrnl.* LVII. 189 Three large masses..recovered from red clay-with-flints filling a solution pipe at Aston Rowant (Oxon.).

IV. 12. Special Combs.: **solution heat treatment** = *solution treatment* below; **solution set** *Math.*, the set of all the solutions of some equation or condition; **solution treatment** *Metallurgy*, a process designed to render an alloy susceptible to age-hardening, by which it is first heated to make a particular constituent enter into solid solution and then quenched; so **solution-treat** *v. trans.*, **solution-treated** *ppl. a.*

1935 G. E. Doan *Princ. Physical Metallurgy* vi. 192 This heating and quenching is called the 'solution heat treatment'. **1979** J. Neely *Pract. Metallurgy & Materials of Industry* xiv. 187/2 Successful solution heat treatment depends on putting the copper into solid solution and trapping it there. **1959** Allendoerfer & Oakley *Fund. Math.* vi. 102 Given a universal set X and an equation $F(x) = G(x)$ involving x, the set $\{x|F(x) = G(x)\}$ is called the solution set of the given equation. **1963** Webber & Brown *Basic Concepts Math.* viii. 166 The truth set of an equation is often called the solution set, and members of the solution set are called solutions of the equation. **1972** A. G. Howson *Handbk. Terms Algebra & Anal.* i. 5 Those objects of a given set which satisfy an open statement form the solution set of the statement relative to that set. Thus the solution set of $x^2 = 2$ relative to the real numbers is the set $\{+\sqrt{2}, -\sqrt{2}\}$. Relative to the rational numbers.., the solution set of $x^2 = 2$ is empty. **1940** J. D. Jevons *Metall. of Deep Drawing & Pressing* xv. 577 Rolled sheet could be 'solution-treated' to place it in a ductile condition. **1977** R. B. Ross *Handbk. Metal Treatment & Testing* 207 Components which have been correctly Solution treated and aged are very often joined by [welding]. **1952** *Jrnl. R. Aeronaut. Soc.* LVI. 235/1 With highly tapered spar booms it is advisable to machine in the solution-treated condition in order that any correction..may be made before precipitation. **1931** *Metallurgist* VII. 12/2 The age hardening..of duralumin and a number of other alloys of similar type requires a previous heat treatment, which is frequently termed the 'solution' treatment. **1970** P. C. Varley *Technol. of Aluminium & its Alloys* v. 62 All solution treatment is carried out in forced air circulation furnaces.

solutioned (sǒlᴵū·ʃənd), *ppl. a.* [f. Solution *v.* + -ed¹.] Treated or covered with solution.
1898 G. L. Hillier *Cycling for Everybody* 81 The solutioned surface should be left alone for some minutes, which allows of the evaporation of the naphtha. **1909** 'R. Andom' *On Tour with Troddles* 264 We strengthened the cover with strips of rubber and solutioned canvas.

solutionist (sǒlᴵū·ʃənist). [f. Solution *sb.* + -ist.] One who solves problems or puzzles; *spec.* an expert solver of crossword puzzles.
1885 *Liverpool Mercury* 24 Oct. 5 A large proportion belong to the party of the Right... M. Paul de Cassagnac calls them 'Solutionists'. **1915** *Competition Prize-Winner* I. 1 Our supporters may rest assured that no solutionist will be permitted to use our advertising space unless.. his methods are absolutely fair. **1926** *Weekly Dispatch* 29 Aug. 3/3 The insertion of clues capable of alternative solutions..gives the ordinary reader an equal chance with the professional solutionist. **1930** *Aberdeen Press & Jrnl.* 3 May 8/2 If you are keen on winning one of the big cash prizes..for competition enthusiasts, why not avail yourself of the help of expert solutionists?

solutionizing (sǒlᴵū·ʃənaiziŋ), *vbl. sb.* [f. Solution *sb.* + -ize + -ing¹.] The process of forming a solution; *spec.* = *solution treatment* s.v. *solution *sb.* 12. So **solu·tionized** *ppl. a.*
1950 *Chambers's Jrnl.* Mar. 189/2 A recently evolved solutionising system, which is used in conjunction with an overhead irrigation plant, allows the soluble plant-foods to be added at any desired rate. **1977** R. B. Ross *Handbk. Metal Treatment & Testing* 206 The hardening of steel, where [solution treatment]..is occasionally referred to as 'Solutionizing' or 'Austenizing'. **1978** *Jrnl. R. Soc. Arts* CXXVI. 690/2 An aluminium alloy containing 6 per cent copper, Hid54, set in a freshly solutionized condition, was proposed as the rivet material.

solutizer (sǫ·liutaizəɹ). [f. L. *solūt-*, ppl. stem of *solvere* Solve *v.* + -ize + -er¹.] = *solubilizer.
1939 *Refiner & Natural Gas Manufacturer* XVIII. 171/2 It was found that the addition of suitable organic solvents or salts ('solutizers') to an aqueous alkaline solution increased the solubility of the resulting solution for mercaptans. **1957** van der Have & Verver *Petroleum & its Products* v. 197 In the so-called solutizer process the extracting medium is an alkaline solution to which substances have been added for promoting the solubility and reducing the hydrolysis of the alkali mercaptides in water. In one form of such a method, the 'solutizers' are the sodium and potassium salts of certain organic acids.

Solutrian, Solutrean, *a.* Add: [ad. F. *solutréen* (G. de Mortillet *c* 1867).] **B.** *sb.* **a.** The Solutrean culture.
1928 C. Dawson *Age of Gods* i. 14 The Aurignacian culture of Europe..was replaced by the sudden intrusion of a new culture—the Solutrean—which originated in Eastern Europe or Asia. **1946** F. E. Zeuner *Dating the Past* ix. 289 When the loess phase of this glacial phase was at its climax, Solutrian appeared, north of the Alpine mountain chains, and in Spain. **1969** G. Clark *World Prehistory* (ed. 2) iii. 61 During the time of the late Solutrean and Early Magdalenian the artists ceased to engrave limbs as though they were hanging from their back-bones. **1975** J. G. Evans *Environment Early Man Brit. Isles* ii. 49 The French Solutrean, an industry of unsurpassed elegance in its flint work.
b. A person of the Solutrean period. *rare.*
1944 H. G. Wells *'42 to '44* 178 The distribution of the Solutreans gives a range quite in accordance with the ideas of W. H. Riddell.

solvable, *a.* Add: **3. b.** *Math.* Of a group: that may be regarded as the last of a finite series of groups of which the first is trivial, each being a normal subgroup of the next and each of the quotients being Abelian.
1892 E. Netto *Theory of Substitutions* xiv. 267 We may carry over the expressions 'transitive', 'primitive' and 'non-primitive', 'simple' and 'compound' from the group to the equation... Conversely, we apply the term 'solvable', which is taken from the theory of equations, also to groups, and speak of solvable groups as those whose equations are solvable. **1898** *Amer. Jrnl. Math.* XX. 277 The necessary and sufficient condition that a group is solvable is that its αth derivative (derived group) is unity. **1929** *Ibid.* LI. 494 The total number of groups of order 72 is 50. Each of these groups is obviously solvable. **1971** D. Gorenstein in Powell & Higman *Finite Simple Groups* ii. 66 The celebrated Feit–Thompson theorem that groups of odd order are solvable implies that every nonabelian simple group has even order. **1982** *Sci. Amer.* Apr. 120/3 An equation is solvable by radicals if and only if the Galois group of the equation is a solvable group.

solvate (sǫ·lvēᴵt), *sb. Chem.* [f. Solv(e *v.* + -ate¹.] A more or less loosely bonded complex formed between a dissolved species and the solvent.
1905 *Amer. Chem. Jrnl.* XXXIV. 489 The existence of hydrates or solvates (in the case of non-aqueous solvents) in one form or another is an old conception. **1922** A. W. Stewart *Physico-Chem. Themes* 269 Another form of explanation is arrived at by assuming that solvates are formed on the addition of the salt. **1969** T. C. Waddington *Non-Aqueous Solvents* ii. 18 Because of its ability to donate its lone pair of electrons to form a covalent coordinate link, as well as by ion-dipole attraction, ammonia forms many solvates.

solvate (sǫlvēᴵ·t), *v. Chem.* [f. Solv(e *v.* + -ate³.] *trans.* To form a solvate with (a dissolved species). Usu. as **solva·ted** *pa. pple.* and *ppl. a.* Also (*rare*) *intr.*, to undergo solvation. Hence **solva·ting** *ppl. a.*
1909 *Publ. Carnegie Inst.* No. 110. 104 When a salt of one of these elements is dissolved in a solvent both the molecules of the salt and the ions formed from them become solvated. **1913** H. C. Jones *New Era Chem.* ix. 165 Non electrolytes solvate very slightly. The electrolytes combine with large amounts of the solvent. *Ibid.*, It is the ions which are the chief solvating agents. **1932** Philbrick & Holmyard *Text Bk. Theoret. & Inorg. Chem.* xi. 312 Hydrogen chloride is a strong acid only in solvents in which it is solvated. **1958** *Proc. Nat. Acad. Sci.* XLIV. 429 Strongly hydrogen-bonding organic solvents have been found to solvate polypeptides and proteins completely. **1968** V. Gutmann *Coordination Chem. in Non-Aqueous Solutions* ii. 34 With increasing donor properties of the solvent its solvating properties become stronger. **1971** *Nature* 1 Jan. 13/2 The study of the solvated electron has its origins in the work begun by Kraus at the beginning of this century on the blue solutions formed by dissolving alkali metals in liquid ammonia and amines. Electrical conductivity and other measurements have established that the solvated electron is responsible for the blue colour of these solutions. **1976** J. Daintith *Dict. Physical Sci.* 4/2 In solution the hydrogen ion, H^+, is solvated by water and often considered to be a hydroxonium ion, H_3O^+. **1977** *Sci. Amer.* July 95/1 The solvated electrons in an alkali-metal-ammonia solution should behave like unpaired electrons in an atom and hence should make the solutions paramagnetic. *Ibid.* 98/3 The entire complex can then readily be solvated by a polar liquid such as an amine or an ether.

solvation (sǫlvēᴵ·ʃən). *Chem.* [f. Solv(e *v.* + -ation.] The process of becoming, or state of being, solvated. Freq. *attrib.*
1909 *Amer. Chem. Jrnl.* XLI. 41 The theory of solvation in solution. **1917** M. H. Fischer tr. *Ostwald's Introd. Theoret. & Appl. Colloid Chem.* 51 These colloids are characterized by their great hydration or solvation. Their particles have taken up a large amount of the dispersion media. **1936** R. W. Gurney *Ions in Solution* i. 4 If a free gaseous ion enters a solvent..energy is liberated. This is known as the Solvation Energy, and has a characteristic value for each ion in each solvent. **1948** *Nature* 31 Jan. 170/1 When the concentration of organic solvent is increased beyond 50 mol. per cent, organic solvent molecules begin to predominate in the medium and eventually replace the aqueous solvation shell. **1967** Margerison & East *Introd. Polymer Chem.* v. 239 The questions associated with solvation, its extent, the type of chemical bonding, the number of solvent molecules involved,..are largely unanswerable. **1976** *Nature* 3 June 435/1 Relaxation rates for ¹⁷O in monosaccharides have been used to study solvation, the revised estimate of the solvation number being 5±1 water molecules for D-glucose and 2·5±1 for D-ribose at 5 °C.

Solvay (sǫ·lvēᴵ). *Chem.* [The name of Ernest *Solvay* (1838–1922), Belgian chemist, who developed the process.] *Solvay* (or † *Solvay's*) *process*: a method of making sodium carbonate using brine, ammonia, and carbon dioxide (which is usu. made as part of the process, by calcining limestone); also called *ammonia-soda process*. Hence *Solvay plant*, etc.
1879 *Sci. Amer. Suppl.* 12 Apr. 2719/3 From these injurious impurities the Solvay soda is almost absolutely free. **1884** *Jrnl. Soc. Chem. Industry* 29 Dec. 633/1 In the Solvay process there have been introduced important improvements. **1888** C. L. Bloxam *Chem.* (ed. 6) 262 (*heading*) Ammonia soda process, or Solvay's process. **1947** Kirk & Othmer *Encycl. Chem. Technol.* I. 402 A 500-ton-per-day soda plant requires gas compressors of about 3000 to 4000 horsepower to handle the suction on the kilns and to compress the gas into the Solvay precipitating towers. **1950** *Thorpe's Dict. Appl. Chem.* (ed. 4) X. 835/1 The earliest Solvay plants used solid salt and dissolved it in a recycled alkaline water obtained from the scrubbing of calciner gases. **1966** *McGraw-Hill Encycl. Sci. & Technol.* II. 399/2 Calcium chloride, obtained as a waste product in the Solvay process. **1974** *Encycl. Brit. Micropædia* IX. 320/1 Its anhydrous form, soda ash (Na_2CO_3),..is now manufactured chiefly by the ammonia–soda (Solvay) process.

solvent, *sb.* **5.** *attrib.* and *Comb.*, as *solvent abuse, recovery, -sniffing*; *solvent-thinned* adj.; **solvent extraction,** the partial removal of a substance from a solution or a mixture of liquids by utilizing its greater solubility in another liquid or its greater permeability through a membrane; so **solvent-extract** *v. trans.*, to purify by means of solvent extraction; also as *sb.*, a fraction extracted from a mixture by this process; a spell of solvent extraction.
1977 Solvent abuse [see *sniffing *vbl. sb.*]. **1949** *Our Industry* (Anglo-Iranian Oil Co.) (ed. 2) iv. 116 The kerosine fraction may..be solvent extracted by means of liquid sulphur dioxide. **1956** *Nature* 4 Feb. 224/2 When insulin was added to an acidic solvent-extract of liver..the recovery of the insulin varied between 92 and 93 per cent. **1963** A. J. Hall *Textile Sci.* iii. 122 Another system of purifying wool is to solvent extract it with an organic solvent such as trichloroethylene. **1978** *Nature* 23 Mar. 298/2 Subsequent solvent extract in mixer settlers will separate plutonium nitrate solution. **1920** *Chem. Trade Jrnl.* LXVI. 103 Quantities of oil for edible purposes were being produced by solvent extraction processes. **1949** *Our Industry* (Anglo-Iranian Oil Co.) (ed. 2) iv. 119 Solvent extraction processes are to-day used extensively in the petroleum industry for refining light distillates, kerosines and lubricating oils. **1978** *Sci. Amer.* July 97/3 Solvent extraction across membranes has been known at least since 1913, when Fritz Haber and Reinhardt Beutner showed that a thin film of oil could be employed as a membrane in two kinds of extraction process. **1947** Kirk & Othmer *Encycl. Chem. Technol.* I. 231 Activated carbon is used on a large scale for solvent recovery. **1977** *Lancet* 8 Jan. 82/1 Cardiac arrhythmias,..neuropsychiatric disorders, and hepatorenal failure have all been reported after solvent sniffing. **1960** *McGraw-Hill Encycl. Sci. & Technol.* IX. 493/1 Solvent-thinned paints, which dry essentially by solvent evaporation, rely on a fairly hard resin as the vehicle.

solventless (sǫ·lvəntlės), *a.* [f. Solvent *sb.* + -less.] Without a solvent.
1936 O. W. Eshbach *Handbk. Engin. Fund.* xii. xxxiii. 70 (*heading*) Solventless varnishes. **1945** *Jrnl. Applied Physics* XVI. 584/2 A thin film of solventless varnish was applied to each sheet. **1970** *Financial Times* 13 Apr. 9/1 A feature of this two-component solventless coating..is that it can be applied cold.

‖ **solvitur ambulando** (sǫ·lvitǒɹ æmbiulæ·ndo). [L. phr., lit. '(the problem) is solved by walking'.] An appeal to practical experience for the solution of a problem or proof of a statement. Also as *sb. phr.* Also in shortened form **ambulando,** by experience; in the course of things.
Originally an allusion to the reported proof by Diogenes the Cynic of the possibility of motion: see Diogenes Laertius VI. 39.
[**1814** *Artis Logicæ Rudimenta. Accessit Solutio Sophismatum* 67 Ineptum est hoc Sophisma 1. Quia solvitur ambulando; quod fecit Diogenes.] **1852** A. H. Clough *Let.* Mar. in *Poems & Prose Remains* (1869) I. 174 It is not..simply one's business in life to 'envisager' the most remarkable problems of humanity... Still we may be assured that only time can work out any sort of answer to them for us. '*Solvitur ambulando.*' **1863** J. Conington *Horace's Odes* p. xxv, How easily the '*solvitur ambulando*' of an artist like Mr. Tennyson may disturb a whole chain of ingenious reasoning on the possibilities of things. **1863** C. Reade *Hard Cash* I. viii. 226 To the à priori reasoners ..he replied by building an engine..hooking on eight carriages, and rattling off up an incline. 'Solvitur ambulando,' quoth Stephenson the stout hearted. **1876** W. James in *Nation* 8 June 369/1 The ultimate decision of which side is right and which wrong shall only be reached *ambulando* or at the final integration of things, if at all. **1906** F. W. Maitland *L. Stephen* xvii. 366 He knew that he would have to proceed empirically. *Solvitur ambulando* —the motto of the philosophic tramp—had also to be the motto of the editor. **1930** J. Laird *Knowl., Belief & Opin.* iv. 103 Perfectly convincing evidence might turn up, so to say, *ambulando*, when we are engaged in something irrelevant. **1934** A. Toynbee *Stud. Hist.* III. 182 A modern Western philosopher applies the historic *solvitur ambulando* to the ancient sophism of the Eleatics. **1955** *Times* 30 Aug. 9/2 To what extent and for what purposes is it justifiable to transform personality by surgery, psychological techniques, or the administration of drugs? In so far as matters of this kind have been the subject of conscious policy in the past solvitur ambulando has been the motto. **1957** G. Ryle in C. A. Mace *Brit. Philos. in Mid-Cent.* 256 The assimilation of language to chess reminds us of what we knew *ambulando* all along.

solvolyse (sǫ·lvǫlaiz), *v. Chem.* Also (*U.S.*) -lyze. [f. *solvolysis after *analyse, analysis*, etc.] **a.** *trans.* To bring about the solvolysis of (a solute); usu. in *pass.* **b.** *intr.* To undergo solvolysis.
1916 *Chem. Abstr.* X. 2349 A table of salts is given which are 'solvolyzed' by C_6H_6. **1921** *Jrnl. Physical Chem.* XXV. 550 Lead chromate solvolyzes. **1965** Phillips & Williams *Inorg. Chem.* I. xv. 569 In sulphuric acid nearly all inorganic salts are solvolysed in the dissolution process. **1974** *Nature* 19 Apr. 671/1 Tosylate..solvolysed in formic acid at 75°C to give an olefin as the major product. **1977** *Lancet* 24–31 Dec. 1298/1 The bile samples were solvolysed (which ensures that bile-acid sulphate esters are included), and individual acids were measured by gas chromatography.

solvolysis (sǫlvǫ·lisis). *Chem.* Pl. -lyses (-lisīz). [ad. G. *solvolyse* (P. Walden 1910, in *Sammlung chem. und chem.-techn. Vorträge* XV. x. 447), f. *solvens* Solvent *sb.*: see *-lysis.] The decomposition or dissociation of a solute brought about by the action of the solvent.
1916 *Chem. Abstr.* X. 2349 Most probably other indifferent, *i.e.*, non-ionizing solvents are also capable of dissociating salts into base and acid. This general property of solvents, which is independent of electrolytic dissociation, is designated 'solvolysis' or solvolytic dissociation. **1922** *Jrnl. Physical Chem.* XXVI. 567 The solvolyses were performed simultaneously in porcelain crucibles in an oven at 225°–230°. **1940** L. P. Hammett *Physical Org. Chem.* vi. 171 The product of the solvolysis is not completely racemized. **1972** R. A. Jackson *Mechanism* iv. 56 Solvolysis of the diester 63 has been postulated to involve the intermediate carbonium ion 64.
So **solvoly·tic** *a.*, pertaining to or of the nature of solvolysis; **solvoly·tically** *adv.*
1916 [see *solvolysis]. **1931** *Chem. Rev.* VIII. 199 Goldschmidt and Mathiesen..report the solvolytic constants for a variety of bases in ethyl and methyl alcohols. **1965** Phillips & Williams *Inorg. Chem.* I. xv. 556 Solvolytic reactions are often of synthetic value, but they also severely limit the range of reactions which may be conducted in a particular solvent. **1974** Gill & Willis *Pericyclic Reactions* vi. 148 With a monohalogeno carbene a mixture of *endo-* and *exo*-halogenobicycloalkanes is formed, only one of which will usually be solvolytically reactive. **1975** *Nature* 21 Aug. 681/1 Overberger and Podsiadly have demonstrated the importance of polymeric chain aggregation on the rate of solvolytic reaction.

solvus (sǫ·lvǒs). Pl. **solvi.** [mod.L., f. L. *solvere* to dissolve, after *liquidus, Solidus².] A line or surface in a phase diagram delimiting the region of stability of a solid solution.
1950 *Jrnl. Geol.* LVIII. 499/2 Now we have a means of establishing the position of the unmixing curve or solvus. **1956** *Amer. Mineralogist* XLIII. 877 In Fig. 4 three solvi are shown. **1963** W. A. Deer et al. *Rock-Forming Minerals* II. 127 If the liquidus minimum passes beyond the intersection with the solvus..the calcium-poor phase would cease to crystallize as a separate phase. **1968** R. A. Higgins *Engin. Metall.* (ed. 2) I. ix. 170 The only new feature of the system..is that we have phase boundaries occurring below the solidus, indicating that phase changes can take place in the solid. On the tin-lead thermal-equilibrium diagram..such phase boundaries.. are indicated by the lines *BC* and *FG*. A line such as *BC* or *FG* is often termed as [*sic*] solvus. **1975** *Nature* 3 Apr. 406/2 At 1,450°C a further solvus was reported between the pigeonite and a more calcic diopside.

‖ **solyanka** (sǒlya·nkā). Also **soljanka.** [Russ.] A soup made of vegetables and meat or fish.
1958 W. Bickel tr. *Hering's Dict. Classical & Mod. Cookery* 113 *Soljanka* is *riba*, fish soup. **1970** 'J. Morris' *Candywine Devel.* xxv. 280 He..heaped the big serving-spoon with *solyanka* from the dish. **1972** M. Glenny tr. *Solzhenitsyn's August 1914* xlii. 422 'Well, now, young men—what is it to be? *Solyanka*, meatballs, scrambled eggs?' said the old man invitingly.

‖ **sol y sombra** (sol i so·mbra). [Sp., lit. 'sun and shade'.] A drink of brandy mixed with anisette or gin.
1930 W. T. Boothby *'Cocktail Bill' Boothby's World Drinks* (rev. ed.) 148 *Sol y Sombra* (*Sun & Shadow*) Brandy (Spanish) ½ jigger Gin ½ jigger. **1966** C. Rougvie *Gredos Reckoning* v. 76 'What's that you're drinking?' '*Sol y sombra*: a mixture of anis and brandy.' **1979** D. Serafin *Saturday of Glory* 17 A *sol y sombra*, an explosive mixture of brandy and anisette.

soma¹. Add: Also 8 **som.**
1. b. In Aldous Huxley's novel *Brave New World*, a narcotic drug which produces euphoria and hallucination, distributed by the state in order to promote content and social harmony. Also *transf.* and *attrib.*
1932 A. Huxley *Brave New World* iii. 66 There is always *soma*, delicious *soma*, half a gramme for a half-holiday, a gramme for a week-end, two grammes for a trip to the gorgeous East, three for a dark eternity on the moon. *Ibid.* v. 95 The *soma* had begun to work. Eyes shone, cheeks were flushed, the inner light of universal benevolence broke out on every face in happy, friendly smiles. *Ibid.* vi. 107 Bernard swallowed four tablets of *soma* at a gulp. *Ibid.* xv. 246 The two Groups..were served by the Deputy Sub-Bursar with their *soma* ration. **1959** *Times Lit. Suppl.* 27 Feb. 105/2 The forecast [in *Brave New World*] of a pain-killing, pleasure-giving pill called *soma*, at once the perfect substitute for alcohol, opium and benzedrine, has not yet been wholly fulfilled. **1968** 'J. le Carré' *Small Town in Germany* ix. 155 If I smoked I'd smoke one of your cigars. I could do with a bit of soma just now. **1976** *Interdisciplinary Sci. Rev.* I. 178/1 Suppose there was national agreement on the need for a 'good' psychoactive drug, one which would be as universally used as the ubiquitous *soma* of *Brave New World*.
2. (Earlier and later examples.)
1785 C. Wilkins tr. *Bhăgvăt Gēētā* ix. 80 The followers of the three *Vēds*, who drink the juice of the *Sōm*.. address me. *Ibid.* 143 *Sōm*—is the name of a creeper, the juice of which is commanded to be drank at the conclusion of a sacrifice. **1972** R. G. Wasson *Soma* ii. 10 My candidate for the identity of soma is *Amanita muscaria*.., in English the fly-agaric.

soma². Add: **1.** Also *fig.*
1914 W. R. Inge *Outspoken Ess.* (1919) 238, I can see no other fate in store for the *soma* of Catholicism; the germ-cells of true Christianity live their own life within it.
2. *Anat.* The compact portion of a nerve-cell, excluding the axon.
1947 R. Lorente de Nó in *Jrnl. Cellular & Compar. Physiol.* XXIX. 211 The word 'soma' will be used to denote the body and all the dendritic branches of the neuron. **1970** *Jrnl. Neurobiol.* I. 340 The somas are likely to be the site of protein synthesis. **1974** *Nature* 8 Nov. 155/1 Studies with a few invertebrate preparations have suggested different ionic mechanisms for spike generation in the soma and axon of the same neurone.
3. The body in contrast to the mind or the soul. Opp. *psyche*.
1958 A. Wilson *Middle Age of Mrs Eliot* 1. 72 He says that between Else fussing with his soma and me with his psyche, all he really wants is a complete rest and he wishes he were travelling with you. **1969** *Listener* 26 June 881/2 A score of youths and girls intent on exploring each other honestly and sincerely in psyche and soma can't help generating some degree of excitement. **1976** *Lancet* 30 Oct. 946/2 No personal physician, at any time, has been able to operate successfully without using substances, as symbols of healing of the soma and the psyche.

somaar, var. *sommer *adv.*

somæsthetic (sōᵘmisþe·tik), *a. Physiol.* Also -esthetic. [f. Gr. σῶμα body + αἰσθητικ-ός perceptive.] Pertaining to or designating those sensations (as of pressure, pain, or warmth) which can occur anywhere in the body, in contrast to those that depend on highly localized sense organs (as sight, balance, or taste).
1899 L. F. Barker *Nerv. System* xlvii. 667 The region of the cortex in which the axones of the general sensory conduction path here considered terminate, I have designated as the somæsthetic area of the cortex. [*Note*] I wish to thank Prof. Gildersleeve, of the Johns Hopkins University, for suggesting this term as a suitable English equivalent for Munk's *Körperfühlsphäre*. **1922** R. S. Woodworth *Psychology* iii. 63 There is a large and important area called the 'somesthetic', connected with the body senses generally, i.e., chiefly with the skin and muscle senses... Destruction of any part of this somesthetic area brings loss of the sensations from the corresponding part of the body. **1949** *Psychosomatic Med.* XI. 338/2 The thalamus probably participates in a crude awareness of somesthetic sensations. **1968** Passmore & Robson *Compan. Med. Stud.* I. xxiv. 51/1 The sensory or postcentral gyrus contains areas 1, 2 and 3 which are often considered together as the somaesthetic cortex, receiving sensations from the opposite side of the body. **1972** *Sci. Amer.* Apr. 82/3 The patient cannot name objects held in his left hand because the somesthetic sensations cannot reach the verbal centers in the left hemisphere. **1977** R. K. Davenport in D. M. Rumbaugh *Language Learning by Chimpanzee* iii. 75 In the brain of nonhuman primates as compared with the human brain, the auditory, visual, and somesthetic association areas are relatively independent.
Hence **somæsthe·sis**, the somæsthetic senses.
1928 *Funk's Stand. Dict.*, Somesthesis. **1950** *Ann. Rev. Psychol.* I. 71 (*heading*) Somesthesis and the chemical senses. **1968** [see *kinæsthesis]. **1977** H. G. Burger in B. Bernardi *Concept & Dynamics of Culture* 423 It [*sc.* the angular gyrus] lies between the association cortices of three non-limbic sensation modalities: vision, audition, and internal sensibility (somesthesis).

Somaj, var. *Samaj.

somal, *a.* Add: **2.** Of or pertaining to the soma of a neurone.
1967 *Jrnl. Neurochem.* XIV. 329 Instead of a somal site of synthesis, an alternative possibility was considered; namely, a local axonal synthesis after entry from Schwann cells or a diffusion of precursor from the soma down into the axon. **1978** *Nature* 27 July 381/1 We compared the values of unit conductances activated by acetylcholine (ACh) applied to the somal membranes with values reported for other molluscan neurones.

Somali (somā·li), *sb.* and *a.* Also **Somal, Somauli,** etc. [Native name.] **A.** *sb.* **1. a.** A Hamitic people, adherents of Islam, living in the Horn of Africa, esp. in the Somali Democratic Republic (Somalia); a member of this people. **b.** The language of this people which belongs to the Cushitic sub-family and existed until recently in spoken form only. **B.** *adj.* Of or pertaining to the Somalis or their country.
1814 Somauli [see Swahili]. **1842** *Jrnl. R. Geogr. Soc.* XII. 241 There are also smaller kalifas that trade to Amin, Ugadin, and other parts of the Somauli country. **1850** *Trans. Bombay Geogr. Soc.* IX. 129 The country inhabited by the Somal is very extensive. *Ibid.* 130 Since the occupation of Aden by the English, the Somaulies have continued to cross over in considerable numbers. *Ibid.* 136 The article in Somauli is expressed by ka, ki or ga, gi. **1856** R. F. Burton *First Footsteps E. Africa* 89 In the latter sense it is the polite address to a Somali. *Ibid.* 644/2 Tale of a Somal chief. **1888** F. L. James *Unknown Horn of Africa* 30 'Hubla' (Somal for 'virgin'). **1878** [see *Galla]. **1921** H. Rayne *Sun, Sand & Somals* 5 The Somali has deeply religious tendencies. **1930** C. G. Seligman *Races of Africa* v. 124 The early history of the Somali is obscure; that they are essentially Hamitic is certain. **1933** [see *Cushite *a.* and *sb.*]. **1937** K. Blixen *Out of Africa* ii. iv. 137 The immigrant Somalis..are severe Mohammedans. *Ibid.* iii. iii. 188 The circle of Somali women in my household. **1948** D. Diringer *Alphabet* ii. iv. 300 This script, called..'the Osmanya script'..has been created recently for the Somali language by 'Isman Yusuf. **1978** *Encounter* Feb. 52/2 It may be that Cubans using Russian weapons saved Ethiopia from defeat by the Somalis in November 1977. **1978** *Language* LIV. 453 The status of Somali (no official orthography; 184, 201) has changed drastically (Romanization adopted).
Hence **Soma·lian** *a.* = *Somali *a.*
1948 D. Diringer *Alphabet* ii. iv. 300 The origin of the Somalian alphabet. **1976** *Guardian* 23 Aug. 7/1 Her father was Christian and disapproved of the general Somalian habit of infibulation. **1981** *Sci. Amer.* Mar. 29/3 The other [skull] found more recently in a fertilizer factory near Mogadishu, where fishermen who had picked it up on a Somalian beach had sent it.

Soman (sōᵘ·mæn). Also **soman.** [Ger., of unknown origin.] The name of an odourless organophosphorus nerve gas.
1951 *Acta Physiol. Scandinavica* Suppl. No. 90. 106 (*table*) Pinacoloxy-methyl-phosphoryl-fluoride (soman). **1953** *Proc. R. Soc. Med.* XLVI. 801 We found..that 3,4 dihydroxyphenylalanine..and other *o*-dihydroxybenzene derivatives were able to protect both true and pseudo ChE's [*sc.* cholinesterases] against, not only Sarin, but also against Tabun and Soman. **1968** [see *Sarin]. **1969** B. Ford *German Secret Weapons* (1970) 109 Soman has the most terrible effects on soldiers: within seconds they would have been reduced to a state of convulsive collapse and death would be relatively certain. **1980** *Sci. Amer.* Apr. 35/3 Soman, first prepared in Germany in 1944, is believed by Western officials to be the standard Russian nerve gas. **1982** *Guardian* 8 Mar. 12/3 The use of Soman and Tabun gases by the Soviet-directed Ethiopians.

somar, var. *sommer *adv.*

Somaschian, *sb.* and *a.* Add: More frequently **Somaschan, Somascan. a.** *sb.* (Further examples.)
1882 T. A. Pope tr. *Capecelatro's Life of St. Philip Neri* II. iii. x. 153 He..co-operated in the erecting of a house of Somaschans for orphan boys. **1936** L. Christiani in E. Eyre *European Civilization* IV. ii. i. 162 The immense achievements of Capuchins,..Somascans, and Barnabites. **1940** E. Graf tr. *Pastor's Hist. of Popes* XXX. iv. 177 In 1647 Innocent dissolved the union between the Doctrinarians and the Somaschans.
So **Soma·schi** *sb. pl.*
1883 *Encycl. Brit.* XVI. 716/2 Clerks regular of St Majolus of Pavia, or 'Somaschi'. **1978** D. H. Farmer *Oxford Dict. Saints* 211/1 [St. Jerome Emiliani] founded orphanages, hospitals, and houses for repentant prostitutes. To look after them he also founded a small congregation of clerks regular, called the Somaschi after their place of origin.

somasteroid (somæ·stěroid), *a.* and *sb. Zool.* Also **Somasteroid.** [ad. mod.L. *Somasteroidea* (W. K. Spencer 1951, in *Phil. Trans. R. Soc.* B. CCXXXV. 87), f. Gr. σῶμα body + mod.L. *Asteroidea*, name of a subclass or class of echinoderms, f. Gr. ἀστεροειδής (see Asteroid *a.* and *sb.*) + L. -*ea*, neut. pl. of -*eus* -eous.] **A.** *adj.* Belonging or pertaining to the subclass or class Somasteroidea, which belongs in the subphylum Asteroidea of star-shaped echinoderms and comprises extinct forms having broad, petal-like arms with a pinnate structure. **B.** *sb.* A somasteroid echinoderm.
1955 L. H. Hyman *Invertebrates* IV. xv. 700 The arrangement recently adopted by palaeontologists.., according to which asteroids and ophiuroids derive from a common somasteroid..ancestor.., must be somehow wrong. **1962** D. Nichols *Echinoderms* iii. 40 The fossilized remains of these ossicles have even been found in the earliest asteroids, the Somasteroids. **1963** *Phil. Trans. R. Soc.* B. CCXLVI. 383 The Asterozoa, or star-shaped echinoderms, may be regarded as a natural taxon, comprising the somasteroids, asteroids and ophiuroids. **1964** [see *phanerozone *a.* (*sb.*)]. **1979** W. D. Russell-Hunter *Life of Invertebrates* xxxiv. 607 The smallest class, Somasteroidea, which arose in the Ordovician, was never an extensively diversified group, being relegated by many neontologists to the status of a subclass of the Asteroidea. (The living genus, *Platasterias*, is no longer regarded as a somasteroid.)

Somastic (sǒmæ·stik). Also **somastic.** [Prob. f. initials of *S*tandard *O*il (see quot. 1930) + Mastic *sb.*] A proprietary term in the U.S. for asphalt-based materials used in coating oil pipelines.
1930 *Official Gaz.* (U.S. Patent Office) 8 July 365/1 Standard Oil Company of California... *Somastic*. For asphalt, asphalt mastic, and pipe coatings. Claims use since May 8, 1930. **1974** *Petroleum Rev.* XXVIII. 634/1 A 40-ft long pipe..emerges from a somastic coating machine at Bredero Price's mill in Leith. **1976** *Offshore Platforms & Pipelining* 131/1 Field joints in the overbend are coated with flexible somastic.

‖ **somaten** (somate·n, somā·ten). Pl. **somatenes.** [a. Catalan (and Sp.) *somatén* an alarm bell; an armed body of citizens.] In Catalonia, a body of civilians armed for the protection of a town or district; a member of this body. Hence **soma·tenist** [Catalan *somatenista*], a member of a somaten.
1845 R. Ford *Hand-bk. Spain* I. 493/1 Thus time was given for the *somaten*, or tocsin, to be rung, and the armed peasantry collected... The Catalan *guerrilleros* were called *Somatenes*, from this bell. **1905** R. Thirlmere *Lett. from Catalonia* I. v. 60 In Olot and other Catalan towns they have what they call *El Somaten*, an armed, local, volunteer force—which closes all the highways at the sound of an alarm bell. **1911** *Encycl. Brit.* XVIII. 566/2 *Miquelets* ..were irregular local troops in Catalonia... They were maintained by the several parishes..and as they had to turn out for duty on sound of the village alarm-bell (*somaten*) they are frequently called *somatenes*. **1928** *Daily Tel.* 29 May 9/5 A member of the 'Somaten' (Militia) to-day succeeded in finding the young man who ran amok with a rifle last Monday... As he refused to surrender the 'Somatenist' fired and killed him. **1929** A. F. G. Bell in E. A. Peers *Spain* 275 A more efficient city-police, together with the Somatenes and the Civil Guard, maintained order throughout Spain.

somatism. For † *Obs.* read *rare* and add later example.
1955 *Mind* LXIV. 495 Reism as such is not yet somatism, but somatism finds its place in the framework of reism as a particular case. Every soul is a body—this is the thesis of somatism.

somatization (sōᵘmătəizēᴵ·ʃən). [f. Gr. σῶμα, σώματ- body, Soma² + -ization.] The occurrence of bodily symptoms in consequence of or as an expression of mental disorder. Hence **so·matizing** *a.*, pertaining to or exhibiting such symptoms.
1925 J. S. van Teslaar tr. *Stekel's Peculiarities of Behav.* II. 341 *Somatization*, conversion of emotional states into physical symptoms. **1943** E. & C. Paul tr. *Stekel's Interpretation of Dreams* I. iii. 25 Cases where the disorder—through somatization—secures bodily expression. **1954** Grinker & Robbins *Psychosomatic Case Bk.* v. 74 In the narrower sense of the word, psychosomatic disturbances are differentiated from those neuroses in which somatization occurs through innervation of the voluntary nervous system. **1966** *McGraw-Hill Encycl. Sci. & Technol.* XII. 501/1 What characterizes the psychosomatic reaction in contrast to the other somatizing patterns is that there is an involvement of specific organs under the control of the autonomic nervous system. **1970** *Psychol. Rep.* XXVII. 756 One aspect of the stress and somatic reaction problem that has not received attention is the relationship between physiological reactions under laboratory-stress conditions and the tendency for individuals to develop somatic complaints under life stresses, as indicated by a personality characteristic of 'somatization'. **1978** *African Jrnl. Med.* VII. 209/2 Somatisation of psychological disorder..is common in the Nigerian population. **1980** I. G. & B. R. Sarason *Abnormal Psychol.* (ed. 3) viii. 202/1 The complaints of somatizing patients often lead to unnecessary surgery.

somato-. Add: **so·matocœl** *Zool.* [a. G. *somatocœl* (K. Heider 1912, in *Verhandl. d. deutsch. zool. Ges.* XXII. 241), f. Gr. κοιλία cavity of the body], each of a pair of cavities in an echinoderm embryo that develop into the main body cavity of the adult; hence **somatocœ·lic** *a.*; **somatome·tric** -me·trical *adjs.*, of or pertaining to the measurement of the body; hence **somato·me·trically** *adv.*; **somatopsy·chic** *a. Psychol.* [ad. G. *somatopsychisch* (C. Wernicke 1892, in *Path. des Nervensystems* (1893) 166], (*a*) of or pertaining to awareness of one's own body (? *obs.*); (*b*) arising from or pertaining to the effects of bodily illness on the mind; **somato-se·nsory** *a. Physiol.* = *somæsthetic *a.*
1955 L. H. Hyman *Invertebrates* IV. xv. 692 The inner walls of the somatocoels meet above and below the intestine to form the primary mesentery. **1962** D. Nichols *Echinoderms* x. 120 Almost as soon as the primary coelomic sacs have been formed, they bud off posteriorly another pair of sacs, the somatocoels, later to form the main coelom of the adult body. **1976** *Nature* 20 May 228/1 All this suggests that the new coelomocytes in the general body cavity can have come only from the somatocoelic epithelium. **1939** *Ibid.* 11 Nov. 807/1 By what somatometric method, which is both reliable and convenient of application, is it possible to assess nutritional status? **1951** *Proc. Sect. Sci. Koninkl. Akad. Wetensch. Amsterdam* C. LIV. 480 (*heading*) The quantitative expression of resemblance in the somatometric study of relationship. *Ibid.*, Somatometrical data of different age-groups are not directly comparable. *Ibid.*, Training and interest will mostly induce the anthropologist to restrict himself to the study of properties than can be demonstrated somatometrically. **1902** *Buck's Handbk. Med. Sci.* (rev. ed.) V. 27/1 Consciousness is a function of the associative mechanism and may be considered in its threefold relationship to the outer world, the body and self—allopsychic, somatopsychic, and autopsychic. **1927** Henderson & Gillespie *Text-bk. Psychiatry* II. 13 His division of concepts into those of the outside world, of the personality, and of the body—'allopsychic', 'autopsychic', and 'somatopsychic'. **1955** *A.M.A. Arch. Neurol. & Psychiatry* LXXIII. 403/2 With the increasing severity in the lesion and the growing disability of the patient, much may be learned about the somatopsychic problem, i.e., about the manner in which the increasingly morbid process affects the state of mind of the patient and his relation to himself and his environment. **1961** *Guardian* 17 May 8/5 We should not allow a preoccupation with psychosomatic illness..to blind us to the advances ..in the treatment of somato-psychic disorders. **1978** F. Mann *Acupuncture* (ed. 3) x. 160 Modern medicine might use the word 'psychosomatic' to describe the diseases considered in this section, as they are physical results of uncontrolled emotion; those in the previous section might be given the label 'somatopsychic', being mental diseases resulting from outer or physical causes. **1952** *Federation Proc.* XI. 5/2 Responses from stimulation of arm and leg subdivisions of somatosensory area I were similar in location but differed in shape. **1975** *Nature* 30 Oct. 738/1 Axons carrying visual, auditory and somatosensory information converge on the tectum and interlace with tectal neurones. **1978** *Sci. Amer.* Sept. 82/2 In the somatosensory area of the cortex the cells in a column respond to the same type of stimulus (pressure, touch, heat, cold) at the same point on the body surface.

somatogamy (sōᵘmătǫ·gămi). *Biol.* [ad. G. *somatogamie* (O. Renner 1916, in *Biol. Centralbl.* XXXVI. 349): see Somato- and *-gamy.] = *pseudogamy b. So **somato-gamous** *a.*
1949 I. F. & W. D. Henderson *Dict. Sci. Terms* (ed. 4) 403/2 *Somatogamy*, pseudogamy; pseudomixis. **1950** W. B. Brierley tr. *Gäumann's Princ. Plant Infection* iii. 20 The nuclear association, initiated by somatogamous fusion, constitutes..the pre-condition for transition from the saprophytic to the parasitic mode of life. **1962** G. Dalldorf *Fungi & Fungous Diseases* x. 136 In fungi that undergo somatogamy, such as yeasts.

somatomedin (sōᵘmătomī·din). *Physiol.* [See quot. 1972.] Any of several peptides present in serum which are thought to act as intermediates in the stimulation of growth by growth hormone.
Quots. 1971, 1972 have no author in common.
1971 J. M. Tanner et al. in *Arch. Dis. Childhood* XLVI. 761/2 A few patients with short stature have the ability to secrete immunoreactive GH..; but their GH fails to stimulate the production of somatomedin, a substance found in the blood, and previously named 'sulphation factor'. **1972** W. H. Daughaday et al. in *Nature* 14 Jan. 107/2 After consideration of many alternatives to the operational terms 'sulphation factor' or 'thymidine factor', we propose the more general term, 'somatomedin'; the prefix, 'somato', connotes both a hormonal relationship to somatotropin and, also, to the soma which is the target tissue of this agent. 'Medin' is included in the name to indicate that it is an intermediary in somatotropin action. **1978** Van Wyk & Underwood in G. Litwack *Biochem. Actions Hormones* V. iii. 102 It is.. generally agreed that to belong in the somatomedin group of peptides, a substance must fulfil three criteria: its concentration in serum must be growth-hormone dependent, it must possess insulin-like actions in skeletal tissues, and it must promote the incorporation of sulfate into proteoglycans of cartilage. **1978** F. L. Strand *Physiology* xix. 360/1 Growth hormone, the somatomedins, insulin, and thyroxine have important anabolic effects that promote wound healing.

somatostatin (sōᵘmătostæ·tin). *Physiol.* [See quot. 1973.] A peptide secreted in the hypothalamus and elsewhere whose actions include the inhibition of the release of various hormones, esp. from the anterior pituitary.
1973 P. Brazeau et al. in *Science* 5 Jan. 79/3 We propose to name the peptide described here *somatostatin*, from somato(tropin), a pituitary factor affecting statural growth, and stat(in), from the Latin 'to halt, to arrest'. **1976** *Nature* 10 June 511/2 Somatostatin inhibits the release of growth hormone, thyrotropin and prolactin from the pituitary, of glucagon and insulin from the pancreas, and also of gastrin. **1977** *Lancet* 23 July 166/2 Somatostatin, a tetradecapeptide first isolated from the hypothalamus.., has been demonstrated in large quantities in the gastrointestinal tract and pancreas.
Hence **so:matostatino·ma** *Path.*, a tumour secreting excessive quantities of somatostatin.
1977 *Lancet* 26 Mar. 668/1 Measurements of insulin and glucagon responses during a glucose-tolerance test in somatostatinoma patients may help to explain this apparent association. **1978** *Jrnl. R. Soc. Med.* LXXI. 173 This somatostatinoma gave rise to the clinical abnormalities of hypochlorhydria, steatorrhoea, and a diabetic glucose tolerance curve.

somatotonic (sōᵘmătǒtǫ·nik), *a.* and *sb.* [f. Somato- + Tonic *a.*] **A.** *adj.* Designating or characteristic of a type of personality which is extroverted and aggressive, classified by Sheldon as being associated with a mesomorphic physique. **B.** *sb.* One having this type of personality. So **somatotonia** (-tōᵘ·niă), somatotonic personality or characteristics.
1937 [see *cerebrotonic *a.* and *sb.*]. **1938** H. G. Wells *Apropos of Dolores* iv. 214 The classification of main human types and temperaments from Hippocrates' down to the cerebrotonics, somatotonics and viscerotonics of today. **1940** W. H. Sheldon *Varieties Human Physique* i. 8 Somatotonia is the motivational pattern dominated by the will to exertion, exercise and vigorous self-expression. It is the drive toward dominance of the functions of the soma. **1950** A. Huxley *Let.* 16 Mar. (1969) 621 We are fortunate to-day in possessing, at long last, a genuinely scientific method for describing physique, temperament and their interrelations..—viscerotonia, somatotonia and cerebrotonia for temperament. **1950** —— *Themes & Variations* 28 The viscerotonic and somatotonic extraverts who are at home in the world. **1969** V. de Sola Pinto *City that Shone* iii. 68, I suppose that in modern psychological jargon, as a child in those distant Edwardian days, I could be described as an introvert and cerebrotonic living in a world of extroverts, somatotonics and viscerotonics.

somatotopic (sōᵘmătotǫ·pik), *a. Neurol.* [f. Somato- + Gr. τοπικ-ός in respect to place (see Topic *a.* and *sb.*).] Characterized by or being a relationship between the locations of neurones in the central nervous system and in the tissues they serve. Also **somatoto·pical** *a.*, in the same sense.
1945 *Federation Proc.* IV. 31/1 Liminal faradic excitation of the cerebellar cortex reveals somatotopic localization in the anterior lobe and lobulus simplex of decerebrate animals. **1958** A. Brodal *Neurol. Anat.* iii. 42 Here again is an example of a somatotopical arrangement of a fibre system. **1965** *Jrnl. Anat.* XCIX. 761 (*heading*) Correlation between nuclear morphology and somatotopic organization in the ventro-basal complex of the raccoon's thalamus. **1969** Truex & Carpenter *Human Neuroanat.* (ed. 6) xvii. 423/1 In the cat there is evidence that corticopontine fibers arising from the sensorimotor cortex project in a somatotopical manner onto two longitudinally oriented cell columns within the pontine nuclei. **1976** *Nature* 20 May 190/3 Physiologists have revealed a strict somatotopic map of the skin in the spinal cord.
Hence **so:matoto·pically** *adv.*, in a manner which preserves such a relationship; **so·matotopy** (*rare*), somatotopic relationship.
1948 A. Brodal *Neurol. Anat.* vi. 161 They have mapped out the areas giving response when different parts of the skin are stimulated, and established a pattern which mainly corresponds to that gained from study of human cases. The body is represented somatotopically, with a certain overlapping between adjacent dermatomes. **1961** *Lancet* 2 Sept. 546/2 Discriminative sensation travels by the lemniscal pathway and is relayed somatotopically via the thalamus to the cortex. **1976** *Progress in Sci. Culture* (E. Majorana Centre) Spring 48 The analysis of the properties, of these two inputs, in terms of somatotopy, frequency of discharge, sensory modality, timing and interaction between them suggests that integration of the information channelled through the two inputs on the same Purkynĕ cell is a basic mechanism of cerebellar operation.

somatotroph (sōᵘ·mătotrǒᵘf). *Physiol.* [Back-formation from next.] A cell of the anterior pituitary which synthesizes somatotrophin (growth hormone).
1968 *Annales de Biologie Animale, Biochimie, Biophysique* VIII. 22 (*heading*) Cytological characteristics of gonadotroph, thyrotroph, corticotroph, somatotroph cells and of prolactin cells in the anterior lobe of the cattle pituitary. **1978** *Jrnl. R. Soc. Med.* LXXI. 434 There are several sera under study where the antibodies react neither with lactotrophs or somatotrophs.

somatotrophic (sōᵘmătotrǒᵘ·fik), **-tropic** (trǒᵘ·pik, -trǫ·pik), *a. Physiol.* [f. Somato- + *-trophic, *-tropic.] Pertaining to or having the property of stimulating body growth; *spec.* applied to the hormone somatotrophin.
1938 H. L. Wieman *Gen. Zool.* (ed. 3) x. 206 The somatotropic hormone is necessary for the normal growth and development of the body. **1952** *Amer. Jrnl. Physiol.* CLXXI. 381 Some of the most striking effects of..cortisone..are directly opposed to the actions of the 'growth hormone' or somatotrophic hormone (STH). **1960** *Biol. Abstr.* XXXV. 1154/2 The somatotropic activity of the plasma was evaluated. **1977** *Lancet* 23 July 198/1, HG.H. secretion from another somatotrophic adenoma..was inhibited..by bromocriptine.
Hence **somatotro·phin, -tro·pin,** a hormone secreted by the anterior pituitary which promotes the release of somatomedin; growth hormone.
1947 H. Selye *Textbk. Endocrinol.* iii. 217/2 Denaturation of somatotrophin by treatment with urea does not influence its growth-promoting potency. **1952** *Endocrinol.* LI. 300 Growth hormone (somatotropin) can produce a growth-promoting and diabetogenic response in the developing chick embryo. **1965** Lee & Knowles *Animal Hormones* ii. 21 The growth hormone, also referred to as somatotrophin (STH), not only controls the rate of growth, but also the metabolism necessary for this growth. **1973** *Sci. Amer.* Sept. 41/3 (*caption*) Somatotrophin, acting with other hormones, regulates the normal growth of children. **1974** D. & M. Webster *Compar. Vertebr. Morphol.* xiii. 307 The growth hormone, somatotropin, produced by acidophilic cells, is a branched protein with about 200 amino acids.

somatotype (sōᵘ·mătotəip), *sb.* [f. Somato- + Type *sb.*¹] The physique of an individual as expressed numerically in terms of the extent to which it exhibits the characteristics of each of three extremes (the endomorph, mesomorph, and ectomorph).
1940 W. H. Sheldon *Varieties Human Physique* i. 7 The patterning of the morphological components, as expressed by the three numerals, is called the somatotype of the individual. After examining..4,000 physiques we were able to describe..76 different somatotypes. **1944** A. Huxley *Let.* 19 July (1969) 508 From these a good draughtsman could turn out true representations of the various somatotypes. **1971** *Nature* 12 Mar. 113/2 An analysis of the somatotype data showed that there was a tendency for the twelve individuals to fall into two groups: a mesomorphic group (seven cases—wide heavy—typical somatotype 4,5,2) and an ectomorphic group (five cases—tall thin—typical somatotype 3,2,6). **1978** G. A. Sheehan *Running & Being* ii. 28, I dug into his [*sc.* Sheldon's] *Atlas of Men* and there I was. Somatotype 235... The number 235 is somatotype shorthand for little or no fat (2); a moderate amount of muscle (3); and a predominance of skin, hair, nervous tissue and thin bones (5). (The limits being one to seven.)

somatotype (sōᵘ·mătotəip), *v.* [f. prec.] *trans.* To assign to a somatotype.
1940 W. H. Sheldon *Varieties Human Physique* vii. 220, 400 northern Negroes whom we have photographed and somatotyped. **1951** Auden *Nones* (1952) 47 Lovers of big numbers go horribly mad, Would have..all of us Well purged, somatotyped, baptised, taught baseball. **1954** W. H. Sheldon *Atlas of Men* 16/2 By the time I had somatotyped the 12,000 men who made up this total series the foundations were well laid.
So **so·matotyping** *vbl. sb.*, the assignation of somatotypes.
1940 W. H. Sheldon *Varieties Human Physique* i. 8 The criteria and the procedure for somatotyping..provide a practicable, objective method for segregating and classifying the varieties of human physique. **1959** A. Huxley *Let.* 15 Oct. (1969) 880 Have you found in your hospital work that somatotyping along Sheldonian lines has helped? **1978** *Nature* 12 Jan. 193/1 Sheldon is the man who invented somatotyping—a classification of variations in human body structure, to which he related variations in temperament, in physical and mental illness, and in patterns of growth and aging.

sombrero. Add: **3.** *Microbiol.* A bacterial plaque in which a ring of partial lysis surrounds a clear central area.
1971 Cunningham & Sercarz in *European Jrnl. Immunol.* I. 414/1 Plaques which were clear in the middle but with a variable concentric zone of partial lysis at the periphery (christened 'sombreros'), were scored as clear. **1975** *Nature* 20 Feb. 639/1 Those which started as sombreros..always grew into larger sombreros.

sombre·roed, *a.* [f. Sombrero + -ed².] Wearing or covered by a sombrero.
1899 F. Remington *Sundown Leflare* 51 Sombreroed and moccasined, Sundown pattered along on his roan pinto. **1906** *Out West* Jan. 49 'Oh, no, you never make it. Too mucho arena!' with an emphatic, disapproving shake of sombreroed head. **1912** D. H. Lawrence *Let.* 17 Sept. (1962) I. 147 Black sombrero'd Italians. **1966** H. W. Yoxall *Fashion of Life* xii. 120 A sombreroed, bearded, cloaked individual.

some, *indef. pron.*, *a.*¹, *adv.*, and *sb.*¹ Add: **A.** *indef. pron.* **I. 4. e.** *to get some*: to have sexual intercourse; to succeed in finding a sexual partner. *U.S. slang.*
1889 W. H. Herndon *Let.* 5 Jan. in E. Hertz *Hidden Lincoln* (1940) 233 Speed about 1839-40 was keeping a pretty woman in this city, and Lincoln, desirous to have *a little*, said to Speed: 'Speed, do you know where I can get *some*?' **1971** E. E. Landy *Underground Dict.* 88 *Get some*.., obtain sexual intercourse with someone. **1978** J. Krantz *Scruples* vi. 173 Since his last visit she was getting some, somewhere, he'd bet his life on it.
f. *and then some*: and (plenty) more in addition. *colloq.* (chiefly *U.S.*).
1908 'Yeslah' *Tenderfoot in Southern California* ii. 22 It rains in sheets, in blankets, and in comforters, and then some. **1914** D. O. Barnett *Lett.* (1915) 19, I picked them out with those glasses, and let them have it, and then some! **1931** T. E. Lawrence *Let.* 10 June (1938) 724 It ..will be 12 guineas and then some! **1958** J. Cannan *And be Villain* i. 24, I waited till the train had gone out and then some. **1976** D. Clark *Dread & Water* ii. 51 People have got to talk... Tell us everything they know and then some.
B. *adj.*¹ **I. 4. e.** (Earlier and later examples.)
1844 *Spirit of Times* 30 Nov. 474/1 Many people have an idea that the 'big mare' will be 'some' in the race. **1876** 'Mark Twain' *Tom Sawyer* i. 8 Smarty! You think you're *some*, now, don't you. **1890** *Dialect Notes* I. 70 To say of a woman that 'she looks *some*', with emphasis on the *some*.., is equivalent to saying that she looks remarkably well.
f. Quite a; a remarkable. Used meiotically, often ironically, to suggest that something or someone is worthy of consideration. *some hope(s)!*: see *hope *sb.*¹ 4a. orig. *U.S.*
1808 J. Mackintosh in R. J. Mackintosh *Mem. Life Sir J. Mackintosh* (1835) I. viii. 448 You know that Bossuet and Arnauld believed their innocence—some authority. **1855** 'Q. K. P. Doesticks' *Doesticks, what he Says* iv. 28 It was 'some' bridge, in fact, a considerable curiosity, and a 'considerable' bridge. **1914** G. Atherton *Perch of Devil* i. 80 They're some geologists, he added with unwilling admiration. *Ibid.* 108 Butte is some education, believe me. **1925** F. Scott Fitzgerald *Great Gatsby* iii. 60 He smiled with jovial condescension, and added 'some sensation!' Whereupon everybody laughed. **1931** Brophy & Partridge *Songs & Slang Brit. Soldier* (ed. 3) 359 *Some hopes!*, it is most unlikely! **1941** W. S. Churchill *Unrelenting Struggle* (1942) 345 When I warned them [*sc.* the French Government] that Britain would fight on alone whatever they did, their Generals told their Prime Minister and his divided Cabinet: 'In three weeks England will have her neck wrung like a chicken.' Some chicken! Some neck! **1958** 'J. Byrom' *Or be he Dead* v. 69 'I gather you have Miss Canning as your assistant sleuth!'..'Some hope!..a good secretary always has to be in love with her boss!' **1976** A. Price *War Game* i. 66 'David has us to console him.'..'Some consolation!' murmured Frances. **1977** J. Wainwright *Do Nothin' till you hear from Me* x. 176 'Some band,' murmurs Ted—and there is suppressed excitement in his voice. I say, 'Ted—believe me—this is going to *be* some band.'
II. 9. a. Also following a numeral. *U.S.*
1968 *Time* 19 Jan. 7/2 Twenty-some years ago, when I was a nurse on the U.S.S. *Hope*. **1971** R. A. Carter *Manhattan Primitive* (1972) xi. 104 He's thirty years old, with..a master's degree and forty-some hours towards a doctorate. **1980** in S. Terkel *Amer. Dreams* 2 There were sixty-some contestants from all over the place.
C. *adv.* **2. b.** (Earlier examples.)
1745 J. Emerson *Jrnl.* 8 Apr. in *Mass. Hist. Soc. Proc.* (1911) XLIV. 74, I read some in Watson. **1785** *Massachusetts Spy* 28 Apr. 2/3 (Advt.), A tall stout looking fellow,..stammers some in his speech. **1817** *Essex Inst. Hist. Coll.* (1866) VIII. 228 The material of which it is built looks some like marble.
d. *to go some*: to go well or fast; to do well; to work hard. *slang* (orig. *U.S.*).
a **1911** D. G. Phillips *Susan Lenox* (1917) II. ii. 24 He had evidently been 'going some' for several days: the sour, worn, haggard face..suggested a moth-eaten jaguar. **1912** J. Sandilands *Western Canad. Dict. & Phrase-Bk.* s.v. *Some*, That's going some may mean great speed or excellence of workmanship, or it may even be used in reference to the speed at which a person races to his ruin. **1915** Wodehouse *Psmith, Journalist* x. 71, I guess we're making a hit. Cosy Moments is going some now. **1966** 'J. Hackston' *Father clears Out* 173 He had the easy movements of the retriever, and for a big dog could go some. **1973** J. Wainwright *Touch of Malice* 8 A uniformed inspector..with..less than five years service under his belt. Jesus—that was going some! **1982** H. Lieberman *Night Call* viii. 47 He'd known the girl for two months; for Daughtry that was going some.
3. (Earlier and later examples.)
c **1780** in *Amer. Speech* 1969 (1973) XLIV. 304 Until it gets some darker. **1913** [see *gun-play]. **1940** W. Faulkner *Hamlet* i. ii. 31 We had done been feeding it [*sc.* a horse] for two–three days now by forced draft..and it looked some better now than when we had brung it home. **1956** G. E. Evans *Ask Fellows who cut Hay* xxv. 231 An old worker..turned the handle and tried it with a few roots. Asked what he thought of it he said with conviction: 'It's some stiff, maaster.' **1976** M. Machlin *Pipeline* iv. 44 He's going to be some pissed off when he finds out about this.

-some, *suffix*⁴, f. Gr. σῶμα body; (*a*) used with this sense, as in *ectosome* s.v. *ecto- *trophosome* s.v. Tropho-; (*b*) used to form words denoting an intracellular particle, as in *acrosome, *chromosome, *lysosome; (*c*) used to repr. *chromosome*, as in *disome, *monosome i.
1921 [see *hexasome* s.v. *hexa-].

somebody, *sb.* Add: **1. b.** Also with sense 'a rival for the affections' in phr. *there* (or *it*) *is somebody else.*
1911 G. B. Shaw *Getting Married* 200 You have never given me a real reason for refusing me yet. I once thought it was somebody else. There were lots of fellows after you. **1935** D. L. Sayers *Gaudy Night* xii. 260 'I suppose,' he said in a savage tone, 'there's somebody else.' **1946** 'Brahms' & 'Simon' *Trottie True* iii. 37 'I know what it is.' Joe tried to stop himself but he couldn't. 'There's somebody else.'
d. Also *somebody-or* (*-the*) *-other.*
1935 D. L. Sayers *Gaudy Night* iii. 255 Mr. Somebody-or-the-other had undertaken..to climb every tree in St. Giles. **1976** 'L. Black' *Healthy Way to Die* iii. 29 The two girls..were the Daughters of Lord Somebody-or-other.
3. (Later example.) *Somebody up there*, God; a supernatural controlling power.
1972 *Dict. Contemp. & Colloq. Usage* 27/3 *Somebody up there loves* (*hates*) *me*, an expression attributing one's good or bad luck..upon being in the good or bad graces of an unseen power above. **1975** J. Updike *Month of Sundays* xxx. 225 In the land of my parish, the shortest day of the year is approaching, and somebody's birthday, I think. **1977** R. Perry *Dead End* i. 9 All I could do was..pray that somebody up there loved me. **1980** *Times* 10 June 10/1 (*heading*) Somebody up there cares for me.

so·meday, *adv.* [f. Some *a.*¹ 2 + Day *sb.*] At some future time. Cf. Day *sb.* 7 b.
Earlier examples of *some day* as a two-word phrase will be found s.v. Some *a.* 2 b (*b*).
1898 G. B. Shaw *Candida* i, in *Plays Pleasant* 94 Theyll ave to give you somethink someday, if it's honly to stop yore mouth. **1902** W. B. Yeats *Cathleen ni Hoolihan* 13 220 We might be put in the way of making Patrick a priest someday, and he so good at his books. **1921** G. B. Shaw *Back to Methuselah* v. 221 Something or other must make an end of you someday. **1939** Joyce *Finnegans Wake* 119 A tea anyway for a tryst someday. **1940** W. Faulkner *Hamlet* ii. i. 110 He could almost see the husband which she would someday have. **1958** [see *Murphy³ 3]. **1966** *Punch* 24 Aug. 294 Won't someday a three-minute montage provide a more significant aesthetic treat than a sonnet? **1978** *Amer. Poetry Rev.* July/Aug. 31/3 It will clutch at a heart it did not seem to know, and someday die.

some one, someone, *pron.* (and *sb.*). Add: (Examples of *pl.*) *someone else*, used pregnantly to mean 'a rival for the affections'. Cf. *somebody else* s.v. *somebody *sb.* 1 b.
1914 'Bartimeus' *Naval Occasions* xxv. 261 It had become necessary to tell Selby that she couldn't love him any longer... Further, by her creed, it was only right that she should tell him about Someone Else as well. **1936** J. Curtis *Gilt Kid* xviii. 178 There's someone on that roof all right. Two someones. **1941** M. Allingham *Traitor's Purse* xi. 129 'She broke the engagement.'..'Why?' 'As she seen someone else?' **1977** A. Hunter *Gently Instrumental* i. 14 Walt half-choked: 'There's—someone else!'.. 'You dirty old queen, that's just what you'd think.' **1978** P. Porter *Cost of Seriousness* 30 To whom someones in the city must pay homage.

somepin, var. *sump'n.

so·meplace, *adv.* and *sb. dial.* and *U.S.* Also **some place.** [f. Some *a.*¹ + Place *sb.*] **A.** *adv.* Somewhere; (at, in, to, etc.) a particular or unspecified place.
1880 *N. & Q.* 24 Apr. 340/2 'No place' in Devonshire is the usual form of 'nowhere', 'no place else' and 'some place else' being the commonest forms of 'nowhere else' and 'somewhere else'. **1896** *Dial. Notes* I. 425 *Some place*, somewhere. **1922** Joyce *Ulysses* 639 Lodging some place about the pit of the stomach. **1933** *Punch* 8 Nov. 511/3 Is there no nicer toon some place else? **1937** J. Steinbeck *Of Mice & Men* i. 18 Some place I'd find a cave. **1948** H. L. Mencken *Amer. Lang.* Suppl. II. 394 The long-awaited grammarian of vulgar American..will have a gaudy time anatomizing such forms as..'It was some place *else*', [*etc.*]. **1959** M. Dolinsky *There is no Silence* v. 78 She's here some-place. **1966** T. Pynchon *Crying of Lot 49* iii. 63 Tony Jaguar decided he could surely unload his harvest of bones on some American someplace... He was right. **1978** *Detroit Free Press* 16 Apr. E.7/4 I've always been able to get a room someplace, but this year I'm struggling.
B. *sb.* Somewhere; a particular or unspecified place.
1922 Joyce *Ulysses* 758 That drunken little barrelly man that bit his tongue off falling down the mens WC drunk in some place or other. **1940** W. Faulkner *Hamlet* iii. ii. 190 There will be some town, some place close where I can live. **1971** B. Patten *Irrelevant Song* 37 (*title*) In someplace further on. **1976** *National Observer* (U.S.) 7 Feb. 15/3 Next the Navy tried Texas, where defense-minded Sen. John Tower said it was a fine idea—for someplace else. **1977** H. Fast *Immigrants* iii. 165 His father came to America from someplace in Poland in eighteen sixty-nine.

somer, var. *sommer *adv.*

som'ers (sɒ·məɹz), *adv. U.S.* Also **somers.** Repr. colloq. pronunc. of *somewheres*.
1876 'Mark Twain' *Tom Sawyer* xxiii. 182 It keeps me in a sweat, constant, so's I want to hide som'ers. **1884** —— *Huck. Finn* xxiv. 241 It's reckoned he left three or four thousand in cash hid up som'ers. **1896** —— in *Harper's Mag.* Aug. 344/2 His aunt Polly wouldn't let him..go traipsing off somers wasting time. **1909** *Dial. Notes* III. 404 *Som'ers*, adv., somewhere. 'He's *som'ers* around.'

Somervillian (sɒməɹvi·lian), *a.* and *sb.* [f. the name of *Somerville* College (founded as a women's college in 1879 and named after Mary Somerville) + -ian.] **A.** *adj.* Of or pertaining to Somerville College, Oxford. **B.** *sb.* A member of the college.
1896 A. D. Godley in *More Echoes from Oxford Mag.* 64 Ye Somervillian students. **1904** R. Bridges *Let.* 26 May (1940) 50 The Somervillians might object. **1922** Byrne & Mansfield *Somerville College 1879–1921* iii. 38 The War Office was granted the use of the Somerville Buildings for the duration of the war..[which] overshadowed the situation of the young Somervillians. **1934** R. Macaulay *Going Abroad* xxix. 254 The two Somervillians..passed most of the afternoon and evening with the Josefs. **1966** E. H. Jones *Margery Fry* v. 35 'They were beautiful, in their Liberty dresses,' said an old Somervillian..remembering the leaders of College society in the 'nineties. **1976** 'A. Cross' *Question of Max* ix. 113 In 1918 Somervillians were still being housed in..Oriel College.

somesthesis, -esthetic varr. *SOMÆSTHESIS, *SOMÆSTHETIC *a.*

something, *sb.*, (*adj.*,) and *adv.* Add: **A.** *sb.* **1. a.** Phrases *something for everybody* (or *everyone*), *something for nothing*. Also used *attrib.*

1869 P. T. BARNUM *Struggles & Triumphs* viii. 132 When people expect to get 'something for nothing' they are sure to be cheated. 1924 G. B. SHAW *Saint Joan* iv. 41 The Jews generally give value. They make you pay; but they deliver the goods. In my experience the men who want something for nothing are invariably Christians. 1938 E. AMBLER *Cause for Alarm* vii. 115 A something-for-nothing proposition always has a string to it. 1955 R. MACAULAY *Let.* 20 Aug. in *Last Lett. to Friend* (1962) 206, I personally think it all to the good, as giving something for every one, however different their minds, backgrounds, and religious temperaments. 1960 *N.Y. Times Bk. Rev.* 17 Jan. 1 There's something for everybody. 1971 *Engineering* Apr. 129/2 Something-for-everyone entertainment. 1976 *Glasgow Herald* 26 Nov. 6/2 But human nature dictates that most people..are liable to take advantage of an opportunity to get something for nothing.

e. (Earlier and later examples.)

c 1863 T. TAYLOR *Ticket-of-Leave Man* II. 32 If Mr. Gibson would only give you employment. He's something in the City. 1907 E. GOSSE *Father & Son* ii. 21 My uncles ..earned a comfortable living, E. by teaching, A. as 'something in the City'. 1951 [see *SLINKY *a.*]. 1962 [see *COMMUTE *v.* 4 b]. 1978 P. FITZGERALD *Bookshop* ii. 20 He was known to drive up to London to work, and to be something in TV. 1979 R. BARNARD *Posthumous Papers* iv. 37 He was something in insurance.

f. *or something* (colloq.), used to express an indistinct or unknown alternative.

1814 JANE AUSTEN *Mansfield Park* I. xi. 223 There were generally delays, a bad passage or *something*. 1899 [see *MITTAGESSEN]. 1913 'S. ROHMER' *Mystery of Dr. Fu-Manchu* i. 4 What, are you moved to London or something? 1926 I. MACKAY *Blencarrow* v. 49 Yet undoubtedly this man was drunk or ill, or something. 1938 *Chatelaine* Oct. 25/3 *Our* things must have tattle-tale gray or somethin' 'cause they never shine like *this*. 1951 M. KENNEDY *Lucy Carmichael* II. iii. 100 'Aren't they engaged or something?' 'I don't know what you mean by *or something*... It's a vulgar, slipshod phrase.' 1958 *N.Z. Listener* 4 July 7/1 Jarden was off the field—had hurt his foot or something—and it seemed that we might be hard up against it. 1969 N. FREELING *Tsing-Boum* viii. 54 She might have a police record or something. 1978 P. MARSH et al. *Rules of Disorder* iii. 69 You have to fight or else people..think you're a bit soft or something.

2. c. (Later examples.)

1931 R. CAMPBELL *Georgiad* iii. 55 Even the devil dwindles to a duiker, Who prides himself as something of a spiker. 1939 R. G. COLLINGWOOD *Autobiogr.* iv. 27, I had become something of a specialist in Aristotle. 1959 *Listener* 17 Dec. 1083/3 It had been, I admit, something of a party. 1978 *Lancashire Life* Sept. 51/1 During the last war he became something of a legend, working incredible hours and doing general and orthopaedic surgery, as well as obstetrics.

3. c. *something* (*good* or *special*), a useful racing tip.

1907 *Racing Expert* 9 July 3 For the benefit of those who care to wait and act upon the best information. 'The Expert' will occasionally wire when he knows Something Special. 1908 *Racing Judge* 6 June 4 Owing to Bank Holiday this Letter will be sent out on Tuesday Evening. ..Something good at Manchester will be given. 1937 PARTRIDGE *Dict. Slang* 800/2 *Something good*, a good racing tip.

4. b. (Later examples.)

a 1902 S. BUTLER *Way of all Flesh* (1903) liv. 249 When Christina pointed out to him that it would be cheap he replied that there was something in that. 1977 B. PYM *Quartet in Autumn* i. 7 'Cheerful, aren't you,' said Edwin, 'but perhaps there's something in it. Four people on the verge of retirement, each one of us living alone.'

d. *something to see* (or *look at*): an impressive sight.

1808 J. MACKINTOSH in R. J. Mackintosh *Mem. Life Sir J. Mackintosh* (1835) I. ix. 501 It was something to see children clinging round the necks of their fathers, and sons carrying their infirm parents in pursuit of health. 1942 T. BAILEY *Pink Camellia* i. 2 In khaki breeches, sitting her horse like a boy, her white shirt open at the throat, she was something to look at. 1957 E. B. WHITE *Let.* June (1976) 440 Martha is really something to see now.

e. *to have* (*got*) *something*, to have an idea or attribute of value or worthy of consideration.

1938 'E. QUEEN' *Four of Hearts* iv. 57 Say..the screwball's got something. Only I got a better idea. 1940 G. GREENE *Power & Glory* I. ii. 25 'I would take..a hostage.'..'You know,' the chief said, 'you've got something there.' 1948 POWYS & BOLTON *Don't listen, Ladies!* in *Plays of Year* (1949) 586 The Crusaders, gentlemen, they *had* something. The husband ordered his clothes from the blacksmith, and his wife's from the locksmith. 1960 *Times* 14 Sept. 12/6 Yet that girl 'had something', as any visitor to the United States will find out. 1973 L. COOPER *Tea on Sunday* i. 20 'I'm not at all the nice little wife she wanted for you.' 'She may have something there.'

f. Used in various phrases expressing admiration, as *isn't* (*that, he*, etc.) *something?*, *to be really something*; *quite something*: see *QUITE *adv.* 5 d.

1958 B. NICHOLS *Sweet & Twenties* ii. 42 The Ritz Bar, in those days, really was something. 1967 M. KENYON *Whole Hog* vii. 81 Isn't that something? So if they [*sc.* pigs] don't know you they're like interested? 1969 WIDDOWSON & HALPERT in Halpert & Story *Christmas Mumming in Newfoundland* 161 Mummers were really something when I was a boy. If you wasn't afraid of them, you wasn't afraid of nothing when you was four or five. 1973 A. CHRISTIE *Postern of Fate* III. x. 213 Perhaps it's something important... And so if they..tried to get whatever it was—that really would be something! 1977 'A. YORK' *Tallant for Trouble* xi. 163 P. C. Abrahams presented arms..decked..out in full-dress white... 'Oh, isn't he something,' Jennie Kamm exclaimed.

g. *to have something going* (with *someone*), to have an 'understanding' or an affectionate relationship (with someone).

1971 V. CANNING *Firecrest* iii. 32 It didn't need any semaphore signals to tell her that there was something going between Mrs. Pilch and Major Cranston. 1973 *Philadelphia Inquirer* (Today Suppl.) 7 Oct. 7/2 Is it true that Sammy Davis Jr. has something going with Linda Lovelace. 1977 E. LEONARD *Unknown Man No. 89* xx. 200 She smiled..like they had something going.

5. a. Also in phr. *a little something*: some food or drink; a snack; refreshments. Cf. sense 1 c in Dict.

1866 GEO. ELIOT *Felix Holt* I. xi. 237 Like the shrill biting talk of a vixenish wife, it..compelled you to 'take a little something' by way of dulling your sensibility. 1926 A. A. MILNE *Winnie-the-Pooh* vi. 77 It was..as if somebody inside him were saying, 'Now then, Pooh, time for a little something.'..So he sat down and took the top off his jar of honey. 1950 J. CANNAN *Murder Included* vii. 146, I wonder if a little something could be provided to pacify the inner man? 1958 WODEHOUSE *Cocktail Time* xix. 159 Butlers always like to keep their strength up with a little something in the middle of the morning. 1977 P. D. JAMES *Death of Expert Witness* IV. iii. 192, I cook a little something for everyone in the evenings.

6. b. Also *slang* (orig. *N. Amer.*), a different matter; an exceptional or extraordinary (person, event, sight, etc.).

1909 R. E. KNOWLES *Attic Guest* 87 But when a lover comes across a couple of states, leaving behind him a big city—and all the girls are sorry to see him go, that's the best of it—that is something else, as we used to say in the South. 1940 W. FAULKNER *Hamlet* I. ii. 33 But when cash money starts changing hands, that's something else. 1949 R. HARVEY *Curtain Time* 67 Getting the small performer dressed for a public appearance was something else again. 1957 E. HORNE in *N.Y. Times Mag.* 18 Aug. 26/3 Something else—A phenomenon so special it defies description. Thus, when asked if the music is great..a cat may reply, 'No, man, not that; it was something else.' 1960 *Melody Maker* 31 Dec. 5/5 *Philly Joe Jones:* Aside from being a fabulous drummer, he could be a great comedian. He has people lying on the floor, he's so funny. Philly's something else. 1968 *Crescendo* Jan. 27/1 The one I rave about more than any other is the band of 1947... That was something else. It was a dream. 1973 R. L. SIMON *Big Fix* (1974) vii. 50 Dillworthy was something else again. 1977 *O.D.* No. 3, 12/1 (*caption*) Oh, wow, these guides are..something else man!

B. *adv.* **2. a.** (Further examples.)

1851 E. RUSKIN *Let.* 25 Nov. in M. Lutyens *Effie in Venice* (1965) II. 218 Nani makes them a great dish of Fish seasoned strongly with Garlic and the smell is something too dreadful if one happens to pass by the door. 1856 G. MEREDITH *Let.* 15 Dec. (1970) I. 38 The *dulness* is something frightful. 1918 C. MACKENZIE *Sylvia Scarlett* I. vii. 208 'These paths are something dreadful, Emmie,' said Mrs. Horne, as the three of them scrambled up through the garden. 1932 L. GOLDING *Magnolia Street* II. ii. 299 The way the razor trembled..now and again was something cruel.

c. Also, with *adj.* used for *adv.* in dial. and colloq. usage.

1898 G. B. SHAW *You never can Tell* in *Plays Pleasant* 211 *Gentleman*: Did you howl? *The Young Lady*: Oh, something awful. 1909 A. WOOLLCOTT *Let.* 24 Sept. (1944) 20 She gads around something fierce, as your friend Bert would say. 1915 J. WEBSTER *Dear Enemy* 300 When he was drunk..he smashed the furniture something awful. 1932 R. LEHMANN *Invitation to Waltz* I. iii. 58 Her husband drinks something shocking. 1963 W. H. MISSILDINE *Your Inner Child of Past* xv. 221, I was taken into the assembly hall. And beat up something terrible. 1978 D. CLARK *Libertines* ii. 41 'I'll put a plaster on that cut for you.'..'Thanks, doctor... It does sting something chronic.'

so·mething, *v.* Hence **so·methinged** *ppl. a. rare.*

1922 E. WALLACE *Valley of Ghosts* xiii. 120 You called me..a fool, and a somethinged fool, almost the first time we met.

so·methingth, *a.* (Earlier example.)

1854 MRS. GASKELL *Lett.* (1966) 302, I am very poor; which eases my cares wonderfully, see somethingth satire of Juvenal.

sometime, *adv.* (and *a.*). Add: **1. d.** Passing into adj. Freq. in phr. *a sometime thing*: something which is occasional or transient.

1935 G. GERSHWIN (*song-title*) A woman is a sometime thing. 1959 *Times Lit. Suppl.* 6 Nov. p. xxxi/1 Going to the movies still entailed leaving home and paying money, so that..the movies were a 'sometime thing' and a reward. 1964 'E. MCBAIN' *Axe* vi. 115 The game..ain't regular, like you said it was. It's a sometime thing, whenever the urge strikes. 1967 *Observer* 19 Nov. 21/2 Money is a sometime thing for Simon Dee, here today and just possibly gone tomorrow. 1969 *Jrnl. Eng. & Gmc. Philol.* LXVIII. 214 Poetic propriety is a sometime thing. 1980 *Newsweek* 17 Nov. 12/2 Political parties are weaker, the Federal bureaucracy has grown unwieldy and party discipline in Congress is a sometime thing.

sometimes, *adv.* Add: **1. c.** Used adjectivally to denote 'occasional'. Cf. *SOMETIME *adv.* (and *a.*) 1 d.

a 1945 E. R. EDDISON *Mezentian Gate* (1958) I. v. 44 Nor did they find wholesome nor comfortable..his sometimes flashings into unforeknowable violence. 1974 *Publishers Weekly* 11 Feb. 56/3 Brando's sometimes generosity, his idealism. 1977 *Rolling Stone* 13 Jan. 36/2 Srouji confessed her entire FBI history, starting with her role in the Sixties as a sometimes informant.

sometimey (sɒ·mtəɪmi), *a.* *U.S. Black* and *Prison slang.* [f. SOMETIME *adv.* (and *a.*) + -Y¹.] Variable, unstable.

1946 MEZZROW & WOLFE *Really the Blues* 378 *Sometimey*, unstable, unpredictable, neurotic. 1969 R. PHARR in A. Chapman *New Black Voices* (1972) 62 She's the evilest and sometime-iest woman I ever shacked up with. 1977 *New Yorker* 24 Oct. 114/2 Men who have transferred to Green Haven from more rigorous but more consistent prisons..say that Green Haven's officers are 'sometimey'.

someway. *adv.* For Now *rare* exc. *dial.* read Now chiefly *U.S. colloq.* and add: **1.** (Later examples.)

1892 B. POTTER *Jrnl.* 8 Oct. (1966) 278 They are related someway to Neil Gow. 1902 W. B. YEATS *Where there is Nothing* (1903) v. 106, I thought he would have had half Ireland with him by this time with his great preaching, but someway when he preaches to the people, they don't seem to mind him much. 1922 JOYCE *Ulysses* 180 Pity of course: but somehow you can't cotton on to them someway. 1930 W. FAULKNER *As I lay Dying* 115 It's like he had got into the inside of you, someway. 1938 M. K. RAWLINGS *Yearling* xiv. 140, I hate the hawks eatin' the quail, but I don't someway mind the 'coons eatin' the grapes. 1978 *Washington Post* 22 July A8, I keep the windows open and run my box fan. I figure the Lord will take care of me somehow, someway.

somewhat, *sb.* and *adv.* Add: **B.** *adv.* **6.** *more than somewhat*, very, extremely; very much.

1930 D. RUNYON in *Collier's* 13 Sept. 7/3, I am now more nervous than somewhat. 1938 D. WHEATLEY *Uncharted Seas* xi. 190, I thought my nerve was pretty good, but this scares me more than somewhat. 1945 *Tee Emm* (Air Ministry) V. 40 Citizens have been known to leave the premises.., being more than somewhat apprehensive of future developments. 1964 WODEHOUSE *Frozen Assets* viii. 143 She said quite a number of things that wounded my sensitive nature more than somewhat. 1974 V. GIELGUD *In such a Night* vii. 67 It burned me up more than somewhat.

somewhen, *adv.* Add: Still found occas., esp. coupled with *somewhere* or *somehow*, but no longer common. (Earlier example in the recent use and later examples alone or coupled with *somehow*.)

1833 J. S. MILL *Let.* 5 July in *Wks.* (1963) XII. 163, I shall write out my thoughts more at length somewhere, and some*when*, probably soon. 1920 H. G. WELLS *Outl. History* (rev. ed.) viii. 37/1 Somewhen about 50,000 years ago..appeared *Homo Neanderthalensis*. 1934 J. L. MYRES in E. Eyre *Europ. Civilization* I. 87 Such accommodation between means and ends, resources and wants, is found to have been achieved, somewhen and somehow, [etc.]. 1975 J. C. MASTERMAN *On Chariot Wheel* v. 40, I cherished the belief that somehow and somewhen I should find my way to Oxford.

somewhere, *adv.* and *sb.* Add: **A.** *adv.* **1. e.** *somewhere in* (*France*, etc.), phr. orig. used during the war of 1914–18 in referring to some locality in the theatre of war without identifying it (because of the restrictions of censorship); hence, in extended use, somewhere unspecified for reasons of security or because one's stay there is temporary.

1915 *Illustr. London News* 20 Feb. 233 (*caption*) The War Area as seen by the Airman: 'somewhere in Flanders' photographed from a reconnoitring Aeroplane. *Ibid.* 241 For the moment 'Victoria' looks like 'Somewhere in France'. 1915 *Daily Sketch* 17 Aug. 12/1 None of these soldiers a year ago expected to be snapped one day..—somewhere in Egypt. 1918 *Wireless World* VI. 390 A Wireless Section 'Somewhere in England'. 1939 JOYCE *Finnegans Wake* 21 There was a brannewail that same sabboath night of falling angles somewhere in Erio. 1939 *War Illustr.* 14 Oct. 144 From 'Somewhere in England' to 'Somewhere in France': 1939 Echoes the Story of 1914. 1943 J. B. PRIESTLEY *Daylight on Saturday* viii. 52 If our lads was fightin' like 'ell somewhere in France, why yer'd see them production figures take a high jump. 1943 *Gramophone* Sept. 63/3 From Mr. Tony Puddy, Somewhere-in-England. 1973 *Jewish Chron.* 9 Feb. 15/2 The girls I visited recently on a training course 'somewhere in Israel' were not all bunched together in one group. 1977 'J. LE CARRÉ' *Honourable Schoolboy* xvii. 400 The Somewhere-in-England sense of makeshift habitation..of every exiled correspondent.

B. *sb.* (Later examples.)

1914 R. BROOKE *Let.* 7 Mar. in *Coll. Poems* (1918) p. cxiii, I shall be glad to be back among you all, and tied to somewhere in England. *Ibid. Let.* Mar. p. cxvii, I want somewhere I needn't always be spick and span in, and somewhere I don't have to pay a vast sum. 1928 W. B. YEATS *Tower* 8 And I myself created Hanrahan And drove him drunk or sober through the dawn From somewhere in the neighbouring cottages. 1930 G. B. SHAW *Apple Cart* II. 69 All their people came from Scotland or Ireland or Wales or Jerusalem or somewhere. 1942 W. FAULKNER *Go down, Moses* 139 A big dog, a hound with a strain of mastiff from somewhere. 1958 B. W. ALDISS *Non-Stop* II. ii. 79 The ship..has come *from* somewhere and is going *to* somewhere. These somewheres are more important than the ship.

somma (sǫ·mā). *Physical Geogr.* Also Somma. [a. It. (*Monte*) *Somma*, proper name of the feature of this kind associated with Vesuvius.] A remnant of an older volcanic cone which partly or wholly encircles a younger cone; the rim of a caldera. Freq. *attrib.*, as *somma ring*.

c 1910 H. R. MILL *Dict. Geogr. Terms* in L. D. Stamp *Gloss. Geogr. Terms* (1961) 426 Somma, originally, the rampart remaining from the old crater of Vesuvius and forming an arc around one side of the new one. The name is sometimes extended to similar formations in other volcanoes. 1917 T. G. BONNEY *T. Anderson's Volcanic Studies* 2nd Ser. vii. 39 This [ridge], according to Dr. Anderson, forms a Somma ring within which rises a smaller and a larger crater, both intermittently active. 1940 *Bull. Volcanologique* VI. 217 The older volcanic mass on the eastern side may be a continuation of the Somma. 1944 C. A. COTTON *Volcanoes as Landscape Forms* xi. 167 An example of a large cumulo-dome which has been built within a wide caldera or 'Somma ring' is the Soufrière of Guadeloupe, in the Lesser Antilles. 1975 FIELDER & WILSON *Volcanoes of Earth, Moon & Mars* vii. 88/1 The so-called somma-type of nested cone..is named after Mount Somma, a residual portion of the encircling volcanic rim unit which was the precursor of the present, centrally disposed, Vesuvius. 1979 *Nature* 1 Nov. 24/1 It [*sc.* Soufrière volcano, St. Vincent] is a strato-volcano 1,220 m high with an open summit crater 1·6 km in diameter located in the southern part of a 2-km wide somma crater.

‖**sommelier** (somǝlye). [Fr.: see SOMLER.] A wine waiter.

1889 *Harper's Mag.* Apr. 698/1 The 'sommelier', or butler, who runs from table to table, laden with bottles, and distributes here and there strange liquids. 1923 E. P. OPPENHEIM *Inevitable Millionaires* xiv. 146 Harold ..making cryptic signs with his fingers which intimated to the sommelier his urgent need of a cocktail. 1955 M. ALLINGHAM *Beckoning Lady* v. 77 He poured the awkward liquid with the skill of a *sommelier*. 1966 *Punch* 20 July 113/2 Although we've still got some cooks to shoot and many sommeliers to educate, our standards have improved beyond all recognition. 1974 *Times* (Wines & Spirits Suppl.) 2 Dec. p. iii/2 An awe-inspiringly stately *sommelier* and long wine lists..can often discourage the sale of wine.

sommer (sǫ·mǝɹ), *adv.* *S. Afr. colloq.* Also **somaar,** 9 **somar, somer.** [a. Afrikaans *somaar, sommer*.] Just, simply, for no specific reason, without further ado.

1835 T. BOWKER *Jrnl.* 12 May (MS.), Bowkers party gone out on patrole *somer* heard from the third Division, they have been shooting some Kafirs, & taking Cattle & Goats. 1850 R. G. CUMMING *Five Yrs. Hunter's Life S. Afr.* I. 27 The Dutch word *somar*..is also a word to which I think I could challenge the most learned schoolmaster in the Colony to attach any definite meaning. It is used by both Boers and Hottentots in almost every sentence. 1920 S. BLACK *Dorp* 174 'Ach, Oom Kaspar, can't I go back to look for my pipe and put it somaar in the box?' 'Nay, kerel, they won't let you go again inside the ballot-box.' 1959 *Cape Times* 15 Aug. (Weekend Mag.) 10/6 Too dangerous to sommer leave it like that with a child about the place. 1969 A. FUGARD *Boesman & Lena* 15 Sannie who? Sommer Sannie Somebody. 1975 *Daily Dispatch* (E. London) 13 June 12 My father was a trader, who sold both blankets and ochre, and to my question as to why the tribesmen didn't sommer buy red blankets instead of going through all the trouble of dyeing white blankets with ochre, he replied that this was for hygienic reasons.

‖**sommité** (somite). [Fr., lit. 'summit, top, tip'.] A person of great eminence or influence.

1856 *Sat. Rev.* 19 Apr. 496/1 We observe the name of numerous *sommités* of the architectural profession. 1859 [see COLLECTIONIZE *v.*]. 1900 W. JAMES *Let.* 2 Apr. (1920) II. 121 We must go in for budding genius, if we seek a European. If an American, we can get a *sommité*! 1938 A. HUXLEY *Let.* 12 Apr. (1969) 434, I thought him [*sc.* Dr. Gerald Webb] a most remarkable old man, and they all say he's one of the *sommités* in the matter of TB. 1965 *Punch* 24 Nov. 752/1 A High Table is an institution; it is not merely a collection of what the French used to call *sommités*.

Somogyi unit (sǭmōᵘgi). *Biochem.* [The name of Michael *Somogyi* (1883–1971), Hungarian-born U.S. biochemist who proposed the unit in 1948 (see quot.).] A unit in terms of which the effectiveness of a solution at catalysing the hydrolysis of starch can be stated (see quot. 1948).

[1948 M. SOMOGYI in *Ann. Surg.* CXXVIII. 676 One unit is that amount of enzyme which hydrolyzes 1 mg. of starch in 1 hour at 37° and pH7 in phosphate buffer, provided that not more than 44% of the starch is hydrolyzed.] 1956 E. KING *Micro-Anal. in Med. Biochem.* (ed. 3) iv. 92 Amylase activity... Somogyi units per 100ml. of plasma. 1961 *Lancet* 12 Aug. 373/1 A new amyloclastic technique which yields results in the equivalent of Somogyi units has been used to measure amylase activity in sera from..patients with diabetes mellitus. 1974 R. M. KIRK et al. *Surgery* vi. 110 The concentration of serum amylase is normally less than 200 Somogyi units.

son, *sb.*¹ Add: **1. a.** (Later example referring to an animal.)

1974 *New Yorker* 29 Apr. 102/2 A son of..a Thoroughbred, and..a quarter-horse brood mare.

3. b. Also, as a term of familiar address without implication of affection. Cf. *old son* s.v. *OLD *a.* 8 a.

1914 G. B. SHAW *Misalliance* 5 *Bentley*:... I should like to wring your damned neck for you. *Johnny* (*with a derisive laugh*): Try it, my son. 1959 E. H. CLEMENTS *High Tension* vii. 121 No good brooding, son. 1967 *Listener* 22 June 807/2 He was then asked to accompany the police to the police-box in order to confirm his identity. He replied, 'Look, son, I am not moving from this spot. If you want me you will have to arrest me.' 1974 S. MARCUS *Minding Store* (1975) ix. 188 Mr. Seeligson said, 'Son, you've done me a great favor. I appreciate all the trouble you've gone to.'

6. d. *son of* ——: at one time a common formula for the title of a sequel to a book or film; hence used *joc.* to designate a programme, product, institution, etc., that is a derivative *of* its predecessor.

1929 E. R. BURROUGHS (*title*) Son of Tarzan. 1934 *Picturegoer* 23 June 20/3 *Son of Kong*... By no means a second *King Kong* this picture, nevertheless, has some clever technical qualities. 1941 'B. GRAEME' (*title*) Son of Blackshirt. 1965 [see *horror film* s.v. *HORROR *sb.* 6]. 1966 'O. MILLS' *Enemies of Bride* iv. 47, I produced a scintillating piece of non-fiction called..*Elizabethan Domestic Drama*... I got a sequel—Son-of-Elizabethan-Domestic-Drama..—into print as well. 1971 R. PETRIE *Thorne in Flesh* iii. 45 We don't want you playing Son of Sexton Blake. You could get hurt. 1976 *Gramophone* Nov. 910/1 The XSV/3000 is recognizably a 'son of' the XUV/4500Q: it has the same slim-line body and lightweight fixing wings. 1979 *Daily Tel.* 29 Mar. 6/3 (*heading*) Cheaper seats likely if 'Son of Concorde' flies. 1981 *Times* 19 Nov. 13/1 President Reagan..has now formally endorsed..negotiations..on strategic arms reductions (now known as 'Start', son of Salt).

7. d. Also used *ellipt.* for *SON OF A BITCH.

1951 W. FAULKNER *Collected Stories* 171 Are you going to sit there and let a black son rape a white woman on the streets of Jefferson?

e. *Son of Heaven* († *Heaven's Son*) [tr. Chinese *tiānzǐ*], the Emperor of China; loosely, any Chinese. Cf. CELESTIAL *a.* 4 and *sb.* 2.

1613 PURCHAS *Pilgrimage* IV. xvi. 369 The King's Title is, *Lord of the world, and Sonne of Heaven*. 1838 GUTZLAFF & REED *China Opened* II. xxvii. 541 To gain such honours as the Mongol princes pay to Heaven's Son, requires a well-stored treasury. 1850 *North-China Herald* 9 Nov. 58/4 One of the common appellations of the Emperors of China has been and still is *T'ien Tsze*, 'the Son of Heaven!' 1923 S. MERWIN *Silk* (1924) 136 It is now my privilege to serve him who is in all but official style the Son of Heaven. 1938 *Foreign Affairs* XVI. 201 A dignity which the 'Sons of Heaven' consider has belonged to them for thousands of years. 1973 J. LEASOR *Mandarin Gold* i. 1 The Emperor, Tao Kuang, the Son of Heaven, who ruled his celestial empire from..Peking, the forbidden city.

f. (*horny-handed*) *son of toil*, a manual labourer. Now often *ironically*.

1873 *Q. Rev.* CXXXV. 543 The peculiar virtues of the horny-handed sons of toil received a severe shock in 1848, and, finally collapsed in 1871. 1902 'MARK TWAIN' in *N. Amer. Rev.* Apr. 441 A crowd of ten thousand.. proud, untamed democrats, horny-handed sons of toil.. and fliers of the eagle. 1933 WODEHOUSE *Heavy Weather* xvii. 298 You look like one of those Sons of Toil Buried by Tons of Soil I once saw in a head-line. 1976 *Times* 23 Mar. 19/4 There won't be any room for your actual horny-handed sons of toil in the TUC; there'll be too many sharp-suited managers.

9. *son-lover*, a son who is his mother's lover.

1913 D. H. LAWRENCE *Let.* Jan. (1932) 102 The old son-lover was Œdipus.

son (sōᵘn), *sb.*² Also **sone.** [a. Sp. *son*, lit. 'sound'.] A slow Cuban dance and song in ⅔ time; *son Afro-Cubano*, a form of the son influenced by Negro dances.

1934 S. R. NELSON *All about Jazz* vii. 167 The Son is far more refined than the Rumba, and when properly danced, is plaintively alluring. The Son is always danced in very slow tempo. This Cuban music is particularly characterized by the continual recurrence of singing. 1939 [see *MACUMBA]. 1954 *Grove's Dict. Mus.* (ed. 5) III. 215/1 When Afro-Cuban dances penetrated the port towns of Cuba they were assimilated by Spanish-Cuban folk music... The hybrid forms are then designated by such titles as *son afro-cubano*. 1956 M. STEARNS *Story of Jazz* (1957) iii. 26 The Rhumba, Conga, Son Afro-Cubano, Mambo, and Cha-Cha are predominantly African. 1958 E. BORNEMAN in P. Gammond *Decca Bk. Jazz* xxi. 270 The sone itself usually consisted of an eight-bar theme for solo voice, followed by an improvised four-bar tag, called montuno, that was sung in choir and repeated twice. Whereas the themes of the sones were usually lilting tunes of obviously Spanish descent, the montunos were unmistakably African. 1964 W. G. RAFFÉ *Dict. Dance* 469/2 As a ballroom dance, the *Son* remained very popular until about 1950, when the *Mamba* began to supersede every other dance in *Cuba*. 1973 [see *rumba dancer* s.v. *RUMBA *sb.* 2.]

‖**so-na** (sōᵘ·na). *Mus.* Also **so na.** [Chinese *suŏnà*.] A Chinese wind instrument (see quots.)

1908 *Jrnl. North-China Branch R. Asiatic Soc.* XXXIX. 87 At Peking the *So Na* is regularly made of two sizes... The similarity of sound between *So Na* and *Zourna*, the name of a Persian oboe, is pointed out by M. Mahillon. 1954 *Grove's Dict. Mus.* (ed. 5) II. 236/1 *So-na* (S.W. *so-la*), conical oboe (wrongly, 'clarinet'). Its name connects it with the Persian *surna*... A wooden pipe with 7+1 holes terminated at one end by a copper bell and at the other end by a small reed mouthpiece. 1968 tr. *Hsia Hsiang* in Gray & Cavendish *Chinese Communism in Crisis* 165 A number of people regularly used to crowd into a small cave and listen to an old poor peasant playing the *So-na*. 1975 C. P. MACKERRAS *Chinese Theatre in Mod. Times* 23 Reed instruments are not of great significance in Chinese drama. The only one that need be mentioned here is the *so-na*, which has eight finger-holes and a double reed. Though in this way similar to the Western oboe, the *so-na* is rather shrill and piercing in tone.

sonagram, -graph, varr. *SONOGRAM, *SONOGRAPH 1.

sonant, *a.* and *sb.* Add: **A.** *adj.* **b.** Syllabic; capable of forming a syllable, or constituting the essential element of a syllable.

1876 [see DIPHTHONG *sb.* a]. 1932 W. L. GRAFF *Language & Languages* 56 The sound that possesses the highest degree of sonority in a syllable is called syllabic or sonant. 1957 C. L. WRENN in *Wiener Beiträge zur Englischen Philologie* LXV. 255 The metrical value of a sonant or vocalic *n* in words like *forbidd'n*..and *heav'n*.

B. *sb.* **1.** (Earlier example.)

1849 J. R. LOGAN in *Jrnl. Ind. Archipel.* III. 229 (*heading*) Surds into sonants differing both in their organic and aspirate classes.

2. A syllabic sound; now usu., a syllabic consonant.

1893 *N.E.D.* s.v. *Consonant* sb., The use of the liquids and nasals as vowels or sonants. 1942 K. MALONE in *Mod. Lang. Q.* Mar. 5 Traces of sonant-consonant opposition may be found in English liquids and nasals: thus *Gardner/Gardiner*. 1949 *Trans. Philol. Soc.* 1948 146 Nasals such as *m, n, ŋ* are often sonants—that is to say, have syllabic function.

3. A consonant that can be either syllabic or non-syllabic, i.e. a liquid, nasal, or semi-vowel.

1933 L. BLOOMFIELD *Language* vii. 121 In most languages there is a third, intermediate group of *sonants*, phonemes which occur in both syllabic and non-syllabic positions. 1976 *Archivum Linguisticum* VII. 93 It [*sc. sted*]..can fall on sonants as well as vowels (for example, [hwal'b], [sdɔr'g]; *hvalpe, storke*).

sonantal (sonæ·ntăl), *a.* *Phonetics.* [f. SONANT *sb.* + -AL.] **1.** Syllabic; = *SONANT *a.* b.

1891 A. L. MAYHEW *Synopsis O.E. Phonol.* 36 The Indg. sonantal liquid l. 1897 *N.E.D.* s.v. *Diphthong*, The combination of a sonantal with a consonantal vowel. *Ibid.*, When these sounds [sc. *y* or *w*]..follow the sonantal vowel, the combination is called a 'falling diphthong'.

2. That is or contains a sonant (*SONANT *sb.* 3). *rare.*

1976 *Archivum Linguisticum* VII. 93 The discussion is divided into sections dealing with developments before geminates, in 'diphthongal groups' (vowel+sonant), with 'sonantal finals' (sonant preceded by another consonant) and before two consonants.

sonar (sōᵘ·nāɹ). orig. *U.S.* [f. initial letters of *so*und *na*vigation (*and*) *r*anging, after *radar*.] A system for use under the sea in which the audible or high-frequency sound reflected or emitted by an object in the sea is used to ascertain its position, nature, or speed; (an) apparatus used for this; also in extended use, a system in which the position of an object in air or water is ascertained by reflected ultrasound, e.g. as used by bats and in diagnostic medicine. Freq. *attrib.* Cf. *ASDIC (see quot. 1963).

1946 *U.S. Navy Press Release* 6 Apr. 1 The word 'sonar' was coined from abbreviations for sound, navigation and ranging, and includes various types of underwater sound devices used in detecting submarines and other submerged objects and in obtaining water depths. 1952 [see *ASDIC]. 1961 W. N. KELLOGG *Porpoises & Sonar* (1962) iv. 48 The idea that sonar is systematically used by the great whales and porpoises is..a new and intriguing thought. 1963 *Times* 2 Feb. 5/2 In order to conform with NATO practice, the name Asdic..has been superseded by the word Sonar, the Admiralty announced yesterday... In future Asdic ratings will be known as Sonar operators. 1968 M. WOODHOUSE *Rock Baby* xiii. 129 Bats swooped past my head, bouncing their tiny high-pitched sonar echoes off me. 1971 *Hi-Fi Sound* Feb. 63 (Advt.), The photograph shows Ferrograph recorders at work in the Sonar Room aboard the Hunter Killer nuclear-powered Submarine H.M.S. 'Churchill'. 1971 *New Scientist* 3 June 568/1 The standard method of detecting submerged submarines is by sonar, either active or passive. 1973 *Brit. Med. Jrnl.* 6 Oct. 28/1 The 'in utero' crown-rump length of the fetus may be determined by sonar in the first trimester of pregnancy. 1974 *Sci. Amer.* Mar. 120/2 Frequency-modulation sonars..can exploit a Doppler signal to measure speed. 1979 *Ibid.* Feb. 67/1 (Advt.), These unique sonar cameras send out an inaudible sound signal that bounces off the subject and returns to the camera, in milliseconds—and the lens automatically rotates to perfect focus.

sonata. Add: Also with pl. ‖ **sonate** (sonā·te).

1. c. *sonata da camera*, a sonata suitable for performance in a room smaller than a concert-hall; *sonata da chiesa*, a sonata suitable for performance in a church.

[1789 C. BURNEY *Gen. Hist. Music* (1935) III. ix. 548 Among the most early of these productions may be ranked the *Suonate per Chiesa*, of Legrenzi, published at Venice 1655; *Suonate da Chiesa e camera*, 1656.] 1801 BUSBY *Dict. Mus.* s.v. Sonata, There are several kinds of *Sonatas*. The Italians, however, reduce them principally to two: the *Sonata da Camera*, or *Chamber Sonata*; and the *Sonata de Chiesa*, or *Church Sonata*. 1883 GROVE *Dict. Mus.* III. 556/2 There are twenty-four 'Sonate da Chiesa' for strings, lute, and organ, twenty-four 'Sonate da Camera' for the same instruments, and twelve Solos or sonatas for violin and violoncello, or 'cembalo'. 1938 *Oxf. Compan. Mus.* 879/2 These may be said to close the period of the Sonata da Chiesa or abstract type of the contrapuntal period. Nearly all these composers also wrote works of the Sonata da Camera (or dance) type. 1968 *Listener* 20 June 813/3 The extent to which purely instrumental pieces were demanded during the services seems altogether extraordinary. This was the cradle of the *sonata da chiesa* and the *concerto grosso*. 1974 *Early Music* July 185/2 Adriano Banchieri was so smitten with the aria..that he set it..as a *sonata da camera* for two violins and bass.

3. *sonata movement.*

1942 *Ann. Reg. 1941* 295 Mr. T. S. Eliot's new poem, *The Dry Salvages*..had a sonata movement, the recurring theme being the timelessness of experience. 1947 A. EINSTEIN *Music in Romantic Era* xviii. 350 The musical construction of a sonata or a sonata-movement does not follow real or idealized feelings.

sondage (sondā·ʒ). *Archæol.* [a. Fr. *sondage* sounding, borehole.] A deep trench dug to investigate the stratigraphy of a site.

[1923 *Glasgow Herald* 13 Sept. 9/5 What is taking place in Berlin now is what the French call 'sondage'... The Chancellor is really endeavouring to see how the land lies.] 1930 *Discovery* Aug. 259/1 Against the south angle of the northern harbour a fairly prominent mound invited a sondage, and it proved to conceal a temple more perfectly preserved than any hitherto found in Mesopotamia. 1952 K. M. KENYON *Beginning in Archaeol.* v. 102 The method of *sondage* is sometimes employed, in which a shaft of comparatively limited area is sunk..with the avowed object of establishing what is the succession of cultures. 1955 L. WOOLLEY *Alalakh* i. 6 In 1948 and 1949 Mr. Sinclair Hood made *sondages* at Tabara al Akrad. 1977 P. BARKER *Techniques Archaeol. Excavation* 76 We have seen that holes dug into extensive layers can be disastrous for their subsequent interpretation, and anyway the information gained by a 'sondage', however small, relates only to the area of the sondage.

sonde (sǫnd). [Fr., 'sounding-line, sounding'.] **a.** A radiosonde or similar device that is sent aloft to transmit or record information on conditions in the atmosphere. Orig. only as the second element in Combs. (as *ballon-sonde, ionosonde, radiosonde*, etc.).

1901 [see *BALLON-SONDE]. 1937 [see *RADIOSONDE]. 1943 P. A. ANDERSON et al. *Captive Radiosonde & Wired Sonde Techniques* (U.S. Nat. Defense Res. Council Project P.D.R.C.-647, Rep. No. 3) 4 When the sonde is moved to a new altitude, a pause of 10–30 seconds is adequate to establish equilibrium readings on the meters. 1949 *Sci. Progr.* XXXVII. 490 A direct method of sounding may provide information on solar spectra from which data on the atmosphere above the highest level reached by the sonde may be derived. 1969 MCINTOSH & THOM *Essent. Meteorol.* vii. 111 More recently, ozone sondes have been used to measure the vertical distribution of ozone. 1975 *Nature* 1 May 20/2 A seventh aircraft was used to drop sondes from 40,000 feet and obtain additional vertical profiles of wind and temperature.

b. An instrument probe for transmitting information about its surroundings underground or under water.

1952 *Bull. Amer. Assoc. Petroleum Geologists* XXXVI. 310 The application of the limestone sonde log to the determination of the porosity profile of wells. 1962 *Research* XV. 298/1 This deflection increases as the sonde continues to enter the [coal] seam. 1962 [see *RE-ENTRY 2 d]. 1975 G. ANDERSON *Coring* vii. 124 The SP log is a measurement of the electrical potential energy in the mud around the sonde as compared with a reference electrode grounded at the surface.

sonder (zǫ·ndǝɹ), *a.* and *sb.* *U.S.* [f. G. *sonderklasse* special class.] **A.** *adj.* Of, pertaining to, or designating a class of small racing yachts. **B.** *sb.* A yacht of this class.

1909 *N.Y. Even. Post* 9 Sept. 3/3 The Sonder yachtsmen, both victors and vanquished, were received by President Taft today. 1913 C. W. ERNST *Letter* (MS.), Our yachtsmen, since 1907, talk of 'sonder-boats', sonder class, sonder race,—meaning certain boats recognised by the International Yacht Racing Union. 1917 *Rudder* Apr. 328/1 Fred M. Hoyt is to have a Marconi rig put upon his sonder boat Skeezix. There are a few sonders left at Marblehead and the innovation will be watched with much interest. 1948 R. DE KERCHOVE *Internat. Maritime Dict.* 695/2 *Sonder*, a class of small yacht originated in Germany, in which the sum of the waterline length, extreme beam, and extreme draft must not be greater than 32 ft.

Sonderbund (zǫ·ndǝɹbunt). [a. G. *Sonderbund* special league, separate association.] A league formed by the R.C. cantons of Switzerland in 1843 and defeated in a civil war in 1847. Also *attrib.*

1847 G. GROTE *Seven Lett. Politics Switzerland* i. 17 The invasion..in..1844..by bands of volunteers..called the Corps Francs—and the separate league of Seven Cantons, called the Sonderbund. 1887 *Encycl. Brit.* XXII. 795/1 The seven Catholic cantons—Uri, Schwyz, Unterwalden, Lucerne, Zug, Freiburg, and Wallis—formed (September 7, 1843) a 'Sonderbund' or separate league... In December 1845 the Sonderbund turned itself into an armed confederation. 1922 E. & C. PAUL tr. *Oechsli's Hist. Switzerland* xxxiv. 386 The Sonderbund war. 1952 E. BONJOUR et al. *Short Hist. Switzerland* x. 262 The founders of this Sonderbund, as it was soon aptly named. 1973 HOWAT & TAYLOR *Dict. World Hist.* 1465/1 The outnumbered Sonderbund expected Austrian and French help, but the swift victory of the Diet's troops under Dufour forestalled its arrival.

‖**Sonderkommando** (zo·ndǝɹkoma:ndo). [G. *Sonderkommando* special detachment.] In Nazi Germany, a detachment of prisoners in a concentration camp responsible for the disposal of the dead; also a member of such a detachment. Also *fig.*

1951 MOYSE & STENHOUSE tr. *Rousset's World Apart* v. 27 The *Sonderkommando*, totally isolated from the world, condemned to live every second of its eternity with tortured and charred bodies. 1960 S. BECKER tr. *Schwarz-Bart's Last of Just* (1961) VIII. 422 The team of *Sonderkommando* responsible for burning the Jews in the crematory ovens. 1976 *Times Lit. Suppl.* 2 July 815/1 A four-legged Sonderkommando who betrays his fellows to starvation. 1979 J. GARDNER *Nostradamus Traitor* xxxv. 165 The womenfolk prepared a great spread of food... 'Enough food to feed the whole of a Sonderkommando.'

sone. Restrict † *Obs.*⁻¹ to sense in Dict. and add: **2.** (sōᵘn). A unit of subjective loudness such that the number of sones is proportional to the loudness of a sound: a tone of frequency 1000 Hz and 40 dB above the listener's audibility threshold produces a loudness of one sone.

1936 S. S. STEVENS in *Psychol. Rev.* XLIII. 416 It is proposed that the unit of the scale be the loudness of 1000 cycle tone 40db above threshold heard with both ears, and that it be called a sone. 1952 *Sci. News Let.* 28 June 411/2 The field studies..used the 'sone' as a unit of loudness. The decibel..measures only the intensity of a noise. It is the loudness that is most objectionable to the human ear. 1958 *Times* 18 July 7/6 The sone scale is..based on the averaged impressions of supposedly normal individuals in identifying the relative loudness of different sounds, *i.e.*, that they are two, three or four times, &c., as loud as a sound of one sone. 1960 *New Scientist* 25 Feb. 454/1 The American Automobile Manufacturers' Association adopted in 1954 a noise specification for new vehicles of 125 sones measured at 50 feet. 1970 [see *NOV].

sone, var. *SON *sb.*²

Soneryl (sǫ·nĕril). *Pharm.* Also **soneryl.** A proprietary name (orig. French) for butobarbitone (5-butyl-5-ethylbarbituric acid, $C_{10}H_{16}N_2O_3$), given in tablet form as a sedative and hypnotic.

1923 *Pharmaceutical Jrnl.* LVI. 170/2 An interesting preparation which Messrs. May and Baker are shortly to issue is 'Soneryl'... Compared with veronal, Soneryl is twice as active. 1923 *Official Gaz.* (U.S. Patent Office) 18 Sept. 395/2 Etablissements Poulenc Frères, Paris... *Soneryl*... Claims use since Apr. 18, 1922. 1933 *Trade Marks Jrnl.* 3 May 510/2 *Soneryl*... A derivative of barbituric acid prepared for use in medicine and pharmacy. May and Baker Ltd... Manufacturing chemists. 1948 J. H. BURN *Lect. Notes Pharmacol.* 44 Tablets of Soneryl (butobarbitone) are used to ensure a good night's sleep either for a patient in hospital or for a normal person travelling by night train. 1959 D. DU MAURIER *Breaking Point* 115, I would sleep on it—if I could sleep, which seemed very doubtful. I took two soneryl tablets, and passed out. 1979 L. BROWN *Encounters With Nature* vi. 79 By taking three Soneryl and four aspirin tablets, I suffered through the night in fitful drowsiness.

‖**son et lumière** (sǫn eˡ lü·miēᵊɹ, son e lümyĕr). Also with capital initials. [Fr., lit. 'sound and light'.] **1.** A form of entertainment using recorded sound and lighting effects, usu. presented at night at a historic building and giving a dramatic narrative of its history. Also *attrib.*

1957 *Times* 1 Aug. 10/1 A dinner party..on the occasion of the preview of the *Son et Lumière* spectacle. *Ibid.* 13 Aug. 11/1 In the five years since M. Paul Robert-Houdin first introduced the technique at..Chambord, *son et lumière* has been adopted at..buildings of historic interest in France. 1959 *Times* 30 June 12/4 Son et lumière performances are being given this week, starting to-night at Greys Court, Henley-on-Thames. 1969 V. ROWE *Loire* 33 At Chambord, in 1952, a new kind of entertainment was born, *son-et-lumière*, a combination of sound and light effects. 1970 N. FLEMING *Czech Point* (1971) xvi. 205, I felt as visible as a statue in a Son et Lumière display. 1977 J. I. M. STEWART *Madonna of Astrolabe* i. 19 A few years ago we had *son et lumière* in aid of some building project.

2. *fig.* Writing or behaviour resembling a *son et lumière* presentation, *esp.* in its dramatic qualities.

1968 *Punch* 16 Oct. 557/3 The effects are produced by a literary *son et lumière*—by liberal direct quotation of anecdotes, poetry, hymns and music, and a series of splendid photographs, rather than words about palaces. 1969 *Listener* 9 Jan. 41/1 Old Mr Seddeby's performance, from 11 o'clock to one daily, as an elder statesman in decline, an illusion which involved a lot of business with a monocle and a cigar-cutter, was pure *son et lumière*. 1978 *Times Lit. Suppl.* 24 Mar. 336/3 It would have been all too easy to end this biography in *son et lumière* and to paint a Tiepolo ceiling on which Forster is floating up to Olympus.

song, *sb.* Add: **2. d.** *song without words*, an instrumental composition in the style of a song (after Mendelssohn's title 'Lieder ohne Worte'); also *transf.*

1871 S. SMILES *Character* viii. 219 Cheerfulness..gives harmony of soul, and is a perpetual song without words. 1883 R. PRENTICE *Musician* II. 95 The second movement [of a Beethoven sonata] is a veritable Song without Words. 1938 *Oxf. Compan. Mus.* 885/1 *Song without words*, a term introduced by Mendelssohn to cover a type of one-movement pianoforte solo, throughout which a well-marked song-like melody progresses, with an accompaniment. 1974 *Encycl. Brit. Macropædia* XI. 902/1 She [*sc.* Fanny Mendelssohn] had herself written some of the *Songs Without Words* attributed to her brother.

4. d. *a song in one's heart*, a feeling of joy or pleasure.

1930 L. HART *With a Song in my Heart* 4 With a song in my heart;—I behold your adorable face. 1946 *Hansard Commons* 9 Apr. 1807, I will find, and find with a song in my heart, whatever money is necessary to finance useful and practical proposals for developing these areas. 1978 *Times* 9 Jan. 13/1 Does the lending rate come down? Then every conservative owner-occupier has a song in his heart.

e. *on* (*full*) *song*, in good form, performing well. *colloq.*

1967 *Autocar* 27 Dec. 10/1 The close and even spacing of the ratios..make it easy to keep the engine 'on full song' during hard driving. 1971 *Daily Tel.* 21 Aug. 16/1 As the table reveals, most of the leading unit trust managers have at least one fund that is 'on song'. 1974 *Observer* 3 Feb. 24/5 Really on song since beating Manchester City in the Cup, Forest won 5–1. 1981 *Radio Times* 11 Apr. 23/2 If you are on song nothing will break your concentration.

5*. *song and dance.* **a.** A form of entertainment (*spec.* a vaudeville act) consisting of singing and dancing. Freq. *attrib.* orig. *U.S.*

[1628 F. DRAKE *World Encompassed* 76 They yet continued their song and dance a reasonable time.] 1872 S. HALE *Let.* 16 Jan. (1918) iii. 78 He did a 'Song and Dance', two, in fact. 1872 *Chicago Tribune* 13 Oct. 5/6 First week of the distinguished song and dance artists. 1895 *N.Y. Dramatic News* 23 Nov. 13/3 The first double song and dance team was comprised of Wash Norton and Ben Cotton. 1940 *Chatelaine* Apr. 36/2, I practiced my song-and-dance act for weeks. 1959 R. LONGRIGG *Wrong Number* iv. 58 So up she pops from hell or wherever, just the time for a bit of song and dance. 1968 *Radio Times* 28 Nov. 53/1 The song-and-dance patter comedian. 1977 *Time Out* 17–23 June 47/2 Pleasant Nilsson-like song 'n' dance numbers.

b. *fig.* A rigmarole, an elaborately contrived story or entreaty, a fuss or outcry. Also *attrib. colloq.* (orig. *U.S. slang*). Cf. sense 4 c in Dict.

1895 E. W. TOWNSEND *Chimmie Fadden* 6 Den, 'is whiskers gives me a song an' dance. 1900 B. MATTHEWS *Confident To-Morrow* 9 And it ain't a song-and-dance I'm giving you either. 1913 KIPLING *Diversity of Creatures* (1917) 292, I don't see how this song and dance helps us any. 1922 S. LEWIS *Babbitt* xxxii. 375 George, what's this I hear about some song and dance you gave Colonel Snow about not wanting to join the G.C.L.? 1949 *Time* 5 Sept. 2/3 Labor Leader Preble..was not impressed by 'the song and dance about [Stefan's] mother and sister being persecuted and murdered'. 1958 'E. DUNDY' *Dud Avocado* III. vi. 266 If only he hadn't felt obliged to make such a song and dance about it. 1967 'S. WOODS' *And Shame Devil* 118 ''Appen tha means well,' he said, his speech suddenly broadened almost out of all recognition, 'and 'appen tha's joost making a song and dance.' 1980 J. DITTON *Copley's Hunch* II. ii. 132 The Prime Minister wants to make a song and dance about it.

6. a. *song-lyric, -sequence, -sheet, -strain, -tune* (earlier and later examples).

1944 C. DAY LEWIS *Poetry for You* vi. 61 The chief thing which poets took over from the song-lyric and preserved in the new lyrical poetry was..'singleness of mind'. 1947 A. EINSTEIN *Music in Romantic Era* xiv. 187 With Op. 24, the Heine song-sequence, he [*sc.* Schumann] began to write lieder. 1930 P. GEDDES et al. (*title*) Song-sheet and welcomes. 1967 A. L. LLOYD *Folk Song in England* i. 29 The countless Sorrowful Lamentations of hanged men did not become anchored in tradition ..perhaps because the song-sheets bearing these effusions are of late appearance. 1876 G. M. HOPKINS *Poems* (1967) 176 So tiny a trickle of song-strain. 1809 E. CUTLER *Diary* 28 Aug. in J. P. Cutler *Life & Times E. Cutler* (1890) v. 98 Very soon a man began to sing a hymn in a familiar song-tune. 1967 A. L. LLOYD *Folk Song in England* iii. 139 As feudal society gives way to capitalism..recitative melodies are replaced by song-tunes.

b. *song-composer, -composition, -singing* vbl. sb. (earlier example), *-writing* vbl. sb. (earlier examples), *-writing* ppl. a.

1947 A. EINSTEIN *Music in Romantic Era* xiv. 184 There were no Italian song-composers. *Ibid.* 191 The procession of musicians who contributed to Romantic song-composition. 1848 W. ALLINGHAM *Diary* 26 Sept. (1907) ii. 43 Dine at Peter Kelly's,..much song-singing afterwards. 1772 J. AIKIN (*title*) Essays on song-writing. 1809 *Belfast Monthly Mag.* Mar. 164/2, I promise.. method in my handling the theory and practice of song-writing. 1947 A. EINSTEIN *Music in Romantic Era* iv. 35 The song-writing Berlin purists.

c. *song-rapt, -wild* adjs.

1885 W. B. YEATS in *Dublin Univ. Rev.* July 137 A wandering song-rapt bird. 1937 BLUNDEN *Elegy* 15 The flight of one small song-wild lark Finds heaven.

7. song-ballet, (*a*) *U.S. dial.*, a ballad; (*b*) a theatrical work combining songs and ballet; *song-cycle* [cf. G. *liederzyklus*], a series of songs intended to form one musical entity, and having words dealing with related subjects; **song-flight,** (a) flight of a characteristic pattern made by a bird as it sings, in a territorial display; **song-form** *Mus.*, a form used in the composition of songs; *spec.* [tr. G. *liedform*] the form of a simple melody with simple accompaniment, or that of a work in three sections of which the third is a repetition of the first; **song-fowl** *poet.* = SONG-BIRD 1; **song-hit** *colloq.*, a song which is a popular success; **song-motet,** a simple type of motet; **song-perch,** a place where a bird perches to sing, so as to establish its territory; **song period,** the part of the year during which the birds of a species sing; **song-plugger** orig. *U.S.*, a person employed to popularize songs, esp. by performing them repeatedly; hence **song-plugging** *vbl. sb.*, **song-plug** *v. trans.*, **song-plugged** *ppl. a.*; **song-post** = *song-perch* above; **song stylist,** a singer admired for his or her style.

1915 *Dialect Notes* IV. 190 *Song-ballet*, *n.*, a song or ballad. 1938 *Sun* (Baltimore) 15 June 6/7 Visitors will join the mountaineers to sing their 'song ballets'. 1962 AUDEN *Dyer's Hand* (1963) 484 We have translated.. Brecht's text for the song-ballet *Die sieben Todsünden* with music by Kurt Weill. 1899 Song-cycle [in Dict., sense 6 a]. 1942 E. BLOM *Music in England* x. 168 Arthur Somervell's settings of poems from Tennyson's 'Maud', which have remained among the world's few great song-cycles. 1978 *Listener* 30 Mar. 412/4 A mature song-cycle by Dallapiccola. 1936 NICHOLSON & KOCH *Songs of Wild Birds* 9 Song-flight is an extra means of making the singer temporarily as conspicuous as possible. 1961 A. J. BERGER *Bird Study* vi. 186 Song, song flights, and other special displays serve an orientation function: they attract a female to the male's territory or to a nest site. 1884 R. PRENTICE *Musician: Grade 3* 4 The simplest song-form is constructed on two or three sentences only. 1902 H. C. BANISTER *Mus. Anal.* i. 2 There is a term now in vogue to designate the simplest of all plans or forms: 'Song-Form' or 'Aria-Form'. 1946 R. BLESH *Shining Trumpets* (1949) v. 109 The blues are essentially a song form. 1954 *Grove's Dict. Mus.* (ed. 5) VII. 962/2 The term 'song form,' derived from the German, has unfortunately been used by different writers with different significations. The vagueness which results and the fact that the term is not happily chosen gives rise to doubts whether it had not better be entirely abandoned. 1877 G. M. HOPKINS *Poems* (1967) 71 Not that the sweet-fowl, song-fowl, needs no rest. 1914 'HIGH JINKS, JR.' *Choice Slang* 18 *Song hit*, a popular song. 1918 [see *HIT *sb.* 4]. 1959 'F. NEWTON' *Jazz Scene* 9 *Pop, pop music*, popular entertainment music as typified by the 'song-hit'. 1942 H. HEWITT *Harmonice Musices Odhecaton* vi. 69 A few 'song-motets' find a place in the Odhecaton. 1974 *Early Music* Oct. 219 Some of his [*sc.* Dufay's] most elegant Latin compositions..are dedicated to the Virgin Mary, and their treble-dominated texture and lyrical charm—they resemble chansons in many ways—explain why they are called song-motets. 1934 *British Birds* XXVIII. 15 If a male is on his song perch when his hen quits her eggs, he usually follows her..to her feeding ground. 1975 I. ROWLEY *Bird Life* v. 61 The kookaburra defends a large area, but in particular a number of song perches. 1908 *British Birds* I. 367 In the middle of the song-period all the individuals of a species found in any locality sing every day. 1961 A. J. BERGER *Bird Study* vi. 171 Many species have a short song period (post-breeding) after the molt has been completed. 1927 *Daily Express* 22 Sept. 9/3 'Clap Yo' Hands' must have been song-plugged for ten minutes right off... 'Do-Do-Do' is another song-plugged number. 1923 *N.Y. Times* 7 Oct. IX. 2/1 *Song plugger*, a retiring representative of a song publisher planted in the audience to call for songs, whistle refrains and applaud. 1927 *Melody Maker* May 437/1 Song pluggers are..vocalists lent by the music publishers to the dance bands just for the nights on which these bands are due to broadcast, and, of course, sing only their employer's numbers. 1976 R. SANDERS in D. Villiers *Next Year in Jerusalem* 208 Gershwin..embarked upon his musical career at sixteen as a Tin Pan Alley song plugger and composer. 1927 *Melody Maker* May 433 (*heading*) Song-plugging thro' the ages. 1972 P. BLACK *Biggest Aspidistra* I. iii. 29 The song-plugging wave did not recede until 1948, when the BBC and the publishers managed to draw up an agreement. 1938 *British Birds* XXXI. 320 The habitat was on open grassy ground with stones and sallow bushes as 'song posts'. 1938 *Sun* (Baltimore) 15 June 6/7 Special guests will be..Miss Florence Clark, of Detroit, noted song stylist. 1973 *Black Panther* 24 Mar. 7/1 Elaine Brown, community activist..is also a musician, composer, lyricist and song stylist.

Song, var. *SUNG.

songfest (sǫ·ŋfest). orig. and chiefly *N. Amer.* [f. SONG *sb.* + *FEST: cf. G. *sängerfest*.] An informal session of group-singing; a festive sing-song.

1912 J. SANDILANDS *Western Canad. Dict.* 9/1 *Songfest*, a feast of song. 1916 *Dialect Notes* IV. 354 *Songfest*. 'There's to be a *songfest* at the church Friday night.' 1953 *Manch. Guardian Weekly* 12 Nov. 9/1 According to the 'New York Times':..Their problem has been greatly complicated by the fact that each time they have tried to speak to the prisoners through the camp public address system the P.O.W. bosses have organised song fests, thrown rocks..or set up a deafening clamour. 1961 'I. Ross' *Old Students Never Die* (1963) viii. 109 The image of Jackie and Alison, holding hands and bowing to us after their little songfest. 1979 *Yale Alumni Mag.* Apr. 30/2 Fifty members of the two groups gathered for a gay, impromptu songfest in Pushkin Square.

Songhai (sǫŋgai·). Also **Songhay, Songhoi,** † **Sungai.** [Native name.] A people of West Africa, living mainly in Niger and Mali; the Nilo-Saharan language of this people. Also *attrib.*

1738 F. MOORE tr. *Leo the African's Geographical Hist. Afr.* in *Trav. Inland Parts Afr.* II. 28 The Negroes have a good many languages amongst them, of which..Sungai is spoken in a great many of their kingdoms. 1858 H. BARTH *Travels & Discoveries in North & Central Afr.* IV. lix. 237 All the huts in these Songhay villages consist merely of reeds. *Ibid.* 238 The Songhay in general are among the most inhospitable people I ever met. 1911 *Encycl. Brit.* XXV. 414/1 According to the *Tarikè Sudan* ..the first king of the Songhoi was called Dialliaman (Arabic *Dia min al Jemen*, 'he is come from Yemen')... The Songhoi emigration must have begun towards the middle of the 7th century... The Songhoi language..is often known as Kissur. 1930 C. G. SELIGMAN *Races of Afr.* iii. 60 Only in 1500, when Melle was captured by the Songhai king, Omar Askia, did the Melestine Empire cease to exist. *Ibid.* 62 The Songhai are moderately tall, with a stature of about 67 inches. 1957 W. M. HAILEY *Afr. Survey 1956* iii. 86 All languages which are known not to be attached to any of the above Units (e.g. Songhai). 1963 [see *SAHARAN *sb.*]. 1966 [see *NILO-]. 1977 H. GREENE *FSO-1* xviii. 161 The Republic of Mali..with Bambara and Fulani tribes in the east, the Songhai and the Malinke and the wandering Tuaregs in the west.

Songish (sǫŋgi·ʃ). Also **Songeesh, Songhees,** 9 **Songhies, Songhish.** [Native name.] An American Indian people of Vancouver Island, British Columbia; the language of this people, a dialect of Straits Salish. Also *attrib.*

1860 [see *KLOOCH]. 1862 R. C. MAYNE *Four Years in Brit. Columbia & Vancouver Island* ii. 30 The road ascends a little hill, on the summit of which lies the Indian village of the Songhies. *Ibid.*, This village of the Songhies presents one of the most squalid pictures of dirt and misery it is possible to conceive. 1865 M. MACFIE *Vancouver Island & Brit. Columbia* xvi. 430 The Songhish tribe, resident near Victoria, hold a general merry-making annually in.. October. 1875 H. H. BANCROFT *Native Races of Pacific States* III. xii. 522 The Songhies said the hunter was transformed into a deer. 1911 *Encycl. Brit.* XXIV. 80/1 *Salishan*, the name of a linguistic family of North American Indian tribes, the more important of which [include] the Salish.., Skokomish, Songeesh, Spokan and Tulalip. 1911 [see *POTLATCH *v.*]. 1973 L. C. THOMPSON in T. A. Sebeok *Current Trends in Linguistics* X. 1010 The rest of the dialects extend from Lummi..across to Songish (Lkungen) around modern Victoria, on Vancouver Island, and on northward to include Saanich.

songket (sǫ·ŋket), *a.* Also 9 **sungkit.** [Malay.] Of cloth: decorated by interweaving short lengths of gold or silver thread into the material. Also *ellipt.* as *sb.*

1894 N. B. DENNYS *Descr. Dict. Brit. Malaya* 107 Silk *sarongs*..woven with silk and gold thread are termed *kain sungkit*. 1953 *Exhib. Malayan Arts & Crafts* (Malayan Agri-Horticultural Assoc.), Hand-woven cloths. The other important category of Malay cloths is that of the *kain songket*. 1963, 1971 [see *KAIN]. 1972 M. SHEPPARD *Taman Indera* 100 (*caption*) The gift is a piece of silk *songket* hand woven cloth, folded in the shape of a bird. *Ibid.* 120 The body of a *songket* sarong is silk, of a single plain colour. The decoration is carried out with gold or silver thread... The songket sarong is worn today by Malay men of all levels of society when attending social functions, and it is therefore in constant demand. 1972 *Malay Mail* 27 May 9/7 (Advt.), We are offering a very wide range of genuine Malaysian songket.

‖**songkok** (sǫ·ŋkǫk). Also 9 **songko.** [Malay.] A cap worn by Malays, resembling a skull-cap.

1894 N. B. DENNYS *Descr. Dict. Brit. Malaya* 108 The skull-cap, *kopia* or *songko*, is worn by some. 1960 S. HARVESTER *Chinese Hammer* i. 8 An affable Indonesian official with his black *songkok* tilted at a rakish angle above his chubby face. 1966 D. FORBES *Heart of Malaya* xiv. 193 The bridegroom wears his *songkok*, the round brimless hat. 1977 P. THEROUX *Consul's File* 38 'This is serious,' he said, glowering and putting on his *songkok*.

songlessness (sǫ·ŋlĕsnĕs). [f. SONGLESS *a.* + -NESS.] The state or condition of being songless.

1924 J. R. HARRIS *As pants Hart* viii. 85, I have often deplored the songlessness of the modern Evangelical and Protestant Churches. 1958 *Times* 9 Sept. 10/5 Another very common bird begins now to sing again after a period of songlessness.

songster. Add: **1. b.** *U.S.* (See quot. 1980.)

1925 ODUM & JOHNSON *Negro & his Songs* i. 1 Anyone hearing him sing day in and day out, together with thousands of others like him, must agree..with the oft-repeated song claim of the 'musicianer', 'music-physicianer' and 'songster', that 'All don't see me goin' to hear me sing.' Not only does he sing, but he sings much and sings long with richness and variety. 1964 *Amer. Folk Music Occasional* i. 61 And if you describe the artist with accuracy, it will be with his own apt word: songster. The term suggests a musician who is both performer and inventor and harks back to the time when every Southern town had its songster, a man who was virtually in charge of the community's social life. 1968 P. OLIVER *Screening Blues* ii. 82 The 'songster' generation of singers. 1970 —— *Savannah Syncopators* 7 The instrumental techniques and traditions of the blues singers, the songsters, the ragtime banjoists and guitarists. *Ibid.* 86 Many of the older blues singers and songsters recall playing for white functions. 1980 *New Grove Dict. Music & Musicians* XVII. 17/1 *Songster*, a black American musician of the post-Reconstruction era who performed a wide variety of ballads, dance-tunes, reels and minstrel songs, singing to his own banjo or guitar accompaniment.

sonic (sǫ·nik), *a.* [f. L. *son-us* sound + -IC.] **1. a.** Employing or operated by sound waves; used esp. with reference to devices and techniques which make use of the reflected echo of a sound pulse.

1923 *Sci. Amer.* May 330/2 Sonic sounding is rendered possible by the fact that sound vibrations, passing through water and striking a solid surface, are returned as an echo to the source from which they originated. 1924 *Telegr. & Telephone Jrnl.* X. 172/2 The United States destroyers *Hull* and *Corry*, equipped with sonic depth sounders..have been ordered to survey the ocean bed. 1933 *Geogr. Jrnl.* LXXXI. 572 The exploration of the Maldive ridge by sonic sounding should be interesting. 1952 *Chambers's Jrnl.* June 364/2 After the War a sonic gun was actually found in a German laboratory. 1961 *Flight* LXXIX. 249/2 In addition to the display system, the RH-1 was fitted with Ryan APN-97 Doppler and the sonic altimeter. 1965 *Punch* 17 Mar. 390/1 'That thing [*sc.* a torpedo] live?' I asked. 'Very,' said the Lieutenant-Commander, 'but it's only an antiquated sonic-homing job.' 1967 *Jane's Surface Skimmer Systems 1967–68* 94/1 It receives craft motion input from a sonic height sensor in the bow. 1976 B. BOVA *Multiple Man* (1977) i. 11 If anyone tried to fire a shot..the scanning lasers would pick up the bullet... Sonic janglers would paralyse everyone in the auditorium.

b. Of or pertaining to sound or sound waves, esp. within the audible range.

1936 *Jrnl. Amer. Chem. Soc.* LVIII. 1071/2 In order to avoid any possible sonic action directly on a test reagent. 1939 [see *sonoluminescence* s.v. *SONO-]. 1942 *Jrnl. R. Aeronaut. Soc.* XLVI. 83 The guiding intention will be, to avoid..the formation of sonic waves, and practical elimination of compression shocks, in order to obtain a minimum of the so-called wave-making resistance. 1947 *Aircraft Engin.* XIX. 180/1 The design of the VG-70 experimental monoplane..was begun as a determined effort in the field of sonic research. 1962 F. I. ORDWAY et al. *Basic Astronautics* viii. 345 Since the flow often speeds up when passing over a body it is possible for sonic flow to exist over parts of a body which is moving at speeds somewhat less than Mach 1. 1969 L. F. YERGES *Sound, Noise, & Vibration Control* I. 1 There is no essential difference between the sonic and vibratory forms of sound energy. 1972 *Observer* (Colour Suppl.) 22 Oct. 53/2 It is this *depth* of sound which stereophony failed to capture, or so the sonic engineers believed when they looked around for their next breakthrough. 1975 *Sci. Amer.* Oct. 135/1 The ultrasonic world is quieter than the sonic, mainly because its sounds are more local. 1977 *Gramophone* Dec. 1016/1 Sonic beauty abounds in *The Triumphs of Oriana*..: 64 minutes of music packed with admirable clarity..on a single disc.

2. Special collocations: *sonic bang* = *sonic boom* below; *sonic barrier* = *sound barrier* s.v. *SOUND *sb.*³ 7 a; *sonic boom*, the sudden loud noise heard when the shock wave from an aircraft travelling faster than sound reaches the ears; *sonic speed, velocity*, the speed at which sound waves travel in a particular medium, esp. air.

1953 *Sci. News* XXX. 118 Subsequently two sonic bangs were heard of the same intensity with a small time interval apparently the same as that between the vapour puffs. 1955 *Times* 20 June 8/7 As the new fighters showed their considerable paces the programme was punctuated with 'sonic bangs' as the speed of sound was exceeded in dives. *a* 1974 R. CROSSMAN *Diaries* (1976) II. 169 Next came the Minister of Aviation, Mr Mulley, who had put in a request for urgent legislation on sonic bangs. 1946 *Jrnl. R. Aeronaut. Soc.* L. 445/2, I do not imagine for one moment that man will be happy until he has conquered the 'sonic barrier'. 1955 *Sci. News Let.* 24 Sept. 195 (*caption*) By giving the aircraft a 'wasp waist', engineers ..made it slip more smoothly through the sonic barrier. 1952 *Times* 2 Sept. 4/5 Aircraft travelling at about the speed of sound cause a loud bang, which has become known as the 'sonic boom'. 1966 *Guardian* 2 May 8/6 Mr Amery, Minister of Aviation, told Parliament that damage from the Concord's sonic booms would be 'negligible'. 1969 *Times* 17 Nov. 4/7 Their home collapsed after another sonic boom. 1977 *New Yorker* 24 Oct. 36/2, I heard the explosion even inside the manuscript room... It sounded like a sonic boom. 1946 *Jrnl. R. Aeronaut. Soc.* L. 436/1 The problems of aerodynamics at speeds well in excess of that of sound are in some respects much simpler than those quite near the sonic speed. 1950 *Sci. News* XV. Plate 7 (*caption*) This model aeroplane..was designed to investigate the forces acting on an aircraft reaching sonic speed. 1962 F. I. ORDWAY et al. *Basic Astronautics* viii. 344 At this point the body is moving at sonic speed and transonic conditions exist. 1942 *Jrnl. R. Aeronaut. Soc.* XLVI. 64 The local 'sonic velocity' *a* is then defined by the expression:—$a^2 = dp/d\rho$. 1949 *Jrnl. Appl. Physics* XX. 638/1 A slowly varying pressure change or signal is transmitted through the mixture with a definite critical or 'sonic' velocity.

Hence **so·nically** *adv.*, by means of sound waves; as regards sound.

1936 *Jrnl. Amer. Chem. Soc.* LVIII. 1070/2 A quantitative study of certain oxidation reactions sonically activated in the audible range. 1959 *Brookhaven Symp. in Biol.* XII. 11 A sonically fragmented sample. 1975 *Gramophone* June 36/1 Sonically the record is obviously to be preferred to Menuhin's earlier recording with Furtwängler. 1981 *Popular Hi-Fi* Mar. 21/2 The Crimson is a better pre-amplifier sonically.

sonicate (sǫ·nikē¹t), *sb.* and *v.* [f. prec. + *-ate*, after *filtrate, precipitate*.] **A.** *sb.* A sample which has been subjected to ultrasound so as to fragment the macromolecules and membranes in it.

1958 M. Litt et al. in *Proc. Nat. Acad. Sci.* XLIV. 144 The production of sonicates (degraded samples) covering more than a tenfold range in molecular weight. **1958** Nishihara & Doty in *Ibid.* 412 Eight 30-cc. aliquots of a stock solution..were exposed to 9-kilocycle sonic waves in a 50-watt Raytheon magnetostriction generator... The samples produced in this way, termed 'sonicates', are listed. **1979** *Nature* 25 Jan. 314/2 (*caption*) Calvaria were washed again, sonicated, and the sonicate ultracentrifuged.

B. *v. trans.* To subject to such treatment.

1960 *Biochem. & Biophys. Res. Communications* III. 471 Tumor mitochondria were sonicated at maximum intensity (1·3 amps) for two 2·5-minute periods. **1974** *Nature* 20 Dec. 655/1 (*caption*) The DNA fragments were prepared by sonicating calf thymus DNA to a molecular weight of 5·4 × 10⁵. **1978** [see below].

Hence **so·nicated** *ppl. a.*; **sonica·tion**, treatment with ultrasound; **so·nicator**, an apparatus for treating samples in this way.

1958 *Proc. Nat. Acad. Sci.* XLIV. 150 The loss in activity due to sonication occurs only as a result of reduction in molecular size below that required for attachment. **1959** *Brookhaven Symp. in Biol.* XII. 11 The variance due to any change in compositional heterogeneity of the sonicated sample, can be obtained. **1964** *Jrnl. Cellular & Compar. Physiol.* LXIV. 153/1 Washed ghosts were sonicated for ten minutes using an MSE 60 W sonicator. **1969** *Nature* 20 Dec. 1164/2 The supernatant contains the polymerase, which may then be assayed simply by adding sonicated calf thymus DNA. **1970** *Sci. Jrnl.* Apr. 42 Cleaning these parts is a major branch of ultraclean technology involving the application of special solvents, the use of 'sonication'—ultrasonic vibration—to jar the dirt loose,..and many other highly specialized procedures. **1978** *Jrnl. Protozool.* XXV. 492/2 Other aliquots of *E*[*ncephalitozoon*] *cuniculi*..were sonicated using a Bronson sonicator.

sonics (sǫ·niks), *sb. pl.* [f. *SONIC *a.*: see -IC 2.] Sonic techniques and equipment generally or collectively.

1955 Hueter & Bolt *Sonics* p. v. The multiplicity of concepts and techniques could be designated by the name *sonics*. *Ibid.*, Sonics encompasses the analysis, testing, and processing of materials and products by the use of mechanical vibratory energy. **1974** 'G. Black' *Golden Cockatrice* vi. 102 All three corvettes were now almost motionless..obviously listening on sonics.

Sonifier (sǫ·nifai,əɹ). Also **sonifer**. [f. as prec. + *-FY + *-ER¹.] The proprietary name of a make of sonicator.

1961 *Official Gaz.* (U.S. Patent Office) 1 Aug. TM 10/1 Branson Instruments, Inc., Stamford, Conn... *Sonifier*. For electrical generators of ultrasonic energy, ultrasonic energy transducers and electrically powered processing systems for the same. First use Oct. 19, 1960. **1963** *Biochim. & Biophys. Acta* LXXI. 232 A suspension of ghosts..was treated with ultrasonic waves generated by a Branson Model S-75 Sonifier for 15 secs. **1966** *Trade Marks Jrnl.* 16 June 809/2 *Sonifier*. 875,310. Ultrasonic apparatus and instruments included in Class 9. Branson Instruments Incorporated..Connecticut, U.S.A.; manufacturers.—8th Feb. 1965. **1980** *European Jrnl. Biochem.* CV. 164/2 Sonication at 4°C with a Branson B-12 sonifer.

Soninke (sǫni·ŋke). [Native name.] A member of a people living in Mali and Senegal; the people itself; the Mandingo language or dialects spoken by this people. Also *attrib.* Cf. *MALINKE.

1886 *Encycl. Brit.* XXI. 662/2 The Mandingoes..comprise the Mandingo proper, occupying Manding, and the Malinkis and Soninkes, scattered along Bambuk, Bure, and Fuladugu. **1911** [see *MALINKE]. **1930** C. G. Seligman *Races Afr.* (rev. ed.) iii. 61 The Mandingo..constitute one of the most important groups of Senegal and West Sudan..and include such large and important tribes as the Soninke, Malinke, and Vei. **1971** N. Levtzion in Ajayi & Crowder *Hist. W. Afr.* I. iv. 124 They are united by..their pride in having once been part of the ancient Soninke kingdom of Wagadu. **1974** *Encycl. Brit. Macropædia* XI. 382/2 Soninke and Dogon are also related to Bambara; Dogon includes many dialects.

Sonne (sǫ·nə). *Bacteriol.* The name of Carl Olaf *Sonne* (1882–1948), Danish bacteriologist, used *attrib.* and formerly in the possessive with reference to the Gram-negative bacterium *Shigella sonnei*, which causes a mild form of dysentery in man. [Described by Sonne in 1915 (*Zentralbl. f. Bakteriol.* LXXV. 408, *Zeitschr. f. klin. Med.* LXXXI. 73).]

1922 *Jrnl. Path. & Bacteriol.* XXV. 393 (*heading*) The Sonne dysentery bacillus in Australia. *Ibid.*, Thjøtta..has recorded the Sonne bacillus from cases of dysentery in Norway. *Ibid.* 394 A culture of Sonne's bacillus. **1927** *Jrnl. Hygiene* XXV. 456 So far, none of the cases of Sonne dysentery confirmed bacteriologically have died. **1930** *Jrnl. Infectious Dis.* XXVII. 468 This experiment was repeated with another strain of the Sonne organism with essentially the same results. **1947** *Ann. Rev. Microbiol.* I. 314 Milder cases are more often caused by the Sonne bacillus than by other *Shigellae*. However, it would be erroneous to conclude from this that infections with *Sh. sonnei* are always light. **1972** *Times* 14 Dec. 2/3 Tests were ordered..after an outbreak in the lines of what was believed to have been Sonne dysentery, a mild form of the disease.

sonnet, *sb.* Add: 3. **sonnet-like** adj., **-thought**, **-writing** (earlier example); **sonnet-sequence**, a set of sonnets with a common theme or subject.

1874 G. M. Hopkins *Jrnls. & Papers* (1959) 259 I looked at some delicate flying shafted ashes—there was one especially of single sonnet-like inscape. **1881** D. G. Rossetti *Ballads & Sonnets* 161 (*title*) The House of Life: a sonnet-sequence. **1920** Wodehouse *Mr. Mulliner Speaking* v. 141 The poet who was spending the summer at the Anglers' Rest had just begun to read us his new sonnet-sequence. **1973** *Listener* 21 June 830/1 Ever since Shakespeare the sonnet-sequence has grouped itself lightly. **1929** Blunden *Near & Far* 59 All that deep-sighing elegy might mourn, Glad lyric hail, and sonnet-thought adorn. **1871** D. G. Rossetti *Let.* 2 Aug. (1967) III. 964 A little sonnet-writing gets done.

sonny. Add: 1. b. A small boy.

1850 *Knickerbocker* XXXVI. 288 'Pa' returned towards the cars; when 'sonny', quickly drawing his pocket-pistol, took a drink. **1939** Joyce *Finnegans Wake* 335 How Holispolis went to Parkland with mabby and sammy and sonny and sissy and mop's varlet de shambles. **1967** [see *MUMMY *sb.*² 2].

2. *Comb.* **sonny boy**, from the title of a popular song, a boy; a man younger than the speaker or writer; freq. as a term of address and with disparaging sense; also *attrib.*; **sonny Jim**: see *sunny Jim* s.v. *SUNNY *a.* 5 c.

1928 A. Jolson et al. *Sonny Boy* (song) 3 Climb up on my knee, Sonny Boy; You are only three, Sonny Boy. **1937** [see *OVER-COMPENSATE *v.*]. **1944** *Gen* 15 June 36/2 So you lay off taking the mike out of women, sonny-boys. **1955** W. Gaddis *Recognitions* iii. iv. 850 Lie back and don't try to remember everything now, sonny boy. **1956** H. Gold *Man who was not with It* (1965) v. 79 A sonny-boy trust that I had to keep. **1970** *Washington Post* 30 Sept. 21/6 When you're the youngest of four children, 'you know, folks call you like you're a sonny boy, and it stuck with me.' **1978** T. Allbeury *Lantern Network* xii. 191 'What do you want, sonny boy?'..'I don't trust you, you English bastard.'

sono- (sōᵘ·no), comb. form of L. *sonus* sound: **sonoche·mistry**, (the study of) the chemical action of sound waves; so **sonoche·mical** *a.*, of or pertaining to sonochemistry; **so:noluminescence** *Chem.*, luminescence excited in a substance by the passage of sound waves through it; hence **so:noluminescent** *a.*; **sono·lysis** *Chem.* [*-LYSIS 1], the decomposition by ultrasound of a liquid, esp. water, as a result of the high temperatures generated within the cavities formed; also, the secondary reactions between the uncombined decomposition products and the liquid itself or compounds dissolved in it; hence **sonoly·tic** *a.*, **sonoly·tically** *adv.*; (as a back-formation) **so·nolyse** *v. trans.*, to subject to sonolysis; **so·nolysed** *ppl. a.*

1953 *Jrnl. Acoustical Soc. Amer.* XXV. 655/1 In view of the frequent attributing of sonochemical reactions to the concomitant heating effect, it is curious that higher temperatures give smaller yields. **1966** *New Scientist* 12 May 367/1 (*caption*) It now seems that a substantial part of the 'fixed' nitrogen available to marine plants and animals is due to sonochemical processes in waves. **1953** A. Weissler in *Jrnl. Acoustical Soc. Amer.* XXV. 651 (*heading*) Sonochemistry: the production of chemical changes with sound waves. [illegible] **1966** *Ibid.* 14 May 367/1 It might well be said that our understanding of sonochemistry is the key to the understanding of numerous chemical changes which take place around us. [illegible] **1969** *Jrnl. Amer. Chem. Soc.* LXI. 1531/1 The luminescence which appears when sound waves pass through liquids has been called acoustic or sonic luminescence, for short, sonoluminescence. [illegible] *Jrnl. Phys. Chem.* LXI. 1531/2 The sonolysis of water.. [illegible] **1964** *Jrnl. Chem. Phys.* XL. 609/1 DCO₂NaO-1*M*, 98%D (pH = 5·5), was sonolyzed in H₂O. **1964** *Jrnl. Physical Chem.* LXVIII. 1460/1 The analogy between the behaviour of radiolyzed and sonolyzed aqueous solutions has been pointed out in several studies. **1966** *New Scientist* 12 May 366/1 We may thus envisage a sonolysed system as a heterogeneous process wherein small centres of very high temperature exist in transient gas bubbles dispersed in a liquid medium.

sonobuoy (sǫ·noboi). Also **sono-buoy**, **Sono-**. [f. *SONO- + BUOY *sb.*] A buoy equipped to detect underwater sounds and transmit them automatically by radio.

1945 *Washington Post* 26 Oct. 1/4 The Sonobuoy, a small buoy tossed overboard from plane or ship in the vicinity of a submarine. This device, in use since 1942, gets the sub's location by sound waves and automatically transmits the information to plane or ship by radio. **1950** *Engineering* 29 Sept. 269/3 There is provision for carrying sono-buoys and military stores on the wings. **1969** *New Scientist* 28 Aug. 420/2 Passive detection, that is, listening for sounds such as sono-buoys, can also be used. **1972** *Sci. Amer.* July 16/3 A hydrophone, dangling from each sonobuoy at a substantial depth, picks up pressure signals that are then relayed by radio to the aircraft. **1973** *Daily Tel.* 24 July 2/7 An RAF Nimrod maritime reconnaissance aircraft will lay Sonobuoys used to detect submarines.

sonofa, sonofer. Colloq. shortening of such phrases as *SON OF A BITCH, *son of a gun* s.v. GUN *sb.* 6 c, etc. Cf. SON *sb.* 7 d.

1951 W. Sanson *Face of Innocence* i. 1 'Why?' Harry said. 'If it isn't you, you old sonofer.' **1968** G. Bagby *Another Day—Another Death* iii. 45 You wouldn't have found one man around here who hadn't threatened..this Clancy sonofa.

son of a bitch. *slang.* Also **son-of-a-bitch**, **sonofabitch**, **sonuvabitch**, etc. In pl., **sons of bitches**. [Cf. SON *sb.* 7 d.]

Now more common in the U.S. than elsewhere.

1. **a.** A despicable or hateful man. Also *attrib.* Cf. *S.O.B.* s.v. *S. I. 4 a.

[c **1330** *Of Arthour & of Merlin* (1973) 333 Abide pou pel malicious! Biche-sone pou dravest amis Pou schalt abigge it ywis! **1605** Shaks. *King Lear* II. ii. One that..art nothing but the composition of a Knaue, Begger, Coward, Pandar, and the Sonne and Heire of a Mungrill Bitch.] **1707** J. Shirley *Triumph of Wit* (ed. 5) 203 There stands Jack Ketch, that Son of a Bitch, that owns us all a grudge. **1748** A. Hamilton *Itinerarium* (1907) 229 It was the landlord ordering his negroes, with an imperious and exalted voice. In his orders the known term of contempt of *son of a bitch* was often repeated. **1762** L. Sterne *Let.* 8 Apr. in *Times Lit. Suppl.* (1955) 8 Apr. 184/4 Phelps is a son of a bitch.. **1776** *Gentl. Mag.* [illegible] Rogue, and—a son of a bitch. **1794** [illegible] **1830** [illegible] **1863** J. Sappington in T. Jefferson *Notes on Virginia* 32 Logan's brother..attempted to strike him, saying, 'White man, son of a bitch'. **1823** Byron *Don Juan* xi. xii. 123 Pray ask of your next neighbour, If he heard not the squawk of fan-born riches, Like lap-dogs, the least civil sons of b—s. **1830** [see BITCH *sb.*¹ 2 a]. **1924** M. Kennedy *Constant Nymph* ii. xvi. 217 You think I ought to want to please every son of a bitch who can pay for a sixpenny ticket. **1930** E. Pound *Let.* 1 Feb. (1971) 224 The stinking son-of-a-bitch who ran the country. **1939** Steinbeck *Grapes of Wrath* xxvi. 388 Most of 'em wouldn't be sons-a-bitches. **1951** R. Chandler *Long Goodbye* iv. 37 Here you fat-bellied son of a bitch, you beat up on me. **1958** *Tatler's Pastor* Hall *Joe* ii. 79 [illegible] **1960** *Spectator* 13 Mar. 397/1 *Ibid.* ii. 68 'Where's that son of a bitch of yours?' **1972** *Listener* 23 Mar. 346/1 A son of a bitch on a sinking ship [illegible]. Saxon et al. *Gumbo Ya-Ya* iii. (1945) 8 Yeah, I'm a son-of-a-bitch. **1953** J. D. Salinger *Catcher in Rye* iii. 30 Boy, I can't stand that sonuvabitch. **1959** *Times* 25 Sept. 9/1 Home..said..that a Scotland Yard report must have been prepared by 'some son-of-a-bitch inspector who wants to blacken my name'. **1963** R. Shaver tr. *Beckett's End in Writers in Revolt* 357 He went hustling along, bowing and scraping and flourishing his hat...The insufferable son of a bitch. **1970** J. Shaw *Beggarman, Thief* ii. iii. 139 Arrogant Hollywood son-ofabitch.

b. With weakened force and neutral or friendly overtones: a fellow, a man.

1951 J. Paul *Springtime in Paris* xv. 294 'The son of a bitch is crazy,' said, in soft English, but completely without malice or disapproval. **1958** J. Carew *Wild Coast* ii. 124 He was a drinking, whoring, kindly savage son-of-a-bitch. **1970** 'A. Hailey' *Overload* iii. xii. 238 Besides, the son-of-a-bitch had guts and was honest.

2. Applied to animals, etc., as a term of abuse.

1771 Smollett *Humph. Cl.* I. 167 Damn the nasty son of a bitch, and them he belongs to! **1954** C. Ogden *Big Knife* in *Famous Plays* 1954 434 Our story is that the dog—a big son of a bitch—yanked free and leaped the fence. **1966** J. Kerouac *On the Road* iv. 24 We were riding this sonofabitch since Des Moines. **1977** *New Yorker* 27 June 26/3 I figured that that goose wouldn't be a meal for me... I went over to the tree, shook it, and yelled, 'Get outta there you son of a bitch!'

3. Applied to a woman as a term of abuse. *rare.*

1963 H. Miller *Black Spring* 250 He got to working overtime in order to lay aside the little bribe which would make the frigid son of a bitch come across like a nymphomaniac.

4. Used as an expletive.

1953 W. Burroughs *Junkie* (1972) ii. 28 'Sonofabitch!' she snarled. 'They can tell when a woman isn't looking for a husband.' **1957** M. Millar *Soft Talkers* iii. 35 Sonuva-bitch, I don't get it. What's the matter? Was that a good deal or..

5. Used in comparisons to suggest strength, ferocity, speed, etc.

1953 W. Burroughs *Junkie* (1972) viii. 75, I hit Philly sick as a sonofabitch. **1976** *New Yorker* 15 Mar. 32/3 Well, I hit that bastard with my white fist and ran out of there like a sonuvabitch.

Hence **son-of-a-bitching** *a.*, a general epithet of abuse.

1930 J. Dos Passos *42nd Parallel* I. 101 Every sonofabitchin yellerleg in the State of Nevada. **1941** E. P. O'Donnell *Great Big Doorstep* vii. 98 Tayo put his bag down, and shed his coat uncertainly. 'Son of a bitchin pole-vaulter,' he muttered. **1960** J. Kirkwood *There must be a Pony!* (1961) ii. 12 The meanest son-of-a-bitching parrot you could ever run up against. **1979** 'A. Hailey' *Overload* iv. xiv. 370 'People!' Paulsen exploded. 'Son-of-a-bitching, stupid people!'

sonogram (sǫ·nǫgræm). Also **sonagram**. [f. *SONO- + -GRAM.] A graphical representation, produced by a sonograph, of the distribution of sound energy among different frequencies, esp. as a function of time.

1959 *New Biol.* XX. 71 (*caption*) Sonograms of songs of two male chaffinches from different 'dialect areas'. **1967** *Jrnl. Acoustical Soc. Amer.* XXIX. 1081/2 (*caption*) Sonagrams of the words 'tack' 'task' 'tact'..illustrating the role that transitions and bursts play in the perception of stops. **1969** R. Stenuit *Dolphin* v. 73 Analysis of the sonogram of a conversation shows click exchanges and whistle exchanges. **1978** *Country Life* 28 Dec. 2225/1 By comparing the sonograms differences can be isolated between the call patterns of individual birds. **1980** *Times* 19 Aug. 12/6 Dr Silver used a hydrophone to record the sounds under water and then obtained a visual record in the form of a sonogram.

sonograph (sǫ·nǫgrɑːf). [f. *SONO- + -GRAPH.] **1.** Also **sonagraph** and with capital initial. An instrument which analyses sound into its component frequencies and produces a graphical record of the results.

Sonagraph is registered as a proprietary term in the U.S.

1951 *Official Gaz.* (U.S. Patent Office) 14 Aug. 345/2 Kay Electric Company, Pine Brook, N.J... *Sona-graph*.. For instrument which is a sound spectrograph which produces permanent visual records showing distribution of energy vs. both frequency and time. Claims use since October 1948. **1953** J. B. Carroll *Study of Lang.* vii. 206 The 'visible speech' machine developed in the Bell Telephone Laboratories..produces the same type of record as does the sonograph, but on a continuous, transitory basis. **1954** *Nature* 13 Mar. 465/1 All this was changed by the invention of the sound spectrograph, now known commercially as the 'Sonograph'. **1956** *New Biol.* XX. 78 The structure of Chaffinch calls has been studied by making recordings on discs or magnetic tapes, and then analysing them on a sound spectrograph, or sonograph. **1976** J. M. Brownjohn tr. *Kirst's Time for Payment* v. 136 The Sonograph can pick up voice frequencies and record them. **1979** *New Scientist* 17 May 537/2 The sonagraph acts as a sonic prism.

2. An image of a tract of seabed obtained by means of side-scan sonar.

1970 *Sci. Jrnl.* Dec. 96/2 By 1964 the geological survey and oceanography of the continental shelf and uppermost continental slope was so evident that the National Institute of Oceanography decided to explore the possibility of adapting the same method for examination of the deeper lying ocean floors. **1974** *Nature* 15 Feb. 433/2 Between this trench and the Italian coast alongslope tectonic trends are seen on sonographs. **1976** *Physics Bull.* Sept. 381/3 Figure 1a..shows a sonograph of a portion of the seabed in the Bristol channel.

Sonoran (sōnō·rān), *a.* [f. *Sonora*, the name of a state in North-west Mexico + -AN.] **1.** Of or pertaining to a biogeographical region including desert areas of the south-western United States and central Mexico.

1880 E. D. Cope in *Proc. Amer. Philos. Soc.* XVIII. 263 This collection..is of interest as serving to fix the extension of the Sonoran fauna to a point further south. **1892** *Proc. Biol. Soc. Washington* VII. 13 The term 'Sonoran Region' has been applied by Cope and others to an important life area which enters the southwestern part of the United States from the table-land of Mexico. **1902** *Nature* 14 Aug. 374/1 It is somewhat of a surprise to find that the author is unable to convince himself of the necessity of a Sonoran region. **1937** *Discovery* July 206/1 This sub-region has much in common with the Sonoran sub-region of the Nearctic Region. **1979** Parse [illegible] *Mag.* Feb. 79/1 There are more than 350 varieties of live animals and plants of the Sonoran desert on exhibit.

2. Of, pertaining to, or characteristic of a grouping of related Indian languages spoken in southern Arizona and northern Mexico.

This classificatory category of some Uto-Aztecan languages into a Sonoran sub-group is now discarded.

[**1875** H. H. Bancroft *Native Races of Pacific States* III. 670 Sound-shifting..has been found by Mr Buschmann in the languages of his Sonora family.] **1891** D. G. Brinton *Amer. Race* 123 The Sonoran branch [of the Uto-Aztecan stock] begins to the north with the Pimas, who occupied the middle valley of the Gila, and the land south of it quite to the Rio Yaqui. I continue for it the name of *Sonoran* given by Buschmann, although it extends far beyond the bounds of that province. **1909** A. F. Chamberlain in *Amer. Anthropologist* XI. 535 A number of Shoshonean languages, from Ute to Nahuatl and some of the Sonoran tongues. **1938** B. L. Whorf in *Ibid.* XXXVIII. 606 In times past some of us hoped that the stock could be classified in such a way that we could summarize the situation by stating generalized reflexes for sub-groups such as 'Shoshonean', 'Piman', 'Sonoran', from which the reflexes of the individual tongues in these groups could be derived as a second step. The hope is vain. No such groups exist. **1964** S. M. Lamb in *Univ. Calif. Publ. Ling.* XXXIV. 121 Further investigation of the northern Mexican languages, which Brinton had..put into a single group called Sonoran, revealed a high degree of diversity among them. **1977** C. F. & F. M. Voegelin in T. A. Sebeok *Native Languages of Americas* I. 482 The traditional major branches of the Uto-Aztecan family (Shoshonean, Sonoran and Aztec).

sonorant (sǫnō³·rănt). *Phonetics.* [f. SONOR(OUS *a.* + *-ANT¹.] A resonant; a sound produced with the vocal tract so positioned that spontaneous voicing is possible; a vowel, a glide, or a liquid or nasal consonant. Also *attrib.* or as *adj.*

1934 in Webster. **1943** [see *RESONANT *a.* 1 b]. **1956** Jakobson & Halle in Sapera & Bastian *Psycholinguistics* (1961) 482 It is advantageous to range these two classes [of phonemes] as 'sonorants' vs. 'obstruents' of consonants. **1963** [see *OBSTRUENT *sb.*]. **1968** Chomsky & Halle *Sound Pattern Eng.* 302 Vowels, glides, nasal consonants, and liquids are sonorant. **1978** *Language* LIV. 327 Among these are..; reduction of palatalized sonorants to simple yod, and of *Ɫ* to *z*. **1981** *Publ. Amer. Dial. Soc.* LXVII. 30 The major sonority distinction is between sonorants (i.e. nasals, liquids and all vocoids) on the one hand and obstruents (stops, affricates and fricatives) on the other.

sonorous, *a.* Add: Also with pronunc. (sǫ·nǒras). (Example concerning pronunc.)

1934 G. B. Shaw *Let.* in *Times* 2 Jan. 11/5 An announcer who pronounced decadent and sonorous as dekkadent and sonnerus would provoke Providence to strike him dumb.

‖ sons bouchés (sǫñ buʃe), *sb. pl.* [Fr., lit. 'blocked sounds'.] In horn-playing, notes stopped by the insertion of the hand into the bell of the instrument; a direction indicating this. Also *attrib.* Cf. *CUIVRÉ, *a.*

1907 T. S. Wotton *Dict. Mus. Terms*, *Sons bouchés*, closed notes on a horn. [**1927** *Grove's Dict. Mus.* (ed. 3) II. 666/1 The composer's intentions should be indicated.. 1959 the placing of a small cross over the note with the word stopped (Fr. *cuivré* or *sons bouchés*).] **1961** R. Pegge in A. Baines *Musical Instruments through Ages* xii. 300 The true stopped notes—referred to as *sons bouchés*..'sons bouchés', 'gestopft', or 'chiuso'. **1977** *Early Music* July 427/1 When one closes the bell fully with the right hand, the *sons bouchés* notes are a *whole tone* and not a semitone above the next lower harmonic.

sont (sǫnt). *v.* *U.S.* Sometimes used to represent *sent* in the written form of Black speech.

1890 *Dialect Notes* I. 71 *Sont*, sent. Used mainly by negroes. **1893** H. A. Shands *Some Peculiarities of Speech in Mississippi* 58 *Sont*..., Negro for *sent*. **1930** W. Faulkner *Sound & Fury* 337, I sont dat boy on dat 'ar hour ago. **1942** —— *Go down, Moses* 16 'What for?' he said. 'She just sont hit to you,' the nigger said. *Ibid.* 106 Ah went to yo house last night, but you want dar. She sont me.

sont, var. SUNT in Dict. and Suppl.

sontag (sǫ·ntæg, zo·ntak). The name of Henriette *Sontag* (1806–54), German singer, used to designate a type of knitted or crocheted jacket or cape, with long ends which are crossed in front of the body and tied behind, worn by women in the second half of the 19th century.

1862 *Harper's Mag.* Oct. 607/1 Constant relays of clothes were successively demanding themselves of 'Clouds', 'Sontags', 'Mariposas', and other dainty feminine wraps. **1863** A. D. T. Whitney *Faith Gartney's Girlhood* xviii. 161 Kate brought quickly, sontag, lady's and cloak. **1900** E. A. Dix *Deacon Bradbury* 45 Did you hear what she said to Mrs Delane about that worsted sontag she brought?

Sonthal, Sonthali, varr. *SANTAL², *SANTALI.

soo·dling, *ppl. a. poet. rare.* [f. SOODLE *v.*] That flows or moves slowly.

1951 Auden *Nones* (1952) 39 The baltering torrent Shrunk to a soodling thread.

soogan, var. SUGGAN in Dict. and Suppl.

soogee (sū·dʒi). Also **soogie, soujie**, etc. = next.

1944 *Time* 10 Jan. 4/3 A bucket of 'soujge'. **1945** *Seafarers' Log* 7 Sept. 4/4 The old soogee bucket and paint brush will be working plenty overtime. **1953** J. Masefield *Conway* (rev. ed.) ii. 124 The recipe for liquid used to clean 'holy ground' was one handful of soft soap..one of 'soojie' (soap powder). **1963** M. Lowry *Ultramarine* iv. 198 Steamflies..are stealthily prowling over..the limp bags of caustic soda and soogee.

So as *v. trans.*, to clean (wood and paintwork) with *SOOGEE-MOOGEE; **soog(ie)ing** *vbl. sb.*

1903 A. Sonnichsen *Deep Sea Vagabonds* iii. 37 There was soodgee-moodgee. Soodgee is a sailor's horror; it means to wash paint-work with a strong solution of soda and water. **1945** *Seafarers' Log* 29 June 12 Soogieing, chipping, painting, etc., shall not be considered an emergency. **1947** *Ibid.* 14 Mar. 5/2 The messhalls were sougeed out, the gallery cleaned. **1950** R. Bissell *Stretch on River* vii. 75, I was sooging down the walls in his [*sc.* the captain's] cabin. **1963** M. Lowry *Ultramarine* v. 217, I had to do a bit of soogeing in the petty officers' washhouse.

soogee-moogee (sū·dʒi,mū·dʒi). *Naut. slang.* Also **soogie-moogie, soojee-moojee, souji-mouji, sugi-mugi**, etc. and without hyphens. [Origin unknown.] **a.** A mixture containing caustic soda used for cleaning paintwork and woodwork on ships and boats. Also *attrib.*

1882 D. Kemp *Yacht & Boat Sailing* (ed. 3) 573 *Soojee Moojee*, a caustic composition sold by yacht fitters for cleaning paintwork. **1883** *Ibid.* 1884 [illegible] A certain caustic composition, known to yachtsmen by the mysterious name of 'skewgy-mewgy'. **1900** F. T. Bullen *Men of Merchant Service* xiii. 120 No tramp second mate can hope to keep his hands out of the paint pot or the soogee-moogee bucket. **1903** [see *SOOGEE *v.*]. **1907** M. Roberts *Flying Cloud* xxiv. 229 She was as clean as any ship with paint and canvas and souji-mouji. **1933** A. W. Nelson *Yankee Swanson* 57 He had a good scrubbing down with 'sudje mudje'. **1934** P. Blundell *Confessions of Seaman* xii. 164 With a bucket containing hot soap and soda water—sougi-mougi sailors call it—he ascended ladders in the swaying engine-room and washed paint-work. **1934** J. Hanley in *Spectator* 26 Jan. 131/1 'There are no sailors to-day,' says [Conrad], 'only Sugi-Mugi men'... Mere washers of paint. Deck-hands whose duty it is to wash and chip paint, morning, noon and night. **1938** W. E. Dexter *Rope Yarns* 82 In all the sailing ships I was in I never came across a long-handled holystone, or 'squilgee', or 'soogy-moogy'. **1939** H. Hodges *Through Mighty Seas* ii. 36 All hands wallowed in soogie moogie. **1962** A. G. Course *Dict. Naut. Terms* 182 *Soogee moogee*, a liquid used for cleaning paintwork and woodwork consisting of soda and water or soap and water.

b. A cleaning operation which involves the use of soogee-moogee.

1935 *Sea Breezes* Jan. 60 The equipment, particularly the sails, would be so perfect that my crew wouldn't mind the 'sojee-moojee' such perfection entailed. **1965** *Time* 31 Dec. 96/3 Soon Sailor Slobodkin..found himself loading garbage, chipping paint and doing soogie moogie (scrubbing paint work) with a crew.

soojee, var. SOOJI. (Later examples.) **sooji halwa** [*HALWA], a kind of dessert made with soojee.

1955 K. P. Jhabvala *To whom she Will* xxi. 137 O she is such a clever girl, kheer she makes, sooji halwa [etc.]. **1971** *Femina* (Bombay) 16 Apr. 57/2 *Sooji Halwa*. 120 gms. sooji (fine semolina); 60 gms. fat; 110 gms. sugar; 30 gms. peanuts; 2 gms. salt.

sook¹ (suk). *Austral.* and *N.Z. slang.* Also **sookie**. [Perh. from Eng. dial. *suck*: see *Eng. Dial. Dict.* s.v. *suck* 19, a 'duffer', a stupid fellow.] A stupid or timid person; a coward; a 'softy'.

1933 N. Scanlan *Tides of Youth* xv. 155 He looked a big sookie and wouldn't say a word. **1941** Baker *Dict. Austral. Slang* 69 *Sook*, a coward, a timid person. **1950** 'B. James' *Advancement of Spencer Button* 9 If he nervously declares he can't fight, and then declares he doesn't want to fight, then he is a 'sook' or a 'sissy'. **1970** G. Greer *Female Eunuch* 79 She may be reviled as a cissy, a sook. **1970** P. White *Vivisector* 12 He wasn't a sook. He could run, shout, play, laugh. **1971** ... [illegible] twice split Billy Abrams's lip, who was two years older. **1975** *Courier-Mail* (Brisbane) 2 Jan. 9/4 The tough speedsters might appear as somewhat of a myth by tearing to different from his mates in case they might think him a bit of a sook.

Hence **soo·key**, **sooky** *adjs.*, cowardly, 'soft', stupid.

1953 D. Cusack *Southern Steel* 328 Get along with you; you're getting real sookey. **1964** *Weekly News* (Auckland) 18 Mar. 38/3 The boys say they feel sooky when the cops hit them. **1970** *N.Z. Listener* 12 Oct. 13/5 Their attitude of tolerant resignation toward the sooky Maoris who are always getting into trouble.

sook² (sŭk). *U.S.* [Origin unknown.] A mature female blue crab, *Callinectes sapidus*, of the eastern coast of the United States.

1955 *Sun* (Baltimore) 30 Oct. 32 'Sooks' are female blue crabs. Their annual migration..has been under way for a month. **1978** J. A. Michener *Chesapeake* 647 He [*sc.* a blue crab] began his own preoccupations in order to mate among the grasses, looking for sooks which were preparing to be shed. [illegible] These overlooked females, on their way south to spend the winter near the entrance of the bay, where fertile sooks traditionally prepared to lay their eggs, sent out pheromonal signals that might be to the vicinity, for this was the final period in which they could be fertilized.

sook, var. *SOUK.

sool (sūl), *v.* Chiefly *Austral.* and *N.Z.* [var. SOWL *v.*¹] **1.** *trans.* Of a dog, to attack or worry; freq. in command *sool* (*him*, etc.) Also *absol.* and *transf.*

1890 [see SOWL *v.*¹ 1]. **1896** Mrs. K. L. Parker *Austral. Legendary Tales* 90 She went quickly towards her camp, calling softly, 'Birree gougou', which meant 'Sool 'em, sool 'em', and was the signal for the dogs to come out. **1904** 'S. Rudd' *Sandy's Selection* 60 'Sool 'im, sool 'im!' the girls were shouting, and the blind dog sooled in moderation. **1935** *Bulletin* (Sydney) 6 Feb. 21/2 A pair of these birds will sometimes 'sool' a hare. **1959** A. Upfield *Bony & Black Virgin* xv. 135 'Sool 'em, Bluey,' Bony pleaded, and the dog nosed about and finally went down under the scrub by a hole made by the foxes. **1960** B. Crump *Good Keen Man* 24 Another young dog would have been better encouragement than a man lumbering about trying to bark like a dog, and noisily skitching and sooling.

2. To urge or goad. Freq. with advbs., esp. *on*.

1898 *Bulletin* (Sydney) 30 Apr. 31/4 The rest of the company came out and 'sooled' on the twelve [arguers]. It was a glorious scrimmage. **1898** E. E. Morris *Austral English* 426/2 Sool.., used colloquially—(1) to excite a dog or set him on. **1911** *Chambers's Jrnl.* Mar. 222/2 Don't 'sool' the dogs to an unwise assault. **1911** L. Stone *Jonah* 31 The Push gathered round, grinning from ear to ear, sooling the women on as if they were dogs. **1916** C. J. Dennis *Moods of Ginger Mick* 39 The bugles East and West sooled on the dawgs o' war. **1921** E. O'Ferrall in Murdoch & Drake-Brockman *Austral. Short Stories* (1951) 158 Served him right if I sooled th' dog on him! **1942** G. Casey *It's Harder for Girls* 52 'Sool the dogs after him,' someone called out. **1960** B. Crump *Good Keen Man* 23, I ran over and sooled the dog after the trotting sow. **1963** B. Pearson *Coal Flat* xii. 227 We'll get an Alsatian and sool it outa you. **1970** G. Greer *Female Eunuch* 190 The hero may..like a lion-tamer sool her on to his enemies. **1977** C. McCullough *Thorn Birds* iv. 80 Father Ralph worked like a man in the grip of some obsession, sooling the dogs after unsuspecting bands of sheep.

Hence **soo·ler**, one who incites; an agitator.

1935 H. R. Williams *Comrades of Great Adventure* 35 Here, as chief 'sooler', he was urging the passing soldiers to patronize the eating-house. **1938** X. Herbert *Capricornia* viii. 98 Then a war-monger, or Sooler, as such people were called in the locality, made his voice heard in the land. **1963** —— *Disturbing Element* 141 She had been sending white feathers round... She had become what her former comrades of the I.W.W. called a Sooler.

soon, *adv.* Add: **I. 4. c.** Also *dial.* and *colloq.*

1907 W. P. Ridge *Name of Garland* ii. 33 Get out of my kitchen soon as you can. **1930** J. B. Priestley *Angel Pavement* ii. 91 We want another man for London and district, soon as we can get one. **1930** W. Faulkner *As I lay Dying* 58, I done put supper on and I'll be there soon as I milk. **1940** —— *Hamlet* I. ii. 26 Jody came in last night. I knowed it soon as I saw him.

8. a, b. (Later examples.) Also *just as soon*; *as soon as look at you*: see *LOOK *v.* 1 a.

1913 W. B. Yeats *Hour-Glass* in *Mask* V. 328 I'd as soon listen to dried peas in a bladder as listen to your thoughts. **1930** A. P. Herbert *Water Gipsies* xvii. 254 I'd as soon have stone-ginger any day. **1966** D. Francis *Flying Finish* x. 133 He waved me to join him, which I would just as soon not have done. **1974** C. Hampton *Savages* (1976) vii. 45 Because Chico well He'd cut your head off soon as say good morning. **1974** M. Hebden *Pride of Dolphins* I. iii. 34 I'd just as soon you dropped me..and let me make my own way home.

9. *soon-arriving*, *-coming*; *soon-come*, *-dropped*, *-finished*, *-forgotten*. With infin., as *soon-to-be*.

a. **1886** E. G. White *Historical Sketches* 164/2 The end so near, the warning of a soon-coming Judgment yet to be given to all nations, tongues, and peoples. **1930** Auden *Poems* 17 Hear something of that soon-arriving day.

b. **1901** G. B. Shaw *3 Plays for Puritans* p. xviii, An hour's soon-forgotten fuss. **1902** W. B. Yeats *Where there is Nothing* (1903) II. 44 Not the fighting of men in red coats, that formal, soon finished fighting, but the endless battle, the endless battle. **1925** Blunden *Eng. Poems* 18 And hyacinth-eyes beneath soon-dropt lids. *Ibid.* 36 That now, this soon-come spring, goes slow and sere.

c. **1961** *Daily Tel.* 30 Aug. 10/2 He wants to appear in Belgrade at the neutralists' conference as the leader at least of a soon-to-be apparently united country. **1975** *Publishers Weekly* 10 Nov. 51/3 Often the dying are 'released' by the knowledge of their soon-to-be end.

II. 11. b. Phr. (*I'd*) *sooner* (*it should be*) *you than me*, and varr. Cf. *RATHER *adv.* 9 f.

1864 Trollope *Can you forgive Her?* I. xiv. 108 You are going down to Cheltenham, are you?..I'd sooner it should be you than me; that's all I can say. **1905** H. A. Vachell *Hill* iii. 53 Phew-w-w!..I'd sooner it was you than me, Verney. **1937** A. Thirkell *Summer Half* vi. 181 'Good old Mr. Lorimer,' said Swan, 'sooner he than I.' **1973** 'J. Sturrock' *Wicked Way to Die* iii. 35 'Talk to her I must.' He shook his head. 'Sooner you than me.'

III. 14. a. (Later examples.) Now freq. (orig. *telegraphese*), as soon as possible.

1815 A. Constable *Let.* 29 Jan. in *J. Constable's Corr.* (1962) I. 113 The picture you request shall be sent per soonest. **1950** C. M. Kornbluth in *Mag. of Fantasy & Sci. Fiction* I. 4 They needed a bright and sparkling little news item..'soonest'. **1962** J. Hay in *E. Queen's 16th Mystery Annual* 163 'Bjornsson and whale to proceed soonest to Regensburg and await further orders,' Twenty-penny cabled Hawker. **1977** 'E. Crispin' *Glimpses of Moon* xiii. 262 Come back to London soonest prepare leave for Libya soonest terrorists blowing up all the oil-wells there. **1977** J. Didion *Book of Common Prayer* II. xiv. 119 I'm getting you together soonest, that's definite.

soon, *a.* Add: **1.** (Later U.S. dial. examples.) Freq. in phr. *a soon start* (*in the morning*).

1913 H. Kephart *Our Southern Highlanders* xiii. 296 Spell is used in the sense of while..and soon for early ('a soon start in the morning'). **1930** G. B. Johnson in B. A. Botkin *Folk-Say* VII. 357 *Soon*, early, quick, alert. 'A soon breakfast', 'a soon man'. **1949** H. Hornsby *Lonesome Valley* ii. 21 People must have got a soon start, because the place was full up already. **1951** L. Craig *Singing Hills* iii. 18 The furrin woman wanted to get a soon start, come morning.

soon (sūn), *sb. rare.* [Substantival use of prec.] The near future.

1940 W. Faulkner *Hamlet* III. i. 166 They are moving not toward a destination in space but a destination in time..; the sleight hand of May shapes them both, not in the immediate, the soon, but in the now.

sooner¹. Add to def.: Chiefly with ref. to the settlement of the territory now known as Oklahoma before the official opening of the area to settlers on 22 April 1889. (Further examples.) Hence (with capital initial), an Oklahoman. Also *attrib.*; **the Sooner State**, Oklahoma. Hence **soo·nerism**, the practice of the unlawful and premature settlement of Oklahoma.

1890 *Congress. Rec.* 17 Jan. 657/2 We have recognized the fact that there are 'sooners' there. **1894** *Columbus* (Ohio) *Dispatch* 19 Mar., An important case growing out of the 'soonerism' at the Oklahoma opening will be given a hearing. **1904** *N.Y. Even. Post* 13 June 7 White 'sooners'..were trying to rob them [*sc.* Indians] of some of the most valuable mineral deposits on their reservation. **1930** *Sat. Rev. Lit.* (U.S.) 22 Mar. 841/3 She has done excellent reporting, has constructed a ripping yarn, has given us novel incidents, novel characters, a fresh setting, has created a strange new Sooner mythology. **1939** *New Yorker* 14 Oct. 73 Oklahoma uses on its road signs a phrase which I first heard in Kansas and never again except in the Sooner State. **1945** J. L. Marshall *Santa Fe* 232 They watched the line through telescopes and, congratulating themselves on their success in checkmating 'Soonerism', figured it would take the first settlers on the fastest horses, about ninety minutes to reach Oklahoma. **1948** *Okla. Cotton Grower* 15 May 2/2 For the Sooner State planter that is perhaps the first major theorem of the business. **1948** [see *OKIE]. **1976** *Billings* (Montana) *Gaz.* 30 June 4-E/2 University of Oklahoma quarterback Joe McReynolds was dropped from the Sooners football roster Tuesday.

sooner². *slang* (chiefly *Austral.*). [perh. as prec.] An idler, shirker; applied as a term of abuse to an ineffectual or obstructive person, object, etc. Also *attrib.* in *sooner dog*.

1892 K. Lentzner *Dict. Slang-English of Australia* 117 *Sooner*, a weak idler, a lazy good-for-nothing. **1919** E. Dyson *Hello, Soldier!* 31 He slugged a tubby Hun, Then choked a Fritzie with his dukes, 'n' pinched the sooner's gun! **1936** F. Clune *Roaming round Darling* xxv. 270 *Onlookers*: Tongue-tied Joe, a sooner dog, a Scotch dog, a dog of all nations, a hungry goat. **1937** Partridge *Dict. Slang* 801/1 *Sooner dog*, one that would sooner feed than fight. **1948** V. Palmer *Golconda* xix. 159 'The dirty sooners!' he burst out. 'They don't know a man when they find one, those heads down south.' **1969** P. A. Smith *Folklore of Austral. Railwaymen* 117 This was an old sooner of an engine. She'd had it. On a stiff climb in a tunnel she began to slip.

soonish, *adv.* (Earlier example.)

1890 S. S. Buckman *John Darke's Sojourn in Cotteswolds & Elsewhere* vii. 58 Good, honest drink..made a man knaow soonish as he wur a-getting nicely forrud.

soop, *v.* Add: **3.** Also without *up*.

a **1822** A. Boswell *Poet. Wks.* (1871) 195 Soop the rink, lads, wide enough, The *hog-scores* mak', and mak' ilk *brough*. **1885** J. Strathesk *More Bits from Blinkbonny* (ed. 2) xiv. 270 Soop weel when I tell ye. **1963** *Times* 17 Jan. 12/5 Rob Roy's country today echoed not with the war cry of the MacGregors but with strange shouts of 'Yes, yes' and 'Soop, soop' as the great bonspiel of the curling game was staged for only the second time in 37 years.

sooping *vbl. sb.* (later examples in sense 3 of the vb.).

1937 T. Henderson *Lockerbie* ix. 57 The ice being keen it required little soopin'. **1976** *Alyn & Deeside Observer* 10 Dec. 5/2 Part of the fun of the game comes in 'sooping'. This is when the players sweep the ice with special brooms in front of a moving stone to help it go further.

soopie, soopje, varr. SOPIE in Dict. and Suppl. **soopolallie**, var. *SOAPOLALLIE.

soor² (sūᵊɹ). *Anglo-Indian slang.* [ad. Hindi *sūar* pig, f. Pali *sūkara-*, Prakrit *sūara-*, *suara-* pig.] A term of abuse for a person: pig, swine.

1848 J. H. Stocqueler *Oriental Interpreter*, Soor, soor-ka-butcha, abusive terms of which the Hindostanee language is fertile. *Soor* is a pig, and *soor-ka-butcha* the offspring of a pig. **1864** Hotten *Slang Dict.* 240 *Soor*, an abusive term... *Anglo-Indian*. **1919** 'Boyd Cable' *Old Contemptibles* vi. 92 'Why don't the *soors* come on an' fight it out?' said Corporal Smedley. **1926** *Blackw. Mag.* July 78/1 '*Soors!*' he cursed. 'They won't give me one of their Union Jacks.' **1936** F. Richards *Old-Soldier Sahib* iv. 74 You black soor, when I order you to do a thing I expect it to be done at once.

soot, *sb.*¹ Add: **5. c.** *soot-blackened*, *-bleakened*, *-clogged*, *-coated*, *-roughened*, *-suffused*.

1894 'Mark Twain' in *Century Mag.* XLVIII. 17/1 It rained all day..apparently trying its best to wash that soot-blackened town [*sc.* St. Louis] white. **1916** Joyce *Portrait of Artist* (1969) 115 The soot-coated packet of pictures which he had hidden in the flue of the fireplace. **1921** W. de la Mare *Veil* 68 Slow wreathed the grease adown from soot-clogged wick. **1932** W. Faulkner *Light in August* vi. 111 Memory..knows remembers believes a corridor in a big long garbled cold echoing building of dark red brick sootbleakened by more chimneys than its own. **1947** W. de la Mare *Coll. Stories for Children* 122 Chapped, soot-roughened hands. **1956** D. Gascoyne *Night Thoughts* 23 The soot-suffused sky-canopy. **1977** H. Fast *Immigrants* I. 29 The still half-naked citizens, soot-blackened and homeless, greeted the ruin as they had always greeted their city.

6. **soot-blower**, a device for clearing soot from the flues of a boiler, furnace, etc.; **soot house** (see quots.).

1930 *Engineering* 16 May 627/1 Steam driers had met with little success while soot-blowers had been widely introduced. **1967** *Trans. Inst. Engineers & Shipbuilders in Scotland* CX. 36 Naval boilers, up to the end of the last war, were not normally provided with sootblowers. **1957** E. E. Evans *Irish Folk Ways* ix. 120 The soot-houses..whose roofs..were stripped in spring in the days when the whole family migrated to the summer pastures. At any rate the soot-house season runs from October to May... In the Hebrides it was customary to strip the soot-laden thatch of the black-houses annually for use as manure, and the roof was left without a smoke-hole to encourage the deposit of soot. **1966** *Daily Tel.* 21 Apr. 4/7 On Achill Island off the west coast of Ireland are the remains of some small buildings... They are called 'soot houses' and were used for the production of soot for fertilising the potato crop.

soot, *v.* Add: **3.** Sometimes with *up*: to fill or choke with a sooty deposit. Also *fig.*

1903 [implied at *sooting* vbl. sb. s.v. SOOT *v.*]. **1925** *Morris Owner's Manual* iv. 46 Bad plug insulation is sometimes caused through sooting. **1929** W. E. Collinson *Spoken Eng.* 84 The engine is knocking. One of the [sparking] plugs is probably sooted up. **1959** M. Pugh *Chancer* i. 9 London was sooting me up and I couldn't shed it, layer by layer, like the plane-trees in the park.

soothe, *v.* Add: **7. c.** With direct speech as object: to say in a soothing manner.

1934 N. Marsh *Man lay Dead* xii. 206 'You shall have every opportunity,' soothed Alleyn. **1976** I. Levin *Boys from Brazil* iii. 77 'I agree, Josef, I agree,' the colonel soothed.

soothe (sūð), *sb. rare.* [Back-formation f. the vb.] A soothing feeling or effect.

1947 *Landfall* (N.Z.) I. iv. 267 They felt the soothe of the darkness. **1971** J. Gardner *Every Night's a Bullfight* vii. 170 Close whispering, kissing; the soothe of flesh against flesh.

soothering, *ppl. a. dial.* and *U.S.* [f. SOOTHER *v.* + *-ING.²] Blandishing, cajoling.

1866 E. L. Linton *Lizzie Lorton* II. i. 13 Thee 'se gitten a soothering tongue in thee head, lad. **1884** [see *SOFTY *a.*]. **1898** G. Bartram *White-Headed Boy* 5 A kind of gentle look in them—a 'soodherin'' look, as we say in Ireland—the way a horse looks at you when he loves you. **1953** *Scots Mag.* Aug. 418 The mither sings a sootherin tune til a waukrif loon.

soothing, *ppl. a.* Add: **2. b.** (Earlier and later examples.) spec. *soothing powder* (in example *fig.*); *soothing syrup*, a medicinal preparation supposed to calm fretful children; freq. *fig.*, flattery; empty reassurance; merely palliative remedies; mawkish or sentimental music, emotion, etc.; hence *soothing-syrupy* adj.

1839 *Spirit of Times* 27 Apr. 87/2 Then comes an ague from Canady, vich can't be cured by American Soothing Syrup nor Durham Mustard. **1861** *N.Y. Tribune* 26 Dec. 2/5 Don't fail to procure Mrs. Winslow's Soothing Syrup for children teething. **1896** [in Dict.]. **1901** W. Churchill *Crisis* II. ix. 193 Senator Bell was their candidate, and they proposed to give the Nation soothing-syrup. *Ibid.* xiv. 246 When the worst comes, the Soothing Syrup men will rally for the Union. **1902** G. H. Lorimer *Lett. Self-Made Merchant* xviii. 261 A lady..in a soothing-sirupy way asked if I would lend it to her. **1914** G. B. Shaw *Translations & Tomfooleries* (1926) 243, I really cannot earn two hundred and fifty guineas by playing soothing syrup to you. **1917** R. Fry *Let.* 20 July (1972) II. 415 My first soothing powder..was to have been a preface but got printed separately. **1926** E. Glyn *Love's Blindness* xvi. 183 If he dispelled..all alarm in Vanessa's imagination, it might possibly be only temporary soothing syrup. **1928** E. O'Neill *Strange Interlude* VIII. 305 Here are passion and hatred and regret and joy and pain and ecstasy, and these are men and women..whose blood is blood and not soothing syrup! **1945** *Richmond* (Va.) *Times-Dispatch* 26 Oct. 5/1 The race between Russia and the United States to obtain strategic bases is fooling no one but the American people—who are still being doped with soothing syrup that differences between ourselves and Russia are those of language and inexperience in foreign affairs. **1963** *Times* 21 Jan. 9/2 Mr. Macmillan and his colleagues must put away the soothing syrup that nothing is really as bad as it seems, that even if all is not quite for the best it automatically will be in the end. **1978** O. White *Silent Reach* xxvi. 268 That's a personal assurance, not official soothing syrup.

sooty, *a.* Add: **2. a.** (Later examples.)

1964 L. Deighton *Funeral in Berlin* xvii. 105 The girl..fluttered her big sooty eyes. **1976** 'A. Hall' *Kobra Manifesto* ii. 23 Black hair and a grey face and sooty bags under his eyes.

e. In the names of plant diseases, as *sooty blotch*, a fungal disease of apples, pears, and citruses which is caused by *Glœodes pomigena* and gives rise to darkish blotches on the skin of the fruit; *sooty mould*, any of several fungal diseases of trees and shrubs which cause a dark discoloration of their fruit.

1901 H. M. Ward *Disease in Plants* xxv. 232 [Honey-dew] serves as nutritive material for various epiphytic fungi—*e.g.* sooty mould, *Capnodium*, *Fumago*, and *Antennaria*. **1902** *Ann. Rep. Secretary Connecticut Board Agric.* 1901 132 Among the diseases in this class which prey upon either the fruit or the foliage of the apple..are the bitter rot..and the sooty blotch. **1939** Sooty blotch [see *fly-speck* s.v. *FLY *sb.*¹ 11]. **1939** *Ann. Bot.* III. 401 The distinction between parasitic and saprophytic 'sooty moulds'..appears to be valid. **1952** E. Ramsden tr. *Gram & Weber's Plant Diseases* ii. 126/2 Sooty mould can be avoided by keeping the tree free from aphides. *Ibid.* 127/1 Associated with *Leptothyrium pomi* is usually the fungus of sooty blotch, *Gloeodes pomigena*. **1969** G. N. Agrios *Plant Path.* ii. 19 Certain fungi, e.g., those causing sooty molds, can cause disease by growing on the surface of the plant and feeding on insect excretions rather than by parasitizing the plant.

5. *sooty-eyed*, *-mossed*.

1874 G. M. Hopkins *Jrnls. & Papers* (1959) 247 Sooty-mossed boulders in foreground. **1964** L. Deighton *Funeral in Berlin* xvii. 106 The sooty-eyed girl laughed.

sop, *v.* Add. **1. d.** With *up*: to soak up, absorb. Also *fig.*

1888 S. O. Addy *Gloss. Words Sheffield* 229 Come sop up that gravy. **1914** J. Galsworthy *Let.* in *Times* 28 Feb. 5/3 The admission or rejection of Tariff Reform, the Disestablishment or preservation of the Welsh Church, I would almost say than the granting or non-granting of Home Rule—questions that sop up *ad infinitum* the energies, the interest, the time of those we elect and pay to manage our business. *a* **1922** T. S. Eliot *Waste Land Drafts* (1971) 5 Blew in to the Opera Exchange, Sopped up some gin. **1951** D. Riesman *Individualism Reconsidered* in A. W. Loos *Religious Faith & World Culture* 73 The everpresent threat of war..used as a rationalization to sop up our 'excessive' comforts. **1962** S. Carpenter in *Into Orbit* 57 The nose [of the capsule] would sop up much of the friction we were running into and would become quite warm. **1973** J. G. Farrell *Siege of Krishnapur* ii. 25 The ladies discovered that while sitting in the boat the hems of their dresses had sopped up a certain amount of bilge water. **1977** A. Carter *Passion of New Eve* x. 158, I sopped up the sauce from the beans with a hunk of bread.

sopelalee, var. *SOAPOLALLIE.

sophianic (sōᵘfiæ·nik), *a. Theol. rare.* [f. SOPHI(A¹ + *-anic* as in MESSIANIC *a.*] Of or pertaining to wisdom.

1936 *Theology* XXXIII. 317 Karl Pfleger says of this Sophianic mysticism that it is 'extraordinarily profound'. **1970** R. Manheim tr. *Corbin's Creative Imagination Ṣūfism* 136 (*heading*) The sophianic poem of a *Fedele d'amore*. *Ibid.* 160 This sophianic intuition is perfectly in keeping with that of the extreme Shi'ites.

sophiology. Add: **1.** (Earlier example.)

1892 J. W. Powell in *Amer. Anthropologist* July 270 For the science of opinions I propose the name *Sophiology*.

2. *Theol.* The doctrine of the Divine Wisdom, as serving to explain the relations between God and the world.

1934 *Theology* XXVIII. 23 In his Christology the author [*sc.* Bulgakov] deliberately and openly relies on Sophiology, the doctrine of the eternal and created Wisdom. **1943** E. L. Mascall *He who Is* x. 135 The 'sophiology', or teaching concerning the Divine Wisdom, which looks back to the fourteenth-century mystic of Mount Athos, St Gregory Palamas, and which became prominent in Russian theology in the last century through ..Vladimir Solovyev. **1970** R. Manheim tr. *Corbin's Creative Imagination Ṣūfism* 98 From this idea of Creation as theophany..arises the idea of a sophiology, the figure of *Sophia aeterna*.

Hence **sophiolo·gical** *a.*; **sophio·logist**.

1933 *Theology* XXVI. 337 This has been related to modern categories of thought by the Russian sophiological school in Paris, especially by Professor S. Bulgakoff. **1937** *Ibid.* XXXV. 92 Such Sophiologists as Bulgakov, Berdyaev, and Solovice.

sophisticate (sǒfi·stikē¹t), *sb.* orig. *U.S.* [Back-formation from the vb.] One who is sophisticated or who has sophisticated tastes. Cf. *SOPHISTICATED *ppl. a.* 2 b.

1923 G. Atherton *Black Oxen* I. 1 All the Sophisticates (as Clavering had named them, abandoning 'Intellectuals' and 'Intelligentsia' to the Parlor Socialists) were present. **1930** H. S. Walpole *Rogue Herries* I. 163 Then Louis of France, making rude gestures, fingers at nose, that he may irritate, polished sophisticate that he is, the barbarian Stanislaus. **1936** 'J. Tey' *Shilling for Candles* xix. 208 Murder and that brittle insincere sophisticate were poles apart. **1942** *Scrutiny* X. 349, I think it is more than an accident that Copland, who started..as a sophisticate of the Big City, should in his mature work have come to express the loneliness..that lies back of all big cities. **1959** *Encounter* Sept. 52/1 For sophisticates, there is a touch of abnormal psychology. **1971** *Hi-Fi Sound* Feb. 25 (Advt.), The simple sophisticate. The Goldring-Lenco GL 69/2 transcription unit..with much more than 'the basics' for enthusiasts who don't require extreme sophistication of design. **1976** *UCT Stud. in English* (Univ. of Cape Town) Oct. 38 To the sophisticate it is a send-up of the genre in the vein of Chaucer's tale of Sir Thopas.

sophisticate, *v.* Add: **1. d.** *trans.* To render sophisticated (in senses in Suppl.; in quot. 1947 with weakened sense). Also *refl.*

1947 C. Morgan *Judge's Story* iv. 17 'But if you have reached the age of twenty-seven without ever having heard of Combined Metallurgical Industries, I am justified.'..'Now sophisticate me.' **1956** M. Stewart *Wildfire at Midnight* iii. 33 Three years of my great friend Nicholas ..would sophisticate a Vestal Virgin. **1978** J. I. M. Stewart *Full Term* xxii. 250, I..took to buying..all the paperbacks I could lay my hands on concerning espionage... It was a field that had sophisticated itself since the distant time when Patullo Minor, the Secret Service Boy, had enthralled his school-fellows with his hazardous escapades.

sophisticated, *ppl. a.* Add: **2. a.** Also, of a literary text: altered in the course of being copied or printed.

1948 *Studies in Bibliogr.* I. 112 This copy is only a sophisticated version of Stow. **1963** *N. & Q.* Mar. 101/1 We know..that F [of *King Lear*] is a sophisticated text, and it seems..possible that we have an example of sophistication here.

b. Of a person: free of naïvety, experienced, worldly-wise; subtle, discriminating, refined, cultured; aware of, versed in, the complexities of a subject or pursuit. Also *transf.* of a play, place, etc., that appeals to a sophisticated person.

Occas. (as in quot. 1952), *Biol.* and *Psychol.* used as opp. *NAIVE *a.* 2.

1895 Hardy *Jude* IV. v. 303 Though so sophisticated in many things she was such a child in others that this satisfied her. **1904** J. C. Lincoln *Cap'n Eri* xii. 230 The only scoffer was the bored Josiah, who, being a sophisticated New Yorker, sat in the best chair and gazed contemptuously upon the entire proceeding. **1915** *New Republic* 13 Feb. 51/2 It is one of those sophisticated melodramas in which a glamor is thrown about the underworld... The dope-fiend, the thief's mistress, the crooked detective, are all exhibited to an audience that apparently prides itself on being 'knowing'. **1933** H. S. Walpole *Vanessa* III. 531 Here in these pages was life, the life that so many polished sophisticated writers missed altogether. **1952** *Arch. Opthalmol.* XLVIII. 607 The sophisticated subject could always distinguish this illusion from the oculogravic illusion. **1954** *Word* X. 236 This conception has cropped up again and again. Even sophisticated thinkers have bent their ingenious efforts to preserving it. **1957** D. Robins *Noble One* vii. 71 She preferred smooth sophisticated young men like Keith who amused and flattered *her*. **1962** P. D. Strevens *Study of Present-Day Eng. Lang.* (1963) 23 The teaching of either language or literature in less educationally and linguistically sophisticated parts of the world. **1969** *Daily Tel.* 18 Oct. 11/5 Its nightclub-restaurant with an 'international' menu and Caribbean band is as sophisticated as you'd find anywhere. **1971** *Ibid.* 17 June 3/3 To the police he showed 'promise' of becoming a sophisticated criminal.

absol. **1952** G. Sarton *Hist. Sci.* I. xvi. 425 It is probable that pederasty was more common in Athens among the aristocrats, the idle rich, and the sophisticated than among the simpler people.

c. Of equipment, techniques, theories, etc.: employing advanced or refined methods or concepts; highly developed or complicated.

1945 C. S. Lewis *That Hideous Strength* xiv. 384 The man was so very allusive and used gesture so extensively that Mark's less sophisticated modes of communication were almost useless. **1952** G. Sarton *Hist. Sci.* I. xi. 289 He represents a second (or third) and more sophisticated stage in the evolution of Pythagorean astronomy. **1956** *N.Y. Times* 1 Apr. 19/1 Navy scientists are virtually exploring multidimensional space in a time machine in the search for what they call 'sophisticated' high-yield weapons. **1960** *Washington Post* 16 June 20/6 Soviet experts are said to have assisted the Peking regime with advanced nuclear reactors of a sophisticated type. **1966** *Times* 28 Mar. (Austral. Suppl.) p. v/4 Victoria now has many sophisticated industrial complexes. **1970** H. Braun *Parish Churches* xvii. 206 The High Gothic font was a sophisticated piece of furniture. **1970** *Daily Tel.* (Colour Suppl.) 28 Aug. 16/4 Laser beams..are useful to scientists as a sophisticated light-source. **1972** L. Alcock *By South Cadbury* viii. 182 The Breiddin had been re-fortified in the late fourth century with a sophisticated timber defence, in the form of a raised fighting platform and look-out towers. **1972** *Sci. Amer.* Sept. 53/2 One of the most sophisticated of all animal communication systems, the celebrated waggle dance of the honeybee. **1979** *Now!* 14 Sept. 78/1 When they raided the flat the police found two-way pocket radios, explosive substances and what were described as 'sophisticated' timing devices.

3. b. Of a printed book, containing alterations in content, binding, etc. which are intended to deceive.

1862 J. H. Burton *Book-Hunter* I. 25 His experience.. rendered him the most merciless detector of sophisticated books. Nothing, it might be supposed on first thought, can be a simpler or more easily recognized thing than a book genuine as printed. But in the old-book trade there are opportunities for the exercise of ingenuity. **1952** J. Carter *ABC for Book-Collectors* 168 Sophisticated..as applied to a book, is simply a polite synonym for doctored or faked-up.

4. *Comb.*, as *sophisticated-looking*.

1925 T. Dreiser *Amer. Tragedy* (1926) I. I. iv. 31 A brisk..and decidedly sophisticated-looking person.

sophi·sticatedly, *adv.* [f. prec. + *-LY².] In a sophisticated manner.

1956 A. Wilson *Anglo-Saxon Attitudes* I. i. 27 He smiled sophisticatedly to show his superiority to congresses. **1960** E. Davies *Beyond Old Bone Trail* i. 2, I was very shy and self-conscious, suffering from what I later found out to be generally and more sophisticatedly known as an inferiority complex. **1971** *Daily Tel.* 1 May 10/2 The work skated sophisticatedly between pastoral musing and a more swinging, jazzy style.

sophistication. Add: **2. a.** Also, the alteration of a literary text in the course of copying or printing.

1956 *Studies in Bibliogr.* VIII. 10 The paucity of '*em's* in the pages set by Compositor B represents the compositor's sophistication of copy. **1963** [see *SOPHISTICATED *ppl. a.* 2 a]. **1981** *Times Lit. Suppl.* 10 July 793/2 It [*sc.* the Folio] also makes numerous minor alterations, many of them literary sophistications.

c. Also, the quality or fact of being sophisticated (sense *2 b): (*a*) worldly wisdom or experience; subtlety, discrimination, refinement; (*b*) knowledge, expertise, in some technical subject.

1915 *New Republic* 16 Jan. 27/1 As to semi-education, the assumption is sound enough, and Dr. Burton's chapters on method and structure, on development and climax and ending, are honest first aids to sophistication. **1934** C. Lambert *Music Ho!* II. 112 In spite of his dazzling and outward sophistication Stravinsky is essentially primitive and naïve. **1951** R. Firth *Elem. Social Organization* v. 163 When we talk..of primitive Greek art.. we are referring..to art that is distinguished primarily by being earlier in time, though it..also bears the character of lack of sophistication. **1964** E. Bach *Introd. Transformational Gram.* vii. 145 The reader of the standard linguistic journals is apt to find articles..that demand considerable mathematical sophistication on his part. **1971** J. B. Carroll et al. *Word Freq. Bk.* p. xxi/1 Complete understanding of the lognormal model requires considerable mathematical sophistication. **1977** R. Williams *Marxism & Lit.* II. iv. 99 Mediation, in this range of use, then seems little more than a sophistication of reflection.

d. (*a*) The property or condition (of a thing) of being highly developed or complicated; technical refinement.

1959 *Time* 12 Oct. 67/3 In the past the usual comment was that Russian space vehicles are big and brawny because of more powerful launching rockets, but that U.S. space vehicles, small and elegant, made up for the Russians' gross size by their sophistication. **1972** L. Alcock *By South Cadbury* viii. 195 Elaborate arrangements to maintain the defences and their garrisons demonstrate the administrative sophistication of Late Saxon England. **1972** *Practical Motorist* Oct. 162/1 On more modern cars, sophistication is now so far advanced that the linkage would virtually require specialist attention!

(*b*) *concr.* An instance of this; a technically advanced characteristic.

1973 *Nature* 9 Nov. 109/2 The range stretches from relatively simple systems such as bacterial flagella and plant viruses..towards bacterial spores and the more complex sophistications of ribosomes, cell walls and mitochondria. **1976** *Early Music* Oct. 451/2 Instamatics cost..over £50 with built-in light meter and other sophistications.

sophomoric, *a.* Add: **b.** Also *ellipt.* as *sb.*

1946 Auden in *Harvard Alumni Bull.* 15 June 707/1 The sophomoric Who face the future's darkest hints With giggles or with prairie squints.

sophrosyne (sōᵘfrǫ·zini). Also **sophrosune**. [ad. Gr. σωφροσύνη prudence, moderation, f. σώφρων of sound mind, prudent.] Soundness of mind, moderation, prudence, self-control.

1889 *Cent. Dict.*, Sophrosyne. **1944** Auden *For Time Being* 89, I am that star most dreaded by the wise, For they are drawn against their will to me, Yet read in my procession through the skies The doom of orthodox sophrosyne [1945 *U.K. ed.* suphrosyne]. **1947** *Mind* LVI. 363 Lord Russell gives us to understand that he has no use for *sophrosyne*. *a* **1963** C. S. Lewis *Poems* (1964) 3 Thus with magistral hand the Puritan Sophrosune Cooled and schooled and tempered our uneasy motions. **1970** J. Gardner *Wreckage of Agathon* 153 Even when his ideas were crazy, the man had sophrosyne, as they used to call it in the old days.

sopie. Add: Also 9 **soopie, soopje, sopi, sopje, soupie, soupje, supje**. (Further examples.)

1790 E. Helme tr. *Le Vaillant's Trav. Afr.* I. v. 90 Those who enter a house are always presented with a *sopi*, that is to say, a glass of rack or gin, or rather of French brandy. **1812** A. Plumptre tr. *Lichtenstein's Trav. S. Afr.* I. II. xii. 167 Whatever Mr. Barrow may say of the *Soopje* as the favourite drink of the colonists, I can very safely affirm, that I never..saw three Africans born, in liquor. **1824** W. J. Burchell *Trav. S. Afr.* II. x. 287 Muchunka..was..stopped from drinking it all off at once as he had seen the others drink their *sopje* (sópy) or dram. **1827** G. Thompson *Trav. & Adventures S. Afr.* I. iii. 33, I alighted and partook of a cup of coffee or a dram (*soopie*) with the hospitable boors. **1835** C. L. Stretch *Jrnl.* 13 May in *Voorloper* (1976) 743 The glass or cup..was presented to the Chief who previously to giving it to the person intended sipped out some portion and as several soupies were given in this way Macamo naturally became quite overcome with the strength of the Brandy. **1849** E. E. Napier *Excursions Southern Afr.* I. vii. 115 A 'Totty', to this day, will share his last sixpence,..or his last 'soupje', with a comrade. **1861** in *Life at Cape* (1973) 37 We ordered a halt to rest a bit, to take a 'soopie', [etc.]. **1862** L. Duff Gordon *Let.* 2 Mar. in *Lett. from Cape* (1925) 131 Though he declined wine or Cape smoke 'soopjes' (drams) with aversion. **1876** F. Boyle *Savage Life* 277 After a while, from his lonely cottage by the ford, came to us the boer farmer in quest of supje (Mercian suppy), of raw spirits and a gossip. **1939** F. B. Young *City of Gold* I. iii. 95 'Come along, Peruvian,' he shouted. 'Just in time for a sopie.' **1981** A. Paton *Towards Mountain* xv. 117 Part of the remuneration of the ['Cape Coloured'] farm workers is the 'sopie', the draught of sour inferior wine that is given them three or four times a day... The sopie has been condemned by generation after generation of social workers, teachers, and ministers of religion.

soppily (sǫ·pili), *adv.* [f. SOPPY *a.* + -LY².] In a soppy or sentimental manner.
1977 *Listener* 21 Apr. 527/1 This soppily indulgent account of..shallow suburbanites. **1980** *N.Y. Times* 15 June VII. 14/1 Caroline thinks Ivan is arrogant, which he is; he thinks she is soppily emotional, which she is.

soppiness. Add: **2.** Mawkish sentiment, facile emotion.
1974 I. MURDOCH *Sacred & Profane Love Machine* 202 Harriet was..fearing tears, a kind of soppiness which might embarrass her dignified friend. **1978** *Times Lit. Suppl.* 1 Dec. 1392/4 Soppiness vies with winsomeness in an exhibition of these postcards at the Bethnal Green Museum.

soppy, *a.* Add: **5.** Full of mawkish sentiment; foolishly affectionate; inane, indulgent; occas. used affectionately. Also *to be soppy on*, to be infatuated with (a person). *colloq.*
1918 H. G. WELLS *Joan & Peter* xi. 369 What Joan knew surely to be lovely, Highmorton denounced as 'soppy'. 'Soppy' was a terrible word in boys' schools and girls' schools alike, a flail for all romance. **1920** H. G. HIBBERT *Playgoer's Mem.* xxxi. 257 The music halls were filled up with the precipitated baseness of pantomime—the puns, the 'unprincipalled' boy, the soppy-sentimental heroine. **1923** C. MACKENZIE *Parson's Progress* x. 121 Everyone will be singing for ever and ever and waving palms and playing harps and all that... I reckon Heaven's soppy, I do. **1929** H. WILLIAMSON *Beautiful Years* xx. 139 'Isn't fair, is it, man?' 'Hush, don't let 'em hear us. They'll think us soppy.' **1930** 'E. BRAMAH' *Little Flutter* xix. 218, I may as well make up my mind that I'm soppy on the blighter. **1935, 1959** [see *DATE *sb.*¹ 1 b]. **1961** *Daily Tel.* 2 Dec. 1/5 Lord Parker, Lord Chief Justice, said yesterday he deplored the tendency towards 'soppy and sentimental' treatment of children in juvenile courts. **1974** J. COOPER *Women & Super Women* 16 Being photographed for the *Tatler* with a soppy expression on her face. **1977** *New Yorker* 8 Aug. 11/1 Side benefits include a Chinese Legionnaire who sings soppy Irish ballads.

sopra bianco: see *BIANCO SOPRA BIANCO.

sopranino (soprani·no), *sb.* and *a.* [a. It. *sopranino*, dim. of SOPRANO *sb.* (and *a.*).] **A.** *sb.* An instrument (usu. wind) of higher pitch than a soprano (see SOPRANO 3 b). **B.** *adj.* Of or pertaining to such an instrument, as a recorder.
1907 T. S. WOTTON *Dict. Mus. Terms* 185 *Sopranino*, the diminutive of Soprano, a term applied to an instrument of higher pitch than that defined as soprano. **1938** *Oxf. Compan. Music* 785/2 The 'Flauto Piccolo' of Handel in *Acis and Galatea* is the little sopranino recorder, an octave above the treble one (or possibly the flageolet was used). *Ibid.* 832/1 *Sopranino Saxhorn in E Flat* (or *F*): also called *Soprano Saxhorn*; also often miscalled *Soprano Flügelhorn* or *Flügelhorn Piccolo*. **1939** A. CARSE *Musical Wind Instruments* xiii. 177 Each group [of saxophones] was in seven sizes... 1. Sopranino..2. Soprano [etc.]. **1954** [see *PARDESSUS 2]. **1964** S. MARCUSE *Mus. Instruments* (1966) 485/2 *Sopranino clarinet*, clarinet pitched between the ordinary C clarinet and the piccolo clarinet in high A♭. *Ibid.*, The sopranino in D was in use in 18th- and 19th-c. orchestras. **1968** *Observer* (Colour Suppl.) 19 May 40/5 His nice Irish wife..warms up the sopraninas (very small recorders) by cradling them inside her blouse. **1977** *Times* 27 May 16/4 A new type of violin, the 11½ in sopranino, with strings of such high tension that when they were first used in the United States, the player wore protective goggles in case they snapped.

soprano, *sb.* (and *a.*). Add: **1. b.** Also *fig.*
1848 GEO. ELIOT *Let.* 8 Mar. (1954) I. 253, I should have written a soprano to your Jubilate.
3. b. *spec.* **soprano saxophone**, the second highest member of the saxophone family, usually pitched in B flat. Abbrev. **soprano sax.**
1859 C. MANDEL *Treat. on Instrumentation Military Bands* 18 The Saxophone. This instrument is made in various keys, viz.:- 1. The B flat soprano-saxophone. **1961** J. A. MACGILLIVRAY in A. Baines *Mus. Instruments* x. 261 The *Tarogato* is a Hungarian instrument resembling a soprano saxophone of wood. **1971** *Guardian* 28 Sept. 10/6 The music is for..flute, trumpet, soprano sax, two basses, and percussion.
5. *ellipt.* = *soprano saxophone*, sense 3 b above.
1876 [see *BARITONE]. **1934** S. R. NELSON *All about Jazz* ii. 57 The other saxophones in common use in the band are the tenor, baritone, soprano and bass. **1967** *Sat. Rev.* (U.S.) 15 Apr. 55/2 Bechet, adjusting the reed of his soprano, looked at him.

sops-in-wine. Add: **1.** (Later examples in revived use.)
1918 'K. MANSFIELD' *Let.* 11 Mar. (1928) I. 152 There is a certain little white pink, striped with dark red, called 'sops-in-wine'. **1972** J. METCALF *Going down Slow* v. 102 What a garden there was at the back of this..inn... Sops in Wine and Floramer, Widow Wail..and deep red damask roses. **1981** T. MCLEAN *Medieval Eng. Gardens* v. 150 Its English name, gillyflower, may be a corruption of July flower, one of its medieval names, others of which were sops-in-wine, queen of delights and..carnation.

sor, sorr, repr. Ir. pronunc. of SIR *sb.* (sense 7 a).
1889 KIPLING *Life's Handicap* (1891) 15 Indirectly, sorr, you have rescued..the peasanthry av a numerous village. **1901** M. FRANKLIN *My Brilliant Career* xix. 162 'How are you enjoying yourself?' 'Treminjous intoirely, sor.' **1914** [see *PHWAT]. **1933** E. O'NEILL *Ah, Wilderness!* II. 66 No harm done. Only careful, Norah, careful... Yes, sorr. **1977** M. KENYON *Rapist* v. 54 A fine figure y'are so, if ye'll pardon the unsolicited estimation, sor.

soralium (sorē¹·liŏm). *Bot.* Pl. **-alia.** [mod.L., f. SORAL *a.* + L. *-ium*, neut. sb. ending.] A well-defined area of the thallus in which soredia occur, characteristic of certain lichens.
1921 A. L. SMITH *Lichens* iii. 144 In lichens of foliose and fruticose structure, and in a few crustaceous forms, the soredia are massed together into the compact bodies called soralia. **1938** G. M. SMITH *Cryptogamic Bot.* I. xiv. 518 They [*sc.* the soredia] may develop over the entire surface of a thallus or in localized pustule-like areas (soralia). **1970** *Sci. Jrnl.* Mar. 32/1 Many lichens have small powdery areas called soralia, the shape and location of which are characteristic for each species. The individual grains of powder in the soralia consist of an algal cell surrounded by fungal hyphae, and it is believed that they have a reproductive function. *Ibid.*, Lichens which have abundant fungal fruit bodies rarely form soralia, and vice versa.

sorb (sǭɹb), *v.* *Physical Chem.* [Back-formation from *SORPTION, after *absorb, absorption*.]
a. *trans.* To collect by sorption. Also *absol.*
1909 J. W. MCBAIN in *Phil. Mag.* XVIII. 918 An idea of the quality of the carbon employed may be obtained from the amount of gas sorbed by it in actual experiment. **1938** *Proc. R. Soc.* A. CLXVII. 407 The two zeolites in the form of three-dimensional networks sorb ammonia copiously without ammoniate formation. **1954** ALEXANDER & HUDSON *Wool* viii. 261 When wool is immersed in hydrogen peroxide, some is initially sorbed by the amino and imino groups without reaction. **1970** *New Scientist* 2 July 9/3 Papers with inked designs sorb best on the inked areas. **1972** *Physics Bull.* Oct. 583/1 This has the advantage that exhausted water vapour is not sorbed by the trap on the fine side of the pump.
b. *intr.* for *pass.*
1970 *New Scientist* 2 July 9/3 The SO_2 sorbs strongly to these sweat patches.
Hence **sorbed, so·rbing** *ppl. adjs.*; also **so·rbate²** [after *distillate, filtrate*, etc.], that which is sorbed.
1909 *Phil. Mag.* XVIII. 923 The total amount of sorbed gas is 67·70 c.c. **1921** *Jrnl. Chem. Soc.* CXIX. 454 Experimental results have always been obtained by shaking a certain volume of a solution of known strength with a known amount of sorbing material such as charcoal, and analysing a sample of the remaining solution. **1928** *Phil. Mag.* V. 749 A revised conception of the mutual relations of sorbent and sorbate in cases where the 'power' time-equation holds. **1946** *Nature* 5 Oct. 475/1 Compact, non-porous sorbing media such as wool. **1949** *Discussions Faraday Soc.* VII. 136 Gmelinite and chabazite occlude a still greater variety of sorbates. **1975** *Nature* 28 Aug. 719/1 The Mg ion of dehydrated offretite should have a strong electrostatic field around it, and sorbed molecules should be strongly attracted to form a complex.

sorbent (sǭ·ɹbĕnt), *sb.* (*a.*) *Physical Chem.* [f. *SORB *v.*, after *absorbent*.] A material having the property of collecting molecules of a substance by sorption; that which sorbs. Also (and orig.) as *adj.*, having this property.
1909 J. W. MCBAIN in *Phil. Mag.* XVIII. 918 My results show that it [*sc.* a specimen of carbon] is highly sorbent towards hydrogen. **1922** *Chem. Abstr.* XVI. 3017 The sorption isotherms were det[ermine]d..with animal charcoal as the sorbent. **1954** *Trans. Faraday Soc.* L. 981 Montmorillonite has already found various large scale uses as a sorbent. **1973** *Nature* 12 Jan. 92/2 The use of a zirconium oxide sorbent may be an economically realistic method of recovering boron from the sea. **1975** *Petroleum Rev.* XXIX. 239/1 In addition CGA maintains equipment to generate foam having a high sorbent capacity.

Sorbian, *a.* and *sb.* Add: (Examples referring to the language.)
1877 [see *LUSATIAN *sb.* and *a.*]. **1889** *Cent. Dict.*, *Sorbian*... *a.* Pertaining to the Sorbs or to their language. **1908** [see *LECHISH *sb.* and *a.*]. **1933, 1972** [see *LUSATIAN *sb.* and *a.*]. **1977** *Language* LIII. 479 Soviet linguists talk about two Sorbian languages (Upper and Lower Sorbian), Western scholars about two Sorbian dialects.

sorbitan (sǭ·ɹbitæn). *Chem.* [f. SORBIT(OL + AN(HYDRIDE.] Any of a number of cyclic ethers which are monoanhydrides of sorbitol; *spec.* the 1,4-anhydride, $CH_2OH{\cdot}CHOH{\cdot}\overline{CH{\cdot}(CHOH)_2{\cdot}CH_2O}$, a colourless crystalline solid. Freq. *attrib.* in names of fatty-acid esters of these compounds, which are used as emulsifiers and surfactants.
1938 *Industr. & Engin. Chem.* Nov. 1222/2 The inner ether hypothesis was further tested by substituting a mixture of preformed sorbitol inner ethers, or sorbitans, for the sorbitol. **1950** KIRK & OTHMER *Encycl. Chem. Technol.* V. 688 Anhydrides of certain hexahydric alcohols and their derivatives, as sorbitan and polyoxyethylene sorbitan esters of fatty acids, are used as emulsifiers and also possess emollient properties. **1958** *Martindale's Extra Pharmacopœia* (ed. 24) 696 The ether-esters, such as sorbitan oleate, are active water-in-oil emulsifying agents, while the polyoxyethylene derivatives mostly produce oil-in-water emulsions. **1969** tr. *Schönfeldt's Surface Active Ethylene Oxide Adducts* iv. 607 A composition containing glycine, phosphoric acid, and, e.g...ethoxylated sorbitan monolaurate as adduct, or sorbitan mono-oleate, is recommended as an antioxidant for fats. **1976** *Nature* 26 Aug. 777/2 By use of surfactants such as sorbitan stearates, water can be emulsified in hydrocarbon solvents and then readily supercooled.

sorbite¹. Substitute for def.: †**1.** A nitride and carbide of titanium found as red microscopic crystals in pig iron. *Obs.*
Quots. 1902, 1904 in Dict. belong with the next sense.
1888 H. M. HOWE in *Engin. & Mining Jrnl.* 18 Aug. 132 (*table*) Minerals which compose iron. Name suggested here..Sorbite. *Ibid.* 1 Sept. 177/1 Sorbite has been detected by Sorby in many cast-irons..as beautiful triangles, rhombs, hexagons and complex crosses. **1919** *Mineral. Mag.* XVIII. 376 It [*sc.* cochranite] is formed under the same conditions [as], and sometimes together with, the copper-red cubes of titanium cyano-nitride, $Ti(CN)_2{\cdot}3Ti_3N_2$. This was named sorbite.., a term afterwards withdrawn, as the same name was given..for one of the transition conditions in carbon-steel.
2. A constituent of steel consisting of microscopic granules of cementite in a ferrite matrix, produced esp. when hardened steel is tempered above about 450°C. [a. F. *sorbite* (F. Osmond 1895, in *Bull. de la Soc. d'Encouragement pour l'Industrie Nationale* X. 491).]
1900 *Metallographist* III. 196 The crystallites represent the solid solution of first solidification, from which, during cooling, plates of cementite separated first, then the eutectic, pearlyte or sorbite (the latter if the melting of the cementite was sufficiently complete). **1924** GREAVES & WRIGHTON *Pract. Micros. Metallogr.* vii. 57 Sorbite is the essential constituent of hardened and tempered steels intended for constructional purposes. **1964** H. HODGES *Artifacts* xix. 218 The effect of gently heating a quenched steel is that at low temperatures any martensite present forms troostite, while at higher temperatures..sorbite is produced. **1967** A. H. COTTRELL *Introd. Metallurgy* xx. 384 Above about 500°C the cementite particles grow competitively..into larger rounded particles dispersed through the B.C.C. iron matrix, giving a spheroidized structure (sorbite).
sorbitic *a.*² (earlier and later examples).
1902 *Jrnl. Iron & Steel Inst.* LXI. 140 Osmond said that probably in future all our steel rails will be made sorbitic. **1927** *Min. Proc. Inst. Civil Engineers* CCXXIV. 319 Photographs indicating the difference between sorbitic and ordinary steel. **1975** *Metals Abstr.* VIII. 1. 298/1 The production of reinforcing wire from grade 80 steel rod with a uniform sorbitic structure is described.

sorbitize (sǭ·ɹbitəiz), *v.* *Metallurgy.* [f. *SORBIT(E² + -IZE.] *trans.* To convert (steel) into a form containing sorbite. Hence **so·rbitized** *ppl. a.*, **so·rbitizing** *vbl. sb.*; **sorbitiza·tion**, the process of sorbitizing.
1918 D. K. BULLENS *Steel* (ed. 2) viii. 167 It required four quenchings to entirely sorbitize the steel. **1927** *Jrnl. Iron & Steel Inst.* CXVI. 582 Conditions of working which tend to keep the oxide content of the bath at a minimum favour the production of sorbitised rails free from hair cracks. **1928** C. J. ALLEN *Steel Highway* I. viii. 97 Another popular means of toughening the steel is by heat treatment, the process known as 'sorbitising' adding considerably to the wearing capacity with but a moderate addition to the cost of the rails. **1930** *Chem. Abstr.* XXIV. 3740 (*heading*) Method of 'sorbitization' of rails used at Nadezhdinsky iron and steel works. **1968** *Metals Abstr.* I. 1266/2 (*heading*) Sorbitizing of rolled wire from the rolling heat. **1975** *Ibid.* VIII. 298/1 (*heading*) Wire-rod sorbitized from its temperature at the end of rolling. **1975** *Chem. Abstr.* July–Aug. 215/1 (*heading*) Sorbitization of carbon steel wire rod during continuous rolling.

sorbitol. Add: It is a hexahydric alcohol, $CH_2OH{\cdot}(CHOH)_4{\cdot}CH_2OH$, found as a dextrorotatory isomer and crystallizing as colourless needles. (Further examples.)
1928 HAAS & HILL *Introd. Chem. Plant Products* (ed. 4) I. 70 Sorbitol occurs in the berries of *Pyrus aucuparia* and also in apple juice. **1955** *Sci. News Let.* 29 Jan. 70/1 A sugary substance called sorbitol..is used to 'entrap' the essential oils and esters that carry the citrus fruit flavor and fragrance. **1964** [see *hexahydric* adj. s.v. *HEXA-]. **1975** *Sci. Amer.* Dec. 80 In response to the raised level of glucose in the aqueous humour (as in the blood) that is characteristic of diabetes..the enzyme catalyzes the reduction of glucose to form the sugar alcohol sorbitol.

Sorbo (sǭ·rbo). Also **sorbo.** [Invented name: cf. ABSORB *v.*] The proprietary name of a make of sponge rubber. Usu. *attrib.*, esp. as *sorbo rubber.* Also *fig.*
1917 *Trade Marks Jrnl.* 15 Aug. 799 Sorbo..Sponge substitutes (India Rubber). Leeson Sponge and Rubber Co. Ltd. **1919** *Ibid.* 7 May 575 Sorbo Rubber-Sponge Products Ltd. **1926–7** *Army & Navy Stores Catal.* 104/2 Sorbo sponges—each 1/3. **1939** *Archit. Rev.* LXXXV. 251 (*caption*) The smaller sketch shows the large rounded sorbo strips fixed to door edges. **1940** *Amat. Radio Handbk.* (ed. 2) 134/1 Most microphones will tend to pick up floor vibrations if rested directly on the table. For this reason they should preferably be placed on a piece of sorbo rubber. **1941** C. KIRKUS *Let's go Climbing* ix. 132, I..put the sorbo mattress underneath my outer clothes, donned gloves and balaclava, and prepared to spend the night. **1941** 'R. WEST' *Black Lamb & Grey Falcon* I. 319 Astra..bounced from the platform like a

great sorbo ball to say good-bye. **1942** M. DICKENS *One Pair of Feet* viii. 159 Toots pursed her sorbo lips over the result. **1963** G. FREEMAN *Campaign* vii. 102 Boat-like red slippers with thick, white sorbo soles. **1976** *Lancet* 9 Oct. 801/2 The endangered areas, trochanters and sacrum, are kept free of pressure by placing the patient on a pillow or, better, sorborubber packs.

sorbose (sǭ·ɹbōᵘz, -s). *Chem.* [f. SORB *sb.*¹ + -OSE².] A ketohexose sugar obtained esp. from rowan berries as a fermentation product of sorbitol.
1889 *Jrnl. Chem. Soc.* LVI. 480 Sorbose ferments more slowly and less completely. **1913** HAAS & HILL *Introd. Chem. Plant Products* 63 Sorbose is a ketanic sugar produced by the fermentative oxidation of the alcohol sorbite contained in the sap of the mountain ash..; this sugar probably does not exist as such in the plant. **1962** *New Scientist* 5 Apr. 804/2 The fermentations already mentioned, with the less difficult fermentations producing ..sorbose (a key intermediate in the synthesis of vitamin C)..., more or less cover the extent of the British fermentation industry. **1970** *Ibid.* 5 Mar. 462/2 The possible non-nutritive sugars include sorbose (found in Mountain Ash berries), xylose (from oat husks), arabinose [etc.].

sorcerer. Add: *sorcerer's apprentice* [tr. F. *l'apprenti sorcier*, the title of a symphonic poem by Paul Dukas (1897), after *der zauberlehrling*, a ballad by Goethe (1797)], one who, like the apprentice in the ballad with his spells, instigates processes which he is unable to control. Also *attrib.*
1952 E. COXHEAD *Play Toward* iii. 24 Of course there was always a sorcerer's-apprentice element in teaching: but of all their creations, Lance was surely not the one who should be getting out of hand. **1966** J. AIKEN *Trouble with Product X* vi. 115, I wondered if his disciples, like the sorceror's [*sic*] apprentice, had not got out of control. **1967** *Sunday Times* 26 Feb. 2/3 The CIA is not a sorcerer's apprentice that has run wild, but..is under strict government control. **1974** HAWKEY & BINGHAM *Wild Card* xiii. 116 Our first priority should be learning to live with the technology we have already, not acquiring more. Because, like the Sorcerer's Apprentice, we just *ain't* going to be able to handle it.

sorcerize, *v.* For *rare*⁻¹ read *rare* and add later U.S. example.
1972 M. MEAD *Blackberry Winter* xv. 202 The threatening sorcerers of the plains..blackmailed our mild-mannered mountain people by promising temporarily not to sorcerize their relatives, providing they were given food and an adequate supply of trade goods.

sordid, *a.* Add: **II. 7.** Also with ppl. adj., as *sordid-seeming*.
1920 D. H. LAWRENCE *Lost Girl* xiv. 329 The dreary, to her sordid-seeming Campagna.
B. *sb.* *rare.* One who is sordid.
1959 C. MACINNES *Absolute Beginners* 13 All the old tax-payers know of this because, of course, for one thing, the poor old sordids recollect their own glorious teenage days. *Ibid.* 184 It doesn't seem possible such sordids as this lot could frighten you. **1960** N. MITFORD *Don't tell Alfred* ix. 97 My children regarded everybody over the age of thirty as old sordids, old weirdies, ruins, hardly human at all.

sordidity. Delete † *Obs.* and add: **a.** (Later examples.)
1917 V. WOOLF *Diary* 22 Nov. (1977) I. 80 Hearing women abuse each other & at the noise others come running with delight—all this sordidity made me think him rather likely to be right. **1923** *Times Lit. Suppl.* 23 Aug. 554/1 His [*sc.* Defoe's] very precise and solid account of licence, adventure, and sordidity [in Roxana]. **1964** A. SWINSON *Six Minutes to Sunset* viii. 150 Carson then switched his attack to the political manœuvring, condemning it for its duplicity and sordidity. **1978** *Times Lit. Suppl.* 1 Dec. 1382/2 Empson seems to insist that one cannot truly admire him without agreeing with him about the sordidity of Englit.
b. (Later examples.)
1927 V. WOOLF *Let.* 7 Feb. (1977) III. 326 Here it is pouring; rain coming through the roof; the sordidity too much for me even. **1936** S. SMITH *Novel on Yellow Paper* 152 The lovely deep rich olive green and brown and yellow sordidity of Sickert's London interiors. **1958** *Times Lit. Suppl.* 30 May 291/1 As the inevitable midnight flittings required less and less transport for the furniture and a large family grew up in an atmosphere of increasing poverty, sordidity and even occasional violence, James observed the domestic scene with the eye of an artist.

sordine, *sb.* and *a.* Add: **A.** *sb.* **2.** (Examples of the form *sordino*, pl. *sordini*.) Also, = DAMPER 2 a (see also quot. 1907). See also *CON SORDINO.
1883 GROVE *Dict. Mus.* III. 636/2 The musical terms 'Senza sordini' and 'Con sordini' applied to the damper-stops were used exclusively by Beethoven in his earlier sonatas. **1888** STAINER & BARRETT *Dict. Mus. Terms* (ed. 3) 406/2 *Sordini*... (3) Dampers of a pianoforte. **1894** G. DU MAURIER *Trilby* I. i. 42 Gecko..played..in minor, in pizzicato, and in sordino. **1907** T. S. WOTTON *Dict. Mus. Terms* 185 *Senza sordini* is..an indication open to misconstruction in piano music of a certain date, since it may mean 'without dampers' i.e. raising the dampers by means of the damper pedal, or it may mean 'without using the mutes'. **1959** *Collins Mus. Encycl.* 617/2 *Sordino*... (2) Damper (in the piano)... (3) Also applied in the late 18th and 19th cent. to a strip of leather (later felt) used to mute the strings of a piano and controlled by a pedal.

sordor. (Later examples.)
1929 *Oxford Poetry* 28 The world..rises and falls On a wave of confetti and funerals And sordor and stinks and stupid faces. **1934** T. S. ELIOT *After Strange Gods* i. 17 The sordor of the half-dead mill towns of southern New Hampshire and Massachusetts. **1982** *Observer* 10 Oct. 30/8 Budd, as a kid, knew the sordor behind the glamour.

sordume (sǭ·ɹdium). *rare.* [Alteration of *SORDUN.] = next.
1955 AUDEN *Shield of Achilles* iii. 76 There I stand in Eden again, welcomed back by the krumhorns, doppions, sordumes of jolly miners.

‖ **sordun** (zordū·n). Also **sordune.** [Ger.] An early form of bassoon, having a cylindrical bore with double reeds.
1876 STAINER & BARRETT *Dict. Mus. Terms* 406/2 *Sordun, Sordono* (It.) (1) An old form of wood wind instrument, having a double reed, with twelve ventages and two keys. **1943** B. MIALL tr. *K. Geiringer's Mus. Instruments* 133 An extremely rare variant of the Bassoon was the *Sordune*... This consisted of a gracefully turned wooden billet containing not merely two but sometimes three parallel cylindrical bores. **1977** *Early Music* Apr. (Advt., recto rear cover), The Sordunes which Praetorius ..illustrates in Syntagma Musicum 1619 show a controlled reed instrument of two parallel bores with a bocal tapering towards the end of its length. The Sordune is a useful bridge between the windcap range of instruments, and the controlled reed instruments. **1978** A. BAINES in J. M. Thomson *Future of Early Music in Britain* 22 Once people have spent money on manufactured sorduns and cornamuses, they want their money's worth from them.

sore, *a.* Add: **II. 9. b.** Also, *a sight for sore eyes*: see *SIGHT *sb.*¹ 1 d.
e. Colloq. phr. *dressed* (or *done*, etc.) *up like a sore finger* (or *toe*) and varr., overdressed. *Austral.* and *N.Z.*
1919 W. H. DOWNING *Digger Dialects* 46 *Sore finger*, an overdressed person (e.g. 'dolled up like a sore finger'). **1939** K. TENNANT *Foveaux* 430 You ought to a seen us in the ole days when we 'ad a procession every year—done up like a sore toe with banners and floats. **1943** J. A. W. BENNETT in *Amer. Speech* XVIII. 91 'All done up like a sore toe' describes someone dressed over-elaborately; many New Zealand children go barefoot much of the time, and it is with this circumstance in mind that we must interpret the simile. **1958** H. D. WILLIAMSON *Sunlit Plain* 10 Get an eyeful of him! Done up like a sore toe. **1965** P. WHITE *Four Plays* 168 I'm gunna get out of this suit. Dressed up like a sore finger.
f. Colloq. phr. *to stick* (or *stand*) *out like a sore thumb*, to be very conspicuous or obvious.
1936 E. S. GARDNER *Case of Sleepwalker's Niece* xiii. 128 'No,' he said, 'that's the one thing in the case that stands out like a sore thumb, now that I stop to think of it.' **1941** —— *Case of Haunted Husband* (1942) xvi. 126 A private detective in that atmosphere would stick out like a sore thumb on a waiter serving soup. **1958** *Spectator* 8 Aug. 187/1 A bad officer will stick out like a sore thumb. **1977** *New Yorker* 15 Aug. 42/3 In the strong late-afternoon light the twelve white houses stood out like twelve sore thumbs.
12. a. Also, angry, resentful. Also const. *about*, *on*, and *at*. Now colloq. (chiefly *N. Amer.*).
1852 DICKENS in *Househ. Words* V. 307/1 The people were very sore about the French marriage. **1884** G. C. DAVIES *Peter Penniless* xxxii. 230 Everybody was greatly amused at the incident, except Quadling, who was sore about it for a long time. **1904** *N.Y. Even. Post* 13 June 1 Kelly denied the charges and said the patrolman was 'sore' on him. **1923** R. D. PAINE *Comr. Rolling Ocean* xiv. 252 All hands were sore on him, but he couldn't take a hint. **1927** [see *HIGH-HAT *v.*]. **1932** WODEHOUSE *Hot Water* xvi. 257 But surely you aren't going to get sore at a little thing like that? **1946** *Sunday Dispatch* 8 Sept. 2/7 They were sore about the decision that had deprived them of complete victory. **1954** [see *NAH¹]. **1975** D. LODGE *Changing Places* iii. 120 Nobody believed him of course, and this made him sore as hell. **1980** *Amer. N. & Q.* Jan. 71/1 Jonson is likely to have been sore about Shakespeare..styling himself gentle.
c. (Earlier example.)
1803 C. WILMOT *Jrnl.* 6 Mar. in T. U. Sadleir *Irish Peer on Continent* (1924) 170 This however is a sore subject, as..there is scarcely any one that one sees who is not a living victim.
13. *sore-footed*, *-hearted* (later examples; hence *sore-heartedness*), *-rimmed*; **sore-back** *attrib.*, (of horses) having a sore back; so **sore-back** *v.* to give (a horse) a sore back, **sore-backed** *a.*
1835 J. E. ALEXANDER *Sketches in Portugal* x. 224 It [*sc.* a hunting-saddle] sore-backs strange horses, is hard and smooth to the rider, and one can't carry any thing on it in the shape of arms or baggage. **1923** in 'Mark Twain' *Speeches* 9 They have always got a sore-back horse lying around somewhere to sell to the stranger. **1901** KIPLING *Five Nations* (1903) 163 A top of a sore-backed Argentine, with a thirst that you couldn't buy. **1933** J. V. ALLEN *Cowboy Lore* iv. 131 But he went to see the gals on a sore-backed hoss. **1814** JANE AUSTEN *Mansfield Park* II. x. 234 She had only to rise and..pass quietly away..sore-footed and fatigued. **1927** T. S. ELIOT *Journey of Magi*, And the camels galled, sore-footed, refractory. **1884** D. BOUCICAULT *Shaughraun* III. i. 20/1 Blessings on your path; it always leads to the poor and to the sore-hearted! **1923** D. H. LAWRENCE *Birds, Beasts & Flowers* 55 Open ..And red at the core with the last sore-heartedness, Sore-hearted-looking. **1915** —— *Rainbow* xiii. 371 He lifted his face, the sore-rimmed eyes half smiling.

sore, *v.*¹ Add: **b.** With *up*. To annoy. *colloq.* (orig. *U.S.*). *rare.*
1929 D. RUNYON in *Hearst's Internat.* July 56/1 It is a sure thing he will get sored up at the second peek. **1963** 'R. EAST' *Pin Men* vi. 162 He sored me up once for all and I left him flat.

sore-head, *a.* and *sb.* Add: Also **sorehead.** **B.** *sb.* **b.** *slang* (chiefly *N. Amer.*). A discontented, dissatisfied person, a malcontent; a mean, niggardly person.
1848 *Weekly Argus* (Albany, N.Y.) 12 Aug. 253/3 As no other selection could be supposed so well to represent such a conventicle of 'sore heads', it is perhaps quite as well it sho'd take that direction as any other. **1912** J. SANDILANDS *Western Canad. Dict. & Phrase-bk.*, *Sore-head*, a person who sees trouble and wickedness in everything. **1916** C. J. DENNIS *Songs of Sentimental Bloke* 130 *Sore-head*, a curmudgeon. **1934** J. T. FARRELL *Young Manhood of Studs Lonigan* iv. 61 'You damn Kike, you got too many horseshoes,' a sore-head said as Davey raked in the pot. **1939** T. WOLFE *Web & Rock* ii. 36 We thought he was a man, but he turns out to be just a little sore-head. **1964** L. NKOSI *Rhythm of Violence* II. i. 23 *Mary*..You two are not to drink until the others arrive! *Jimmy* Rubbish! We have priority claims! *Mary*.. Soreheads! What makes you think you have priorities over others! **1978** M. PUZO *Fools Die* xx. 226, I was holding court with a lot of my customers, who were all telling me what a bunch of shit the whole business was, caused by a few soreheads.
sore-headedness (earlier example).
1860 *Marysville* (Calif.) *Appeal* 31 Mar. 2/2 The patriots of the Customs House [are] suffering from the sore-headedness which so often follows an unsuccessful attempt at ascendency in the political scale.

Sorelian (sŏre·liăn, -ī·liăn), *a.* [f. the name of Georges *Sorel* (1847–1922), French political philosopher + -IAN.] Of, pertaining to, or characteristic of Sorel or his views on the regeneration of society through proletarian or syndicalist violence.
1921 N. ANGELL *Fruits of Victory* v. 165 The Sorelian philosophy of violence and instinctive pugnacity..gives us the tendency to an infinite splitting of the Labour movement. **1931** R. SOLTAU *French Pol. Thought in Nineteenth Cent.* xiv. 460 Sorelian philosophy can be examined from two aspects. **1936** WIRTH & SHILS tr. *Mannheim's Ideology & Utopia* iii. 125 This attitude takes many forms—appearing first in the anarchism of Bakunin and Proudhon, then in the Sorelien [*sic*] syndicalism, and finally in the fascism of Mussolini. **1970** H. ARENDT *On Violence* iii. 73 The strange revival of the life philosophies of Bergson and Nietzsche in their Sorelian version. **1979** *Dædalus* Winter 19 Pragmatic proposals have replaced the Sorelian myth.

sorely, *adv.* Add: **5.** *sorely-battered*, *-needed*, *-sweated*.
1900 W. S. CHURCHILL in *Morning Post* 1 Jan. 6/1 The engine was soon crowded and began to steam homewards —a mournful, sorely-battered locomotive. **1917** —— in M. Gilbert *Winston Churchill* (1977) IV. Compan. I. 87 A mere bluff designed to induce him to dissipate sorely-needed forces on coastal defence. **1952** R. CAMPBELL tr. *Baudelaire's Poems* 129 Who come to waste their sorely-sweated pittance.

Sörensen (sö·rənsən). *Chem.* Also **Sorensen.** The name of Søren P. L. *Sørensen* (1868–1939), Danish biochemist, used *attrib.* and in the possessive to denote a titration method employed in the estimation of amino-acids, consisting in treating the sample with formaldehyde, which combines with the amino groups, and then titrating the carboxylic acid groups against base.
Sørensen first described the method in 1907 (*Compt.-Rend. des Travaux du Lab. de Carlsberg* VII. 1).
1914 J. A. MANDEL tr. *Hammarsten's Text-bk. Physiol. Chem.* (ed. 7) ii. 166 On this behaviour is based Sörensen's formoltitration which serves for the estimation of amino-acids in the urine. **1916** A. P. MATHEWS *Physiol. Chem.* ix. 363 The capacity of the digestive products of combining with formaldehyde steadily increases, as is shown by the Sorensen titration. **1934** W. R. FEARON *Introd. Biochem.* vii. 100 Sörensen's method of formaldehyde titration..is applicable to many amino compounds, including ammonium salts in urine. **1973** BIGGS & WOODSON *Clin. Biochem.* iii. 40 The Sörensen formol titration technique can be used..to obtain accurate titration values.

Soret (so·re). *Chem.* [The name of Jacques-Louis *Soret* (1827–90), Swiss physicist, who first detected the absorption in a study of blood (*Bibliothèque univ. et Rev. suisse: Arch. des Sci. phys. et nat.* (1878) LXI. 347).] *Soret* (also †*Soret's*) *band*, a characteristic intense band at a wavelength of approximately 400 nanometres which occurs in the ultraviolet absorption spectra of porphyrins and their derivatives.
1899 W. D. HALLIBURTON *Essent. Chem. Physiol.* (ed. 3) 148 Oxyhæmoglobin shows a band (Soret's band) between the lines G and H. **1959** *Lancet* 2 May 912/2 The

most effective wavelength would be expected to lie in the region of the Soret band, 390–410 mμ, where light absorption is maximal. **1977** *Jrnl. Amer. Chem. Soc.* XCIX. 4191/1 The wavelength of irradiation corresponded to a region of intense Soret band absorption (in tetraarylporphine dications) or to the region of near minimal.. absorbance on the long wavelength side of the Soret band (in other species).

‖ **soroban** (sǫ·roban). [Jap., f. Chinese *suàn-pán* SWAN-PAN.] A kind of abacus used in Japan, adapted from the swan-pan.
1891 A. M. BAKER *Japanese Girls & Women* x. 266 Crowds of clerks sitting upon the matted floors, each with his *soroban*, or adding machine, by his side. **1903** W. DEL MAR *Around World through Japan* xiii. 133 The addition of seven and six presented difficulties unless he had an abacus (*soroban*). **1958** HOSOI Sō in *Japan* xxii. 612/1 In the beginning of the Edo Era, the *soroban* (abacus) was used by Japanese merchants and engineers for their calculations. The *soroban* was introduced from China. **1965** *Australian* 23 Nov. 12 The appearance of a low-cost, small-type, locally-made electronic computer..has driven a wedge under the four-century reign of the soroban.

‖ **soroche** (sorōᵘ·tʃe). [Sp., ad. Quechuan *suruči*, name of some mineral to which mountain sickness was attributed, and hence 'mountain sickness'.] A name in the Andes for mountain sickness.
1878 I. L. BIRD *Lady's Life in Rocky Mountains* (1879) iv. 48, I feel a singular lassitude... This is said to be the milder form of the affection known on higher altitudes as *soroche*, or 'mountain sickness'. **1891** E. B. CLARK *Twelve Months in Peru* 104 A headache with a weighty feeling on the brow, vomiting, and breathlessness are the usual symptoms of *sorroche*. **1922** *Glasgow Herald* 4 Jan. 4 In order to minimise the effects of soroche, or mountain sickness, on persons suffering from weak heart the company provide cars with compartments equipped with oxygen. **1963** *Times* 7 June 13/6 Faced with another presidential election this Sunday, the Peruvian people have apparently succumbed to some political equivalent of the *soroche*, the mountain sickness that shortens the breath, softens the will, and enfeebles the body. **1970** *Sci. Amer.* Feb. 56/1 Even mountain natives sometimes lose their acclimatization to high altitude and incur *soroche* (chronic mountain sickness), which is characterized by extreme elevation of the relative number and mass of red cells in the blood..and ultimately congestive heart failure if the victim remains at high altitude. **1981** L. LEAMER *Assignment* ix. 129 It's great to breathe some real air again... No more *soroche*.

Soroptimist (sŏrǫ·ptimist), *a.* and *sb.* [f. L. *sor-or* sister + OPTIMIST *sb.* (*a.*) (prob. after the *Optimist* Club, founded in 1911).] **A.** *adj.* Chiefly in *Soroptimist Club*, an international service club for professional and business women, founded in California in 1921. Also *Soroptimist International*. **B.** *sb.* A member of a Soroptimist club.
1921 *Charter* (Soroptimist Club, Oakland, California), Whereas, the persons who [*sic*] names appear at the foot hereof..having enrolled themselves as Charter Members of the Soroptimist Club (Oakland), [etc.]. **1924** *Glasgow Herald* 8 Mar. 6 The woman publisher who represents her profession in the Soroptimist Club (i.e., women Rotarians). **1924** *British Weekly* 14 Aug. 431/4, I know now what a Soroptimist is. **1930** M. BRADBURY *Cook Bk.* 2 The Soroptimist Club Code of Ethics... As a Soroptimist I pledge my untiring efforts to understand life, my true and right relation to sister Soroptimists and humanity at large, [etc.]. **1955** L. P. HARTLEY *Perfect Woman* iv. 39 Would it be easy, with possibly some Tablers and Soroptimists listening in? **1977** *Times* 6 Apr. 18/7 Among her many activities, Marina devoted a great deal of time to the Soroptimist International organisation. She was president of the National Union of Soroptimist Clubs (Italy) from 1969–1971, when their international golden jubilee was celebrated in Rome.

sororal, *a.* Add: **3. b.** *sororal polygyny* (Anthrop.), in some polygamous kinship systems, the custom whereby the first wife's sister(s) are preferred as secondary wives.
1952 A. R. RADCLIFFE-BROWN *Struct. & Funct. in Primitive Soc.* iii. 67 Sororal polygyny (marriage with two or more sisters). **1971** P. B. HAMMOND *Introd. Cultural & Soc. Anthrop.* vi. 159/2 Sometimes the desirability of a man's marrying his wife's sister is stressed, a usage referred to as sororal polygyny. **1977** C. R. & M. EMBER *Cultural Anthrop.* (ed. 2) viii. 154/2 Jealousy.. seems to be lessened if one man is married to two or more sisters (sororal polygyny). *Ibid.*, The Crow Indians practice sororal polygyny, and co-wives usually share the same tepee.

sororate (sorōᵊ·rĕt). *Anthrop.* [f. L. *soror* sister + -ATE¹, after LEVIRATE.] In some kinship systems a custom whereby, on the death of his wife, a man is expected to marry her (unmarried) sister; also occas. = *sororal polygyny* s.v. *SORORAL *a.* 3 b; also *attrib.* Hence **sorora·tic** *a.*, characterized by such a custom.
1910 J. G. FRAZER *Totemism & Exogamy* IV. 140 The other [custom] is the rule which allows or requires a man to marry the younger sisters either of his living or of his deceased wife... The latter custom..has no distinctive name, but on analogy I propose to call it the *sororate*. **1921** E. WESTERMARCK *Hist. Human Marriage* (ed. 5) xxxi. 263 The sororate, like the levirate, can be..interpreted as the outcome of existing conditions. **1947** CHAPPLE & COON *Princ. Anthrop.* xiii. 311 The sororate occurs when a man marries two or more sisters. **1952** M. N. SRINIVAS *Relig. & Society among Coorgs* v. 148 As sororatic unions are preferred among the Coorgs, a mother's younger sister steps into the mother's shoes in the event of the mother's death. **1963** W. N. STEPHENS *Family in Cross-Cultural Perspective* i. 27 In many primitive societies broken homes are automatically 'mended' by means of the sororate and levirate. **1970** *Internat. & Compar. Law Q.* XIX. 1. 139 The levirate and sororate institutions could perpetuate the inter-kin tie.

sorority. **2.** (Further *attrib.* examples.)
1911 *Daily Colonist* (Victoria, B.C.) 1 Apr. 2/5 All of the girls lived at a sorority house, 4522 Eighteenth avenue, Northeast. **1948** A. N. KEITH *Three came Home* xiii. 229, I had my sorority pin, which had been specially set in my college days with a diamond. **1977** [see *RAT *sb.*¹ 2 e].

sorosilicate (sorosi·likĕ·t). *Min.* [a. F. *sorosilicate* (V. Billiet 1945, in *Bull. de la Soc. belge de Géol.* LIII. 182), f. Gr. σωρός heap: see SILICATE.] Any of the group of silicates characterized by isolated pairs of SiO^4 tetrahedra that share an oxygen atom at a common apex.
1947 [see *PHYLLOSILICATE]. **1960** *Amer. Mineralogist* XLV. 1 Perrierite is a sorosilicate, with a high number of O-atoms not bonded to silicon. **1972** R. W. FAIRBRIDGE *Encycl. Geochem. & Environmental Sci.* 101/1 In the sorosilicates (two isolated tetrahedra which share one oxygen atom) calcium is present mainly in the epidote group.

sorosis. **2.** (Examples.)
1869 GEO. ELIOT *Jrnl.* 16 Feb. in *Lett.* (1956) V. 14 The Ladies of the 'Sorosis' at New York..proposed to make me an honorary member of their society—I declined. **1902** *Out West* May 557 The founding of the first woman's club, Sorosis of New York, was almost simultaneous with the union of the Atlantic and Pacific by the completion of the first transcontinental railway in 1869. **1942** C. MORLEY *Thorofare* xxiii. 98 I've promised to read a paper on my trip abroad to our literary sorosis.

sorption (sǭ·ɹpʃən). *Physical Chem.* [Extracted from *absorption* and *adsorption*.] The combined or undifferentiated action of absorption and adsorption. Freq. *attrib.*
1909 J. W. MCBAIN in *Phil. Mag.* XVIII. 916 The non-committal name 'sorption' may be coined to designate the sum of the phenomena, while 'absorption' and 'adsorption' should be restricted to proven cases of the solution or surface condensation respectively. **1932** *Trans. Faraday Soc.* XXVIII. 182 The rate of sorption at constant temperature was very nearly directly proportional to the pressure. **1945** *Electronic Engin.* XVII. 325/1 M. Francis—dealing mainly with the sorption effects of the walls—gives a detailed account of the use of the McLeod gauge with non-permanent gases. **1954** *Thorpe's Dict. Appl. Chem.* (ed. 4) XI. 33/1 In its practical aspects, the term 'sorption' is usually applied to the concentration at solid surfaces of gases and vapours, and of dissolved substances and colloids, and is therefore largely synonymous with adsorption. **1969** *McGraw-Hill Yearbk. Sci. & Technol.* 342/1 Sorption pumping, a process for producing a vacuum in a closed system by the capture and holding of gases by certain solids or liquids, has recently been developed into a practical technique. **1971** [see *ROUGH *v.*¹ 7 d]. **1973** *Nature* 12 Jan. 92/1 New work on the sorption of boron compounds may provide a way of isolating the element from seawater. **1978** *Ibid.* 16 Mar. 230/1 It is important in the field of forest products science to know the sorption mechanism of water in wood.
Hence **so·rptive** *a.*, of, pertaining to, or exhibiting sorption.
1921 *Jrnl. Chem. Soc.* CXIX. 928 After each heating, the density was determined and then the sorptive capacity at each of the three temperatures. **1938** *Proc. R. Soc.* A. CLXVII. 392 The paper describes an attempt to relate the sorptive properties [of zeolites] towards permanent gases..and towards ammonia..to the structure and properties revealed by the X-ray method.

sorr: see *SOR.

sorrel, *sb.*¹ Add: **7. b.** *sorrel soup* (earlier example).
1797 J. WOODFORDE *Diary* 20 Apr. (1931) V. 28 We had for Dinner to day, some Haddocks..Sorrell Soup, a boiled Tongue & Veal Cutlets.

sorrel, *a.* and *sb.*² Add: **A.** *adj.* **d.** *Comb.*, as *sorrel-coloured* adj.; **sorrel-top** *colloq.* (orig. *U.S.*), a red-haired person.
1887 W. B. YEATS *Let.* 13 Aug. in *Lett. W. B. Yeats to Katherine Tynan* (1953) 37, I enclose these trivial verses... The Fairy Doctor. The fairy doctor comes our way Over the sorrel-coloured wold. **1863** 'E. KIRKE' *My Southern Friends* iv. 58 'Har, you lousy sorrel-top,' said the trader to the red-faced and red-headed bar tender. **1904** 'O. HENRY' in *Everybody's Mag.* Feb. 187/1, I guess they don't raise 74-inch sorrel-tops with romping ways down in his precinct. **1918** G. FRANKAU *One of Them* xix. 145 Once more released to lavish wealth and name On head or blonde or sorrel-top or raven.

sorrow, *sb.* Add: **8. sorrow song**, a lament; *spec.* a song expressing the sorrows of the American Black people.
1903 W. E. B. DU BOIS *Souls of Black Folk* xiv. 250 (*heading*) The sorrow songs. *Ibid.*, They that walked in darkness sang songs in the olden days—Sorrow Songs—for they were weary at heart. **1936** A. LOCKE *Negro & his Music* iii. 25 These 'sorrow songs' are more than a priceless heritage from the racial past, they are promising material for the Negro music of the future. **1943** J. COLERIDGE-TAYLOR *Memory Sketch* v. 35 On her return from a South African tour, she [*sc.* Ada Crasby] gave a recital,..asking my husband to accompany her in his 'Six Sorrow Songs'. **1962** R. E. POOL *Beyond Blues* 18 Marcus Garvey headed a movement of a 'back to Africa' Zionism which was a symptom of race consciousness and of the Negro's awakening discovery of the land of long, long ago: 'All the way from Africa to Georgia, I carried my sorrow songs.'

sorry (sǫ·ri, sɒ·ri), *sb.* *Midlands* and *north dial.* Also **sorrey.** [Var. SIRRAH: cf. SIRREE.] A term of address (now expressing familiarity) for a man or boy.
a **1796** S. PEGGE *Derbycisms* (1896) 65 *Sorry*,..sirrah; in speaking to a boy or lad. **1913** D. H. LAWRENCE *Sons & Lovers* ii. 31 'Shall ter finish, Sorry?' cried Barker, his fellow butty. **1965** BROPHY & PARTRIDGE *Long Trail* 183 *Sorry*, mate, pal, chum. Usually in vocative and chiefly among Yorkshire and Lancashire troops. **1977** SCOLLINS & TITFORD *Ey up, mi Duck!* II. 56 *Sorrey*, the local version of the traditional term 'sirrah'... Nowadays, a term of familiarity, as in: 'Eh up, *sorrey*! Aah's it gooin?' Towards Nottingham the pronunciation sometimes approximates more to 'Surrey'.

sorry, *a.* Add: **7.** *sorry-looking* (earlier example); **sorry-go-round** [after MERRY-GO-ROUND: cf. †MERRY-GO-SORRY], a depressing cycle of events.
1872 'MARK TWAIN' *Roughing It* v. 48 The cayote is a..sorry-looking skeleton. **1959** V. NABOKOV *Nabokov's Dozen* 1 The blurred Mount St. George..on the picture postcards which since 1910..have been courting the tourist from the sorry-go-round of their prop, among.. lumps of rock and..sea shells. **1964** *Punch* 29 Apr. 630/1 It was time to stop the 'sorry-go-round' of inflation.

sort, *sb.*² Add: **I. 4. b.** Also without marked disparagement: of some (untypical or unusual) kind, not having the usual characteristics, equipment, facilities, etc.
1946 D. GWYNN *Bishop Challoner* x. 155 In the Midlands the Franciscans had a school of sorts at Edgbaston. **1959** N. MARSH *False Scent* (1960) i. 21 There's a party of sorts at half-past which I hope may amuse you. **1972** *Times Lit. Suppl.* 4 Aug. 909/4 (Advt.), He is a poetic eye, a visionary of sorts. **1973** *Times* 18 May 22/7 During the Second World War he was a soldier of sorts (he even rose from the ranks).
8. c. (*b*) Hence passing into use as a parenthetic qualifier expressing hesitation, diffidence, or the like, on the speaker's part; also (only in the full form *sort of*) following the statement it qualifies.
1903 G. B. SHAW *Man & Superman* II. 67 I'll sort of borrow the money from my dad until I get on my own feet. **1930** A. BENNETT *Imperial Palace* lxiii. 509, I don't believe they sort of understand English people, Italians don't. **1949** *Granta* Christmas 43/2 One of us had to do a big strong man to sort of separate them. **1958** [see *BUGGER *v.* 2 c]. **1973** *Art Internat.* Mar. 68/1, I sort of use music as a connection to more of the things I want to be about. **1976** *National Observer* (U.S.) 25 Sept. 17/1 He calls it the 'Icarus Human-Powered Aircraft.' 'It's sort of a cumbersome name.'
1923 J. MANCHON *Le Slang* 283 They hung back in their breeching sort-of, ils s'appuyaient sur l'avaloir, si je puis dire. **1952** M. LASKI *Village* xv. 206 It just happened, sort of, and we couldn't either of us help it. **1959** *Psychiatry* XXII. 293/1 Except I feel like, well, what you're doing anyway is just sitting here and saying all these things just to tease me and to taunt me, sort of. **1976** *National Observer* (U.S.) 28 Feb. 21/4 And it is all those things, sort of; and yet it is a really fine book.
10. b. *sort of thing*, used adverbially to indicate the inexactness or indefiniteness of the preceding words. Cf. sense 8 c (*b*) above. *colloq.*
1935 E. RAYMOND *We, the Accused* v. ii. 572 What he doesn't know about the law isn't worth knowing, sort of thing. **1968** *Guardian* 24 Apr. 9/2, I don't just give him a sharp slap in temper, sort of thing. **1979** A. FOX *Threat Warning Red* xi. 161 A dummy run. Only go through the motions, sort of thing.
11. b. Also (in phr. *the right sort*) with ref. to one person.
c **1863** T. TAYLOR *Ticket-of-Leave Man* III. 43 But don't look glum, Bob, you're the right sort, you are. **1914** G. B. SHAW *Fanny's First Play* I. 173 But hes the right sort: I can see that. **1936** [see *RIGHT *a.* 8 d].
c. (Earlier examples.)
c **1869** TAYLOR & DUBOURG *New Men & Old Acres* I. 10 Fanny Bunter—in spite of her Ruskinism-run-mad—isn't half a bad sort. **1875** W. READE *Outcast* 202 Our host told us the old woman was his mother, and we musn't [*sic*] mind her being cross, she being a real good sort all the same.
d. Proverb. *it takes all sorts to make a world*; also ellipt. *it takes all sorts*.
[**1620** T. SHELTON tr. *Cervantes' Hist. Don Quixote* II. vi. 34 In the world there must bee of all sorts. **1767** JOHNSON *Let.* 17 Nov. (1952) I. 194 The World, says Locke, has people of all sorts.] **1844** D. W. JERROLD *Story of Feather* xxviii. 161 Click can't get off this time?..Well, it takes all sorts to make a world. **1908** K. GRAHAME *Wind in Willows* iv. 89 The Wild Wood is pretty well populated.. with all the usual lot, good, bad, and indifferent... It takes all sorts to make a world. **1940** [see *COIN *v.*¹ 5 d].

1951 E. COXHEAD *One Green Bottle* i. 35 'I daresay it takes all sorts—' conceded Harry vaguely. **1965** J. FLEMING *Nothing is Number when you Die* II. iii. 68 She shrugged. 'It takes all sorts, you know.' **1975** J. I. M. STEWART *Young Pattullo* iii. 71 'My father's a banker during the week and a country gent at week-ends. Takes all sorts you know.' 'Takes all sorts?' 'To make a world.'
e. A girl or young woman; a girl-friend. (Predominantly in male use.) *slang* (orig. *Austral.*).
1933 F. CLUNE *Try Anything Once* 93 'Look here, George,' I said. 'Lend me a suit of civvies. I've got to meet a great little sort, and her father has a dead nark on soldiers.' **1953** T. A. G. HUNGERFORD *Riverslake* 144 Felix came in after tea and said that his sort could come. **1968** K. DENTON *Walk around my Cluttered Mind* 137 They'd told me, 'Don't worry about bringing anything except a bottle. The sorts are laid on.' Even after only ten months I understood this to mean that there would be feminine company. **1970** *Daily Progress* (Charlottesville, Va.) 7-c/3 He [*sc.* a skinhead in the U.K.] wants only to drink, go out with 'sorts', another word for girls, perhaps take pep pills or marijuana. **1972** A. DRAPER *Death Penalty* ii. 13 Ben..drove..to pick up Jeannie—his 'sort' or 'gimpy'. For that was how he described his girl friend.
12. b. *all-sorts*, as *sb.*: see *ALL E. 13.

sort (sǭɹt), *sb.*⁴ *Computers.* [f. SORT *v.*¹ II.]
a. The action of arranging items of data in a prescribed sequence.
1956 *Jrnl. Assoc. Computing Machinery* III. 156 Seven passes will effect a complete sort but an eighth pass will be required to collect the items back on to one reel of tape. **1964** C. DENT *Quantity Surveying by Computer* vi. 79 The effect of this sort..will be to arrange the narrative items under their proper headings. **1973** *Computers & Humanities* VII. 202 We decided to remove the umlauts before performing the sort.
b. Special Comb.: **sort key**, a characteristic feature of items of data according to which the data may be arranged; **sort program**, a program written to perform a sort; **sort routine**, a routine written to perform a sort.
1967 D. G. HAYS *Introd. Computational Linguistics* x. 171 The sort has brought together all the contextual spans with a common sort key. Print the first sort key in some obvious place... Then begin printing the contextual spans with that sort key, one after another. **1969** *Computers & Humanities* III. 137 Each letter group, plus the word from which it was derived and the frequency of occurrence of that word..is then sorted in alphabetical order using the letter group as the first sortkey and the position of the first letter of each group.. as the second sortkey. **1963** *Communications Assoc. Computing Machinery* VI. 266/2 The tape merging program initially determines an optimum number of merge passes for merging the sorted files produced by the sort program. **1973** *Computers & Humanities* VII. 203 The IBM sort program..handled eight fields of character or numerical data. **1964** C. DENT *Quantity Surveying by Computer* vi. 79 If we rearrange the keyword so that the trade and heading numbers occupy the most significant position, a standard sort routine with four decks will sort the blocks into the order specified by the trade number plus the heading number. **1969** *Computers & Humanities* III. 137 The letter concordance program accepts the keyword output of the concordance program and generates sort records. These are then sorted by a standard sort routine.

sort, *v.*¹ Add: **II. 11. e.** To reprimand (a person); to deal with (a person) by means of force, repression, etc. *colloq.*
1941 BAKER *Dict. Austral. Slang* 69 *Sort someone out, to*, to reprove a person, put him in his place. **1943** HUNT & PRINGLE *Service Slang* 61 To pick a quarrel and use force is to 'sort out' someone. **1958** J. BLACKBURN *Scent of New-Mown Hay* ii. 24 Get this fellow Kirk sorted out and don't overdo it. **1965** P. ARROWSMITH *Jericho* ix. 92 Let's all go down and sort out that peace pickets' camp. **1974** *Times* 25 Feb. 10/7 Richards came in to sort Willis out and, although Willis prevailed in the end, it was not before Richards had hit him several times for four.
f. To separate out and resolve the complexities of (a problem); to clear up (a confusion or difficulty); to put to rights, deal with. Also, with a person as object: to solve the problems of (someone), 'put (him) straight'. Also *refl.*
1948 'N. SHUTE' *No Highway* v. 128 Will you see if you can get that one sorted out? **1954** J. MASTERS *Bhowani Junction* vi. 51 If I send a message, sort it out between Macaulay and that depilated Sikh assistant of Taylor's. **1962** *Woman's Own* 15 Sept. 69/2 Perhaps you haven't yet decided, assuming this is something which will sort itself out when the time comes. **1963** [see *FAVOUR *sb.* 2 b]. **1963** A. ROSS *Australia 63* 17 The two Perth matches ..allow the visiting side to sort themselves out. **1973** M. AMIS *Rachel Papers* 46 No, don't tell me she's the very girl to show me what egotistical folly it is to compartmentalize people in this sad way; don't tell me she's going to sort me out, take me on, supply the *cognitio* and comic resolution.
12. a. (Later example, of a thing.)
1925 E. F. NORTON *Fight for Everest, 1924* 51 Nearly 300 yak-loads of provision boxes, rolls of bedding and stores of all sorts, dumped higgledy-piggledy off the yaks, began to sort themselves into orderly lines and piles.
16. a. For *Sc.* and *north.* read orig. and chiefly *Sc.* and *north.* and add later examples (corresponding in meaning to sense *11 f).
1950 'D. DIVINE' *King of Fassarai* x. 82 Take her along to Sergeant Marker. Let him sort it. **1975** M. BRADBURY *History Man* xiii. 220 'Tomorrow will sort itself, Barbara,' says Felicity, 'you'll manage.' **1976** L. HENDERSON *Major Enquiry* ix. 57 They're the boys to get it sorted, all young, keen, and raring to go.

sorta, sorter (sǭ·ɹtə), repr. a colloq. pronunc. of the phr. *sort of* (see SORT *sb.*² 6, 8).
Formerly also *sort a*, *sort o'*.
1790, etc. [see SORT *sb.*² 8 c]. **1853** [see *JACK-LEG, JACKLEG *a.* and *sb.*]. **1898** J. D. BRAYSHAW *Slum Silhouettes* 1 Bloomfiel'—never yeard the nime. Wot sorter covey? **1952** B. MALAMUD *Natural* 126, I hoped she would straighten him out and sorta hold him in the team. **1967** E. & M. A. RADFORD *No Reason for Murder* xvii. 115 'He ain't a'goin' to come to life agin, guv'nor.'..'I dunno... Mebbe he might—in a sorta way.' **1978** J. WAINWRIGHT *Jury People* xxxv. 106 A funny sorta bloody joke. **1981** H. R. F. KEATING *Go West, Inspector Ghote* vii. 85 Come on in here. We got some sorta kook.

sortable, *a.* Restrict *Obs.* exc. *dial.* to senses in Dict. and add: **4.** Capable of being sorted or arranged.
1972 *Computers & Humanities* VI. 188 *Description*: Creates sortable tape from card or card-image input, sorts tape, adds..other information to sorted tape.

sortal, *a.* Delete † *Obs.*⁻¹ and add later examples. Also as *sb.*
1959 P. F. STRAWSON *Individuals* v. 168, I shall draw a rough distinction..between two kinds of non-relational tie which bind particulars and universals. This is the distinction between *sortal* and *characterizing* universals. **1975** *Times Lit. Suppl.* 28 Feb. 215/3 The orthodox Fregean and Russellian view that the sortals 'man', 'woman', 'cat' are predicables and not names. **1980** D. WIGGINS *Sameness & Substance* 7 Any predicate whose extension consists..of all the particular things or substances of one particular kind, say horses, or sheep, or pruning knives, will be called here a sortal predicate.

sorted, *ppl. a.* Add: **2.** Also *sorted-out*.
1927 HALDANE & HUXLEY *Animal Biol.* ii. 64 Each sorted-out pack will be complete in having one card of each kind.
3. *Physical Geogr.* Said of shapes and other features displayed on patterned ground, where the stones forming the patterns are distributed in a way suggesting their having been sorted according to size.
1950 A. L. WASHBURN in *Rev. Canad. de Géogr.* IV. 9 In order to standardize the terminology for the purpose of this paper, the broad classification of patterned ground indicated below has been adopted... Patterns on horizontal ground. Sorted circles. Sorted polygons... Patterns on sloping ground. Sorted stripes. *Ibid.* 13 The stones of sorted stripes range in size from gravel in the narrow stripes to boulders in the largest ones. **1956** —— in *Bull. Geol. Soc. Amer.* LXVII. 830/1 A sorted net is patterned ground whose mesh is intermediate between that of a sorted circle and a sorted polygon and has a sorted appearance commonly due to a border of stones surrounding finer material. **1968** R. W. FAIRBRIDGE *Encycl. Geomorphol.* 374/2 In contrast to circles, sorted polygons.. apparently never develop singly. **1977** A. HALLAM *Planet Earth* 89 The forces generated by freezing and thawing tend to segregate particles of different grain size in the soil to produce sorted polygons, whose margins are outlined by the coarser soil fragments.

sorter. Add: **3.** A machine that can sort punched cards into a prescribed order by means of a code punched on selected columns of the cards.
1917 L. R. DICKSEE *Office Machinery* ii. 16 Beyond all question, the most notable..Sorting Machine is the Hollerith Sorter. **1940** W. J. ECKERT *Punched Card Methods Sci. Computation* 10 The sorter automatically sorts the cards into groups according to the information punched in any chosen column of the cards. The cards to be sorted are placed in the hopper..and the sorted cards fall into the various receptacles. **1949** [see *INTERPRETER 5 a]. **1956** G. A. MONTGOMERIE *Digital Calculating Machines* viii. 146 The second basic piece of equipment is the sorter... A pack of punched cards can be fed into this machine and each card will be automatically examined in turn. **1970** O. DOPPING *Computers & Data Processing* iv. 68 If a deck of cards is to be ordered according to a multi-digit number..by means of a sorter, it is necessary to let the cards pass through the sorter as many times as the number field has digits.

‖ **sortes** (sǭ·ɹtīz, sǭ·ɹtēⁱz), *sb. pl.* [L., pl. of *sors* lot, chance.] In phrases *sortes Virgilianae*, *Homericae*, *Biblicae*: divination, or the seeking of guidance, by chance selection of a passage in Virgil, Homer, or the Bible. Also *ellipt.* and *transf.*
a **1586** SIDNEY *Apol. Poet.* (1595) sig. B4, Whereupon grew the worde of *Sortes Virgilianæ*, when by suddaine opening *Virgils* booke, they lighted vpon any verse of hys making. **1646** T. BROWNE *Pseud. Ep.* v. xxi. 272 The first an imitation of *sortes Homericæ*, or *Virgilianæ*, drawing determinations from verses casually occurring. **1700** J. WELWOOD *Memoirs* 100 Lord Falkland, to divert the king, would have his Majesty make a trial of his fortune by the *Sortes Virgilianæ*, which..was an usual kind of augury some ages past. **1740** H. WALPOLE *Let.* 25 Sept. (1974) XXXVII. 79 In three words I will give you her picture as we drew it in the *Sortes Virgilianæ*—Insanam vatem aspicies. I give you my honour, we did not choose it. **1801** M. EDGEWORTH *Belinda* II. xiii. 25 Several volumes of French plays and novels were lying there, and Clarence Hervey raking up one of them, cried: 'Come, let us try our fate by the sortes Virgilianæ.' **1845** G. E. JEWSBURY in A. Ireland *Sel. Lett. G. E. Jewsbury to J. Welsh Carlyle* (1892) 179, I send it you by way of a 'sortes', and the Bible has as much virtue—that way—as Virgil! **1886** D. C. MURRAY *Cynic Fortune* xv. 183 In the practice of the *sortes* (which was a favourite occupation of his) [he] was elevated or depressed by the text he fell upon. **1897** A. C. BENSON *Diary* June in D. Newsome *On Edge of Paradise* (1980) ii. 63, I took a Sortes Biblicae before refusing. **1947** H. NICOLSON *Diary* 11 Dec. (1968) 118, I consult *sortes Biblicas*. My Bible opens at Ezekiel XL 22. **1969** G. GREENE *Travels with my Aunt* I. xvi. 170 The Sortes Virgilianae—a game my mother considered a little blasphemous unless it was played with the Bible. **1975** V. CANNING *Kingsford Mark* vi. 105 He acknowledged the encouragement of the *sortes*. All the omens were right.

sortie, *sb.* Delete ‖ and add: **2. a.** (Earlier example.)
1778 H. WALPOLE *Let.* 8 Oct. (1967) XXIV. 413 Before their last *sortie*, one heard nothing but *What news of the fleets?*
d. An operational flight by a military aircraft.
1918 B. HALL *En l'Air* viii. 76 My machine was a single-seated Nieuport biplane... I carried 1,000 rounds of ammunition... An air sortie at dawn! **1941** *Hutchinson's Pictorial Hist. War* 19 Mar.–13 May 64/2 The main target of the R.A.F.'s night sorties is the industrial centre of Emden. **1955** *Times* 29 June 10/3 In the five active days of the test the two sides flew 12,347 sorties and simulated the dropping of over 300 atomic bombs. **1969** G. MACBETH *War Quartet* 37 Then Waking..we were up.. for New sorties. **1977** *R.A.F. News* 27 Apr.–10 May 11/3 The Phantom has an average sortie capability of more than 1½ hours.
e. *Photogr.* A series of aerial photographs taken during one flight; *transf.*, a photographic session. Also *attrib.*
1953 R. J. C. ATKINSON *Field Archaeology* (ed. 2) i. 24 The unit of classification of prints is the *sortie*, that is, a series of prints taken on a single flight. Each sortie consists of one or more strips of prints, running approximately in an East–West direction, each print overlapping the next to East and West by about two-thirds of its width. *Ibid.*, Each print..is labelled at the top with the sortie-number. *Ibid.* 25 Sortie plots may be bought separately at the same charge per plot. **1959** N. MAILER *Advts. for Myself* (1961) 229 One sortie when she was photographed sipping a soda she shaped the second straw into a heart. **1963** W. K. KILFORD *Elem. Air Survey* x. 226 When the sortie is complete the film is processed... The relative position of each photograph.. of the sortie is plotted. **1969** G. C. DICKINSON *Maps & Air Photographs* xv. 246 A set of photographs taken during one flight is usually known as a *sortie*. *Ibid.* 247 More detailed information about coverage is contained on index diagrams, known as sortie plots.
3. b. (Earlier example.)
1809 D. THOMPSON *Jrnl.* 18 Aug. (1950) II. 31 The Sortie of the [stream] that falls into the [stream] at Deer's Horns Plains.
4. *attrib.* in *Astronautics*, designating spacecraft designed to return to earth after a period; so *sortie mission*.
1972 *National Observer* (U.S.) 27 May 6/4 NASA also plans a 'sortie module', a laboratory for six scientists and engineers that would be carried in the orbiter's cargo bay for earth-orbit missions lasting from one to four weeks. **1972** *New Scientist* 6 July 3 NASA has offered Europe the 'sortie can'—a pressurised laboratory module that is to swing out from the cargo bay of the orbiting shuttle. **1973** *Times* 15 June 27/7 Here the proposal is that Europe would develop a 'sortie lab' or 'spacelab' module. **1976** LOHMAN & LEE in L. G. Napolitano *Space Activity Impact on Sci. & Technol.* 108 In a series of flights, these aircraft duplicate the observations which would be conducted in a single sortie mission.

‖ **sortie de bal** (sorti də bal). [Fr., lit. 'departure from (the) ball'.] A woman's evening cloak with a quilted lining, popular in the late nineteenth century.
1864 M. B. CHESNUT *Diary* 29 Dec. in C. V. Woodward *M. Chesnut's Civil War* (1981) xxxix. 696 Mary was wrapped in a snowy swansdown *sortie de bal*. **1895** [see *DOVE *sb.* 1 d]. **1908** A. BENNETT *Old Wives' Tale* III. ii. 296 It was a tall and mature woman who wore over a dress of purplish-black silk a vast flowing *sortie de bal* of vermilion velvet, looped and tasselled with gold.

sorting, *vbl. sb.* Add: **4. a.** *sorting-out operation, process*; **sorting code** *Banking*, a code number which identifies a branch office in the banking system, and is used to facilitate the processing of cheques and credit transfers.
1959 D. S. TRAVERS in *Electronics in Banking* (Institute of Bankers) 50 The cheque of the future... The code appears in five blocks... Block 2—Bank and branch sorting code. **1965** PERRY & RYDER *Thomson's Dict. Banking* (ed. 11) 524/2 *Sorting code numbers*, a system devised to assist customers of banks who make considerable use of the Credit Transfer system... The number is placed in the box provided on the credit by the customer before passing the credit to the bank. It consists of three groups of two figures each, eg. 20-03-92, the first digits denoting the bank, and the remainder identifying the branch. **1977** *Canad. Jrnl. Linguistics* 1976 XXI. II. 149 In (44) the criterion of the sorting-out operation is P_1. **1960** E. H. GOMBRICH *Art & Illusion* II. iv. 144 Classical art also underwent an evolution, a sorting-out process after its heroic period.
b. *sorting table*; with *adv.*, as *sorting-out centre*.
1952 S. SPENDER *Learning Laughter* vii. 97 The reader should be able to visualize them going to the sorting-out

centre near Haifa. 1881 A. Bathgate *Waitaruna* xii. 172 The 'pickers-up' were busy gathering the fleeces..and carrying them to the sorting table.

sort-out. [f. vbl. phr. *to sort out*: see Sort *v.*[1] 11 in Dict. and Suppl.] **a.** The action or an instance of sorting out (of things or situations that are in disarray). **b.** A fight or dispute.

1937 Partridge *Dict. Slang* 802/1 *Sort-out*, a fight, a mellay. 1961 *John o' London's* 25 May 591/2 An enlightened attempt at a black-and-white sort-out. 1964 C. Willock *Enormous Zoo* i. 14 Beaton himself had a sort-out with a buffalo bull that tried to overturn his Land-Rover. 1972 *Tel.* (Brisbane) 18 May 18/4 He was the most cantankerous character I have met. I had only been here two days when we had our first sort-out. 1981 G. Hammond *Revenge Game* vii. 65 You folk finish the first sort-out downstairs.

S.O.S. (e:s,ōu,e·s), *sb.* Also **SOS.** [The letters *s*, *o* and *s*, chosen because easily transmitted in Morse code.] **1. a.** The international wireless code-signal of extreme distress, used esp. by ships at sea.

1910 J. A. Fleming *Princ. Electr. Wave Telegr. & Teleph.* (ed. 2) 882 This signal, S,O,S, has superseded the Marconi Company's original high sea cry for help, which was C,Q,D. 1910 E. Lawton *Boy Aviators in Nicaragua* 263 S.O.S. is now the wireless distress call. 1924 *Mod. Wireless* III. 310/3 The famous signal 'SOS' was adopted officially by the International Radio Telegraph Convention in July, 1908. 1930 'Sapper' *Finger of Fate* 245 The S.O.S. had been picked up by three other boats... The last S.O.S. had broken off abruptly in the middle of the message. 1942 [see *dash *sb.*[1] 7 f].

b. *transf.* An urgent message or appeal for help.

1918 *Punch* 13 Mar. 176 S.O.S. at Suburban Pictures. (In cases of emergency affecting any of the audience messages are sometimes thrown on the screen by the courtesy of the management.) 1931 E. F. Benson *Mapp & Lucia* iii. 80 There was nothing but helpful sunny cordiality in response to this S.O.S. 1965 D. Francis *Odds Against* xix. 235 You'd sent him to me as a sort of S.O.S.

2. As an abbrev. of var. jocular phrases. **a.** 'same old story, stuff', etc. **b.** 'shit on a shingle': chipped beef on toast. *U.S. Mil. slang.*

1918 *Sat. Even. Post* 27 Apr. 62 'What have you got this morning, Thompson? SOS?' 'Yes sir,' returned the striker. 'Same old slum.' 1926 Maines & Grant *Wise-Crack Dict.* 14/2 *S.O.S.*, same old story. 1959 I. & P. Opie *Lore & Lang. Schoolch.* ix. 162 The possessors of young and healthy appetites are lyrical about their food. School dinners are 'muck', 'pig swill'..'S.O.S.' (Same Old Slush). 1963 *Amer. Speech* XXXVII. 271 Haskell students do employ the standard American slang initials *S.O.S.*, but..Girls use them to mean 'same old stuff'... Boys..refer more specifically to creamed beef on toast.. and this abbreviation stands for either 'same old shit' or 'shit on a shingle'. 1974 *News & Observer* (Raleigh, N. Carolina) 8 July 23/2 The troops still sometimes get S.O.S. for breakfast, whether they want it or not.

3. *attrib.* and *Comb.*, as *S.O.S. call*, *message*, *signal*; **S.O.S. redouble** *Bridge* (see quot. 1926[2]); also *ellipt.*

1915 *Daily Mail* 10 May 4/5 The Marconi operators showed magnificent coolness in that hour of trial. They made the 'S.O.S.' call, and they had to make it quickly. 1938 *Encycl. Brit. Bk. of Year* 122/1 S.O.S. and police messages broadcast from all transmitters during the year reached a total of 1,213. 1978 J. Thomson *Question of Identity* vii. 70 I'd've heard if he was dead... They'd've put it on the radio; an S.O.S. message. 1926 M. C. Work *Auction Bridge Compl.* x. 134 The No Trumper resorts to a recently invented signal of acute distress and makes the 'S.O.S. Redouble'. *Ibid.* 498 *S.O.S.* (*redouble*), redouble made to indicate weakness rather than strength. 1939 N. de V. Hart *Bridge Players' Bedside Bk.* 135 If the strong opening No Trumps..is being played, a redouble by a player who has opened with One No-Trump is not an SOS. 1967 R. L. Frey *Bridge Players' Encycl.* 470/2 In rare circumstances a player may redouble his partner's bid as an SOS instead of his own bid. 1977 C. H. Goren *Bridge Compl.* (ed. 3) 684 *Redouble*, a call that further increases the scoring value of tricks and penalties after an opposing double; sometimes also used as a request for partner to rescue (S.O.S. redouble). 1917 'Sapper' *No Man's Land* 71 A row of grey-painted rockets with a red top, which in case of emergency send up the coloured flares that give the S.O.S. signals to those behind. 1927 H. A. Vachell *Dew of Sea* 257 He received a letter from the gentleman, regarded (and rightly) by the McCullough as an S.O.S. signal. 1929 C. H. Smith *Bridge of Life* iv. 83 When one gave an SOS signal all the rest came to his assistance.

S.O.S., *v.* Also **SOS.** [f. prec.] **1.** *intr.* **a.** To make an S.O.S. signal or signals. **b.** *Bridge.* To execute an S.O.S. redouble.

1918 Kipling *Land & Sea Tales* (1923) 114 Then..this Baxter-man got busy with his wireless and SOS'ed like winkie. 1926 M. C. Work *Auction Bridge Compl.* x. 135 In the event of a business pass, the No Trumper can S.O.S. if in need of help. 1975 G. Howell *In Vogue* 75/1 We have been passing through the awkward age, when instead of conversing we 'S.O.S.ed' in monosyllabic slang.

2. *trans.* To send an urgent message requesting (someone) *to* do something.

a 1936 Kipling *Something of Myself* (1937) viii. 221 One of the Captains S.O.S.-ed me to give him 'something to tell these somethinged tourists about it'.

so's: see *'s 5.

‖ **sosatie** (sǫsā·ti). *S. Afr.* Also **sas(s)atie, sas(s)atje, sosaartje,** etc. [Afrikaans, f. S. Afr. Du. *sasaatje*, f. Mal. (Javanese) *sesate* skewered meat.] Curried or spiced meat grilled on a skewer.

1833 *Cape Good Hope Lit. Gaz.* 2 Sept. 138 (Pettman), *Sasaitie*, or cabobs, is really no despicable eating. 1870 *Cape Monthly Mag.* Oct. 224 '*Sosaartjes*', '*smoor-picklaar*', and all sorts of vegetable '*breedies*', are importations from India. 1883 'R. Iron' *Story Afr. Farm* II. ii. iv. 67, I got the Hottentot girl to show me how to make 'sar-sarties' this morning. 1885 L. H. Meurant *Sixty Years Ago* 29 There existed in those days what we termed 'Sasaatje and Rice' houses, places where a favourite Dutch dish called 'Sasaatjes' was served in the evenings...two sassaatjes (diamond-shaped inch-sized pieces of mutton, curried and about half a dozen stuck upon a bamboo skewer, and then roasted upon a gridiron). 1894 *Cape Argus* 22 Dec. (Pettman), A Hittite..with a long spear and a very pronounced intention to spit you on it, like a sassatje. *a* 1920 [see *bobotie]. 1939 S. Cloete *Watch for Dawn* xxi. 310 A woman had so much to do with food. Sassaties, bobotee, snysels. 1948 H. V. Morton *In Search of S. Afr.* 293 Sosaties..are a popular and delicious grill which can be as simple as veal or mutton cutlets sprinkled with curry powder and roasted on a skewer over a clear wood fire, or as complex as pieces of mutton or pork soaked in wine or vinegar, spiced with coriander, turmeric, pepper, tamarind and grilled in the same way. 1953 *Cape Times* 21 Mar. 3/3 These [South African recipes] are destined for Holland, where they will be needed to prepare *braaivleis* and *sasaties* when the South African *volkspele* team visits Culemborg and holds a *braaivleisaand* there. 1973 *Farmer's Weekly* (S. Afr.) 25 Apr., At a braaivleis the sosaties are eaten straight from the grill.

‖ **soshi** (sōu·ʃi). Pl. **soshi.** [Jap., lit. 'strong man', a Chinese *suōshì*, f. *suō* lusty, valiant + *shì* warrior.] A mercenary political agitator or intimidator, a terrorist; a bodyguard.

1891 B. H. Chamberlain *Things Japanese* (ed. 2) 128 Since 1888, there has sprung up a class of rowdy youths, called *soshi* in Japanese—juvenile agitators who have taken all politics to be their province. 1894 G. N. Curzon *Probl. Far East* ii. 33 The *soshi* or professional rowdies, who are ready, for a consideration, to let out their services to either party in Japan. 1896 L. Hearn *Kokoro* vi. 95 Soshi form one of the modern curses of Japan. They are mostly ex-students who earn a living by hiring themselves out as rowdy terrorists. 1910 Lady Lawson *Highways & Homes of Japan* xxv. 284 At one time this extraordinary man was a *soshi* or political bully, one of the turbulent class who suffer from too much education and too little to eat. 1930 M. D. Kennedy *Changing Fabric of Japan* vii. 121 In 1923, meetings in favour of Manhood Suffrage were broken up by gangs of *soshi*. 1977 G. M. Berger *Parties out of Power in Japan 1931–1941* iv. 147 Nakamizo Tamakichi, a former seiyūkai bodyguard (*sōshi*), organized a group of several hundred toughs.

so-soish, *a.* (Earlier examples.)

1819 Keats *Let.* 15 Apr. (1931) II. 353 Do you and the miss Birkbecks get groggy on any thing—a little so so ish so as to be obliged to be seen home with a Lantern[?]. 1826 M. Wilmot *Let.* 13 Jan. (1935) 231 Major F[alconer] goes on so, so, ish, and poor dear Bessy out of spirits.

soss, *v.*[1] **2.** (Later *poet.* example.)

1951 Auden *Nones* (1952) 39 The three wise Maries come, Sossing through seamless waters.

sostenuto, *a.* and *sb.* Add: **A.** *adj.* **1.** (Later examples.)

1955 *Times* 27 July 5/6 Mr. Thomas Matthews, the soloist,..missed some of its finest effects, chiefly because he under-emphasized its sostenuto spaciousness. 1976 *Gramophone* Sept. 410/1 The *sostenuto* opening and eager staccato continuation are both beautifully judged.

2. (Earlier example.)

1826 M. Kelly *Reminisc.* I. 114 He had been the first cantabile singer of his time, and his sostenuto singing was still admirable.

B. *sb.* Delete *rare*[−1] and add later examples. Also *fig.*

1933 N. Douglas *Looking Back* II. 460 It was September, that wonderful month when the melodies of summer linger on..in a tranquil *sostenuto*. 1934 C. Lambert *Music Ho!* v. 309 His [*sc.* Sibelius's] use of the brass is sparing, concentrating far more on its sostenuto than on its percussive qualities. 1943 D. Gascoyne *Poems 1937–42* 14 The heart's receptive chalice in pure hands upheld Towards the sostenuto of the sky.

sot, *sb.*[1] **3. sot-weed**: for † read *Obs.* exc. *Hist.* and add earlier and later examples.

1698 E. Ward *London Spy* I. 9 We had each of us Stuck in our Mouths a lighted Pipe of Sotweed. 1708 E. Cook (*title*) The sot-weed factor; or, a voyage to Maryland. 1961 J. Barth (*title*) The sot-weed factor. 1965 E. Tunis *Colonial Craftsmen* iii. 52 Early Americans not only grew a lot of 'sot weed' they also consumed a lot of it.

sotalol (sōu·tălǫl). *Pharm.* [Etym. unkn.: see -ol.] The compound, $CH_3.NH(SO_2).C_6H_4.CH(OH).CH_2.NH.CH(CH_3)_2$, a beta-adrenergic blocking agent used (in the form of its hydrochloride) in the treatment of cardiac arrhythmias.

1968 *Jrnl. Pharmacol. & Exper. Therapeutics* CLX. 231 The antiinflammatory action of morphine is reversed by nalorphine but not by a *beta* adrenergic blocking agent, sotalol (MJ 1999; 4-(2-isopropylamino-1-hydroxylethyl)-methanesulphonanilide hydrochloride). 1974 R. G. Shanks et al. in A. G. Snart *Adv. in Beta-Adrenergic Blocking Therapy—Sotalol* I. 24 Sotalol is the only drug which blocks all adrenergic β-receptors and has no membrane stabilizing activity or intrinsic sympathomimetic activity. 1978 *Nature* 19 Oct. 596/2 In contrast, β-adrenergic antagonists lacking the hydrophobicity of propranolol, such as atenolol, practolol and sotalol, are not able to induce meiosis.

sotch (sǫtʃ). *Physical Geogr.* Pl. **sotchs, sotches.** [Fr. dialectal word of pre-Latin origin (Robert).] A doline, esp. one in the Causses region of France.

1910 H. R. Mill *Dict. Geogr. Terms* in L. D. Stamp *Gloss. Geogr. Terms* (1961) 426/2 *Sotch*, dolines (*q.v.*) (Causses, France). The term 'cloup' is used in Aquitaine. 1922 *Geol. Mag.* LIX. 394 The only fertile and habitable regions on these barren plateaus are the 'sotches' or hollows, where the red earth formed by the denudation of the limestone is preserved in large funnel-shaped hollows. 1937 *Geogr. Jrnl.* LXXXIX. 63 A detailed study of the karst landscape, showing the important part played by the dolinas or 'sotchs' in the dissection of the surface. 1972 M. M. Sweeting *Karst Landforms* iv. 48/2 The sotchs are both circular and elongated, their diameters and depths variable and their topographic situations diverse.

Sotho (sū·tu). Also **Suthu, Suto,** etc. [Native name.] **A.** *sb.* A subdivision of the Bantu people which includes the Basuto and various other tribes chiefly found in Botswana and the Transvaal; also, the languages spoken by these people. **B.** *adj.* Of or pertaining to this group of peoples or their languages. Cf. *Sesotho.

1928 G. P. Lestrade *Bantu Tribes S. Afr.* I. i. 14 There has..been..intermarriage between the Bavenda and the Basotho (there was a Sotho invasion of Bavendaland). 1928 *Africa* I. 481 In reality there are two main branches of this group, Chwana spoken in Bechwanaland and the Western Transvaal, and Suto with its dialects. 1929 [see *Nguni *sb.* and *a.*]. 1936 [see *hill-culture* s.v. *hill *sb.* 4 d]. 1948 M. Guthrie *Classification Bantu Lang.* iv. 69 These radicals from Sutho. 1957 C. G. Seligman *Races Afr.* (ed. 2) viii. 120 The Sotho..includes the Southern Sotho of Lesotho..the Tswana..of Botswana and western Transvaal, and the Northern Sotho..of central and northern Transvaal. 1977 *Times* 7 Sept. 7/4 [The South African] Secretary for Information..gives the population..as..whites,..Coloureds,..Indians,..Zulus, ..Xhosas,..Tswanas,..North Sothos,..South Sothos, [etc.]. 1979 *Listener* 25 Oct. 545/1 The text Modimo o Lerabu—'God is Love' in the Sotho language. Both Zulu and Sotho are used in the worship.

‖ **Soto** (sōu·tǫu). [Jap.] One of the three sects of Zen Buddhism. Freq. *attrib.*

1893 S. Kuroda *Outlines Mahâyâna* vi. 24 Dôgen introduced the Sôtô sect, 2176 after Buddha, or 1227 A.D. 1894 *Trans. Asiatic Soc. Japan* XXII. 430 The Zen sects.. are divided..into three divisions. The *Rinzai*..from 1168 A.D., the *Sôtô* from 1223 A.D. and the *Obaku* from 1650 A.D. 1917 A. K. Reischauer *Stud. in Japanese Buddhism* iii. 117 The chief difference between the Soto and the Rinzai branches of the Zen Sect is that the former puts more weight upon book learning as a subsidiary aid to silent meditation. 1949 C. Humphreys *Zen Buddhism* v. 91 It is in Japan that Zen can be best studied, and although there are three sects of Zen, the Rinzai, Soto and Obaku, there is little difference between them. 1977 'E. V. Cunningham' *Case of One-Penny Orange* (1978) iii. 29, I am Zen. The Soto School.

sotol (so·tol). [Amer. Sp., f. Nahuatl *tzotolli*.] A plant of dry regions belonging to the genus *Dasylirion* of the family Agavaceæ, native to south-western North America and bearing linear leaves and small white flowers; also, the fibre from the leaves of this plant or the beverage made from the sap.

1881 *Amer. Naturalist* XV. 874 The home of the sotol is Western Texas, Southeastern New Mexico and Northern Chihuahua. 1908 D. T. MacDougal *Bot. Features N. Amer. Deserts* 9 This is the typical sotol region. 1942 Castetter & Bell *Pima & Papago Agric.* 213 The mature leaves were dried by the Papago in the sun on a sotol mat. 1964 F. O'Rourke *Mule for Marquesa* 72 The scorched grey of candelilla and guayule, Spanish daggers, sotol with the crowns burned a dirty white by sun and wind. 1976 *Hortus Third* (L. H. Bailey Hortorium) 364/1 An alcoholic beverage, sotol, is extracted from the trunks. 1976 F. A. & D. L. Latorre *Mexican Kickapoo Indians* iii. 44 Two benches..are built and covered with dyed sotol or cattail mats. *Ibid.* v. 64 The favored material used today for basketmaking is sotol.

sotter, *v.* Add: **b.** To bubble. Also *poet.*

1834 J. Galt *Lit. Life* III. 51 The blood was sottering out of his shoe mouth. 1886 W. S. McIntosh in D. H. Edwards *Mod. Sc. Poets* IX. 70 This were burnie sae sottered an' sang. 1951 Auden *Nones* (1952) 64 The sharp streams and sottering springs of A commuter's wish.

sottise. Add: (Later examples.) Also *transf.*

1952 J. Biggs-Davison *Tory Lives* ii. 48 Tact and flexibility were not often exhibited by the Monarch of whose *sottise* his predecessor was wont to speak. 1963 N. Macrae *Sunshades in October* iii. 39 The Conservative Government..dissected two-and-a-half of these three *sottises*. 1977 *Times* 23 Mar. 16/8 The *Daily Mail* Diary.. is not slow to criticize errors and *sottises* in rival newspapers.

‖ **sottisier** (sotīzye). [Fr.] A collection of *sottises*; esp. a list of written stupidities. Also *transf.* and *fig.*

1929 E. Pound *Let.* 1 Feb. (1971) 224 The simplest and briefest form of attack is by a sottisier... Make your sottisier from *Poetry* and the main literary reviews, Sunday supplements, etc. 1944 *Horizon* Sept. 187 Nor is it [*sc.* *Finnegans Wake*] a mere *sottisier*. 1959 *Observer* 30 Aug. 13/7 A cast-iron certainty for the September Sottisier Stakes. 1969 J. Gross *Rise & Fall of Man of Letters* ix. 238 It would be easy to compile a fair-sized sottisier of such remarks, but not very amusing. 1976 *Times Lit. Suppl.* 26 Mar. 345/5 The sloppy editing and the *sottisier* of an index, with its general confusion of names, styles, titles and sexes. 1978 *Times* 22 Mar. 11/8 Ginger Rogers..under-rehearsed commentary is a *sottisier* of all the awful things Americans are supposed to say.

soubrette. Add: **1.** (Later examples.) In extended use, a woman playing a role or roles in light entertainment, e.g. on television or at a seaside variety show, with implications of pertness, coquetry, intrigue, etc.

1905 R. Beach *Pardners* (1912) i. 33 Variety house, with..two-ton soubrettes, with Barrios diamonds and hand-painted socks. 1951 *Sunday Pictorial* 21 Jan. 6/3 Harriet Cohen has the vivacity and femininity of a soubrette. 1956 *Ann. Reg. 1955* 360 Adele Leigh scored a special success in the soubrette part of Bella. 1958 *Time* 8 Oct. 6/3 His 'wife', Miss Rosita Segovia, takes too soubrette-like a view of this role. 1977 *Listener* 27 Jan. 111/1 By the time she was 18, she was a soubrette at the Gaiety. 1978 *TV Times* 28 Jan.–3 Feb. 26/2 Television's most resilient soubrette, Una Stubbs, can talk about love, marriage and happiness with a twinkle in her eye.

soubrettish *a.* (later example).

1979 J. Crosby *Party of Year* (1980) xiv. 156 The soubrettish face darkened.

soucar: see also *sahukar (now the more usual spelling).

Soudan. Add: Also = Soudanese *sb.* and *a.*

1867 'Ouida' *Under Two Flags* I. xiii. 297 Chasseurs, Zouaves..mingled with jet-black Soudans. 1889 W. F. Butler *Charles George Gordon* iii. 58 Some of his old Soudan soldiers.

Soudanese *a.* (earlier example).

1884 E. W. Hamilton *Diary* 13 May (1972) II. 615 In the House of Commons yesterday there was an abnormal display of excitement..on the occasion of the Soudanese or Gordonese vote of censure moved by Sir M. H. Beach.

soufflé, *sb.* and *a.* Add: **A.** *sb.* **a.** (Earlier and later *fig.* examples.)

1888 [in Dict., sense b]. 1916 E. Pound *Let.* 27 July (1971) 89 Sometime, certainly, you must have the soufflé of contemporary French poets. 1964 *New Statesman* 1 May 675/3 A girl, her yellow hair in a sort of lacquered soufflé, ran from the police court, crying black tears from the cosmetics of her child's eyes. 1980 T. Morgan *Somerset Maugham* viii. 202 [The play] *Caroline* was a soufflé whipped up..out of boredom.

b. (Earlier example.)

1845 E. Acton *Mod. Cookery* xix. 491 A common soufflé-pan may be purchased for four or five shillings.

B. *adj.* (Later example.)

1972 *Trans. Oriental Ceramics Soc.* XXXVIII. 141 The entire vessel covered with *soufflé* gilding.

‖ **souffrante** (sūfrant), *a.* [Fr., fem. sing. pres. pple. of *souffrir* to suffer.] Of women: delicate, indisposed or ill; prone to anxiety or depression.

1827 E. Grosvenor in G. Huxley *Lady Elizabeth & Grosvenors* (1965) vii. 141 She is a very interesting person, very handsome, but pale and 'souffrante'. 1877 Geo. Eliot *Let.* 10 July (1956) VI. 392, I have been *souffrante*, but am content to bear my share of such trouble. 1938 A. Thirkell *Let.* in M. Strickland *A. Thirkell: Portrait of a Lady Novelist* (1977) vii. 120, I have been so *souffrante* this summer with perpetual headaches. 1977 *Times Lit. Suppl.* 28 Jan. 94/4 A tearful, tortured female, forever *souffrante*, forever worried, forever hoarding and cherishing her sufferings.

‖ **souffre-douleur** (sūfr,dūlör). Also **souffre douleur.** [Fr., lit. 'suffer sorrow'.] One who is in a subservient position and must listen to or share another's troubles; also *spec.* a woman who acts as a paid companion to an older woman.

1845 M. Gardiner *Strathern* II. xvii. 17 The woman on whose arm she leans is her *dame de compagnie*, her *souffre douleur*. 1864 C. M. Yonge *Trial* II. iii. 55 A younger brother and legitimate *souffre douleur*. 1907 M. E. Braddon *Dead Love has Chains* i. 3 She had her maid and her *souffre-douleur*, a dowerless kinswoman of six-and-twenty. 1962 *Punch* 7 Nov. 684/1 Josephine, employed as Aunt March's *souffre-douleur*. 1981 W. Gérin *Anne Thackeray Ritchie* xx. 240 When away from home he always wrote regularly to his 'souffre douleur'.

soufrière (sūfriēr). [Fr., f. *soufre* Sulphur *sb.* + *-ière* -ier.] = Solfatara.

1879 J. W. Boddam-Whetham *Roraima & Brit. Guiana* vi. 53 How pleasant it would be to spend a few days on each of these West Indian islands! To visit their *souffrières*, their mountain forests, their wild hills, and their cultivated estates! 1902 *Pop. Sci. Monthly* July 273 The warm springs and solfataras (or souffrieres [*sic*]) on Martinique and other islands displayed unwonted activity. 1939 *Nature* 14 Oct. 677/1 The soufrières of Montserrat are considered as having been caused mainly by the penetration of meteoric water to a subterrain of sulphided limestone that had been reheated intensively by uprising magma. 1972 *Whitaker's Almanack 1973* 786/2 Montserrat..contains three active soufrières and several hot springs.

sougee, var. *soogee.

sough, *v.*[1] **1. b.** (Later example.)

1975 W. McIlvanney *Docherty* i. xviii. 124 Trees were brooding presences, soughing incantations. Every bush hid an invisible force, frequently malevolent.

sought, *ppl. a.* Add: **b.** *sought-after* (later examples).

1944 Auden *For Time Being* (1945) 32 Her famous, memorable, sought-after evenings. 1978 *Lancashire Life* Apr. 45/2 Foulridge took umbrage,..continuing to build on its [*sc.* Foulridge's] reputation of being a sought-after spot to live.

sougi-mougi, var. *soogee-moogee. **soujge, souji-mouji,** varr. *soogee, *soogee-moogee.

‖ **souk** (sūk). Also **sok, sook, soug, suk(h, suq.** [Fr., ad. Arab. *sūk* market-place.] An Arab market or market-place, a bazaar (sense 1 a).

1826 Denham & Clapperton *Narr. Trav. N. & Cent. Afr.* ii. 51 The soug, or market, is well supplied with every necessary and luxury. 1829 J. L. Burckhardt *Trav. Arabia* 54 In a row of eight or ten shops are sold rice, onions, butter, dates, and coffee-beans... This is what the Arabs call a *souk*, or market. 1855 R. F. Burton *Personal Narr. Pilgrimage to El-Medinah* I. 333 There is a large 'Suk', or market-place in the usual form, a long narrow lane darkened by a covering of palm leaves, with little shops let into the walls of the houses on both sides. 1899 A. E. W. Mason *Miranda of Balcony* ii. 24 Every evening he comes down to the Sôk, buys milk and bread. 1909 G. W. Furlong *Gateway to Sahara* i. 21 One afternoon, I was passing here from the *Suk*..when a ragged, unkempt fellow appeared in the caravan road there, acting most strangely. 1921 *Glasgow Herald* 20 Oct. 4 The suqs or covered streets, which, being screened from the glare of the sun, afford fine shelter for shops and markets. 1926 D. Byrne *Brother Saul* v. 64 When Anna went abroad.. to the sook of the perfumers. 1931 *Observer* 6 Sept. 13 The sun-smitten pavement of the *sukh*. 1959 W. Thesiger *Arabian Sands* xiii. 258 Behind the..houses which lined the water-front were the *suqs*, covered passageways, where merchants sat in the gloom, cross-legged in narrow alcoves among their piled merchandise. 1968 R. Hargreaves *Bloodybacks* iii. 74 The importation of assorted consignments of *filles du roi*, nubile but dowerless wenches willing to marry anyone in a position to provide them with a home... Their distribution was reminiscent of nothing so much as the disposal of Christian slaves in an Oriental *Sok*. 1978 L. Heren *Growing up on The Times* iii. 91 The assassins escaped into the souks of the city [*sc.* Damascus], and no more was heard of them. 1981 *Financial Times* 12 Dec. 7/2 The lust for gold..grabs most visitors to Sharjah and Dubai, where the gold souks gleam with the stuff.

soul, *sb.* Add: **I. 2. c.** Also in Naut. phr. *soul and body lashing* (see quot. 1962).

1903 C. Protheroe *Life in Mercantile Marine* 150 The best method of arranging his oil-skins to keep the water out,..known as a 'soul and body lashing'. 1936 B. Adams *Ships & Women* iv. 87 All wore rope yarns tightly tied about wrists and ankles... We call those rope yarns 'soul and body lashings'. 1962 A. G. Course *Dict. Naut. Terms* 182 *Soul and body lashings*, rope yarns tied round the waist and sleeves of oilskin jackets, and round the bottom of oilskin trousers, to prevent the water, from seas crashing on board, getting under the oilskins. They also prevented the wind from ballooning up inside the oilskins.

d. *soul and conscience*: in *Sc. Law*, the formula by which medical testimony in writing is authenticated; also *attrib.* (see quot. 1976).

1892 A. M. Anderson *Criminal Law of Scotland* v. xiii. 252 Medical reports are made on soul and conscience, read at the trial, and sworn to as true. 1925 W. J. Lewis *Manual of Law of Evidence in Scotland* iii. ii. 84 Medical certificates on soul and conscience, apparently holograph, appear, in non-contentious matters, to be generally accepted without further evidence. 1976 L. Kennedy *Presumption of Innocence* iii. 147 There was a soul and conscience certificate in relation to Mrs Carmichael; this meant that a doctor had sworn on his soul and conscience that she was unfit to attend the court.

3. b. Also in somewhat weakened use, deep feeling, sensitivity.

1823 Byron *Don Juan* xiv. lxxi. 150 But there was something wanting on the whole—I don't know what, and therefore cannot tell—Which pretty women—the sweet souls!—call *Soul*. 1853 Lytton *My Novel* III. ix. iii. 22 Oh, no! no picture of miserable, vicious, Parisian life. This is beautiful; there is *soul* here. 1904 F. H. Jackson *Mural Painting* 21 Benozzo Gozzoli..filled his long life with the production of the most charming wall-paintings, which, if he had had what is often called 'soul', would have placed him very near the summit of the Palace of Art.

3*. a. The emotional or spiritual quality of Black American life and culture, manifested esp. in music (see quot. 1973).

1946 *Ebony* Sept. 34/2 He uses a bewildering, unorthodox technique and his playing is full of what jazzmen refer to as 'soul'. 1954 *Grove's Dict. Mus.* (ed. 5) IV. 600/2 Louis Armstrong declared that 'Anything played with beat and soul is jazz.' 1964 *Amer. Folk Music Occasional* i. 17 It's just really rough what the colored entertainers have to go through sometimes... That's why the colored people sing the blues; that's why they sing with soul. 1973 S. Henderson *Understanding New Black Poetry* 74 In the late 1950's the word 'Soul' surfaced in the musical community and quickly spread to the wider Black Community, where it came to mean not only a special kind of popular music..but also..'racial spirit' and 'racial flavor'... The word is losing some of its popularity now.

b. *ellipt.* for *soul music*, see sense 25 below.

1961 [see *funk *sb.*[3] 2]. 1968 P. Oliver *Screening Blues* ii. 46 The distinction between gospel music and the most recent development of blues and rock 'n roll—soul—is one of content rather than style. 1975 *New Yorker* 28 Apr. 6/3 She's lately been branching out from a strict regimen of blues and folk songs..to include some rock, soul, and Nashville-inspired ditties. 1979 *Radio Times* 19 July 60/1 The word 'soul' probably originated with Ray Charles... Soul is the music of experience... It's one person's heart speaking to another person's.

c. *attrib.* passing into *adj.* (*a*) Characteristic of or pertaining to Black people or culture; (*b*) of or pertaining to soul music.

1962 *John o' London's* 1 Feb. 113/3 Feldman is not really a soul-merchant. 1968 *N.Y. Times* 17 June 46 Sonny Charles, the organist, took over, singing with a soul appeal that caught up even this predominantly white audience. 1969 C. Himes *Blind Man with Pistol* xxi. 231 The big white man thought they were talking about him in a secret language known only to soul people. 1971 B. Malamud *Tenants* 63, I swear to myself I will be the best writer, the best Soul Writer. *Ibid.* 121 From across the street..Bill spied him and whooped, 'Lesser, man, for Christ's sake, cross on over here. I got some soul people with me.' 1972 *Sat. Rev.* (U.S.) 27 May 18/1 You'll be surprised how many soul folk speak Dutch and work and play in surprising Amsterdam. 1975 D. Pitts *Target Manhattan* (1976) xxvi. 105 They had.. listened to a group of black soul singers. 1976 *Drum* (E. Afr. ed.) June 10/2 Soul language is a language of protest, a language of self-assertion, a language that rejects the white man's values. 1981 *Westindian World* 28 Aug. 5/6 The Crusaders are among the finest exponents of the art of making a good listenable soul record.

4. f. *to have no soul*: to be lacking in sensibility or right feeling.

1704 Swift *T. Tub* ii. 64 That Fellow, cries one, has no Soul; where is his Shoulder-knot? 1850 'L. Limner' *Christmas Comes* 9 He seeks refuge in his organ, much to the annoyance of a little tailor in the attic who has no soul in him. 1919 G. B. Shaw *Inca of Perusalem* in *Heartbreak House* 209 You have no soul for fine art.

g. *to have a soul above* (something): to be superior to or have higher aspirations than (something).

1795 G. Colman *New Hay at Old Market* 10 My father was an eminent Button-maker..but I had a soul above buttons... I panted for a liberal profession. 1834 F. Marryat *Peter Simple* I. i. 2 My father, who was a clergyman..had..a 'soul above buttons'. 1889 E. Dowson *Let.* 27 Oct. (1967) 112, I have still a soul above tractlets. 1899 G. B. Burgin *Bread of Tears* i. iii. 51 Miss Mercy Tressock evidently wrote a very bad hand, and she hadn't a soul above blots: they were dotted copiously about on every page. 1909 'O. Henry' *Rus in Urbe* in *Hampton's Mag.* Aug. 160/1 She had a soul above ducks—above nightingales.

h. *to make one's soul*: see Make *v.*[1] 47.

6. b. (Later example.)

a 1902 S. Butler *Way of All Flesh* (1903) xiii. 56 He had stuck to his post... He had said to himself: 'I..am the very soul of honour. 1976 R. Lehmann *Sea-Grape Tree* 30 He's the soul of courtesy but he can be a wee bit difficult.

III. 12. d. In Tsarist Russia, a serf. Also *transf.*

1806 M. Wilmot *Jrnl.* 17 Aug. in *Russ. Jrnls.* (1934) iii. 271 One..often hears two Ladies..talking to each other about the sale of Lands, purchase of *Souls* (slaves). 1895 C. Garnett tr. *Turgenev's Fathers & Children* i. 2 Nikolai Petrovitch Kirsanov..had..a fine property of two hundred souls, or, as he expressed it—since he had arranged the division of his land with the peasants..of nearly five thousand acres. 1943 E. M. Almedingen *Frossia* iv. 169 The good Boyarin made it known that her dowry would be..five hundred souls, all under the age of fifty. 1969 V. G. Kiernan *Lords of Human Kind* vi. 225 Africans were being disposed of as Europeans were by their princes not long before, when the Congress of Vienna ..distributed them in lots of so many thousand 'souls'. 1977 V. S. Pritchett *Gentle Barbarian* i. 5 Spasskoye was..a self-sufficient feudal community..an empire numbering 5,000 'souls'.

14. b. *The Souls*, a late nineteenth-century aristocratic coterie with predominantly cultural and intellectual interests.

1890 B. Potter *Jrnl.* 31 Dec. (1982) I. 349 Balfour.. would crush them in the intervals between a flirtation with one of the 'Souls' and the reading of a French novel. 1934 H. G. Wells *Exper. Autobiogr.* II. ix. 766 The 'Souls', the Balfour set. 1980 D. Newsome *On Edge of Paradise* ii. 47 The young and wealthy aspirants to public eminence and the eligible daughters of leading families... The group to be christened by Lord Charles Beresford in 1888 'the Souls'.

V. 19. *soul-food*, *-minster*, *-music*, *-power*, *-work* (later example).

1920 W. R. Lethaby in *London Mercury* Mar. 575 The history that can be seen and touched is a strong and stimulating soul-food, entirely different from vague and wearying written history. 1937 Blunden *Elegy* 16 Foremost of all a matin hymn From these soul-minsters leaps aloft. 1900 W. James *Let.* 20 July (1920) II. 133, I..sit *thinking of letters*, and of the soul-music with which they might be filled if my tongue could only utter the thoughts that arise in me to youward. *a* 1930 D. H. Lawrence *Phoenix* (1936) v. 607 They combine with their soul-power some great technical skill. 1927 —— *Let.* 9 Jan. (1932) 679 Painting is more fun and less soul-work than writing.

20. *soul-brother* (later examples), *-mate* (later examples), *-thief*, *-twister*.

1970 R. Lowell *Notebook* 151 We were an empire, soul-brothers To Babylon and China. 1978 *Listener* 20 July 90/1 Baudelaire recognised in Poe a soul brother and mirror image. 1915 F. M. Hueffer *Good Soldier* iii. v. 202 He thought that Mrs Basil had been his soulmate. 1976 Botham & Donnelly *Valentino* ix. 71 Convinced that he had found the woman who would be his life's soulmate. 1889 W. B. Yeats *Lett. to New Island* (1934) 195 Perhaps they are evil-spirits, these soul-thiefs, and not fairies at all. 1928 D. H. Lawrence *Phoenix II* (1968) 284 Dmitri Karamazov doesn't go half the lengths of the other Russian soul-twisters. 1956 D. Gascoyne *Night Thoughts* 33 This soultwister blisters the paint of the set.

21. *soul-making*, *-mating*, *-prompting*, *-transfiguring*.

1819 Keats *Lett.* (1958) II. 102 Call the world if you Please 'The vale of Soul-making'. 1922 Joyce *Ulysses* 138 If aught that the..hand of sculptor has wrought in marble of soultransfigured and of soul-transfiguring deserves to live. *a* 1930 D. H. Lawrence *Phoenix* (1936) v. 605 Man just doesn't know how to interpret his own soul-promptings. 1939 A. Huxley *After Many a Summer* x. 140 Love, Passion, Soul-mating—all in upper-case letters. 1958 *Times Lit. Suppl.* 21 Feb. 101/2 The mid-Victorian novelists..thought of it [*sc.* Oxford] as a moral testing-ground or 'a vale of soul-making'.

22. *soul-awakening*, *-deadening*, *-destroying* (later examples), *-inspiring*, *-satisfying* (later examples), *-searing*, *-shattering*, *-stirring* (later example), *-testing*.

a 1822 Shelley *Posthumous Poems* (1824) 320 Soul-awakening music, sweet and strong. 1926 C. Barry *Detective's Holiday* iv. 33 Suddenly a soul-awakening boom behind him smote his ears. 1909 Mrs. H. Ward *Daphne* viii. 186 This dull, soul-deadening English life. 1937 *Atlantic Monthly* CLIX. 57/1 Exact information which really can be taught is despised as soul-deadening. 1898 G. B. Shaw *Candida* ii, in *Plays Pleasant* 123 What dreadful—what soul-destroying cynicism! 1930 *Engineering* 25 July 111/3 A common indictment against modern conditions is that machine tending is 'soul-destroying'. 1976 J. Snow *Cricket Rebel* 40 It was often soul destroying. On wet wickets or slow ones, I was expected to charge up and down and let it go when I knew I had no earthly chance of getting anything out of the wicket. 1794 J. Trumbull in *Columbian Muse* 58 And damp'd, alas! thy soul-inspiring ray, Where Virtue prompted and where Genius soar'd. 1979 'A. Hailey' *Overload* ii. i. 106, I guess it's real soul-inspiring to work in a ritzy layout like this. 1890 Kipling *Life's Handicap* (1891) 151 He was afraid for the sake of another—which is the most soul-satisfying fear known to man. 1939 F. Scott Fitzgerald *Lett.* (1964) 48 It is not very soul-satisfying because it [*sc.* the cinema] is a business of telling stories fit for children. 1936 *Times Lit. Suppl.* 21 Mar. 242/3 We are..given a soul-searing account of a Russian pogrom. 1979 'A. Hailey' *Overload* iii. xii. 253 A week and a half had passed since the soul-searing night when he learned that Ruth's life was endangered by cancerous cells at large in her body. 1899 Kipling *From Sea to Sea* II. xxv. 5 The result is soul-shattering. 1974 R. Harris *Double Snare* xi. 73 She and I had a soul-shattering row, and weren't on speaking terms. 1927 *Granta* 14 Oct. 9/1 He rapidly composed and delivered a few soul-stirring orations. 1932 Wodehouse *Louder & Funnier* 212 The unmistakable look of a man who has passed through some soul-testing experience. 1965 J. A. Michener *Source* (1966) 192 Captain Epher's plan of battle required daring from all the Hebrews and soul-testing courage from a few.

23. *soul-struck*, *-transfigured*.

1949 Blunden *After Bombing* 3 The child soul-struck with the yellow flag's new fire. 1922 Soul-transfigured [see *soul-transfiguring*, sense 21 above].

25. soul-bearer, among the Akan peoples of West Africa, a person deemed to carry within him the external soul of a ruler or important person; **soul-body** *Spiritualism*, a spiritual body (see Spiritual *a.* 4 a); **soul-bolts** *pl.*, 'the bolts which fasten the soul in place', used in var. slang phrases expressive of surprise or shock; cf. *soul-case* below; **soul brother** orig. *U.S. Blacks*, a fellow Black man; cf. *soul sister* below; **soul-candle,** (*b*) [tr. Yiddish *neshome licht*, f. *neshome* soul (Heb. *nešāmā*) + *licht* Light *sb.* (G. *licht*)] in Judaism, a candle lit on the eve of the anniversary of a parent's death, and also on the eve of Yom Kippur (the custom is said to derive from Prov. xx. 27); **soul-case**: restrict † to sense in Dict. and add: (*b*) *U.S.* and *Austral. slang*, 'the casing of the soul', chiefly used in slang phrases expressive of hardship or suffering; cf. *soul-bolts* above; **soul-catcher,** among various tribes of North American Indians, a hollowed bone tube used by a medicine man to contain the soul of a sick person (see also quot. 1976); **Soul City,** an epithet applied to the Harlem area of New York city; also *transf.*; **soul-doctor** *slang*, (*a*) a clergyman; (*b*) a psychiatrist; † **soul-driver,** (*a*) a clergyman; (*b*) *U.S.*, a person who trades the services of convicts, indentured servants, or slaves; **soul food** orig. *U.S. Blacks*, the kind of food typically eaten by Black people, *spec.* those foodstuffs originating in the southern states of America; **soul-force** = *satyagraha; **soul-friend**: also in extended use (see quots.); **soul kiss** = *deep kiss* s.v. *deep *a.* IV. c; so **soul-kiss** *v. trans.*; hence **soul-kissing** *vbl. sb.*; **soul music,** a type of music popularized by Black singers which incorporates elements of rhythm and blues and gospel music; also *ellipt.*: see sense *3 b; **soul sister** orig. *U.S. Blacks*, a fellow Black woman; cf. *soul-brother* above; **soul-sleeper** (later U.S. examples); **soul stuff** = next; **soul-substance,** a hypothetical immaterial substance believed to form the 'spirit' or 'self' of each person (in some cultures also of animals and objects), and which is independent of the material body and outlives it.

1951 E. L. R. Meyerowitz *Sacred State of Akan* ii. 51 Like every king.., the queenmother has her elders, among whom are several spokeswomen, female *akrafo* or soul-bearers. 1967 *Times* 14 Nov. 17/2 (Advt.), A Baga wood nimba shoulder mask,..an important large Ashanti gold soul-bearer's disc,..a large New Ireland Uli. 1961 Soul body [see *exteriorize *v.*]. 1971 *Spiritualist* Oct.–Dec. 6/2 Help each other that your soul-body may rise in beauty and can be admired when you reach the World of Spirit. 1850 H. Melville *White Jacket* II. xliv. 296 Start my soul-bolts, maties, if any more Blue Peters and sailing signals fly at my fore! 1902 J. Burgess *Some Shetland Folk* 77 If du has, I'll knock the bloomin' sowl-bolts out of him. 1903 'T. Collins' *Such is Life* vi. 234 'Wouldn't think that horse had a devil in him as big as a bull-dog,' observed the horse-driver. 'Shake the soul-bolt out of a man, s'posen you *do* stick to him.' 1959 *Jazz* Fall 291 It's one of those type LPs. I had all 'soul brothers'. 1969 *Listener* 4 Sept. 319/3 And if you think the main feeling being expressed is self-pity (and the self-generated violence and frustrations of self-pity), then so what? This is strictly for soul-brothers. 1973 H. Nielsen *Severed Key* xiv. 144 I've got some soul brothers hustling baggage at LAX. 1978 I. B. Singer *Shosha* viii. 141 Toward the evening meal, she lit a large candle stuck in a pot of sand—a 'soul candle'—and put on a silk holiday dress. 1835 A. B. Longstreet *Georgia Scenes* 109 When you come to the last half mile of each heat, run his heart, liver, lights, and soul-case out of him. 1896 J. C. Harris *Sister Jane* 277 The way that hoss flung around wi' you was enough to jolt your soul-case loose. 1901 F. J. Gillen *Diary* 15 Apr. (1968) 34 Flies were celebrating some festival all night and worried the very soul cases out of us. 1962 R. Tullipan *March into Morning* 13 Then he got the bright idea of bringin' in orphan kids and working the soulcase off them until they turn eighteen and have to be paid more money. 1932 D. Jenness *Indians of Canada* 333 Peculiar to the medicine-men of the Haida, Tlinkit, and Tsimshian was the use of a special 'soul-catcher', a bone tube, generally carved, for capturing the wandering souls of the sick and restoring them to their bodies. 1969 *Times* 22 Sept. 14/2 One invariably sees a face in the centre of a soul-catcher, a tube of hollowed bone into which the shaman [of the Tsimshian Indians] sucked the soul of a sick man—to keep it safe from harm while the illness lasted. 1976 *Times* 10 Nov. 18/4 A nine-inch bone soul catcher of the Tsimshian tribe reached £12,000... A soul-catcher is a tube within which a medicine man would catch the imp that caused a sickness. 1964 *N.Y. Times Mag.* 23 Aug. 62/3 Soul City, Harlem. 1971 B. Malamud *Tenants* 89 Lesser descended..into Soul City by himself. 1977 M. Herr *Dispatches* (1978) 196 Danang was Soul City for many of us, it had showers and drinks. 1785 Grose *Dict. Vulgar Tongue*, *Soul doctor*, or *driver*, a parson. 1962 D. Lessing *Golden Notebook* i. 202 Anna Wulf is sitting on a chair in front of a soul-doctor. 1699 B. E. *New Dict. Canting Crew*, *Soul-driver*, a Parson. 1774 in *Amer. Hist. Rev.* (1900) VI. 77 Among them there was two Soul drivers. They are men who..drive them [*sc.* servants and convicts] through the Country..untill they can sell them to advantage. 1818 *Massachusetts Spy* 4 Nov. (Th.), Two men, in the character of soul drivers, lodged in the jail for safe keeping, five negros. 1846 *Swell's Night Guide* 132/2 *Soul driver*, a methodist parson. 1973 A. Dundes *Mother Wit* 230 Individuals who speculated in the purchase and sale of slaves were called 'Negro-drivers' or 'soul-drivers'. 1964 *N.Y. Times Mag.* 23 Aug. 62/3 Soul *food*, chitterlings, collard greens, ham hocks, grits, black-eyed peas and rice, and the like. 1969 L. Sanders *Anderson Tapes* (1970) xxviii. 71 This soul food crap—knuckles and hocks and greens. 1972 *Times* 15 Nov. 10/5 *Soul food.* Professional chef with knowledge of American Southern food..wanted for a new restaurant..in Chelsea. 1978 *Broadcast's Programme Edinburgh TV Festival* 8/3 The social centre of the series is a soul-food grocery owned by a West Indian entrepreneur. 1969 *It* 4–17 July 10/4 With soul force we'll look to the needs of our brother In a world that's our universal home. 1977 *Arab Times* 14 Dec. 2/5 'The voice of women..is a special soul-force in the struggle for a non-violent world,' the 36-year-old pacificist leader from strife-torn Northern Ireland declared. 1929 I. M. Clark *Church Discipline in Scotland* i. 29 Columba had a method of entrusting those who had sinned to the spiritual care of individual monks of his community, who were termed soul-friends and whose duty it was to restore the souls of those penitents. 1979 *Church Times* 11 May 2/3 A special sort of job is being offered to spiritually gifted women in the diocese of Truro. The Bishop..wants them to train to be 'soul friends'—so that they may give spiritual guidance and direction with his formal backing and recognition. 1953 H. Waugh *Last seen Wearing* 55 She calls him exciting and lets him soul-kiss her. 1960 Wentworth & Flexner *Dict. Amer. Slang* 504/1 *Soul kiss*, a long passionate open-mouthed kiss, during which a lover's tongue licks, caresses, or explores the tongue and mouth of the beloved. 1970 R. Davies *Fifth Business* ii. 130 Some of them were experts in what were then called French kisses or soul kisses, which the irreverent called 'swapping spits'. 1973 E. Jong *Fear of Flying* 82, I had the distinct sensation of kissing my own mouth—like when I was nine and used to wet a piece of my pillow with saliva and then kiss it or try to imagine what 'soul-kissing' was like. 1961 *Commonweal* 24 Mar. 658 It's called 'soul music' because its practitioners have incorporated some of the backbeat, rhythms, and exclamatory melodic lines of Negro gospel music. 1968 P. Oliver *Screening Blues* 6 Soul music, which exploits the intensity of expression of religious song, the form and instrumental character of the blues and the maudlin sentiments of pop music. 1974 *Black World* Mar. 57/2 Soul music belched from windows where Black women wearing tired faces gazed impassively down at the hopeless street. 1980 *Oxford Times* 8 Feb. 15/1 They get equally close to the style and spirit of American soul music. 1967 Wentworth & Flexner *Dict. Amer. Slang* Suppl. 705/1 *Soul sister*, a female Negro. *Negro use only*. 1968 *N.Y. Times* 17 June 24 Plate glass in Negro-owned establishments remained intact and displayed the words, 'Soul Brother' or 'Soul Sister'. 1976 *Drum* (E. Afr. ed.) June 10/3 African girls have always plaited their hair, and it was the soul sisters in America who were copying the girls in Africa. 1860 *Southern Enterprise* (Thomasville, Georgia) 13 June 2/5 Soul Sleepers is the name of a new religious sect which has recently made its appearance at Fairfield, Iowa... They.. think that the soul is a mortal substance, and sleeps with in the body until resurrection. 1887 J. Kirkland *Zury* 65 He and Peddicomb had both been connected with the little sect of Christians called 'Soul-sleepers'. 1880 *Cent. Dict.*, *Soul-stuff*..the hypothetical substance of the soul; psychoplasm. 1909 W. James *Mem. & Stud.* (1911) viii. 202 If there were in the universe a lot of diffuse soul-stuff, unable of itself to get into consistent personal form.. it might get its head into the air, parasitically, so to speak, by profiting by weak spots in the armour of human minds. 1972 D. Davies *Dict. Anthropol.* 165/2 *Soul-stuff*, mana. The spiritual power with which every man is endowed... Many societies seek to enhance this power by standing in the tribe. It can only be gained by special feats... It is also thought to be found in the hair. 1890 W. James *Princ. Psychol.* I. x. 318 But what is this abstract numerical principle of identity..? May it be the indivisible Soul-Substance, in which, according to the orthodox tradition, my faculties inhere? 1924 W. B. Selbie *Psychol. Relig.* ii. 28 Anthropologists are..fairly generally agreed that underlying all religions is what they call animism, or belief in a soul substance diffused over the whole world, but in things as well as in persons. 1972 H. J. Eysenck *Encycl. Psychol.* II. 57/2 Heraclitus..considered fire as the primary force and 'soul-substance' because it moved and transformed matter.

‖ **soulagement** (sūlaʒmaŋ). [Fr. Cf. Solace *sb.*[1] and Solacement.] Solace, relief.

1777 Earl of Carlisle *Let.* 18 Feb. in J. H. Jesse *Geo. Selwyn & Contemporaries* (1844) III. 171, I know our house might be a *soulagement* to you. 1949 H. Nicolson *Diary* 17 Apr. (1968) 169 All this calm reflects my deep *soulagement* at Vita's improvement in health. 1968 A. Coates *Myself Mandarin* xvii. 245 One of my Secretariat colleagues telephoned me..to remind me..to return my instrument... With a certain *soulagement* I learned that he was referring merely to a piece of paper—the frightening document on the desk.

soulful, *a.* Add: **1.** (Earlier example, and later examples of extended sense.)

1860 Princess Victoria *Let.* 6 Feb. in R. Fulford *Dearest Child* (1964) 234 Her Princess said to Fritz Karl while they were singing something of Mozart 'You are looking soulful.' 1931 A. Huxley *Music at Night* 18 When the great obvious truth is affirmed..in a series of soulful close-ups..the sensitive can only wince. 1951 *Sunday Pictorial* 21 Jan. 6/4 His eyes become deceptively soulful.

3. Expressive of Black feeling; characteristic of Black music. Also as quasi-*adv.*

1964 *Amer. Folk Music Occasional* i. 46, I sing my song more soulful. 1973 *Black Panther* 15 Sept. 15/1 The audience is encouraged to join hands with their neighbors and all standing sway with the soulful music. 1979 *Black World* Apr. 9/1 'Soul' is a highly valued concept among Afro-Americans. Soulful behavior may be called something else by some Afro-Americans, but its value goes...

soulfully, *adv.* (Later examples.)

1922 Joyce *Ulysses* 279 Bronze, listening by the beerpull, gazed far away. Soulfully. 1973 *Daily Tel.* 23 Apr. 15/2 Her two daughters,..one bitter and promiscuous, one soulfully virginal, appeared as remote from ordinary living as folk-tale princesses.

soulhood. (In Dict. s.v. Soul *sb.*) (Later examples.)

1933 S. Sassoon *Traveller to his Soul* in *Satirical Poems* (ed. 2) 68 The problem which concerns me most..Is, bluntly stated, 'Have I got a soul?' And, soulhood granted, while millenniums roll, Will it inhabit some congenial clime..Whereon to meet what the tame 'the Whole'? 1940 C. S. Lewis *Problem of Pain* ix. 129 Supposing, as I do, that the personality of the tame animals is largely the gift of man—that their mere sentience is reborn to soulhood in us as our mere soulhood is reborn to spirituality in Christ—I naturally suppose that very few animals indeed, in their wild state, attain to a 'self' or ego.

soul-search, *v.* [Back-formation f. Soul-searching *ppl. a.*] **a.** *trans.* To examine penetratingly and thoroughly; to make a soul-searching analysis of. *nonce-use.* **b.** *intr.* To engage in examination of one's thoughts, to reflect deeply.

1966 *Economist* 29 Jan. 421/2 Trotsky writing in exile, Mr Alistair Cooke soul-searching the trial of Mr Alger Hiss, Adlai Stevenson eulogising Robert Frost. 1968 A. Marin *Clash of Distant Thunder* (1969) xv. 114 When we tried Eichmann people used to argue and discuss and soul-search. We Jews are very good at that. 1972 *Daily Tel.* 30 Aug. 11/4 Last week there was the heart-rending case of a man in public life soul-searching about the reasons his son took to drugs.

Also as *sb.*, an act of soul-searching; hence **soul-searched** *ppl. a.*

1966 *Guardian* 8 July 3/5 (*heading*) Soul-search by the Liberals. 1970 E. M. Brecher *Sex Researchers* ix. 255 More prolonged eye-contacts, or 'soul-searches'. 1978 *Times* 9 Oct. 18/4 The most honest and soul-searched response of the conclave.

soul-searching, *ppl. a.* Also in extended and trivial use. Hence **soul-searchingly** *adv.*

1938 Partridge *World of Words* ix. 309 The following sextette, ultimately selected from a much longer list of words that had been soul-searchingly gathered for your delectation. 1959 *Economist* 17 Jan. 203/1 After a 'soul-searching' parliamentary debate, an openly penitent Belgian government has announced its plan for dealing with the suddenly agitated Congo. 1960 *News Chron.* 13 Oct. 3/2 The deeply soul-searching: 'What do you think of modern jazz?'

soul-searching, *vbl. sb.* [f. prec.] Examination of one's soul or conscience, deep reflection; penetrating consideration of a state of affairs.

1948 'E. Crispin' *Buried for Pleasure* xvii. 145 The announcement..is not made without a good deal of preliminary soul-searching. 1953 *Time* 6 July 6/3, I would suggest some honest soul-searching for these unrealistic fathers, and orchids to Father Gerald Murphy and the increasing number of his prototypes among the younger Jesuits. 1974 *Country Life* 18 Apr. 926/2 What has been the subject of..agitated soul-searching is whether rector and churchwardens..had any right whatsoever to sell it. 1978 P. G. Winslow *Coppergold* 39 It's causing a bit of soul-searching at the Yard.

souly, *a.* Restrict † *Obs.* to senses in Dict. and add: **1. b.** = Soulful *a.* 1 in Dict. and Suppl. *colloq. rare.*

1911 D. H. Lawrence *White Peacock* vii. 131 I'm not one of your souly sort.

Soumak (sū·mak). Also **Soumac, Sumac,** etc. [Orig. uncertain, perh. a corruption of the Azerbaijan place-name *Shemakha*.] A type of rug or carpet made in the neighbourhood of Shemakha in Azerbaijan, distinguished by a flat, napless surface and loose threads at the back. Freq. *attrib.* Cf. *Kashmir 2.

1904, etc. [see *Kashmir 2]. 1932 D. C. Minter *Mod. Needlecraft* 228/2 *Knitting Stitch*..sometimes called Sumac or Kelim Stitch..is..a long diagonal stitch. It gives a surface effect like those of various traditional woven rugs. 1959 *Chambers's Encycl.* III. 135/2 There are two main varieties of smooth-faced carpets: Kilim, woven like tapestry, and Soumak, where the pattern weft passes over four and under two warp threads. 1960 H. Hayward *Antique Coll.* 261/1 *Soumac rugs*, Caucasian rugs in a tapestry weave with conventional designs. 1974 *Times* 9 Mar. 23/5 Specialists in Kelims, Soumaks and Tribal Rugs.

sou markee (sū mā·kī). Also **sou marquee, sumarkee,** etc. [ad. Fr. *sou marqué*, lit. 'marked sou'.] A small French coin of the eighteenth century issued for the colonies and circulating esp. in the West Indies and North America. Hence *loosely*, something of little value.

1665–7 J. Lauder *Jrnls.* in *Publ. Scottish Hist. Soc.* (1900) XXXVI. 92 On the place wheir they make it its sold for a sous marky la livre. 1826 *Massachusetts Spy* 5 July (Th.), Who the d—l would give a sumarkee to read the newspapers after breakfast? 1896 H. G. Parker *Pomp of Lavilettes* 54 I'll bet he's got nothing more than what he went away with, and that wasn't a sou markee! 1903 A. H. Lewis *Boss* 181, I don't pony for a sou markee. 1936 J. A. McKenna *Black Range Tales* 268 Marshall drifted from one settlement to another, and he likewise died without a sou marquee. 1952 R. M. Hamilton *Canad. Quotations* 138/2 It's not worth a sou marquee. Phrase common in the Maritime provinces meaning, of trifling value; a reference to French Guiana sous which, counterstamped by other West Indian colonies, were sometimes carried north to Canada. 1977 F. Leiber *Swords & Ice Magic* 80 'By the beats of Titchubi,' the former breathed, 'this is no *sou marque*, black dog no *chien noir*.'

sound, *sb.*[3] Add: **1. a.** Also, pressure waves that differ from audible sound only in being of a lower or a higher frequency. Cf. *infrasound, *ultrasound.

1967 I. M. Freeman *All about Sound & Ultrasonics* xiii. 99 Sonar is just one of the many uses that engineers and scientists have found for ultrasonic sound, which is often called ultrasound. These are names for sound that is too high in frequency to be heard. 1973 D. Ensminger *Ultrasonics* i. 6 Perhaps the animal that is best known for its use of ultrasonics is the bat. Many scientists have studied these interesting animals and their use of sound to find food. 1978 R. B. Minnix in Lipscomb & Taylor *Noise Control* i. 30 Infrasound is concerned with very low frequency (below about 20 Hz) longitudinal mechanical waves where sound is felt rather than heard.

c. *sound-on-film* (Cinemat.), the incorporation of the sound track with the film. Freq. *attrib.* Cf. *married print* s.v. *married *ppl. a.* 3.

1931 B. Brown *Talking Pictures* 270 Sound-on-film recording may be monitored direct from a photo-electric cell in the recording machine. 1957 Manvell & Huntley *Technique Film Music* ii. 27 The true arrival of the sound film was the arrival of sound-on-film. 1976 *Oxf. Compan. Film* 450/1 In the Vitaphone process the sound came from a disc precariously synchronized with the picture. The limitations of this system were quickly recognized and 'sound-on-film' became standard.

d. *Physics.* Applied to various kinds of wave motion (designated *zero*, *second*, *third*, etc., *sound*) that are predicted or observed to occur in superfluids and physically bear some resemblance to ordinary ('first') sound.

1944, etc. [see *second sound* s.v. *second *a.* 7 a]. 1957 tr. L. D. Landau in *Soviet Physics JETP* V. 102/1 It is shown..that in a Fermi liquid at absolute zero other waves can be propagated; these differ in nature from ordinary sound, and we shall call them waves of 'zero sound'. 1959 K. R. Atkins in *Physical Rev.* CXIII. 962 This article discusses the possible existence of two hitherto undetected types of wave propagation in liquid helium II. Third sound is a surface wave of long wavelength on a liquid helium film... Fourth sound may exist in narrow two-sided channels. *Ibid.*, To discuss wave propagation in liquid helium II, it is necessary to write down two separate hydrodynamical equations, one for the superfluid component and the other for the normal component. In first sound the two components move in the same direction in phase, and there is a first-order oscillation of the density but only a second-order oscillation of the temperature. In second sound the two components move in opposite directions out of phase, and the temperature oscillation is then first-order while the density oscillation is only second-order. 1969 W. E. Keller *Helium-3 & Helium-4* vi. 203 (*caption*) Attenuation and propagation velocity of sound in liquid He[4] showing the characteristics associated with the transition from first sound to zero sound for two frequencies. 1974 D. J. Bergman in K. D. Timmerhaus et al. *Low Temperature Physics—LT 13* I. 507 Following our experience with third sound, we may expect that in fourth sound, too, when the channels that hold the helium are sufficiently small so that the normal fluid motion is completely locked out, the only important source of attenuation will be conduction of heat into the walls of the helium channels. 1974 *Nature* 15 Mar. 194/3 The report..that they have observed the propagation of fourth sound in the two newly discovered phases of liquid [3]He amounts to the first unequivocal evidence that both of these new phases are superfluids. 1976 *Physics Bull.* Aug. 351/2 'Zero sound'..corresponds to oscillations in shape of the Fermi surface. 1981 *Nature* 2 Apr. 359/2 Second sound is an unusual type of propagating wave mode, which can occur in superfluids, involving fluctuations in the local temperature and entropy of a medium rather than in the local density and pressure as found in a conventional sound wave.

e. *sound and light* = *son et lumière 1. Used *attrib.*

1960 *Woman* 23 Jan. 35/3 The pretty little town of Buxton, one of the first in England to stage a 'sound and light' production for summer visitors. 1966 J. Philips *Wings of Madness* (1967) i. i. 9 The Sound and Light program put on..every night..kept tourists in town. 1979 *United States 1980/81* (Penguin Travel Guides) 548 On weekend evenings, a multimedia sound-and-light show using laser beams.

2. a. (Examples with reference to a television set: cf. *picture *sb.* 2 f.)

1928 *Television* Oct. 10/2 A one-act play was..televised ..and receiving televisors within a range of four miles tuned in both sight and sound. 1960 [see *picture *sb.* 2 f]. 1979 R. Jaffe *Class Reunion* (1980) iii. iii. 322 Emma was..watching television, but she had the sound..low.

3. a. Also, a phenomenon identical to an audible sound except that it is inaudible by reason of its frequency (cf. sense *1 a).

1950 *Sci. Amer.* Aug. 52/2 The English physiologist H. Hartridge..watched bats flying through darkened rooms and advanced the theory that they might be orienting themselves by means of ultrasonic sounds too high in frequency for human ears to hear. 1976 L. H. Schaudinischky *Sound, Man, & Building* i. 8 Above 20 000 Hz extends the 'infinite' supersonic range, the ultrasound. Man is not equipped with an organ capable of directly responding to sounds in that range, but where infrasound is concerned it may be picked up with the aid of a special sense of touch. 1978 J. Goldstein in P. M. Lipscomb *Noise & Audiol.* i. 6 In order to be heard, a sound must be within a certain frequency range because there are limitations in the frequencies the human ear can perceive.

d. *pl.* Popular music; also in *sing.*, a tune or record. *slang* (orig. *U.S.*).

1955 *Amer. Speech* XXX. 304 Kenton's music is round sounds. 1961 Rigney & Smith *Real Bohemia* p. xvii, *Sounds*, music, mainly jazz. 1968 *Daily Mirror* 27 Aug. 7/5 Together cats don't buy records, they buy *sounds*, and they never blow their cool.

e. A characteristic style of (usu. popular) music indicated by a defining word or words. Cf. *Mersey sound* s.v. *Mersey.

1963 [see *gear *sb.* 5 c]. 1967 *Radio Times* 21 Dec. 35/4 *The Greek Sound*... Tonight's programme is about the new genre, which in the last eight years has given a new impetus and vitality to Greek popular music. 1970 *Guardian* 15 June 9/5 Steel Bands and the Reggae Sound beloved of skinheads. 1974 *Listener* 13 June 767/1 In 1927, there was an inimitable Ellington sound, and so there was at the end.

4. f. The impression produced by a statement or report, freq. in phr. *to like the sound of* (some person or thing). (See note at sense 4 d in Dict.)

1859 Mrs. Gaskell *Let.* 21 Mar. (1966) 543, I like the 'sound' of him extremely, and I hope he will like me when we come to know each other. 1965 R. Sheckley *Game of X* (1966) xxii. 155 'You take care of the piloting, and we will handle the navigating.' Somehow I didn't like the sound of that.

5*. In elliptical uses. **a.** *Cinemat.* and *Broadcasting.* The department in charge of recording sound. Also, an engineer in this department; the equipment used by him.

a 1940 F. Scott Fitzgerald *Last Tycoon* (1941) iii. 30 Call sound, and if he's been heard from, call him. 1969 M. Steinbeck *On Stage* 165 The voice track on a film is called the sound track. The engineer in charge and the whole unit is referred to simply as 'sound'. The director may call out before a take, 'Is sound ready?' 1972 *Listener* 21 Dec. 852/1 Sequence of calls before a shot. Production Assistant: 'Quiet. Going for a take. Standing by.' Director: 'Right.' Sound: 'Sound running.'

b. = *radio *sb.* 2 b. Cf. *sound radio*, sense 7 b below. Also *attrib.*

1949 *Times* 17 Feb. 5/3 The first hundred thousand mark is about to be reached in..television licences..compared with the 11 m. for sound. 1955 *Times* 29 June 11/2 So far not even B.B.C. television has found the way to transfer the aura of the 9 p.m. sound news to television. 1967 'M. Hunter' *Cambridgeshire Disaster* iv. 28 If necessary he would give up television, ask for a transfer to Sound, anything to get more time at home. 1972 P. Black *Biggest Aspidistra* iii. iii. 171 The most obvious effect of the Coronation for television was the demand for sets... Though the BBC still regarded sound as the senior service..the sound audience never again exceeded television's.

6. a. *sound-aspect*, *-association*, *-change* (later examples), *-clause*, *-colour*, *-combination*, *-complex*, *-development*, *energy*, *event*, *-feature*, *-gesture*, *-group*, *-history* [tr. G. *lautgeschichte*], *-image*, *-intensity*, *-language*, *level*, *-mark*, *-output*, *-pattern*, *-picture*, *poem*, *power*, *-quality*, *-sentence*, *-sequence*, *-structure*, *-symbol*, *-system* (earlier and later examples), *-type*, *-unit*, *-value*, *-wave* (earlier example), *-word*, *-world*.

1936 H. Mulder *Cognition & Volition in Lang.* 46 The life of the language as regards its sound-aspect. 1954 A. H. Gardiner *Theory of Proper Names* (ed. 2) 73 Even logicians..overlooked the importance of the sound-aspect. 1924 Mawer & Stenton *Introd. to Survey of Eng. Place-Names* ix. 174 Its chief weakness is the remoteness of the sound-association between the original compound name and the suggested simple derivative. 1912 L. Bloomfield in *Jrnl. Eng. & Gmc. Philol.* XI. 623 S[heffield] confuses the factors—sound-change and analogy—that constitute change in language. 1939 [see *palatal *a.* 2 b]. 1962 W. Nowottny *Lang. Poets Use* i. 5 Calling in alliteration's aid and that of a sound-change. *a* 1889 G. M. Hopkins *Jrnls. & Papers* (1959) 273 We may now say of rhythm i.e. verse that it is the recasting of speech into sound-words, sound-clauses and sound-sentences. 1890 G. B. Shaw in *Star* 9 May 2/5 Marlowe's line was not 'mighty'..but it was tuneful, exquisitely emphasised, and sometimes gorgeous in its sound color. 1962 *Listener* 9 Aug. 225/1 Schönberg's 'melody of sound-colours' (*Klangfarbenmelodie*). 1924 Mawer & Stenton *Introd. to Survey of Eng. Place-Names* v. 100 An unfamiliar English sound or sound-combination was altered to suit the Norman pronunciation. 1965 *Language* XLI. 93 First a child learns a sound-combination and then he attaches meaning to it! 1931 G. Stern *Meaning & Change of Meaning* 31 If the sound-complex is to be apprehended as meaning something..a mental content must accrue to it. 1900 E. Björkman *Scand. Loan-Words in M.E.* I. 30 There are some tests of form which are not based on differences of sound-development between Scandinavian and English. 1965 *English Studies* XLVI. 141 Surnames, like Johe *Le Roper*..reflect the spoken dialect, but do not necessarily prove indigenous sound-developments. 1931 G. O. Russell *Speech & Voice* iv. 21 (*heading*) Sound-energy not air motion. 1962 A. Nisbett *Technique Sound Studio* 239 They readily remove sound energy from the air at their resonant frequency, and this is then mopped up within the absorber. 1962 P. Strevens *Papers in Lang.* (1965) xii. 146 When sound events are recorded, the technical standard of recording is important. 1939 *Word Study* Mar. 2/1 Linguistics..deals with the use of a limited number of definable events—the significant sound-features of a language—occurring in certain definable sequences. 1964 W. R. Lee in D. Abercrombie et al. *Daniel Jones* 292 Sounds and sound-features which belong to neither language. 1938 I. Goldberg *Wonder of Words* iv. 55 Sound-gesture, such as Paget draws upon in this etymology, is precisely what it is called. 1956 J. Lotz in L. White *Frontiers of Knowl.* xiv. 219 Marginal sound-gestures like the bilabial trill used when shivering: *Brrr!* 1928 O. Jespersen in *Proc. Brit. Acad.* XIV. 352 There are no other words than *switch* and *stretch* beginning and ending with exactly these sound-groups. 1964 J. Vachek in D. Abercrombie et al. *Daniel Jones* 199 If followed by a vowel, the same sound-group was preserved unimpaired. 1933 O. Jespersen *Essentials Eng. Gram.* vi. 62 The sound-history of French also serves to explain some striking peculiarities concerning the use of the letter *g* in English spelling. 1964 *English Studies* XLV. 422 A detailed knowledge of sound-history..and sound-substitutions. 1943 tr. M. Buber in H. Read *Educ. through Art* ix. 279 Sound-image after sound-image.. emerges from vibrating throat..into the surrounding air. 1951 A. Gardiner *Theory of Speech & Lang.* 70 It is only the sound-image connected with the words which can be reproduced in a physical copy. 1973 S. Heath in *Screen* Spring/Summer 108 A *langue* is defined by Saussure as a system of signs, a sign being the union of

SOUND

signifiant ('sound-image') and *signifié* ('concept'). **1982** *Listener* 16 Dec. 26/3 There's something wrong with the way a taped sound-image remains fixed in eternity. **1934** *Discovery* Dec. 346/1 Noise is a subjective phenomenon and cannot be directly measured. The stimulus causing this impression of sound is a sound-intensity which can be defined and measured objectively. **1952** *Mind* LXI. 215 It is impossible to imagine a sound-intensity divorced from any definite sound-pitch. **1969** *Gloss. Acoustical Terms* (*B.S.I.*) 16 *Sound intensity*,..the sound energy flux through unit area. **1918** *Amer. Jrnl. Philol.* XXXIX. 89 A Dakota Indian..would not understand a Neapolitan, even though he would sooner understand the gestures than the sound-language. **1937** R. A. Wilson *Birth of Lang.* 160 The twenty-six already differentiated elements of sound-language. **1931** S. K. Wolf in L. Cowan *Recording Sound for Motion Pictures* xx. 301 It is necessary to have some means of varying sound levels in theatres. **1974** *Physics Bull.* June 227/1 Leeds City Council, decided to use its licensing laws to limit sound levels in ballrooms, discotheques and similar places of entertainment. **a 1892** W. Whitman *Daybks. & Notebks.* (1978) III. 671 One of the first desiderata..is a set of..sound-marks attached to letters..each mark belonging to that specific sound. **1978** *Sci. Amer.* Jan. 29/3 We cannot shut our earlids; awake, we are always open to..the old soundmarks we remember and cherish. **1937** Sound output [see *control engineer* s.v. *control *sb.* 5]. **1947** Crowther & Whiddington *Science at War* 175 It was found that the sound-output was mainly due to propellers. **1925** *Language* I. 41 One must ascertain if the sound is a typical form or one of the points in its sound pattern, or is merely a variant of such a form. **1977** P. Strevens *New Orientations Teaching of English* xii. 154 Accent features are manifested in sound-patterns of various kinds. **1903** A. W. Patterson *Schumann* xvi. 186 The whole forms a kind of sound-picture representing the various personages in the dance. *Ibid.* xvii. 203 What if the tone poet..knew infinitely better than his..advisers what was or was not fitting in the great sound-poem to which his genius gave birth? **1971** *Guardian* 18 Feb. 10/6 Artaud wrote sound poems. **1947** Crowther & Whiddington *Science at War* iii. 155 A transmitter producing about 50 watts of sound-power in water was adequate. **1950** D. Jones *Phoneme* 12 An alphabetic system of phonetic transcription consists of letters representing sound-qualities. **1977** *Broadcast* 28 Nov. 14/2 The singles we get are so badly pressed that we get complaints from listeners about the sound quality. **a 1889** Sound-sentence [see *sound-clause* above]. **1914** L. Bloomfield in *Trans. Amer. Philol. Assoc.* XLV. 69 The various parts of this sound-sequence..have been heard and uttered by the speaker (or the hearer). **1962** F. Behre *Contrib. Eng. Syntax* 134 The sound-sequence..must correlate with certain extra-lingual elements to be inferred from the context. **1959** D. Cooke *Lang. Mus.* v. 234 Music..has now become pure sound-structure, an intellectual and aesthetic delight. **1936** *Science & Society* I. 38 Certain sound-symbols are universally attached to the same referent by all members of the community. **1975** *Language for Life* (Dept. Educ. & Sci.) xxvi. 521 The learning of sound-symbol correspondences should take place in the context of whole word recognition and reading for meaning. **1879** H. Sweet in *Trans. Philol. Soc. 1877–9* 544, I am fully conscious that mine is a very inadequate study of an exceptionally difficult sound-system. **1897** *Mod. Lang. Notes* XII. 244 Least understood..is the historical development of the sound-systems of modern dialects. **1949** J. R. Firth in *Trans. Philol. Soc. 1948* 132 More detailed notice of 'h' and the *glottal stop* in a variety of languages will reveal the scientific convenience of regarding them as belonging to the prosodic systems of certain languages rather than to the sound systems. **1977** *Canad. Jrnl. Linguistics 1976* XXI. 177 No information about how they work in the sound system of a language is gained. **1941** Sound-type [see *allophone]. **1964** J. C. Catford in D. Abercrombie et al. *Daniel Jones* 29 The laryngologists have no tradition of systematic..description of phonologically pertinent sound-types. **1934** J. J. Logan *Outl. Eng. Philol.* 24 A syllable, thus, is a sound-unit. **1920** T. S. Eliot *Sacred Wood* 133 It is an arrangement and choice of words which has a sound-value and at the same time a coherent comprehensible meaning. **1964** W. R. Lee in D. Abercrombie et al. *Daniel Jones* 288 There is a tendency to give them [*sc.* letters] the sound-values they possess in the learner's mother tongue. **1848** *Trans. R. Irish Acad.* XXI. 65, I proceed now to explain..the circumstances of the great sea wave and of the aërial sound wave, attending most great earthquakes. **a 1889** Sound-word [see *sound-clause* above]. **1953** H. Read *True Voice of Feeling* I. viii. 144 The caesura is..the breaking of the rhythm into sense words of different length from the sound marks. **1961** *Times* 19 June 9/6 Its sound-world is the old sound-world—parts of it exult in the manner of Richard Strauss. **1976** *Gramophone* Aug. 319/3 Decca and DG engineers help their artists to create a much more limpid and crystalline soundworld.

b. (Also in other objective combinations.) *sound-detector, -locator; sound-deadening, -reflecting; sound-absorption, -production; sound-absorbent, -absorptive, -imitative* adjs.

1961 P. Strevens *Papers in Lang.* (1965) xi. 137 The upper surface..is hard, and therefore probably less sound-absorbent. **1935** *Discovery* May 126/2 The latest designs and materials for sound-proofing and sound absorption. **1972** *Lebende Sprachen* XVII. 37/1 *Sound absorption*, 1) the process of dissipating..sound energy. 2) The property possessed by materials..of absorbing sound energy. **1937** *Archit. Rev.* LXXXI. p. lxxii/1 The complete unit is also lined with sound-absorptive material. **1977** *Chicago Tribune* 2 Oct. vi. 9/2 Rehearsals with empty seats are one thing, performances with every seat..filled with sound-absorptive bodies quite another. **1945** Nelson & Wright *Tomorrow's House* iii. 16/2 The existence of walls lined with books constitutes an excellent sound-deadening treatment. **1962** A. Nisbett *Technique Sound Studio* ii. 48 Of the various possible sound-deadening systems, it is best to try to avoid those which give a padded-cell effect. **1878** *Chambers's Jrnl.* 29 June 413/1 An extremely delicate sound detector. **1942** W. Simpson *One of our Pilots is Safe* 54 Chances of escaping detection would be good, either by enemy fighters high above or sound detectors on the ground. **1921** E. Sapir *Language* 4 The interjections and sound-imitative words. **1956** J. Lotz in L. White *Frontiers of Knowl.* xiv. 223 Even sound-imitative words vary: thus the English *splash* corresponds to Hungarian *loccsan*. **1919** Sound-locator [see *locator 4]. **1941** D. Masters *So Few* ix. 106 Human ears listening at the sound locators to detect the course [of the aircraft]. **1977** *Jrnl. R. Soc. Arts* CXXV. 419/2 The Sound Locator..greatly assisted the anti-aircraft personnel to plan their defences in advance. **1925** P. Radin tr. *Vendryès's Language* 20 The study of sound-production, that is to say,..phonation. **1933** *Archit. Rev.* LXXIII. 232 Only a small area of the walls has a sound-reflecting surface. **1962** A. Nisbett *Technique Sound Studio* 44 A 'bathroom' acoustic would be provided by a small room with strongly sound-reflecting walls.

7. a. sound-attribute *Linguistics*, a prosodic feature; **sound barrier,** the obstacle to supersonic flight posed by such factors as increased drag and reduced controllability, which occur when aircraft not specially designed for such flight approach the speed of sound; also *fig.*; *to break the sound barrier*, to travel faster than sound; **sound channel** *Oceanogr.*, a layer of water in which sound is propagated over long distances with minimum energy loss, usu. because of refraction back into this layer from above owing to the temperature gradient, and from below owing to the pressure gradient; **sound-conditioned** *a.* [*condition *v.* 9], sound-insulated; having improved acoustic qualities; hence **sound conditioning; sound effect,** (*a*) orig. *U.S.* (usu. in *pl.*), a sound typical of an event or evocative of an atmosphere, produced artificially in a play, film, etc. (cf. *effect *sb.* 3 c); also *attrib.* and *transf.*; (*b*) the effect produced by the sound of a word; **sound-insulated** *a.*, insulated against sound; also **sound insulation; sound-law** *Philol.* [tr. G. *lautgesetz*], a rule stating the regular occurrence of a phonetic change in the history of a language or language family; **sound meter,** an instrument for measuring the intensity of sound; **sound moderator,** a device fitted to a firearm which reduces the noise of report, a silencer; **sound pressure,** the difference between the instantaneous pressure at a point in the presence of a sound wave and the static pressure of the medium; **sound print** = *sonogram; **sound-proofed** *a.*, that has been made sound-proof; **sound-ranging** *Mil.* (see quot. 1973); hence **sound-ranger,** one trained in sound-ranging; **soundscape** [Scape *sb.*[2]], (*a*) a musical composition consisting of a texture of sounds; (*b*) the sounds which form an auditory environment; **sound-shift,** *Philol.* = *shift *sb.* 14 c; **sound-shifting** [tr. G. *lautverschiebung*]; **sound spectrogram** = *sonogram; **sound spectrograph** = *sonograph 1; hence **sound-spectrographic** *a.*; **sound spectrography; sound-substitution** *Linguistics*, the replacement of one phoneme by another; hence (as back-formation) **sound-substitute** *v. trans.*, to replace (one phoneme) by another (*rare*); **sound-symbolism** *Linguistics*, the (partial) natural representation of the sense of a word by its sound; hence **sound-symbolic** *a.*, pertaining to or manifesting such symbolism; **sound-tight** *a.* = *sound-proof* adj.

1932 D. Jones *Outl. Eng. Phonetics* (ed. 3) i. 2 The student of spoken English..must learn the proper usage in the matter of the 'sound-attributes' (length, stress, and voice-pitch). **1945–9** *Acta Linguistica* V. 88 The phonemes of a given language are realized in concrete sounds and sound-attributes. **1939** *Jrnl. R. Aeronaut. Soc.* XLIII. 818 It is noteworthy that the curve, which at first is flat, rises gradually for a while, without the enormous increases which other experimenters have found between M.n. 0·6 and 0·8, and which have made them speak of a concrete 'sound barrier'. **1952** *Times* 8 Sept. 5/2 Their moment of triumph after breaking once more through the sound barrier. **a 1955** in T. H. Pear *Eng. Social Differences* (1955) iii. 112 Is there a Sound Barrier against your Son? **1955** *Times* 7 July 8/3 The bang that shook London early on Tuesday morning was caused..by a Gloster Javelin breaking the sound barrier. **1963** *Listener* 14 Mar. 457/1 The African rhythmic element is not part of the Asian musical heritage, and there are totally different tonal systems which constitute a kind of 'sound-barrier' which jazz has had to crash. **1973** A. Price *October Men* xvi. 231 When the General whispered, people moved..when he growled, they broke the sound barrier. **1976** *Lancs. Evening Post* 7 Dec. 1/4 When we went through the sound barrier I only felt a very slight judder. **1946** *Bull. Geol. Soc. Amer.* LVII. 928 The velocity of propagation of sound decreases, due to the temperature decrease, from the surface to 4000 feet and then increases, due to pressure increase, from there to bottom. This type of velocity pattern is known as a sound channel. **1972** M. G. Gross *Oceanogr.* vii. 200 This sound channel is a typical feature of the open ocean at depths of around 1000 meters at mid-latitudes to near the surface in polar regions. **1947** *Sun* (Baltimore) 5 Aug. 6 (Advt.), You travel all the way by the same luxurious Panagra DC-6..air-conditioned and sound-conditioned for your comfort! **1972** *Fortune* Jan. 81/2 Sound conditioning assures privacy in these garden apartments. **1909** *Moving Picture World* 10 July 56/1 (Advt.), Yerkes & Co... Manufacturers of high grade sound effects for moving pictures. **1911** D. S. Hulfish *Cycl. Motion-Picture Work* II. 191 The orchestra comprises pianist and drummer, and a 'sound effect' man. **1928** *Exhibitor's Herald & Moving Picture World* 28 Apr. 21/2 The experts of Victor..will..arrange for the synchronized orchestration and sound effects for this picture, in which airplane battles will have an important part. **1942** Partridge *Usage & Abusage* 298/1 Passing over such obviousness as *bang, crash, hiss*..we see that imitation is most effective when the echoism and sound-effects extend over a succession of words. **1951** W. Empson *Structure of Complex Words* 412 *Rebuke* is prim, apparently from the sound-effect. **1958** *Listener* 25 Dec. 1091/3 The studio managers who twiddle the knobs and the sound-effects engineers. **1966** *Ibid.* 24 Feb. 284/1 A meteorite passed across the sky and produced a brilliant light, together with sound effects. **1972** P. Black *Biggest Aspidistra* i. iv. 36 Producers deplored the attention their ingenuities received, but the public was and is fascinated by sound effects. **1933** Sound-insulated [see *air-conditioning *vbl. sb.*]. **1970** Sound-insulated [see *insulated *ppl. a.* 3]. **1932** *B.B.C. Year Bk. 1933* 365 (*caption*) Eel grass for sound-insulation sandwiched in walls of pumice concrete. **1969** *Gloss. Acoustical Terms* (*B.S.I.*) 49 *Sound insulation*, means taken to reduce the transmission of sound. **1874** H. Bendall tr. *Schleicher's Compar. Gram.* 12 *Vowel sound-laws* (i.e. influence of vowels and consonants on vowels) were not existent in the original Indo-European language. **1911** L. Bloomfield in *Jrnl. Eng. & Gmc. Philol.* X. 619 Synonymous words might be collected to prove almost any desired sound-law. **1974** R. Quirk *Linguist & Eng. Lang.* i. 3 There are good historical reasons..for our firmly associating it [*sc.* 'language'] with..'sound-laws'. **1928** *Sci. Abstr.* A. XXXI. 39 Discusses the differences between physical and physiological intensity of sound and describes a form of sound meter for technical use. **1974** *Physics Bull.* Oct. 481/2 Dawe Instruments..has introduced the type 1400H sound-meter which uses a ceramic microphone... Sound levels as low as 24 dB can be measured. **1934** *Rep. Departmental Comm. Statutory Definition & Classification of Firearms & Ammunition* 44 in *Parl. Papers 1934–5* (Cmd. 4758) VIII. 871 There is procurable an appliance known as a silencer or sound moderator which can be fitted to almost all types of firearms for the purpose of reducing the noise of the explosion of the cartridge. **1953** W. G. B. Allen *Pistols, Rifles & Machine Guns* xiii. 172 Silencers are not permitted by law on privately owned weapons, but a 'sound moderator' may be used providing the appropriate endorsement is made on the..Certificate... The only sound moderators on sale are for .22 in. weapons. **1976** *Shooting Times & Country Mag.* 16–22 Dec. 47/1 (Advt.), Erma Emi semi-automatic carbine,..sound moderator, 'scope. **1916** *Sci. Abstr.* B. XIX. 514 (*heading*) Sound pressure. **1930** *Jrnl. Sci. Instruments* VII. 113 The response at a particular frequency is measured by the E.M.F. developed by the microphone per unit sound pressure per unit area. **1976** *Acustica* XXXV. 255/1 The transfer function is subtracted from the harmonic analysis of sound pressure to produce the source spectrum. **1969** R. Petrie *Despatch of Dove* i. iv. 64 Have you ever seen a soundprint of your own voice? **1932** *Times Educ. Suppl.* 20 Aug. 321/4 The divisions between class-rooms are soundproofed with eelgrass quilting. **1956** N. Mailer *Man who studied Yoga* in *New Short Novels* II. iv. 19 Scream my little one. It will do you no good. The walls are soundproofed. **1978** C. Tomlinson *Shaft* 39 The sighs that in a giant building rise up trapped between its sound-proofed surfaces. **1978** J. B. Hilton *Some run Crooked* ii. 17 They've learned to be radio mechanics, asdic operators, sound-rangers and flash-spotters. **1919** *Sci. Amer.* 17 May 509/1 Both parties to the late conflict excited their ingenuity..to improve methods of sound-ranging, on land and in the air and at sea. **1934** T. E. Lawrence *Let.* 19 Mar. (1938) 793 Research..to develop the art of sound-ranging, and anti-aircraft gunnery. **1973** J. Quick *Dict. Weapons & Mil. Terms* 407/3 *Sound ranging*, a method of locating the source of a sound, such as that of a gun report or a projectile burst, by calculations based on the intervals between the reception of the sound at various previously oriented microphone stations. **1968** *Time* 4 Oct. 6 In this collection, he proved his mastery of the subtle colors, treacherous rhythms, and delicate contrapuntal lines that fashioned Debussy's impressionistic soundscapes. **1973** *Daily Colonist* (Victoria, B.C.) 22 Sept. 5/5 The world soundscape project..counted horn blasts at intersections around the world. **1977** *Times Lit. Suppl.* 11 Feb. 144/3 A small number of jazz musicians have..gravitated towards the soundscapes of Varèse and Stockhausen. **1977** *Guardian Weekly* 18 Sept. 18/1 The 'soundscape', Schafer's word to describe our sonic environment, the day-to-day background of our auditory experience. **1911** L. Armitage *Introd. Study Old High German* ii. iii. 57 OHG is distinguished from all other W. Gmc. languages by a series of Sound-changes affecting its consonant system, which are usually grouped together under the name of the Second or HG. Sound-Shift. **1922** O. Jespersen *Language* ii. 43 The first book in the 1822 volume [of Grimm's *Grammatik*] contains..his exposition of the 'sound shift' (lautverschiebung), which it has been customary in England since Max Müller to term 'Grimm's Law'. **1965** C. F. Hockett *Sound Change* 192 What then of the neat discrete 'speech sounds' of the comparativists? Even more, what of their 'sound shifts'? **1880** A. H. Sayce *Introd. Sci. Lang.* I. iv. 324 Practically the sound shiftings [in the Semitic dialects] are confined to the sibilants. **1908** J. Wright *O.E. Grammar* 100 The first sound-shifting, popularly called Grimm's Law, refers to the changes which the Indo-Germanic explosives underwent in the period of the Germanic primitive community. **1945** R. K. Potter in *Science* 9 Nov. 470/2 The beat of the heart may be recorded slowly and converted to the sound spectrogram form by high speed reproduction. **1974** *Sci. Amer.* Mar. 86/3 The sound spectrograms of Infant A's cries looked exactly like what we have come to regard as being typical of a normal infant. **1945** *Science* 9 Nov. 465/1 The patterns..were made by an instrument that we have called the sound spectrograph. **1977** *Time* 21 Mar. 64/3 The most striking evidence came from a sound spectrograph, a machine that reduces speech to electronic 'pictures' called spectrograms or voiceprints. **1947** R. K. Potter et al. *Visible Speech* i. 4 A sound spectrographic record of the words '*Visible Speech*' is shown in Fig. 3. **1976** *Word 1971* XXVII. 57 Sound-spectrographic and cineradiographic analysis of neonatal cry and crysound. **1948** *Language* XXIV. 4 That we have reached a crucial point in the development of phonemics is clear from the first published results of sound spectrography. **1962** *Amer. Speech* XXXVII. 67 Surgical study..using synchronized cineradiography and sound spectrography. **1953** K. Jackson *Lang. & Hist. in Early Britain* ii. 558 A possible case of pre-lenition *b* sound-substituted by AS. *b*. **1898** *Trans. Amer. Philol. Assoc.* XXIX. 38 It is not always easy to say where sound-substitution ceases and natural speech begins. **1926** L. Bloomfield in *Language* II. 164 Whoever speaks a foreign language or dialect may in it substitute resemblant features of his native speech... Linguistic substitution of phonemes is *sound-substitution*. **1959** A. Campbell *O.E. Gram.* 200 In early loan-words this would arise by the operation of native sound-changes, but in later ones sound-substitution might produce similar results. **1964** R. H. Robins *Gen. Linguistics* 14 The onomatopoeic and 'sound-symbolic' part of language is of great significance. **1977** *Word 1972* XXVIII. 318 A new polar response pair with no relevance to the sound to be considered for membership in one of the sound-symbolic semantic clusters in a phonetic-symbolism experiment. **1901** H. Oertel *Lect. Study Lang.* 328 It would..embrace the attempts at word-painting and sound-symbolism. **1922** O. Jespersen *Language* 396 The idea that there is a natural correspondence between sound and sense, and that words acquire their contents and value through a certain sound symbolism, has at all times been a favourite one with linguistic dilettanti. **1957** R. W. Zandvoort *Handbk. Eng. Gram.* ii. ii. 111 Thus *a man's club* by the side of *a men's club*; *a woman's college* by the side of *a women's college*. This seems to be to some extent a matter of 'sound-symbolism': the singular forms are preferred because they have a more 'manly' sound. **1977** G. W. Hewes in D. M. Rumbaugh *Language Learning by Chimpanzee* i. 48 Sound-symbolism may be explicable on the basis of mouth-gesture. **1932** Kipling *Limits & Renewals* 81 The door was shut; and it's sound-tight for reasons connected with the last nights of the condemned.

b. In combinations referring to the mechanical or electrical transmission, broadcasting, or reproduction of sound, as *sound boom, broadcasting, -crew, engineer, man, negative, programme, radio, record, recorder, recordist, source, studio, system, transmission*; *sound-recording* vbl. sb. and ppl. adj.; *sound-reproducing* ppl. adj.; **sound archive,** a library in which sound recordings are preserved; **sound-book** *disused*, a book supplied with gramophone records to supplement the text; **sound camera** *Cinemat.* (see quot. 1959); **sound check** *colloq.*, a test of sound equipment before a musical performance to ensure that the sound production is correct; **sound-film** *Cinemat.*, a cinematic film with accompanying recorded sound (see also quots. 1923, 1929); **sound gate** *Cinemat.*, the part of a sound head where the sound track is scanned as the film passes through it; **sound head** *Cinemat.*, the part of a film projector concerned with producing an electrical signal from the sound track (see also quot. 1959); **sound-mix:** see *mix *sb.*[2] 2; **sound mixer** see *mixer 1 c; hence **sound-mixing** *vbl. sb.*; **sound picture** = *sound-film* above; also, any recording of an auditory event; **Soundscriber,** a machine for the recording and subsequent reproduction of the spoken word (a proprietary term in the U.S.); **sound shop,** a shop which sells equipment for playing, reproducing, or recording music; **sound stage,** a stage having acoustic properties suitable for the recording of sound (*spec.* one used for filming); **sound stripe** *Cinemat.*, a narrow band of magnetic material on the edge of a film, which contains the sound track; **sound-thief** *slang*, an expert in 'bugging' or the installation and operation of concealed microphones; **sound track** *Cinemat.*, the sound constituent of a film, recorded on the edge of the film stock as either an optical or a magnetic band; also, such a record independent of the film; freq. *attrib.*; also *fig.*; hence **sound-track** *v. trans.*, to provide with a sound track; to serve as a sound track for; **sound truck,** (*a*) = *loud-speaker van* s.v. *loudspeaker 2; (*b*) (see quot. 1959[2]).

1962 (*title*) BBC sound archives recorded programmes library World War 1939–1945. **1977** *Times* 16 May 7/5 In July the Sound Records Department of the Imperial War Museum will be opening to the public... Some have been acquired from..the..BBC sound archives. **1937** *Discovery* Feb. 61/2 *Songs of Wild Birds*. By *E. M. Nicholson* and *L. Koch*. With gramophone records... It is the first sound-book published in Britain. **1938** *Times Lit. Suppl.* 17 Dec. 805/2 The sound-book..seems to be catching on. **1975** *Country Life* 13 Feb. 390/2 Ludwig Koch..conceived the idea of a sound-book—'a combination of text, picture and sound, the last supplied by gramophone records attached to the book'. **1961** G. Millerson *Technique Television Production* i. 14 Another camera and sound boom have taken over. **1929** *Television* Jan. 10/3 (*caption*) The Baird Company's Concert Party and Engineers, photographed in the sight and sound broadcasting studio in Long Acre. **1940** R. S. Lambert *Ariel & all his Quality* vii. 183 The coming of War, which would make sound broadcasting..indispensable..would sound the death-knell of television. **1958** *Listener* 21 Aug. 260/1 One must not imagine that sound broadcasting will fail to be of value to the community for many years to come. **1977** Sound broadcasting [see *television broadcasting* s.v. *television 3 b]. **1904** *Science Siftings* 26 Mar. 353/1 A wonderful camera that will photograph noises... With this sound camera, all noises..can be realistically reproduced. **1958** *New Statesman* 26 July 106/1 ITN's roving reporter, Robin Day, roved as far as Egypt with sound-cameras. **1959** W. S. Sharps *Dict. Cinematogr.* 130/2 *Sound camera*. (1) A film picture camera that makes no external noise in operation and is therefore suitable for use when sound is being recorded. (2) A camera that records sound on film. **1976** *Oxf. Compan. Film* 646/1 Optical sound cameras are now used only to produce negatives for making married prints of finished films. **1977** *Rolling Stone* 13 Jan. 10/1 He runs his hands through his straw-thatched hair as his new band kicks off the sound check with 'You Wear It Well'. **1961** K. Reisz *Technique Film Editing* xii. 185 Having chosen his topics, the producer must get together his unit—cameraman, editor, script-writer and the sound-crew. **1974** A. Morice *Killing with Kindness* ii. 14 It was some American production they were recording over here... He and the rest of the sound crew had been given Tuesday off. **1937** *Amer. Speech* XII. 101 Sound effect..refers to the diabolical work of the sound man or, with greater dignity, the sound engineer. **1973** J. Porter *It's Murder with Dover* iv. 34 The TV cameraman..lowered his camera... A nearby sound engineer agreed. **1923** *Mod. Wireless* I. 418/2 The successful production of such a sound record upon a separate film, the sound-film and the picture-film being run simultaneously. **1927** *Daily Mail* 2 July 8/2 The sound-film of the Walker-Milligan fight which was made by the British Phonofilm Co. **1929** *Times* 30 July 13/2 Contrasting 'dialogue films', which, in imitation of the stage, depend principally on dialogue to tell their story, with 'sound films', which use sound as a supplement to silent technique. **1957** Manvell & Huntley *Film Music* 9 We have tried to show how the first principles of sound film music composition were developed through the imaginative collaboration of composers and film-makers. **1964** N. Marsh *Dead Water* vi. 162 A badly-synchronised sound-film. **1975** G. Howell *In Vogue* 65/2 Sound came in 1927, and by the end of 1928 the worst sound film could outdraw the best silent movie. **1931** G. F. Jones *Sound-Film Reprod.* 12 The film must pass through the sound gate at a uniform speed, in order that the pitch of the music or speech shall not vary. **1960** O. Skilbeck *ABC of Film & TV* 61 The sound gate is the corresponding point—though here the film is in constant flow—in a sound camera or head. **1931** S. K. Wolf in L. Cowan *Recording Sound for Motion Pictures* xx. 289 (*caption*) Schematic diagram of Western Electric sound head. **1959** W. S. Sharps *Dict. Cinematogr.* 131/1 *Sound head*,..the mechanism in a film printing machine that is concerned with the printing of the sound track. **1979** *Amat. Photographer* 10 Jan. 88/1 The sound heads are well screened to reduce hum level and are only brought into contact with the film when the projector is set to 'forward, sound'. **1929** *N.Y. Times* 20 Oct. ix. 8/5 *Playback*..provides a means for the director, the actors and the sound men to determine in general how a scene will sound immediately after it has been taken. **1935** S. W. Pring tr. *L. Sabaneev's Music for Films* vi. 93 The volume of sound emitted is regulated, not by the conductor, but by the soundman in the monitor room. **1971** D. E. Westlake *I gave at Office* (1972) 12 At noon the engineer and the sound man and the director and I would all leave The Hub. **1932** Sound-mix [see *mix *sb.*[2] 2]. **1971** *Sat. Rev.* (U.S.) 25 Dec. 44/1 It was necessary to add quite a lot of traffic noise on the final sound mix. **1938** Sound mixer [see *cut *v.* 21 e]. **1972** D. Francis *Smokescreen* i. 9 The sound mixer took off his ear-phones..and fiddled ..with the knobs on his Nagra recorder. **1977** *Times* 18 Apr. (Gramophone Suppl.) p. iv/6 The controls of the sound-mixing console. **1929** *N.Y. Times* 20 Oct. ix. 8/6 *Soup*, the developing bath in which a sound negative is developed. **1928** *Times* 24 Dec. 28/1 Either British acoustics or the Anglo-German mechanism will presumably be installed in the Gaumont houses, to the exclusion of U.S. sound pictures. **1955** *Radio Times* 22 Apr. 47/3 A recorded sound picture of the Dutch people's struggle to win new land from the sea. **1979** J. Gardner *Nostradamus Traitor* xi. 37 There was a clean sound picture from almost every part of the flat. **1955** *Radio Times* 22 Apr. 1 Radio Times..BBC Sound and Television Programmes. **1966** R. Williams *Communications* (ed. 2) iii. 68 There are more emphatic differences in the distribution of interests in the various BBC sound programmes. **1938** K. Baily in *Radio Times* 21 Oct. 12/2 With a sound radio system that is chiefly a utility service, and in which listeners are participating, the ultimate fusion of vision with sound will be easily achieved. **1952** *Times* 1 Jan. (Rev. of 1951) p. v/2 Sound radio (wireless declined farther towards archaism) has done much during the year. **1971** M. Lee *Dying for Fun* xxiii. 107 The sound radio producer was supervising the recording of an interview. **1900** R. S. Baker *Boy's Bk. Inventions* vii. 258 The cylinder on which the sound pictures or records were to be made was covered with tin foil. **1977** Sound record [see *sound archive* above]. **1957** J. S. Huxley *Relig. without Revelation* (rev. ed.) vii. 171 The invention of the gramophone and the sound-recorder. **1961** L. van der Post *Heart of Hunter* i. i. 30 Charles Leonard, the mechanic who was also our sound recorder..would like nothing better than to go on recording Bushman music and folklore. **1871** *Eng. Mechanic* 17 Nov. 233/1 In sound-recording, I do not think that electro-magnetism would be of much service. **1931** *Electronics* Apr. 587/1 (*heading*) Effects of optical slits in variable area sound recording. **1933** *Chem. Abstr.* XXVII. 50/3 (*title*) Discharge lamp for use with sound-recording apparatus. **1967** A. L. Lloyd *Folk-Song in Eng.* i. 64 [Cecil] Sharp made the notations by ear without the controlling help of sound-recording. **1975** *Language for Life* (Dept. Educ. & Sci.) xv. 234 Another facility of value to the English department..is a sound recording studio. **1958** *Times* 18 Feb. 5/2 Thomas Arthur Howell.., sound recordist.., Twickenham. **1977** *Broadcast* 4 Apr. 25/3 BBC contract news cameramen and sound recordists spelled out their growing concern over pay and conditions. **1931** L. Cowan *Recording Sound for Motion Pict.* 387 *Sound head*, compartment on the projector which contains sound-reproducing systems and mechanisms for guiding and driving film. **1958** M. Kelly *Christmas Egg* iii. 105 Displays of perfectionists' sound-reproducing equipment. **1969** *Gloss. Acoustical Terms* (*B.S.I.*) 41 *Sound reproducing system*, an apparatus for re-creating sound which has been recorded. **1946** *Sun* (Baltimore) 6 Feb. 13/1 The Soundscriber is a recording device which enables observers to describe the position and actions of their assigned horses during a race. The description can be played back immediately..and compared with the pictures of the race. **1950** *Official Gaz.* (U.S. Patent Office) 24 Oct. 1011/2 *Sound Scriber*...for electric sound recording and reproducing machines... Claims use since Feb. 15, 1936. **1968** C. M. Vines *Little Nut-Brown Man* iv. 73 He dictated into the soundscriber, and handed to me the papers referred to in his dictation. **1972** *Daily Colonist* (Victoria, B.C.) 24 May 21/8 Phil Barker tuning a hi-fi set (he's a salesman in a sound shop). **1962** A. Nisbett *Technique Sound Studio* 247 *Crossfade*, a gradual mix from one sound source or group of sources to another. **1931** L. Cowan *Recording Sound for Motion Pict.* 243 Special buildings—sound stages—had to be constructed in which recording could be carried on. **1958** [see *bank *sb.*[3] 10 a]. **1978** S. Sheldon *Bloodline* xvii. 205 Rhys brought Elizabeth to a sound stage, where they made motion pictures for research and for their world-wide advertising and products divisions. **1965** *Focal Encycl. Photogr.* (rev. ed.) I. 1418/1 Recording live sound effects or commentary..may be done on the film actually exposed in the camera (usually containing a magnetic sound stripe) or on a tape recorder. **1979** *Amat. Photographer* 10 Jan. 88/1 One of the main criticisms of sound stripe reproduction has been background hiss and hum picked up at the recording stage. **1929** *Morning Post* 24 May 12/7 There are now 17 sound-studios in New York and Long Island. **1962** A. Nisbett *Technique Sound Studio* 272 *Sound studio*,..any room or hall which is primarily used for microphone work. Its most important properties lie in its size and its acoustics—the way in which sound is diffused and absorbed, and the reverberation time. **1964** M. McLuhan *Understanding Media* xxix. 296 Everyone has at some time wished he were equipped with his own sound system during a movie performance. **1977** 'J. le Carré' *Hon. Schoolboy* iii. 56 Where it was operable, he ran moles and sound-thieves in tandem..[that is], Karla had liked to back up his agent operations with microphones. **1929** *Photoplay* Apr. 31/2 *Sound track*, the narrow band of space along the left side of picture film on which is printed the ribbon-like strip of light and dark lines which constitute the record from which sound is projected. **1946** G. Millar *Horned Pigeon* xiv. 191 He made rude sucking noises with his lips, an exaggerated sound-track for the scene he witnessed through the window. **1949** Sound-track *v.* [see *kinescope *v.*]. **1957** Wodehouse *Over Seventy* xvi. 154 This is not always the laughter of a real studio audience. Frequently, it is tinned or bottled. They preserve it on sound tracks, often dating back for years. **1968** *Radio Times* 28 Nov. 57/5 Excerpts from the sound-track album of Finian's Rainbow. **1977** *New Statesman* 2 Sept. 314/1 The mindlessly self-pitying lyrics were just about swallowable if used to soundtrack shots of Kingston's corrugated iron shanty towns. **1982** *London Review Bks.* IV. xxiv. 8/1 When M. Hulot's author balances a soundtrack, the human voice plays a small and outclassed part in the din of the inanimate. **1935** *Discovery* Sept. 278/2 The ultra-short wave sound transmissions will stimulate further perfection of sound-reproducing apparatus. **1969** *Gloss. Acoustical Terms* (*B.S.I.*) 11 *Sound transmission*, the transfer of sound energy from one medium to another. **1936** P. Rotha *Documentary Film* iv. ii. 208 Sound-trucks are essentially large and cumbersome objects. **1940** *Nation* 30 Mar. 432/3 Forbidding..the operation of their own sound trucks, and the presentation of their own movie. **1959** *Economist* 2 May 433/1 In the cities, towns and villages of Japan over the past three weeks, the days..have been rendered hideous by 'sound-trucks' rumbling through the streets. **1959** W. S. Sharps *Dict. Cinematogr.* 132/1 *Sound truck*, a mobile sound recording unit, usually with its own power supply. **1971** *Black Scholar* Dec. 56/1 The first time we went out on the soundtrucks, I was on the soundtrucks, the first leaflet we put out, I wrote, the first demonstration, I made up the pamphlets.

sound, *a.* Add: **I. 3. c.** Also *spec.* of currency: having a fixed or stable value, esp. based on gold. Freq. as *sound money*. orig. *U.S.*

1841 J. Tyler in J. D. Richardson *Messages & Papers of Presidents 1789–1897* (1897) IV. 85 The idea..of furnishing a sound paper medium of exchange may be entirely abandoned. **1895** *Nation* 19 Dec. 438/1 He has astonished the friends of sound money. **1903** R. T. Ely *Studies in Evol. Industrial Society* 482 The Fabians have been in favour of what is called with us sound currency. **1938** H. V. Hodson *Slump & Recovery* vii. 217 The 'sound-money' provision that only unquestionably strong banks should be allowed to reopen. **1958** *Spectator* 8 Aug. 198/2 Are they now Sound Money men, after thirteen years of Tory-Socialist inflation?

sound, *v.*[1] Add: **I. 2. c.** *to sound off.* (*a*) Of a band: to strike up (see also quot. 1909). Also *imp. U.S. Mil.*

1909 Webster s.v. *sound, to sound off. Mil.*, at a certain point in the ceremony of parade or guard mounting in the United States army, to play, usually marching in quick time from right to left of the line and back;—said of the band or field music. **1919** *Review* (N.Y.) 30 Aug. 350/3 The organization of all possible 'errors' in the use of language into categories and hierarchies, and parading them before classes with all the pomp of 'Sound off!' and 'Pass in review!'. **1936** *Amer. Speech* XI. 61 The adjutant commands, 'Sound off!' and the band marches, playing, back and forth before the stationary troops... And so,..when a man talks loud and long, playing the tune of his own thoughts before uninterested comrades, he is said to be *sounding off*.

(*b*) to speak out, to speak loudly; to complain, protest; to brag; to put forward one's opinion, esp. forcefully and at length. *colloq.* (orig. *U.S.*).

1918 G. E. Griffin *Ballads of Regiment* 39 You low-down, dirty rookey! What in blazes do you mean By sounding off and beefing, not a rag upon you clean. **1920** *Amer. Legion Weekly* 13 Aug. 28 (*caption*) *Sounding off.* But he is sounding off before inspection. You can't blame him because he has been hoping and waiting for the *Weekly*..but it hasn't come. **1935** C. G. Finney *Circus of Dr. Lao* 63 Kate, don't go sounding off that way in front of all these people. **1939** J. Steinbeck *Grapes of Wrath* xiii. 174, I didn' mean to sound off at ya, mister. It's the heat. **1943** *Amer. Mercury* Nov. 554 A guy who *sounds off* (talks too much) is told to *knock it off*. **1951** *Sunday Pictorial* 21 Jan. 10/2 The 'Pic' cites a few examples with the sincere wish that someone will shut them up the next time they sound off. **1960** L. Cooper *Accomplices* II. i. 80 He used to sound off about the chap and blackguard him all ends up. **1972** 'E. Lathen' *Murder without Icing* (1973) xxii. 195 We thought he was just sounding off. **1979** 'A. Hailey' *Overload* iv. ix. 340 It adds up to him being an exhibitionist with a need to 'sound off' constantly, even in small ways.

3. d. *Black English.* = *to play the dozens* s.v. *play *v.* 16 e; *to sound on* (someone): to taunt, to criticize (someone). Cf. sense 13 below.

1962 R. D. Abrahams in *Jrnl. Amer. Folklore* LXXV. 215 When men do 'sound'..it provides a very different kind of release than when adolescents do. **1971** B. Malamud *Tenants* 73 I'm not soundin on you, Lesser, but how can you be so whiteass sure of what you sayin if my book turns out to be two different things than you thought? **1972** W. Labov *Language in Inner City* p. xxii, Johnny..had a curious bald spot on the top of his head several months ago, since grown over, and he is still sounded on regularly by reference to this bald spot. **1973** E. Bullins *Theme is Blackness* 107 Hey..baby..why you got to sound on me like that? **1974** H. L. Foster *Ribbin', Jivin', & Playin' Dozens* iv. 160 He knows how to 'run a game', to 'signify', to 'woof'..and to 'sound'.

6. (Later examples of legal use with reference to other types of action.)

1918 *Law Rep. Appeal Cases* 289 Whether it sounds in debt or in damages such a cause of action implies a present obligation to pay simultaneous with its coming into existence. **1947** *Law Rep.* 23 Aug. 466 An action against a salvor for negligence or misconduct sounds in tort. **1964** *Mod. Law Rev.* XXVII. iii. 264 To juggle with the language of the forms of action and say that the plaintiff's action sounds in tort not contract, cannot alter the fact that the line between liability and non-liability is drawn by seeing whether the act..is, or is not, a breach of contract between two other persons. **1972** *N.Y. Law Jrnl.* 24 Oct. 20/4 While the action sounds in contract, the complaint sets forth two causes of action for unliquidated amounts.

II. 12. (Example.)

1817 Ld. Sefton *Let.* 30 Dec. in *Creevey Papers* (1903) I. xii. 268 It was put into my hand while a surgeon was sounding my bladder..to ascertain whether I had a stone or not.

13. To taunt. Cf. sense 3 d above. *U.S. slang. rare.*

1958 H. Salisbury *Shook-up Generation* iv. 63 He had heart. He would do things no other boy would dare. He would sound a cop on the beat and run away laughing.

sound, *v.*[2] Add: **2. a.** (Later *transf.* example.)

1972 *Science* 5 May 464/1 Lightweight ionosondes have been placed in satellites, and these sound from the height of the satellite..*down* to the peak of the F layer.

6. a. Also with *out*.

1944 E. S. Gardner *Case of Careless Kitten* ii. 21 He..wants someone to sound out Aunt Matilda on how she'll feel. **1956** A. H. Compton *Atomic Quest* 230 Japan was sounding out Russia for her help in negotiating a conditional surrender. **1960** *News Chron.* 25 Feb. 1/2 He had sent a three-man mission to Madrid to sound-out the Spanish.

sound-alike. [f. Sound *v.*[1], after *look-alike*.] A person or thing that closely resembles another (or others) in sound or name. Also *attrib.* or as *adj.*

1970 *Sat. Rev.* (U.S.) 31 Oct. 59/1 It is another of those more-of-the-same pieces, like the sound-alikes of another era, written by all those minor contemporary imitators of Haydn and Mozart. **1972** *Eye, Ear, Nose & Throat Monthly* Apr. 142/1 When a pharmacist takes a prescription over the telephone..there is always the possibility that a drug not intended by the prescriber is likely to be dispensed. Such an error could be the result of a 'sound-alike' or a 'read-alike' drug. **1975** *Verbatim* May 12/1 The lost *r* may one day ruin my professional reputation—by converting the word that's meant into an unrelated sound-alike. **1977** *Rolling Stone* 5 May 66/2 Hire a Pendergrass soundalike—David Ebo. **1979** *Logophile* II. v. 10/2 *The Encyclopedia of Homonyms*..claims to be 'the only complete comprehensive collection of 'sound-alike' words ever published'.

Soundex (sau·ndeks). *Information Sci.* [f. Sound *sb.*[3]] Used, usu. *attrib.*, with reference to a phonetic coding system intended to suppress spelling variations, used esp. in *Med.* to encode surnames for record linkage. Also *absol.*, material encoded using the Soundex code. Hence **Soundex-code** *v. trans.*, to encode (a name or other data) using this code; **Soundex-coded** *ppl. a.*; **Soundex coding** *vbl. sb.*

1959 *Science* 16 Oct. 955/3 The surnames were first reduced to phonetic codes, consisting in each case of the first letter of the name followed by three numeric digits and known as the Russell Soundex Code. *Ibid.* 958/1 The family names on all cards in both files were Soundex coded by means of the computer. **1967** *Amer. Jrnl. Human Genetics* XIX. 340 The usefulness of surname information in its Soundex coded form can be shown to be considerably greater than that of the full alphabetic surnames. **1968** *Brit. Med. Bull.* XXIV. 207/2 The master file of identities of patients is held on magnetic tape in a quasi-alphabetical sequence, based on the Soundex code of surnames. **1970** *Jrnl. Interdisciplinary Hist.* I. 116 Our data are apparently twice as resistant to Soundex coding as Nitzberg's. *Ibid.* 117 No simple strategy such as 'search one Soundex pocket before and one after' will enable us to effect more linkages. *Ibid.*, We..first applied our pretreatments to all surnames in our test file ..and then applied the files of Soundex to the pretreated surnames. **1972** *Computers & Humanities* VI. 190 In the first attempt a quasi-alphabetic sequence known as the Soundex code, which has been used in the field of medical record linkage, was utilized. The Soundex system sets aside the least reliable portions of Anglo-Saxon surnames, thus Smith/Smyth/Smythe all are codified the same.

sounding, *vbl. sb.*[1] Add: **2. b.** *Black English.* Playing the dozens (*play *v.* 16 e).

1962 R. D. Abrahams in *Jrnl. Amer. Folklore* LXXV. 209 The dozens are commonly called 'playing' or 'sounding'. **1965** *Ibid.* LXXVIII. 343 Sounding, especially Mother-Sounding, demonstrates the second place given to the mother-son bond in comparison to the primary place assigned the clique. **1972** W. Labov *Language in Inner City* p. xviii, The setting was essentially that of a party..with card games, eating and drinking, singing and sounding. **1974** H. L. Foster *Ribbin', Jivin', & Playin' Dozens* v. 183 In Pottsville, Pennsylvania, the term is sounding... In a Brooklyn, New York, secondary school the terms ranking and sounding are still used.

sounding, *vbl. sb.*[2] Add: **1. a.** Now usu. by means of echo.

1966 *McGraw-Hill Encycl. Sci. & Technol.* XIII. 216/1 Since about the middle of the 1920s, virtually all deep-sea soundings have been made by echo sounding.

b. Also with *out*.

1969 *Daily Tel.* 11 Nov. 21/4 The secret sounding-out by Plessey and BSR was an effort to clear the hurdle before breathing a word, but a share jump precipitated events.

c. *transf.* The determination of any physical property at a depth in the sea or at a height in the atmosphere; an instance of this.

1875 *Proc. R. Soc.* XXIII. 249 Temperature-soundings were taken on the 28th of September and on the 3rd of October, at depths of 2800 and 1420 fathoms respectively. **1947** *Sci. Progr.* XXXV. 88 These soundings have also shown exceedingly dry layers..to exist from time to time in the troposphere. **1955** E. Burgess *Frontier to Space* iii. 24 The use of the rocket for altitude sounding is by no means a new idea. **1974** *Physics Bull.* Jan. 11/3 Further instrumental developments are bound to follow and it may be that balloon and rocket soundings of the atmosphere will soon become obsolete.

d. *Archæol.* A trial boring made on a site to gain preliminary information. Cf. *sondage.

1957 K. Kenyon *Digging up Jericho* 170 Our excavation at the highest point on the central ridge..was only a restricted sounding. **1967** *Amer. Anthropologist* LXIX. 401/2 At Chagar Bazar, Huwaish (at which only a seven-day sounding was carried out), and in the rest of Mesopotamia 'religious responsibilities rested with the local secular chiefs'.

5. *sounding-pole*; **sounding balloon** = *ballon-sonde; **sounding rocket,** a rocket designed to carry scientific instruments into the upper atmosphere in order to make measurements during its flight.

1902 Sounding balloon [see *ballon-sonde]. **1937** C. G. Philp *Stratosphere & Rocket Flight* v. 33 By the aid of 'sounding' balloons..data has been obtained of the earth's atmosphere at a height of over 23 miles. **1965** R. A. Craig *Upper Atmosphere* ii. 17 Sounding balloons are most commonly made of neoprene and inflated with helium or hydrogen. **1875** 'Mark Twain' in *Atlantic Monthly* May 569/1 You can go and get the sounding-pole. **1975** *Islander* (Victoria, B.C.) 23 Feb. 13/2 Tying themselves together like mountain climbers, sounding poles in hand, they forded the river. **1947** *Amer. Jrnl. Physics* XV. 139/1 (*caption*) The WAC Corporal sounding rocket, which reached 43 mi altitude. **1962** F. I. Ordway et al. *Basic Astronautics* iv. 121 Data obtained from the sounding rockets were correlated with readings from the Explorer 4 artificial satellite. **1978** Pasachoff & Kutner *University Astron.* xxviii. 710 (*caption*) The ultraviolet spectrum of 3C 273, taken with a 40-cm telescope, the largest ever flown on a sounding rocket.

sounding, *ppl. a.*[1] Add: **1. d.** *sounding sand* = *singing sand* s.v. *singing *ppl. a.* 4 b.

1884 *Proc. Amer. Assoc. Adv. Sci. 1883* 251 The sounding sand is near the surface only, at the depth of one or two feet the acoustic properties disappear. **1897** G. P. Merrill *Treat. Rocks, Rock-Weathering & Soils* ii. ii. 143 On certain Hawaiian beaches, such sands [*sc.* shell sands] give out a distinct note..when walked over, or even when shaken in a closed vessel, and are popularly known as sounding, or singing, sands. **1976** *Nature* 5 Feb. 368/2 Hardly surprisingly in view of their weird effects, sounding sands are incorporated into folklore and legends going back at least 1,500 years.

soup, *sb.* Add: **1.** (Further *fig.* examples.) Phr. (*from*) *soup to nuts* (U.S. colloq.), from beginning to end, completely; everything.

1876 Geo. Eliot *Let.* 2 May (1956) VI. 244 Are you not sometimes made rather desponding by the reading of newspapers and periodicals?..All information is given in a soup of comment. **1910** C. Mathewson *Won in Ninth* 143 He knew the game from 'soup to nuts'. **1938** H. Asbury *Sucker's Progress* 16 For many years a common expression was 'from soda to hock', meaning the whole thing, from soup to nuts. **1946** E. O'Neill *Iceman Cometh* i. 79, I know all about that game from soup to nuts. **1964** F. O'Rourke *Mule for Marquesa* 42 'Everything here we asked for?' 'Soup to nuts... Nothing but the best.' **1977** *Undercurrents* June–July 9/1 The twelve page Corruption Supplement is a rich soup of sex, planning scandals, corruption trials, housing fiddles, [etc.].

b. *Biol.* A solution rich in organic compounds which, it is believed, formerly made up the oceans or lakes of the earth and was the environment in which cellular life originated. Freq. as *primordial soup*.

[**1929** J. B. S. Haldane in *Rationalist Ann.* 8 When ultra-violet light acts on a mixture of water, carbon dioxide, and ammonia, a vast variety of organic substances are made... Before the origin of life they must have accumulated till the primitive oceans reached the consistency of hot dilute soup.] **1956** *Amer. Scientist* XLIV. 356 One plausible explanation is that spontaneous resolution of an *early* biosynthetic intermediate from the primordial nutritional 'soup' of the first organisms led to a monoconfigurational world. **1971** I. G. Gass et al. *Understanding Earth* ix. 126/1 This primitive soup provided a nutrient 'broth' for the first living organisms which finally arose within it. **1976** R. Dawkins *Selfish Gene* xi. 211 Floating chaotically free in the primeval soup. **1977** *Vole* No. 4. 13/2 We both [*sc.* humans and plants] have common ancestors..in that pool of organic nutrients known as the primordial soup.

2. b. For *U.S.* read 'orig. *U.S.*' and add further examples.

1915 J. Buchan *Thirty-Nine Steps* ii. 37, I was in the soup—that was pretty clear. **1917** Lloyd George *Let.* 31 July (1973) 184 Henderson has now put us into the soup & there is no knowing what will happen. **1925** [see *eyebrow 1 d]. **1939** H. G. Wells *Holy Terror* i. ii. 38 We're in the soup... We've got to do 1914 over again. **1968** *Listener* 23 May 660/3 You find you may want to move a group of pictures..to a different part of the building, and if the rooms over there are designed for quite a different kind of picture, you're rather in the soup. **1977** C. McCullough *Thorn Birds* xvii. 455, I do feel very sorry for her, and it makes me more determined than ever not to land in the same soup she did.

d. Fog; thick cloud. Cf. *pea-soup a.

1901 [see sense 2 c]. **1941** F. H. Joseph *Let.* 7 Apr. in *Britain at War* (1942) 4 It wasn't long..before we were in the soup again. **1966** E. West *Night is Time for Listening* iii. 107 Over the North Sea the soup was dense and threatening; turbulence was marked. **1972** J. Gores *Dead Skip* (1973) xxiii. 161 Ballard watched the taillights recede into the soup.

e. Nitro-glycerine or gelignite.

1902 *N.Y. Tribune* 22 Oct. 8/4 Dynamite or nitro-glycerine is called 'soup'. **1903** I. K. Friedman *Autobiogr. of Beggar* vii. 218 Louis learned how ter make de 'soup' from a gang of 'yeagers' dat used ter blow de doors off country banks. **1905** [see sense 2 c]. **1920** 'Sapper' *Bull-Dog Drummond* x. 265 I've got the soup here—gelignite. **1930** D. L. Sayers *Strong Poison* xiii. 169 Sam put the soup in at the 'inges and it blowed the 'ole front clean off. **1960** *Observer* 24 Jan. 5/1 The American petermen had started it long before the First World War by using soup, nitro-glycerine in liquid form, to pour through a little plasticine channel to blow the fashionable combination lock safes.

f. *Photogr.* and *Cinemat.* A processing chemical, esp. the developer.

1929 *N.Y. Times* 20 Oct. ix. 8 *Soup*, the developing bath in which a sound negative is developed. **1934** *Tit-Bits* 31 Mar. 12/3 The chemicals in which the film is developed are known as 'soup'. **1969** Gish & Pinchot *Lillian Gish* ix. 102 Joe showed me how film was developed in the 'soup'. **1978** L. Deighton *SS-GB* xxiii. 220 Any special instructions? Over or under development? Fine grain soup? **1979** *SLR Camera* Dec. 60/1 When you've mixed the soup remember to keep it in a well stoppered dark bottle which has been thoroughly cleaned.

g. *Surfing.* (See quot. 1962.)

1962 T. Masters *Surfing made Easy* 65 *Soup*, the foam or broken portion of a wave. **1966** *Weekly News* (N.Z.) 19 Jan. 6/2 When going through waves, point the board directly into the oncoming 'soup'. **1968** *Surfer Mag.* Jan. 24/3 By standing feet parallel, you can float over breaking soup. **1977** *Surfing World* (Austral.) XVII. ii. 88 Plow through miles of soup.

3. *soup-plate* (earlier example; also *fig.*), *-pot* (earlier example), *-tureen* (earlier example).

1726 D. Eaton *Let.* 16 Feb. (1971) 46, I..left directions in writing..what to pack up. I wrote down all manner of herbes, and the soop plates, &c. **1924** E. M. Forster *Passage to India* i. iii. 28 A sunk soup plate of a lawn. **1939** H. Hodge *Cab, Sir?* 217 The badge itself..is called a 'soup-plate'. **1964** C. Willock *Enormous Zoo* v. 80, I shone my torch..and found a couple of large pink soup-plates glaring back at me—a hippo. **1751** H. Glasse *Art of Cookery* (ed. 4) App. 331 Put them with the Fins and Head in a Soop-pot. **1834** Dickens *Sk. Boz* (1836) 1st Ser. I. 160 Delighted to screen himself behind a soup tureen.

b. soup-and-fish *slang*, men's evening dress, a dinner suit.

1918 Wodehouse *Piccadilly Jim* i. 26 He took me to supper at some swell joint where they all had the soup-and-fish on but me. **1945** 'A. Gilbert' *Black Stage* xi. 149 What do you do about dinner here? Soup-and-fish or just a clean collar? **1970** H. McLeave *Question of Negligence* (1973) xviii. 141 Get him to take off his soup-and-fish and show us his scar.

4. soup bunch *U.S. dial.* (see quot. 1923); **soup-fin (shark),** a brown or grey shark with large teeth, *Galeorhinus zygopterus*, found off the Pacific coast of North America and once hunted for the value of its liver and fins; **soup gun** *U.S. Mil. slang*, a mobile army kitchen (? *obs.*); **soup-kitchen** (earlier example); hence **soup-kitchener,** one who accepts food from a soup-kitchen; **soup line** *U.S.*, a queue of people waiting to be fed at a soup-kitchen; **soup man** *Criminals' slang*, an expert user of nitro-glycerine, etc.; **soup-shop,** (*a*) (earlier example); **soup-strainer (moustache)** *colloq.*, a long moustache; **soup-ticket** (earlier example).

1923 *Dialect Notes* V. 244 *Soup bunch*, a small bundle of vegetables for soup. **1938** *Mississippi* (Federal Writers' Project) 286 The grocery stores and the fruit and vegetable stands sell 'soup bunches' which provide the base for home-cooked vegetable soup. **1905** D. S. Jordan *Guide to Study of Fishes* I. xxx. 541 The soup-fin shark..is found on the coast of California, where its fins are highly valued by the Chinese. **1941** *Sun* (Baltimore) 25 Nov. 5/3 Tales of big profits in soupfin shark liver fishing sent E. Smith..hustling..to get his share. **1961** E. S. Herald *Living Fishes of World* 27/2 In 1942 and 1943 about five thousand soupfins were caught..west of Los Angeles. **1975** *Daily Colonist* (Victoria, B.C.) 22 Aug. 16/5 In San Diego the markets call it shark or 'soupfin'. **1975** *Islander* (Victoria, B.C.) 30 Nov. 10/2 One shark hunted to near extinction because of its liver..is the soup fin shark. **1918** C. J. Swan *My Company* 72 The cooks took the 'soup gun', as they immediately nicknamed the kitchen, all apart. **1928** A. C. Havlin *Hist. Company A* 37 In spite of being accompanied by our 'soup gun', we frequently charged the trenches assisted only by coffee and a strip of bacon between two slices of bread. **1839** C. Sinclair *Holiday House* xi. 255 We never had a drop of broth from the soup-kitchen all winter. **1907** G. B. Shaw *Major Barbara* ii. 220 You lie, you old soup-kitchener, you. **1938** C. Himes *Black on Black* (1973) 167 The panic which he had prophesied was on hand and already soup lines had come into existence. **1980** *TWA Ambassador* Oct. 69/3 We had soup lines and the Depression because men lost confidence in themselves. **1961** B. Knox *Die for Big Betsy* ii. 44 'Denby's a "soup" man,' he said. 'Specializes in second-rate safe-blowings.' **1799** *Manch. Mercury* 8 Jan. 4/5 The plan of the soup shops at Birmingham might be advantageously followed at Manchester. **1932** Wodehouse *Hot Water* viii. 153 He did not propose to have a valet hanging around him festooned with fungus and snorting at him all the time from behind a great beastly soupstrainer. **1962** E. Lucia *Klondike Kate* iii. 86 A soulfully humming male quartet in soup-strainers and sideburns. **1968** *Listener* 1 Aug. 140/1 At the telegraph office we aroused with great difficulty an elderly man with a large grey soup-strainer moustache. **1839** E. Hall *Diary* 29 Jan. in O. A. Sherrard *Two Victorian Girls* (1966) i. 11 Our poor house was besieged by a host of people come for soup tickets.

soup, *v.* Add: **2.** [cf. Soup *sb.* 2 b.] To place in difficulties, to bring to grief. Usu. in pa. pple. *colloq.*

1895 W. C. Gore in *Inlander* Dec. 114 *Soup.* v., to cause to fail; to bring to grief. **1922** Joyce *Ulysses* 160 Luck I had the presence of mind to dive into Manning's or I was souped. **1964** *Daily Tel.* 16 Jan. 26/4 Admitting that he earned £3,000 a year, Lord Taylor said that if he accepted a junior Ministry he would be 'souped'.

3. [cf. quot. 1911 s.v. Soup *sb.* 2 c; perh. infl. by Super-.] Orig. and chiefly with *up*. To modify (an engine, aircraft, motor vehicle, etc.) to increase its power and efficiency. Also *transf.* and *fig. colloq.* (orig. *U.S.*).

1931 [implied at *souped(-up) *ppl. a.*] **1933** C. K. Stewart *Speech Amer. Airman* 92 *Soup Up*, to supercharge. **1939** *Sun* (Baltimore) 3 Aug. 1/6 We have done this without 'souping up' our engines, without putting alcohol in our gasoline,..or flying with motors which last only five hours. **1949** A. Hynd *We are Public Enemies* i. 22 Dillinger..bought two new Fords. He souped up the motors... Now he was ready to act as his own getaway driver. *Ibid.* 29 John Dillinger and five other public enemies arrived in three souped-up Ford cars. **1959** *Spectator* 17 Apr. 557/1 The collection is souped up with frantic editorial comments. **1962** *John o' London's* 8 Feb. 140/2, I don't think Mr. Hauser was at his most perceptive in souping-up what was already very funny. **1965** L. Whitten *Progeny of Adder* (1966) 31 The quintet, souped up on sets—tranquilizers and pep pills taken together. **1972** F. Warner *Lying Figures* iii. 35 The coffee soups her up so that she has to take a tranquillizer. **1976** K. Benton *Single Monstrous Act* v. 152 He had lovingly souped up the Escort's engine, and now gave it full throttle. **1979** J. Gardner *Nostradamus Traitor* xxxix. 188 A German car: Opel Kadett, souped, and probably reinforced.

So **sou·ped(-up)** *ppl. a.*; **sou·ping(-up)** *vbl. sb.*

1931 *Automotive Industries* 30 May 826/1 Ray Keech's run at Daytona Beach in the White Triplex powered with three 'souped-up' Liberty engines. **1941** *Time* 18 Aug. 76/2 Its hero, Slave Trader Matthew Flood, is built like a souped-up Abraham Lincoln. **1949** [see *soup *v.* 3]. **1956** D. Walker *Harry Black* xiii. 196 You're like a souped-up version of my mother. **1957** *New Yorker* 2 Nov. 95/2 Their superb High Fidelity components reproduce all the sounds of the original..with no 'souped-up' tones, squeaks or other distortions. **1960** *News Chron.* 16 June 4/6 Without any souping at all, the Mini-Minor..produces a very useful performance. *Ibid.* 4/7 A specially cast manifold for the souped version of the Mini-Austin. **1961** *Times* 7 Nov. 19/1 In Britain a thriving business has grown up in tuning and modifying the engines of existing models to give more performance. So widespread has this practice (referred to by enthusiasts as 'brewing up', 'souping up', or merely 'hotting up') become, particularly with 'Minis' that the B.M.C. introduced an 'officially hotted up' version last September. **1965** *Listener* 18 Nov. 795/1 As if lacking confidence in his own directorial inventiveness, Visconti takes recourse during one sequence to a modulated version of Fellini's style, and at another juncture provides his audience with souped-up..Antonioni. **1975** B. Garfield *Death Sentence* (1976) ii. 11 A souped-up car with enormous rear tires growled past him. **1980** *SLR Camera* July 7/1 News from the Colonies tells us that Ilford have introduced a 'souped-up' 1D-11 for processing black and white film in the USA.

‖ **soupe** (sūp). The French word for Soup *sb.*, usu. used with defining addition, as *soupe à l'oignon* (onion soup), etc.

1767 'Coriat Junior' *Another Traveller!* I. ii. xx. 196, I well remember that the good father supt up his *Soupe à l'Oignon*. **1777** P. Thicknesse *Year's Journey* II. xliv. 98 A Frenchman eats his *soupe* and *bouille* at twelve o'clock. **1794** [see Soup maigre *attrib.*]. **1863** G. Meredith *Let.* 9 Feb. (1970) I. 210 Soupe, à la sage femme. **1865** M. Eyre *Lady's Walks* xxxi. 330 They..dine, except on fête days, on *soupe aux choux*, or *aux haricots*. **1883** 'Wyvern' *Culinary Jottings for Madras* (ed. 4) xxx. 321 '*Soupe à l'oignon*':—Slice a couple of Bombay onions; powder them well with flour... Grated Parmesan should accompany. **1904** A. Bennett *Great Man* xxvi. 306 The host's first spoonful of *soupe aux moules*. **1952** D. Ames *Murder, Maestro, Please* xvii. 126 She insisted on making *soupe à l'oignon* to go with the contents of the bar. **1975** *New Yorker* 6 Oct. 135/1 Good pastries are still sold, along with such popular dishes as *soupe Chinoise*, a pungent concoction containing shrimp, meat, and vegetables. **1980** E. Leather *Duveen Letter* xiv. 163 The waiter brought the *soupe à l'oignon*.

soupé, var. *souper *sb.*[2]

souper, *sb.*[1] (Earlier example.)

1854 *Tablet* 11 Nov. 713/4 Every Souper, every tyrannical agent, every sworn Orangeman, will be a friend to this new order of things.

‖ **souper** (supe), *sb.*[2] Also **soupé.** [Fr.] Esp. in France: an evening meal, supper; *souper intime* (æ̃ntīm) ['intimate']: for two, in privacy.

1787 H. Maty tr. *Riesbeck's Trav. Germany* I. xx. 238 But, alas! so soon as the body is satisfied here, so soon does the mind long for the friendly *dines* and *soupés* of Paris. **1834** *Baboo* I. i. 10 Those soupers are inestimable, and must not cease. **1851** E. Ruskin *Let.* in W. James *Order of Release* (1948) ix. 169 The monk received us very kindly and soon after we had souper, milk, macaroni, salt fish, fritters and very bad wine. **1967** J. Richardson *Courtesans* iii. 37/2 (*caption*) A courtesan bound for a *souper intime*. **1970** *Guardian* 30 Jan. 9/1 The *souper intime* with..low lights, and music. **1976** *New Yorker* 26 Jan. 40/3 By the time she returned, to give Joey a *souper* of boiled-liver chunks, macaroni, pig's maw, beef kidney, and rice, it was almost midnight.

soupie varr. *sopie, *soupy *sb.*

‖ **soupirant** (sūpirañ). [Fr., = sighing (lover), pres. pple. of *soupirer* to sigh.] A male admirer, a suitor.

1849 Thackeray *Pendennis* I. x. 93 And is Sir Derby Oaks..another *soupirant*?.. Another admirer of Miss Fotheringay? **1969** J. Fowles *French Lieutenant's Woman* lxi. 443 But he no sooner saw that than he saw the reality of such an arrangement—how he would become the secret butt of this corrupt house, the starched *soupirant*, the pet donkey.

soupje, var. *sopie.

souple. Add: Also *souple silk*. Hence **sou·pling** *vbl. sb.*, the act or process of partially degumming raw silk which is to be made into souple silk.

1888 A. Sansone *Dyeing* I. viii. 159 *Souple Silk*, silk which only loses 5 to 8 per cent of its weight, and is consequently not completely deprived of its gum. *Ibid.* 160 *Soupling*: Work 1½ hours in water, containing 3 to 4 grains of cream of tartar per litre. **1927** Horsfall & Lawrie *Dyeing Textile Fibres* x. 283 In the case of degumming for souple silk, the object is to remove only a portion of the gum... An operation called 'soupling' then follows, when the silk is treated for one and a half hours in a solution containing three to four parts tartar per 1,000. **1964** S. R. Crockett *Dyeing & Printing* ii. 23 Varying grades of degumming are exemplified by ecru silk..souple silk..and boiled-off silk.

soupy, *a.* Add: **1.** (Earlier example.)

1869 Dickens *Uncomm. Traveller* (1958) xxxiii. 333 The dirty table-cloths, the stuffy, soupy, airless atmosphere.

2. Sentimental; mawkish.

1953 R. Crawshay-Williams *Let.* 1 Aug. in B. Russell *Autobiogr.* (1969) III. ii. 92, I was glad to see..your emphasis..upon the role of power politics rather than ideologies—and also your re-emphasis upon the way in which science and scientific method have conditioned (all that is 'best' in) Western Values. It is maddening the way in which the opposite 'soupy' belief is accepted even by most unsoupy people. **1976** *National Observer* (U.S.) 20 Nov. 24/5 He has included them in the autobiography, along with..a series of sincerely affectionate, if soupy, tributes to Daddy from family and friends. **1977** *New Yorker* 4 July 82/3 There is the silliness of the movie's plangency: hard to feel soupy about a talented couple giving up their love because of the stardust in their eyes.

Hence **sou·piness,** sentimentality.

1963 Wodehouse *Stiff Upper Lip, Jeeves* v. 40 That squashy soupiness of hers, that subtle air she had of being on the point of talking baby-talk. **1977** *Gramophone* Jan. 1153/1 The slow movement brings a hint of soupiness in the tone.

soupy (sū·pi), *sb.* *U.S. Mil. slang.* Also **soupie.** [f. Soup *sb.* + -y[6], -ie.] (A summons to) a meal.

1899 J. R. Skinner *Hist. Fourth Illinois Volunteers* 26 Answered the familiar call of 'soupy, soupy, soupy,' at 5:30 o'clock. **1918** *Stars & Stripes* 5 July 4, I say 'Yum yum' when 'soupie' blows. **1939** *Amer. Speech* XIV. 30/2 *Soupy*, n., mess call (general U.S. Army usage).

sour, *a.* and *sb.*[1] Add: **A.** *adj.* **I. 1. b.** (Later examples.)

1865 C. F. Browne *Artemus Ward: his Travels* 151 A Vigilance Committee, which hangs the more vicious of the pestiferous crowd to a sour apple-tree. **1922** Joyce *Ulysses* 160 We'll hang Joe Chamberlain on a sourapple tree.

2. b. Esp. in *to go* (or *turn*) *sour* (*on* a person).

1928 *Daily Tel.* 20 Mar. 11/5 Sir Victor Sassoon..advised the House to pass the bill, as there was a danger of the Government, in racing parlance, 'going sour'. **1952** C. Day Lewis tr. *Virgil's Aeneid* ix. 194 Let only my luck stay good And not turn sour on me. **1957** A. Macnab *Bulls of Iberia* xv. 214 He cannot afford to ease up in one or two bulls, or the whole afternoon may go sour on him. **1964** L. Nkosi *Rhythm of Violence* 50 What is a cynic but a romanticist turned sour? **1971** A. Sampson *New Anatomy of Britain* 278 It is at the meetings with Treasury men that so many political ideals have been defeated, so many bold promises gone sour. **1981** P. Niesewand *Word of Gentleman* i. 14 Moorhouse and his party had wiped the floor with the opposition... Then suddenly everything went sour.

3*. Of petroleum, natural gas, etc.: containing a relatively high proportion of sulphur. Opp. *sweet*.

1919 E. W. Dean *Motor Gasoline Properties* (U.S. Bureau of Mines Techn. Paper No. 214) 24 There is a possibility that gasoline 'sour' to the doctor test may have been the cause of certain reported corrosion of metal parts of carburetors. **1925** *Petroleum Age* 1 Jan. 16/2 Sour oils also have a distinctively unpleasant odor which is absent in sweet oils. **1936** W. L. Nelson *Petroleum Refinery Engin.* xxiv. 527 For 'sour' sulfur-bearing light distillates, the doctor treatment must be used. **1967** *Wall St. Jrnl.* 31 Jan. 32/2 Recovery of elemental sulphur from 'sour' gas is expected to materially increase available supplies. **1979** *Economist* 11 Aug. 67/1 There is a sour gas formation under the country's best oil field, Yibal.

II. 4. b. Of music: out of tune.

[**1593** Shaks. *Richard II* v. iv. 42 How sowre sweet Musicke is, When Time is broke, and no Proportion kept?] **1937** *Amer. Speech* XII. 48/2 *Sour*, out-of-tune playing. **1976** *Gramophone* Feb. 1356/1 String tone is wirey, even a bit sour in the G minor, especially during loud passages.

III. 9. a. *sour-favoured*, *-tongued* (later example).

1916 Joyce *Portrait of Artist* iv. 187 The face was eyeless and sourfavoured. **1930** Blunden *Summer's Fancy* 22 And black-capped and gowned The sour-tongued master stared and hovered nigh.

10. sourball, sour-ball *U.S.*, (*a*) a peevish or sour-tempered person; also *attrib.* or as *adj.*; (*b*) a boiled sweet with an acid taste; **sour beef** *U.S. local* = *sauerbraten; **sour bread,** restrict † to sense in Dict. and add: (*b*) *U.S.*, sourdough bread; **sour cream,** *spec.* fresh cream soured by the addition of lactic acid; **sour crop** *Vet. Sci.*, oidiomycosis of chickens, turkeys, or other poultry, producing a crop filled with foul-smelling liquid and often thickened and ulcerated; **sour grapes:** see Grape *sb.*[1] 1 a in Dict. and Suppl.; hence **sour-grapeism,** the action or practice of disparaging something because it is out of one's reach; **sour-grapey** *a.*, disparaging because something is out of reach; **sour-grapiness;** **sour-mash** *U.S.*, (whisky made from) fermenting grain mash; also *attrib.*; **sour orange,** the Seville orange, *Citrus aurantium* distinguished by its thick skin and bitter pulp; also, the tree bearing this fruit; also *attrib.*; **sourpuss, sour-puss** *slang* (orig. *U.S.*) [*puss *sb.*[2]], a sour-faced person; a grumbler; a killjoy; also *attrib.*; so **sour-pussed** *a.*, sour-faced, miserable; **sour veld(t)** *S. Afr.*, grassland covered with coarse grass lacking nutritive value.

1900 *Dialect Notes* II. 62 *Sour-ball*, a chronic grumbler. 1933 *Manufacturing Chemist* Nov. 41/1 Assorted Sour Balls (purchased in a railroad depot, Boston, Mass.)... Balls had a coating of grain. 1935 J. O'Hara *Appointment in Samarra* iv. 123 My God, you're sourball tonight. 1962 E. Lacy *Freeloaders* vi. 113 You think Gil is nuts? He's been acting the sourball all day. 1964 [see *halva]. 1976 *N.Y. Rev. Bks.* 15 Apr. 33/1 The witness from those years is overwhelming, and not just from snobbish intellectuals and sourball novelists. 1935 *Evening Sun* (Baltimore) 2 Mar. 18/3 Mrs. Haberkorn was 'a world champion' sour beef cooker. 1947 *Sun* (Baltimore) 3 Nov. 11/8 (Advt.), Old fashioned sour beef & dumplings. 1968 E. Staebler *Food that really Schmecks* 36 Sauerbraten (Sour Beef Pot Roast). 1884 H. A. Dwight *Bread-Making* 46 Sour bread is such a common evil that a special chapter should be given to it... Sour bread follows..as a consequence of sour yeast. 1902 W. Faulkner *Go down, Moses* 196 Then for two weeks he ate the coarse, rapid food—the shapeless sour bread, the wild strange meat. 1977 H. Fast *Immigrants* iii. 201 Lunch was homemade sausage meat..and fresh milk as thick as cream, and with it Mary Gallagher's home-baked sour bread and home-churned butter. 1855 E. Acton *Mod. Cookery* (rev. ed.) vi. 143 'Sour cream' is an ingredient not much approved by English taste, but it enters largely into German cookery. 1961 'E. Lathen' *Banking on Death* (1962) iii. 22 Roast beef, baked potato—'For God's sake, no sour cream!' 1978 D. Francis *Trial Run* iii. 45 The object of her curiosity..spooned sour cream into his borsch. 1951, 1975 Sour crop [see *oidiomycosis]. 1975 B. Meyrick *Behind Light* xv. 199 'Sour crop,' he announced..as he gently felt the chicken's full crop. 1853 Mrs. Gaskell *Cranford* i. 5 There, economy was always 'elegant', and money-spending always 'vulgar and ostentatious'; a sort of sour grapeism which made us very peaceful and satisfied. 1957 R. W. Zandvoort *Handbk. Eng. Gram.* ix. ii. 307 The suffix is added to syntactic word groups..in such formations as *sour-grapeism*, [etc.]. 1962 *Punch* 11 Apr. 579/1 It may have sounded a silly and sour-grapey sort of thing to say. 1980 *Good Housekeeping* Nov. 15/3 Perhaps I'm being a tiny bit sour grapey. 1970 *Guardian* 30 July 9/4 One Amsterdam camp site owner who..almost moulded away with sour grapiness. 1885 'C. E. Craddock' *Prophet of Gt. Smoky Mountains* 150 Him an' me run a sour mash still on the top o' the mounting. 1892 'Mark Twain' *Amer. Claimant* i. 23 Over-confidence and gaiety induced by overplus of sour-mash. 1958 'W. Henry' *Seven Men at Mimbres Springs* 216 The reservation doctor..was definitely given to a rigorous regimen of sourmash Kentucky bond taken internally for pain as self-directed. 1976 T. Stoppard *Dirty Linen* 65 Big bellied, red-eyed men in white crumpled suits swig from medicine bottles of two-year-old sour mash bourbon. [1890 E. Bonavia *Cultivated Oranges & Lemons* pl. vi, The Seville Orange of Kandy..known there by the name of *Amool Dódan* (sour round orange).] 1920 H. J. Webber in *Bull. Calif. Agric. Exper. Station* No. 317. 268 An examination of sweet and sour orange seedling stock..showed the presence of many widely different types. 1926 H. H. Hume *Cultivation of Citrus Fruits* iv. 45 Sour oranges, or bigarades, are distinguished from the sweet varieties by their broadly winged petioles. 1938 M. K. Rawlings *Yearling* i. 12 There were..sour orange biscuits. 1973 *Advocate-News* (Barbados) 26 Feb. 5/1 A virus of unknown nature..was found to be infecting sour orange seedlings. 1937 *Sun* (Baltimore) 28 May 14/7 Hadley doesn't look like the kind of sour-puss who would do that. 1942 *Penguin New Writing* XV. 92 He pretends to be more interested in the antics of his birds than in the puffings an' blowings of a sourpuss of a council clerk. 1960 *Guardian* 15 Mar. 7/3 It's about time we got away from sourpuss champions. 1966 'H. MacDiarmid' *Company I've Kept* i. 34 All the Moral Rearmers and other sour-pusses in Scotland. 1980 *Logophile* IV. i. 45/2 He had always been henpecked by his wife, a sourpuss with a waspish temper. 1952 J. Steinbeck *East of Eden* xlvii. 520 Henry was a man who liked fun—needed it. A sour-pussed associate could make him sick. [1801 Sour veldt: see *sour grass 3.] 1863 J. S. Dobie *S. Afr. Jrnl.* (1945) 76 On across the Little Tugela ..over rank sour-veldt grass. 1894 T. R. Sim *Sk. Flora Kaffraria* 14 The sour veld..is composed of rank strong growing grasses. 1948 *Star* (Johannesburg) 20 Oct. 3/7 Sourveld management presents formidable problems. 1978 *Jrnl. Afr. Hist.* XIX. 479 Seasonal loss of nutrition of the plateau grasses (i.e. the presence of sourveld).

B. *sb.*[1] **3.** (Earlier example.)

1862 J. Thomas *How to mix Drinks* 59 The brandy sour is made with the same ingredients as the brandy fix, omitting all fruits.

sour, *v.* Add: **4. b.** In pa. pple., also const. *on* (the source of embitterment, etc.). *U.S.* and *Austral. colloq.*

1898 E. N. Westcott *David Harum* xli. 346 He's kind o' soured on the hull thing. 1906 E. Dyson *Fact'ry 'Ands* xvii. 225 'Fact is,' said the packer, 'we're gettin' er bit soured on wimmin.' 1907 *St. Nicholas* XXXIV. 601/2 Maybe if I get any more soured on Hammond I'll skate over with my trunk and try Ferry Hill.

source, *sb.* Add: **4. e.** Also, a person supplying information, an informant, a spokesman.

1934 Webster s.v. *source n.*, one who or that which supplies information. 1940 W. Faulkner *Hamlet* ii. i. 131 The Varners would know by now from the one incontrovertible source, the girl herself, that two of them were not guilty. a 1961 E. M. Mills in *Webster* s.v. [1]*source*, Sources close to the chief executive report he is planning to request the Legislature to approve state purchase. 1973 *Atlanta* (Georgia) *Jrnl.* 19 Apr. 17A/1 Deputy White House press secretary Gerald Warren issued the following statement: 'The White House is not prepared to react to a story based on sources.' 1979 E. Newman *Sunday Punch* i. 3 He had pointed me in the direction of a couple of stories—he was a kindly man and, as a source, needed no special motivation.

f. *attrib.*, as (sense 4 e) *source book, data, document, material, study.*

1899 A. B. Hart (*title*) Source-book of American history. *Ibid.* p. xvii, The Source Book is meant to supplement, not to supplant the text-book. 1900 [in Dict., sense 4 e]. 1948 L. MacNeice *Holes in Sky* 43 We rarely read their poems, Mere source-books now. 1961 J. D. Rosenberg *Darkening Glass* (1963) v. 101 'The Nature of Gothic'..is the source book for *Unto This Last.* 1974 *Education & Community Relations* Jan. 3 The researchers also included a question on what support teachers would welcome from external sources and seven ideas were suggested, i.e. in-service courses, teachers guides or source books, pupils books, films, TV lessons, radio lessons and visiting speakers. 1982 *N. & Q.* Dec. 535/2 *Le Menagier de Paris* has long been known as an invaluable sourcebook for practical details of everyday life in a reasonably prosperous middle-class household in France in the 1390s. 1971 J. Howlett in B. de Ferranti *Living with Computer* ii. 17 The general principle is to..use it..as source data for a whole series of studies. 1920 A. J. Grieve in A. S. Peake *Commentary on Bible* 725 It has therefore been surmised that the writer has here incorporated an Aramaic (possibly Greek) source-document. 1977 *New Yorker* 29 Aug. 35/2 Source documents, once put into computer-readable form, tend to become relatively inaccessible, and in some computer systems are even eliminated. 1936 *Time* 21 Sept. 47/1 For most of their source material the editors relied on second-rate writers. 1955 W. Moore *Bring Jubilee* xix. 182 It is not easy to see behind source material, to visualise state papers, reports, letters, diaries as written by men. 1978 *Early Music* Oct. 597/3 The discussion of the music combines a flair for words with great attention to stylistic interactions and the lessons to be learned from study of the source material. 1964 *English Studies* XLV. 252 Even those readers least interested in source-study are likely to have their notions of Shakespear's work made altogether more accurate. 1979 *Studies in Eng. Lit.: Eng. Number* (Tokyo) 3 Source study—by this is meant here not a mere source-hunting but a comparative study between words and their sources—is certainly rewarding so far as *Confessio Amantis* is concerned.

5. a. More widely, any point where, or process by which, energy or some material component enters a physical system; opp. Sink *sb.*[1] 8 in Dict. and Suppl. (Earlier and further examples.)

Freq. without const., but otherwise not really distinct from sense 4 d in Dict.

1855, etc. [see *sink *sb.*[1] 8]. 1926 H. Glauert *Elem. Aerofoil & Airscrew Theory* iii. 21 A sink is a negative source or a point at which fluid is disappearing. 1956 E. H. Hutten *Lang. Mod. Physics* iv. 139 The engine is in contact with two heat reservoirs (the boiler and condenser, or the source and sink of energy) at different temperatures. 1971 I. G. Gass et al. *Understanding Earth* xix. 263/1 Boundaries at which the net effect of motion is to generate surface area are here termed sources.

b. *Electronics.* (The material forming) the part of a unipolar transistor which corresponds in function to the cathode of a thermionic valve.

1952 [see *gate *sb.*[1] 8 h]. 1962 Simpson & Richards *Physical Princ. Junction Transistors* viii. 173 The source and drain..are ohmic electrodes attached to the *n*-type body of the device [*sc.* a field-effect transistor]. 1977 *Sci. Amer.* Sept. 74/3 The inversion creates a continuous *n*-type channel from source to drain and large currents can flow.

6. *Comb.*, as (sense 4 e) *source-hunter, -hunting*; **source-criticism** *Theol.*, analysis and study of the sources used by the authors of the biblical text; hence **source-critical** *a.*; **source program** *Computers*, a program written in a language other than machine code, usu. a high-level language (cf. *object program* s.v. *object *sb.* 10); **source rock** *Geol.*, a rock formation in which a particular mineral material originates; *spec.* a deposit in which petroleum is formed.

1977 J. L. Houlden *Patterns of Faith* iii. 26 This is particularly true of the gospels of Matthew and Luke, where, according to the source-critical orthodoxy.., visible alterations of the Markan basis could be observed. 1901 J. Moffatt *Historical New Testament* App. 677 No method which neglects source-criticism can satisfactorily explain the doublets [in the Apocalypse]. 1931 K. E. Kirk *Vision of God* 498 An elaborate source-criticism which must be adjudged..to be in the main based upon the theory, and therefore to involve a vicious circle. 1977 G. W. H. Lampe *God as Spirit* iv. 102 The Jesus whom historical research tries to reconstruct through the laborious processes of source criticism, form criticism, and redaction criticism. 1964 D. Daiches *Eng. Lit.* iv. 82 An attempt to rescue literary study from the philologists and source-hunters. 1956 *Canad. Forum* June 67/1 His treatment of sources and analogues lacks the rigorous testing which we require of source-hunting in literary studies after twenty years of sniping by new critics and old scholars alike. 1979 Source-hunting [see *source study*, sense *4 f]. 1959 M. H. Wrubel *Primer of Programming for Digital Computers* vi. 129 The program can be corrected or modified at the source program stage and reassembled. 1970 O. Dopping *Computers & Data Processing* xix. 304 Instead of machine instructions, the source program contains statements or symbolic instructions, which the computer then translates to an object program by means of a special program. 1973 C. W. Gear *Introd. Computer Sci.* iv. 158 A language compiler accepts as input a set of statements called a source program. 1931 *Bull. Amer. Assoc. Petroleum Geologists* XV. 161 Source rocks of petroleum include carbonaceous or 'bituminous' sedimentary deposits, containing aquatic plant and animal remains..and the products of their biochemical and geochemical alterations. 1965 G. J. Williams *Econ. Geol. N.Z.* xix. 352/1 Early borings at Kotuku..left little doubt that the oil at Kotuku comes from the Oligocene Cobden Limestone—though this does not necessarily mean that the Limestone is the source rock. 1971 I. G. Gass et al. *Understanding Earth* xxii. 323/2 The concentration of diamond source-rocks in the older cratons.

source, *v.*[1] Restrict † *Obs.* to senses in Dict. and add: **4.** *trans.* **a.** In pass., *to be sourced in*, to originate in, to be based in; to mention as a source.

1941 W. C. Handy *Father of Blues* xxii. 298 Affinities that may be sourced in a common ultimate Oriental origin. 1972 'J. Godey' *Three Worlds of Johnny Handsome* ii. 23 Mitchell became aware of a rumbling sound vaguely sourced in the floor. 1978 *Maledicta 1977* I. 326 Over twenty of Mr. Tamony's scripts are sourced in H. L. Mencken's *American Language, Supplements* 1 and 2. 1982 *Times Lit. Suppl.* 17 Dec. 1394/1, I also drew on a scientific paper discussing Vittoz's work by the Chicago psychoanalyst, Dr Harry Trosman, which is sourced in my notes.

b. To obtain from a specified source; *spec.* of components (for a vehicle). Chiefly in pa. pple.

1972 *Wall St. Jrnl.* 24 Feb. 1/5 Ford works on stripped-down cars, called 'Asian Model Ts' that could be sourced and assembled anywhere in Asia. 1980 *Times* 20 Mar. 27/5 One component manufacturer said last night: 'Our indications are that less than 10 per cent of the Bounty [*sc.* a new car] will be sourced in the United Kingdom.' 1981 *Times* 6 Feb. 18/2 Counterfeited goods, largely sourced from south east Asia..have mainly been finding their way into British export markets. *Ibid.* 15 Aug. 15/4 British manufacturers..have stressed that Nissan should source at least 80 per cent of the contents of the cars it plans to make in the United Kingdom in Europe.

Hence **sou·rcing** *vbl. sb.*, *spec.* the obtaining of goods and components from a specified or understood source.

1960 *Business Week* 2 Jan. 67/3 Businessmen now refer to imports from foreign plants as 'sourcing'—a term that until recently referred to company purchases from a domestic supplier. 1960 *Wall St. Jrnl.* 15 Mar. 14/5 There is a growing tendency toward foreign 'sourcing', the purchase or production of finished goods or components abroad. 1970 *Daily Tel.* 1 Oct. 2/2 Ford, British Leyland and other manufacturing companies has had to resort to 'dual sourcing' for some components because of this year's unprecedented run of major supplier strikes. 1972 *Wall St. Jrnl.* 24 Feb. 1/5 Experience under the U.S.-Canadian auto pact and in the Common Market has emboldened Detroit to expand multi-national sourcing of parts and components.

source language. [f. Source *sb.* + Language *sb.*] **1.** A language from which a translation is made.

1953 *Philos. Sci.* XX. 217 One of the decisive steps in certain methods of machine translation is the determination of the syntactic structure of any given sentence in the source-language (i.e., the language from which we translate). 1964 M. A. K. Halliday et al. *Linguistic Sciences* 123 Translation as activity faces only one way; the translator observes an event in one language, the 'source' language, and performs a related event in another, the 'target' language. 1974 R. Quirk *Linguist & Eng. Lang.* vi. 97 The difficulties vary profoundly according to the manifold combinations of source- and target-languages involved.

2. *Computers.* The programming language in which a program or procedure is written. Cf. *object language 3.

1959 *Communications Assoc. Computing Machinery* Feb. 9/2 Sections 2 and 3 give a formal description of the Fortran source language, insofar as arithmetic type statements are concerned. 1963 *Ibid.* VI. 430/1 The debugging system has been implemented for Fortran as the source language and could be easily adapted to other problem-oriented languages. 1975 T. Bartee *Introd. Computer Sci.* xiii. 377 For high-level compiler languages such as Fortran, PL/I, and Algol, there is an attempt to make the source language machine-independent.

sourceless, *a.* (Later examples.)

1908 *Punch* 25 Sept. 436/1 The unnoticed, sourceless wound. 1981 *Washington Post* 8 July c1 Talent—this sourceless asset that one cannot command to be born.

sourdine, *sb.* Add: **a.** *Early Mus.* (Later examples.)

1891 C. R. Day *Descr. Catal. Mus. Instruments recently exhibited at R. Mil. Exhibition* v. 96 The combination of a double reed with a cylindrical bore..presents certain theoretical difficulties... During the 14th, 15th, and 16th centuries instruments thus constructed were in common use. Those most generally met with were the *krumhorn* ..and the *sourdine.* 1941 N. Bessaraboff *Anc. Europ. Mus. Instruments* 80 There were several families of instruments, such as the cromornes, the racketts ('the sausage bassoons'), the sourdines. 1976 D. Munrow *Instruments Middle Ages & Renaissance* vi. 45/3 The tone is much softer and more muffled, however, and very similar to that of a rackett; hence the other name given to the instrument..*sourdine* (French).

b. A mute or damper.

c 1779 W. Waring tr. *Rousseau's Dict. Mus.* 384 *Sourdine*, a small instrument of copper or silver, which is applied to the bridge of a violin or violincello, to render the sounds weaker. 1959 *Collins Mus. Encycl.* 618/2 *Sourdine*... (1) Mute (of a string or wind instrument). (2) Damper (in the piano). (3) A device for muting the strings of the piano.

sour-dough, *sb.* Add: **1. c.** *N. Amer.* Fermenting dough, esp. that left over from a previous baking, used as leaven; bread made from this. Also *attrib.*

1868 J. Anderson *Sawney's Lett.*, Gie my respecks to your guid wife; If ever I get hame to Fife, I'll teach her hoo to mak loaf bread, Wi' sour dough. 1922 G. C. F. Pringle *Tillicums* 156 In every old prospector's cabin.. you would see a bowl which contained sour dough from the previous making. This was used as yeast to be mixed in with the dough at the next baking. 1931 *Sun* (Baltimore) 5 Mar. 12/3 [Maryland] might be thought wanting in pride in the culinary art for which the State enjoyed well-earned fame when Indiana was eating sour dough and jerked meat. 1959 M. Shand *Summit & Beyond* xi. 197 There was good old sourdough bread, too. 1966 Mrs. L. B. Johnson *White House Diary* 3 Apr. (1970) 380 They had sour-dough biscuits too—although they had trouble making them rise in this high altitude. 1971 *Daily Colonist* (Victoria, B.C.) 23 Apr. 4/7 In early America, a pot of sourdough starter was part of a bride's dowry. 1976 *National Observer* (U.S.) 26 June 8/3 San Francisco sourdough has a character all its own, because there are microscopic organisms in the air here that do funny things to the starter. Some say it's the same wild yeasts that work on the nearby vineyards. 1979 *United States 1980/1* (Penguin Travel Guides) 370 Two-pound Porterhouses..served with sourdough bread and pinto beans.

2. Substitute for def.: An experienced prospector in Alaska, the Yukon, or the Northwest territories. (Earlier example.)

In quot. 1904 in Dict. for '[of frost]' read '[below zero]'.

1898 *Klondike Nugget* (Dawson, Yukon Terr.) 20 July 1/4 The usual strong expletives had been used expressive of their meeting and Mr. Chee Chaco was not looking for information from his old friend Mr. Sour Dough.

sour grass. Add: **3.** *S. Afr.* The coarse grass characteristic of sour veld (see *sour *a.* 10).

1801 J. Barrow *Acct. Trav. S. Afr.* I. iii. 110 That division of the district called the *Zuure-veldt*, or Sour Grass plains. 1852 M. B. Hudson *S. Afr. Frontier Life* I. 236 About Graham's Town and Lower Albany is 'the Zuurveld or Sour Grass Country'.

soursob (sau·rsɒb). *Austral.* [Alteration of Sour-sop, perh. in reference to the acid sap.] A bulbous plant, *Oxalis cernua*, of the family Oxalidaceæ, native to South Africa and widely naturalized as a weed elsewhere, bearing divided leaves and clusters of bright yellow bell-shaped flowers; also called the Bermuda buttercup.

1907 *Jrnl. Dept. Agric. S. Austral.* X. 802 Can anyone tell me to destroy a weed called by some 'Soursob'? 1909 J. M. Black *Naturalised Flora S. Austral.* 41 *Oxalis cernua*, Thunb. Soursob. 1961 *Times* 23 May 3/1 (Advt.), Biological control of several important weeds and pests of South African origin, such as cape tulip, spiny emex, soursob,..white wax scale and black beetle. 1972 *Advertiser* (Adelaide) 2 Sept. 5/7 Never have I met an Englishman who actually grew soursobs or oxalis or whatever they are properly called at home.

sour-sop. Add: **2*.** = *soursob.

1885 *Garden & Field* (Adelaide) Aug. 29/3 'Amateur' wants to know a remedy for 'soursops', and if gas lime would do. *Ibid.* Sept. 41/1 Now there's a fellow who wants £500 to tell farmers how to kill the Soursops or oxalis. 1930 A. J. Ewart *Flora of Victoria* 687 O[*xalis*] *cernua*..Soursop. A perennial with a brown bulb and a tapering root... A troublesome weed, native to South Africa, originally a garden escape.

sous-, *prefix.* Add: **sous-chef, -officier** (further examples), **-préfecture, -préfet.**

1825–40 F. Burney *Jrnls. & Lett.* (1980) VIII. 49 M. d'Arblay continued in his humbler office, of *sous chef* to one of the *Bureaux de l'Interieur.* 1902 A. Bennett *Grand Babylon Hotel* xxi. 238 We have found in our second sous-chef an artist. 1973 *New Society* 15 Nov. 412/2, I gradually pieced together the hierarchy. After the chef..came Jacko, the sous-chef. 1980 N. Freeling *Castang's City* i. 1 Lasserre was a Commissaire... He was the sous-chef, the Chief of Staff. 1895 E. Dowson *Let.* 2 Oct. (1967) 316 Owing to a great demand for lodgings by officiers and sous-officiers, we found it..impossible to find what we wanted. 1972 G. Bell *Villains Galore* vi. 68 He had been a *sous-officier* in the [Foreign] Legion. 1940 J. Joyce *Let.* 16 Sept. (1966) III. 487 Our applications.. passed through the sous-préfecture of this department this morning. 1976 N. Freeling *Lake Isle* viii. 43 Soulay was a sous-préfecture, and sub-prefects are small beer. 1865 L. Booth in S. Pakenham *Sixty Miles from England* (1967) v. 67 Etiquette expects yearly residents to leave cards on the Sous-Préfet. 1944 H. Nicolson *Let.* 27 June (1967) 382 They went to the *souspréfecture.* The *sous-préfet*..had been secretary to Pucheu. 1974 E. Ambler *Dr Frigo* iii. 186 The sous-Préfet assured me of his most distinguished sentiments.

sousaphone (sū·zăfōun). Also **Sousaphone.** [f. the name of John Philip *Sousa* (1854–1932), American bandmaster and composer, after *saxophone*, etc. (see also quot. 1939).] A large bass wind instrument of the helicon type. Also *attrib.*

1925 *Punch* 27 May 561/3 An instrument called the Sousaphone weighs eighteen pounds and is twenty feet long. 1935 *Ibid.* 23 Oct. 470/2 As it was removed from its case he saw that it was a sousaphone—the instrument that plays the 'oom' in Viennese waltzes while the rest of the orchestra follows with the 'wump wump'. 1939 *Internat. Cycl. Mus.* 1772/2 The first sousaphone was made by C. G. Conn in 1899 expressly for Sousa's band and its bell opened directly upward. The present bell-front type was first made in 1908. 1958 [see *exotica]. 1974 P. De Vries *Glory of Hummingbird* i. 4 The sousaphone tuba he played in the local marching band.

souse, *sb.*[1] Add: **1.** (Examples in W. Indies.)

1929 W. J. Locke *Ancestor Jorico* viii. 108 We were given..souse, which is the gelatinous parts of a pig pickled in lime-juice. 1952 S. Selvon *Brighter Sun* ii. 23 They make souse—boiled pork, seasoned with lime and pepper and cucumber. 1958 B. Hamilton *Too Much of Water* iv. 74 A real Barbadian breakfast. 'Maan,' he said, 'I give you flying fish an' pepper-pot, an' pudding and souse.' 1974 *Sunday Advocate-News* (Barbados) 10 Mar. 8/1 She is selling a popular Barbadian delicacy—pudding and souse.

3. *souse meat.*

1972 E. Wigginton *Foxfire Bk.* 20 Souse meat. Boy, that's the best stuff I ever eat. 1976 *Washington Post* 7 Nov. xi/5 We will try to re-create the atmosphere of a country store. Sardines,..souse meat and soda crackers.

souse, *sb.*[5] Add: **1. b.** A heavy drinking-bout. *U.S. slang.*

1903 G. Ade *People you Know* 13 (*heading*) The periodical souse. 1930 E. Wallace *Calendar* xviii. 244 If ever a man had an excuse for a souse, you've got it. 1946 E. O'Neill *Iceman Cometh* iii. 199 Bejees, we'll go on a grand old souse together.

3. A drunkard. *slang* (chiefly *U.S.*).

1915 J. London *Jacket* 213, I remember you mentioned playing chess with that royal souse of an emperor's brother. 1936 Wodehouse *Laughing Gas* i. 11 The lad.., who is pretty generally recognized as London W.1's most prominent souse. 1953 R. Chandler *Long Good-Bye* v. 27 Sylvia is not a souse. When she does get over the edge it's pretty drastic.

souse, *v.*[1] Add: **I. 3. d.** To intoxicate thoroughly. Chiefly in pa. pple. Now *slang.*

1613 [see Soused *ppl. a.* 2]. 1902 H. L. Wilson *Spenders* ix. 87, I could see then that he was good and soused. 1953 K. Tennant *Joyful Condemned* xxxi. 306 Grandma used to get a bit soused sometimes, but she fed me O.K. 1976 M. Russell *Double Deal* vi. 46 Ralph's a pro. He's soused every night, and I don't recall an edition going astray yet.

II. 6. c. To drink so as to become intoxicated, to carouse. *slang.*

1921 E. O'Neill in *Theatre Arts Mag.* V. 32 Ain't you sousin' with 'em most every day? 1923 M. Watts *Luther Nichols* 43 Just as they're middling honest and don't souse.

soused, *ppl. a.* **2.** (Later example.)

1932 J. T. Farrell *Young Lonigan* i. 34 Coming home, he had almost gotten into a mixup with some soused mick.

‖ **sous-entendu** (suzɑ̃tɑ̃dü). [Fr.] Something not expressed but left to be understood by the hearer or reader.

1865 Mill *Exam. Hamilton's Philos.* xxii. 442 No shadow of justification is shown for thus..adopting into logic a mere *sous-entendu* of common conversation in its most unprecise form. 1907 W. De Morgan *Alice-for-Short* xxxii. 332 She knew well enough that the unheard portions of the conversation were worse than what had reached her ears, and the *sous-entendus* probably still worse than they. 1972 *Times Lit. Suppl.* 31 Mar. 374/3 It is satisfyingly like other such duets, though embellished at least once with an erotic *sous-entendu.*

‖ **souteneur** (sūtənör). [Fr., = protector, f. *soutenir* to sustain.] A man who lives on the earnings of a prostitute or prostitutes under his protection. Also *transf.*

1906 tr. *Weininger's Sex & Character* ii. x. 234 The souteneur is always a criminal, a thief, a fraudulent person, or sometimes even a murderer. 1920 A. Huxley *Limbo* 38 After midnight he would write novels with a feminine pen, earning the money that would make his unproductive male labours possible. A kind of spiritual *souteneur.* 1927 *Observer* 11 Dec. 20/2 The souteneur.. extends some shadowy kind of patronage and protection over the girl in return for a 50 per cent or still higher share of her earnings as prostitute. 1969 *Punch* 26 Mar. 469/1 The death-knell of revolution in Britain was the football match between strikers and the police in 1926. It was the moment when the Russian *souteneurs* of the British Communist Party realised with disgust that revolutions can't be made, they have to happen. 1977 M. T. Bloom *13th Man* viii. 147 The underworld of traffickers, *souteneurs*, and prostitutes..at Buenos Aires.

‖ **soutenu** (sūtənü), *a.* and *sb. Ballet.* [Fr., pa. pple. of *soutenir* to sustain.] **A.** *adj.* Of a movement: sustained, performed slowly. **B.** *sb.* A sustained or slowly-performed movement, *spec.* a complete turn on point or half point.

1930 [see *plié]. 1947 N. Nicolaeva-Legat *Ballet Educ.* iv. 63 After *en dedans*, the opposite *soutenu* is performed and the *port-de-bras* executed with the active foot extended in front. 1978 *N.Y. Times* 29 Mar. c19/1 The faint sounds of a tune to dance by are drifting out, then a louder shout. 'All you have there is a little soutenu... You did six counts, she did seven.'

south, *adv.*, *prep.*, *sb.*, and *a.* Add: **A.** *adv.* **1. b.** Also, in or into the southern states. *U.S.*

1852 Mrs. Stowe *Uncle Tom's Cabin* i. x. 142 To appreciate the sufferings of the negroes sold south. 1936 M. Mitchell *Gone with Wind* iii. 51 The air was always thick with threats of selling slaves south and of direful whippings. 1976 M. G. Eberhart *Family Fortune* (1977) vii. 77 'Suppose Mr Jeff sell me south?'.. 'He can't sell you... You and all the slaves..were set free.'

B. *sb.* **2. a.** (Further examples, of Ireland.) *spec.* The Republic of Ireland.

1913 R. Kipling *Let.* Dec. in Ld. Birkenhead *Rudyard Kipling* (1978) xvi. 257 Which is the most [*sic*] dangerous enemy? The South playing a game it has not got its heart in, or the North in a blind rage? 1974 D. Seaman *Bomb that could Lip-Read* vii. 60 The whole attitude of the South baffled and angered him. Irish politics were beyond him. 1978 D. Murphy *Place Apart* ii. 33 In Northern Ireland one has a wide choice of names for the rest of the island: the Twenty-six Counties, the Free State, Southern Ireland, the South, Eire and the Republic of Ireland.

3. *South of England*, also (freq. with hyphens) *attrib.* (earlier and later examples); *South of France, spec.* the French Riviera; also *attrib.*

1741 M. W. Montagu *Let.* 29 July (1966) II. 245 They are gone to Marseilles and design passing some months in the South of France. 1815 E. Fremantle *Diary* 8 Sept. in *Wynne Diaries* (1952) xxxi. 533 We determined to take the former Road, particularly as the South of France is not quite quiet. 1847 C. Brontë *Jane Eyre* III. iv. 106 A large, fashionable, south-of-England city. 1872 C. M. Yonge *P's & Q's* ii. 10 At the time of her death, Elspeth and Persis had been in the South of France. 1922 M. Arlen *Piracy* iii. v. 183 We are going to the south of France to-morrow. 1922 E. Sitwell in *New Age* 6 July 120/1, I liked..the warm, South-of-France feeling about her, and her faded hair that was like dry, powdery mimosa. 1940 'G. Orwell' *Inside Whale* 42 Dickens..is a south-of-England man, and a cockney at that. 1971 J. Brunner *Honky in Woodpile* ii. 14 I'll start on the Sunday papers and catch their South of England editions. 1980 I. Murdoch *Nuns & Soldiers* i. 83 Soho in summer was his South of France.

6. *Bridge.* The player sitting opposite and partnering north: occas. in conventional printed representations of the game, the player who wins the bidding and plays the hand.

1926 [see *east *sb.* 4]. 1933 C. Vandyck *Contract Contracted* iii. 31 South deals and bids 1 Diamond. North seeing the possibilities of a Slam gives a Slam Invitation by bidding 5 Spades. 1958 *Listener* 2 Oct. 541/1 One would expect South to pass. 1964 Frey & Truscott *Official Encycl. Bridge* 514/2 In bridge writing for general reading, South is, conventionally, the declarer... However, in reporting International Matches, the actual positions at the table are used. 1978 *Country Life* 14 Dec. 2098/2 Most Souths without further thought would bid Three No Trumps.

7. A collective name for the industrially and economically less advanced countries of the world, typically situated to the south of the industrialized nations.

1975 *Economist* 18 Oct. 103/2 North-south dialogue... This week's preliminary get-together [between] the west and the oil and non-oil developing nations..illuminated the snake pit ahead. 1977 *N.Y. Times* 22 Sept. 43 Today, any regional struggle over who is to become managing director of the I.M.F. is far less likely to be one between the United States and Western Europe as between the 'North' and the 'South'—that is, the developed, industrial countries and the so-called developing countries, some oil-rich and others oil-poor; some well on the way to industrialization, and others desperately poor. 1978 *New Internationalist* May 6/1 Present patterns of technology transfers from North to South vindicate Bertrand Russell's view..that 'I am compelled to fear that science will be used to promote the power of dominant groups rather than to make men happy.' 1979 *Newsweek* 19 Nov. 144 The turbulent years of the 1970s have witnessed an uneasy confrontation between the North and the South, and a largely unresolved debate on a whole series of specific economic problems.

C. *adj.* **1. b.** (Further examples.) Also, denoting the southern part of a city, town, district, ocean, etc.; *South Kensington* (colloq. *Ken*), a district of London noted esp. for the museums and other cultural and scientific institutions located there; also *ellipt.*, any of these institutions; *South-Spain* attrib., designating or pertaining to a South Spainer (see sense 1 c below).

1835 *Penny Cycl.* III. 25/2 The South Atlantic Ocean does not offer any other peculiarity in its formation, but the Northern is distinguished by several. 1845 *Encycl. Metrop.* VIII. 600 (*heading*) South London. 1862 A. J. Munby *Diary* 18 Jan. in D. Hudson *Munby* (1972) 114, I walked past the South Kensington Museum and along the Cromwell Road. 1882 *Girl's Own Paper* 1 Apr. 432/2 We advise your going direct to the British Museum or South Kensington, and make a study of one gallery after another. 1885 A. Edwardes *Girton Girl* I. iii. 59 There was no South Kensington, and we never called ourselves art students. 1885 *Encycl. Brit.* XVIII. 119/1 When the 'Challenger' was cruising in the South Pacific..the water was found to be uniformly warmer than the air. 1924 J. Buchan *Three Hostages* iii. 50 He was M.P. for a South London division. 1924 W. Runciman *Before Mast* iii. ii. 78 A visit was paid to a very fine South Spain barque. *Ibid.*, As it was a Friday we had presented to us a real South Spain meal, pea soup and pork. 1933 M. Allingham *Sweet Danger* xiv. 172 If we get away with this we

might start on the South Ken. There's a large-size model of a flea there I've always had my eye on. 1933 J. Masefield *Bird of Dawning* 98 All South-Spain ships pass where we pass, going or coming. 1944 W. Temple *Church looks Forward* ii. 17 That, broadly speaking, is the aim of the South India Scheme. 1946 *Whitaker's Almanack 1947* 783/2 South Georgia is permanently inhabited and is an important seat of the whaling industry. 1948 *Times* 11 May 3/3 The Communists continued with their campaign to keep the population of South Korea from voting in today's elections. 1949 O. Hammerstein (*title of musical*) South Pacific. 1950 *Times* 28 Apr. 5/4 The Federal Government has decided to send a mission.. to the island of Amboyna, where the revolt which resulted in the proclamation yesterday of the 'independent Republic of the South Moluccas' originated. 1954 *Times* 21 July 6/1 There will probably be two other enclaves for Viet-minh troops now in south Viet Nam. 1965 M. Allingham *Mind Readers* i. 22 Her drawl, which Peggie had known when it was pure South London, was now very Mayfair. 1966 *Times* 22 Jan. 7/1 Both the Yemen Government and the Arab League have welcomed the N.L.F.–O.L.O.S. merger into the new anti-British militant 'Front for the Liberation of Occupied South Yemen' (F.L.O.S.Y.). 1972 F. MacCarthy *All Things Bright & Beautiful* ii. 44 Lethaby..retired from the Central [School] to concentrate on the South Kensington professorship. 1972 H. Kurath *Studies in Area Linguistics* 54 The South Atlantic States—the Southern and South Midland dialect areas. 1975 *Times* 11 Mar. 13/2 Little girls with South Ken accents. 1982 *Daily Tel.* 11 Oct. 18/4 Spinks of St. James's tell me..that the South Atlantic Medal will be seen.

c. (Further examples.) *South Spainer*, formerly, a ship engaged in trade with Spain; a sailor on such a ship. See also *South African *sb.* and *a.*, *South American *sb.* and *a.*

1821 A. Royall *Lett. from Alabama* (1830) 137 She married a South Carolinian. 1839 *Dublin Rev.* May 449 'Read Mr. James's book,' said a South Australian colonist to the writer of this article. 1842 *Penny Cycl.* XXII. 270/2 South Polar countries. 1856 C. Nordhoff *Merchant Vessel* viii. 97 They hold all manner of foreign vessels, or 'south Spainers', in supreme contempt. 1877 *Encycl. Brit.* VII. 188/2 *South Slavic* [dictionary]—Richter and Ballman, Wien, 1839–40. 1881 *Ibid.* XII. 755/2 The South Indian [railway line] (the only one on the narrow gauge), in the extreme south, from Cape Comorin to Madras city. 1889 *N.Y. Semi-Weekly Tribune* 6 Dec. 13/4 Three ballots were put in the box for the South Dakotans to draw from. 1903 H. Holmes *Life & Adv. on Oceans* 8 There may be truth in the saying that a South Spainer, bound for a warm climate, can put his clothes in a stocking. 1924 W. Runciman *Before Mast* ii. iii. 46 Never a cargo vessel looked cleaner or better cared for than this little South-Spainer. 1936 A. W. Clapham *Romanesque Archit.* iv. 81 The system of barrel-vaults without direct lighting of the aisled nave is general in the south-French school. 1939 South Caucasian [see *Laz]. 1941 C. S. Forester *Captain from Connecticut* iii. 57 Hubbard's South Carolinian speech. 1949 *Britannica Bk. of Year* 380/1 An interim agreement was signed calling for some U.S. troops to remain in Korea until South Korean military forces could be well organized. 1950 'P. Woodruff' *Island of Chamba* ii. 43 There will be food in the Persian style as well as Moghul and South Indian. 1951 *Britannica Bk. of Year* 342/2 In East Indonesia some prominent Ambonnese people..on April 25 [1950] seceded and proclaimed a South Moluccan independent republic. 1958 T. Hickinbotham *Aden* xii. 196 The South Arabian League which originally advocated union between the Colony and the Protectorate. 1963 E. Humphreys *Gift* ii. i. 207 'No thanks, Stel, love.' With her I tended to talk as if I were a South Walian. 1966 J. Cleary *High Commissioner* i. 20 The Americans will accuse the Chinese and vice versa. The same with the South Vietnamese and the Viet Cong. 1971 *Guardian* 12 July 7/2 Guiding bewildered Irishmen and rooted South Londoners through the intricacies of Notting Hill Gate. 1972 'E. Peters' *Death to Landlords!* ix. 135 You are a South Indian yourself, Mr. Narayanan. 1975 *Amer. Speech 1973* XLVIII. 60 A South Dakotan holds that a wet bituminous road is slippery but an icy road is slick. 1975 *Times* 3 Dec. 1/6 The hijackers, all South Moluccans who demanded the return of the islands. 1977 *Trans. Philol. Soc. 1975* 176 *Cuillem* looks possible, *c* presumably representing South Walian phonetics. 1978 *Amer. Speech* LIII. 41 Three areas of the eastern states provided the bulk of the English-speaking settlers for the North Central states: South Midlanders from the Upland South, [etc.]. 1979 *Guardian* 5 Nov. 11/6 The marvellous Tipi cover by South Dakotan Sioux braves.

2. c. *South Bank, spec.* the southern bank of the Thames and the areas adjacent to it, (*a*) noted esp. for the cultural complexes and public gardens developed between Westminster and Blackfriars bridges for and since the Festival of Britain in 1951; also *attrib.* and *ellipt.*, any of these complexes; (*b*) used (freq. *attrib.* and as *adj.*) with reference to the policy of the Anglican diocese of Southwark to re-express traditional beliefs and practices in ways that would make them better suited to contemporary life.

1951 H. Nicolson *Diary* 4 May (1968) 206 Viti and I go to the South Bank Exhibition. We are entranced. 1961 *Guardian* 19 May 22/5 A glass-and-wood pavilion in best South Bank style. 1963 M. Frayn in Sissons & French *Age of Austerity* xv. 329 The South Bank site—a derelict slum, low-lying, marshy, and heavily blitzed. *Ibid.* 330 A model of the South Bank made out of toilet rolls. 1963 *Guardian* 8 July 14/3 The problems of the South Bank parishes which this so-called South Bank religion is trying to tackle. 1965 Lunn & Lean *Cult of Softness* iv. 44 The new predestination exercises a certain attraction on South Bank theologians. 1967 A. Laski *Seven Other Years* xiii. 179 It may be that I am going a little far in calling it atheism; it might, I suppose, be regarded as an effusion of so-called South Bank Christianity. 1968 J. Turner *Requiem for Two Sisters* i. 5 The vicar..had never thought it part of his duty to dress in a black suit and dog-collar all the time. Indeed, his opinions were to a great extent South Bank. 1977 'E. Crispin' *Glimpses of Moon* vi. 88 No one takes any notice of the clergy nowadays, except for Humanists waiting to welcome South-Bank bishops into the fold. 1977 *Skateboard Special* Sept. 3/4, I do a lot of riding on the South Bank. 1980 *Times* 22 Oct. 13/7 The National have given us many fine productions... This one lapse of quality does not justify the witch-hunt on the South Bank.

3. (Later examples.) Also *fig.*

1917 N. Douglas *South Wind* ii. 16 For Nepenthe was famous not only for its girls and lobsters, but also for its south wind. 1937 C. Mackenzie (*title*) The south wind of love. 1946 L. B. Lyon *Rough Walk Home* 28 Ask that for these may blow The hot south rage of life again.

4. (Later example.)

1973 *Express* (Trinidad & Tobago) 1 Feb. 17/1 Behind the move to promote female calypsonians is well-known south businessman, Mr. Lall Parsotan.

south-[1]. Add: **1.** *south-facing.*

1961 *Times* 23 Dec. 3/4 The aspect is right, that is south-facing. 1978 'J. Bell' *Swan-Song Betrayed* ii. 16 Her workroom, small but south-facing.

2. Also in *Comb.* with advbs., as **south-about** *adv.*, by a southerly route; also *attrib.*; **south-away** quasi-*sb.*, somewhere to the south.

1958 *Times* 20 Dec. 3/3 Bass frequent the coast from Suffolk south-about to Cheshire. 1961 *Times* 24 Nov. 14/6 So it was decided..to send the ship by the south-about route via Cape Horn. 1954 J. R. R. Tolkien *Fellowship of Ring* 16 The folk of the Marish..came..up from south-away.

South African, *sb.* and *a.* [f. *South Africa*: see South *a.* 1, -an.] **A.** *sb.* **1.** A native or inhabitant of South Africa (see below).

This area of southernmost Africa consisted in the 19th century of a group of British and Boer territories; in 1910, following the British victory in the Boer war (1899–1902), these united to form the Union of South Africa; in 1961 the Republic of South Africa was established.

1806 J. W. Janssens *Let.* 17 Jan. in G. M. Theal *Rec. Cape Colony* (1899) V. 298 The interest of the few unfortunate Men who I have the Honor to command, that of the brave and good Dutch South Africans.., put me under the necessity to accept the painful conditions. 1871 J. Mackenzie *Ten Years North of Orange River* p. v, I would specially direct attention to that part.. which describes the results of the past contact of Europeans with South Africans. 1897 G. A. Parker *S. Afr. Sports* p. xiii, I am glad of this opportunity of dedicating this, the first compilation relating to South African sports, to the foremost South African [*sc.* Cecil Rhodes], and one who has for many years encouraged outdoor games with characteristic liberality. 1913 C. Pettman *Africanderisms* 3 As South Africans our lot is cast in a country which..is quite young. 1949 E. Pound *Pisan Cantos* lxxx. 92 And persuaded an Aussie or Zealander or S. African To kneel with him in prayer. 1978 J. Branford *Dict. S. Afr. Eng.* p. xiii, This text..has been written for South Africans of all racial groups.

2. *absol.* uses of the adj.

1930 *Economist* 8 Nov. 866/2 South Africans [*sc.* shares] remained firm. 1969 *Guardian* 24 Oct. 9/5 You will need some medium-dry sherry... You could go for a good South African at about £1.

B. *adj.* Of or pertaining to South Africa or its inhabitants.

1824 (*title of newspaper*) The South African Commercial Advertiser. 1838 W. B. Boyce *Notes on S. Afr. Affairs from 1834 to 1838* p. xv, Much of the misrepresentation of South African affairs, arises from the fact, that the Colonists labour under the..disadvantage of being..unconnected with the powerful interests, which in England press the claims of the Colonists of other British dependencies upon the attention of Parliament and the Public. 1876 in J. Flint *Cecil Rhodes* (1976) iii. 39 Lord Carnarvon's South African policy. 1894 [see South *a.* 1 c (*b*)]. 1913 C. Pettman *Africanderisms* 8 The words of Portuguese, Indian, and Malay origin, still current in South African Dutch. 1921, 1925 [see *Afrikaans *sb.*]. 1949 A. Wilson *Wrong Set* 41 Everyone was anxious to know what Harry thought of the South African hock. 1967 L. Meynell *Mauve Front Door* v. 64 South African sherry was absolutely right on that evening. 1978 J. Branford *Dict. S. Afr. Eng.* p. xi, South African English ..is in every sense, culturally, lexically, grammatically and phonologically, a 'mixed bag'.

Hence **South Africanism,** (*a*) distinctive South African quality; (*b*) a word or idiom peculiar to or characteristic of South Africa.

1959 *Listener* 29 Oct. 714/2 South Africans..were lulled into thinking the essential South Africanism of General Smuts permanently secure. 1961 *Personality* 16 May 27, I think my favourite South Africanism is 'Bioscope'. 1978 J. Branford *Dict. S. Afr. Eng.* p. xiii, These South Africanisms include..a number of English phrases or usages.

South American, *sb.* and *a.* [f. *South America*, the name of the southern part of the continent of America, excluding Central America: see South *a.* 1, -an.] **A.** *sb.* A native or inhabitant of South America.

1775 [see South *a.* 1 c (*a*)]. 1826 F. B. Head *Rough Notes Journeys Pampas & Andes* 306 The Spanish South Americans have certainly become independent of the government of Spain. 1863 T. W. Hinchliff *S. Amer. Sk.* viii. 173 Spaniards, Italians and South Americans have a vile habit of using the knife. 1913 A. S. Peck *S. Amer. Tour* xxxiii. 361 The manufacturer..should understand that the South Americans in general are not eager to trade with us. 1943 H. F. Artucio *Nazi Octopus in S. Amer.* i. i. 7 This book is a summing up of experiences and observations which might have been those of any South American who believed in the principles of democracy. 1979 *Guardian* 23 Oct. 28/5 The race got off to a bad start, blamed on Europeans and South Americans surging forward before the signal went off.

B. *adj.* Of or pertaining to South America or its inhabitants.

1820 *Times* 11 Mar. 3/5 (*heading*) South-American affairs. 1833 [see South *a.* 1 c (*b*)]. 1869 *Month* Jan. 82 The same operation could be carried on in the South American plains. 1950 T. D. McCown in J. H. Steward *Handbk. S. Amer. Indians* VI. i. 2 The osseous human remains and the artifacts of human manufacture..have been accumulated mainly by European and South American scientists over a period of about 100 years. 1976 'A. Hall' *Kobra Manifesto* xi. 153 Half the people in the queue were South American Indians.

sou·thbound, *a.* (*sb.*) Also **south-bound.** [f. South-[1] + Bound *ppl. a.*[1]] **A.** *adj.* **1.** Bound or directed southwards; travelling south.

1885 H. M. Jackson *Zeph.* vi. 227, I am going on the south-bound train. 1960 'E. McBain' *Killer's Payoff* ix. 88 A southbound trip that eventually led back to the city. 1976 W. Greatorex *Crossover* 170 He ran wildly across the south-bound traffic. 1980 K. Follett *Key to Rebecca* xxvi. 276 Southbound trains are less in demand.

2. Intended for traffic travelling south.

1971 *Daily Tel.* 29 Dec. 1/5 Most of the pile-ups were on the southbound carriageway near the junction of the M1 and M10. 1980 *West Lancs. Even. Gaz.* 1 Mar. 3 Sections of north and southbound carriageways of the M6 near Preston will be closed for at least four hours.

B. *ellipt.* as *sb.* A southbound train.

1903 'O. Henry' in *Everybody's Mag.* Feb. 173/1 Passengers on the south-bound saw them seated together. 1932 W. Faulkner *Light in August* xv. 340 Folks.. begun to come in and buy tickers for the southbound.

South Devon. [f. South *a.* 1 + *Devon.] A bull or cow belonging to the breed so called, characterized by its large size and light red or fawn colour, and used for both milk and beef production; also, the breed itself.

1897 W. Housman *Cattle* ii. 59 A breed of cattle, now called the South Devon breed, but long known by the local name of 'South Hams', has sprung up. *Ibid.* 61 The 250 animals of 1893..give the aggregate of 2,961 entries of South Devons. 1946 F. H. Garner *Brit. Dairying* iii. 161 At maturity the South Devon is..heavier than any other breed. Quite commonly South Devon cows weigh 15 or 16 cwt. 1977 'E. Crispin' *Glimpses of Moon* v. 69 Then came the cows, fourteen-hundredweight yearling South Devons.

southdown. 3. (Earlier example.)

1826 B. R. Haydon *Let.* 18 Aug. in *Autobiogr. & Mem.* (1927) xxiv. 331 His sly hints as I passed his shop that he had 'a bit of South Down, very fine'.

south-east, *adv.*, *sb.*, and *a.* Add: **B.** *sb.* **3.** (Later examples.)

1968 *Radio Times* 28 Nov. 20/5 A look at some non-broadcast music events taking place in London and the South East. 1972 P. Johnson *Offshore Islanders* iii. 146 There was an enormous bias in favour of the south-east. Clergymen did not want to serve in the wilder and poorer districts of the north and west.

C. *adj.* **1. b.** With proper names, denoting the south-eastern division of a continent, race, etc., and with sbs. and adjs. derived from them.

1893 *Geogr. Jrnl.* Nov. 474 Travel and Adventure in South-east Africa: being the narrative of the last eleven years spent..on the Zambesi. 1909 *Prospectus* (South-East Borneo Rubber Plantations Ltd.) 2 This Company has been formed for the purpose of acquiring and working the Rubber Plantation known as Tanah-Intan..situated in South-East Borneo. 1946 F. Owen *Campaign in Burma* vii. 44 To complete the picture of South-East Asia Command we must reintroduce Lieutenant-General Joseph Stilwell. 1959 'M. Derby' *Tigress* ii. 86 Young South-East Asians at play. 1964 *Whitaker's Almanack 1965* 71/1 South-East England Development Proposals. 1968 O. Wynd *Sumatra Seven Zero* v. 57 The South-east Asian male is never turned out burly. 1971 H. Trevelyan *Worlds Apart* xvii. 193 The Deputy Minister in charge of South-East Asian affairs spoke to me..about the responsibilities of the co-chairman for Vietnam.

south-easter. (Earlier example.)

1797 A. Barnard *Let.* 15 Oct. in *S. Afr. a Century Ago* (1901) 100 What a bold south-easter we have had these two days!

south-ea·sterner. [f. South-eastern *a.* + -er[1].] An inhabitant or native of the south-eastern part of a country.

1960 G. Ashe *From Caesar to Arthur* v. 115 The resulting cleavage between Vortigern and the south-easterners whom he aspired to govern was widened by the Pelagian heresy. 1964 *New Society* 26 Mar. 3/2 At present two out of three southeasterners live within 40 miles of Charing Cross.

souther, *sb.* (Earlier example.)

1851 *Austral. & N.Z. Gaz.* XXX. 483 During the night a 'stiff souther' put [the *Pauline*] again on shore.

southerly, *a.* Add: **2.** For '*Southerly buster* (see Buster 3)' read '*southerly burster, buster* (see Burster 2, Buster 3 a in Dict. and Suppl.)'. (Earlier example.)

1850 B. C. Peck *Recollections of Sydney* viii. 132 It is almost a corollary, that the evening of a hot-wind day brings up a 'southerly buster', as we have heard the vulgar call it, very chill indeed..as this wind comes from the southerly region of the Australian Alps.

B. *sb.* A wind blowing from the south; a southerly buster. *Austral.* and *N.Z.*

1943 K. Tennant *Ride on Stranger* viii. 79 When the Southerly blew, the stiff leaves..twisted rim-on to the blast. 1964 R. Braddon *Year Angry Rabbit* xiv. 123 What use is it being able to guarantee fine weather, or rain, or a cool southerly only on the coast? 1973 P. White *Eye of Storm* i. 65 How exotic, how naked her body felt when the southerly began to blow at the end of a sticky summer's day, caressing her inside her dresses.

southern, *a.* and *sb.* Add: **A.** *adj.* **1. b.** (Earlier and further examples.) *Southern Baptist*, a Baptist who is a member of a church belonging to the Southern Baptist Convention, first organized in 1845; also *attrib.*

1789 *Deb. Congress U.S.* 28 Apr. (1834) 215 Suppose a member from Massachusetts was to propose an impost on negroes, what would you hear from the Southern gentlemen, if fifty dollars was the sum to be laid? 1846 J. Soule in *Jrnls. Gen. Conf. Methodist Episcopal Church, South* (1851) I. 105 Southern Methodists were able so far to conciliate public opinion, and quiet popular apprehension, as to carry on..the ordinary operations of church enterprise and discipline. 1866 in W. L. Fleming *Documentary Hist. Reconstruction* (1907) II. 247 In 1845, when the Southern Baptist Convention was organized,..in proportion to the population there were more negroes than white people who were members of our churches. 1932 *N.Y. Times* 3 Nov. 19/3 The Southern Baptist handbook for 1932 declares Southern Baptists are 'still wasting money in riotous living'. 1936 M. Mitchell *Gone with Wind* ix. 195 A delicately nurtured Southern belle with her Irish up. 1964 'E. McBain' *Ax* ii. 32 A simpering smile on her lips, as though she were a Southern belle waiting to be asked for a dance. 1978 N. Longmate *Hungry Mills* i. 21 The 'Southern gentlemen', a breed already famous for their independence and arrogance, owned the large plantations. 1979 *Arizona Daily Star* 1 Apr. E7/4 Television station WFAA, which aired his program, said Robison's remarks crossed over from religious to political proselytizing and then canceled the half-hour program. The result has been a wave of support for the popular Southern Baptist evangelist.

3. b. *Southern Cross*: also *transf.*, the Australian national flag.

1855 R. Carboni *Eureka Stockade* xxxvii. 50 There is no flag in old Europe half so beautiful as the 'Southern Cross' of the Ballaarat miners. 1917 'H. H. Richardson' *Richard Mahoney* ii. i. 96 The 'Southern Cross' hoisted—a blue bunting that bore the silver stars of the constellation after which it was named.

4. c. *U.S. spec.* Pertaining to or belonging to the southern States of America (cf. South *sb.* 2 c). *Southern Comfort*, the proprietary name of a brand of alcoholic drink, based on whisky and orig. manufactured in the U.S.

1819 D. Thomas *Trav. through Western Country* 100 The mistress..treated us to milk, in the true spirit of southern hospitality. 1836 *Southern Lit. Messenger* II. 111/2 We have known a New Englander laugh at the Southern use of the word clever. 1860 *Charleston* (S. Carolina) *Mercury* 15 Nov. 2/5 The 'Lone Star' was very suggestive of the additions which may hereafter be made to the Independent Southern Confederacy. 1877 C. Hallock *Sportsman's Gazetteer* 96 The Southern Fox Squirrel inhabits the Southern States from North Carolina to Texas. 1880 'Mark Twain' *Tramp Abroad* xlix. 574 Hot wheat-bread, Southern style. 1925 G. P. Krapp *Eng. Lang. in Amer.* I. 40 It is..much easier for an American to call up in his mind a kind of image of the Eastern and Southern types of American speech than of the Western or General type. 1934 *Official Gaz.* (U.S. Patent Office) 24 July 786/2 Midland Distilleries, Incorporated, St. Louis, Mo... *Southern Comfort.* For Cordial. 1947 *Trade Marks Jrnl.* 12 Nov. 707/1 *Southern Comfort* B647,105. Wines, spirits (beverages) and liqueurs. Southern Comfort Corporation.., 2121, Olive Street, St. Louis,..Missouri, United States of America; Manufacturers. 1962 A. Lurie *Love & Friendship* iv. 60 He had a slight Southern accent. 1978 M. G. Eberhart *Nine O'Clock Tide* i. 21 I'll take a little drink..Southern Comfort. 1979 *United States 1980/81* (Penguin Travel Guides) 280 The corn bread, fried chicken..and the like are served buffet-style in a big Southern-style mansion. 1980 *Blair & Ketchum's Country Jrnl.* Oct. 142 (Advt.), Grandma Johnston's southern fruit cake.

5. b. southern beech = *nothofagus.

1914 W. J. Bean *Trees & Shrubs Hardy in Brit. Isles* II. 97 The southern beeches are only adapted for the milder parts of the country. 1957 M. Hadfield *Brit. Trees* 189 The so-called 'southern beeches'..represent the beech family in South America, south-east Australia, and New Zealand. 1974 [see *nothofagus].

9. *southern-shaped*; **southern-fried** *U.S.*, cooked in a manner characteristic of the southern states; also *fig.*

1972 *St. Louis Post-Dispatch* 5 Nov. c1/4 'Ah sure thought we were gonna win,' he said in his Southern-fried drawl. 'Ah really did.' 1973 *Sat. Rev. World* (U.S.) 6 Nov. 43/1 Southern-fried chicken..with bananas, sweet corn, and tomatoes. 1976 N. Thornburg *Cutter & Bone* vi. 142 Not hillbilly really. Just a good ole boy, southern fried. 1982 *Times* 19 June 7/5 Southern fried chicken really is a speciality. 1922 D. H. Lawrence *Aaron's Rod* xiv. 195 There was..something inhuman and possessed-looking in their foreign, southern-shaped faces, so much more formed and demon-looking than northern faces.

B. *sb.* **3.** *U.S.* The dialect of English spoken in the southern states.

1935, 1951 [see *New England a]. 1975 *New Yorker* 21 Apr. 33/3, I listened to the Governor's lady talking for some minutes to some of her South Delaware friends, and they *were* talking Southern. 1981 J. Scott *Distant View of Death* x. 147 Saying in her comedy Southern: 'Why, Colonel..you jest *spoil* lil' ol' me.'

southernism. Add: **1. b.** An idiom, expression, or word peculiar to a more southerly part of Britain, esp. to the South of England.

1967 P. J. Bawcutt *Shorter Poems of Gavin Douglas* p. lxxv, The absence of the southernisms and archaic verbal inflections that are found in Douglas, side by side with the regular Middle Scots forms. 1978 *Trans. Yorks. Dial. Soc.* LXXVIII. 9 More definitely attributable to error is the intrusion of Southernisms such as *hond* for hand, *darter* for daughter and, grammatically, *she* for her as object pronoun.

2. (Earlier example.)

1861 *N.Y. Tribune* 15 July 6/4 Southernism has raised the standard and gage of social condition absolutely; and those who are so unfortunate as not to be high-born—i.e., born at the South—are given to feel that they must eke out their shortcomings with an extra amount of Southern ardor and Pro-Slavery talk.

southernization (sʌ·ðənaizēi·ʃən). [f. Southernize *v.*: see -ation.] The act of making southern in respect of character.

1976 *Time* 27 Sept. 98/2 These developments helped to modify the old stereotypes and mitigate fear of Southernization in the North. 1976 *National Observer* (U.S.) 6 Nov. 14/2 Great pains were taken to ensure the success of this historic occasion, otherwise known as 'The Southernization of Central Park'.

sou·thernness. Delete *rare* and add later examples.

1931 Blunden *Votive Tablets* 210 The weaknesses of his verse,..the exuberant southernness of so much of it might be effeminate. 1966 *Listener* 10 Mar. 357/3 Cleanth Brooks's recent book on the Southernness of Faulkner. 1973 D. Aaron *Unwritten War* vii. 118 Charleston, the hatchery of rebellion and quintessence of Southernness. 1980 T. Holme *Neapolitan Streak* 10 He had never shown ..prejudice against Peroni's southern-ness as most northerners invariably did.

Southeyan (sɒ·ðiăn), *a.* [f. the name of the English poet and prose writer Robert *Southey* (1774–1843) + -an.] Of, pertaining to or characteristic of the writings of Southey. Hence **Southeya·na** [Ana *suff.*], writings, etc., relating to Southey.

1817 Keats *Let.* 11 May (1931) I. 32, I am very near agreeing with Hazlitt that Shakespeare is enough for us. By the by what a tremendous Southean [*sic*] article his last was. 1931 Blunden *Votive Tablets* 191 A modern house would not contain a collection of Southeyana. *Ibid.* 198 The last sentences of the extract show the Southeyan good sense. 1974 R. Holmes *Shelley* ix. 207 The prose notes are constantly more powerful and effective than the long-drawn Miltonic or Southeyan rhetoric of the verse.

southland. Restrict 'Now *arch.* or *poet.*' to sense 1 and add: **2.** (Further examples.)

1849 J. G. Whittier in *National Era* 1 Nov. 174/4 The South land hath its fields of cane, The Prairie boasts its heavy grain. 1905 *Florida Times-Union* 7 May ii. 8/1 Yet is this place rich in its treasured holdings of art, its clustered memories and traditions of the Old South or the southland of ante-bellum days. 1956 G. P. Kurath in A. F. C. Wallace *Men & Cultures* (1960) 153 The Charleston, after seething in the Southland as a Negro round dance, was discovered in 1923. 1974 P. McCutchan *Call for Simon Shard* iv. 32 He entered the Southland.. unofficially, no fanfares, no men from Canberra. 1978 *Guardian Weekly* 1 Oct. 9/1, I believe in the Southland.. I believe in the people from South Carolina..and I believe in the people of Georgia.

3. Southland beech = *silver beech* s.v. *silver *sb.* and *a.* 21 e, in reference to the region of New Zealand on the west coast of the South Island.

1947 J. C. S. Brough *Timbers for Woodwork* xvi. 138 Imported beeches are..Red beech..and Southland, or 'Silver' beech. 1966 G. W. Turner *Eng. Lang. Austral. & N.Z.* viii. 166 The Southland beech is not really a beech.

southpaw (sau·θpǭ:), *sb.* (*a.*) *colloq.* (orig. *U.S.*, in *Baseball*). [f. South *a.* + Paw *sb.*[1]]

1. A person's left hand. (In quot. 1848, a punch or blow with the left hand.)

1848 *Democratic B-hoy*, 'I say, Lewy, give him a sockdologer!' 'Curse the Old Hoss, what a south-paw he has given me!' 1885 *Sporting Life* 14 Jan. 4/3 They had always been accustomed to having their opponents hug their bases pretty close, out of respect for Morris' quick throw over to first with that south-paw of his. 1942 Berrey & Van den Bark *Amer. Thes. Slang* § 121/53 *Southpaw, wrong hand* or *fist*, the left hand or fist. 1948 *Chicago Tribune* 20 Apr. i. 20/5 He waved his big south-paw and ducked under the roof.

2. One who pitches or throws with the left hand; a left-handed person.

In *Boxing*, a southpaw leads with his right hand.

1891 *Chicago Herald* 24 July 6/1 The new south-paw.. came to town yesterday. 1911 *Daily Colonist* (Victoria, B.C.) 15 Apr. 8/5 Davis came up to bat... He faced the twirler right-handed. He always does with southpaws. 1932 *Ring* Apr. 5/2 McCoy was a slow southpaw who had proved just a good workout for Joe Chip. 1942 Berrey & Van den Bark *Amer. Thes. Slang* § 430/10 Left-handed person,..*south-paw.* 1947 J. Gunther *Inside U.S.A.* xl. 657 Ah won't even go to the Polo Grounds unless a southpaw's pitchin'. 1951 *Sport* 6–12 Apr. 8/2 On the same bill, Joe Lucy, the young southpaw, meets South African lightweight Gerald Dreyer. 1955 *Sci. News Let.* 14 May 310/2 The family cat may have a preferred paw.., and pussy is most often a southpaw when she is not ambidextrous. 1959 *Sunday Times* 8 Nov. 32/6 In the ball parks all over the United States the so-called 'diamond', formed by the track between the bases, is always oriented to the same points of the compass, so that in whatever park a team is playing the pitcher on his mound will always have his right hand on the north side of his body; hence a left-hander is a 'southpaw'. 1967 *Boston Sunday Herald* 26 Mar. ii. 7/1 Rocket Rod Laver leads the greatest tennis show on earth into Boston Garden Monday night... The freckle-faced southpaw is the top-seeded player. 1970 H. McLeave *Question of Negligence* (1973) vi. 48 'Nobody told me he was a southpaw.' Even the psychiatrist had.. forgotten that the surgeon cut with his left hand. 1976 *Billings* (Montana) *Gaz.* 26 June 1-B/2 The 6-0 lefthander, the only southpaw listed on the Angels' roster, struck out six and walked the same number. 1976 'A. Burgess' *Beard's Roman Women* (1977) v. 110 Donatella, a southpaw, animated this [*sc.* her left shoulder-blade] while lifting the one remaining chair from the front room. 1978 M. Kenyon *Deep Pocket* ix. 103 He wore shorts and boxing gloves. ''E's a southpaw,' Peckover said.

3. *attrib.* or as *adj.* Left-handed; also *transf.*, left-footed, and *fig.*

1891 *Cricket* 29 Oct. 463/1 The Germantown man returned the ball like a flash to the wicket, and the 'south-paw' batsman was run out. 1932 J. T. Farrell *Young Lonigan* iii. 126 It was swell for Studs to play,.. knowing he had made that good kick,..to run back and pick one of Helen's southpaw kicks out of the air. 1949 *Sun* (Baltimore) 3 June 18/8 They would have been bunched against southpaw pitching. 1957 R. Watson-Watt *Three Steps to Victory* xliii. 245 This was, however, a south-paw kind of compliment. 1969 *New Scientist* 6 Nov. 277/2 Jack Bodell has just become the first southpaw heavyweight champion in British boxing history.

Hence **sou·thpaw** *v. trans.*, to pitch with the left hand; **sou·thpawing** *vbl. sb.*, the action of pitching with the left hand.

1928 *Daily Ardmoreite* (Ardmore, Okla.) 12 Apr. 8/1 Herb Pennock southpawed his way the route for the Yankees. 1938 *Chicago Tribune* 4 Apr. 21/1 The White Sox positively refused to be awed today by the southpawing of Larry French. 1951 *Sun* (Baltimore) 23 Aug. 20/1 Jim Burns southpawed his eighth straight triumph.

southron, *sb.* Add: **2. b.** (Earlier examples.) Also *transf.*

1828 *Free Press* (Tarboro, N. Carolina) 9 Nov., I am a Republican in principle, and a Southron in feeling. 1831 J. J. Audubon *Ornith. Biogr.* I. 110 When those [mocking birds] which had gone to the Eastern States..have returned, they are instantly known by the 'southrons' who attack them on all occasions.

South Sea. Add: **5. c.** *South Sea Islander.*

1832 A. Earle *Narr. Residence N.Z.* (1966) 58, I am persuaded that these South Sea islanders, though so nearly of the same complexion, still are not of the same race. 1913 C. Mackenzie *Sinister Street* I. ii. iii. 173 That belt of yours, Michael, would give a South Sea Islander a headache. 1974 J. Pope-Hennessy *R. L. Stevenson* xi. 211 Louis..was as brown as a South Sea Islander.

south-side. Add: **a.** Also *attrib.*

1869 *Bradshaw's Railway Man.* XXI. 58 The railway.. joins the Glasgow and South Western and the Caledonian (southside branches), on the south side of the city. 1937 H. G. Wells *Star Begotten* vi. 105 The Punic Wars..he presented as a gigantic necessary struggle between noble north-side soldiers and revengeful, obdurate but extremely competent south-side loanmongers. 1955 Keepnews & Grauer *Pict. Hist. of Jazz* iii. 43 Pianist Blythe, who died in 1930, was..in the forefront of the 'South Side style', which meant small-band recordings and jobs at small, rough clubs. 1979 *Irish Times* 28 Sept. 12/6 If the northside area described here is compared with a southside suburb the differences become more marked.

South Suffolk. [f. South(down + Suffolk.] A sheep belonging to the breed so called, developed in New Zealand by crossing Suffolk and Southdown sheep and used to produce lean meat and short, fine wool.

1950 *N.Z. Jrnl. Agric.* Feb. 144/3 Two instances of fresh development with sheep breeds may be mentioned: First, the aim of some breeders in the South Island '..to establish a new breed by crossing Suffolk and Southdown with a view to producing a sheep having Southdown characteristics with a maximum of lean meat'—the South Suffolk. 1956 G. Bowen *Wool Away!* (ed. 2) xii. 148 South Suffolk. This is a New Zealand type derived from crossing Southdowns with Suffolks, and is being very widely used..in Canterbury and is spreading rapidly to other parts of New Zealand. 1977 *N.Z. Herald* 8 Jan. 4-8/6 (Advt.), 23 South Suffolk ewes.

south-west, *adv.*, *sb.*, and *a.* Add: **B.** *sb.* **1. b.** *U.S.* The south-western states.

1835 J. H. Otey in W. M. Green *Mem. J. H. Otey* (1885) 15 Connected with my journey to the South-West, was an ardent desire to forward..a projected plan of a *Literary and Theological Seminary*, to meet the wants of Episcopalians in Tennessee and Mississippi and Louisiana. 1897 *Sears, Roebuck Catal.* 342/2 The best specimens of broad burlesque for which the Southwest is so distinguished. **1925** C. F. LUMMIS (*title*) Mesa, cañon and pueblo; our wonderland of the Southwest. **1935** A. G. MACDONELL *Visit to America* ix. 142 It must have been from the Spanish south-west..that the first prospectors came..into Montana. **1976** 'R. MACDONALD' *Blue Hammer* xv. 83 Mildred was the most beautiful woman in the South-west. **1979** G. MACDONALD *Camera* ix. 130/2 In the South-West the pueblo Indians..were..much photographed.

c. *ellipt.* for South West Africa (now Namibia).

1976 J. MCCLURE *Rogue Eagle* i. 22 Ma..said she'd write to her family in South West, but he said that was just another Bantustan these days. **1978** S. NAIPAUL *North of South* II. vi. 245 Abraham had been in Namibia ('South-West')... It was in South-West..that he and Tessa had met.

C. *adj.* **2. b.** With proper names, denoting the south-western division of a city, country, continent, etc.

1858 *1st Rep. South-West London Protestant Inst.* 3 *Rules.* 1. That this Society be called 'The South-West London Protestant Institute'. **1899** *Geogr. Jrnl.* May 563 A notice of Dr. Rehbock's work on irrigation in German South-West Africa. **1946** *Whitaker's Almanack 1947* 789 The *Orange*..is the principal river of the south,..flowing into the Atlantic between the Protectorate of South-West Africa and the Cape of Good Hope. **1950** H. BRIERCLIFFE *Southern England* p. iii, There is necessarily some overlapping with..South-west England and the Midlands. **1968** *Ann. Reg. 1967* 326 The South African Government remained completely unmoved by United Nations' efforts to plan the implementation of the 1966 General Assembly resolution that South West Africa be removed from South African control. **1980** *Times* 14 June 1/8 The Middle East, or south-west Asia as the Americans now call it.

south-wester, *sb.* Add: **2.** (Earlier example.)

1836 R. S. SURTEES *Let.* in A. Mathews *Mem. Charles Mathews* (1839) IV. ix. 193 Throwing aside his hat, he put on one of the boatman's 'south-westers'.

3. A (white) inhabitant of Namibia (formerly known as South West Africa).

1976 *Times* 21 Aug. 13/3 South Africa's original plans to break up South West Africa into eight or nine mini-states, of which the only viable ones would be a 'ranchistan' containing the 99,000 rich German, Afrikaans and English-speaking 'Southwesters', and Ovamboland. **1978** *Guardian Weekly* 7 May 6/2 The Teutonic calm of Windhoek... Like their White Rhodesian counterparts most 'Southwesters' as they call themselves have yet to grasp the dimension of the change.

south-western, *a.* Add: **2. b.** *U.S.* Of, pertaining to, or characteristic of the south-western states.

1806 *New Eng. Palladium* (Boston) 30 July 2/1 The President appoints the Legislative Councils in our South-western Territories. **1832** *Jrnl. Gen. Convention Prot. Episc. Church* 51 Delegates have been chosen to co-operate with Alabama and Louisiana in organizing the contemplated South-Western Diocese. **1973** J. M. WHITE *Garden Game* 150 The walls were whitewashed in simple, South-Western style.

south-westward, *a.* (Later example.)

1972 *Science* 19 May 791/1 The southwestward flow of cold drier air is suggested to account for two aspects.

souvenir, *sb.* Delete ‖ and add: **2. c.** *Mil. slang.* In the 1914–18 war, a jocular term for a bullet or shell.

1915 D. O. BARNETT *Let.* 17 May 140 They kept sending their big black souvenirs over. **1929** *Papers Mich. Acad. Sci. Arts & Lett.* X. 324 *Souvenirs*, shells.

3. *souvenir-hunter, programme, shop.*

1923 KIPLING *Irish Guards in Gt. War* I. 131 Being a hardened souvenir-hunter, he is reported to have removed the official German name-board of the establishment. **1976** B. JACKSON *Flameout* (1977) iii. 40 Souvenir hunters were a menace..stealing bits of metal that could, if left in position, help determine the cause of the crash. **1962** L. DEIGHTON *Ipcress File* ii. 22 A cigarette-girl..tried to sell me a souvenir programme. **1950** J. FLANNER in *New Yorker* 25 Feb. 84/2 The best souvenir shops for the pilgrims are in the Via della Conciliazione. **1980** J. GARDNER *Garden of Weapons* III. iii. 246 From the souvenir shop the coach party went ..for a drink.

souvenir (sūvĕnīə·ɹ), *v.* [f. the sb.] **1.** *trans.* To pierce with a bullet or shell. Cf. sense *2 c of the sb. *Mil. slang* (in the war of 1914–18).

1915 *Chambers's Jrnl.* Oct. 663/1 Our periscope was 'souvenired' later on with a rifle-bullet clean through the tin sides.

2. To provide with or constitute a souvenir of (something). *rare.*

1917 W. OWEN *Let.* 25 Nov. (1967) 510 How much better than a photograph does it [*sc.* a poem] souvenir that day! **1976** *Vogue* Jan. 7/2 The Tate..is issuing a special Constable diary..and a Constable paper-weight. So the exhibition will be fully souvenired.

3. To take as a 'souvenir'; to appropriate; to pilfer, steal. Also *absol. slang* (orig. *Mil.*).

1919 W. H. DOWNING *Digger Dialects* 46 *Souvenir*,..to steal, find, capture, etc. **1920** *Punch* 28 Jan. 65/1 The Major..set the ladies souveniring among old water-tin stoppers, which he alleged to be the plugs of hand-grenades. **1944** F. CLUNE *Red Heart* 19, I dug up his body, souvenired his false teeth. **1956** S. HOPE *Diggers' Paradise* ix. 83 But early, too, numbers of youngsters show that tendency to 'souvenir' which is the euphonious term for pilfering. **1969** I. BROWN *Rhapsody of Words* 120 Silver spoons and jewellery souvenired from rooms with open windows. **1975** J. I. M. STEWART *Gaudy* vii. 116 It's possible that people sometimes souvenir such things.

Hence **souveni·ring** *vbl. sb.*

1969 'M. INNES' *Family Affair* xiii. 145 It had been lifted much as somebody might lift a china gnome..from a suburban garden. Souveniring, as they say. **1972** *Guardian* 15 May 1/4 The House of Commons is determined to end the 'souveniring' of cutlery... All crests will be removed from the cutlery in some of the visitors' cafeterias.

‖**souvlaki** (su-, suvlā·ki). Pl. **souvlakia.** [mod. Gr. σουβλάκι, f. σούβλα skewer.] A Greek dish consisting of pieces of meat grilled on a skewer. Also *attrib.* Cf. *KEBAB.

1958 R. LIDDELL *Morea* II. vii. 171 They had bought *souvlakia* or bread and cheese at every halt. **1963** J. M. STUBBS *Home Bk. Greek Cookery* vii. 97 The best *souvlakia* I have eaten in Greece are those sold..at Antirhion. **1972** J. AIKEN *Butterfly Picnic* i. 19 Lamb grilled on skewers, souvlakia, and all the different forms of mince that the Greeks delight in. **1979** M. A. SHARP *Sunflower* iii. 30 The building..was..sandwiched between a *souvlaki* stand and a tiny hotel.

souzalite (sū·zăləit). *Min.* [f. the name of A. J. A. de *Souza*, 20th-c. Brazilian mining administrator: see -ITE[1].] A hydrated basic phosphate of aluminium, iron, and magnesium, found in association with scorzalite as fibrous masses of green crystals.

1947 PECORA & FAHEY in *Bull. Geol. Soc. Amer.* LVIII. 1217 The new minerals are named in honor of Dr. Evaristo Scorza and Dr. Antonio José Alves de Souza... Souzalite is a fibrous, green hydrous iron magnesium aluminum phosphate. **1951** C. PALACHE et al. *Dana's Syst. Min.* (ed. 7) II. 911 Scorzalite is known from the Corrego Frio pegmatite, Minas Geraes, Brazil, where it occurs with souzalite and brazilianite. **1970** *Amer. Mineralogist* LV. 152 Souzalite is a hydrothermally reworked product of scorzalite and occurs as bluish-green aggregates of poly-synthetically twinned prismatic crystals.

sov. (Earlier examples.)

1829 P. EGAN *Boxiana* 2nd Ser. II. 492 'Come,' said an old ring-goer, 'here's my *sov.* to begin.' **1846** *Swell's Night Guide* 101 But of whatever size if a known gentleman and some liberality (as nothing less than five sovs would be any inducement) you may perhaps get admittance in the absence of her keeper.

Sov[2] (sǫv), colloq. abbrev. of *SOVIET *sb.* 2. Usu. in *pl.* with *the.*

1967 [see *CHICOM *sb.* and *a.*]. **1969** H. H. COOPER *Cave with Two Exits* II. 163 Their only worry would be that Washington might think the Sovs had done it. **1977** *Time* 28 Feb. 23/3 Certainly in every case I know of, the opposition—usually the Sovs but sometimes the Chicoms —were involved up to their eyeballs on the other side. **1981** P. FOX *Satan's Messenger* ix. 68 'Comeback from the Sovs?' 'Never heard of you.'

sovereign, *sb.* and *a.* Add: **I. 2. δ. suffering** (later dial. and slang examples). Now *rare.*

1836 [see *BLANK *sb.* 12]. **1914** E. PUGH *Cockney at Home* 221 I've played..till twelve at night, and then not made half a suffering.

II. A. *sb.* **1.** † **d.** A free citizen or voter of America. *U.S. Obs.*

1846 ... *Indiana Hist. Soc. Publ.* (1905) III. 412 Thousands of children in our state have not received even the trifling aid which these [public] funds afford. This fact illustrates the situation of thousands of the future sovereigns of our beloved State. **1861** *Harper's Mag.* Mar. 570/1 Deacon E— lived out West. 'The sovereigns' of that section met in caucus to appoint delegates to a County Convention. **1866** 'MARK TWAIN' *Lett. fr. Hawaii* (1967) xi. 110, I am a free-born sovereign, sir, an American.

4. b. *attrib.* Also *sovereign purse.*

1907 ... *Yesterday's Shopping* (1969) 492/1 Gentlemen's sovereign purses, Russia leather. **1976** *Lancashire Life* Dec. 59/1 Years afterwards I showed him a sovereign purse containing a solitary half sovereign.

B. *adj.* **5. d.** *Banking.* Of or pertaining to a commercial loan made to a sovereign state.

1977 *47th Ann. Rep. Bank Internat. Settlements* 102 This..meant a notable increase in the sovereign loan portfolio; but what about the corresponding rise in the country of 'sovereign risks'? **1982** *Daily Tel.* 8 Dec. 21/4 Only £26 million was set aside as a general provision, which is where the bank is believed to take account of sovereign loans. **1983** *Times* 31 Mar. 17/3 The report calls for much greater availability of information about sovereign lending.

Soviet (sǭu·viet, sǫ·viet, -v-, -ĕt), *sb.* and *a.* Also **soviet.** [a. Russ. *sovét* council.] **A.** *sb.* **1. a.** In the U.S.S.R.: one of a number of elected councils which operate at all levels of government, having legislative and executive functions.

The term was also applied to various revolutionary councils set up prior to the establishment of socialist rule in 1917.

1917 *Times* 27 July 6/4 (*heading*) Hostile vote against the Soviet. *Ibid.* 8 Sept. 6/4 A meeting of the Central Committee of the Soviet was held..at which the situation on the front was considered. **1920** *Edin. Rev.* July 59 Soviets, *i.e.*, councils or committees of workmen's and soldiers' delegates, are elected in every township, village or rural district for the purpose of local administration. **1930** *Times Lit. Suppl.* 30 Oct. 880/1 The chairman of the village soviet..may in theory be master in his own limited sphere; in practice he is the servant of a Communist 'cell'. **1941** E. STRAUSS *Soviet Russia* iv. 33 Workers and soldiers ..organized their own Councils or Soviets. **1953** B. MIALL tr. *Delbars's Real Stalin* vii. 48 The first Soviets of working-class deputies were formed. The president of the Soviet of St. Petersburg was a Menshevik. **1965** B. PEARCE tr. *Preobrazhensky's New Economics* 191 No more workers and office-workers are employed by the state, the local soviets, and the co-operatives than are employed in private industry, private trade, and agriculture. **1979** O. SELA *Petrograd Consignment* 20 During the 1905 uprising in St. Petersburg, together with Rakovsky and Trotsky he [*sc.* Helphand] led the Soviet.

b. In other countries: a similar council organized on socialist principles.

1918 *Daily Mirror* 12 Nov. 2/4 (*heading*) Berlin Soviet Meets. The first sitting of the Workers' and Soldiers' Council in Berlin was held..this evening in..the Reichstag. **1934** *Fundamental Laws Chinese Soviet Republic* vi. 79 The First All-China Congress of Soviets..calls upon the Chinese workers and peasants..to fight resolutely against [illegible]. **1977** J. CLEARY *High Road to China* ix. 136 The Bolshevists..in Saxony..have taken over some of the towns, declared soviets.

c. *transf.* and *fig.*

1945 *Tee Emm* (Air Ministry) V. 40 Pistons, connecting rods, and other vitals cease to follow the paths their designer intended and form a sort of Soviet of miscellaneous salvage. **1947** CROWTHER & WHIDDINGTON *Science at War* 86 Owing to their character of complete equality and outspokenness, these meetings were called 'Sunday Soviets'. **1972** *History Workshop Pamphlet* No. 6. 26 The cavilling system..was an embryo of workers' control... It was a little Soviet which had grown up within the capitalist system.

2. A citizen of the U.S.S.R. Chiefly in *pl.* (hence loosely = Soviet Union or its leaders).

[illegible] **1926** *Commercial & Financial Chron.* 24 Jan. 288/1 He [*sc.* Clemenceau] insisted upon writing the final paragraph, 'affirming that the Allies had not changed their attitude towards the Soviets'. **1930** *Amer. Speech* VI. 122 [illegible] Soviets go on hunger strike. **1948** W. S. CHURCHILL *End of Beginning* 221 The Soviets had to repel the terrific onslaught of Germany. **1959** *Daily Tel.* 7 Feb. 11/4 President Eisenhower, seeking one word to cover citizens of the Soviet Union, has turned the criticism of purists and adopted the term 'Soviets'. **1964** R. A. BUTLER in *Listener* 13 Aug. 222/2, I am sure that the Soviets are not going to attack us, or anything like that, at the present time. **1977** C. MCCARRY *Secret Lovers* iii. 34 'Who did Bulow meet in Dresden?' 'A Soviet, an Army captain named Kalmyk.'

B. *adj.* **1.** Of, pertaining to, or having, a system of government based on soviets; *Soviet Union*: the Union of Soviet Socialist Republics.

1918 *Decrees issued by Revolutionary Peoples Govt.* I. 11 The Soviet Government does not look backward, but forward. **1920** *Glasgow Herald* 9 Mar. 8 The [American] Government has virtually decided to permit the resumption of trade relations with Soviet Russia. **1922** A. J. TOYNBEE *Survey Internat. Affairs 1920–3* 369 The new 'Red' Army of Soviet Armenia. **1928** H. N. BRAILSFORD *How Soviets Work* vii. 99 What the Soviet Union has done on a small scale for backward races like the Tartars and Bashkirs may one day have immense significance for..Central Asia. **1946** *Ann. Reg. 1945* 83 It was agreed that the Soviet Union's claims for reparations should be met by removals from the Russian zone in Germany. **1965** M. HEYWARD tr. *Myrdal's Rep. Chinese Village* x. 114 The rest of the Soviet regiments and northern Shensi set up their own soviet republic. **1974** tr. *Sniečkus's Soviet Lithuania* 16 The Communists called for a socialist revolution in Lithuania and the establishment of Soviet power.

2. Of, pertaining to, under the influence of, or living in the U.S.S.R.

1928 *Russian Economist* I. 64 This is the secret of 'bourgeois' diplomacy, and this riddle is being solved by Soviet diplomacy and with it by all the Russian-speaking people. **1932** *Sun* (Baltimore) 27 Jan. 12/7 If what is Russia is now known as the Soviet Republic, should we not have some adjective similar to 'French', 'American', etc.?..'Soviet' has been regularly used—Soviet literature, Soviet morals, and so on. **1935** A. HUXLEY *Let.* June (1969) 397 The thing simply turned out to be a series of public meetings organized by the French Communist writers, and by the Russians as a piece of Soviet propaganda. **1961** *Ann. Reg. 1960* 409 New trade agreements were negotiated also with several countries in the 'Soviet block'. **1965** *Listener* 22 Sept. 431/1 The more enterprising..read Soviet magazines and newspapers. **1977** *Times* 14 June 16/7 He is a Soviet Jew whose family has been refused an exit visa to go to Israel. [illegible] ship. **1964** T. B. BOTTOMORE *Elites & Society* vi. 111 The unified elite in Soviet-type societies is contrasted with the plurality of elites in Western-type societies. **1978** *Detroit Free Press* 5 Mar. (Parade Suppl.) 14/4 Romanov would crack down on the mishmash of more than 100 government ministries and independent agencies that create confusion in Sovietland.

3. In combination with adjs. designating another country or people in the sense 'Soviet and..', as *Soviet-American, -Chinese, -German,* etc.

1939 W. S. CHURCHILL in *Daily Mirror* 24 Aug. 14/2 In view of the Soviet–German intrigue and all other information to hand it is becoming increasingly difficult to see how war can be averted. **1958** *Listener* 28 Aug. 295/2 The theme of Soviet–Arab friendship. **1965** H. KAHN *On Escalation* xiii. 249 The U.S. in fact was carefully concerned to limit, if not avoid direct Soviet–American confrontations. **1971** H. TREVELYAN *Worlds Apart* xvi. 177 In Moscow we saw little prospect of any new initiative being successful at that moment when Soviet–Chinese relations were in an uncertain phase of manœuvre. **1978** F. MACLEAN *Take Nine Spies* iv. 158 The Soviet–German Pact of August 1939.

Hence **Sovie·tic** *a.* (now *rare*), of or pertaining to the (Russian) Soviet system; **So·vietism,** the (Russian) Soviet system; **So·vietist** *rare* an adherent of the Soviet system; **Sovie·tophile** *a.*, that loves the Soviet Union; **So:vietopho·bia,** fear of the Soviet Union (cf. *Russophobia* s.v. Russo- b in Dict. and Suppl.); hence **Sovie·tophobe.**

1919 E. E. CUMMINGS *Let.* 7 Nov. (1969) 62 All N.Y.'s radicals are throwing up their hats in celebration of the anniversary of Sovietism. **1920** W. T. GOODE *Bolshevism at Work* 68 The order existing in Sovietic Moscow. **1920** *Glasgow Herald* 19 Aug. 7 All Russia, apart from the Sovietists, bears no ill to Poland. **1934** Sovietic [see *dope-dream* s.v. *DOPE *sb.* 5]. **1950** *Sun* (Baltimore) 4 Jan. 1/8 Controversy over what the Truman Administration..can do to keep Sovietism in China from engulfing Formosa, the last refuge of the Nationalists. **1955** *Bull. Atomic Sci.* Jan. 35/3 The strong wine of Sovietophobia on which most of the contributors had dined was just milk for babes at the Burnham table. **1957** V. NABOKOV *Pnin* iii. 71 Only another Russian could understand the reactionary and Sovietophile blend presented by the pseudo-colorful Komarovs. **1966** *Listener* 3 Mar. 325/1 This bloody love..which must go on vitiating all our attempts at Sovietophobia. **1976** *Survey* Summer-Autumn 237 After 1968 Sartre discovered that ultimately his philosophy was more likely to culminate in anarchy than in Sovietism. **1980** *Daily Tel.* 8 July 14 Should not the British media sort out this phobia? Otherwise 'Sovietophobes' might well be in danger of alienating the most convinced of their potential allies, i.e. the Russians.

Sovietize (sǭu-, sǫ·viĕtəiz), *v.* Also **sovietize.** [f. *SOVIET *sb.* and *a.* + -IZE.] *trans.* To convert to a Soviet system of government; to bring into conformity with soviet, communist or Marxist principles; to subject to the influence or control of the Soviet Union.

1920 *Glasgow Herald* 18 Aug. 7 Lenin's attempt to Sovietise the..countries possessed by the Cossacks of the Don, Terek, and Kuban. **1922** *Ibid.* 29 July 8 Not long since the Bolshevists succeeded in Sovietising Bokhara. **1928** *Daily Express* 1 Nov. 9 The izvoschiks (cabdrivers) of Moscow are to be organised, their hours of work regulated, and..their cabs equipped with meters. The task of 'Sovietising' the izvoschik will not be an easy one. **1936** *Sun* (Baltimore) 15 Sept. 13/8 Owners of Lille textile mills, fearful lest their factories 'be Sovietized', tonight defied Government efforts to grant a forty-hour week to labor. Owners of the mills..demanded the right to prevent 'establishment of Soviets' in their plants. **1954** A. KOESTLER *Invisible Writing* 138 Bokhara the noble was being sovietised more ruthlessly than any other Asiatic town. **1968** J. M. WHITE *Nightclimber* xvi. 108 [He] assisted Panchevski, the Defence Minister, to Sovietize the army. **1980** *English World-Wide* I. 1. 20 Yiddish spelling was proletarianized, declericalized, and 'Sovietized' in the U.S.S.R.

Hence **So:vietiza·tion;** also, more *rarely*, **So·vietizer; So·vietized** *ppl. a.*; **So·vietizing** *vbl. sb.* and *ppl. a.*

1920 *Glasgow Herald* 12 Aug. 7 A Sovietised Poland subject to the Moscow Government. **1921** *Ibid.* 17 Jan. 11 It remains to be seen how the Persian Court will take to the idea of Sovietisation. **1922** *Ibid.* 29 July 8 The Amir..cannot but view the Sovietising of this region with great disfavour. **1925** *Ibid.* 26 Mar. 8 The principal virtue of wireless is its sovietising power. **1939** *Sun* (Baltimore) 29 May 13/2 An effort by the medical profession to brand socialized medicine, 'Sovietized medicine' has failed. **1939** *War Illustr.* 9 Dec. 392/3 In Russian Poland conditions were on the whole rather better... The process of Sovietization was carried out gradually. **1948** J. TOWSTER *Political Power in U.S.S.R.* I. iv. 70 The People's Commissariat for the Affairs of the Nationalities..operated as watchdog, organizer, sovietizer, and protector of the nationalities. **1949** F. MACLEAN *Eastern Approaches* I. vi. 70 At Talgar we boarded a lorry full of highly Sovietized Kazakh girl students. **1955** *Times* 16 Aug. 9/7 As constituent units of the U.S.S.R., the process of sovietization has been applied to the Baltic States without mercy. **1968** V. V. ASPATURIAN in A. Kassof *Prospects for Soviet Society* vii. 159 Sovietization is here defined as the process of modernization and industrialization within the Marxist-Leninist norms of social, economic, and political behavior. **1974** V. NABOKOV *Look at Harlequins* III. i. 132 The President of Quirn..timorously sympathized with the fashionable Sovietizers. **1982** *Daily Tel.* 30 July 12 If the Western alliance splits and the Russians establish military supremacy in Europe,..we should fear..'Sovietisation'.

Sovietology (sǭu-, sǫ:viĕtǫ·lŏdʒi). Also **sovietology.** [f. *SOVIET *sb.* and *a.* + -OLOGY.] The study and analysis of affairs and events in the U.S.S.R. So **So:vieto·logist,** a student of Soviet affairs; **So:vietolo·gical** *a.*; **So:vieto·lo·gically** *adv.*

1958 *Spectator* 3 Jan. 10/2 The Sovietologist really can help his listeners by explaining what has happened. *Ibid.* 13/3 A complete service with serious Sovietological analysis. **1958** *Times Lit. Suppl.* 17 Oct. 595/2 Many works of fuller and more detailed scholarship are..already becoming available as 'Sovietology' develops more and more into a major industry. **1963** *Ibid.* 4 Jan. 3/1 Mr. Dudinstev's brief excursion into fantasy may be sovietologically significant. **1968** *Soviet Studies* XIX. 467 The change in the function of Soviet ideology..could then be matched by a change in the function of Sovietology, at least in the field of theoretical-ideological controversy. **1971** H. TREVELYAN *Worlds Apart* xx. 237 The Sovietologists of the Western press, working on the documents in London or Washington, were forced by the nature of their occupation to draw conclusions, not always justified by the facts. **1976** *Daily Tel.* 21 Oct. 16/6 A newspaper photograph showing the arrival of President Tsedenbal of Mongolia at Moscow Airport has been examined in detail by the 'Way of the World' Sovietological department. **1979** *Dædalus* Winter 121 There are cracks in sovietology as well as in the Soviet monolith itself.

‖**sovkhoz** (sǫ·vkǫz). Also **sovhoz, sovkhos,** etc. Pl. **sovkhoz, sovkhozes, sovkhozy.** [Russ., f. *sov(étskoe khoz(yáĭstvo* Soviet farm.] In the U.S.S.R.: a state-owned farm. Also *attrib.* Cf. *KOLKHOZ.

1921 *Russian Economist* I. 385 Sovkhoses, i.e., Soviet farms, that include agriculture of industrial workmen as well as the States' farms proper. **1926** *Spectator* 29 May 922/1 In Soviet Russia any estate is liable to be turned into a *Sovkhos*, a government model farm. **1932** H. G. WELLS *Work, Wealth & Happiness of Mankind* iv. 184 The Sovkhoz is a state plantation, a really scientifically planned and directed modern large-scale organization of production... The Sovkhozy have to take up lands hitherto uncultivated. **1938** *Nature* Mar. 453/2 The 'organization of the sovkhozes' (large-scale State agricultural enterprises). **1943** E. M. ALMEDINGEN *Frossia* v. 219 New tractors..meant help for thousands of our Sovhoz farms. **1953** O. CAROE *Soviet Empire* xi. 175 Land and water had been nationalized and, after taking a large share for the *Sovkhozes* or state farms, redistributed in an arbitrary fashion. **1955** H. HODGKINSON *Doubletalk* 28 The collective farms are not to be confused with the State farms or *sovkhoz*..which are owned by the State and worked by government employees. **1967** *Bull. Inst. Study USSR* (Munich) June 15 A wave of sovkhoz development followed which, beginning in 1954, did not recede until 1964. **1977** tr. *Le Monde* in *Guardian Weekly* 27 Nov. 12/5 They..lay siege to the stores in search of rare or common articles which cannot be found in their *kolkhoz* or *sovkhoz* (state farm) general stores.

‖**sovnarkhoz** (sǫ-vnāɹkǫz). Also **Sovnarkhoz.** Pl. **sovnarkhozy, sovnarkhozie.** [Russ. *sovnarkhóz*, abbrev. of *sovét naródnovo khozyáĭstva*, council of national economy.] In the U.S.S.R.: a regional council for the local regulation of the economy.

These councils were introduced in 1957 and abandoned in 1965.

1958 *Ann. Reg. 1957* 207 The country was split into 105 'economic regions', in each of which an 'economic council' (*sovnarkhoz*) was established, responsible to the Republican Government. **1962** *Economist* 3 Mar. 788/3 The 107 smaller economic sub-divisions and their councils, or *sovnarkhozy*. **1964** *Ann. Reg. 1963* 211 The most important decisions were now taken by the Party's new Central Asian Bureau or the single regional *sovnarkhoz*. **1964** *Times Rev. Industry* Mar. 90/1 The programme included visits to the Moscow and Leningrad Sovnarkhozie (Councils of National Economy). **1964** *Economist* 12 Dec. 1242/2 The system of regional councils, or Sovnarkhozy, introduced by Mr Khrushchev in 1957.

Sovnarkom (sǫ-vnāɹkǫm). [a. Russ. *sovnarkóm*, abbrev. of *sovét naródnykh komissárov*, council of people's commissars.] The highest executive and administrative organ of government of the U.S.S.R. (renamed the Council of Ministers in 1946). Also, a council having analogous functions in one of the republics of the U.S.S.R.

1938 *Ann. Reg. 1937* 196 The Sovnarkom ordered the Gosplan to finish the schedule for the third five-year period. **1939** G. B. SHAW *Geneva* I. 21, I am Commissar Posky of the Sovnarkom and Politbureau, Soviet delegate to the League Council. **1948** J. TOWSTER *Political Power in U.S.S.R.* 248 The central executive committees and sovnarkoms of the constituent republics. **1959** *Times Lit. Suppl.* 2 Oct. 553/3 His [*sc.* Trotsky's] refusal of an offer by Lenin in 1922, twice repeated, that he should be appointed a vice-president of Sovnarkom. **1959** E. H. CARR *Socialism in One Country* II. IV. xx. 244 Even in the domain of treaty-making Sovnarkom acquired independent constitutional powers.

sow, *sb.*[1] Add: **8. a.** *sowcunt* (coarse *nonce-wd.*), *-feeder*; *sow-dugged adj.*; **sow-belly** *U.S. slang*, (salted) side of pork.

1867 W. L. GOSS *Soldier's Story of his Captivity* 205 My captor presented to me a generous slice of 'sow-belly'. **1945** B. MACDONALD *Egg & I* (1946) III. viii. 97 Tits fed this baby pickles, beer, sow-belly and cabbage. **1976** G. EWART *No Fool* III. 69 To go into your South, a different life. Sow-belly and cornbread with syrup poured over it. **1922** JOYCE *Ulysses* 541 (*Her sowcunt barks.*) Fohracht! **1960** AUDEN *Homage to Clio* 55 Steatopygous, sow-dugged and owl-headed. **1960** *Farmer & Stock-breeder* 16 Feb. (Suppl.) 37/2 Such an arrangement with individual sow-feeders, allows for better attention to each sow.

c. Also *sow beaver, grizzly bear.*

1959 E. COLLIER *Three against Wilderness* xxi. 210 She was an old sow beaver who could be reckoned upon to give birth to four or five sturdy kits. **1976** *Telegraph-Journal* (St. John, New Brunswick) 12 Aug. 12/4 A sow grizzly bear that..mauled him..was only trying to protect her young.

sow, *v.*[1] Add: **2. a.** Also, *to sow* (land) *to* (a crop). Cf. PUT *v.*[1] 26 b, *PLANT *v.* 6 a.

1939 *Sun* (Baltimore) 4 July 16/2 There will be no possibility of spreading the galls to land that is sown to wheat or rye. **1972** *Morning Star* 4 Jan. 4/1 This was cattle-breeding country, with a dairy produce industry and with only about 75,000 acres sown to grain.

4. c. *transf.* in *Mil.* To lay or 'plant' (an explosive mine); *spec.* to drop (mines, etc.) by aircraft into the sea or otherwise. Also *absol.*

1939 *Sun* (Baltimore) 20 Nov. 8/2 In the last conflict the Germans sowed 44,000 mines, 11,000 of them in British home waters. **1943** *Ibid.* 26 Nov. 1/5 After they have dropped their first flares they remain over the target area, keeping it marked by sowing more flares. **1944** K. DOUGLAS *Alamein to Zem Zem* (1946) xiv. 82 Mines were sown in the tracks of vehicles, where other vehicles might be expected to follow. **1974** *Times* 18 Apr. 1/3 A lot of anti-personnel mines sown on the canal banks have slipped into the water. **1979** J. BARNETT *Backfire is Hostile!* xiii. 135 Twenty-four Tu-16 Badgers began.. sowing at forty-two thousand feet.

sowlth (sault). [ad. Ir. *samhailt* likeness, apparition.] A formless, luminous spectre. Chiefly in the writings of W. B. Yeats.

1829 G. GRIFFIN *Collegians* II. xxviii. 289 The Sowlth was seen upon the Black Lake last week. **1892** W. B. YEATS *Countess Kathleen* iii. 54 Call hither now the sowlths and tevishes. **1895** —— *Poems* 79 Pooka, sowlth, or demon of the pit. **1963** *Times Lit. Suppl.* 1 Feb. 78/4 In the first version [of *The Countess Kathleen*]..there is a naive elaboration, in which 'sheogues', 'tevishies', 'sowlths', and other rustic spirits appear.

sown, *ppl. a.* Add: **2.** (Later *absol.* examples contrasted with *desert.*)

1926 T. E. LAWRENCE *Seven Pillars* (1935) v. lviii. 328 The difference between Hejaz and Syria was the difference between the desert and the sown. **1940** J. BUCHAN *Memory Hold-the-Door* i. 22 We had for our playground both the desert and the sown. **1957** K. KENYON *Digging up Jericho* 29 The age-long struggle of the Desert and the Sown.

sox (sǫks), commercial and informal spelling of *socks*, pl. of SOCK *sb.*[1]

Also used as the final element in the names of some sports teams, esp. in *U.S. Baseball.*

1905 H. G. WELLS *Kipps* I. ii. 37 He abbreviated every word he could; he would have considered himself the laughing-stock of Wood Street if he had chanced to spell *socks* in any way but 'sox'. **1912** G. FRANKAU *One of Us* v. 41 To dollars deaf, impervious to invective, They plunged profaning hands in shirts and sox. **1942** Z. N. HURSTON in *Amer. Mercury* July 88 Dat broad couldn't make the down payment on a pair of sox. **1948** *Richmond* (Va.) *Times-Dispatch* 11 June 27/1 The Boston Red Sox today socked the Cleveland Indians 15–7. **1965** *Liberator* Aug. 20/1 And Sweet Mac was there; legs crossed, showing his..two-fifty sox. **1977** *West Briton* 25 Aug. 17/7 (Advt.), Assorted nylon plain & fancy short sox.

Soxhlet (sǫ·kslĕt). *Chem.* Also **soxhlet.** The name of Franz *Soxhlet* (1848–1926), Belgian chemist, used *attrib.* (and † in the possessive) to denote an apparatus and method which he devised for the continuous solvent extraction of a solid.

1889 *Jrnl. Chem. Soc.* LV. 359 When using the ordinary form of Soxhlet extractor, there is always a doubt as to the exact time when the substance is completely extracted, unless the whole apparatus is taken to pieces. **1899** *Jrnl. Physiol.* XXIV. 319, I used casein which had been extracted for a week in Soxhlet's apparatus. **1945** M. F. GLAESSNER *Princ. Micropalaeont.* x. 239 Space and equipment for..simple porosity and permeability tests, soxhlet extraction of bituminous rock samples, and gas analysis may be provided. **1950** *Jrnl. Org. Chem.* XV. 256 The salt-like product was extracted with butyl alcohol in a Soxhlet apparatus for 6 hours. **1968** R. O. C. NORMAN *Princ. Org. Synthesis* vii. 224 Barium hydroxide is placed in the thimble of a Soxhlet extractor over a flask of boiling acetone.

soy[1]. **1, 2.** For Latin names substitute *Glycine max.*

3. Equivalent to *SOYA 2: *soy bottle, jam, oil, protein*; **soy frame,** an ornamental stand with a ring frame used for holding a soy bottle; **soy-sauce** = sense 1 (earlier and later examples). See also *SOY-BEAN.

1960 Soy bottle [see *soy frame* below]. **1970** *Canadian Antiques Collector* Oct. 18 (*caption*) Four 12-sided soy bottles with shoulder bands of diamonds cut in high relief. **1788–1815** *Watson & Bradbury Pattern Bk.* in F. Bradbury *Hist. Old Sheffield Plate* (1912) IX. 197 Soy frames.. Sugar Tongs... Snuffers. **1912** F. BRADBURY *Hist. Old Sheffield Plate* IX. 271 The soy frames and cruet frames.. are sufficiently illustrated here to give a very fair general idea of the different fashions. **1931** E. WENHAM *Domestic Silver* x. 142 Snuffer-trays and soy-frames can be..made into attractive inkstands. **1960** H. HAYWARD *Antique Coll.* 261/2 *Soy frame*, silver or plated oblong or oval stand with ring frame for holding soy or sauce bottles. **1956** B. Y. CHAO *How to cook & eat in Chinese* I. iii. 50 Similar to soy sauce is a *soy jam*, which is much thicker in consistency. **1976** *Billings* (Montana) *Gaz.* 30 June 4-B/1 Soy oil lost some 25 points. **1974** *Sci. Amer.* Feb. 19/1 Soy protein is nutritionally somewhat less complete than meat. **1795** Soy sauce [see *SOY-BEAN a]. **1959** R. KIRKBRIDE *Tamiko* vii. 54 They had hors d'oeuvres of raw wild vegetables, sashimi, thin slices of raw tuna and sea bream with soy sauce. **1978** *Nagel's Encycl.-Guide: China* 380 The seven sauces used in the stew are replaced by soy sauce, into which the meat is dipped before being cooked.

soya. Add: **1.** (Later examples.)

1905 [see *MISO]. **1970** *Times* 20 Apr. 4/5 Liveweight gains in turkeys and pigs..were as good as those with fishmeal or soya.

2. *attrib.*, as *soya flour, meal, milk, oil*; **soyaburger,** a hamburger made with (beef and) soya beans; also, a mixture of minced beef and soya beans; **soya link** [LINK *sb.*[2] 2 c] = next; **soya sausage,** a sausage made with minced soya beans. Cf. *SOYA BEAN.

1953 POHL & KORNBLUTH *Space Merchants* ii. 14 When real meat got scarce, we had soyaburgers ready. **1974** *Globe & Mail* (Toronto) 29 Oct. 15/9 The federal government decided to allow soyaburger products to be marketed. **1930** *Times Lit. Suppl.* 27 Feb. 167/2 Soya flour prepared by the ordinary methods soon turns rancid. **1951** *Good Housek. Home Encycl.* 333/1 Stir in the soya flour. **1965** B. SWEET-ESCOTT *Baker Street Irregular* vii. 200 That unspeakable dish, the soya link, the staple diet of the British in the Mediterranean campaign. **1968** J. W. PURSEGLOVE *Trop. Crops: Dicotyledons* I. 265 Soya meal, the residue after the extraction of the oil, is a very rich protein feeding stuff for livestock. **1977** C. MCFADDEN *Serial* (1978) viii. 22/2 Harvey..drank his soya milk without complaint. **1917** H. A. GARDNER *Paint Researches* xxiii. 316 A series of tests were conducted to determine the rate of drying of soya oil. **1982** *Times* 14 Apr. 11/1 Sunoil, linoil and soyaoil prices appeared to be rising. **1971** D. MEIRING *Wall of Glass* xxiii. 195 They went to the counter. It was soya sausages, potatoes, cabbage.

soya bean. Also † **soja bean; soyabean.** [f. SOYA + BEAN.] **a.** = SOY[1] 2. Cf. also next entry.

1897 *Publ. Georgia Dept. Agric. 1896* 64 A display of soja beans..a legume of exceptionally fine quality for stock feed. **1905** *Chambers's Jrnl.* Mar. 220/2 Soya beans ..are grown all over Japan and in Manchuria. **1930** *Times Lit. Suppl.* 27 Feb. 167/2 The high nutritive value of the soya bean has long been recognized. **1958** *Times Rev. Industry* Apr. 92/2 The 1957 increase was well spread; there were..good crops of soyabeans in the United States. **1968** J. W. PURSEGLOVE *Trop. Crops: Dicotyledons* I. 265 Soya beans are one of the world's most important sources of oil and protein. **1973** *Saint Croix Courier* (St. Stephen, New Brunswick) 26 July 1 Canada is not a big user of soya beans.

b. *attrib.*

1911 *Daily Colonist* (Victoria, B.C.) 6 Apr. 14/2 In the cargo for Victoria was a shipment of 500 tubs of soya bean oil. **1944** V. HODGSON *Diary* 7 May in C. Driver *British at Table* (1983) ii. 16, I have an order with the Dairy for a pound of sausage..of soya bean flour. **1966** GETTENS & STOUT *Painting Materials* 62 Soya Bean Oil... A typical analysis of soya bean oil gives 14 per cent of palmitic acid. **1977** G. SCOTT *Hot Pursuit* iii. 26 Drinking glass after glass of..soya bean milk bought from little carts at every corner.

soy-bean. (In Dict. s.v. SOY[1] 3.) Add: **a.** (Earlier and later examples.)

1795 tr. *C. P. Thunberg's Trav.* IV. 121 Soy-sauce, which is every where and every day used throughout the whole empire,..is prepared from Soy Beans..and salt, mixed with barley or wheat. **1970** *N.Y. Times Encycl. Almanac* 259/2 [Iowa] ranks first in popcorn and oats, and is second in soybeans. **1975** *New Yorker* 26 May 50/3 He said that in the nineteen-sixties the United States 'had excess capacity in corn and soybeans'. **1978** J. IRVING *World according to Garp* xv. 300 A field of corn and a field of soybeans.

b. *attrib.*, esp. as *soybean oil.*

1935 *Cereal Chemistry* XII. 442 It is only within the last decade that the use of soybean flour for food purposes in this country has been seriously considered. **1938** A. A. HORVATH *Soybean Industry* xiv. 97 Quantities of soybean oil have been used in the manufacture of foundry cores. **1956** B. Y. CHAO *How to cook & eat in Chinese* I. iii. 49 The most important flavourer of Chinese food is *soy-bean sauce* or soy sauce for short. **1967** D. & E. T. RIESMAN *Conversations in Japan* 58 Delicacies such as fresh ginger root, soybean soup, meat and vegetables. **1973** P. THEROUX *Saint Jack* v. 52 Over in an armchair drinking soy-bean milk..sat old Mr. Tan Lim Hock. **1979** C. MACLEOD *Luck runs Out* vii. 76 Her successful campaign to have soybean cutlets put on the menu.

Soyer (soi·əɹ). The name of Alexis Benoît *Soyer* (1809–58), French-born cook to fashionable society in England, subsequently working for Irish famine-relief and with the British army in the Crimea, used *attrib.* and formerly in the possessive, as **Soyer('s) stove,** to designate a table-top cooking-range (the *Magic stove*) developed by him in 1849, or his Field stove invented in 1857. Also *absol.*

1856 A. SOYER *Mod. Housewife* (new ed.) 513 (Advt.), Soyer's Magic Stove and Lilliputian Apparatus, specially adapted for out-door cooking. **1857** —— *Culinary Campaign* 524 These receipts are also applicable for barracks, in camp, or while on the march, by the use of Soyer's New Field Stove, now adopted by the military authorities. **1858** VOLANT & WARREN *Mem. Alexis Soyer* xiii. 157 M. Soyer at once perceived the importance of this little apparatus, and..it was brought out..as 'Soyer's Magic Stove', to which was added the 'Camp Kitchen'. **1878** *Instructions to Mil. Cooks in Prep. Dinners* 9 The field cooking apparatus in use in the Army are Feetham's and Soyer's stoves... Soyer's stove..is fitted with a boiler only, and will cook for 50 men. **1941** *Jrnl. R. Army Med. Corps* LXXVII. 274 Soyer stoves or other improvised water heaters. **1981** J. BARNETT *Firing Squad* II. 116 Blackened cauldrons, known in the Army as Soyer stoves, spew thick smoke from tall thin chimneys... The men are offered soup from the Soyers.

sozzle, *sb.* Add: Also † **sossle. 2.** (Earlier example.)

1848 BARTLETT *Dict. Americanisms* 321 *Sossle*, or *sozzle*, a lazy or sluttish woman.

sozzle, *v.* Add: **3.** *intr.* [Back-formation f. *SOZZLED *ppl. a.*] To imbibe intoxicating drink. *slang.*

1937 G. FRANKAU *More of Us* xv. 160 Then Sophie called; and brooding, 'Nice schemozzle If that lot stays to feed as well as sozzle.' **1953** N. FITZGERALD *Midsummer Malice* xx. 242 We can sit here and sozzle gently and enjoy ourselves.

sozzled (sǫ·z'ld), *ppl. a. slang.* Also † **sosselled.** [f. SOZZLE *v.*] Intoxicated, drunk; drunken.

1886–96 in Farmer & Henley *Slang* (1903) VI. 301/2 She was thick in the clear, Fairly sosselled on beer. **1904** ADE *True Bills* 26 It was customary to mix Tea,..Egg-nog,..and Straight Goods until..the last Caller was Sozzled. **1921** *Blackw. Mag.* Feb. 157/1, I wasn't what you'd call sozzled. I might have been lit up a bit, but sozzled—no. **1935** D. L. SAYERS *Gaudy Night* xx. 414 He was beautifully sozzled last night. **1951** 'J. WYNDHAM' *Day of Triffids* i. 23 'Gin, blast it! T'hell with gin!'.. The voice gave a sozzled chuckle. **1963** N. MARSH *Dead Water* (1964) i. 13 'She'm sozzled,' said Wally, and indeed, it was so. **1972** E. STAEBLER *Cape Breton Harbour* xiv. 130 With a sozzled smile he began to sing about a little yellow dory.

spa, *sb.* Add: **2. b.** Also *spa bath, pool.* A health bath containing hot, aerated water. *U.S.*

1974 *Los Angeles Times* 13 Oct. III. 8 (Advt.), The Original Santa Barbara Hotub is a superbly-engineered spa that is beautiful furniture. **1976** *Outdoor Living* (N.Z.) I. II. 59/1 A spa pool is a large, hot bath with aerated water, bubbling softly around, massaging your body. **1977** *Times* 29 Oct. 11/5 The latest craze [in Los Angeles] is bathing with your friends..in a jacuzzi or spa-bath. **1979** *Arizona Daily Star* 8 Apr. J1/5 Spas in or next to the pools are also a hot item. Ragel says he sells spas with about 50 percent of his pools.

3. b. A commercial establishment which offers health and beauty treatment (esp. for women) through steam baths, exercise equipment, massage, and the like. *U.S.*

1960 *Life* 8 Feb. 111/1 The submerged specter above.. is getting a hydraulic underwater massage at a plush health spa near San Diego called the Golden Gate beauty resort whose customers are usually female. **1976** *Vogue* Dec. 214/1 Most American spas are designed exclusively for women. **1981** W. SAFIRE in *N.Y. Times Mag.* 21 June 10/2 Only fuddy-duddies go to the *gym*,..the upscale..crowd goes to the *spa*.

space, *sb.*[1] Add: **II. 6. d.** For *U.S.* read orig. *U.S.* and add earlier and later examples.

1894 E. L. SHUMAN *Steps into Journalism* 83 Articles by the beginner are nearly always submitted 'on space'. **1933** E. WAUGH *Scoop* III. i. 259 I've only been on the paper three weeks... It is the first time I've drawn any money... I'm 'on space', you see. **1971** D. AYERST *Guardian* xxv. 357 Williams represented the *Guardian* in St Petersburg at first on a small salary..and then..on space.

e. Room in a newspaper, periodical, etc., or on some other medium, which may be acquired for a specific purpose, esp. advertising.

1930 *Economist* 29 Nov. 1003/2 In advertising Britain is far behind America in buying space. **1940** R. S. LAMBERT *Ariel & All his Quality* vii. 168 Selling 'space'.. breeds a very different outlook from providing programmes. **1950** *Times* 7 Feb. 5/5 In the last election, one company gave space to the Communist Party and the Commonwealth Party, but the main newsreels adhered to the general agreement that space should be given only to the main parties.

8. a. (Later examples.)

1901, etc. [see *outer space* s.v. *OUTER *a.* 3]. **1924** R. GRAVES *Mock Beggar Hall* 40 May not Space be housing and sheltering millions of other beings like us, or different from us? **1959** *Daily Tel.* 23 Feb. 11/6 For the human body, space begins about 12 miles up, where there is not enough air left to burn a candle. **1961** 'C. E. MAINE' *Man who owned World* vii. 86 Such is human psychology that a living man returning from space attracts less attention than a dead man not returning. **1962** F. I. ORDWAY et al. *Basic Astronautics* ii. 26 On April 12, 1961, a 27-year old Russian air force pilot Yuri Gagarin..whirled once around the earth in an orbit at an average altitude of 158 miles. Some 108 minutes after launching, he had returned to earth the first man to travel in space.

9. e. = *living space* (a) s.v. *LIVING *vbl. sb.* 7 a. *slang* (chiefly *N. Amer.*).

1976 *New Times* 19 Mar. 36 Werner Erhard through est, has created the 'space' for them to 'be' and given them the 'opportunity' to 'take responsibility' for their lives. **1977** C. MCFADDEN *Serial* (1978) iii. 13/2 Leonard had a lot going for him otherwise, and Kate liked the space he was in. **1980** G. B. TRUDEAU *Tad Overweight*, Seriously, I think I know where you're coming from, and I'd like to share that space. **1981** *Gossip* (Holiday Special) 31/3 The reason why I can say that so boldly is because they give me my space. They let me be me.

10. d. A portion of a page (in a newspaper, etc.) available for a specific purpose, esp. advertising; a period or interval of broadcasting time available to or occupied by a particular programme or advertising 'slot'. Esp. in injunction *watch this space!* (freq. *transf.*). Cf. sense 6 e above.

1917 *B.E.F. Times* 20 Jan. 15/2 (Advt.), Watch this Space. **1956** *B.B.C. Handbk. 1957* 78 Plays from the West End..are often heard in the more 'popular' programme spaces. **1972** *Sci. Amer.* Feb. 114/1 Kant's own book was discussed in this space a couple of years ago from the paper-back edition issued by the University of Michigan Press. **1979** J. RATHBONE *Euro-Killers* iv. 44 Where is he? Watch this space for exciting revelations in the next few days.

15. c. *Telecommunications.* An interval between consecutive marks in a mark–space signalling system such as telegraphy. Opp. *MARK *sb.*[1] 13 e.

1859, etc. [see *MARK *sb.*[1] 13 e]. **1906** A. E. KENNELLY *Wireless Telegr.* xi. 153 A dash has the length of three dots, and the space separating dots or dashes in a letter are [*sic*] of dot length... The space separating adjacent letters is three dots long and the space separating words, six dots long. **1954** *Electronic Engin.* XXVI. 230/1 The principle..is to explore the centre of each received signal element..to determine whether it is 'mark' or 'space', and use the information so obtained to initiate new signals of correct length. **1968** D. C. GREEN *Radio & Line Transmission (A)* xvi. 292 [In the Murray code] each character is represented by a combination of five signal elements that may be either a mark or a space. In Great Britain a mark is represented by a negative potential or the presence of a tone and a space is represented by a positive potential or the absence of a tone.

16*. *Math.* An instance of any of various mathematical concepts, usu. regarded as a set of points having some specified structure; cf. **metric space*, **topological space*, **vector space.*

1911 *Trans. Amer. Math. Soc.* XII. 287 It is not always necessary to set up a definition of distance for the Hilbert space; for other domains of objects to do so might be very difficult or even impossible. **1927** *Bull. Amer. Math. Soc.* XXXIII. 14 The Hilbert space of infinitely many dimensions in which the coordinates $x_1, x_2, x_3, \ldots, x_n, \ldots$ of each point are subject to the condition that the sum of their squares be a convergent series is a metric space in which distance is defined by the formula [etc.]. **1932** M. H. STONE *Linear Transformations in Hilbert Space* i. 1 The word 'space' has gradually acquired a mathematical significance so broad that it is virtually equivalent to the word 'class', as used in logic. **1964** A. P. & W. ROBERTSON *Topological Vector Spaces* i. 5 A topological space is a set provided with a structure that enables convergence and continuity to be considered. **1968** P. A. P. MORAN *Introd. Probability Theory* i. 2 The experiment can turn out in one of a..number of exclusive ways which we denote as $E_1, E_2, E_3, \ldots$, and which we call the 'elementary events'. The set of all such events is called the 'space' of elementary events.

III. 17. a. *space-band, -bar, -gauge.*

1888 J. HARRISON *Man. Type-Writer* 18 In front of the four banks of keys there is a narrow strip of wood which is called the 'space-bar'. *Ibid.* 25 The 'space-gauge' is a little thumb-piece at the extreme right of the carriage. **1895** E. COLLYNS *Typists' Man.* 17 The space between the lines is regulated by the 'Space Gauge'. **1904** C. T. JACOBI *Printing* (ed. 3) x. 132 A stationary box..contains a series of space-bands. **1919** B. DE BEAR *Typewriting* 22 You depress the space-bar whenever you want to leave a space in a line of the work. **1930** *Daily Express* 23 May 4/6 Spacebands are pushed up to fill out the line to the required width, and then the whole line is automatically conveyed to the face of a mould and filled with molten metal. **1957** *Encycl. Brit.* XVIII. 502/2 By touching another key, a double wedge spaceband is placed between the words. **1962** *Which?* Dec. 359/2 The space bar.. moved the carriage exactly half a space when depressed, the other half when released.

b. *space continuum, harmony, music, -occupancy* (earlier example), *-perception, -relation* (earlier example), *-sensation, -sense, -symmetry, -value.*

1865 S. HODGSON *Time & Space* ii. 65 Their space-relations are not capable of analysis into relations of time. *Ibid.* 75 The space-senses sight and touch..are brought into play simultaneously with the other senses. **1875** G. H. LEWES *Probl. Life & Mind* II. 278 What is signified in speaking of material extension is space-occupancy. **1886** W. JAMES *Let.* 12 Sept. in R. B. Perry *Tht. & Char. W. James* (1935) I. 604 Of already written things I have a long-finished article on space-perception, [etc.]. **1890** —— *Princ. Psychol.* II. xx. 195 Let the movement *bc*, of a certain joint, derive its absolute space-value from the cutaneous feeling it is always capable of engendering. *Ibid.* 219 We must..seek to discover *by what means* the circumstances can so have transformed a space-sensation. **1911** —— *Some Probl. Philos.* xi. 182 God, as the orthodox believe, created the space-continuum, with its infinite parts already standing in it, by an instantaneous *fiat.* **1924** R. M. OGDEN tr. *Koffka's Growth of Mind* 72 Psychology of space-perception. **1932** F. L. WRIGHT *Autobiogr.* II. 145 Freedom of floorspace and elimination of useless heights worked a miracle in the new dwelling place... An entirely new sense of space values in architecture came home. **1933** H. READ *Art Now* ii. 78 At our period the artist..had to infer the extension of plane surfaces..the placing of all objects in a space continuum. **1957** —— *Tenth Muse* xxxi. 279 A distinction between an aesthetic consciousness determined by time-sense (music and poetry) and an aesthetic consciousness determined by space-sense (the plastic arts). **1963** *Times* 30 Apr. 15/1 A 'poème plastique', written in 1918, uses the very up-to-date idea of instrumental units separated in space; a kind of early stereophonic, space music. **1965** W. LAMB *Posture & Gesture* iv. 56 The process of variation, sometimes under the heading Narrow-Wide, is recognized and figures as a component in 'Space Harmonies'. **1977** J. LE CARRÉ *Hon. Schoolboy* xix. 467 Jerry..walked into the reception room... Space music was playing and there was even conversation under it. **1979** *Nature* 11 Oct. 433/1 Depending on the symmetry of the lattice and of the arrangement of the atoms within each cell, a crystal is assigned to one of the 230 possible space-symmetry groups.

c. *space-coordinate, -derivative, -inversion.*

1888 *Rep. Brit. Assoc. Adv. Sci. 1887* 507 Certain relations which held between the fluid velocities *u, v, w*, and their space-derivatives at any point of a rigid boundary. **1967** CONDON & ODISHAW *Handbk. Physics* (ed. 2) II. vi. 41/1 The Minkowski matrix η is a Lorentz matrix which defines the space-inversion transformation $x' = -x, y' = -y, z' = -z, t' = -t$. **1968** M. S. LIVINGSTON *Particle Physics* vii. 137 Scientists, philosophers, and others have been interested in the significance of space-inversion invariance. **1970** G. K. WOODGATE *Elem. Atomic Struct.* iii. 41 And r_i is the space-co-ordinate of the incident wave at the position of the *i*th electron.

d. For *U.S.* read orig. *U.S.* *space-writer* (later example); *space rate* (examples); relating to the purchase of (advertising, etc.) space, as *space-buyer, salesman.*

1902 E. BANKS *Autobiogr. Newspaper Girl* 207 [By] the 'guarantee space' system..a member of the staff is guaranteed a stipulated sum of money every week, and as much over that amount as he or she makes by writing at ordinary or special space-rates. **1934** S. BECKETT *More Pricks than Kicks* 92 'Well' insisted the space-writer. **1939** F. M. FORD *Let.* 14 Mar. (1965) 316 You can be certain of occupying a certain space in the pages of the Review and being paid at the usual..space rates. **1948** G. V. GALWEY *Lift & Drop* i. 11 Mrs Lawson, the space-buyer of Rooster's. **1954** KOESTLER *Invisible Writing* IV. xxxv. 377 Dr. Magnus..was now space-salesman for an obscure little Polish gazette. **1972** G. BROMLEY *In Absence of Body* iii. 41 He's the chief space buyer. **1979** *Amer. Film* July–Aug. 55/1 'Now who in the audience will know what a space salesman is?' quizzed Jaffe; a space salesman sells advertising space in a magazine.

e. In sense 8 a, outer space regarded as a field for human activity; (many of these formations are modelled on analogous uses of *air, air-*): *space agency, biology, bus, conquest, -crew, doctor, exploration, explorer, journey, law, lifeboat, liner, museum, navigation, navigator, pilot, relay, research, science, scientist, taxi, technology, travel, traveller, tug,* etc.

1958 *Science* 11 Apr. 807/2 Herbert F. York..has been named chief scientist of the Defense Department's new space agency. **1970** *Times* 15 Apr. 1/5 This firm decision was taken today by the space agency in preference to the much riskier feat of attempting a landing a day earlier after a faster return. **1960** *IRE Trans. Military Electronics* IV. 284/2 To gain some insight into the problems.. ultimately to be studied in space biology, using these missiles as experimental tools. **1977** J. TODD in S. Brand *Space Colonies* 49/2 During the hey-day of interest in space exploration (summer 1962) a symposium on the ecological aspects of space biology was convened. **1961** *New Scientist* 27 July 216 Ultimately, the space bus named *Ranger* will find its way to the moon, running on electric power drawn from the sun. **1967** *Boston Sunday Globe* 23 Apr. 1/4 Komarov, 40, spent 24 hours 17 minutes in space Oct. 12–13, 1964, aboard the Voshkod 1 'space bus'. **1949** E. F. RUSSELL in 'E. Crispin' *Best SF* (1955) 209 The biped tribes..need all their unity to cope with space-conquest. **1951** A. C. CLARKE *Sands of Mars* x. 126 Visiting space-crews..soon got bored if they had nothing to do between trips. **1974** *Sci. & Technical Aerospace Rep.* XII. 1600/2 Aeromedical problems of weightlessness and the transfer of spacecrews between Soyuz and Apollo spacecrafts are discussed. **1953** J. N. LEONARD *Flight into Space* xi. 103 Using the scientific method of dissecting a many-sided problem into its separate parts, the space doctors discuss and study the dangers of space individually. **1964** *Skylights* Mar. 1 Pulmonary atelectasis—collapse of the small lung sacs—was feared by space doctors, but did not appear during the 17-day test span. **1957** *IRE Trans. Mil. Electronics* I. 43 (*heading*) Space exploration—the new challenge to the electronics industry. **1969** *Guardian* 7 June 2/8 Russia was spending a significantly higher percentage..on space exploration. **1959** K. VONNEGUT *Sirens of Titan* i. 30 The state of mind on Earth with regard to space exploration was much like the state of mind in Europe..before Christopher Columbus set out... The monsters between space explorers and their goals were not imaginary. **1975** *New Yorker* 21 Apr. 108/2 It has turned out that our real space explorers have necessarily been practical men. **1901** H. G. WELLS *First Men in Moon* xx. 248 All through the major portion of that vast space journey I hung thinking of such immaterial things. **1961** *Daily Tel.* 6 May 8/2 Cdr. Shepard's successful space journey is an immense relief not only to the Americans but the entire free world. **1955** A. G. HALEY in *Jet Propulsion* (1956) XXVI. 951/1 We have about as clear a vision of the space law that will prevail one or two centuries from now as Hammurabi in the 22nd century B.C. **1960** *Daily Tel.* 17 Aug. 13/3 Mr Shawcross also announced that he would resign as chairman of the organising committee on space law, recently set up in London. **1980** *Oxf. Compan. Law* 1165/2 *Space law*, principles of law accepted by nations as binding on them and their nationals in engaging in activities in outer space..and in relation to celestial bodies. **1966** *Observer* 4 Dec. 2/7 American scientists are planning 'space lifeboats' to rescue the crews of disabled spaceships. **1944** E. COLLINS *Mariners of Space* i. 14 Earth's new Space Liner..leaves Croydon to-day at noon. **1982** A. HEMINGWAY *Psyche* i. 16 The castaway..was a former waiter on a spaceliner. **1977** SACHS & JAHN *Celestial Passengers* xxxii. 190 A new space museum is being developed a mile from Disneyland in Anaheim, California. **1931** J. M. WALSH *Vandals of Void* iv. 40 A ticklish job..is this of space navigation. **1976** *Internat. Aerospace Abstr.* XVI. 23/2 Time intervals in problems of space navigation and communication are often obtained by determining the phase of binary signals. **1936** *Forum & Century* July 36/2 Suppose that a breed of space navigators has begun to appear on earth. **1951** A. C. CLARKE *Exploration of Space* 82 His position is, clearly, only one of the things a space-navigator would want to know. **1962** *Amer. Speech* XXXVII. 43 Before April 12, 1961, the concept expressed by *cosmonaut*, in reference to both American and Soviet space flights, was rendered in the American press by such terms as *astronaut, spaceman,..space navigator.* **1944** E. COLLINS *Mariners of Space* iii. 21 Space pilots and their mechanics buzzed bee-like in and out of their quarters. **1978** *Space Picture Library Holiday Special* 6 The man who rushed forward.. could do little more than break the space-pilot's fall. **1958** *Listener* 4 Dec. 910/1 The result implied that the moon could be used as a space relay for transatlantic radio communication. **1957** *IRE Trans. Mil. Electronics* I. 43/1 The development of a system to control remotely a space-research vehicle. **1982** M. DUKE *Flashpoint* xv. 108 We've already had benefits from space research. **1957** D. J. ENRIGHT *Apothecary's Shop* 232 At least one writer, Robert Conquest, is exploring space-science as a subject for his poetry. **1978** *Nature* 16 Feb. 599/1 If the future facing space science 20 years ago lay full of hope and promise, the symposium indicated how confused and uncertain the picture is today. **1953** M. O. HYDE *Flight Today & Tomorrow* 100 Space scientists look to the rocket to carry them beyond the earth. **1969** *Times* 2 May 16/4 Space scientists have discovered six concentrations of dense material below the surface of the moon. **1952** W. LEY in C. Ryan *Across Space Frontier* 114 The space station..is always spinning, and obviously it cannot be stopped just to enable a space taxi to enter one of the turrets. **1970** N. ARMSTRONG et al. *First on Moon* xiv. 369 An orbiting space station and the 'space taxi'..to take astronauts there and back. **1958** *Science* 11 Apr. 803/1 To be strong and bold in space technology will enhance the prestige of the United States among the peoples of the world. **1972** *Guardian* 10 July 11/7 France..and West Germany [are] eager to embrace the most advanced of space technologies. [**1929** Space travel: see *space station*, sense 19 below]. **1931** J. M. WALSH *Vandals of Void* i. 23 In the early days of space travel more than one ship was pirated. **1951** 'J. WYNDHAM' in *Best of John Wyndham* (1973) 196 The question of continued space-travel ships of the present types becomes grave. **1978** I. WATSON in C. Priest *Anticipations* 13 What kind of space travel..? Well, they can only be going to the stars. **1930** *Science Wonder Q.* Spring 342/2 (*caption*) Illustrating the journey of the space travelers from Astropol to Venus. **1949** 'M. INNES' *Journeying Boy* iv. 39 So might the earth's first space-traveller exclaim as his rocket took off for the moon. **1976** *Listener* 22 July 83/3 A journey of merely five light years would take about 500,000 years... 15,000 generations of men and women..would successively replace the original crew of stellar space-travellers en route. **1961** *Aeroplane* C. 184/1 The one-man 'space tug' would be used for assembling a large space-station in orbit. **1970** *Physics Bull.* Apr. 145/2 A manned moon station is foreseen, as are..a 'space shuttle' for commuting between the earth and vehicles in low earth orbit, and a 'space tug' for transport to Mars in the 1980s.

f. Applied to sprays designed to produce droplets that will remain suspended in the air for a long period.

1956 *Aerosol Age* June 70/2 This opens up some interesting possibilities in the lower cost, non-toxic pressurized space sprays. **1958** HERZKA & PICKTHALL *Pressurized Packaging* xiv. 271 Although it is possible to produce a space deodorant which employs only deodorant perfume and propellant..it is better to use the perfume in conjunction with glycols. **1973** J. B. WILKINSON et al. *Harry's Cosmeticology* (ed. 6) xliv. 764 A good example of the functional use of space sprays is the aerosol room deodorant. **1974** M. O. JOHNSON in Sciarra & Stoller *Sci. & Technol. Aerosol Packaging* xx. 541 (*heading*) Air fresheners and space bactericides.

18. a. *space-based, -dependent, -spanned.*

1931 C. DAY LEWIS *From Feathers to Iron* 45 Space-spanned, God-girdled, love will keep its form, being planned of bone. **1958** I. ASIMOV *Naked Sun* viii. 107 Try getting rid of me against my will and you'll be looking down the throats of space-based artillery. **1962** CORSON & LORRAIN *Introd. Electromagn. Fields* 534 Space-dependent functions can also be represented with the exponential notation. **1972** *Guardian* 9 Feb. 3/8 The US is evolving..an entire space-based defence network.

b. *space-devouring, -travelling, -wasting.* Also with (formally identical) vbl. sbs.

1907 W. JAMES *Let.* 14 Feb. (1920) II. 265 The magnificent space-devouring Subway roaring me back and forth. **1934** C. LAMBERT *Music Ho!* II. 108 His time travelling is like the space travelling of a character like Douglas Fairbanks. **1938** *Times Lit. Suppl.* 1 Oct. 625/3 The space-travelling itself forces a more direct comparison with..'The First Men in the Moon'. **1949** E. MUIR *Coll. Poems* (1960) 177 Its space-devouring eyes Pass me and hurry on. **1962** F. W. HOUSEHOLDER in Householder & Saporta *Probl. Lexicogr.* 281 Others objected to them as (a) space-wasting, (b) often irrelevant and unhelpful. **1979** J. PATON *Sea of Rings* xv. 122 We've never encountered any other space-travelling civilisation.

c. In adjectival phr., as *space-to-ground.*

1958 C. C. ADAMS *Space Flight* 144 Whether bombing or space-to-ground missile attacks would be any more effective from such a [space] station has not been established. **1967** *Economist* 29 Apr. 479/1 A Soviet space ship that, according to some reports, was having a variety of troubles with attitude control, power consumption and space-to-ground communication.

19. space age, the period of human exploration and exploitation of space; freq. *attrib.*, applied to products supposed to be characteristic of this age; hence (*nonce-wds.*) **space-ager,** one living in this age; **space-agey** *a.*, characteristic of this age; **space-averaged** *a.* *Physics*, averaged over a region of space; **space blanket,** a light metal-coated plastic sheet designed to retain heat; **space-borne** *a.*, carried through space; also, carried out in space or by means of instruments in space; **space-bound** *a.*, bound or limited by the properties of space; **space cabin,** a chamber designed to support human life in space; **space cadet,** a trainee spaceman; also *transf.*, esp. a (young) enthusiast for space travel; **space capsule,** a small spacecraft containing the instruments or crew relating to the purpose of a space flight; **space chamber,** a chamber in which conditions in space or a spacecraft can be simulated; **space charge** *Electronics*, a collection of particles with a net electric charge occupying a volume, either in free space or in a device; freq. *attrib.* and in *Comb.*, as *space-charge-limited* adj.; **space club,** a group of nations that has launched or intends to launch spacecraft; *spec.* a consortium of European nations formed to co-operate in space research and development; **space colony,** a large group of people imagined as living and working in a space station or on another planet; **space curve** *Geom.*, a curve that is not confined to any one plane; **space density** *Astr.*, frequency of occurrence per specified volume of space; **space fiction,** science fiction set in space or on other worlds or involving space travel; so **space-fictional** *a.*; **space-filler,** something that serves to occupy an otherwise vacant space; *spec.* a brief or insignificant item in a newspaper or magazine; **space fleet** *Science Fiction*, a fleet of spacecraft; **space flight,** a journey or travel through space; **space flyer,** (*a*) a spacecraft; (*b*) an astronaut; **space frame** *Engin.*, a three-dimensional structural framework which is designed to behave as an integral unit and to withstand loads applied at any point; **space gun,** (*a*) a large gun which projects a spacecraft into space; (*b*) a hand-held gun whose recoil is used by an astronaut or spaceman to propel himself; **space heater,** any self-contained appliance for heating an enclosed space within a building; also **space heating; space helmet,** a helmet worn in space to protect the head and provide air; also *transf.*; also **space-helmeted** *a.*; **space industry,** the sector of industry which manufactures goods and materials in connection with space flight; **Space Invaders,** the name of an animated computer game in which a player attempts to defend himself against a fleet of enemy spaceships; also, the attacking force itself; **space lab, spacelab** = next; *spec.* (with capital initial(s)) as a proper name (see quot. 1980); **space laboratory,** a laboratory in space, esp. a spacecraft equipped as a laboratory; **space lattice** *Cryst.*, substitute for def.: a regular, indefinitely repeated array of points in three dimensions in which the points lie at the intersections of three sets of parallel equidistant planes and every point is surrounded by the same pattern of points in the same orientation; a three-dimensional Bravais lattice; (examples); **space launcher,** a rocket used to lift spacecraft into space; **space lift, spacelift** [after *AIR-LIFT 2], an act of transporting goods or personnel in space; **space medicine,** the branch of science concerned with the medical effects of being in space; **space myopia** (see quot. 1973); **space needle,** a small rod or fibre of conducting material in orbit about a planet; **space observatory,** an astronomical observatory in space; **space-occupying lesion** *Path.*, a mass, freq. a tumour, which has displaced brain tissue; **space opera** chiefly *U.S.* [cf. **horse opera*, *SOAP OPERA], space fiction, esp. of a primitive and extravagant kind; an example of this genre; **space-order,** an ordering of points or events in space; **space physics,** the physics of extraterrestrial phenomena and bodies, esp. within the solar system; **space-plane,** (*a*) (see quot. 1961); (*b*) = *SHUTTLE *sb.*[1] 8 c; **space platform** = *space station* below; **space-port,** a base from which spacecraft a launched; (in fiction) a base at which spaceships take off and land; **space probe,** an unmanned spacecraft for research or reconnaissance; **space programme,** a programme of exploration of space and development of space technology; **space race,** the competition between nations to be first to achieve various objectives in the exploration of space; **space-reddening** *vbl. sb. Astr.*, the reddening of starlight as a result of wavelength-dependent absorption and scattering by interstellar dust; also **space-reddened** *ppl. a.*; **space rocket,** a rocket designed to travel beyond the earth's atmosphere; **space satellite** = *SATELLITE *sb.* 2 c; **space-saving** *a.*, that uses space economically or tends to the better use of available room; also as *sb.* and **space saver,** a device or appliance designed to this end; **space shot,** the launch of a spacecraft and its subsequent progress in space; **space shuttle** see *SHUTTLE *sb.*^{1} 8 c; **space sick, space-sick** *a.*, sick from the effects of space flight; hence **space sickness**; **space simulator,** a device which simulates the conditions of space, or of the interior of a spacecraft; **space-speak** [*-SPEAK], the jargon of space technologists, considered as a corruption of standard English; **space stage** *Theatr.*, a modern stage on which the significant action alone is lighted, the rest remaining in darkness; hence **space staging**; **space station,** a large artificial satellite used as a base for operations in space; **space suit,** a garment designed to protect the wearer against the conditions of space; so **space-suited** *a.*, wearing such clothing; **space vehicle,** a spacecraft, esp. a large one; **space velocity** *Astr.*, the velocity in space of a star relative to the sun, equal to the vector sum of its proper motion and its radial velocity; **spacewalk, space walk,** an act or spell of physical activity undertaken in space outside a spacecraft; also as *v. intr.*; hence **spacewalking** *vbl. sb.* and *ppl. a.*; also **spacewalker**; **space warp,** an imaginary distortion of space-time that is conceived as enabling space travellers to make journeys that would otherwise be contrary to the known laws of nature; **space wave** *Radio* [tr. G. *raumwelle* (A. Sommerfeld 1911, in *Jahrb. der drahtl. Telegr.* IV. 166)], the radio wave that passes from a transmitter to a receiver either directly through space without reflection or with reflection from the ground; **spaceway** *Science Fiction*, an established route of space travellers; usu. *pl.*

1946 H. HARPER *Dawn of Space Age* I. i. 5 We have had an age of steam-power, an age of electricity and of the petrol engine, and an age of the air, and now with the coming of atomic power the world should, in due course, find itself in the space age. 1960 K. AMIS *New Maps of Hell* iii. 80 The outset of the space age and the immense technological effort involved in it are obviously the propelling force of much science fiction today. 1963 *New Yorker* 8 June 96 The space-age, space-tested material that makes possible this smart, new look in luggage. 1980 *Times Lit. Suppl.* 7 Nov. 1258/4 Our space-age Palace of History—the new computerized Public Record Office at Kew. 1959 *Times* 9 Mar. 13/5 One of the rockets, the space-ager firmly believes, will have him aboard. 1962 *Punch* 28 Nov. 781/1 A modern caravan..trying to look zippy and space-agey. 1946 *Nature* 26 Oct. 582/2 Such time- or space-averaged statistical structures are becoming increasingly familiar to X-ray crystallographers. 1962 CORSON & LORRAIN *Introd. Electromagn. Fields* iii. 91 (*caption*) To find the space-averaged field intensity produced by the dipoles, we calculate the field intensity at *O*..and then repeat this calculation for many other points *O′*. 1972 *Brit. Med. Jrnl.* 29 Jan. 293/2 The body temperature should be slowly raised to normal, using a 'space blanket' and heating pads if necessary, in a warm room. 1953 J. N. LEONARD *Flight into Space* 157 Undoubtedly one of the great preoccupations of the space-borne astronomers will be to study the moon and the planets. 1965 *New Scientist* 26 Aug. 485/1 One would have thought that the bugs could have been eliminated from the fuel cell system before it ever became space-borne. 1968 *Ibid.* 28 Mar. 680 In spite of considerable lobbying to make optical astronomy a space-borne science, many practising observers show little enthusiasm for the idea. 1975 *Nature* 22 May 287/1 With the shuttle taking up an ever-increasing share of the space budget, there is likely to be little money to spare for expanding space-borne astronomy. 1960 *Analog Science Fact & Fiction* Nov. 14/2 He banged it shut behind him and, feeling that he might as well continue with his spacebound existence, walked all the way to the elevator. 1958 D. G. SIMONS in M. Alperin et al. *Vistas in Astronautics* I. vi. 301 The systems and controls required to establish a space cabin capsule. 1961 *Guardian* 10 Mar. 1/5 As on previous occasions in this series of Russian experiments, the space cabin, as it is called, weighed 4·5 tons. 1974 *Sci. & Technical Aerospace Rep.* XII. 1013/1 (*heading*) Survival of infectious microorganisms in space cabin environments. 1952 *Newsweek* 13 Oct. 39/2 (*caption*) Test pilot A. M. 'Tex' Johnston..resembles a space cadet in the new high-altitude helmet and suit designed to protect pilots in the upper air. 1957 P. MOORE *Science & Fiction* i. 18 Lucian's seamen are the logical ancestors of the rocketeers and space-cadets of to-day. 1958 C. C. ADAMS *Space Flight* p. vii, There have been space books for children—our present space cadets and future rocket pilots. 1979 *Harvard Mag.* May–June 15 How can I be one of the first to see these new worlds in detail and not be crawling with gooseflesh? Me, an original space cadet? 1959 *Listener* 15 Jan. 118/1 An American firm is given a contract to build a space-capsule designed to put a man into orbit round the earth. 1963 *Ann. Reg. 1962* 398 The larger two-man Gemini space capsules, orbiting for a fortnight on end. 1977 G. SCOTT *Hot Pursuit* x. 88 It starts..with the space capsule... A Russian satellite, one of the Cosmos series. 1959 *Daily Tel.* 23 Feb. 11/7 Col. Steinkamp and his colleagues have been carrying out interesting tests, lasting from four hours to a week, in a sealed 'space chamber'. 1966 *Science World* 7 Jan. 10 The Air Force has been testing the ability of men to live for long periods in a new gas mixture that may be used in space ships... There have been two short tests in 'space chambers'. 1913 *Physical Rev.* II. 450 (*heading*) The effect of space charge and residual gases on thermionic currents in high vacuum. 1921 *Ibid.* XVIII. 56 The maximum space-charge limited current was the same for each [tube]. 1956 *Nature* 11 Feb. 285/2 The corresponding current..for maximum space-charged-limited pulsed emission from the surface of the oxide coat is 8 amp. per sq. cm. 1962 SIMPSON & RICHARDS *Physical Princ. Junction Transistors* iv. 54 For this reason the region AB is often referred to as the space-charge or depletion region. 1980 J. W. HILL *Intermediate Physics* xxii. 210 In the Maltese cross tube and the deflection tube, the space charge formed by the filament is attracted towards the positively-charged anode and accelerated. 1961 *Economist* 14 Jan. 116/2 West Germany agreed by the 1954 treaties not to manufacture long-range missiles. Participation in a space club with a military potential would take the erosion of the treaties a stage further. 1970 *Daily Tel.* 3 Sept. 5 Britain failed to join the 'space club' yesterday because of a fault in the second stage of the Black Arrow rocket fired from Woomera, Australia. 1974 *N.Y. Times* 19 May IV. 6 The space colonies..would provide an alternative to earth if the earth's resources ever reach the point of depletion. 1971 *Proc. Nat. Acad. Sci.* LXVIII. 815/1 Our interest here is in space curves that are the central curves of elastic rods. 1931 *Astrophysical Jrnl.* LXXIV. 268 (*heading*) A numerical method of determining the space density of stars. 1978 *Nature* 10 Aug. 569/1 *RS Canum venaticorum* systems are the most plentiful binary stars known, having a space density of at least 10^{-6} systems pc^{-3}. 1952 *Space Science Fiction* May 2/1 We like good space fiction, and we intend to bring you the best of it... The space-opera of flashing rayguns and invincible heroes has long since been overdone. 1960 *Guardian* 19 Aug. 5/4 A frantic urge for escape, but where to? Astrology, necromancy, space-fiction? 1979 *Daily Tel.* 14 Dec. 13/3 Star Trek is the latest in an increasing number of space fiction films which..tend to find individuality. 1963 V. GIELGUD *Goggle-Box Affair* xviii. 191 Space-fictional horrors. 1911 H. S. HARRISON *Queed* xviii. 232 There's a little squib about the college that may serve as a space-filler. 1956 *Nature* 17 Mar. 530/2 Into the large centre well 3–4 drops of anti-serum are deposited and a glass or aluminium plug... The latter serves merely as a space-filler to spare anti-serum. 1972 *Sci. Amer.* July 13/3, I have written some poetry, mostly nonserious, that has found its way into medical journals as space-filler. 1944 E. COLLINS *Mariners of Space* iii. 25 Space-Captain Jan Marthus of the Martian Space Fleet steered his friend into the restaurant. 1979 J. PATON *Sea of Rings* 23 William Robert Mahony, ex-Captain, Space Fleet, aged 46. 1931 *Wonder Stories* Jan. 900/1 We know now what conditions are necessary for a space flight... After all space flying is too great a matter to be limited by national pride and jealousy. 1949 A. C. CLARKE *Across Sea of Stars* (1959) 76 There is a timelessness about space-flight..unmatched by any other experience of man. 1978 J. UPDIKE *Coup* (1979) v. 183 The hollow head with which a mummified Pharaoh is helmeted for *his* space-flight. 1911 *Mod. Electrics* Nov. 516/1 He knew now that Fernand 600 10 had carried off his sweetheart in a space-flyer and that the machine by this time was probably far out from the earth's boundary. 1931 *Wonder Stories* Feb. 958 To old and seasoned space-fliers like Professor Galloway and myself, there was something ludicrous in all this emotional bustle..over a little hop to the Moon. 1962 M. V. GLENNY tr. *Gartmann's Space Travel* 130/1 The space flyer will..encounter two opposed physical conditions: pressure and weightlessness. 1962 *Listener* 1 Mar. 368/2 All three American space-fliers had had to be landed in the sea. 1912 A. MORLEY *Theory of Structures* xiii. 380 (*heading*) Space frames. 1967 *Jane's Surface Skimmer Systems 1967–68* 48/1 The craft is of lightweight space-frame construction in marine aluminium. 1974 *Times Lit. Suppl.* 4 Jan. 14/4 The debased standards of theatre design today and..artists who think 'empty space, lighting, and maybe an aluminium space frame' are enough. 1935 H. G. WELLS *Things to Come* 12 The stormy victory of the new ideas as the Space Gun fires and the moon cylinder starts on its momentous journey. 1954 K. W. GATLAND *Devel. Guided Missile* (ed. 2) 197 All the propellant could be consumed in the first second of take-off—as Jules Verne proposed in his famous 'space-gun'. 1968 *Amer. Speech* XLIII. 166 *Space gun*, a handheld instrument used to propel an astronaut outside the capsule. 1970 N. ARMSTRONG et al. *First on Moon* viii. 180 This was where I had to use the little space gun. 1976 P. MOORE *Next Fifty Years in Space* i. 16 It is not impossible that the space-gun principle may have its uses in the future, but it will be confined to firing non-fragile payloads off airless worlds. 1925 *Sci. Amer.* Mar. 162/3 Space heaters. 1951 *Good Housek. Home Encycl.* 203/1 The stove..is primarily designed as a boiling ring but it will also serve as a space heater. 1980 *Amat. Gardening* 25 Oct. 9/1 Electrical space heaters are extremely expensive to run at high temperatures. 1934 *Jrnl. Inst. Heating & Ventilating Engineers* XIII. 234 The open fire is still..the most widely-used domestic space-heating appliance. 1973 *Guardian* 17 Mar. 12/5 What is especially intolerable..is that..electricity with a starting efficiency of only 20 per cent or so is allowed to be sold for space heating, a role particularly suitable for the 80 per cent of low grade heat that has been thrown away. 1954 *Newsweek* 6 Dec. 108/1 It is significant, too, that the American kid of 1954, dazzled by space helmets and death-ray guns, has time in his daily make-believe for the mustang and the six-shooter. 1973 *Times* 29 Aug. 3/2 Scientists are developing a 'space helmet' respirator to protect miners against dust. 1979 R. JAFFE *Class Reunion* (1980) II. ii. 191 Her blonde, teased, sprayed bubble hairdo..looked like a space helmet. 1957 *Time* 22 July 52/1 From a sealed chamber like the cabin of a rocket ship, and from space-helmeted human guinea pigs who live in it, medical researchers..hope to learn answers to some fundamental questions about the body's consumption of fuel and oxygen. 1982 D. MACKENZIE *Raven's Revenge* x. 94 A space-helmeted motorcyclist. 1962 M. V. GLENNY tr. *Gartmann's Space Travel* 9 The space industry ..has plans for huge multi-purpose earth satellites. 1972 *Guardian* 10 July 11/2 The unhappy fragmental European space industry. 1979 *Los Angeles Times* 23 Sept. VII. 17/1 Nobody likes to be a loser, but when playing Space Invaders, most gamesters don't seem to mind. 1980 *Guardian* 2 Feb. 8/3 Driven out of the BR station buffet by bleeping Space Invaders. 1980 *Washington Post* 2 Sept. B1 A world-class Space Invaders player can keep the machine going for an hour. 1982 *London Rev. Bks.* IV. xxiv. 7/1 The advent of the Space Invaders can't mean anything except that new inventions bring new possibilities. 1966 *Electronics* 31 Oct. 134 Although the Gemini computers are highly flexible.., they are not versatile enough for space labs. 1975 K. GATLAND *Missiles & Rockets* xv. 246 Hatches on top of the cargo compartment will open to permit Space Lab to be hinged out into space. 1979 *Fortune* 29 Jan. 77 In the microgravity of an orbiting spacelab, NASA will make crystals, alloys, and medicines never seen on earth. 1980 T. FURNISS *Space Satellites* 30/2 One of the payloads the Shuttle carries is the Spacelab research station. Spacelab is built in Europe by the member countries of the European Space Agency. 1960 Space-laboratory [see *meteor bumper* s.v. *METEOR]. 1973 *Guardian* 28 May 2/2 The battered American space laboratory, Skylab, cooling parasol now clutched tightly over her gold-foiled head. 1895 Space-lattice [see *LATTICE *sb.* 3* a]. 1923 GLAZEBROOK *Dict. Appl. Physics* IV. 18/2 In the crystals of very simple chemical compounds..the space-lattice is directly formed by the chemical atoms. In the more complicated crystalline substances..the space-lattice points are surrounded or replaced by groups of atoms. 1973 J. G. TWEEDDALE *Materials Technol.* I. iii. 59 Although there are countless varieties of crystals, there can be only 14 types of space lattice. 1961 *Daily Tel.* 1 Feb. 16/4 (*heading*) Joint effort for space 'launcher'. 1963 *Guardian* 9 Nov. 7/7 The Soviet Union has now apparently agreed that Governments should be allowed to license private space-launchers. 1954 'J. CHRISTOPHER' *22nd Cent.* 65 As many as possible would be got away to those planets by a full space lift. 1964 *Yearbk. Astron. 1965* 142 What then might be realized is a joint use of the nationally developed space hardware in a space-lift of supplies to obtain the first firm footholds on the Moon in the shape of a scientific base. 1949 *Time* 12 Sept. 29/2 The U.S. Air Force's School of Aviation Medicine..has set up an interplanetary research section, [and] named it the Department of Space Medicine. 1962 F. I. ORDWAY et al. *Basic Astronautics* xiii. 539 This centrifuge is unique among those used in space medicine research. *Ibid.* xii. 474 Other visual phenomena associated with space flight include space myopia... Looking out into the darkness of space, the astronaut would not know whether his eyes were focused at infinity or only a few feet from his ship. 1973 *Gloss. Aeronaut. & Astronaut. Terms* (*B.S.I.*) XVII. 1 *Space myopia*, the tendency of the human eye to accommodate for a distance commonly of the order of six feet when in a featureless environment, resulting in potential failure to perceive objects at considerably greater distance. 1961 *Daily Tel.* 23 Oct. 1/3 (*heading*) Space needles begin to form radio band. 1964 Space needle [see *DIPOLE 2]. 1952 F. L. WHIPPLE in C. Ryan *Across Space Frontier* 136 Our space observatory can give us vital information as to how some stars die in a spectacular blaze of glory. 1972 *Guardian* 22 Aug. 2/4 A space observatory, Copernicus, was launched here today, the fourth to be put in orbit. 1961 *Lancet* 9 Sept. 570/1 The patients reported in this paper had small pretectal space-occupying lesions demonstrated both clinically and radiologically. 1978 *Jrnl. R. Soc. Med.* LXXI. 226 These cases illustrate some atypical presentations of tuberculosis—as epilepsy, cranial nerve palsy,..or space-occupying lesion. 1949 *Sat. Rev. Lit.* (U.S.) 24 Dec. 7/3 No less than eight of this year's crop of science-fiction novels are what is known in the trade as 'space operas'—books built round the theme of interplanetary travel. 1952 [see *space fiction* above]. 1960 K. AMIS *New Maps of Hell* ii. 44 In space-opera, Mars takes the place of Arizona with a few physical alterations, the hero totes a blaster instead of a six-gun. 1969 H. WARNER *All our Yesterdays* ii. 41 [Wilson] Tucker is responsible for the use of 'space opera', which he proposed in the January, 1941, *Le Zombie* as a name for the 'hacky, grinding, stinking, outworn spaceship yarn'. 1978 *Broadcast* 31 July 24/1 The Seven-Up [TV advertisement] series..is meant as a space opera send up. 1890 W. JAMES *Princ. Psychol.* II. xx. 276 The obvious objection is that mere serial order is a *genus*, and space-order a very peculiar species of that *genus*. 1927 B. RUSSELL *Outl. Philos.* iv. 50 A written word is a series of pieces of matter, having an essential space-order. 1961 *Adv. Astronaut. Sci.* VI. 779 Space physics. 1962 F. I. ORDWAY et al. *Basic Astronautics* iv. 117 An important characteristic of space physics is that it is closely related to two aspects of geophysics, namely atmospheric physics and ionospherics. 1980 *Jrnl. R. Soc. Arts* May 357/1 The College has now built more instruments..making thereby vast additions to space-physics knowledge. 1961 *Aeroplane* C. 597 A spaceplane is an aircraft capable of entry into orbit, using for propulsion purposes the atmosphere through which it has passed. 1978 *N.Y. Times Mag.* 29 Jan. 26 Designed to take off like a rocket, fly in orbit like a spacecraft, and return to a runway landing like a glider, these huge spaceplanes are expected to make the near reaches of space more accessible than ever before.

1958 F. A. WARREN *Rocket Propellants* xi. 196 Thought of space platforms, space ships, satellite stations, and high-altitude exploratory rockets overshadows consideration of other rocket uses. 1980 M. BABSON *Dangerous to Know* vii. 47 She'd crashed like a chunk of rubble from an abandoned space platform. 1935 *Amer. Speech* X. 54/1 Spaceport. 1943 *Astounding Science-Fiction* Feb. 9/1 Carew had landed him at one of the less expensive spaceports. 1962 *Daily Progress* (Charlottesville, Va.) 23 Feb. 11 After these few words, Glenn set out for the ride through brilliant sunshine to this space port—where it all began—and his meeting with President Kennedy. 1977 *Time* 30 May 45/2 For one scene, set in a brawling space-port bar, the casting director went to a London firm called Uglies, Ltd. 1977 *Daily Tel.* 28 July 1/6 The small spaceport at Kagoshima, at the southern tip of Japan, looked more like a station for amateur rocketry than a serious rival to Cape Canaveral. [1955 E. BURGESS *Frontier to Space* viii. 152 We would then have the deep-space probe.] 1958 *Listener* 20 Nov. 822/1 Direct contact between some form of space probe and the moon..must be close at hand. 1977 *Whitaker's Almanack 1978* 158/2 A Russian space probe has revealed that the lower layers [of Venus] are extremely dense. 1958 *New Statesman* 6 Sept. 263/2 It was Congress, rather than the President, that took the initiative in pushing a space programme. 1977 B. LANGLEY *Death Stalk* ii. 23 A number of senators had a vested interest in seeing that the space programme continued. 1959 *Listener* 29 Jan. 226/2 The possible nature of Britain's contribution to the space-race. 1967 M. KENYON *Whole Hog* iii. 31 If you've got something which could keep the same men..in the capsules..the space race would be won. 1978 *Nature* 9 Mar. 119/2 Czechoslovakia has won the 'little space race' for the third nation to put a citizen into orbit. 1959 *Listener* 24 Dec. 1111/2 The stars are said to be space-reddened. In the same way atmospheric dust causes the sun to appear red at sunset. 1931 *Astrophysical Jrnl.* LXXV. 392 The differential absorption or space reddening at 1000 parsecs in the galactic plane is unquestionably real. 1937 Space reddening [see *INTERSTELLAR *a.*]. 1928 *Discovery* June 190/1 The arguments used in regard to objections so far raised meet the case in respect of space-rocket machines. 1936 'J. BEYNON' *Planet Plane* iv. 36 You can be sure that if they were building a space rocket anywhere we'd have heard of it. 1958 *Listener* 16 Oct. 606/1 A space-rocket, aimed towards the moon, is successfully launched from Cape Canaveral. 1977 'J. FRASER' *Hearts Ease in Death* vii. 76 Mark Dunton's bed..looked like some sort of space rocket, with tubes connecting various parts of Mark's body to pieces of apparatus. 1954 *Jrnl. Brit. Interplanetary Soc.* XIII. 165 Since it is theoretically possible today to design and build instrumented rocket vehicles for both orbital and escape missions, one is often asked whether there is not justification for a space-satellite programme, such as the Americans suggested in 1948. 1962 E. SNOW *Red China Today* (1963) lxxxvi. 723 The development of space satellite espionage and electronic detection devices had so far advanced that all essential information needed to guarantee against surprise attacks would soon be in possession of both sides. 1974 P. CATTERMOLE *All about Space Exploration* vi. 74 Unmanned space satellites had been approved by the U.S. Government. 1970 *Toronto Daily Star* 24 Sept. 5/2 (Advt.), Just out on sale!..New..Space-saver Consolette [television]. 1921 *Sci. Amer.* 30 July 79/1 Home Building..Many unique space-saving devices are now being used. 1934 WEBSTER, *Spacesaving*, n. 1936 *Punch* 11 Mar. 287/1 For space-saving reasons, [I] have exchanged the old edition of the D.N.B. for its India paper form. 1964 E. BACH *Introd. Transformational Gram.* ii. 17 The latter notation is useful as a space-saving device. 1978 M. & N. WARD *Home in Twenties & Thirties* 23 Cost saving was one necessity, space saving another. 1961 *Guardian* 29 Mar. 1/6 Russia did not give advance notice of space shots. 1969 *Times* 19 Feb. 13/4 Between now and April, Mars lies in a favourable position for space shots. 1977 D. BAGLEY *Enemy* xxxii. 259 Designing a trajectory for a space shot to Pluto. 1949 A. C. CLARKE *Across Sea of Stars* (1959) 93, I was sure I'd never be space sick. 1971 *New Yorker* 27 Feb. 32 If an astronaut were made to move his hand repeatedly in the wrong direction in relation to the spin, he could easily get spacesick. 1951 A. C. CLARKE *Sands of Mars* i. 3 Space-sickness was a thing of the past. 1969 *New Scientist* 2 Oct. 28/1 The Russian cosmonaut Titov was the first to complain of space sickness in 1961. 1959 *IRE Trans. Mil. Electronics* III. 96/1 For the first space simulator, it is proposed to use a combination of visual references and cabin motion to give a hint or illusion of 'g' forces. 1974 *Sci. & Technical Aerospace Rep.* XII. 37 (*heading*) Radiometer for measuring a wide range of irradiances in space simulators. 1966 *Science* 13 May 875/1 We read of 'space speak' on every hand. Newspapers and magazines discuss it in their science columns, and popular fancy seems to have been captured by it. The belief is that the space effort has given us, in addition to the possibility of going to the moon, a new linguistic phenomenon. 1982 M. LEAPMAN *Yankee Doodles* III. 175 Transiting is a typical piece of spacespeak in that it makes a verb out of a noun. 1928 J. DOLMAN *Art of Play Production* xviii. 397 Three types of modern stages have..been reasonably successful in accomplishing the true purpose of formalism. One is..the so-called 'space stage', the essential feature of which is light, so controlled as to reveal only the significant action and to suppress the background altogether in a void of darkness. The methods of the space stage are..adaptable to the purposes of expressionism. 1961 *Twentieth Cent.* Feb. 121 A space stage..a broad platform with no barrier between audience and performer. 1959 *Listener* 9 July 73/1 A television equivalent to Brechtian 'space staging'. [1929 *Science Wonder Stories* Sept. 365 (*heading*) The spatial station as a basis for spatial travel.] 1936 P. E. CLEATOR *Rockets through Space* vi. 141 So great are the possibilities of the space-station that von Pirquet is of the opinion that the achievement of interplanetary travel..must depend upon the construction of such a station. 1956 J. G. PORTER in A. Pryce-Jones *New Outl. Mod. Knowl.* 135 We are to visualize a space station, a sort of artificial satellite of the earth which is to act as a landing stage for all space ships. 1969 *Sci. Jrnl.* Feb. 66/1 Although any manned satellite might literally be called a manned space station, the term is usually restricted to spacecraft which remain in orbit for long periods and which carry relatively large crews. 1929 *Science Wonder Stories* July 175/1 Normal communication by speech would be impossible. Of course, this is not true of enclosed, air-filled rooms... But it is true when one is out 'in the open' (in the space suit). 1962 J. GLENN et al. in *Into Orbit* 244 G-suits are not to be confused with pressure suits (or, now, spacesuits) which the Astronaut wears during space flight to maintain atmospheric pressure at high altitudes. 1979 D. ADAMS *Hitch-Hiker's Guide to Galaxy* iii. 25 He will automatically assume he is also in possession of a toothbrush,..space suit etc., etc. 1951 COGGINS & PRATT *Rockets, Jets, Guided Missiles & Spaceships* v. (*caption*) Switching off the electromagnets in his boots, the space-suited chief engineer kicks off for a look at his project. 1977 *Daily Mirror* 10 May 19/2 A Staffordshire housewife..saw two Space-suited people with long blonde hair looking down at her from a craft above her house. 1946 *N.Y. Times* 29 July 1/2 They are to serve as pioneers for the long-range guided missiles and 'space' vehicles. 1959 *Daily Tel.* 23 Feb. 11/7 Protective clothing..will be needed to help guard a man against the heavy G forces imposed on him because of the great thrust upwards that a space vehicle will develop. 1959 *Times* 15 Sept. 11/3 In putting a space vehicle on to the moon the Russians have provided the most complete..proof of the length of the lead that they now hold. 1977 'M. UNDERWOOD' *Fatal Trip* xx. 117 We can often track a load of porn..as successfully as the Americans track a space vehicle..to Mars. 1921 *Bull. Nat. Res. Council Nat. Acad. Sci.* II. 196 Average space velocities vary from 10 to 30km/sec., there being a well-marked increase in average space velocity as one proceeds from the blue to the redder stars. 1927 H. N. RUSSELL et al. *Astronomy* II. xix. 652 The space velocity, being found from the proper motion, parallax, and radial velocity, demands for its determination the combination of observations made in very different ways. 1978 PASACHOFF & KUTNER *University Astron.* iii. 62 We can combine the values of velocity in the plane of the sky and in the radial direction to find the actual velocity of the star in space, the space velocity. 1965 *Newsweek* 14 June 30/3 Thirteen new layers had been added [to his spacesuit]..to protect his torso and legs against micro-meteorites and the extreme temperatures on his spacewalk. 1969 *Daily Tel.* 17 Jan. 1/2 The link-up was controlled manually by the cosmonauts. The two who made the space 'walk' reported 'feeling fine' after their feat. 1970 *Guardian* 8 July 20/7 Astronaut figures, 12 ft high, will 'space walk' over a quarter-mile section of Blackpool promenade in this autumn's illuminations. 1978 *Nature* 19 Jan. 201/3 The work aboard Salyut-6 is more interesting than his previous mission on Salyut-4, since the cosmonauts can now go for spacewalks. 1979 M. COLLINS *Flying to Moon* vi. 44 We would fly alongside the second Agena, and I would space walk over to it. 1965 *Newsweek* 14 June 30/1 Spacewalker White had trouble sleeping, due to the excitement. 1965 *Ibid.* 21 June 24/1 He was firmly convinced that spacewalking is an easily mastered art. 1970 *Daily Tel.* (Colour Suppl.) 10 Apr. 9/4 Theoretically we could now send a space-walking astronaut as an electric repair man to one of these satellites if anything went wrong. 1947 *Jrnl. Brit. Interplanetary Soc.* VI. 138 The next step is to explain that matter and energy are mutually convertible..and from this it is an easy process to pass to the 'invention' of a plausible device which produces a 'space-warp' in the opposite sense to the normal one, by means of the expenditure of energy in some form. Bodies in the region of this artificial 'space-warp' therefore acquire a negative weight—what could be simpler? 1953 *Galaxy Sci. Fiction* Nov. 53/2 He read the next one ..about a star-ship that hit a space warp and got hurled into another universe. 1974 G. BUTLER *Coffin for Canary* ix. 108 Don't unnerve me, boss. *You* don't believe in space-warps and voyagers back from another age. 1913 *Rep. Brit. Assoc. Adv. Sci. 1912* 403 His theory leads to the conclusion that there are not only space-waves (*Raumwellen*) in these media, but also surface-waves (*Oberflächenwellen*) at the boundary surface. 1943 F. E. TERMAN *Radio Engineers' Handbk.* x. 674 The ground wave can conveniently be divided into two components, a surface wave and a space wave. 1974 HARVEY & BOHLMAN *Stereo F.M. Radio Handbk.* vii. 145 The range of the space-wave is chiefly determined by the height of the transmitting and receiving aerials. 1947 *Astounding Sci. Fiction* Sept. 171 You can't believe..your eyes... It was a platitude of the spaceways. 1979 *Daily Tel.* 1 Feb. 15/2 Everyone loves Harry Harrison; the genre would be poorer without this Monty Python of the spaceways.

space, *v.* Add: **4. d.** *intr.* With *out.* To experience a drug-induced state of euphoria; to become disoriented by the use of narcotic stimulus. Cf. *SPACED *ppl. a.* 4. *U.S. slang.*

1968–70 *Current Slang* (Univ. S. Dakota) III–IV. 116 *Space out*, *v.* To achieve an euphoric state because of drug use... To lose one's train of thought while under the influence of a drug. 1970 *New York* 16 Nov. 48/3 Karenga ..looks like he's going crazy or spacing out on dope.

spacearium (spē^{i}sē^ə·riŏm). [f. SPACE *sb.*^{1} + PLANET)ARIUM.] A large room arranged so that scenes representing space may be projected on to its interior; a planetarium.

1962 *Times* 18 Apr. 11/7 Five separate exhibits..will include such things as the 'Spacearium'. 1976 *Nature* 16 Sept. 180/1 At his disposal is a team of 250, a vast archive, a huge 'spacearium', an enormous theatre with a 55-foot high screen, scores of audio-visual aids and, of course, hundreds of artefacts—which means aircraft and spacecraft.

spacecraft (spē^i·skrɑft). Also **space-craft, space craft.** Pl. **spacecraft.** [SPACE *sb.*^{1}] Any vehicle designed to travel in space.

1930 *Sci. Amer.* Aug. 142/1 Valier was the principal proponent of working toward the space craft from the known forms of surface or air craft. 1932 D. LASSER *Conquest of Space* xvii. 279 Our experience with cosmic speeds and distances is not equal to the task of guiding a space-craft on its perilous journey. 1946 [see *ARTIFICIAL *a.* 1 c]. 1959 *Observer* 2 Aug. 11/5 In the next few days, 40,000 ft. above the Californian desert, the world's first manned spacecraft will switch on its rocket engines for the first time. 1960 *New Statesman* 13 Feb. 214/2 One..complains bitterly of the 'world-wide conspiracy in operation throughout this planet which has succeeded in suppressing *almost all* knowledge of, and discussion of the space-craft known to terrestrials as "Flying Saucers"'. 1960 *Guardian* 15 Dec. 9/5 An American attempt to put a 388 lb spacecraft into orbit around the moon. 1968 *Times* 12 Oct. 18/6 They suggest that the altitude measurement made by the Russian space craft as it approached the surface of Venus was almost exactly half what it should have been. 1976 A. DAVIS *Television: First Forty Years* 47 There are graphics departments..and props departments that can build models of, say, spacecraft.

spaced, *ppl. a.* Add: **2. a.** (Examples in *techn.* contexts.)

1920 *Wireless World* 12 June 200 (*heading*) Spaced loop aerial. 1949 G. A. BRIGGS *Sound Reproduction* iii. 35 Sound is picked up by 3 spaced microphones, recorded on separate channels, and played back through 3 spaced speakers. 1959 K. HENNEY *Radio Engin. Handbk.* (ed. 5) xxvi. 27 Analysis..indicates that, regardless of the value of the incident and polarization angles, the spaced-loop output voltage goes through a null when the directions of propagation are contained in the planes of the loop antennas. 1968 *Times* 29 Nov. (Sound of Leisure Suppl.) p. vi/5 The spaced microphone technique (using microphones a few inches or several feet apart, according to taste) relies both on volume and time of arrival differences. 1978 *Geophysical Research Lett.* V. 917/1 Observations of the troposphere with a VHF radar using a large antenna for transmission and small spaced antennas for reception.

b. With *out.*

1937 'M. INNES' *Hamlet, Revenge!* III. v. 295 The number of spaced-out acts committed by the criminal. 1952 C. DAY LEWIS tr. *Virgil's Aeneid* IX. 200 Where the ring of defenders was thinnest and daylight showed Between their spaced-out bodies. 1976 'J. ROSS' *I know what it's like to Die* xxxi. 203 The spaced-out drops of water from the tap.

c. Of children: born or conceived at intervals (of a kind denoted by qualifying advb.).

1939 [see *EXPECT *v.* 4 e (b)]. 1965 M. STEWART *Airs above Ground* i. 10 She acquired a wealthy London banker..the kind of man..safely ensconced in the Jaguar belt with three carefully spaced children away at carefully chosen schools. 1976 *Times* 25 Feb. 16/5 The advantages of a smaller or adequately spaced family.

4. In a state of drug-induced euphoria, 'high'; removed from actuality or disoriented, by narcotic stimulus. Also *transf.* Freq. with *out. slang* (orig. *U.S.*).

1968–70 *Current Slang* (Univ. S. Dakota) III–IV. 116 *Spaced out*, adj., mentally deficient; strange; absent-minded. 1969 *Negro Digest* Sept. 10 Spaced poems say that our ancestors are in the air and will communicate with us. 1969 *Time* 26 Sept. 41 The culture has its own in-group argot:..'straights' (everyone else), 'heat' (the police)..and being 'spaced out' (in a drug daze). 1971 J. MANDELKAU *Buttons* v. 68, I remember being really spaced out and someone handing me a ladybird—telling me how nice they tasted. 1974 A. LURIE *War between Tates* vi. 131 'You look exhausted.' 'I am sorta spaced out.' 1975 *New Yorker* 5 May 6/1 Vibraharpist Mike Mainieri and pianist Warren Bernhardt play some heavily lyrical and spaced-out duets. 1977 C. MCFADDEN *Serial* (1978) xx. 47/1 She just sort of stood there, feeling totally spaced, because the whole number was nothing short of *unreal*.

spacefaring (spē^i·sfɛ^{ə}riŋ), *vbl. sb.* and *ppl. a.* [f. SPACE *sb.*^{1}, after *seafaring*.] **A.** *vbl. sb.* Space travel. **B.** *ppl. a.* That engages in space travel.

1959 P. ANDERSON *Virgin Planet* (1969) iii. 20 Who had ever begun the idea that spacefaring was one long wild adventure? 1962 *New Scientist* 1 Mar. 489 The leaders of the two spacefaring countries. 1973 *Sci. News* 3 Nov. 283 My own view..was that they [sc. pulsars] were perfect interstellar navigation beacons, the sort of markers that an interstellar spacefaring society would want to place throughout the galaxy. 1978 D. R. MASON *Mission to Pactolus R* i. 11 A long session in a Fingalnan brothel would set them up for another stint of spacefaring.

Also **spa·cefarer,** one who travels in space.

1974 *New Scientist* 27 June 772 Such an asteroid-ark might one day encounter spacefarers of another civilisation, members of an intellectual brotherhood of the stars.

space group. *Cryst.* [f. SPACE *sb.*^{1} + GROUP *sb.*, tr. G. *raumgruppe* (A. Schoenflies 1891, in *Krystallsyst. und Krystallstruct.* II. iv. 359).] Any of the 230 sets of symmetry operations, derived from the point groups by the inclusion of translations, glide planes, and screw axes, which are used to classify crystal structures.

1901 *Rep. Brit. Assoc. Adv. Sci. 1901* 322 [Schoenflies] calls the groups of operations, whether those of Jordan or those added by himself, 'space-groups' (*Raumgruppen*). 1903 H. HILTON *Math. Crystallogr.* xvii. 159 Every space-group is isomorphous with one of the thirty-two point-groups consistent with the law of rational indices. 1934 W. P. DAVEY *Study of Crystal Struct.* viii. 240 The 230 space-groups are listed..giving..the Schoenflies code for the derivation of the space-group. 1974 *Nature* 11 Jan. 85/2 Dark-field observation could establish the presence of centres of symmetry, glide planes and screw axes, which could lead to the establishment of the space group.

spaceless, *a.* Add: Hence **spa·celessly** *adv.*; **spa·celessness,** the quality or condition of being unbounded by space, or of lacking space.

1895 G. MACDONALD *Lilith* xxxix. 286 We were not in the outer darkness; had we been, we could not have been *with* her; we should have been timelessly, spacelessly, absolutely apart. 1920 S. ALEXANDER *Space, Time, & Deity* I. 342 To possess spacelessness or timelessness (eternity). 1981 M. SPARK *Listening with Intent* v. 91 The spacelessness of this room where I lived..with..a bed for sitting and sleeping on.

spa·ce-like, *a. Physics.* [f. SPACE *sb.*^{1} + -LIKE.] Resembling or having the properties of space; *spec.* being or related to an interval between two points in space-time that lie outside one another's light cones (so that no signal or observer can pass from one to the other).

1914 [see *four-vector* s.v. *FOUR C. 2]. 1920 A. S. EDDINGTON *Space, Time & Gravit.* iii. 60 Even if the discovery of a new ray led us to modify the reckoning of time and space, it would still be necessary in the study of material systems to preserve the *present* absolute distinction of time-like and space-like intervals. 1934 A. HUXLEY *Beyond Mexique Bay* 215 It may be possible to think of vast durations as composed of space-like fragments. 1955 L. ROSENFELD in W. Pauli *Niels Bohr* 87 An arbitrary family of space-like surfaces. 1964 *Cambr. Rev.* 24 Oct. 51/2 The commutativity..of the field operators at spacelike distances. 1978 PASACHOFF & KUTNER *University Astron.* xxvii. 694 The distance between the events is greater than the distance that light can travel in the time between the events. Such an interval is called spacelike.

spaceman (spē^i·smæn). Also **space-man.** [SPACE *sb.*^{1}] **1.** (In Dict. s.v. SPACE *sb.*^{1} 17 d.) (Later example.)

1935 G. GREENE *England made Me* II. 80 He's News... I'm a space man... I can't afford to miss a thing.

2. One who travels in space or comes from another planet.

1942 *Thrilling Wonder Stories* Apr. 110/1 Maybe Lambert was a spaceman. Maybe he wasn't, but if he knew anything at all about spaceman's lingo he'd have to give now. 1944 F. BROWN in B. Aldiss *Introducing SF* (1964) 69 Earth's armada, all ten thousand ships and half-million fighting spacemen. 1956 'J. WYNDHAM' *Seeds of Time* 81 Whither thou goest, I will go... It was all right for a tribe of nomads, but nowadays the wives of soldiers, sailors, pilots, spacemen—. 1962 A. LURIE *Love & Friendship* xiii. 250 Amateur experts..Visitors from another world... They think they're space men. 1977 P. CARTER *Under Goliath* iv. 21, I was reading a good book all about spacemen.

Hence **spa·cemanship.**

1957 *Economist* 2 Mar. 400/1 Dr von Braun has always entertained, and shared with the public, the most ambitious ideals of spacemanship. 1966 *Life* 7 Jan. 32 (*caption*) Down from adventure and spacemanship, Stafford and Schirra pop out for an eye-level look at the sea.

spacer^1. Add: **1. a.** (Later examples.)

1916 *Chambers's Jrnl.* July 480/1 A new unit in roofing called a spacer allows of ordinary plain tiles or slates being laid wider apart than usual. 1943 *Sun* (Baltimore) 15 Apr. 26/3 She figured out a way to save countless man-hours in the manufacture of spacers—small rubber rings used in Rolls Royce airplane engines. 1954 *Automobile Engineer* XLIV. 508/3 A tubular spacer and two thrust washers round the shaft separate the roller bearings. 1962 *Which?* July 197/2 The second type [of life-jacket]..used *sealed-in air* between inner and outer coverings..which are held apart by 'spacers' of absorbent foam plastic. 1976 *Shooting Mag.* Dec. 12 (Advt.), It's a quality production gun, with one major difference. You can adjust the stock and insert various 'spacers' to achieve the perfect cast and drop.

b. *Archæol.* In full, *spacer plate.* A flat bead perforated with several holes in the same plane, by which the threads of a primitive multi-strand necklace are held apart.

1924 D. RANDALL-MACIVER *Villanovans & Early Etruscans* vi. 173 Fusiform and spherical gold beads, white and blue glass cylinders and small perforated spacers of amber formed the other elements of the necklace. 1940 V. G. CHILDE *Prehist. Communities of British Isles* vii. 124 In Ireland..the crescentic necklace was translated into a crescentic collar, cut out of a sheet of native gold and relieved with designs in panels, imitating the spacer-plates of the necklaces. 1958 *Antiquity* XXXII. 209 The complex-bored amber spacers from Hagenau. 1980 J. J. TAYLOR *Bronze Age Gold Work* iii. 41 The spacer-plate motif can be found in the Beaker repertoire pre-dating Wessex, to which the earliest dated spacer-plate is likely to belong.

3. *Biol.* A section of DNA which is not represented in the final RNA transcript, separating two sections which are.

1970 O. L. MILLER et al. in *Cold Spring Harbor Symp. Quantitative Biol.* XXXV. 505/2 Non-matrix segments of varying lengths (spacers) are intercalated between matrix units. 1974 *New Scientist* 9 May 329 Chemists..have succeeded in synthesising a nucleotide complementary to a definite part of a small bacteriophage.., a part regarded as a spacer. 1982 W. I. P. MAINWARING et al. *Nucleic Acid Biochem. & Molecular Biol.* vii. 258 There are spacer regions between genes which *may or may not* be transcribed. If transcribed, their contribution to the RNA transcript is removed by an excision process during the maturation of the RNA.

4. Special Comb.: **spacer gel,** a part of the gel used in electrophoresis (see quot. 1975).

1968 J. R. SARGENT *Methods in Zone Electrophoresis* vii. 90 It is possible to eliminate the sample gel and the spacer gels..and to apply the sample directly on top of the main gel. 1975 DAVIS & SIMPKINS in Williams & Wilson *Biologist's Guide to Princ. & Techniques Pract. Biochem.* iv. 115 The upper third of the gel consists of a large pore 'stacking' or 'spacer' gel, and the lower portion of smaller pore 'running' or 'separating' gel. The function of the stacking gel is to concentrate the sample as it moves through the gel, so that on entering the running gel it is an extremely narrow band.

spacer^2 (spē^i·səɹ). *Science Fiction.* [f. SPACE *sb.*^{1} + -ER^1.] **1.** = *SPACEMAN 2.

1955 C. M. KORNBLUTH *Mindworm* 59 'I'm a spacer,' he said... 'Venus.' 1958 I. ASIMOV *Naked Sun* i. 11 The Galaxy was closed to Earthmen. It was pre-empted by the Spacers, whose ancestors had been Earthmen centuries before.

2. A spaceship, spacecraft.

1962 *New Worlds Sci. Fiction* Sept. 50 The spacer broke through the low hanging layer of clouds above. 1978 D. R. MASON *Mission to Pactolus R* i. 5 A Fingalnan voice spoke into the quiet command cabin of the hurrying spacer.

spaceship (spē^i·sʃip). Also **space ship, space-ship.** [SPACE *sb.*^{1}] **1.** A spacecraft; esp. a manned one under the control of its crew.

1894 J. J. ASTOR *Journey in Other Worlds* I. vi. 93 'What sort of space-ship do you propose to have?' asked the vice-president. 1929 *Sci. Wonder Q.* I. 65/2 Since the space ship..was circling freely in space..the proximity of the moon was no hindrance to leaving the ship. 1936 *Jrnl. Brit. Interplanetary Soc.* III. 26 It is essential that the space-ship should be in constant communication with its base station. 1938 C. S. LEWIS *Out of Silent Planet* iv. 38 All he ever remembered of his first meal in the space-ship was the tyranny of heat and light. 1948 'N. SHUTE' *No Highway* ix. 226 Designs for a rocket-propelled Space Ship, I think he called it, for a projected journey to the moon. 1951 A. C. CLARKE *Sands of Mars* i. 9 Sleek, streamlined spaceships..had been the dream of the early twentieth century. 1962 F. I. ORDWAY et al. *Basic Astronautics* i. 2 The spaceship is stimulating the study of space navigation. 1970 N. ARMSTRONG et al. *First on Moon* xiv. 338 If you're going to run a spaceship you've got to be pretty cautious about how you use your resources. 1977 P. CARTER *Under Goliath* vi. 29 It's from Mars... It's a spaceship from Mars.

2. *spaceship earth* (also with capital initials), a phr. used to draw attention to the finite nature of the earth's resources.

1966 B. WARD (*title*) Space ship earth. 1969 R. B. FULLER *Operating Man. Spaceship Earth* iv. 52 We have not been seeing our Spaceship Earth as an integrally-designed machine which to be persistently successful must be comprehended and serviced in total. 1976 *Conservation News* Sept./Oct. 12/1 There is a growing awareness of the need to conserve the life-support systems of spaceship earth. 1981 *Nature* 5 Nov. 41/3 All species on 'Spaceship Earth' have an equal right to exist.

space-time (spē^i·s,tai·m), *adj. phr.* and *sb.* Also **spacetime, space time.** [SPACE *sb.*^{1}] **A.** *adj. phr.* Pertaining to or situated in both space and time.

1905 R. B. PERRY *Approach to Philos.* 131 An experimentally verifiable system must contain space-time variables. 1938 *Harper's Mag.* Jan. 151/2 If you denounce him as a 'Jew', apart from his space-time characteristics, you perform a monstrous act. 1965 *Math. in Biol. & Med.* (Med. Res. Council) v. 227 For many years leukaemia has been alleged to occur in clusters, though the evidence has been largely anecdotal and its interpretation intuitive because of the difficulties..of formulating the hypothesis of space-time clustering in precise and testable terms. 1973 *Nature* 27 July 214/2 The projection of the most recent seismic migration wave to the latitude of maximum recurrence of earthquakes provides the basis for setting the space-time bounds on the next earthquake greater than magnitude 5.

B. *sb. Physics.* [tr. G. *raumzeit.*] Time and three-dimensional space regarded as fused in a four-dimensional continuum containing all events. Also *attrib.*

1915 E. CUNNINGHAM *Relativity* 9 (*heading*) The space-time transformations. *Ibid.* 10 Newtonian Relativity consists in the fact that either (x, y, z, t) or (x', y', z', t') are equally valid as space-time coordinates. 1927 S. ERTZ *Now East, now West* viii. 117 I've got quite drunk on theories about the space-time continuum. 1930 *Morning Post* 17 June 13/1 The 'metrical structure' of space-time now adopted by Einstein is based on the assumption that there is a meaning in saying that two short lines are equal in length, that they are parallel, and that the angle between them is not altered when they are moved so as to remain parallel to themselves. 1934 E. MUIR *Coll. Poems* (1960) 44 Lengthening league by league The ghastly thin anatomy of Space Time Stripped to the nerve. 1940 W. EMPSON *Gathering Storm* 18 But if it parts Into uncommunicable spacetimes. 1956 *Nature* 10 Mar. 458/1 He [sc. H. Weyl] discovered the first 'unified field theory' in which the Maxwell field appears along with the gravitational field as a geometrical property of space-time. 1967 G. STEINER *Lang. Silence* 283 We are imprisoned in the unexamined assumption or unconscious illusion of a homogeneous, forward-flowing space-time continuum. 1978 PASACHOFF & KUTNER *University Astron.* xxvii. 695 The basic conclusion of the general theory [of relativity] is that gravitational fields change the geometry of spacetime. We say that in the presence of gravitational fields, spacetime becomes curved.

spacewoman (spē^i·swumăn). [f. SPACE *sb.*^{1}, after *SPACEMAN.] A female traveller in space; a woman who comes from another planet.

1962 N. MITCHISON (*title*) Memoirs of a spacewoman. 1963 *Daily Tel.* 18 Oct. 23/3 (*caption*) Russia's first space woman, Valentina Tereshkova, 26, waving at Prestwick yesterday when her plane made a two-hour refuelling stop. 1978 G. HOUSEHOLD *Last Two Weeks of Georges Rivac* i. 14 It doesn't matter which way up I'm put. I ought to be a spacewoman.

spaceworthy (spē^i·swɔ̄ɹði), *a.* [f. SPACE *sb.*^{1} + WORTHY *a.*] In a fit condition for space travel. Also **spa·ceworthiness.**

1959 'WYNDHAM' & PARKES *Outward Urge* iv. 156 She reclaimed the damaged Satellites, and made three of them spaceworthy again. 1959 J. BLISH *Clash of Cymbals* ii. 39 The unending demands of the city's spaceworthiness. 1960 *New Statesman* 30 Jan. 146/2 Apparently no other proofs were needed to demonstrate the efficacy of their measures to ensure the space-worthiness of the cosmic vehicle. 1968 *Daily Tel.* 16 Dec. 1/8 Russia's unmanned Zond 5 and 6 made successful 'boomerang' flights round the moon, and Soyuz 3 has been proved space-worthy for four days.

spacey, var. SPACY *a.* in Dict. and Suppl.

spacing, *vbl. sb.* Add: **1. b.** *attrib.* Also, *spacing collar, washer.*

1916 *Automobile Engineer* VI. 92/2 (*caption*) Adjustable spacing collar for milling cutters. 1923 *Popular Wireless* 27 Oct. 294/1 (Advt.), Spacing washers, large ... per doz. 2½d. 1971 *Buck & Hickman Tool Buyer's Guide* 96 Spacing Washers also supplied 1⅛ in. bore × 2¼ in. dia. *Ibid.*, Spacing collars and washers for milling machine arbors.

spacistor (spē^i·sistəɹ). *Electronics.* [f. SPACE *sb.*^{1} + *TRANS)ISTOR.] A kind of semiconductor device (see quot. 1957).

1957 STATZ & PUCEL in *Proc. IRE* XLV. 317/1 New devices are considered in which electrons or holes are injected directly into space-charge regions of reverse-biased junctions avoiding the diffusion of carriers through field-free regions. The case considered is one in which the junction is biased at a voltage such that the injected carriers are multiplied by the avalanche process. A device of this type shall be called a spacistor. 1962 SIMPSON & RICHARDS *Physical Princ. Junction Transistors* viii. 175 The spacistor can have input and output impedances of the order of tens of megohms and the coupling between input and output terminals is almost purely capacitative. 1969 A. MARCUS *Electronics for Technicians* viii. 160 There is not, as yet, general agreement upon a symbol for the spacistor transistor.

Spackle (spæ·k'l). *N. Amer.* Also **spackle.** [Cf. SPARKLE *v.*^{2} 4 b and G. *spachtel* putty knife, mastic, filler.] A proprietary name for a compound used to fill cracks in plaster and produce a smooth surface before decoration.

1928 *Official Gaz.* (U.S. Patent Office) 7 Feb. 17/1 *Spackle*... A surfacing compound for filling imperfection so as to bring up to a smooth and level surface areas that are to be painted or decorated. Claims use since Aug. 1, 1927. 1951 *Home Painting, Wallpapering & Decorating* ii. 34 You must be sure there is paint on the surface upon which you are going to put the spackle. 1971 *Black World* Mar. 65/2 After he did that he mixed the spackle with water and spread it as evenly as he could over the crack. 1975 *Amer. Speech 1969* XLIV. 26 *Spackle, n.*, a prepared paste that, when mixed with water, is used to repair cracks and holes in plaster or gypsum wallboard.

Hence as *v. trans.*, to repair or fill with Spackle (see also quot. 1940); **spa·ckling** *vbl. sb.*

1940 R. MAYER *Artist's Hand-bk. Materials & Techniques* xii. 506 *Spackling* or *Sparkling* (probably from the German *spachteln*, to putty up). The rectifying of a defect in plaster or a mural painting by digging out the defective spot and filling it in with a plastic gesso, plaster of Paris, Keene's cement, or other similar material. 1941 *Pop. Science* July 149 In addition to patching plaster, other prepared powders or so-called 'spachtling [*sic*] compounds' are sold for patching, filling, and smoothing purposes. 1971 H. SMITH *View from Chivo* vii. 69 Then he coated each wire with a spackling compound. 1971 *Black World* Mar. 67/1 People who could watch for him if Death came suddenly sweeping through the crack in the kitchen wall that he hadn't spackled properly. 1980 *Redbook* Oct. 58/3 She cooked better than I and it was a fact that I Spackled the walls better than she.

spacy, *a.* Add: Also **spacey.** **2.** In a heightened state of consciousness; disoriented from reality, = *SPACED *ppl. a.* 4. *slang* (chiefly *U.S.*).

1968–70 *Current Slang* (Univ. S. Dakota) III–IV. 117 *Spacey*, adj., appearing to be functioning on a level higher than normal. 1972 *Last Whole Earth Catalog* (Portola Inst.) 67/3 Her expression was very spacy... Her eyes were so weird it was hard to read anything very definite in her face... She..sat on the curb in the lotus posture and stared off into the east. 1975 *Globe & Mail* (Toronto) 12 July 32/6 'You get pretty spacey after doing a lot of interviews,' she said. 'One time someone asked me if Dylan had changed a lot and I was so spacey I said, "Well, he doesn't bum drinks any more."' 1977 C. MCFADDEN *Serial* (1978) ii. 10/2 She spotted Martha's ex-husband-once-removed with his spacy new old lady. 1980 J. A. CARVER *Panglor* (1981) viii. 138 His head felt large, and a little spacey, and he felt a heightened sense of geometry, of perspective.

3. Of, pertaining to, or purportedly characteristic of (conditions in) outer space; esp. used in connection with electronic music.

1971 *It* 2–16 June 19/2 A good studio-produced album sounding very live and certainly spacey. 1977 *Rolling Stone* 13 Jan. 62/5 Terje Rypdal,..spun spacey riffs clothed in lots of distortion. 1980 *Radio Times* 27 Sept.–3 Oct. 89/3 For 18 years Ron Grainger's tune gave the programme [*sc.* 'Dr. Who'] an eerie, spacey feel.

spad^1 (spæd). orig. *U.S.* [Var. of SPUD *sb.*] A spike with a flattened end containing a hole to support a plumb-line, so that it can be driven into a tunnel roof to provide a marker.

1908 L. W. TRUMBULL *Man. Underground Surveying* iii. 93 A majority of the mines of the United States are using punched horseshoe nails known as 'spads'. The head of the horseshoe nail..is hammered flat and a hole punched through it. 1957 R. C. A. HOOPER *Winiberg's Metalliferous Mine Surveying* (ed. 4) ix. 186 Some surveyors prefer to have the width of the spad along the longitudinal axis of a drive rather than at right angles. 1973 B. A. BARRY *Construction Measurements* xvi. 220 Line is carried forward in tunneling by tape and theodolite to marks placed either on the floor or side or roof of the tunnel. Driving plugs and spads into the roof to serve as horizontal control (traverse) points keeps them free from traffic damage or disturbance.

Spad^2 (spæd). [f. the initials of *Société pour Aviation et ses Dérivés*, the designers.] Any of several types of French aircraft, esp. a biplane fighter much used in the war of 1914–18. Also *attrib.*

1917 *Sci. Amer.* 6 Oct. 249/1 Still faster is the 'Spad', a tiny biplane. 1918 J. M. GRIDER *War Birds* (1927) 60 They have Pups and Spads and Avros. *Ibid.* 88 There were three Spads so Capt. Foggin asked for Spad pilots. 1920 A. J. L. SCOTT *Hist. Sixty Squadron, R.A.F.* ii. 15 Foot..was given a 'Spad', on which he did great execution during the autumn. 1926 J. L. PRITCHARD *Bk. Aeroplane* xii. 230 Spad machines were recognised as among the fastest..used by the Allies during the war. 1929 HALL & NILES *One Man's War* 170 In very high, thin air the Nieuports manœuvred very well, better perhaps than the Spads did, because the Spad was a heavier machine. 1970 *Word Watching* Apr. 7/1 *Spad*, A-1 fighter-bomber escort for rescue helicopters. 1973 D. LEES *Rape of Quiet Town* vi. 101 Vague memories of anti-aircraft drill from the days of Spads and Fokkers.

spadaite (spā·dă,ait). *Min.* [ad. G. *spadait* (F. von Kobell 1843, in *Jrnl. für prakt. Chem.* XXX. 467), f. the name of Lavinio *Spada* de Medici (1801–63), Italian writer, politician, and mineralogist: see -ITE^1.] A hydrated magnesium silicate, $MgSiO_3.2H_2O$, found as an alteration product in lavas, and occurring as amorphous masses coloured red by impurities.

1848 J. D. DANA *Man. Mineral.* 149 Spadaite. A flesh-red mineral, near Schiller spar. 1931 *Amer. Mineralogist* XVI. 232 Spadaite occurs only in the ore shoots, associated with the silicates mentioned, quartz, various sulphides (chiefly chalcopyrite and bornite) and their oxidation products, and small quantities of native gold. 1968 I. KOSTOV *Mineral.* 334 Spadaite, $MgSiO_3.2H_2O$,..corresponds to hydrated enstatite.

spaddle. For *Obs.*$^{-1}$ read *Obs. rare* and add: **a.** (Later example.) **b.** = SPATTLE *sb.*^{2} 1.

1835 H. C. TODD *Notes upon Canada & U.S.A.* 8 Jonathan..[uses] *spaddle*, for spade. 1861 MRS. BEETON *Bk. Housek. Managem.* 760 The principal utensils required for making ice-creams are ice-tubs, freezing-pots, spaddles, and a cellaret. The use of the spaddle is to stir up..the cream.

spade, *sb.*^{1} Add: **2. c.** More forcefully, in colloq. phr. *to call a spade a* (*bloody*) *shovel*: to speak with great or unnecessary bluntness.

1919 W. S. MAUGHAM *Moon & Sixpence* iii. 12 We did not think it hypocritical to draw over our vagaries the curtain of a decent silence. The spade was not invariably called a bloody shovel. 1945 N. BALCHIN *Mine Own Executioner* ii. 34 Sometimes..I get so fed up with all the mumbo-jumbo and abracadabra and making of holy mysteries about simple things that I like to call a spade a shovel. 1978 CADOGAN & CRAIG *Women & Children First* ii. 48 As a literary starting-off point, the determination to call a spade a bloody shovel has imposed a fundamental limitation... Outspokenness..is simply not enough.

4. a. *spade-type* adj.

1960 *Farmer & Stockbreeder* 29 Mar. 57/2 Most impressive do I find these new spade-type rotary cultivators which have been imported from the Continent. 1967 *Jane's Surface Skimmer Systems 1967–68* 31/1 The steering gear comprises twin balanced spade-type rudders.

b. *spade-armed, -footed* (earlier example), *-handed, -proud.*

1782 J. TRUMBULL *M'Fingal* III. 61 Till looking back he spied in rear The spade-arm'd chief. 1962 E. SNOW *Red China Today* (1963) lxvii. 508 Battalions, whole divisions, of spade-armed peasants in this general area have been working on a plan. 1867 *Amer. Naturalist* I. 108 The Spade-footed Toads..are more uncertain in their appearance, being governed entirely by the dampness or dryness of the season. 1934 DYLAN THOMAS *18 Poems* 25 When blood, spade-handed, and the logic time Drive children up like bruises to the thumb. 1941 L. B. LYON *Tomorrow is Revealing* 24 It hurt me, the efficient, spade-proud hole, That earth-room with its tapestry of boughs.

5. spade lug *Agric.*, each of a number of metal lugs that are bolted to the rim of a tractor wheel so as to project radially outwards and give an improved grip; hence **spade-lugged** *a.*; **spade terminal** *Electr.*, a flat, spade-shaped piece of metal having a slot or hole in it for fixing under a nut or bolt to make an electrical connection; **spade-work,** (*a*) work done with a spade for the preparation of ground; (*b*) *fig.*, preliminary work, difficult or laborious preparation, pioneering research; hence (*rarely*) **spade-worker.**

1921 *Trans. Amer. Soc. Agric. Engin.* XV. 175 The tri-spade lugs were cast with three spades on each casting.. and were staggered on the wheel when in place similar to the bolt on spade lugs. 1950 *Engineering* 5 May 506/1 The use of spade lugs in place of plain steel rims and strakes gradually reduced the tractor weight per drawbar-horse-power. 1967 J. OATES *Farm Machinery* xi. 77 Spade lugs and strake bars are used to bite more deeply into the ground. 1945 H. J. HINE *Tractors on Farm* (ed. 2) iv. 38 With spade-lugged steel wheels the spaces between the lugs must be cleaned out from time to time with a paddle. 1968 *Wireless World* Feb. 133/1 (Advt.), Heavy duty terminals... Black only will take spade terminals and wander plug. 1976 *Gramophone* May 1841/1 A twin phono-plug low capacitance signal cable is supplied, plus a green spade terminal earth wire. 1778, etc. Spade-work [in Dict., sense 4 a]. 1912 H. G. ALDIS in *Cambr. Hist. Eng. Lit.* IX. xiii. 346 Brian Twyne, a diligent Oxford antiquary who had done much pioneer spade-work in the same field. 1927 *Daily Tel.* 12 July 9/1 The discovery of a helpful blood-test for cancer may be placed among the important advances the near future may give us as a result of spade-work already carried out. 1931 E. F. BENSON *Mapp & Lucia* iv. 93 Lucia..had insisted that all the credit was due to Drake's wife, who had planned everything (or nearly) and had done all the spade-work. 1951 *Sport* 27 Apr.–3 May 8/3 Ernie's crafty spadework has been responsible for many of the goals netted by Jack Milburn and George Robledo. 1977 A. CLARKE *Letter from Dead* ix. 105, I did a bit of spadework on him yesterday..and he seemed to be thawing a little. 1912 'SAKI' *Chron. Clovis* 267 'Where I think you political spade-workers are so silly,' said the Duke, 'is in the misdirection of your efforts.'

spade, *sb.*^{2} Add: **1. b.** *attrib.* in *sing.*

1904 'O. HENRY' *Cabbages & Kings* vi. 105 The invitations to the musicale came sliding in by pairs and threes and spade flushes. 1973 *Country Life* 10 May 1331/1 West led the Spade Knave, which I took with dummy's Ace.

c. *fig.* in advb. phr. *in spades*, very much, in abundance, extremely. (Spades is the highest ranking suit in Bridge.) *colloq.* (chiefly *U.S.*).

1929 D. RUNYON in *Hearst's International* Oct. 62/2, I always hear the same thing about every bum on Broadway, male and female, including some I know are bums, in spades, right from taw. 1964 WODEHOUSE *Frozen Assets* i. 19 'It's the law I'm beefing about. You didn't make the law.' 'But I administer it.' 'I'll say you do. In spades.' 1972 R. NIXON *Diary* in *Mem.* (1978) 619 Anybody who gets to the top in the Communist hierarchy and stays at the top has to have a great deal of political ability and a great deal of toughness. All three of the Soviet leaders have this in spades.

3. a. A Black person, a Negro, esp. male: freq. in White use, as a term of contempt or casual reference. Formerly among U.S. Blacks, a very dark-skinned Negro. *slang* (orig. *U.S.*).

1928 C. MCKAY *Home to Harlem* vi. 56 Jake is such a fool spade. Don't know how to handle the womens. 1929 W. THURMAN *Blacker the Berry* I. 34 Wonder where all the spades keep themselves? 1945 L. SHELLY *Jive Talk Dict.* 17 *Spade*, colored person. 1957 C. MACINNES *City of Spades* II. ii. 118 A British lady with a wild love of Spades, and a horrid habit of touching you on the shoulder because she says 'to stroke a darkie brings you luck.' 1969 [see *DINGE *sb.*^{2}]. 1971 N. SAUNDERS *Alternative London* xxviii. 263 On Saturdays try Brixton market—nearly as big, more genuine, lots of spades. 1978 J. A. MICHENER *Chesapeake* 678 The four Turlocks hated Negroes and never hesitated in voicing their disgust. 'Goddamned spades killed my cousin Captain Matt—one of them gets out of line with me, he's dead.'

b. *attrib.* or as *adj.*

1928 C. MCKAY *Home to Harlem* vi. 56 She was of the complexion known among Negroes as spade or chocolate-to-the-bone. 1952 C. BROSSARD *Who walk in Darkness* x. 61 These spade intellectuals really think they've made it when they get a white girl. 1964 *Negro Digest* Feb. 55/1, I can't see why no colored man'd want to marry no white chick... Not when there's so many fine spade chicks around. 1978 M. PUZO *Fools Die* liv. 568 Two spade hookers went gliding by arm in arm.

Hence **spa·delet** *nonce-wd.* [-LET], a Black child.

1959 C. MACINNES *Absolute Beginners* 62, I passed a crocodile of infants, and among them a number of little Spadelets.

spade, *v.*^{1} **2. b.** (Earlier example.)

1785 G. WASHINGTON *Diary* 5 Sept. (1925) II. 410 Began to spade up the Lawn in front of the Court yard.

spadger. Add: **2.** *transf.* A boy. *colloq. rare.*

1899 *Captain* II. 273/2 If we've got to take these three young spadgers..we shall want something bigger'n this here gig. 1978 K. BONFIGLIOLI *All Tea in China* I. iii. 29 'See here, young spadger,' he said..'if you should be a little short of tin..come and spend a night or two at Great Coram Street.'

Hence as *v. intr.* (*rare*), (*a*) to catch sparrows or other small birds, to go sparrow-netting; (*b*) *fig.*, to play or frolic *about* in the manner of a sparrow.

1939 F. THOMPSON *Lark Rise* ix. 171 In winter in the 'eighties the youths and big boys of the hamlet would go out on dark nights 'spadgering'. For this a large net upon four poles was carried... When they came to a spot where a flock of sparrows or other small birds was roosting, the net was dropped over the hedge..and the birds enclosed were slaughtered. 1967 *Listener* 7 Sept. 315/2 Tommy Steele spadgered larkily about among the zoomorphs.

spadgers, var. *SPAGGERS.

spadona (spădō^u·nă). *S. Afr.* [ad. It. *spadone* large sword.] An imperfectly developed feather taken from a young ostrich in its first year.

1881 in A. Douglass *Ostrich Farming in S. Afr.* xiv. 91 Light Spadona..white Boos. 1896 R. WALLACE *Farming Industries of Cape Colony* xi. 235 Spadonas..are pointed like a sword, hence the name, which comes from the Italian. 1955 G. ARCHMAN in Saron & Hotz *Jews in S. Afr.* vii. 130 The different types of feathers that the ostrich produces—chicks, wings, bodies, tails, spadonas.. bloods, and female bodies.

Spaetlese, var. *SPÄTLESE.

spag (spæg). Slang abbrev. of *SPAGHETTI 1 a. Also **spag bol** (bɒl), spaghetti Bolognese.

1948 PARTRIDGE *Dict. Forces' Slang* 176 *Spag*, spaghetti. (Italian front.) 1969 *Southerly* XXIX. 308 'I'll shout you a plate of steak and spag,' I said. 'It's only a buck.' 1970 D. CLARK *Deadly Pattern* iv. 86 To eat oxtail and spag bol properly, you've got to be stripped to the waist.

spaggers (spæ·gəɹz). Also **spadgers.** Slang abbrev. of *SPAGHETTI 1 a. Cf. *-ER^6 and prec.

1960 I. JEFFERIES *Dignity & Purity* v. 96 Thinking of the spaggers I would have rustled up, left to myself. 1980 I. MURDOCH *Nuns & Soldiers* ii. 127 'You said you were tired of spaghetti and potatoes and —' 'Spuds and spadgers fill you up at least.'

spaghetti (spăge·ti). [It., pl. of *spaghetto* thin string, twine.] **1. a.** A variety of pasta made in strings intermediate in thickness between macaroni and vermicelli. Occas., a dish of spaghetti.

1849 E. ACTON *Mod. Cookery* (ed. 8) p. xxxii, *Sparghetti* [sic]—Naples vermicelli. 1888 MRS. BEETON *Bk. Househ. Managem.* § 2952 Maccheroni, or Spaghetti, a smaller kind of macaroni,..generally follows the soup. 1921 F. SWINNERTON *Coquette* II. xvi. 175 The waitress approached, bearing two large plates piled high with spaghetti. 1931 B. STARKE *Touch & Go* iv. 51 A..schoolteacher took us to lunch in Avon and showed us how to manage yards of spaghetti by rolling it up on our forks. 1949 J. MORE *Land of Italy* 4 A capacity to eat..*pasta* is an essential requirement... The principal varieties.. are the rope-like spaghetti of Naples, the pipe-like macaroni,..and the omelet-like cannelloni. 1956 N. DE LA FÈRE *Italian Bouquet* vi. 66 He slipped the menu under my eyes... I..refused to admit that the only recognisable word was 'spaghetti'. 1965 C. K. STEAD in *N.Z. Short Stories* (1966) 337 Julian came into Gomeo's and asked for a spaghetti. 1981 M. NABB *Death of Englishman* II. iii. 97 Old men..munching slowly at their *spaghetti* with toothless gums.

b. *fig.* and *transf.*

1935 A. HUXLEY *Let.* 17 Feb. (1969) 391 Orlo Williams ..has read every inch of spaghetti that has ever emerged from the Italian presses. 1940 O. NASH *Face is Familiar* 195 And they give you a look that implies that your spine is spaghetti and your soul is lard. 1946 F. HAMANN *Air Words* 50 *Spaghetti*, (1) electric wiring; (2) strings of sealing compound. 1960 COOKE & MARKUS *Electronics & Nucleonics Dict.* 446/2 *Spaghetti*, insulating tubing used over bare wires or as a sleeve for holding two or more insulated wires together. The tubing is usually made of varnished cloth or a plastic. 1973 C. WILLIAMS *Man on Leash* (1974) viii. 119 He was always..experimenting and lashing up nutty pieces of electronic spaghetti. 1981 T. BARLING *Bikini Red North* xiii. 276 His vasectomy could be reversed. 'New techniques... We can join up all that miniature spaghetti with incredible accuracy.'

2. An Italian: usu. contemptuous. *slang.*

1931 'D. STIFF' *Milk & Honey Route* iii. 38 Italian hobos are equally rare. They are the 'wops' or 'spaghettis'. 1977 *New Society* 29 Sept. 651/1 When the war criminal, Herbert Kappler, who served a life sentence for having commanded an execution squad that massacred 335 Italian civilians, was spirited out of a military hospital in Rome by his wife, this was generally welcomed [in Germany] with an almost racialist feeling of superiority vis-a-vis those doting spaghettis.

3. Complex roadways forming a multi-level junction, esp. on a motorway. *colloq.*

1963 *Lebende Sprachen* VIII. 166/1 Driver talk..Plenty of spaghetti. 1966 *Guardian* 4 June 14/2 Details of one of the biggest pieces of motorway spaghetti so far designed in Britain were published... It is the Gravelly Hill interchange on the M6. 1972 *Daily Tel.* 19 June 19/3 The transport profession has produced its own jargon... Multi-level motorway-junctions are 'plates of spaghetti'. 1976 *Times Lit. Suppl.* 20 Feb. 184/5 The best-known examples [of three-dimensional mazes] are motorway spaghetti, aesthetically unpleasing and disturbing to drive through.

4. *attrib.* and *Comb.* *spaghetti house, joint,*

sauce, tongs; spaghetti-like adj.; **spaghetti Bolognese,** spaghetti served with a sauce of which the principal ingredients are beef and tomato; **spaghetti bowl,** a network of pipelines constructed to carry materials between petrochemical companies on the Gulf Coast of the U.S.; also *transf.*; **spaghetti junction** *colloq.*, a complex junction of roads at different levels; applied *spec.* to a major interchange on the M6 near Birmingham; also *fig.*; **spaghetti (shoulder) strap,** a thin cord-like shoulder strap for a dress or the like; **spaghetti tubing** *colloq.*, tubular insulation for electrical wire; **spaghetti Western,** a 'Western' (*WESTERN *sb.* 4) or film set in the U.S. 'old west', but made in Italy or by Italians, esp. cheaply.

[1947 L. P. de Gouy *Gold Cookery Bk.* 760 (*heading*) Spaghetti alla Bolognese.] 1950 E. David *Bk. Mediterranean Food* 93 If you are serving the classic spaghetti *Bolognese*..see that it is highly flavoured. 1973 'M. Underwood' *Reward for Defector* vii. 56 [He] reached out for everything he needed to make himself a plateful of spaghetti bolognese. 1958 *Houston* Apr. 16/1 The mighty petrochemical industry today is feeding from an overflowing 'Spaghetti Bowl'. *Ibid.* 16/3 Industrial Sales Engineer G. R. Walton of the Houston Pipe Line Company (the man who coined the term 'Spaghetti Bowl' about two years ago). 1970 *Chem. & Engin. News* 26 Oct. 30 Mention 'the spaghetti bowl' and most West European petrochemical producers would probably think of the pipeline system in the Gulf Coast of the U.S. rather than the maze being developed in their own back yard. 1962 I. Fleming *Spy who loved Me* I. ii. 34 Derek took me right across London to a spaghetti house called 'The Bamboo'. 1900 Ade *Fables in Slang* 158 He knew his Works were good, because all the Free and Untrammeled Souls in the Spaghetti Joint told him so. 1982 H. Engel *Murder on Location* (1983) ix. 87, I could see a spaghetti joint across the street. 1971 *Evening News* (Worcester) 15 Nov. 7/4 Worcester will have its own 'spaghetti junction' if the big multi-level interchange is ever constructed in the Arboretum. 1978 *Listener* 5 Jan. 25/4 We pass abruptly from the *proprium* to the *ordinarium*—from the winding country road into a great spaghetti junction of criss-cross melodies. 1980 S. Brett *Dead Side of Mike* xi. 125 He got held up..under the spaghetti junction between the M23 and M25 because of road works. 1979 K. Bonfiglioli *After You with Pistol* vi. 25 We parted in a spaghetti-like tangle of insincere matiness. 1953 A. Boni *Talisman Italian Cook Bk.* xiii. 222 (*heading*) Spaghetti sauce home style. 1968 C. Drummond *Death & Leaping Ladies* v. 111 He was neatly avoiding spilling spaghetti sauce over a very snappy jacket. 1980 B. Freemantle *Charlie Muffin's Uncle Sam* vii. 73 The man hadn't changed his shirt... There was spaghetti sauce on the collar. 1972 *New Yorker* 30 Sept. 81/1 Black cocktail dresses with sparkling spaghetti straps. 1977 *Lancashire Life* Dec. 104/3 The demure look from Elizabeth Hayes consists of a lace-trimmed polyester satin bedjacket which, when taken off, reveals a sexy nightdress with spaghetti shoulder straps. 1980 *Times* 12 Feb. 7/5 A natty little camisole top with spaghetti shoulder straps topped with an amazing sort of opened up tube of knitting which seals your arms and then blossoms into a shrug for the back and shoulders. Do not ask me how it is done. 1972 *House & Garden* Feb. 78/1 Spaghetti tongs, 85p. 1977 'E. McBain' *Long Time no See* iii. 32 Knives and forks piled haphazardly..paper napkins, spaghetti tongs, a corkscrew. 1922 *Science & Invention Mag.* May (Advt., rear cover), Spaghetti Tubing, black or yellow in 2½-ft. lengths..18 cents per length. 1969 M. Pei *Words in Sheep's Clothing* (1970) iii. 22 'Spaghetti Western' and 'Sukiyaki Western' are terms applied to cheap Westerns produced in Italy and Japan. 1973 J. Susann *Once is not Enough* i. 34 It started with the flop of Melba's picture... When your kid is busted into pieces, you can't worry about a spaghetti western. 1977 G. Marton *Alarum* 56, I wanted to see a Spaghetti Western movie.

‖ **spaghettini** (spagetiˑni). [It.] A thin variety of spaghetti.

1953 A. Boni *Talisman Italian Cook Bk.* ix. 156 Bring water and salt to boil, add spaghettini and cook 12 minutes. 1974 *Encycl. Brit. Micropædia* VII. 789/3 Among the popular cord forms [of pasta] are spaghetti ('little string'), a finer type called spaghettini, and the very fine vermicelli. 1977 D. E. Westlake *Nobody's Perfect* (1978) II. v. 120 You can buy me spaghettini with clam sauce.

spake (spēˡk), *sb. S. Wales.* [Of unknown origin: perh. var. Spoke *sb.*] A string of rail-mounted wagons or trolleys used in coal mines to transport men.

1935 *Trans. Inst. Mining Engineers* LXXXVIII. 384 In mines where the men are brought out in spakes, provision should be made for their protection from the cold intake air. *Ibid.* XC. 140 A spake is composed of a number of flat trolleys fitted with wooden seats placed at intervals at right angles to the length of the trolley, and consequently at right angles to the slant. 1971 *Guardian* 6 July 1/5 A colliery accident in which an underground train ran out of control... It was remarkable the train—known as a spake—did not leave the rails. 1979 *Times* 29 Dec. 12/5 The seams lay, in places, three miles below ground. To reach them by 'spake', the man-carrying wagons, entailed a journey of half an hour.

spall, *v.*[1] Add: **1. b.** (Later example.)

a 1925 F. S. Anthony *Follow Call* (1936) ii. 22, I landed..on a patch of broken stone I had spalled up..for metalling in front of the stand.

2. a. (Further examples.) Also, to detach as small fragments or particles.

1971 *Sci. Amer.* June 29/2 As the shock wave traversed a gas bubble some of its energy would go into spalling liquid from the inner surface of the bubble and projecting it through the void to strike the bubble wall at the other side. 1973 J. G. Tweeddale *Materials Technol.* II. vi. 134 Thermal fracturing is a somewhat crude way in which material may be 'spalled' off the surface of a brittle material which has low thermal-shock resistance. 1980 M. Napier *Blind Chance* xii. 109 She looked at the huge chip of stone spalled off by a bullet.

b. *Nuclear Physics.* To cause spallation of (a nucleus).

1976 *Nature* 16 Sept. 201/1 Stronger shock waves..spall nearly all the nuclei to free nucleons.

3. (Later examples.) Also without *off.*

1940 K. Rexroth *In what Hour* 33 Novelty emerges after centuries, a rock spalls from the cliff. 1968 *Engineering* 26 July 171/3 These alloys would spall after enamelling. 1977 *Sci. Amer.* Feb. 35/1 High-energy impacts cause large pieces of the target to spall off. 1980 *National Trust* Autumn 14/2 Damp has penetrated the stone..and the corners have broken off or spalled.

spallation (spǭlēˡ·ʃən). [f. Spall *v.*[1] + -ation.] **1.** *Nuclear Physics.* The detachment of a number of nucleons or small nuclei from a larger nucleus, esp. as a result of the impact of an energetic nucleon or other particle. Freq. *attrib.*, as *spallation product.*

1948 Lapp & Andrews *Nuclear Radiation Physics* xiii. 305 When a high-energy neutron strikes a target nucleus, a reaction known as spallation may occur. 1950 Glasstone *Sourcebk. Atomic Energy* x. 239/2 They do not undergo fission and break up into two, more or less equal, parts. Instead, such nuclei emit various numbers..of nucleons, leaving a series of products of lower mass and atomic number. The name *spallation* has been proposed for reactions of this kind. [*Note*] This term is based on the verb 'to spall', meaning to break up by chipping off small fragments, which was suggested as appropriate by W. H. Sullivan. 1953 *Ann. Rev. Nuclear Sci.* II. 401 O'Connor & Seaborg..bombarded uranium with 380-Mev α-particles and then analyzed radiochemically for spallation and fission products. 1962 B. G. Harvey *Introd. Nuclear Physics & Chem.* ix. 183 Reactions in which several particles are emitted are often called 'spallation', regardless of the mechanism by which they take place. 1979 *Nature* 15 Feb. 519/1, ^{14}C can be produced fairly readily in heavy ion collisions, or as a spallation product of heavy nuclei.

2. *gen.* Spalling.

1971 *New Scientist* 8 Apr. 72/3 An outer spallation zone due to the interaction of shock waves with the rock surface. 1974 Hawkey & Bingham *Wild Card* xv. 133 The crash guys..look for recrystallization at micro-promontories, Neumann banding, parallel banding, spallation.

spalled, *ppl. a.* Add: **1.** More widely, broken off or chipped by spalling. Also *spalled-off.*

1931 G. W. Tyrrell *Volcanoes* ii. 67 Such rocks..may be distinguished from agglomerate by the..occurrence of fragments of spalled-off scoria in the interstices between the blocks. 1971 *Ann. Rep. Delegates Sci. Area Univ. Oxford* 67 Coke ash slag on spalled ends of Steetley tarred dolomite blocks. 1975 *Country Life* 9 Oct. 924/3 Spalled and fractured brickwork was cut out and replaced.

spalliard, dial. form of Espalier *sb.* (Simple and attrib. examples.)

1886 F. T. Elworthy *W. Somerset Word-Bk.* 696 *Spalliard* (spaal·yurd), *sb.*, espalier, a trained fruit tree. I think, sir, we must dig up that *spalliard* plum. 1920 [see *pleach *sb.*].

Spam (spæm). Also **spam.** [App. blend of Sp(iced *ppl. a.* + H)am *sb.*[1], but see also quot. 1937[1].] **1.** The proprietary name of a type of tinned meat consisting chiefly of pork; also (with small initial) applied *loosely* to other types of tinned luncheon meat.

1937 *Squeal* 1 July 1/2 In the last month Geo. A. Hormel & Co..launched the product *Spam*... The 'think-up' of the name [is] credited to Kenneth Daigneau, New York actor... Seems as if he had considered the word a good memorable trade-name for some time, had only waited for a product to attach it to. 1937 *Official Gaz.* (U.S. Patent Office) 26 Oct. 750/2 Geo. A. Hormel & Company, Austin, Minn.. Spam..For Canned Meats—Namely, Spiced Ham. Claims use since May 11, 1937. 1939 J. Steinbeck *Grapes of Wrath* v. 49 The tractor driver stopped..and opened his lunch: sandwiches wrapped in waxed paper, white bread, pickle, cheese, Spam. 1942 *Yank* 28 Oct. 8 There, arrayed in all their glory, were slices of ham, spam, bologna and potato salad. 1951 'A. Garve' *Murder in Moscow* xiii. 127, I received..four tins of meat—spam, I think it was called. 1957 H. Roosenburg *Walls came tumbling Down* ix. 199 We were offered Spam sandwiches. 1971 C. Bonington *Annapurna South Face* xi. 134 That night he made supper, a magnificent concoction of fried Spam and fried new potatoes. 1981 G. MacBeth *Kind of Treason* iv. 41 A plate of Molly's best Spam sandwiches.

fig. 1958 *Listener* 6 Nov. 750/2 An actor can only turn the quite unconscious richness of it [*sc.* the Hoxton voice] into—spam.

2. *Comb.* **Spam can** *slang*, a streamlined steam locomotive formerly used on the Southern Region of British Rail; **Spam Medal** *Mil. slang*, a medal awarded to all the members of a force (see also quot. 1962).

1967 G. F. Fiennes *I tried to run Railway* v. 54 We borrowed from the Southern for trials two Battle of Britain class engines. We took these Spam Cans out. 1971 D. J. Smith *Discovery Railwayana* x. 59 *Spam can*, streamlined locomotive of the SR. 1945 Partridge *Dict. R.A.F. Slang* 40 Naffy gong 1939–45 star (medal). Since late 1943... It is also called *the spam medal.* 1959 *Legionary* Mar. 11/1 As all of us overseas at the time were volunteers, it meant that everybody wore one and so, in patronizing fashion, we tagged it [*sc.* the Canadian Volunteer Service Medal] the Spam Medal. 1962 Granville *Dict. Sailors' Slang* 110 *Spam medal*, 1939–45 star whose ribbon has the same colours as the NAAFI girls' arm flash. As *spam*, a kind of *spiced-ham*, was sold in the NAAFI canteen, what more obvious term could suggest itself?

Hence **spa·mmy** *a.*, consisting or tasting chiefly of (bland) luncheon meat; also *fig.*, commonplace, mediocre, unexciting.

1959 *Observer* 11 Jan. 18/3 Skipton is toned down to scale with our spammy age. 1960 J. Stroud *Shorn Lamb* i. 13 We got a spammy sort of meal.

span, *sb.*[1] Add: **5. c.** *Psychol.* Mental extent; the amount of information that the mind can be conscious of at a given moment, or the number of items it can reproduce after one presentation; esp. const. *of*, as *span of apprehension, attention, consciousness,* etc. Cf. *memory span* s.v. *memory 12.

1887 *Mind* XII. 76 The highest number correctly reproduced is to be regarded as the limit which we wish to find, and which we term here the *span.* *Ibid.* 79 We might expect that 'span of prehension' should be an important factor in determining mental grasp. 1890 W. James *Princ. Psychol.* I. xi. 405 The question of *the 'span' of consciousness* [*sc.* to how many things can we attend at once] has often been asked and answered. 1922 R. S. Woodworth *Psychol.* xi. 262 The 'span of attention' for objects..is measured by discovering how many such objects can be clearly seen, or heard, or felt, in a single instant of time. Measurement of this 'span' is one of the oldest experiments in psychology. 1945 *Mind* LIV. 165 She reduced the number of choices to a range lying within or just within the 'span of attention' of her subjects. 1971 *Jrnl. Gen. Psychol.* Jan. 129 No matter how high the level of luminance, the span of apprehensions will not exceed eight items during the critical interval. 1979 R. Jaffe *Class Reunion* (1980) III. ii. 312 The child was very bright and had a long attention span.

d. The maximum lateral dimension of an aircraft, or of a wing, from wing tip to wing tip.

1909 R. Kennedy *Flying Machines* ii. 34 It will be difficult to get a monoplane of sufficient span for heavy lifts, together with a strong construction. 1927 C. L. M. Brown *Conquest of Air* vii. 98 It was a biplane, the two main wings being 32 ft. in frontal width (span) and 5 ft. in depth (chord). 1953 *New Biol.* XIV. 73 A vulture has a broad wing of large surface area as well as large span. 1968 Miller & Sawers *Technical Devel. Mod. Aviation* v. 137 For the first time Douglas made a basic change in the wing..and increased the span by 10 feet to give greater lift and fuel capacity.

e. A range of numerical values; the difference between the highest and lowest values in a range.

1962 *Gloss. Terms Automatic Data Processing (B.S.I.)* 13 *Range.* 1. All the values which a quantity may have. 2. The difference between the highest and lowest of these values (in mathematics often called the span). 1974 *Physics Bull.* Jan. 31/3 The series covers the temperature range 223 to 573 K in eight spans.

f. *span of control*: in Business Studies, the area of activity, number of functions or subordinates, etc., for which an individual or organization is responsible.

1937 L. Gulick in Gulick & Urwick *Papers on Sci. of Administration* i. 7 Span of control. Just as the hand of man can span only a limited number of notes on the piano, so the mind and will of man can span but a limited number of immediate managerial contacts. 1956 E. Bridges in A. Dunsire *Making of Administrator* 5 He will look to see that too many people are not reporting to any one head—that the 'span of control' or of management as they call it, is not too wide. 1962 *Rep. Comm. Broadcasting 1960* 226 in *Parl. Papers 1961–2* (Cmnd. 1753) IX. 259 Since there might well be some scores of companies, it is highly unlikely that a corporation could exercise its responsibilities effectively; the 'span of control' would be too great. 1976 P. R. White *Planning for Public Transport* i. 20 Operation of private party coach hire or excursions may require frequent decisions on pricing and scheduling, and hence a small span of control for each manager is desirable.

6*. *Math.* That which is generated by the elements of some set. Cf. *span *v.*[1] 5*.

1968 D. Passman *Permutation Groups* ii. 155 We define the span of A, Span A, to be the subspace of FG spanned by all the functions a_{ij}. We list some basic properties of the span. 1981 *Sci. Amer.* Oct. 153 (*caption*) Their [*sc.* vectors'] span is a plane because any point in the plane can be reached by vectorially adding some scalar multiple of *A* to some scalar multiple of *B*.

7. (sense 5 c) *span test*; (poet.) *spanlong, -wide* adjs.; **span loading** *Aeronaut.*, the gross weight of an aircraft divided by its wing span or, more commonly, by the square of the wing span; also *transf.*, of a bird; **span wire,** each of the series of wires suspended across the route of a tram or trolleybus to carry the overhead electric wire; **span-worm** *U.S.* = Looper[1] in Dict. and Suppl.

1929 *Jrnl. R. Aeronaut. Soc.* XXXIII. 359 This..depends primarily on 'span loading' that is weight/(span)². 1953 *New Biol.* XIV. 72 The sinking speed is a function of the 'span loading', i.e. weight/wing span. A low sinking speed can only be achieved by having a wing span which is large relative to the weight of the bird. 1983 D. Stinton *Design of Aeroplane* iv. 120 In straight and level flight the lift loading across the span..is equal to..the weight of the aircraft divided by the wingspan. Both are referred to collectively as the span loading. 1957 Blunden *Poems of Many Years* 281 Spanlong rabbits quite forget danger's eye. 1971 *Jrnl. Gen. Psychol.* Apr. 238 We have here a well-defined figural-span measure which does *not* link with other span tests. 1943 E. Muir *Narrow Place* 7 Was this the ground That stretched beyond the span-wide world-wide ditch. 1891 *Electr. World* 21 Mar. 225/1 In the case of side pole and span wire, construction poles should be placed at the points represented and a span wire run between them. 1963 A. T. Dover *Electric Traction* (ed. 4) xxiii. 358 At curves the span wire must be on a level with the trolley wire, otherwise the hangar will be pulled out of the vertical. 1820 *Amer. Farmer* I. 375/3 What can our obliging correspondents tell us about the..best method of destroying that dreadful plague of our orchards, the span worm. 1903 [see *measuring *ppl. a.* b]. 1972 Swan & Papp *Common Insects N. Amer.* xix. 293 The Bruce spanworm is a major defoliator of aspen in the prairie regions.

span, *v.*[1] Add: **I. 5*.** *Math.* To generate. Cf. *span *sb.*[1] 6*.

1941 Birkhoff & MacLane *Survey Mod. Algebra* vii. 170 This subspace is evidently the smallest subspace containing all the given vectors; hence it is called the subspace generated or spanned by them. 1964 A. P. & W. Robertson *Topological Vector Spaces* i. 3 If L is a linearly independent subset of a vector space E and if S is a subset containing L and spanning E, there is a base B of E with L ⊂ B ⊂ S. 1981 *Sci. Amer.* Oct. 156/2 For example, the *u*–*v* plane is a vector space that can be spanned by two vectors directed along the positive *u* and *v* axes.

II. 6. (Further example.)

1976 *Offshore Platforms & Pipelining* 151/1 Any relatively stiff pipeline laid in a wavy seabed will span in places.

span, *v.*[2] Add: **1. a.** Also *absol.* (Earlier example.)

1793 J. Baxter *Jrnl.* 2 Feb. in *Amer. Speech* XL. (1965) 199, I and John Schenck spaned in together.

spanandry (spæ·nændri). *Zool.* [ad. F. *spanandrie* (P. Marchal 1913, in *Ann. des Sci. Nat.: Zool.* XVIII. 268), ad. Patristic Gr. σπανανδρία scarcity of population, f. Gr. σπάνις scarcity + ἀνήρ, ἀνδρ- man: see -y[3].] Lack or extreme scarcity of males in a population. Hence **spana·ndric** *a.*

1924 *Nature* 14 June 880/1 A. Vandel: Geographical spanandry in a Branchiopod Crustacean, *Lepidurus apus.* 1967 *Science* 28 Apr. 483/3 Polygamy is extreme, but, since this follows from the fact that males are produced in much smaller numbers than females, *spanandry* seems a better term, and is used here. 1976 *Bull. Entomol. Res.* LXVI. 179 (*heading*) The identity of the greenhouse thrips *Heliothrips haemorrhoidalis* (Bouché) (Thysanoptera) and the taxonomic significance of spanandric males. 1976 *Entomol. Meddelelser* XLIV. 60 The major types of parthenogenesis: thelytoky, cyclical parthenogenesis,..and spanandry are reviewed.

spancelled, *ppl. a.* and *pa. pple.* Add: Also *fig.*

1910 J. M. Synge *Deirdre of Sorrows* II. 35 You're seven years spancelled with Naisi and the Pair. 1965 *New Statesman* 26 Nov. 848/1 In England it is the wives who regard themselves as the spancelled party. 1980 *Times Lit. Suppl.* 20 June 688/4 He was..spancelled by a lack of money that meant he could never be apprenticed to a leading London surgeon.

Spandau (spæ·ndau). Also **spandau.** [f. *Spandau*, name of a district of (West) Berlin.] **a.** A German machine-gun used during the war of 1914–18 (see quot. 1966). **b.** Applied to other machine-guns of German design, esp. the MG34 and MG42 of the war of 1939–45. Freq. *attrib.* as *Spandau (machine) gun.*

a 1918 J. T. B. McCudden *Five Years in Royal Flying Corps* (1919) v. xii. 241, I distinctly noticed the red-yellow flashes from his parallel Spandau guns. 1929 E. W. Springs *Above Bright Blue Sky* 221 He was suddenly cold as he awaited the crack of the spandaus. 1938 G. S. Hutchinson *Machine Guns* xii. 333 The British were armed with Vickers, the Germans with Spandaus. 1944 K. Douglas *Alamein to Zem Zem* (1946) iv. 29 In their pit lay a Spandau machine-gun. 1966 T. R. Funderburk *Fighters* 27 The Maxim gun was manufactured in England..as a Vickers gun. The same gun was manufactured under license by the German Weapons and Munitions Factory, a state arsenal at Spandau, Berlin, and was known as a Spandau gun. A light version ..was developed by the Germans for the use of aerial gunners and known as a Parabellum. 1968 A. Diment *Bang Bang Birds* vi. 97 Vitconne is crouched behind a Spandau lent to him by a friendly German officer. 1971 F. W. A. Hobart *Pictorial Hist. Machine Gun* 228 (*caption*) First German Airborne Corps defending Cassino... The MG-42 was a very effective dual purpose machine gun. The Germans used it a lot on a tripod to produce fixed line fire. The British troops called it the 'Spandau'.

Spandex (spæ·ndeks). orig. *U.S.* Also **spandex.** [Arbitrarily f. Expand *v.*] **a.** A synthetic elastomeric fibre composed largely of polyurethane. **b.** A proprietary name for certain fabrics made from this fibre.

1959 *Federal Register* 10 Feb. 981/3 The following generic names for manufactured fibers..are hereby established... *Spandex*, a manufactured fiber in which the fiber-forming substance is a long chain synthetic elastomer comprised of at least 85 percent of a segmented polyurethane. 1962 *New Scientist* 22 Mar. 697/1 As materials classified as spandex may have 15 per cent of some other compound to be introduced with the polyurethane, wide variations in fibre properties are possible. 1968 *Trade Marks Jrnl.* 7 Feb. 176/2 *Spandex*... All goods included in Class 22 made wholly or substantially of spandex fibres. Monsanto Chemical Company..U.S.A.; manufacturers.—4th Sept. 1964. 1972 *Brazilian Bull.* (Brazilian Trade Centre, London) Apr. 3 The Brazilian subsidiary of the U.S. Du Pont company is to build a factory at Paulínia..for the production of Lycra Spandex fibre. 1973 *Materials & Technology* VI. iv. 334 The illustration shows cross-sections through three well-known brands of spandex fibre, these being 'Lycra'.., 'Spanzelle'.., and 'Vyrene'. 1978 *Neiman-Marcus Christmas Bk.* 26 Nylon and spandex maillot.

spandite (spæ·ndəit). *Min.* [Blend of Spessartite and Andradite.] A manganiferous garnet resembling spessartine but containing more iron and calcium.

1907 *Rec. Geol. Survey India* XXXV. 22 The garnet is intermediate in composition between spessartite and andradite, and..Mr. Fermor proposes to introduce the term spessart-andradite, which may be shortened for convenience in general use to spandite. 1955 Brown & Dey *India's Mineral Wealth* (ed. 3) 606 Spandite, a garnet intermediate in composition between spessartite and andradite, occurs in the Kodurite series of the Srikakulam district of Andhra. It varies in colour from deep orange to brown orange-red and blood-red and never displays the lighter orange shades of the true spessartite. 1972 *Mineral. Abstr.* XXXIII. 121/1 X-ray data for hiddenite diopside, [etc.]..and a probable spandite are given.

spandrel. Add: **1. c.** On oriental patterned rugs or carpets: one of the spaces between the central field and the border, or between an arch motif and its frame.

1900 J. K. Mumford *Oriental Rugs* x. 147 In the spandrels over the arch of the prayer rugs there is a repetition of the pear patterns. 1931 A. U. Dilley *Oriental Rugs & Carpets* iii. 63 Large central medallions, brilliant pendants, traceried spandrels and graceful floral scrolls. 1962 C. W. Jacobsen *Oriental Rugs* II. 264 Mudjars have a definite prayer arch design... Many of these have panels above the spandrel with van dykes designs. 1967 *Times* 26 Jan. 22 (Advt.), A *Silk Heriz* with all over blue spandrels on an old mellowed field, classical rust medallion, floral border.

spandrelled, *a.* (Earlier example.)

1813 M. Edgeworth *Let.* 1 May (1971) 37 The beauty of the spandrilled ceiling with all its rich, and light ornaments.

spandy, *a.* Delete *rare* and add earlier example. Also *spandy-bright, spandy new.*

1838 'T. Titterwell' *Yankee Notions* 116, I have heard of a ghost that always came in a new coat..and a spandy clean dickey. 1903 K. D. Wiggin *Rebecca of Sunnybrook Farm* i. 14 The trouble is to get the shoes... These are spandy new I've got on. 1968 J. Updike *Couples* iii. 227 Ben's lank hairs ran together to make black seams, like sores downrunning into the tops of his comically new topsiders, cup-soled, spandy-bright. 1973 E.-J. Bahr *Nice Neighbourhood* ii. 23 Don has this very definite fixation that I am going to bang up our spandy new car.

spang, *v.*[2] For '*Sc.* and *north.*' read 'orig. *Sc.* and *north.*' and add: **1.** (Later examples.) Also *fig.*

1966 *If, Worlds of Sci. Fiction* Dec. 39/2 [He] kneed the screen door open so that it spanged against the outside wall. 1976 L. Sanders *Hamlet Warning* (1977) xix. 165 A stream of bullets spanged off the metal around him. 1979 *Observer* 4 Feb. 4/7 We shared champagne and Coke with the nurses. Thank God spanged out on all sides.

spang, *adv.* For *U.S.* read 'orig. and chiefly *U.S.*' and add later examples.

1921 D. Canfield *Brimming Cup* I. v. 65 The brooks were..all running spang full to the very edge with snow-water. 1925 Wodehouse *Sam the Sudden* xiii. 91 If he thinks a young bride's going to stand for that sort of conduct right plumb spang in the middle of what you might call the honeymoon, [etc.]. 1936 M. Mitchell *Gone with Wind* xxxii. 547 So you needs a spang new pretty dress. 1962 *Punch* 26 Dec. 943/1 With its superb flair for the nostalgic and traditional the BBC has put this year's Christmas spang in the middle of its dramatised version of The Old Curiosity Shop. 1971 'D. Cory' *Sunburst* xiii. 212 Will you *look* at that?.. Spang in the middle of the bloody road.

spangle, *sb.*[1] Add: **4. c.** A fowl or pigeon belonging to a variety distinguished by speckled plumage.

1854 *Poultry Chron.* II. 66/1 Eighteen pens of beautiful silver spangles added to the old laurels of Mr. Vivian. 1855 *Ibid.* III. 175/2 The Toys [*sc.* pigeons] are as follows: Suabians or Spangles, [etc.]. *Ibid.* 355/2 The third kind is what the old breeders of Game fowls call 'Spangles'. The cock is red and white in the hackle and saddle, and black and white in the tail and breast. The hens are partridge-colour, spotted with white. 1948 G. O. Rickwood *Constable's Country* 14 The..'King's Arms'..was a rendezvous of cock-fighters in days when the 'feeders' of the birds—Shropshire reds, Staffordshire jet-blacks,..spangles and other noted breeds—were..important personages in the hierarchy of the sport.

spangled, *ppl. a.* **1.** (Later *fig.* examples.)

1920 E. Sitwell *Wooden Pegasus* 13 Flickered down the street together In the spangled weather. 1943 L. B. Lyon *Evening in Stepney* 17 By spangled weirs we pass the gas-works. 1977 [see *rhinestone b].

Spanglish (spæ·ŋgliʃ). [Blend of Spanish *sb.*[1] + English *sb.*] A type of Spanish contaminated by English words and forms of expression, spoken in Latin America.

[1954 S. Tió *A Fuego Lento* 62 Esta lengua nueva se llamará el 'Espanglish'. La etimología es clara. Viene de español y de english.] 1967 *Time* 7 Apr. 12 A historical pageant known as a 'Texas Fandangle'—border-country Spanglish for fandango, the frenetic Mexican dance. 1972 *Daily Tel.* 28 Nov. 13/8 Argentina is not alone in falling victim to 'Spanglish'. Chilean housewives who have problems with the plumbing call for el gasfitter. 1974 *Amer. Speech 1970* XLV. 230 Spanglish may be characterized as a gradual relexification of Puerto Rican Spanish through borrowings, adaptations, and innovations. 1976 M. Millar *Ask for me Tomorrow* (1977) vi. 43 A mixture of Spanish and English slang sometimes called Spanglish.

spangolite (spæ·ŋgŏləit). *Min.* [f. the name of Norman *Spang*, 19th-cent. U.S. mineral collector + *-ol-* + -ite[1].] A hydrated basic sulphate and chloride of copper and aluminium, $Cu_6AlSO_4(OH)_{12}Cl{\cdot}3H_2O$, which is a secondary mineral found as pale green, prismatic or tabular, rhombohedral crystals.

1890 S. L. Penfield in *Amer. Jrnl. Sci.* CXXXIX. 371, I take pleasure in not only expressing at this time my thanks to Mr. Spang for his kindness but also in naming the mineral..Spangolite after him. 1949 *Amer. Mineralogist* XXXIV. 182 A few months ago, Mr. Hatfield Goudey..found a new occurrence of spangolite at Majuba Hill, Pershing County, Nevada. 1968 I. Kostov *Mineralogy* 514 Spangolite is ditrigonal-pyramidal.., also with perfect {001} cleavage.

Spaniard, *sb.* (and *a.*). Add: Hence **Spaniarde·ss,** a Spanish woman; **Spa·niardly** *adv.*, in a manner typical of the Spanish.

1909 Kipling *Rewards & Fairies* (1910) 42 De Avila..very Spaniardly hung them all for heretics. 1931 O. Nash *Hard Lines* 93 The bashful Spaniardess apparently finds the amorous Spaniard..menacing to her virtue.

spaniel, *sb.*[1] (and *a.*). Add: **3. a.** *spaniel eye; spaniel-eyed* adj.

1958 M. Kelly *Christmas Egg* III. 125 Brett could well imagine his assumed spaniel eyes of reproach. 1975 T. Allbeury *Special Collection* xi. 79 Felinski wondered what women would make of those liquid, spaniel eyes. 1963 P. Fleming *Kolchak* vii. 84 A plump, spaniel-eyed, ineffective little man.

spaniel, *v.* **a.** (Later examples.)

1924 Galsworthy *Forest* II. i. 33 Devoted to him; spaniels round him all the time. 1958 'W. Henry' *Seven Men at Mimbres Springs* vii. 80 With that reference to the gun, Sparhawk began spanielling again, obsequious as ever.

Spanish, *a.* (*adv.*) and *sb.*[1] Add: **A.** *adj.* **1. d.** Of a Jew or Jewish institution: of or belonging to the Sephardim. Also *Spanish-Hebrew, Spanish (and) Portuguese* (cf. *Portuguese *a.* a.) Hence **Spanish-Jewish** *a.*

1817 M. Edgeworth *Harrington* iv. 88 This Spanish Jew must..be a most accomplished and amiable person. 1851 [see Sephardi]. 1876 Geo. Eliot *D. Deronda* VII. lv. 238 His mind went to the synagogue..and heard the Spanish-Hebrew liturgy. 1892 [see *Ashkenazim]. 1894 I. Zangwill *King of Schnorrers* v. 105 The Mahamad..administered the affairs of the Spanish-Portuguese community. 1902 G. E. Mitton *Hampstead & Marylebone* 80 In Bryanston Street there is a synagogue which was built for the Spanish and Portuguese Jews. 1932 C. Roth *Hist. of Marranos* xii. 315 On his death in 1762, he left..a legacy of £1,000 to the Spanish and Portuguese community. 1949 'R. West' *Meaning of Treason* I. vi. 122 That slender and distinguished old gentleman of Spanish Jewish descent, Mr Salzedo. 1977 *Early Music* Apr. 262 Basil Douglas Ltd. presents..Sephardic Romances from before the expulsion of the Spanish Jews (1492). 1981 *Times* 16 Oct. 9/1 Elias Canetti..was born in Bulgaria, of Spanish-Jewish descent. 1982 *Times* 26 Jan. 11/5 Sir Moses Montefiore ..was an intensely loyal Englishman. The Spanish Portuguese Jewish Congregation..refused a request to transfer the remains to Israel.

2. a. *Spanish saddle.*

1821 S. F. Austin *Jrnl.* 8 July in *Texas Hist. Assoc. Q.* (1904) VII. 287 Swapped away Wilsons Horse & an old Grey..for a mule, & exchanged a french saddle for a Spanish one. 1897 E. Hough *Story of Cowboy* 67 The Spanish saddles of the Southwest were often heavily decorated with silver. 1945 *Elk Mountain Pilot* (Crested Butte, Colorado) 19 July 3/1 (Advt.), For Sale... A Spanish Saddle, excellent condition.

c. *Spanish shawl.*

1925 G. Greene *Babbling April* 17 And the night was so hot, And no one can see in the dark, And a rent in the Spanish shawl. 1967 A. Wilson *No Laughing Matter* II. 92 A grand piano on which [is] a white Spanish shawl with red and green embroidered roses. 1975 'R. Player' *Let's talk of Graves* iii. 77 The year '54..the year of the largest crinolines... Thrown over them..were..large Spanish shawls.

e. (*a*) Denoting a style of art or architecture native to or characteristic of Spain; (*b*) denoting a style of decoration or architecture imitative of that of Spain. Also *Spanish-style* adj.

1927 *Sunset Mag.* May 87/1 Many builders and real estate men are masquerading whole city blocks of houses under the name of Spanish. 1931 S. Sitwell (*title*) Spanish Baroque art. 1937 J. Laver *Taste & Fashion* xviii. 258 The style of his interior decoration may be shortly described as Spanish 'baroque'. 1950 A. Wilson *Such Darling Dodos* 159 There were Regency bedrooms, a Spanish Baroque dining room. 1953 S. Bedford *Sudden View* I. i. 20 The dining-car..turned out to be..decorated with machine-carved Spanish Renaissance woodwork of astonishing gloom. 1960 *Encounter* Apr. 3/2 There is something ancient and unfamiliar about its [*sc.* a skyscraper's] situation among Spanish-style San Francisco homes. 1970 H. Braun *Parish Churches* xi. 148 It [*sc.* the arch] is often four-centred or, in some of the more opulent examples, the three-centred 'Spanish' arch. 1976 *Liverpool Echo* 22 Nov. 14/2 (Advt.), Hall, lounge, Spanish arch to dining room, [etc.]. 1977 *N.Z. Herald* 8 Jan. 4-3/1 (Advt.), $5000 deposit will secure this charming Spanish bungalow in New Windsor. 1979 *Arizona Daily Star* 1 Apr. (Advt. Section) 19/5 Enter this custom Spanish territorial home over a wooden bridge. 1979 N. Hartley *Quicksilver* vi. 77 Several Spanish-style interior patios.

3. b. *an old Spanish custom*: phr. used *joc.* to justify a long-standing practice which is unauthorized or otherwise irregular.

1932 *N. & Q.* 13 Feb. 122/1 Could any reader tell me the origin of the phrase, 'An old Spanish custom,' as applied, in a jocular sense, to any unauthorised practice? 1966 M. Torrie *Heavy as Lead* x. 115 Giving Sir Ganymede lunch at the pub..appeared by this time to have become an old Spanish custom. 1982 *Listener* 25 Nov. 13/2 The December issue of *Encounter*..lifts some lids on the 'old Spanish customs' of Fleet Street print unions.

6. *Spanish-American* (earlier example), *-Indian* (earlier example), *-Mexican.*

1705 R. Beverley *Virginia* 51 By their Accounts, we suppose him to have come from the Spanish Indians, some-where near Mexico, or the Mines of St. Barbe. 1727 E. Dorrington *Hermit* I. i. 1, I accidentally fell into Discourse with a Spanish Mexican Inhabitant, named Alvarado. 1811 *Niles' Reg.* I. 14/2 The Creoles—Spanish Americans—i.e. the descendants of Spaniards born in this country. 1935 E. Farjeon *Nursery in Nineties* II. iii. 91 A beautiful Spanish-Mexican girl who smoked cigarettes. 1980 *Amer. Speech* LV. 39 Geographic names of Spanish linguistic origin [are] utilized..to determine the sphere of Spanish-Mexican influence in California.

7. Spanish bowline (see quot. 1968); **Spanish Civil War,** the civil war in Spain (1936–9), espoused on both sides as a popular 'cause' throughout Europe and America, in which Nationalist rebel forces, with Fascist support, overcame the Republican Government and its anti-Fascist allies (cf. *International Brigade* s.v. *international *a.* 2); **Spanish-Colonial** *a.*, designating a style of architecture characteristic of the former Spanish colonies in the Americas; also *absol.*; **Spanish comb,** a decorative comb having a deep top, worn in the hair; **Spanish dance,** the traditional dance form of Spain, of gypsy origin and characterized by elaborate heel-work and freq. involving the use of castanets; hence **Spanish dancer, dancing; Spanish flu** *colloq.* = *Spanish influenza* below; **Spanish foot,** a foot (of a chair or other piece of furniture) of a scroll form with vertical ribs; **Spanish guitar,** the standard six-stringed (orig. five-stringed) non-electric guitar, used for both folk and classical music; **Spanish hat** (see quot. 1960); **Spanish influenza,** a popular name for influenza caused by an influenza virus of type A; *esp.* that of the pandemic which began in 1918; **Spanish omelette,** an omelette containing a selection of tomatoes, onions, potatoes, and other vegetables; **Spanish tile** *Building*, (*a*) a roofing tile that is curved cylindrically and slightly tapered, to be laid alternately convex and concave so as to overlap at both sides; (*b*) *U.S.*, a curved roofing tile that is laid convex upwards and overlaps at one side only by means of a straight projection; **Spanish tummy** *colloq.*, a stomach upset of a type freq. experienced by visitors to Spain; **Spanish War** = *Spanish Civil War* above.

1968 E. Franklin *Dict. Knots* 26 *Spanish bowline*, a double loop knot that is tied in the bight in which the two loops are splayed. 1974 *Maclean's Mag.* May 10/2 Spanish bowlines to make slings for scaffolding. 1936 C. Prieto *Spanish Front* xi. 80 It is hardly necessary in this book to give a detailed description of the Spanish Civil War. 1981 A. Price *Soldier no More* xi. 144 He'd subscribed to all sorts of causes, from the Spanish Civil War onwards. 1927 *Sunset Mag.* May 13/2 The popular desire seems to be to call this architecture Spanish. Architects are inclined to call it Spanish-Colonial. 1937 R. Newcomb *Spanish-Colonial Archit. in U.S.* viii. 37 Many in these states have wished to build in the Spanish Colonial rather than in the American Colonial. 1973 G. Sims *Hunters Point* xiii. 114 It's what they call Spanish Colonial style with wooden columns and a roof of heavily twisted tiles. 1977 H. Fast *Immigrants* I. 39 There was much substance if little taste all through the dining room,..a curious and uninspiring marriage of Spanish Colonial and Victorian. 1873 *Young Englishwoman* Mar. 131/2 Diadem plaits or torsade, fastened with a Spanish comb. 1923 M. Kennedy *Ladies of Lyndon* iv. 234 Could you find my Spanish comb? And..that black lace shawl. 1975 *Times* 23 Aug. 7/2 She hoped her mother would let her wear a Spanish comb she had rashly bought. 1931 K. Boyle *Plagued by Nightingale* xvi. 137 He was doing a lively Spanish dance to the piercing screams of their laughter. 1974 W. Foley *Child in Forest* 20 Our elder sister..once did a 'Spanish' dance on the end of the bed, with a cracked soap-dish for a castanet. 1948 'La Meri' *Spanish Dancing* i. 1 There are great Spanish dancers, male and female, who are not Spanish. *Ibid.* 4 Hermetic Spain is that promised land to which we go to find..the greatest..Spanish dancing in the world. 1980 A. Cornelisen *Torregreca* vi. 129 At *Carnevale*..Maria made "Spanish dancers" dresses for herself and her little brother. 1918 W. Owen *Let.* 24 June (1967) 560 About 30 officers are smitten with the Spanish Flu. 1937 K. Blixen *Out of Africa* II. v. 161 When we had the Spanish Flu on the farm, Farah was..shivering with fever. 1979 D. Williams *Genesis & Exodus* xi. 213 Those who had survived 1914–18 and the plague of Spanish flu that followed. 1902 F. C. Morse *Furnit. of Olden Time* vi. 151 The chair..is of the style called Queen Anne. It has Spanish feet. 1923 J. C. Rogers *English Furnit.* II. ii. 56 Legs also were given a sudden broadening like an inverted cup... In some cases there was the carved 'Spanish' foot. 1975 *Country Life* 29 May (Suppl.) 40h (Advt.), Early 16th century Virginia walnut gateleg table with..Spanish feet. 1862 G. Borrow *Wild Wales* i. 10 Playing remarkably well on the guitar—not the trumpery German thing so called—but the real Spanish guitar. 1934 S. R. Nelson *All about Jazz* ii. 54 The Spanish Guitar, with its resonant tone and range, has created a minor revolution in the ranks of the fretted instrument players. 1961 A. Birch in A. Baines *Mus. Instruments through Ages* vii. 168 At some point during the sixteenth century a fifth course became standard for guitar in Spain... It was this new five-course instrument which was to carry the name of 'Spanish guitar'. 1784 E. Sheridan *Jrnl.* 1 Oct. (1960) 26, I found a Spanish hat was what I must bye... Even silk Balloons are almost out—I have not seen a Cap since I came. 1840 Thackeray in *Fraser's Mag.* XXI. 688/1 A ricketty lay-figure, in a Spanish hat and cloak. 1882 C. M. Yonge *Unknown to History* II. iv. 47 Captain Fortescue..a long plume in his Spanish hat. 1960 C. W. Cunnington et al. *Dict. Eng. Costume* 201/2 *Spanish hat*,..a large hat of velvet, satin or sarcenet, the brim evasé, trimmed with feathers. 1980 A. Crawley *Dial 200–200* xi. 119 Maria, looking fabulous in her white satin bolero, tight trousers and black Spanish hat. 1918 *Policeman's Monthly* Oct. 4/2 The members of the Hartford department comprise the motor-cycle squad, one of whom was a victim of Spanish influenza this week. 1940 *N. & Q.* 30 Mar. 218/1 The ravages of the 'Spanish influenza' which, between 1918 and 1922, caused four times as many deaths as those caused by the last war. 1976 Botham & Donnelly *Valentino* viii. 64 The worst Spanish influenza epidemic since the turn of the century had swept through California. 1886 S. T. Rorer *Mrs Rorer's Philadelphia Cook Bk.* 260 (*heading*) Spanish omelet. 1935 S. Lewis *It can't happen Here* xxv. 270 One eye was..so surrounded with bruised flesh that..it looked like a Spanish omelet. 1974 A. Williams *Gentleman Traitor* xv. 248 He..had the African cook prepare him a Spanish omelette. [1904 F. E. Kidder *Architect's & Builder's Pocket-bk.* (ed. 14) III. 1430 Galvanized-iron tiles of the 'Spanish' pattern.] 1913 —— *Building-Construction & Superintendence* (ed. 9) II. iv. 278 The rafters..are covered with Ludovici interlocking Spanish tiles set on the 'shiplap' roof-boarding. 1956 E. Molloy *Builders' & Decorators' Ref. Bk.* XVII. 3 The English pantile..is a descendant of the Spanish tile.., also called the over-and-under tile. 1979 *Tucson* (Arizona) *Citizen* (Advt. Suppl.) 28 Apr. 16/4 High beamed ceilings, brick floors, French doors, and Spanish tile all help in adding southwestern flavor to a contemporary look. 1967 *Sunday Times* (Colour Suppl.) 21 May 39 They never seem to get Spanish tummy, and their children are never overtired brats. 1968 A. Brown *Slay me Suddenly* ix. 129 'Where's Herbert?' 'Vomiting... Spanish tummy, I suppose.' 1937 H. Nicolson *Diary* 27 July (1966) 310 He [*sc.* Anthony Eden] thinks that the Spanish War will last another year. 1977 C. McCarry *Secret Lovers* x. 130 Is it a good book? You said you were in the Spanish war.

8. a. (*c*) **Spanish Merino** = Merino 1; **Spanish sheep,** (i) = Merino 1 (earlier example); (ii) = *Jacob 4.

1788 W. B. Conyngham *Let.* 28 June in H. B. Carter *His Majesty's Spanish Flock* (1964) iii. 60 In answer to what Evidence I have relative to the success of my Cross from the Spanish sheep I have..the greatest Reason to believe that the Breed may be greatly improved. 1802 F. L. Humphreys *Life D. Humphreys* II. 346 A Gold Medal..is presented to you..for your patriotic exertions in introducing into New-England one hundred of the Spanish Merino breed of Sheep. 1891 R. Wallace *Rural Economy Austral. & N.Z.* xxvi. 357 The Spanish Merino is a sheep of large size, producing a superior quality of strong combing wool. 1913 Spanish sheep [see *Jacob 4]. 1964 H. B. Carter *His Majesty's Spanish Flock* p. x, These were the men who..transformed the Spanish Merino from an envied monopoly of one nation into the essential foundation of the modern world trade in wool. 1974 *Times* 25 Nov. 3/3 There are now about 150 registered flocks [of Jacob sheep] in Britain, comprising 3,000 spotty sheep, also known as Spanish or piebald.

9. Spanish beard (earlier example); **Spanish bluebell** = *Spanish squill* below; **Spanish cedar,** a species of Central American cedar, esp. *Cedrela mexicana*, or its timber; **Spanish dagger**: also more generally, one of several species of *Yucca*, esp. *Y. gloriosa*; (earlier and later examples); **Spanish harebell** = *Spanish squill* below; **Spanish moss** (earlier example); **Spanish squill,** a bulbous plant, *Endymion hispanicus* (formerly *Scilla hispanica*), bearing loose racemes of blue, pink, or white bell-shaped flowers; (earlier and later examples); **Spanish stopper** = *gurgeon stopper* s.v. *gurgeon; cf. Stopper *sb.* 8.

1763 tr. *La Page du Pratz's Hist. Louisiana* II. iv. 37 The other excrescence is commonly found upon trees near the banks of rivers and lakes. It is called Spanish beard. 1924 L. H. Bailey *Man. Cultivated Plants* 164 Spanish Bluebell..fl[ower]s blue to rose-purple, usually a dozen or more, ascending or nodding in an open raceme. 1979 *Guardian* 5 June 10/1 Large white butterflies..thrusting their long tongues into the Spanish bluebells. 1907 Spanish cedar [see *cigar-box* s.v. *cigar 2]. 1947 J. C. Rich *Materials & Methods of Sculpture* x. 287 Spanish cedar is not a true cedar variety, but the wood is favored by some sculptors for carving. 1972 *Handbk. of Hardwoods* (Forest Prod. Res. Lab.) (ed. 2) 53 'Central American cedar' is sometimes called 'Spanish cedar' in reference to the former Spanish colonies. 1859 A. Van Buren *Sojourn in South* 108 A tall 'Spanish dagger' stood leaning its crested head. 1939 G. B. Pickwell *Deserts* 25/1 Spanish daggers bloom in deserts. 1975 *Islander* (Victoria, B.C.) 3 Aug. 3/4 The Spanish Dagger, with fruits that are eaten raw, baked on hot stones, made into jelly or dried for winter use. 1808 *Curtis's Bot. Mag.* XXVIII. 1102 (*heading*) Spanish Harebell. 1823 E. James *Acct. Exped. Rocky Mts.* III. 220 The Spanish moss disappears northwardly of the 33d degree of north latitude. 1790 *Curtis's Bot. Mag.* IV. 128 (*heading*) Spanish Squill. 1977 *Chicago Tribune* 2 Oct. XI. 13/2 Late—Spanish squill..and double late tulips. 1883 G. O. Shields *Rustlings in Rockies* xxi. 195 Within the space of this five acres may be found ..Spanish stoppor [*sic*]. 1908, 1921 Spanish stopper [see *gurgeon].

10. *Spanish-looking, -surnamed* adjs.

1944 N. Coward *Star Quality* (1951) 85 Rather a good Spanish-looking sideboard. 1976 *Billings* (Montana) *Gaz.* 20 June 2-A/1 In what was called the first tabulation of Latino 'segregation trends', the study also said segregation of Spanish-surnamed pupils increased in the 1970s in all regions of the nation. 1981 R. Rendell *Put on by Cunning* vii. 65 A dark, Spanish-looking girl.

B. *sb.* or *ellipt.* **2. a.** Delete *rare*⁻¹ and add later examples.

1832 W. Irving in P. M. Irving *Life & Lett. Washington Irving* (1863) III. 43 The levee..presents the most whimsical groups of people of all nations, castes, and colors—French, Spanish, half-breeds, [etc.]. 1880 *News & Press* (Cimarron, New Mexico) 24 June 2/2 The famous Pecos Church, built by the Spanish in 1680. 1932 [see *moment *sb.* 1 d]. 1962 *Amer. Speech* XXXVII. 207 English speakers [in northern Colorado] refer to Spanish speakers as *Spanish.* The word *Spaniard* is not used.

Hence **Spa·nishness, Spa·nishry,** the quality of being Spanish; **Spa·nishy** *a.*, of a Spanish type or character.

1922 Joyce *Ulysses* 273 Big Spanishy eyes goggling at nothing. 1957 *American Anthropologist* LIX. 818 Spanish-speaking Venezuelans and 'Spanishy' local-born people. 1960 L. Durrell in *N.Y. Times Bk. Rev.* 12 June VII. 1/1 We travel really to try and get to grips with this mysterious quality of 'Greekness' or 'Spanishness'. 1963 *Times Lit. Suppl.* 17 May 356/5 This intense 'Spanishry' of Unamuno's..may be a reason..for his comparative neglect in this country. 1965 *Listener* 25 Nov. 873/2 In the company of the desolate *Gigues*, even Iberia takes on a less obvious Spanishry. 1977 V. S. Pritchett *Gentle Barbarian* iii. 41 Her Spanishness had its Islamic roots. 1979 B. Malamud *Dubin's Lives* vii. 228 She's been reading Spanish love poems... Her voice sounds Spanishy.

Spanisher (spæ·niʃəɹ). *colloq. rare.* [f. Spanish *a.* + -er[1].] A Spaniard.

1910 Kipling *Rewards & Fairies* 293 The Spanishers had shut down all their Dutch ports against us English. 1940 [see *aim *v.* 5 b].

spank, *v.*[1] Add: **1. d.** *N.Z. colloq.* To milk (a cow).

1897 I. Scott *How I stole 10,000 Sheep* ii. 8 We got on pretty well and spanked, *i.e.* milked, his cows for him night and morning. 1948 D. W. Ballantyne *Cunninghams* II. ix. 199 They spanked cows and built fences.

spanker[1]. Add: **3. b.** In full *spanker mast.* The fourth (or fifth) mast of a ship with four or more masts.

1853 D. McLean *Great Republic* 17 The ship has 4 masts, the after one named the spanker mast. 1892 *Nautical Mag.* June 546 During the first four years of the four-posters, the 'after mizzen' and 'spanker mast' were sometimes heard of. 1902 [see *pusher 2 b]. 1946 *Amer. Neptune* VI. 138/1 Mr. Bush called the fourth mast the 'jigger' and the fifth the 'spanker', the reverse of the usual nomenclature. 1970 H. Chevalier *Last Voyage of Rosamond* 37 We soon not only learned the names of the four masts—the foremast, the mainmast, the mizzen and the spanker.

spanker[2]. (Later example.)

1931 K. Boyle *Plagued by Nightingale* iv. 28 Washerwomen bending over the stream..and with wooden spankers beating the life out of dish-cloths.

spanking, *vbl. sb.* Add: Also *fig.*

1922 H. Crane *Let.* 25 Feb. (1965) 80 Your translations amuse me without interesting me as thoroughly as your nicely administered spanking of McAlmon. 1978 *Times Lit. Suppl.* 1 Dec. 1393/1 A. A. Milne's Winnie-the-Pooh stories receive a monstrous spanking for their promotion of snobbery and imperialist values.

spanking, *ppl. a.* Add: **5.** Used as *adv.* Very, exceedingly; esp. as *spanking new*, brand-new. Cf. Span-new *a.* *colloq.*

1886 H. Baumann *Londinismen* 188/2 A spanking fine dinner. 1905 *Dial. Notes* III. 71 *Bran spankin' new*,..absolutely new. 1925 F. Scott Fitzgerald *Great Gatsby* i. 6 The [house]..on my right was a colossal affair..spanking new under a thin beard of raw ivy. 1959 *Weekly Times* (Melbourne) 30 Sept. 1 (Advt.), Imagine that great day..when you take delivery of your new working partner: a spanking new Ferguson 35. 1972 *Newsweek* 10 Jan. 34/2 The spanking new city of Brasília was carved painfully out of the wilderness. 1977 *Time* 19 Dec. 12/1 The spanking-white train chugged into Reading station. 1979 *Radio Times* 5–11 May 4/1 Luxurious surroundings at Lewisham's spanking new leisure centre. 1979 W. Styron *Sophie's Choice* vii. 164 On the driveway there now rested a spanking clean and polished Cadillac sedan.

spankled, *ppl. a.* (Later example.)

1866 G. M. Hopkins *Jrnls. & Papers* (1959) 143 Carnations..have their tongue-shaped petals powdered with spankled red glister.

spanned (spænd), *a.* *Biol.* [f. Span *sb.*[1] + -ed[2].] Of a culture of cells or micro-organisms: having a restricted lifespan; unable to propagate asexually without limit.

1968 *Exper. Cell Res.* XLIX. 116 To produce 'spanned' amoebae the nutritional state of amoebae must be altered for a protracted period. 1971 J. Z. Young *Introd. Study Man* xxii. 293 Transfer of a normal nucleus into spanned cytoplasm produced resumed logarithmic growth, and a spanned nucleus in normal cytoplasm produced the spanned condition. 1973 *Nature* 22 June 444/1 The switch from a 'spanned' culture to one that is 'immortal' does not necessarily signify a qualitative change in other cellular properties.

spanner[1]. Add: **2. b.** Colloq. phr. *to throw a spanner in the works* and varr.: to cause disruption, to interfere with the smooth running of something. Cf. *monkey-wrench* s.v. *monkey *sb.* 17 a.

1934 Wodehouse *Right Ho, Jeeves* xi. 142 He should have had sense enough to see that he was throwing a spanner into the works. 1939 A. Ransome *Secret Water* i. 18 We can't go. It's all off. The First Lord's chucked a spanner in the works. 1946 D. L. Sayers *Unpopular Opinions* 111 She was in love with Leicester—why didn't she marry him? Well, for the very same reason that numberless kings have not married their lovers—because it would have thrown a spanner into the wheels of the State machine. 1959 *News Chron.* 10 July 4/1 Mr. Cousins has thrown a spanner into the Labour Party's works. 1960 R. East *Kingston Black* ix. 90 My department might be able to throw a spanner into the works—if necessary. 1977 *Time Out* 17–23 June 5/4 Either way, the 60 workers occupying the factory have put a spanner in the works.

4. *attrib.* and *Comb.*, as **spanner tight** *a.*, of a nut: as tight as can be secured manually with a spanner; **spanner wrench** *U.S.*, a non-adjustable spanner.

1925 *Morris Owner's Man.* 53 The nuts should always be kept (small) spanner tight. 1931 *Daily Express* 31 Jan. 3/6 Even where the nut was absolutely spanner tight. 1940 *Sun* (Baltimore) 30 Mar. 20/1 The fuel door..was bolted closed the night before the ship sank, but was found open with a spanner wrench beside it when salvagers examined the sunken vessel. 1969 *Publ. Amer. Dial Soc.* LII. 35 *Spanner wrench*,..a wrench having a fixed distance between its jaws which fits on the hose couplings and is used to tighten or loosen connections.

spanners (spæ·nəɹz), *sb. pl.* Chiefly *dial.* [Pl. of *spanner*, f. Span *sb.*[1] or *v.*[1] + -er[1]: cf. Spanner[3].] A game of marbles (see quot. 1881). Also **spa·nnims.**

1847 J. O. Halliwell *Dict. Archaic & Provinc. Words* II. 779/1 *Spannims*, a game at marbles played in the eastern parts of England. 1881 *Cassell's Bk. Sports & Pastimes* 250 Spanners. This is a good simple game for two players. Player No. 1 shoots off his taw, player No. 2 following suit, it being his object either to hit his opponent's taw with his own, or to place his own within a span of it. 1903 *J. Strutt's Sports & Pastimes of People of England* (new ed.) IV. iv. 304 The following is a list of the present (1902) best known marble games:—Bounce About,..Handers,..Pyramid,..Spanners. 1948 J. L. Bailes *Vocab. Marbles* in *Trans. Yorks. Dial. Soc.* VIII. 23 *Spannims*, sb., E. Ang., the game of hundreds.

spanning, *ppl. a.* Add: **2.** *Math.* Of a subgraph, esp. one that is a tree: that includes and connects every vertex of a graph.

1956 *Proc. Amer. Math. Soc.* VII. 49 The set of edges eventually chosen must form a spanning tree of *G*, and in fact it forms a shortest spanning tree. 1965 Busacker & Saaty *Finite Graphs & Networks* i. 20 A spanning tree is a maximal subgraph of a connected graph which contains no circuits and is a minimal subgraph which joins all vertices. 1972 R. J. Wilson *Introd. Graph Theory* iv. 46 Given any connected graph *G*, we can choose a circuit and remove one of its edges, the resulting graph remaining connected... We repeat this procedure with one of the remaining circuits, continuing until there are no circuits left. The graph which remains will be a tree which connects all the vertices of *G*; it is called a spanning tree of *G*... More generally, if *G* now denotes an arbitrary graph with..*k* components, we can carry out the above procedure on each component of *G*, the result being called a spanning forest (or skeleton).

spanspek (spa·nspek). *S. Afr.* Also **spanspec, sponspe(c)k.** [ad. Afrikaans *spaanspek* sweet melon, f. Du. *Spaans(ch)e* Spanish + *spek* bacon.] = Cantaloup in Dict. and Suppl.

[1731 G. Medley tr. *Kolb's Descr. Cape Good Hope* II. 277 *Melo Hispanicus*, i.e. The Spanish or Musk-Melon. The Musk Melons, produc'd at the Cape, are as good as those produc'd in Spain. The Cape-Europeans call 'em Spanish Bacon.] 1886 G. A. Farini *Through Kalahari Desert* v. 61 Mr. Barlett came and asked if I would like to buy any musk-melons (*spauspeck*). 1913 J. J. Doke *Secret City* 78 The ground was cumbered by water-melons and spanspeks. 1951 R. Campbell *Light on Dark Horse* iv. 70 The district is..scarcely excelled for the size and succulence of its..spanspec, watermelons, strawberries, [etc.]. 1958 *Cape Times* 12 Dec. 11/4 The always-popular sweet melon—spanspek—is not expected in quantities till the New Year. 1971 *Fair Lady* 10 Nov. 111 Baked egg custard/or Medium slice sponspek melon. 1975 *Cape Times* 13 Jan. 2 A Paarl spanspek exporter said in an interview at the weekend that these prices..were paid only at the beginning of the season.

Spansule (spæ·nsiūl). Also **spansule.** [f. Span *sb.*[1] + Cap)sule.] A proprietary name for a capsule that when swallowed releases a drug steadily for several hours, or that releases various drugs at prearranged times. Cf. *sustained-release* s.v. *sustained *ppl. a.* 1 c.

1954 *Official Gaz.* (U.S. Patent Office) 2 Mar. 18/2 Smith Kline & French Laboratories, Philadelphia... *Spansule*... For capsules containing multiples of specially coated globules..and providing for the gradual release of a medicament in the gastro-intestinal tract. 1954 *Trade Marks Jrnl.* 6 Oct. 1000/1 *Spansule*... Pharmaceutical preparations... Smith, Kline & French International Co. 1955 Osol & Farrar *Dispensatory of U.S.A.* (ed. 25) 89/1 Dexedrine Sulfate Spansules contain 15 mg. of the dextro salt in the form of tiny pellets, of which there are over a hundred in each capsule, having varying disintegration times so as to release the drug uniformly over a period of 8 to 10 hours. 1971 C. Williams *And Deep Blue Sea* (1972) ix. 102 Our labs came up several years ago with a timed-release spansule; the opiate takes effect in about twenty minutes and then an aphrodisiac eight hours later. 1974 'J. Black' *Oil* (1975) IV. viii. 349 The two men had stayed up the remainder of the night, swallowing Dexamyl spansules while they meticulously went over all that Needham would have to do.

spanwise (spæ·nwəiz), *a.* *Aeronaut.* [f. Span *sb.*[1] + -wise.] Following the direction of the span of a wing or other aerofoil.

1946 *Jrnl. Brit. Interplanetary Soc.* VI. 95 The phenomenon of tip stall is brought about by spanwise drift in the boundary layer over a swept wing. 1955 J. Shapiro *Princ. Helicopter Engin.* vii. 360 Spanwise mass distribution in uniform section blades..gives rise to considerable bending moments when the blade is stiff. 1967 *Jane's Surface Skimmer Systems 1967–8* 121/2 A drive pad..enables the engines of a multi-engined aircraft to be coupled together by spanwise shafting. 1977 *Jrnl. R. Soc. Arts* CXXV. 350/1 A swept wing was made with narrow spanwise slots on the surface.

spar, *sb.*[1] Add: **4. b.** *Aeronaut.* Each of the main members of a wing on older aircraft, which run transversely to the fuselage and carry the ribs.

1866 *1st Rep. Aëronaut. Soc.* 35 But with all such arrangements the apparatus must fail—*length of wing is indispensable!* and a spar thirty feet long must be strong, heavy, and cumbrous. 1895 *Amer. Engineer & Railroad Jrnl.* Aug. 387/2 Being caught by a side puff, the machine was blown over, and the front starboard spar was too much broken to mend on the field. 1919, etc. [see *rib *sb.*[1] 11 e]. 1930 Nayler & Ower *Aviation of To-Day* vii. 154 The modern steel spar of an aeroplane wing..can be treated by calculation. 1960 C. H. Gibbs-Smith *Aeroplane* xiii. 96 In 1919, he [*sc.* Adolph Rohrbach] started building smooth-skinned metal surfaces, combined with metal box-spar construction in the wings, thus allowing more stresses to be borne by the surfaces.

4*. *Oil Industry.* Also **Spar, SPAR.** An installation intended to float above a submarine well-head and provide large storage tanks and various service facilities, esp. for loading tankers.

1973 *Times* 31 Oct. (Offshore Supply Suppl.) p. iii/2 The concrete spar is anchored above a submarine manifold with pipelines from production platforms. The lower part of the spar is a 300,000-barrel capacity storage chamber. 1975 *Offshore Progress—Technol. & Costs* (Shell Briefing Service) 18 One of the newest deep water concepts is the Spar—floating storage and loading terminal all in one. 1976 *Offshore Platforms & Pipelining* 218 The SPAR has been designed to maintain a constant draft in both the loaded and ballasted conditions. 1979 *North Sea Progress* (Shell Internat. Petroleum Co.) 8 Other methods involve ..the use of custom-built SPAR-type semi-submersible production units.

6. spar tree *Forestry*, a tree or other tall structure to which cables are attached for hauling logs.

1925 [see *high lead* s.v. *high *a.* 21]. 1965 [see *high line* s.v. *high *a.* 21]. 1980 *Beautiful Brit. Columbia* Fall 40 Atop the cliff stands a long-disused spar tree; close by the bay are the broken bricks, cement and tile that testify to a long-abandoned project here.

spar, *sb.*[3] Add: **5.** *Comb.*, as **sparmate** *U.S.*, a sparring partner.

1937 *Sun* (Baltimore) 30 Aug. 15/2 Ten days ago one of his sparmates opened up a deep cut under his eye with a punch. 1950 J. Dempsey *Championship Fighting* 15, I found plenty of kid sparmates. 1974 *Los Angeles Times* 13 Oct. III. 13/3 Chartchai, three-time world flyweight champion, has been staying at a Tokyo hotel,..with his wife and his manager and his wife, sparmate and trainer.

spar, *v.*[3] Add: **2. c.** (Earlier examples.) Also with *off.*

1843 T. Talbot *Jrnls.* (1931) 4 [We became] finally the prey of an insidious sand-bar, where after hours of sparring,..we again resume the slow ascent. 1875 'Mark Twain' in *Atlantic Mag.* May 568/1 Maybe she 'strikes and swings'. Then she has to while away several hours (or days) sparring herself off.

sparagmite (spæ·rāgməit). *Geol.* [ad. G. *sparagmit* (J. Esmark *Reise von Christiania nach Drontheim*, etc. (1829) 25), f. Gr. σπάραγμα fragment, piece torn off: see -ite[1].] A generic term for the feldspathic sandstones (chiefly arkoses), conglomerates, and other fragmental rocks which occur in late Pre-Cambrian formations in Scandinavia.

[1866 *Geol. Mag.* III. 384 (*table*) 'Spragmitic' strata—conglomerates, Schists, Dolomitic beds.] 1882 A. Geikie *Text-bk. Geol.* 657 In central and northern Norway the Archæan gneiss is overlaid by reddish and grey sandstones and conglomerates (Sparagmite), with schists, quartzites, and limestones. 1930 Peach & Horne *Geol. Scotland* iii. 72 The Sparagmite of Scandinavia is an arkose resembling the dominant type of the Torridon Sandstone. 1979 *Nature* 25 Jan. 290/1 Micaceous meta-arkoses (sparagmites) have, in places, a basal conglomerate resting unconformably on the Basal gneisses.

‖ **sparagmos** (spăræ·gmǫs). [a. Gr. σπαραγμός tearing, rending.] The tearing apart of a hero, or his ritualized death in an analogous manner as represented in some tragedies or myths when it symbolizes part of the cycle of death and rebirth. Also *fig.*

1949 F. Fergusson *Idea of Theater* i. 32 The ritual had ..its *Sparagmos*, in which the royal victim was literally or symbolically torn asunder. 1957 N. Frye *Anat. of Criticism* iii. 192 The disappearance of the hero, a theme which often takes the form of *sparagmos* or tearing to pieces. Sometimes the hero's body is divided among his followers. 1961 J. Holloway *Story of Night* vii. 130 The death of Coriolanus is almost as much a *sparagmos* of the ritual victim by the whole social group as was possible on the stage. 1977 *Dædalus* Summer 62 Not that conferences alone can do this, but they are signals that the reconstitution of anthropology at a higher level under the aegis of processualism is under way. Otherwise the centrifugal drift, indeed, the suicidal sparagmos, will go on and on.

spare, *sb.*[1] Add: **3. a.** (Later examples.) Also *spec.* in *pl.*, spare parts.

1907 C. W. Brown *Petrol Engine* i. 11 The manufacture is simplified and the number of 'spares' which the owner of a car is called upon to carry considerably reduced. 1914 *Vanity Fair* Jan. 95/1 Some ingenious modifications have been devised for taking care of the 'spare'. 1930 A. P. Herbert *Water Gipsies* xxiii. 334 I'll put you to bed in the spare, and let nobody come near you. 1957 *Practical Wireless* XXXIII. 701/2 A suitable piece of aluminium can probably be found in almost any spares box. 1976 M. Maguire *Scratchproof* v. 66 Did it usually take him an hour to put on the spare? Would he mind if I looked at the punctured tyre? 1979 B. Parvin *Deadly Dyke* v. 23 A small room with a single bed..had never been used... It would have been thought of as the spare.

b. *slang.* An unattached woman, esp. one available for casual sex. Freq. in phr. *a bit of spare.*

1969 J. Boland *Shakespeare Curse* xxi. 169 Kelley was a man whose wife was in an advanced state of pregnancy. You think he'd turn down a bit of spare if it was offered to him? 1974 P. Cave *Dirtiest Picture Postcard* x. 61 The men would not have to bother with the married girls anyway. There's plenty of spare about. 1978 R. Busby *Garvey's Code* iv. 44, I..got the impression Maurice was ..on the look-out for a bit of spare... Some of the girls we get in here..don't leave much to the imagination.

4. For *U.S.* read orig. *U.S.* and add earlier and later examples. Also in skittles.

1843 *Knickerbocker* XXII. 327 His bowling at nine-pins was the very perfection of carelessness. He was never guilty of a 'spare'. 1976 *Bridgwater Mercury* 21 Dec., Keith Pollard, whose top-of-the-board 84 included four spares, led Alleycats to a runaway home win..in a first division Puriton and District Skittles League Game.

spare, *a.* and *adv.* Add: **I. 1. a.** (*a*) *spare tyre*, (i) an extra tyre carried in a motor vehicle for emergencies; similarly, *spare wheel*; also (ii) *transf.*, a roll of fat around the stomach (*colloq.*).

1917 Kipling *Diversity of Creatures* 162 The policeman laid his hand on the rim of the right driving-door (Woodhouse carries his spare tyres aft). 1920 'O. Douglas' *Penny Plain* xxiii. 267 It was a tyre gone... Stark put on the spare wheel and they started again. **1961** *Harper's Bazaar* Dec. 43/1 The deep diaphragm section slims you... That 'spare-tyre' has vanished! **1971** D. DEVINE *Dead Trouble* v. 48 My spare tyre keeps me warm. You're too skinny. **1972** *Country Life* 7 Dec. 1592/3 The luggage boot is..fairly well filled by the spare wheel. **1977** *Lancashire Life* Nov. 153/1 There is no need for a spare tyre to clutter up the Mini's limited boot space.

(*d*) *spare room*, a room not regularly used, esp. a bedroom reserved for visitors.

1814 JANE AUSTEN *Mansfield Park* I. iii. 54 The absolute necessity of a spare-room for a friend was now never forgotten. **1837** *Southern Lit. Messenger* III. 333 One of the third-story rooms we must keep for a spare room. **1855** *Knickerbocker* XLVI. 380 They have stolen away into the spare-room, otherwise, parlor. **1904** A. DALE *Wanted: a Cook* 332 The wine-cellar was under the bed in the spare-room. **1928** GALSWORTHY *Swan Song* III. ix. 280 He spied a spare-room window open at the top. **1953** E. SIMON *Past Masters* III. 169 The spare room, newly done up, was frequently inhabited by..distinguished visitors. **1977** J. PORTER *Who the Heck is Sylvia?* ix. 79 Her habit of knocking on the spare-room door before entering.

e. *colloq.* Of persons: off-duty, idle (cf. sense 1 a (*c*) in Dict.). Also, useless, superfluous.

1919 *Athenæum* 1 Aug. 695/2 'To be spare' is to be temporarily off duty. **1925** FRASER & GIBBONS *Soldier & Sailor Words & Phrases* 266 *Spare, to look*, to be idle: not engaged on any particular job. **1936** J. CURTIS *Gilt Kid* xv. 154 We can't stand around here spare... Come on. **1970** 'D. HALLIDAY' *Dolly & Cookie Bird* viii. 117 Janey stayed there with her manicured hand on his brow..and I felt a bit spare.

f. Phr. *to go spare*: (*a*) to be unemployed; (*b*) to become infuriated or distraught. *colloq.*

c **1942** R. DIMBLEBY *Let.* in J. Dimbleby *Richard Dimbleby* (1975) vii. 163 I'd be grateful if your team would remember an at least practised broadcaster who appears to be 'going spare'! **1958** F. NORMAN *Bang to Rights* 169 When he saw what I had done he went spare. **1969** J. N. SMITH *Is he Dead, Miss ffinch?* xv. 95 The train had just gone. His lordship nearly went spare. **1975** T. HEALD *Deadline* iv. 68 What's the time? Monica will be going spare.

II. 4. e. Of style: unadorned, bare, simple.

1965 *Listener* 7 Oct. 552/2 The narrative..was spare, precise, almost a little cold, and made its tale of muddle and butchery thereby the more devastating. **1966** *Ibid.* 12 May 702/2 We feel the participants to be in agony and it is impossible to remain indifferent to them. This achievement has something to do with the spare, angular dialogue. **1977** *Times* 23 May 25/1 Tom Courtenay gives a frighteningly spare performance in One Day in the Life of Ivan Denisovich.

III. 8. *spare-bodied* (later example); **spare-time** *a.*, that is done in one's spare time; operating in or occupying spare time.

1936 'M. INNES' *Death at President's Lodging* ii. 39 The spare-bodied man that he was. **1931** Spare-time [see *PART-TIME *a.*]. **1955** BLUNDEN *Addresses on General Subjects* 24 This poet [*sc.* Shelley] almost achieved, as one of his spare-time labours, the establishment of the first steamship service in the Mediterranean. **1973** A. HOLDEN *Girl on Beach* 143 He really is a professional lawyer after all, and merely a spare-time amateur art critic. **1978** *Nagel's Encycl.-Guide: China* 320 The 'Spare Time Industrial University' at Shanghai.

[illegible]

2. *transf.* A visceral organ or other bodily part from a donor, or a prosthetic device, which is to be used to replace a defective organ, etc., in a person. Freq. *attrib.*, as *spare-part(s) surgery*. *colloq.*

1944 [see *BANK *sb.*[3] 7 f]. **1960** S. PLATH *Colossus* 87 The storerooms are full of hearts. This is the city of spare parts. **1963** *Daily Tel.* 21 Sept. 9/5 Spare part surgery is still in its infancy. **1967** *New Scientist* 25 May 449/1 With the technique of kidney transplantation now firmly established..spare-part surgeons are now turning their attention to other organs in the body. **1968** *Guardian* 6 May 8/1 Spare parts surgery is too important to be left to the surgeons alone. **1970** D. J. MARLOWE in *Mystery Writers' Choice 1977* (1977) 4 'Well, what about your body?' I still didn't care for the idea about being used for spare parts. **1977** B. PYM *Quartet in Autumn* ii. 20 She could donate certain organs to assist in research or spare-part surgery.

sparganosis. Add: **3.** Infection with larval tapeworms of the genus *Sparganum*. (The only current sense.)

1928 *Proc. Soc. Exper. Biol. & Med.* XXVI. 254 Sparganosis results from applying frogs infected with these spargana to inflamed and ulcerated areas of the body. **1954** *Amer. Jrnl. Tropical Med. & Hygiene* III. 123 It is probable that human sparganosis is acquired in Korea by the consumption of raw snake, which is a fairly common practice in this area. **1976** EDINGTON & GILLES *Path. in Tropics* (ed. 2) iii. 210 Sparganosis has been reported in man from many tropical countries such as South East Asia, Africa, Madagascar, China, and South America.

sparge, *sb.* Add: **3.** *attrib.*, as **sparge arm** *Brewing*, a sparge pipe used to sprinkle hot water over the malt, usu. arranged to rotate above the tun; **sparge pipe,** a horizontal perforated pipe used to sprinkle or spray water or other liquids, esp. one used to flush a slab urinal.

1947 *Brewing* (ed. 2) ii. 23 The holes in the actual sparge arms have to be made much larger. **1971** J. S. HOUGH et al. *Malting & Brewing Sci.* x. 261 Other rotating machinery within the mash tun comprises sparge arms (which are moved by water pressure) for spraying liquor evenly over the goods. **1910** MAXWELL & BROWN *Encycl. Municipal & Sanitary Engin.* 444/2 *Sparge pipe*, a pipe having fine holes drilled throughout its length so as to deliver a spray of water as is required for flushing. **1948** *Archit. Rev.* CIV. 289 A sparge pipe on the sky-light ridge cools the glass by water spray in summer. *c* **1972** *Shanks* (Barrhead) *Catal.* 93 Chromium plated flush and sparge pipes with clips.

sparge, *v.* Add: **4. b.** To aerate (a liquid) with air (in quot. *absol.*).

1973 *Nature* 23 Feb. 534/1 The *p*H was adjusted to 3·8 with concentrated HCl and aeration and mixing accomplished by sparging with air.

sparger. Add: Also, an appliance for sprinkling a liquid other than water, or for aerating a liquid with air.

1949 E. CHAIN in H. W. Florey et al. *Antibiotics* II. xvii. 700 The fermenters..were made of steel; they contained a sparger for air distribution. *Ibid.*, The air sparger consisted of a ¾-in. stainless steel pipe bent to form a 12-in. square. **1963** ECKENFELDER & MCCABE *Biol. Waste Treatment* 183 Spargers, each with four 5/16 in. diameter holes radiating horizontally from the center were checked for oxygen absorption efficiency. **1976** *IEEE Spectrum* May 54/2 An early example of..fatigue failure was that of the liquid boron, or 'poison', sparger at the Garigliano Plant of Ente Nazionale per l'Energia Electrica..in Cessa Aurunca, Italy, in 1964.

spargosis. Add: Also *fig.*

1934 'H. MACDIARMID' *Stony Limits & Other Poems* 23 In open country..watching an aching spargosis of stars.

[illegible]

where growing more and more out of the thing that had been Man. **1920** H. J. LASKI *Let.* 15 May in *Holmes-Laski Letters* (1953) I. 263 They may be stupid, lazy, what you will; but ninety-nine out of every hundred have a divine spark in them somewhere which sympathy and enthusiasm is sure to light. **1932** *Week-End Rev.* 9 Jan. 46/2 She felt she had not got into touch, had not given herself, had not transmitted the divine spark. **1957** *Oxf. Dict. Chr. Ch.* 1218/1 As originally formed, man was a powerless entity who wriggled on the ground like a worm ..until a Divine spark set him on his feet. **1968** F. LUNDBERG *Rich & Super-Rich* xv. 632 (*heading*) The divine spark among the rich.

6. c. *pl.* usu. const. as *sing.* One who works with electrical equipment: a radio operator, an electrician, etc. *slang.*

1914 *Dialect Notes* IV. 151 *Sparks*, wireless operator. **1917** *Wireless World* V. 37 In the Service the regular nickname for wireless telegraphists is graphically expressed as 'Sparks'. **1922** P. F. WESTERMAN *Wireless Officer* iii. 25 A burly, jovial-featured man..greeted Mostyn as he stepped off the gang-plank. 'Hello, you're our Sparks, aren't you?' **1934** *Sun* (Baltimore) 31 Jan. 20/3 'Sparks', the radio operator, was busy at his key. **1938** H. BORUST *In Plain Clothes* xii. 184 Scene shifters, 'sparks' (light men), wardrobe-keepers. **1971** *Guardian* 24 Sept. 12/4, I went to Manchester as a spark's mate—an electrician's mate. **1975** *Listener* 10 Apr. 461/3 Lord Sneaker tells his sparks to wrap up the lights. **1977** M. BABSON *Murder, murder, Little Star* viii. 56 The Technical Crew were called by the names of the jobs they did... Sparks was the electrician, Props was the property master, Camera the cameraman. **1980** R. MITTON *Master & Son* i. 9 Meet Ulrica Halsted..the sexiest Spark that ever went to sea.

d. Short for *spark telegraphy* (cf. senses *7 b, d).

1921 *Wireless World* 2 Apr. 21/1 Commencing by pointing out the advantages and disadvantages of various circuits for the reception of Spark, C. W. and Telephony, Captain Tingey gave many useful hints. **1922** *Ibid.* 15 Apr. 76/2 One ought to..switch that connection on to different places when one is receiving spark or telephony. **1925** *Weekly Dispatch* 22 Nov. 8/2 If the Government were to replace spark by continuous wave the loss on old apparatus could be set aside by the revenue from wireless licences.

7. b. More widely, as *spark discharge(r)*, *frequency*, *-gap* (later examples; also *attrib.*), *source*, *station*, *tester*, *transmitter*.

1848 *Patent Jrnl.* 5 Aug. 266/1 In frictional spark discharges, the consequent shock, light, and other peculiarities are in part owing to waves of..polarization. **1973** L. R. LENTZ et al. in *Automotive Electr. Equipment* (Inst. Mech. Engineers) 63/2 The fuel mixture is ignited prior to the occurrence of the normal spark discharge. **1921** E. E. BUCHER *Pract. Wireless Telegr.* 101 Spark dischargers for radio-telegraphy. **1906** J. A. FLEMING *Princ. Electric Wave Telegr.* ii. 157 The author has..devised the following appliances for measuring spark frequency. **1925** W. GREENWOOD *Wireless Telegr. & Teleph.* iv. 75 If a low spark frequency is required the alternator circuit can be tuned to the alternator frequency, and the spark gap lengthened. **1905** *Electrician* Feb. 614/1 Measurements of spark-gap resistance in wireless telegraph senders. **1935** *Discovery* Aug. 226/1 There are two different types of short-wave generator in actual use, the valve and the spark-gap oscillators. **1967** *New Scientist* 14 Dec. 671/1 The operator closes the discharge switch, and the electric charge in the capacitors leaps across the spark gap. **1944** *Jrnl. Optical Soc. Amer.* XXXIV. 773/2 A type of interrupted spark source involving no mechanical parts has been developed for use in quantitative spectrographic analysis. **1956** *Nature* 4 Feb. 222/1 A high-precision spark source and an optical arrangement containing a rotating mirror are needed. **1913** *Year-Bk. Wireless Telegr. & Teleph.* 401 It does not follow..that a continuous-wave station is immune from interference by a spark station. **1925** *Sci. Abstr.* B. XXVIII. 232 (*heading*) International measurements of the wave-lengths of spark stations. **1925** *Morris Owner's Man.* 84 The motorist should try each plug in turn with the aid of a 'spark tester'. **1916** J. A. FLEMING *Princ. Electric Wave Telegr.* (ed. 3) 671 The types of transmitter employing such condenser discharges are called spark transmitters. **1934** A. L. ALBERT

[illegible]

for charged particles consisting of two charged electrodes separated by a gas that is ionized by the passage of the particle; **spark erosion** *Engin.*, a method of machining metal in which a series of electric sparks is used to remove droplets from the piece; freq. *attrib.*; hence **spark-erode** *v. trans.*, to machine (a piece) by spark erosion; **spark line,** a spectral line corresponding to an atom in a given state of ionization; **spark machining** *Engin.* = *spark erosion* above; **spark-prop** *Criminals'*

slang, a diamond pin, a tie-pin; **spark spectrum,** a spectrum produced by an atom in a given state of ionization, commonly excited under laboratory conditions by an electric spark; **spark telegraphy,** an early method of radio-telegraphy in which high-frequency oscillations are set up in a transmitting aerial by the discharge of a highly charged capacitor through a spark gap in series with an inductance connected to the aerial; hence **spark telegraph.**

1902 *Encycl. Brit.* XXXIII. 230/2 The distance at which the effects of the oscillatory spark could be perceived by the aid of the coherer was closely connected with the height of this air-wire or aerial connected to the spark balls and coherer. **1924** O. LODGE *Harmsworth's Wireless Encycl.* III. 1864/2 The object of replacing the pointed ends of the wires by spark balls is to prevent the gradual leaking discharge. **1961** *Rev. Sci. Instruments* XXXII. 482/1 The spark chamber is a direct outgrowth of an older detector called the spark counter. **1974** FRAUENFELDER & HENLEY *Subatomic Physics* iv. 56 Spark chambers have many of the advantages of bubble chambers, and they can be triggered. [**1868** Spark coil: see *SPRENGEL.] **1900** G. D. HISCOX *Horseless Vehicles* vii. 121 The Edison spark coil..is a short, thick coil, which will give a hot, bright spark, and yet will have an instantaneous discharge. **1902** [in Dict., sense 7 b]. **1922** [see *jump-spark, jump spark* s.v. *JUMP-]. **1971** *Sci. Amer.* May 86/2 The ordinary automobile spark coil..is the commonest version of the induction coil. **1935** *Sci. Abstr.* XXXVIII. 718 The general name 'spark counter' is proposed for the new counters devised by the author. **1970** *Nucl. Instruments & Methods* LXXXVII. 181/1 Cylindrical spark counters have been designed that employ boron nitride disks as converters. **1980** J. W. HILL *Intermediate Physics* xxiii. 220 (*heading*) The spark counter. **1960** *Metal Treatment* XXVII. 206/1 When a die showed signs of wear, one must be able to take it out, spark-erode it and replace it. **1955** *Aircraft Production* XVII. 421/1 The process of spark-erosion machining is becoming widely used for special purposes, such as machining holes in tungsten-carbide. **1977** R. B. ROSS *Handbk. Metal Treatments & Testing* 360 Spark erosion is extremely useful where shapes are required in hardened or difficult to machine materials. **1980** *West Lancs. Evening Gaz.* 4 Jan. 10 (Advt.), Familiarity with close tolerance machining and bench work is essential and some experience of spark erosion machining would be an advantage. **1879** *Proc. R. Soc.* XXX. 27 The spark lines are in the sun, but the less refrangible member of the wide triplet and the blue line seen in the flame are absent. **1932** *Ibid.* CXXXIV. 611 It is proposed to give a complete catalogue of the spark lines of arsenic. **1950** *Jrnl. Optical Soc. Amer.* XL. 180/1 By adjusting the conditions of discharge either arc or spark lines may be made to predominate. **1954** *Engineer* 2 July 12/2 Although spark machining may, in theory, be carried out with electrode and workpiece separated only by air, in practice a liquid dielectric is used. **1973** J. G. TWEEDDALE *Materials Technol.* II. vi. 152 Spark machining is applicable only to electrically conducting materials. **1879** *Macmillan's Mag.* Oct. 506/1 My pal said, 'Pipe his spark prop' (diamond pin). **1923** J. C. GOODWIN *Sidelights on Criminal Matters* iii. 32 To steal a tie-pin, or 'spark prop' as it is termed in the slang of thieves, [etc.]. **1873** *Phil. Trans. R. Soc.* CLXIII. 266 (*table*) The spark-spectrum of the chloride. **1905** E. C. C. BALY *Spectroscopy* 374 The induction coil is used..for the production of the so-called spark spectra of substances. **1970** G. K. WOODGATE *Elem. Atomic Struct.* vi. 105 Na I is also called the arc spectrum of sodium, Mg II the first spark spectrum of magnesium and Al III the second spark spectrum of aluminium. **1934** A. L. ALBERT *Electr. Communication* xv. 426/1 (*caption*) The generation of damped waves with a spark telegraph set. **1898** *Ludgate* VII. 78/1 Mr. Marconi, in July, 1897, came to England to introduce his new plan of 'Spark Telegraphy'. **1908** *Rep. Brit. Assoc. Adv. Sci. 1907* 730 A movement which much more nearly corresponds with the actual current in the vertical wire as used in spark telegraphy.

spark, *v.*[1] Add: **1. a.** *spec.* in *Electr.* To produce or emit an electric spark or sparks by ionization of the medium separating two conductors at different potentials. Also *fig.* in phr. *to spark on all cylinders* = to *function* (etc.) *on all cylinders* s.v. *CYLINDER *sb.* 6.

1884 S. P. THOMPSON *Dynamo-Electr. Machinery* 60 Any dynamo in which the curve of potentials at the commutator presented such irregularities..would probably spark excessively at the collector. **1905** T. H. HAWLEY *Motor Ignition Appliances* iv. 20 If our charge fails to explode or the plug to spark, the fault must usually be sought for elsewhere. **1926** R. W. HUTCHINSON *Wireless* 112 The spark gap consists of two small spheres sparking across the diameters of two larger ones. **1967** L. BACON in L. Holmes *Odhams New Motor Man.* iii. 81 One simple check can clear the whole of the ignition circuit—are the plugs sparking? **1977** M. HINXMAN *One-Way Cemetery* viii. 55 John realized his inspector was sparking on all cylinders. He looked a damned sight fresher than Waller felt.

c. With *over*. To be crossed or connected by a spark as a result of a breakdown in insulation. Also *fig.*

1915 *Standardization Rules Amer. Inst. Electr. Engineers* 48 The voltage at which a given gap sparks over is found by taking the voltage corresponding to the spacing..and multiplying by the correction factor. **1966** R. ARDREY *Territorial Imperative* iii. 88 When antagonists face each other..inhibited from further attack..their energy.. 'sparks over'—another ethologist's term—into a third instinctual channel which will cause no damage. **1974** *Sci. Amer.* Feb. 78/2 Since the magnets were not designed to work under water many of them sparked over and failed when power was fed into them.

3. d. *fig.* To fire, to inspire; to kindle, to set in motion; *to spark off*, to be the immediate cause of (something hard to control). orig. *U.S.*

1912 L. J. VANCE *Destroying Angel* ii. 21 Abrupt inspiration sparked the imagination of Peter Stark, and he began to sputter with enthusiasm. **1941** *Sun* (Baltimore) 24 Apr. 15/2 He is the type [of ballplayer] that sparks an infield and hustles all of the time. **1947** *Richmond* (Va.) *Times-Dispatch* 1 Dec. 10/1 We hope that circumstances which might well spark another world conflict will not rise in Palestine. **1957** *Economist* 26 Oct. 287/2 Any encroachment on vital western interests is liable to spark off the sort of war that would incinerate communism along with communists. **1962** *Listener* 4 Oct. 501/1 A story has come out of California and sparked the November election campaign, which is now beginning to smoulder and crackle. **1964** *Ann. Reg. 1963* 100 The final decision to end Federation sparked off immediately a new constitutional controversy. **1970** S. L. BARRACLOUGH in I. L. Horowitz *Masses in Lat. Amer.* iv. 158 The more widely distributed post-reform incomes help spark development by changing propensities to invest. **1978** *Dumfries Courier* 20 Oct. 5/1 The club has taken no stand against this sort of behaviour, which can spark off bad behaviour among the spectators. **1979** *IEEE Trans. Professional Communication* XXII. 70/1 Man has always been intrigued by the elusive nature of the brain mechanisms which spark new and unexpected ideas to solve problems. **1981** *Times* 24 July 23/1 Stocks on the New York Stock Exchange closed higher due to a late afternoon rally, sparked by bargain hunting among oil stocks and blue chip issues.

sparker. Add: **3.** A powerful form of sonar apparatus used to investigate solid structures underlying sediment on the sea bed.

1961 *Ann. Reg. 1960* 406 The whole area was surveyed by the newly developed 'sparker' technique, an extension of the echo-sounding method used for charting the sea-bottom. The sparker used an intense sound source (a spark) which enabled the sound wave to penetrate the sea bottom. The sound waves were reflected from the interfaces between different kinds of rock. **1969** J. MAVOR *Voyage to Atlantis* II. iii. 84 The 'sparker' and line hydrophone for listening to the reflected sound signals were the most powerful in the world but required a depth of 60 feet and 150 yards of towline astern. **1974** *Nature* 26 Apr. 745/1 An 8 kJ sparker was also used.

sparking, *vbl. sb.*[1] Add: **b.** Special Combs.: **sparking coil** = *spark coil* s.v. *SPARK *sb.*[1] 7 d; **sparking plug,** a device that is fitted to the cylinder head of an internal-combustion engine and used to ignite the explosive mixture in the cylinder by the discharge of a spark between two electrodes at its end.

1897 G. D. HISCOX *Gas, Gasoline, & Oil Vapor Engines* x. 74 The sparking coil..consists of a bundle of iron wire, insulated and wrapped with insulated copper wire. **1929** *Proc. Inst. Automobile Engineers* XXIII. 252 If the sparking coil and plug gap be so balanced that the spark is just able to pass without a following arc, then the least widening of the gap will result in a failure of the spark to pass. **1902** Sparking plug [in Dict.]. **1908** tr. *Lieckfeld's Oil Motors* v. 85 A further improvement in electric ignition was introduced by the French firm [*sc.* de Dion et Bouton] when they brought out the sparking plug. **1929** E. LINKLATER *Poet's Pub* xxvi. 279 The life of a sparking-plug is a fierce tropical existence. **1970** K. BALL *Fiat 600, 600 D Autobook* iii. 35/2 For the best performance, sparking plugs should be renewed every 10,000 miles.

sparkle, *v.*[1] **3. b.** (Later example.)

1981 *Times* 10 June 20 Banks and insurances sparkle. An initial flurry of activity was seen in banks and insurances.

sparkler. 5. (Later examples.)

1975 *Country Life* 2 Oct. 839/3 Other sparklers made by the champagne method can be excellent: notably Sparkling Saumur and Seyssel from Savoy. **1981** *Times* 6 Mar. (Bride & Home Suppl.) p. vi/4 There are lots of good sparkling wines..Blanquette de Limoux or the Spanish sparklers.

sparklessly, *adv.* (Example.)

1906 H. M. HOBART *Elem. Princ. Continuous-Current Dynamo Design* iv. 106 Prior to coming into the position of short-circuit under the brush, it has been carrying a current of, say, 100 ampères in one direction. Immediately after emerging from the position of short-circuit, it will be carrying a current of 100 ampères in the opposite position. The change must take place sparklessly.

sparklet. Add: **3.** Properly **Sparklet,** with capital initial. The proprietary name of a metal capsule containing carbon dioxide under pressure, used to carbonate the water in a siphon; the siphon or bottle containing such a capsule; the carbonated water produced by this device. Freq. *attrib.*

1896 *Trade Marks Jrnl.* 23 Dec. 1102 *Sparklets.* 198134. Metal capsules to contain compressed gases. Aerators, Limited,..London; Manufacturers. **1902** KIPLING in *Collier's* 6 Dec. 8/1 Captain Mankelton..fed me canned beef and biscuits and give me a cigar—a Henry Clay—and a whiskey and sparklet. **1903** M. A. STEIN *Sand-Buried Ruins Khotan* ix. 158 The water from the well.. tasted extremely brackish, and neither filtering nor the lavish use of 'Sparklets' could make it palatable. **1905** [see *AERATE *v.* 3]. **1911** *T. Eaton & Co. Catal.* Spring & Summer 175/2 Fresh soda water at a minimum cost with a Sparklet Bottle and a Sparklet Bulb. **1920** *Blackw. Mag.* Apr. 542/2 His particular fancy in drinks—usually whisky and sparklet. **1929** J. BUCHAN *Courts of Morning* II. xii. 271 A mess-servant brought him a long drink of lime-juice and sparklets. **1933** J. CARY *American Visitor* v. 51 He bustled to fetch glass and sparklet bottle. Bewsher was famous for his hospitality. **1971** R. DENTRY *Encounter at Kharmel* iii. 54 He squirted soda water from the sparklet syphon into the generous measure of whisky.

sparkly (spɑ̄·ɹkli), *a.* [f. SPARKLE *sb.* or *v.*[1] + -Y[1].] Sparkling; that sparkles. Also *fig.*, lively.

1922 *Glasgow Herald* 18 Dec. 8 Among the shiny, sparkly rings. **1957** J. FRAME *Owls do Cry* 124 Evening bags covered with sparkly beads. **1978** *Detroit Free Press* 5 Mar. 14 (Parade Suppl.) 14E/3 (Advt.), A wee bouquet—jewel-bright & sparkly as springtime! **1979** *Rescue News* Sept. 8/3 The shop needs a good facelift and more positive selling policy with a wider range of goods and a sparklier range of staff.

spark out, *predic. a.* (and *adv.*) *dial.* and *colloq.* [perh. f. SPARK *sb.*[1] + OUT *adv.*; cf. SPARK *v.*[1] 2 c.] **a.** Utterly extinguished. **b.** *fig.* Forgotten; (completely) unconscious. Also as *adv.*

1880 J. HARTLEY *Halifax Clock Almanack* 17 Th' fire wor spark aght. **1882** NODAL & MILNER *Gloss. Lancs. Dial.* 249 He'll goo spark-out—*i.e.* be entirely lost or forgotten. **1936** J. CURTIS *Gilt Kid* v. 51 Eileen..was drunk, proper lit up and had passed spark out. **1952** M. ALLINGHAM *Tiger in Smoke* xi. 172 He's spark out, only just breathin'. Bin like that two days. **1966** A. PRIOR *Operators* xvi. 264 When she goes out, she's spark out. For three hours at least. **1977** *Zigzag* Aug. 5/1 If she tried to physically stop me doing something I would knock her spark out. **1971** F. NORMAN *Dodgem-Greaser* iii. 32 He keeled over like a felled tree and crashed to the canvas, spark out.

spa·rkover. Also **spark-over.** [f. vbl. phr. *to spark over*: see *SPARK *v.*[1] 1 c.] = *FLASH-OVER.

1915 *Standardization Rules Amer. Inst. Electr. Engineers* 48 The Spark-over voltage, for a given gap, decreases with decreasing barometric pressure and increasing temperature. **1936** *Nature* 19 Sept. 509/1 The crackling sound of more or less prolonged duration finishes with a loud crack coincident with the final sparkover. **1942** [see *ACCELERATING *vbl. sb.*]. **1946** *Nature* 2 Nov. 603/2 The sparkover voltage of different electrode arrangements in air.

spark plug, *sb.* Also (chiefly in *fig.* use) **spark-plug, sparkplug.** [f. SPARK *sb.*[1] or *v.*[1] + PLUG *sb.*] **1.** = *sparking plug* s.v. SPARKING *vbl. sb.*[1] b in Dict. and Suppl. Cf. *PLUG *sb.* 2 m.

1903 *Motor* 3 June 376/1 A small stiff bristle tooth brush..is very convenient for cleaning the porcelain of a spark plug. **1914** F. STRICKLAND *Petrol Motors* (ed. 2) 55 The high-tension leads and spark plug. **1920** J. D. MORGAN *Electr. Spark Ignition* iv. 43 A natural starting point for the study of spark plugs..is the voltage required to produce sparking. **1931** J. A. POLSON *Internal Combustion Engines* xii. 250 Deposits of carbon on spark plugs may cause failure by providing a path from the central electrode to the shell. **1973** V. CANNING *Flight of Grey Goose* ix. 171 He can't use ours [*sc.* a boat]... I got the spark plug in me pocket.

2. *fig.* One who or that which initiates or is the driving force behind any activity or undertaking. *colloq.* (chiefly *U.S.*).

1941 *Sun* (Baltimore) 24 Aug. 15/2 Introducing Hal Sieling... He's sparkplug of infield. **1953** M. MCCARTHY *Groves of Academe* viii. 159 In an experimental college like our own..a teacher's excitement is the spark-plug behind the whole system. **1958** [see *LIQUIDITY 1 d]. **1977** *Time* 14 Mar. 53/3 Lillee is the Australian sparkplug.

Hence **spa·rk-plug** *v. trans.*, to inspire, encourage, or lead (some activity). *U.S. colloq.*

1945 *Sun* (Baltimore) 22 Sept. 4-0/2 Our own reconversion directors have just taken the lid off the construction industry with the hope..that a boom here will help sparkplug a general re-employment drive over the reconversion badlands. **1956** A. H. COMPTON *Atomic Quest* 322 It is important that our government spark-plug the advance of nuclear power. **1961** *Time* 7 July 63/2 Ray Stevens, 67, who sparkplugged the company's diversification, was president from 1956 to 1960.

sparky, *a.* **2.** (Further *fig.* examples.)

c **1865** G. M. HOPKINS *Poems* (1967) 147 Thus he.. Gilds with some sparky fancies his black night. **1928** D. H. LAWRENCE *Lady Chatterley's Lover* x. 134 Pure, sparky, fearless new life! **1979** *Daily Tel.* 6 Apr. 15/2 If they [*sc.* the films] are all as sparky as these larks they deserve support.

sparmannia (spɑ̄ɹmæ·niă). [mod.L. (Linnæus filius *Supplementum Plantarum* (1782) 41), f. the name of Andres *Sparrman* (1748–1820), Swedish traveller + -IA[1].] A large hairy shrub of the genus *Sparmannia* (family Tiliaceæ) native to southern Africa, and bearing toothed heart-shaped leaves and clusters of white flowers, esp. *S. africana*.

1801 *Curtis's Bot. Mag.* XV. 516 African Sparmannia... This beautiful shrub is a native of the Cape of Good Hope. **1818** M. EDGEWORTH *Let.* Sept. (1971) 101, I wish I could get a Sparmania [*sic*] for Lady Lansdowne. **1885** T. BAINES *Greenhouse & Stove Plants* 317/2 Most of the

insects that are troublesome on indoor plants will live on this Sparmannia. **1956** X. FIELD *Housewife Bk. House Plants* III. 84 The sparmannia likes plenty of elbow room and is inclined to swamp its neighbours. **1976** *Homes & Gardens* Aug. 75/3 Bold foliage plants that will fill out your glass-enclosed jungle include..sparmannia.

sparrer² (spæ·rəɹ), repr. dial. pronunc. of SPARROW; also as first element of SPARROW-GRASS.

1884 [see NEGRO 1 d]. **1935** in Z. N. Hurston *Mules & Men* I. vii. 153 He seen a sparrer sittin' on a dead limb of a tree. **1961** *John o' London's* 19 Oct. 447/1 Cloak-and-dagger intellectuals, or game little cockney sparrers. **1970** N. STREATFEILD *Thursday's Child* xxiii. 156 That Ebeneezer we 'ave to 'elp..'asn't got no more brains than a sparrer. **1979** R. CASSILIS *Arrow of God* III. ix. 78 A dark-skinned..dhoti-clad cockney sparrer.

sparring, *vbl. sb.* Add: **3. sparring partner,** a boxer employed to practise with another in training for a contest; also *transf.* and *fig.*, esp. a person with whom one enjoys arguing.

1908 *Captain* Sept. 530/2 Jack was the best sparring partner he could have. **1930** *Daily Express* 8 Sept. 2/3 Cal Barton (Birmingham), Jack Hood's sparring partner. **1958** *Victorian Studies* I. 248 Mill..is used with the utmost disrespect as a sort of punch-drunk sparring-partner. **1971** H. WILSON *Labour Govt.* xxv. 493 Mr Mikoyan, my old sparring-partner and friend of the 1947 trade talks, now in honourable retirement, was waiting for us. **1977** 'E. MCBAIN' *Long Time No See* iv. 60, I was workin [*sic*] with Warren and a sparring partner... Warren's got a fight Tuesday night.

sparrow. Add: **1. d.** A chirpy, quick-witted person; used *spec.* of a Londoner, in *cockney sparrow*, etc.

1861 C. M. YONGE *Stokesley Secret* iv. 62 'A cock-sparrow for her London manners.'..'A London-bred sparrow; a pert forward chit.' **1892** —— *That Stick* I. ix. 95, I care about my neighbours..after a sort, but the jolly city sparrows of the slums for me! **1961** [see *SPARRER²]. **1969** J. GARDNER *Compl. State of Death* vi. 116 'Your mother's English?' 'Yes. Ma's a little cockney sparrow.' **1969** G. BUTLER *Coffin's Dark Number* xii. 151 She recognized her for what she was, one of those sharp little London sparrows. **1977** *Zigzag* Aug. 4/1 There are never any low class twits, only cockney sparrows and 'characters'.

3. *sparrow-shot*; *sparrow-legged* adj.

1965 F. SARGESON *Memoirs of Peon* vi. 173 Two young sparrow-legged ruffians. **1761** STERNE *Tr. Shandy* III. x. 33 Small curses..upon great occasions..are but so much waste of our strength... They are like sparrow shot.. fired against a bastion. **1955** L. DURRELL *Tree of Idleness* 48 Under rain, that rattles down the leaves like sparrow-shot.

4. sparrow-brain *colloq.*, (a person with) a tiny brain, (a person of) limited intelligence or perception; **sparrow-bub** *dial.*, a fledgling sparrow; **sparrow cop** *U.S. slang* (see quots.); **sparrow-fall** *poet.* (cf. SHAKES. *Ham.* v. ii. 233), a sparrow's death; hence *gen.* mortality; also *attrib.*; **sparrow-fart,** (*a*) *dial.* and *slang*, break of day, very early morning; also *pl.*; (*b*) *rare*, a person of no consequence.

1930 V. SACKVILLE-WEST *Edwardians* v. 241, I don't suppose it satisfies anyone, except perhaps a sparrow-brain like mother. **1975** H. FLEETWOOD *Picture of Innocence* iii. 45 She didn't actually *care* about her, and even, with her sparrow brain, despised her. **1917** D. H. LAWRENCE *Look! We have come Through!* 159 Curious long-legged foals, and wide-eared calves, and naked sparrow-bubs. **1896** *Harper's Mag.* June 104/1 The boys do call a park policeman a sparrow-cop, don't they? **1935** A. J. POLLOCK *Underworld Speaks* 111/1 *Sparrow cop*, police or motorcycle officer who patrols parks and boulevards (petty thefts and petting parties). **1960** WENTWORTH & FLEXNER *Dict. Amer. Slang* 506/2 *Sparrow cop*, a policeman in disfavor with his superiors and assigned to a park to guard the grass. **1946** DYLAN THOMAS *Deaths & Entrances* 40 Mammoth and sparrowfall Everybody's earth. **1970** T. HUGHES *Crow* 32 An old man..Gazed towards the nearby polished shoes And slowly forgot the deaths in Homer The sparrowfall natural economy Of the dark simple curtain. **1886** R. HOLLAND *Cheshire Gloss.* 331 *Sparrowfarts*, very early morning 'Tha mun be up by *sparrowfarts* or tha'll be too late.' **1922** JOYCE *Ulysses* 747 Miss This Miss That Miss Theother lot of sparrowfarts skitting around talking about politics they know as much about as my backside. **1947** D. M. DAVIN *Gorse blooms Pale* 194 There we were as usual this morning at Sparrowfart, me with the jeep pulled up in front of the General's caravan. **1974** H. MCLEAVE *Only Gentlemen can Play* (1975) II. i. 92 It was important enough to bring you out here at sparrow fart.

sparrowy *a.*, (*b*) characteristic of or resembling a sparrow; so *sparrowy-looking* adj.

1926 S. SITWELL *All Summer in Day* I. i. 18 The sun, also, was very sparrowy in voice, and both sun and birds seemed satisfied for the moment. **1948** I. BROWN *No Idle Words* 31 Were they [*sc.* buntings] bunters in the perky.. sparrowy sense. **1953** D. A. BANNERMAN *Birds Brit. Isles* I. 300 The female is a sparrowy-looking little bird at all seasons. **1955** R. GRAVES *Greek Myths* I. lxxiv. 252 Castor and Polydeuces..come darting on 'sparrowy wings' through the upper air. **1976** A. J. RUSSELL *Pour Hemlock* vii. 63 She clutched steno pad to sparrowy chest.

sparse, *a.* **4.** *Comb.* (earlier and later examples).

1870 J. R. LOWELL in *Atlantic Monthly* Jan. 8 And thrust far off The Heaven..To voids sparse-sown with alienated stars. **1923** D. H. LAWRENCE *Birds, Beasts, & Flowers* 172 What would they do, those..sparse-haired elephants slowly following? **1924** J. MASEFIELD *Sard Harker* III. 201 He held on across the foothills through the sparse-growing sage.

Spartacist (spɑ̄·ɹtăsist, -kist). Also **Spartakist.** [ad. G. *Spartakist*, f. *Spartakus* *SPARTACUS, the name of a Thracian slave-leader in the Gladiatorial War (73–71 B.C.) against Rome, adopted as a pseudonym by K. Liebknecht (see below) in his political tracts: see -IST.] A member of a German radical socialist group formed in 1916 by Karl Liebknecht (1871–1919), Rosa Luxemburg (1870–1919), and Franz Mehring (1846–1919) and dedicated to ending the war of 1914–18 through revolution and to establishing a socialist government. Also *attrib.* or as *adj.*

1919 W. R. INGE *Outspoken Ess.* i. 18 The 'Spartacist' scoundrels who have betrayed and ruined their country. **1920** *19th Cent.* Mar. 560 The extreme Left wing of the Independents, known as Spartacists. **1925** *Contemp. Rev.* Dec. 715 The movement which a few Spartakists originated in the hope of establishing Soviet rule in Germany. **1965** *Listener* 4 Nov. 700/2 Otto Neurath..had been a member of the short-lived revolutionary Spartacist government in Munich. **1974** J. WHITE tr. *Poulantzas's Fascism & Dictatorship* IV. ii. 168 The process followed particular steps... 1918–19. Failure of the German revolution and defeat of the Spartakist militants.

So **Spa·rtacan** (-kăn) *rare* = *SPARTACIST; **Spa·rtacism** (-siz'm, -kiz'm), the policy and principles of the Spartacists.

1918 *N.Y. Times* 15 Dec. I. 3/5 Spartacism appeared in Munich openly..when 'the bloody events in Berlin and the guilt of the Government' were discussed. *Ibid.*, Premier Eisner visited the meeting and defied the Spartacan leaders. **1919** *Nation* (N.Y.) 19 Apr. 632 The Programme of the Spartacans. **1919** J. M. KEYNES *Econ. Conseq. Peace* 271 A victory of Spartacism in Germany might well be the prelude to Revolution everywhere. **1920** *Glasgow Herald* 9 Apr. 9 Spartacism in Germany is a domestic matter for the German Government to deal with.

Spartacus (spɑ̄·ɹtăkŏs). [See prec.] Used *attrib.* in *Spartacus group, league* [tr. G. *Spartakusbund*], the Spartacists.

1918 *Spectator* 30 Nov. 607/1 The Spartacus group—the wild adherents to Liebknecht and Rosa Luxemburg. **1974** *Encycl. Brit. Macropædia* XI. 205/2 Rosa Luxemburg..in an alliance with Karl Liebknecht and other like-minded radicals..formed the..Spartacus League.

Spartakist, var. *SPARTACIST.

spartina (spɑ̄ɹtəi·nă, -ī·nă). [mod.L. (J. C. D. Schreber *Linnæus' Genera Plantarum* (ed. 8, 1789) I. 43), ad. Gr. σπαρτίνη rope.] = *RICE-GRASS c.

1836 W. J. HOOKER *Compan. Bot. Mag.* II. 258 Distinguishing characters of both our British Spartinas. **1867** M. PLUES *Brit. Grasses* vii. 192 *Spartina stricta*, Smith. Cord Spartina. **1907** *Bull. Misc. Inf. R. Bot. Gardens Kew* 191 Others regard the three recognisable English Spartinas as varieties. *Ibid.* 193 These Spartina-swamps extend along each side of the river [Beaulieu]. **1925** *Jrnl. Ecol.* XIII. 83 *Spartina* is eagerly devoured by beasts of all kinds. **1934** [see *GLYCERIA]. **1943** J. W. DAY *Farming Adventure* xxi. 241 Spartina grass will live and grow in salt water on mud where no other plant could exist. **1965** *Times* 31 Aug. 10/6 Spartina grass.. thrives on tidal marshes. **1977** *Birds* Autumn 42/3 The fields of spartina have shrunk to spiky islands which are vanishing fast.

spartum. (Later example.)

1841 H. H. WILSON *Trav. Moorcroft & Trebeck in Himalayan Prov.* I. I. i. 10 A..swinging bridge..of the construction common in these mountains. The ropes used in its formation are made of a variety of spartum or star-three grass.

spas, var. *SPAZ.

spasm, *sb.* Add: **4.** Special Combs., as **spasm band** *U.S.* (now *Hist.*), a group, freq. of children, playing jazz on home-made musical instruments; **spasm music** *U.S.* (now *Hist.*), music played by spasm bands; **spasm war,** a war in which the combatants use their complete thermo-nuclear capabilities.

1926 WHITEMAN & MCBRIDE *Jazz* xiii. 267 When the last fearful note died, he turned to the leader. 'Stale Bread,' said he [*sc.* a judge], 'you may be a band, but you're a spasm band. Discharged.' The name stuck and the spasm band went on playing. **1943** I. LANG *Background of Blues* 4 Then there were the spasm bands... The saloons and sidewalks where spasm music and street singers found their most generous audiences were those of the French Quarter and particularly those in Storyville. **1964** HALL & WHANNEL *Pop. Arts* x. 297 In the early decades of the century the 'spasm bands' played a kind of home-made jazz on improvised instruments. **1943** Spasm music [see *spasm band* above]. **1965** H. KAHN *On Escalation* i. 14 But we need alternatives other than all-out spasm war or peace at any price. **1967** M. H. HALPERIN *Contemp. Mil. Strategy* (1968) ii. 15 The assumption that a general nuclear war between the United States and the Soviet Union would be an all-out, or 'spasm' war.

spasm (spæ·z'm), *v.* [f. the sb.] **1.** *trans.* To afflict with spasms (only in *pass.*); to cause to move convulsively. *rare.*

1790 J. BYNG *Jrnl.* 27 Aug. in *Torrington Diaries* (1935) II. 284 Otherwise I feel myself very low, and enfeebled, relaxed by day, and spasm'd by night. **1962** J. D. MACDONALD *Key to Suite* (1968) ix. 152 He spasmed his body inward, dropped the few remaining inches and landed on the railing.

2. *intr.* To twitch convulsively; to suffer a spasm.

1900 W. D. HOWELLS *Lit. Friends & Acquaintances* viii. 269 Of a person who had a nervous twitching of the face..he [*sc.* Henry James Sr.] said 'He *spasmed* to the fellow across the room and introduced him.' **1958** 'W. HENRY' *Seven Men at Mimbres Springs* xii. 142 Frank's gun fell from his spasming fingers. **1970** J. HANSEN *Fadeout* ix. 74 The boy's fine head did its slow, neck-straining roll while the unexpectedly deep voice spasmed and his mouth labored. **1978** J. IRVING *World according to Garp* xv. 304 The prostrate pig..squealed, its short legs spasmed.

spasmodic, *a.* **5.** (Earlier example.)

1832 CARLYLE *Let.* 28 Aug. (1977) VI. 211 Were I of the spasmodic School, I could gnash my teeth, now and then, over such a banishment: but..I reflect rather what deluges of Folly and Falsehood I stand safe from.

spasmogen (spæ·zmodʒĕn). *Pharm.* [Back-formation from next.] A spasmogenic drug.

1952 *Brit. Jrnl. Pharmacol. & Chemotherapy* VII. 91 There was little differentiation between the degrees of inhibition of..the 'direct' acting spasmogens, as compared with that shown by atropine. **1961** *Jrnl. Pharm. & Pharmacol.* XIII. 446 Common spasmogens such as histamine and 5-HT. **1974** *Nature* 4 Oct. 427/2 The response of the ileum to spasmogens varied widely at different times of the year.

spasmogenic (spæzmodʒe·nik), *a.* [f. SPASM + -O + *-GENIC.] † **1.** *Med.* (See quot.) *Obs.*

1899 CHURCH & PETERSON *Nervous & Mental Dis.* VII. vi. 546 If such a hyperesthetic zone arises from or becomes associated with some mental storm, pressure upon it may serve to revive the memories in question and provoke a hysterical fit. It is then denominated a spasmogenic or hysterogenic point or zone.

2. *Pharm.* Of a drug or other substance: promoting the contraction of smooth muscle.

1913 *Chem. Abstr.* VII. 3163 KCl has definit[e] spasmogenic properties in children. **1942** *Jrnl. Pharmacol. & Exper. Therapeutics* LXXIV. 275 Papaverine antagonizes the spasmogenic agents equally. **1975** *Nature* 13 Mar. 151/2 It can block the spasmogenic effects of 5-HT on the smooth muscle of isolated guinea pig ileum.

spasmolytic (spæzmŏli·tik), *a.* and *sb.* *Pharm.* [f. SPASM + -O + *-LYTIC.] **A.** *adj.* That relieves spasm of smooth muscle.

1937 *Jrnl. Pharmacol. & Exper. Therapeutics* LX. 13 The spasmolytic action of Syntropan exerted through the nerve endings on the isolated rabbit intestine is only about 20 times less than that of Atropine. **1953** [see *SERPASIL]. **1964** W. G. SMITH *Allergy & Tissue Metabolism* iv. 56 The effects of high dosage include ganglionic blockage and spasmolytic activity. **1982** *Jrnl. Med. Chem.* XXV. 1358/2 Some of these compounds have intensive spasmolytic activity for colonic motility.

B. *sb.* A spasmolytic drug.

1937 *Brit. Med. Jrnl.* 11 Sept. 560/1 The employment of spasmolytics in obstetrics is useful. **1978** *Acta Neurol. Scandinavica* LVII. 65 The drug is..capable of depressing distressing clonus, and it is concluded that it deserves further testing as a spasmolytic.

Hence **spasmo·lysis,** the action of such drugs.

1946 *Biol. Abstr.* XX. 1122/1 Although the compounds ..have some papaverine-like action on smooth muscle, spasmolysis results largely from their anticholinergic action. **1964** *Jrnl. Amer. Geriatric Soc.* XII. 1083 (*heading*) Spasmolysis: a new technique for treatment of spasticity. **1977** *Scandinavian Jrnl. Respiratory Dis.* Suppl. No. 98. 47 (*heading*) Sensitization of contracted tracheal smooth muscle to β-adrenergic spasmolysis by subthreshold doses of papaverine.

spasmoneme (spæ·zmonīm). *Zool.* [ad. G. *spasmonem* (G. Entz 1892, in *Math. und naturwissensch. Berichte aus Ungarn* X. 27), f. Gr. σπασμό-ς pulling, convulsion + νῆμ-α thread.] One of the three strands in the stalk of a vorticellid whose sudden contraction causes the stalk to coil tightly, withdrawing the animal from possible danger.

1901 [see *AXONEME]. **1941** BEAMS & KING in Calkins & Summers *Protozoa in Biol. Res.* ii. 94 The spasmoneme does not function like a true muscle but like a modified flagellum. **1978** *Jrnl. Cell Biol.* LXXVII. 358/2 Within the stalk is the contractile organelle, often referred to as the myoneme, though the term spasmoneme is preferable since recent studies indicate that the organelle is biochemically distinct from the myosin- and actin-based contractile systems.

spasmophilia (spæzmofi·liă). *Path.* [f. SPASM + -O + *-PHILIA.] Undue tendency of the muscles to contract, *esp.* as caused by a deficiency of systemic calcium.

1859 R. G. MAYNE *Expos. Lex. Med. Sci.* 1177/2 *Spasmophilia*, an epithet by Jos. Frank applied to erratic, spasmodic affections. **1892** D. H. TUKE *Dict. Psychol.*

Med. II. 843/1 Hyperæsthesia, with the corresponding hyperkinesis, spasmophilia or convulsions, is the principal symptom of spinal irritation. **1907** *Brit. Jrnl. Children's Dis.* IV. 448 J. Zahorsky includes tetany, carpal spasm, carpopedal spasm, laryngospasm and eclampsia, under

[illegible]

years 1922–24 were failures. **1963** *Washington Post* 2 Oct. B-2 The 1963 fall of spat (young oysters) in the James River seed area has been inadequate to maintain seed stocks for the third successive year. This contrasts sharply with the above average spatfall throughout the

[illegible]

spatial; to invest with spatial qualities or relations. So **spa·tialized** *ppl. a.*, **spa·tializing** *vbl. sb.*

1906 S. S. LAURIE *Synthetica* I. 24 There could be no

[illegible]

SPATTEE

able three-seater had fully cantilevered wings..and spatted' undercarriage.

spattee (spæti·). [f. Spat *sb.*⁵, after *puttee.*] Formerly, an outer stocking or legging worn by women for protection against wet and cold.

1926 *Bulletin* 17 Aug. 4/3 The invention is called the 'Highland spattee' and is a Highlander's stocking made spat-fashion to allow the wearer to slip it easily over her shoes and silk stocking. **1928** *Daily Tel.* 10 Jan. 6 The knitted spattee. **1939** Joyce *Finnegans Wake* i. 11 All spoiled goods go into her nabsack: curtrages and rattlin buttins, nappy spattees and flasks of all nations.

spatter, *sb.*² Add: **2.** *Geol.* Magmatic material emitted as small fluid fragments by a vent or fissure associated with a volcano; also, a fragment of this.

1953 *Bull. U.S. Geol. Survey* No. 994. 23 Small steep-sided cones composed very largely of spatter are common on the rift zones of Hualalai volcano. **1969** [see **driblet sb.* 4]. **1971** *New Scientist* 10 June 611/1 Irregular explosions threw bright orange lumps of spatter as high as 20 metres. **1976** G. B. Oakeshott *Volcanoes & Earthquakes* vi. 49 Activity had concentrated on the building of a combined cinder-and-spatter cone made up of hardened lava fragments and congealed blobs or spatters.

spatter, *v.* Add: **7. spatter rampart** *Geol.*, a wall or ridge formed of spatter along the edge of a fissure in a volcanic area; **spatterware, spatter ware** (see quots. 1959, 1977²).

1953 *Bull. U.S. Geol. Survey* No. 994. 6 The common basalts of Hawaii erupt quietly, building only low spatter ramparts and diminutive spatter-and-cinder cones. **1967** G. Macdonald in Hess & Poldervaart *Basalts* I. 53 Spatter ramparts may be several kilometers long, although generally they are not continuous over the entire distance. **1977** A. Hallam *Planet Earth* 96 Spatter cones and spatter ramparts form around parasitic vents and fissures when eruptions are less violent and the magma more fluid. **1935** *N.Y. Times* 9 June x. 9/2 'Gaudy Dutch' was the name given to some of the spatterware made in the shape of peacocks and tulips for this area many years ago. **1959** L. Gross *Guide to Antiques* viii. 97 Spatterware is a fairly heavy earthenware with characteristic decoration of color applied with a sponge or spattered on to give a stippled effect. **1977** Fleming & Honour *Penguin Dict. Decorative Arts* 751/1 *Spatter ware*, C19 wares decorated with bright colours applied with a sponge through a stencil, usually giving a blotchy effect. **1980** *Times* 4 Oct. 14/3 On the Staffordshire 'spatterware', in particular, Sotheby's American experts were out by a factor of 10 on some of their estimates.

spatulate (spæ·tiŭlēⁱt), *v.* [f. Spatul(a + -ate³.] *trans.* To stir or mix with a spatula. Also *absol.*

1923 in L. P. Anthony *Dict. Dental Sci.* 269/1. **1954** O. C. Applegate *Essentials of Removable Partial Denture Prosthesis* 261 Spatulate thoroughly to free the mix of contained air. **1956** J. N. Anderson *Appl. Dental Materials* xviii. 216 The difficulty of spatulating such a mixture..usually results in overspatulation. **1957** *Jrnl. Amer. Ceramic Soc.* XL. 254/1 The fine powders were spatulated intensively with either pine oil or a commercial screen oil. **1981** *Scand. Jrnl. Dental Res.* LXXXIX. 100/2 Dentists often spatulate for too short a time.

2. *Surg.* To give a spatulate form to (a tubular vessel).

1976 *Ann. Thoracic Surg.* XXII. 235/2 The end of the divided subclavian artery was then spatulated. **1977** *Urologia Internat.* XXXII. 369 The..terminal ureter is amputated as needed and the ureter is spatulated for about 10 mm on its anterior surface.

Hence **spa·tulating** *vbl. sb.*

1966 *Brit. Jrnl. Urol.* XXXVIII. 525 An atraumatic catgut suture is then inserted..at the apex of the spatulating incision. **1977** *Urologia Internat.* XXXII. 368 The main differences are the way the submucosal tunnel is prepared, the resection and spatulating of the terminal ureter, [etc.].

spatulated, *a.* Add: Also as *ppl. a.*, given a spatulate form.

1957 *Jrnl. Urol.* LXXVII. 413 The technique of anastomosing the spatulated ends of the ureter was used without the ureterotomy above the site of anastomosis. **1977** *Urologia Internat.* XXXII. 369 The spatulated end of the ureter is spread over the trigonal mucosa.

spatulation. Restrict *rare* to sense in Dict. and add: **III.** [f. *spatulate *v.*] **2.** Chiefly *Dentistry.* The process of stirring or mixing with a spatula.

1939 J. Osborne *Dental Mechanics* i. 5 (*heading*) Spatulation time. **1940** E. Covington *Efficient Dental Assistant* iii. 71 Rapid and longer spatulation accelerate [*sic*] setting time with most plasters. **1957** V. J. Kehoe *Technique Film & Television Make-Up* xii. 147 Mix by spatulation until a smooth cream is obtained. **1977** *Jrnl. Biomed. Materials Res.* XI. 863 The spatulation was conducted at approximately 150 rpm, but that rate was difficult to maintain throughout the mixing interval. **1981** *Scand. Jrnl. Dental Res.* LXXXIX. 97/2 Twelve fillings with diameters of 6 mm were made.., varying the spatulation time from 5 to 60 s at 5 s intervals.

3. *Surg.* The procedure of spatulating a tubular vessel; also, the spatulated portion.

1977 *Urologia Internat.* XXXII. 372 We advocate spatulation of the terminal end of the ureter. **1979** *Brit. Jrnl. Urol.* LI. 105/2 The stitch is continued around the upper ends of the spatulations.

‖ **Spätzle** (ʃpe·tslĕ, ʃpe·ts'l), *sb. pl.* Also **Spatzle, spätzle,** and (anglicized) **-s.** [Ger. dial., lit. 'little sparrows'.] Noodles of a type made in southern Germany.

1933 E. E. Amiet *Palmer House Cook Bk.* 247 Spatzles. Into 3 soup-spoonfuls of flour, beat 1 egg and 3 soup-spoonfuls of milk. Add a pinch of salt, a pinch of pepper and a little ground nutmeg. Set a colander over boiling water and squeeze this mixture through, boil for a minute and pour into a sieve and serve. **1959** M. Crosland tr. *Rovan's Germany* 179 The *Spätzle*, a kind of fresh pasta from Swabia. **1978** J. Irving *World according to Garp* viii. 155 Garp had cooked an elegant Paprika Chicken and spätzle.

spaug (spǭg). *Anglo-Irish.* [Ir. *spág.*] A clumsy, awkward foot.

1910 P. W. Joyce *English as we speak it in Ireland* 331 *Spaug*, a big clumsy foot:— 'You put your ugly spaug down on my handkerchief. **1922** Joyce *Ulysses* 128 Taking off his flat spaugs and the walk.

spaulty (spǭ·lti), *a. dial.* Also **spoalty, spoulty.** [f. Spalt *a.* + -y¹.] Dry and brittle.

1895 J. J. Raven *Hist. Suffolk* xix. 266 When turnips are hard and brittle..they are said to be *spoalty*, which William Ellis (1750) spells *spalt*, and Professor Skeat notes as a Cambridge-Shire word. **1904** *Eng. Dial. Dict.* V. 643/1 Them turnips is spoulty. **1906** Kipling *Puck of Pook's Hill* 238 Did he promise me a set of iron cramps or ties for the roof? They never came to hand, or else they were spaulty or cracked.

spawn, *sb.* Add: **3.** (Later *fig.* example.)

1920 D. H. Lawrence *Touch & Go* 6 The plays of A People's Theatre are—oh heaven, what are they?—not popular nor populous nor plebeian nor proletarian nor folk nor parish plays. None of that adjectival spawn.

8. *spawn-like* adj.

1938 S. Beckett *Murphy* 249 The iris was reduced to a thin glaucous rim of spawnlike consistency.

spawn, *v.* Add: **2. b.** (Later example.)

a 1930 D. H. Lawrence *Last Poems* (1932) 198 Oh I have loved the working class Where I was born, And lived to see them spawn into machine-robots.

6. (Further examples, *spec.* of tornadoes or the like.)

1955 *Sci. News Let.* 18 June 394/1 The general atmospheric conditions in which hurricanes are spawned are known. **1965** *Listener* 21 Oct. 610/1 Every summer the violent climate of the central United States spawns a series of devastating storms—tornadoes—twisters they call them here. **1976** *Bay City* (Michigan) *Times* 12 July 1/5 Powerful thunderstorms roamed much of the East Sunday spawning tornadoes and flash floods.

spawned *ppl. a.* **b.** Also with *out.*

1972 *Trout & Salmon* June 41/1 Unripe fish as well as spawned-out carcasses and unused roe..are sold in the market. **1972** L. Hancock *There's a Seal in my Sleeping Bag* vi. 132 Birds congregated on the salmon river to feed on the spawned-out salmon.

spaz (spæz). *slang.* Also **spas.** [Abbrev. of *spastic *sb.*] = *spastic *sb.* b (see also quot. 1977¹).

1965 P. Kael *I lost it at Movies* iii. 259 The term that American teen-agers now use as the opposite of 'tough' is 'spaz'. A spaz is a person who is courteous to teachers, plans for a career..and believes in official values. A spaz is something like what adults still call a square. **1975** M. Amis *Dead Babies* viii. 47, I know how long, you little spaz. **1977** *Amer. Speech* 1975 L. 67 *Spas*,..1: Uncoordinated, clumsy person 2: Person regarded as dull, foolish, or stupid 3: Friend (jocular use) 'Hey spas, do we have any homework?' **1977** J. Wambaugh *Black Marble* (1978) viii. 117 The man's a spaz! A total spaz! **1982** *Guardian* 26 Oct. 8/4 Come onnnnn—bag your face, you geek, you grody totally shanky spaz.

speak, *sb.* Add: **4.** (Earlier example.)

1811 *Lexicon Balatronicum* s.v. *Speak*, He has made a good speak; he has stolen something considerable.

5. = *speakeasy.

1930 *Bookman* (U.S.) LXXII. 398/1 Better grade speaks in Times Square are dispensing with femme shills and hangers-out. **1952** [see *open *a.* 2 c]. **1977** H. Fast *Immigrants* iv. 242 We're just lucky it happened in a speak, because maybe no one will bring any charges.

speak, *v.* Add: **I. 1. c.** *spec.* To speak to another by means of a telephone; —— *speaking* (where —— is a speaker on a telephone), phr. used by the speaker to announce his identity.

1885 *List of Subscribers* (United Telephone Co.) p. xiv, 'Who speaks?' came distinctly from the wires into the office. '2577,' was the reply—it was the hotel number. **1925** F. Scott Fitzgerald *Great Gatsby* ix. 200 But the connection came through as a man's voice, very thin and far away. 'This is Slagle speaking.'.. 'Yes?' The name was unfamiliar. **1933** 'Sapper' *Knock-Out* i. 9 Standish.. took the receiver from the other's hand. 'Hullo! Sanderson,' he said. 'Yes—Standish speaking. What now?' **1973** J. Wainwright *Pride of Pigs* 166 'Quince?' said the voice. 'Speaking.' Quince hooked his fingers through the carrying handle of the Trimphone, telephone hand set... He said: 'Who's that?.. Who's speaking?' **1977** L. Meynell *Hooky gets Wooden Spoon* xiii. 151 C.I.D. here..who is it speaking, please?

g. *spec.* To propose marriage. Cf. sense *13 d.

1604 Shaks. *Othello* i. iii. 166 She thank'd me, And bad me, if I had a Friend that lou'd her, I should but teach him how to tell my Story, And that would wooe her. Vpon this hint I spake. **1803** G. Colman *John Bull* iii. i. 31 *Lady Caroline.* Lard, Mr. Shuffleton!.. You never spoke anything to—that is—to justify such a—. *Shuffleton.* (Aside.) That's as much as to say, speak now. **1858** Trollope *Doctor Thorne* ii. vi. 81, I think you may speak now, Frank... She is very fond of you. **1904** H. James *Golden Bowl* I. i. xii. 213 He liked..to feel that he should be able to 'speak'..the word itself being romantic. **1932** S. Gibbons *Cold Comfort Farm* xiv. 195 Flora did not dare to imagine what would happen if they returned from the ball and he had not spoken. He *must* speak! **1964** M. Laski in S. Nowell-Smith *Edwardian England* iv. 198 An interval might have been found—perhaps in the conservatory, perhaps on a sofa in a dark nook under the stairs—when *he* had spoken and *she* accepted.

4. a. (*b*). (Earlier example.)

1824 Longfellow *Let.* 2 Mar. in S. Longfellow *Life H. W. Longfellow* (1886) I. iii. 37 If this were not another building, I should have imagined I occupied the same chamber that you did in former times, for it seemed to be the very highest point of the dwelling, the very *apogee*, so to speak.

II. 10. speak for —. a. *speak for yourself*: expressing a desire to dissociate oneself from what another has just said or the assumptions behind it.

1738 Swift *Polite Conversation* i. 16 Pray, sir, speak for yourself. **1778** Boswell in *London Mag.* Feb. 58/1, I remember hearing a late celebrated infidel tell that he was not at all pleased when the infidel wife of his friend, a poet of some eminence, addressed him in a company in London, 'we Deists'.—Speak for yourself, Madam, said he abruptly. **1824** Scott *Redgauntlet* III. x. 292 'Speak for yourself, friend,' said Peter, scornfully. **1916** G. B. Shaw *Androcles & Lion* Prologue 3 *Megaera*: Everybody knows that the Christians are the very lowest of the low. *Androcles*: Just like us, dear. *Megaera*: Speak for yourself. **1946** L. P. Hartley *Sixth Heaven* ii. 46 'Speak for yourself,' said Barbara defiantly. **1976** *Times Lit. Suppl.* 30 Apr. 508/4 We learn that 'when viewing serials..we feel we know these characters well enough, for example, to say hello to in the street.'.. My reaction to this is to say: speak for yourself!

c. (Later examples.)

1943 *Sun* (Baltimore) 25 Feb. 6/1 (Advt.), We hope to preserve even more food this year. But well over half of this season's pack is already spoken for by the Government. **1971** *Petticoat* 17 July 29/2 He's not married, but he's involved, as they say, spoken for, and has lived with his girlfriend in London for the last few years.

12*. speak past —. To talk at cross-purposes with; to speak incomprehensibly to. Cf. G. *vorbeireden.*

1952 G. Himmelfarb *Lord Acton* i. 2 This historian.. ended his life..as a lecturer doomed to speak past his audience. **1975** *United Church Observer* Nov. 15/1 The most important issue..is whether we can speak together and converse together, not *at* each other or *past* each other.

13. speak to —. d. *spec.* To propose marriage to. Cf. sense *1 g.

1809 Ld. Granville *Let.* 14 Nov. in B. Askwith *Piety & Wit* (1982) iv. 64, I spoke to Harriet last night; she was very nervous and so was I... She consented to my speaking to the Duke [her father]. **1840** Thackeray in *Fraser's Mag.* XXII. 230/1 'Will you marry me?' In fact, this very speech had been taught him by cunning Gann, who saw well enough that Swigby would speak to one or other of his daughters. **1863** Mrs. Gaskell *Dark Night's Work* iv. 46 He had some discussion with himself as to whether he should speak to her, and so secure her promise. **1977** G. Butler *Brides of Friedberg* i. 12 Next day someone I would *much* rather have accepted spoke to me riding in the Row. But it was too late.

III. 17. speak out. b. *to speak out in meeting* (examples).

1830 *Mass. Spy* 23 June 4/1 O dear, I spoke out in meeting. **1906** *Springfield* (Mass.) *Weekly Republic* 13 Sept. 8, I do not think the president will think any the less of me for speaking right out in meeting and saying that I am not for it.

IV. 21. c. *to speak one's piece*: see *piece *sb.* 17 g.

27. c. For † *Obs.*⁻¹ read *rare* and add later example.

1936 J. Steinbeck *In Dubious Battle* vi. 97 If the bitch ever whelps, I'd like to speak a pup.

VI. 36. speak-back = **talk-back*; **speak-box,** an intercom device by a (usu. outside) door which allows a caller to speak to someone elsewhere in the building (cf. *voice-box* (*b*) s.v. *voice *sb.* 14); **speak-easy**: delete and see *speakeasy; **speak-out,** an occasion on which people can speak freely and unreservedly; **speak-your-weight machine,** a weighing machine which announces one's weight in spoken words.

1940 *Chambers's Techn. Dict.* 788/1 *Speak-back*, the subsidiary microphone-amplifier-reproducer in a motion-picture studio, by which the remote recordist can speak to the director on the sound-stage. **1960** *Guardian* 14 Apr. 8/3 His most recent hero..attempts to teach speak-your-weight machines to sing. **1962** N. Freeling *Love in Amsterdam* ii. 106 The buzzer went, and he heard Sophia's voice on the speakbox. **1966** *Illustr. London News* 30 July 11 Kidbrooke has its own T.V. studio which incorporates a 'speak-back'. **1966** L. Deighton *Billion-Dollar Brain* xv. 146 His voice was..like a speak-your-weight machine. **1968** *Guardian* 19 Aug. 14/5 Americans Abroad for McCarthy held a 'speak out' near the Speakers' Corner in a sunlit but swampy Hyde Park yesterday. **1970** W. J. Burley *To kill a Cat* ix. 152 He rang the bell. Almost at once he was startled by a woman's voice from close at hand. A speak-box which he hadn't expected. He found the little metal grille and spoke into it. 'Come up, please.' **1977** *Daily Tel.* 25 Oct. 11/4 At the New York 'speak out', women were invited to tell anonymously of abuse suffered from promiscuous or sexually taunting male employers or superiors.

-speak (spīk), *suffix.* The verb Speak used, after Orwell's *Newspeak* and *Oldspeak*, as a substantival suffix (cf. Speak *sb.* 1) to denote a particular variety of language or characteristic mode of speaking.

1949, etc. [see *Newspeak]. **1949**, etc. [see *Oldspeak]. **1960** K. Amis *Take Girl Like You* xi. 140 Charlton, his creep-speak effectively silenced, had departed in protest-march style. **1966** *Science* 13 May 875/1 We read of 'space speak' on every hand. Newspapers and magazines discuss it in their science columns... The belief is that the space effort has given us, in addition to the possibility of going to the moon, a new linguistic phenomenon. **1972** *College English* Jan. 439 (*heading*) Doublespeak: dialectology in the service of Big Brother. **1980** *Times* 27 Feb. 14/2 Such emphasizers as *undoubtedly* (Ponderoso Speak: *indubitably*)..diminish, if they do not actually destroy the assurance of a statement. **1981** *Times* 28 Jan. 7/1 (*heading*) Haigspeak rewrites the grammar. **1981** *Guardian* 1 May 2/4 'I am very sorry that I cannot be with you today... I am most grateful and touched that you have decided to name a locomotive after me,' it [*sc.* a telegram] said in classic royalspeak.

speakable, *a.* Add: **3.** Able or fit to be spoken *to.*

1956 'C. Blackstock' *Dewey Death* x. 227 I'm simply not speakable to this morning. **1979** *Daily Tel.* 20 Sept. 18/3, I don't think he is speakable to at the moment... Could you ring back in 20 minutes?

speakeasy (spī·kīzi). *slang* (orig. and chiefly *U.S.*). Also **speak-easy.** [f. Speak *v.* + Easy *adv.*] A shop or bar where alcoholic liquor is sold illegally. Also *attrib.*

1889 [see Speak *v.* 36]. **1895** L. Pendleton *Corona of Nantahalas* iv. 45 A sort of rural 'speak easy', where the colourless liquid was poured into the purchasers' bottles from a new and innocent-looking kerosene can. **1903** A. H. Lewis *Boss* xiii. 162 That..no side-doors or speak-easy racket [should be] stood for. **1922** Joyce *Ulysses* 418 In the speakeasy. Tight. I shee you, shir. **1946** [see *creep joint* s.v. *creep *sb.* 6]. **1958** S. Traill in P. Gammond *Decca Bk. Jazz* vi. 75 Every cheap speakeasy had its resident piano player. **1961** W. Vaughan-Thomas *Anzio* vii. 138 Inevitably some of these underground caves became 'speak-easy' dens where the local black-marketeers sold *vino* to the troops. **1968** [see *prohibition era* s.v. *prohibition 6]. **1982** *Age* (Melbourne) 3 Feb. 6/6 Unable to find a respectable job, she first became a bootlegger during the Prohibition era and ran a speakeasy.

speaker. Add: **7.** (Earlier example.)

1875 W. D. Whitney *Life & Growth of Lang.* iv. 72 The difficulty is one which English-speakers can hardly realize.

8. = *loud-speaker.

1926 *Jrnl. Franklin Inst.* CCII. 436 This speaker employs a six-inch cone driven by an electromagnetic power unit. **1954** R. Dahl *Someone like You* 66 Maybe the great radio engineer doesn't know how to connect the mike to the speaker? **1978** *Hi-Fi News* Sept. 15 (Advt.), High Fidelity speakers for the discerning ear.

9. *attrib.* and *Comb.*, as (sense *8) *speaker grille, system*; **speaker-hearer,** a person regarded as a user of language; **speaker-key,** a key fitted to a wind instrument to enable the playing of notes an octave or a twelfth higher (cf. *octave key* s.v. *octave *sb.* (*a.*) 8); **speaker-listener** = *speaker-hearer* above; **speaker-phone** *U.S.*, a telephone receiver which need not be held in the hand; **Speaker's Conference,** a conference, first set up in 1916, whose purpose is to examine electoral law and reform under the chairmanship of the Speaker of the House of Commons; **Speakers' Corner,** the north-east corner of Hyde Park, near Marble Arch, noted as a place where soap-box orators traditionally air their views; also *transf.*

1979 P. Way *Sunrise* iv. 44 There was a bell push and a speaker-grille just above it. **1965** N. Chomsky *Aspects of Theory of Syntax* i. 4 To study actual linguistic performance, we must consider the interaction of a variety of factors, of which the underlying competence of the speaker-hearer is only one. **1982** *Amer. Speech* LVII. 16 He must ultimately be willing to make claims about this base with respect to a speaker-hearer's capabilities. **1890** D. J. Blaikley *Acoustics in Relation to Wind Instruments* 31 In one direction advantage is taken..to aid the player in producing certain notes, notably on the clarinet; the thumb,—or speaker-key of which is designedly used to open a small air-way, thereby introducing a weak place, by which means certain sub-divisions of the air column are aided, and certain others are hindered. **1972** S. Richmond *Clarinet & Saxophone Experience* vi. 104 By pressing the speaker key for the second harmonic all notes will sound an octave higher. **1965** N. Chomsky *Aspects of Theory of Syntax* i. 3 Linguistic theory is concerned primarily with an ideal speaker-listener, in a completely homogeneous speech-community. **1978** *Archivum Linguisticum* IX. 10 The ideal speaker-listener of generative theory. **1955** *Sun* (Baltimore) 29 July 8/6 The new 'hands-free' Speakerphones enable you to take notes, refer to records, have others in the room with you join in the telephone conversation. **1968** *Time* 5 Apr. 54 Emerson, a municipal court judge in Downey, Calif., finds the speakerphone invaluable for getting a brief piece of testimony from a policeman, parole officer or technical expert. [**1916** *Times* 16 Dec. 9/5 The Speaker's Electoral Reform Conference is for the moment in suspense.] **1917** *Times* 18 Jan. 9/5 The recommendations of the Speaker's Conference on Electoral Reform. **1974** *Times* 5 Mar. 2/3 Electoral reform has always been a matter for Parliament itself, expressing its view by means of free votes on recommendations by a Speaker's Conference. **1980** *Guardian* 18 Feb. 3/7 The last Speaker's Conference on electoral law, which met between 1972 and 1974, recommended.. that the minimum age for standing for Parliament should be reduced. **1936** J. C. Goodwin *One of Crowd* xix. 271 Speakers' Corner is a mixed grill of apostles and propagators, of oddities and crudities, of fanatics and eccentrics. **1953** Earl Winterton *Orders of Day* xxii. 313 'Speakers' Corner' in Hyde Park. **1982** *Times* 16 Mar. 10/2 The crypt of St Paul's cathedral is regaining some of its historical reputation as an ecclesiastical Speakers' Corner. **1938** C. Himes *Black on Black* (1973) 172 A speaker system was installed to throw his powerful voice even farther. **1974** *Times* 4 Mar. 1/8 At about 4.40 pm the hijacker in the cockpit announced over the speaker system that the aircraft would be landing at Amsterdam.

speakerine (spī·kərīn). [a. Fr. *speakerine*, f. *speaker* announcer (f. Speaker) + fem. suff. *-ine.*] A woman announcer on radio or television; a television hostess.

1957 *Manch. Guardian* 19 Dec. 5/3 B.B.C. television viewers will see a programme..in which ten European countries will take part... The British contribution is to be the 'finale', in which Sylvia Peters is to introduce the women announcers—known as 'speakerines'—of Europe. **1973** D. May *Laughter in Djakarta* v. 97 We're going to hear the news from this lovely speakerine? **1979** J. Rathbone *Euro-Killers* ix. 96 He..turned on the television. A speakerine was announcing the programme.

speakie (spī·ki). *temporary.* [f. Speak *v.* + -ie, after *movie, *talkie.] A stage play in contrast to a (silent) film; rarely, = *talkie. Usu. *pl.*

1921 J. M. Barrie *Let.* 18 Sept. (1942) 192 The ordinary stage drama he [*sc.* Charlie Chaplin] called the 'Speakies'. **1927** *Observer* 24 Apr. 15/1 She prophesied the downfall of the 'speakies' and the triumphant survival of the 'movies'. **1927** *Daily Tel.* 11 Oct. 6 An innovation last night was the introduction of a real stage set in the middle of the film... Whether this mixture of 'movie' and 'speakie' is desirable may be questioned. **1928** *Sunday Dispatch* 15 July 14 'Talkies' or 'speakies' as they are calling them in Hollywood, have very definitely arrived.

speaking, *vbl. sb.* Add: **2. b.** Now chiefly *U.S.* (Further examples.) *at this* (or *the*) *present speaking*, at this moment.

1835 J. P. Kennedy *Horse-Shoe Robinson* I. vi. 78 If I suspicioned a bamboozlement, which I am not far from at this present speaking. **1863** C. C. Hopley *Life in South* I. 57 Then came the 'speaking', as the sermon was called. **1891** M. E. Ryan *Told in Hills* iii. v. 205 At the present speaking the days are not picnic days. **1895** 'C. E. Craddock' *Mystery Witch-Face Mt.* 206 Thar war a big crowd at the cross roads ter hear the speakin'. **1942** J. Thomas *Blue Ridge* v. 155 Men..will travel miles to a speaking—which may be a political gathering or one for.. discussing road building.

3. b. (Later example.)

1920 D. H. Lawrence *Lost Girl* iv. 54 She began to hate outspokenness and direct speaking-forth of the whole mind.

5. a. *speaking engagement, order, part* (later examples), *tour, trip, voice* (examples).

1870 O. Logan *Before Footlights & behind Scenes* iii. 37 By and by I got into 'speaking parts', such as the Duke of York in Richard the Third. **1879** *Law Rep. Appeal Cases* IV. 40 If the Court of Quarter Sessions stated upon the face of the order, by way of recital, that the facts were so and so, and the grounds of its decision were such as were so stated, then the order became upon the face of it, a speaking order. **1908** E. Terry *Story of My Life* xiv. 355 Melba..had a bad cold, and therefore a frightful *speaking* voice for the moment. **1924** W. Holtby *Crowded Street* xxxvi. 270 Delia..departed northwards on a speaking tour. **1931** F. L. Allen *Only Yesterday* ii. 32 He would win them to his cause, making a speaking trip through the West. **1944** N. Streatfeild *Curtain Up* viii. 99 Mime..she loved. Then there were her speaking parts. **1973** 'E. McBain' *Hail to Chief* vi. 109, I have a speaking engagement... I'm talking at a women's college. **1977** *Rolling Stone* 5 May 15/4 White, a big man with a rich, resonant speaking voice which turned into a tough growl when he sang. **1978** M. Dickens *Open Book* xix. 173 On this speaking tour, my engagements fell roughly into two main categories. **1979** Ld. Denning *Discipline of Law* ii. i. 66 It was possible to extend it to include not only the order of the Tribunal itself—when it was a 'speaking order'—but in addition all the documents properly before the Tribunal and considered by them.

c. (Example with *upon.*)

1801 M. Edgeworth *Belinda* I. xii. 370 Lady Delacour is not upon speaking terms with this Mrs. Margaret Delacour; she cannot endure her.

speaking, *ppl. a.* Add: **1.** *speaking clock*, a telephone service giving the correct time in words (cf. *talking clock* s.v. *talking *ppl. a.* 2); *speaking stop*, a stop key on an organ which permits or prevents the sounding of a rank of pipes.

1934 *P.O. Electr. Engineers' Jrnl.* XXVII. 142/1 For some time past a speaking clock has been installed in Paris. **1938** *Oxf. Compan. Mus.* 660/2 An organ of 168 actual 'speaking stops' (we so call the stops which really sound, as distinct from other devices.) **1977** *Gramophone* Mar. 1444/2 This is a very large Compton organ indeed, with 37 speaking stops on the pedal. **1978** 'H. Carmichael' *Life Cycle* xiii. 139 If nobody at Scotland Yard has a watch you could've dialled the Speaking Clock.

speakingness. (Later example.)

1957 E. Bowen in K. Mansfield *34 Short Stories* 15 Words had but one appeal for her, that of speakingness.

speako (spī·ko). *U.S. slang.* [f. *speak(easy + *-o².] = *speakeasy.

1931 *Amer. Mercury* Dec. 415/2 There's only one thing in the world that puts a speako over, and that's good will. **1932** *Ibid.* Jan. 11 A brewery which supplied every other speako between Fourteenth and Canal streets. **1941** J. M. Cain *Mildred Pierce* (1943) i. 29 Making the grand tour of all the speako's he knows.

spear, *sb.*¹ Add: **I. 2. a.** (Later examples.)

1930 R. Campbell *Adamastor* 60 A starved mongrel.. From where he crouched, a thrilling spear of pain, Hurled forth his Alleluia to the sky. **1934** T. S. Eliot *Rock* ii. 78 Encompassed with enemies armed with the spears of mistaken ideals.

b. The 'sack'; dismissal. *Austral. slang.*

1912 in Stewart & Keesing *Old Bush Songs* (1957) 273 I've been many years a shearer and I fancied I could shear, I've shore for Rouse of Guntawung and always missed the spear. **1941** Baker *Dict. Austral. Slang* 69 *Get the spear*, to be dismissed from a job.

II. 8. *spear-tip*; *spear-forest, -print, surge.*

1946 S. Spender *European Witness* xiv. 217 A country of clustered spear-forests and gloomy heaths. **1911** E. Pound *Canzoni* 4 Deep in my heart that spear-print stays, That wound I got beyond the waters. **1900** Chesterton *Wild Knight & Other Poems* 103 The crest of the spear-surge. **1930** T. S. Eliot tr. *St.-J. Perse's Anabasis* 43 The horsemen..feeding on their spear-tips the pure disasters of sunshine.

9. a. *spear-fisher, -fisherman.*

1951 T. C. Roughley *Fish & Fisheries of Australia* ix. 309 Members will not..seek quarrels with line-fishermen or other spear-fishermen regarding priority rights of fishing at any place. **1962** *Underwater Swimming* ('Know the Game' Ser.) 19/1 The spearfisherman should always carry a knife. *Ibid.*, An added safety device is to have a float anchored in the diving area to which the spearfisher can go to rest or leave his catch. **1982** *Times* 21 July 3/3 Dr Paul Cragg, a biologist, was in favour of resuming grants for spearfishermen.

b. *spear-fishing* (later examples; hence, as a back-formation, *spear-fish* vb. intr. and trans.); also similative, as *spear-flashing.*

1962 *Times* 6 Apr. 7/2 Sail, snorkel, skin-dive, spear-fish in tropical Florida. **1963** *Harper's Bazaar* Jan. 65/1 On the Côte d'Azur, many of the big fish have been.. spear-fished out of sight. **1973** J. Jones *Touch of Danger* xix. 106, I spearfished... Sonny..was no adept with flippers or speargun. **1945** Spear-fishing [see *flipper *sb.*² 1 b]. **1960** M. A. Gabrielsen et al. *Aquatics Handbk.* xiv. 102/1 Spear fishing is becoming a popular competitive as well as recreational sport. **1973** J. Jones *Touch of Danger* xix. 107 Sonny was against spearfishing for sport. **1937** Blunden *Elegy* 78 Against high blue Spear-flashing white the spire.

c. *spear-shaken, -stuck, -tipped.*

1947 S. Spender *Poems of Dedication* iv. 56 Above the destroyed city reborn city..Tower of wings climbing spear-shaken skies. **1943** D. Gascoyne *Poems 1937–42* 5 Whose are these hollow red-filmed eyes And thorn-spiked head and spear-stuck side? **1954** W. Faulkner *Fable* 132 The spear-tipped iron fence beyond which the three sentries flanked the blank door beneath the three morning-windy flags.

d. *spear-straight* adj.

1848 J. R. Lowell *Poems* 2nd Ser. 69 A stem.. Standing spear-straight in the waist-deep moss. **1919** J. Masefield *Reynard the Fox* ii. 107 With spear-straight stern.

10. spear-carrier, a carrier of a spear, a spearman: used *transf.* as (*a*) orig. *Theatr. slang*, an actor with a walk-on part; hence, an unimportant participant; (*b*) *U.S. colloq.*, a proponent or 'standard-bearer' (cf. *spear-head 1 b); **spear gun,** a type of weapon used in spearfishing which operates by firing a detachable harpoon; also *attrib.*; hence **spear-gunner**; **spear tackle** *Austral.*, an illegal tackle in rugby football in which a player is lifted and thrust to the ground head first; hence as *v. trans.*

1960 *New Yorker* 13 Aug. 97/1 The 'Quartet' is full of characters who in one novel may seem irritatingly superfluous spear-carriers,..but who in the 'Quartet' turn out to be members of a literary repertory company. **1963** *Times* 20 May 12/5 Most of those spear-carriers not only don't know where the United States is but they don't know where they are themselves. **1967** *N.Y. Times* 21 May 26/1 Dr. King had 'emerged as the public spear-carrier of a civil disobedience program'. **1976** *Times* 18 Mar. 10 In Wisconsin on the same day Representative Morris Udall, the 'liberal-progressive' spear carrier, will have to win to stay in the race. **1981** N. Marsh *Black Beech & Honeydew* (rev. ed.) x. 215 The students.. would begin to accept the enormous challenge of a Shakespeare play and their own real importance, if only as spear-carriers, in doing so. **1982** *Sunday Sun-Times* (Chicago) 20 June 100/1 By the time Breakfast at Wimbledon telecasts are beamed into the United States on Fourth of July weekend, American tennis pros Davis, Dunk and Hardie will have vacated their present lodging and be long gone from the venerable tournament that they graced momentarily as spear-carriers. **1951** T. C. Roughley *Fish & Fisheries Austral.* ix. 303 The sport of fishing with spears or spear-guns under water..has had only a brief history. *Ibid.* 304 Most Australian spear-gun fishermen use a gun with rubber as the motive power. **1979** J. Leasor *Love & Land Beyond* i. 7 The..five-pronged fork of an underwater spear gun. **1951** T. C. Roughley *Fish & Fisheries Austral.* ix. 308 Those responsible for such an attitude know little of the spear-gunner's activities. **1969** *Sun-Herald* (Sydney) 13 July 36/2 Canterbury were penalised for a spear tackle on Cavanagh. **1977** *Telegraph* (Brisbane) 8 Nov. 3/3 He was injured after he was allegedly spear-tackled... A player is spear-tackled when an opponent tackles low, lifts the man with the ball high, turns him over and thrusts him into the ground head first.

spear, *sb.*² Add: **2. d.** The edible shoot, including stem and tip or head, of asparagus or of sprouting broccoli (esp. calabrese).

1952 *Quick Freezing* Jan. 9/1 At a Birds Eye press conference held recently it was stated that three new products have been added to that range. These are: (1) Chicken livers... (2) Broccoli spears, broccoli cuts... (3) Pineapple slices in syrup. **1966** *Harrod's Food News* Sept. 2/1 Brocolli Spears—8 oz. 2/11 *Ibid.* 5/1 Asparagus spears Spanish (5½ in. long)—17 oz. tin 4/9. **1969** *Oxf. Bk. Food Plants* 162/1 Asparagus..is usually considered to be a luxury vegetable. The part eaten is the young shoot or 'spear'. **1974** P. Westland *Taste of Country* ii. 37/2 Cover with the cooked broccoli spears and then the cheese sauce. **1979** *Sunset* Apr. 178/2 A light entrée, it's especially good when accompanied with sliced ham..and additional spears of freshly cooked asparagus.

spear, *v.*³ Add: **1. b.** To dismiss. *Austral. slang.* Cf. *spear *sb.*¹ 2 b.

1911 'S. Rudd' *Dashwoods* 13 If I was the boss here I would. I'd spear him without warnin'.

c. To cause to move like a spear; to spear-head.

1920 W. Camp *Football without Coach* v. 85 The ball should be held in the hand and speared through the air by giving the hand a twist as the ball leaves it. **1951** *Daily Progress* (Charlottesville, Va.) 11 Feb. 1/6 The Second Division at the end of its eight-day battle had set the pattern of the Eighth Army's new hunt-and-kill offensive with aggressive tank forces spearing the way. **1969** G. Macbeth *War Quartet* 46 So when I Speared the first squadron in the dawn assault Over the cliffs, that wool..Warmed the heart's beating.

d. To beg; to obtain by begging. *U.S. slang.*

1912 *Railroad Man's Mag.* Apr. 493/1 They had mooched the stem and threw their feet, And speared four-bits on which to eat. **1926** *Amer. Speech* II. 390/1 To *make the grade* or *connect* is to get the amount of money one is after. *Spear* is another word for connecting. **1942** Berrey & Van den Bark *Amer. Thes. Slang* § 370/3 Beg; request a loan or gift,..*spear*.

2. b. To move like a spear. *rare.*

1944 *Times* 14 Apr. 4/2 Yesterday the tanks handed over to the infantry, and speared south-east and south-west towards the Crimean mountains.

spear-head, *sb.* Add: **1. b.** The leading part or element (of a thrust, movement, etc.); a person or group leading an attack.

1929 *Times* 12 Nov. 17/3 The Belfast members of the society..were made the spearhead of a thrust for an advance of wages for shipyard joiners alone. **1932** *Times* 12 July 6/3 Afterwards they [*sc.* Yorkshire] broke down before the Nottinghamshire attack, with Larwood as its spearhead, and lost seven wickets for 53 runs. **1940** *Hutchinson's Pictorial Hist. War* 10 Apr.–11 June 176 Synchronizing their Blitzkrieg attack with dive-bombers, German tanks acted as the spearhead in the rapid Nazi advance through the Low Countries and Northern France. **1945** L. Mumford *City Development* xii. 130 The park system is thus the very spearhead of comprehensive urban planning. **1946** *R.A.F. Jrnl.* May 168 He was called upon to organize the new spearhead for Bomber Command. **1951** 'J. Wyndham' *Day of Triffids* xii. 223 When they found that we represented only a group similar to their own, and were not the spearhead of a rescue party on the grand scale their interest would lapse. **1958** *Manch. Guardian* 20 Aug. 4/2 Lord Cameron and his colleagues do not think that the dockers' claim can be fairly regarded as the spearhead of a new national wage movement. **1962** *Times* 27 Nov. 13/2 Mr. W. P. Tapley..told me of.. preparations to..keep traffic flowing regardless of the weather. 'Salt,' he said, 'is still the spearhead.' **1970** A. Toffler *Future Shock* iii. 40 America, as the spearhead of superindustrialism, represents a new, quicker, and very much unwanted tempo. **1977** 'J. le Carré' *Honourable Schoolboy* xviii. 442 The spearhead of the operation will be handled by ourselves. If supportive action is required, Martello will supply it.

3. *spear-head army, forces, group.*

1931 W. S. Churchill *World Crisis* VI. xx. 301 Amid these varying schemes one plain question stood forth. Should Mackensen with the 'spear-head army' go on or stop? **1978** R. V. Jones *Most Secret War* xlix. 486 The enthusiasm of its members was such that they were sometimes ahead of our spearhead forces. **1977** M. Walker *National Front* ii. 39 During 1961, the Special Branch had been aware that the Spearhead group had been formed.

Hence **spea·r-head** *v. trans.*, to act as the spear-head of, to lead (a movement, attack, etc.).

1938 *Daily News* (Los Angeles) 27 July 8/1 Liberal leader who spearheaded the debate. **1943** *Sun* (Baltimore) 9 July 1/6 He also disclosed that the Marines spearheaded the direct invasion of New Georgia by landing at Segi. **1944** *Times* 19 Feb. 4/7 Spearheaded by our fleet, we have been able to drive the enemy from these bases. **1957** K. A. Wittfogel *Oriental Despotism* 8 Those nonbureaucratic groups and strata which, in feudal Europe and Japan, spearheaded the rise of a commercial and industrial society. **1968** P. Oliver *Screening Blues* ii. 88 Spearheaded by singers like Little Richard Penniman, Ray Charles or B. B. King.., the musical forms which had been held at a distance for so long were allowed to merge. **1980** M. Fonteyn *Magic of Dance* 289 She was such an intelligent artist and fine dancer that she was able to spearhead the ballet reforms, still advancing rather slowly. **1983** *Times* 21 Jan. 16/5 Furniture sales appear to be spearheading the upturn in consumer spending.

Spearman¹ (spiə·ɹmæn). *Statistics.* The name of Charles Edward *Spearman* (1863–1943), English psychologist, used *attrib.* and in the possessive to designate a coefficient he devised as a measure of the degree of agreement between two rankings, being their product-moment correlation coefficient; symbol ρ or *R.*

1907 *Drapers' Company Res. Mem.* (Biometric Ser.) iv. 22 There is a further very serious indictment to be made against Spearman's R... R retains a constant value for wide variations in ρ_{12}. **1942** *Biometrika* XXXII. 277 The Spearman coefficient ρ may be regarded as a sample grade correlation. *Ibid.* 278 The error involved in using Spearman's ρ from a small sample. **1970** *Jrnl. Gen. Psychol.* LXXXIII. 91 A Spearman rank correlation coefficient.. between the median latency scores on the last day of training and the mean of the median latencies over the three days of testing produced a value of $-·3$.

spearmint. Add: **1. c.** *ellipt.* A piece of chewing-gum flavoured with the oil extracted from this plant.

1920 'Sapper' *Bull-Dog Drummond* ix. 240 A grim-faced man at the wheel..had apparently felt the seriousness of the occasion so acutely, as to deposit his third piece of spearmint on the underneath side of the steering-wheel for greater safety. **1945** J. Steinbeck *Cannery Row* i. 9 No Abbeville child..knew the lack of a stick of spearmint ever afterward.

spear-point. Add: **1.** (Later *Comb.* example.)

1930 Blunden *Poems* 55 Like stars in frost as spear-point-bright.

2. Also *fig.*

1937 R. Warner in C. Day Lewis *Mind in Chains* 37 Let us utilise this weight without blunting the spearpoint of our advances. **1963** *Times* 15 Jan. 8/4 The spearpoint of aggression has been blunted in Vietnam.

spec, *sb.*¹ Add: **1. b.** In recent use more generally, as a gamble, on the off chance.

1928 R. Campbell *Wayzgoose* ii. 36 Some came on spec and others came on bikes. **1938** F. D. Sharpe *Sharpe of Flying Squad* xxv. 256 He never tries anywhere on spec., and never does more than two houses a night. **1970** G. Greer *Female Eunuch* 178 The third book bought on that same day was bought on spec. **1978** *New York* 3 Apr. 36/3 Franklin Thomas remembers starting the center on spec with an analysis that projected a 50–50 chance of succeeding. **1981** B. Hines *Looks & Smiles* 197 'Is he expecting you?' 'No, we just came on spec.'

spec, *sb.*² *U.S. slang.* [Short for Spectacle *sb.*¹ or Spectacular *sb.*] **a.** In a circus: (see quot. 1926). **b.** An elaborate and expensive television show.

1926 *Amer. Speech* I. 283/1 *Spec.*, the opening spectacle, or grand entry. **1949** *New Yorker* 5 Nov. 61 Mrs. Webster rode an elephant in the 'spec'. **1959** G. Marx *Let.* 7 Dec. in *Groucho Lett.* (1967) 268 Our little play..could possibly be done as a TV spec.

spec, *sb.*³ *colloq.* [Short for Specification.] A detailed working description; a standard of manufacture or construction. Also *transf.* Freq. *pl.*

1956 *Mag. Fantasy & Sci. Fiction* Oct. 20/2 Belle was not only a perfect secretary.. she also had personal specs which would have delighted Praxiteles. **1966** *Aviation Week & Space Technology* 5 Dec. 5/3 (Advt.), The reliability requirements are, in many cases, virtually unbelievable. Seemingly, a unit built to these kinds of specs would almost have to work perfectly forever. **1976** J. Carroll *Madonna Red* (1977) i. 30 The 707 spec sheets she had memorised..at the Black September training camp. **1979** *Amat. Photographer* 30 May 99/1 The basic specs of these two new OMs remain the same. **1979** *Truck & Bus Transportation* July 20/2 Leyland Australia is basically working within the U.K. spec. to keep costs down.

spec, *sb.*⁴ Colloq. abbrev. of *specialist 2 c. *U.S.*

1958 *Army–Navy–Air Force Register* 14 June 7 Grade E7. Title, Old: Master Spec. New: Specialist-7. **1969** I. Kemp *Brit. G.I. in Vietnam* iii. 43 Specialist Fourth Class—or Spec. Four—Much, a small man..who talked in a continuous monotone. **1977** 'E. McBain' *Long Time no See* xii. 200 These are designations of rank. An E-3 is a Pfc., a Spec 4 is Specialist 4th Class, a corporal. An E-5 is a three-striper, and so on.

spec, *a. colloq.* [Short for Speculative *a.*] Of or pertaining to the practice of building houses without prior guarantee of sale, esp. in estate developments. Also as *adv.* Cf. Speculative *a.* 7 a in Dict. and Suppl.

1958 *Observer* 2 Mar. 8/4 As 'developers' and L.P.T.B. leap-frogged over each other..far into the Home Counties, we created the vast paradise of spec' building and hire-purchase. **1958** *Spectator* 4 July 13/1 Builder-designed 'spec' houses. *Ibid.* 8 Aug. 193/2 A film snippet of 'spec.' housing. **1962** D. Tench *Law for Consumers* vii. 100 Where the buyer buys a completed new house from the builder or developer (what is sometimes called a 'spec' built house). **1965** *New Society* 11 Nov. 6/3 How seldom local authorities and spec builders use any research at all. **1970** J. Betjeman *Ghastly Good Taste* (new ed.) p. xxv, Spec. builders and advertisement hoardings and litter droppers. **1978** J. Wainwright *Jury People* iii. 191 That estate..contains..'high density' housing... 'Spec building'—that is, or was, the term used for such estates. **1978** *Listener* 14 Sept. 337/3 The new, spec-built, often neo-Georgian houses.

special, *a.*, *adv.*, and *sb.* Add: **A.** *adj.* **3. a.** (Further examples.)

1854 Geo. Eliot tr. *Feuerbach's Essence Christianity* iii. 44 God as God..has no more significance for religion than a fundamental general principle has for a special science. **1859** Bartlett *Dict. Amer.* (ed. 2) 432 *Special deposit*, a deposit made in a bank subject to the control of the depositor, and which is not made a part of the funds of the bank to be used by it in its business. **1866** *Weekly New Mexican* (Santa Fe) 22 Dec. 2/2 One of the senators from San Miguel county having resigned, a special election was held. **1886** I. M. Rittenhouse *Jrnl.* 3 Feb. in *Maud* (1939) xi. 368 Eliza brought me a special delivery letter from my good boy. **1904** *Harper's Mag.* Feb. 462/2 And what a lucky chance that brought me a 'special delivery'. **1907** R. Hermon-Hodge *Let.* 20 Jan. in R. S. Churchill *Winston S. Churchill* (1969) II. Compan. i. 641, I am sure you will agree that while the class of men which provides the Yeomanry with officers must often present 'special cases' such as yours, it would never do to make them precedents. **1908** Special school [see *defective *sb.* 2 c]. **1911** W. James *Some Probl. Philos.* i. 4 Limited by the omission of the special sciences, the name of philosophy has come more and more to denote ideas of universal scope exclusively. **1921** *Handicapped Children* (Connecticut Board of Educ.) 6 Let us consider those children who are in need of special education because of some degree of mental abnormality. **1929** E. C. Thomas *Lay Folks' Hist. Liturgy* i. xiv. 62 In addition to the missions mentioned above, there seems to have been a special relationship between Britain and Galicia. **1944** *Act 7 & 8 Geo. VI* c. 31 § 228 In fulfilling their duties under this section, a local education authority shall, in particular, have regard..to the need for securing that provision is made for pupils who suffer from any disability of mind or body by providing, either in special schools or otherwise, special educational treatment. **1944** H. A. Hodges *Wilhelm Dilthey* 110 At the close of the Middle Age the emancipation of the special sciences began. **1945** *Hansard Commons* 7 Nov. 1299 We should not abandon our special relationship with the United States and Canada about the atomic bomb. **1957** Ld. Hailey *African Survey 1956* xvii. 1160 The academic standards to be adopted in the new University College in the first instance were to be those of the University of London, and an application to enter into 'special relationship' with the University of London was accepted. **1958** *Times* 4 July 15/1 The new scheme of 'special deposits' to be introduced by the Bank of England will in effect be kept in reserve in case of need. **1961** *Yale Review* LI. 21 If Britain enters 'Europe', the possibly illusory 'special relationship' with the United States will lose much of its plausibility. **1963** *Sunday Express* 7 July 1 There is a faction in Washington which does not share the official United States policy of friendship for Britain, and which dislikes the 'special relationship' between the two countries. **1968** Special class [see *preemie]. **1974** *State* (Columbia, S. Carolina) 15 Feb. 22-A/6 The beleaguered Republican candidate in the February 5 special election to replace the late Rep. John Saylor of Pennsylvania. **1974** Special vote [see *overall, over-all *adv.* 1 c]. **1976** *Pacifist* Jan. 4/1 The Secretary of State for Defence insists that defence is a special case—housing, social services, education are not? **1977** Special delivery [see *recorded *ppl. a.* 1 b]. **1977** *Times Educ. Suppl.* 21 Oct. 47/3 A teacher with experience in Special Education to teach basic subjects to groups of slow-learning children. **1978** *London Clearing Banks* (Committee London Clearing Bankers) 68 From 1958 onwards the Bank was able to supplement its open-market operations with calls for special deposits from the clearing banks; when a call was made, each bank had to lodge at the Bank of England a cash deposit of a specified percentage of its total deposits. **1979** R. Jaffe *Class Reunion* (1980) iii. iv. 333 The baby had been sent to a special school because she had been born with a kind of sickness. **1980** *Encounter* May 41/2 The encounters took on a peculiarly symbolic significance at a time when various signs point to the possible demise of Israel's 'special relationship' with West Germany and the United States.

c. *special edition* (earlier and later examples).

1872 B. Jerrold *London* xx. 177 The news-boy would deliver the special edition. **1939** L. MacNeice *Autumn Jrnl.* v. 23 They are selling and buying the late Special editions snatched and read abruptly. **1957** V. Brittain *Testament of Experience* iv. 162 Newsvendors calling special editions ran up and down.

d. Special collocations: *Special Air Service*, a special section of the armed forces trained in commando techniques of warfare; cf. *S.A.S.* s.v. *S 4 a; *special area*, a depressed area of high unemployment designated in 1934 (see quot.) for development and improvement; cf. *development area* s.v. *development ii; *Special Branch*, a section of the C.I.D. which deals with police matters relating to political security; *Special Drawing Right*, an additional drawing right allocated to member countries of the International Monetary Fund, allowing them extra powers to purchase foreign currency from the Fund, and so increase their foreign exchange reserves; usu. *pl.*; cf. SDR s.v. *S 4 a; *special effect* (Cinematogr.), a scenic illusion created by props and camera-work (see quot. 1951); usu. *pl.*; *special interest* (U.S.), a group or corporation which seeks special advantages for itself, usu. by political means; *special paper*, an extra advanced examination of the General Certificate of Education, formerly known as a scholarship paper; *Special Reserve*, special units formed in peacetime to furnish a reserve for the regular army in time of war; *special stage* (see quot. 1967).

1942 *We speak from the Air* (Ministry of Information) xvii. 53 There is no better example of co-operation between the services than in the organisation and training of the Special Air Service troops that has been quietly taking place for some time. **1977** *Proc. R. Soc. Med.* LXX. 504/1 Major I. T. Houghton discussed first aid in the Special Air Service Regiment. This was formed in 1941 by David Stirling for operations behind the enemy lines in North Africa. **1980** *Jrnl. R. Soc. Arts* July 491/1 The Special Air Service..is trained for special tasks and it therefore represents the ready-made anti-hijacking force. **1934** *Act 25 Geo. V* c. 1 § 8 (1) This Act may be cited as the Special Areas (Development and Improvement) Act, 1934. **1945** *Archit. Rev.* XCVII. 109/1 The former Special Areas, now renamed (with adjustments) the Development Areas, are not necessarily economic and social entities. **1979** G. Pottinger *Secretaries of State for Scotland 1926–76* xiv. 148 The modest incentives dating originally from the Special Areas Acts of 1934 and 1937, and modified by post-war legislation, to persuade firms to move to depressed districts were no longer considered adequate. **1894** J. G. Littlechild *Reminiscences* i. 8 The later years of my service at Scotland Yard were spent in connection with the 'special' branch of the Criminal Investigation Department. **1936** 'N. Blake' *Thin Shell of Death* ix. 159 Scotland Yard had got into touch with the Special Branch in Dublin. **1962** *Listener* 15 Mar. 459/2 Warrants of this kind would be normally executed by special branch officers acting under the directions of the Director of Public Prosecutions. **1979** J. Wainwright *Duty Elsewhere* vii. 30 The Special Branch... Originally, it was the Special *Irish* Branch. Terror met by terror. **1982** *Listener* 16 Dec. 4/1 It is held by those in Special Branch that contact with Security Service is already so close that SB ought to take over primary responsibility for counter-espionage. **1967** *Bankers' Mag.* CCIV. 216 The scheme for creating new 'reserve units'.. had already been abandoned for a plan for establishing new special automatic drawing rights with the imf... The special drawing rights will be shown as part of the reserves in Britain and the us... The special drawing rights are to be administered by the imf, but kept separate from its other assets. **1971** H. Wilson *Labour Govt.* xxxv. 726 The IMF Special Drawing Rights scheme..had come into effect on 28 July [1969]. **1978** *Internat. Relations Dict.* (U.S. Dept. State Library) 23/2 The IMF has created its own international unit of account—the Special Drawing Right (or SDR or 'paper gold'). **1937** Special effects [see *editor *sb.* 5]. **1951** *Jrnl. Soc. Motion Picture & Television Engineers* LVII. 53/1 Studio terminology combines under 'special effects' a large variety of items, materials, equipment and processes which..aid in realistically imitating natural phenomena..which otherwise could be considered only at prohibitive expense or with impossible hazards. **1978** R. Hill *Pinch of Snuff* vii. 65 The special effects department must be getting better. **1910** G. Pinchot *Fight for Conservation* 134 The people of the United States believe that..the Senate and the House no longer represent the voters by whom they were elected, but the special interests by whom they are controlled. **1980** *Outdoor Life* (U.S.) (Northeast ed.) Oct. 21/2 It is a measure of the arrogance of the special interests that 53 of the new lease applications are for tracts inside the proposed wilderness area. **1963** *Times Educ. Suppl.* 1 Feb. 197/4 In your issue of November 23 a letter was published concerning the restriction imposed by the S.S.E.C. that no A level candidate may offer more than two of the special papers which are to replace the scholarship papers. **1976** *Bridgwater Mercury* 21 Dec. 1/5 Karen..passed all of her A level subjects, English, Economics and History at Grade A, as well as gaining a Distinction on the Special Paper in the last subject. **1908** *Times* 11 Apr. 8/2 The second order details the precise conditions of the transfer of all Militia units, except those disbanded, to the Special Reserve. **1908** *Regs. for Officers of Special Reserve* 1 The Special Reserve of Officers is a branch of the Reserve of Officers, established by Royal Warrant dated the 3rd April, 1908. *Ibid.*, The Special Reserve. Special reservists form part of the first class of the army reserve. **1909** *Army & Navy Gaz.* 19 June 589/2 Mr. Haldane stated that bayonets are now being issued to all infantry units of the Special Reserve which are being armed with the short rifle. **1931** [see *cadre 2]. **1935** *Regs. for Officers & Airmen of Special Reserve serving in R.A.F. Squadrons* (Amendment 5) 1 Applications for the attendance of Special Reserve personnel to give evidence in private lawsuits will be reported at once through the usual channels. **1967** P. Moss *Story so Far* vi. 65 A lot of international rallies..have special stages, or tests, over the worst bits of road they can find. Not all big rallies have to do this but on those which do it is on these special stages that the rally is won and lost. **1977** *Belfast Tel.* 27 Jan. 25/4 The other two members of the team have elected him to drive all the competitive sections—or special stages—along the 18,000 mile route.

4. a. (Further examples.)

1843 N. Simons *Statutes at Large, U.K.* XVI. 131 Commissioners for General Purposes to execute all Matters with respect to the Duties under all the Schedules except such as are directed to be executed by Special or other commissioners. **1926** *Daily Chron.* 13 May 1/3 Measures must be taken to demobilise the special constabulary.. and to ensure the speedy return to normal conditions. **1932** Auden *Orators* i. 36 Jokes about special constables and conscientious objectors. **1956** in J. Biggs-Davison *Hand is Red* (1973) x. 134 Members of the R.U.C. and B Special Constabulary. **1970** *Britain 1970: an Official Handbk.* (Central Office of Information) 105 All police forces have an attachment of special constables, who are volunteers willing to perform police duties without pay in their spare time. In England and Wales the function of special constables is to act as auxiliary to the regular force when required. In Scotland they are employed only in emergencies, although they may be assigned for duty for training purposes. **1970** P. Laurie *Scotland Yard* 294 *Special patrol group*, a unit of uniform policemen, about 130 strong, attached to no particular area. They are available for intensive patrolling, searches, guards, raids, etc. **1971** *Halsbury's Statutes England* (ed. 3) XXXIV. 1252 Unlike the General Commissioners..the Special Commissioners are civil servants, and their powers and duties are not limited to any particular area... They do, however, act independently of the Crown. **1980** P. G. Winslow *Counsellor Heart* xx. 215 Men were sent to saturate the area. The Special Patrol Group was called in.

c. *special buyer*, the bill-broker of the Bank of England in the discount market; *special partner* (U.S.) = *limited partner* s.v. *limited *ppl. a.* 2 b.

1941 *Economist* 1 Feb. 149/1 The help given to the market last week by the special buyer has made itself more felt. **1965** Seldon & Pennance *Dict. Econ.* 311 In Britain open market operations in Treasury Bills are conducted through a firm of discount brokers (the 'Special Buyer') on instructions from the Bank of England. **1822** *Laws of State of New-York* ccxliv. 259 That partnerships, to be formed under this act, shall consist of one or more partners, jointly and severally responsible..who shall be called general partners; and one or more partners, who furnish certain funds or capital to the common stock, whose liability shall extend no further than the fund which he or they have furnished to the partnership stock, and who shall be called special partners. **1839** Marryat *Diary Amer.* 1st Ser. II. 251 In America, if a person wishes to become a special partner (a sleeping partner) in any concern, he may do so to any extent he pleases. **1889** *Cent. Dict.* 4309/1 If the statute governing partnerships is violated the special partner becomes liable as a general partner.

5. a. (Further examples.)

1790 F. Burney *Diary* Jan. (1842) V. iii. 85 Mr Fairly's marriage..was by special licence, and at the house of Sir R— F—. **1825** Special licence [see *droit¹ 1 b]. **1849** *Rep. Sel. Comm. Public Libraries* 138 in *Parl. Papers* XVII. 1 The second description of libraries would be those destined to represent special branches of literature; for instance the Commercial Library, or the Library of Fine Arts, at Hamburgh, would be a most perfect instance of those special libraries. **1908** H. Cecil *Let.* 20 Aug. in R. S. Churchill *Winston S. Churchill* (1969) II. Compan. ii. 808 A special licence costs a lot of money..so you had better go for an ordinary one or for banns. **1937** *Discovery* June 192/2 At the request of the British Council, the Association of Special Libraries and Information Bureaux has undertaken the compilation of a short list of standard and technical books by British authors. **1954** K. Amis *Lucky Jim* iii. 27 Have you got that syllabus together.. the list of stuff for your special subject next year? **1966** A. Battersby *Math. in Management* ix. 232 The appeal to elasticity is evident in any retail outlet with price reductions, 'special' offers' or trading stamps. **1971** P. D. James *Shroud for Nightingale* vii. 243 He might well have told his special nurse or blurted it out in his delirium. **1973** *Scotsman* 7 Aug. 6/2 The 'wedding' in the registrar's office at Dumfries took place after the couple were granted a special licence in the courthouse only 50 yards away. **1976** *Gloss. Documentation Terms* (*B.S.I.*) 62 *Special library*, a library maintained by an association, government agency, parliament, research institution, learned society, professional association, museum, business firm, industrial organization, chamber of commerce, etc. or other organized group, the greater part of its collection being in a specific field or subject. **1977** *Western Morning News* 1 Sept. 1/1 Instant coffee is to be on 'special offer' in various shops and supermarkets for the next six months.

c. *special* (*theory of*) *relativity*: see *relativity 2.

7. *special magistrate, order, prosecutor, term.*

1852 *Alabama Reports* XX. 446 The Court of Commissioners of Roads and Revenue in this State has no power to hold special terms, except in cases expressly authorized by law. **1859** *Indiana Reports* XI. 562 The indictment was signed by James F. Suit, as prosecuting attorney; but it appears that he was a special prosecutor elected by the people. **1889** *Cent. Dict.*, *Special orders*, in *law*, those orders which are made only in view of the peculiar circumstances of the case, and require notice to the adversary and a hearing by the court. **1919** *Act 9 & 10 Geo. V* c. 100 § 26 Anything which under the Electric Lighting Acts may be effected by a provisional order confirmed by Parliament may be effected by a special order made by the Electricity Commissioners and confirmed by the Board of Trade. **1934** *Northeastern Reporter* CXC. 270/2 'Special term of court' is one not fixed by general statute establishing court terms, but ordinarily called pursuant to power granted by statute. **1936** *Panel* Mar. 3/3 Several thousand additional witnesses have been examined in the office of Mr. Dewey, the Special Prosecutor. **1966** Tacheron & Udall *Job of Congressman* vii. 198 Special orders are designed primarily to authorize the House to disregard the regular rules of procedure so that a particular matter can be handled with dispatch. **1973** *N.Y. Law Jrnl.* 25 July 12/8 The sole question raised on this appeal is whether or not Special Term erred when it concluded that a 1960 Pontiac convertible ..was covered under an automobile insurance policy issued with respect to a 1960 Buick sedan. **1974** *Greenville* (S. Carolina) *News* 22 Apr. 1/3 'The special prosecutor has already indicted and is preparing to try' seven former Nixon campaign or administration aides. **1975** J. P. Morgan *House of Lords & Labour Govt.* ii. 64 Orders are known in the Lords as Special Orders. **1979** *Internat. Jrnl. Sociol. of Law* VII. 279 The function of the Special Magistrate was 'to represent in his own person the whole policy of Britain'.

7*. *Math.* Of a group: that can be represented by matrices of unit determinant.

1903 J. E. Campbell *Theory Continuous Groups* i. 17 This group is called the special linear homogeneous group; it is a sub-group of the general linear homogeneous group. **1955** B. Higman *Appl. Group-Theoretic & Matrix Methods* xii. 181 The rotation group in three dimensions is isomorphic with a factor group of the special unitary group in two dimensions containing half the number of elements. **1955, 1967** [see *SU* s.v. *S 4 a]. **1968** M. S. Livingston *Particle Physics* xii. 212 Special unitary group theory based on two fundamental states (known by the symbol SU(2)) leads to the prediction of a multiplet structure in the spectrum of substates. **1971** D. Gorenstein in Powell & Higman *Finite Simple Groups* ii. 68 The latter group is the projective special unitary group.

8. *special agreement, -interest, -occasion, -procedure, -purpose, -range.*

1944 *Act 7 & 8 Geo. VI* c. 31 § 236 If at any time the managers or governors of an aided school or a special agreement school are unable or unwilling to carry out their obligations..it shall be their duty to apply to the Minister for an order revoking the order by virtue of which the school is an aided school or special agreement school. **1945** Nelson & Wright *Tomorrow's House* vii. 80/1 The living-room..is..turning into a special-purpose room like the study. **1952** *Times* 4 Feb. 2/3, £250,000 for grants in aid towards a new 'special agreement' secondary school, and for works at four aided polytechnics. **1956** Abraham & Hawtrey *Parliamentary Dict.* 188 *Special procedure order*, a term applied by the standing orders of both Houses to an order, made or confirmed by a minister, in relation to which the Statutory Orders (Special Procedure) Act, 1945, applies. **1957** *Times Lit. Suppl.* 8 Nov. 678/5 Instruction covers everything from the regular features of the dinner table to special-occasion items for the buffet or the hors-d'œuvre tray. **1959** W. K. Richmond *British Birds of Prey* p. xi, How far, if at all, was it safe to rely on a phraseology which enjoyed no common currency, however acceptable it might be to this or that special-interest group? **1961** *Act 9 & 10 Eliz. II* c. 62 2nd Schedule § 2 Where a trust fund includes special-range property,..the special-range property shall be carried to a separate part of the fund. **1964** D. Foulkes *Introd. Admin. Law* i. 6 The first part of the procedure provides for the proper publication of the order (known as a 'special procedure order') and the making of objections, and the second part for parliamentary scrutiny. **1982** *Time* 8 Nov. 92/2 The special-interest slangs generated then were interminably publicized.

C. *sb.* **5. a.** (Further examples.)

1833 W. H. Breton *Excursions in New South Wales* ii. 51 A Government Establishment formerly existed at Wellington Valley..and it was to this place that all the principal convicts, or those called *specials* were sent: that is to say, those of good connections. **1867** J. Morison *Australia* ix. 220 A laudable consideration was shown by the Government to a class of convicts belonging to..'the upper classes of society', and who were known by the name 'specials'. **1939** *War Illustr.* 9 Dec. 393/3 Accompanied by German police, bands of Sudeten 'specials' raided the University and the Czech societies, and many persons were taken to the Gestapo headquarters. **1955** *Times* 20 July 4/3 There were only two police constables on St. Mary's, reinforced for the occasion by four 'specials'. **1972** *Times* 7 Apr. 5/2 (*heading*) Specials platoon fired into crowd.

b. Also, a special article, dish, edition, offer, programme.

1867 *Oregon State Jrnl.* 19 Jan. 2/3 A Washington special says that [etc.]. **1899** J. F. Fraser *Round World on a Wheel* xiv. 162, I described..the London evening papers..the 'specials', 'extra specials', and 'second extra specials'. **1914** Joyce *Dubliners* 115 They went into the parlour at the back and O'Halloran ordered small hot specials all round. **1933** E. Waugh *Scoop* ii. iv. 212 Make up the Irish edition with his morning cable... If the follow-up comes in..run a special. **1939** J. B. Priestley *Let People Sing* ii. 44 What about ordering the *Ninepenny Special*—steak-and-kidney pudding and peas? **1952** *Hist. Times* IV. i. iv. 415 The leading articles and 'specials' exposing the exaggerations of the 'yellow peril' school published during the year. **1958** *Punch* 1 Jan. 60/3 Accepting one of his hostess's 'specials' at one of her now-famous 'get-together' parties. **1961** *Listener* 28 Dec. 1135/2 There have, however, been single performances and 'specials' worthy to be recalled. **1966** T. Pynchon *Crying of Lot 49* i. 14 Pink flyers advertising specials at the markets. **1977** B. Pym *Quartet in Autumn* ii. 24 I'll do one of my specials—baked beans on toast with a poached egg on top. **1979** *Tucson* (Arizona) *Citizen* 20 Sept. 7B/6 NBC made a deal with the studio to make a few two-hour 'Buck Rogers' movies, to be shown occasionally as specials. **1980** *Redbook* Oct. 46/1 Before you go to the supermarket, check your newspaper for sales. Watch for patterns: 'specials' usually occur toward the beginning of the month, when supplies are highest.

spe·cial, *v. slang.* [f. the sb.] **1.** *intr.* To work as a special correspondent for a newspaper.

1915 T. Burke *Nights in Town* 318, I have worked on six newspapers... I have done everything, from subbing to specialling.

2. *trans.* Of a member of the staff of a hospital: to attend continuously to (a single patient).

1961 'K. Norway' *Waterfront Hospital* vii. 130 She seems determined to special Emlyn Roberts. **1967** *Nursing Times* 27 Jan. 111/1 A nurse will have to 'special' the patient to make the necessary observations. **1978** L. Isenberg in D. Abse *My Medical School* 207 One night we were asked if someone would 'special' a lady who might require surgery during the night for suspected abdominal trauma.

specialism. Add: **2.** Also, a specialized area of knowledge or work; a professional or academic field.

1937 L. MUMFORD in F. Mackenzie *Planned Society* p. vii, Specialists..cannot plan: for planning involves the job of coordinating specialisms, focussing them in common fields of knowledge, and canalizing them in appropriate channels of common action. **1950** [see *EXPERTISE]. **1967** *Times Rev. Industry* May 125/2 These [*sc.* technical specifications] can be prepared in a fashion likely to impress a group of people of different specialisms, including operating managers, engineers, finance officers and scientists. **1977** *Antiquaries Jrnl.* LVII. 342 But why did the Americanist W. M. Bray find it so difficult to extract a subject from his own specialism?

specialist. Add: **2. b.** [tr. Russ. *spetsialist.*] In Communist parlance, a person with a specialist knowledge in some area of science, engineering, or culture; an engineer, scientist.

1929 V. M. MOLOTOV *Communist Party Soviet Union* 39 Of course there cannot be many among the ranks of the old specialists who could be taken into the Party... The Shakhty case revealed clearly enough that we have some of the most bitter enemies among the specialists, whose skill we must nevertheless use. **1974** T. P. WHITNEY tr. *Solzhenitsyn's Gulag Archipelago* I. I. ix. 334 The Case of Glavtop—May, 1921. This case was important because it involved *engineers*—or, as they had been christened in the terminology of the times, 'specialists', or spetsy. **1977** 'S. LEYS' *Chinese Shadows* (1978) ii. 101 It [*sc.* the Tower of the Six Harmonies] is such a sturdy building that an army of 'specialists' would have been necessary to demolish it.

c. An enlisted man in the U.S. army employed on specialized duties. Cf. *SPEC *sb.*[4]

Freq. prefixed to the name of a soldier. The grades of specialists are modelled on the ranks, from corporal to master sergeant, but do not correspond to them in other respects.

1955 *Army–Navy–Air Force Register* 14 May 1/1 Army personnel in the top four enlisted grades will be separated into two groups, non-commissioned officers and specialists... Those who perform non-leadership duties of a technical or administrative nature will be designated 'Specialists'.. and will rank among themselves as Master Specialists (E-7), Specialist First Class (E-6), Specialist Second Class (E-5), and Specialist (E-4). **1969** I. KEMP *Brit. G.I. in Vietnam* iii. 67, I got on particularly well with the new crew chief, Specialist Fifth Class Jaycelon. **1974** *Encycl. Brit. Micropædia* VII. 406/2 *Specialist*, military, any of four enlisted ranks in the U.S. Army corresponding to the grades of corporal (Specialist 4) through sergeant first class (Specialist 7).

2*. *Ecol.* A species which is closely adapted, and largely restricted, to a particular mode of life.

1966 *Amer. Naturalist* C. 607 When the gain to a jack-of-all-trades in reduced travelling time makes up for his lower hunting efficiency compared to the patch specialists, then the jack-of-all-trades will outcompete both specialists. **1973** P. A. COLINVAUX *Introd. Ecol.* xli. 576 Animals become specialists in eating particular plants; the plants, on an evolutionary time scale, respond with new poisons. **1978** *Nature* 5 Jan. 56/2 *K*-selected species (or 'specialists') exhibit conservative strategies for survival in predictable environments, and are adapted against pressures from predators and interspecific competition.

‖**spécialité** (spesialite). Also **specialité.** [Fr.: see SPECIALITY.] A special or distinguishing feature, characteristic, etc.; a speciality. Also *specialité de la maison*, a dish which a restaurant considers its speciality, or one for which it is particularly noted. Cf. SPECIALITY 5 c in Dict. and Suppl.

1839 W. A. SHEE *Jrnl.* Sept. in *My Contemporaries* (1893) iii. 79 The jousts..are..the *specialité* of such a meeting. **1857** GEO. ELIOT *Let.* 1 Sept. (1954) II. 379 Every new or renovated journal should have a specialité—do something not yet done, fill up a gap. **1861** K. STANLEY *Let.* 1 Feb. in B. & P. Russell *Amberley Papers* (1937) I. iii. 114 Mr. Oliphant..is very anxious that Japan should not be thought his spécialité so he will not learn the language. **1893** SOMERVILLE & 'ROSS' *Vine Country* ix. 167 The *spécialité* of the town, ordered expressly for us by Monsieur A., the macaroons made at the convent according to an ancient recipe known to the nuns. **1914** 'SAKI' *When William Came* xiv. 242 Iced mulberry salad, my dear, it's a *spécialité de la maison*, so to speak; they say the roving husband brought the recipe from Astrakhan, or Seville, or some such outlandish place. **1935** A. E. W. MASON *They wouldn't be Chessmen* xii. 137 He would take his luncheon at a famous restaurant on the Quai which had a specialité de la maison. **1956** R. MACAULAY *Towers of Trebizond* xi. 119 We got several good trout, and two odd-shaped fishes which must have been a specialité of that lake, as we had never seen them before. **1957** 'P. PORTOBELLO' *Deb's Mum* 18 Old Lady Dolgelly, who made a *spécialité* of housekeeping. **1962** 'R. SIMONS' *Killing Chase* v. 68 'How about *Specialité de la Maison*?' 'What's that?' 'Cod roes on toast.' **1976** 'M. INNES' *Gay Phoenix* iii. 42 These Paphians, like the restaurants, had their *spécialités*.

speciality. Add: **5. c.** (Later examples.) Cf. *SPÉCIALITÉ.

1977 T. HEALD *Just Desserts* i. 16 Bognor chose..a speciality of the house called Pollo Sophia Loren. **1980** F. KING *Indirect Method* 161 You must try these profiteroles. Speciality of the *maison*. **1982** J. MIDGLEY *Stone Killer* iii. 23 They dined on the restaurant speciality, Sicilian-style macaroni.

6. b. *attrib.*

1930 E. FERBER *Cimarron* xxiii. 373 Fascinating little speciality shops..just like those..on Madison Avenue. **1938** [see *BACK *sb.*[1] 25 a]. **1959** *Times* 9 Mar. (Britain's Food Suppl.) p. xiv/4 Dishes produced in the speciality restaurant often surpass..those of establishments that attempt to run the full range. **1967** *Times Rev. Industry* Feb. 31/1 The newly-created group was thus firmly based in three different markets—household and pharmaceutical products and speciality foods. **1979** N. FREELING *Widow* iii. 10 A gigantic supermarket, with a covered gallery of speciality shops all around.

7. *Theatr.* Used *attrib.* to designate a performer or performance, esp. in variety entertainments, of an unusual or specialized character.

1933 P. GODFREY *Back-Stage* xv. 187 The next important subdivision of the chorus consists of the speciality dancing troupe. *Ibid.*, A few speciality solo steps. **1952** W. GRANVILLE *Dict. Theatrical Terms* 166 *Speciality artiste*, a vaudeville artiste who specializes in impersonations, juggling, etc. **1962** A. NISBETT *Technique Sound Studio* iii. 69 It is also possible to balance most dance bands and speciality groups on a single microphone provided that sufficient trouble is taken. **1967** *Stage* 2 Mar. 21/4 (Advt.), Wanted for long summer season speciality act, strong male vocal..and two girl dancers. **1970** W. J. BURLEY *To kill Cat* iii. 58 We had a cabaret artiste of that name... She was a speciality dancer and her art possessed great artistic merit.

specialization. Add: **2.** *Biol.* A specialized character or adaptive feature in an organism.

1918 F. W. JONES *Probl. Man's Ancestry* 30 Pithecoid specializations vary so much in their manifestations in the different groups of monkeys. **1978** *Nature* 26 Jan. 353/1 It seems that both intermuscular and subcutaneous lipid accumulation and a reduction in skeletal ossification have evolved as specialisations to reduce density in *Pleuragramma*, a pelagic Antarctic fish without a swim bladder.

specialty. Add: **9.** *N. Amer.* = SPECIALITY 6, *7, and 8.

1888 G. O. SEILHAMER *Hist. Amer. Theatre* II. 118 Dr. Bayley, a specialty performer..gave entertainments in this country as early as 1752. **1901** *Daily Colonist* (Victoria, B.C.) 3 Nov. 12/3 These expensive attractions in conjunction with a big specialty company of no mean pretensions will go to make up an excellent show. **1915** in S. Marcus *Minding Store* (1974) ii. 23 Twenty-five specialty shops in one building. **1919** [see *FOLLY *sb.*[1] 5 b]. **1936** J. STEINBECK *In Dubious Battle* i. 3 Department stores and specialty shops. **1961** BOWMAN & BALL *Theatre Lang.* 338 *Specialty number*, a song not intended to be fully integrated with the rest of the entertainment—Musical comedy. **1968** *Globe & Mail* (Toronto) 17 Feb. B4 Roblin Steel Corp..., maker of specialty steel forgings and castings.., plans to buy privately owned Washburn Co., maker of specialty household products. **1978** *Language* LIV. 189 M & M apparently prefer to leave these rich but page-devouring matters to specialty works.

speciate (spī·ʃiēⁱt), *v.* *Biol.* [Back-formation from *SPECIATION.] *intr.* Of a population of plants or animals: to exhibit evolutionary development leading to the recognition of a new species. Hence **spe·ciating** *ppl. a.*, showing or inducing speciation.

1964 *Oceanogr. & Marine Biol.* II. 206 C[*odium*] *fragile* appears to be speciating as it moves into waters where these two subspecies are present. **1970** *Watsonia* VIII. 67 The lack of flower-constancy and of other speciating factors in the visitors. **1973** P. A. COLINVAUX *Introd. Ecol.* xli. 579 Man alone can change his niche without speciating. **1981** *Nature* 26 Feb. 743/2 Only rarely will members of the small speciating population be fossilized, so that the lineage will show what are known as 'punctuated equilibria'—a long series of unaltered fossil forms followed by a rapid shift to a new type.

speciation (spīʃiēⁱ·ʃən). *Biol.* [f. SPECI(ES + -ATION.] The formation of new and distinct species in the course of evolution.

1906 O. F. COOK in *Science* 30 Mar. 506/2 Speciation.. is the origination or multiplication of species by subdivision, usually..as a result of environmental incidents. **1926** *Nature* 21 Aug. 271/1 Thus *speciation* through continuity stands in contrast with *mutation* through discontinuity. *Ibid.* 272/1 Isolation is the most important factor in the speciation of birds. **1953** J. S. HUXLEY *Evolution in Action* iii. 71 Much of speciation represents a frill of mere diversity. **1978** *Nature* 21 Sept. 255/1 Speciation involves the splitting of a single evolutionary lineage into two or more genetically independent ones.

specie. Add: **7. b.** For 'Now *rare* or *Obs.*' read 'Now *Obs.* exc. as erron. sing. of SPECIES *sb.* 10'. (Later examples.)

1974 A. SCOTT-JAMES *Sissinghurst* vi. 74 There are.. thickets of specie roses in many odd corners. **1974** *Country Life* 25 Apr. 1033/2 (Advt.), We specialise in Roses (specie & old-fashioned). **1980** *Daily Tel.* 22 Jan. 11/2 Castrated rats and other animals live longer than normal creatures of the specie. **1980** *Pan Am Clipper* Oct. 48/1 Is he [*sc.* man] descended directly from apes, or is he a specie that evolved from an entirely new.. branch of the primate tree?

8. specie jar, a large glass or china jar formerly used for storage in chemists' shops and now used only for display; **specie point** = *gold point* s.v. *GOLD[1] 10; **specie-room**, a strong-room on a ship in which gold coin was deposited.

1914 *N. & Q.* 14 Feb. 127/1 The well-known 'specie jars' of chemists' shops. **1929** C. J. S. THOMPSON *Myst. Apothecary* xix. 257 The large cylindrical vessels called 'specie jars', with metal or gilded lids, that still decorate some of the chemists' shop-windows. **1861** G. J. GOSCHEN *Theory Foreign Exchanges* iv. 50 There would have existed a certain competition..to buy at a heavy discount, much below specie point, in order subsequently to realize at least the specie value. **1966** A. GILPIN *Dict. Econ. Terms* 190 *Specie points or Gold points*, the extreme points of variation in a rate of exchange under the gold standard. **1891** *Scribner's Mag.* Nov. 603/2 In these days of heavy gold shipments, the specie-room on the steamship is a very important institution. *a* **1929** R. BEDFORD in Murdoch & Drake-Brockman *Austral. Short Stories* (1951) 100 I'll bring dynamite..and blow the specie-room open..and haul out the gold-boxes.

species, *sb.* Add: **9. f.** *Chem.* and *Physics.* A particular kind of molecule, ion, free radical, etc.; a distinct kind of atom (esp. a radioactive one) or sub-atomic particle.

[**1857** W. A. MILLER *Elem. Chem.* III. 45 Other remarkable species of compounds which are obtained by substitution, are those in which a portion of the hydrogen of the original body is displaced by chlorine.] **1895** C. S. PALMER tr. *Nernst's Theoret. Chem.* IV. ii. 521 We will select as a further..example of complete heterogeneous equilibrium, a system composed of H_2O and SO_3, i.e. two molecular species. **1948** *Nature* 28 Feb. 291/2 The use of tracer materials, radioactive species, and radiations. **1962** COTTON & WILKINSON *Adv. Inorg. Chem.* xiv. 427 Pure sulfuric acid contains a number of species in equilibrium. **1967** M. CHANDLER *Ceramics in Mod. World* vi. 179 Natural uranium is not a single nuclear species. It contains two isotopes. **1971** *Physics Bull.* Dec. 720/3 Using the techniques of flash photolysis..he has identified the spectra of many new species, like CH_2, CH_3 and NH_2. **1974** *Nature* 13 Dec. 538/1 Whether *n* is considered to be the total number of particles in the Universe or the number of a given species, such as electrons or nucleons, is not important within the accuracy considered here.

14. *species-cross, diversity, -evolution, -formation, group*; **species being**, a term [tr. G. *Gattungswesen* (P. C. Reinhard, 1797)] used by Marx to denote man's objective consciousness of life and the mastery of the natural world through work which characterize the human species; man considered in respect to these qualities; **species pair**, a pair of species which are similar, sympatric, and closely related, but distinct; **species-poor, -rich** *adjs.*, having a small, or large, number of species; so **species richness**; **species rose**, a rose belonging to a distinct species and not to one of the many varieties produced by hybridization; cf. *OLD ROSE; **species sanitation** *Med.*, measures taken against a particular species of mosquito in order to reduce the incidence of malaria; **species-specific** *a.*, found in or characteristic of the members of one species only; hence **species specificity**; **species-uniform** *a.*, consistent throughout a species.

1959 M. MILLIGAN tr. *Marx's Econ. & Philos. MSS of 1844* 75 Conscious life-activity directly distinguishes man from animal life-activity. It is just because of this that he is a species being. *Ibid.* 76 It is just in the working-up of the objective world, therefore, that man first really proves himself to be a *species being*. **1979** GLASSNER & FREEDMAN *Clinical Sociol.* iv. 95 Workers are alienated from..their 'species being', or from their human capacity for conscious and creative activity. **1926** J. S. HUXLEY *Ess. Pop. Sci.* 25 Where the offspring of species-crosses are perfectly fertile, [etc.]. **1967** *Oceanogr. & Marine Biol.* V. 257 An intuitive character referable to any natural population or assemblage of individuals is species diversity or biotic diversity. **1972** Species diversity [see *species richness* below]. **1946** F. E. ZEUNER *Dating Past* xii. 355 This genus experienced an episode of abundant species-evolution from the Eocene to the Miocene, or roughly for 50 million years. **1977** R. HOLLAND *Self & Social Context* viii. 246 Mead's concept of the social-self.. never clarifies the relation between species-evolution and individual development. **1941** J. S. HUXLEY *Uniqueness of Man* vi. 155 Chromosome-doubling after crossing is a method of species-formation in which the isolation is not spatial but genetic. **1975** *Nature* 24 Jan. 290/3 Some hakes, especially the western South American species-group (*M*[*erluccius*] *gayi*), might support a greater fishery than at present. **1942** J. S. HUXLEY *Evolution* vi. 284 Overlapping Species Pairs. Numerous puzzling cases are presented by extremely similar species which overlap over much of their range and yet remain distinct. **1959** *New Biol.* XXVIII. 70 Of some sympatric species-pairs, one member releases pollen in the morning and the other in the evening, and the stigmas of each species are receptive only at the appropriate times, so considerably reducing the chances of receiving foreign pollen. **1964** V. J. CHAPMAN *Coastal Veg.* ix. 225 In the absence of grazing a luxuriant and species-poor Festucetum rubrae develops. **1976** *Jrnl. R. Soc. Arts* CXXIV. 640/1 The Southern Ocean is characterized by a species-rich, productive ecosystem which contrasts sharply with the species-poor, relatively barren terrestrial and freshwater ecosystems of the islands and continental landmass. **1973** *Nature* 30 Mar. 344/2 In maintaining or reconstructing types of herbaceous vegetation in which the density of flowering plants exceeds 20 species/m²—the so-called 'species-rich' communities, success is often frustrated by competitive exclusion. **1981** *Country Life* 12 Feb. 376/3 Species-rich hedges were treated sympathetically. **1972**

Ecology LIII. 279/2 The pattern of bird species richness (mean number of species per census) was quite similar to the pattern of bird species diversity. **1930** J. N. HART *Rose Growing* ix. 57 The species roses are actually wild roses, either native..or imported. **1935** N. MITCHISON *We have been Warned* II. 196 There were Penzance briars, and species roses growing unpruned. **1976** LD. HOME *Way Wind Blows* xiv. 200 A very attractive garden to the south and front of the house..had been filled with a wide variety of species roses. **1930** M. F. BOYD *Introd. Malariol.* vi. 418 'Selective control', 'species control' or 'species sanitation', as it is variously designated, *i.e.*, limiting efforts to the control of one species. **1945** *New Biol.* I. 107 The disease can be controlled by applying anti-mosquito measures to these kinds [*sc.* malarial vectors] only. This method of control, species sanitation, was first used, with spectacular results, by Watson in Malaya. **1959** A. A. SANDOSHAM *Malariology* i. 19 The increased knowledge of the systematics and bionomics of local anopheline fauna made it possible to evolve the more scientific and more economic method of mosquito control referred to as 'species-sanitation'. **1924** *Jrnl. Exper. Med.* XL. 106 The question arises whether these antigens are simple species-specific proteins. **1956** *Nature* 21 Jan. 133/1 The reduction in oxygen consumption appears to be caused by a species-specific antibody in the N'Dama serum. **1980** A. KENNY *Aquinas* iii. 76 Chomsky has argued that it is impossible to explain the rapidity with which children acquire the grammar of a language from the..utterances of their parents unless we postulate a species-specific innate language-learning ability. **1925** *Jrnl. Exper. Med.* XLII. 141 Species specificity of cells is of a different order as opposed to species specificity of proteins. **1964** M. HYNES *Med. Bacteriol.* (ed. 8) vii. 76 The differences that make a protein molecule a specific antibody are only minor; in chemical structure and species-specificity..it is still a γ-globulin molecule. **1968** R. W. LANGACKER *Lang. & its Structure* ix. 247 The view that linguistic experience serves more to activate language than to shape it accounts for the fact that language is species uniform and species specific. **1976** *Word 1971* XXVII. 225 If a sign language is treated on a par with an oral language, then language is neither species-specific nor species-uniform, because other species are now known to be capable of learning a sign language.

speciesism (spī·ʃīziz'm). [f. SPECIES *sb.* + -ISM.] Discrimination against or exploitation of certain animal species by human beings, based on an assumption of mankind's superiority.

1975 R. D. RYDER *Victims of Sci.* 16, I use the word 'speciesism' to describe the widespread discrimination that is practised by man against other species... Speciesism and racism both overlook or underestimate the similarities between the discriminator and those discriminated against. **1976** *New Society* 3 June 544/1 If racism and sexism are wrong, what about speciesism? Peter Singer apologises for using this graceless term. **1979** *Listener* 7 June 777/1 'Speciesism' is intended to convey the idea that one animal species, human beings, supposes that it has the right to exploit other species. **1982** *Times* 8 Apr. 13/1 Animals have rights... There are forms of 'speciesism' as corrupting as 'racism' or 'sexism'.

Hence **spe·ciesist** *a.* and *sb.*

1975 R. D. RYDER *Victims of Sci.* 21 The main speciesist defence of cruelty to animals is that mankind benefits—in terms of knowledge, economy or sport, for example. **1977** *Daily Tel.* 5 May 16 A demonstrating mob of anti-speciesists, perhaps accompanied by species even odder than themselves, may soon be..cawing, roaring and singing..at your door. **1978** *Nature* 9 Nov. 122/2 Gould is, in general, rather good at puncturing human speciesist vanity, and in particular he will have nothing to do with the myth that evolution represents progress toward man.

specific, *a.* and *sb.* Add: **A.** *adj.* **2. d.** *Physics.* (i) Of or designating a dimensionless number equal to the ratio of the value of a property of a given substance to the value of the same property of some reference substance (as water) or of vacuum under the same conditions, so providing a relative value for comparison with different substances, as *specific gravity* (see GRAVITY 4 c); *specific heat* (see HEAT *sb.* 2 d); *specific inductive capacity* = *dielectric constant* s.v. *DIELECTRIC *a.* 2 b; *specific viscosity*, the difference between the viscosity of a solution of a given concentration and that of the pure solvent, divided by the viscosity of the pure solvent.

1838 M. FARADAY in *Phil. Trans. R. Soc.* CXXVIII. 33, I feel satisfied that the experiments altogether fully prove the existence of a difference between dielectrics as to their power of favouring an inductive action through them; which difference may..be expressed by the term specific inductive capacity. **1918** *Physical Rev.* XII. 50 One arm was then filled with water, and the other with a mixture of water and ethyl alcohol, the specific inductive capacity of which was known. **1935** *Jrnl. Physical Chem.* XXXIX. 157 Staudinger later adopted the term 'specific viscosity', for the quantity $\eta_r - 1$. **1944** [see *PERMITTIVITY]. **1959** K. HENNEY *Radio Engin. Handbk.* (ed. 5) iv. 2 The dielectric constant *K* of an insulating material is the ratio of the capacitance C_x of a capacitor using the material as the dielectric to the capacitance C_a using air as the dielectric... This property of the material is sometimes called inductivity or specific inductive capacity. **1966** M. L. MILLER *Struct. Polymers* v. 206 At the highest rate of shear used in these experiments, the reduced specific viscosity was independent of concentration.

(ii) Of or designating a physical quantity that is referred to a unit of mass, volume, or other measure in order to form a number independent of the properties of the particular system studied, and so measuring an inherent property or characteristic that can be scaled to describe a given system or used as an indicator of the effect of an action or process, as *specific acoustic impedance* (see *IMPEDANCE 2); *specific activity*, the activity of a given radioisotope per unit mass; *specific charge*, the ratio of the charge of an ion or sub-atomic particle to its mass; *specific conductance, conductivity*, the conductivity of unit length of a material of unit cross-sectional area; the reciprocal of resistivity (see *RESISTIVITY 1); *specific (fuel) consumption*, the weight of fuel consumed by an engine per unit time per unit of power or thrust developed; the reciprocal of specific impulse; *specific impulse* (see *IMPULSE *sb.* 2 c); *specific ionization*, the number of ion pairs produced by an ionizing particle per unit path length; *specific refraction, refractive constant*, a constant relating the refractive index (*n*) of a material to its density (ρ), given by $(n^2 - 1)/(\rho(n^2 + 2))$; *specific resistance, resistivity* = *RESISTIVITY 1; *specific rotation, rotatory power*, the angle through which the plane of polarization of light of a specified wavelength is rotated by passage through a column of an optically active substance of given length (usu. 10 cm.) and at unit concentration; *specific surface*, the surface area per unit volume of a finely-divided substance; *specific thrust* = *specific impulse* above; *specific volume*, the volume of a substance per unit mass; the reciprocal of density.

In mod. use, there is a tendency to restrict the application of *specific* to quantities that are referred to unit mass. Accordingly, alternative terms are being advocated to replace those that do not conform to this narrow definition, as *conductivity* for *specific conductance* and *specific conductivity*, *relative density* for *specific gravity*, *specific heat capacity* (which is referred to unit mass) for *specific heat*, *dielectric constant* or *relative permittivity* for *specific inductive capacity*, and *resistivity* for *specific resistance* and *specific resistivity*.

The examples follow in alphabetical order.

1938 *Nature* 18 June 1098/1 The specific activity of phosphatide P extracted from human blood corpuscles 24 hours after administration of labelled sodium phosphate was found to be 40 times less than that of plasma inorganic P. **1961** G. R. CHOPPIN *Exper. Nuclear Chem.* xi. 186 Much higher levels of specific activity may be counted with no resolving time losses. **1926** R. W. LAWSON tr. *Hevesy's Man. Radioactivity* i. 6 From the magnitude of the deflexion, combined with similar deflexion experiments in an electric field, we can determine the magnitude of the specific charge (*e/m*). **1971** D. F. JACKSON *Concepts of Atomic Physics* ii. 18 The specific charge of the lightest known ion, that of hydrogen, is $9{\cdot}579 \times 10^7$ C. kg⁻¹. **1885, 1886** Specific conductance [see *CONDUCTANCE]. **1924** J. R. PARTINGTON in H. S. Taylor *Treat. Physical Chem.* I. xi. 517 In such cells conductivity water with a specific conductance of $0{\cdot}21 \times 10^{-6}$ ohm⁻¹ can be kept 12 hours without change. **1958** CONDON & ODISHAW *Handbk. Physics* IV. ix. 141/1 In terms of Ohm's law, the defining equation..for specific conductance reduces to $\kappa = I/E$..where *I* is the current and *E* the potential applied to a centimeter-cube sample of the conductor. **1898** C. L. SPEYERS *Text-bk. Physical Chem.* ix. 166 A 5% aqueous solution of KCl at 0° has a specific conductivity of 0·06617 mhos. **1957** G. E. HUTCHINSON *Treat. Limnol.* I. viii. 558 (*caption*) Composition of standard bicarbonate waters of varying specific conductivities. **1931** *Automotive Industries* 9 May 726/2 Fig. 10 shows the variation in specific consumption, power and head temperature of the A70 cylinder, with change in fuel flow. **1946** J. W. VALE *Aviation Mechanic's Engine Man.* ix. 271 At any altitude the specific fuel consumption increases with the increase of power output. **1966** J. H. HORLOCK *Axial Flow Turbines* viii. 214 For the high by-pass ratios..the specific fuel consumption drops rapidly with increasing turbine temperature. **1932** *Physical Rev.* XXXIX. 884 The specific ionization thus determined does not exceed 32 ion-pairs per cm, in water-saturated air at 68 cm pressure. **1961** G. R. CHOPPIN *Exper. Nuclear Chem.* iii. 23 This specific ionization is a measure of the rate of energy loss. **1940** GLASSTONE *Textbk. Physical Chem.* viii. 524 The difference between the specific refractions for two wave lengths, e.g., the Hα and Hγ lines, is called the specific dispersivity. **1899** J. WALKER *Introd. Physical Chem.* xiv. 138 Another specific refractive constant is given by [etc.]. **1899** J. MCCRAE tr. *Reychler's Outl. Physical Chem.* III. ii. 197 In order to determine the specific resistance of a solution in ohms, we consider a cubical mass of the solution, the length of whose side is 1 centimetre. **1935** WILSON & DOWSE tr. *Holzer's Foundations Short Wave Therapy* 74 The specific resistance is not..a constant, independent of frequency. **1978** P. W. ATKINS *Physical Chem.* xxv. 820 The resistance of a material increases with its length *l* but decreases with its cross-section *A*... The proportionality coefficient is called the resistivity, or specific resistance. **1958** *Chambers's Techn. Dict.* 716/1 *Resistivity*, a term denoting volume resistivity, i.e. the resistance of a block of the material in question having unit length and unit cross-sectional area; also called specific resistivity. **1964** R. F. FICCHI *Electrical Interference* viii. 133, ρ is the specific resistivity of the conductor. **1899** J. WALKER *Introd. Physical Chem.* xv. 150 The specific rotation of lævorotatory oil of turpentine is 37·01°. **1940** GLASSTONE *Textbk. Physical Chem.* viii. 585 The optical rotatory power of a pure substance, particularly in the liquid state, is generally expressed in terms of its specific rotation or specific rotatory power. **1958** CONDON & ODISHAW *Handbk. Physics* VI. vi. 120/1 The molecular rotation is the product of the specific rotation by the molecular weight *M* of the active material. **1876** *Jrnl. Chem. Soc.* I. 667 It is proposed to substitute a_D for the specific rotatory power obtained by means of the sodium ray. **1899** J. WALKER *Introd. Physical Chem.* xiv. 139 The specific rotatory power is usually denoted by the symbol [α]. **1924** *Chem. Abstr.* XVIII. 3507 The specific surfaces of several varieties of charcoal were measured. **1951** A. E. ALEXANDER *Surface Chem.* i. 4 Colloidal materials such as charcoal and clays show adsorption phenomena very markedly owing to their large specific surfaces.. which arises [*sic*] from their fine state of subdivision. **1977** ROWELL & FARINATO in L.-H. Lee *Characterization of Metal & Polymer Surfaces* II. 399 The specific surface of a monodisperse colloid becomes independent of the number concentration and refractive index of the spherical particles. **1949** D. G. SHEPHERD *Introd. Gas Turbine* iii. 78 The curves for specific thrust are in general of similar shape.., increasing with T_{max} and having in an optimum value at a certain pressure ratio. **1966** J. H. HORLOCK *Axial Flow Turbines* viii. 211 A high value of specific thrust means that small engine weight is required. **1868** JONES & WATTS *Fownes's Man. Chem.* (ed. 10) II. 250 The numbers obtained.., representing the specific volumes of the various solid and liquid elementary substances, present far more cases of discrepancy than of agreement. **1957** *Amer. Inst. Physics Handbk.* II. 117 Many tables and other aids have been prepared for the routine calculation of density and specific volume of sea water.

4. c. Of a duty or tax: assessed on an article or goods according to quantity or amount without reference to value.

1789 *Deb. Congress U.S.* 9 Apr. 107, I shall not pretend to say that there ought not to be specific duties laid upon every one of the articles enumerated. **1845** J. K. POLK *Diary* 1 Nov. (1929) 23, I had recommended..the abolition of the minimum principle and specific duties. **1901** J. S. NICHOLSON *Pol. Econ.* III. 348 If the tax is specific and not *ad valorem*. **1930** M. CLARK *Home Trade* III. xxii. 187 Specific duties are those which are based on the quantity of the imported produce, i.e. they are so much per lb. or so much per gallon, etc. **1959** *Chambers's Encycl.* V. 512/2 Specific duties are expressed as an amount of money on the unit amount of the product while *ad valorem* duties are expressed as a percentage addition to the selling price.

5. *specific epithet* (chiefly *Bot.* and *Microbiology*), the second (adjectival) element in the Latin name of a species according to the binomial system, which follows the generic name and serves to distinguish a species from others in the same genus; *specific name*, (*a*) (now chiefly *Zool.*) = *specific epithet* above; (*b*) (now chiefly *Bot.* and *Microbiol.*), the Latin name of a species, which in the binomial system comprises a generic name and a specific epithet.

1753, etc. Specific name [in Dict.]. **1871** *Nature* 20 July 221/1 The mistake Cotteau is accused of making of assigning to Desor instead of Agassiz the specific name of *Pseudodiadema hemisphæricum* is entirely unfounded. **1905** *Règles Internat. Nomancl. Zool.* (Congrès Internat. de Zool.) 31 A specific name becomes a subspecific name when the species so named becomes a subspecies. **1906** *Internat. Rules Bot. Nomencl.* 1905 47 When a..species is moved into another genus..the first specific epithet.. must be retained. **1926** *Rep. Brit. Assoc. Adv. Sci. 1925* 75 A species [of animal] is a community, or a number of related communities, whose distinctive morphological characters are..sufficiently definite to entitle it, or them, to a specific name. **1945** *Rhodora* XLVII. 274 Binomial nomenclature was not intended by Linnaeus to supersede the polynomial specific name. **1964** *Internat. Code Zool. Nomencl.* ii. 7 The name of a species consists of two words (binomen)..;..the first word is the generic name, the second word is the specific name. **1966** *Internat. Code Bot. Nomencl.* iii. 27 *Tuber*..was accompanied by binary specific names, e.g. *Tuber cibarium*, and is therefore admissible. *Ibid.* 30 The name of a species is a binary combination consisting of the name of the genus followed by a single specific epithet. **1970** *Watsonia* VIII. 156 The specific epithet *racemosa* is..not applicable in the genus *Amelanchier*. **1974** *Encycl. Brit. Macropædia* II. 1019/2 Gaspard Bauhin, a Swiss botanist of the late 16th and early 17th centuries, designated plants by a generic and a specific name. **1982** ARMS & CAMP *Biology* (ed. 2) xx. 310 Specific epithets..are adjectives, and the same one may be combined with different generic names and used for a number of unrelated organisms; for example, *Erythronium americanum*, the trout lily; *Euarctos americanus*, the American black bear.

B. *sb.* **2.** Usu. *pl.* Also *loosely*, details, particulars.

1966 *New Statesman* 9 Sept. 350/2 The latter [*sc.* journalism]..considers the specifics of an event, using implicit general principles of behaviour out of necessity. **1972** G. BROMLEY *In Absence of Body* iii. 30 Let's get down to specifics. What can we actually do to help? **1975** *N.Y. Times* 11 Sept. 8/1 Placing this tragedy of a woman's sexual obsession with her stepson within the arresting specifics of this strange setting does at least remove it from the fury-bestrewn never-never-land of the antique Greek drama. **1977** F. BRANSTON *Up & Coming Man* xii. 125 He told us he had been investing in property in London, but he was a bit vague about the specifics. **1980** *Jrnl. R. Soc. Arts* Feb. 152/1 Planning should start..with specifics rather than concepts.

3. A specific word, name, etc., *spec.* in taxonomy or toponymy.

1962 BURRILL & BONSACK in Householder & Saporta *Probl. Lexicogr.* 195 The elements in geographic names that indicate the class of the entity, e.g., in *Red Hill*..

or *Lake Erie*.., *Hill* and *Lake*, are the generic elements (or 'generics'). The elements that identify the particular entity, in the above instances *Red* and *Erie*, are called the specific elements (or 'specifics'). **1969** J. FOWLES *French Lieutenant's Woman* viii. 50 Although many scientists of the day gratefully used her [*sc.* Mary Anning's] finds to establish their own reputation, not one native type bears the specific *anningii*. **1977** *Word 1972* XXVIII. 133 Most of the specifics, or second elements, in such names are demonstrably of Gaelic and not of Pictish origin.

-specific. The adj. used as the final element in combination with sbs. to designate phenomena peculiar to the category indicated by the precedent element: specific to or that specifies (something).

1924, etc. [see *species-specific* adj. s.v. *SPECIES *sb.* 14]. **1936**, etc. [see *organ-specific* adj. s.v. *ORGAN *sb.*[1] 8]. **1949** K. DAVIS *Human Society* xxi. 600 Age-specific fertility trends show that a stationary or a declining population will soon eventuate. **1965**, etc. [see *language-specific* adj. s.v. *LANGUAGE *sb.* 6 b]. **1971** J. Z. YOUNG *Introd. Study Man* v. 84 Study of some brain-specific proteins. **1978** *Dædalus* Fall 7 Herodotus' report..refracts into intensely age-specific opinions. **1979** *Nature* 25 Jan. 251/1 But there is a far less trivial, and far more nation-specific way in which intellectual resources may be used or squandered.

specificity. Delete 'Chiefly *Med.* and *Path.*' and add: **1. a.** (Later examples.)

1922 [see *DIRECTEDNESS]. **1946** *Scrutiny* XIV. 109 George Eliot's genius appears in the specificity with which she exhibits the accomplishments in Gwendolen of the kind of conscious advantage she resembles Isabel in enjoying. **1977** J. F. FIXX *Compl. Bk. Running* vii. 89 Few runners had supposed it could possibly be so hot on a mid-April day in Massachusetts, so practically nobody had trained properly. As a result, most people's times were terrible. The same specificity principle applies to terrain.

b. The narrowness of the range of substances with which an antibody or other agent acts or is effective.

[**1896**: in Dict., sense 1.] **1904** G. H. F. NUTTALL *Blood Immunity & Blood Relationship* ix. 381 Wassermann.. brought the question of specificity into greater prominence. *Ibid.* 443/1 (Index), Specificity of precipitins. **1935** N. P. SHERWOOD *Immunol.* xii. 274 Two kinds of specificity can be demonstrated by immune reactions, one that applies to species and the second to type variation within a species. **1971** *Sci. Amer.* July 26/1 The specificity, or narrow spectrum, of vaccines is a limitation; it means that a different vaccine is required for each virus or strain of virus.

c. *Biol.* The degree to which a parasite or symbiote is restricted in its range of hosts.

1924 *Rep. Brit. Assoc. Adv. Sci. 1923* 453 They [*sc.* parasitic nematodes] may be divided broadly into a section with more or less strict 'specificity' and a section with members occurring in various hosts, often of quite distantly related groups. **1955** *Sci. Amer.* July 77/1 It is also known that the protein coat determines the specificity of the virus, i.e., whether or not it will attack a certain bacterium. **1965** B. E. FREEMAN tr. *Vandel's Biospeleology* xv. 245 This is more a case of parasitic or symbiotic specificity than cavernicolous specialisation.

2. (Later examples.)

1928 HARTSHORNE & MAY *Studies in Deceit* II. xii. 221 (*heading*) Specificity of attitudes. **1958** R. WILLIAMS *Culture & Society* III. iii. 231 A principal virtue was always the specificity, not only of definition, but of illustration. **1981** *Times* 16 July 6/1 They also want to avoid 'specificity' because they have not yet formulated fully-fledged policies.

specificness. (Later examples.)

1905 W. STEVENS *Let.* 31 Dec. (1967) 85 Reflections.. on Japanese life, on specificness,..on my future. **1966** J. ELLIS in C. E. Bazell *In Memory of J. R. Firth* 83 By *register-choice* is meant the particular register out of the performer's range to which the utterance may be assigned ..the specificness of the assignment depending on the delicacy of the analysis.

specifier. Add: **b.** That which specifies.

1954 A. J. AYER *Philos. Ess.* iii. 61 Let us say of any two singular referential statements S and S'..that S is a specifier of S' if and only if S' is not a component of S, S entails S', and S' does not entail S. **1964** E. A. NIDA *Toward Sci. Transl.* xii. 258 As the data are analyzed both semantically and structurally, certain 'specifiers' can be associated with the elements in question, as tags of identification. In the comparison stage, the specifiers of the source-language text are matched with specifiers of the receptor language, so that the corresponding semantic and structural elements are properly identified. **1971** [see *INTENSIFIER].

specimen. Add: **8. b. specimen-book** (earlier example); **specimen page**, a page submitted by a printer as a sample setting for a book; **specimen tree**, a tree planted on its own, away from other plants of a similar size.

1871 W. BLACKWOOD *Let.* 19 Sept. in *Geo. Eliot Lett.* (1956) V. 190, I also send you by Book Post a specimen Book from which you will..be able to select the colour for the paper cover. **1835** DICKENS *Let.* 9 Dec. (1965) I. 102, I have received neither specimen page nor proofs. **1877** W. PATER *Let.* 30 Jan. (1970) 27 Dear Mr. Macmillan. Of the two specimen pages, I enclose the one I think preferable. **1926** S. UNWIN *Truth about Publishing* ii. 37 As soon as the printers' estimate and specimen page are received and have been checked, the estimate has to be completed by the addition of the cost of paper, binding and other items. **1975** J. BUTCHER *Copy-Editing* ii. 13 Specimen pages are intended to show solutions to all the general typographical problems that the printers will meet in the book. **1933** A. OSBORN *Shrubs & Trees for Garden* xxiv. 119 Weeping trees are unsuitable for grouping, but very valuable for planting as specimen trees on the lawn. **1961** E. WAUGH *Unconditional Surrender* III. ii. 230 Guy took to walking..in the public gardens... There were winding paths, specimen trees, statuary. **1980** *Amat. Gardening* 25 Oct. 15/1 It makes a shapely specimen tree as well as being good for making a hedge or screen.

speck, *sb.*[1] Add: **2. c.** Also *the Speck* (Austral. colloq.), Tasmania.

1930 *Bulletin* (Sydney) 11 June 21 N.S.W., V., Q., S.A., W.A. and the Speck. **1949** *Geogr. Mag.* Feb. 373 *Tassie* and *The Speck*, meaning Tasmania. **1963** *Times* 12 Mar. (Austral. Suppl.) p. v/4 Tasmania—affectionately known as 'the speck'.

d. (Later example.)

1936 M. MITCHELL *Gone with Wind* xl. 719 You're smart enough about dollars and cents... But you..aren't a speck smart about folks.

5. *Comb.*, as *speck-like* adj.

1917 J. MASEFIELD *Lollingdon Downs* 56 No spark of him is specklike in his glass. **1965** E. BISHOP *Questions of Travel* I. 11 A specklike girl and boy, Alone, but near a specklike house.

speck, *v.*[1] Add: **5.** *Austral.* [Both this and sense 1 of the vbl. sb. below may properly repr. abbrev. of SPECULATE *v.*: cf. SPEC *sb.*] **a.** *intr.* To search for small particles of gold or opal on the surface. **b.** *trans.* To search the surface of (the ground) for traces of gold or opal; to discover (particles of gold, etc.) in this manner.

1888 H. LAWSON *His Father's Mate* in *Stories* (1964) I. 139 A pick and shovel, and a gold dish..with which he used to go 'a-speckin'' and 'fossickin'' amongst the old mullock heaps. **1903** R. BEDFORD *True Eyes* lviii. 305 With little cries of delight he 'specked' a four-ounce slug of the red gold. *Ibid.* lx. 315 They had sieved and dry-blown and 'specked' the little tongue of auriferous soil. **1926** *Spectator* 14 Aug. 240/2 Went 'specking' in nearby creeks. Got colours of gold but no nuggets. **1936** I. L. IDRIESS *Cattle King* xxiv. 211 Next morning they picked up gold. In trembling excitement they 'specked' piece after piece. **1969** E. WALLER *And there's Opal out There* 116 A couple of tourists specking for bits of potch and opal.

specking, *vbl. sb.* [f. SPECK *v.*[1] + -ING[1].] **1.** *Austral.* The action of searching for surface gold or opal. Cf. sense 5 of the vb. above.

1901 M. VIVIENNE *Travels in W. Australia* 171 Almost everyone in the camp went out for an afternoon's specking (looking on the ground for nuggets). **1945** *Walkabout* (Melbourne) 1 Mar. 14 Most of the residents of Lightning Ridge are experts at the art of 'specking'.

2. The discoloration of pottery through contamination of the glaze, or this effect contrived for decoration.

1967 M. CHANDLER *Ceramics in Mod. World* iii. 96 In the case of a glaze mix, it is particularly important to avoid contamination and specking. *a* **1977** *Harrison Mayer Ltd. Catal.* 22/3 Ilmenite...can be used in glazes to obtain specking effects.

speckle, *sb.* Add: **2. b.** A granular appearance seen in images formed by originally coherent light as a result of the interference of waves that have been reflected at a rough surface or have passed through an inhomogeneous medium; also, each of the light or dark areas giving rise to this appearance. Freq. *attrib.*

1965 *Jrnl. Optical Soc. Amer.* LV. 247 Exposing photographic film directly to the backscattered radiation confirms the independent existence of the speckles. *Ibid.* 252/2 Both speckle pattern and diffraction pattern were recorded (photographically) at the same distance from the aperture. **1970** A. LABEYRIE in *Astron. & Astrophysics* VI. 85/1 'Speckle' refers to the grainy structure observed when a laser beam is reflected from a diffusing surface... In large telescopes, the image of point stars also features a speckle pattern, due to seeing induced phase fluctuations on the wavefront. **1975** T. S. MCKECHNIE in J. C. Dainty *Laser Speckle & Related Phenomena* iv. 126 We may reduce speckle by simply reducing the coherence of the illumination. **1976** *Physics Bull.* Aug. 357/2 Objects viewed in highly coherent light acquire a peculiar granular appearance. This is the laser speckle phenomenon. **1977** *McGraw-Hill Encycl. Sci. & Technol.* 397/1 The size of the speckles is equal to the diffraction-limited resolution limit of the telescope, regardless of the resolution limit determined by the turbulent atmosphere. **1979** *Nature* 5 July p. vii/2 A double laser speckle camera which is used for non-destructive stress, vibration, and flaw analysis of engineering components.

3. speckle interferometry, the analysis of speckle in two or more images, differing only in the instant of exposure, as a means of obtaining information about the source of light or the agent that caused the speckle; so **speckle-interferometric** *a.*

1973 *Astrophysical Jrnl.* CLXXXII. L139 Speckle interferometric techniques are an effective way of obtaining information about small solar features without the problems of lifting large telescopes above the Earth's atmosphere. **1970** A. LABEYRIE in *Astron. & Astrophysics* VI. 85 Key words: speckle interferometry. **1972** *Sci. Amer.* Feb. 106 The technique, known as speckle interferometry, can also be used to map local deformations in stressed mechanical parts. **1973** *Astrophysical Jrnl.* CLXXXII. L139 Speckle interferometry is potentially more powerful than two-aperture Michelson stellar interferometry because the entire aperture is used. **1978** PASACHOFF & KUTNER *University Astron.* II. vi. 148 The speckle interferometry technique involves taking photographs of the speckle pattern with very short exposures—on the order of 1/100 second—or using electronic detection devices and then using mathematical techniques and computer assistance to deduce the properties of the starlight that entered the telescope.

speckle, *v.* Add: **2.** (Earlier and later examples.) Also, to become speckled.

1703 tr. *H. van Oosten's Dutch Gardener* IV. ix. 218 If you water them in the Heat of the Sun, the leaves will speckle, and so often lose their Spindel. **1973** R. ADAMS *Watership Down* ix. 36 As the plants moved in the breeze, the sunlight dappled and speckled back and forth over the brown soil.

speckled, (*ppl.*) *a.* and *pa. pple.* **1.** (Later examples.)

1965 *Jrnl. Optical Soc. Amer.* LV. 247 When a diffuse surface is illuminated by a coherent monochromatic source such as a laser, the illuminated area appears speckled. **1978** PASACHOFF & KUTNER *University Astron.* II. vi. 148 At any one instant, the image of a star through a large telescope looks speckled because different parts of the image are affected by different small turbulent areas in the earth's atmosphere.

3. a. (*b*) *speckled hen, perch*: (earlier examples).

1877 C. HALLOCK *Sportsman's Gazetteer* 276 Locally they are..severally known as yellow perch,..speckled hen, etc. *Ibid.* 378 Silver Perch, or Speckled Perch.

b. speckled wood, (*b*) a brown butterfly with yellowish spots, *Pararge egeria*, found in lightly shaded places in Britain, much of continental Europe, and North Africa.

1766 M. HARRIS *Aurelian* 132 Speckled wood... It flies in woods. The caterpillar feeds on grass. **1974** [see *meadow brown* (*butterfly*) s.v. *MEADOW *sb.* 4 b].

c. speckled yellows, a disease of sugar beet characterized by distorted and discoloured leaves, caused by a deficiency of manganese.

1938 *Brit. Sugar Beet Rev.* XII. 77/2 Fields affected with 'Speckled Yellows' can be recognised from a considerable distance. **1959** *New Biol.* XXX. 91 Diseases such as 'grey speck' of oats, 'speckled yellows' of sugar beet and 'marsh spot' of peas are caused by the low availability of manganese in the soil. **1960** *Farmer & Stockbreeder* 15 Mar. 149/2 (*caption*) Sugar Beet leaf—illustrating deficiency of manganese ('speckled yellows').

speckling, *vbl. sb.* (Later examples.)

1965 *Jrnl. Optical Soc. Amer.* LV. 247/1 Speckling has ..been observed when viewing a translucent material which is backlighted by a cw [*sc.* continuous wave] laser. **1973** *Astrophysical Jrnl.* CLXXXII. L139 Small, high-contrast features such as umbral dots and faculae near the limb [of the sun] show the speckling most clearly.

spe·cky, *a.*[1] *colloq.* [f. SPEC(S + -Y[1].] Bespectacled.

1956 R. JENKINS *Guests of War* IV. i. 167 The unbraw unlovable puke married to yon specky gasping smout of a barber. **1959** I. & P. OPIE *Lore & Lang. Schoolch.* ix. 172 A girl or boy with spectacles is known as 'Four-eyes', 'Specky four-eyes', '*Annie* four eyes'... Occasionally he is 'Eye balls'..and 'Specky Jock' (Scotland).

Speclette (spekle·t). [f. SPEC(S + *-lette* (refashioned on Fr. model after -LET): cf. -ETTE.] A pair of spectacles that folds at the bridge (see quot. 1962[1]).

1931 *Illustr. London News* 31 Oct. 707/3 (Advt.), Speclettes the delightful and fashionable folding spectacles. **1962** L. S. SASIENI *Optical Dispensing* viii. 182 In the 'Speclette' the sides are folded in halves on to the lenses and the two halves of the front folded over each other in the plane of the lenses, the one sliding across the other. *Ibid.* 183 (*caption*) 'Speclette' folding spectacles.

spect, *v.* Also spec, speck, 'spect, etc. Repr. (chiefly U.S.) non-standard pronunc. of (*I*) *expect* or *suspect*.

1839 F. A. KEMBLE *Jrnl. Residence Georgian Plantation* (1863) xii. 118 Good for colored folks, missis; me 'spect not good enough for white people. **1852** MRS. STOWE *Uncle Tom's Cabin* II. xx. 38, I spect I grow'd. Don't think nobody never made me. **1893** H. A. SHANDS *Some Peculiarities of Speech in Mississippi* 59 Speck (spec). Used by negroes for both *expect* and *suspect*. **1914** 'BARTIMEUS' *Naval Occasions* xx. 182 'Spect's you wants yer breakfus'—same's me! **1927** A. P. RANDOLPH in A. Dundes *Mother Wit* (1973) 200 What's th' matter wid you? 'Specks you got dat Randolph fever, too, eh? **1976** *Washington Post* 7 Nov. K2/2 We'll teach them..how to say: I SPEC (as in 'I spec I will do that.'). **1976** C. DEXTER *Last seen Wearing* xix. 151 'Has the wife got the chips on, Lewis?' 'I 'spect so.' **1977** F. PARRISH *Fire in Barley* x. 106, I 'speck you want me to feed the zoo.

spectacle, *sb.*[1] Add: **II. 7. c.** (Earlier and later examples.) † Also (*rarely*) in sing., a score of zero in one innings. Cf. *PAIR *sb.*[1] 2 b.

1835 *Bell's Life* 13 Sept. 3/4 Good put a spectacle on him first ball. **1865** *F. Lillywhite's Guide to Cricketers* 27 The ominous 'spectacles' have been worn by the best sighted men. **1907** E. V. LUCAS *Hambledon Men* 230 It is believed he never made two noughts, or 'a pair of spectacles', in any match of note! **1979** *Wisden Cricket Monthly* Dec. 21/3 Who got a 'pair of spectacles' for Yorkshire on his first appearance?

9. a. *spectacle-blurred* adj.

1932 W. FAULKNER *Light in August* (1933) xiii. 291 Misshapen, with his gray stubble and his dark spectacle-blurred eyes.

spectacular, *a.* and *sb.* Add: **1.** Also *fig.*

1934 J. B. PRIESTLEY *Eng. Journey* ix. 316 Both my companions knew about this yard, which had been a spectacular failure in which over a million of money had been lost.

4. (Later examples.) Also *spec.* a radio or television programme, entertainment, etc., produced on a lavish or spectacular scale.

1953 *N.Y. Times* 3 Jan. 8/5 Thirteen 'spectaculars' will be affected, including the giant British Overseas Airways sign, Cunard and Canadian Pacific Lines displays and advertisements for gin, wine, radio and television. **1954** *Ibid.* 28 Mar. x13/1 Its [*sc.* NBC's] big feature..will be a series of costly and lavish ninety-minute 'spectaculars'—opera, drama, musical comedy, circuses, ice shows, etc. **1958** *Times* 28 Mar. 3/4 A television 'spectacular' transmitted by the National Broadcasting Company. **1966** *Punch* 8 June 858/2 *The Disorderly Knights*, a historical novel of the sixteenth century by Dorothy Dunnett, is a five hundred page spectacular: enormous in every possible way. **1969** *Listener* 20 Feb. 249/3 Radio drama may miss its former purse-power, and the multi-studio 'spectacular' is a fashion of the past. **1971** *Scope* (S. Afr.) 19 Mar. 4/2 It was a golfing spectacular the old pros will talk about for years. **1978** S. BRILL *Teamsters* x. 391 The ceremony and dinner party were followed by an entertainment spectacular put on by.. Barbara McNair, Billy Daniels, Ed McMahon and Frank Sinatra.

specta·cularism. [f. SPECTACULAR *a.* and *sb.* + -ISM.] Spectacular character or quality.

1888 G. B. SHAW in *Star* 18 July 2/4 These are but details, only to be noted at Covent Garden, because so much is there sacrificed to spectacularism on the highest scale. **1931** *Aberdeen Press & Jrnl.* 14 Oct. 6/4 The spectacularism of Noel Coward.

spectate, *v.* Add: **2.** *intr.* [Back-formation from SPECTATOR.] To be a spectator rather than a participant, esp. at a sporting event.

1929 *Amer. Speech* IV. 501 An advertisement in an Iowa paper reads: 'Ladies Only at the Promenade Roller Rink Thursday afternoon, 2:30 to 5. No men will be allowed to skate or to spectate.' **1971** H. C. RAE *Marksman* II. iii. 118 On Sunday afternoons..the boys meet and play soccer... Occasionally Gordon went there, to play or to spectate. I'm not sure which. **1974** 'P. B. YUILL' *Bornless Keeper* xiii. 125 He enjoyed spectating at a good snarling-match. **1980** L. BIRNBACH et al. *Official Preppy Handbk.* 90/2 They..provide entertainment for the Prep women who spectate when they're not flinging a ball about themselves.

spectating *ppl. a.* (later example). Also as *vbl. sb.*

1942 *Amer. Speech* XVII. 24 'Tackle the shoe problem —shoes for *spectating*' was the heading of an advertisement of a Department Store, Oct. 11, 1941. **1966** *Listener* 12 May 697/2 Only a gigantic comic talent..could.. communicate the comic grandeur to a spectating, as opposed to a reading, audience. **1978** D. FRANCIS *Trial Run* xix. 234 I'd ridden in races in that state..so why fret at some gentle spectating.

spectation. Delete † *Obs.* and add later example.

1940 R. G. COLLINGWOOD *Ess. Metaphysics* xxxi. 307 The mere act of spectation could in time generate the idea of a cause.

spectator. Add: **2. b.** *spectator sport*, a sport which affords good entertainment for spectators as well as for participants. Also *transf.* and *fig.*

1943 *Amer. Speech* XVIII. 95 The American 'spectator-sports' (of clothes) has been mistranslated in at least one advertisement [in New Zealand] as 'spectacular-sports'. **1944** M. LASKI *Love on Supertax* xi. 109 Burn all those clothes you've got on..and get back into a decent unpretentious spectator-sports-suit. **1954** *Encounter* Feb. 57/1 The fascination of the great spectator-sports—soccer, athletics, cricket, lawn tennis—is partly due to the effect of the game on its audience. **1969** A. GLYN *Dragon Variation* ix. 268 Well, call this [*sc.* chess] a spectator sport, twenty goddam minutes and nobody's moved a thing. Give me tennis! **1975** *New Yorker* 10 Feb. 110/3 With hard times upon us there may be a question in the minds of even the least dedicated office-holders both here and in Albany about how long their hard-pressed constituents will let them get away with treating representative government as a minor spectator sport. **1979** *Guardian* 12 June 8/1 Watching election coverage all through the night is a great spectator sport.

spectatorial, *a.* Add: Hence **spectato·rially** *adv.*

1930 A. HUXLEY *Brief Candles* 14 People think I'm an excellent psychologist. And I suppose I am. Spectatorially. But I'm a bad experiencer. **1973** *Times Lit. Suppl.* 14 Dec. 1536/5 Our Londoner applauds not a murder, but a representation; and he himself is spectatorially present at a representation, not at a murder. **1980** D. NEWSOME *On Edge of Paradise* vi. 211 Arthur was invited to a number of spectatorially promising occasions.

spectatory. **2.** (Earlier and later examples.)

1829 H. FOOTE *Compan. to Theatres* 135 Such as do not wish to mix in their frolics, may witness them from the spectatory of the theatre. **1881** P. FITZGERALD *World behind Scenes* I. ii. 21 One of the most difficult questions is how is the *salle* or 'spectatory', as old writers call it—not a bad word either—to be effectively lighted.

spectinomycin (spe:ktinǒmai·sin). *Pharm.* [f. mod.L. *spect-ābilis* visible, remarkable (see SPECTABLE *a.*), + *-MYCIN.] An antimicrobial substance obtained from the fungus *Streptomyces spectabilis* and used esp. to treat gonorrhœa that is resistant to penicillin.

1964 *Antimicrobial Agents & Chemotherapy* 438/1 The in vitro activity of spectinomycin..was tested against routine gonococcal isolates and against resistant strains isolated from penicillin therapy failure cases of gonorrhea. **1976** *Lancet* 25 Dec. 1379/2 Spectinomycin succeeded in 21 (95%) of 22 patients treated. **1977** *Maclean's Mag.* Feb. 54 The only cure so far for the new strain is spectinomycin, a drug four times as costly as penicillin and hence not widely applicable in the Far East.

spectral, *a.* Add: **5. a.** Also applied to a property or parameter which is being considered as a function of frequency or wavelength, or which pertains to a given frequency range or value within the spectrum. (Later examples, which also illustrate the extended use of 'spectrum'; cf. *SPECTRUM 3 a, b.)

1919 *Sci. Abstr.* A. XXII. 563 Section 2 considers the definition of temperature. This is based upon thermal radiation and spectral distribution. **1950** *Audio Engin.* Aug. 14/2 A knowledge of the spectral characteristics of sound sources will indicate the regions in the frequency scale to which particular attention must be paid in assessing the effect of response changes in the sound system. **1951** *New Biol.* XI. 34 A spectral absorption curve, in which the proportion of light absorbed is plotted against wavelength. **1957** G. E. HUTCHINSON *Treat. Limnol.* I. vi. 376 The spectral composition of the total reflected light when the sun is high is little different from that of the incident. **1964** *Oceanogr. & Marine Biol.* II. 13 The turbulent fluctuations of velocity or of some other property of the water..may be observed directly, and the spectral distribution of energy..derived from the observations. **1966** *McGraw-Hill Encycl. Sci. & Technol.* IV. 582/2 Any spectral emissivity value is valid only for a narrow wavelength interval. **1971** *Physics Bull.* July 385/2 Spectral intensity is intensity per unit bandwidth (W sr⁻¹ Hz⁻¹). *Ibid.* Nov. 653/3 An investigation of the spectral content, vibrato, attack and sound pressure of vowels sung by male and female students under technical and performing conditions. **1977** I. M. CAMPBELL *Energy & Atmosphere* viii. 272 The spectral absorption characteristics of alkyl nitrates are rather similar to those of nitric acid.

d. *Math.* Of or pertaining to the spectrum of a transformation (*SPECTRUM 5*).

1948 P. R. HALMOS *Finite Dimensional Vector Spaces* ii. 80 The characteristic equation, and consequently every other spectral concept such as the proper values and their multiplicities, is invariant under replacing A by BAB⁻¹. **1968** P. A. P. MORAN *Introd. Probability Theory* iii. 118 Hence *U* and *V* are non-singular matrices, and we have $P = U^{-1}\Lambda U = V\Lambda V^{-1}$, which is a 'spectral' decomposition of *P*.

6. Special collocations: *spectral analysis*, chemical analysis of substances by means of their spectra; analysis of light or another oscillating system into a spectrum; *spectral index* (see quot. 1956); *spectral series* = *SERIES *sb.* 16; *spectral term*: see *TERM *sb.* 11 d; *spectral type* (Astr.), any of the types used to classify stellar spectra, each being associated with stars of a characteristic range of temperatures and compositions and designated by a letter or letters.

1862 Spectral analysis [in Dict., sense 5 b]. **1888** *Phil. Mag.* XXV. 343 (*heading*) Mathematical spectral analysis of magnesium and carbon. **1930** *Proc. IRE* XVIII. 1199 Expression (9) lends itself to spectral analysis into its component frequencies by the following process. **1978** *Nature* 16 Mar. 232/2 As a further step, we carried out a spectral analysis according to the techniques of Blackman and Tukey on the time series for each of our latitude bands. **1956** *Observatory* LXXVI. 181 The usual terminology is adopted, where the flux density *S* from a discrete source refers to the flux in both polarizations, and the spectral index *x* refers to the index in the relationship $S \propto (\text{frequency})^x$. **1967** *Astrophysical Jrnl.* CL. 5 The average spectral index of twenty-six spirals between 40 and 21 cm is −0·83. **1974** *Nature* 4 Oct. 398/2 The spectral shape of the pulsed emission..can be approximated by a power law with an energy spectral index of $\alpha \sim 1$ to about a GeV. **1900** *Sci. Abstr.* III. 465 A most useful review of the present state of knowledge respecting spectral series. **1974** G. REECE tr. *Hund's Hist. Quantum Theory* iv. 61 Several attempts were made to give a theoretical interpretation of the spectral series. **1890** A. M. CLERKE *System of Stars* iii. 37 About eleven-twelfths of all the stars show linear spectra of absorption. They fall into two great divisions, corresponding to Father Secchi's first and second spectral types. **1924** *Proc. Amer. Acad. Arts & Sci.* LIX. 217 If the stellar density for any spectral type were uniform throughout space, the number of stars visible should double with every increase of half a magnitude in brightness. **1973** SMITH & JACOBS *Introd. Astron. & Astrophysics* x. 268 Since we will have occasion to refer to specific spectral types in the following paragraphs, we give the spectral sequence from hot to cool stars (40,000 K to 3000 K) here: O, B, A, F, G, K, and M. Each spectral class is further divided into ten subclasses 0 to 9.

spectrally, *adv.* Add: **2.** In the form of a spectrum; as regards, or in terms of, the spectrum.

1914 S. E. SHEPPARD *Photo-Chem.* viii. 325 Spectrally dispersed rays. **1971** *Jrnl. Gen. Psychol.* LXXXIV. 95 Spectrally pure stimuli are being produced by passing the output of a xenon arc through narrow-band interference filters. **1973** *Sci. Amer.* Nov. 77/1 The results of this experiment suggest that noise spectrally adjacent to the signal is most effective in masking recognition. **1980** *Ibid.* Oct. 123/1 The light..is spectrally pure and monochromatic.

spectre, *sb.* Add: For '7–8 **specter**' read '7– **specter** (now *U.S.*)' and add: **6. b.** *spectre-faint, -pale.*

1924 R. GRAVES *Mock Beggar Hall* 5 The exiled Alcibiades Beheld him in the Chersonese, Yet spectre-faint. **1928** V. WOOLF *Orlando* v. 233 The spectre-pale beech trees.

spectrin (spe·ktrin). *Biochem.* [f. SPECTR(E *sb.* + -IN[1]; so called because it was isolated from 'ghosts' (sense *11 e) of red blood cells.] A fibrous protein constituent of the membranes of red blood cells, forming a network on the inside of the plasma membrane.

1968 MARCHESI & STEERS in *Science* 12 Jan. 204/2 Because this protein appears to be a new molecular species and is extractable from erythrocyte ghost membranes, we suggest that it be called Spectrin. **1974** *Sci. Amer.* Mar. 27/3 The two heaviest polypeptide components, with molecular weights of 255,000 and 220,000, are collectively known as spectrin. **1978** *Bio Systems* X. 98/1 A variety of proteins other than actin can form filaments (e.g., flagellin, spectrin, spasmoneme filaments, skeletin).

spectro-. Add: **spectrohe·liogram**, a photograph obtained with a spectroheliograph; **spectroheliograph**, add def.: (*a*) an instrument which photographs the sun using light of a particular wavelength, *esp.* that of the Balmer α emission line of hydrogen; † (*b*) a spectroheliogram; (later examples); hence **spe:ctroheliogra·phic** *a.*; **spe:ctrohelio·meter**, a spectrophotometer for use in studying the sun; **spectrohe·lioscope**, an instrument which provides a directly observable monochromatic image of the sun by means of a rapidly scanning device which transmits light of only one wavelength (which may be modified so that Doppler shifts can be observed); **spectrophone**, add def.: a device in which a body of gas may be caused to emit sound waves when illuminated by a periodically interrupted beam of electromagnetic radiation (usu. visible or infra-red); (later examples); **spe:ctrophosphori·metry**, the spectrometric study of phosphorescence; so **spe:ctrophosphori·meter**, a spectrometer designed for this; **spe:ctrophosphorime·tric** *a.*; **spe:ctrophotofluoro·meter**, a spectrophotometer designed for the study of fluorescence; hence **spe:ctropho:tofluorome·tric** *a.*, **-fluorome·trically** *adv.*; **spe:ctropolari·meter**, an instrument designed to measure rotation of the plane of polarized light as a function of wavelength; so **spe:ctropolarime·tric** *a.*; **spe:ctropolari·metry**; **spe:ctroradio·meter**, a combination of a spectroscope and a radiometer, designed to measure the intensity of electromagnetic radiation over a range of wavelengths; so **spe:ctroradiome·tric** *a.*; **spe·ctroradio·metry**; **spe·ctrotype** *Immunol.*, the range of antigens to which a given antibody is reactive.

1905 *Astrophysical Jrnl.* XXI. 354 Spectroheliograms were obtained showing detail in the centre of the disk. **1968** *New Scientist* 11 Jan. 97/1 (*caption*) X-ray spectroheliograms of solar plages obtained with OSO-4. **1973** Spectroheliogram [see *RASTER *sb.*[2] a]. **1903, 1915** Spectroheliograph [see *FLOCCULUS 3 b]. **1965** P. WYLIE *They both were Naked* iv. 152 Big gadgets—telescopes, spectroheliographs, particle accelerators and the like. **1905** *Astrophysical Jrnl.* XXI. 279 Our new explanation of the spectroheliographic results will be founded on the hypothesis that the sun is an unlimited mass of gas in which convection currents..are continually forming. **1973** *Sci. Amer.* Oct. 76/2 The major experiments of Skylab include a spectroheliometer from the Harvard College Observatory that is mapping the sun at wavelengths of from 300 to 1,350 angstroms with a resolution of five seconds of arc. **1976** *New Yorker* 6 Sept. 40 The spectroheliometer showed that they extended up from the chromosphere..into the transition region between the chromosphere and the corona. **1906** *Astrophysical Jrnl.* XXIV. 42 This instrument..constitutes a spectrohelioscope, and was intended for the visual study of prominences. **1929** G. E. HALE in *Encycl. Brit.* XXI. 179/2 The spectrohelioscope renders visible to the eye many of the phenomena of the solar atmosphere photographed with the spectroheliograph and also permits their velocities in the line of sight to be measured. **1955** *Sci. Amer.* Sept. 194/2 In 1890 George Ellery Hale and Henri Deslandres independently invented the spectrohelioscope. This instrument utilized the red light of hydrogen to produce an image of the entire disk of the sun. **1948** *Chem. Abstr.* XLII. 1467 An app. was constructed for the detn. of gases absorbing in either the infrared or the visible region, on the principle of Bell's spectrophone. **1965** *New Scientist* 21 Oct. 199/3 The collisional phenomenon has recently been investigated.. using a device named the 'spectrophone', in which a sample of gas is subjected to a sequence of pulses of infrared (140 a second) during which the molecules acquire vibrational energy. If they lose that energy by collision, the increase in molecular velocities, and therefore in pressure, is detected by a microphone in the gas, which accordingly registers a 140 c/s note. **1961** *Nature* 8 Apr. 166/1 The measurements with our spectrophosphorimeter set an upper limit to the lifetime of the pyrene dimer. **1978** *Ibid.* 19 Jan. 236/1 The weak afterglow spectrum is also shown in Fig. 1*a*, at a gain factor of 20 for 100% solid crystalline carbazole at an identical setting of the spectrophosphorimeter. **1968** M. Zander *Phosphorimetry* iii. 136 The quantitative spectrophosphorimetric analysis of mixtures. *Ibid.* 138 The qualitative and quantitative analysis of a mixture is possible by spectrophosphorimetry. **1974** *Nature* 30 Aug. 763/1 Analysis of the fluorescent material (by chromatography on the adsorbent *in situ* by spectrofluorimetry *in situ*, and by the latter technique and by spectrophosphorimetry after solvent extraction from the adsorbent). **1956** *Rev. Sci. Instruments* XXVII. 664/2 Spectrophotofluorometer. Continuous activation of compounds and measurement of the resulting fluorescence throughout the visible and ultraviolet regions is provided by this instrument. **1974** *Nature* 1 Feb. 291/1 The continuous formation of NADH:NADH was measured fluorometrically at 37°C with an Aminco Bowman spectrophotofluorometer using an excitation wavelength of 340 nm and an emission wavelength of 460 nm. **1964** *Jrnl. Exper. Med.* CXX. 509 The spectrophotofluorometric technique of Shore *et al.*.. was used with slight alterations. **1975** *Nature* 17 Apr. 636/1 Cells were centrifuged and the histamine released into the supernatant..was determined spectrophotofluorometrically. **1926** *Sci. Abstr.* A. XXIX. 310 (*heading*) A spectro-polarimeter for the ultra-violet. **1971** *Nature* 16 July 192/1 Circular dichroism was measured at 25°C with a JASCO ORD/UV-5 spectropolarimeter equipped with a CD attachment. **1960** C. Djerassi *Optical Rotatory Dispersion* iii. 28 These refractive index gradients.. account for the 'blanking-out' phenomenon noted in the spectropolarimetric examination of ketal formation. *Ibid.* 18 The single most important factor responsible for the renewed interest in rotatory dispersion has been recent advances in ultraviolet spectropolarimetry. **1927** *Jrnl. Optical Soc. Amer.* VII. 439 The essential parts of a spectroradiometer consist of (1) a suitable spectrometer for dispersing thermal radiation into a spectrum, and (2) suitable radiometric instruments for measuring the spectral radiation intensities. **1975** *Nature* 10 Apr. 512/2 Measurements of the spectral energy distributions..of natural radiation between 400 and 800 nm were made using a spectroradiometer. **1922** *Jrnl. Optical Soc. Amer.* VI. 1021 The transmission screen method should prove useful in supplementing the spectroradiometric measurements..on fainter stars. **1951** *Electronics* Jan. 81/3 This is a spectroradiometric curve of a particular color, obtained by measuring..the number of watts radiated by the source at each wavelength. **1921** *Jrnl. Optical Soc. Amer.* V. 133 The fiducial line in this Bureau's spectroradiometry is the yellow helium line. **1945** R. A. Sawyer *Exper. Spectroscopy* xi. 277 The methods used in the investigation of the infrared radiation..are essentially the methods of spectroradiometry. **1974** *Nature* 16 Aug. 532/2 The A5A idiotype has been found to be associated with a particular antibody spectrotype. **1981** *Exper. Parasitol.* LII. 216/2 When a clone of S1 spectrotype was allowed to establish a relapsing infection on two separate occasions, two variants of different spectrotypes were produced.

spectrochemical (spektroke·mikăl), *a.* [f. Spectro- + Chemical *a.*] Of or pertaining to spectrochemistry; *spectrochemical series*, a series of ligands arranged in order of magnitude of the ligand field splitting that they cause in the electronic orbitals of a central atom.

1896 *Jrnl. Chem. Soc.* LXX. i. 553 The results of a spectrochemical investigation..establish the formula. **1938** R. Tsuchida in *Bull. Chem. Soc. Japan* XIII. 393 Arranging in the ascending order of *P* [*sc.* 'the work which would be done by the system if the ligand were to approach from infinity to the seat of co-ordination'], we obtain a spectrochemical series for the first band: viz., NH_3, H_2O, F^-, Cl^-, Br^-, I^-. **1960** *Jrnl. Iron & Steel Inst.* CXCV. 375/1 The most satisfactory boron line for spectrochemical use is at 2497·73 Å. **1966** Phillips & Williams *Inorg. Chem.* II. xxviii. 396 Now when the [crystal-field splitting] energies, Δ, obtained from spectra ..are compared for a large number of different cations, it is found that there is an approximately constant ligand series, the Fajans–Tsuchida spectrochemical series: $CO > CN^- > NO^- > .. S^{2-} > Br^- > I^-$. **1975** P. S. Braterman *Metal Carbonyl Spectra* i. 3 Electronic spectra contain *d-d* bands that place CO in the spectrochemical and nephelauxetic series.

Hence **spectroche·mically** *adv.*, as regards spectrochemistry; by spectrochemical methods.

1905 *Nature* 15 June 160/2 Several simple molecules of this kind may be combined into one crystallised particle of the spectrochemically normal diamond. **1966** P. W. J. M. Boumans *Spectrochemical Excitation* i. 4 Alloys containing more than two elements..were investigated spectrochemically.

spectrochemistry (spektroke·mistri). [f. Spectro- + Chemistry.] The branch of chemistry dealing with the chemical application of spectroscopy, esp. in analysis, and with the interpretation of spectra in chemical terms.

1893 *Jrnl. Chem. Soc.* LXIV. i. 254 (*heading*) Spectrochemistry of nitrogen. **1905** *Nature* 15 June 162/1 It is of course of the greatest value to be able to examine the constitution of the bodies without affecting them chemically; and spectrochemistry..gives us the means of doing so. **1957** *Technology* Sept. 258/2 In the department of applied science..radio-chemistry, spectro-chemistry and metallurgy are all prominent. **1966** P. W. J. M. Boumans *Spectrochemical Excitation* p. viii, This book will..contribute to further progress in the field of theoretical spectrochemistry.

spe:ctrofluori·metry. Also -fluoro-. [f. Spectro- + *fluoro-, fluorimetry.] The spectrometric study of fluorescence. So **spe:ctrofluori·meter** (-fluoro-), a spectrometer designed for this; **spe:ctrofluorime·tric** (-fluoro-) *a.*, **-me·trically** *adv.*

1957 *Analyst* LXXXII. 611 The sensitivity of the spectrofluorimeter depends not only on the sensitivity of the detector and the intensity of the exciting light, but also on the slit width and aperture of the analysing monochromator. **1961** *Analytical Chem.* XXXIII. 1362/1 The spectrofluorometer has a sensitivity equivalent to 0·01 p.p.m. of quinine sulphate. **1962** *Lancet* 26 May 1130/1 The plasma was analysed spectrofluorometrically for the pharmacologically active spirolactone. **1965** *Analytical Chem.* XXXVII. 137/1 (*heading*) Spectrofluorometric determination of submicrogram amounts of aluminium and beryllium. **1965** *Jrnl. Clin. Path.* XVIII. 375/1 (*heading*) A device for spectrofluorimetry. **1971** *Nature* 2 July 24/2 All the chromosomes in the human complement can be paired by these patterns, which can be analysed most accurately by a quantitative spectrofluorimetric method. **1975** D. H. Burrin in Williams & Wilson *Biologist's Guide to Princ. & Techniques Pract. Biochem.* v. 147 Spectrofluorimetry is most accurate at very low concentrations. *Ibid.*, Spectrofluorimeters enable the utilization of great spectral selectivity since.. two monochromators may be used. **1957** *N.Y. Times* 8 Nov. 33/4 In the meantime, using spectrofluorometry, thin-layer chromatography and a third technique with iodine, the Coast Guard and the E.P.A. conducted 741 tests at its laboratories in Georgia and Connecticut. **1976** *Nature* 16 Sept. 242/2 The amine was isolated by ion-exchange chromatography..and estimated spectrofluorimetrically.

spectrogram. Add: More widely, a visual representation of a spectrum of any kind. (Later examples.)

1939 *Amer. Speech* XIV. 313/2 Examination of eleven subjects singing the vowel 'ah' normally and after hydrogen inhalation... Illustrated with acoustic spectrograms. **1961** *Brit. Birds* LIV. 388 The original spectrograms have a frequency scale from 0 to 10 kilocycles per second. **1975** *Nature* 25 Sept. 295/1 Spectrograms obtained with a Carnegie image tube on the 2·1-m telescope at the Kitt Peak National Observatory have been used to complete the determination of radial velocities for 50 of the 52 galaxies. **1978** *Hi-Fi News* Sept. 175/1 None of these, however, exceeded −70dB ref. 36 V RMS across the load (shown in spectrogram *fig. 2*). **1981** *Amer. Speech 1977* LII. 237 The degree of centralization was determined by the use of a bilogarithmic scale in measuring the spectrograms of eighty instances of /aɪ/.

spectrograph. Add: **1.** (Later examples.) More widely, any apparatus for producing a visual record of a spectrum (optical or otherwise). Cf. *mass spectrograph* s.v. *mass *sb.*[2] 10 d, *sound spectrograph* s.v. *sound *sb.*[3] 7.

1940 *Geogr. Jrnl.* XCV. 276 My own work was research on the Ozone Layer with a Dobson Spectrograph. **1955** *Times* 15 June 5/5 Such spectrographs are used in laboratories to examine the ions existing in ionized gases, and an American version has been used in rockets. **1967** M. Schlauch *Language* v. 102 There is an instrument called the spectrograph which can record and photograph the complex series of vibrations of the air which convey a sequence of spoken sounds. **1974** *Encycl. Brit. Macropædia* XI. 605/1 The most noteworthy observation made with the parabola spectrograph was the spectrum of rare gases present in the atmosphere... There was a line corresponding to an ion of mass 22 that could not be attributed to any known gas.

spectrographic *a.*, **-graphically** *adv.* (examples relating to spectra other than of light); also **spectro·grapher**, one who uses a spectrograph.

1946 *Nature* 20 July 79/1 The result of this selection is a volume from which spectrographers may derive helpful guidance in the correct choice of analytical methods. **1959** E. Pulgram *Introd. Spectrography of Speech* i. 20 The position, which I have at least in part adopted in consequence of learning to view language spectrographically. *Ibid.* ii. 27 The experimental linguist.., like the spectrographer of speech, is interested in the cultural *and* the physical values of the phenomena at the same time. **1962** *Amer. Speech* XXXVII. 62 Phonemic analysis of the stress-intonation system based on spectrographic analysis of a specific corpus. **1974** *Sci. Amer.* Mar. 86/3 Another promising diagnostic technique for detecting abnormalities in infants, in addition to the spectrographic analysis of the cry, is the measurement of..changes in brain waves in response to sounds. **1981** *Amer. Speech 1977* LII. 269 A carefully developed argument, supported with spectrographic evidence.

spectrology. **2.** (Later example; still *rare*.)

1969 James & Sternberg *Design of Optical Spectrometers* p. x, Spectroscopy (or *spectrology*, to introduce here a word coined by Fellgett) belongs to this branch of physics.

spectrometer. Add: Now in extended use, any of a wide range of instruments for producing spectra and measuring the positions, etc., of spectral features. Cf. *mass spectrometer* s.v. *mass *sb.*[2] 10 d.

1901 *Sci. Abstr.* IV. 930 Universal spectrometer... This instrument can be used for studying any portion of the spectrum, either visually, photographically, or photometrically, with two different dispersions. **1950** *Sci. News* XV. 19 Monochromatic light of chosen wavelength, from a spectrometer, is now used in order to stimulate the retinal receptors. **1958** *Antiquity* XXXII. 124 The characteristics of the trace-elements present are detected by a special Gamma-ray Spectrometer. **1978** P. W. Atkins *Physical Chem.* xix. 616 The centre of the e.s.r. spectrum of the methyl radical occurs at 329·4 mT when the spectrometer is using 9·233 GHz microwaves.

spectrometric *a.* (later example); **spectrometry** (examples); also **spectrome·trically** *adv.*; **spectro·metrist**, an expert in spectrometry; a person employed to operate a spectrometer.

1948 *Physical Rev.* LXXIV. 1222/1 By extending the method..a higher order mass resolution is obtained which is generally usable in the mass spectrometry of positive ions. **1953** G. P. Barnard *Mod. Mass Spectrometry* 286 A different method was used..to determine the half-life of ^{85}Kr mass spectrometrically. **1954** Powell & Ross in *Appl. Mass Spectrometry* (Inst. Petroleum) 7 In Raman spectrometry, the material is irradiated with monochromatic light..and the light scattered at right angles to the incident beam is examined spectrometrically. **1958** *Antiquity* XXXII. 124 The staff consists now of two Senior Scientific Officers, two Research Assistants, three technicians, a secretary and two part-time workers (optical spectrometrist and statistician). *Ibid.*, In Optical Spectrometry, a programme of analysis of prehistoric European bronzes..is being carried out. **1962** *Oxf. Univ. Gaz.* 9 Mar. 775/2 Continuing his research on the history of British Bronze Age metallurgy, he completed a programme of spectrometric analyses of Early Bronze Age material. **1971** *Nature* 27 Aug. 646/2 Some 3,3-diaryl oxetanes which could not be isolated but only detected spectrometrically in solution. **1973** *Physics Bull.* Feb. 108/1 The term chemi-ionization has unfortunately taken on a different meaning amongst mass spectrometrists. **1975** *McGraw-Hill Yearbk. Sci. & Technol.* 193/1 Flame spectrometry, traditionally a single-element technique, has been advanced recently to include multielement analysis.

spectrophotometer (spe:ktrǒfotǫ·mĭtəɹ). [f. Spectro- + Photometer.] An instrument designed to measure the relative intensity of light (usu. transmitted or emitted by a substance under study) at different wavelengths in a particular region of the spectrum.

1881 [in Dict. s.v. Spectro-]. **1938** *Jrnl. Optical Soc. Amer.* XXVIII. 18 This spectrophotometer was designed specifically to measure in the visible region the absorption spectrum of visual purple. **1948** *Nature* 21 Feb. 285/2 Total penicillin may be assayed non-biologically with the aid of the..ultra-violet spectrophotometer. **1966** *McGraw-Hill Encycl. Sci. & Technol.* II. 594 d/1 The split-beam spectrophotometer..measures the difference in absorption at any given wavelength between two nearly identical cell suspensions. **1976** C. Skegg tr. *Welz's Atomic Absorption Spectroscopy* v. 84 In modern spectrophotometers a logarithmization of the signal takes place within the instrument, so that the readout on the display is already in absorbance and no conversion is required.

Hence **spe:ctrophotome·tric** *a.*, of, pertaining to, or employing a spectrophotometer; **spe:ctrophotome·trically** *adv.*; **spectrophotometry**, the technique of using a spectrophotometer.

1884 Spectrophotometric [in Dict. s.v. Spectro-]. **1897** Spectrophotometrically [in Dict. s.v. Spectro-]. **1899** *Phil. Mag.* XLVIII. 421 In direct comparisons of white light this latter principle has been successfully applied; but in spectrophotometry the optical difficulties in the way of its use have prevented the obtaining of corresponding results. **1942** *Astrophysical Jrnl.* XCVI. 451 Spectrophotometric observations of the light of the night sky have been made for the stronger lines of the visual region of the spectrum. **1961** *Times* 9 Jan. 2/3 (Advt.), The Physical Assay Division..is concerned..with..application of industrial methods of analysis including ultraviolet and infra-red spectrophotometry. **1972** R. A. Jackson *Mechanism* iv. 76 If the reaction is followed spectrophotometrically the presence or absence of stable intermediates is often indicated from the changes in the spectrum as the reaction proceeds. **1976** G. Thurston *Coronership* i. 13 Methods of analysis by chromatography and spectrophotometry enable accurate results to be produced with speed. **1977** I. M. Campbell *Energy & Atmosphere* ix. 307 (*caption*) Concentration profile of O_3.. versus altitude in the upper mesosphere, measured by rocket-borne spectrophotometric detection.

spectroscopic, *a.* Add: **2. b.** *spectroscopic binary* (Astr.), a star whose binary nature is revealed only by a study of its spectrum.

1896 *Circular Harvard Univ. Observatory* No. 11, Professor Solon I. Bailey has found that the star μ[1] Scorpii is a spectroscopic binary. **1930** R. H. Baker *Astron.* viii. 338 Some examples of spectroscopic binaries among the bright stars are: Capella, Spica, Castor, β Aurigae and Algol. **1975** *Sci. Amer.* Mar. 28/2 The normal star associated with Cygnus X-1 is a single-line spectroscopic binary with a period of 5·6 days. The term spectroscopic here means that the presence of two stars is indicated by a periodic Doppler shift of the spectral lines of at least one of the stars as they revolve around a common center of gravity. **1978** Pasachoff & Kutner *University Astron.*

vi. 134 (*caption*) Alcor and Mizar provide examples of both visual and spectroscopic binaries.

spectroscopy. Add: In mod. use, the investigation of spectra by any of various instruments.

1955 *Sci. Amer.* Sept. 144/2 The faintness of the aurora and the rapidity with which it changes make spectroscopy difficult, but development of the technique..has produced beautiful spectrograms of auroras extending..into the near ultraviolet and the near infrared. **1966** *McGraw-Hill Encycl. Sci. & Technol.* XII. 589/1 The important instruments that are used in spectroscopy include spectroscopes, spectrometers, spectrographs, interferometers, and spectrophotometers. **1978** *Nature* 21 Sept. 199/2 We intend carrying out further spectroscopy on the object, providing that it does not fade too quickly, to monitor spectral variations.

spectrum. Add: **3. a.** Also, a dark band containing bright lines produced similarly; such a (coloured or dark) band, or the pattern of lines in it, as characteristic of the light source; hence, the pattern of absorption or emission of light or other electromagnetic radiation over any range of wavelengths exhibited by a body or substance.

1824 *Edin. Philos. Jrnl.* X. 39 Lines are also seen in the spectrum of other fixed stars of the first magnitude. **1879** *Encycl. Brit.* X. 215/1 When the light of a burning metal is examined with a properly-arranged prism, it is seen to give a dark band or spectrum which is traversed by certain vertical bright lines. **1900** *Proc. R. Soc.* LXVI. 45 The expected argon spectrum was almost entirely absent. **1925** G. A. Lindsay tr. *Siegbahn's Spectroscopy of X-Rays* vi. 195 X-ray spectra afford one of the most direct sources of information concerning the inner structure of the atom. **1966** *McGraw-Hill Encycl. Sci. & Technol.* VIII. 420/2 Microwave spectra of atoms can be used to measure.. nuclear electric and magnetic moments. **1971** *Physics Bull.* July 401/1 The laser spectrum extends from the vacuum ultraviolet to the far infrared. **1978** Pasachoff & Kutner *University Astron.* xxviii. 709 All the lines in the spectrum of 3C 48 were shifted by 37 per cent, a still more astounding redshift.

b. The entire range of wavelengths (or frequencies) of electromagnetic radiation, from the longest radio waves to the shortest gamma rays of which the range of visible light is only a small part; any one part of this larger range.

1888 *Encycl. Brit.* XXIII. 142/1 When a telescope is to be constructed for photographic purposes the aim should be to unite..the rays near that portion of the spectrum which act most powerfully on the photographic plate. **1923** Glazebrook *Dict. Appl. Physics* IV. 891/1 Beyond the photographic limit, investigation of the infra-red spectrum by means of the heating effect of the rays has been carried on. **1947** *Sci. News* IV. 54 The wave lengths of the visible spectrum, from red over yellow, green, blue to violet, lie between 700 and 350 millionths of a millimetre. **1962** *Rep. Comm. Broadcasting 1960* i. 5 in *Parl. Papers 1961–2* (Cmnd. 1753) IX. 259 The division by international agreement of the frequency spectrum into bands allocated to particular services forms part of the International Radio Regulations. **1978** Pasachoff & Kutner *University Astron.* ii. 21 The new ability that astronomers have to study parts of the electromagnetic spectrum other than light waves enables us to increase our knowledge of celestial objects manyfold.

c. An actual or notional arrangement of the component parts of any phenomenon according to frequency, energy, mass, or the like. Cf. *mass spectrum* s.v. *mass *sb.*[2] 10 d, *power spectrum* s.v. *power *sb.*[1] 18 f.

1887 *Science* 11 Mar. 238/1 It is proposed to analyze a composition by forming what may be called a 'word-spectrum', or 'characteristic curve', which shall be a graphic representation of an arrangement of words according to their length and to the relative frequency of their occurrence. **1897** J. J. Thomson in *Phil. Mag.* XLIV. 297 When the cathode rays are deflected by the electrostatic field, the phosphorescent band breaks up into several bright bands separated by comparatively dark spaces; the phenomena are exactly analogous to those observed by Birkeland when the cathode rays are deflected by a magnet, and called by him the magnetic spectrum. **1933** *Proc. R. Soc.* A. CXLII. 347 A large number of experiments was made to determine the distribution of α-particles over the whole of the spectrum, which includes α-particles of ranges between 7 cm. and 12 cm. **1939** *Psychol. Rec.* III. 60 Fig. 1 shows the acoustic spectra of three tones of the same singer and vowel, sung at the three different intensity levels at approximately the same frequency. **1962** A. C. Gimson *Introd. Pronunc. Eng.* iii. 21 The spectrum above 4,000 cps would appear to be largely irrelevant to the recognition of our vowels. **1971** *Nature* 3 Sept. 2/2 The idea..is that even quite small explosions can be distinguished from earthquakes of comparable size by the high frequency parts of their seismic spectra. **1973** Williams & Fleming *Spectrosc. Methods in Inorg. Chem.* (ed. 2) iv. 181 In many cases, convenient starting points for counting the spectrum are the peaks at *m/e* 28 (N_2^+) and *m/e* 32 (O_2^+).

d. *fig.* The entire range or extent of something, arranged by degree, quality, etc.

1936 R. Campbell *Mithraic Emblems* 20 Their sistered stridences ignite The spectrum of the poets' lyre. **1952**, etc. [see *broad-spectrum* a. s.v. *broad D. 2]. **1958** *Listener* 28 Aug. 308/2 At the other end of the political spectrum Lloyd Warner has used similar methods in his nostalgic account of the status system of old New England. **1964** G. L. Cohen *What's Wrong with Hospitals?* i. 18 Theoretically, students remain long enough on each type of ward to give them a spectrum of experience. **1971** *Sci. Amer.* July 25/2 At the polar ends of the age spectrum—children and 'senior citizens'—the trends at the moment are following different courses. **1979** *Practical Woodworking* Mar. 42 P— hand tools embrace a wide spectrum of products.

5*. *Math.* (See quots.)

1948 P. R. Halmos *Finite Dimensional Vector Spaces* ii. 79 The set on n proper values of A, with multiplicities properly counted, is the spectrum of A. **1972** A. G. Howson *Handbk. Terms Algebra & Anal.* xvii. 83 The set of all eigenvalues of a linear transformation *t* of a finite-dimensional vector space *V* is known as the spectrum of *t*.

6. b. *spectrum allocation.*

1960 *McGraw-Hill Encycl. Sci. & Technol.* XI. 260/1 (*caption*) Radio spectrum allocations. **1980** *Sci. Amer.* Feb. 32/1 The control of interference lies at the heart of spectrum allocation, which entails the development of systematic plans for the use of frequencies in radio communication.

7. Special Comb.: **spectrum analyser**, a device which analyses a system of oscillations into its spectral components.

1942 *Radiation Lab.* (Mass. Inst. Technol.) *Man.* No. M-115 (*title*) Spectrum analyzer (Type 103) for pulsed oscillators at 3,000 Mc/sec. **1973** *Times* 14 Dec. 8/8 So far they had spent 150 man-hours in preliminary work, setting up 'spectrum analysers, computers, graphical displays and other advanced equipment'.

specular, *a.* **3. c.** Add def.: Designating or pertaining to reflection by a surface in which incident light is reflected as in a mirror. (Later examples.)

1927 *Jrnl. Optical Soc. Amer.* XIV. 371 The instrument is so constructed that the identical areas employed in the measurement of specular density may be used without change for the measurement of the diffuse density. **1940** *Chambers's Techn. Dict.* 789/2 *Specular density*, the photographic density in an image measured with parallel light, as contrasted with diffuse density, when the total light passed is measured, including that dispersed. **1967** E. Chambers *Photolitho-Offset* vii. 76 Some surfaces have both specular and diffuse reflection—such as varnished wood. **1974** *Jrnl. Optical Soc. Amer.* LXIV. 546/2 Conventional definitions of reflectance treat the road as a *Ultrasound Diagnosis* II. 10 Only echoes specularly component of reflectance.

specularite (spe·kiŭlărəit). *Min.* [f. Specular *a.* + -ite[1].] = *specular iron ore* s.v. Specular *a.* 3 b.

1892 E. S. Dana *Dana's Syst. Min.* (ed. 6) 213 Specular iron; Red Hematite, Red Ocher. Specularite. **1959** Berry & Mason *Mineralogy* x. 364 In the variety called specular hematite or specularite the color is black and the luster metallic and splendent. **1959** J. D. Clark *Prehist. Southern Africa* ix. 244 Specularite (specular iron) was apparently an often sought-after medium for paint in the Later Stone Age. **1970** *Prof. Papers U.S. Geol. Survey* No. 700-c. 103/1 A specularite-bearing conglomeratic sandstone containing minor malachite was noted about 1 mile southeast of the copper locality.

specularly, *adv.* Delete *rare*$^{-1}$ and add later examples.

1930 *Jrnl. Optical Soc. Amer.* XX. 23 The smoother the surface, the greater is the intensity of the specularly reflected light and the greater..is the lustre or gloss. **1978** D. N. Lapedes *McGraw-Hill Dict. Physics & Maths.* 920 *Specular reflection factor*, the ratio of the specularly reflected light to the incident light. **1980** *Rec. Advances Ultrasound Diagnosis* II. 10 Only echoes specularly reflected could be identified with any certainty.

speculative, *a.* and *sb.* Add: **A.** *adj.* **7. a.** *speculative builder*, a builder who has houses erected without securing buyers in advance. Hence *speculative-built* adj. Cf. *spec *a.*

1868 *1st Rep. Comm. Employment Children, Young Persons, & Women in Agric.* 35 in *Parl. Papers 1867–8* XVII. 95 Cottages..have been put up by speculative builders of the flimsiest materials. **1902** G. K. Chesterton *Twelve Types* 13 The colossal diagram of streets and houses is..the opium dream of a speculative builder. **1933** *Archit. Rev.* LXXIV. 120 There is a possibility, of course, that the speculative builder who has bought this estate is an intelligent man. **1960** Pidgeon & Crosby *Anthology of Houses* 94, 2-storey, speculative-built terrace houses. **1973** *Listener* 25 Jan. 118/1 The idiocies and crudities permitted to the developer—or, as I prefer to call him, the speculative builder.

8. Special collocations: *speculative fiction* (see quot. 1953); *speculative grammar*, a late medieval scholastic grammatical system in which the structure of language is interpreted through scholastic philosophy in terms of our perception and representation of the world by the 'modes of signification' (*modi significandi*) (cf. *modistæ); any one of the grammatical theories arising from this analysis.

1953 R. A. Heinlein in *Library Jrnl.* July 1188/1 The term 'speculative fiction' may be defined negatively as being fiction about things that have not happened. **1978** Speculative fiction [see *sci-fi]. **1951** R. H. Robins *Anc. & Mediaeval Gram. Theory in Europe* iii. 80 Most of these philosophical or 'speculative' grammars were entitled *De Modis Significandi* (whence the name 'Modistae'), or, as we might put it to-day, 'On Semantics'; they covered a great deal more ground than would now be included in 'grammar' narrowly considered. **1968** J. Lyons *Introd. Theoretical Linguistics* i. 15 It was the task of scientific, or 'speculative', grammar to discover the principles whereby the word, as a 'sign', was related on the one hand to the human intellect and on the other to the thing it represented, or 'signified'. **1972** Hartmann & Stork *Dict. Lang. & Linguistics* 139/1 New in these universal speculative grammars [were]..the refinements in syntactic analysis, e.g. the function of prepositions, the formal criteria of grammatical acceptability, and the concepts of dependency, government and transitivity. **1975** *Canad. Jrnl. Linguistics* XX. 134 Speculative grammar attempted to show how the 'modes of existence' of objects were apprehended by the 'modes of understanding' of the human intellect.

speculum. Add: **4*.** = *speculum metal* (in Dict., sense 5 a).

1912 *Phil. Mag.* XXIV. 321 The gold surface was brought into closer proximity to the speculum surface. **1929** *Bureau of Standards Jrnl. Res.* (U.S.) II. 343 Data are presented on the ultra-violet reflecting power of various metals—beryllium, chromium,..speculum, stellite, and stain-less steel. **1941** *Proc. Physical Soc.* LIII. 263 It is not without interest to note that speculum has not such a good reflecting power as the three former materials. **1966** *McGraw-Hill Encycl. Sci. & Technol.* XIII. 650/1 Among special cast bronzes are bell metal.. and speculum.

speech, *sb.*[1] Add: **I. 1. d.** *spec.* in *Linguistics*. = *parole *sb.* 2*.

[**1924**: see *speech-utterance*, sense 12 a below.] **1935** [see *parole *sb.* 2*]. **1937** J. R. Firth *Tongues of Men* 16 De Saussure's famous lectures..in which the speech-language distinction is regarded as fundamental. **1953** U. Weinreich *Languages in Contact* ii. 9 The question of merging vs. unmerged coexistence is a problem *par excellence* in speech-language relations. **1964** *English Studies* XLV. (Suppl.) 35 A second aspect of no less importance is the distinction between 'language' (*langue*) and 'speech' (*parole*). **1974** M. Taylor in *Metz's Film Lang.* p. ix, *Speech* (parole) is the antithesis, or, rather, correlative, of language system: language system is the social aspect of language, whereas speech is the utterance, the actual practice, of a language system.

III. 8. d. Also † *His Majesty's Speech*, *Speech from the Throne*, *King's* (or *Queen's*) *Speech*: a speech delivered by the sovereign (in person or by commission) at the opening or prorogation of Parliament; now *spec.* the speech delivered by the sovereign at the opening of Parliament, written by his or her ministers and setting forth the policies and legislative programme of the Government. Also, a speech delivered by the representative of the sovereign at the opening of the legislative assembly of a member of the Commonwealth.

1603 in *Jrnls. House of Commons 1547–1628* I. 146/2 His Majesty's Speech ended, Mr. Speaker..presented himself to his Majesty. **1641** *Diurnall Occurrences of Parliament 20th Jan.–10th Mar. 1628* 1 M. Selden reported to the House that his Majesties Speech made the last day of the Parliament, in the upper House, is also entred by his Majesties command. *Ibid.* 5 (*heading*) The Kings Speech. **1751** *Parliamentary Hist.* V. 279 Nor, like the former Speech from the Throne, is it mention'd by any Historian. **1771** [in Dict.]. **1792** J. Woodforde *Diary* 15 Dec. (1927) III. 395 The Kings Speech in the House of Lords, a very long one. **1844** Erskine May *Law of Parl.* vii. 142 The session is opened at once by the Queen's speech. *Ibid.* xxi. 326 On the opening of Parliament, the Queen, in her speech from the throne, addresses the commons. **1897** [in Dict.]. **1906** *Daily Colonist* (Victoria, B.C.) 11 Jan. 4/1 The opening [of the Legislative Assembly] will be attended with the usual ceremonies, and in the King's Speech will be indicated some of the salient points of the government policy. **1923** J. C. W. Reith in *Radio Times* 23 Nov. 290/3 At the opening of Parliament..our proposal to broadcast the King's Speech was..declined. **1964** Abraham & Hawtrey *Parliamentary Dict.* 165 A Queen's speech is read by the Lord Chancellor on proroguing Parliament, but this is never debated... This speech reviews the session which it concludes. **1971** *Guardian* 17 Aug. 2/1 The traditional Speech from the Throne read in Maltese by the new Governor-General. *a* **1974** R. Crossman *Diaries* (1975) I. 508 As a backbencher I never dreamed of attending the Queen's Speech debates, regarding them as the most boring occasions.

IV. 12. a. *speech-acoustics, act, area, -behaviour, -breathing, -correctionist, -energy, -event, -feeling, -group, -habit, -material, -melody, -movement, -organ, -pattern, -response, -rhythm, science, -situation, -sound* (earlier example), *-stuff, -style, -system, -unit, -utterance, -way.*

1949 *Archivum Linguisticum* I. 1. 42 Philologists are beginning to turn away from phonetics to speech-acoustics. **1961** *Amer. Speech* XXXVI. 222 A treatment of speech acoustics up to spectrography. **1946** C. Morris *Signs, Lang. & Behavior* ii. 37 There is no language.. without the production of sign-vehicles, and it is such production which constitutes a speech-act. **1955** J. L. Austin *How to do Things with Words* (1962) iv. 40 Here there is an obvious parallel with one element in *lying*, in performing a speech-act of an *assertive* kind. **1974** D. Hymes *Foundations of Socio-Linguistics* ii. 52 A party (speech situation), a conversation during the party (speech event), a joke within the conversation (speech act). **1982** *Papers Dict. Soc. N. Amer. 1977* 86 Speech acts are not predictable from code characteristics either. **1933** Speech area [see *dialect 2 b]. **1961** *Amer. Speech*

XXXVI. 95 Speech areas can be delineated and subdivided on the basis of heteroglosses. **1931** T. H. Pear *Voice & Personality* ii. 22 There is the person whose speech-behaviour adumbrates what would develop if at this point the speaker received encouragement. **1980** *English World-Wide* I. 283 Seven of the essays are by German-writing authors on linguistic problems of German, ranging from urban speech (Vienna) to the speech behaviour of accused in court. **1955** *Brit. Jrnl. Psychol.* XLVI. 54 Measures of speech-breathing activity promised to be more immediately relevant to the changing states of tension and affect during interview. **1977** D. Fry *Homo Loquens* iii. 23 An interesting feature of speech breathing is that the moments at which we breathe in are far from being arranged haphazard. **1972** J. L. Dillard *Black English* vii. 267 Speech correctionists and educators... One Negro speech correctionist-psychologist..went so far as to indulge in a little too-elementary learning theory: language, being a learned activity, can be learned badly. **1943** Speech energy [see *audio frequency* s.v. *audio-]. **1933** L. Bloomfield *Language* ii. 24 We have yet to examine B, the speech-event in our story. **1948** J. R. Firth in *Lingua* I. 400 A *speech event* in a context of situation is therefore a technical abstraction from utterances and occurrences. A *speech event* may be sub-divided into *speech items*. **1976** *Word 1971* XXVII. 197 In my analysis, the communicative process is divided into the threefold gradation proposed by Hymes of *speech situation, speech event*, and *speech act*. **1916** L. Bloomfield in *Trans. Amer. Philol. Assoc.* XLVII. 13 Our speech-feeling seems to distinguish quite clearly between predicating and non-predicating utterances. **1979** *Amer. Speech 1976* LI. 135 The double negative is both a part of our speech-feeling and a sensible way to strengthen a negative statement. **1925** L. P. Smith *Words & Idioms* 245 Linguistically considered, England, the Dominions, and the United States may almost be regarded as one speech-group. **1964** C. Barber *Present-Day Eng.* v. 124 There are phase-differences between different speech-groups, and it would be unsafe to assume that the words currently fashionable in a Birmingham rock-and-roll club were simultaneously fashionable in a West End night-club, or that the picturesque phrases used by schoolboys were still fashionable in R.A.F. messes. **1928** O. Jespersen *Internat. Lang.* 1. 26 Everybody will necessarily transfer some of his speech-habits to the international language. **1979** M. Millar *Murder of Miranda* ii. 72 It's a speech habit I picked up from all the teenagers. **1912** A. D. Sheffield *Grammar & Thinking* vii. 188 Sentence-study..can profitably keep in view the diverse speech-material that the pupil meets in his work with foreign languages. **1962** A. J. Bliss in Davis & Wrenn *Eng. & Med. Studies presented to J. R. R. Tolkien* 29 Either a fragment of speech-material has one of the rhythms which are acceptable, in which case *ictus* and stress inevitably coincide; or else it cannot be used in verse at all. **1934** *Essays & Studies* XIX. 141 This speech-melody of ordinary intercourse. **1970** *English Studies* LI. 278 This latter feature, also known as intonation or speech melody, is of course almost a subject in itself. **1918** R. Bridges in *Poems of G. M. Hopkins* 96 It was at one time the author's practice to use a very elaborate system of marks, all indicating the speech-movement. **1957** C. E. Osgood et al. *Measurement of Meaning* i. 12 Little or no correspondence between thought-movements and speech-movements was found. **1925** Grattan & Gurrey *Our Living Lang.* p. xxi, Sounds are produced and modified by the position of speech-organs. **1961** *Amer. Speech* XXXVI. 217 Theory of the syllable must be based on the articulatory movements of speech organs. **1936** G. K. Zipf *Psycho-Biology of Lang.* v. 195 One infers the nature of speech-patterns from the exemplifications of the patterns, i.e. the configurations of speech-elements. **1969** M. Pugh *Last Place Left* xxix. 207 His speech patterns were as elaborate as ever, but his voice was no longer so well modulated. **1974** *Howard Jrnl.* XIV. 80 The restricted and elaborate codes which characterize the speech patterns of the lower and middle classes respectively. **1927** G. A. de Laguna *Speech Development* ii. 36 The correlation between the speech-response and its objective conditions is a correlation between independently variable elements of response and independently variable elements of the external situation. **1910** G. Saintsbury *Hist. Man. Eng. Prosody* iv. iii. 316 The presence of closely allied forms [of the alliterative line], in the different Scandinavian and Teutonic languages, assumes..a natural rise from some speech-rhythm or tune-rhythm proper to the race and tongue. **1976** J. Lee *Ninth Man* 275 His poor father with the snicker-provoking Germanic speech rhythms. **1933** *Amer. Speech* VIII. 37/1 Graduate curricula in speech science, phonetics, speech psychology, and rhetoric. **1977** *Whitaker's Almanack 1978* 530 First degrees..are awarded..in *Speech Science* by the University of Sheffield. **1953** *English Studies* XXXIV. 258 'Man' and 'garden' in this context denote..definite, individualized concepts... Taken out of the context (or of the speech-situation) they are semantically colourless. **1980** *English World-Wide* I. 1. 99 This is still used..if at least one of the participants in a speech situation has not been educated in English or..Bahasa Malaysia. **1840** Geo. Eliot *Let.* 21 Dec. (1954) I. 77 Pray bring Phonarthron—speech-sound is a boon that I often need—I shall expect from it..a key to the classic Oriental and Sclavonic tongues. **1934** Joyce *Let.* 9 Aug. (1966) III. 316 Also why for you make me big speechstuff about Frankee Doodles? **1936** J. Kantor *Objective Psychol. Gram.* xi. 156 As a test of the validity of the speech-style conception we may inquire into its applicability to speech studies. **1978** *Amer. Speech* LIII. 66 Samarin notes that speech styles of glossolalia are socioculturally determined, as are speech styles of English prayers. **1946** H. Jacob *On Choice of Common Lang.* ii. ii. 96 Inflected systems are highly resistant to simplification... However, most of the European speech-systems have progressed considerably in the right direction. **1964** *English Studies* XLV. (Suppl.) 37 Whereas Aristotle started from..'speech' (*parole*), these philosophers had the speech-system (*langue*) in mind. **1936** G. K. Zipf *Psycho-Biology of Lang.* 233 A sentence is a speech-element, or a speech-unit. **1949** C. E. Bazell in *Travaux du Cercle Linguistique de Copenhague* V. 77 This succession of speech-units need not answer to anything in the system. **1924** L. Bloomfield in *Mod. Lang. Jrnl.* Feb. 319 Actual speech-utterance, *la parole*, varies not only as to matters not fixed by the system..but also as to the system itself. **1956** J. Whatmough *Language* 42 There is an inverse relationship between frequency of occurrence and the comparative perspicuity that accompanies the utilization of a speech-utterance. **1931** H. Shenton *Internat. Communication* 1. 46 This approach to the problem might well be called a study of the speech-ways of mankind. **1955** *English Studies* XXXVI. 17 Current Elizabethan usage, harking back to much older, popular speechways. **1972** H. Kurath *Studies in Area Linguistics* i. 12 Any native speaker's usage is in a large measure representative of the speechways of a social or age group in his community.

b. *speech analyser, -trainer, -writer; speech-training.*

1973 *Canad. Jrnl. Linguistics* XVIII. 90 Using the speech analyzer..we then recorded an intonation curve (in Hz) for each sentence on photo-sensitive paper. **1955** T. H. Pear *Eng. Social Differences* 99 Many people.. flare up at any suggestion (except from a speech-trainer consulted voluntarily) of possible improvements in their speech. **1933** *Amer. Speech* VIII. 11/1 To complete any gaps there may be in the speech training of the members. **1977** P. Strevens *New Orientations Teaching of Eng.* vii. 86 Young learners will learn best through mimicry with speech training games for interest and for special points of difficulty, but with little or no use of phonetics. **1834** J. S. Mill in *Monthly Repos.* VIII. 419 One of our politicians..reproached him..with being a λογογράφος, or speech-writer. **1976** H. Wilson *Governance of Britain* iv. 88 The transatlantic custom of using speech-writers, recently imported into Britain for the use of certain eminent politicians and others, is only to be deplored.

13. speech area, a region of the brain involved in the comprehension or production of speech, a speech-centre; **speech chain** *Linguistics*, an utterance regarded as a sequence of elements; **speech clinic**, a centre for the treatment of speech defects; **speech code** *Cryptology*, a simple verbal code formed by the regular substitution of secret words; **speech coil**, a coil that drives the cone of a loud-speaker according to the signal current flowing in it; **speech-community** *Linguistics*, a group of persons sharing a language or variety of a language; **speech island** *Linguistics*, a small area inhabited by speakers of a language or dialect other than that spoken in the surrounding areas; **speech pathology**, the study and treatment of defective speech; hence **speech pathologist**; **speech physiology**, the study of the physical production of speech sounds; **speech-prefix**, in the text of a play: the name or description of the speaker(s) of a line or lines, set at the head of each speech; **speech psychology**, the study or application of psychological methods and techniques useful in learning to speak a language; also **speech psychologist**; **speech recognition**, the process of identifying and interpreting or responding to the sounds produced in human speech; **speech recognizer**, a machine capable of responding to the content of speech; **speech-song** = *Sprechgesang; **speech stretcher** *Phonetics* (see quot. 1972); **speech synthesizer**, a machine designed to generate sounds imitative of the human voice and recognizable as meaningful speech; **speech therapy**, the training of patients in the production of a full range of speech sounds; hence **speech therapist**, one who practises this; **speech-to-noise ratio**, the signal-to-noise ratio of speech.

1885 *Harper's Mag.* Mar. 638/2 (*in figure*) Speech area. **1913** *Q. Rev.* Jan. 124 Over a large portion of the highest level of the brain the special work of each group of cells or 'area' is now known. If our speech-areas are diseased we cannot speak. **1968** Passmore & Robson *Compan. Med. Stud.* I. xxiv. 52/1 (*caption*) The three speech areas [*sc.* Broca's area, superior area, Wernicke's area] shown on the left cerebral cortex. **1950** D. Jones *Phoneme* 1 Nearly every utterance, or 'speech chain', is made up of a large number of small elements. **1953** C. E. Bazell *Linguistic Form* i. 5 But the smaller the number of choices, at any one point of the speech-chain, the smaller the probability of open juncture. **1963** Denes & Pinson (*title*) The speech chain: the physics and biology of spoken language. **1963** R. I. McDavid *Mencken's Amer. Lang.* 320 The spread of technical medical terminology to education, as *clinic* (yielding *reading clinic* and *speech clinic*). **1976** *New Yorker* 15 Nov. 146/2 A young man.. who took a Ph.D. in speech pathology at Iowa in 1936 and then left to set up a speech clinic and research centre. **1973** 'A. Hall' *Tango Briefing* xviii. 221 Fred was the standard speech-code name for any third member of an active cell. **1928** *Wireless World* 6 June 603/2 (*caption*) Various gauges of wire for speech coil. **1934** *Discovery* Oct. 301/2 The 2-in. speech coil attached to the 11-in. cone works in a flux density of 11,500. **1975** G. J. King *Audio Handbk.* vi. 132 The speech coil is composed of inductance, distributed capacitance and resistance. **1894** G. E. Karsten in *Publ. Mod. Lang. Assoc.* IX. 327 It is pre-eminently the speech-community which moulds the individual's language. **1911** L. Bloomfield in *Jrnl. Eng. & Gmc. Philol.* X. 629 A language is formed (i.e., a new speech-community is segregated) by definite changes in the outer surroundings of a group of people. **1950** R. A. Hall *Leave your Language Alone!* x. 153 Theoretically, it might be possible to keep an otherwise-normal speech-community hermetically sealed off from all outside sources of borrowing. **1978** K. Hudson *Jargon of Professions* 10 Speech communities are no longer as self-contained as they were. **1888** M. D. Learned in *Amer. Jrnl. Philol.* IX. 65 We are to seek the causes which have contributed to the formation of this important speech-island in the domain of German dialects. **1933** [see *Lusatian *sb.* and *a.*]. **1957** *Publ. Amer. Dial. Soc.* xxvii. 5 In one instance, that of *stone boat*, there is an additional speech island along the Mississippi, opposite St. Louis. **1978** *Amer. Speech* LIII. 44 A large speech island appears in the German-settlement area of Missouri and Illinois. **1972** J. L. Dillard *Black English* vii. 267 A linguistically sophisticated speech pathologist like Joan Baratz. **1982** *Amer. Speech* LVII. 213 Speech pathologists, audiologists,..and many others have had some introduction to formal linguistic analysis. **1931** L. E. Travis *Speech Path.* p. vii, Speech pathology is in its growing pains. **1976** *New Yorker* 15 Nov. 148/2 One of his advisers suggested that he sign up instead for the graduate program in speech pathology at the University of Iowa. **1936** G. K. Zipf *Psycho-Biology of Lang.* iii. 96 The experimental phoneticist..attempts to determine by his laboratory study of speech-physiology what changes [in a language] are possible. **1961** *Amer. Speech* XXXVI. 222 A treatment..of speech physiology by means of X-ray stills and films. **1959** *N. & Q.* June 213/1 It is therefore recommended that for the last speech of 'Elder Worthy' on I, 60..the speech-prefix 'Young Worthy' be substituted. **1978** *Studies in Eng. Lit.: Eng. Number* (Tokyo) 12 There are eleven Latin speech-prefixes. The first five are '*Omn.*' for all the characters on the stage. **1937** Palmer & Hornby *Thousand-Word Eng.* i. 21 It is in the nature of a designed *plateau* (as speech-psychologists call it), that is..a given stage..at which the learner may pause. **1921** H. E. Palmer *Princ. Lang.-Study* 19 A logical order of progression in accordance with principles of speech-psychology. **1933** *Amer. Speech* VIII. iv. 37/1 There are graduate curricula in speech science, phonetics, speech psychology, and rhetoric. **1953** Fry & Denes in W. Jackson *Communication Theory* xxx. 426 (*heading*) Mechanical speech recognition. *Ibid.* 427 The reasons for the failure of these..systems becomes clear when the mechanism of human speech recognition is considered. **1970** *New Scientist* 30 Apr. 216/2 Research on speech-recognition devices is still in its extreme infancy. **1980** *TWA Ambassador* Oct. 25/1 A second went to Bell Telephone Laboratories for a computerized speech-recognition system that can respond to human sentences. **1953** W. Jackson *Communication Theory* 431 Any mechanical speech recognizer requires for its operation a considerable amount of linguistic information. **1970** O. Dopping *Computers & Data Processing* xi. 162 A 'normal' speech recognizer would recognise words regardless of the speaker. **1909** *Cent. Dict. Suppl.*, Speech-song. [**1925** Song-speech: see *Sprechgesang.] **1946** E. Blom *Everyman's Dict. Mus.* 580/1 *Speech-song*,..a term for a kind of singing that approximates to speech and touches the notes, indicated by special signs, without intoning them clearly at the proper pitch. **1959** *Listener* 17 Dec. 1093/2 The Roman practice of narrating during Holy Week the Evangelists' accounts of the Passion in a stylized speech-song (*tonus lectionis*). **1976** P. Stadlen in D. Villiers *Next Year in Jerusalem* 324 The Bible's casual hint at Moses' 'heavy tongue'..[is] realized, by having Moses engage in speech song while..Aaron is made to sing. **1948** M. Joos *Acoustic Phonetics* (Lang. Monogr. No. 23) 129 The usefulness of the speech stretcher for phonetic demonstration is immense. **1972** Hartmann & Stork *Dict. Lang. & Linguistics* 216/2 *Speech stretcher*, a device used in phonetic research to slow down recorded speech without changing the pitch or distorting it in any other way. **1953** *Jrnl. Acoustical Soc. Amer.* XXV. 735/1 A speech synthesizer would be required to simulate..closely the actual dimensions of the vocal tract. **1970** *Times Lit. Suppl.* 23 July 787/4 If the zealous phoneticist is dissatisfied with the acoustics of a real human voice he can nowadays, it seems, ring down for a speech synthesizer, couple it up to his computer, and manufacture ideal vowel sounds. **1933** S. M. Stinchfield *Speech Disorders* i. 10 A clinician, psychologist or speech therapist might suspect that one of the following conditions would be found in such a case. The child may be deaf, or some childhood illness may have slowed up his rate of development. **1975** M. Kenyon *Mr Big* ii. 21 A speech therapist..who'd insisted that correct speech being a matter of breath control for six months he would simply have to learn to breathe. **1933** S. M. Stinchfield *Speech Disorders* vii. 141 It is worth-while to spend some time in reviewing the more important types of nervous disorders, in order to better understand their implications, in undertaking speech therapy. **1976** E. Ward *Hanged Man* xl. 267, I took the speech therapy and the office-boy jobs. **1951** *Engineering* 23 Feb. 226/3 Those concerned with the telephone apparatus..have paid much attention to questions of speech intelligibility, but in its broad aspects the matter is of direct interest to most shop executives. It is not their business to design loud speakers, but they are certainly concerned with the speech-to-noise ratio in workshops. **1981** *Amer. Speech* XXXVI. 221 Monosyllabic, bisyllabic, and trisyllabic words presented for identification in seven different speech-to-noise ratios.

speechful, *a.* **b.** (Earlier example.)

1820 C. R. Maturin *Melmoth the Wanderer* IV. xxviii. 228 No form of Guido's, hovering in exquisite and speechful undulation between earth and heaven.

speechless, *a.* Add: **2. b.** Also (*dial.*) as *adv.*

1915 D. H. Lawrence *Rainbow* i. 34 The woman's not speechless dumb. She's not clutterin' at the nipple. She's got the right to please herself.

speechment. (Earlier example.)

1826 J. O'Keefe *Recoll.* I. ix. 341 After this grand speechment, it was thought advisable not to let her have any thing at all to say.

speed, *sb.* Add: **II. 5. c.** (Later examples.) Freq. in sense d below; also *ellipt.*, a bicycle having the number of gears indicated.

1955 *Radio Times* 22 Apr. 30/1 (Advt.), 6-valve, 3-speed, autochange table radiogram. **1970** 'D. Halliday' *Dolly & Cookie Bird* iii. 26 A runaway self-propelled two-speed-gear lawnmower. **1975** *Sat. Rev.* (U.S.) 3 May 44/1 Then they'll hop onto their sleek 3-, 5-, or 10-speeds and ..explore America's city streets.

d. Any of the possible gear ratios of a machine, esp. a bicycle or motor vehicle; the equipment associated with this; = *gear *sb.* 7 b.

1866 *English Mechanic* 22 June 263/1, I think in velocipede construction it is necessary to have two speeds... By pulling the lever A it throws the small wheels out of gear, and the driver can use the large wheels for level ground and the small ones for ascending hills. **1904** Kipling *Traffics & Discoveries* 304, I was on the point of reversing and working my way back on the second speed ere I ended in some swamp, when I saw sunshine through the tangle ahead and lifted the brake. **1907** G. B. Shaw *John Bull's Other Island* iv. 80 The pig..put in the fourth speed with its right crubeen. **1926** T. E. Lawrence *Seven Pillars* (1935) ix. cii. 560 The armoured car was too heavy for the flints, and always she sank in a little, making heavy going on third speed. **1951** N. Mitford *Blessing* ii. ii. 170 I'll get you a bike with three speeds. **1974** *Encycl. Brit. Macropædia* II. 522/2 The simplest automobile transmission is the sliding-spur-gear type with three or four forward speeds and a reverse. **1980** P. Lively *Judgement Day* x. 129 A shiny new bike with three speeds.

6. b. (Later examples.)

1935 S. C. Johnson *Foulsham's Compl. Photographer* iii. 26 The last of the four factors which must be considered in determining the length of an exposure is the speed of the plate or film. **1977** 'J. le Carré' *Hon. Schoolboy* vi. 123 Loading a cassette into the camera, he set the film speed.

c. *Photogr.* = *shutter speed* s.v. *shutter *sb.* 3 b.

1917 P. L. Anderson *Pictorial Photogr.* v. 95 Of course these high speeds are not necessary for indoor work, where an exposure of less than 1/8 second is practically never desired. **1947** A. Ransome *Great Northern?* xxiii. 286 He set the aperture at f.11, the speed at a twenty-fifth of a second. **1977** J. Hedgecoe *Photographer's Handbk.* 162 Speeds of 1/500–1/2000 sec allow you to freeze subject action beyond the perception of the eye.

d. The rate, measured in words per minute, at which a person can write shorthand or can type; *spec.* (freq. in *pl.* of both skills) applied to the capacity of a particular person.

1886 *Encycl. Brit.* XXI. 840/2, 180 or 200 words a minute is no uncommon speed in certain styles of speech such as the conversational,—a speed which many [shorthand writers]..would never acquire. **1933** Smith & Munro *Guide to High Speed Writing in Pitman's Shorthand* i. 10 It is well within the capacity of the majority.. to reach a speed of, say, 160 words a minute, and..all writers of the system should make up their minds that that is to be their *minimum* speed. **1957** C. Smith *Case of Torches* xi. 137 He..said there wouldn't be much shorthand or typing. So I thought—well, I don't want to lose my speeds, then I heard about this job. **1976** H. Tracy *Death in Reserve* i. 14 I'm a sort of secretary.. I've got high speeds in shorthand and typing.

6*. An amphetamine drug, esp. methamphetamine, freq. taken intravenously. Cf. *speedball 1 a. *slang* (orig. *U.S.*).

1967 [see *freak *sb.*[1] 4 c]. **1967** [see *Methedrine]. **1969** Fabian & Byrne *Groupie* viii. 66 Now he was on speed the paranoid fantasies were really beginning. **1970** N. Saunders *Alternative London* xxii. 175 Amphetamines ('speed') are stimulants which can temporarily reduce fatigue, increase mental activity and give you a general feeling of well-being. **1975** J. Symons *Three Pipe Problem* ix. 65 'What was he on?'.. 'Speed mostly. Sometimes acid.' **1978** G. Vidal *Kalki* v. 100 Dr Lowell produced a hypodermic needle. I let him shoot me up... I assumed that he had given me speed.

9. c. *to be one's speed*: to suit one's tastes, interests, or abilities; to be one's 'cup of tea' (*cup *sb.* 12 b (ii)). *colloq.* (chiefly *U.S.*).

1923 E. L. Rice *Adding Machine* v. 77 'Did you ever carve a leg of lamb?'.. 'No, corned beef was our speed.' **1954** R. Bissell *High Water* iii. 36 'I'm gonna buy you an Uncle Wiggly book,' I said. 'He'd be just your speed.' **1970** J. Sangster *Touchfeather, Too* i. 11 Lesbianism.. isn't really my speed at all. I'm a normal type girl.

11. a. *speed-meter.*

1938 *Times* 14 Oct. 15/4 The driver of the omnibus.. traversed the evidence of the speed-meter by pleading that his omnibus was so constructed as to be incapable of travelling at the alleged illegal speed. **1958** *Manch. Guardian* 21 Jan. 6/3 The use of radar speedmeters to enforce the law on the roads promises to be more successful than most efforts to reduce..road accidents.

11. b. *speed-bike, bowler, launch.*

1904 *Trans. Inst. Naval Archit.* XLVI. 161 The American speed launches. **1950** W. Hammond *Cricketers' School* v. 52 No one exemplified better than Larwood the true speed-bowler's action. **1955** S. Spender *Coll. Poems 1928–1953* ii. 74 Speed-bikes and tracks are real. **1977** *Arab Times* 3 Dec. 10/5 West Indian speed bowlers Andy Roberts and Michael Holding helped bring about the collapse of the Australian 'super test' cricket team.

11. c. *speed-law, record, trial* (earlier example); *speed-time, -torque* adjs.; **speed-boat**, a high-speed motor boat; hence **speed-boating**, riding in a speed-boat; **speed bug** [*bug *sb.*[2] 4 a] *slang*, one who enjoys travelling at high speed; an addict of speed sports; **speed bump** *colloq.* = *sleeping policeman* s.v. *policeman 1 e; **speed cop** *slang*, a policeman or official detailed to enforce traffic laws, esp. a motorcycle patrolman; **speed demon** [*demon 2 e] *slang*, one who likes to travel at great speed, a 'speed bug'; also *transf.*; **speed freak** (orig. *U.S.*), a person addicted to an amphetamine drug (cf. sense 6* above); **speed gun**, a hand-held device for estimating the speed of a moving vehicle (proprietary name in U.S.); **speed hog** [*hog *sb.*[1] 7 c] *slang*, one who causes annoyance by exceeding the normal or legal speed limit; **speed king** *slang* (orig. *U.S.*), a motor-racing champion; **speed limit**, (*a*) the maximum speed a vehicle is capable of achieving (quot. in Dict.); (*b*) the maximum speed permitted by law on certain types of road or to specified classes of vehicle; **speed merchant** *colloq.*, one whose 'business' concerns the use of speed; *spec.* (*a*) *Cricket*, a fast bowler; (*b*) one who enjoys driving or riding at high speed (cf. *merchant *sb.* 3); **speed-read** *v. trans.* and *intr.*, to read rapidly by assimilating several phrases or sentences at once; so **speed-reader**, one who speed-reads; **speed-reading** *vbl. sb.*; **speed shop** *slang* (see quot. 1954); **speed trap**, a system operated by the police for detecting motorists exceeding the speed-limit; **Speed-walk** *U.S.*, a proprietary term for a moving walkway for conveying passengers; also (with small initial) in general use; cf. *Travolator; **Speedwriting** orig. *U.S.*, the proprietary name of a form of shorthand which uses the letters of the alphabet; hence **speedwriter**. See also *speedball, *speedway.

1911 *New Fry's Mag.* May 224/1 The number of speedboats, pure and simple, has grown greatly. **1929** 'Seamark' *Down River* i. 6 Hillary Kittredge..had fallen under the lure of speed-boat racing. **1940** R. Chandler *Farewell, my Lovely* xxxvii. 280 The speedboat scuffed the *Montecito's* ancient sides. **1976** H. Kemelman *Wednesday the Rabbi got Wet* liii. 311 There are houses all around this little lake and each..must have a speedboat or an outboard. **1928** *Sunday Dispatch* 5 Aug. 11/3 The heavy demand for petrol, created by the new sport of 'speed' boating. **1975** *Country Life* 30 Oct. 1129/1 A useful side effect of the speed-boating..was the weed being cut by the propellers of the boats. **1914** *Automobile Topics* XXXIV. 191/1 The trials..were crowded with excitement for the speed bugs who filled the grandstands. **1928** Speedbug [see *aquaplane]. **1975** *Public Works* Aug. 73/1 Speed bumps had been installed in many apartment complexes and shopping center parking lots. **1978** T. L. Smith *Money War* (1979) i. 59 As he approached the speed bumps, Hogan slowed... They took the bumps gently and then pulled off the road. **1924** *Cape Argus* 12 Jan. 20 These 'speed cops', however, wear uniform and are not got up to appear like ordinary motor-cyclists. **1933** *Amer. Speech* VIII. 72/2 His Grace, on being stopped, demanded 'Are you a speed-cop?' The patriotic magistrates fined him £10. 10s. and suspended his license for three months. **1948** *Sat. Even. Post* 3 July 77/3 Speed cops still speak politely to me. **1941** Speed demon [see *dice *v.* 1 a]. **1962** A. Lurie *Love & Friendship* vii. 113 Helen, the regular cashier, was a speed demon. **1971** *Black Scholar* Jan. 35/1 Jack Johnson.. has been called many things, e.g. show off, fool, speed demon. **1967** Speed-freak [see *freak *sb.*[1] 4 c]. **1973** R. Mills *Young Outsiders* ii. 63 In the summer of 1970 Jimmy would have been labelled a 'speedfreak' within the circles in which he moved. **1972** *Official Gaz.* (U.S. Patent Office) 1 Aug. TM 29/1 CMI, Incorporated, Minturn, Colo. Filed Feb. 1, 1971. *Speed Gun* for Traffic Radar..First use Dec. 10, 1970. **1972** *Tuscaloosa* (Ala.) *News* 14 Dec. 32/4 The hand-held Digital Doppler, a new tool police are using to catch speeders... The police purchased eight of the devices—also known as 'speed guns'. **1976** *Tel.* (Brisbane) 24 June 1 The speed guns are being used by traffic police mainly in areas of potential traffic hazard. **1928** *Daily Express* 23 Aug. 9/4 The scheme to limit 'speed-hogs'. **1974** *Country Life* 30 May 1360/1, I join issue with the RYA on their objection to..boat registration... Without it the speed hog..will often continue to go undetected. **1913** *Illustr. Technical World Mag.* June 493/1 (*caption*) Ralph De Palma. The 'speed king' of 1912. **1938** C. Graves *Swiss Summer* 108 German princes, English speed-kings..are usually to be found here. **1976** *Western Mail* (Cardiff) 22 Nov. 3/2 Yesterday's event was held to raise money for the memorial fund to the Welsh speed king who died in a rally incident in the summer. **1926** *Scribner's Mag.* Aug. 152/1, I hear you break speed-laws as recklessly as hearts. **1902** *To-day* 13 Mar. 648 The Automobile Club urges that the existing speed limit should be abolished. **1909** *Chambers' Jrnl.* Mar. 225/1 A speed-limit sometimes as low as..four and three-quarter miles per hour. **1926** *Motor* 26 Oct. 561/1 (*heading*) The speed limit. 35–40 m.p.h. or none at all? **1970** P. Laurie *Scotland Yard* v. 113 Fifty mph..is only ten mph above the speed limit there. **1973** H. McCloy *Change of Heart* i. 6 As for my mother, I can't make her drive ten miles over the speed limit. **1913** J. B. Hobbs *How to make Century* xi. 80 The speed merchant was now sending up such hot samples that his every delivery was more likely to take a wicket than offer a chance of runs. **1923** *Daily Mail* 15 Feb. 6 The goggled 'speed merchant' cannot see so well as usual. **1951** *Ithaca* (N.Y.) *Jrnl.* 9 Aug. 21/1 [He] had a glass arm and he certainly was no speed merchant. **1982** J. B. Hilton *Sunset Law* i. 11 A County Court judge..drove at reckless speed about the highways, menacing speed-merchants with a Smith & Wesson. **1960** *Time* 24 Nov. 21 Bagwell taught himself to speed-read, gulping whole paragraphs at a glance. **1973** *Sat. Rev. World* (U.S.) 20 Nov. 64/2 (*caption*), I speed-read your last book in two minutes and twenty seconds. **1965** *N.Y. Times* 5 June 22 The Evelyn Wood Reading Dynamics Institute..teaches prospective speed-readers to see every word on the page—but to read three words at once, not one word out of three. **1975** 'S. Marlowe' *Cawthorn Jrnls.* xix. 162 He wished he had something to read... He was a speed-reader. He had an IQ of 140. **1983** *Listener* 20 Jan. 25/2 A library for an extra-terrestrial speed-reader cruising above the British Isles. **1965** G. Jackson *Let.* 3 Oct. in *Soledad Brother* (1971) 89 Be sure to look into the course on speed reading. **1977** *Time* 3 Jan. 13/3 Having promised to read every bill brought up before the state senate, he ran into 2,300 of them. So he took a speed-reading course and read them all. **1935** P. Frankau *I find Four People* iv. 235 About this time the copywriter accomplished a speed-record for journalism. **1942** E. Paul *Narrow St.* xvii. 132 The *Bremen* made her maiden trip to New York and set a speed record. **1954** *Amer. Speech* XXIX. 102 *Speed shop*, a parts house, where engine parts and equipment are sold, and sometimes where hot rods are built. **1962** *Punch* 17 Oct. 561/1 Engineering firms and speed shops supply every beefed or stripped refinement. **1977** *Hot Car* Oct. 49/3 You can often pick up reasonable headers off the shelf from a good speed shop. **1917** A. T. Dover *Electric Traction* ii. 8 A speed-time curve, for a run between two stations, is usually made up of periods of—(1) acceleration; (2) constant speed, or 'free running'..; (3) coasting..; and (4) retardation or braking. **1976** P. R. White *Planning for Public Transport* x. 216 In urban areas, higher rates of acceleration followed by a period of coasting can enable a given schedule to be covered using less energy (this is illustrated with reference to speed-time curves in Chapter 4). **1920** *Whittaker's Electr. Engineer's Pocket-bk.* (ed. 4) 544 For electric traction on railways and tramways, a motor possessing a variable speed-torque characteristic is preferable to one possessing a constant speed-torque characteristic. **1962** G. A. T. Burdett *Automatic Control Handbk.* I. 2 The motor must often have a speed/torque characteristic to match that of the driven machine. **1927** U. Sinclair *Oil!* 9 'Sit still,' said the man. 'Don't look round. A speed-trap!' **1980** G. M. Fraser *Mr American* xviii. 344 She had been caught in a police speed trap on the way to Brighton. **1883** W. H. Maw *Rec. Pract. Marine Engineering* I. 286/2 It was not found possible to take indicator diagrams on the full speed trials, but a considerable number of trials were run at lower speeds progressing up to 11·74 knots per hour. **1955** *Britannica Bk. of Year* 490/1 *Speedwalk*, a moving sidewalk. **1956** *Official Gaz.* (U.S. Patent Office) 4 Sept. TM 9 Passenger Belt Conveyors, Inc., Akron, Ohio... Speedwalk..For Passenger Belt Conveying Apparatus. First use Mar. 29, 1954. **1960** *Times Rev. Industry* Apr. 28/1 A twin-track passenger-carrying conveyor belt..is being constructed in London... Known as speed-walks..and travolators, at least two dozen installations are already in use in the United States. **1978** *Jrnl. R. Soc. Arts* CXXVI. 431/1 We can expect to see improvements in short-haul feeder and ancillary systems such as speedwalks, travolators and escalators. **1955** W. Gaddis *Recognitions* iii. ii. 740 You didn't send me a speedwriter down. **1925** *Speedwriting* (Brief Eng. Systems, Inc.) 7 (*caption*) Speedwriting, the wonderfully efficient new system of shorthand, developed by Miss Emma B. Dearborn, can be written after a few weeks' study, either by pencil or on any make of typewriter. **1927** *Official Gaz.* (U.S. Patent Office) 8 Aug. 242/1 *Speedwriting*... Printed Lessons and Examination Sheets Issued from Time to Time. Claims use since Dec. 29, 1924. **1929** *Radio Times* 8 Nov. 439/2 Speedwriting (The Universal Shorthand). **1938** E. B. White *Quo Vadimus?* 24 They sell a new kind of shorthand course, called the Quigley Method of Intensive Speedwriting. **1962** *New Scientist* 5 July 11/1 Part of the Speedwriting method is to cut out most of the vowels and unnecessary movement in making other letters. **1976** T. Stoppard *Dirty Linen* 24 You do speedwriting I suppose?.. Yes, if I'm given enough time. **1982** Barr & York *Official Sloane Ranger Handbk.* 84/1 Every Sloane secretarial college has a nimbus of girls with their shorthand books or red speed-writing books.

e. *Photogr.* In names for an electronic flash-gun, as *speed flash, gun, lamp, -light.*

1940 A. L. M. Sowerby *Wall's Dict. Photogr.* (ed. 15) 315 Externally, a speed-gun takes the form of a box or casing that can be attached to the camera to make the whole a single unit. **1950** W. F. Berg *Exposure* 339 A special kind of flash lamp is the speed flash, also known as multiple or electronic flash... One of the most important advantages of speed lamps is their exceedingly short flash time. **1953** *Sun* (Baltimore) 10 Jan. 5 (*caption*) A photo snapped of Prime Minister Winston Churchill..showed five-sided 'spots'. It was raining at the time and was attributed to reflections from a speedlight. **1969** A. Feininger *Compl. Colour Photographer* ii. 45 Electronic flash or speedlight has..advantages over conventional flashbulbs. **1979** *SLR Camera* Jan. 36/1 The AE-1 is.. an automated system camera. It includes a power winder for auto film transport, a speedlite for auto flash photography and an automatic exposure control.

speed, *v.* Add: Also *Pa. pple.* 8— **speeded.**

II. 11. d. Also *fig.*

1909 C. F. G. Masterman *Condition of England* ii. 23 When life has become 'speeded up' to the motor-car level. **1931** *Times Lit. Suppl.* 3 Dec. 979/3 He brought youth to the theatre; he speeded it up. **1974** A. Lurie *War between Tates* xiv. 277 He has the sense of being slowed down and speeded up alternatively.

13. e. To drive a motor vehicle fast, esp. at an illegal speed; to break the speed limit in a motor vehicle.

1931 Galsworthy *Maid-in-Waiting* xiii. 118 'I'm going to speed,' said Jean, looking back. The speedometer rose rapidly. **1941** M. Halsey *Traffic Accidents & Congestion* v. 41 The very word *speed* has come to mean *speeding* or going too fast. **1949** R. A. Byrd *Driving to Live* xi. 156 It becomes ridiculous, in the light of this new discovery [*sc.* road-stare phenomenon], for any driver to brag about his ability to speed. **1954** B. Preston *Focus on Road Accidents* ii. 63 If the motorist continued to speed and was caught, several times a day, every day, then he would soon stop speeding. **1969** *New Yorker* 14 June 29/2 If you speed, we'll charge you the same amount we charge anyone. **1979** D. Anthony *Long Hard Cure* viii. 72, I..went back to my car. I sped a little on the way back to town.

f. To be addicted to or under the influence of an amphetamine drug. Also *fig.* Usu. as *pres. pple.* Cf. sense *6* of the sb. *slang.*

1973 R. Mills *Young Outsiders* ii. 60 If you are speeding you go out and do more things. **1977** *Rolling Stone* 7 Apr. 47/1 'The best diet for the road,' he says, 'is soup for lunch and candy for supper. It keeps the weight off and you're speeding on all that sugar by show time.' **1978** S. George *Screen Test* ii. 18 'You speeding?' He shrugged. 'Yes. Cancels the alcohol.'

spee·dball. Also **speed-ball, speed ball.** [Speed *sb.*] **1. a.** A dose of a drug, esp. a mixture of cocaine and morphine or of cocaine and heroin. *slang* (orig. *U.S.*).

1909 R. Beach *Silver Horde* v. 58 You must have fed him a speed-ball, for I never saw a guy gear up so fast... He's developed a remarkable burst of speed. **1935** N. Ersine *Underworld & Prison Slang* 69 *Speedball*, a mixture of *coke* and *morph* which is much in demand among drug addicts. **1953** W. Burroughs *Junkie* xiv. 139 A shot of morphine would be nice later when I was ready to sleep, or, better, a speedball, half cocaine, half morphine. **1964** *Guardian* 18 Apr. 6/3 To give a nobbler eighteen months for slipping a speedball in a greyhound's breakfast is like tossing up a ball and then punishing it for having the effrontery to bounce. **1974** M. C. Gerald *Pharmacol.* xv. 291 The 'speedball', a mixture of cocaine and heroin, is frequently employed to modulate the extreme feelings of excitement.

b. A glass of wine, *spec.* when strengthened by additional alcohol or spirits. *U.S. slang.*

1926 *Amer. Speech* I. 653 *Speed-balls*, wine. **1931** 'D. Stiff' *Milk & Honey Route* 215 *Speed balls*, sherry wine. These days it may be any wine, even *dago red.* **1931** G. Irwin *Amer. Tramp & Underworld Slang* 178 *Speedball*, a glass of wine, more especially when 'doped' or made stronger by the addition of some alcohol, ether or strong spirits.

2. a. A ball-game resembling soccer, but in which a ball caught in the air may be passed by hand (see quot. 1976). *U.S.*

1923 *Amer. Boy* Jan. 24 (*heading*) Speedball—a new boys' game. *Ibid.* 25/1 Rychener..called out to Coach Mitchell, 'This is certainly a speedy game. Let's call it speedball.' **1928** *Sportswoman* Oct. 22/2 Speedball furnishes exercise for the whole body—legs, trunk, arms. **1933** *Jrnl. Health & Phys. Educ.* Oct. 38/1 Speedball has gained popularity until it is now used by high schools, private schools, teachers' colleges, and universities from coast to coast. **1946** E. D. Mitchell in Fox & Davies *Official Soccer-Speedball Guide* It is fitting that this particular year should be a reminder to me to see what has been happening to Speedball... It was twenty-five years ago in the Fall of 1921, at the University of Michigan, that it was first played. **1961** J. S. Salak *Dict. Amer. Sport* 413 The conversion of a ground ball to an aerial ball is the essential difference between speedball soccer. **1976** *Webster's Sports Dict.* 410/1 *Speedball*, a game played between 2 teams of 11 players on a football field with the object of kicking a ball between the uprights of the goal or passing the ball over the end line to score and preventing the opposing team from scoring.

b. *Boxing.* A type of small, fast punch-ball.

1955 F. Mills *Learn Boxing with Me* v. 72 The platform pear ball is not an easy ball to punch... The ball to use is the small-type speedball, weighing about nine to ten ounces. The heavier kinds are too slow to be beneficial. **1975** Oliver & Riley *Boxing* ix. 46 Gyms vary in..the amount of equipment..available. Essentially what is needed is a ring, punch-balls, speed-balls, heavy bags and punching-pads.

speeded-up, *ppl. a.* [f. Speed *v.* + -ed[1] + Up *adv.*]

a. Of a film: giving the appearance of rapid motion usu. as a result of being projected at a speed greater than that of original shooting. Cf. Speed *v.* 11 d.

1931 J. S. Huxley *What dare I Think?* ii. 54 It is one of the most astonishing spectacles to see, on the speeded-up film, the processes of cell-division, of organization of growth. **1951** A. C. Clarke *Sounds of Mars* xi. 143 You've all seen the films we made—especially the speeded-up colour one showing a complete cycle of Saturn's phases. **1968** A. Diment *Bang Bang Birds* viii. 146 Everything was happening like a speeded up holiday movie. **1976** *Listener* 25 Mar. 381/2 A speeded-up film of a flower opening and closing.

b. *transf.* Having an increased speed.

1962 M. McLuhan *Gutenberg Galaxy* 47 The new institutes for speeded-up reading. **1980** U. Curtiss *Poisoned Orchard* vii. 65 Sarah went to meet her at a speeded-up saunter.

speeder. Add: **3.** (Earlier example.)

1847 *Knickerbocker* XXX. 517 A few [girls] tend the 'warpers', the 'spoolers', and the 'speeders'.

4. Also *spec.* one who exceeds the speed limit in a vehicle.

1974 R. B. Parker *God save Child* i. 6 [The police] have to arrest drunks and flag down speeders and break up fights.

5. *N. Amer.* A small vehicle running on railway tracks used for line maintenance, etc., orig. only manually propelled.

1905 J. Outram *In Heart of Canad. Rockies* 152 The top of a box car is the choicest of propelling methods, unless one can get a ride on a hand-car or a speeder, with opportunity to slacken speed or stop whenever one desires. **1934** *Sun* (Baltimore) 6 Aug. 2/1 One hour in advance of the pilot train section men went over the President's route in gasoline 'speeders'. **1947** A. Saunders *Algonquin Story* xi. 128 The husband, apparently a section man, had rushed up the tracks on a three-wheeled 'speeder' to bring Molly back to help. **1960** J. J. Rowlands *Spindrift from House by Sea* ii. 107 They had taken him out with pneumonia..fifty miles on a gasoline speeder..at twenty below zero. **1970** R. & J. Paterson *Cranberry Portage* xxiii. 152 For days I travelled by locomotive, trolley, speeder, anything moving my way.

speeding, *vbl. sb.* Add: **4. b.** The act of driving a motor vehicle fast, esp. at an illegal speed. Also *transf.*

1908 *Evening Star* (Washington) in *Daily Chron.* 7 Oct. 4/7 Baby carriages are required to carry lights at night in Chicago. That rapid city may yet find it necessary to provide special police to keep the baby carriages from speeding. **1911** *Daily Colonist* (Victoria, B.C.) 2 Apr. 2/4 It was decided that everything possible would be done to assist the authorities in enforcing reasonable regulations regarding speeding. **1922** J. A. Dunn *Man Trap* ii. 24 Driving a car about the countryside at the expense of many fines for speeding. **1977** E. W. Hildick *Loop* xvii. 115 The copper... He's been trying to get you all afternoon... I hope you haven't been doing no speeding.

5. *Comb.* **speeding citation, ticket** *U.S.*, a summons given to a motorist who has violated a speed regulation.

1974 *Marlboro Herald-Advocate* (Bennettsville, S. Carolina) 18 Apr. 1/4 Out of the 228 speeding citations..there was only one request for a jury trial. **1960** 'E. McBain' *Heckler* v. 70 He had flatly refused to square any raps for them, raps ranging from speeding tickets to disorderly conduct. **1978** R. Ludlum *Holcroft Covenant* ix. 114 If it'll get you out of Rio..I'll go like a greased pig in a slaughterhouse and pay the speeding tickets from my per diem.

speeding, *ppl. a.* Add: **2. d.** Of motor vehicles, motorists, etc.: travelling fast, esp. at an illegal speed.

1957 R. Matheson *Shores of Space* 156 People who had ..been struck down by speeding cars. **1960** *Daily Tel.* 16 Jan. 1/5 Now the speeding motorist will find things more difficult. **1978** 'M. Yorke' *Point of Murder* xiii. 125 A speeding Mini, in the small hours, might attract notice if a police patrol car were around.

speedo (spī·do), colloq. abbrev. of Speedometer.

1934 *Passing Show* 21 July 19/3 The puzzled eyes of his scarlet face went hopelessly to the speedo. The needle was steady at fifty-five. **1938** L. MacNeice *I crossed Minch* vi. 81 Lever..stepped on the gas, He said,..'Just you watch the speedo climb.' **1960** *News Chron.* 10 Oct. 6/6 The speedo needle does not obscure the mileage indicator. **1969** M. Calthorpe *Defectors* 140 He glanced at the speedo... He'd driven almost five miles. **1976** P. Hill *Hunters* vii. 93 The car [was as] steady as a rock.. as the speedo reached up towards its limit.

speedometer. Add: **1. a.** (Later examples.)

1912 *Motor Man.* (ed. 14) 115 (*caption*) Details of system of friction drive for shaft of speedometer. **1932** *World Today* Feb. 261/1 The noisiest thing about this car was its speedometer, which maintained a noise something between a click and a swish. **1955** L. G. Pringle *Driving Lessons* 35 Chancing to glance at his speedometer, he discovered he was travelling at 45 m.p.h. through a built-up area! **1978** K. Jolly *Driving made Easy* 35 The speedometer shows your speed in miles per hour... It should be checked frequently quickly when on the move.

b. The distance-indicator that is often incorporated in the speed-indicator of a motor vehicle; an odometer. *U.S.*

1929 *Collier's* 12 Jan. 8/2 Don't believe the speedometer of that magnificent limousine. It could have been fixed for sale purposes in twenty minutes. **1938** Palmer & Crooks *Millions on Wheels* v. 87 Several courts have decided..that setting speedometers to indicate lower than actual mileage is a fraudulent practice... The buyer of a used car should therefore not even bother to look at the speedometer mileage. **1942** R. W. Riis *Repairmen will get You* ii. 19 The car has gone at least 20,000 miles, although I think the speedometer doesn't register all the miles.

2. *attrib.*, as *speedometer needle.*

1909 *Quarterly Rev.* Jan. 148 Men are tempted to feel as if they must be acting rightly when the speedometer needle hovers below the figures 10. **1977** N. Slater *Crossfire* iv. 77 The speedometer needle [was] jumping erratically around the 50 mph mark.

speedster (spī·dstəɹ). [f. Speed *sb.* + -ster, after *roadster.*] **1.** A fast motor vehicle; a speed-boat.

1918 Webster Addenda, *Speedster*,..a high-speed roadster, usually with low seats, accommodating two persons seated side by side. **1928** *Daily Express* 3 Sept. 5/5 But I'm not telling everyone who comes up in a swell speedster [*sc.* a speed-boat]. **1977** 'J. Gash' *Judas Pair* vi. 72 He lived about fifteen miles off the trunk road... I patted my speedster and swung the handle.

2. a. A motorist who drives fast.

1921 *Daily Colonist* (Victoria, B.C.) 16 Oct. 25/3 As a means of checking up on the speedster, a device has been invented. **1930** *Aberdeen Press & Jrnl.* 17 May 8/5 Lochiel criticised the making of trunk roads for the use of south speedsters. **1974** J. Cleary *Peter's Pence* vii. 197 McBride drove fast: in Rome slow drivers got more attention..than speedsters.

b. More generally, a person who moves or acts very quickly, a fast runner. Also applied to animals.

1947 *Richmond* (Va.) *Times-Dispatch* 9 Nov. B7/7 The 175-pound speedster from Bridgeton, N.J., never quit running until he was across the goalline. **1966** D. Varaday *Gara-Yaka's Domain* iv. 48 The four speedsters made a sudden concentrated rush at the meat-rich tycoon. **1977** *Time* 13 June 21/2 An American track-meet promoter, anxious to lure a top dash man to his indoor meet to increase the gate, called a speedster and promised him $800 plus expenses for joining the field.

speed-up. [f. vbl. phr. *to speed up*: see Speed *v.* 11 d.] **a.** The act or process of increasing the speed or working rate of a thing. Also *attrib.*

1923 *Daily Mail* 3 Mar. 7 Train Speed-up..The speeding up of all services. **1944** *Sun* (Baltimore) 18 May 18/6 (*heading*) Assembly speedup law is planned. **1953** J. S. Huxley *Evolution in Action* i. 33 With a hundred-thousand-fold speed-up [of the film]..the overall processes of evolution became visible. **1970** R. Lowell *Notebook* 158, I too in the end will see the things like this, Whatever I've lived assumed in one bright glance Like speed-up reading. **1977** P. Somerville-Large *Eagles near Carcase* v. 102 'Go on,' I said, 'not too fast.'.. The last thing I wanted was a speed-up.

b. The act or process of increasing productivity, esp. without raising rates of pay. orig. and chiefly *U.S.*

1935 *Sun* (Baltimore) 9 Feb. 2/1 Labor unrest..flows from..inequitable hiring and rehiring methods, espionage, speed-up and displacement of workers at an extremely early age. **1938** *Reader's Digest* May 55/2 There is no piecework, no speedup. **1943** *Sun* (Baltimore) 14 June 10/7 The extreme manifestation of management is the ruthless speedup or stretch-out, where production is increased without thought of the health of the workingman. **1959** N. Mailer *Advts. for Myself* (1961) 184 The productive speed-up tends to be replaced by..'featherbedding'. **1962** E. Snow *Other Side of River* xxxi. 235 In 1955, for example, the *Workers Daily* revealed that alleged gains made in 'speed-up' drives by shocking over-use of labor were more than offset by heavily increased losses in man-power output. **1974** *Time Out* 22 Nov. 64/3 It's a depressing and probably accurate picture of a failure of consciousness in a barely unionised community where speed-ups, piece-work and cheap labour are rife.

spee·dway. [Speed *sb.*] **1. a.** *U.S.* A track or road prepared for fast horse-driving. Now *Hist.*

1894 [in Dict. s.v. Speed *sb.* 11 c]. **1968** *N.Y. City* (Michelin Tire Corp.) 97 Trotters were in fashion, and they whipped through the Park to the speedways of Harlem.

b. A road on which motor vehicles may travel fast. Also *gen.*, a motorway. Also *fig.* orig. *U.S.*

1903 *N.Y. Times* 16 Aug. 24/7 The numerous owners of rapid roadsters are devoting no inconsiderable portion of their Summer leisure to spirited brushes on the new speedway. **1919** *Public Opinion* 14 Nov. 471 (*heading*) The Speedway to Prosperity. **1927** [see *autostrada]. **1946** J. W. Day *Harvest Adventure* xiv. 247 The traffic on that concrete speed-way, the Great West Road. **1976** J. van de Wetering *Tumbleweed* xx. 176 Their car..was joining the main traffic on the speedway.

2. *Sport.* **a.** A racing track for motor vehicles. Also *transf.*

1925 *Kansas City* (Missouri) *Star* 31 May 1/1 Here is how they finished..in the thirteenth annual 500-mile motor classic on the Indianapolis speedway yesterday. **1968** D. Braithwaite *Fairground Architecture* 170/1 *Speedway*, a loose term which could refer either to an Ark —mounted with *motor-cycles*, or to a Monte Carlo Rally. **1979** *Arizona Daily Star* 1 Apr. c 12/2 USAC's 200-mile race at the Ontario Motor Speedway last weekend turned out better than a lot of people figured.

b. *transf.* A sport in which motorcyclists race several laps about a short oval dirt track. Freq. *attrib.*

1930 S. Elder *Romance of Speedway* i. 9 Now that Speedway Racing has taken its place as one of our national pastimes. *Ibid.* ii. 27 In the first rank among Speedway riders. *Ibid.* 33 Habitual frequenters of Speedway gatherings. **1937** D. W. Harding in *Scrutiny* VI. 111. 254 When we have passed..from the state of idly watching men..and have accepted the invitation to form part of an audience for jugglers, acrobats..or speedway riders..we may speak of display entertainment. **1950** *Sport* 22–8 Sept. 21/2 The Wembley crowds shot up and speedway in general benefited. **1961** K. Reisz *Technique Film Editing* (ed. 9) ii. 77 Bill Fox (Dirk Bogarde) had been an ace speedway rider before the war. **1977** *Western Morning News* 30 Aug. 12/3 Exeter gave themselves a tremendous boost in their challenge for the Gulf British League speedway title when they snatched a precious point from title rivals Reading.

speedy, *a.* Add: **5. b.** Also *speedy trial* (U.S. Law), a criminal trial held after a minimum of delay, considered to be a citizen's right. Also *attrib.*

[**1776** G. Mason *Virginia Decl. Rights* in *Virginia Gaz.* (Williamsburg) 1 June 2/2 In all capital or criminal prosecutions a man hath a right to demand the cause and nature of his accusation, to be confronted with the accusers or witnesses, to call for evidence in his favour,

and a speedy trial by an impartial jury. **1789** J. Madison in T. Lloyd *Congress. Reg.* 4 June 428 In all criminal prosecutions, the accused shall enjoy the right to a speedy and public trial. **1807** J. Overton *Tennessee Reports* (1813) I. 253 The 9th section of the bill of rights secures to the citizen a speedy, public trial, and to demand the cause of accusation against him. **1878** *Michigan Reports* XXXVIII. 739 The right to a speedy and public trial in criminal cases by an impartial jury cannot be taken away by legislation.] **1893** *Southeastern Reporter* (1894) XVIII. 284/1 The 'speedy trial', and the policy of the law to expedite the trial of criminal cases, forbid that the person accused of crime shall be detained in prison beyond any term of the court at which he can lawfully be tried. **1901** *Daily Colonist* (Victoria, B.C.) 22 Oct. 1/6 Robt. Cameron, a penitentiary convict, before Judge Bole in the Speedy Trial court today, pleaded guilty to attempting to escape. **1926** J. Black *You can't Win* xviii. 262, I..decided to go before the court under the Speedy Trials Act. A defendant electing for a speedy trial dispenses with a jury and saves time and money for the community. **1951** F. H. Heller *Sixth Amendment* iv. 60 The right to a speedy trial may not be asserted merely in order to forestall the ends of public justice. **1976** *Billings* (Montana) *Gaz.* 30 June 1-c/3 Judge Sorte rejected defense motions to dismiss the charges against the couple on grounds of prejudicial pretrial news coverage and lack of a speedy trial.

speel, *v.*[2] For entry substitute:

† **speel,** *v.*[2] *dial.* and *slang. Obs.* Also **spiel.** [Origin obscure: perh. f. Speel *v.*[1]] **a.** *intr.* To go fast; to run *away*, make off.

a **1818** W. Midford *Cappy* in T. Thompson *Coll. Songs* (1827) 49 Owre his airm hung a basket—thus onward he speels, And enter'd Newcassel wi' Cap at his heels. **1829** *Sessions Papers at Old Bailey* 9 Apr. 329 The blake come, i spelld away. **1882** *Sydney Slang Dict.* 9/2 Shake this mob, Bill, and speel to the den, and let our lushy shicksters bring the ruin in. **1905** in A. B. Paterson *Old Bush Songs* 56 No more shall we muster the river for fats, Or spiel on the Fifteen-mile plain. *a* **1921** G. H. Gibson in *Penguin Bk. Austral. Ballads* (1964) 207 With a turn o' speed..As can spiel like a four-year brumby. **1945** S. J. Baker *Austral. Lang.* 70 Horses are said..*to speel* when they travel fast.

b. *trans.* Phr. *to speel the drum* (see quots.). Cf. *drum *sb.*[1] 9 d.

1839 H. Brandon in W. A. Miles *Poverty, Mendicity & Crime* 167/2 *To speel the drum*, to run away with the stolen property. **1859** [in Dict.].

Speenhamland (spī·nămlænd). The name of a village near Newbury, Berks., used *attrib.*, esp. as *Speenhamland system*, and (rarely) *absol.* of a system of poor relief adopted by the magistrates there in 1795, and subsequently established throughout most of rural England.

[**1797** F. M. Eden *State of Poor* I. ii. 577 This shews.. what should be the weekly Income of the industrious poor, as settled by the Magistrates for the County of Berks, at a meeting held at Speenhamland, May the 6th, 1795.] **1835** R. Hall *Let.* 10 July in *1st Ann. Rep. Poor Law Commissioners* 208 It is singular that the provisions of the Speenhamland Act of Parliament, as the Berkshire table is called,..contemplate only the case of the *industrious* poor. **1854** G. Nicholls *Hist. Eng. Poor Law* II. xii. 139 The famous Berkshire bread-scale, locally known as the 'Speenhamland Act of Parliament'. **1902** *Encycl. Brit.* XXVI. 665/2 The well-known Speenhamland scale (1795), by which a larger or lesser allowance was given to a family according to its size and the prevailing price of corn. **1925** C. M. Waters *Econ. Hist. England 1066–1874* vi. i. 332 The enclosures and Speenhamland had quadrupled the poor-rate. **1934** B. Russell *Freedom & Organization 1814–1914* vi. 78 An important step in the development of the Poor Law was taken by the inauguration of what is called the 'Speenhamland' system in 1795. **1959** *Chambers's Encycl.* VII. 550 The 'Speenhamland' system of poor relief..remained in force until the Poor Law Amendment Act of 1834. **1973** *Listener* 19 Apr. 500/3 Bill Jordan, a sociologist,..sees the social security system..as a modern Speenhamland system, deepening the division between the rich middle classes and the 'pauperised' poor.

Speeton (spī·tən). *Geol.* The name of a village on the North Yorkshire coast, used *attrib.* with reference to a series of clays of Lower Cretaceous age which outcrop there.

1829 J. Phillips *Geol. Yorkshire* i. iii. 74 From the termination of the white cliffs the coast bends to the northward, and exhibits in succession, rising from beneath the chalk, the Speeton clay and the coralline oolite series. **1882** A. Geikie *Text-bk. Geol.* 816 The marine Neocomian strata of England are well exposed on the cliffs of the Yorkshire coast at Filey, where they occur in a deposit long known as the 'Speeton Clay'. **1946** L. D. Stamp *Britain's Structure & Scenery* xviii. 204 In the northern basin the Lower Greensand is represented by the Speeton Clays, with 'carstone', succeeded by marls and the famous Red Chalk of Hunstanton and Lincolnshire. **1969** Bennison & Wright *Geol. Hist. Brit. Isles* xiv. 323 On the north side of the Yorkshire Wolds a narrow outcrop of clays is found and it extends to the coast near Speeton. These clays, some 300 feet thick and known as the Speeton Series, range in age from the Ryazanian to the Albian.

‖ **Speisesaal** (ʃpai·zəzāl). [Ger.] In German-speaking countries: a dining room, dining hall.

1871 *Monthly Packet* Christmas 57 The gentlemen betook themselves to the Speise saal, about which a cheerful bustle prevailed. **1929** E. M. Brent-Dyer *Rivals of Chalet School* v. 61 'No walk *this* day!' said that young person, looking sadly out of the window of the *Speisesaal* while they were having hot coffee and biscuits. **1969** J. Eastwood *Come die with Me* (1970) ii. 20 She..went down to the Speisesal [*sic*].

spelæology. Add: The form **speleo-** is now usual. Also, the hobby of exploring caves.

1937 *Caves & Caving* I. 7/1 The objectives of the British Speleological Association... 1. The promotion of friendship and the exchange of ideas amongst those who are interested in speleology. 2. The study of the geological, hydrological, archaeological and 'sporting' aspects of speleology. **1974** *Islander* (Victoria, B.C.) 8 Dec. 15/4 Press reports..have promoted public interest in the sport and science of speleology.

spelæologist (later examples); also, an explorer of caves.

1955 *Times* 20 Aug. 5/3 A team of Spanish speleologists from Pamplona..have completed a 10-day exploration of the Larra region of the Pyrenees. **1971** *Daily Tel.* 1 May 11/4 Speleologists (or cavemen) should explore the Dove's Nest in Borrowdale.

speleothem (spī·li,oþem). *Geol.* [f. Gr. σπήλαιον cave + θέμα that which is laid down, deposit.] Any structure which is formed in a cave by the deposition of minerals from water, e.g. a stalactite, stalagmite, etc.

1952 G. W. Moore in *Nat. Speleol. Soc. News* June 2/1 In an effort to relieve the ambiguities of 'formation', the term speleothem is proposed... It is suggested that the word be used as a general term for secondary mineral deposits formed from water in caves, such as stalactites, helictites, rimstone terraces. **1968** *Nature* 6 July 49/1 Temperature dependent fluctuations in the ^{18}O/^{16}O composition of calcite deposited on speleothems could be used to fill this gap in the present understanding of past climates. **1976** W. B. White in Ford & Cullingford *Sci. of Speleology* viii. 271 There is a competition between shapes guided by the flow path of the solution and shapes guided by the particular mineral and its crystal habit. This gives rise to two broad classes of speleothems, dripstone and flowstone forms and erratic forms. **1980** *Cambr. Encycl. Archaeol.* 53/1 Methods involving the measurement of different isotopes are being widely applied to derive long climatic records from ice cores, cave deposits like stalagmites (speleothems) and tree rings.

spell, *sb.*[1] Add: **4. a.** *spell-enslaved, -soaked.*

1938 Dylan Thomas *Let.* 14 Oct. (1966) 210 A still room in a spellsoaked house. **1955** J. R. R. Tolkien *Return of King* vi. iv. 227 The creatures of Sauron, orc or troll or beast spell-enslaved, ran hither and thither.

spell, *sb.*[3] Add: **3. b.** For *Austr.* read *dial.* and *Austral.* and add earlier and further examples.

Examples from 1863 to 1975 may be found in *Dict. Newfoundland English* (1982).

c **1845** J. Tucker *Ralph Rashleigh* (1929) xi. 146 Both men took a hoe and gave the children a spell. **1847** J. O. Halliwell *Dict. Archaic & Provincial Words* 781/2 *Spell*,..pleasure; relaxation. *Somerset.* **1862** J. S. Dobie *Jrnl.* 10 Sept. in *S. Afr. Jrnl.* (1945) 23 Invited to stay and give my horse a day's spell. **1867** M. A. Barker *Station Life in N.Z.* 128 We were all so breathless that a 'spell' (do you know that means 'rest'?) would have been most acceptable. **1931** G. L. Nute *Voyageur* 96 Every five miles or so a halt was made to rest the dogs and to allow the men to smoke. These stops were termed 'spells' or 'pipes', and the voyageurs spoke of a day's journey as being so many spells or pipes. **1940** F. Sargeson *Man & his Wife* (1944) 30 About half-way back to the shore he took a spell. **1954** B. Miles *Stars my Blanket* xix. 141 We stopped for a 'spell' in the sandy bed, spreading out ground-sheets in the shade of the trees that lined the centre of it. **1972** *Regional Lang. Studies Newfoundland* May 9 *Spell*, a period of rest or a short sleep. Used..in Newfoundland and in Anglo-Irish. **1977** *N.Z. Herald* 8 Jan. 1-9/2 Marie Kay has been back racing only a little over a month following a long, enforced spell.

4. d. (*b*) (Earlier example.)

1745 D. Giddings *Jrnl.* 27 May in *Essex Inst. Hist. Coll.* (1912) XLVIII. 299, I..continued in ye Trench a Spell.

e. *N.Z.* One of the periods into which a game of rugby is divided.

1900 *N.Z. Illustr. Mag.* III. 237/1 Usually in the second spell, when play is getting more exciting. **1934** A. E. Mulgan *Spur of Morning* i. iii. 23 The school won a hard game by six to nil, a try in each spell.

f. *Cricket.* A period or series of overs during a session of play in which a bowler bowls unchanged.

1976 J. Snow *Cricket Rebel* 77 On that last afternoon at Colchester I took three quick wickets in my first spell. **1977** *World of Cricket Monthly* June 27/3 He was brilliantly caught by Richards at second-slip off Croft's first ball of a new spell.

spell, *v.*[2] Add: **3. e.** *U.S. to spell* (someone) *down*: to defeat (someone) in a spelling-contest. † **f.** *U.S.* To put to the test in spelling. *Obs.*

1854 B. F. Taylor *Jan. & June* 259 They all stand in solid phalanx by schools, and the struggle is, to spell each other down. **1866** C. H. Smith *Bill Arp* 171 He then spelt him right straight along on all sorts of big words, and little words. **1871** E. Eggleston *Hoosier Schoolm.* (1872) iv. 47 Ralph dreaded the loss of influence..if he should be easily spelled down. **1932** *Randolph Enterprise* (Elkins, W. Va.) 18 Feb. 4/5 He was fairly spelled down at close of my school at Job, spring 1882, by a little girl not 7 years old. **1952** T. Pyles *Words & Ways of Amer. Eng.* (1954) v. 82 He who misspelled had to take his seat; the master's purpose was to 'spell down' the pupils.

g. *fig. to spell out* (orig. *U.S.*): to explain (something) step by step; to state explicitly or in detail.

1940 *San Francisco News* 31 Dec. 11 In the interest of clarifying public opinion, these opponents should spell out their position fully. **1952** B. Wolfe *Limbo '90* (1953) xxiv. 390 Spelling out the bald verities to a retarded child. **1956** J. Potts *Diehard* xiii. 196 If you weren't such a fool you'd know it too. You want me to spell it out in words of one syllable for you? **1960** *Guardian* 1 Nov. 8/5 Dr. Kaldor would apparently have us spell all these things out. But..there really is a limit to the amount of detail ..to which a party in Opposition can commit itself. **1973** G. W. Turner *Stylistics* vi. 171 An electrician asking which of three wires is 'the earth' clearly means 'the earth(ed) wire' or 'the wire connected with the earth', but hardly needs to spell it out. **1978** J. A. Michener *Chesapeake* v. 239 With studied care Thomas Janney spelled out the terms of the deal he had arranged.

spellbinder. For *U.S.* read orig. *U.S.* and add later examples. Also *gen.*

1906 *Calgary* (Alberta) *Eye Opener* 6 Oct. 1/3 He is a spell-binder all right. **1932** [see *flesh-creeper* s.v. *flesh *sb.* 12 b]. **1939** *English* II. 245 Coleridge..would seem to have been something of a spell-binder: we have many witnesses..to the potency of the fascination he exerted over those with whom he came in contact. **1969** *N.Y. Rev. Books* 2 Jan. 15/2 'The books of all three are products of men who do not necessarily believe in their thoughts.' The risk is obvious of course—the false prophet, the spellbinder. **1974** P. de Vries *Glory of Hummingbird* (1975) ii. 11 A ministry replete enough with intimations..that he was less than a spellbinder.

spe·llbinding *ppl. a.*; **spe·ll-bindingly** *adv.*

1977 *Gramophone* Dec. 1121/3 A group of songs with orchestra..sounded so spellbindingly lovely. **1978** K. J. Dover *Greek Homosexuality* iii. 164 Philosophy..was not the product of solitary meditation, to be communicated by a spell-binding orator.

spelldown (spe·ldaun). *U.S.* [f. vbl. phr. *to spell down*: see *spell *v.*[2] 3 f.] An eliminating contest in spelling.

1943 *Nat. Geogr. Mag.* Dec. 755 Among cherished memories of 'the days of real sport', the old-fashioned 'spelldown' takes high rank. **1951** E. Graham *My Window looks down East* xv. 131 Mrs. Rowen decided to give him another chance and started the 'spelldown' over, but the children were consulted first. They agreed. The spelling started again. **1958** *Time* (Atlantic ed.) 23 June 38/1 In the..19th round of the spelldown, 13-year-old Betty Morgan,..choked up on *chiaus.*

spelling, *vbl. sb.*[2] Add: **1. b.** *U.S.* A spelling-bee, spelling-test. *rare.*

1860 O. L. Jackson *Colonel's Diary* (1922) 23 The boys were anxious for a spelling in the evening. **1889** J. W. Riley *Pipes o' Pan at Zekesbury* 45 How her face used to look in the twilight As I tuck her to spellin'. **1975** *Budget* (Sugarcreek, Ohio) 20 Mar. 14/3 The young folk are having german spellings once a week.

3. *spelling card, match, mistake, reform* (earlier example), *reformer* (later examples); **spelling pronunciation**, the pronunciation of a word according to its written form; **spelling school**, † (*a*) a building in which spelling is taught; (*b*) *U.S.*, a contest in spelling.

1850 C. M. Yonge *Langley School* xviii. 166 'Nobody' left the gate open,..tore the spelling cards, scratched the slates. **1974** 'J. le Carré' *Tinker, Tailor, Soldier, Spy* xxii. 186 Spelling cards lay spread over the floor. **1845** H. Greeley in *Publ. Mod. Lang. Soc. Amer.* (1941) LVI. 501 It used to be the custom that the head of the first class and the next should choose sides for a 'spelling match'. **1967** B. Banfill *Pioneer Nurse* xiii. 146 An invitation for all of us to an old-fashioned Spelling Match. **1966** N. Marsh *Black Beech & Honeydew* iii. 74, I won a Navy League Empire Prize..with an essay containing thirty-one spelling mistakes. **1901** E. Koeppel (*title*) Spelling-pronunciations. **1927** L. Bloomfield in *Amer. Speech* II. 438/1 This last feature is a fairly close parallel to our 'spelling pronunciations', such as the full form *fore-head* for *forrid* and the now perhaps accepted *waistcoat* and *seam-stress* for *weskit* and *semstress.* **1944** [see *pronunciation-spelling* s.v. *pronunciation 6]. **1977** P. Strevens *New Orientations Teaching of English* xii. 153 One further characteristic of American pronunciation that contrasts with British speech is the frequency of 'spelling pronunciations' in both place-names and proper names. **1848** A. J. Ellis (*title*) A plain statement of the objects and advantages of the spelling reform. **1908** G. K. Chesterton *All Things Considered* 220 Some spelling-reformers..do spell his name phonetically. **1936** *Discovery* May 164/2 Unlike many spelling reformers, he respects tradition and the 'look' of a word. **1704** Swift *Tale of Tub* 16 There is also, the Spelling School, a very spacious Building. **1832** E. M. Chamberlain *Jrnl.* 25 July in *Indiana Mag. Hist.* (1919) XV. 241 In the evening I appointed a spelling school at which I invited all the parents to attend. **1948** E. N. Dick *Dixie Frontier* 138 Backwoods debating societies, spelling schools, story telling, and singing helped to while away the time.

spelling, *vbl. sb.*[3] Add: (Later *attrib.* example.) *rare.*

1939 Dylan Thomas *Map of Love* 15 Endure burial under the spelling wall.

spelling (spe·liŋ), *vbl. sb.*[4] [f. Spell *v.*[3]] **1.** *Sc.* The practice of acting as a substitute for another or taking turns at some work or labour.

1920 *Glasgow Herald* 16 Apr. 7 The Sheriff finds that the custom known as 'spelling' is recognised in the West

of Scotland, and in this case the defender's stevedore assented to the pursuer 'spelling' for the regular employee. **1955** *Times* 17 Aug. 5/4 Organized 'spelling'—that is to say, arrangements between members of a gang that they shall take it in turns to leave their work—is still to be found in Glasgow and Liverpool, in spite of many attempts to stamp it out. **1965** *Daily Express* 6 Aug. 4/5 'Spelling' in Glasgow is the same as 'welting' in Liverpool.

2. *Austral.* Resting from work. Also *attrib.*

1911 *Chambers's Jrnl.* Aug. 591/2 Old Davy..settled down on a selection near Grassmere which the Cornet Scrubber,..used as a spelling-place in his spare hours. **1926** B. Cronin *Red Dawson* ii. 36 There was need for their spelling before they sat in on the game in real earnest.

spelter, *sb.* Add: **1.** Also *locally* applied to various ores; (later example).

1912 *Trans. R. Geol. Soc. Cornwall* XIV. 153 In some of the deeper levels..an intimate mixture of chalcopyrite and garnet, which sometimes contains cassiterite also, locally known as 'spelter', has been met with in considerable quantities.

speltoid (spe·ltoid), *a.* and *sb.* *Bot.* [a. G. *speltoid* (H. Nilsson-Ehle 1917, in *Botaniska Notiser* 305), f. *spelt*: see Spelt *sb.*[1] and -oid.] **A.** *adj.* Of a wheat: resembling or having certain characteristics of spelt. **B.** *sb.* A speltoid wheat.

1920 *Hereditas* I. 116 A distinctive character of these speltoids..is the short outer glumes, abruptly cut off at the top as in *Triticum spelta*... Speltoid heterozygotes are also characterized by longer straws and longer and more lax spikes than the mother sort. **1932** *Proc. 6th Internat. Congr. Genetics* I. 286 Speltoid and dwarf types ..are found in the progeny of treated plants of *Triticum vulgare.* **1939** Sansome & Philp *Recent Adv. Plant Genetics* vii. 212 The speltoids differ from the parental type in having lax ears, thick keeled glumes which can only be pulled away from the grain with difficulty, and by the presence of awns. **1975** *Proc. R. Soc. B.* CLXXXVIII. 149 Reasonably uniform lines were available for test crosses with *vulgare* and speltoid tester stocks. *Ibid.* 156 *Spelta* and speltoid cannot be distinguished phenotypically... The low frequency of defectives in F_2's with base-sterile speltoids indicates a high degree of buffering by the polygenic system.

Hence **spe·ltoidy**, speltoid character.

1944 *Genetics* XXIX. 233 Speltoidy (so called because of the resemblance to *T. spelta*) has been shown..frequently to be due to the effects of a particular monosome. **1958** *Nature* 28 Jan. 1813/1 Okamato..has shown that chromosome IX is in the A genome, and has thus reduced the historical importance of speltoidy and compactoidy.

spelunker (spelʌ·ŋkəɹ). *N. Amer. slang.* [f. as Spelunk + -er[1].] One who explores caves, esp. as a hobby; a caver, a speleologist.

1942 A. F. Harlow *Weep no More, my Lady* xxiii. 407 A young man named Floyd Collins, a native of the hills a few miles north of the Mammoth, had long been a confirmed 'spelunker', or cave bug. **1946** *Life* 4 Nov. 143 Cave exploring,..which bears the scientific name of speleology, has a group of amateur followers who like to call themselves 'spelunkers'. **1965** P. Ordway *Night of Reckoning* (1967) 131 Not being a spelunker at heart, let's hope..there aren't too many bats. **1971** G. G. Luce *Body Time* ii. 54 A courageous spelunker and geologist, Michel Siffre, who spent two months on a subterranean glacier in the French Alps in 1962. **1980** 'E. McBain' *Ghosts* viii. 155 The cave seemed not in the least bit inviting. He had always considered spelunkers the choicest sorts of maniacs.

Hence (as a back-formation) **spelu·nk** *v. intr.*, to explore caves as a hobby; **spelu·nking** *vbl. sb.*, the practice or hobby of exploring caves.

1946 *Life* 4 Nov. 143 (*heading*) Life goes spelunking. **1959** *Weekend* 12 Sept. 77/2 To take a second look at Cadomin caves *Weekend* organized another spelunking expedition. **1965** R. McDowell *Hound's Tooth* vii. 105 They had no legitimate complaint if he chose to take Jenny spelunking in Hunter's Cave. **1974** *Islander* (Victoria, B.C.) 8 Dec. 15/3 Spelunking..is a sport which has become popular within our time. **1979** *United States 1980/81* (Penguin Travel Guides) 563 You can see these on..walks through the cave,..and on spelunking tours where you do your own locomoting on all fours.

Spencer, *sb.*[2] Add: [In sense *5, from the name of Christopher Miner Spencer (1833–1922), U.S. inventor and manufacturer.]

2. a. (Earlier examples.)

1795 *Sporting Mag.* V. 324/2 Spencers. These fashionable coatlets. *Ibid.* VI. 41/2 A young gentleman then approaching, dressed in a light coat, and a blue spencer.

b. (Earlier example.)

1799 J. Woodforde *Diary* 19 June (1931) V. 200 Very cold indeed again today, so cold that Mrs. Custance came walking in her Spenser with a Bosom-friend.

c. (Earlier example.)

1831 J. Brown *Let.* 26 Oct. (1912) 27 Then there is the odd dress of the sailors, with bright yellow worsted spencers and large slouched hats.

d. A type of under-bodice (usu. made of wool) worn esp. by women and girls to provide extra warmth in winter.

1881 in A. Adburgham *Shops & Shopping* (1964) xvii. 189 Light, elastic, inexpensive [hand-knitted clothes], including Jerseys, Cardigans, Vests, Spencers, Combinations, etc. **1924** [see *hug-me-tight 1]. **1953** 'P. Wentworth' *Ivory Dagger* xvi. 69 Miss Silver stood revealed in a slip petticoat of grey artificial silk and a neat white spencer whose high neck and long sleeves had..a narrow crochet edging. **1972** L. Hancock *There's a Seal in my Sleeping Bag* viii. 204, I wore two spencers (Australian item of thermal underwear).

5. A type of rim-fire repeating rifle or carbine used esp. during the U.S. Civil War. Freq. *attrib.*

1866 'F. Kirkland' *Pictorial Bk. Anecdotes & Incidents* 660/1 Harris ordered the skirmish line forward,..with orders to silence the troublesome battery..with the aid of the Spencer rifle. **1873** J. H. Beadle *Undeveloped West* 545 My horse, bridle, saddle, lariat, gun (a Spencer) and two Navajo blankets cost me two hundred dollars. **1904** *Kynoch Jrnl.* Apr.–June 96 The second repeater—and the one most prominent in the war—was the Spencer, having a magazine in the butt containing seven cartridges. **1923** J. H. Cook *Fifty Yrs. on Old Frontier* 5, I had traded a pistol..for a Spencer carbine. **1949** *Exciting Western* May 36/2 The .52 Spencer he kept under the bunk was a souvenir of Malvern Hill. **1974** *Encycl. Brit. Micropædia* IX. 412/1 *Spencer carbine*, military rifle with a magazine in the buttstock that contained seven cartridges. The cartridges were fed into the chamber by means of a trigger-guard operating lever.

Spencerian, *a.* and *sb.* Add: Also † **Spencerean. A.** *adj.* **1.** (Earlier examples.)

1865 *N.Y. Social Science Rev.* I. 70 Spencerean philosophy will permeate and penetrate the world of thought. **1878** W. James in *Jrnl. Speculative Philos.* XII. 9 This explicit acknowledgement..seems, after all, to bring back unity and simplicity into the Spencerian formula.

2. [f. the name of the U.S. calligrapher Platt Rogers *Spencer* (1800–64).] **a.** Designating a system of handwriting developed by Spencer.

1863 *Amer. Jrnl. Educ.* XIII. 876 (Advt.), Spencerian system of penmanship. **1865** *21st Ann. Rep. Board of Trustees Public Schools Washington D.C.* (1866) 19 The committee recommend that the Spencerian System of Penmanship be substituted for that of Potter and Hammond. They deem the Spencerian superior. **1906** *Mod. Writing Master* Sept. & Oct. 5/1 The condemning of the old Spencerian penmanship by the educators..marks the first point at which one can detect sober and serious thought on the subject. The old Spencerian had many faults as a system.

b. Of, pertaining to, or characteristic of the cursive script developed by Spencer and used in his system.

1883 G. A. Gaskell *Penman's Handbk.* 21/2 Spencerian, by P. R. Spencer, 1848. **1892** Howard & Brown *Lessons in Rapid Writing* 8 Use an elastic steel pen with fine and well tempered point. Gillott's No. 604..and Spencerian No. 1, are as good as any pens manufactured. **1895** *Montgomery Ward Catal.* Spring & Summer 118/1 Spencerian Script Ruler, a heavy maple ruler.., copies of penmanship printed on both sides of ruler. **1911** G. S. Porter *Harvester* xix. 479 That is where Uncle Henry showed his fine Spencerian hand. **1966** H. Nielsen *After Midnight* (1967) iv. 60 On the sheet of paper one line was written in a severe Spencerian script. **1972** J. Mosedale *Football* ii. 16 A document in Spencerian script faded by the years. **1982** 'W. R. Duncan' *Queen's Messenger* xii. 153 His script was close to Spencerian at times.

B. *sb.* (Earlier example.)

1878 W. James in *Jrnl. Speculative Philos.* XII. 7 Whether any Spencerian would hail with hearty joy their advent is another matter.

Spencerite. (Earlier example.)

1871 J. McCosh *Christianity & Positivism* vii. 182 There is a..set of youths in our day who will become Comtists, or Millites, or Spencerites, or even Huxleyites.

spencerite[2] (spe·nsərəit). *Min.* [f. the name of L. J. *Spencer* (1870–1959), British mineralogist + -ite[1].] A hydrated basic zinc phosphate, $Zn_4(PO_4)_2(OH)_2.3H_2O$, found as colourless or white monoclinic crystals, freq. in stalactitic masses, in a zinc mine near Salmo, British Columbia.

1916 *Nature* 29 June 375/1 Prof. T. L. Walker: Spencerite, a new zinc phosphate from British Columbia. **1918** *Univ. of Toronto Studies (Geol. Ser.)* No. 10. 13 The spencerite forms the central part of the stalactitic growths and would appear to have been sealed up by the later shell of calamine. **1972** *Mineral. Mag.* XXXVIII. 690 Spencerite is an example of a mineral containing Zn in both octahedral and tetrahedral co-ordination.

spencite (spe·nsəit). *Min.* [f. the name of Hugh S. *Spence* (1885–1978), Canadian mineralogist + -ite[1].] A metamict borosilicate of rare-earth and other elements (chiefly yttrium and calcium), which is closely related to tritomite and occurs as brownish- or greenish-black grains and irregular masses.

1961 C. Frondel in *Canad. Mineralogist* VI. 576 Spencite is a new borate-silicate of calcium and yttrium, $(Ca,Fe)_2(Y,La)_3(B_2Si_{4-5}Al_1)_4(O,OH,F,Cl)_{26}$ ... This new mineral was collected by Hugh S. Spence in 1934 from a prospect pit in Cardiff township..and was tentatively identified as thalenite. **1968** I. Kostov *Mineral.* 319 Spencite, $Y_3(CeTh)CaBSi_3O_{13}$, is another member [of the tourmaline group] occurring as black grains. **1973** *Canad. Mineralogist* XII. 71/1 Spencite varies widely in its physical properties and chemical composition. **1976** *Mineral. Abstr.* XXVII. 246/2 Spencite (tritomite-Y) is described from nepheline syenite pegmatites in the Dugdu and Kadyros Palaeozoic alkaline massifs.

spend, *sb.*[1] Add: **1. a.** (Further examples.) Also, the amount spent.

No longer restricted to the phrases in Dict.

1976 *Computers in Higher Education & Research: Next Decade* (Dept. Educ. & Sci.) 22 About £21 million should be spent for universities on new machines, buildings and operating costs..with a spend of about £10 million for research councils for similar purposes. **1982** *ICL News* Oct. 4/1 On the hardware side customer spends are relatively high—typically around £3 million for an installation. **1983** *Observer* 16 Jan. 8/4 The battle for advertising spend.

b. *ellipt.* for 'spending money'. Freq. in *pl.* *colloq.* and *dial.* (chiefly northern).

1970 *Guardian* 9 Dec. 9/2, I can remember when Lancashire children..turned over their unopened wage packet to Mum, who gave them back very modest 'spends'. **1976** *West Lancs. Even. Gaz.* 13 Dec. 6/2 What do other OAPs get for 'spends'. **1977** P. Carter *Under Goliath* xxiii. 127 Nearly everyone I knew got their spends on Friday night so they would all be at the pictures.

† **2.** *Obs. slang.* Semen, vaginal secretion; ejaculation. Cf. *spend *v.*[1] 15 c.

1879–80 *Pearl* (1970) 13, I felt her crack deluged with a warm, creamy spend whilst my own juice spurted..in loving sympathy. *Ibid.* 217 You dissembling, bleeding, rotten..lump of shit, rubbed over with a little spend. *c* **1888–94** *My Secret Life* III. 143, I could always go on pushing after a spend in those days, my prick would not loose its stiffness for minutes afterwards. **1891** *Simple Tale of Suzan Aked* (1898) iii. 100 Then, of course, not a drop of spend can get into me, because it is all caught by the letter.

spend, *v.*[1] Add: **I. 1. b.** (Later examples with *for.*)

1971 *Publishers' Weekly* 4 Oct. 42/2 Countless people.. have longed to own the Oxford dictionary and could not afford to spend $300 for it. **1977** H. Fast *Immigrants* 6 Anna persuaded him..to spend two dollars for a heavy jacket.

II. 15. c. To ejaculate; to have an orgasm. *slang.*

1662 Pepys *Diary* 7 Sept. (1970) 191, I went up to her and played and talked with her and, God forgive me, did feel her; which I am much ashamed of, but I did no more, though I had so much a mind to it that I spent in my breeches. **1714** *Cabinet of Love* 19 For at one instant both together spent. **1763** Wilkes & Potter *Essay on Woman*, Oft when we spend we propagate unknown. **1868** *Index Expurgatorius of Martial* 1 When you say, Hedylus, 'I shall spend, finish if you mean to finish', my flame languishes. **1922** Joyce *Ulysses* 739 He made me spend the 2nd time tickling me behind with his finger. **1980** R. L. Duncan *Brimstone* vii. 163 He felt himself spending at the very moment she contracted around him.

spending, *vbl. sb.* Add: **4.** † **b.** *Obs. slang.* An orgasm; an ejaculation. Cf. *spend *v.*[1] 15 c.

1856 W. Whitman *Leaves of Grass* 242 The babes I beget upon you are to beget babes in their turn, I shall demand perfect men and women out of my love-spendings. **1879–80** *Pearl* (1970) 15 She came again in another luscious flood of spendings.

7. b. In the sense 'of or pertaining to expenditure or disbursement', as *spending -power, spree, tax.*

1930 *Economist* 13 Sept. 477/1 Economy appeared to be denounced as a 'policy of stagnation' and the dangerous line of 'increased spending-power' once more held out as a remedial policy. **1942** *Time* 7 Sept. 94/3 The Senate Finance committee..pondered a new Treasury proposal for a 'spending tax' designed to encourage war savings. **1956** A. Huxley *Adonis & Alphabet* 140 We are now squandering the capital of metallic ores and fossil fuels... How long can this spending spree go on? **1968** *Punch* 23 Oct. 563/1 At the back of my mind there is a tiny doubt about the feasibility of the Spending Tax. **1980** 'R. B. Dominic' *Attending Physician* xvi. 144 You're not going on a spending spree with all that loot, are you?

spe·ndsavour, *a.* *nonce-wd.* [f. Spend *v.*[1] + Savour *sb.*] That has lost its savour.

1879 G. M. Hopkins *Poems* (1967) 81 Are you beamblind, yet to a fault in a neighbour deft-handed? are you that liar And, cast by conscience out, spendsavour salt?

spe·ndthri:ftiness. *rare.* [f. Spendthrifty *a.* + -ness.] The act or quality of being spendthrifty.

1950 H. J. Massingham *Curious Traveller* iv. 76 For such spendthriftiness no honest word can be said. **1959** *Times* 29 June 12/6 The feeling of pleasurable spendthriftiness.

Spenglerian (ʃpeŋglɪə·riən), *a.* [f. the name of the German philosopher Oswald *Spengler* (1880–1936) + -ian.] Of, pertaining to, or characteristic of the philosophy of Spengler, esp. as expressed in his work *Der Untergang des Abendlandes* ('The Decline of the West').

1922 *Contemp. Rev.* July 52 Here the Spenglerian antithesis between Apollonian and Faustian is apparent. **1933** *PMLA* XLVIII. 608 It is hardly necessary to refer to Spenglerian pessimism. **1948** A. Toynbee *Civilization on Trial* 12 We do here seem to find a certain measure of Spenglerian uniformity. **1964** P. Worsley in I. L. Horowitz *New Sociology* 373 Spenglerian-cycle schemas. **1978** H. Wouk *War & Remembrance* xvi. 165 They sometimes played chess, and Hesse gloomed over the board in Spenglerian tones about the collapse of European man.

Also (now *rarely*) **Spe·nglerism**, the philosophy of Spengler or views in accordance with his; **Spe·nglerist** *a.* and *sb.*

1922 *Contemp. Rev.* July 50 There is no doubt in the minds of the Spenglerists that..from the ruins of Russia will arise a new culture. *Ibid.* 51 He will not have 'Spenglerism' dismissed as mere materialist determinism. *Ibid.* 52 The Spenglerist philosophy has not so far been exploited in the interests of German chauvinism. **1931** *Times Lit. Suppl.* 5 Feb. 93/2 His criticism of Pragmatism, Behaviourism, Spenglerism..are..intellectually acute.

Spenserian, *a.* (Earlier example.)

1817 Coleridge *Biog. Lit.* I. iv. 84 The Spencerian stanza, which always..recalls to the reader's mind Spencer's own style.

spent, *pa. pple.* and *ppl. a.* Add: **II. 5. b.** Also *ellipt.* as *sb.*, a spent herring.

1957 W. C. Hodgson *Herring & its Fishery* ii. 18 In this area [of the North Sea] there are large shoals of small immature herrings of low fat-content mixed with recovered spents that are just beginning to fatten. **1975** *New Yorker* 22 Dec. 55/1 April and May are the slack months, since the spring spawning has by then completed itself, and the herring are what are known as 'spents'— thin, indolent, not worth the catching.

spergula. (Earlier example.)

1829 [see *cerastium].

sperm, *sb.* Add: **I. 4. a. sperm bank** = *semen bank* s.v. *semen 2; hence **sperm banking**; **sperm count**, the number of spermatozoa in the ejaculate or in one millilitre of it; **sperm morula**, a ball of spermatozoa.

1963 B. Russell *Let.* 10 Mar. in *Autobiogr.* (1969) III. iv. 173 If a sperm-bank, such as you envisage, had existed during the régime of Hitler, Hitler would have been the sire of all babies born in his time in Germany. **1977** M. Sokolinsky tr. *Merle's Virility Factor* vi. 130 Suppose you've had a vasectomy done... The men kept the sperm..in a sperm bank, in case their wives wanted to be fertilized. **1972** *Science* 7 Apr. 32 Biologist Mark Lappé..is disturbed that commercial outfits are the first to introduce large-scale sperm banking. **1941** *Endocrinology* XXVIII. 783 For approximately 18 days after the fever, the human total sperm counts remain at a relatively normal level. **1979** *Daily Tel.* 29 Nov. 19/1 Her twin was told she was unlikely to have children because of her husband's low sperm count. **1889** *Cent. Dict.*, Sperm morula. **1921** *Nature* 7 July 586/1 The occurrence of developing sperm-morulæ in microscopic sections of.. oysters has..been already observed. **1936** *Mém. Musée Royal d'Histoire Naturelle de Belgique* (2nd Ser.) III. 1003 Since females on English [oyster] beds nearly always carry some sperm morulae amongst the eggs, self-fertilisation will nearly always be possible.

b. *sperm-producing* adj.

1927 Haldane & Huxley *Animal Biol.* ii. 52 Their walls are composed of germ-cells (sperm-producing cells).

II. 5. b. (Earlier examples.)

1775 in *Essex Inst. Hist. Coll.* (1887) XIII. 202 He heard sperm candles were 3s. **1839** C. F. Briggs *Adventures Harry Franco* II. 143 Mrs. Brown's house was brilliantly illuminated with a sperm candle in each side light.

spermacetic (spɜ̄ɹmăsī·tik), *a.* *rare*[−1]. [f. Spermacet(i + -ic.] Of or pertaining to spermaceti; *spermacetic oil* = *spermaceti oil* s.v. Spermaceti 2 a.

1922 Joyce *Ulysses* 393 They..rubbed him all over with spermacetic oil.

spermacide (spɜ̄·ɹmăsəid). *rare*[−1]. [f Sperma- + -cide.] = *spermicide.

1908 *Chem. Abstr.* II. 3128 Reports upon Dr. Bamberger's lupina powder, spermacide, and choleysin are given.

spermalege (spɜ̄·ɹmălēdʒ). *Ent.* [ad. F. *spermalège* (J. Carayon 1959, in *Rev. de Zool. et de Bot. Africaines* LX. 82), f. L. *sperma* Sperm *sb.* + *legere* to gather, collect.] In female bedbugs, an organ in which sperm are received and stored.

1964 H. E. Hinton in K. C. Highnam *Insect Reproduction* (Symp. R. Entomol. Soc. Lond. No. 2) 101 These cells constitute the mesodermal part of the organ of Ribaga (or Berlese), or, as it is now aptly called by Carayon, the mesodermal part of the spermalege. **1978** H. V. Daly *Introd. Insect Biol.* iv. 62/2 In some bedbugs a spermalege, or specialized organ of the integument, is developed where the puncture normally is made.

sperm-aster (spɜ̄·ɹm,æstəɹ). *Biol.* Also **spermaster, sperm aster.** [f. Sperm *sb.* + *aster 4.] A star-shaped configuration which in some species forms ahead of the sperm nucleus as it enters the egg in fertilization, and which develops into the amphiaster.

1904 *Biol. Bull.* VI. 226 One is the growth and division of the sperm aster, the other the growth of the nucleus. **1925** E. B. Wilson *Cell* (ed. 3) v. 400 In almost all cases the sperm-nucleus, as it advances within the egg, is typically preceded by a sperm-aster which sooner or later divides to form an amphiaster that is the forerunner of the cleavage-amphiaster. **1946** *Nature* 17 Aug. 239/1 Colchicine treatment..inhibits spindle formation as well as the appearance of the spermaster in fertilized eggs and of the monaster in parthenogenesis. **1977** *Jrnl. Embryol. & Exper. Zool.* XL. 194 It would be of the utmost interest to establish whether a causal relationship does also exist between sperm aster formation and the movement and directionality of the female pronucleus.

spermatic, *a.* **5.** (Earlier example.)

1685 J. Locke *Jrnl.* 22 June in Ld. King *Life & Lett. John Locke* (1830) I. 309, I saw, at Mr. Lewenhook's, several microscopical observations..but the best of all his glasses, and those by which he describes his spermatic animals, we did not see.

spermaticide (spɜɹmæ·tisəid). Now *rare.* [f. Spermato-: see -cide.] = *spermicide. Hence **sperma·ticidal** *a.* = *spermicidal *a.*

1923 M. C. Stopes *Contraception* v. 104 Of all the chemical substances used as spermaticides, undoubtedly quinine is in the most general use. **1931** Spermaticide, spermaticidal [see *spermicide]. **1934** J. Ellison et al. *Sex Ethics* v. 93 The spermaticidal effect of these pessaries is not very high, and many failures have resulted. *Ibid.*, The use of lactic acid jelly ointment, as a lubricant, will assist in ensuring that all parts of the vagina are covered with spermaticide.

spermatid. Add: *spec.* one that is formed by the meiotic division of a secondary spermatocyte and develops into a spermatozoon without dividing again. (Later examples.)

1924 E. W. MacBride *Study of Heredity* ii. 46 In the case of the male cell each secondary spermatocyte divides into two precisely equal cells called spermatids. **1959** [see *spermatogonium* s.v. *spermato- 1]. **1968** *New Scientist* 2 May 218/2 It now appears that the X and the Y chromosomes become genetically inert before the spermatids (the cells from which the spermatozoa develop) are formed.

Hence **sperma·tidal** *a.*

1975 *Biochem. & Biophysical Res. Communications* LXVII. 183 Proteins extracted from elongated spermatids..reveal the presence of a new spermatidal basic protein fraction.

spermatize, *v.* Restrict † *Obs.* to sense in Dict. and add: **2.** [Back-formation f. *spermatization.] *trans. Mycology.* To effect spermatization upon. *rare.*

1932 *Mycologia* XXIV. 347 Apothecia are produced only when the isolate from the crocus is used as a source of microconidia to spermatize any of the other six.

Also **spermatiza·tion** (see quot. 1932); **spe·rmatizing** *ppl. a.*

1932 F. L. Drayton in *Mycologia* XXIV. 346 The development of apothecia by *Sclerotium Gladioli*..has been induced by means of placing the microconidia of one thallus on certain structures which develop on another thallus. This process may be called 'spermatization', and it will be so designated throughout this paper. **1945** *Ibid.* XXXVII. 635 He attempted to intercross these four races by the spermatization method. **1967** M. E. Hale *Biol. Lichens* iii. 42 The spermatizing element may be any hypha that can contact the ascogonium. **1970** *Cytologia* XXXV. 427 The mechanism of spermatization in this fungus..is of a type not so far reported.

spermato-. Add: **1. spermatocyte**, a cell which gives rise to spermatids by meiosis; the *primary spermatocyte* gives rise in meiosis I to two *secondary spermatocytes*, which each give rise in meiosis II to two spermatids; (later examples); **spermatogonial** *a.* (examples); **spermatogonium**, (*a*) *Bot.* (see quot. 1957); (*b*) *Zool.*, a primordial male reproductive cell which undergoes mitosis and gives rise to primary spermatocytes; (later examples).

1920 Spermatocyte [see *oogonium 2]. **1959** [see *spermatogonium* below]. **1970** Spermatocyte [see *oocyte]. **1902** *Biol. Bull.* III. 44 In the case of the accessory chromosome, it has been found advantageous to trace its course through the spermatogonial divisions in *Brachystola*, and through the spermatocyte changes in *Hippiscus.* **1976** *Nature* 22 Jan. 209/1 After 600 rad of X rays, the mean spermatogonial mutation rate was one-third that of the 7-locus mean. **1957** Snell & Dick *Gloss. Mycol.* 144/2 *Spermatogonium*, the proper form for what is called 'spermogonium'.., but seldom used. **1959** W. Andrew *Textbk. Compar. Histol.* xii. 490 [In fishes] spermatogenesis consists of the same general cytological stages as in the invertebrates: namely, spermatogonia, spermatocytes, spermatids, and spermatozoa. **1968** Passmore & Robson *Compan. Med. Stud.* I. xxxvii. 27/2 Spermatogonia show little sign of activity until puberty when they begin to multiply rapidly. **1970** [see *oogonium]. **1974** *Sci. Amer.* Sept. 56/3 In the course of spermatogenesis the two important objectives are reduction of the chromosome number from the diploid number (46) of the spermatogonium to the haploid number (23) of the spermatozoon and the preparation of the spermatozoon for its role in fertilization.

2. spermatoci·dal *a.* = *spermicidal *a.*; **spe·rmatocide** = *spermicide; **spermatoge·nically** *adv.*, as regards the production of sperm; **spermatogenesis**, the formation and development of spermatozoa (cf. *spermiogenesis); (later examples); **spermatology**, † (*a*) a treatise on sperm; (*b*) the scientific study of sperm; (examples).

1928 *Funk's Stand. Dict.* 2337/3 *Spermatocidal*, checking or killing the motor-power of spermatozoa. **1937** Spermatocidal [see *chinosol]. **1949** *New Gould Med. Dict.*, Spermatocide. **1980** *Nature* 10 Apr. 548/1 Spermatogenically active testes. **1978** P. J. Hogarth *Biol. Reproduction* ii. 11 The final phase of spermatogenesis in the testes is known as spermiogenesis.., and consists of the progressive conversion, without division, of the spermatids into functional spermatozoa. **1980** Spermatogenesis [see *spermiogenesis]. **1833** Dunglison *Dict. Med. Sci.* II. 312/2 *Spermatology*, a treatise on sperm. **1970** B. A. Afzelius in B. Baccetti *Compar. Spermatol.* 568 In spite of the great efforts to solve the problems in comparative spermatology we find ourselves confronted with more questions than answers. **1976** T. Sharpe *Wilt* iv. 32 He got the Nobel prize for spermatology.

spermatophore. Add: Hence **spermatopho·ric** *a.*

1959 W. Andrew *Textbk. Compar. Histol.* xii. 475 [In the common squid] peculiarly complicated organs, the spermatophoric organs and sacs, receive the sperm from the vas deferens, pack them into bundles, and cover them with a tunic..before their ejection as spermatophores by the penis. **1977** M. J. & J. Wells in Giese & Pearse *Reproduction Marine Invertebrates* IV. vi. 297 In..the wider initial section of the first spermatophoric gland, the sperm are embedded in a mucin secretion.

spermatozoan, *a.* (Example.)

1954 G. I. M. Swyer *Reproduction & Sex* ii. 24 The spermatozoan content of the ejaculate.

spermicide (spɜ̄·ɹmisəid). [f. Spermi- + -cide.] Any substance which kills spermatozoa, esp. one used as a contraceptive. So **spermici·dal** *a.* Cf. *spermacide, *spermaticide, and *spermatocide* s.v. *spermato-.

1929 *Jrnl. Hygiene* XXIX. 323 The spermicidal powers of the various chemical contraceptives sold to the public have never previously been compared, and no one has been in a position to say that one spermicide is preferable to another. **1931** *Ibid.* XXXI. 190 The words 'spermicide' and 'spermicidal' are preferred, on grounds of euphony, to 'spermaticide' and 'spermaticidal'. **1936** *Discovery* Sept. 277/2 Baker has shown that soap solution is more spermicidal than most chemical contraceptives. **1962** A. Huxley *Island* vi. 84 Before the age of rubber and spermicides. **1977** V. Coleman *Paper Doctors* ix. 93 There were 15 different types of spermicidal contraceptive waiting to be prescribed. **1977** E. J. Trimmer et al. *Visual Dict. Sex* (1978) xiv. 135/3 Spermicides..should not be used alone as a contraceptive method.

spermidine (spɜ̄·ɹmidīn). *Biochem.* [f. Sperm(in + *-idine.] A colourless liquid triamine, $H_2N(CH_2)_3NH(CH_2)_4NH_2$, which is found very widely in living tissues, in company with spermine, and has a number of metabolic functions.

1927 H. W. Dudley et al. in *Biochem. Jrnl.* XXI. 97 This base has been given the name spermidine since it not only occurs in association with spermine but also has been found to be structurally related to the latter substance. **1964** A. White et al. *Princ. Biochem.* (ed. 3) xxvi. 538 Spermine and spermidine are present in significant amounts in ribosomes and appear to be essential to their structure and function. **1974** *Nature* 17 May 250/1 The polyamines spermine spermidine and their precursor putrescine occur ubiquitously and in high concentrations in animals and plants.

spermin. Add: The spelling *spermine* (-īn) is now usual. (Later examples.)

1927 [see *spermidine]. **1937** *Thorpe's Dict. Appl. Chem.* (ed. 4) I. 315/1 The polyamino-compounds are of little interest with the exception of spermine, a tetra-amino derivative of the constitution $NH_2 \cdot [CH_2]_3 \cdot NH \cdot [CH_2]_4 \cdot NH \cdot [CH_2]_3 \cdot NH_2$, first isolated from human semen by Schreiner, and occurring in many mammalian organs. **1964, 1974** [see *spermidine].

spermiogenesis (spɜ̄ɹmi,odʒe·nėsis). *Biol.* [f. Spermi- + -o + -genesis.] The development of spermatozoa; *spec.* the maturation of spermatids into spermatozoa (the last phase of spermatogenesis).

1920 *Biol. Bull.* XXXIX. 333 The general topography of spermiogenesis has been worked out on *Murgantia.* **1956** *Nature* 25 Feb. 387/2 A study of spermiogenesis in the bandicoot..has shown that an acrosome is indeed present. **1980** D. J. Begley et al. *Human Reproduction* iii. 30/2 Following spermatogenesis and spermiogenesis, which take about 64 days in man, sperm pass through the rete into the epididymis.

spe·rmy, *a.* For *rare*[−1] read *rare* and add later examples. Also, full of sperm. (In quot. 1851, with reference to spermaceti.)

1851 H. Melville *Moby Dick* II. xxxviii. 253 If you unload the skull of its spermy heaps..you will be struck by its resemblance to the human skull. **1973** M. Amis *Rachel Papers* 31, I decided I would ring her when I got back. It would be intelligent to do it while I still felt tolerably spermy and Joycean after my night with Gloria.

speromagnetic (spiᵊ:rōmægne·tik), *a.* *Physics.* [f. Gr. διασπείρειν to scatter + *antiferr)omagnetic *a.*] Applied to an amorphous magnetic material in which the individual electron spins are aligned more or less anti-parallel to their closest neighbours but overall there is a statistical distribution of orientations with no preferred direction. Also **speroma·gnetism.**

The form *sperimagnetic* occurred throughout the 1973 coinage paper. This was corrected by Coey et al. in *Physical Rev. Lett.* (1976) XXXVI. 1062.
1973 *Nature* 21–28 Dec. 445/2 The new phenomenon of speromagnetism (or amorphous superparamagnetism) adds a further delectable layer of complexity to what is already a highly elaborate network of phenomena. **1973** Coey & Readman in *Ibid.* 477/2 Such a configuration, with the spins fixed relative to each other but possessing no overall preferred direction, is illustrated in Fig. 2c. It is termed 'sperimagnetic' [*sic*]..and..is probably a fairly common phenomenon in amorphous compounds. **1976** *Physical Rev. Lett.* XXXVI. 1062/1 Speromagnetism may also occur where the anisotropy dominates the exchange, in pure amorphous rare-earth metals, for example. **1976** *Physics Bull.* July 294/3 Speromagnetic ordering may also result when the exchange is distributed about zero with a range of positive and negative values.

sperrylite (spe·riləit). *Min.* [f. the name of Francis L. *Sperry* (*c* 1862–1906), U.S. chemist + -ITE[1].] Platinum arsenide, $PtAs_2$, found as opaque greyish-white isometric crystals.
1889 H. L. Wells in *Amer. Jrnl. Sci.* CXXXVII. 67 (*heading*) Sperrylite, a new mineral. **1902** H. A. Miers *Mineral.* II. 332 Sperrylite, the only natural compound of platinum, occurs as brilliant microscopic crystals..in a nickeliferous ore consisting of various sulphides, at Sudbury in Ontario. **1926** *Mineral Mag.* XXI. 94 A very fine and unusual crystal of sperrylite..came from a new adit on the Tweefontein farm..about 10 miles NNW. of Potgietersrust, Waterberg district, Transvaal. **1975** *Canad. Mineralogist* XIII. 327/2 Sperrylite is a common mineral in many platinum-bearing deposits.

‖ **spes** (spē[i]z). *Law.* [L.] A hope or expectancy, esp. of some future benefit. Also in various Latin phrases (see quots. 1959).
1815 *Decisions First & Second Divisions Court of Session* (Faculty of Advocates) 28 Feb. 258 There is a *jus crediti* conferred on the heir of a marriage-contract. It is not a mere *spes successionis*, but a *jus crediti*. **1945** *Law Rep.* 10 May 363 *Sine spe recuperandi*, in the sense of abandoning all hope of his owner recovering the ship. **1952** *Ibid.* 10 July 486 The taxpayer's answer to that is, first of all, that this is not property at all but a mere *spes* or hope of getting something. **1959** Jowitt *Dict. Eng. Law* II. 1664/2 *Spes recuperandi*, the hope of recovery. If it is entertained by a person in danger of death, it makes a declaration by him inadmissible in evidence... *Ibid.*, *Spes successionis*, an expectation of succession, as distinct from a vested right. **1977** Johns & Greenfield *Dymond's Capital Transfer Tax* ii. 40 The fiction of *Re Scott* does not go so far as to require one to suppose that the beneficiary had more than an interest in expectancy ..or indeed a mere *spes*, when he died.

speshul (spe·ʃəl), repr. supposed colloq. pronunc. of SPECIAL *a.* and *sb.*
1900 *Times* 7 July 10/1 Boys..trample continually on your toes and screech everlastingly into your ears 'Cigarettes, cigars, chocolates'..'Speshul 'dition—latest cricket scores.' **1979** *Jrnl. Lancs. Dial. Soc.* Jan. 5 His first *buzz* was a *football speshul* to Maine Road.

spessartine. Now usu. with pronunc. (-īn). Substitute for def.: Garnet containing manganese in place of calcium; manganese garnet. (Later examples.)
1910 *Encycl. Brit.* XI. 471/1 Spessartite, or spessartine, ..is a fine aurora-red garnet, cut for jewelry when sufficiently clear. **1977** A. Hallam *Planet Earth* 128 An individual [garnet] can be regarded as an intimate mixture of two or more of the following end-members: pyrope (magnesium-aluminum garnet), almandine (iron-aluminum garnet), spessartine (manganese-aluminum garnet), grossularite (calcium-aluminum garnet), andradite (calcium-iron garnet), and uvarovite (calcium-chromium garnet).

spessartite (spe·sǎːtəit). [f. as SPESSARTINE: see -ITE[1].] **1.** *Min.* = SPESSARTINE in Dict. and Suppl. Chiefly *N. Amer.*
1868 J. D. Dana *Syst. Mineral.* (ed. 5) 268 Manganese-Aluminagarnet: Spessartite. **1910** [see *SPESSARTINE]. **1934** G. L. English *Getting acquainted with Minerals* II. 228 Some of the fine garnets of Delaware County, Pennsylvania are spessartite. **1953** F. H. Pough *Field Guide to Rocks & Minerals* II. 296 Spessartite is less common and not often properly identified when it is in a schist. **1971** R. Purvis *Treasure Hunting in Brit. Columbia* i. 21/2 Spessartite is pink to dark brown.
2. *Petrogr.* [ad. G. *spessartit* (H. Rosenbusch *Mikrosk. Physiogr.* (ed. 3) (1896) II. 529).] A porphyritic lamprophyre in which the feldspar is sodic plagioclase and the phenocrysts consist of an amphibole or pyroxene, usu. green hornblende.
1908 P. MacNair *Geol. & Scenery of Grampians* II. x. 64 Dykes and sills of lamprophyre, including kersantites, vogesites, and spessartites, are..to be met with throughout the Highlands. **1930** Peach & Horne *Chapters Geol. Scotland* iv. 108 The dark basic sills in the dolomites of the Assynt region..include representations of vogesites and spessartites. **1966** [see *lamprophyre* s.v. *LAMPRO-]. **1970** *Jrnl. Geol.* LXXVIII. 742/1 The dikelets probably formed during later stages of crystallization of the spessartite.

spew, *sb.* Add: **1. c.** Surplus material exuded between the halves of a mould during the manufacture of plastic objects. Freq. *attrib.*
1933 *Industr. & Engin. Chem.* June 647/1 The degree of flow can be readily controlled by the location, size and placement of the spew hole. **1945** A. T. Birkby *Phenolic Plastics* iv. 43 Provision should be made for air venting, and, in the case of compression moulds, for spew-ways. *Ibid.* viii. 93 'Flash' or 'spew' soon begins to appear round each press, and unless swept up and removed becomes a nuisance. **1964** Wordingham & Reboul *Dict. Plastics* 165 *Spew groove*, in moulding operations, the groove in a mould which permits the escape of surplus material.
4. Special Comb.: **spew frost** = *needle ice* s.v. *NEEDLE *sb.* 14.
1938 C. F. S. Sharpe *Landslides & Related Phenomena* iii. 27 Growths of frost crystal of this sort are known as spew frost.., feather-ice, or needle-ice..and on the European continent as Pipkrake or Kammeis. **1939** [see *needle ice* s.v. *NEEDLE *sb.* 14].

sphacelia. Add: Hence **sphace·lial** *a.*
1909 B. M. Duggar *Fungous Dis. Plants* xi. 245 The surface mycelial areas are thrown into folds and numerous short conidiophores arise, bearing small ovate conidia. This is known as the sphacelial stage. **1938** G. M. Smith *Cryptogamic Bot.* I. xii. 453 This is followed by a progressively upward metamorphosis into compact tissue until the whole mycelium has been changed into a sclerotium that is capped with remnants of the sphacelial tissue. **1976** G. C. Ainsworth *Introd. Hist. Mycol.* vii. 187 The major contribution of L. R. Tulasne in his classic paper in 1853..was to demonstrate that the sphacelial phase, the sclerotia, and the ascocarps were all stages of one fungus, *Claviceps purpurea*.

sphaerite (sfiə·rəit). *Min.* [ad. G. *sphärit* (V. von Zepharovich 1867, in *Sitzungsber. der K. Akad. der Wissensch.* (*Math.-Nat. Classe*) LVI. 24), f. *sphärisch* spherical: see -ITE[1].] A hydrous aluminium phosphate now identified with variscite.
1886 J. D. Dana *Syst. Mineral.* (ed. 5) 587 Sphærite... In globular concretions with a drusy faceted surface, without a distinct fibrous or concentric structure. **1921** *Bull. U.S. Geol. Survey* No. 679. 135 Sphaerite... Spherical masses of white fibers. **1950** *Amer. Mineralogist* XXXV. 1059 It appears that sphaerite is wholly identical with variscite.

sphæro-. Add: **sphærocobaltite** (earlier example); cf. *spherocobaltite* s.v. *SPHERO-; **sphæ·rocone** *Palæont.*, an ammonoid with a very involute shell in which the outer whorl conceals the inner one and the whole has a globular form; **sphærosome**, var. *SPHEROSOME; **sphærolite**, obs. var. SPHERULITE in Dict. and Suppl.
1877 *Mineral. Mag.* I. 267 Sphœrocobaltite [*sic*].. occurs in spheroidal forms with roselite at Schneeberg, Saxony. **1923** Sphærocone [see *SERPENTICONE]. **1970** R. M. Black *Elements Palaeont.* viii. 89 Sphaerocones occur repeatedly during ammonoid history.

sphagnum. For 'mod.L.' read 'mod.L. (J. J. Dillenius *Hist. Muscorum* (1741) 240)' and add: **1.** (Earlier example.)
1741 J. J. Dillenius *Hist. Muscorum* 240 The larger soft and hollow-leaf'd Bog Sphagnum.
3. Special Comb.: **sphagnum bog, swamp,** a bog in which the plant-life consists chiefly of mosses of the genus *Sphagnum*.
1911 C. B. Crampton *Vegetation Caithness* iv. 51 In recent times..*Sphagnum* bogs have been reduced to their present small proportion in the moorland associations. **1951** V. Nabokov *Conclusive Evidence* iv. 47 Beyond the lower course of the river..the vast expanse of a misty-blue sphagnum bog. **1972** Dean & Smith *Wisconsin* 78/2 Sphagnum bogs are one stage in the succession from open-water lake to conifer swamp. **1890** Sphagnum swamp [in Dict., sense 1]. **1941** J. Buchan *Sick Heart River* II. 124 A bottomless half-frozen sphagnum swamp ..heaved under his tread.

sphairistike (sfē·ri-stikī). *Obs.* exc. *Hist.* [ad. Gr. σφαιριστική (τέχνη) (skill) in playing at ball, f. σφαιριστικός: see SPHAIRISTIC *a.* Cf. STICKÉ.] A type of tennis first played in 1873 which was later developed into and renamed lawn-tennis.
1874 G. D. Fitzgerald in *Field* 21 Mar. 270/2 Sir, I have lately seen a new game played which will be a great acquisition as an out-of-door amusement at country houses. The game is called Sphairistike, or Lawn Tennis. **1874** F. Kilvert *Diary* 27 July (1969) III. 55 We began to play 'sphairistike' or lawn tennis. **1927** *Times* 10 June 10/4 Badminton quickly got a certain popularity in England, and from it Major Wingfield invented sphairistike. The net of sphairistike was practically the badminton net, 5 ft. high at the sides, with the same, if a broader, red tape along the top. *Ibid.*, The name 'sphairistike', however, was impossible (if only because people would pronounce it as a word of three syllables to rhyme with 'pike'), and it was soon rechristened. **1965** *Punch* 16 June 895/2 The British invented lawn tennis, though the original name for it, 'sphairistike', didn't catch on. *Ibid.* 896/3 Arthur Balfour..mildly suggested that a better name for 'sphairistike' would be 'lawn tennis'! **1982** 'J. Gash' *Firefly Gadroon* v. 62 My eyes lit upon a genuine old Sphairistike racquet.

sphalerite. (Examples.)
1871 *Jrnl. Chem. Soc.* XXIV. 312 (*heading*) On the occurrence of thallium in sphalerite from Geroldseck (Breisgau). **1925** A. S. Alexander *Tramps across Watersheds* v. 235 Some of these minerals and ores are:—gold, silver bloom, galena or lead ore, blende or sphalerite or zinc ore..and a great variety of quartz. **1952** [see *ROBINSONITE]. **1969** Bennison & Wright *Geol. Hist. Brit. Isles* x. 247 The later, mesothermal, lodes..carry such minerals as galena and sphalerite with a little uranium ore as well. **1973** *Times* 1 Dec. 11/3 'This piece', he said, indicating a toffee-coloured chunk of sphalerite from Picos de Europa, Santander, Spain, 'was brought up from the mine for me by a local dealer'.

sphecid (sfe·sid), *sb.* and *a.* *Ent.* [f. mod.L. family name *Sphecidæ*, f. *Sphex* SPHEX (Linnæus *Systema Naturæ* (ed. 10, 1758) I. 569): see -ID[3].]
A. *sb.* A fossorial wasp of the family Sphecidæ. **B.** *adj.* Of or pertaining to an insect of this kind.
1895 J. H. & A. B. Comstock *Man. Study of Insects* xxii. 650 The Sphecids or the Thread-waisted Wasps.. are the most commonly observed of all our digger-wasps as certain species build their mud nests in the attics of our houses. **1926** E. O. Essig *Insects Western N. Amer.* xxviii. 871 Sphecid or Thread-Waisted Wasps..are solitary, nest in the soil, and provision the nests with spiders or insects. **1966** C. Sweeney *Scurrying Bush* v. 75, I saw a handsome, inch long, metallic blue-black wasp fly out from behind the row of books. It was a sphecid or mud dauber. **1976** R. M. Bohart et al. (*title*) Sphecid wasps of the world. *Ibid.* 1 Adult sphecids feed on a variety of food.

spheno-. Add: **1. spheno-mandi·bular.**
1893 H. Morris *Human Anat.* ii. 190 The sphenomandibular ligament (long internal lateral)..is a thin, loose band, situated some little distance from the joint. **1967** G. M. Wyburn et al. *Conc. Anat.* iv. 112/2 The sphenomandibular ligament runs from the spine of the sphenoid to the lingula at the mandibular foramen.

sphenochasm. (sfī·nokæz'm). *Geol.* [f. SPHENO- + Gr. χάσμ-α CHASM.] (See quot. 1958.)
1958 S. W. Carey in *Continental Drift* (Geol. Dept., Univ. of Tasmania, Hobart) 193 A Sphenochasm..is the triangular gap of oceanic crust separating two cratonic blocks with fault margins converging to a point, and interpreted as having originated by the rotation of one of the blocks with respect to the other. **1965** A. Holmes *Princ. Physical Geol.* (ed. 2) xxix. 1086 The Gulf [of California] sphenochasm appears to have been filled with rheid mantle material in much the same way as the Red Sea. **1971** *Nature* 2 July 21/2 As the Gulf of Honduras sphenochasm completed its opening, the Yucatan and Nicaraguan cratons assumed their modern positions relative to North America.

sphere, *sb.* Add: **I. 7. d.** Read 'orig. esp. in Africa or Asia' and add later examples.
1950 L. Fischer in R. H. S. Crossman *God that Failed* 223 It provided for a spheres-of-influence division of the areas accessible to Soviet-Nazi aggression. **1973** A. Broinowski *Take One Ambassador* iv. 43 The Japanese themselves are told they can't resort to force, even in what they see as their own sphere of influence. **1981** *Times* 21 Feb. 13/5 A programme of reform [in Poland] sufficiently limited to reassure the Russians that their sphere of influence is safe.
II. 8. a. Esp. in *Math.*, the set of all points at a specified distance from a specified point. (Later examples.)
1934 C. C. Krieger tr. *Sierpinski's Introd. Gen Topology* vi. 77 The set $K(p,r)$ (where $p \in M$, and $r > 0$) is called an open sphere of centre *p* and radius *r*. **1959** E. M. Patterson *Topology* i. 3 Since all spheres are homeomorphic, we speak of *the* sphere, rather than *a* sphere. **1968** E. T. Copson *Metric Spaces* iii. 32 If we impose on the set..of all ordered pairs of real numbers the metric $\rho(\mathbf{x},\mathbf{y}) = \max\{|x_1 - y_1|, |x_2 - y_2|\}$ the spheres are squares.
12. sphere gap *Electr.*, a form of spark gap with two spherical electrodes, used esp. in devices for measuring high voltages.
1913 *Trans. Amer. Inst. Electr. Engineers* XXXII. 739 The sphere gap has been suggested as a standard instrument to be used in the measurement of high voltage. **1962** *Newnes Conc. Encycl. Electr. Engin.* 767/2 The measurement and recording of testing voltages requires either a voltage divider..or a sphere gap..capable of measuring the peak voltage.

-sphere, *suffix.* [f. the sb., after *atmosphere.*] Used in names of more or less spherical structures or regions forming part of or associated with the earth (or any celestial object), as *BARYSPHERE, *BIOSPHERE, *IONOSPHERE, *MAGNETOSPHERE, etc.

spherical, *a.* and *sb.* Add: **A.** *adj.* **4.** *spherical wave*: a wave in which the wave fronts are concentric spheres.
1907 *Chem. Abstr.* I. 1470 The spherical wave of explosion is propagated in a fluid medium according to formulas analogous to those derived for plane waves. **1976** D. Ross *Mechanics of Underwater Noise* ii. 35 The intensity of a spherical wave is proportional to the square of the pressure.
5. Forming parasynthetic adjs., as *spherical-bodied, -roofed, -surfaced.*
1804 Spherical-bodied [in Dict., sense 1 α *Comb.*]. **1946** *Nature* 26 Oct. 26 583/2 The aberrations of a spherical mirror are corrected by a single spherical-surfaced meniscus lens. **1977** *N.Y. Rev. Bks.* 13 Oct. 16/1 The spherical-roofed auditorium at CIA headquarters in Langley.

spherics, occas. var. *SFERICS *sb. pl.*

spherify, *v.* For *rare*[−1] read *rare* and add: Also, to turn *into* a spherical body. Also **spherifica·tion.**
1848 Poe *Eureka* 74 Several fragments..were..spherified into a moon. *Ibid.* 75 Three moons..having been formed..by the rupture and general spherification of as many distinct ununiform rings.

spheristerion. For *rare*[−1] read *rare* and add earlier example in literal sense (in the Latinized form **spheristerium**).
1764 R. Adam *Ruins Palace Spalatro* 11 On the other side of the Cella Media was a Spheristerium.., a room alotted for the different exercises of the ball.

sphero-. Add: **spheroco·baltite** *Min.* = *sphærocobaltite* s.v. SPHÆRO-; **sphero-cyli·ndrical** *a.*, (of a lens) having a spherical and a cylindrical surface; **sphe·roplast** *Biol.* [-PLAST], a bacterium or plant cell bound by its plasma membrane, the cell wall being deficient or lacking and the whole having a spherical form; hence **sphe·roplasting** *vbl. sb.*, treatment (as with an enzyme) that converts cells to spheroplasts; **sphe·roplasted** *a.*; **spherosymme·trical** *a.*, spherically symmetrical.
1889 *Cent. Dict.*, Spherocobaltite. **1924** *Amer. Mineralogist* IX. 61 Spherocobaltite. **1975** *Soviet Physics—Crystallogr.* XX. 74/1 The reaction involved in the synthesis of $CoCO_3$ has been studied in connection with the problem of growing single crystals of spherocobaltite. **1881** Sphero-cylindrical [see *ASTIGMATIC *a.* 2]. **1962** L. S. Sasieni *Optical Dispensing* x. 264 A toric lens is a curved sphero-cylindrical lens. **1958** C. Hurwitz et al. *Jrnl. Bacteriol.* LXXVI. 612/2 Lederberg (1956) has referred to these forms as protoplasts. In view of current uncertainty as to the fate of the cell wall in this case, we shall use 'spheroplasts' (McQuillen, 1958, personal communication) as a neutral term. *Ibid.* 613/2 Spheroplasts are osmotically fragile in a hypotonic environment. **1976** *Ann. Rev. Microbiol.* XXX. 42 Conversion of these cells to spheroplasts results in the release of periplasmic proteins into the supernatant. **1973** *Jrnl. Bacteriol.* CXVI. 491/1 Spheroplasted cells were found to lose endonuclease I from periplasm to surrounding medium. **1965** *Acta Path. Microbiol. Scand.* LXIII. 412 The mechanism of the ..spheroplasting effect of the incubation temperature is not obvious. **1973** *Jrnl. Bacteriol.* CXVI. 493/2 Spheroplasting does not affect gamma-ray induced DNA degradation in *E. coli* CRthy⁻. **1964** *Progress Reaction Kinetics* II. 299 The ions SO_4^{2-} and F^-..both can be considered as fairly spherosymmetrical.

spherocyte (sfe·rosəit, sfiə·ro-). *Med.* [f. SPHERO- + -CYTE.] A red blood cell which is biconvex instead of biconcave.
1908 Christophers & Bentley in *Sci. Mem. Officers Med. & Sanitary Dept. Govt. India* XXXV. 77 As such cells..are a sign of most important pathological changes, we have thought it desirable to have a name to designate the condition and have termed them spherocytes. **1947** *Amer. Jrnl. Med. Sci.* CCXIV. 255/1 The increased hemolysis seems due to the fact that spherocytes and ovalocytes are removed by the spleen more readily than are normal biconcave disks. **1977** *Lancet* 12 Nov. 1025/2 The blood smear showed many fragmented red blood-cells and some spherocytes.
Hence **spherocy·tic** *a.*, of, pertaining to, or characterized by the presence of spherocytes; **spherocyto·sis**, any spherocytic condition.
1933 F. W. Price et al. *Textbk. Pract. Med.* (ed. 4) xi. 775 Acholuric jaundice. Synonyms.—Spherocytosis; congenital hæmolytic anæmia. **1937** E. B. Krumbhaar in R. L. Cecil *Textbk. Med.* (ed. 4) 1266 Hemolytic Jaundice. (Congenital, Familial, Spherocytic Anemia, [etc.]). **1947** *Amer. Jrnl. Med. Sci.* CCXIV. 255/1, 2 families which have a characteristic history and the clinical findings of hereditary spherocytic hemolytic icterus but show no spherocytosis. **1970** Passmore & Robson *Compan. Med. Stud.* II. xxxi. 3/2 The specific abnormality underlying spherocytic haemolytic anaemia is unknown. **1977** *Lancet* 6 Aug. 305/1 Repeated phagocytosis in lymphoreticular tissues is a feature of many hæmatological disorders, of which hereditary spherocytosis is an example *par excellence*.

spheroidal, *a.* Add: **2. c.** *Metallurgy.* = *NODULAR *a.* 4.
1920 *Jrnl. Iron & Steel Inst.* CII. 261 The present experiments were undertaken to clear up the cause of the formation of the spheroidal cementite. **1957** *Technology* July 172/3 To-day, nodular or spheroidal graphite cast iron is an article of commercial manufacture. **1977** *Metals Abstr.* X. 1801/1 Tests were made on boronizing spheroidal graphite cast iron (SGCI) to improve surface hardness and wear resistance.

spheroidize (sfiə·roidəiz), *v.* *Metallurgy.* [f. SPHEROID *sb.* + -IZE.] **a.** *intr.* Of grains, esp. of graphite in cast iron or steel: to undergo conversion into spheroids.
1912 in A. Sauveur *Metallogr. Iron & Steel* App. II. 6 On long heating the pro-eutectoid and pearlitic cementite spheroidize slowly, and neighboring particles merge. **1936** G. F. C. Gordon *Elem. Metallurgy* (ed. 2) ix. 125 The effect of annealing hyper-eutectoid steel, if done with great care, is to cause the network of free cementite to 'spheroidise'. **1939** E. C. Rollason *Metallurgy for Engineers* viii. 112 Prolonged annealing induces greater ductility at the expense of strength, owing to the tendency of the cementite in the strained pearlite to 'ball-up' or spheroidize.
b. *trans.* To convert (grains) into spheroids.
1918 *Jrnl. Iron & Steel Inst.* XCVII. 433 The cementite is completely spherodised [*sic*] into round-shaped globules. **1958** *Times Rev. Industry* July 48/3 The high carbon chrome tubes..have to be 'spheroidized' to get them into a good machinable condition. **1978** *Nature* 20 July 237/1 One hypothesis is that the BeO particles were spheroidised (grain refining agents should be angular and sharp-cornered) either during the hot isostatic pressing or during the time when molten.
Hence **sphe·roidized** *ppl. a.*, **sphe·roidizing** *vbl. sb.*; **spheroidiza·tion**, the process of converting to spheroids.
1912 A. Sauveur *Metallogr. Iron & Steel* xii. 15 It may well be asked whether..the spheroidizing of pearlite is due to excessively slow cooling. *Ibid.*, Spheroidized pearlite is softer, less tenacious, and more ductile than lamellar pearlite. **1920** *Jrnl. Iron & Steel Inst.* CII. 267 The temperature interval of spheroidisation in low carbon steels is very small, extending only to about 20°. **1923** Glazebrook *Dict. Appl. Physics* V. 341/1 Pearlite which has undergone spheroidising is known as granular pearlite. **1939** E. C. Rollason *Metall. Engineers* viii. 118 During subsequent hardening operations the time required to dissolve fine spheroidised cementite is less than for the lamellar type. **1944** Gregory & Simons *Heat-Treatment of Steel* xv. 181 The reheating temperature is below the lower critical or transformation point, as in the spheroidization of the carbide particles in high carbon tool steels. **1958** *Times Rev. Industry* July 48/3 To soften them they are heat treated in spheroidizing furnaces. **1967** A. H. Cottrell *Introd. Metallurgy* xx. 384 Above about 500°C the cementite particles grow competitively ..into larger rounded particles dispersed through the B.C.C. iron matrix, giving a spheroidized structure (Sorbite). **1977** *Industr. & Engin. Chem. Process Design & Development* XVI. 108/1 The process involves the melting and spheroidization of 44–125-μ magnetite grit.

spherosome (sfiə·rosōᵘm). *Biol.* Also **sphæro-.** [ad. F. *sphérome* (P.-A. Dangeard 1919, in *Compt. Rend.* CLXIX. 1008): see SPHERO- and *-SOME[4].] A cytoplasmic liquid droplet or cell organelle found in plant tissues, often associated with hydrolytic enzymic activity; the plant structure answering to the lysosome in animal tissues. Hence **spheroso·mal** *a.*
1954 *Biol. Abstr.* XXVIII. 2398/1 The sphaerosomes are real and permanent cytoplasmic structures, and..are neither artifacts nor temporary storage droplets of fat or other lipoids. **1958** *Exper. Cell Res.* XV. 611 The spherosomes are seen as opaque white globules 0·7 μ in diameter, or as dark-rimmed spheres. **1966** *Protoplasma* LXII. 220 In the cells studied..spherosomal motion is rapid during the first 24 hr. **1976** Bell & Coombe tr. *Strasburger's Textbk. Bot.* (rev. ed.) 16 The nucleus, plastids, mitochondria, spherosomes and golgi bodies remain throughout within the cytoplasm.

spherule. Add: **3.** Special Comb. **spherule cell** *Ent.*, a kind of cell in the hæmolymph of certain insects (see quot. 1969).
1935 R. E. Snodgrass *Princ. Insect Morphol.* xiv. 394 The spherule cells of caterpillars differ in many ways from those of Coleoptera, but they appear to be cells of the same type. **1969** R. F. Chapman *Insects* xxxiii. 677 Spherule cells, found in Lepidoptera and Diptera, are round or oval cells with large, non-refringent, usually acidophilic inclusions filling the whole cell. **1974** Spherule cell [see *œnocytoid* s.v. *ŒNO-].

spherulite. Add: Formerly also **sphærolite, spherolite.** [ad. G. *sphærolit* (now *sphärolit*) (A. G. Werner, *a* 1816: see W. G. W. Becker *Journal einer Bergmännischen Reise* (1816) II. p. vi).]
1. a. Now *Obs.*
b. Substitute for def.: *Geol.* A small spheroidal mass found in rock; *spec.* one consisting of many crystals which have grown radially from a point. (Earlier and later examples.)
1829 *Trans. Geol. Soc.* II. 202 A previous separation had taken place of the feldspathose spherolites. **1844** C. Darwin *Geol. Observations on Volcanic Islands* iii. 58 The sphærulites are either white and translucent, or brown and opaque. **1856** *Q. Jrnl. Geol. Soc.* XII. 340 The multiplication and confusion of these crystallites or sphærulites ultimately destroy the glassy character of the substance altogether. **1868** *Mem. California Acad. Sci.* I. 52 Small globular grains, from the size of a pin-head to that of a rifle bullet, called 'sphærolites' by Beudant. **1909** J. P. Iddings *Igneous Rocks* I. vi. 230 Such spherulites..consist of radiating prisms of alkali feldspar with submicroscopic graphic intergrowths of quartz. **1931** [see *ORBICULE]. **1939** W. H. Twenhofel *Princ. Sedimentation* xv. 570 Spherulites with radiate structure are evidently a form of crystal aggregate, whereas those with amorphous internal structure are either excremental particles or colloidal aggregates as glauconite, greenalite, or chamoisite. **1974** A. C. Tennissen *Nature of Earth Materials* v. 280 Spherulites have a radial type of internal structure.
c. A small spheroidal mass of crystals or fibrils of any substance, esp. a polymer.
1893 *Mineral. Mag.* X. 97 (*table*) $LiNaSO_4$... Spherulitic... Some of the spherulites have a radial structure. **1911** *Chem. Abstr.* V. 1052 The author describes the peculiar structure of spherulites obtained in cholesterol and β-naphthalene benzoate. **1933** *Bureau of Standards Jrnl. Res.* (U.S.) X. 488 The microscopical examination of the crystals of rubber hydrocarbon indicates that they are spherulites composed of many fine needles. **1963** *New Scientist* 7 Nov. 334/1 Some polymers, like nylon and polystyrene are highly crystalline. They do not normally form single crystals, however, but produce so-called spherulites which are clusters of plate-like or needle-like individual crystals. **1975** *Sci. Amer.* Dec. 99/1 The spherulite, or 'sunburst' microstructure normally results from crystallization under quiescent conditions.

spherulitic, *a.* Add: Also † **spherolitic. 1.** Also of other substances. Cf. *SPHERULITE 1 c. (Later examples.)
1893 [see *SPHERULITE 1 c]. **1975** *Sci. Amer.* Dec. 100/1 The microstructure of spherulitic polymers is also significantly affected by the phenomenon of secondary crystallization.
2. (Earlier and later examples.)
1829 *Trans. Geol. Soc.* II. 202 Even where no traces of spherolitic structure are visible, the ribboned arrangement of the rock is owing to the mass of matter having been drawn out in the direction of the zones while still liquid. **1945** *Trans. Faraday Soc.* XLI. 321 One of the clearest demonstrations of the spherulitic structure was provided by a specimen of polythene. **1962** F. W. Billmeyer *Textbk. Polymer Sci.* v. 149 The point of initiation of spherulitic growth, its nucleus, may be a foreign particle..or may arise spontaneously in the melt.
3. (Earlier and later examples.)
1844 C. Darwin *Geol. Observations on Volcanic Islands* iii. 64 The little brown sphærulitic globules of the rocks of Ascension. **1933** *Bureau of Standards Jrnl. Res.* (U.S.) X. 486 The appearance and properties of the spherulitic clusters did not depend upon the sample of ammoniated latex from which they had been prepared.
Hence **spheruli·tically** *adv.*
1975 *Sci. Amer.* Dec. 104/2 A spherulitically crystallized polymer.

sphincter. Add: **2. a.** *sphincter control.*
1949 M. Mead *Male & Female* v. 115 They [sc. Samoan children] do not need to fear that they themselves, by their unsteady sphincter control,..will endanger the normal order of existence. **1957** *Psychoanal. Rev.* XLIV. 121 The attainment of anal sphincter control in childhood is so fundamental in human socialization that the surgical destruction of anal sphincter control must result in a severe emotional and social disruption.
Hence **sphincte·rial** *a.*; also **sphi·nctered** *a.*, possessing a sphincter (of a specified kind).
1889 *Cent. Dict.*, Sphincterial. **1963** R. P. Dales *Annelids* i. 32 A terminal bladder or vesicle closed by means of a sphinctered nephridiopore. **1965** Auden *About House* (1966) 27 A second childhood, petulant, weak-sphinctered In a cheap hotel. **1976** R. Pound *A. P. Herbert* xxvi. 298 Its equability, with or without the reinforcement of vitamins, deep breathing, and the eccentric sphincterial discipline, may have added to his length of days.

sphincterotomy (sfiŋktĕrǫ·tŏmi). *Surg.* [f. SPHINCTER: see -TOMY.] The surgical cutting of or into a sphincter.
1890 in Billings *Med. Dict.* **1892** A. Duane tr. *Fuchs's Textbk. Ophthalm.* iv. iii 726 Stellwag's operation is called oblique blepharotomy or sphincterotomy. **1922** J. Joyce *Let.* 4 Oct. (1966) III. 67 He says that..a slight operation —not iridectomy or iridotomy but sphincterotomy will probably restore a great part of former vision. **1977** *Proc. R. Soc. Med.* LXX. 160/2, I do not think this particular technique is suitable for simple procedures such as cholecystectomy or sphincterotomy.

sphingid (sfi·ndʒid), *sb.* and *a.* *Ent.* [f. mod.L. family name *Sphingidæ*, f. *Sphinx* SPHINX, adopted as a generic name by Linnæus (*Systema Naturæ* (ed. 6, 1748) 63): see -ID[3].] **A.** *sb.* A hawk-moth of the family Sphingidæ. **B.** *adj.* Of or pertaining to an insect of this kind.
1911 *Trans. Zool. Soc. London* XX. 85 Almost the entire surface is thinly coated with fine, short, white hair, an exceptional feature with Sphingid larvæ. *Ibid.* 95 *Erinnyis ello*... The commonest Sphingid of tropical America. **1930** *Proc. Entomol. Soc.* V. 24 The caudal horn was movable in many Sphingid caterpillars. **1933** *Discovery* Apr. 125/1 Eye-spots..occur particularly in sphingids. **1973** *Nature* 16 Mar. 205/2 Recently internal ocelli have been reported in several adult sphingids. **1981** G. Keynes *Gates of Memory* iii. 41 A boy in another House had been fortunate enough to catch a specimen of an extremely rare immigrant sphingid moth.

Sphingine (sfi·ndʒəin), *a.* *rare*[−1]. [f. *sphing-* stem of SPHINX + -INE[1].] Sphinx-like; characteristic of a Sphinx; enigmatic, inscrutable.
1925 A. Huxley *Those Barren Leaves* II. i. 86, I would put on..my most Sphingine smile.

sphingo- (sfi·ŋgo). *Biochem.* Comb. form of Gr. Σφίγξ, stem Σφιγγ- SPHINX (see quot. 1881 for *sphingosin*), used in the names of a number of related compounds isolated from the brain and nervous tissue, as **sphingoli·pid(e**, any naturally occurring fatty acid derivative of a sphingosine; hence **sphi:ngolipido·sis** (see quot. 1962); **sphingomy·elin** [MYELO-], any of a number of complex phos-

pholipids which are phosphoryl choline derivatives of *N*-acyl sphingosines; **sphi·ngosine** (formerly -in), a colourless crystalline base, $C_{18}H_{37}NO_2$, or any of various homologues and derivatives of this, which combined as sphingolipids occur widely in brain and nervous tissue.
1947 H. E. Carter et al. in *Jrnl. Biol. Chem.* CLXIX. 77 Among the lipide constituents [of nerve tissue] there are at least three, the cerebrosides.., sphingomyelins.., and gangliosides.., which are derivatives of the organic base sphingosine. Sphingosine may also be present in other compounds... As a matter of convenience it is proposed that the term sphingolipide be used to designate these substances. **1978** J. R. Holum *Org. & Biol. Chem.* xi. 225 The acyl units in the acylamido parts of the sphingolipids are not the usual fatty acids found in neutral fats. **1962** Knudson & Kaplan in Aronson & Volk *Cerebral Sphingolipidoses* 395 The sphingolipidoses are hereditary diseases in which there is an accumulation of sphingolipids in one or more tissues of the body... There are at least three enzyme defects among the sphingolipidoses. **1976** *Adv. Exper. Med. & Biol.* LXVIII. 9 (*caption*) Examples of early prenatal diagnoses carried out with the present microtechniques in pregnancies at risk for sphingolipidoses. **1883** J. L. W. Thudichum in *12th Ann. Rep. Local Govt. Board 1882–3: Rep. Med. Officer 1882* App. B. No. 3. 221, I have for the purposes of the present research isolated and analysed two representatives of this remarkable class of bodies, one *amidomyelin*..another *sphingomyelin*, which was found to be ..a genuine educt and principle of the brain. **1920** [see *CEPHALIN[2]]. **1946** *Biol. Rev.* XXVI. 285 Sphingomyelins are phosphatides in which the sphingosine or a closely related base is bound by an NH—CO linkage to a fatty acid..and by an ester linkage to choline phosphoric acid. **1973** Sphingomyelin [see *LECITHIN]. **1881** J. L. W. Thudichum in *Ann. Chem. Med.* II. 18 A body remained insoluble which was of an alkaloidal nature, and to which, in commemoration of the many enigmas which it presented to the inquirer, I have given the name of *Sphingosin*. **1908** Hall & Defren tr. *Abderhalden's Text-bk. Physiol. Chem.* ii. 20 On being subjected to hydrolysis this substance took up two molecules of water and formed one molecule of cerebronic acid, one of sphingosine and one of galactose. **1957** [see *CEREBROSIDE]. **1968** A. White et al. *Princ. Biochem.* (ed. 4) iv. 73 Although the above C_{18} sphingosines are most abundant in sphingolipids, other homologous C_{16}, C_{17}, C_{19}, and C_{20} sphingosines also are found among the naturally occurring sphingolipids.

Sphinx. Add: **5. a.** *sphinx-look.*
1923 D. H. Lawrence *Ladybird* 230 The queer, blank, sphinx-look with which he gazed out beyond himself.
c. sphinx moth (later examples).
1939 Duncan & Pickwell *World of Insects* x. 168 The caterpillars of the family of sphinx moths..have earned their name of 'sphinx' by their habit of rearing up their front ends, drawing in their heads, and thus assuming a threatening attitude. **1972** *Sci. Amer.* June 73/1 The larger sphinx moths weigh from two to six grams.

sphygmo-. Add: **sphygmographic** *a.* (earlier example); **sphygmogra·phically** *adv.*; **sphygmomanometer**: delete 'or rate'; **sphy:gmomano·metry**, the use of a sphygmomanometer; **sphy:gmomanome·tric** *a.*; **sphygmometer**, substitute for def.: an instrument for exhibiting or measuring the force or rate of the pulse (earlier and later examples); **sphygmo·metry; sphygmoscope** (earlier example).
1867 *Brit. Med. Jrnl.* 20 July 40/1, I refer to pulse No. 10 principally for the purpose of shewing how completely the sphygmographic form may be modified by merely functional, that is to say nervous, disorder. *Ibid.* 13 July 20/1 The full pulse (sphygmographically, that in which the second event is well marked or developed). **1885** J. B. Yeo tr. *M. J. Oertel's Respiratory Therapeutics* II. 472 When..inspiration is slow and cautious,.. sphygmographically the pulse waves altered by the rise of blood pressure immediately succeed to the average normal ones. **1902** *Amer. Jrnl. Physiol.* V. 205 For sphygmomanometric work it was found necessary to pack the small space between this collar and the forearm with soft muslin to prevent a distention of the reflected bands when the pressure within was raised. **1905** *Johns Hopkins Hosp. Rep.* XII. 69 Points of interest in sphygmomanometry. **1962** *Lancet* 15 Dec. 1225/2 Many of the difficulties inherent in clinical sphygmomanometry of the new-born infant have been overcome by the latest development in photoelectric methods. **1834** *Ibid.* 20 Sept. 936/2 At the meeting of the French Academy of Sciences on the 1st inst, M. Magendie read a report on an instrument invented by a Dr. Herisson, called the 'sphygmometer', and intended to measure the state of the pulse... The bottom of the instrument is placed over the radial artery, each pulsation of which elevates the mercury, and thus discloses to the eye the minutest variation of the circulation. **1890** *Practitioner* June 421 (*caption*) Upper curve, radial pulse obtained from healthy adult male by air modified sphygmograph (sphygmometer). **1867** *Med. Rec.* (N.Y.) 15 July 243/2 Herrisson's and Blundell's ideas on sphygmometry were sunk in oblivion. **1908** G. Oliver *Studies in Blood-Pressure* (ed. 2) ii. 42 Writers on sphygmometry have always grouped together all the instruments which derive their readings of the arterial pressure from a single artery. **1856** *Lancet* 8 Nov. 510/1 The numerous cases of disease of the heart which have come under the care of Dr. Scott Alison..have afforded abundant means of applying..the new sphygmoscope, or cardioscope, (contrived by that physician).

sphygmology (sfigmǫ·lŏdʒi). *Med.* [See SPHYGMO- and -LOGY.] The study of the pulse. Hence **sphygmolo·gical** *a.*
1890 Billings *Med. Dict.* II. 570/1 Sphygmology. **1931** G. Sarton *Introd. Hist. Sci.* II. 75 Sphygmology is even older than urology and more universal. *Ibid.* 87 The Shang-han-lun is divided into ten books of which the first is a sort of scientific (sphygmological) introduction to the others. **1941** *Bull. Hist. Med.* X. 210 It seems probable that some Aegimius may be credited with being the founder of sphygmology. **1973** C. R. S. Harris *Heart & Vascular System Anc. Greek Med.* v. 255 The supposed perception of fullness is a typical feature of post-Herophilian sphygmology from which only the Empirics apparently dissented. *Ibid.* vii. 409 But enough of these sphygmological conundrums.

spic (spik), *sb.* and *a.* *U.S. slang.* Also **spick, spig, spik,** and with capital initial. [Shortened f. *SPIGGOTY *sb.* (and *a.*)] **A.** *sb.* **a.** A contemptuous and offensive name for a Spanish-speaking native of Central or South America or the Caribbean; a spiggoty.
1913 H. A. Franck *Zone Policeman 88* i. 10 It was my first entrance into the land of the *panameños*, technically known on the Zone as 'Spigoties', and familiarly, with a tinge of despite, as 'Spigs'. **1916** E. Peixotto *Our Hispanic Southwest* 102 The Mexican men they despise and call 'spicks'. **1928** S. Lewis *Man who knew Coolidge* ii. 116 We need a supply of cheap labour, and where get it better than by encouraging these Wops and Hunks and Spigs and so on to raise as many brats as they can? **1936** *Opportunity* Aug. 239/1 Frank was just a 'huerco' to his mother, 'spick' to his white schoolmates in Queensville, Texas. **1949** W. Faulkner *Knight's Gambit* 137, I don't intend that a fortune-hunting Spick shall marry my mother. **1953** F. Scott Fitzgerald *Tender is Night* ix. 275 'He's a spic!' he said. He was frantic with jealousy. **1964** E. Lacy *Pity Honest* iii. 48 This is becoming a tough neighbourhood, full of Spics. **1977** D. E. Westlake *Nobody's Perfect* (1978) 39 You'd put your kid in a school with a lotta niggers and kikes and wops and spics?
b. The Spanish language; *spec.* Spanish-American.
1933 E. Hemingway *Winner take Nothing* (1934) 200, I wish I could talk spik... I don't get any fun out of asking that spik questions. **1977** *Amer. Speech 1975* L. 67 *Spic n*, 1: Spanish language 2: course in the Spanish language 'I've had two years of Spic.'
B. *adj.* = *SPIGGOTY *sb.* (and *a.*) 3. *derog.*
1919 *Ladies' Home Jrnl.* Sept. 27 The Marines had been..silencing the elusive 'spick' bandit in Santo Domingo. **1950** R. Moore *Candlemas Bay* 29 Jerry Canneri. Or Carnoodle. Some such damn spik name. **1976** N. Thornburg *Cutter & Bone* iii. 74 A nigger fag and two spic girls with a pet monkey.

‖ **spiccato** (spikā·to), *a.* (*adv.*, *sb.*) *Mus.* [It., detached, distinct.] Of a style of staccato bowing, esp. on the violin: detached, *i.e.* with short breaks between notes caused by controlled bouncing of the bow; of music played in this manner. Also used *advb.*, as a direction to the performer and as *sb.*, an instance of spiccato playing; a passage in this style.
1724 [see STACCATO *a.* (*adv.*, *sb.*)]. **1740** J. Grassineau *Mus. Dict.*, *Spiccato*, signifies to separate, divide, part,.. that is, to give every note its distinct sound, and is the contrary of what we call slurring. **1883** Grove *Dict. Mus.* III. 650/2 *Spicato* [sic].., a term applied in violin-playing to a particular vibratory style of bowing. *Ibid.* 682/2 The Spicato [sic] is marked by dots over the notes. **1938** *Oxf. Compan. Mus.* 891/2 *Spiccato*,..in music..implying staccato effect. **1955** *Times* 12 May 5/6 The scherzo, which incidentally allowed Mr. Campoli to unfold long stretches of the most prodigious *spiccato* bowing. **1964** *Listener* 25 June 1043/1 Paganini's innovations in violin technique—the big bouncing *spiccato*, for instance. **1978** *Gramophone* Aug. 355/3 The lyrical grace of Norbert Brainin's playing in the Adagio and his *spiccato* triplets in the finale.

spice, *sb.* Add: **2. e.** A medicated preparation added to cattle or horse feed. ? Now only *Hist.*
1707 J. Mortimer *Whole Art of Husbandry* 157 Take a quart of Ale, half an ounce of Diapente.., Horse-spice two Ounces. **1928** E. P. Oppenheim *Chron. Melhampton* 143 A retired dealer in cattle spices. **1961** M. W. Barley *Eng. Farmhouse & Cottage* v. i. 253 Thomas Morrison kept a much more interesting shop, in the last years of the seventeenth century... There was ironmongery.. 'horse spice and jollop' for the farmer.
7. a. *spice-cabinet, jar; spice-bazaar.*
1924 R. Graves *Mock Beggar Hall* 6 Scepticos heard this popular Figment in the spice-bazaar. **1893–4** *T. Eaton & Co. Catal.* Fall & Winter 120/3 Spice cabinets, 90c. **1908** *Sears, Roebuck Catal.* 359/3 German china cereal and spice jars..with names of spices or cereals on each jar. **1977** C. Watson *One Man's Meat* iii. 25 The dining enclosure..was screened from cook top and sluice unit by rubber plants and rows of spice jars.
b. *spice-enrichened, -sweet* adjs.
1940 C. Day Lewis tr. *Virgil's Georgics* II. 39 Nor all Arabia's acres of spice-enrichened soil. **1953** W. de la Mare *O Lovely England* 51 The spice-sweet gorse.
8. spice-berry: for *U.S.* read *N. Amer.* and add earlier example; **spice-bush** (earlier example); **spice-isle**, one of the spice-islands (*poet. rare*); **spice-wood**, (*a*) (earlier example).
1792 G. Imlay *Topogr. Descr. N. Amer.* 216 There is a variety of shrubs in every part of the country, the principal of which are the myrtle and spice berry. **1770** G. Washington *Jrnl.* 15 Oct. (1925) I. 409 The Soil.. being as black as Coal and the Growth, Walnut, Cherry, Spice Bushes. **1885** W. B. Yeats in *Dublin Univ. Rev.* Sept. 121 Where spice-isles nestle on the star-trod seas. **1756** P. Kalm *Resa til N. Amer.* II. 204 Spicewood. (Laurus æstivalis. *Spec.* 370).

spice, *v.* Add: **1. b.** Also with *up*, to enliven, to make more interesting or racy. *colloq.*
1927 *Scribner's Mag.* Apr. 390/1 The brazen forgery in *The Gentlemen's Magazine* seems to have been a facetious attempt to spice up a sober-toned, political news-letter. **1979** *Arizona Daily Star* 8 Apr. (Wedding Suppl.) 15/2 One Tucson couple spiced up a wedding with circus performances, complete with a juggler and unicyclist.
spicing *vbl. sb.*: also with *up*.
1934 C. Lambert *Music Ho!* II. 127 There is no instance ..of the spicing up of a simple harmonic basis.

spick, *sb.*[4] *dial.* Also **spic.** [Var. of dial. *speak*: see *E.D.D.*] A withy or rod, usu. pointed and doubled, used to secure thatch; a spar (SPAR *sb.*[4]).
1890 J. D. Robertson *Gloss. Words County of Gloucester* 147 *Speeks* or *spicks*, the pieces of wood used for holding together the thatch on a rick. **1893** Dartnell & Goddard *Gloss. Words Wiltshire* 152 *Spick*.., in thatching, the same as *spar*. **1934** *Times Lit. Suppl.* 1 Feb. 71/3 The thatcher's tackle of 'spicks' or 'spars' (pointed hazel-rods) which fix the bands to hold down the thatch. **1939** D. Hartley *Made in England* ii. 57 A bundle of the doubled withies split and bent into a twisted hook or double prong... These, when used in stack work, are called 'spics' or 'speks'. **1949** K. S. Woods *Rural Crafts Eng.* III. vii. 109 *Thatching Spics.* The riving, or rending of the wood, which is the basis of a number of crafts, is illustrated.

spick, *a.* (Later example.)
1920 D. H. Lawrence *Lost Girl* vi. 99 He liked to have his clothes neat and spick.

spick, var. *SPIC *sb.* and *a.*

spick-and-spanness. (Examples.)
1911 Mrs. H. Ward *Case of Richard Meynell* II. viii. 174 [He] was himself a model of spick-and-span-ness. **1931** *Times Lit. Suppl.* 21 May 407/1 The ancient houses..had a touch of mysterious romance that the bright spick-and-spanness of the new architecture misses. **1975** *New Yorker* 28 Apr. 52/2 A life that, in spite of its inferior neatness and spick-and-spanness, was essentially more serious and finer than our own.

spiculation. (Earlier example.)
1868 G. M. Hopkins *Jrnls. & Papers* (1959) 185 Spiculation in a dry blot in a smooth inkstand.

spicule. Add: **5.** (Earlier examples.)
1835 J. Paget *Let.* 16 Apr. in S. Paget *Mem. & Lett. Sir J. Paget* (1901) 57 The same appearances have been noticed in our dissecting-rooms, where they have been attributed to the deposition of small spicules of bone (which, indeed, they somewhat resemble). **1865** Milton & Cheadle *N.-W. Passage by Land* xv. 301 The fallen timber lay as thickly and entangled as the spiculæ in the children's game of spelicans.
6. *Astr.* Any of numerous short-lived, relatively small radial jets of gas observed to occur in the sun's atmosphere in the chromosphere and lower corona.
1945 W. O. Roberts in *Astrophysical Jrnl.* CI. 136 Small spikes of chromospheric material, observed in Hα with the coronagraph and quartz-polaroid monochromator are described. These spicules, seen in polar regions of the sun, have very brief lifetimes, amounting on the average to 4 or 5 minutes. **1948**, etc. [see *JET *sb.*[2] 4 c (ii)]. **1974** Bray & Loughhead *Solar Chromosphere* ii. 60 Spicules in the polar regions of the sun tend to follow the direction of the overlying coronal rays. **1978** Pasachoff & Kutner *University Astron.* viii. 191 Spicules are best seen when we are looking off the edge of the sun, beyond the limb.

spider, *sb.* Add: **3. a.** (Earlier example.)
1807 in *Austin Papers* (1924) I. I. 132, 2 Spiders with Covers.
4. Also, an alcoholic drink of similar ingredients (see quots. 1861, 1888); a soft drink with ice-cream floating in it.
1861 H. Earle *Ups & Downs* 283 They are..up to unlimited 'spiders', or lemonade and sherry. **1888** E. Finn *Chron. Early Melbourne* ii. 548 The favourite tipple of the bushman was mixed brandy and ginger beer—a 'spider', as it was called. **1941** *Coast to Coast* 229 'You've had your drink, so now you've got to buy us all a spider at Smith's'... I didn't want to go back and sit in Smith's and drink silly coloured muck with ice-cream floating in it. **1965** G. McInnes *Road to Gundagai* 14 She reached for a thick yellow glass and poured in the ginger beer.. an enormous dollop of ice-cream which she dropped into the ginger beer. 'There's your spider.' **1974** Buckley & Hamilton *Festival* 127 You used to strut into the milk bar as though you owned the place. 'A lime spider, Harry.'
6. a. Also, similar parts of instruments and other apparatus. (Earlier and later examples.)
1860 Clark & Colburn *Recent Practice Locomotive Engine* 52/1 In driving wheels, the centre, or 'spider', for a 5-feet wheel to carry 4½ tons, will weigh 1800 pounds and upwards. **1935** A. G. Ingalls *Amateur Telescope Making* (ed. 4) 371 Another interesting diffraction phenomena [*sic*]..known to most able telescope designers, is the fact that there will be fewer diffraction lines from a four-legged diagonal support spider than from one having only three legs. **1961** Miczaika & Sinton *Tools of Astronomer* iii. 75 Diffraction of light by the spider supporting the secondary mirror is a frequent complaint. **1966** L. A. H. Eastman tr. *Schenkel's Plastics Extrusion Technol. & Theory* xi. 326 The melt is..fed via a 90° bend into a

distributor spider, which may have four to eight symmetrically arranged radial channels.
d. *Austral. Opal-mining.* (See quot. 1912.)
1912 *Empire Mag.* Nov. 281/2 *Spider*, a small iron instrument which serves the double purpose of holding the candle, and 'lifting' the seam of opal. **1940** I. L. Idriess *Lightning Ridge* xxiii. 158, I gouged around and under, then pryed it out with the spider point. **1958** M. D. Berrington *Stones of Fire* iii. 33 A candle in a 'spider' that queer, spiked holder that is used below ground. **1967** *Sunday Mail Mag.* (Brisbane) 10 Dec. 3/5 The candle in its spider dropped to the floor and went out.
e. *Engin.* A metal sleeve within which an object may be gripped by screws or wedges.
1920 *Bull. U.S. Bureau of Mines* No. 182. 7 *Spider*, tool that encircles and holds the pipe by means of steel wedges. *Ibid.* 17 The swinging spider..is probably one of the most useful inventions..for the handling of casing in drilling oil wells. **1940** [see *CAT-HEAD *sb.* 3 c]. **1950** A. W. Judge *Centre, Capstan & Automatic Lathes* II. iii. 135 The short end of the hub faces outwards, and the spider is gripped between the arms by three chuck jaws. **1977** R. D. Langenkamp *Handbk. Oil Industry Terms & Phrases* (ed. 2) 159 The spider is manually locked around a length of tubing just below the tool joint. Some advanced types of elevator spiders are air operated.
f. *Electronics.* A flexible linkage formerly placed between the moving cone and the fixed magnet assembly of a loudspeaker.
1928 *Wireless World* 6 June 608/1 A centring device in the form of a brass spider attached to the pin is supplied. **1948** G. A. Briggs *Loudspeakers* vi. 20 The bakelised spider gives a sharply defined bass response to the cone, resulting in a crispness in the tone. **1959** N. H. Crowhurst *Basic Audio* I. 49 To prevent the coil rubbing against the magnet poles, a centering 'spider' or suspension is used, which allows free movement in the direction of vibration, while preventing the coil from moving against the pole faces.
7. a. Also, a trotting gig. *Austral.*
1945 S. J. Baker *Austral. Lang.* ix. 175 *Spider* or *junker*, a trotting gig. **1955** A. Ross *Australia 55* 34 The drivers, dressed in silks like jockeys, sit behind their animals in tiny carriages known as spiders. **1969** *West Australian* 5 July 32/5 Causing Pyraket to strike and badly buckle the inside wheel of Master Flame's spider.
b. An early bicycle with the benefit of steel wheels, as opp. to those of wood. Cf. *spider-wheel*, sense 10 a below. Now only *Hist.*
1874 *Bicycling* 4 Had he lived in the days of the 'Coventry Spiders'. **1908** E. M. Sneyd-Kynnersley *H.M.I.* ix. 82 Safety bicycles were not yet: the Boneshaker was not tempting, and the Spider was perilous.
8. d. A spider-nævus.
1942 *Amer. Jrnl. Med. Sci.* CCIV. 251 The well known increase in excretion of estrogenic substances in pregnancy coincides with the period during which vascular spiders and palmar erythema tend to appear. **1948** D. Ballantyne *Cunninghams* iii. 15 Winter weather gave her the blue spiders. **1969** *Daily Colonist* (Victoria, B.C.) 9 Jan. 2/3 Does wearing a 'pants-type' girdle cause broken blood vessels in the thighs?..These little vessels are called 'spiders'... These spiders are commonest in women; hence the hormone (estrogen) level is thought to have a bearing. **1974** R. M. Kirk et al. *Surgery* vi. 107 Hepatic failure causes weakness,..vascular spiders (named from the spider-like appearance of dilated arterioles), palmar erythema..and encephalopathy.
8*. *Cards.* A variety of patience played with two packs.
1890 'Cavendish' *Patience Games* 186 The Spider.. requires quite sufficient exercise of thought to render it very interesting. **1901** 'Tarbart' *Patience* 49 Spider. Played with two full packs of cards. **1930** W. S. Maugham *Gent. in Parlour* xv. 78, I knew seventeen varieties of patience. I tried the Spider and never by any chance got it out.
9. a. *spider-cloth, floss, -form.*
1916 D. H. Lawrence *Amores* 86 Great grey spider-cloths hanging Low from the roof. **1978** C. Tomlinson *Shaft* 13 Finer than the lines Of spider floss. **1954** J. R. R. Tolkien *Two Towers* 332 There agelong she had dwelt, an evil thing in spider-form.
d. *spider-tongued; spider-spruce, -thin.*
1948 C. Day Lewis *Poems 1943–1947* 82 But look at her parlour, all tidied and spider-spruce! **1928** V. Woolf *Orlando* vi. 257 The Serpentine..was a bronze colour; spider-thin boats were skimming from side to side. **1939** G. Greene *Confidential Agent* I. ii. 48 A..cotton bed-spread, clean and faded and spider-thin. **1934** Dylan Thomas *18 Poems* 24 Some let me make you of autumnal spells, The spider-tongued, and the loud hill of Wales.
e. Instrumental, as *spider-curtained* adj.
1925 Blunden *English Poems* 40 The spider-curtained darkness in the attic of black Jacob's farm.
10. a. spider angioma *Path.*, a spider-nævus; **spider-man**, one employed to work on high structures; a steeple-jack; **spiderveil**, a kind of veil; **spider veins**, small dilated superficial veins around varicosities on a leg; **spider-wheel**, (*c*) a metal wheel with wire spokes (formerly applied *spec.* to a bicycle-wheel); hence **spider-wheeled**, *a.*, fitted with spider-wheels; **spider-work**, also *spec.* in *Needlework* = *OPUS ARANEUM.
1956 *New Gould Med. Dict.* (ed. 2) 77/1 Spider angioma. **1961** R. D. Baker *Essent. Path.* xvi. 412 At autopsy one notes ascites and subcutaneous edema of the legs, often with hydrothorax. Spider angiomata are frequently observed on the skin. **1955** *Britannica Bk. of Year* 489/2 *Spiderman*, an erector of building structures. **1958** *Radio Times* 25 July 3/1 These spider-men and steel-erectors work at great heights, often where there are no means of protection. They walk along girders at dizzy heights as though they were strolling along Piccadilly. **1962** *B.S.I. News* July 11/1 Safety harness worn by window-cleaners and spidermen. **1972** J. Wainwright *Night is Time to Die* 8 They used an expression familiar to all working policemen. 'Sudden Death.' It covers everything: from.. the spiderman who takes that one chance too many..to the hippy who spins into permanent orbit. **1922** Joyce *Ulysses* 441 In smart Saxe tailormade, white velours hat and spiderveil. **1976** *Vogue* Jan. 20/4 The treatment of broken and spider veins on legs. **1875** *Eng. Mechanic* 23 Apr. 146/2 With the spider-wheels I found that there was rather a tendency to get loose. **1882** *Bicycle* 15 The Spider-Wheel, invented by the Coventry Machinists Company and now almost universal. **1906** *Chambers's Jrnl.* Oct. 735/1 The introduction of the free spider-wheel, pneumatic-tired cycle. **1969** *West Australian* 5 July 32/5 On the turn out of the back straight in the last lap Majestic Scott's spider wheel was badly buckled. **1977** *Weekly Times* (Melbourne) 19 Jan. 57/4 (Advt.), Semi-tipper, 10–1 spread bogey... Hercules body and hoists.. 900 × 20 tyres, spider wheels. **1886** *Century Mag.* July 338/2 There may be a crowd of onlookers in every kind of trap, from a four-in-hand drag to a spider-wheeled buggy drawn by a pair of long-tailed trotters. **1943** J. W. Day *Farming Adventure* iii. 40 A high spider-wheeled dogcart. **1865** Spider-work [see *OPUS ARANEUM]. **1874, 1883** Spider-work [in Dict.].
b. spider beetle, a long-legged beetle of the family Ptinidæ.
1954 Borror & DeLong *Introd. Study of Insects* xxii. 379 The Ptinidae, or spider beetles, are small long-legged beetles..somewhat spiderlike in appearance. **1979** P. L. G. Bateman *Household Pests* II. 106 Spider beetles are basically scavengers and infestations often originate in old birds' nests.
c. spider flower, an annual herb of the genus *Cleome* of the family Capparaceæ, esp. *C. hasslerana*, which has clusters of pink or white flowers with long stamens (cf. *CLEOME); **spider lily,** a bulbous plant belonging to the genus *Hymenocallis*, native to North and South America, or *Crinum*, native to tropical regions, both of the family Amaryllidaceæ, and bearing clusters of white or pink flowers, often fragrant; **spider plant,** (*b*) a perennial herb, *Chlorophytum comosum*, of the family Liliaceæ, native to South Africa, of which forms bearing variegated linear leaves and clusters of white flowers are much cultivated as house plants.
1861 A. Wood *Class-Bk. Bot.* (ed. 10) 240 Spider Flower...Herbs or shrubs. **1909** A. E. Mack *Bush Calendar* 4 Faded by the excessive rain were the red spider-flowers. **1931** W. N. Clute *Common Names Plants* 101 The spiderflower (*Cleome*) named from the long and sprawling stamens like spider's legs. **1968** Peterson & McKenny *Field Guide Wildflowers* 230 Spider-flower... Note the extraordinarily long stamens projecting beyond the 4 narrow-stalked pink or white petals. **1887** *Harper's Mag.* Feb. 351/1 The exquisite white spider-lily, nodding in clusters on long stalks. **1908** E. J. Banfield *Confessions of Beachcomber* I. i. 21 Along the deltas of the creeks are fragrant, gigantic 'spider lilies' (*Crinum*). **1946** D. C. Peattie *Road of Naturalist* v. 58 The cypress woods around Charleston with sudden spider lilies. **1980** A. Desai *Clear Light of Day* i. 7 They went slowly up the wide stairs between the massed pots of spider lilies and asparagus fern. **1946** M. Free *All about House Plants* xv. 126 That plant with striped leaves known to many as spider-plant..increases by means of plantlets produced on the ends of its flower stalks. **1979** S. Rifkin *McQuaid in August* vii. 48 Three enormous spider plants hung..in front of the window.
spi·derish *a.*, resembling a spider; hence **spi·derishness.**
1935 O. Stapledon *Odd John* i. 3 Strangers were often revolted by his uncouth proportions. They called him spiderish. **1944** G. B. Shaw *Everybody's Political What's What?* xxxvi. 320 Commercial ability is often really mere spiderishness.

spider, *v.* Add: **2. a.** *intr.* To move in a manner suggestive of a spider. **b.** *trans.* To cause to move or appear thus.
1938 G. Greene *Brighton Rock* vi. i. 236 Ida Arnold had been trained by the Board. Queerer things than that had spidered out under her fingers and old Crow's. **1975** *New Yorker* 26 May 39/3 It is impossible to resist a postscript at the bottom of that august form, though no doubt it would have to be spidered up the margin. **1976** 'F. Clifford' *Drummer in Dark* vi. 27 His fingers spidered over the map, stressing a detail here, a field of fire there.
Hence **spi·dering** *ppl. a.* and *vbl. sb.*
1973 T. Pynchon *Gravity's Rainbow* I. 55 His little bureau is dominated now by a glimmering map,..written names and spidering streets. **1975** *New Yorker* 12 May 141/1 He wishes only, with his nimble, sinister spidering amid the complexities of our cultural situation, to give us —one of his favorite words—*frissons*.

spider-web, *sb.* Add: **2. a.** (Later examples.)
1923 G. H. McKnight *English Words* iv. 53 The early years of the war were productive in slang language among the British troops... Wire entanglements were *fly traps* and *spider webs*. **1957** O. Nash *You can't get there from Here* 61 This is the most enticing spiderweb of a tarradiddle ever spun. **1978** J. Carroll *Mortal Friends* II. v. 187 He was running his finger along the cold pane of the window, making spiderwebs in the fog.
b. A variety of turquoise characterized by a network of fine dark lines running through it. Also *spider-web turquoise*.
1936 M. Bedinger *Navajo Indian Silver Work* 18 There is another famous variety, of exceptionally good quality... It is a dark blue with black tracery all through it, which gives it the name of 'Spider Web' turquoise. **1968** J. Sinkankas *Van Nostrand's Standard Catalog Gems* iv. 202 Persian spiderweb is similar except that a network of fine black lines divides the surface of the gem into a mosaic of even patches. **1975** R. Webster *Gems in Jewellery* xiv. 79 The turquoise and the matrix of dark brown limonite or fawn-coloured sandstone are then cut together to give stones known as turquoise matrix or, if the veins are fine, spider web turquoise.
spider-web *v.*, (*b*) *intr.* in various fig. uses (see quots.); **spider-webby** *a.*, resembling a spider-web.
1823 L. Hunt in *Liberal* II. 369 'What the devil's here? To bring my stockings home at last undone?'.. 'Undone... They so spider-web, it's a despair.' **1864** Spider-webby [in Dict. s.v. SPIDER *sb.* 9 d]. **1884** 'Mark Twain' *Huck. Finn* ix. 74 The rain would thrash along by so thick that the trees off a little ways looked dim and spider-webby. **1936** I. L. Idriess *Cattle King* iv. 34 He and others have spider-webbed from around the Australian coasts far into the inland. **1974** M. Hoyt *Thirty Miles for Ice Cream* x. 110 The deciduous trees, without their leaves, show at that distance as gray or a spider-webby black.

spidery, *a.* **1. b.** (Earlier example.)
1825 Coleridge *Let.* 21 Feb. (1971) V. 414 As we advance in years, the World, that spidery Witch, spins it's threads narrower and narrower, still closing in on us.

spiel (spīl, ʃpīl), *sb.*[2] *slang* (orig. *U.S.*). [a. G. *spiel* play, game; see also next.] **1.** Talk, a story; a speech intended to persuade or advertise, patter. Also *transf.*
1896 Ade *Artie* xi. 100 There was a long spiel by the high guy in the pulpit. **1906** *Daily Colonist* (Victoria, B.C.) 25 Jan. 6/1 We appointed him mayor at five minutes' notice and gave him the job of giving the Chinks the right kind of a spiel. **1925** Wodehouse *Sam the Sudden* xiii. 92 He pulled this long spiel about having had a letter from a guy he used to know named Finglass. **1926** J. Black *You can't Win* ii. 9 Your capable beggar on the street does not say 'please'. He rips off his spiel in such exact and precise language that you get your dime without it. **1931** 'D. Stiff' *Milk & Honey Route* 189 In the missions, to make him lonely and the more susceptible to the 'righteous spiel', they sometimes sing *Where is my wandering Boy Tonight?* **1937** *Printers' Ink Monthly* May 42/2 *Spiel*, the advertising copy. **1944** D. Burley *Orig. Handbk. Harlem Jive* 59, I latched on to this hard, mad spiel. **1953** H. Miller *Plexus* I. iii. 137 Flatter the pants off him! Then go into a little spiel—you know what I mean. Give him some pointers on how to launch the magazine. **1959** *New Statesman* 19 Sept. 344/1 At the end of these flights the poor bored hostess is still compelled to repeat her antique *spiel*; 'We hope you have enjoyed your flight.' **1962** J. Wain *Strike Father Dead* 59, I gave her just the Christian name, and she gave me the spiel about never having met anybody called that before, and its being a nice name, and so forth. *a* **1974** R. Crossman *Diaries* (1975) I. 82 Then Callaghan started off with a long spiel which he read aloud from a Treasury brief on pale blue paper, describing the extreme gravity of the economic situation. **1980** *Listener* 13 Nov. 665/3 A long spiel..from a tart about how much horrider Soho has become.
2. A swindle, a dishonest line of business.
1901 'J. Flynt' *World of Graft* iv. 169 I've been shut up a number of years..but I didn't mind them as much as you would; I took them as part of the spiel. **1921** P. & T. Casey *Gay Cat* 303/1 'What's your spiel?' asks one hobo of another. **1932** W. Hatfield *Ginger Murdoch* 175, I reckon you were thinking you had shaken me off, and could go about your spiel, whatever it is. **1954** T. A. G. Hungerford *Sowers of Wind* 174 This isn't a spiel, Colonel... I know this bloke, and he's on the level.

spiel (spīl, ʃpīl), *v.* *slang* (orig. *U.S.*). Also † **speel, speil.** [ad. G. *spielen* to play, gamble].
1. a. *intr.* To gamble. Also rarely *trans.*
1859 [implied at *SPIELING *vbl. sb.*]. **1882** *Sydney Slang Dict.* 8/1 *Speel*, to gamble. **1892** I. Zangwill *Childr. Ghetto* I. iv. 124 They played loo, 'klobbiyos', napoleon... Old Hyams did not *spiel*, because he could not afford to. **1931** [see *HALF *sb.* 6 i]. **1953** W. Mankowitz *Bespoke Overcoat* xiv. 21 You go to the dog tracks in the evening? Not for me... Horses? No horses, neither. You must spiel something. Poker, shemmy?
b. To play music.
1870 *Territorial Enterprise* (Virginia City, Nevada) 16 July 3/1 The new 'circus' is to be seen at the corner of D street and Sutton avenue—down var der orkan goes a spielin'. **1871** *Nassau Lit. Mag.* Feb. 179 Come now, old fellow, 'speil'. **1947** G. S. Perry *Cities of Amer.* 187 Denver's Symphony chooses to spiel only when winter's winds doth blow.
2. To talk, esp. volubly or glibly; to patter. Also with *away*.
1894 *Mid-Winter Appeal* (San Francisco) 10 Mar. 1/3 Tell [the barker] to stop spieling now and then. **1904** 'O. Henry' *Cabbages & Kings* xiii. 220 If you can borrow some gent's hat in the audience, and make a lot of customers for an idle stock of shoes come out of it, you'd better spiel. **1914** [see *BALLYHOO *sb.*]. **1920** Wodehouse *Coming of Bill* I. v. 60 Spiel away, ma'am... The floor's yours. **1946** Mezzrow & Wolfe *Really Blues* (1957) vi. 70 One of the funniest things I ever heard was Mac spieling in Yiddish. **1966** R. Sheckley *Mindswap* vii. 49 Silent and disdainful, scorning to spiel, the little man stood with arms folded as Flynn walked up to the booth.
3. *trans.* To tell, to reel *off*; to announce; to perform.
1904 'O. Henry' *Cabbages & Kings* iii. 58 I'll come right back and hear you spiel the rest before bedtime. **1936** W. A. Gape *Half a Million Tramps* v. 139 When my

turn came I was not ready to 'spiel' off the answers. 1962 *Coast to Coast 1961–62* 81 Garish neons had spieled, in Latin letters, the delights of innumerable honkeytonks. 1970 A. TOFFLER *Future Shock* xviii. 378 Each participant spieled off his reason for attending. 1977 *Time* 28 Nov. 64/1 In a few hours he would be on a..stage singing his songs and spieling his narrative jazz poetry to an audience of college kids.

Hence **spie·ling** *vbl. sb.*

1859 G. MATSELL *Vocabulum* 84 *Speiling*, gambling. 1898 A. M. BINSTEAD *Pink 'Un & Pelican* ix. 190 A raid upon a 'spieling' club by the police. 1904 'O. HENRY' in *McClure's Mag.* July 353/2 It was just what Buck wanted—a regular business at a permanent stand, with no open air spieling..on the street corners every evening. 1937 G. FRANKAU *More of Us* vii. 78 Nor think this spieling shames our British blood. 1959 T. H. WHITE *Godstone & Blackymor* 47 There was no patter now, no fair ground spieling. 1981 *Observer* 9 Aug. 3/2 'Pitching', or spieling, is how traders sell by a kind of inverted auction: prices start out sky-high, and buyers leap into the breach as the pitcher brings them tumbling down.

spiel, var. *SPEEL *v.*[2]

spieler. For *Austr. slang* read *slang* (orig. *U.S.*) and add: Also † **speiler.** **1.** (Earlier and later examples.) Now chiefly *Austral.*

1859 G. MATSELL *Vocabulum* 83 *Speiler*, a gambler. 1885 [see *MURRUMBIDGEE]. 1911 W. H. KOEBEL *In Maoriland Bush* xxii. 283 The professional sharper or 'spieler'..wings his periodical flights from Sydney or Melbourne. 1929 *Detective Fiction Weekly* 13 July 731/2 Hard on their trail would come all the 'magsmen', the 'spielers', the dips, the 'broadsmen', and the 'pickers up'. 1935 H. R. WILLIAMS *Comrades of Great Adventure* 234 The spielers worked the three-card trick on the 'mugs'. 1957 J. WATEN *Shares in Murder* 156 You could match your wits against smart con-men and spielers.

2. One who spiels (*SPIEL *v.* 2); a 'barker'; a voluble speaker.

1894 *Mid-Winter Appeal* (San Francisco) 19 May 15/1 Some spielers for the Midway who attempted to lick the Camp gate keeper were sent up for 24 hours. 1901 [see *BALLYHOOER]. 1936 J. L. HODSON *Our Two Englands* vi. 109 The public..are as fascinated by a good spieler or barker on the show front as they were then. 1937 *Printer's Ink Monthly* May 42/3 *Spieler*, a radio commentator. 1956 *Time* 11 June 42 Mary Costa..agrees that a girl spieler should be 'good-looking but not too flashy to detract from the product'. 1976 *National Observer* (U.S.) 8 May 16/1 His style is all style, a curious amalgam that incorporates at its corniest Dare to Be Great spieler Glenn Turner.

3. A gambling club.

1931 *Police Jrnl.* IV. 502 A cardsharper (broadsman) met a confidence trickster (con-head) and a thief (tea-leaf) in a gaming house (speiler). 1945 P. CHEYNEY *I'll say she Does* iii. 57 It's a gamblin' *spieler*... There's big play there every night. 1955 D. WEBB *Deadline for Crime* iv. 88 Throughout Soho and Mayfair there are a number of what are known as speilers, illicit gambling dens run by the underworld mainly for the underworld, or wealthier mugs of the racecourses. 1962 [see *KITE *sb.* 4 c]. 1976 J. O'CONNOR *Eleventh Commandment* iv. 62 A well-known boxing referee who used to run a dirty low-down dive of a spieler.

‖ **Spielraum** (ʃpī·lraum). *Philos.* Pl. **Spielräume.** [Ger. (J. von Kries *Die Principien der Wahrscheinlichkeitsrechnung*, 1886), f. *spiel* play + *raum* room.] The range of possibilities (orig. in probability theory) within which the probability of an outcome or likelihood of a hypothesis is to be assessed.

1921 J. M. KEYNES *Treat. Probability* vii. 88 Briefly..he [*sc.* von Kries] may be said to hold that the hypotheses for the probabilities of which we wish to obtain a numerical comparison, must refer to 'fields' (Spielräume). 1937 *Mind* XLVI. 486 In this connexion he criticises von Kries's '*Spielraum*'-theory of probability. 1944 *Horizon* IX. 235, I propose to borrow from a theory of probability the concept of a *spielraum*. In connection with the judgments we make about freedom, this metaphor of a playing-space can be applied in three ways. 1961 J. N. FINDLAY *Values & Intentions* iii. 111 The image of a *Spielraum* or space of possibilities seems..indispensable to the full development of partial belief. 1969 P. GEACH *God & Soul* vii. 97 There has to be this element of chance in things if human choices are to have any *Spielraum*.

spiff, *v.*[2] Add: Also with *up*.

1979 *Arizona Daily Star* 22 July J 3/2 The man doing it was an interior decorator, not an art conservator, and he did what he felt was best—he went in and spiffed up the church.

spiffy, *a.* (Earlier and later examples.)

1853 D. G. ROSSETTI *Let.* 2 Nov. (1965) I. 161 The frame for my water-colour has just come in and is spiffy cheesy jammy nobby [etc.]. 1958 T. STANWELL-FLETCHER *Clear Lands* 24 This was with the comparatively new, very spiffy Canadian government's *C. D. Howe*. 1978 H. WOUK *War & Remembrance* viii. 86 She's turned into quite the spiffy New York gal.

spiflicate, *v.* (Earlier example.)

1749 *Gentl. Mag.* Dec. 563 Whence the term *spifflicated*?

spiflicating *ppl. a.* (earlier example).

1852 [see *ROWDY-DOW *sb.*].

spiflicated (spi·flikēⁱtėd), *ppl. a. slang* (orig. *U.S.*). Also **spifflicated.** [f. prec. + -ED[1].] Intoxicated, drunk.

1906 'O. HENRY' *Four Million* 114 He uses Nature's Own Remedy. He gets spifflicated. 1910 [see *PIE-EYED *a.*]. 1927 *New Republic* 9 Mar. 71/2 The following is a partial list of words denoting drunkenness now in common use in the United States..spifflicated. 1931 *Sun* (Baltimore) 6 Jan. 6/7 Almost every name you could think of to describe the state of being drunk was given, but one splendid one that I know of was omitted—'spifflicated'. *a* 1966 'M. NA GOPALEEN' *Best of Myles* (1968) 338 Drunk; jarred;..spiflicated. 1971 H. A. SMITH *View from Chivo* i. 5, I do not believe..that I was spifflicated last night.

spig, var. *SPIC *sb.* and *a.*

spiggoty (spi·gǫti), *sb.* (and *a.*). *U.S. slang.* Also **spiggity, spigotti, spigoty.** [Orig. uncertain: prob. repr. broken English (see sense 1, quot. 1938): now generally superseded by *SPIC *sb* and *a.*] **1.** A contemptuous name for a Spanish-speaking native of Central or South America.

1910 'W. LAWTON' *Boy Aviators in Nicaragua* 331 Ring..steamed down here on his gunboat just in time to fire that shell and throw a scare into the spiggotys at the very physicky moment. 1913 H. A. FRANCK *Zone Policeman 88* 10 It was my first entrance into the land of the panameños, technically known on the Zone as 'spigoties'. 1916 *Recruiter's Bull.* (U.S. Marine Corps, N.Y.) Jan. 10/1 Well, we did not get to Haut de Cap—we ran into a 'nest' of Spiggoties and things were pretty warm for about three hours. 1934 R. STOUT *Fer-de-Lance* xv. 248 'He's a dirty spiggoty.' 'No, Archie, Mr Manuel Kimball is an Argentine.' 1938 *Amer. Speech* XIII. 311/1 'Spiggoty' originated in Panama during Construction Days, and is assumed to be a corruption of 'spikee de' in the sentence 'No spikee de English', which was then the most common response of Panamanians to any question in English. 1959 R. CAMPBELL *I would do it Again* xxvi. 196, I learned that the young ladies were mostly Spigoties, a name applied to those of mixed Spanish and Indian blood.

2. Spanish-American; more generally, a foreign language.

1914 E. O'NEILL *Movie Man* in *Lost Plays* (1950) 37 Say, you're getting to be a regular talker of spigoty! Slip me the answer to that word 'basta', will you? 1922 H. L. FOSTER *Adventures Trop. Tramp* ix. 132 Just stood around the dock and jabbered a lot of spiggoty talk at me, like I could understand spiggoty! I don't know a word of this damned Spanish, and I'm glad of it! 1923 *Amer. Legion Weekly* 27 Apr. 12 Yer S.O.L. kid!! All I can dope out is the date—I can't get this frog spiggity a 'tall!!

3. *attrib.* passing into *adj.* Central or South American; foreign.

1918 W. S. POAGUE *Diary* 1 Feb. (? 1919) 28 We repaired to the town and had a white man's dinner with some spigotti liquor. 1937 H. KLEMMER *Harbor Nights* 197 *Marijuana* is a popular spiggoty drug which has spread rapidly into the North during recent years.

spigot, *v.* Add: Also (*lit.*), to insert in the manner of a spigot.

1910 *Automobile Engineering 1911* 19/3 Single separate cylinders..are spiggotted deeply into the crankcase. 1954 *Automobile Engineer* XLIV. 507/2 This cover is spigoted into the housing and bolted to the front wall of the box.

spik, var. *SPIC *sb.* and *a.*

spike, *sb.*[2] Add: **2. d.** Usu. in *pl.* One of a number of sharp-pointed metal studs driven into the sole of a cricket boot, running shoe, etc., to give a surer foothold. Also (*pl.*) by metonymy, a pair of spiked shoes.

1832 P. EGAN *Bk. Sports* 348/2 And all in spikes and jackets clad, Elate for vict'ry came. 1898 *N.Y. Tribune* 23 Apr. 9/3 He was in collision with Jennings and McGann and his foot and legs were injured with their spikes. 1955 R. BANNISTER *First Four Minutes* ii. 16, I suddenly noticed that my best pair of spikes had split along the side. 1976 J. WAINWRIGHT *Who goes Next?* 29 'Footprints.. Spiked. Now *he* isn't wearing spikes.' Enfield nodded towards the corpse. 'The two Herberts who found him—I doubt if *they'll* have spiked shoes.'

e. *fig.* A prickly resentment; anger, venom. Freq. in phr. *to have* (or *get*) *the spike*, to be (or become) angry or offended.

1890 J. D. ROBERTSON *Gloss. Words County of Gloucester* 147 'To have the *spike*' is to be out of temper, or offended. 1895 *Daily News* 4 Jan. 3/7 Of course Chris gets the spike (in a temper) because Sullivan had shopped him. 1922 JOYCE *Ulysses* 388 He had in his bosom a spike named Bitterness. 1960 N. HILLIARD *Maori Girl* II. xi. 141 But you don't have to get the spike with me just for that. 1978 *Chicago* June 166/3, I had located the spike inside him, the one that Arabs get hooked on when they detect a Jewish émigré.

f. (*a*) A quantity of alcohol, esp. spirits, added to a drink. *U.S. slang.*

1906 *Dialect Notes* III. 157 *Spike*, *n.*, alcohol, an alcoholic beverage. 'This punch has a good big spike in it.' 1969 J. CHEEVER *Bullet Park* xiv. 189 She..returned with a bottle of whiskey and spiked her coffee... The spike steadied her hand. 1974 *Times-Picayune* (New Orleans) 14 Aug. III. 2 It's like chips without dips, or punch without the spike.

(*b*) A small quantity *of* a radioisotope or other substance added to a material in order to act as a tracer, reference, etc.

1959 R. E. TATE in Hausner & Schumar *Nucl. Fuel Elements* viii. 110 Spike enrichment, in which some of the fuel elements..contain plutonium dispersed in an inert matrix, requires a high through-put of the enriched elements. 1962 *Analytical Chem.* XXXIV. 709/2 The U[233] plus U[238] isotopic dilution spike is added to an unknown uranium sample. 1965 *Jrnl. Geophysical Res.* LXX. 1844/1 Five to twenty grams of sample was dissolved in 1 *N* HCl, TH[230] and U[233] spike added, and the solution evaporated to dryness. 1976 *Nature* 24 June 685/2 After ensuring that the spike and sample were well mixed, the cadmium was chemically extracted by ion exchange.

g. A bayonet. *Mil. slang.*

1928 E. BLUNDEN *Undertones of War* 270 The cowman now turned warrior mentioned out His up-and-down *sans* fierce 'bundook and spike'.

h. A hypodermic needle or syringe used for the injection of an intoxicating drug; hence, the drug itself or an injection of this. *slang* (orig. *U.S.*).

[1923 J. MANCHON *Le Slang* 285 *Spike*, *s*... 30 *V*[*ulgaire*] une aiguille.] 1934 *Detective Fiction Weekly* 21 Apr. 107/2 Both me and the twist was on junk and when they fanned us they found a spike on me but no stuff. 1953 ANSLINGER & TOMPKINS *Traffic in Narcotics* 315 *Spike*, a drug. *Also* a hypodermic needle, an injection of a drug. 1959 'E. MCBAIN' *Pusher* viii. 383 'You say you shot up together? Did you both use the same syringe?' 'No, Annabelle had his spike, and I had mine.' 1964 *Daily Tel.* 25 Nov. 22/6 Among the terms used by addicts are..'blast parties', for groups of marijuana smokers, and 'spikes' for hypodermic needles. 1974 J. WAINWRIGHT *Evidence I shall Give* xxxvii. 211 It was a mounting yearning. A craving... He needed a spike—badly! 1979 P. DRISCOLL *Pangolin* xvii. 139 This punk kid, shooting amphetamines, can't find enough spikes.

i. (*a*) *Electr.* A pulse of very short duration in which a rapid increase in voltage is followed immediately by a rapid decrease; (*b*) a burst of electromagnetic radiation marked by short duration or great intensity, esp. one from space.

1935 *Arch. Neurol. & Psychiatry* XXXIV. 1140 Sharp negative spikes in the record often seem to be associated with motor movements of a clonic sort. 1957 *Wireless World* Jan. 10/2 Some of the output-signals have the form of sharp spikes, each pulse of ignition interference producing two spikes of the same polarity. 1969 J. J. SPARKES *Transistor Switching* iv. 109 Noise-voltage spikes in the earth line or at the input have to exceed about 600 mV before significant signals appear at the output. 1973 T. PYNCHON *Gravity's Rainbow* I. 146 He's been under Rollo Groast's EEG countless times since first he came to 'The White Visitation', and all's normal-adult except for, oh once or twice perhaps a stray 50-millivolt spike off a temporal lobe. 1969 *Astrophysical Jrnl. Lett.* CLVII. L73 Four out of five consecutive optical spikes, each reaching in some 10 days a peak of luminosity that is a factor of 2–3 above a varying background level,..have been observed for the single QSS 3C 345. 1974 *Nature* 8 Nov. 113/1 Well known transient phenomena such as supernovae, galactic radio noise spikes. 1975 D. G. FINK *Electronics Engineers' Handbk.* XI. 13 The typical output of an optical laser consists of a series of spikes. 1977 *Sci. Amer.* Oct. 53/2 Grindlay and Gursky suggested that the X-ray photons of a burst are released in a two-second 'spike'. 1980 *Nature* 7 Feb. 551/1 There are no clear spikes in the Kavalur bursts.

j. *Journalism.* A spindle on which recent newspaper stories are filed, *spec.* when rejected for publication.

1936 B. BROOKER *Think of Earth* II. iii. 141 The editor picked up a spike-file from the top of the desk. 1942 W. FAULKNER *Go down, Moses* 374 He took the press association flimsy from its spike and handed it to Stevens. 1962 [see *copy-taster* s.v. *COPY *sb.* C]. 1974 D. SEAMAN *Bomb that could Lip-Read* vii. 58 The P.A. copy was neatly pierced by a spike, Fleet Street's time-honoured way of giving the thumbs-down to a story.

4. Also *spec.* the casual ward of a workhouse (see CASUAL *a.* 9) (earlier and later examples); an institution affording more or less temporary accommodation for the homeless.

1866 *Temple Bar* XVII. 184 Let the 'spikes' be what they may they were a great deal better than the 'paddingkens'. 1903 J. LONDON *People of Abyss* viii. 78 On asking him what the 'spike' was, he answered, 'The casual ward. It's a cant word.' 1933 'G. ORWELL' *Down & Out* xxvi. 189 D'you come out o' one o' de London spikes (casual wards), eh? 1949 C. GRAVES *Ireland Revisited* viii. 125 At first we did not understand thieves' slang, or that a 'spike' meant a workhouse. (We were told to avoid the Portsmouth 'spike'.) 1972 *Times* 27 Dec. 2/8 'If this place was not here,' a proud articulate Glaswegian 'dosser' said, 'we'd be on the road or in the reception centre, the spike. I have been in the spike for the past 11 months.' 1980 *Guardian* 2 Oct. 18/1 A generation ago there were half a dozen lodging houses in the town..as well as the 'spike' or casual ward of the workhouse.

5. c. *spike-heeled, -helmeted, -tailed* (earlier example).

1953 D. DODGE *To catch Thief* i. 11 She was dressed for the evening; a long gown, fragile, spike-heeled slippers, a fur wrap. 1981 A. LURIE *Language of Clothes* iv. 106 French-speaking Canadians..negotiating the icy snow-heaped streets..in nyloned legs and spike-heeled boots. 1916 R. GRAVES *Goliath & David* 6 And look, spike-helmeted, grey, grim, Goliath straddles over him. 1870 *North Alabamian* (Tuscaloosa, Ala.) 12 Jan. 1/2 'Who's that gentleman, my little man?' was asked of an urchin. 'That one with the spike-tailed coat?'

6. spike-fiddle *Mus.* = *REBAB; **spike heel,** a fashionable narrow high heel of a woman's shoe, tapering towards a point (cf. *stiletto heel* s.v. *STILETTO *sb.* 5); hence, a spike-heeled shoe; **spike microphone,** (*colloq.*) **mike** (see quot. 1962).

1940 C. SACHS *Hist. Mus. Instruments* (1942) xii. 242 Most Islamic instruments..have no place in art music, with the exception of the Persian *spike fiddle* (called *rabāb* or *kamānga a'gūz* in the Near East)... Malay fiddles, which have preserved the old Persian name in the form *rebab*, are much simpler. 1974 SCHACHT & BOSWORTH *Legacy of Islam* (ed. 2) x. 500 Al-Fārābī also provides the first description of a bowed instrument, the *rabāb*. This was later known in two forms, one with a separate neck, the other a spike-fiddle with a hemispherical sound-chest. 1929 D. L. MOORE *Pandora's Letter Box* xi. 205 The 'spike' heel now popular is disgustingly difficult to balance on. 1950 'S. RANSOME' *Deadly Miss Ashley* xi. 136 A pair of sandals..featured spike heels decorated with brilliants. 1971 D. C. BROWN *Yukon Trophy Trails* ii. 31, I wanted to live in a log cabin, shoot my own steaks and never wear another pair of spike heels. 1962 *Symposia of Zool. Soc.* VII. 8 There is also the 'spike' microphone, which is extremely small, and can be driven through hard material to record sounds in an inner chamber. 1966 *Economist* 3 Dec. 1029/2 Though there is no federal law against bugging or wire-tapping *per se*, the Supreme Court held that the spike microphone driven through the wall of Mr Black's hotel room constituted physical trespass. 1950 *Washington Post* 20 Mar. B1/4 Police had slipped the 'spike mike' into a wall common to the adjoining premises. 1973 'D. HALLIDAY' *Dolly & Starry Bird* xvi. 248 *No one* in sneakers with spike mikes and tapes and transmitters?

spike, *sb.*[4] *slang.* [Back-formation f. *SPIKY *a.*[2] 4: a use of SPIKE *sb.*[2]] An Anglican who advocates or practises Anglo-Catholic ritual and observances.

1902 *Church Times* 14 Mar. 320/2 A priest is wanted for this parish. A hard-working Catholic. Not a 'spike'. 1914 J. W. LEGG *Eng. Church Life* 159 It would seem that there were spikes (as Dr. Bright of Christ-church used to call them) in 1768. 1922 E. RAYMOND *Tell England* II. iii. 204 My altar has generally been two ration boxes, marked 'Unsweetened Milk', but the spike has surrounded it. And look here.., the spike knows how to die. He just asks for his absolution and his last sacrament, and—and dies. 1930 SAYERS & 'EUSTACE' *Docs. in Case* i. 37 He turned out to be an earnest and cultivated middle-aged spike from Keble. 1952 R. MACAULAY *Let.* 23 May (1961) 318 Is she one who would be shocked at seeing communicants at High Mass? There have always been those there, I gather; and 'spikes' don't like it. 1963 C. MACKENZIE *My Life & Times* II. 203 In that summer of 1897 Sandys Wason was still a deacon... It was he who started using 'spike' for an extremist... One can still be spiked up; one can still talk of spikiness, and in Anglican circles be understood. 1980 A. N. WILSON *Healing Art* iv. 47 There were several other effigies of famous spikes, including the legendary Father Tooth.

Hence as *v.*[2] *trans.* with *up*: to make (more) 'spiky' or High Church; to enliven with ritual; also **spi·kery,** 'spiky' character or behaviour.

1923 C. MACKENZIE *Parson's Progress* xvi. 214 Was it really worth while trying to spike up the Rector and his services and his flock? 1958 B. PYM *Glass of Blessings* iii. 48 A new vicar trying to spike things up a bit. 1965 C. E. POCKNEE *Parson's Handbk.* (ed. 13) vii. 85 There is no ancient authority for the custom of sitting for the psalms. This is a slovenly piece of modern 'spikery'. 1972 C. STEPHENSON *Merrily on High* ii. 35, I was encouraged by the high church ladies who would listen with amusement and interest to my plans for 'spiking' up the church. 1980 A. N. WILSON *Healing Art* ix. 110 For all her spikery, there would always be a part of herself which found it impossible to shake off the freedoms of scepticism.

spike, *v.*[1] Add: **2.** *fig.* (Earlier and later examples, esp. in phr. *to spike* (*some*)*one's guns*.)

1823 T. CREEVEY *Let.* 11 Mar. in *Creevey Papers* (1903) II. iii. 66 He has himself entirely spiked his guns in the House of Commons. 1862 E. HALL *Jrnl.* 9 Mar. in O. A. Sherrard *Two Victorian Girls* (1966) II. 291 He proceeded to kiss her forehead... She should have spiked the first gun instead of leaving it to clear the way for the advance of others. 1927 *New Republic* 21 Sept. 122/2 They have flitted from one foolish suggestion to the other. The silliest of these was that, to spike the third-term objection, Mr. Coolidge would agree, if elected in 1928, to resign at the expiration of his eighth year of continued occupancy. 1953 L. P. HARTLEY *Go-Between* 16 My enemies would be off their guard, they would never suspect danger from a gun they had so thoroughly spiked. 1971 S. E. MORISON *European Discovery Amer.: Northern Voy.* xiv. 469 It remained for Samuel de Champlain to spike the legend of a City of Norumbega, storied like a New Jerusalem.

5. c. To lace (a drink) with alcohol; to fortify (beer, etc.) by the addition of spirits. Also *transf. slang* (orig. *U.S.*).

1889 L. PENDLETON *In Wiregrass* xviii. 201 Water from biled hops an' poke root, an' 's sweetened wi' 'lasses and spiked wi' good strong whiskey. 1900 *Dialect Notes* II. 63 *Spike*, to fortify a drink by adding wine or spirits. 1915 *Ibid.* IV. 229 *Spike*, to flavor with wine or whiskey, as 'She spikes her cakes.' 1941 J. SMILEY *Hash House Lingo* 52 *Spike*, (bar) add liquor to a drink. 1952 B. MALAMUD *Natural* 24 A crushed cocoanut [*sic*] drink which he privately spiked with a shot from a new bottle. 1962 *Sunday Times* (Colour Suppl.) 14 Oct. 24/1 Spike a *béchamel* sauce with Parmesan cheese. 1980 G. THOMPSON *Murder Mystery* (1981) xxii. 175 She made tea, which he spiked with bourbon.

d. Of a newspaper editor: to reject (a story or part of one) as by filing it on a spike (*SPIKE *sb.*[2] 2 j).

1908 A. S. M. HUTCHINSON *Once aboard Lugger* v. vii. 263 Tiny little scrap of news..copied out a dozen times by Mr. Issy Jago and left..at the offices of as many newspapers. Seven sub-editors 'spiked' it, [etc.]. 1940 W. P. CROZIER *Jrnl.* 9 Dec. in D. Ayerst *Guardian* (1971) xxxiv. 539 E.A.M. disgruntled because I spiked pars (for London letter) on last night's bombing. 1950 C. M. KORNBLUTH in *Astounding Science Fiction* July 150/2 The M[anaging] E[ditor]..decided nobody would believe it. He spiked the story on the 'dead' hook. 1961 B. WELLS *Day Earth caught Fire* vii. 115 This is the newsroom, the place where all the best stories are spiked. 1978 L. HEREN *Growing up on The Times* viii. 283, I discovered that my story had been spiked. It was the first and only time the paper had questioned my judgment, and I felt badly about it.

e. *intr.* To inject another or (for *refl.*) oneself with an intoxicating drug. Also *trans.* and *fig.*

1935 N. ERSINE *Underworld & Prison Slang* 69 *Spike*, to take a shot of dope. 'He spiked about an hour ago.' 1971 J. MANDELKAU *Buttons* v. 68 Almost immediately I was spiked with wine and acid. 1973 T. PYNCHON *Gravity's Rainbow* I. 47 Kevin Spectro will take his syringe and spike away a dozen times tonight..to sedate Fox (his generic term for any patient). 1974 *Guardian* 28 Jan. 11/5 The addicts..'ll sometime try and spike you, try and get you mainlining too. 1977 'J. LE CARRÉ' *Honourable Schoolboy* ii. 46 The girl..alone and spiked with tiredness.

f. To enrich (a nuclear reactor or its fuel) *with* a particular isotope; to add a small proportion of some distinctive material to.

1956 *Ann. Rev. Nucl. Sci.* VI. 330 A natural uranium power reactor of this size might not become critical, so that the reactor will be 'spiked' with a few enriched elements. 1959 F. G. FOOTE in Hausner & Schumar *Nucl. Fuel Elements* v. 78 The uranium can be spiked with either plutonium or U[233]. 1971 *New Scientist* 13 May 386/2 Early work in Britain on spiking enriched uranium with plutonium as a possible fuel for Advanced Gas-cooled Reactors. 1974 *Nature* 1 Feb. 310/2 The homogenised mixture was spiked with a known volume of a solution containing 1 μg ml[-1] of the N-nitroso compound. 1976 *Lancet* 4 Dec. 1223/1 Concentrations were measured against cadmium standards..prepared from blood spiked with cadmium chloride standard solution. 1977 *Rolling Stone* 13 Jan. 31/2 Silkwood had slipped a vial of plutonium into her vagina or rectum, then used a syringe to spike her samples at home.

7. *intr.* To rise in a spike; to protrude angularly.

1958 *Listener* 18 Sept. 418/2 High mounds of rubble and tangled, bombed machinery which spiked into the air like the legs of dead animals. 1975 N. NICHOLSON *Wednesday Early Closing* I. 21 St. George's steeple spikes up against the sky, graceful as a larch tree and bold as Blackpool Tower.

spike-bozzle (spəikbǫ·z'l), *v. slang* (orig. *Mil.*). Now *rare.* Also **spike-boozle.** [f. SPIKE *v.*[1] 2: the second element is obscure (but see quot. 1962).] To render (an enemy plane, etc.) unserviceable; to destroy; to upset. Also *transf.*

1915 H. ROSHER *In R. Naval Air Service* (1916) 123 Last night 'old man Zepp' came over here... Two machines went up to spike-bozzle him, but..never even saw him. 1923 *Blackw. Mag.* July 5/1 Piffers used to wear them..until the clothing department spike-bozzled them. 1942 BERREY & VAN DEN BARK *Amer. Thes. Slang* § 759/10 Spike bozzle, to chase an enemy plane. 1962 W. GRANVILLE *Dict. Sailors' Slang* 110/2 *Spike boozle*, render ineffective; sabotage, blow up or otherwise destroy a ship or upset a plan. A merging of *spike*, a gun and bam-*boozle*, cheat, mystify, bemuse.

spiked, *a.*[2] Add: **3. a.** Laced or fortified with alcohol. Also *transf.* and *fig.*

In some examples parsable as pa.pple. of the verb.

1909 *Dialect Notes* III. 374 *Spike*, *v.tr.* 1. To mix an alcoholic with a non-alcoholic beverage. Chiefly in the *pp.* 'This lemonade is heavily spiked.' 1929 *Detective Fiction Weekly* 23 Mar. 161/2 'I can ditch a drink that I suspect of being spiked.' This word 'spiked' was that year [*sc. c* 1899] the very newest slang, signifying 'doped'. 1942 E. PAUL *Narrow St.* i. 2 Some of the early risers huddled round the counter to swallow their coffee, often spiked with cheap, watered rum or cognac. 1960 J. MCNAMEE *Florencia Bay* 52 Then the promoter who, on the strength of spiked assays, had floated a company. 1974 P. CAVE *Mama* (new ed.) XI. 95 The babble of conversation was liberally spiked with laughter, merriment and enthusiasm.

b. *transf.* Containing a small addition of a radioactive or otherwise distinctive material; enriched.

1959 F. G. FOOTE in Hausner & Schumar *Nucl. Fuel Elem.* v. 78 The best use of plutonium-spiked uranium fuels would be in fast neutron reactors. 1962 *Analytical Chem.* XXXIV. 709/2 Separate spiked and unspiked analyses must be made of the unknown uranium sample. 1975 *Sci. Amer.* Oct. 27/2 The idea is to add about half a gram of plutonium 239 to each kilogram of uranium. Since a kilogram of natural uranium contains 7·2 grams of U-235, the 'spiked' fuel would contain a total of 7·7 grams of fissionable material.

spikey: delete 'obs.'

spikily, *adv.* (Later examples.)

1959 *Times* 18 June 15/4 Higo Harada's Music for Piano and Violin, Op. 12,..was spikily intellectual music, rarely visited by any spirit of delight. 1976 J. COOPER *Harriet* (1977) xiii. 81 Her spikily mascara'd eyes softened.

spikiness. (Later examples.)

1962 *Listener* 3 May 788/3 The curious spikiness of the quartet writing. 1977 D. BAGLEY *Enemy* i. 10 Her comments were acute... I found her spikiness of mind very agreeable. 1979 E. H. GOMBRICH *Sense of Order* viii. 200 Does it make sense, therefore, to diagnose the spikiness of Gothic shapes by the same standards we might adopt when seeing such spiky forms in a contemporary hand?

spiking, *vbl. sb.* Add: **3.** The action of adding a spike (*SPIKE *sb.*[2] 2 f (*b*)).

1962 *Newnes Conc. Encycl. Nucl. Energy* 772/2 Spiking was first carried out in the NRX reactor in 1951. 1974 *Sci. Amer.* Feb. 121/1 The use of isotope-spiking and mass spectrometry. 1979 *Liquefied Petroleum Gas* (Shell Internat. Petroleum Co.) 4 In some instances LPG can also be transported by 'spiking'—that is, enriching crude oil with small quantities of LPG.

spiky, *a.*[2] For 8 **spikey** read 8— **spikey** and add: **3.** (Later examples.)

1930 M. KENNEDY *Fool of Family* xv. 147 'How spikey you are!' protested Fenella mildly. 'Oh yes. Keep your temper when I'm rude. You would.' 1955 *Times* 4 Aug. 10/5 What matter that the melodic line is as seductively curved as that of any of the great operatic romanticists of the near past, instead of being spiky and angular in the contemporary fashion? 1964 CRYSTAL & QUIRK *Prosodic & Paralinguist. Features in Eng.* iv. 47 Pitch variation, with extremes in a 'spiky' movement on the one hand..and in a 'glissando' movement..on the other. 1981 N. J. CRISP *Festival* vii. 176 He seemed more relaxed ..not as spiky and difficult as he had been.

4. Of a particularly ritualistic or High-Church Anglican character. *slang.*

1893 W. BRIGHT *Let.* 20 Oct. (1903) 348 The ultras, as they might be called, on the Catholic side, present Church ideas, too often, in a form altogether too hard to be attractive; I believe I am said to have called it 'spiky', in a letter to my friend the Principal of Ely College. 1921 *Church Times* 12 Aug. 147/3 We wonder what would be thought of some of his sayings if they were uttered by a spiky young curate to-day. 1950 A. WILSON *Such Darling Dodos* 34 She became a daily communicant and delighted the more 'spikey' of her neighbours. 1962 *Times Lit. Suppl.* 13 July 505/1 Her story is of the American priest, Charles Phillips, whose churchmanship would in England be rated high-to-spiky. 1977 B. PYM *Quartet in Autumn* xxiv. 212 He had been a server at the spikiest Anglo-Catholic church.

spile, *sb.*[2] Add: **2. b.** Also *Canad.* (Earlier and later examples.)

1844 *Knickerbocker* XXIII. 444 The spiles you see sticking from sugar-holes in every maple. 1868 *Amer. Naturalist* Mar. 39 He remembers very distinctly making 'spiles' of its [*sc.* elder] stems when tapping sugar-trees. 1947 K. M. WELLS *Owl Pen* 89 It is time..to get the rusty spiles and sap buckets down from the beams in the woodshed. It is maple syrup time. 1973 L. RUSSELL *Everyday Life Colonial Canada* xi. 144 The operator drilled a hole into the side of the [maple tree] trunk..and set into this a small wooden spout called a spile.

spile, *sb.*[3] **1. d.** *spile-driver* (earlier example).

1865 *Atlantic Monthly* Apr. 393/1 By means of a spile-driver, an iron pipe..is driven down until it rests upon the solid rock.

spilikin, var. SPILLIKIN, SPELLICAN in Dict. and Suppl.

spilite. Add: Also with pronunc. (spi·ləit). [orig. formed as F. *spillite* (Al. Brongniart: see A. H. de Bonnard in *Nouveau Dict. d'Hist. nat. appliquée aux Arts* (rev. ed.) (1819) XXIX. 371), f. Gr. σπίλος spot, stain.] In mod. use, an altered basalt, commonly amygdaloidal in structure, which is characterized by the albitization of its constituent feldspars and the presence of numerous secondary minerals, and is exemplified by many pillow-lavas. (Earlier and later examples.)

Brongniart did not use *spilite* in its usual modern sense, applying it to amygdaloidal diabase, and much variation in usage has occurred subsequently. There is a comprehensive review by T. G. Vallance in *Proc. Linnean Soc. New South Wales* (1960) LXXXV. 8–52.

1834 *Jrnl. Geol. Soc. Dublin* I. 119 In adopting the term spilites in preference to that of amygdaloid,..I have been induced to do so because the English names were.. irrelevant to the variable nature of the mineral substances. 1893 *Geol. Mag.* Decade III. X. 59 Prof. de Lapparent.. refers to these 'spilites' as amydaloidal melaphyres. 1911 *Ibid.* Decade V. VIII. 203 In addition to the pillow-structure there are certain characteristics that mark the spilites of Great Britain. The first of these is that they are as a rule very completely decomposed, and the second that their felspars are always rich in soda. 1937 A. JOHANNSEN *Descriptive Petrogr. Igneous Rocks* III. 300 The British Geological Survey uses spilite 'to designate the Carboniferous, Devonian, and Ordovician pillow-lavas of Devon and Cornwall..' This variation from the original usage apparently still persists, for Holmes in 1920, defined them as basaltic rocks whose feldspars have been albitized. The term should be dropped. 1960 *Proc. Linnean Soc. New South Wales* LXXXV. 10 English-speaking geologists made little or no use of the term during the 19th century. Rocks which would have been called spilites by the French were usually known as greenstones. *Ibid.* 12 There appear to be four main criteria used in the recogniton of spilites, namely, fabric, mineralogy, chemical composition and geological occurrence. Much of the confusion associated with the name spilite derives from the fact that various workers have emphasized different features. 1974 F. FIALA in G. C. Amstutz *Spilites & Spilitic Rocks* 18 Spilites are abundant in pillow lavas. Nevertheless, both terms are not synonymous, as has been assumed by many workers. 1981 F. J. TURNER *Metamorphic Petrol.* (ed. 2) 223 (*heading*) Spilites as metamorphic rocks.

Hence **spilitiza·tion,** alteration into spilite; **spi·litized** *a.*

1946 *Amer. Jrnl. Sci.* CCXLIV. 313 During spilitization the ferro-magnesian minerals were broken down, and..the manganese, with other constituents, migrated to the tops of the flows. 1960 *Q. Jrnl. Geol. Soc.* CXVI. 388 Dewey and Flett..came to the right conclusion in attributing a large proportion of spilitization of pillow-lavas to reactions quickly achieved under submarine volcanic conditions. 1962 E. A. VINCENT tr. *Rittmann's Volcanoes & their Activity* viii. 225 In thick submarine basalt flows.. it is often to be observed that only their upper layers consist of spilitized pillow lavas. 1974 F. FIALA in G. C. Amstutz *Spilites & Spilitic Rocks* 19 The association of spilitized diabases and spilites with keratophyres is.. marked in the initial Devonian volcanism of northern Moravia. 1978 *Nature* 8 June 459/1 Due to the vitric nature and widespread secondary alteration (incipient spilitisation and weathering) the exact petrologic character and petrogenetic affinity of these rocks have not been clearly defined.

spilitic (spəil-, spili·tik), *a. Geol.* [f. SPILITE in Dict. and Suppl. + -IC.] Of, pertaining to, or of the nature of spilite; characterized by the presence of spilite.

1911 *Geol. Mag.* Decade V. VIII. 205 The basic felspar can be traced in every stage of development as we follow the spilitic lavas from the outer precincts of the aureoles to the actual contacts with the granite. *Ibid.* 243 The spilitic suite of igneous rocks. 1965 A. HOLMES *Princ. Physical Geol.* (ed. 2) xxx. 1126 The volcanic rocks interbedded with eugeosynclinal sediments are characteristically submarine lavas of the spilitic suite.., including pillow lavas. 1974 D. S. COOMBS in G. C. Amstutz *Spilites & Spilitic Rocks* 377 Augite was thus *unstable* relative to chlorite..during the production of the diagnostic spilitic mineralogy. 1977 A. HALLAM *Planet Earth* 201 The geosynclinal belt across Europe, with its associated zones of spilitic volcanics and serpentines, has been interpreted as site of an old ocean.

spill, *sb.*[1] Add: **1. b.** (Later examples.)

1904 *Kynoch Jrnl.* Oct.–Dec. 204 The reader will observe..a flaw technically termed a 'spill', the result of a small hollow or depression near the surface of the ingot, which, in the process of rolling has been closed, but which, in the extending process of pressing the metal into a case has again been opened & made manifest. 1954 A. R. BAILEY *Text-bk. Metallurgy* xi. 377 Cavities near the surface..are likely to become opened up during working, become oxidised and fail to weld up, thus forming surface oxide laminations, sometimes known as spills.

2. a. (Earlier example.)

1821 M. EDGEWORTH *Let.* 14 Nov. (1971) 268 Harriet performed to admiration as Fire eater—I held lamp..and lighted spills which she seemed to devour.

b. *spill-holder* (earlier example), *-jar, vase.*

1851 *Parker's Jrnl.* 5 Apr. 185/2 The mantelpiece is probably painted to imitate marble, and on it are placed two 'spill-holders' of perforated card, with bouquets worked in silk on each. 1868 C. L. EASTLAKE *Hints on Household Taste* v. 134 Chimneypiece spill-vases, made of brass,..decorated with a pattern in encaustic colour. 1903 A. BENNETT *Leonora* x. 282 Reaching a second spill from the spill-jar on the mantlepiece. 1978 *Country Life* 13 Apr. (Suppl.) 38/1 (*caption*) A Pair of English Porcelain Spill Vases.

spill, *sb.*[4] Add: **2. a.** *spec.* = *oil spill* s.v. *OIL *sb.*[1] 6 e.

1972 L. M. HARRIS *Introd. Deepwater Floating Drilling Operations* xvii. 178 Equipment and practices designed for safety and reliability are the first line of defense against oil spills and pollution. Should a spill occur, however, advance planning can reduce its severity. 1975 *Petroleum Rev.* XXIX. 237/3 The ability of present-day booms to contain a spill is limited to good weather conditions.

4. A diffusion of light, esp. beyond the area intended to be illuminated.

1952 [see *BAFFLE *sb.*[1] 5]. 1972 T. COE *Don't lie to Me* (1974) i. 4 Her features..hard to read in the dim spill from a nearby streetlight. 1977 P. SCUPHAM *Hinterland* 9 A spill Of light poured off rough drapery As blacks and whites and ochre tones Work shifts about the curtain wall.

5. *Austral. Pol.* A vacating of other posts after one important change of office.

1956 J. T. LANG *I Remember* 311 There had to be an annual election of leader. That made it inevitable that some members would intrigue against the leader hoping for a Cabinet spill. 1974 *Courier-Mail* (Brisbane) 6 July 3/1 After Mr. Tucker's election as new party leader, Mr. Dean (Sandgate) moved for a 'spill' of all other Opposition front bench positions. 1975 *Australian* 18 Mar. 1 It will be left to Mr Fraser's supporters to force the issue and move against Mr Snedden through either a spill of leadership positions or a motion of no confidence.

6. *attrib.* and *Comb.*, as *spill-proof* adj.; **spill burner,** a form of burner used in some gas turbines which allows excess fuel to be recirculated; **spill valve,** a valve which serves to allow the escape of surplus fluid.

1945 *Proc. Inst. Mech. Engineers* CLIII. 464/2 (*caption*) o Indicates the condition giving a mean particle size of 200μ: On the spill burner this limit is not attained. 1972 H. COHEN et al. *Gas Turbine Theory* (ed. 2) vi. 177 A second practical method of obtaining good atomization over a wide range of fuel flow: the spill burner. It is virtually a simplex burner with a passage from the vortex chamber through which excess fuel can be spilled off. 1920 E. BUTLER *Internal Combustion Engine Design & Pract.* (ed. 2) viii. 138 When used on an automobile, owing to vibration, they [*sc.* accumulator cells] should be occasionally examined for scaling or other damage, and also for loss of solution, if not spill proof. 1944 W. A. KOEHLER *Princ. & Applications Electrochem.* (ed. 2) II. iv. 69 A portable radio battery with a transparent plastic case and spill-proof cover. 1963 *Glamour* Oct. 12/1 (Advt.), New spray mist! Unbreakable. Spill-proof... Intimate by Revlon. 1922 *Trans. Inst. Engineers & Shipbuilders Scotland* LXV. 421 This spill valve opens from the discharge chamber of the pump and is worked by the same lever as actuates the pump plunger. 1959 *Motor Manual* (ed. 36) ii. 30 At a predetermined moment, this outlet is closed and fuel is then forced to the engine cylinder until a second outlet (commonly called a spill valve) is opened so as to release the pressure and return the surplus oil to the supply side of the system. 1975 T. D. MORTON *Reed's Motor Engin. Knowledge for Marine Engineers* iii. 77 A fuel spill valve (pneumatically loaded) maintains rail pressure as decided at the controls.

spill, *v.* Add: **II. 12. a.** (Further example.)

1918 D. H. LAWRENCE *New Poems* 30 In the street spilled over splendidly With wet, flat lights.

13. b. Also from a parachute.

1925 *Literary Digest* 11 July 25/1 Used as a dividing mark in folding the parachute, and also to 'spill' the wind out of it after a landing. 1942 A. M. LOW *Parachutes in Peace & War* iii. 48 The parachute might 'spill air' & drop faster. 1976 A. WHITE *Long Silence* vii. 59 Spill air, ride the motion down as rapidly as is safe. Look around ..trying to make out the other parachutes.

15. a. (Later *transf.* and *fig.* examples.)

1920 H. CRANE *Let.* 14 Apr. (1965) 37 A mood which rose and spilled over in a slightly cruder form than what you see. 1962 *Amer. Speech* XXXVII. 17 The metropolitan area of New York City..has spilled into the surrounding countryside, engulfing cities and communities which had their own economic and cultural centers. 1962 A. NISBETT *Technique of Sound Studio* iv. 84 Where a tape has spilled and wound round the spindle and become crumpled, speech recordings are often still playable. 1972 B. MOORE *Catholics* i. 17 An anchor spilled like entrails from its bow, falling deep into the sea. 1976 *Times* 10 Sept. 1/6 Cape riots spill into white zones. The anti-apartheid rioting spilled out of Cape Town and spread into white areas of Cape Province. 1976 P. & W. PROCTOR *Women in Pulpit* vii. 124 Although Connie Parvey is primarily a campus chaplain, her work in this case spilled over into the role of hospital chaplain.

16. a. *trans.* To utter (words); to confess or divulge (facts). *slang* (orig. *U.S.*).

[1574: in Dict., sense 10 b.] 1917 R. W. LARDNER *Gullible's Travels* 213 'Go ahead and spill it,' I says. 1920 C. SANDBURG *Smoke & Steel* 44 Men at tables spill Peloponnesian syllables. 1923 'B. M. BOWER' *Parowan Bonanza* iv. 47 Maybe he taught the parrot that lingo just to have her spill it in town and start a rush. 1925 E. WALLACE *King by Night* xxxi. 143 Spill it quick, Goldy. 1944 [see *OIL *sb.*[1] 3 j]. 1953 K. TENNANT *Joyful Condemned* xxiv. 233 She was going to spill everything to him.. She would have pooled you, too. 1973 'B. MATHER' *Snowline* vi. 70 You didn't come down here..just to tempt me..with a beer. Spill it. 1977 I. SHAW *Beggarman, Thief* II. i. 119 He picked up the phone to call the Colonel, spill everything.

b. *to spill the beans*: to reveal a secret. *slang* (orig. *U.S.*).

1919 T. K. HOLMES *Man from Tall Timber* xxviii. 355 'Mother certainly has spilled the beans!' thought Stafford in vast amusement. 1921 R. D. PAINE *Comr. Rolling Ocean* viii. 136 The beans are spilled, and that is what Maddigan guessed the moment he set eyes on you. 1928 [see *CLEAN *a.* 3 d]. 1929 E. LINKLATER *Poet's Pub* vii. 91 'Tell me the truth,' she says. 'Spill the beans, Holly, old man!' 1945 *Sun* (Baltimore) 28 Nov. 1/1 A Government publication in this country spilled the beans concerning our urgent interest in experiments with uranium. 1958 E. DUNDY *Dud Avocado* I. vi. 93 Spilling beans of shattering truths or equally shattering lies. 1966 D. VARADAY *Gara-Yaka's Domain* vii. 82 Wilson in an indulgent moment of weakness 'spilt the beans'. 1979 G. HAMMOND *Dead Game* vii. 83 You asked me to trust you... So now I think you'd better spill the beans. 1982 *Listener* 23–30 Dec. 3/1 Julian Critchley spills the beans about El Vino and says why he likes it.

c. *to spill one's guts* (*out*): to divulge as much as one can, to confess. *slang* (chiefly *U.S.*).

1927 C. F. COE *Me—Gangster* iv. 78 'Throw him out, eh?' the old man snarled... 'Throw him out an' have him spill his guts about the whole gang?' 1945 S. J. BAKER *Austral. Lang.* vii. 140 *To hold one's guts*, to be silent, and *to spill one's guts*, to talk, reveal a secret. 1973 *Black Panther* 8 Sept. 10/3 Mistakenly believing that Haldeman and another assistant had told the truth during previous questioning, Butterfield spilled his guts out. 1979 'A. HAILEY' *Overload* III. viii. 226 The kid—he was eighteen, by the way, and not long out of trade school—broke down and spilled his guts.

spill-. Add: **3. spillbank** (see quot. 1961).

1909 H. M. WILSON *Irrigation Engin.* (ed. 6) xix. 560 The material brought up by these [buckets] is deposited on one of two endless belt-carriers running on booms which dump it on either spillbank. 1961 L. D. STAMP *Gloss. Geogr. Terms* 427/2 Spill-bank is the term used by British and Indian engineers engaged in river training for the bank of coarse alluvium spilled over by a river in flood (C. C. Inglis).

4. In the sense 'that empties by spilling', as *spillbucket.*

1938 L. MACNEICE *Earth Compels* 20 Bric-a-brac Pick-a-back Spillbucket Splits.

spillage (spi·ėldʒ). [f. SPILL *v.* + -AGE.] The action or fact of spilling; that which spills or is spilt. Also *attrib.*

1934 WEBSTER, *Spillage*.., the act of spilling; that which spills or is spilled over; as, a trough to divert any *spillage*. 1937 *Times* 9 Oct. 9/6 A miner..was trapped underground by a fall of rock and completely buried. Dr. Saunders arrived at 3.15 p.m., by which time the rescue party had removed most of the spillage from the imprisoned man's body. 1947 *Sun* (Baltimore) 26 June 8/1 Recent oil spillages in and near Baltimore harbor have resulted in great annoyance to bathers. 1962 A. NISBETT *Technique of Sound Studio* vii. 128 Remember also when handling music tapes that any damage due to spillage or stretching, etc., will be more noticeable than on speech tapes. 1963 *Daily Mail* 15 May 4/4 On every bottle of brandy sold in the Hilton, two and two-thirds ounces are written off for 'spillage'. 1969 *Gloss. Aeronaut. & Astronaut. Terms* (*B.S.I.*) IV. 3 *Spillage drag*, the difference between the drag at a given intake flow and the drag at some specified intake flow. *Ibid.* 4 *Spillage*, the amount by which the intake flow is less than some specified intake flow. 1969 *Daily Tel.* 30 Aug. 12/6 A pesticide spillage has killed 11,000 fish in the South Holland Drain. 1972 *Ibid.* 14 Apr. 19/3 A glass is often placed beneath the tap as a substitute for the drip tray and used for topping up... The patron should ensure that he rejects any beer which contains a mixture of this spillage. 1973 *Times* 30 July 11/2 If you spill anything you simply move your saucepan to another burner, lift off the support, and underneath there is a spillage bowl which has collected what has boiled over and which has only to be taken out and rinsed. 1976 *Washington Post* 23 Jan. A 14/4 Accidents are caused by drivers trying to avoid rock spillage..on the Beltway. 1976 *Offshore Platforms & Pipelining* 222/1 The utmost attention was given to reducing the risk of oil spillage during operations. 1976 *Cambridge Independent Press* 16 Dec. II. 7 (Advt.), Family-size oven with 'Cook Clean' panels... Hob light, auto-timer, slide-out spillage trays. 1977 *R.A.F. News* 27 Apr.–10 May 3/4 Allegations that excessive 'spillage' drag from the engine intakes will seriously impair Tornado's performance, particularly at high angles of incidence in the transonic speed range, were refuted.

spiller, *sb.*[3] **2.** (Earlier example.)

1884 *Bull. U.S. Nat. Museum* No. 27. 998 Mackerel pocket or spiller... The pocket was introduced into the mackerel-seine fishery in 1878 for holding the surplus catch which would otherwise spoil before being cleaned and salted.

spiller (spi·lə ɹ), *sb.*[4] [f. SPILL *sb.*[1] + -ER[1].] = SPILL *sb.*[1] 2 a.

1936 M. MITCHELL *Gone with Wind* 71 Pork took a long spiller from the mantelpiece, lit it from the lamp flame and went into the hall.

spillikin, spellican. Add: **3.** (Further examples.) Also in sing.

1907 E. GOSSE *Father & Son* ii. 50 My nerves were a packet of spilikins. 1940 W. DE LA MARE *Pleasures & Speculations* 71 No fine shades of psychology, or ethical spellicans are here. 1945 —— *Burning-Glass* 44 To ponder upon a moth..A spelican from his palm.

spilling, *vbl. sb.*[1] Add: **3. c.** The action of causing air to escape from a parachute; also such an escape of air.

1930 *Flight* 11 July 784/2 As usual, he judged his distance perfectly, and by judicious 'spilling' landed right in front of the club enclosure. 1951 *Gloss. Aeronaut. Terms* (*B.S.I.*) III. 15 *Spilling*, the escape of air, with local partial collapse, at the periphery of a parachute canopy, caused either by the instability of the parachute or by side-slipping.

spillover (spi·lōᵘvəɹ), *sb.* and *a.* Also **spill-over.** [f. SPILL- + OVER *adv.*] **A.** *sb.* That which spills over; the process of spilling over; (an) incidental development; a consequence, a repercussion, a by-product.

1940 [see *KAROK]. 1949 *Richmond* (Va.) *Times-Dispatch* 6 Oct. 26/1 A rush to buy got under way as soon as the opening bell sounded. This was evidently a spillover from yesterday when the market established a new high for the year. 1957 J. I. M. STEWART *James Joyce* 10 This has no relevance to the action, and is a spill-over from Joyce's more openly autobiographical writing in the history of Dedalus. 1957 P. WORSLEY *Trumpet shall Sound* 269 Weber..looks for the source of change in social tensions..; the danger of resentment of disciplinary authority..; or the spill-over into irrational channels of affect which is not absorbed by the rational order. 1962 *Lancet* 12 May 1009/1, 32 patients had pulmonary disease preoperatively, presumably owing to 'spill-over', and only in those with chronic pulmonary suppuration did this fail to clear up. 1970 *Daily Colonist* (Victoria, B.C.) 11 Nov. 3/5 The threat to Canadian security possibly being greater from a spillover of violence and the potential of anarchy in the streets. 1971 *Gloss. Electrotechnical, Power Terms* (*B.S.I.*) III. ii. 27 *Spill-over*, in a.c. signalling on multi-link connections, that part of a signal which passes from one section to another before the connection between the sections is split. 1973 'D. HALLIDAY' *Dolly & Starry Bird* vii. 97 The revolver was pointed straight at my head... I wasn't keen on the spillover into small arms. 1977 *New Yorker* 26 Sept. 66/3 The continuing rise in crime, its increasing spillover into the white community, and the failure of our criminal justice system..have created apprehension. 1979 *United States 1980/81* (Penguin Travel Guides) 309 Soho is a very livable combination of 19th-century cast-iron buildings, spillovers from Little Italy. 1980 *Times Lit. Suppl.* 5 Sept. 965/5 What economists call 'spillovers'—those unwanted side-effects, incidental to the legitimate production and use of man-made goods, that are familiar to the public as pollution, noise, congestion and other pervasive hazards and disadvantages.

B. *attrib.* or as *adj.* That results from spilling over; incidentally developed.

1953 J. S. HUXLEY *Evolution in Action* iv. 96 The nervous excitation spills over and is discharged into another channel, that of digging a nest hole. Such irrelevant spill-over activities are called displacement activities. 1961 *Listener* 2 Nov. 692/2 It is always experimental (and exciting) to work on the basis of 'spillover' audiences, and place a known 'difficult' programme immediately after a known winner. 1967 *Spectator* 14 July 44/1 With the postwar growth of technology and population these disservices or 'spillover effects'..have become too conspicuous to be ignored. 1971 *Physics Bull.* Nov. 654/2 The tilting of the subreflection of a Cassegrain antenna with the object of redirecting spillover radiation away from warm earth regions, has considerable geometric optical consequences. 1981 *Times* 31 July 19/1 If the United States slips into recession the spill-over effect could hamper what appears to be a slow recovery in the world chemical industry.

spillspilling (spi·lspiliŋ), *ppl. a. nonce-wd.* [Redupl. f. SPILL *v.*: see -ING[2].] Repeatedly spilling.

1922 JOYCE *Ulysses* 439 Advances with a tilted dish of spillspilling gravy.

spillway (spi·lwēⁱ). Also **spill-way.** [f. SPILL- + WAY *sb.*[1]] **1. a.** A channel or slope built to carry away surplus water from a reservoir.

1889, 1892 [in Dict. s.v. SPILL- 3]. 1943 J. S. HUXLEY *TVA* xi. 83 The road bridge across the spillway..became part of a bold composition finishing on a strong horizontal line and broken only by the small scale tower of the lift shaft. 1962 R. B. FULLER *Epic Poem on Industrialization* 133 It was but a warning trickle down the spillway Of the cosmic power dam. 1979 *United States 1980/81* (Penguin Travel Guides) 611 The world's largest concrete dam..has a spillway twice as high as Niagara Falls.

b. *Physical Geogr.* A natural feature providing a channel for the overflow or escape of water from a lake.

1914 DAVID & PRIESTLEY *Brit. Antarctic Exped. 1907–9 Rep. on Sci. Investigations* I. i. 5 At intervals the horst is breached by valleys..which form outlets, 'by-washes', or 'spillways' for the surplus inland ice and snow. 1929 C. R. LONGWELL *Pirsson's Textbk. Geol.* (ed. 3) I. vii. 156 Each of the lakes..has streams or springs flowing into it, and an outlet or spillway determined by the lowest point in the rim of the containing basin. 1957 G. E. HUTCHINSON *Treat. Limnol.* I. i. 80 When the ice had fully retreated, spillways to the west were established. 1971 *Nature* 3 Sept. 34/2 The latter [*sc.* a gorge] may have operated for a time as a spillway from a glacial lake covering the upper part of the Chew valley.

2. *attrib.*, as *spillway dam.*

1913 P. A. M. PARKER *Control of Water* vii. 415 We find in spillway dams, that such damage as occurs is apparently due to a partial vacuum induced by the flowing water. 1951 W. F. HEALD *Scenic Guide Oregon* 11 The spillway dam, 1450 feet long crosses the north channel.

spilt, *ppl. a.* Add: **2. a.** (Further examples.) Also with *out.*

1881 O. WILDE *Poems* 149 The heart of the lotus drenched..With the spilt-out blood of the rose-red wine. 1939 DYLAN THOMAS *Map of Love* 11 Carved birds blunt their striking throats on the salt gravel, Pierce the spilt sky with diving wing in weed and heel An inch in froth.

b. (Earlier examples.)

1738 SWIFT *Polite Conv.* I. 27 'Tis a Folly to cry for spilt Milk. 1828 J. NEAL *Rachel Dyer* xx. 248, I pity you both, but there's no help for you now—never cry for spilt milk.

spin, *sb.*[1] Add: **1. b.** The product of a machine which rotates and twists toffee.

1913 D. H. LAWRENCE *Sons & Lovers* I. 4 She..went to get Annie a spin of toffee.

2. a. (Examples with reference to spin-drying.)

1966 D. V. DAVIS *New Domestic Encycl.* (ed. 2) iii. 127 Each spin should be restricted to about 15 seconds. 1969 K. J. MILLS *Washing Wisdom* vi. 97 If a spin has been selected, then this will happen for the required time.

c. (Earlier and further examples.) Also, the ability to impart such a motion to the ball; spin-bowling.

1851 J. PYCROFT *Cricket Field* ix. 174 The more spin you give the ball, the better the delivery; because then the ball will twist, rise quickly, or cut variously, the instant it touches the ground. 1855 *F. Lillywhite's Guide to Cricketers* 81 A bowler with a great spin. 1861 *Bell's Life* 10 Nov. 6/3 But however good their trundling, pitch or pace, or break or spin, Still the monarch of all bowlers, to my mind was Alfred Mynn. 1951 *Sport* 27 Apr.–3 May 12/1 Wardle's left-arm slows and the rapid advance of Eddie Leadbetter, the 23-year-old leg-break and googly bowler, offer spin enough to check the best opponents. 1955 *Times* 5 July 4/1 Silk was probably right in thinking that his seamers would get more out of it than his spinners, but what he did not know was that Oxford were likely to lose their composure against spin. 1958 *Times* 19 July 3/5 Even if we are to be short of spin in Australia our pace attack will be as formidable as it was four years ago. 1977 *World of Cricket Monthly* June 32/3 It was left to the spin of Holford to separate the dangerous pair.

d. *Aeronaut.* A steep descent in which an aircraft describes a helix at an angle of attack greater than the stalling angle; *flat spin*: see s.v. *FLAT *a.* 15.

1915 *Aeroplane* 10 Nov. 578/2 Several times their aeroplane got into a 'spin'. 1930 NAYLER & OWER *Aviation To-Day* 324 To come out of the spin the pilot pushes his stick forward so as to unstall the wings. 1939 [see *SPIRAL *a.*[1] 1 b]. 1953 *Aeroplane* 30 Jan. 140/2 Apart from spins that resulted in fatal accidents, there were also quite a number before 1914 which had less drastic consequences. 1977 *New Yorker* 27 June 62/1 The Liberator..went into a spin, dived toward the earth.

e. The continued revolving of the clutch of a motor after being disengaged.

1919 B. H. DAVIES *Motor Driving* 129 The disc A is pressed against the disc B, which damps out the 'spin'. 1948 A. W. JUDGE *Mod. Motor Engineer* (ed. 4) II. 305 If the car is fitted with a clutch stop, or brake, this will effectively obviate clutch spin on disengagement. 1977 J. H. HAYNES *Ford Fiesta Owners Workshop Man.* v. 95/2 Clutch spin is a condition that occurs when there is an obstruction in the clutch.

f. *colloq.* The act of playing a gramophone record, esp. on the air; a session of playing gramophone records.

1977 *Broadcast* 28 Nov. 10/1 Records not receiving maxiplay but likely to get at least one daytime spin. 1977 *R.A.F. News* 11–24 May 20/6 Disc jockey SAC 'Duke' Bedford set off on a record non-stop spin of 72¼ hours.

4. a. Now freq. a ride or run in a motor vehicle or aircraft.

1907 G. MEREDITH *Let.* 7 Sept. (1970) III. 1606 To vary my growls..I hire a motor and have a spin of 100 miles, a way of ensuring appetite and prolonged sleep. 1942 A. CHRISTIE *Body in Library* xi. 102 He took his car and went for a spin down to the front. 1960 M. SPARK *Ballad of Peckham Rye* vii. 164 'I called for you last Saturday,' Mr. Druce said. 'I thought you would care for a spin.' 1976 *Southern Even. Echo* (Southampton) 12 Nov. 14/5 A 15-year-old boy..took his father's car, and went in it for a spin around Basingstoke. 1978 G. VIDAL *Kalki* iii. 59 'We'll go for a spin,' said Kalki... First we had to get through to the Katmandu airport. That took an hour... It was five o'clock before we were able to take off.

c. *Austral.* and *N.Z. slang.* A (good, bad, etc.) experience or piece of luck.

1919 in W. H. DOWNING *Digger Dialects* 47. 1929 K. S. PRICHARD *Coonardoo* xix. 188 Mollie had had a crook spin when the children were little. 1934 T. WOOD *Cobbers* xi. 134 People generally said they were having a bad spin. 1948 D. BALLANTYNE *Cunninghams* 132 What a miserable bloody spin he was having. 1960 N. HILLIARD *Maori Girl* IV. iii. 261 Give her a decent spin now and she'll turn out all right. 1964 H. P. TRITTON *Time means Tucker* 113 When I remarked that he'd had a tough spin he grinned, 'Served me right for being such a blanky fool.'

6. *Math.* The local rotation of a continuous medium, as expressed by the curl of the local velocity; vorticity.

1878 W. K. CLIFFORD *Elements of Dynamic* I. II. ii. 123 The velocity-system due to a definite angular velocity about a definite axis is spoken of as the rotation-velocity. To specify it completely we must assign its magnitude and the position of the axis... A rotation-velocity, so denoted, shall be called a spin. 1878, 1882 [in Dict., sense 2 a]. 1937 S. L. GREEN *Hydro- & Aero-Dynamics* ii. 14 There is an essential difference between motions with and without spin, i.e. between rotational and irrotational motion. 1958 *Science* 4 Apr. 731/3 The arrow, which points along the axis of spin and has length equal to the angular speed, is the 'vorticity'... The case of no spin is appropriate to some applications, particularly for waves on water and for aeronautics.

7. *Physics.* An intrinsic property of certain elementary particles which is a form of angular momentum and is usu. pictured as a rotation (it is distinct from angular momentum possessed by virtue of occupation of an orbital); a vector representing this in the case of a particular particle.

1926 UHLENBECK & GOUDSMIT in *Nature* 20 Feb. 264/1 To start with, we shall consider the effect of the spin on the manifold of stationary states which corresponds to motion of an electron round a nucleus. 1938 R. W. LAWSON tr. *Hevesy & Paneth's Man. Radioactivity* (ed. 2) viii. 83 The electrons exhibit a rotation proper to themselves and generally referred to as 'spin', and this confers upon them the property of small magnets. 1943 *Ann. Reg. 1942* II. 365 Mesons are of two types—the more usual with the longer life and zero spin and another of shorter life having a spin 1*h*. 1955 *Sci. News Let.* 19 Feb. 117/3 They have passed proton beams through two hydrogen filled chambers to get atoms whose spins are all in the same direction. 1966 C. R. TOTTLE *Sci. Engin. Materials* i. 13 Beryllium..can have two electrons in the 2s state, since they can have opposite spin. 1974 P. W. ATKINS *Quanta* 223/1 The spin is the intrinsic, characteristic, and irremovable angular momentum of a particle. A convenient fiction is to suppose that the spin is the angular momentum arising from the rotation of a body about its own axis.

8. *Austral. slang.* [Perhaps a different word.] Five pounds in money.

1941 *Coast to Coast 1941* 225 'How'd you go at the two-up?' I asked. 'Aw, I got a spin,' said Tom. 1949 L. GLASSOP *Lucky Palmer* 15 'Not five bob. A spin,' said the carpenter, fishing a five pound note out. 1962 S. GORE *Down Golden Mile* 261 Backed Sweet Friday for a spin... But it never run a drum.

9. *attrib.* and *Comb.*, as *spin axis*; (sense *7) *spin angular momentum, multiplicity* (= *MULTIPLICITY 1 e (*a*)), *quantum number, state*; **spin-allowed** *a. Physics*, consistent with the selection rules describing changes in spin quantum number; **spin bowler** *Cricket*, a slow bowler who imparts spin to the ball on delivery; also **spin-bowling; spin-dye** *v. trans. Textiles*, to dye (textiles) by a process which incorporates the colouring matter before the filament is formed; so **spin-dyed** *ppl. a.*, **spin-dyeing** *vbl. sb.*; **spin echo** *Physics*, a radio-frequency signal induced in a coil surrounding a system of (esp. nuclear) spins in a static magnetic field in the plane of the coil following the application of two radio-frequency pulses to the coil; freq. *attrib.*; **spin flip** *Physics*, the quantum jump of a particle from one spin state to another; **spin glass,** a dilute solid solution of a magnetic substance in a non-magnetic host; **spin-labelling** *Chem.*, the technique of labelling (*LABEL *v.* 2) with stable paramagnetic radicals which can be studied using electron spin resonance techniques; so **spin-label** *v. trans.*; also as *sb.*, a radical or compound used in spin-labelling; **spin-labelled** *a.*; **spin-lattice** *Physics*, used *attrib.* with reference to the interaction between a crystal lattice and a particle possessing spin; **spin-orbit** *Physics*, used *attrib.* with reference to the interaction between spin and orbital motion, esp. of an electron in an atom; **spin polarization** *Physics* = *POLARIZATION 1*; so **spin-polarized** *a.*; **spin-spin** *Physics*, used *attrib.* with reference to the interaction between two or more particles possessing spin; **spin-stabilized** *a. Astr.*, (of a rocket, spacecraft, etc.) stabilized in a desired orientation by being made to rotate about an axis; so **spin-stabilization**; **spin tunnel** *Aeronaut.* = *spinning tunnel* s.v. *SPINNING *vbl. sb.* 8 c; **spin vector** *Math.* and *Physics*, a vector representing rotation; *spec.* one which by its magnitude and direction represents the intrinsic angular momentum of a particle; **spin wave** *Physics*, a cooperative oscillation in the alignment of electron spins, propagated through a magnetic material in the form of a wave.

1972 DEPUY & CHAPMAN *Molec. Reactions & Photochem.* iii. 34 Electronic transitions between states of the same multiplicity, i.e., singlet-singlet and triplet-triplet transitions, are spin-allowed. 1973 Spin-allowed [see *PHOSPHORESCENCE]. 1928 *Proc. R. Soc.* A. CXVII. 610 Goudsmit and Uhleenbeck have introduced the idea of an electron with a spin angular momentum of half a quantum and a magnetic moment of one Bohr magneton. 1977 *Dædalus* Summer 27 Some quantum numbers, such as electric charge and spin angular momentum, refer to physical, measurable attributes of the particle. 1922 GLAZEBROOK *Dict. Appl. Physics* I. 421/2 A true circular disc, mounted eccentrically on a spin-axis normal to its plane, illustrates the state of bad static balance. 1926 *Nature* 20 Feb. 264/1 This couple will cause a slow precession of the spin axis. 1977 *Dædalus* Fall 48 The spin axis of this satellite could be oriented at will by command from the Earth. 1920 D. J. KNIGHT in P. F. Warner *Cricket* 42 Let any player who does not believe in this dictum go and face such spin bowlers as Barnes, Hearne, [etc.]..on a sticky wicket. 1955 *Times* 14 July 3/5 One has doubts as to just how much the spin bowlers approve of these pitches. 1976 DEXTER & MAKINS *Testkill* 90 A book he was planning on the history of the great spin bowlers. 1955 *Times* 5 July 4/6 They are clearly a useful team, well equipped with stroke-players but perhaps a little short of spin bowling. 1963 *Times* 7 June 4/3 Mushtaq Mohammad, Pakistan's Test batsman, amply confirmed his ability against spin bowling. 1975 *Cricketer* May 20/1 Critics of the English cricket scene often bemoan the lack of spin bowling. 1948 *Jrnl. Soc. Dyers & Colourists* LXIV. 291/2 Dyes suitable for spin-dyeing..are those of nitrodiarylamine, azo, and anthraquinone types. 1961 F. D. LEWIS *Chem. & Technol. Rayon Manufacture* xii. 148 Lack of space precludes more than a cursory consideration here of the spindyeing of viscose rayon yarns. *Ibid.* 149 It would not seem possible to spindye thread with a single organic pigment. 1963 A. J. HALL *Textile Sci.* iii. 132 Generally, by this use of pigments instead of dyes it is possible to produce coloured fibres and yarns having the maximum fastness properties; they are often designated 'spin-dyed' yarns. 1949 *Bull. Amer. Physical Soc.* XXIV. VII. 13/2 Spin Echoes. E. L. Hahn, University of Illinois. 1963 *Times* 8 May 2/7 (Advt.), Post-Doctoral Research Fellowship..for work on relaxation time measurements by spin-echo techniques. 1979 *Nature* 22 Nov. 367/1 Experiments using the weakly inelastic scattering of neutrons have also become feasible; they give similar information [about the motion of polymer chains in a liquid], but for higher frequencies: from $10^{11}s^{-1}$ with 'back scattering spectrometers' down to $2 \times 10^{8}s^{-1}$ with the most recent 'spin echo' method. 1955 *Physical Rev.* C. 1505/1 The special form of the dispersion relations depends on the high-frequency behavior of the spin-flip amplitude. 1971 *New Scientist* 1 Apr. 6/2 Light.. from a Q-switched carbon-dioxide laser is the 'pumping' source for the spin-flip laser. 1975 *Sci. Amer.* May 89/2 Units of time and distance are specified in terms of the frequency of the hydrogen spin-flip at 1,420 megahertz. 1970 P. W. ANDERSON in *Materials Res. Bull.* V. 549 (*heading*) Localisation theory and the Cu-Mn problem: spin glasses. 1976 *New Scientist* 2 Dec. 533/1 The essential feature of the spin glass transition is that below some critical temperature, the magnetic atoms are locked or frozen into random orientations (hence the analogy with a glass). 1979 *McGraw-Hill Yearbk. Sci. & Technol.* 247/2 The transition from the paramagnetic to spin glass phase occurs as the material is cooled through a certain temperature point. 1965 T. J. STONE et al. in *Proc. Nat. Acad. Sci.* LIV. 1010 (*heading*) Spin-labeled biomolecules.

Ibid. 1785 Stone..obtained a measure of the rotational mobility of the region of the macromolecule to which the spin labels were bonded. *Ibid.* 1791 It appears likely that the spin-labeling method will be useful in studies of the interaction of haptens, coenzymes, inhibitors, and substrates with proteins and other macromolecules. **1969** *New Scientist* 30 Oct. 224/2 The use of spin-labelled substrate analogues shows clearly..the subtle changes in protein conformation that occur during the enzymic process. **1974** *Nature* 12 Apr. (verso rear cover), The most commonly used spin labels are molecules which contain a nitroxide moiety. **1975** D. H. Burrin in Williams & Wilson *Biologist's Guide to Princ. & Techniques Pract. Biochem.* v. 161 By spin labelling glycerophosphatides with a stable nitroxide free radical, the lateral diffusion of the labelled molecules in a membrane..may be studied. **1976** *Tetrahedron Lett.* June 2180 Spin labelling with carbonyl compounds was carried out for 1–2 days in pH 7·5 aqueous phosphate buffer. **1938** *Physica* V. 502 The period of the alternating field must be of the order of magnitude of the relaxation time τ of the spin-lattice equilibrium. **1978** P. W. Atkins *Physical Chem.* xix. 629 The motion of nuclei can affect the shapes and widths of lines in n.m.r. just as it does in e.s.r., and the spin-lattice and spin-spin relaxation times can be discussed in precisely the same way. **1956** *Nature* 18 Feb. 306/1 The 'electronegativity' of all the cations is not that of their ground states, for several of them are in states of lower spin-multiplicity. **1977** I. M. Campbell *Energy & Atmosphere* viii. 219 Radiative transitions between states of the same spin multiplicity are easy in the absence of contravention of other selection rules. **1932** Bacher & Goudsmit *Atomic Energy States* 14 The interaction between the two electrons is smaller than the spin-orbit interaction of the 3*p* electron. **1963** G. Troup *Masers & Lasers* (ed. 2) 185 Because of spin-orbit coupling..the 'spin' S in the Hamiltonian is not necessarily equal to the true spin of the ion, but is rather an 'effective spin' related to the multiplicity of levels actually found. **1978** W. J. Kaufmann *Exploration of Solar System* xi. 396 This means that Mercury rotates three times about its axis while circling the sun twice. This phenomenon is known as spin-orbit coupling. **1966** *Proc. Physical Soc.* LXXXIX. 587 The fundamental spin polarization (**P**) is typically represented by $\mathbf{P}(x) = P_0 \hat{\xi} \cos Qx$ where **Q** is the spin-density wave vector and $\hat{\xi}$ is a unit vector along the direction of polarization. **1970** I. E. McCarthy *Nuclear Reactions* i. i. 7 They [*sc.* the proton and the neutron] may be identified separately by measuring the spin polarization of the beams. **1977** *New Scientist* 24 Feb. 455/2 The process is called 'spin polarisation by optical pumping'. **1968** M. S. Livingston *Particle Physics* vii. 138 If parity is not conserved in other weak interactions, it could lead to an asymmetry in the direction of emission of β rays from spin-polarized radioactive nuclei. **1980** *Nature* 17 Jan. 248/1 One way of producing spin-polarised electrons is to take a storage ring, fill it with electrons or positrons, and leave for an hour or so. **1930** Pauling & Goudsmit *Struct. Line Spectra* iv. 53 It seems to be sufficient to give all electrons the same rotation, so that they have the angular momentum $sh/2\pi$, with *s*, the spin quantum number, always 1/2. **1964** J. W. Linnett *Electronic Struct. Molecules* i. 9 Since each spatial orbital is defined by the three quantum numbers *n*, *l* and *m*, this is equivalent to saying that each orbital can accommodate two electrons, and these only if they have different spin quantum numbers. **1934** H. E. White *Introd. Atomic Spectra* xii. 186 (*heading*) Spin-spin-, or ss-coupling. **1936** *Proc. R. Soc.* A. CLV. 641 This degeneracy can only be removed by an external magnetic field or by the spin-spin interaction between the ions. **1978** Spin-spin [see *spin-lattice* above]. **1961** *Planetary & Space Sci.* IV. 262/2 Spin stabilization turned out to be one of the more difficult design problems encountered in the program. **1976** M. H. Kaplan *Mod. Spacecraft Dynamics & Control* iv. 124 The use of spin stabilization in this orbit..was seen as a means of achieving the..mission at an early date. **1956** *Spaceflight* I. 19/1 The satellite..was intended to be spin-stabilized so that one hemisphere..always faced the Sun. **1976** M. H. Kaplan *Mod. Spacecraft Dynamics & Control* iv. 124 The concept of a spin-stabilized, 24-hour satellite was first proposed by the Hughes Aircraft Company in the fall of 1959. **1956** *Physical Rev.* CIV. 488/1 If the two spin states are equally abundant, this indication implies that the average neutron width is not the same in each state. **1947** A. Pope *Wind-Tunnel Testing* i. 11 The vertical tunnels for testing parachutes differ from the spin tunnels in that they require an even velocity front instead of the dish-shaped front required for the spin tunnel. **1959** F. D. Adams *Aeronaut. Dict.* 158/1 The air speed in the spin tunnel may be kept equal to the rate of descent of a tested model, causing the model, while spinning, to remain at a given height relative to the observer. **1882** Spin-vector [in Dict., sense 2]. **1899** C. J. Joly in W. R. Hamilton *Elem. Quaternions* (ed. 2) I. III. ii. 492 This vector γ has been called the spin-vector of the function φ. **1948** *Physical Rev.* LXXIII. 415/1 The integral Lorentz transformations are represented by exactly those spin transformations which, together with their inverse, map integral spinvectors into integrals pinvectors. **1981** *Sci. Amer.* Apr. 47/3 A particle with one-half quantum of intrinsic spin can have only two possible orientations; in the simplest case, where the particle is in motion, the spin vector can point either in the same direction the particle is moving or in the opposite direction. **1936** *Proc. R. Soc.* A. CLV. 644 For long waves one can carry out the transition to a continuum in the same way as Bloch did in his theory of 'spin waves'. **1953** *Rev. Mod. Physics* XXV. 235/2 The rf magnetic field excites spin waves with wave numbers in the range 1 to 10⁸ cm⁻¹. **1973** *Sci. Amer.* Jan. 88/3 Waves produced by oscillations of magnetic moment in ferromagnetic and antiferromagnetic materials (spin waves) generate quasiparticles called magnons.

spin, *sb.*² (Earlier example.)

1842 C. Ridley *Let.* Mar. in *Cecilia* (1958) vii. 90 Mrs. Dixon, a good lady..who was sitting in a very tidy, very hot room with two old spins as companions.

spin, *v.* Add: **I. 2. a.** (Examples relating to man-made fibres.)

1899 *Jrnl. Soc. Arts* 8 Dec. 62/2 The solution of gelatine must be coloured to the required shade before being spun. **1921** T. Woodhouse tr. *Foltzer's Artif. Silk* 23 This solution..is conducted or spun through special capillary tubes. **1974** *Encycl. Brit. Macropædia* VII. 258/2 Acetate is dry spun by extruding acetone solutions of cellulose acetate into hot air.

c. (Examples relating to man-made fibres.)

1899 *Jrnl. Soc. Arts* 8 Dec. 63/2 If too much water is present the collodion will not be tenacious and therefore will not spin. **1963** A. J. Hall *Textile Sci.* ii. 50 If a suitable proportion of opaque white titanium dioxide pigment is added to the viscose solution just before it is spun into filaments, these can be produced with a lower lustre or even be matt.

3. a. (Examples relating to man-made fibres.)

1891 *Jrnl. Soc. Chem. Industry* 30 Apr. 359/1 This invention relates to the process and apparatus for spinning artificial silk from nitro-cellulose. **1973** *Materials & Technol.* VI. iv. 292 Dry spinning. This method is used to spin filaments from syrups which can be prepared by dissolving the fibre-forming materials in a suitable solvent.

II. 10. c. *Cricket.* Of a ball: to travel through the air with spin (Spin *sb.*¹ 2 c).

1851 J. Pycroft *Cricket Field* ix. 174 Clarke is not conscious of any attempt to make his ball spin or twist: a certain action has become habitual to him. **1970** N. Cardus *Full Score* 120 Mailey would tell me how much he revelled in the 'feel' of a ball spinning from his fingers. **1980** *Cricketer International* Apr. 83/3 Hollies..came up to bowl the next ball. It neither spun nor twisted in the air but drifted up and then down in a graceful parabola.

d. *Aeronaut.* Of an aircraft or its pilot: to perform or undergo a spin (*spin *sb.* 2 d).

1914 *Aeroplane* 1 July 17/2 If a 'scout' started to spin round its own nose it would never come into control again. **1918** J. M. Grider *War Birds* (1927) 66, I am going up to ten thousand [feet] and shut off and spin down and see what happens. **1931** C. D. Barnard *Learning to Fly* 151 Only a stalled aeroplane will spin—in other words when the machine is no longer airborne. **1952** *Technical Rep. Aeronaut. Res. Committee 1943* II. 755 The model usually spins more steeply and recovers from the spin more easily than the aeroplane.

e. Of a motor clutch: to continue to revolve after being disengaged.

1918 *Dyke's Automobile Encycl.* (ed. 7) 662/1 When a clutch spins, when thrown out of engagement, it is difficult to shift gears. **1928** *Motor Man.* (ed. 27) 57 The flange..is brought into contact with the stop, which acts as a brake and prevents the cone or plate spinning. **1965** D. Kaberry *Ford Corsair* viii. 53/1 If the clutch is spinning, difficulty will be experienced in engaging gear, particularly from a standstill.

11. a. Fig. phr. *to spin one's wheels* (U.S. colloq.), to mark time, to do nothing productive.

1960 *Wall St. Jrnl.* 15 Mar. 1 'We're just sitting here spinning our wheels,' says a disgruntled Naval aviator in California. **1974** *Evening Herald* (Rock Hill, S. Carolina) 19 Apr. 4/1 The Selective Service System has, in fact, done little but spin its wheels for the past 14 months.

e. *Cricket.* Of a bowler: to impart spin to (a ball) on delivery; to cause (the ball) to break after pitching.

1904 *Westm. Gaz.* 21 May 3/1 He can spin the ordinary left-hander's break-back. **1920** D. J. Knight in P. F. Warner *Cricket* 49 Rhodes and Woolley..seem to spin the ball in such a way that it gets straight up from the pitch in quite a different manner. **1960** [see *cut *v.* 31 a].

f. To make (an aircraft) perform a spin (*spin *sb.* 2 d).

1918 J. M. Grider *War Birds* (1927) 87 Then Ortmeyer ..spun a Camel into the ground and killed himself. **1928** O. Stewart *Aerobatics* 13 Machines fitted with Handley-Page automatic slots are extremely difficult to spin. **1952** *Technical Rep. Aeronaut. Res. Committee 1943* II. 766 The aircraft was spun from 28,000 ft., one 2½-turn spin to the left being carried out and two 2½-turn spins to the right.

g. = *spin-dry *v.* Also *absol.*

1959 *Which?* Aug. 92/1 Tumbler driers are intended to take washing that has either been wrung or spun to a half-dry state. **1966** D. V. Davis *New Domestic Encycl.* (ed. 2) iii. 127 If you spin a minimum-iron fabric while it is hot you are much more likely to press in creases. **1969** K. J. Mills *Washing Wisdom* vi. 98 Hook the outlet pipe over the sink. Spin for 15–20 seconds.

h. To play (a gramophone record).

1965 *Listener* 23 Dec. 1036/3 If he will dust off his old Plum Label HMV 78..and spin it, he will hear Gertrude Lawrence very distinctly saying to Noël Coward: 'Strange how potent cheap music is!' **1966** T. Pynchon *Crying of Lot 49* v. 140 She sat alone..listening to Mucho's colleague Rabbit Warren spin records.

13*. spin down: a. *trans. Biol.* To centrifuge so as to cause the separation of components.

1947 Ackerman & Regato *Cancer* xiv. 817 The fluid is spun down and sectioned. **1965** *Proc. Nat. Acad. Sci.* LIV. 400 A unit of blood was drawn into a heparin receiving pack and spun down slowly for 15–20 min. **1978** *Sci. Amer.* Dec. 30/2 If the unbonded cells are then spun down into a mixed-up mass and given fluid containing antibiotics and the right salts, new hydras form in a few days.

b. *intr. Astr.* Of a rotating body, esp. a star: to rotate more slowly, usu. because of decreasing angular momentum. Also *trans.*

1967 *Nature* 24 June 1297/2 The boundary layer suction causes a slow internal motion which stretches vortex lines and spins down the solar interior. *Ibid.* 1299/1 The region itself spins down quickly. **1976** *Sci. News* 30 Oct. 280 This pulsar is spinning down so slowly that when the effects of its motion across the sky on its apparent (to us) spin rate are considered, it may actually be spinning up from the point of view of someone riding along with it. **1979** *Nature* 29 Feb. 602/2 Any loss of radiation from the region will transport vast quantities of angular momentum away from the rotating body. The reaction back on the body itself will cause it to spin down.

13. spin off: a.** *trans.* To throw off by or as if by centrifugal force in spinning; freq. *fig.*, esp. (*a*) *U.S. Comm.*, to distribute (stock of a new company) to shareholders of a parent company; to create (a company) in this way; (*b*) to produce as a by-product, side-effect, or indirect benefit.

1957 *N.Y. Times* 9 June F1/3 Right now, there is considerable speculation that du Pont will 'spin off' its G.M. stock—that is, give it to its own stockholders in the form of a dividend. **1959** *Wall St. Jrnl.* 20 May 1/5 A subsidiary set up in early 1950 to rent out space in an engineering firm's new building was spun off in December, 1954. **1964** *Science* 29 May 1113 The Systems Development Corporation (SDC) was 'spun off' by RAND in 1956 to help specifically with design and programming for the first computerized air defense system. **1969** *Physics Bull.* June 215/2 If..pulsars are neutron stars,..then these could rotate at up to 10³ Hz; they would 'spin off' electrons (or plasma) which would be accelerated in the star's magnetic field. **1969** *Wall St. Jrnl.* 3 July 4/2 The publicly owned company then 'spins off' those shares to its holders, who, in turn, often sell the shares to other traders. **1972** *Real Estate Rev.* Winter 5/2 A black who has invested his savings in a commercial or residential venture is not permitted..to parlay small investments into big ones by spinning them off through refinancing vehicles. **1972** *Publishers Weekly* 4 Dec. 35/1 From the file, the publisher can now produce updated editions with minimum effort, and can spin off subsidiary products as well. **1974** *Nature* 29 Mar. 459/2 Several papers have already been spun off from discussions and presentations at the CETI conference. **1977** *Time* 10 Jan. 47/3 Tandy Corp. has spun off most of its other businesses into separate companies chaired by Charles Tandy. **1979** *Daily Tel.* 15 Jan. 5/6 'Softly, Softly' was spun off 'Z Cars' and no one complained about that.

b. *intr.* To be thrown or move off by or as if by centrifugal force in spinning; usu. *fig.*

1969 *N.Y. Rev. Books* 16 Jan. 33/2 (Advt.), More interesting, however, is the way in which the material is organized to allow the class or individuals to spin-off into the study of related problems. **1969** *Sci. Jrnl.* Nov. 74 People have claimed..that space is a great thing because all the technology evolved will ultimately spin-off to the commercial market place. **1971** *New Scientist* 4 Mar. 488/1 These small companies specialising in technologically advanced products have 'spun off' largely from powerful local universities. **1971** *Daily Tel.* (Colour Suppl.) 6 Aug. 12/4 This prestige spins off to make Europeans more and more rail conscious.

13*. spin out:** *intr.* Of a vehicle: to skid round out of control. *N. Amer. slang.*

1954 *Amer. Speech* XXIX. 102 Last Sunday some car spun out and hit five parked cars. **1971** M. Tak *Truck Talk* 150 *Spin out*, to lose traction on a slippery road.

13**. spin up:** *Astr.* Of a rotating body, esp. a star: to rotate more quickly because of a gain in angular momentum or a redistribution of matter. Also *trans.*

1967 *Nature* 24 June 1299/1 If the interior were rotating at the rapid rate suggested by Dicke, the convective envelope would be spun up to this rate in a very short time. **1974** *Ibid.* 29 Nov. 366/1 As the Sun evolves, its rotating core gets smaller while its outer part spins up. **1976** [see sense 13* b above]. **1978** *Nature* 16 Feb. 635/1 Before the neutron star starts to accrete and spin up..it spends a comparatively long time being spun down in the weak stellar wind of its companion. **1979** *Ibid.* 11 Jan. 116/1 The neutron star would be spun up in the process.

spina. 1. *spina bifida*: substitute for def.: A congenital malformation of widely varying severity in which there is a failure of one or more vertebræ to surround completely the meninges and spinal cord, usu. with effects on spinal cord function. (Further examples.) [mod.L. (N. Tulpius *Observationes Medicæ* (1641) III. 233, 235).]

1853 J. Erichsen *Science & Art of Surgery* xliv. 631 Spina bifida may be met with in any part of the vertebral column. **1885** *Trans. Clin. Soc. London* XVIII. 361 Lastly, this examination serves to complete the refutation of the view..that spina bifida in the great majority of cases is due to a dropsy of the central canal of the cord. **1887** *Lancet* 2 July 4/2 The term 'spina bifida occulta' is applied to a defect in the arches of the vertebræ such as occurs in the commoner forms of spina bifida, but no tumour or cyst appears externally. The deformity is, however, accompanied by some remarkable external conditions. **1965** E. D. Smith *Spina Bifida* i. 5/1 In 1875, Virchow, who introduced the term 'spina bifida occulta', used it to describe a spina bifida in association with lumbosacral hypertrichosis. **1966** [see *meningo-myelocele* s.v. *meningo-]. **1974** Passmore & Robson *Compan. Med. Stud.* III. xxxvi. 9/1 Spina bifida is an incomplete vertebral arch. When this bony defect is covered with skin and fascia it is usually symptomless, and is called spina bifida occulta.

spinach. Add: Now also with pronunc. (spi·nɛtʃ). **3. b.** Nonsense, rubbish. *U.S. colloq.* (now *rare*).

[**1928** C. Rose in *New Yorker* 8 Dec. 27/2 (*caption*) 'It's broccoli dear.' 'I say it's spinach, and I say the hell with it.'] **1929** J. P. McEvoy *Hollywood Girl* xiii. 205 It's a flop and then I says to him, in other words I say it's spinach and I say to hell with it. **1933** E. Hawes (*title*) Fashion is spinach. **1934** A. Woollcott *While Rome Burns* 304 This..reticence..will..be described by certain temperaments as..good taste... I say it's spinach. **1950** R. Bissell *Stretch on River* xxi. 207 'It's a transferral of intent. It's a result of childhood trauma. It's Oedipus denial,' said my sister-in-law, who was beautiful, thank god, so you could put up with this spinach.

4. spinach beet: = *silver beet* s.v. *silver *sb.* and *a.* 21 c; (later examples); **spinach-green**, (*a*) a dark green vegetable dye made from spinach; (*b*) a dark green colour; also *attrib.* or as *adj.*; **spinach jade** (see quot. 1964); also *attrib.*

1885 W. Robinson tr. *Vilmorin-Andrieux's Veg. Garden* 379 The leaves of the Common White Leaf-Beet, or Spinach Beet, may be cut for use even earlier. **1978** *Times* 17 July 14/3 The experts recommend growing Swiss chard, otherwise known as spinach beet. **1845** E. Acton *Mod. Cookery* xx. 508 (*heading*) Spinach green, for colouring sweet dishes. **1861** Mrs. Beeton *Bk. Househ. Managem.* 250 (*heading*) Spinach green for colouring various dishes... Pick and wash the spinach free from dirt, and pound the leaves in a mortar to extract the juice. [etc.] **1896** [in Dict.]. **1937** *Burlington Mag.* June 300/1 A fine example of the spinach green jade K'ang-hsi period. **1969** R. Gordon *Kungli Kunglish* a The engine..was painted a spinach green. **1975** *Times* 3 July 14/3 A figured silk confection in spinach green. **1980** *Catal. Fine Chinese Ceramics* (Sotheby, Hong Kong) 220 A spinach-green jade covered censer, raised on tripod supports. **1958** W. Willetts *Chinese Art* I. ii. 61 Siberian jade has a rather distinctive appearance owing to the presence of small particles of black graphite embedded in the stone, which leads the Chinese to call it 'spinach jade'. **1964** M. Medley *Handbk. Chinese Art* 108/2 Spinach jade, the Chinese *po-ts'ai-yü*, a nephrite from Siberia characterised by black flecks of graphite. **1976** 'M. Delving' *China Expert* xiii. 158 Men was wearing a shirt of dark grey silk..fastened up to the neck with spinach jade buttons.

spinachy (spi·nɛdʒi, -ɛtʃi), *a.* [f. Spinach + -y¹.] Characteristic or suggestive of spinach.

1950 O. Nash *Family Reunion* (1951) 18 So spinach was spinach, I had no call to be spinachy. **1981** P. Theroux *Mosquito Coast* xv. 191 Tiptoeing through this spinachy swamp on duckboards.

spinal, *a.* Add: **I. b.** *spinal puncture* or *tap*: the insertion of a needle into the subarachnoid space of the spine, usu. in the lumbar region, so that cerebrospinal fluid may be withdrawn or something introduced.

1896 *Brit. Med. Jrnl.* Suppl. 4 Jan. 1/3 Only a few drops of fluid could be obtained by spinal puncture. **1919** A. Levinson *Cerebrospinal Fluid* i. 25 Corning, who was the first to use spinal puncture, employed an operation that was fraught with danger to the cord. **1973** Norack & Tharaken *Nervous System* iv. 35 Some CSF is withdrawn and replaced by air which acts as a contrast medium. The air is introduced by passing a needle either directly into the ventricle or between the lower two lumbar vertebræ (spinal tap) into the lumbar cistern. **1979** *Sci. Amer.* Aug. 66/3 He underwent at least 48 spinal taps, three air encephalograms and numerous myelograms. **1980** K. E. Moyer *Neuroanatomy* xxi. 265/2 In a spinal tap, or spinal puncture, the needle is always introduced into the subarachnoid space below the termination of the spinal cord itself.

3. (Further examples.)

1884 *Queen* 9 Feb. 132/2 (Advt.), Partial paralysis, Spinal curvature, constipation, congestion, &c. **1958** J. Blackburn *Face of Lion* viii. 54 A cripple with one leg longer than the other and a pronounced spinal curvature.

5. b. *spinal reflex*, a reflex involving the spinal cord but not the brain.

1868 *Phil. Trans. R. Soc.* B. CXC. 141 In the Dog and Cat the spinal reflex movements are more forcible. **1924** *Jrnl. Physiol.* LVIII. 411 Shivering to cold cannot be produced as a spinal reflex. **1978** *Brain Res.* CXLII. 431 Stimulation of all three segmental nerves simultaneously produced up to a 100% increase in size of the spinal reflex.

6. Also of a seat or carriage: designed to support the spine. Now *Hist.*

1884 *Queen* 16 Feb. 189/2 (Advt.), Every requisite for Invalids' Carriage.. Adjustable Spinal Couches. **1900** *Illustr. London News* 25 Aug. 293/2 (Advt.), Adjustable Bath Chair or Spinal Carriage. **1917** *Harrods Gen. Catal.* 1038/2 The Adjustable Spinal Bath Chair and Spinal Carriage Combined, enabling a person to take outdoor exercise in a sitting, reclining, or horizontal position. **1973** *Times* 7 May 17/7 Many years ago I travelled in a spinal carriage every few weeks from Selby to Leeds and back in the guard's van.

6*. *Physiol.* **a.** Involving the spine as containing a major part of the central nervous system: *spinal anæsthesia, analgesia*, anæsthesia, analgesia induced by an injection into the spine (see quot. 1938); *spinal block* (*a*) an obstruction to the flow of the cerebro-spinal fluid; (*b*) spinal anæsthesia or analgesia; *spinal shock*, a temporary flaccid paralysis and loss of reflexes in some muscles that may follow an injury to the spine, the ones affected being those whose nerves come from a point in the spinal cord below the site of the injury.

1885 *N.Y. Med. Jrnl.* XLII. 483/2 (*heading*) Spinal anæsthesia and local medication of the cord. **1912** *Jrnl. Amer. Med. Assoc.* 23 Nov. 1859/1 There are practically no contraindications to the employment of spinal analgesia. **1938** Maxson & Babcock *Spinal Anesthesia* i. 1 Spinal anesthesia is the term most commonly used... It is technically correct when all the sensory faculties—touch, temperature and muscle sense, as well as pain—are abolished in the affected region. Spinal analgesia is the correct technical term when pain sense alone is abolished without the loss of the epicritical faculties. The distinction between the two terms, however, is rarely made. **1974** Lichtiger & Moya *Introd. Pract. Anesthesia* xv. 152 Tetracaine is the most commonly used drug for spinal anesthesia. **1976** D. D. Moir *Obstetric Anaesthesia & Analgesia* xiii. 209 Low spinal analgesia is commonly used to administer forceps delivery. **1928** *Arch. Neurol. & Psychiatry* XIX. 613 The compression of the veins of the neck..is used most often in spinal lesions where it is commonly used to determine whether a 'spinal block' is present. **1976** D. D. Moir *Obstetric Anaesthesia & Analgesia* vii. 115 A low spinal block creates a tranquil patient, free of all pain. **1898** *Phil. Trans. R. Soc.* B. CXC. 134 Goltz's descriptions of spinal shock are masterly, but they refer entirely to the Dog, and to transsection of the cord in the lumbar region. **1978** *Jrnl. Neurol. Sci.* XXXVI. 335 A patient in spinal shock has..

b. Used to describe an animal whose spine has been severed from its brain.

1906 C. S. Sherrington in E. A. Schäfer *Text-bk. Physiol.* II. 818 The spinal frog, when placed on its back, does not, as a rule, right itself. **1977** *Brain* XL. 230 'Spinal man' cannot stand, and shows no primary extensor reflex. **1962** [see prec. sense]. **1975** *Sci. Amer.* Aug. 75/2 Sherrington found for example, that a spinal dog whose hind leg..would simultaneously brace the opposite leg to assume the weight removed from the withdrawn leg.

spinar (spi·nāɹ). *Astr.* [f. *spin*(*ning st*)*ar*, after *Quasar*, *Pulsar*.] A hypothetical, supermassive, rapidly rotating celestial object which may be located in the nuclei of some active galaxies and quasars, and which could help to account for the huge energy output of quasars.

1971 Morrison & Cavaliere in D. J. K. O'Connell *Nuclei of Galaxies* 485 By spinars we mean a class of bodies storing a considerable fraction of their total energy in a single, continuous degree of freedom—rotation... Pulsars are a very special spinar, the spinars of galaxies. **1971** *Astron. & Astrophysics* LVI. 166/2 We will consider..a supermassive oblique rotator (SOR) having solid-body rotation and quasi-dipole magnetic field. In the 'cold' version such a rotator coincides with a 'spinar' or a 'magnetoid' proposed by Morrison. **1978** *Nature* 28 Sept. 282/1 The nonthermal radiation of a spinar..in some circumstances far exceed the thermal (Eddington-limited) component.

spindle, *sb.* Add: **I. 4. b.** Delete latter part of def., and transfer quot. 1899 to sense 4 d below.

c. *Cytology.* A bipolar configuration of fibres to which the chromosomes become attached by their centromeres at metaphase of mitosis before being pulled towards the poles; cf. *spindle fibre*, sense 17 below.

1878 *Q. Jrnl. Microsc. Sci.* XVIII. 114 The portion of the spindle which remains in the egg after the formation of the second polar cell reconstitutes itself in a nucleus. **1937** Hackman & Hulett *Animal Biol.* iii. 38 A star-shaped figure of radiating fibres is seen in the cell. This divides into two, forming a spindle-shaped structure which, with a radiating 'star' at each end, and the chromosomes arrange themselves where the fibres from the two stars meet, in the centre of the spindle. **1973** *Sci. Amer.* Oct. 77/3 The action of the mitotic spindle in pulling the chromosomes apart when a cell divides.

d. *Anat.* [tr. G. *spindel* (W. Kühne 1863, in *Arch. f. path. Anat. u. Physiol.* XXVII. 520).] = *muscle spindle* s.v. *muscle *sb.* 3 d.

1894 *Jrnl. Physiol.* XVII. 238 The spindles have been studied by Golgi (1880). Golgi's definition of them is: bundles of completely developed muscle-fibres, surrounded by a special sheath, and to be found in muscles at every period of growth. **1940** Maximow & Bloom *Text-bk. Histol.* xiv. 276 The muscle fibers of the spindle are approximately half as thick as the ordinary muscle fibers. **1962** *Phil. Trans. R. Soc.* B. CCXLV. 82 All spindles contain two distinct types of intrafusal muscle fibre, 'nuclear bag fibres' and 'nuclear chain fibres', which differ in structure and innervation. **1974** [see *muscle spindle*].

e. *Anat.* Any of numerous small sensory organs within tendons and aponeuroses which consist of a spindle-shaped bundle of tendon fibres containing the branching endings of a nerve and enclosed in a capsule; a neurotendinous spindle; = *tendon organ, spindle* s.v. *tendon c.

1896 E. L. Bilstein tr. *Stöhr's Text-bk. Histol.* II. 115 The medullated nerves of tendons terminate in part in a close plexus of gray nerve-fibers, and in part in tendon-spindles. **1901**, etc. [see *neurotendinous* adj. s.v. *neuro-]. **1905** J. S. Ferguson *Normal Histol.* ix. 138 Nerve fibres enter the spindle and give off several medullated branches which run between the tendon bundles near the axis of the spindle. **1954** T. L. Peele *Neuroanat. Basis Clinical Neurol.* xix. 420/1 Neurotendinous spindles are usually present near the musculotendinous junction. **1966** T. S. & C. R. Leeson *Histology* xx. 440/2 Neuromuscular spindles lie in muscle... Neurotendinous spindles are similar and are located in tendons and aponeuroses near their junctions with muscle.

f. *Med.* A configuration seen in an electroencephalogram (see quot. 1935).

1935 A. L. Loomis et al. in *Science* 14 June 597/2 The amplitude [of the waves] builds regularly to a maximum and then falls regularly so that we have designated these 'spindles', because of their appearance in the record. **1952** *Confinia Neurologica* XII. 73 Spindles are most prominent in the thalamus of cats under barbiturates. **1965** *Math. in Biol. & Med.* (Med. Res. Council) vi. 175 Often exact identification of the maxima and minima of the spindles was rather difficult. **1977** *Lancet* 12 Nov. 1021/1 A slowing of the alpha, concurrent alpha and theta, and beta spindles are found during relaxation and on the borders of sleep.

II. 7. c. A spindle moulder (see sense *17).

1920 F. T. Hill *Pract. Aeroplane Constr.* 108 This is known as a French spindle, and its cutting action, in order to form the recesses, is shown in the enlarged view on the right. **1925** W. J. Blackburn *How to work Spindle Moulder* iii. 29 In working a spindle there are three kinds of cutters—those used on the square block, those used with the slotted cutters, and those used on the French spindle.

9. a. (Further examples.)

1824 'A. Singleton' *Lett. from South & West* 82 They [*sc.* Virginians] also call, what we [*sc.* New Englanders] call the spindle, the tassel. **1871** *Amer. Naturalist* V. 245 The corn..sent forth a new tassel or spindle.

12. a. (Later example.)

1904 *Hartford* (Connecticut) *Courant* 19 Aug. 13 What this man was really doing was simply placing a spindle on Magazine Rock.

III. 16. *spindle-shaped.*

1849 D. Bunce *Travels with Dr. Leichhardt* 15 Blue spindle-shaped fruits or berries. **1927** [see sense 4 c above]. **1965** J. S. Cox *Illustr. Dict. Hairdressing & Wigmaking* 14/1 Spindle-shaped curler, a rounded rod, tapering towards each end.

17. spindle-back, used *attrib.* to designate a chair with a back consisting of framed cylindrical bars; **spindle-berry**, the bright red fruit of the spindle-tree, *Euonymus europæus*; **spindle cell** *Med.* and *Biol.*, a narrow, elongated cell; *spec.* one in the blood of lower vertebrates analogous to the platelet in mammals; freq. *attrib.* (cf. *spindle-celled*, sense 16 in Dict.); **spindle fibre** *Cytology*, any of the microtubular strands which form the visible structure of a spindle (sense 4 c above); **spindle hour**, an hour during which a spindle is involved in spinning, used as a unit of measurement; **spindle machine**, moulder, a woodworking machine used to shape mouldings, in which one or more cutters are carried on a spindle; so **spindle moulding**; **spindle oil**, a light distillation product of petroleum, used for lubrication esp. of high-speed machinery.

1896 *Heal & Son Catal. Bedroom Furnit.* 156 Solid Oak Spindle-back Rush-Seat Chair. 12/9 to 4. **1928** *Ibid.* Cottage Furnit. 18 Spindle Back' Arm Chair, stained oak colour. 28/-. **1937** *Times* 15 Nov. 19/4 One could repeat this story in respect of the spindleback chairs made in the West of England from the Selsley Forth to Herefordshire. **1959** G. Savage *Antique Collector's Handbk.* 126 The Lancashire spindle-back chair is similar in many ways to the ladder-back. **1931** 1958 *Cent. Dict.* 1039 The dying glory of bracken, oak, birch, mountain-ash and spindle-berry. **1925** *Daily Mail* 21 Sept. 15/4 Pink spindle berries are lovely in a pewter mug. **1950** P. J. Brooker *Goose Cathedral* viii. 167 The hedges were hung with a sapphire of spindleberries—lurid purple bursting into fiery orange. **1878** Spindle-cell [in Dict., sense 16 b]. **1901** A. P. Ohlmacher in Hektoen & Riesman *Text-bk. Path.* I. 200 It [*sc.* round-celled sarcoma] grows more rapidly than the spindle-cell type and is more malignant than the spindle-cell sarcoma. **1905** J. S. Ferguson *Normal Histol.* iii. 34 In the denser forms of mature connective tissue..the connective tissue cells lose their typical embryonal stellate form and become somewhat fusiform; they are then known as the spindle cells of connective tissue. *Ibid.* vi. 81 Controversy has arisen from the supposed analogy of the true blood platelets of human blood with certain other structures found in the blood of the lower vertebrates, especially the 'spindle-cells' of birds and amphibians. **1949** A. S. Romer *Vertebrate Body* xiii. 427 In most nonmammalian vertebrates the thrombocytes take the form of spindle cells—small, oval, pointed structures with a central nucleus. **1959** W. Andrew *Textbk. Compar. Histol.* ix. 371 Spindle cells are conspicuous and probably should be thought of as fusiform lymphocytoid cells rather than platelet-forming elements. **1971** T. J. Hara in Hoar & Randall *Fish Physiol.* V. iv. 88 Aside from the taste buds, specialized epidermal 'spindle' cells were found on the head and body of minnows and various teleost fishes. **1976** *Path. Ann.* XI. 214 (*caption*) Spindle cell variant of thymic carcinoid tumor. **1878** *Q. Jrnl. Microsc. Sci.* XVIII. 229 The spindle-fibres are identical with the stellate rays. **1974** *Encycl. Brit. Micropædia* VI. 946/1 In anaphase the chromatid pairs separate and are pulled to opposite ends of the cell by the spindle fibres. **1930** *Times* 24 Mar. 23/5 Mill activity in the cotton growing states, measured by spindle hours, established a high record. **1970** P. R. Lord *Spinning in '70's* 11 Production per spindle hour has been increased by raising spindle speeds. **1902** G. Ellis *Mod. Pract. Joinery* xxiii. 364 *Spindle machine*, an irregular moulding machine in which the

cutters are fixed at the end of a vertical spindle which projects through the table. **1915** *Machine Woodworker* 15 Nov. 15/2 Running moulds from thicknessed boards on spindle machine may do if it is a special type of mould. **1912** *Ibid.* 15 July 17/1 A spindle moulder being a machine that has to do a large variety of work, the stock of cutters should be large. **1925** W. J. Blackburn *How to work Spindle Moulder* i. 11 As the spindle moulder is used for working regular sections upon edges of various shaped pieces of wood it is best to place it close to the band saw. **1965** F. L. Dunsmore *Technique Woodworking Machinery* II. i. 14 No machine has a greater selection of cutter heads than the spindle moulder. **1979** *Building* Dec. 105/3 Planing, thicknessing, sawing, routing and spindle moulding. **1887** B. J. Crew *Pract. Treat. Petroleum* iv. 316 Spindle oils. Distillers of residuum usually divide their products into three classes. The third..product constitutes the stock for spindle and machinery oils. **1931** *Engineering* 2 Jan. 1/2 Oil Pt. will be recognised as a 'spindle oil', used only for lightly loaded high speed journals. **1977** *Lubricants Business* (Shell International Petroleum Co.) 1 Refining of the distillates, which removes unsuitable components, produces lubricating oil fractions of the desired properties ranging from thin spindle oil to heavy cylinder oil.

Hence **spi·ndleless** *a.*, having no spindle or spindles.

1964 *Gloss. Letterpress Rotary Printing Terms* (B.S.I.) 7 *Spindleless reel stand*, a reel stand supporting the reel on free-running cones at each side of the reel. **1967** *Economist* 29 Apr. 459/3 There [*sc.* in Czechoslovakia] the BD 200 spindleless spinning unit has been developed; by next year it will be modified to run at 40,000 rpm as against the conventional spindle's 10,000.

spindle, *v.* Add: **5.** To recess and taper (a spar for an aeroplane's wing); to cut *out* (a recess) in a spar.

1918 *Aeronaut. Jrnl.* Feb. 44 Jigs for small parts should be so constructed that several pieces may be spindled at the same time. **1919** Pippard & Pritchard *Aeroplane Struct.* 201 Questions of strength determine the amount which can safely be spindled out. **1920** F. T. Hill *Pract. Aeroplane Constr.* v. 106 Having drilled the spar, the next operation will be to spindle out the recess. **1928** *Technical Rep. Aeronaut. Res. Committee 1926–7* 466 These specimens were first formed..with a length of 2 inches spindled to give a cross section geometrically similar to the fractured portion of the spar.

spindleage (spi·ndlɛdʒ). [f. Spindle *sb.* + -age.] The total number of cotton spindles in use at a given time and in a specified area.

1921 A. S. Wade *Cotton Spinning* 15 The loss of spindleage due to the reduction of the working week. **1954** R. Robson et al. in A. F. W. Coulson *Man. Cotton Spinning* I. i. 8 By 1928 the number of spindles had risen to just over 165 million but this was to represent the peak world spindleage. **1972** C. H. Lee *Cotton Enterprise* ii. 16 The spindleage of the trade grew from 1·7 million to four or five million between the early 1780's and 1812.

spindled, *ppl. a.* Add: **3. b.** Of a spar or strut for an aeroplane wing: having been recessed and tapered. Also with *out*.

1919 S. Camm *Aeroplane Construction* iii. 23 This [steel] strut..is superior to the solid spindled strut, which possesses a tendency to buckle laterally. **1928** *Technical Rep. Aeronaut. Res. Committee 1926–7* 467 The fracture [in through part of the bolt hole and tended to follow a 'zig-zag' course on its way into the spindled portion. **1930** *Flight* 17 Jan. 119/1 The wing is of equally simple construction with two spindled-out spruce spars and light girder ribs.

spindling, (*vbl.*) *sb.* Add: **4.** The process of recessing and tapering a spar for an aeroplane wing. Also *attrib.*, as *spindling jig, machine*.

1918 *Aeronaut. Jrnl.* Feb. 40 In the spindling of spars the job should be placed in a jig having stops set. **1920** F. T. Hill *Pract. Aeroplane Constr.* v. 108 The base of the spindling jig. *Ibid.* 110 The only marking out required for spindling is the end of the recess. *Ibid.* 116 Flanges are grooved on the spindling machine with a jig.

5. *Med.* The occurrence of fairly regular, alternating increases and decreases of amplitude in an electroencephalogram.

1963 *Electroencephalogr. & Clin. Neurophysiol.* XV. 766/2 This randomness can account for such characteristics as spindling without involving concepts of beating between alpha patterns. **1968** *Brit. Med. Bull.* XXIV. 257/1 Barnet and Lodge (1966) have averaged responses to ton-clicks or pure-tone bursts during deep sleep (high-voltage spindling in the EEG) in 22 infants.

spindly, *a.* Add: **3.** *Comb.*, as *spindly-legged, -stemmed* adjs.

1951 J. Cleary in Murdoch & Drake-Brockman *Austral. Short Stories* 434 The other boy and the two girls, the freckle-faced, spindly-legged..girls, are all working now. **1971** *Flying* Apr. 71/2 It was a big, spindly-legged beast. **1897** Spindly-stemmed [in Dict., see sense 1 a].

spin-down (spi·ndaun), *sb.* [f. Spin *v.* + Down *adv.*] A decrease in the speed of rotation of something.

1963 *Jrnl. Fluid Mechanics* XVII. 402 The initial angular velocity Ω_0 = 3·668 sec⁻¹, and the final angular velocity Ω_1 = 3·933 sec⁻¹. (This was a case of 'spin-down'.) **1967** *Astrophysical Jrnl.* CXLIX. L121 The 'spin-down' time of a cup of tea is approximately ($R^2/\nu\Omega$)^½, where *R* is the radius of the cup and Ω is the angular velocity of the tea. **1971** *New Scientist* 26 Aug. 452/2 The favourite explanation that decreases in period, or 'spin-ups', occur when starquakes take place in the deformed crust of the star seems to fall flat on its face when confronted with the discovery that the Crab pulsar, at least, also undergoes 'spin-downs'. **1977** *Astron. & Astrophysics* LX. 851 Pulsar braking index..and possibly magnetospheric friction can account for a spin-down of Venus within some 10⁹ years.

spin-down (spindau·n), *a.* *Physics.* [f. Spin *sb.*¹ + Down *adv.*] Being or pertaining to a particle whose spin points downwards.

1965, etc. [see *spin-up *a.*].

spin-drier, -dryer (spindrai·ǝɹ). Also without hyphen and as two words. [f. *spin-dry *v.*, *Drier, Dryer.*] A machine which removes excess water from washing by spinning it rapidly in a rotating perforated drum.

1959 *Archit. Rev.* LXXV. 76/3 The laundry is all electric, and is equipped with a Rotary Washer and Spin Dryer, and Rotary Ironer, in which all the laundry of the house can be done without resort to clothes lines. **1948** *Sun* (Baltimore) 15 July 8 (Advt.), Be among the first to see the amazing new easy spindrier with automatic spin-rinse! **1955** *Observer* 10 Jan. 10/3 Spin-driers liberate housewives from the effort of wringing and greatly reduce the time of drying clothes. **1959** G. Freeman *Jack would be Gentleman* vi. 114 We didn't have a spin dryer like you've got, only reached in the kitchen. **1961** *Spectator* 30 June 738/1 A canny Scot who salvaged Vat 69 filter pads and extracted whisky from them in his spin-drier was taken to court. **1970** J. Aiken *Died on Rainy Sunday* 12 The spin-dryer whirred to a halt.

spindrift. Add: **a.** Also *fig.* (example *attrib.*).

1946 Dylan Thomas *Deaths & Entrances* 16 Not for the proud man apart From the raging moon I write On these spindrift pages.

b. *transf.* Driving snow, sand, etc. Also *attrib.*

1961 Webster, *Spindrift*, sand, dust, or snow driven before the wind like sea spray. **1971** C. Bonington *Annapurna South Face* xii. 147 At Camp V, the spindrift avalanches swished down all night, sweeping the slope, burying the tent. *Ibid.*, Cataracts of spindrift kept pouring over the cliffs above. **1973** D. Haston *In High Places* vii. 60 A spindrift avalanche had come right down the chute, filling the tent..back and covering everything. This wasn't the first time we'd been spindrifted.

c. *Comb.*, as *spindrift-laden, -streaming* adjs.

1975 D. Haston in C. Bonington *Everest the Hard Way* xiv. 214 Once again it was cold with snapping spindrift-laden wind. **1976** B. Bridges *Harbingers* 7 Or green and spindrift-streaming shelf of stone.

spin-dry (spindrai·), *v.* Also **spin dry.** Inflected both as for Spin *v.* (*dry* taken as adj.) and Dry *v.* [f. Spin *v.* + Dry *v.* or Dry *a.*] *trans.* To remove excess water from (washing) by spinning it rapidly in a rotating perforated drum; to dry partially in a spin-drier. Also *absol.*

1957 *Sat. Even. Post* 19 Mar. 135 It takes the Savage [Washer & Dryer] just one-tenth the time to *spin-dry* the entire load as it does to *turn tub*. **1961** *Good Housek. Home Encycl.* 304/1 The automatic type..will..spin-dry the clothes at a turn of the switch. **1968** *Observer* 2 Mar. 10/3 If clothes are spin-dried and dried immediately they save on the after-effects of ironing. **1968** *Times* 6 June p. 1x/5 Thus clothes can be boiled and then spun-dry ready for ironing. **1969** *Guardian* 15 Feb. 4/4 Finally, it [*sc.* a washing-machine] spin-dries the clothes. **1975** *Country Life* 27 Jan. 212/2 Give it a wash, however and spin dry if you like.

So **spin-drying** *vbl. sb.*; **spin-dry-ing** *vbl. sb.* also.

1958 *Home Beautiful* Jan. 50/2 The newest models [of washing machines] include a drying feature known as the 'spin dry' or 'extractor'. *Ibid.* 51/1 If the 'spin dry' is to work efficiently, it must not be overloaded and clothes must be evenly distributed. **1966** *Archit. Rev.* CXX. 346/1 Spin-drying considerably reduces the line-drying time as opposed to wringing. **1962** *Which?* Aug. 238/1 Too much spin drying can set the creases into the clothes. **1978** P. Porter *Cost of Seriousness* 36 When I start an allegory It's plunged Into Spin-drying programmes Right through to the spin-dry.

spine, *sb.*¹ Add: **I. 5. c.** A tall mass of lava projecting upwards from the mouth of a volcano.

1903 *Amer. Jrnl. Sci.* CLXVI. 270 The ancient summit of the mountain [*sc.* Mont Pelée] has lost most of its former prominence above the rim of the crater, but within the old Caldera a cone has risen which overtops the surrounding walls and terminates in a spine rising hundreds of feet above the main mass of the new cone. **1954** W. D. Thornbury *Princ. Geomorphol.* xix. 500 The rapid growth of plug domes is further illustrated by one on Santa Maria volcano on Guatemala... It had a spine which attained a height of 66 meters. **1976** P. Francis *Volcanoes* iii. 124 Evidence of just how viscous the lavas were comes from the spine which was pushed up out of the vent of Mt Pelée, reaching over 300 metres in height.

II. 6. b. Also *fig.*

1912 Galsworthy *Inn of Tranquillity* 189 The vice of drawing these distorted morals has permeated the Drama to its spine. **1977** *New Yorker* 9 May 116/2 A beauty called Laura..who is the spine of the place. **1977** H. Fast *Immigrants* 111. 308 The great Jewish conductor, Schuman-Heinsland 44 Doors shake on their jambs: the spine of the house Thrills as the sprung wood quivers, and goes still. **1979** *Times Lit. Suppl.* 26 Mar. 83/1 No one can be the spine..of a book..but it is the spine of the book that takes some strain. **1981** F. Inglis *Promise of Happiness* i. 3 We try to say what some of the best books are like, so that we can hand them on... This expression of the gift relationship..gives spine and structure to this account.

9. b. The back of a book, that is, the part bearing the title, etc., which is visible when the book is standing on a shelf; also, the corresponding part of a dust-jacket or a shallow box.

1922 M. Sadleir *Excurs. in Victorian Bibliogr.* 14 Be wary of books in cloth which bear no publisher's imprint on the spine. **1931** *Publishers' & Booksellers' Gloss.* 17 The practice of having the selling price printed on the spine of a book jacket. **1952** 'C. Canning' *Spider* 14 A The trade..rival. **1961** J. Nisbett *Technique Sound Studio* vi. 113 Programme details are written..on the spine of the tape box. **1966** *Listener* 10 Oct. 571/1 The collected spines of the books in the rack. **1967** P. Brett *So much Blood* i. 16 Brown velvet upholstery and the leather spines of books gave the room a comfortable look. **1976** 'E. Lathen' *Double, Double* i. 14 She had put up..a row of books with their spines outward.

III. 10. a. *spine-chiller*; *spine-ache* adj.; **spine-bashing**.

1851 H. Melville *Moby Dick* III. xlviii. 281 Launch me, spine-wise, on the sea. **1947** Dylan Thomas *Let.* 13 June (1966) 314 The hill to the village is a spinebreaker.

11. spine-basher *Austral. slang*, a loafer; so **spine-bashing** *vbl. sb.* (and *pres. pple.*); **spine-bill**, insert after 'honey-eaters': of the genus *Acanthorhynchus*; (later examples); **spine-chiller, -tingler**, something (rarely someone) that inspires excitement and terror; *esp.* a horror or suspense story, film, etc.; **spine-chilling** *ppl. a.* and *vbl. sb.*; **-freezing** *ppl. a.*, inspiring excitement and terror, horrifying; **spine road**, a major road linking other important routes or points; **spine-thriller, -tingler**, something pleasurably frightening; *esp.* an exciting story, etc.; **spine-tingling** *ppl. a.*, pleasurably frightening or disturbing; **spine-chilling**; **spine wall** *Building* (see quot. 1969).

1946 R. Rivett *Behind Bamboo* 399 *Spinebasher*, one who lies on his back, always resting. **1951** *Sydney Morning Herald* 20 Mar. 14 The elbow-benders, spine-bashers, eternal babblers keep one ear to the loudspeakers, an ear to the ground. **1941** *Argus* (Melbourne) *Week-End Mag.* 15 Nov. 1/4 Spine bashing, having a rest; loafing. **1945** L. Glassop *We were Rats* 168 'She's sweet,' I said. 'Go and do some spine bashing.' **1966** G. W. Turner *Eng. Lang. in Austral. & N.Z.* vi. 135 Lying down and doing nothing, what the Australian calls *spinebashing*, is not confined to the lower ranks. **1977** A. Nash *Bush Days* 34 Spine-bills flashed by. **1938** *Mod. Lang. Notes* Mar. 177 Horror stories and spine-chillers. **1977** *Daily News* (Perth, Austral.) 19 Jan. 21/2 Weighing in with just the right amount of horror, this spine chiller will be published. **1965** L. Deighton *Ipcress File* xx. 201 His spine could be spine-chilling. **1973** Mackenzie *Speech* XV. 205/1 *Spine-chiller*, a mystery film or play. **1948** Berrey & Van den Bark *Amer. Thes. Slang* § 504/4 *Sensational story or book*..spine chiller or tingler. **1957** *London Mag.* Aug. 55 *Elective Affinities* is a far more important novel than Walpole's *Castle of Otranto*, and yet there has only been one edition of it in the past hundred years, compared to the dozen or so of Walpole's unreadable spine-chiller. **1969** E. Lemarchand *Alibi for Corpse* xv. 189 Even now the idea of her..gives me the willies. And Twentyman was a spine-chiller. **1976** *Listener* 21 Oct. 500/2 The show..is an extremely able piece of work, a Grand Guignol frisk..child, suicide. This was a spine-chiller, and no mistake. **1946** 'M. Innes' *From London Far* II. 20 A sweet and—as he hoped—wholly spine-chilling smile. **1958** *New Statesman* 25 Jan. 103/1 With unerring instinct the Chaucer..the spine-chilling specialist of *The Wages of Fear* and *The Fiends*, has bought a new tension in Picasso. **1958** *Times* 20 Dec. 9/2 The new Greek film is a spine-chilling and blood-curdling. *Cases of Night* is masterly. **1969** H. Agar *Saving Remnant* v. 127 'What..will so many Jews?'—that spine-chilling comment. **1983** *Listener* 10 Feb. 16/3 She fascinating with her antiquarian donnish history instead of the mildly spine-chilling ghost story she expected. **1960** Woodhouse *Jeeves in Offing* iii. 42 Mrs. Cream..worked in her most spine-freezing manner..to her spine-freezer. **1961** ——— *Ice in Bedroom* i. 9, I dipped into one of her products once..spine-freezing. **1935** J. Noranmi *N.Z. Cel* xii. 134 Tracers of a spine-tingling kind ran to turn this one. **1937** *Discovery* Dec. 393/1 The spine-tingling of dry heat. **1961** *Guardian* 18 Apr. 3/4 The spine-tingling road from Bonnyrigg to Pathhead. **1967** *Times* 3 Dec. 2/8 Spine roads have been suggested to interconnect the town. **1975** *Guardian* 11 Oct. 28/4 Oxford city's consultants drew up an alternative plan..the spine road. **1976** *Times* 20 Sept. 2/4 As the national..spine roads. **1912** *Maclean's Mag.* Nov. 135 (*heading*) Producing spine thrillers. How successful melodramas are furnished—some criticisms about art of capitalizing spines. **1962** *Listener* 21 June 1085/1 It requires a spine-tingler. **1948** *Times* 21 June 3/5 *Spine-tingler*, a thriller or horror story. **1941** *Radio Times* 6 Jan. 31/4 *Cop Spine-tingling* 'Bazan (see *spine-chiller* above). **1978** *TV Times* 1 Feb. 1 Feb. 19/1 This week's *The South Bank Show* homes in on the best Shakespearian action alive today. Also..the two spine-tingling historical romance. **1968** *Blues Unlimited* Dec. 4 John Lee's spine-tingling guitar. **1978** J. B. Hilton *Some run Crooked* xii. 118 He spoke the words quietly..and there was something spine-tingling about them. **1949** *Archit. Rev.* CV. 236 Construction is box-frame with continuous reinforced concrete slabs, and with load-bearing cross walls and a spine wall. **1963** *Gloss. Gen. Building Terms* (*B.S.I.*) 22 *Spine wall*, an internal loadbearing wall running in the direction of the main axis of a building or structure.

spinel. Add: **2. a.** The formula of the typical species is $MgAl_2O_4$. (Further examples.)

1963 F. H. Pough *Field Guide Rocks & Minerals* II. 141 Spinel, like corundum, is a mineral of metamorphosed limestones and low-silica pegmatites, and consequently

it is commonly associated with corundum. **1977** A. Hallam *Planet Earth* 137 The name spinel broadly refers to a group which includes the minerals magnetite and chromite. In a narrower sense, spinel is the magnesium aluminum oxide member of the group.

b. The general formula of the minerals is $A^{2+}B_2^{3+}O_4$, *spec.* where *B* is Al. The name is also applied to any of a large number of artificial minerals having similar structures. (Further examples.)

1944 C. Palache et al. *Dana's Syst. Min.* (ed. 7) I. 688 In addition to the natural spinels a considerable number of artificial spinels have been described, where *A* = Co, Cd, Cu and *B* = Co, Ti, Sn, V, Ga, In, Mn. One of the further complexities of composition in the artificial spinels is a tendency for substitution of *A* atoms for part of the *B* atoms of the formula, i.e., the formula may be written $BABX_4$ (as in the compounds $GaMgGaO_4$, etc.). **1967** D. H. Martin *Magnetism in Solids* i. 39 The two crystal types to which the most fully investigated ferrimagnetics belong are those known as 'spinels' and 'garnets'. **1974** D. M. Adams *Inorg. Solids* iii. 69 The structure of a spinel is conveniently described by a parameter γ which equals the fraction of A^{2+} in octahedral sites. Thus, for a 'normal' spinel, γ = 0, for an 'inverse' spinel, γ = 1. A random distribution of A^{2+} amongst the two groups of sites gives γ = 2/3.

c. *spinel structure*.

1944 C. Palache et al. *Dana's Syst. Min.* (ed. 7) I. 688 Another chemical feature of interest is the ability of artificial Mg, Al spinel to contain considerable excess of Al_2O_3 without impairment of the spinel structure. **1967** M. Chandler *Ceramics in Mod. World* vi. 178 Today most ferrites have spinel structures.

spineless, *a.* Add: Hence **spi·nelessly** *adv.*, **spi·nelessness.**

1920 *Chambers's Jrnl.* 18 Dec. 35/2 His spinelessness and low tastes. **1948** A. L. Rowse *End of Epoch* 186 The spinelessness, the disastrous disunity of the liberal elements in Germany. **1960** *Middlesex Hospital Jrnl.* Jan.–Feb. 12/1 At interviews, the salesman applicant has to steer a course between spinelessness and cockiness. **1977** J. Porter *Who the Heck is Sylvia?* xii. 110 Capitulating as spinelessly as usual, Miss Jones..unfastened her seat belt. **1981** Craig & Cadogan *Lady Investigates* xi. 214 Gregory's vacillation..seems to be an example of the kind of spinelessness that he usually condemns in others.

spinet¹. Add: **2.** *U.S.* In full *spinet piano*. A type of small piano.

1936 *Arts & Decoration* Jan. 30/1 The spinet pianos are inviting musical instruments. Their design, suitable to the modern apartment or small house, delights the decorator and the owner. **1937** *Étude* Dec. 803/3 In a few months manufacturers had worked out many refinements in the popular vertical pianos, which came to be known under the generic terms of 'consoles' or 'spinets'. **1970** W. Apel *Harvard Dict. Mus.* (ed. 2) 804/2 *Spinet*,..a piano with strings perpendicular to the keyboard (as in an upright piano) but with an indirect action. **1977** J. Philips *Five Roads to Death* (1978) I. ii. 31 There was a little spinet piano in one corner.

spinifex. Insert in etym. after mod.L.: (Linnæus *Mantissa Plantarum* (1771) II. 163). **1.** For *Tricuspis* substitute '*Triodia* or *Plectrachne*'.

spinkie (spi·ŋki). *dial.* [f. Spink *sb.*¹ + -ie.] = Spink *sb.*¹ 1.

1911 D. H. Lawrence *White Peacock* II. vii. 316 There are two or three robins' nests, and a spinkie's.

spinks (spiŋks). *Austral.* [f. Spink *sb.*¹ + *-s².] = *Jacky Winter* s.v. *Jacky 3.

1945 Baker *Austral. Lang.* xii. 211 The Brown Flycatcher is called..spinks. **1969** [see *Jacky Winter* s.v. *Jacky 3].

spinless (spi·nlės), *a.* [f. Spin *sb.*¹ + -less.] Having no spin, or no tendency to spin.

1936 *Aeroplane* 22 Jan. 116/2 The plane must be totally different from anything on the market to-day. It must..be spinless and viceless. **1951** *Physical Rev.* LXXXI. 282/2 (*heading*) On the Lamb shift for spinless electrons. **1966** *Nucl. Sci. Abstr.* XX. 1931/1 (*heading*) A fortran code for computing angular distributions, polarizations and transport cross sections of neutrons scattered by spinless targets. **1981** M. Gell-Mann in J. H. Mulvey *Nature of Matter* viii. 184 That theory should include..the spinless particles we believe we need for spontaneous symmetry breaking.

spinnable, *a.* Add: (Further examples.)

1926 *Chambers's Jrnl.* 16 Oct. 726/1 The cotton fibres of spinnable length are removed..by machines called 'gins'. **1927** M. H. Avram *Rayon Industry* 186 Not every solution which possesses sufficient viscosity and is capable of being drawn into filaments is spinnable. **1967** A. Ziabicki in H. F. Mark et al. *Man-Made Fibers* 16 There are known many spinnable liquids, such as honey and high-mineral oils, which nobody would consider fiber-forming.

Hence **spinnabi·lity**, the capacity for being spun; applied esp. to a solution from which a synthetic fibre may be drawn.

1939 *Chem. Abstr.* XXXIII. 4493 (*heading*) The relation between elasticity, streaming anomalies and spinability of solutions with particular reference to fibrinogen. **1953** R. J. W. Reynolds in R. Hill *Fibres from Synthetic Polymers* v. 91 Spinnability occurs at only a moderately high molecular weight. **1979** Tadmor & Gogos *Princ. Polymer Processing* xv. 637 The problems of draw resonance and spinnability are not fully understood.

spinnaker. Add: Also carried on other sailing vessels. (Further examples.) Also *fig.*

1912 E. K. Chatterton *Fore & Aft* vii. 233 She [*sc.* a Thames barge] has all her canvas tanned with the exception of her jib staysail, which her skipper calls a 'spinnaker'. **1925** H. W. Smyth *Sea-Wake* 179 The spinnaker hand, young and unspoiled in eye or thought, kept looking round him. *Ibid.* 191 'Hooroo,' began the spinnaker man, and no one interrupted him now; he had done the best day's spinnaker work of his life. **1927** G. Bradford *Gloss. Sea Terms* 168/1 Large jibs on English coast craft are sometimes called spinnakers. **1949** *Sun* (Baltimore) 9 Aug. 8/5 The big China-built cutter..carried a huge sky-blue Nylon spinnaker across the finish line. **1957** R. Campbell *Coll. Poems* II. 253 The spring with rosy spinnaker outfanned Comes curling silver fleeces through the land. **1959** *Yachting World Annual* 9/2 *Columbia's* combination of a small, flat spinnaker and small short-luffed spinnaker staysail seemed to be pulling more effectively. **1963** G. Millar *Oyster River* 60 We put the nine-foot dinghy in the river (easily done with a short boom fitted into a spinnaker cup set low on the aft side of the mainmast). **1967** J. Howard-Williams *Sails* xii. 164 The spinnaker-staysail..sets under the spinnaker on a reach, in order to fill a slight gap forward. **1972** *Sail* May 100 A typical low aspect ratio spinnaker staysail.

‖**spinnbar** (ʃpi·nbāɹ), *a.* [Ger.] Of a viscous liquid: capable of being drawn into strands; spinnable.

1944 G. W. S. Blair *Survey Gen. & Appl. Rheol.* v. 68 Many spinnbar materials show flow-elasticity. **1945** *Proc. R. Soc. Med.* XXXIX. 8 Curves cannot be directly obtained for spinnbar materials because the ratio of pressure (stress) to rate of flow not only falls with rising stress but increases with increasing deformation (work-hardening).

‖**spinnbarkeit** (ʃpi·nbāɹkait). Also **Spinnbarkeit.** [a. G. *spinnbarkeit* (H. Erbring 1936, in *Kolloid-Beihefte* XLIV. 173), f. *spinnbar* (see prec.).] The capacity of a viscous liquid, esp. the cervical mucus, for being drawn into strands; spinnability.

[**1938** G. W. S. Blair *Introd. Industr. Rheol.* x. 108 Erbring, studying this property in detail, calls it 'Spinnbarkeit', which perhaps is best translated as 'fibrosity'.] **1944** —— *Survey Gen. & Appl. Rheol.* v. 69 Both tack and 'Spinnbarkeit' are often associated with adhesive properties. **1954** G. I. M. Swyer *Reproduction & Sex* iv. 47 During the two or three days immediately prior to ovulation, the healthy cervix pours forth a copious flow of clear, almost watery mucus which flows out over the vaginal surface of the cervix... At this time it possesses in marked degree the rheological property variously called 'spinnbarkeit', 'fibrosity' or, perhaps better, 'ductility', that is, the ability to be drawn out into a fine thread (or to be blown up into a sizeable bubble). **1975** *Acta Endocrinologia Suppl.* CXCIX. 242 The amount, spinnbarkeit and ferning of the cervical mucus correlated well with the estradiol concentration.

spinner. Add: **II. 7. c.** The person who tosses the coins in the game of two-up; *come in spinner*: the cry commonly used to start the game. Chiefly *Austral.*

1911 L. Stone *Jonah* II. vi. 215 The spinner placed the two pennies face down on the kip, and then, with a turn of the wrist, the coins flew twenty feet into the air. **1945** T. Ronan *Strangers on Ophir* 119 Cries of 'Another quid to see him go. Get set on the side. All set, come in Spinner.' **1948** V. Palmer *Golconda* iv. 25 Step in, spinner... I'm backing heads... They're up. Fair go. **1958** 'N. Culotta' *They're a Weird Mob* viii. 117 He began to play... 'Come in, spinner'..I sang softly. **1964** [see *kip *sb.*⁸]. **1975** L. Ryan *Shearers* 97 'All set?' Lofty asked. 'Set!' Sandy said. 'Come in, spinner!'

d. *Agric.* A rotating device for lifting potatoes out of the ground.

1923 J. R. Bond *Farm Implements & Machinery* xi. 164 The old high-speed spinner did considerable damage to the tubers. **1943** J. W. Day *Farming Adventure* xvii. 197 Another machine with a great future is the potato digger or spinner. **1960** *Times* 5 July (Agric. Suppl.) p. viii/2 Spinners are still by far the most universal lifting appliances. **1973** M. Partridge *Farm Tools through Ages* v. 154/2 An iron share..preceded the spinner as the machine was hauled forward, serving to break down the soil.

e. *Cricket.* A spin bowler.

1951 *People* 3 June 7/5 Jim Sims, Middlesex spinner, tells me that he's never felt fitter than he does this season. **1963** A. Ross *Australia 63* 13 Since the departure of Laker and Wardle no class spinner of any kind had emerged. **1976** Dexter & Makins *Testkill* 78 The spinner, Flinders,..hit Ackroyd for two staggering sixes into the Mound Stand.

f. *Surfing.* (See quot. 1970.)

1962 *Surfer* (1962–3) Dec.–Jan. 52/2 (Advt.), One beautiful turn can't make a good movie, but...1286 beautiful turns, nose rides, spinners..can. **1968** *Surfer Mag.* Jan. 73/2 He showed them skeg-first takeoffs, spinners. **1970** *Studies in English* (Univ. of Cape Town) I. 31 Another popular hot dogging stunt is the *spinner*, in which the surfer turns himself in a full circle on the board, preferably while he is on the nose.

10. *Aeronaut.* A metal fairing that is attached to and revolves with the propeller boss of an aircraft in order to streamline it.

1918 *Flight* 11 July 767/2 The airscrew..had its boss enclosed in the usual 'spinner'. **1944** *R.A.F. Jrnl.* Aug. 287 Peter looked at..the starboard engine cowling, and the spinner revolving at high speed, yet so perfectly made that it seemed to be a motionless dome. **1969** K. Munson *Pioneer Aircraft 1903–14* 159/1 The 2-blade propeller had a shallow, bowl-shaped spinner similar to that fitted to the Bristol M.1C.

11. Comb.: **spinner magnetometer**, a magnetometer used to measure the remanent magnetism of rocks, baked clay, etc., in which a sample is spun between coils and induces in them a current dependent on the strength and direction of the magnetic field; see also *spinning magnetometer* s.v. *spinning *vbl. sb.* 8 d.

1955 *Jrnl. Geophysical Res.* LX. 332 Spinner magnetometer. This magnetometer is a further development of the type first pioneered by Johnson. **1963** *Jrnl. Sci. Instruments* XL. 162/1 The sensitivity of the spinner magnetometer is proportional to the speed at which the specimen is rotated. **1973** *Nature* 4 May 28/2 The remanent magnetizations were measured with a spinner magnetometer.

spinneret. Add: Also **spinnerette. 1.** (Further example.)

1926 T. H. Savory *Brit. Spiders* 8 Now it raises its abdomen..and secretes a drop of silk from its spinnerettes.

2. A cap or plate having a number of small holes through which a spinnable solution is forced in the production of man-made fibres; an individual hole or channel in such a plate.

1894 *Work* 7 July 391/2 The resultant..substance..is next forced..through minute holes in a glass spinneret. **1927** M. H. Avram *Rayon Industry* viii. 197 Glass spinnerettes are now produced with 66 holes. **1957** D. C. Hague *Economics of Man-Made Fibres* iii. 93 Nylon is a truly synthetic fibre... In fact, it is a plastic which has been forced through a spinnerette to form a textile fibre. **1973** *Materials & Technology* VI. iv. 291 The number of holes in a spinneret will determine the number of filaments which will be present in the yarns; this usually varies between 15 and 100.

spinning, *vbl. sb.* Add: **I. 1. c.** The process or action of drawing into a thread; *spec.* the process of forming a man-made fibre by drawing or extruding a melt or viscous solution of a polymer through a spinneret; *dry, melt, wet spinning*: (see quots. 1974).

1883 [in Dict., sense 1 b]. **1896** *Jrnl. Soc. Chem. Industry* 30 May 317/2 The production of a lustrous thread of cellulose in continuous length, by the process of drawing or 'spinning' is..an accomplished fact. **1910** A. F. Barker *Textiles* iii. 59 Vanduara silk is obtained by using gelatine as a basis, the threads, after spinning, being treated with formaldehyde to render them insoluble in water. **1927** T. Woodhouse *Artificial Silk* v. 37 Coagulation may be effected in warm air by so called 'dry-spinning', when the solvents can be vaporized by such air. **1963** A. J. Hall *Textile Sci.* ii. 75 With the introduction of nylon an entirely new method of fibre spinning was established—so-called melt-spinning in which the polymer..is melted in a novel device above the spinneret so that it can..be extruded through the multi-holed spinneret into cold air. **1974** *Encycl. Brit. Macropædia* VII. 258/2 In wet spinning, the solution of fibre-forming material is extruded into a coagulating bath that causes the jets to harden. *Ibid.*, In dry spinning the fibre-forming substance is dissolved in a solvent before the solution is extruded. As the jets of solution emerge from the spinneret, a stream of hot air causes the solvent to evaporate from the spinning solution, leaving solid filaments. *Ibid.*, In melt spinning the fibre-forming material is melted and extruded through spinnerets, and the jets harden into solid filaments as they cool on emerging from the spinneret.

4. b. Of a motor clutch: the fault of continuing to revolve after being disengaged.

1913 W. E. Dommett *Motor Car Mech.* 125 The clutch shaft has a coned brake which prevents 'spinning' when gear changing. **1948** A. W. Judge *Mod. Motor Engineer* (ed. 4) II. 305 In some cases the use of a thicker lubricant in the gear-box will prevent clutch spinning.

c. *Aeronaut.* The action of an aircraft when in a spin (*spin *sb.*¹ 2 d).

1915 *Aeroplane* 10 Nov. 578/2 It is always possible to avoid spinning or side-slipping in fog or cloud. **1930** Nayler & Ower *Aviation To-day* 324 Spinning..was first started in the War as a means to bewilder, or escape from, the enemy. **1977** *R.A.F. News* 27 Apr.–10 May 11/4 The Phantom pilots go up with an instructor for a twice-yearly check-out in the trials and tribulations of spinning.

6. (Further examples.) Also *concr.*

1927 *Daily Tel.* 11 May 18/6 To place orders for general metal spinnings. **1964** H. Hodges *Artifacts* iv. 74 The method of shaping bronze vessels known as spinning is virtually a mechanical form of raising. **1973** J. G. Tweeddale *Materials Technol.* II. iv. 86 Spinning has certain similarities to panel-beating.

II. 7. b. *spinning-machine* (later examples), *nozzle*; also in terms denoting substances that are spun to form man-made fibres, as *spinning dope, solution, syrup*.

1959 *Times Rev. Industry* Sept. 5/3 There has been an expansion in the production of man-made fibres, already coloured during their spinning by the addition of pigments to the spinning dope. **1899** *Jrnl. Soc. Arts* 8 Dec. 63/2 The filtering is to eliminate every particle of suspended matter which may exist in the collodion before it arrives at the spinning machines. **1975** J. Kasparek in E. Dyson *Rotor Spinning* x. 161 (*heading*) Processing of man-made staple fibres on the..rotor spinning machine. **1914** *Chem. Abstr.* VIII. 258 A process and device for perforating and cleaning the capillary tube of spinning nozzles for artificial silk manufacture. **1921** T. Woodhouse tr. *J. Foltzer's Artificial Silk* xix. 192 The spinning nozzles or spinnerets, from which separate threads..issue. **1931** S. E. & E. R. Trotman *Artificial Silks* 49 The spinning nozzle consists of a head or rose containing a number of capillary apertures through which the spinning solution enters the coagulating bath or evaporating chamber. **1921** T. Woodhouse tr. *J. Foltzer's Artificial Silk* vi. 40 When the solution of the cotton is complete, the spinning solution begins to decompose, unless it is kept at a low temperature. **1973** O. Steinerová tr. *B. Piller's Bulked Yarns* xi. 434 The latter [*sc.* viscose staple fibres] were made dyeable by acid wool dyes due to addition of protein particles to the spinning solution. **1973** *Materials & Technology* VI. iv. 290 The spinning syrup has to be extruded through very tiny holes in the spinneret.

c. *spinning gallery*.

1956 R. W. McDowell in W. A. Singleton *Stud. Archit. History* II. 133 Reference must be made to the 'spinning galleries'..an attractive feature of some.. ..Lakeland villages. **1976** G. Moffat *Short Time to Live* xi. 115 'What's brought you to Sandale?'..'Vernacular architecture, sir... Interiors too: spice cupboards, stone stairways, spinning galleries.'

8. c. *spinning tunnel*, a wind tunnel with a vertical air flow for testing the behaviour of model aircraft in simulated spins. Also *free-spinning tunnel*.

1934 *Rep. & Mem. Aeronaut. Res. Committee* No. 1578. 2 Tests in the Free Spinning Tunnel were accordingly projected as a check upon the validity of the results which could be obtained with small dynamical models. *Ibid.* 12 In the spinning tunnel the models are usually about 1/25 scale, and thus the rate of rotation is about five times that of the full scale spin. **1937** *Technical Rep. Aeronaut. Res. Committee 1936* I. 452 The R.A.E. Free Spinning Tunnel was brought into use in 1932 to examine the spinning properties of various existing and projected designs of aeroplanes. **1939** *Ibid. 1937* I. 552 The effect of mass distribution has been explored as a matter of routine on all designs tested in the spinning tunnel. **1947** A. Pope *Wind-Tunnel Testing* i. 10 The NACA has two free-spinning tunnels, one 15 ft in diameter, the other 20 ft.

d. *spinning magnetometer* = *spinner magnetometer* s.v. *SPINNER 11.

1960 *Archaeometry* III. 47 The great advantage that a spinning magnetometer has over the astatic type is that it can be used in a normal laboratory in the presence of a relatively large amount of local magnetic interference. **1963** R. M. Cook in Brothwell & Briggs *Science in Archaeol.* I. v. 64 In the spinning magnetometer the sample is rotated continuously to produce an alternating current.

spinning, *ppl. a.* Add: **3.** *spinning reserve* (Electr. Engin.), reserve power-generating capacity which is available to meet sudden increases in load.

1932 *Rep. Proposed Amer. Stand. Defs. Electr. Terms* (Amer. Inst. Electr. Engineers) 62/2 Spinning reserve is that reserve generating capacity connected to the bus and ready to take load. **1974** *Times* 21 Jan. 15/6 These 'spinning reserves' are carried on plant which is generating power, but not fully loaded. **1979** 'A. Hailey' *Overload* I. i. 5 GSP & L's last spinning reserve had been brought to full load.

Hence **spi·nningly** *adv.*

1923 *Daily Mail* 19 May 6 The ball is cracked spinningly through the gap between point and third man.

spinning-jenny. Add: **2.** Also, the whole of such an apparatus. (Earlier example.)

1879 H. A. Simmons *Ernest Struggles* iv. 72 To the ceiling of the taproom was fixed what the men called a 'spinning jenny', which was a revolving hand, like that on a clock, with a number of figures round it. It was with this that the customers won and lost pots of beer.

spino-. Add: **spino-cerebe·llar, -tha·lamic** *adjs.*

1900 E. A. Schäfer *Text-bk. Physiol.* II. 806 Another spino-cerebellar system, mainly crossed (heteromeric), lies in the ventro-lateral edge of the lateral column. **1975** *Sci. Amer.* Jan. 71/2 They found that one of the main afferent tracts leading to the cerebellum, the ventral spinocerebellar tract, conveys information not about the state of the body or the external environment but about the activity of inhibitory interneurons in the spinal cord. **1900** E. A. Schäfer *Text-bk. Physiol.* II. 807 Others [*sc.* spinal cells], it is said..enter the diencephalon, ending in the ventro-lateral nucleus of the optic thalamus, forming a spino-thalamic system. **1974** D. & M. Webster *Compar. Vertebr. Morphol.* xii. 278 Like the ventral spinothalamic tract, with which it is confluent, the lateral spinothalamic tract joins the medial lemniscus in the brainstem.

spinodal (spəinōᵘ·dăl), *a.* and *sb.* *Physical Chem.* [f. Spinod(e + -al.] **A.** *adj.* Being or pertaining to a spinodal; involving a metastable condition described by such a curve. **B.** *sb.* A curve which is the locus of stationary points in a system of curves; *spec.* such a curve in a pressure-volume diagram or the like which delimits a region of thermodynamic metastability from one of instability.

1956 *Nature* 3 Mar. 419/2 A single kinetic law did not apply at all temperatures, which suggests that the spinodal curve may have some influence. **1961** *Acta Metallurgica* IX. 536/2 The spinodal is the locus of points within a miscibility gap where $\partial^2F/\partial x^2 = 0$. F is the free energy of mixing and x the composition. *Ibid.* 801/2 Au–Ni thin foils decompose by what is almost certainly a spinodal mechanism. **1969** *Trans. Metallurg. Soc. AIME* CCXLV. 1707 Just above and just below the spinodal, we expect the reaction to proceed along a path which is not necessarily the energetically most favorable one. **1973** *Physics Bull.* Apr. 230 Spinodal decomposition of equimolar NaCl–KCl mixed crystals. **1974** *Nature* 29 Nov. 381/2 The locus of all the tangent points *c* defines a metastability limit (spinodal).

spin-off (spi·nǫ̈f), *sb.* and *a.* orig. *U.S.* Also **spinoff.** [f. vbl. phr. *to spin off*: see spin *v.* 13**.]

A. *sb.* **1.** *Comm.* A distribution of stock of a new company to shareholders of a parent company; a company so created.

1951 Stanley & Kilcullen *Federal Income Tax* 182 Sec. 112 (*b*) (11), added by the 1951 Act, permits the distribution of stock in a spin-off without recognition of gain to the stock-holders, subject to certain restrictions designed to prevent the use of spin-offs to distribute earnings and profits. **1956** *Sun* (Baltimore) 30 May 15/1 The proposed 'spin-off' was to be on the basis of three shares of Bestwall Gypsum for each share of Certain-Teed. **1969** *Daily Tel.* 4 June 3 Many had been anticipating a complete spin-off by B P of its United States subsidiary with a United States quotation and a chance of more direct public participation in the group's Alaskan activities. **1974** *Telegraph* (Brisbane) 8 May 46/4 Spea is a subsidiary of the Italstrade Company. Italstrade, in turn, is a subsidiary, or spin-off, of Italstat. **1981** *Observer* 4 Oct. 21/1 A growing phenomenon in British business life: the hive-off, spin-off or demerger—the management buy-out, in fact. **1981** *Times* 28 Oct. 19/5 Even split into four separate companies, the spin-offs would be equal fifteenth in the league table.

2. A by-product, an incidental development, side-effect, or benefit; the production or accrual of side-effects or indirect benefits; *spec.* (*a*) a business, organization, etc., developed out of or by (former) members of another larger business, etc.; (*b*) a show, television programme, etc., developed from an idea or character in another.

1959 *Wall St. Jrnl.* 12 May 1/4 Numerous firms have been organized by M.I.T. scientists who decided to strike out on their own—'spin-offs from M.I.T.', one research official terms them. **1961** *Guardian* 10 Oct. 6/6 'Technological fall-out' or 'technological spin-off'..are the terms used to denote the desirable social byproducts of the plan to send men to the moon. **1963** *Listener* 7 Nov. 735/2 The development out of (or 'spin off', as the Americans call it) magazines [*sc.* magazine programmes] must not be interpreted as any lack of conviction in their continuing role. **1967** *Technology Week* 23 Jan. 75/2 There were to be spin-offs in the form of a series of assist devices for emergency, temporary or permanent assistance to cardiac function. **1967** *Daily Tel.* 15 May 9/8 The close season is also the signal for another series of BBC Comedy Playhouse 'try-outs'. It produced 'Steptoe'; and last season, to use the current jargon, the 'spin-offs' in series form were 'The Whitehall Worrier', [etc.]. **1968** *Economist* 13 Jan. 55/2 It was flatly denied that the huge military and space programmes had been of any advantage, in terms of technological spin-off, to industry. **1968** P. McKellar *Experience & Behav.* xv. 398 In this connexion we encounter the notion of 'spin off', the term used for other applications of findings that have emerged from space research. **1969** *Daily Tel.* (Colour Suppl.) 28 Mar. 7/3 A car..is a means of transport with a horrifying spin-off of death and injury. **1975** *Lady* 17 July 97/1 One of the spin-offs of our affluent society is that more people can afford to keep dogs. **1976** *TV Times* (Brisbane) 22 May 7/2 There is a tradition in American TV—if a show is a success, do a spin-off. In other words, take one or two characters from the parent series and build another series around them. **1977** Sachs & Jahn *Celestial Passengers* xxxii. 198 Space spinoffs have resulted in many new products to improve the quality of our recreational activities. *Ibid.* 193 Probably the best-known space spinoff to health is the cardiac pacemaker. **1979** *Jrnl. R. Soc. Arts* CXXVII. 626/1 If we can improve our productivity ..then there will be all sorts of spin-offs from this in the way of leisure industries and service industries.

B. *attrib.* or as *adj.* That develops or is created as a spin-off.

1966 *National Observer* (U.S.) 18 July 7 Although the column hasn't been as successful churning up front-page news stories as some editors had hoped, it does develop an occasional 'spin-off' story. **1967** *Boston Globe* 18 May 35/6 Shares of the spin-off company will be distributed tax free to United Fruit stockholders. **1969** *Physics Bull.* July 268/2 Many of the successful 'spin-off' firms in the United States were based on a transfer of technology by individuals from large and continuing programmes in government and university laboratories. **1974** *Financial Times* 8 Apr. 23/7 Hardly anyone earns less than the proposed new minima, which would therefore raise only a few earnings and so would hit employers' wage bills and eat into Stage Three allowances only through the spin-off effect on holiday and sick pay. **1979** *Amer. Jrnl. Trop. Med. & Hygiene* XXVIII. 1043/2 No attempt was made to document 'spinoff' costs, notably losses of tourism revenue. **1980** J. Wainwright *Man of Law* xi. 64 With hindsight..I *knew*... But honesty demands that I ask spin-off questions. How much did I know?

spinone (spinō̆ᵘ·ni). [It.] A wire-haired gun-dog of an Italian breed, usually white with tan or brown markings, drooping ears, and a docked tail.

1945 C. L. B. Hubbard *Observer's Bk. Dogs* 188 The Spinone..has for centuries been the Italian all-purpose shooting dog. **1964** E. F. Dalglish tr. *Schneider-Leyer's Dogs of World* xvi. 219 Spinone... A big, squarely built, rough-coated dog with long ears and docked tail. **1972** *Daily Tel.* 5 Dec. 7/7 Breeds of dog whose docking is permitted by the Kennel Club are: sporting spaniels,.. Italian spinones,..Weimaraners, [etc.].

spinor (spi·nɔɹ). *Physics.* [a. G. *spinor* (B. L. van der Waerden 1929, in *Nachr. von d. Ges. d. Wissensch. zu Göttingen* 100), f. Spin *v.* + *-or*, after Tensor, Vector.] Any quantity existing in a space and having the property that rotation through 360° reverses its sign and leaves it otherwise unchanged; also, applied to quantities constructed from two or more of these in the way that tensors may be constructed from vectors.

1931 *Physical Rev.* XXXVII. 1022 With the spinor analysis developed by B. Van der Waerden.., which comprises all representations of the Lorentz group, even those not contained in ordinary tensor calculus, one is able to write all derivations and equations in an automatically covariant form. **1952** *Amer. Jrnl. Physics* XX. 253/2 The available literature on spinors is exceedingly hard to understand. **1974** P. A. M. Dirac *Spinors in Hilbert Space* ii. 4 Spinors, like tensors, are geometrical objects embedded in a space and have components that transform linearly under transformations of the coordinates of the space. Spinors differ from tensors in that they change sign when one applies a complete revolution about an axis, while tensors are unchanged. **1974** G. Reece tr. *Hund's Hist. Quantum Theory* xv. 200 It was soon seen that the four-component Dirac ψ was a particular kind of entity... It was called a spinor. **1977** K. O. May tr. *Iyanaga & Kawada's Encycl. Dict. Math.* I. 216/2 Although $SO(n)$ ($n \geq 3$) is a connected Lie group, it is not simply connected. The simply connected Lie group which is locally isomorphic to $SO(n)$ is called the spinor group and is denoted by $Spin(n)$.

Hence **spino·rial** *a.*, involving spinors, decribed by means of spinors.

1968 *Physics Bull.* Nov. 381/1 The wavefunction must have four components which transform according to the 'spinorial' transformation encountered by Pauli. **1978** *McGraw-Hill Yearbk. Sci. & Technol.* 356/1 A spinorial gauge field with spin 3/2.

spin-out (spi·naut). *N. Amer. slang.* [f. vbl. phr. *to spin out*: see spin *v.* 13***.] A skidding spin by a vehicle out of control.

1957 *Daily Progress* (Charlottesville, Va.) 28 Oct. 14/6 A spin-out in the last 10 minutes of the race may have cost Tony Briggs of Charlottesville top honors in the first preliminary race before the President's Cup automobile race. **1971** *Maclean's Mag.* Oct. 39/3 He had trouble eliminating his spin-outs, those heart-stopping moments when the car slithers in circles and semi-circles on the track.

Spinozism, Spinozist. Add: Also **Spino·zaism, Spino·zaist.**

1852 H. Melville *Pierre* xx. ii. 381 Why, lad, I have received propositions from the Editors of the Spinozaist to contribute a weekly column to their paper, and you know how very few can understand the Spinozaist. **1912** *Q. Rev.* Oct. 393 Modern Spinozaism is inclined to identify ethics with religion. *Ibid.* 398 He no longer speaks as a Spinozaist.

Spinozite: for *rare*⁻¹ read *rare* and add later example.

1946 *Mind* LV. 101 Some of these assumptions..are often confidently made by others than professed Spinozites.

spin-rinse (spi·nrins). [f. Spin *v.* + Rinse *sb.*] A rinsing of washing in a rotating perforated drum which draws off water; a combined rinse and partial spin-dry.

1948 [see *spin-drier, -dryer]. **1961** *Guardian* 22 Mar. 8/3 A powerful spin-dryer.., to which a crafty bit of design has given a built-in spin-rinse percolating downwards through the clothes to give an effective rinse, hitherto lacking in the conventional spinner.

spin-scan (spi·nskæn), *a.* and *sb.* [f. Spin *v.* + Scan *sb.*] **A.** *adj.* Applied to devices whose scanning motion is provided by the rotation of the craft carrying them.

1967 *Electronics* 6 Mar. 73 (Advt.), SB RC also built the spin-scan camera that has been sending back high-resolution black-and-white photos of the earth's cloud cover. **1979** *Sci. Amer.* Nov. 21 (Advt.), The satellite's 'camera' is a visible-infrared spin-scan radiometer.

B. *sb.* A scan performed or produced by such a device. Also *attrib.*

1972 *Space Research* XII. 1768 Television would have taken a shorter exposure (a fraction of a second, compared with several minutes for a spin scan). **1974** *Science* 25 Jan. 318/2 A 2·5 centimeter telescope aboard Pioneer 10 is capable of making two-dimensional spin-scan maps..at high resolution. **1974** *Nature* 6 Sept. 18/1 The most striking early results were the hundreds of photographs that resulted from the spin-scans of the planet, giving higher resolution pictures of the planet than had previously been possible.

So **spi·n-scanning** *a.*

1972 *Space Research* XII. 1765 A summary description is made of spin-scanning devices for various space missions.

spinster. Add: **3.** (Sense 2 b) *spinster-baiting*.

1938 L. MacNeice *Mod. Poetry* 191 Merely a piece of rather cruel spinster-baiting.

spi·nsterishness.

1913 R. West *Let.* June in G. N. Ray *H. G. Wells & Rebecca West* (1974) 23 Your spinsterishness makes you feel that a woman desperately and hopelessly in love with a man is an indecent spectacle. **1930** R. Macaulay *Staying with Relations* iii. 44 The elegant spinsterishness

of Claudia and Benet had turned, in Julia,..to something more sensuous.

spinstry. 2. For *Obs.*⁻¹ read *rare* and add later example.

1894 *Wales* Aug. 192/1 The local gossips..watched the progress of events from the heights of maternity and spinstry.

spintrian. For † *Obs.* read *rare* and add later examples.

1887 L. C. Smithers tr. *Forberg's Man. Class. Erotol.* viii. 166 More than three may enjoy themselves together; this is what we call after Tiberius, the spintrian kind. **1913** C. MacKenzie *Sinister St.* I. ii. ix. 287 My library ..holds as many secrets as the Spintrian books of Elephantis, long ago lost and purified by the sea.

spinulosin (spəiniŭlōᵘ·sin). *Biochem.* [f. mod.L. *spīnulōs-us*, specific epithet of the fungus from which it was first isolated (see Spinulose *a.*): see -in¹.] A purple-black crystalline benzoquinone derivative, $C_8H_8O_5$, which is produced in cultures of the moulds *Penicillium spinulosum* and *Aspergillus fumigatus*.

1938 Anslow & Raistrick in *Biochem. Jrnl.* XXXII. 689 Birkinshaw & Raistrick (1931) reported the isolation from cultures of strains in the *Penicillium spinulosum* Thom series of a new mould metabolic product which was unnamed at that time, but for which the name *spinulosin* is now proposed. **1946** [see *fumigatin]. **1950** L. F. & M. Fieser *Org. Chem.* (ed. 2) xxxii. 770 Spinulosin..has been prepared from fumigatin by Thiele addition of acetic anhydride, hydrolysis of the resulting tetraacetate, and oxidation. **1979** *Biol. Abstr.* LXVII. 6332/2 A strain of *A. fumigatus* (Fresenius)..was grown on a culture medium with added saccharose. The synthesis of..spinulosin was studied under many different temperature, pH and inoculum conditions.

spin-up (spi·nʌp), *sb.* [f. Spin *v.* + Up *adv.*¹] An increase in the speed of rotation.

1960 *Jrnl. Geophysical Res.* LXV. 2994 (*caption*) Increased stability of both the theoretical and observed motion is seen after spinup. **1968** *New Scientist* 3 Oct. 8/2 The particular merit of the autogyro is that its rotor does not depend on engine power, except for the spin-up before take-off. **1969** *Nature* 29 Nov. 873/1 The time scale for the cracking and initial spin up is of the order of 0·01 to 0·1 s, the time for a shear wave in the crust to propagate a stellar radius. **1971** [see *spin-down sb.] **1977** *Astron. & Astrophysics* LX. 85/1 Subsequent retrograde spinup via thermally driven atmospheric tides can explain the present slow retrograde motion of the planet.

spin-up (spinʌ·p), *a.* *Physics.* [f. Spin *sb.*¹ + Up *adv.*²] Being or pertaining to a particle with a spin pointing upwards.

1965 *Sci. Amer.* Jan. 106/3 The electrons' magnetic field at the nucleus is produced by..the net spin density, or the 'spin up' density minus the 'spin down' density. **1970** *Nature* 25 July 372/2 Four spin-up electrons at the corners of a tetrahedron and four spin-down electrons at the corners of an interpenetrating tetrahedron pointing the opposite way are arranged around each oxygen centre. **1973** *McGraw-Hill Yearbk. Sci. & Technol.* 287/1 With this technique, it is possible to calculate the difference between the spin-up and the spin-down populations in each electron shell. **1980** Rudden & Wilson *Elem. Solid State Physics* ii. 41 The first available electron is placed in the 1s spin-up state, the second in the 1s spin-down state.

spiny, *a.* Add: **3. b. spiny rat,** a rodent of the family Echimyidæ, found in tropical South and Central America and distinguished by bristly fur.

1876 A. R. Wallace *Geogr. Distrib. Animals* II. xvii. 238 The Echimyidæ, or spiny rats, are a family, chiefly South American. **1924** W. B. Scott *Hist. Land Mammals Western Hemisphere* v. 184 The spiny rats (*Echimys* and *Loncheres*) are so called from their appearance, not because they are related to the true rats. **1974** H. MacInnes *Climb to Lost World* ix. 147 A spiny rat, a rare creature: Mike Atherley and I both saw it more than once and thought that it had no tail.

Spion Kop: see *kop 2.

spiral, *sb.* Add: **2. c.** *U.S. Football.* A kick or pass in which the ball in flight spins round its long axis.

1896 Camp & Deland *Football* vi. 61 *Spiral*, a kick similar to the twister, in which the ball maintains a true course while revolving on its long axis. **1910** W. Camp *Book of Foot-Ball* viii. 308 Long passes are best made by holding the ball like a spear and sending a spiral. **1972** J. Mosedale *Football* v. 69 Fourteen times his soft spirals connected, for 216 yards.

d. *fig.* A progressive movement in one direction (esp. upwards or downwards, and marking a relentlessly deteriorating state of affairs), considered to take the form of a spiral; *spec.* one caused by the interaction or alternate overtaking of interdependent quantities; *vicious spiral*: see *vicious *a.* 9.

1897 P. Geddes *Let.* 10 Feb. in P. Boardman *Worlds of P. Geddes* (1978) vi. 155 New money = new crime = new report.. = new police and so on, in downward spiral. **1931**, etc. [see *inflationary *a.*]. **1939** *Economist* 16 Dec. 405 The fear of an accelerating spiral of wages and prices. **1958** *Spectator* 15 Aug. 211/1 Steps to avoid a new arms race spiral. **1965** *Listener* 23 Sept. 439/2 We have got to get rid of the endless spiral, price increases and pay claims. **1975** *Physics Bull.* Aug. 345/1 The tendency for departments with the highest demand for undergraduate places to take the largest slice of the UGC cake has meant that engineering and physical science departments have fallen into a spiral of fewer students resulting in lower grants. **1980** *Times* 15 Jan. 14 Price of antique silver follows the gold spiral.

e. *Aeronaut.* A descent (or, rarely, a climb) made by an aircraft in the form of a helix; a continuous banking turn accompanying a descent or ascent.

1910 *Sphere* 30 July 103/2 This descent..was composed of several high-speed dives and short spirals. **1918** E. M. Roberts *Flying Fighter* 272 When I came out of my spiral,..my engine would not start again. *Ibid.* 301 The Hun machine started upward in a spiral. **1941** Pope & Otis *Elements of Aeronaut.* ix. 85 The spiral, usually done with power off, is merely a gliding turn, continued to make several complete turns, gradually descending. **1975** G. H. Saunders *Dynamics of Helicopter Flight* v. 178 A neutrally stable spiral would require the pilot to take back out his cyclic input once his desired bank angle has been achieved.

3. c. (Earlier example.)

1850 Ld. Rosse in *Phil. Trans. R. Soc.* CXL. 511 Night excellent, a spiral seen in an oblique direction, resolved well, particularly towards the centre, where it is very bright.

spiral, *a.*¹ and *adv.* Add: **1. b.** Also, of a descending course or path. (Further examples, from *Aeronaut.*)

1908 F. W. Lanchester *Aerodonetics* vii. 180 The aerodrone..loses its equilibrium and comes rapidly to earth with a kind of spiral dive. **1912** *Flight* 31 Aug. 787/1 The machine at once started a spiral nose-dive. **1939** *Aircraft Engin.* XI. 40/3 There are many references to spiral or corkscrew descent in the literature..but this always implies a fully controllable motion at an angle of incidence below the critical angle; very different to the spin proper. **1961** C. B. Smith *Testing Time* 53 To their horror, the men on the ground saw the aircraft drop out of control into a whirling dive, the 'spiral dive' which they knew meant almost certain death.

d. *Surg.* Of a fracture: curving round a long bone lengthwise.

1897 *Lippincott's Med. Dict.* 955/2 Spiral fracture. **1934** *Practitioners Library Med. & Surg.* V. ii. 301 Fractures of the Shaft of the Femur... Nonunion may occur... This is likely to happen in the oblique or spiral types. **1950** *Brit. Encycl. Med. Pract.* (ed. 2) IV. 369 (*caption*) Spiral fracture of tibia in boy aged 8 years, three weeks after accident; no clinical signs or symptoms except his refusal to use his leg. **1976** M. Machlin *Pipeline* xlii. 455 At the Medical Center Hospitals in Houston they had told him that the spiral break would take at least six months to heal to the point where the cast could be removed.

3. a. *spiral bevel gear*, a bevel gear that is also a spiral gear; *spiral binding*, a book binding in which a helical wire passes through a closely spaced row of holes near the inside edge of each leaf; so *spiral-bound* adj.; *spiral divergence* (Aeronaut.) = *spiral instability* below; *spiral gear*, a gear wheel whose teeth are cut obliquely to the axis of the wheel and are curved to form part of what is approximately a spiral or helix; *esp.* a skew gear of this kind; so *spiral gearing*; *spiral instability* (Aeronaut.), an instability in which an aeroplane undergoing a banked turn tends to enter a descending spiral as a result of sideslipping and reduction of the radius of turn; *spiral stability* (Aeronaut.), the capacity of an aeroplane not to enter a spiral while executing a banked turn, or to recover from a steeply-banked spiral path.

1915 V. W. Pagé *Automobile Repairing* ix. 767 The advantages of the spiral bevel gear are mainly due to the shape of the teeth which roll into engagement more smoothly than the ordinary form of bevel gears. **1973** *Transmission & Rear Axle—Bedford Trucks & Coaches* (Vauxhall Motors) 168 The differential, spiral bevel gear and pinion can be serviced without removing axle from vehicle. **1949** Melcher & Larrick *Printing & Promotion Handbk.* 280/2 *Spiral binding*, see *mechanical bindings*. **1968** F. H. Holliday *Man. Stationery, Office Machines & Equipment* ii. viii. 440 The sheets..are fed on to a conveyor to be trimmed, knocked up and punched to receive the spiral binding. **1961** *Lebende Sprachen* VI. 104/1 Spiral-bound stenographer's notebook. **1969** D. Francis *Enquiry* iii. 37 He stood..holding a spiral bound notebook. **1976** E. Ward *Hanged Man* xxxi. 200 A thick spiral-bound document. **1949** *Jrnl. R. Aeronaut. Soc.* LIII. 541/1 There are two common kinds of aircraft instability which usually would be judged tolerable according to the above criterion. These are the phugoid oscillation (longitudinal-symmetric motion) and the spiral divergence (lateral-antisymmetric motion). **1970** T. Hacker *Flight Stability & Control* vii. 159 The possibility of eliminating spiral divergence by design can be made to stand out by means of stability diagrams. **1888** J. G. Horner *Lockwood's Dict. Mech. Engin.* 338 *Spiral gear*, includes helical, stepped, and worm gearing. **1914** —— *Gear Cutting* v. 48 Since the special gear-cutting machines have come into general use, the manufacture of spiral gears has been established on a better commercial basis than hitherto. **1971** B. Scharf *Engin. & its Lang.* xii. 155 In some types of multiple gear train, the same gear drives a helical gear (with parallel axis) as well as a spiral gear (with skew axis). *a* **1877** Knight *Dict. Mech.* III. 2275/1 *Spiral gearing*, a gear-wheel having meshing spiral ribs and grooves. The teeth run around the periphery of the gear-wheel, and meet in an angle on a line midway from either edge of the wheel. **1930** *Engineering* 2 May 559/2 Spiral gearing has been found to have many valuable applications. **1914** L. Bairstow et al. *Rep. & Mem. Advisory Comm. for Aeronautics* No. 77. 168 A machine which is liable to spiral instability when gyroscopic actions are eliminated, cannot become stable owing to gyroscopic action of the propeller and engine. **1970** T. Hacker *Flight Stability & Control* vii. 162 The elimination of spiral instability..is achieved by the constraint of the angle of bank ϕ, obviously in addition to correcting the heading. **1947** C. F. Toms *Introd. Aeronautics* v. 222 The use of a certain amount of dihedral..is essential for both directional and spiral stability. **1978** M. Simons *Model Aircraft Aerodynamics* xi. 126 If the model is primarily a thermal soarer..effort should be concentrated on spiral stability.

b. *spiral thickening*.

1933 *Trop. Woods* XXXVI. 4 Spiral Thickenings.—Helical ridges on the inner face of, and a part of, the secondary wall. **1953** K. Esau *Plant Anat.* xi. 228 Such secondary thickenings [of the xylem] are called, respectively, annular, spiral or helical, and reticulate.

c. Also *spiral cleavage*, a pattern of embryogenesis characteristic of certain invertebrate groups, in which the third cell division is asymmetrical and destroys all but four-fold rotational symmetry.

1892 E. B. Wilson in *Jrnl. Morphol.* VI. 377 The events of the cleavage fall into three very marked periods which I shall designate respectively as the (1) spiral, (2) transitional, and (3) bilateral periods. *Ibid.* 441 The third spiral cleavage of the primary micromeres gives rise to four apical cells. **1948** *New Biol.* V. 113 The same fundamental pattern (called spiral cleavage because of the oblique direction of many of the divisions; the actual pattern of cells resulting is not a spiral) is found also in other groups, such as the flatworms. **1967** L. A. Borradaile et al. *Invertebrata* i. 2 The annelid superphylum has eggs that develop by means of spiral cleavage. When the blastula divides from the four-cell stage to the eight-cell stage the second quartet lie on top of and between the cells of the first quartet.

e. *Astr.* Special collocations in sense 2, as: *spiral arm*, an arm of a spiral galaxy; *spiral galaxy*, a galaxy in which bright stars and gas clouds tend to be located along arms that appear to spiral from a central nucleus; *spiral nebula*, a spiral galaxy (now chiefly *Hist.*).

1914 A. S. Eddington *Stellar Movements & Struct. of Universe* xi. 243 The star clouds of the Milky Way form its [*sc.* our galaxy's] spiral arms. **1978** Pasachoff & Kutner *University Astron.* xxiii. 582 The interstellar extinction prevents us from studying parts of the spiral arms farther away from the sun. **1913** *Astrophysical Jrnl.* XXXVII. 112 The stellar accumulations might be arranged so as to produce the phenomenon of the Milky Way—on the supposition of a spiral galaxy. **1944** H. Shapley *Galaxies* i. 26 The subclassifications of bright spheroidal and spiral galaxies..are possible only for those systems near enough for large-scale photography. **1980** *Sky & Telescope* July 25 By studying star formation in spiral galaxies, we can do more than test theories of spiral structure. **1850** Ld. Rosse in *Phil. Trans. R. Soc.* CXL. 505 The other spiral nebulæ discovered up to the present time are comparatively difficult to be seen. **1920** A. S. Eddington *Space, Time & Gravitation* x. 160 The most remote objects known are the spiral nebulae, whose distances may perhaps be of the order a million light years. **1978** Pasachoff & Kutner *University Astron.* xxiii. 572 Another class of objects was once known as 'spiral nebulae'... However, these spiral nebulae are now known to be galaxies in their own right.

4*. = *spiral-bound* adj., sense 3 a above.

1977 H. Greene *FSO-1* ix. 83 [He] whipped out large black reading glasses, and peered downward at a spiral pad. **1978** R. Thomas *Chinaman's Chance* xiii. 135 Durant was seated on the couch..a secretary's spiral notebook in his lap.

spiral, *v.* Add: **1. b.** To fly an aircraft in a spiral path. Also with *down, downwards*.

1916 E. C. Middleton *Aircraft* iii. 33 The pilot either 'spirals' or glides down, until he is able to ascertain the direction of the prevailing wind. **1922** H. L. Foster *Adventures of Tropical Tramp* xi. 173 The aviator spiraled downwards towards his landing place. **1941** Pope & Otis *Elements of Aeronautics* ix. 85 Practice is required.. to spiral over a fixed spot when the wind is blowing. **1978** M. Simons *Model Aircraft Aerodynamics* iv. 37 Even better rates of climb would result if the model did not have to spiral.

c. *fig.* To move rapidly in one direction (usu. upwards), in a manner considered to resemble a spiral; to increase or decrease in response to the same movement of another quantity or other quantities. Cf. sense 2 d of the sb. above.

1922 H. Crane *Let.* c 18 June (1965) 91 Under the influence of aether and *amnesia* my mind spiraled to a kind of seventh heaven of consciousness. **1941** *Time* 20 Oct. 35/1 Even if import and farm prices resist all controls, processors' and retailers' prices will rise but not spiral with them. **1942** E. W. Kemmerer *ABC of Inflation* 156 If wages and the prices of farm products are not adequately restricted but are permitted to spiral upward..the whole price situation will get out of control. **1959** *Listener* 18 June 1052/1 The cost of living has spiralled. **1977** *Milestones* Summer 19/1 A similar table published in the Autumn 1974 issue of *Milestones* shows

how much the cost of spare parts has spiralled. **1979** *Tucson* (Arizona) *Citizen* 20 Sept. 1A/4 The dollar spiraled downward on European money markets today. **1979** *Daily Tel.* 5 Dec. 21 The..risks of sending demand, output and tax revenue spiralling down again while the cost of spending programmes such as unemployment benefit..spiral up.

Hence **spi·ralling** *ppl. a.* and *vbl. sb.*

1944 Auden *For Time Being* (1945) 90 Even the problems of Trade Cycles And Spiralling Prices are regarded by the experts As practically solved. **1958** *Spectator* 18 July 92/3 The twin threat of a renewed world currency crisis and a spiralling of trade restriction. **1965** [see *on-so *adv.*]. **1969** H. Perkin *Key Profession* iv. 138 The post-war situation of spiralling prices and incomes. **1979** E. H. Gombrich *Sense of Order* vi. 160 Spiralling terminations suggest the curling of elastic matter. **1979** *Nature* 14 June 622/1 Viscous drag on the planet's orbital motion would then lead to a spiralling into the stellar core. **1980** *Oxf. Diocesan Mag.* June 7/2 The failure of extrinsic motivation to fill the need is evident from the spiralling demands of frustrated materialists.

spiralism (spəiᵊ·rălizʼm). *Sociol.* [f. Spiral *sb.* + -ism.] A term for mobility in career and place of residence as part of individual success in an industrial economy. So **spi·ralist** *sb.* and *a.*

1957 W. Watson in M. Gluckman *Closed Systems & Open Minds* (1964) vi. 147 The progressive ascent of the specialists of different skills..forms a characteristic combination of social and spatial mobility which may be called 'spiralism'. *Ibid.* 148 A spiralist's possible range of activity is the total number of fields within which his education qualifies him to compete. *Ibid.* 149 The generic spiralist culture. **1962** A. Sampson *Anat. of Britain* xxviii. 461 The career of a corporation manager cuts across traditional societies and local communities; he is a 'spiralist'—moving towards the top in narrowing circles, from one community or country to another, gathering local experience before he settles in the head office as a senior executive. **1969** C. Bell *Middle Class Families* ii. 40 There were 13 'blocked' spiralists who admitted they were not at the top. **1971** —— & Newby *Community Stud.* v. 158 Social mobility in Franza proceeds by a kind of spiralism. Individuals must 'get out' in order to 'get on'.

spiralization (spəiᵊ:rălaizēⁱ·ʃən). [f. Spiralize *v.* + -ation.] The acquisition of spiral form; (in quot. 1851, a spiral coil).

1851 H. Melville *Moby Dick* II. xviii. 141 The whale-line..is spirally coiled away in the tub..so as to form.. layers of concentric spiralizations. **1910** J. A. Fleming *Princ. Electr. Wave Telegr.* (ed. 2) 127 The increase in resistance of a solenoid due to the spiralization. **1928** T. C. Chamberlin *Two Solar Families* xvii. 159 The propulsion of solar bolts into planetary space and the drawing of these into orbits by the joint action of the sun and the passing star (the spiralization) was only the beginning of the process of..nebularization. **1941** *Cold Spring Harbor Symp. Quantitative Biol.* IX. 15/2 The outstanding question is whether or not this amount of 'contraction' can all be accounted for by spiralization. **1969** I. B. Raikov in T. T. Chen *Res. in Protozool.* III. 75 Shortly before nuclear division, the chromosomes undergo spiralization and become longitudinally split. **1980** *Macromolecules* XIII. 799/2 Since..spiralization of the chain cannot be invoked for steric control, the stereoregularity can only result from the asymmetric spatial arrangement around the metal atom of the catalytic complexes.

spiralize, *v.* Add: Also *intr.*, to move in a spiral. Hence **spi·ralized** *ppl. a.*, formed into a spiral shape; **spi·ralizing** *ppl. a.*

1851 H. Melville *Moby Dick* I. xxxv. 265 Drink and pass!.. Short draughts—long swallows, men; 'tis hot as Satan's hoof. So, so; it goes round excellently. It spiralizes in ye. **1928** T. C. Chamberlin *Two Solar Families* 145 A group of bodies already separated may respond to the spiralizing whirl of two centers of gravity even more freely than an eruptive body from which the spiralized matter must be shot forth. **1940** *Jrnl. Genetics* XL. 67 The suggestion has been made..that whereas all non-heteropycnotic regions have a spiral structure the heteropycnotic regions are not spiralized at all. **1977** *Nucl. Instruments & Methods* CXLV. 258/2 A new extrapolation is computed to associate track segments. In this way a spiralizing track can be followed. **1980** *Macromolecules* XIII. 798 Neither a chiral carbon nor a spiralized chain participates in the first two addition steps.

spiramycin (spəirămai·sin). *Pharm.* [ad. F. *spiramycine*, f. *spira-* (of unkn. origin): see *-mycin.] A mixture of macrolide antibiotics obtained from the fungus *Streptomyces ambofaciens*.

1955 *Antibiotics Ann.* 1954–5 724 A new antibiotic, spiramycin, has been extracted from the culture filtrates of a previously undescribed species of streptomyces. **1974** *New England Jrnl. Med.* CCXC. 1110/2 Treatment with spiramycin during pregnancy reduced the overall frequency of the fetal infections.

spirane (spəiᵊ·rēⁱn). *Chem.* Also **spiran** (-æn). [ad. G. *spiran*: see *spiro- 2 and -ane.] Any organic compound having two rings in the molecule with a single atom (usu. of carbon or nitrogen) common to both; a spiro-compound.

In quot. 1919 used *attrib.* of the shared atom.

1911 *Jrnl. Chem. Soc.* C. i. 497 Spirans or dispirans are homo- or hetero-cyclic compounds containing two rings with a carbon atom common to both. **1919** *Ibid.* CXV. i. 323 Strains existing in one ring of a *spiro*-compound could not possibly make themselves felt in the second ring unless communicated by a spirane carbon atom, which itself is in a state of strain. **1950** E. de B. Barnett *Stereochem.* iv. 55 Any spirane in which the rings are identical but unsymmetrical should be enantiomorphic. **1972** Natta & Farina *Stereochem.* iv. 113 A class very similar to the allenes consists of the spirans—bicyclic compounds whose rings are attached to only one atom. **1978** R. L. Baumgarten *Org. Chem.* xvii. 289 Biphenyls substituted with bulky ortho substituents, substituted allenes, and spiranes are among the classes of compounds that exhibit optical activity although there is no asymmetric carbon atom present.

spirant, *sb.* (Earlier example.)

1862 W. D. Whitney in *Jrnl. Amer. Oriental Soc.* VII. vii. 319 The labial series has no sibilants; for its pair of fricatives, surd and sonant..so lack the hissing quality which distinguishes the..sibilants, that it seems preferable to put them in another class; which..we will call the spirants.

spirantal (spəiræ·ntăl), *a.* [f. Spirant *sb.* + -al.] = Spirant *a.*

1893 J. Clark *Man. Linguistics* iii. 41, kj, ghj, k j, ghʷj give for result a sort of geminated spirantal sound. **1928** J. & E. M. Wright *Elem. Middle English Gram.* (ed. 2) vi. 131 The palatal spirantal element began to disappear in pronunciation from about the end of the fourteenth century in the south Midland and southern dialects. **1965** [see *Chipewyan].

spirantization (spəiᵊ:răntaizēⁱ·ʃən). [f. next: see -ization.] Making into a spirant; development of a spirantal pronunciation. Also *attrib.*

1911 *Amer. Jrnl. Philol.* XXXII. 37 Breath stress was weaker here than in Germanic, where spirantization took place in spite of the escape of breath in a preceding nasal. **1939** *Language* XV. 64 The second point..is the problem of the spirantization of ungeminated non-emphatic stopped consonants. **1973** J. M. Anderson *Structural Aspects Lang. Change* 209 A second spirantization rule converted [g̃] → [ɣ] through loss of occlusion. **1977** *Trans. Philol. Soc. 1975* 11 There is no principled way to separate a rule like the spirantization of intervocalic /b d g/ in Spanish from German obstruent devoicing or OE verbal ablaut.

spirantize, *v.* (In Dict. s.v. Spirant *sb.* and *a.*) Add: *trans.* To pronounce as a spirant, to make into a spirant. So **spi·rantizing** *ppl. a.* and *vbl. sb.* (Further examples.)

1911 *Amer. Jrnl. Philol.* XXXII. 36 The former [*sc.* Iranian] is a spirantizing language, like the old Germanic. *Ibid.*, The spirantizing of voiceless aspirates in Iranian. **1936** *Language* XII. 247 If with the vowels we may include the semivowels, we note here the Balto-Slavonic palatalizations,..[and] the spirantizing of Latin *c* (that is, [k]) as it develops into the Romance languages. **1964** S. K. Chatterji in D. Abercrombie et al. *Daniel Jones* 411 In Bengali, the aspirates *ph, bh*, are normally spirantized to [ɸ, β]. **1977** *Word 1972* XXVIII. 252 It cannot be argued that early Guarani might have had an occlusive [d] in initial position and that it would have been a simple matter to spirantize it.

spire, *sb.*¹ Add: **10. b.** *spire-passion*.

1944 Blunden *Cricket Country* v. 64 There is only one person known to me who quite equals my spire-passion.

c. *spire-straight* adj.

1933 C. Day Lewis *Magnetic Mountain* 26 There, as a candle's beam Stands firm and will not waver Spire-straight in a close chamber.

† spireme. *Cytology. Obs.* Also **spirem.** [ad. G. *spirem* (W. Flemming *Zellsubstanz, Kern und Zelltheilung* (1882) xx. 195), f. Ionic Gr. σπείρημ-α coil, convolution.] The tangled strands of chromosomal material seen in the early stages of cell division, formerly believed to be a single continuous strand (or two in a diploid cell, etc.); = *skein *sb.*¹ 2 c.

1889 *Q. Jrnl. Microsc. Sci.* XXX. 171 We call this stage, with Flemming, the 'Knäuel-Stadium' (skein stage), or 'spirem', or 'mother-skein'. **1905** *Amer. Naturalist* XXXIX. 484 During synapsis the reticulum becomes transformed into a definite spirem. **1910** *Encycl. Brit.* VII. 714/1 As the spireme thread contracts, it segments into a number of short, and usually U-shaped, segments —the 'chromosomes'. **1936** *Discovery* May 161/1 The hypothesis of the continuous spireme, long given up by cytologists, is resurrected, and entirely inaccurate statements are made.

spirit, *sb.* Add: **II. 6. c.** (Further examples.) In Christian charismatic groups: *baptism in* (*of*, etc.) *the Spirit* (and similar phrases), an experience subsequent to conversion and water-baptism, usually evidenced by speaking in tongues (see quot. 1972²): in allusion to Mark i. 8 and parallel passages; *to receive the Spirit*, to experience conversion, evidenced by speaking in tongues; *to sing in the Spirit*, to sing in a language apparently unknown to the singer (cf. I Cor. xiv. 15); so *song in the Spirit*; *the spirit moves me*: see Move *v.* 11.

1865 T. T. Carter in *Oxford Lent Sermons* xiii. 198 Within this new dispensation of the Spirit there is a specially sacred Presence of our Lord. **1896** R. A. Torrey *Baptism with Holy Spirit* i. 20 The Baptism with the Holy Spirit is the Spirit of God coming upon the believer,..imparting to him gifts not naturally his own, but which qualify him for the service to which God has called him. **1902** *Christian* 11 Dec. 13/1 If we are born again, we have received the Spirit once for all. **1903** R. C. Morgan *Outpoured Spirit* viii. 53 By coming unto Jesus ..as members of the Spirit-baptized Church. **1909** T. B. Barratt *In Days of Latter Rain* 98 One who is filled with the Spirit..gets visions, prophesies, sings in the Spirit, staggers under the overwhelming weight of the Glory. **1932** C. E. Raven *Creator Spirit* iv. 113 It is just as much within this one order that the Spirit works in creation as in those fuller manifestations which we call the Incarnation and Atonement. **1941** D. Gee *Pentecostal Movement* i. 7 For the individual recipient of the baptism in the Spirit it is subsequent to..regeneration. **1948** L. W. Brown in M. Warren *Triumph of God* vi. 154 There are.. those who maintain that no man has truly received the Spirit unless he has spoken with tongues. **1964** N. Bloch-Hoell *Pentecostal Movement* ii. 28 Mrs. Arthur.. gives a striking account of angel visions and singing in the Spirit in perfect Latin followed by correct translation into English. *Ibid.*, With regard to the 'song in the Spirit' and the testimony of the nuns it must be remembered that the latter was no first-hand evidence. **1972** S. Tugwell (*title*) Did you receive the Spirit? **1972** R. A. Wilson tr. *Hollenweger's Pentecostals* ii. 22 In a Bible school in Topeka, Kansas, run by Charles Parham, speaking in tongues was recognized as a distinguishing characteristic of the baptism of the Spirit. **1977** G. W. H. Lampe *God as Spirit* iii. 91 Perhaps the most original and significant insights of Paul are that the Spirit's inspiration makes men Christlike and, ideally, makes the community a visible re-embodiment of Christ.

8. b. *that's the spirit:* exclamatory phr. used in commendation of someone's courage, determination, etc.

1923 *Spectator* 21 Apr. 657/2 'That's the spirit!' has been the immediate comment of the country. **1930** M. Allingham *Mystery Mile* xxvii. 272 'I was ashamed that he should be..in my service.' 'That's the spirit,' said Campion. **1963** *Mad Mag.* July 47/1 Good! Good! *That's* the spirit! **1974** G. F. Newman *Price* ii. 70 'I'm absolutely certain I'll be completely vindicated.' 'That's the spirit.'

9. a. *moving spirit* (Moving *ppl. a.* 2 b).

1902 [see Moving *ppl. a.* 2 b]. **1926** S. Baldwin *On England* 66 In a city like this, where the intelligence and the moving spirits of industry meet round this table, what an opportunity you have! **1954** N. Mitford *Madame de Pompadour* xix. 242 The state papers still passed through her hands and the work was done in her room, but she ceased to be the moving spirit.

10. a. *spirit of place* [tr. L. *genius loci*], the characteristic atmosphere and influence of a particular place.

1918 D. H. Lawrence in *Eng. Rev.* Nov. 319 All art partakes of the Spirit of Place in which it is produced. **1955** G. Murchie (*title*) The Spirit of Place in Keats. **1980** P. Lively *Judgement Day* iii. 26 It was as though the spirit of place, nowadays, exerted its power only over the young. **1981** *Sunday Tel.* 25 Jan. 30/2 Of even the great cathedral of Canterbury Hilaire Belloc wrote with deep sadness for a spirit of place sought but not found.

c. (Further examples.)

1880 *Church Times* 7 May 295/1 Anything in more flagrant violation of the spirit and letter of the Prayer Book can hardly be conceived than to send the Elect out of the choir to put on 'the rest of the Episcopal habit'. **1926** G. M. Trevelyan *History of England* ii. iii. 172 Radicals appealed to the letter and the spirit of 'Magna Charta' against gagging acts, packed juries and restrictions of the franchise. **1961** [see *protectorate law* s.v. *protectorate *sb.* 4]. **1982** *Church Times* 15 Jan. 20/2 It is..neither in the letter nor spirit of these resolutions for anyone..to act in a way which has the effect of forcing one view upon those who hold the other.

III. 11. b. (Further examples, with reference to Matt. xxvi. 41.)

1666 G. Torriano *Piazza Universale di Proverbi Italiani* 268/2 The *spirit* is ready, but the body is lame. **1925** A. Huxley *Those Barren Leaves* iii. xiii. 269 The spirit is willing, but the flesh is weak. Weak in pain, but weaker still.., more inexcusably weak, in pleasure. **1935** C. Isherwood *Mr. Norris changes Trains* ix. 138, I really must apologize for my shortcomings as a correspondent... The spirit was willing, dear boy. I hope you'll believe that. **1938** W. S. Maugham *Summing Up* x. 30 Though ceasing my methodical study of the old masters (for though the spirit is willing, the flesh is weak), I have continued with increasing assiduity to try to write better.

c. Also (with reference to John iv. 23), *in spirit and in truth*, spiritually and sincerely.

1920 Galsworthy *In Chancery* iii. xiv. 328 At this moment he knew with certainty that he would never be near to her in spirit and in truth, nor she to him. **1953** A. J. Toynbee *World & West* vi. 96 Our hearts are hungry for a divinity that we can worship in spirit and in truth.

IV. 17. d. *out of spirits* (later examples).

1907 G. B. Shaw *John Bull's Other Island* iv. 94 You seem rather out of spirits... You havnt got neuralgia, have you? **1966** G. Greene *Comedians* i. v. 140 He seemed tired and out of spirits.

VI. 23. a. *spirit-call, -communication, -doctor, -harmony, -house, -land* (earlier example), *-medium, -mischief, -nature, -possession, -stream, thing, tower, -visit, -voice* (earlier example), *-wall, -world* (earlier example).

Some of these are used in sense 23 g but now have a wider use in Anthropology.

1949 Blunden *After Bombing* 17 For young, for old, a spirit-call Even now, bright music, stirs the air. **1869** Geo. Eliot *Let.* 11 July (1956) V. 48 'Spiritualism' (by

which I mean, of course, spirit-communications). **1960** W. Naylor *Silver Birch Anthol.* 7 The uncertain art of continuous spirit communication. **1936** *Discovery* June 187/1 The would-be spirit-doctor, who must have what the African regards as a faculty for 'seeing spirits'—we might term it clairvoyancy. **1928** Blunden *Undertones of War* xvii. 177 The flush and abundance of antique life and memorial and achievement, such as blend into the great spirit-harmony of the cities in that part of Europe. **1856** Spirit-house [see *circle *sb.* 21 b]. **1966** Mrs. L. B. Johnson *White House Diary* 27 Oct. (1970) 437 There were 'spirit houses' everywhere—brightly decorated little houses..about the size of a bird house, standing on stilts —one beside almost every residence. These houses are to entice the spirits so they will leave *your* house to you. **1973** P. Berton *Drifting Home* vi. 89 In Little Salmon, the graves are as numerous as the houses. Indeed they are like small cabins—a village of spirit houses with sloping roofs, glass windows and curtains, containing dead flowers and teapots and plates for the use of the deceased. **1841** E. A. Poe in *Graham's Mag.* May 241/2 A traveller from the spirit-land. **1853** *Theatr. Jrnl.* 22 June 190/1 A celebrated 'spirit-medium'..advertises a *seance* for the demonstration of spiritual communications. **1979** P. Niesewand *Member of Club* vii. 50 We've..got.. to enlist the support of spirit mediums. The ancestral spirits are supposed to speak through them. **1915** W. B. Yeats *Reveries over Childhood & Youth* xiv. 70 Somnambulistic country girls..become mediums for some genuine spirit-mischief, surrendering to their desire for the marvellous. **1962** Auden *Dyer's Hand* (1963) 135 The greatest of spirit-nature pairs and the most orthodox is, of course, Don Quixote-Sancho Panza. **1909** W. James in *Proc. Soc. Psychical Research* XXIII. lviii. 35 The whole record of spirit-possession in human history. **1950** Blesh & Janis *They all played Ragtime* x. 188 The incalculable power generated by the ring-shout rhythms brings about 'spirit-possession', referred to in jazz as being 'sent out of this world'. **1964** Gould & Kolb *Dict. Soc. Sci.* 638/2 *Spirit possession* seems too broad a term to specify the social functions and active controlling role characteristic of the shaman. **1925** Blunden *English Poems* 93 The time will come when, at the point to die I'll wish a spirit-stream as cool and clear. **1886** W. B. Yeats in *Dublin Univ. Rev.* Jan. 75 Two spirit things a man hath for his friends: Sorrow that gives for guerdon liberty, And joy. *a* **1957** Spirit-thing [see *root-sensation* s.v. *root *sb.*¹ 20]. **1955** E. Pound *Classic Anthol.* iii. 158 When he planned to begin a spirit tower Folk rushed to the work-camp and overran All the leisure of King Wen's plan. **1936** E. Sitwell *Victoria of England* i. 20 According to Mr Owen's account, the Duke returned.., after death,..in order to confide matters of importance to him... These spirit-visits must, one imagines, have been the result of the Duke's interest in minor details. **1844** E. A. Poe in *Graham's Mag.* Mar. 142/1 During the swell of the organ, the spirit-voice of the deceased addresses itself to the murderer. **1937** *Burlington Mag.* Oct. 162/1 These walls are now generally called *ying pi*—shadow- or spirit-walls... They are often found in front of the street-door, but sometimes also inside the enclosure behind the entrance in order to shut out a view of the interior. They served too as a protection against demons. The idea is current that demons are stupid and therefore only walk straight forward. **1948** G. H. Johnston *Death takes Small Bites* ii. 47 Alongside the road, protecting the big houses from wandering devils, were spirit-walls, blank and forbidding. **1847** W. Smith tr. *Fichte's Present Age* 62 So does the Spirit-World not indeed *flow* together at the breath of Love, for in it there is no Winter, but there all *is* and *abides* in eternal communion with the mighty Whole.

b. *spirit-bride, -wind* (further examples).

1904 W. H. Hudson *Green Mansions* xxi. 306 Nor did my mournful spirit-bride come to me. **1917** G. Frankau *City of Fear* 26 My Soul would bide with its spirit-bride At the Inn of a Thousand Dreams. **1894** *Wales* May 38/2 Lord, round me then with weeping clouds, And let my mind In quick blasts sigh beneath those shrouds, A spirit-wind. **1934** Blunden *Choice or Chance* 3 (*heading*) Spirit-Wind.

d. *spirit-dazzling, -giving, -lifting, quelling* (earlier example), *-quenching, -strangling, -uplifting*.

1946 R. S. Thomas *Stones of Field* 12 Deadly as a falcon brooding over its prey In a tower of spirit-dazzling and splendid light. **1946** Blunden *Shelley* xvi. 199 The spirit-giving blue sky of spring. **1841** E. A. Poe in *Gift 1842* 161 Spirit-lifting ecstasy of adoration. **1812** Spirit quelling [see *nerve-racking* s.v. *nerve *sb.* 11 b]. **1601** Donne *Progresse of Soule* in *Poems* (1633) 3 Spirit-quenching sicknesse, dull captivitie. **1948** F. R. Leavis *Great Tradition* v. 232 Coketown (the spirit-quenching hideousness of which is hauntingly evoked). **1930** Blunden *Summer's Fancy* 49 Of sloughs, of sieges, of Golgothas, of spirit-strangling. **1951** L. MacNeice tr. *Goethe's Faust* 299 Here is the prospect free, Spirit-uplifting.

e. (*a*) *spirit-baptized, -filled*; (*b*) *spirit-rotten*.

1903 R. C. Morgan *Outpoured Spirit & Pentecost* 53 How are we to be filled with the Spirit? By drinking; by coming unto Jesus, and as members of the Spirit-baptized Church. **1974** Spirit-baptized [see *Pentecostal *sb.* b]. **1895** J. MacNeil (*title*) Spirit-filled life. **1936** J. Brice *Pentecost* xiii. 226 The prayer-life of the Spirit-filled believer is transformed through his new apprehension of the Father. **1977** *Belfast Tel.* 17 Jan. 2/8 (Advt.), Full Gospel Bible College at Millmount, Randalstown... We can promise you dynamic spirit-filled Ministry with David Hathaway. **1922** D. H. Lawrence *Fantasia of Unconscious* v. 75 We are sympathy-rotten, and spirit-rotten, and idea-rotten.

g. *spirit guide, healer, healing, photograph, photography* (earlier example), *-writing* (earlier example).

1856 *Spiritual Herald* Feb. 3 Spirit hands, spirit voices, spirit healing. **1864** Browning *Dram. Pers.* 179 David.. peeps in the glass ball, Sets to the spirit-writing, hears the raps. **1872** *Spiritualist* 15 Apr. 25/1 Within the last six weeks in London, spirit photography has set in like a flood. **1877** H. P. Blavatsky *Isis Unveiled* II. ii. 118 A medium must be *passive*; and if a firm believer in his 'spirit-guide' he will allow himself to be ruled by the latter. **1909** W. James in *Amer. Mag.* Oct. 585/2 Rappings, apparitions, poltergeists, spirit-photographs, and materializations. **1956** R. M. Lester *Towards Hereafter* x. 124 My husband and I visited a trance medium, and a spirit healer we now know as Dr Light spoke through her. **1960** *Spectator* 28 Oct. 649 One of their [*sc.* Spiritualists'] chief interests is 'spirit healing'; that is, any healing that is brought about by a non-human agency. **1982** E. Jenkins *Shadow & Light* xxii. 172 A matter between him and his spirit-guides, not one with which human beings had anything to do.

i. Special *Comb.*: **Spirit baptism** = *baptism in the Spirit*, sense 6 c above.

1964 N. Bloch-Hoell *Pentecostal Movement* vii. 142 Glossolalia, in connection with the Spirit baptism, was generally believed to be a permanent Gift of Grace. **1977** G. W. H. Lampe *God as Spirit* vii. 199 It [*sc.* Pentecostalism] is very emphatic in its assertion of a two-stage relationship to God, reception of the Spirit, or Spirit-baptism, being sharply differentiated from the state of being a Christian without that added gift of God.

24. b. *spirit-flask* (earlier and later examples); *spirit-shop* (earlier example), *vault* (earlier example).

1829 G. Griffin *Collegians* II. xxix. 302 The assault made by Danny on her spirit flask, which she now.. discovered to be empty. **1961** L. G. G. Ramsey *Connoisseur New Guide to Antique English Pott., Porc. & Glass* 73 Other manifestations of Early Victorian work..were the ..spirit flasks formed into the shapes of human figures. **1835** *Mirror of Parliament* 20 May 983/2, I would willingly, if there was any chance of succeeding, include spirit-shops. **1848** Mrs. Gaskell *Mary Barton* II. iv. 57 The rest was taken in a spirit vault, and the refreshment was a glass of gin.

c. *spirit-merchant* (earlier example).

1822 *Sunday Times* 20 Oct. 3/4 *Bankrupts*... James Cayne, jun. and Thomas Bullock Watts, of Yeovil, Somersetshire, spirit-merchants.

e. *spirit duplicator, kettle*.

1958 *Daily Mail* 24 July 6/6 We came to the spirit duplicators... These machines *were* rather magical. They duplicated in seven different colours. **1890** *Girl's Own Paper* 4 Jan. 213/3 An Etna or small spirit kettle, and a bottle of spirits; some tea and a little sugar. **1923** M. Beerbohm *Peep into Past* 10 His portly form..may be seen bending over the little spirit-kettle. **1960** C. Day Lewis *Buried Day* vii. 137 Cake-stand,..heavy silver teapot, spirit-kettle and all.

spiriting, *vbl. sb.* Add: **3. b.** The application of a spirit, as a finishing process in french polishing. Also with *off*.

1933 C. Crampton *Canework* (ed. 6) iii. 36 The last operation in french polishing, which gives the final gloss, is known as 'spiriting off'. **1960** C. H. Hayward *Staining & Polishing* ix. 55 Purpose of spiriting. The idea of this process is gradually to burnish the surface of semi-soft shellac.

spiritism. (Earlier example.)

1856 *Spiritual Herald* July 203 Witchcraft was an unfriendly form of spiritism, and calculated to produce fear.

spirit-lamp. (Earlier example.)

1767 J. Priestley *Hist. & Present State of Electricity* viii. xii. 700 The bar was heated by a spirit lamp placed underneath it.

spirit-rapper. (Earlier example.)

1853 H. Paul *Rappings & Table Movings* 9 A rival spirit rapper! A genuine Medium.

spirit-rapping. Add: **1.** (Earlier example.)

1852 *Harper's Mag.* Dec. 129/1 The spirit-rappings are again engrossing a lion's share of the talk.

2. (Earlier example.)

1853 H. Mattison (*title*) Spirit rapping unveiled!

spiritual, *a.* and *sb.* Add: **A.** *adj.* **I. 1. a.** (Further *Comb.* examples, in apposition.)

1916 Joyce *Portrait of Artist* (1969) v. 252 Turned off that valve at once and opened the spiritual-heroic refrigerating apparatus, invented and patented in all countries by Dante Alighieri. **1929** D. H. Lawrence *Paintings of D. H. Lawrence* A4ᵛ This no doubt is all in the course of the growth of the 'spiritual-mental' consciousness.

c. (Later examples.) See sense 5 of the sb. in Dict. and Suppl.

1905 [see *gospel song* s.v. *gospel *sb.* 9]. **1964** *Amer. Folk Music Occasional* i. 15 Q. What kind of music did you like best at this time, spiritual music or blues or what? R. I always liked the spiritual music the best.

e. *spiritual home* (with no religious connotation), a place or milieu, other than one's home, which seems especially congenial or in harmony with one's nature, or to which one feels a sense of belonging or indebtedness.

1932 *Week-End Rev.* 7 May 586/1 If they write about it at all they make it clear that Europe is their spiritual home. **1941** A. Christie *Evil under Sun* iii. 52 A man like you would be at Deauville or Le Touquet... That's your —what's the phrase?—spiritual home. **1962** Wodehouse *Service with Smile* xi. 164 He disliked Lord Ickenham, considering him a potty sort of feller whose spiritual home was a padded cell in some not too choosy lunatic asylum. **1967** [see *paper *sb.* 8 b]. **1977** [see *Rastafarian *a.* and *sb.*].

3. a. (Further examples.)

1843 *Quincy* (Illinois) *Herald* 15 Dec. 3/1 Hyram Smith has had a revelation confirming the spiritual wife system. **1901** Kipling *Kim* v. 111 The comfort..of being properly ..respected as her spiritual adviser by a well-born woman. **1925** *Ladies' Home Jrnl.* Apr. 38/1 It is considered by the elderly women of Utah a great and sacred privilege to be the spiritual wife of Brigham Young or the Prophet Joseph Smith in the world to come. **1951** J. T. McNeill *History of Cure of Souls* iii. 42 Among ancient civilizations, that of India seems to have given the greatest prominence to the spiritual director. **1980** *Tablet* 26 Jan. 8/2 He.. cited..divorced Catholics in second marriages who, after prayer and consultation with a spiritual director, have returned to the sacraments. *Ibid.* 95/3 A letter I wrote in reply to my spiritual children was produced.

4. c. *spiritual healing*, cure or healing attributed to the agency of (a) spirit; faith-healing.

1899 H. W. Dresser *Spiritual Healing* vi. 54 How can I live according to the principles of spiritual healing among people who have no sympathy with the new principles. **1926** W. T. Walsh *Sci. Spiritual Healing* iii. 30, I announced a service of spiritual healing to be held in the church every Thursday morning after the celebration of the Holy Communion. **1980** *Spiritual Healer* XXVIII. 131 We have known many instances of migraine..being overcome through spiritual healing.

B. *sb.* **5.** Substitute for def.: = *Negro spiritual* s.v. *Negro 7. (Earlier and later examples.)

1866 *Harper's Mag.* May 775/1 Maum Rina flavored all her dishes with these 'spirituals', as they are called among the negroes. **1926** A. Niles in W. C. Handy *Blues* 9 These songs [*sc.* the blues] were woven of the same stuff as the other overlapping items in the long list,—the work-songs, love-songs..; yes, and decidedly the spirituals. **1947** S. Bellow *Victim* iv. 39 Harkavy and a girl he had brought to the party were singing spirituals and old ballads. **1981** M. Doody in Martin & Mullen *No Alternative* iv. 53 'Oh! Sometimes it causes me to tremble, tremble, tremble' as the spiritual goes.

spiritualism. Add: **2.** (Earlier example.)

1796 [see *ideal *sb.* 1].

3. (Earlier example.)

1853 J. Dix *Transatlantic Tracings* xiv. 244 Every two or three years the Americans have a paroxysm of humbug—..at the present time it is Spiritual-ism.

4. (Earlier examples.)

1850 J. R. Logan in *Jrnl. Indian Archipelago* IV. 552, I would proceed at once to facts illustrative of the different forms of spiritualism which prevail in Eastern Asia and Asianesia. **1867** E. B. Tylor in *Proc. R. Inst.* V. 90 A slight acquaintance with the spiritualism of the savage has sometimes led to its being considered as the result of a degeneration from the opinions of more cultured races.

spiritualist. Add: **3.** (Earlier example.)

1796 F. A. Nitsch *Kant's Princ. concerning Man* 153 The Spiritualists having discovered immaterial objects convert the mind into a spirit, and so on.

4. (Earlier example.)

1852 E. B. Browning *Let.* 13–14 May in *Lett. Brownings to George Barrett* (1958) 181 Lady Elgin is a great spiritualist with a leaning to Irvingism & a belief in every sort of incredible thing.

5. (Earlier example.)

1837 J. S. Mill in *Westm. Rev.* XXVII. 17 Many sterling thoughts are..disguised in phraseology borrowed from the spiritualist school of German poets and metaphysicians.

spiritus (spi·ritus). [L., = breath, aspiration, spirit.] **1.** *Gr. Gram.* = Breathing *vbl. sb.* 9; *spiritus asper*: see Asper *sb.*¹; *spiritus lenis* (lī·nis), smooth breathing; the sign (') placed over an initial vowel to show that it is not aspirated.

[**1470** Priscianus *Grammatici* (1855) I. ii. 51 In dictione tenor certus, absque ea incertus, non potest tamen sine eo esse. Similiter spiritus asper vel lenis.] **1867** Max Müller *Chips from German Workshop* II. 87 Sanskrit v is generally represented in Greek by the Digamma, or the spiritus lenis. **1878** W. G. Rutherford *First Gr. Gram.* 3 This sign (') is called *spiritus asper*, or rough breathing. **1888** King & Cookson *Princ. Sound & Inflexion Gr. & Latin* viii. 171 An initial *j* becomes the spiritus asper, sometimes changed to spiritus lenis. **1936** G. K. Zipf *Psycho-Biol. of Lang.* iii. 84 In Greek..the aspiration of the first aspirate was lost, even though the first aspirate was only an *h* (spiritus asper), that is, a phoneme which consisted exclusively of the aspiration. **1948** D. Diringer *Alphabet* viii. 458 By adopting this system of rough and smooth breathing (*spiritus asper* and *spiritus lenis*) for the vowel sounds,..the Greek alphabet helped to preserve flexibility in the Greek speech. **1976** *Archivum Linguisticum* VII. 170 Thus Schmalstieg does not give a clear explanation of..other initial aspirations like the *spiritus asper* in Greek.

2. *spiritus rector* (re·ktǭɹ), a ruling or directing spirit.

1911 W. James *Some Probl. Philos.* i. 16 Thus a '*spiritus rector*' would be a metaphysical,—a 'principle of attraction' a theological,—and a 'law of the squares' would be a positive theory of the planetary movements. **1935** A. Revusky *Jews in Palestine* xii. 202 The leader and *spiritus rector* of this movement, Vladimir Jabotinsky. **1942** D. D. Runes *Dict. Philos.* 300/1 *Spiritus rector*,.. some sort of subtle natural force in corporeal beings. The alchemists applied the expression to some substance..said to be capable of transmuting metals into gold, and also an elixir which was supposed to prolong life indefinitely. **1959** R. F. C. Hull tr. *Jung's Aion* in *Coll. Wks.* IX. 167 It [*sc.* the self]..is older than the ego, and..is the *spiritus rector* of our fate. **1977** *Times* 20 Dec. 13/6 It has often been reported that the international oil companies were the real *spiritus rector* for the explosion of oil prices during the winter of 1973–74. **1980** *Encounter* May 35/1 More than fifteen years ago he was the *spiritus rector* of the *European Journal of Sociology*.

spiro-. Add: **1. spirochæte** (also **-chet-** in this word and its derivatives): now with pronunc. (spəi·rokīt); in mod. use, any bacterium of the order Spirochætales, comprising actively motile non-spore-forming organisms having a helical form; (later examples); hence **spirochæ·tal** *a.*, that is a spirochæte; caused by spirochætes; **spi:rochætici·dal** *a.*, lethal to spirochætes; **spirochæ·ticide** a spirochæticidal substance; **spirochæto·sis,** infection with or a disease caused by spirochætes.

1916 *Jrnl. Exper. Med.* XXIII. 377 We discovered a spirochætal microorganism which is now believed to be the cause [of Weil's disease]. **1922** *Encycl. Brit.* XXXI. 906/1 It is best described as spirochætal jaundice rather than by the older name of Weil's disease. **1969** Edington & Gilles *Path. in Tropics* vii. 298 Leptospirosis is caused by infection with spirochaetal organisms. **1908** *Practitioner* Oct. 549 The treponema, or, as they are usually designated in this country, the spirochaetes. **1919** *Chambers's Jrnl.* June 415/2 This organism belongs to the class known as spirochaetes, of which the spirochaete of syphilis and that of relapsing fever are other members. **1939** R. Campbell *Flowering Rifle* ii. 47 The fierce spirochete. **1973** R. G. Krueger et al. *Introd. Microbiol.* iii. 59/1 The order is subdivided into two families: Spirochaetaceae for the larger..spirochaetes, which are free-living or parasitic in shellfish; and Treponemataceae for spirochetes that are only about 0·2–0·3 μm in diameter and do not exceed 16 μm in length. **1913** *Jrnl. Exper. Med.* XVIII. 435 (*heading*) A study of the spirochæticidal action of the serum of patients treated with Salvarsan. **1949** M. A. Jennings in H. W. Florey et al. *Antibiotics* II. xxxi. 1037 Penicillin possessed some spirocheticidal activity. **1920** *Jrnl. Amer. Med. Assoc.* 25 Dec. 1768/1 The efficacy of..spirocheticides in the so-called chemical prophylaxis of syphilis is limited to a period of not more than eight hours after the spirochete has had the opportunity of invading the healthy person. **1906** *Jrnl. Hygiene* VI. 580 (*heading*) Spirochaetosis of mice due to *Spirochaeta muris* n.sp. in the blood. **1951** Whitby & Hynes *Med. Bacteriol.* (ed. 5) xxi. 350 Spirochætes and fusiform bacilli of Vincent's type may be found in large numbers in various conditions, particularly of the lung, known as spirochætosis. **1981** *Brit. Med. Jrnl.* 21 Nov. 1362/1 A review of 100 consecutive rectal biopsy specimens obtained from patients with rectal bleeding or diarrhoea showed that 10 had spirochaetosis.

2. *Chem.* Formative element used in the names of organic compounds whose molecular structure includes two rings with a single atom (usu. of carbon or nitrogen) common to both. Also in *Comb.* or as quasi-*adj.*, as *spiro-compound*. [Introduced in Ger. (A. Baeyer 1900, in *Ber. d. Deut. Chem. Ges.* XXXIII. 3771).]

1908 *Chem. Abstr.* II. 75 (*heading*) Spirocyclane. *Ibid.*, The term 'spiro' is applied by the author to systems of two cycloids, having only one carbon atom in common. **1909** *Jrnl. Chem. Soc.* XCVI. i. 652 The product, decomposed by cold water, yields the spirocyclane derivative. **1915** *Ibid.* CVII. ii. 1080 (*heading*) The formation and stability of *spiro*-compounds. **1932** *Proc. R. Soc.* A. CXXXIV. 359 The anhydrous crystals of racemic *spiro*-dihydantoin have a density of 1·94. **1960** *Jrnl. Amer. Chem. Soc.* LXXXII. 5560/1 According to the number of spiro atoms present, the compounds are distinguished as monospiro, dispiro, trispiro compounds *etc.* **1976** A. L. Ternay *Contemp. Org. Chem.* vii. 185 A spiro compound results when two rings share one atom. **1978** *Nature* 5 Jan. 44/2 During irradiation of spiropyrans, in non-polar solvents dimer and charge-transfer complexes are formed. **1978** *Further Perspectives Org. Chem.* (CIBA) 40 Evidence for the possible spiro intermediate would give enormous support and we are seeking it.

spirograph (spəiᵊ·rogrɑf). *Med.* [f. Spiro- + -graph.] An instrument which provides a continuous tracing of the movements of the lungs during respiration.

1890 in Gould *New Med. Dict.* **1934** *Publ. Mod. Lang. Assoc. Amer.* XLIX. 1168 The air pressure and vibrations taken back of the point of contact with the spirograph. **1977** *Bull. Exper. Biol. & Med.* LXIV. 1662 The investigation is stopped when changes in the helium concentration in the spirograph decrease to a certain small value.

Hence **spi·rogram,** the tracing produced by a spirograph; **spirogra·phic** *a.*, pertaining to or observed by means of a spirograph; **spirogra·phically** *adv.*; **spiro·graphy,** the study of respiration by means of a spirograph.

1946 L. N. Gay *Bronchial Asthma* i. 7 (*caption*) Spirographic tracings of Hofbauer. **1956** Hinshaw & Garland *Dis. of Chest* vi. 89 Tracings of respiration, sometimes called spirograms, are readily made by means of a calibrated spirometer equipped with a writing device. **1959** *Presse Médicale* No. 47. Suppl. 2/1 Having studied clinically and spirographically a population of 232 patients, authors make first a distinction between three groups according to the intensity of dyspnea. *Ibid.*, Spirography allows effectively to make a distinction between three groups of patients. **1977** *Lancet* 22 Jan. 156/1 (*caption*) Spirogram of habitual hyperventilator showing irregular amplitude and frequency of respiration and sighing tendency. *Ibid.* 26 Mar. 709/1 Their diagnosis of hyperventilation is based on..a spirographic pattern for which criteria of abnormality are lacking. **1977** *Bull. Exper. Biol. & Med.* LXXXIII. 3 On the basis of indices obtained by the motion-picture spirography method, a definite idea can thus be obtained of the functional state of the apparatus for controlling respiration under different conditions.

spirogyra (spəirodʒəi·rā). Also Spiro-. [mod.L. (H. F. Link in C. G. D. Nees von Esenbeck *Horæ Physicæ Berolinensis* (1820) i. 5). f. Spiro- + Gr. γῦρος, γυρά round.] A green filamentous fresh-water alga of the genus of this name.

1875 Bennett & Dyer tr. *Sachs's Text-bk. Bot.* II. 220 Each single cell of Spirogyra, like an isolated cell of Closterium, &c., constitutes an individual. **1906** G. F. S. Elliot *First Course Pract. Bot.* 57 The joint protoplasm, called the 'zygospore',..begins to divide and form a new Spirogyra. **1937** E. E. Stanford *Gen. & Econ. Bot.* xiii. 346 Spirogyra..is so abundant, so characteristic, and so readily studied that it commonly passes as a 'characteristic' chlorophyte, which decidedly it is not. **1962** *Listener* 16 Aug. 240/1 There are many other kinds of microscopic organisms,..the desmids, the green strands of Spirogyra, and the rest. **1971** P. Crampton tr. *Heyerdahl's Ra Expeditions* iii. 50 A shallow pool about as big as a wash-basin, but completely covered with green spirogyra.

spiroidal (spəiroi·dăl), *a.* [f. Spiroid *a.* + -al.] = Spiroid *a.*

1928 T. C. Chamberlin *Two Solar Families* xiv. 143 If a series of projected bolts were shot forth from a sun in succession toward the star while all were being pulled forward by the star but in different degrees, the train they formed would obviously be curved in a spiroidal manner. **1972** *Nature: Physical Sci.* 6 Nov. 22/1 The majority of the crystals were acicular or whisker-like although a few were spiroidal.

spirolactone (spəirolæ·ktōᵘn). *Chem.* and *Pharm.* [f. *spiro- 2 + *lactone 2.] Any spirane in which one of the rings is a lactone; *spec.* any of the series of steroid derivatives to which spironolactone belongs.

1958 *Canad. Med. Assoc. Jrnl.* LXXIX. 883/1 The administration of Spirolactone results in a persistent and significant increase in sodium excretion. **1959** *Jrnl. Org. Chem.* XXIV. 743/1 The spirolactone side chain was built onto a steroid nucleus containing an aromatic A ring. **1961** *Lancet* 16 Sept. 618/2 With the development of the 17-spirolactones, which are aldosterone antagonists, a new method has become available for investigating possible cases of primary aldosteronism. **1970** Passmore & Robson *Compan. Med. Stud.* II. xi. 10/1 The main reason for believing that spirolactones are aldosterone antagonists is that they reverse all the renal effects of aldosterone.

spironolactone (spəiᵊ:rǫnolæ·ktōᵘn). *Pharm.* [f. *spirolactone.] A steroid spirolactone derivative, $C_{24}H_{32}O_4S$, which is an aldosterone antagonist, increasing sodium excretion, and is used esp. in the treatment of œdema and hypertension associated with hyperaldosteronism.

1960 *Austral. Ann. Med.* IX. 188/1 In the present paper we describe the use of a similar drug, spironolactone ('SC 9420', 'Aldactone', Searle). **1961** *Lancet* 2 Sept. 528/2 Spironolactone 100 mg. four times daily definitely improved six out of seven myasthenic patients. **1974** R. M. Kirk et al. *Surgery* vii. 146 Treatment with diuretics or spironolactone, and dietary salt restriction, help to reduce the accumulation of fluid. **1977** *Lancet* 17 Sept. 600/2 Spironolactone relieved muscle fatigue in paramyotonia congenita—an effect possibly mediated by the accompanying hyperkalæmia.

spiroplasma (spəiᵊ·roplæzmă). *Biol.* [mod. L., f. Spiro + Plasma.] Any of a group of pathogenic prokaryotes lacking a cell wall and related to the mycoplasmas, but characterized by their helical structure and rotatory movement.

1973 Davis & Worley in *Phytopathology* LXIII. 407/1 In order to distinguish the organism from previously described species and to avoid possible inappropriate use of the term 'mycoplasmalike',..we are coining the term spiroplasma. **1976** *Ann. Rev. Microbiol.* XXX. 181 More recently, spiroplasmas have been cultivated from corn plants infected with corn stunt disease.

spit, *sb.*[1] Add: **4. b.** An instrument used by Customs officers for probing and examining cargo.

1925 *Chambers's Jrnl.* 19 Sept. 668/1 A barbed 'spit' is used for examining cotton, wool, and various coarse goods, so that a portion of the interior may be drawn out. A flat, wooden 'spit' is used in the examination of rolls of carpet, cloth, linoleum... A short, fine 'spit' is used for probing the stuffing and packing in and around furniture. **1970** M. Greener *Penguin Dict. Commerce* 310 *Spit*, a weapon used by Customs authorities to discover whether dutiable goods are hidden in other cargo.

7. spit-jack, a spit with a turning mechanism (see quot. 1967).

1967 *Antique Finder* Aug. 11/3 *Spit jacks*... These mechanisms were fitted on the wall at the side of the fireplace in order to rotate the carcass in front of the fire. A heavy weight was suspended from a cord and wound over the cylinder. The power was conveyed by a series of cogged wheels to another cylinder connected by a cord with a grooved disc on the end of the spit itself, which it slowly turned. **1971** R. Howe *Mrs Groundes-Peace's Old Cookery Notebk.* 65 There was the 'spit-jack', a weight-driven spit, considered in the sixteenth century as a labour-saving device.

spit, *sb.*[2] Add: **1. c.** *spit and sawdust*: the floor covering (esp. formerly) typical of the general bar of a public house (see quot. 1937); hence, the bar itself. Freq. *attrib.* (also *transf.*). Cf. *sawdust *sb.* 3 b.

1937 Partridge *Dict. Slang* Add. 995/1 *Spit and sawdust*, a general saloon in a public-house: C. 20. Ex the sawdust sprinkled on the floor and the spitting on to the sawdust. **1969** D. Clark *Death after Evensong* ii. 40 Where shall I start? In the spit and sawdust? I take it you'll do the saloon bar yourself. **1971** R. Busby *Deadlock* x. 158 The Porter's Arms, a spit and sawdust public house near the markets. **1972** *Guardian* 24 Feb. 10/2 Pub styles polarize into 'Spit and Sawdust'..and 'Architects' Fanciful'. **1976** *Rhyl Jrnl. & Advertiser* 9 Dec. 1/1 He thought that a 'rough and ready, spit and sawdust affair' could be produced and made available for consideration at the council's estimates meetings. **1977** *Lancashire Life* Mar. 58/2 They also convert buildings into billiard clubs, where the decor is attractive and the spit-and-sawdust era is just a memory.

2. c. *to go for the big spit*: to vomit. *Austral. slang.*

1967 F. Hardy *Billy Borker yarns Again* 40 Don't tell me the Gargler went for the big spit. **1970** *Private Eye* 10 Apr. 16 He goes for the big spit and accidentally entombs a nice old lady and her dog in tepid chuck. **1975** R. Beilby *Brown Land Crying* 225 Goin' for the big spit, was I? I don't remember.

3. a. Also, *the* (*dead*) *spit of. colloq.*

1885 Hall Caine *Shadow of Crime* II. xxvi. 129 A brother..the spit of hissel'. **1901** E. W. Hornung *Black Mask* 37 I'll chance you having another ring..the dead spit of mine. **1921** 'K. Mansfield' *Let.* Sept. (1977) 232 One of his [sc. Cézanne's] men gave me quite a shock. He's the *spit* of a man I've just written about, one Jonathan Trout. **1936** M. de la Roche *Whiteoak Harvest* v. 98 Easy for a boy to look like his grandmother. There was Renny—the spit of old Gran! **1953** A. Upfield *Murder must Wait* xvii. 154 The son's the dead spit of the old man. **1966** [see *gramp].

b. *spit and image*: see also *image *sb.* 4 a. Also *spit image* and absol. *spit*. Cf. the (later) forms *spitten image*, *spitting image* s.v. *spitten *a.*, *spitting *ppl. a.* 3 respectively.

1929 J. B. Priestley *Good Companions* i. v. 166 That's theirs... It's the spit image o' yours, too. **1949** *Penguin New Writing* XXXVI. 35 My husband saw a man that was the spit-image of King no further away than Jackson. **1966** *New Statesman* 18 Feb. 235/1 For most of the last two acts he's catapulted in and out of doors, changing on the way into the cheerfully sodden porter of the bordel who happens to be his spit.

5. spit-curl: for *U.S.* read orig. *U.S.*; (earlier and later examples); **spit-insect,** restrict † to sense in Dict. and add: (*b*) the cuckoo-spit insect, *Philænus spumarius*; cf. *spittle-bug* s.v. *spittle *sb.*[2] 4.

1831 *Boston* (Mass.) *Transcript* 9 Sept. 2/1 What would the reverend Doctor say of the 'spit curls', and Chinese precision of a modern dandyzette's head gear? **1857** [see *beau-catcher* s.v. *beau *sb.* 3]. **1903** Farmer & Henley *Slang* VI. 314/1 *Spit-curl, subs. phr.* (costers').—A curl lying flat on the temple; a *soap-curl*. **1957** L. Durrell *Justine* i. 37 A spitcurl at each temple. **1968** J. Ironside *Fashion Alphabet* 198 *Spit curls*, small curls brought forward on to the cheek, often moistened with 'spit'. **1950** J. Brooke *Goose Cathedral* iv. 82 Little green spit-insects dropped out of the tamarisks into one's cup. **1917** Kipling *Divers. Creatures* 9 There's a nice little spit-kitten for you!

Spit, *sb.*[4] Colloq. abbrev. of *Spitfire* (in full *Supermarine Spitfire*), a British fighter aeroplane produced between 1936 and 1947.

1941 *Saturday Rev. Lit.* (U.S.) 4 Oct. 9/3 A pilot 'aviates'..a Wimpy or Spit (Spitfire). **1948** G. V. Galwey *Lift & Drop* ii. 25 From time to time Miss Procter..said 'Here come the jet jobs,' or 'The Spits are lovely.' **1965** *New Statesman* 7 May 718/2 'Look, Ron,' he cried, 'my old Spit!' **1980** J. McClure *Blood of an Englishman* vi. 64, I was flying Spits, Hurricanes, while Bonzo..was in Bomber Command.

spit, *v.*[1] Add: **1. c.** Of a Customs officer: to examine with a 'spit'.

1925 *Chambers's Jrnl.* 19 Sept. 668/1 The officer.. selects a number of bales for inspection. These he 'spits' —that is, he inserts a special iron instrument, which is capable of penetrating to the interior of a large bale and extracting a small piece of the commodity.

spit, *v.*[2] Add: **I. 2. b.** (Later *fig.* example.)

1915 T. S. Eliot *Love Song J. Alfred Prufrock* in *Poetry* (Chicago) June 132 How should I begin to spit out all the butt-ends of my days and ways.

c. Also, *to spit blood*, (*a*) to express vehement anger, to rage (*colloq.*); (*b*) *slang*, of a spy, etc.: to fear exposure; *to spit chips* (Austral. slang), (*a*) to feel extreme thirst; (*b*) to manifest acute anger or vexation.

1901 *Bulletin Reciter* (Sydney) 108 While you're spitting chips like thunder... And the streams of sweat near blind you. **1946** A. Marshall *Tell us about the Turkey, Jo* 142, I was spitting chips. God, I was dry! **1947** J. Morrison *Sailors belong Ships* 189 Old Mick Doyle's with them. He's spitting chips because they're not using sea

water. **1954** P. Gladwin *Long Beat Home* 17 It's enough to make you spit chips when you think of Sydney—movies and vaudeville comedies and a decent musician once in two years. **1963** J. Joesten *They call it Intelligence* i. iv. 45 When a resident agent is forced to lie low, because counter-espionage is on his trail, it is said of him that he is 'spitting blood'. **1963** Wodehouse *Stiff Upper Lip, Jeeves* ix. 72 If there's one thing that makes a collector spit blood, it's hearing about another collector getting a bargain. **1965** I. Southall *Ash Road* 77 Not when I saw Mr Fairhall last. He was spittin' chips because Peter had gone away. **1966** L. Deighton *Billion-Dollar Brain* x. 90 A man tailed or suspected is said to be 'spitting blood'. **1966** 'L. Lane' *ABZ of Scouse* 102 When I think of it I could spit blood.

3. a. (Further Amer. examples.)

1773 *Let.* 26 Aug. in *Maryland Hist. Mag.* (1920) XV. 280 The wind is strong at N:E. & it spits Rain. **1803** B. Hunt *Diary* 16 Apr. in *Chester Co.* (Pa.) *Hist. Soc. Bull.* (1898) 7 Continues cold and Spiting snow. **1835** N. Wyeth *Jrnl.* 12 Jan. in F. G. Young *Sources Hist. Oregon* (1899) I. 243 Spit snow all day at night set in to snow.

4. Freq. in phr. *to spit it out*: to reveal, confess, disclose something.

1855 Mrs. Gaskell *North & South* I. xxv. 313 I'm easier in my mind for having spit it out. **1904** [in Dict.]. **1920** Galsworthy *Skin Game* i. 31 Don't be so mysterious, mother. If you know something, do spit it out! **1935** Auden & Isherwood *Dog beneath Skin* ii. v. 115 Go on, then, spit it aht! **1950** J. Cannan *Murder Included* vi. 123 'I've gotta clue.'.. 'Spit it out, for mercy's sake, boy.' **1981** A. Price *Soldier No More* 43 'Well—spit it out, man! Don't just stand there,' Willis exhorted him.

II. 7. b. Also *spec.* with *out*, of ceramic glazes: to form blisters which burst during firing.

1904 *Trans. Eng. Ceramic Soc.* IV. 30 The china vase.. with the Seger porcelain glaze..has no sign of blister.., and does not spit out in the kiln. *Ibid.* 32 The earthenware trials that have spit out are not verified by the china trials.

c. (Later examples with *on*.)

1949 H. L. Mencken *Mencken Chrestomathy* xxx. 626 Is it hot in the rolling-mill? Are the hours long? Is $15 a day not enough? Then escape is very easy. Simply throw up your job, spit on your hands, and write another 'Rosenkavalier'. **a 1975** Wodehouse *Sunset at Blandings* (1977) ii. 20 There was a time when you had to employ wild horses to drag me from London, and they had to spit on their hands and make a special effort.

10. *spit and polish*: also in extended use, precise correctness, smartness; freq. as a derogatory expression in contrast with purposeful work or utility. Also *attrib.*, smart in appearance; hence *spit-and-polished* adj.

1914 C. Beresford *Mem.* I. 120 From that day onwards I set myself steadily against bright-work and spit-and-polish. **1920** *Q. Rev.* Jan. 196 Gunnery had been neglected in pursuit of 'spit and polish'. **1949** 'J. Tey' *Brat Farrar* xiii. 116 He had understood Brat's distrust of the [stables'] spit and polish. **1950** *Daily Progress* (Charlottesville, Va.) 23 Sept. 1/7 Officially Pretzer was.. part of the spit and polish First Infantry Division. **1958** *Times* 12 Nov. 3/3 Holst's suite, 'The Planets', was very, very much better..thanks to..the orchestra's ever-increasing spit-and-polish. **1977** L. Meynell *Hooky gets Wooden Spoon* xiii. 172 Some nice chubby-faced spit and polish Sandhurst type. **1977** *Times* 8 Nov. 4/3 The spit-and-polished toecaps of his boots. **1979** *United States 1980/81* (Penguin Travel Guides) 614 The 4,300 spit-and-polished midshipmen have a 3:45 pm dress parade on Warden Field.

III. 11. The vb. stem in combination as **spit-cat, kitten** *colloq.* = Spitfire *sb.* 2.

1898 A. Ollivant *Owd Bob* v. 51 Eh, but art' a tearin' spit-cat surely! **1912** Kipling *Divers. Creatures* (1917) 9 There's a nice little spit-kitten for you!

Spitalfield(s). Add: Also applied to silk and velvet made up there into furnishings, etc., or to the weavers (orig. Huguenot refugees) involved in this trade. Occas. *absol.*

1812 H. & J. Smith *Rejected Addresses* 120 Spital-fields with real India vies... Old calico, torn silk, and muslin new. **1819** M. Edgeworth *Let.* 24 Mar. (1971) 185 Mr. Buxton very plain sense and admirable facts about Newgate and Spitalfield weavers and all that. **1861** C. M. Yonge *Young Step-Mother* iii. 26 He was..of the Protestant French sort, that..ran away from the Sicilian vespers, or the Edict of Nantes, I don't remember which; only the Spitalfields weavers have something to do with it. **1894** *Country Gentleman's Catal.* 155/2 Umbrellas.. very best Spitalfields twill silk. **1952** M. Cost *Hour Awaits* 89 The small red drawing-room, its damask hangings of Spitalfields silk. **1955** R. Fastnedge *Eng. Furniture Styles* iii. 86 Queen Anne's state bed, at Hampton Court, is hung, with Spitalfields velvet, patterned in rich colours on a cream ground. **1972** P. Rogers *Grub Street* ii. 115 Such persistant troublemakers as the Spitalfields weavers. **1976** T. Jeal *Until Colours Fade* ii. 30 The Red Drawing Room..its walls were lined with faded crimson Spitalfields' silk. **1980** J. Rose *Elizabeth Fry* iv. 68 The Spitalfields weavers, thrown out of work by the powerlooms.

spi·tball. *N. Amer.* Also **spit-ball, spit ball.** [f. Spit *sb.*[2] + Ball *sb.*[1]] **1.** A spittle-ball (see Spittle *sb.*[2] 4), esp. one thrown as a missile by a schoolchild.

1846 *Knickerbocker* XXVII. 410 They..crooked pins, made pop-guns, ejected spit-balls. **1899** W. James *Talks to Teachers* 92 The spitballs that Tommy is ready to throw. **1939** L. M. Montgomery *Anne of Ingleside* xi. 71 Bertie Shakespeare Drew threw a spit-ball at her. **1956** M. Stearns *Story of Jazz* (1957) xix. 252 Gillespie was also perfecting his spitball-throwing technique while in the Calloway band. **1977** I. Shaw *Beggarman, Thief* ii. iii. 143 They [sc. the teachers] spend most of the time trying to keep the kids from..throwing spitballs.

2. *Baseball.* A ball moistened on one side with saliva or sweat before pitching, so that it acquires a swerve. (Illegal in the official game.)

1905 J. J. McGraw *Official Baseball Guide* 13 The perfect 'spit' ball drops from a batter's hips to his knees or below in perhaps two feet of forward motion. **1912** C. Mathewson *Pitching in Pinch* 20 Some spit-ball pitchers announce when they are going to throw a moist one by looking at the ball as they dampen it. **1928** G. H. Ruth *Baseball* vi. 75 All spit balls break down, but by turning the wet spot one way or the other the pitcher can make the ball break in or out as he desires. **1946** Mezzrow & Wolfe *Really the Blues* (1957) viii. 125 His arms [were] pumping like he was a pitcher winding up to shoot a spitball over the plate. **1976** *Billings* (Montana) *Gaz.* 7 July 4-A/2 That's what I can do for the Cubs. I guarantee that they will be snarling and swearing, gouging, spiking, mauling. They'll be throwing beanballs and spitballs.

3. *transf.* and *fig.*

1888 *Judge* 10 Nov. 68/1 All statements to the opposite effect are spit-balls at the moon. **1925** Fraser & Gibbons *Soldier & Sailor Words* 266 *Spit ball*, hand-grenade. (U.S. Army.) **1931** W. G. McAdoo *Crowded Years* xv. 225 Their vitriolic comments..consisted chiefly of mere verbal spitballs thrown in a..mood of..sabotage. **1933** E. O'Neill *Days without End* i. 33 They'd turned naughty schoolboys and were throwing spitballs at Almighty God. **1960** I. Wallach *Absence of Cello* 58 She also thought that their impulse to outrage was over-developed. 'You all sit around and throw spitballs at the world,' she said. **1981** T. Barling *Bikini Red North* xii. 263 It doesn't make a spitball of difference. The deadline stands.

Hence **spi·tball** *v. intr.*, to throw out suggestions for discussion; **spi·tballer,** one who throws spitballs.

1928 *Chicago Tribune* 7 June 19/4 The Giants..made only three hits off..Clarence Mitchell, the southpaw spitballer. **1955** H. Kurnitz *Invasion of Privacy* (1956) viii. 64 I'm just thinking out loud... Spitballing we call it in the movie business. **1961** J. B. Priestley *Saturn over Water* iv. 52 No, don't tell me we're not talking about painting. I'm only spitballing while I try to think. **1976** C. Larson *Muir's Blood* xvii. 98 'Are you serious?' Blixen asked. 'I'm spitballing,' Schreiber replied. **1977** *New Yorker* 2 May 100/3 The spitballer won't grow into His father's jacket.

spitchered (spi·tʃəɹd), *a. slang* (orig. *Naut.*). [f. Maltese *spicca* finished, ended, perh. ult. f. It. *spezzare* to break into pieces, fragment.] Rendered inoperative, ruined.

1920 *Blackw. Mag.* Mar. 340/2 'Cease firing!' screamed the C.O. 'He's spitchered.' **1946** J. Irving *Royal Navalese* 163 *Spitcher'd*, finished, done. **1970** P. Dickinson *Seals* i. 14 That damned gadget might..be functioning right as rain in thirty seconds, or it might be spitchered for ever.

spite, *sb.* Add: **2. d.** *spite fence*, a wall, fence, etc., erected with the intention of causing annoyance. orig. and chiefly *U.S.*

1899 *Everybody's Mag.* I. 70/2 Meanwhile an ordinance was passed making the building of spite fences illegal. **1928** *Hearst's Internat.* Aug. 89/2 For the full depth of the boundary ran a tall fence of unpainted boards set upright. This fence was fully eighteen feet-high. It was what is known as a spite-fence. **1957** R. F. V. Heuston *Salmond on Torts* (ed. 12) v. 201 An occupier of land is free to erect a wall or fence with the sole object of blocking his neighbour's view, or preventing him from acquiring an easement of light—a 'spite-fence' as it is sometimes called. **1977** *New Yorker* 23 May 50/3 To gain privacy from the street, he put up a nine-foot wall, which one of his neighbors, interpreting it as a spite fence, sued unsuccessfully to have removed.

spitfire, *a.* **2.** (Earlier example.)

1858 Mayne Reid *Ran away to Sea* (1859) xii. 93 Even under such a wind she still continued to carry most of her sail..while her storm, spit-fire, and third jibs were still kept bent to the breeze.

Spi·thead. The name of an anchorage and a strait which lie off Portsmouth, used *attrib.* in *Spithead nightingale* (see quots.) and *Spithead pheasant* (*a*) a bloater; (*b*) a kipper. *Naut. slang.*

1890 Barrère & Leland *Dict. Slang* II. 283/1 *Spithead nightingales* (naval), boatswains, and boatswains' mates, on account of their calls. **1948** Partridge *Dict. Forces' Slang* 177 *Spithead pheasant*, a, a bloater. (Navy.) **1954** *News Chron.* 4 Nov. 1/4 He had to produce: 1. a cat; 2. a Spithead Pheasant (a kipper); 3. a Sick Bay Shackle (a safety-pin). **1963** *Times* 26 Feb. 12/6 Spithead pheasant—slang for a kipper—still appears on the Navy's menu, but Spithead nightingales—bo'sun's mates and call boys are 'off'.

spi·t-out. *Ceramics.* [f. vbl. phr. *to spit out*: see *spit *v.*[2] 7 b.] Accidental blistering of a glaze when fired, caused by air or gas bubbles; the blisters so caused.

1909 *Trans. Engl. Ceramic Soc.* VII. 1 'Spit-out' is one of a number of 'mysterious' faults which occur in the manufacture of pottery. **1930** A. B. Searle *Encycl. Ceramic Industries* III. 199/1 'Spit-out' is most abundant when the atmosphere in the kiln is alternately reducing and oxidising. **1964** H. Hodges *Artifacts* ii. 52 Pin-holes ..and spit-out (fine, broken blisters) are most commonly due to the evolution of gases from the body during firing puncturing the molten glaze. **a 1977** *Harrison Mayer Ltd. Catal.* 19/2 Spit-out. This is a common fault which occurs during the decorating fire and consists of the formation of minute blisters or bubbles in the glaze... Spit-out occurs at the higher enamel kiln temperature.

spi·t-roast, *v.* [f. Spit *sb.*[1] + Roast *v.*] *trans.* To cook (meat or fish) on a spit, either over a fire or in an oven. Freq. as *vbl. sb.* and *ppl. a.*

1954 J. A. Beard *Compl. Bk. Barbecue & Rotisserie Cooking* 39 (*heading*) Rotisserie & spit roasting. **1959** *Good Food Guide* 27 The new landlord has already succeeded in astonishing the locality by spit-roasting a whole calf in the open for a Church festival. **1961** *House & Garden* Feb. 45 A spit-roasted chicken bought from the supermarket. **1961** S. Beck et al. *Mastering Art French Cooking* vi. 240 Spit roasting, when the chicken is wrapped in fat and continually rotated, is far less exacting than oven roasting. **1973** *Country Life* 13 Dec. 2024/2 Pike is particularly delicious when spit-roasted whole. **1976** *Times* 14 Feb. 11/5 Spit roasted is how these birds are served... Baste them frequently while spit roasting..you can check whether your spit roasted chicken is ready. **1978** *Lancashire Life* Nov. 150/2 John explains how the spring chicken will be spit roasted.

‖**spitskop** (spi·tskǫp). *S. Afr.* Also **spitzkop,** and in dim. forms **spitzkopje, -koppie.** [Afrikaans *spitskop*, f. Du. *spits* pointed + *kop* *kop.] A sharp-pointed hill; also (with capital initial) used as a proper name.

1872 C. A. Payton *Diamond-Diggings S. Afr.* 3 He has the tabular mountains and 'spitzkops' (sharp pointed hills and peaks)..till he arrives within a few hours' distance of the Vaal. **1889** [see *randjie]. **1905** G. W. Stow *Native Races S. Afr.* xx. 396 The headquarters of 'Kousopp were at the two spitzkopjes to the left of the 'Gumaap. **1937** A. J. H. Goodwin in I. Schapera *Bantu-Speaking Tribes S. Afr.* ii. 33 The flat-topped and 'spitzkop' hills so typical of the Karroo country and other parts. **1974** *Standard Encycl. S. Afr.* X. 221/2 The Spitskoppies group is a collection of gigantic granite rocks or 'island' mountains rising out of the Namib plain... It is a broken range dominated by the Great Spitskop (.. 1742 metres).

spit-sticker. Add: Also in shortened form **spit-stick.**

1909 in Webster. **1920** R. J. Beedham *Wood Engraving* 3 The spit-sticks..are tools by which the cut can be easily broadened by gradually pressing the point deeper into the wood. **1960** G. Lewis *Handbk. of Crafts* 187 The Spitstick, like the graver, is made in several thicknesses. Its section is triangular, with sides tapering gently down to a point.

spitten (spit'n), *a. dial.* Repr. corrupted pronunc. *spit and* (*image, picture*) s.v. Spit *sb.*[2] 3 b in Dict. and Suppl. Cf. *spitting image* s.v. *spitting *ppl. a.* 3; *splitting image* s.v. *splitting *ppl. a.* 5.

1878 W. Dickinson *Gloss. Dial. Cumberland* (ed. 2) 92/1 *Spitten picter*.., a strong likeness. **1887** Hall Caine *Deemster* II. xxvi. 233 He looked the spitten picture of my ould father. **1910** A. Bennett *Clayhanger* i. ii. 17 A nice-behaved young gentleman, and the spitten image of his poor mother. **1928** *Eng. Jrnl.* Dec. 819 Hit's the spitten image of you, Jilson. Your likeness, pint blank... Thar you are a-settin' in your cheer, a-holdin' your fiddle; thar's your basket. **1936** W. Holtby *South Riding* v. iv. 313 Spitten image of his dad, little Alf is, isn't he, Reg? **1941** J. Faulkner *Men Working* vi. 105 Ifen hit ain't the spitten image of that church acrost the street.

spitter[2]**.** Add: **3.** *U.S. Baseball.* = *spitball 2.

1908 *Baseball Mag.* July 7/2, I found by holding the ball with my finger tips and steadying it with my thumb alone I could get a peculiar break to it... It is not a 'spitter'. **1975** *New Yorker* 14 Apr. 98/2 The next pitch broke down sharply over the plate, and everyone cried, 'Spitter! Hey, a spitter!'

spitting, *vbl. sb.*[2] Add: **4. c.** Also *spitting distance*.

1959 P. Bull *I know Face, but*.. iv. 70 One of the reasons I had closed with the Chatham offer was that it was within spitting distance of London. **1965** *Listener* 3 June 841/3 All this the spectators can see at spitting distance. **1977** *Western Mail* (Cardiff) 5 Mar. 6/1 More than doubled pre-tax profits has taken the Midland Bank to within spitting distance of its two giant High Street rivals—National Westminster and Barclays.

spitting, *ppl. a.* Add: **1.** (Later *poet.* examples.)

1910 W. B. Yeats *Green Helmet* 33 With my spitting-cat heads, my frenzied moon-bred band. **1953** Dylan Thomas *Under Milk Wood* (1954) 48 And in Willy Nilly the Postman's dark and sizzling damp tea-coated misty pygmy kitchen where the spittingcat kettles throb and hop on the range.

2. spitting cobra, the African black-necked cobra, *Naja nigricollis*.

1931 R. L. Ditmars *Snakes of World* xiii. 167 The Spitting Cobra or Black-necked Cobra..comes close to being the most dangerous snake of Africa. **1976** G. Durrell *Stationary Ark* ii. 28 This [word] was expectorated with a venom that would have done credit to a spitting cobra.

3. *spitting image*, alteration of *spitten image* (see *spitten *a.*). Cf. *splitting image* s.v. *splitting *ppl. a.* 5.

1901 A. H. Rice *Mrs. Wiggs* vii. 94 He's jes' like his pa —the very spittin' image of him! **1917** A. W. Blue *Quay Head Tryst* 70 He's the spittin image o' a thrawn fechter. **1929** H. S. Walpole *Hans Frost* iii. v. 370 In another twenty years..she would be her mother's spitting image. **1938** N. Coward *Operette* i. vii. 58 Believe it or not, she was the spitting image of Princess Ena! **1960** H. Pinter *Caretaker* ii. 33 Your spitting image he was. **a 1974** R. Crossman *Diaries* (1975) I. 243 Far from adapting himself to his new position he is adapting his new position to himself (No. 10, as I saw the other day, is the spitting image of his little house in Hampstead).

spittle, *sb.*[2] Add: **4. spittle bug** *U.S.* = *frog-hopper* s.v. Frog[1] 8; cf. Cuckoo-spit[2] 1.

1882 *Vermont Agric. Rep.* VII. 77 Dr. Cutting spoke of the frog hopper, usually known as the spittle bug on grass. **1948** *Sun* (Baltimore) 9 June 26/8 Spittle-bug infection has damaged alfalfa fields. **1972** Swan & Papp *Common Insects N. Amer.* xiii. 136 Spittlebugs are named for the sticky, bubbly mass of froth with which the nymphs surround themselves.

spi·ttled, *a.* [f. Spittle *sb.*[2] + -ed[2]: cf. Spittle *v.*[1]] Covered with spittle.

1926 *Scots Observer* 1 Jan. 5/1 The tangled, insect-spittled vetches. **1939** Dylan Thomas *Map of Love* 12 The spittled eyes, the salt ponds in the sleeves.

spittly, *a.* Delete † *Obs. rare*⁻⁰ and add: Also **spittley.** (Examples.)

a 1861 [see *boxy *a.*[2] 1]. **1935** E. R. Eddison *Mistress* xxi. 434 Spittly and slimy..from the beast's mouth I plucked it out. **1949** L. A. G. Strong *Maud Cherrill* ix. 47 Once again would come the sly leer and the spittly whisper. **1980** J. O'Faolain *No Country for Young Men* iv. 71 'God blosht the wet turf,' she whispered and blasted it herself with spittly breath.

spitty (spi·ti), *a.* [f. Spit *sb.*[2] and *v.*[2] + -y[1].] Resembling spit; spitting or inclined to spit. Also *fig.*

1742 W. Ellis *Mod. Husb.* June x. 73 They believe the Grass-hopper will breed from this spitty Matter. **1865** H. J. Daniel *Muse in Motley* 43 He wud spet, Iss, spetty like a toad. **1909** *Chambers's Jrnl.* June 388/1 It took longer to get acquainted with the Major, who..spoke in such a spitty, hissy, foreign kind of way. **1945** B. MacDonald *Egg & I* vi. 99 Great winds came bounding down..; blew rain at us in spitty gusts. **1950** *Audio Engin.* Sept. 32/2 The program material sounds *ragged* and *spitty*, and the *whisker* at the main peak can sometimes be detected. **1976** U. Holden *String Horses* ii. 26 Lavender..smiled her spitty smile.

Spitz. Substitute for def.: A dog belonging to one of a group of northern breeds distinguished by thick fur, a pointed muzzle, pricked ears, and a tail curled over the back. Also *attrib.*, as *Spitz dog*. (Earlier and further examples.)

1842 H. Stanley *Let.* 18 July in N. Mitford *Ladies of Alderley* (1938) 38 In this town [sc. Ems] wherever you go you see a white Spitz dog with a curly tail & black eyes. **1897** *Private Life of Queen* xviii. 147 The perky little tan-coloured German Spitz-dog, 'Marco'. **a 1910** 'O. Henry' *Rolling Stones* (1912) 181 Six burglars..bore away..a five-hundred-dollar prize Spitz dog. **1948** [see *laika]. **1978** *Times* 11 Feb. 2/4 The Japanese spitz, a small, white prick-eared creature like an undernourished Samoyed, appeared at Crufts for the first time.

Spitzenberg (spi·tsənbɜɹg). Also **spitzenberg, Spitzenbergh, Spitzenburg.** [Origin unkn.; cf. quot. 1795.] An apple with a red and yellow skin, belonging to the North American variety so called, developed from a seedling first found at Esopus, N.Y.; also, the tree bearing this fruit.

1795 J. Jay *Let.* 12 Dec. in *Columbia Library Columns* (1970) XIX. 43 Ten are what we call Spitzenberghs, from the Name of the Man in whose orchard the first tree of the kind was found. **1809** 'D. Knickerbocker' *Hist. N.Y.* iii. i. 122 Mottled and streaked with dusty red, like a spitzenberg apple. **a 1817** T. Dwight *Trav. New-Eng.* (1821) I. 45 The varieties of apple-trees are:.. Spitzenberg, Holden Sweeting, Fall pippin. **1869** *Rep. Comm. Agric. 1868* (U.S. Dept. Agric.) 482 [He] names the Baldwin for dessert and cooking, the Spitzenburg for cooking. **1894** H. Frederic *Copperhead* 71 He..picked out another apple—a spitzenberg this time. **1921** *Daily Colonist* (Victoria, B.C.) 8 Oct. 19/1 (Advt.), 320 Boxes of Apples..from our own ranch..Spitzenberg at $3.23. **1949** J. B. Herrick *Memories* 5 Father laid in..a barrel each of Spitzenburgs and Baldwins.

Spitzflöte (spi·tsflötə). Also with small initial and anglicized **Spitz-flute, spitz flute,** etc. [Ger., f. *spitz* pointed, acute + *flöte* flute.] An organ stop of the type of the gemshorn, yielding a tone resembling that of the flute.

1855 E. J. Hopkins *Organ* 118 The pipes of the Spitzflute are slightly conical. **1884** *Encycl. Brit.* XVII. 831/1 The 8 spitzflöte may be regarded as a variety of open diapason. **1894** T. Elliston *Organs* 119 *Spitz flute*, 8 ft. pitch. **1923** N. A. Bonavia-Hunt *Mod. Organ Stops* 31 The spitzflöte differs in having a more graduated taper. **1925** W. L. Sumner *Organ* x. 301 In British organs the stop is usually called spitzflöte or spitz flute... The spitzflöte has an interesting..and attractive tone with more harmonic development than the usual English gemshorn. **1966** H. & J. Norman *Organ Today* xi. 120 The Spitzflute (or more accurately Spitzflöte) is a large-scale open flute, the taper giving it blending qualities that would otherwise be lacking.

spitzkop, var. *spitskop.

spitzy (spi·tsi), *a.* [f. Spitz + -y[1].] Resembling or pertaining to a Spitz dog. Also *fig.*, pointed.

1937 Auden & MacNeice *Lett. from Iceland* xii. 185 A sort of little sheepdog, black and white with a thick but not very long coat, a broad forehead and a spitzy foreface. **1952** C. L. B. Hubbard *Pembrokeshire Corgi Handbk.* i. 14 The extremely sharp-pointed tips to the ears prevalent in the more Spitzy Pembrokeshires. **1968** *Sunday Truth* (Brisbane) 3 Nov. 58/4 He had a 'spitzy nose'—a sharp nose—a scar on his left arm where he had been hurt by machinery.

spiv (spiv), *sb. slang.* [Origin obscure: perh. from Spiff *v.*[2], Spiffy *a.*] A man who lives by his wits and has no regular employment; one engaging in petty blackmarket dealings and freq. characterized by flashy dress.

1934 A. Bracey *School for Scoundrels* 336 *Spiv*, petty crook who will turn his hand to anything so long as it does not involve honest work. **1937** *Even. News* 12 Mar. 15/6 (Advt.), With the Lincoln and the Grand National in the offing, the twisters, the welshers, the 'spivs' and the 'boys' are getting ready for a profitable session of the gentle sport of rooking the racegoer. **1939** [see *barrow *sb.*[2] 4]. **1945** [see *lairy *a.*[2] 1]. **1947** *Times* 13 Aug. 6/1 If spivs meant men living by their wits, the nets could be thrown very wide indeed. **1948** I. Brown *No Idle Words* 107 It is queer that its opposite [of *deep*] wide, should have replaced it in Spiv-English. **1952** 'J. Henry' *Who lie in Gaol* viii. 123 In appearance, he resembled the typical spiv; with coat-hanger shoulders, and pointed shoes, and a smile that would have been an asset to any confidence man. **1958** *People* 4 May 1/4 Who was responsible for letting the spivs hi-jack the crowds at Wembley? **1965** *New Statesman* 26 Nov. 851/1 The emergence of Robert Stephens, once a type-cast West End performer of small, seedy spiv-roles, as a major heroic actor in the making. **1978** *Cornish Guardian* 27 Apr. 3 Metrication will be an open invitation for every spiv and racketeer to cheat the British public.

Hence as *vb.*, (*a*) *intr.*, to make one's living as a spiv; (*b*) *trans.*, to spiff, to spruce (oneself) *up*; **spivved** (spivd), *ppl. a.*; **spi·v(v)ery,** behaviour characteristic of a spiv or the state of being a spiv; **spi·vvish, spi·v(v)y** *adjs.*, characteristic of a spiv; **spi·vishly** *adv.*

1945 B. Naughton in C. Madge *Pilot Papers* 99 See his Spivy coat—the width of the lapels, the padded-out shoulders? **1947** *Times* 18 Nov. 2/4 Instead of that brave new Britain all they had left was a land fit for bookies to spiv in. **1948** *Chambers's Jrnl.* Oct. 547/1 That is their focus, their touchstone—recognition of a unique quality in him, not to be obliterated by the sordid elements in his story, his obvious shiftiness of character, his spivery attributes or his too frequent lapses. **1948** C. Day Lewis *Otterbury Incident* ii. 18 Tilting his hat at an even more spivvish angle. **1951** Koestler *Age of Longing* i. i. 7 Surrounded by relatives and friends, a spivvy son and an insipid daughter,..he would have to meet the ultimate ordeal. **1952** A. Wilson *Hemlock & After* i. v. 96 I've spivved along on my own steam as far as I can go. **1956** D. M. Davin *Sullen Bell* 87 There were few people in the restaurant, spivvish-looking and absorbed. **1957** R. Hoggart *Uses of Literacy* ix. 225 Where domestic or personal roots are weak or have been forcibly broken, these attitudes can quickly lead to an extensive moral 'spivvery'. **1959** I. Jefferies *Thirteen Days* iii. 37 Scruffy lot. They'll get spivved up quick enough for a Naafi girl. **1965** *New Statesman* 19 Mar. 463/1 The spivvish businessman. **1966** J. Gloag *Sentence of Life* ii. 30 The inspector's jacket was sharply, spivishly waisted. **1971** B. W. Aldiss *Soldier Erect* 177 We spivved ourselves up, put on clean shirts, and strolled out of camp. **1976** *Times* 5 Feb. 21/5 Willott..has no time for spivvery within or without the law. **1976** *Listener* 26 Feb. 245/4 The pelvic lead singer, spivved, moustached, sharp. **1978** *Time Out* 18 Aug. 55/5 Hoffman, spivvy and moustached for maximum seediness, is an ex-con on parole who can't go straight.

splack, *adv. rare*⁻¹. [Echoic.] With a sound suggestive of splashing and smacking.

a 1960 E. M. Forster *Maurice* (1971) xxxviii. 182 Mr London and Mr Featherstonhaugh dived splack into the water lilies.

splake (splēⁱk). *N. Amer.* [f. Sp(eckled (*ppl.*) *a.* + Lake *sb.*[4]] A trout produced by crossing the lake trout, *Salvelinus namaycush*, and the speckled trout, *S. fontinalis*.

1954 *Sun* (Baltimore) 27 May 2/7 Biologists of the Ontario Department of Lands and Forests have completed the first step of an experiment in producing a new variety of game fish. The 'splake' has resulted. It is a hybrid, the result of crossing speckled and lake trout. **1968** *Globe & Mail* (Toronto) 5 Feb. 6/4 A hardier type of trout called splake..is being introduced. **1979** *Whig-Standard* (Kingston, Ontario) 9 Feb. 26/1 Splake from inland waters are highly colored compared to those from the great lakes which are usually silvery.

splanchno-. Add: **splanchnocra·nium,** those lower and anterior bones and cartilages of the head and face that are derived from visceral arch elements; opp. *neurocranium* s.v. *neuro-; hence **splanchnocra·nial** *a.*; **splanchnome·galy** [Gr. μεγάλ-, μέγας large], an enlarged condition of the viscera.

1907 W. N. Parker tr. *Wiedersheim's Compar. Anat. Vertebrates* (ed. 3) 75 A series of cartilaginous arches arise in serial order on the ventral side of the brain-case; these encircle the anterior part of the alimentary tract like hoops, incomplete dorsally, and are distinguished as the visceral portion of the skull (splanchnocranium). **1931** Saunders & Manton *Man. Pract. Vertebr. Morphol.* xii. 121 The brain-box and sense capsules form the neurocranium, while the jaws and visceral arches form the splanchnocranium. **1974** D. & M. Webster *Compar. Vertebr. Morphol.* iv. 58 Of the splanchnocranial cartilages, the most anterior pair is the mandibular arch. *Ibid.* 59 (*caption*) Lateral view of the splanchnocranium of an acanthodian. [**1900** Dorland *Med. Dict.*, Splanchnomegalia.] **1910** *Practitioner* Jan. 34 There is sometimes also a general increase in the size of the various internal organs, a condition of splanchnomegaly. **1970** N. G. Schneeberg *Clin. Endocrinol.* v. 70/2 Generalized marked splanchnomegaly occurs, though enlargement of the liver is most common.

splash, *sb.*[1] Add: **1. e.** A small quantity of liquid, *spec.* a dash of soda-water or tonic, etc., added to spirits as a drink.

1922 Joyce *Ulysses* 551 Here, to buy yourself a gin and splash. **1935** G. Greene *England made Me* v. 243 The atmosphere of..the week-end jaunt, the whisky and splash. **1965** 'J. le Carré' *Looking-Glass War* xv. 173 Woodford's wife added a little soda to her Scotch, a splash: it was habit rather than taste. **1977** *Rolling Stone* 30 June 81/3, I walked to the bar, ordered a double vodka with a splash of tonic no fruit.

2. a. *to make a splash* (earlier example).

1804 G. Colman *Let.* 9 Jan. in A. Mathews *Mem. Charles Mathews* (1838) I. xx. 434 A vile part, surely, for the *début* of a man who is to make a *splash*.

b. Also *attrib.*

1832 J. Romilly *Diary* 6 May (1967) 12 Missed Hearing Roze's Sermon..:—hear it was poor: tho with a splash passage against the wickedness of manufactures.

d. *colloq.* The prominent display in a newspaper of an advertisement, headline, or story; the material so displayed, usu. of a dramatic or sensational character. Freq. *attrib.*

1922 Joyce *Ulysses* 638 The usual splash page of letterpress about the same old matrimonial tangle. **1932** H. Nicolson *Public Faces* xi. 288 We must get the news back here before to-night—time for full splash in to-morrow's papers. **1933** W. Moss *Polit. Parties Irish Free State* iii. 127 'Splashes', i.e. full-page advertisements using the most effective appeals and backed with extensive and detailed argument. **1948** *Manch. Guardian Weekly* 1 July 2 The perfect 'splash' story in the lull after the Republican Convention. **1960** J. Robb *Front Page Story* 46 Percy tossed the latest copy over to Bert, the 'Splash' sub. 'Splash' sub was the title given to the man who normally handled only the page-one lead. He was usually..the best sub-editor in the office. **1966** *New Statesman* 13 May 680/3 The paper had reverted to its old jittery habit of lifting other people's splashes and exclusives. **1974** *Globe & Mail* (Toronto) 23 Oct. 12/7 'Anything could happen,' said..Michael Wilson after the London papers gave splash play to his announcement. **1977** *New Yorker* 24 Oct. 55/2 The violent crimes.., if they occur in New York City, get a one-day splash in the *News*. **1979** P. Niesewand *Member of Club* iv. 31 Courtney turned back to the paper ..to look at the front page. The splash story was the row over the death of a young South African commando. **1982** *Chicago Sun-Times* 2 Dec. 67 Speakes gave a thorough briefing on the Cabinet Council study, providing the rest of the nation a Page 1 splash the next day.

4. b. *Med.* Also *splash sound*. A sound produced by a mixture of air and liquid in the stomach or other cavity when it is sharply disturbed. Cf. *splashing *vbl. sb.*[1] 1 b.

1890 F. Taylor *Pract. Med.* 435 The presence of air and liquid together in the pleural cavity may be demonstrated by the test known as Hippocratic succussion, or splash sound. **1908** Hutchison & Rainy *Clin. Methods* (ed. 4) 66 It should be remembered..that a splash may be elicited over even a normal stomach shortly after a meal containing much fluid,..and care should also be taken not to mistake a splash produced in the transverse colon for a stomach splash. **1938** N. L. Eckhoff in H. Rolleston *Brit. Encycl. Med. Pract.* VII. 226 A splash may be heard in cases of pyloric obstruction, and in some cases of obstruction in the colon. **1971** [see *splashing *vbl. sb.*[1] 1 b].

6. b. Amphetamines. *U.S. slang.*

1969 J. Gardner *Complete State of Death* ix. 166 The American..addicts..call most of the amphetamines 'speed', in the same way as they talk of others as 'bennies', 'splash', 'cartwheels'. **1974** M. C. Gerald *Pharmacol.* xvii. 332 Amphetamines (Benzedrine),..*Bennies, Peaches, Splash*.

splash, *v.*[1] Add: **I. 2. a.** Also, to pour *out* with a splash.

1908 E. F. Benson *Blotting Bk.* ii. 51 Mills..splashed himself out a liberal allowance of brandy into his glass.

c. *colloq.* To present (news, etc.) ostentatiously, or as a 'splash' (*splash *sb.*[1] 2 d).

1930 *London Mercury* Feb. 317 All the evening papers have 'splashed' the story. **1934** A. P. Herbert *Holy Deadlock* 210 She was 'splashed' in the newspapers as a sort of modern St. Joan. **1946** J. W. Day *Harvest Adventure* xvi. 272 London and provincial newspapers 'splashed' this first organized revolt against a tyranny and waste of public money. **1958** *Punch* 1 Jan. 59/3 The story was splashed over the front page. **1969** [see *par *sb.*[4]]. **1979** A. Brink *Dry White Season* ii. vi. 138 The photograph of Emily embracing Ben was splashed on the front page of an English newspaper.

d. *colloq.* To spend (money) extravagantly or ostentatiously. Freq. const. advbs., esp.

in phr. *to splash* (money) *out on* (something). Also *intr.*

1934 *Times* 7 Mar. 7/5 Public money ought not to be splashed about in this manner without grave and searching examination by the House of Commons. **1938** S. Beckett *Murphy* 79 He thought for a second of splashing the fourpence. **1946** F. Sargeson *That Summer* 82 After we'd splashed on a talkie we went home. **1960** S. Barstow *Kind of Loving* ii. ii. 170, I splash eight-and-six on a pound box of chocolates and send them with a little note. **1973** *Courier & Advertiser* (Dundee) 1 Mar. 2/2 Allied now plan to splash out an extra £150,000 on advertising. **1973** E. Lemarchand *Let or Hindrance* xiv. 170 They..splashed the lolly around when the pay packets came in. **1978** *Morecambe Guardian* 14 Mar. 17/2 Splash out on something new to wear; the result will be worthwhile.

II. 5. d. Const. *down*. Of a spacecraft: to alight on the sea after a space flight. Also *transf.* Cf. *splashdown.

1962 *Daily Tel.* 4 Oct. 1/7 Cdr. Walter Schirra.. 'splashed down' safely in the Pacific at 10.28 (BST) last night. **1965** *Economist* 4 Sept. 873/1 By the time they splashed down on Sunday Lieutenant-Colonel Cooper and Lieutenant-Commander Conrad had completed the longest-ever manned flight in space. **1969** *Times* 3 June (Suppl.) p. iii/4 Apollo 11 is due to splash down at 5.52 on July 24. **1978** *Times* 1 Aug. 2/1 The ill-fated [balloon] Zanussi in which they splashed down.

splash-. Add: **1. splashback,** a panel fastened to and protecting the wall behind a sink, cooker, etc., from splashes; **splash cymbal,** a small, light cymbal; **splashguard,** a guard fitted to an object to prevent splashing; **splash party** *U.S.*, a party at which the guests engage in swimming and other water sports; **splashplate,** a metal splashback on a cooker; **splash pool,** a shallow paddling pool for children; **splash-proof,** *a.*, impermeable to splashes; **splash zone,** that area adjacent to the sea, a waterfall, etc., that is continually splashed by water.

1926–7 *Army & Navy Stores Catal.* 295/3 *Bathroom accessories*... White opal Splash Back with fitted 5-in. shelf. **1961** *Which?* Dec. 317/1 Oven, grill and hotplate controls were either on the front of the cooker or on the splash-back. **1971** *Ideal Home* Apr. 52/1 Plain blue tiles form a splashback panel round the back. **1982** *Daily Tel.* 3 Aug. 9/1 Cast iron washstand.., with tiled splashback and mirror, is rare and costly at £1,000. **1961** Splash cymbal [see *crash cymbal* s.v. *crash *sb.*[1] 7]. **1964** Splash cymbal [see *choke *v.* 10 b]. **1917** *Harrods Gen. Catal.* 516/1 The 'Reliance' Splash Guard. Solid Red Rubber... When ordering please state outside diameter of tap. **1962** *Which?* Mar. 87/1 All the continuous feed models [of waste-disposers]..had a splashguard, a slotted rubber cover across the entrance to the grinding chamber which allows you to push waste through but prevents water and scraps from being flung out. **1930** *Amer. Speech* VI. 121 *Splash*.., swimming: Splash Party of Girls' Club Lively Affair. **1956** W. H. Whyte *Organization Man* xxii. 291 'Splash' parties (all you can eat for a dollar). **1967** *Gloss. Terms Gas Industry* (B.S.I.) 89 *Splashplate*, a panel above and behind a hotplate. **1970** *Which?* June 175/1 Many of the stoves had splash plates—usually of painted metal. **1971** *Sunday Express* (Johannesburg) 28 Mar. 16/3 (Advt.), Beautiful 3 bedroomed home with really lovely garden... Splash pool for children. **1976** *Billings* (Montana) *Gaz.* 16 June 5-D/4 (Advt.), Keep kids cool all summer. 8' × 18" splash pool with vinyl liner. **1929** *Radio Times* 8 Nov. 451/2 A 20 amp-hour accumulator... Vent is large, splash-proof and spray-tight and screws in. **1965** *Wireless World* July 36 (Advt.), Dustproof and splashproof. **1933** *Jrnl. Marine Biol. Assoc.* XVIII. 453 During the winter, continued storms will keep the height of the splash zone at five feet or more. **1956** F. W. Adams in D. L. Linton *Sheffield* 330 [*Scapania undulata*.] Common in waterfalls, 'splash zones', reservoir-overflows, and just below the water line on stones in non-calcareous streams. **1976** *Offshore Platforms & Pipelining* 30/1 A central shaft..will permit divers to go 50 m below sea level when there is rough water in the splash zone.

2. Used (usu. without following hyphen) with reference to a system of lubrication in which oil is distributed throughout an engine in the form of drops initially splashed from a reservoir by the working of certain moving parts. So *splash lubrication*.

1906 *Evening Express* (Liverpool) 9 Mar. 7/4 If the splash system of lubrication were more generally adopted by the makers there would be no trouble on this score. **1907** F. Strickland *Man. Petrol Motors & Motor Cars* xvi. 247 The difficulty in splash lubrication is to provide enough oil for the various parts, without having so much in the crank chamber that it works up into the cylinder and produces smoke. **1924** A. W. Judge *Mod. Motor Cars* I. iv. 119 A low speed engine, an over-cooled one, or a splash-lubricated one would require a low-viscosity or 'thin' grade [of oil]. **1939** J. I. Clower *Lubricants & Lubrication* xiv. 358 Splash oiling is widely used on small, vertical and horizontal single-acting machines [sc. compressors]. **1963** [see *fly-wheel]. **1966** *McGraw-Hill Encycl. Sci. & Technol.* VII. 607/1 In splash lubrication, the oil entering the engine flows to troughs into which extensions of the connecting rods dip.

splash-board. Add: **1.** (Earlier example.)

1826 Disraeli *Vivian Grey* II. iii. 121 By dint of rattling the whip against the splash-board.

4. A splashback; a protective panel attached to a wall.

1868 C. L. Eastlake *Hints on Household Taste* viii. 183 The common bed-room wash-stand..has a splash-board to protect the wall against which it is placed. **1976** *Billings* (Montana) *Gaz.* 16 June 11-A/4 Even a minimal remodeling job can cost $2,500 before the flower-print wallpaper is stuck on the splashboard.

splashdown (splæ·ʃdaun). Also **splash-down.** [f. Splash- + Down *adv.*] The alighting of a spacecraft on the sea. Also *transf.*

1961 *Washington Post* 5 May B2 They [sc. several warships] are strung out about 60 miles apart, and their mission is to retrieve Shepard after 'splashdown'. *Ibid.* 6 May A1 The perfect flight, from lift-off to splash-down, lasted only 15 minutes. **1968** [see *recovery fleet* s.v. *recovery 10]. **1968** *Guardian* 28 Dec. 8/2 From launch to splashdown the Apollo-8 mission went entirely without hitch. **1970** N. Armstrong et al. *First on Moon* xiv. 347 The Navy's weathermen had predicted acceptable splashdown conditions. **1978** *Times* 1 Aug. 2/2 (*caption*) He [sc. a member of the crew of a balloon] hears the recording of their splashdown transmission.

splasher[1]**.** Add: **3.** A piece of cloth or the like hung behind a washstand to protect the wall from splashes. *U.S. Obs.* exc. *Hist.*

1895 *Montgomery Ward Catal.* Spring & Summer 123/1 Stamped Splashers, white cotton duck; size, 18 × 28 inches. *Ibid.* 577/2 Toilet Splasher for protecting wall above washstand, etc., waterproof, made of sewed wood splints, to roll up. **1905** D. Belasco in M. J. Moses *Representative Amer. Dramas* (1925) ii. 73/1 A washstand, backed by a 'splasher' of white oilcloth, is near the bed. **1945** *Sun* (Baltimore) 20 June 4-0/3 The furnishings of the Victorian style bedroom were simple linen scarfs... They were used on the flat surfaces, and 'splashers' of linen were fastened to the wall back of the washstand.

splashily, *adv.* (Later example.)

1978 D. Francis *Trial Run* vi. 87, I watched some horses..their feet plopping splashily in the wet.

splashiness. For *rare*⁻⁰ read *rare* and add: Also, the quality of being splashy.

1978 *Listener* 30 Mar. 412/4 There is a certain wayward splashiness about some of its more 'texturising' sections —all bells, string glissandi and chattering winds.

splashing, *vbl. sb.*[1] Add: **1. b.** *Med.* Noisy motion of air and liquid inside the body.

1890 F. Taylor *Pract. Med.* 344 In cases of hydro- or pyo-pneumothorax, shaking the patient will elicit a splashing sound. **1897** Hutchison & Rainy *Clin. Methods* 62 If 'splashing' be elicited it will be partly heard and partly felt. Distinct splashing elicited three hours after a meal..is very suggestive of a dilated stomach. **1971** R. B. Cole *Essent. Respiratory Dis.* iii. 29 The presence of air and fluid in the pleural space (hydro-pneumothorax) is indicated if a splash can be heard on auscultation when the thoracic cage is gently shaken. It should not be confused with the splashing of gastric contents.

3. splashing-board = Splash-board 1.

a 1817 Jane Austen *Northanger Abbey* (1818) I. vii. 87 Seat, trunk, sword-case, splashing-board, lamps, silver moulding, all you see complete; the iron work as good as new, or better. He asked fifty guineas.

splashy, *a.*[2] Add: **3.** (Later examples.) Cf. *splash *sb.*[1] 2 d.

1972 'T. Cox' *Don't lie to Me* ix. 87 The killing had originally gotten a pretty splashy play in the newspapers. **1976** *Scotsman* 25 Nov. 14/2 A preposterous miracle has rescued the 'Observer'. It was the sort of last-minute, cliff-hanging rescue which the 'Observer' would hesitate to make a big, splashy story out of.

splat, *sb.*[2] [f. *splat *v.*[2]] **1.** *colloq.* A slapping and splashing sound; a smack.

1958 S. A. Grau *Hard Blue Sky* vi. 375 Annie was throwing the mud by handfuls and listening to the splat and sizzle. **1974** *Publishers Weekly* 11 Nov. 48/1 She tosses her head as she thinks how superior she'll be and, of course, the basket of eggs falls with a 'Splatt!' [*sic*]. **1976** 'Trevanian' *Main* (1977) iv. 72 A spoiled child dangles from her free hand... She gives it a good shake and a splat on the bottom. **1979** D. Gurr *Troika* xxiv. 176 The dough hit him across the face with a vicious slapping *splat*.

2. *Metallurgy.* **a.** Used *attrib.* and in *Comb.* with reference to a method of cooling hot liquid metal extremely rapidly by causing droplets of the metal, propelled by a shock wave, to strike and spread out upon a (usu. rotating) metal surface; so *splat-cool* vb., *-cooled* ppl. adj., *-cooling* vbl. sb., *splat-quench* vb., etc.

1960 *New Scientist* 28 July 286/2 Known as 'splat cooling', it consists in melting about 25 mg of metal in a shock tube by inductive heating. The resultant liquid drops are shot out of the tube by the shock wave and, travelling at high speed, impinge on the inside rim of a fast-revolving copper wheel. Because the splattering drops spread into a very thin film and the centrifugal force imparted by the spinning wheel ensures good thermal contact, the metal or alloy cools at something like a million degrees a second. **1965** *Trans. Metall. Soc. AIME* CCXXXIII. 1584/1 Heat-transfer coefficients.. for aluminum and silver splat cooled essentially on a nickel substrate. **1968** *Acta Metallurgica* XVI. 1204/2 Alloys from 5 to 95% Sn in approximately 5% steps were splat cooled. **1968** *Jrnl. Chem. Physics* XLVIII. 1911/1 No stacking disorder was observed in splat-quenched two-phase alloys between α-Pb and ε [sc. a Pb–Bi alloy]. **1972** *Materials Sci. & Engin.* X. 343/1 Small pieces of the homogenized ingots were splat quenched at a rate of approximately 10⁸ deg per sec. **1974** *Nature* 8 Nov. 100/2 He [sc. Pol Duwez] conceived the idea of blasting a small molten drop of alloy by means of a gaseous shock wave against a sloping piece of copper: the technique soon acquired..the onomatopoeic designation 'splat-cooling'. **1976** *Ibid.* 29 Jan. 271/3 A technique for splat-quenching and compaction of Al–Fe alloys.

b. A thin, localized film of metal produced in splat-cooling.

1965 *Trans. Metall. Soc. AIME* CCXXXIII. 1581/1 Upon impact the metal spread into a thin nonuniform film called a splat, about 10⁻⁴ cm thick. **1976** *Materials Sci. & Engin.* XXIII. 101/2 Chemical analyses of splats ..indicated that some Cr was lost.

splat, *v.*[2] (*int.* and *adv.*) *colloq.* [Shortening of Splatter *v.*]

1. *intr.* To land with a sharp smacking sound, or with a sound as of slapping and splashing. Also as *adv.*, as *to go splat*, and *int.*

1897 'H. S. Merriman' *In Kedar's Tents* v. 55 A bullet went 'splat' against a rock. **1922** J. A. Dunn *Man Trap* xvi. 226 A bullet whistled by Jimmy's head, *splatting* on the lava crust. **1937** J. Steinbeck in *Esquire* Sept. 200/3 His fist splatted into Johnny Bear's smiling mouth. **1970** R. D. Abrahams *Positively Black* iv. 101 This here bird flies over head and really lets go, right splat on her head. **1976** J. Grady *Great Pebble Affair* (1977) 96 I'll jump out of the window... I hope I splat and spoil all the upholstery on those goddamn jalopies. **1977** *Undercurrents* June–July 34/3 Picked up the tap and splat! It leapt across the room.

2. *trans. Metallurgy.* To cause droplets of (a metal) to strike a surface and form a film in the process of splat-cooling.

1965 *Trans. Metall. Soc. AIME* CCXXXIII. 1583/1 Two metals, aluminum and silver, were selected to be splatted because of their high thermal conductivities and lack of oxidation problems.

Hence **spla·tted** *ppl. a.*, **spla·tting** *vbl. sb.*

1976 *Phil. Mag.* XXXIV. 236 The average energy loss ..was determined..from specimens immediately after splatting. *Ibid.*, Prior to performing microanalysis it was necessary to calibrate the variation of energy loss E_p with copper concentration for splatted material.

splather (splæ·ðəɹ), *v. dial.* [Perh. var. of Splatter *v.*] *intr.* To splash; to sprawl.

1877 F. Ross et al. *Gloss. Words Holderness* 134/2 *Splather*,..to splash water or mud. **1886** R. Holland *Gloss. County of Chester* 333 A procumbent plant which spreads over the ground would be said to '*splather* about'. **1887** T. Darlington *Folk-Speech S. Cheshire* 365 *Splather*, ..to sprawl. 'He had bu' just spokken th' word, an' o'er he went *splatherin*' i' th' middle o' th' bruk.' **1936** F. Clune *Roaming round Darling* xiv. 118 The closed Experiment Farm, now profusely splathered with flora and fauna animate and inanimate.

Also as *sb.* (see quot. 1877); hence **spla·thering** *ppl. a.*

1877 E. Peacock *Gloss. Words Manley & Corringham, Lincolnshire* 235/1 *Splather*, (1) a splash. (2) Noisy talk. **1879** *Dewsbre Olmenac* in *Eng. Dial. Dict.* (1904) V. 673/2 Matters hev gotten squared up efter a deeal o' conflab an splather. **1929** J. B. Priestley *Good Companions* i. v. 188 She sends a long splathering telegram and never puts her address in it. **1937** Partridge *Dict. Slang* 811/2 *Hold your splathers!* be silent!

splatter, *v.* Add: **1. c.** Of objects.

1931 W. G. Carr *By Guess & by God* 91 A salvo of shells splattered around the periscope. **1976** 'E. McBain' *Guns* (1977) iii. 66 It had certainly been traumatic pulling the trigger of the .38 and watching the back of that cop's head come off and splatter onto the Seagrim's poster. **1978** *Amer. Poetry Rev.* July/Aug. 4/1 His deep brown feces splatter over Queen Anne's Lace and the waving sedge Of the pond.

2. b. To beat or batter.

1881 J. Sargisson *Joe Scoap's Jurneh through Three Wardles* 31 If it wasn't for that oald grey heid o' thine Ah wud splatter thee. **1897** W. Beatty *Secretar* xiv. 110, I would have splattered his harns on the causey. **1959** I. & P. Opie *Lore & Lang. Schoolch.* x. 198 Recommendations by boys in East and South-East London: 'Bash him up.' 'Beat him up.'.. 'Splatter him.'

splattered, *ppl. a.* (Later example.)

1979 E. Newman *Sunday Punch* i. 1 The prize being, for some of the fighters, fame and fortune, often accompanied by splattered noses.

splay, *sb.* Add: **1. d.** A tapered widening of the carriageway at a road intersection or corner provided in order to increase visibility for motorists.

1956 *Proc. Inst. Civil Engineers* V. ii. 356 The splay provided in the boundary line of the minor road approaching the intersection should remove any possible restriction on vision at the point where the vehicle turning right-in has to cross the right-out stream from the minor road. **1963** W. F. Cassie in E. Davies *Traffic Engin. Practice* x. 280 A minor road intersecting a major road. There are three possibilities..: (*a*) right-angled intersection; (*b*) right-hand splay; and (*c*) left-hand splay. **1968** J. L. Paisley in *Ibid.* (ed. 2) iv. 99 At uncontrolled junctions visibility splays should be provided to give full visibility to right and left. **1977** *Cornish Times* 19 Aug. 5/2 The Cornwall Surveyor..offered no objections subject to the formation of a suitable access with visibility splays of 30 by 500 feet in each direction along the adjoining highway.

splay, *a.* Add: **1.** (Later examples.)
1895 R. L. Stevenson *Amateur Emigrant* 38 We had a fellow on board, an Irish-American, for all the world like a beggar in a print by Callot; one-eyed, with great, splay crow's-feet round the sockets. **1952** Dylan Thomas *Coll. Poems* p. viii, Though song Is a burning and crested act, The fire of birds in The world's turning wood, For my sawn, splay sounds.

2. splay fault *Geol.*, a subsidiary fault diverging at an acute angle from a larger dislocation.
1942 E. M. Anderson *Dynamics of Faulting* vii. 150 Splay faults may..be expected to diverge from the main fracture at about this angle [*sc.* 22½°]. **1969** Bennison & Wright *Geol. Hist. Brit. Isles* vii. 147 These structures continue across into Northern Ireland but are much less strongly developed there and are replaced by series of en-echelon and splay faults. **1971** *Nature* 19 Feb. 538/1 North and south of the Gregory rift..the periclinal ends of the uplifted area are broad transverse depressions traversed by splay-faults.

splay foot, splay-foot. 2. For † *Obs.* read '*Obs.* or *rare*' and add later example.
1922 Joyce *Ulysses* 690 A slender splayfoot chair of glossy cane curves.

spleen, *sb.* Add: **9. c. spleen index, rate,** the proportion of the population having enlarged spleens (as determined by palpation), useful as indicating the incidence of malaria.
1969 Edington & Gilles *Path. in Tropics* ii. 13 The former [methods] determine parasite rates in random blood samples and spleen indices... A close correlation exists between parasite and spleen rates. **1903** Stephens & Christophers *Pract. Study Malaria* xxiii. 261 Above ten years, the spleen rate is usually considerably in excess of the parasite rate. **1935** *Discovery* Jan. 11/1 In these districts today the spleen rate, indicating the incidence of malaria amongst the inhabitants, is very low. **1963** E. Pampana *Textbk. Malaria Eradication* iv. 72 The spleen rate underestimates the true percentage of enlarged spleens.

spleen, *v.* Add: **1. c.** (Later examples.) *U.S.*
1889 R. T. Cooke *Steadfast* xviii. 198 [It] makes me spleen to think on't! **1902** H. L. Wilson *Spenders* x. 110 Well, I knew Dan'l J. purty well, and I spleened against some of his ways, but that's done fur.

spleenical (splī·nikăl), *a.* *rare*$^{-1}$. [f. Spleen *sb.* + -ical.] Spleenful.
1818 Keats *Let.* Oct. (1931) I. 262 You see there is nothing spleenical in all this. The only thing that can ever affect me personally..is any doubt about my powers for poetry.

‖ **splendeurs et misères** (splaṅdœ̄r e mizę̄r'). [Fr., lit. 'splendours and miseries', from the title of Balzac's novel sequence *Splendeurs et Misères des Courtisanes* (1839–47).] The glories and degradation of life set side by side; hence, applied to other co-existent extremes of conditions. Cf. *splendours and miseries* s.v. *splendour *sb.* 5.
1952 *Observer* 13 Jan. 7/5, I don't seem to have read any piece of criticism in the past month that doesn't speak of *splendeurs et misères*, so I'd better jump on the band-waggon. **1977** *N.Y. Rev. Bks.* 15 Sept. 42/4 *Mario Puzo Inside Las Vegas*..seems to me one of the liveliest testimonials to the *splendeurs et misères* of gaming fever I've read.

splendid, *a.* Add: **6.** *splendid isolation*, phr. used with reference to the political and commercial uniqueness or isolation of Great Britain; also *transf.* Cf. Isolation a (quots. 1896[1,2]).
[**1896** G. E. Foster in *Official Rep. Deb. H. Com. Canada* 16 Jan. 176 The great mother Empire stands splendidly isolated in Europe.] **1896** *Times* 22 Jan. 10/1 Splendid isolation... A few weeks ago England appeared to stand alone in the world, surrounded by jealous competitors and..unexpected hostility. **1898** [in Dict.]. **1902** J. Chamberlain in *Times* 7 Jan. 4/4 It is the duty of the British people to count upon themselves alone... I say alone, yes, in a splendid isolation, surrounded and supported by our kinsfolk. **1912** *Review of Reviews* July 63/1 The abandonment by Great Britain of her splendid isolation. **1933** *Times* 21 Feb. 12/3 A Nazi band played in splendid isolation in the square. **1976** *National Observer* (U.S.) 13 Nov. 1/2 In a little while Williams and his bodyguard will be leaving this splendid isolation (we are in a hotel suite 30-odd floors above Central Park South) and going to the Ambassador Theatre.

splendidness. (Later example.)
1980 *N.Y. Times* 5 Oct. vii. 33/5 Miss Leffland at her best is extraordinarily good; indeed, there is not a contemporary writer of short stories from whom truth of feeling, splendidness of insight, and a human beauty both aching and real can more confidently be expected.

splendi·ferously, *adv.* *joc.* [f. Splendiferous *a.* + -ly².] Splendidly; magnificently.
1900 J. Vaizey *About Peggy Saville* viii. 57 If you weren't so fat, you would do splendiforously for Ophelia. **1930** D. H. Lawrence *Apocalypse* (1932) i. 22 And nowhere does this happen so splendiferously as in Revelation. **1959** *Good Food Guide* 292 This splendiferously-named hotel. **1982** *Country Life* 28 Jan. 251/4 It is almost impossible to use *Brewer* to look up what you want to find..and this goes as splendiferously for the new edition as ever in the past.

splendi·ferousness. *joc.* [f. Splendiferous *a.* + -ness.] Splendour; magnificence.
1934 in Webster. **1971** J. H. Gray *Red Lights on Prairies* v. 103 The splendiferousness of the new schools, warehouses.

splendour, *sb.* Add: **5.** *splendours and miseries* = *splendeurs et misères.
1943 S. Sitwell (*title*) Splendours and miseries. **1971** A. Morice *Death of Gay Dog* iv. 44, I took in a survey of the room, to see what further splendours and miseries were in store for us. **1981** *Times* 1 Apr. 11/2 *Omnibus* looked at the splendours and miseries of acting.

spleniculus (splīni·kiulŏs). *Med.* [mod.L., f. L. *splen* Spleen *sb.* + *-iculus*, diminutive ending.] A detached portion of the spleen, a small accessory spleen.
[**1848** Quain & Sharpey *Elem. Anat.* (ed. 5) II. 1089 These are commonly named accessory or supplementary spleens (splenculi [*sic, bis*], lienculi).] **1897** *Brit. Med. Jrnl.* 16 Jan. 145/2 A spleniculus was left [after splenectomy], and that patient did not suffer from any of these symptoms. **1939** McNee & McMichael in H. Rolleston *Brit. Encycl. Med. Pract.* XI. 402 In the neighbourhood of the spleen there are usually some small haemolymph glands or spleniculi... Their enlargement may cause persistence of clinical symptoms after splenectomy. **1973** A. I. S. MacPherson et al. *Spleen* iii. 97 Spleniculi or accessory spleens are found in 10 to 20 percent of all post-mortem examinations.

splenium (splī·niŏm). *Anat.* [a. L. *splēnium* (see Splenial *a.*¹ and *sb.*).] The thick posterior part of the corpus callosum of the brain.
1845 W. J. E. Wilson *Anatomist's Vade Mecum* (ed. 3) viii. 411 Posteriorly [the corpus callosum]..forms a thick rounded fold (splenium), which is continuous with the fornix. **1902** D. J. Cunningham *Text-Bk. Anat.* 529 The massive posterior end [of the corpus callosum]..lies over the mesencephalon... It is called the splenium, and it consists of an upper and a lower part. **1968** Passmore & Robson *Compan. Med. Stud.* I. xxiv. 44/2 The corpus callosum forms a prominent feature and its divisions into the rostrum, genu, body and splenium can be readily recognized. **1977** *Neurology* XXVII. 688/2 If we infer that inferior splenium has a significant functional role in relaying visual verbal codes from the right to left cerebral hemisphere, a plausible corollary may be that color-valuing tends to use superior or dorsal splenium.

spleno-. Add: **sple·nocyte** [-cyte], one of the mononuclear leucocytes formerly thought to be characteristic of the spleen; **sple:nohepato·me·galy** [Hepato-; cf. next] = *hepatosplenomegaly* s.v. *hepato-; **splenomega·lia, -me·galy** [Gr. μεγάλ-, μέγας large], enlargement of the spleen; hence **splenome·galic** *a.*; **sple·nopexy** [-*pexy], surgical fixation of a wandering spleen; **sple:noporto·graphy** [ad. F. *splénoportographie* (G. Sotgiu et al. 1952, in *La Presse Médicale* LX. 1295/1)], radiography of the hepatic portal system following the introduction of a contrast medium into the spleen to make the system detectable; so **splenopo·rtogram,** a radiograph obtained in this way; **splenore·nal** *a.*, connecting the spleen and a kidney; **splenotomy** (earlier example).
1900 Dorland *Med. Dict.* 621/1 Splenocyte. **1925** Strong & Elwyn *Bailey's Text-bk. Histol.* (ed. 7) iv. 82 They are..free mesenchymal elements..aggregated in enormous numbers in the mammalian splenic pulp (splenocytes). **1979** *Nature* 18 Jan. 218/2 Some of these IFC were..cultured..in the presence of irradiated syngeneic splenocytes or thymocytes as feeder layer. **1900** Dorland *Med. Dict.* 621/1 Splenohepatomegaly. **1936** Splenohepatomegaly [see *lipoidosis]. [**1898** *Allbutt's Syst. Med.* V. 539 Splenic anæmia is the name by which the disease is best known in this country; but it has also been called..*spleno-megalie primitive*; under the last name chiefly it is described in French literature.] **1900** Dorland *Med. Dict.* 621/2 *Splenomegalia, splenomegaly*, enlargement of the spleen. **1952** Cole & Elman *Textbk. Gen. Surg.* (ed. 6) xxiv. 679 The typical case of hemolytic jaundice or anemia is diagnosed without difficulty. Cardinal manifestations are slight to moderate jaundice, splenomegalia, and anemia of the spherocytic, reticulocytic type. **1900** *Guy's Hosp. Rep.* LIV. 1 Splenomegalic cirrhosis of the liver. **1900** Splenomegaly [see *splenomegalia* above]. **1903** T. K. Monro *Man. Med.* 316 (*heading*) Splenic anæmia (primary splenomegaly). **1974** *Trypanosomiasis & Leishmaniasis* (Ciba Foundation) 207 Symptoms like marked general lymphadenopathy and splenomegaly should be considered as signs of activation of the immune apparatus. **1897** *Brit. Med. Jrnl.* 16 Jan. 133/2 The great difference between splenectomy and splenopexy is that in the former case the individual is deprived of an organ, and, although we do not know with any degree of precision what its real use is, we are nevertheless bound to display caution in dealing with it. **1923** Pool & Stillman *Surg. Spleen* v. 88 Although the immediate results of splenopexy are claimed to be favorable, the late results are uncertain. **1955** R. T. Shackelford *Surg. Alimentary Tract* II. vi. 996 The fixation (splenopexy) of a 'wandering spleen'..has been abandoned. Splenectomy rather than splenopexy is indicated. **1953** *Ann. Surg.* CXXXVIII. 583/2 The splenoportogram for this patient..demonstrated a well functioning portacaval shunt in the three-second film which was taken during the injection of the last 5 cc. of the opaque media. **1969** S. I. Schwartz *Princ. Surg.* xxix. 1052/1 (*caption*) Normal splenoportogram. Note site of injection in spleen and diffusion of radiopaque material through organ. **1953** *Ann. Surg.* CXXXVIII. 582/2 It employs the visualization of the splenic and portal veins by roentgenographic examination. This is accomplished by injecting contrast media directly into the spleen... Sotgiu, Cacciari and Frassineti, working in Bologne, reported favorable results in four patients. It is from these authors that we have borrowed the term 'Splenoportography'. **1980** *Rec. Adv. Surg.* X. 5 Pre-operative determination by means of..splenoportography..and selective arteriography..are important. **1945** *Ann. Surg.* CXXII. 479 It is undesirable..to resort to the use of a vein graft in the performance of a splenorenal anastomosis. **1974** R. M. Kirk et al. *Surgery* vi. 108/1 (*caption*) Portal venous hypertension is reduced by portacaval or splenorenal shunt. **1976** *Lancet* 11 Dec. 1300/2 Our results suggest that hypergastrinæmia is more likely to follow portacaval than splenorenal anastomosis. **1868** *Guy's Hosp. Rep.* XIII. 416 The operation of splenotomy as a means of cure [for leucocythæmia] is physiologically unsound.

splenosis (splīnōu·sis). *Med.* [f. Splen(o- + -osis.] The presence in the body of numerous separate pieces of living splenic tissue.
1939 Buchbinder & Lipkoff in *Surgery* VI. 933 A case is reported of autotransplantation of splenic tissue throughout the abdominal cavity following trauma of the spleen... We offer the term splenosis to describe this condition. **1947** *Canad. Med. Assoc. Jrnl.* LVI. 376/2 A case of splenosis is reported, in which seeding occurred after splenectomy for non-traumatic reasons. **1973** A. I. S. MacPherson et al. *Spleen* ix. 224 The seeding of viable cells from the ruptured spleen throughout the peritoneal cavity and their subsequent growth to form numerous tiny nodules of histologically normal splenic tissue has been described and has been given the name 'Splenosis'.

splenunculus (splīnʊ·ŋkiŭlŏs). *Med.* [mod.L., f. L. *splēn* Spleen *sb.* + *-unculus*, diminutive ending.] = *spleniculus.
[**1848**: see *splendiculus.] **1897** [see Lienculus]. **1909** Adami & Nicholls *Princ. Path.* II. x. 222 Nearly 400 of these splenunculi were found scattered throughout the abdominal cavity. **1962** *Lancet* 26 May 1104/1 After the removal of the splenunculus there was remission. **1974** W. A. Sodeman *Pathologic Physiol.* (ed. 5) xxiii. 680/2 Such a hemolytic relapse has been documented..in a splenectomized patient with hereditary spherocytosis; at surgery a splenunculus was found to weigh 217 grams.

splib. *U.S. Blacks.* [Origin obscure.] A Negro.
1964 *N.Y. Times Mag.* 23 Aug. 62/2 *Soul brother*, Negro; also referred to as *scobe, blue, splib.* **1969** A. Young *Dancing* 5 Nobody want no nice nigger no more... They want an angry splib A furious nigrah.. **1970** *Atlantic Monthly* Jan. 38 Any other terms such as 'boy', 'spook', 'splib',..or 'colored' carry connotations of prejudice. **1976** *Amer. Speech* 1974 XLIX. 184 *Splib*,..liberal black who looks angry but will not upset the status quo.

splice, *sb.* Add: **1. b.** Also, a joining of metal girders or rails, concrete beams, etc. (Examples.)
1877 *Rep. & Awards Group XVIII U.S. Centennial Comm. Internat. Exhib. 1876* 66 (*caption*) Rails and splices used on Pennsylvania Railroad and branches. **1892** L. de C. Berg *Safe Building* II. x. 113 In locating the rivets of a splice care should be taken not to weaken the original plate. **1934** Relf & Johansen *Handbk. Aerodynamics* (ed. 2) I. ix. 638 In splicing built-up or box spars the splices in the flanges should come at the points of inflection. **1951** R. D. Chellis *Pile Foundations* xi. 271 If piles are restrained throughout their full length in firm soil..a splice consisting of a full butt weld or single web and flange plates welded on in the field suffices. **1977** J. P. Cook *Composite Construction Methods* x. 209 In the usual steel frame, column sizes are normally kept constant for a height of two stories, with the column splice about 3 or 4 ft above the floor level.

c. *Cricket.* The v-shaped tang of a bat-handle, which forms a joint with the blade; the joint itself. Also in slang phr. *to sit* (*on* or *upon*) *the splice*, to play a cautious defensive game.
1906 A. E. Knight *Compl. Cricketer* 44 The spliced handle was a later development, and some of the earliest splices were of ash. *Ibid.* 352 To sit upon the splice.—To play with too much caution, to deliberately refrain from scoring. **1912** *Daily Mirror* 9 July 14/2 Vidler sat icily on the splice, playing the right game and keeping up an end while Crutchley got runs. **1927** *Punch* 26 Jan. 108/1 'I don't think you're helping the score at all,' I protested. 'You're just sitting on your splice and leaving it all to me.' **1935** *Encycl. Sports* 195/1 Cricketers who bought a new bat treated it almost with reverence... They oiled every part of it, except the handle and splice. **1963** A. Ross *Australia 63* iii. 79 Dexter, hooking at him, nearly returned a gentle catch off the splice. **1968** L. Frewin *Best of Cricket's Fiction* II. 267 There was a business-like look about him, the air of one who without being the least downhearted or inclined to sit upon the splice, was yet determined to take no foolish risks.

d. A joint made in editing or repairing film or magnetic or paper tape.
1923 F. A. Talbot *Moving Pictures* (ed. 3) viii. 100 In making the splice care has to be observed to introduce the two pieces true to frame-line, gauge and marginal perforations. **1931, 1936** [see *bloop *v.*]. **1949** Frayne & Wolfe *Elem. Sound Recording* xxix. 603 A program can be edited in advance by cutting out portions or inserting

other portions, since splices in the tape cannot be heard. **1973** *Screen* Spring/Summer 43 Montage in the narrow sense (ie as an editing splice) has a diminishing importance in certain modern films.

splice, *v.* Add: **I. 2. a.** Also, to fasten together (metal girders and rails, concrete beams, etc.).
1892 L. de C. Berg *Safe Building* II. x. 113 If any part of a girder..is spliced, made of two parts, the number of rivets each side of splice..should be made sufficient to transfer the full strength of original plate across the joint. **1913** W. H. Sellew *Steel Rails* iv. 263 The stiffness of the rail that is to be spliced. **1951** R. D. Chellis *Pile Foundations* ix. 201 (*caption*) Exposing reinforcing preparatory to splicing a large precast concrete pile. **1976** R. Chudley *Construction Technol.* III. vii. 75 Where members of the frame are joined or spliced together the connections are generally mechanical (nut and bolt).

c. Also *spec.* in *Biol.*, to join or insert (a gene or gene fragment).
1975 *Nature* 18 Dec. 563/1 The genes to be cloned would first be spliced on to either a bacterial plasmid.. or on to the DNA of bacteriophage lambda which would then infect the bacterium. **1977** *Sci. News* 29 Jan. 70 The controversial research in question is a class of experiments that..include splicing the genes of a virus or bacteria to partially purified DNA from mammals or birds..known to produce potent toxins or pathogens. **1979** *Newsweek* 4 June 64 One valuable product has already resulted from the work: human insulin, manufactured by splicing fragments of DNA that manufacture the hormone in humans into an intestinal bacterium.

e. To make a splice or joint in (a length of film or magnetic or paper tape); to join (film or tape) *in, on* or *up*. Occas. *intr.* (or *absol.*).
1912 F. A. Talbot *Moving Pictures* xii. 137 Occasionally when a film is being run through the projector it becomes severed by some means or other. Before it can be used again the break must be repaired by splicing the two parts together. **1931** Wilkinson & Reis in L. Cowan *Recording Sound for Motion Pictures* xiv. 200 When film is spliced on a standard splicing machine, the splice crosses the sound track at right angles to its length. **1958** W. E. Stewart *Magn. Recording Techn.* ii. 42 The wire, still useful in some applications, cannot be spliced so easily as tape. **1962** L. Deighton *Ipcress File* xxiv. 156 The film lab had been very thorough, they had spliced on to the end of the film the incident of my arrest. **1973** A. Broinowski *Take One Ambassador* x. 140 The odd *faux pas*..would have to be cut; splice in a bedroom scene there. **1974** N. Freeling *Dressing of Diamond* 174 Snip a bit off this [tape] and splice it up. **1978** L. Davidson *Chelsea Murders* xxiii. 141 He put in six solid hours at the editing... He compared and cut and spliced till two in the morning. **1978** [see *splicer 2 b]. **1980** S. Hockey *Guide Computer Applic. Humanities* ii. 25 Sections of corrected [paper] tape can be spliced or glued into the original.

f. *Cricket.* To strike (the ball) with the splice of the bat, as a mishit.
1982 *Guardian* 19 Feb. 22/2 Botham went for a swinging pull shot, and spliced it tamely but safely to mid-wicket.

3. b. *intr.* To get married. Also const. *with*.
1874 E. Eggleston *Circuit Rider* xxiii. 216, I heerd say as he was goin' to splice with a gal that could pray like a angel afire. **1875** J. G. Holland *Sevenoaks* xii. 155, Jim, be ye goin' to splice? **1981** T. Heald *Murder at Moose Jaw* xii. 114 If the old flapper spliced with the colonel she stood to lose a million dollars.

spliced, *ppl. a.* (Earlier and later examples.) Also, reinforced by splicing.
1859 T. P. Shaffner *Telegraph Man.* xli. 597 Fig. 13 is the two ends spliced, having first been cleaned. **1867** G. H. Selkirk *Guide to Cricket Ground* iii. 44 A new handle can be inserted..and the 'spliced bat' will be quite as good as before—indeed, many players have their bats spliced at first, thinking it a great improvement. **1891** W. G. Grace *Cricket* ii. 42 This one [*sc.* a bat] had a spliced handle with a strip of whalebone down the centre of it, and was very much prized. **1897** *Sears, Roebuck & Co. Catal.* 243/3 Men's Seamless Cotton Half Hose... Spliced heels and toes. **1931** Wilkinson & Reis in L. Cowan *Recording Sound for Motion Pictures* xiv. 201 A similar section of silent track, matching the average density of the spliced tracks, is..cemented into the hole. **1968** J. Ironside *Fashion Alphabet* 69 *Spliced heel*, heel reinforced with the same fabric as the stocking. **1970** E.-O. Libuda tr. *Heinhold's Power Cables & Application* xlvi. 457 Brazing is necessary for spliced connections [in copper conductors].

splicer. Add: **2. a.** A tool or implement used in splicing.
1923 J. H. Cook *Fifty Years on Old Frontier* 114 A wire cutter, splicer, and staple-puller combined.

b. A mechanical device used to splice film or tape.
1927 E. G. Lutz *Motion-Picture Cameraman* xi. 243 (*caption*) B. & H. 16 mm. rewinder and splicer. **1931** Wilkinson & Reis in L. Cowan *Recording Sound for Motion Pictures* xiv. 202 A diagonal splicer has been evolved to produce a silent splice. **1953** K. Reisz *Technique Film Editing* 274 The two reels, one of sound track, one of picture, both provisionally held together by paper clips, are joined on a splicer. **1978** G. McDonald *Fletch's Fortune* xxx. 201 'I need one of those cassette tape recorders. You know, with a tape splicer. I need to splice some tape.' 'Mine doesn't have a splicer.'

splicing, *vbl. sb.* Add: **I. 1. a.** (Later examples.)
1912 F. A. Talbot *Moving Pictures* xii. 138 To facilitate..splicing a small clamping device is used. **1931** L. E. Clark in L. Cowan *Recording Sound for Motion Pictures* ix. 144 In splicing a certain amount is lost at each end. **1973** *Screen* Spring/Summer 42 The montage in question is not..necessarily montage in the narrow sense (ie splicing). **1980** *Daily Tel.* 18 Jan. 14/6 The scientists.. have modified common laboratory bacteria by 'gene splicing' to produce human interferon.

b. *splicing-block, machine, -tape.*
1927 E. G. Lutz *Motion-Picture Cameraman* xi. 241 Joining films is made with the aid of a splicing-block which has as its main feature series of teeth over which the film-perforations fit. **1931** Splicing machine [see *splice *v.* 2 e]. **1958** J. Tall *Techniques of Magnetic Recording* x. 161 The splicing tape should be laid down.. at an angle. **1974** H. Burstein *Questions & Answers about Tape Recording* xviii. 246 The leading edge of the splicing tape presents a resistance.

spliff (splif). *slang* (orig. *W. Indies*). Also splif. [Origin unknown.] A cannabis cigarette, *spec.* one rolled in a conical form; a smoke of cannabis.
1936 *Daily Gleaner* (Kingston, Jamaica) 3 Oct. 35 Here is the hot-bed of ganja smoking..and even the children may be seen at times taking what is better known as their 'spliff'. **1953** R. Mais *Hills were Joyful Together* i. xi. 111 He took the spliff and lit up, dragged long at it, drawing the smoke deep down into his lungs. **1959** C. MacInnes *Absolute Beginners* 76 A third just said, 'Great', with a soft dream in his eyes—but that may have been because he'd just been dragging on a splif inside the toilet. **1969** Fabian & Byrne *Groupie* (1970) i. 9 He showed me how to roll spliffs—as he called them—so that I could roll for him. **1972** J. Brown *Chancer* ii. 30 They might be going ..to meet a pusher or they might be going just for a few spliffs and to catch the party feeling. **1975** *High Times* Dec. 137/1 Like Marley, he's a spliff-toking Rastafarian. **1977** *Transatlantic Rev.* lx. 192 Coon rolled a massive splif and blew clouds of ganja gremlins through the hatch.

spline, *sb.* Add: **1. b.** (Examples.)
1904 *Drapers' Co. Res. Memoirs* XIII. 12 The curves.. were plotted with our coordinatograph for a series of values of *k* or *r* on a large scale, drawn in with a spline and integrated with a Coradi compensating planimeter. **1953** A. H. Robinson *Elements Cartogr.* v. 83/1 For larger curves, the defining points of which are far apart, a flexible curve or a spline with weights is more satisfactory.

2. Now *esp.* such a key that is formed integrally with the shaft; also, a corresponding recess in a hub along which the key may slide.
1909 Kimball & Barr *Elements Machine Design* viii. 196 Sometimes it is desirable to have the hub free to slide axially along the shaft, but constrained to rotate with it. In such cases a feather or spline is used. **1932** R. C. H. Heck in C. E. O'Rourke *Gen. Engin. Handbk.* xviii. 532 Often a gear has to slide along its shaft. The key in this service is called a spline or feather. **1952** F. J. Camm *Newnes Engineer's Ref. Bk.* (ed. 5) 854 The efficiency of a spline for driving purposes is measured by the amount of contact made by the male and female splines. **1966** G. W. Michalec *Precision Gearing* vii. 324 Generally, involute internal and external teeth are mated, but non-involute splines are also suitable. **1979** *Industrial Fasteners Handbk.* i. 318 There are two basic forms of spline—straight-sided splines which may number 4, 6, 10 or up to 16 splines equally distributed around the circumference of a shaft, and serrated splines which are in the form of adjacent triangular teeth.

3. *Math.* Also *spline curve*. A continuous curve constructed so as to pass through a given set of points and have continuous first and second derivatives.
1946 I. J. Schoenberg in *Q. Appl. Math.* IV. 48 For $k = 4$ they represent approximately the curves drawn by means of a spline and for this reason we propose to call them spline curves of order *k*. **1966** *Notices Amer. Math. Soc.* XIII. 140 This paper extends and strengthens convergence properties previously published..for periodic splines and for nonperiodic splines satisfying general end conditions. **1978** *Nature* 1 June 407/2 Cubic spline interpolation was applied at standard depths to 800 m.

spline, *v.* Add: (Examples.) Also, to secure (a part) by means of a spline. **splining** *vbl. sb.* (examples).
1901 *Shop & Foundry Practice* II. § 15. 14 Fig. 13 shows a jig designed for holding shafts for key-seating or splining, plain cutters being used for the purpose. *Ibid.* 15 The shafts..which are to be splined or key-seated, are laid into these grooves and are clamped. **1920** *Autocar Handbk.* (ed. 10) x. 125 In order to allow the gear wheels to drive the shaft on which they are placed, or vice versa, the latter is often grooved or 'splined'. **1926** *Motor Man.* (ed. 26) v. 82 The worm-wheel spindle emerges from the off side of the steering box, and splined thereto is a short lever, generally known as the drop arm. **1953** *Straight-Sided Splines & Serrations* (*B.S.I.*) 5/2 If two shafts are splined to different fits by the same cutter.. the radius of the arc at the root..will vary slightly from the designed arc.
Hence **splined** *ppl. a.*, provided with a spline or splines.
1909 N. Hawkins *Mech. Dict.* 529/2 *Splined shaft*, a shaft provided with a long feather way; as a splined feed rod on a lathe. **1952** P. S. Houghton *Gears* ix. 156 The length of bearing in an involute splined hub will depend upon the chosen materials. **1967** [see *knock-off *a.* 2 b]. **1979** *Industrial Fasteners Handbk.* i. 318 British Standard 3550:1963 specifies dimensions of involuted splined shafts and splined holes with a 30° pressure angle.

splint, *sb.* Add: **3. b.** *S. Afr.* A fragment or broken piece of diamond.
1872 C. Rhodes *Let.* in B. Williams *Cecil Rhodes* (1921) iv. 29 You must not however think that every diamond one finds is a beauty, the great proportion are nothing but splints. **1887** J. W. Matthews *Incwadi Yami* xxvii. 415 Faithfully carrying out their master's behests, and never robbing him of a single splint. **1903** W. R. Cattelle *Precious Stones* 79 Beyond the small pieces resulting from cleavages, other fragments are saved which cannot be cut to jewels. Some of these are called 'splints', and are used for mechanical purposes or ground to powder.

4. a. Also, an object used to fasten or immobilize teeth or the jaws.
1948 *Brit. Dental Jrnl.* LXXXV. 223/1 Removable acrylic bite blocks or splints with clasps were constructed.., so that the bite was opened..and the mandible moved forward. **1962** Blake & Trott *Periodontology* xiv. 144 The main object of periodontal splint design is to make use of sound teeth to give stability to mobile teeth.

9. b. (sense 2) *splint-bottom; splint-bottomed* adj.
1850 *Knickerbocker* XXXVI. 73 She wiped out the seats of some splint-bottomed chairs with her calico apron. **1876** 'Mark Twain' *Tom Sawyer* vi. 68 The master, throned on high in his great splint-bottom armchair, was dozing. **1919** T. K. Holmes *Man from Tall Timber* iv. 36 A comfortable armchair with splint-bottom. *Ibid.* v. 46 Their splint-bottomed armchairs.

splintage. Delete *rare*$^{-0}$ and add examples. For 'surgical' read 'surgical or dental'.
1956 H. & A. J. Kazis *Complete Mouth Rehabilitation* vii. 88 As a result of splintage there is a reduction of the stress load on the supporting bone. **1970** R. D. Muckart in G. Murdoch *Prosthetic & Orthotic Pract.* xi. 471 Maintenance of mobility and the prevention of deformity or contracture is by far the most important function of upper-limb splintage. **1978** *Jrnl. R. Soc. Med.* LXXI. 186 The wrist fixation device is designed to control Colles' fracture by internal splintage.

splinter, *sb.* Add: **1. f.** A splinter group (see sense 7 b below). orig. and chiefly *U.S.*
1948 *Sun* (Baltimore) 20 Aug. 1/2 The Republican party and its Dewey-Warren ticket, without 'leftist or extreme right splinters', is the nation's only hope 'to put an end to disunity'. **1972** D. E. Westlake *Bank Shot* viii. 56 Probably a new splinter... They keep fractionalizing, makes it extremely difficult to keep proper surveillance. **1977** *New Yorker* 9 May 67/2 The old-guard splinter of the Congress which Mrs. Gandhi had routed in 1969. **1981** *Listener* 1 Jan. 24/1 A newly imaginative use of a Red Brigade splinter.

7. a. splinter bid *Bridge*, an unusual jump bid showing a singleton or void in the suit bid; **splinter-deck,** an armour-plated deck on a ship (see also quot. 1909); **splinter hæmorrhage,** a narrow, elongated hæmorrhage resembling one produced by a splinter.
1977 *Oxf. Times* 11 Feb. 8/7 The bidding went: One Heart—pass—Three Spades (splinter bid showing a singleton or void together with a heart fit). **1978** *N.Y. Times* 29 Mar. c 25/2 For slam purposes, the splinter bid, or unusual jump to show a singleton or a void, solves many problems. **1909** *Cent. Dict. Suppl.* 346/2 A deck worked for protective purposes below a protective deck is called the splinter-deck. **1933** *Jane's Fighting Ships* 171 Above again is a 1$\frac{1}{16}$" splinter deck against aerial attack. **1973** J. Quick *Dict. Weapons & Mil. Terms* 416/3 *Splinter deck*, a deck fitted with armor. **1931** W. Boyd *Path. Internal Dis.* i. 40 There may be small 'splinter hemorrhages' under the nail—a linear track as if a sliver had been run in. **1971** Robbins & Angell *Basic Path.* ix. 274/2 Seeding of the nail beds and of the skin produces small petechial hemorrhages known as 'splinter hemorrhages' or microabscesses.

b. *attrib.* or as *adj.* Of or pertaining to a group, party, etc., which splits itself off as an independent entity from a larger political or social group, esp. as *splinter group*. orig. *N. Amer.*
1935 *Economist* 19 Oct. 742/2 The new 'splinter parties', it will be observed—the Co-operative Commonwealth Federation (farmer-labour), Mr. H. H. Stevens's 'Reconstructionists', and the Social Credit League—have hardly succeeded in making a scratch on the traditional surface of Canadian politics. **1948** *Manch. Guardian Weekly* 15 Jan. 5/2 He challenges the Republicans to dodge the stigma of reaction and encourages the Wallace 'splinter groups' to hold firm. **1948** *Sun* (Baltimore) 23 Feb. 8/3 While splinter minorities may have a voice in the legislature, they cannot extend that voice beyond their own minority base. **1950** *Times* 27 Feb. 5/3 The Cabinet is made up of an uneasy coalition of splinter parties. **1958** *Spectator* 17 Jan. 73/2 The formation of superior-minded splinter groups which have no wish to become part of the main body of the Church. **1898** Middleton & Tait *Tribes without Rulers* 209 Splinter-segments of a clan do not form cores of tribes. **1964** R. Braddon *Year Angry Rabbit* v. 43 The fourteen new splinter nations now beginning to flake off the edges of a ripely rich Soviet Russia. **1968** *Guardian* 17 June 8/1 The ineffective splinter-group politics of the Fourth Republic. **1975** *N.Y. Times* 8 Nov. 26/2 The women's movement has increasingly allowed itself to be dominated..by radical splinter groups and issues which lack support among the majority of women. **1978** L. Heren *Growing up on The Times* ix. 293 Malcolm X formed a splinter movement, the organization of Afro-American unity. **1979** D. Sanders *Queen sends for Mrs Chadwick* 82 Supposing..the two main parties are deadlocked... They would have to turn to whatever splinter party happened to be closest to their own line.

splinter, *v.* Add: **3. c.** *fig.* To break off to

form a splinter group or groups; *loosely*, to divide or split. Also with *off*.
1967 M. L. King *Trumpet of Conscience* iii. 49 Under the impact of social forces unique to their times, young people have splintered into three principal groups, though of course there is some overlap among the three. **1972** *Guardian* 11 Jan. 9/1 Later Frank Ashbourn joined them ..and in May 1970 he and Mersh splintered off to form South Sea Bubble. **1976** *Oxford Diocesan Mag.* July 11/1 But the village's young people, distressed at seeing the parishioners splinter off to other towns for church, asked to hold the new prayer assembly in place of Mass.

splintering, *vbl. sb.* Add: Also *fig.*
1958 *Listener* 11 Sept. 366/2 He [*sc.* Ben-Gurion] is.. haunted by a well-nigh perverse splintering of political parties at home. **1968** P. Oliver *Screening Blues* ii. 62 The distinctions in the names [of Negro Churches] were indicative of the splintering of the churches into separate factions.

splinterless (spli·ntəɹlės), *a.* [f. Splinter *v.* + -less.] Of glass: guaranteed not to break into splinters; = *splinter-proof *a.* 2 (now the more usual term).
1928 *Sunday Express* 15 Jan. 6 The manufacture of splinterless glass. **1928** *Daily Express* 19 June 3 A splinterless shock-proof safety glass.

splinter-proof, *sb.* and *a.* Restrict *Mil.* to senses in Dict. and add: **B.** *adj.* **2.** That does not break into splinters; designed so as not to splinter.
1941 *AWA Techn. Rev.* V. 214 The control panel is mounted..below a re-entrant window of splinter-proof glass. **1964** *Economist* 19 Sept. 1124/3 A perfect mirror. It is unbreakable, splinterproof, lightweight.

splinting, *vbl. sb.* **2.** For 'surgical' read 'surgical or dental', and add later example.
1960 W. L. McCracken *Partial Denture Construction* xiv. 270 Splinting should not be used for the purpose of retaining a tooth that would otherwise be condemned for periodontal reasons.

split, *sb.*¹ Add: **2. d.** *Canad.* (chiefly *Newfoundland*). A piece of kindling-wood. Usu. in *pl.*
1858 R. T. S. Lowell *New Priest in Conception Bay* I. 74 The fire, where the round bake-pot stood, covered with its blazing 'splits'. **1919** W. T. Grenfell *Labrador Days* 198 'Get a few more splits, then, boy,' she replied, 'and I'll be cutting t' pork t' while.' **1976** Taylor & Horwood *Beyond Road* 55 Well, one time I was only a small boy gettin' in the splits—that's kindling.

e. *Anglo-Irish.* A piece of bogwood burned for illumination.
1892 *Ballymena Observer* 29 Apr. 6/1 *Splits*, long thin pieces of bogwood used for giving light. **1957** E. E. Evans *Irish Folk Ways* xiv. 185 Considerable use was made of buried timber dug from the bogs, of oak for roofing beams and..resinous 'splits' to give light.

4. d. *U.S.* = *split-up* s.v. *split-.
1972 *N.Y. Law Jrnl.* 10 Oct. 3/2 Tacking is permitted for stock dividends and splits, recapitalizations, [etc.]. **1976** [see *split-down* s.v. *split-].

6. Also, a policeman.
1932 'G. Orwell' *Coll. Essays* (1968) I. 89 He would.. exclaim 'Fucking toe-rag!'..meaning the 'split' who had arrested him. **1935** G. Ingram *Cockney Cavalcade* xiii. 202 'Here's the 'splits', boys!' A young lad who had been at the entrance with some others, had seen a police-car draw up and risked his liberty by dashing in to warn the hall occupants. **1966** W. Merrilees *Short Arm of Law* 140 At this point a destination board attendant asked another railway employee what the splits were after.

7. b. Also *spec.*, a half-bottle of champagne.
1973 T. Pynchon *Gravity's Rainbow* i. 5 All that's keeping him up there is an empty champagne split in his hip pocket, that's got hooked somehow. **1980** *N.Y. Times* 6 Nov. c2/3 To uncork a split of Champagne, some of which froths to the ground.

e. A sweet dish consisting of sliced fruit (esp. banana, split open lengthways), with ice-cream, syrup, etc. orig. *U.S.*
1920, etc. [see *banana split* s.v. *banana 3]. **1936** [see *parfait]. **1938** G. Greene *Brighton Rock* i. i. 17 That's what I want, a sundae. Delia likes splits best. **1939** A. Huxley *After Many a Summer* i. x. 135 Virginia was at the soda-counter, pensively eating a chocolate-and-banana split. **1979** M. Denny *Fruit in Season* 33 Banana splits... Place one banana per person in a dish with a portion of ice-cream in the centre... Pour a little chocolate sauce over.

f. *N. Amer.* A split-level house.
1970 *Toronto Daily Star* 24 Sept. 28/7 Back splits, side splits, bungalows. **1976** *Billings* (Montana) *Gaz.* 6 July 5-d/7 (Advt.), This gorgeously decorated 4 level split. **1980** *Times* 7 Apr. 5/6 French-speakers [in Montreal] would buy 'side halls, split levels, back splits'.

g. A split shift (see *split *ppl. a.* 3 a).
1973 R. Busby *Pattern of Violence* iii. 41 I'm working the split today. Get that boss of yours to give you a couple of hours off. **1977** P. Carter *Under Goliath* xxvi. 145 She..went moaning on... They were still at it at nine o'clock when Mr Black came back from his split.

8. *slang.* **a.** A division or share of the proceeds of a legal or illegal undertaking.
1889 Clarkson & Richardson *Police!* xxiii. 321 A share... Regular, split, drop. **1916** *Variety* 27 Oct. 12/1 W. S. Campbell..would not accept the 55–45 division of the receipts offered by the management, Campbell wanting a 50–50 split. **1934** J. T. Farrell *Young Manhood of Studs Lonigan* xiii. 206, I wasn't working for a long time, and then I got me this job, and now I'm also lined up with a can-house, and get my split on anybody I bring there. **1964** J. P. Clark *Three Plays* 121 Both thieves Will certainly be content to settle For an even split. **1973** J. Leasor *Host of Extras* i. 24 'I'll give you five thousand cash, the pair.' I must know *someone* who could advance this on the promise of a fifty per cent split down the middle of the selling price?

b. *N. Amer.* A girl, a woman.
1935 A. J. Pollock *Underworld Speaks* 111/2 *Split*, a girl. **1975** *Globe & Mail* (Toronto) 16 Dec. 9/5 An announcement was posted that the force's first female officer Constable Jacqueline Hall, had been hired. 'He's gone and hired another split, as if we don't have enough whores and splits in the department already,' Mrs. Nesbitt quoted the sergeant as saying.

9. *Croquet.* (See quot. 1961.)
1896 *Cassell's Bk. Sports & Pastimes* 305 The Split is a stroke used when you desire in taking croquet to move both balls some distance. **1961** *Croquet* ('Know the Game' Ser.) 36/1 *Split*, a croquet stroke in which the balls go in different directions.

10. *U.S. Sports.* A draw; a drawn series of matches.
1967 [see *double-header c]. **1974** *Cleveland* (Ohio) *Plain Dealer* 13 Oct. c1/1 The loss evened the C's exhibition slate to 2–2 and gave them a split in the two-game series with the Toros. **1976** *Springfield* (Mass.) *Daily News* 22 Apr. 40/2 With the VL getting only a split in six battles.

split, *v.* Add: **I. 5. e.** *to split one's* (or *the*) *ticket* or *ballot*: to vote for candidates of more than one party in an election. Also *ellipt.* *U.S.*
1842 *Spirit of Times* (Philadelphia) 14 July 2/1 The cry is raised of 'Vote the whole ticket! Don't split your ticket!' **1848** J. J. Hooper *Widow Rugby's Husb.* (1851) 23 Never split in my life. **1905** *N.Y. Even. Post* 17 Oct. 1 Plenty of talk is heard about intentions to split ballots. **1946** *Chicago Daily News* 20 Nov. 18/5 Democrats..decided the country did need a change, and split their ticket. **1975** R. Stout *Family Affair* (1976) xiii. 141 He asked if I had split the ticket, and I said yes, I had voted for Carey but not for Clark. **1980** *Times* 8 Oct. 8/4 To persuade electors to 'split the ticket'—to vote for a Republican President and for a Democratic senator.

f. *to split the atom*, to cause atomic nuclei to undergo fission. Also *fig.*
1909 *Busy Man's Mag.* Oct. 44/2 He [*sc.* Professor J. J. Thomson] is known both as 'The Man of Ion', and as the man 'who split the atom'. **1930** Sayers & 'Eustace' *Documents in Case* ii. 262 If anyone goes quietly away into a corner to experiment with high-voltage electric currents, they start a lot of ill-informed rubbish about splitting the atom. **1932** *Discovery* Mar. 69/2 The problem of splitting the atom is briefly this: given..that at the centre of every atom there is a minute nucleus whose electrical charge fixes the elementary nature of the atom, can we by any agency detach a part of this charge? **1935** J. Guthrie *Little Country* xxi. 335 With the blast of his cornet, Archibald Packer had split the Temmian atom. **1964** M. Gowing *Britain & Atomic Energy 1939–1945* 18 They bombarded a foil of the metal lithium, disrupting the lithium nuclei which, after combining with incident protons, split into two alpha particles. The experimenters had 'split' atoms by artificial means. **1981** *Daily Tel.* 24 Sept. 16/4 The first scientists to work on 'splitting the atom'.

6*. *slang* (orig. *U.S.*). To depart from, to leave. Freq. in phr. *to split the scene*: cf. *scene 8 e.
1956 O. Duke *Sideman* iii. vii. 272 Naw, man—I split that scene. **1963** *Freedomways* III. 522 Evil Indians sink feathered arrows into the good guys, who kicked a couple of times and then split the scene. **1968** Busby & Holtham *Main Line Kill* vi. 66 Where you bin? We thought you split the scene without giving us the word. **1971** *Sunday Sun* (Brisbane) 26 Sept. 3/3 When he split the Brisbane scene he left behind documents that could be incriminating to the drug gangsters. **1973** *Black Panther* 27 Oct. 17/2 We'll be splitting this place soon and once the book is written we won't have to come back. **1978** S. Wilson *Dealer's Move* i. 12 He and Miranda split Scotland for good and came down to London.

II. 10. c. *U.S. Sports.* To draw, to tie; *spec.* in *Baseball*, to win one game of a double-header, or to win half of the games in a series. Also *trans.*
1975 *Cleveland* (Ohio) *Plain Dealer* 31 Mar. 1-d/5 If Houston loses both of its remaining games and the Cavs split, the Cavs..have a better record against other division teams than Houston does. **1979** *Tucson* (Arizona) *Citizen* 20 Sept. 10d/1 He split two decisions this season in hookups with Gaylord Perry.

11. c. *slang* (orig. and chiefly *U.S.*). Of a couple: to become divorced; to separate.
1942 Berrey & Van den Bark *Amer. Thes. Slang* § 360/2 *Divorce*...split. **1951** E. Coxhead *One Green Bottle* x. 267 'Why did Chris go off early? Is anything wrong?' 'We've split,' Cathy answered. **1976** *National Observer* (U.S.) 14 Aug. 1/4 They had to split. If they don't love each other, what else can they do? **1978** *Detroit Free Press* 2 Apr. 19a/2 The [divorce] suit ended months of speculation that the TV sportscaster and film producer were splitting. **1982** 'J. Gash' *Firefly Gadroon* i. 13 Women are always unreasonable... We split after a terrible fight.

13*. *slang* (orig. *U.S.*). To depart, to take one's leave.
1954 *Time* 8 Nov. 42 *Split*,..depart. **1956** O. Duke *Sideman* iii. ix. 294 But that's why the cat split. **1956** B. Holiday *Lady sings Blues* (1973) iii. 38, I grabbed him and told him to do something because I had to split for the bathroom again. **1962** *Radio Times* 17 May 43 After the gig, dad, let's split to your pad for some suds. **1967** W. Murray *Sweet Ride* viii. 128 Since nobody asked you over, why don't you just split so we can finish our lunch? **1977** *Sounds* 1 Jan. 21/4 In the main hall Roger Scott from London's Capital Radio arrived, took one look at the wasteland and split.

III. 14. a. With *out*: also *slang* (now *Obs.* or *rare*), to separate or disentangle from another.
1924 G. C. Henderson *Keys to Crookdom* 419 *Splitting out*, separating pickpocket from his victim in case of trouble. The stall splits out the wire. **1931** *Collier's* 16 May 66/2 Everybody else is busy trying to split out Regret and the bloodhounds.

b. With *out*: also *slang*, to quarrel; to part company; to take one's leave (cf. sense 13* above). With *up*: also *colloq.*, to break up a relationship (esp. of a couple); *spec.* to become divorced.
1850 G. W. Matsell *Vocabulum* 84 *Split out*, no longer friends; quarrelled; dissolved partnership. **1879** *Macmillan's Mag.* Oct. 505/2 There is a reeler over there which knows me, we had better split out. **1903** G. H. Lorimer *Lett. from Self-Made Merchant to his Son* viii. 104 He and his father split up, temporarily, over it, and, of course, it cost me the old man's trade and friendship. **1927** J. Black *You can't Win* x. 132 'Where are you going, kid?'..'If you are going to split out, I'll go to San Francisco for a while.' **1942** Berrey & Van den Bark *Amer. Thes. Slang* § 360/2 *Divorce*...split up. **1956** B. Holiday *Lady sings Blues* (1973) xxii. 176 Even if I could have split out I'd have been in a snowstorm of lawsuits. **1959** 'E. Peters' *Death Mask* i. 15 When we split up..I felt it was all my fault. I had to be free of him. **1976** M. Machlin *Pipeline* xlvii. 491 You just split out like a streak of blue lightning, without saying nothing to nobody. **1976** W. Corlett *Dark Side of Moon* i. i. 29 'He thought his parents were..splitting up?' 'Divorce?.. he thought it was on the cards.'

split-. Add: **split-down** *U.S. Stock Exchange* (see quot. 1976); cf. *split-up* below; **split-off,** an act of splitting off; something that splits off or that has become split off; **split-up,** an act of splitting up; *spec.* in the U.S. Stock Exchange: the division of a stock into two or more stocks of the same total value; cf. *split-down* above.
1932 *Sun* (Baltimore) 16 Apr. 15/8 The whys and wherefores of the 'split-down' movement in the capital structures of various corporations, in contrast with the stock 'split-ups' popular in the boom days of 1928–29 are now being explained in Wall Street. **1976** D. W. Moffat *Econ. Dict.* 257/2 The reverse split, or split-down in which a corporation reduces the number of shares into which its ownership is divided. The single word split usually refers to a split-up. **1935** Z. N. Hurston *Mules & Men* i. ii. 40 Ah knowed one preacher dat was called to preach at one of dese split-off churches. De members had done split off from a big church... He come and preached at dis split-off for two whole weeks. **1964** *New Statesman* 14 Feb. 248/1 The split-off of science into a separate culture. **1878** H. Sweet tr. H. Paul in *Trans. Philol. Soc.* (1879) 390 Even in the parent Indogermanic language long before its split-up, there were no longer any roots, stems and suffixes, but only ready-made *words*. **1908** G. H. Lorimer *Jack Spurlock* iv. 76, I should have told her then about my split-up with the Governor. **1928** E. S. Mead *Corporation Finance* (ed. 6) I. xxx. 358 We may note finally the difference between a stock 'split-up' and a stock dividend. In the first case two or three shares are issued for one share of existing stock, and when this is par-value stock the par value is reduced. **1944** *Sun* (Baltimore) 9 Sept. 11/1 Pepsi-Cola added ⅞ to its brisk upturn of the previous day in response to the three-for-one splitup proposal. **1975** *High Times* Dec. 51/2 Last year I was still spinning from my split-up with a man. **1976** Split-up [see *split-down* above].

split, *ppl. a.* Add: **2. a.** *split-ring* (earlier example); *split bearing* (Mech.), a bearing for a shaft in which the housing and bush are each split laterally into two parts for ease of assembly; *split flap* (Aeronaut.), a flap occupying only the lower part of the wing thickness.
1902 R. Grimshaw *Mod. Workshop Hints* xiv. 268 (*heading*) Filling split bearings with babbitt. **1973** O. S. Nock *Gresley Pacifics* I. vii. 91/2 The inside big end.. necessarily had split bearings. **1929** *Techn. Notes U.S. Nat. Advisory Comm. Aeronaut.* No. 324. 1 It is known that..a suction exists between the parts of a split flap located at the trailing edge. **1968** Miller & Sawers *Technical Devel. Mod. Aviation* iii. 84 The adoption of retractable undercarriages, which increased drag when they were lowered for landing, made it less important to use flaps which increased drag as greatly as the split flap. **1853** C. M. Yonge *Heir of Redclyffe* II. xxi. 340 It was locked, but the key was on her own split-ring.

b. *split-face, skirt, stitch* (earlier example); *split baluster* (see quot. 1969); *split end* (Amer. and Canad. Football), an end (*end *sb.* 3 g) positioned at some distance from the rest of the formation; *split falls* (see quot. 1960); *split graft* (Med.) = *split-skin graft*, sense 5 a below; *split jump* (Figure-skating), a jump during which the legs are momentarily kicked out into the splits position; *split pea*, rhyming slang for 'tea' (*obs.*); cf. *Rosy Lee; *split shot, stroke* (Croquet) = *split *sb.*¹ 9; *split-turn*, a sharp turn.

1904 P. Macquoid *Hist. Eng. Furnit.* ix. 228 The split baluster ornament..has been variously named split baluser, cannon, or mace decoration. **1934** *Burlington Mag.* Sept. 125/1 An extensive use of relief decoration in the form of turned 'split balusters' is also rather characteristic of many of these pieces. **1969** J. Gloag *Short Dict. Furnit.* 635 *Split baluster*, a turned baluster split centrally, and applied ornamentally to a surface. **1955** C. V. Mather *Winning High School Football* vii. 187 (caption) The halfback splits half the distance with the split end. **1968** *Globe & Mail* (Toronto) 10 July 27/5 Adkins will be the split end with underrated Jay Roberts, a tough blocker, remaining at tight end. **1977** *New Yorker* 10 Oct. 177/2 Using only two backs..and sending four split ends..downfield, Restic had Harvard throw fifty-seven passes that afternoon, thirty-one of them valid. **1923** D. H. Lawrence *Birds, Beasts & Flowers* 182 And white teeth showing in your dragon-grin as you race, you split-face. **1939** *Country Life* 11 Feb. p. xxxiii/1 (Advt.), Made in Cavalry Twills..Sheppards Checks, Split falls or fly front. **1960** C. W. Cunnington et al. *Dict. Eng. Costume* 75/2 *Falls*, a buttoned flap to the front of breeches and..of pantaloons and trousers... 'Small' or 'Split Falls' was a narrow central flap. **1929** *Surg., Gynecol. & Obstetrics* XLIX. 96/2 In lining a contractile cavity with a split graft allowance should always be made for contraction. **1958** *New Biol.* XXVII. 40 Split-grafts are prepared by enzyme digestion of the fibres joining the epidermis to the dermis, which frees the epidermis for use as the graft. **1961** J. S. Salak *Dict. Amer. Sports* 416 *Split jump*,..a variation of the jump from the back edge with the free toe assisting. **1968** Split jump [see *lay-back 2 b]. **1857** 'Ducange Anglicus' *Vulgar Tongue* 20 *Split-Pea*, tea. **1931** S. Kaye-Smith *Hist. Susan Spray* iii. 296 I'll make you a nice cup of split pea. **1975** *Oxf. Compan. Sports & Games* 218/2 When the balls travel in different directions the stroke is also known as a split shot. **1976** *Scotsman* 20 Nov. (Weekend Suppl.) 4/1 The look of clothes today suggests country more than town... Capes and ponchos, loose knits and split skirts, are more at home on town birds than country cousins. **1880** L. Higgin *Handbk. Embroidery* iii. 22 *Split Stitch* is worked like ordinary 'stem', except that the needle is always brought up *through* the crewel or silk, which it splits. **1897** *Encycl. Sport* I. 254/1 *Split stroke*, taking croquet so as to drive the balls on courses nearly at right angles to one another. **1932** W. Faulkner *Sartoris* iii. 252 The damn thing zoomed past and did a split-turn and came back at me again.

c. Phr. *to keep on a split yarn* and varr.: to keep in a state of alert. *Naut. slang.*
1929 *Papers Mich. Acad. Sci., Arts & Lett.* X. 298/2 *Having everything on a split yarn*, ready to start at once. **1958** W. King *Stick & Stars* 73 All submariners had to be kept on a split yarn in case England was invaded.

3. a. *split beam*, a beam (of radiation, etc.) that has been split into two or more components, *spec.* as used in a radar technique in which a single aerial transmits alternately two beams slightly displaced from each other in order accurately to obtain the direction of a target; freq. *attrib.*; *split beaver* (slang) (see quots.); *split decision* (Boxing), a decision made on points in which the judges and referee are not unanimous in their choice of a winner; *split entrance, entry* adj. (N. Amer.), designating a house in which the entrance is half-way between the levels of the two floors; also *absol.* as *sb.*; cf. *split-level *a.*; *split-field* = next; usu. *attrib.*; *split-image*, (*a*) an image in a rangefinder or focusing system that has been bisected by optical means, the halves of which are displaced when the system is out of focus, used esp. in various types of camera; usu. *attrib.*; (*b*) = *splitting image* s.v. *splitting *ppl. a.* 5; *split instant, minute, moment*, an extremely small space of time; cf. *split-second *a.* and *sb.*; *split page* (U.S. Journalism) (see quot. 1970); *split-phase* (Electr.), used *attrib.* with reference to devices, esp. induction motors, that utilize two or more voltages at different phases produced from a single-phase supply; also *transf.*; *split rail* (orig. U.S.), a fence rail split from a log; freq. *attrib.*, as *split-rail fence*; *split run*, a press run of a newspaper in which some portions contain certain copy, advertisements, etc., not carried by other portions; *split screen* (Cinemat., Television, and Computing), a screen on which are projected simultaneously two or more images; *split shift*, (*a*) a working shift split into two or more periods separated by an interval or intervals of several hours; (*b*) a shift overlapping the times of two other shifts; *split ticket* (earlier and later examples); *split trial* (U.S. Law), a trial conducted in two stages of which the first will establish facts necessary to the impartial or swift conducting of the second; *split week*, (*a*) *Theatr.* (see quot. 1948); (*b*) a working week in which days off occur other than at the weekend.
1947 Crowther & Whiddington *Science at War* 25 This 'split-beam' method of direction-finding gives very accurate results. **1966** D. Taylor *Radar* ii. 24 Special stations..with facilities for 'split-beam' d.f. [*sc.* direction finding] were provided for accurate tracking of ships and fire-control purposes. **1966** [see *spectrophotometer]. **1978** R. V. Jones *Most Secret War* xlii. 397 The method was to set a Freya station on the coast of France so that its split-beam pointed over the target in London. **1972** *New Society* 7 Dec. 591/1 The business has evolved its own jargon; full frontals are 'beavers', becoming 'split-beavers' if the legs are parted. **1976** Lieberman & Rhodes *Compl. CB Handbk.* vi. 137 *Split beaver*, stripper. **1977** E. J. Trimmer et al. *Visual Dict. Sex* (1978) xxiv. 270 In the further stages of frankness 'beaver' and 'split beaver' shots show the hairy vulva. **1978** J. Irving *World according to Garp* xiii. 241 Pictures of naked women... If you could see the sex parts..that was a beaver... If the parts were *open*, that was called a *split* beaver. **1970** *Times* 28 Sept. 13/4 Buchanan, the British lightweight champion, gained a 15 round split points decision over Ismael Laguna of Panama. **1976** *Daily Times* (Lagos) 8 Oct. 30/3 The 29-year-old Panther..then said he had already petitioned the Nigerian Boxing Board of Control over the decision which gave Billy Savage the title by a split decision on September 24. **1968** *Globe & Mail* (Toronto) 13 Feb. 31/2 (Advt.), Beautiful split entrance bungalow. **1967** *Boston Sunday Globe* 23 Apr. b42/5 (Advt.), Keep that city job and enjoy country living in these unusually attractive split entry ranches. **1976** *Laurel* (Montana) *Outlook* 23 June 19/1 (Advt.), You will never regret buying this new 4 bdrm split entry. **1941** *Amateur Photographer's Handbk.* (ed. 2) vi. 121 Some people find this split-field type of range finder difficult to use. **1976** C. Reynolds *Photoguide to Filters* 156 One special accessory is the split-field lens. **1950** R. A. McCoy *Pract. Photogr.* ii. 15 To operate the split image type [of rangefinder] it is necessary to look through the finder and observe that the image seems to be broken in the center and offset. **1960** *Focal Encycl. Photogr.* (rev. ed.) 946/2 The parallax effect appears as a split image which joins up across a dividing line when the lens is set to maximum sharpness. **1977** J. Hedgecoe *Photographer's Handbk.* 15 As a focusing aid a 'split image' or focusing screen rangefinder may be sunk into the center of the underside of the screen. **1981** 'M. Innes' *Lord Mullion's Secret* 179 He was by a strange freak of heredity the split image of one commemorated by Nicholas Hilliard some centuries ago. **1936** M. Mitchell *Gone with Wind* xx. 348 'Rain,' she thought... But, in a split instant: 'Rain? No!... Cannon!' **1931** W. G. Carr *By Guess & by God* 27 Using his one periscope for split-minute looks. **1957** I. Asimov *Naked Sun* ii. 31 For one fleeting split moment he bent his head back and stared directly at Solaria's sun. **1953** B. Westley *News Editing* 419/1 *Split page*, same as 'second front page'. **1957** J. Steinbeck *Pippin IV* 58 Colour photographs filled the split-page of every newspaper. **1970** R. K. Kent *Lang. Journalism* 26 *Split page*, the front page of a newspaper's second section; second front page. **1895** S. P. Thompson *Polyphase Electric Currents* ix. 175 This is a form of split-phase motor having two or more sets of coils placed at different angles. **1921** W. S. Ibbetson *Motor & Dynamo Control* vi. 174 This split-phase winding has a very high resistance and induction, so that the current in it lays nearly 90° behind that in the running coils. **1953** *Pedestrian* Summer 26 Sometimes the policeman is operating what is known as a split phase; pedestrians may cross half the road in front of halted traffic and not realize that the traffic on the other half has the right of way. **1976** C. G. Grolle *Compl. Guide Electr. Repairs* viii. 120 All split-phase motors have a centrifugal switch that drops out the contacts on the start winding after full speed is attained. **1826** T. Flint *Recollections of Last Ten Years* 206 Scarcely has a family fixed itself, and enclosed a plantation with the universal fence,—split rails [etc.]. **1897** *Essex Antiq.* (Salem, Mass.) Feb. 27/2 The split-rail fence is also old. Logs, generally of ash, about nine feet in length, and a foot or more in diameter, split the entire length into about sixteen equal parts, formed the rails, which were chamfered at each end. Of such split sections posts were also made, having holes cut in them in the proper places to receive the ends of the rails. **1934** *Bulletin* (Sydney) 14 Feb. 30/2 A new post or a new set of rails has to be put in a split-rail fence. **1976** Scott & Koski *Walk-In* (1977) xxx. 216 A weathered split-rail fence..announced the boundary. **1961** Webster, Split run. **1963** D. Ogilvy *Confessions Advert. Man* (1964) vi. 110 In split-run tests, long copy invariably outsells short copy. **1977** D. Grossman *Samson Management Lexicon* ii. 19 Split-run copy testing. **1979** *Austral. Financial Rev.* 15 Aug. 22/6 The commission's investigations cover practices known in some sectors of the trade as 'split runs' and 'blowing'. A split run involves several print runs of the same editorial content, but with different advertising content. **1953** R. Bretz *Techniques Television Production* xi. 206 In the case of the phone conversation the split screen might appear in a direct cut after a single shot of the person making the call. **1958** *Times* 20 Jan. 3/2 Attempts to quicken the action [of a film] by a split screen device fail lamentably in their object. **1970** W. Wager *Sledgehammer* (1971) xv. 91 As if in some film.. Williston's neatly typed dossiers..jumped into focus... Actually they appeared side by side in a split-screen effect, hung there for a long moment and vanished. **1977** *Time* 26 Sept. 42/2 Alvy and Annie, on a split screen, talking to their shrinks about the frequency with which they have sex. **1955** M. Reifer *Dict. New Words* 196/1 *Split shift*,..a work schedule or shift in which there is a break in the working hours. **1960** *Guardian* 30 June 10/4 Split shifts (e.g. 4–9 a.m. and 5–8 p.m.) and split days off. **1964** G. L. Cohen *What's Wrong with Hospitals?* ii. 39 Wards operating a three-shift system, instead of the generally abhorred 'split shift' which gave nurses a useless afternoon break. **1970** F. McKenna *Gloss. Railwaymen's Talk* p. v, The footplate crew has an even worse cycle—what is called the 'split shift' system. **1978** *Detroit Free Press* 16 Apr. f5/3 (Advt.), We have psychiatric nursing positions available on all shifts. No split shifts. **1836** J. Hoyt *Let.* 21 Nov. in W. L. Mackenzie *Life M. Van Buren* (1846) 262, I was reproached by you for having voted a 'split ticket'. **1964** *Economist* 31 Oct. 482/2 A 'split-ticket' group..to urge voters to support Mr Johnson and Mr Keating, the Republican senatorial candidate. **1960** *Annals Amer. Acad. Pol. & Soc. Sci.* CCCXXVIII. 52/1 Of all the time-saving remedies, the split trial should prove the most powerful. **1967** *North-Western Reporter* 2nd. Ser. CL. 323/1 In that year [*sc.* 1878] secs. 4697–98–99, R.S. 1878, were enacted and provided a split trial in which the insanity issue was tried first and if the accused was found sane he was then tried on his plea of not guilty before the same jury. **1926** Split week [see *ham *sb.*¹ 5]. **1948** H. L. Mencken *Amer. Lang.* Suppl. II. 691 *Split week*, a week on the road divided between two or more towns. **1974** P. Wright *Lang. Brit. Industry* ix. 81 *Split weeks* have also become fashionable instead of unvarying Saturday-to-Saturday weeks.

d. *fig.* With reference to division or dissociation affecting a person's mental life or the self. In special collocations, as *split consciousness, man, mind, -mindedness, personality*; *split-minded* adj.
1958 R. F. C. Hull tr. *Jung's Undiscovered Self* v. 74 The rupture between faith and knowledge is a symptom of the split consciousness which is so characteristic of the mental disorder of our day. **1944** H. Read *Educ. Free Men* x. 32 We divide the intelligence from the sensibility of our children, create split-men (schizophrenics, to give them a psychological name), and then discover that we have no social unity. **1962** M. McLuhan *Gutenberg Galaxy* 51 (*heading*) The Homeric hero becomes a split-man as he assumes an individual ego. **1938** Split mind [see *schizoid *a.* a]. **1945** Koestler *Yogi & Commissar* iii. i. 121 Typical examples of socially approved split-mind patterns are the Astronomer who believes both in his instruments and in Christian dogma [etc.]. *a* **1974** R. Crossman *Diaries* (1977) III. 372 The fact is that Jim is absolutely split-minded. **1976** *Jrnl. R. Soc. Arts* CXXIV. 630/2, I must admit to being rather split-minded on this subject. **1947** S. O'Faolain *Irish* i. 23 A delightful dualism—the moderns would call it splitmindedness. **1963** R. F. C. Hull tr. *Jung's Mysterium Coniunctionis* in *Coll. Wks.* XIV. iii. 248 The surprisingly common phenomenon of masculine split-mindedness, when the right hand mustn't know what the left is doing. **1919** M. K. Bradby *Psycho-Anal.* x. 129 The split personalities of hysterics and mediums..have a subjective meaning. **1931** E. Wilson *Axel's Castle* ii. 40 A theory which makes one's poetic self figure as one of the halves of a split personality. **1966** 'H. MacDiarmid' *Company I've Kept* xiii. 259 In my view, he [*sc.* Walter Elliott] was a split personality. **1974** *Listener* 31 Jan. 131/1 Every nation becomes a bundle of contradictions and paradoxes—practically a split personality.

5. a. *split-site comprehensive* (*school*), *school*; *split-brain*, used with reference to a person or animal whose corpus callosum has been severed or is lacking, so that there is no direct connection between the two halves of the brain; *split-dose* (Med.), applied to the technique of administering a given quantity of ionizing radiation in several exposures so as to reduce its harmful effects in relation to its therapeutic ones; *split-half* (Statistics), used *attrib.* with reference to the technique of splitting a body of supposedly homogeneous data into two halves and calculating the results separately from each to assess their reliability; also *absol.*; *split-skin graft* (Med.), a skin graft which involves only the superficial portion of the thickness of the skin; cf. *split graft*, sense 2 b above.
1958 R. W. Sperry in Harlow & Woolsey *Biol. & Biochem. Bases Behavior* 418 In recent efforts to learn more about connectivity principles in perceptual integration, we have been putting to use the demonstrated functional independence of the two hemispheres in what we have come to call the 'split-brain' preparation... In these split-brain animals one can leave intact a whole hemisphere to maintain generalized background function. **1968** Passmore & Robson *Compan. Med. Stud.* I. xxiv. 94/2 At times, the non-dominant hand may 'go off on its own' and have to be restricted by the dominant hand. One begins to doubt whether a split brain man is singular or plural. But in no sense does he resemble a schizophrenic, in spite of the layman's interpretation of that word. **1972** R. E. Ornstein *Psychol. of Consciousness* ii. 55 In day-to-day living, these 'split-brain' people exhibit almost no abnormality. **1947** *Radiology* XLIX. 321/1 In this study the split-dose technic was applied to recovery as tested by lethal effects. **1968** *Brit. Med. Bull.* XXIV. 246/2 Young..has written a program to synthesize the results of split-dose experiments from survival curves at various phases of the cycle. **1935** *Psychol. Rev.* XLII. 158 This conception of split-half or comparable-form reliability as simply inter-item correlation can and should be brought into relationship with Kelley's concept of reliability as adequacy of sampling. **1946** *Jrnl. Educational Psychol.* XXXVII. 473 Since any test may be split in a large number of ways, the split-half method of estimating test reliability fails to give a unique result. **1971** *Computers & Humanities* V. 260 Because 0101 and 0410 had high internal reliability (split half), we did not cut, Xerox, and translate further samples from these books. **1972** *Guardian* 8 Mar. 12/6 Since they are..formed from a merger of two or three existing schools, the split-site comprehensive schools have some attraction for local educational authorities. **1973** *Times* 11 Apr. 8/6 A mini-bus is used in one of our split-site comprehensives. **1975** *Times* 30 Dec. 3/2 The survey of 18 split-site schools, most of them divided by one or two miles across cities and industrial roads, concludes that they are 'the unfortunate by-products of imposing a comprehensive system too quickly'. **1981** *West Lancs. Evening Gaz.* 11 Nov. 9/8 But in 24 of the 63 cases, the authority would have to create either a split-site school, or a school with more than 490 pupils. **1929** *Surg., Gynecol. & Obstetrics* XLIX. 82 (*heading*) The use and uses of large split skin

grafts of intermediate thickness. 1977 *Proc. R. Soc. Med.* LXX. 480/1 Excision and split-skin graft undertaken in 5 patients was successful in the 3 who were traced.

split-arse, -ass, *a. slang.* [f. SPLIT *ppl. a.* + ARSE *sb.*] **1.** *split-arse mechanic* (see quot.).

1903 FARMER & HENLEY *Slang* VI. 317/1 *Split-arse mechanic*, ..a harlot.

2. *Forces' Slang* (orig. *Air Force*). **a.** Classy, showy; (of an airman) reckless, that performs stunts; (of aircraft) having good manœuvrability.

1917 F. T. NETTLEINGHAM *Tommy's Tunes* 49 The expression 'a splitass merchant' is applied indiscriminately to a reckless individual or to a really good flyer capable of executing stunts with a modicum of safety. 1919 W. H. DOWNING *Digger Dialects* 47 *Split-ass*, unusual. 1934 V. M. YEATES *Winged Victory* II. xii. 288 They were sufficiently splitarse and did all the stunts, but there was nothing like a Camel for lightness of touch. 1946 J. IRVING *Royal Navalese* 163 Split-ass, showy. 1966 A. LA BERN *Goodbye Piccadilly* vi. 67 The Royal Air Force and the Fleet Air Arm used to describe certain flyers as 'split-arse types'. This coarse expression was reserved for outstandingly reckless airmen.

b. *split-arse cap* (see quots.).

1931 BROPHY & PARTRIDGE *Songs & Slang Brit. Soldier* (ed. 3) 360 *Split-arse cap*, the old R.F.C. cap, rather like a Glengarry. 1945 C. H. WARD-JACKSON *It's a Piece of Cake* (ed. 2) 56 *Split-arse cap*, the field service cap as distinct from the peaked dress service cap.

Hence as *sb.*, a flying stunt (see also quot. 1919); as *v. intr.*, to make a sudden turn in an aircraft; to perform stunt flying; **split-arsing** *vbl. sb.* and *ppl. a.*

1917 F. T. NETTLEINGHAM *Tommy's Tunes* 49 So won't you splitass back along the track To my dear old Omer Town. 1919 W. H. DOWNING *Digger Dialects* 47 *Split-ass*, an aeroplane on its side in banking for a sharp turn. 1929 *Papers Mich. Acad. Sci., Arts & Lett.* X. 325/2 *Splitass*, to do stunt flying, or to fly in a reckless manner. 1931 BROPHY & PARTRIDGE *Songs & Slang Brit. Soldier* (ed. 3) 360 *Split-arsing*, stunting low and flying near the roofs of..huts. 1934 V. M. YEATES *Winged Victory* III. iv. 306 The triplanes had come down, and did some diving at the splitarsing Camels but didn't hit anyone. *Ibid.* vi. 347 Something fired at him. He splitarsed and nearly hit an SE. 1945 C. H. WARD-JACKSON *It's a Piece of Cake* (ed. 2) 56 *Split-arse*, stunt.

split-level, *a.* [f. SPLIT *ppl. a.* + LEVEL *sb.*] **1. a.** Designating a house or other building which has a floor between the floor-levels of two adjoining storeys (see quot. 1957); applied to this type of design. Also, designating a room having a floor on two levels. Also *absol.* orig. *U.S.*

1952 *N.Y. Times* 7 Sept. VIII. 1/1 A community of 129 split-level houses. 1955 *New Yorker* 1 Oct. 109/2 In the majority of the ads for new houses, 'split-level' is the big word, the selling word. There are, it appears, a good many kinds of split-levels. 1957 *Times* 12 Nov. (Canada Suppl.) p. xv/2 A visit to one of the new planned communities will convince the stranger that Canada has gone 'split-level' mad, provided he knows what split-level means. This is the house originally designed for a sloping site, with entrance midway between upper and lower floors. 1958 *Spectator* 18 July 116/2 It is the first split-level church building in America. 1959 [see *MULTILEVEL *a.*]. 1960 *Guardian* 5 Feb. 6/6 The split-level £2,000 room. *Ibid.* 24 Feb. 12/5 The house is on a sloping site, so this split-level treatment makes sense. 1963 J. N. HARRIS *Weird World Wes Beattie* (1964) ii. 20 One split level was much like another, but the general arrangements of Mrs. Ledley's house showed good taste. 1968 *P. AMERMAN* *Vac* xvii. 97 There was a spiral staircase leading to split-level bedrooms. 1980 *Times* 29 Feb. 13/1 (Advt.), Luxury individually designed modern split level detached house.

b. Of a cooker: having the oven and hob in separately installed units.

1960 *Times* 4 Jan. 13/1 British manufacturers are at last taking advantage of the experience of the Americans.. Now, in England, a number of manufacturers are producing standard units incorporating split level cookers with the oven at waist level, the hob fixed into the working surface, [etc.]. 1976 *S. Wales Echo* 25 Nov. 27/4 (Advt.), Extended kitchen, including Creda split-level oven and hob. 1978 *Cornish Guardian* 27 Apr. 16/3 (Advt.), Large kitchen, split level cooker etc.

2. *transf.* and *fig.*

1955 *N.Y. Times Mag.* 30 Oct. 4/4 Mr. Spectorsky has looked on a community.. and seen only what he planned to see—the fieldstone houses and the split-level personalities, the couples who cannot make ends meet on forty or fifty thousand a year. 1959 *N.Y. Times* 23 Aug. 1. 66/4 (Advt.), Urban and suburban fashion-seekers find the split-level dress a good investment. 1960 *Washington Post* 14 Jan. D 1/2 The migrant worker leads a split-level life.. One state may provide adequate housing and other services. But as soon as he steps across a state line he may find himself in an area with no program to meet his needs. 1965 *New Statesman* 19 Mar. 456/3 George is the natural man drawn to pubs and barmaids, Arbuthnot the conscience that matters at him. This sort of split-level Englishman seems to belong with Mr. Pooter. 1968 *Language* XLIV. 501 Morphologically conditioned sound change (analogical change) is normally of such a nature that no new phoneme arises in a 'split-level' analysis. 1973 *Irish Times* 2 Mar. 13/4 We Irish are now inured to a split-level morality: we react in low key and circumspection when one of 'our own people' murder or maim; but vehemently when the slayer is 'one of them'.

split-second, *a.* and *sb.* Also split second(s).

[Abbrev. of *split second(s) hands*: see SPLIT *ppl. a.* 2 a and SECOND *sb.*[1] 3.] **A.** *adj.* **1.** Designating a stop-watch having two seconds hands, one underneath the other and one of which may be stopped independently of the other. Also *absol.*

1884, 1888 [in Dict. s.v. SPLIT *ppl. a.* 2 a]. 1897 *Sears, Roebuck Catal.* 377/1 This is the best cheap Split Second Horse Timer made. 1916 H. L. WILSON *Somewhere in Red Gap* ix. 379 When I left 'em Jake was holding a split-second watch on the waiter he'd just given an order to. 1977 T. C. COLLOCOTT *Dict. Sci. & Technol.* 1106/1 *Split-seconds chronograph*, a chronograph with two independent centre seconds hands, one underneath the other.

2. Occurring, executed in, or lasting a fraction of a second; timed or calculated to a fraction of a second; sudden, instantaneous.

1946 'M. INNES' *From London Far* III. v. 206 The issue of modern naval conflicts depends upon split-second decisions. 1951 *People* 3 June 2/4 Watch for the two split-second appearances of twenty-one-years-old Audrey Hepburn. 1957 *New Yorker* 2 Nov. 68/2 He pictures himself as being locked in a death struggle with ringing telephones, split-second appointments, and traffic jams. 1959 W. GOLDING *Free Fall* iv. 82, I was..sitting up like, willing them to die..because this demanded split-second timing. 1975 R. L. DUNCAN *Dragons at Gate* iv. 125 An instant pressure on the trigger, sound dispersed by a silencer, split-second work. 1978 *Poland* May 50/2 A player who cannot respect punctuality..cannot be counted on to carry out complex tactical maneuvers requiring split-second decisions when he is under a great mental strain.

B. *sb.* A fraction of a second.

1912 CHESTERTON *Manalive* I. iv. 97 Mr. Moon stood for one split second astonished. 1935 S. LEWIS *It can't Happen Here* iv. 46 Typed revelations timed to the split-second. 1940 [see *RECEIVER[1] 1 c]. 1950 *Sport* 22–28 Sept. 12/1 (caption) Racing in, a split second too late for the tackle, is Jones. 1966 *Times* 24 Mar. 5/4 Whose split second..had been gone. 1978 J. WAINWRIGHT *Jury People* xi. 141 For a split second, she'd known pain she'd never dreamed existed..then she'd passed out.

splitter, *sb.*[2] Add: **1. d.** An auxiliary set of gears that provides a set of ratios between those of the main gearbox. Also *attrib.*

1963 *Automobile Engineer* LIII. 228/2 An overdrive in front of the gearbox can be used as a splitter gear..to provide intermediate ratios between any two of the gearbox. These can be engaged without operating the clutch. 1967 *Economist* 8 July (Commercial Vehicles Suppl.) p. xx/1, Even when there are enough speeds in the splitter box (which may be either in front or behind the normal box) rear axles are getting more complicated. 1977 *Belfast Tel.* 21 Jan. 21/8 (Advt.), 1969 Leyland Super Commer, 6-wheel Tipper, fitted with 9 s.s., splitter box, 401 engine. 1977 D. RUTHERFORD *Return Load* ii. 32 'How many gears have you got?' 'Thirteen forward and two reverse.' 'It's got a splitter.' Sally said.. 'You don't have to move the gear lever thirteen times.'

5. *Hunting slang.* A first-rate hunt.

1843 *Ainsworth's Mag.* III. 219 'What a fine country before us!' observes a third; 'what a splitter we shall have!' says a fourth. 1976 *Shooting Times & Country Mag.* 16–22 Dec. 25/2 There was more than a holding scent and..we were in for a splitter.

splitting, *vbl. sb.* Add: **2. c.** *spec.* in *Psychol.* The process of division or dissociation affecting the mind or self. Also with *off.*

1890 W. JAMES *Princ. Psychol.* I. x. 227 According to M. Janet these secondary personalities are always abnormal, and result from the splitting of what ought to be a single complete self into two parts. 1910 tr. *Freud's Orig. & Devel. Psycho-Anal.* in *Amer. Jrnl. Psychol.* XXI. 191 We followed his (sc. Janet's) example when we made the mental splitting and the dissociation of personality the central points of our theory. *Ibid.*, I soon came to another view of the origin of hysterical dissociation (or splitting of consciousness). 1927 HENDERSON & GILLESPIE *Text-Bk. Psychiatry* v. 101 The most extensive, and at the same time the most profound, of all personality changes is the 'splitting of the personality' that occurs in advanced schizophrenia. 1946 M. KLEIN *Contributions to Psycho-Analysis* (1948) 346 The early splitting of the breast into a good and bad 'breast mother' as a way of dealing with ambivalence had been very marked. 1961 J. A. HADFIELD *Introd. Psychotherapy* xviii. 174 A severe shock such as a car accident or even a severe illness can result in a splitting off of a certain phase of life.

5. *splitting mill* (earlier example); **splitting field** *Math.*, the least field which includes all roots of a specified polynomial.

1841 H. SCRIVENOR *Hist. Iron Trade* vi. 120 All below that size were cut in the splitting-mill. 1942 E. ARTIN *Galois Theory* ii. 22 A splitting field is of finite degree since it is constructed by a finite number of adjunctions of algebraic elements, each defining a field of finite degree. 1971 E. C. DADE in Powell & Higman *Finite Simple Groups* viii. 256 If F is algebraically closed, then it is a splitting field for any simple F-algebra.

splitting, *ppl. a.* Add: **4.** (Earlier examples.)

1828 *Oxonian* I. 461 Felt a splitting head-ache under my night-cap. 1836 DICKENS *Let.* 18 Dec. (1965) I. 209 The head ache continues here..is so great that my head is actually splitting.

5. *splitting image*, an exact likeness. Also (*dial.*) *splitten image*. Cf. *spitting image* s.v. *SPITTING *ppl. a.* 3.

1880 E. W. BARRETT *Specimens Westmoreland Dial.* II. 36 Soa t'kersmas up t'fells Et just be t'splitten image Ov a kersmas 'mang yersells. 1930 D. HARTLEY *Made in England* I. 3 Evenness and symmetry are got by pairing the two split halves of the same tree, or branch. [Hence the country saying: he's the 'splitting image'—an exact likeness.)

splittism (spli·tiz'm). [tr. Chinese *fēnliè zhǔyì*: see SPLIT *v.* and -ISM.] In Communist use: the pursuance of factional interests in opposition to official party policy. Also *transf.* Hence **spli·ttist**, one who practises splittism; also as *adj.*

1962 *Guardian* 25 Dec. 7/1 That dread word, 'Splittism', which has never before darkened a page of the Sino-Soviet polemic, broke through to the surface of the Peking 'People's Daily' yesterday in the first open discussion of the possibility of a split. This term, taken from the translation of the Chinese document into English by the official New China News Agency, appears from the context to be identical with the Russian Communist concept of 'fractionalism'. 1964 *Economist* 8 Feb. 490/3 In Peking's view splittism means opposing and betraying Marxism-Leninism, usually in the interests of the bourgeoisie. 1968 *Ibid.* 5 Oct. 24/3 It is no longer the 'splittists' in Peking who are on trial before the majority of the world's communist parties, but the hard-liners in the Kremlin. 1969 *Daily Tel.* 11 Feb. 19/7 Prelimo and the Organisation of African Unity Liberation Committee are aware of the dangers of 'splittist'—as Peking calls it—tendencies. 1976 *Times* 20 Oct. 10 We must..repudiate all those who betray Marxist-Leninist-Mao Tse-tung thought, tamper with Chairman Mao's directives, practise revisionism and splittism, [etc.]. 1978 *Peking Rev.* 10 Mar. 9/2 The 'gang of four'..once again indulged in splittist activities and devised all kinds of schemes and plots. 1980 *Economist* 31 May 9/1 One challenge to western unity..had been compounded by another..the danger of splittism was evident.

splodgily (splɒ·dʒili), *adv.* [f. SPLODGY *a.* + -LY[2].] In a splodgy manner. Also *fig.*

1963 *Times* 16 Feb. 4/7 This matches the closing not wisely but too well, and her imaginative splodgily-organized supporters were splodgily built. 1978 R. HILL *Pinch of Snuff* xiv. 141 'Your girl's got two minutes, missus,' he said splodgily.

splosh (splɒʃ), *sb. colloq.* and *dial.* [Echoic: cf. SPLASH *sb.*[1] and SLOSH *sb.*] **1.** The dull, splashing sound of the impact of a hard object striking or struck by something wet and soft; the impact itself. Also, a quantity of liquid suddenly dashed or dropped.

1894 C. E. DE LEONG in *Calif. Hist. Soc. Q.* (1930) IX. 133 Storming like fury..a mixture of snow and rain..a splosh almost knee deep. 1899 E. M. STOOKS *Not Exactly* xi. 180 After all 'tis but drew minutes or so of accurst laziness. 1908 E. NESBIT *Enchanted Castle* ii. 38 Gerald..heard the splosh of the fish of that one. 1916 J. E. BANGS *From Pillar to Post* xii. 233 The committee bundled me into the back..and it landed with a splosh. 1925 the soft was heard.. the hall with no more damage than one rather slush splash of snow. 1929 W. DEEPING *Second Youth* xxi. 111 The road had dropped a splosh of water on Uncle Reginald's new hat. 1930 A. G. STREET *Gentleman of Party* i. 11 An hour of steady splosh, splosh, splosh, their horses' feet in thick batter of mud which churned over the road. 1954 L. M. BOSTON *Children of Green Knowe* ix. 94 A snow flake landed on his cheek.. He felt the splosh. 1983 'F. PARRISH' *Bait on the Hook* ii. 35 'At splosh o' paint on the mudguard. Got reason to remember 'at, I 'ave.

2. *slang.* Money.

1893 G. ELEN 'E *dunno where 'e Are* (song) Since Jack Jones come into that little bit o' splosh. 1916 'TAFFRAIL' *Pincher Martin* vi. 100 The others spent two pounds of some splosh some-ow. 1934 *Week. Gaz.* (weekly ed.) 30 Aug. 1/3 Poetry in the Old Kent Road who can come into a little bit of splosh. 1950 WODEHOUSE *Nothing Serious* 216 The pillows of having all that splosh in the old sock. 1967 A. WILSON *No Laughing Matter* II. 82 Intentions need a bit of splosh to back em up.

splosh, *v.* (*int.* and *adv.*) *colloq.* and *dial.* [f. prec.] **a.** *trans.* To splash (something); to cause (something) to move with a splashing sound. **b.** *intr.* To move with a splash. Also *int.* and as *adv.*

1890 *Harry Fludyer* 47 Such larks when you heard the ball go splosh on a man's hat. 1901 J. PRIOR *Forest Folk* III. 114 Nell continued to turn the handle of her old barrel churn. 'Splosh! splosh! went the cream. 1904 *Eng. Dial. Dict.* V. 678/2 In Sc. 1913 'Hd. You are scorching water about for?' 1923 WODEHOUSE *Inimit. Jeeves* iii. 30, I began to sing like a lanky nightingale as I sploshed the sponge away. 1924 GALSWORTHY *White Monkey* ii. iii. 194 Down came death—splosh!—and a creature wiped out, like a fly on a wall. 1926 *Somerset Year Bk.* 54 She brought down th' liddle shutter again splosh-bang. 1930 R. CAMPBELL *Poems* 10 Jack Squire through his own teardrops sploshes. 1931 W. FAULKNER in *Amer. Mercury* Mar. 261/2 She made the sound into the cup and the coffee sploshed out on to her hands and her dress. 1941 W. FORTESCUE *Trampled Lilies* x. 61, I sploshed through the morass of mud and manure. 1966 M. WOODHOUSE *Tree Frog* iii. 24 He was sploshing around in a quicksand. 1978 S. RADLEY *Death & Maiden* x. 6 He sploshed on down the muddy track.

Hence **splo·shing** *vbl. sb.*

1929 W. DEEPING *Roper's Row* xxx. 334 The sound of debate would cease suddenly. Boards creaked. Sploshings and sandpaperings recommenced. 1974 P. DICKINSON *Poison Oracle* v. 115 The awkward sploshings of his paddle..had been the loudest sounds in the marsh.

sploshy, *a.* Add: **a.** (Earlier example.) Also, characterized by splashing or sploshing.

1838 *N.Y. Advertiser & Express* 28 Mar. 1/4 On Tuesday it was muggy and sploshy. 1972 [see *PLASTICIZED *ppl. a.* 2].

b. = SPLOTCHY *a.*

1942 E. WAUGH *Put out More Flags* iii. 199 Cedric turned from the portrait of Angela.. "Is it finished?" 'Yes. It was very hard to make the man finish it, though.' 'It hardly looks finished now, does it Daddy? It's all sploshy.'

c. = SPLASHY *a.*[2] 3.

1966 [see *coffee-table book* s.v. *COFFEE *sb.* 5 b].

spludge (splɒdʒ). *U.S. dial.* [Echoic: cf. next.] A 'splurge' or ostentatious display; *to cut a spludge* = *to cut a splurge* s.v. *SPLURGE *sb.* 1 a.

1831 *Essayist* II. 80/2, I was naturally anxious to see a little more of Tennessee life, and inasmuch as it was said there was to be a great *spludge* at the shooting, I went with him. 1856 F. M. WHITCHER *Widow Bedott Papers* ix. 89 She tries to cut a spludge and make folk think she's a lady. 1933 *Dialect Notes* IV. 113 [Kansas words] *Spludge*, splurge.

splurge, *sb.* For *U.S.* read orig. *U.S.* and add: **1. a.** (Later examples.) Also in phr. *to cut a splurge*, to make an ostentatious display.

The first quot. in Dict. should be cited from 'F. Herbert,' *The Talisman for 1829* (1828), 11. 107.

1860 BARTLETT *Dict. Amer.* (ed. 3) 112 *Cut a splurge*, ..to make a show or display in dress. 1895 *Punch* 1 June 258/2 My anti-gambling old Gaffer 'Sa], is for spludge splurge. 1897 *Chicago Tribune* 19 Sept. 37/1 Two shrewd young Hoosiers..came to Chicago in 1892 and cut a big splurge in monetary and real estate circles. 1977 *Daily Tel.* 19 Jan. 4/3 Presidential inauguration ceremonies..will go on for the rest of the week in a $5 million (about £1,700,000) splurge of dances, parties and celebrations.

b. A sudden extravagant indulgence, esp. in spending.

1928 *Publishers' Weekly* 16 June 2429 The Sales Force hadn't the courage to urge big splurges. 1929 W. DEEPING *Roper's Row* xxx. 342 He was not to be the slave of other people's animal appetites, their sex splurges. 1951 M. HILLS *Orchids on your Budget* iv. 74 For years she has been putting something aside—not for a rainy day, but for a spring splurge. 1957 M. MILLAR *Soft Talkers* 193 A chicken salad which she looked at cavernously but barely touched, as if she knew too well the penalties of such a splurge. 1976 *National Observer* (U.S.) 25 Dec. 2/1 The Air Transport Association predicts a 6 per cent increase in air travel this year. 'We don't see any tremendous splurge,' a spokesman said.

c. *spec.* in *Journalism*, a large or showy advertisement, feature, etc.

1906 E. PARTRIDGE *Charm of Words* 31 A full-page advertisement..is usually called either a *splash* or, if showy, a *splurge*. 1971 'L. BLACK' *Death has Green Fingers* x. 109 We've got..quite a splurge about the rose industry generally.

splurge, *v.* For *U.S.* read orig. *U.S.* and add: **b.** *intr.* To spend money extravagantly. Freq. *const. on*. Also *trans.*

1934 in WEBSTER. 1947 *Chicago Sun* 28 Jan. 17/2 When I got around to furnishing my office, I thought I'd splurge on a good 18th Century English armchair. 1961 *Observer* 19 Mar. 3/3 The cigarette manufacturers splurged [on advertising], in spite of or because of cancer fears. 1976 *High Times* Dec. 159/1 If you really get into omelettes, you should splurge and purchase a good copper or stainless steel omelette pan. 1977 *Spare Rib* June 6 So Twinlock bought the ad.. I was sure it didn't look good when Bhopal an offer which is so compelling that feminists will splurge this sort of money? 1979 R. JAFFE *Class Reunion* (1980) II. v. 129 You don't have to splurge on Maxim's now—we can go somewhere simple.

splurgy, *a.* For *U.S.* read orig. *U.S.* and add earlier and later examples.

1843 *Yale Tomahawk* May 4/3 They even pronounce his speeches splurgy. 1937 *Sunday Times* 20 Feb. 6/1 Diana is equally obviously one of those 'big, splurgy' actresses who have been successes up to the word Gone.

splutteringly (splʌ·təriŋli), *adv.* [f. SPLUTTERING *ppl. a.* + -LY[2].] In a spluttering manner.

1941 *Penguin New Writing* X. 21 The rifles volleyed spluttering three times and the funeral was over. 1969 *Listener* 16 Jan. 85/1 This is often a splutteringly funny book, but the humour seems to me to spring from an underlying sadness and spiritual discomfort. 1978 P. MASON *Shaft of Sunlight* iii. 47 He..made me try—splutteringly—to explain why the whole of the Rhondda valley could not be sentenced to transportation for life without having even poached a hare.

spoalty, var. *SPAULTY *a.*

Spock-marked (spɒ·kmāːkt), *a. joc.* [f. the name of Benjamin McLane *Spock* (b. 1903), U.S. physician and specialist in child care + MARKED *ppl. a.*, after *pock-marked*.] (Adversely) affected by an upbringing held to be in accordance with the principles of Dr. Spock, esp. as set forth in his *Common Sense Book of Baby and Child Care* (1946).

1967 A. COMFORT *Anxiety Makers* vi. 186 Liberal and sane advice can still leave the children of anxious parents permanently Spock-marked. 1974 J. COOPER *Women & Super Women* 10 Permanently Spock-marked, they believe the world owes them a living.

Spode[2]. (Earlier examples.)

1869 C. SCHREIBER *Jrnl.* 16 Sept. (1911) I. 37 The only thing the small shops at Exeter presented was a little Spode basket. 1875 L. TROUBRIDGE *Life amongst Troubridges* (1966) 106 We fished out several things—a blue Spode plate, for which we gave a shilling, [etc.].

spodic (spɒ·dik), *a. Soil Science.* [f. Gr. σποδ-ός wood ashes, embers + -IC.] Of a soil horizon: that is illuvial deposit rich in aluminium oxide and organic matter and usu. also contains iron oxide.

1960 *Soil Classification (7th Approximation)* (U.S. Dept. Agric., Soil Survey Staff) v. 46/1 The spodic horizon is an illuvial horizon that is characterized by the accumulation of appreciable amounts of organic carbon, an illuvial accumulation of free iron not accompanied by roughly equivalent amounts of illuvial crystalline clay, or an illuvial accumulation of organic carbon usually if not always accompanied by an accumulation of aluminum combined in some form other than that of the crystalline silicate clays. 1973 J. G. CRUICKSHANK *Soil Geogr.* iii. 91 Podsolisation involves the translocation downward of laterally of iron and aluminium compounds from silicate clays, and their concentration in a specific lower horizon (B₂ Bs, or spodic horizon). 1979 *Nature* 4 Oct. 366/2 Soils in this forest are histosols (peat soils) with some spodic (podzolic) characteristics.

spodiosite (spɒ·d-, spəᵘ·diǝsəit). *Min.* [ad. Sw. *spodiosit* (H. V. Tiberg 1872, in *Geol. Fören. Stockholm Förhandl.* I. 84), f. Gr. σπόδιος ash-grey: see -ITE[1].] A phosphate and fluoride of calcium, Ca_2PO_4F, found as orthorhombic crystals; also, any of a series of artificial minerals having similar crystal structures but containing other elements.

1887 *Jrnl. Chem. Soc.* LII. 1. 346 In consequence of the chemical and crystallographical resemblance between spodiosite and kjerulfine, the author suggests that the two minerals may be isomorphous compounds. 1911 *Jrnl. Physical Chem.* XV. 469 The spodiosite found at Nordmark in Sweden occurs in orthorhombic prisms. 1953 *Mineral. Mag.* XXX. 167 Those orthorhombic cell dimensions provide some justification for the inclusion by Dana of spodiosite in the wagnerite group. 1970 *Inorg. Chem.* IX. 2264/1 Spodiosites with the general formula Ca_2XO_4Cl, with X = P, V, Cr, and As have been prepared. 1974 *Jrnl. Amer. Ceramic Soc.* LVII. 102/1 Spodiosite analogues, a class of compounds related to apatites, are used in phosphors and have possible future industrial applications as laser hosts and bioceramic materials.

Spodosol (spɒ·dəsɒl). *Soil Science.* [f. Gr. σποδός wood ashes, embers + *-SOL.] Any soil belonging to an order characterized by the presence of a spodic horizon and including most podzols and podzolic soils.

1960 *Soil Classification (7th Approximation)* (U.S. Dept. Agric., Soil Survey Staff) xiii. 192/1 The Spodosols include primarily the soils that have been called Podzols, Brown Podzolic soils, and Ground-Water Podzols. Not all soils called Podzols, however, are in this order. 1969 BUCKMAN & BRADY *Nature & Properties of Soils* (ed. 7) xi. 317 The low native fertility of most Spodosols makes them uncompetitive for tilled crops. 1978 R. W. SIMONSON in V. M. Chaney *Quaternary Soils* 13 Direct evidence for transfers of organic matter within profiles has been obtained by measuring amounts moving down from A horizons of Spodosols.

spoewslang, var. *SPUUGSLANG.

spoil, *sb.* Add: **III. 11. a.** (sense 10) *spoil heap* (later examples), *tip*.

1927 F. B. YOUNG *Portrait of Clare* I. ii. 34 The black dome of the Mawne Road spoil-heap fell away on her left into the tree-softened contours of Mawne Bank. 1973 *Times* 5 May 3/8 About three hundred of the 3008 Eppleston spoilheap, started in the 1820s and at its peak containing well over 1,500,000 tons of red shale, dwarfing the houses. 1980 *Times Rev. Industry* May 58/1 Around it [sc. Doncaster] stretches a flat, ill-drained corner of Yorkshire's West Riding, pockmarked with colliery spoil tips. 1972 *Times* 4 July 3/2 A devastated landscape of abandoned quarry and spoil heaps.

b. spoils system (earlier example).

1838 *N.Y. Mavo Political Sk. Eight Years in Washington* xi. 51 Jefferson..authorized a friend to compromise 'spoils system' to a certain extent. 1866 *Chambers's Jrnl.* 20 Jan. 33/1 Come on, you beggar!..I'll spile your pretty face for you.

spoil, *v.*[1] Add: **III. 10. d.** (Later examples.) Also *transf.*

1847 *Sporting Life* 16 Oct. 166/2 Hudson returned some heavy hitting; but Cannon would not be denied, although he met with a stopper that would have spoiled the upper works of the best chancery lawyer in the kingdom. 1866 *Chambers's Jrnl.* 20 Jan. 33/1 Come on, you beggar!..I'll spile your pretty face for you.

11. e. To render (a ballot paper) invalid, as by improper marking, deliberate defacement, etc.; to invalidate (a vote) in this manner.

1872 *Act* 35 & 36 *Vict.* §33 If the voter inadvertently spoils a ballot paper, he can return it to the officer, who will, if satisfied of such inadvertence, give him another paper. ?1886 *Truth About Irish Election 1885* (Irish Loyal & Patriotic Union) 24 He clearly informed him that he would *spoil his vote*. 1953 *Ann. Reg. 1952* 235 The pro-Germans had been urged to abstain or to spoil their ballot papers. 1978 G. HERMET et al. *Elections without Choice* i. 3 The difference between free and controlled elections is indicated by the opportunity a voter has..to have his franchise recognised through registration..to decide how to vote, even to spoil his ballot, without external pressure.

14. b. Also const. *in*.

a 1960 E. M. FORSTER *Maurice* (1971) vii. 42 Durham.. would be leaned to at all hours curled up in his room and spoiling to argue.

spoilage. Add: **2. b.** The deterioration or decay of foodstuffs and perishable goods. orig. *U.S.*

1928 *Mineral Water Trade Rev.* 18 Jan. 16/1 The question of spoilage is not thoroughly dealt with in this country. Spoilage is an American term denoting any kind of deterioration found in a bottled carbonated beverage. 1958 *New Scientist* 24 July 483/1 The time that elapses between the killing of a whale and its arrival at the processing plant is often long enough for serious bacterial spoilage to develop, impairing both the yield and quality of the oil and the flesh itself. 1976 *National Observer* (U.S.) 27 Mar. 3/1 Israel has relied on subtler tactics to control the West Bankers. These have included detaining farmers' produce trucks to cause spoilage.

spoiled, *ppl. a.* Add: **3. c.** Of a vote or ballot paper: rendered invalid.

1944 *Federal Reporter* (U.S.) CXXXVIII. 248/1 In the election..201 [employees] cast ballots, with the result 39 unchallenged votes for the Union, 51 for International .. 7 votes were challenged spoiled, or blank. 1958 W. MACKENZIE *Free Elections* xv. 131 Even if there is compulsory voting, this general dissent may express itself through the proportion of 'spoiled papers' handed in. *Ibid.* 136 Administrative difficulties arise not over papers that are clearly 'spoiled' but over marginal cases. 1973 *Irish Times* 2 Mar. 8/1 Electorate, 37,299.. Spoiled votes, 488. 1976 *Guardian* 17 Apr. 24/5 There was only one spoiled paper in the 94 per cent poll.

d. *spoiled nun, priest*, a nun or priest who has repudiated her or his vocation.

1914 S. JOYCE *Dublin Diary* (1962) 26 He is the spoiled priest to his finger tips. 1916 J. JOYCE *Portrait of Artist* (1969) v. 182 In any case..the spoiled priest..is a gentleman. 1932 F. SCOTT FITZGERALD in A. Mizener *Far Side of Paradise* (1951) 307 The story should do this. Show a man who is a natural idealist, a spoiled priest. 1977 *Times Lit. Suppl.* 13 May 593/1 Romantic wickedness (or, in O'Flaherty's own spoiled-priest phrase, 'romantic sin').

spoiler. Add: **2. b.** *Boxing.* An inferior fighter who sets out to disrupt his opponent's style.

1948 *Sun* (Baltimore) 29 June 15/2 Walcot is only a timid defensive heavyweight to wit the words 'offense' and 'attack' are abhorrent. In the language of the ring he is known as a 'spoiler', the type of fighter who can make an opponent look bad but who can never look good himself. 1959 *Times* 14 Oct. 16/5 Carroll, a 'spoiler' when at close quarters, has a good left jab.

c. *U.S.* One who mars the chance of victory for an opponent, while not being a potential winner. Also, applied to a thing. Esp. in *Sport* and *Pol.*

1950 *Sun* (Baltimore) 9 June 20/4 In meeting San Francisco here..the Colts will get a crack at their biggest 'spoiler' of the past two seasons. 1963 *N.Y. Times* 5 Nov. 40 It seems clear that William Buckley will poll enough votes to be a 'spoiler', though it is not yet certain for whom. 1967 CONGRESSE & SHANNAN *Tiger of Tammany* iv. 117 Van Buren was the 'spoiler' in a three-way race. 1968 *N.Y. Times* 25 Jan. 19 Mr. Javits..described the six-year-old Conservative party as 'a faction' that had been set up as 'a spoiler' party. 1972 *Publ.* 22 Aug. 11/2 3/4 What he [sc. George Wallace] is doing these days is laying the premise for another 'spoiler' campaign in 1972. 1976 J. V. O'BRIEN *William O'Brien* viii. 164 It was felt that the Protestant vote had gone not to O'Brien but to the spoiler candidate and ex-Lord Mayor. 1979 *N.Y. Post* 20 Aug. 44 It looks like we're out of it now, so we'll just have to be spoilers from here on. *Ibid.* 46, I suppose you can't have a 'spoiler' in soccer. 1980 *Washington Post* 11 Sept. 1/1 Not that Anderson is yet seen as a serious challenger for the presidency.. But he is now perceived as a genuine factor in the presidential election, a certified political spoiler.

3. a. *Aeronaut.* A flap or the like that can be made to project from the upper surface of an aircraft wing to break up a smooth airflow and so increase drag.

1928 DE PORT & BODEN *Rep. U.S. Air Corps Material Div. Airplane Branch* No. 2876, 2 The 'spoiler' or 'lift destroyer' described in this report causes a decrease in lift by a 'burbling' action. 1932 *Rep. U.S. Nat. Advisory Comm. for Aeronautics* No. 439. 3/1 When ailerons and spoilers are used together the full effect of both is not obtained if the spoilers are located directly in front of the ailerons. 1947 A. C. DOUGLAS *Gliding & Advanced Soaring* iii. 24 When the spoilers are raised a certain amount of the lift of the wing is destroyed and the machine must glide at a steeper angle in order to keep the speed constant. 1978 R. JANSSON *Ness Caper* xiii. 115 Thackeray trimmed once more, and briefly pulled on the spoilers to reduce our air speed.

b. A structure on a motor vehicle intended to reduce lift and so increase the pressure on the wheels at high speed; also, one located so as to reduce the drag caused by components on the underside of a car.

1963 *Times* 8 Mar. 3/7 A special feature..is the tail, which is sharply cut-back in Ferrari style, the roof sweeping down to end in a 'spoiler' or raised lip. This treatment eliminates turbulence and wind drag by killing low-pressure over the tail, and keeping weights on the

rear wheels at high speed. 1965 *Observer* 10 Oct. 5/1 It has retractable spoilers which can be pushed up to increase high-speed stability. 1974 L. DEIGHTON *Spy Story* i. 10 His car..was..all dressed up in black vinyl, Lamborghini-style rear-window slats, and even a spoiler. 1979 *Time* 2 Apr. 6 (Advt.), The front spoiler is there to give you increased tyre adhesion at high speeds.

spoiling, *vbl. sb.*[1] Add: **3.** *Rugby Football.* The act or process of disrupting the opposing side's play; usu. *attrib.* Also *transf.*

1937 C. W. JONES *Rugby Football* i. 1 Experiments are being made with a new rule affecting the spoiling around the scrummages. 1959 *Times* 19 Oct. 14/2 Some dazzling moves started by breaks by Mulligan, who throughout found Jeeps in fine spoiling form. 1967 J. POTTER *Foul Play* ix. 111 A finer display of spoiling tactics and spectacular bouleversements had probably never disgraced a hockey field. 1978 *Rugby World* Apr. 41/3 It was a very interesting game, but I thought the criticism by the Barbarians of spoiling tactics after the game was unjustified.

spoil-sport. Add: (Earlier example.)

1801 M. EDGEWORTH *Belinda* I. iii. 97 Harriet swore at the colonel, for the veriest *spoil-sport* she had ever seen.

Hence **spoil sport** *v. intr.*, to behave as a spoil-sport.

1869 TAYLOR & DUBOURG *New Men & Old Acres* III. 72 *Brown.* I'm locked in.. Let me out. *Lilian.* (Unlocking door.)..What was that for? *Brown.* (Laughing.) For fear I might spoil sport. 1946 R. TENNANT *Lost Haven* (1947) xvii. 285 He did not want to spoil sport; he wanted the whole gang.

spoilt, *ppl. a.* Add: (Earlier and later examples.)

1816 JANE AUSTEN *Emma* III. xvii. 312, I am losing all my bitterness against spoilt children. 1935 W. C. GILLINGS *Handbk. for Presiding Officers* 25 The spoilt ballot paper shall be at once cancelled by the presiding officer. 1983 *Times* 5 July 10/7 We are not protecting like spoilt children.

spoilure (spoi·liūr). *rare.* [f. SPOIL *v.*[1], after *failure.*] The act of plundering or despoiling.

a 1918 W. OWEN *Poems* (1931) 77 Be slowly lifted up.. Great gun..about to curse.. Be not withdrawn, dark arm, thy spoilure done. 1928 *Daily Sketch* 7 Aug. 12/5 Lion, my Lord, this camp was left in your charge. You made a big spoilure of it, so I am taking your work away.

Spokane (spōᵘkæ·n), *sb.* (and *a.*) Also 9 Spokain, Spokein. [Native name.] A tribe of the Salish group of North American Indians; the language of this people. Also *attrib.* or as *adj.*

1831 R. COX *Adventures on Columbia River* I. ix. 197 The Spokans we found to be a quiet, honest, inoffensive tribe. 1838 S. PARKER *Jrnl. Exploring Tour beyond Rocky Mts.* 284 We passed to-day several small villages of the Nez Percé and Spokein nations. 1875 H. H. BANCROFT *Native Races Pacific States* III. iii. 615 The northernmost Salish language is the Shushwap.., then there are the.. Pend d'Oreille, the Spokane, the Soaiatipi, and the Okanagan, which with others spoken on the Columbia show close affinities. 1894 *Messenger Sacred Heart* June Apr. 273 Colonel George Wright gained a general victory over the..Spokanes. 1933 V. F. RAY *Sanpoil & Nespelem* 9 The Sanpoil spoke a dialect of Interior Salish and were surrounded by other Salish speaking peoples. All of the neighboring dialects were intelligible to them except Columbia and Spokane. 1944 N. W. ROSS *Westward Women* 71 She had to learn the Spokane language... [This] was a difficult task, as the sounds the Spokanes made in speech were like nothing so much as the sounds of breaking corn. 1965 *Canad. Jrnl. Linguistics* Spring 78 Languages of sure affiliation.. Spokan. *Ibid.* 159 In Spokane, one of the Interior Salish languages showing the greatest contrast, there are four terms that are generation-reciprocal. 1974 J. FAHEY *Flathead Indians* ii. 22 Flatheads married freely among contiguous Pend Oreilles, Nez Percés, and Spokanes.

spoke, *sb.* Add: **1. d.** *Basket-Making* = STAKE *sb.*[1] 5 d.

1897 A. FIRTH *Cane Basket Work* ii. 17 Spokes, the coarser canes used as the foundation, and round which the weavers are placed. 1925 A. A. GILL *Practical Basketry* 39 After the spokes are arranged for weaving, take a short strand. 1958 O. R. SCOTT *Basketry Step by Step* 8 The uprights of a basket are called stakes or spokes.

b. spoke-wood (earlier example).

1851 J. BROWN *Forester* (ed. 2) iv. 362 Young oaks, of the size generally termed spoke-wood, sell well.

spoken, *ppl. a.* Add: **2. d.** *the spoken word*, speech (as opp. to written language, etc.), esp. in the context of radio broadcasting.

1832 CARLYLE in *Fraser's Mag.* Apr. 257/1 Whether man can any longer be so interested by the spoken Word, as he often was in those primeval days. 1929 *Radio Times* 29 Nov. 432 Poetry..has its roots in the spoken word: the written word is only a means of saving poetry from the oblivion of time. 1946 R. S. LAMBERT *Ariel & All his Quality* iii. 60 The 'spoken word' is the most contentious and most closely scrutinised part of the broadcast programme. 1954 *Ann. Reg. 1953* 62 Training..in understanding of the written and spoken word. 1961 *Listener* 28 Sept. 456/1 One criticism that has been made of spoken-word material in the Third is that it has sometimes been too esoteric. 1973 *Ibid.* 1 Nov. 574/1 It was the treatment of the spoken word that seemed to fox the early fathers of broadcasting.

spokenness (later example).

1947 L. MACNEICE *Dark Tower* 20 But when no character can be presented except through spoken words, without the characters being fully, that very *spokenness* makes this distinction between subjective and objective futile.

spokeslady (spōᵘ·ksléⁱdi). [Cf. SPOKESWOMAN.] = SPOKESWOMAN.

1976 *Richmond* (Va.) *Times-Dispatch* 23 June 7/4 'Don't you quote us, though,' the spokeslady hastened. 1969 *Guardian* 31 July 9/5 'It came out like that, and they decided to leave it like that,' a spokeslady said. 1972 *Daily Tel.* 25 Sept. 2/1 Our most clamorous spokeslady for conscious opinion.

spokesman. 2. (Later examples.)

1976 *Times* 21 May 2/4 Mary Whitehouse, spokesman for the campaign. 1976 *Daily Tel.* 20 Jan. 1/1 A spokesman for the British Medical Association.

spokesmanship. Add: **b.** [Cf. *-MANSHIP.] Skilful use of the position of spokesman.

1960 *Encounter* May 27 Such spokesmanship underplays the potential excitement of the work itself. 1963 *Times* 19 Apr. 8/5 Official explanations about why there is no such sign at the chief bookshop of all in Kingsway are simply an exercise in spokesmanship.

spokespeople (spōᵘ·ks,pī:p'l). [Cf. SPOKESMAN.] Two or more 'spokespersons'.

1972 *Guardian* 19 Apr. 8/4 Spokespeople for several leftist groups feel the murders are being carried out by a terrorist organization called 'The Black Hand'. 1977 *Undercurrents* June–July 5/2 Each group of ten or twenty had elected a 'spokesperson' who met in plenary with other 'spokespeople'. 1983 *Listener* 18 Aug. 27/2 The BBC spokespeople were obliged to come back with efforts to explain that the comparison..was not as clear as it might seem.

spokesperson (spōᵘ·ks,pɜ̄ːs'n). [f. SPOKESMAN, after *chairperson* (see *PERSON *sb.* 2 f).] A manufactured substitute for 'spokeswoman' or 'spokesman'.

One of numerous words formed to avoid alleged sexual discrimination in terminology.

1972 *Guardian* 18 Feb. 11/3 The spokesperson (non-sexist term) for UCWE complained that she had been physically assaulted by a university administrator'. 1976 *New Yorker* 29 Nov. 172/2 One's heart and imagination.. were bound by the ascetic, sexual, Christian woman who recurs in Mrs. Spark's novels so often as to suggest, if not an alter ego, a spokesperson. 1978 J. IRVING *World according to Garp* xviii. 402 A 'spokesperson' for the Ellen Jamesians. 1981 *Economist* 28 Nov. 39/1 As a feminist Illip Miss Joan Lester..has been made spokesperson for women's rights and welfare.

spokeswoman. (Later examples.)

1974 *Sunday* (Charleston, S. Carolina) 28 Apr. 5-A/8 A Union spokeswoman said Mrs. Agnew had scheduled her visit and was receiving routine care. 1976 *Film & Television Technician* Nov. 7/4 'The arguments against racism have to take place at a local level, they can't be imposed from above,' said a spokeswoman from Transport House. 1979 *Guardian* 23 Jan. 1/8 A spokeswoman..said the response from members yesterday was 'phenomenal'.

spondulicks. Add: (Further examples.) Also as quasi-*sing.* **spondoolick, spondulick,** etc. Also, a piece of money, a coin.

1864 [see *SINGLE *a.*]. 1868 *All Year Round* 19 Sept. 334/1 A restaurant..where the tallest sort of feeding may be had at all hours at the lowest possible cost to the man of spondulics. 1869 E. SAMPSON *Tales of Fancy* 36, I was instructed to give him up the sum of two in advance in order to draw out the spondulicks. 1902 W. N. HARBEN *Abner Daniel* 58 The one with the spondoolix wonders harder than the one who has none. 1912 W. OWEN *Let.* 25 Apr. (1967) 131, I shall do nothing until further brings me some 'spondoolicks'. 1913 P. OPPENHEIM *Inevitable Millionaires* xi. 121 [Do I understand that the young man..has dissipated the whole of his patrimony, in twelve months?' he inquired. 'Every bean,' Harold assented. 'Not a spondulik left'. 1966 *Courier-Mail* (Brisbane) 9 Aug. 1/10 (Advt.), *Spondoolicks*. Dollars, pounds or call it what you like, you'll save plenty at Maclay's. 1980 *Private Eye* 29 Feb. 13/1 No one seemed very anxious to come up with the spondulicks.

spondylitic (spɒndili·tik), *sb.* and *a.* *Path.* [f. SPONDYLIT(IS + -IC.] **A.** *sb.* A person suffering from spondylitis.

1900 in DORLAND *Med. Dict.* 1949 F. HERNAMAN-JOHNSON et al. *Ankylosing Spondylitis* 1. ii. 16 Russian surgeons, and later, American surgeons, began to operate on spondylitics. *Ibid.* iii. 26 The patient with a commencing spondylitic arthritis of the hip may show some of the classical early signs of tuberculosis. 1977 *Lancet* 17 Sept. 592/1 Clinical investigation of first-degree relatives of patients with classic ankylosing spondylitis with irradiated siblings more in approximately 15%. 1980 *Jrnl. R. Soc. Med.* LXXIII. 65 Views of the cervical spine showed minor spondylitic changes.

B. *adj.* Of spondylitis; caused by or associated with spondylitis.

spondylolisthesis (spɒ·ndilɒlisþī·sis). *Path.* [mod.L. (H. F. Kilian *Spondylolisthesi* (1853) 33), f. Gr. σπόνδυλ-ος, var. σφόνδυλος vertebra + ὀλίσθησις dislocation, slipping.] The forward displacement of the body of the lowest lumbar vertebra (but not the posterior lamina and spine) relative to the bones of the pelvis, or of any other lumbar vertebra relative to the one below it.

1858 *Brit. & Foreign Med.-Chir. Rev.* XXII. 177 (heading) The nature and origin of spondylolisthesis. 1885 *Trans. Obstetr. Soc.* XXVI. 187 Neugebauer considers that the predisposing cause of spondylolisthesis is either a congenital deficiency between the superior and inferior articular process on both sides, or a fracture of the same parts. 1932 *Brit. Jrnl. Surg.* XIX. 374 There are two types of spondylolisthesis... In [the first]..the entire 5th lumbar vertebra slips forwards upon the sacrum and carries the rest of the spine with it. The second type.. consists in the separation (or spondylolysis) of the 5th lumbar vertebra into two portions.., in such a way that the part bearing the spinous process and inferior articular surfaces moves backwards and the rest of the vertebra slips forwards upon the sacrum. 1977 *Proc. R. Soc. Med.* LXX. 421/1 In a proportion of patients the spondylolysis progresses to spondylolisthesis and may then cause symptoms and signs. 1981 *Church Times* 4 Sept. 12/4 As one born with congenital spondylolisthesis (a dangerous defect of the spine linkage requiring bone grafts chiselled off my hip), I am delighted with my restricted life.

Hence **spo:ndylolisthe·tic** *a.*

1884 *Amer. Jrnl. Med. Sci.* LXXXVIII. 600 Neugebauer's description of spondylolisthetic subjects applied exactly to this case. 1932 *Brit. Jrnl. Surg.* XIX. 380 That the supports of the spondylolisthetic 5th lumbar vertebra are very insecure is obvious.

spondylolysis (spɒndilɒ·lisis). *Path.* [f. as prec. + *-LYSIS.] The splitting or partial disintegration of a vertebra.

1885 *Trans. Obstetr. Soc.* XXVI. 188 A specimen of spondylolisthesis upon one side, a hemiolisthesis of the fourth lumbar vertebra caused by a similar spondylolysis. 1932 [see *SPONDYLOLISTHESIS]. 1959 *Jrnl. Bone & Joint Surg.* XLI. A. 311 A preceding bone abnormality or defect of the isthmus need not be necessarily present for the development of spondylolysis. 1977 [see *SPONDYLOLISTHESIS].

Hence **spondyloly·tic** *a.*

1959 *Jrnl. Bone & Joint Surg.* XLI. A. 312 In the spondylolytic cases..the articular processes above and below the affected vertebra were free to reach and compress the isthmus.

spondylosis (spɒndilōᵘ·sis). *Path.* [ad. F. *spondylose* (P. Marie 1898, in *Bull. de la Soc. Méd. des Hôp.* 11 Feb. 121), f. as prec.: see -OSIS.] Ankylosis of the spine, or of a vertebral joint.

1900 *Jrnl. Nerv. & Mental Dis.* XXVII. 558 (heading) A case of rhizomelic spondylosis. 1942 *Jrnl. Bone & Joint Surg.* XXIV. 827 The apparatus and routine used in twenty-two cases of rhizomelic spondylosis treated by roentgenotherapy combined with well-established orthopaedic measures. 1949 F. HERNAMAN-JOHNSON et al. *Ankylosing Spondylitis* 1. iv. 40 (heading) 'Old man's spine' (spondylosis). *Ibid.*, The radiograph of a case of spondylosis shows a general curvature of the spine. 1977 *Brit. Med. Jrnl.* 26 Mar. 822/1 The local and root symptoms of cervical spondylosis are usually well controlled by non-operative methods.

Hence **spondylo·tic** *a.*

1964 *Amer. Jrnl. Roentgenol.* XCII. 1268/1 Spondylotic changes in the cervical and lower thoracic spinal column sometimes affect the spinal cord. 1972 *Lancet* 8 July 71/2 Patients with spondylotic myelopathy.

spondylus. Add: Also *attrib.*, as *spondylus shell*.

1932 R. F. FORTUNE *Sorcerers of Dobu* v. 189 The groom and his kin..must accumulate..a spondylus shell necklace or two. 1979 *Archaeology* July 11/2 Many Precolumbian ornaments and beads were manufactured from the rosy-colored rims of spondylus shells.

sponge, *sb.*[1] Add: **I. 1. c.** (Further examples.)

1872 'MARK TWAIN' *Roughing It* xlvii. 333 One of the boys has gone up the flume..throwed up the sponge.. kicked the bucket..he's dead! 1874 TROLLOPE *Phineas Redux* I. xxxix. 325 When..Thursday afternoon came, Mr. Daubeny 'threw up the sponge'.

II. 6. d. (Earlier examples.)

1747 H. GLASSE *Art of Cookery* (ed. 2) xvii. 297 To make White Bread..when your Spunge has stood its proper Time clear the Oven, and begin to make your Bread. 1796 A. SIMMONS *Amer. Cookery* 38 *Butter biscuit.* One pint each milk and emptins, laid into flour, in sponge.

g. With defining word: a type of thick jelly eaten as a dessert.

1859 J. H. WALSH *Eng. Cookery Bk.* 275 Lemon sponge. .. Take half an ounce of isinglass..the juice of eight lemons. 1907 *Yesterday's Shopping* (1969) 55/1 Sweets.. Lemon Sponge, Raspberry Sponge (in copper moulds, 10/0 extra, returnable). 1978 E. LOTHIAN *Country House Cookery from West* 19 Orange sponge. 1 oz (30 g) gelatine. 4 oranges.

h. A sponge-cake; the mixture from which such a cake is made.

1877 *Cassell's Dict. Cookery* 920/1 *Sponge, Savoy,*.. pour the batter gently into a mould. 1907 *Yesterday's Shopping* (1969) 55/2 Golf Sponge, iced chocolate, coffee, pink or white..each of 0/11. 1917 *Harrods Gen. Catal.* 1269/2 Sponge Swiss Roll..per lb. 1/8... Caracas Roll (Rich Chocolate Sponge)..per lb. 1/10. 1948 *Good Housek. Cookery Bk.* II. 447 Steamed sponge puddings. *Ibid.* 579 Genoese Sponge (basic recipe) 3 oz. butter 2½ oz.

flour ½ oz. cornflour 3 large eggs 4 oz. caster sugar. 1960 R. DANIEL *Death by Drowning* v. 54 A jam sponge, please. 1975 *Times* 10 May 13/4 The mixture can be baked..as a sponge flan.

IV. 11. a. *sponge box.*

1895 *Army & Navy Co-op. Soc. Price List* 191/2 Sponge box for travelling, patent aluminium. 1970 *Canadian Antiques Collector* Oct. 18/2 Similar trifles for feminine use included snuff boxes, sponge boxes and bodkin cases.

13. sponge-bag trousers, a pair of men's checked trousers, patterned in the style of many sponge-bags; **sponge biscuit** (later examples); **sponge cloth,** (*b*) a thin piece of spongy material used for cleaning; (*c*) a type of cotton fabric (see quot. 1957); **sponge mixture,** (*a*) a packet of prepared dry ingredients for making a sponge-cake; (*b*) the ingredients of a sponge-cake mixed together ready for baking; **sponge rubber** liquid rubber latex processed into a sponge-like substance; freq. *attrib.*; **sponge sandwich** a sponge-cake consisting of two halves sandwiched together with a filling; in earlier use, covered with custard and eaten as a pudding.

1915 V. WOOLF *Voyage Out* xxiii. 376 Can't you imagine him—bald as a coot with a pair of sponge-bag trousers? 1977 A. J. AYER *Part of My Life* ii. 35 The members of Pop also had the privileges..of wearing coloured waistcoats, sponge-bag trousers, braid on their tail-coats, flowers in their button-holes and sealing wax on their top-hats. 1837 MRS. GASKELL *Let.* 18 Mar. (1966) 10 Aunt L. has..expressed a strong wish to hear 'her dear little voice once again' and has a spunge biscuit behind her pillow this 4 days to give her. 1892 T. F. GARRETT *Encycl. Pract. Cookery* 147/1 *Sponge Biscuit.*—Beat ten eggs very thick and smooth. 1954 D. HARTLEY *Food in England* vii. 218 (heading) Egg and lemon jelly (using sponge biscuits). 1902 D. SALOMONS in A. C. Harmsworth *Motors & Motor Driving* vi. 93 Sponge cloths are a desirable accessory for cleaning and for polishing up. 1919 *Queen* 26 July 138 White sponge cloth is the thing for this new coat and skirt. 1957 M. B. PICKEN *Fashion Dict.* 318/2 *Sponge cloth*... Cotton fabric of coarse yarn woven in honeycomb weave to produce open spongy effect. 1976 W. TREVOR *Children of Dynmouth* iii. 58 Timothy rinsed the sponge-cloth he was using, squeezing it out in his bowl of dirty water. He wiped the inside of the oven..and closed the door. 1926–7 *Army & Navy Stores Catal.* 50/1 Sponge mixture..pkt. -/5½. 1962 'O. MILLS' *Headlines make Murder* x. 119 She..poured boiling water on her sponge mixture. 1975 *Times* 10 May 13/4 A sponge mixture that you bake yourself tastes very much nicer than a shop bought one. 1932 *New Yorker* 9 Apr. 56/3 A luxurious soft pile combined with a sponge rubber back. 1934 G. F. CHARNOCK *Mech. Technol.* (ed. 2) xxii. 278 Sponge rubber, such as is sometimes used for upholstery, and in which the pores are many times larger than the cells of expanded rubber, is not such an effective insulator. 1951 *Archit. Rev.* CIX. 164 (*caption*) A sponge rubber overlay is fitted over the springs. 1967 N. FREELING *Strike out where not Applicable* 20 Metal furniture, upholstered in sponge rubber, covered with grey plastic. 1884 *Myra's Cookery Bk.* xiv. 309 Sponge sandwiches... Sponge cakes 6—cut in half lengthways. 1917 *Harrods Gen. Catal.* 1269/2 Sponge Sandwiches..each 1/6. 1967 A. LASKI *Seven Other Years* iii. 30, I want you to go..and get a sponge sandwich for tomorrow. A chocolate sponge with cream.

sponge, *sb.*[2] **2.** (Later examples with *down.*)

1954 M. STEWART *Madam, will you Talk?* vii. 60, I hadn't time for a bath, but I took a quick cool sponge down. 1960 *House & Garden* Mar. 63/1 All the paper will need will be a sponge down. 1977 W. GOLDING *Moving Target* (1982) 66 Ann has just had a sponge-down in the beastly bath.

sponge, *v.* Add: **I. 4.** Also *intr.*

1962 M. E. MURIE *Two in Far North* II. vii. 171 The [bread] sponge didn't sponge in spite of red damask tablecloth and fur parka I had lovingly wrapped it in.

Hence **spo·ngeable** *a.*, able to be wiped with a sponge.

1971 *Ideal Home* Apr. 75/2 Spongeable wallpaper. 1976 *Milton Keynes Express* 23 July 22/5 (Advt.), Roller blind kits and dozens of fabulous spongeable fabrics at Bedford Wednesday Market behind statue.

sponge-cake. Add: Now usu. made without milk. (Earlier example.) Also *fig.*

1808 JANE AUSTEN *Let.* 15 June (1952) 191 You know how interesting the purchase of a sponge-cake is to me. 1902 W. JAMES *Var. Relig. Exper.* xiv. 364 Naturalistic optimism is mere syllabub and flattery and sponge-cake in comparison. 1909 W. S. CHURCHILL *Let.* 15 Sept. in R. S. Churchill *Winston S. Churchill* (1969) II. *Compan.* II. xii. 911 It appears to me to belong to the whipped cream and sponge cake style of painting.

Hence **spo·nge-cakey** *a.*

1858 MRS. GASKELL *Let.* 1 Oct. (1966) 896 Some little sponge cakey puddings. 1971 B. W. ALDISS *Soldier Erect* 15, I flung my sexual emotions into gear by imagining sponge-cakey vulvas.

sponged, *ppl. a.* Add: **3. b.** Of colour, paint, design, etc.: applied with a sponge. Hence *sponged ware*, pottery decorated by being dabbed with a sponge impregnated with colour. Cf. *spatterware* s.v. *SPATTER *v.* 7.

1925 *Heal & Son Catal.: Table Wares* (1972), A sponged design in clear blue under-glaze on white ground. 1957 MANKOWITZ & HAGGAR *Conc. Encycl. Eng. Pott. & Porc.* 101/1 Mocha was made here [at Greens] as well as sponged and lined wares. 1971 L. A. BOGER *Dict. World Pott. & Porcelain* 323/1 *Sponged Ware.* In American ceramics; decorating the surface of pottery by dabbing with a sponge or something of the sort.

sponging, *vbl. sb.* Add: **4.** *Cookery.* The action or process of setting a sponge of flour, yeast, water, and salt.

1895 J. GOODFELLOW *Elem. Princ. Breadmaking* xiv. 93 The golden rules to follow in sponging are..Work at as low a temperature as possible... Use as little yeast as possible. 1929 E. B. BENNION *Breadmaking* 250 Sponging and doughing. 1949 A. R. DANIEL *Baker's Dict.*, *Sponging*, the baker's term for setting a sponge of flour, yeast, or barm, water, and salt.

spongio-. Add: **spongioplasm** (later examples); now † *Obs.*; [ad. G. *spongioplasma* (F. Leydig *Zelle und Gewebe* (1885) vii. 173)].

1891 [see *RETICULUM 4 b]. 1896 [see *ENCHYLEMA]. 1933 M. FERNÁN-NÚÑEZ tr. *Ramón-Cajal's Histology* xvii. 299 The chromatic granules offer in their interior a vacuolated spongioplasm. 1936 W. SEIFRIZ *Protoplasm* xv. 266 The older workers in cytology held similar opinions, expressed in the 'spongioplasm' (framework) and 'hyaloplasm' (intervening fluid) of Leydig and the 'ground substance' and 'reticulum' of Carnoy and others.

spongioblastoma (spɒ:ndʒiɒblæstōᵘ·mă). *Path.* [ad. G. *spongioblastom*: see *spongioblast* s.v. SPONGIO- and *-OMA.] A malignant tumour, usu. of the brain or optic nerve, believed to be derived from spongioblasts.

1918 *Neurol. Bull.* I. 276/2 He [*sc.* Kaufmann] suggests the name *columnar cell glioma* and provisionally the term *spongioblastoma*. 1925 *Arch. Neurol. & Psychiatry* XIV. 145 They are both—neuroblastomas and spongioblastomas—richly cellular, very proliferative and rapidly growing. 1967 *Nursing Times* 27 Jan. 108/2 The highly malignant spongioblastoma multiforme which can kill within a month or two of its first symptom.

spongiosa (spɒndʒiōᵘ·zā). *Anat.* [L. *spongiōsa* (sc. *substantia*): see SPONGIOSE *a.*] **a.** The tissue constituting the bulk of the posterior grey column of the spinal cord. Now *rare.* **b.** Cancellous or spongy bone tissue, such as that found within the ends of the long bones.

1947 H. C. ELLIOTT *Textbk. Nerv. System* xiii. 157/1 Like the dorsal gray column, the spinal V nucleus has a dorsolateral substantia gelatinosa, the relay nucleus for pain fibers, and a more central spongiosa for touch fibers. 1949 *New Biol.* VI. 73 If we take a bone such as the thigh-bone (femur) and saw it in two lengthways, we find that the two ends near the joints consist of a spongy lattice (spongiosa) while the shaft is a hollow tube of dense bone (compacta). 1954 T. L. PEELE *Neuroanat. Basis Clinica Neurol.* v. 79/1 The 'body', or spongiosa, is subdivided into a ventrally placed basal part adjoining the intermediate gray columns, a dorsally placed 'cervix' or neck, and a 'caput' or head. 1966 *Lancet* 31 Dec. 1430/1 Histological sections..confirmed the increase in trabecular bone in experiment 11. The increase was striking in both primary spongiosa..and secondary spongiosa. 1982 *Calcified Tissue Internat.* XXXIV. 425/2 Osteoclasts in the primary spongiosa near the growth plate were the first to incorporate ³H-TdR.

spongiosis (spɒndʒiōᵘ·sis). *Path.* [f. SPONGIO- + -OSIS.] † **a.** (See quot. 1907.) *Obs.* **b.** Accumulation of fluid between the cells of the epidermis.

1907 A. WHITFIELD *Skin Dis. & their Treatm.* xi. 172 Some of the epithelial cells become swollen from the imbibition of the plasma, and the condition thus produced is known as spongiosis. 1932 R. L. SUTTON *Introd. Dermatol.* iv. 30 Intercellular edema, or spongiosis, is shown by widening of intercellular spaces. 1966 WRIGHT & SYMMERS *Systemic Path.* II. xxxix. 1461/2 Spongiosis, separation of cells of the prickle cell layer by an accumulation of fluid between them—in other words, oedema of the epidermis. Sometimes referred to tautologically as 'intercellular oedema' to distinguish it from so-called 'intracellular oedema' (which is hydropic degeneration of the cytoplasm of epidermal cells, as seen, for example, in certain viral infections). 1976 *Lancet* 27 Nov. 1168/2 The epidermis showed several areas of basal vacuolation and spongiosis with many eosiniphilic bodies.

spongo-. Add: **spongolite** *Geol.* [ad. F. *spongolithe* (L. Cayeux 1897, in *Mém. Soc. géol. du Nord* IV. 99)], a rock formed almost entirely of sponge spicules.

1945 M. F. GLAESSNER *Princ. Micropalaeont.* ii. 23 Some rocks known as spongolites are largely formed of siliceous sponge spicules. 1963 *Geol. Mag.* C. 296 The Ardagh spongolites appear to have been laid down as a bioherma1 deposit. 1968 R. W. FAIRBRIDGE *Encycl. Geomorphol.* 780 In rare cases, sponge banks accumulate quite extensive deposits of siliceous spicules (spongolite or sometimes 'spiculite').

spongy, *a.* Add: **1. e.** Of suspension and braking systems in motor vehicles: deficient in firmness.

1952 FRADZEE & BEDELL *Automotive Maintenance & Trouble Shooting* x. 396 A spongy pedal on hydraulic brake systems may be due to excessive clearance between the shoes and the drum. 1954 I. FLEMING *Live & let Die* xiii. 134 All the fun of driving had been taken out of them ..with hydraulic-assisted steering and spongy suspension. 1962 *Which? Car Suppl.* Oct. 123/1 The Fiat 1500..had one disconcerting point [*sc.* in its braking system]—the long pedal travel necessary, which always felt 'spongy' and gave the impression that there was less power available in the brakes than in fact was the case. 1967 B. C. MACDONALD *Car Doctor A to Z* iii. 19 (*heading*) Pedal has 'spongy' feel.

sponson, *sb.* Add: **3. a.** *Canad.* An air-filled buoyancy chamber in a canoe, intended to reduce the risk of sinking even if the canoe becomes filled with water; so *sponson canoe.*

1911 *Daily Colonist* (Victoria, B.C.) 14 Apr. 19/1 (Advt.), Don't fail to see the safety sponson canoes that cannot sink. 1917 P. L. HAWORTH *On Headwaters of Peace River* i. 6 The craft in question were really Chestnut sponson canoes, seventeen feet long... It had never been my intention to take a sponson canoe on the trip. 1968 R. M. PATTERSON *Finlay's River* 109 This proved to be a seventeen-foot Chestnut 'Pleasure Model' canoe fitted with sponsons—that is, with air-chambers along the gunwales so designed that the canoe, with an average load in it, would float even when swamped and full of water.

b. A projection from the hull or body of some kinds of aircraft, intended to increase its lateral stability in the water; also, a stabilizer in the form of a float at the end of a wing.

1928 V. W. PAGÉ *Mod. Aircraft* xvi. 668 It is a braced monoplane type, the hull being supplemented by sponsons of aerofoil section at each side. 1930 *Flight* 31 Oct. (Aircraft Engineer Suppl.) 1192b/2 Design of wing-tip floats or sponsons. 1935 [see *AERODYNAMICS]. 1965 [see *HYDROPLANE *sb.* 2]. 1971 *Maclean's Mag.* Oct. 70/2 Only part of the propeller, some of the rudder, and pieces of the sponsons at the front actually cut into the water. 1983 *Times* 5 Aug. 2/8 The helicopter hit the sea... The impact ripped open the bottom of the fuselage and removed the sponsons containing emergency flotation gear.

sponsor, *sb.* Add: **4.** One who pays, or contributes towards, the cost of a broadcast programme or other spectacle, *spec.* in return for commercial advertisement.

1931 P. DIXON *Radio Writing* 18 The sponsor wants a dramatic type of program and is willing to spend one thousand dollars a week for the program. 1953 *Manch. Guardian Weekly* 2 July 15/2 United States broadcasting started as a service of information and entertainment for the family accompanied by restrained acknowledgements to the sponsor before and after each programme. 1956 AUDEN & KALLMAN *Magic Flute* (1957) 57 To name a sponsor or to praise a brand. 1972 'E. LATHEN' *Murder without Icing* (1973) vii. 67 It is axiomatic in all sports coverage that the sponsor's time should never intrude on the action.

sponsor, *v.* Add: **1. b.** To promote and support (a resolution, bill, etc.), esp. in a legislative assembly.

1961 *Time* 14 July 25/3 The U.A.R. forthwith sponsored a Security Council resolution. 1964 *Ann. Reg. 1963* III. ii. 152 Virtually the same resolution as had been vetoed in the Security Council was sponsored in the General Assembly by 44 nations. 1973 *Daily Tel.* 19 Nov. 3/3 Sir Gerald Nabarro..successfully sponsored legislation on clean air, coroners, oil burners, and thermal insulation.

2. To pay, or contribute towards, the expenses of a radio or television programme, a performance or other event or work, *spec.* in return for advertising space or rights.

1931 F. A. ARNOLD *Fourth Dimension* x. 78 The travelogue type of program, sponsored by a tourist agency or a steamship company. 1931 P. DIXON *Radio Writing* 25 When an advertiser decides that the program's worth sponsoring. 1963 *Amer. N. & Q.* I. 67/1 Ohio State University and its Press deserve great credit for sponsoring the work through the English Department, the University Libraries, and the Graduate School. 1976 *Jrnl. R. Soc. Arts* June 364/2 But a lot of them are sponsoring these concerts now, which are being broadcast regularly with the names of the sponsors.

3. To support (someone) in a fund-raising activity by pledging a certain sum for each unit completed. Cf. *SPONSORED *ppl. a.* 2.

1967 *Oxfam News* Jan. 2/3 Over £35,000 was raised by young people..sponsored by friends at a penny a mile or more. *Ibid.* June 6/2 Marathon marches..are a big money-spinner... Everyone who walks collects threepence per completed mile from a friend who has sponsored them.

spo·nsored, *ppl. a.* [f. SPONSOR *v.* + -ED[1].] **1.** Financially supported or promoted; freq. of radio or television programmes, etc., having (a portion of) their expenses paid by a commercial interest in return for granting advertising space or rights.

1931 F. A. ARNOLD *Fourth Dimension* 31 Sponsored programs are those that are prepared for advertisers or organizations that pay for their time on the air and also pay for the program that is broadcast. 1932 *B.B.C. Yearbk. 1933* 35 The danger to Press interests that would be involved in changing over from public-service to sponsored broadcasting. 1953 O. LANCASTER in *Daily Express* 28 Nov. 1/4 (*caption*) Will sponsored TV help Miss Cheesecake in her career? 1973 *Ann. Rep. Curators Bodl. Libr. 1971–2* 54 The OSTI-sponsored experiment with MARC tapes was completed in 1971.

2. Of a fund-raising activity (orig. a walk), usu. organized on behalf of a charity, in which each participant obtains pledges from sponsors to donate a certain sum for each unit completed.

[1966 *Oxfam News* Apr. 1/4 There were Oxfam-sponsored walkers..in the..annual Margate to Maidstone 50-mile walk. Sponsors paid 3d. for each completed mile.] 1967 *Ibid.* Jan. 2/3 Teenage support for Oxfam increased. ..The 'sponsored walk' caught on. 1970 *Times* 11 May 10/4 (*caption*) People taking part in a sponsored walk along the Grand Union Canal..to raise funds for the British Council for Rehabilitation of the Disabled. 1973 [see *SLIM-IN]. 1977 *R.A.F. News* 11–24 May 8/4 Flt Lt O'Doherty..organised a sponsored team ride from Lands End to John O'Groats.

sponsorship. (Later examples.)
1931 P. Dixon *Radio Writing* 25 The sponsorship of the Philadelphia Symphony Orchestra by Philco. *Ibid.* 26 The men who do 'Amos 'n' Andy'..worked for years without sponsorship and at very small salaries. 1955 *Bull. Atomic Sci.* Mar. 104/3 Will sponsorship, however, be forthcoming soon enough and on a sufficient scale? 1966 *Oxfam News* Oct. 4/1 A contingent..from Oxfam House took part in the walk and faces..became..winning and chatty as sponsorships were sought. 1978 'D. Rutherford' *Collision Course* 20 Regent have decided to put some sponsorship money into motor racing.

spontaneity. Now freq. with pronunc. (spɒntănē¹·ĭti).

spoof, *sb.* Add: **1.** (Earlier example.)
1884 *Topical Times* 13 Dec. 3/2 The revival of Spoof in Great Britain and America is..wholly due to private enterprise, and..Monsieur Arthur Roberts.

2. a. (Earlier example.)
1889 E. Dowson *Let.* 19 May (1967) 80 We sat dejectedly in the office but were obliged to admit finally that it was a case of 'spoof'.

b. A skit or 'send-up'; *spec.* a film, play, or other work that satirizes a particular genre.
1958 *Oxf. Mag.* 13 Mar. 374/2 This..programme..proved to be an experimental double-dose of theatrical spoof. 1958 *Films in Review* May 254/1 There had been a few films which foreshadowed the screwball pattern, particularly the hilarious Hollywood spoof *Bombshell*. 1975 D. Lodge *Changing Places* ii. 61 Even the weather forecast seemed to be some kind of spoof. 1977 *Time* 3 Jan. 71/1 The prolific Gardner sets a spoof of pulp fiction inside a philosophical monologue on good and evil.

3. (Earlier and later examples.) Also as *adj.*
1884 *Topical Times* 13 Dec. 3/2 At Byzantium many Spouf Courts were; but when it became the capital of the Greek empire the game fell into disuse. 1914 G. B. Shaw *Fanny's First Play* 229 How am I to know how to take it? Is it serious, or is it spoof? If the author knows what his play is, let him tell us what it is. 1927 [see *SPOOF *v.* b]. 1946 V. Tempest *Near Sun* iii. 27 The spoof-feint raid that has been put on to make Jerry think that we are going to central Germany. 1978 K. Gregory *First Cuckoo* 25 [Readers] are invited to spot the 'spoof' letter whose apparent erudition hoaxed the editor into printing it.

spoof, *v.* Add: **a.** (Earlier and later examples.) Also, † to avoid by means of a ruse. Also *absol.*
1889 E. Dowson *Let.* 10 Mar. (1967) 48 It is the 'après' wh. spoofs us. *Ibid.* 11 Nov. 115 The Lord Mayorlet's Tom Foolery was a nuisance. I spoofed it successfully by going from Limehouse to Bloomsbury by tram. 1933 *Sun* (Baltimore) 8 July 6/7 It will be found necessary to handle our Doctor with much circumspection if he is not spoofing. 1977 'J. le Carré' *Honourable Schoolboy* i. 31 The story had everything... It spoofed the British.

b. To make (something) appear foolish by means of parody; to 'send up'. Also *absol.*
1927 *Observer* 20 Nov. 20/5 This is a spoof piece which fails, through sheer clumsiness, to spoof. 1953 [see *GOOGOL]. 1981 W. Safire in *N.Y. Times Mag.* 29 Mar. 10/3 'Urbababble'..spoofs the lingo of those urbane people in the city business.

c. To render (a radar system, etc.) useless by providing it with false information.
1972 *Sci. Amer.* July 18/3 The opposition need not even destroy the installation; by sampling the pulses emitted by the system it can contrive to 'spoof' or jam it. 1977 *R.A.F. News* 27 Apr.–10 May 20/4 They were, however, 'soft spots' in a defensive system. They would have to be attacked again and again and they would have to be spoofed by the considerable number of methods now available.

Hence **spoofed** (spŭft) *ppl. a.*; **spoo·fer,** one who spoofs; **spoo·fery,** † (*a*) *pl.*, a low sporting club; also *spec.* the Adelphi (cf. Spoof *sb.*, quot. 1889) (*obs. slang*); (*b*) trickery, hoaxing; **spoo·fing** *vbl. sb.* and *ppl. a.*
1895 *People* 6 Jan. 13/4 About half-past one this morning I was in the 'Spooferies'—Where? In the 'Spooferies'—the Trafalgar Club they call it now—in Maiden-Lane. 1903 A. M. Binstead *Pitcher in Paradise* x. 227 'And when it comes to comparing the Spooferies with the House of Lords,' the missive concluded, 'it is a million to one on the Spooferies.' The other name of the Spooferies was the Adelphi Club. 1914 *Conc. Oxf. Dict.* Addenda, *Spoofer.* 1920 *Quill* Dec. 9 The after-dinner speeches..were brilliant impromptu spoofings directed at the guest of honor. 1926 A. H. Godwin *Gilbert & Sullivan* 219 Bunthorne..is, in common language, a spoofer. 1926 K. Grahame in G. Sanger *Seventy Years a Showman* 20 The whole thing was unabashed 'spoofery'—clumsy fakes, dried fish, abortions in bottles,..and so on. *a* 1936 Kipling *Something of Myself* (1937) iii. 48 There were..'spoofing'-letters from subalterns to be guarded against. 1958 P. Scott *Mark of Warrior* ii. 176 He's got me spoofed in terms of the exercise. 1965 H. Kahn *On Escalation* v. 86 If the super-ready status is accompanied by limited 'spoofing' or 'jamming' or other hostile acts. 1975 *New Yorker* 19 May 22/3 It's an atrocity, of course, and one of the most spoofed of all the Jeanette MacDonald–Nelson Eddy operettas. 1976 *Time* 20 Dec. 53/3 Gardner has set himself two roles..: the hilarious spoofer of pulp fiction [etc.]. 1978 G. Mitchell *Wraiths & Changelings* xv. 147 Ghosts are part of my stock-in-trade..so..I'm up to most of the dodges... The plainsong..was the only artistic effort..attempted in any of the spoofery.

spoogslang, var. *SPUUGSLANG.

spook, *sb.* Add: **2.** *slang* (orig. and chiefly *U.S.*). An undercover agent; a spy.
1942 Berrey & Van den Bark *Amer. Thes. Slang* § 458/16 'Spotter.' (One who spys upon employees.).. *Silent eye, spook, spotter.* *Ibid.* § 765/7 *Rat, rubber heel, spook, spotter,* a person employed to detect irregularities. 1954 *People* (Austral.) 3 Nov. 24/1 The *spooks* were senior constables who wore no uniform, worked in pairs and followed constables about the city and suburbs to see if they did their work properly. 1961 *John o' London's* 20 Apr. 434/1 The idea of making a living as a spy—'spook' in current Washington slang—is repugnant to most of us. 1966 R. Thomas *Spy in Vodka* (1967) vi. 50 I'd like him to get out of the spook business. 1979 L. Pryor *Viper* i. 9 'My training was also in espionage at the CIA farm.'..'A spook,' I said in wonder.

3. *slang* (orig. and chiefly *U.S.*). A derogatory term for a Black person.
1945 L. Shelly *Hepcats Jive Talk Dict.* 17/2 Spook (n), frightened negro. 1953 K. Tennant *Joyful Condemned* xxvii. 262 The boss of the ward..was doing time for going with 'spooks'—negroes. 1966 *New Statesman* 25 Nov. 778/1, I find a disturbing minority of my English contemporaries..pointedly tossing off inconsequential remarks about spades and spooks in my company. 1977 E. Leonard *Unknown Man, No. 89* xxiii. 235 We almost had another riot... The bar-owner..shoots a spook in his parking lot.

spook, *v.* Add: **1. a.** (Later example.)
1976 *Publishers Weekly* 21 June 88/1 The ghost of the highwayman Black Charlie who spooks Flora with regular visitations.

b. To frighten or unnerve; *spec.* (of a hunter, etc.) to alarm (a wild animal). *slang* (chiefly *N. Amer.*).
1935 E. Hemingway *Green Hills Afr.* i. i. 13 We spooked one [kudu]... No chance of a shot. 1944 *Nat. Geogr. Mag.* June 669/1 To get photographs of the herds Williams took to the saddle, since a man on foot is liable to 'spook', or stampede them. 1959 W. Faulkner *Mansion* ix. 222 Pupils and teacher both who were already spooked..by the sudden presence of the unexplained white woman. 1973 A. Garner *Red Shift* 12 You're spooking me. You're too quiet. 1980 M. Gordon *Company of Women* (1981) ii. iv. 187 You always act like you're waiting for something... It spooks me.

2. a. (Earlier examples.) Also *fig.*
1871 *N.Y. Tribune* 24 Feb. 1/5 Once he saw Toussaint L'Ouverture spooking about with an air of mournful majesty. 1886 [see *ASTRALLY *adv.*]. 1973 E.-J. Bahr *Nice Neighbourhood* xviii. 190 A free-wheeling teen-ager..[who] seems to be spooking around half-shot all the time.

b. To take fright; to become alarmed. *N. Amer. slang.*
1928 R. Santee *Cowboy* xvii. 250 As luck would have it I got a throw, for the cattle spooked an' run. 1941 E. Hemingway *For whom Bell Tolls* xxii. 272 He'll probably leave tracks like an old bull elk spooking out of the country. 1957 W. Faulkner *Town* i. 14 The old dug-in city fathers..spooked to the desperate expedient of..exhuming..the story of the Cuban dice game. 1974 R. M. Pirsig *Zen & Art of Motorcycle Maintenance* iii. xx. 245, I spook very easily these days... *He* never spooked at anything.

spooked, *ppl. a.* *U.S. slang.* [f. Spook *v.* + -ed[1].] **a.** Frightened; nervy; dogged by ill-fortune.
1937 E. Hemingway *To have & have Not* i. iii. 50 He would get to worrying and get so spooked he wouldn't be any use. 1952 B. Harwin *Home is Upriver* xiii. 128 The cattle backed away to the far side, a skittish, spooked mass of bristling horns and rolling eyes. 1969 L. Sanders *Anderson Tapes* (1970) li. 137 Don't get spooked when something comes up you didn't figure on. 1970 E. Tidyman *Shaft* (1971) x. 132 There were still some people.. staring into the night. The spooked, the stoned and the sleepless. 1977 E. Leonard *Unknown Man, No. 89* xxi. 214 He was running for town, spooked good now, in a panic.

b. *spooked up*, excited, pepped up.
1939 Wodehouse *Uncle Fred in Springtime* iii. 43, I saw one of those Western pictures at our local cinema last night, in which a character described himself as being all spooked up with zip and vinegar. That is precisely how I feel. The yeast of spring is fermenting in my veins, and I am ready for anything. 1969 C. Burke *God is Beautiful, Man* (1970) 85 Well this makes her pretty happy and she gets so spooked up about it that she ran into the city and forgot her jar of water.

spookery. (In Dict. s.v. Spook *sb.*) (Later examples.)
1927 [see *PSYCH *sb.* 2 a]. 1935 *Times Lit. Suppl.* 3 Jan. 10/4 The author does not exclude provincial and foreign spookeries. 1960 *Times* 20 Aug. 3/7 *Der Freischütz*..the quintessence of early romanticism with its naive spookery. 1973 E. Pace *Any War will Do* (1974) i. 23 Taking corners too quickly seemed to be another sign of the immaturity..that makes a man opt for the profitless thrills of 'spookery' in the first place.

spoo·king, *vbl. sb.* [f. Spook *v.* + -ing[1].] **1. a.** The action of calling spirits; a séance.
1919 E. H. Jones *Road to En-Dor* i. 2 'What's the suggestion?' Alec asked. 'Spooking,' said I. 'Cripes!' said Alec... Matthews brought in the..table... Little wrote a letter of the alphabet on [squares] and arranged them in a circle... I polished the tumbler... We had constructed our first 'Ouija'. 1930 H. G. Wells *Autocracy of Mr. Parham* ii. iv. 135, I will not relax one jot or one tittle in these precautions until I have demonstrated forever the farcical fraudulence of all this solemn spooking.

b. Haunting, frightening.
1961 *Amer. Speech* XXXVI. 224 The use of *spook* can extend to the frightening of people; it is not limited to physical *spooking* but can indicate mental frightening. 1966 T. Pynchon *Crying of Lot 49* v. 106 Oedipa wondered what hangups, crises, spookings in the middle of the night might be developed from the shadowed subtleties of his mouth, hidden under a full beard. 1979 *Angling* July 19/3 The spooking of fish by striking at line bites.

2. The action of spying.
1977 M. Herr *Dispatches* (1978) 51 The romance of spooking started to fall away..leaving the spooks on the beach.

spoo·kist. [f. Spook *sb.* + -ist.] A spiritualist or medium.
1902 G. W. E. Russell *For Better? For Worse?* xi. 171 A man whom I knew well was taken suddenly and seriously ill, and his relations, who were enthusiastic spookists, telegraphed for the celebrated clairvoyante Mrs Endor. 1920 M. Asquith *Autobiogr.* I. v. 72 If he could 'get through'—to use the orthodox expression of the spookists—he would find all his opinions on this subject more than justified. 1959 *Listener* 9 Apr. 645/1 This eminent spookist had such a look of the slim operator.

spooky, *a.* (In Dict. s.v. Spook *sb.*) Add: **1. a.** Of, pertaining to, or characteristic of spirits or the supernatural; frightening, eerie. (Earlier and later examples.)
1854 *Wide West* (San Francisco) 16 July 1/5 After treading many dark passages, the guide, having unlocked all sorts of 'spooky' looking iron doors,..ushered us before the tomb. 1906 E. Dyson *Fact'ry 'Ands* xviii. 239 There was somethin' spooky 'n' soopernatural erbout er pertickler weird 'n' unaccountable erfluvium. 1929 T. Wolfe *Look Homeward, Angel* xxvii. 378 Don't start that ..spooky stuff! It makes my flesh crawl. 1948 *Time* 1 Nov. 90/2 Shakespeare's *Macbeth* is a turbulent melodrama, full of spooky claptrap. 1960 R. Dahl *Kiss, Kiss* 187 This..is really beginning to get interesting—a trifle spooky, too. 1977 J. F. Fixx *Compl. Bk. Running* viii. 104, I had the spooky feeling that I was dressed up in somebody else's body. 1980 G. Mitchell *Whispering Knights* ix. 98 'It's a spooky-looking place,' said Capella nervously.

b. *Surfing.* Of a wave: dangerous or frightening.
1966 *Surfer* VII. iv. 48 Morne Plage features a left breaking over a coral reef and tapering into a big black deep spooky pass a half a mile from the coast. 1970 *Studies in English* (Univ. Cape Town) I. 34 Waves, especially the bigger and more powerful ones, are often dangerous and frightening, and can sometimes be referred to as *spooky*... *Spooky* might indicate the difficult or the unpredictable, as in: 'Things get a little spooky when you're faced with fifteen feet of soup.'

2. Of a person (or animal): nervous; easily frightened; superstitious. *N. Amer. slang.*
1926 D. Branch *Cowboy & his Interpreters* 12 There were times when the steer would get spooky and mad. 1932 L. Golding *Magnolia St.* ii. v. 354 I'm not a spooky person but I sometimes think he was the Devil. 1947 *Westerners Brand Bk.* (Denver Posse) 51 Range cattle.. were too 'spooky' in those days for man-made bridges. 1962 G. MacEwan *Blazing Old Cattle Trails* i. 4 Attendants knew that the nervous and spooky longhorns were easily alarmed and would stampede at the slightest provocation. 1979 *Fortune* 26 Mar. 24/2 Even those spooky about coping with Italian traffic can easily find the well-marked way to Monza, about ten miles northeast of Milan.

3. Of or pertaining to spies or espionage. *U.S. slang.*
1975 *Times* 12 June 18/4 The Central Intelligence Agency spooks are the most spooky spooks. 1979 L. Pryor *Viper* iv. 79 They're tough, crusading terrorists.. That isn't going to make your job..any easier... Keep your spooky friends out of my life. 1980 J. Melville *Chrysanthemum Chain* 120 Somebody on the spooky side of the Embassy might have a view.

Hence **spoo·kily** *adv.*, in a spooky manner.
1955 R. Hobson *Nothing too Good for Cowboy* xviii. 186 These wild ones [*sc.* steers] were held in one bunch and slithered spookily along immediately behind my saddle horse. 1959 'P. Quentin' *Shadow of Guilt* xiv. 126 We both stood looking at the shirt... It had a spookily human quality.

spool, *sb.*[1] Add: **1. d.** (Later examples.)
1955 R. Hobson *Noting Too Good for Cowboy* xviii. 186 Two spools are attached in a light-tight box to the top of the instrument. 1936 *Discovery* Aug. 238/2 The film..is mounted in a supply spool on the top part of the camera. It is carried around guide spools and across the focal plane to the take-up spool. 1967 *Tape Recording Mag.* July 260/1 The..conventional, spool-to-spool machines. 1977 W. Marshall *Thin Air* v. 58 Feiffer watched the twin spools on the tape recorder winding in Number Two's words.

2*. The sliding member of a spool valve.
1960 Lee & Blackburn in J. F. Blackburn et al. *Fluid Power Control* ix. 239 The spool valve must be very

accurately made since it depends upon closeness of fit between spool and sleeve to hold the leakage down. 1974 *Encycl. Brit. Micropædia* X. 344/3 In the mid or neutral position of the spool, ports A and B are blocked. The movement of the spool may be manually or electrically controlled.

3. a. spool valve, a valve in which a shaft with channels in its surface slides inside a sleeve with ports in it, the flow between which depends on the position of the shaft in the sleeve.
1908 R. Peele *Compressed Air Plant for Mines* xx. 248 The throw of the spool valve..is produced indirectly by the introduction of a system of small, auxiliary ports, connecting the ends of the valve chest with the cylinder. 1926 R. M. Evans in J. Roberts *Mining Educator* I. 618/2 There are two forms..: the air-thrown or spool-valve drill for use with compressed air, and the tappet drill for use with steam. 1971 *Farmers Weekly* 19 Mar. 82/4 Three hydraulic spool valves were fitted on the side of the trailer. One lifted and lowered the trailer. 1976 D. E. Turnbull *Fluid Power Engin.* vi. 172 The spool valve was originally developed..more than a century ago when it was used for controlling the flow of steam to steam engine cylinders at the famous Gorton locomotive works in Manchester.

b. Designating (an article of) furniture popular in N. Amer. during the second half of the 19th cent. and decorated with spool-shaped turnery.
1928 Johnson & Sironen *Man. Furnit. Arts & Crafts* i. xiii. 79 The first and most noticeable product of the [powered wood-working] machine was a type [of furniture made up almost entirely of turned posts, frames, legs and arms and became known as spool furniture. 1931 S. Glaspell *Ambrose Holt & Family* xxvii. 263 She went up to the bed, a walnut spool bed. 1946 *Sun* (Baltimore) 16 July 10/3 The spool rack..had nice roomy shelves, which were used to store many back copies of the *Sunpapers*. 1969 J. Gloag *Short Dict. Furnit.* 636 *Spool Furniture*, an American style, introduced in the 1850s, and based on the use of spool turning on the members of chairs, beds, tables, and wash stands. This form of decoration with its string of spools was an attenuated variation of the more robust bobbin turning..of the mid- and late 17th century. 1974 R. B. Parker *God save Child* (1975) xxii. 153 There was a spool bed with a gold-patterned spread. 1981 *Times Lit. Suppl.* 20 Feb. 210/1 Jane Carlyle sitting in her spool chair.

spool, *v.* Delete *rare* and add: **a.** (Later examples.) Also with advbs. and *transf.*
1976 S. Brett *So Much Blood* x. 126 I'll spool through and see if there's anything relevant.'.. Gerald busied himself spooling on and playing snatches of the tape. 1976 B. Lecomber *Dead Weight* iii. 46 The Boeing's engines spooled up to a keening howl. 1977 *Time* 10 Jan. 6/1 Poised at the gate. Once more the course spools through the mind. Now the Starter's signal. 1978 J. Carroll *Mortal Friends* iv. iv. 433 His fear unraveled itself and then spooled quickly around his winch of a throat.

b. Also, to wind (film, etc.) on (to) a spool. Also *transf.*
1979 G. Scott *Hot Pursuit* ix. 80, I began to spool the film of what had happened through my mind. 1980 S. Brett *Dead Side of Mike* x. 116 Shall I spool it [*sc.* a cassette] back and see if there's something we missed?

spooled *ppl. a.* (later example); **spoo·ling** *vbl. sb.*
1897 *Sears, Roebuck Catal.* 333/1 Spooled Wire for Tissue Paper Work. 1969 *Listener* 13 Feb. 193/3 There is no lacing or spooling. The films in the cassette are sealed. 1976 *Offshore Platforms & Pipelining* 209/1 Studies show that this method of spooling between flanges can be used on lines and risers.

spoon, *sb.* Add: **2. c.** A spoonful of sugar or other substance.
1922 Joyce *Ulysses* 69 He scalded and rinsed out the teapot and put in four full spoons of tea. 1944 C. Himes *Black on Black* (1973) 200 By the time I find they was gettin' up for breakfast all the breakfast was gone but a spoon of grits. 1968 P. Dickinson *Skin Deep* vii. 142 How many spoons shall I put in?.. Bob likes six. 1980 R. Hill *Spy's Wife* iv. 22 Aspinall came in with a tea tray. 'Three spoons for me,' said Monk.

d. A dose or measure of an intoxicating drug, *spec.* two grammes of heroin. *U.S.*
1968–70 *Current Slang* (Univ. S. Dakota) III–IV. 118 *Spoon*, a level teaspoon of heroin. (Drug users' jargon). 1977 J. Cheever *Falconer* 46 Two spoons had been found, hidden in Farragut's toilet bowl.

e. *pl.* A pair of spoons held in the hand and beaten together as a simple percussion instrument.
1972 *Jazz & Blues* Nov. 27/1 'Main line' has added interest in that Shorty accompanies himself on spoons. 1977 P. Carter *Under Goliath* xxvii. 147 It was a terrific party... Mr Mitchell played the spoons and Mr Gannon brought out his accordion and we had a singsong.

4. c. Delete quot. 1814 and add earlier and later examples. Also, a lofted stroke played with this club.
1790 C. Jones *Hoyle's Games Impr.* 288 The Spoon.. [is used] when in a Hollow. 1858 [see *GUTTA[2] 1 b]. 1927 *Daily Mail* 8 July 14/4 The 3 at the difficult twelfth, where he was five yards from the pin with a drive and a spoon and holed the putt, would have shaken any youthful rival. 1962 *Times* 9 June 3/2 He struck a spoon, which ran pleasingly up on to the ninth green. 1971 L. Koppett *N.Y. Times Guide Spectator Sports* vi. 128 The No. 3 wood, or the 'spoon', provides distance with more loft.

f. *Cricket.* A ball lofted by a soft or weak shot; a stroke which 'spoons' the ball. Cf. Spoon *v.*[2] 2 b.
1871 'Thomsonby' *Cricketers in Council* 3 A ball hit into the air is a 'spoon', unless it goes a long way, when it becomes a 'skyer'. 1906 A. E. Knight *Compl. Cricketer* 353 *Spoon*, a badly mistimed hit.

g. *Surfing.* = *ROCKER[1] 5 e.
1963 *Pix* 28 Sept. 62/4 *Spoon*, the slight upward slope in a surfboard. 1970 *Studies in English* (Univ. Cape Town) I. 28 *Rocker*, or *banana*,..indicates the curvature of the surfboard along its length; in other words, the surfboard, when viewed from the side, is higher at both ends than in the centre. More specific is the word *spoon*, which applies to the upturn of the nose of the surfboard.

8. a. (Earlier example with *about*.)
c 1859 J. S. Coyne *Everybody's Friend* i. i. 7 It was one of my nonsensical effusions, when I was spoons about you... Mrs. F. Spoons! Feath. Well, when I was dying in love with you, my dear.

b. (Earlier and later examples.) Also in *sing.* (*rarely*), an instance of sentimental love-play; a fond lover.
1846 *Spirit of Times* 18 Apr. 92/2 The girls are beautiful, with a very liberal allowance of 'the spoons', as our friend Smith would say. *c* 1921 D. H. Lawrence *Mod. Lover* (1934) 188 A young chap goes out on Sunday night for a bit of a spoon. What is it but natural? *Ibid.* 195 Yes, his reputation as a spoon would not belie him. He had lovely lips for kissing. 1939 Joyce *Finnegans Wake* 115 Some softnosed peruser might mayhem take it up erogenously as the usual case of spoons, *prostituta in herba*.

9. a. *spoon-case* (later example), *-tray*, *-victuals* (earlier and *fig.* examples).
1922 Joyce *Ulysses* 30 And snug in their spooncase of purple plush, faded, the twelve apostles. 1765 J. Wedgwood *Let.* 17 June (1965) 34 The articles are..teapot and stand, spoon-tray, Coffeepot, [etc.]. 1977 Fleming & Honour *Penguin Dict. Decorative Arts* 751/1 *Spoon tray*, a small oval or oblong dish used in mid-c18 England to hold tea-spoons. 1777 *Pennsylvania Even. Post* 11 Feb. 73/2 Philip Clark..has a remarkable way of throwing his head back when he eats spoon victuals. 1877 C. Box *Eng. Game Cricket* 461 *Spooning*, getting under the ball. In derision, it is called 'spoon victuals', especially at Cambridge.

b. *spoon-seat.*
1922 Joyce *Ulysses* 653 He..drew two spoonseat deal chairs to the hearthstone.

11. spoon-back, the back of a chair (of a type esp. popular in the late-18th and 19th cent.) curved concavely to fit the shape of the occupant; a chair of this style; hence *attrib.*, as *spoon-back chair*; also **spoon-backed** *a.*; **spoon-bait** (earlier *fig.* example); **spoon-bending,** the distortion of a spoon-handle by apparently psychokinetic means; also **spoon-bend** *v. intr.* and **spoon-bender; spoon bow,** a ship's bow having full round sections reminiscent of the bowl of a spoon; so **spoon-bowed** *a.*; **spoon bread** *U.S.* (chiefly *Southern*) = *egg bread* s.v. *EGG *sb.* 7 (of such a consistency that it is usu. served with a spoon); **spoon canoe** *Canad.*, a spoon-bowed canoe; **spoon drain** *Austral.*, a shallow drain across a street.
1909 G. O. Wheeler *Old Eng. Furnit.* (ed. 2) v. 167 In our illustration..we see one of those transitional specimens with cane-panels and spoon-back. *Ibid.* v. 156 (*caption*) Queen Anne cabinet, and spoon-backed chair showing early cabriole legs. 1936 *Burlington Mag.* July 42/2 Half-way between the spoon-back chair and the strapwork back of the George II period. 1969 J. Gloag *Short Dict. Furnit.* (ed. 2) 636 *Spoon back*, sometimes used in America for the banded back chair, of the Queen Anne period, the term may have been suggested because the profile of the back resembles the curve of a spoon. In England a comparatively rare early 19th century chair with an open concave back and semi-circular top rail is called a spoon-back... Mid-Victorian single chairs with oval, waisted backs are also described as spoon-backed. 1979 A. Scholefield *Point of Honour* 40 We sat down in two Victorian spoon-back chairs. *c* 1878 J. Albery *Crisis* in *Dram. Works* (1939) II. 321 She has thrown away her heart..on..young Denham. Any cold, glittering thing does for spoon-bait. 1975 *Nature* 2 Oct. 354/3 Some of the children still claimed they could spoon-bend without cheating. 1977 *Times* 3 Nov. 6/1 Britain and Japan have a higher proportion of spoon-benders a head than any other country..people who can distort cutlery simply by thinking about it. *Ibid.* 6/4 The military implications of spoon-bending. 1979 J. Wainwright *Duty Elsewhere* i. 7 He was ready to give E.S.P. the benefit of a man-sized doubt. He even claimed to have an open mind concerning the spoon-bending gag. 1909 *Cent. Dict.* Suppl., Spoon-bow. 1927 [see *bulb-keel* s.v. *BULB *sb.* 5]. 1969 *Islander* (Victoria, B.C.) 23 Nov. 11/2 She is still under construction, a 54-foot fishing schooner with the same spoon bow as the famous Bluenose and the Lunenburg schooners. 1900 Spoon-bowed [in Dict., sense 10 a]. 1932 *Scribner's Mag.* June 364/3 It was time for me to speed back to the spoon bread and young broiled turkey that were being prepared for me now in Edith's kitchen. 1941 W. A. Percy *Lanterns on Levee* i. 11 Oh, the poor little boys..never put a lump of butter into steaming batter-bread (spoon-bread is the same thing). 1960 J. J. Rowlands *Spindrift* iii. 176 Spoon-bread made from coarse water-ground corn-meal. 1979 M. G. Eberhart *Bayou Rd.* i. 17 We can have some flour and spoon bread and chicken. 1907 T. Crosby *Among An-ko-me-nums of Pacific Coast* 141 The canoes of the Pacific coast are of the type usually called 'dugouts'..[including] a 'spoon canoe', flat-bottomed and nearly straight with hardly any bow or stern. 1976 *Islander* (Victoria, B.C.) 13 June 5/3 Their graceful spoon canoes, hand hewn, 30 feet long, 3 feet 6 inches wide..were only used on the shallow draft, northern rivers. 1934 *Bulletin* (Sydney) 7 Mar. 21/1, I saw your ropes fly off when you went over the spoon drain. 1972 *Advertiser* (Adelaide) 13 June 5/8 To lessen level crossing accidents, a double spoon drain at the approaches to all level crossings may help.

spoon, *v.*[2] Add: **I. 1.** *fig.* and *transf.* (Earlier example.)
1840 M. Edgeworth *Let.* 30 Dec. (1971) 574 At Coffee time she spooned out a fine compliment to Miss E about Frank and Rosamond.

2. b. (Earlier example.) Also, to hit (a simple catch).
1836 E. Jesse *Angler's Rambles* 296 She had a perfect knowledge of what was a *bad hit*; and when her lover *spooned* a ball up into the air, which was of course caught, he generally walked off to a distant part of the field. 1912 A. Brazil *New Girl at St. Chad's* vii. 115 She played too soon at a short-pitched ball, and spooned a catch to mid-on. 1976 J. Snow *Cricket Rebel* 113 Soon afterwards he spooned a simple catch from a stroke ringsiders described as a 'protective jab'.

spooner[2]. (Later examples.)
c 1921 D. H. Lawrence *Mod. Lover* (1934) 195 Emmie ..cuddled into his arms. He was famous as a spooner, and she was famous as a sport. 1976 *Times* 24 Mar. 13/3 A wilderness of forms for the convenience of spooners.

spoonerism. (Later examples.)
1923 [see *MARROWSKY b]. 1976 *Oxford Diocesan Mag.* July 15/1, I am *not* going to put on any weight until I'm fifty, when I shall allow myself to become matronly, ready to be a follower of 'soda and gobbly matrons', as enjoined by the marriage service. (A good Spoonerism that, created quite involuntarily by my mother some years ago.)

Hence **Spooneri·smus** [nonce mock-German], a spoonerism; **spoo·nerize** *v. trans.*, to alter (a word or phrase) by a spoonerism; hence **spoo·nerized** *ppl. a.*
1923 A. Huxley *Antic Hay* xx. 284 When pain and anguish wring the brow, an interesting mangle thou, as we used to say in the good old days when the pun and the Spoonerismus were in fashion. 1927 *Daily Express* 22 July 7 Zoojolical Gardens... But why not let the misprint stick? The Zoo gardens are 'jolical' gardens, and probably the London Zoological Society would have no objection to them being spoonerised as such. 1972 D. W. Bahlman in E. W. Hamilton *Diary* p. xxv, The Herbert family, Hamilton, and other friends..called themselves the Bilton Waggers, a Spoonerized version of Wilton Baggers. 1974 V. Nabokov *Look at Harlequins* (1975) ii. v. 101 Only a lunatic would have chosen a pair of third-rate publicists to write about—spoonerizing their names in addition!

spoon-feed, *v.* Add: Hence **spoon-feeding** *vbl. sb.* (in quots., *fig.*).
1905 M. F. Reany *Medical Profession* i. 21 'Spoon-feeding'..is excellent for obtaining good examination results, but is it quite so productive of good practioners? 1978 *Listener* 19 Jan. 87/2 The mixture of spoonfeeding and stimulation is..the essential quality of the lectures on scientific themes given..to an audience of children.

spooniness. 2. (Earlier example.)
1856 D. G. Rossetti *Let.* June (1965) I. 303, I fear tin is out of question, as I think *all* contributors write for love or spooniness.

spoony, *a.* **1. b.** (Earlier example.)
1834 C. Mathews *Let.* 13 Mar. in A. Mathews *Mem. Charles Mathews* (1839) IV. xiii. 280 It [*sc.* misreporting a speech] has such a *spoony* appearance, breaking out in a fresh place with such a phrase.

sporadic, *a.* Add: **2. a.** (Earlier example.)
1813 T. Young in *Quarterly Rev.* Oct. 255 All the Asiatic and European languages..which may be subdivided into five orders, Sporadic, Caucasian, Tartarian, Siberian, and Insular.

d. *Astr.* Applied to a meteor that is isolated and does not appear to belong to a shower. Also *absol.*, a sporadic meteor.
1929 C. Hoffmeister in *Astrophysical Jrnl.* LXIX. 167, I have developed some methods..to find out..which portion of meteors..may be regarded as 'sporadic', i.e., the directions of motions being distributed in an irregular way without indication of radiation in currents... Meteors emanating from a real radiant are mostly of cometary character and sporadic meteors are mostly of interstellar origin. 1954 A. C. B. Lovell *Meteor Astron.* xxi. 429 A good deal of attention has been given..to the problem of the velocity of sporadic meteors, and the conclusion now seems inescapable that they must be contained in the solar system as distinct from the interstellar view which has prevailed for so long. 1961 D. W. R. McKinley *Meteor Sci. & Engin.* i. 3 The number of sporadics greatly outweighs the total number of meteors belonging to the well-known showers. 1978 V. F. Buchwald *Iron Meteorites* I. i. 11/1 It is commonly inferred that all meteors—including the sporadic—are fragments of comets.

3. b. *sporadic E-layer, -region*: a discontinuous region of ionization that occurs from time to time in the E-layer of the ionosphere and results in the anomalous reflection back to earth of VHF radio waves. Also *ellipt.* as *sporadic E, sporadic-E,* used *attrib.* and *absol.*

1937 *Terrestrial Magnetism* XLII. 76 No relation between the sporadic *E*-region ionizations and magnetic disturbances of widespread character is apparent so far. 1949 *Nature* 2 Apr. 528/2 The usual measure of sporadic *E*-layer ionization is the so-called 'critical frequency', that is, the highest reflected frequency for this layer. 1955 *Sci. Amer.* Sept. 129/1 One of the outstanding mysteries of the ionosphere is a type of irregularity called 'sporadic E'. 1967 *Economist* 22 July 280/1 UHF is virtually immune to sporadic-E problems which so beset VHF. 1974 E. Harnischmacher in F. Verniani *Struct. & Dynamics of Upper Atmosphere* 272 The Sporadic E layer, Es, is the most irregular and the thinnest layer of the ionosphere.

4. *Math.* Being a finite simple group that does not fall into any of the infinite classes into which most finite simple groups fall. Also *absol.* as *sb.*
1968 *Proc. Nat. Acad. Sci.* LXI. 398 (*heading*) A perfect group of order 8,315,553,613,086,720,000 and the sporadic simple groups. 1968 *Manifold* I. 12 Apart from the Mathieu groups, no other sporadic simple groups were known until Zvonomir Janko discovered one in Australia ..in 1965. 1973 *Amer. Math. Monthly* LXXX. 1028 Still, some hardy souls felt a thorn in their side. For the five groups of Mathieu all reason defied; Not A_n, not twisted, and not Chevalley, They called them sporadic and filed them away. 1980 *Sci. Amer.* May 68/1 There is good reason to believe the number of sporadic groups is finite, and indeed many mathematicians believe the sporadics that have already been identified, a total of 26, complete the list of finite simple groups.

sporangiole: see Sporangiolum in Dict. and Suppl.

sporangiolum. (Examples of the form *sporangiole*.)
1895 F. W. Oliver tr. *Kerner's Nat. Hist. Plants* II. 673 In *Thamnidium* the sporangial branch ends in a large sporangium, and in addition bears laterally a number of tiny sporangia (sporangioles) containing four spores each. 1928 C. W. Dodge tr. *Gäuman's Compar. Morphol. Fungi* ix. 105 Under favorable conditions of nourishment, however, they are continued through several generations when the sporangioles become as large and multispored as the sporangia. 1953 J. Ramsbottom *Mushrooms & Toadstools* xviii. 209 In this common form..the fungus usually shows finely-branched endings (arbuscules) with characteristic granular masses (sporangioles) when they are being absorbed by the host plant. 1969 F. E. Round *Introd. Lower Plants* iv. 67 No flagellate cells have ever been found in the class [*sc.* Zygomycetes]..; the spores germinate directly into a germ tube and are formed either in a large spherical sporangium or in various reduced few-spored 'sporangioles' or even singly.

sporangiospore (sporæ·ndʒiŏspō°ɹ). *Bot.* [f. Sporang(ium + -o + Spore.] A spore produced in a sporangium.
1889 in *Cent. Dict.* 1899 *Ann. Bot.* XIII. 477 Léger [in 1897] distinguishes three types of spores in the Mucorineae: sporangiospores, chlamydospores and zygospores. 1930 H. M. Fitzpatrick *Lower Fungi* vii. 149 In *Geolegnia* the encysted sporangiospores are thick-walled and remain quiescent in the sporangium until freed by the disintegration of its wall. 1974 *Encycl. Brit. Macropædia* XII. 761/2 The lower fungi (*i.e.*, more primitive) produce spores in sporangia, which are saclike sporophores whose entire cytoplasmic contents cleave into spores, called sporangiospores. Sporangiospores are either naked and flagellated (zoospores) or walled and nonmotile (aplanospores).

spore. Add: **3. a. spore print,** a permanent image of the spore-producing structures of a fungal fruiting body, made by allowing spores to fall a short distance on to a surface where they adhere.
1900 McIlvaine & Macadam *Thousand Amer. Fungi* p. xxx, When a spore-print is to be taken, select a fully-grown specimen, remove the stem, place the spore-bearing surface upon the gummed paper, cover tightly with an inverted bowl or saucer, and allow to stand undisturbed for eight or ten hours. 1969 F. E. Round *Introd. Lower Plants* vi. 85 Spore colour is a valuable aid to identification and is easily detected by placing a mature 'mushroom' head on a piece of white paper and leaving overnight to produce a 'spore print'.

sporeling (spō°·ɹliŋ). [f. Spore + -ling[1], after *seedling*.] A young plant (e.g. a fern) developed from a spore.
1910 C. T. Druery *Brit. Ferns* iv. 27 Given improved seedlings (we prefer seedling to sporeling, even in Ferns, since practically a seed precedes the young Fern) of this class, the probability is that their offspring will vary still more. 1914 *Plant World* Feb. 31 Both tetraspores and carpospores..were sown on oyster shells... In about twenty-four hours, when the resulting sporelings were firmly attached, the shells were clamped to boards. 1946 *Nature* 19 Oct. 536/1 Ontogeny was also illustrated by living sporelings and beautiful microscope preparations from fronds of different ages of the fern-royal. 1970 R. Carson *Rocky Coast* i. 38 The algal spores swarming in the water, ready to settle down and become sporelings.

Spörer (ʃpō·rəɹ). *Astr.* Also **Spoerer, Sporer.** The name of Gustav-Friedrich Wilhelm *Spörer* (1822–96), German astronomer, used *attrib.* and in the possessive, as **Spörer's law,** an empirical relationship noticed by Spörer according to which the mean latitude of sunspots tends to decrease as a sunspot cycle progresses; **Spörer minimum,** the interval between about A.D. 1400 and 1510 during which little activity is thought to have taken place on the sun.
1922 *Monthly Notices R. Astron. Soc.* LXXXII. 536 Each cycle begins with an activity in high latitude; each cycle ends with the last remnants of activity transferred to a low one. This is what is known as 'Spoerer's Law of Zones'. 1968 P. Moore *Sun* v. 52 Spörer's law is undoubtedly significant, and if we could explain it we should obtain a valuable clue as to the cause of the solar cycle. 1976 J. A. Eddy in *Science* 18 June 1196/1 The earlier minimum, which we might call the Spörer Minimum, persisted by our 10-parts-per-mil criterion from about 1460 through 1550... We can presume that the Spörer Minimum was probably as pronounced as the Maunder Minimum and that during those years there were few sunspots indeed. *Ibid.* 1199/3 The Spörer Minimum of the 16th century is coincident with the other severe temperature dip of the Little Ice Age. 1978 *Daily Tel.* 8 May 11/8 The 15th century..marked the Sporer Minimum of sunspots..and was cold.

sporidesmin (spɒride·zmin). *Biochem.* Also **sporo-.** [f. mod.L. *Sporidesmium*, generic name of the fungus from which the first such substance was isolated (H. F. Link 1825, in D. C. L. Willdenow *Linnæus' Species Plantarum* (ed. 4) VI. ii. 120), f. Sporid(ium (or *spori-*, comb. form of Gr. σπορά Spore) + Gr. δεσμ-ός band, bundle: see -in[1].] Any of a class of toxins that are produced by fungi and cause various diseases of animals, esp. sheep, when ingested.
1959 Synge & White in *Chem. & Industry* 5 Dec. 1547/1 We propose the name 'sporidesmin' for the substance now described. 1963 *Jrnl. Chem. Soc.* 3172 On alkaline degradation ammonia, methylamine, and a red ketone were isolated from both sporidesmins. 1969 Edington & Gilles *Path. in Tropics* xi. 486 Various toxic substances have been isolated from moulds contaminating animal foodstuffs during this century, sporodesmin isolated from *Pithymyces chartarum* and causing liver necrosis and cirrhosis in sheep being an example. 1975 *N.Z. Jrnl. Agric.* Sept. 17/1 Very high zinc intakes protected against sporidesmin, the fungal toxin causing facial eczema. 1981–2 *Deer Farmer* (N.Z.) Summer 11/1 The susceptibility of both Red and Fallow deer to sporidesmin, the facial eczema toxin.

Spork (spǭɹk). Also **spork.** [Blend of Sp(oon *sb.* + F)ork *sb.*] A proprietary name for a piece of cutlery combining the features of a spoon, fork, (and sometimes, knife).
1909 in *Cent. Dict.* Suppl. 1970 *Official Gaz.* (U.S. Patent Office) 11 Aug. TM 65 Van Brode Milling Co., Inc., Clinton, Mass... *Spork* for Combination Plastic Spoon, Fork and Knife. 1971 P. J. R. Nichols *Rehabil. Severely Disabled* II. iii. 117 Spoons or spoons with fork ends (sporks), either fitted with a swivel or shaped to a child's needs, are the commonest aids supplied. 1975 *Equipment for Disabled: Home Managem.* (ed. 4) 50/1 The spork can be adapted by bending and/or lengthened by rivetting on an extension. 1976 *Trade Marks Jrnl.* 22 Dec. 2628/1 *Spork*... Cutlery, forks and spoons, all included in Class 8. D. Green and Company.., Sutton, Surrey..; manufacturers and merchants.

sporo-. Add: Also with pronunc. (spǭ°ro). **sporoge·nesis,** the formation of spores; **sporophy·tically** *adv.*; **spor·oplasm,** the protoplasm of a spore.
1890 Webster, *Sporogenesis*, reproduction by spores. 1905 *Bot. Gaz.* XL. 93 The events of sporogenesis in *Pallavincia Lyellii* present..no fundamental differences from those of other liverworts and higher plants. 1969 A. M Campbell *Episomes* xiv. 166 It was suggested.. that activation of an episome might play a causative role in sporogenesis. 1970 *Bot. Gaz.* CXXXI. 139/2 The incompatibility system of the family is of the homomorphic, sporophytically controlled type. 1893 R. R. Gurley in *Bull. U.S. Fish Comm.* 1891 413 *Cystodiscidae*... A bivalve shell..: condition of sporoplasm unknown. [*Note*] *Sporoplasm.* Protoplasm of the spore. 1947 *Ann. Rev. Microbiol.* I. 6 Typically each spore results from the cooperative activity of six cells, two giving rise to the valves of the sporocyst, two producing the polar capsules, ..and two being sporoplasm cells each with a gamete nucleus. 1979 *Jrnl. Protozool.* XXVI. 448/2 Mature spores are short-lived and within 12–24 h begin to extrude their sporoplasms in all directions.

sporodesmin, var. *SPORIDESMIN.

sporodochium (-dōᵘ·kiǒm). *Bot.* Pl. **-dochia.** [ad. mod.L. *sporidochium* (H. F. Link 1824, in D. C. L. Willdenow *Linnæus' Species Plantarum* (ed. 4) VI. ii. 1), f. Gr. σπορά Spore + δοχή receptacle + L. *-ium* neut. suffix.] A fungal fructification composed of an exposed mass of conidia overlying a cushion-like layer of short conidiophores.
1913 *Phytopathology* III. 24 The chief contrast to pionnotes lies in the Tubercularia-like hemispherical form of the sporodochia. 1947 *Ann. Rev. Microbiol.* I. 66 Miller.. presented good evidence that the sporodochia produced by isolates of *Fusarium* growing on agar media are in reality patch mutants. 1971 P. H. B. Talbot *Princ. Fungal Taxonomy* x. 143 A sporodochium..is a compound sporophore of pulvinate (cushion-shaped) form, composed of a stromatic base giving rise to closely grouped erect conidiophores.

Hence **sporodo·chial** *a.*, of or pertaining to a sporodochium.
1913 *Phytopathology* III. 36, I find it very easy to produce cotton, tomato, and cowpea wilt with the young pionnotes and sporodochial stage of the 3-septate conidia which the vascular parasites freely form in pure culture. 1973 *Ibid.* LXIII. 831/1 Sporodochial development counts were made at full bloom.

sporonin (spɒ·rǭnin). *Biochem.* [a. G. *sporonin* (Zetsche & Huggler 1928, in *Ann. der Chem.* CDLXI. 94): see Sporo- and -in[1].] An inert substance forming the resistant outer covering of spores. Cf. *SPOROPOLLENIN.
1928 *Chem. Abstr.* XXII. 2949 This, after prolonged digestion with HCl, followed by boiling with 5% alkali, affords a new brownish yellow substance, sporonin, $(C_{10}H_{14}O_3)_x$. 1964 *Grana Palynologica* V. 247 According to Zetsche the membranes consisted of cellulose and a specific substance called sporonin, or pollenin if derived from pollen, which was responsible for the chemical stability of the membranes. 1974 Stanley & Linskens *Pollen* ix. 138 Because of the similarity of spore wall sporonin to pollen wall pollenin they [*sc.* Zetsche et al.] subsequently used the word '*sporopollenin*'.

sporont (spɒ·rɒnt, spŏr-). *Zool.* [f. Sporo- + Gr. ὀντ-, ὤν, pres. pple. of εἶναι to be, exist.] A protozoan cell at a stage of the life cycle following syngenesis and preceding the formation of spores.
1885 [see *CEPHALONT]. 1900 *Jrnl. R. Microsc. Soc.* 337 The macrogametes are fertilised by microgametes, and in each case an oocyst is formed round the conjugates. The new nucleus (sporont-nucleus) divides, the daughter-nuclei also divide, and later the protoplasm, so that four sporoblasts are formed within the oocyst. 1912 [see *GAMONT]. 1947 *Ann. Rev. Microbiol.* I. 7 Fusion of these gametes in pairs produces eight zygotes which become sporonts. 1977 *Jrnl. Protozool.* XXIV. 55/1 Binucleate and quadrinucleate sporonts..were observed.

Hence **sporontici·dal** *a.* [-cide], lethal to sporonts.
1970 W. Peters *Chemotherapy & Drug Resistance in Malaria* iv. 123 Pyrimethamine alone at a dose of 0·078 mg/kg had no sporontocidal action. 1977 *Martindale's Extra Pharmacopoeia* (ed. 27) 343/2 The biguanides.. have..a rapid sporonticidal action against some strains.

sporopollenin (-pɒ·lĕnin). *Biochem.* [a. G. *sporopollenin* (Zetsche & Vicari 1931, in *Helv. Chim. Acta* XIV. 64): see *SPORONIN and Pollenin (in Dict. and Suppl.).] An inert substance, consisting largely of polysaccharides, that forms the resistant outer coating of spores and pollen grains.
1931 *Chem. Abstr.* XXV. 2455 The crude sporopollenin, *i.e.* mixt. of spore and pollen membranes, obtained was purified as previously described. 1966 *Jrnl. Chem. Soc.: C* 16/2 A major problem in the study of sporopollenins is their degradation by sufficiently mild methods to give relatively large molecular species. 1978 Hoyle & Wickramasinghe *Life Cloud* x. 91 The first possibility..was that it might be a rather exotic material, sporopollenin. This substance forms a major component of the protective coatings in pollens and spores and is chemically and thermally extremely stable.

sporotrichosis (-trikōᵘ·sis). *Med.* [f. mod.L. *Sporotrich-um* (orig. *Sporothricum*), former name of the genus *Sporothrix* (H. F. Link 1809, in *Mag. für die neuesten Entdeckungen in der ges. Naturkunde* III. 12), f. Gr. σπορά Spore + τριχ-, θρίξ hair: see -osis.] A disease caused by the chronic presence in the tissues of the fungus *Sporothrix schenckii*, common in soil and wood and freq. introduced by a superficial scratch, typically producing nodules and ulcers in the lymph nodes and skin.
1908 *Brit. Jrnl. Dermatol.* XX. 301 Sporotrichosis is not incompatible with the presence of syphilis or tuberculosis. 1939 *Yearbk. Dermatol. & Syphilol.* *1938* 57 The most common type of sporotrichosis reported is the cutaneous or localized form with secondary regional lymphangitis. 1976 Edington & Gilles *Path. in Tropics* (ed. 2) v. 284 Sporotrichosis..assumed almost epidemic proportions in miners in South Africa due to splinters infecting the skin.

sport, *sb.*[1] Add: **I. 1. e.** In proverbial phr. *the sport of kings* (latterly, influenced by sense 5), orig. applied to war-making, but later extended to hunting and horse-racing (also surf-riding).
[*a* 1668 Davenant *Soldier going to Field* in *Works* (1673) 322 For I must go where lazy Peace, Will hide her drouzy head; And, for the sport of Kings, encrease The number of the Dead. 1691 Dryden *King Arthur* ii. 19, I count not War a Wrong: War is the Trade of Kings, that fight for Empire; And better be a Lyon, than a Sheep.] 1735 W. Somervile *Chace* i. 14 My hoarse-sounding Horn Invites thee to the Chace, the Sport of Kings, Image of War, without its Guilt. 1843 R. S. Surtees *Handley Cross* I. xiii. 253 'Untin', as I have often said, is the sport of kings—the image of war without its guilt, and

only five-and-twenty per cent. of its danger. 1918 G. Frankau *Poet. Works* (1923) II. xxi. 130 Weep for the King of Sports, the Sport of Kings;..On thousand tracks, unridden, desolate, Hay waves from winning-post to starting-gate. 1935 T. Blake *Hawaiian Surfboard* iii. 66 News reels and still cameramen will be on hand to shoot the thrilling rides that always accompany big surf, so the rest of the world may see the 'sport of kings' by picture. 1961 L. Mumford *City in History* ii. 44 With concentration on war as the supreme 'sport of kings', an ever larger portion of the city's new resources..went into the manufacture of new weapons. 1968 W. Warwick *Surfriding in N.Z.* 1 Surfriding was practised almost exclusively by members of Hawaiian royal families: hence surfriding's now anachronistic title, 'Sport of Kings'. 1978 *N.Y. Times* 29 Mar. B4/7 The track plans to feature rhinoceroses thundering down the homestretch in the next novelty of the sport of kings.

II. 6. b. *fig.* (Later examples.)
1954 N. R. Ker in R. M. Wilson *Ancr. Riwle* p. xii, If the whole method of writing and the orthography of the Caius manuscript were typically English the aberrant *r* could be explained..as a 'sport' by an English scribe in a period of experiment and change. 1971 *Lancet* 9 Oct. 811/1, I found myself immersed in a tangle of phys- words... Except for a few sports like 'physbuttock' (from fizz) they all come from two roots.

8. d. A good fellow, a lively, sociable person (applied to men or women); one who behaves in a 'sportsmanlike' fashion. Also *good sport*; *old sport* (freq. as a familiar term of address, more usually of men than women); *be a sport*, act in a generous and sportsmanlike spirit.
1881 Ld. Rhondda *Let.* 30 Oct. in *D. A. Thomas Viscount Rhondda* ii. 24 X— didn't make herself particularly unpleasant to me, though no doubt she was annoyed about something. I think she is rather a sport because she is such a good type of a certain class of character. 1905 *Punch* 22 Mar. 199, I shouldn't mind, Old Sport. 1913 *Ibid.* 21 May 405, I say, old chap, I've not had a smoke for half-an-hour, so I think I'll go on top. Be a sport and go inside with the women, will you? 1915 R. H. Davis *With Allies* viii. 159 All that was asked of the stranded Americans was to keep cool and, like true sports, suffer inconvenience. 1917 M. B. Owen *Secret of Typewriting Speed* 85 It is better to be known by everyone as a 'fine girl' than a 'good sport'. 1918 C. Mackenzie *Sylvia Scarlett* i. ii. 64 You're no sport, Maudie. You've got the chance of your life and you're turning it down. 1920 W. J. Locke *House of Baltazar* xvii. 208 The old man must be a good sport. *a* 1922 T. S. Eliot *Waste Land Drafts* (1971) 5 Myrtle was always a good sport. 1923 Galsworthy *Captures* 145 'Let me go, mister!' came the hoarse voice again. 'Be a sport!' 1925 F. Scott Fitzgerald *Great Gatsby* iii. 57 Want to go with me, old sport? 1931 W. Faulkner *Sanctuary* vi. 57 'Come on,' Temple said. 'Be a sport. It wont take you any time in that Packard.' 1932 J. Farrell *Young Lonigan* ii. 65 He told himself Bertha was a pretty good sport, all things considered. 1933 E. O'Neill *Ah, Wilderness!* iii. i. 83 He's a hot sport, can't you tell it? 1942 A. Christie *Body in Library* ix. 85, I did like her. I thought she was a good sport. 1952 M. Laski *Village* ii. 48 People turned to slap Roy on the back and say 'Well done, old sport.' 1973 *Guardian* 2 Mar. 15/8 The North Vietnamese.. walked out of a drafting committee in protest against.. one of Waldheim's men... Waldheim is being a good sport and going round explaining that he is not the least offended. 1982 P. McGinley *Goosefoot* viii. 114 Will you come out..this evening?..Come on, be a sport.

e. Chiefly *Austral.* A familiar form of address, esp. used to a stranger. Occas. *sports.*
1935 [see *JOLT *sb.* 4 a]. 1943 K. Tennant *Ride on Stranger* v. 48 The small boy rose and said rapidly: 'Fair go, sports.' 1952 E. Lambert *Twenty Thousand Thieves* i. 29 'Have a swig, sport.' He took the bottle..and helped himself to a mouthful. 'Thanks, sport.' He handed the bottle back and idly he noted that he had never called a man 'sport' before. 1962 L. Davidson *Rose of Tibet* ii. 48 Houston introduced himself. 'Glad to know you, sport. You've caught me at a busy moment.' 1975 R. Beilby *Brown Land Crying* 80 'Come on, sport,' the doorman was saying patiently. 'You can't stop here. You've had a skinful.'

8*. *colloq.* † **a.** A film about athletic sports (*obs.*). **b.** *ellipt.* or in *pl.* The sports section of a newspaper. *U.S.*
1913 [see *feature picture* s.v. *FEATURE *sb.* 4 c]. 1923 *Nation* (N.Y.) 17 Oct. 25/1 Crime and comic strips, sports and 'columns'—the *Leader* provides them all. 1955 W. Tucker *Wild Talent* v. 65 He asked, 'Can I have the sports?' Conklin pulled the section from the paper.

c. A sports car; a sports model of a motor car.
1948 M. Laski *Tory Heaven* ix. 123 Rupert..was driving her down in his own super-charged super-sports Bentley. 1952 A. R. D. Fairburn *Strange Rendezvous* 50 Epicene Sir Giles..plays at Walton Heath, and drives a sports. 1974 R. Rendell *Face of Trespass* xviii. 168 The powerful purr of a Jaguar sports.

III. 9. a. *sport-lover; sport-minded* adj.
1929 *Daily Express* 12 Jan. 3 To-day the thoughts of sport-lovers will be spread over thirty-two battle-grounds, where the third round of the F.A. Cup competition will be fought. 1960 V. Jenkins *Lions down Under* vii. 103 The people of Christchurch are extremely sport-minded.

b. *sports centre, club, complex, day, deck, department, desk, edition, editor* (later example), *equipment, field, girl, ground, hall, news, page, pavilion, programme, section, shop, stadium, -writer; sports-mad, -minded* adjs.
1973 *Times* 27 July (Leisure Suppl.) p. i/2 The Sports Council..claimed that England and Wales need 842 indoor multi-purpose sports centres built by 1981 to supply the leisure demands of the public. 1965 'J. le Carré' *Looking-Glass War* xii. 145, I got the knife at cost.. through the sports club. 1976 *Globe & Mail* (Toronto) 30 Jan. 4 (*caption*) Metro says university to pay 50%-plus to use sports complex. 1940 F. Sargeson *Man & his Wife* (1944) 54 When the last war ended I was at the High School. We got the news of the armistice on our annual sports day. *a* 1936 Kipling *Something of Myself* (1937) viii. 229 The new three-deckers..hellishly noisy from the sports' deck to the barber's shop. 1981 J. M. Brinnin *Beau Voyage* (1982) 59 The sports deck echoing with the click of the discs used in shuffleboard. 1975 G. Howell *In Vogue* 80/1 Jean Patou's new sports department sells jersey and marocain bathing suits. 1968 D. Francis *Forfeit* viii. 101 The sports desk is a big asset to the paper. 1959 M. Shadbolt *New Zealanders* 75 Mr Jackson lay belly-upwards on an unshaded part of the baked-brittle back lawn. The Saturday sports edition covered his face. 1969 Sports editor [see *night-editor* s.v. *NIGHT *sb.* 13 c]. 1969 H. MacInnes *Salzburg Connection* iv. 50 He had opened a sports-equipment shop in Salzburg. 1931 *What is Fascism & Why?* 174 You must give houses, schools, baths, gardens, and sports fields to the working Fascist people. 1938 J. Betjeman in *New Statesman* 12 Nov. 777/1, I adore you, Pam, you great big mountainous sports girl. 1933 J. Buchan *Prince of Captivity* iv. i. 325 The aeroplane..had landed in the sports ground of the factory. 1972 G. Bromley *In Absence of Body* ix. 119 'Where's the match played?' 'At their sports ground.' 1943 Sports hall [see *SIPOREX]. 1976 *Dumfries & Galloway Standard* 25 Dec. 18/6 (Advt.), Sports Hall Area £3.00 per hour. 1963 *Times Lit. Suppl.* 24 May 370/2 Australians..are sports-mad. 1960 I. Cross *Backward Sex* iii. 76 She was not exactly sports-minded. 1928 *Radio Times* 27 Apr. 149/2 (*caption*) Bert simply puts up with [the radio set]..for the sake of the sports news. 1967 Sports news [see *city page* s.v. *CITY 9]. 1930 J. B. Priestley *Angel Pavement* ii. 57 Mr. Smeeth.. arrived at the sports page, where the prospects of certain women golfers were discussed at considerable length. 1976 L. Henderson *Major Enquiry* ii. 11 Milton glanced only briefly at the headlines of the newspaper before he turned to the sports pages. 1931 'G. Trevor' *Murder at School* ii. 42 The pair had reached the sports pavilion. 1938 J. Joyce *Let.* 6 June (1966) iii. 424 The director who has charge of the singing almost resents even a friendly introduction from the director, say, who controls the sports programme. 1973 J. Porter *It's Murder with Dover* vi. 60 He sat in front of the television watching that sports programme. 1940 G. Marx *Let.* in *Groucho Lett.* (1967) 46, I picked up the paper Tuesday morning, nervously turned to the sports section. 1926 S. Lewis *Mantrap* xxv. 288 Joe and he would..found a sports-shop of their own. 1971 C. Whitman *Death Suspended* iv. 80 Alec was at his sports shop. 1973 *Times* 5 Dec. 16/3 Those responsible for the conduct and management of any designated sports stadium. 1932 Sportswriter [see *QUARTERBACK *sb.* 2 b]. 1972 J. Mosedale *Football* ii. 21 The principal address at the Philadelphia Sportswriters Association banquet.

c. Used (chiefly *pl.* in U.K. and *sing.* in U.S.) to designate articles of attire suitable for outdoor sports or for informal wear, as *sport(s) clothes, coat, jacket, shirt, shoe, skirt, suit, wear,* etc.; also *sport(s)-coated, -jacketed, -shirted* adjs.
(*a*) 1912 *Sphere* 17 Aug. p. ii/1 The shirts I can also specially commend for holiday as well as for..sports wear. 1914 *Queen* 4 July 15 (Advt.), Smart fitting mercerised Sports Coat. 1922 *Moving Picture Stories* 23 June 23/2 A great many of the new sports clothes are shown with the divided skirt and pantalette cuff. 1925 *Eaton's News Weekly* 2 May 10 The sports shirt above, of fine white Oxford material. 1927 'C. Barry' *Mouls House Mystery* xx. 178 In an hour, Gilmartin was at his home, clad in an old sports jacket. 1930 J. Buchan *Castle Gay* xii. 191 At a small draper's..a jacket of rough tweed was purchased—what is known in the trade as a 'sports' line. 1948 'J. Tey' *Franchise Affair* x. 110 The male with the sports jacket and the pin-striped trousers. 1952 B. Hamilton *So Sad, so Fresh* ii. 22 These sports-coated, bepiped, sophisticated young bloods. 1955 N. Fitzgerald *House is Falling* vii. 107 Brigadier Poodle Poole-Casey, hatless and sports-jacketed. 1962 J. D. Salinger *Franny & Zooey* 132 And there was old Dick..Sitting at a table in blue jeans and a gruesome sports jacket. 1967 N. Freeling *Strike out where not Applicable* 70 She was the kind of woman that would not look her best in sports clothes. 1973 'S. Harvester' *Corner of Playground* iii. i. 173 The sports-shirted wrestler. 1975 G. V. Higgins *City on Hill* ii. 34 His silver-blue double-knit sportscoat was too large. 1982 'M. Innes' *Sheiks & Adders* vii. 61 An English youth..in a very commonplace sports-shirt and dark trousers.
(*b*) 1916 H. L. Wilson *Somewhere in Red Gap* v. 188 Beryl Mae Macomber in her sport shirt. 1925 *Eaton's News Weekly* 2 May 10 Sport shoes of smoked elk and tan calf, with crepe rubber sole and heel. 1927 P. Bottome *Belated Reckoning* iv. 57 The happy princess.. beautiful in her English 'sport suit' and crushed felt hat. 1946 *Reader's Digest* July 85/1 Found in addition were an Army blouse..and a sport jacket. 1946 *Chicago Daily News* 17 May 35/8 And to think I've been afraid to be seen outside in my new sport coat! 1950 M. Huxley *Let.* 21 June in A. Huxley *Lett.* (1969) 625 When you order shirts please ask whether they have an Airtex blue with sport collar, long sleeves; that is a collar which at will is worn with or without tie. 1966 H. Kemelman *Saturday Rabbi went Hungry* (1967) xiv. 89 The local chief of police was wearing a sport shirt and chinos. 1972 J. Mosedale *Football* ii. 21 He arrived resplendent in a new sport jacket. 1977 R. E. Harrington *Quintain* xi. 120 Atlas had changed from his plaid dinner jacket into a sport coat.

d. *pl.* Applied to fast, low-built motor cars of a racing type. Freq. as sports car. Also *transf.* and occas. in *sing.*
(*a*) 1925 *Correct Lubrication* 52 Lea-Francis (sports models). 1928 *Proc. Inst. Automobile Engineers* XXII. 316 With regard to the acceleration of 'sports' cars, I agree that rough running and noise are taken for power development. 1932 G. Winn *Unequal Conflict* xviii. 350 She nourished a wild hope that..she would find Derek's silver sports car standing outside. 1936 'N. Blake' *Thou Shell of Death* iii. 46 A Lagonda sports-tourer. 1955 E. Pound *Classic Anthol.* i. 60 His sports-car leads with the iron-grays, Six reins are in his hand. 1963 Bird & Hutton-Stott *Veteran Motor Car* 73 One of the pleasantest sports-touring cars of the pre-war era. 1967 N. Freeling *Strike out where not Applicable* 39 It is not far.. especially for the fast sports coupé. 1971 'D. Shannon' *Murder with Love* iv. 68, I don't somehow think Mrs. Franks drives a Mercedes sports model. 1977 C. McCullough *Thorn Birds* xviii. 484 He pushed on, gunned the red sports car up..the Domokos Pass.
(*b*) 1927 U. Sinclair *Oil!* iv. 80 But that didn't trouble Mr. Bankside, who had already..bought himself..a big new limousine, also a 'sport-car'. 1927 *Scribner's Mag.* Feb. 159/1 Laban, furious, mounts his Sport-model Camel and takes after the elopers. 1955 *Amer. Speech* XXX. 238 The best known writer on the subject, Tom McCahill, calls his most recent book *The Modern Sports Car* (New York, 1954); but a volume by Raymond F. Yates and Brock W. Yates..is named *Sport and Racing Cars* (New York, 1954). Of the two, *sports car* is unquestionably the more commonly used. 1978 J. Irving *World according to Garp* xiv. 280 The man's sport car still chugged like an animal.

10. a. sportcast(er, sportcasting = *sportscaster*, etc., sense 10 b below; **sportfest** [*FEST], a festival of sport; a meeting for athletics or other competitive sports; **sport-fisherman,** a sea-going boat equipped for sportfishing; **sportfishery; sportfishing** orig. *U.S.*, fishing by rod and line for sport or recreation; hence (as a back-formation) **sportfish,** a fish caught thus for sport rather than (primarily) for eating.
1939 *Amer. Speech* XIV. 6 Newspaper and magazine columns..'Sportorial', 'Sportcast', 'Sportlight'. 1938 *Variety* 28 Dec. 30/4 Jim Britt, WBEN sportcaster, promised a copy of a set of health rules to any listeners who would write in. 1941 *Time* 31 Mar. 50/1 Adam has sponsored the sportcasting of big-time bouts. 1937 Auden in Auden & MacNeice *Lett. from Iceland* xi. 147 The sport-fest was a primitive affair. Some part singing ..and a swimming race. 1950 E. Bradner *Northwest Angling* iii. i. 172 As a sport fish, the salmon is highly valued by the anglers of the Pacific Coast. 1971 *Nature* 18 June 422/3 The billfishes of the family Istiophoridae, which include several well known sportfish such as the sailfish, the marlins, and spearfishes. 1954 *Field & Stream* Jan. 72 (*caption*) *Miss Chevy IV*, unique 34-foot, V-bottom, twin-engine sport fisherman. 1967 *Jane's Surface Skimmer Systems 1967–68* 107/2 Sportfisherman is a luxury fishing yacht with living and sleeping accommodation for four. 1955 (*title*) Sport fishery abstracts. I. 1. 1910 C. G. Holder in *Proc. 4th International Fishery Congress* 1. 201 (*heading*) Sport fishing in California and Florida. 1926 *Daily Colonist* (Victoria, B.C.) 15 July 5/3 Something had to be done in order to save the sport fishing of Vancouver Island. 1978 A. Gilchrist *Cod Wars* v. 36 The Icelandic government (well aware of the amount of money which sport-fishing brings into the country) takes care of the availability of salmon in two important ways.

b. *pl.* **sportscaster** *N. Amer.* [after *broadcaster*: cf. *newscaster*, etc.] one who presents a sports broadcast on radio or television; a broadcasting sports commentator; hence **sportscasting;** also **sportscast,** a sports broadcast; **sportsfest** = *sportfest*, sense 10 a above; **sports finder** *Photogr.*, a direct-vision viewfinder usu. consisting of a simple frame which allows action outside the field of view of the camera to be seen, fitted esp. to twin-lens reflex cameras; **sports medicine,** the branch of medicine dealing especially with the consequences of engaging in sports.
1961 Webster, Sportscast. 1976 N. Nielsen *Brink of Murder* i. 11 The sportscast on the portable TV was in progress. 1938 *Amer. Speech* XIII. 239 Note that newscaster and sportscaster are now common terms in *Variety*. 1952 B. Wolfe *Limbo* xx. 323 With each passing day the sportscaster's voice lost a few more decibels of its professional bounce. 1964 M. McLuhan *Understanding Media* xxx. 303 A sportscaster had just begun his fifteen-minute reading from a script. 1981 'E. McBain' *Heat* iii. 51 The sportscaster read off the baseball scores. 1969 C. Armstrong *Seven Seats to Moon* xii. 125 J sat all the way through the sportscasting. 1976 *Listener* 5 Aug. 151/1 Television caught both aspects of this mammoth *sportsfest* [*sc. The Olympic Games*] very well. 1953 A. Matheson *Leica Way* 100 The sports finders are an earlier version of the brilliant finders. 1977 J. Hedgecoe *Photographer's Handbk.* 165 Most twin lens reflexes have a 'sports finder' which folds out of the hood. 1961 (*title*) Journal of sports medicine and physical fitness. 1973 *Times* 13 Dec. 7/2 Britain is far behind some European countries, America and Australia in her attitude to sports medicine, Dr Ian Adams, medical officer to Leeds United Football Club, said. 1977 J. F. Fixx *Compl. Bk. Running* xxii. 248 The general physician is not interested in sportsmedicine.

sport, *v.* Add: **I. 2. e.** (Earlier example.)
1789 *Loiterer* 20 June 9 The Squire of the Parish..will ..give him unlimited leave to sport over his Manor.

II. 11. a. Restrict ? *Obs.* to *to sport timber*; *to sport* (usually *one's* or *the*) *oak*: (later examples).
1911 Beerbohm *Zuleika D.* xii. 187 The man who now occupied my room had sported his oak—my oak. 1932 R. Aldington *Soft Answers* 72 My bell rang again...

Had they come back to make my flat the battleground for another Idiot scene? It was no use sporting the oak with the lights in my windows. **1951** S. Spender *World within World* ii. 50 If one arrived early one was liable to find the heavy outer door of his room, called 'the oak', sported as a sign that he was not to be disturbed. **1974** J. I. M. Stewart *Gaudy* viii. 145 The light on the little landing was extinguished, but Ivo Mumford's oak had not been sported.

13. (Later example.)
1894 A. Morrison *Tales of Mean Streets* 41 There was a milliner's window, with a show of..hats... 'Which d'yer like, Lizer?—..I'll sport yer one.'

‖ **sportif** (sportif), *a.* (*sb.*) [Fr.] Of a sporting character, sportive; interested in or pursuing athletic sports. Also, of a garment: suitable for sporting or informal wear. Also *absol.* and as *sb.*
1934 C. Lambert *Music Ho!* iv. 242 The musical equivalent of this obsession with the mechanical, the *sportif* and the soi-disant contemporary, is provided by the naïvely realistic orchestral pieces of Honegger, such as *Pacific 231* and *Rugby*. **1938** E. Hemingway *Fifth Column* (1939) 171 There were two kinds: the drunkards and the sportifs... The sportifs took it out in exercise. **1958** *Manch. Guardian* 4 Aug. 3/6 Last year it was..the man who played boogie-woogie on a gramophone all night in the next room and when I complained said I wasn't *sportif* and made a bigger row than ever. **1963** *Times* 1 May 9/6 Elegance is the word that immediately comes to mind—elegance with more than a touch of the *sportif*, for the sportswoman is a good bit better than you might expect from a car of less than 1-litre capacity. **1966** *Listener* 30 June 936/1 All nations are now *sportif*, although they do not necessarily acclaim the same sports as we do. **1977** *Times* 30 July 10/1 The hooded sweat-shirt is..casual and sportif yet real and warm.

sporting, *vbl. sb.* Add: **3. b.** *sporting column, page, paper; sporting event; sporting goods, jacket.*
1901 *Bookman* Oct. 123/2 Americans..have noted the peculiarities of the diction of the writers of the sporting columns. **1920** *Times* 14 June 15/3 Among the sporting events which follow each other in..June, Royal Ascot..is unique. **1869** *Boyd's Business Directory* 500 John H. Mann, importer and dealer in guns, fishing tackle, gun powder, and all sporting goods. **1978** R. B. Parker *Judas Goat* xxii. 135 'Picked up a new shotgun at a sporting goods store,' he said. **1837** Dickens *Sk. Boz* 2nd Ser. 53 A brown coat, something between a great-coat and a 'sporting' jacket, on his back. **1915** *Lit. Digest* 21 Aug. 360/3 Bozeman Bulger..contributes to the sporting page of the New York *Evening World*. **1961** P. White *Riders in Chariot* ix. 239 He would..go away, or reach for the sporting page. **1849** C. Brontë *Shirley* III. iv. 75 He reads only a sporting paper. **1907** M. E. Braddon *Dead Love has Chains* vi. 139 Game has to be forgiven in a man, like smoking, and sporting papers, and mistresses.

c. sporting editor *U.S.*, a sports editor; **sporting print,** a print of a scene taken from the field-sports; also *transf.*
1857 *Spirit of Times* 1 Aug. 340/2 We see exactly, where the 'sporting editor' of *The Times* has made his fatal mistake about handicaps and handicappers. **1899** T. Hall *Tales* 128 The somewhat intellectual-looking sporting editor of the aforesaid *Universe*. **1849** Thackeray *Pendennis* I. xviii. 168 Six sporting prints, and four groups of opera-dancers..formed the late occupant's pictorial collection. **1964** R. Jeffries *Embarrassing Death* viii. 84 Two thousand men eager to pay five or ten shillings for a 'sporting print' every month. **1973** G. Greene *Honorary Consul* ii. iii. 81 A corridor hung with Victorian sporting prints: riders falling into a stream, checked at a bullfinch, rebuked by the master.

sporting, *ppl. a.* Add: **2. a.** Also *spec.* in phr. *sporting parson.*
1826 F. Reynolds *Life & Times* I. iii. 99 The family consisted of the Dowager Lady Grandison,..an old Irish Major—a sporting parson—the house apothecary—my father, my aunt, and myself. **1901** *Daily Tel.* 23 July 10/6 Those who imagined that the last 'sporting parson' had disappeared from the Church of England are quite mistaken. **1982** M. Young *Elmhirsts of Dartington* ii. 21 His mother..meant him to be a priest, not a sporting parson..but a proper God-fearing priest.

b. *sporting man* (earlier example). Also used in other collocations referring to low gaming and betting.
1824 R. Humphreys *Mem. J. Decastro* 206 Bob Todrington, a sporting man (caricatured by Old Dighton). **1857** *Household Words* 12 Sept. 264/1 With its sparring snobs, and flashing satins, and sporting gents and painted cheeks. **1902** G. B. Shaw *Mrs Warren's Profession* p. xxix, Well, does anybody who knows the sporting world really believe that bookmakers are worse than their neighbors? **1946** K. Tennant *Lost Haven* (1947) ii. 40 Her mother was entertaining some sporting friends who had dropped in to settle up certain transactions. **1967** S. Beckett *Stories & Texts for Nothing* iii. 84 Thronged already with sporting men fevering to get their bets out of harm's way before the bars open.

c. *N. Amer.* Used *spec.* to denote a prostitute or loose woman, as *sporting girl, woman.* Cf. Sportswoman b.
1925 *Amer. Speech* I. 151/2 The woman of the underworld is spoken of as a 'sporting woman'. **1938** A. J. Liebling *Back where I came From* 152 Most of the women..go out by the day as house-workers. There may be a few sporting girls, but if so they don't work their own block. **1951** E. Paul *Springtime in Paris* iv. 89 The place Xavier Privas, where the former sporting girls and their male friends congregate. **1971** J. Gray *Red Lights on Prairies* vi. 143 The existence of a colony of sporting women at Nose Creek was prejudicially affecting the morals and welfare of the Community.

3. a. (Earlier and later examples.) Also, of or characterized by conduct consonant with that of a sportsman or 'good sport'.
1799 *Times* 1 June 4/3 Hunting Box, pleasantly situate in a sporting part of the Country. **1920** 'O. Douglas' *Penny Plain* xi. 115 'Isn't it awful..about our minister marrying..a girl twenty years younger than himself.' 'But how sporting of him,' Pamela said. **1923** Wodehouse *Inimitable Jeeves* x. 112, I had..got as far as the lift before I remembered what it was that I had meant to do to reward Jeeves for his really sporting behaviour in this matter of the chump Cyril. **1962** S. Raven *Close of Play* iii. xv. 186 By declaring when they did, they left Baron's Lodge with three-hundred and twenty-two runs to make in two hours... It was, on the whole, a sporting declaration.

b. Also, an opportunity that a sportsman might consider. (Later examples.)
1913 *Granta* 7 Mar. 255/2 If bad shows are booked for this theatre the *actors* are not the people to be blamed; they are, naturally, trying to do their best—give them a sporting chance. **1977** M. Allen *Spence in Petal Park* xxxi. 146 All that rubbish they learnt on the rugger field about giving the other fellow a sporting chance... The world just doesn't work like that.

sportive, *a.* Add: **6. a.** (Later example.)
1969 *Daily Tel.* 13 Mar. 18 Sportive readers of this paper's report yesterday on the pay talks of 90 Tonbridge cricket ball makers are puzzled..that these craftsmen should be represented by the National Union of Furniture Trade Operatives.

b. Of clothes: suitable for sporting or informal wear. Cf. *Sportif *a.*
1935 *Amer. Speech* X. 193/2 Combinations like *smoothly sportive, fetchingly feminine*..are numberless. **1963** C. Beaton *Diary* 15 Feb. (1979) 358 In his yachting jacket and sportive shoes, he has something about him of the swash-buckling style that reminds me of Douglas Fairbanks, Senior.

sportster (spōᵊ·ɹtstəɹ). [f. Sport *sb.*¹ + -ster.] **a.** A sports coat. *rare.* **b.** A sports car. Cf. *Roadster 2 d.
1963 *N.Y. Times* 20 Dec. 11 Sportster. Made of cotton Safari cloth, lined with wool alpaca. **1971** *New Scientist* 6 May 324/1 (*heading*) Spark speeds sportsters. **1974** P. J. Filby *Specialist Sports Cars* 90/1 The 803, which was an open sportster with a ladder frame chassis bonded to its fibreglass body.

sportswoman. Add: **1.** Also *transf.* in recent use, a woman who displays the typical good qualities of a sportswoman. Cf. note at Sportsman 1 a.
1906 Kipling *Puck of Pook's Hill* 148 She'd say she'd get us whipped. She never did, though... Aglaia was a thorough sportswoman. **1925** F. Scott Fitzgerald *Great Gatsby* iv. 86 Miss Baker's a great sportswoman... She'd never do anything that wasn't all right. **1948** F. Thompson *Still glides Stream* ii. 42 If I can't bear a bit of pain at my time of life I'm no sportswoman.

sporty, *a.* Add: **2.** Of a motor car: of a racing type; resembling a sports car in appearance or performance.
1962 A. Lurie *Love & Friendship* xii. 228 Traded in his Pontiac got him one of those new Valiants,..kind of sporty for a man his age. **1966** *Economist* 9 July p. xvii/2 Chevrolet's Corvair only began to sell when the Monza version was introduced after a few months, with bucket seats and a 'sporty' reputation, but precious little else for the extra $250. **1972** *Sunday Express* 9 Jan. 5/1 Performance is distinctly sporty. Its top gear acceleration of 40–60 in 9 seconds beats anything else in the class.

‖ **sposa** (spǭ·sā). Now *rare.* [It.] A wife; a bride. Also as **cara** (cā·rā) **sposa,** 'dear wife'; a devoted female companion. Cf. *Sposo.
1624 J. Chamberlain *Let.* 4 Dec. (1939) II. 589 Tom Cary a privado of the Princes bedchamber was dispatcht..into Fraunce with a love letter and some rich and rare jewell to the Sposa. **1781** N. Mundy *Let.* 15 Oct. in A. E. Newdigate-Newdegate *Cheverels* (1898) iii. 48 Adieu my dear Sr Roger, may you and yr Cara Sposa meet in health & spirits. **1793** W. B. Stevens *Jrnl.* 13 Mar. (1965) i. 72 His Wife..handsome enough for a Cara Sposa. **1797** E. Wynne *Jrnl.* 22 Jan. in A. Fremantle *Wynne Diaries* (1952) xx. 265 The *sposa* received so many fine compliments of congratulation that she was quite at a loss. **1821** Shelley *Let.* 22 Oct. (1964) II. 363 La Guiccioli his [*sc.* Byron's] cara sposa who attends him impatiently.

s'pose (spōᵘz), *v.* Also **spose, 'spose, spoze.** Repr. an informal pronunc. of Suppose *v.*
1852 R. S. Surtees *Mr Sponge's Sporting Tour* xvii. 84 Law, ma! but you don't 'spose pa would ever allow such a thing. **1873** Trollope *Harry Heathcote* (1874) i. 12 'I s'pose the poor must live somewheres,'..said Mrs. Growler, the old maid-servant. ? **1912** R. Fry *Let.* (1972) I. 358, I 'spose women weren't artists enough. **1936** L. Durrell *Let.* in *Spirit of Place* (1969) 40, I no longer care... A new phase I spoze. **1945** A. Kober *Parm Me* 44 If you're not enjoying yesself, Max, spose you sit with them. **1968** D. O'Grady *Bottle of Sandwiches* (1969) i. 7 'And s'posin' it rains in the meantime?' 'See what you mean. Small marquee, eh?' **1971** *Black World* June 72/1, I paid them out of receipts I was sposed to turn in. **1980** H. R. F. Keating *Murder of Maharajah* iii. 37, I s'pose we're not strictly his guests. We're the Maharajah's guests.

sposhy, *a.* (Earlier example.)
1842 *Yale Lit. Mag.* VIII. 96, I can't always decipher quail tracks—specially in *sposhy* weather.

‖ **sposo** (spǭ·so). Now *rare.* [It.] A husband. Also as **caro** (cā·ro) **sposo,** 'dear husband'. Cf. *Sposa.
1778 F. Burney *Diary* Aug. (1940) i. 12 Hetty, who, with her *sposo*, was here to receive us. **1792** —— *Let.* 2 Oct. (1972) I. 229 Her caro sposo has continued very tolerably well. **1816** Jane Austen *Emma* III. vi. 88 The thing would be for us all to come on donkies, Jane, Miss Bates, and me—and my caro sposo walking by. **1858** Trollope *Three Clerks* III. xviii. 328 The gentleman who has the honour of being her intended sposo. **1887** *Athenæum* 21 May 670/2 Italian girlhood..has two sole points of interest, the *sposo* and the fashion plate. **1976** *Times Lit. Suppl.* 30 Jan. 103/1 Nor was her husband..a mere *caro sposo*: their devotion was total.

spot, *sb.*¹ Add: **II. 4. c.** (Earlier and later examples.) Now usu. without the definite article.
1856 *Spirit of Times* 22 Nov. 196/1 Addison County leads the van (or 'knocks the spots off', as we say here) in Vermont, and is celebrated over the world for its fine horses. **1861** *Atlantic Monthly* June 747/1, I wish I had control of chain-lightning for a few minutes;..I'd make it come thick and heavy and knock spots out of Secession. **1903** A. Bennett *Truth about Author* xiii. 171 'We will write a play together... We can do something that will knock spots off—' etc., etc. We determined upon a grand drawing-room melodrama which should unite style with those qualities which make for financial success on the British stage. **1943** A. L. Rowse *Cornish Childhood* 186 They [*sc.* the Nazis]..have at any rate been intelligent, and knocked spots off those public-school gentlemen.

d. A pip on a playing-card. Also, a playing-card having a specified number of pips (cf. sense 5 d). In recent use *U.S.*
1578 J. Stockwood *Sermon preached..24 Aug.* 142 They perfectlye can tell howe manye spottes there be in a payre of Cardes..when as they scarce reade a leafe of the Bible twice in a Moneth. **1844** 'J. Slick' *High Life N.Y.* II. xxx. 215 'Jest so,' sez I, a fli[n]gin down the ten spot o' clubs and the ace-o' diamonds. **1873** J. H. Beadle *Undevel. West* iv. 92 The ace is your winning card. The eight and ten spot win for me. **1976** *Washington Post* 19 Apr. B8/3 South won the opening trump lead with his ten-spot. **1977** D. Anthony *Stud Game* xix. 118, I had a poker game... I played mechanically, counting spots and backing the odds.

5. d. (Earlier example.)
1846 Durivage & Burnham *Stray Subjects* 135, I moved towards the money, but he prevented my raising it, by covering it with a *twenty-spot*.

e. Substitute for def.: Either of two marine food fishes of the family Sciænidæ, *Leiostomus xanthurus* or the channel bass, *Sciænops ocellata*. (Earlier and later examples.)
1877 C. Hallock *Sportsman's Gazetteer* 396 Spot.—*Liostomus obliquus.* **1961** E. S. Herald *Living Fishes of World* 192/1 The spot, *Leiostomus xanthurus*, found from Cape Cod to Texas, is easily recognized by the spot above the base of the pectoral fin.

f. = *spot board* in sense 14 below.
1922 E. J. Evans *Building Contracts* xviii. 81 It is essential that all plant, such as Derby's sieves, cornice moulds..spots, scaffolding, etc.,..be on site in readiness. **1927** A. H. Telling *ABC of Plastering* 195 It is part of the labourer's business to see that the 'spot' is supplied with 'stuff' for the plasterer's use. **1964** [see *spot board*, sense 14 below].

III. 7. a. (Further examples, esp. *colloq.* with *of* and abstract nouns.)
1915 H. Rosher *In R.N.A.S.* (1916) 117 Pity I'm not due for another spot of leave yet. **1924** D. B. W. Lewis *At Sign of Blue Moon* 272 What about a spot of lunch? **1933** M. Allingham *Sweet Danger* ii. 21 Since the spot of bother last year, the tunnel is no longer a tunnel. **1951** *People* 3 June 5/6 A dead-broke barrister who turns burglar following a spot of light coaching from a retired cracksman. **1959** H. Hamilton *Answer in Negative* iv. 52 The police will be coming..to deal with a spot of trouble. **1976** F. Muir *Frank Muir Book* 43 The first major French ballet company visited New York in 1827. They had a spot of bother with their tights.

d. Also preceded by a defining word and *absol.*
1924 Lawrence & Skinner *Boy in Bush* vi. 85 Y' slog t' th' nearest pub t'cadge a beer spot. **1930** J. Devanny *Butcher Shop* 200 The 'spot' she craved, or failing that a cup of tea. **1936** Wodehouse *Laughing Gas* ix. 90 May I offer you a spot?.. I can recommend the Scotch. **1950** G. Greene *Third Man* viii. 63 'Have another drink, Mr. Martins?' 'No, I don't think I will.' 'Well, I'd like another spot.'

e. *slang* (orig. and chiefly *U.S.*). A term of imprisonment. Usu. preceded by numeral, designating a term of the specified number of years. Cf. sense 5 d.
1901 'J. Flynt' *World of Graft* 220/2 *Spot*, term in prison. A 'one spot' means a sentence of one year. **1907** J. London *Road* 84 He had never been in the particular penitentiary to which we were going, but he had done 'one-', 'two-', and 'five-spots' in various other penitentiaries (a 'spot' is a year). **1966** M. Brewer *Man against Fear* x. 105 He was serving a three spot for cunning... He got into a row with one of the warders.

8. d. = *Slot *sb.*² 5*, esp. in a performance, show, or programme; *spec.* in *Broadcasting*, a short interval for an advertisement or

announcement; an advertisement or announcement occupying such an interval. orig. *U.S.*
1923 *N.Y. Times* 7 Oct. IX. 2 *Spot*, an act's position on a bill. **1926** *Amer. Mercury* Dec. 465/1 Fields and Fink moved in to a homer in the next to closing spot. **1937** *Amer. Speech* XII. 101 *Spot* or *spot announcement* refers to a brief announcement, usually commercial, spotted at various times. **1950** *Sport* 22–28 Sept. 18/2 Jack..continued to fill the centre-half spot in the Celtic line-up. **1958** *Listener* 25 Sept. 462/2 Occasional spectacular successes may please advertisers who happen to have 'spots' next to them. **1958** *Manch. Guardian* 29 Sept. 4/4 Spots, unlike full-length programmes, would reach people not already prejudiced in favour of the candidate. **1960** *Twentieth Cent.* Apr. 357 Dons..will do anything for a television spot. **1962** H. E. Beecheno *Introd. Business Stud.* ix. 81 They [*sc.* TAM ratings] also help to determine, on the commercial channel, the prices of advertising 'spots'. **1967** *Technology Week* XX. 95/2 (Advt.), Your spot? Perhaps operating a ground computer complex. **1972** *Newsweek* 10 Jan. 9/2 Among these super-rich, two families contend for the No. 1 spot in the financial pecking order. **1972** G. Durrell *Catch me a Colobus* vii. 139 He appeared on the local television as 'Uncle Ambrose', doing a children's spot in which he always had an animal of some sort to show them and talk to them about. **1976** *Southern Even. Echo* (Southampton) 17 Nov. 21/3 The 125cc solo class provides a championship medal for yet another Italian, Pietro Bianchi, with Spain's Angel Nieto in second spot. **1977** J. Wainwright *Do Nothin' till you hear from Me* x. 177 Tricks to get the tempo moving... Tricks to introduce a solo spot. **1978** S. Brill *Teamsters* iii. 98 What pervaded Fitzsimmons' union..was a generally low level of competence in the top-paying spots. **1980** *Times* 5 Feb. 19/4 At present rates £44,300 would buy 49 30-second spots on each [radio] station during day-time shows.

e. *colloq.* With preceding adj.: a situation of the (unpleasant or difficult) kind specified; *in a spot*, in difficulties, in trouble.
1929 C. F. Coe *Hooch* v. 103 Jimmie Daust is in a bad spot, too. **1932** Wodehouse *Hot Water* xvi. 257 If that guy's in a spot, I'm glad of it. **1940** A. Christie *One, Two, buckle my Shoe* 160 It's the sort of business that might land him in a tight spot. **1967** E. S. Gardner *Case of Queenly Contestant* vii. 84 He was afraid his father would find out. He was in a spot. So he turned to the troubleshooter. **1978** A. Price *'44 Vintage* xviii. 200 She'd probably only been humouring him, like any nice girl in an awkward spot might do.

f. *colloq.* A place of entertainment; *spec.* = *night spot* s.v. *Night *sb.* 14.
1954 G. Greene *Twenty-One Stories* 205 I should be taken to plenty of Spots if I wasn't with a husband. **1956** B. Holiday *Lady sings Blues* (1973) iii. 32 It was jumping with after-hour spots, regular-hour joints, restaurants, cafés, a dozen to a block. **1970** C. Major *Dict. Afro-Amer. Slang* 108 *Spot*, usually a nightclub but also any popular place.

9. b. In phr. *the man on the spot*. See also *on-the-spot *a.*
1897 I. Malcolm in R. S. Churchill *W. S. Churchill* (1967) I. Compan. ii. xii. 848, I write like the 'man on the spot' The most inconceivable rot. **1922** M. Asquith *Autobiogr.* II. i. 21, I took my host aside and asked him if 'the man on the spot'—generally a favourite with the stupid—had given him his views on South Africa. **1955** G. Greene *Quiet American* i. ii. 21, I always like to know what the man on the spot has to say. **1978** *Listener* 26 Jan. 119/3 If the man on the spot senses paralysis on the field, he ought to be able to abdicate.

d. *to put* (a person) *on the spot*: (*a*) *colloq.* (orig. *U.S.*), to place (someone) in a particular location; to put in a difficult or embarrassing position; (*b*) *U.S. slang*, to arrange for the murder of (someone), to kill; also *fig.* Cf. sense 8 e above.
1928 *Detective Fiction Weekly* 11 Aug. 735/2 We learned that the State still had one reliable witness, who could 'put us on the spot'. **1929** *Amer. Speech* IV. 343 *Put-on-the-spot*, left waiting at an appointed meeting place. **1929** M. A. Gill *Underworld Slang* 9/2 *Put on the spot*, killed. **1930** *Punch* 16 Apr. 442 You get rid of inconvenient subordinates..by 'putting them on the spot'—that is deliberately sending them to their death. **1930** *Sun* (Baltimore) 31 Oct. 1/3 Confident that America's prohibition law is about to be 'put on the spot', enterprising English vintners are already preparing for a resumption of their happy relations with their American clientele. **1934** G. Ade *Let.* 24 June (1973) 184 The Democrats have put every independent voter on the spot by nominating [Sherman] Minton at the dictation of Paul McNutt. **1937** *Sunday Express* 21 Feb. 1/3 (*heading*) Englishman 'put on the spot'. **1951** M. Lowry *Let.* 5 June (1967) 244 The last thing we'd want to do is to put you on the spot or embarrass you. **1960** H. Innes *Doomed Oasis* ii. iii. 145, I couldn't exactly say it in my report of the search. It would have put the Company on the spot, if you see what I mean.

10. c. *Cricket.* (*a*) The point at which a ball should pitch for optimum length and direction; (*b*) an irregular place on the pitch from which the ball, when bowled, may move in an unexpected direction.
1855 F. Lillywhite *Guide to Cricketers* 29 The second day, however, Dean and Nixon found out the 'spot', and seemingly deposited every ball..on the precise place. **1859** *All Year Round* 23 July 305/2 The wicket..had no dead spots, no lively ones; no chance for 'shooters', none for 'bumpers'. **1901** *Encycl. Sport* I. 247/1 To bowl 'on the spot' is to bowl a good 'length'. **1908** W. E. W. Collins *Leaves from Old Country Cricketer's Diary* ix. 145 Their fast bowler found a spot on which the ball shot dead instead of bumping. **1950** F. N. S. Creek *Teach yourself Cricket* vi. 111 A mere mechanical ability to pitch the ball on some regular spot called a 'good length' is of little value. **1980** *Wisden Cricket Monthly* Apr. 31/1 He was heartened by the presence of an awkward spot—a pretty large one, right on his length—from which the ball flew viciously from the very start.

d. Phr. *to hit* (or *go to* or *touch*) *the spot*: to be exactly what is required, 'to fit the bill' (said esp. of food or drink). *colloq.* (chiefly *U.S.*).
1868 *Putnam's Mag.* I. 670/1 'I hope that last corjul set you up?' 'Yes, Mr. Plunkitt, it went right to the spot.' **1897** *Strand Mag.* May 500/2 Then percussion or detonation was tried, and that 'touched the spot'! **1908** 'O. Henry' *Voice of City* 235 Oh, pass the bottle... That hits the spot... My first drink in three months. **1923** W. Nutting *Massachusetts Beautiful* 241 Did ever a dish of apple dowdy go to the spot like that? **1949** F. P. Keyes *Dinner at Antoine's* xvii. 268 That hot chocolate and those big chunks of roast beef certainly hit the spot. **1974** P. de Vries *Glory of Hummingbird* xviii. 275 'They haven't got a name for it [*sc.* a pancake] yet.'.. 'Batter Up. It hits the spot.' **1976** 'E. McBain' *Guns* (1977) vii. 194 Colley would love a drink... A gin and tonic would hit the spot.

e. *Stock Exch.* With preceding adj.: a share in which dealings are of the specified kind.
1928 *Daily Sketch* 10 Aug. 20/3 Courtaulds' shares remain a firm spot at 4½. **1981** *Times* 23 Apr. 18/3 Another dull spot was Danish Bacon, which gave up 2p to 104p after results.

f. *Assoc. Football.* With *the*: the penalty spot (see *Penalty 5).
1970 *Liverpool Daily Post* 26 Sept. 5/5 *Smith made no mistake from the spot*: Smith scored a penalty. **1976** *Ilkeston Advertiser* 10 Dec. 19/5 Penalty expert Alan Crisp gave the goalkeeper no chance from the spot. **1977** *Guardian* 9 Mar. 20/3 Amid strong protests from several Forest players, the referee pointed to the spot.

11. b. (Earlier example.) Also in *sing.*
1881 *Harper's Mag.* Apr. 734/2 'Spots', 'futures', 'longs' and 'shorts' were unknown terms. **1930** *Sunday Times* 12 Oct. 2/4 Raw Rubber prices became easier, and spot was dealt in down to 3 11-16d. **1976** *Scotsman* 25 Nov. 2/8 Silver fluctuated with sterling and ended about 2p down at 262.50p for spot and 273p for three-month.

11*. *ellipt.* for *Spotlight *sb.* (*a*).
1920 Wodehouse *Jill the Reckless* xvi. 231 Another debate on the subject of blues, ambers, and the management of the 'spot'. **1930** [see *Flood *sb.* 6*]. **1960** J. Symons *Progress of Crime* vi. 218 We cut off our engines after putting the spots on first to make sure this was the right show. **1968** K. Weatherly *Roo Shooter* 40 While the kangaroo shooter swings the spot in wide-reaching sweeps, the rabbit shooter uses it at about thirty-five yards and shoots into the sidelight. **1977** *Times* 24 Sept. 22/4 Wall spots are just over £6, table and standard lamps at reasonable prices.

IV. 12. b. *spot-development, -group, -zone.*
1909 G. Forbes *Hist. Astron.* 105 During the sun-spot maximum the corona seems most developed over the spot-zones—*i.e.*, neither at the equator nor the poles. **1926** H. Macpherson *Mod. Astron.* iii. 41 The regions where the bombs are likely to appear are around and among active spot-groups. *Ibid.* 45 A relation between high-rotation speeds and spot-development.

d. *spot market, rate.*
1939 *Sun* (Baltimore) 12 Dec. 17/6 As the price of tin for future delivery advanced..local cash or spot market prices declined slightly. **1956** *Ann. Reg. 1955* 436 Dealings in lire were restricted to the spot market. The Bank of England began the publication of official spot T.T. rates for lire. **1982** *Daily Tel.* 3 Aug. 13/2 Saudi oil is being traded on the spot market at up to $3 a barrel discount. **1933, 1971** Spot rate [see *Premium 3 c].

e. In sense 9: = 'made on the spot', as *spot decision, fine*, etc. See also *spot test* (*b*), sense 14 below, and *Spot check *sb.*
1921 Z. Cope *Early Diagnosis Acute Abdomen* i. 3 Spot-diagnosis is impressive but unsafe. **1934** *Amer. Speech* IX. 113/1 One man may make a *spot count*, for eight or nine hours. **1953** *Times* 29 July 6/6 He said that the 'spot scrutiny' of vehicles such as has been carried out in Bedfordshire was one of the best ways of identifying those which should not be on the roads. **1959** *Economist* 21 Mar. 1047/1 He may now be obliged to make spot decisions with inadequate notice in highly controversial circumstances. **1976** *Wymondham & Attleborough Express* 17 Dec. 2/1 Prosecution is a waste of time of already crowded courts, so what about spot fines for these moronic motor maniacs. **1979** H. S. Kent *In on Act* xxii. 247 Although..I would often be glad of a 'spot opinion' on one of my new problems, it was never quite the same again.

14. spot ad, advertisement *Broadcasting*, an advertisement occupying a short break during or between programmes; so **spot advertising** *vbl. sb.*; **spot announcement** *Broadcasting*, an announcement occupying a short break during or between programmes; **spot board** *Plastering* (see quot. 1964); **spot cash** orig. *U.S.*, money paid on the spot; **spot commercial** = **spot advertisement*; **spot dance,** a competitive dance in which a spotlight plays on the dancers until the music stops, at which time the couple on whom the spotlight rests wins; **spot effect** *Broadcasting*, a sound effect created in the studio (see quots. 1941 and 1976); also *attrib.* in *pl.*; **spot-fish** = sense 5 e in Dict. and Suppl.; **spot height** *Surveying*, the height of a point above mean sea level, usu. as marked on a map; **spot kick** *Assoc. Football*, a penalty kick; also *attrib.*; **spot lamp** = *Spotlight *sb.*; **spot level** *Surveying* = **spot height*; also *fig.*; **spot-list** *v. trans.*, to place (a building) on a list (*List *sb.* 6 a (*b*)) as the result of special consideration; so **spot-listing** *vbl. sb.*; **spot map,** a map in which spots or points indicate individual locations or occurrences of something; **spot meter,** a photometer that measures the intensity of light received within a cone of small angle, usu. 2° or less; **spot news** *Journalism* (orig. *U.S.*), news reported of events as they occur; also *attrib.*; **spot plate** *Chem.*, a plate having several small depressions in which spot tests can be performed; **spot-proof** *a.*, (of a fabric) that is not susceptible to small stains; **spot-reducing** *vbl. sb.*, reduction of fat in selected areas of the body; **spot test,** (*a*) a chemical test performed using a single drop of sample; (*b*) = ***spot check** *sb.*; also **spot-test** *v. trans.*, to subject to a spot test; **spot testing** *vbl. sb.*; **spot welding** *Engin.*, a form of resistance welding in which one or more small, localized welds are produced on the overlapping surfaces to be joined; so **spot weld** *sb.*, a weld of this kind; also (with hyphen) as *v. trans.*, to join by spot welding; **spot-welded** *ppl. a.*; **spot welder,** an apparatus that carries out spot welding; **spot wobble** *Television* (now *disused*), small periodic oscillation of the scanning spot, formerly used to render the scanning lines less noticeable; **spot zoning** *U.S.*, the special rezoning (see *Zoning *vbl. sb.*) of an area to meet a particular interest.
1934 J. Rorty *Our Master's Voice Advertising* 269 Canada prohibits the broadcasting of 'spot ads'. **1962** *Rep. Comm. Broadcasting 1960* 82 in *Parl. Papers 1961–2* (Cmnd. 1753) IX. 259 The principle of the control of advertising time applied only to 'spot' advertisements, and not in any analogous way to advertising magazines. **1961** *Ann. Reg. 1960* 452 The amount of 'spot' advertising in the clock hour was to be reduced to 7½ minutes. **1979** *Guardian* 4 Aug. 8/7 This danger [of competition] can..be avoided..if only a different kind of advertising is permitted on the new commercial channel: if it may show only *block* not *spot* advertising. **1937** Spot announcement [see sense 8 d above]. **1949** *Consumer Reports* May 236/2 A 'spot announcement'..can..be inserted between the close of one commercial program and the opening of the next. **1976** *National Observer* (U.S.) 3 July 13 To defend his position, Mr. Reagan taped a 1½ minute 'spot' announcement. **1931** W. Verrall in T. Corkhill *Brickwork, Concrete & Masonry* VIII. xxxi. 1940 The spot board, as it is sometimes called, is made of floor boards, about 3 ft. long, nailed together with cleats at the back. **1939** *Archit. Rev.* LXXXV. 213 Where hand-mixing [of plaster] is carried out the mess and waste that are inevitable when plaster is mixed on 'banker-boards' and then transferred to 'spot-boards' should be avoided where possible. **1964** J. S. Scott *Dict. Building* 307 *Spot board* or *gauge board* or *spot*, a plasterer's board about 3 ft square on which he works up the plaster before he puts it on. It rests on a stand about 27 in. high. **1879** *Bradstreet's* 8 Oct. 4/3 A business Utopia where credit shall be unknown and 'spot cash' an unvarying rule. **1909** A. N. Lyons *Sixpenny Pieces* ii. 12 The spot-cash practitioners of Mile End Road are rather strange..to us. **1957** V. Packard *Hidden Persuaders* xii. 131 After the car went on sale reports from dealers stated that 90 per cent of the people buying paid spot cash. **1978** *Dumfries & Galloway Standard* 21 Oct. 21/3 (Advt.), Caravans wanted, spot cash, any size, any make. **1955** *Times* 28 July 9/7 Television uses many old pictures to fill less valuable time, mainly as vehicles for strings of 'spot' commercials that effectively break any thread they may possess. **1962** A. Nisbett *Technique Sound Studio* i. 18 He [*sc.* the announcer] also presents programme trailers and (in many countries outside Britain) spot commercials. **1976** *National Observer* (U.S.) 10 Apr. 2/5 The networks have said they will make regular prime-time spot-commercial time available. **1944** M. Laski *Love on Supertax* iii. 40 Such refinements as 'spot-dances' and..jazz superbly played. **1947** *Daily Gleaner* (Kingston, Jamaica) 4 Nov. 15/1 Spot dances, raffles, games and a grand floor show. **1941** *B.B.C. Gloss. Broadcasting Terms* 31 *Spot effects*, effects created in a studio where the performance of which they form part is taking place. **1944** L. MacNeice *Christopher Columbus* 17 'Effects'..are of two kinds—records on a gramophone or spot effects in the studio. **1961** G. Millerson *Telev. Production* i. 14 A spot-effects man sounds the horn of an 'approaching car'. **1976** B. Armstrong *Gloss. TV Terms* 84 *Spot effect*, a brief sound—door closing, bell ring—added during the dub. **1875** *Fur, Fin & Feather* 122 You are always welcome to a seat in his boat, if disposed for snipe or duck, or spot-fish. **1913** A. R. Hinks *Maps & Survey* i. 23 There is a strong tendency to give spot heights for summits, and not for the bottoms of depressions. **1928** E. D. Laborde *Pop. Map Reading* iii. 65 In actual practice, it is often difficult to find the exact height of ground: small areas of high ground are not always marked with spot heights. **1977** D. Beaty *Excellency* vi. 74 He drew his track on the maps and charts..and memorized the spot-heights over the Alps. **1950** *Sport* 22–28 Sept. 4/4 He has handed over the responsibility of taking penalty kicks to left-back 'Jock' Ferrier, who converted a spot-kick last Saturday. **1971** *Post* (S. Afr., Cape ed.) 9 May 23/5 Dumile Melane never faltered with the spot-kick amid cheers from jubilant Cubs supporters. **1977** *Belfast Tel.* 22 Feb. 30/6

Magee remains the Blues' spot kick expert and, if a penalty is awarded.., he will be the man to take it. **1937** *Motor Catal.* (East London Rubber Co.) 151 'Raydyot' Spot Lamp... Universal movement. **1962** H. C. Weston *Sight, Light & Work* (ed. 2) vi. 197 A low voltage 50 watt spot-lamp unit..can be mounted at a distance of 6 ft. from the work and yet provide an illumination upwards of 300 lm/ft.² **1976** *Liverpool Echo* 22 Nov. 14/3 (Advt.), Daimler Sovereign 1973, automatic, power steering,..spotlamps, [etc.]. **1908** N. F. Mackenzie *Methods of Surveying* x. 118 From these 'spot levels' the actual contour lines are sketched in by interpolation. **1920** J. K. Finch *Topographic Maps & Sketch Mapping* i. i. 19 The exact height of the ground is only shown at important points on hachure maps. This is done by means of 'spot levels' or 'spot heights' scattered over the map. **1958** *Listener* 20 Nov. 813/1 As the polls attempt a nation-wide cross-section one would expect them to reflect the gradual development and movement of political opinion..more steadily and consistently than the spot-levels, so to speak, which are taken at by-elections, with all their variations of locality, candidate, and so on. **1974** W. H. Irvine *Surveying for Construction* iii. 12/1 At present spot levels are shown to the nearest foot but with metrication the levels will be to the nearest 100 mm. **1974** *Country Life* 31 Jan. 197/3 After representations by the Victorian Society, the building was recently spot-listed. **1977** M. Binney in Binney & Burman *Change & Decay* 188/1 If a church is of any merit yet still not listed it is worth asking the Department of the Environment to consider spot-listing it. **1973** *Times* 8 Jan. 3/3 Widespread 'spot listing' to save buildings threatened by development may be counter-productive. **1901** Parkes & Kenwood *Hygiene & Public Health* xi. 666 'Spot maps'—maps of a district, on which the deaths or cases of various infectious diseases are spotted out—furnish valuable graphic expressions of any grouping. **1973** J. J. McKelvey *Man against Tsetse* ii. 81 In three months the commission acquired 460 collections of flies and made a spot map of tsetse locations. **1979** *Dictionaries* I. 36 The spot maps will be effectively utilized to show the distribution of a single [dialectal] term. **1955** J. Lipinski *Miniature & Precision Cameras* vi. 240 The incident light meter may measure the light incident upon the scene. Or the spot meter may measure it from a white card. **1976** *Physics Bull.* Sept. 395/3 A range of luminance (brightness) spotmeters and illuminance meters will be displayed. **1912** S. Washburn *Cable Game* 9 The other type, the 'cable men', are collectors of what might be called 'spot' news. **1936** E. Waugh *Waugh in Abyssinia* 82 There were demands from Fleet Street for daily items of 'spot news'. **1976** *National Observer* (U.S.) 15 May 2/3 Other Pulitzer winners were: Sydney Schanberg of the New York Times, international reporting;..and the staff of the Chicago Tribune, spot-news reporting. **1928** *Chemist-Analyst* XVII. i. 18/2 Spot plate for outside indications. **1937** J. W. Matthews tr. *Feigl's Spot Tests* ii. 5 Spot plates are made from glazed porcelain, and usually contain 6–12 adjacent depressions of equal size that hold 0·5–1 c.c. of liquid. **1981** *Sci. Amer.* Feb. 128/1 The aluminum compound responsible for the buffering effect of buffered aspirin can be detected with filter paper or a spot plate. **1950** Spotproof [see *Habutai]. **1958** *Times* 6 Oct. 13/1 Leather and suède jackets and coats many of which are treated to be spot-proof against rain. **1960** *Sunday Express* 27 Nov. 14/3 Gymnasia with spot-reducing and general slimming equipment. **1970** *Harrod's Catal.* May 14/1 Spot reducing 'Faradic' and 'G.5' from £2. 2. 0. **1921** *Chem. Abstr.* XV. 2599 Spot tests which depend upon the formation of $PbCrO_4$ or of $AgCrO_4$ are fairly sensitive. **1948** *Ann. N.Y. Acad. Sci.* XLIX. 268 Prior to quantitative analysis..the fractions are spot-tested on paper impregnated with ninhydrin. **1955** *Times* 16 June 9/3 'Spot tests' could be extended, and it would be well if the Minister were to declare his intention with regard to them. **1960** *Jrnl. Iron & Steel Inst.* CXCIV. 285/3 For Cu, a spot test is suggested which allows estimation in the range 0·1–0·5 p.p.m. **1969** *Listener* 28 Aug. 287/1 Miss Atwood's Toronto heroine works for Seymour Surveys, spot-testing consumer products. **1942** *Industr. & Engin. Chem.* (Analytical Ed.) 15 Apr. 278/2 (*heading*) Application of infrared radiation to spot-testing. **1908** *Engineering* 9 Oct. 486/1 The pedal of one of the machines on view had been made..by attaching a piece of angle iron on each side of the lever by two 'spot welds'. **1951** *Ibid.* 13 Apr. 439/1 The shear strength of an individual spot weld was nearly twice that of a rivet. **1977** *Modern Railways* Dec. 488/1 The main handicap..was the high labour cost which resulted from the 50,000 rivets and 60,000 spot-welds involved in the construction of one vehicle. **1909** *Engineering* 15 Jan. 69/3 We have seen three thicknesses of 3/16-in. iron spot-welded together, the spot being about ¼ in. in diameter. **1958** *Ibid.* 14 Mar. 344/2 The panels are..automatically seam-welded before being spot-welded to the framing. **1973** A. Parrish *Mech. Engineer's Ref. Bk.* vi. 34 Aluminium is readily spot welded but the welding parameters must be closely controlled because the metal has a short plastic temperature range. **1921** *Automobile Engineer* Mar. 106/3 (*caption*) A spot welded bonnet. **1969** J. G. Tweeddale *Welding Fabrication* II. iii. 87 In a spot welded assembly the joint is comparatively rigid. **1914** *Proc. Inst. Mech. Engineers* 167 An attempt was made to put down one small machine, a spot welder or a contact welder. **1963** H. R. Clauser *Encycl. Engin. Materials & Processes* 573/2 Semi-portable spot welders are..available for assemblies such as automotive structures, housings and cases that cannot be handled by permanently placed..machines. **1976** Spot welder [see *seam welder* s.v. *Seam *sb.*¹ 10]. **1908** *Engineering* 9 Oct. 486/1 The method of spot-welding appears cheaper, quicker, and as good as riveting. **1912** *Automobile Engineer* 30 Oct. 387/3 In the spot-welding machine the sheets are joined at spots instead of rivets. **1970** K. Ball *Fiat 600, 600 D Autobook* xii. 141/1 The body structure is made up from nine separate assemblies joined together by spot welding. **1951** *Sun* (Baltimore) 19 Oct. (B ed.) 5/7 First sets to incorporate 'spot wobble' are 15-inch home receivers. Focusing is by the normal line-pattern method, but the flick of a switch converts to 'spot wobble'. **1956** *B.B.C. Engin. Monogr.* No. 1. 5/2 The obliteration of the line structure by carefully adjusted spot wobble is of course essential in any telerecording system if beat patterns are to be avoided when the film is scanned. **1932** *Sun* (Baltimore) 6 May 14/2 All cities which have zoning laws go through a period of attacks that aim to break them down for private as opposed to public interests. The practice is called 'spot' zoning. **1933** *Ibid.* 16 Nov. 12/1 The Board of Zoning Appeals has disapproved a proposed 'spot-zoning' ordinance in the interest of the promoters of a distillery. **1946** *Amer. Home* Sept. 96/3 There is..one very deplorable habit among the creators and administrators of Zoning Laws which was, to permit what is known as 'spot zoning'. This may allow a small 'spot' of business to establish itself in an exclusively residential area. **1976** *Tulane Law Rev.* L. 357 The term 'spot zoning'..may mean rezoning not in accordance with a comprehensive plan, rezoning in the absence of a mistake or change, or rezoning which is arbitrary and improper.

spot, *v.* Add: **I. 3.** (Earlier example.)
1850 *Rep. Comm. Patents: Agric. 1849* (U.S.) 456 These varieties spot better, and produce a finer leaf than any I have ever seen.

II. 4. d. (Earlier examples.)
1856 E. M. Curr *Waste Lands of Province of Wellington* 30 The practice of which I speak is called in New Zealand 'spotting' or 'spoiling' country. **1864** E. Muter *Trav. & Adv. Officer's Wife* II. xiii. 260 'Cockatoos', as the station-owners call the other [class of land-purchasers], who 'spotted' his run all over with fifty and one hundred acre sections.

e. To moisten *with* a drop of liquid; to place a drop of (liquid) *on* (*to*) a surface, etc.
1954 R. E. Oesper tr. *Feigl's Spot Tests* (ed. 3) II. iv. 158 The moist reagent paper is spotted with the test solution. **1972** *Sci. Amer.* June 34 (*caption*) The mixture is spotted at one corner (*X*) of a square of filter paper; the fragments are separated by chromatography in one direction and by electrophoresis in another. **1977** *Lancet* 26 Nov. 1140/2 Blood is spotted on to filterpaper. **1978** *Nature* 9 Feb. 577/2 Even concentrated lysates did not result in killing or lysogenisation when spotted on a lawn of Mu-sensitive bacteria.

7. Also without *out*.
1915 B. E. Jones *Cinematograph Bk.* 176 Having cleaned and spotted the film, attention may be given to any torn portions or broken perforations. **1937** *Discovery* Feb. p. xiv/1 This book [*sc.* J. Deschin *New Ways in Photography*]..contains many suggestions and practical methods for getting better results in..spotting prints. **1979** *Amateur Photographer* 7 Feb. 4 (Advt.), A custom hand print, cropped, hand printed, spotted, dry mounted and heat sealed.

7*. *N. Amer.* To place (something) in a particular location; *esp.* to place (a railway car) in the proper place for loading or unloading.
1917 *Dial. Notes* IV. 329 *Spot*, to stop (a car) at the proper place on a railroad track. **1936** B. Brooker *Think of Earth* i. iii. 38 It was a wilderness of rusting steel and rotten ties where old dump-cars were 'spotted' for repairs. **1937** *Liberty* 25 Dec. 20/1 Already strategically spotted throughout..our population are sergeants and lieutenants and chiefs of police of the new 'supertrained' school. **1947** *Sun* (Baltimore) 13 May 6/3 Passengers would walk..under cover to their planes spotted alongside the pier. **1956** T. Raddall *Wings of Night* (1957) xii. 106 You might phone the railway people and tell 'em to spot four or five boxcars on the Hall's Creek siding not later than Tuesday. **1962** J. Onslow *Bowler-Hatted Cowboy* xiv. 136 When I arrived at the yards the cars were ready spotted opposite the loading-chutes. **1970** J. H. Gray *Boy from Winnipeg* 55 The rest of us spotted our lunch kits, towels, and swimming things at a convenient table. **1979** *Arizona Daily Star* 22 July 1. 6/4 If you're playing a Nevada audience, you can't give them all new stuff, and you have to be careful how you spot it.

III. 9. b. (Earlier examples.)
1848 E. Z. C. Judson *Mysteries & Miseries of New York* i. 116 To spot is to recognize—to mark. **1849** G. G. Foster *New York in Slices* 15 The expertness acquired by the keepers of these shops in 'spotting' their man is truly wonderful.

d. *Mil.* To locate (an enemy position). Also *intr.* and const. *for.*
1914 *Aeroplane* 11 Nov. 425/2 He poised..for a spell to spot the lurking place of the battery. **1915** D. O. Barnett *Let.* 23–25 Jan. 51, I had a man with a periscope spotting for me, and he registered some near things for the Bosch's face. **1916** 'Boyd Cable' *Action Front* 135 'Stand by for trouble. That brute is spotting for his gun.' The aeroplane dropped a light, turned, and circled round to the left. **1918** C. Bright *Telegraphy, Aeronautics & War* 51 While he is spotting he is continually subjected to tremendous shelling. **1942** *Hutchinson's Pictorial Hist. War* 18 Mar.–9 June 62 (*caption*) Men of a Cypriot company..spotting for enemy aircraft. **1973** R. Dentry *Encounter at Kharmel* ix. 152 I'll come in low. You spot. I'll be busy.

e. To watch out for and observe (a certain class of objects) as a hobby. See also *Train-spot *v.*
1919 G. B. Shaw *Inca of Perusalem* in *Heartbreak House* 212 Chips keeps owls and rabbits. Spots motor bicycles. **1957** *Times Lit. Suppl.* 8 Nov. 675/2 As other boys spot railway engines, Alan Villiers as a child in Melbourne spotted and studied the big sailing ships lying at anchor in Port Phillip bay.

spot check, *sb.* orig. *U.S.* Also hyphened. [f. Spot *sb.*¹ + Check *sb.*] A check made on the spot; a quick check made on a random sample. Also *attrib.*
1933 *Sun* (Baltimore) 15 Aug. 8/1 The spot check made on bread prices from February 15 to August 2 in sixteen representative cities. **1946** *N.Y. Times* 30 June L-25/2 Spot checks are made but many of these slaughter houses go as long as five weeks without a check. **1947** *Sun* (Baltimore) 3 Apr. 5/1 Troops and police made a surprise spot-check of identities in the center of Jerusalem today, hastily erecting roadblocks while they examined the papers of pedestrians and persons in trucks and automobiles. **1963** *Language* XXXIX. 465 A survey of the vast..literature of ethology, supplemented by repeated spot checks of ongoing research projects, reveals that the study of signaling behavior in animals has, by and large, been taxonomically parochial. **1972** *Daily Tel.* 1 Mar. 2 A spot check by the Inner London Education Authority last autumn showed that 17,000 out of 165,000 secondary school pupils were absent that day. **1978** L. Deighton *SS-GB* iv. 37 They saw the soldiers doing the spot-check. Parked across the roadway there were three Bedford lorries. **1980** *Sunday Times* 24 Aug. 1/1 Six French paratroopers have also been charged after spot-checks on kitbags revealed stolen electrical goods.

Hence **spot-che·ck** *v. trans.*, to subject to a spot check; also *absol.*; **spot-che·cking** *vbl. sb.*
1944 U. Sinclair *Presidential Agent* i. ii. 43 Robbie Budd lived and breathed and ate and talked aeroplanes: cat-walks and bulkhead segments, stabilizers and de-icers, sub-assemblies and spot-checking—a whole new vocabulary. **1955** *Times* 5 July 10/7 They had decided..not to make vehicle tests compulsory and they were considering falling back on the proposal..to extend 'spot checking' to private vehicles. **1962** *Guardian* 15 Jan. 3/4, I recently saw vehicles being spot-checked by the Ministry of Transport. **1962** E. Snow *Red China Today* (1963) xxxviii. 277 We spent another forty days spot-checking all over the county. **1972** *Accountant* 14 Sept. 314/1 He'd spot-checked them against actual orders. **1976** C. Weston *Rouse Demon* (1977) xix. 90 By spot-checking instead of monitoring each complete tape, Adrian had managed to go through seven. **1980** P. Moyes *Angel Death* i. 9 Once in a while, a Customs Officer will..do a certain amount of spot-checking on visiting private yachts.

spo·tlight, *sb.* Also **spot light, spot-light.** [f. Spot *sb.*¹ + Light *sb.*] A source of artificial light casting a narrow and intense beam; *esp.* (*a*) *Theatr.*, a lamp used to cast intense illumination on a small area of a stage; also, the light cast by such a lamp; also *fig.* (cf. *Limelight); (*b*) an auxiliary lamp on a motor vehicle.
1904 *Minneapolis Times* 1 Aug. 4 In the drama that was enacted Mr. Galvin was not in the spotlight at any time, but it cannot be denied that his was an important part. **1916** *Lit. Digest* (N.Y.) 8 Jan. 89/1 It will put the magnates and the self-styled fighters..into the background and give the players the spot-light. **1920** *T. Eaton Catal.* Spring & Summer 395/5 Spotlights are very necessary accessories, especially in country driving. **1921** *Daily Colonist* (Victoria, B.C.) 11 Oct. 7/4 (Advt.), These new Spot Lights constitute the most remarkable improvement in the history of the flashlight business. This flashlight is capable of throwing a beam 300 feet or more. For motoring, picking out road signs, boating,..or for night flash signalling, nothing can equal them. **1922** C. Aiken *Jig of Forslin* 25 While in the warm dark seats, we watch the spot-light Dazzle upon the singer's hair and eyes. **1922** M. B. Houston *Witch Man* xiv. 193 He has a nice baritone, but he's the sort of man who gravitates to the centre of the floor and deliberately absorbs all the spot-light. **1928** *Daily Express* 7 July 5/2 Ahead of him was the professor's spot-light, making a little glow upon the ground... Charlie..dared not approach within serviceable distance of the professor's torch. **1930** J. B. Priestley *Good Companions* ii. ii. 295 He comes to me and asks all about curtains and footlights and spot-lights and props, and he goes about looking so important when he's anything to do. **1931** H. F. Pringle *Theodore Roosevelt* i. xi. 135 The spotlight focused on Roosevelt, a spotlight so white and continuous that the other..commissioners found themselves in..shadow. **1937** *Motor Catal.* (East London Rubber Co.) 151/1 'Bosch' Spotlight... Gives a brilliant, long beam, which can be trained upon the kerb, on the road ahead, or signposts. **1946** D. C. Peattie *Road of Naturalist* ii. 24 Some sixty years ago the spotlight fell upon Pronuba. **1949** 'P. Wentworth' *Spotlight* xxviii. 175 He wanted the hall to be dark, with the single lamp on the mantelpiece arranged to be as much like a spotlight as possible. **1957** *Encycl. Brit.* XXI. 285/2 The spot-light—in contrast to the flood light—which can be controlled and focussed accurately upon one particular spot. **1964** L. Deighton *Funeral in Berlin* xix. 113 A V.W. saloon with blue flasher and spotlight full on. **1967** N. Freeling *Strike out where not Applicable* 129 It was a kind of altar. No spotlights, thank heaven, but a decided feeling of stiffly-bunched-madonna-lilies. **1972** *Times* 30 Nov. 23/8 A strong consumer and governmental spotlight on food prices. **1973** 'R. MacLeod' *Nest of Vultures* ii. 40 A half-dozen blinding spotlights were trained on a postage-stamp stage. **1978** R. Ludlum *Holcroft Covenant* vi. 79 Several passengers were reading under the beams of tiny spotlights, but most were asleep. **1980** M. Booth *Bad Track* i. 18 On the front of the radiator grille was mounted a pair of very large Cibie spotlights that dwarfed the standard headlamps.

spo·tlight, *v.* Also **spot-light.** [f. the sb.] *trans.* **a.** To illuminate with a spotlight. Also *fig.*
1923 [implied at *Spotlighted *ppl. a.*]. **1926** H. T. Wilkins *Marvels Mod. Mechanics* 234 These panorama lamps can be swung round in a circle, and, along with flood lights, be concentrated in a beam of rays to 'spotlight' the stage stars. **1935** *Amer. Speech* X. 192/2 A dress that scarfs and sleeves itself with stripes will *spotlight* her and *outsmart* her neighbors. **1942** *Sun* (Baltimore) 1 Jan. 22/8 He hoped that the publicity in connection with the measure has 'spotlighted' the abuse of charity racketeers. **1955** *Radio Times* 22 Apr. 15/3 Each programme in the series will spotlight a particular breed of dog. **1963** *Listener* 21 Feb. 353/3 His understanding of the jazz idiom spotlighted the 'bluesy' features of the Concerto beautifully. **1973** R. Perry *Ticket to Ride* i. 11 The

standard lamp..was softly spot-lighting her from behind. **1975** J. Cleary *Safe House* ii. 73 An elderly motorbike..the weak beam of its headlamp spotlighting them as they stood at the back of the truck. **1976** C. Blackwood *Stepdaughter* i. 56, I feel that my future career as a painter stands spot-lit and exposed. **1977** F. Young in J. Hick *Myth of God Incarnate* ii. 29 Too often the so-called outdated substance-categories have been spotlighted and criticized.

b. Chiefly *U.S.* To hunt (game) by spotlight. Cf. Jack-light *v.*
1934 in Webster. **1968** *Daily Progress* (Charlottesville, Va.) 26 Jan. 3/4 After illegally spotlighting and killing a deer [etc.].

So **spo·tlighted, -lit** *ppl. a.*; **spo·tlighter,** one who hunts by spotlight; **spo·tlighting** *vbl. sb.*, (*a*) illumination by spotlight; (*b*) hunting by spotlight; also *attrib.*
1923 B. Hecht *Florentine Dagger* v. 72 A crowd..stood watching officials and the spotlighted figures of mystery enter. **1932** *Observer* 26 June 13/4 The spotlit nocturne that follows..is starred by Spessiva's solos. **1934** Webster, Spotlighter, Spotlighting. **1952** *New Yorker* 4 Oct. 119/1 With insufficient spotlighting, the spectacle was like a succession of obscure museum paintings in need of cleaning. **1953** E. Simon *Past Masters* ii. 110 Everyone returning from the spotlighted centre was surrounded. **1961** *Daily Progress* (Charlottesville, Va.) 27 Jan. 3 (*heading*) Deer 'spotlighting' convictions hit 14. *Ibid.*, The deer 'spotlighters' were ordered to pay a total of $3,250 in fines. **1972** L. Hancock *There's a Seal in my Sleeping Bag* i. 13 He was surprised to learn in Australia that pit-lamping or spotlighting not only is legal but the accepted way of hunting.

spot on, -on, *adv.* and *a.* (*phr.*) [f. Spot *sb.*¹ + On *adv.*] Completely accurate(ly), precise(ly), exactly right.
1920 E. H. Young *By Sea & Land* i. 56 All these X-chasing instruments make my head go round. You can overdo all that, I say; shoot quick and spot on—that is the best way in the long run. **1956** C. Willcock *Death at Flight* iii. 34 'What a spot for a front tyre to choose to burst,' he said. 'Funny thing, sir, the outer casing looks as though it was in spot-on condition.' **1961** *Punch* 18 Jan. 130/3 'Self-discipline and self-restraint', to use the P.M.'s spot-on terms, become minimal requirements in the ordinary citizen. **1962** L. Davidson *Rose of Tibet* vii. 134 Our oracle here warned that you would be coming... The time was fixed fairly spot-on. **1966** M. Woodhouse *Tree Frog* xxv. 182 Neither the altimeter nor the compass is one hundred per cent spot on, because we didn't have time for an instrument check. **1969** W. Garner *Us or Them War* i. 14 'I'm still getting things right?' The plain-clothes man said, 'Spot on, Mr Morton.' **1973** *Times* 14 May 9/1, I moved my run-up back from 109ft to 114ft and fitted in 18 strides so that I was hitting the board spot on. **1976** *Courier-Mail* (Brisbane) 2 June 4/5 It's about time somebody with influence said the sort of things Mr. Colin Lamont..said about education. He is spot on. **1977** *Sounds* 9 July 8/1 Pristine production and spot on playing, A.R.S. could be up there alongside the L.A. crowd if they concentrate more on albums full of songs of this quality. **1978** R. Perry *Dutch Courage* vii. 82 Although the sarcasm was overdone, the cynicism was spot on. **1981** T. Barling *Bikini Red North* ix. 194 Getting here..with a day to spare is spot-on timing. **1982** *N. & Q.* Oct. 472/2 His thesis is provocative, its evidences spot-on, and his conclusions pretty convincing.

spotted, *a.* and *ppl. a.* Add: **3. a.** (Further examples.) Also, = *Rocky Mountain* (*spotted*) *fever* s.v. *Rocky *a.*¹ 1 b.
1902 Wilson & Chowning in *First Biennial Rep. Montana State Board of Health* 27 Enough was accomplished to warrant the formation of a working hypothesis ..that the so-called 'Spotted Fever' is due to the presence in the patient's blood of the above mentioned haematozoan..; and that the parasite is conveyed to man through the bite of a tick. **1903** *U.S. Hygienic Lab. Bull.* XIV. 7 (*heading*) Spotted fever (tick fever) of the Rocky Mountains. *Ibid.*, I have suggested as a name for the disease 'Tick Fever', as there are already two diseases sometimes called 'spotted fever'.

6. spotted dog, (*a*) a white or light-coloured dog with black or dark spots, esp. a Dalmatian; (*b*) *fig.* = *Spotted Dick* in Dict.; also *attrib.*; **spotted wilt,** a virus disease of herbaceous plants, esp. tomatoes, in which it causes curling and necrotic spotting of the leaves.
1854 C. M. Smith *Working-Men's Way in World* xii. 288 For supper come smoking sheep's-heads..and 'spotted dog', a very marly species of plum-pudding. **1910** F. W. Hackwood *Inns, Ales, & Drinking Customs* 288 The 'Talbot' readily became known among the vulgar as the 'Spotted Dog'. **1930** E. C. Ash *Pract. Dog Bk.* 47 The Spotted Dog [*sc.* the Dalmatian] became the dog of the Circus. *a* **1936** Kipling *Something of Myself* (1937) i. 18 An enormous currant roly-poly—a 'spotted dog' a foot long. **1974** *Country Life* 25 Apr. 990/1 The other hound ..reminded me of a spotted dog pudding at school. **1919** C. C. Brittlebank in *Jrnl. Dept. Agric. Victoria* XVII. 231 It is well to have a common name for every tomato disease, and I propose that of 'Spotted Wilt' for this latest one, from the spotting and subsequent wilting of the attacked plants. **1950** *N.Z. Jrnl. Agric.* June 587/1 Iceland poppies should be inspected regularly from now on for symptoms of the virus disease known as spotted wilt. **1979** *Ann. Appl. Biol.* XCIII. 173 Chlorotic ring-spots.., leaf specking, terminal bud necrosis,..and severe stunting of groundnut (*Arachis hypogaea*) were shown to be caused by tomato spotted wilt virus.

spotter. Add: **2. a.** (Earlier example.)
1876 *Scribner's Monthly* Apr. 911/2 The stockholders and directors, the 'car-starters' and 'spotters'.

c. A look-out, an observer; a scout; one who identifies or looks out a target; *spec.* (Mil.), an aviator who locates enemy positions and directs fire.
1903 A. M. Binstead *Pitcher in Paradise* v. 127 So vigilant and observant was that solitary optic that its owner was the chosen 'spotter' of a most relentless band of racecourse pickpockets. **1914** *Illustr. London News* 29 Aug. 320/3 Electric contrivances for communicating messages between the 'spotter' aloft and the gun-layer below. **1918** E. M. Roberts *Flying Fighter* 108 The fire was being directed from the ground from what the battery commander called the O.P., or observation post. He sent me up to that post with one of the spotters. **1925** J. C. Goodwin *Queer Fish* xvi. 153, I surmise that they are 'spotters', posted where they are to warn the proprietor of the card-room should the police or their informers put in an appearance. **1935** A. J. Cronin *Stars look Down* i. ix. 70, I heard the Tynecassel spotter was coming down to watch ye at the next Sleescale match. **1935** A. J. Pollock *Underworld Speaks* 112/1 *Spotter*, an accomplice who sizes up the victim and surroundings for the purpose of robbery, murder, kidnaping or a beating. **1939** *Illustrated* 16 Dec. 28 The 'spotter' sweeps the sky with his powerful telescope in search of aircraft. **1940** *Sun* (Baltimore) 13 Aug. 1/7 The Germans..have mounted long-range guns between Calais and Boulogne..but they will be simply shooting into the blue unless they have effective spotters aloft. **1940** *Times* 13 Nov. 3/1 Spotters with machine-guns man the roofs. **1943** *Times* 9 Oct. 2/5 The spotters 'tell' every outgoing or incoming aircraft by direct telephone line to the nearest R.O.C. centre. **1949** *Times* 5 Jan. 4/5 Aircraft from Kuala Lumpur carried spotters for the naval gunners. **1955** *Times* 21 June 9/4 One spotter aircraft to each forest operating company is the minimum requirement for efficiency. Such spotters can ask for..'bulldozer' rights for a patrol on a hot scent to pass through the areas of other units without being shot in error. **1959** I. & P. Opie *Lore & Lang. Schoolch.* xvii. 373 The sentinel may..be called a guard, spotter, scout, [etc.]. **1963** T. & P. Morris *Pentonville* 353 Crime or crimes are to a great extent planned in gaol..then the necessary men (i.e. safe blower, drivers, spotters, con men, fence,) are found in here and given the word. **1976** *Globe & Mail* (Toronto) 11 June 9/8 Once Mr. Drmbie or his spotter sight a fire, they circle, swoop in at treetop level and note the size of the blaze. **1976** *Eastern Even. News* (Norwich) 13 Dec. 4/3 Chic is heavily involved in all forms of entertainment and is the local 'spotter' for television's 'New Faces'. **1977** 'O. Jacks' *Autumn Heroes* vii. 96 'Spotter plane, boss.'..'I don't like the idea of a spotter.' **1980** J. O'Faolain *No Country for Young Men* i. 23 The Tans came in shooting... At the barracks, 'spotters' with bags over their heads and slits for their eyes identified known activists.

d. One who spots (*Spot *v.* 9 e) trains, etc. See also *Train-spotter.
1950 *Oxf. Jun. Encycl.* IX. 400/1 The main activity of spotters' clubs is the collecting of locomotive numbers. **1957** *Railway Mag.* Dec. 887/2 (*heading*) Necktie for spotters. **1958** *Times* 6 Dec. 7/4 As a young 'spotter' [of car registration numbers in Egypt] I quickly noticed that the Arabic face was always legible at a considerably greater distance than the Western face. **1973** *Times* 27 July (Leisure Suppl.) p. iv/4 Steam railway enthusiasts seem to grow in number each year as spotters young and old rue the passing of steam.

3. *attrib.*, as (sense *2 c) *spotter aircraft, pilot, plane.*
1942 *Daily Tel.* 1 Jan. 1/8 Three enemy fighters..and a 'spotter' aircraft were shot down. **1959** *Times* 18 May 6/1 From Marks Tey, Essex, on the main Colchester road, spotter aircraft yesterday radioed a report of a 20-mile queue of close-packed traffic. **1976** *Globe & Mail* (Toronto) 11 June 9/8 A second source of fire information—three Cessna spotter aircraft flying 3,000 feet overhead signalling radio reports into the fire control room. **1967** *Listener* 21 Sept. 355/3 The radio in the company HQ chatters with constant news from the ground control and the spotter pilots. **1958** *Observer* 3 Aug. 1/4 Floating 1,000 ft. over Southern England in the Automobile Association's spotter plane..I watched the holiday traffic streaming out of London yesterday. **1977** *Time* 22 Aug. 10/1 Israeli batteries—sometimes directed by observers in spotter planes—fired 16 times at Palestinian forces. **1981** E. Clark *Send in Lions* v. 53 Spotter planes searched the vastness of the Sahara.

spotting, *vbl. sb.* Add: **1. c.** Also without *out*.
1940 *Amer. Speech* XV. 360/1 *Spotting*, eliminating spots from negatives. **1951** *Good Housek. Home Encycl.* 269/2 *Spotting*,..the term applied to a process in the cleaning of a garment such as a suit. **1958** *Spectator* 4 July 15/3 The store or small cleaner will deal with finishing such as spotting and pressing. **1977** J. Hedgecoe *Photographer's Handbk.* 267 Spotting is the filling in of clear specks (which would otherwise print black) until they match the surrounding tone, and so vanish.

4*. A slight discharge of blood via the vagina, *esp.* as a side-effect of oral contraceptives; light staining due to this.
1900 Dorland *Med. Dict.* 624/2 *Spotting*, a slight menstrual show upon a woman's napkin. **1944** Miller & Bryant *Gynecol. & Gynecol. Nursing* iv. 57 A faint spotting noted on the underclothing or nightclothing may be the only symptom of an early carcinoma of the cervix. **1962** *Lancet* 22 Dec. 1315/2 The clinical criterion of adequate dosage most commonly used was absence of break-through bleeding and spotting. **1977** *Ibid.* 5 Nov. 947/1 According to Islamic precepts, intercourse is prohibited during menstruation, and spotting is interpreted as a prolonged menstrual period.

4.** The action of observing, acting as a look-out; *spec.* (Mil.), the action of locating enemy positions from an aircraft and directing fire.
1914 A. Hurd *Fleets at War* iv. 136 Jellicoe did all that was possible..to instal a fire-control set of instruments in each ship for 'spotting' and controlling at long-range shooting. **1918** W. H. Berry *Aircraft in War & Commerce* vi. 77 At the Dardanelles..some of the 'spotting' [by seaplanes] for the battleship guns..was not beaten later. **1928** C. F. S. Gamble *Story of North Sea Air Station* viii. 113 The duties of this air station were:..(*f*) Spotting for coastal batteries.

5. *spotting scope* [*Scope *sb.*³ b].
1960 C. E. Chapel *Art of Shooting* xv. 147 The shooter may have a spotting scope. This is a telescope of comparatively high power, which may be mounted on a tripod and placed near the shooter..and used to observe the location of shot holes on the target. **1970** R. A. Steindler *Firearms Dict.* 238/2 With the help of a spotting scope.., the spotter informs the shooter as to the exact point of bullet impact on the target. **1980** *Hunting Ann. 1981* 62/2 If you are serious about trying for a trophy, use a spotting scope.

spotty, *a.* Add: **2. c.** *gen.* Unsteady, uneven; patchy; sporadic, intermittent. orig. and chiefly *U.S.*
1932 *Sun* (Baltimore) 25 Feb. 19/6 The advance [of the curb market] was somewhat spotty, but on the whole, the list developed a firm tone. **1934** *Ibid.* 17 Sept. 8/1 Business conditions..are..'spotty'. One section may flourish,..while another suffers. **1937** R. S. Morton *Woman Surgeon* i. 21 My grown brothers played whist with my father before dinner... His luck was spotty. **1937** E. B. White *Let.* 31 May (1976) 155 My attendance at meals may be a little spotty—for a twelvemonth I shall not adjust my steps to a soufflé. **1957** K. A. Wittfogel *Oriental Despotism* 55 The spotty distribution of his administrative centers. **1970** A. Toffler *Future Shock* v. 72 Available statistics, unfortunately, are spotty. **1977** H. Fast *Immigrants* i. 57 Clair's schooling was spotty, but she learned to read. **1979** *Dictionaries* I. 97 Philosophers' general interest in dictionaries has been spotty.

spoulty, var. *Spaulty *a.*

spouse-breach¹. Add: *transf.* (Later *arch.* example.)
1922 Joyce *Ulysses* 47 He [*sc.* a dog] rooted in the sand,..a pard, a panther, got in spousebreach.

spout, *sb.* Add: **I. 4. a.** (Earlier example.)
1834 W. H. Ainsworth *Rookwood* II. iii. v. 345 To the spout with the sneezers in grand array.

b. *up the spout* (earlier *fig.* examples).
1829 P. Egan *Boxiana* 2nd Ser. II. 351 At the expiration of thirty-five minutes, and seventeen rounds, the *flue faker* acknowledged he was '*up the spout*'. **1846** *Swell's Night Guide* 64 And when she saw all hope was up the spout, She spouted everything a spout would take.

c. *to put* (someone) *up the spout*: to make pregnant (esp. out of wedlock); also simply *up the spout*, in the womb; pregnant. Cf. *up the pole* s.v. *Pole *sb.*¹ 1 b. *slang.*
1937 Partridge *Dict. Slang* 815/2 *Spout, up the*,..pregnant with child... Often in form, *to have been put up the spout*. **1949** *Landfall* III. 234 Well, they say he put her up the spout. **1956** P. Scott *Male Child* ii. i. 95 All these years taking every possible care and suddenly there's one up the spout. **1970** 'S. Troy' *Blind Man's Garden* viii. 100 Up the spout, isn't she? I thought Michel would have had more bloody savvy.

d. A gun barrel. Usu. in phr. *up the spout*, of a bullet or cartridge: in the barrel and ready to be fired. *slang.*
1943 C. H. Ward-Jackson *Piece of Cake* 55 *Spout*, a gun barrel. **1943** *R.A.F. Jrnl.* Aug. 12 He'd a round up the spout from being on this guard..and when he pulled the trigger there was a bang and a flash. **1969** M. Gilbert *Blood & Judgement* vii. 70, I can count six here in the clip... There's probably one up the spout. **1966** D. Varaday *Gara-Yaka's Domain* xi. 128 The pin failed to fire the dud cartridge in the chamber. There was no time to pump another into the spout. **1976** G. Seymour *Glory Boys* xv. 191 The Uzi was concealed there. Wonder if he's put the catch off, thought Jimmy, put one up the spout.

III. 11. spout-bath *N.Z.*, a natural douche-bath; **spout cup,** (*a*): for † in Dict. read 'Now only *Hist.*' and add later examples.
1929 C. C. Martindale *Risen Sun* 164 There are things in New Zealand that they call 'spout-baths'. A solid shining stream thuds on to your back from a height, and you feel as strong as it does. **1977** *N.Z. Herald* 5 Jan. 2-20/11 (Advt.), Rotorua Boulevard Motel, New units, sleep 1–6, restaurant adj, TV, putting green, hot swimming pool, 6 pvte spa and spout baths. **1956** G. Taylor *Silver* iv. 80 *Spout Cups*. Found also in faience, they are no more than a tankard or two-handled cup..ending in a curved spout. **1970** *Canadian Antiques Collector* June 28/2 What were spout cups? Used to feed invalids and children. Popular in the 18th century. Bulbous body, domed cover, duck-neck spout with handle generally at the right angle to the spout.

spouting, *ppl. a.* Add: **2. c.** *spouting well* = Spouter 3 b.
1861 *Chem. News* 28 Sept. 164/2 At about 325 feet in depth, there is about 40 feet of white sandstone, near the top of which 'spouting wells' are found. **1898** *Knowledge* 1 June 124/2 The 'spouting' wells of Russia entirely eclipse those of America in output. **1912** Tower & Roberts *Petroleum* v. 83 On the basis of the general character of the yield wells are divided into two classes, the flowing, spouting, or gushing wells, and the pumping wells.

spoze, var. *S'pose *v.*

‖ **Sprachgefühl** (ʃprā·χgĕfül). [Ger., f. *sprache* speech + *gefühl* feeling.] The intuitive feeling of a speaker for the essential character of a language; linguistic instinct. Also *loosely*, the character or genius of a language.

1902 GREENOUGH & KITTREDGE *Worlds & their Ways* ix. 127 Men of genius may take great liberties with their mother tongue without offence; but..let them violate its *Sprachgefühl*, and their mannerism becomes, as it were, a foreign language. **1914** *Pedagogical Seminary* June 256 If there is to be instruction in modern foreign languages the age of about ten is for most children the best to begin it. If it comes before this the child has not developed sufficient *Sprachgefühl* for his *own* language, has not developed sufficient idiomatic sense of it to avoid the danger of interference. **1938** E. PARTRIDGE *World of Words* II. vi. 163 The *Sprachgefühl*, feeling for speech, exercises a pervasive influence in a language so long cultivated as English. **1953** K. JACKSON *Lang. & Hist. in Early Britain* 682 It may be thought that the language was in some way committed by its *Sprachgefühl* to a penultimate accent. **1976** *Amer. Speech 1973* XLVIII. 224 Whether this development is associated with the loss of sprachgefühl in a community where the use of a language is declining is a possibility to be considered.

spraddle, *v.* Add: **2.** *trans.* To spread or stretch (one's legs, etc.) wide apart. Also *transf.*

1913 J. LONDON *Son of Sun* vii. 241 He stood with legs spraddled over a large grass basket. **1928** 'BRENT OF BIN BIN' *Up Country* xv. 264 As she walked she spraddled the off hind hoof. **1929** 'SEAMARK' *Down River* i. 6 Let this decrepit tub of yours spraddle her old legs a little faster.

3. The vb.-stem in combination with ppl. adjs., to form adjs., as *spraddle-footed, -hipped, -legged. U.S.*

1935 Z. N. HURSTON *Mules & Men* I. viii. 177 Don't set there all spraddle-legged. **1974** D. SEARS *Lark in Clear Air* ii. 33 They were big, old, spraddle-hipped, spavined bays with a lot of Clyde in them. **1975** J. GORES *Hammett* (1976) xiv. 102 The boy stood spraddle-footed on the porch.

spra·ddled *ppl. a.* (also with *out*: see also quot. 1927); **spraddling** *ppl. a.* (later examples).

1898 H. S. CANFIELD *Maid of Frontier* 89 He rode with the spraddling seat of a man more accustomed to the plow than to the scout's saddle. **1926** E. HEMINGWAY *Sun also Rises* xviii. 230 They held him and lifted him. It was uncomfortable and his legs were spraddled. **1927** *Amer. Speech* Dec. 169 To put on airs or be dudishly dressed was to be 'all spraddled out'. **1930** D. RUNYON in *Collier's* 1 Feb. 46/2 Down he comes all spraddled out. **1935** L. A. G. STRONG *Tuesday Afternoon* 83 Between his spraddling knees he had spread a clean white napkin. **1940** *Harper's Mag.* Oct. 513/2 Every now and then a nigger would come flying out and go sailing through the air..spraddled like a flying squirrel. **1975** J. GORES *Hammett* (1976) xxxii. 221 Laverty's right knee pumped, twice, up between Lynch's spraddled legs.

sprag, *sb.*[3] Add: **2. a.** More widely, any of several devices formerly fitted to motor vehicles to prevent them from running backwards down a hill (see quots.).

1902 A. C. HARMSWORTH et al. *Motor & Motor-Driving* xv. 332 The Sprag..is an adjunct fitted to most cars... The sprag should be dropped before the car actually starts to run backwards. *c* **1915** *Autocar Handbk.* (ed. 6) xiv. 216 The sprag is normally held up clear of the ground by a cord, but when the car is likely to stop on a hill the driver should release the cord in good time, and let the 'devil' drag on the ground. Then directly the car stops, the pointed end of the bar digs into the ground. *Ibid.* 217 Another good form of sprag consists of a strong pawl, which is allowed to trip over ratchet teeth out on a revolving part of the gear... Even with good brakes the one great advantage of a sprag is that it permits the car to be restarted on the steepest hill with both brakes off. **1933** *Motoring Encycl.* 723/2 The sprags often fitted on horse-drawn vehicles..have been revived in a new form on motor vehicles. *Ibid.*, The sprag is arranged at the rear of the gear-box, and consists of a roller and wedge adapted to produce a gripping motion on a drum..which turns with the transmission shaft.

b. *fig.* Cf. SPOKE *sb.* 4 a.

1914 'HIGH JINKS, JR.' *Choice Slang* 21 A sprag in the wheel of progress. **1973** R. DENTRY *Encounter at Kharmel* vii. 111 If you were in the President's shoes, how would you put a sprag in Ziauddin's wheel?

sprag, *v.* Add: **3.** To accost truculently. *Austral. slang.*

1916 C. J. DENNIS *Songs of Sentimental Bloke* 130 *Sprag*, to accost truculently. **1935** L. LUARD *Conquering Seas* iii. 41 'Twas only to save you from getting spragged. *a* **1938** C. J. DENNIS in *Penguin Bk. Austral. Ballads* (1964) 236 A tug named Tyball..Sprags 'em an' makes a start to sling off dirt.

Sprague–Dawley (sprēᶦg,dǭ·li). The names of R. W. *Dawley* (1897–1949) U.S. physical chemist who established the strain, and of his wife, née *Sprague*, used *attrib.* to designate an inbred strain of rat much used in laboratories.

A proprietary name in the U.S.A.

1951 *Proc. Soc. Exper. Biol. & Med.* LXXVII. 635/2 Weanling rats of the Sprague-Dawley (SD) strain have a lower choline requirement..than was observed in rats of the Alabama Experiment Station (AES) strain. **1967** B. S. WOSTMANN et al. in M. L. Conalty *Husbandry Lab. Animals* 195 Germfree rats were of the Sprague-Dawley or Lobund Wistar strains. **1970** *Jrnl. Gen. Psychol.* LXXXIII. 88 The *Ss* were 70 male albino rats of the Sprague Dawley strain. **1978** *Official Gaz.* (U.S. Patent Office) 3 Jan. TM 67/1 Sprague-Dawley. The Mogul Corporation... Pub. 10-11-77. Filed 6-9-77.

sprain, *v.*[1] Add: **2.** *to sprain one's ankle*: (of a woman) to be seduced (and become pregnant); to lose one's virginity. *euphem.* Obs. exc. *Hist.*

1785 F. GROSE *Classical Dict. Vulgar Tongue* s.v. *Ankle*, a girl who is got with child, is said to have sprained her ankle. **1940** M. SADLEIR *Fanny by Gaslight* I. 286 'I suppose you are still—still a ——.'..'I suppose you mean, have I sprained my ankle yet?'

spraing, *sb.* Restrict *Sc.* to sense in Dict. and add: **2.** Also † **sprain.** A disease of potatoes in which they appear sound on the outside but show curved lines of discoloration when cut.

1909 *Jrnl. Board Agric.* XVI. 33 Several complaints have reached the Board respecting the losses caused to potato growers from the ailment known as sprain in potatoes, and it is asserted that cases of it..are causing some anxiety in Scotland and elsewhere. **1929** *Trans. Brit. Mycol. Soc.* XIV. 150 Spraing. (Sprain, Internal Rust Spot.)... Internal brown spot. **1980** F. HOPE in E. Gram et al. *Recognition & Control of Pests & Diseases of Farm Crops* (ed. 3) 155/1 Spraing is thought to be caused by a virus. **1980** *Amat. Gardening* 4 Oct. 12/1 The [potato] crop is affected with a physiological disorder known as spraing.

spraints, *sb. pl.* **b.** Delete *rare* and add: in *sing.* or *collect.* sense. (Later examples.)

1960 G. MAXWELL *Ring of Bright Water* xi. 155 There is a lavatory at every other holt, and the excrement (which is known as 'spraint', and has no offensive odour..) often forms a high pyramidal pile. **1979** *Guardian* 16 Mar. 11/3 A spraint is a blackish smear of digested fishbone and otter-lunch which the animal tends to leave.

sprangly, *a.* (Earlier examples.)

1840 C. F. HOFFMAN *Greyslaer* II. III. i. 103 His great sprangly beard. **1886** *Leslie's Pop. Monthly* XXII. 503/1 We can command a view through their sprangly branches.

spraser, sprasy, varr. *SPRAZER.

sprauncy (sprǭn·(t)si), *a. slang.* Also **sprauntsy, spronсy.** [Of uncertain origin: perh. related to dial. *sprouncey* cheerful (*Eng. Dial. Dict.*).] Smart or showy in appearance or sound of voice.

1957 L. P. HARTLEY *Hireling* xi. 90 She's bought new sprauncy clothes for the children. **1959** H. HOBSON *Mission House Murder* viii. 57 That sprauncy Girton voice of hers. **1969** *Guardian* 4 Feb. 7/2 The 'sprauntsy' (showy) antique dealers. **1971** G. EWART *Gavin Ewart Show* I. 21 Two sprauncy birds inhibit the parkway. **1976** 'P. B. YUILL' *Hazell & Menacing Jester* xiii. 142 They pay sixty grand to live in a spronсy little street like this.

sprawl, *sb.* Add: **1. d.** The straggling expansion of an indeterminate urban or industrial environment into the adjoining countryside; the area of this advancement. Freq. with defining adj. (see *suburban* and *urban sprawl* at first element in Suppl.).

1955 *Times* 23 Aug. 10/2 It is sad to think that London's great sprawl will inevitably engulf us sooner or later, no matter how many 'green belts' are interposed in the meantime between the colossus and ourselves. **1958** *Listener* 23 Oct. 641/1 As the new industrial zones came to life on the edges of the built-up areas, they frequently appeared to be no more than an extension of the old industrial sprawl. **1967** *Ibid.* 3 Aug. 147/2 Planning so far has failed to contain sprawl. **1971** P. GRESSWELL *Environment* 122 Green Belt policy stopped sprawl in crucial places at a crucial time. **1977** *Listener* (N.Z.) 15 Jan. 9/2 It's a sad reflection on our society that 'suburban' has become a dirty word, synonymous with 'subtopia' and responsible for many urban problems from neurosis to sprawl.

spraw·lingly, *adv.* [f. SPRAWLING *ppl. a.* + -LY[2].] In a sprawling manner.

1921 *Spectator* 7 May 585/2 Gauntly outlined, white and still, Three haystacks peer above the hill; Three agèd rakes thrust sprawlingly Fantastic tendons to the sky. **1980** *Economist* 15 Mar. 104/1 Several books of sociology, history, politics and literary criticism uneasily brought together into one weightily erudite but sprawlingly inconclusive survey.

spray, *sb.*[1] **1. c.** Delete † and add later examples.

1870 HARDY *Satires of Circumstance* (1914) 20 When I set out for Lyonnesse, A hundred miles away, The rime was on the spray. **1893** F. THOMPSON *Poems* 66 Oh, there were flowers in Storrington On the turf and on the spray. **1921** W. DE LA MARE *Veil* 19 Thrush and robin perched mute on spray.

2. d. A brooch or clip fashioned in imitation of a bouquet of flowers, or of a twig with fruit or foliage. In full, *spray brooch.*

1803 C. WILMOT *Jrnl.* 6 Mar. in T. U. Sadleir *Irish Peer on Continent* (1920) 142 Necklaces, sprays of brilliants, towering on the head like feathers, diamond nets, combs. **1863** MRS. GASKELL *Dark Night's Work* v. 64 Your pearls ..were..handsome..but we must have them re-set; the sprays are old-fashioned. **1939** P. WENTWORTH *Lonesome Road* viii. 50, I should love to give her a diamond spray from Woolworth's. *Ibid.* xii. 73 The oak spray..two diamond oak-leaves and three acorns. **1951** *Catal. of Exhibits, South Bank Exhib., Festival of Britain* 39/1 Ruby and diamond spray brooch. **1966** *Harper's Bazaar* Sept. 76 A pair of Victorian diamond sprays in the hair.. £900. **1979** *Country Life* 11 Oct. Suppl. p.v., Modern diamond and emerald spray brooch.

spray, *sb.*[2] **2.** For *Med.* read orig. *Med.* and add: **a.** (Earlier and further examples.) Now applied to any jet of (esp. liquid) particles emitted by an atomizer or similar device.

1870 *Brit. Med. Jrnl.* 2 July 21/1 The value of sprays in the treatment of affections of the throat and windpipe ..is much hindered by the inconvenience attending the use of the apparatus... Clarke's [apparatus] produces a fine spray. **1891** KIPLING *Light that Failed* viii. 146, I haven't any spray, and I never leave charcoal unfixed overnight. **1963** R. CARSON *Silent Spring* iii. 23 There were cockroaches in the house..and..a spray containing endrin was used. **1974** A. J. HUXLEY *Plant & Planet* xxviii. 337 Temperate fruit growers may apply fifteen different sprays in a season.

3. b. *spray-based, -clouded, -dabbled, -haired.*

1895 W. B. YEATS *Poems* 210 With blown, spray-dabbled hair. **1930** R. CAMPBELL *Adamastor* 74 The rocks, spray-clouded, are your signal guns. **1939** DYLAN THOMAS *Map of Love* 4 Spray-based and rock-chested sea. **1956** P. LARKIN *Less Deceived* 38 Rain patters on a floor that tilts and sighs. Fast-running floors, collapsing into hollows, Tower suddenly, spray-haired.

c. *spray booth, can, job, nozzle, pump*; **spray-gun,** an apparatus for applying a liquid substance in the form of spray; an atomizer; **spray irrigation,** a form of irrigation in which water is sprayed from pipes running along or above the ground and reaches the surface in the form of droplets; **spray line,** a perforated pipe used in spray irrigation; **spray pond,** a pool over which water is sprayed in a chamber through which air is passed, so as to cool the air or humidify it; **spray refining, steelmaking,** a continuous method of making steel in which molten iron falling in a stream is atomized by jets of oxygen and flux that combine with impurities in the droplets; hence **spray steelmaker,** an installation in which spray steelmaking is carried out; **spray tower,** a hollow tower in which a liquid is made to fall as a spray, e.g. to cool it or to bring it into contact with a gas.

1959 *Times* 17 Feb. 2/3 The manufacture of Industrial Equipment such as Ovens, Spraybooths. **1975** 'R. BUTLER' *Where all Girls are Sweeter* vi. 63 A man was spraying a car in a spray booth. **1972** *Times* 14 Dec. 9/5 A higgledy-piggledy mixture of colours..all laid on with spray cans and felt-tipped marker pens. **1980** *Times Lit. Suppl.* 5 Sept. 972/4 Not only are felt-pens and spray-cans rare in Africa, but so is literacy. **1920** *Brass World* Dec. 343 (*caption*) A new spray gun invention. **1944** R. CHANDLER *Five Murderers* 31 A woman..was popping at aphis [*sic*] with a spray-gun. **1956** 'B. BUCKINGHAM' *Three Bad Nights* xiv. 122 Battered old spray-guns with which they were sprinkling the plants. **1974** 'S. WOODS' *Done to Death* 29 Mr Gillespie was pottering about in his garden with a green spray-gun in his hand. **1931** *Circular U.S. Dept. Agric.* No. 195. 1 Recently spray-irrigation equipment, designed for the irrigation of general farm crops, has been developed in Germany. **1950** *N.Z. Jrnl. Agric.* LXXX. 520/1 Centrifugal pumps are most commonly used for spray irrigation. **1974** WITHERS & VIPOND *Irrigation* ii. 55 Spray irrigation..is suitable for use in temperatures down to −9°C (16°F). **1963** L. DEIGHTON *Horse under Water* xvii. 69 The last rays of the sun did a spray job on one side of da Cunha's bony head. **1971** *Engineering* Apr. 95/2 (Advt.), Your problem will be solved by an enthusiastic team well versed in the study of Spray Characteristics and Patterns for any particular Sprayjob. **1961** H. J. HINE *Dict. Agric. Engin.* 130 It is possible to divide up the methods of overhead application to field soil into spray lines, sprinklers and rainers. **1974** WITHERS & VIPOND *Irrigation* ii. 54 A more elaborate spray line consists of a water pipe, about 30 mm diameter, perforated in one quadrant along its length, and mounted on light tressles. **1919** FRASER & JONES *Motor Vehicles & their Engines* viii. 65 The suction created by the rush of air past the spray nozzle causes the gasoline to be delivered to the mixing chamber in a fine spray. **1966** GURNEY & COTTER *Cooling Towers* iii. 62 (*caption*) Typical sections of cooling tower timber distribution system showing header, lateral and porcelain spray nozzle. **1924** L. C. LICHTY *Measurement Natural Gas* III. iv. 349 Two general methods are used to cool the water, namely, the spray pond or cooling tower method. **1951** J. JACKSON *Cooling Towers* ii. 7 Spray ponds are only suitable where a close approach to the wet-bulb temperature is not required; their main use lies in cooling water prior to discharge into a river. **1913** *Chambers's Jrnl.* Jan. 61/2 The stamps are moistened within the machine by a tiny spray pump. **1950** *N.Z. Jrnl. Agric.* May 470/1 A spray pump [for a garden]..may be of bucket or knapsack type according to personal preference. **1967** A. H. COTTRELL *Introd. Metallurgy* xi. 141 Very recently a new steelmaking process, spray-refining, has been developed by the British Iron and Steel Research Association. A stream of molten iron from a blast furnace falls through a ring of jets, which breaks it up into droplets,..and a second ring of jets which inject powdered lime and other fluxes. **1975** *Steel USSR* V. 493 The optimum working pressure in the

chamber for carrying out spray refining of pig iron in a vacuum is 150 mmHg. **1965** *Economist* 25 Dec. 1437/2 (*heading*) Spray steelmaking. **1966** *Ibid.* 15 Oct. 293/3 As operated at present, the spray steelmaker is far from continuous: it only runs during part of a blast furnace tap and its capacity is limited by the size of the ladle that is pushed underneath. **1966** *Observer* 16 Oct. 9/1 The revolutionary 'spray' steel-making process, developed by the British Iron and Steel Research Association in Sheffield,..has all the makings of a major technological advance. **1976** *Metals Abstr.* IX. 1750/2 (*heading*) An investigation into oxidation processes occurring during spray steelmaking using naturally alloyed pig iron. **1937** T. K. SHERWOOD *Absorption & Extraction* vi. 168 Spray towers are best suited to the absorption of very soluble gases. **1951** J. JACKSON *Cooling Towers* ii. 7 Like the cooling pond and the spray pond, the spray tower is probably suitable only for easy cooling duties. **1979** A. L. LYDERSEN *Fluid Flow & Heat Transfer* vi. 155 Spray towers are the simplest type of gas scrubbers.

d. Special Comb. relating to natural spraying or the production of spray, as **spray region,** the region at the top of the atmosphere where molecules are so far apart that their paths are determined by gravity and upward moving ones are likely to escape into space; **spray zone,** that area near the sea or a waterfall that is often moistened by spray.

1949 *Jrnl. Brit. Interplanetary Soc.* VIII. 255A While it is generally believed that the Zodiacal Light is due to extra-terrestrial particles, it is possible that the scattering particles are molecules in the 'spray' region, very high up in the Earth's atmosphere. **1963** G. M. B. DOBSON *Exploring Atmosphere* i. 8 The top of this spray region, which is the top of the atmosphere, is quite indefinite. **1946** *Ecology* XXVII. 321/1 These communities of the spray zone are present only in situations either exposed directly to spray or very well protected from desiccation. **1976** *New Phytologist* LXXVI. 361 Two contrasting coastal habitats which create salinity problems for species occupying them are the spray zone and the salt marsh.

spray, *v.*[4] Add: **1. b.** *fig.*

1923 H. G. WELLS *Men Like Gods* III. iv. 284 From some slope above a lark had gone heavenward, spraying sweet notes. **1924** R. CAMPBELL *Flaming Terrapin* ii. 23 The trees came crashing down lengthwise, And sprayed their flustered birds into the skies. **1976** E. DUNPHY *Only a Game?* iv. 125 They were giving us space to come further forward, and giving us room to turn and spray it about.

2. b. *transf.* To subject to a rapid succession or shower of bullets, shot, etc.

1926 T. E. LAWRENCE *Seven Pillars* VI. lxxvii. 404 Wood..got the Indians ready to spray the guard tent. **1977** C. FORBES *Avalanche Express* xv. 152 The man behind the gun...sprayed the Wagon-Lit... A hail of bullets thudded into their compartment.

3. b. Of a male cat: to mark its environment with the smell of its urine, as an attractant to the female.

[**1949** M. C. GAY *How to live with a Cat* xi. 175 A tomcat sprays whiffs of cat smell about in the hope of luring a maiden feline to his moon-bespattered lair.] **1954** L. F. WHITNEY *Compl. Bk. Cat Care* xvi. 199 The males specifically lose their potent and..obnoxious odour, stop spraying and use their pans or boxes. **1967** A. LEWIN *Unaltered Cat* II. iv. 117 If he's frustrated, he'll begin to spray. It's an unpleasant odour, except, of course to a female cat.

sprayable (sprēᶦ·ăb'l), *a.* [f. SPRAY *v.*[4] + -ABLE.] Capable of being sprayed.

1957 B. A. DOMBROW *Polyurethanes* iv. 67 Heaters.. are necessary to reduce the initial viscosity of the prepolymer to a sprayable consistency. **1972** D. G. SHEPHERD *Aerospace Propulsion* iv. 108 Another possible fuel is powdered metal, for example magnesium or aluminium, mixed with a liquid hydrocarbon to form a pumpable and sprayable slurry.

spray drying (stress variable). Also **spray-drying.** [SPRAY *sb.*[2] or *v.*[4]] A method of drying foodstuffs, ceramic materials, etc., by spraying finely-divided particles of the substance into a current of hot air or another gas, the water in the particles being rapidly evaporated.

1921 *Jrnl. Industr. & Engin. Chem.* May 448/2 Spray drying is comparatively expensive, mainly because it is impossible to utilize all the heat going through the drying chamber. **1946** [see *flash-drying* vbl. sb. s.v. *FLASH *sb.*[1] 14 b]. **1968** *Economist* 15 June 64/3 The dried milk market, so far dominated by Cadbury's 'Marvel' which is the product of an improved spray drying technique, is worth around £3 million a year.

Hence (as a back-formation) **spray-dry** *v. trans.*; **spray-dried** *ppl. a.*; **spray drier,** a machine or installation that performs spray drying.

1921 *Jrnl. Industr. & Engin. Chem.* May 448/2 It is very difficult to give the cost of operating a spray dryer either in dollars or in heat units. **1932** *Bull. Hannah Dairy Res. Inst.* No. 3. 134 Rancidity appeared less frequently in roller-dried than in spray-dried powders. **1945** *Biochem. Jrnl.* XXXIX. p. xxvi, We have examined ..the biological value of the proteins (nitrogen) of a sample of spray dried milk manufactured in 1939. **1950** *Thorpe's Dict. Applied Chem.* (ed. 4) x. 197/2 The essential features in spray-drying eggs are the avoidance of scorching..and the production of a powder containing not more, and preferably less, than 5% moisture. **1961** M. HYNES *Med. Bacteriol.* (ed. 7) xii. 196 More than 10 per cent of imported spray-dried egg contain live *Salmonellæ*. **1967** M. CHANDLER *Ceramics in Mod. World* ii. 62 The spray dryer consists of a cylindrical drum..with a conical outlet at the bottom. **1971** *Sci. Amer.* Oct. 5/3 (Advt.), Our engineers solved the..problem..by spray-drying it [*sc.* powdered ceramic] in a blast of hot air. **1979** A. L. LYDERSEN *Fluid Flow & Heat Transfer* vi. 167 A spray dryer for detergent containing 45 weight% water.

sprayman (sprēᶦ·mæn). Also **spray-man.** [f. SPRAY *v.*[4] + MAN *sb.*[1]] A person who sprays crops with insecticide or the like.

1959 *Bull. World Health Organization* XX. 23 No clinical symptoms of dieldrin poisoning have been observed among the teams of spray-men. **1969** W. F. DURHAM in Miller & Berg *Chem. Fallout* xxiii. 439 Both respiratory and dermal exposure of spraymen were greater for day than for nocturnal spraying. **1973** J. J. MCKELVEY *Man against Tsetse* iii. 190 The spraymen walked along the outer margins of the vegetation on each river bank.

spray-on (sprēᶦ·ǫn), *a.* [f. SPRAY *v.*[4] + ON *adv.*] Of a liquid substance: that is applied in the form of spray. Also of the container.

1959 *Wall St. Jrnl.* 3 July (Eastern ed.) 13/4 The company also reports brisk sales of its new aerosol package spray-on cans of enamel. **1965** *Harper's Bazaar* Dec. 89/1 Spray-on fragrance flacons combine luxury with efficiency of application. **1976** R. RENDELL *Demon in my View* ii. 19 Arthur still felt guilty about using spray-on polish instead of the old-fashioned wax kind.

spray-painting (stress variable). [SPRAY *sb.*[2] or *v.*[4]] The application of paint in the form of spray.

[**1902**: see *painting-machine* s.v. *PAINTING *vbl. sb.* 6.] **1921** C. W. TERRY *Practical Motor Body Building* xliii. 310 Spray painting by compressed air is a system followed by many of the large modern motor car builders. **1934** *Archit. Rev.* LXXV. 12/2 As a constructivist painter Moholy-Nagy was the first to..employ spray-painting to secure more exact chromatic effects. **1968** J. IRONSIDE *Fashion Alphabet* 229 *Flocking*. This is a method of applying patterns or dots to fabric... The method is rather similar to spray-painting. **1978** *N.Y. Times* 30 Mar. C13/3 (Advt.), French polishing and spray painting for piano refinishing.

Hence (as a back-formation) **spray-paint** *v. trans.*, to paint (a surface) by means of a spray; also as *sb.*, paint that is applied thus; **spray-painted** *ppl. a.*

1928 *Daily Express* 5 Oct. 2/1 A new base-material with which the car is covered, and which can be spray-painted to any finish desired by the customer. **1973** 'E. MCBAIN' *Hail to Chief* iv. 51 The decrepit brick building..had been spray-painted with graffiti. **1975** B. GARFIELD *Hopscotch* x. 97 He spent the next hour buying..automobile spray-paint. **1975** *Listener* 16 Oct. 491/3 Political posters and spray-painted slogans on every square inch of stone.

sprazer (sprā·zəɹ). *slang.* Also **spraser, sprasy, sprazey,** etc. [ad. Shelta *sprazi*.] Sixpence; a sixpenny piece. Cf. *SPROWSIE.

1931 [see *HALF *sb.* 6 i]. **1934** [see *HOLE *sb.* 2 d]. **1936** J. CURTIS *Gilt Kid* xvii. 175 I'll..let you have the odd twenty-seven [pounds] and a sprazer. **1939** J. B. PRIESTLEY *Let People Sing* x. 257 See if we can't take another spraser or two from the punters. **1961** *John o' London's* 30 Nov. 610/3 Sixpence—*tanner, spraf, sprarsy.*

spread, *sb.* Add: **I. 2. e.** *Diamond-cutting.* The width of a stone considered in proportion to its depth.

1813 J. MAWE *Treat. Diamonds* i. 47 The artist..has to examine carefully, in what direction the stone may be cut, so as to afford the greatest breadth, or *spread* as it is technically termed. **1930** W. R. CATTELLE *Precious Stones* 62 Since the trade have found how important it is to have a proper 'spread' to the stone..there has been a tendency to demand stones too shallow for the best results.

† **f.** *Aeronaut.* = *SPAN *sb.*[1] 5 d. *Obs.*

1894 *To-Day* II. 171/2 The wings have a spread of twenty yards square. **1909** A. BERGET *Conquest of Air* 188 The spread of the wings is 10·20 metres. **1912** C. B. HAYWARD *Pract. Aeronaut.* 262 The ratio of spread to depth (aspect ratio) of the monoplanes is usually less than that of biplanes.

g. *Econ.* The difference between two rates or prices.

1919 A. C. WHITAKER *Foreign Exchange* xii. 369 The spread between the local and the foreign money rates. **1928** *Britain's Industr. Future* (*Liberal Industr. Inquiry*) IV. xxiv. § 7.331 The Linlithgow Committee..came to the conclusion that 'the spread between the producer's and the consumer's prices is unjustifiably wide'. **1938** *Sun* (Baltimore) 8 June 6/1 Charges of price-cutting came into the open at a hearing May 19 on distributors' request for a wider 'spread' between the prices they paid for milk and those for which they sold it. **1978** *N.Y. Times* 29 Mar. B2/1 But adding to the rise in farm beef prices is the widening spread between what a cattle raiser gets for a steer and what a roast costs in the supermarket.

h. The degree or manner of variation of a quantity among the members of a population or sample.

1929 *Jrnl. du Conseil* IV. 219 It has hitherto been the practice to employ such statistical methods as finding the mean length, median length, or semi-inter-quartile range in dealing with the 'spread' of length frequency groups. **1957** *Practical Wireless* XXXIII. 696/2 The current gain of the first stage transistor has been quoted as about 50 times, but it could be as low as 30 times, due to manufacturing spread. **1974** *Listener* 7 Nov. 595/2 Now there is a more even spread of intelligence and the skew in the graph is at both ends.

i. The expansion of a person's girth, esp. at middle age; paunchiness. Also, an example of this. Usu. in phr. *middle-aged spread*: see *MIDDLE-AGED *a.* 1 b. *colloq.*

1930 *Field* 29 Nov. 775/2 An older woman..middle-aged, with, possibly, a 'spread'. **1931**, etc. [see *MIDDLE-AGED *a.* 1 b]. **1976** N. THORNBURG *Cutter & Bone* iv. 86 Swanson could have passed for a decade older, having already achieved a comfortable middle-aged spread.

3. c. A ranch, esp. for raising cattle; a large farm. Also *fig.* orig. and chiefly *U.S.*

1927 W. JAMES *Cow Country* 67 He'd paid a big price for the said spread, and he was lord and master there sure enough. **1947** *Trail Riders Bull.* Feb. 20/1, I wuz top bronc buster for the Tumblin' L spread. **1963** R. D. SYMONS *Many Trails* iii. 25 So he had a few dollars coming in, and could see his way, as he would have said, to running a 'spread'. **1966** G. DURRELL *Two in Bush* vi. 186 Harry, together with Bevan Bowan, took us out to a 'spread' not far from Canberra (a tiny little smallholding of some 200,000 acres) on which they were investigating another facet of the kangaroo's biology. **1973** J. WAINWRIGHT *Devil you Don't* 30 'The Ponderosa' was his spread and no cheap, jumped up, fiddle-foot was gonna muscle in. **1981** J. BEECHING *Death of Terrorist* iii. 34 He thought of his ranch up in Texas. 'Not a big spread,' he said modestly.

d. *Geol.* A relatively thin sedimentary deposit.

1956 A. L. ARMSTRONG in D. L. Linton *Sheffield* vi. 90 It is from these beds that the sands and gravels of the lower terraces and valley spreads were mainly derived. **1977** *Antiquaries Jrnl.* LVII. 187 Some sarsens..could be derived from chalk or Greensand as could the soli-flucted spread in the Vale of Pewsey.

e. *Cytology.* A microscopic preparation (as a smear or a squash) in which material is spread for observation rather than thin-sectioned, esp. for the purpose of showing chromosomes at metaphase.

1963 *Stain Technol.* XXXVIII. 284 Heteroploidy was not observed in any of the spreads, unlike those observed in long term cultures of rabbit endothelial cells. **1968** *Brit. Med. Bull.* XXIV. 261/2 Typical spreads produced in her laboratory show metaphase material in about one of every 50 fields of 100 × 100 μ. **1978** *Nature* 23 Mar. 325/1 Figure 2 shows the distribution of silver grains in autoradiograms of human metaphase chromosome spreads.

4. d. *Billiards.* A rebound of a cue ball from the object ball at a considerable angle from its former course. *U.S.*

1858 M. PHELAN *Game of Billiards* (ed. 3) 102 To effect a 'spread' it is not necessary to hit the object-ball so far off from the centre as would appear at the first glance. **1913** M. DALY *Daly's Billiard Bk.* iii. 46 Try the same plan for the dead follow and the dead 'spread' (wide angle carom).

II. 6. b. Any substance suitable for spreading on bread to make it tasty, such as paste or jam. orig. *U.S.*

1866 *Hours at Home* III. 507/2 A late rebel told me that, while with Lee upon the Gettysburgh campaign, he went to a farm-house one day and demanded some 'spread', as they call marmalade in that matter-of-fact country. **1886** F. R. STOCKTON *Casting away of Mrs. Lecks* 40 There was some sort of jam left at the bottom, so that the one who gets the last biscuit will have somethin' of a little spread on it. *c* **1938**, etc. [see *sandwich spread* s.v. *SANDWICH *sb.*[2] 3]. **1962** M. DUFFY *That's how it Was* xiii. 107, I had to..mix up some chocolate spread from cocoa, sugar and a little milk. **1972** *Daily Tel.* 11 Nov. 2/7 There were increases of 3·41 per cent in the prices of jams, honey, and spreads.

9. An article or advertisement displayed prominently in a newspaper or periodical; *spec.* printed matter occupying two facing pages. Also *fig.* orig. *U.S.*

[**1858** O. W. HOLMES *Autocrat of Breakf.-Table* 131 One gives a 'spread' on lines, and the other on paper—that is all.] **1877** *Harper's Mag.* Dec. 50/1 His remarkable ability is best seen when occasion arises for a 'spread'. **1924** in WEBSTER. **1931** *Week-End Rev.* 7 Nov. 563/2 The inclusion of a four-page 'spread', printed in two colours. **1940** [see *centre spread* s.v. *CENTRE *sb.* 19]. **1951** M. MCLUHAN *Mech. Bride* (1967) 61/1 Another full-page spread shows a threesome in a panic. **1956** H. KURNITZ *Invasion of Privacy* xviii. 117 The afternoon papers, and the late radio broadcasts..had given the Morley case a big spread. **1969** A. GLYN *Dragon Variation* viii. 238 You'll give the match a good spread in your papers, won't you, Paul? **1972** D. LEES *Zodiac* 7 Get a centre page spread with pictures.

10. *U.S. Stock Exchange.* A contract combining the option of buying shares of stock within a specified time, at a specified price above that prevailing when the contract is signed, and the option of selling shares of the same stock within the same time, at a specified price below that prevailing when the contract is signed. Cf. *SPREAD EAGLE *sb.* 3* b, STRADDLE *sb.* 2 a.

1879 in WEBSTER. **1885** *Harper's Mag.* Nov. 844/1 A 'straddle'..differs from the 'spread' in that the market price at the time of purchase is filled into the latter, while in the 'straddle' the price may vary from that of the market, by agreement or otherwise. **1900** S. A. NELSON *ABC of Wall St.* 160 *Spread.* This is a double stock privilege which entitles the holder to the right to deliver or demand a certain amount of stock on specified terms, or grain price differences between different options, or between the same option in different cities, or between the put and call price. **1957** CLARK & GOTTFRIED *University Dict. Business & Finance* (1967) 332/1 If a stock is selling at 100, a speculator may buy a *spread* option for $5 per share, with a *spread* in the buying and selling prices of 5 points up or down. Thus, if the stock later goes below 90 (including the 5 point *spread* and the 5 point cost of the option), it can be sold at a profit. **1970** SLOAN & ZURCHER *Dict. Econ.* (ed. 5) 412 *Spread*, as applied to security trading, two separate options, a *put*..specifying a price below the prevailing market, and a *call*..specifying a price above the prevailing market, both options applying to the same security and expiring on the same date.

11. *Bridge.* (See quot. 1929.) Cf. *SPREAD *v.* 19.

1929 M. C. WORK *Compl. Contract Bridge* 245 *Spread*, a hand which Declarer can show in proof of his ability to win all thirteen tricks. **1977** *Field* 13 Jan. 65/1 Only the duplication of values prevented the contract from being a spread.

12. a. *Geol.* An array of seismometers used simultaneously to detect and record disturbances resulting from a single shot in a geophysical survey.

1942 *Geophysics* VII. 138 The problem of the wide shot spread is much more complicated than that of the vertical shot. **1945** *Ibid.* X. 351 To correct for these weathering variations a series of short refraction shots was taken at each recording spread. **1962** *Jrnl. Geophysical Res.* LXVII. 2852/2 The method used was to keep the geophone spread fixed and to move the shot away from the spread. **1977** A. HALLAM *Planet Earth* 29/1 Large spreads of seismometers have recently been set up, one of the largest being in Norway.

b. *Oil Industry.* The total assemblage of men and equipment needed for a particular job, esp. laying a pipeline.

1974 *Petroleum Rev.* XXVIII. 765/1 The prospect of lay barge spreads capable of operating in depths of up to 1,000 feet certainly posed problems. **1975** *North Sea Background Notes* (Brit. Petroleum Co.) 32 The land line was laid in three spreads, two working between the Tay and Cruden Bay and the third from the Tay to Grangemouth. **1976** *Offshore Platforms & Pipelining* 175/3 Two of the pipeline spreads will use automatic welding.

spread, *v.* Add: **I. 1. g.** *Mus.* To play (the notes of a chord or a chord itself) in rapid succession instead of simultaneously.

1873 [in Dict., sense 1 a]. **1938** *Oxf. Compan. Mus.* 48/1 *Arpeggiare* (It.), to play harpwise, i.e. (on the piano, etc.) to spread the notes of a chord, from the bottom up. **1953** W. EMERY *Bach's Ornaments* 102 On the harpsichord or clavichord, a chord is often very much more effective when spread than when played *sec*. **1977** *Gramophone* June 68/2 The violinist also spreads some of his two-note chords.

4. e. To record or enter *on* a documentary record. *U.S.*

1845 J. F. COOPER *Chainbearer* II. iv. 44 It will greatly aid the reader..if I spread on the record the language that passed between my late agent and..his confidant. **1894** T. F. ROBLEY *Hist. Bourbon County, Kansas* 184 Councilmen Dimon, White and Drake caused the following order to be spread upon the minutes. **1910** *Atlantic Monthly* Feb. 231 Achievements in that field are naturally not spread on the record as are exploits in railway financiering. **1931** *Randolph Enterprise* (Elkins, W. Va.) 26 Mar. 1/5 Resolved that a copy of these resolutions be sent to the family, a copy spread on the minutes of the lodge, [etc.].

7. b. (Earlier *absol.* examples.)

1860 J. G. HOLLAND *Miss Gilbert's Career* x. 173 He sort o' stands round, and spreads, and lets off all the big talk he hears. **1884** 'MARK TWAIN' *Huck. Finn* xxi. 206 He howled, and spread around, and swelled up his chest.

19. *Bridge.* (See quot. 1964.) Cf. *SPREAD *sb.* 11.

1929 M. C. WORK *Compl. Contract Bridge* 245 *Spread*, to 'claim the rest'. **1964** FREY & TRUSCOTT *Official Encycl. Bridge* 520/1 *Spread*,..verb: to spread the hand, either as a claim or as a concession of the remaining tricks.

spread, *ppl. a.* Add: **1. d.** *Phonetics.* Pronounced with the lips drawn out rather than rounded; unrounded.

1902 [see *fan consonant* s.v. *FAN *sb.*[1] 11]. **1965** *Language* XLI. 26 The spread phonemes /i/ and /ê/ are realized in the back allophones.

3. spread head *U.S. Journalism*, a display heading; hence **spread-headed** *a.*

1907 *Everybody's Mag.* XVI. 321 There is no such thing as fashion in dogs, despite the frequent 'spread heads'. **1923** O. G. VILLARD *Some Newspapers & Newspapermen* 152 It has not been able to rise without the vulgar comic section or the Sunday pictorial, and the usual spread-headed Sunday features.

4. In parasynthetic combinations, as *spread-kneed, -legged, -lipped, -winged* adjs.

1932 W. FAULKNER *Light in August* vii. 140 McEachern lowered himself stiffly to the top of a feed box, spread-kneed, one hand on his knee and the silver watch in the other palm. **1969** L. MICHAELS *Going Places* 21, I stood spread-legged, bolt naked. **1973** G. W. TURNER *Stylistics* ii. 63 If Australians tend to use spread-lipped vowels.., they perhaps seem matey fellows. **1936** R. CAMPBELL *Mithraic Emblems* 89 A spread-winged phœnix from its ash The Cross remained against the sky. **1972** R. ADAMS *Watership Down* xlv. 375 Four partridges..sailed down, spread-winged, into the field.

Hence **sprea:dou·tness** *rare*, the quality or condition of being spread out.

1879 W. JAMES in *Mind* IV. 339 Since the essences of things are as a matter of fact spread out and disseminated through the whole extent of time and space, it is in their spread-outness and alternation that he [*sc.* the entire man] will enjoy them. **1915** G. F. STOUT *Man. Psychol.* (ed. 3) II. i. 214 There is another inseparable character belonging to many kinds of sensation, though, probably, not to all, which may be called Extensity or extensiveness or diffusion or 'spreadoutness'.

spreadable (spre·dăb'l), *a.* [f. SPREAD *v.* + -ABLE.] That can be easily spread: esp. of butter, etc. So **spreadabi·lity.**

1940 'PLASTES' *Plastics in Industry* xi. 165 The easy spreadability of animal and vegetable glues is a very important property. **1962** P. BRACKEN *I hate to housekeep Bk.* vi. 65 The moisture in the onion turns this into a good spreadable spread for sandwiches or canapés. **1971** *Nature* 26 Nov. 238/1 The causes of cracks in cheese, poor spreadability in butter. **1979** J. RATHBONE *Euro-Killers* i. 10 The butter was..cool but spreadable.

spread eagle, *sb.* Add: **3*. a.** *U.S. Stock Exchange.* An operation by which a broker agrees to buy shares of stock within a specified time at a specified price, and sells the option of buying shares of the same stock within the same time at a higher price. **b.** = *SPREAD *sb.* 10. Cf. STRADDLE *sb.* 2 a.

1857 *Merchants' Mag.* XXXVII. 136 The buyer can call when he pleases, which would compel the 'spread eagle' operator to deliver. **1870** J. K. MEDBERY *Men & Myst. Wall St.* 86 One modification of this is the Spread Eagle, formerly a highly popular style of speculation with capitalists who had plenty of money and a wide-awake broker. **1910** *Encycl. Brit.* V. 55/1 A combined option of either calling or putting is termed a 'straddle', and sometimes on the American stock exchange a 'spread-eagle'.

6. quasi-*adv.* Like a spread eagle, with arms and legs outstretched.

1922 JOYCE *Ulysses* 492 A blond feeble goosefat whore ..lolls spreadeagle in the sofacorner. **1954** *Sun* (Baltimore) 16 Jan. 12/2 He told the two men to stand 'spread eagle' against the wall of a building while Joseph L. Klingenberg..frisked them. **1973** W. H. HALLAHAN *Ross Forgery* iv. 59 Ross stripped the klutz of all his clothes and staked him spread-eagle to the ground under a tropical sun.

spread-eagle, *v.* Add: Also **spreadeagle.**

2. b. (Earlier and later examples.)

1887 *N. & Q.* 1 Oct. 278/2 Cod—as well as haddock and ling..may be seen spread-eagled across transverse sticks to dry. **1955** J. P. DONLEAVY *Ginger Man* xxvi. 252 The kangaroo fell spread-eagled to the floor. **1976** 'D. HALLIDAY' *Dolly & Nanny Bird* v. 65 He spreadeagled a broad, powerful hand and clenched Johnson's in it.

c. (Earlier and later examples in *Cricket.*)

1887 *Cricket* 24 Nov. 460/2 A high one from Miss Tompkins spread-eagled Miss White's wicket. **1955** *Times* 15 June 6/3 McDonald added only 12 runs in 65 minutes before his stumps were spreadeagled by Worrell.

Hence **spread-eagled** *ppl. a.*

1940 L. MACNEICE *Last Ditch* 21 Light on her feet and gentle with her fingers; Put on a little flesh, became an easy Spreadeagled beauty for Renaissance painters. **1982** B. FANTONI *Stickman* xxiii. 161 Stepping over Lonnie's spread-eagled body, I lit a cigarette.

spreader. Add: **I. 4. d.** = *lifting beam* s.v. *LIFTING *vbl. sb.* 2 b. Also *spreader beam.*

1960 S. P. OPPENHEIMER *Erecting Structural Steel* vii. 88 Slings having a center ring and known as spreaders or bridle slings are used. *Ibid.*, When a long, thin piece, which might buckle under compression, is to be lifted, an additional heavy spreader beam is used above separate single slings placed at the ends. **1968** W. G. RAPP *Construction of Structural Steel Building Frames* vi. 175 To choke the load with a sling around it one eye is passed through the other, the free eye is then hooked or shackled into the hoisting spreaders on the main load hook. **1970** *Specification for Flat Lifting Slings* (*B.S.I.*) 20 Use a spreader beam or other suitable arrangement to ensure that the slings are as nearly as practicable vertical.

II. 7. A surfactant.

1918 *Jrnl. Econ. Entomol.* XI. 67 The surface tension and specific gravity are probably factors of importance in determining the value of a spreader. **1941** *Nature* 12 Apr. 438/2 The efficiency of..'spreaders' in a given oil may be gauged by the direct measurement of their spreading power against surface contamination in the Adam-Langmuir surface pressure trough. **1963** H. MARTIN *Insecticide & Fungicide Handbk.* iii. 57 Where high or medium volume is used, the addition to the spray of surface-active 'wetters' or 'spreaders' may be recommended in order to increase the cover obtained.

III. 8. *Comb.*, as **spreader-bar** = SPREADER 3; **spreader light,** a light attached to the spreader of a yacht.

1927 C. A. LINDBERGH *We* iii. 44 The wheels touched earth..rolled.., sank into the spreader bar and we nosed over. **1954** W. FAULKNER *Fable* 222 When they sat up it was together as though a spreader bar connected them. **1968** J. ARNOLD *Shell Bk. Country Crafts* xi. 166 Between the two horses, the trace-chains were kept steady by a spreader-bar. **1947** *Sun* (Baltimore) 3 July 15/5 The practice of a flashlight on sails as something approaches in the darkness is a good one, the authorities say, and the spreader lights now carried by many sailing yachts is [*sic*] an even better idea. **1977** G. V. HIGGINS *Dreamland* ix. 96 With the spreader lights on the mainmast on, I took down the mizzen sail.

spreading, *vbl. sb.* Add: **2. spreading-board,** (*a*) a board on which sheep are laid while being shorn (*rare*); (*b*) a setting-board for insect specimens; **spreading factor** *Biol.* = *HYALURONIDASE.

1874 HARDY *Far from Madding Crowd* I. xxii. 247 The issue of their dialogue was the taking of her hand by the courteous farmer to help her over the spreading-board into the bright May sunlight outside. **1909** in WEBSTER. **1963** V. NABOKOV *Gift* ii. 111 To drive a pin smoothly through the insect's thorax, stick it in the cork groove of the spreading board. [**1930** *Science* 14 Nov. 508/2 The ink particles were..spread through a wider area under the influence of the factor.] **1939** *Nature* 9 Dec. 978/2 The occurrence of spreading factor in bacterial filtrates has so far been found to go parallel with their mucinase content. **1978** *Exper. Cell Res.* CXV. 227 Techniques previously developed for the purification of the fetal calf adhesion and spreading factor were directly applied to human serum.

spreading, *ppl. a.* **2. b.** *spreading adder*, substitute for def.: the hog-nosed snake, *Heterodon contortrix*, a harmless snake which characteristically inflates its body, flattens its head, and hisses loudly. Cf. HOG-NOSE. (Examples.)

1842 J. E. DEKAY *Zool. N.Y.* III. 52 The Hog-nosed Snake..is also called Dead Adder, Spreading Adder, Hog-nose and Buckwheat-nose. **1904** R. STUART *River's Children* 91 Rattlers and copperheads, spreading-adders, moccasins, and conger-eels come up to the island. **1931** W. FAULKNER in *Amer. Mercury* Mar. 261/2 Her mouth pursed out like a spreading adder's, like a rubber mouth. **1972** E. WIGGINTON *Foxfire Bk.* 295 Spreading adder. They tell me they're an awful poisonous snake.

sprea·d-over. Also **spreadover.** [f. SPREAD *v.* 4 c + OVER *prep.*] A system of distributing work or holidays over a period of time; *spec.* an arrangement by which a fixed number of work-hours may be performed at varying times within a given period. Also *attrib.* Cf. *split shift* (*a*) s.v. *SPLIT *ppl. a.* 3 a.

1923 *Westm. Gaz.* 14 Apr. 1/2 A 'spread-over' of 44 hours. **1924** *Ibid.* 7 Mar. 1/2 The 'spread-over' system—the performance of eight hours' work any time within twelve hours. **1930** *New Statesman* 27 Dec. 350/2 The Lancashire miners resolved..to work the spreadover. **1963** *Times* 1 June 12/2 There is in most Coventry factories no observance of the Whitsun holiday, the period having been moved to September to allow a better spread-over of holidays. **1976** P. R. WHITE *Planning for Public Transport* v. 104 A substantial number of bus crews.. work 'split' or 'spreadover' shifts, covering both morning and evening peaks. **1978** *Daily Tel.* 21 Mar. 2/8 Supplementary payments bus crews may qualify for, such as spread-over duties, unsocial hours, [etc.].

spreadsheet (spre·dʃīt). *Computers.* [f. SPREAD *ppl. a.* + SHEET *sb.*[1]] A program that allows any part of a rectangular array of positions or cells to be displayed on a VDU screen, with the contents of any cell able to be specified either independently or in terms of the contents of other cells.

1982 *Micro Software Mag.* May-June 4/1 Software designers have used two distinct methods in their attempt to provide the perfect package for financial modelling on a microcomputer: dynamic on-screen spreadsheet calculations, and the more traditionally processed logic file. **1983** *Daily Tel.* 13 Aug. 18 A good spreadsheet will let you put in all your figures, then just press one specified key and do all the calculations at once producing your completed accounts.

spready, *a.* For † *Obs.*[−1] read *rare* and add:

1. (Later example.)

1924 R. HALL *Unlit Lamp* v. 30 She looks like a tree... A beech tree? No, that's too spready—a larch tree.

2. Of a meal: plenteous, lavish. Cf. SPREAD *sb.* 7.

1962 G. AVERY *Greatest Gresham* ix. 165 Nice to have a spready tea for once.

3. Spreadable, easily spread.

1974 J. E. UNDERHILL *Wild Berries Pacific Northwest* 19 With a jelly..we..have a product that is very smooth and 'spready'.

‖ **Sprechgesang** (ʃpre·χʸgĕza:ŋ). *Mus.* Also **sprechgesang.** [Ger., lit. 'speech song'.] A style of dramatic vocalization intermediate between speech and song. See next.

1925 W. H. KERRIDGE tr. *Wellesz' Arnold Schönberg* 138 The *Dreimal sieben Gedichte* (*Thrice seven songs*), from Albert Giraud's *Pierrot Lunaire*..are written for a *Sprechgesang* (song-speech), piano, flute.., clarinet.., violin.., and violoncello. **1938** *Oxf. Compan. Mus.* 883/1 'Sprechgesang', the older term, is properly singing tinged with a speaking quality, whereas 'Sprechstimme' is rather speech tinged with a singing quality. **1947** E. NEWMAN *Life Richard Wagner* IV. xxii. 437 There used to be much talk, a generation or so ago, about the so-called *Sprechgesang*—a term apparently designed to make more acceptable to us the canine sounds emitted by some Wagnerian singers. **1959** *Listener* 11 June 1039/3 The Stravinsky sounded tense and its Sprechgesang (or was it perhaps just ordinary muttering) very dramatic. **1978** P. GRIFFITHS *Conc. Hist. Mod. Music* iii. 36 *Pierrot lunaire* is an ambiguous work in many senses. The soloist is required to use Sprechgesang, a mode of vocalization lying between song and speech.

‖ **Sprechstimme** (ʃpre·χʸʃti:mə). *Mus.* Also **sprechstimme.** [Ger. (Grimm, 1871), lit. 'speech voice'.] A term used by Schoenberg to describe the voice of a performer singing according to the principles of *Sprechgesang*; also *loosely*, = prec.

1922 *Music & Lett.* III. 83 *Pierrot Lunaire*..represents the zenith of Schönberg's powers... In a foreword the composer explains that the voice part is to be what he calls a 'Sprechstimme', neither song nor speech, but something in between. **1938** [see *SPRECHGESANG]. **1954** *Grove's Dict. Mus.* (ed. 5) VIII. 26/2 *Sprechstimme*.., a human voice written for by a composer who requires from the performer a delivery according to the principles of *Sprechgesang*. **1961** *Listener* 17 Aug. 257/3 The *Gurrelieder* provided Schönberg with the first occasion to make use of a speaking part that anticipates the *Sprechstimme* of *Pierrot Lunaire* in being in strict rhythms and with a fixed pitch. **1968** *Ibid.* 31 Oct. 591/2 Berg's strict distinction between Schoenbergian *sprechstimme* and half-sung phrases frequently receives cavalier treatment. **1982** *Times Lit. Suppl.* 25 June 693/2 His words are shared out among the whole company, cowled and carrying candles, like a coven going in for *sprechstimme*.

spree, *sb.* Add: **1. a.** *transf.* (Later examples.) Cf. also *shopping, spending spree* under the first element in Suppl.

1955 *Times* 29 Aug. 11/1 What has been described as a profit boom and a dividend spree is being used to foster the idea that wage claims can be advanced one after the other without respite. **1976** J. SNOW *Cricket Rebel* 21 A six-hitting spree by Jim Pressdee who..made 115. **1983** *Times* 22 Feb. 17/2 (*heading*) Kellock joins own shares buying spree.

spree *v.* (earlier example); also *trans.*, to spend (money) recklessly; **spreeing** *vbl. sb.* (earlier examples).

1845 in N. E. Eliason *Tarheel Talk* (1956) iv. 137 They both had been spreeing it the evening before with some members of Congress. **1860** Spreeing [see *BUMMING *vbl. sb.*]. **1885** 'MARK TWAIN' *Let.* 11 Sept. (1917) II. 457 The drunkenness (and sometimes pretty reckless spreeing) ceased before he came East. **1897** —— *Following Equator* i. 33 It was the remittance-man's custom to..spree away the rest of his money in a single night. **1907** G. B. SHAW *Let.* 27 July (1972) II. 703 The guarantee fund shall not be drawn upon for current expenses at the Savoy (Barker would spree it on a single scene in Peer Gynt). **1928** —— *Intelligent Woman's Guide Socialism* lxiv. 296 They destroy the sense of security which induces the possessors of spare money to invest it instead of spreeing it.

spreeu, var. *SPREW[2].

‖ **spreite** (ʃprəi·tə). *Palæont.* Also **Spreite.** Pl. **spreiten.** [Ger., a layer or lamina, esp. something extending between two supports.] A banded pattern of uncertain origin found in the infill of the burrows of certain fossil invertebrates.

1962 W. HÄNTZSCHEL in R. C. Moore *Treat. Invertebrate Paleont.* W. 182/2 The vast majority of trace fossils remain unchanged through the geologic eras. This is true for nondescript, smooth furrow-like crawling trails and cylindrical burrows, as well as for more distinctive U-shaped burrows with *Spreite*. *Ibid.* 200/1 *Gyrophyllites*.., vertical shaft from which rosettes of short, simple (feeding) tunnels radiate at different levels, as in a mine; 'leaves' with sculpture of *Spreiten* burrows. **1976** *Nature* 17 June 576/2 The burrow walls are laminated parallel to the long axis of the burrow, and the infill exhibits a cuspate ('spreite') structure.., probably indicating active backfilling by the animal responsible.

sprekelia (sprekī·liă). [mod.L. (L. Heister *Systema Plantarum Generale* (1748) 5), f. the name of J. H. von *Sprekelsen* (d. 1764) who sent bulbs of the plant to Linnæus + -IA[1].] A bulbous plant of the monotypic genus so called, belonging to the family Amaryllidaceæ, native to Mexico, and bearing linear leaves and large crimson flowers; = *jacobæa lily* s.v. *JACOBÆA 2.

1840 *Edwards's Bot. Reg.* XXVI. 33 (*heading*) The Tumbler Sprekelia, shorter-flowered variety. **1962** *Amateur Gardening* 10 Mar. 33/2 Sprekelia..takes us to.. Mexico... The oddly shaped flowers with three erect and three drooping petals are deep crimson. **1975** *Country Life* 16 Jan. 158/3 Dormant sprekelia bulbs will be.. replanted.

Sprengel (ʃpre·ŋ'l). The name of Hermann Johannes Philip *Sprengel* (1834–1906), German-born English chemist, used *attrib.* and in the possessive to denote devices developed by him, as **Sprengel('s) (air, mercury) pump,** a vacuum pump in which exhaustion is produced by trapping bubbles of gas between short columns of mercury falling down a narrow vertical tube; also *absol.*; **Sprengel('s) tube,** a glass U-tube that narrows to a capillary at each end, used to determine the specific gravity of a liquid by weighing the tube when filled with the liquid and then with water at the same temperature.

1868 *Chem. News* 7 Feb. 73/2 Could any one give me information about the dimensions for a Sprengel air pump, and the quantity of mercury required. I wish to use it for exhausting vacuum tubes for a 4-inch spark induction coil. **1868** E. ATKINSON tr. *Ganot's Physics* (ed. 3) 144 Sprengel's air pump. **1879** *Jrnl. Chem. Soc.* XXXVI. 197 The use of Sprengel's tube..has given much better results [than a specific gravity bottle]. **1883** *Encycl. Brit.* XVI. 31/1 One [improvement]..is to pass the mercury, before it enters the 'falling' tube, through a bulb in which a good vacuum is maintained, by means of an ordinary air-pump or a second 'Sprengel'. **1890** *Rep. Brit. Assoc. Adv. Sci. 1889* 232 The density of ice and of water at various temperatures may then be determined, a Sprengel tube—which is easily made—being used for warm water. **1906** *Chem. News* 19 Jan. 28/2 It is by the well-known Sprengel mercury pump that his name will be best remembered. **1932** *Proc. R. Soc.* A. CXXXV. 514 A continuously operating Sprengel pump..removed the gas from that part of the apparatus..and delivered it, without loss, into the Töpler pump. **1934** *Analyst* LIX. 172 The undersurface of the arms of the Sprengel tube is made of white enamel glass. **1962** J. THEWLIS *Encycl. Dict. Physics* V. 715/1 The Sprengel pump was used extensively in the early days of vacuum experimentation. It is now obsolete, because its rate of working is very low and the ultimate vacuum reached only of the order of 10^{-3} mm mercury.

sprew[2]. Add: Also **spreeu, spreo, spreuw.** (Earlier and later examples.)

1795 C. R. HOPSON tr. *C. P. Thunberg's Trav. Europe, Africa & Asia* (ed. 2) II. 48 A kind of *Corvus*, (or crow) called *Spreuw*, was found..in great plenty. **1801** [see *FISCAL *sb.* 3]. **1867** E. L. LAYARD *Birds S. Afr.* 173 The red-wing spreo, aided by finches,..would soon pick the crop [of grapes]. **1913** J. J. DOKE *Secret City* 229 'Jolly fine birds, spreeuws,' he said. **1939** S. CLOETE *Watch for Dawn* 106 A spreeu dug its beak into some half-dried dung for the fly-worms that were in it. **1961** B. C. TAIT *Durban Story* 69 Blue-black sprews flash above incredibly green lawns.

‖ **sprezzatura** (spre:tsatū·ra). [It.] Ease of manner, studied carelessness; the appearance of acting or being done without effort; *spec.* of literary style or performance.

1957 N. FRYE *Anat. Criticism* 93 The quality that the Italian critics called *sprezzatura*. **1960** E. H. GOMBRICH *Art & Illusion* III. vi. 193 *Sprezzatura*, the nonchalance which marks the perfect courtier and the perfect artist. **1960** *Spectator* 14 Oct. 569 The style governed by *sprezzatura*, dash and mandarin neoclassicism. **1973** *Times Lit. Suppl.* 14 Sept. 1063/2 Literary fashion and his own aristocratic *sprezzatura* demanded that he affect an unconcern.

sprig, *sb.*[1] Add: **2.** Also *spec.* (*N.Z.*), a stud or spike attached to the sole of a boot, esp. in *Sport.*

1930 [implied at *SPRIG *v.*[1] 1]. **1949** D. M. DAVIN *Roads from Home* I. ii. 27 John hammered the last tack into a sprig of his football boots. **1972** *Guardian* 11 Nov. 21/5 You look at Sid [Going] when we're changing, he's got sprig (stud) marks all over him. **1981** I. A. GORDON in *N.Z. Listener* 2–8 May, The *sprig* (though it has acquired a new meaning on the football field), was originally a short headless nail.

sprig, *sb.*[2] Add: **6.** *sprig mould, -muslin.*

1922 HARDY *Late Lyrics & Earlier* 1 And maids come forth sprig-muslin drest. **1951** J. B. KENNY *Compl. Bk. Pottery Making* vii. 123 A sprig mold is a block of plaster with a depression shaped like the ornament in reverse. **1956** G. HEYER (*title*) Sprig muslin. **1976** *Canadian Antiques Collector* (Toronto) Mar.–Apr. 20/1 There were over 22 sprig moulds for making applied decorative relief to pots.

sprig, *v.*[1] Add: **1.** Also *spec.* (*N.Z.*), to equip (a boot) with sprigs. Cf. *SPRIG *sb.*[1] 2.

1930 C. V. GRIMMETT *Getting Wickets* v. 105 Had his boots been properly sprigged, it is probable that he would easily have taken the catch.

sprig, *v.*[2] Add: **2.** Also *absol.*

1960 H. POWELL *Beginner's Bk. Pottery* II. ii. 21 When sprigging, you may find that the small, thin sprigged shapes dry too quickly.

sprigged, *ppl. a.* Add: **1. c.** Of ceramic ware: adorned with, or forming, sprigs or other ornaments in applied relief. Cf. SPRIG *sb.*[2] 4 b.

1756 J. BOWCOCKE *Acct. Bk.* in L. Jewitt *Ceramic Art Gt. Brit.* (1878) I. vii. 209 Mr. White: 1 imag'd cup and 7 sprig'd chocolates. **1906** R. L. HOBSON *Porcelain of All Countries* xx. 190 The 'sprigged' pattern..consists of sprays of Chinese plum. **1960** [see *SPRIG *v.*[2] 2]. **1971** L. A. BOGER *Dict. World Pott. & Porc.* 323/2 *Sprigged ware*, in English ceramics; a contemporary English name given to 18th century wares decorated in applied reliefs principally of flowers, foliage, and stems.

sprigging, *vbl. sb.* Add: **3.** The process of decorating ceramic ware with sprigs or other ornaments in applied relief.

1928 W. B. HONEY *Old Eng. Porc.* iii. 63 'Sprigging' is the Staffordshire name for the process of applying these reliefs. **1961** L. G. G. RAMSEY *Connoisseur New Guide Antique Eng. Pott., Porc. & Glass* 35 The ornament was obtained by 'luting' to the surface of the pot previously moulded relief-motifs, and by connecting these with 'stems' formed of threads of clay rolled out thin between the hands. This whole process was known as 'sprigging'. **1969** G. WILLS *Eng. Pott. & Porc.* 102 What is known to potters as 'sprigging'..was done by using a patterned mould made of plaster. The soft clay was pressed into it and the surplus scraped carefully away.

sprightle (sprəi·t'l), *v. Midland dial.* [Back-formation f. SPRIGHTLY *a.*] *intr.* With *up*: to become lively or alert.

1896 G. F. NORTHALL *Warwickshire Word-bk.* 224 *Sprightle up! excl.*, be lively, alert (sprightly). **1920** D. H. LAWRENCE *Lost Girl* ix. 206 Oh but she sprightles up a bit sometimes.

spring, *sb.*[1] Add: **I. 4. spring branch** (earlier example); **spring hole, house**: for *U.S.* read *N. Amer.* (later examples).

1823 *Amer. State Papers: Public Lands* (U.S. Congr.) (1834) III. 811 One hundred [acres] on Sweet spring branch, St. Mary's river. **1956** K. M. WELLS *By Jumping Cat Bridge* xiv. 83 We followed it, up to the spring hole on the edge of the flat land, a no-good bit of bog hole on the edge of arable land. **1933** H. ALLEN *Anthony Adverse* ii. 26 A fresh cheese cool from the spring house, and a firm, white loaf caught in a silver clamp provided with a small steel saw in the shape of a dragon's head with teeth amused Maria. **1972** E. WIGGINTON *Foxfire Bk.* 207 Roll the sausage into balls, pack them in a churn jar..and set in the water trough in your spring house. **1980** *Knoxville* (Tennessee) *News-Sentinel* 6 Apr. C4/5 He also said, 'Milk kept in the *springhouse* will blink and sometimes clabber during a severe thunderstorm or toad-strangler.'

II. 6. h. *ellipt.* A spring salmon (see sense 7 c below).

1913 *Chambers's Jrnl.* Oct. 729/2 Next in value comes the 'spring', the largest fish often weighing sixty-five pounds. **1921** *Daily Colonist* (Victoria, B.C.) 12 Mar. 9/4 In May it [*sc.* the largest cannery in the North] will commence to pack springs. **1975** H. WHITE *Raincoast Chron.* (1976) 222/1 Some men hand trolled springs during the winter.

7. a. *spring-broidery, crop, fever* (earlier example), *fret, shoot* (later example), *-song*; **spring cabbage,** a variety of cabbage that matures in the spring; a cabbage of this variety; **spring greens,** the leaves of young cabbage plants, used as a vegetable; also, a cabbage of a variety that matures in spring; **spring roll,** a Chinese snack consisting of a pancake filled with vegetables and fried in the shape of a roll; **spring training** *Baseball*, pre-season fitness and skills training taking place in spring.

1862 G. M. HOPKINS *Vision of Mermaids* (1929), Until it seem'd their father Sea Had gotten him a wreath of sweet Spring-broidery. **1842** J. C. LOUDON *Suburban Horticulturist* III. v. 629 This [*sc.* tying up the leaves] may be usefully practised with the earliest spring cabbages. **1900** W. D. DRURY *Bk. Gardening* 1196/1 (Index), Spring cabbages. **1979** P. J. SALTER et al. *Know & grow Veg.* iv. 104 Spring cabbage..may be short of nitrogen. **1834** *Chambers's Edin. Jrnl.* III. 255/1 The remainder may be ready for spring crop with very little labour. **1962** E. SNOW *Red China Today* (1963) xxiii. 174 In July, on a visit to the Agricultural Exhibition Building, I was told by the director that, the spring crop having failed, only phenomenally good weather during August could save the autumn harvest. **1978** *Biol. Abstr.* LXVI. 2091/2 (*heading*) Effect of gibberellic acid..sprays on a spring crop of artichoke. **1843** LYTTON *Last of Barons* II. vi. vii. 256 The love of Sibyll was no common girl's spring-fever of sighs and blushes. **1897** KIPLING in *Scribner's Mag.* Dec. 679 'Send your road is clear before you when the old Spring-fret comes o'er you. **1969** in *Current Trends in Linguistics* (1972) X. 155 *Spring fret*, animals' feeling of restlessness in the spring. (Ga.). **1919** C. S. PEEL *Daily Mail Cookery Bk.* (ed. 2) vii. 134 Vegetables which can be cooked in this manner—cauliflowers, brussels sprouts, spring greens, broccoli tops, [etc.]. **1937** MIDDLETON & HEATH *From Garden to Kitchen* iii. 52 Every alternate one can be pulled out and used as spring greens. **1972** Y. LOVELOCK *Veg. Bk.* I. 69 Spring greens, either hearted or leafy, are gradually superseding the tougher British colewort. **1943** M. P. LEE *Chinese Cookery* i. 21 Spring rolls... After having made all the rolls, fry them in..hot lard. **1972** D. BLOODWORTH *Any Number can Play* xxii. 219 Helping himself to a spring roll from a trolley of hot Chinese delicacies. **1946** DYLAN THOMAS *Deaths & Entrances* 41 Tell his street on its back he stopped a sun And the craters of his eyes grew spring-shoots and fire. **1857** J. W. DAVIDSON *Mendelssohn's Six Bks. Songs without Words* p. ii/1 The *Allegretto Grazioso*, in A major..to which he is believed to have given the name of *Frühlingslied* (Spring Song). **1889** O. WILDE in *Woman's World* II. 111/2 However, if the 'Songs of the Inner Life' are not very successful, the 'Spring Songs' are delightful. **1958** *Anglia* LXXVI. 55 The well-known Spring-song..'Lencten is come wiþ love to toune'. **1897** *Sporting News* 27 Mar. 5/3, I am on my way to join the Boston team at Savannah where the players have all been ordered to report for spring training. **1928** G. H. RUTH *Babe Ruth's Own Book of Baseball* i. 17 In 1925 when I collapsed in Asheville, during the spring training trip, a lot of people figured I'd never put on a uniform again. **1976** *National Observer* (U.S.) 12 June 14/3 Randolph was hardly awed at becoming a Yankee... I came to spring training looking for a job.

c. spring peeper, a very small tree frog, *Hyla crucifer*, of eastern North America; **spring salmon** *N. Amer.*, a Pacific coast salmon that returns from the sea to the river in spring, esp. the chinook, *Oncorhynchus tshawytscha.*

1906, etc. Spring peeper [see *PEEPER[1] 2 b]. **1950** *Chicago Tribune* 28 Mar. 14/3 Then there are those who listen for choruses of spring peepers. **1977** *Globe & Mail*

(Toronto) 11 Apr. 8/3, I don't know how many people other than those who live in the country are familiar with one of our prime sounds of spring: the spring peeper. **1850** G. Hines *Voy. round World* 331 In this country [*sc.* America] they are generally distinguished by the names of spring-salmon and fall-salmon. **1905** D. S. Jordan *Introd. Study of Fishes* II. 80 The economic value of any species depends in great part on its being a 'spring salmon'. **1964** G. C. Carl *Some Common Fishes Brit. Columbia* 28 A few mature [Chinook salmon] may enter the larger rivers in late spring or early summer (hence the name 'spring salmon').

8. *spring-filled, -impassioned, -scented.*

1932 W. Faulkner *Light in August* (1933) vii. 138 Into the bleak, clean room the springfilled air blew in fainting gusts. **1881** O. Wilde *Poems* 178 Each spring-impassioned tree Flames into green. **1925** G. Greene *Babbling April* 25 She waits without fear that spring-scented day.

IV. 14. e. *slang* (orig. *U.S.*). An escape or rescue from prison.

1901 'J. Flynt' *World of Graft* ii. 32 It is comparatively easy to make a 'spring' out of the clutches of the law when there is sufficient money to hand around to the various persons with 'pull'. **1923** A. Stringer *Diamond Thieves* xx. 385, I *did* wait. But his swell friends didn't come across wit' any spring. **1968** 'B. Mather' *Springers* xv. 161 We just..waited for what we knew would eventually come. A spring. **1977** F. Ross *Dead Runner* i. 41 Springing some bugger from the Scrubs—O.K. Not easy. ..You can't pull a spring like that without help on the inside.

21. c. *Bootmaking.* The raising or rise of the toe of a last above the ground-line. Also, arch or curvature in the instep.

In quot. 1885 *spec.* applied to the steel support which gives the desired curvature to the boot or shoe.

1885 J. B. Leno *Art of Boot & Shoemaking* viii. 54 The patterns for the springs of large size boots should have a good quarter of an inch left on at sides and bottom. *Ibid.* xix. 145 The spring is attached to the inner sole... Care should be taken in..fixing the metal shield... Instances have occurred in which the spring, through being inadequately shielded, has pierced through the inner sole to the foot. **1902** F. Y. Golding *Manuf. Boots & Shoes* 107 *Spring* is the term used to denote the elevation of the toe of the last... If the substance of the sole be light very little spring is required. *Ibid.*, Sometimes the term 'spring' is used to describe the hollowness or arch of the waist. **1905** E. J. C. Swaysland *Boot & Shoe Design* 20 For light dress work the spring of the toe should be half an inch. **1916** F. Plucknett *Boot & Shoe Manuf.* ii. 17 It is advisable to put spring into the forepart of the last, equal to the amount which the boot would probably acquire in wear. **1935** J. Ball in F. Y. Golding *Boots & Shoes* VI. xi. vi. 39 Spring is the amount of *curve* from the joint to the point of the toe, and whereas a light flexible shoe requires very little, the stouter and stiffer the sole the greater the amount of spring.

V. 24. b. (Later *attrib.* examples.)

1916 *Daily Colonist* (Victoria, B.C.) 23 July 4/6 The mate was on the forecastle head, but outside the rail, when the spring line parted. One end of the big hawser flicked up and smashed the mate's leg. **1975** H. White *Raincoast Chron.* (1976) 156/1 A deck-hand expertly heaves a throwing-line ashore. Somebody catches it, pulls in the steel spring-line from amidships.

25. a. *spring bell* (earlier example), *blind, clip* (earlier example), *heel, mattress* (earlier example), *roller, seat* (earlier example), *stay* (earlier example), *tab, tine, washer.*

1847 *Illustr. London News* 13 Mar. 165/1 There is..a spring-bell, which the Prisoner is to sound when he requires the attendance of an officer. **1835** Dickens in *Evening Chron.* 18 June 3/4 Spring blinds were fitted to the windows. **1889** *Cassell's Bk. Househ.* I. 250/2 Spring blinds, though very nice when they work well, are subject, as an old housekeeper once said, to 'tempers'. **1934** *Cassell's Home Encycl.* 86/2 Automatic or spring blinds are made with a hollow roller. **1871** Spring clip [see Release *sb.*[1] 6 b]. **1897** *Sears, Roebuck Catal.* 194d/2 Chocolate Goat Button, made with spring heel, new opera toe with tip. **1952** M. Allingham *Tiger in Smoke* viii. 140 'He's got nerve.'..'Likewise spring heels and rubber bones.' **1850** Spring mattress [see Mattress[1] 1]. **1821** P. Egan *Life in London* (1822) II. iv. 266 A fine collection of maps, concealed by the cornices of the book-cases, on spring rollers, can be referred to without the least trouble. **1971** *Habitat Catal.* 86/2 Roller blinds ..come complete with spring roller and fixing brackets. **1853** *Heal & Son Catal.: Bedsteads,* Sofas and Couches.. with squab or spring seats. **1745** in J. S. McLennan *Louisbourg* (1918) 177 Ye fore Spring Stay was Shott away. **1942** *Rep. & Memoranda Aeronaut. Res. Committee* No. 2029. 9 Notes on the Spring Tab. R.A.E. Report No. B.A. 1665. (5058.) April, 1941. (Unpublished.) **1969** *Gloss. Aeronaut. & Astronaut. Terms (B.S.I.)* v. 10 *Spring tab*, a balance tab, the angular movement of which is geared to the compression or extension of a spring embedded in the main control circuit. **1923** J. R. Bond *Farm Implements & Machinery* v. 49 The simple spring tine as first introduced on the Canadian cultivator had not sufficient strength. **1971** *Farmers Weekly* 19 Mar. 84 The RCM firm offers a choice of tines to fit one basic frame—a normal spring tine..or subsoiler tines. **1912** *Motor Man.* (ed. 14) 206 Spring washers are less effective, but answer well enough for the less vital parts of the mechanism. **1929** *Ibid.* (ed. 26) 20 *Spring washer*, a tempered steel washer cut through at one place and given a 'set' to provide a certain amount of spring. Used under a nut to keep it secure.

c. (Further examples.)

1948 *Penguin Music Mag.* June 54 Sir Henry Wood.. recommended the use of spring-clip mutes which fastened on the music desks. **1972** *Times* 23 Sept. 9/1 These spring-clip pans..spring open when a clip at the side is released. **1897** *Sears, Roebuck Catal.* 192/1 The most handsome and up-to-date spring heel shoe ever placed on the market. **1934** *Heal & Son Catal.: Better Furnit.* 2 'Chassis'..(to fix to spring-mattress frame). **1973** *Sun* 4 Sept. 20/2 (Advt.), Spring roller blinds from £2.94... Inc. all fittings, fringe and pull cords. **1907** Spring-side safety iron [see *Iron *sb.*[1] 4 j]. **1921** W. de la Mare *Memoirs of Midget* viii. 48 She, too, was in black, with a long, springside boot. **1923** J. R. Bond *Farm Implements & Machinery* v. 54 (*caption*) Canadian spring-tine cultivator with corn and grass seed boxes. **1960** *Farmer & Stockbreeder* 15 Mar. 98/3 The spring-tine rake to the rear..is provided to meet necessary safety regulations. *Ibid.* 134/1 It's a Spring-tine Cultivator—Spring-tine Harrow..all in one.

26. c. *spring-doored, -driven, -heeled, -mattressed, -mounted, -tailed, -tined.*

1978 J. Irving *World according to Garp* xv. 311 In a Laundromat... They had opened one of the big spring-doored dryers. **1928** J. E. Haswell *Horology* vii. 78 In the category 'spring-driven' clocks are placed the numerous types which derive their motive power from the energy of a coiled mainspring. **1958** *Times* 20 Aug. 2/6 Zhukov, Pudov, and Ozog all flashed past the now weary Eldon in pursuit of the fleeing, spring-heeled Pole. **1977** *Vogue* Dec. 11/2 'Colleagues' has something inert and passive about it, compared with the spring-heeled activity of 'work mates'. **1920** Galsworthy *In Chancery* ii. xiv. 236 His spring-mattressed bed. **1960** E. L. Delmar-Morgan *Cruising Yacht Equipm. & Navigation* 37 If the compass is of the spring-mounted type [etc.]. **1952** Dylan Thomas *Coll. Poems* 174 No springtailed tom in the red hot town. **1938** C. Culpin *Farm Machinery* vii. 95 The spring-tined harrow is a really light cultivator that can be adjusted to produce very variable effects.

d. Special Combs.: **spring bows** = Bow-compass 1; **spring collet** *Engin.*, a tapered collet that is slotted along much of its length, so that when moved in a similarly tapered seat the separate parts are pressed against the stock inside the collet; **spring line**, a line where the water table reaches the surface and along which springs are numerous; **spring-loaded** *a.*, containing a compressed or a stretched spring pressing one part against another; **spring rate** = *Rate *sb.*[1] 8 c; **spring sail** (see quot. 1931); **spring sweep** = prec.; **spring-tree**, a saddle-tree with two springs.

1895 *Army & Navy Co-op. Soc. Price List* 686 Spring Bows, set of three, Ink, Pencil, and Divider, in case. **1964** G. Baker *Scale Drawing* 31 Spring-bows do not have knee-joints because the angle of working is so small as not to matter. **1932** W. P. Turner *Machine Tool Work* xiii. 290 The clamping lever..is pushed back, which moves the hollow push sleeve..against the end of the spring collet, forcing it into the nose cap, thus closing the collet on the stock and holding it firmly. **1971** B. Scharf *Engin. & its Lang.* viii. 61 The bore (internal diameter) of the spring collet matches the size and shape of the bar to be gripped. **1932** *Geol. Mag.* LXIX. 407 West of the spring line, thin outcrops of hard ferruginous sandstone occur in the clays. **1963** L. F. Chitty in Foster & Alcock *Culture & Environment* vii. 181 Other finds suggest an alternative route by the present hillside road over Clee Hill, with its far-extending outlook, keeping above the spring-line to Doddington. **1975** J. G. Evans *Environment Early Man Brit. Isles* vii. 159 Stream courses have a similar effect.., and the well known phenomenon of 'spring-line settlement' is an extreme example of this form of geographical determinism. **1904** *Engin. Rev.* XI. 418/2 Delivery valves..of the mushroom spring-loaded type. **1938** *Brit. Jrnl. Psychol.* XXIX. 40 A spring-loaded pencil..made a record on a sheet of graph paper. **1979** J. Wainwright *Tension* v. 21 The door is spring-loaded. **1957, 1959** Spring rate [see *Rate *sb.*[1] 8 c]. **1931** S. P. B. Mais *England of Windmills* p. xxiv, [Andrew] Meikle was ..the inventor of the spring sail. *Ibid.*, The spring sail is made up of wooden shutters, or a canvas-covered wire frame, hinged at one edge and connected to a common sail-rod so that they all open or shut simultaneously. **1971** S. Freese *Windmills & Millwrighting* iv. 58 The framework of spring sails consists of eight or ten sail bars, spaced 3 ft. apart and forming a series of bays, in each of which are three shutters 4 or 5 ft. by 11 in., wood or metal framed, and covered by wood or canvas. **1919** *Jrnl. Franklin Inst.* CLXXXVII. 178 In some cases these shutters worked against the tension of a spring instead of the pull of a weight, and such were known as Meikle's 'spring sweeps'. **1924** *Trans. Newcomen Soc. 1922–3* III. 50 It was desired to fit patent sail regulation instead of 'spring' or 'sail' sweeps. **1963** E. H. Edwards *Saddlery* xiv. 97 In the modern spring tree of correct design..the necessity for a cut-back head is largely obviated by the fact that the head itself is set back at an angle of 45 degrees. **1976** *Horse & Hound* 3 Dec. 31 (Advt.), Parzival all purpose/jumping [saddle]. Again built on the same spring tree, finished in Stubben's special dark brown Sella leather.

spring, *v.*[1] Add: **I. 3. e.** Phr. *Where did you spring from?* and varr., used when someone appears unexpectedly. *colloq.*

1892 I. Zangwill *Childr. Ghetto* III. ii. vi. 109 'Hullo! where did *you* spring from?' It was Raphael who had elicited the exclamation. He suddenly loomed upon the party. **1924** Wodehouse *Leave it to Psmith* ix. 181 'Wherever,' she inquired, 'did you spring from, Ed?' **1932** G. Heyer *Devil's Cub* ix. 141 Several persons hailed him, demanding to know whence he had sprung. **1971** 'S. Woods' *Serpent's Tooth* 154 She was, perhaps, the last person he had expected.. 'Where did you spring from?'

f. *U.S. slang.* To escape or be released from arrest or imprisonment.

1904 H. Hapgood *Autobiogr. of Thief* ix. 188 Soon after I was transferred from Sing Sing..a friend..said '..If you can get on Keeper Riley's gallery I think you can spring (escape).' **1926** *Clues* Nov. 162/2 *Spring*, to be released, as from jail or prison. **1955** *Publ. Amer. Dial. Soc.* xxiv. 184 You get snatched in the neck and it costs you twelve hundred to spring. **1962** 'K. Orvis' *Damned & Destroyed* xii. 82 When I sprung..Moss was standing by the prison door.

g. *Austral.* and *U.S. slang.* To pay *for* a treat. Also without const.

1906 E. Dyson *Fact'ry 'Ands* xviii. 250 Feathers..said reproachfully to the Aberdeen engineer..whose turn it was to 'spring': 'Blime, cobber, er yer givin' ther barmaid er perpetual 'oliday 'r what?' **1973** W. McCarthy *Detail* iii. 163 I'm springing for chow tonight and charging it to the Service. **1976** M. Machlin *Pipeline* ix. 107 We'll spring for the booze.

IV. 21. d. Also const. *on* and without following prep.

1876 'Mark Twain' *Tom Sawyer* xxxiv. 265 Old Mr. Jones is going to try to spring something on the people here to-night. **1895** Roberts & Morton *Adventures of Arthur Roberts* xi. 145 'Dinner!' ejaculated Johnson. 'Yes, we shall have to spring the landlady for that at once. That's where she will want a bit of the ready money on account.' **1922** *Ladies' Home Jrnl.* July 72/1 But one day she sprang a surprise; she sprinkled the salad with Dromedary [Shredded Coconut]—*father actually smiled.* **1943** K. Tennant *Ride on Stranger* i. 1 'One I didn't mind,' he admitted. 'Two's plenty. But to spring it on a man like this.' His tone was one of outrage. **1969** *Listener* 14 Aug. 204/1 The official French government spokesman sprang the wholly unexpected news: the franc was to be devalued. **1979** D. Eden *Storrington Papers* xiii. 149 She's a bit upset. I did rather spring it on her.

22. c. Also *fig.*

1930 D. Hammett *Dain Curse* xv. 165 'All right, spring it,' I said, as we sat down in his..living room. 'Any trace of Gabrielle yet?' he asked. 'No. But spring the puzzle. Don't be literary with me, building up to climaxes and the like... Just spread it out for me.' **1980** J. Barnes *Metroland* iii. iv. 161, I started again, more seriously this time, masochistically trying to spring that familiar trigger for panic and terror.

22*. To release (a person) from custody or imprisonment, esp. to contrive such a release by means of bail. Also, to contrive an unlawful escape from prison. *slang* (orig. *U.S.*).

1900 'Flynt' & 'Walton' *Powers that Prey* 62 It cost his push a thousand plunks to spring him from the coppers. **1911** C. B. Chrysler *White Slavery* ix. 70 If you get nicked..you can 'raspberry' the 'bull' and get 'sprung'. **1929** *Sun* (Baltimore) 15 Nov. 1/6, I got sick of him [*sc.* the lawyer] and I started examining the jury myself. They sprung me in five minutes. **1936** J. Steinbeck *In Dubious Battle* iii. 24 They'll give him the works if George doesn't get busy. Tell George to try to spring him for a drunk... If a sanity board ever gets hold of that poor devil, he's in for life. **1963** *Security Gaz.* V. 187/3 Those who may be preparing to 'spring' an inmate. **1967** *Punch* 15 Mar. 375 In the main the British bail system works better than the American one. Over there a professional bondsman..'springs' the accused man for roughly ten per cent of the bail fixed. **1974** *Daily Tel.* 2 Sept. 3 Miss Mary Tyler, the English school-teacher who has spent more than four years in Indian jails awaiting trial, is to be returned to a high security prison this week in case militant Maoists try to 'spring' her. **1977** J. Cheever *Falconer* 13 I'm in cellblock F... Last Tuesday they forgot to spring us for supper. **1980** *Observer* 3 Apr. 9/5 What the Minister has in mind is the 'springing' of the dockers in 1972 following the intervention of that hitherto mysterious figure, the Official Solicitor.

23. c. *Naut.* To move, haul, or swing (a vessel) by means of a spring or cable. Cf. Spring *sb.*[1] 24.

a **1865** W. H. Smyth *Sailor's Word-bk.* (1867) 646 *Sprung*..the ship slued round by means of guys. **1898** S. B. Luce *Text-bk. Seamanship* 217 Ships may be sprung broadside to the wind..for the purpose of better ventilation; or in engagements at anchor, to bring the guns to bear on various points. **1922** C. C. Soule *Naval Terms & Definitions* 69/2 *Spring*, to turn a vessel with a line.

d. *Bootmaking.* To raise (the toe or waist of a last) above the ground-line (see also quot. 1953).

1905 E. J. C. Swaysland *Boot & Shoe Design* iii. 21 An upward curve in the waist of about an eighth of an inch... This is very much less than lasts are usually sprung in the waist. **1916** F. Plucknett *Boot & Shoe Manuf.* ii. 16 Provision should be made for alteration in shape which would be likely to take place in wear.., *e.g.* springing the toe of the last. **1953** A. V. Goodfellow in J. H. Thornton *Textbk. Footwear Manuf.* ii. vii. 89 The effect of springing a pattern, or changing the relationship between one part and another of the same pattern is to lengthen one line and shorten another.

springal(d[1]**.** **β.** (Later example.)

1909 W. Deeping *Red Saint* xl. 340 The King's men who still held the castle, had thrown springalds of fire down upon the houses, setting the thatch ablaze.

‖ **springar** (spri·ŋgăɹ). [Norw.] A Norwegian country-dance in three-four time; also, the music for this dance.

1947 [see *halling]. **1959** *Listener* 2 July 37/1 To achieve the truly national in Norwegian music they should not rest content to quote *springars* and *hallings*, and other dances. **1964** W. G. Raffé *Dict. of Dance* 474/1 *Springar* has nine different steps, and is a dance in which partners show their individual qualities—the man his virility and accomplishment, the girl her skill and grace.

spri·ng-back. Also spring back, springback. **1.** [f. Spring *sb.*[1] 22.] **a.** A folder with a spring clip incorporated in the spine to hold the papers placed in it; also, the clip itself. **b.** = *loose back* s.v. *Loose *a.* 9. Also *attrib.*

1895 *Montgomery Ward. Catal.* Spring & Summer 247/1 Music Folio, imitation leather, spring back. **1919** 'Etienne' *Strange Tales from Fleet* 64 The stupendous and well-filled 'spring-back', replete with bills, a few receipts, and reams of official correspondence. **1923**, etc. [see *loose back* s.v. *Loose *a.* 9]. **1948** G. V. Galwey *Lift & Drop* vi. 162 Bourne opened the spring-back correspondence file. **1982** P. D. James *Skull beneath Skin* xli. 332 The archives were bound in springback folders.

2. [f. Spring *v.*[1]] The capacity to spring flexibly back into position after subjection to pressure. Freq. *attrib.* or as *adj.* Also *fig.*

1945 *Richmond* (Va.) *Times-Dispatch* 2 Mar. 19/3 It [*sc.* a type of compressed wood] can be made..with negligible springback tendencies. **1970** *Motoring Which?* July 116/1 All the arm-type mirrors we included were either spring-back or collapsible. **1972** A. Amin tr. *Ahmad's No Harvest but Thorn* x. 98 She leapt over the spring-back wooden fence like a young mouse-deer startled by a *pulut* fox. **1973** J. G. Tweeddale *Materials Technol.* II. iv. 87 Whenever a material possesses significant elasticity, as in cold working, there is difficulty with spring back. **1975** *N.Z. Jrnl. Agric.* Sept. 55/2 The finished fence..has a feeling of controlled flexibility and springback to true that augurs well for its stock-holding qualities. **1977** *Time* 31 Jan. 40/1 Charles Schultze agrees that the first quarter 'will show a pretty good springback' from the recent economic pause.

spring-board. Add: **2.** (Earlier examples.)

1799 *Times* 1 June 3/4 He positively leaps over a large tilted waggon and four horses..and does not make use of a spring board or trampoline. **1841** Thoreau *Jrnl.* 23 Jan. (1906) I. v. 174 Like the spring-board on which tumblers perform and develop their elasticity.

3. Also *Austral.* and *Canad.*

1934 *Bulletin* (Sydney) 19 Sept. 20/1 Nerves of steel are needed day by day By those who on the springboard stand, while forest monarchs sway. **1972** *Daily Colonist* (Victoria, B.C.) 16 July 6/5 He captured first place in the springboard chop open competition to become Canadian champion. **1975** H. White *Raincoast Chron.* (1976) 102/2 Metal-tipped springboards set in notches were used by the early fallers to climb above the butt-swell [of a tree].

springbok. Add: **2.** *pl.* (usu. with capital initial). **a.** A nickname for a South African national sporting team or touring party. Also *sing.*, one who represents South Africa in international sport.

1906 *S. African News Weekly* 3 Oct. 24/1 A crowd of 9,000..accorded the springboks a great reception as they walked on the field. **1932** *Grocott's Mail* (Grahamstown, S. Afr.) 2 Jan. 3 It cannot be said that many English rugby critics strongly favour England's chance against the Springboks on Saturday. **1959** *Cape Argus* 23 Jan. 19/6 (*heading*) Hockey Springboks train. **1970** *Times* 5 Feb. 9/3 Such an admission would be a natural sequel to our recent failure to protect our Springbok guests from repeated insult and persecution. **1975** *Country Life* 16 Jan. 137/3 The French, who lost both their matches against the Springboks..are often underestimated.

b. A nickname for a contingent of South African troops. Also *sing.*, a South African soldier.

1916 'Captain' (*title*) With the Springboks in Egypt. **1925** Fraser & Gibbons *Soldier & Sailor Words* 267 *Springboks*.., the South African contingent in the war. From their badge, the Colonial emblem of a springbok antelope. **1943** J. Burger *Black Man's Burden* 236 The Springboks (Union troops) are spoken of as the successors to the pioneer Voortrekkers. **1944** *Eastern Province Herald* 25 Aug. 1 The company is doing urgent tank maintenance and repair work for the Fifth and Eighth armies; and the Marshall said that he was impressed by the efficiency of the Springbok 'backroom boys'.

spring chicken. [f. Spring *sb.*[1] 6 b.] **1.** A small chicken (esp. a roasting bird); *spec.* one aged between eleven and fourteen weeks and weighing around one and a half kilograms (see also quot. 1958).

1780 J. Woodforde *Diary* 13 June (1924) I. 285 We had for dinner..three nice Spring Chicken rosted. **1849** [in Dict. s.v. Spring *sb.*[1] 7 a]. **1892** [see *On *prep.* 20 l]. **1917** *Harrods Gen. Catal.* 1267 Poultry... Chickens (Spring). **1943** L. I. Wilder *These Happy Golden Years* xxi. 191 Ma finished frying a spring chicken while the new potatoes and the peas were cooked. **1958** *Times* 18 Aug. 11/2 The term 'spring chicken' has come to be something of a misnomer for the small bird now available at poulterers..all the year round. It is the product of a specialist industry which has developed this particular type of bird to come to maturity in about 10 weeks only. **1974** *Times* 7 Mar. 13/5 Spring chickens are also in the shops, these are a little older [than *poussins*]—from eight weeks to four months.

2. *fig.* A young person. Freq. in phr. *to be no spring chicken* and varr., to be no longer young. *colloq.* (orig. *U.S.*).

1910 *National Police Gaz.* (U.S.) 6 Aug. 3/1 She wasn't a Spring chicken, by any means, yet she wasn't old. *a* **1911** D. G. Phillips *Susan Lenox* (1917) II. ii. 20 Miss Hinkle was showing her age—and she was 'no spring chicken'. **1964** 'E. McBain' *Axe* vii. 125 I'm not a spring chicken any more... I'm fifty-two years old. **1973** G. Beare *Snake on Grave* iv. 23 She was no spring chicken any more, she was pushing forty. **1977** *New Yorker* 10 Oct. 131/1 Americans were impressed by Dr. White's prescription whenever they saw a photo of President Eisenhower swinging a golf club or of Dr. White, no spring chicken himself, bent low over the handlebars of his bicycle. **1981** D. M. Thomas *White Hotel* iv. i. 145 You're just a spring chicken, Lisa dear!

spring-cleaning. (Earlier examples.)

1857 E. M. Stone *Life J. Howland* i. 28 At the annual 'spring cleaning', they discovered a bundle of manuscripts. **1873** H. Martineau *Let.* 6 Mar. in *Autobiogr.* (1897) III. 416 This will be a busy month, with the spring cleaning and whitewashing.

spring-clean *v.*: also *intr.* and as *sb.*, the act of spring-cleaning. Also *transf.* and *fig.*

1849 C. Brontë *Shirley* I. xi. 289 Very handsome.. these shining brown panels are..but—if you know what a 'Spring-clean' is—very execrable and inhuman. **1908** in *Englische Studien* (1935) Apr. 119 He was helping his wife to 'spring-clean'. **1926** *Socialist Rev.* Dec. 14 House to house inspections [should be] made in the worst areas; in fact, a regular spring-clean of the whole town organised. **1957** L. Durrell *Justine* i. 23 It was thrown away by Hamid in the course of a spring-clean. **1961** *Countryman* Autumn 515 The exhausted hound awoke and started a relentless spring-clean of his dusty coat. **1978** N. Marsh *Grave Mistake* i. 36, I was helping springclean at the time. **1979** *Guardian* 19 July 14/5 If, early next week, the Carter White House has truly been spring-cleaned..then the Presidency will be much strengthened.

springer[1]**.** Add: **3. a.** (Earlier example.)

1727 J. G. Scheuchzer tr. *Kæmpfer's Hist. Japan* I. 137 Tobiwo is what the Dutch call a Springer (Flying-fish) because it leaps out of the water.

4. c. *Naut. slang.* A physical-training instructor in the navy.

1935 *N. & Q.* 29 June 465/1 The officer in charge of physical training was known in my ship..as 'Bunje'. The modern term is 'Springer', both words being descriptive of the acrobatic nature of these duties. **1964** J. Hale *Grudge Fight* vi. 93 The springers all fancy their chance in the training line.

11. *slang.* A racehorse on which the betting odds suddenly shorten (see quot. 1961).

1922 E. Wallace *Flying Fifty-Five* xi. 67 The 'springer' in the market, the horse that opened at ten to one and came rapidly to five to two. **1961** J. Prescot *Case for Hearing* iv. 61 Plenty of punters like to know how the market's moving so that they can go for the 'springer', the horse that suddenly shortens in price because someone in the know slaps a lot of money on at the last possible moment.

Springfield (spri·ŋfīld). The name of *Springfield*, Mass., used *attrib.* and *absol.* to designate firearms produced at the U.S. government armoury located there.

1813 *Niles' Reg.* IV. 87/2 [The] gun..is but one pound and a half heavier than the common Springfield gun. **1849** *Knickerbocker* XXXIII. 3 They carried slung from their saddles the excellent Springfield carbines, loading at the breech. **1866** *Harper's Mag.* Aug. 407/1 'Puss' had been doing his best with his 'Springfield', but all at once he stopped firing. **1904** *Kynoch Jrnl.* Apr.–June 95 The Government of the United States had adopted in 1855 the Springfield rifle—a muzzle-loader, calibre .58 bore, with which a 500 grain bullet & 60 grains of powder were used. **1933** F. B. Willoughby *Alaskans All* 3 His flannel shirt, high laced boots, the hunting-knife in his belt, and the Springfield by his side, made it difficult for me to realize he was a priest. **1964** M. McLuhan *Understanding Media* xxxii. 341 The old-timers who backed the Springfield rifle against perimeter fire. **1976** 'O. Jacks' *Assassination Day* (1978) v. 81 The rifle..was an old World War II American Springfield.

spring garden. Restrict † *Obs.* to senses in Dict. and add: **d.** A garden containing many plants that bloom early in the year.

1883 W. Robinson *Eng. Flower Garden* i. p. l/1 What we should counsel those who care for their spring gardens is this: To begin and always to work with a series of nursery beds. **1914** E. A. Bowles *My Garden in Spring* xvii. 295 If you have enjoyed strolling round the spring garden with me..I hope later on you will accompany me on a second journey to review the summer aspect of the place and plants. **1947** H. Nicolson *Diary* May (1968) 97 Sissinghurst is lovely. The spring-garden has lost its early bloom.

‖ **springhaas** (spri·ŋhās). *S. Afr.* Also (with hyphen) **spring-haas.** [Afrikaans, f. Du. *spring* jump, leap + *haas* hare.] The jumping hare, *Pedetes capensis*, a nocturnal rodent with large and powerful hind legs and a hopping gait.

1785 G. Forster tr. *Sparrman's Voy. Cape Good Hope* II. 195 By the colonists it [*sc.* the *jerbua Capensis*] is called *berg-haas*, or *spring-haas* (the mountain or bounding hare). **1837** J. E. Alexander *Narr. Voy. W. Afr.* I. 347 One of the party shot a *spring-haas*, or jumping hare, formed like the kangaroo, with very short fore-legs and long hind ones. **1890** A. Martin *Home Life on Ostrich Farm* xi. 225 Most uncanny of all the hares is the springhaas. This creature..is never seen in the daytime, and can only be shot on moonlight nights. **1942** S. Cloete *Hill of Doves* 145 His pulse..is fast and jumping like a springhaas.

spring-hare (spri·ŋhēᵊɹ). Also springhare, spring hare. [Partial tr. Afrikaans *spring-haas*: see prec.] = *Springhaas.

1822 W. J. Burchell *Trav. Interior S. Afr.* I. 343 Besides the holes of the Aardvark, those of the Springhaas (Springhare)..were very frequent. **1853** F. Galton *Trop. S. Afr.* ix. 281 The spring-hare..is a creature about two feet long, shaped like a kangaroo in body and in tail, but with a different head; it burrows and lives in holes all day, but at night frisks about and grazes. **1912** J. Stevenson-Hamilton *Anim. Life Afr.* xvi. 253 The Spring Hare..is found locally through Africa south of the Equator. **1937** K. Blixen *Out of Africa* ii. 96 The little springhares were out on the roads..jumping along. **1966** E. Palmer *Plains of Camdeboo* x. 179 All around us in their roomy burrows were springhares, those kangaroo-like creatures with their enormous hind legs that go hoppity-hoppity over the veld at night.

springing, *vbl. sb.*[1] **III. 13. c.** *springing-board* (earlier example).

1846 Geo. Eliot *Let.* 29 Oct. (1954) I. 224, I do not know whether I can get up any steam again in the subject. ..I must have the book as a springing-board.

springing, *vbl. sb.*[2] Add: Also, the state or quality of a set of springs; hence, the suspension of a vehicle.

1952 *Times* 19 Aug. 11/6 The springing gives a comfortable ride for the passengers. **1978** *Lancashire Life* Apr. 141/1 The springing is quite soft, so it soaks up the bumps.

springing, *ppl. a.* **10.** (Earlier example.)

1725 D. Eaton *Let.* 13 July (1971) 30 Mr. Joseph Lynwood..considering..that he had better mind other business than shooting (for they were all springing spaniels), hang'd three of them of his own accord.

springless, *a.* **5.** (Literal example.)

1927 E. Sitwell *Rustic Elegies* 82 Delivered from that springless frost.

springy, *a.* Add: **1. b.** Characteristic of the season of spring; spring-like.

1860 S. Warner *Say & Seal* II. xviii. 229 It was April now, and a soft springy day. **1936** N. Coward *To-night at 8.30* III. 85 Quite Springy out, isn't it?

sprinkler. Add: **4.** *attrib.* and *Comb.*, as (sense 1 c) *sprinkler installation, system;* **sprinkler irrigation** = *spray irrigation* s.v. *Spray *sb.*[2] 3 c.

1935 *Jrnl. R. Aeronaut. Soc.* XXXIX. 164 Sprinkler installation for air defence. **1967** *Guardian* 24 May 1/6 The store had no sprinkler installation. **1950** O. W. Israelsen *Irrigation Princ. & Pract.* (ed. 2) vi. 142 Sprinkler irrigation systems may be divided into three groups according to the purpose for which they are used. **1955** McColly & Martin *Introd. Agric. Engin.* xxxvii. 531 Horizontal centrifugal pumps and turbine pumps are the common types used for sprinkler irrigation, as they give a steady flow of water. **1980** *Jrnl. R. Soc. Arts* Mar. 175/1 There are sophisticated forms of sprinkler irrigation..whereby the optimal nutrient content is fed to the crops at just the right moment. **1930** *Engineering* 28 Feb. 279/2 Sprinkler systems in various store rooms. **1973** *Scottish Sunday Express* 5 Aug. 1/2 Why was no sprinkler system installed in the Isle of Man Summerland entertainment centre in which 51 people died in one of Britain's worst-ever fires?

Hence **spri·nklered** *a.*, of or pertaining to a sprinkler system; *spec.* (chiefly *N. Amer.*), provided with or watered by a sprinkler system.

1930 F. E. Wolfe *Princ. Property Insurance* ix. 151 In Great Britain the principle is commonly applied to plural risks, to sprinklered and other fire-resisting risks. **1968** *Globe & Mail* (Toronto) 13 Feb. B11/6 (Advt.), 16,500 sq. ft. on second floor, vinyl tile floors, fully sprinklered, oil and steam heated,..air-conditioned offices. **1975** J. Montague *Slow Dance* 48 Our true Catholic world, a graveyard..water sprinklered grass, collapsing wreaths. **1978** *N.Y. Times* 29 Mar. D12/6 (Advt.), Over 50,000 sq. ft. now available in prime industrial location Complete burglar & fire alarm systems ..Totally sprinklered.

sprinkling, *vbl. sb.*[1] Add: **4. sprinkling irrigation** = *spray irrigation* s.v. *Spray *sb.*[2] 3 c.

1927 *Bull. New Jersey Agric. Exper. Station* No. 453. 3 Three general methods of irrigation, known as surface, sub-, and sprinkling irrigation have been developed. *Ibid.*, Sprinkling irrigation, also known as overhead or Skinner irrigation, from the name of the inventor, is the method of watering crops by means of spray from small nozzles.

sprint, *sb.*[1] Add: **3. sprint car** orig. *U.S.*, a type of racing car (see quot. 1969).

1954 *Motorsport* June 19/2 Sprint cars were drawing big crowds at the half-mile speedways. **1955** *Sun* (Baltimore) 14 Apr. (B ed.) 18/7 Drivers for the latest entries are two Speedway veterans..and two rookies from the sprint-car circuit. **1961** [see *Dragster]. **1969** *Britannica Bk. of Year 1968* 801/1 *Sprint car*, a rugged racing automobile that is midway in size between midget racers and ordinary racers, has about the same horsepower as the larger racers, and is usually raced on a dirt track. **1981** *Telegraph* (Brisbane) 27 Nov. 18/3 The Queensland Bomber and National Sprintcar champion, Bob Kelly also will be having a crack at tonight's title.

sprit, *v.*[1] Add: Hence **spri·tted** *ppl. a.*

1911 D. H. Lawrence *White Peacock* ii. viii. 335 We.. went along the wet furrows, sticking the spritted tubers in the cold ground.

sprite, *sb.* **1. a.** (Later *arch.* example.)

1928 R. Campbell *Wayzgoose* ii. 53 As she bent above her chosen Knight A lovely fragrance ravished all his sprite.

spritely, *a.* and *adv.* Add: Still found as a variant of Sprightly *a.* and *adv.*

1898 G. B. Shaw in *Sat. Rev.* 21 May 683/1 The younger generation is knocking at the door; and as I open it there steps spritely in the incomparable Max. **1927** *Daily Tel.*

21 June 14/7 The programme was the usual compound of items serious and spritely—mostly spritely of course—and from time to time dancers and comedy cyclists flashed across the stage. **1977** *Monitor* (McAllen, Texas) 9 Jan. 5c/1 Instantly the room is filled with the spritely, bubbling sound of ragtime. **1981** *N.Y. Times* 10 May xi. 23/1 The clams Casino..covered with a spritely seasoned well-blended onion-garlic topping.

spritty (spri·ti), *sb. colloq.* Also **sprittie.** [f. Sprit(sail + -y[6].] A spritsail barge.

1948 H. Benham *Last Stronghold of Sail* ii. 27 The ketch barge skipper had no doubt of his superior status as compared with the sprittie sailorman... He admitted that in their own weather the spritties were faster. **1952** W. G. Arnott *Alde Estuary* vi. 46 The last spritties to visit Snape were the *Una* and *Beatrice Maud* about 1928. **1960** *Guardian* 1 July 7/5 The Cambria, the last 'spritty' to trade under sail as a coaster.

spritz (sprits, ‖ʃprits), *v. U.S.* [f. G. *spritzen* to squirt.] *trans.* To sprinkle, squirt, or spray.

1917 *Dialect Notes* IV. 339 *Spritz*,..to sprinkle. 'Look out, I'll *spritz* you.' **1948** Mencken *Amer. Lang.* Suppl. II. 202 *Spritz*, to sprinkle or squirt. **1950** T. Shane *Bar Guide* 156 Spritz seltzer over lemon juice, lime juice, grape juice, and sugar. **1976** N. Thornburg *Cutter & Bone* i. 15 The dog, a male, was spritzing a dwarf palm tree. **1978** *Detroit Free Press* 16 Apr. D10/1 Pierre de France..maintains the stars tell which fragrances men and women should be spritzing on their bodies.

Hence **spritz** *sb.*; **spri·tzing** *vbl. sb.*

1935 H. R. Rudd *Hocks & Moselles* x. 125 Light wine.. often..possesses that natural sparkle or spritz that is so attractive in the Moselle wines. **1975** *High Times* Dec. 89/3 (Advt.), This spray will make any room smell like a candy store... A one-second spritz will mask any odour. **1976** *National Observer* (U.S.) 21 Aug. 12/3, I found myself inexplicably caught up by the nonsense..the inevitable spritzing of the seltzer bottle.

spritzer (spri·tsəɹ, ‖ʃpri·tsəɹ). Chiefly *N. Amer.* [a. G. *spritzer* a splash.] A mixture of wine and soda water; a drink of this mixture.

1961 in Webster. **1964** *Vogue* Apr. 71/1 Drink *Spritzer* (dry white wine and soda). **1972** G. Baxt *Burning Sappho* iii. 50 Flo's [drink] was a white wine spritzer. **1979** *Toronto Star* 7 Apr. G1/6 Order a spritzer and never look at who you're talking to—something better could be walking in the door.

‖ **spritzig** (ʃpri·tsig), *a.* [Ger.] Of wines: sparkling, *pétillant*. Also as *sb.*

1949 T. A. Layton *Choose your Wine* (ed. 2) vii. 66 Sometimes the wines are shipped so that there is a delightful natural sparkle in them..called by the Germans *spritzig*. **1959** *Times* 21 Sept. 13/3 This is a wine that is *spritzig*, as the Germans say, or what the French call *pétillant*—that is, it has a slight prickle or sparkle. **1968** *Vogue* Dec. 141/1 A young moselle is a perfect aperitif, especially if it has that rare characteristic *spritzig*: an almost imperceptible touch of effervescence. **1972** *House & Garden* Feb. 100/4 KWV Late Vintage..is slightly *spritzig*, and has a very fresh taste. **1980** *Times* 9 Aug. 8/8 The wine is very big, with..a slight touch of spritzig on the palate.

s-process: see *S II. 9.

sprocket, *sb.* Add: **2. b.** *sprocket hole*, each of a line of holes along film or paper tape with which sprockets can engage to propel it or keep it correctly aligned.

1910 F. H. Richardson *Motion Picture Handbk.* 101 Old, dry films jump because..the sprocket holes are shrunken. **1931** B. Brown *Talking Pictures* viii. 182 Bad tracking produces a loud hum due to either the picture or the sprocket holes running across the light beam. **1960** M. G. Say et al. *Analogue & Digital Computers* ix. 263 A row of holes across the tape consists of one sprocket hole, 0·043 in. diameter.., and up to five information holes, 0·072 in. diameter. **1971** *Southerly* XXXI. 138, I could print those four or five with the sprocket holes still on the film. **1980** S. Hockey *Guide Computer Applic. Humanities* ii. 25 It is normally necessary to leave a foot or so of blank tape with only the sprocket holes punched at each end of a paper tape.

c. (Later examples.) Also *Cinemat.*, a sprocket-wheel that propels film by engaging with perforations along its edge.

1910 F. H. Richardson *Motion Picture Handbk.* 110 The intermittent sprocket..should be watched and promptly renewed when there are signs of serious wear. **1925** R. V. Johnson *Mod. Picture Theatre Electrical Equipment & Projection* 67 The intermittent sprocket pulls the film through the gate, and the bottom sprocket carries the film to the lower spool box. **1953** L. J. Wheeler *Princ. Cinemat.* vi. 181 It is reasonable to expect the sprocket dimensions to be different to those in the camera mechanism. **1977** D. MacKenzie *Raven & Kamikaze* vi. 77 Slade unwound the film from the sprockets and put it back in the can.

attrib. **1910** F. H. Richardson *Motion Picture Handbk.* 102 A machine with worn intermittent sprocket teeth. **1935** *Discovery* Nov. 325/2 Disturbances such as sprocket-tooth ripple.

sprocketed, *a.* Add: Also, furnished with sprocket holes.

1967 P. Grosset *Compl. Bk. Amateur Film Making* viii. 143/2 You will need to use..sprocketed tape (which cannot slip, shrink or stretch). **1978** [see next].

sprocketless (sprǫ·ketles), *a.* [f. Sprocket *sb.* + -less.] Not employing or requiring sprockets.

1967 *Boston Sunday Herald* 26 Mar. vi. 2/8 The smooth, quick operation results from the projector's sprocketless film drive. **1978** *Broadcast* 23 Apr. 24/2 Synchronising sprocketed film..with sprocketless video tape.

sprog (sprǫg). [Cf. Sprag *sb.*[2]] **1.** *Services' slang.* A new recruit; a trainee; a novice. Also *occas.*, one of inferior or ordinary rank. Freq. *attrib.*

1941 *New Statesman* 30 Aug. 218/3 *Sprog* (R.A.F.), a tyro. **1942** *Word Study* Dec. 6/2 'Hey, *sprog*,' said the corporal, 'how about us gettin' another *cliner*?' **1943** J. Hillier in *Penguin New Writing* XVI. 23 Never mind, Wendy, you sprogs of 'B' flight will learn to fly yet—if you live long enough! **1946** J. Irving *Royal Navalese* 164 *Sprog*, a new entry. **1949** J. R. Cole *It was so Late* 62 Each time a new course of pilots arrived the sprog officers among them used to fall over themselves after her. **1970** C. Wood *'Terrible Hard' says Alice* ii. 31 Pasty-faced sprog subalterns. **1978** F. Branston *Sergeant Ritchie's Conscience* i. 12 Some sprog copper, so new he did not even recognize him.

2. *slang* (orig. *Naut.*). A youngster; a child, a baby.

1945 'Tackline' *Holiday Sailor* vii. 75, I can't deny him nothing. Always giving the sprog a tanner to nip off and buy himself some nutty. **1949** in Partridge *Dict. Slang* (ed. 3) Add. 1181/2 Nobby Clark's gone on leave, his wife's just had a sprog. **1968** 'O. Mills' *Sundry Fell Designs* x. 115 All those sprogs, and that ghastly earth-mother missus. **1973** M. Amis *Rachel Papers* 64 Here I attempted a few minutes' work, not easy because the fifty bawling sprogs had classes there in the afternoon. **1981** D. Clark *Longest Pleasure* ii. 14, I don't think he's been really with us since the sprog came along.

sproncy, var. *Sprauncy *a.*

sprosser (sprǫ·səɹ). [Ger.] The thrush-nightingale, *Luscinia luscinia*, of the family Turdidæ, found in eastern Europe and Asia.

1871 Newton & Saunders *Yarrell's Hist. Brit. Birds* (ed. 4) I. 320 A second species of Nightingale occurs.. long known to German bird-fanciers as the sprosser. **1912** Baxter & Rintoul *Rep. Scott. Ornithol. 1911* 3 Curiously enough, both the Common Nightingale..and the Northern Nightingale or Sprosser..were added in spring to the Scottish list. **1954** D. A. Bannerman *Birds Brit. Isles* III. 304 This [*sc.* the thrush-nightingale] is the 'sprosser' of Germany, inhabiting the same type of cover as its better known relative.

sprottle (sprǫ·t'l), *v. dial.* [Cf. Spartle *v.*[2], Sprattle *v.*[2]] *intr.* To sprawl, to struggle helplessly. Hence **spro·ttling** *ppl. a.*

1829 J. Hunter *Hallamshire Gloss.* 85 *Sprottle*, to struggle with inefficacious vehemence. **1839** A. Bywater *Sheffield Dial.* 15 An we sprottlin abaht to ger up. **1917** D. H. Lawrence *Look! We have come Through!* 97 Why do you spurt and sprottle like that, bunny? **1921** —— *Tortoises* 19 It goes right through him, the sprottling insect. **1937** A. Uttley *Ambush of Young Days* iv. 65 The lamb suddenly gave a tiny wailing 'Baa', and tried to struggle to its feet. 'It's sprottling. It'll be all right,' cried Josiah, our old manservant.

sprount (spraunt), *v. rare*[−1]. [App. altered form of Sprout *v.*[1]; but cf. Sprounting *vbl. sb.*] *intr.* To sprout.

1939 Dylan Thomas *Map of Love* 64 The half-liquid plants sprounting from the bog.

sprouse, var. *Sprowsie.

sprout, *sb.*[1] Add: **3. b.** *U.S. colloq.* and *slang.* A young person, a child.

1934 *Jrnl. Amer. Folk-Lore* Jan./Mar. 51 One time she was getting ready to go to a play-party. Some of the young sprouts were waiting for her. **1942** Berrey & Van den Bark *Amer. Thes. Slang* § 383/2 Child..(little or young) sprout. **1950** R. Moore *Candlemas Bay* 24 I'm going to beat the living pickle out of this goddam sprout of mine. **1951** *Harper's Mag.* July 36/1 A girl out your way has married..and is coming home with a sprout. **1983** *Verbatim* IX. iii. 23/2 The young sprouts and broths of lads who feel their oats and are full of beans.

6. sprout-land (earlier and later examples).

1851 Thoreau *Jrnl.* 12 Feb. in *Writings* (1906) VIII. 156 It is refreshing to walk over sprout-lands, where oak and chestnut sprouts are mounting swiftly up again into the sky. **1914** R. Frost *North of Boston* 87 A rock-strewn town where farming has fallen off, And sprout-lands flourish where the axe has gone.

sprouter. Restrict *rare* to sense in Dict. and add: **2.** A container in which seeds (esp. of mung beans) are sprouted.

1971 *Health Food Age* June 33/2 (Advt.), Get more vitamins daily with a seed sprouter. *Ibid.*, Be one of the first to get a Miracle Seed Sprouter. **1977** C. McFadden *Serial* (1978) xvii. 41/2 She was starting mung beans in her sprouter.

sprowsie (sprau·zi). *slang.* Also **sprouse, sprowser.** [Prob. var. of *Sprazer.] Sixpence; a sixpenny piece.

1931 'G. Orwell' in *Coll. Essays* (1968) I. 70 *Sprowsie, a*, a sixpence. **1933** —— *Down & Out in Paris & London* xxxii. 236 These..are some of the cant words now used in London... A hog—a shilling. A sprowsie—sixpence. **1960** A. Prior in *Pick of Today's Short Stories* XI. 180, I walked across to the record player and took some silver out of my pocket... 'Half-Nelson, do me a favour and put a sprouse in there for me... I've got no change.' **1966** F. Shaw et al. *Lern Yerself Scouse* 34/2 *Sprowser*, sixpenny piece.

spruce, *sb.* Add. **5.** *spruce green;* **spruce budworm**, the brown larva of a tortricid moth, *Choristoneura fumiferana*, which damages the foliage of spruce trees in North America; **spruce grouse**, also Franklin's grouse, *Canachites canadensis*; (earlier and later examples); **spruce hen**, a female spruce grouse; **spruce partridge** (earlier and later examples); **spruce pine**, one of several North American conifers, formerly esp. the bog or black spruce, *Picea mariana*, but now usually *Pinus glabra*; **spruce tea**, an infusion of tender spruce shoots.

1884 *Rep. Comm. Agric.* (U.S. Dept. Agric.) 378 The Reddish-Yellow Spruce-Bud Worm..was found to be very injurious to the white spruce. **1925** [see *Biocœnosis]. **1976** *Maclean's Mag.* 3 May 56/3 Except for its larva..the spruce budworm is an unexceptional creature. **1939** *Sun* (Baltimore) 11 Feb. 20/2 Enrollees of the Civilian Conservation Corps will be attired in new uniforms.. of 'forest' or 'spruce' green. **1844** J. E. DeKay *Zool. N.Y.* ii. 206 The Spruce Grouse... The flesh is bitter, and has a peculiar taste as if boiled in turpentine. **1946** T. M. Stanwell-Fletcher *Driftwood Valley* 13 Several times we've scared up coveys of spruce grouse along the trail. **1966** *Kingston* (Ontario) *Whig-Standard* 21 Jan. 11/1 The spruce grouse is so retiring..that few people get to see one. **1902** W. D. Hulbert *Forest Neighbors* 87 Spruce hens and partridges. **1959** W. A. Leising *Arctic Wings* 31, I had not seen the spruce hen until the shot bird came tumbling out of the tree. **1971** A. Fry *Long Journey* vii. 41 We came on half a dozen spruce hens. **1771** W. Richardson *Jrnl.* in *Canad. Hist. Rev.* (1935) XVI. 57 There are more spruce Partridges in the woods than I have seen anywhere in this country. **1917** T. G. Pearson *Birds Amer.* II. 15/1 There is no such bird as the Spruce Partridge. It is the Spruce Grouse. **1963** *Calgary Herald* 4 Oct. 27/7 Spruce partridge and Franklin's grouse season has been open since Aug. 31. **1765** J. Bartram *Jrnl.* 25 Sept. in *Trans. Amer. Philos. Soc.* (1942) XXXIII. 30/2 Ye 2 leaved or spruice pine grows very large in swamps. **1842** M. Crawford *Jrnl.* 10 Sept. (1897) 19, I noticed the White Pine and the Spruce Pine. **1886** *Outing* VIII. 60/2 One morning I entered a clump of bushes near a spruce-pine thicket. **1913** H. Kephart *Our Southern Highlanders* xiii. 295 The hemlock tree is named spruce-pine. **1949** *Boston Globe Mag.* 4 Dec. 11/2 He went up to the white 'Honor Roll' board nailed to a big spruce pine. **1967** N. T. Mirov *Genus Pinus* iii. 183 *Pinus glabra* (spruce pine) is the least common pine of the southeastern United States. **1783** in *New Brunswick Mag.* (1899) II. 320 Some chocolate is wanted for our Masting Camp for at present we use Spruce Tea which causes sum murmuring. **1936** *Discovery* Jan. 31/1 That unpalatable beverage, spruce tea. **1956** *Beaver* Summer 18, I lashed out strong laxatives all round and ordered spruce tea to be brewed and administered constantly.

spruce, *v.*[2] *slang* (orig. *Mil.*). [Of unknown origin.] **a.** *intr.* To lie, practise deception; to evade a duty, malinger. Also with quasi-obj. **b.** *trans.* To deceive.

1917 W. Muir *Observ. Orderly* xiv. 230 To spruce is to dodge duty or to deceive. A man who contrived to slip out of the ranks of a squad when they were performing some distasteful task would be said to 'spruce off'. **1919** *Athenæum* 8 Aug. 728/1 *Spruce*, deceive. **1925** Fraser & Gibbons *Soldier & Sailor Words* 267 *Spruce*.., to lie. To deceive. **1951** A. Baron *Rosie Hogarth* 222 Write your own cheque. It's yours for the asking..Go on! I ain't sprucing. **1967** G. M. Wilson *Cake for Caroline* vi. 71 Dr. Meunier's no fool, he'd have known if she was sprucing..Malingering. Faking tummy trouble. **1969** H. Carvic *Miss Seeton draws Line* ix. 172 Them two old tarts at the Nut House, they spruced you proper. **1970** A. Hunter *Gently with Innocents* xiii. 166 The coin is damning. He can't spruce his way round that. **1978** *Daily Tel.* 26 May 16 A kipper..by inference, should cost more than the untreated fish. Who is sprucing whom?

Hence **spru·cer**, one who tells tall stories, a trickster.

1917 W. Muir *Observ. Orderly* xiv. 230 He would be denounced as a 'sprucer' if he managed to arrive late for his meal and yet, by a trick, to secure a front place in the waiting queue at the canteen. **1919** *Athenæum* 15 Aug. 759/1 'A sprucer' is a man who tells tall stories. A man who is 'ticked off' for wrong doing by his officer may escape further punishment by 'sprucing him up a yarn'. **1930** P. MacDonald *Link* ix. 194 This is where the G.D. begins to show up for the sprucer that he is. **1968** *Listener* 25 Jan. 111/2, I suspect Peter Eckersley was pulling Cutforth's leg. He was a good 'sprucer', as they used to say in Swadlincote.

sprue, *sb.*[2] Add: **a.** A channel through which molten metal or plastic flows into a mould cavity or the runner supplying it. (Further examples.)

1939 J. Osborne *Dental Mechanics* xvi. 173 Sprues are necessary to convey the gold to the denture base. In order that this may be accomplished as quickly as possible it is usual to use three, four or even five sprues. **1943** Simonds & Ellis *Handbk. Plastics* xv. 593 In a single-cavity mold, the sprue usually leads directly to the cavity, and there is no runner. **1979** L. J. Boucher *Comprehensive Rev. Dentistry* xxiii. 640/2 The sprues should be large enough that the molten metal in them will not solidify until after the metal in the casting has solidified.

b. A piece of metal or plastic attached to a casting, having solidified in the mould channel; *spec.* a stem joining a number of toys or other small items of moulded plastic.

1875 [in Dict.]. **1911** G. H. Wilson *Dental Prosthetics* ix. 353 The investment is broken and brushed away from the casting... The sprues are snipped away and the edges filed. **1940** M. G. Swenson *Complete Dentures* xxxiii. 576 The sprues..are removed by cutting with a carborundum disc or a jeweler's saw. **1966** *Airfix Mag.* June 308/1 Stretched sprue comes in very useful for numerous jobs. It is exceptionally rigid as compared with wire. **1971** W. R. Matthews *Plastic Scale Model Aircraft* 72 Some practice is needed to master the technique and initially the failure rate is high—the plastic sprue just pulls in two. **1977** *Vole* I. 16/1 The sprue is the little plastic tree to which the functional bits of plastic are attached as in an aeroplane modelling kit.

c. An object used to form such a channel, e.g. by being withdrawn or by melting.

a **1877** Knight *Dict. Mech.* III. 2292/2 *Sprue*, a piece of metal or wood used by a molder in making the ingate through the sand. **1908** G. V. Black *Operative Dentistry* II. 353 The hole left connecting the mold with the little crucible by the removal of the sprue is too small for even the very highly heated gold to run through by its own weight. **1979** L. J. Boucher *Comprehensive Rev. Dentistry* xiv. 388/2 The wax sprue melts at the same time as the wax in the pattern.

2. *Comb.* **sprue former** *Dentistry*, (*a*) a wax channel for guiding moulding material into a mould; (*b*) = sense 1 c above; (*c*) a tube or syringe for producing a wax channel.

1911 G. H. Wilson *Dental Prosthetics* x. 344 The completed wax model is prepared for flasking by attaching gate or sprue formers. **1920** *Ibid.* (ed. 4) x. 406 To operate the sprue former remove the screw plunger and fill the barrel with the desired wax. **1930** S. S. Jaffe in I. G. Nichols *Prosthetic Dentistry* xxx. 537 Make five sprue formers about the size of matchsticks. **1936** E. W. Skinner *Sci. Dental Materials* xxix. 252 After obtaining the wax pattern, it is used for preparing a casting mold. A small length of wire, called the sprue former, is attached. **1961** J. N. Anderson *Applied Dental Materials* (ed. 2) xv. 141 The pattern may be removed by attaching a metal sprue former and then withdrawing the wax in a straight line away from the cavity. **1975** Brewer & Morrow *Overdentures* xiv. 197/1 The stainless steel sprue former is placed in the impression in the hydrocolloid formed by the cone attached to the master cast.

sprue, *v. Dentistry.* [f. Sprue *sb.*[2]] *trans.* To furnish with a sprue or sprues.

1943 F. A. Peyton *Restorative Dental Materials* vii. 246/2 (*heading*) Spruing the pattern. **1949** *Brit. Dental Jrnl.* LXXXVI. 114/1 The wax pattern is thoroughly cleansed... It is sprued, the reservoir added and mounted on a rubber crucible former. **1963** J. Osborne *Dental Mechanics* (ed. 5) xiv. 319 If the pattern has been sprued from above then the mixed investment is coated over the pattern and sprues. **1975** Brewer & Morrow *Overdentures* xiv. 201/1 After completion of the wax-up, the case is sprued.

So **sprue·ing** *vbl. sb.*, provision of sprues.

1930 S. S. Jaffe in I. G. Nichols *Prosthetic Dentistry* xxx. 537 This method of 'sprueing' is only suitable for a centrifugal casting machine. **1979** L. J. Boucher *Comprehensive Rev. Dentistry* xxiii. 640/2, 8 to 12 gauge round wax forms are usually used for multiple sprueing of removable partial denture castings.

sprued (sprūd), *a. Dentistry.* [f. Sprue *sb.*[2] + -ed[2].] Furnished with a sprue or sprues.

1930 S. S. Jaffe in I. G. Nichols *Prosthetic Dentistry* xxx. 538 Place the waxed and sprued model on the mix.. and allow it to set. **1963** C. R. Cowell et al. *Inlays, Crowns, & Bridges* v. 46 The sprued pattern must be invested immediately because delay increases the chances of distortion due to the release of strains in the wax. **1979** L. J. Boucher *Comprehensive Rev. Dentistry* xxii. 579/1 The sprued pattern should be placed one-quarter inch from the end of the casting ring.

spruik (sprŭk), *v. Austral.* and *N.Z. slang.* [Of unknown origin.] *intr.* Esp. of a showman: to deliver a speech, hold forth, speak in public.

1916 C. J. Dennis *Songs of Sentimental Bloke* 42 'E'll sigh and spruik, an' 'owl a love-sick vow. **1934** V. Palmer *Swayne Family* 250 Wonder you didn't get a job spruiking for the pictures down there in town. **1941** K. Tennant *Battlers* xiii. 143 The ampster's is an easy job. He stands in the front row of the listening crowd registering intense interest and enthusiasm while the showman 'spruiks'. **1955** D. Niland *Shiralee* 106 Kelly was chanting the count, banging the big drum, lining the fighters up on the board again, spruiking to the crowd. **1963** *Truth* (Wellington, N.Z.) 15 Oct., Announcers who spruik just for the sake of hearing themselves need to be mighty sure of their facts. **1975** H. Porter *Extra* 244 Hollow-chested men.. who sell agitated toys on street corners or spruik outside strip-tease joints.

Hence **sprui·ker**, a speaker employed to attract custom to a sideshow, a barker; a public speaker.

1924 *Truth* (Sydney) 27 Apr. 6 *Spruiker*, a speaker. **1933** *Bulletin* (Sydney) 2 Aug. 10/4 General Blamey's somewhat ill-advised efforts to prevent political spruikers addressing the Friday-night crowds. **1952** J. Cleary *Sundowners* 154 Rupe sounds like a spruiker from the Domain. **1959** J. Wright *Generations of Men* xvi. 203 Spruikers dressed in red flannel shirts..shouted and gestured in hoarse invitation. **1977** C. McCullough *Thorn Birds* v. 93 'Come on, chaps, who'll take a glove?' the spruiker was braying. 'Who wants to have a go?'

spruit. (Earlier examples.)

1832 *Graham's Town Jrnl.* 13 Apr. 62 They were joined by Lieut. Warden and his party.., who had been ordered to come over the back of the mountains and rendezvous at the *Spruits* of the Keiskamma. **1850** N. J. Merriman *Jrnl.* in *Kafir, Hottentot & Frontier Farmer* (1854) 85 In two days more we found ourselves, just after sundown, at Carl Spruit, (a small stream,) about ten miles this side of Bloemfontein.

sprung, *ppl. a.*[1] Add: **5.** *sprung rhythm*, a term coined by Gerard Manley Hopkins (1844–89) for a poetic metre used by him which approximates to the rhythm of speech and in which each foot consists of one stressed syllable either alone or followed by a varying number of unstressed syllables; hence applied to verse, etc., using this metre.

1877 G. M. Hopkins *Lett. to R. Bridges* (1955) 43 The *Deutschland*, though in sprung rhythm, is marked with accents. *Ibid.* 45, I do not of course claim to have invented *sprung rhythms* but only *sprung rhythm*; I mean that single lines and instances of it are not uncommon in English. *Ibid.*, The choruses in *Samson Agonistes* are intermediate between counterpointed and sprung rhythm. In reality they are sprung. *Ibid.* 46 Why do I employ sprung rhythm at all? Because it is the nearest to the rhythm of prose,..the least forced, the most rhetorical and emphatic of all possible rhythms. **1879** —— *Let.* 27 Feb./10 Mar. in Hopkins & Dixon *Corr.* (1935) 23, I shd. add that the word Sprung which I use for this rhythm means something like *abrupt* and applies by rights only where one stress follows another running, without syllable between. *c* **1883** —— *Poems* 3 Sprung Rhythm, as used in this book, is measured by feet of from one to four syllables, regularly, and for particular effects any number of weak or slack syllables may be used. *Ibid.* 6 Sprung rhythm..has..ceased to be used since the Elizabethan age, Greene being the last writer who can be said to have recognized it. **1935** W. B. Yeats *Let.* 16 Dec. (1954) 844, I think of writing for the first time in sprung verse (four stresses) with a certain amount of rhyme. **1940** H. S. Walpole *Roman Fountain* xii. 169 'But the metre —' 'This is sprung verse and you can run as many syllables into a line as you please.' **1961** *Listener* 23 Nov. 863/1 The myth is created in strikingly personal terms—a kind of sprung verse, for example, developed [by Isaac Rosenberg] quite independently of Hopkins. **1977** J. Milroy *Lang. of G. M. Hopkins* v. 116 Many 'sprung' lines can be quoted in which some of the metric feet have only one strongly stressed syllable.

b. *transf.* in *Mus.*

1944 M. Tippett *String Quartet No. 2 in F sharp* (verso front cover), The 4th *movement* needs a decisively sprung rhythm on which virtually the whole movement is based. **1959** *Listener* 8 Jan. 80/1 *The Heart's Assurance* was the first work in which Tippett's long-breathed melody and sprung rhythm flowered into what one might call creative ornamentation. **1975** *Gramophone* Jan. 1313/3 Even an early work like the Double Concerto.. whose characteristic 'sprung-rhythm' style, derived from the Elizabethan and Jacobean madrigal,..demands from the string orchestra a litheness and an emancipation from the bar-line that were not common in those days.

6. Produced unexpectedly in order to disconcert (in quots., of an alibi which has been 'sprung' upon the prosecution in a court of law). Cf. Spring *v.*[1] 21 d.

1966 *Listener* 1 Sept. 301/1 The sprung alibi in which the defence suddenly at the last minute..produces an alibi which has not been heard of before. **1973** Smith & Keenan *Eng. Law* (ed. 4) ii. 22 The abuse of 'sprung' or late alibis had been so widespread in criminal trials that Sect. 11 of the Criminal Justice Act, 1967, provides that, in general, notice of alibi must be given in advance of a trial on indictment. This is not required in summary trials because of the ease with which the prosecution can ask for an adjournment where the defendant 'springs' an alibi on the prosecution at the last moment.

sprung, *ppl. a.*[2] [Irreg. f. Spring *v.*[2] by analogy with Spring *v.*[1], Sprung *ppl. a.*[1] The expected regular form, *springed*, is rare.] **1.** Provided with a spring or springs. Also *Comb.*, as *sprung-edge(d)* adj.

[**1884**: see Spring *v.*[2] 4.] **1909** *Westm. Gaz.* 11 Nov. 5/1 An entirely new form of suspension is to be introduced which is claimed not only to render the best sprung car extremely coarse in comparison, but which does away with springs altogether. **1916** 'Boyd Cable' *Action Front* 146 The longer but smoother journey in the sweetly-sprung motor ambulance. **1932** *Daily Tel.* 8 Oct. 19/6 (Advt.), 50 luxuriously sprung easy chairs in various coverings. **1948**, etc. [see *interior-sprung mattress* s.v. *Interior *a.* 3]. **1976** *Star* (Sheffield) 29 Oct. 3/3 (Advt.), 3 piece 10 cushioned traditional suite in high quality Dralon, incorporating sprung-edged seating. **1976** *Evening Standard* 29 Dec. 6 (Advt.), Pocketed spring mattress. Sprung-edge base.

2. Of a floor, esp. a dance floor: constructed so as to be resilient. Cf. *spring floor* s.v. Spring *sb.*[1] 25 a.

1966 J. B. Priestley *Salt is Leaving* xv. 202 Lights all colours. Spring floor. Five-piece band. **1972** G. Hale *Floors* ix. 91 *Spring floors*, floors intended for dancing should have a degree of resilience and for this reason ballroom floors..are often fully sprung. **1978** *Daily Tel.* 31 Oct. 19/5 The hotel's sprung dance floor still exists.

spud, *sb.* Add: **3. f.** *Forestry.* A chisel-like implement used to remove the bark from timber.

1914 Moon & Brown *Elem. Forestry* 383 *Spud*, a tool for removing bark. **1919** N. C. Brown *Forest Products* iii. 68 The spudder..proceeds to peel off the bark by inserting the spud between the bark and the wood, and gradually pries it off. **1966** A. E. Wackerman et al. *Harvesting Timber Crops* (ed. 2) viii. 198 Some [Redwood bark] is still removed by hand with spuds, but hydraulic methods are also in use. Mechanical power spuds are also used, as are tractors with spud attachments similar to dozer blades.

g. = *spade lug* s.v. *Spade *sb.*[1] 5.

1917 *Proc. Inst. Automobile Engineers* XII. 80 A machine with a 30 cwt. axle loading and short spuds pulls partly by adhesion and partly by grip. **1933** Watson & More *Agriculture* (ed. 3) i. v. 157 A disadvantage of wheeled tractors is that the spuds or bars that are needed on the wheels for soft land are damaging to roads. **1950** C. Davies *Mechanized Agric.* vi. 39 There are many designs of this [skeleton wheel]—as, in fact, there are of wheel lugs, spuds and cleats.

h. *Thatching.* (see quots.)

1939 H. J. Massingham *Country Relics* i. 13 The equipment is the same when a roof is rethatched, except for the addition of a thatching spud still used by the older thatchers... It is a light, squarish slab of wood..with one end chamfered on both sides to an edge and an oblong slit cut out at the other end through which the fingers pass when the tool is grasped. It is thrust up into the old thatch at the eaves in order to make an opening for the new yealm to be made fast at its thin end. **1968** J. Arnold *Shell Book of Country Crafts* 185, I watched one man.. driving home his pegs with a flat oblong board: this had a spade-like handle and was called a 'spud'. **1972** [see *Legget].

5. (Earlier example.)

1845 E. J. Wakefield *Adventure in N.Z.* I. xi. 319 Pigs and potatoes were respectively represented by '*grunters*' and '*spuds*'.

b. = *Potato *sb.* 5 d.

1960 I. MacCormick *Small Victory* 86 Are you any good at darning socks, Sister? I got a *spud* in my spare pair. **1978** M. de Larrabeiti *Rose beyond Thames* 34 There were huge spuds in the heels of their socks.

6. Each of a number of poles that can be put out from a dredger and stuck into the bed or bank of the river so as to keep the vessel stationary.

1891 F. Wyatt *Phosphates of America* iv. 51 The boats are held in position at the four corners by 'spuds' or strong square poles with iron points, which are dropped into the water before dredging is begun. **1912** C. Prelini *Dredges & Dredging* xx. 166 The bow spuds are of great size, 43 in. square and 52 ft. long. **1925** G. W. Pickels *Drainage & Flood-Control Engin.* xiv. 397 Spuds are devices for holding the hull still while the dredge is working. There are three of them, one at the center of the stern and one on each side near the front end. The rear spud is always vertical and consists of a heavy timber with a pointed iron shoe at its lower end which moves in a box or guide frame attached to the hull. When lowered, the pointed end is forced into the bottom of the ditch or river and prevents the rear end of the hull from moving. The spud is raised by a cable operated by the spud engine. The side spuds may be either vertical or inclined. *Ibid.* 398 Inclined spuds are fastened to the A-frame and their lower ends rest on the banks, hence they are called bank spuds. **1969** R. Hammond *Mod. Dredging Pract.* iv. 136 The spuds are provided with sliding collars in order to allow for the wide range of dredging depths.

7. *Plumbing.* A short length of pipe used as a connecting piece between two components or taking the form of a projection from a fitting to which a pipe may be screwed.

1905 *Internat. Libr. of Technol.* LXXI. §15. 68 The end..of the spud is threaded iron-pipe size for screwing into the tapping of an ordinary iron kitchen boiler. **1907** R. M. Starbuck *Mod. Plumbing Illustr.* 123 Ventilation.. consists in connecting a pipe from the local vent spud on the water-closet bowl to a heated flue. **1908** A. G. King *Practical Steam & Hot Water Heating & Ventilation* xi. 105 The branches should be one size larger than the vertical pipe or 'spud' supplying the radiator valve, or one size larger than the riser which they feed. **1939** [see *spud wrench* in sense 8 below]. **1972** J. Hastings *Plumber's Compan.* 184 The 'spud connection' is a brass fitting comprising a socket on one end for receiving the nozzle and on the other end a male thread for connecting a pipe.

8. *attrib.* and *Comb.*, as **spud barber** *slang*, one who peels potatoes; **spud-bashing** *slang* (orig. *Mil.*), the peeling of potatoes; **spud can** *Oil Industry*, a structure that can be sunk into a soft sea-bottom by temporary ballasting and then used as the base of a tower platform extending above the surface of the water; **Spud Islander** *Canad. slang*, a native or inhabitant of Prince Edward Island, which is noted for its fine potatoes; **spud line**: in slang phr. *in the spud line*, pregnant; **spud wrench** (see quots. 1939, 1960[1]).

1935 A. J. Pollock *Underworld Speaks* 112/2 *Spud barber*, a potato peeler (prison). **1961** G. Foulser *Seaman's Voice* 48 The galley-boy [was] just a spudbarber after all. **1940**, etc. Spud-bashing [see *Bashing *vbl. sb.* 3]. **1980** *Times* 13 Nov. 10/1 Between dashing home from the office..and having a bath, there is not much time for spud bashing. **1975** *Offshore* Sept. 49-04/3 Spud cans are designed for soft seabeds, giving minimum penetration. **1976** *Offshore Platforms & Pipelining* 53/3 The tower base consists of a truss-reinforced stiffened shell called a spud can... After the tower is uprighted, the spud can is artificially forced into the ocean bottom until the desired load-carrying capability is reached. **1957** *Globe Mag.* (Toronto) 29 June 4/2 (*caption*) Spud Islanders are known throughout Canada for the quality of their potatoes. **1962** G. MacEwan *Blazing Old Cattle Trail* xxxiii. 222 In 1900, the twenty-four-year-old 'Spud Islander' journeyed to the far Northwest. 1937 J. Curtis *There ain't no Justice* xxiv. 245 You mean she's in the spud line? 1967 H. W. Sutherland *Magnie* vi. 80 It couldn't have been himself that put Kathleen Ertall in the spud line. 1939 W. T. Walters *Steam & Hot Water Fitting* x. 145 Spud Wrenches. This type of wrench is another handy tool. It is made to fit the spuds of the different sizes of the union radiator valves and traps. 1960 S. P. Oppenheimer *Erecting Structural Steel* vii. 93 The open-end spud wrench ..is the most common... It is so called because the handle is formed into a long, heavy pin (or spud) that is thrust through and used by the erection men for matching up holes in connections to be fastened together. 1960 D. A. Halperin *Building with Steel* xii. 163/1 It is permissible to tighten the bolts with a long handled socket or spud wrench.

spud, *v.* **3. a.** Substitute for def.: To begin to drill (a hole for an oil well) by imparting an up-and-down motion to the drilling bit. Now usu. more widely, to drill (a well) through the upper part of the overburden; also *absol.* Freq. const. *in* and occas. written *spud-in*. Also with *out*. (Further examples.)

1913 V. B. Lewes *Oil Fuel* 64 If the hole is not deep enough, it has to be 'spudded out' to the necessary depth. 1924 *Bull. Amer. Assoc. Petroleum Geologists* VIII. 643 The driller, with his hand on the brake and familiar with the action of his machinery and pump since the well was 'spudded' in, is by far the best judge of the formation in which he is drilling. 1928 *Publ. Texas Folk-Lore Soc.* VII. 59 He had a 100,000 barrel gusher and was spudding in on another location. 1948 *Sun* (Baltimore) 16 Apr. 12/1 Drillers spudded in the first well of the big Leduc field in November, 1946. 1966 *Southern Reporter* CLXXX. 746/2 Substantial surface preparations to drill are sufficient to be considered 'commencement' of drilling operations for lease-clause purposes,..provided that such preliminary operations are continued..until well is actually spudded in. 1967 *Economist* 18 Nov. 788/2 The company has a world-wide business instrumentation for well-drilling.., whenever 'wildcats are spudded'. 1975 *BP Shield Internat.* May 1/2 BP's drilling contractors.. will spud and drill to completion the first Forties well. 1977 *Irish Times* 8 June 10/3 The Deminex consortium yesterday disclosed that it spudded-in its exploratory well in block 34/15, off the west coast, on Friday last.

spudder² (spʌ·dəɹ). [f. Spud *v.* + -er¹.] A small drilling rig used for spudding.

1922 L. C. Sands in D. T. Day *Handbk. Petroleum Industry* I. 250 These lighter outfits generally manipulate the drilling tools by the spudding principle and are often called 'spudders'. 1935 *Discovery* Apr. 119/1 The see-saw motion of the [rocking] 'beam'..transmits to the tool a motion very similar to that of the spudder. 1960 *Pacific Reporter* CCCLIV. 444/2 Plaintiff..purchased four small standard drilling rigs (commonly referred to as 'spudders') at an aggregate cost of $15,750. 1977 R. D. Langenkamp *Handbk. Oil Industry Terms & Phrases* (ed. 2) 160 Spudders are used in shallow-well workovers, for spudding in, or bringing in a rotary-drilled well.

spudding, *vbl. sb.* (In Dict. s.v. Spud *v.*) Add: **1.** *spec.* The preliminary drilling of a well-hole through the upper part of the overburden (see *spud *v.* 3 a). Freq. as *spudding in*, *spudding-in*. (Earlier and further examples.)

1885 *Engineering* 26 June 708/2 The bore is usually started by what is termed spudding—*i.e.*, working the tools direct from the Bull wheel, lifting and dropping by slacking or surging the rope on the wheel. 1922 A. Blum *Petroleum* III. i. 162 The hole is started with a bit, eight to twenty inches across the cutting edge, the size depending upon the proposed depth of the hole... This is called 'spudding in'. 1929 Babbitt & Doland *Water Supply Engin.* vii. 160 The well hole is usually started by 'spudding', which consists in drilling without the walking beam because of the shorter length of tools and the lack of 'stretch' in the rope. 1967 *Sun* 22 Sept. 8/5 Exploratory drilling starts this weekend with the 'spudding-in' of a well 15 miles off the Lincolnshire coast. 1974 *Bull. Amer. Assoc. Petroleum Geologists* LVIII. 1124/2 The present drilling program began in 1971 with the spudding of the Eider well.

2. Special Comb.: **spudding bit,** a drilling bit used in spudding.

1907 *Internat. Libr. Technol.: Rock Boring* 16 The spudding bit in Fig. 16 is for drilling earth down to solid rock. 1924 L. C. Uren *Textbk. Petroleum Production Engin.* v. 125 A special spudding bit is used which is shorter than the usual pattern. 1939 D. Hager *Fund. Petroleum Industry* ix. 225 With the Lucy spudding bit a cement plug is set at the bottom of the hole, and the bit, by being raised or lowered, digs a hole several feet deep into the side of the hole.

spugslang, var. *spuugslang.

spumante (spumæ·nti). Also erron. **spumanti.** [It., = 'sparkling'.] A sparkling (usu. sweet) white wine from Asti in Piedmont; in full, *Asti Spumante*.

1908 C. E. Hawker *Chats about Wine* 89 The Frenchman with his Sillery..The Italian with his Asti Spumante, all favour sweetness. 1938 C. Isherwood *Lions & Shadows* i. 37 To celebrate the achievement, Mr. Holmes has ordered Asti Spumanti. What a marvellous drink!..So cool and fizzy and sweet. 1945 *Comment from Italy* (Three Arts Club) 22 It was beer issue night, and most of the groups had reinforced it with bottles of vino, vermouth, and spumanti as well. 1958 L. Durrell *Balthazar* III. x. 206 Pursewarden was inveighing somewhat incoherently against the Cervonis for serving Spumante instead of champagne. 1971 *Country Life* 25 Nov. 1466/2 A dry spumante wine. 1982 *Sunday Tel.* (Colour Suppl.) 7 Mar. 38/3 The Spumanti was flowing in the officers' mess, everyone cheering.

spume, *sb.* Add: **3.** *spume-bow*.

1921 W. de la Mare *Veil* 15 With myriad spume-bows roaring ocean swills The cold profuse abundance of the rain.

spumed (spiūmd), *ppl. a.* [f. Spume *sb.* + -ed².] Flecked as with spume.

1923 D. H. Lawrence *Birds, Beasts & Flowers* 60 Cyclamen leaves..Frost-filigreed Spumed with mud.

spumoni (spŭmōᵘ·ni). *U.S.* [ad. It. *spumone* (used in the same sense). f. *spuma* foam, Spume.] A kind of ice-cream dessert (see quot. 1950).

1929 M. Lief *Hangover* v. 83 Mogador picked up a spoon and helped himself to some of the star's spumoni ice-cream. 1936 J. dos Passos *Big Money* 126 After they'd had their spumoni Mr. Barrow ordered himself brandy. 1950 Fransden & Nelson *Ice Cream & Other Frozen Desserts* xvii. 177 A special spumoni cup should be used. Press one-fourth of a spumoni cupful of vanilla ice cream around the sides and bottom to line the cup. Add chocolate ice cream... Finish filling with a mixture of fruit and whipped cream. 1968 P. Durst *Badge of Infamy* iv. 37 In addition to the minestrone and pizza there was a green salad..spumoni and coffee. 1977 J. Wambaugh *Black Marble* (1978) xi. 246 Had her little two-year-old spumoni sucker *in the car* when I picked her up.

spun, *ppl. a.* Add: **1. a.** Also of normally rigid materials, as glass. Also Comb. *spun-dyed* adj., of materials coloured during spinning. (Further examples.)

1779 Spun glass [in Dict.]. 1899 *Jrnl. Soc. Arts* XLVIII. 62/1 Spun glass is probably the earliest production which resembles natural silk. 1936 H. Miller *Black Spring* 192 Suddenly like spun-glass under a blue flame, the street quickens into tongues of fire. 1940 *Engineering* 1 Nov. 343/3 The advantages of such spun-concrete products as pipes and poles are well known. 1947 J. C. Rich *Materials & Methods of Sculpture* iv. 64 Spun-copper bowls are highly recommended for use in mixing plaster of Paris. 1980 *Nature* 14 Feb. p. xiii/2 The body is a spun steel bowl and lid on die cast iron base.

Comb. 1955 Cockett & Hilton *Basic Chem. Textile Colouring & Finishing* i. 34 Spun-coloured or pigmented fibres. These are sometimes known as spun-dyed fibres, although the method of colouration is not a dyeing process in the usual sense. 1975 R. H. Peters *Textile Chem.* III. i. 4 Other ways of colouring with insoluble compounds are to introduce them in fibre manufacture, i.e. in the polymer solution prior to extrusion giving 'spun-dyed' filaments.

b. *spun gold* (later *attrib.* example). Hence *spun-golden* adj.

1966 T. Pynchon *Crying of Lot 49* i. 21 Frail girls with heart-shaped faces, huge eyes, spun-gold hair. 1978 S. Sheldon *Bloodline* ii. 40 She had spun-golden hair and skin as delicate as porcelain.

c*. Applied to vegetable protein, esp. soya, that has been spun into fibres so as to resemble meat and to the meat substitutes made from it.

1973 *Guardian* 21 Apr. 12/3 'Spun steak.' Looks like steak..only it is made of spun soya. 1976 T. Heald *Let Sleeping Dogs Die* v. 81 Canine comestibles... Doggy buckwheat..spun protein tripe.

2. (Earlier example.)

1869 K. H. Digby *Little Low Bushes* 244 Carheil is like a long and spun-out speech.

spunk, *sb.* Add: **5. c.** *coarse slang.* Seminal fluid.

For the sense development, compare the obs. slang *mettle*, which had the same meaning.

c 1888–94 *My Secret Life* I. 87 It seemed to me scarcely possible, that the sweet, well dressed, smooth-spoken ladies..could let men put the spunk up their cunts. 1896 A. Beardsley *Let.* 21 Dec. (1970) 236 She played dirty tricks... Till the spunk trickled right down her femur. 1923 J. Manchon *Le Slang* 289 *Spunk*,..2° courage, vivacité, feu; on dit de préférence *mettle*, parce que 3° O[bscène] (= *come*, s.). 1971 B. W. Aldiss *Soldier Erect* 196 By sprawling right back on the seat and ignoring the stink of the shit-pit below me, I managed in no time to lob some spurts of spunk over my stomach, with some relief. 1978 J. Irving *World according to Garp* ii. 32 The boys were beating off, in turn, and rushing with their hot spunk in their hands to the microscopes in the infirmary lab—to see if they were sterile.

6. spunk-water *U.S.*, rain-water that collects in hollow tree-stumps, popularly thought to be a cure for warts.

1876 [see *blame *v.* 7 d]. 1949 *Time* 29 Aug. 7/2 Spunk-water, spunk-water, wash away my warts!

spunkily, *adv.* Add: Also spiritedly, courageously. *U.S.*

1949 J. Flanner in *New Yorker* 9 Apr. 58/2 A hundred artists, from all over Italy,..are spunkily giving their first show this month. 1976 *Publishers Weekly* 21 June 88/2 Judith hurries out to be with her sister and spunkily unravels the mystery.

spun-yarn, spunyarn. Add: **2.** (Later *Comb.* examples.) *spun-yarn major*, *spun-yarn trick* (slang): see quots. 1929, 1925.

1916 *In Northern Mists* xvi. 63 The practice has been known of getting everything ready the night before and the proper fastenings replaced with pieces of spunyarn which can be cut with a sailor knife as soon as the signal is made. Of course, it is not playing the game: that's a 'spunyarn trick'. 1925 Fraser & Gibbons *Soldier & Sailor Words* 267 *Spun yarn tricks*, underhand dealing. 1929 F. Bowen *Sea Slang* 132 *Spun-yarn major*, a lieutenant-commander in the Navy. 1942 'Sea Wrack' *Six Bells* 163 Above her head sounded the drumming footsteps of Tommy and his crew of spunyarn majors as they clattered round the decks securing stays.

spur, *sb.*¹ Add: **I. 2. d.** (Earlier example of (*a*).) Hence *spur-of-the-moment* attrib. phr.

1801 *Ann. Reg. 1799* II. 27/1 Volunteers, with a party of the Surrey cavalry, attended and prevented the populace in general from taking that step, which, perhaps, the best feelings of human nature had, upon the spur of the moment dictated. 1948 C. Day Lewis *Otterbury Incident* viii. 94 Toppy is tops at spur-of-the-moment tactics. 1958 C. Williams *Man in Motion* (1959) vii. 77 There's no such thing as a spur-of-the-moment suicide. 1978 M. Puzo *Fools Die* xv. 161 Junkies, alcoholics, amateur pimps, small-scale thieves and spur-of-the-moment rapists.

II. 6. d. (Later examples in calligraphy and typography.)

1912 E. M. Thompson *Introd. Greek & Latin Palaeography* xix. 521 Long strokes [of letters]..are occasionally provided with an ornamental spur near the top of the vertical stems. 1976 *Visible Language* X. 44 *Spur*, a small projection, usually pointed, from a stroke or terminal.

IV. 12. a. *spur-clink*.

1911 E. Pound *Canzoni* 43 The silken trains go rustling, The spur-clinks sound between.

14. spur blight, a fungus disease of raspberries and loganberries causing discoloured patches on the stems and the death of buds at the nodes, and weakening the laterals; **spur line,** a railway branch-line; **spur mark,** one of the marks left on the base of a glazed pot by the spurs (sense 8 c) on which it has rested during firing (see also quot. 1933); **spur road** (later examples); now applied to a connecting road that branches off from a motorway or main highway.

1915 *Bull. Colorado Agric. Exper. Station* No. 206 (*title*) Spur blight of the red raspberry caused by Sphaerella rubina. 1941 *Sun* (Baltimore) 18 Feb. 6/4 Anthracnose, spur blight on red raspberries and cane blight can be controlled by promptly removing and burning all old fruiting canes after harvest and spraying the new canes one to four times in the growing season. 1979 Scopes & Ledieu *Pest & Disease Control Handbk.* vi. 55 Some of the fungicides for spur blight will also control cane spot. 1924 Kipling *Debits & Credits* (1926) 166 'E 'ad us all screened in over in a cuttin' on a little spur-line. 1977 H. Fast *Immigrants* 8 This great railroad..has begun the construction of a spur line to connect its main line with the City of San Francisco. 1895 R. Mills *Catal. Blue & White Orient. Porc.* 52 'Spur marks' are little projections of the paste, apparently to prevent the bottom of the vessel touching the oven. They are peculiar to Japanese porcelain. 1933 *Burlington Mag.* Oct. 160/2 On the bottom of the interior of the bowl will be found five spur marks where the pontil was broken off. 1972 *Trans. Oriental Ceramics Soc.* XXXVIII. 23 A ring of spur marks usually is to be seen on the base of proto-Yüeh pieces. 1958 *Times* 19 Mar. 10/3 The spur road to London Airport will be served by a flyover junction. 1963 *Listener* 31 Jan. 198/2 Heavy trucks rumble along the spur roads from the Alaska Highway to mining camps deep in the frozen interior. 1977 *Jrnl. R. Soc. Arts* CXXV. 359/1 These needs could be met by a six-lane spur road connecting with adjacent motorways.

14. b. spur-dog, a small spiny shark, *Squalus acanthias*, found in the Atlantic and the Mediterranean; (later examples).

1921 [see *nurse-hound* s.v. *nurse *sb.*² b]. 1959 A. C. Hardy *Open Sea* II. ix. 179 The spur-dog can easily be distinguished by the prominent spine immediately in front of each of its dorsal fins. 1976 *Evening Post* (Nottingham) 13 Dec. 7/2 Other trophies for the best specimen fish went to..Eric Rawson for an 8½ lb. spur dog, [etc.].

spur, *v.*¹ Add: **I. 2.** Also occas., with an action or activity as object.

1951 *Newsweek* 27 Sept. 74/3 Much of this expansion has been spurred by the government. 1976 *National Observer* (U.S.) 17 Apr. 7/2 Mae Craig,..Liz Carpenter, and I spurred a move for the survivors of the Spruce Goose week end to entertain our millionaire host with an appreciation party.

spuria (spiūᵊ·riă). *sb. pl.* [L., neut. pl. of *spurius* spurious: cf. *trivia *sb. pl.*] Spurious works, words, etc.

1918 E. Marsh *Rupert Brooke* 110, I hope this note will not start a vain hunt for *spuria* among the published poems. 1935 *Mind* XLIV. 105 There is a record of posthumously published remains and an interesting section on spuria. 1977 *Times Lit. Suppl.* 22 Apr. 496/4 The Platonic corpus (including spuria and doubtful works).

spurling² (spɒ̈·ɹliŋ). *Naut.* [Origin unknown.] Only in Combs., as **spurling gate** (see quot. 1927); **spurling pipe** (see quot. 1962). Cf. Spurling-line.

1927 G. Bradford *Gloss. Sea Terms* 169/2 *Spurling gate*, the iron casting set in the deck through which the anchor chain passes. 1938 F. A. Worsley *First Voyage* 217 The spurlingate pipes were now uncovered and the ends of the chain cables roused up. 1962 A. G. Course *Dict. Nautical Terms* 186 *Spurling pipe*, the pipe that encloses the anchor cable from windlass to cable locker.

‖ **spurlos versenkt** (ʃpū·rlōs fɛrze·ŋkt), *adj. phr.* [Ger., = sunk without trace.] Sunk

without trace; usu. *fig.*, done for, lost from sight. Hence occas. (with partial translation) *sunk spurlos*. Cf. *to sink without trace* s.v. *sink *v.* 1 a.

The phrase became widely known as a result of the publication in September 1917 of a secret telegram sent in May of that year by Count Luxburg, the German minister in Buenos Aires, to Berlin, advising that Argentine shipping should be either turned back or sunk without trace.

1918 W. H. Allen in *Stories of Americans in World War* 24 He was merely seeking an excuse for the inhuman conduct he planned. The order 'spurlos versenkt'—to be sunk without leaving a trace—was to be obeyed. 1922 A. Huxley *Mortal Coils* 92 You're done for..sunk—spurlos. 1928 H. Crane *Let.* 23 Oct. (1965) 329, I haven't had a creative thought for so long that I feel quite lost and *spurlos versenkt*. 1930 W. F. Sands *Undiplomatic Mem.* xv. 225 My plan for neutralization of Korea..was sunk without trace, completely *spurlos*. 1946 W. S. Churchill in *Compl. Speeches* (1974) VII. 7337 He has departed '*spurlos versengkt*' [*sic*] as the German expression says—sunk without leaving a trace behind. 1963 *Oxf. Med. School Gaz.* XV. 23 Many distinguished research workers have ceased production when they have become professors and like submarines in dangerous waters have been *spurlos versenkt*.

spur-of-the-moment *attrib. phr.*: see *spur *sb.*¹ 2 d.

spurrite (spɒ̈·rəit). *Min.* [f. the name of Josiah E. *Spurr* (1870–1950), U.S. geologist + -ite¹.] A monoclinic carbonate and silicate of calcium, $(Ca_2SiO_4)_2.CaCO_3$, which is a product of the contact metamorphism of limestone and usu. occurs as pale grey crystalline masses.

1908 F. E. Wright in *Amer. Jrnl. Sci.* CLXXVI. 551 Spurrite occurs in the hand specimens either in pure, unaltered state, except for minute inclusions of magnetite, or together with yellow garnet, calcite and gehlenite. 1938 *Mineral. Mag.* XXV. 38 Spurrite is..a constituent not only of the silicated zone of the contact but also occurs isolated in calcite in the marbles derived from the chalk. 1968 I. Kostov *Mineralogy* 304 The minerals of the group are contact-metamorphic products in limestones, spurrite being a characteristic high-temperature mineral.

spurt, *v.*³ Add: **1. b.** Of stocks and shares: to rise suddenly in price or value. Cf. Spurt *sb.*¹ 2 d.

1931 *Economist* 27 June 1385/2 Dunlops and Imperial Chemicals spurted on bear closing. 1977 *Belfast Tel.* 19 Jan. 4/1 Beecham 406p spurted 10p to 15p among top industrials. 1982 *Times* 27 Apr. 15/2 Building contractor J. Jarvis spurted 41p to 341p in response to a dawn raid.

spurtlet (spɒ̈·ɹtlėt). *rare*⁻¹. [f. Spurt *sb.*¹ or *sb.*³ + -let.] A little spurt.

1921 E. M. Forster *Let.* 28 Sept. in *Hill of Devi* (1953) 132, I annoyed him.., and he had a spurtlet of temper and said I was not his friend.

sputnik (spu·tnik). Also **Sputnik.** [a. Russ. *spútnik*, lit. 'travelling companion', f. *s* with + *put'* way, journey + *-nik*, agent suffix (cf. *-nik).] An unmanned artificial earth satellite, esp. a Russian one; *spec.* (usu. with capital initial) the proper name of a series of such satellites launched by the Soviet Union between 1957 and 1961.

The first Sputnik, launched on 4 October 1957, was the first artificial satellite.

1957 *Times* 9 Oct. 10/6 Pride in the launching of the *sputnik* ('fellow-traveller'), as the satellite is called, as well as the guided missile, were reflected in a speech by Mr. Krushchev..last night. *Ibid.* 30 Oct. 10/2 Mr. Khrushchev replied: 'To peace and to the *sputnik* as a symbol of peace!' *Ibid.* 4 Nov. 11/2 The régime which sends a second Sputnik girdling the earth has just emerged from another of its secretly contrived shifts of political power. 1958 A. Huxley *Let.* 15 Feb. (1969) 846 The technical advances in these psychological, physiological and bio-chemical fields are probably far more important.. than the physical and engineering advances which have put sputniks into the heavens. 1964 M. McLuhan *Understanding Media* iii. 44 When Sputnik had first gone into orbit a schoolteacher asked her second-graders to write some verse on the subject. 1971 *New Scientist* 10 June 638/1 China's remarkable progress in the field is underscored by the weight of its first sputnik (unmatched ..by any satellite launched by France or Japan). 1983 *N.Y. Times* 7 Jan. A1/4 It is not a dangerous situation.. and we have no worries about the fate of this sputnik.

b. *transf.* and *fig.*

1958 *Newsweek* 10 Feb. 25/1 We may find ourselves confronted with a sputnik in the chemical, biological, and radiological field, as we did in missiles. 1959 *Daily Tel.* 10 Dec. 16/7 Internal 'sputniks', pills containing miniature radio transmitters, which can travel around the intestines. 1963 *Punch* 17 Apr. 549/1 Such Hollywood *sputniks* as Frank Sinatra and Sammy Davis Jnr. 1968 [see *loop *v.*¹ 6].

2. *attrib.* and *Comb.*, as *sputnik diplomacy*, *race*, *town*; **Sputnik double** *Bridge*, a take-out double of a suit overcall of one's partner's opening bid; also *absol.* as *Sputnik*.

1957 *N.Y. Times* 20 Oct. IV. 4/1 Since the Soviet space satellite has been in its orbit, Moscow has been showing what many are now calling sputnik diplomacy. 1959 *Listener* 15 Jan. 96/2 The rocket would set the stage for a diplomatic offensive [in Russia] similar to the sputnik diplomacy a year ago. 1958 *Bridge World* July 33/2 We noticed that the negative double ('Sputnik') would fit. 1976 *Country Life* 1 Apr. 846/2 A Sputnik double, if played, will lead to the same final contract. 1957 *Times Lit. Suppl.* 27 Dec. 782/4 America's defeat in the sputnik race. 1959 *Daily Tel.* 27 Apr. 10/2 The political sensationalism of the sputnik race. 1958 *Daily Mail* 7 June 5/5 Russian planners will deal with overspill population from big cities... They are planning 'sputnik' towns. 1966 *Listener* 19 May 729/3 In preparation for the creation of Moscow's own ring of 'Sputnik towns'—though this development may not happen until after 1980.

Hence **spu·tnik(e)ry, sputniki·tis** (*nonce-wds.*).

1957 *Observer* 20 Oct. 14/2 We rang up Hamley's to see how Sputnikitis was hitting them... 'No, I'd not say our space toys were on the up... It's always in competition with cowboys and Red Indians, you see.' 1957 *Economist* 30 Nov. 762/1 The United States and the Soviet Union are pouring in money and scientists in an Antarctic form of sputnikitis. 1960 *Spectator* 10 June 826 The abnormal concentration of effort in such fields as 'sputnikery'. 1961 *New Scientist* 6 July 38/1 The narrower field of sputnikry.

sputter, *sb.* Add: **4.** Special Comb.: **sputter ion pump** *Physics* [perh. f. the vb.], a pump in which the gas is absorbed by a getter that is deposited by sputtering it from a cathode.

1962 *Sci. Amer* Mar. 82/1 (*caption*) Sputter ion pump works by ionizing gas molecules and removing them from the chamber..to be evacuated. 1980 J. F. O'Hanlon *User's Guide Vacuum Technol.* ix. 221 The sputter ion pump has the advantage of freedom from hydrocarbon contamination and ease of fault protecting but does suffer from the reemission of previously pumped gases.

sputter, *v.* Add: **5. a.** Also with adv., to move *away*, come *in*, etc., with a sputtering sound; also *fig.*

1936 L. C. Douglas *White Banners* ix. 200 The taxi sputtered away. 1977 N. Sahgal *Situation in New Delhi* xii. 123 Thank God no one was around as they sputtered in. 1977 *Time* 11 Apr. 36/3 Terrorism sputters on, but Argentines have learned to cope with it, even ignore it.

b. Also with *out*, to sputter and die out (in quots., *fig.*).

1964 D. MacArthur *Reminiscences* vi. 162, I was certain that..his advance would sputter out as it ran ahead of its supply line. 1974 H. L. Foster *Ribbin'* v. 228 In most cases, if teachers would not interfere, these incidents would sputter out.

6. *Physics.* **a.** To remove atoms of (a metal) *from* a cathode by bombarding it with fast positive ions; to deposit (metal removed in this way) *on* another surface.

1910 *Phil. Mag.* XX. 337 A relatively thick film was sputtered on a 1 mm. quartz plate. 1924 *Science* 31 Oct. 392/2 The cathode drop sputters tungsten from the cathode in an amount..between 10^{-8} and 10^{-7} grams. 1949 S. Frankel in J. F. Blackburn *Components Handbk.* v. 182 A gold plating is sputtered on and is baked for at least one hour at 500°C. 1961 *Proc. IRE* XLIX. 1148/2 The gas is found in the metal which has been sputtered from the cathode. 1965 *Wireless World* Aug. 409/1 The positive plasma ions..impinge with sufficient energy to sputter atoms onto an adjacent substrate. 1974 *Sci. Amer.* Apr. 35/1 The film can be sputtered onto the substrate in a vacuum chamber.

b. To cover (a surface) with metal by sputtering.

1910 *Phil. Mag.* XX. 331 Two plates were sputtered simultaneously so as to insure the same thickness for both films. 1962 *Sci. Amer.* Mar. 86/3 The steady hail of ions 'sputters' the surface. 1971 *Physics Bull.* Sept. 554/2 Pure silica glass surfaces have been sputtered by 20 keV argon ions at 0° incidence to the general plane.

sputtered (spɒ·təɹd), *ppl. a.* [f. prec. + -ed¹.] Formed by or resulting from sputtering (see *sputtering *vbl. sb.* 2).

1910 *Phil. Mag.* XX. 337 A second blank quartz plate.. was now rigidly fastened to the blank side of the sputtered plate. 1921 *Physical Rev.* XVIII. 215 Sputtered films are crystalline. 1971 *Physics Bull.* Sept. 554/2 Redeposition of low energy sputtered particles on to adjacent surfaces.

sputterer. For *rare*⁻⁰ read *rare* and add examples.

1881 *Punch* 15 Oct. 174/1 Whilst facts *be* facts, whilst truth its cold *douche* showers, Hydrants & squirts can ne'er be equal powers: One sturdy spout will spoil the sputterers all. 1935 A. J. Pollock *Underworld Speaks* 112/2 *Sputterer*, a person who has just begun opium smoking and smokes a small quantity daily, getting much satisfaction from the drug with the belief that he will never become addicted.

sputtering, *vbl. sb.* Add: **2.** The removal of atoms from a substance subject to bombardment, esp. from a metallic cathode bombarded by positive ions, and usu. with subsequent deposition on an adjacent surface.

1902 *Phil. Mag.* IV. 653 The metallic films..were obtained by sputtering from a cathode *in vacuo* on glass strips. 1930 *Rev. Mod. Physics* II. 186 'Sputtering', or disintegration of an electrode subjected to positive ion bombardment is a well known and often troublesome phenomenon. 1948 L. D. Smullin in Smullin & Montgomery *Microwave Duplexers* v. 210 Sputtering is a process in which the cathode is heated by positive-ion bombardment to the point where particles are boiled out of the cathode and finally condense on the anode or on the tube walls. 1952 [see *ion bombardment* s.v. *ion 2]. 1976 *Sci. Amer.* May 115/3 By the process known as sputtering, the impact of electrons and protons on the surface could chip away atoms and release them into the atmosphere, from which they would quickly escape.

sputum. Add: (See quot. 1973.)

1973 *Lancet* 24 Feb. 420/1 Bacteriologists and exfoliative cytologists often report on a specimen sent as sputum: 'No sputum present, saliva only.' *Ibid.*, Sputum is material brought up from the trachea by the actions of coughing or hawking, or the same material spat out with (usually) an admixture of saliva. (Material hawked up into the pharynx and swallowed is generally referred to as swallowed sputum, so the first half of this definition is necessary—sputum is sputum even if it is not spat out.)

‖**spuugslang** (spū·χslaŋ). *S. Afr.* Also **spoew-, spoog-, spug-, spuw-,** etc. [Afrikaans *spuugslang*, *spoegslang*, f. *spu(ug)*, *spoe(g)* to spit + *slang* snake.] = *ringhals.

1789 W. Paterson *Narr. Four Journeys Country of Hottentots* 165 The Spoog Slang, or Spitting Snake. 1812 A. Plumptre tr. *Lichtenstein's Trav. S. Afr.* I. I. vii. 95 A very rare sort of serpent, called here the *spugslang* (the spurting snake)..is from three to four feet long, of a black colour, and..when attacked it will spurt out its venom. 1911 *Encycl. Brit.* XXV. 299/1 It shares with the cobra a third Dutch name, that of 'spuw slang' (spitting snake). 1923 Kipling *Land & Sea Tales* 34 He gave us half-a-crown for a spuugh-slange—a kind of snake. 1931 R. L. Ditmars *Snakes of World* xiii. 172 Another name [for the ringhals] is Spoew-slang, applied from the reptile's 'spitting' its poison.

spy, *sb.* Add: **1.** Joc. phr. *one's spies*: one's private or unofficial sources of information.

1955 E. H. Clements *Discord in Air* iii. 32 He designed the engine... I got that from my spies. There's no secret about it, anyway. 1965 M. Frayn *Tin Men* v. 31 The Queen's going to pay an official visit... So I'm told by my spies, anyway.

5. *spy-catcher*, *-catching*, *fever*, *-fiction*, *film*, *-government*, *mania* (earlier example), *movie*, *-net*, *network*, *novel*, *play*, *scandal*, *scare*, *school*, *series*, *story*, *system* (earlier example), *thriller*, *trial*, *work* (later example); **spy in the cab** *colloq.* = *tachograph; **spy in the sky,** a satellite or aircraft used to gather intelligence; freq. *attrib.* (with hyphens); **spy-master,** the head of an organization of spies; **spy plane,** an aircraft used to gather intelligence; **spy ring** = *ring *sb.*¹ 11 c; **spy satellite,** a space satellite used to gather intelligence, usu. military; **spy-ship** (later examples).

1952 O. Pinto *Spy-Catcher* i. 7 There is indeed excitement at times in the life of a real spy-catcher. 1976 *Eastern Evening News* (Norwich) 13 Dec. 1/1 Spycatcher.. 'Jock' Wilson, head of London's C.I.D. 1978 Cadogan & Craig *Women & Children First* x. 223 The most irresistible wartime subjects were evacuation and spy-catching. *Ibid.* 225 The spy-catching tales of Dorita Fairlie Bruce. 1973 'D. Halliday' *Dolly & Starry Bird* xvi. 234 The spy fever had spread like foot-and-mouth disease. 1963 *Times Lit. Suppl.* 8 Feb. 92/2 The famous heroes of British spy-fiction..have seldom been professionals. 1942 *N.Y. Times* 13 June 11/2 A tautly intriguing and sometimes hair-raising spy film. 1972 'E. Ferrars' *Breath of Suspicion* iii. 46 It struck Richard, that this must be a spy film. 1929 D. H. Lawrence *Pansies* 127 The big, flamboyant Russia Might have been saved if a pair Of rebels like Anna and Vronsky Had blasted the sickly air Of Dostoevsky and Tchekov, And spy-government everywhere. 1968 J. Kelson in *Headlight* Sept. 15/3 The tachograph is a gadget that has been hotly argued over.. but..against all..obstacles to productivity and efficiency, the Spy in the Cab is the crowning irrelevancy of the age. 1980 *Daily Tel.* 30 Dec. 1/1 The new rules on drivers' hours and the use of tachographs, the so-called 'Spy in the cab', will come into force on Thursday. 1960 *N.Y. Times* 12 June IV. 6/2 The U-2 reconnaissance 'overflights' provided, by aerial photography and tape recording of Soviet radio and radar emissions, the most important intelligence gathered by the C.I.A. The 'spy in the sky' more than compensated for the very few spies on the ground that the United States has been able to infiltrate into Russia. 1961 *Time* 7 July 16 The spy-in-the-sky warning system. 1969 *Daily Tel.* 18 Apr. 24/8 The NSA..is responsible for ship, aircraft and 'spy-in-the-sky' satellite espionage. 1976 *Daily News* (N.Y.) 11 June (CB & Sound Suppl.) 2/1 *Spy in the Sky*, police aircraft. 1977 *Time* 23 May 33/2 A lot of the information is picked up by those spy-in-the-sky satellites. 1892 *Englishman in Paris* II. xii. 275 The spy mania..became positively contagious. 1943 *Time* 18 Jan. 38/2 The Nazi spymaster and Naval Attaché, Captain Dietrich Niebuhr. 1978 A. Neave *Nuremberg* xiii. 148 The older generation remembered him for his incompetence as a spy-master in the United States during the First World War. 1969 G. Lyall *Venus with Pistol* iii. 20 It's all a bit like something out of a bad spy movie. 1955 W. Tucker *Wild Talent* xvi. 215 A master spy-net, efficiently directed. 1977 G. Markstein *Chance Awakening* xxv. 76 Hentoff is..a key figure in a British-based spy network. 1919 C. Mackenzie *Sylvia & Michael* vi. 222 Our spy-novels and spy-plays must have been of priceless assistance to the Germans. 1979 *Guardian* 1 Mar. 7/2 Literaturnaya Gazeta..cast doubt on the value of spy novels. 1960 *Aeroplane* XCVIII. 627/1 That the U-2 'Spy-plane' operation is nothing new is evident from what I heard in America about three years ago. 1962 *Spectator* 14 Sept. 351 The Chinese..brought down an American spyplane. 1976 J. Pover *Day of Reckoning* iv. 21 Wasn't there something about a spy plane—yes..the U-2 incident. 1919 Spy-play [see *spy novel* above]. 1943 Spy ring [see *ring *sb.*¹ 11 c]. 1960 *Guardian* 22 July 1/2 The spy-ring members began checking on the habits of people working

in radar..establishments. 1980 *Times* 22 Jan. 15/1 General Miyanaga, the leader of the spy ring, had been a Russian-speaking specialist working in intelligence for many years. 1960 *Washington Post* 29 Mar. A16 Much of the money..will go for additional Atlas intercontinental ballistic missiles in hardened sites and for acceleration of the Midas, or 'spy' satellite development. 1976 W. H. Canaway *Willow-Pattern War* xx. 204 There's a camouflaged installation... You couldn't see it from a spy satellite. 1977 *Arab Times* 14 Dec. 1/1 Georg Leber, defending his role in a burgeoning spy scandal. 1923 R. Macaulay *Told by Idiot* III. xxi. 257 The German spy scare, the British spy scare, these fevers were worked up in the jingo press. 1976 J. Lee *Ninth Man* 74, I like an informed citizenry. Spy scares keep them alert. 1968 *Observer* 22 Dec. 11/2 New recruits..are sent for training to Group Four's headquarters... A large country house.. it conjures up all the spy-school images. 1975 *Listener* 28 Aug. 281/4 Drama..is supplied by *The Eiger Sanction* ..but only so far as the most standard of television spy series. 1962 *Daily Tel.* 13 June 1/4 (*heading*) Fourth Soviet spyship watching U.S. tests. 1977 C. Forbes *Avalanche Express* xxiii. 242 The 17,000-ton Soviet freighter..the new pride of Soviet..spy ships. 1923 Spy story [see *daffy *a.*]. 1978 F. Maclean *Take Nine Spies* 333 The well-known story of the Watchmaker of Orkney ..as neat a spy story as one could wish. 1823 T. Bewick *Memoir* (1975) xi. 103 This happy society was however, at length broken up, at the time when the War on behalf [of] despotism was raging & the Spy system was set afloat. 1952 I. Asimov *Stars, like Dust* iv. 48 All young fools who get their notions of interstellar intrigue from the video spy thrillers are easily handled. 1977 *Amer. N. & Q.* XV. 76/2 Somehow, the categorization 'spy thriller'.. seems to diminish the 'classic' quality. 1972 K. Benton *Spy in Chancery* xxi. 249, I don't hanker after another spy trial, thank you. 1915 J. Buchan *Thirty-Nine Steps* vii. 184 It is ordinary spy work.

spy, *v.* Add: **I. 1. b.** Phr. *to spy out the land*; freq. *fig.*

1535 [in Dict.] 1913 Galsworthy *Dark Flower* III. v. 320 What had Dromore come for? To spy out the land, discover why Lennan and his wife thought nothing of the world 'outside'. 1936 A. Christie *ABC Murders* xv. 112 This man must have been spying out the land beforehand and discovered your brother's habit of taking an evening stroll. 1958 P. H. Newby *Ten Miles from Anywhere* 124 Maybe you think I'm up to no good? A poultry thief spying out the land. 1979 A. Boyle *Climate of Treason* i. 32 An unofficial representative of the new Soviet régime arrived in London to spy out the land not long after the 1918 Armistice.

4. d. In the names of children's games: (*a*) *hy-spy*, *I spy*: see Hy-spy in Dict. and Suppl.; (*b*) *I spy (with my little eye)*, a game in which one player selects an object (visible to all) for the others to guess, giving them its colour or its initial letter with the words 'I spy with my little eye something (blue, etc.) *or* beginning with —'.

1946 R. Lehmann *Gipsy's Baby* 80 We remained below and played *I Spy*—with colours, not the alphabet, so that my brother could join in. 1969 I. & P. Opie *Children's Games* x. 275 Their participation in intellectual guessing games, even of the humble order of..'I Spy With My Little Eye', is apt to be limited to occasions when they are restricted and unable to play anything else. 1975 *Language for Life* (Dept. Educ. & Sci.) vi. 85 Stories, and such verbal games as 'I-spy' and 'Knock-knock', encourage children to explore speech sounds and help them develop a better intuitive understanding of these sounds.

spy-. Add: *spy-camera*, *-microphone*, *-mike*, *-window* (earlier example).

1968 *Punch* 2 Oct. 495/1 Television spy-cameras, hooked to machines which will translate a visual image into patterns and templates, hooked to high-speed rotary presses, will start stamping out copies in all sizes. 1972 *Times* 14 Dec. 2/8 A Henry Moore bronze..disappeared from the Lefevre Gallery..although a television 'spy' camera was operating. 1977 J. Hedgecoe *Photographer's Handbk.* 21 Made in Czechoslovakia the Mikroma 'spy camera'..has a seven speed shutter and a f3.5 lens. 1960 *Sunday Express* 17 July 1 Tiny hidden spy-microphones planted by foreign agents. 1955 Pohl & Kornbluth *Space Merchants* i. 4 Nothing but the usual State Department and House of Representatives spy-mikes. 1960 *Sunday Express* 17 July 1 (*heading*) Spy-mikes sensation. Arms firms checked for 'listening walls'. 1755 S. Richardson *Let.* 15 Aug. in *Sel. Lett.* (1964) 321 My spy-window—Ay, that is the window of my vexation —workmen are—workmen... Only, that workmen I wish they were.

spying, *vbl. sb.* Add: **b.** *spyingpoint*.

1922 Joyce *Ulysses* 254 What is it?..—Find out, Miss Douce retorted, leaving her spyingpoint.

sqn, abbrev. of Squadron *sb.* in Dict. and Suppl.

1914 W. S. Churchill *Let.* 26 Aug. in M. Gilbert *Winston S. Churchill* (1972) III. *Compan.* I. 55, I am trying to get the 3rd Sqn (yours I hope) for the Infantry division I am forming from Marines and Naval reservists. 1942 Partridge *Dict. Abbrev.* 91/2 *Sqn.*, squadron, whether naval or aerial. *Sqn Ldr.*, Squadron Leader. 1977 *R.A.F. News* 22 June–5 July 5/1 Sqn Ldr Sean Maffett..has been a commentator for nearly ten years.

squab, *sb.* Add: **5. b.** In mod. use, the padded back or side of a car-seat.

1904 *Car* 15 June 114/2 Two extra seats..fold up underneath the back squab. 1924 *Motor* 28 Oct. 700/1 Several different methods have been invented to enable the angle of the squab to be varied. 1966 'A. Hall' *9th Directive* vii. 62, I slid back and rested my head on the rear squab... It was a big car, comfortable. 1972 *Drive* Spring 147/3 The height, legroom and squab level of the driver's seat can be adjusted.

squad, *sb.*¹ Add: **3. b.** Also *transf.* and *fig.*

1914 Joyce *Dubliners* 244 Three squads of bottles of stout and ale and minerals, drawn up according to the colours of their uniforms. 1940 T. S. Eliot *East Coker* v. 14 The general mess of imprecision of feeling Undisciplined squads of emotion. 1979 *N. & Q.* Feb. 83/2 The only unforgiveable fault is the abbreviation of titles to unpronounceable squads of initials.

4. b. *Sport* (orig. *U.S.*). A group of players forming a team or from which a team is chosen.

1902 *Harvard Bull.* 19 Mar. 2/2 The rest of the squad will leave the cage as soon as the ground is dry enough. 1920 W. Camp *Football without Coach* i. 17 A player should take the ball in his hands,..release it and pass it to the next player. This next man repeats the performance, and so it goes through the squad. 1950 *Daily Ardmoreite* (Ardmore, Okla.) 15 Jan. 14/3 It was the second loss of the season for the Tiger 'A' squad as the unbeaten Cougars took them 47 to 40. 1975 *Cricketer* May 4/1 Intikhab Alam ..has been omitted from the squad of 19 from which the 14 will finally be chosen. 1981 G. Boycott *In Fast Lane* i. 7 Like everyone who has ever played or watched a cricket match I have my own views about the tour party —and I wouldn't have picked quite this squad.

c. A unit within a police force, organized to investigate or prevent a particular type of crime; freq. in ellipt. use for *flying squad* s.v. *flying *ppl. a.* 4 e (*b*). See also *fraud*, *murder*, *riot*, *vice squad* at first element in Suppl.

1905 *N.Y. Times* 22 June 8/6 Commissioner McAdoo selected yesterday the men for the special squad which will arrest women in the streets. 1928 E. Wallace *Flying Squad* xv. 132 You do your best, eh? You did your best to put Bradley away, and draw the attention of the Squad to you and me! 1938 [see *guy *v.*⁴]. 1939 'N. West' *Day of Locust* xxvii. 221 A big squad of policemen was trying to keep a lane open between the front rank of the crowd and the façade of the theatre. 1962 *Daily Tel.* 15 June 22/5 Three detectives, two of them drug squad officers, flew to Gibraltar from London yesterday to investigate the haul of illegal drugs found in the cruiser Belfast. 1980 *Times* 24 Jan. 5/3 Serious crime squads throughout Britain are searching for the inventor and manufacturer of a black box which contains a device that can reverse electricity meter readings.

5. *squad car*, *drill* (earlier example), *leader*, *room*.

1938 F. D. Sharpe *Sharpe of Flying Squad* x. 120 A Squad car went down at once. 1956 D. G. Browne *Rise of Scotl. Yard* III. xxv. 339 Squad cars, area cars, and 'Q' cars..kept in touch by wireless with the Information Room at Scotland Yard. 1970 G. F. Newman *Sir, You Bastard* iii. 109 Sneed parked behind the squad car, which had been directed via the information room. 1891 *Macm. Mag.* Oct. 466/1 The best thing for them to do would be to go back to squad drill. 1953 *Amer. Speech* XXVIII. 23 A special group is formed by these military words and phrases..squad, squad leader, and training. 1967 tr. *Quotations from Chairman Mao Tsetung* (ed. 2) 106 The secretary of a Party committee must be good at being a 'squad leader'. 1946 *Sun* (Baltimore) 3 July 15/4 John Moynahan, lockup keeper at the town hall police station, was puzzled as he watched Midnight, the seven-year-old station cat, leap madly around the squadroom. 1981 W. Marshall *Perfect End* 5 [He] went through the lobby and into the squadroom.

squaddie (skwǫ·di). *Services' slang.* Also **squaddy.** [f. Squad *sb.*¹ + -ie, perh. influenced by Swaddy *sb.*] A member of a squad; a private soldier; a recruit. Also *transf.*

1933 G. Ingram *Stir* xvi. 254 You get the screws and squaddies to shoot at us! Why, you're—well mad, you are. 1943 Hunt & Pringle *Service Slang* 61 *Squaddie*, recruit—new to the squad. 1959 J. Jefferies *Thirteen Days* viii. 105, I had a motley but effective army of luckless squaddies who had been selected by orderly sergeants. 1970 *Daily Tel.* (Colour Suppl.) 30 Oct. 25/4 Most of the Beatles have been seen several times, as have Danny Cohn-Bendit, Jean-Jacques Lebel, the CIA, of course, and drug squaddies with joss sticks and beards. 1978 J. B. Hilton *Some run Crooked* xi. 114 'It needn't have been a squaddy who'd lost the knife.' 'It was a soldier's knife.'

‖ **squadra** (skwa·dra). *Hist.* Pl. **squadre.** [It.; cf. Squadron *sb.*] In Italy: a paramilitary squad organized to support and promulgate Fascism; a Fascist cadre.

1922 *English Rev.* XXXV. 558 When the Fascista army is mobilised..full authority is given to a secret military command. Their smallest unit is the 'manipolo'; then come the 'squadra', 'centuria', 'coörte', and 'legione'. 1924 K. L. Roberts *Black Magic* ii. 55 The smallest military unit of the Fascisti was the squadra of eight or ten or fifteen men—any small number. Three squadre made a manipolo. 1967 C. Seton-Watson *Italy from Liberalism to Fascism* xiii. 571 As the *squadre* grew in strength and boldness, they held up socialist local councils at the pistol point [*sic*] and forced them to resign. 1974 J. White tr. *Poulantzas's Fascism & Dictatorship* III. iii. 126 Para-military *squadre* were also formed outside the *fasci*, even if most fascists took part in them.

† **squadrilla** (skwǫdri·lă). *temporary.* [Blend of Squadron *sb.* and Flotilla. Cf. F. *escadrille*.] = *squadron *sb.* 3 b.

1914 *Daily Mail* 28 Dec. 5/1 A squadrilla of five German aeroplanes caused a hundred casualties in the suburbs of Warsaw. 1916 *Glasgow Herald* 18 Aug. 8 Squadrillas of aeroplanes were sent forward to bring down or drive back the enemy aviators. 1917 *Daily Chron.* 11 Dec. 2/3 The strength of the German aviation services..rather more than 200 squadrillas.

squadrism (skwǫ·driz'm). [ad. It. *squadrismo* (also used), f. *squadra.] The organization and activities of the *squadre*.

1926 B. B. Carter tr. *L. Sturzo's Italy & Fascismo* ii. 46 The phenomenon of *arditismo* and *squadrismo*, or the use of armed irregular bands. 1932 H. R. Spencer *Govt. & Politics of Italy* xii. 115 After the March on Rome the existence of the *squadra* and the guerilla habit known as *squadrismo* constituted a serious difficulty for a régime that was seeking to regularize itself. 1940 *Ann. Reg. 1939* 195 All the new office-holders are old Fascists from the period of 'squadrism' (the earliest Fascist formation). 1954 B. & R. North tr. *Duverger's Pol. Parties* I. i. 39 Hitler reacted violently against the tendencies of Roehm, Mussolini against the excesses of squadrism. 1967 C. Seton-Watson *Italy from Liberalism to Fascism* xiii. 571 At the end of 1920 *squadrismo* spread from its training ground in Venezia Giulia to Emilia.

squadrist (skwǫ·drist). Pl. ‖ **squadristi** (skwadri·sti); **squadrists.** [ad. It. *squadrista*, f. as prec.] A member of a *squadra*. Also *attrib.*

1938 E. Ambler *Cause for Alarm* iv. 62 His father had been killed by the Squadristi in nineteen-twenty-three. 1957 *Encycl. Brit.* XII. 802/1 In the early years the *squadristi* had freely used rubber truncheons and castor oil. 1967 C. Seton-Watson *Italy from Liberalism to Fascism* xiii. 571 In July 1920..youths of the Trieste Fascio..burnt down the headquarters of the Slovene organisations... This was the first appearance of *squadristi*. 1973 *Times Lit. Suppl.* 2 Mar. 226/1 The *ras* of Cremona could claim the support of the 'squadrists'. *Ibid.*, The ultimate limit to Farinacci's squadrist ethic was its dependence upon Fascism's capacity to retain power. 1977 M. Walker *National Front* i. 19 They [*sc.* Mosley's army] had neither the numbers nor the organization of Hitler's *Sturmabteilung* or Mussolini's *Squadristi*.

squadrol (skwǫ·drōᵘl). *U.S. slang.* [f. Squad *sb.* + Pat)rol *sb.*] A small police van.

1961 in Webster. 1965 J. McCormick *Bravo* ii. 44, I sat between them in the front seat of the squadrol, stone faced, my arms crossed, as we drove off under the thin pealing of noontime bells from Loyalty Chapel. 1972 B. Garfield *Line of Succession* (1974) I. 38 The Plymouth..was a block ahead when the squadrol, its red and blue lights flashing, came in sight on a collision course. 1976 *Tel.* (Brisbane) 21 Dec. 36/5 No one got excited when the small van, called a 'squadrol', pulled up in front of the Starr Hotel at 617 West Madison.

squadron, *sb.* Add: **I. 3. b.** *Air Force.* A small operational unit in an air force, consisting of aircraft and the personnel necessary to fly them.

1912 *Times* 9 May 14/5 A party of officers and non-commissioned officers..are to leave the Aviation School at Farnborough on May 15... These will form the nucleus of two flying squadrons of the new Royal Flying Corps. 1919 *Daily Mail Year Bk.* 46/2 The range of such raiding squadrons..tends to grow constantly from day to day. 1939 [see *group *sb.* 3 e]. 1942 T. Rattigan *Flare Path* I. 26 He was on a week's leave, and we were married before he went back to his Squadron. 1959 [see *flight *sb.*¹ 1 h]. 1978 R. V. Jones *Most Secret War* xliv. 420 The main point, though, was the esprit de corps, and this was what Hartley had meant with his 'last *squadron* in the Air Force'.

5. b. (Later examples.)

1930 T. S. Eliot tr. *St.-J. Perse's Anabasis* 39 Squadrons of stars pass the edge of the world. 1978 J. A. Michener *Chesapeake* 759 What Steed did next, in the late 1950s, was to pension off his field hands and purchase a squadron of gigantic automatic corn harvesters.

8. *squadron commander*, *leader* (later examples), *officer*.

1907 R. Hermon-Hodge *Let.* 12 Jan. in R. S. Churchill *Winston S. Churchill* (1969) II. Compan. I. 640 In reporting on the Regiment in 1905 the GOC remarks 'I thought the Squadron Commanders exceptionally well qualified for their positions'. 1976 *Southern Even. Echo* (Southampton) 15 Nov. 9/2 If the junior NCO's are not doing their job properly it does not matter how good the squadron commander is. 1919 W. S. Churchill *Let.* 8 Feb. in M. Gilbert *Winston S. Churchill* (1977) IV. Compan. I. 517 The ranks in contemplation are as follows:—Air Marshal: Air Commodore: Wing Commander: Squadron Leader: Flight Leader: Flying Officer or Observer. *a* 1944 K. Douglas *Alamein to Zem Zem* (1946) ii. 13 He now found himself second in command of a squadron whose squadron leader had been a subaltern under him before. 1943 J. R. Williams *Aircraftwoman Grey* xiii. 146, I have asked Squadron Officer Hedley, the Group Officer, to call and see you. 1972 A. Price *Col. Butler's Wolf* i. 9 Squadron Leader Roskill is a colleague of mine at the Ministry of Defence. 1971 K. B. Beauman *Partners in Blue* iv. 72 The Squadron Officer was one of six of this rank appointed on September 27th... She had joined the Directorate in August.

squalene (skwē̆i·lī̆n). *Chem.* [f. Squal(us + -ene.] A colourless, oily, liquid, triterpenoid hydrocarbon, $C_{30}H_{50}$, which in animals is an intermediate in the biosynthesis of cholesterol and occurs esp. in the liver oils of sharks and other elasmobranch fishes.

1916 M. Tsujimoto in *Jrnl. Industr. & Engin. Chem.* VIII. 896/2 From..the fact that the author has discovered the hydrocarbon first in the liver oils of the squaloid sharks, he proposes the name '*Squalene*' for the hydrocarbon. 1938 [see *œstrin]. 1964 *New Scientist* 22 Oct.

221/1 These rearrangements occur as part of a marvellously complicated, concerted enzymic process by which the long molecule of squalene is folded and formed into rings. 1978 *Sci. Amer.* Sept. 94/1 Sterols, such as cholesterol and the steroid hormones, are flat, platelike molecules derived from the compound squalene.

squall, *sb.*³ Add: **3.** Special Comb.: **squall line,** a line along which high winds and storms are occurring (see also quot. 1950).

1906 *Q. Jrnl. Meteorol. Soc.* XXXII. 264 From the Kew curves we might be led to suppose that the velocity in the squall was approximately of the same magnitude as the velocity of the squall line. 1923 N. Shaw *Air & its Ways* 75 The surface boundary of the polar front in this region is called the 'squall line'. 1950 *Jrnl. Meteorol.* VII. 21/1 The term *squall line* is among the oldest in meteorology and is perhaps the least clearly defined. Prior to the general adoption..of the frontal theory of cyclones, it was customary to designate as a squall line any line of storms projecting in a general southerly and easterly direction from a depression... With the advent of the frontal theory, some of these lines of storms were redesignated more descriptively as *cold fronts*... There remained the lines of storms which appear in general in the warm sector of cyclones, roughly parallel to the cold front, and along which there is intense convective activity. 1979 L. J. Battan *Fund. Meteorol.* ix. 187 Most often the storms regarded as being in the organized class are those that form in lines or bands of thunderstorms, sometimes called squall lines... They commonly are initiated along a cold front, or ahead of, and nearly parallel to it.

squally (skwǭ·li), *a.*¹ [f. Squall *sb.*² + -y¹.] Of a child, etc.: that screams discordantly or shrilly; squalling, noisy.

1861 [see *mother's blessing* s.v. *mother *sb.*¹ 15 a]. 1947 D. Riesman in *University Observer* Winter 24/2, I prefer sensitive and cultivated people to squally brats on trains. 1958 *Listener* 25 Dec. 1093/3 Maria Callas seemed right out of the picture, squally and weak at climaxes.

squalmish (skwā·miʃ, skwǫ·miʃ), *a.* *U.S. colloq.* Also **squamish, squawmish.** [Var. Qualmish *a.*, perh. influenced by Squeamish *a.*] Nauseous, qualmish, queasy.

1867 'Mark Twain' *Notebk.* (1935) vi. 59, I am..very tired of being seasick... All I take an interest in is being squalmish and getting to shore again. 1902 S. Clapin *New Dict. Americanisms* 381 *Squawmish*, in parts of New England, said for queasy. 1944 H. Wentworth *Amer. Dial. Dict.* 590/1 *Squalmish, squamish*.., squeamish, qualmish. 1948 *Daily Progress* (Charlottesville, Va.) 24 Jan. 4/4, I am not only not interested in food but maybe a little squalmish at the very thought of it.

squalor. Add: **1. b.** Also, the quality or state of being mentally squalid, or of lacking intellectual sensitivity and order.

1933 E. A. Robertson *Ordinary Families* v. 88 The whole family is much the same. Lives Dangerously in mental squalor. 1941 A. Huxley *Grey Eminence* vi. 139 This strength and magnificence, so very different from our own weakness and mental squalor.

squaloro·logy. [f. Squalor + -ology.] The study of squalor, esp. as a supposed science. So **squalorolo·gical** *a.*; **squaloro·logist,** a student of or a person particularly interested in squalor.

1956 *Observer* 7 Oct. 17/4 Toughish thriller about gigolo-adventurer smuggling between Venice and Yugoslavia. Distinguished by some sociological and squalorological realism. 1957 *Brittanica Bk. of Year* 814/1 *Squalorologist*, a writer stressing unpleasant matters. (1955). 1958 *Observer* 27 Apr. 14/7 To congratulate Dan Farson on two vivid squalorological [*sic*] scoops; one of methylated-spirit drinkers on a bomb-site.., the other of Soho's afternoon nudist shows. 1961 *John o' London's* 6 July 57/4 This [*sc.* a film] is a piece of glamourised squalorology. 1973 *Observer* 29 July 31/7 Himself an Ohio policeman and expert squalorologist.

squame. Add: **4.** *Med.* **a.** A small flake of dead tissue shed from the surface of the skin in some disorders.

1911 M. Morris *Dis. Skin* (ed. 5) i. 16 The scale, or squame, is a dry and usually laminated exfoliation of the epidermis. 1953 S. Beckett *Watt* 170 A..constipated man, covered with squames. 1975 *Sci. Amer.* Nov. 70/3 Finally they fuse into the flakes called squames, which are eventually shed from the surface.

b. A squamous cell (see *squamous *a.* 8).

1949 in *New Gould Med. Dict.* 1954 *Jrnl. Obstetr. & Gynecol.* LXI. 156/2 Smears of patients with erratic or inactive curves consist mainly of intermediate squames and have a high proportion of basal cells. 1973 *Gray's Anat.* (ed. 35) 27/2 Squamous (pavement) epithelium. This is composed of flattened, interlocking, polygonal cells (squames).

Squamish (skwǭ·miʃ), *sb.* and *a.* Also † **Skwamish; Squawmish.** [a Comox cognate of Squamish *sqxʷúʔmiš*.] **A.** *sb.* **a.** A (member of a) North American Indian people of the Coast Salish group of southwestern British Columbia. **b.** The language of this people. **B.** *adj.* Of, pertaining to, or characteristic of this people or their language.

1846 R. B. Sage *Scenes in Rocky Mountains* 221 The Indians principally consist of the following tribes: the Snakes,..Squamishes,[etc.]. 1884 *Ann. Rep. Canad. Dept. Indian Affairs 1883* 189 (*table*) Fraser River Agency..Squamish..Hou Sound..367. 1907 T. Crosby *Among An-ko-me-nums* i. 10 The An-ko-me-nums..inhabit the valley of the Fraser River..and include the.. Songees, Skwamish, Sumats. 1928 J. Mooney *Aboriginal Population Amer. North of Mexico* 29 Squawmish tribes. 1934 Webster, Squawmish. 1955 H. G. Barnett *Coast Salish of British Columbia* 3 Aboriginally the Squamish made their homes on the river named after them at the head of Howe Sound. 1957 *Encycl. Brit.* XIX. 888/1 The principal [Salishan] groups or tribes are:..2, Coast Salish, ..Nanaimo, Squawmish, Lummi, [etc.]. 1967 A. H. Kuipers (*title*) The Squamish language. 1969 P. Sedlak in *Working Papers on Language Universals* I. 24 *Squamish* ..must be analyzed differently since the phonetic specification is not that indicated for the cases just discussed *Ibid.* 26 *Mandarin Chinese* involves two more features than those necessary to describe *Squamish*. 1977 H. Landar in T. A. Sebeok *Native Langs. of Americas* II. III. 356 Squamish. South Georgia Salish; 100 to 200 in British Columbia (1962).

squamish, var. *squalmish *a.*

squamous, *a.* Add: **6.** (Earlier example.)

1829 *Glasgow Med. Jrnl.* II. 327 These eruptions may be exanthematic, vesicular..or squamous.

8. *Med.* Designating epithelium that contains (or consists of) a layer of very thin, flattened cells, and the cells themselves. Also *squamous-celled* adj.

1860, 1872 [in Dict., sense 5]. 1891 Squamous-celled [in Dict., sense 6]. 1908 *Practitioner* Sept. 355 Wherever the disease occurs, seeing that it is commonly a squamous-celled epitheliomatous ulcer, it is reasonable to suppose that it starts as the result of some previous lesion causing a denudation of the epithelium at one of these spots. 1937 E. E. Hewer *Text-bk. Histol.* ii. 9 Three types can be distinguished:—(1) Simple squamous epithelium...(2) Simple columnar and cubical epithelium...(3) Stratified epithelium. 1947 *Radiology* XLIX. 281/1 A variety of other tumors, such as..squamous-cell carcinomas of the vagina and fore-stomach..were observed. 1968 Passmore & Robson *Compan. Med. Stud.* I. xvii. 2/1 There are two varieties of stratified squamous epithelium, cornified (keratinized) and non-cornified (non-keratinized). 1978 *Jrnl. R. Soc. Med.* LXXI. 726 Metaplasia into stratified squamous-cell epithelium is very common in the middle ear.

squander-bug. *colloq.* Also **squander bug, squanderbug.** [f. Squander *v.* + *bug *sb.*² 4 a, after *jitterbug*, etc.] A symbol of reckless extravagance and waste, first used in government publicity campaigns to promote economy during the war of 1939–45 and represented as a devilish insect; a likeness of this. Also, one who is profligate with money or resources. Hence **squa·nderbugging.**

Introduced in 1943 by the National Savings Committee.

1943 *Times* 8 Jan. 8/1 (Advt.), Beware the treacherous Squander Bug! He's the prince of fifth-columnists—doesn't believe in a nest-egg for the future—doesn't believe in making money fight for Britain... Join a Savings Group to defeat the Squander Bug! 1943 O. Lancaster *More Pocket Cartoons* 55 (*caption*) Oh, just look at this sweet little squander-bug in platinum and rubies that darling Boysy's just sent me. 1944 *Convoy* Feb. 26 A Government so dexterous, through its armies of Public Relations Depts., in persuading us to squash the squander bug or grow more onions or shave in tepid water. 1946 J. W. Day *Harvest Adventure* xvi. 268 That is the question to ask these rural squander-bugs. 1966 *Punch* 26 Jan. 108/2 The more favoured nations continue to outdo each other in squanderbugging. The new Russian satellite will have showers, electric razors, fresh linen and a load of other luxury items. What a blow to American prestige. 1976 *Listener* 23 Dec. 843/3 No initiative could be broached..for fear of earning the name of Squander-bugs.

squa·nderlust. *U.S. slang.* [f. Squander *v.*, after *wanderlust.] A strong desire to spend money or to waste assets; also *loosely* = next.

1935 *Amer. Speech* X. 155/1 Louis Ludlow, member of Congress and former Washington correspondent..is the author of *America Go Bust*, which 'focuses attention on the bureaucratic *squanderlust*' of our times. 1960 *Wall St. Jrnl.* 5 May 8/2 Politicians... moved..by an any-year philosophy of squanderlust, lead the country down the road to economic trouble. 1977 *Time* 18 July 72/3 No longer the ultimate expression of corporate and personal squanderlust, the private plane is now a ubiquitous.. means of air travel to smaller cities.

squanderma·nia. *colloq.* [f. Squander *v.* + -mania.] An insane desire or obsession to spend money recklessly or to waste assets. Hence **squanderma·niac** *a.* and *sb.*

1920 *Public Opinion* 2 July 3/1 The public are deeply roused upon the Squandermania issue. 1921 *Glasgow Herald* 2 Apr. 7 The burden imposed upon him [*sc.* the tax-payer] by a 'Squandermaniac' Government. 1922 *Public Opinion* 10 Mar. 228/1 The real squandermaniac would be revealed in the man with an infinite capacity for standing at street corners. 1931 *Punch* 4 Nov. 481/1 The triumph of Reason over Squandermania. 1933 C. Mackenzie *Water on Brain* vii. 85 There cheek by jowl sat the squandermaniac and the suicidal junior clerk. 1956 *Sun* (Baltimore) 14 June (ed. B) 16/2 They think that people who support the provision of some military aid to our allies are 'soft' and afflicted with 'squandermania'. 1976 *Times* 16 Aug. 11/5 Sheer 'squander-mania'..has currently invaded our society.

squantum (skwǫ·ntɒm). *U.S. local.* Also **Squantum.** [f. *Squantum*, the name of a sea-coast village (now part of Quincy), in Norfolk Co., E. Mass: see also M. Mathews *Dict. Americanisms*.] In Massachusetts: a picnic, a 'clambake'; *spec.* an annual feast formerly held on the sea-shore at which sea-food was eaten.

1812 *Boston Gaz.* 24 Aug. 2/5 The Squantum Celebration Will be this day, (Monday) at the old celebrated spot... The Feast was postponed on Saturday on account of the weather... The antient Celebrators of the Squantum Feast, will be honored..with the presence of their illustrious friends. 1832 S. G. Goodrich *Syst. Universal Geogr.* 106 The feast of *Squantum* is held annually on the shore to the E. of Neponset Bridge, at a rocky point projecting into Boston Bay. 1855 H. A. Wise *Tales for Marines* i. 21, I wish to all fired smash I was..hazin' round with Charity Bunker and the rest o' the gals at a squantum. 1890 E. Bellamy *Six to One* vi. 60 The squantum was to be held at a point on the narrow peninsula.. that divides the ocean from the broad lagoon.

square, *sb.* Add: **II. 8. b.** *method* (or *principle*) *of least squares* [tr. F. *méthode des moindres quarrés* (A.-M. Legendre *Nouvelles Méthodes pour la Détermination des Orbites des Comètes* (1806) 74)], the technique of estimating a quantity, fitting a graph to a set of experimental values, etc., so as to minimize the sum of the squares of the differences between the observed data and their estimated true values; so *least square(s)* attrib., denoting estimates, regression lines, etc., obtained by this method, the method itself, and the processes which it involves.

[1812 *Phil. Mag.* XXXIX. 242 M. Legendre had not in any way mentioned the method which he has denominated that of small squares, (*moindres carrées*,). *Ibid.* 243 The principle of the small squares.] 1825 *Ibid.* LXV. 10 The principle of least squares will hold good, whatever law of probability be adopted. 1830 Poisson in *Q. Jrnl. Sci., Lit. & Arts* VI. 96 This embarrassment..remained to the period when M. Legendre proposed a direct and uniform method of forming the final equations, which was generally adopted under the name of Method of least squares of the errors, which was assigned to it by its author. 1872 Thomson & Tait *Elem. Nat. Philos.* I. iii. 115 *A* and *B* are to be found by the method of least squares from values of *l* observed for different given values of *t*. 1916 L. D. Weld *Theory of Errors & Least Squares* v. 87 If more [observations] are made, least-square reduction may be applied to their adjustment. 1939 A. E. Treloar *Elem. Statistical Reasoning* iv. 59 'Least squares' solutions, or the minimizing of squared deviations to reach representative values, may be made with facility, whereas the minimizing of absolute deviations becomes so involved that the problems can rarely be solved that way. 1950 A. McF. Mood *Introd. Theory Statistics* xiii. 311 The primary reason that the method of least squares is commonly used for curve fitting is merely that it leads to a simple linear system of equations for determining the coefficients. 1957 G. E. Hutchinson *Treat. Limnol.* I. vii. 469 *C* is determined from the distribution of temperatures by the method of least squares. 1970 *Jrnl. Gen. Psychol.* July 127 The principle of the method is to rotate one matrix (usually a principal-components matrix) as close as possible to a hypothesized factor matrix, in a least-square solution. 1974 *Encycl. Brit. Macropædia* VII. 967/1 Gauss developed a technique for calculating its orbital components with such accuracy that several astronomers late in 1801 and early in 1802 were able to locate Ceres again without difficulty. As part of his technique, Gauss used his method of least squares, developed about 1794.

10. e. A square piece of material used as a scarf or cravat.

1882 *Queen* 7 Oct. 334/2 Lace Bows... Lace Sets... Indian Muslin and Lace Squares, from 1s. 11d. to 12s. 6d. 1926 in C. W. Cunnington *Eng. Women's Clothing in Present Cent.* (1952) vi. 190 The latest scarf conceit is a square of chiffon caught round the neck. 1960 C. Dale *Spring of Love* iv. 94 Miss Burroughs [wearing]..a silk square over her shoulders against the draught. 1966 B. Kimenye *Kalasanda Revisited* 60 Removing the georgette square which had been tightly binding her head. 1979 A. Scholefield *Point of Honour* 142 Yellow knotted silk square at the throat.

11. b. *Cricket.* A closer-cut area at the centre of a ground, any strip of which may be prepared as a wicket.

1899 *Lawns* (Sutton & Sons, Reading) 32 The club purse must determine the extent of ground to be treated in the manner we recommend, but while the work is in progress it is worth while to strain the point to make the playing square sufficiently large, say, at the very least, forty yards in the line of the wickets, by thirty yards in width. 1924 H. de Sélincourt *Cricket Match* v. 110 As they reached the square, five Raveley men emerged, running, from the Pavilion, and called loudly for the ball. 1950 F. J. Reed *Lawns & Playing Fields* xvii. 174 On established cricket squares mowing should commence as early as possible, setting the machine high and gradually lowering the cut as the season advances. 1976 J. Snow *Cricket Rebel* 22 We were not to meet up as a side until we got on to the square.

c. *Mil. slang.* A parade ground.

1915 F. H. Lawrence in *Home Lett. T. E. Lawrence* (1954) 644 There were 10 officers on the square when I joined in September, and four of them are now dead, four wounded and one missing. 1925 Fraser & Gibbons *Soldier & Sailor Words* 268 *Square*,..an army term for the drill or parade ground. In general, the Barrack Square.

1962 A. WESKER *Chips with Everything* I. iii. 17 This is the Square. We call it a square-bashing square. 1982 'W. HAGGARD' *Mischief-Makers* ii. 21 He had failed to pass Sandhurst. He had failed to pass off the square and had been put back a term.

14. Delete † *Obs.* and add later examples.

1819 J. WILSON *Compl. Dict. Astrol.* 379 *Square*, the quartile aspect, containing a quadrant or right angle. 1861 R. J. MORRISON *Hand-bk. Astrol.* I. I. iii. 8 When a sextile aspect or distance of *sixty* degrees falls in the latter, Ptolemy intimates that it has the effect of a square, or *ninety* degrees. And when a trine falls in signs of short ascension, he says that the effect is also that of a square aspect. 1929 V. E. ROBSON *Alan Leo's Dict. Astrol.* 188 The Square is the most critical and conflicting of aspects. a 1963 L. MACNEICE *Astrol.* (1964) viii. 258 Sextiles..are supposed to be 'good'..aspects, while the square (90°) is considered 'bad'. 1975 I. M. HICKEY *Astrol.* viii. 72 Squares represent the lessons we have failed to learn.

15. g. *Bookbinding.* Usu. in *pl.* The portion of the cover of a bound book which projects beyond the leaves.

1835 'J. A. ARNETT' *Bibliopegia* 207 Squares.—That portion of the boards of a volume which projects over the edges. 1876 *Encycl. Brit.* IV. 43/2 The same processes are followed with the sides and the 'squares' when any ornamentation is tooled upon them. 1901 D. COCKERELL *Bookbinding* ix. 131 If the book has been trimmed, or is to remain uncut, a little more must be allowed for the 'squares'. 1946 E. DIEHL *Bookbinding* II. xi. 148 If the squares are too large when the boards have been laced on, it is a simple matter to cut them down.

h. A given space on the page of a newspaper, etc., considered as a unit of measurement for advertisements. *U.S.* (now *Hist.*).

1800 *Impartial Observer* (Natchez, Mississippi) 5 May 1/1 Advertisements..which exceed a square will be inserted at the same proportionate price. 1877 *Harper's Mag.* Dec. 111/1 These newspaper people set an extraordinary value on their squares, as they call them. 1943 C. CROW *Great Amer. Customer* 122 The standard space measurement [for advertising] was the 'square', which meant a space equal in depth to the width of a column—approximately two column inches.

15*. Slang uses. **a.** One who is square (*SQUARE *a.* 9 d); a person considered to hold conventional or old-fashioned views. orig. *U.S. Jazz.*

1944 *Sun* (Baltimore) 27 Jan. 10/5 *Square*, in musician's jargon, anyone who is not cognizant of the beauties of true jazz. 1944 D. BURLEY *Orig. Handbk. Harlem Jive* 70 Are you going to be a square all you days? 1947 [see *HIP *v.*[4]]. 1952 'E. BOX' *Death in Fifth Position* (1954) i. 23 Though I might not be entirely a square I was..hopelessly ignorant of all that..mattered. 1959 H. HOBSON *Mission House Murder* ii. 15 The odd fifty million citizens who don't dig them are dead-beats—squares. 1965 G. HACKFORTH-JONES *Storm in Harbour* ix. 142 You and I are what the up and coming generation call squares. We live in the past and we don't like what we see of the present. 1968 T. WOLFE *Electric Kool-Aid Acid Test* xxvii. 386 We're in two different worlds. You're a hippie and I'm a square. 1974 *Howard Jrnl.* XIV. 101 The 'square' are women who are basically pro-authority, in favour of law and order, and share the values of 'respectable' society. 1977 J. D. DOUGLAS in Douglas & Johnson *Existential Sociol.* i. 42 Marihuana has been widely used for decades by artists and other groups, probably also as a way of expressing feelings against the squares.

b. A cigarette containing tobacco, rather than marijuana. *U.S.* (chiefly *Blacks'*).

1970 H. E. ROBERTS *Third Ear* 13/1 *Square*, a cigarette. 1971 *Black Scholar* Sept. 36/2 Why, why, he kept asking himself, as he lit a square.., why do I keep having that dream. 1974 *Black World* Nov. 57 Light me up a square, baby.

16. Also, a square meal. orig. and chiefly *U.S.*

1882 O. MERDIAN *Let.* 20 Sept. in *Frontier* (1930) X. 252/1, I went in..and had some dinner..ate a square & talked awhile. 1927 J. BARBICAN *Confess. Rum-Runner* xxiii. 260 We sure was hungry for the dough, for it was weeks since we had roped in our three squares a day. 1962 'E. MCBAIN' *Like Love* ii. 21 But he had had a clean bed to sleep in, and three squares a day, as the saying goes. 1979 'H. HOWARD' *Sealed Envelope* x. 135 Mine was a lousy job. There must be a better way of making three squares a day.

IV. 20. *square one*: the beginning, the starting-point. Freq. as *back to* (also *in, on*) *square one.*

Often said to derive from the notional division of an association football pitch into eight numbered sections for the purposes of early radio commentaries (see *Radio Times*, 1927, 28 Jan.). This suggestion cannot be upheld with any certainty, and the phrase may simply come from a board-game such as Snakes and Ladders.

1960 *Times* 21 May 9/2 As far as building up a basis for profitable negotiations is concerned the two sides are back in square one. 1965 *Listener* 24 June 930/2 Let us drop the logical knot that twin studies have tied us in and go back for a moment to square one. 1965 *Guardian* 13 Oct. 2/7 The city's medical officer..said they were still in 'square one', and would stay there till they got some real facts. 1966 J. I. M. STEWART *Aylwins* x. 126 That he had seized a chance to break off our interview at that point seemed to argue a refusal to abide by this judgement of the matter. We were back, so to speak, in Square One. 1970 G. F. NEWMAN *Sir, You Bastard* 279 A couple of wrong answers and Sneed knew he'd be right back on square one. 1973 G. TALBOT *Ten Seconds from Now* (1974) viii. 111 After each of those successful essays it was 'back to Square One'. 1977 'M. INNES' *Honeybath's Haven* x. 98 Honeybath broke off in these bold proposals, suddenly aware that Edwin was weeping. It was like being back on square one. 1980 D. BOGARDE *Gentle Occupation* xii. 332 'Black is black, Moluccans.'..'Are coloured people. They are dark,' said Emmie with force. 'Well, don't let's *have* any blasted children'... 'But I do. I want.'..'Oh for God's sake. We're back to square one again.'

V. 21. *attrib.* and *Comb.*, as **square-bashing** *vbl. sb.* and *ppl. a. Mil. slang* [*BASHING *vbl. sb.* 3], drilling; hence **square-basher**; **square-free** *a. Math.* [tr. G. *quadratfrei*], (of an integer) equal to the product of a set of different primes; not divisible by a perfect square; **square-pusher** *slang*, (*a*) a respectable girl; (*b*) a boyfriend; hence **square-pushing**, the act or practice of walking out with a girl (popularly associated with accompanying nursemaids, etc., about town squares); love-making; also as *pres. pple.*

1959 *Spectator* 21 Aug. 212/2 The transition away from the era of the square-basher and the char-and-wadder is painfully slow. 1943 Square-bashing [see *BASHING *vbl. sb.* 3]. 1946 C. FRY *Phoenix too Frequent* 28 There, do you see her, you acorn-chewing infantryman? You've made her cry, you square-bashing barbarian. 1962 [see sense 11 c above]. 1975 'G. BLACK' *Big Wind for Summer* ii. 20 Attached to a Malay regiment, supervising weapon training and square bashing. 1960 NIVEN & ZUCKERMAN *Introd. Theory Numbers* xi. 226 The set of square-free integers has natural density $6/\pi^2$. 1971 G. HIGMAN in Powell & Higman *Finite Simple Groups* vi. 209 All elements have square-free order. 1890 BARRÈRE & LELAND *Dict. Slang* II. 158/1 A square pusher is a girl of good reputation. 1922 JOYCE *Ulysses* 425, I seen you up Faithful place with your squarepusher, the greaser off the railway, in his cometobed hat. 1918 W. J. LOCKE *Rough Road* x. 116 'Go "square-pushing"?' said Doggie contemptuously, using the soldiers' slang for walking about with a young woman. 1922 JOYCE *Ulysses* 161 Squarepushing up against a backdoor. Maul her a bit. 1928 F. E. BAILY *Golden Vanity* xii. 178 Left me cold in a strange place to go square-pushing with some forward young woman. 1930 J. B. PRIESTLEY *Good Companions* I. iv. 134 'E wouldn't bother, though, too busy square-pushing, taking the girls out, see.

square, *a.* Add: **I. 1. b.** *square mile*: also *spec.* a familiar term for the (heart of the) City of London.

1966 L. SOUTHWORTH *Felon in Disguise* i. 13 Being a non-residential area, murders seldom occur in the square mile. 1971 *Guardian* 3 Mar. 18/4 Prince Charles was made a Freeman of the City of London yesterday... It was the kind of traditional occasion that the square mile does so well. 1975 *Times* 1 Mar. 12/2 The City Corporation hopes to have redeveloped 90 per cent of the square mile by 1980.

6. d. *Assoc. Football*, etc. Of a group of players: positioned in a line at right angles to the direction of play (*spec.* as a defensive weakness).

1972 G. GREEN *Great Moments in Sport: Soccer* x. 97 Often, too, Mullen and Hancocks would find each other with long, cross-field passes which travelled from one touchline to the other during the course of an attack, with the result that opposing defences were often caught square offering vital openings to the forwards in the middle. 1977 *Times* 28 Feb. 8/3 They were goals Middlesbrough always looked like taking against Arsenal's soft, square defence.

7. c. Also in other sports.

1955 *Times* 14 May 3/4 (*heading*) All square in Davis Cup.

7*. *Mus.* Of rhythm: simple, straightforward.

1958 *Times* 27 Oct. 12/3 Attempts have been made.. in recent years to get away from the square style of playing that arose out of the hymns and other purely vocal music. 1967 A. L. LLOYD *Folk Song in England* iv. 170 The earlier melodies are more vigorous, squarer, franker in cast. 1976 *Early Music* July 270 The opening sinfonia for strings and trombones is remarkably like several opera overtures of the time, with square rhythms [etc.].

II. 8. a. *square deal*: see *DEAL *sb.*[2] 4 c.

9. d. Designating one who is out of touch with the ideas and conventions of a particular popular contemporary movement (orig. *Jazz*); conventional; old-fashioned. Formerly opp. *HEP *a.* Also of things. *slang* (orig. *U.S.*).

1946 B. TREADWELL *Big Bk. Swing* 125/2 *Square*, not versed in Swing, puritanical. 1950 J. VEDEY *Band Leaders* 175 Consummate performer that Ellington is, he put these numbers over to the delight of all types of audience, young and old, sophisticated and 'square'. 1953 W. BURROUGHS *Junkie* (1972) x. 110 The other patients were a pretty square and sorry lot. 1959 N. MAILER *Advts. for Myself* (1961) 264 They wish this newspaper to be more conservative, more Square—I wish it to be more Hip. 1959 *Punch* 2 Sept. 103/1, I..told her that the bang-opening was old-hat and a completely square method of writing these days. 1965 F. RAPHAEL *Darling* i. 7 You know books. Those things with pages very square people still occasionally read. 1971 B. MALAMUD *Tenants* 80, I didn't expect it to be that good, not from the square dude you are. 1977 P. G. WINSLOW *Witch Hill Murder* II. xvii. 219 He wants to be in with the law. Square at heart.

10. c. (Earlier example.)

1825 H. WILSON *Mem.* (ed. 2) III. 360 As though I had been the Duchess's chosen daughter-in-law, for whom he was making all square.

III. 11. g. Having membership of the Freemasons; in accordance with the Masonic code.

1888 KIPLING *Man who would be King* in *Phantom 'Rickshaw* 73, I am hoping you will give him the message on the Square. 1927 —— *Limits & Renewals* (1932) 172, I told him I was something else besides a G.P... From then on he told the tale on the Square. 1974 *Times Educ. Suppl.* 21 June 2/4 How many local councils..are riddled with freemasonry? At how many appointments are the best men..passed over because they are not on the square?

IV. 12. a. *square-bracketed, -ended, -faced* (earlier example), *-pupilled, -sectioned, -shaled; square-hung, -pied; square-looking* (earlier example); *square footage.*

1970 *Guardian* 26 Nov. 13/4 Attention is focussed on the heavy-typed and square-bracketed passages. 1923 *Trans. Scottish Ecclesiological Soc.* VII. 65 The Lady Chapel was ..removed and a long square-ended one substituted. 1936 W. FAULKNER *Absalom, Absalom!* 369 He saw then the square-ended saw chunk beside the wall. 1978 A. & G. RITCHIE *Anc. Monuments Orkney* 71 The church consisted of a rectangular nave with a porch at its W end and a square-ended chancel at the E end. 1872 GEO. ELIOT *Middlemarch* II. III. xxiv. 32 She was of the same curly-haired, square-faced type as Mary. 1963 A. SMITH *Throw out Two Hands* i. 16 One man sitting down in a thing 3 feet 11 inches by 2 feet 11 inches tends to occupy the bulk of the available square footage. 1874 G. M. HOPKINS *Jrnls. & Papers* (1959) 255, I am not so sure of the tiles being squarehung—they may have been lozenges. 1845 POE in *Godey's Lady's Bk.* Feb. 63/1 On the very tips of their heads were certain square-looking boxes. 1868 G. M. HOPKINS *Jrnls. & Papers* (1959) 178 Like the skin of a white snake square-pied with black. 1957 T. HUGHES *Hawk in Rain* 39 And looked down A square-pupilled yellow-eyed look. 1964 W. L. GOODMAN *Hist. Woodworking Tools* 99 Another jack..with a similar casting screwed to a square-sectioned wooden stock. 1917 E. POUND *Lustra* 183 Breaking the riven waves On square-shaled rocks.

14. a. square capital *Palæogr.*, a form of rectilinear capital letter, *spec.* characterizing a script used in early Latin manuscripts (cf. RUSTIC *a.* 5 b); **square cut** *Cricket*, a cut hit square on the off-side; hence **square-cut** *v. trans.*, to cut (a ball) thus; **square cutter; square dance,** a dance in which four couples face inwards from four sides; also *loosely*, a country-dance; hence **square-dance** *v. intr.*; **square dancer, -dancing** *vbl. sb.*; **square dinkum**: see *DINKUM *a.*; **square drive** *Cricket*, a drive hit square on the off-side; hence **square-drive** *v. trans.*, to drive (a ball or bowler) in this manner; **square engine,** an internal-combustion engine in which the length of the stroke is approximately equal to the bore of the cylinders; **square-eyed** *a. joc.*, affected by or given to excessive viewing of television; **square Hebrew** *Palæogr.*, the standard Hebrew script which displaced the Aramaic form towards the end of the Biblical period, and has been adopted for use in printed texts; **square John** *N. Amer. slang*, an upright, respectable person; *spec.* one who is not a drug-addict; **square law** *Physics*, a law relating two variables one of which varies either directly or inversely as the square of the other (cf. *inverse square* (*law*) s.v. INVERSE *a.* 3 a); also used *attrib.* of a device whose action obeys the square law; **square-lipped rhino(ceros)** = *white rhino(ceros)* s.v. *WHITE *a.* 11 a; **square motor** = *square engine* above; **square-mouthed rhino(ceros)** = *white rhino(ceros)* s.v. *WHITE *a.* 11 a; **square piano** = *square pianoforte*; **square pianoforte** (earlier example); **square pin**: on an electrical plug, a pin with a rectangular rather than a circular cross-section; **square-rig,** (*b*) *Naut. slang*, the uniform of a naval rating (see also quot. 1942); **square-rigger** (earlier example); **square serif** *Typogr.*, a type-face distinguished by straight serifs as thick as the other parts of the letters; **square-shooter** *slang* (orig. and chiefly *U.S.*), an honest, dependable, sound person; hence **square-shooting** *a.*, honest, respectable; **square thread** *Mech.*, a screw thread which in cross-section is castellated in form, with the width and height of the thread equal to the width of the valley between threads; **square wave** *Electronics*, (a voltage represented by) a periodic wave that varies abruptly in amplitude between two fixed values, spending equal times at each; **square well** *Nucl. Physics*, a potential well of square section; **square wheels,** used *joc.* of a set of wheels which give a jolting ride, as if they were square; also *fig.*

1699 M. LISTER *Journey to Paris* 108 The same MS... is written in Square Capitals and very short Lines. 1883 [see RUSTIC *a.* 5 b]. 1906 E. JOHNSTON *Writing & Illumin. & Lettering* I. i. 37 Square Capitals were formed, pen-made Roman Capitals, of the monumental type. 1897 K. S. RANJITSINHJI *Jubilee Bk. Cricket* iv. 164 A square-cut travels somewhere between point and third-man. It is

the commonest form of cut. 1956 N. CARDUS *Close of Play* 150 Those who saw him will cherish memories of his vehement hooking..his square-cuts. 1976 DEXTER & MAKINS *Testkill* 167 Hunt let the first ball go by, then square-cut the second with great majesty. 1920 D. J. KNIGHT in P. F. Warner *Cricket* 32 Another beautiful square cutter is J. T. Tyldesley. 1870 L. M. ALCOTT *Old-Fashioned Girl* vii. 132 I'm going to begin with a redowa, because..it's better fun than square dances. 1902 *Encycl. Brit.* XXVII. 375/2 'Dull Sir John' and 'Faine I would' were square dances popular in England three hundred years ago. 1955 *Times* 28 June 11/4 The term 'square dance' is the American equivalent of the English 'country dance'. 1959 *Manch. Guardian* 7 Aug. 6/4 The entire population turns out to square-dance in the main streets. 1976 *West Lancs. Evening Gaz.* 15 Dec. 1. 8/2 One woman square dancer had fallen on the polished floor of the hall. 1977 *Times* 24 Jan. 4/7 The mass free admission square dance..bore witness that square dancing is alive and well across the continent. 1900 W. J. FORD *Cricketer on Cricket* xii. 140 His strokes are limited to the off-side, chiefly cuts and square drives. 1954 J. FINGLETON *Ashes crown Year* 271 May brilliantly square-drove him for 4. 1977 *Guardian* 3 Jan. 11/7 Amiss began the afternoon by square driving Bedi's first ball for four. 1930 *Engineering* 7 Mar. 303/2 He then analyses the main differences between a square engine and one with a stroke:bore ratio of 2. 1964 J. BRAINE *Jealous God* viii. 136 'Square-eyed sods,' he said. 1976 *Listener* 8 July 2/2 He called the television set 'the Devil's Box', claimed..that it would turn the bronzed, outdoor-loving youngster into a round-backed, square-eyed weakling. 1915 Square Hebrew [see note in Dict., sense 14 a]. 1948 D. DIRINGER *Alphabet* 261 The Aramaic script therefore became the parent of the 'square Hebrew'. 1974 *Encycl. Brit. Micropædia* IV. 983/1 Between the 6th and 2nd centuries BC, Classical, or Square, Hebrew gradually displaced the Aramaic alphabet. 1934 *Detective Fiction Weekly* 21 Apr. 113/1 The man who works for a living..is generally referred to in terms of contempt such as working stiff, Honest John, Square John, sucker or scissor bill. 1935 A. J. POLLOCK *Underworld Speaks* 112/2 *Square John*, a dope peddler who is not addicted. 1962 'K. ORVIS' *Damned & Destroyed* ix. 62, I played it even safer with those uptown Square Johns. 1968 *Daily Colonist* (Victoria, B.C.) 2 Nov. 8/1 He kept saying that McWhirter was a 'square John'. 'What does a "square John" mean? Does it mean an ordinary law-abiding citizen?' Mr. Owen-Flood asked. 'As far as I know,' Porter replied. 1921 *Physical Rev.* XVIII. 263 For the weaker [magnetic] fields there is a decided curvature in the lines which gradually smooth out into practically straight lines. It is in this lower region that the 'square law', proposed by Sir J. J. Thomson, holds. 1945 *Electronic Engin.* XVII. 734/2 The valve will work satisfactorily as a square law rectifier. 1958 W. T. O'DEA *Social Hist. Lighting* i. 8 A room forty feet by thirty could be lit quite cheerfully by the candles and would be dismal by the light of a single bulb. The reason is the 'square law' so well known to anyone who has studied physics. Ten feet away from a light source the illumination per unit area is only one-hundredth of what it is one foot away, and so on. 1976 *Nature* 29 Jan. 294/2 The vector voltmeters had a band-width of 1 kHz and employed a square-law detector following the narrow-band filter. [1931 C. R. S. PITMAN *Game Warden among his Charges* i. 3 There is the huge, square mouth—from which it derives its sobriquet of 'square-lipped'.] 1961 *New Scientist* 9 Nov. 340/2 A rare animal, the white or square-lipped rhino, is threatened by extinction. 1970 *Nature* 28 Mar. 1180/1 (*caption*) By 1966 ..there were about 800 southern square-lipped rhinoceroses in the Hluhluwe and Umfolozi reserves in Natal. 1912 C. B. HAYWARD *Pract. Aeronaut.* 349 Design in this field [*sc.* aircraft engines] has..gone back to automobile standards of several years ago when it was customary to build what are known as square motors, i.e. those in which the bore and stroke are the same. 1881 *Proc. Zool. Soc.* 726 The square-mouthed rhinoceros is a huge ungainly-looking beast. 1915 ROOSEVELT & HELLER *Life-Histories Afr. Game Animals* II. xxi. 662 The square-mouthed rhinos..seemed to be of a perceptibly lighter gray. 1853 DICKENS *Bleak Ho.* xxxviii. 379 A little jingling square piano. 1938 R. FIELD *All this & Heaven Too* (1939) I. xvi. 208 One of the girls went to a square piano and began trying the keys. 1980 *Early Music* July 377/2 Included in the sale was the Zumpe square piano of 1766 (£3,200) the earliest known piano to have been made in England. 1787 *Brit. Patent* specn. 1596 (*caption*) This figure represents the movement of a square Piano Forte. 1965 P. HONEY *Planning Electricity in House* iii. 75 The 13-amp. plug has square pins (which are superior to round pins) and will not fit any other size of socket. 1942 BERREY & VAN DEN BARK *Amer. Thes. Slang* § 791/10 *Square rig* or *rigger*, a double-breasted uniform. 1951 N. COWARD *Star Quality* 24 Attired as they were in the usual 'Square-Rig' of British Ordinary Seamen, they caused a mild sensation. 1962 W. GRANVILLE *Dict. Sailors' Slang* 112/1 *Square rig*, uniform worn by 'men dressed as seamen', the jumper, flannel, jean collar and bell-bottomed trousers. 1979 *Navy News* Feb. 4/1 It is a once-only increase to enable these ratings to complete the replacement of old-pattern suits of square rig, without being out of pocket. 1855 C. NORDHOFF *Merchant Vessel* 285 Our mate..had never before been in a 'square-rigger'. 1940 H. F. LOCK *Basic Typogr.* iii. 30 The modern square-serif letters are derived from the 'Antiques' and 'Egyptians' of a century ago. 1967 E. CHAMBERS *Photolitho-Offset* ii. 13 Leading the contemporary field are the square serif and the sans serif. 1978 S. RICE *Bk. Design* 214 *Sans serif*, typeface design not having the small bracketing accents at the ends of the letter strokes... Square serif letters do not properly belong to this grouping. 1914 JACKSON & HELLYER *Vocab. Criminal Slang* 79 *Square-shooter*,..a dependable person; a reliable, compact-keeping person. 1928 S. LEWIS *Man who knew Coolidge* i. 51 There's a man that it's a pleasure to do business with, a square-shooter if ever there was one. 1937 WYNDHAM LEWIS *Blasting & Bombardiering* I. vi. 63 My friend..was somehow treacherous and not at all the good sport and 'square-shooter' I had supposed him to be. 1962 E. LUCIA *Klondike Kate* ii. 24 Kitty was looked upon as a 'square shooter' in the rough give-and-take game of the Dawson gambling joints. 1922 Square shooting [see *ANIMAL A. 6]. 1932 J. DOS PASSOS *1919* 428 One of them made a speech in English and another one in Sicilian saying that this was a squareshooting concern that had always treated laborers square. 1908 E. OBERG *Handbk. Small Tools* i. 29 The Acme thread..has of late become widely used, having in most instances taken the place of the square thread on account of its better wearing qualities. 1939 S. E. WINSTON *Machine Design* iii. 72 The Square Thread..is probably the most typical transmission screw thread, as its mechanical efficiency is considerably higher than that of such threads as the V thread. 1975 BRAM & DOWNS *Manuf. Technol.* iv. 120 In the square thread the sides are parallel and normal to the axis of the screw. 1944 *Jrnl. Scientific Instruments* XXI. 64 (*heading*) A simple variable 'square-wave' stimulator for biological work. 1965 *Wireless World* Sept. 460/1 One popular method of amplifier stability assessment is 'square-wave testing' in which a suitable square wave is applied to the input, and the output inspected on an oscilloscope screen. 1975 G. J. KING *Audio Handbk.* v. 120 It is assumed that the input waveform is a true square wave of very small rise time. 1939 *Physical Rev.* LVI. 890/1 The square well fitting proton-proton scattering is not as deep as that fitting proton-neutron scattering. 1954 *Ibid.* XCVI. 461/1 The smoothing of the edges of the square-well potential was of significance for the interpretation of the elastic proton scattering with heavy nuclei. 1975 W. F. HORNYAK *Nuclear Structure* iv. 242 Another simple potential well used to represent the average potential of the single-particle model is the three-dimensional square well. 1924 *Radio Times* 19 Dec. 594/1 (Advt.), Thousands of Wireless enthusiasts..are running their broadcasting reception on square wheels, enduring..distortion... Are *you* getting square wheel reception? 1977 'O. JACKS' *Autumn Heroes* viii. 111 It was almost impossible to stand upright. The truck was operating on square wheels.

b. square flipper [app. a folk etymologizing of Newfoundland dial. *fipper*, *fripper*, etc. (also used), of uncertain origin], the bearded seal, *Erignathus barbatus*, native to Arctic regions (earlier and later examples); **square-tail**, (*b*) the char or brook trout, *Salvelinus fontinalis*, native to eastern North America.

1774 G. CARTWRIGHT *Jrnl.* 12 Dec. (1792) II. 38 A squarephripper was caught in a net to-day. 1784 T. PENNANT *Arctic Zool.* 161 The Seal-hunters in Newfoundland have a large kind, which they call the Square Phipper, and say weighs five hundred pounds. 1832 J. MCGREGOR *Hist. & Descr. Sk. Brit. Amer.* I. 108 The harp seal..the hooded seal..the square flipper, the blue seal, and the jar seal. 1842 J. B. JUKES *Excursions in Newfoundland* I. 312 The 'square fipper'..is, however, very rare. 1861 L. DE BOILIEU *Recoll. Labrador Life* 91 These seals are not like the Square Frippers. 1911 D. M. LINDSAY *Voyage to Arctic* iii. 39 Square flippers..are also found on the coast [of Newfoundland]. 1957 *Beaver* Spring 49/1 They were immensely strong and could carry off a square-flipper seal single-handed. 1935 B. PERRY *And gladly Teach* vii. 151 The lake was perfect for bathing and boating. There were big 'square-tails' in it then. 1972 *Trout & Salmon* Feb. 14/2 It brought to mind an experience I had in Labrador this past summer when I was up there fishing for squaretail trout.

square, *adv.* Add: **3. d.** *U.S. colloq.* Completely, exactly.

1862 E. S. PHILBRICK *Let.* 2 Nov. in E. W. Pearson *Lett. from Port Royal* (1906) 103 His heart failed him and he backed square out. 1880 'MARK TWAIN' *Tramp Abr.* xvii. 152 He..shot the dragon square in the center of his cavernous mouth. 1903 A. D. MCFAUL *Ike Glidden* xxvi. 236 Hain't I bin a-runnin' my legs right square off this four days? 1921 R. D. PAINE *Comr. Rolling Ocean* i. 8 It surely did hit me square between the eyes.

e. *Mus.* In a square fashion (*SQUARE *a.* 7*).

1960 L. BERNSTEIN *Joy of Music* 105 Now that you've heard what syncopation is like, let's see what that same Blues we heard before would sound like without it... *Played 'square' by sax, no vibrato.*

4. b. *square on*: (*a*) *Cricket*, of a bowler: having one's body square to the batsman; (*b*) *fig.*, directly, honestly. Also as *adj. phr.*, straightforward.

1963 A. ROSS *Australia 63* ii. 65 His actual delivery, a shade reminiscent of George Tribe's, is made more square-on than is classical. 1968 K. WEATHERLY *Roo Shooter* 145 Jim, we've got to look at this square on. We haven't given the game away. It's given us away. 1977 *Lancashire Life* Dec. 60/2 Some of his more ambitious attempts, however, don't quite 'come off', while others are 'square-on' to the point of suggesting a lack of imagination.

5. b. *Cricket.* At right angles to the line of the delivery.

1851 J. PYCROFT *Cricket Field* vii. 154 Practise diligently with leg-balls, till balls..a little wide of leg-stump go nearly square. 1891 [see *FINE *a.* 7 f]. 1909 W. CAFFYN *Seventy-One not Out* viii. 90 He was a fine leg-hitter, generally hitting square. 1963 T. E. BAILEY *Improve your Cricket* i. 38 (*caption*) The ball has been played behind square, but it can be played in front.

square, *v.* Add: **I. 1. d.** Also *fig.*

1930 T. S. ELIOT tr. *St.-J. Perse's Anabasis* 53 The boundless unreckoned year, squared out with dawns and fires.

3. e. *refl.* To put oneself into a posture of defence. Also *fig.* Cf. sense 11 in Dict. and Suppl. orig. *U.S.*

1823 J. F. COOPER *Pioneers* II. xv. 223 Square yourself, you lubber,..and we'll soon know who's the better man. 1864 'MARK TWAIN' *Celebrated Jumping Frog* (1867) 107 Caesar..squared himself to receive his assailants. 1893 *Harper's Mag.* Mar. 643/1 With a look of determination on his face, [he] squared himself to write. 1977 J. I. M. STEWART *Madonna of Astrolabe* i. 28, I was squaring myself to the necessity of telling him that I was no good for the purpose he had in mind when our walk came suddenly to an end.

f. *Assoc. Football*, etc. To pass (the ball) across the pitch, esp. towards the centre.

1972 G. GREEN *Great Moments in Sport: Soccer* viii. 85 Bloomfield made it 2–3 from a squared header by Groves. 1976 E. DUNPHY *Only a Game?* iv. 113 Having to go round the back of the goal and square it back before you could score. 1978 *Times* 12 Jan. 10/3 Tait..squared the ball into the stride of Rafferty who hit in a first-time shot.

II. 4. e. *colloq.* With *away*: to put in proper order, to tidy up, to 'sort out'.

1909 R. A. WASON *Happy Hawkins* (1912) xvi. 203 She had a head on her, Barbie had, an' when she got squared away, she made 'em all get down an' scratch. 1947 *Seafarers' Log* 19 Dec. 10/1 Motion carried that Ship's Delegate contact all tripcard men who have not acted in a way becoming to a Union man and get them squared away. 1956 H. KURNITZ *Invasion of Privacy* xiv. 87 Let's get you squared away, Mr. Jarrold. You knew damn well what you were doing..didn't you? 1966 T. PYNCHON *Crying of Lot 49* i. 20 He outlined what she was in for.. decide what to liquidate and what to hold on to, pay off claims, square away taxes, distribute legacies. 1980 'R. B. DOMINIC' *Attending Physician* ii. 12 We've got Mrs Bertilucci squared away... Sorry about her yelling at you.

5. d. (Earlier examples.)

1821 J. A. QUITMAN *Let.* 4 Dec. in J. F. H. Claiborne *Life & Corr. J. A. Quitman* (1860) I. 69, I paid my $25, squared my bill, and departed. 1855 'Q. K. P. DOESTICKS' *Doesticks, what he Says* xvii. 141 Damphool squared up his broad bill, and paid his washerwoman, which left him dead broke.

e. *Sport.* To make the scores of (a match, etc.) equal. Also *absol.*

1923 *Daily Mail* 8 May 12 The American captain missed his chance to square at the 17th, where he had a putt of 4 feet to win the hole. 1926 WODEHOUSE *Heart of Goof* ii. 66 Bradbury, driving another long ball, won the fifteenth, squaring the match. 1955 *Times* 26 July 3/1 England, in fact, if they are to win..must make 366 in six hours to-day, while South Africa, to square the rubber, have eight wickets yet to take. 1976 J. SNOW *Cricket Rebel* 54 The last few tense and dramatic overs later on in the final Test in Georgetown when the West Indies were pressing for victory to square the series.

6. c. *Austral.* With *off*: to placate or conciliate (a person). Also *intr.*

1945 BAKER *Austral. Lang.* vi. 134 *Square off, to*, to apologize, to produce a glib explanation for some lapse or misdeameanour. 1969 *Courier-Mail* (Brisbane) 15 Jan. 7/3 Moloney said..that he had been drinking at hotels... As they were driving home, he..decided..to buy beer to 'square off' with his wife. 1976 *Nature* 19 Feb. 519/2 Squaring off the proprietors of the three national chains of newspapers, whose unquestioning support he [*sc.* Mr. Fraser] enjoyed throughout the campaign.

III. 11. d. With *off*: to assume a fighting attitude. Also *fig.* orig. *U.S.*

1838 J. C. NEAL *Charcoal Sketches* 41 If he 'squares off' at a big fellow, he is obliged..to hit his antagonist on the knee. 1856 'Q. K. P. DOESTICKS' *Plu-ri-bus-tah* xi. 126 Then, at once, squared off at Cuffee, Instantly 'sailed into' Cuffee. 1873 J. H. BEADLE *Undevel. West* xxxvi. 773 The bow appeared to be rearing up to square off at the midday sun. 1942 E. PAUL *Narrow St.* xxix. 267 The rest of the world were squaring off for a life-and-death struggle. 1960 T. MCLEAN *Kings of Rugby* 170 Wellington was still ahead, 6 to 3, as the two teams squared off for the second half. 1974 'E. LATHEN' *Sweet & Low* iv. 49 The yelling started... It sounded to me like they were going to square off.

12. c. (Earlier and later examples.) Also *fig.*, to get moving; to put oneself into shape; to make ready.

1849 N. KINGSLEY *Diary* 3 Sept. (1914) 57 The wind died away and soon sprung out from the South and [we] squared away before it. 1868 H. WOODRUFF *Trotting Horse Amer.* ix. 101 They must be wakened up from time to time, so as to make them get out of their sluggish habit and square away. 1889 'MARK TWAIN' *Connecticut Yankee* xxxvii. 479, I didn't waste any time..but squared away for business. 1961 'E. LATHEN' *Banking on Death* (1962) xiii. 103 Miss Todd wasted no more time on idle chatter but squared away to her typewriter.

14. In Comb., as **square-up** *colloq.*, a quarrel.

?1949 DYLAN THOMAS *Sel. Lett.* (1966) 339 Bert and I had a regular square-up, but he came over to my way of thinking.

squared, *ppl. a.* Add: **1. a.** Also with *off.*

1965 J. A. MICHENER *Source* (1966) 580 No square corners would be allowed, no neatly squared-off towers. 1973 D. MAY *Laughter in Djakarta* v. 83 Tapping some magazines into a squared-off pile.

c. (Earlier example.)

1887 *Science* 11 Mar. 240/1 The graphic representation of the results will be readily understood. It is only necessary to take a sheet of 'squared' paper, or paper ruled in two directions at right angles to each other.

2. (Later examples.) Also *fig.*

1964 J. W. LINNETT *Electronic Struct. Molecules* vi. 92 There will be five squared-terms..and twenty cross-terms ..in the square of the total function. 1979 *Tucson* (Arizona) *Citizen* 20 Sept. 11A/1, I don't know if tarantulas are considered insects or just some form of spider.. squared.

squa·rehead. *slang.* Also **square-head.** [SQUARE *a.* 9 c.] **1.** An honest person; one who is not a criminal (see quots.).

1890 BARRÈRE & LELAND *Dict. Slang* II. 293/1 'Honesty among thieves' is undoubtedly the production of a square-head or sham thief; a good thief will rob anybody. 1950 *Austral. Police Jrnl.* Apr. 119 *Square-head* (or *squarey*), one who has no convictions. 1977 *Courier-Mail* (Brisbane) 16 Feb. 5/1 Half an hour or so later a squarehead (criminal slang for someone who is going straight), who was minding the money in a bag for Blair, handed it over to one of the Toecutters.

2. A foreigner of Germanic extraction, esp. a German (spec. *Army slang* in the war of 1914–18) or Scandinavian.

1903 FARMER & HENLEY *Slang* VI. 333/1 *Squarehead*,.. a German or Scandinavian. 1906 *Soldier Slang* in C. McGovern *Sarjint Larry an' Frinds*, *Square-head*, a soldier of German birth and addicted to the use of German idioms. 1918 H. C. WITWER in *Collier's* 16 Mar. 21/2 The English call 'em 'Uns..we call 'em squareheads. 1932 J. DOS PASSOS *1919* 61 She started firing across the North Star's bows with a small gun that the squareheads manned. 1942 C. BARRETT *On Wallaby* v. 95 A bunch of Germans with this mob behaved very differently: the square-heads were sullen and surly. 1953 *Times Lit. Suppl.* 23 Oct. 678/2 Those of Swedish extraction in America, known as squareheads, suffer the reputation of being dull, heavy, stupid, simple. 1973 D. JONES *Let.* 15 Apr. in R. Hague *Dai Greatcoat* (1980) iv. 244 In the ranks we often used 'Squarehead' of old Jerry—'That poor old "squarehead" they brought in from last night's raid looked pretty far gone to me.'

3. In misc. *transf.* and *fig.* uses.

1919 *Dialect Notes* V. 62 *Square-head*, a dull, stupid person. 1949 *Sun* (Baltimore) 26 Oct. 10/3 A real 'square-head' seaman, who knows what it is like to drift for days without food. 1958 'CASTLE' & 'HAILEY' *Flight into Danger* i. 13 This plane is crammed with squareheads who are going to Vancouver..to root like hell for their boys.

square-headed, *a.* Add: **d.** Level-headed, sensible.

1896 H. G. WELLS *Let.* 24 Jan. in *Exper. Autobiogr.* (1934) I. vi. 402 He's a first rate, square headed, thoroughly honest man. 1922 JOYCE *Ulysses* 82 Squareheaded chaps those must be in Rome: they work the whole show.

e. Stolid, dull.

1936 C. S. LEWIS *Allegory of Love* iv. 172 To believe thus is to attribute to Chaucer a square-headed vulgarity of thought and feeling.

f. Describing one of a Germanic race.

1942 E. PAUL *Narrow St.* xxix. 268 The square-headed, owl-like Nazis. 1979 O. SELA *Petrograd Consignment* 35 A big, square-headed man.

Hence **square-hea·dedness.**

1930 *Times Lit. Suppl.* 10 Apr. 321/1 An underlying sanity and square-headedness about Huneker's critical judgments.

squareness. Add: **3. b.** *Assoc. Football*, etc. Of a defence: the condition of being square (*SQUARE *a.* 6 d) and lacking in depth.

1978 *Guardian Weekly* 19 Nov. 23/5 Taking full advantage of Manchester United's inadequacy in the air and punishing the squareness of their defence on the ground.

4. Conventionality, dullness. Cf. *SQUARE *a.* 9 d.

1961 *John o' London's* 16 Nov. 548/2 Where Squareness is the ultimate low, anything can get by if it proclaims itself Hip loudly enough. 1972 M. J. BOSSE *Incident at Naha* ii. 119, I was beginning to care about our Bostonian, even though he was capable of, like, ultimate squareness. 1977 P. USTINOV *Dear Me* xiii. 173 It is out of fear of what is known as squareness that we rarely say what we really think or feel.

squarer. Add: **4.** *Electronics.* A device that converts a sinusoidal or other periodic wave into a square wave of the same period.

1965 *Wireless World* Aug. 402/1 If the square wave output is integrated and used to bias the first transistor, and if the squarer is d.c. coupled throughout and has an odd number of stages, then the feedback maintains the mark-to-space ratio constant. 1971 *Physics Bull.* July 427/1 The basic module was an inverter and was demonstrated as a squarer, then two or more were joined together to produce various logic gates and multivibrators.

Squaresville (skwɛ̄ᵊ·ɹzvil). orig. *U.S.* Also **Squareville.** [f. SQUARE *sb.* or *a.*: see *-VILLE.] An imaginary town characterized by dullness and conventionality. Also *attrib.* or as *adj.*

1956 'E. MCBAIN' *Cop Hater* (1958) viii. 75 This guy is from Squaresville, fellas, I'm telling you. He wouldn't know a ·45 from a cement mixer. 1961 *Times* 27 Apr. 17/2 To look round a crowded room and murmur 'Squareville' or 'Hicksville' may be elliptical, but is certainly more effective than the full form. 1967 P. WELLES *Babyhip* (1968) ii. 31 It's such squaresville talk. 1968 *Listener* 11 July 51/1 And they went away, more than ever convinced that the war between the generations was for real. And through the window there floated a querulous, puzzled voice. 'A queer fish, real squaresville.'

square-toed, *a.* Add: **1. b.** *transf.* in *U.S. Naut.* use. Now only *Hist.*

1851 H. MELVILLE *Moby Dick* I. xvi. 110 You may have seen many a quaint craft in your day, for aught I know: square-toed luggers, mountainous Japanese junks. 1886 *Forest & Stream* 13 May 316/3 Even the regular 'square-toed' schooner, supposed to be original to San Francisco, flourishes up the James. 1948 R. DE KERCHOVE *Internat. Maritime Dict.* (1958) 771/1 *Square-toed frigate*,..local name given in Quincy, Mass., to the scow sloops engaged in the granite trade. (Obsolete.)

squaring, *ppl. a.* Add: **3.** That multiplies a quantity by itself.

1961 *Technology* Feb. 44/2 An automatic neon overload indicator..detects signal peaks exceeding the range of the squaring unit. 1979 *Proc. London Math. Soc.* XXXVIII. 514 The squaring map $a \to a^2$, determined by this Jordan product, is the same as that determined by functional calculus.

squarish, *a.* Add: Occas. **square-ish.** (Later examples.)

1921 S. COLVIN *Memories & Notes* viii. 130 In spite of her squareish build she was supple and elastic in all her movements. 1950 M. PEAKE *Gormenghast* lxxvii. 417 A largish, square-ish room. 1981 *Times* 21 Apr. 14/5 Later, the mausoleum was superseded by a small squarish chapel.

squash, *sb.*[1] Add: **I. 3. a.** *squash-ball* (example), *-racket* (= the bat used in the game: examples), *-rackets* (= the game: later examples); **squash tennis** *U.S.*, a game similar to squash rackets, played with a lawn-tennis ball.

1901 E. MILES *Game of Squash* i. 16 Americans generally use a Lawn-Tennis ball..and a Squash-Tennis racket, which is like a miniature Lawn-Tennis racket. 1905 H. A. VACHELL *Hill* ii. 26 He bought..a 'squash' racquet, 'squash' balls, and a yard ball. 1917 *National Squash Tennis Assoc. Rules* 18 Description and specifications of a Squash Tennis Court, as adopted by the National Squash Tennis Association... The dimensions of a Standard Court shall be..Length 32 feet 6 inches Width 17 feet. 1928 *N.Y. Times* 12 Dec. 32/6 The success of the American players in the English squash racquets tournament... At the [American] colleges..squash racquets became more popular. 1930 A. DANZIG *Racquet Game* III. i. 157 The two varieties of squash—squash tennis and squash racquets—have so much in common that they may be called first cousins. 1973 *Times* 28 Sept. 5/6 There is a wide disparity between the international and American versions of squash rackets. 1975 *Oxf. Compan. Sports & Games* 985/1 The squash racket is not as strong as the rackets racket. *Ibid.* 992/2 Squash tennis was born in a school in Concord, New Hampshire,..but..was refined by Feron, of New York, who first wrapped netting round the ball.

b. *ellipt.* for *squash rackets* or occas. (*U.S.*), *squash tennis.*

1899 *N.Y. World* 8 Aug. 14/4 'Squash'..is a variation of the time-honored court tennis. 1902 E. MILES (*title*) Racquets, tennis and squash. 1930 [see *squash tennis*, sense 3 a above]. 1952 J. B. PICK *Phoenix Dict. Games* 183 Squash is played with a rubber ball on a four-walled court. *Ibid.* 185 Service in squash is not the deadly weapon it is in rackets. 1975 *Oxf. Compan. Sports & Games* 986/1 Squash is derived from, and has much in common with, the much older game of rackets, and originated at Harrow School.

II. 5. c. A social gathering; an informal religious or literary meeting.

1904 H. JAMES *Golden Bowl* I. III. xiv. 252 The intrinsic oddity of the London 'squash', a thing of vague, slow, senseless eddies. 1916 L. EINSTEIN *Let.* 31 July in *Holmes-Einstein Letters* (1964) II. 134 The season, however, which I loathe, was rendered nicer by the war ..no more fat squashes but agreeable dinners and luncheons. 1938 M. WHITLOW *J. Taylor Smith* xi. 114 The Inter-varsity Christian Fellowship have a 'date' with him for a series of 'Squashes' at Oxford and Cambridge. 1977 L. GORDON *Eliot's Early Years* iii. 47 In 1912, Conrad Aiken took 'Prufrock' to a 'poetry squash' in London and showed it to Harold Monro. 1979 *PN Review 13* 19/1 Lord, you know that next week is the Freshers' Squash.

6. b. *Biol.* A preparation of softened tissue that has been made thin for microscopic examination by pressing or tapping.

1942 DARLINGTON & LA COUR *Handling of Chromosomes* v. 40 Sections have now been largely replaced by smears and squashes for all but the smallest masses of material. 1971 *Nature* 18 June 452/2 Fifty cells of three females and thirty-two cells of four males were examined. A testis squash was also made to give meiotic figures. 1981 *Japanese Jrnl. Genetics* LVI. 529 This method enables the observation of both C-banding patterns and the karyotypes by aceto-orcein squash technique in the same chromosome complement.

7. Also, a drink made from the juice of crushed fruit other than lemons; = *CRUSH *sb.* 4 e. Freq. as second element of combinations: see *lime-squash* s.v. *LIME *sb.*[2] 2, *orange squash* s.v. *ORANGE *sb.*[1] 7 a.

1914 C. MACKENZIE *Sinister Street* II. III. vii. 644 Will you have a squash and a biscuit? 1936 *Discovery* June 192/1 Fruit Squashes, containing the pulp of the fruit, were analogous to the well-known orange and lemon squashes. The blackcurrant squash was remarkable for its delicate flavour. 1939 A. P. HERBERT *Water Gipsies* (rev. ed.) x. 99 Jane suggested that they should..have a lemon-squash... Ernest did not want a squash. 1967 *Coast to Coast 1965–6* 185 'Come on in and I'll make you a squash.'..Meg squeezed a lemon for his drink. 1980 *Brit. Med. Jrnl.* 29 Mar. 913 Most fruit squashes are unsuitable for babies.

8. *attrib.* as **squash bite** *Dentistry*, an impression of the teeth made by biting the jaws together on a piece of plastic material.

1914 N. G. BENNETT *Sci. & Pract. Dental Surg.* xxxvii. 607/1 A 'squash bite' impression, or even one taken in the ordinary way in a tray, is very easily distorted in removing from the mouth. 1940 J. OSBORNE *Dental Mechanics* v. 47 It is usual if this type of block is used to have taken a 'squash bite' at the impression stage. 1963 C. R. COWELL et al. *Inlays, Crowns, & Bridges* ii. 10 The relationship of the prepared tooth to adjacent and opposing teeth must be recorded in the indirect technique with a wax or an alginate squash bite.

squash, *sb.*[2] Add: **4. squash-berry,** the red berry of *Viburnum pauciflorum*, a deciduous North American shrub; = *moose-berry* s.v. MOOSE[1] b; **squash blossom,** the flower of the plant on which squashes grow, applied *attrib.* to jewellery made by the Navajo which is characterized by designs (of Spanish, and ult. Moorish, derivation, representing pomegranates) resembling this flower.

1937 LADY ROCKLEY *Some Canadian Wild Flowers* 77 Its [*sc. Viburnum pauciflorum's*] berries known as the 'Squash-berry' are gathered and make an excellent preserve. 1966 A. R. SCAMMELL *My Newfoundland* 32 He even shook his head at bakeapple jam, squashberry jelly and 'meshberries'. 1974 J. E. UNDERHILL *Wild Berries Pacific Northwest* 23 Huckleberries, Blueberries, Squash-berries, and many others, may be made into delicious jams. 1923 D. H. LAWRENCE *Birds, Beasts & Flowers* 19 The fig, the horseshoe, the squash-blossom. Symbols. 1930 D. & M. R. COOLIDGE *Navajo Indians* xvi. 115 The beautiful squash-blossom pendants which the Hopis like so much. 1944 J. ADAIR *Navajo & Pueblo Silversmiths* v. 83, I watched Charie make other pieces. One of them was a squash-blossom bead. 1950 S. H. BABINGTON *Navajos, Gods & Tom-Toms* xv. 170 The pronged pieces in the beautiful so-called squash blossom necklace are the buttons which were sewed along the outside seams..of Spanish army officers' pants. 1977 C. MCFADDEN *Serial* (1978) ii. 10/2 Carol..had embellished it with her trade-mark jewelry: an authentic squash-blossom necklace.

squashed, *ppl. a.* (In Dict. s.v. SQUASH *v.*[1]) Add: **2.** Special collocations: **squashed fly (biscuit)** *colloq.* = *GARIBALDI 3; **squashed tomato** *slang*, a name given in different localities to various children's games (see quots.).

1900 J. S. FARMER *Public School Word-bk.* 85 *Squashed flies*,..biscuits with currants. 1909, etc. [see *GARIBALDI 3]. 1931 C. LITHGOW *Simple Sailor* v. 49 In 'the break', they grappled for their milk and bun, or 'squashed-fly' biscuit. 1977 K. M. E. MURRAY *Caught in Web of Words* xvii. 321 Gwyneth remembered her anguish as a little girl at finding nothing in her parcel but a Garibaldi ('squashed fly') biscuit. 1959 I. & P. OPIE *Lore & Lang. Schoolch.* xviii. 381 There are more than sixty established names for the pursuit of illegally knocking at doors... *Squashed tomato.* Wolverhampton. 1963 S. MARSHALL *Exper. in Educ.* ii. 56 The new look given to the age-old playground game which for some unknown reason has become 'Squashed Tomato' in the language of today... A voice was giving orders to the players. 'John, three scissors south towards Cambridge. Carol, two pigeon steps towards Newmarket [on a map painted on a school playground].' 1969 I. & P. OPIE *Children's Games* iv. 157 The game is usually known as 'Sardines', but also.. 'Squashed Sardines', and 'Squashed Tomatoes'. *Ibid.* vi. 189 *Squashed Tomato.* Both caller and called run towards each other, with arms crossed in front of them. The one advancing remains at the spot where they squash into each other. The caller returns to his place in front.

squashily (skwǫ·ʃili), *adv.* [f. SQUASHY *a.* + -LY[2].] In a squashy or squelchy manner.

1922 *Blackw. Mag.* Oct. 485/2 A small damp object.. struck the ground squashily near where I was standing. 1924 'L. MALET' *Dogs of Want* viii. 262 Upon the seat.. Mr. Noakes heavily, not to say squashily, subsided.

squat, *sb.*[1] Add: **6. a.** *spec.* in Gymnastics and Weight-lifting.

Earlier called *crouch*.

1954 M. FALLON *Muscle Building for Beginners* x. 56 Keep the head up and the back flat, and resist any temptation to lean forward, particularly at the lowest point of the squat. 1959 LOKEN & WILLOUGHBY *Compl. Bk. Gymnastics* iv. 35 *Squat Head Balance.* Start this stunt from a squat position with the hands on the mat and the inside of the knees resting on the elbows. 1964 G. C. KUNZLE *Parallel Bars* ii. 42 Simple squats on one bar. *Ibid.* 44 Complete the squat off by pushing away strongly with the arms and drawing the shoulders forwards. 1977 J. F. FIXX *Compl. Bk. Running* vii. 91 Some runners and coaches think weightlifting is essential to good performances. Emil Zatopek..used to do squats while holding his wife, Dana, on his shoulders.

b. *hot squat*: see *HOT *a.* 1 e.

8. a. The illegal occupation of an uninhabited building (esp. by a group of homeless people organized for this purpose); the period of such an occupation.

1946 *Daily Mail* 20 Sept. 2/3 The Great Squat is over... Today at 1200 hours the rearguards of Squat-Force will retire. 1963 S. COOPER in Sissons & French *Age of Austerity* 44 Early in September 1946 Londoners were startled by what was christened the Great Sunday Squat. 1969 *Guardian* 27 Sept. 9/2 The Diggers have decided not to take part in any more hippie squats. 1970 N. SAUNDERS *Alternative London* xvii. 122 They then organised the squatters in East London, which has developed into the longest squat ever, lasting over ten months, in Arbour Square. 1975 *Times* 8 Jan. 3/6 This is the biggest squat ever, a serious attempt to house homeless people. 1981 *Daily Tel.* 3 Mar. 2/1 This squat cost the ratepayer £46,700—money we need not have spent had the squat not taken place.

b. A house, flat, or building occupied by squatters; a squatter's place of residence.

1975 *Guardian* 26 Sept. 5/8 He's at 14 Algernon Road. It's a squat. 1977 M. DRABBLE *Ice Age* II. 211 They'd been hosed out of their last squat. 1980 *Daily Tel.* 28 Oct. 17/3 A whipround among punks from a squat near the police station raised £12.50.

squat (skwǫt), *sb.*[4] *U.S. slang.* [Prob. f. slang *to squat* to void excrement.] Nothing at all; (following a negative construction) anything. Orig. as second element of phr. *doodly-squat* [prob. f. U.S. slang *doodle* excrement].

1934 Z. N. HURSTON *Jonah's Gourd Vine* xviii. 217 She ain't never had nothin'—not eben doodly-squat, and when she gits uh chance tuh git holt uh sumpin de ole buzzard is gone on uh rampage. 1946 MEZZROW & WOLFE *Really Blues* viii. 107 These cats weren't from doodlely-squat. *Ibid.* 373 *Doodlely-squat*, nothing, no more than the product of a child who squats to do his duty. 1967 WENTWORTH & FLEXNER *Dict. Amer. Slang* Suppl. 705/2 *Squat*,.. = zot. *Ibid.* 712/2 *Zot*,..a grade or a score of zero. 1975 G. V. HIGGINS *City on Hill* i. 18 A lot of people that didn't care squat about the war went with us on that point. 1977 *Rolling Stone* 30 June 82/1 Under no circumstances would I ask those..judges down in Oswego to give him back his shingle on the condition it doesn't mean doodly-squat. 1979 P. BENCHLEY *Island* ii. 26 It'll be another forecast-of-Armageddon cover that won't amount to squat.

squat, *pa. pple.* and (*ppl.*) *a.* Add: **2.** (Later examples.)

1956 G. E. EVANS *Ask Fellows who cut Hay* xxv. 228 Another feature of the dialect is the expressive vigour of many of the words and phrases:..*squat* (pronounced with a very broad *a*) hidden or quiet. 1962 M. PROCTER *Devil in Moonlight* xv. 155 We'll keep it squat and take a chance on having trouble later.

4. *squat-lobster*: for (see quot. 1902) read: a crab-like marine animal belonging to the family Galatheidæ, and add later examples.

1928 RUSSELL & YONGE *Seas* iii. 67 There are also squat-lobsters, which have long claws and broad, flattened bodies. 1978 *Sci. Amer.* Dec. 99/2 The galatheids (the squat-lobsters, a group intermediate between the macrurans and the true crabs) have reflecting superposition eyes with square facets.

squat (skwǫt), *adv. rare*[−1]. [f. the (*ppl.*) *a.*] In a direct and straightforward manner, 'flat'.

1909 KIPLING *Songs from Books* (1913) 24 Tell old Winter, if he doubt, Tell him squat and square-a!

squat, *v.* Add: **II. 9. c.** To occupy an uninhabited building illegally (esp. said of a group of homeless people organized for this purpose); to live as a squatter (*SQUATTER *sb.*[1] 1 d).

An isolated early example of this sense is entered as *transf.* under sense 9 a in Dict. (its correct date, however, is 1880, not 1879 as given there).

1937 'G. ORWELL' *Road to Wigan Pier* v. 81 In one town I remember a whole colony of them who were squatting, more or less illicitly, in a derelict house which was practically falling down. 1946 *Daily Worker* 9 Sept. 4/3 We..decided to assist homeless people to squat in certain of these buildings. 1969 *Listener* 15 May 665/1 No one expects to see 40,000 people squatting this year as there were 23 years ago. 1969 *Peace News* 13 June 5/1 One startling realisation..is how few is the number of families that have had the courage to squat. 1980 *Oxf. Compan. Law* 1171/2 Persons may squat in buildings by reason of inability to find other accommodation and may do so deliberately as a protest against shortage of housing in the area.

d. *trans.* (*a*) To install (someone) as a squatter. (*b*) To occupy (a building) as a squatter.

1973 *Guardian* 23 Mar. 9/5 Shelter, the campaign for the homeless, has squatted a homeless family of six people in an Ealing council house..reserved by the council for a homeless widow with four children. 1975 *Daily Tel.* 22 July 12 Much has been made of cases in which occupied privately-owned property has been squatted in the temporary absence of the owner. 1976 *Milton Keynes Express* 28 May 11/7 He added that he squats dozens of homeless people in corporation or council houses. 1977 *It* June 5/1 By January '76 the place was squatted by Enrique Ahriman, self-styled Demon of Confusion.

squa·t board, squatboard. *Naut.* [f. SQUAT *v.* + BOARD *sb.*] (See quots.)

1905 *Rudder* Feb. 62/2 There is one institution on the St. Lawrence which, it is believed, is peculiar to it and but little known of elsewhere; the 'squat board'. This is an appendage in the form of a horizontal plane attached to the stern. 1953 M. V. BREWINGTON *Chesapeake Bay: Pictorial Maritime Hist.* vi. 167 The stern settled so badly when underway 'squatboards' were necessary. 1955 J.-O. TRAUNG *Fishing Boats of World* I. 8/2 To overcome the resulting 'squatting' or settling aft, 'squatboards' have been added. These are flat wooden fins or planes placed at and nearly parallel to the waterline under and abaft the stern, to hold the stern up when the boat is driven hard.

squat tag. *U.S.* [f. SQUAT *v.* + TAG *sb.*[2]] A version of the game of tag in which a player may gain temporary immunity by squatting on the ground.

1883 W. W. NEWELL *Games & Songs Amer. Children* xi. 159 In *squat-tag*, the fugitive is safe while in that position. 1960 V. WILLIAMS *Walk Egypt* 95 Half a dozen children played Squat Tag around a wagon.

squatted, *ppl. a.* Add: **4.** Occupied by squatters. Also with *in*.

1963 S. COOPER in Sissons & French *Age of Austerity* 46 The squatted-in camps were..a useful temporary stop-gap for the housing problem. 1973 *Guardian* 5 Mar. 6/4 Everyone dropped in on another squatted house. 1975 *Time Out* 25 July 5/4 Islington Council is currently threatening to have electricity and gas supplies cut off to dozens of squatted houses in Finsbury Park. 1981 *Times* 14 July 2/4 According to Mrs Jean Styles..only two flats are empty and two 'squatted'.

squatter, *sb.*[1] Add: **1.** (Examples of the early Austral. use indicated by the note to sense 2 in Dict.)

1830 J. BETTS in *Occas. Papers Univ. Sydney Austral. Lang. Res. Cent.* (1965) No. 4. 13 A clan of people called 'Squatters'. These were generally emancipated convicts, or ticket-of-leave men, who, having obtained a small grant, under the old system, or without any grant at all, sat themselves down in remote situations, and maintained large flocks, obtained generally, in very nefarious ways, by having the run of all the surrounding country. 1833 W. H. BRETON *Excursions in N.S.W.* 442 There are likewise in the colony certain persons called 'squatters' (the term is American) who are commonly..of the lowest grade. 1835 *Sydney Gaz.* 28 Apr. 2 In every part of the country squatters without any reasonable means of maintaining themselves by honesty, have formed stations, and evidently pursued a predatory warfare against the flocks and herds in the vicinity.

d. One who occupies an uninhabited building illegally (esp. as a member of an organized group).

1880 W. H. DIXON *Royal Windsor* IV. xxix. 269 The King's house was a wreck; the fanatic, the pilferer, and the squatter, having been at work. 1946 *Times* 12 Aug. 2/3 Doncaster Rural District Council has turned on the water supply for a colony of its 'squatters' in military huts at Sprotborough. 1952 M. LASKI *Village* xiii. 185 The London squatters had moved into their flats and their hotels, and triumphantly held the police and all the authorities at bay. 1968 *Guardian* 2 Dec. 1/3 The London Squatters Campaign—formed three weeks ago. 1973 LD. DENNING in *All England Law Reports* III. 395 [McPhail v. Persons unknown]. What is a squatter? He is one who, without any colour of right, enters on an unoccupied house or land, intending to stay there as long as he can. 1980 *Oxf. Compan. Law* 1171/2 A squatter is a trespasser and liable to criminal penalties if he forces entry against the opposition of the lawful occupier or if, having been warned, he fails to leave.

2. Also *N.Z.*

1872 M. A. BARKER *Christmas Cake in Four Quarters* IV. ii. 260 Amongst our most constant guests were the Scotch shepherds of a neighbouring 'squatter'. 1891, 1911, 1933 [see *COCKY *sb.*[2] 2]. 1959 P. R. STEPHENS in A. McLintock *Descr. Atlas N.Z.* 38 The squatters soon became the dominant political force in the new country.

3. b. Substitute for def.: A bronze-wing pigeon of the genus *Phaps*, either *P. elegans* or *P. chalcoptera*.

c. *Cricket.* A ball which remains low on pitching; a shooter.

1955 I. PEEBLES *On Ashes* 109 In Statham's first over to Miller there were three 'squatters'. 1959 *Times* 7 Aug. 4/4 Phelan failed by only a whisker to bowl Pataudi with a squatter.

4. squatter's (or **squatters'**; occas. †**squatter**) **right(s)** orig. *U.S.*, the right of a squatter to the land on which he has settled; also in extended and *fig.* use.

1854 H. D. THOREAU *Walden* i. 54 These are all the materials excepting the timber, stones and sand, which I claimed by squatter's right. 1857 T. H. GLADSTONE *Englishman in Kansas* 168 The 'squatter-right' to a lot of ground is bought and sold on the strength of the law.. which asserts its power by rifle and tomahawk. 1883 *Brandon* (Manitoba) *Daily Mail* 24 Feb. 4/2 The infernal row you are all making up there about grievances, monopolies, squatters' rights, etc. 1944 N. STREATFEILD *Curtain Up* xvi. 222 A talent once accepted acquired squatter's rights, as it were. 1958 B. HAMILTON *Too Much of Water* x. 209 They had, by constant use.., almost acquired squatters' rights over a small table in the aft corner. 1968 E. S. RUSSENHOLT *Heart of Continent* II. v. 76 Families already living along the Assiniboine, exercise 'squatter's rights', and lay claim to the newly-surveyed River Lots. 1973 'TREVANIAN' *Loo Sanction* (1974) 207 The lone painter..had come to assume over the years that the space, the stove, and the tea were his by squatter's right.

‖ **squattez-vous** (skwǫ·tē¹vū:), *imp. phr. slang.* [Joc. f. SQUAT *v.*, after Fr. *asseyez-vous* sit down.] Sit down.

1899 KIPLING *Stalky & Co.* 177 *Squattez-vous* on the floor, then!... I swear you aren't going to sit on *my* bed! 1959 R. POSTGATE *Every Man is God* xxiii. 216 'Squattez-vous' was an invitation to sit down—Captain Roddman offered it to tenants who called on him in his office.

squatting, *vbl. sb.* Add: **3.** Also in sense 9 c of *SQUAT *v.*

1946 *Times* 17 Aug. 2/5 A new form of 'squatting' is reported from Gravesend, Kent. 1969 *Guardian* 27 Sept. 9/2 The Diggers will continue to support squatting by genuinely homeless families. 1976 *Times* 21 May 2/3 We publicly would not yield to the left..on other issues, such as squatting.

squatting, *ppl. a.* Add: **1. b.** Occupying an empty building as a squatter.

1963 S. COOPER in Sissons & French *Age of Austerity* 44 One of the squatting families. 1970 *Guardian* 12 Sept. 5/7 Last October the first four squatting families moved into flats which..had been empty for up to two years.

squaw, *sb.* (and *a.*). Add: **5. a.** *squaw dance* (earlier example), *hitch* (earlier example); **squaw boot** (see quot. 1975); **squawman** (earlier examples); **squaw winter** (earlier examples); **squaw wood** (see quot. 1944).

1952 J. K. HOWARD *Strange Empire* 336 All the women had beaded ornaments and lavishly embroidered 'squaw boots'. 1975 C. CALASIBETTA *Fairchild's Dict. Fashion* 50/2 *Squaw boot*, below-the-knee boot made of buckskin with fringed turned down cuff at top, soft sole, no heel, worn by American Indian women and popular with young people in the 1960's. 1864 in *Beaver* (Winnipeg) (1963) Autumn 52/2 Oregon Jack gave a squaw dance at which everybody got very drunk, I believe. 1887 LEES & CLUTTERBUCK *British Columbia* 232 Other hitches there are of less fame than this, notably the 'Squaw Hitch', a comparatively simple affair. 1866 *Rep. Indian Affairs* (U.S.) 91 White men, who have located in the vicinity of the reservation, and are known as squaw men. 1877 R. I. DODGE *Hunting Grounds Gt. West* xliii. 427 Squaw men. This is the name given by Indians to those men, not of their tribe, who, by purchase of squaws (marriage), have been adopted by or are tolerated in it. 1861 *Amer. Agriculturist* XX. 321/2 The best authorities put them immediately after Squaw Winter, which is the first cold snap that destroys tender vegetation. 1871 *Lakeside Monthly* V. 4/2 Those single-minded, grand old fellows.. kicked the light snow of 'squaw winter' from their Spanish-leather boots. 1914 *Outing* June 191/2 The cooking fire is only the beginning of the possibilities of 'squaw wood'. 1944 R. F. ADAMS *Western Words* 153/1 *Squaw wood*, a slang name for dried cow chips; also used in speaking of small, dry, easily broken sticks when used for fuel. 1968 C. HELMERICKS *Down Wild River North* I. vi. 100 Anything is squaw wood that you don't have to chop.

b. squaw-berry, the edible berry of one of several shrubs, esp. the bear-berry, *Arctostaphylos uva-ursi*, an evergreen prostrate creeper; **squaw corn,** a variety of maize having soft grains of various colours.

1852 *Anglo-Amer. Mag.* I. 418/2 The partridge leads her young brood forth to feed upon the soft luscious fruits of the huckleberry and squaw-berry. 1884 M. G. C. HALL *Lady's Life on Farm in Manitoba* 162 We have had jelly made of squawberries. 1956 V. FISHER *Pemmican* (1957) 161 Dried elderberry, or squawberry or wild currant he did not care for. 1824 J. DODDRIDGE *Notes Settlement Virginia* 90 How widely different is the large squaw corn, in its size, and the period of its growth. 1914 E. STEWART *Lett. Woman Homesteader* 151 They had a small patch of land..on which was raised the squaw corn that hung in bunches from the rafters. 1975 *Daily Colonist* (Victoria, B.C.) 5 Oct. 22/5 Nowadays, squaw corn is grown purely for its highly ornamental, variegated ears with kernels in purple, mauve, red, yellow, cream and black, very nice for Thanksgiving..displays.

squawk, *sb.* Add: **1. b.** *fig.* A complaint, a protest, esp. in phr. *to set* (or *put*) *up a squawk. slang* (orig. and chiefly *U.S.*).

1909 C. B. CHRYSLER *White Slavery* ix. 70 'Snatchin' simps' is good enough for Little Willie, there is no 'fall', no squawk, all you have to do is to stall. 1914 JACKSON & HELLYER *Vocab. Criminal Slang* 79 *Squawk*,..a protest; a vociferous demonstration, as an indignant repudiation of an injustice..'If you don't put up a squawk they'll trim you.' 1948 M. LASKI *Tory Heaven* ix. 126 They was just told to shut down and shut down they did..there wasn't a squawk out of none of them. 1973 'B. MATHER' *Snowline* ii. 25 Our starry-eyed bleeding-hearts and permissives at home set up a squawk. 1976 'R. B. DOMINIC' *Murder out of Commission* xvi. 147 How in God's name can we set up a squawk? We don't know what's going on.

squawk, *v.* Add: **1. c.** *U.S. slang.* To turn informer, to 'squeal'.

1872 G. P. BURNHAM *Memoirs U.S. Secret Service* p. vii, *Play baby*, to whine; 'squawk'; or assume innocence. 1929 W. R. BURNETT *Little Caesar* VI. v. 251 You know.. Joe squawked. 1935 *Amer. Speech* X. 12/2 *Belch*, to confess or carry information to the police. Modern *to squawk*. 1937 *Times Lit. Suppl.* 25 Dec. 974/4 The thief who 'squawks' is expelled as professionally infamous; his occupation's gone.

d. *U.S. slang.* To complain, protest.

1875 J. G. HOLLAND *Sevenoaks* xvii. 239 He mustn't squawk an' try to git another feller to help 'im out of 'is bargain. 1926 J. BLACK *You can't Win* iv. 41 Usually the sucker is a married man and can't squawk. But when he does squawk..the only thing to do is to blow back his money. 1939 *Time* 23 Jan. 30/2 Since most Hummert ghosts are glad to add caviar to bread-&-butter from other jobs, they have seldom squawked. 1948 *Sun* (Baltimore) 7 Jan. 13/1 When you pass a law and hire somebody to enforce it, you can't squawk if your kids get pinched for violating it. 1951 E. PAUL *Springtime in Paris* vi. 121 The contractor had been getting away with plenty..and would not dare squawk, no matter how high a bill was presented. 1976 M. MACHLIN *Pipeline* xxvii. 319 If the EPA ever finds out and squawks, they'll just fight it out in the courts.

squaw·k box. *U.S. slang.* Also with hyphen or as one word. [f. SQUAWK *sb.* or *v.* + BOX *sb.*[2]] **a.** A loud-speaker or public-address system.

1945 *New Yorker* 17 Mar. 28/3 The squawk box became alive for that second before any message comes over. 1950 'D. DIVINE' *King of Fassarai* xxv. 219 There was a radio blaring..its music was punctuated with the hoarse stridencies of the squawk-box. 1973 *New Scientist* 20 Sept. 684/2 The squawk box emits two marginally different frequencies, almost out of audible range, through separate speakers. 1978 W. F. BUCKLEY *Stained Glass* xi. 104 The whistle stopped blowing, the passengers were all in their seats, the coordinators stopped speaking into their squawk boxes.

b. A speaker or receiving device which forms part of an intercommunication system, esp. in an office.

1954 *Sun* (Baltimore) 16 Apr. 27/2 Office buildings were deserted or skeleton staffed as interoffice 'squawk-boxes' carried in the doings on Thirty-third street by radio. **1962** L. DEIGHTON *Ipcress File* i. 15 Even over the squawk-box I could hear the lift in Alice's voice. **1964** P. GALLICO *Hand of Mary Constable* 48 To the right of the desk, on a small table, there was a squawk box with some dozen switches for inter-office communications. **1970** N. ARMSTRONG et al. *First on Moon* p. xiii, *Squawk box*, a small speaker, connected by telephone line to Mission Control, for home or office reception of live conversation between the spacecraft and earth. **1976** P. HARCOURT *Dance for Diplomats* ix. 96 The buzzer on my intercom made frantic noises...'Coming,' I said into the squawk box.

squawker. Add: **2.** (Examples.)

1896 G. B. SHAW *Let.* 8 Dec. (1965) I. 712 Yes, Lena is a fascinating squawker. **1923** H. C. WITWER in *Collier's* 29 July 26/3 To show you what a cheap squawker this Rags is, Spence tells me he has just welshed on a bet with him.

3. A loudspeaker designed to reproduce accurately sounds in the middle of the audible range.

1959 N. H. CROWHURST *Basic Audio* I. 67 Many installations use two or more speakers of different sizes..: large speakers (woofers) for the low frequencies, medium-sized speakers (squawkers) for the mid-range, and small speakers (tweeters) for the high frequencies. **1975** G. J. KING *Audio Handbk.* vi. 150 The output..drives the middle-frequency unit (squawker) and the high-frequency unit (tweeter) via a 4 kHz frequency divider.

squawmish, var. *SQUALMISH.

squdge (skwɒdʒ), *v.* [Perh. blend of SQUEEZE *v.* or SQUASH *v.*[1] and PUDGE[1]; cf. next and *SQUSH *v.*] *trans.* To squash or squeeze; to hug tightly. Also as *sb.*

1870 R. H. D. BARHAM *Life & Lett. R. H. Barham* II. vii. 27 At last he got so terrible bad surely nothing would ease him, so that we was forced to *squdge* him under the blankets. **1909** KIPLING *Rewards & Fairies* (1910) 5 They've put us into boots..and my toes are squdged together awfully. **1928** J. M. BARRIE *Peter Pan* v. ii. 159 Oh, Peter, how I wish I could take you up and squdge you! **1975** *New Society* 13 Nov. 388/1 Why does a large tube of toothpaste have a large hole? It is by length we judge the squdge of our brush, not by girth.

squdgy (skwɒ·dʒi), *a.* [Perh. f. prec. + -Y[1], or blend of SQUASHY *a.* and PUDGY *a.*[1]] Soft and moist or yielding, squashy.

1892 KIPLING *Barrack-Room Ballads* 51 Elephints a-pilin' teak In the sludgy, squdgy creek. **1919** W. DEEPING *Second Youth* xvii. 145 He made haste to shake Joseph Bluett's squdgy hand and escape. **1959** M. STEEN *Woman in Back Seat* I. v. 97 'Don't you like babies?' Lavinia shook her head... 'They're so squdgy, and they haven't got any shape!'

squeak, *sb.* Add: **2. a.** Also *fig.* in neg. contexts (*colloq.*). Cf. *PEEP *sb.*[1] 2 d.

1977 *Spare Rib* July 10/1 We've hardly heard a squeak out of them since. **1982** S. BRETT *Murder Unprompted* iv. 41 'I'm surprised you haven't heard anything about it..' You sure you haven't heard anything?' 'Not a squeak.'

3. b. (Later examples without qualifying adjs.)

1880 MRS. H. WOOD in *Argosy* XXIX. 191 At the last moment, when the ship was getting away, and I had given the captain up, he came on board... 'I've had a squeak for it, Johnny,' he laughed, as he shook my hand. **1939** A. RANSOME *Secret Water* x. 121 You oughtn't to have waited. It's going to be a squeak getting home across the Wade.

4. Restrict † *Obs.* to sense in Dict. and add: **b.** A piece of incriminating information offered to the police; *to put in the* (or *a*) *squeak*: to turn informer, to inform *against*.

1922 [see *FLATTY[2] 3]. **1936** J. CURTIS *Gilt Kid* ii. 22 You'll..turn grass and put in the bleeding squeak against me. **1955** D. WEBB *Deadline for Crime* i. 14 Then the squeak goes in. A bent buyer grasses to the law. **1973** A. HUNTER *Gently French* iv. 33, I can see another villain putting a squeak in but knocking off Freddy would be just stupid.

squeak, *v.* Add: **4. b.** To cause (something) to squeak.

1913 C. MACKENZIE *Sinister St.* I. II. xv. 401 Michael solemnly regarded the fair-haired boy of two who was squeaking an indiarubber horse. **1977** 'J. GASH' *Judas Pair* xv. 177 Could he see the curtain? I'd moved it without squeaking its noisy runners.

5. *intr.* **a.** *to squeak through*: to get through by a narrow shave, to scrape through.

1938 H. NICOLSON *Diary* 1 Sept. (1966) 358 We may just squeak through. On the other hand, we may get into the same mess as in 1914. **1943** *Sun* (Baltimore) 22 Nov. 14/2 The Irish squeaked through to a 14-to-13 verdict over Iowa Pre-Flight. **1971** J. BISHOP *Days of Martin Luther King, Jr.* iv. 329 The President..said he was not optimistic about the passage of the civil rights bill. It would require strong bipartisan support to squeak through. **1977** *Time* 7 Mar. 24/2 Rabin only squeaked through by sweeping the votes allotted to Israel's conservative kibbutzim.

b. With preps.: to make one's way by a narrow shave, to scrape *by*, *into*, etc. Chiefly *U.S.*

a **1961** H. H. MARTIN in *Webster* s.v., By six months of hard cramming..he squeaked by the finals. **1966** *Economist* 27 Aug. 810/3 His Progressive Conservatives squeaked back into power with only 39 per cent of the votes. **1968** *Ibid.* 20 Apr. 20/3 The Bill squeaked out of the Rules Committee on a single vote. **1974** *Union* (S. Carolina) *Daily Times* 24 Apr. 1/4 Texaco..squeaked by Mobil last year to become the country's second biggest oil firm. **1977** *Monitor* (McAllen, Texas) 26 June 1B/5 Jimmy Connors squeaked past a valiant Stan Smith in five sets.

squeaker. Add: **1. d.** *Criminals' slang.* An informer. Cf. SQUEALER 2 b.

1903 FARMER & HENLEY *Slang* VI. 335/1 *Squeaker*, (1) a blab.., and (2) an informer. **1924** E. WALLACE *Room 13* vii. 171, I want to talk with you, you dirty squeaker! You're the fellow that told the deputy I was getting tobacco in through a screw. **1930** *Observer* 19 Oct. 17 The recent attempt to murder him..was not due to..the impulse to remove rivals or 'squeakers'. **1950** [see *GRASSER[1]]. **1973** A. HUNTER *Gently French* ii. 14 Dutt had been brooding over the tip-off mystery... The squeaker must have been Rampant.

2. e. Also, esp. = CICADA.

1897 'NATALIAN' *S. Afr. Boy* viii. 76 The youthful genius who brought two squeakers—tree cicadæ—before school hours, and released one in each room. **1959** S. J. BAKER *Drum* 147 Squeaker: A type of cicada.

3. b. (Later examples.) Esp. as a party toy (see quot. 1980).

1930 [see *paper streamer* s.v. *PAPER *sb.* 12]. **1939** N. COWARD *To step Aside* 49 She was seated on the knee of an Argentine with a paper fireman's hat on her head, blowing a squeaker. **1980** J. W. HILL *Intermediate Physics* v. 45 The party 'squeaker', a paper tube containing a weak coiled spring which uncoils when it is blown, works on the same principle.

4. b. *N. Amer. colloq.* A game won by a very narrow margin.

1961 in WEBSTER. **1969** *Eugene* (Oregon) *Register-Guard* 3 Dec. 1D/1 Remember that 26–24 squeaker over Stanford, when the Indians kicked a field goal with 63 seconds left to 'win' the game. **1970** *Globe & Mail* (Toronto) 28 Sept. 18/8 Ottawa Rough Riders..lost a squeaker to Montreal.

squeaking, *ppl. a.* Add: **2. b.** *squeaking sand*, sand that gives out a short, high-pitched sound when disturbed.

1966 *Sedimentology* VI. 136 A 'squeaking' sand was found by the writer in Gower, S. Wales and this has been used for comparison with the 'booming' sand of the desert. **1976** *Nature* 5 Feb. 368/2 The most common of the musical sediments is probably squeaking (otherwise known as singing, barking or whistling) sand which produces a high frequency note in the range 500–2,500 Hz.

squeaky, *a.* Add: **a.** Proverbial phr. *the squeaky wheel gets the grease* (and varr.): the person who makes the most fuss or trouble gets the attention.

[**1937** in J. Bartlett *Familiar Quotations* 518/2 The wheel that squeaks the loudest is the one that gets the grease. **1948** in B. Stevenson *Home Bk. Prov.* 2883/2, I hate to be a kicker, I always long for peace, But the wheel that does the squeaking is the one that gets the grease.] **1969** J. LA MARSH *Mem. Bird in Gilded Cage* x. 291 Mrs. Kinnear did..put on a most relentless campaign over the years..in line with the best of political principles—'the squeaky wheel got the grease'. **1974** *Hansard* (Canada) 17 Oct. 502/1 It is the old story: the squeaky wheel gets the grease.

c. *Comb.* **squeaky clean** (also with hyphen), (of hair, etc.) washed and rinsed so clean that it squeaks; completely clean; freq. *fig.*, above criticism, beyond reproach.

1975 *Country Life* 8 May 1176/2 *No one*..is in a position to criticise... No one is, in the current idiom, that squeaky-clean. **1976** N. THORNBURG *Cutter & Bone* i. 24 Still towelling his hair, Bone returned to the living room. ..'Behold, the squeaky clean Richard Bone.' **1978** *Guardian Weekly* 13 Aug. 16/2 The [Ford Motor] company has denied making any illegal payments, claiming that it is 'squeaky clean' in this area. **1980** [see *SET-ASIDE *sb. phr.*]. **1981** L. DEIGHTON *XPD* xvii. 155 His..long dark hair was wavy and squeaky clean.

squeal, *sb.* **3. a.** *U.S. slang.* An act of informing against another.

1872 G. P. BURNHAM *Memoirs of U.S. Secret Service* 152 This 'squeal' among the 'queersmen' brings this foul business straight home to *you*. **1903** *N.Y. Sun* 5 Nov. 3 Ever since his so-called 'squeal' at the Lexow investigation he has been a marked man. **1907** 'O. HENRY' *Trimmed Lamp* 185, I always thought that Kike's squeal on his boss was about the lowest-down play that ever happened.

b. *U.S. Police slang.* A call for police assistance or investigation; a report of a case investigated by the police.

1949 S. KINGSLEY *Detective Story* I. 14 'This is Jim's squeal, ain't it?'..'Yeah, I'll take it... This is my partner's case.' *Ibid.* II. 86 Get me the old files on that Cottsworth squeal! **1960** 'E. MCBAIN' *See them Die* vii. 79 Parker's on the prowl, Hernandez is answering a squeal. **1972** B. GARFIELD *Line of Succession* (1974) I. 3 The first cop said, 'Do it. Send in a squeal—we'll want the wagon.' **1973** 'E. MCBAIN' *Hail to Chief* i. 6 The appearance of Homicide cops at the scene of a murder was mandatory, even though the subsequent investigation was handled by the precinct detectives catching the squeal.

squeal, *v.* **3.** (Earlier example.)

1846 *National Police Gaz.* (U.S.) 413/2 Some dozen of the infamous rogues, well known to them, who infest that city, will be 'pulled' until they find one that will 'squeal'.

squeamy (skwī·mi), *a.* orig. *U.S.* [f. SQUEAM(ISH *a.* + -Y[1].] = SQUEAMISH *a.*

1838 C. GILMAN *Recoll. Southern Matron* v. 44, I feel so squeamy-like at my stomach. **1863** 'E. KIRKE' *My Southern Friends* v. 76 Doan't be squeamy, gal; out with it. **1880** *Harper's Mag.* Sept. 582/1, I expect they'd eet so much sweet it kinder made 'em squeamy. **1908** H. G. WELLS *War in Air* v. 174 They're a bit squeamy now, but you wait till they've got their hands in.

squeegee, *sb.* Add: **1. b.** A similar implement for cleaning windows, windscreens, etc., or for other purposes requiring smooth application of pressure.

1918 L. E. RUGGLES *Navy Explained* 130 Squeegee—..used in civil life to clean windows. **1955** *Sci. News Let.* 2 Apr. 224/1 Resembling a window cleaner's squeegee, the sweeper is ribbed to fold gently back and forth when it is moved across the floor. **1955** *Sci. Amer.* Oct. 126/3 A small squeegee or an automobile windshield wiper will help in the cleaning job. **1962** *Which?* Oct. 314/1 The *Gnomist transfer* had to be stuck on with water and a squeegee, which the manufacturer provided. *a* **1977** *Harrison Mayer Ltd. Catal.* 15/1 The colour in paste form is forced by means of a squeegee through a fine mesh mounted on a frame.

3. *Comb.* **squeegee band** *Naut. colloq.*, an improvised band (see quots.). Cf. *WASH-BOARD 3 d.

1916 'TAFFRAIL' *Pincher Martin* x. 176 Before very long the 'squeegee band', composed of two drums, a dozen fifes, many mouth-organs, and an unholy number of mess kettles and other noisy utensils, was marching round the deck. **1979** A. C. HAMPSHIRE *Just Old Navy Custom* 225 *Squeegee band*, composed of instruments not usually found in a band, e.g. Jew's harp; mouth-organ; comb and paper, etc.

squeegee, *v.* Add: **3.** *intr.* To use a squeegee.

1972 *Times* 20 Sept. 3/3 He washes, then squeegees. **1977** *Centuryan* (Office Cleaning Services) Christmas 17/2 Eva squeegeed off as first of the guest shiners.

squeeze, *sb.* Add: **1. b.** (Further examples, esp. in *Pol.* contexts.) Colloq. phr. (orig. *U.S.*) *to put the squeeze on* (someone): to exert influence on (someone) to act in a particular way, to 'pressurize' someone. Also without indirect obj.

1777 J. WEDGWOOD *Let.* 13 Apr. (1965) 204 Mrs. Du Burk's assurance in asking us to pay her debts is very great... It is another squeeze, and I would not pay a stiver. **1941** E. B. WHITE *Let.* 24 June (1967) 210, I am writing you direct to put the old personal squeeze on you. **1942** R. CHANDLER *High Window* (1943) xxxiii. 220 She hired me to..put the squeeze on Linda for a divorce. **1949** *Ann. Reg. 1948* 6 The tension was not relaxed, for there now began the Russian 'squeeze' in Berlin. **1954** 'J. CHRISTOPHER' *22nd Century* 21 They're putting the squeeze on. But there's got to be an excuse before they can swallow us. **1969** A. G. FRANK *Latin Amer.* xxv. 394 The imperialist squeeze obliges them to react by squeezing their workers. **1978** S. BRILL *Teamsters* vi. 217 Spilotro's army of enforcers..put the squeeze on hard-pressed loan-shark victims.

d. (Earlier and later examples.) Also, a restriction in the supply of money, credit, goods, etc.; *spec.* in Stock Market usage, pressure applied to dealers in shorts to cause them to settle at a loss. Cf. *credit squeeze* s.v. *CREDIT *sb.* 14.

1872 *Chicago Tribune* 23 Oct. 1/5 The Gold Room was treated to a slight sensation to-day in the shape of a 'squeeze' in cash gold, which was made as high as 3/8 per cent per diem for borrowing. **1924** G. G. MUNN *Encycl. Banking & Finance* 523/1 A money squeeze refers to a temporary shortage in the supply of loanable funds accompanied by difficulty in borrowing and marking up of interest rates. **1927** W. H. HUBBARD *Cotton & Cotton Market* (ed. 2) 396 While we have never had a corner since 1910, we have had in recent years a succession of annoying premiums on the near deliveries. The trade calls these minor corners a 'squeeze'. **1937** *Sun* (Baltimore) 25 Sept. 15/8 'Longs' are traders who in recent months have been accumulating contracts specifying delivery of corn to them in September. 'Shorts' are those who have sold these contracts, many of them reportedly without having possession of the corn to deliver. The attempt to make these 'shorts' pay a comparatively high price either to buy back their contracts or to buy the corn to deliver on them is known as a 'squeeze' in market parlance. **1943** [see *ROLL BACK, ROLL-BACK, ROLLBACK 2]. **1958** J. K. GALBRAITH *Affluent Society* xvi. 184 A severe squeeze will ordinarily be placed on the capital requirements of smaller-scale firms. **1979** B. HINES *Price of Coal* I. 48 I'm talking about spending thousands of pounds of public money... I thought there was a squeeze on?

e. *Bridge.* A tactic used to force an opponent to discard or unguard a potentially winning card.

1926 *Work-Whitehead Auction Bridge Bull.* Jan. 117/1 The Squeeze is unquestionably the least understood of the several more or less rare plays arising from time to time in the proper play of Auction hands. **1933** *Sunday Times* 5 Feb. 5/1 The coup formerly only known as the 'Vienna Coup', but now, more appropriately, also termed the 'Squeeze'. **1959** *Listener* 5 Nov. 802/3 He played for a squeeze. **1976** *National Observer* (U.S.) 20 Nov. 21-A/2 Another 'cooked' story is behind this week's hand. It involved a refusal to finesse and ended with a very fancy squeeze.

6. c. (Earlier example in the sense 'plan' or 'work'.)

c **1863** T. TAYLOR *Ticket-of-Leave Man* III. 59, I owe him one for spoiling my squeeze.

7. b. *slang.* An impression of an object made for criminal purposes.

1882 *Sydney Slang Dict.* 8/2 *Squeeze*, an impression of a keyhole in wax. **1930** G. D. H. & M. COLE *Burglars in Bucks.* III. xxxiv. 135 Where did the dummy keys..come from?.. If they were forgeries it would be simpler, for Sir Hiram might remember if anyone had handled his keys long enough to take a squeeze. **1941** BAKER *Dict. Austral. Slang* 71 *Squeeze*, an impression of a keyhole in wax.

9. a. (Earlier example.)

1848 J. F. COOPER *Oak Openings* II. v. 78 In one instance, however, a young Indian had a still narrower 'squeeze' for his life.

c. A difficult situation.

1905 *Dialect Notes* III. 22 *Tight squeeze*,..a difficulty. **1972** *National Observer* (U.S.) 27 May 1/1 The safest drivers are those who know what their cars can do and how to make them do it in a squeeze.

9*. *Baseball.* The use of squeeze play (*SQUEEZE PLAY 1 a); a bunt made to try to bring home a runner from third base.

1908 *Spalding's Official Base Ball Guide* 279 Under Ned Hanlon the Cincinnati team worked the 'squeeze' nearly as well as the New York Americans. **1942** L. FONSECA *How to pitch Baseball* III. i. 93 Another play for which the pitcher—and catcher, too—must always be on guard is the squeeze, one of baseball's most spectacular plays. **1976** *Billings* (Montana) *Gaz.* 30 June 2-E/6 In the eighth inning, the Royals tried to salvage the game with a squeeze with the bases loaded.

10. a. **squeeze bunt** *Baseball*, the bunt (*BUNT *sb.*[3] 2) made in squeeze play (*SQUEEZE PLAY 1 a); also as *v. intr.*; **squeeze-pidgin** *slang*, a bribe.

1952 B. FELLER *Pitching to Win* viii. 108 During the 1951 season, we had a number of squeeze bunts, those which score a runner from third base, beat us in several important ball games. **1955** P. RICHARDS *Mod. Baseball Strategy* xi. 130 Many managers make a big mistake asking pitchers to squeeze-bunt. **1974** *Cleveland* (Ohio) *Plain Dealer* 13 Oct. C. 1/2 The big run for the defending world champions came on a two-strike, suicide squeeze bunt by Bert Campaneris in the fifth inning. **1946** J. IRVING *Royal Navalese* 165 *Squeeze-pidgin*, a tip: a bribe. **1970** 'B. MATHER' *Break in Line* i. 11 'What's a squeeze-pidgin?'..'A bribe... Something you *squeeze* out of somebody.'

b. *attrib.* uses in *Bridge* (sense 1 e above).

1936 E. H. DOWNES *Squeezes, Coups & End Plays* 10 The Squeeze trick must always be won in the hand opposite the final entry card. **1947** J. BROWN *Winning Tricks* xxi. 233 A long suit is not necessarily for a squeeze, although long suits have come to be associated with squeeze positions. **1954** G. S. COFFIN *Bridge Play from A to Z* 328 Many types of preparatory squeezes occur..such as the squeeze long-suit..; the squeeze finesse, the squeeze strip, etc. **1964** FREY & TRUSCOTT *Official Encycl. Bridge* 526/1 The squeeze finesse is characterized by the presence of a symmetric menace which must be guarded with an equal number of cards by both opponents. **1974** *Times* 5 Jan. 8/7 The counter-attack by the declarer to keep one move ahead of the squeeze-breaking defence.

squeeze, *v.* Add: Also with *dial. preterite* and *pa. pple.* 9– **squoze**, *pa. pple.* 9 **squozen**.

1844 'J. SLICK' *High Life N.Y.* II. 195, I sot down on a bench runnin over with harnsome gals, that squoze close together and squinched themselves up to make room for me. **1866** J. T. STATON *Rays fro' Loominany* 107 Awd welly as lief they'd squozzen my guts eawt as speighlt my bonnet. **1878** *Roger Plowman's Excursion to London* ii. 21, I gently squoze hur 'and. **1928** A. A. MILNE *House at Pooh Corner* viii. 143 He squeezed and he sqoze [*sic*], and then with one last squze he was out. **1931** *Sun* (Baltimore) 1 Sept. 8/7 'Orange?' repeated Waitress No. 1. 'Do you want it squoze?' **1933** M. LOWRY *Ultramarine* vi. 237 He just sort of *squoze* the rabbit. **1938** J. AGATE *Diary* 17 Dec. in *Selective Ego* (1976) 115 Before going down to correct my proofs at the *S.T.* squeezed in a performance of the *Messiah*. Or, rather, Beecham squoze it in for me.

1. f. To fire *off* (a round, shot, etc.) from a gun. *colloq.*

1956 *Amer. Speech* XXXI. 192 A rifleman never fires a shot, he *squeezes off a round*. **1975** A. PRICE *Our Man in Camelot* vii. 136 He..got his gun clear just as Harry squeezed off his first shot.

g. To approach or 'push' (a certain age). *colloq.*

1976 *National Observer* (U.S.) 20 Nov. 24/1 But that takes a lot out of a man, particularly when he's squeezing 70. So sometimes he's pretty tired. **1978** *Guardian Weekly* 30 July 21/1 The original heroine, now squeezing forty.

3. e. To exert commercial or financial pressure on (someone); to restrict a supply of money, credit, goods, etc.; *spec.* in Stock Market usage, to force dealers in shorts to settle at a loss.

1885 *Harper's Mag.* Nov. 842/1 The bulls get a 'twist on the shorts' by artificially raising prices, and 'squeezing', or compelling the bears to settle at ruinous rates. **1900** S. A. NELSON *ABC Wall St.* 160 When shorts become frightened after having oversold and then are forced to violently bid up prices in competition with the owners of stocks they are said to have been squeezed. **1902** L. L. BELL *Hope Loring* xiv. 272 You squeezed me badly in '93. **1951** *Times* 3 Jan. 7/2 The domestic consumer of coal and coke is already being squeezed, but he often uses more gas and more electricity as a result. **1970** *Daily Tel.* 3 Sept. 3/2 The Government can scarcely ask banks to squeeze their customers when a State Corporation is advertising loans to attract business. **1978** *Jrnl. R. Soc. Arts* CXXVI. 390/2 Manufacturing industry's profits have been greatly squeezed.

5. e. *Bridge.* To force (an opponent) to discard a guarding or potentially winning card.

1926 *Work-Whitehead Auction Bridge Bull.* Jan. 118/2, I will give the three cards remaining in each of the four hands to show how South was squeezed by the lead of the Queen of Clubs. **1934** G. F. HERVEY *Mod. Contract Bridge* xxi. 223 The Americans now call this coup [*sc.* the Vienna coup] Squeezing or Squeeze Play. The name is apt as the play of the declarer is such that he squeezes the opponents and forces them to discard and unguard a suit. **1949** H. G. FREEHILL *Squeeze at Bridge* i. 19 The essential features of the squeeze are three. First: there must be a squeeze-card. That is, a card to the lead of which the player who is squeezed has to discard. **1959** T. REESE *Bridge Player's Dict.* 209 Playing no trump, South lays down the ten of hearts and West is squeezed; he must either unguard spades or throw away the winning diamond. **1979** N. SQUIRE *Squeeze Play Simplified* i. 4 The two menaces are on the left of the player to be squeezed, the squeeze card on his right.

6. a. (Later example.)

1974 J. GARDNER *Corner Men* xiii. 188 The man in the rear of the Merc fired once... Wright squeezed twice.

squeeze-. Add: **squeeze bottle**, a bottle made of flexible plastic, squeezed to expel the contents; **squeeze-box** *slang*, † (*a*) *Naut.*, a ship's harmonium (*obs.*); (*b*) an accordion or concertina; **squeeze cementing** *Oil Industry* (see quot. 1938); **squeeze lens** *Cinemat.* (orig. *U.S.*), an anamorphic lens attachment (cf. *ANAMORPHIC *a.* 2); **squeeze toy**, a child's doll or similar toy which sounds when pressed; **squeeze tube** = TUBE *sb.* 2 d; a tube-shaped container which yields its contents when squeezed.

1953 *Wall St. Jrnl.* 16 Sept. 23/4 The principal addition to the product line will be polyethyline, best known in its form of 'squeeze bottles' and transparent packaging material. **1964** V. E. YARSLEY et al. *Cellulosic Plastics* i. 5 To-day 'squeeze-bottles' in polythene are used in increasing quantities all over the world. **1976** 'E. MCBAIN' *Guns* (1977) vii. 177 The doctor..takes a squeeze-bottle.. and then wets a piece of gauze. **1909** J. R. WARE *Passing Eng.* 232/1 *Squeeze-box*,..the ship harmonium—used in the hasty Sunday service. From the action of the feet. **1936** *Amer. Speech* XI. 280/1 *Squeeze-box*, an accordion. **1938** AUDEN & ISHERWOOD *On Frontier* III. i. 99 Get yer squeeze-box. [*First Soldier begins to play the accordion.*] **1942** E. PAUL *Narrow St.* x. 77, I told my Marseilles accordion teacher that the experience had disabled me from practising the squeeze-box for a week. **1963** *Times* 26 Jan. 11/1 They imagine everyone spends their time dancing to squeezebox music and living on Knackwurst and Apfelstrudel. **1973** C. BONINGTON *Next Horizon* vii. 105 He was already ensconced in the bar at the Clachaig, his squeeze box out, a dram of whisky at his side and a cigarette in his mouth. **1938** *Oil Weekly* 28 Feb. 36/1 Squeeze cementing means that cement slurry is forced, or 'squeezed', by pressure into or against a permeable formation or through perforations in casing and liners for the purpose of shutting off water or for reducing gas/oil ratios. **1974** D. K. SMITH in P. L. Moore et al. *Drilling Practices Man.* xvi. 400 Squeeze cementing is necessary for many reasons, but probably the most important use is to segregate hydrocarbon producing zones from those formations producing other fluids. **1957** *Amer. Cinematographer* Mar. 149/1 Focal distortion is practically eliminated by a technique that adds a squeeze (anamorphic) lens to the system that produces partial scene compression in the camera, with the remainder being effected in the printing process. **1977** J. HEDGECOE *Photographer's Handbk.* 31 *Squeeze lens.* This is an 'anamorphic' supplementary lens system continuing cylindrical lens elements. When mounted on a normal lens it squeezes the image by reducing either its vertical or its horizontal size without affecting the other dimension. **1954** *Toys & Novelties* Mar. 421 (Advt.), A complete new line of unique squeeze toys by Alan Jay. **1976** *National Observer* (U.S.) 25 Dec. 6/4 He would also like to see a uniform law covering sound levels of toys... 'I have been responsible for manufacturing more than 35 million squeeze toys in my career, and I have never run into anything like this,' Young says. **1872** S. HALE *Let.* 1 Oct. (1919) iv. 86 We went to Rowney's,—delicious! —and I bought two squeeze tubes..; there were water-colours there. **1962** J. GLENN in *Into Orbit* 201, I pulled a squeeze-tube of apple sauce out of its receptacle and parked it in the air in front of me.

squeezed, *ppl. a.* Add: **1. b.** Also with *out*, *up* (earlier example).

1831 M. EDGEWORTH *Let.* 30 Apr. (1971) 533 A squeezed up poke bonnetted old mother. **1880** in I. M. Tarbell *Hist. Standard Oil Co.* (1904) II. 325 This constitutes the only foundation for the oft-repeated expressions 'crushed out', 'squeezed out', and 'bulldozing'. **1951** W. FAULKNER *Requiem for Nun* 225 The long invincible arm of Progress ..released him, not flung him away like a squeezed-out tube of paint, but rather..merely opening its fingers.

squee·ze play. Also **squeeze-play.** [f. SQUEEZE *v.* + PLAY *sb.*] **1. a.** *Baseball.* A tactic whereby the batter bunts so that a runner at third base can attempt to reach home safely and score.

1905 *St. Louis Globe-Democrat* 2 July II. 8/5 Like the spit ball, the squeeze play is not strictly an invention of the present season. **1909** *Collier's* 15 May 29/2 The 'squeeze' play, of which we have heard so much in the past two or three years, is the method of scoring a man from third base on the hit-and-run system. **1949** *Chicago Tribune* 27 Sept. II. 1/8 The mighty sox..had to resort to the squeeze play to score their all-important victory. **1979** *Arizona Daily Star* 22 July C 4/4 Not many pitchers throw inside anymore. It's like the squeeze play. It's a lost art.

b. *transf.* and *fig.* An act of coercion or 'pressurizing'; the application of force or pressure to obtain an end. *colloq.* (chiefly *U.S.*).

1916 *Independent* (N.Y.) 5 June 410 (*heading*) The 'squeeze-play' at the capital. **1932** *Sun* (Baltimore) 14 Sept. 10/1 Mostly employers of relatively small numbers of workers, they appear to have been caught in an economic 'squeeze play' in which 'contract' shops..are played against each other in pursuit of the lowest bids. **1944** D. WECTER *When Johnny comes marching Home* IV. vii. 557 You perhaps mentioned the fact that Hitler was putting the squeeze play on Hindenburg a few years later. **1966** *Economist* 10 Sept. 1009/1 The real victims of the Franco government's squeeze-play—the Spaniards whose prospects of being allowed to go on crossing the border daily to work in Gibraltar are now in jeopardy. **1977** D. ANTHONY *Stud Game* xx. 122 She hinted that her partner had worked this squeeze-play once before, with..a pregnant girl..and a widow who paid off.

2. *Bridge.* The employment of a squeeze. Cf. *SQUEEZE *sb.* 1 e.

1926 *Work-Whitehead Auction Bridge Bull.* May 50 The squeeze play is made when running a suit, by forcing an opponent to discard a number of times. **1934** G. F. HERVEY *Contract Bridge Dict.* 139 Squeeze Play is the basis of practically all double-dummy problems. **1959** T. REESE *Bridge Player's Dict.* 209 The technique of squeeze play is complicated, and even in the simplest three-card ending..certain basic elements have to be present, such as threat cards, entries, a squeeze card and correct timing. **1976** *Country Life* 19 Feb. 440/2 South did not understand squeeze play... When the Nine of Clubs falls from East, the squeeze against West is marked.

squeezer. Add: **4.** Usu. *pl.* A playing-card which has its value indicated in one or two corners, so that a player may ascertain his hand while holding the cards closely arranged. (App. orig. used in Poker, but now standard.)

1876 *Paper & Printing Trades Jrnl.* Sept. 8/1 The 'Squeezers' Playing Cards introduced by Messrs. Lawrence Brothers..are rapidly rising in popular estimation. **1888** *Amer. Humorist* 15 Sept. 3/1 The editor picked up his hand, slid the squeezers past his good eye, and began to softly whisper the 'Pirate King'. **1906** G. FRANKAU *X.Y.Z. of Bridge* 47 Horatius in a long frock-coat Rending two 'squeezer' packs. **1930** C. P. HARGRAVE *Hist. Playing Cards* xiii. 345 The New York Consolidated Card Company..issued indexed cards..under the name of 'Squeezers'.

squeezy, *a.* Add: Also, capable of being squeezed, esp. as *squeezy bottle*, squeeze bottle.

1971 R. A. PERCY *Charlesworth on Negligence* (ed. 5) xviii. 732 Using a squeezy bottle to squirt coolant liquid into the interior of a workpiece on a centre lathe. **1972** *Guardian* 5 Dec. 16/6 Strong colours ready mixed in squeezy bottles, a boon for..young children.

squegger (skwe·gəɹ). *Electronics.* [Of uncertain origin: perh. a shortened respelling of *self-quench(ing* + -ER[1].] An oscillator whose oscillations build up to a certain amplitude and then cease for a time before beginning again; also, the production of such oscillations.

1921 *Radio Rev.* May 248 The production of notes by a high-frequency oscillating valve making use of a grid leak and condenser, and the factors governing the pitch of this note, were discussed. The application of this arrangement (termed by Major Prince a 'squegger') to a receiver..was also described. **1932** *Admiralty Handbk. Wireless Telegr. 1931* xvi. 727 Receivers in which..self-oscillations alternately build up and are damped out without the necessity of a separate circuit to vary the reaction. These are called self-quenching receivers or squeggers. **1932** LADNER & STONER *Short Wave Wireless Communications* viii. 147 The presence of audible modulation..may also be brought about by a high resistance in the grid circuit... This effect..was called originally a 'squegger'. **1932** *Wireless Engineer* IX. 186 In the super-regenerative class, two receivers were available, one of the self-quench (or squegger) type. **1946** C. A. QUARRINGTON *Mod. Pract. Radio & Television* II. x. 156 Time bases employing the squegger principle are inclined to be affected unduly by working conditions.

Hence (as a back-formation) **squeg** *v. intr.*, (of an electrical circuit) to oscillate intermittently, to be self-quenching; † **sque·gger-ing** *vbl. sb.*; **sque·gging** *vbl. sb.* and *ppl. a.*

1933 *QST* Sept. 66/2 Coupling between this 'squegging' oscillator and the receiver is obtained. *Ibid.* Nov. 17/1 The receiver may be made..to operate as a super-regenerator, by increasing the screen voltage to the point where 'squegging' (low frequency oscillation) starts. **1938** *Television* XI. 540/3 The valve is allowed to oscillate continuously with pauses of quiescence ('squeggering'). **1939** *Radio Amateur's Handbk. 1940* (ed. 17) iv. 61/1 Too much feed-back will cause the oscillator to 'squeg', or operate at several frequencies simultaneously. **1942** *Electronic Engin.* XIV. 629/3 The great difficulty in ultra high frequency oscillators is to obtain sufficient oscillation amplitude without squeggering or dead-spots in the turning range. **1947** *Ibid.* XIX. 162 Increase of R_2 to successively greater values results in the oscillator squegging.

1959 K. HENNEY *Radio Engin. Handbk.* (ed. 5) xii. 22 The adjustment of the feedback to give oscillation without pulsing or squegging is very critical. **1973** *Simulation* XXI. 77/2 There are circumstances under which squegging is desirable, such as in a superregenerative receiver. *Ibid.* 79/2 A squegging oscillator is highly sensitive to the settings of certain parameters.

squelch, *sb.* Add: **1. c.** A devastating argument or retort; a crushing blow. *slang* (orig. *U.S.*).

1942 BERREY & VAN DEN BARK *Amer. Thes. Slang* § 11/3 Something conclusive or decisive (as an argument, answer, blow, or the like),..squasher, squelch, squelcher. **1964** *Pix* (Australia) 4 Aug., *Pix* has added to its large collection of famous squelches with contributions from readers. **1977** *Amer. Speech 1975* L. 155 If I use *humdinger*, I have to face the derisive remark of my elder son, 'Nobody used that word since Theodore Roosevelt died' .., or the equally crushing squelch of the younger one, 'Papa, that is not even funny.'

4. *Electronics.* A circuit that suppresses all input signals except ones of a predetermined character; *spec.* in *Radio*, a circuit that suppresses the noise output of a receiver when the signal strength falls below a predetermined level. Freq. *attrib.*

1937 [see *QUIETING *vbl. sb.* 2 a]. **1945** *FM & Television* Apr. 31/1 The feature of this squelch circuit is that sharp pulses of interference do not open the squelch... It has been found that a 1-microvolt signal can open the squelch, but noise pulses of considerable amplitude do not. **1950** J. K. HENNEY *Radio Engin. Handbk.* (ed. 4) xvii. 822 A signal of 140 db (0·1 μu) is sufficient to open the squelch of a police receiver. **1959** [see *MUTING *vbl. sb.* b]. **1976** *CB Mag.* June 64/1 (Advt.), the transceiver is combined through a switchable standby circuit to interrupt the music when the adjustable squelch level is broken. **1981** *Daily Tel.* 5 Nov. 2 (Advt.), Ultra compact, yet has variable squelch to cut signal 'chopping'.

squelch, *v.* Add: **1. b.** (Later examples.) Now chiefly *U.S.*

1910 *Dialect Notes* III. 455 *Squelch*, *v. tr.*, to snub, to turn down. 'She *squelched* him.' **1936** L. C. DOUGLAS *White Banners* vi. 135 An inquisitive maid-of-all-work who might try to be chummy unless promptly squelched. **1960** *Wall St. Jrnl.* 15 Mar. 14 Recent attempts by other domestic unions to hurtle national boundaries have been squelched. **1978** H. WOUK *War & Remembrance* xxv. 252 That'll squelch him, I assure you, and he'll be as quiet as a mouse.

3. *trans. Electronics.* = *QUIET *v.* 2 d.

1950 J. K. HENNEY *Radio Engin. Handbk.* (ed. 4) xvii. 822 This voltage drop biases the first audio amplifier beyond cutoff, thus squelching the set. **1976** *S9* (N.Y.) May/June 107/1 The light..then remains on (although the receiver may be squelched, it will then pick up all calls on the channel until it is reset by the firefighter).

squelched *ppl. a.* (*fig.* examples).

1914 'HIGH JINKS, JR.' *Choice Slang* 18 *Squelched*, ignored, insulted, 'sat upon'. **1928** *Sat. Even. Post* 7 Jan. 20/2 Apparently squelched, she made no reply.

squelchy, *a.* **1.** (Examples in more general use.)

1914 [see *PAPAIA]. **1928** *Daily Express* 12 June 10/5 What Will Woman Do Next?.. Start an agitation, I shouldn't wonder, for the suppression of the word 'women'. Sober and dignified once.., the over-work of the last ten years has made a sickening, squelchy mess of it.

squench, *v.* Add: **2.** (Later examples.) *rare.*

1865 *Punch* 20 May 200/2 Mr. Newdegate had a plan, whereof not much need be said, as it was squenched by 126 to 42. **1923** U. L. SILBERRAD *Lett. Jean Armiter* iv. 100 You are not easily squenching Art, with a capital A, when it is once fairly talking.

squib, *sb.* **4. a.** Delete † *Obs.* and add later examples in similar use; also, a short or thin person.

1898 *Leeds Mercury Suppl.* 26 Mar. (E.D.D.), Ah'll knock thee dahn, yo' little squib, if tha doesn't shut thi gob. **1979** *Courier-Mail* (Brisbane) 28 Apr. 17/2 We have numerous utility expressions for people such as..sparrow squib, nugget and streak, for men of varying sizes.

d. For def. read: *erron.* for *SQUIL.

e. A horse lacking courage or endurance; hence, a coward. *Austral. slang.*

1908 [see *MONTY]. **1924** *Truth* (Sydney) 27 Apr. 6 *Squib*, a coward. **1933** *Bulletin* (Sydney) 15 Nov. 8/4 In the result stayers reproducing the old-time qualities of the Australian thoroughbred became rare; but speedy squibs abounded. **1936** A. RUSSELL *Gone Nomad* 55 There's no place in this town for squibs. **1947** *Coast to Coast 1946* 217 'You're a bloody lot of squibs,' said Darky disgustedly. **1951** F. HARDY *Power without Glory* 160, I know these johnnys; they're all squibs. **1978** *Telegraph* (Brisbane) 8 Feb. 20/1 Don Ash is the sort of bloke who makes you feel a squib for crying off with some minor ache or pain from that daily canter around the block.

squib, *v.* Add: **I. 3. b.** With *on*: to betray or let down (someone). Also without const., to funk, to behave in a cowardly manner; to wriggle or squirm. *Austral. slang.*

1938 S. J. BAKER in *Observer* 13 Nov. 11/3 Rat on = to betray, let down. Squib on = ditto. **1941** —— *Dict. Austral. Slang* 71 *Squib, to*, to funk, to be afraid of something. **1945** *Coast to Coast 1944* 5 I'll let him through; it's better to squib than to wreck old Dutch and myself, too. **1947** K. TENNANT *Lost Haven* xvii. 292 You'll probably squib out of it at thel ast moment. **1962** *Coast to Coast 1961–62* 83 He could finish on a good wicket in anything. And never squib on a bloke.

II. 8. To avoid (a difficulty or responsibility); to funk, to shirk through fear or cowardice. *Austral. slang.*

1934 *Bulletin* (Sydney) 1 Aug. 46/4 He judges the dodger's done just what he'd do himself; squibbed it and dropped his load. **1936** F. CLUNE *Roaming round Darling* xv. 138 Oxley and Evans attempted to explore the Macquarie, but floundering about the marshes Oxley squibbed the job once more. **1940** N. MONKS *Squadrons Up* i. 26 The censor..said that we should not let the enemy Command know their fighters were squibbing it [*sc.* avoiding combat]. **1955** D. NILAND *Shiralee* 50 The rough-and-tumble doesn't worry me. I'm not squibbing the issue.

squibber. (Earlier example.)

1810 *Irish Mag.* Aug. 377/1 Nothing like a squibber writing of his own acquirements.

squid, *sb.* Add: **1. d.** Also with capital initial. A ship-mounted anti-submarine mortar with three barrels, developed in the war of 1939–45.

1947 CROWTHER & WHIDDINGTON *Science at War* iv. 160 A three-barrelled mortar for throwing three projectiles each was developed, and named the Squid... The expenditure of ammunition required to sink a U-boat with the Squid was very much less than with depth charges. The production of the Squid was ordered direct from the drawing board in the urgency of the situation in 1943. **1962** W. GRANVILLE *Dict. Sailors' Slang* 112/2 *Squid box*, housing of a *squid*, a triple-barrelled mortar for firing depth charges. It is placed on the sterns of destroyers and frigates. **1973** J. QUICK *Dict. Weapons* 418/3 *Squid*, a British shipborne surface-to-sub-surface medium-range antisubmarine mortar system. A triple-barreled mortar fires a pattern of three mortar bombs which are programmed to give a three dimensional explosive pattern ahead of the target.

3*. A stable configuration of a parachute which is only partially extended.

1947 *Techn. Rep. Aeronaut. Res. Comm. 1946* II. 1465 If a parachute is released into an airstream and its speed relative to the air is greater than a certain critical speed, it will not open fully. Instead it will take up a dynamically-stable partly-open shape known as a 'squid'. **1949** *Jrnl. R. Aeronaut. Soc.* LIII. 1055/1 For a parachute in a steady squid state, the inflow to the canopy equals the outflow.

4. squid-hound (earlier example); also *attrib.*

1794 A. THOMAS *Newfoundland Jrnl.* (1968) xiii. 183 Whenever Squids are found is also found a Fish called Jumpers, or Squid Hounds, from the avidity with which they pursue and eat squids. **1934** E. REYNARD *Narrow Land* v. 250 Hut moved fast, almost as fast as the squidhound bass.

SQUID (skwid), *sb.*[2] *Physics.* Also **Squid**, **squid**. [Acronym f. the initials of *s*uper-conducting *qu*antum *i*nterference *d*evice.] A device consisting essentially of a superconducting ring containing one or more Josephson junctions, made the basis of a very sensitive magnetometer by utilizing the fact that a change in the magnetic flux linkage of the ring by one flux quantum produces a sharp change in the ring's impedance.

1967 W. S. GOREE in *Proc. Symp. Physics of Superconducting Devices* (U.S. Naval Research Lab.) 9 One embodiment of this device is the SQUID originated by Mercerau and others, which is two Josephson junctions in a superconducting ring. **1972** *Cryogenics* XII. 28/1 SQUID magnetometers and related devices have been used to make quantitative measurements of thermal fluctuations at temperatures as low as 0·023 K. **1979** *Nature* 22 Feb. 643/1 The magnetisation was measured with a Squid magnetometer. **1979** *McGraw-Hill Yearbk. Sci. & Technol.* 379/2 The dc SQUID has recently been developed for fast switching applications, that is, as an electronic logic or memory element. **1982** *Economist* 3 Apr. 120/3 At the frontiers of R and D now being done into superfast computers..is a device that is known as a squid—which is short for superconducting quantum interference device.

squid, *v.* Add: **2.** Of a parachute: to achieve a stable configuration when only partially extended.

1943 *Rep. & Mem. Aeronaut. Res. Council* No. 2119. 4 A non-porous parachute will not squid. **1951** W. D. BROWN *Parachutes* vii. 68 The mouth of the canopy will begin to collapse inward, and the parachute will squid. **1956** W. A. HEFLIN *U.S.A.F. Dict.* 148/1 *Critical closing speed*, in wind-tunnel tests, the airspeed at which an open parachute begins to squid, i.e. close or collapse into longitudinal shape.

Hence **squi·dded** *ppl. a.*; **squi·dding** *vbl. sb.*

1894 Squidding [in Dict., s.v. SQUID *v.*]. **1943** *Rep. & Mem. Aeronaut. Res. Council* No. 2119. 1 Squidding obviously may be a source of trouble in applications of parachutes. *Ibid.* 5 If a parachute is taken in the squidded condition and the air-speed reduced, the effective permeability of the canopy falls. **1969** J. GARDNER *Founder Member* viii. 132 A parachute training instructor went over all the elementary lessons..recalling things like critical speeds, oscillation and squidding.

squidge[2]. *U.S. slang. rare.* [Origin unknown.] One charged with performing troublesome duties for another (see quots.).

1907 ADE *Slim Princess* vii. 80 The squidge—that means the fellow who does all the worrying and gets nothing out of it. **1942** —— *Let.* 16 June (1973) 230 When Mr. and Mrs. Al Laflin and I traveled in distant countries, we always hired a 'squidge' the moment we arrived in a new town. His job was to stay with us and accept all the hardships and worries. **1953** [see *PATSY *sb.*]

squidge, *v.* [Cf. SQUIDGE *sb.*; sense 2 may represent an independent imitative formation.)] *trans.* **1.** To squeeze; to squelch or to mix roughly; to press together, so as to make a sucking noise. Also *intr. dial.* and *colloq.*

1881 H. & C. R. SMITH *Isle of Wight Words* 34 *Squidge*, to squeeze. **1939** J. STEINBECK *Grapes of Wrath* x. 136 The children squidged their toes in the red dust. **1947** —— *Wayward Bus* iii. 33 His shoes squidged sloppily on the floor. **1952** —— *East of Eden* xxxvi. 427 Now spit in your palm... Squidge the spit all together. **1969** *Guardian* 16 Aug. 3/8 Waiting for the day when lunch will be a blob of something squidged out of a tube. **1971** B. W. ALDISS *Soldier Erect* 242 The leeches were at me... I squidged them underfoot. **1977** *Basildon Yellow Advertiser* 19 May 5/2 To clean very dirty nails make a thick lather in a sponge and then 'squidge' your nails in the sponge—the suction will draw out the dirt.

2. *Tiddlywinks.* To play (a wink) by snapping it with a larger counter.

1955 *Sci. of Tiddlywinks* (Cambridge Univ. Tiddlywinks Club) 8 If the tiddlywink is squidged on the floor it will not rise more than an inch or two. **1962** *Time* 14 Sept. 56 A squopped wink cannot be squidged again until it is de-squopped. **1977** SHARP & PIGGOTT *Bk. Games* 165 No player may squidge another's wink.

squidger (skwi·dʒəɹ). *Tiddlywinks.* [f. prec. + -ER[1].] **a.** The larger wink used to propel or flip a player's winks. **b.** One who propels winks with another wink of larger size. (*rare*).

1955 *Sci. of Tiddlywinks* (Cambridge Univ. Tiddlywinks Club) 6 The game is played with five variable factors..(3) The squidger (the instrument used to flick with; this is usually another round counter in normal games). **1958** *Sports Illustr.* 7 Apr. M5 Each tiddlywinker plays with two large and four medium-size winks and, to hoist them, uses a wink of suitable size. It is called a squidger. **1962** [see *SQUOPPER]. **1975** *Sunday Mail Color Mag.* (Brisbane) 28 Sept. 21/2 The *squidger* is the tiddly-winker's most valuable tool—the large wink, with which he flips the smaller counters. **1977** [see *SQUOPPED *ppl. a.*].

squidgy, *a.* Add: **1.** (Earlier and later examples.)

1891 KIPLING *Life's Handicap* 30 There's a dale more in nature than your squidgy little legs have iver taken you to, Orth'ris, me son. **1942** G. KERSH *Nine Lives Bill Nelson* xi. 68 He..liked to get about with a few nice squidgy blondes. **1978** *Morecambe Guardian* 14 Mar. 6/2 (Advt.), Upholstery does not now rely on spectacular outlines to make its impact. Gone are the way-out 'squidgy' shapes that the Italians launched so successfully a few years ago, instead the line is neat and classical, with a strong accent on comfort.

2. Moist and pliant; squashy, soggy. Esp. of food. Cf. *SQUDGY *a.*

1973 *Times* 17 May 18/6 We settled for sandwiches. This was not a good idea. They were made of squidgy and tasteless bread of the wrapped and sliced variety. **1976** U. HOLDEN *String Horses* vi. 69 Her kiddies spread ketchup on their bread, liking squidgy food. **1977** *Times* 28 May 11/6 A nice squidgy fruit cake.

Hence **squi·dgily** *adv.*

1933 E. A. ROBERTSON *Ordinary Families* ix. 193 He turned out to be as squidgily repulsive as only an undersized Italian can be.

squiff[2]. *slang. rare.* [Of uncertain origin: perh. f. SQUIFFY *a.*] **1.** A contemptible person.

1939 E. POUND *Let.* Feb. (1971) 322 Yes, do better than that squiff, that femme ouistiti and lowest degree of animal life (apart from Cambridge Eng. profs).

2. The head. In phr. *off* (one's) *squiff* = *off* (one's) *head* s.v. HEAD *sb.* 34.

1960 H. PINTER *Room* 115 Mr. Kidd: I don't know what I think. (He sits.) I think I'm going off my squiff.

squiffed, *a. slang.* [var. SQUIFFY *a.*] Intoxicated, drunk.

a **1890** in Barrère & Leland *Dict. Slang* (1890) II. 295/1 He rolls home rather squiffed, just as the day is dawning. **1926** S. HOWARD *Lucky Sam McCarver* I. 36, I think I did meet you once before, though I was a bit squiffed at the time. **1961** [see *CHUMBLE *v.*]. **1977** B. GARFIELD *Recoil* ii. 28 I'm already a little squiffed. Ought to go on the wagon.

squiffer (skwi·fəɹ). *slang.* [Origin obscure.] A concertina; also, an organ-bellows or organ. Also *transf.* (see quot. 1934).

1914 G. B. SHAW *Fanny's First Play* I. 175 If you can get Holy Joe to sprint a hundred yards, I'll stand you that squiffer with the gold keys... What's a squiffer?..a concertina. **1934** 'I. HAY' *David & Destiny* v. 78 What exactly is a Squiffer... Strictly speaking, a concertina. By a process of excusable exaggeration, an organ-bellows, or even the organ itself. By a characteristic confusion of ideas, a person who blows an organ.

squiffy, *a.* Add: **1.** (Earlier example.)

?**1855** MRS. GASKELL *Lett.* (1966) 375 Curious enough there is a Lady Erskine, wife of Lord E, her husband's eldest brother living at Bollington, who tipples & 'gets squiffy' just like *this* Mrs E.

2. Askew, skew-whiff.

1941 BAKER *Dict. Austral. Slang* 71 *Squiffy*, askew. **1977** G. MELLY *Rum, Bum & Concertina* vii. 85, I never associated it with an orgy, a term I felt to imply a Roman profusion of grapes, wine, buttocks, breasts, marble *chaises-longues*, and squiffy laurel crowns.

squiggle, *sb.*[2] Add: **1.** (Later examples.) Esp. a wavy or twisting line drawn on a surface.

1928 *Daily Express* 9 July 10/3 No hieroglyphics in timetables to be looked up, no particular squiggle to be remembered lest one is unhooked from the express at Dijon or goes to Rome instead of Florence. **1934** R. FRY *Let.* 3 Aug. (1972) II. 692 The squiggles in the foreground represent runnels of water flowing everywhere. **1957** V. S. NAIPAUL *Mystic Masseur* v. 78 The handwriting became a hasty, tired squiggle, and the note-book was abandoned. **1979** *Listener* 3 May 603/3 From time to time, squiggles appeared on the screen which were interpreted by some as subliminal instructions to vote labour.

2. *Comb.*, as **squiggle-eyed** *adv.*: in phr. *to look squiggle-eyed* (*at* someone), to view askance or unfavourably. Also as *adj.* (Only in P. G. Wodehouse.)

1927 WODEHOUSE *Meet Mr Mulliner* vi. 178 There is a certain stage in the progress of a man's love when he feels like curling up in a ball and making little bleating noises if the object of his affections so much as looks squiggle-eyed at him. **1941** —— *Berlin Broadcasts* in *Performing Flea* (1961) i. 265 The internee is always being told to show his passport, and if he has not got one, the authorities tend to look squiggle-eyed. **1960** —— *Jeeves in Offing* xii. 136 She's always been a bit squiggle-eyed about Phyllis, because in Switzerland she held the view that we were a shade too matey. **1972** —— *Pearls, Girls & Monty Bodkin* i. 9 He was very fond of her—in a brotherly way, of course, to which even Gertrude Butterwick, always inclined to look squiggle-eyed at his female friends, could not have taken exception.

squiggle, *v.* Add: **2.** (Later U.S. examples.) Also *fig.*

1922 W. STEVENS in *Dial* July 91 Our bawdiness..Is equally converted into palms, Squiggling like saxophones. **1978** *Detroit Free Press* 16 Apr. D11/1 Snitman..drew from the water a thin, blackish-brown worm that squiggled in his palm, trying to sink tiny teeth into his flesh.

3. To write (something) in a squiggly manner, to scrawl. Also, to squeeze or daub (paint) from a tube thus.

1942 N. BALCHIN *Darkness falls from Air* vi. 108, I signed it. Lennox squiggled underneath, 'I foresee difficulties but the scheme is right in principle. J.L.' **1969** *Observer* 12 Jan. 8/5, I watched a little girl squiggle yellow and red paint on a piece of white paper. **1973** *Daily Tel.* 24 Oct. 17/2 It was the 15th tube [of paint], so I squiggled it on.

† **squil** (skwil). *Oxford University Slang. Obs.* exc. *Hist.* Also **squill**. [Curtailed form of L. *Esquilīnus* of or pertaining to the Esquiline Hill, one of the seven hills of Rome (see quots.). The Esquiline was largely inhabited by low persons; it was used as a cemetery, ult. for paupers (cf. Hor. *Sat.* I. 8).] A name formerly used by Christ Church men to designate opprobriously any member of the university not a member of Christ Church (see also quot. 1970).

1721 [see HODMAN 3]. **1874** C. WORDSWORTH *Soc. Life Eng. Univ. 18th Cent.* 304 Ch. Ch. was unpopular:..the men gave themselves airs, with wonderful ignorance and conceit they claimed to belong to an House, not to a College; those of other Colleges were 'squils' and 'hodmen'. **1900** H. L. THOMPSON *Christ Church* vii. 151 The phrase 'Squils and Hodmen' needs some explanation. The first word is now happily forgotten, but was in use within the last twenty years, as a colloquial designation of members of other colleges. It was supposed to be a corruption of 'Ex-Collegees', or 'Esquilini'. **1921** S. PAGET *H. Scott Holland* iv. 82 He belonged to us, not to 'out-college men', whom some of us called Squills, *i.e.* dwellers on the Esquiline. **1970** BILL & MASON *Christ Church & Reform* p. ix, We are both non-gremial members of Christ Church, and in the abusive word of an earlier day are therefore 'squills'.

squilgee, *sb.* Add: (Earlier example.)

1851 H. MELVILLE *Moby Dick* III. viii. 66 Edgewise moved along the oily deck, it operates like a leathern squilgee.

squilgee *v.*: also *fig.*

1905 J. C. LINCOLN *Partners of Tide* vii. 147 That doctor squilgeein' my maintop with his physic stuff has made me feel A1 again.

squill. 1. c. (Examples of *red squill* as a substance.)

1947 *Federation Proc.* VI. 333/1 The identity of the cardiac and the convulsant properties of red squill receives support from an experiment. **1963** *Times* 25 Jan. 13/7 The poisons are principally red squill, sodium fluoro-acetate, and rodenticides containing elemental phosphorus. **1974** M. C. GERALD *Pharmacol.* iii. 57 Strychnine and red squill have been used as rat poisons.

squinch, *v.* Add: **1.** (Later examples.)

1939 *Real Detective Mag.* Aug. 89 She squinched and twisted her too prominent nose in a way that was not at all becoming. **1956** R. ELLISON in W. King *Black Short Story Anthol.* (1972) 263 Buster stopped and looked at me, squinching up his eyes with his head cocked to one side. **1974** *Gen. Systems* XIX. 65/2 Within a few hours, the palsy paralyzes the seventh cranial nerve, squinching half the victim's face. The eye cannot close and it waters excessively. The lips displace, and the mouth corner sags.

2. *intr.* To squeeze up so as to occupy less place; to crouch. Also with advbs., as *down*, *over*, etc.

1843 'J. SLICK' *High Life in N.Y.* II. 195 Wal, she squinched a trifle and gin a leetle start. **1942** W. FAULKNER *Go down, Moses* ii. 158 The old woman was kind of squinched down in one corner. **1972** M. J. BOSSE *Incident at Naha* III. 144 Virgil..waited for me to move from the doorway, which I did not by rising but by squinching over.

squinch-owl. *U.S. local.* [SQUINCH *v.*] = *SCREECH-OWL 1 b.

1880 J. C. HARRIS *Uncle Remus* xix. 89 Word went roun' dat de man Squinch Owl done kotch nudder watzizname. **1929** W. FAULKNER *Sound & Fury* 48, I heard a squinch owl that night. **1966** *Publ. Amer. Dial. Soc. 1964* XLII. 23 *Screech* (or *scritch*, or *squeach*, or *squinch*, or *scroochi*) *owl*, *Otus asio*, each name of which suggests some characteristic of the little owl.

squint, *sb.* Add: **1. d.** *Radar.* Lack of alignment between the axis of a transmitting aerial and the direction of maximum radiation, deliberately introduced in some systems. Freq. *attrib.*

1947 L. N. RIDENOUR *Radar Syst. Engin.* vi. 197 'Squint', which results from improper installation or trimming of antennas, has the same operational effect as crabbing of the aircraft in a cross-wind. **1969** BARTON & WARD *Handbk. Radar Measurement* ii. 31 The fall-off in energy ratio restricts the practical squint angles to about half the individual beamwidth. **1969** C. A. WILEY in Kayton & Fried *Avionics Navigation Syst.* viii. 370 (*caption*) Squint-mode window display.

squint, *v.* **6. b.** (Earlier example.)

1844 QUEEN VICTORIA *Jrnl.* 21 Sept. in D. Duff *Victoria in Highlands* (1968) I. 58 We then began our descent, 'squinting' the hill, the ponies going as safely..as possible.

squinted, *ppl. a.* Add: **2.** *Radar.* Of an aerial: having squint (sense *1 d).

1966 R. W. BICKMORE in R. C. Hansen *Microwave Scanning Antennas* iv. 297 Resolution will suffer if a linear phase term is programmed into the processor to cause 'squinting' of the synthetic beam. This is due to the projected aperture effect and is true for all electrically squinted antennas. **1969** BARTON & WARD *Handbk. Radar Measurement* ii. 31 The feeds required for generation of slightly squinted beams must overlap in the focal plane.

squinty, *a.* Restrict *Obs.*[—0] to sense in Dict. and add: **2. a.** Of persons: = SQUINT-EYED *a.* 1 a, b.

[**1819** KEATS *Let.* 20 Sept. (1958) II. 204, I am like a squinti(n)g [squinty: 1931 ed.] gentleman who saying soft things to one Lady ogles another.] **1922** JOYCE *Ulysses* 343 She knew right well..what made squinty Edy say that because of him cooling in his attentions. **1957** *Numbers* (N.Z.) Mar. 16, I was only a squinty little third-former.

b. Of the eyes: characterized by squint or oblique vision.

1925 T. DREISER *Amer. Tragedy* II. I. ii. 174 The young girl with the yellow hair and squinty blue eyes. **1978** J. IRVING *World according to Garp* ii. 37 The eyes..as red-lidded and squinty as a pig's.

squire, *sb.* Add: **5. d.** As a term of polite address to a gentleman not formally a squire. More recently, a jocular or familiar address to another man, not necessarily of different status.

1828 J. F. COOPER *Notions* I. 102 His usual address is 'friend', or sometimes he contemplates a stranger of a gentlemanly appearance, with the title of 'squire'. **1864** TROLLOPE *Can you forgive Her?* (1865) II. i. 6 'Well, Squire,' said Scruby, 'how is it to be?' **1959** [see *LONG *a.*[1] 7 c]. **1962** C. WATSON *Hopjoy was Here* x. 111 You see, squire, it's reasonable. **1968** *Listener* 22 Feb. 255/3 At a garage in Paddington I overheard a very sleek young man..ask the attendant to 'fill both tanks up'. 'Blimey, squire,' was the reply, 'you going all the way to Marble Arch?' **1977** N. J. CRISP *Odd Job Man* i. 6 'Good-night then, squire,' he said, to the barman. **1982** *Times* 15 May 14/6 (*caption*) Tell you what, squire—keep the pension and I'll take the cash!

6. Also, applied more widely to any local dignitary.

1848 *Knickerbocker* XVIII. 379 Every body is a squire in these days. **1873** 'MARK TWAIN' & WARNER *Gilded Age* 17 'Squire' Hawkins got his title from being postmaster of Obedstown. **1935** H. W. HORWILL *Dict. Mod. Amer. Usage* 301/1 In Am. the *squire* is primarily a justice of the peace, but the name is loosely given, most commonly as a title, to any prominent resident in a village. **1948** W. FAULKNER *Intruder in Dust* ii. 34 He had already telephoned Squire Fraser.

7. (Earlier example.)

1874 [see *count-fish* s.v. *COUNT *sb.*[1] 9].

squire, *v.* Add: **1.** Now more freq. in U.S. use. **a.** (Later examples.)

1901 G. B. SHAW *Cæsar & Cleopatra* III, in *Three Plays for Puritans* 148 Ftatateeta comes to her. Apollonius offers to squire them into the palace. **1949** N. MITFORD *Love in Cold Climate* I. ix. 97 Squiring royal old ladies to the supper-room. **1967** *Boston Herald* 1 Apr. 20/5 You will squire Machree wherever she wants to go. **1977** I. SHAW *Beggarman, Thief* III. x. 334 The sight of his mother ..being squired..off the plane by a man who seemed not much older than himself had disturbed him.

b. (Later examples.)

1962 D. LESSING *Golden Notebk.* I. 120 That second night of dancing she was squired by Stanley while her husband drank in the bar until it closed. **1977** *Time* 31 Jan. 31/1 At least one of Carter's aides turned up: Pollster Pat Caddell, who squired Hugh Hefner's daughter Christie.

squirearchal, *a.* (Earlier example.)

a. **1833** J. S. MILL in *Jurist* IV. 15 Some stupid younger son of a squirearchal house.

squirearchy[1]. **1. a.** (Earlier example.)

a. **1796** W. TONE *Autobiogr.* (1893) II. iii. 55 Such is the honesty of the Squirearchy of Ireland.

Squi·rearchy[2]. [f. the name of Sir John Collings *Squire* (1884–1958), Eng. poet and man of letters, punningly after prec.] The influential literary circle, composed principally of critics and poets, which surrounded Squire, esp. during his editorship of the *London Mercury* (1919–34).

1930 J. ARROW *J. C. Squire v. D. H. Lawrence* 5 Mr. J. C. Squire has pronounced... The magnanimous verdict is that the work of which the 'Squirearchy' does not approve will be 'easily forgiven and even forgotten'. **1959** *Times Lit. Suppl.* 10 Apr. 210/1 When Dr. Leavis founded *Scrutiny* it was with the laudable intention of cleaning out the Augean Stables of the Squirearchy, and an academic Hercules indeed he proved to be. **1978** J. PEARSON *Façades* viii. 146 The next encounter [of] the Sitwells.. was the battle with the group who called themselves..the 'Squirearchy'.

squirl (skwɜ̄ɹl). *dial.* and *colloq.* [Perh. blend of SQUIGGLE *sb.*[2] and TWIRL *sb.* or WHIRL *sb.*] A flourish or twirl, esp. in handwriting.

1843 J. BALLANTINE *Gaberlunzie's Wallet* ix. 212 Look at the lang turns o' his L's, and the squirls o' his B's; he's been weel brought up. **1922** *Sunday at Home* Oct. 59/2 'What a squiggly handwriting,' she said... 'They would not let us write that way at school. But when I am grown up I shall make lots of squirls.' **1955** E. BOWEN *World of Love* iii. 58 They are simply signed with a squirl. **1971** H. R. LOYN *Wulfstan MS.* 22, S2 three times uses a distinctive squirl, rather like a reversed *c*, to show the end of a section before a blank half-line. **1979** *PN Rev.* 10 7/1 The anterior reinforcing board..has a similar but not identical configuration of interlaced and interlocking squirls.

squirm, *sb.* Add: **c.** A twisting or curving form of decoration characteristic of *art nouveau*; hence *colloq.* (with *the*), the style itself.

1909 [see *art nouveau* s.v. *ART *sb.* VI. e]. **1972** F. MACCARTHY *All Things Bright & Beautiful* ii. 46 Art nouveau was called 'the squirm'. When 'the squirm' arrived in Britain..Lewis F. Day and Walter Crane were outraged.

squirm, *v.* Add: **2. b.** *trans.* To twist or contort (something) *into* a new form. *rare.*

1876 'MARK TWAIN' *Tom Sawyer* xxi. 171 A brain-racking effort was made to squirm it into some aspect or other that the moral and religious mind could contemplate with edification.

squirmer (skwɜ̄·ɹməɹ). [f. SQUIRM *v.* + -ER[1].] One who squirms or writhes, esp. with embarrassment; an evasive person.

1951 O. NASH *Family Reunion* 109 Ho, squirmers and writhers, how long will ye suffer The medical tyrant, the social rebuffer! **1977** P. DICKINSON *Walking Dead* iv. 59 He was anxious to pin the little squirmer down.

squirmy, *a.* **2.** (Later examples.)

1922 E. RAYMOND *Tell England* I. ii. 46, I felt a pleasing, squirmy excitement to think that we were to walk on to the..field in the company of the great Middlesex amateur. **1943** D. POWELL *Time to be Born* viii. 182 It was a squirmy feeling and reminded Ken..that..the carapace should be a little thicker. **1982** BARR & YORK *Official Sloane Ranger Handbk.* 57/1 One would never talk about true feelings—excessive piety is rather squirmy.

squirrel, *sb.* Add: **1. b.** Also, squirrel-fur in fashionable use (in the 19th and 20th cent.). Also *ellipt.*, a coat of squirrel fur.

1827 *Manch. Guardian* 27 Oct. 1/2 (Advt.), A perfectly new, elegant..and fashionable assortment of Furs, consisting of Chinchilla, Russia Fitch,..Squirrel and Sable. **1895** *Army & Navy Co-op. Soc. Price List* 15 Sept. 1038 Cape, lined throughout with grey and white Squirrel. **1930** P. HAMBLEDON *Straight Flame* iv. 169 A fur coat.. fashioned of the softest silvery squirrel from Manchuria. **1978** *Country Life* 31 Aug. 598/3 There are..squirrels and durable musquashes for the less well heeled.

6. b. Also, made of squirrel skin or fur, as *squirrel coat, hat, lining.*

1936 E. M. DELAFIELD *Provincial Lady in Amer.* 235 Ella, elegant..in..grey squirrel coat. **1969** L. HELLMAN *Unfinished Woman* xi. 153 A little girl in a fine squirrel coat and hat. **1974** *Selfridge Xmas Catal.* 14 Leather coat with silvered racoon collar and squirrel lining, £330.00.

7. a. squirrel-cage: spec. *transf.* in *Electr.*, a form of rotor resembling a squirrel-cage, used in small electric motors (usu. *attrib.*); **squirrel-headed** *a.* (later U.S. example); also **squirrel-headedness.**

1895 S. P. Thompson *Polyphase Electr. Currents* iv. 117 A solid cylinder of iron is improved [as a rotor] by surrounding it with a mantle of copper, or by a squirrel-cage of copper bars. **1920** *Whittaker's Electr. Engineer's Pocket-bk.* (ed. 4) 552 Motors with squirrel-cage rotors require special features in the rotor to obtain large starting torque without large losses at normal load. **1972** *Sci. Amer.* Oct. 114/2 The air is circulated and exhausted by a system of squirrel-cage blowers. **1953** *Sun* (Baltimore) 30 Jan. B4/1 Former President Harry S. Truman was quoted as saying..that action..to curb President Eisenhower's Government reorganisation powers 'was a squirrel-headed thing to do'. **1955** E. Pound *Section: Rock-Drill* lxxxv. 6 Not to pamper this squirrel-headedness.

b. **squirrel briar**, a name used in New England for either of two local species of *Smilax*, *S. glauca* or *S. rotundifolia*, deciduous prickly vines bearing blue-black berries.

1910 C. B. Graves et al. *Catal. Flowering Plants & Ferns Connecticut* 125 Smilax glauca..Saw, Cat or Squirrel Briar. Common. Dry or moist open woods and thickets. *Ibid.*, Smilax rotundifolia..Common Green Briar. Horse, Cat, Bull or Squirrel Briar. **1979** C. MacLeod *Family Vault* viii. 53 Something grabbed her..but it was only a squirrel briar.

squirrel, *v.* Restrict † *Obs.* to senses in Dict. and add: **1. c.** *intr.* To go round in circles like a caged squirrel; to run or scurry (*round*) like a squirrel.

1921 *Sat. Westm. Gaz.* 29 Jan. 21/1 His deeply affectionate but explosive father inexhaustibly squirrelling round the cage of conventional ideas. **1953** A. Clarke *Coll. Plays* (1963) 374 My soul was waiting for me, so small, It wriggled and squirrelled to my shoulder. **1966** D. Francis *Flying Finish* xvii. 200 The useless thoughts squirrelled round and round, achieving nothing. **1983** 'M. Yorke' *Find me Villain* v. 39 Nina's mind went on squirrelling round unhappily.

2. *trans.* To store *away* in the manner of a squirrel; to save, hoard; to cache. Rarely with *up* and without advb.

1939 J. Steinbeck *Grapes of Wrath* xxviii. 569, I been squirrelin' money away. **1949** M. Mead *Male & Female* vii. 145 Mouths first suck and later bite, and are capable of spitting, of squirrelling food in the cheek all night. **1953** J. Blish *Case of Conscience* in *If. Worlds of Sci. Fiction* I. iv. 40 I've got more plans to that effect squirrelled away. **1959** J. Braine *Vodi* vi. 95 Dick felt somehow provident, squirrelling up at least one good event to remember in the winter. **1965** G. Holden *Don't go it Alone* (1966) xvii. 141 He wasn't the type of person to squirrel away something and forget it. **1974** D. Smith *Look Back with Love* viii. 68, I started a bottom drawer, but it petered out after I had squirrelled one tea-cosy and a pen-wiper. **1981** N. Freeling *One Damn Thing after Another* viii. 58 Arlette..was good at squirrelling away things in obscure places.

squirrelling, *vbl. sb.* Restrict † to sense in Dict. and add: **2.** Hoarding, saving up; storing *away*.

1960 *Listener* 31 Mar. 564/2 Surplus hours..used either to offset the effect of tight times, or as a safeguard against situations when..work became short, and time had to be booked at the 'lower' time rate. I have heard this practice described as 'squirrelling' in other industries. **1965** *Wall St. Jrnl.* 17 Feb. 1 Presumably, the worsening could result from public hoarding of the new coins or the squirreling away of old ones as keepsakes. **1969** *Daily Tel.* 31 July 15/1 We resolved to resist all temptations to lay in 1969 supplies of anything which remained in bulk from 1968's feverish squirrelling. **1972** *Guardian* 25 Sept. 8/1 The specialist's squirrelling of facts.

squirrelly (skwi·rĕli), *a.* Also **squirrely.** [f. Squirrel *sb.* + -y[1].] **1.** Resembling or characteristic of a squirrel.

1925 D. H. Lawrence *Reflections on Death of Porcupine* 204 The chipmunks..were little squirrelly things with stripes down their backs. **1965** H. Gold *Man who was not with It* ii. 11, I saw that squirrelly look which says: 'Win it all back with the rent money.'

2. Inclined to rush this way and that, unpredictable. Of a person: demented, crazy; jumpy, nervy.

1928 O. Elton *Survey Eng. Lit. (1730–80)* I. iii. 70 Lady Sarah Lennox's letters are 'at first squirrelly and girlish'. **1934** M. H. Weseen *Dict. Amer. Slang* 198 *Squirrely*, abnormal; queer; crazy. **1960** *Spectator* 25 Mar. 438 Her description of contestants going 'squirrelly' (demented) with exhaustion..makes an unpleasant sociological document. **1970** K. Platt *Pushbutton Butterfly* (1971) x. 117, I got all squirrelly after you left. I just couldn't sit around doing nothing. **1977** J. Wambaugh *Black Marble* (1978) viii. 108 Squirrelly dog, Pattie Mae... Squirrelly! Just like his goddamn owner. I never shoulda said I'd bring him today.

squirt, *sb.* Add: **2. d.** A jet-propelled aeroplane, punningly after *jet*. *Air Force slang. temporary.*

1945 L. R. Gribble *Battle Stories of R.A.F.* xxiii. 59 To fly the squirts in combat meant the development of a new technique. **1948** *N.Y. World-Telegram* 30 Dec. 11/7 The plane itself is called a 'blow torch', a 'flame thrower', a 'squirt', [etc.].

3. c. *transf. spec.* in *Air Force slang*, a burst of gun-fire.

1942 Brennan & Hesselyn *Spitfires over Malta* i. 28, I gave him a squirt with the scatter guns. **1948** *Welsh Rev.* Winter 287 There was another squirt of song and then silence. **1978** K. Amis *Jake's Thing* iv. 34 To be given the choice of two [buses]..was..certainly welcome in the increasing rain and squirts of cold wind.

d. A compressed radio signal transmitted at high speed. Cf. sense *9 of the vb.

1968 M. Woodhouse *Rock Baby* vii. 64 It can't transmit continuously, though, surely... What does it do, transmit in squirts? **1974** 'J. le Carré' *Tinker, Tailor, Soldier, Spy* vi. 50 Irina..boiled down the microdots and played radio for him on a high-speed squirt to beat the listeners.

5. a. For (Chiefly *U.S.* and *dial.*) read orig. *U.S.* and add: (Later examples.) Also *spec.* a child or young person.

1914 G. Atherton *Perch of Devil* i. 39 She had 'sized him up' as a 'squirt'.., but he was 'a long sight better than nothing'. **1924** Kipling *Debits & Credits* (1926) 153 They both shook 'ands with the young squirt across the table. **1935** 'N. Blake' *Question of Proof* i. 17 It's about time that squirt Wemyss was suppressed. **1955** M. Gilbert *Sky High* iii. 46 Most people who send letters like that are cowardly little squirts. **1958** B. Malamud *Magic Barrel* (1960) 138 George..remembered him giving him nickels..when he was a squirt. **1967** [see *half-portion* s.v. *half- II. n]. **1977** J. Bingham *Marriage Bureau Murders* xi. 134 Sidney Shaw, the little squirt..ought to be able to intercept a letter.

squirt, *v.* Add: **II. 4. †d.** Phr. *to squirt a mouldy*, to fire a torpedo. *Naut. slang. Obs.*

1916 [see *mouldy *sb.*].

9. To transmit (information) in highly compressed or speeded-up form. Also *absol.*

1971 C. Egleton *Last Post for Partisan* xii. 113, I should have been given the means to squirt... You pre-record the message and then push it through in a second, before they have a chance to find your frequency. **1977** P. van Rjndt *Blueprint* xiii. 231 The message is recorded on a separate tape which is then treated electronically in order to compress the message... The normal procedure is to 'squirt' it over radio transmission. **1979** C. McCarry *Better Angels* ii. i. 106 Radio equipment..could squirt a million words from one continent to another via satellite in a droplet of electric energy that required less than a millisecond to send or receive.

squirter. Add: **4.** (Later examples.)

1966 *Punch* 3 Aug. 193 The advertisements inviting operators to take up space at forthcoming fairs, feasts, wakes and well-dressings are full of cavillings like..'No squirters or confetti.' **1983** 'F. Parrish' *Bait on Hook* ii. 29 You squirt it [*sc.* paint] out of a tin... You can't do a thin line easily with a squirter.

5. *slang.* A revolver.

1935 G. Ingram *Cockney Cavalcade* xvi. 260 There've been reports..of some of you..doing that film stuff with squirters and handkerchiefs over your faces.

squish, *sb.* Add: **3.** *Engin.* In some internal-combustion engines, the forced radial flow of mixture from the cylinder into the combustion chamber as the piston approaches the cylinder head at the end of a stroke. Freq. *attrib.*

1934 *Proc. Inst. Mech. Engineers* CXXVIII. 155 Satisfactory mixing was brought about largely by the 'squish'. **1953** *Proc. Inst. Mech. Engineers: Automobile Division 1951–2* 103/2 Since squish occurs late in the compression stroke, it is not likely to affect maximum-power spark timing. **1957** *Encycl. Brit.* XII. 505/1 The squish turbulence..feeds most of the charge that would otherwise have been last to burn into the flame front. **1979** R. H. Warring *Know Your Model Aero Engines* xii. 72 With a squish head, the outer section of the head is flat, with a smaller hemispherical combustion chamber in the middle.

squish, *v.* Add: **1.** Also *colloq.*

1976 D. Heffron *Crusty Crossed* ix. 69 We squished our teeth into the berries in our mouths to stop giggling. **1977** G. Durrell *Golden Bats & Pink Pigeons* v. 129 He [*sc.* an octopus] had wedged himself, or rather squished himself, into a small crevice.

2. b. Of a person, etc.: to proceed or make *one's way* with a squishing sound. *colloq.*

1952 *Sun* (Baltimore) 9 July 30/4 (*caption*) Soaked to the point of not caring, this waterlogged pedestrian squishes his way across a downtown street. **1963** M. Beadle *These Ruins are Inhabited* iii. 40 We squished down a rutted lane. **1965** F. Knebel *Night of Camp David* ii. 50 Tires squishing through the slush and spraying muddy water from little pools at the edge of the pavement. **1978** *Chicago* June 72/2 The highlight of Day is to squish around in the foam spread in a contained area by the Village Fire Department.

squishy, *a.* Add: Also *fig.*

1953 *Freckles* in *News & Observer* (Raleigh, N.C.) 9 Mar. 15/7 Everyone knows you're squishy about Miss Springtime! **1973** *Times Lit. Suppl.* 26 Oct. 1310/5 The very delicacy and respect for words in themselves which are radio's chief strengths do tend to make it rather a squishy medium. **1981** *Sunday Tel.* 8 Mar. 13/3 All the lovely squishy feeling flowed over us when we waved and the driver waved and tooted.

squit, *sb.*[1] For *dial.* read *dial.* and *slang* and add: **1.** (Later examples.)

1909 G. B. Shaw *Let.* 4 Nov. in *Lett. to Granville Barker* (1956) 160 Some little squit of a nervous boy who can cry and scream like a burlesque of Eugene. **1928** A. Huxley *Point Counter Point* xxi. 389 Miserable scrofulous little squit! **1947** E. Coxhead *Play Toward* iv. 102 It's impossible, darling. That—that little squit—and Peggy Jacques! **1976** J. Snow *Cricket Rebel* 11, I was left in the hands of a second year prep boy, my 'nursemaid', to be introduced to the way of life of a new boy or 'squitt' as he was called.

2. (Later examples.)

1959 A. Wesker *Roots* i. 26 Love? I don't believe in any of that squit—we just got married. **1976** *Norwich Mercury* 19 Nov. 6/8 Dont talk squit.

squitter. **1.** For Now *dial.* read Now *dial.* and *colloq.* and add later examples. Cf. *skitter *sb.*[1] 1.

1958 P. Scott *Mark of Warrior* ii. 131 'Aren't you sleeping?' 'I get the squitters pretty regularly.' **1976** A. Price *War Game* 13, I reckon squitters was queen [of the battlefield]. More of the poor bastards crapped themselves to death than killed each other. **1981** Ld. Harewood *Tongs & Bones* ii. 37 We went incessantly to those over-public latrines... My squitters were at their worst.

2. *Radar.* Random pulses produced by a transponder in the absence of interrogating signals.

1958 *Proc. Inst. Electr. Engineers* CV. B. Suppl. No. 8. 299/2 An average of 2700 pulses/sec are produced by the trigger circuit. These pulses are of constant amplitude and shape, but entirely random in time spacing and are known as 'squitter'. **1976** P. Honold *Secondary Radar* i. 67 If the dynamic characteristics of the receiver are too low, the message transmitted may be falsified.., should the signal level be very high. This..will also cause additional dead times—if these are triggered by interference pulses (squitter).

squiz (skwiz), *sb.* *Austral.* and *N.Z. slang.* [f. Quiz *sb.*[2], prob. blended with Squint *sb.*] A look or glance.

1916 C. J. Dennis *Songs Sentimental Bloke* 130 *Squiz*, a brief glance. **1922** K. Mansfield *Garden Party* 72, I say Laura..you might just give a squiz at my coat before this afternoon. See if it wants pressing. **1948** V. Palmer *Golconda* xxx. 254 With one squiz at it [*sc.* the ore] he can tell how much it would crush to the ton. **1961** R. Park *Hole in Hill* (1962) viii. 60 'All right, let's have a squiz.' They looked out very cautiously. **1972** J. McClure *Caterpillar Cop* iv. 62 You won't find it there... I had a good squiz over the whole area.

squiz, *v.* *Austral.* and *N.Z. slang.* [f. prec.] *trans.* and *intr.* To look (at) or glance; to peer.

1941 Baker *Dict. Austral. Slang* 71 *Squiz*, to look at, inspect. **1947** *Coast to Coast* 132 Kids, who used to slip up the lane behind our place and squiz through the cracks in the fence. **1948** D. W. Ballantyne *Cunninghams* (1963) xvi. 73 They walked down the passage and squizzed in the dining-room.

squizzed (skwizd), *a.* *U.S. slang. rare.* [Orig. unknown.] Drunk, tipsy.

1845 *Knickerbocker* XXV. 75 He was rescued from the pyre..looking like a squizzed cat. **1941** *Sat. Rev. Lit.* 22 Mar. 5/1 A judge of good whiskey, who is, for the purpose of this narrative, slightly squizzed.

squizzle (skwi·z'l), *v.* *colloq.* and *dial.* Now *rare.* [Imitative.] *intr.* To squirt *out*, to squish. Hence **squi·zzling** *ppl. a.*

1856 'J. Phœnix' *Phœnixiana* (ed. 11) xviii. 130 When the mouth is filled with the luscious fruit, and the..sweet though embarrassing juice is squizzling out all over the chin, and shirt-bosom. **1872** Hardy *Under Greenw. Tree* I. ii. 19 Such a squizzling and squirting job as 'tis in your hands. **1901** *Westm. Gaz.* 27 Feb. 1/3 Boots squeak and squizzle in the mud.

squodgy (skwǫ·dʒi), *a.* *colloq.* Also **squodgey.** [Imitative: cf. Squashy *a.*] Soft and soggy; squelchy.

1970 *Daily Tel.* 15 June 11 Paper cups..tend to go squodgy and superimpose their own faint flavour to hot drinks. **1981** *Church of England Newspaper* 28 Aug. 10/3 One buries by pouring earth upon the deceased, rather than searching for squodgey ground in which to submerge the body.

squoggy (skwǫ·gi), *a.* *colloq.* [Prob. blend of Quaggy *a.* and Soggy *a.*] Wet and miry.

1950 J. Cannan *Murder Included* vi. 121 'These heel-marks.'..'The ground's too squoggy to tell whether they're male or female.'

squop (skwǫp), *v.* *Tiddlywinks.* Also † **squallop** (pa. t. and pa. pple. † **squapt**). [Of unknown origin.]

1. *trans.* To cover and immobilize (another's wink) with one's own; also with another player as obj.

1956 *Minutes Cambr. Univ. Tiddlywinks Club* Mar. 12 in G. Consterdine *On Mat* (1967) 8 If both members of the team have all their winks either cupped or covered, then it shall be said that they are *squapt*. The verb shall be declined thus: *Present* I squallop..*Past* I squapt..*Past participle*—Squapt..*Noun*—squap. **1958** *Sunday Times* 2 Mar. 16/3 Mr. Harry Secombe squopped his own captain, who observed irritably that he was a Charlie. **1962** *Christian Sci. Monitor* 1 Oct. 13 In tournament play games are generally limited to 25 minutes, and a scoring system is instigated which takes account of how many winks a team has potted, squopped, etc. **1971** *Ottawa Citizen* 6 Feb. (Canadian Mag.) 24 You can't get something going for you if too many of your winks get squopped by an opponent's wink. **1978** *Boston Herald American* 25 Apr. 6

A gentle shot which either frees one of your winks or squops one of your opponent's winks is referred to as a 'piddle'.

2. *intr.* To cover an opponent's wink with one's own.

1963 *Observer* 22 Dec. 11/8 To squop is to cover an opponent's wink. **1971** *Philadelphia Evening Bull.* 13 Feb. 8 Squidgers in hand, the Hark Yon Tree Hath No Leaves But They Will Out Club is off to Toronto to squop and counter-squop for the North American Tiddlywinks Association Championship. **1977** *Cornell Alumni News* July 25 Sunshine tried to defuse the potential blitz by coming inside Drix's zone and threatening to squop.

Hence **squopped** *ppl. a.*, **squo·pping** *vbl. sb.* and *ppl. a.*; **squo·pper**, a player who squops.

1962 *N.Y. Times* 5 Aug. 23 It was in squopping that the British excelled. **1962** *Time* 14 Sept. 56 A squopped wink cannot be squidged again until it is de-squopped, either by the original squopper or by a squopped player's partner. **1962** *Life* 14 Dec. 122 Two-man units each with a powerful offensive squidger and a canny defensive squopper. **1971** *Ottawa Citizen* 22 Jan. 24/1 When it comes to squidging and squopping, Rosemary Wain and Andy Tomaszeuski are two of the best in the business. **1975** *Sunday Mail Color Mag.* (Brisbane) 28 Sept. 21/2 Squopping is the ability to flip one of your winks on top of your opponent's wink—and prevent him playing it. Then the squopped player tries to free his captive counter by squopping the counter on top. **1977** Sharp & Piggott *Bk. of Games* 165 The owner of a squopped wink must wait until it is freed... The owner of a squopping wink..must ensure that his squidger first touched his own wink.

squop (skwǫp), *sb.* *Tiddlywinks.* [f. prec.] The act or achievement of covering an opponent's wink with one's own; a wink squopped in this way.

1962 *Time* 14 Sept. 56 The squop shot is entirely new to them. **1962** *N.Y. Times* 21 Oct. 83 The high point of the match was a septenary squop by Mr. Stein in which he immobilized seven winks with a single shot. **1979** *Harvard Mag.* May–June 40 The wink is shot into a key position, from which it can defend friendly squops, attack enemy squops, or set up a strategic zone.

squoze, squozen: see *squeeze *v.*

sqush (skwɒʃ), *v.* *U.S. colloq.* and *dial. rare.* [Imitative: cf. Squash *v.*[1], *squdge *v.*, Squish *v.*] **1.** *intr.* To collapse into a soft, pulpy mass.

1884 'Mark Twain' *Huck. Finn* xxix. 303 He'd a squshed down like a bluff bank that the river has cut under. **1955** E. Pound *Section: Rock-Drill* lxxxviii. 41 Belascio or Topaze, and not have it sqush, a 'throne', something God can sit on without having it sqush.

b. To squelch, squeeze messily.

1929 in H. Wentworth *Amer. Dial. Dict.* (1944) 592/1, I..'squshed' through many a weary mile of mud. **1949** H. Hornsby *Lonesome Valley* xix. 249 He could take off his shoes and walk around in the mud and let the mud ooze and sqush between his toes.

Sr., abbrev. of: **1.** Sir *sb.* (q.v.).

‖ **2.** Señor 1 a, Signor 1 a.

a **1912** W. T. Rogers *Dict. Abbrev.* (1913) 182/1 *Sr.*,..Señor. **1973** *Listener* 3 May 582/3 Sr Bassetti felt that the Italian Communist Party is not the ogre it used to be. **1977** 'D. Cory' *Bennett* ii. 43 My instructions with regard to the search for Sr Bennett are..unequivocal. **1982** *Listener* 14 Jan. 8/3 Sr Fraga won the recent elections in his native Galicia.

3. Senior *a.* 1 a. Chiefly *U.S.*

1936 *N.Y. Herald Tribune* 5 June 2/2 (*heading*) Extradition of Ellis Parker Sr. to await Republican Convention. **1973** D. Aaron *Unwritten War* (1975) iii. vii. 108 James Sr., for all his geniality and charm, was an opinionated and strong-minded man. **1977** D. Delman *Nice Murderers* iv. 53 Horowitz felt..a stab of sympathy for Kennicutt, Sr. Fathers and sons, he thought.

Sranan (srā·nan). [Prob. Taki-Taki: in full *Sranan Tongo*, Surinam tongue (also used).] = *talkee-talkee. Also *attrib.*

1953 L. L. E. Rens *Hist. & Social Background of Surinam's Negro-Eng.* x. 135 The term NE was used simply because none of the other appellations (takki-takki, Krioro, sranang tongo) was adequate. **1957** *Lingua* VI. 374 (*heading*) The verbal system of Sranan. **1964** *English Studies* XLV. 423 'Phonetic Interpretation', by which term he [*sc.* Echteld] means the substitution of Sranan phonemes for the original English ones. **1974** R. A. Hall *External Hist. Romance Languages* 45 The Portuguese element in Sranan. **1980** *English World-Wide* I. i. 6 This argument is valid even for such cases as Krio, Tok Pisin or Sranan.

Sri: see *Shri, Sri. **'sright**: see *'s 1 c.

Sri Lankan (ʃrī læ·ŋkăn, ʃrilæ·ŋkăn, s-), *sb.* and *a.* [f. Skr. *Sri Laṅkā* name of the island known in English until 1972 as Ceylon (f. *śri* *Shri, Sri, honorific prefix + *Laṅkā* name of (the chief town of) the island) + -an.] **A.** *sb.* A native or inhabitant of Sri Lanka. **B.** *adj.* Of or pertaining to Sri Lanka or its people.

1973 *Advocate-News* (Barbados) 29 Dec. 7/2 (*heading*) Sri Lankan is world billiards champion. **1974** *Times* 9 Sept. 12/8 At the Edinburgh military tattoo, the Sri Lankan police reserve band have been giving a display. **1976** *Scotsman* 25 Nov. 3/5 BOC International are fighting the Sri Lankan plans to nationalise their Ceylon Oxygen offshoot. **1977** J. Laker *One-Day Cricket* 66 The Sri Lankans rattled the score along. **1982** *Church Times* 29 Jan. 11/4 In all its life the Church there was both Anglican and totally Sri Lankan.

S-rope: see *S 2 c.

‖ **sruti** (ʃru·ti). *Mus.* [Skr. *śruti*, lit. hearing, listening.] A microtonal interval in Indian music.

1792 W. Jones in *Asiatick Researches* III. 69 They unanimously reckon twenty-two *srutis*, or quarters and thirds of a tone, in their octave. **1891** [see *grama[2]]. **1954** *Grove's Dict. Mus.* (ed. 5) IV. 457/1 That disposes of the widely accepted view of the 'quartertone'—that it is half a semitone, and that Hindus have 22 of these *srutis* to make a melody with! What they have is five, six or seven notes to the octave, but 60 or more ways of disposing them. **1972** R. Shankar *Indian Music* iv. 135 The scale of twenty-two srutis is never sung chromatically, and the intervals are not important to one another but only as *groups* of intervals.

S-scroll(ed), -shaped, -sofa: see *S 2 c.

St. Add: **e.** (with small initial) for *stumped (by)* in *Cricket*. Cf. Stump *v.*[1] 8.

stab, *sb.*[1] Add: **2. f.** In colloq. phr. *to make* (or *have*, etc.) *a stab at* (something), to try, attempt; to make a shot at. orig. *U.S.*

1895 W. C. Gore in *Inlander* Dec. 115 *Stab, to make a*, to make a blind attempt to answer a question. **1908** K. McGaffey *Show Girl* 235, I..made a stab for the rail. **1915** Wodehouse *Something Fresh* xi. 315 'I *do* wish that this time you would endeavour..not to make a fool of yourself.'..'I'll have a jolly good stab at it, governor.' **1930** Galsworthy *Roof* vi. 96 D'you think you'll be able to travel the day after to-morrow?..I'll have a good stab at it, as my more genial colleagues say. **1940** 'N. Shute' *Old Captivity* x. 294 We may have to come back again... But I think we'll have a stab at it. **1961** *Press Jrnl.* Apr. 10/3 Let's say you're going to take a stab at writing up the annual office picnic. **1973** 'S. Woods' *Enter Corpse* 113 'Now that,' said Nelson, 'I can't believe.' 'You might have a stab at it,' Maitland suggested, 'It happens to be true.' **1980** W. Maxwell *So Long, see You Tomorrow* (1981) ii. 7 She may have made a stab at being a mother to my older brother and me.

g. In fig. phr. *stab in the back*, a treacherous deed. Cf. sense *1 g of the vb.

1922 Joyce *Ulysses* 621 That stab in the back touch was quite in keeping with those Italianos. **1934** R. H. Lutz *Causes of German Collapse in 1918* v. 132 (*heading*) The 'stab-in-the-back' question. **1934** —— tr. von Kuhl in *Ibid.*, Some maintain that we lost the war owing to the stab in the back administered to the Army by those at home... On the other side the 'stab-in-the-back legend' is rejected as 'one of the most malignant and..stupidest legends'. **1953** J. W. Wheeler-Bennett *Nemesis of Power* i. i. 67 For several days before his actual appearance..Hindenburg was closeted with..the extreme Nationalist leaders. In this brief period was crystallized the legend of the 'stab-in-the-back', in justification of which many innocent Germans were to suffer when the National Socialists came to power. *Ibid.*, The Marshal [*sc.* Hindenburg] testified..on November 18... He..addressed himself to the German people. Their defeat, he told them, was not attributable to the Army but to the civilian demoralization and disunion. The irreproachable Army had received a 'stab-in-the-back' (*Dolchstoss*) from the Revolution. **1959** *Times* 21 Oct. 13/4 Professor Nordhoff, the managing director of Volkswagen, reacted as if he were the victim of another stab-in-the-back legend. **1971** A. Bullock *Twentieth Cent.* 25/1 Our knowledge of the recent past..will be based on hear-say, myths ('the Stab in the Back', for instance).

3*. *Oil Industry.* (See quot. 1975.)

1972 L. M. Harris *Introd. Deepwater Floating Drilling Operations* xii. 133 The integral marine-riser system has the choke and kill lines installed on the riser joints so that they are simultaneously stabbed and made up during the stab and make up of the marine-riser connector. **1975** L. Crook *Oil Terms* 106 *Stab*, the operation of guiding one end of a pipe into the connection of another pipe to 'make up' a connection.

4. **stab-and-drag** *Archæol.*, a technique of ceramic decoration whereby a point is drawn along the surface of a pot and pushed in deeper at intervals (usu. *attrib.*); **stab-stitch** *Needlework* (see quot. 1964); hence as *v. trans.* and **stab-stitching** *vbl. sb.*

1931 V. G. Childe in *Archæol. Jrnl.* LXXXVIII. 47 Handles and flat bases first make their appearance in Fort Harrouard II. In the same levels the fine incision of the early Chassey style gives place to deep incision, or stab-and-drag. **1931** S. Piggott in *Ibid.* 78 *Stab-and-drag* lines—made with a point that is drawn along the clay and pushed in deeper at intervals—occur only on certain Scottish pots, e.g., from Unstan. **1954** —— *Neolithic Cultures* vii. 204 Certain vessels with stab-and-drag motifs. **1978** *Proc. Prehist. Soc.* XLIV. 276 Decorated wall-sherd; fine vertical stab and drag lines. **1917** E. R. Hambridge *Simple Dressmaking* i. 7/2 Back stitches should be frequent in stab tacking... Cf. Fig. 28 for stab stitch, which is similarly worked. *Ibid.* 10/1 Stab-stitching..is back-stitching, but worked with the needle placed vertically through the material..instead of horizontally. *Ibid.*, Running, run-stitching, and half back-stitching can also be stab-stitched, but strong thread or silk should be employed. **1932** D. C. Minter *Mod. Needlecraft* 159/2 Stab stitching..close to the fold of the felt..may be employed. *c* **1951** *Glovemaking* (Dryad Leaflet 31) 6 A hand-made glove..is almost always sewn on the right side, and various stitches can be used. The most popular of these is the 'prix' or stab-stitch. **1964** *McCall's Sewing* ii. 32/2 *Stab-stitch* a stitch in which the needle is brought in and out of the fabric at right angles. **1976** *Woman's Weekly* 6 Nov. 42/3 Stab-stitch the boots together in pairs.

stab (stæb), *sb.*[5] *Med.* [a. G. *stab* rod, after V. Schilling's use of G. *stabförmig* rod-shaped, *stabkern* rod-nucleus (*Zeitschr. f. exper. Path. u. Therapie* (1911) IX. 691, 692): cf. Staff *sb.*[1]] Used *attrib.* and *absol.* to designate white blood cells characterized by a nucleus in the form of a single bent or twisted rod (orig. regarded as abnormal forms).

1929 R. B. H. Gradwohl tr. *Schilling's Blood Picture* ii. 128 The neutrophilic degenerative stab or staff forms are not present in the normal blood picture. *Ibid.* 135 They [*sc.* degenerative forms] are practically insignificant, with the exception of the stabs..which deserve special mention as a degenerative phenomenon in the nuclear shift. **1938** W. Magner *Textbk. Hematol.* v. 79 Schilling divides the neutrophile leukocytes into the following classes. (1) Myelocytes. (2) Juvenile leukocytes or metamyelocytes. (3) Stab, staff or rod-nuclear cells. **1972** W. J. Williams et al. *Hematol.* iii. 27/2 (*heading*) Band form or stab cell. *Ibid.* lxvi. 562/2 The stab is the least mature cell of the granulocytic series found in the peripheral blood of normal persons.

stab, *v.* Add: **1. g.** In fig. phr. *to stab* (a person, etc.) *in the back*, to harm or damage in a treacherous manner. Cf. *back-stabber.

1916 G. B. Shaw in *N.Y. Times Mag.* 9 Apr. 2/2 The cry that 'England's Difficulty Is Ireland's Opportunity' is raised in the old senseless, spiteful way as a recommendation to stab England in the back when she is fighting some one else. **1932** Kipling *Limits & Renewals* 384 He makes my job ten times more difficult than it need be..stabs me in the back with his crazy schemes for betterment. **1956** N. Nicolson *Diary* 4 Nov. in *Diaries & Lett. H. Nicolson* (1968) 315, I did not want to publish any letter until the crisis in Egypt had ended, as otherwise I might be accused of stabbing the troops in the back. **1979** F. Olbrich *Sweet & Deadly* viii. 91 All these years with me he's been completely honest and now he stabs me in the back.

2. d. Delete *nonce-use* and add: Also *fig.*, to pierce like a pointed weapon.

1920 R. Macaulay *Potterism* vi. v. 259 Gideon's fate pilloried on that placard had stabbed through him. **1946** D. C. Peattie *Road of Naturalist* ii. 33 Outside our mortal dusty sphere, Canopus must be a horrible, blinding searchlight stabbing through a black and icy void.

7. *Oil Industry.* To guide (a length of pipe) so as to connect it properly to another member.

[**1922**: implied at *stabber 1 d.] **1932** *Amer. Speech* VII. 271 *Stab*, to guide (pipe) in making connections so that the threads engage properly. **1948** *Petroleum Handbk.* (ed. 3) v. 85 As the empty elevator hook is hoisted the derrick man latches in a stand as it passes his level. The stand is picked up and 'stabbed' into the tool joint at the rotary table. **1976** *Offshore Platforms & Pipelining* 6/3 Only one pile add-on will have to be stabbed and welded to drive the sleeve piles.

stabber. Add: **1. d.** *Oil Industry.* One who stabs pipes (see *stab *v.* 7).

1922 F. M. Towl in D. T. Day *Handbk. Petroleum Industry* I. 410 The joint is lifted into place, and a man, the 'stabber', standing by the end of the joint with a handspike, moves the joint until it is straight with the last joint laid. **1976** M. Machlin *Pipeline* iv. 50, I worked as a stabber—that's the man who sits on the end of the walking beam and holds the pipe straight.

stabbing, *vbl. sb.* Add: **1. b.** **stabbing board** *Oil Industry* (see quot. 1932).

1932 *Amer. Speech* VII. 271 *Stabbing-board*, a board in the derrick at the height of one joint, from which the derrick man stabs pipe. **1974** R. D. Langenkamp *Handbk. Oil Industry Terms & Phrases* 126 The derrick man stands on the stabbing board and assists in guiding the threaded end of the casing into the collar of the preceding joint that is hanging in the slips in the rotary table.

stabilate (stēⁱ·bilēⁱt). *Biol.* [f. L. *stabil-is* Stable *a.* + *-ate*, after *filtrate*, *precipitate*, etc.] A sample of biological material from a homogeneous source which is preserved by freezing on a single occasion to serve as a standard, portions of it being restored as desired.

1965 Lunsden & Hardy in *Nature* 6 Mar. 1032/2 A separate term is now required to convey this implication of the stabilization of biological characters in preserved material and to distinguish preserved from 'strain' material. The word 'stabilate' is now proposed for this purpose. A 'stabilate' may be defined as a population of an organism preserved in viable condition on a unique occasion. **1970** *Ibid.* 19 Sept. 1255/1 A stabilate of infected blood was frozen to −70° at the start of each experiment to ensure that a strict homologous challenge of vaccinated animals was subsequently made. **1976** *Ibid.* 24 June 698/2 All eight surviving cattle were challenged with a similar injection of stabilate.

Hence as *v. trans.*, to make into a stabilate.

1978 *Tropenmedizin und Parasitologie* XXIX. 443/2 The isolate was needle-passaged once in a goat and then stabilated.

stabile (stēⁱ·bəil). [f. L. *stabil-is* Stable *a.*, after *mobile *sb.*[3]] A rigid sculpture or similar construction of wire, sheet metal, etc., of a type first developed by Alexander Calder (1898–1976), Amer. sculptor and painter, opp. *mobile *sb.*[3] 1 a. (The name was suggested by Jean Arp.)

1943 J. J. Sweeney *Alexander Calder* 30 The sculpture which was exhibited at the Galerie Percier shows a full assimilation of his experiences... It was to this type of stationary abstract sculpture that Arp, a few months later, gave the name 'stabiles'. **1949** [see *mobile *sb.*[3] 1 a]. **1957** [see *kinetic *a.* 6]. **1967** *Times* 28 Feb. (Canada Suppl.) 31 This 46-ton 'stabile' is conceived and created in nickel stainless steel. **1977** P. Leach *Baby & Child* iii. 179 Her stabile will still be useful when she is in her cot.

stabilimentum (stăbilime·ntə̆m). *Ent.* Pl. **-menta.** [See Stabiliment.] A conspicuous broad band of silk running across the web of certain kinds of spider.

1912 J. H. Comstock *Spider Bk.* iv. 203 Some of the orb-weavers strengthen their webs by spinning a zigzag ribbon across the centre or below the hub (Fig. 188). This ribbon has been termed the *stabilimentum*. **1952** T. H. Savory *Spider's Web* iv. In addition to radii and spirals there are certain accessory structures which most webs possess. Of these the most conspicuous, when present, is the stabilimentum. **1983** *New Scientist* 10 Feb. 366/1 It seems that stabilimenta are really early warning landmarks for birds.

stabilimeter. Restrict *Aeronautics* to sense in Dict. and add: **2.** *Biol.* A device attached to or forming an animal's cage, the movement of which is recorded as a measure of the animal's activity.

1940 *Jrnl. Exper. Psychol.* XXVII. 91 A similar but smaller stabilimeter has been constructed for studies on spontaneous activity and conditioning in guinea pigs. **1954** *Jrnl. Compar. & Physiol. Psychol.* XLVII. 97 The basic apparatus unit was a stabilimeter-type activity recorder... It involves a circular living cage resting on a central pivot about which it can tilt slightly in any direction. **1972** A. K. Bartoshuk in Kling & Riggs *Woodworth & Schlosberg's Exper. Psychol.* (ed. 3) xviii. 822/2 Stabilimeter activity increased markedly for the younger rats..but scarcely at all for 100-day rats.

stability. Add: **1. f.** Also of an atomic nucleus or sub-atomic particle.

1955 J. A. Wheeler in W. Pauli *Niels Bohr* 163 (*heading*) Nuclear fission and nuclear stability.

2. e. Also, of a physical property or the system possessing it.

1925 W. Greenwood *Text-bk. Wireless Telegr. & Teleph.* vi. 121 Stability is essential [in a receiver]. By this is meant that the sensitivity must be capable of being controlled by the operator, and must not be liable to be upset by the reception of very strong signals. **1962** J. H. & P. J. Reyner *Radio Communication* iv. 153 Modern conditions call for high stability, the capacitance being required to remain constant with time. **1963** B. Fozard *Instrumentation & Control Nucl. Reactors* x. 122 For high quality equipments..requiring a gain stability of 0·1% both H.T. and heater supplies must be stabilised. **1975** D. G. Fink *Electronics Engineers' Handbk.* xvii. 50 The stability of feedback systems is of great importance since an unstable system will not be effective in maintaining the controlled variable at approximately the desired value.

stabilization. (In Dict. s.v. Stabilize *v.*) Add: **a.** (Later examples.)

1922 *World's Paper Trade Rev.* 10 Mar. 767/1 Stabilisation of wages is an urgent necessity in order that the industry might enjoy continued peace. **1945** *Electronic Industries* Sept. 226 *Stabilization*, a system for maintaining a radar beam in a desired direction in space despite the roll and pitch of the ship or aircraft. **1963** B. Fozard *Instrumentation & Control Nucl. Reactors* x. 122 Stabilisation of heater current is conveniently arranged by joining the valve heaters in series and supplying them with direct current from a regulated supply. **1975** *Petroleum Rev.* XXIX. 92/1 The stabilisation consists of the separation of light hydrocarbon components from the raw crude so that the 'stabilised' product is 'stable' in the sense that it can be stored and handled under atmospheric conditions.

b. *attrib.*

1940 *Times* 5 June 7/4 Inasmuch as the wage structure of the country has reached a stabilization point, the only justification for wage adjustments now will be in conditions common to all industry. **1958** *Spectator* 6 June 752/2 Sir Oliver Franks..called for a 'stabilisation' loan [from the U.S.] and the reinforcement of the IMF resources. **1969** *Advancement of Sci.* XXVI. 65/2 Management was thus seen as concerned primarily with stabilization policy and with the short-run, since stability is by definition a short-term affair. **1974** *Daily Tel.* 11 Mar. 18/5 The South African government may introduce a gold stabilisation fund.

stabilize, *v.* Add: **1.** Also with other objs. and *refl.*

1916 H. Barber *Aeroplane Speaks* 71 If an aeroplane was not stabilized in this way. **1938** G. H. Richter *Textbk. Org. Chem.* xxix. 595 The majority of the reactions of naphthalene are..the formation of unstable 1,4 addition products..which then stabilize themselves by the elimination of two atoms. **1958** *Times* 30 Aug. 6/1 The rocket during the whole of its flight was stabilized, to prevent rotation. **1975** *Petroleum Rev.* XXIX. 89/3 The raw crude oil will be stabilised and the natural gas liquids recovered.

3. *intr.* To become stable; *esp.* to cease varying in value.

1961 in Webster. **1971** *Timber Trades Jrnl.* 21 Aug. 16/1 Lumber stabilises at approximately 15% moisture content under normal use conditions. **1974** *State* (Columbia, S. Carolina) 26 Apr. 19-A/2 The Nixon administration contends the economy will stabilize in the second quarter. **1981** *Times* 4 Feb. 16/7 He was obliged with his pilot to eject from a Javelin which had stabilized in an unrecoverable spin.

stabilized, *ppl. a.* Made stable. **1. a.** (Later examples.) Prevented from oscillating or moving.

1958 *Times Rev. Industry* Sept. 96/1 The country's [*sc.* South Africa's] highways consist of gravelled or stabilized roads. **1968** R. W. Fairbridge *Encycl. Geomorphol.* 126/1 Sand-sized pellets from drying crusts of saline flats with 8% or more of clay..form stabilized dunes. **1977** 'D. Rutherford' *Return Load* ii. 40 The modern, stabilized vessel provided good accommodation for the drivers.

b. Prevented from fluctuating in value or quality.

1918 *Nation* (N.Y.) Feb. 129/2 To give every farmer just returns and stabilized prices.

2. Of cloth: treated in order to prevent stretching or shrinking.

1960 *Textile Terms & Definitions* (Textile Inst.) (ed. 4) 140 *Stabilized finish*, treatment applied to a textile material to increase its resistance to dimensional changes in laundering and use. **1977** *Austral. Sailing* Jan. 41/1 Cheret, Banks and Fogh spinnakers..are made from either ¾ oz or ½ oz stabilised or zephylite cloth.

3. Made stable in character or behaviour; *spec.* of a drug addict, able to live more or less normally on a repeated constant dose.

1961 *Drug Addiction* (Min. of Health) 11 Arising also..is the conception of a 'stabilised addict'... A careful scrutiny of the histories of more than a hundred persons classified as addicts reveals that many of them who have been taking small and regular doses for years show little evidence of tolerance and are often leading reasonably satisfactory lives. **1964** G. L. Cohen *What's Wrong with Hospitals?* vii. 152 They sit in one charabanc, and the stabilized male chronics in another. **1967** M. M. Glatt et al. *Drug Scene* vii. 81 The 'stabilized addict', i.e. a person who was able to maintain a fixed dose of the narcotic drug without the need for ever-increasing doses and who, at the same time, was able to follow his occupation. **1976** J. I. M. Stewart *Young Pattullo* viii. 188 Fish looked stabilized, at least until the following day, when I could take stock of his state again.

stabilizer. Add: **1. a.** (Later examples.)

1911 G. C. Loening *Monoplanes & Biplanes* x. 136 A horizontal surface placed at the rear acts as a longitudinal stabilizer. **1918** J. M. Grider *War Birds* (1927) 221 He managed to land with his stabilizer wheel. **1920** *Discovery* Mar. 77/2 Captain Caquot solved the problem by fitting the peculiar large stabilisers, or tails, which give the balloon a conspicuous, almost an uncanny appearance. **1948** W. Ley *Rockets & Space Travel* vii. 180 Schaefer and I are inclined to believe that even a 1 *g* rocket might be able to do without a stabilizer. **1962** F. I. Ordway et al. *Basic Astronautics* ii. 17 The Dutch tried using small fins attached to the body of the rocket in order to do away with the unwieldy stabilizer stick. **1973** *Times* 21 Feb. 7/1 The stabilizer of the Soviet airliner..had jammed.

b. *Naut.* A device intended to reduce the rolling of a ship in heavy seas. Cf. *gyro-stabilizer* s.v. *gyro- b.

1913 *Engineering* 13 June 820/1 The perfect stabiliser must act against the forces which are acting on the ship in such a way as to always resist the effect of the sea in producing motion. **1946** *Nature* 24 Aug. 250/2 A chapter follows on the sperry gyro ship-stabilizer, in which each gyro rotor weighs 100 tons and spins at 800 revolutions a minute. **1958** *Engineering* 28 Mar. 386/3 The installation of anti-roll Denny-Brown stabilisers in the Cunard Liner Queen Mary has now been finished. **1968** W. J. Fox *Marine Auxiliary Machinery* (ed. 4) xvi. 461 It is in the provision of this form of [flume] control that the main differences lie between the flume stabilization system and the older types of passive tank stabilizer. **1983** *Listener* 14 July 12/1 We abruptly lurched... 'Technical proving of the stabiliser,' said the announcement.

c. Chiefly *N. Amer.* A stabilizing device on a motor vehicle or tractor; *spec.* = *sway bar* (*b*) s.v. *sway-. Also *stabilizer bar*.

1931 *Automobile Engineer* Nov. 481/3 The stabiliser bar on the Panhard-Levassor..is fitted transversely at the rear end of the chassis. **1939** *Audel's New Automotive Guide* xxxii. 827 Practically all modern cars are provided with stabilizers as standard equipment. **1949** Landon & Hafferkamp in Frazee & Bedell *Automotive Fundamentals* viii. 445 The stabilizer may be thought of as a third spring which connects the two individual suspensions. **1962** *Which? Car Suppl.* Oct. 138/1 Clamp bolts of stabiliser bar [were] slightly loose. **1966** Shippen & Turner *Basic Farm Machinery* I. viii. 94 *Stabilizers*. These are also used to prevent side to side movement of certain farm implements. They usually take the form of metal bars slightly cranked at each end and they fit between an anchorage point beneath the rear axle and the ends of the linkage bars. **1977** 'D. Rutherford' *Return Load* ii. 32 She's got optional six-wheel drive with stabilizers operating on the front and rear axle.

d. *pl.* A pair of small supporting wheels fitted one at each side of the rear wheel of a child's bicycle, to keep it upright.

1960 A. L. Pullen *Cycling Handbk.* (ed. 3) i. 10 (*caption*) Child's lightweight cycle with stabilizers. **1970** *Kay & Co.* (Worcester) *Catal.* 1970–71 Autumn/Winter 982/1 RSW 11 Cycle... It has a 12 inch frame, 11 inch pneumatic tyres, a spacious rear hold-all, and stabilizers. **1978** *Dumfries & Galloway Standard* 21 Oct. 21/7 (Advt.), Child's bicycle, with stabilisers, as new, to suit 3–6 years.

2. (Examples.) More widely, an additive which inhibits chemical or physical change in a substance, esp. one used to prevent the breaking of an emulsion.

1909 *Chem. Abstr.* III. 1342 Vaseline which was added originally to protect the gun has turned out to be a valuable stabilizer for the powder. **1932** *Discovery* Aug. 240/1 The coal was prevented from separating from the oil by the addition of a 'stabilizer'. **1940** *Thorpe's Dict. Appl. Chem.* (ed. 4) IV. 520/1 In the early days of smokeless powders a small amount of alkali was added as a stabiliser. **1948** J. Osborne *Dental Mech.* (ed 2) xi. 183 Since the monomer will polymerize by the action of heat or light, it is necessary to prevent this occurring... An inhibitor, or stabilizer, is therefore added to all dental monomers. **1963** W. J. Woolgar *Plastics in Plumbing* v. 57 A small proportion of stabilizer is added, to prevent the P.V.C. from decomposing at processing temperatures. **1966** *Punch* 5 Jan. 9/1 Mrs. Joyce Butler..has introduced a Labelling of Food Bill which calls for a detailed specification..of anti-oxidants, emulsifiers and stabilisers in foods. **1977** *Lancet* 15 Oct. 780/2 Because it increases the viscosity of solutions guar gum is widely used in the food industry as a thickener and stabiliser of fat emulsions.

3. *Electronics.* **a.** A circuit or device for preventing unwanted feedback. **b.** A circuit that holds the output voltage of a power supply at a constant level despite changes in supply voltage or load, by comparison with a fixed reference voltage. Also *stabilizer circuit*.

1924 Moyer & Wostrel *Practical Radio* viii. 119 A common stabilizer for all the radio-frequency stages is generally sufficient. **1936** *Physical Rev.* L. 1094/2 (*heading*) Electronic voltage stabilizers. **1958** D. G. A. Thomas in O. R. Frisch *Nuclear Handbk.* xv. 24 The principle of..stabilisers is to compare the output voltage (or fraction of it) with that of a voltage reference source, amplify the difference, and use it to control the output. **1974** Harvey & Bohlman *Stereo F.M. Radio Handbk.* v. 82 An amplifier is not essential but its use improves the regulation of the stabilizer circuit.

4. *gen.* Something that reduces variation in the condition or behaviour of anything.

1955 *Times* 15 July 9/2 These stabilizers included increased social service payments, agricultural price supports..an easy credit policy..and a tax system which made tax liabilities fall faster than income during the recession. **1958** *Manch. Guardian* 21 Jan. 6/1 The 'built-in stabilisers' of the American economy are beginning to grip.

5. *Math.* A subgroup of a permutation group, being the group of elements that map some subset of the permuted elements on to itself.

1965 J. J. Rotman *Theory of Groups* iii. 49 If $x \in X$, then the set of all $t \in G$ that fix x forms a subgroup H_x of G (called the stabilizer of x). **1979** *Proc. London Math. Soc.* XXXVIII. 200 Hence the stabilizer of e, Stab $(e) = \{g \mid ge = e\}$ is finite.

stabilizing, *ppl. a.* (Later examples.)

1935 H. A. L. Fisher *Hist. Europe* I. xii. 137 The Arabs were poets, dreamers, fighters, traders; they were not politicians. Nor had they found in religion a stabilizing or unifying power. **1954** P. Mason *Ess. Racial Tension* 141 Stabilizing devices are introduced to preserve the dominant position and genetic purity of the dominating race. **1960** C. Day Lewis *Buried Day* i. 18 For a rather rootless person like myself, there is something reassuring, if not stabilising, in the mere known existence of individual ancestors. **1981** *Sci. Amer.* Feb. 36/3 The ability to monitor Russian missile-carrying submarines could..be considered stabilizing in that it deters surprise attack.

Stabit (stæ·bit). *Engin.* [Arbitrary, suggested by Stability.] A large mass of concrete so shaped that when large numbers are placed together they tend to interlock and form a strong barrier that will act as a breakwater.

1962 *Dock & Harbour Authority* May 29/2 The Stabit is basically a hollow tetrahedron and, besides its use for breakwater construction, it is considered to be particularly suitable for beach and coast protection works. **1963** *Guardian* 30 Aug. 3/5 (*caption*) Six hundred concrete 'teeth' have been fixed in position to protect the root of Shoreham harbour's east breakwater from the pounding of heavy seas. The 'teeth'..are called Stabits, and are interlocking in a sloping, double layer. **1968** *Proc. 11th Conf. Coastal Engin.* II. 797 The number of 29 ton Stabits which have been used on the reconstructed moles at Benghazi is about 10,000.

stable, *a.* Add: **3. d.** *Nucl. Physics.* Of an isotope: not subject to spontaneous radioactive decay, or decaying only very slowly.

1904 F. Soddy *Radio-Activity* viii. 126 The elements known to the chemist are stable *because* they exist and have survived. **1924** O. Lodge *Atoms & Rays* ii. 30 Even uranium is not quite stable... The element with 82 active pairs [of electrons] would be fairly or perhaps quite stable, and would be indistinguishable from lead. **1942** J. D. Stranathan *'Particles' of Mod. Physics* v. 167 Those elements with the greatest number of stable isotopes are Sn with 10 and Xe with 9. **1956** *Nature* 28 Jan. 159/2 Long-lived aluminium-26 may easily be confused with the stable aluminium-27 as product nucleus. **1977** J. L. Harper *Population Biol. of Plants* xi. 365 Small

amounts of ^{15}N, a stable isotope, had been applied to the soil.

stable, *sb.*[1] Add: **1. d.** *transf.* and *fig.* Also in phr. *straight from the stable* = *straight from the horse's mouth* s.v. *horse *sb.* 25 c.

1907 Lady Monkswell *Jrnl.* 13 July in *Victorian Diarist* (1946) II. 188 The beautiful Surrey landscape looks down into this purgatory of motor stables & everything that motors require. **1922** Joyce *Ulysses* 145 Tell him that straight from the stable. **1949** A. Christie *Crooked House* iii. 14 My information..came from the stable itself... She dined with me.

2. b. *slang.* A group of prostitutes working for the same person or organization.

1937 [see *knock-down *sb.* 5]. **1940** 'J. Crad' *Traders in Women* i. 26 He..now runs a 'stable' of white women for coloured seamen in Cardiff. **1973** C. & R. Milner *Black Players* ii. 35 Many players [*sc.* pimps] have several ladies, who constitute their stable. **1979** N. Hynd *False Flags* xi. 91 The consulate maintained a stable of young women..whose only purposes were those of sexual entrapment.

c. *transf.* (In quots., of motor vehicles.)

1949 *Sun* (Baltimore) 29 July 12/3 The man who owned a stable of bulldozers laughed and said, 'We'll do it tomorrow morning.' **1974** *Spartanburg* (S. Carolina) *Herald* 18 Apr. c1/2 John Greenwood, a self-made millionaire, announced Wednesday that his stable of Corvettes would be running out of John Green Automotive garage on highway 221 at I-85.

3. b. *transf.* An establishment where boxers are trained; a group of boxers under the same management.

1897 *Nat. Police Gaz.* (U.S.) 26 May 11/4 His boxing academy at..Dale End, Birmingham, is being largely supported, and some likely lads will shortly emerge from Anthony's stable. **1936** *Sun* (Baltimore) 16 Mar. 3/3 He insisted his occupation was manager of prizefighters, but the arresting detectives failed to recognize the pugilists he mentioned as his 'stable'. **1953** *Chicago Daily Sun-Times* 29 Dec. 40/5 Some years ago Rocky's hometown pal, Al Columbo, sent Marciano to New York as a candidate for membership in the Weill stable. **1976** *Scotsman* 24 Dec. 15/1 Maurice Hope..becomes the third member of the East London stable managed by Terry Lawless.

c. A team or organization which prepares motor cars for racing; a group of racing cars owned by the same enterprise.

1935 Eyston & Lynden *Motor Racing & Record Breaking* ii. 20 The Italian term [*scuderia*] finds its equivalent in..our own 'stable', and the formation of such terms among racing men is a development of very recent years. *Ibid.* vii. 65 Every important Continental racing stable was represented. **1957** S. Moss *In Track of Speed* v. 61 He returned to Europe at the head of a racing stable of mechanics. **1966** *Publ. Amer. Dial. Soc.* 1964 xlii. 8 *Scuderia*.., a stable of cars, usually privately entered.

d. More widely, an establishment which trains or produces persons, etc., esp. of a characteristic quality or type. Also, a group *of* persons (*spec.* in publishing) under the same management or trained at the same place.

1942 *Tee Emm* (Air Ministry) II. 137 Robert..takes the lead in the latest offering from the Air Ministry Instructional Film stables. **1951** *Sun* (Baltimore) 19 June 7/2 Best known in the SRP's stable of führers is Remer. **1963** *Listener* 14 Mar. 458/1 The now-famous group centred on George Webb..which used to meet at the 'Red Barn', Bexleyheath, the stable out of which Humphrey Lyttelton came, played, typically, for a small audience of each other. **1970** C. L. Cline *Lett. George Meredith* I. 37 Lucas recruited the best staff of artists in the business..mainly from the *Punch* stable. **1977** 'J. le Carré' *Hon. Schoolboy* i. 17 Luke..had been the star turn in his magazine's Saigon stable of war reporters.

e. In colloq. phr. *from the same stable*, from the same source.

1950 *Sun* (Baltimore) 6 Jan. 5 (*caption*) This jet plane is from the same stable as the Comet. **1959** *Listener* 2 July 36/1, I would be interested to hear a play from the same stable. **1962** H. E. Beecheno *Introd. Business Stud.* v. 42 We find many competing products are coming from the same 'stables'. **1972** G. Bell *Villains Galore* xiv. 213 No one was quite sure who Boote actually was, except that he was from the same stable as Stallion.

5. a. *stable-dog*, *-girl*, *-jacket* (earlier example), *jockey*, *slang*, *-theatre*.

1865 *Our Young Folks* I. 461 It began to be remarked that this was a stable-dog, educated for the coach-boy and stable. **1967** N. Freeling *Strike out where not Applicable* 36 He blusters..shouting at the stable-girls. **1852** C. M. Yonge *Two Guardians* i. 12 Edmund..seeing a boy in a stable jacket, asked Marian if he should not let him lead the ponies round by the drive. **1971** D. Francis *Bonecrack* xvi. 215 'You can stay on..if you like.'..'What as?' he said apprehensively... 'Stable jockey,' I said. **1894** Stable slang [see *lad *sb.*[1] 2 c]. **1903** Somerville & 'Ross' *All on Irish Shore* 269 His speech, what there is of it, is ungarnished with stable slang. **1928** T. S. Eliot *Dialogue on Poetic Drama* in *Dryden's Ess. Dram. Poesie* p. xxvi, We shall end with a..cosmopolitan little-theatre ... What is..more likely is that nothing will be done at all. We are all too busy..to prance about in a stable-theatre.

b. **stable block**, a building designed to house stables; **stable companion**, also *transf.*; **stable-craft**, the knowledge and skill involved in the proper maintenance of stables and stable animals; **stable manure** = *stable-dung*; **stable-mate** = *stable companion* in Dict. and Suppl.

1977 P. D. James *Death of Expert Witness* ii. i. 56 As she had come round the corner of the house from putting her bicycle in the old stable block, Inspector Blakelock had been standing at the front door. **1892** *Strand Mag.* July 36/1 Its stable companion was the Challenge tricycle. **1920** Galsworthy *Tatterdemalion* xii. 179, I used to like very much his attitude to the young 'stable-companion' who had arrived with him. **1968** 'J. le Carré' *Small Town in Germany* xiii. 214 'Praschko was up there, was he? In Berlin? With the Russians and Aickman?' 'Stable companions.' **1931** *Times Lit. Suppl.* 25 June 502/3 His remarks on stablecraft are also thoroughly sound. **1953** G. Brooke (*title*) Introduction to riding and stablecraft. **1629** J. Parkinson *Parad.* i. i. 2 Sandy loame..may soone be helped with old stable manure of horses. **1864** Trollope *Small House at Allington* II. xxx. 306 There was..a vexed question between Hopkins and Joliffe the bailiff on the matter of—stable manure. **1973** 'I. Drummond' *Jaws of Watchdog* vii. 95 Give it plenty of muck... Stable manure is the best. **1941** Wyndham Lewis *Let.* May (1963) 288 The 'practical politicians' and ..their stable-mates the hardboiled business-men' have somehow or other to be tamed. **1958** *Daily Sketch* 2 June 15/3 Guersillus's stablemate, Paridel, will not now run in the Derby. **1979** J. Leasor *Love & Land Beyond* i. 13, I run a stable mate of the Cord, an Auburn 851 cabriolet.

stable (stēⁱ·b'l), *sb.*[2] *Coal Mining.* [Of uncertain origin: perh. the same word as prec.; cf. also Staple *sb.*[4]] An excavation in a face to accommodate a coal-cutting machine or loader working into it. Also *stable hole*.

1906 *Trans. Inst. Mining Engin.* XXXI. 401 Coal-cutters, which have a disc or a chain-jib fixed to their longitudinal centre, cannot cut close to the ends of the face, and require a stable-hole or heading to be driven in advance at each end. **1914** G. L. Kerr *Practical Coal-Mining* (ed. 5) vi. 148 This avoids the necessity of making jib-holes or 'stables' at the start and finish of the cut. **1945** *Trans. Inst. Mining Engin.* CIV. 209 Stables to a depth of at least twice the depth of the cut are required at each end of the face. **1968** *Economist* 8 June 85/3 The great benefit of the machine is in avoiding the present concentration of men in the 'stable hole' the most dangerous part of the pit. **1973** L. J. Thomas *Introd. Mining* vii. 278 A longwall face stable was necessary so that the face cutting and loading machinery could stand in it.. while the face conveyor was pushed over.

stable, *v.* Add: **1.** Also *transf.*

c **1957–8** E. M. Forster *Life to Come* (1972) 175 British officers are never stabled with dagoes, never, it was too damn awkward for words. **1962** *Daily Tel.* 8 Jan. 15/7 The possibility of 'stabling' Underground trains in the tunnels instead of in the open at depots during periods of severe icy weather..unfortunately is unworkable.

stabled, *ppl. a.*[2] Add: Also *transf.*

1962 *Daily Tel.* 8 Jan. 15/8 No passenger service could start until the last stabled train had left.

stably, *a.*[2] Add: (Earlier and later examples.) Also *Comb.*, as *stably-smelling* adj.

1851 E. Ward *Jrnl.* 20 June (1951) iii. 196 Slept at Wortley's on the floor—rather stably. **1930** J. Dos Passos *42nd Parallel* i. 12 The musty stably smelling herdic cab. **1968** E. McGirr *Lead-Lined Coffin* iii. 100 The next [shed] had been a stable years ago: there was still a faint stably whiff attached to it.

stachybotryotoxicosis (stækibǫ:tri,otǫksikōᵘ·sis). *Vet. Sci.* [f. mod.L. *Stachybotrys* (A. C. J. Corda *Icones Fungorum* (1837) I. 21/1), generic name (f. Gr. στάχυς ear of corn + βότρυς cluster) + Toxicosis.] Toxicosis caused by toxins of the graminivorous fungus *Stachybotrys alternans*, affecting esp. horses and characterized by hæmorrhage and necrosis.

1945 V. G. Drobotko in *Amer. Rev. Soviet Med.* II. 238 *Stachybotryotoxicosis* is the name we have given to a new disease of horses, of unknown etiology, which appeared in the Ukraine several years ago. **1970** Jubb & Kennedy *Path. Domestic Animals* (ed. 2) I. iv. 352/2 Stachybotryotoxicosis is caused by preformed toxins of *Stachybotrys alter[n]ans* (*atra*) growing on substances rich in cellulose. **1978** *Bio Systems* X. 193/1 A group of toxins ..produced by various species of *Fusarium*, *Trichoderma*,..and other genera [of fungi] are implicated in a number of mycotoxicoses. Among these are alimentary toxic aleukia and stachybotryotoxicosis.

stachyose (stæ·ki,ōᵘz). *Chem.* [a. G. *stachyose* (Von Planta & Schulze 1890, in *Ber. d. Deut. Chem. Ges.* XXIII. 1693): see Stachys and -ose[2].] A tetrasaccharide first isolated from the Chinese artichoke, *Stachys affinis*, and found in a number of plants of this genus and in the seeds of most leguminous plants.

1890 *Jrnl. Chem. Soc.* LVIII. ii. 1089 The carbohydrate, for which the authors propose the name *stachyose*, has a slightly sweet taste, and forms a neutral solution. **1937** *Thorpe's Dict. Appl. Chem.* (ed. 4) I. 497/1 The characteristic carbohydrate is stachyose, $C_{24}H_{42}O_{21}$, $4H_2O$, yielding on hydrolysis one molecule of glucose, one molecule of fructose, and two molecules of galactose from one molecule of the carbohydrate. **1977** *Jrnl. Nutrition* CVII. 1861/1 A 1:1 mixture of soybean flour and sucrose would contain 2% stachyose.

stachys. Add: **b.** (Earlier example.) Cf. *rabbit's ear 1.

1789 W. Aiton *Hortus Kewensis* II. 300 Wood Stachys, or Hedge-nettle..Fl[owers] July and August.

stack, *sb.* Add: **1. b.** (Earlier and later examples.) Also in *pl.* and as *advb. ellipt.*, a pile of money. orig. *U.S.*

1870 'Mark Twain' in *Galaxy* Sept. 425/2 Never saw 'such a stack of them on one establishment'. **1892** —— *Amer. Claim.* xxiv. 236 Stacks of money had been placed in bank [*sic*] for him and Hawkins by the Yankee. **1904** W. N. Harben *Georgians* xxiii. 222 My boy, I had stacks an' stacks of fun on that trip. **1919** Wodehouse *My Man Jeeves* 15 I'm a bit foggy as to what jute is, but..Mr. Worple had made quite an indecently large stack out of it. **1952** E. F. Davies *Illyrian Venture* vii. 127 Chesshire had stacks of letters from a girl friend and decided to read one a day for a month. **1968** B. Hines *Kestrel for Knave* (1972) 81 I'm not that bad, I'm no worse than stacks o' kids, but they just seem to get away with it. *Ibid.* 83 It's stacks better than roamin' t'streets doin' nowt.

c. *to swear on a stack of Bibles* (see quot. 1909). *U.S. colloq.*

1866 M. Reid *Headless Horseman* II. lvii. 287 I'll sware it on the crass—or a whole stack av Bibles if yez say so. **1909** *Dialect Notes* III. 378 *Swear on a stack of Bibles (a mile high)*,..an exaggerated or emphatic form of oath. **1926** M. J. Atkinson in J. F. Dobie *Rainbow in Morning* (1965) 82, I would not believe him if he swore to it on a stack of Bibles as high as his head. **1956** B. Holiday *Lady sings Blues* (1973) ii. 24 Mom..swore on a stack of Bibles I was eighteen.

d. A set of shelving on which books are arranged for storage, esp. in a library (also *book-stack* s.v. *book *sb.* 18); hence, a part of a library designed for the storage of books, and to which access by readers may be restricted. Freq. *attrib.* and in *pl.*

1879 C. A. Cutter in *Library Jrnl.* IV. 235 The new wing..consists of a perfectly uniform series of book stacks arranged like a gridiron. **1884** Stack-room [in Dict., sense 8 b]. **1900** *Library Jrnl.* Nov. 679/2 Electric signals are also a part of the apparatus, and convenient elevators for passengers and freight are provided in the book-stacks. **1910** A. E. Bostwick *Amer. Public Library* 284 The relation of reading room to stack must be such as to make these [carriers] easily operable. **1933** *Times* 9 Nov. 9/2 Before leaving the building they paused to visit one of the new two-tier book-stacks on the ground floor. **1946** *Library Quarterly* Apr. 128/2 It is a modern brick building, five stories high, and contains, in addition to the stack space, a small reading room. **1956** 'C. Blackstock' *Dewey Death* i. 5 There were some seven thousand books. ..Barbara spent most of her spare time in the history section, wandering from stack to stack. **1966** C. Potok *Chosen* (1967) ii. viii. 152 Its stacks were filled mostly with bound volumes of scholarly journals and pamphlets. **1980** *Cosmopolitan* Dec. 221/1, I located a promising title for my Proust researches. 'Not on the open shelves.' I would have to order it to be fetched for me from the stacks of the library.

e. *Aeronaut.* A series of aeroplanes circling at different altitudes and awaiting landing instructions.

1947 *Britannica Bk. of Year* 841/2 As the bottom plane lands, each member of the stack drops 1,000 ft. and a new plane can then be brought in on top **1952** *Jrnl. R. Aeronaut. Soc.* LVI. 615/2 Once an aircraft was in a stack it was difficult to bring it forward. **1965** *Observer* 31 Oct. 1/1 He came in for a third attempt after circling for a further 40 minutes in the Watford 'stack'. **1976** L. Deighton *Twinkle, twinkle, Little Spy* x. 94 We joined the stack.. and circled to await landing permission.

f. In a computer or calculator, a set of registers or storage locations which store data in such a way that the most recently stored item is the first to be retrieved; also, a list of items so stored, a push-down list.

1960 E. W. Dijkstra in *Numerische Math.* II. 312 The basic concept of the method is the so-called stack. One uses a stack for storing a sequence of information units that increases and decreases at one end only. **1963** *Ann. Rev. Automatic Programming* IV. 183 A stack is merely an area of storage with an associated administrative quantity, the 'stack pointer', which controls the addressing of the stack. **1973** C. W. Gear *Introd. Computer Sci.* viii. 359 We can choose a section of memory at execution time to store this stack. **1976** *Sci. Amer.* June 88/1 (Advt.), HP's special logic system with four-register stack almost completely eliminates the need to re-enter data.

5. c. In *fig.* phr. *to blow one's stack* = *to blow one's top* s.v. *blow *v.*[1] 24 i. *slang* (orig. *U.S.*).

1947 Berrey & Van den Bark *Amer. Thes. Slang* Suppl. § 48/1 *Blow one's stack*, to become angry or excited. **1952** R. Bissell *Monongahela* 189 When Andrew [Carnegie] received the minutes and read them he blew his stack a mile high. **1965** F. Knebel *Night of Camp David* xi. 173 O'Malley looked startled. 'Well, he..he was goddam mad. Frankly, he blew his stack.' **1979** W. H. Canaway *Solid Gold Buddha* xxiv. 156, I ain't whingeing, honest... I'm sorry I blew me stack.

7. (Later examples.) No longer *dial.* Cf. *sea-stack* s.v. *sea *sb.* 23 a.

1944 A. Holmes *Princ. Physical Geol.* xiv. 287 Later the arch falls in, and the seaward portion of the headland then remains as an isolated stack. **1957** G. E. Hutchinson *Treat. Limnol.* I. ii. 173 Coastal islands formed by cutting behind promontories, so producing isolated stacks, occur along the margins of large lakes. **1975** [see *severe *a.* 9 b].

8. a. (sense 2) *stack-shaped* adj.

1864 J. A. Grant *Walk across Afr.* 62 Grain is housed under the eaves of stack-shaped huts. **1921** *Glasgow*

Herald 26 Mar. 7 About a dozen 18-pounder shell cases, some of which contained curious stack-shaped bombs.

b. stack gas, gas emitted by a chimney-stack.

1945 H. D. SMYTH *Gen. Acct. Devel. Atomic Energy Mil. Purposes* viii. 91 It became essential to know whether the stack gases (at Clinton and at Hanford) would be likely to spread radioactive fission products in dangerous concentrations. **1973** H. GRUPPE *Truxton Cipher* xiii. 135 The smell of stack gas lay heavy upon the destroyer's upperworks.

stack, *v.*¹ Add: **1. b.** *Aeronaut.* To order (aircraft waiting to land) at different flight levels and in landing sequence above an airport; to place (an aeroplane) in a waiting stack (freq. with *up*). Also *intr.* (of aircraft), to form a stack.

1941 K. HENNEY *Radio Engin. Handbk.* (ed. 3) xvii. 616 The present practice of 'stacking' airplanes..limits the number of landings..to about 4 per hour. **1943** M. FEIGEN *Pocket Aviation Quiz Bk.* 55 Planes cruising above an airport at varying assigned altitudes in order not to collide while awaiting their turns to land are said to be: stacked up. **1949** *Sun* (Baltimore) 4 Nov. 2/6 Planes 'stack up' over the range station near Mount Vernon. **1965** P. WYLIE *They both were Naked* I. i. 4 We'd spent that interval..'stacked up' and waiting for planes..to be called down for landings. **1975** D. LODGE *Changing Places* vi. 218, I hope to hell we aren't stacked for hours over Kennedy.

2. b. To pile *up* one's chips at poker. Now usu. *fig.*, to present oneself, measure up; to arise, build up. *colloq.* (chiefly *U.S.*).

1896 ADE *Artie* ii. 10 He'd stack up, you know, an' feel in his pockets and then he'd say: 'I'm forty-seven cents loser.' **1911** R. D. SAUNDERS *Col. Todhunter* xiii. 198 Old Bill Strickland, of Nineveh, somehow don't seem to stack up the right way against the Honourable Stephen K. Vancey. **1921** R. D. PAINE *Comr. Rolling Ocean* iv. 71, I wish this trouble hadn't stacked up between us. **1938** *Sun* (Baltimore) 6 Apr. 11/5, I think every one will agree my record stacks up favourably enough with that of any other pro. past or present. **1951** M. MCLUHAN *Mech. Bride* (1967) 48/1 See how you stack up with your fellow men on the following issues. **1965** WODEHOUSE *Galahad at Blandings* x. 169 I've never been a brainy sort of guy, and what I want is a wife with about the same amount of grey matter I have, and that's how Vee stacks up. **1977** R. E. MEGILL *Introd. Risk Analysis* xv. 170 Dougherty and Nozaki ranked knowing your competitors as of equal value to knowing how well your own estimate stacks up.

6. a. To shuffle or arrange (playing-cards) dishonestly. In *fig.* phr. *to stack the cards* (etc.) *against*: to reduce (a person or thing's) chance of success. Cf. PACK *v.*² 5, STOCK *v.*¹ 23 b. orig. *U.S.*

1825 in M. Bayard Smith *Forty Yrs. Washington Soc.* (1906) 186 John Randolph observed after counting the ballots, 'It was impossible to win the game, gentlemen, the cards were stacked.' **1896** J. F. LILLARD *Poker Stories* 54 The stranger got skinned right and left. The cards were stacked and marked on the back, so that he didn't have any chance at all to win. *c* **1926** 'MIXER' *Transport Workers' Song Bk.* 31 He'll know this when I stack the cards. **1941** B. SCHULBERG *What makes Sammy Run?* v. 81 You read the papers, you know how the cards are stacked against this nut. **1977** *New Yorker* 24 Oct. 37/1 He..confirmed our worst fear: the deck is stacked... He picked up a cardboard box containing several packs of cards. **1978** G. VAUGHAN *Belgrade Drop* x. 67 'Pin your ears back,' Yardly murmured. 'We've got a lot stacked against us.'

b. = PACK *v.*² 4 in Dict. and Suppl. Also *fig.*

1948 *Durant* (Okla.) *Daily Democrat* 2 July 1/5 His young polltaker detected no signs of 'stacking' the poll for any candidate. **1963** *Wall St. Jrnl.* 25 Jan. 16/3 The government is now stacked from top to bottom with men who reflect their President's prejudices. **1970** *New Yorker* 28 Nov. 104/2 Legally, marriage is still stacked in favor of the man. **1976** *N.Y. Times* 10 Apr. 29/1 Gov. George C. Wallace of Alabama charged today that..efforts were being made..to stack delegate-selection procedures against him.

stackable (stæ·kăb'l), *a.* [f. STACK *v.*¹ + -ABLE.] Able to be stacked or piled up: esp. of chairs and other furniture. Hence **stackability.**

1958 *Archit. Rev.* CXXIII. 255 (caption) The chairs are stackable. **1960** *House & Garden* Oct. 80/2 Stackable Italian table..for use in large works canteens. **1969** *Jane's Freight Containers 1968–69* 490/3 Stackability... When the ends are folded down, five stacked together stand 8 ft high. **1973** J. ELSOM *Erotic Theatre* x. 209 Surrounded by a Brighton beach of stackable chairs. **1977** *Design Engin.* July 41/1 Stackable valves often introduce unacceptable pressure losses and fault finding often means dismantling the whole stack.

stacked, *ppl. a.* Add: **4.** Of a female figure: well-rounded and attractively shaped. Also of a woman: having a prominent bosom. Similarly, *stacked up*, *well-stacked*. *U.S. slang* (as a term of male approbation).

1942 BERREY & VAN DEN BARK *Amer. Thes. Slang* § 37/14 *Shapely*,..stacked up nicely. **1952** *Esquire* June 131/2 The singer ain't a bad-looking broad, she's well-stacked and sort of young. **1954** *Amer. Speech* XXIX. 234 There is a present fashion of referring to an attractively proportioned woman as well stacked, or simply, stacked, period. **1965** *Liberator* (N.Y.) Aug. 22/3 Two stacked broads approached. Everyone attained a hip position. It consisted of pulling pants high, rolling the Hi-Lo collar, [etc.]. **1975** D. LODGE *Changing Places* v. 154 What about your wife?.. Is she well-stacked? **1981** 'D. SHANNON' *Murder Most Strange* vii. 147 A cute little blond chick.. really stacked.

5. *stacked head*, in a tape recorder: a head (*HEAD *sb.* 11 g) in which the gaps corresponding to the tracks in multi-channel recording are located one above another.

1954 C. A. TUTHILL *How to service Tape Recorders* iv. 45 Typical individual or stacked heads, currently being released by The Brush Development Company, are cast in synthetic resin. **1962** A. NISBETT *Technique Sound Studio* 267 For stereo, the first and third tracks are recorded at the same time, using a stacked head (one with both gaps in line). *a* **1975** *Tape Talk* (Sony Corporation) 3 *In-line heads*, arrangement of stereophonic heads on a tape recorder in which the head gaps are mounted one directly above the other. Also called 'stacked heads'.

6. *stacked heel*, a heel built by stacking or appearing to stack thin layers of wood or other material. Also *stacked-heeled* adj.

[**1960** *Harper's Bazaar* Aug. 24 Stack-heeled moccasin in deep brown leather.] **1960** *Sunday Express* 11 Sept. 15/4 In Paris Dior..showed Cuban 'stacked' heels. **1964** in Hamblett & Deverson *Generation X* 13 Mod girls have a short back-and-sides hair cut like men used to, wear shift-style dresses with round collars... Stacked heeled shoes, white stockings. **1966** T. PYNCHON *Crying of Lot 49* vi. 177 She turned, pivoting on one stacked heel. **1977** J. WAMBAUGH *Black Marble* (1978) v. 62 Comfortable loafers with low stacked heels in colors to match wool-knit pants and jackets.

7. *stacked deck*, a pack of cards that has cheatingly been set in a prearranged order; also *fig.* Hence more widely (and without the implication of dishonesty), of odds. Cf. *STACK *v.*¹ 6 a. Chiefly *U.S.*

1964 A. WYKES *Gambling* 329/1 *Stacked deck*, a pack of cards that a cheat has prearranged for his own benefit. **1971** P. O'DONNELL *Impossible Virgin* vii. 144 They worked alone, without hirelings, and so often went in against stacked odds. **1976** *Honolulu Star-Bull.* 21 Dec. H-3/1 The Minnesota Vikings, shooting to become the first National Football League team to make four Super Bowl appearances, will be playing with a stacked deck Sunday. **1978** E. TIDYMAN *Table Stakes* II. vii. 345 You played the game with a stacked deck. You had his daughter, his trust, and his face.

stacker, *sb.*¹ Add: **2. a.** More widely, any machine for raising individual items or bulk material and depositing them on a stack or pile; also, a stacker crane. (Later examples.)

1922 G. F. ZIMMER *Mech. Handling & Storing* (ed. 3) xxxviii. 654 A similar stacker is illustrated in Fig. 954; it is composed of a slat conveyor, and is for handling cases. **1950** W. STANIAR *Plant Engin. Handbk.* xx. 1417 A number of different types of portable elevators, stackers, or tiering machines are made with the lifting mechanism either motor or hand operated... Portable stackers are made for the handling of smaller units. **1979** *Belt Conveyors for Bulk Materials* (Conveyor Equipment Manufacturers Assoc., U.S.) (ed. 2) i. 7 Belt conveyors, with their stackers and reclaimers, have become the only practical means for large-scale stockpiling and reclaiming of such bulk materials as coal, ore, and taconite pellets.

b. A part of a data-processing machine in which punched cards are deposited in a stack after having passed through the machine.

1962 in *Gloss. Terms Automatic Data Processing* (*B.S.I.*) 90. **1969** P. B. JORDAIN *Condensed Computer Encycl.* 82 A card stacker ensures the correct sequencing of emerging cards. **1971** J. T. MURRAY *Introd. Computing* vii. 126 The input hopper provides the cards which are sorted into any of the six available stacker pockets.

3. Special Comb.: **stacker crane**, a hoist running on a fixed horizontal track for stacking and retrieving pallets or the like.

1959 W. STANIAR *Plant Engin. Handbk.* (ed. 2) XXVIII. 32 The makers of the stacker crane claim that it may be employed for safe stacking of materials to greater heights than with other forms of equipment. **1979** *Computers in Shell* (Shell Internat. Petroleum Co.) 6 A computer-controlled stacker crane takes the pallet and places it in one of the thousands of pallet spaces in the racks, the location being recorded by the computer.

stacking, *vbl. sb.* Add: **a.** (Examples in sense *1 b of the vb.)

1942 H. L. SMITH *Airways* 365 Since speed is the commodity sold by the air-line operator, he is interested in any system that can solve the problem of stacking. **1969** *Daily Tel.* 14 Nov. 1/7 Stacking over a 'holding area' while waiting a turn to land, is not uncommon. **1979** C. WOOD *James Bond & Moonraker* i. 13 If there were no stacking positions at Heathrow..he would be home in time to..eat supper with the family.

b. *stacking chair*; **stacking fault** *Cryst.*, a break in the regular order of stacking of layers of atoms in a crystal.

1939 MARTIN & SPEIGHT *Flat Bk.* 101 Stacking chair, by Alvar Aalto in natural birch or lacquered..20s od. **1951** *Catal. of Exhibits, South Bank Exhib., Festival of Britain* 87/2 Cantilevered all-purpose stacking chairs. **1982** E. DEWHURST *Whoever I Am* vii. 90 Motioning her niece to fetch up one of the stacking chairs. **1951** *Phil. Mag.* XLII. 815 The best-known examples of translation twinning are 'stacking faults' in the sphere-packing lattices, *i.e.* breaches of the stacking rules which lead to face-centred cubic or hexagonal close-packing. **1976** *Sci. Amer.* Nov. 105/2 In brass, bronze and certain stainless steels, for example, stacking faults extend over distances equivalent to many atomic diameters.

stack-up. [f. vbl. phr. *to stack up*: see *STACK *v.*¹ 1 b.] The arrangement of objects in a stack or pile; a build-up. *spec.* in *Aeronaut.* = *STACK *sb.* 1 e.

1945 F. WALTON *Airman's Almanac* 403 *Stack-up*, airplanes flying at different altitudes over an airport awaiting signal to land. **1948** *Jrnl. R. Aeronaut. Soc.* LII. 681/2 Once the diversion and 'stack up' problem is eliminated the plain jet..can become almost as cheap to operate as propeller-powered aircraft. **1950** *Daily Progress* (Charlottesville, Va.) 14 Dec. 1/8 The department has under study a plan for shifting first class surface mail to the air lines in order to avoid a stack-up of such mail. **1970** *Sci. Amer.* Mar. 86/1 Unsafe weather conditions at airports, forcing the stack-up and diversion of aircraft,..take an immense annual toll in inconvenience. **1976** *Lebende Sprachen* XXI. 153/2 The failure was caused by a build-up/stack-up of adverse manufacturing tolerances.

stade¹. Add: **3.** *Geol.* (See quot. 1961); = *STADIAL *sb.*, *STADIUM 5.

1961 *Bull. Amer. Assoc. Petroleum Geologists* XLV. 660/2 A stade was a climatic episode within a glaciation during which a secondary advance of glaciers took place. **1964** *Prof. Papers U.S. Geol. Survey* No. 501-D. 104/2 Currently, the oldest group of moraines define the early stade of Bull Lake Glaciation, and the middle and youngest groups together define the late stade. **1974** *Nature* 26 Apr. 752/2 The 'cold' water fauna [belongs] to the period of the Loch Lomond readvance stade.

stadial, *a.* Restrict † *Obs. rare*⁻¹ to sense in Dict. and add: **2. a.** *Geol.* Of or pertaining to a glacial stadium or stade; *stadial moraine* = *recessional moraine* s.v. *RECESSIONAL *a.* 3.

1937 WOOLDRIDGE & MORGAN *Physical Basis of Geogr.* xxii. 378 Retreatal stages of the ice in the valleys are marked by 'stadial moraines'. **1970** R. J. SMALL *Study of Landforms* xi. 388 If the ice is affected by episodes of retreat, separated by stillstands, a number of smaller, sub-parallel ridges ('recessional' or 'stadial' moraines) will be formed. **1971** S. C. PORTER in K. K. Turekian *Late Cenozoic Glacial Ages* xi. 320 In the Puget Lowland only two stadial episodes have been determined for the Puget lobe. **1979** *Nature* 19 July 199/2 The landforms produced by the stadial glaciers in the Highlands are often clear and abundant.

b. *Archæol.* Pertaining to or expressed in terms of a series of successive stages into which a culture or period can be divided.

1959 *Science* 6 Feb. 305/2 Willey and Phillips' stadial conception of New World prehistory. **1968** L. R. BINFORD in S. R. & L. R. Binford *New Perspectives in Archeol.* I. i. 15 The application of such a scale to innumerable empirical cases..can never provide us with an understanding of the processes operative in the past which resulted in the stadial sequence. **1977** G. CLARK *World Prehistory* (ed. 3) ii. 41 The use of the term 'Neolithic' in this connection in itself poses a problem. If stadial terms are rarely used in contemporary archaeological discourse, it seems all the more important to be clear about their meaning.

B. *sb. Geol.* = *STADE¹ 3.

1954 *Antiquity* XXVIII. 8 The tundra-like phase at the beginning of the third Würm stadial. **1964** K. W. BUTZER *Environment & Archeol.* xxvii. 395 It seems improbable that food-collectors at the Neanderthaler level would venture through thousands of kilometers of harsh environment at the height of any of the Würm stadials. **1979** SEYFERT & SIRKIN *Earth Hist. & Plate Tectonics* (ed. 2) xiv. 515/2 Glacial advances occurred during stadials.

stadiometer. Add: **2.** An apparatus for measuring a person's height.

1972 *Sunday Times* 22 Oct. 65/2 His product range does not end at anthropometers. There is a stadiometer for measuring height to a high degree of accuracy. **1980** *Brit. Med. Jrnl.* 29 Mar. 915 When the child is old enough to stand a special stadiometer can be used to measure height.

stadium. 2. Add to final sentence: *spec.* an enclosed area for sporting events equipped with tiers of seats for spectators. (The pl. *stadiums* is usual in this sense.)

1928 *Times* 20 Apr. 6/6 It would be difficult to imagine a more impressive enclosure than the Stadium [*sc.* Wembley Stadium] for the holding of the greatest football festival of the year. **1938** L. MACNEICE *Earth Compels* 59 It's no go the picture palace, it's no go the stadium. **1972** G. GREEN *Great Moments in Sport: Soccer* iii. 44 As for Lenin Stadium itself, it was very much like Wembley before the cover went on, an elliptical concrete monster, liberally dotted with exits and entrances, and steeply tiered.

5. *Geol.* = *STADE¹ 3. Now *Obs.* or *rare.*

1910 *Zeitschr. für Gletscherkunde* IV. 246 The later stages of glaciation in the scheme proposed by Geikie.. may prove to correspond to the *stadia* of the Alpine region which are so clearly brought out by Penck and Brückner in their recent great work. **1914** W. B. WRIGHT *Quaternary Ice Age* vii. 156 An investigation of the stages of retreat in the Etsch district has shown that three main stadia can also be distinguished here. **1937** *Ibid.* (ed. 2) xi. 185 This stadium must..be dated back before the Swiss pile-dwellings, or in round numbers 7000 years ago.

staff, *sb.*¹ Add: **I. 9. f.** = *train staff* (*a*) s.v. TRAIN *sb.*¹ 22 b. Cf. *TOKEN *sb.* 7 b.

1885 E. B. IVATTS *Railway Managem. at Stations* 559 *Staff* (*train*), a piece of wood or metal used on single lines, which..confers the 'right of road' for an engine. **1902**

Encycl. Brit. XXXII. 147/1 The staff..is delivered to the engine-driver at station A, and constitutes his authority to occupy the main track between that station and stations B. On reaching B he surrenders the staff, and receives another one which gives him the right to the road between B and C. **1931** D. L. SAYERS *Five Red Herrings* xxvii. 330 The station-master marched across, carrying the staff under his arm. **1974** J. S. HOLDEN *Wallington Branch* v. 87 One instance is recorded of the branch train being snowed up.. A porter, despatched with the staff, had to make his way on foot.

III. 21. a. *General Staff*: hence *Chief of the General Staff*, *Chief of Staff*, the senior staff officer of a service or commander.

1904 *Rep. War Office (Reconstruction) Committee* II. 22 in *Parl. Papers* (Cd. 1968) VIII. 121 In accordance with our recommendations, a Chief of the General Staff of the Army has been appointed. **1937** H. NICOLSON *Diary* 21 Apr. (1966) 299 It is a man's dinner and consists mainly of the Cabinet and a few Chiefs-of-Staff. [illegible] *a* **1974** R. CROSSMAN *Diaries* (1975) I. 595 Couldn't we have officials there as well, like the Chiefs of Staff?

b. *ellipt.* (chiefly with initial capital) = *staff sergeant*, sense 25 in Dict.

1925 FRASER & GIBBONS *Soldier & Sailor Words & Phrases* 269 *Staff*, staff-sergeant. **1943** *Yank* 8 Oct. 15 [illegible] **1958** P. SCOTT *Mark of Warrior* I. i. 23 A sergeant..addressed them as 'Gentlemen' and told them to address him as 'Staff'. **1965** 'J. LE CARRÉ' *Looking-Glass War* xii. 156 'You call me "Staff",' the instructor said. **1978** J. BARNETT *Head of Force* x. 88 [illegible]

22. b. Construed as *pl.*: members of a staff, employees. Rarely, an employee.

1951 E. M. BRENT-DYER *Chalet School & Jo* xvii. 220 You're almost a Staff, so that'll be all right. **1955** *Times* 10 May 17/4 A continuing supply of higher technical staff. **1970** *Daily Tel.* 4 May 14 A Yorkshire factory with 250 staff, nearly all women. [illegible]

c. *spec.* in a business organization: (*a*) the employees responsible for providing advisory and ancillary services to line managers and their subordinates; (*b*) salaried (as opp. to wage-earning) employees.

1915 C. E. KNOEPPEL *Installing Efficiency Methods* viii. 18 He [*sc.* the engineer] decides to create an organization to be known as the 'staff'—advisory in nature and without jurisdiction over any of the line officials. **1923** O. SHELDON *Philos. Managem.* iv. 114 A 'Staff and Line' organization..is based upon a strict demarcation between thinking and doing; between the actual execution of production, which is the 'Line', and the business of analysing, testing, comparing, recording, making researches, co-ordinating information, and advising, which is the 'Staff'. [illegible] **1980** *Daily Tel.* 16 Feb. 25/5 Almost all the major companies pay more [in redundancy payments] than the statutory minimum and generally staff will enjoy more generous benefits than shop-floor workers.

IV. 25. a. *staff appointment* (earlier example), *car*, *duty* (example), *job*; **staff college** (later examples); **staff-wallah** *slang* [cf. WALLAH *b*], a disparaging term for a non-combatant army officer; **staff-work**, the supportive work of planning, organization, etc., done by staff officers for the commander; also in civilian contexts.

1802 J. ORROK *Let.* 22 July (1927) [illegible]

b. Also in passenger ships.

1936 S. G. MCNEIL *In Great Waters* 136, I was to be the first holder of the newly created post of Staff-Captain. **1944** *New Yorker* 23 Dec. 26/1 One man who had this second feeling to a degree was Lemuel Watkins, staff captain and second in command to the *Marquette*'s master. **1957** D. G. O. BAILLIE *Sea Affair* 111 In ships carrying anything over a thousand passengers, there is usually a Staff Captain who devotes his entire activities to their proper entertainment.

c. (Later examples.) Similarly, *staff member*. Also in sense 'of, provided or reserved for, arranged by, the staff of an establishment', as *staff canteen*, *dance*, etc.

1940 R. S. LAMBERT *Ariel & all his Quality* i. 33 He fought shy of..social gatherings, except for appearances at occasional staff dances. **1942** WYNDHAM LEWIS *Let.* 27 Jan. (1963) 313 Your staff-member who has ended up a prisoner in Athens had an interesting journey. **1966** *Economist* 3 Sept. 888/1 The employers' habit of describing concessions of better pay and conditions to manual workers as creating 'staff conditions' for all has often meant that yet further concessions have been made to 'staff' employees, in order to preserve their differentials over the men on the shop floor. **1972** K. BENTON *Spy in Chancery* xvii. 192 She probably had lunch there. There's a staff canteen. **1977** F. BRANSTON *Up & Coming Man* xiv. 150 Freelancing... [illegible]

26. staff and ticket (system) an elaboration of the staff system (below) allowing for the movement of several trains in one direction along a single line, whereby the last train carried the staff (sense 9 f above) and the preceding trains carried tickets pertaining to this (*Obs.* exc. *Hist.*); **staffholster** *nonce-wd.*, a holster for a watchman's staff; **staff-man**, a member of a staff; **staff nurse**, a trained nurse in a hospital, ranking above a registered nurse and below a ward sister; **staff photographer**, a photographer on the staff of a newspaper or journal; **staff-room**, a common room for the use of the staff, as in a school; also *transf.*, the staff itself; **staff-student** *a.*, designating the relation between students and teaching staff, esp. in phr. *staff-student ratio* (cf. *pupil-teacher ad*), s.v. *PUPIL *sb.*¹ 5); **staff system**, a block system on railways according to which an engine-driver may not proceed along a single line without carrying the staff (sense 9 f above) authorizing him to do so; **staff ticket**, a ticket used on railways to operate the staff and ticket system (see above); **staff writer**, a writer employed on the staff of a newspaper, radio or television station, or the like.

1887 C. E. STRETTON *Safe Railway Working* iii. 64 All other single lines should be worked under the train staff and ticket system. **1889** W. M. ACWORTH *Railways of Eng.* x. 392 Single lines are commonly worked on what is known as the 'staff and ticket' system. **1927** E. T. MACDERMOT *Hist. G. W. Railway* I. ii. xx. 611 The Wycombe Branch, on its extension to Thame in 1862, seems to have been the first on which the Train Staff and Ticket system was introduced. **1969** *Railway Mag.* Feb. 67 Up to 1892, all movement of trains was controlled by the 'staff and ticket', which was replaced by the Webb-Thompson electric train staff. **1922** JOYCE *Ulysses* 423 At a corner two night watch in shouldercapes, their hands upon their staffholsters, loom tall. **1940** *Chambers's Techn. Dict.* 799/2 *Staffman* (*Surv.*), the surveyor's assistant whose duty it is to hold the levelling staff while the instrument is sighted upon it and readings are being taken. **1976** *Times* 27 Sept. 12/3 [illegible] **1977** P. POLLACK *Pict. Hist. Photogr.* (rev. ed.) [illegible] **1888** Staff nurse [in Dict., sense 25 c]. **1973** J. MCCLURE *Caterpillar Cop* viii. 116 [illegible] **1978** *Jrnl. R. Soc. Med.* LXXI. 401 [illegible] **1941** *Times* (weekly ed.) 30 July 6/4 This special study was taken in the garden of 10, Downing Street by a staff photographer. **1977** P. POLLACK *Pict. Hist. Photogr.* (rev. ed.) [illegible] **1925** [illegible] **1953** K. TENNANT *Joyful Condemned* [illegible] **1977** P. KEMP in [illegible] **1974** *Howard Jrnl.* XIV. [illegible] **1887** C. E. STRETTON *Safe Railway Working* iii. 64 (*heading*) The train staff system. **1902** *Encycl. Brit.* XXXII. 147/1 In the United Kingdom and in Australia the means for preventing collisions between trains running towards each other is the 'staff system'. **1966** K. MØLLER *Amer. & Brit. Railway Eng.* 39 The staff system is still in use in Britain; every engine running over a certain section of a line must carry the corresponding token or staff. **1885** E. B. IVATTS *Railway Managem. at Stations* 559 Two or three trains may be passed forward by 'staff ticket' in this way, but ultimately either a train, an engine, or a messenger must convey the staff to enable trains being sent in the opposite direction. **1914** Staff writer [see *CREDIT *sb.* 13 f]. **1941** W. ABBOT *Handbk. Broadcasting* (ed. 2) xxvi. 322 Staff writers prepare commercial continuity, talks, announcements, interviews, special-occasion scripts, original plays, adaptations, and often station publicity. **1967** E. E. WILLIS *Writing Television & Radio Programs* i. 11 Writers who are permanently attached to the staff of a broadcasting organization and are paid on a regular weekly or monthly basis instead of being paid per script are known as *staff writers*. **1977** D. GREENE in Bond & McLeod *Newslett. to Newspapers* ii. 89 [illegible]

staffelite (stæ·fĕlait). *Min.* [ad. G. *staffelit* (C. A. Stein 1866, in *Jahrb. des Vereins für Naturkunde im Herzogthum Nassau* XIX–XX. 57), f. *Staffel*, name of a locality in Hesse, W. Germany: see -ITE¹.] A carbonate-containing variety of apatite found as colourless or yellow masses; carbonate-fluorapatite; = FRANCOLITE.

1868 J. D. DANA *Syst. Mineral.* (ed. 5) 534 Staffelite of Stein..occurs incrusting the phosphorite of Staffel, in botryoidal, reniform, or stalactitic masses. **1939** *Mineral. Mag.* XXV. 407 Francolite and staffelite are identical, and the name francolite has priority. [illegible] Staffelite..is a stalactitic apatite and occurs in the Staffel deposits near Limburg, West Germany.

staffer² (sta·fəɹ). orig. and chiefly *U.S.* [f. STAFF *sb.*¹ + -ER¹.] A member of a staff.

a. Of a newspaper or journal: a staff writer.

1949 *Cavalier Daily* (Univ. of Virginia) 22 Oct. 4/2 Staffers of the Daily Pennsylvanian visited the Princetonian offices following the Penn-Princeton football game. **1952** G. REINHARDT *Crime without Punishment* 290 He knew..what confidential memos had been passed between the managing editor and publisher of the *New York Times*..a fact doubtless not known to most *Times* staffers. **1973** E. B. WHITE *Let.* 24 May (1976) 645 The story of *The New Yorker* has yet to be well told. Many staffers were indignant about parts of the Thurber book.

b. More widely, of a business or other organization.

1950 in WEBSTER *Add.* **1962** *Housewife* (Ceylon) Apr. 34 (*caption*) Mr. Neale talks to staffers of a local advertising agency. [illegible] **1980** *Information Retrieval & Library Automation* XVI. 11/3 The issue of April 1980 for example contains a brief but informative report on 'interlending' in Czechoslovakia, prepared by a BLL staffer who spent two weeks surveying the Czech scene.

c. *spec.* of the President of the U.S.; a member of the President's White House staff.

1969 R. NEUSTADT in A. King *Brit. Prime Minister* 143 The functional equivalents between a British Cabinet and our set of influentials—whether Secretaries, Senators, White House staffers, Congressmen or others. **1976** *National Observer* (U.S.) 14 Feb. 16/6 [illegible] **1978** W. SAFIRE in *N.Y. Times Mag.* [illegible]

staffless, *a.* Add: **2.** Having no business or domestic staff; without employees. Also *transf.*

1957 'P. PORTOBELLO' *How to be Deb's Mum* 38 Her husband nursed the chef, a foreign refugee, back to health from influenza in an otherwise staffless but courtly castle. **1965** L. MEYNELL *Double Fault* ii. iii. 126 In these almost staff-less days I can't tell you how comforting it is to be awake in the morning and hear the gravel of the drive being raked. **1976** T. HEALD *Let Sleeping Dogs Die* iii. 46 You'll have to take pot luck... I'm staffless now..but I shall be able to rustle something up.

Staffordshire. Add: **a. Staffordshire bull terrier**, a small stocky terrier of the breed so called, first developed by crossing bulldogs and terriers, characterized by a fawn, blue, or brindle coat, often with white markings, and a short, broad head with dropped ears; **Staffordshire cone**, a kind of pyrometric cone; **Staffordshire knot**: also, a Stafford knot or half-hitch used as a craftsman's device or motif; **Staffordshire ware** (earlier example).

1765 J. WEDGWOOD *Let.* 17 June (1965) 34 An order from St. James's for a service of Staffordshire ware. **1904** *Our Dogs* 13 July 42/2 (Advt.), Old Staffordshire red brindled Bull-Terrier Dog Pup, five weeks, makes 20lbs. **1904** *Ibid.* 2 Jan. 24/2 (Advt.), Grand Litter Staffordshire Bull-terrier Pups, gamest strain. **1908** B. W. WATSON *Old Silver Platers & their Marks* 9/2 (*table*) 1808 Feb. 4 Josh Gibbs... Staffordshire Knot. **1950** A. C. SMITH *Dogs since 1900* xiv. 194 [illegible] Staffordshire Bull Terrier Club was founded. **1967** M. CHANDLER *Ceramics in Mod. World* i. 38 Cones used in continental Europe are still known as Seger cones, the slightly modified cones used elsewhere [illegible] Staffordshire cones. **1968** J. ARNOLD *Shell Bk. Country Crafts* 280 The figures represented by [Corn] Dollies cover a considerable variety.... The Horn of Plenty, ..the Staffordshire Knot, [etc.]. **1971** *Vogue* 15 Oct. [illegible] Staffordshire knots. [illegible] 70/1 Staffordshire Cones..are slender, trihedral pyramids made of ceramic materials and are so constituted as to deform when subjected to elevated temperature for a period of time. *Ibid.*, 'Staffordshire Cones', have been manufactured in England for nearly fifty years. **1977** J.

WAMBAUGH *Black Marble* (1978) viii. 126 Pattie Mae didn't do so well with the Staffordshire bull terrier.

b. (Earlier example in sense 'Staffordshire ware'.)

[**1770** H. WALPOLE *Let.* 6 May (1967) XXIII. 211 We have Etruscan vases, make of earthen ware in Staffordshire, from two to five guineas.] **1866** QUEEN VICTORIA *Let.* 16 May in R. Fulford *Your Dear Letter* (1971) 75, I send her a tea-set..which I hope she will often use. It is Staffordshire.

c. *ellipt.* = *Staffordshire bull terrier*, sense a above. Also *attrib.*

1903 *Our Dogs* 27 June 944/1 (Advt.), Pure-bred Staffordshire Bitch, good guard, and very game. **1943** H. N. BEILBY *Staffordshire Bull Terrier* i. 3 The Staffordshire will, by his intelligence and general usefulness, make a very strong appeal. **1968** [see *FILLED *ppl. a.* 2].

stag, *sb.*¹ Add: **1. b.** (Later example.)

1935 T. S. ELIOT *Murder in Cathedral* i. 29 Cabined in Canterbury, realmless ruler, Self-bound servant of a powerless Pope, The old stag, circled with hounds.

5*. *dial.* and *colloq.* A big, romping girl; a bold woman.

1684 G. MERITON *Yorks. Dialogue* 55 Nea, nea, great stags, what a durdum thou macks! **1790** GROSE *Provinc. Gloss.* (ed. 2), *Stag*,..a romping girl. **1877** F. ROSS et al. *Gloss. Words Holderness* 135/2 *Stag*,..a rude, romping girl. **1922** JOYCE *Ulysses* 425 The likes of her! Stag that one is. Stubborn as a mule! **1922** D. H. LAWRENCE *Aaron's Rod* iii. 33 She too was a tall stag of a thing.

6. d. A spell of duty. (See also quot. 1881.)

1881 S. EVANS *A.B.E. Evans's Leicestershire Words* (new ed.) 255 A 'stag' is also one set to watch while his fellows are engaged in anything in which they wish not to be caught. **1931** BROPHY & PARTRIDGE *Songs & Slang Brit. Soldier: 1914–18* (ed. 3) 361 *Stag*, sentry-go. **1958** R. STOREY *Touch it Light* in J. C. Trewin *Plays of Year* XVIII. 341 There's seven stags in the hours o' darkness and only five of you to do 'em. Somebody has to do two. **1975** A. BEEVOR *Violent Brink* iv. 97 The films would be handed in for processing when they were relieved at the end of their two hour 'stag'.

e. *ellipt.* for *stag-dinner*, *-party*, etc. (sense 8 c in Dict. and Suppl.). *N. Amer.*

1904 *Brooklyn Eagle* 28 May 3 The Myrtle Fishing Club will have a stag at Hurman Hub's Park this evening. **1947** *Chicago Tribune* 19 Oct. (Comic Suppl.) 6 The marchin' and chowder club's throwin' a stag tonight. **1971** R. LEWIS *Fenokee Project* viii. 148 He's getting married tomorrow. Tonight he's holding his stag, and most of the men from the dam are going along.

f. *U.S.* A man who attends a social function without a female partner. Also quasi-*adv.* in phr. *to go stag*.

1905 N. DAVIS *Northerner* 213 'No man not escorting a lady'—a stag, you know—could go upon the floor. **1905** *Dialect Notes* III. 21 Are you going to the dance stag? **1924** P. MARKS *Plastic Age* xix. 210 True, he was not 'dragging a woman', but several of the brothers were going 'stag'; so he felt completely at ease. **1928** *Daily Express* 14 Dec. 19 A needy or avaricious 'stag'—as male dancers are called in the United States. **1948** *This Week Mag.* 1 May 16/3 The sign read: 'No Stags Allowed'. **1979** R. JAFFE *Class Reunion* (1980) I. viii. 117 A lot of boys went to the parties stag. Social life was easy for them, not the way it was for girls, who had to wait..until someone called. **1980** R. L. DUNCAN *Brimstone* iii. 59 They're not going to let you in by yourself. They have a rule against stags.

8. c. For *U.S. slang* read *slang* (orig. *U.S.*). *stag-night*, *-party* (later examples): freq. applied *spec.* to a celebration held on the eve of a man's marriage. Cf. *hen-party* s.v. HEN *sb.* 8.

1965 *Listener* 9 Sept. 373/2 On 'stag nights' it [*sc.* the entertainment] is pretty blue. **1973** in E. Dunphy *Only a Game?* (1976) iv. 110 We went out this evening for his stag night. **1923** 'BARTIMEUS' *Seaways* xii. 234 We don't want any women. We'll just have a stag party and talk Service shop and play pool afterwards. **1978** J. WAINWRIGHT *Thief of Time* 83, I know people... Class strippers. Stag-party hostesses. There's a real market.

9. a. stag film orig. *U.S.*, a pornographic film made for a male audience; **stag line** *U.S.*, the group of unattached young men at a social function; **stag movie** orig. *U.S.* = *stag film* above.

1968 *Wall St. Jrnl.* 11 Sept. 18/1 Pornography is not one of the nation's truly burning issues, and showing stag films is not our idea of how to run the world's greatest deliberative body. **1977** *Gay News* 7–20 Apr. 23/2 She.. made these very tame, anodyne stag films that she's always denying. **1934** J. O'HARA *Appointment in Samarra* i. 16 She would get twice around the dance floor with the same partner, then someone would step out of the stag line and cut in. **1977** G. V. HIGGINS *Dreamland* v. 47 As a member of Porcellian I had been invited to the stag line at a gathering on Beacon Hill. **1960** *Christian Herald* July 14/2 Teen-agers bought 'stag movies' for as much as $50 a reel. **1971** *Ink* 12 June 3/1 What he found was a hundred men having their mid-shift tea break and enjoying a stag movie.

stag, *v.*³ Add: **2. b.** *trans.* To deal in (shares) as a stag.

1935 *Times* 27 Nov. 19/2 The loan was heavily stagged, for the total applications exceeded £14,000,000. **1966** *New Statesman* 23 Sept. 456/3 The gilt-edged market has now improved to the point where the new issue of ICI loan stock seems likely to be stagged even more heavily than the last. **1981** *Daily Tel.* 20 July 15/2 The offer is likely to be subscribed although the opportunities for stagging the issue will be limited.

4. *intr.* To go to or attend a social occasion unaccompanied. Also const. *it*. *U.S. slang.*

1900 *Dialect Notes* II. 64 *To stag it*, to go to a party without escorting a lady. **1941** *Sat. Even. Post* 10 May 74/3 If you won't go with me to the picnic, I'll stag. **1973** *Lebende Sprachen* XVIII. 38/1 He had planned to stag at the class dance.

5. *trans.* To cut (trousers or other articles of clothing) off short. Also with *off*. *N. Amer.*

1902 [implied at *stagged* ppl. a. below]. **1905** *Terms Forestry & Logging* (U.S. Dept. Agric. Bureau Forestry) 49 *Stag*, to cut off trousers at the knee, or boots at the ankle. **1942** L. RICH *We took to Woods* vii. 188 One stags one's pants, one's shirt sleeves, anything that needs to be abbreviated quickly, even one's hair. **1953** R. MOON *This is Saskatchewan* 215 They [*sc.* the lumberjacks] wore pants stagged off or rolled half way to the knee so as not to be confused with mere city dwellers. **1972** *Islander* (Victoria, B.C.) 30 Apr. 16/1 He was always dressed in the same way..heavy..underwear, tin pants stagged to the proper working length.

Hence **stagged** *ppl. a.*, (of trousers) cut off short; also with *off*. *N. Amer.*

1902 S. E. WHITE *Blazed Trail* xxvii. 190 A gigantic young riverman in the conventional stagged (*i.e.* chopped off) trousers. **1933** E. HEMINGWAY *Winner take Nothing* 29 He wore stagged trousers and lumbermen's rubbers and a mackinaw shirt. **1956** H. S. M. KEMP *Northern Trader* 114 He had the mackinaw shirt and stagged-off pants, [etc.].

stage, *sb.* Add: **I. 1. i.** (*a*) (Further examples.) In mod. use, a division of a stratigraphic series, composed of a number of zones and corresponding to an age in time; the rocks deposited during any particular age. [tr. F. *étage* (introduced in this sense by A. d'Orbigny 1841, in *Paléont. Française: Terrains Crétacés* I. 417).]

1898 [see *GROUP *sb.* 4 b (iii)]. **1915** C. SCHUCHERT *Text-bk. Geol.* II. xxx. 582 The epochs and series are further divided into ages (time) and stages (rocks), but these divisions have as yet no scientific precision. **1931** [see *SERIES 11 a (ii)]. **1960** J. M. WELLER *Stratigr. Princ. & Pract.* 443 Biostratigraphic zones are combined to form larger units termed stages. **1966** D. T. DONOVAN *Stratigraphy* vii. 160 Thus we speak of Ordovician System or Ordovician Period, Albian Stage or Albian Age, according to whether we are referring to the rocks themselves or the time occupied by their accumulation. **1976** H. D. HEDBERG *Internat. Stratigraphic Guide* vii. 71 Currently recognized stages are variable in time span, but on the average they range from 3 to 10 million years as indicated by isotopic age determinations.

(*b*) A glacial or interglacial period.

1895 *Jrnl. Geol.* III. 247 These [flood-loams] cannot always be separated from the similar deposits of later glacial stages which must obviously have been deposited over the same tracts. **1939** A. K. LOBECK *Geomorphology* ix. 314 The following names have been given to the several stages of glaciation [in America]: Nebraskan, Kansan, Illinoian, Iowan, and Wisconsin... The world may now be in an interglacial period, to be followed by another glacial stage. **1969** R. V. RUHE *Quaternary Landscapes in Iowa* ii. 26 Four of the major units of the standard glacial and interglacial stages were established in Iowa.

4. g. (Later example in sense 'landing-place'.)

1969 F. MOWAT *Boat who wouldn't Float* iii. 24 We emerged at the base of a spindly and unbelievably rickety stage (as fishermen's wharves are called) made of peeled spruce poles.

i. A boxing ring. Now *Hist.*

1829 P. EGAN *Boxiana* 2nd Ser. II. 44 He was carried upon the shoulders of several men, from the stage to a private room in the Stand. **1954** F. C. AVIS *Boxing Ref. Dict.* 106 *Stage*, the old name for the ring. **1982** S. B. FLEXNER *Listening to America* 105 Broughton's rules called for a 'stage' with a one-yard square chalked or scratched in the middle.

j. *Canad.* An erection on which meat is kept out of the reach of animals, or on which meat is dried.

1715 J. KNIGHT *Let.* 30 June in *Lett. from Hudson Bay* (1965) 51 Wee were all forcd to leave the factory & to take our Selves to ye Woods & to gett on trees & Stages for Six Days. **1800** A. HENRY *Jrnl.* 9 Sept. in E. Coues *New Light on Greater Northwest* (1897) I. iii. 91 We then arranged camp..and made a suitable stage near by, to hold fresh meat, etc. **1922** *Beaver* Mar. 39/2 Passing a considerable amount of jerked meat on a stage, I entered the wigwam. **1940** F. NIVEN *Mine Inheritance* 45 The erection of stages, platforms raised high on poles above the prairies on which food could be left beyond the reach of leaping wolves.

5. a. *to hold the stage*: see *HOLD *v.* 6 g; *to set the stage*: see SET *v.* 74.

f. *stage left* (or *right*): (on) the left (or right) side of a stage (as considered from a position facing the audience). Similarly, *stage centre*. Also *fig.*

1947 *Gloss. Techn. Theatr. Terms* (Strand Electr. & Engin. Co.) 28 *Stage left*, that half of the stage on the actor's left when facing the audience. **1961** BOWMAN & BALL *Theatre Lang.* 351 *Stage right*,..right stage, or right of stage. **1972** F. WARNER *Lying Figures* III. 16 Epigyne ..sits stage left in hanging basket chair. **1977** 'C. AIRD' *Parting Breath* vi. 79 The Devil always enters stage left. **1979** *Internat. Jrnl. Sociol. of Law* Feb. 26 There are three things that are very evident already: one is that he brought to a 20th century anthropological stage-center classic legal problems that preoccupied Maine in the 19th century.

IV. 8. a. More recently, as *fare stage*, one of the principal stops on an omnibus or tram route, which marks the start of a new step in the fare structure: see *FARE *sb.*¹ 9.

9. c. (Examples of sense 'an omnibus'.)

1853 'MARK TWAIN' *Let.* 26 Oct. (1917) I. 28 The Phila. 'bus drivers cannot cheat. In the front of the stage is a thing like an office clock. **1912** J. MILNE *John Jonathan & Co.* 92 A fleet of motor-buses, which the New Yorkers call 'stages', short for stage-coaches, meanders up and down it [*sc.* Fifth Avenue]. **1939** *Nat. Geogr. Mag.* Feb. 133/2 Mammoth sleeper buses (which they still call 'stages'). **1973** R. HAYES *Hungarian Game* iii. 30 Nearly a dozen standbys had taken the stage back to Mammoth village.

10. b. *stage-by-stage* adj. phr., that proceeds by stages; step-by-step.

1956 *Nature* 25 Feb. 391/1 Using the Townsend electron avalanche process in a gas in a stage-by-stage system. **1959** *Daily Tel.* 14 Apr. 22/3 (*heading*) Stage-by-stage atomic offer to Russia. **1962** E. SNOW *Red China Today* (1963) xix. 139 An accurate stage-by-stage itinerary prepared for me by the First Army Corps showed a main trek of some 6,000 miles.

11. a. (Later examples.)

1908 E. M. FORSTER *Room with View* x. 170 She was too great for all society, and had reached the stage where personal intercourse would alone satisfy her. **1940** J. BUCHAN *Memory Hold-the-Door* ii. 39, I have used the word politics, but at this stage I was no politician, being interested only to a small degree in theories, and not at all in parties. **1956** W. S. CHURCHILL *Hist. Eng.-Speaking Peoples* II. xx. 247 It was at this stage that a group of lawyers and gentry decided to offer Cromwell the crown. **1966** *Oxf. Univ. Gaz.* 23 Dec. 429/1 This legislation will.. come forward next term or in the following term, and it is at that stage that members of the House will be asked to take the responsibility of deciding [etc.]. **1977** J. THOMSON *Case Closed* i. 15 I'm not risking you making a balls-up of it at this stage in the game.

c. (Examples in *Ent.*)

1925, 1932 [see *INSTAR *sb.*]. **1974** *Nature* 18 Jan. 154/2 The term 'stage' is here used as equivalent to the French term *étape* and is composed of several 'stadia' separated by a moult.

d. *slang.* A period of imprisonment during which privileges are allowed.

1932 'JOCK OF DARTMOOR' *Dartmoor from Within* ii. 56 In his fourth year he [*sc.* the convict] enters the highest stage. In this stage he is permitted a tobacco and cigarette ration. **1958** F. NORMAN *Bang to Rights* 30 My punishment was three days bread and water..and twenty eight days stage.

11*. a. *Electronics.* A part of a circuit usu. comprising one transistor or valve, or two or more functioning as a single unit, and the associated resistors, capacitors, etc.

1920 *Jrnl. Inst. Electr. Engineers* LVIII. 65/1 This is the first attempt to deal comprehensively with the problems of the multiple-stage amplifier. **1930** *Proc. IRE* XVIII. 1715 It will be seen that there are two stages of push-pull high-frequency amplification. **1944** *Electronic Engin.* XVI. 392 A multivibrator functioning as a divider requires three valves per dividing stage. **1961** *Listener* 9 Nov. 776/2 The relatively poor amplifying stages..in even the most costly television sets [are] incapable of providing the full and almost distortion-free sound of a genuine 'high-fidelity' system. **1975** D. G. FINK *Electronics Engineers' Handbk.* XIII. 18 An amplifier may take the form of a single stage or a complex single stage or it may employ an interconnection of several stages... For a multistage amplifier, the individual stages may be essentially identical or radically different.

b. *Astronautics.* Each of two or more sections of a rocket that have their own engines and propellant and fall away in turn as their propellant becomes exhausted.

1935 C. S. PHILP *Stratosphere & Rocket Flight* xii. 62 The first method consists of a rocket built in two or more stages, the first stage being a relatively low-power engine for use in the lower parts of the earth's atmosphere, and the second stage, or subsequent multiple stages, of increased power for use in the higher and more rarefied regions. **1948** *Jrnl. Brit. Interplanetary Soc.* VII. 168 Into this section propellant from the 1st, 2nd and 3rd stage tanks is automatically transferred..so that as its own propellant is drawn off by the motor it is replaced and both the 4th and Final stages achieve 'release velocity' with tanks at capacity level. **1955** *Times* 4 Aug 6/2 We shall be limited to one-stage rockets at first, but afterwards we may work on two-stage rockets which will reach greater heights. **1963** *Ann. Reg. 1962* 445 The satellite, weighing 170 lb., was placed in orbit by a three-stage Delta rocket. **1975** K. GATLAND *Missiles & Rockets* viii. 185 The first stage engines burnt for about 2½ minutes, boosting the Apollo astronauts to an altitude of 36 miles.

12. a. *stage-carpenter* (earlier example), *-carpentering*, *crew*, *design*, *designer*, *-hand* (earlier example), *lighting*, *-picture*, *show*, *trick*, *version*; *stage army*, *aside*, *-dialect*, *-villain* (earlier example); similarly *stage Australian*, *Frenchman*, *Irishman*, etc.; *stage Irish* sb. and adj.

1922 C. S. CHURCHILL *Let.* 4 Jan. in M. Soames *Clementine Churchill* (1979) xiii. 203 All the sad events of last year culminating in Marigold passing and re-passing like a stage Army through my sad heart. **1957** A. C. L. DAY *Outl. Monetary Economics* xiii. 177 There was, therefore, a stage army of cash moving from bank to bank through each week, helping improve appearances. **1813** M. EDGE-

WORTH *Let.* 16 May (1971) 53 Lady Derby he says is always *acting*—that there is continually a stage *aside* which betrays her. **1945** *Essays & Studies 1944* XXX. 35 A dry unsympathetic comment delivered curtly like a stage-aside, 'Would he had blotted a thousand!' **1965** *Times Lit. Suppl.* 16 Sept. 812/1 Only the baggy trousers and the wide brimmed hat anchor him to the image of the 'stage' Australian, or the Boy from the Bush. **1826** O'KEEFFE *Recoll.* I. iv. 146 Years after, some such enthusiastic spirit possessed the stage carpenters at Cork. **1899** 'MARK TWAIN' in *Cosmopolitan* Oct. 593/2 He had to retire from his profession of stage-carpentering. **1959** *Guardian* 13 Nov. 9/3 The Banana Boat song was booming on the telly in the stage-crew's room. **1975** *New Yorker* 21 Apr. 112/1 The theatre can devote all its resources—orchestra, singers, coaches, stage crews, lighting team—to the preparation and performance of the work. **1943** J. LEYDA tr. *S. Eisenstein's Film Sense* ii. 77 This is an important law which can be found in painting, in stage design..of this period. **1977** J. AIKEN *Last Movement* vi. 116 Is that your profession—stage design? **1938** L. BEMELMANS *Life Class* II. iii. 141 An energetic hostess.. will often arrive with a squadron of orchestra leaders, architects..and stage designers. **1978** R. LUDLUM *Holcroft Covenant* xiii. 151 There was a man in London, a stage designer, who'd had a brief vogue as a decorator among the wealthy on both sides of the Atlantic. **1927** *New Republic* 12 Oct. 218/2 Mr. Wiley has a further advantage over his fellow craftsmen in being master of two stage-dialects—pidgin English and Negro. **1966** G. N. LEECH *Eng. in Advertising* viii. 78 A music-hall comedian adopts a 'stage-dialect'. **1824** in A. Mathews *Mem. Charles Mathews* (1839) III. xx. 453 Talbot is the stock Morbleu, which he makes a monkey—a ballet-master—in short, a stage Frenchman. **1885** F. LESLIE in *Entr'acte Ann.* 22/1 The stage hands were non-expectorants, and the ladies were quite vexed at the clean condition of the stage. **1962** *Listener* 1 Mar. 387/3 These pages are littered with wild stage-Irish cries of 'Jasez', 'begod', and the like. **1962** A. LURIE *Love & Friendship* xi. 208 'That's a gra-and idea,' Charley said, stage Irish. **1977** A. J. BLISS in D. O'Muirithe *Eng. Lang. in Ireland* 9 At this early date a conventional 'stage Irish' had been established. **1980** J. O'FAOLAIN *No Country for Young Men* v. 96 He must think he'd fallen into a stage-Irish household. **1860** *Players* I. 131 The dialect he assumed, though it may not have been so productive of laughter as that in which the 'stage Irishman' usually delivers himself. **1911** G. B. SHAW in *Evening Sun* (N.Y.) 9 Dec. 4/6 The stage Irishman of the nineteenth century, generous, drunken, thriftless, with a joke always on his lips and a sentimental tear always in his eye. **1973** J. ELSOM *Erotic Theatre* ii. 33 A stage actress—as recognizable a type as a stage Irishman and more frequently seen. **1895** *New Budget* 4 Apr. 21/1 One would have practically to invent new methods of scene-painting and stage-lighting. **1908** G. B. SHAW *Let.* 2 Aug. (1972) II. 804 There were some very clever tricks of stage lighting in the second act of Siegfried. **1983** 'J. LE CARRÉ' *Little Drummer Girl* iii. 51 The stage lighting was too good, she couldn't penetrate the haze. **1920** W. B. YEATS *Poems* p. vii, When our stage-pictures were made out of poor conventional scenery and hired costumes. **1949** F. FERGUSSON *Idea of Theatre* i. 28 The contemplation of the final stage-picture or epiphany. **1980** *Times* 29 Feb. 13/2 The WNO *Onegin* is..a procession of stage-pictures far beyond the everyday purview of opera production. **1895** G. B. SHAW in *Sat. Rev.* 16 Feb. 217/1 Stage shows with nothing to redeem their obvious silliness but a promise of as much lewdness as the audience will stand. **1982** N. FRYE *Great Code* v. 117 The Puritan and Jansenist prejudice against 'stage shows'. **1776** *St. James's Chron.* 19 Oct., Allowing reasonably for stage trick, this appears to us to be extravagantly over-done. **1895** G. B. SHAW *Let.* 28 Nov. (1965) I. 572 This is not one of my great plays..: it is only a display of my knowledge of stage tricks. **1856** A. C. RITCHIE *Mimic Life* 105 Desdemona, according to the stage version (which omits her during the midnight brawl). **1955** *Radio Times* 22 Apr. 31/2 '*A Woman of No Importance*'.. Adapted for radio from the stage version. **1885** A. EDWARDES *Girton Girl* II. xvii. 281 Dismissed as one occasionally sees the frustrated stage villain, long before the final falling of the curtain!

b. *stage-line* (earlier example), *-office*.

1830 *Williams's N.Y. Ann. Reg.* 115 Other principal Stage lines from Albany. **1812** J. MCNAB *Jrnl.* 4 Mar. in *Beaver* (1973) Summer 9/2 We..inquired at the stage office when the sleigh sets out for New York on Friday. **1872** 'MARK TWAIN' *Roughing It* 22 The first thing we did..was to hunt up the stage-office, and pay..for tickets per overland coach.

13. stage box (later examples); **stage direction,** (*b*) stage-management (also *fig.*); **stage director** orig. *U.S.*, a stage-manager; also, more recently, a director (sense *1 g); **stage-door,** (*a*) (earlier example of *stage-door keeper*); † (*b*) a door at the side of the proscenium arch (*obs.*); **stage-door Johnny** *slang* (chiefly *U.S.*), a (young) gentleman who frequents stage-doors for the company of actresses; **stage-entrance** = *stage-door*; **stage-fright** (earlier example); **stage name** (later examples); **stage presence,** the (forceful) impression made by a performer on an audience; **stage school,** an academy of drama; **stage-set** = SET *sb.*¹ 28 in Dict. and Suppl. (also *transf.*); **stage-setting** (earlier and later examples); also *fig.*; **stage-struck** *a.* (later examples); **stage-whisper** (earlier and later examples); hence as *v. trans.*, (*a*) to address (a person) in a stage-whisper; (*b*) with obj. as direct speech: to say (something) in a stage-whisper; **stage-whispered** *ppl. a.*, spoken in a stage-whisper; **stage-whispering** *ppl. a.*; **stage-worthy** *a.* (later examples); hence **stage-worthiness.**

1857 H. MARTINEAU *Autobiogr.* I. iv. 388 [Mr. Macready] gave us the stage box, whenever we chose to ask for it. **1982** C. CASTLE *Folies Bergère* ii. 64 He was to be seen..accepting congratulations in a stage box. **1833** R. DYER *Nine Years of Actor's Life* 78, I began a correspondence with the well-known Henry Lee, and finally agreed to take the stage direction of his theatres. **1962** V. NABOKOV *Pale Fire* 55 When morning finds us marching to the wall Under the stage direction of some goon Political, some uniformed baboon. **1782** T. HALL in G. O. Seilhamer *Hist. Amer. Theatre* (1889) II. v. 55 Before you see one of your stage directors Or, if you please, one of those strange projectors. **1849** *Theatrical Mirror* 27 Aug. 101/1 Mr A. Harris, stage-director of the Royal Italian Opera, of Covent Garden, has been presented with a piece of plate. **1908** E. TERRY *Story of my Life* xiv. 326 It was not as an actor but as a stage director that he wanted to work. **1979** A. WILLIAMSON *Funeral March for Siegfried* ix. 42 'We'll have to chase up the stage staff,'.. 'I can give you the stage director's address.' **1761** A. MURPHY *Way to keep Him* (ed. 4) v. 101 Enter Lady Constant. *Lovemore* No way to escape?—[Attempts both stage doors, and is prevented]. **1776** R. Y. WALSINGHAM *Let.* 6 Feb. in J. Boaden *Private Corresp. David Garrick* (1832) II. 134 That you will be so good as to pardon the stage-door keeper for admitting me last night. **1829** H. FOOTE *Companion to Theatres* 33 At this time, the proscenium was altered; stage doors were introduced, there having been none in the original building. **1883** D. COOK *On Stage* I. ix. 187 Of such stage-doors as are here described there is no London theatre in possession. **1912** *Out West* Feb. 139/1 No theater can hope to do business without stage door Johnnies. **1922** [see *RICH *adv.* and *Comb.* 10 b]. **1952** GRANVILLE *Dict. Theatrical Terms* 169 *Stage-door Johnny*, the Victorian buck..who haunted the stage door of the Gaiety Theatre, London,..when some of the most beautiful women of the day were members of the chorus. **1976** BOTHAM & DONNELLY *Valentino* iv. 35 Two Ziegfeld Follies girls who were doing the town with a pair of wealthy stage-door Johnnies. **1830** J. BERNARD *Retrospections of Stage* II. ix. 273 He got the carpenter to fix a bucket on a swivel, over the stage-entrance of the Theatre. **1956** B. HOLIDAY *Lady sings Blues* (1973) xii. 112 We were supposed to pick up all the cats at the Braddock Motel.., near the stage entrance of the Apollo. **1876** 'MARK TWAIN' *Tom Sawyer* xxi. 169 A ghastly stage-fright seized him. **1941** A. CHRISTIE *Evil under Sun* iv. 66 He doubted if Arlena Stuart, to give her her stage name, had ever wanted to be alone in her life. **1959** T. S. ELIOT *Elder Statesman* II. 51 You know I meant my stage name. The name by which you knew me. **1977** *Sounds* 9 July 28/1 *Stooges*..headed by James Osterburg—stage name Iggy Stooge. **1929** *Melody Maker* Feb. 195/2 There was a wide gap between the stage presence of the cornet soloist and the stage presence of the other artists. **1959** R. LONGRIGG *Wrong Number* iv. 49 Mrs. Proctor, in the soubrette part of the kitchen-maid, made up for her vocal uncertainty with a racy and convincing stage presence. **1977** *Zigzag* Aug. 16/1 Syl's got a lot of what directors call 'stage presence'. **1936** N. STREATFEILD *Ballet Shoes* iv. 55 She..ran an ordinary stage school where the children learnt all kinds of dancing. **1977** S. BRETT *Star Trap* ii. 24 He came out of one of the stage schools... He may have been a child star in films. **1861** Stage set [see SET *sb.*¹ 28]. **1947** J. C. RICH *Materials & Methods of Sculpture* i. 13 In designing for a garden..an excellent and highly recommended procedure is first to make a small three-dimensional scale model of the garden and its immediate environs, and to use this small 'stage-set' actively as an aid in determining the nature and placing of the garden ornament. **1958** S. SPENDER *Engaged in Writing* i. 13 A large, bare room with faded nineteenth-century murals, like the back of an operatic stage-set. **1977** *Proc. R. Soc. Med.* LXX. 427/2 A collaborative effort in which satirical comedy is fused with the sort of music, dance, lavish costumes and stage-sets used in court ballets. **1881** C. C. HARRISON *Woman's Handiwork* III. 152 All the little invisible wires that control the scenery and 'stage setting' of a home-interior. **1929** *Oxford Poetry* 10 For three-and-twenty years, he curled And drooped, on this stage-setting of the world. **1982** P. RABY '*Fair Ophelia*' iv. 48 Ciceri extended into the sphere of stage settings the reforms which Talma himself had introduced so far as historical accuracy of costume was concerned. **1911** Stage struck [see *RUMBLE *sb.* 6]. **1976** *Southern Even. Echo* (Southampton) 6 Nov., Despite warnings of financial trouble in the theatre,..she has remained stage-struck throughout her life after deciding at the age of four to become an actress. **1864** H. MORLEY *Jrnl.* 17 Dec. (1866) 355 His bedroom scene, spoken throughout in an oppressively ostentatious stage whisper, is an intolerable blunder. **1927** J. N. MCILWRAITH *Kinsmen at War* xx. 198 Mrs. Secord spoke in a stage whisper. **1932** KIPLING *Limits & Renewals* 80 Private Gillock, who poses as a wit, was stage-whispering me for leave to 'put a shot into his radiator'. **1941** B. SCHULBERG *What makes Sammy Run?* iii. 49 'What are you thinking about, honey?' Billie stage-whispered. **1960** J. RAE *Custard Boys* I. v. 52 'Who's your German friend?' he asked in a stage whisper. **1977** W. M. SPACKMAN *Armful of Warm Girl* 38 A huge handsome white-haired classmate flung himself jovially upon them..to beg in a whooping and waggish stage-whisper. **1979** R. LITTELL *Debriefing* iii. 33 'Do you have the pouch?' she stage-whispers. **1978** G. SIMS *Rex Mundi* iv. 26 The stage-whispered duet started again... I made out only odd words. **1883** 'MARK TWAIN' *Life on Miss.* xxxi. 342 'The captain's voice, by G—!' said the stage-whispering ruffian. **1973** *Times Lit. Suppl.* 19 Oct. 1272/4 The recent Jonathan Miller production of *The Malcontent* has demonstrated that play's stageworthiness. **1959** *Times* 4 Dec. 15/1 None of his [*sc.* Mussorgsky's] operatic undertakings is stageworthy. **1979** *Amer. N. & Q.* Nov. 40/1 In an effective and stageworthy central scene, the old miser's secrets..are discovered.

stage, *v.* Add: **3. d.** *transf.* To mount or put on (a spectacle). Also, to effect (a recovery); *to stage a comeback*: see *COME-BACK *sb.*² 2.

1924 F. J. HASKIN *Amer. Govt.* (rev. ed.) xxxvii. 437 In combating..bootlegging, Federal agents..staged raids that revealed..the widespread extent of Volstead Law violations. **1951** *Sport* 27 Apr.–3 May 5/1 It's hard to think that the event can be staged at Wembley. **1958** A. H. COMPTON *Atomic Quest* 122 His Nazi-trained students staged a protest. **1967** [illegible] More than 500 students staged a sit-in at Cambridge University yesterday. **1981** *Times* 9 Apr. [illegible] A gradual return of confidence saw equities and gilts stage a rally yesterday.

6. b. Of a pilot or aircraft: to make a brief landing in the course of a long journey.

1971 P. PURSER *Holy Father's Navy* [illegible] staged in Iceland and was on his way to Norway. **1973** D. KYLE *Raft of Swords* (1974) [illegible] flight was staged at Vancouver.

7. *trans.* *Astronautics.* To separate (a section or stage) from the upper or remaining part of a rocket. Also *intr.* of the stage.

1957 [illegible]

8. b. To cause (a person) to pass through stages; to bring about (something) in stages.

1957 A. C. CLARKE *Deep Range* [illegible] **1962** E. SNOW *Red China Today* (1963) [illegible] We staged them through quick courses of training and refresher training in the Ningtu technique. **1980** *Daily Tel.* 15 Mar. 1/7 The Government will 'stage' the payment of the increases to key staff this year.

stage-coach. b. (Earlier examples.)

1831 *Boston Transcript* 2 Aug. 2/3 The entertainment happened to be the 'Stage Coach,' which was acted so wretchedly that it was impossible to make head or tail of it. **1873** 'COOLIDGE' *What Katy Did* (1872) [illegible]

staged, *ppl. a.* Add: **4.** That proceeds by stages: = *PHASED *ppl. a.* 2.

1960 *Economist* 15 Oct. 255/2 A group of commissioners who disagree with those proponents of a more balanced policy of staged development. **1969** *Daily Tel.* 16 May 1 A mutual and staged withdrawal of all foreign troops from South Vietnam. **1973** *Ibid.* 13 Dec. 28 New wage rates which would be staged in over 2½ years.

stage-manage, *v.* (In Dict. s.v. STAGE-MANAGER.) Add: (Later *fig.* examples.) Also *absol.*

1910 WODEHOUSE *Psmith in City* i. 3 My pater wants to jump in and stage-manage. **1924** J. BUCHAN *Three Hostages* [illegible] **1958** *Washington Post* 16 June [illegible] U.S. Embassy in Moscow. **1964** 'E. MCBAIN' *Ax* ii. 36 We did it [illegible] **1980** D. LODGE *How Far can you Go?* [illegible] she was now convinced, was being stage-managed [illegible] them all naked together.

Hence **stage-managed** *ppl. a.*

1930 *Times Lit. Suppl.* 1 May 359/3 In 1891 France and Russia had signalized their close relations by the enthusiastic, though carefully stage-managed, reception given to the French fleet at Kronstadt. **1974** *Listener* 10 Oct. 461/3 These are not press conferences at all, but carefully stage-managed events, put on..for the audiences of BBC and ITV news.

stage-management. (In Dict. s.v. STAGE-MANAGER.) Add: (Earlier example.) Also *fig.*

1812 C. MATHEWS *Let.* 15 Jan. in A. Mathews *Mem. Charles Mathews* (1838) II. [illegible] **1949** M. STEEN *Twilight on Floods* [illegible] **1972** A. GIDDENS [illegible] *Social & Polit. Theory* [illegible] Durkheim [illegible] largely due to Durkheim's own stage-management.

stage-manager. Add: (Earlier and later examples.) Also, in mod. usage, one who is in charge of the technical side of a production (see quot. 1961). Also *transf.* and *fig.*

1805 R. W. ELLISTON *Let.* 10 July in G. Raymond *Life & Enterprises of Robert William Elliston, Comedian* (1857) I. ii. [illegible] **1906** [illegible] **1926** E. DYSON [illegible] astonishing frock is at heart a stage-manager. **1961** BOWMAN & BALL *Theatre Lang.* 349 *Stage manager*, the head of the production staff, who runs the stage-show.. during rehearsals, in technical matters.. and who, once the production opens, takes complete charge of the stage, the actors.., and the crews. **1982** S. Brett *Murder Unprompted* ii. 21 The Stage Manager's calming voice came over the loudspeaker, 'Beginners, Act One, please.'

Hence as *v. trans.*, to stage-manage (*rare*); **stage-manageress,** a female stage-manager.

1900 M. Beerbohm *Let.* 11 Dec. (1964) 138 As stage-manageress she has been absolutely intelligent and sweet and charming. **1902** G. B. Shaw *Let.* 9 Aug. (1972) II. 281 It is..very important to get the last scene well stage managered, with a big surging crowd. **1926** E. F. Crosse *Let.* in *Times* 13 Apr. 15/5 There was no shyness and the stage manageress explained to me all details.

stager. Add: **5.** One who erects scaffolding in a shipyard. Cf. Stage *sb.* 4 e.

1927 *Dict. Occupational Terms* §668 *Stager*,..erects staging on which workmen stand to work. **1974** *Socialist Worker* 26 Oct. 16/5 Management agreed that stagers in the Society of Boilermakers be made a skilled section with parity with packers and sheeters.

stagflation (stægflēɪ·ʃən). *Econ.* [Blend of Stag(nation and In)flation: cf. *slumpflation.] A state of the economy in which stagnant demand is accompanied by severe inflation.

1965 I. Macleod *Hansard Commons* 17 Nov. 1165/1 We now have the worst of both worlds—not just inflation on the one side or stagnation on the other, but both of them together. We have a sort of 'stagflation' situation. **1971** R. Boyson *Down with Poor* 5 The result of all this extra state interference..has been..what might be called rampant stagflation, that is to say stagnation in production and raging inflation which..destroys belief in the future. **1974** W. Rees-Mogg *Reigning Error* iv. 75 So-called stagflation and slumpflation are the inevitable reflection of the progressive divergence between a rising nominal and a falling real supply of money. **1976** F. Zweig *New Acquisitive Society* II. xi. 137 Even recession, stagnation, 'stag-flation' or steeply rising unemployment do not rule out wage explosions. **1979** *N.Y. Rev. Bks.* 25 Oct. 44/2 Stagflation ate away at prosperity. In 1975 industrial production fell.

staggeen. (Earlier example.)

1829 G. Griffin *Collegians* I. vii. 135 A parcel of ould staggeens, sir, that's running for a saddle, that's all the races they'll have.

stagger, *sb.*¹ Add: **3.** (Earlier example.)

1865 'Mark Twain' *Screamers* (1871) 149 He would make one more stagger at it anyway.

3*. A staggered arrangement or disposition; *spec.* in *Aeronaut.*, an arrangement of the wings of a biplane such that the leading edge of the upper wing is in front of or behind that of the lower wing.

1915 W. E. Dommett *Aeroplanes & Airships* 104 When the wings of a biplane are set with the upper one slightly ahead of, or abaft of the other, they are said to be staggered. The stagger is measured by the angle made by the line joining the leading edges with the normal to the fore and aft axis of the aeroplane. It is convenient to call the stagger positive if the upper wing is ahead of the lower. **1919** H. Shaw *Text-Bk. Aeronaut.* iii. 58 The effect of positive stagger on a biplane is to cause the lower plane to work in a slight downdraught from the upper plane, and so decreases its angle of incidence, giving the machine greater stability. **1937** *Times* 13 Nov. 13/5 All the footmarks lay on the same straight line with no 'stagger' right or left of it. **1950** R. G. Batson *Roads* v. 83 A stagger interrupts the passage of traffic from one branch of the minor road to the other. **1957** L. L. Beckford *A.B.C. of Aeronaut.* 95/2 When the upper wing is placed behind the arrangement is known as Back Stagger. **1972** *Times* 14 Nov. 15/7 The method of arranging the 'stagger' was one of the matters on which Customs and Excise consulted a wide range of trade bodies. **1980** M. Booth *Bad Track* i. 17 At the head of the small approach road was a T-junction... At a stagger from the slip-road was another, going down the other side to join the motorway.

4. a. stagger-juice: for *Austral. slang* read *slang* and add earlier example; **stagger tuning** *Electronics*, the tuning of different stages of an amplifier to slightly different frequencies so as to broaden the overall frequency response; so **stagger-tuned** *a.*

1905 *Chambers's Jrnl.* Oct. 730/1 A liquor labelled Scotch whisky, but commonly known as 'stagger juice'. **1947** F. E. Terman *Radio Engin.* (ed. 3) vii. 360 The behavior of such stagger-tuned pairs under conditions corresponding to maximal flatness..is of particular interest. **1975** D. G. Fink *Electronics Engineers' Handbk.* xiii. 57 The simple shunt-compensated stage has found extensive use in stagger-tuned pulse-amplifier applications. **1953** Fowler & Lippert *Television Fund.* vi. 121 Some of the later-model TV receivers make use of a single tuned inductance between tube sections of the i-f amplifier and obtain the necessary bandwidth by means of stagger-tuning. **1979** G. M. Miller *Handbk. Electronic Communication* ix. 285 The problem here is..how to get a wide-enough bandwidth but still have relatively sharp falloff at the pass-band edges. Most TV IF amps solve this problem through the use of stagger tuning.

b. That involves the implementation of a staggered arrangement, as *stagger hours, schedule, plan.* Cf. sense *9 b of the vb. orig. *U.S.*

1918 *Dial* 2 Nov. 369/1 'Stagger' hours have been instituted—whereby one department goes to its work a half hour earlier than another, thus relieving the congestion of the street cars. **1933** *Sun* (Baltimore) 16 Aug. 4/2 The animals would be marketed..under a system of 'stagger' shipments to prevent swamping market and packing facilities. **1943** *Ibid.* 16 Nov. 11/3 State liquor stores went on a new 'stagger' schedule of hours..with doors opening at 9 A.M. and closing at 3 P.M. **1947** *News Chron.* 11 Apr. 4/6 (*heading*) Firms favour 'stagger' plan. **1960** *Wall St. Jrnl.* 7 Apr. 8 The stagger system is election of directors in classes for various terms of office.

stagger, *v.* Add: **II. 9. a.** Delete *Mech.* and add: Also, to position (things) at successively greater distances from the straight line they would otherwise form. (Later examples.)

1916 H. Barber *Aeroplane Speaks* 63 By staggering the top surface forward..it is removed from the action of the lower surface and engages undisturbed air. **1937** *Memo. Lay-out & Constr. Roads* (Min. of Transport) No. 483. 13 Where a minor road crosses a major road constructed with a single carriageway the minor road should be staggered, preferably to the left. **1959** *Listener* 6 Aug. 208/2 Then the road engineers got to work and staggered the cross-roads.

b. To arrange (holidays, times, etc.) so that they do not coincide; to arrange (an event or action) so that its implementation is spread over a period of time, or so that it is performed by different persons at different times. orig. *U.S.*

1918 *Daily Chron.* 23 Apr. 4/2 In order to maintain efficient service for industrial traffic, it is urged by the Tramways..Committee that factories and other places of business should 'stagger' their times of opening and closing. **1929** *Sun* (Baltimore) 27 Sept. 12/3 The days of rest will be staggered, one-fifth of the workers presumably laying off each twenty-four hours. **1934** *Ibid.* 3 Nov. 2/2 Omitting their national shows and staggering the introduction of new models. **1946** in P. N. S. Mansergh *Const. Rel. between Brit. & India* (1979) VIII. 211 Though the departure of officers on proportionate pension could be 'staggered', the right to retire could not be taken away. **1951** *Engineering* 7 Sept. 302/2 A scheme for staggering working hours, to ease the..load at..peak hours. **1962** E. Godfrey *Retail Selling & Organ.* xvii. 172 Many firms have found it better to stagger stock-taking, so that all departments are not disrupted at the same time. **1978** L. Davidson *Chelsea Murders* III. xxi. 122 Lunch was being staggered, the six of them dashing out individually for a sandwich.

staggered, *ppl. a.* Add: **2.** *spec.* **a.** Positioned alternately on one side and the other of a line, or obliquely at successively greater distances from it; also, composed of parts so placed.

1875, 1905 [in Dict.]. **1909** V. Lougheed *Vehicles of Air* iv. 205 (*heading*) Staggered surfaces. **1930** *Engineering* 20 June 787/2 The removal of completely assembled portions of [railway] track would be awkward in the face of the general use of staggered joints. **1937** *Times* 13 Apr. (Motor Suppl.) p. x/1 Where this is not practicable staggered crossing or roundabouts are recommended. **1947** *Highway Engineers' Ref. Bk.* iv. 122/2 On straight roads side-mounted lanterns may be arranged either opposite to each other or in a staggered arrangement. **1960** C. H. Gibbs-Smith *Aeroplane* I. xi. 69 The Goupy had staggered biplane wings with pivoting wing-tip ailerons. **1961** F. C. Avis *Sportsman's Gloss.* 62/2 *Staggered start*, one in a race on an oval track..where the runner in the outside lane starts apparently in front of the runner in the inside lane, but only to compensate for the greater distance of the outside lane. **1968** Batson & Proudlove *Roads* (ed. 2) iv. 67 (*caption*) Layouts of T-junctions and staggered cross-roads on 2-way and divided roads. **1978** J. A. Michener *Chesapeake* xi. 654 From the woods came six doves, flying low in their wonderfully staggered fashion.

b. Arranged not to coincide in time; having starting and finishing times that overlap. Cf. *stagger *sb.* 4 b.

1932 *Sun* (Baltimore) 8 Oct. 13/7 Severely curtailed appropriations, which necessitated recourse to a 'staggered' schedule of openings. **1940** *Ann. Reg. 1939* 358 A scheme of 'staggered' hours was evolved whereby some playhouses would be allowed to keep open till ten on condition that others still kept to the six o'clock rule. **1955** *Times* 12 Aug. 5/2 'Staggered' working hours for nearly 10,000 office workers in the West End of London may be introduced to ease the peak hour travelling problem. **1980** M. Babson *Dangerous to Know* iii. 15 The rest of the staff arrived at staggered hours.

staggering, *vbl. sb.* Add: **2. c.** See *stagger *v.* 9 b.

1955 *Times* 19 May 7/3 Difficulties had arisen by all the pool mail being received in the same week and as a result 'staggering arrangements' were made to spread the traffic. **1959** *Ann. Reg. 1958* 507 He undertook to consider whether the staggering of hours should be made compulsory. **1971** *Timber Trades Jrnl.* 14 Aug. 53/1 The reduction in productivity..due to the staggering of holidays. **1979** *Daily Tel.* 15 Dec. 2/3 The inquest had been told that the staggering procedure under which the 32 [parachute] troops left opposite doors of their Hercules had gone out of synchronisation.

staggering, *ppl. a.* Add: **1. d.** (Later examples.)

1922 Joyce *Ulysses* 413 Staggering bob in the vile parlance of our lower class licensed victuallers signifies the cookable and eatable flesh of a calf newly dropped from its mother. **1966** W. S. Ramson *Austral. Eng.* iv. 70 *Staggering bob*, a widespread dialect phrase used either of a very young calf which has not yet found its legs or of the veal from such a calf.

2. Also in trivial use: amazing, astounding; enormous.

1934 J. B. Priestley *Eng. Journey* vi. 202 He still controlled this staggering array of properties, extending from remote industrial villages in Yorkshire to Shaftesbury Avenue. **1939** [see *bally *a.* and *adv.*]. **1951** *Sport* 30 Mar.–5 Apr. 9/1 Staggering offers have been made for Twentyman, a half-back of immense promise. **1958** P. H. Gibbs *Curtains of Yesterday* iv. 53 Summoning two of his men he had the lorry filled up with a staggering amount of food—cheese, bread, bully beef, hams, tinned food. **1978** *Lancashire Life* July 55/4 Rich man though he was ..the High Sheriff must have faced a staggering bill.

staggeringly, *adv.* Add: Also in trivial use, amazingly; exceedingly.

1976 J. Cooper *Harriet* xvi. 131 She felt staggeringly untogether... She had a blinding headache. **1979** *PN Rev.* 9 60/2 It is the mark of staggeringly few of the great pulpiteers. **1979** *Guardian* 3 Nov. 11/3 What he is going to say in his six lectures..is staggeringly banal.

sta·ggerment. *nonce-wd.* [f. Stagger *v.* + -ment.] Great amazement, astonishment.

1933 J. R. R. Tolkien *Let.* 16 Mar., (MS.) I have actually been presented by a well-wishing old gentleman with a complete N.E.D., to my staggerment, as I had quite given up hope of possessing one. **1937** —— *Hobbit* xii. 221 To say that Bilbo's breath was taken away is no description at all. There are no words left to express his staggerment. **1975** *Church Times* 31 Jan. 11/1 As I lifted my arms for the quick frisk, I was conscious of unjournalistic feelings of staggerment. I had been through this process often enough at airports across the world, but never before, ever in a cathedral for goodness' sake.

staggy (stæ·gi), *a.* [f. Stag *sb.*¹ + -y¹.]

1. a. *N.Z.* and *N. Amer.* Of an animal or its meat: having the characteristics or appearance of a mature male (see quots.).

1933 *Press* (Christchurch, N.Z.) 9 Dec. 17/8 *Stag*, imperfectly, or late castrated male sheep or steer. Hence *staggy*. **1934** Webster, *Staggy*, having the appearance of a mature male animal;—said of female domestic animals. **1950** *N.Z. Jrnl. Agric.* Sept. 201/3 Complaints about the proportion of 'staggy' New Zealand wether lamb carcasses have been received recently from the United Kingdom. **1974** *Globe & Mail* (Toronto) 12 Dec. 8/2 The quality of beef within a grade can vary, depending on whether an animal is staggy or whether it weighs less than 550 pounds or more than 700. A staggy steer exhibits bull-like characteristics and has tougher meat.

b. *Austral.* (See quot.)

1891 R. Wallace *Rural Econ. Austral. & N.Z.* i. 30 Sometimes the [potato] sets remain fresh and do not decay in the soil after the haulms have developed; they remain 'staggy' or hard and woody.

2. a. Abounding in stags. *rare.*

1921 *Blackw. Mag.* July 34/2 Very staggy ground this. Indeed, a sambur grunted in covert which we watched.

b. *fig.* Of a tree, etc.: having bare branches. Cf. Stag-headed *a.* 2.

1933 R. Campbell *Flowering Reeds* 21 Bare trees.. Down the long avenue in staggy flight Are hunted by the hungers of the gale. **1961** P. White *Riders in Chariot* i. 13 Slapped by a staggy elder-bush..Whipped by the little sarsaparilla vine. **1973** —— *Eye of Storm* vii. 322 Under the staggy orange trees, amongst the hummocks, in the green haze of Noamurra.

stag-horn. Add: **3. b. stag('s) horn coral,** a branching coral of the genus *Acropora*.

[**1785** in G. H. Millar *New Syst. Nat. Hist.* iv. ii. 286 The coral plants..sometimes shoot out like trees without leaves in winter;..sometimes they are found to resemble ..the antlers of a stag, with great exactness and regularity.] **1884** R. Rathbun in G. B. Goode *Fisheries U.S.: Nat. Hist. Aquatic Animals* I. v. 841 Among the true stony corals are the Stag-horn Corals..and many others. **1928** Russell & Yonge *Seas* vii. pl. 59 (*caption*) Stag's Horn Coral. **1977** G. Durrell *Golden Bats & Pink Pigeons* v. 123 The predominant coral was Stag's horn,.. like a great graveyard of all the finest Victorian deer trophies, decked out in white and electric blue.

4. *Naut.* (See quot. 1961.)

1923 *Man. Seamanship* (Admiralty) II. 87 The 15-in. cordite whips..are either taken to the motor..or to the special staghorn for lowering. **1961** F. H. Burgess *Dict. Sailing* 196 *Staghorn*, a metal bollard with two horizontal arms.

5. *Path.* Used *attrib.* to designate a large calculus of the kidney having the branched form of the renal pelvis that it occupies.

1910 *Lippincott's New Med. Dict.* 924/1 Stag-horn calculus. **1926** Young & Davis *Young's Pract. Urol.* I. vi. 377 In some extreme cases the kidney is only a thin sheet of dense scar tissue in which no trace of tubules or glomeruli can be found, overlying a large stag-horn calculus. **1961** R. D. Baker *Essent. Path.* xvii. 441 (*caption*) The large staghorn calculus is a cast of the renal pelvis and of renal calyces. **1974** J. D. Maynard in R. M. Kirk et al. *Surgery* viii. 161 Renal calculi may be entirely symptomless, particularly the very large staghorn type filling most of the pelvicalyceal system.

staging, *vbl. sb.* Add: **1. a.** Also, *spec.* shelving for plants in a greenhouse.

1886 *Bk. Garden Managem.* 437 The great desideratum in the arrangement of staging of any kind, as far as the plants themselves are concerned, is to bring them as close to the light as possible. **1929** *Radio Times* 8 Nov. 417/2 Chrysanthemums are now displaying..bloom under glass. ..Be careful not to spill water on the floor and staging. **1974** *Country Life* 24 Jan. 150/2 *Tulipa humilis* and crocus ..have been moved to staging in an airy, unheated greenhouse.

3. To Chiefly *Anglo-Indian* add: and *U.S.* (Earlier and further examples.) Now *rare.*

1840 *Southern Lit. Messenger* VI. 381/2 He does not follow the sea nor staging. **1864** *Harper's Mag.* Oct. 563/1 In an ancient adobe building,..Mr. Banning carried on his staging and teaming operations. **1894** *Outing* XXIV. 399/2 Stagin' in them days, stranger, was *stagin'*.

4. *Astronautics.* The arrangement of stages (*stage *sb.* 11* b) in a rocket; the separation and falling away of a stage from the remainder of the rocket when its propellant is spent.

1959 H. S. Seifert *Space Technol.* xix. 19 In the staging operation the uncoupling of the two sections must take place smoothly so as not to put asymmetric loads on the continuing part of the structure. **1962** D. Slayton in J. Glenn et al. *Into Orbit* 26 About thirty seconds after staging..the Atlas goes over into about a 14° pitch to get you headed into the precise angle for a good orbit. **1966** H. O. Ruppe *Introd. Astronaut.* I. iii. 88 In conventional 'tandem staging' or 'series staging' each complete stage burns and separates before the next stage ignites. In parallel staging, components (engines, tanks, or stages) operate simultaneously.

5. Used *attrib.* to designate a stopping-place or assembly-point at a place intermediate between a base and a destination, as *staging-area, -point, -post.* orig. and chiefly *Mil.*

1945 *Amer. Speech* XX. 259 Other terms are used metaphorically: some which are particularly adaptable are..*ration allowance, staging area.* **1971** *Fremdsprachen* XV. 45 The states in the U.S. Coastal Zone contain most of the nation's population and industry, the gateway for maritime trade of about $40 billion, the staging area for the $500 million fish and crustacean industry. **1976** *New Yorker* 24 May 29/1 Almost all the American athletes will be 'processed' at a staging area on the premises of the State University College at Plattsburgh. **1955** *Times* 25 Aug. 9/2 Egypt is the vital staging point in air routes for the reinforcing of the Far East. **1969** H. Horwood *Newfoundland* i. 2 The point that I had reached had once been a meeting place for Indian bands, a summer staging-point. **1952** *Times* 21 Aug. 5/6 The American base in England is one of a number of strategic strongholds developed by the S.S.A.F. since the war. There are air bases, staging posts, and other installations in widely scattered places all over the world. **1959** *Listener* 1 Jan. 8/1 Explorers have met and drunk tea at the South Pole, as if it were just any other staging post. **1970** H. Trevelyan *Middle East in Revolution* 154 All that was left at the revolution was an R.A.F. staging post and technicians helping to train the Iraqi Forces.

‖ **stagione** (stadʒɪō·nĕ). *Opera* and *Ballet.* [It., lit. 'season'.] (See quot. 1978.) Freq. *attrib.* Also in *Comb.*, as ‖ **stagione lirica** [lit. 'lyrical season'], the opera season of an Italian theatre.

1946 E. Blom *Everyman's Dict. Music, Stagione*,..the term is used esp. for an opera season. **1963** *Times* 19 Feb. 15/3 Como's Teatro Sociale..has opened its *stagione lirica* for 1963 with a work by Verdi. *Ibid.* 20 Feb. 16/7 The *stagione* and *abonnement* system (necessitating unvaried programmes on a subscription throughout the season), while working well for opera, is hard on the ballet. **1970** *Daily Tel.* 12 Oct. 11/5 He..examined the problems of opera in English and the 'stagione' versus repertory systems. **1978** Ld. Drogheda *Double Harness* xxi. 289 The repertory system meant running a large number of different operas on consecutive nights, whereas *stagione* (literally 'season' in Italian) meant grouping a series of performances of one opera fairly closely together, performed with basically identical casts and conductor.

stagnant, *a.* Add: **2. a.** Also of (the ice of) a glacier or ice sheet.

1902 *Geol. Survey New Jersey* V. iii. 86 Such isolated bodies of ice doubtless preserved their motion..for a time. But when they became small, or when the local topography was unfavorable to motion, they became stagnant, and all the drift they held was let down on the surface as the ice melted. **1949** *Amer. Jrnl. Sci.* CCXLVII. 291 The lower 9 miles of this glacier are stagnant, but the upper part and most of the tributary ice streams are active; some receding and some advancing. **1973** R. J. Price *Glacial & Fluvioglacial Landforms* viii. 206 Examples of downwasting, stagnant valley glaciers are known to have existed in the Scandinavian mountains.. during the last of the Pleistocene glaciations.

stagnate, *v.* Add: **1. a.** Also of (the ice of) a glacier or ice sheet.

1924 *Bull. N.Y. State Mus.* No. 251. 159 If any general cause were to operate to deprive the whole glacier of a part of its pressure head, this part would be more likely to respond by stagnating. **1968** R. W. Fairbridge *Encycl. Geomorphol.* 1045/1 A variety of landforms have been interpreted as evidence that ice stagnated and melted over large areas during its retreat from various parts of Europe and North America.

stagnation. Add: **1. a.** Also of ice.

1929 *Geogr. Rev.* XIX. 256 (*heading*) The stagnation and dissipation of the last ice sheet. **1943** *Amer. Jrnl. Sci.* CCXLI. 97 When the ice finally disappeared, the gravel blanket..would be let down to form kames and hummocky gravel deposits such as have commonly been taken as indicators of general ice stagnation. **1973** R. J. Price *Glacial & Fluvioglacial Landforms* viii. 207 When stagnation of a valley glacier occurs, glacial erosion ceases.

2. Also *spec.* in *Econ.*, an absence or low rate of growth.

1938 A. H. Hansen *Full Recovery or Stagnation* xx. 319 It ought to appear incongruous..to follow a chapter on secular stagnation with one on inflation. *Ibid.*, Paradoxical though it be, the more we sink into deep stagnation with vast unemployment of labor and resources, the more imminent is the danger of inflation. **1965** J. L. Hanson *Dict. Econ. & Commerce* 362/2 *Stagnation thesis*, the belief that in advanced economies saving might be so great as to make the maintenance of full employment difficult. **1972** *Oxf. Univ. Gaz.* CII. Suppl. No. 7 p. 4 Whereas the adoption of the I.M.F. prescriptions had apparently led to stagnation in Argentina. **1974** M. B. Brown *Economics of Imperialism* ix. 224 Concentration..would be discouraged in periods of rapid economic growth and encouraged during stagnation or slump.

3. *Comb.* **stagnation point** *Aeronaut.*, a point on the leading edge of a moving aerofoil at which the air is at rest relative to the aerofoil.

1926 H. Glauert *Elements Aerofoil & Airscrew Theory* ii. 14 Consider first the pressure which occurs at a stagnation point, where the fluid is brought to rest at the nose of the body. **1955** *Sci. Amer.* Oct. 126/3 At what is called the 'stagnation' point, just in front of the model, the streamline splits in two, one half flowing around each side of the obstruction. **1979** Bertin & Smith *Aerodynamics for Engin.* vii. 269 We see that the temperature of the air at the stagnation point is sufficiently high that we could not use an aluminum structure.

stagnationist (stægnēɪ·ʃənist), *a.* and *sb.* Chiefly *Econ.* [f. Stagnation + -ist.] **A.** *adj.* Characterized by stagnation; promoting stagnation. **B.** *sb.* One who advocates or forecasts stagnation.

1951 A. Koestler *Age of Longing* v. 103 After some twenty million factory workers..had sent in resolutions calling for death to the 'stagnationist vermin', Edwards published another book. **1958** *Listener* 10 July 57/2 Mr. Nicholson..joins the ranks of the 'stagnationists'—those who would preserve forms when the functions have decayed. **1964** *Economist* 25 July 332/1 Consumption has moved into a stagnationist stage of the cycle. **1972** Hunt & Sherman *Economics* i. vi. 84 Marx..was not a 'stagnationist'—that is, he did not believe capitalism would suffer one long depression or that mass unemployment at high levels would last forever.

‖ **Stahlhelm** (ʃtā·lhelm). [Ger., lit. 'steel helmet'.] The Steel Helmet organization. Also *attrib.* Hence **Sta·hlhelmer,** a member of this organization. See *steel helmet.

1927 *Daily Tel.* 16 Aug. 10/6 The Stahlhelm's boast that it mustered 80,000 members was grossly exaggerated. **1927** *Times* 29 Nov. 15/5 The Stahlhelm organization of ex-soldiers. **1928** *Times* 4 June 13/2 It is the Stahlhelmers' boast that they embody the traditions of the old Army. **1930** [see *minority 3 b]. **1934** *New Republic* 18 July 249/2 The *Stahlhelm*, the Steel Helmets, is a voluntary organization of khaki-uniformed veterans..tending ..to be brought into semi-official relationship with the Nazi party. **1978** W. Fest *Dict. German Hist.* 150 *Stahlhelm*, association of ex-servicemen founded on 29 Dec. 1918... The Stahlhelm was anti-republican and from the late twenties it became militant in its demand for an authoritarian government... In 1934 it was converted into a 'National Socialist front-line fighters' union', but dissensions with the new Nazi members led to its dissolution in 1935.

stain, *sb.* Add: **7. stain painting,** a style of painting in which diluted acrylic paints are applied to unsized canvas; a painting executed in this style; hence **stain painter,** an exponent of this style; **stain-resistance,** resistance to staining; hence **stain-resistant** *a.*

1965 *New Statesman* 30 Apr. 693/2 Some of the hard-edge and stain painters are making matters worse by panicking themselves into the optical movement. **1965** Stain painting [see *hard-edge* s.v. *hard *a.* (*sb.*) 22]. **1974** *Globe & Mail* (Toronto) 23 Oct. 14/8 Her big canvases.. are stain paintings (staining is a technique using acrylics mixed with water, on unsized canvas). **1959** *Times* 12 Jan. 11/5 For use on wool, cotton, or synthetic fibres to improve oil, grease, and water stain-resistance. **1960** *Farmer & Stockbreeder* Suppl. 12 Jan. 3/3 Casual coat by Salbry is in waterproof, stain-resistant Norzon and has a fleecy wool lining.

stainierite (stēɪ·niərait). *Min.* [ad. Du. *stainierit* (V. Cuvelier 1929, in *Natuurwetensch. Tijdschr.* XI. 177), f. the name of Xavier *Stainier* (b. 1865), Belgian geologist: see -ite¹.] A hydrous oxide of cobalt which is usu. found as black needles forming microcrystalline crusts on cobalt ores, and is now regarded as the same as heterogenite.

1930 *Mineral. Abstr.* IV. 248 This is regarded as the crystalline equivalent of the colloidal heterogenite, and is named stainierite. **1941** *Ibid.* VIII. 86 X-ray powder photographs of heterogenite, stainierite, mindigite, and trieuite from Katanga..show identical patterns, and these minerals are to be regarded as varieties of the heterogenite group, differing in the relative amounts of Co_2O_3 and CuO. **1962** [see *heterogenite]. **1968** I. Kostov *Mineralogy* 228 The last mineral [*sc.* heterogenite] is also known as stainierite.

stainless, *a.* (*sb.*). Add: **A.** *adj.* **2.** Highly resistant to staining or corrosion. See also *stainless steel.

1897 *Sears, Roebuck Catal.* 244/1 Ladies' Fast Black Cotton Hose... Fast color and stainless. **1921** *Engineer* 11 Nov. 504/2 Stainless iron may be easily forged by hand into such difficult objects as spurs. **1932** *Discovery* May 145/1 Low carbon stainless steel, frequently called stainless iron, contains less than 0·12 per cent of carbon and from 11 to 12 per cent of chromium. **1945** *ABC of Cookery* (Ministry of Food) i. 8 Any cutlery which is not stainless should be kept well polished. **1960** *Good Housek. Cookery Bk.* (rev. ed.) 574/1 A sharp-pointed, stainless vegetable knife. **1979** T. Motogoshi in M. Nurse *Stainless Steel* 19/1 The use of stainless tubes has been boosted by the increase in the erection of high rise buildings.

B. *ellipt.* for *stainless steel; articles made of this.

1971 G. V. Higgins *Friends of Eddie Coyle* (1972) x. 64, I don't care whether it's the stainless or not... I got to have the stuff. **1975** *New Yorker* 26 May 111 (Advt.), Galax, a unique combination of solid brass and 18-8 stainless, is light-years ahead in function and design. **1977** C. McFadden *Serial* (1978) i. 8/2 They spent it rapidly on ..Dansk stainless and Rosenthal china.

stainless steel. [f. Stainless *a.* + Steel *sb.*¹] **1.** A chromium-steel alloy, usu. containing about 14 per cent of chromium when used for cutlery, etc., that does not rust or tarnish under oxidizing conditions because of the formation of a film of oxide on its surface. Freq. *attrib.*

1917 *Sci. Amer.* 31 Mar. 329/3 A steel that does not stain or tarnish is one of the latest new materials... It is called 'stainless steel'. **1920** *Glasgow Herald* 4 Aug. 9 Since the Armistice there has been an enormous sale of stainless steel for cutlery purposes. **1926** J. H. G. Monypenny *Stainless Iron & Steel* i. 6 Stainless steel was first introduced to the public in 1914 in the form of table cutlery. *Ibid.* 13 The discovery of the non-corrosive properties of stainless steel by Brearley, in 1913, was something entirely new in the history of chromium steels. **1932** *Discovery* May 145/1 There are three principal types of stainless steel. Low carbon stainless steel, frequently called stainless iron, contains less than 0·12 per cent of carbon and from 11 to 12 per cent of chromium... Cutlery steel, so called because it was first used for knives, contains 0·3 per cent of carbon and from 12 to 15 per cent of chromium... Austenitic stainless steel has the highest resistance of all to corrosion and is the most generally useful of the stainless steels. **1958** *Times Rev. Industry* Apr. 73/1 There were..many worthy examples of..the application of newer materials, especially stainless steel and aluminium, to domestic requirements. **1979** V. S. Naipaul *Bend in River* v. 83 The stainless steel jug..had only a stale-looking trickle of powdered milk.

2. *fig.* or in fig. context: chiefly used *attrib.* with allusion to the qualities of brightness, hardness, coldness, etc., associated with stainless steel.

1963 *Times* 10 June 6/5 Miss Astrid Varney's Isolde has the hallmark of true greatness upon it, but sometimes the voice takes on a stainless steel quality that would be more suited to the role of Ortrud. **1964** G. Lyall *Most Dangerous Game* xxiii. 192, I was..searching for the stainless-steel glint of a river to fix my position. **1974** *Publishers Weekly* 25 Mar. 6/2 She is..married to Sir Alfred Ayer, the celebrated English philosopher, whom she describes affectionately as 'the man with the stainless steel mind'. **1978** R. Doliner *On the Edge* (1979) ix. 142 His ego was gleaming hard, made of stainless steel.

stair, *sb.* Add: **1. d.** (Later examples.)

1928 T. S. Eliot *Song for Simeon* 2 They shall praise Thee and suffer in every generation..Light upon light, mounting the saints' stair. **1930** R. Campbell *Poems* 4 All the gifts my faith has brought Along the secret stair of thought.

5. a. *stair-carpet* (earlier example).

1817 M. Holyoke in G. F. Dow *Holyoke Diaries* (1911) 167 Began to mend Stair carpet.

b. stair dancer *slang*, a thief who steals from open buildings; cf. *dancer 6; **stairlift,** a device that can be built into a domestic staircase for the conveyance of disabled or infirm people up and down stairs; **stair-rod** (earlier example); also (in *pl.*) a proverbial comparison for heavy rainfall), **stair-step** *sb.* (earlier and later examples); also *fig.* and as *adj.*, resembling a stair-step; **stair-step** *v.*: also *intr.*, to resemble stair-steps; hence **stair-stepper, stairstepping** *ppl. a.*; **stair-tread** = Tread *sb.* 11; **stairwell,** the shaft containing a flight of stairs, a well (Well *sb.*¹ 8 a).

1958 *Times* 10 Feb. 4/5 'Stair dancer' is..the name given by the police to the thief who walks in and out of City offices, looking for something to steal. **1977** 'E. Crispin' *Glimpses of Moon* xii. 235 Since he was a stair dancer, a walk-in thief, judges had been inclined to be lenient until the last occasion, when his offence had been said..to have been aggravated by his having broken a window to 'effect an entrance'. **1977** *Hansard Commons* 24 Jan. 472 Stairlift and personal passenger vertical lifts for the disabled. **1980** *BSI News* Aug. 13/1 Stairlifts and homelifts are now extensively used in domestic situations, where they can be an invaluable aid to the disabled or infirm person. **1843** Dickens *Christmas Carol* iv. 130 The old man raked the fire together with an old stair-rod. **1963** *Times* 22 Apr. 4/6 During the morning the rain came down like stair-rods. During the match it turned to a swirling drizzle. **1977** D. MacKenzie *Raven & Ratcatcher* i. 14 The rain was falling in stair rods. **1794** N. Parry *Jrnl.* 26 June in *Kentucky Hist. Soc. Reg.* (1936) XXXIV. 386 Ky. hill on the south shore is exceeding bad, being long, steep, & broken with Limestone, somewhat resembling stair-steps. **1925** C. R. Cooper *Lions 'n' Tigers* iv. 76 This was the district of 'stair-steps', of thin, narrow-shouldered women, trailed by processions of children, five and six in a line. **1944** S. Putnam tr. *E. da Cunha's Rebellion in Backlands* ix. 415 This position.. occupied a broad stairstep on the slope of the hill between Mount Mario..and the Vasa-Barris. **1959** *Wall St. Jrnl.* (Eastern ed.) 12 Aug. 19/2 The 'stairstep' plan in use in Nashville calls for integration classes starting with the first grades and adding one grade each year. **1961** Webster, Stair-step *v. intr.* **1963** *Amer. Speech* XXXVIII. 203 Some of the English terms are literal equivalents of terms used by German-speaking skiers and might be called loan translations:..*stair step*, Treppenschritt. **1976** *Billings* (Montana) *Gaz.* 11 July 1-c/3 Yellow clover which carpets the banks smells sweet, grassy rims stairstep against the distant horizon. **1925** C. R. Cooper *Lions 'n' Tigers* iv. 77 Don't need many ladders aroun' this country... All they have t'do is line up the kids and walk on their heads. Ever see so many stair-steppers? **1972** *Nat. Geographic* Sept. 335 (*caption*) Stairstepping head-waters of the..river cascade out of a..bog. **1919** *Brit. Manufacturer* Nov. 34/1 Sections..such as are utilised for stair-treads, cornices, etc. **1931** Dougherty & Kearney *Fire* vii. 99 Stair wells and other shafts extending from the first floor to the roof. **1958** 'W. Henry' *Seven Men at Mimbres Springs* xiv. 164 Shortly, they heard his step on the stationhouse stairs, then saw his shadow rise out of the stairwell's greater blackness. **1977** B. Bainbridge *Injury Time* xiii. 109 Lights burned on the stairwells of the flats and along the deserted balconies.

staircase. Add: **1. a.** *spec.*, at Oxford and Cambridge, a college staircase and the rooms accessible from it; in *transf.* use, the people living in those rooms. *moving staircase*: see *moving *ppl. a.* 3.

1762 [in Dict.]. **1861** T. Hughes *Tom Brown at Oxford* I. i. 13 The rest is divided into staircases, on each of which are six or eight sets of rooms. **1914** C. Mackenzie *Sinister St.* II. iii. x. 699 'Alan, who are these mysterious creatures that come down for cocoa at ten?'.. 'They'd bore you rather... They're people who live on this staircase. I don't see them any other time.' **1974** J. I. M. Stewart *Gaudy* ii. 26 The corridor is not an Oxford institution. One lives on a staircase: commonly one set of rooms on either hand, storey by storey, from ground floor to attics. A hospitable man will give a party for the whole staircase. **1977** K. Benton *Red Hen Conspiracy* xvi. 130 He was one of my pupils... Leader of a very rowdy staircase.

d. *spirit* (also *bravery*) *of the staircase*, phrases rendering Fr. *esprit de l'escalier* (see *esprit 2 c). See also sense 4 below.

a **1906** J. Morley in H. W. & F. G. Fowler *King's English* (1906) i. 32, I thought afterwards, but it was *the spirit of the staircase*, what a pity it was that I did not stand at the door with a hat, saying, 'Give an obol to Belisarius.' **1906** [see *esprit 2 c]. **1976** L. Hellman *Scoundrel Time* 110 Ah, the bravery you tell yourself was possible when it's all over, the bravery of the staircase.

e. *Electronics.* A voltage that alters in equal steps to a maximum or minimum value.

1956 *Electronics* Feb. 192/2 A..staircase generator which generates a negative-going staircase is shown in Fig. 1. **1959** *Ibid.* 23 Jan. 36/3 Each pulse in the train causes one step of the staircase. **1965** *Wireless World* Sept. 425/1 This signal..consists of a 12·5 μs bar, a sine-squared pulse..and a five-step staircase.

4. (sense 1 d) *staircase afterthought, thought, wit*; **staircase generator** *Electronics*, a signal generator whose output is a staircase (sense 1 e above).

1964 *Guardian* 17 Jan. 11/6 The Senator, having written this last passage, decided not to deliver it, a staircase afterthought that will cost him dear. **1956** Staircase generator [see sense *1 e]. **1976** *Pract. Electronics* Oct. 812/2 The output from the staircase generator is then fed into one input of a comparator. **1958** J. Lodwick *Bid Soldiers Shoot* ii. vi. 210 'At what time did it start to snow that night, Lodwick?' He was referring..to the night of the parachutage... Staircase thought makes me wish that I had suggested that they have a look at my companion's boots, to which the virginal blanket fallen from Heaven might still..be clinging. **1920** A. Dobson in *National Rev.* July 654 Staircase-wit. If you fail to understand a joke within twenty-four hours, your symptoms indicate sluggish apprehension... This is what the French call '*L'esprit de l'escalier*'.

stair-foot. b. (Later example.)

1914 D. H. Lawrence *Widowing of Mrs. Holroyd* i. ii. 22 At that instant the stairfoot door opens slowly, revealing the children.

stairway. Add. **b.** *spec.* in *Geomorphol.*, a series of abrupt changes of level in the floor of a glaciated valley.

1904 *Jrnl. Geol.* XII. 570 The tread of the steps in the long stairway..greatly lengthened in down-canyon order. **1957** J. K. Charlesworth *Quaternary Era* I. xiii. 297 Cirques in almost all glaciated regions frequently occur in tiers, the 'tandem cirque' or 'cirque stairway'.., each step often with its rock-basin and tarn. **1974** [see *riser 7 b].

stake, *sb.*¹ Add: **1. e:** *to pull up stakes* (earlier and later examples); also, *to pick up stakes*; *to set stakes* = *to drive stakes*; *to tie* (*someone*) *to the stake*: see Tie *v.* 2 in Dict. and Suppl.

1703 S. Sewall *Diary* 15 Apr. (1879) II. 76 Went to my Bounds, asserted them,..then ordered Kibbe to pull up the Stakes. Told Mr. Lynde's Tenants what my Bounds were..; forwarn'd them of coming there to set any Stakes. **1817** J. K. Paulding *Lett. from South* I. 83 When they have exhausted one hunting-ground, [the Indians] pull up stakes, and incontinently march off to another. **1924** 'R. Daly' *Outpost* xvii. 165 I've sometimes thought of pulling up stakes and pushing further into the mountains. **1949** *Boston Globe* 15 May (Fiction Mag.) 6/2 We'll set our stakes, an' I'll slip down to Dawson an' record the claim. **1974** M. Allen *Super Tour* (1975) i. 23 'I'm assuming you're in a position to pick up stakes in a hurry.' 'As long as it will take to pack two bags.' **1980** *Dallas Times Herald* 10 May (Week End Suppl.) 6/1 The economic incentive that Europeans once had to pull up stakes and move to America.

2. c. *stake and rider* (later *attrib.* examples); *stake and ridered* adj. (earlier example).

1846 *Knickerbocker* XXVII. 208 Already the 'stake and ridered' fence was beginning to enclose the cleared land. **1884** G. W. Cable *Dr. Sevier* II. liv. 175 Again they followed him along a line of stake-and-rider fence. **1950** *Pennsylvania Dutchman* Jan. 3/3 He could do nothing better than to quickly place his gun behind him in a corner of a stake-and-rider fence.

5. c. Also *ellipt.* for *stake-body truck* (sense 7 below). *N. Amer. colloq.*

1968 *Globe & Mail* (Toronto) 17 Feb. 50/3 (Advt.), Immediate delivery on new pick-ups, panels, vans and stakes. *Ibid.*, Ford 2½ ton stake with covered compartment for carrying personnel. **1978** *Detroit Free Press* 16 Apr. F13/7 (Advt.), 1978 Chevy truck sale. Pickups. Elcaminos. Stakes. Stepvans.

d. (Examples.)

1903, 1910 [see *league *sb.*² 3*]. **1959** D. Wright *Baskets & Basketry* vi. 136 *Stakes*, rods driven in with the bottom sticks to form the foundation of the sides of a basket. **1964** H. Hodges *Artifacts* x. 146 Most baskets were made by first weaving a base, although solid wooden bases drilled to take the uprights, or stakes, were occasionally used.

6. (Further examples.) Also *Stake* (*in* or *of*) *Zion*.

1843 H. Caswall *Prophet of 19th Cent.* 90 Other 'churches' established by 'revelations' given to Smith, are called 'Stakes of Zion', or simply 'Stakes'. **1857** *Southern Illinoisian* (Shawneetown) 1 May 1/3 Throughout the States and Territories, at various and convenient locations, the Mormons have what are termed 'stakes in Zion', and each stake is governed by a Presidency. **1870** J. H. Beadle *Life in Utah* 124 All the wealthy members were to follow him to western Pennsylvania, and establish a new 'stake' for the others to gather to! **1905** *Out West* Sept. 246 The Stakes of Zion, I will explain, are those gathering places of the Saints that are outside of Zion proper—Jackson county, Missouri, where the holy city it is believed will yet be built. **1961** *Guardian* 23 Jan. 2/3 The share of the Manchester 'stake' (stake is roughly the Mormon for diocese) is £2 million. **1976** *Times* 18 June (Spec. Rep. Mormons) p. iii/7 A stake is administered by local lay members.

7. *stake-fence, -hole*; (sense 6) *stake centre, house, president*; **stake-boat** (earlier and later examples); also, a fixed boat to which other boats may be moored; **stake-body** *U.S.*, a body for a lorry, etc., which has an open, flat platform fitted with removable stakes (sense 5 c in Dict.) along the sides in order to retain the load; also *attrib.* in *stake-body truck*, a lorry fitted with such a body; **stake-net** (later examples); **stake-truck** = *stake-body truck* above.

1839 *Spirit of Times* 13 July 217/2 After a smart pull for it, she [*sc.* a boat] overhauled them one by one, passing the Washington about half way to the stake-boat. **1902** *Federal Reports* CXIII. 926 The tug left the tow in order to engage in other work, picking up light boats, and towing them down to a stake boat off Liberty Island. **1943** A. Gibbs *U-Boat Prisoner* 121 We went over to Jersey and took a motorboat to what he called a stake boat, where we found the barge. **1907** *Cycle & Auto. Trade Jrnl.* 1 Feb. 390 (*caption*) Studebaker 3½ ton paying load truck with stake body. **1913** *Hub* July 123 (*caption*) Universal Chain Drive Stake Body Truck. **1976** *Washington Post* 19 Apr. D3/5 (Advt.), Discount center for stake bodies, step vans, medium duty trucks, [etc.]. **1978** J. Gores *Gone, no Forwarding* (1979) 4 He..opened the door of his stake-body truck. **1976** *Times* 18 June (Spec. Rep. Mormons) p. iii/7 Twenty stakes are now functioning, most of them headquartered in new stake centre church buildings. **1882** W. D. Hay *Brighter Britain!* I. vii. 190 A stake-fence ought to be proof against both pigs and cattle. **1913** J. Masefield *Daffodil Fields* 46 Beside the stake-fence Lion stopped. **1977** Stake-hole [see *post-hole 1]. **1930** L. Foster *Larry* 131 Then we all paraded down to the Stake House (Mormon Districts are called 'Stakes'), where there was a pioneer's meeting. **1936** *Sun* (Baltimore) 17 Feb. 7/1 A number of Rock Hall fishermen walked from their homes over the ice in the Chesapeake Bay to their stake nets near Tolchester Beach the past week, cut out their nets, and landed three hundred pounds of rock fish. **1973** W. Elmer *Terminol. Fishing* ii. 73 The stake net is a fixed net now only found on the south-west coast of Scotland. **1947** G. S. Perry *Cities of Amer.* iii. 39 Each stake president will parcel out the acreages he has agreed to accept among his bishops, and each bishop will divide his commitment among the Saints in his ward. **1907** *Cycle & Auto. Trade Jrnl.* 1 Jan. 346a (*caption*) Five-ton Imperial Stake Truck. **1975** C. Weston *Susannah Screaming* xxvi. 137 An old stake truck passed him, the back jammed with long-haired college-age kids.

stake, *sb.*² Add: **1. c.** Hence *spec.*, a shareholding (*in* a company).

1955 *Times* 4 Aug. 12/1 Pilkington Brothers, famous the world over for plate and other kinds of glass, is proposing to increase its stake in another hardly less famous glass business. **1969** *Listener* 31 July 137/2 If the local paper has a stake in a local commercial station, this would tend to perpetuate the monopoly in local news and comment which has existed for too long in many provincial cities. **1981** *Times* 6 May 24/7 (*heading*) Trafalgar House buys 14·9 pc stake in French Kier.

3. b. *colloq.* Used *fig.* with defining words to denote a particular business or way of life in which success is attained through competition.

c **1885** A. W. Pinero *Magistrate* (1892) i. 24 You nominated yourself for the Matrimonial Stakes. Mr. Farringdon's The Widow, by Bereavement, out of Mourning, ten pounds extra. **1885** *Sat. Rev.* 7 Feb. 181/2 The hothouse kind of life..enabled this nervous, delicate, and curiously constituted competitor to win the Novel Stakes time after time. **1901** G. B. Shaw *Admirable Bashville* ii. i. 309 Yet so threadbare as to accept these consolation stakes. **1926** Galsworthy *Silver Spoon* ii. v. 145 He was not going to enter for the slander stakes. **1936** J. Curtis *Gilt Kid* iii. 34 Both the men looked as if they might be on the Jo Roncing [i.e. poncing] stakes. **1969** *Listener* 3 Apr. 470/1 No music is more recuperative than Mozart's and, in the therapy stakes, none runs it as close as Webern's. **1977** *Spare Rib* May 37/1 Energy and money were spent outdoing other girls in the beauty stakes.

4. (Earlier and later examples.) Also *N. Amer.*, a grub-stake; a sum of money earned or saved; a store of provisions or sum of money necessary for survival during a certain period.

1738 W. Byrd *Hist. Dividing Line* (1901) 178 [We] recommended to the men to manage this, their last stake, to best advantage. **1853** 'P. Paxton' *Yankee in Texas* 204 The horse is his last resource... When lost, the quondam owner is said to be flat broke or flat footed, and must beg, borrow, or steal, for a stake. **1873** J. H. Beadle *Undeveloped West* 510 It is a splendid country to travel through; a miserable poor one to stop in to make a 'stake'. **1899** 'J. Flynt' *Tramping* i. i. 20 It is usually immaterial to him what happens to society as such so long as he [*sc.* a thief] can make a 'stake'. **1931** 'D. Stiff' *Milk & Honey Route* xi. 117 A hobo may go to town with a stake and blow it in during two or three nights of slapping it up. **1946** E. O'Neill *Iceman Cometh* i. 53 I'll make my stake and get my new gamblin' house open before you boys leave. **1966** *Islander* (Victoria, B.C.) 21 Aug. 5/1 They [*sc.* loggers] seldom worked more than a few months in one place, just long enough to gather a stake, which they spent in a few days in town. **1978** J. Updike *Coup* (1979) vii. 279, I worked in that oil town in the Rift..and when I had a little stake I hitched back to Istiqlal.

5. *stake-race*; **stake-man** *U.S. slang*, a hobo, a tramp.

1899 'J. Flynt' *Tramping* ii. v. 310 He learns to travel merely for travel's sake, and develops into a 'stake-man', who only works long enough to get a 'stake' and then go off on a trip again. **1901** [see *bindle¹ a]. **1896** H. M. Blossom *Checkers* ii. 20 Y' see, take a big stake-race like this, where every horse is a 'cracker-jack',..and they've all got a chance. **1968** *Globe & Mail* (Toronto) 17 Feb. 1/5 Deep in the bush you can have just about all the comforts of home as the great uranium stake race enters its final two days.

stake, *v.*¹ Add: **1. a.** Also *N. Amer.*, to claim (land) by marking it with stakes; also *absol.*

1908 M. A. Grainger *Woodsmen of West* 78 Now Billy Hewlitt was a 'timber-cruiser'—a man who sought for forest timber, to stake it. **1916** *Yukon Territory* (Canada Dept. Interior) 12 In a short time Bonanza was staked from end to end. **1945** *Clarke County Democrat* (Grove Hill, Alabama) 24 May 1/6 The California Oil Company.. has staked a location for the drilling of a test well. **1959** M. Shand *Summit & Beyond* vii. 113 There was a report of a [gold] strike up White River. The men were talking of going to 'stake'. **1968** *Globe & Mail* (Toronto) 17 Feb. 1/7 It is land that was staked before in the madness of 1954, when it became known that one of the world's greatest concentrations of uranium ores had been uncovered.

c. Phr. *to stake* (*out*, † *off*) *a claim*, to make or register a claim (to land) by marking it with stakes; freq. *fig.* orig. *U.S.*

1851 *State Jrnl.* (San José, Calif.) 15 Mar. 2/1 It was estimated that ten thousand people were on the ground staking off 'claims'. **1876** [in Dict., sense 1 b]. **1904** J. London *Daughter of Snows* xiii. 140 You staked that claim before he was dry behind the ears. **1928** H. Crane *Let.* 17 Apr. (1965) 324 Skepticism may stop there.., but I am not exactly satisfied by that... I still stake some claims on the pertinence of the intuitions. **1939** Wodehouse *Uncle Fred in Springtime* xiv. 199 Up till now, he had regarded Lord Emsworth as the most promising claim that any prospector for ore could hope to stake out. **1949** *Nat. Hist.* Apr. 189/3 Sometimes when the bee hunter finds a nest that has not yet reached its peak of honey production, he will 'stake his claim' by marking the tree so that other hunters will know of his prior discovery.

8. *colloq.* (orig. *U.S.*). **a.** Usu. with adv. *out.* To maintain surveillance of (a place, etc.) in order to detect criminal activity or apprehend a suspect. Cf. *stake-out.

1942 Berrey & Van den Bark *Amer. Thes. Slang* § 499/4 *Stake out*,..to surround a criminal retreat to spy upon or prevent escape. **1943** R. Chandler *Lady in Lake* (1944) xxix. 157 They had the house staked. **1962** L. Deighton *Ipcress File* 221 When..the French police staked out the courier routes, they found..50,000 dollars of forged signed travellers' cheques. **1967** M. Procter *Exercise Hoodwink* xiii. 91 The house was 'staked out.'.. A man called Whipper Slade emerged. 'Coo, he's a real deadleg,' said a detective who recognised him. **1974** *Black World* June 28/1 Places that are so staked out with doormen and electronic gadgets that only god can enter the lobby. **1981** *Daily Tel.* 6 July 2/2 Police were tipped off that trouble might occur with skinheads at nearby Greenford. They staked out a likely disco there.

b. *to be staked out*: to be set, or to set oneself, to maintain surveillance of a place.

1951 M. Spillane *Big Kill* vi. 122 He's been a cop a long time. He's been staked out often enough to spot it when he's being watched himself. **1974** J. A. Michener *Centennial* iv. 162 The Pawnee reacted as had been expected, with a countercharge of their own, and their leaders had covered only a short distance when they spotted Lame Beaver staked out, his rifle at the ready. **1979** H. Kissinger *White House Years* xix. 756 David Bruce..came to the Embassy through the front door where the press was staked out.

stake, *v.*³ Add: **5.** *colloq.* (orig. *U.S.*). To furnish with money or supplies, etc.; to grub-stake. Also with compl. introduced by *with* or *to* indicating the commodity, etc., supplied.

1853 'P. Paxton' *Stray Yankee in Texas* 219 The jo-fired mean whelp wouldn't stake me. **1894** 'Mark Twain' *Pudd'n-head Wilson* iv. 58 Tom staked him with marbles to play 'keeps' with. **1917** G. B. McCutcheon *Green Fancy* 25 He staked her to a ticket to New York. **1934** R. Graves *I, Claudius* xxiii. 333 He gave me a purse of money and muttered in my ear: 'Tell nobody that I'm staking you, but put this on Scarlet.' **1942** Z. N. Hurston in *Amer. Mercury* LV. 88 If Jelly really had had some money, he might have staked him..to a hot. Good Southern cornbread with a piano on a platter. **1969** *Coast to Coast 1967–68* 138 They would stake him, buy his grub, supply him with horses and packs, pay him a hundred pounds for her scalp. **1978** M. Puzo *Fools Die* ii. 38 He felt their happiness for him, and to repay it he said, 'Now let me stake you guys, you too, Diane. Twenty grand apiece.'

staked, *ppl. a.* Add: *staked and ridered* = *stake-and-ridered* adj. s.v. Stake *sb.*¹ 2 c.

1901 W. N. Harben *Westerfelt* xx. 271 The scarecrow in the cornfield beyond the staked-and-ridered rail fence looked like the corpse of a human being flattened against the yellow sky. **1852,** etc. [see *ridered *a.* 2].

stakement (stēɪ·kmĕnt). *Hist.* [f. Stake *v.*¹ + -ment.] The entitlement of tenants whose rents are in arrears to have their eviction delayed.

1904 M. Bateson *Borough Customs* I. ii. 297 The 'stakement' appears to be the equivalent of the Germanic 'wiffatio'. **1927** J. S. Furley *Ancient Usages City of Winchester* 22 The *Usages* closes with an account of 'stakement'. When a tenant was in arrears with his rent the landlord could take no steps till after year and day. *Ibid.*, Stakement is symbolic... At Exeter the custom was to carry away a stone..but the stake was used at Reading, at Fordwich, and the other Cinque Ports. **1927** *Contemp. Rev.* Aug. 261 The procedure of 'Stakement', a court process that delayed adverse possession, is very early and full of interest. **1935** K. M. E. Murray *Constitutional Hist. Cinque Ports* ii. 16 Such is the practice of 'stakement' for rent arrears.

sta·ke-out. *colloq.* (orig. *U.S.*). Also as one word. [f. vbl. phr. *to stake out*: see *stake *v.*¹ 8.] An act or period of surveillance of a place by police or investigative agents. Also *transf.* and *attrib.*

1942 Berrey & Van den Bark *Amer. Thes. Slang* § 499/2 *Stake-out*, a surrounding of a criminal retreat to spy upon or prevent the escape of criminals. **1943** R. Chandler *Lady in Lake* (1944) xl. 208 Somebody stood behind that green curtain..as silently as only a cop on a stake-out knows how to stand. **1955** *Sun* (Baltimore) (B ed.) 11 Jan. 26/8 Bergen was arrested..by..rookie policemen who were part of an extensive stakeout in the area which had been set up to trap the person or persons responsible for the large number of burglaries. **1960** *Washington Post* 15 Jan. A12 The success of the police in these instances has been largely due to the 'stake-out' system. **1966** J. Gardner *Amber Nine* viii. 95 He had enjoyed that time as stake-out man at London Central: lurking, watching (and watching for) people. **1972** *New Yorker* 16 Jan. 27/1 A group of agents..who were conducting a stakeout of a house in suburban Wilmette. **1979** *Daily Tel.* 19 Sept 36/5 Father Frederick Linale.. said police asked him last week if they could use the mission on a 'stake out'. **1983** *Times* 29 Jan. 6/8 President Reagan's impromptu visit to a Boston Irish pub for a quick beer broke up a stake-out by armed FBI agents.

staker¹. Add: **c.** *Canad.* One who stakes a (mining) claim.

1898 *Yukon Midnight Sun* (Dawson, Yukon Territory) 15 Aug. 1/3 He also said that until after September 1st no one but the original staker has any business upon these claims. **1921** *Daily Colonist* (Victoria, B.C.) 13 Oct. 2/4 The original stakers are reported to be taking out $35 a day in coarse gold, some very large nuggets being found. **1954** *North Star* (Yellowknife, N.W. Territory) Aug. 2/1 Though a few stakers were 'made', it was generally the brokers and speculators who skimmed off the cream.

staker². (Later examples.)

1975 J. Symons *Three Pipe Problem* vii. 45 Willie was a great staker of his reputation. **1976** *Sunday Times* (Lagos) 7 Nov. 21/3 (Advt.), Pools Magnet have won over ten million Naira for Stakers all over the world.

stakey (stēɪ·ki), *a.* *slang* (chiefly *Canad.*). Also **staky.** [f. Stake *sb.*² + -y¹.] Well provided with money; 'flush'.

1919 *Camp Worker* (Vancouver) 28 June 7/1 If they hold a job for a month they have done something out of the ordinary, and as a consequence, they are never very staky, and the question of funds comes first in most cases. **1927** *Amer. Speech* II. 392/1 When a *staky* worker comes to town, his giving money to his impecunious acquaintances is called the *pay-off*. **1960** *Weekend Mag.* (Montreal) 8 Oct. 28/1, I wasn't going to go in, because at the time I wasn't stakey. **1970** P. St. Pierre *Chilcotin Holiday* 31 On being assured that the prisoner was stakey, the mayor imposed a ten-dollar fine. **1973** B. Broadfoot *Ten Lost Years* xv. 172 Why, we was making 15 cents a glass... Both of us were getting stakey as hell.

Stakhanovite (stăkā·novəit), *sb.* and *a.* [f. the name of the Soviet coal-miner Alekséĭ Grigór'evich *Stakhánov* (1906–77) + -ite¹; cf. Russ. *stakhánovets* sb., *stakhánovskiĭ* adj.

The Soviet authorities publicized the prodigious output achieved by Stakhanov in 1935 as part of a campaign to increase industrial output.]

A. *sb.* In the U.S.S.R. during the 1930s and 1940s, a worker whose productivity exceeded the norms and who thus earned special privileges and rewards; *transf.*, one who is exceptionally hard-working and productive. **B.** *adj.* Designating, pertaining to, or characteristic of such a worker or such workers collectively.

1935 *Time* 16 Dec. 25/3 In the coal mine at Stalino two assistant foremen, a checkweighter and an electrician were arrested for the murder of a fast-working Stakhanovite. **1936** W. Citrine *I search for Truth in Russia* 349 In the best building works the Stakhanovite builders have shown examples of high productivity of labour in bricklaying, beton work, plastering and excavating work. **1938** *Times Lit. Suppl.* 8 Jan. 18/4 The *udarnik* (or Stakhanovite, as he now is called) was 'curiously like the pep-it-up-team-work-factory-spirit fellows I knew at home, and equally detested by his clock-watching fellow-workers'. **1949** [see *shock *sb.*² 7 b]. **1952** [see *norm c]. **1959** *Times* 29 May 8/5 This conferring with East Anglian trade union leaders and stakhanovites of his party..has probably been the most useful thing to come out of his [*sc.* Hugh Gaitskell's] tour. **1961** C. T. Hsia *Hist. Mod. Chinese Fiction* xviii. 495 The application of Stakhanovite methods to literary production could only mean further deterioration in quality. **1977** *Time* 1 Aug. 5/3 Though U.S. workers have been regularly chided at home for goofing off on the job, they are veritable Stakhanovites compared with some of their European counterparts.

Hence **Stakha·novism,** a movement in the U.S.S.R. aimed at encouraging hard work and maximum output, following the example of Stakhanov; also *transf.*; **Stakha·novist** *a.* and *sb.* = *Stakhanovite *sb.* and *a.*

1936 V. M. Molotov (*title*) What is Stakhanovism? **1937** *Nature* 27 Feb. 364/2 He outlined the development of the principles of scientific management from the pioneer work..to such recent manifestations as 'Stakhanovism' in Russia. **1938** *Downside Rev.* LVI. 370 Before summing up the authors insert a valuable chapter on the Stakhanovist movement in Russia. **1940** *Manch. Guardian Weekly* 16 Feb. 130 News of what in the Soviet Union is called 'Socialist emulation' confirms again and again that certain 'Stakhanovists' accomplish 'norms' (their own normal production) of 260 to 400 per cent of normal. **1949** E. Fitzgerald tr. *Labin's Stalin's Russia* ii. 53 Stakhanovism (with all its serious industrial accidents) was introduced in Novosibirsk. **1954** *Encounter* Feb. 39/2 The tense, driving Stakhanovist atmosphere of Soviet Europe. **1957** *Times Lit. Suppl.* 25 Oct. 641/2 The romantic legend by which democracy lives, on the other hand, venerates Stakhanovism in high places. **1970** G. Greer *Female Eunuch* 123 Many others [*sc.* working wives] pride themselves on the way they manage to run a home and hold their own in a job at the same time, accepting the patronizing title of 'working wonders' in a kind of unofficial Stakhanovism. *Ibid.* 275 Women's literature is full of the trumpeting of female Stakhanovists.

staking, *vbl. sb.* Add: **a.** Also *N. Amer.*, the action of *stake *v.*¹ 1 a.

1952 *North Star* (Yellowknife, N.W. Territory) Nov. 3/2 There, the mineral claims map sheet is completely filled, so far as staking is concerned. **1968** *Globe & Mail* (Toronto) 17 Feb. 1/7 At noon on Monday, 100,000 acres of land beyond the geological formation that gave Elliot Lake 13 uranium mines in the 1950s, will be opened to staking. **1979** *Arizona Daily Star* 5 Aug. (Advt. Section) 4/3 You will be responsible for the acquisition of uranium mineral properties required through staking, bids,..joint ventures and/or purchase.

d. staking rush *Canad.*, a rush (Rush *sb.*² 4) to stake (mining) claims.

1953 *North Star* (Yellowknife, N.W. Territory) July 1/1 There are possibilities of an additional poll in the Marion River area, where a staking rush is now under way. **1964** *North* (Ottawa) May–June 14 A miniature 'staking rush' to the Pine Point area took place during the following winter.

staky, var. *stakey *a.*

stalactite. Add: **3.** (Earlier and later examples.)

1851 W. Irving *Alhambra* (rev. ed.) 76 The lofty ceiling was originally of the same favorite material, with the usual frostwork and pensile ornaments or stalactites. **1931** C. Hill *Moorish Towns in Spain* 147 We stand amid a small labyrinth of columns, under a roof of stalactites in stone. **1974** *Encycl. Brit. Micropædia* IX. 517/1 A peculiar type of faceted, crystal-shaped stalactite is found in Turkey; this form became the most common Turkish capital decoration.

4. (Earlier example in sense 3.)

1855 J. Fergusson *Illustr. Handbk. Archit.* I. ix. v. 463 Instead of the simple curves of the dome, the roofs are made up of honeycombed or stalactite patterns, which look more like natural rock-work than the forms of an art.

Stalag (stā·læg, ‖ ʃtā·lag). [a. Ger., abbrev. of *stammlager* main camp.] In Nazi Germany: a prison-camp primarily for captured enemy private soldiers and non-commissioned officers. **Stalag Luft, Stalagluft** (-luft) [G. *luft* air], such a camp for Air Force personnel.

1940 *Times* 30 July 7/3 There are three types of camps for British prisoners known officially as Oflag, Stalag, and Dulag (contractions for Offizierslager, Stammlager, and Durchgangslager). **1941** [see *Ilag]. **1944** V. G. Garvin tr. *Gary's Forest of Anger* xxix. 141 The old man was sent to a Stalag in Poland. **1945** [see *Oflag]. **1947** *News Chron.* 24 Jan. 1/4 Scharpwinkel, who is believed to have ordered the murder of many of the 50 R.A.F., Commonwealth and Allied airmen after their escape from Stalagluft III in March 1944. **1959** W. Faulkner *Mansion* xiii. 295 If it had been a..book instead of a war..they would have escaped. But he..never knew anyone who ever actually escaped from a genuine authentic stalag, so they had to wait for regular routine liberation. **1974** *Times* 18 Oct. 16/6 On at least first sight Long Kesh *looks* like one of the Stalags. **1978** R. V. Jones *Most Secret War* xxxi. 265 While in Stalag Luft 3 he not only concealed his knowledge of our new radar devices, but built a radio transmitter.

stalagmometer. Add: (Further examples.) Hence **sta:lagmome·tric** *a.*

1910 *Chem. Abstr.* IV. 1765 (*heading*) The significance of the stalagmometric method. **1920** *Jrnl. Biol. Chem.* XLIV. 378 The amount of lipase in blood serum, determined by..the stalagmometric method. **1940** Glasstone *Text-bk. Physical Chem.* vii. 482 The weight of a definite number of drops is determined, or the number obtained from a given volume of liquid, as it flows between two marks on a stalagmometer, is counted. **1976** *Coll. Czech. Chem. Commun.* XLI. 1845 The dependence between the mass of the detached drop and the rate of its formation was determined by means of three stalagmometers with different sizes of the dripping area and with a different angle of the capillary edge.

stale (stēɪl), *sb.*⁷ *colloq.* [Absol. use of Stale *a.*¹] A stale cake or loaf of bread, etc.

1874 Hardy *Madding Crowd* II. iii. 39, I went to Riggs's batty-cake shop, and asked 'em for a penneth of the cheapest and nicest stales, that were all but blue-mouldy, but not quite. **1937** DeArmond & Graf *Route Sales Managem.* 4 The man who sells and delivers bread to the grocer must remove the stales each day. **1937** 'G. Orwell' *Road to Wigan Pier* i. 15 Frayed-looking sweet-cakes..bought as 'stales' from the baker.

stale, *a.*¹ Add: **3. c.** *Comm.* That has remained inactive for a considerable time; (of a cheque) out-of-date.

1889 Barrère & Leland *Dict. Slang* II. 297/2 *Stale bear*,..a man who has sold stock which he does not possess, and has not bought it back. A bear who has been short of stock for a considerable period... *Stale bull*, a man who has held stock for a long period without profit. **1901** C. Duguid *How to read Money Article* viii. 37 The time comes when the 'bull campaign' turns into a 'stale bull account', that is, when the bulls are anxious to sell, even at a loss. **1901** *Business Terms & Phrases* (ed. 2) 199 *Stale cheque*,..a cheque which has remained unpaid for some considerable time. **1930** M. Clark *Home Trade* 271 Stale bulls are those who come to the conclusion that they have waited long enough for a rise in price and who, therefore, sell out. **1939** F. Lee *City Page* iii. 61 Eventually every crossed cheque has to be paid into a banker's account, usually within a time limit of six months; otherwise the cheque is 'stale'. **1957** Clark & Gottfried *Dict. Business & Finance* (1967) 332/2 *Stale*,..in business, out of date, or outstanding for a long time. **1979** F. E. Perry *Dict. Banking* 11/2 A banker receiving a cheque antedated by six months or more for payment would regard it as 'stale'.

6. Also *stale-smelling* adj.; (in sense 3 c above) *stale-dated* adj.

1963 J. N. Harris *Weird World Wes Beattie* (1964) xv. 179 This check came along... It was stale-dated, see? I mean it had been drawn the previous September, and this was about May. **1936** E. Wilson *Travels in Two Democracies* 286 They carried me into a stale-smelling building. **1973** T. Pynchon *Gravity's Rainbow* i. 24 The city around them at once a big desolate ice-box, stale-smelling.

stately, *adv.* Delete *rare* and add later examples.

1920 [see *class-jealousy* s.v. *class *sb.* 9]. **1938** G. Greene *Brighton Rock* vii. i. 279 She smelt faintly, stately, of Californian Poppy. **1957** 'M. M. Kaye' *Shadow of Moon* xxi. 326 The drawing-room smelt stately of cigar smoke.

stalemated, *ppl. a.* Add: Also *fig.*

1952 D. MacArthur *Reminiscences* (1964) x. 409 The terrible blood tribute exacted by this type of stalemated attrition. **1965** *Economist* 9 Jan. 104/1 The United States accepted a stalemated substitute for victory in Korea. **1977** *Time* 31 Jan. 7/3 The Vice President also wants to know just what allied or U.S. initiatives Europeans would welcome to get the stalemated talks..going once again.

Stalin (stā·lin). The name of Joseph *Stalin* (see *Stalinism) used *attrib.* to designate things instituted by him or developed during his leadership of the Soviet Union, as *Stalin Line, Prize, tank* (see quots.); **Stalin organ** *Mil. slang*, a type of Soviet multi-barrelled mobile rocket launcher.

1942 *Foreign Affairs* XX. 317 The Stalin Line, with Kiev as a strong base behind it, runs along the Dniester and then through very easily defensible country in the direction of Zhitomir and Korosten. 1942 *Nature* 25 Apr. 475/1 The following awards of Stalin Prizes for outstanding scientific work in 1941 have been made. 1945 *Times* 1 Jan. 4/1 The Russians are using an entirely new type of tank... It is the super-heavy Stalin tank and has the biggest tank gun in the world. 1953 *Sun* (Baltimore) 16 May (B° ed.) 1/1 Eastern Germany, under Russian directions, is to help build a vast fortified 'Stalin line' defense belt across Eastern Europe, including V-2 rocket bases and bomber fields. 1954 Koestler *Invisible Writing* ii. 23 Those who survived have..become Stalinite prize court jesters. 1955 *Britannica Bk. of Year* 490/1 *Stalin organ*, a device for firing multiple rockets. 1977 *Time* 21 Feb. 21/1 Sakharov..won the Stalin Prize and was thrice awarded the country's highest civilian medal. 1982 *Daily Tel.* 5 Aug. 28/6 There was the characteristic, terrifying noise of 'Stalin organs' being fired—40 rockets going off within 10 seconds.

Stalinesque (stāline·sk), *a.* [f. as prec. + -ESQUE.] Of, pertaining to, or characteristic of Joseph Stalin, his policies, activities, etc.; Stalinist.

1943 L. Adamic *My Native Land* 129 When I quoted to him a rumor that he was engaging in Stalinesque hold-ups to procure money for revolutionary purposes, he did not deny it. 1979 *Times* 14 Nov. 12/4 Stalin rebuilt the city [sc. Minsk] in Stalinesque style: a grandiose central avenue with a trade union palace.

staling (stēⁱ·liŋ), *ppl. a. Bot.* [f. Stale *v.*² + -ING².] Of fungal products: diverting or inhibiting fungal growth.

1916 A. H. Graves in *Mem. N.Y. Bot. Gard.* VI. 326 The problem was now to obtain..this substance, which we may for convenience call the 'staling substance', free from mycelium. 1948 *Nature* Mar. 422/1 A similar effect is obtained when the metabolism of the latter [sc. a second parasite of a plant] leads to the formation of substances ('staling substances') which by and by inhibit growth; hence the formation of lesions of limited size. 1979 *Experientia* XXXV. 200/2 This may possibly be due to reduction of growth of both interacting colonies caused by diffusing staling products.

Stalinism (stā·liniz'm). [f. Joseph *Stalin* (Russ. Iosif *Stálin*), the assumed name of Iosif Vissariónovich Dzhugashvíli (1879–1953), leader of the Soviet Communist Party and head of state of the Soviet Union + -ISM.] The policies pursued by Stalin, based on but later deviating from Leninism, esp. the formation of a centralized, totalitarian, objectivist government.

1927 *Daily Tel.* 21 Nov. 10/3 A violent denunciation of 'Stalinism' and its 'terrorising of the party'. 1941 Koestler *Scum of Earth* 23 We had realised that Stalinism had soiled and compromised the Socialist Utopia. 1947 C. Malamuth tr. *L. Trotski's Stalin* 422 On the theoretical plane every bit of 'Stalinism' has issued from the criticism of the theory of permanent revolution as it was formulated in 1905. 1955 *Times* 5 May 15/4 Indirectly the tenets of Stalinism and Trotskyism are being subjected in new tests at the conference tables of international diplomacy. 1958 *Times Lit. Suppl.* 7 Mar. 123/1 The Yugoslav system has relapsed into totalitarianism after a promising breakaway from Stalinism. 1968 *Russian Review* XXVII. 309 A formal definition of Stalinism would run something like this: a one-man dictatorship in which a single dictator ruling arbitrarily, uncontrolled by any party organs, is the sole interpreter of the Marxist-Leninist dogma, and is surrounded by the cult of his personality. A revised definition..would be the same as above, but the single personality clause would be replaced by the CC Politburo collective leadership. 1971 I. Deutscher *Marxism in our Time* (1972) 86 in the end the nonviolent meaning of Marxism was suppressed under the massive, crushing weight of Stalinism. 1977 *Time* 21 Mar. 12/2 In a bitter statement, Gui accused the Communists of practicing 'Stalinism', calling himself the victim of the chamber's 'will for my political execution'.

Stalinist (stā·linist), *sb.* and *a.* [f. as prec. + -IST.] **A.** *sb.* A follower or supporter of Stalin or his policies. **B.** *adj.* Of, pertaining to, or characteristic of Stalin, his followers, or his policies.

1928 *Observer* 22 Jan. 14/5 Open calculations measured in advance by the Stalinists. 1930 *Times* 27 Mar. 13/3 The Stalinist group of Communist leaders. 1939 G. Greene *Lawless Roads* iii. 107 The papers were full of a Stalinist plot. 1941 Koestler *Scum of Earth* 113 The sectarian hatred between Stalinists, Trotskists, and Reformists still existed. 1955 *Times* 3 Aug. 9/6 Khrushchev, the reputed Stalinist, goes on a mission to China. 1974 tr. *Wertheim's Evolution & Revolution* i. 70 In Stalinist Russia..there was no sign whatever that the Russian way was considered to be a deviation from the usual succession of phases. 1977 *Socialist Press* 2 Mar. 11/4 Their series of Conferences..were 'bureaucratically controlled' as was that called by the Stalinist-led BLMC Combine Committee. 1978 I. B. Singer *Shosha* xiii. 229 The designation..as Fascist lackeys and agents of Hitler evoked protests even from sworn Stalinists.

Stalinite (stā·linait), *sb.* and *a. rare.* [f. as prec. + -ITE¹.] = *Stalinist *sb.* and *a.*

1927 *Daily Tel.* 6 Dec. 11 The struggle between the Trotskists and the Stalinites. 1938 R. H. S. Crossman in *New Statesman* 7 May 780/2 Unlike the Stalinites, he has learnt nothing from the experience of Fascism, and so can greet it as a positive advance towards the millennium. 1945 Koestler *Yogi & Commissar* III. i. 134 The *New Statesman and Nation's* interpretations of Stalinite policy display all the ingeniousness of the official Apologist. 1973 *Times* 19 Jan. 4/5 (*heading*) 'Almost Stalinite' to deny effective power to local party bosses.

Stalinize (stā·linəiz), *v.* [f. as prec. + -IZE.] *trans.* To transform, etc., in accordance with Stalin's policies and practices. Hence **Stalinization.**

1949 I. Deutscher *Stalin* 366 Even the spoken language became 'Stalinised' to a fantastic extent. 1954 Koestler *Invisible Writing* x. 120 The natives were drawn into the towns, educated, Russified and Stalinised. 1956 *Washington Post* 19 Nov. 42/1 Even today, after the unspeakable horror of the blood bath in Hungary, the betting is still somewhat against a 're-Stalinization'. 1959 *Manch. Guardian* 5 Aug. 10/3 Returning to Hungary..in 1944 to take part in the 'Stalinisation' of his country. 1971 I. Deutscher *Marxism in our Time* (1972) vi. 124 This was the first dangerous attack on the autonomy of the [Polish] Communist Party, the first act, as it turned out, of 'Stalinization'. 1979 *Dædalus* Winter 135 Mutual contamination or interpenetration between the Communist party (which had not been Stalinized in the first place...) and Italian society. 1981 *Guardian* 16 Feb. 5/4 Mr Gerald Kaufman..said he did not regard 'the Stalinisation of the Labour Party as democracy'.

Stalinoid (stā·linoid), *a.* (and *sb.*). [f. as prec. + -OID.] Resembling or having some characteristics of Stalinism; loosely Stalinist. Also as *sb.*

1940 *Amer. Mercury* Apr. 498/2 Years of careful reading of the Stalinoid literature of hallucination. 1947 *Sun* (Baltimore) 4 Nov. 10/3 Food prices have come down—to the consternation of the Stalinoid cohorts. 1952 *Manch. Guardian Weekly* 16 Oct. 7/3 It is Marxist, hostile to Stalinoids and to modern Trotskyites. 1952 P. Selznick *Organizational Weapon* vii. 297 The term 'stalinoid' is usually employed as a rough synonym for 'fellow traveler'. 1966 *Guardian* 26 May 6/5 The Stalinist intellectuals and the Stalinoid liberals. 1977 *Daily Tel.* 12 Oct. 18 The Nobel Peace Prize has gone to some curious candidates in its time, including Le Duc Tho, the villainous Stalinoid secretary of the North Vietnam Communist party.

stalk, *sb.*¹ Add: **4. d.** *coarse slang.* A penis, esp. one that is erect.

1597 Shaks. *Lover's Compl.* (1609) l. 147 My woeful self..Threw my affections in his charmed power, Reserv'd the stalke and gave him al my flower. 1608 — *Per.* iv. vi. 46 *Bawd*. Heere comes that which growes to the stalke, neuer pluckt yet I can assure you. Is shee not a faire creature? 1939 Joyce *Finnegans Wake* (1964) ii. 236 Just so stylled with the nattes are their flowerheads now and each of all has a lovestalk unto herself. 1961 Partridge *Dict. Slang Suppl.* 1203/2 *Stalk*,..an erection. 1976 A. White *Long Silence* iv. 37, I had a stalk on me long as my arm. 1978 J. Updike *Coup* (1979) ii. 79 My stalk verged upon erection. 1979 W. Styron *Sophie's Choice* ii. 45 She..prepared to take between those lips unkissed by my own the bone-rigid stalk of my passion.

5. c. *colloq.* A lever mounted on the floor or on the steering column of a motor vehicle, which controls the gears or such devices as horn, indicators, lights, windscreen wipers, etc.

1964 *Road & Track* Jan. 22/2 The gears are selected by an odd curved floor stalk which looks the passenger seat but they go in all right. 1973 *Country Life* 15 June 1577/2 Steering-column stalks look after the windscreen washer and wiper. 1977 *Daily Tel.* 19 Jan. 12/6, I liked the cloth-trimmed seats and the control stalks on the steering column in place of the old rocker switches on the fascia.

10. stalk switch, a switch in the form of a stalk or lever mounted on the steering column of a motor vehicle (see sense 5 e above).

1976 *Evening Post* (Nottingham) 25 Dec. 10/8 Stalk switches, within easy reach of fingers on the wheel, control the lights, wipers, wash and flasher, also horn and direction indicators.

stalking-horse. 2. a. Delete ? *Obs.* and add later examples.

1965 *Times* 12 Jan. 6/2 This meant that the Europeans would regard us as the stalking horse or paid hand of Uncle Sam and would not wish us to participate fully in European affairs. 1977 J. M. Harrison in *Bond & McLeod Newslett. to Newspapers* iii. 208 Zenger was actually a stalking horse for the group of wealthy politicians who owned the *New York Journal*. 1980 *Jewish Chron.* 15 Feb. 1/1 It raises the fear that the Irish may be acting as a stalking horse for the whole European Economic Community.

stall, *sb.*¹ Add: **3. c.** A parking space for a motor vehicle, usu. marked out but not partitioned off. *U.S.*

1940 *College Topics* (Univ. of Virginia) 4 Nov., These stalls will be painted to facilitate parking and a time limit of one hour has been ordered for both sides of the street in this area between 8 a.m. and 6 p.m. 1955 J. H. Schmitz in Aldiss & Harrison *Decade the 1950s* (1976) 17 Cord hurriedly flew the skipboat round the station and rolled it back into its stall. 1976 C. Weston *Rouse Demon* (1977) xxiii. 110 Her car was in its stall in the subterranean garage.

d. One of a series of urinals separated by divisions, in a men's lavatory; also, a compartment in a wash-room. Also *urinal stall, toilet stall.*

1967 *Gloss. Sanitation Terms (B.S.I.)* 58 *Stall urinal*, a urinal having a back curved on plan to form a stall for the user... When stall urinals are fixed in ranges, divisions or cloaking pieces are provided between each stall. 1969 C. Logue *New Numbers* (1971) 62 Mechanical faucets drench a line of porcelain stalls. 1977 P. D. James *Death of Expert Witness* ii. vii. 90 The male washroom, apart from the urinal stalls, differed very little from the women's. 1978 R. Ludlum *Holcroft Covenant* xxiii. 265 If they let the weapon through, he was to reassemble it immediately, in the toilet stall of a men's room.

5. d. *transf.* pl. Those who occupy the stalls in a theatre.

1901 G. B. Shaw *Three Plays for Puritans* Pref. p. viii, English indiennce on the theatre, as far as the stalls are concerned, does not exist. 1920 *Daily Mail* 27 Sept. 4/5 'I wonder whether we shall ever get our 'stalls' back,' a West End box-office manager remarked to me; the 'stalls' in the front-of-the-house vernacular signifying a particular class of playgoer. 1927 *Sunday Express* 10 Apr. 5/4 'Why should the stalls stand to oblige the pit?' asked a satellite near me.

12*. [f. *Stall *v.*² 9 d, e.] **a.** *Aeronaut.* The condition of an aircraft when the streamline flow over its wings breaks down, usu. owing to a low air speed or a high angle of attack; the sudden loss of lift (and height) associated with this.

1918 J. M. Grider *War Birds* (1927) 88 He went straight up three hundred feet and stalled and fell out of the stall right into the middle of the field. 1927 *Glasgow Herald* 31 Aug. 10 There is only one issue for the airman near the ground—a spin and a crash. 1928 [see *lever v.*² 1]. 1966 M. Woodhouse *Tree Frog* xxv. 191 He couldn't slow down to my speed without..stalling and nobody..would risk a stall this close to the ground. 1976 W. Garatoi *Gossover* 204 The big jet fell ten thousand feet. ..that stall would have turned into a spin..with a less-experienced captain.

b. The sudden stopping of an internal-combustion engine at low revolutions (see also quot. 1959).

1959 *Motor Manual* (ed. 36) iv. 83 The very simple torque converter..would work well only at one speed. It could, for instance, be designed to give quite high multiplication at 'stall' (the moment when the car is on the point of moving). 1973 R. Rosenblum *Mushroom Cave* (1974) 3 The [boat's] motor was finicky; tying off the cord had precipitated a coughing fit in the carburetor, followed by a stall.

13. a. (sense 12*) *stall warning* (usu. *attrib.*).

1958 *Chambers's Techn. Dict.* Add. 1016/1 *Stall-warning indicator*, a device fitted to aeroplanes which do not provide positive warning of the approach of a stall by buffeting. 1976 B. Lecomber *Dead Weight* ii. 32 The stall-warning light blazed urgently as I tried to haul the shuddering nose up.

13. b. stall-holder, (*c*) one who occupies a seat in the stalls of a theatre, concert-hall, etc.; **stall-keeper,** (*c*) (earlier example); **stall plate:** also, a similar plate bearing the arms of a knight of another Order (earlier and later examples); **stall seat,** a seat in the stalls of a theatre; **stall shower,** a shower-bath enclosed in a cubicle; **stall turn** *Aeronaut.*, a turn achieved by stalling one wing of an aircraft, causing increased drag on that wing and reduction of the radius of the turn; hence **stall-turn** *v. intr.*

1849 *Theatr. Programme* 34/2 The Committee have the honor to announce the following Stall Holders:—The Duchess of Leeds, [etc.]. 1965 *Times* 10 Jan. 4/1 Last night's performance in the Albert Hall allowed stall-holders, at least, to hear more of that detail than ever before. 1842 *Ainsworth's Mag.* II. 157 Went to keep a stall at the fancy fair. Threw all the other stall-keepers into the shade. 1841 N. H. Nicolas *Hist. Order of Bath* 206 A copy of the Inscriptions on some of the Stall Plates of Knights of the Bath, will be found in the appendix. 1867 J. Brooke-Little *Royal Ceremonies of State* vii. 102/1 Henry VII's magnificent chapel at the East end of Westminster Abbey..makes a splendid setting for the installation of Knights Grand Cross who, like Garter Knights display their banners and have stall plates. 1920 *Daily Mail* 17 Sept. 4/5 Before the war approximately 90 per cent. of the occupants of stall seats in a West End theatre of any repute were in evening dress. 1939 R. Chandler in *Dime Detective Mag.* Aug. 65/1 A glass stall shower, monogrammed towels on a rack. 1978 S. Sheldon *Bloodline* xvii. 202 She walked through a tiled bathroom that included a marble bathtub and a stall shower. 1963 *Beaver* (Winnipeg) Summer 25/2 The plane went into a stall turn. 1942 *R.A.F. Jrnl.* 3 Oct. 30 How insecure the safety belt seemed when called upon to do stall turns in the back seat of a Hart. 1949 *Times* 9 Feb. 2/3 The aircraft..climbed steeply, stall-turned, and..burst into flames. 1963 A. V. Bramble *Air Plane Flight* 105 The 'stall turn' is a useful manoeuvre for changing to a reciprocal course (heading altered through 180°) in less time and space than by normal turning.

stall, *sb.*³ Add: **4. a.** More recently, without *off*.

1926 *Swell's Night Guide* 41 They are never at a loss for a stall. 1931 W. Faulkner *Sanctuary* xvii. 156 If it was a stall, dont common sense tell you I'd have invented a better one? 1939 E. S. Gardner *D. A. draws Circle* (1940) vii. 98 'Sometimes when he'd be working, I'd take meals up to him. I think that was just a stall.' 'You mean the meals were for someone else?' 'Yes.'

b. An act of stalling (for time) or prevarication. Cf. *Stall *v.*² 3 a. *colloq.*

1945 *Sun* (Baltimore) 21 Nov. 1/1 The 200 delegates termed the company reply 'a stall pure and simple'. 1963 J. N. Harris *Weird World Wes Beattie* (1964) iv. 49 It was a very good stall, if you know Edgar... So he left the whole matter in abeyance. 1977 D. E. Westlake *Nobody's Perfect* (1978) 141 Tell him to give me a while to get the cash together... This isn't a stall... I do have the money.

stall, *v.*¹ **III. 9.** Restrict † *Obs.* to senses in Dict. and add: **c.** Of a draught animal: To come to a halt because of mud or other impediment.

1807 C. W. Janson *Stranger in Amer.* 172 The last time he passed, his horses *stalled*, that is, they were for some time unable to drag the wagon through the worst places. 1857 W. Chandless *Visit to Salt Lake* II. vi. 233 His teams were none too strong, and twice he 'stalled' hopelessly, and had to send to the nearest farm for a yoke of cattle.

d. Of an aircraft or its pilot: to enter a stall.

1910 R. Loraine *Jrnl.* Apr. in W. Loraine *Robert Loraine* (1938) vi. 100 The machine leapt higher, so did my heart, higher still—then—puff!—I came to earth, having stalled and crashed. 1917 D. F. Pitt. 1951 An aeroplane can only reduce its flying speed to a certain minimum, after which it will stall. 1931 *Statesman* (Calcutta) 5 Dec., It is claimed that the autogiro cannot 'stall', or lose flying-speed. 1958 D. Piggott *Gliding* vii. 94 The actual speed at which the glider stalls will be raised when the glider is being turned or accelerated..or if a heavier load is being carried. 1975 L. J. Clancy *Aerodynamics* v. 98 If a particular wing is such that it stalls too suddenly, it may be necessary to provide some artificial pre-stall warning device.

e. Of an engine or motor vehicle: to stop suddenly as if of its own accord. Also with the driver or the occupants of the vehicle as subj.

1914 R. & E. Shackleton *Four on Tour in Eng.* 204 A few miles beyond Chipping Norton we stalled near the foot of a hill—and found that it was because of an inexcusable forgetting of gasoline! 1932 *Birmingham Post* 17 Dec. 16/2 The men drove off in the van. A few minutes later the engine stalled. 1962 'C. Blackstock' *Dewey Death* ix. 207, I share..a car with a friend... She once stalled in the middle of Piccadilly Circus. 1973 R. Rosenblum *Mushroom Cave* (1974) 5 He is go'd off the throttle, his engine died, the boat's motor stalled.

f. To trifle or linger *around* (also *along*); to 'hang about'. *U.S. colloq.*

1916 H. M. Bower 'Phantom Herd' i. 5 I've been stalling along and keeping the best of the bucks in the foreground. *Ibid.* xii. 194, I stalled around out there till my money gave out. 1976 P. G. Winslow *Witch Hill Murder* (1977) ii. xv. 207, I hoped he might answer sort of friendly..and I've been kind of stalling around.

g. *transf.* and *fig.*

1932 E. D. Paine *Comrades of Rolling Ocean* ii. 22 When things happened too fast, his mind stalled on a dead centre. 1953 N. Tinbergen *Herring Gull's World* i. 11 It stalls, makes a sharp turn and dives down. 1971 C. Bonington *Annapurna South Face* xvi. 199, I immediately noticed the lack of oxygen; once again my progress stalled into a crawl with rests at almost every step.

11. b. Also *fig.*, of an assembly, plan, etc.: to be held up. Chiefly *U.S.*

1910 *Outlook* 2 July 473 Congress would have been stalled in its efforts to prepare certain legislation without the leaders. 1953 *Times* 31 Oct. 3/6 General Thimayya said that 'explanations', which have been stalled for the past fortnight, were to be resumed to-morrow. 1970 G. F. Newman *Sir, you Bastard* v. 143 Both their requests to make phone calls were stalled. 1978 *N.Y. Times* 29 Mar. a 8/3 the seventh session of the..conference was stalled at its beginning here today over the question of who would preside.

c. Add *def.*: To cause (an aircraft, vehicle, engine, etc.) to stall.

1904 W. Wright in M. McFarland *Papers Wilbur & Orville Wright* (1953) I. 441 We allowed the machine to turn up a little too much and it stalled it. 1918 Bookman & Shark *Putnam's Automobile Handbk.* xiv. 159 Stalling the motor is the result of feeding too little gas, or..the accelerator. 1930 J. Dos Passos *42nd Parallel* iv. 266 Being careful to get out to crank the car as he had stalled the motor. 1947 F. S. Holliday *Driving Test Fully Explained* iii. 10 A timely check of gears will prevent you from stalling the engine. 1965 A. Christie *Autobiog.* (1977) vii. iii. 331, I stalled the engine once or twice..and I was rather chary about passing things. 1973 *Daily Tel.* 11 July 21/5 There was no structural failure in the Russian TU-144 supersonic airliner until after the pilot had stalled it.

12. c. To weary or tire; to fatigue. Usu. in *pass.* Chiefly *Sc.* and *north dial.*

1816, etc. in *Eng. Dial. Dict.* s.v. Stall v. 19. 1948 I. Brown *No Idle Words* 108 Stalled, applied to depressed human beings, is a good old usage. 'You look stalled' is 'you're bored'..

stall, *v.*² Add: **1. b.** *intr.* To screen a pickpocket's operation; to act as a look-out during a robbery or burglary.

1846 in Dict., sense 1]. 1882 *Sydney Slang Dict.* 9/2, I pinched a swell of a fawney and fenced it for a double finnip and a cooter. My jomer stalled. 1926 J. Black *You can't Win* xxi. 338 Coppers located 'work' for burglars and stalled for them while they worked.

3. a. *trans.* To put (someone) *off* for the time being. Now usu. without advb.

1829 P. Egan *Boxiana* 2nd Ser. II. 345 He would not be '*stalled off*' by the most knowing of the knowing. 1930 *Sat. Even. Post* 26 July 26/1 We might be able to stall them for two or three days with the idea that Tony is in Washington tryin' to fix the rap against them. 1948 A. Hynd *Pinkerton Case Bk.* 56 He kept stalling the woman off with one excuse or another. 1963 J. N. Harris *Weird World Wes Beattie* (1964) iii. 36 So I stalled him off. I said I couldn't remember meeting anyone at Mac's. 1977 J. Crosby *Company of Friends* xx. 128 Elaine is in Paris. To bargain some more. It's the only way left to stall him. It's important to stall him.

b. *intr.* To prevaricate; to be evasive. Also, to play for time or temporize. Freq. in U.S. colloq. phr. *quit stalling* (usu. imp.). orig. *U.S.*

1903 A. H. Lewis *Boss* 23 [If] Big Kennedy shows up to stall ag'inst you, why I should say (etc.). 1932 W. Faulkner *Light in August* ix. 202 'Quit stalling,' the stranger said. 'If you crooked the guy, say so.' 1934 D. Runyon in *Collier's* 24 Nov. 52/2 All she can think of..is to stall for time. 1953 J. Hilton *Time & Time Again* iii. 220 He was stalling for time. 1959 F. Hobson *Death on Back-Bench* viii. 101 Just quit stalling, and come clean. 1969 *Listener* 9 Jan. 41/2 Suppose Mr Ransom suddenly seized Miss Cold and flung her down on the office table with a snarl of 'Quit stalling,' like one of her favorite film actors. 1980 S. Saipaul *Black & White* ii. 34 For more than a year the Guyanese courts had been stalling on the custody suit his parents had brought.

stalled, *ppl. a.* Add: **4. b.** *Archæol.* *stalled cairn*: on the Orkneys, a Neolithic cairn covering a burial-chamber which was divided into lateral cells by stone slabs projecting from the wall.

1934 J. G. Callander in *Proc. Soc. Antiq. Scotl.* LXVIII. 320 (*heading*) A long stalled chambered cairn or mausoleum (Rousay type) near Midhowe, Rousay, Orkney. 1937 C. S. T. Calder in *Ibid.* LXXI. 117 (*caption*) Stalled cairn, Calf of Eday: plan and sections. 1954 S. Piggott *Neolithic Cultures Brit. Isles* viii. 234 The *Camster* tombs..are..from the northern shores of the Moray Firth..and there occur on the Orkneys, which include the ultimate form the long 'Stalled Cairn' type ...are mostly concentrated in the islands of Rousay and Eday. 1963 *Field Archæol.* (Ordnance Survey) (ed. 4) 31 Other peculiar features in Orkney are the 'stalled' cairns in which what is really a long and carefully-constructed cist is divided into as many as ten or more compartments by upright stones. 1973 C. Renfrew and his team found on the island of Rousay with its remarkable number of stalled cairns.

5. *a.* Of an aircraft or aerofoil: in an air-flow that has ceased to be streamline. Also applied to flight in this condition.

1912 O. Wright in M. McFarland *Papers Wilbur & Orville Wright* (1953) II. 1050 There is only one explanation I can give and that is the one that you suggest, that the machine was 'stalled'. 1932 *Jrnl. R. Aeronaut. Soc.* XXXVI. 312 Two sensations are characteristic of stalled flying: first of all, a very noticeable decrease in the effect of the elevator control and secondly the reversed effect of elevator control on the inclination of the gliding path. 1966 *McGraw-Hill Encycl. Sci. & Technol.* XIII. 228/2 For some wings, if the root section is placed at an angle near stall, the tip sections will probably be stalled. 1969 W. Thomson *Threat for Flight* 8 If the angle of attack is made too large..flow becomes turbulent and the blade is said to be stalled.

c. Of a vehicle or engine: stopped through having stalled.

1966 B. H. Deal *Fancy's Knell* (1967) x. 153 A man in the middle stood up..the stalled traffic..in the middle of the street. 1977 'D. Rutherford' *Return Load* i. 13 Sally restarted the stalled engine, engaged first gear and tentatively let in the clutch.

staller². Add: **1.** (Recent example without *up*.)

1977 *New Society* 7 July 6/2 At first he was a staller. 'You get a woman trying to get on a train, someone goes in front of her, stalls her for ten to 15 seconds, and she's forgot about her bag.'

2. One who stalls or prevaricates; a person who gains time by obstructionism.

1937 Partridge *Dict. Slang* 823/1 *Staller*, a person constantly, or very good at, making excuses or playing for time. 1977 G. Greene *FSO-*1 ix. 76 You are..one of the world's most transparent stallers..are convened to talk about Larry. 1981 'W. Haggard' *Money Men* vii. 74 He was an experienced staller. Postponing action was called good judgement.

sta·ll-in. *U.S.* [f. Stall *v.*¹ 11 c + *-IN³.] A form of protest in which participants block the roads with immobilized vehicles.

1964 [see lie-in s.v. *Lie *sb.*¹ 6]. 1977 *Time* 14 Nov. 21/3 They need hardly fear the kind of traffic stall-ins staged by New York City residents to protest the arrival of the Concorde.

stalling, *vbl. sb.* Add: **5.** The event of coming to an unintended halt or stalling (*Stall *v.*¹ 9). *b. Aeronaut.* freq. *attrib.*, as *stalling point, speed.*

1808 M. L. Weems *Let.* 17 May in E. E. F. Skeel *M. L. Weems* (1929) II. 377 Stalling of Waggons, sweeping away of Stages, drowning of Horses &c. &c. are dreadful. 1888 J. Kirkland *McVeys* 220 The occurrence—the 'stalling' of a wagon and team, was common enough in those early days. 1912 O. Wright in M. McFarland *Papers Wilbur & Orville Wright* (1953) II. 1052 The liability of the machine to dive in case of 'stalling' is present in every one. 1916 H. Barker *Aeroplane Speaks* 89 If it[*sc.* the engine] is throttled down, then the course must be one of a steeper angle than B, or there will be danger of stalling. 1917 'Contact' *Airman's Outings* iv. 100 At times he varied this method by lifting the machine almost to stalling point, letting her down again, and repeating the process. 1928 C. F. S. Gamble *North Sea Air Station* v. 84 An inherently stable machine was.. found to be of little value for aerial fighting, as the possession of a high degree of manœuvrability, combined with a low 'stalling' speed, are some of the essentials for this work. 1932 H. H. Price *Perception* vi. 149 A short-eared owl flying at just above stalling-point. 1952 L. Nathan *Car Driving in Two Weeks* iii. 21 To prevent stalling of engine the following procedure is invaluable. 1966 D. Francis *Flying Finish* 129 We touched down.. at a fraction above stalling speed. 1976 'A. Hall' *Kobra Manifesto* i. 15 He reached the stalling-point and dropped tail first and bounced and tilted and..then bucked forward.

6. *Surfing.* (See quots.)

1962 T. Masters *Surfing made Easy* 65 *Stalling a board*, stepping or leaning back on a board to slow it down. 1968 W. Warwick *Surfriding in N.Z.* 10/2 Step back on your board and put it out of trim, this will slow your board down... This manoeuvre is called stalling.

stalling (stǭ·liŋ), *vbl. sb.*² [f. Stall *v.*² + -ING¹.] **1.** The action of helping a pickpocket by distracting or jostling his victim. *Criminals' slang.*

1908 J. M. Sullivan *Criminal Slang* 24 *Stalling for a dip*, arranging [a pickpocket's] victims so that they can be successfully robbed. 1926 *Flynn's* 16 Jan. 638/1 The gay cat and spotters got 'em on location; then it was a case of palin', stick up, stallin' or rollin' in any way you please.

2. Prevaricating, temporizing; the action of being evasive or devious.

1927 *Vanity Fair* XXIX. 132/4 'The run-around' is stalling or failing to keep a promise. 1952 *Landfall* Sept. 227 Walk into a Saturday-afternoon bar and hear the noise; do you get the impression of *stalling*?

stallion. Add: **1. a.** Also *fig.*

1924 R. Campbell *Flaming Terrapin* ii. 27 He would hear the whinnying stallions of the wind career. 1940 L. MacNeice *Last Ditch* 15 The stallions of the soul—Eager to take the fences That fence about my soul.

2. b. Delete † *Obs.* and add: Now only in former sense.

1933 D. Parker *After Such Pleasures* 138 Go answer it, you damned—you damned *stallion*! 1978 L. Meynell *Papersnake* v. 70 Barton amused himself by keeping a tally of Lasting's women; 'that insatiable stallion' he called him.

3. b. Among U.S. Blacks, a tall, good-looking girl or woman. *colloq.*

1970 C. Major *Dict. Afro-Amer. Slang* 108 *Stallion*, a good-looking black woman. 1975 R. H. Rimmer *Premar Experiments* (1976) i. 148, I love you Samantha Brown. In black ghetto language, you're a lovely stallion.

Stamford (stæ·mfǝ̈ɹd). The name of a town in Cambridgeshire, used *attrib.*, as **Stamford ware** *Archæol.*, a kind of Saxo-Norman lead-glazed pottery made of estuarine clay from the vicinity of Stamford.

1956 G. C. Dunning in D. B. Harden *Dark-Age Britain* III. 230 The term 'Stamford ware' may be proposed for the lead-glazed pottery of fine quality of the late Anglo-Saxon and Norman periods, provided it is clearly understood that this is a generic term for pottery made of.. specific clays but at more than one centre within the area. 1956 *Proc. Cambridge Antiquarian Soc.* XLIX. 48 Saxo-Norman glaze only occurs on this Stamford ware. 1962 [see *lead-glazing* vbl. sb. s.v. *Lead *sb.*¹ 12]. 1974 M. Ingate *Sound of Weir* xx. 172 Stamford ware made from 900 to the 13th century from middle Jurassic esturine clay.

staminoid (stæ·minoid), *a. Bot.* [f. L. *stāmin-*, Stamen + -OID.] Of the nature of or resembling a stamen.

1869 M. T. Masters *Vegetable Teratology* 301 The scales that are met with in some plants, either as excrescences from the petals, or as imperfect representatives of stamens or other organs, are occasionally staminoid. 1930 *Jrnl. Genetics* XXIII. 107 A collection..in which the petals had become metamorphosed into stamens, showed every type of staminoid petal. 1974 *Flora* CLXIII. 405 (*heading*) Staminoid petals in *Geranium pratense* L. and their inheritance.

‖ **Stammbaum** (ʃta·mbaum). *Linguistics.* [Ger., family tree: the sense was introduced by A. Schleicher in *Darwinische Theorie u. die Sprachwissenschaft* (1863) 13.] A family tree of languages: see *family-tree* s.v. *Family *sb.* 11. Hence **Sta·mmbaumtheorie** (-te:ori) (see quot. 1954).

1939 E. Prokosch *Compar. Gmc. Gram.* 21 August Schleicher conceived the Indo-European primitive language as the trunk of a linguistic 'Stammbaum'. 1954 M. A. Pei *Dict. Linguistics* 162 *Pedigree theory*, the theory (*Stammbaumtheorie*) formulated by August Schleicher in 1866, according to which a parent language split into two branches, each of which again bifurcated into two languages, etc. 1965 *Language* XLI. 106 Rodríguez stresses ..the ultimate reconcilability of the Stammbaumtheorie and the Wellentheorie. 1971 W. Labov in W. O. Dingwall *Survey of Linguistic Science* 426 One of the classic unresolved dichotomies of historical linguistics is the opposition of the *Stammbaum* and *wave theories* of linguistic differentiation.

‖ **Stammtisch** (ʃta·mtiʃ). [Ger., f. *stamm* tree trunk, cadre + *tisch* table.] A table reserved for regular customers in a German restaurant, beer-hall, etc.

1938 L. Bemelmans *Life Class* (1939) II. vii. 197 Political discourses at his *Stammtisch*, the table regularly reserved for him and his group of friends, at the Löwenbräu. 1940 G. Frankau *Self-Portrait* xi. 62 We lunched, with others of the same ilk, at his 'Stammtisch' (perpetually reserved table) in the Hamburg Rathskeller. 1964 I. Fleming *You only live Twice* iv. 54 The quiet corner table that appeared to be his *Stammtisch*. 1970 L. Deighton *Bomber* xxi. 312 The three best tables were marked with *Stammtisch* flags, so that only regulars would dare to sit there.

stamp, *sb.*³ Add: **II. 9. b.** Maize that has been crushed or pounded with a wooden pestle. *S. Afr.* Cf. *stamp mealies*, sense 20 below, and Samp.

1923 *S. Afr. Pioneer* Dec. 143/2 All partook freely of the feast of meat and stamp. 1976 J. McClure *Rogue Eagle* vi. 112 The price of mealie *stamp* in Maseru.

III. 14. e. = *insurance stamp* s.v. *Insurance 5.

1912 [see *insurance stamp* s.v. *Insurance 5]. 1946 *Act* 9 & 10 *Geo. VI* c. 67. 720 Contributions..are payable by means of adhesive stamps. 1974 *Times* 6 Feb. 14/2 The qualification test has been simplified for the emergency: a declaration that 26 stamps have been paid within the previous 12 months is enough to entitle people to the full rate for a year.

f. = *trading stamp* s.v. *Trading *vbl. sb.* b.

1933 in *Parl. Papers 1932–33, Rep. Cttee on Gift Coupons & Trading Stamps* 12 (Cmd. 4385) XII. 387 The stamps are given to the customer in proportion to the amount spent and are stuck by him into a collecting book... When the book is full..the stamps may be tendered in exchange for a gift. 1963 J. T. Story *Something for Nothing* iii. 89 You get a grocer, you get a baker, you get a hairdresser and a chemist and a garage and a draper all giving stamps. 1976 A. Grey *Bulgarian Exclusive* I. i. 17 The two psychedelic gift mugs the garage.. had given them in exchange for seven Heron stamps.

IV. 19. *stamp-licker; stamp-licking* vbl. sb. and ppl. adj. (freq. with reference to menial office work).

1928 F. Le Gros Clark *Apparition* xiii. 176 You've never even held a commission. Bloody stamp-licker in an office. 1978 J. Updike *Coup* (1979) vii. 269 In the bureaucracy of Kush Amid so many posts for stamp-lickers and boot-lickers. 1913 *Punch* 14 May 382/3 There is something after all to be said for the Stamp-licking Act. 1973 G. Talbot *Ten Seconds from Now* (1974) ii. 22 At first the job was stamp-licking office boy on the commercial side. 1979 *Nature* 4 Jan. 7/1 A total paid staff of two people who do everything from typing, stamp-licking and driving to..producing scholarly catalogues.

20. stamp collection, a philatelist's collection of postage stamps; also *fig.*; **stamp machine,** (*b*) a vending machine which supplies postage stamps; **stamp mealies** *S. Afr.* [ad. Afrikaans *stampmielies*] = sense 9 b above; cf. *Stamped *ppl. a.* 1 b; **stamp war,** competition amongst retailers to attract custom by providing the best trading-stamp offer; an instance of this.

[1865 *Stamp-Collector's Mag.* 1 Jan. 2/1 When we first saw a postage-stamp collection, more than ten years past, it contained about a hundred and fifty specimens.] 1884 *Stamp Collectors' Jrnl.* 15 Jan. 19/1 The value of a stamp collection does not depend entirely upon the amount of money expended for the album and the stamps. 1926 J. S. Huxley *Ess. Pop. Sci.* 164 This *corpus* of fact..is only a vast stamp-collection, no more than a lumber-room, unless each generation in its turn will make it live. 1978 S. Sheldon *Bloodline* iv. 62 He was basically a retiring man, content to make a modest living, reside in a little apartment in Passy and tend to his small stamp collection. 1944 J. D. Carr *Till Death do us Part* xviii. 191 There's no stamp machine at the post office... Anyone who wants stamps must buy 'em..over the counter. 1969 R. Thomas *Singapore Wink* xxvi. 250 Trippet and I went in search of a stamp machine. We fed dimes and nickels and quarters into it until we had almost three dollars' worth. He helped me lick them. 1952 L. Green *Lords of Last Frontier* 79 We now live well and keep strong on stamp mealies from Oorlog's place. 1963 *Daily Tel.* 14 Oct. 1/4 (*heading*) Stamp war challenge to Garfield Weston. 1972 *Guardian* 16 Oct. 9/5 In the early sixties the stamp war broke out. Different supermarket chains started offering different stamps, each one claiming to give better value and better gifts.

stamp and go. Add: **1.** Also, a shanty sung to accompany this action.

1929 F. Bowen *Sea Slang* 132 Stamp and Go. The shanty sung for a straight pull along the deck. 'What shall we do with the drunken sailor?' is probably the best known.

2. (Usu. with hyphens.) In the West Indies: a simple, quickly prepared codfish fritter (see also quot. 1893).

1893 C. Sullivan *Jamaica Cookery Bk.* 87 Stamp and go. These are rough cakes made with cornmeal and flour ..salt fish and a little butter... The country people as they travel stop at the way-side shops and buy them... Hence the name. 1953 *Caribbean Q.* III. i. 11 *Stamp-and-go*,..a kind of codfish fritter that is quickly made. 1970 *New Yorker* 10 Jan. 54/3 (Advt.), A chief who can water your mouth with a..native Stamp-and-Go.

stamped, *ppl. a.* Add: **1. b.** *stamped mealies* = *Stamp *sb.*³ 9 b. *S. Afr.*

1911 A. B. Lamont *Rural Reader for S. Afr.* 189 'Stamped mealies'—that is mealies from which the outer husk has been removed—are made by stamping the grain with a heavy wooden stick in a large bowl. 1937 C. R. Prance *Tante Rebella's Saga* 82 Oom Fanie and his family had just asked a blessing on their supper of pumpkin and stamped mealies.

2. c. Also *stamped* (*and*) *addressed envelope*: a self-addressed envelope with a postage stamp affixed, enclosed with a letter so that the recipient may reply at the sender's expense. Freq. required by an organization of a private enquirer and often abbrev. *S.A.E., s.a.e.* (cf. *S 4 a).

In this sense the word may now be considered f. Stamp *sb.*³ + -ED².

1873 *Young Englishwoman* Apr. 208/1 We..will return your edging if you will send a stamped addressed envelope. 1949 N. Mitford *Love in Cold Climate* I. xiii. 131 The fella says here to enclose a stamped and addressed envelope but I don't think I shall pander to him. *a* 1953 Dylan Thomas *Under Milk Wood* (1954) 39 Here's a letter for you with stamped and addressed envelope enclosed. 1977 *Graduate* 9 Dec. 14/2 (Advt.), Enquiries and applications, with stamped addressed envelope, to Mr P Austin, DoE Archaeological Services, Carlisle Castle, Carlisle, Cumbria. 1981 M. Spark *Loitering with Intent* ii. 52 An unpublished poem..which..had been rejected eight times, returning to roost in my own stamped addressed envelope.

stampede, *sb.* For **a, b** read **1, 2** and add: Also † **stampado, stampido. 1. a.** *α.* (Earlier example.)

[1826 T. Flint *Francis Berrian* I. ii. 46 Instantly this prodigious multitude..took what the Spanish call the 'stompado'. With a trampling like the noise of thunder,.. they [*sc.* the horses] took to their heels.] 1828 in *Missouri Hist. Rev.* (1914) VIII. 187 A little before daylight, the mules made an abortive attempt to raise a stampido.

b. In N. Amer., an exhibition of cowboy skills, a rodeo; *spec.* that held at Calgary, Alberta (usu. *Calgary Stampede*), for the first time in 1912 and annually since 1919.

1912 *Calgary* (Alta.) *Daily Herald* 31 Aug. 6/1 Calgary is on the eve of its Stampede festival. 1919 *Eye Opener* (Calgary, Alta.) 9 Aug. 4 Come to Calgary Stampede Week and have the time of your life. 1923 C. M. Barbeau *Indian Days* 5 Picturesque stampedes take place every summer in the July celebrations at Banff. 1948 *Ada* (Okla.) *Even. News* 2 July 1/5 A capacity crowd was on hand for the opening performance of the Hereford Heaven Stampede. 1950 B. Hutchison *Fraser* xvii. 251 Here the ranchers and Indians gather once a year for the innocent fun of the stampede. 1974 *Sat. Rev. World* 2 Nov. 30/2 The Calgary Stampede during the first two weeks of each July..offer[s] competition in matches ranging from wild-cow milking to buffalo riding.

2. a. Also *spec.* (N. Amer. Hist.), a concerted rush of prospectors to the goldfields.

1872 R. W. Raymond *Statistics of Mines* III. iv. 202 Rocky Bar..has suffered somewhat from the stampede to the bars of the Snake River. 1916 *Yukon Territory* (Canada Dept. Interior) 11 In the autumn of 1886 coarse gold was discovered in the Fortymile River, and ..the usual stampede occurred. 1937 C. L. Andrews *Pioneers & Nuggets of Verse they Panned* 17 The stampede to the gold fields of the Tanana Valley..caused an exodus from Dawson. 1965 *Canad. Geogr. Jrnl.* Apr. 119/1 Not only was it the last of the old-fashioned stampedes in which dog teams and men vied for space along the narrow trail, but it also ushered in the air age of prospecting.

b. (With initial capital.) An uproarious kind of dance. Also *Stampede Dance* (in quot., a dancing-party).

1856 *Spirit of Times* 13 Dec. 238/2 The following was the programme of dancing: Part the Fourth—Scotch Reel,..French Four, General Stampede. 1870 J. C. Duval *Adventures Big-Foot Wallace* xlii. 263, I see you haven't yet introduced the Texas national dance—the Stampede. 1950 *Chicago Daily News* 10 May 10/1 The annual 'Stampede Dance' of the Order of the Builders, State of Illinois, will be held May 20.

stampede, *v.* Add: Also † **stompede. 1. a.** *β.* (Earlier examples.)

1838 *Hesperian* Nov. 37/2 When we awoke, we found that the flies had *stompeded* our horses, to use the expression of the country, which means that they made them so restive that they broke loose from the hopples. 1844 J. Gregg *Commerce of Prairies* II. 35 A party of Mexicans..*stampeded* and carried away, not only their own horses, but those of the Texans.

b. Also, to cause (an individual) to take precipitate action.

1890 C. King *Sunset Pass* 56 Don't get stampeded. Just keep cool; watch and listen. 1912 R. Pocock *Man in Open* 104 The lady attracted attention by screaming, so the third shot stampeded poor Jones. 1924 *Machinists' Bull.* (Winnipeg) Oct. 3/2 Efforts are being made by various agencies to use the present condition as a club to stampede the men and disgust them with their Organization. 1950 *Time* 3 Apr. 20/2 A solid, grey, calm man, never rushed to a conclusion, impossible to stampede.

2. a. (Earlier example.)

1823 S. Williams in E. C. Barker *Austin Papers* (1924) I. 699 On the way..the Cavallada Stampeded and a part of the horses and mules were not recovered.

b. Also *spec.* of a prospector: to rush to the goldfields.

1877 R. W. Raymond *Statistics of Mines* VIII. v. 263 Among the miners who had 'stampeded' to Cedar were many of the best prospectors in the Territory. 1898 M. Landreville *Appeal of Yukon Miners* 23 Miners are prone to stampede to any district which has the appearance of greater richness than the one wherein they are at work. 1951 V. B. Angier *At Home in Woods* 41 The prospectors who stampeded through here around '98 on their way to the Yukon had a pretty good trick.

stampeder. (In Dict. s.v. Stampede *v.*) Add: *N. Amer.* **1.** One who takes part in a sudden or unreasoning rush of persons, esp. for gold. (Further examples.)

1859 E. H. N. Patterson *Jrnl.* 22 May in L. Hafen *Overland Routes to Gold Fields* (1942) 143 In anything that I may have heretofore written that might be deemed disparaging of the 'stampeders', I do not wish to be understood as intentionally unjust or uncharitable. 1884 *Century Mag.* Oct. 844/2 In the days of the stampeders and the toboggan trains, this was the only house on the trail. 1936 N. A. D. Armstrong *Yukon Yesterdays* 13 No wonder that hundreds of the weaker stampeders returned to the base to take the first steamer back home! 1963 *Beaver* (Winnipeg) Summer 42/2 Before the turn of the century, hordes of rushing stampeders..had reason to curse the savage stretch of water.

2. One who causes a stampede amongst cattle or horses.

1862 *Harper's Mag.* Sept. 450/1 Horses..which, having been 'hobbled' beside the fires of their respective owners, had..escaped the notice of the stampeders.

stamper, *sb.* Add: **3. e.** A matrix or copy of an original disc recording used to press other copies of a gramophone record.

1918 [see *Matrix 4 c]. 1935 [see *Master *sb.*¹ 9* a]. 1952 [see *Mother *sb.*¹ 10*]. 1975 G. J. King *Audio Handbk.* vii. 154 Most gramophone records start as very high quality tape recordings, the edited material then being recorded in disc form on to a lacquer blank, from which the stamper is ultimately derived.

stamping, *vbl. sb.* **3. stamping ground:** for *U.S.* read orig. *U.S.* and add earlier and further examples.

1821 D. Dunklin *Let.* 25 Dec. in E. C. Barker *Austin Papers* (1924) I. 456 It is unnecessary to undertake to give you any details of affairs in your old Stamping-ground. 1862 *Harper's Mag.* June 34/1, I found myself near one of these 'stamping grounds', and a simultaneous roar from five hundred infuriated animals gave notice of my danger. 1915 E. R. Burroughs *Return of Tarzan* xxiii. 326 The woman he loved was within a short journey of the stamping-ground of his tribe. 1955 *Times* 19 Aug. 2/5 Henry Wood used the Proms as a stamping ground for new music because there was little opportunity otherwise for making it known. 1977 J. I. M. Stewart *Madonna of Astrolabe* iv. 74 Tell Charles Atlas I'd have a go if asked —I suppose as Tutor for Admissions it's his stamping-ground.

stance, *sb.*² Add: **1. c.** Also *spec.* in *Mountaineering*, a ledge or foothold on which a climber can secure a belay.

1920 G. W. Young *Mountain Craft* v. 218 It is..vital for a leader to know what character of stance he requires in order to bring up his following safely. 1933 G. D. Abraham *Mod. Mountaineering* viii. 154 The second man looped the rope behind him around the projection from a stance a few feet along the ledge. 1956 [see *Intilted *ppl. a.*]. 1971 C. Bonington *Annapurna South Face* xi. 127, I had to wait another hour while he safeguarded his stance with carefully placed pitons.

d. Also *transf.*, the position of the player's body in readiness or in playing a stroke. Similarly *gen.*, a standing attitude or way of positioning.

1929 M. Lief *Hangover* 234 At dinner Mogador's young bride was plainly worried about her fork-stance. 1936 M. Allingham *Flowers for Judge* v. 84 No one who saw him could have dreamed for a moment that he regarded himself as anything else but the Head of the Firm. His poise and stance proclaimed it. 1965 'W. Trevor' *Boarding-House* ii. 19 Galleity and Mrs Slape stood close together..humble in their stance. 1970 J. G. Farrell *Troubles* II. 220 Driscoll instantly dropped into a boxing stance, right fist guarding his chin, left fist pumping exaggeratedly back and forth.

e. *fig.* An attitude adopted in relation to a particular object of contemplation; a policy, 'posture'.

1960 *Amer. Speech* XXXV. 215 An 'unlinguistic' stance is evidenced in the view that some variants embody language 'corruption'. 1964 *Ann. Reg. 1963* 216 In general those Parties in economically more advanced countries adopted a pro-Soviet stance, although several had dissident pro-Chinese minorities. 1972 [see *Rhetoricize *v.*]. 1977 J. I. M. Stewart *Madonna of Astrolabe* xx. 277 Moderate regret and underlying unconcern established itself as our public stance.

3. b. *Sc.* The pitch of a showman or street-trader; a location for a fair or market. Cf. Stand *sb.*¹ 15.

1814 *Farmer's Mag.* Nov. 466 If they are not in the market the night before, it is not often that a stance can be got after day-light in the morning. 1924 *Kelso Chron.* 25 July 4 This old-established Border fair was held on the usual stance on St Boswells Green on Friday. 1933 *Cases Court of Session* (Scotland) 65 A street trader shall not carry on business on any stance..unless he holds a permit from the chief constable for such stance. 1964 M. Banton *Policeman in Community* ii. 31 He..can attend to less pressing matters such as an application for a news-vendor's stance.

c. *Sc.* A standing-place for (a row of) public vehicles; a bus-stop or taxi-rank. Cf. Stand *sb.*¹ 17.

1926 *Edinburgh Corp.* (*General Powers*) *Order Confirmation* 25 in *Bills Public* I. 461 'Stance' means a place where omnibuses may stop a longer time than is necessary for the taking up and setting down of passengers. 1931 A. A. MacGregor *Last Voyage* 24 The erection of stance poles [for tram-cars] along Princes St. 1978 *Dumfries Courier* 13 Oct. 6/2 Travellers will find that early buses and taxis are temporarily sitting in different stances than is usual.

stand, *sb.*¹ Add: **I. 2. e.** Also, the place at which a halt is made; the performance itself; *transf.*, esp. in *one-night stand*: see *One *numeral a., pron.*, etc. 33.

1895 *N.Y. Dramatic News* 19 Oct. 11/1 Denver was the second stand of the week. 1931 *Amer. Speech* VI. 336 *Stand*, n., a town or city where a show stops to give performances. 1938 D. Baker *Young Man with Horn* III. 149 He'd been making stands at moving-picture houses all over the country. 1959 *Times* 16 Dec. 3/2 A number of travelling road-shows do one or two-night stands at such unlikely places as the Constitution Hall. 1964 Mrs. L. B. Johnson *White House Diary* 22 Apr. (1970) 115 Mrs. Eisenhower invited them in 1953, and had entertained around four thousand guests, about a four-hour stand. 1973 G. Beare *Snake on Grave* ii. 12 When the Sands in Vegas offered her a stand she took Latch with her to play for her.

f. The mean sea-level at a given epoch in the past; also, the level of the sea at high or low tide.

1934 Webster, *Stand*, the state of the tide at high or low water when there is no vertical movement. 1966 *Gloss. Oceanogr. Terms* (U.S. Naval Oceanographic Office) (ed. 2) 156/1 Where a double tide occurs, the stand may last for several hours even with a large range of tide. 1972 *Science* 13 Oct. 190/3 On Barbados, sedimentological considerations suggest that the high stand associated with the last interglacial (terrace III, 124,000 years ago) lasted no longer than about 5,000 years. 1978 *Nature* 18 May 185/3 At that time [*sc.* 22,000 years ago], sea-level was about 300 foot below its present stand because so much of the Earth's water was locked in glaciers.

4. c. (Earlier and later examples in *Cricket.*) Now *spec.* a partnership between two batsmen at the crease.

1851 J. Pycroft *Cricket Field* x. 189 Then comes the time when your great gun tumbles down his men: and that is the time that some sure, judgmatic batsman.. comes calmly and composedly to the wicket and makes a stand. 1912 P. F. Warner *Eng. v. Austral.* iv. 29 Barnes and Strudwick made a capital last wicket stand. 1980 *Wisden Cricket Monthly* Mar. 6/3 Charlie Davis..was then joined by..Garfield Sobers, in a stand of 254 in 363 minutes.

5. c. (The performance of) a stallion or bull at stud. Also, a stud or stud-farm. *U.S.*

1797 E. Chambers *Let.* 29 Nov. in J. Steele *Papers* (1924) I. 151 As a covering horse I am of Opinion he would make a very good Stand. 1836 *Russellville* (Kentucky) *Weekly Advertiser* 21 Jan. 3/3 (Advt.), Merlin is now at this stand in Elkton... Books are opened for those who may wish to enter their mares. 1959 W. Faulkner *Mansion* i. 9 He had to lead the cow the three miles back..to claim a second stand from the bull.

d. = Erection 4. *slang.*

1867, etc. [see *cock-stand* s.v. *Cock *sb.*¹ 23]. 1868 *Index Expurgatorius of Martial* 88 Maevius who while sleeping only gets A piss-proud stand that melts away on waking. 1903 Farmer & Henley *Slang* VI. 346/1 *Stand*,.. (venery).—1. An *erectio penis*.

9. (Later examples in *Gymnastics.*)

1956 Kunzle & Thomas *Freestanding* i. 26 Also try jumps from stand, both from half knee bend and with very little knee bend. 1964 G. C. Kunzle *Parallel Bars* ii. 50 Push away from the bars with the left hand to land in side stand.

II. 11. †**e.** A degree of proficiency measured by achievement in school-work; a mark or grade awarded in assessment. *U.S. Educ. Obs.*

1900 *Dialect Notes* II. 64 *Stand*, degree of proficiency in college studies, as evidenced by marks and honors. 1904 *N.Y. Even. Post* 17 Mar. 7 The highest stand man of the non-elective scholastic period was Dean Wright of 1868, who attained a stand of 3·71 on a scale of 4·00. 1921 R. D. Paine *Comrades of Rolling Ocean* i. 11, I had a rotten stand in your course.

12. b. The post or station of a sheep-shearer. *Austral.* and *N.Z.*

1893 S. Newland *Paving Way* II. xviii. 339 As applicants for a 'stand' on the shearing-floor began to camp about, his audiences became more numerous. 1901 'R. Boldrewood' *In Bad Company* 21 It's hard on a chap, when he comes to a shed..to be told that all the stands is took up. 1922 C. G. Turner *Happy Wanderer* 143 Four hundred men might answer the roll-call where only one hundred could 'get a stand'—i.e. a chance to shear. 1933 *Bulletin* (Sydney) 6 Sept. 20/2 They run a record long shearing, six stands being considered sufficient to cope with 60,000 sheep. 1949 D. Walker *We went to Australia* 97 We watched them [*sc.* shearers] in the 'eight-stand' shed at Nareeb. 1956 G. Bowen *Wool Away!* (ed. 2) x. 110, I have set out the plan of a three-stand shed..which should meet the needs of a large cross-section of sheep farmers.

15. b. (Later U.S. example.)

1929 [see *Meet *sb.* 1 b].

18. (Further examples.)

1885 *Daily Tel.* 11 Nov. 3/7 Many changes have taken place at Aintree, and, if the weather had permitted, the new stands would have been finished off. 1902 *Daily Mail* 7 Apr. 5/1 (*heading*) Stand collapses at a football match. 1977 *Evening Post* (Nottingham) 27 Jan. 20/4 There will be other rises with best seats in the stands going up to £1.40.

IV. 23. c. *Metallurgy.* A set of rolls and their auxiliary fittings which during any one pass provide one gap for the metal being rolled.

1874 *Jrnl. Iron & Steel Inst.* I. 349 Space is left at the end of the train for two stands of merchant roughing and finishing rolls. 1958 A. D. Merriman *Dict. Metallurgy* 338/2 A stand is usually..described as 2-high, 3-high, [etc.].. A rolling-mill may consist of a single stand or of several stands in series. 1973 G. F. Bryant *Automation of Tandem Mills* ix. 160 In 1968, a fifth stand was added to the BSC Abbey Works tandem cold-rolling mill.

d. *Oil Industry.* A number of lengths of drill pipe (usually from one to four) joined together, esp. when being unscrewed from a string or racked in a derrick.

1913 B. Redwood *Petroleum* (ed. 3) I. v. 317 The casing with which it is desired to shut off the water must admit of being moved quite freely in the bore-hole, so that it may be raised or lowered the full length of a 'stand', that is, for a distance of, say, three lengths or joints. 1949 [see *Rack *v.*³ 4 a]. 1960 C. Gatlin *Petroleum Engin.* v. 52/2 Only two or three joints per stand will be pulled when using shorter derricks. 1973 J. W. Jenner in Hobson & Pohl *Mod. Petroleum Technol.* (ed. 4) iv. 120 The drawworks then hauls the blocks up the derrick until a stand of three joints of drill pipe is above the rotary table.

26. (Earlier U.S. example.)

1851 W. Kelly *Excursion to California* I. v. 83 A stand of prairie plover most opportunely made their appearance as we pulled up.

29. For *U.S.* read orig. *U.S.* and add: *spec.* one of trees; now also applied to natural assemblages of plants, esp. when only one species is present or considered. (Later examples).

1905 *Terms Forestry & Logging* (U.S. Dept. Agric. Bureau Forestry) 22 *Stand*, all growing trees in a forest or in part of a forest. Syn.: growing stock. 1912 Hawley & Hawes *Forestry in New England* i. 8 The term 'stand' is the unit of description applied to any definite portion of a forest having a definite distinguishing characteristic. Thus in a certain type we may have a stand of young growth; a stand of diseased and damaged trees; a stand of exceptionally tall specimens, etc. These stands may be extensive, covering many acres or they may be confined each to a small part of an acre. 1947 K. Tennant *Lost Haven* xvii. 284 Nice stand of trees, brushwood, coach-wood, white soapy box. 1967 M. J. Coe *Ecol. Alpine Zone Mt. Kenya* 28 All the genera cited..are..obvious components of the alpine zone, with the megaphytic Senecios and Alchemilla scrub forming almost pure stands under suitable conditions. 1975 P. Lively *Going Back* i. 8 Old wooden chicken-houses half-submerged in grass and cow-parsley and stands of nettles. 1979 H. W. Hocker *Introd. Forest Biol.* iv. 99 The stand is a basic unit of management and is important to planning silvicultural operations... In the terminology of plant ecology, a forest stand would be designated as a community, or an ecosystem... Foresters use the term stand to designate the tree portion of the ecosystem.

stand, *v.* Add: **I. 3. b.** Of a stallion: to be available as a stud-horse to serve mares (esp. *at* a certain place). Also, *to stand at stud* (cf. Stud *sb.*²). orig. *U.S.*

1766 *Virginia Gaz.* 4 Apr. 3/3 Merry Tom Stands at my house, and covers mares at a guinea the leap. 1788 W. Lenoir in N. E. Eliason *Tarheel Talk* (1956) 297 Whirligig will stand this season..& will cover Mares at Forty Shillings per season Twenty Shillings per cover. 1846 *Spirit of Times* 18 Apr. 94/1 Young Dread will stand this season at Watertown, Jefferson County, N.Y. 1891 J. L. Kipling *Beast & Man in India* viii. 207 Importing English thoroughbreds, Arabs, and Norfolk trotters who stand as sires at the service of farmers. 1959 *Times* 31 Aug. 13/5 (*heading*) Shantung to stand in England. 1974 D. Francis *Knock Down* xiv. 171 Nestegg is standing at stud in Ireland. 1974 *New Yorker* 29 Apr. 102/2 Go Man Go stands at stud at Buena Suerte Ranch. *Ibid.* 102/3 Tony B Deck's father, who stood in Perry, Oklahoma, was murdered in his stall. 1977 *N.Z. Herald* 5 Jan. 1–12/5 Aristoi is by Sir Gaylord from Attlea, by Mt Trouble from Athenia, by Pharamond II, and stands at the Preston Farm Stud.

6. c. Of the penis: to become or remain erect.

? *a* 1508 Dunbar *Tua Mariit Wemen* sig. b ii, And a stif standand thing staiffis in mi neiff. *c* 1593 Nashe *Choise of Valentines* (1899) 12 'Unhappie me,' quoth shee, and wilt' not stand? Com, lett me rubb and chafe it with my hand!' 1762 T. Bridges *Homer Travestie* I. iv. 189 She guides his weapon where she list;..a touch of her soft hand, If fallen down, will make him stand. 1868 *Index Expurgatorius of Martial* 82 That's the way to make your Martial stand. 1903 Farmer & Henley *Slang* VI. 346/2 Also (proverbial) 'Stand always, as the girl said.'

7. Also *to stand from under*: see *Under *adv.* 4 c.

14. to stand pat. For *U.S.* read Chiefly *U.S.* and add: (*b*) (earlier example). **stand-pat** *a.* (later examples); **stand-patter** (later examples); **stand-pattism** (later examples).

1890 *Stock Grower & Farmer* 29 Mar. 7/1 When it came to them two accomplishments he stood pat. 1920 W. A. White *Let.* 4 Feb. (1947) 204 He also has on [the Republican Platform Committee]..a lot of old high-binder standpatters who haven't had an idea since the fall of Babylon. 1922 *Nation* 18 Nov. 271/1 The Bonar Law Government..remains in power, but probably on a minority vote. This deprives it of the right to pursue the standpat Toryism on which it made the elections. 1952 W. D. Jacobs *William Barnes* v. 79 A sobering realization is that a certain antagonism to a new credo of language is plain stand-pattism. 1975 G. V. Higgins *City on Hill* vii. 202 The mood of the country may be such that a stand-patter is the only candidate who can be elected. 1977 *Daily Tel.* 17 Feb. 14/7 He is highly critical of stand-pat, counter-reformation Catholicism.

II. 24. *all standing. to be brought up all standing*: also *actively*; *to gybe all standing* (see quot. 1976); (earlier *transf.* examples); also *fig.*

1837 *Southern Lit. Messenger* III. 178 This reflection brought me up, as the sailors say, 'all standing'. 1840 R. H. Dana *Two Years before Mast* xxxi. 231 The mate ..turned in 'all standing', and was always on deck the moment he was called. 1879 Hartigan & Walker *Stray Leaves* 2nd Ser. 198 [They] gained their respective domiciles, and turned in 'regimental', or, as Jack has it, 'all standing', for their..last night's rest in Old England. 1884 [see Gybe *sb.*²]. 1903 A. Bennett *Let.* 27 Mar. (1966) 35 The close of the book, as it stands, will 'bring him up all standing'. 1924 A. J. Small *Frozen Gold* i. 13 A spring..would take him from his chair, all-standing, sheer to the throat of the swaggering giant who held the gun. 1976 *Oxf. Compan. Ships & Sea* 365/2 To gybe without attending the runners, or to do so involuntarily, is known as to 'gybe all-standing', and is dangerous.

29. Also, of (a pot of) tea: to be left to draw.

1933 E. A. Robertson *Ordinary Families* vii. 162 Well, I'll get Olive to bring some fresh tea, then. This has been standing rather a long time. 1935 G. Santayana *Last Puritan* III. v. 333 Don't keep this tea standing any longer... It will be poison. Make some fresh for the boys when they come down. 1976 L. Henderson *Major Enquiry* xii. 76 It will be ready soon. I like tea to stand properly, don't you?

V. 59. b. (Later examples.)

1919 Conrad *Arrow of Gold* iv. ii. 162 Captain Blunt jumped up. 'My mother can't stand tobacco smoke.' 1949 'G. Orwell' *1984* II. iii. 134 'I could have stood it if it hadn't been for one thing,' he said. 1964 I. Murdoch *Italian Girl* iii. 39 Do turn that music off, would you? I can't stand music in the background. 1981 E. A. Taylor *Cable Car Murder* (1983) xxiii. 170 Don't get me started on her; I can't stand her.

61. b. Also, to put up or make a present of (a sum of money), esp. as part of a larger amount sought.

1844 Dickens *Let.* 22 July (1977) IV. 157 If you should decide to come, I will very gladly stand £10 of this Thirty. 1914 G. B. Shaw *Fanny's First Play* I. 177, I cant pay the fine and get him out; but if youll stand 3 pounds I'll stand one; and thatll do it. 1970 G. F. Newman *Sir, you Bastard* viii. 214 Friends able to stand five-thousand pounds surety for his bail.

VI. 71. stand for—. m. To endure, put up with, tolerate. Cf. sense 59 in Dict. *colloq.* (orig. *U.S.*).

1896 Ade *Artie* xii. 107 They say they can't stand for that kind o' work. 1911 R. W. Chambers *Common Law* x. 282 It's going to be hard for her. She can't stand for a mutt—and it's the only sort that will marry her. 1916 E. V. Lucas *Vermilion Box* lxvi. 72 So crabbed and odd and disagreeable that the store let him go... Two weeks ago he lost his position in the country store. Even that place could not stand for him. 1927 *Punch* 20 Apr. 428/1 The English public, it appears, will only stand for American films. 1952 M. Laski *Village* xix. 265 Me and Dad have stood for a lot of things..but there's one thing we won't stand for and that's any hole-and-corner business. 1967 N. Freeling *Strike out where not Applicable* 77 Marguerite wouldn't have stood for being humiliated. 1973 E.-J. Bahr *Nice Neighbourhood* i. 6 He was a man who just purely couldn't stand for anyone..to be asleep when he was awake.

74. stand on—. m. In (chiefly imp.) phr. *stand on me*, (you may) rely on me, believe me. Cf. *stand upon* sense 78 c in Dict. *slang.*

1933 *Cornh. Mag.* June 697 'E'll finish like a crab—stand on me fer that. 1935 Wallace & Curtis *Mouthpiece* i. 17 If any of your clients ever want to go abroad.. in a hurry—never mind about passports, eh? Just stand on me. 1959 F. Norman (*title*) Stand on me, a true story of Soho. 1970 G. F. Newman *Sir, you Bastard* i. 35 You'll be all right, stand on me.

75. stand over—. Also *transf.* in extended use; *Austral. slang*, to intimidate or threaten; to extort money from (someone).

1932 V. Woolf *Pargiters* (1978) 31 Miss Edwards, the small dressmaker,..could cut out quite well, but one had to stand over her. 1939 K. Tennant *Foveaux* 173, I just had Thompson in here and he stood over me for three quid. 1940 *Punch* 24 Apr. 449/2, I could *occasionally* leave her to wash up a few cups or something like that without actually standing over her the whole time. 1953 K. Tennant *Joyful Condemned* iii. 21 There's many a man thought he was going to stand over some little lowie and now he's..looking through the bars. 1967 K. Giles *Death & Mr Prettyman* ii. 58 [Australian loq.] You could stand over—pardon, persecute me. 1978 D. Francis *Trial Run* vii. 105, I should stand over 'them... Make yourself a bit of a nuisance, so they send it [*sc.* a Telex] to get rid of you.

76. stand to—. f. (Later *arch.* example.)

1935 T. S. Eliot *Murder in Cathedral* i. 20 Do not ask us To stand to the doom on the house, the doom on the Archbishop.

s. [tr. Ir. *seasamh do.*] To be to one's advantage, to sustain. *Anglo-Irish.*

1907 Yeats *Deirdre* 34 Women, if I die, If Naoise die this night, how will you praise? What words seek out?

for that will stand to you; For being but dead we shall have many friends. **1914** JOYCE *Dubliners* 11 Why, when I was a nipper, every morning of my life I had a cold bath... That's what stands to me now. **1922** —— *Ulysses* 622 Through all those perils of the deep..there was one thing, he declared, stood to him..a pious medal he had that saved him.

VII. 91. stand by. c. Now, of a juror: to withdraw from the jury, esp. at the challenge of the prosecution. Also *trans.* with juror as obj. Cf. CHALLENGE *sb.* 3 a.

1828 *Act 9 Geo. IV* c. 54. 500 Nothing herein contained shall affect..the power of any court in Ireland to order any juror to stand by. **1890** T. BRETT *Comm. Present Laws of Eng.* II. XIII. viii. 1162 The names of those ordered to 'stand by' are called again. **1923** W. J. BYRNE *Dict. Eng. Law* 167/1 The Crown, although it can challenge for cause, has no peremptory challenge, but it may order any person to 'stand by', and need not show cause. **1969** *Sunday Tel.* 12 Jan. 4/6 The defence..could have called 'Challenge!' 70 times without offering reasons. The prosecution called 'Stand by for the Crown!' nine times. (In a British court the phrase 'I object' does not arise in this context.)

trans. **1927** A. M. SULLIVAN *Old Ireland* ii. 41 If this infamous creature attempts to 'stand by' a single Nationalist juror, you will ram it down his throat! **1960** V. T. H. DELANY *Christopher Palles* xi. 101 The Crown 'stood by' 96 jurors, while the defence challenged 36. **1979** *Criminal Law Rev.* May 273 The Crown's right to 'stand-by' jurors is a term which refers to the Crown's power of challenge.

e. (Examples of *gen.* use.)

1866 'MARK TWAIN' *Lett. from Hawaii* (1967) 117 Just as you take a sustaining breath and 'stand by' for the crash, his poor little rocket fizzes faintly in the zenith. **1917** 'CONTACT' *Airman's Outings* iv. 84 For thirty hours the flight had 'stood by' for a long reconnaissance... A slight but steady rain washed away all chance of an immediate job. **1943** *Ann. Reg. 1942* 28 The wastefulness of keeping so many men merely 'standing-by'. **1972** *Listener* 21 Dec. 852/1 Sequence of calls before a shot. Production Assistant: 'Quiet. Going for a take. Standing by.'

92. stand down. d. *Mil.* To come off duty; to relax after a state of alert (also *trans.*).

1916 I. GURNEY *Let.* 25 Oct. in *PN Review* 29 (1982) 32/1 Our last orders were as follows.—From Stand to 5.30. Stand Down, clean rifles... Stand to 5–5.30. Stand Down. **1918** E. S. FARROW *Dict. Mil. Terms* s.v. *Stand to*, Stand down is the order countermanding 'stand-to'. **1919** W. H. DOWNING *Digger Dialects* 47 *Stand-down*, the order by which the period of intense armed vigilance is ended at daybreak, nightfall, or after the alarm of a threatened enemy attack has passed over. **1931** F. TILSLEY *Other Ranks* 108 They religiously stood-to and stood-down every dawn and dusk. **1973** *Daily Tel.* 29 Oct. 30/3 Pres. Nixon ordered..troops in Europe..to remain on the alert..but elsewhere round the world American forces were stood down. **1983** *Times* 12 Feb. 1/1 Acas officials were fighting to keep alive the proposal for a third-party intervention to settle the water workers' strike. But a lull in the peace process is expected over the weekend after the Acas conciliation team was stood down.

94. stand forward. (Earlier example.)

1790 *Loiterer* 9 Jan. 7, I shall be happy to contribute my mite... I dare say his Lordship would stand forward [i.e. with a donation].

95. stand in. d. Also *rarely*, to fall in *with* (a proposal).

1911 M. BEERBOHM *Zuleika Dobson* viii. 138 'Dorset,' he said huskily, 'I shall die too.'..'I stand in with that,' said Mr. Oover [an American]. 'So do I!' said Lord Sayes.

f. To fill the place of another (usu. temporarily); to deputize *for* (a person). *spec.* in *Cinemat.*, to act as a substitute *for* a principal actor. Cf. *STAND-IN 2.

1904 in *Eng. Dial. Dict.* V. 725/2 *e. Ken.* Mrs. —— will stand in while Mrs. —— is ill. **1943** HUNT & PRINGLE *Service Slang* 62 *Stand-in*, a deputy; one who 'stands in' for you, or does your duty while you go out. **1955** *Times* 6 June 7/6 There is always a way, especially in Russia where queueing has had to be carried to a fine art. You can employ the willing services of an Armenian or Georgian or other agile-minded person who will stand in for you or will, in turn, get another to stand in for him. **1958** *Listener* 7 Aug. 210/2 The people who stand-in for the film stars when the rough stuff begins. **1978** 'B. GRAEME' *Double Trouble* ii. 20 She has to stand-in for the star while they are working out lighting, camera angles and so on... They try to have a stand-in as much like the star as possible.

96. stand off. f. (Earlier examples.)

1878 J. H. BEADLE *Western Wilds* ii. 38 He offered him fifty thousand for it, and the feller stood him off for seventy-five thousand. **1883** J. HAY *Bread-Winners* xvii. 274 Come, come, Sam, don't stand me off that way.

g. To lay (an employee) off temporarily. Also *intr.* of an employee. Cf. LAY *v.*[1] 54 f in Dict. and Suppl.

1918 [implied at *STAND-OFF *sb.* 5]. **1927** CARR-SAUNDERS & JONES *Soc. Struct. Eng. & Wales* 135 It is not uncommon for indentures to contain a clause enabling the employer to 'stand off' the apprentice without pay if there is no work for him. **1930** *Daily Express* 8 Sept. 11/4 Thirteen hundred Chislet miners..went on strike..as a protest against the standing-off of six men. **1940** H. G. WELLS *New World Order* §5. 58 A state of five million people with half a million of useless hands, will be twice as unstable as forty million with two million standing off. **1952** M. LASKI *Village* i. 17 It wasn't very nice me having to go out to work when Mr. Wilson was stood off, and him staying at home to keep an eye on the children. **1960** G. E. EVANS *Horse in Furrow* i. 22 The day-men were liable to be *stood-off* on wet days. **1976** *Eastern Daily Press* (Norwich) 16 Dec., He was later stood off but went back there to work from 1962 to 1975.

100. stand over. b. (Earlier example.)

1822 M. EDGEWORTH *Let.* 30 May (1971) 404 A beef and pigeon pie that had stood over from the preceding week.

101. stand to. c. *Mil. ellipt.* for *to stand to one's arms*, sense 76 d in Dict. Hence, to come or remain on duty. Cf. *stand-to*, sense 104 below.

1915 F. H. LAWRENCE *Let.* 7 Mar. in *Home Lett. T. E. Lawrence* (1954) 671, I thought the Germans were attacking us, so I passed the word along for all my men to stand to, as we call it. *a* **1918** W. OWEN *Poems* (1963) 60 Eh? What the 'ell! Stand to? Stand to! Jim, Give's a hand with pack on, lad. **1942** E. WAUGH *Put out More Flags* i. 22 She saw him as Siegfried Sassoon, an infantry subaltern in a mud-bogged trench, standing to at dawn,.. waiting for zero hour. **1977** J. B. HILTON *Dead-Nettle* xv. 120 After days of pointless standing-to in dew-drenched hedge-bottoms, there was a cleaning-up of uniforms.

103. stand up. e. Also, † to present oneself for marriage.

1842 *Amer. Pioneer* I. 314 They were married without any previous preparation..he standing up in a hunting dress, and she in a short gown and petticoat of homespun.

f. (Earlier example.)

1836 DICKENS *Sk. Boz* (1st Ser.) I. 252 Nobody thought of 'standing up' under doorways or arches.

g. (Later examples.) Also, freq. in extended phr. (*only*) *the clothes one stands up in.*

1937 'G. ORWELL' *Road to Wigan Pier* ix. 182, I.. planned..how one could..start out with no money and nothing but the clothes one stood up in. **1944** N. STREATFEILD *Curtain Up* vi. 71 Monsieur Manoff and most of his pupils..escaped to America... They had, of course, nothing but what they stood up in. **1981** E. LONGFORD *Queen Mother* v. 83 Queen Wilhelmina of the Netherlands ..brought only the clothes she stood up in plus a tin hat.

j. (Later *poet.* example.)

1896 A. E. HOUSMAN *Shropshire Lad* vii. 11 When smoke stood up from Ludlow And mist blew off the Teme.

p. Also with impersonal subj.: to endure or withstand.

1921 G. B. SHAW *Back to Methuselah* p. lxxxv, I had seen Bible fetichism, after standing up to all the rationalistic batteries of Hume, Voltaire, and the rest, collapse before the onslaught of much less gifted Evolutionists. **1940** *Punch* 11 Dec. p. xiii, Nylon tufts will stand up to an incredible amount of hard use.

q. To fail to keep an appointment with (someone), esp. a social engagement or 'date' with a member of the opposite sex. *colloq.* (orig. *U.S.*).

1902 O. V. LIMERICK *Billy Burgundy's Opinions* 57, I am awfully sorry I had to stand you up last night. **1906** 'O. HENRY' *Four Million* 122 Rosy's stuck to the affirmative this time for two whole days. But it's five hours yet till the time, and I'm afraid she'll stand me up when it comes to the scratch. **1936** J. CURTIS *Gilt Kid* xxv. 244 It must be getting along. He didn't want Maisie to think that he was standing her up. **1940** R. CHANDLER *Farewell, my Lovely* xxxii. 246 You stood me up for an hour the other night. **1952** J. CANNAN *Body in Beck* ix. 186 Time and again..I stood up the chaps so as to climb with him. **1959** *New Statesman* 11 Apr. 519/2 It isn't long, however, before she discovers the canker in the bud, in the person of William, a young smoothie from Madison Avenue, who stands her up. **1978** L. THOMAS *Ormerod's Landing* iii. 43 'What about the other agent, the lady?'..'Stood you up, I shouldn't wonder,' laughed Charles.

r. In colloq. phr. *to stand up and* (also *to*) *be counted*, to show one's political colours; also more widely, to display one's conviction or sympathy, esp. when this requires courage. orig. *U.S.*

1904 *Hartford* (Connecticut) *Courant* 12 Aug. 10 Another democratic paper, the 'Sacramento Bee', follows the example of the 'Chicago Chronicle' and stands up to be counted for Roosevelt. **1945** *Somerset News* (Princess Anne, Maryland) 22 Mar. 1/3 Why then weren't Shore delegates men enough to stand up and be counted? **1968** *Listener* 1 Aug. 134/2, I suppose in the end it was having to stand up and be counted as part of 'The New Establishment'; being forced to own up that I earn my living and have my being in that world. **1969** *Ibid.* 13 Nov. 658/3, I was a great one for demos in my youth... I can remember a good deal of self-righteousness, in standing up to be counted with the saints. **1973** 'M. INNES' *Appleby's Answer* iv. 38 A mild-mannered man. But he felt he must stand up and be counted. **1977** D. JAMES *Spy at Evening* xxiv. 199 People like you haven't stood up to be counted.

s. To sustain close examination, to be tenable: esp. of a charge or theory.

1948 in M. McLuhan *Mech. Bride* (1967) 6/3 Authorities here voiced doubt..whether such a charge would 'stand up'. **1962** *Listener* 10 May 814/3 It will be interesting to see if this conclusion stands up when more results become available. **1976** N. FREELING *Lake Isle* xxxii. 232 It won't stand up in law, I tell you.

t. *will the real — please stand up?*, a catch-phrase which requests that a person clarify his position or make himself known (often rhetorical). orig. *U.S.*

1971 *Black Scholar* June 32 (*heading*) Will the real black man please stand up? **1973** *Illustr. London News* July 76/3 Will the real Kate Brown please stand up and show herself? **1981** *Nature* 12 Mar. 89 (*heading*) Will the real Grenville Orogeny please stand up.

VIII. 104. stand-alone *a. Computers*, designating a part of a computer system that can be used independently; **stand-away** *a.*, (*a*) of a person: reserved, chilly, 'standoffish'; (*b*) of a collar, etc.: that lies or rises away from the neck of the wearer; also *absol.* as *sb.*; **stand-back** *rare*, (*a*) a source of reassurance or support; a dependable person; (*b*) one who holds back; **stand-down** *Mil.* (now esp. *Air Force*), the action or state of coming or remaining off duty or of relaxing from a period of vigilance; the end of a spell of duty; **stand-over**, (*b*) *Austral. slang*, used *attrib.* to designate the perpetrator of extortion by threat, a protection-racketeer, as *stand-over man*; or the process of such extortion; occas. *absol.* and *transf.*; **stand-to** *Mil.*, *ellipt.* for *stand-to-arms*; also, the time of coming on duty, as at dawn or dusk, or in preparation for an attack; also *attrib.*; **stand-to-arms**: also, the period of standing to arms.

1966 C. J. SIPPL *Computer Dict. & Handbk.* 295/2 *Stand-alone capability*, a multiplexor designed to function independently of a host or master computer, either some of the time or all of the time. **1969** *Computers & Humanities* III. 137 A system for typing concordance output on an IBM Selectric typewriter is being developed for use with a stand-alone device which consists of a magnetic tape unit, keyboard, and heavy-duty selectric typewriter. **1977** *Sci. Amer.* Sept. 147/2 At the next level in the hierarchy of capability and function are the small computer systems that are prepackaged as stand-alone units... They have a self-contained power supply. **1938** J. CARY *Castle Corner* iii. 155 Rifty had Bridget for his partner but Bridget was even more standaway than usual. **1955** *Sun* (Baltimore) 3 Feb. (B ed.) 3/1 Most of Balenciaga's collars are simple turnover standaways. **1964** *McCall's Sewing* i. 12/1 Stand-away collars, collarless styles and V- or U-shaped necklines will make the neck seem longer. **1971** *Woman's Own* 27 Mar. 24/1 A stand-away collar is good for a thick neck. **1915** D. H. LAWRENCE *Rainbow* xii. 313 Gudrun was..a great comfort and shield to her... This was a great stand-back to Ursula, who suffered agonies when she thought a person disliked her. **1922** —— *Aaron's Rod* x. 122, I had a corporal called Wallace... He was my stand-back. **1946** J. W. DAY *Harvest Adventure* vi. 89 We've got a fine lot o' stand-backs to tell others to git forrad. **1919** W. H. DOWNING *Digger Dialects* 47 *Stand-down*, the hours at which the above orders [to go off duty] are regularly given each day. **1925** FRASER & GIBBONS *Soldier & Sailor Words* s.v. *Stand-to*, 'Stand-down' was the corresponding order at the end of the Danger Period, used in like manner as an expression for a definite point of time. **1945** C. H. WARD-JACKSON *It's Piece of Cake* (ed. 2) 57 *Stand-down*, time-off when flying is cancelled. **1949** W. S. CHURCHILL *Second World War* II. II. xvi. 297 On February 13, 1942, Admiral Raeder had his final interview on 'Sea Lion' and got Hitler to agree to a complete 'stand-down'. **1953** EARL WINTERTON *Orders of Day* xix. 270 Next morning at 'stand down', on a cold grey day, I went to have a look at where the place had been where I had spent so many years of my life [engaged in fire-watching]. **1978** H. WOUK *War & Remembrance* xxviii. 284 Then would come the bored wait, the stand-down, the recovery of aircraft, and the resumed plan of the day. **1939** K. TENNANT *Foveaux* 174 He didn't deserve to be a 'standover man' if he couldn't move quicker. **1941** BAKER *Dict. Austral. Slang* 71 *Standover (man)*, a criminal who exacts toll from other lawbreakers or innocents. *Ibid.*, *Standover, work the*, to act as a 'stand-over man'. Also, a door-to-door hawker's term for bullying tactics adopted to intimidate housewives. **1954** L. H. EVERS *Pattern of Conquest* 198 Don't come the stand-over tactics you used with Charlie. **1977** *Daily News* (Perth, Austral.) 19 Jan. 16/5 Detectives believe the shooting is part of a war that has broken out amongst stand-over men in the massage parlour industry in Melbourne. **1915** in W. Wood *In Line of Battle* (1916) 217 At stand-to, 6 a.m. Much shelling. **1928** E. BLUNDEN *Undertones of War* ii. 21 Let me take you..back now..into the stand-to billets in Festubert village. **1942** *R.A.F. Jrnl.* 3 Oct. 31 There was that stand-to in the small hours one day. **1954** W. K. HANCOCK *Country & Calling* vii. 195 In the quiet middle years of the war we were allowed to lie down after our training until stand-to at 6 a.m. **1975** T. ALLBEURY *Special Collection* xi. 78 They made a permanent stand-to situation with just six of their anti-missile missiles. **1915** A. D. GILLESPIE *Let.* 11 Mar. in *Lett. from Flanders* (1916) 43 As I write, during the evening 'stand to arms', the birds are all singing in spite of the sniping.

standard, *sb.* Add: **A. II. 10. b.** Also *double standard*: see *DOUBLE *a.* 6.

12. a. *standard of comfort* (examples); also *standard of life*.

1879 A. & M. P. MARSHALL *Econ. Industry* II. vii. 102 The Standard of Comfort which young people are prudent enough to secure for themselves before they marry, varies from place to place and from time to time. **1898** B. BOSANQUET (*title*) Standard of life, and other studies. **1907** G. B. SHAW *John Bull's Other Island* II. 41 He guesses Broadbent's standard of comfort a little more accurately than his sister does. **1936** J. M. KEYNES *Gen. Theory Employment, Interest & Money* xvi. 218 The position of equilibrium, under conditions of *laissez-faire*, will be one in which employment is low enough and the standard of life sufficiently miserable to bring savings to zero.

b. Also *transf.*, the form or class in which pupils are prepared for a particular standard.

1878 F. KILVERT *Jrnl.* 16 Jan. (1977) 300 Gave the upper standards at the school questions on paper on the Catechism. **1915** D. H. LAWRENCE *Rainbow* xiii. 353 She made friends with the Standard Three teacher. **1934** G. B. SHAW *On Rocks* Pref. 168 The likeliest outcome is an elaborate creed of useful illusions, to be discarded bit by bit as the child is promoted from standard to standard or from form to form. **1966** C. ACHEBE *Man of People* i. 2

Sixteen years or so he had been my teacher in standard three. **1973** *Express* (Trinidad & Tobago) 17 Mar. 7/1 He left Trinidad a Seventh standard pupil of the St. Helena C.M. School.

16. Short for: **c.** *standard lamp*, sense 30 in Dict.

1910 H. G. WELLS *New Machiavelli* III. iii, in *English Rev.* Sept. 292 The light of the big electric standard in the corner. **1939** O. LANCASTER *Homes, Sweet Homes* 10 All over Europe the lights are going out..olde Tudor lanthorns, standards and wall-brackets. **1974** M. INGATE *Sound of Weir* ix. 74 Tall lamps, 'they're Standards,' said Iris, had large coloured shades.

d. A standard form of a language (see *STANDARD *a.* 3 e). *Modified* (also *Received*) *Standard*: see the first element in Suppl.

1913 *Mod. Lang. Teaching* Dec. 262/2 While within the London sphere of influence..Received Standard goes on quite gaily, the London type of Modified Standard has won the day in this area, among those sections of the community who might otherwise speak a Kentish..or Surrey type of Modified Standard. **1972** HARTMANN & STORK *Dict. Lang. & Linguistics* 218/1 Deviations from the respective established standards are called non-standard or sub-standard.

III. 26. c. A tune or song of established popularity, esp. in *Jazz*.

1937 *Amer. Speech* XII. 184/1 *Standard*, a number whose popularity has withstood the test of time. **1938** 'JELLY ROLL' MORTON in *Downbeat* Aug. 31/1, I also transformed..*After the Ball*, *Back Home in Indiana*, etc., and all standards that I saw fit. **1947** R. DE TOLEDANO *Frontiers of Jazz* p. xii, Half a dozen ancient Melrose stock arrangements of jazz standards 'as played by King Oliver'. **1959** 'F. NEWTON' *Jazz Scene* ii. 29 The repertoire [of jazz] consists of so-called 'standards'—themes which for one reason or another, lend themselves to profitable jazz playing. They may be drawn from any source, the traditional blues and the current popular song being the most important. **1971** C. FICK *Danziger Transcript* 95 There was a Cuban quartet in the garden playing light standards. **1980** M. BOOTH *Bad Track* v. 84 For an hour or so, the band jammed,..before going into standards that they knew and admired.

IV. 30. standard-bred *a.* (earlier example); also *N. Amer.* as *sb.*, a horse of this breed, developed esp. for harness racing (contrasted with *thoroughbred*); **standard lamp** (earlier and later examples).

1888 G. W. CURTIS *Horses, Cattle, Sheep, & Swine* x. 56 When an animal meets the requirements of admission, and is duly registered, it should be accepted as a standard bred trotting animal. **1948** *Sun* (Baltimore) 20 Aug. 18/1 The standardbreds..make running horses look like sissies. **1976** *National Observer* (U.S.) 2 Oct. 7/2, I grew up in.. the western part of Illinois..and every farmer..had a standardbred, either trotter or pacer... They just tied sulky bikes to the backs of the nags and drove them around the track themselves. **1894** *Country Gentlemen's Catal.* 115/1 Wrought Iron Standard Lamp, with copper Oil Container, 70/-. **1932** R. LEHMANN *Invitation to Waltz* I. xi. 110 The light was very bright and white, coming from three brass standard lamps with white silk shades. **1980** A. N. WILSON *Healing Art* xvi. 195 By the fireplace..gas blazed, and a standard lamp shone dimly.

B. *adj.* **I. 1. a.** Also freq. in special scientific collocations, as *standard atmosphere*, (*a*) a unit of atmospheric pressure, equal to 760 torr or 1013·25 millibars; (*b*) a hypothetical atmosphere with defined surface temperature and pressure and specified profile of temperature with altitude, used esp. in aviation and space research; *standard cable*, a unit of attenuation formerly used in telephone engineering (see quot. 1963), now replaced by the *BEL; *standard candle*, a disused unit of luminous intensity, defined as the intensity of the flame of a spermaceti candle of specified properties (see quots.), now replaced by the *CANDELA; also *transf.*; *standard cell*, any of several forms of voltaic cell designed to produce a constant and reproducible electromotive force as long as the current drawn is not too large; *standard deviation*: see *DEVIATION 2 d; *standard error*, a measure of the statistical accuracy of an estimate, equal to the standard deviation which a large population of such estimates would have; *standard wire gauge*, one of the series of standard thicknesses for wire and metal plates in the United Kingdom; any specific measure in this series; abbrev. *s.w.g.*, *S.W.G.* s.v. *S 4 a.

1911 W. N. SHAW *Forecasting Weather* p. xi, The accepted normal pressure of the atmosphere, or 'standard atmosphere', is that of a column of mercury 76 centimetres high at the freezing-point of water under the conditions as to gravitation which are to be found in latitude 45°N. or S. **1924** *Official Bull. Internat. Comm. Air Navigation* VII. 34 The Commission decides: To adopt the Regulations set out hereunder concerning the definition of an international standard atmosphere. **1930** *Meteorol. Gloss.* (Met. Office) (ed. 2) 162 The International Standard Atmosphere which is used as the basis of graduation of altimeters assumes at mean sea level a temperature of 15°C., a pressure of 1,013·2 mb., and a lapse rate of 6·5°C. per kilometre from sea level up to 11 km., above which the temperature is assumed constant at −56·5°C. **1963** JERRARD & MCNEILL *Dict. Sci. Units* 130 One standard atmosphere corresponds to a barometric pressure of 29·9213 inches or 760 mm of mercury of density 13·595 g cm⁻³, where the acceleration due to gravity if 980·665 cm sec⁻²... The ICAO [*sc.* International Civil Aviation Organisation] standard atmosphere was introduced about 1940. **1977** I. M. CAMPBELL *Energy & Atmosphere* iii. 47 The origin of the rising temperature from 11 to 50 km in the standard atmosphere is the degradation of a portion of the solar irradiance to thermal energy through the agency of primary absorption by ozone. **1906** J. POOLE *Pract. Telephone Handbk.* (ed. 3) xxvi. 413 Standard Cable and Equivalents.—In the agreement entered into in February 1905 between the British Post Office and the National Telephone Co. certain standards of telephonic transmission were stipulated, and these were to be measured by comparison with the transmission results obtained with standard telephone instruments through certain lengths of standard test cable. **1924** *Trans. Amer. Inst. Electr. Engineers* XLIII. 797/1 The 'mile of standard cable' has been used in telephone engineering in this country for over twenty years..as the unit for expressing the transmission efficiency of telephone circuits and apparatus. **1963** JERRARD & MCNEILL *Dict. Sci. Units* 131 The unit compared the attenuation produced in the circuit under test with that in a standard cable which was defined as a theoretical cable one mile in length, resistance 88 ohms, capacitance 0·054 microfarad, inductance one millihenry and leakance 5×10^{-6} mho. The standard cable produced an attenuation of about 20% for a 800 c/s input. **1879** Standard candle [in Dict.]. **1937** G. S. MONK *Light* v. 36 The standard candle was originally of sperm wax, weighing ⅙ lb., ⅞ in. diameter, and burning 120 grains per hr. **1959** *Listener* 2 July 14/2 By studying the period of a Cepheid, we can.. find out its real luminosity; its apparent magnitude is easy to measure, and hence its distance may be determined, so that these convenient variables act as our standard candles in space. **1976** *New Scientist* 2 Dec. 530/1 The new finding opens the way to calibrate a new standard candle, namely, the absolute brightness of a galaxy by means of an easy measurement in radio astronomy. **1872** L. CLARK in *Proc. R. Soc.* XX. 447 We have therefore the mean value of the electromotive force of the standard-cells, as determined by the electrodynamometer, 18 observations..1·45735 Volt. **1920** *Whittaker's Electr. Engineer's Pocket-bk.* (ed. 4) 100 The original standard cell devised by Latimer Clark is a mercury-zinc cell using zinc and mercurous sulphates as electrolyte and depolarizer respectively. **1980** J. P. BROMBERG *Physical Chem.* xvii. 315 The voltage of the unknown cell..can be determined from the calibrated slide wire and the known voltage of the standard cell. **1897** G. U. YULE in *Jrnl. R. Statist. Soc.* LX. 821 We see that $\sigma_1\sqrt{(1-r^2)}$ is the standard error made in estimating *x*. **1956** J. H. BURN *Lect. Notes Pharmacol.* (ed. 4) 129 The standard error of the mean is proportional to the standard deviation and inversely proportional to the number of animals used. **1962** J. H. KINOSHITA et al. in A. Pirie *Lens Metabolism Rel. Cataract* 409 The results are given as the mean ± standard error of the mean of 12 determinations. **1858** J. WHITWORTH *Misc. Papers Mech. Subjects* 68 (*caption*) Standard wire-gauge. **1884** *Rep. Board of Trade on Proc. & Business Weights & Measures Act, 1878* 3 in *Parl. Papers* XXVIII. 851 The new standard wire-gauge has been adopted by the War, Admiralty, and India Departments. **1941** *Trans Newcomen Soc.* XXI. 94 He [*sc.* Sir Joseph Whitworth] named it the 'Standard Wire Gauge'; it covered a range of 38 sizes of from 0·3 in. to 0·0018 in. diam. **1963** JERRARD & MCNEILL *Dict. Sci. Units* 151 The Standard Wire Gauge..classifies wire diameters in geometrical progression. **1971** B. SCHARF *Engin. & its Lang.* ix. 75 In Great Britain the main gauges are the Standard Wire Gauge (S.W.G., s.w.g.), and the Birmingham Gauge (B.G.).

b. Also *standard costs* Accountancy (see quot. 1959); freq. *attrib.*, as *standard cost card* etc. Hence *standard costing*. Also as parasynthetic adj., *standard-rated*.

1917 W. N. POLAKOV in *Trans. Amer. Soc. Mech. Engineers* XXXVIII. 587 Carrying out the analysis of the economy limit to its logical conclusion, the standard cost of the product is arrived at. **1918** G. C. HARRISON in *Industrial Managem.* LVI. 393/2 In the 'Standard Cost Card' shown in Form I it will be seen that the estimated or standard cost of the bolts is $11·079 per thousand. **1921** (*title*) Standard costing principles & practices for the plywood industry. **1935** C. M. GILLESPIE *Accounting Procedure for Standard Costs* iv. 62 The structure of standard cost cards becomes complex. **1959** *Chambers's Encycl.* I. 39/2 Two major developments in cost accounting must be briefly mentioned. Firstly the introduction of standard costing. Standard costs are estimates made in detail for operations, processes of articles on the basis of predetermined standards. **1974** *Terminol. Managem. & Financial Accountancy* (Inst. Cost and Managem. Accountants) 15 *Standard cost rate*, a rate calculated by dividing the expected overhead cost attributable to a cost centre by the predetermined quantity of the base to which the rate is applied.

Comb. **1972** *V.A.T.: Gen. Guide* (H.M. Customs) 30 Where an amount payable covers both standard-rated and zero-rated goods or services, the amount must be split in fair proportion. **1977** *Jrnl. R. Soc. Arts* CXXV. 440/1 The most sensitive systems might have legislation applied at a luxury rate whilst others could be standard rated.

e. Bridge. *Standard American*, the commonest system of bidding in the U.S.

1961 A. SHEINWOLD *Short Cut to winning Bridge* 13/1 Most of the hands in this book are bid according to the principles of 'Standard American'. **1963** H. SCHENKEN *Better Bidding* i. 16 The most popular system in this country is called Standard American... They all have one thing in common with Standard American: the opening bid of one in a suit is almost unlimited. **1968** ROTH & RUBENS *Mod. Bridge Bidding* p. xvi, Standard American is the most widely used although not..the best approach to bidding. **1976** *National Observer* (U.S.) 24 July 15/3 If you are playing under Precision's rules or by agreement in Standard American and you open light.., you can pass third hand with a subminimum count.

3. b. Also *standard work*.

1821 HAZLITT in *London Mag.* Aug. 179/2 Pictures are scattered like stray gifts through the world... There are plenty of standard works still to be found in this country. **1849** C. BRONTË *Shirley* I. v. 94 One should not be apathetic in studying standard works. **1922** JOYCE *Ulysses* 628 You know the standard works on the subject. **1969** J. GROSS *Rise & Fall Man of Lett.* iv. 115 He goes on to supply a list of standard works, English and European,..which would certainly keep most ordinary readers tied up for as far ahead as they could plan.

e. Applied to that variety of a spoken or written language of a country or other linguistic area which is generally considered the most correct and acceptable form, as *Standard English*, *American*, etc.; Received Standard; also, *standard pronunciation* = *received pronunciation* s.v. *RECEIVED *ppl. a.* 1 b.

1836 *Q. Rev.* Feb. 356 It is, however, certain that there were in his [*sc.* Higden's] time, and probably long before, five distinctly marked forms, which may be classed as follows:—1. Southern or standard English, which in the fourteenth century was perhaps best spoken in Kent and Surrey by the body of the inhabitants. **1859** *Proposal Publ. New Eng. Dict.* 3 As soon as a standard language has been formed, which in England was the case after the Reformation, the lexicographer is bound to deal with that alone. **1878** J. A. H. MURRAY in *Encycl. Brit.* VIII. 396/2 Chaucer's language is well known to be more southern than standard English eventually became. **1908** H. SWEET *Sounds Eng.* 7 Standard English, like Standard French, is now a class-dialect more than a local dialect: it is the language of the educated all over Great Britain. **1909** D. JONES *Pronunc. Eng.* i. 1 Standard Pronunciation. **1919** G. P. KRAPP (*title*) The pronunciation of standard English in America. **1925** —— *Eng. Lang. in Amer.* I. v. 296 The informal or local speech will often seem more penetrating, more genuine than the standard speech. **1947** PARTRIDGE *Usage & Abusage* 304/1 Standard English and Standard American are the speech of the educated classes in the British Empire and the United States. **1962** P. H. JOHNSON *Error of Judgment* ii. 5 Had spoken standard English varied by a few fancies such as 'crorss' for 'cross' and 'poyt' for 'poet'. **1975** *Times Lit. Suppl.* 7 Feb. 136/2 Local names pronounced in [Chinese] dialects widely different from 'Mandarin' or, as it must now be called, Standard Speech. **1978** *English Jrnl.* Dec. 7/1 There is also a kind of 'standard standard'. Some people call it 'broadcast' or 'publications' standard, because most newspapers and television news shows use it.

f. *standard time*: see *TIME *sb.* 27.

3*. *Math.* That does not involve infinitesimal quantities.

1961 A. ROBINSON in *Proc. K. Nederlandse Akad. v. Wetensch.* A. LXIV. 434 We consider in the first instance functions, relations, sets, etc. which are defined already in R_0 [*sc.* the set of all real numbers]... Such concepts will be called standard (functions, relations, sets, etc.). **1972** [see *NON-STANDARD *a.* b].

Hence **sta·ndardless** *a.* [-LESS], having no standard or standards; unprincipled; **sta·ndardness** [-NESS].

1912 GALSWORTHY *Inn of Tranquillity* 217 How can we help it, seeing that we are undisciplined and standardless, seeing that we started without the backbone that schooling gives? **1944** AUDEN *Sea & Mirror* in *For Time Being* iii. 29 On the shuddering edge of the bohemian standardless abyss. **1972** J. L. DILLARD *Black English* vi. 238 A combination of age-grading, status-grading, and peer group influence causes a special feature to operate among young males at about the age of puberty. At that time, their graph of standardness will actually swing 'downward' a bit. **1973** *N.Y. Law Jrnl.* 2 Aug. 4/8 Such a subjective determination as is proposed here lacks the necessary standards to insure a nondiscriminatory result. The danger of discrimination which inheres in such a standardless approval is..evidenced by the determination in question here.

standardizable (stæ·ndǎrdəizāb'l), *a.* [f. STANDARDIZE *v.* + -ABLE.] That may be standardized.

1922 *19th Cent.* Feb. 185 The sale of the more standardisable profits. **1980** *U.S. News & World Rep.* 1 Dec. 45/1 Legal clinics offering adoptions, divorces, wills and other frequently cut-and-dried standardizable services at low cost are flourishing.

standardly (stæ·ndǎrdli), *adv.* [f. STANDARD *a.* + -LY[2].] In a standard manner; according to common practice; normally, generally.

1957 *Proc. Aristotelian Soc.* LVII. 237 Knowledge of a particular event will standardly give one knowledge only of the location of its witnesses. **1978** A. RYAN in Hookway & Pettit *Action & Interpretation* 75 A game in which we are standardly beset by something like Prisoners' Dilemma problems.

‖**Standartenführer** (ʃta·ndāɹtenfü:rəɹ). Ger., lit. 'leader of the standard'.] Under the Third Reich, a commanding officer of a unit of the Schutzstaffel or Sturmabteilung.

1943 W. NECKER *German Army of To-day* iv. 168 *Staff Officers*:.. *SS-Standartenführer* = colonel. **1971** BENDER & TAYLOR *Uniforms, Organization & Hist. Waffen-SS* II. 72 SS-Standartenführer, promoted on Nov. 4, 1944 to SS Oberführer, Wilhelm Mohnke. **1973** 'D. JORDAN' *Nile Green* xliv. 225 He was on the staff of the SS Hauptamt. He ended the war as Standartenfuhrer. **1978** *Detroit Free Press* 16 Apr. (Record) 12/2 These strips had been collected by a German doctor who was writing a treatise on tatooes, and also by the 28-year-old wife of the Standartenfuehrer or commanding officer.

stand-by. Add: **1. c.** The state of being immediately available to come on duty if required; readiness for duty. Also *transf.* Usu. in phr. *on stand-by*. orig. *Naut.*

1946 R. E. HIGGINBOTHAM *Wine for my Brothers* iv. 75 I'm on stand-by—take the wheel in fifteen minutes. **1959** WALLIS & BLAIR *Thunder Above* (ed. 2) ii. 13 With a full load we could use an extra stewardess. You're on stand-by, aren't you? **1960** *Times* 22 Mar. 12/2 Darkness brought peace to the riot areas, but an active citizen force was put on standby. **1971** W. KEENAN *Murder on Melancholy* v. 48 We'll use two of the vans every night, the other will be on standby. **1974** D. FRANCIS *Knock Down* iv. 45 I'm on stand-by from four this afternoon for twelve hours... Most stand-bys are just a bore.

d. *spec.* in civilian aviation, a stand-by passenger; *on stand-by*, waiting for a stand-by seat; in possession of a stand-by ticket. See sense 9 below.

1961 'E. LATHEN' *Banking on Death* (1962) xix. 136 Four stand-bys who were convinced that by keeping in motion their chances of getting on a plane were improved. **1962** J. D. MACDONALD *Key to Suite* (1968) ii. 14 'You got in real early.' 'Earlier than I wanted to. But all they could do for me on anything later was to put me on standby.' **1970** *Guardian* 8 Sept. 11/4 All the flights are full. They're likely to be standby half the night for seats on any plane. **1973** [see *STAGE *sb.* 9 e].

4. *ellipt.* for: **a.** *stand-by credit*, *loan*, etc.; **b.** a stand-by fare or ticket.

1959 *Daily Tel.* 18 Dec. 20/5 It was in December, 1956, after the Suez crisis, that Britain drew $561 millions from the Fund and arranged for a 'stand-by' of $738 millions.. to be drawn upon if necessary. **1975** *Offshore* Aug. 4/1 The Snell tug was lounging around, tagging up a stand-by at a thousand dollars a day. **1980** *Daily Tel.* 29 Aug. 7/2 The £90 Scottish standby, the airline claims, is more than £25 cheaper than the second-class rail fare.

II. *attrib.* or as *adj.* **5.** Of a charge for electricity: remaining constant, fixed; levied for the availability of an electrical supply in a given period, irrespective of the amount used; *standby issues*: (see quot. 1940). Also *transf.*

1900 *Jrnl. Inst. Electr. Engineers* May 680 What are called the Standing or Stand-by Charges. Other items which might fairly be added to the stand-by charges are the rent, rates and taxes, and part of the management expenses. **1907** *Chambers's Jrnl.* 1 June 423/2 What are called the 'stand-by' costs are also much reduced in the gas system. **1931** *Discovery* Feb. 55/1 The 'standby charges' of internal combustion cars are negligible, and therefore running costs are extremely low. **1940** *Chambers's Techn. Dict.* 802/2 Stand-by losses, that part of the power expended in a generating station in order to maintain plant in instant readiness to take a load. **1961** *Gloss. Electrotechnical, Power Terms* (*B.S.I.*) 11. vii. 7 *Standby charge*, a financial charge for the availability of a supply under the conditions of a standby tariff, to be paid by the consumer irrespective of whether or not he makes use of the standby supply.

6. a. Of machinery or equipment: kept in a position of reserve, *spec.* in case of failure of a primary device or supply.

1908 *Sears, Roebuck Catal.* 205/2 The Stand-By Dry Batteries..will produce more current and last longer. **1930** *Engineering* 28 Feb. 295/1 A standby machine of this type can be useful. **1942** *R.A.F. Jrnl.* 3 Oct. 25 From the..roof, electric lights and stand-by hurricane lamps are suspended. **1954** 'J. CHRISTOPHER' *Twenty-Second Cent.* 122, I knew they would have the stand-by generators on in a minute or two. **1969** *Gloss. Terms Magnetic Components & Pinnacles* (*B.S.I.*) 4 *Stand-by starting coil*.. a magnetic compass which provides a secondary heading reference for steering a ship. **1973** *Daily Tel.* 8 Mar. 2/6 Stand-by electric generators kept for emergencies have been packed ready for transport.

b. Of (a body of) persons: on stand-by; available to come on duty. More generally, ready to stand in for another if required. orig. *Naut.*

1923 J. H. MCCULLOCH *Million Miles* iii. 59 It was the stand-by man from the other watch, dragging us out again for our four-hour battle on the deck. **1937** *Amer. Speech* XII. 100 The public is..not so accustomed to stand-by passenger or pianist, an artist who remains on call for emergency work. **1946** R. E. HIGGINBOTHAM *Wine for my Brothers* 178 Dane was stand-by man, and he started over the catwalk for the bridge. **1958** *Economist* 16 Aug. 507/2 When he asked the assembly to create a United Nations 'standby' peace force, he was making a gesture of conciliation. **1974** *Senator* (S. Carolina) *Daily Item* 18 Apr. 6a/1 The White House has secretly appointed wealthy campaign contributors to the standby corps, which would take over if Congress were to.... **1977** G. CLARE *Last Waltz in Vienna* (1982) ii. 144 My role was that of stand-by boy-friend very much playing second fiddle.

c. Similarly, of a vehicle or craft held in reserve.

1959 *Economist* 14 Mar. 992/2 In case of breakdown a standby vehicle [etc.] can be hired from the manufacturers when needed. **1974** *BP Shield Internat.* Oct. 20/1 I'm keeping in touch with our standby vessel, *Ottorburn*. **1976** P. R. WHITE *Planning for Public Transport* viii. 185 It may be cheaper to move to direct sale coupled with retention of standby vehicles and crews to provide extra bookings if required. **1982** *Times* 5 June 8/5 Several of the smaller frigates have been pulled out of standby fleet.

7. Designating a state, condition, or position of readiness. Also *stand-by duty*.

1924 *Western World* X. 355/1 The receiving telephones are hung on a special rest, this automatically putting the call-receiver in a stand-by position. **1944** *Daily Progress* (Charlottesville, Va.) 25 May 4/1 Ammunition plants which were closed or placed on a standby basis..have been ordered reopened. **1959** *Times* 11 Sept. 7/2 This new cell is specially designed for standby duties. **1977** *Cornish Times* 19 Aug. 15/3 The present Saltash ambulance station is manned from 7 a.m. to 11 p.m. each day with a stand-by-basis operating during the remaining eight hours.

8. Applied to an economic or financial measure prepared for implementation should certain conditions obtain; spec. *stand-by credit*: an additional credit facility reserved at low interest which may be drawn upon at standard rates if needed; cf. *line of credit* s.v. *LINE *sb.*[2] 30 c. Hence of loan arrangements, etc.

1947 *Sun* (Baltimore) 25 Nov. 14/3 The President's proposals for stand-by price and wage controls. **1957** *Encycl. Brit.* III. 56/2 It is a common practice for banks to grant their regular customers a 'line of credit' under which the bank agrees to extend loans up to a certain maximum... Borrowers sometimes pay a small interest charge, amounting to perhaps ¼% of the unused part of the line of credit, under so-called 'stand-by' agreements. **1957** *Times* 27 Dec. 13/1 The decision to ask for an extension of the $739 standby credit with the International Monetary Fund..is wise. **1966** *Economist* 4 June 996/1 Ways to stimulate business..preferably by winning from Congress standby authority to cut taxes across the board. **1973** *Times* 29 Nov. 6/5 We'd have to have some sort of standby tax authority..during the remaining time. **1977** *Time* 8 Aug. 18/2 Italy has repaid on schedule an International Monetary Fund stand-by loan.

9. In civilian aviation: designating a system of seat allocation whereby a passenger does not book in advance, but may board at a cheaper rate the next flight with spare unbooked capacity; also *stand-by fare*, *passenger*, *ticket*, etc. Also as *adv.* Cf. sense 1 d above.

1963 *Guardian* 9 Feb. 12/2 BEA had proposed a one third cut in fares for stand-by passengers, namely, people prepared to chance obtaining a seat after booked passengers had boarded the plane. **1963** *Daily Tel.* 19 Feb. 1/8 Stand-by night tourist fares in April and May represent a rate of less than 2d a mile... The full cost of the stand-by ticket will be refunded if the passenger fails to get a flight. **1968** 'A. YORK' *Predator* viii. 124, I had to come in on a standby tourist. But they've had a last-minute cancellation. **1970** D. HARRIS *Hijacked* (1971) 6 We've been notified not to accept standby passengers for Flight 901. **1977** *Daily Tel.* 16 Sept. 1/3 The cost of travelling stand-by from Gatwick is £149 as opposed to..the normal return fare between London and New York of £392. **1978** *Times* 18 July 2/6 The argument over where to buy stand-by tickets was causing much confusion... Pan Am and TWA handled many..stand-by passengers at their London offices.

standee. For *U.S.* read orig. and chiefly *U.S.* and add: **1.** (Earlier examples.) Also *transf.* in *Theatr.* U.S.

1831 *American* (Harrodsburg, Kentucky) 25 Mar. 1/3 'I say Captin, we've landed,' said a 'standee', roared a dozen voices.. 'Captain, I demand a berth.' **1849** G. G. FOSTER *N.Y. in Slices* xxiv. 99 Some public reporter.. thus nobly earns the privilege of a seat in the dress circle.. or a standee in the lobby on full nights.

2. a. (Earlier and later examples.) Also *spec.* a standing passenger in a public vehicle.

1856 *Knickerbocker Mag.* 278 Occasionally one of the standees, perched at the top, and the 'standees' are thrown against each other like alley-pins by a 'ten-strike'. **1934** WODEHOUSE *Right Ho, Jeeves* xvii. 211, I wedged myself in among the standees at the back. **1942** *Sun* (Baltimore) 22 Jan. 40/5 Mr. Maxwell admitted that the bus left the terminal at Overlea with fourteen standees. **1954** E. E. CUMMINGS *Lett.* 1 Dec. (1969) 238 A 90'clock, all 'standees' had to be given the seats of all ticketowners who..hadn't yet occupied theirs; so..the Brattle Theatre were not enough chairs... Major Barrimore, Superintendent of..some and Dr. Maine formed a rather ill-assorted group of standees. **1962** 'C. MARSH' *Dead Water* iii. 81 They were not enough chairs. **1967** J. LEE *Ninth Man* 70 No seats were vacant, and standees stared angrily at the fighters taking up so much space. **1982** R. B. FLEXNER *Listening to America* 109 The stands..seated 4,000 fans,..with over 300 more added standees also watching.

b. *attrib.*, esp. of public transport vehicles, as *standee bus*.

1937 *Sun* (Baltimore) 27 Oct. 5/2 Spectators began storming the playhouse long before curtain time, and by time the show began, there were a standee audience. **1952** *Public Transport Assoc. Jrnl.* July 336 Large capacity single deck motor buses of a special construction... The Minister had under review the whole question of standing passengers (including standing passengers on 'Standee' type vehicles). **1959** KITCHIN & WESLOCK *Road Transport Law* (ed. 12) 124 Laden weight of p.s.v. is total of vehicle with water, oil and fuel, plus 140 lb per seat (and, in the case of 'standee' vehicles registered after December 31, 1954, an additional 140 lb for each standing passenger in excess of eight). **1962** *Daily Tel.* 5 Nov. 20/5 The introduction of 'standee' buses, carrying 36 standing passengers and 35 seated. **1970** *Commercial Motor* 25 Sept. 102/2 The stanchions which one would expect in a 'standee' bus are not fitted, since passengers in general may have handrails and stanchions would best be fitted. **1976** P. R. WHITE *Planning for Public Transport* iv. 80 The pattern of three or four sets of sliding doors per car...is related to the much more limited seating capacity, often arranged longitudinally, and a high proportion of standee space. **1985** *Dallas Morning News* 1 Feb. 56/3 This [bus] would have a dual exit and a standee area on the lower deck.

stander. Add: **I. 1. d.** One who 'stands' another a drink: see STAND *v.* 61 b. *nonce-use.*

1922 JOYCE *Ulysses* 419 Will immensely splendiferous stander permit one stooder..to terminate one expensive ..libation.

stand-in. [f. vbl. phr. *to stand in*: see STAND *v.* 95.] **1.** A friendly or profitable understanding (*with* another), esp. a corrupt arrangement or 'put-up job'. *U.S. colloq. Obs.*

1870 *Food Jrnl.* 1 Nov. 523 The affair is settled amicably by a 'stand in', which means that the purchaser shall buy the other, or others, a certain sum not to bid against him. **1908** H. E. MCCARTY *Scrapes of Snow Girl* 89 My heart went out to him the minute he said he had a stand in with the right people. **1926** J. BLACK *You can't Win* vi. 91 The whole thing was a 'stand in' from the captain down. Everybody's satisfied. The sucker has his money, and the city gets the fine.

2. a. *Cinemat.* One who substitutes for a principal film actor while the cameras and lighting for a scene are set. Formerly, *stand-in man*. Chiefly *U.S.*

1928 *N.Y. Times* 25 Mar. VIII. 6/2 *Stand-in man*, substitute for the star used by the director when rehearsing actions and testing lights on a scene. **1935** *Evening Sun* (Baltimore) 26 Sept. 46/1 Gene Dorothy Granger, actress, and George Lollier, actor and 'stand-in' for Richard Dix, had a long-planned wedding last year. **1937** *Daily Mirror* 8 Mar. 5/2 Frances is often described as Olmsted's double, but it's the first time she has been his stand-in. 'Being a 'stand-in' does not necessarily mean that one is an understudy or even requires any acting. **1948** 'N. SHUTE' *No Highway* vi. 91 A Hollywood beauty.. In twenty years, and it must be years before she got an inkling what it was that differentiated her from all the stand-ins and walkers-on. **1958** *People* 4 May 15/1 He won't use a stand-in or double for his fights. **1976** M. MACHLIN *Scratchproof* ii. 22 The stand-ins were called for. The shot was lined up.

b. *gen.* One who fills the place of or substitutes for another. Also *transf.*

1937 D. RUNYON in *Collier's* 21 Aug. 22/1 Nobody cares much about having a stand-in for Nicely-Nicely..and many citizens are in favor of pulling out of the contest altogether. **1940** *Punch* 7 Feb. 156/1 Easily bored by the public functions that our even officials are forced to do.. the German ruler has made..use of the 'stand-in' since he came to power. **1944** *Ann. Reg. 1943* 1 Apr. (B ed.) 3/1 An absorbable gelatin sponge..may serve..as a stand-in for the clot where it becomes necessary to remove part of a clot in the tissues.. **1958** S. ELLIN *Eighth Circle* (1959) II. 3, I like arrest was a fake; he was just a stand-in for the real culprit. **1968** T. STODDARD *Real Inspector Hound* (1970) 11 An army of assorted stand-ins, understudies, tools or the rumours-up, the right-hand men.. **1980** *Daily Tel.* 24 Nov. 17/2 He has also a stand-in to. **1980** *Times* 18 Nov. 1/4 As a decoy, a scapegoat, a stand-in for us at the time.

c. *attrib.*

1938 *N.Y. Times* 28 Dec. 11/1 'Stand-in' ruse jails 3... Police game collectors, with previous convictions on which they might have been convicted, were using 'stand-ins' to receive new sentences for them. **1958** *Engineering* 11 Apr. 457/3 Preliminary experiments were made on 'stand-in' compounds, which it was hoped would simulate the behaviour of polonium. **1976** *Southern Even. Echo* (Southampton) 25 Nov. 15/4 Stand-in goalkeeper, Les Northcote, stood between Tonbridge and a hammering at the hands of unbeaten Salisbury.

standing, *vbl. sb.* Add: **4. h.** *Law* (orig. *U.S.*). A position from which one has the right to prosecute a claim or seek legal redress; the right itself; = *locus standi* s.v. LOCUS *sb.*[1] 4. (See also quot. 1962.)

1924 *Chicago Junction Case* in *U.S. Reports* CCLXIV. 271 Mr. Justice George Sutherland, dissenting... The complainants have no standing, to vindicate the rights of the public. **1962** *Stanford Law Rev.* May 433 Defined generally, if not very helpfully, in the context of this article 'standing' is the word of art used to express the right to invoke the power of a court. **1967** H. W. R. WADE *Administrative Law* (ed. 2) iv. 126 These remedies are matters of policy, and in which the federal courts hold worthy of legal protection from the effects of unconstitutional governmental action. **1967** H. W. R. WADE *Administrative Law* (ed. 2) iv. 126 These remedies..show..for the purpose of conferring standing..upon the prerogative orders. **1972** *N.Y. Law Jrnl.* 24 Oct. 13/1 It is sufficient for purposes of standing that plaintiff establish a causal connection between the violations alleged, by they fraud or breach of fiduciary duty, and plaintiff's loss. **1984** *Law Reports* (Appeal Cases) June 639 The rules as to 'standing' for the purpose of applying for prerogative orders..are not the same as ..

9. b. The position of a person or organization in a graduated table, esp. in *Sport* and *Educ.* Also, a score indicating this. Freq. *pl.*

1881 *N.Y. Herald* 12 Sept. 11/5 The appended table will show the standing of the clubs to date. **1904** *Spalding's Official Base Ball Guide* (ed. 28) 108 To find the Standing of the Clubs—Divide the number of games won by games played. Example: Pittsburgh, in 1903, played 140 games and won 91; 91 divided by 140 equals .650. **1917** E. EARLE *Life of U.S. Naval Academy* v. 111 Class standing is affected in some measure by conduct. **1938** R. BANNING *Annapolis Today* vii. 99 Poor standing in studies and conduct records.. in combination with the aptitude marks, was given to determine their final class standings, for which the courses are computed. **1968** *Globe & Mail* (Toronto) 13 Feb. 8/1 The appointment leaves 10

vacancies in the 102-seat Senate. Standings now are: Liberal 60; Conservative 29; Independent 2; Independent Liberal 1. **1977** *Belfast Tel.* 24 Jan. 17/7 He now leads the world drivers' championship standings with 13 points.

11. standing point (earlier example); **standing room** (earlier *Theatr.* and later examples); also in phr. *standing room only*, esp. in a theatre or similar place of resort (abbrev. *S.R.O.* s.v. *S 4 a).

1847 W. SMITH tr. *Fichte's Characteristics Present Age* xvi. 254 A new Standing point in life, or, as we may say, a new sense is thereby unfolded. **1807** *Cabinet* I. 54 To be rewarded at last..by finding—a little standing room! **1889** G. B. SHAW *Fabian Ess. Socialism* 11 The board is at the door, inscribed 'Only standing room left'. **1910** *National Police Gaz.* (U.S.) 5 Nov. 2/2 The Davenport Lady Minstrels opened their season in Asheville, N.C., recently, to standing room only. **1934** WODEHOUSE *Right Ho, Jeeves* xi. 100 The place being loaded down above the Plimsoll mark and standing room only. **1973** *Times* 1 June 7/2 Lord's became..standing room only for Gooch's first hundred. **1956** MRS. L. B. JOHNSON *White House Diary* 5 Jun. (1970) 154 A crowded affair that would have been standing room only. **1965** S. THOMPSON *Murder Reported* xix. 146 That bar is mobbed. Standing room only.

standing, *ppl. a.* Add: **I. 1. c.** *spec.* in *Sport* (esp. *Athletics*): performed from a standing position (cf. *CROUCH *sb.*[2] b). Also *standing start*, of a motor car, etc.: a start, esp. of a race or performance trial, from a stationary state.

1875 *Encycl. Brit.* III. 13/1 The running hop-step-and-jump, standing high-jump, standing long jump and high jumps are fairly exercised. **1906** *Daily Chron.* 14 Sept. 1/6 The standing-jump. **1908** H. GRIFFIN *Athletics* 65 The standing long and high jumps are fairly exercised. **1907** *Daily Chron.* 2 Oct. 9/2 Jenatzy..covered..the first kilometre in 1 min. **1932** *World Ot.* 9/2 ..the first kilometre, from a standing start, in 3 min. 1 sec. **1953** *Illustr. London News* 9 Dec. 962/1 (Advt.), Speed up Brooklands Test hill from a standing start, 18.76 m.p.h. **1965** *Times* 29 Nov. 4/1 It had been timed at.., from standing start, a type of start in which the horses line up exactly at the pole marking the distance they are to run and break at the starter's command... No starting gate is used. **1960** E. S. & W. J. HOGAN *High Speed Rugby* vii. 64 When a long pass is needed, the 'dive' pass, done properly, has advantages—even from a standing position. **1973** D. RUTHERFORD *Kick Start* vii. 139, I was making a standing start... The Norton Commando accelerates from 0 to 100 m.p.h. in 15 seconds.

f. *standing ovation*: a rousing ovation conferred by an audience standing as a mark of enthusiastic approval, esp. after a speech.

1960 J. RUSSELL *Autobiogr.* III. ii. 87, I was deeply touched by being given a standing ovation when I rose to speak. **1968** D. WITTON *Labour Govt.* xxvii. 564 At the end there was a spontaneous and wild standing ovation. **1981** S. JACKMAN *Game of Soldiers* ix. iii. 175 The men gave him a standing ovation, whistling and stamping as he stood there grinning and bowing.

2. c. *standing crop*, a growing crop; now used *spec.* in *Ecol.* to denote the total quantity of living things in an (esp. planktonic) eco-system, or in some component of one.

1861 J. BROWN *Forester* (ed. 3) vii. 477 There is great danger of having it [sc. fire] among the standing crop, and very often to the damage of a considerable portion of the standing crop. **1935** P. S. WELCH *Limnol.* ix. 215 Various schemes..available for the production of a lake can be expressed in terms of (1) standing crop—the total amount of plankton present in the water on a selected date—and (2) annual crop. **1946** *Ecol. Monogr.* XVI. 324/1 The size of the standing crop at any given time is the result of the summation of the excess of production over destruction from the beginning of the growth of the population to the moment of observation. **1957** G. E. HUTCHINSON *Treat. Limnol.* I. xii. 746 The addition of a large quantity of phosphorus in 1938 undoubtedly led to a considerable increase in the standing crop of the lake. **1979** R. BREWER *Princ. Ecol.* iv. 129 At any one time an area has a certain amount of living energy stored in biomass, often referred to as standing crop.

3. c. *standing iron*, a metal spike on the collar of a sledge dog, to which a ribbon or similar decoration may be attached. *Canad.*

1934 P. H. GODSELL *Arctic Trader* 59 Bells jangled as the dogs proudly tossed their heads and displayed their bells and standing irons. **1939** *Beaver* Sept. 23 Fox tails and coloured ribbons decorated the leather collars and standing irons. **1960** J. W. GOSSELL *I want no Lady* iv. 64 The fluttering ribbons of ribbons on the standing irons of the team.

4. b. *standing salt*: in medieval and later times, a large, often ornate, salt-cellar placed in the middle of a dining-table. Cf. SALT *sb.*[1] 7 a, b. *Hist.*

1878 W. J. CRIPPS *Old Eng. Plate* x. 255 A cylindrical standing salt, of the year 1554. **1931** E. WENHAM *Domestic Silver* v. 42 The imposing standing-salts.., from the Middle Ages to the third quarter of the seventeenth century, were the symbols of social distinction. **1956** G. TAYLOR *Silver* iii. 43 The most important, and often the most elaborate piece of table plate during the Middle Ages was the standing salt. **1972** *Times* 18 Nov. 24/1 (Advt.), An Elizabeth I silver-gilt standing salt.

II. 11. d. Also *standing matter, type.*

1875 J. SOUTHWARD *Dict. Typogr.* (ed. 2) 129 *Standing matter*, composed matter which remains standing for some future use. **1916** *Estimating for Printers* 24 If a job repeats and the printer has kept it standing without ..rent, the job should be estimated as though it were reset, and the advantage of standing type kept by the printer. **1964** F. BOWERS *Bibliogr. & Textual Crit.* i. v. 31 This hypothesis could be supported by other variants in the standing type..but not to their pages of reset type.

e. Physics. *standing wave*, a wave in which the positions of maximum and minimum oscillation remain stationary; = *stationary wave* s.v. *STATIONARY *a.* 1 e.

1896 *Knowledge* 1 June 136/1 Each wave crest maintains its position relative to the stone, and from this comes the term *standing wave*. **1905** *Trans. R. Soc. Edin.* XLI. 592 The remarkable analogy between the sound-vibrations of an elastic body and the light-vibrations of a radiating atom is at least suggestive. It is not, for instance, conceivable that the latter are caused by 'standing waves' in the elastic system which conveys them to the atom? **1947** A. E. SLATER & A. C. DOUGLAS *Gliding & Adv. Soaring* i. 29 Similar 'standing waves' have now been found, and soared in, elsewhere. For instance, in an experimental flight to leeward of the Alps a sailplane has reached 30,000 feet above sea level. **1962** A. NISBETT *Technique Sound Studio* ii. 40 Standing waves may arise between opposite sides of the microphone a standing wave pattern will be set up. **1977** A. HALLAM *Planet Earth* 55/2 Such features, called antidunes or standing waves, are often seen in streams running across beaches at low tide.

III. 15. a. Also *standing order*, a written directive to a banker instructing that a regular payment be made from an account, usu. to another party; similarly, *transf.* in *Commerce*.

1913 BAGSHAW & HANNAFORD *Pract. Banking* ix. 144 Executing Standing Orders... A customer may instruct his banker to pay a certain sum at a stated time, and the main instances of this are in the payment of annual subscriptions to clubs. **1937** A. F. FRANCIS *Pract. Branch Banking* xv. 207 All those regular periodical payments, such as rent, insurance premiums, club subscriptions.. which are made by the Branch on the instructions of customers, come under the general description of 'Standing Orders'. **1962** D. FRANCIS *Dead Cert* xi. 134 I'll place a standing order with Interflora, for lilies. **1972** L. LYALL *Blame Dead* ix. 67 The yearly premium is paid by standing order, so the payments to the life insurance companies and probably the rest of them...

17. b. *standing committee* (earlier and later examples).

a 1676 H. ELSYNGE *Expedicio Billarum Antiquitus* (1954) 23 Even this doth prove that there were standing Committees. **1857** *N.Y. Times* 15 Jan. 4/4 The House of Commons are divided into standing committees. **1868** H. SCOBELL *Memorials* 19 In Parliament there have usually been Five Standing Committees appointed at the beginning of the Parliament, and remaining during all the Session. **1868** ERSKINE MAY *Law of Parl.* (ed. 6) xiv. 379 There is further an exceptional class of committees, called standing committees. The only committee properly so termed is one to whose appointment, being by standing order, is permanent, the nomination only being renewed from session to session. **1930** ...the committee of public accounts, under a standing order of the 23rd April 1861. **1942** *Legislative Assembly Deb.* (Delhi) 17 Mar. 1461 A Standing Finance Committee of this Assembly has been appointed. It is not proposed—at present at any rate—to appoint any other Standing Committee of the Legislature. **1967** J. D. LEES *Committee System U.S. Congr.* ii. 18 Standing committees..are permanent committees that continue from Congress to Congress... The U.S. Congress and the..Committee system used in the House of Commons which are not specialised and whose main purpose is to save time of the House. **1978** *National People's Congress* appoints a standing Committee. Theoretically at least, that committee (made up of a Chairman, 13 Vice-chairmen, a Secretary-general and 35 ordinary members) is all-powerful.

IV. 19. Naut. *all standing*: see STAND *v.* 24.

attitude. no. [illegible]

stand-off, *attrib. phr.*, *a.* and *sb.* Add: **A.** *attrib. phr.* and *adj.* **3.** Of an object: that projects or is positioned a short distance away from a surface or another object; that serves to hold something in such a position.

1932 *Chambers's Jrnl.* Feb. 148/2 The well at the base of the shower-unit has a stand-off waste-pipe and the well can be used as a foot-bath. **1962** *Air-Cushion Vehicles* I. 28/2 On top of the trusses are secured box-section full-length longerons, known as 'stand-off booms', which are braced together and complete the primary structure. **1964** R. F. FICCHI *Electr. Interference* v. 52 Component manufacturers have developed the ceramic stand-off, or stand-up capacitor... **1977** *Gramophone* Oct. 743/1 The GC300..is finished with a..matt black aluminium plate with four stand-off feet.

4. *Mil.* Of a guided missile: designed to be launched against its target from an aircraft at long range, esp. as *stand-off bomb*. Also *stand-off range.*

1957 *Times Survey British Aviation* Sept. 2/1 Improved marks of the V-class bombers will carry a powered guided bomb (the so-called 'stand-off bomb'), the first weapon ..carried for the striking force for many years. **1960** A. BUCHAN *Nato Guided Missile*..by the so-called 'stand-off bombs'. **1969** *Times* 30 Apr. 2/3 The Swingfire anti-tank weapon is now being delivered to the British Army, and we hope that the Division will shortly obtain a contract to develop Swingfire further by exploiting its ability to 'kill' armour from the air and at long stand-off range. **1971** E. LUTTWAK *Dict. Mod. War* 181/1 *Stand-off missile*, a missile, generally fitted with a nuclear warhead, which is launched by a bomber and substitutes for the latter in the final phase of the attack. **1978** R. V. JONES *Most Secret War* xlvi. 463 A long-range glider bomb, the BV 246.. an example of a 'stand-off' missile.

B. *sb.* For *U.S.* read Chiefly *U.S.* (Earlier U.K. example.)

1865 TROLLOPE *Can You forgive Her?* II. xxiv. 183 There's a lot of talk about some women—what the men call a 'nollimy tangere'.

b. *Mexican stand-off*, no chance to benefit (or *spec.* to defend oneself); hence, a general stalemate (cf. sense 3 in Dict. and Suppl.).

1891 *N.Y. Sporting Times* 10 Sept. 4/3 'Monk' Cline, who got a Mexican stand-off from Dave Rowe last season with Louisville. **1929** HOSTETTER & BEESLEY *It's a Racket!* 231 *Mexican stand-off*, to kill and escape. **1950** J. O'HARA *Appointment in Samarra* vii. 202 That's the ..Mexican stand-off as we called it. **1953** *Time* 14 Sept. 8 It's the..in Chicago... when seven men were given the Mexican stand-off against the ..in the middle of a gang garage. **1958** 'W. HENRY' *Seven Men at Mimbres Springs* xvi. 189, I rightly and firmly believe we've taken some of the ... **1979** D. MACKENZIE *Raven settles Score* 16 It was a Mexican standoff. Neither could go to the law but..nor could the Koreans.

c. In *gen.* use, any uneasy stalemate or deadlock; an impasse. Freq. in *Pol.* contexts.

1958 *Spectator* 31 Oct. 585/1 On the Fuchs-Hillary stand-off whether to continue to Scott Base or not. Sir Vivian tactfully writes: 'Unfortunately this exchange became known publicly.' **1971** T. W. ROBINSON *Cultural Revolution in China* i. 15 The standoff with Cambodia was precipitated by the propaganda activities of the Chinese embassy and the New China News Agency. **1974** *Spartanburg* (S. Carolina) *Herald* 19 Apr. A1/3 A 19-year-old Colombian..died in a hail of police bullets after a stand-off all day. **1982** W. CUNNINGHAM *Case of Sliding Pool* (1982) xii. 145 We can't do anything, neither can he. It's a standoff.

3. (Examples.)

1843 J. H. GREENE *Exposure Arts Gambling* 287 Thus, if a man bets on the ace and deuce, and if the ace comes to his side, and the deuce to the dealer's side, it is called a stand-off, and no one wins. **1891** 'MARK TWAIN' in *St. Nicholas* Nov. 21/2 It was about a stand-off; so both of them had to whoop up their dangerous adventures, and try to get ahead that way. **1958** *Sun* (Baltimore) 17 May 44/5 Supreme Court decisions were about a standoff. There was a ruling against the General Electric in a patent suit... The High Court agreed to consider the validity of the TVA on an appeal by eighteen southeastern utilities. **1964** F. BOWERS *Bibliogr. & Textual Crit.* iii. 115 But in *Old Fortunatus* the evidence is rather a stand-off. **1968** *Times* 11 Oct. 5 contain the uncorrected state of A and 6 the uncorrected state of B. **1974** *Cleveland* (Ohio) *Plain Dealer* 26 Oct. 7-D/7 Cleveland Benedictine and Cambridge battled to a 0-0 standoff in a non-league football game here Friday night.

4. (Examples.) Also *fig.* *U.S.* Now *rare*.

1883 B. HARTE *Carquinez Woods* 63, I reckon you'd better make it [*sc.* a bet] a stand-off. **1885** ... that'll give us both a stand-off..I'll find out and let you know. **1891** R. E. RYAN *Tod on Hills* xv. 186 ... I'd suggest a stand off on the bootlities—till your return. **1906** 'O. HENRY' in *Everybody's Mag.* Aug. 165/2 I've suggested a stand-off at a delicatessen but down-town.

5. A rest; a temporary cessation from work. Cf. *STAND *v.* 96 g. *rare.*

1918 *Jrnl. R. Naval Med. Service* IV. 181 He should have had two stand-offs. **1930** C. R. SAMSON *Fights & Flights* 100 He told me..to give my cars a stand-off for the rest of the day.

6. Rugby Football. *ellipt.* for *stand-off half.* Cf. *fly-half* s.v. FLY *sb.*[1] 8.

1922 *Daily Mail* 11 Nov. 11 Cassels at stand-off seeming to be able to take any sort of pass. **1939** *Daily Tel.* 18 Dec. 11/3 Their backs, with the exception of P. Hodgson, at stand-off, are untried... **1959** *Listener* 14 May 700/3 Castleford..are captained by their stand-off Alan Hardisty. **1980** *Sunday Times* 7 Sept. 35 Even now, too days later, it still has something of a whirl for the Lions stand-off.

7. Something serving to hold an object clear of a surface or another object.

1967 *Boston Sunday Herald* 24 Mar. 1 22/1 (Advt.), Two may strap stand-offs. **1974** ... *Phys. Bull.* Dec. 592/2 Two quartz blocks are positioned adjacent to the device, and a preformed gold tape is bonded across the quartz standoffs and the back contact of the diode. **1979** *Engin. Materials & Design* Aug. 56/2 The brochure details a selection of fasteners, from a plain self-clinching nut or stud, to standoffs, self-locking, floating and miniature types, having thread sizes from M2 to M12.

stand oil. Also **sta-ndöl.** [tr. G. *standöl*: see STAND *sb.*[3], OIL *sb.*[1] So called from its formerly being prepared by allowing linseed oil to stand.] Linseed oil or another drying oil that has been thickened by heating, used in paints, varnishes, and printing inks.

1908 LIVACHE & MCINTOSH *Manuf. Varnishes & Kindred Industries* (ed. 2) III. vi. 153 Reh's Stock Mixing Varnishes.—... (For instance, raw) linseed oil [illegible] follows. (1) Linseed oil, 'stand oil'.—Raw oil is heated without driers until it deposits mucilage and clarifies, after which it is tanked for three to four weeks before use. **1934** H. Hiler *Notes on Technique of Painting* iii. 234 A good varnish containing some standoil (and a hard resin). **1955** W. Gaddis *Recognitions* i. iii. 124 This Van Eyck,..how sharp the lines are, look how smoothly they flow, it's perfect painting in stand oil, isn't it. **1981** *National Gallery Report 1980–81* 69 The problem of distinguishing, in the dried state, between raw and heat-treated oils (stand oils) remains obstinately unresolved.

stand out. Add: Now usu. **standout** or with hyphen. **2. a.** One who stands out from the crowd; an outstanding or conspicuous person or thing. *N. Amer. colloq.*

1928 *Collier's* 29 Dec. 26/2 When the show opened, this girl had improved in her dancing so amazingly that she was a distinct 'standout'. **1936** *Sun* (Baltimore) 13 July 4/1 The smash-ups are staged realistically, the standout being that in which an apparently new..car hurtles over a parapet and arcs one hundred feet or so through the air. **1938** D. Baker *Young Man with Horn* iii. ii. 145 He was blessed above the run of band leaders. It isn't given to all organizers to have a stand-out in their organizations. **1944** *Sat. Rev. Lit.* (U.S.) 8 Jan. 14/3 John Hersey's novel about the AMG in Italy, 'A Bell for Adano', is likely to be one of the action stand-outs of the new season. **1958** *Times* 24 Nov. (Canad. Suppl.) p. xii/6, If they see a boy who is a standout,..they can't go wrong by putting him down as a good prospect. **1979** *Time* 8 Jan. 46/2 Another campus standout, Phyllis Wallace, now in her 50s, has been burdened by prejudice against blacks as well as women.

b. *attrib. phr.* or as *adj.*

1932 *Sun* (Baltimore) 15 Aug. 9/1 California's rowing triumph and Japan's first swimming conquest were the stand-out performances. **1955** Keepnews & Graver *Pictorial Hist. Jazz* xi. 117/2 Red Nichols and Miff Mole were the stand-out figures in this phase. **1978** J. Irving *World according to Garp* viii. 162 Robert Muldoon, a standout tight end for the Philadelphia Eagles.

standpoint. 2. (Earlier example.)

1854 Geo. Eliot tr. *Feuerbach's Essence Christianity* ix. 96 This abstraction..is determined by the essential standpoint of man.

stand-still, standstill, *a.* Add: **2. a.** Characterized by the absence or restriction of movement.

1829 P. Egan *Boxiana* 2nd Ser. II. 233 Jem had now reduced the 'big one' to his own weight, and had also placed him upon the stand-still system. **1852** J. Reynolds *Hist. Illinois* 266 The cotillions, or stand still dances, were not then known. **1927** *Daily Express* 27 Dec. 1 A 'standstill' order prohibiting all movements of cattle, sheep, pigs, or goats in thirteen counties. **1975** J. G. Evans *Environment Early Man Brit. Isles* ii. 46 Intervening organic horizons represent standstill phases when the climate ameliorated sufficiently for a continuous vegetation cover to form.

b. *Econ.* Of an agreement, etc.: that seeks to maintain the present state of affairs, esp. by deferring the necessity to repay an international debt. Also *transf.* and of the debt itself.

The original *standstill agreement* was concluded in 1931 between German and U.S. and other Allied banking and commercial concerns, and allowed for the postponement of German short-term credit repayments in the light of the country's severe economic plight.

1931 *Times* 17 Aug. 9/2 The agreements negotiated between the national 'standstill' committees and the German bankers will..be mentioned in the report. *Ibid.* 20 Aug. 9/1 The 'standstill' agreement for the prolongation of short-term credits to Germany was signed..today. **1932** *Sun* (Baltimore) 12 Sept. 7/4 The standstill agreement..was designed to lift the pressure of world creditors on Germany until that nation could nurse its financial affairs back into something like normal vigor. **1938** H. V. Hodson *Slump & Recovery* ix. 331 The story of Germany's 'standstill' debt in this period. **1940** *Economist* 10 Aug. 191/2 The new standstill agreement was signed on behalf of the British, American and Swiss Banking Committees. **1955** *Ann. Reg. 1954* 425 Although further 'reactivation' of standstill debts was effected during the war, the amount of standstill debts declined substantially. **1970** *Internat. & Compar. Law Q.* 4th Ser. XIX. ii. 270 Even an obligation of member States merely to abstain—a so-called standstill clause..must be clear enough as to permit its direct application. **1980** *Wall St. Jrnl.* 6 June 1/2 Diamond International disclosed that it is discussing several courses of action that include a 'standstill agreement' with Cavenham Holdings.

stand-up, *a.* and *sb.* Add: **A.** *adj.* **1. d. stand-up comic** (also **comedian**), a comedian whose act consists of standing before an audience and telling a succession of jokes.

1966 *Listener* 11 Aug. 194/1 In television complex sentences need to be eschewed, especially by stand-up comics. **1969** *Ibid.* 24 Apr. 588/2 The audience, used to jazz singers, satire companies or stand-up comedians, could make nothing of his work. **1969** *New Yorker* 14 June 92/2 Bruce had just begun to surface as a standup comic. *a* **1975** Wodehouse *Sunset at Blandings* (1977) vi. 44 I've known Home Secretaries who were as cheerful as stand-up comics. **1980** D. MacKenzie *Raven & Paperhangers* i. 9 Playing straight man to a stand-up comic in a Vegas night club.

2. a. Also, designating a cafeteria or other establishment at which patrons stand at a counter to eat or drink; also applied to the counter itself.

1920 H. G. Wells *Outl. Hist.* xviii. 130/2 A stand-up buffet for light refreshments. **1971** R. J. White *Second-Hand Tomb* v. 59 Jasper..took a peep into the stand-up bar. **1981** M. C. Smith *Gorky Park* i. vii. 87 A stand-up cafeteria next to the children's store.

B. *sb.* **4.** Also *rarely*, a counter at which stand-up refreshment can be obtained and consumed.

1897 'Mark Twain' *Following Equator* xiii. 143 He halted in front of the best restaurant, then glanced at his clothes and passed on, and got his breakfast at a 'stand-up'.

6. An act of failing to keep an engagement or 'date' with another; also, *to give* (someone) *the stand-up*. Cf. *Stand *v.* 103 q. *slang* (orig. *U.S.*).

1921 A. G. Empey *Madonna of Hills* vii. 56, I wonder if that jane's double-crossed me and and given me the stand-up? **1940** R. Chandler *Farewell, my Lovely* xxxix. 297 'It's a little late, but I've had a lot to do.' 'Another stand-up?' Her voice got cool. **1961** J. MacLaren-Ross *Doomsday Bk.* ii. vii. 172 Find out where he's ringing from if it's another stand-up. **1977** D. Ramsay *You can't call it Murder* i. 15 We made a dinner date... He didn't show up... I'd write it off as an ordinary stand-up, except that he left that ten-gallon job [sc. hat] behind.

7. A police identification parade. *U.S. slang.*

1935 A. J. Pollock *Underworld Speaks* 113/2 *Stand-up*, the police line-up; show-up. **1949** *Philadelphia Even. Bull.* 14 Apr. 2/4 Jackson was brought to City Hall last night to take a look at Norman in a police standup, but he could not positively identify the prisoner.

Stanford–Binet (stæ·nfǒɹd bi·ne). *Psychol.* The names of Stanford University and Alfred Binet (1857–1911), used *attrib.* and *absol.* to designate the revision and extension of the Binet-Simon intelligence tests (see *Binet–Simon) undertaken by L. M. Terman and first published in 1916, which established the concept of an intelligence quotient.

1918 Cuneo & Terman in *Pedagogical Seminary* Dec. 414 The purposes of this study were..to correlate the results of Stanford–Binet tests with school marks. *Ibid.* 428 The Stanford–Binet scale was given to 112 representative school children. **1937** Terman & Merrill *Measuring Intelligence* iii. 51 The old Stanford–Binet..yields mental ages slightly too high at the younger ages. **1939** *Brit. Jrnl. Psychol.* XXX. 8 The Stanford–Binet test gives a good estimate of general intelligence though it gives verbal ability undue weight. **1954** A. Anastasi *Psychol. Testing* i. 11 In America, a number of revisions were prepared, the most famous of which is..known as the Stanford–Binet. **1974** R. M. Pirsig *Zen & Art of Motorcycle Maintenance* (1976) i. vii. 80 His Stanford–Binet IQ, which is essentially a record of skill at analytic manipulation, was recorded at 170.

Stanhope. Add: **4.** The name of the historian Philip Henry Stanhope (1805–75), 5th Earl Stanhope, used *attrib.* and *absol.* to designate the historical essay prize founded by him at Oxford University in 1855, or essays associated with this prize.

[**1856** *Oxford Univ. Calendar* 273 Earl Stanhope's Prize. In a Convocation holden on Friday, December 14, 1855, the following Regulations for a Prize founded by the Right Hon. Earl Stanhope, with a view to the full development and the continued growth and welfare of the School of Law and Modern History, and accepted by Convocation on the 20th day of June, 1855, were agreed upon.] **1861** J. A. Symonds *Let.* 17 Apr. (1967) I. 285 Conington..declares the Stanhope to be about as much too long. *Ibid.* 1 June 296, I confess to being a little anxious myself about the Stanhope, though one ought never to be disturbed on account of Prizes. **1863** *Oxford Univ. Calendar* 105 (*heading*) Stanhope historical essay. **1896** J. Buchan in W. Buchan *John Buchan* (1982) iv. 81 Get the Newdigate. Get the Stanhope. **1915** A. Huxley *Let.* 8 June (1969) 71, I may come up here in September to do some reading in Bodley for the Stanhope Prize. **1916** *Ibid.* 18 June 101, I trust I don't have to read the Stanhope at the Encaenia. **1934** *Oxford Univ. Gaz.* 21 Nov. 140/2 Stanhope Essay Prize, 1935. The subject for 1935 is 'Edward, the Black Prince.' **1976** M. Green *Children of Sun* ii. 51 John Buchan..went to Oxford, where he won the Stanhope Essay Prize, the Newdigate Poetry Prize..and a first-class degree.

stanhopea (stænhōᵘ·piă). [mod.L. (J. Frost 1829, in *Curtis's Bot. Mag.* LVI. 2948), f. the name of Philip Henry Stanhope (1781–1855), 4th Earl Stanhope, President of the Medico-Botanical Society.] An epiphytic orchid of the genus of that name, native to tropical America and bearing large, often fragrant, flowers.

1829 *Curtis's Bot. Mag.* LVI. 2948 (*heading*) Splendid Stanhopea. **1890** W. Watson *Orchids* lxi. 465 Some cultivators use nothing but sphagnum for Stanhopeas. **1917** L. H. Bailey *Stand. Cycl. Hort.* VI. 3222/2 Stanhopeas enjoy a shady, moist location. **1962** *Amateur Gardening* 7 Apr. 6/1 The stanhopeas have large highly scented wax-like flowers.

stanine (stæ·nəin). *Psychol.* [Blend of Standard *sb.* and Nine *sb.*] A nine-point scale on which test scores can be grouped in descending order of achievement, first developed by the United States Air Force in 1942 (see quot. 1968); also, a score on such a scale. Freq. *attrib.*

1945 *Sun* (Baltimore) 1 Oct. 6/5 The result..was a 'stanine' rating (stanine being an invented word, from 'standard of nine'). **1957** E. R. Hilgard *Introd. Psychol.* (ed. 2) xvi. 400/2 A score of..650 on a Graduate Record Examination, or 8 on the Air Force stanine scale is above that achieved by 93 per cent of those on whom the test was calibrated. **1961** A. Anastasi *Psychol. Testing* (ed. 2) iv. 93 Raw scores can readily be converted to stanines by arranging the original scores in order of size and then assigning stanines in accordance with the normal curve percentages. **1968** *Internat. Encycl. Social Sci.* I. 38/1 Single-digit standard scores, ranging from 1 to 9, with a mean of 5 and a standard deviation of 2 are called *stanines* (standard nines). **1976** *Word 1971* XXVII. 330 The performance of the Experimental children placed them at the second stanine with respect to French-speaking children in Montreal. **1976** *Woman's Day* (U.S.) Nov. 62/2 But few educators use such simple terms anymore. Instead, we may be unnecessarily confused by a *stanine* score.

Stanislavsky (stænislæ·vski). *Theatr.* Also **Stanislavski.** The name of the Russian actor and director Konstantin *Stanislavsky* (1863–1938), used *attrib.* to designate the style and technique of acting practised and taught by him: see *method *sb.* 2 e.

1924 J. J. Robins tr. *Stanislavsky's My Life in Art* xxxiii. 351 Chekhov gave that inner truth to the art of the stage which served as the foundation for what was later called the Stanislavsky System. **1958** *Spectator* 22 Aug. 248/3 The success of the Bolshoi Theatre Ballet rests on the application to dancing of the Stanislavsky technique. **1972** *Village Voice* (N.Y.) 1 June 55/4 The Stanislavski approach may meet one of its greatest challenges in a Stein play. **1974** T. P. Whitney tr. *Solzhenitsyn's Gulag Archipelago* I. i. x. 387 Krylenko abandoned the Stanislavsky method, didn't assign the roles, relied on improvisation.

Hence **Stanisla·vskian, -yan** *a.*

1958 *Spectator* 20 June 813/1 Stanislavskyan acting was being practised by Schepkin fifty years before the genesis of the System. **1965** *New Statesman* 8 Oct. 537/1 Gielgud..is the outstanding Stanislavskian actor in the West. **1979** A. Williamson *Funeral March for Siegfried* x. 50 He would have thought most [singers]..incapable of expressing so lucid a Stanislavskian principle.

Stanley (stæ·nli). The name of Edward Smith *Stanley* (1775–1851), 13th Earl of Derby, zoologist, used *attrib.* to denote birds named in his honour, as **Stanley bustard,** a large black and brown bustard, *Neotis denhami* (formerly *Otis stanleyi*), native to south-eastern Africa; **Stanley crane** = *paradise crane* s.v. *Paradise *sb.* 8; (*Tetrapteryx paradisea*, formerly *Anthropoides stanleyanus*).

1831 J. E. Gray *Zool. Misc.* 12 Stanley Bustard. *Otis Stanleyi.* Above vermiculated, black and white. **1884** R. B. Sharpe *Layard's Birds S. Afr.* (ed. 2) 634 Stanley Bustard..is common in the northern portions of the colony. **1912** J. Stevenson-Hamilton *Anim. Life Afr.* xvii. 273 The Stanley bustard..is partial to..the high country of East and South Africa. **1979** K. B. Newman *Birdlife in S. Afr.* xii. 134/2 Both the Kori Bustard and the second largest, the Stanley Bustard, perform rather grand courtship displays. **1867** E. L. Layard *Birds S. Afr.* 303 The 'Stanley' or 'Blue' Crane is not abundant in any locality. **1937** *N. & Q.* 19 June 450/2 A Stanley Crane, elegant in dove-like grey. **1958** [see *paradise crane* s.v. *Paradise *sb.* 8]. **1966** E. Palmer *Plains of Camdeboo* xi. 187 Our blue crane is also known as the Stanley crane. It belongs only to South Africa.

stannite. Add: **2.** (Earlier and later examples.) [ad. F. *stannine* (F. S. Beudant *Traité élémentaire de Minéralogie* (ed. 2, 1832) II. 416).]

1868 J. D. Dana *Syst. Mineral.* (ed. 5) 68 Stannite... Probably tetragonal, and hemihedral like chalcopyrite. **1976** [see next].

stannoidite (stæ·noidəit). *Min.* [f. Stann(ite + -oid + -ite¹.] A sulphide of copper, tin, iron, and zinc, $Cu_8(Fe,Zn)_3Sn_2S_{12}$, found as yellow-brown orthorhombic crystals having a metallic lustre, superficially resembling those of stannite.

1969 A. Kato in *Bull. Nat. Sci. Mus. Japan* XII. 165 (*heading*) Stannoidite,..a new stannite-like mineral from the Konjo Mine, Okayama Prefecture, Japan. **1975** *Econ. Geol.* LXX. 834 Mawsonite and stannoidite characteristically occur with bornite and chalcopyrite in xenothermal ore deposits in Japan. **1976** *Nature* 10 June 482/2 The Japanese ores, however, contain tin minerals other than cassiterite (for instance, stannite, stannoidite, and mawsonite).

stanol (stēⁱ·nǫl). *Chem.* [f. *st(er)ol about -an(e, after *cholestanol*, *ergostanol*, etc. (the respective saturated derivatives of *cholesterol, *ergosterol, etc.).] Any fully saturated sterol.

1949 L. F. & M. Fieser *Nat. Products related to Phenanthrene* (ed. 3) iii. 93 Members of the sterol series that are fully saturated, like cholestanol, are conveniently described as stanols; those that contain one double bond are stenols. **1958** C. W. Shoppee *Chem. Steroids* ii. 78 The test is positive with stenols, negative with stanols. **1972** *Nature* 21 July 149/1 Although stanols have been shown to occur in *Sphagnum* moss and in a pollen mixture they are extremely rare in the rest of the contemporary plant kingdom. **1979** *Experientia* XXXV. 186/1 Saturated sterols (stanols) comprise 50–75% of the total sterols of this insect.

Stanton (stæ·ntən). [Name of Sir Thomas Edward *Stanton* (1865–1931), English engineer.] **Stanton number,** a dimensionless measure of heat transfer used in forced convection studies, equivalent to the ratio of the Nusselt number to the product of the Reynolds and Prandtl numbers, viz. $h/c_p\rho v$, where h is the heat transfer coefficient of the fluid, c_p is its heat capacity at constant pressure, ρ is its density, and v is its velocity.

Defined as the reciprocal of the Prandtl number in some dictionaries, but no textual evidence for this has been found.

1942 W. H. McAdams *Heat Transmission* (ed. 2) iv. 95 Stanton number. **1966** W. M. Kays *Convective Heat & Mass Transfer* xi. 248 The decrease in Stanton number noted here occurs in what still seems to be essentially a turbulent boundary layer. **1978** *Internat. Jrnl. Heat & Mass Transfer* XXI. 282/2 From the results for the Nusselt number, the dimensionless heat transfer, expressed as the Stanton number, can be determined: $St = Nu/Re\,Pr$.

stanza. Add: **4.** *Sport.* A half or other session of a game.

1945 S. J. Baker *Austral. Lang.* xvii. 299 In football news we find..*first* or *second stanza*, the first or second half of a game. **1974** *News & Reporter* (Chester, S. Carolina) 22 Apr. 10-a/5 Comfort Control warmed their bats up in the final three stanzas to gain the victory. **1981** *National Times* (Austral.) 25–31 Jan. 23/3 There is also a growing habit of describing the next half or quarter of a football match as a 'stanza'—perhaps because it is so poetic.

5. *Comb.*, as **stanza-form,** the form of a stanza (sense 1); arrangement in stanzas.

1927 E. V. Gordon *Introd. Old Norse* 293 There were variants of the normal stanza-forms. **1957** E. T. Cone in N. Frye *Sound & Poetry* i. 6 Zelter, keeping close to the unusual stanza-form..produces a top-heavy musical period. **1976** *Classical Q.* XXVI. 16 The catalectic effect in English or German is a function of the particular stanza-form. **1978** *Early Music* Oct. 629/1 A large number of 'wasted pages'..are taken up with setting out the poetic texts in stanza form.

stapelia. Add: Hence **stape·liad,** a plant belonging to one of a group of closely related genera including *Stapelia* and others formerly considered part of it.

1933 White & Sloane *Stapelieae* i The corolla of Stapeliads is usually fleshy. **1966** E. Palmer *Plains of Camdeboo* xvi. 261 Among the great plant travellers are the Stapeliads of the family Asclepiadaceæ. **1977** W. P. U. Jackson *Wild Flowers Table Mt.* 48/2 *Orbea variegata*..is the only Stapeliad in the Cape Peninsula.

staph (stæf). Colloq. abbrev. of mod.L. *Staphylococcus*, name of a genus of pathogenic bacteria.

1933 Partridge *Slang To-day & Yesterday* iii. iii. 190 Staph, staphylococcus, one of the commonest types of bacteria. **1956** A. Huxley *Let.* 13 Aug. (1969) 804 The average mortality after surgery was twenty-nine per cent, with peaks, during epidemics of streps and staphs, of over fifty per cent. **1978** J. Irving *World according to Garp* ii. 47 He later went through Harvard Business School, a staph infection, and a divorce.

staphylinid, *sb.* and *a.* (Examples.)

1925 A. D. Imms *Gen. Textbk. Entomol.* iii. 481 Staphylinid larvæ are typically campodeiform. **1965** B. E. Freeman tr. *Vandel's Biospeleology* xiii. 208 The staphylinids found in caves belong to three ecological categories. **1978** *Nature* 16 Mar. 209/1 Among the many other examples are various species of staphylinid beetles that march with army ants.

staphylorrhaphy. (Earlier example.)

1835 *Lancet* 31 Jan. 648/1 In the operation of staphylorraphy,..if care be not taken to leave the ends of the sutures rather long..the knots will readily untie.

staple, *sb.*¹ Add: **2. d.** A piece of thin wire (characteristically shaped in the form of three sides of a rectangle), driven through papers, etc., and clinched to bind them.

1895 *Army & Navy Co-op. Soc. Price List* 540/2 Patent Staple Presses (For Fastening Papers &c.)..Wire Staples. Size ½ inch, per 1000 0/6. **1898** G. B. Shaw *Let.* 4 Mar. (1972) II. 11 Come along & bring some long staples (½" will do) with you. **1907** *Yesterday's Shopping* (1969) 349/1 The Self-Feeding Automatic Staple Press. Holds 25 staples which travel automatically. Staples are in strips of 25, as shown in illustration. **1926–7** *Army & Navy Stores Catal.* 369/3 The 'Longdon' Combined Stapler..only one staple can pass at a time. **1940** *Brit. Stationer Ann.* 2 (Advt.), Ace Stapling Machines..Staples: ¼ in... 7/2 per box of 5,000. **1967** *New Yorker* 15 July 26/3, I..dropped into a stationery store..looking for a mechanism of sorts with which to run a staple into my thumb. **1981** H. Engel *Ransom Game* (1982) iii. 21, I wondered how could I possibly discuss anything with a man who called staples Bostitch pins.

3. b. *Mus.* A metal tube on to which the double reed of a wind instrument is tied.

1880 Grove *Dict. Mus.* II. 486/2 It [sc. the oboe] is usually made in three pieces, a top, bottom, and bell joints, to which is added a short metal tube, the staple, on which the reed..is attached by means of silk. **1908** *Ibid.* (ed. 2) IV. 42/2 The bassoon reed is placed directly upon the 'crook' of the instrument, but the oboe reed is built up upon a small tube or 'staple'. **1953** E. Rothwell *Oboe Technique* 49 The part of the cane from the tip downwards which has been scraped and thinned with the knife after tying the cane on the staple, is known as the 'scrape' (or sometimes as the 'lay'). **1976** D. Munrow *Instr. Middle Ages & Renaissance* i. 8/4 The [shawm] player presses his lips against a metal disc at the base of the staple, taking the entire reed inside his mouth. *Ibid.* vi. 39/3 (*caption*) Set of modern rackett reeds mounted on their staples.

4. staple-driver (later example, in sense *2 d); **staple gun,** a hand-held device for driving staples home; hence (hyphened) as *v. trans.*; † **staple press** = *stapler².

1895 *Montgomery Ward Catal.* Spring & Summer 116/3 Staple Driver, for binding books, papers, pamphlets, etc. ..staple is placed in holder, driven to place. **1960** G. Lewis *Handbk. Crafts* 350 A staple gun can be useful, but is not essential. **1975** *Harpers & Queen* May 128/3 Looks like some faggot decorator went nuts in here with a staple gun. **1977** J. Fraser *Hearts Ease* ix. 107 The heavy squad..had staple-gunned plastic sheets to cover the hole. **1895, 1907** Staple press [see sense 2 d above].

staple, *sb.*² Add: **3. b.** Also, the principal or basic food on which a community lives.

1970 C. Furtado in I. L. Horowitz *Masses in Lat. Amer.* ii. 33 In the case of exported staples, there occasionally appeared competitive productive areas which were better situated geographically or which had access to protected markets. **1971** *Sci. Amer.* Sept. 113/1 Except for the staples and tea, tobacco and candy, there is no strong desire for non-Eskimo foods. *Ibid.* Oct. 21/1 It is only recently that human populations have come to depend heavily on a single cultivated plant staple for food. **1977** *N.Y. Rev. Bks.* 23 June 16/1 Much of this inflation can be attributed to a rise in the world price of oil and food staples.

staple, *sb.*³ Add: **4. staple fibre, yarn.**

1928 E. Fyleman tr. *Hottenroth's Artificial Silk* i. 16 Staple fibre..consists of artificial silk, the thread of which is cut into sections of about the length of cotton or worsted staple. **1974** *Sci. Amer.* Apr. 57/3 The viscose-rayon process..regenerates pure cellulose..as a continuous filament or a staple fiber. **1955** *Times* 10 May 18/3 More inquiry continues to come forward for filament rayon and staple yarns. **1968** J. Ironside *Fashion Alphabet* 209 The filaments are bunched together, cut into short lengths, combed, drawn and spun into spun or staple yarn, which is fuzzier and is made into fabrics.

staple, *v.*¹ Add: **b.** (Later examples in sense *2 d of the sb.).

1964 R. Petrie *Murder by Precedent* vi. 87 At no time after stapling the copies did I move any sheets apart. **1975** D. Ramsay *Descent into Dark* iv. 130 You were stapling that article on abortion together.

stapled, *a.*² Add: Also *spec.* of papers, fastened together with a staple or staples.

1956 S. Bellow *Seize Day* (1957) i. 32 Several pages of blue hectographed script, stapled together. **1958** *N. & Q.* Jan. 46/2 The cost of this book by English standards ($1.50 for a text in stapled paper cover) is prohibitive. **1974** 'P. B. Yuill' *Bornless Keeper* xii. 113 He picked up the wad of stapled foolscap sheets.

stapler² (stēⁱ·pləɹ). [f. *staple *sb.*¹ 2 d + -er¹.] A device for fastening together papers, etc., with a staple or staples.

1951 *Catal. of Exhibits, South Bank Exhib., Festival of Britain* 64 Office equipment..'Velos' plier stapler. **1969** 'C. Keith' *Missing Book-Keeper* iv. 48 The shallow centre drawer..contained..some extra staples for the stapler on top of the desk. **1973** *Sci. Amer.* July 113/2 Edges of the foil can be fastened together with tape or with a stapler. **1978** *Detroit Free Press* 16 Apr. (Advt. Suppl.) 1/4 Full-size standard stapler with open-channel loading.

stapling (stēⁱ·pliŋ), *vbl. sb.* [f. Staple *v.*¹ + -ing¹.] The action of the verb, esp. that of fastening papers together with staples; the fastening so made.

1898 G. B. Shaw *Let.* 4 Mar. (1972) II. 11 The plays..require endless cutting & folding and stapling into brown paper covers. **1940** [see *staple *sb.*¹ 2 d]. **1961** *Lebende Sprachen* VI. 104/1 Stapling machine, stapler. **1970** *Guardian* 5 May 4/5 He looked at his report again and noticed the stapling had been interfered with. **1973** D. Miller *Chinese Jade Affair* xix. 190 The secretary came in with..official seals, a stapling gadget, sealing-wax.

star, *sb.*¹ Add: **I. 1. e.** (Earlier and later examples of phr. *to see stars.*)

1839 *Spirit of Times* 16 Nov. 434/1 She fetched me a slap in the face that made me see stars. **1924** Galsworthy *White Monkey* ii. xi. 206 '*Per ardua ad astra*,' 'Through hard knocks we shall see stars.' **1966** D. Varaday *Gara-Yaka's Domain* xiv. 160 Had it been daylight I would still have seen the stars caused by the searing pain I felt in my thumb! **1977** *Cleethorpes News* 27 May 32/3 Already three fighters have seen stars as they have been sent crashing to the canvas.

3. b. (Further examples.) *to thank* (or *praise*) *one's lucky stars*: see Lucky *a.* 3.

1819 M. Wilmot *Let.* 24 Oct. (1935) 24 O ye stars and garters how often do I wish for Mary and a green Lawn!!! **1913** [see *bobby-dazzler]. **1931** F. L. Allen *Only Yesterday* ix. 237 In any café in Paris one might find an American expatriate thanking his stars that he was free from standardization at last. **1976** M. Millar *Ask for me Tomorrow* (1977) ix. 77 My stars, you needn't shout.

5. a. Also applied to celebrities in other spheres of entertainment. **film star**: see *film *sb.* 7 c; **movie star**: see *movie b.

1919 G. B. Shaw *Annajanska* in *Heartbreak House* 265 You still want to be a circus star. **1941** *Picturegoer & Film Weekly* 6 Sept. 3/1 Barbara Mullen is no longer unknown. It would be an exaggeration to say *Jeannie* makes her a star. It needs more than one part to do that. **1946** *R.A.F. Jrnl.* May 175 The British Forces Network finds new stars of radio from the ranks of the R.A.F. **1976** *Oxf. Compan. Film* 468/2 Hayley..has been the more prominent in films, scoring a success as a teenage star, notably in *Tiger Bay*.

b. (Earlier and later examples.)

1829 G. Griffin *Collegians* II. xx. 103 Anne Chute..was, beyond all competition, the star of the evening. **1881** C. E. Pascoe *Everyday Life in our Public Schools* 218 The Torpid game is reserved as a sphere for young 'stars' who come up to Harrow with a reputation. **1973** *Art Internat.* Mar. 55/1 George Washington Wilson's *The Brig and Cliffs, Filey, From the North Landing*..was for me the star of the group [of photographs]. **1975** *Nature* 16 Oct. 531/3 It was supported by a galaxy of scientific stars, including 14 Nobel Prizewinners.

c. *Sport.* An outstanding performer.

1916 [see *clay court* s.v. *clay *sb.* 9]. **1928** E. O'Neill *Strange Interlude* vi. 210 I'm going to start in training him..so he'll be a crack athlete when he goes to college...I want him to..be a bigger star than Gordon ever was. **1930** *Sun* (Baltimore) 26 Dec. 11/7 No player is now on the field more than half of that, so even a star averages only about a half hour's real work throughout the year. **1964** G. C. Kunzle *Parallel Bars* ix. 407 The content of five difficulties and one superior difficulty was more than most international stars had at that time. **1979** R. Jaffe *Class Reunion* (1980) iii. iii. 318 Emma..was the star of her gymnastics class.

6. c. A badge of rank, authority, or military service.

1890 [see *pack *v.*¹ 9 a]. **1895** *Montgomery Ward Catal.* Spring & Summer 296/3 Policemen's regulation rubber coats..pocket for billy and shield for star. **1908** W. H. Davies *Autobiogr. Super-Tramp* 48 With that the marshal of the town stood before the open door, showing the star of his authority on his dark clothes. **1924** C. J. Tolley *Mod. Golfer* 6, I passed on from there in '16 with a second lieutenant's star in that regiment. **1942** [see *Sam Browne]. **1946** W. S. Churchill *Victory* 143 Two new stars for operations in the East will also mark the service of those who have gone out..to finish the war against the Japanese. **1977** R. Ludlum *Chancellor Manuscript* xxix. 312 My permanent rank is brigadier-general. I will undoubtedly receive my second star in June.

d. A small star of coloured paper, awarded to a (usu. primary) schoolchild for a good piece of work. (The star is often stuck alongside the work in an exercise book, or displayed on a wall-chart.)

1977 *Cleethorpes News* 6 May 17/2 The books were all very neat and dotted with gold stars on nearly every page. The stars are worth house points which build up over the term. **1978** *Jrnl. R. Soc. Arts* CXXVI. 351/1 If they get 80 per cent or over, they get a blue star, 90 per cent or over a gold star. And ten lessons with stars will win them a prize.

10. c. In guidebooks, one of a number of stars or asterisks against the name of a hotel, restaurant, resort, etc., indicating its rank in a grading system. Cf. *four-star, *five-star, etc., under the first elements.

[**1886** S. Coolidge *What Katy did Next* vi. 138 'Following a star', in their choice of a hotel..they had decided upon one of those thus distinguished in Baedeker's guide-book.] **1905** E. M. Forster *Where Angels fear to Tread* vi. 172 Giotto..has painted two frescoes... That is why Baedeker gives the place a star. **1939** J. B. Priestley *Let People Sing* v. 99 A bad inn that is given two stars in the Automobile Association's handbook. **1963** R. Carrier *Great Dishes of World* 128 If one gave stars to the regions of France—as well as to their better restaurants—for the excellence of their cooking, Burgundy would have an unchallenged 'three'. **1974** *Guardian* 20 Mar. 1/1 The Michelin Guide..said that no [British] restaurants had been awarded two or three stars.

d. Used in various other grading systems, as for cognac, refrigerators, petrol, etc.

1922 Joyce *Ulysses* 498 It was in consequence of a portwine beverage on top of Hennessy's three stars. **1951** R. Postgate *Plain Man's Guide to Wine* ix. 125 'Three Stars' indicates the standard and satisfactory degree of distillation... Five Stars should indicate a good brandy. **1968** S. E. Ellacott *Everyday Things in England 1914–68* ii. 42 Frozen foods and refrigerators were graded with stars according to the length of time they could be kept fresh. One star denoted a week, two stars a month, and three stars three months. **1971** *Homes & Gardens* Aug. 86/1 Two stars indicate that the temperature is about 10 deg. F. (−12 deg. C.) and frozen food will last for up to four weeks. **1982** *Sunday Tel.* 1 Aug. 6/1 My correspondent objects to the use of star petrol ratings in place of octane numbers.

11. d. *Physics.* A photographic image consisting of a number of lines emanating from a central point, which represent the paths of sub-atomic particles produced by the impact of a cosmic ray or other energetic particle.

1938 *Nature* 1 Oct. 613/1 After an exposure of five months the plates have now been developed and examined. They present singular tracks and stars like those reported in previous papers. **1948** *Science* 26 Nov. 588/2 Approximately 75% of the heavy negative mesons give rise to stars when they come to rest in the emulsion. **1957** G. E. Hutchinson *Treat. Limnol.* I. iii. 212 Reactions in which a number of particles are produced.., when they are recorded in a photographic emulsion, are recognized as cosmic ray 'stars'. **1974** *Encycl. Brit. Macropædia* V. 201/2 Interaction of the cosmic-ray particle with the constituents of the emulsion can often be observed as a 'star' in the emulsion; *i.e.*, a spot from which secondary particles are emitted.

12. i. *Electr. Engin.* A star-connected set of windings; *in star*, by means of a star connection.

1907 [see *neutral *a.* 4 f]. **1924** A. L. Cook *Elem. Electr. Engin.* xviii. 509 When starting, the three groups of coils are connected in star (or Y). **1962** *Newnes Conc. Encycl. Electr. Engin.* 709/1 A 4-branch star converts to a 6-branch mesh. **1974** Howatson & Lund *Princ. Heavy Current Engin.* 109 It is possible to connect star to delta but..no fourth wire can be used and so the calculations become tedious if the system is not balanced. **1976** F. de la C. Chard *Electricity Supply* v. 147 Transformers for 3-phase duty..may have both primary and secondary windings connected in delta or star.

14. (Later examples of sense 'a prisoner of star-class'.)

1928 *Notes on Imprisonment* (Home Office) 11 Promotion to the Third Stage may be earned after 12 months, or by 'Stars' after six months. **1945** *Prisons & Borstals* (Home Office) ii. 18 An Ordinary Class prisoner comes into the Second Stage after 12 weeks, a Star after 4 weeks. **1962** 'J. Bell' *Crime in Our Time* vi. ii. 170 Nowadays, the 'stars' as a general rule are by no means first offenders. **1976** A. Miller *Inside Outside* iv. 48 Several..said that if that was what one-time Stars became, they were cured of returning.

II. 15. a. *star-crowd, -distance, -field, -glimmer, -image, -rise.*

1870 T. W. Higginson *Army Life in Black Regiment* ix. 209, I know moon-rise, I know star-rise, Lay dis body down. **1904** *Nature* 9 June 135/2 The error inherent in the star-images. *a* **1918** W. Owen *Poems* (1963) 95 And tiring after beauty through star-crowds, Dared I go side by side with you. **1920** A. S. Eddington *Space, Time & Gravitation* viii. 127 The measurement of the displacement of the star-image on the photographic plate. **1952** C. Day Lewis tr. *Virgil's Aeneid* iv. 82 Often as star-rise, the troubled ghost of my father, Anchises, Comes to me in my dreams. **1954** J. R. R. Tolkien *Fellowship of Ring* ii. ix. 402 In the star-glimmer they must have offered their cunning foes some mark. **1961** Webster, *Star field*, a region of the sky containing stars either as seen in a telescope or recorded on a photograph. *a* **1963** S. Plath *Crossing Water* (1971) 63 The pale, star-distance faces. **1968** P. Moore tr. *E. L. Schatzman's Struct. Universe* i. 11 In 1838 F. W. Bessel..measured the first star-distance. **1976** *Houston* (Texas) *Chron.* 22 Sept. vii. 1/2 Sumner's first lecture covers the starfield, with the movements of the sun, moon, planets illustrated and explained.

c. *star-dusted, -filled, -guided, -powdered, -sown, -strewn, -studded, -taught* (earlier and later examples) adjs.

1786 T. Dwight *Amer. Poems* (1793) I. 39 Let every sage and seer, Dreamer of dreams, and star-taught prophet hear! **1868** J. R. Lowell in *Atlantic Monthly* May 627 It is wider Than the star-sown vague of space. **1886** W. B. Yeats *Mosada* 2 For Azolar The star-taught Moor said thus it was decreed. **1895** M. H. Kingsley *Diary* 5 June in *Trav. W. Afr.* (1897) vii. 124 The hills silhouetted against the star-powdered purple sky. **1896** A. E. Housman *Shropshire Lad* 87 The star-filled seas are smooth to-night. **1901** H. G. Wells *First Men in Moon* v. 62 That airless, star-dusted sky! **1915** G. Frankau *Tid'apa* v. 25 Blue-dark against star-strewn turquoise, rose the ramparts of Lallong Ridge. **1920** J. Gregory *Man to Man* xxiv. 284 The field of star-strewn sky. **1930** J. Masefield *Wanderer of Liverpool* 87 Star-lighted, star-guided, the sea-gleaming beautiful thing. *a* **1936** A. E. Housman *More Poems* (1936) 166 No star is lost at all From all the star-sown sky. **1945** W. de la Mare *Burning-Glass* 11 Ev'n happier in watch of..A star-strewn nightfall. **1958** *People* 4 May 19/5 Star-studded with ex-League players. **1962** *Daily Tel.* 28 Apr. 20 (*heading*) U.S. seeking 'star-guided' missile for NATO. **1977** *New Yorker* 29 Aug. 20/2 It's Mickey Mouse, imprisoned inside a star-dusted transparent balloon. **1978** *Listener* 16 Mar. 339/1 In an otherwise star-studded cast, the lead characters are not quite up to their role.

d. *star-eyed* (later example), *-glimmering, -tall* adjs.

1878 G. M. Hopkins *Poems* (1967) 77 Star-eyed strawberry-breasted Throstle. **1900** W. B. Yeats *Shadowy Waters* 9 More shining winds, more star-glimmering ponds? **1943** S. Spender *Spiritual Exercises* 6 Outside, the eternal star-tall mountains gleam.

19. (Earlier and further examples.) Also in sense *5 c.

1839 F. Marryat *Diary in Amer.* II. ii. xiii. 121 They look for importations of star actors from this country. **1876** 'Mark Twain' *Tom Sawyer* iv. 50 He had been around among the star pupils inquiring. **1879** C. E. Pascoe *Dramatic Notes* 68 What is known as 'star-acting' usually forms the principal feature of the bills of this theatre. **1890** W. James *Princ. Psychol.* I. xi. 453 The laws of stimulation and association..*may* at times simply form the background for a 'star-performer', who is no more their 'inert accompaniment' or their 'incidental product' than Hamlet is Horatio's or Ophelia's. **1909** *Times* 24 Aug. 15/6 The engaging of outside 'star' players to strengthen a county [cricket] side. **1917** T. E. Lawrence *Lett.* (1938) 219 The star film showed the Pyramids. **1927** *Melody Maker* Sept. 923/1 The all too few star performers in this country. **1933** J. Cary *Amer. Visitor* xiii. 169 Uli turned out one of the star pupils. **1943** N. Coward *Middle East Diary* (1944) 23 He must possess ..what is described in the theatre as 'star quality'... Complete authority, a direct eye, and a compelling economy of gesture. **1950** *Sport* 22–28 Sept. 12/4 Have too many star players been allowed to drift away from the Swansea Town fold? **1961** *Ann. Reg. 1960* 446 Star acting was also the chief merit of *The Last Angry Man*. **1972** *Vogue* Feb. 89/3 Twiggy's got this rather mysterious presence. Star quality. **1977** *New Yorker* 19 Sept. 75/1 I'm sure Mme. Dorfmann is furious with the review, because he's her star pupil. **1978** J. Gardner *Dancing Dodo* xxxii. 256 You're playing the star role—Hercule Poirot—and I've got to coach you. **1982** *London Review Bks.* IV. xxiv. 20/2 Star quality, however, was not at all what was looked for in those who played opposite a superstar like Kean.

20. star-back *slang*, an expensive, reserved seat at a circus; **star bill,** a poster advertising a theatrical star; **star billing** = *top billing* s.v. *billing *vbl. sb.*³ b; also *transf.* and *fig.*; **star boarder** *U.S.*, a boarder, usu. of long standing, having or regarded as having special privileges; also used euphemistically of more complicated relationships; **starburst,** (*a*) an explosion of a star or stars, or an explosion producing an appearance of stars; (*b*) *Photogr.*, a lens attachment which causes a bright light source to appear in a photograph with added star-like rays; also, the effect so produced; usu. *attrib.*; **star cloud,** a region of the sky where stars appear to be especially numerous and close together; **star connection** *Electr. Engin.*, an arrangement in a polyphase system in motors and the like by which one end of each phase winding is connected to a common point; so **star-connected** *a.*; **star count,** a statistical survey of the stars in various directions in space to ascertain the numerical distribution across the sky of stars brighter than some given magnitude; **star-crossed**: delete † and add later examples, after quot. 1592 (Shakes.) in Dict.; **star-cut** *a.*: also in extended uses (see quots.); **star-delta** *Electr. Engin.*, used *attrib.* with reference to the use of star connection when an induction motor is started with a change to delta connection for continuous running; **star drag** *Angling* (see quot. 1960); **star network,** a data or communication network in which all terminals are independently connected to one central unit; **Star of David** = *Magen David; **star point** *Electr. Engin.*, the common junction of the windings in a star-connected system; **star quad** *Telecommunications*, a quad (*quad *sb.*⁶) in which the four conductors are all twisted around a common axis, with members of each pair being diametrically opposite each other; usu. *attrib.*; **star stream,** † (*a*) a narrow band of the sky that is rich in stars; (*b*) each of the two groups of stars in star streaming; **star streaming,** the phenomenon (explained by the rotation of the galaxy) in which stars show a broad tendency to have proper motions in one or other of two opposite directions, thus falling approximately into two intermingled groups ('streams'); **star system,** (*a*) in the world of film and theatrical entertainment, the practice of promoting an eminent artiste in leading roles; (*b*) a large structured collection of stars, a galaxy; **star-tracker,** a self-regulating device which maintains its orientation relative to a star, used in the control systems of spacecraft; so **star-tracking** *vbl. sb.*; **star turn,** the principal or most important item in an entertainment; also *fig.*; **star-vehicle,** a play or film designed especially to show off the talents of a particular actor or actress; **star witness,** the principal or most important witness in a trial.

1931 *Amer. Mercury* Nov. 354/1 *Starbacks*, the reserved seat section. **1933** E. Seago *Circus Company* iii. 23 He sat with me in the 'star backs'. **1965** M. Stewart *Airs above Ground* vi. 72 Tim had no difficulty in getting what he called 'starback' seats... These, the best seats, were rather comfortable portable chairs..right at the ringside. **1901** A. Chevalier *Before I Forget* 157 Just ordered fresh stock of special printing, star bills, &c. **1956** H. Kurnitz *Invasion of Privacy* iii. 27 One picture a year..: Star billing. **1959** *Listener* 5 Feb. 238/1 The Chinese still give credit to the Russians for their technical aid, but there has been a change in the star billing. Now they say these things were done by Chinese engineers with Soviet assistance. **1967** *Amer. N. & Q.* June 156/1 *Behind Spanish American Footlights* is the key reference tool in its field, and as such rates star billing. **1979** *Listener* 25 Oct. 547/3 The *Radio Times*..gives Ted Heath (1970–74) star billing. **1877** in H. Asbury *Gem of Prairie* (1940) iv. 135 Jessie Curtis, star boarder, is still at 519 State Street. **1897** *Boston Jrnl.* 16 Jan. 6/5 'I'm afraid you are about to be dethroned.' The Star Boarder—'Why?' **1908** J. M. Sullivan *Criminal Slang* 24 *Star boarder or lodger*, a boarder..in good financial standing who has all the privileges of a husband. **1922** N. B. Tarkington *Gentle Julia* 113 The pill-boxes [for insects] ..evidently contained star boarders, for they were pierced with 'breathing holes'. **1935** J. Hargan *Gloss. Prison Lang.* 8 *Star-boarder*, a lifer. **1935** A. J. Pollock *Underworld Speaks* 113/2 *Star boarder*, the inmate of a house of prostitution, who earns the most money. **1976** 'O. Bleeck' *No Questions Asked* xii. 137 He lived here and we split expenses... He was sort of a star boarder. **1965** G. McInnes *Road to Gundagai* iii. 56 Rockets whooshed skyward ending in great parabolic starbursts. **1977** G. Michanowsky *Once & Future Star* i. 3 At the dawn of human history,..this starburst occurred and briefly became visible as a..second sun in our sky. **1977** J. Hedgecoe *Photographer's Handbk.* 181 A clear sky can be filled with diverging lines if the sun is spread by means of a starburst filter. **1978** *Amateur Photographer* 2 Aug. 110/1 Sparkling highlights are emphasized with a cross screen (star-burst). **1979** *Ibid.* 30 May 92/3 A very special filter which not only gives a star-burst effect, but rays of the starburst are tinged with the colours of the spectrum. **1924** *Proc. Amer. Acad. Arts & Sciences* LIX. 217 Fainter stars of division A reveal through their distribution the relative nearness of the star clouds in Cygnus. **1947** *Astrophysical Jrnl.* CV. 257 The regions of the star clouds in Sagittarius and Ophiuchus are rich in clearly marked globules. **1896** D. C. & J. P. Jackson *Alternating Currents* II. viii. 395 If the armature is star connected, the pressure between rings is equal to the vector sum of the pressure developed in the two coils. **1976** F. de la C. Chard *Electr. Supply* v. 145 A star-connected winding has only $1/\sqrt{3}$ times the line voltage across each phase but carries the full line current. **1894** G. Kapp *Electric Transmission of Energy* (ed. 4) xii. 418 The armature is drum wound with star connection. **1969** *Power System Protection* (Electr. Council) II. xi. 404 Busbar reactors. (b) Tie bar connected. Two methods are commonly adopted and may be classified as (i) the star connection.., and (ii) the ring connection. **1889** *Nature* 8 Aug. 345/1 For simple star-counts, we have only to substitute star-counts by magnitudes over selected areas of the sky. **1933** *Discovery* Feb. 41/2 The modern method employed for thus studying the distribution of stars is the statistical one of 'star-counts'. **1947** *Astrophysical Jrnl.* CV. 257 Stoddard has made star counts according to photographic and photovisual magnitudes for four large globules. **1962** C. Oman *Mary of Modena* v. 172 Abbé Armand Jean de Bouthillier de Rancé of La Trappe had, according to rumour, well known star-crost love before he suddenly renounced the world at the age of thirty-six. **1973** *Alberta Hist. Rev.* Winter 12/1 But if Uncle Charlie's first motivation was star-crossed love, his second was certainly horses. **1977** W. M. Spackman *Armful of Warm Girl* 18 She wailed in star-crossed despair. **1967** Wodehouse *Company for Henry* iii. 45 Clichy double overlay weight,..the top flattened by a large window, star-cut base. **1972** J. Howard-Williams *Sails* (ed. 3) xii. 172 The genniker (or spanker, or star-cut spinnaker) is..a cross between a genoa and a spinnaker... It is primarily a racing sail... It is cut rather like a storm spinnaker, with full width foot tapering to the rule minimum 75 per cent at mid-girth, with narrow shoulders from there up. **1976** *Yachting World* Oct. 72/1 Spinnakers are sewn in a special corner of Hood's loft with a five-step machine. This sews a unique five-stitch seam on star-cut spinnakers. **1908** W. B. Hird *Elem. Dynamo Design* ix. 260 If the star-delta method of connection is used [in the starting of a squirrel-cage motor], the ratio must be 1 to 1·73, and no other choice can be made. **1962** *Newnes Conc. Encycl. Electr. Engin.* 517/1 Star-delta starting is one method of reducing starting current of 3-phase motors. **1976** F. de la C. Chard *Electr. Supply* v. 148 A star-delta connection for a step-down transformer is normally used. **1950** *N.Y. Times* 30 Dec. 18/8 While you could stand a heavier bait rod and a reel with star drag, the regulation bait rod will suffice. **1960** C. Willock *Angler's Encycl.*, *Star-drag*, adjustable tension device often built into big sea multiplying reels. Tension is varied by means of a star-shaped nut. **1979** *Angling* July 33/1 (Advt.), Rugged star drag design with white-oak leather washers for smooth line control. **1977** *Financial Times* 21 Feb. 13/1 Modern data networks have come a long way since the banks first started installing their enormous star networks, consisting of one or two very big computer centres serving several thousand simple terminals installed in branch bank offices. **1979** T. Housley *Data Communications & Teleprocessing Systems* ii. 52 There are four basic network configurations that can be used: the star network, the ring network, the mesh network, and the hierarchical network. The star network..is probably the most common. **1941** M. Samuel tr. *Bein's Theodore Herzl* ii. vii. 229 At one side hung a flag: a white field with two blue stripes and the Star of David. **1979** *N.Y. Rev. Bks.* 25 Oct. 3/2 Jan Friedländer's book plates displayed a score by Chopin set within a Star of David. **1908** W. B. Hird *Elem. Dynamo Design* vii. 186 The star point..is usually connected to earth, both at the generator and at the load end. **1969** *Power System Protection* (Electr. Council) II. xi. 404 Each section of busbar is connected via a reactor to a common star point, and if the feeders and generators are suitably arranged little or no current need flow through the reactors. **1930** *Gloss. Terms Telegraphs & Telephones (B.S.I.)* 31 *Star quad cable*, a cable containing a number of quads, each quad formed by twisting together four insulated conductors about a common axis. **1958** J. R. G. Smith *Elem. Telecommunications Pract.* v. 73 Other features of star-quad cables are that in some cables certain of the pairs are screened by being wrapped in metallized paper..; they are placed in the centre of the core and used for music. **1970** P. Norman in T. L. Squires *Telecomm. Pocket Bk.* v. 53 Some 4,000 miles of these [balanced pair] systems are installed in this country, consisting mainly of 24-pair,

40 lb per mile conductors in star-quad formation. **1894** *Knowledge* 1 June 133/1 The streams are in most cases accompanied by narrow black channels in the general nebulosity, which run parallel to and alongside of the star streams. **1904** J. C. Kapteyn in H. J. Rogers *Congress of Arts & Science* (1906) IV. 418 Here we have a clear indication that we have to do with two star-streams. **1925** *Ark. f. Mat., Astr. och Fysik* XIXa. xxi. 1, I have tried to find a clue to a possible connection between the star-streaming discovered by Kapteyn, 'the two star-streams', and the asymmetrical drift of high stellar velocities. **1968** W. M. Smart *Riddle of Universe* vi. 106 Kapteyn's star-streams are consistent with the phenomenon of galactic rotation and are indeed explained by it. **1906** *Rep. Brit. Assoc. Adv. Sci. 1905* 257 (*heading*) Star streaming. **1921** *Discovery* Feb. 36/1 In 1904 Professor Kapteyn..read at the Astronomical Congress at St. Louis, U.S.A., a paper of far-reaching importance, in which he announced the discovery of star-streaming. **1979** Lang & Gingerich *Source Bk. Astron. & Astrophysics* lxxvii. 514 Karl Schwarzschild showed that it was unnecessary to think of two star streams and that the phenomenon of star streaming could be explained by assuming that the individual motions of the stars are distributed in an ellipsoid with the long axis in the direction of motion of Kapteyn's two star streams. **1832** *Rep. Sel. Comm. on Dramatic Lit.* in *Parl. Papers 1831–32* VII. 30 The star system that has been adopted by the two great theatres. **1870** R. A. Proctor *Other Worlds than Ours* 256 To return for a moment to fig. 2, it will be seen at once that an aperture extending laterally through a star system so shaped must have a particular direction and be perfectly straight in order to be visible to observers placed, as we are supposed to be, in the central opening. **1890** G. B. Shaw in *Star* 18 Apr. 2/3 The familiar star system trick of making the minor characters slur their work in order to leave plenty of time for the mock pregnant pauses..of the leading actor. **1928** A. S. Eddington *Nature Physical World* viii. 167 The first partitions [of the gaseous nebulae] are the star-systems such as our galactic system. **1937** A. Calder-Marshall in C. Day Lewis *Mind in Chains* 71 This tendency..has led to the star-system deplored by film critics almost without exception. **1967** P. Moore *Amat. Astronomer's Gloss.* 89 *Nubeculæ* (or Magellanic Clouds), the nearest of the external star-systems, and so the brightest as seen with the naked eye. **1971** *Guardian* 5 June 9/4 Hollywood was devising the 'star system', the big solo buildup, the personality cult of the silent screen. **1962** *Aeroplane & Commercial Aviation News* CIII. 32/1 The main outstanding problem in the provision of such a system is the development of a suitable daylight star-tracker. **1978** *Nature* 5 Oct. 378/2 The star tracker in the scientific instrument can identify and guide on stars brighter than 14 mag. **1964** *Discovery* Oct. 7/3 A star-tracking ability can readily be developed from the same system. **1898** A. M. Binstead *Pink 'Un & Pelican* ii. 44 The 'star turns' in the entertainment, which took place twice every day, were the then unknown Paul Cinquevalli, Batty, the natty horseman..the 'Beautiful Geraldine', and two savage and sullen brown bears. **1906** E. Dyson *Fact'ry 'Ands* vii. 78 In fact the packer soon found that waiting was Eric's 'star turn'. **1909** *Flight* 3 July 398/1 M. Blériot..is the 'star turn' at the Brayelle aerodrome at Douai just now. **1915** H. G. Wells *Boon* 328 'Inevitably,' said the Bishop, 'this theatricalism, this star-turn business, with its extreme spiritual excitements, ..leads to such a breakdown as afflicts you.' **1951** J. G. Fennessy *Sonnet in Bottle* iv. iv. 116 You're our star turn, it's up to you. **1977** 'E. Crispin' *Glimpses of Moon* xii. 232 His normal chief recreation..is watching other people at work. And in this..it is the two workmen who are the star turns. **1932** *New Yorker* 9 Jan. 43/1 As a spectacle it is a typical star vehicle hitched up to a singer with..a weak larynx. **1953** K. Reisz *Technique Film Editing* i. 60 Many of the films built round the personality of Greta Garbo..are little more than ingeniously contrived star-vehicles, yet they cannot be dismissed as worthless. **1974** *Listener* 2 May 580/2 Few self-respecting dramatists want to construct star-vehicles nowadays, and so Robert Morley..has written one for himself. **1924** W. M. Raine *Troubled Waters* xvii. 183 Haight [*sc.* a district attorney] was very gentle and considerate of his star witness. **1978** P. Moyes *Who is Simon Warwick?* xi. 137 The prosecutors were..satisfied with the evidence. Susan Benedict was to be their star witness.

22. a. star-anise, also, *Illicium verum*, a small evergreen tree found in southern China; the fruit of this tree, or the oil or spice obtained from it; (later examples); **star tulip,** a glabrous perennial bulbous plant belonging to any of several species of the genus *Calochortus* (family Liliaceæ), native to temperate western N. America; (formerly distinguished from Mariposa lilies, but now sometimes used synonymously for the whole genus).

1883 Star anise [see *coffin-wood* s.v. *coffin *sb.* 13]. **1972** K. Lo *Chinese Food* i. 12 One or two pieces of star-anise and a sprinkling of cinnamon. **1895** W. Robinson *English Flower Garden* (ed. 4) ii. 347/2 One of the most experienced growers of the Calochorti, Dr. Wallace, of Colchester, writes of the family thus:..'Other dwarf forms among the Star Tulips lasted well up to the end of June, when the beautiful Mariposa Lilies continued the display with their tall spikes.' **1921** M. Hampden *Bulb Gardening* xvii. 185 Every garden should contain Calochorti... They consist of three groups, but I consider two only fit for the amateur gardener's patronage; these are known as Mariposa, or Butterfly Tulips, and Star Tulips. **1925** [see *mariposa lily]. **1974** H. G. W. Fogg *Compl. Handbk. Bulbs* vii. 49/2 Of those described as star tulips, *C[alochortus] benthami*, clear yellow with dark central blotch, and *C. maweanus* 'Major'..are among the finest.

23. (Later example.)

1976 *Sci. Amer.* Apr. 94/2 A type of opal new to us has recently been mined near Spencer, Idaho. It is a star opal that shows streaks of colour in symmetrically arranged angular patterns similar to the rays of a star sapphire.

star, *v.* Add: **7.** (Earlier example.)

1827 *Gardener's Mag.* II. 105, I wonder indeed that members of a (professedly) liberal society should quietly submit to be classed and regulated, and starred and scheduled, like the items in a paper of assessed taxes.

8. a. (Later examples.) Also in sport, to play a star role in a team.

1933 *Radio Times* 14 Apr. 73/1 There may have been two Zazels, one of whom..retired before 1890, when the other took over the name and 'starred' at the Westminster Aquarium. **1972** J. Mosedale *Football* viii. 116 Turner.. starred from 1940 through 1952. **1976** *Oxf. Compan. Film* 633/2 She returned to Britain to star in *Say Hello to Yesterday*. **1978** *Dumfries Courier* 20 Oct. 5/3 Carson starred on the right wing, and was the mainspring of a lively Queens side.

d. To advertise as a film or theatrical star; to give a star part to (an actor or actress); (with a film, etc., as subject) to present in a leading role.

1895 G. B. Shaw *Let.* 27 Mar. (1965) I. 508 It is good business to star Janet. **1922** *Encycl. Brit.* XXX. 699/2 He appeared in a minor rôle on the New York stage in 1901; later he was 'starred' in several comedies and musical pieces. **1929** A. C. & C. Edington *Studio Murder Mystery* viii. 102 Already we are going to star her! Already we've half bought a story, just for her. **1936** [see *feature *v.* 4 b]. **1951** 'N. Shute' *Lonely Road* (ed. 2) p. v, In 1936 a film was made..starring Clive Brook. **1962** E. Albee *Who's afraid of Virginia Woolf?* (1964) i. 5 *Chicago* was a 'thirties musical, starring little Miss Alice Faye. Don't you know *anything*? **1980** *Sunday Times* 21 Sept. 14 Since then he has earned an international reputation with plays..and The Faith Healer, which starred James Mason last year on Broadway.

starboard, *sb.* (and *a.*). Add: **A.** *sb.* **a.** Also used with reference to aircraft.

1909 F. T. Jane *All World's Air-Ships* 142 Motor–3-cylinder 10–12 h.p. Buchet, mounted directly on the lower plane, a little to starboard of centre line. **1977** J. Cleary *High Road to China* ii. 65 The Bristol slid to starboard, kept sliding and I let it go, feeling I was getting it under control.

B. *attrib.* or as *adj.* Also used with reference to aircraft.

1917 [see *port *sb.*⁴ (*a.*) 2 a]. **1948** 'N. Shute' *No Highway* iii. 61 The second pilot..came down into the cabin... Then he said: 'Which is the boffin?'..'Sitting on the starboard side, near the front.' **1976** B. Lecomber *Dead Weight* iii. 45 The tailwheel lock had broken... The starboard prop had run away.

star-bright. Add: Also **star bright. b.** Delete '? *technical*' and add def.: Of wine and cider: perfectly clear and free from sediment (see quot. 1979). (Later examples.)

1923 A. L. Simon *Supply, Care & Sale of Wine* xvi. 109 Before bottling any wine, you must make sure:–..That the wine is star-bright. *Ibid.* 110 These..conditions are, to my mind, rules without exception, but there are bottlers who make an exception to the 'star-bright bottling' rule when they bottle Vintage Ports. **1945** *Wine & Spirit Trade Rec.* 18 June 650/1 A device that helps the examination of Wines is to paint sheets of tin..with black pigment..and to fix it in an angle of the cellar walls where a short gas jet can be lit behind the bottle so that the Wine can be seen..under the best conditions for discovering whether the Wine is 'star bright' or not. **1973** *Times* 20 Oct. 14/2 Modern vinification makes the majority of wines 'star bright' but the presence of a little deposit in many fine wines means quality. **1979** Turner & Roycroft *Winemaker's Encycl.* 167 Wines which are brilliantly clear and reflect highlights when in the bottle are said to be 'star-bright'. All wines that are entered in competitions or are otherwise exhibited publicly should be star-bright.

starch, *sb.* Add: **4.** Freq. in phr. *to take the starch out of* (a person or thing): to remove the stiffness, formality, or pompousness from (someone, etc.), esp. by ridicule; to deflate.

1840 *Spirit of Times* 25 Apr. 90/1 There is something in training in these parts that will be very apt to 'take the starch out' of any 'conceit' they may have. **1846** [in Dict.] **1889** G. B. Shaw in *Hawk* 13 Aug. 172/1 This is the sort of thing that takes the starch out of the most bumptious critic. **1922** *Daily Mail* 20 Nov. 12 The home forwards were unquestioned masters, and on the day's play would have taken the starch out of any other pack in the country. **1936** V. W. Brooks *Flowering of New England* vi. 113 The British reviews were cold and formal. ..The great Romantic critics had not appeared, to take the starch out of their pompous manners.

5. a. *starch-free* adj.

1939 A. Thirkell *Before Lunch* iv. 88 Miss Starter is on a diet and has to have a special bread called Kornog, which is practically starch-free. **1971** *Guardian* 29 June 13/2 Anyone on a starch-free diet should seek some other country [than Scotland].

b. starch blocker, a dietary preparation that supposedly affects the metabolism of starch so that it does not contribute to a gain in weight; **starch gel,** a gel made from starch and an aqueous buffer solution, used as the supporting medium in a method of zone electrophoresis; so *starch-gel electrophoresis*; **starch hyacinth,** a small bulbous plant, *Muscari neglectum*, belonging to the family Liliaceæ, native to Europe and western Asia, and bearing spikes of dark blue, strongly scented flowers; **starch-reduced** *a.*, processed so as to contain less than the normal proportion of starch.

1981 *Arizona Med.* XXXVIII. 848/1 A new product has appeared on the market... It is not a drug, but a processed food made from a certain type of bean. It is also called NBE (Northern Bean Extract) or The Starch Blocker. **1983** *Daily Tel.* 14 Apr. 6/5 Slimmers who use starch blockers..are wasting their money... Experts.. say they do not affect the quantity of starch digested and could have unpleasant effects if they did work. **1955** O. Smithies in *Biochem. Jrnl.* 630/1 A starch gel containing the desired buffer is prepared in a suitable plastic tray. *Ibid.* 635/1 α_1-Globulin does not appear as a definite band between α_2-globulin and albumin in starch gel electrophoresis. **1961** *Lancet* 5 Aug. 291/1 On electrophoresis in starch gel, no macroglobulin moved out from the point of insertion. **1978** *Jrnl. R. Soc. Med.* LXXI. 192 Investigations into this isoenzyme composition of the circulating CK using starch-gel electrophoresis. **1790** *Curtis's Bot. Mag.* IV. 122 We have thought it better to call this species the Starch Hyacinth, the smell of the flower in the general opinion resembling that substance. **1808** J. E. Smith *Eng. Bot.* XXVII. 1931 Starch Hyacinth..is so abundantly wild in many places. **1900** G. Bell *Let.* 11 Jan. (1927) I. v. 61, I went there yesterday afternoon for starch hyacinths and cyclamen and had a tremendous scramble. **1927** F. B. Young *Portrait of Claire* IV. iii. 372 In sheltered crevices gentian, starch-hyacinth and chionodoxa mocked with their living blue the surly Midland winter. **1939** A. Thirkell *Before Lunch* iv. 89 Pepso is only starch-reduced. **1972** 'G. North' *Sgt. Cluff rings True* ix. 76 Sugar substitutes, starch-reduced biscuits and breads, low-calorie soups.

Star-chamber, † **starred chamber.** Add: **2. b.** (Later examples.) Also *fig.*

1896 W. C. Gore in *Inlander* Jan. 150 *Star chamber*, an oral examination given to a student privately. **1934** [see *Gay-Pay-Oo]. **1958** M. Kennedy *Outlaws on Parnassus* xiii. 208 A large group of [novelists]..will never allow that the novel can be subjected to legislation. If some Star Chamber has been set up, they..will have none of it. **1973** *Times* 4 Dec. 17/6 The constitutional propriety of the present industrial star chamber is dubious. **1975** 'S. Marlowe' *Cawthorn Jrnls.* (1976) xviii. 150 Mexican law's based on the Code Napoleon... You could say the courtroom's a Star Chamber instead of a field of forensic battle.

starchy, *a.* Add: **2. b.** Of food: containing much starch.

1948 *Good Housek. Cookery Bk.* i. 12 *Carbohydrates.* These comprise starchy foods and sugars. **1977** C. McCullough *Thorn Birds* i. 13 No one carried a pound of superfluous flesh, in spite of the vast quantities of starchy food.

4. Also *transf.*

1874 'Mark Twain' *Lett. to Publishers* (1967) 81 It will thus be a mighty starchy book. **1880** [in Dict.]. **1977** E. W. Hildick *Loop* xxiv. 164 Maybe that's putting it a bit too formal, Ralph, just a bit starchy.

stardom. Delete *nonce-wd.* and add: (Later examples.) Also, the status of a celebrity or star performer in other spheres of activity. Also *fig.*

1901 C. Morris *Life on Stage* xxv. 203 At the curtain's fall her stardom was over. **1927** *Radio Times* 21 Oct. 143/2 From a clerkship in an insurance office to 'stardom' at a Royal Variety Performance is a long step. **1941** 'N. Blake' *Case of Abominable Snowman* xvi. 177 'I fancy lady Macbeth had put in a good deal of quiet work before the play opens, telling her husband what a worm he was.' 'Grooming him for evil stardom?' **1951** *Sport* 16–22 Mar. 13/3 Quietly plodding the hard road to stardom is Terry Webster, goalkeeper of Derby County. **1977** *New Yorker* 9 May 143/2 The media..impel a Prime Minister to seek stardom at the expense of the Cabinet. **1979** *Tucson* (Arizona) *Citizen* 20 Sept. 1D/3 Giangardella has been tackling opponents since he was a fourth grader at Mt. Carmel Elementary School back in Niles. He went on to stardom at Niles McKinley High School.

star-dust. Add: Also **stardust. 3.** *fig.* That which is illusory or insubstantial. Freq. in phr. *to have stardust in one's eyes* and varr.

1933 E. Sitwell *English Eccentrics* iv. 116 Here comes the star-dust, the noisy chattering crowd of tipsters and smaller racing men. **1934** V. Woolf *Walter Sickert* 20 His [*sc.* Sickert's] paint has a tangible quality; it is made not of air and star-dust but of oil and earth. **1967** Wodehouse *Company for Henry* iii. 45 Clichy double overlay weight..centred by a pink rose within a ring of white stardust canes. **1975** A. Hunter *Gently with Love* xxii. 81 'You had a different opinion of her once.'..'I must have had some stardust in my eyes.' **1977** *New Yorker* 4 July 82/3 There is the silliness of the movie's plangency: hard to feel soupy about a talented couple giving up their love because of the stardust in their eyes.

stare, *v.* Add: **2. b.** Also (in *fig.* sense) to be apparently obvious but nevertheless overlooked.

1966 J. B. Priestley *Salt is Leaving* iii. 43 She found the Mahler album..'It must have been staring you in the face,' she added. **1972** 'G. North' *Sgt. Cluff rings True* xvi. 120 'It was staring me in the face,' the Sergeant said, and he should have seen it sooner. **1979** A. Boyle *Climate of Treason* viii. 236 The over-caution of Cowgill in fumbling for conclusions that stared him in the face demanded some patience.

d. *to stare* (someone) *down, out*: to stare at someone without being first to blink or lower

STARE DECISIS

one's gaze, usu. as an expression of resistance or hostility; to outstare. Also *fig.*

1856 Dickens *Little Dorrit* (1857) i. xxiv. 215 'She looked at the Princess, and the Princess looked at her.' 'Like trying to stare one another out,' said Maggy. **1946** T. H. White *Mistress Masham's Repose* xiv. 115 Miss Brown searched out her pupil's eyes and fixed them with her own. She had a..trick of staring Maria down. **1965** 'T. Hinde' *Games of Chance* i. iv. 110 That made me shout at Kenny a lot, and mimic him, and stare him out. **1972** R. Thomas *Porkchoppers* (1974) xii. 107 He spent nearly a minute staring at Goff. Goff had stared back, thinking that he was damned if he'd let any pal of Cloke's stare him down. **1979** *Guardian* 12 Jan. 8/5 Some measure of fiscal 'mid-term adjustment'..is called for. So is a serious attempt to stare down the local government workers. **1979** G. Seymour *Red Fox* iv. 56 The maid in the starched apron stared him out.

6. stare-you-out, the activity of staring someone out (see sense 2 d above), esp. as a children's game; also *attrib.*

1962 E. O'Brien *Lonely Girl* ix. 107 In the village.. people stopped to look..with savage stare-you-out eyes. **1972** J. Quartermain *Rock of Diamonds* xxvi. 140 She held her expression... I grinned and played 'Stare-you-out'. But I blinked first. **1977** D. Morris *Manwatching* 75 (*caption*) Such is the impact of the close-quarters gaze that the schoolboy game of stare-you-out is extremely difficult to maintain over a long period of time.

‖ **stare decisis** (stēᵊri dèsəizis, stāre dèsīsis). *Law.* [L., lit. to stand by things decided.] The legal principle of determining points in litigation according to precedent; properly as *vbl. phr.*, to be bound by precedents.

1782 F. Buller in E. H. East *Rep. King's Bench* (1801) I. 495 The rule *stare decisis* is one of the most sacred in the law. **1811** *Q. Rev.* Dec. 434 The learned judge..professes his anxiety 'stare decisis', and to abide by authorities. **1845** H. Broom *Legal Maxims* ii. 61 It is..an established rule to abide by former precedents, *stare decisis*, where the same points come again into litigation. **1936** J. C. Gardner *Judicial Precedent Scots Law* iii. 41 It probably cannot be said that the *stare decisis* doctrine had no part in the operation of precedent on the inferior Courts in Scotland. **1970** *Internat. & Compar. Law Q.* XIX. i. 145 A rigid adherence to *stare decisis*..is of little assistance where the law hopes to offer an answer to the people it serves. **1982** *Sci. Amer.* June 20/2 The doctrine of *stare decisis* (namely that precedents should be followed) may be imposed at the players' option, or it may arise without explicit amendment, as successive judges feel impelled to treat 'similarly situated' persons 'similarly'.

‖ **starets, staretz** (stā·ryets). Pl. **startsy, startzy** (stā·rtsi). [Russ., = (venerable) old man, elder.] In the Russian Orthodox Church, a spiritual leader or counsellor. Also *transf.*

1923 G. Buchanan *My Mission to Russia* I. xviii. 240 Rasputin..thus gradually acquired the reputation of a holy man, or elder (*starets*), and was credited with the gifts of healing and prophecy. **1955** J. D. Salinger in *New Yorker* 29 Jan. 36/2 He meets this person called a starets—some sort of terribly advanced religious person—and the starets tells him about a book called the 'Philokalia'. **1966** A. Bloom *Living Prayer* v. 73 The Staretz Ambrose of Optina had the kind of vision which allowed him to see a person's real good. **1975** *Christian* III. 91 The Startsy became enormously influential in the last hundred years of Tsarism. **1976** *Ibid.* III. 150 She was meeting..a man who was himself the spiritual soul of one of the greatest *startzy* (spiritual fathers) of the nineteenth century West, the Abbé Huvelin. **1981** A. Edwards *Sonya* 492 Rasputin..'cured' Tsarevich Alexis's hemophilia in 1904. Thereafter, with each recurrence of her son's illness, the Tsarina grew more dependent upon this starets. **1983** *Church Times* 4 Feb. 7/1 They tell us of the hidden work of the *Startsy*, the 'elders' or spiritual fathers, whose counsel and prayer is an inspiration to many.

starey (stēᵊ·ri), *a.* Also **stary.** [f. Stare *v.* + -y[1].] **1.** Inclined to stare; giving the appearance of staring.

1924 *Chambers's Jrnl.* Aug. 557 A bit flushed and starey about the eyes but still breathing. **1950** T. E. Lawrence *Mint* xiv. 50 Her eyes were starey, like a haddock's. **1960** P. Coleridge *Running Footsteps* 170 Did he go blind at the end? He was getting very stary, very vacant, like. **1975** *Listener* 9 Oct. 479/1 Vanessa Redgrave, very starey and earnest. **1980** S. King *Firestarter* i. 57 There was something starey in her eyes that made him think about those combat-fatigue stories you heard during wartime.

2. = Staring *ppl. a.* 3.

1955 W. W. Denlinger *Complete Boston* i. 94 They [*sc.* internal parasites] reflect themselves in a dull starey coat.

star-fish, starfish. Add: **3.** Special Comb.: **starfish bed** *Geol.*, a stratum rich in starfish fossils; (usu. with capital initial(s) as a proper name).

1863 *Q. Jrnl. Geol. Soc.* XIX. 289 Capping these, in Down Cliffs, is the Starfish-bed, above which begin the Middle Lias Sands. **1935** J. Pringle *South of Scotland* ii. 34 Near the head of the Lady Burn the mudstones are underlain by a hard greenish-grey calcareous sandstone, the well-known Starfish Bed. **1970** R. M. Black *Elements Palaeont.* x. 141 While, in general, they are rare fossils, in some horizons (referred to as 'starfish' beds) they are relatively numerous, e.g. in the Upper Ordovician, Girvan and in the Lower Ludlow in Herefordshire. **1977** J. W. Perkins *Geol. explained in Dorset* xi. 138 The base of the Starfish Bed forms a strong spring-line, and hillsides below this level are often cut by deep wooded gullies as a result.

star-gazy, stargazy, stargazey (stā·ɹgē[i]zi), *a.* Also **star-a-gaze, starry-gazey, starry-gazy.** [f. Star-gaze *v.* + -y[1].] *star-gazy pie*, a kind of fish pie traditionally made in Cornwall (see quots.).

1847 J. O. Halliwell *Dict. Archaic & Provinc. Words* II. 799/1 Starry-gazy-pie. A pie made of pilchards and leeks, the heads of the pilchards appearing through the crust as if they were studying the stars. *Cornw.* **1864** F. T. O'Donoghue *St. Knighton's Keive: a Cornish Tale* Gloss. 303 *Star-a-gaze pie*, a mackerel pie with the heads above the paste, gazing upwards, as it were. **1954** D. Hartley *Food in England* x. 246 *Stargazey pies.* These are properly made of pilchards... The cooks covered the body of the fish—but left the head sticking out. **1966** *Punch* 14 Sept. 385/1 To provide the dishes that one's forbears ate—roast saddle of hare,..or stargazy pie, or syllabub—would be to proclaim oneself madly affected. **1970** A. Pascoe *Cornish Recipes Old & New* 30 (*heading*) Stargazy pie. **1980** *ABMR* Feb. 75/1, I now believe that heavy cake, like starry-gazey pie, was originally made from pilchards.

stark, *adv.* Add: **2. b.** (Later examples of *stark raving mad.*) Also *stark ravers* (slang): see *ravers *a.*

1958 E. Dundy *Dud Avocado* iii. vi. 270 My first thought was that I had gone stark raving mad..and that I was now hallucinating in a looney bin. **1968** [see *cool *v.* 5 b].

Stark (stāɹk). *Physics.* The name of Johannes Stark (1874–1957), German physicist, used *attrib.* with reference to an effect observed by him (*Sitzungsber. der k. Preuss. Akad. der Wissensch.* (1913) 20 Nov. 932) in which spectral lines of a gas are broadened, split, or shifted when the source is in an electrostatic field (either applied externally or due to charged particles within it). Cf. *Zeeman.

1914 *Science* 2 Oct. 493/2 The Stark-electric effect differs from the Zeeman effect in that the various lines of the same series are not equally affected. **1939** *Nature* 20 May 834/1 The spectroscopy team is drawn from an even wider area,..dealing..with atomic spectra, hyperfine structure, Stark effect and related phenomena. **1962** G. M. Barrow *Introd. Molecular Spectrosc.* v. 92 Some dipole-moment determinations that have been made using the Stark effect in rotational spectra. **1979** *Physical Rev.* A. XX. 504/2 Stark-broadening parameters for the brighter isolated tin lines are given in Table VI.

starkers (stā·ɹkəɹz), *a. slang.* [f. Stark *adv.*: cf. *-er[6].] **1.** = Stark naked *a.*

1923 J. Manchon *Le Slang* 292 *Starko*,..= *stark naked* = *starkers*, tout nu. **1952** M. Allingham *Tiger in Smoke* vi. 105 We was all starkers and painted black. **1963** *Guardian* 20 Apr. 12/6 There was no stripping... The girls were starkers all the time. **1972** 'D. Devine' *Three Green Bottles* 12, I dare you to go in starkers. **1981** *Times Lit. Suppl.* 11 Sept. 1046/3 There is also Rosemary Martin as the Art School model in David Storey's *Life Class*, and Cathy Kessler, similarly starkers in *Lay-By*.

2. Stark raving mad. Cf. Stark *adv.* 2 b in Dict. and Suppl.

1962 E. Grierson *Massingham Affair* v. i. 245 She's.. a bit starkers, poor soul, and driving Matron..up the wall. **1972** L. P. Davies *What did I do Tomorrow?* iii. 38 You belted out of that room... They thought you were starkers.

starko (stā·ɹko), *a. slang.* [f. Stark (naked + -o[2].] = Stark naked *a.*

1923 [see *starkers *a.* 1]. **1935** L. Luard *Conquering Seas* iii. 40 They'll be stripping us starko next. **1946** B. Marshall *George Brown's Schooldays* iv. 16 Doing a bundle means getting dressed from starko in five minutes. **1961** J. Pudney *Thin Air* xv. 197 Leave him in his birthday suit. Miss bloody Garth can walk back to Midsomer starko and explain to the folks that she's been a man all the time. **1979** C. Brand *Rose in Darkness* iv. 25 She opened the door, practically starko, and couldn't think what the lady was so surprised about.

starlet. Add: **3.** A young promising performer (usu. an actress) in the world of entertainment; also used in sport, etc. Cf. Star *sb.*[1] 5 a, c in Dict. and Suppl.

1920 J. Ferguson in *Northern Numbers* 97 Some 'starlet' sings, Into the footlights' glare. **1938** *Life* 29 Aug. 37 When Hollywood uncovers a starlet with a unique personality, like Marie Wilson, they turn her over to the still-picture department for a publicity 'build-up'. **1951** M. McLuhan *Mech. Bride* (1967) 24/1 As much time goes into the search for a title of some indigestible cold lard as in launching a starlet. **1959** Anon. *Streetwalker* iii. 48 A paragraph about a film starlet catches my eye. **1976** *Star* (Sheffield) 29 Oct. 28/1 Sheffield United manager Jimmy Sirrel produced another starlet from his club's youth policy today when he named Gary Hamson in his side. **1978** J. Wainwright *Thief of Time* 105 Very few 'starlets' end up as stars. They stay there—chocolate-box perfection with nil personality.

star-light, starlight, *sb.* and *a.* Add: **3. b.** Special Comb.: **starlight scope** *Mil.*, a device incorporating an image intensifier for use as a gun sight or telescope when there is little light.

1969 I. Kemp *Brit. G.I. in Vietnam* vii. 146 The North Vietnamese found two very valuable, expensive, and secret items of equipment, known as 'Starlightscopes'. The Starlightscope is an infra-red telescope for observation at night, which can also be fixed to a rifle for shooting in the dark. **1973** T. O'Brien *If I die in Combat Zone* iv. 28 Look at this... It's a starlight scope... Supposed to let you see in the dark. **1977** *Time* 23 May 33/2 There's the 90-mm. recoilless rifle with a 'starlight' scope for enhanced visibility.

‖ **starover** (stārǒvyḗ·r). Pl. **starovers, starovery.** [Russ.] = *Old Believer.

1861 A. P. Stanley *Lectures on Hist. Eastern Church* xii. 471 The real force, the permanent interest, of the Rascolniks lies in the eight millions of souls who call themselves *Starovers*; that is, 'the Old Believers'. **1957** *Oxf. Dict. Chr. Ch.* 1287/1 *Starovery*, another name for the Russian sect of the Old Believers. **1963** N. V. Riasanovsky *Hist. Russia* xix. 220 The Old Believers or Old Ritualists—*starovery* or *staroobriadtsy*—rejected the new sign of the cross, the corrected spelling of the name of Jesus, the tripling instead of the doubling of the 'Hallelujah' and other similar emendations.

starquake (stā·ɹkwē[i]k). *Astr.* [f. Star *sb.*[1] after Earthquake.] A sudden change of shape or structure undergone by a neutron star, pulsar, etc.

1969 *Nature* 9 Aug. 598/1 Neutron stars which satisfy the criterion, $\omega \gtrsim 10^3\ s^{-1}$, should have starquakes as they slow up. **1970** *New Scientist* 4 June 465/3 After the starquake, the pulsar readjusts its shape and speeds up. **1976** *Sci. Amer.* Oct. 78/3 In still other neutron-star models the source of the gamma rays is ascribed to 'starquakes', volcanic activity or other sudden changes in size or shape. **1978** Pasachoff & Kutner *University Astron.* xi. 314 As a result of this 'starquake', the matter would then be distributed slightly closer to the center of the star.

starred, *ppl. a.* Add: **2. e.** *spec.* thus marked in order to indicate some special category or merit. (Later examples.)

1914 *Hansard Lords* 24 Nov. 459 My original arrangement with Lord Kitchener was that a starred man should neither be solicited for recruitment nor accepted for the Army if he offered himself. **1927, 1937** [see *nap *sb.*[4] 2 c]. **1940** *Hansard Lords* 6 Aug. 148 As far as I remember, the starred question was introduced at the instance of the late Lord Curzon about twenty years ago, and the object was to enable noble Lords to put down questions which they would wish to see mentioned in the House, rather than dealt with by a written reply, but upon which no debate should take place. **1964** F. White *West of Rhone* xxii. 233, I stopped at a starred hotel... It deserved its star, for it was very good. **1970** R. Lowell *Notebk.* 104 Four stone inkfish, thrice stepped on, lifting the spout—Not starred in any guidebook. **1971** *Guardian* 19 July 8/2 Margaret Drabble..whom he much admires for..her starred first. **1974** R. Quirk *Linguist & Eng. Lang.* xi. 158 American English items [in a dictionary] are prefixed by a warning asterisk (an unhappy emblem when we consider what a starred form means in linguistics). **1976** N. Roberts *Face of France* xix. 185 The starred items on the menu.

f. Also *spec.* of a shattered vehicle windscreen of splinterproof glass.

1960 *B.S.I. News* Mar. 8/2 A windscreen of laminated glass may crack under impact but it will hold together, though 'starred', and remain in one place except in the most violent of collisions. **1979** R. Perry *Bishop's Pawn* ix. 153, I continued driving blind, unable to see through the starred windscreen.

starrer. Restrict *local* to sense in Dict. and add: **2.** *slang.* A play or film which provides an impressive leading role for an actor or actress.

1951 Green & Laurie *Show Biz* 571/2 *Starrer*, starring vehicle. **1978** M. Puzo *Fools Die* xxvii. 311 A Kellino starrer would get the studio's two million back.

starrigan (stæ·rigăn). *Newfoundland.* Also **starigan.** [Perh. ad. Ir. *stairricín* stump, stick.] A young, stunted, or decayed evergreen tree (chiefly fir), esp. cut for firewood; a stick, branch, or stump of this.

1895 *Jrnl. Amer. Folk-Lore* VIII. 39 *Starrigan*, a young fir-tree, which is neither good for firewood nor large enough to be used for timber, hence applied with contempt to anything constructed of unsuitable materials. *c* **1900** in *Regional Lang. Stud. Newfoundland* (1978) VIII. 26 *Starrigan*, a green stick, especially a var of small dimensions. **1903** *Newfoundland Q.* Dec. 5 He could get nothin' there but a few green var starrigans, or dun boughs. **1920** Grenfell & Spalding *Le Petit Nord* 94 Light snow has fallen during the night, and every 'starrigan', every patch of 'tuckamore' is 'decked in sparkling raiment white'. **1964** *Newfoundland Herald* 26 Jan., *Starrigans*, actually dry tree stumps which formed an important source of fuel in the depression days. **1977** *Decks Awash* Sept. 64 De bull would get de tacklin An off 'e'd go for starrigans Wid frosty snow acracklin. **1981** *Publ. Amer. Dial. Soc.* LXVIII. 43 *Starrigan*.., very sappy fir, this name usually applied only when cut for firewood.

starring, *vbl. sb.* Add: **1.** (Earlier and later non-*attrib.* examples.)

1841 *Punch* 17 July 12/1 We consider Mr. Phelps' opposition to this ruinous system of 'starring' as commendable and manly. **1849** *Theatrical Programme* 25 June 38 Queen's Theatre, Hull... Starring is the order of the day here. **1940** *Illustr. London News* CXCVI. 188/2 This kind of 'starring' I regard as a foolish departure at a theatre which has built up a reputation as a repertory with a first-rate team.

starry, *a.* Add: **2. d.** Of or pertaining to stars in the world of entertainment.

1907 G. B. Shaw *Let.* 24 May (1972) II. 690, I have.. utterly rejected the starry part of it [*sc.* a production of *Man & Superman*]. **1918** R. Wagner *Film Folk* 8 The starry firmament of Los Angeles. **1929** H. G. Wells *King who was King* i. 30 The film *entrepreneur*..has to secure the services of a starry lady. **1959** *Manch. Guardian* 30 Jan. 7/1 One of the most persistent critical complaints about Miss Dresdel has been her tendency to shine too brightly even in the starriest company. **1981** *Times* 24 Jan. 7/2 Gertrude Lawrence..needed something very big, something very starry.

starry-eyed, *a.* [f. Starry *a.* + Eyed *ppl. a.*] **a.** Of persons: idealistic, uplifted, romantic; ingenuous, naïve.

1936 M. Mitchell *Gone with Wind* xxix. 489 She had never stood starry-eyed when the Stars and Bars ran up a pole. **1945** Nelson & Wright *Tomorrow's House* xviii. 206/2 Starry-eyed prospective home builders. **1955** M. Dickens *Winds of Heaven* v. 130 Holding hands and being starry-eyed over bottles of Chianti in romantic little cafés. **1958** *Spectator* 10 Jan. 37/3 He can hardly be so starry-eyed as to suppose that a similar enterprise on a national scale would be blessed by the whole hierarchy. **1964** S. Brittan *Treasury under Tories* ii. vii. 208 He understood both businessmen and bureaucrats too well to be starry-eyed about either. **1979** *Dædalus* Summer 107 A starry-eyed bachelor woos the kitchen maid.

b. *transf.* and *fig.*

[**1928** 'Brent of Bin Bin' *Up Country* x. 163 She expressed starry-eyed sympathy with his loss, having a struggle to keep back the tears.] **1947** A. L. Rowse *End of Epoch* i. 14 The starry-eyed vacuity of the unteachable and the uneducable. **1958** *Listener* 25 Sept. 476/2 All too often United States policy-makers..tend to hover between a starry-eyed idealism and a dangerous brinkmanship. **1973** *Times Lit. Suppl.* 3 Aug. 891/4 Mr Grunberger's account is balanced, but..his conclusion is starry-eyed. **1978** *Jrnl. R. Soc. Arts* CXXVI. 345/2 It is just not possible to accomplish anything like so starry-eyed a concept.

starscape (stā·ɹskēɪp). [f. Star *sb.*[1], after Landscape.] A view or prospect of a sky filled with stars. Also *fig.*

1926 S. Leslie *Cantab* (ed. 2) xiii. 163 He stared through the window at the Sussex star-scape. **1947** *Horizon* Aug. 100 He [*sc.* Paalen]..leads us into 'starscapes', a happy nomenclature invented by him. **1958** L. Durrell *Balthazar* i. i. 13 Winter: freezing snow, cool sand, clear sky panels...magnificent starscapes. **1965** *Spectator* 1 Jan. 16/1 One of those golden, glittering Covent Garden starscapes which..are so fascinating to diagnose through opera glasses. **1968** A. C. Clarke *2001: Space Odyssey* xliv. 210 The scurrying flecks of light no longer moved across the redly-glowing starscape still thousands of miles below.

star-ship. Also **star ship, starship.** [f. Star *sb.*[1] + Ship *sb.*[1]] **1.** (In Dict. s.v. Star *sb.*[1] 20.)

2. *Sci. Fiction.* A large manned spacecraft designed for interstellar travel.

1934 *Astounding Stories* Dec. 9 To start the year we offer you *Star Ship Invincible*, by Frank K. Kelly. **1956** F. Pohl *Alternating Currents* 77 A particle of meteoric matter slammed into *Starship Terra II* in hyperspace. **1967** J. Blish *Star Trek* 40 Capt. James Kirk of the starship *Enterprise*..had seen more planets than most men knew existed. **1972** H. C. Rae *Shooting Gallery* iv. 242 Now he was to be..flung, like one of those plastic starships, out into a galaxy of worlds which didn't really exist. **1978** *Listener* 30 Mar. 405/1 Smaller monitor screens were scattered about. It could have been the bridge of a star-ship. **1980** *Daily Tel.* 14 Jan. 8/1 Could star ships ever be propelled by the violent mutual annihilation of matter and anti-matter?

star-spangled, *ppl. a.* Add: **2.** *star-spangled banner.* **a.** The national flag of the United States of America. Also *fig.* Cf. *Stars and stripes* s.v. Star *sb.*[1] 6 b.

1814 F. S. Key in *Baltimore Patriot* 20 Sept. 2/1 The star-spangled banner in triumph shall wave. **1843** Dickens *Martin Chuzzlewit* (1844) xxi. 261, I thank you, sir, in the name of the star-spangled banner of the Great United States. **1869** 'Mark Twain' *Innoc. Abr.* xlix. 515 A robe..that was a very star-spangled banner of curved and sinuous bars of black and white. **1887** —— *Let.* 18 Dec. (1917) II. 480 It had bought them of the star spangled banner Masterthief. **1949** *Chicago Daily News* 15 Dec. 3 Was the Star-Spangled Banner made in a brewery?

b. With capital initials: the name of the U.S. national anthem.

1814 [in Dict.] **1843** *Quincy* (Illinois) *Herald* 3 Mar. 1/1 It is a most beautiful history of that national ballad, 'The Star Spangled Banner'. **1899** Kipling *From Sea to Sea* II. xxv. 15 When we had chanted 'The Star-Spangled Banner' not more than eight times, we adjourned. **1936** *Nation* 10 Oct. 419/2 As the band played the Star Spangled Banner at the end,..the President stood erect. **1977** G. Markstein *Chance Awakening* xliii. 128 He was on his feet..with 17,000 other people in the stadium as the Star Spangled Banner thundered out of the loudspeakers.

start, *sb.*[2] Add: **5. b.** Also *start-to-stop*, used (usu. *attrib.*) with reference to train journeys or their schedules.

1899 *Railway Mag.* IV. 375/1 They comprise one of the best start-to-stop runs I have ever had on a British line. **1931** *Times Educ. Suppl.* 19 Sept. (Home & Classroom Section) p. ii/2 (*caption*) The Great Western Railway Company regained this week the record for the fastest start-to-stop journey in the world. **1936** *Discovery* Nov. 356/1 Two or three runs booked, start-to-stop, at over 80 miles an hour. **1968** O. S. Nock *Railway Enthusiast's Encycl.* 279 The run of the 'Silver Jubilee' from King's Cross to Darlington is also tabulated; and finally that of the 'Coronation', introduced in 1937, with the fastest start-to-stop schedule ever tabled with steam in Great Britain.

g. *Sport.* By synedoche, a contest, race, or game. Chiefly *N. Amer.*

1944 *Sun* (Baltimore) 23 Feb. 12/3 Davis is a welterweight... Davis isn't that good. At least he never has been in most of his previous starts. **1949** *Richmond* (Va.) *Times-Despatch* 10 Oct. 13/2 The Rebels, in gaining their third win in four league starts..won it as convincingly as the score would indicate. **1966** *Telegraph* (Brisbane) 22 Jan. 5/2 He [*sc.* a horse] started racing in November, and in five starts has tallied a win, second, and a third. **1970** *Globe & Mail* (Toronto) 25 Sept. 32/3 Winless in 14 previous starts this season, Miss Ella Cinders had little trouble with Sandy Hawley up last Saturday as she galloped to a 12-length win. **1970** *Washington Post* 30 Sept. D1/7 The hapless, helpless Nats..couldn't hold off the East Division champions, who have captured nine consecutive starts.

h. The act of beginning to build a house. Also *housing start.*

1946 *Sun* (Baltimore) 20 Aug. 8/2 The Wyatt office claims about 406,000 'starts' of dwelling units in the first five months of the year. **1955** *Times* 30 May 11/1 New housing 'starts' rose in April but by less than they usually do over March. **1966** *New Statesman* 25 Nov. 769/2 What is worrying is that the starts are falling in the private sector and, as a house takes an average of about a year to build, the effects will be projected into next year's figures. **1976** *National Observer* (U.S.) 22 May 8/3 Around 28 to 29 per cent of all single-family housing starts.

i. Phr. *for a start*: to begin with. *colloq.* Cf. *starter 2*.

1951 E. Paul *Springtime in Paris* iii. 56, I..found Montherlant's *Les Célibataires*... 'That's a good one for a start.' **1971** *Radio Times* 21 Aug. 47/3 What makes Raven unusual? *For a start* he's 46, and..he was a ballet dancer a lieutenant of infantry, a classical actor and a television producer. **1978** L. Thomas *Ormerod's Landing* iii. 48 Everybody else knows... The submarine crew know for a start.

12. start button, a switch that is pressed in order to set a machine or process in action; **start-line** = *starting-line* s.v. *starting *vbl. sb.* 2 b; chiefly *transf.* and *fig.*, esp. in *Mil.* use (see quot. 1961).

1964 Start button [see *control register* s.v. *control *sb.* 5]. **1968** *Brit. Med. Bull.* XXIV. 190/1 When the start-button of the machine is pressed, it simply causes the programmed procedure to operate on the data, giving rise to an action which will depend entirely on the data and the procedure. **1977** D. MacKenzie *Raven & Kamikaze* iii. 40 He..plugged the cable into a wall-socket and thumbed the start button. **1945** E. Waugh *Brideshead Revisited* 224, I would..think at such and such a time.. I shall cross the start-line and open my attack for better or worse... With Julia there were no phases, no start-line, no tactics at all. **1946** G. Millar *Horned Pigeon* ii. 32 Rommel's Afrika Korps were on the start line of their long advance. **1961** W. Vaughan-Thomas *Anzio* v. 69 The concept of a start line—an essential part of infantry tactics—is simply a matter of applied common sense... Just as in a race in athletics all competitors must line up at a starting-point..so, in the infinitely more exacting race of an infantry attack, the unit—be it battalion, brigade or division—needs some feature on the ground along which the troops can be lined for the take off. **1982** J. Wainwright *Anatomy of Riot* 15 In Army parlance he was going to be the field commander when the war left the start-line.

start, *v.* Add: **I. 3. e.** Also with *forth* (*arch.* and *poet.*). Cf. sense 4 b in Dict.

1916 Joyce *Portrait of Artist* ii. 86 All day the stream of gloomy tenderness within him had started forth and returned upon itself in dark courses and eddies. *Ibid.* iv. 171 He seemed..to see the amount of his purchase start forth immediately in heaven..as a frail column of incense. **1939** C. S. Lewis in M. Black *Importance of Lang.* (1962) 37 A new metaphor simply starts forth, under the pressure of composition or argument.

11. a. Also const. *out.* orig. *U.S.* Cf. sense 12 d in Dict. and Suppl.

1925 E. O'Neill *Desire under Elms* in *Compl. Wks.* II. 170 We're free, old man..an' we're startin' out for the gold fields of Californi—a! **1933** —— *Ah, Wilderness!* ii. 71 He dared me to race him... So I said all right and we started out. We swam and swam and were pretty evenly matched. **1978** W. F. Buckley *Stained Glass* xii. 117 There are escort vessels, and it is quite a muddle if every boat decides for itself when to start out.

c. Of a motor vehicle or its engine: to begin to operate. Also with *up.*

1902 A. C. Harmsworth *Motors & Motor-Driving* ix. 165 A petrol engine will generally start most easily with all the cold-air inlets closed. **1904** J. F. Gill *How to build Petrol Motor* 57 When a speed of about four miles per hour has been obtained, the current may be switched on, after which, the motor should start. **1932** D. L. Sayers *Have his Carcase* xii. 145 The Morgan wouldn't start, not for toffee... On their..putting in a new one [*sc.* lead], the engine had started up at once. **1971** *Daily Tel.* 27 Oct. 13/5 The engine would start instantly from cold without using the choke control. **1973** L. Cooper *Tea on Sunday* iii. 44, I left early..because of the bad weather. I was afraid my car might not start.

12. d. *to start in*: for (U.S. colloq.) read colloq. (orig. *U.S.*); (later examples); also *to start in on*; *to start out*: also const. *prep.*; *to start over* (U.S.): to begin again.

1892 'Mark Twain' *Amer. Claimant* 138 He had started out on a high emprise. **1912** Wodehouse *Prince & Betty* iv. 53 Then we start in. **1924** F. Scott Fitzgerald *Let.* 27 Oct. (1964) 168 I'm tired of being the author of *This Side of Paradise* and I want to start over. **1925** J. Buchan *John Macnab* vii. 144 In this country, once you start in on politics you're fixed in a class and members of a hierarchy. **1929** D. H. Lawrence *Pansies* 20 Then I am willing to fight, I will roll my sleeves up And start in. **1951** M. McLuhan *Mech. Bride* (1967) 68/2 She started out as a Tillie the Toiler. **1957** O. Nash *You can't get there from Here* 84 Once they start in on ideas and ideals they'll end up spouting ideologies and isms. **1965** M. Bradbury *Stepping Westward* ii. 112 Now go back to the beginning and start over. **1976** *National Observer* (U.S.) 26 June 8/6 If it doesn't work, throw the whole mess out and start over, adding a teaspoon of active dry yeast and a few drops of vinegar at the beginning. **1978** T. Allbeury *Lantern Network* vii. 87 Chaland had started in straight-away. 'Bonnier your group is far too big.'

f. To begin to go *to* (school). *U.S.*

1836 W. Sewall *Diary* 10 Aug. (1930) 172/1 Henry and Catherine started to school. **1898** C. A. Bates *Clothing Bk.* No. 1279 That boy..will have to start to school soon. **1931** *Amer. Speech* VII. 20 *Start*, begin to go. Used mainly in the one expression, start to school. 'I started to school when I was five.' (Widespread.)

g. *to start* (*in*) *on*: to attack; to nag or bully. Cf. sense 12 d above. *colloq.*

1907 G. B. Shaw *Major Barbara* ii. 214 When trade is bad..and the employers az to sack arf their men, they generally start on me. **1953** K. Tennant *Joyful Condemned* iii. 23 I'll give you five minutes, and then I'll start in on you. So hand over the two quid. **1967** J. Morrison in *Coast to Coast 1965–6* 140 The minute I mentioned it she started on me. **1968** Bethell & Burg tr. *Solzhenitsyn's Cancer Ward* I. xxi. 332 The critics may start it on you.

h. Colloq. phr. *don't you start*, expressing exasperation at hearing sentiments (of praise, criticism, advice, etc.) repeated by another speaker.

1934 N. Marsh *Man lay Dead* x. 167 'You're a—a wonder,' finished Nigel seriously. 'Don't you start!' said Mrs North. **1956** P. Scott *Male Child* ii. v. 157 'She ought to get out more.' He grinned. 'Now don't you start.' **1974** 'S. Woods' *Done to Death* 129 'She's a spinster.'.. 'Don't you start!' said Hugh explosively.

13. d. Also *loosely*, to begin.

1979 *Southern Star* (Eire) 29 Sept. 3/4 Mr. McCarthy also told Mr. Coleman that £160,000 had been allocated for further work on the Cork–Brandon road and the Viaduct aided by the EEC Regional Development Fund and would start up in two weeks after the meeting. **1981** E. North *Dames* viii. 138 The music started up and played..a tango.

II. 18. Also const. *forth. rare.*

a **1817** Jane Austen *Northanger Abbey* (1818) II. xiii. 245 She took the first opportunity..to start forth her obligation of going away very soon.

23. c. Also with *up* (occas. *absol.*).

1910 *Marine Oil Engine Handbk.* 14 It is possible to start up from cold on petrol. *Ibid.* 57 There is little difficulty about starting up small engines. **1945** C. S. Lewis *That Hideous Strength* xiv. 383 He started his engine up and they drove away. **1979** D. Clark *Heberden's Seat* i. 8 A car with a set of jump leads to start me up would do it.

e. Also, to set *up* a business (occas. *absol.*). Cf. sense 12 c in Dict.

1974 McArthur & Atkins *Dict. Eng. Phrasal Verbs & their Idioms* 215 He has started up a new business. **1975** Cowie & Mackin *Oxf. Dict. Current Idiomatic Eng.* I. 305/2 He *started up* a successful car hire firm. *Ibid.*, They were thinking of *starting up* in the fruit and vegetable trade.

g. To conceive (a baby), to succeed in conceiving (a child). Also, *to start a family.*

1931 N. Mitchison *Corn King & Spring Queen* iii. vii. 306 She wanted to start another baby at once, but he was very anxious that she should be as strong and well as possible at Plowing Eve. **1934** H. G. Wells *Exper. Autobiogr.* II. viii. 638 We were now justified in starting a family. **1938** E. Bowen *Death of Heart* i. 24 Irene had started Portia. **1956** *Mademoiselle* Sept. 185/1 After the apple farm was started we were going to start a child. **1973** G. Greene *Honorary Consul* ii. iii. 96 'He wanted to marry... And if there's a child—' 'Have you started one?' 'No.' **1975** M. Babson *There must be some Mistake* xiv. 107 We would have married immediately, perhaps have started a family. **1977** *Listener* 25 Aug. 227/1 When a couple have become used to a two-wage standard of living, how do they give it up to start a family?

h. To begin to suffer from or succumb to an illness, esp. a cold.

1932 E. M. Delafield *Thank Heaven Fasting* i. i. 14, I think Cecily's starting a cold. **1943** J. B. Priestley *Daylight on Saturday* xi. 68 The very sight of her streaming face..had made him feel that at any minute he would start a cold too. **1958** P. Kemp *No Colours or Crest* vii. 147 He himself was recovering from the malaria he had started at Arborie, but was still very weak.

24. b. Phr. *to start something*: to cause some trouble, agitation, etc. *colloq.* (orig. *U.S.*).

1917 U. B. Sinclair *King Coal* 78 Either the man was an agitator, seeking to 'start something', or else he was a detective sent in by the company. **1917** Wodehouse *Uneasy Money* xvi. 179 You certainly did the wrong thing. You started something! **1924** E. O'Neill *Welded* i. 97 We're not 'starting something' now, are we—after our promise? **1943** F. J. Bell *Condition Red* 59 The Japs.. slunk by without starting anything.

III. 28. start-stop, used *attrib.* with reference to an electric telegraph system in which each group of elements transmitted is respectively begun and ended with signals activating and deactivating the receiving mechanism.

1922 *Electrician* 8 Sept. 265/2 The teletype..is a 'start-stop' printer. **1937** *Sci. Abstr.* B. XL. 48 A machine for correcting start-stop 5-unit signals. **1974** R. N. Renton *Internat. Telex Service* iii. 12/2 In the start-stop system, although the driving motors may be running, the sending and receiving devices are normally held at rest in a zero-phase position.

startability (stāɹtăbi·liti). [f. Start *v.* + -Ability.] **a.** Of a fuel: the degree to which it facilitates the starting of an engine. **b.** Of an engine: the degree to which it can be readily started.

1933 *Petroleum Handbk.* (R. Dutch-Shell Group) viii. 131 The exact point in the distillation curve which controls 'startability' varies somewhat with the atmospheric temperature and the particular engine in which the gasoline is used. **1935** *Jrnl. R. Aeronaut. Soc.* XXXIX. 907 Experimental evidence has shown that until very low temperatures are reached the startability of a gasoline is roughly dependent upon the percentage distilling to 100°C. **1971** *Good Motoring* Sept. 4/3 Checking startability on steep hills. **1976** *Drive* Sept.–Oct. 114/2 The Beetle has a deserved reputation for..unfailing startability in all weathers.

starter. Add: **I. 2. a.** (Later *fig.* examples.) Also *transf. spec.* an idea that deserves initial consideration (only in neg. contexts: cf. *non-starter).

1917 'I. Hay' *Carrying On* iv. 93 That exasperating race of bad starters but great stayers, the British people. **1947** G. B. Shaw in *Musical Times* Jan. 10/1 They can all sing passably in tune and are selected..because they are good readers and good starters. **1948** M. Allingham *More Work for Undertaker* xiii. 166 The tip about Brownies [*sc.* shares] was never even a likely starter. **1960** *Times* 7 July 13/2 'Bevanism' was never a starter as a political philosophy or programme. **1976** *Listener* 18 Nov. 641/2 The objections to it are so strong that it isn't a starter.

b. With qualifying adj.: a motor vehicle or engine which starts (well, slowly, etc.). Cf. sense *11 c of the vb.

1952 M. Steen *Phoenix Rising* viii. 179 That's my car..; it's the easiest starter... Get in and start the engine. **1975** *Country Life* 27 Feb. 508/3, I found the Lancia a good starter with no need for choke.

2*. a. Phr. *as* or *for a starter, for starters*: to begin with, for a start. *colloq.* (orig. *U.S.*).

1873 J. H. Beadle *Undeveloped West* xxii. 450 He gave me twenty drops of laudanum as a starter. **1902** G. H. Lorimer *Lett. Self-Made Merchant* v. 64 All that he ever needed was a few hundred for a starter. **1947** *Chicago Tribune* 3 Sept. 6/3 As a starter, agents have begun a canvass of small independent food wholesalers. **1950** *Manch. Guardian Weekly* 9 Nov. 7 He wired how many frogs' legs did they think they could handle. They told him ten thousand as a starter. **1969** D. Francis *Enquiry* xii. 168, I fell with a crash. 'That's for starters,' he said. **1970** *New Yorker* 5 Dec. 142/1 (Advt.), For starters, here's the line-up of Knicks and Rangers games for the rest of the month. **1973** *Listener* 6 Dec. 767/3 The vehicles for enlargement could be found in new stations or wavelengths..but what happens is likely to require as a starter some change in current assumptions. **1978** *Globe & Mail* (Toronto) 11 Jan. 8/1 For starters, do not call us scalpers. We are ticket hosts.

b. A dish eaten as the first course of a meal, before the main course (also in *pl.*). *colloq.*

1966 [see *dessert 1 b]. **1966** *Sunday Express* 16 Oct. 18/3 You get a three-course dinner, with four 'starter' courses and seven main dishes to choose from, and a sweet. **1966** *Vogue* Nov. 154/3 Starters include fish soup, cock-a-leekie, duck-liver pâté. **1968** *New Society* 22 Aug. 266/1 The first course of a meal is sometimes called a 'starter', which is perhaps not so much non-U as jargon. **1969** P. Highsmith *Tremor of Forgery* xvii. 155 'Try this Tunisian starter. Turns up on every menu.' He meant the antipasto of tuna, olives, and tomatoes. **1975** *Reveille* 20 June 2/2 They are equally good as a 'starter' or served with salad. **1979** V. Canning *Satan Sampler* ii. 30 There was avocado pear for what some people disgustingly called 'starters'.

II. 4. b. *N. Amer. Sport.* The player in a team game who starts the game; in Baseball *spec.* the pitcher.

1967 *Boston Herald* 8 May 16/6 The victory gave Atlanta starter Pat Jarvis a 3–0 record. **1968** *Globe & Mail* (Toronto) 3 Feb. 35/2 Two of our starters are in Quebec City on an exchange visit, one player is away sick and Bill Edwards is still injured. **1974** *State* (Columbia, S. Carolina) 3 Mar. 3-D/4 Collins..led the way with 20 points as again all starters hit in double figures. **1976** *Billings* (Montana) *Gaz.* 16 June 3-C/7 The Phillies jumped on starter Ed Halicki and reliever Dave Heaverlo for an eight-run lead after two innings.

5. a. Freq. in phr. *under starter's orders* (Horseracing): subject to the instruction of the starter, ready to begin a race. Also *transf.* and *fig.* Cf. Order *sb.* 23 a in Dict. and Suppl.

1965 M. Spark *Mandelbaum Gate* vi. 197 Freddy has said to tell you we are under starter's orders; what is starter's orders? **1973** P. Malloch *Kirkback* i. 10 'Drink it up, chum. I forgot we were under starter's orders.' Gilchrist drank it. Five minutes later they were on their way. **1974** *Times* 27 Jan. 10/5 With the first day of the £44,000 Philadelphia indoor tennis tournament only half over, six of the 16 seeds were already out of the running. Nastase, Newcombe and Orantes were all injured and, like five other entrants, did not even come under starter's orders. **1976** *Milton Keynes Express* 11 June 3/2 Show jumping commentator Dorian Williams put the horses under starter's orders. **1979** Reese & Flint *Trick 13* 82 They're under starter's orders... They're off!

b. *U.S.* (*a*) One who directs the operation of lifts in a large building; (*b*) an employee of a hotel, station, etc., who organizes transport for patrons (see also quot. 1917).

1909 *Pacific Monthly* Feb. 123/1 Thanks to the crowd in the lobby, the uniformed 'starter' had not seen the bum and come over from the elevators to order him away. **1917** *Street Railway Employment in U.S.* (U.S. Bureau of Labor Statistics) 14 *Starters.*—See that cars leave terminal points on scheduled time..reroute cars to straighten schedules and perform duties of inspectors. **1922** S. Lewis *Babbitt* 32 The little unknown people who inhabited the Reeves Building corridors—elevator-runners, starter..—were in no way city-dwellers. **1931** Wodehouse *Let.* 19 May in *Performing Flea* (1961) 66 A man standing in the crowd outside a movie theatre here after a big opening hears the carriage starter calling for 'Mr Warner's automobile'. **1932** *Making Bus Operations Pay* iv. 77 The station is kept open 24 hours a day by a force of eleven paid employees and three redcaps to handle baggage. Included in the paid force is a station manager, three ticket managers..and a special officer who also acts as a taxi starter. **1978** R. Ludlum *Holcroft Covenant* 541 She was given a number on the twelfth floor, the top floor, but as it was the lunch hour, the starter doubted anyone was there. **1981** *Washington Post Mag.* 22 Mar. 8/4 'A *starter*,' she said. 'You know, the man who gets a guest a taxi.'

6. a. (Later examples.)

1926 W. Faulkner *Soldiers' Pay* ii. 86 She turned the switch and tried to reach the starter with her foot. **1934** *Discovery* Nov. 324/2 A hand starter is provided on the engine, or it may be started from a car battery. **1969** G. Macbeth *War Quartet* 30 When we kicked the starters, drove On across other countries..I remembered all Who fought. **1970** K. Ball *Fiat 600, 600D Autobook* xi. 135/2 Dismantling of the starter is a simple task and is similar to that for the generator. **1977** [see *shock *sb.*[3] 6*].

b. An automatic switch forming part of the auxiliary circuit of some fluorescent lamps, the purpose of which is to enable the electrodes to become hot enough for a discharge to occur after it becomes inactive. Also *starter switch.*

1942 C. L. Amick *Fluorescent Lighting Man.* ii. 22 Each lamp requires a separate starter and a separate ballast. *Ibid.* 23 The heat from the discharge itself keeps the cathodes hot during normal operation, hence the starter switch can remain open. **1962** *Newnes Conc. Encycl. Electr. Engin.* 433/2 In place of the starter switch..a cathode-heating transformer can be used, so avoiding the need for replacement parts. **1967** P. Honey *Household Electricity* 44 Fluorescent lamps..generally need a special circuit with what is called a starter, a device which heats up the electrodes for a second or two after the current is switched on.

7. Delete (See quot.) and substitute: A culture containing bacteria, yeast, or the like, used to initiate souring or fermentation in the making of butter, cheese, dough, etc. (Earlier and later examples.)

1896 *Vermont Agric. Rep. 1895* 67 This may be done.. by using a 'starter' made from cream. **1935** *Discovery* Nov. 340/2 In the manufacture of butter 'starters' are employed to sour the cream under controlled conditions. **1939** K. Pinkerton *Wilderness Wife* vii. 74 At least sour dough bread was sure... Robert had sponsored it by making the 'starter', a mixture of flour and water which grabbed its yeast germs from the air. **1950** *N.Z. Jrnl. Agric.* Jan. 27/2 Quick granulation is achieved by the addition of a 'starter' and the storing of the packed honey immediately in a consistent temperature of about 56 degrees F. **1973** W. H. T. Tayleur *Home Brewing & Wine-Making* vii. 55 The optimum amount of starter in wine-making is about 1 to 15.

8. *attrib.* and *Comb.*, as (sense 6) *starter button, cord, motor, switch*; **starter home** orig. *U.S.*, a first home, usu. one bought by a young couple; also *ellipt.*; **starter set** orig. *U.S.*, a small set of china intended to be the basis of a larger collection; also *transf.*

1971 Starter button [see *starter motor* below]. **1977** D. Beatty *Excellency* vi. 80 He primed the engines, pressed the starter button, heard the propeller creak round. **1971** *Scope* (S. Afr.) 19 Mar. 65/2 Damned engine doesn't have a starting handle, just a starter cord. **1976** *Billings* (Montana) *Gaz.* 27 June 5-D/7 (Advt.), Also a fenced back yard makes it a great starter home. **1979** *Arizona Daily Star* 5 Aug. (Advt. Section) 21/2 This 3 bedroom, 1 bath home will make a super starter for a young family. **1980** *Times* 6 Feb. 6/6 Local authorities and builders should provide more 'starter' homes to meet the demand for lower priced small homes for sale, Mr John Stanley, Minister for Housing and Construction, said yesterday. **1928** *Correct Lubrication* 39 Starter motor spins without turning engine. **1971** R. Dentry *Encounter at Kharmel* vi. 100 Pepper..leaned hard on the starter button. The starter motor set up a pulsating, piercing shriek. **1946** *Time* 29 July 64/2 It was inexpensive—a 20-piece 'starter set' sold for about $6. **1970** *Guardian* 17 Dec. 9/5 A starter set consisting of four soup bowls, four 10½-inch plates, four eight-inch plates, [etc.]. **1977** *Montgomery Ward Catal.* Spring–Summer 510/3 Men's starter set [of golf clubs]. **1925** *Morris Owner's Man.* 9 The starter switch should immediately be released. **1965** Priestley & Wisdom *Good Driving* ii. 19 The starter switch usually takes the form of a knob or button which is either pressed or pulled out.

starting, *vbl. sb.* Add: **1. a.** Also with *off* and *up* in some senses. Freq. *attrib.*

1821 M. Edgeworth *Let.* 29 Jan. (1971) 235, I hope this can be arranged as I found it a starting off point and I could not conclude the agreement without it. **1895** Kipling in *Century Mag.* LI. 265/2 There was the same 'starting-off place'—a pile of brushwood. **1912** *Motor Man.* (ed. 14) iii. 108 Cars having compressed air starting-up devices are always equipped for rapid tyre inflation from the air pressure cylinder. **1927** [see *activation]. **1946** *Happy Landings* July 1/1 The engine was not turned by hand through one cycle before the starting-up operation. **1959** *Times* 12 June 17/5 Extraordinary starting-up costs at the Company's new borate refinery. **1967** E. Short *Embroidery & Fabric Collage* i. 28 There is a definite starting-off point, namely the subject, which then has to be translated into a suitable flat pattern.

2. a. *starting cord.*

1977 D. MacKenzie *Raven & Ratcatcher* viii. 114 He whipped the starting cord on the small outboard motor. It caught at the second attempt.

b. starting block, (usu. *pl.*) a shaped rigid block for bracing the feet of a runner at the start of a race; also *fig.*; **starting gate** *Sport*, (*a*) a barrier device used at the beginning of a race (esp. of horses) to ensure a simultaneous start for all competitors; (*b*) a point from which individual runs are timed, as in skiing etc.; also *transf.* and *fig.*; **starting grid** = *grid 6 c; **starting handle,** a handle used to start a machine; *spec.* a detachable one that is turned to start the engine of a motor vehicle; **starting line,** a real or imaginary line used to mark the place from which a race starts; also *fig.*; **starting pistol,** a pistol used to give the signal at the start of a race; **starting salary,** the salary (on a pay-scale) earned at first by an employee taking up a new post.

1937 Bresnahan & Tuttle *Track & Field Athletics* iii. 54 There are two opinions on the matter of support for the feet at the start of the sprints. One is that holes in the track be used as a means of foot support, while the other is that starting blocks on top of the track be used. *Ibid.* xvi. 455 Those who prefer to employ starting blocks rather than holes in the track claim that blocks have the advantage in that they protect the track and make adjustments..easier. **1961** F. C. Avis *Sportsman's Gloss.* 63 *Starting blocks*, triangular pieces of wood fixed to the track and against which the sprinter's heels are pushed at the start of a race. **1977** J. Wainwright *Pool of Tears* 205 He crouched, like a sprinter on starting-blocks. **1983** *Listener* 29 Sept. 4/2 All the Opposition parties are poised nervously on the starting-blocks of the new political year. **1898** Starting-gate [in Dict., sense 2 a]. **1930** *Times* 24 Mar. 4/2 Numbered saddle cloths, the starting-gate, and the totalisator are among the reforms which came from them. **1940** C. Day Lewis tr. *Virgil's Georgics* i. 31 When racing chariots have rushed from the starting-gate. **1950** *Sun* (Baltimore) 18 Feb. 12/1 The competitors fear for life and limb when they are encased in the gate, which resembles the starting gate used for horses, and is designed to eliminate false starts. **1968** *Globe & Mail* (Toronto) 5 Feb. 19/3 The 18-year-old student was assessed a one-second penalty by the jury panel, which reviewed a claim by two officials that Debbie was too quick out of the starting gate. **1971** *Language* XLVII. 5 Acceptance of this very general position, however, does little more than put one in the proper starting gate. **1976** *New Yorker* 8 Mar. 105/1 Mount Sterling stumbled coming out of the starting gate, unseating his rider, Maple. **1957** Starting grid [see *grid 6 c]. **1978** 'D. Rutherford' *Collision Course* 100 Positions on the starting grid are allocated on practice times. **1876** Starting handle [in Dict., sense 2 a]. **1886** D. Clerk *Gas Engine* ix. 242 The starting handle is then let go, and the motor piston runs over its ports. **1932** D. L. Sayers *Have his Carcase* xii. 145 After..exercise on the starting-handle, they had diagnosed trouble with the ignition. **1973** P. Audemars *Delicate Dust of Death* x. 136 They left the bus... Perhaps they could not start it... I tried the starting handle myself. It was jammed. **1906** Starting-line [see *repair station* s.v. *repair *sb.*[1]]. **1920** [see *baton *sb.* 2 b]. **1962** A. Nisbett *Technique Sound Studio* x. 173 Anything but sheer surrealism was doomed to remain stickily on the starting-line. **1974** *Howard Jrnl.* XIV. 84 In a meritocratic society those whose powers of oral and written expression are extremely limited are left at the starting line in the race for social status. **1982** J. Jarver *Athletics Fundamentals* 20/1 The aim of the sprint start is to get the athlete away from the starting line as fast as possible. **1935** 'N. Blake' *Question of Proof* ii. 30 Get set! Go!! That was the starting-pistol you heard. **1962** A. Nisbett *Technique Sound Studio* x. 178 A modified revolver is better than a starting pistol. **1969** H. Perkin *Key Profession* iv. 138 New assistant lecturers and junior lecturers..came in..at starting salaries much lower than those in schoolteaching. **1973** *Nature* 7 Dec. p. i, Starting salary would be in the vicinity of £3,000.

startle, *v.* **3. a.** Delete † *Obs.* and add later example.

1972 R. Adams *Watership Down* v. 18 To rabbits, everything unknown is dangerous. The first reaction is to startle, the second to bolt. Again and again they startled, until they were close to exhaustion.

b. (Later example.)

1961 R. Graves *More Poems* 42 In the course of travel, you must have startled at Some coign of true felicity.

Hence **sta·rtlement,** the state or condition of being startled, alarm; hence, something that gives rise to this.

1927 *Chambers's Jrnl.* Feb. 92/2 No startlement was in her face by now. **1960** 'S. Harvester' *Chinese Hammer* i. 10 A strange expression of startlement came into her cool dark eyes. **1975** *Islander* (Victoria, B.C.) 4 May 3/3 Even so he [*sc.* a mouse] was dreadfully nervous and would leap like a miniature kangaroo at the least startlement.

start-up, *sb.*[2] Also **startup.** [f. vbl. phr. *to start up*: see *start *v.* 23 c, e.] The action or process of starting up a series of operations, a piece of machinery, a business, etc. Also *attrib.*

1945 H. D. Smyth *Gen. Acct. Devel. Atomic Energy Mil. Purposes* ix. 94 In estimating time schedules this 'start-up' or 'equilibrium' time must be added to the time of construction of the plant. **1954** *Trans. IRE on Nuclear Sci.* I. 3/1 During startup the behavior of the reactor is similar to that of an amplifier with positive feedback. **1959** *Wall St. Jrnl.* (Eastern ed.) 18 Dec. 9/1 Start-up expenses connected with new store openings have been heavy. **1968** *Globe & Mail* (Toronto) 13 Jan. B 5/4 Production target is late 1969 with startup to have cost more than $90-million. **1975** P. R. F. Mathijsen *Guide to European Community Law* ii. iv. 87 It is proposed that modern agricultural enterprises be set up with special aids such as start-up grants, investment aid and guaranteed credits. **1976** B. Jackson *Flameout* (1977) iv. 63 The flight data recorder..tape-recorded engine start-up, takeoff, climb-out, [etc.]. **1980** *Daily Tel.* 20 Mar. 28 (Advt.), They will assume responsibility for the start-up, management, organisation, and profitability of large work sites. **1983** *Sunday Times* 23 Jan. 56/1 The fund is designed to take advantage of the tax relief introduced in 1981 for investors putting up to £20,000 a year into start-ups. **1983** *Times* 19 Feb. 15 (*heading*) Business start-up funds worth the risk for top taxpayers.

starty, *a.* (Earlier example.)

1861 Mrs. Stowe *Pearl of Orr's Island* I. ix. 70 The little feller was very starty and fretful in his sleep last night.

starvation. Add: **2. b.** *starvation diet, line, point, rations.*

1848 Mill *Pol. Econ.* I. ii. xii. 433 Wages may fall below starvation point. **1869** J. Greenwood *Seven Curses of London* iii. 46 The child is welcome to live on starvation diet just as long as it may. **1915** Mrs Belloc Lowndes *Diary* 3 Mar. (1971) 56 British prisoners..have one cup of coffee with no milk or sugar..a 2 lb. loaf of bread for two days..Practically starvation rations. **1925** Joyce *Let.* 25 Mar. (1966) III. 117, I have now been put on a starvation diet by way of adding to my present pleasures. **1937** 'G. Orwell' *Road to Wigan Pier* v. 80 A man and wife on twenty-three shillings a week are not far from the starvation line. **1957** P. Worsley *Trumpets shall Sound* 15 This does not mean that Melanesians always live near the starvation-line. **1977** D. Williams *Treasure by Degrees* i. 16 The production of pure food.. unadulterated by mechanical or chemical intervention ..would have put the nation on a starvation diet. **1977** D. Francis *Risk* v. 56 If he caught me..he'd..leave me in the dark on starvation rations.

3. Also *colloq. ellipt.* for *starvation cold.*

1967 E. Grierson *Crime of One's Own* xi. 89 Queer day to leave it open. It's starvation. **1977** *Lancashire Life* Mar. 68/3 They mostly had open top decks and open staircases, and it was starvation in the winter.

starve, *v.* Add: **II. 7. e.** Phr. *starve the crows* and varr. = *stone the crows* s.v. *crow *sb.*[1] 3 d. *Austral. slang.*

1918 H. Matthews *Saints & Soldiers* 116 'Starve the crows,' howled Bluey in that agonised screech of his. **1936** A. Russell *Gone Nomad* vi. 46 Starve the crows! I laugh ev'ry time I think of it. **1966** G. W. Turner *Eng. Lang. Austral. & N.Z.* vi. 118 The well-known *stone the crows*.. occurs in such forms as *starve the wombats*..*starve the ninnies* and several more. **1966** 'J. Hackston' *Father clears Out* 156 Trooper Newbigun turned his horse's head and rode off with such dignity that Albert Horne said, just like an Australian, 'Gawd starve the crows!' **1968** *Courier-Mail* (Brisbane) 18 Nov. 10/7 Though the Dad and Dave expression 'Starve the lizards' is well enough known it has taken the spring of 1968 to bring about a similar but far from fanciful phrase for the flying foxes.

starved, *ppl. a.* Add: **d.** *Pottery.* Of a glaze: lacking the expected brilliance after firing.

1964 H. Hodges *Artifacts* ii. 52 Under-firing may result in starved glazes which have a dull appearance. **1968** H. Powell *Pottery Handbk. Clay, Glaze & Colour* ii. 56 A starved glaze is lacking in shine. *a* **1977** *Harrison Mayer Ltd. Catal.* 18/2 Starved glaze. The glaze surface is dull in areas which have been adjacent to porous refractories during firing. As the term implies glaze volatiles are sucked away from the surface of the glaze by the porous refractory.

starver. Add: **2.** *Austral. slang.* A saveloy.

1941 Baker *Dict. Austral. Slang* 71 *Starver*, a saveloy. **1959** D. Niland *Big Smoke* 211, I know what the things I eat cost me. Starvers, crumpets, stale cakes, speckled fruit, pies.

stash, *v.* Add: Also **stach. 1.** (Earlier example.) Also *absol.*

1794 *Sessions Papers* 17 Sept. 1200/2 He says, Miller, it is, *stash*, I am satisfied.

2. To conceal, to hide; to put aside for safe keeping; to stow or store. Freq. with *away.* Formerly *Criminals' slang*; orig. *U.S.* in revived mod. use.

1797 *Humphry Potter's Dict. Cant & Flash Lang.* (ed. 3) 55 *Stach*, to conceal a robbery. **1821** *Sessions Papers* 14 Dec. 66/1 He begged of me to *stash* it, which means say nothing about it. **1914** Jackson & Hellyer *Vocab. Criminal Slang* 80 *Stash*,..to hide;..to cease talking; to 'plant'. **1927** *Dialect Notes* V. 477 Billy he done stashed the jug in th' brush, an' now the damned ol' fool caint find hit! **1937** C. R. Cooper *Here's to Crime* v. 102 A friend of mine had it stached in his cellar, in a fruit jar. **1937** D. Runyon in *Collier's* 16 Jan. 9/4 She must have some scratch of her own stashed away somewhere. **1944** *Daily Progress* (Charlottesville, Va.) 25 May 6/6 A customer at least has a sporting chance to pick up a bottle of brandy, gin, or rum if the dealer doesn't have a bottle of old Kentucky corn julep stashed away under the counter for him. **1952** *Manch. Guardian Weekly* 20 Mar. 4/3 The big gift already stashed away in the farmers' bank accounts. **1962** J. Heller *Catch-22* vi. 51 Just when I was all set to really start stashing it away they had to manufacture fascism and start a war. **1970** R. Price *Howling Arctic* i. 15 Travel proved too difficult after a while so they stashed the sledges and walked on. **1974** F. Forsyth *Dogs of War* (1975) i. i. 39 With all fees paid, he netted a cool £500,000, which was still stashed in the Zwingli Bank. **1978** J. A. Michener *Chesapeake* 670 The watermen ferried dead birds to the ice shelf, stashed them and returned to fetch others.

stash (stæʃ), *sb. slang* (orig. and chiefly *U.S.*). Also **stach.** [f. the vb.] **1. a.** Something, or a collection of things, stashed away; a hoard, stock; a cache.

1914 Jackson & Hellyer *Vocab. Criminal Slang* 80 *Stash*,..used as a noun in the sense of something cached. **1942** Berrey & Van den Bark *Amer. Thes. Slang* § 207/5 *Cache*..stach, stash, stash-away. **1954** Webster, *Stash*,.. something stashed away or the place where it is stashed. **1969** *New Yorker* 31 May 90/1, I liked..the stash of Pucci shifts. **1970** G. Jackson *Let.* 26 Mar. in *Soledad Brother* (1971) 199, I *want* my food and drink from the people's stash. **1975** B. Garfield *Hopscotch* xxv. 257 If he told Oakly the truth about going to ground then he'll want to clean out his stash..he's..got to have money. **1979** *Daily Tel.* 10 Apr. 3/2 Chief Insp. Newark said he was satisfied Barnes had no stashes of money hidden away. **1980** *Encounter* May 37 Even crane-crews angle For a share of the stash, Their lines urging up A grey, enormous catch.

b. A cache of an (illegal) drug; a quantity (of a drug); the drug itself. (See also quot. 1942.)

1942 Berrey & Van den Bark *Amer. Thes. Slang* § 509/12 *Stash*, concealed equipment for taking narcotics. **1953** W. Burroughs *Junkie* iii. 36 Taking junk hidden by another junkie is known as 'making him for his stash'. It is difficult to guard against this form of theft because junkies know where to look for a stash. **1959** [see *bust *sb.*[3] f]. **1967** *Trans-Action* Apr. 11/1 Someone cruises by in a car and brings a nice 'stash' of 'weed'. **1968** T. Wolfe *Electric Kool-Aid Acid Test* xi. 133 The Hermit ..was..keeper of the communal acid stash down there in the cave. **1975** *High Times* Dec. 11/1 Anyone who turns stash knows that most people will pay any price to get high. **1978** *N.Y. Times* 30 Mar. B1/2 A number of dubious substances, such as 'African Yohimbe Smokestuff'. This, the label said, should be added 'to your regular stash to turn your domestic green into African Red'. **1982** *Guardian* 14 Dec. 11/6 The hairy young man in Lee Cooper jeans..asking 'Anyone seen my stash?'

2. *slang* (orig. *Criminals'*). A hiding-place, a hide-out; a rendezvous; a dwelling, 'pad'.

1927 *Amer. Speech* II. 390/2 A *stash* is a hiding-place. **1930** R. Chadwick in *Liberty* 23 Aug. 33/2 If we were on a bank job in a strange city the stash would be in a room we had rented several weeks in advance. In a small town, though, you don't have any stash, because an hour after you moved in everybody in the burg would be checking in. **1946** Mezzrow & Wolfe *Really Blues* viii. 132 No Hotel Ritz for us this time; our stash was over some kind of feed store. **1963** L. Deighton *Horse under Water* xviii. 77 We set up 'Art for the Average Guy, Inc.', just a little stash on East 12th. **1965** *Listener* 7 Jan. 31/2 Susan Sontag went to see Philip Johnson, the New York architect, or rather she 'moseyed along to his stash on Park'.

stasiology (stæsiǫ·lŏdʒi). *rare.* [tr. F. *stasiologie* (see quot. 1954), f. Gr. στάσι-ς party, faction (see Stasis in Dict. and Suppl.): see -ology.] The science or study of political parties.

1954 B. & R. North tr. *M. Duverger's Pol. Parties* 422 The development of the science of political parties (it could perhaps be called *stasiology*). **1966** K. West *Power in Liberal Party* p. vii, There is now a field of academic enterprise known as analytical stasiology whose practitioners believe that by constructing a series of conceptual frameworks one may arrive at a 'science' of political parties.

stasipatric (stæsipæ·trik), *a.* *Biol.* [f. Gr. στάσις Stasis + πάτρα fatherland (f. πατήρ father) + -ic.] Applied to a form of speciation in which new taxa are considered to arise within the geographical range of the parent species, each part of which comes to be occupied by one of the new taxa. Hence **stasipa·trically** *adv.*

1967 M. J. D. White et al. in *Austral. Jrnl. Zool.* XV. 298 If a term, equivalent to allopatric and sympatric, is needed to describe the process of direct conversion of an essentially continuous population into a number of contiguous taxa (races, semispecies, or species) by the spread of chromosomal rearrangements around which isolating mechanisms develop, one might perhaps choose the adjective *stasipatric*, which is intended to indicate the essentially unchanging geographic range of the superspecies. **1968** *Science* 8 Mar. 1069/1 In the case of a chromosomal rearrangement which first establishes itself near the edge of a species distribution, one can imagine it spreading both inwards through the range of the species (stasipatrically) and outwards into previously unoccupied territory (allopatrically). **1973** L. S. Dillon *Evolution* x. 138/2 The stasipatric model is not designed to displace the established ones but to supplement them.

stasis. Restrict *Path.* to sense in Dict. and add: Also pronounced (stæ·sis). **1. b.** *gen.* Inactivity; stagnation; a state of motionless or unchanging equilibrium.

1920 *Glasgow Herald* 30 Nov. 9 The prevailing mood of Labour is indefinite; a condition of stasis has been caused by the coal strike and the dread of unemployment. **1930** W. Empson *Seven Types of Ambiguity* vii. 245 He is drawn taut between the two similar impulses into the stasis of appreciation. **1933** T. S. Eliot *Use of Poetry & Use of Criticism* vi. 103 Arnold represents a period of stasis; of relative and precarious stability, it is true, a brief halt in the endless march of humanity in some, or in any direction. **1940** E. Muir *Story & Fable* v. 186 This could be done by so controlling the chemical processes of the body as to produce a self-subsistent balance, an everlasting, living stasis. **1943** *Sewanee Review* LI. ii. 337 Art, according to Dedalus-Joyce, tends toward the achievement of *stasis*, which implies a state of contemplation, of detachment from the *kinesis* of life. **1972** *Times Lit. Suppl.* 1 Sept. 1020/3 We see him in the moment of stasis before action. **1978** J. Updike *Coup* (1979) iii. 91 A religion whose antipodes are motion and stasis.

c. In the psychoanalytical theory of Wilhelm Reich (see *Reichian *sb.* and *a.*), a hypothetical accumulation of unused or repressed sexual energy.

1942 T. P. Wolfe tr. *Reich's Function of Orgasm* iii. 58 The role of sexual stasis in increasing antisocial and perverse sexual impulses. *Ibid.* 361 *Stasis*, the damming-up of sexual energy in the organism, thus the source of energy for the neuroses. **1953** in *Wilhelm Reich: Sel. Writings* (1961) 12 *Stasis neurosis*, all somatic disturbances which are the immediate result of the stasis of sexual energy, with stasis anxiety at its core. **1973** D. Boadella *W. Reich* vii. 194 There are two fundamental biological responses to sexual stasis, or any other blockage to emotional functioning.

2. [Gr. στάσις in sense 'faction, discord'.] Party faction, civil strife.

1933 R. J. Bonner *Aspects of Athenian Democracy* v. 91 Solon tried to strengthen the government against sedition, or *stasis*, as the Greeks called it, by requiring every citizen to take one side or another in case of serious party strife. **1956** A. W. Gomme *Commentary on Thucydides* II. 374 From *stasis* in Kerkyra to *stasis* in the Greek world generally..to universal conditions of *stasis* and war as its stimulus. **1963** M. I. Finley *Anc. Greeks* 51 The dividing-line between politics and sedition (*stasis* the Greeks called it) was a thin one in classical Greece, and often enough *stasis* grew into ruthless war. **1975** N. G. L. Hammond *Classical Age of Greece* 166 The weakening of traditional obligations and the revolution in the economy which arose from the war were among the factors which led to the outbreak of *stasis*, civil war, in 411 and 410 at Athens.

stat[1] (stæt). Colloq. abbrev. of *Photostat b.

1960 'E. McBain' *Give Boys Great Big Hand* v. 40 Here are the stats, kid. **1971** C. Fick *Danziger Transcript* (1973) 81 Make stats of all the graduation and group pictures. **1977** J. Aiken *Last Movement* i. 33 Wonderful reviews... I'll send you stats of all the notices.

stat[2] (stæt). Colloq. abbrev. of Statistic *sb.*, Statistics. Cf. *math[3]. orig. and chiefly *U.S.*

1961 Webster, *Stat*,..statistics. **1973** *Amer. Speech* 1970 XLV. 86 *Stat*, Elementary Statistics. **1976** *Springfield* (Mass.) *Daily News* 22 Apr. 39/1 A key stat: Marcel Dionne had only two shots Tuesday night. He scored on one, had another in the second period. **1977** *Rolling Stone* 5 May 46/3 Certainly, his stats in spring training were not impressive; his earned run average was over 4.00.

stat. (stæt), *adv.* *Pharm.* [Abbrev. of L. *statim.*] On a prescription: immediately.

1875 W. H. Griffiths *Lessons on Prescriptions* iv. 18 *Stat.*, immediately. **1971** *Lancet* 25 Sept. 700/2 *Stat.*, to be given at once.

stat- (stæt), *prefix.* [f. Stat(ic *a.* and *sb.*] Used in combination with the names of the practical electrical units to form the C.G.S. electrostatic system of units, as *statampere, statcoulomb, statfarad, statgauss, statohm, statvolt.*

[**1903** A. E. Kennelly in *Trans. Amer. Inst. Electr. Engin.* XXII. 534 In a comprehensive system of electromagnetic terminology, the electric C.G.S. units should also be christened... They might be denoted by the prefix *abstat*. Thus the C.G.S. electric unit of e.m.f. would be the abstatvolt.] **1920** *Proc. Amer. Philos. Soc.* LIX. 365 B_e is the electric flux density in 'statgausses'. **1925** W. H. Timbie *Elements of Electricity* xvi. 584 A statcoulomb is a small unit of charge, one coulomb being equal to 3×10^9 statcoulombs. **1932** S. S. Attwood *Electric & Magnetic Fields* i. 19 The unit of potential is an erg per statcoulomb and is called the statvolt in the electrostatic C.G.S. system. **1937** H. L. Curtis *Electr. Measurements* ii. 11 The prefixes *ab* and *abstat* were proposed by Kennelly ... The latter has generally been shortened to *stat*. No one of the proposed names has been adopted by any international organization. **1939** J. B. Whitehead *Electricity & Magnetism* vi. 62 The name of the electrostatic unit of capacitance is the statfarad. One farad is equal to 9×10^{11} statfarads. **1958** Condon & Odishaw *Handbk. Physics* A-9/2 The cgs unit of charge is called the esu or the statcoulomb. It is the unit defined by using dynes of force and cm of distance and adopting $4\pi\epsilon_0 = 1$. The corresponding unit of current is called the statampere, defined as the flow of one statcoulomb per second. *Ibid.* A-11/1, 1 statohm $= c^2/10^9$ ohms..with $c = 2{\cdot}997930 \times 10^{10}$ cm/sec. **1963** Jerrard & McNeill *Dict. Sci. Units* 12 In recent years the prefix stat has sometimes been used to denote electrostatic units, thus 1 stat volt = 300 practical volts, 1 stat ampere = $(1/3) \times 10^{-9}$ ampere. This prefix is an abbreviation for abstat which was proposed for electrostatic units at the same time as ab was suggested for electromagnetic units.

statal, *a.* Restrict *rare* to sense in Dict. For *a* **1864** Edw. Bates (W.) read **1862** E. Bates in *Official Opinions Attorneys Gen.* X. 388, and add later examples. Also of a state in other federations.

1880 A. Tourgée *Fool's Errand & Invisible Empire* II. xi. 489 Public education flourished as a part of the statal economy. **1949** *Times* 7 Feb. 5/3 All the states outside this special category have already been merged with provinces or have joined one or another of the six great statal groups.

2. *Linguistics.* Of a passive verbal form: expressing a state or condition rather than an action (opp. *actional *a.*).

1935 [see *actional *a.*]. **1961** R. B. Long *Sentence & its Parts* v. 119 Thus *have* has no passive when it is statal. **1968** *Language* XLIV. 236 The 'statal passives' are quite different in their internal representations from the true passives, and actually are not passives at all. **1975** *Ibid.* LI. 362 The ambiguity is between the 'statal' and 'dynamic passive' meaning.

statalon, var. *statolon.

statary, *a.* Restrict † *Obs.* to senses in Dict. and add: **3.** *Ent.* (stēⁱ·tări). Pertaining to or designating army ants during that phase of their life cycle when they return to a fixed colony each night.

1933 *Jrnl. Compar. & Physiol. Psychol.* XV. 297 When in the 'statary' condition, [ant] colonies do not appear as susceptible to the atmospheric changes that apparently furnish the stimuli for the bivouac-change movements of nomad colonies. **1940** *Jrnl. Compar. Psychol.* XXIX. 434 Swarm division is exceptional in a statary colony, but is a regular morning event in a nomadic colony. *Ibid.* 435 Smaller swarms..characterize the statary period. **1972** *Sci. Amer.* Nov. 73/3 The actual regulator of the ants' nomadic and statary behavior, as Schneirla eventually demonstrated, was not some external influence but the breeding cycle within the colony.

state, *sb.* Add: **I. 4. a.** (Further examples.) Also in generalized or abstract sense: each of the possible modes of existence of a system; the condition of a device that determines what output it produces for a given input.

1937 A. M. Turing in *Proc. London Math. Soc.* XLII. 250 We know the state of the system if we know the sequence of symbols on the tape, which of these are observed by the computer.., and the state of mind of the computer. **1942**, etc. [see *Markov]. **1954** *Jrnl. Franklin Inst.* CCLVII. 170 Once a stable state has been reached for all secondary relays, then further circuit changes can occur only if modification is made in the input state. **1961** F. M. Reza *Introd. Information Theory* ii. 54 Let an experiment have a finite number of *n* possible outcomes, $a_1, a_2, \ldots,$ and a_n, called states. We assume the process to be of the finite Markov type and initially in the state *k*. **1962** A. Gill *Introd. Theory of Finite-State Machines* i. 6 Roughly, the state of a finite-state machine at any given sampling time is that variable which, together with the input symbol, enables one to predict the output symbol. *Ibid.* iv. 130 Let *M* be an *n*-state machine with the input alphabet $X = (\xi_1, \xi_2, \ldots, \xi_p)$. **1964** F. L. Westwater *Electronic Computers* ii. 21 The reliable and stable electronic devices are so far two-state devices... If we had a three-state device we could..develop a ternary system of arithmetic. **1979** J. R. Gibson *Electronic Logic Circuits* i. 5 Logic elements may be combined to form multiple-state systems and the states of such systems may be used to represent numbers in systems other than the binary one.

b. *spec.* in *Physics*, a condition of an atom or other quantized system described by a particular set of quantum numbers; *esp.* one characterized by the quantum numbers *n*, *L*, *S*, *J*, and *m*. Cf. *level *sb.* 3 e.

1913 N. Bohr in *Phil. Mag.* XXVI. 5 If in these expressions we give *τ*, different values, we get a series of values for W, *ω*, and *a* corresponding to a series of configurations of the system. According to the above considerations, we are led to assume that these configurations will correspond to states of the system in which there is no radiation of energy; states which consequently will be stationary as long as the system is not disturbed from outside. **1925** *Astrophysical Jrnl.* LXI. 39 Every spectral line is now believed to be emitted (or absorbed) in connection with the transition of an atom (or molecule) between two definite (quantized) states, of different energy-content—the frequency of the radiation being exactly proportional to the change of energy. **1929** N. V. Sidgwick *Electronic Theory of Valency* ii. 18 These states are distinguished by the fact that in them the electron possesses an integral number..of quanta of energy. **1935** P. A. M. Dirac *Princ. Quantum Mech.* (ed. 2) i. 11 A state of a system may be defined as an undisturbed motion that is restricted by as many conditions or data as are theoretically possible without mutual interference or contradiction. **1935** Condon & Shortley *Theory Atomic Spectra* iv. 122 The terms are designated as ¹*S* (doublet *S*), ²*P*, ³*D*, ... according to the *P* value of the configuration from which they arise. The separate levels are designated by adding the value of *j* as a superscript, thus ²*S*½ ... To specify an individual state the value of *m* is given as a superscript. **1955** E. B. Wilson et al. *Molecular Vibrations* x. 246 The selection rules for overtone frequencies will next be considered. These are transitions between the ground vibrational state and an excited state in which one quantum number is greater than one and all other quantum numbers are zero. **1970** [see *level *sb.* 3 e]. **1972** DePuy & Chapman *Molec. Reactions & Photochem.* iii. 43 Formaldehyde, which is planar in its ground state, distorts slightly toward a pyramidal structure in the S_1 state. **1978** P. W. Atkins *Physical Chem.* xix. 631 The isotope ⁵⁷Co decays slowly..and forms an excited nuclear state of ⁵⁷Fe.

5. a. (Examples of *the* (or *a*) *state of affairs.*) *spec.* in the philosophy of L. Wittgenstein (1889–1951): see quots. 1922, 1962.

1776 *St. James's Chron.* 23–25 May 4/1 Administration had, to the Scandal of all good Government, suppressed every Thing relative to the true State of Affairs in America. **1909** G. B. Shaw *Press Cuttings* 31 We [women] should lose our influence completely under such a state of affairs. **1911** E. L. Thorndike *Animal Intelligence* vi. 245 By a satisfying state of affairs is meant one which the animal does nothing to avoid. **1922** tr. *Wittgenstein's Tractatus* 31 It would..appear as an accident, when to a thing that could exist alone on its own account, subsequently a state of affairs could be made to fit. *Ibid.* 35 Objects contain the possibility of all states of affairs. *Ibid.* 43 The picture contains the possibility of the state of affairs which it represents. **1958** R. L. Green *Land of Lord High Tiger* ii. 33 Sad affairs of State! Sad state of affairs! Affairs in a sad state! **1962** M. Cranston tr. *Hartnack's Wittgenstein & Mod. Philos.* ii. 13 A 'state of affairs' is a fact that in itself does not consist of facts... A state of affairs is a combination of possible facts. *Ibid.* 14 If an elementary sentence, or, better, an elementary proposition is true, then the state of affairs which is spoken of exists. **1973** A. Kenny *Wittgenstein* v. 73 States of affairs, we are told, are independent of one another.

e. *state of the art*: the current stage of development of a practical or technological subject; freq. (esp. in *attrib.* use) implying the use of the latest techniques in a product or activity.

[**1889**: see Status 4.] **1910** H. H. Suplee *Gas Turbine* 6 It has therefore been thought desirable to gather under one cover the most important papers... In the present state of the art this is all that can be done. **1955** *Jrnl. R. Aeronaut. Soc.* LIX. 471/1 Flight instruments and flight techniques of human pilots had to be brought up to a state where automatic flying could be fitted into a consistent state-of-the-art picture. **1957** R. A. Heinlein *Door into Summer* (1960) v. 69 Engineering is the art of the practical and depends more on the total state of the art than it does on the individual engineer. **1967** *Technology Week* 23 Jan. 18/1 (Advt.), Poseidon's requirements range from weapon effects on electronics to the design and use of state-of-the-art test checkout equipment. **1970** J. Earl *Tuners & Amplifiers* iv. 79 An average magnetic cartridge at the current state of the art produces an output of 1 mV per channel for each velocity unit of 1 cm/S. **1970** *Daily Tel.* (Colour Suppl.) 9 Oct. 7 Our highly sophisticated, technological defence establishment is advancing the state-of-the-art weaponry into new and therefore secret areas. **1975** *Language* LI. 1009 What emerges about the state of the art in linguistics is incidental to the presentation of L[abov]'s analysis of language use in various speech communities. **1976** *National Observer* (U.S.) 27 Mar. 15/2 State-of-the-art solar research is well advanced in New Mexico. **1976** *Offshore Platforms & Pipelining* 16/3 Except for an innovation or two, the platform is state-of-the-art and no more than an extension of existing capabilities. **1978** *Sci. Amer.* Apr. 64/3 In the 1950's C. Miller Fisher.. proposed that anastomosis, or joining, of cerebral arteries beyond the point of occlusion might be appropriate in some stroke cases, but such manipulations were then still beyond the state of the art. **1978** *SLR Camera* Aug. 54/1 It is still not so easy to produce decent pictures at such a venue, even when one is replete with 'state of the art' cameras, long lenses and fast film.

f. *State of the Union message*: a yearly address delivered by the President of the U.S. to Congress, giving the Administration's view of the state of the nation and its plans for legislation.

[**1787** *Constitution as formed for U.S.* II. iii. 10 He shall from time to time give to the Congress information of the state of the union.] **1945** *Newsweek* 15 Jan. 30 (*caption*) Three days before the President's State of the Union message, the 79th Congress takes its oath of office. **1959** *Ann. Reg.* 1958 182 The President appeared in person before both Houses of Congress on 9 January to deliver his State of the Union Message. **1968** W. Safire *New Lang. Politics* 427 State of the Union *messages* (preferred over speeches, addresses, or reports..) have inclined to be lengthy statements of legislative intent. **1974** *Guardian* 31 Jan. 2/7 President Nixon's sixth—and very possibly his last—State of the Union message.

g. *the state of play*: the position in which a matter or business stands at a particular time.

1966 *Rep. Comm. Inquiry Univ. Oxf.* II. 56 Receiving applications..and keeping colleges informed on the 'state of play' on each candidate. **1971** A. Garve *Late Bill Smith* i. 38 You know the state of play, you can handle everything. **1979** R. Mutch *Gemstone* v. 56 Write me a short report... I want the state of play. I want it accurate and I want it now.

14. b. *Bibliography.* (See quots. 1931.)

1931 P. H. Muir *Points* 12 A word is still needed to describe changes made before any publication takes place. These changes may be made while the entire edition is still in the publisher's hands, they may take place at the printer's, at the binder's, or even at a stage intermediate between the issue of some of the review copies and the actual date of publication. Any differences that may arise before that time will be referred to as 'states'. *Ibid.* 13 It is probable that copies of the book in all three states will be issued on the same day. They will all be 'first issues'; but some will be first, some second, and others again third 'states' of the first edition. **1931** G. Worthington *Bibliogr. Waverley Novels* p. viii, I have.. decided to use the word 'State' whenever between two copies of a first edition there are differences of sufficient importance to be noticed; and my 'First State' is the variety which I believe to be preferable..to any other. *Ibid.* 37 Guy Mannering... There are four States of the first edition. **1949**, etc. [see *issue *sb.* 14 b (ii)]. **1972** *Scholarly Publishing* III. 123 The text was published almost simultaneously in English, French, and German, all with the same title. To distinguish between the various states and issues is nearly hopeless.

IV. 31. c. (Earlier Amer. example.)

1634 *Mass. Bay Rec.* (1853) I. 117 When I shalbe called to give my voice touching any such matter of this state, wherein ffreemen are to deale, [etc.].

33. b. Short for *State Department* (sense 41 a below). *U.S.*

1955 W. Tucker *Wild Talent* xi. 151 Somebody down at State had to soothe his ruffled feelings. **1971** R. Ludlum *Scarlatti Inheritance* I. ii. 18 It took an executive order from the President to get it out of State. **1979** H. Kissinger *White House Years* ii. 29 The new President wanted to change the negotiating instructions on Vietnam drafted at State that reflected the approach of the previous Administration.

c. Preceded by the name of a State: short for *state university* (sense 38 e below).

1928 *Time* 29 Oct. 28/3 Penn mauled Penn State, 14–0. **1975** J. Wyllie *Butterfly Flood* (1977) xxiv. 177, I can outswim you any time at all. I used to swim for Penn State. **1980** 'R. B. Dominic' *Attending Physician* ii. 11 Two students in Ohio State sweatshirts.

VII. 38. a. (Further examples.)

1881 E. W. Hamilton *Diary* 5 June (1972) I. 144 It is certainly a great misfortune that the three foremost men on the front Opposition bench should be dependent on State aid. **1885** State school [see Real *a.*¹ 10]. **1917** A. S. Neill *Dominie Dismissed* ii. 27 Our rulers..send the rest of the sons of the community to State schools where they are trained to be disciplined and content with their lot. **1937** *Discovery* Oct. 305/2 State aid..might ultimately be the only hope of British agriculture. **1943** Koestler *Arrival & Departure* IV. 165 State-bureaucracies and managers establish themselves in vital hedgehog positions. **1944** Wyndham Lewis *Let.* 20 Aug. (1963) 378 But Stalin has a working state-system, with the air purged of humbug. **1948** J. Towster *Political Theory in U.S.S.R.* iii. 42 The Party..was not yet equipped to substitute collective-farm and state-farm production for kulak production. **1972** *Listener* 23 Nov. 690/2 With the inevitable rise in food prices..the Government..[should] introduce state subsidies to stabilise prices. *a* **1974** R. Crossman *Diaries* (1975) I. 351 If he stays in the state system—in the village school and then in Banbury. **1977** B. Freemantle *Charlie Muffin* iv. 43 Had she been born in a council house..and attended a state school..Janet would have been a slag.

e. (Further examples.)

For numerous other *attrib.* uses see *D.A.E.* and *D.A.*

1831 J. M. Peck *Guide for Emigrants* 256 One sixth part is to be..bestowed on a state college or university. *a* **1857** *Michigan Gen. Statutes* (1882) I. 171 The board of state auditors are hereby authorized..to procure plans, drawings and estimates for a state capitol. **1885** State park [see *land-jobbing* s.v. *land *sb.*¹ 12]. **1925** L. S. Dunaway *What Preacher Saw* 81 Jeff Davis declared that the ground on which the new state capitol was constructed was 'too poor for two Irishmen to raise a row on!' **1941** *N.Y. Times Mag.* 26 Jan. 19/3 Police work forms only part of the State trooper's duties. **1952** *Mind* LXI. 471 Suppose we then see him in earnest conversation with a scout from the state university. **1964** M. Banton *Policeman in Community* iv. 94 Many of the State police forces again are very different. The men wear big hats like the Canadian mounties and concentrate upon highway patrol. **1969** C. Burke *God is Beautiful, Man* (1970) 30 They took a trip to a place that looked like a state park. **1976** *National Observer* (U.S.) 1 May B5/4 The bulk of the newly admitted enjoyed a style of training, in junior and state colleges, little superior to that of the high school. **1976** A. Price *War Game* I. ix. 173 He's a New Englander... He left his state university nine years ago.

40. b. (Further examples.)

1833 J. S. Mill in *Jurist* IV. 14 Let the State endowments be once withdrawn from the church of England. **1842** —— in *Morning Chron.* 13 Jan. 3/6 If an established church is not to be independent of state control, no established church ought to exist. **1848** in A. Prentice *Tour in U.S.* viii. 76 A real, state-paid bishop, whilome a minister of the Scotch Relief Kirk. **1856** Geo. Eliot in *Westm. Rev.* X. 64 Its patent machinery of state-appointed functionaries. **1876** *Fortn. Rev.* 1 Apr. 630 Proof that, as the *Katheder-Socialisten* maintain in Germany, the part of state-intervention will go on steadily increasing. **1887** G. B. Shaw *Let.* 8 June (1965) I. 173 The sweeping away of our..wicked workhouse prisons in favour of State-owned farms and factories. **1891** H. Spencer *Justice* 270 Under the existing system of ownership, those who manage the land, experience a direct connexion between effort and benefit; while, were it under State-ownership, those who managed it would experience no such direct connexion. **1899** W. S. Churchill in *Morning Post* 6 Dec. 5/7 There will be those who will dream another dream of a brave system of State-aided—almost State-compelled—emigration. **1901** *Edin. Rev.* Apr. 453 In Germany we have seen a State-paid clergy help to create and keep on foot the great parliamentary party of the Centre. **1926** H. Sheehy-Skeffington *Let.* 15 Feb. in *Lett. of Sean O'Casey* (1975) I. 168 In no country save in Ireland could a State-subsidized theatre presume on popular patience. **1927** Carr-Saunders & Jones *Social Structure Eng. & Wales* 148 To complete the tale of State-provided

benefits that school children may receive. **1927** A. Huxley *Proper Studies* 220 In most modern countries the only state-supported orthodoxy is a sexual orthodoxy. **1942** *Contemp. Jewish Rec.* V. 274 The state-controlled *Kolkhoz* or *Sovkhoz* systems in Soviet Russia. **1946** J. W. Day *Harvest Adventure* xviii. 301 No National Park or State-planned 'lung' could offer lovelier scenery so near a manufacturing area. **1946** State-run [see *private *a.* 4 d]. **1952** C. P. Blacker *Eugenics: Galton & After* 185 France has a Minister for Population and a state-sponsored National Institute of Demographic Studies. **1958** A. Quinton in *Victorian Stud.* I. 254 [T. H.] Green's defence of state intervention is now generally recognised as an important intellectual preparation for the British Labour Party. **1973** *Listener* 17 May 635/1 That section of the Left which wants greater equality without a massive increase in state ownership. **1977** *N.Y. Rev. Bks.* 31 Mar. 22/3 On state control of industry, the pivotal socialist issue, he had no consistent views. **1978** S. Sheldon *Bloodline* iv. 69 He borrowed money on the real jewelry from the Crédit Municipal, the state-owned pawnshop. **1979** *Jrnl. R. Soc. Arts* CXXVII. 125/2 There was now state-aided scientific research.

41. a. state capitalism, a system of socialism whereby the State exerts exclusive control over a substantial proportion of the means of production, and over the deployment of capital created by this; hence **state-capitalist, -capitalistic** *adjs.*; **state-centred** *a.*, that centres on the State; **State Council,** the highest administrative and executive body of the People's Republic of China; **State Department** *U.S.*, the federal department for foreign affairs, presided over by the Secretary of State, = *Department of State* s.v. Department *sb.* 3 b; **State Enrolled Nurse,** a nurse enrolled on a State register and having a qualification lower than that of a State Registered Nurse; **state line** *U.S.*, the boundary line of a State; **state-oriented** *a.*, directed towards the State; **State Registered Nurse,** a nurse enrolled on a State register, and better qualified than a State Enrolled Nurse, **State rights:** also of States composing other federal nations; **State Scholarship,** a scholarship awarded by the State for study at a university; **state socialism,** socialism achieved by State ownership of public utilities and industry; hence **state socialist** *a.* and *sb.*; **state-socialistic** *a.*; **state vector** *Physics*, a vector in a space whose dimensions correspond to all the independent wave-functions of a system, the instantaneous value of the vector conveying all possible information about the state of the system at that instant; **state visit,** a visit by a head of state to a foreign country for ceremonial rather than official purposes; also *fig.*; **state-wide** *a.*, of, pertaining to, or extending over a whole state (usu. in sense 31 c; occas. sense 30); also as *adv.*

1903 D. Modell tr. *P. Kropotkin's Mod. Science & Anarchism* x. 94 Anarchism cannot see in the next coming revolution a mere exchange of monetary symbols for labor-checks, or an exchange of present Capitalism for State-capitalism. **1928** E. & C. Paul tr. *Stalin's Leninism* 436 The left-wing communists were of opinion that State capitalism is incompatible with the system of the proletarian dictatorship. **1978** Ld. Hailsham *Dilemma of Democracy* xvii. 110 Modern capitalism is the state capitalism of the Coal Board, British Rail or the Post Office. **1965** B. Pearce tr. *Preobrazhensky's New Economics* 153 In the Entente countries the economic system of the war period was state-capitalist to a considerably smaller extent. **1945** Koestler *Yogi & Commissar* III. iii. 201 Soviet Russia is a State-Capitalistic totalitarian autocracy. **1957** K. A. Wittfogel *Oriental Despotism* 33 The state-centered system of land grants as it prevailed in early China. **1977** *Dædalus* Summer 53 Is the state-centered concept of international politics..still relevant to the age of interdependence? **1969** Plano & Olton *Internat. Relations Dict.* 179 *State Council*, the highest executive decision-making body of the state apparatus, comparable to the Council of Ministers in the Soviet Union. **1978** tr. *Documents 1st Sess. 5th National Congr. People's Republic of China* 153 The State Council is the Central People's Government and the executive organ of the highest organ of state power. **1790** *Deb. Congress U.S.* 1 Apr. (1834) 1505 The resolution laid on the table yesterday, respecting the State Department. **1836** in *Ann. Rep. Amer. Hist. Assoc.* 1907 (1908) II. 117 This morning..I went to the State Department, to have a conference with the present Acting Secretary of State. **1930** *Times* (Weekly ed.) 9 Jan. 35/1 On the eve of its meeting the State Department published the New Year greetings exchanged between King George and President Hoover. **1970** E. Snow *Red China Today* (new ed.) 26 Under pressure he had himself made a State Department-accredited 'correspondent'. **1978** R. Ludlum *Holcroft Covenant* iii. 41 'It's the State Department. For you.' 'State? This is Lieutenant Miles, NYPA police.' **1961** *Nursing Mirror & Midwives' Jrnl.* 23 June 1143/2 The lettering on badges and certificates of existing State Enrolled Assistant Nurses will not be altered when the title is changed to State Enrolled Nurse as from June 28 this year. **1974** *Encycl. Brit. Macropædia* XIII. 399/1 Legislation in 1960 changed the title of a less qualified grade from state-enrolled assistant nurse to state-enrolled nurse. **1783** *Virginia Gaz.* 20 Dec. 2/3 George R. Clark, Surveyor State Line. **1868** *Harper's Mag.* June 123/2 He pronounced 'good-by' to the Prairie State, at the State line. **1973** R. L. Simon *Big Fix* (1974) xviii. 132, I was..driving into a desert dawn. I had already crossed the state line. **1961** *Encounter* Sept. 24/1 The political self-consciousness of the individual citizen is State-oriented. **1920** *Nursing Mirror & Midwives' Jrnl.* 17 July 270/1 A Bill for the State Registration of Nurses was drafted by the College in which it was provided that ..all the nurses on the College Register would automatically become State-registered nurses without further fee. **1965** D. Edwards-Rees *Story of Nursing* vi. 66 From 1925 onwards the only way for a new-comer to become a state-registered nurse was to take a three-year course and to pass the state examination at the end of it. **1977** P. Hill *Liars* ix. 117 On the mantelshelf were pictures.. one..of Rose in the full uniform of a State Registered Nurse. **1907** W. S. Churchill in R. S. Churchill *Winston S. Churchill* (1969) II. Compan. I. 627 A more practical explanation of any inconsistency in the attitude of the States is to be found in the growing cleavage upon the question of State rights *versus* Federal authority. **1930** W. K. Hancock *Australia* vi. 121 In truth, Australian public opinion (even that section of it which is disposed to favour 'State rights') has only a fitful, ineffective interest in maintaining the Federal balance. **1944** *Ann. Reg.* 1943 63 State Scholarships to universities should be of such a value as to enable the holder to take full part in the life of the university. **1966** *Rep. Comm. Inquiry Univ. Oxf.* II. 27 Source of financial assistance... A college scholarship or exhibition and a State Scholarship or LEA award. **1879** G. J. Holyoake in *19th Cent.* June 1114 State socialism is one of the diseases of despotism. *Ibid.*, State Socialism, so far as any taste for it exists in England, is a growth of Toryism. *Ibid.* 1115 The only persons in this country likely to be suspected of the State Socialiste craze are the working class co-operators. *Ibid.* 1116 True co-operators are no State Socialists. **1912** State socialist [see *radical *sb.* 5 a]. **1930** W. K. Hancock *Australia* vii. 137 Mr. Eggleston says that the Victorians are destroying their State socialism by political sabotage. **1968** H. W. Laidler *Hist. Socialism* xliii. 739 The social legislation of Bismarck has often been referred to as state socialism. *Ibid.* 742 Mr. Hillquit's distinction between government ownership and collective ownership..was.. the distinction made between the state socialist and the democratic socialist approach. **1929** Condon & Morse *Quantum Mechanics* vi. 205 The components of the state vector along the different principal axes multiplied by their complex conjugates give the probabilities of each value for the state in question. **1951** *Physical Rev.* LXXXII. 914/1 Quantum mechanics involves two distinct sets of hypotheses—the general mathematical scheme of linear operators and state vectors with its associated probability interpretation and the commutation relations and equations of motion for specific dynamical systems. **1970** D. T. Gillespie *Quantum Mechanics Primer* iv. 41 The state of the system is completely described by the state vector in the sense that anything which is in principle knowable about the system at time *t* can be learned from the function $\Psi_t(x)$. **1794** H. Wilson *Let.* 12 Apr. in *Fingall Papers* (Nat. Libr. Ireland MS. 8023(5)), I think your resolution as to a certain complimentory State Visit is perfectly correct, both as to yourself and your Lady. **1857** C. M. Yonge *Dynevor Terrace* I. xiv. 218 On Monday we go to Leffingham... After that, more state visits, unless I can escape to Oxford. **1914** W. Owen *Let.* 1 Jan. (1967) 225, I had to buy a waistcoat today, to complete the black suit: for the State Visit to Mr. Aumont which French Politeness imposes. **1966** *Listener* 23 June 900/2 President de Gaulle arrives in Moscow for twelve-day state visit. **1979** G. St. Aubyn *Edward VII* vii. 320 He [*sc.* the King]..decided that he wished to pay a State Visit abroad. **1911** C. E. Persons et al. *Labor Laws* 62 A state-wide organization was sufficient to bring the necessary pressure to bear. **1927** [see *rodeo 3 b]. **1948** J. Towster *Political Power in U.S.S.R.* iv. 86 Maximum unification of all economic activity by one state-wide plan. **1954** W. Faulkner *Fable* 189 A horde of Federal agents and sheriffs and special officers like the converging packs of a state- or nation-wide foxhunt. **1974** *Hartsville* (S. Carolina) *Messenger* 22 Apr. 2-A/5 The American Patriot Reading Club will offer youngsters statewide the opportunity to become better acquainted with the men, women, places and events, which have been prominent in the history of our state and the nation. **1980** *News & Observer* (Raleigh, N. Carolina) 28 Oct. 22/3 That immunization of cats against rabies be required statewide, and that the N.C. Wildlife Resources Commission be permitted to hunt and trap foxes for rabies control.

b. state's evidence: see Evidence *sb.* 7 c; **States-rights:** hence **States-righter,** an advocate of States-rights; also in other federal nations.

1948 *States' Rights* (Birmingham, Alabama) 26 July 1/1 States' Righters is all right, but the term, 'Dixiecrats', leaves the wrong impression. **1959** *Times Lit. Suppl.* 28 Aug. 491/3 It undoubtedly rallied every states-righter and virtually every Southerner, however progressive, behind Governor Faubus. **1972** *Accountant* 30 Mar. 405/2 Mr Macaw is regarded as a devout 'States-righter', who is firmly opposed to the creation of any Federal body to control the securities industry [in Australia].

stated, *ppl. a.* **2. b.** *stated meeting* (earlier example).

1866 *Harper's Mag.* Apr. 676/2 At one of their stated weekly meetings recently there was a large attendance.

statehood. For 'Chiefly with reference to the U.S.' read 'Orig. with reference to the U.S.' and add later examples.

1952 *Times* 9 Dec. 9 (*heading*) Solid achievements [by Israel] after four years of Statehood. **1971** I. Deutscher *Marxism in Our Time* (1972) v. 99 In a vast part of the continent a struggle was indeed going on for the achievement of independent statehood and nationhood. **1976** *Time* 27 Dec. 14/3 What the Palestinians want most of all is the sense of national identity that would arise from statehood.

State house. Add: **4.** *N.Z.* A house owned and let by the government.

1941 W. B. Sutch *Poverty & Progress in N.Z.* vii. 153 In the building of State houses of high quality a suggestion as to necessary standards has been made. **1964** B. Crump in *N.Z. Listener* 11 Dec. 5/3 It was a state house with shrubs and flowers in a street full of state houses with flowers and shrubs. **1978** B. Mason in *Islands* Aug. 15 After years of waiting, they got a State house last summer.

stateless, *a.* Add: **c.** Not being a citizen of any state; having no nationality.

1930 E. F. W. Gey van Pittius *Nationality within Brit. Commonw.* xiii. 133 A person of dual nationality can at least enter more than one country and seek protection from both while abroad. Not so the stateless person. **1957** C. Day Lewis *Pegasus* 36 He travels on, not only blind But a stateless person. **1968** J. Lock *Lady Policeman* xv. 132 The alien—a forty-seven year old Jugoslav-born stateless person—looked at her blankly. **1981** G. Markstein *Ultimate Issue* 153 In the absence of valid documents, we must grade a suspect stateless.

Hence **sta·telessness,** the condition of having no nationality.

1930 E. F. W. Gey van Pittius *Nationality within Brit. Commonw.* xiii. 132 In case of statelessness, which is the reverse of dual nationality or multiple nationality, an unfortunate individual is placed in the unenviable position of being without any country at all. **1959** *Times* 24 Mar. 9/5 The fundamental theory that statelessness should be overcome or reduced by the application of the principle of *jus soli* (the law of the soil) under which the nationality of a State is acquired by birth on its territory. **1963** H. Arendt *Eichmann in Jerusalem* x. 155 This was one of the few cases in which statelessness turned out to be an asset.

stately, *a.* Add: **4. a.** *stately home*: originally in allusion to quot. 1827; now a fixed phrase designating a great country-house. Also *attrib.* and *Comb.*

1827 F. Hemans in *Blackw. Mag.* Apr. 392 The stately Homes of England, How beautiful they stand! **1831** J. Brown *Let.* 26 Oct. (1912) 26 There is certainly something unapproachable in the 'stately homes of England'. **1848** A. Brontë *Tenant of Wildfell Hall* III. xv. 311, I.. looked back, for one last view of her stately home. **1874** Jewitt & Hall (*title*) The stately homes of England.. illustrated with 210 engravings on wood. **1920** D. H. Lawrence *Touch & Go* III. i. 68 This is what happens to the stately homes of England—they buzz with inky clerks, or their equivalent. **1934** Wodehouse *Right Ho, Jeeves* xiv. 167 Some stately-home owners of the name of Stretchley-Budd, hanging out in a joint called Kingham Manor. **1938** N. Coward *Operette* I. vii. 54 The Stately Homes of England How beautiful they stand, To prove the upper classes Have still the upper hand. **1945** B. Goolden *Ichabod* ix. 50 A group of persons restricted to the Services, the stately homes of England, and a line running roughly from Stanhope Gate to Eaton Square. **1959** Duke of Bedford *Silver-Plated Spoon* xi. 215 We had jumped straight into the front rank of the stately homes business. **1977** B. Pym *Quartet in Autumn* v. 45 There would be visits to a safari park and to the stately homes that offered the best attractions.

statement. Add: **2. a.** Also *transf.* and *fig.*

1953 W. M. Ivins *Prints & Visual Communication* i. 3 Since the invention of writing there has been no more important invention than that of the exactly repeatable pictorial statement. **1958** *Listener* 18 Sept. 437/3 Apart from some inherent nastiness in the story there is a fundamental weakness in its theatrical statement. **1970** *New Yorker* 26 Sept. 29/3 We had in mind thirty or forty people making a quiet statement by riding together. **1977** C. McFadden *Serial* (1978) xl. 86/2, I wish you'd stop shaving your goddamn legs. You might not know it, but you're making a *statement*.

b. *Computers.* An expression in a program language that specifies some operation or task, corresponding to one or more instructions according to the context and the level of the language.

1957 *Proc. Western Joint Computer Conf.* 188/2 He.. programmed the job in four hours, using 47 FORTRAN statements. **1957** D. D. McCracken *Digital Computer Programming* xviii. 215 The automatic coding systems directed toward scientific calculations accept problem statements in a form similar to ordinary mathematical language. **1959** M. H. Wrubel *Primer of Programming for Digital Computers* vi. 126 There are statements for performing arithmetic; statements for branching and looping; and input and output statements. **1967** W. F. Sharpe *Introd. Computer Programming using BASIC Lang.* i. 6 Every statement must begin with a legal command (after the line number, if any). **1973** J. K. Hughes *PL/I Programming* i. 8 The statement to input the data.. should be specified. This could be accomplished with the statement get list (A, B, C, D, E); Notice how all PL/I statements are ended with a semicolon. **1981** Monds & McLaughlin *Introd. Mini & Micro Computers* vii. 99/1 An assembler translates a machine oriented language, ie, a language where, in general, one statement gives rise to one machine instruction... A compiler translates a problem-oriented language..into machine code. Each statement in the language usually gives rise to more than one, and sometimes many, machine instructions.

3. b. (Earlier example.)

1885 W. Whitman *Daybks. & Notebks.* (1978) II. 375 Half-annual 'Statement' from D Mckay \$20.71 '¹⁸ for 6 mo's preceding Dec 1, '85.

c. statement of affairs *Accounting*, a list of assets and liabilities not expressed as a formal balance sheet.

1895 *Reports Tax Cases* (1907) III. 456 Every man and every company having foreign or colonial investments, of course, knows of the interest arising from them, takes note of it and enters it in any statement of affairs which may require to be made up. **1928** R. G. Williams *Elem. Bk.-Keeping* xiv. 240 The preparation of the statement of affairs may possibly depend upon information which is not supplied by the books of account, and often undue reliance must be placed upon the trader's memory. **1978** J. Kellock *Elem. Accounting* xi. 184 The term 'statement of affairs' is one commonly used in the subject of incomplete records and may be defined as a list of assets and liabilities. **1981** *Daily Tel.* 19 Dec. 17/6 In the statement of affairs presented at the creditors' meeting £31,000 of client account balances appeared as part of the net assets of the company. **1983** *Ibid.* 12 Mar. 19/1 Not only was there a complete absence of a statement of affairs—the receiver said that he had been unable to lay his hands on any of the company's books—but it was apparent from the scant information he was able to provide that the liabilities are far in excess of known assets.

5. Special Comb. (in sense 2): **statement-form** *Logic* = *propositional function* s.v. *propositional *a.* b.

1942 J. Findlay in *Mind* LI. 261 A statement-form is an expression containing variables such as x, v, etc., which gives rise to statements when expressions with a constant meaning are substituted for those variables. Thus 'x is long' is a statement-form. **1950** M. G. White in M. Farber *Philosophic Thought in France & U.S.* 711 How shall we define or make clear the meaning of the statement-form 'x is a history of y'? **1961** E. Nagel *Structure of Sci.* v. 95 Instead of being statement-forms the postulates of the theory appear to be statements.

statemental (stēⁱtme·ntăl), *a.* [f. Statement + -al.] That makes, consists of, or is characterized by a statement or statements.

1939 C. W. Morris in *Kenyon Rev.* I. 413 Scientific discourse is, in summary, statemental or predictive in character. **1955** J. L. Austin *How to do Things with Words* (1962) vi. 72 The primary or primitive use of sentences must be..statemental or constative. **1976** [see *sentential *a.* 2 b].

state-monger, states-monger. For *Obs.* exc. *arch.* read Now *rare* or *arch.*

1891 F. W. Bain *Antichrist* I. 53 The Revolutionary statesmongers were far too much in a hurry. **1965** C. Campos *View of France* x. 241 Perhaps..France is the source of light, as a certain tall statemonger would have it. Or perhaps Arnold was right after all and the English are a more earnest nation.

Stater² (stēⁱ·tər). *Irish Hist.* [f. State *sb.* + -er¹.] A member of the Irish Free State army.

1925 S. O'Casey *Juno & Paycock* III. 111 Ah, why didn't I remember that then he wasn't a Die-hard or a Stater, but only a poor dead son! **1936** 'N. Blake' *Thou Shell of Death* xiii. 239 During the Civil War..the Staters and the Republicans had a battle in the garden. **1965** L. Fleming *Head or Harp* x. 83 The streets themselves were full of 'Staters'.

state-room. Add: **3. a.** (Earlier examples.)

1774 J. Schaw *Jrnl. Lady of Quality* (1921) i. 22 Our Bed chamber which is dignified with the title of State Room, is about five foot wide and six long. **1832** F. Trollope *Domestic Manners of Americans* (ed. 2) I. xvii. 259 They occupied a state-room which Captain Hall had secured for his party.

b. (Earlier example.)

1853 *Southern Standard* (Charleston, S. Carolina) 31 Aug. 2/5 Messrs. Eaton & Gilbert..have built a beautiful car for the Hudson River Railroad which is divided into state rooms of eight feet square.

stateship (stēⁱ·tʃip). *Irish Hist.* [f. State *sb.* + -ship.] = Tuath.

1917 D. Figgis *Gaelic State Past & Future* 11 To make more easy the general administration of the country, he [*sc.* Cormac] regrouped the administrative units of the country. Until then the nation had consisted of a number of separate stateships. **1918** —— in *Studies* VII. 259 For the unit of the Irish polity—sometimes spoken of as the Gaelic State—was the Tuath. It was at once a political and an economic unit; a stateship of a state. **1918** A. de Blacam *Towards Republic* iii. 24 Along the western counties, where the Irish language still predominates, traces of the life of the stateships linger to this day.

Sta·teside, stateside, *a.* and *adv. colloq.* (orig. and chiefly *U.S.*). Also **states side** and hyphened. [f. *State-s* (State *sb.* 31 d) + Side *sb.*¹ 15 b.] **A.** *adj.* Of, in, or pertaining to the continental United States of America.

1944 Karig & Kelley *Battle Report* I. vii. 151 Hearing that several United States news correspondents were in Soerabaja it occurred to him that Stateside newspapers might carry stories of the air attack. **1952** *Chambers's Jrnl.* Aug. 486/1 After the biennial yacht-race to Honolulu is over..the problem of how to return the yachts to their stateside ports can no longer be postponed. **1960** *Encounter* Feb. 31/2 The kids keep up with the latest Stateside fad. **1967** A. Diment *Dolly Dolly Spy* vi. 80 She was tall and slim, as most of those Stateside career women are. **1975** tr. *Melchoir's Sleeper Agent* II. 148 Larry numbly examined his mangled, disfigured hands. 'I.. guess I bought myself a Stateside ticket.' **1979** *Sci. Amer.* Feb. 27/2 Agent Orange and its stateside analogues, much used defoliants for weeds and forest cover.

B. *adv.* Towards or in the continental United States of America.

1945 *Sun* (Baltimore) 12 Mar. 9-0/6 'Stateside' is a mighty popular word out here [*sc.* in Guam] because a service man going 'stateside' is going home. **1950** 'D. Divine' *King of Fassarai* xvii. 137 You're just fresh from States-side. You'll see what I mean when you've been on the island for a week. **1963** L. Deighton *Horse under Water* xliv. 176 Fernie fixed the consignment to a ship heading stateside. I notified my contacts in New York. **1966** E. McGirr *Funeral was in Spain* 74 I'm going back states side in a few days. **1969** *Oz* Apr. 27/2 This guy knew some of the Angels State-side, or at least he thinks he knows. **1976** M. Machlin *Pipeline* xxxix. 426 I'm willing to cash in my chips, drag up stateside, and get going on something else.

statesman¹. Add: **1.** *elder statesman*: see *elder *a.* 1 c.

stateswoman. Add: Hence **sta·teswomanship; sta·teswomanlike, -ly** *adjs.*

1850 Thackeray *Pendennis* II. vi. 62 She..discharges I don't know what more duties of British stateswomanship. **1959** C. Pankhurst *Unshackled* xii. 205 The signatories of this stateswomanly letter included Viscountess Acheson, the Hon. Mrs. Guy Baring. **1966** *Punch* 30 Mar. 442/1 Mrs. Indira Gandhi..will 'meet and talk with the Prime Minister', either at the airport or a nearby hotel. With stateswomanlike caution, of course, she'll be rehearsing her formal greeting with a blank after the Mister.

stathmokinesis (stæːþmokinī·sis). *Biol.* and *Med.* [ad. F. *stathmocinèse* (A.-P. Dustin 1938, in *Compt. Rend. de l'Assoc. des Anatomistes* XXXIII. 209), f. Gr. σταθμό-ς station, stage + κίνησις motion.] The type of cell division produced by substances such as colchicine, characterized by a halt or long delay at metaphase. Hence **sta:thmokine·tic** *a.*, (of a drug) that produces stathmokinesis; applied also to the method of measuring rates of cell division by means of such a drug.

1945 *Bot. Rev.* XI. 148 While some cells showed complete pycnosis, normal telophase stages and cytoses occurred. Some cells assumed normal telophase stages and formed giant or polyploid nuclei. Dustin proposed the name of 'stathmocinesis' for this type of indirect division and applied the name of 'stathmocinetic poison' to colchicine. He contended that the arrest of the nuclear division in the metaphase was preceded by a phase of excitation which distinguished this poison from those which merely inhibited division. **1971** *Brit. Jrnl. Cancer* XXV. 692 Since the stathmokinetic method we used cannot provide information about the proportion of cells actually in the proliferative cycle, the proportion or index of arrested metaphases..can only be used to find the potential doubling time..of the tumour cell population. **1977** W. A. Aherne et al. *Introd. Cell Population Kinetics* iii. 18 (*heading*) Metaphase arrest (stathmokinesis). *Ibid.*, The stathmokinetic agents in use today are colchicine, its derivative Colcemid, and the increasingly popular vinca alkaloids vinblastine and vincristine.

static, *a.* and *sb.* Add: **A.** *adj.* **3. b.** *static friction* (see quot. 1878).

1878 *Phil. Trans. R. Soc.* CLXVII. 509 Coulomb pointed out the necessity of distinguishing between the friction which resists the relative movement of surfaces already in motion, or what is now called kinetic friction, and the friction which tends to prevent surfaces at rest from being set in motion, or what is now called static friction. **1922** Glazebrook *Dict. Appl. Physics* I. 389/2 The subject of static friction is of considerable importance in the theory of the stability of engineering structures. **1980** J. W. Hill *Intermediate Physics* iv. 29 The block should, in absolutely frictionless conditions, begin to move however small the applied force but in practice static friction opposes the movement.

c. *Econ.* Of or pertaining to an economic system in a state of equilibrium. Cf. *statics 1 d.

1899 J. B. Clark *Distrib. of Wealth* iii. 31 What a static theory openly and intentionally puts out of sight—namely, changes that alter the mode of production. *Ibid.* v. 60 The study of the unreal static state is a heroic..use of the isolating method of study. **1904** J. N. Keynes *Scope & Method Pol. Econ.* (ed. 3) iv. 147 An economic theory is termed *static* if it is based on the assumption of ..a state in which there occurs no essential modification of the general conditions under which production and consumption, distribution and exchange, are carried on. **1947** P. A. Samuelson *Foundations Econ. Analysis* xi. 312 Gustav Cassel..considers Economic Dynamics to be a third stage of analysis, following a pure Static Economy and a 'quasi-static' Uniformly Progressive Economy. **1974** A. S. Campagna *Macroeconomics* iii. 50 Static analysis is concerned with states of equilibrium, and static models inquire into the forces leading to, maintaining, and reestablishing, if necessary, the equilibrium condition.

4. c. *Phonetics.* Of consonants, = Continuant *a.* 2; of tones, not changing pitch during utterance.

1931 [see *kinetic *a.* 5]. **1939** L. H. Gray *Foundations Lang.* 53 All consonants can be uttered by themselves whether they are *static* (or *continuant*), i.e., can be held continuously without changing quality (notably nasal, lateral, rolled, fricative, and sibilant sounds..); or are *kinetic*, i.e., cannot be so held. **1958** R. Kingdon *Groundwork Eng. Intonation* p. xxii, *Static tone*, a tone in which the voice remains steady on a given pitch throughout its duration. *Ibid.* 4 The Static Tones are the level tones, accompanied by stress, which are used on the words to which it is desired to give prominence in the sentence, but to which no particular feeling is attached. **1961** [see *kinetic *a.* 5]. **1968** B. M. H. Strang *Mod. Eng. Structure* (ed. 2) vi. 92 Kingdon divides the notes and tunes of English speech into two kinds, the level or Static Tones, and the moving or Kinetic Tones. **1969** D. Crystal *Prosodic Systems & Intonation in English* iv. 158 *Stress.* Perceivable increase in loudness accompanied by unmarked pitch movement (the norm in the pitch-range system, static tone).

6. c. *Computers.* Applied to a store in which the data are held at fixed positions in the device, and any location can be accessed at any moment.

1947 A. W. Burks et al. in *Coll. Wks. J. von Neumann* V. 44 We distinguish two broad types of such devices: static, and dynamic or pulse-type accumulators. **1970** O. Dopping *Computers & Data Processing* x. 134 A ferrite core memory..is an example of a static memory, while a drum memory..is a dynamic memory in which a particular memory can be read or written only when the continuously rotating drum is at a certain angular position. **1977** J. C. Boyce *Digital Computer Fund.* viii. 212 The basic building block of the static memory discussed in this section is a 16-pin integrated circuit.

7. Special collocations (mostly in sense 3): *static characteristic* (*curve*) (Electronics), a graph showing the relationship between two parameters of a valve, transistor, etc., measured under steady conditions (strictly at zero frequency and with no load impedance); *static line*, a line connecting a parachute to the body of an aircraft and serving to open and release it automatically when tensioned by the movement of the parachutist away from the aircraft; *static pressure*, the pressure of a fluid on a body when the latter is at rest relative to it; *static-pressure tube* = *static tube* below; *static test*, a test of a device or object in a stationary position, or under conditions that are constant or change only gradually; so *static-test* vb. trans.; *static testing* vbl. sb.; *static thrust*, thrust generated by a stationary aero-engine or rocket engine; *static tube* (Aeronaut.), the part of the pitot-static head that registers static pressure, consisting of a tube aligned parallel to the airflow, closed at the forward end, and having holes along its length; *static water*: during the war of 1939–45, a store of water with no pressure of its own intended for use as an emergency supply, esp. in fighting fires.; freq. *attrib.*, as *static-water tank*.

1919 *Wireless World* May 77/2 From the static characteristic of a valve..it is clear that if the grid voltage varies between sufficiently small limits, the law of variation will be exactly reproduced in the anode circuit. **1939** H. J. Reich *Theory & Applications Electron Tubes* iii. 43 Strictly a static characteristic is one obtained with steady voltages, whereas a dynamic characteristic is one obtained with alternating voltages. **1975** D. G. Fink *Electronics Engineers' Handbk.* vii. 39 The performance of a transistor over wide ranges of current and voltage is determined from static characteristic curves. **1930** C. Dixon *Parachutes for Airmen* ii. 20 The 'Guardian Angel' was a typical example of the 'automatic' design. It was fitted into a tube fixed at the side of the aeroplane, with a static line attached to the harness on the airman. **1966** M. R. D. Foot *SOE in France* iv. 78 All a parachutist has to do is to jump through the hole; his parachute is opened automatically by a thin wire called a 'static line' which his own weight breaks. **1977** *R.A.F. News* 30 Mar.–12 Apr. 3/3 If new members decide they want to move on to free-fall parachuting, they must first make six static line jumps. **1915** W. E. Dommett *Aeroplanes & Airships* 103 *Pressure head*, a combination of pitot tube and static pressure or suction tube. **1923** Glazebrook *Dict. Appl. Physics* V. 66/2 The air thus blows tangentially past these holes, and the pressure within the tube is equal to the static pressure of the surrounding air. **1948** C. E. Chapel *Aircraft Basic Sci.* i. 21/1 The center of pressure can be located for each angle of attack by installing parallel rows of static pressure tubes at right angles to the leading edge of a wing flush with the upper and lower surfaces. **1978** J. D. Anderson *Introd. Flight* iv. 95 Static pressure is a consequence of just the purely random motion of the molecules. **1905** *Jrnl. Iron & Steel Inst.* LXVIII. 486 Overheating..will produce brittleness under shock or vibratory stress, although static tests may have given fairly satisfactory results. **1918** Cowley & Levy *Aeronautics* ix. 176 Fig. 97 gives thrust and torque required to drive the propeller for a 'static test', i.e. when the forward speed is zero. **1961** *Time* (Atlantic ed.) 24 Feb. 12/2 The Saturn has been static-tested, but will not be operational until 1965. **1962** *B.S.I. News* July 9/1 The 'static' test used..in approving well over a million Kite-marked safety belts is extremely effective in finding out any weakness. **1966** *McGraw-Hill Encycl. Sci. & Technol.* XI. 610a/2 The static test facilities must be capable of handling and disposing of the hot gases..that are expelled from the rocket. **1968** A. E. Roy tr. *T. de Galiana's Conc. Encycl. Astronautics* 259/1 The complete rocket, held in a test-stand, is then static tested. **1950** G. G. Smith *Gas Turbines & Jet Propulsion* (ed. 5) xii. 205 (*heading*) Static testing. **1958** F. A. Warren *Rocket Propellants* x. 173 Because of the dangers involved in static testing, special precautions must be taken to protect personnel. **1962** F. I. Ordway et al. *Basic Astronautics* vii. 319 Static testing may be divided into two categories: powerplant static testing and vehicle capture firing. **1916** M. A. S. Riach *Air-Screws* vii. 89 The 'Static' thrust of an air-screw on an aeroplane is usually measured by attaching a spring balance to the rear portion of the machine and attaching a rope from the spring balance to some fixed support. **1952** A. Y. Bramble *Air-Plane Flight* x. 153 The problem of providing the essential static thrust for take-off found its solution in the variable-pitch propeller. **1962** F. I. Ordway et al. *Basic Astronautics* vii. 319 Some of the common measurements made..are given below.

(1) Thrust (static thrust developed by the engine) [etc.]. **1923** Glazebrook *Dict. Appl. Physics* V. 66/1 A 'head', consisting of two parts—the 'pressure' or 'Pitot' tube and the 'static' tube—is fixed on some exposed part of the aircraft. **1940** A. C. Kermode *Mechanics of Flight* (ed. 4) ii. 29 In modern types the Pitot tube is connected to the inside of a capsule..while the Static tube is connected to the casing of the instrument. **1965** V. H. Brix tr. *Martynov's Pract. Aerodynamics* vi. 141 The most widely used instrument for measuring static pressure is the static tube or sonde. **1944** *Ourselves in Wartime* 108/2 Thousands of static water tanks were established in parks, squares and in the basements of bombed buildings, so that never again would there be a shortage of water through the destruction of the mains by the enemy. **1958** L. W. Tancock in *Aspects of Translation* 31 You may remember those tanks of water stored in our towns against fires caused by enemy action. We had none in North Wales, but..near where we lived a torrent came tumbling down the valley and a large notice stood by indicating that this was *static water*. **1976** 'J. Charlton' *Remington Set* xxiii. 119 This place used to be an airfield in the war... Under here there's a big static water tank.

B. *sb.* **4. a.** Atmospherics; radio noise.

1913 *Wireless World* Nov. 508/2 Communication will also be had with New Orleans, which the static formerly prevented. **1938** D. Canfield *Fables for Parents* II. 124 That woman who talks about cooking is coming on splendidly. Not a speck of static. Wouldn't you like to bring your sewing over and listen? **1950** 'N. Shute' *Town like Alice* v. 156 He wanted to see a live broadcast of 'Much-Binding-in-the-Marsh' which he listened to on short wave from Brisbane when the static permitted. **1960** *Practical Wireless* XXXVI. 413/1 The background noise caused by external static can be troublesome. **1978** R. Ludlum *Holcroft Covenant* xvii. 196 There was a sudden burst of static from a radio speaker beneath the dashboard.

b. In *fig.* use, aggravation or interference; confusion, fuss, trouble, criticism. *slang* (orig. *U.S.*).

1926 Maines & Grant *Wise-Crack Dict.* 8/1 *Full of static*, not worth listening to. **1953** W. Burroughs *Junkie* (1972) viii. 77 An Sol said, 'Hell, I love junk... But if I can't use it without I get static all the time from the law, I'll get off junk and stay off.' **1969** C. Young *Todd Dossier* 154 If I notified Security it would just mean a lot of fuss, ..where was the requisition, all that static. **1974** L. Deighton *Spy Story* xvii. 186 Spare me the static... Why didn't you lay it on for me, about working for the goddam Brits? **1979** D. Anthony *Long Hard Cure* v. 42 The words are full of static, a reaction to that attack.

5. Static electricity. Freq. *attrib.*

1916 'B. M. Bower' *Phantom Herd* xiv. 233 All that negative I took to-day is chock full of 'static'. **1951** Koestler *Age of Longing* 195 The umbrellas..had a tendency to charge the people who carried them with static, which sometimes discharged itself in crackling sparks at a handshake, kiss or other bodily contact. **1956** *Planning* XXII. 128 Static elimination is the removal of the harmful electric charges which accumulate on fibres and thin sheets during manufacturing processes. **1978** *Hi-Fi News* Sept. 130/1 (Advt.), Not unlike a common magnet attracting iron particles, static scavenges and draws dust particles onto the record surface. **1979** *Globe & Mail* (Toronto) 5 May 8/1 The winter smell of the settlement has become that of laundry aids—fluffers and static eradicators—whose smells are wafted into the icy air.

-static, formative element (f. Gr. στατικός causing to stand, stopping: see Static *a.* and *sb.*) used in the senses (*a*) 'inhibiting flow', as in Hæmostatic *a.* and *sb.*; (*b*) 'inhibiting growth', as in *bacteriostatic *a.*, *fungistatic *a.*

staticisor (stæ·tisəizəɹ). *Computers.* [f. Static *a.* + -ise¹ + -or.] A device which converts a succession of bits into an array of simultaneous states, thereby storing them.

1949 Williams & Kilburn in *Proc. Inst. Electr. Engineers* XCVI. III. 82 Information may be represented 'dynamically' by pulses, which only exist transiently, or 'statically' by d.c. coupled flip-flop circuits, which retain the information until purposely reset to a standard condition... The set of flip-flop circuits is called a 'staticisor'. **1956** *Electronic Engin.* XXVIII. 154/2 The diodes in one horizontal row store the binary digits of the same significance in each of the words, and all have access to the trigger V_1 in the input-output staticizor. **1960** Haley & Scott *Analogue & Digital Computers* vii. 191 The digits are transient and must therefore be stored or converted into a static form by setting a row of five staticisors.

So **sta·ticize** *v. trans.*, to store by means of a staticisor.

1952 *Electronic Engin.* XXIV. 30/1 In almost every design, a modest number of electronic triggers is retained since they allow words to be set up ('staticized') or discharged ('dynamicized') at different rates. **1960** Haley & Scott *Analogue & Digital Computers* vii. 193 Once the function digits have been 'staticised' there comes the problem of decoding.

statics. Add: **1. d.** *Econ.* That part of economic theory which examines the forces and conditions obtaining at a state of equilibrium in an economic system, without consideration of changes through time; esp. as *comparative statics* (see quot. 1974).

1871 W. S. Jevons *Theory Pol. Econ.* p. viii, The nature of Wealth and Value is explained by the consideration of indefinitely small amounts of pleasure and pain, just as the theory of Statics is made to rest upon the equality of indefinitely small amounts of energy. **1891** J. N. Keynes *Scope & Method in Pol. Econ.* iv. 141 In so-called economic statics we are frequently engaged in examining the effects of particular changes. **1920** A. Marshall *Princ. Econ.* (ed. 8) v. v. 366 The problem of normal value belongs to economic Dynamics: partly because Statics is really but a branch of Dynamics. **1947** P. A. Samuelson *Foundations Econ. Analysis* ii. 8 This method of comparative statics is but one special application of the more general practice of scientific deduction in which the behavior of a system (possibly through time) is defined in terms of a given set of functional equations and initial conditions. **1954** W. Jaffé tr. *Walras's Elem. Pure Economics* viii. 117 Had I supposed utility to be a *variable* functionally related to time, then time would have had to figure explicitly in the problem. And we should then have passed from economic *statics* to economic *dynamics*. **1963** R. E. Kuenne *Theory of Gen. Econ. Equilibrium* i. 15 Comparative statics is a *method* of employing static models analytically by imposing changes upon the data of the model. **1974** A. S. Campagna *Macroeconomics* iii. 50 With a model in equilibrium any change in the variables will cause the model to react until a new equilibrium is reached. The comparison of these two equilibrium states is called comparative statics.

† 2. = *static *sb.* 4 a. *Obs.*

1918 in Webster Add. **1921** *Sci. Abstr.* B. XXIV. 156 On the North Atlantic coast the statics come in large proportion from the S.W. **1926** *Glasgow Herald* 15 May 4 A wall of 'statics' may be responsible for the fact that no wireless messages have been received from the airship for some time.

station, *sb.* Add: **II. 7. b.** *to keep station, on station.*

1923 *Man. Seamanship* (Admiralty) (ed. 2) II. x. 176 The leading ship should therefore at once reduce speed, even though the other is keeping station on her. **1939** *War Illustr.* 2 Dec. 372/3 The absolute necessity of maintaining the order in which their ships are placed in the convoy, i.e. to keep station, and not to alter course except at the order of the commodore of the convoy. **1972** *Lebende Sprachen* XVII. 149/1 *On station*, the status of an ocean station vessel when within the limits of the assigned ocean station. **1979** *Courier-Mail* (Brisbane) 1 Jan. 15/4 Quentin Fogarty managed to get film on the return flight of a large glowing orange object keeping station with the aircraft about 8 km away. **1982** *Times* 31 Mar. 4/4 HMS Endurance will remain on station as long as is necessary.

11. b. (Earlier example.)

1848 *Alfred in India* 12 There are also numbers in the civil service, and they reside at what are called *stations*.

c. *Air Force.* An aerodrome where personnel are employed or garrisoned.

1911, etc. [see *air-station* s.v. air *sb.*¹ B. III. 7]. **1922** *Encycl. Brit.* XXXI. 85/1 At the outbreak of the World War the stations on the organized east coast system of aerial patrol were as follows:—Eastchurch, [etc.]. **1942** T. Rattigan *Flare Path* I. 5, I suppose you came to see someone up at the station. **1977** *Daily Tel.* 7 Nov. 2 The Service's Ground Branch is most seriously affected, with one in three group captains nominated for command making it known at pre-selection stage that they are not interested in taking over their own station.

13. d. A mission station; a mission (Mission *sb.* 6).

1834 J. A. Wilson *Jrnl.* 2 Jan. in *Missionary Life & Work in N.Z.* (1889) I. 7 We are..satisfied with the central position of the station. The large tribes are within a circle of twenty miles. **1851** H. R. Richmond *Jrnl.* in *Richmond-Atkinson Papers* (1960) I. ii. 85 On Monday evening we arrived at Mr Ashwell's station at Tukapoto. **1883** C. F. Wilder *Sister Ridnour's Sacrifice* 229 The converts..have been for many weeks at the station. **1923** O. Schreiner *Thoughts on S. Afr.* 16 A minister of the Dutch Reformed Church..came to spend a night at our station. The accommodation of an up-country mission house is limited.

e. *U.S.* A branch post office.

1896 *Ann. Rep. U.S. Postmaster General* 14 All detail matters relating to the establishment and discontinuance of post offices, the establishment of stations,..would be superintended personally by the district supervisors. **1939** J. L. Floherty *Make Way for Mail* x. 158 Twenty-four sub-postal stations are connected by tube system with the general post office. **1960** E. K. Meador *Billion Dollar Pork Barrel* i. 20, I..was appointed a regular clerk..at the same station where I had been working as a sub-carrier. **1977** *Times Lit. Suppl.* 18 Feb. 187/4 (Advt.), P.O. Box 307, Times Square Station, New York.

f. A broadcasting station; an establishment or organization transmitting radio or television signals. Cf. *radio station* s.v. *radio sb. 7; *television station* s.v. *television 3 c.

1912: see *radio station* s.v. *radio *sb.* 7. [**1913** *Wireless World* Apr. 8 (*caption*) Aden Wireless Station.] **1922** *Science & Invention* Feb. 937/2 We amateurs and experimenters sit by our cozy fireside..and enjoy wireless telephone music sent out by the now famous Westinghouse broadcasting station at Newark, or from one of the dozen other stations. **1923** *Daily Mail* 13 May 8/2 You turn the handle a quarter of an inch, 'tuning'-in to the Cardiff station. **1944** S. S. McKay *W. Lee O'Daniel & Texas Politics, 1938–1942* i. 22 In addition to the WBAP programs the Doughboys were taken to Dublin every Wednesday night for a program over a small station there. **1959** A. Webster *Roots* II. i. 37 She turns the radio on, turning the dial knob through all manner of stations and back again until she finds some very loud dance music. **1969** *Listener* 6 Feb. 177/2 The daily output of all these stations is remarkable considering their limited resources in money, staff and equipment. **1978** J. Irving *World according to Garp* xix. 432 The Vermont station carried the game..from Philadelphia.

g. A location in an automated system (e.g. for data processing or a manufacturing process) where a particular operation takes place.

1948 *Math. Tables & Other Aids to Computation* III. 150 The card reading unit is similar to a standard IBM Reproducing Punch, except that a full reading station has been inserted ahead of the punching station. **1949** B. L. Davies *Technol. Plastics* xvii. 317 Each one of a number of moulds is placed at a station round the table and a cam device is provided for opening it and ejecting the moulding automatically as the table rotates. **1976** *Sci. Amer.* Feb. 77 (*caption*) The three components..are mounted on an underbody shuttle and carried rear end first past five welding stations in sequence.

h. *colloq.* The headquarters of an intelligence service.

1973 A. Mann *Tiara* vii. 57 Would you ask the station to let me know each day what they hear? **1975** N. Luard *Robespierre Serial* iv. 18 A member of the Paris station, a young cypher clerk. **1978** R. Cassilis *Winding Sheet* I. xiv. 44 He was a good man... A good Head of Station.. who hadn't forgotten his tradecraft.

14. (Earlier example.) Also *N.Z.*

1822 J. Dixon *Voyage to N.S.W.* 47 There have, however, been instances of stock-keepers, at distant stations, having been murdered. **1845** W. Deans *Let.* 25 Nov. in J. Deans *Pioneers of Canterbury* (1937) 100 There will be no stations so far inland here as there are in Australia. **1851** E. Shortland *Southern Districts N.Z.* xiii. 245 We arrived early in the afternoon at the native station near Lake Wairewa. **1930** [see *homestead *sb.* 2 b]. **1950** *N.Z. Jrnl. Agric.* May 463/1 Years previously, when the present farm was part of a large station.

V. 28. (sense *11 c, 13 b) *station commander*; (sense *13 f) *station director, manager, operation*; (sense *13 h) *station chief*; (sense 14) *station holder, horse, man, manager, owner*; (sense 20) *station announcer, buffet, manager, platform* (earlier example).

1964 'J. H. Roberts' '*Q*' *Document* (1965) ii. 54 The blaring voice of the station announcer calling the trains. **1939** N. Coward *To step Aside* 61, I had some tea with her in the station buffet at Dieppe. **1941** J. D. Carr *Case of Constant Suicides* i. 15 Drinking slopped tea..in a steamy station-buffet. **1978** D. Kyle *Black Camelot* viii. 119 He went into the station buffet and bought a cup of tea and a Banbury cake. **1974** W. Garner *Big Enough Wreath* iv. 41, I have a station chief in London. **1943** C. H. Ward-Jackson *It's a Piece of Cake* 8 We've a new Station Commander who is keen on physical training, and life is one long parade from six in the morning till ten at night. **1972** *Police Rev.* 1 Dec. 1569/2 Why should not the station commander himself take action? **1978** R. V. Jones *Most Secret War* xliv. 419 The Station Commander told Hartley that he had been watching all three squadrons. **1923** J. Reith *Diary* 25 Oct. (1975) ii. 132 Very busy on new regulations for SB... I am leaving it more to the station directors. **1878** E. S. Elwell *Boy Colonists* 27 When the men..wanted any of these articles for *personal use*..the station-holder sold them to them. **1911** C. E. W. Bean '*Dreadnought*' *of Darling* vii. 75 Leagues away even from the homestead cows or the station horses. **1966** 'J. Hackston' *Father clears Out* 14 No station horse of any standing would have approved of our little farm. *Ibid.* 174 Fearfully I looked round as if at that moment one of the station men would come riding up and find me with the dead sheep. **1911** C. E. W. Bean '*Dreadnought*' *of Darling* xvii. 167 One station manager told us that he had found scores of them [*sc.* rabbits.] dead around their burrows. **1962** *B.B.C. Handbk.* 30 The station managers must aim to build a partnership between the broadcaster and the community. **1965** *Guardian* 11 Feb. 16/4 Stationmasters in the London-Midland region of British Rail..would be replaced by station managers, who would have wider powers. **1977** D. L. Altheide in Douglas & Johnson *Existential Sociol.* iv. 142 Secondly, there is the relation of the news to the overall station operation, including sales, production, and programming. **1911** E. M. Clowes *On Wallaby* iii. 69 A station-owner's life, even in these days, is not all beer and skittles. **1968** K. Weatherly *Roo Shooter* 42 Three weeks later it rained... The station owners smiled. **1886** Kipling *Plain Tales from Hills* (1888) 117 Nobody ever dreamed of seeing him handcuffed on a station platform.

29. station agent, (*a*) chiefly *U.S.*, a person in charge of a stage-coach or railway station; (*b*) a person working for an intelligence service; **station break** *U.S.*, a break (*break *sb.*¹ 8 k) between radio or television items or programmes, during which the station identifies or advertises itself; **station hand:** also *N.Z.*; **station head,** the chief of an intelligence service headquarters; **station-keeping:** also *transf.* and as *ppl. a.*; **stationman,** a person employed on the (underground) railways, as a platform attendant, porter, etc.

1857 *Trans. Illinois Agric. Soc.* II. 25 The active co-operation of this company, through its station agents, ..in bringing forward an interesting show. **1879** E. J. Simmons *Mem. Station Master* (1974) ix. 131 The Long and Narrow Railway paid their station-agent better than the Great Smash Company. **1910** J. Hart *Vigilante Girl* xv. 203 He stopped at the stage station... When the station agent looked to see what the man had written [etc.]. **1948** *Westerners Brand Bk.* (Denver Posse) 21 Louis J. F. Jaeger..was the Butterfield station agent at Fort Yuma. **1974** J. Grady *Six Days of Condor* 109 Who do you suppose was station agent out of Taiwan? **1949** *Consumer Rep.* May 236/2 A 'station-break' can..be inserted between the close of one commercial program and the opening of the next. **1971** D. E. Westlake *I gave at Office* (1972) 133 I've been in this business long enough to know a lead-in for a station break when I see one, and that finish was a natural, so I took a station break. **1930** L. G. D. Acland *Early Canterbury Runs* 1st Ser. i. 8 When an old fashioned squatter or station hand used the word 'homestead' he used it to signify the owner's residence. **1974** J. Grady *Six Days of Condor* 108 He worked his way..from special field agent to station head. **1962**

Punch 18 Apr. 597/3 A British scheme 'envisages'.. 'twelve station-keeping active-relay satellites'. **1971** *Daily Tel.* 14 Sept. 2/6 The 'station-keeping' device would be useful in fog.. to indicate the correct distance behind the vehicle in front. **1952** *Britannica Bk. of Year* 667/1 *Stationman*, a platform attendant of the London Transport (Underground) railways. **1963** *Times* 24 May (London Underground Transport Suppl.) p. vii/4 Passengers refused to leave trains at the haughty ordering of stationmen who refused to give any explanation of why the journey could not continue. **1968** *Daily Tel.* (Colour Suppl.) 13 Dec. 14/1 The wild adventures of train controllers and station men, or signalmen and porters as they used to be called in the dull old days. **1974** P. Wright *Lang. Brit. Industry* i. 19 Notice that on the London tubes, the 'degrading' rank of *porter* has been engulfed in *stationman*.

stationarity. Add: More widely, the state of being stationary or unvarying; stationariness; constancy.

1955 M. Loève *Probability Theory* ix. 418 In the integral stationarity case, the basic ergodic inequality takes very simple forms. **1972** *Nature* 8 Dec. 338/1 Hot spots have been hypothetically identified as mantle plumes and their stationarity with respect to the Earth's spin axis demonstrated. **1973** *Sci. Amer.* July 82 The demographic behavior of these nations to pointing in the direction of a rough equilibrium of deaths and births, that is, stationarity. **1973** *Animal Behaviour* XXI. 161/1 The assumption of stationarity of the processes in equation (7) seems justified if one bears in mind that the animals observed lived under constant conditions.

stationary, *a.* Add: **1. d.** *stationary bicycle, bike* (N. Amer.), a fixed machine, resembling a bicycle, used in fitness exercises.

1962 E. Lucia *Klondike Kate* ii. 53 And pedalled for hours on the stationary bicycle to keep her figure. **1969** *Sears Catal.* Spring/Summer 400/2 Stationary Bike.. Pedal for miles without leaving the comfort of your own home. Chain-driven pedal action gives you the same exercise as regular bicycle. **1976** *Woman's Day* (U.S.) Nov. 154/2 If you don't want to be on public display, try a stationary bicycle or running in place in your bedroom.

e. Physics. *stationary wave* = *standing wave* s.v. *Standing *ppl. a.* 11 c.

1856 D. Lardner *Hand-bk. Nat. Philos.* IV. iv. i. 350 (*heading*) Stationary waves. **1867** J. Tyndall *Sound* 101 The step of a water-carrier is sometimes so timed as to throw the surface of the water in his vessel into stationary waves, which may augment in height until the water splashes over the brim. **1905** *Brit. Pat.* 8,290 In consequence of the interference of the impressed and reflected oscillations, the phenomenon of 'stationary waves' ..is produced. **1962** Walshaw & Jobson *Mechanics of Fluids* xii. 387 By passing light through the divergent part of a supersonic nozzle, the existence of stationary waves inclined to the stream at the appropriate Mach angle may be confirmed on a shadowgraph.

f. Of an artificial satellite: geostationary.

1970 *Gloss. Aeronaut. & Astronaut. Terms* (*B.S.I.*) xviii. 3 *Stationary satellite*, a synchronous satellite in a circular, equatorial orbit, moving in the direction of rotation of the primary body. **1979** *Sci. Amer.* Feb. 58/1 Data on the expanses not accounted for by the World Weather Watch are provided by five geosynchronous ('stationary') weather satellites, [etc.].

2. b. *stationary state* (Physics), a steady state; *spec.* any of the stable orbits of the electrons in the Bohr model of the atom.

1909 *Rep. Brit. Assoc. Adv. Sci.* 634 If we are given the probability that the coordinates of the system may be between given limits, then a condition for the stationary state is that the mean values of the accelerations of ... are zero. **1913** N. Bohr in *Phil. Mag.* XXVI. 7 The dynamical equilibrium of the systems in the stationary states can be discussed by help of the ordinary mechanics, while the passing of the systems between different stationary states cannot be treated on that basis. **1934** *Jrnl. Chem. Soc.* 359 Each spectral term, multiplied by Planck's constant, may be taken to represent, for the corresponding stationary state of the atom, the work necessary to remove the electron to an infinite distance from the proton. **1974** G. Reece tr. *Hund's Hist. Quantum Theory* v. 67 Bohr now sums up his results as follows: in the 'stationary states' classical mechanics is used. **1977** I. M. Campbell *Energy & Atmosphere* iv. 65 Under normal conditions, E:C* will be the type of reactive intermediate to which the Stationary State Approximation can be applied, i.e. d[E:C*]/dt = 0.

c. *Math.* That is not instantaneously changing; associated with a derivative whose value is zero.

1901 G. A. Gibson *Elem. Treat. Calculus* vi. 105 Since f'(x) measures the rate of change of the function it is usual to class those values of the function for which f'(x) is zero as stationary values. **1902** Snyder & Hutchinson *Differential & Integral Calculus* xiii. 153 A point at which the direction of bending changes from positive to negative, or *vice versa*..is called a point of inflection, and the tangent at such a point is called a stationary tangent. **1954** H. R. Cooley *First Course Calculus* v. 89 Points at which the derivative is equal to 0 are sometimes called stationary points. They are points at which, if a particle were progressing along a curve from left to right, its vertical motion would be momentarily stopped. **1978** K. Ahmad *Trad. & Mod. Math.* 85 A stationary point is a point at which dy/dx = 0. Maxima, minima and points of inflection are stationary points.

d. *Statistics.* Applied to a series of observations that has attained equilibrium, so that the expected value of any function of a section of it is independent of the time for which it has been running.

1938 H. Wold *Stud. in Analysis of Stationary Time Series* 1 Observational series which describe phenomena changing with time may be roughly classified in two broad categories, viz. evolutive and stationary. *Ibid.*, Stationary time series are unchanging in respect to their general structure. The fluctuations up and down in such a series may seem random or show tendencies to regularity—in any case, the character of the series is, on the whole, the same in different sections. **1968** P. A. P. Moran *Introd. Probability Theory* iii. 151 A set of numbers, p₀, p₁, ... such that ... will be called a 'stationary distribution' of the process. **1975** *Nature* 11 Dec. 490/1 Two assumptions are commonly made about earthquakes: first, that their occurrence has a stationary random Poisson distribution.

station-house. Add: **2.** Also *U.S.*, the police-station itself.

1870 *Galaxy* Feb. 272 An headquarters of police..is called in New York a station-house, though in many other places this word is more correctly used to indicate a stopping house on railroads. **1931** H. F. Pringle *Theodore Roosevelt* i. xi. 138 He arrived..at a station house in the lower part of the city, and interrupted the meditations of the sergeant. **1963** *Listener* 4 Apr. 585/1, I began my police career in June, 1951. After three months' training I was assigned to the W.54th Street station house. **1979** *Honolulu Advertiser* 8 Jan. D-1/4 ..1000 Hasidic Jews stormed a Brooklyn police station-house.

5. Also *N.Z.*

1888 A. McKay in *Bull. N.Z. Geol. Survey* No. 1. 4 The station-house was so far wrecked. **1933** L. G. D. Acland in *Press* (Christchurch) 4 Nov. 15/7 *Men's hut*, house where the station hands live. On some stations it is called the station house.

stationnaire (stɛ̄ʃɔnɛ̄ᵊ·ɹ). *Obs.* exc. *Hist.* [Fr.] A naval guard-ship, stationed at a foreign port for the use of an ambassador.

1895 *F.O.* 64/1351 (Public Record Office) No. 14 He feared the arrival of the second stationnaires would inevitably excite ill-feeling amongst the Mahomedans. **1914** R. Rankin *Inner Hist. Balkan War* xii. 364 As yet no step has been taken at hand except the weakly-armed 'stationnaires' or European guard-ships which at all times lie in the Bosphorus for the use of the Ambassadors. **1922** *Glasgow Herald* 21 Dec. 9/3 As regards the foreign Embassy stationnaires, Lord Curzon said that he had never thought they could be detrimental to Turkey's sovereignty. **1958** H. Buchanan *Ambassador's Daughter* ..the 66 Ambassadors accredited to the Sublime Porte before the Turkish Revolution, were given a summer residence on the shores of the Bosphorus,..with an armed sloop or stationnaire always anchored in the vicinity in case of a rising of the Turks.

station-wagon. Also station wagon, (-)waggon. **†1.** *U.S.* **a.** A type of horse-drawn covered carriage, used for conveying passengers. *Obs.*

1894 *Hub News* p. 3 May 27/1 Business has been fairly good this spring.. Traps are in most demand, next come buggies, cutunders, and business rockaways or station wagons. **1901** *Varnish* 13 July 253/1 Then we would all know the difference between a cabriolet and an extension-top phaeton; a station wagon and a rockaway.

b. A similar type of motorized carriage.

1904 *Motor World* 21 Jan. 680/1 The station wagon is a new model exhibited the first time this year. **1904** *Sci. Amer.* 30 Jan. 99/1 Manufacturers of Electric Broughams, Landaus,.. Station Wagons, Surreys.

2. orig. *U.S.* An estate car; a saloon motor car with a rear door or doors, and capable of carrying goods as well as passengers.

1929 *N.Y. Times* 6 Jan. xi. 1/5 The Ford Motor Company is having its own exhibition.. The three new models added to the line are on display there. These are the town sedan, the convertible cabriolet, and the station wagon. **1930** *Amer. Speech* V. 376 The other commercial bodies have usually retained old names—*delivery wagon, station wagon.* **1942** *E. African Ann.* 1941–2 11/1 Safari cars..have box-bodies, of the station-wagon model. **1949** P. McLean *Eastern Approaches* ii. iii. 205 This was a new cut-down Ford station wagon, with room in it for six people and a certain amount of kit. **1951** 'J. Wyndham' *Day of Triffids* v. 88 We kept on foot for a while, looking out for a suitable car. We found a large, almost new station-waggon. **1969** *Sydney Morning Herald* 24 May 55/4 (Advt.), Datsun station wagon. **1976** *Encounter* June 9/1 They advanced, the distinguished party,..down the steps, into Halpert's Dodge station wagon, where, along with two sacks of lawn fertiliser, they fitted comfortably.

statism. Restrict *rare* to senses in Dict. and add: **3. b.** = *Étatisme.

1919 *Social. Rev.* XI. 62 Traditional phrases such as 'The Appeal to Democracy', 'Freedom for Little Nations', etc., have been used so often, with so poor a result during the last eighteen months, that all the time 'individualism' and 'statism' have been struggling together for supremacy and power under their cover. **1940** *Sun* (Baltimore) 5 Nov. 5/7 Republican Senator Charles L. McNary concluded his Vice-Presidential campaign tonight with the charge the New Deal is taking deeper and deeper refuge in paternalism and statism'. **1945** A. Huxley *Let. 8 Aug.* (1969) 531 Men and women.. brought up under Statism.. have been taught to believe that the State is more important than the individual. **1962** *Times Lit. Suppl.* 23 Nov. 919/1 Anarchic egocentricity thus merges with the high-minded, patriotic. This has caused schizophrenia in British Labour. **1970** *Daily Tel.* 1 Dec. 1/4 In South America today..various forms of Marxist-inspired Statism are establishing themselves. **1979** *Times* 2 Apr. 54/2 The shortfall itself is rooted in policies that have led to too much statism and not enough private initiative.

statist, *sb.*¹ Add: **3.** (With capital initial.) A member of a conservative Belgian nationalist party which sought to maintain the power of the provincial assemblies or States in the late eighteenth century.

1909 *Camb. Mod. Hist.* VI. xviii. 653 Only a short time was, however, to elapse before they [sc. Estates General] split asunder into two irreconcilable parties—the Statists of van der Noot..and the Democrats of Vonck. **1921** E. Cammaerts *Belgium* 11 'Statists' and 'Vonckists'. **1966** V. R. Lorwin in R. A. Dahl *Pol. Oppositions in Western Democracies* v. 149 The Statists, or Van der Nootists, opposed the reforms of Joseph II; they sought to maintain the customs and privileges of the established Catholic Church and the narrowly based oligarchy of landowners, masters of urban crafts, and nobles who dominated the sclerosed provincial assemblies or Estates. **1974** *Encycl. Brit. Macropædia* XI. 157/1 Van Der Noot made a triumphant entry into Brussels, where he and his 'Statists' were supported by the Estates of Brabant.

4. A supporter of statism.

1929 *National Observer* (U.S.) 4 Apr. 17/2 So much for rent control, just one of many well-meant disasters visited upon us by statists. **1970** *N.Y. Rev. Bks.* 21 Oct. 49/1 McCagg sees Stalin as a 'statist', more interested, that is, in building the Soviet state than in the Communist Party.

statist (stēᵊ·tist), *a.* [f. Statism: see -ist.] Of, pertaining to, advocating, or based on statism.

1926 *New Statesman* 9 Jan. 26/2 Tory propagandists and Labour re-thinkers share in the admiration of the new 'statist' economic system we have in Britain. **1960** *New Left Rev.* May–June 6/2 Communist and Labour fundamentalism are the 'statist' variety. **1973** *Advocate-News* (Barbados) 8 May 3/1 in other systems, the co-operatives has been the peripheral to the main sector, whether capitalist or statist. **1976** K. Joseph *Monetarism is not Enough* 17 But the whole economy is not private. Nearly two-thirds is statist, and insensitive in itself to contraction of the money supply. **1977** *N.Y. Rev. Bks.* 27 Oct. 26/3 They include not only members of the socialists but also working people who have experienced statist authoritarianism, rigidity, narrowness, and felt impelled to speak up, however discreetly. **1979** *New Society* 9 Aug. 293/2 The commitment to democracy was real in the later Marcuse—including a commitment to defend existing democratic liberties against statist and fascist incursions.

statistic, *sb.* **1. b.** Delete ¶ and substitute for entry: A quantitative fact or statement. (Examples.)

1880 'Mark Twain' *Tramp Abroad* xvi. 148 There is not a statistic wanting. It is as succinct as an invoice. This is what a translation ought to be. **1928** H. Belloc *Hist. England* III. iii. 244 Before the siege-piece had been developed as a fairly reliable arm, a city or a castle wall was attacked in one of five ways.. There was direct attack.. There was a starving out of the garrison.. Probably, if a statistic could be made, the latter would be found the most commonly successful method of the true Middle Ages. **1934** *Punch* 14 Mar. 294/1 Few citizens realise that there *is* any river traffic other than the boat race. Let me give them a statistic:—At least 3,000 craft of all sizes pass under Waterloo Bridge every week. **1949** E. Hyams *Not in our Stars* xvi. 198 Although the first dead was a horror and a tragedy, the ten thousandth was a statistic. **1973** *Times* 24 Apr. 12/2 The statistic of 22·2 unemployed to every notified vacancy in Scotland. **1975** *Nature* 11 Sept. 82/1 A more pertinent statistic is that about 98% of all light taxes with second family cars, lie within the 50-mile range of present battery technology.

c. *Statistics.* Any of the numerical characteristics of a sample (as opposed to one of the population from which it is drawn). Cf. *Parameter 2 f.

1922, etc. [see *Parameter 2 f]. **1925** R. A. Fisher *Statistical Methods for Research Workers* iii. 43 The utility of any particular statistic, and the nature of its distribution, both depend on the original distribution. **1976** *Biometrika* LXIII. 438 The sample sizes are enormous, the smallest being 23517, so that under the null hypothesis this statistic should be distributed as a χ² variate with 4 degrees of freedom.

statistical, *a.* Add: **2. c.** *statistical significance* = *significance 3.

1938 *Jrnl. Parapsychol.* II. 210 The primary requirement of statistical significance is met by the results of this investigation. **1971** *Jrnl. Gen. Psychol.* LXXXV. 68 None of the..interactions reached statistical significance. **1974** *Jrnl. Dental Res.* LIII. 763/2 Whether to use the standard test for statistical significance or the Bechhofer test depends on the problem that is posed.

d. Of a branch of science, or a physical process or condition: not absolutely precise but dependent on the probable outcome of a large number of small events, and so predictable; *statistical mechanics*, the description of physical phenomena in terms of a statistical treatment of the behaviour of large numbers of atoms, molecules, etc., esp. as regards the distribution of energy among them; hence *statistical-mechanical* adj.

1885 J. W. Gibbs in *Proc. Amer. Assoc. Adv. Sci.* XXXIII. 57 (*heading*) On the fundamental formula of statistical mechanics, with applications to astronomy and thermodynamics. **1900** *Rep. Brit. Assoc. Adv. Sci.* 617 The statistical dynamics of the distribution of the molecules. **1902** J. W. Gibbs (*title*) Elementary principles in statistical mechanics. **1917** *Proc. K. Akad. van Wetensch. te Amsterdam* XIX. 578 The statistical mechanical explanation Boltzmann gave of it [sc. the second law of thermodynamics] rests on statistical foundations which are destroyed by introduction of the quanta. **1927** [see *Fermi-Dirac]. **1945** H. D. Smyth *Gen. Acct. Devel. Atomic Energy Mil. Purposes* ix. 102 Such effects are used in six 'statistical' separation methods: (1) gaseous diffusion (2) distillation.. In all these 'statistical' methods the separation factor is small so that many stages are required. **1955** H. B. G. Casimir in W. Pauli *Niels Bohr* 132 As long as this phenomenon [sc. superconductivity] is not understood an essential element is lacking in our comprehension of statistical mechanics and of the nature of the solid state. **1955** Friedman & Weisskopf in *Ibid.* 138 The statistical method of determining the yield of nuclear reactions gives a reasonable account of their most important features. **1956** E. H. Hutten *Lang. Mod. Physics* iv. 149 This H-function may be taken as the statistical-mechanical analogue of entropy, since it exhibits a one-sided change in time. **1962** J. Riordan *Stochastic Service Systems* iii. 28 The system is said to be in statistical equilibrium.. It is not without changes, for probabilities are still in question, but the probabilistic description of its behavior in time is invariant with time. **1964** J. D. Bernal in *Proc. R. Soc.* A. CCLXXX. 302, I found..that we did not know much about heaps and that to understand heaps we had to open a new subject, that of statistical geometry. *Ibid.* 307 It would be very well worth while to examine the purely geometrical properties of random close-packed aggregates without holes, particularly in relation to kinds of lines that can be drawn through neighbouring points. This should be one of the first tasks of the proposed statistical geometry. **1977** *Nature* 13 Aug. 459/1, I believe that the most widespread quantum statistical mechanics can be 'grafted' onto the classical statistical mechanics of Boltzmann and Gibbs and that therefore quantum mechanics does not require anything essentially new. **1974** G. Reece tr. *Hund's Hist. Quantum Theory* i. 14 The great achievement of statistical physics was that of deriving thermodynamics from mechanical principles. *Ibid.* iii. 46 In the years 1900–7 Einstein completed a basis for statistical thermodynamics.

statistically, *adv.* Add: *statistically significant*: see *Significant *a.* 5.

statisticize, *v.* Add: Also *intr.*, to collect or employ statistics. Hence **stati·sticizing** *vbl. sb.*

1927 *Sunday Express* 17 July 11 Miss Inderrieden has reduced desertions to a percentage basis. She says that the wife is to blame in fifty percent of desertions.. This is a dreadful specimen of statisticizing. It means absolutely nothing. **1977** *Nature* 10 Apr. 602/3 The objective ..is not to help budding statisticians to statisticize but to help experimental scientists to experiment properly.

statistics. Add: **1. a.** (Earlier example.)

1770 W. Hooper tr. *Bielfeld's Elem. Universal Educ.* III. xiii. 269 The science, that is called statistics, teaches us what is the political arrangement of all the modern states of the known world.

2. a*. = *vital statistics* s.v. *Vital *a.* Also *transf.* *colloq.*

1958 *Times* 24 Feb. 11/3 It is a pretty thought to contemplate all those statistics hipping and swaying and shimmering. *Ibid.* 13 Nov. 9/4 To-day, except for the slightly high waist-lines and one or two modified trapezes, feminine statistics were where nature intended them to be. **1960** *Punch* 3 Aug. 148/2 An enticing girl with yellow hair and sound statistics. **1978** O. White *Silent Reach* vi. 61 Next time you get into position, try squeezing her statistics.

3. *Physics.* The statistical description appropriate to the behaviour and properties of an ensemble of many atoms, molecules, etc., esp. as regards the distribution of energy among them; *spec.* = *quantum statistics* s.v. *Quantum 7 a.

1873 J. C. Maxwell in *Nature* 25 Sept. 440/1 The modern atomists have therefore adopted a method which is I believe new in the department of mathematical physics, though it has long been in use in the Section of Statistics. **1920** *Phil. Mag.* XLIX. 124 In the case of a system of gases, to which the statistics assumed to be regular, the potential energy remains approximately constant. **1903** [in Dict., sense 2 a]. **1909** *Proc. R. Soc.* A LXXXIII. 86 The general thesis of which a development is here attempted is thus the molecular statistics of distributions of energy. **1927** [see *Fermi]. **1928** [see *Bose-Einstein]. **1950** W. J. Moore *Physical Chem.* xii. 396 In deriving the Boltzmann statistics, we assumed that the individual particles were distinguishable and that any number of particles could be assigned to one energy level. **1979** *Sci. Amer.* Feb. 80/1 Two fundamental categories of particles, the fermions and the bosons. These categories are distinguished by the intrinsic angular momentum, or spin, of the particles, and by their statistics, or behavior in groups.

stative, *a.* and *sb.* Add: **A.** *adj.* **3.** Now also applied to languages other than Hebrew, to sentences, and to words other than verbs.

1932 E. H. Blake in J. T. Hatfield et al. *Curme Vol. Linguistic Stud.* 38 All temporal and locative relations have theoretically three aspects, a stative indicating existence or rest at a time or place (time when, place where), an ablative indicating continuance or motion from a time or place,.. and a terminal indicating continuance or motion to a time or place. **1939** L. H. Gray *Foundations of Lang.* vii. 202 The stative verb expresses 'the state of being in a certain condition'. **1964** *Language* XL. 78 Stative verbs occur with the nasal simulfix and typically with stative prefixes. **1967** Fenn & Tewkesbury *Speak Mandarin* 3 The stative sentence consists of a subject and a stative verb. **1970** *Language* XLVI. 830 English adjectives, much like English verbs, seem to divide into *stative* and *active* ones. **1979** *Trans. Philol. Soc.* 130 A fundamental distinction exists in Abkhaz between stative and non-stative (or dynamic) verbs.

B. *sb.* Now also applied to languages other than Hebrew, and to words other than verbs.

1966 J. Buse in C. E. Bazell *In Memory of J. R. Firth* 54 These constraints may be used to divide full-words into the four sub-classes of verb, stative, noun and negative. **1973** *Canad. Jrnl. Linguistics* XVIII. 103 In order to explain the possessive function of statives and the limitations of this function in Sechelt, it is necessary to consider a cross-section of occurrences of this verbal aspect and to define its semantic range.

stato-. Add: **sta:to-acou·stic** *a.* *Anat.*, pertaining to the senses or faculties of both equilibration and hearing; *spec.* the epithet of the eighth cranial nerve (the vestibulocochlear nerve); **sta·tocone** [Gr. κῶνος dust], each of the large number of granules in the statocyst of some animals, similar to a statolith, but smaller; also, an otolith in a vertebrate; also **statoco·nia** *sb. pl.*; **statolith** (later examples); also, an otolith in a vertebrate.

1958 *Gray's Anat.* (ed. 32) 144 The ganglia of the vagus, glossopharyngeal, stato-acoustic, facial and trigeminal nerves are derived from the ganglion-crest, but they migrate ventrally and soon come to lie on the ventrilateral aspect of the hind-brain. **1964** J. Z. Young *Model of Brain* vii. 122 There is thus the possibility of correct interaction of visual and stato-acoustic information, these being the two chief systems that project to the mid-brain roof. **1974** D. & M. Webster *Compar. Vertebr. Morphol.* xi. 255 (*caption*) Cross section through the labyrinth. **1923** Borradaile *Animal* ... **1939** Parker & Haswell *Text-bk. Zool.* I. xii. 707 Each statocyst [in molluscs] may contain a number of minute statoconia or, more usually, a single, larger statolith. **1969** *Biol. Bull.* CXXXV. 441 In the labyrinth of teleosts there are generally three large statoliths.. Most other vertebrates, however, have otolith masses consisting of a very great number of small statoconia held more or less firmly together by an organic gel. **1970** *Nature* 29 Aug. 832/1 Inner ear sensory surfaces from the Pacific herring..were prepared.. After removal of the statoconia (otoliths) the tissues were dehydrated. **1980** *Gray's Anat.* (ed. 36) 1190/1 The gelatinous membrane into which the cilia project is flattened and is termed an otolithic membrane.. because it contains numerous minute crystalline bodies called otoliths, otoconia, or statoconia. **1955** *Sci. News* XXXVI. 92 Even plants which have no statoliths respond to gravitational stimuli. **1962** D. Nichols *Echinoderms* vi. 85 These are tiny fluid-filled spheres containing calcareous statoliths, the differential movement of which is registered by special nerves. **1969** *Nature* 28 June 1323/1 Starch statoliths are mobile starch grains found in almost all plant organs that respond to gravity. **1975** *Ibid.* 2 Oct. 380/2 Statoliths of at least one species of cephalopod are composed of aragonite, a stable form of calcium carbonate.

statolon (stæ·tŏlɒn). *Biochem.* Also *erron.* statalon. [Prob. a blend of mod.L. *stoloniferum* (see below) and Stoloniferous *a.* and *-static.] A complex polysaccharide obtained from the mould *Penicillium stoloniferum*, now known to contain a fungal virus which is an antiviral agent, stimulating the release of interferon.

1961 Probst & Kleinschmidt in *Federation Proc.* XX. 441/2 A fermentation broth of *Pen. stoloniferum* [sic] prophylactically inhibits a number of viruses both in animals and in tissue culture.. The active principle, designated as statolon, has been subjected to a variety of purification procedures. **1964** *Proc. Nat. Acad. Sci.* LII. 741 Statolon is a complex anionic polysaccharide with a relatively high content of galacturonic acid. **1967** *New Scientist* 10 Aug. 300/2 W. J. Kleinschmidt and L. F. Ellis have now refined the technique of fractionating statalon, and have discovered, to their surprise, that one of the active fractions contained numerous virus-like hexagonal particles.. They have confirmed that it is these particles that are responsible for stimulating interferon activity in mice. **1971** *Sci. Amer.* July 26/3 Eli Lilly and Company demonstrated the presence of a fungal virus in statolon, an extract of another penicillium species. **1976** P. Collard *Devel. Microbiol.* v. 64 There are a number of compounds that either stimulate the production of interferon in the host cells, as does Statolon, originally believed to be a fungal product but now shown to be a virus, or release interferon which is already pre-formed in the cells. **1978** *Nature* 29 June 760/1 Tilorone, statolon and Newcastle disease virus (NDV), all potent inducers of interferon in mice, induced a marked increase in spleen cell cytotoxicity.

stator¹. Add: **1.** (Earlier example.)

1895 S. P. Thompson *Polyphase Electr. Currents* v. 113 In describing the parts of polyphase motors..[we] shall call the rotating part the *rotor*, and the stationary part the *stator*.

2. More widely, an immovable part of any turbine; *spec.* a stator blade (see sense 3 below), or a row of such blades.

1916 J. W. M. Sothern *Marine Steam Turbine* (ed. 4) viii. 424 The turbine consists of two principal parts, the rotor or moving part, and the cylinder (sometimes called the 'stator') or stationary part. **1951** Cohen & Rogers *Gas Turbine Theory* vi. 133 Each stage will consist of one rotating row followed by a stator but it is usual to provide an additional stator row at entry to guide the air correctly into the first rotor. **1962** F. J. Ordway et al. *Basic Astronautics* x. 404 The rotor turns within the stator, which is a series of blades arranged in a circle around the inside wall of the compressor housing. **1967** N. E. Borden *Jet-Engine Fundamentals* 80 The first four stages of the high-pressure compressor are provided with variable stators, automatically positioned by hydraulic actuators. **1971** P. J. McMahon *Aircraft Propulsion* iv. 124 Figure 4.2 shows a view in the radial direction of the blading of a compressor stage—rotor plus stator—with the stators of the previous stage included for completeness.

3. *attrib.* and *Comb.*, as *stator coil, winding*; **stator blade,** a small stationary aerofoil fixed to the casing of an axial-flow turbine, rows of which are positioned between the rows of rotor blades.

1946 G. G. Smith *Gas Turbine & Jet Propulsion* (ed. 4) iv. 58 Stator blades to direct the air flow between stages will also be of either steel or light alloy. **1977** J. L. Kerrebrock *Aircraft Engines & Gas Turbines* v. 120 There are normally three types of blade row in an axial compressor: the inlet guide vanes, the rotor blades, and the stator blades. **1904** W. R. Bowker *Dynamo, Motor & Switchboard Circuits* 109 The 'stator' coils are arranged in a six-pole grouping. **1895** S. P. Thompson *Polyphase Electr. Currents* v. 115 The stator winding is usually the primary, the rotor winding secondary.

stats (stæts). Colloq. abbrev. of Statistics.

1. (With capital initial.) A department responsible for collecting or recording numerical facts or data.

1942 N. Balchin *Darkness falls from Air* vi. 102 Why not give Stats. a new man and leave Giles where he is? **1953** D. Parry *Going up—going Down* iv. 130 He visited the Stats. Branch. **1966** M. Woodhouse *Tree Frog* i. 2 Do you know what they are in Stats?

2. = Statistics 2 a.

1962 L. Deighton *Ipcress File* x. 64, I am not a statistician.. I was getting pretty fed up with his house-breaking stats. **1974** *Cleveland* (Ohio) *Plain Dealer* 26 Oct. 4-D/1 The position statistics have been impressive even though individual stats for many have not reached expectations. **1976** *New Yorker* 12 Apr. 90/1 Drove to Fort Lauderdale.. found the ballpark, parked, climbed to the press box, said hello, picked up stats and a scorecard, took the last empty seat, filled out my card. **1981** J. Wainwright *Tainted Man* 169 The trumpet.. This never-ending paper crap. Stolen cars, crime stats.

3. = Statistics 1 a.

1970 *Guardian* 31 Jan. 13/3, I..went on to stress the importance of what..I have decided to dub stats education.. It's..vital that people should know more about statistics.

statue, *sb.* Add: **1. c.** *pl.* The name of various children's games which involve the players standing still in different postures.

1906 *Dialect Notes* III. 158 *Statues*, n., the name of a game in which children pose. **1916** N. Douglas *London Street Games* 41 Catch-in-the-Rope is also for boys and girls, and so is.. Statues.. When you play this game you have to line yourselves up against a wall or a house; then the judge comes along and pulls one of your forwards and in that moment you have to make a posture and a face, and pretend to be a statue. **1935** E. Farjeon *Nursery in Nineties* v. 240 She quickly suggests a game, Magical Music, or Forfeits, or Statues. She.. thumps the only tune to which Statues can be played. **1950** R. Pym *Some Tame Gazelle* xxi. 236 When they realised that a prayer was being said, they stood stiffly with the urn, like children playing a game of 'statues'. **1981** L. Deighton *XPD* xiv. 356 'You're moving,' called Stein loudly. It was a good-natured complaint, the sort that children might use when playing the game of 'statues'.

2. *statues game.*

1975 'D. Jordan' *Black Account* ii. xxxv. 175 Guy stood by the door like a child playing the statues game.

statued, *ppl. a.* Add: **2.** (Earlier example.)

1731 L. Theobald *Orestes* iv. iii. 57 The Statued Goddess born in solemn Pomp, Guarded, as 'twere the Spoils of hostile Bands?

status. Add: Now usu. with *pl.* statuses (stēᵊ·təsēz). **1. b.** *status asthmaticus*, the condition of a patient during a prolonged severe asthmatic attack; also *ellipt.*, = status epilepticus.

1947 Dorland & Miller *Med. Dict.* (ed. 21) 1392/1 Status asthmaticus. **1962** J. H. Burn *Drugs, Med. & Man* xviii. 179 Sometimes the attacks [of asthma] follow one another so steadily that the patient's life is in danger. He is then said to be in 'status asthmaticus' and in the past patients have died. **1971** *Where* Dec. 360/1 There was no sign of epilepsy until she was two and a half, when she went into status (a condition where the fits follow one another without pause).

3. a. (Examples of pl. use.) Also *social status*.

1854 Mill *Pol. Econ.* (ed. 3) II. iv. vii. 331 As to civil and social status, in a new reform bill, the opportunity was not taken. **1873** in *Dict.*). **1901** G. B. Shaw *Socialism for Millionaires* in *Fabian Tract* No. 107. 15 A millionaire does not really care whether his money does good or not, provided he finds his conscience eased and his social status improved by giving it away. **1936** R. Linton *Study of Man* viii. 114 There are no rôles without statuses or statuses without rôles. **1946** D. L. Sayers *Unpopular Opinions* 120 The commercial middle classes acquired the plutocratic and aristocratic notion that the keeping of an idle woman was a badge of superior social status.

1955 T. H. Pear *English Social Differences* i. 25 Each individual..can have many statuses. **1960** D. Potter *Glittering Coffin* iv. 54 The links between job and social status..are too often taken for granted. **1960** P. C. Reisterro in I. L. Horowitz *Masses in Lat. Amer.* xiv. 517 Cultural ascent in society refers to the acquisition of those cultural forms corresponding to a higher class or social status. **1977** R. Holland *Self & Social Context* v. 89 Occupants of similar statuses may support each other against threats from members of a rôle-set, as when teachers support each other against parents of their students.

5. a. *attrib.* or as *adj.* That confers prestige on its possessor; having a high social status; superior.

1969 M. Mead *Male & Female* xiii. 266 A caustic critic has labelled the one child of middle-class families as a 'status child', a child that merely gives the parents the status of *having had a child*. **1956** C. W. Mills *Power Elite* iv. 79 As a status model the debutante declined. **1960** S. Kauffmann *If it be Love* i. viii. 113 A status-car, matched to your income, and a station wagon. **1961** A. Smith *East-Enders* vii. 122 One of the status cafes on the edge of the East End, beautifully kept. **1964** A. Wykes *Gambling* viii. 107 This was a status-club. **1967** *Listener* 6 July.. **1971** *Guardian*.. **1974** B. Rook *Hist.* xvi. 153 It's the most status-y kind of thing you can do. **1977** D. Morris *Manwatching* 129 What the rapist wants is.. the total object subjection of his victim.. This is Status Sex, and it is not by any means an exclusively human pattern of activity.

b. *Comb.*, as (sense 3) *status-conscious, -dissenting, -ridden* adjs.; **status anxiety,** anxiety about one's social status, esp. the fear of losing it; **status group(ing),** any group of people who have similar social standing (see quot. 1978); **status-seeker,** one who is concerned with improving or demonstrating his social status; hence **status-seeking** *vbl. sb.* and *ppl. a.*; **status symbol,** a possession or asset sought or acquired as a symbol of social prestige; hence **status-symbolism;** **status system,** a social structure or organization in which status derives from one's position or achievement in some aspect of the group's activity; **status-trophy** *rare* = *status symbol* above.

1959 *Encounter* Sept. 58/1 The political conflicts of the 'fifties..were..explained by sociological concepts such as 'status anxiety'. **1971** Halsey & Trow *Brit. Academics* xiii. 309 This may be due to the status anxiety associated with marked social mobility. **1959** V. Packard *Status Seekers* (1960) i. 6 Wives..tend to be more status-conscious than their husbands. **1974** tr. *Wertheim's Evolution & Revolution* 234 Aslan peasants are generally status conscious and..the more prosperous ones often recognize mutual social obligations only towards those whom they consider to be their peers. **1963** J. M. Mogey *Family & Neighbourhood* viii. 140 The remainder of St. Ebbe's people we may call, by contrast, status-dissenting. **1969** *New Society* 4 Feb. 26/3 On the newer suburban housing estates..a new species has developed, the status-dissenting working class person. **1920** *Status group* [in Dict., sense 2 a]. **1978** *Listener* 19 Jan. 77/3 Status groups—for example, peers of the realm or vagrants—are social networks of those who share similar social prestige. **1964** *Punch* 29 July 173/2 We hear a lot about status-groupings in the U.S. **1953** *Observer* 16 Aug. 7/5 Miss Thurburn can move us to admiring recognition when she deals with.. the status-ridden society. **1962** *Guardian* 14 Aug. 5/3 The prestige motions of a Pakistani civil servant are almost as inflexible as those found in our own status-ridden society. **1959** V. Packard *Status Seekers* (1960) i. 7 The status seekers..are people who are continually straining to surround themselves with visible evidence of the superior rank they are claiming. **1979** *United States 1980/81* (Penguin Travel Guides) 224 There are no phone calls to take, except for those status-seekers foolish enough to have telephones in their cars. **1981** R. Sears in Parsons & Shils *Toward General Theory of Action* 477 Secondary motivational systems..between mother and child..include aggression, dependency, self-reliance..and status-seeking. **1960** *Guardian* 23 Dec. 6/6 The status-seeking educational system. **1962** *Punch* 31 Jan. 227/3 Status-seeking, broken marriages, intrigue among research staff of Midland firm. **1962** E. Snow *Red China Today* (1963) xi. 300 It should not be supposed that they are any less outer-directed than the status-seeking sons of Madison Avenue. **1976** J. Wainwright *Who goes Next?* 52 This over-expensive, status-seeking building. **1955** Status symbol [see sign-vehicle s.v. *Sign *sb.* 12 b]. **1957** *Wall St. Jrnl.* 29 Oct. 1/2 The most common sources of interoffice rivalry over status symbols involve such obvious executive trappings as the size of the desk, the quality of drapes and carpets in private offices, [etc.]. **1965** G. Maxwell *House of Elrig* v. 77, I was singled out as a target for jealousy by those whose parents would have liked 'a handle to their names' and whose status-symbol cars compared spectacularly with the modest and practical conveyances my mother chose. **1981** *Church Times* 23 Oct. 10/2 A fragment of the true Cross—a status symbol if ever there was one. **1957** *Wall St. Jrnl.* 29 Oct. 9/2 In their place he introduced a highly formalized system of status symbolism. **1968** *Punch* 7 Aug. 183/2 The main design criteria [sic] of motor-car manufacture are now neither comfort nor status-symbolism, but simple safety. **1942** Warner & Lunt *Status System Mod. Community* ii. 16 We shall hereafter refer to the eighty-nine behavioral structures (or statuses) as social positions or statuses, and the total social system of Yankee City as the positional or status system. **1978** *Listener* 26 Jan. 35/2 Our Martian would quickly conceptualise pair-bonding in what we call marriage, scientific organisations in the social relations of discovery, status systems in the relations of dominance and submission, and so on. **1964** Auden in *Listener* 1 Oct. 515/2 This unpopular art which cannot be..hung as a status-trophy by rising executives.

statusful (stēᵊ·təsfŭl), *a.* [f. Status + -ful.] Having or conferring (high) social status or distinction.

1969 *Daily Tel.* (Colour Suppl.) 27 Oct. 57/1 Ordering a 22-carat Dunsburgy with ocelot seats and rhinestone headlamps is a bid for statusful individuality. **1975** *Times* 22 Aug. 10/1 'Statusful' speakers, like BBC newsreaders. **1977** J. A. Fishman in H. Giles *Lang., Ethnicity & Intergroup Relations* i. 37 It is not ethnicity *per se* that is of concern but rather the recognition of less statusful, narrower, 'peripheral' identities. **1978** *Amer. Speech* LIII. 14 To imitate the user's special familiarity with that less statusful or less responsible class of people who have such special familiarity and use the term.

status quo. Add: (Later *attrib.* examples.) *status quo ante*: the state of affairs previously existing (also *absol.*).

1877 in [Dict.]. **1955** M. I. Stevens *Let.* 9 Mar. (1967) 709 Everything is now proceeding in status quo ante. **1965** H. Kahn *On Escalation* xiii. 244 It is normal that status quo nations will tend to be crisis-avoiding nations. **1967** *Listener* 27 July 103/1 Nasser was outbargained in view, and that the Arabs were re-supplied with arms, we should be back at a sort of status quo ante. **1971** *Guardian* 6 Dec. 1/2 The TUC puts its money firmly on the need for a 'status quo' declaration by the employer. In other words, the unions insist that the employer must..abandon his authority to change a work practice or agreement without the blessing of his employees.. Workers can be expected to observe procedures only if they contain a 'status quo' clause and the employer observes it. **1976** *New Yorker* 24 May 115/2 There is a great move to the status quo ante. **1978** R. Lewis *Inevitable Fatality* vi. 161 The attempts by Sir Henry Monroe to return QWARTA [sc. a company] to its status quo.

statusy (stēᵊ·təsi), *a.* *colloq.* [f. Status + -y¹.] Possessing, indicating, or imparting a high status.

1962 *Guardian* 29 Nov. 7/1 No matter how many irrelevant if status pieces the 'New Yorker' publishes. **1964** *Punch* 22 Jan. 125/1 As statusy a combination as you can get. **1969** *Time* 26 Sept. 72/2 Pretty soon about 95% of the kids smoked. It got to be a really statusy thing. **1971** *Daily Colonist* (Victoria, B.C.) 5 Sept. 21/6 Katherine Gibbs, who in 1911 founded the statusy secretarial school that bears her name.

statute, *sb.* Add: **9.** *statute labour* (further Canad. examples).

1820 J. M. O'Brien *Jrnl.* 16 Feb. (1968) xvi. 155 We shall have a statute law.. **1873** *Canad. Bk.* .. **1968** *Canad.* .. **1962** .. *Heart of Continent* iv. xii. 223 The Council of Assiniboia for 1901..abolishes statute labour, that time-honoured method of doing road-work.

statute-book. 2. *on the statute-book* (later examples).

1906 G. B. Shaw *On Rocks* ii. 253 You will have to wait two years and go through the whole job again before you can get your Bill on the statute book as an Act of Parliament. **1972** *Times* 2 Feb. 16/6 His Lordship thought as the Act was on the statute book the judge was fully entitled to have regard to it.

statutory, *a.* Add: **2. b.** *statutory company*, a company created by statute, as distinguished from a chartered company or a joint-stock company; also, any company other than one incorporated by royal charter; *statutory holiday*, a holiday established by statute; *statutory instrument*, a common type of subordinate legislation (see quot. 1946); *statutory meeting*, a general meeting of the members of a company, held in accordance with a statute; *spec.* the first such meeting, held between one and three months after the company is entitled to commence business; *statutory rape* (*U.S.*), sexual intercourse with a female who is below the age of consent (whether it occurs against her will or not); *statutory tenant*, a person who is legally entitled to remain in possession of premises although his tenancy of them has expired; so *statutory tenancy*.

1915 *Act* 5 & 6 *Geo. V* c. 44 § 2 The expression 'statutory company' means any railway company, canal company, dock company, water company, or other company incorporated by special Act, who are for the time being authorised under such an Act to construct, work, own, or carry on any railway, canal, dock, water, or other public undertaking, and includes any person or body of persons so authorised. **1970** M. Gregory *Penguin Dict. Commerce* 106 Certain contracts are not binding unless made by deed. There are: (1) gratuitous promises, (2) transfers of shares in statutory companies. **1911** *Daily Colonist* (Victoria, B.C.) 28 Apr. (Mag. section) 7/1 Yesterday [sc. Easter Monday] was a statutory holiday and many Victorians took advantage of the fact to drop their daily business cares. **1979** *Globe & Mail* (Toronto) 2 June 5/7 He also promised to provide time-and-a-half pay for policemen working on statutory holidays. **1976** *Alyn & Deeside Observer* 10 Dec. 28/4 (Advt.), Refuse Collection.. Due to the incidence of Christmas and New Year statutory holidays it has been necessary to re-arrange certain collection days. **1946** *Act* 9 & 10 *Geo. VI* c. 36 § 1(1) Where by this Act or any Act passed after the commencement of this Act power to make, confirm or approve orders, rules, regulations or other subordinate legislation is conferred on His Majesty in Council or on any Minister of the Crown then, if the power is expressed ..in the case of a power conferred on His Majesty, to be exercisable by Order in Council; (b) in the case of a power conferred on a Minister of the Crown, to be exercisable by statutory instrument, any document by which that power is exercised shall be known as a 'statutory instrument'. **1974** R. Crossman *Diaries* (1976) II. 549 The issue the journalists were taking concerned a statutory instrument. **1853** *Bradshaw's Railway Manual, Shareholder's Guide, & Official Directory* V. 213 Which are to be held in March and September. **1900** *Act* 63 & 64 *Vict.* c. 48 § 12 Every company limited by shares and registered after the commencement of this Act shall, within a period of not less than one month nor more than three months from the date at which the company is entitled to commence business, hold a general meeting of the members of the company, which shall be called the statutory meeting. **1962** G. Cummings .. **1894** *Act* .. **1898** *Northeastern Reporter* LXXX. 439 The respondent was convicted of statutory rape. **1959** *Times* 26 Oct. (Canadian ed.) 99/2 His taste for young flesh led to three statutory rape scandals..but the older he got, the more he [sc. Errol Flynn] seemed a cardboard sinner. **1977** J. Crosby *Sunset Beach* .. **1920** *Act* 10 & 11 *Geo. V* c. 17 § 15 Conditions of a statutory tenancy. **1948** *Daily Mail* 25 July 7/3 Mr. Hunt's case was that he owned the freehold of 104, Upper Richmond-road and occupied two rooms. Mr. Sullivan was the statutory tenant. **1972** *N.Y. Law Jrnl.* 20 Oct. 2/2 The parties concede that the defendants' occupancy of the apartment after the expiration of the lease was that of 'statutory tenants', since the premises were rent controlled under the City Rent and Rehabilitation Law. **1973** *Country Life* 3 May 1 (Suppl.) 17 (Advt.), 3-bedroomed flat let unfurnished to statutory tenant.

Staunton (stǭ·ntən). The name of Howard *Staunton* (1810–74), English chess-player and writer, used *attrib.* and *absol.* to designate chess-men of a design now accepted as standard.

[1891 R. B. Swinton *Chess for Beginners* ii. 5 A chess-board and a set of men are necessary. The latter alone can be purchased at all prices, from sets in 'African ivory', after Staunton's pattern, down to flimsy wooden pieces. *Ibid.* 7 Your wooden men..will..be easily identified with the drawings of Staunton's men which form headpieces to the..chapters.] **1898** *Dict. Nat. Biogr.* LIV. 121/1 Staunton's name was conferred on the set of chessmen which are recognised as the standard type among English-speaking people. His 'Chess Player's Text-book' was issued in 1849, without date, to be sold with the Staunton chessmen. **1913** H. J. Murray *Hist. Chess* ii. x. 773 Chessmen of fanciful shapes, however, are often made as curiosities. For actual play, most players would prefer to use the 'Staunton chessmen', the pattern of which Howard Staunton designed in 1849. **1951** G. Frankau *Oliver Trenton* ix. 77 The set of ivory Stauntons I won from him. **1959** I. Jardine *Chess* ix. 113 The pieces of a 'Staunton design' are..the most popular nowadays.. The photograph shows what they look like in your Staunton-type set. **1977** M. Kelly in D. Marcus *Best Irish Short Stories* II. 66 The Staunton with its austere lack of pretension. **1979** P. Alexander *Show me Hero* iv. 55 A chess-board and one of the earliest Staunton sets.

stave, *sb.*¹ Add: **I. 2. i.** [Cf. Norw. *stav.*] A vertical wooden post forming part of the framework of a building, usu. a stave church (see sense 8 c below); also, a plank used in the walls of such a construction.

1915 H. G. Leach *Scandinavia of Scandinavians* ii. 162 In architecture, the most distinctive survivals from the Middle Ages in Norway are the so-called 'stave' churches, tepee-like structures, built of wooden staves, rising roof above roof. *Ibid.*, Stave churches often contain elaborate wooden carvings which have served as models for modern Norwegian decoration.. This is of course the earliest type of the church of Urnaes..one of the earliest existing stave structures, with its intricate animal and vegetable motives. **1974** *Encycl. Brit. Micropædia* IX. 539/1 The chief foundation of the stave church consists of four horizontal wooden beams, from which rise four corner posts, or staves.

7. b. An alliterating letter in a line of Old English verse. Also *head-stave*. Cf. G. *stab, hauptstab*.

1891 H. Sweet *Anglo-Saxon Reader* (ed. 7) p. lxxxv, In our texts..the letters or staves are in italics. *Ibid.*, We denote the first or chief one of the first half-line the head-stave, while I has either one or two called under-staves.

1959 R. B. Le Page in *Jrnl. Eng. & Gmc. Philol.* LVIII. 434 The two alliterating staves, one in each half-line, have a definite structural function. **1964** K. Malone in *Anglia* 67 Grammatically *fela* is *momma* makes a unit but because of the m-stave that binds the two halves of the line together the on-verse must be classified as D in spite of the *ic* clause.

IV. 8. c. **stave church** [tr. Norw. (Bokmål) *stavkirke*], a church built with walls of upright planks or staves, of a type mainly built in Norway from the eleventh to the thirteenth century; **stave mill** *N. Amer.*, a mill making cask-staves.

1915, etc. [see sense 2 i above]. **1933** T. Linstrom *This is Norway* facing p. 32 (*caption*) Borgund Stave Church in Laerdal. **1936** A. W. Clapham *Romanesque Archit.* viii. 164 A highly remarkable class of buildings in Norway, which includes the celebrated mast or stave-churches of Norway. **1968** G. Jones *Hist. Vikings* ii. vii. 116 An oak-built stave church at some time before the year 1050. **1932** R. Flannagan *County Court* 188 Widowed five years before by the death of her husband, who went to jail for the later husband's chain of stave-mills and had prospered. **1964** *Daily Progress* (Charlottesville, Va.) 7 May 13/8 Blades and other stave mill operators say they will measure long as the industry as it is profitable. **1968** E. R. Buckler *Ox Bells & Fireflies* vi. 102 If there was urgent need of ready money..you worked off and on in the stave mill.

stavies, var. *stovies *sb. pl.*

Stavka (sta·fkă, sta·vkă). [Russ., f. *stavít'* to put, place.] The general headquarters of the Russian army.

1928 *Daily Express* 16 July 8 The mutual relations between the Stavka (G.H.Q.) and the Duma are revealed clearly for the first time. **1951** W. S. Churchill *World Crisis* VI. ix. 142 By August 6 the Russian General Headquarters—in future called the Stavka—learned definitely that the German main forces..were entraining for the French front. **1949** I. Deutscher *Stalin* 460 The *Stavka*, the Red Army's G.H.Q., was in his offices in the Kremlin. **1963** P. Fleming *Siege at Peking* .. **1977** *Stavka*, besides embodying all the worst technical vices of Russian military bureaucracy, was rotten to the core with dishonesty, nepotism and intrigue.

stay, *sb.*¹ Add: **1. b.** Also, a supporting wire or cable on an aircraft.

1894 O. Chanute *Progress in Flying Machines* 237 This main aeroplane..is trussed and stiffened in every direction by wire stays. **1908** H. G. Wells *War in Air* x. 317 It had taken only an hour or so to substitute wing stays from the second flying machine and to replace the nuts he had himself removed. **1919** S. Camm *Aeroplane Construction* xiii. 108 The various wires used in construction may be classified into four distinct types, namely, cable wire, stranded cable..solid wire, stay, the stranding type of cable..are employed for stay wires, the extra flexible cable used for controls, and the swaged rods in plane or streamline form.

3. *stay-wire*, (*b*) a supporting wire on an aircraft.

1919 [see sense 1 b above]. **1969** K. Munson *Pioneer Aircraft 1903–14* p. 9 The supersonic horizontal surfaces, a method of support..formed by stretching cloth over frames of wood and wire, constitute the 'wings' or supporting part of the apparatus. They are connected to each other through hinge-joints by upright standards and lateral stay-wires.

stay, *sb.*² Add: **5. a.** *stay-making* (earlier example).

1843 Dickens *Let.* 2 Nov. (1974) III. 589 Trades... I think of..stay-making [etc.].

b. *stay braid* (example).

1759 *Newport* (Rhode I.) *Mercury* 26 June 4/3 To be sold by Jacob Richardson,..Stay Braid and Cord.

stay, *v.*¹ Add: **4. c.** (Earlier example with *down.*) Also *to stay over* (orig. U.S.): to stop overnight; *to stay with*: to remain in the mind or memory.

1884, etc. [see *over *adv.* 9 b]. **1895** Kipling *Day's Work* (1898) 175 She had 'stayed down there but weathers', as the saying is, because her brother..could not afford the expense of her keep at even a cheap hill-station. **1922** A. Woollcott *Let.* 26 Mar. (1944) 35 want to tell you that seldom has anything I have heard stayed with me like the hymn and half of first prose in the *Spoon River Anthology*. **1973** *Christian Science Monitor* 12 July 19/4 On the way home that stayed with me. **1977** *Jrnl. Home Econ.* .. **1981** E. A. Taylor *Cable Car Murder* (1981) xvii. 130 We had a rather long visit. I stayed over, and she took me to the train the next morning.

6. b. (Earlier example.) Also used outside U.S.

1843 *New Mirror* 23 Sept. 385/2 And now we have put her in black and white, where she will 'stay put'. **1891** 'L. Malet' *Wages of Sin* iv. v. 217 It takes a lot of talent strength to sit, either mentally or physically, really still. Not to fidget. To 'stay put', in short. **1924** J. Buchan *Three Hostages* vii. 100 He's able enough, but he won't stay put, and that makes him pretty well useless. **1948** F. Clune *Roaming round Darling* xx. 159 Here, for the time being, Sturt must 'stay put', while the Poet and I begin rolling down the Darling. **1969** *Globe & Mail* (Toronto) 17 Mar. 3/8 Fire Chief Dawson told him to stay put until the car could be pulled away safely. **1978** R. Busby *Garvey's Code* xii. 159, I keep the gun.. And you stay put.

8. b. (Later examples.) Also *U.S.*

1951 *Amer. Speech* XXVI. 75/1 'Do you stay here?' In common Negro usance stay is used for 'live' but is heard otherwise. **1959** A. Fullerton *Yellow Ford* v. 45 'Would you care to stay round here, man?' I had not caught on at first, to her meaning: the verb 'stay' is used in South Africa when in England we'd say 'live'. **1962** W. Faulkner *Reivers* i. 19 Mr Wimbush stays a solid eight miles from town. **1980** D. Morris *Mrs Gandhi* p. xiii, In March 1977, my wife..and I went to see her in New Delhi, at 1 Safdarjang Road, the house where she had stayed since she had first become Prime Minister in 1966.

c. Similarly, *to be here to stay*.

1936 M. Mitchell *Gone with Wind* xlii. 739 Everyone knew hard times were here to stay. **1960** *Jrnl. post-industrial* v.* ... **1966** *Listener* 5 May 661/3 It's a small thought, perhaps, but I have a..though, when viewed against a much more important fact that Mahler is indubitably here, and here to stay. **1969** ... uncertainties about the future of radio, one thing seems certain: local radio is here to stay, and we shall have more of it. **1973** J. Wainwright *Dig Grave* 83 'I don't go for this union stuff,' he said. 'We'll kill that in other ways..' 'They're here to stay, mate, whether you go for 'em, or not.' **1976** *Guardian* 17 Apr. 11/8 Thatcherism is here to stay.

12. b. For *U.S. colloq.* read *colloq.* (orig. *U.S.*) and add: Also *fig.*, to concentrate on, to apply oneself to, to continue with.

1956 H. Surmelian *Invasion of Privacy* .. I gave up on water. Stay with it. **1961** 'A. A. Fair' *Stop at Red Light* (1962) vii. 108 That adds up, Donald. Stay with it. You're doing fine. **1969** *Guardian* 15 July 2/1 These astronauts..have an amazing capability to stay with their work. **1976** J. Rossi *I know what it's like to Die* xv. 158 'I've got to stay with it,' he thought, 'because I can't just drop it.' **1982** *Times* 6 Feb. 15 (*heading*) Fed stays with its tight money policy.

II. 17. b. *to stay the course*: to hold out to the end of a race. Freq. *fig.*

1889 *Daily Tel.* 21 Nov. 5/7 Doubts are also entertained..concerning her [sc. a horse's] ability to stay the course. **1916** *Times* 8 May 9/1 If we are to 'stay the course' better for us, other sections must be prepared for greater sacrifices. **1925** A. Huxley *Those Barren Leaves* I. viii. 103 'Do you suppose you'd shall be a scholar and a gentleman?'.. 'One will certainly have stopped being a gentleman,' he answered. 'One's begun to stop even now, thank heaven.' 'But the scholar will stay the course?' **1966** *Listener* 10 Mar. 363/1 There was much to be learnt from this programme..about metal fatigue, for instance..or those who could stay the course. **1985** *Forbes Sun* iv. 16/2 When President Reagan exhorted Senators and Congressmen to stay the course, the actual meaning of his words was the opposite of his intended meaning.

31. *stay-away*, (*b*) an act or process of staying away, esp. from work; also as *adj.*; **stay-down** *a.*, of, pertaining to, or designating a strike staged by miners staying down a mine; **stay-in** *a.*, of, pertaining to, or designating a strike in which the strikers remain in their place of work; also *absol.* as *sb.*, (one who participates in) a strike of this kind; **stay-up** *a.*, of stockings: remaining in place without garters or suspenders; also *absol.* as *sb.*

1940 *Sun* (Baltimore) 10 Sept. 7/2 A 'stay-away' strike by hundreds of Allegany county school children continued into its second week today, with parents' support. Students have stayed out of school in a protest against the consolidation program of the Allegany County Board of Education. **1963** *Listener* 28 Feb. 354/1 The calls for general work-stoppages during recent years have been 'stay-aways' directed against political and other restrictions rather than against employers. **1976** *Times* 24 Aug. 2/4 Thousands of people in Soweto heeded a call not to go to work.. Whether the mass stay-away..reflected widespread support for the strike call is unclear. **1948** *Sun* (Baltimore) 11 Feb. 2/1 (*heading*) 'Stay-down' strikes occupy British mine. *Ibid.*, Three hundred miners are staging a 'stay-down' strike in the Wales-wood mine. **1958** *Times* 23 Feb. 5/1 More than 400 delegates from all parts of South Wales conferred at Cardiff on Saturday discussed measures to stop the wave of stay-down strikes. **1980** *Listener* 29 May 686/2 Miners in Hungary were winning themselves better conditions with a new tactic, the stay-down hunger strike. **1935** *Political Q.* May 95 It was the threat of the 'stay-in' strike. **1936** *Times* 29 Apr. 5/7 (*heading*) Lock-out of 'stay-in' strikers. **1937** *Amer. Speech* XII. 319 'The stay-in' type of action takes place during working hours. It is a sit-down, folded arms or crossed arms striker. When it is prolonged beyond that period, it becomes, in addition, a stay-in. **1944** *Times* 12 June 14/1 Some of the stay-ins conceded on the day of the offer he the evening, but the national executive were not satisfied. **1950** Mills & Brown *Wagner Act to Taft-Hartley* viii. 278 In the 1936–1937 'stay-in strike' management was locked out and kept off the job. **1968** *Amer. Speech* XLIII. 63 Drop-outs and stay-ins have not been honored. **1959** *Guardian* 20 Aug. 4 No roll—stay up tops. **1953** *Ibid.* 20 July (E ed.) 3 No supporters are necessary for they have their own stay-up tops! **1969** J. Gardner *Complete State of Death* ix. 174 Her woollen kaftan riding up to display the dark elasticated top of her stay-up stockings. **1973** *Nation* (Barbados) 25 Nov. 3 (Advt.), Nylon Stay-ups 99 c.

stay-at-home, *a.* and *sb.* Add: **A.** *adj.* Also *spec.* avoiding going abroad on military service.

1946 W. S. Churchill *Victory* 78 You hear all this talk by the stay-at-home Left Wing intelligentsia that the soldiers will hold us guilty if we do not have a new world waiting for them on their return.

B. *sb.* **a.** (Earlier example.) Also *spec.* one who avoids going abroad on military service. Also *transf.*

1836 *Dickens Let.* 27 Nov. (1965) I. 200 Mrs. Dickens is a great stay-at-home just now. **1883** A. Pinkerton *Spy of Rebellion* xxv. 499 Extravagant ideas of a struggle which should be 'short, sharp and decisive', were the only ones entertained by the great army of 'stay at homes'. **1918** *Nation* (N.Y.) 7 Feb. 131/1 Students.. accused as stay-at-homes, shirking, and fail to suffer for the Fatherland. **1949** E. Hyams *Grape Vine in England* 143 Not by any means all the root-dwelling insects change into winged insects. The stay-at-homes winter among the roots of the vine. **1951** A. Cooke in *N.Y. Times Mag.* 19 July 6/4 We..sent in 'technicians' and followed them with stay-at-home sacrifice, but burdened their stay-at-homes with no extra taxes.

b. A staying at home, *spec.* a strike.

1959 [see *moon world* s.v. *Moon *sb.* 9]. **1962** *Guardian* 20 Apr. 18/1 They disagreed on whether the stay-at-home should be for a year or two. **1971** *Times* 14 Aug. 1/4' If today's stay-at-home is maintained..the organizers will have demonstrated that they are capable of arranging mass protests among urban blacks.

Staybrite (stēᵊ·brait). [f. Stay *v.*¹ + an arbitrary respelling of Bright *a.*] A proprietary name for a make of stainless steel.

1925 *Metallurgist* I. 155/2 Sir Wm..prepared photograms of this chromium-nickel steel—known as 'Staybrite'—in both the fully quenched and softened condition. **1930** *Engineering* 11 Apr. 486/1 Some trouble had been experienced with Staybrite jackets. **1937** G. Frankau *More of Us* ii. 26 In hall of mouse-marble, guy with Staybrite, stood old Dorkins, butler hugely. **1937** *Trade Marks Jrnl.* 7 July 853 Staybrite.. Stainless steels. Firth-Vickers Stainless Steels Ltd., Sheffield. **1964** Bird & Hallows *Rolls-Royce Motor Car* xi. 244 Staybrite radiator shell, single-point suspension and glass-petrol-intake cap.

Stayman (stēᵊ·mən). The name of Samuel M. Stayman (b. 1909), an American authority on contract bridge, used *attrib.* and *absol.* to designate a convention used in bidding at contract bridge.

1952 S. M. Stayman *Expert Bidding at Contract Bridge* vi. 53 The bidding method that has become known as the 'Stayman' Convention was not named by me. Since I was the first to describe it in print [sc. ..1945], and bridge players ..called it by my name. **1955** *Compl. Stayman System of Contract Bidding* viii. 83 The Stayman system takes a position in the centre of the two popular standards. **1962** *Times* 4 Apr. 6/7 The use of artificial 'asking' bids such as the Neapolitan One Club, the Stayman Two Clubs and the Blackwood conventions. **1972** *Guardian* 26 June 12/4 When asked if I play Stayman, I find it difficult to answer. There are many versions of this convention. **1976** *National Observer* (U.S.) 10 July 18/3 The partnership also was playing 'double barrel' Stayman in which two clubs shows a weak hand and two diamonds forces to game.

stay-put (stēᵊ·pŭt), *a.* and *sb.* *colloq.* Also stay put, stayput. [f. vbl. phr. *to stay put*: see Stay *v.*¹ 6 b.] **A.** *sb.* A refraining from movement or travel. (Stress variable.)

1947 Koestler *Sons of Earth* 194 It is a sort of general stay put. *Ibid.* 204 Crowd of smart civilians queuing up to get petrol for their cars—and actually getting it in spite of stay-put order etc.

b. One who refuses to move, one who stays at home. Chiefly *Austral.*

1967 Partridge *Dict. Slang* Suppl. 1388/2 *Stay-put*, one who holds his ground ('stays put'): coll., esp. Australian: since ca. 1930. **1977** C. McCulloch *Thorn Birds* vi. 120 Things were more amicable between vagabonds and stay-puts.

B. *adj.* Remaining where or as placed; during which one remains in one place. Also, refusing to move, refraining from travel.

1947 *N.Y. Times Mag.* 9 Feb. 9/5 Drive—and..Other important details you'll like: the gay-fun-feature for extra long wear, our new stay-put moccasin foot. **1963** *Guardian* 23 July 4/5 A stay-put holiday at home. **1974** *Daily Tel.* (Colour Suppl.) 13 Dec. 31/1 In the United States, a mobile home can have more than one owner. Some cost almost as much as a stay-put house of comparable size. **1976** *Sears Catal.* Spring/Summer 22 Fashion-back jeans and shorts with soil release.. Bar tacked. Stay-put zipper. Band waist, belt loops. **1973** *Times* 13 Jan. 12/8 A centre for touring as well as for stay-put holidaymakers. **1977** *Bornes Dail.* 7 May 2/1 Thirty four of Miri's stay-put squatter families who have been defying eviction deadlines for two years have now lost their battle for right.

stay-pu·tter. *colloq.* [f. as prec. + -er¹.] One who refrains from moving.

1927 J. Adams *Errors in School* ii. 41 Leaving this suggestion of a restricted meaning of the term *idea* to the tender mercies of the stay-putters. **1971** 'E. Fenwick' *Impeccable People* xix. 105, I don't understand you stay-putters, but you're probably wise people.

stchi. Add: Also shchi (now the usual form), shtchee, shtchi, shtshi, etc. (Earlier and further examples.)

1824 J. D. Cochrane *Narr. Pedestrian Journey through Russia* iv. 105, I never entered a cottage, but shtshee (a cabbage soup), with meat, milk and bread, were placed before me unasked. **1908** *Daily Chron.* 14 June 3/4 The delicious and universal cabbage soup is not in English letters *Shchi*,..but—as nearly as maybe—*Sshchi*. **1904** *Westm. Gaz.* 21 Apr. 7/1 The dish of shtchee—fermented cabbage and meat—with vodka, he greatly enjoyed. *Ibid.* 21 Jan. 16/1 That extraordinary mixture of pickled cabbage, meat broth, and about a hundred other ingredients which goes by the name of *shtshi*. **1941** A. L. Simon *Conc. Encycl. Gastron.* III. 18/2 The most popular form of cabbage soup in France is called *Garbure*; the Russian edition of it is known as *Stschi*. **1958** W. Bickel tr. *Hering's Dict. Classical & Mod. Cookery* 113 *Schtschi*, Russian Cabbage Soup. **1971** *Times* 9 Aug. 5/8 [Soup made from wild sorrel] is a welcome change in the summer and is called *green schchi* (*schchi* being the traditional cabbage soup). **1973** R. Martin *Internat. Dict. Food & Cooking* 259/2 *Shchi*, (*Russian*), green vegetable soup of meat *bouillon* and cabbage. **1977** *N.Y. Rev. Bks.* 14 Apr. 10/4 In exchange for a few dissident intellectuals the Ibanskians import from America tons of *shchi*, the Russian national dish of cabbage soup.

stead, *sb.* **III. 12. f.** For *dial.* read *dial.* and *colloq.* and add later examples. Now usu. considered (also as *'stead*) to represent Instead *phrasal comb.*

1903 K. D. Wiggin *Rebecca of Sunnybrook Farm* xxvi. 279 Rebecca's fifty dollars had to be swallowed up in a mortgage, 'stead of goin' towards school expenses. **1916** G. B. Shaw *Pygmalion* II. 121, I want to be a lady in a flower shop stead of selling at the corner of Tottenham Court Road. **1939** Joyce *Finnegans Wake* 283 They ought to told you every last word first stead of trying every which way to kinder smear it out poison long. **1971** *Black World* Oct. 62/1 The sweet-potato bread was a dollar quarter this time stead of dollar regular. **1978** J. Thomson *Question of Identity* x. 100 He'd've been all right with me... 'Stead of which..he marries her.

steadite (ste·dəit). [f. the name of J. E. *Stead* (1851–1923), English metallurgist + -ite¹.] **1.** *Metallurgy.* A constituent of phosphorus-rich irons and steels which is a eutectic of austenite and iron phosphide (and sometimes also cementite) and contains dissolved phosphorus.

1902 *Jrnl. Iron & Steel Inst.* LXI. 118 Mr. A. Sauveur (Boston)..suggested [in correspondence]..'Steadite', to designate the eutectic alloy of iron and the phosphide Fe_3P formed in iron rich in phosphorus. **1925** *Machinery* XXVI. 501/1 Steadite is distributed throughout the metal and gives the abrasive action previously mentioned. **1964** S. H. Avner *Introd. Physical Metall.* xi. 327 Steadite is relatively brittle, and with high phosphorus content, the steadite areas tend to form a continuous network outlining the primary austenite dendrites. **1975** *Brit. Foundryman* LXVIII. 106/1 Phosphorus..segregates in the last portion to solidify, known as steadite, which is composed of iron, iron carbide and iron phosphide.

2. *Min.* A siliceous variety of apatite, usu. containing iron, found as yellowish needles in basic slag.

1911 V. A. Kroll in *Jrnl. Iron & Steel Inst.* LXXXIV. 130 In the new mineral, which..he [*sc.* the author] distinguishes by the name of Steadite, measurement of the angles reveals the occurrence of the hexagonal system. **1950** *Mineral. Mag.* XXIX. 184 Steadite is frequently a late constituent to crystallize; its crystals are the commonest in the vugs of basic slags. **1951** *Science* 29 June 755/2 Metallographic studies..prove the presence of cohenite, steadite, and schreibersite, as well as troilite. The percentage of nickel is about double that of the meteorite fragments found in the same area.

steady, *sb.* Add: **1. a.** *spec.* in Newfoundland, a part of a river which has little or no perceptible current. (Later examples.)

1842 J. B. Jukes *Excursions in & about Newfoundland* II. 241, I understood from a salmon-fisher, the only person inhabiting the neighbourhood, that a succession of 'steadies', with occasional rapids, may be met with for twelve miles farther. **1907** J. G. Millais *Newfoundland* xi. 206 At noon we entered a beautiful 'steady'. **1969** H. Horwood *Newfoundland* i. 4 The canoe..bounced joyously past the white water into the still and foam-flecked steady below.

b. A regular boyfriend or girlfriend. *colloq.* (orig. *U.S.*).

1897 F. Moss *Amer. Metropolis* III. ix. 172 Her 'steady' is Jim Clarke. **1927** *Vanity Fair* Nov. 132/3 His steady has quit him for another or he is lonesome for her. **1950** 'N. Shute' *Town like Alice* 313, I suppose he's turning into Rose's steady. **1960** Auden *Homage to Clio* 56 You won't find a steady in *that* museum Unless you prefer Tea with a shapeless angel to bedtime With a lovely monster. **1978** *Daily Mirror* 12 Jan. 3/4 Meanwhile Rod was flying off to Rio—with a farewell kiss for his latest 'steady', 23-year-old model Bebe Bluell.

steady, *a.* (and *adv.*) Add: **A.** *adj.* **4. e.** (Earlier example.)

1857 A. Mathews *Tea-Table Talk* II. 343 Ceylon coffee, heretofore steady and pressing for immediate sale, is now inactive.

f. *Cricket.* Of a batsman or his play: consistent, safe, cautious.

1826 S. Maunder in R. Dagley *Death's Doings* 54 A steady Player, careful of his fame, May have a *good long Innings*. **1833** *New Sporting Mag.* V. (Cricketers' Reg.) 13 This style [of bowling] Pilch met by steady play. **1857** T. Hughes *Tom Brown's School Days* II. viii. 398 To the suggestions that Winter is the best bat left, Tom only replies, 'Arthur is the steadiest, and Johnson will make the runs if the wicket is only kept up.' **1890** J. Lillywhite *Cricketers' Annual* 143 W. G. Turnbull: A steady bat, but lacking power; should hit more at loose balls. **1924** H. De Sélincourt *Cricket Match* iv. 104 As a matter of sad fact there was no steady and reliable batsman upon the side. *Ibid.* v. 159 He felt a batsman, pure and simple; and decided that he was..in for a good, steady display of batting.

6. c. Of a boyfriend or girlfriend: regular or constant. *colloq.* (orig. *U.S.*).

1887 *Lantern* (New Orleans) 23 July 2/2, I expect my steady company at the house this evening. **1922** S. Benson *Poor Man* v. 127 She had just mislaid her last steady beau, so she was at the moment a little susceptible. **1932** J. Dos Passos *1919* 43 Della let Joe kiss her when they said good night and he began kinder planning that she'd be his steady girl. **1975** D. Lodge *Changing Places* iii. 126 She's become Charles Boon's steady girl friend. **1977** *Rolling Stone* 5 May 47/4 He has no steady girl.

B. *adv.* **a.** Also *to go steady*, to keep regular company (*with* someone) as a boyfriend or girlfriend. *colloq.* (orig. *U.S.*).

1905 E. Wharton *House of Mirth* II. xiii. 493, I thought we were to be married: he'd gone steady with me six months and given me his mother's wedding ring. **1923** *Saucy Stories* 1 Mar. 78/1 Puzzled, she asked him, Well, wasn't we... Didn't you go with me steady? **1946** *Coast to Coast 1945* 136 Are you going steady with anyone, Billy? **1962** M. Urquhart *Frail on North Circular* xii. 70 Noticing a huge, cheap ring on Joan's finger. 'Where'd you get that?' 'It's a going steady ring.' **1978** F. Weldon *Praxis* xx. 163 I'm going steady with one of the young doctors.

b. Also in other contexts or *transf.* Freq. in colloq. phrases expressing caution, as *steady as she goes, steady on* (*with* something), *steady there*, etc.; *steady the Buffs* (Buff *sb.*² 6), hold on! keep calm! be careful!

1825 H. Wilson *Mem.* II. 162 Here the men, forgetful of the caution..began to draw [their swords]. Steady there!! Never a finger or a high to move. **1853** C. Brontë *Villette* I. ix. 168 She was going to bestow on me a kiss.. but I said, 'Steady! Let us be steady, and know what we are about.' **1888** Kipling *Story of Gadsbys* 6 I'd like to see Mr Khan being rude to that girl! Hullo! Steady the Buffs! **1893** *Illustr. London News* 18 Feb. 222/2 (Advt.), Steady there, Spencer with the milk, Rosebery here has not had a drop yet. **1903** G. B. Shaw *Man & Superman* II. 70 Here! Mister! arf a mo! steady on! **1936** A. Ransome *Pigeon Post* xiii. 140 'Sorry,' sobbed Titty. 'Awfully sorry. I didn't mean to.'..'All right, Titty... All right... Steady on.' **1953** N. Jacob *Morning will Come* xiii. 241 He was growing nervous, and kept saying, 'Steady, Charles, steady the Buffs!' **1959** J. Verney *Friday's Tunnel* i. 12 Here, steady on with the sugar, greedy guts. **1971** *Time* 30 Aug. 4/2 No changes were contemplated in the Administration's approach. 'Steady as she goes was the watchword,' said Shultz. **1972** J. Wainwright *Night is Time to Die* 155 Steady the Buffs! (thought Ripley). He's goading you. **1976** in R. Crossman *Diaries* II. 307 The 'steady as she goes' budget was welcomed by foreign bankers. **1976** *Shooting Times & Country Mag.* 18 Nov. 28/2 Opening gates, and holding them open without a hound rushing into the field, at the words 'steady there!' **1979** A. Williamson *Funeral March for Siegfried* xxxiii. 167 'She had last year been Andersson's mistress—' 'Here, steady on!' cried Von Wolstenholm..purple with indignation.

steady, *v.* Add: **1. c.** (Further example.)

1876 G. M. Hopkins *Wreck of Deutschland* iv, in *Poems* (1967) 52, I steady as a water in a well, to a poise, to a pane.

5. Also with *up*.

1932 E. Waugh *Black Mischief* iii. 104 When you're convinced he's steadied up a bit, let him have chambers of his own in one of the Inns of Court. **1963** *Times* 2 Feb. 5/1 The port's Trawler Officers' Guild asked the owners to co-operate in steadying up the men who go on board from the public houses and take bottles with them on late-night sailing.

steadyish, *a.* (Later example.)

1924 H. De Sélincourt *Cricket Match* iv. 104 Gauvinier tried to arrange for a steadyish man to go in at No. 7.

steady state. [f. Steady *a.* + State *sb.*] **1.** An unvarying condition; a state of equilibrium.

1885 *Electrician* 10 Jan. 180/1 With special arrangements (solenoidal) of impressed force, there is no transmission of energy in the steady state. **1905** *Nature* 27 July 293/2 His [*sc.* Planck's] ensemble of systems has not yet reached a statistical 'steady state'. **1930** Ruark & Urey *Atoms, Molecules & Quanta* ix. 272 These five quantum numbers may be taken as those required to fix the steady states of an atom in a strong magnetic field. **1963** *Wall St. Jrnl.* 19 Aug., It long had been held by many scientists that the universe is in a steady state, that, in its general features, it is the same now as it always has been. **1971** *Nature* 8 Jan. 75/1 If present patterns..continue, the annual death rate from lung cancer will increase from about 15,000 to more than 45,000 in the steady state that will be attained in the 1980s. **1977** *Lancet* 3 Sept. 509/2 Plasma-carbamazepine concentrations were in steady state in all patients.

2. *attrib.* (freq. with hyphen). **a.** *gen.*

1909 *Phil. Mag.* XVII. 251 There is no *a priori* reason why there should not be different 'steady state' formulæ corresponding to different kinds of matter. **1942** *Jrnl. Appl. Physics* XIII. 710/2 The steady state current is then the difference of the expressions for the total current and for the transient current. **1965** H. I. Ansoff *Corporate Strategy* iii. 32 The microeconomic theory of the firm..is basically a steady-state theory concerned with successive equilibrium conditions. **1970** J. Earl *Tuners & Amplifiers* iii. 66 For good quality listening we should not go in for anything less than..8W steady-state power per channel. **1976** *Conservation News* Nov./Dec. 22/1 The author is sceptical that steady-state economics will ever find practical application, since people's desires for consumption show no evidence of having a reasonable upper limit. **1977** A. Hallam *Planet Earth* 296 Hutton and Lyell were postulating a steadystate Earth—ceaseless piecemeal local change was occurring. But the consequence of such change was to maintain from the indefinite past to the indefinite future an overall, constant equilibrium in the terrestrial economy. **1978** P. Marsh et al. *Rules of Disorder* iv. 112 Evidence for the existence of a 'steady-state' system of rules and a sense of social propriety.

b. *Astr.* Used with reference to any cosmological theory which embraces the principle that on a large scale the universe is essentially unchanging in time and space; *spec.* the theory propounded by Bondi, Gold, and Hoyle of an isotropic universe expanding at a constant rate, with matter being continuously created so that the mean density of the universe remains constant.

1948 Bondi & Gold in *Monthly Notices R. Astron. Soc.* CVIII. 252 (*heading*) The steady-state theory of the expanding universe. **1955** *Sci. News* XXXVII. 22 The steady-state theory was designed to overcome a difficulty in the world-models of general relativity which no longer exists. Until a few years ago it was thought that these gave a value for the age of the universe..which was too small to satisfy other astrophysical evidence. **1969** *Times* 23 June 6/7 The steady state theory..says that the universe looks roughly the same from any position and at any time in the past, present or future. **1971** J. Z. Young *Introd. Study Man* xxvi. 363 There are other versions of steady-state cosmologies according to which new galaxies are continuously being formed by newly created matter. **1973** [see *radio astronomy]. **1977** J. Narlikar *Struct. Universe* iv. 132 With the help of the PCP [Perfect Cosmological Principle], Bondi and Gold were able to deduce a number of important properties of the steady-state Universe. As we shall see.., they were able to deduce that such a Universe must continually expand. **1978** Pasachoff & Kutner *University Astron.* xxix. 732 If we have indeed discovered radiation from the big bang itself then clearly the steady state theory is discredited.

Hence **steady-stater,** an advocate of a steady-state theory of the universe.

1966 *Time* 11 Mar. 51 As the galaxies move farther away from each other, steady-staters believe, new galaxies are constantly being formed. **1973** 'D. Halliday' *Dolly & Starry Bird* xiii. 188 Black holes..according to theory are nonluminous stars..concealing whole galaxies..to the satisfaction of all the Steady Staters.

steak. 1. a. Add to def.: or specifying how it should be cooked, as *stewing steak* (meat from a less tender cut: see *stewing *vbl. sb.* b).

2. c. (Earlier and later examples.) Cf. *Hamburger 2.

1884 *Boston Jrnl.* 16 Feb. 2/2 We take a chicken and boil it. When it is cold we cut it up as they do meat to make Hamburg steak. **1951** *Good Housek. Home Encycl.* 502/2 *Hamburg steak*, a fried or baked flat cake of freshly minced seasoned steak, very popular in the United States.

3. *steak dinner, pie* (examples), *sandwich*; *steak hammer, tongs* (earlier example); in names of restaurants or other eating-places serving mainly beefsteak, as *steak bar, house, restaurant*; **steak and kidney,** used *attrib.* to designate a pie or pudding containing a mixture of beefsteak and kidney; also *ellipt.*; **steak au poivre** (*o* pwāvrə), beefsteak flavoured with coarsely crushed peppercorns before cooking; = *pepper steak* s.v. *pepper *sb.* 5; **steak Diane** (‖ dɪˌaˑn), a dish consisting of thin slices of beefsteak fried with seasonings, esp. Worcestershire sauce; **steak knife,** (*a*) a butcher's knife; (*b*) a serrated table knife; **steak tartare,** a dish consisting of raw minced beefsteak mixed with egg and seasonings.

1910 Steak and kidney [see *say *v.*¹ B. 1 a]. **1930** H. Burke *Cookery Bk.* 103 Steak and kidney pudding... Put in the beef and kidney (see Steak and Kidney Pie recipe). **1960** I. Jefferies *Dignity & Purity* iv. 59 Cobb..spirited us off to a nearby pub for steak and kidney pud. **1965** L. Sands *Something to Hide* ix. 154 The sight of the steak-and-kidney glistening succulently between them. **1977** C. McCullough *Thorn Birds* ix. 206 The seven of them sat in the small dining room eating steak-and-kidney pie. **1953** Beard & Watt *Paris Cuisine* 132 Steak au poivre. **1976** 'F. Clifford' *Drummer in Dark* iv. 17 Avocado vinaigrette and steak au poivre... Choosing didn't take long. **1971** *Guardian* 10 June 7/8 Fire damaged a kitchen, restaurant, and steak bar in..Hull yesterday. **1957** *Gourmet Cookbk.* II. 270 (*heading*) Steak Diane. **1974** W. Garner *Big Enough Wreath* viii. 103 The waiter [was] serving his steak Diane. *a* **1964** C. Whitney in D. Macarthur *Reminiscences* viii. 271 We were seated and served a steak dinner. **1979** *Tucson* (Arizona) *Citizen* 20 Sept. 8 c/1 About 2,000 steak dinners will be served. **1934** Webster, Steak hammer. **1974** *Habitat Catal.* 81/3 Wooden steak hammer with a square toothed head knocks coarser cuts of steak into succulent shape. **1762** J. Boswell *Jrnl.* 15 Dec. (1950) 86, I went into the City to Dolly's Steak-house in Paternoster Row and swallowed my dinner by myself. **1954** I. Levin *Kiss before Dying* III. v. 178 They went to a steak house on Fifty-second street. **1977** B. Roueché *Fago* (1978) II. iii. 101 There was a steakhouse restaurant across the street. **1895** *Montgomery Ward Catal.* Spring & Summer 447/2 Steak knives,..12 inch blade... No butcher shop would be without them after a trial. **1951** *Catal. of Exhibits, South Bank Exhib., Festival of Britain* 59/2 Steak knife, hollow-ground blade. **1959** L. Smith *One Hour* xxv. 321 Even now, I see those steak knives cutting through the meat. **1979** N. Hynd *False Flags* v. 43 The barman gave Mason a steak knife. The sandwich arrived. **1723** J. Nott *Cook's & Confectioner's Dict.* sig. Kk2, To make a Stake-Pye. **1791** J. Woodforde *Diary* 8 Aug. (1927) III. 291 We did our best and gave them some Beans and Bacon..Stake Pye and a Codlin Pudding. **1930** H. Burke *Cookery Bk.* 103 Steak Pie. Follow the Steak and Kidney Pie recipe, omitting the kidney. **1981** 'M. Yorke' *Hand of Death* xiv. 121 He went home to the steak pie Nancy had prepared. **1970** J. Updike *Bech: a Book* 199 We..made all those steak restaurants in the East Fifties light up like seraglios under bombardment. **1941** B. Schulberg *What makes Sammy Run?* v. 85 Sammy had his mouth full of..steak sandwich. **1979** J. van de Wetering *Maine Massacre* ix. 123 Would you like a sandwich? A steak sandwich? **1911** A. Filippini *Internat. Cook Bk.* 676 (*heading*) Steaks, Tartare. **1958** *Observer* 26 Jan. 5/6 A steak tartare. **1969** 'M. Underwood' *Shadow Game* ii. 22 Peacock was having Steak Tartare...rather appropriate food for one who dwelt in a ruthless world of sophisticated gangsterism. Raw meat! **1845** E. Acton *Mod. Cookery* (ed. 3) vii. 161 If..it should be necessary, for want of steak-tongs, to use a fork, it should be passed through the outer skin..of the steak.

steal, *sb.*² Add: **1. b.** (Earlier example.)

1872 *Daily Gaz.* (Little Rock, Arkansas) 1 Apr., Of all the swindles and steals that have ever been proposed or carried out in our State, this is the largest and boldest.

c. *colloq.* (orig. *U.S.*). A bargain.

1942 Berrey & Van den Bark *Amer. Thes. Slang* § 546/2 Advantageous purchase; a bargain,..*steal*. **1951** *N.Y. Herald-Tribune* 14 Dec. 6 The asking price is $45,000, but I'm pretty sure you could get it for 43,000, and at that price it's a *steal*. **1960** *News Chron.* 2 May 3/1 At £30,000 it was a steal. I think it's worth £75,000. **1969** C. Drummond *Odds on Death* vi. 142 A car like this..is a steal at three thousand quid. **1979** *Fortune* 15 Jan. 67 A sentimental gesture, but it was a steal—a quarter of a million acres for less than $10 an acre!

3. b. (Earlier and later examples.)

1867 *Chicago Times* 26 July 5/2 Norton made first base, but, on essaying Berthrong's steal, he was similarly ousted. **1908** [see *run down *sb.* 1]. **1949** *Oregonian* (Portland) 10 Aug. III. 4/1 Davis overthrew second in an attempt to nail Hale on a steal. **1976** *National Observer* (U.S.) 12 June 14/3 Don't worry. I give the steal sign, and if you're thrown out, I'll take the blame. **1976** *Washington Post* 19 Apr. D 4/4 Washington's glamour boy also struck out when he labeled the 'Yanks' Roy White 'an excellent steal man, 15 for 16 last year' only to correct himself a minute later by giving White 16 steals in 31 attempts last season.

c. *Basketball.* An act of obtaining possession of the ball from an opponent.

1974 *State* (Columbia S. Carolina) 15 Feb. 3-B/4 Then, on a steal, Iona tied it up 62–62.

steal, *v.*¹ Add: **I. 1. g.** With a person as quasi-obj., in phr. *to steal* (someone) *blind*, to rob or cheat (someone) totally or mercilessly. *colloq.* (orig. *N. Amer.*).

1974 *Times* 28 Feb. 9/5 Mr. Howard Hughes, the eccentric multimillionaire..replied: 'Because he's a no-good, dishonest son of a bitch, and he stole me blind.' **1975** *Citizen* (Ottawa) 29 Oct. 21/2 Trustee Dalton McGuinty..said there was no other way to keep students from 'stealing us blind'. **1977** I. Shaw *Beggarman, Thief* I. ii. 21 We'd've been stolen blind without him. **1978** D. Bagley *Flyaway* xxi. 182 These people are Fulani... We're not staying here—they'd steal us blind.

4. h. *to steal* (*the*) *picture, scene, show*: (*colloq.* (orig. *U.S.*)) in theatrical contexts, to outshine unexpectedly the rest of the cast; also *transf.*, to become or make oneself the centre of attention; *to steal* (one's) *thunder*: see *thunder *sb.* 3 c.

1928 *Amer. Speech* III. 368 If a 'part' actor leaves a better impression on the audience and critics than the 'star',..the 'part' actor or actress 'steals' the picture. **1934** *Everyman* 24 Aug. 201/2 (*caption*) It seems we've stolen the show, Aussie. **1937** H. G. Wells *Brynhild* ix. 143 He appeared in bright new flannels,..the best-looking author in the bunch. He stole the picture. **1941** F. Scott Fitzgerald *Last Tycoon* iii. 37 'Somebody been catching flies on him?' she asked—a phrase for stealing scenes. **1942** E. Waugh *Put out More Flags* iii. 189 They came to the little party.., and stole the scene. **1963** Auden *Dyer's Hand* 185 Short of cutting him [*sc.* Falstaff] out of the play altogether, no producer can prevent him from stealing the show. **1979** *Tucson Mag.* Jan. 55/3 Kate Gardiner could well steal the show in the delectable role of Dorine.

5. g. (Earlier examples.) Also in Basketball and Ice Hockey, and *fig.*

1836 *New Sporting Mag.* Oct. 361 [The batsmen's scores] added to the byes they stole, and the wide balls bowled, sufficed to make a hands of eighty-six runs. **1851** J. Pycroft *Cricket Field* x. 196 A sharp runner..will often try a long-stop's temper by stealing runs. **1862** *Sunday Mercury* (N.Y.) 13 July 6/2 Creighton..made his base by a missed fly-catch of Sawyer's; Brainard and Young getting their runs by stealing in on the pitcher and catcher. **1936** *Philadelphia Rec.* 31 July 15/1 No Landon speech is likely to startle anybody. You know in advance that he will never take a full cut at the ball, try to steal a base or catch a line drive with one hand. **1938** M. Dutton *Hockey* vi. 110 It is hard enough to steal the puck in your own end zone, without trying to regain a lost puck in the other fellow's. **1942** C. Bee *Basketball Library* IV. ii. 7 An attempt to 'steal' the ball from a good dribbler often leaves the defensive player out of position. **1968** *Globe & Mail* (Toronto) 5 Feb. 18/3 Ballantyne was ahead 5–3 going into the sixth end, but Lawrie tied it up in the seventh and stole one in each of eighth and ninth for the victory. **1978** *Boston Globe* 4 Jan. 42/2 Hollins stole the ball with seven seconds to play and scored.

6. c. (Earlier example.)

1743 W. Ellis *Mod. Husbandman* July xvi. 77 One of my Hen Pheasants..got Abroad, and stole her Nest.

stealage, *a.* **a.** (Earlier example.)

1769 J. Wedgwood *Let.* 7 Apr. (1965) 72 On calculating the expence, breakage, and stealage by Sea, and comparing it with Land Carriage, I find there is about half saved.

steam, *sb.* Add: **I. 6. c.** Cheap wine laced with methylated spirits; methylated spirits as an intoxicant. *Austral.* and *N.Z. slang.*

1941 Baker *Dict. Austral. Slang* 71 *Steam*, cheap wine, esp. laced with methylated spirits. **1963** A. Lubbock *Austral. Roundabout* 59 To my regret, I never got a chance to sample either 'plonk', or 'steam'! **1966** J. K. Baxter *Pig Island Lett.* 36 I'd give old Rose the go-by For a bottle of steam tonight.

7. d. Also, *under one's own steam*; *like steam* (Austral.), furiously; *to let off steam*: also *fig.*, to relieve one's pent-up energy by vigorous activity; to give vent to one's feelings, esp. harmlessly; *to run out of steam*: see *run *v.* 66 c.

1863 *Blackw. Mag.* Feb. 249/1 This is a free country, and a few eloquent or blustering Radicals serve to 'let off the steam' of their class. **1869** H. James *Let.* 16 Apr. in J. Strouse *Alice James* (1980) viii. 138, I feel an irresistible need to let off steam periodically & to confide to a sympathetic ear the impressions which the week has generated in my soul. **1905** H. Lawson *Coll. Verse* (1968) II. 4 We was draftin' 'em out for the homeward track and sharin' 'em round like steam. **1912** Conrad in *English Rev.* XI. 311 We are not allowed to bring them in under their own steam. **1916** H. J. Laski in *Holmes–Laski Lett.* (1953) I. 25, I intend to write you a weekly letter to Washington—for I must let off steam somewhere. **1949** J. Symons *Bland Beginnings* 142 'Would you be kind enough to..see Miss Cleverly home.' 'That's not necessary... I can move under my own steam.' **1976** J. I. M. Stewart *Young Pattullo* iii. 72 It's just a dining club letting off steam. **1979** B. Hardy *World owes Me Nothing* 102, I hammered at the door like steam and over he came and opened it.

8. b. Short for *steam radio*, sense 17 below.

1959 C. MacInnes *Absolute Beginners* 112, I heard one of your arias on the steam, last evening. **1960** *Spectator* 15 July 103 John Arlott over on steam is still the best of the commentators. **1973** G. Talbot *Ten Seconds from Now* (1974) v. 83 Frank Gillard..crowned his Corporation career by becoming Managing Director of Radio, our 'Head of Steam'.

12. *steam-bakery, bath* (earlier and later examples), *heater, radiator.*

1794 J. B. S. Morritt *Let.* 24 June (1914) ii. 47 After a violent steam bath, they would run out and roll in the snow. **1879** *Bradstreet's* 22 Nov. 2/1 In close rooms close stoves are better than steam radiators. *a* **1884** Knight *Dict. Mech.* Suppl. 861/1 *Steam Heater*,..a low pressure steam-heating apparatus. **1903** Ade *In Babel* 29 For ten years it had braced itself against the onsweeping rush of big machine-shops and steam-bakeries. **1962** A. Lurie *Love & Friendship* xii. 228 That..asthma kind of like a steam radiator. **1967** G. F. Fiennes *I tried to run Railway* ii. 7 The second [mistake was] to report myself for pulling off the steam-heater pipe. **1977** C. McCullough *Thorn Birds* xi. 251 Hot nights in Gilly were bearable compared to this steam bath.

13. *steam-stack, trap.*

1877 Steam trap [see Trap *sb.*¹ 8 b]. **1935** Joyce *Let.* 28 Aug. (1957) 381, I would also like a pleasure yacht with a steamstack. **1955** *Times* 12 July 1/6 Before ordering any steam trap ask for its expectation of life. It is no use saving on steam equipment to pay it out later servicing traps.

14. *steam dredge, dredger, drill, -dryer, elevator, -mill* (examples), *press* (later examples), *shovel* (hence *-shovelful*), *trowel, -trumpet.*

1815 D. Drake *Nat. & Statist. View Cincinnati* iii. 137 The most capacious..building in this place is the Steam Mill. **1861** *Mitchell's Maritime Reg.* 1651/3 The launch of the Ancona, a very fine steam dredger, of 300 tons, recently took place at Southampton. **1873** H. James *Let.* 25 Apr. (1974) I. v. 373 It was once a goodly old palace and though pitifully inconvenient as a hotel, is charminger to stay in than if it had a steam elevator. **1877** *Encycl. Brit.* VII. 464/1 The construction of large river steam dredges is now carried on by many engineering firms. **1879** R. J. Burdette *Hawk-Eyes* 25 The depot policeman looked in to say to him that if he was tired out, he would send in a section hand or the steam shovel to give him a spell. **1880** *Harper's Mag.* Aug. 344/2 The grist from it [*sc.* the tide mill] is said to be of a better quality than from the steam-mills, as being less heated in the process. **1904** *Jrnl. Inst. Electr. Engin.* XXXIII. 965 Steam-dryers are fitted in the flues of two of the boilers. **1906** T. Roosevelt *Let.* 20 Nov. (1919) 182 There the huge steam-shovels are hard at it; scooping huge masses of rock and gravel and dirt. **1906** W. De Morgan *Joseph Vance* xli. 367 He told how she and he were awakened by the sudden stoppage of the screw, followed by the roar of the steam-trumpet. **1925** L. R. Harris in *Messenger* VII. 387/1 A so-called 'steam-drill'..guaranteed to drill a hole faster than any ten men could drill one in the old way with sledge hammer and steel. **1928** *Observer* 15 Apr. 5/4 The people in the restaurants shovel food into their mouths as the steam-trowel takes up its load of earth. **1937** *Discovery* Dec. 362/2 All advances in technique such as the steam press and the linotype, had been developed by the news-printer and later used by the book-printer. **1966** T. Pynchon *Crying of Lot 49* iii. 65 A wildcat transistor outfit that..was underselling even the Japanese and hauling in loot by the steamshovelful. **1972** J. Mosedale *Football* vii. 95 After a trip through the steam presses, caps and uniforms were either too large or too small. **1978** J. Irving *World according to Garp* iv. 79 His mouth still reminded Garp of a steam shovel's power.

15. *steam barge, bus, ferry* (examples), *ferry-boat, hopper* (Hopper¹ 6), *launch* (earlier example), *lawn-mower, locomotion, locomotive, lorry, omnibus, railway, schooner, yacht* (later examples).

1819 *N.Y. Even. Post* 4 Jan. 2/5 Steam sch[oone]r Ramapo, Reid, New Orleans. **1834** J. B. Purcell *Jrnl.* 21 Mar. in *Catholic Hist. Rev.* (1919) V. 253 Mr. Mtgomery an hour & ½ in crossing the River in Steam-ferry boat. **1842** J. McDonogh *Papers* (1898) 65 The steam ferry which runs from one side of the river to the other lands a short distance below my house. **1866** *Mitchell's Maritime Reg.* 18 Aug. 1033/3 Messrs. C. and W. Earle launched from their yard a steam barge [named *Lion*] the first of its class built in Hull. **1869** *Bradshaw's Railway Man.* XXI. 34 A steam ferry across the river Severn. **1872** F. Trevithick *Life Richard Trevithick* II. xxi. 207 Cast-iron wheels were ordered with a view to steam locomotion in the Cordilleras. *Ibid.* xvii. 26 The high-pressure steam-puffer ..moved..towards the broken mass..and..changing its powers from steam-crane to steam-locomotive, conveyed it to the port. **1877** *Encycl. Brit.* VII. 464/2 The steam hoppers employed to receive and remove the dredgings carry about 500 tons of excavations. **1878** C. Schreiber *Jrnl.* 30 June (1911) II. 155 The Embassy steam launch met us. **1890** G. Meredith *Let.* 14 Apr. (1970) II. 997, I am promised a steam-yacht to take me up at Oban. **1902** H. C. Moore *Omnibuses & Cabs* i. iv. 38 The first real steam omnibuses, the 'Era' and 'Autopsy', were invented by Walter Hancock, of Stratford, and placed on the London roads in 1833. **1915** *Naut. Gaz.* 31 Mar. 4/1 The Panama Canal has brought us the steam schooner and other Pacific curiosities. **1916** *Law Rep. I King's Bench* 148 The defendants, who were brewers, used a Steam lorry weighing five tons, for the purpose of delivering beer from their premises to various public-houses served by them. **1923** *Blackw. Mag.* Nov. 681/1 In the harbour.. there were lying odd craft... The one romance of life for these steam-hoppers..had been quenched. **1928** J. Mason *Before Mast in Sailing Ships* 174 He was picked up by a steam barge which happened to be passing. **1933** V. Sommerfield *London's Buses* 5 (*caption*) Three of Hancock's Steam Buses, 1832 to 1836. **1946** G. Foreman *Last Trek of Indians* 116 A steam ferryboat was in service. **1946** Noble & Junner *Vital to Life of Nation* vi. 88 Sumner..began experimenting in the design of a steam wagon in 1889, a year or two later producing a steam lawn mower. *Ibid.* 96 (*caption*) A London steam omnibus of 1902. **1958** *Listener* 11 Sept. 379/2 He was knocked down and killed by a steam lorry. **1965** D. Arundell *Sadler's Wells* xi. 144 Mrs. Warner and Phelps were shown arriving..in a first-class steam-railway carriage. **1970** F. McKenna *Gloss. Railwaymen's Talk* 1 Most steam locomotive depots in England are embedded in the older parts of our towns and cities. **1971** *Sat. Rev.* (U.S.) 6 Nov. 86/3 The purpose of the California steam-bus project is to demonstrate how effectively city buses can operate at low levels of exhaust emission. **1977** D. Jack *Leyland Bus* i. 11 The back-bone of the business was production of steam lawn-mowers selling with 30-inch roller at £85 each. **1977** H. Fast *Immigrants* i. 72, I got two steam schooners, wooden ships, six hundred tons each. **1980** *Times* 25 June 4/2 The inaugural voyage of the National Trust's restored 1859 steam yacht the Gondola took place on Coniston Water yesterday.

16. a. *steam-bent, -hauled, -heated, -operated, -set.* Also with vbl. sbs., as *steam-bending, cleaning, -heating*; and vbs., as *steam-bend, -clean* (trans.).

a **1884** Knight *Dict. Mech.* Suppl. 861/1 In Campbell & Pryor's method of steam heating for dwellings, the steam boiler and radiators are inclosed in a heating room in the cellar. **1890** *Harper's Mag.* Sept. 576/1 The Kents lived in a steam-heated flat. **1917** S. Graham *Priest of Ideal* iv. 53 This mansion, with its good roof and closed windows and doors, and probably steam-heating to keep out the damp. **1934** *Discovery* Nov. 314/1 In 1934 the German railways made some striking accelerations in the schedules of their steam-hauled expresses. **1936** *Ibid.* Nov. 357/1 Acceleration of steam-operated trains in Great Britain continues. **1946** *Nature* 5 Oct. 474/1 This at once prompted Rudall to examine the effect of 50 per cent urea on steam-set β-keratin. **1949** *Sun* (Baltimore) 14 Dec. 5/2 Already, automobile 'laundries' and firms that steam-clean buildings have been told to cease operation. **1956** *Handbk. Hardwoods* (Forest Prod. Res. Lab.) 2 Classification of timbers according to their steam-bending properties is..based mainly on the minimum bending radius of sound, clear specimens 1 in. thick at a moisture content of about 25 per cent. **1956** *Amer. Speech* XXXI. 86 Advertisement of Adelaide Steam Cleaning Service. **1962** J. T. Marsh *Self-Smoothing Fabrics* ix. 122 It is not necessary to employ a special chamber for curing, but only the usual steam-heated cylinders or a stenter. *Ibid.* ii. 9 The steam-set fabric has a pleasant handle, and good crease recovery. **1966** A. W. Lewis *Gloss. Woodworking Terms* 106 *Oak, Japanese*..easier to work than European oak. Steam-bends excellently. **1973** *Times* 4 Oct. 24/5 The chair is demountable and consists of eight wooden staves..steam-bent into a soft, flowing outline. **1977** *Modern Railways* Dec. 493/3 Steam-hauled excursions would be operated over this short length. **1978** *Detroit Free Press* 2 Apr. 2F/4 The natives claim 'the diggers', which is even more impolite than calling them Aussies, come to this spectacular thermal display to get their suits steam-cleaned for free.

b. Objective, as *steam-raising.*

1923 *Engineering* 26 Jan. 101/2 The boilers, furnaces ..economisers, coal bunkers and other details of the steam-raising equipment are carried by the steel framework of the building. **1979** *Improved Energy Efficiency* (Shell Internat. Petroleum Co.) 5 Substantial savings are possible in steam-raising.

17. steam age, the era when trains were drawn by steam locomotives; also *attrib.* or

as *adj.*, belonging to this era; *fig.* out-of-date; **steam beer,** a Californian effervescent beer; **steam calliope** *U.S.* = *steam-organ*; **steam car** (earlier U.S. and later examples); **steam-carriage** (earlier example); **steam cracking,** the thermal cracking of petroleum using steam as an inert diluent which reduces polymerization and increases the yield of olefins; hence **steam-cracked** *a.*; **steam cracker,** an installation for this process; **steam curing,** the curing or hardening of a material by treatment with steam; hence **steam-cure** *v. trans.*; **steam-cured** *ppl. a.*; **steam distillation** *Physical Chem.*, distillation of a liquid in a current of steam, used esp. to purify at temperatures below their normal boiling points liquids that are not very volatile and are immiscible with water; hence (as a back-formation) **steam-distil** *v. trans.* and *intr.*; **steam-doctor** (earlier example); **steamfitter,** one who installs the pipes of a steam-heating system; a steam-heating engineer; **steam fly,** the small brown cockroach, *Blattella germanica*, commonly found in kitchens and bathrooms; **steam heat:** see sense 12 in Dict.; now *spec.* (the heat produced by) a steam-generating central heating system, used in passenger trains and buildings; **steam iron,** an electric iron containing water which is heated and emitted as steam from its flat surface to assist in the pressing of clothes; † **steam-kettle,** a kettle used in sick-rooms to create a moist warm atmosphere (*obs.*); **steam line,** a line in a phase diagram representing the conditions of temperature and pressure at which water and water vapour are in equilibrium in the absence of ice; **steam organ** (later examples); **steam point,** (*a*) a temperature at which liquid water and water vapour are in equilibrium; *spec.* the boiling point of water under standard atmospheric pressure; (*b*) *N. Amer.*, a metal pipe which is driven into frozen earth and down which steam can be passed in order to thaw the ground for mining; **steam radio,** colloq. name for sound radio, considered outmoded by television; hence, a radio receiver; **steam-raiser,** a person employed in an engine shed to light the fires of locomotives and raise steam; **steam room,** (*b*) *U.S.*, a vapour bath; **steam table** *U.S.*, a table in a cafetaria, etc., slotted to hold containers of cooked food kept hot by steam circulating beneath them; **steam turbine:** see Turbine 1 b.

1941 Auden *New Year Letter* III. 66 The genius of the loud Steam Age. **1961** *Spectator* 4 Aug. 181 The jet-age author gets the same sort of romance out of beaten-up old Dakotas..as steam-age Robert Louis Stevenson did from a schooner. **1978** W. Garner *Möbius Trip* (1979) I. i. 34 Suppose you're..a bullion dealer. You're not happy with your present security. It's a little bit steam age. **1898** *Western Brewer* XXVIII. 278/1 Steam beer.. is bottom fermenting... The steam beer mash is made according to the English downward mashing method. **1941** *American Neptune* Oct. 402 Claus Spreckels is the reputed inventor of the great San Francisco speciality, steam beer. **1959** *San Francisco Chron.* 28 June 1 There won't be a drop of steam beer in Northern California after a few more days. **1974** W. R. Hunt *North of 53°* xv. 102 Many saloons served the 'choicest goods' and steam beer at two bits a glass. **1868** *Daily Territorial Enterprise* (Virginia City, Nevada) 29 Aug. 3/1 Even a steam calliope would not cause our firm nerves to tremble as vigorously as this worst of all combinations of unsweet tones. **1936** J. Dos Passos *Big Money* 164 The clanking roar of the rollercoaster and the steam-calliope. **1976** *St. Louis Post-Dispatch* 16 Sept. 1/2 A steam calliope is the ransom for the return of Nipper. **1833** *Amer. Rail Road Jrnl.* II. 225/2 The Steam Car accomplished the distance. **1903** J. Fox *Little Shepherd of Kingdom Come* v. 65 'Steam cars!' they cried. **1962** E. Lucia *Klondike Kate* 7 They clambered aboard..the steam cars. **1969** *Listener* 3 July 31/3 Mr Donald Healey, developer of Austin-Healey sports cars, is rumoured to be building..a 140 m.p.h. steam car... The current spate of steam-car development projects. **1788** in *Rep. U.S. Comm. Patents 1849* (1850) 581 If any person..shall make..any elevator, hopper-boy, or any steam-carriage..without the consent of the said Oliver Evans. **1962** Murphree & Ciprios in *Mod. Petroleum Technol.* (ed. 3) ix. 318 The octane number of this steam-cracked naphtha ranges from about 80 to 100 research method (unleaded). **1968** *Economist* 2 Nov. 73/1 When finished, the plant will include a new steam cracker, with a production capacity of 340,000 tons a year of ethylene, 200,000 tons of propylene, [etc.]. **1959** *Petroleum Times* 25 Sept. 602/1 No. 3 olefine plant is..based on the steam cracking process. **1962** Murphree & Ciprios in *Mod. Petroleum Technol.* (ed. 3) ix. 318 Although the primary purpose of steam cracking is the production of light hydrocarbons, the process also produces material in the gasoline boiling range. **1977** *Shell in Base Chemicals* (Shell Internat. Petroleum Co.) 4 Benzene, toluene and mixed xylenes coming from oil are extracted in special plants from reformate and pyrolysis gasoline, formed when lower olefins are manufactured by the steam-cracking of liquid feedstocks. **1910** *Cement Era* VIII. 169/1 Blocks of 1 part cement to 8 parts sand and [*sic*] steam cured at 80 pounds pressure showed a crusting strength of 2,100 pounds per square inch. **1962** J. T. Marsh *Self-Smoothing Fabrics* xi. 177 It sometimes happens that, with cotton goods which have been steam-cured, the crease recovery is very slightly below that obtained in an atmosphere free from steam. **1909** *Chem. Abstr.* III. 1210 Steam-cured blocks may be made all winter. **1962** J. T. Marsh *Self-Smoothing Fabrics* xi. 176 It was possible to show a linear relation between the improvement in resistance to abrasion of steam-cured fabrics over dry-cured fabrics and the amount of steam present. **1907** *Engin. News* 5 Sept. 249/1 (*heading*) Effect of steam curing on the crushing strength of concrete. **1921** Hatt & Voss *Concrete Work* I. viii. 181 Steam curing is accomplished in curing tunnels with a roof of such a shape that it will drain the condensed moisture to the sides of the tunnel. **1967** M. Chandler *Ceramics in Mod. World* iv. 128 The whole assembly is..put into a steam-curing cabinet. **1923** W. M. Cumming et al. *Systematic Org. Chem.* ii. 24 When the liquid to be steam-distilled is lighter than water, the small glass tube E is extended to the bottom of the receiver. **1964** *Oceanogr. & Marine Biol.* II. 152 The compound is reduced to trimethylamine with $TiCl_3$, which is then steam-distilled into an excess of standardized acid. **1974** Rosser & Williams *Mod. Org. Chem. for 'A' Level* xiii. 253 If care is not taken to dry organic liquids thoroughly, the water/liquid mixture will often steam-distil over at a temperature lower than the actual boiling point of the pure liquid. **1904** *Analyst* XXIX. 385 (*heading*) Laboratory apparatus for steam distillation. **1954** *Thorpe's Dict. Appl. Chem.* (ed. 4) XI. 85/2 The chief advantage offered by steam distillation is that a substance of fairly low volatility can be separated from non-volatile impurities at a temperature much below its normal boiling-point. **1972** P. R. S. Murray *Princ. Org. Chem.* ix. 58 Steam distillation is most effective when one of the components to be separated..has a high molecular weight. **1830** *Cincinnati Chron.* 6 Feb. 2/3 The Mayor was induced..to issue his warrant for the apprehension of a black man calling himself Caesar Gimsoun, and practising in this city as a *steam doctor*. **1906** *Daily Colonist* (Victoria, B.C.) 12 Jan. 7/7 (Advt.), To plumbers, steam fitters etc. We have just received two carloads of iron pipes in all sizes. **1977** J. Crosby *Company of Friends* xvii. 114 They both laughed...talking about the problems of the trade like steamfitters discussing occupational hazards. **1933** M. Lowry *Ultramarine* iv. 184 'You shuffle them—' '—King of the steamflies, eh—'. **1944** *Jrnl. R. Army Med. Corps* LXXXIII. 188 The steam fly or German cockroach. **1962** *New Scientist* 11 Oct. 75/1 The German cockroach, commonly referred to as 'the steam fly', is dark brown to tan in colour and is also very widely distributed. **1904** *Railway Mag.* XIV. 169/2 Since the general introduction of steam-heat..it appears to be an easy matter for the guard to simply turn a valve to supply sufficient steam to heat the cars comfortably. **1941** J. Masefield *In Mill* 36 The winter steam-heat made it impossible to wear a coat while at work. **1974** *Times* 1 Apr. 14/5 Hugh Lawson, deputy city engineer of Nottingham, gave a speedy talk on his city's steam-heat system. **1951** *Good Housek. Home Encycl.* 154/2 Electronic steam irons are now available, but..are not always so effective as they might appear. **1962** *Which?* Sept. 270/1 If you want to avoid using a damp cloth or damping the clothes, then your choice may well be a steam iron. **1972** *Guardian* 30 Aug. 9/5 The opening for filling steam irons with distilled water is usually mingy, and the thing overflows. **1890** F. Taylor *Man. Pract. Med.* (1891) 356 In the intervals, the laryngitis is to be treated by a moist warm atmosphere (steam-kettle) and mild opiates as in other cases. **1879** Steam line [see *hoar-frost b]. **1937** M. W. Zemansky *Heat & Thermodynamics* xi. 175 In the particular case of water..the vaporization curve is called the steam line. **1910** 'I. Hay' in *Granta* 11 June 12 Even the steam organs seemed to have stopped of their own accord. **1962** L. Davidson *Rose of Tibet* vi. 106 It was as though he had been pushed into a steam organ at a fair. The stupefying blare of sound. **1895** *Jrnl. Chem. Soc.* LXVII. 196 It is absolutely necessary that the surface of the pyrometer should be free from all soluble salts when the steam point is being taken. **1909** *Yukon Territory* (Canada Dept. Interior) iv. 38 A steam point is an iron pipe of about 5½ feet in length,..connected to a boiler supplying the steam... The miner drives the steam point into the ground, where it is left..until a hole is thawed. **1965** *Jrnl. Chem. Physics* XLII. 274/2 In 1954 the size of the Kelvin degree was fixed by assigning the value 273·16°K to the triple point of water, so that the value of the steam point is now subject to experimental determination. **1974** W. R. Hunt *North of 53°* iv. 15 Steam points replaced wood fires for thawing and this greatly speeded the mining work. **1957** V. Gielgud *Brit. Radio Drama 1922–1956* The flight from 'steam-radio' to television has become an admitted rout. **1961** *Radio Times* 6 Apr. 41/4, I am the proud possessor of 'square eyeballs', but still feel that the good old 'steam' radio has a winner in the *Scrapbook* series. **1976** J. Snow *Cricket Rebel* 7 Overseas tours were followed equally avidly on the old 'steam' radio in the lounge. **1925** Paterson & Webster *Man. Locomotive Running Shed Management* viii. 114 The usual method is for the steam raiser to 'line' the grate along the firebox sides and well into the corners with coal, leaving the centre of the grate bare. **1947** H. Webster *Locomotive Running Shed Pract.* 177 Following an interval of 30 minutes or so the fire is inspected by the steam-raiser who breaks it up and adds fresh fuel. **1972** *Village Voice* (N.Y.) 1 June 73/1 A steamroom in which vapors rise and good men fall. **1861** Steam table [see *hot plate* s.v. *hot *a.* 12 c]. **1944** S. Bellow *Dangling Man* 32, I looked around at the steam tables and the posters of foundering ships. **1976** M. Machlin *Pipeline* xiii. 154 Next to the sandwiches was a steam table with several containers of soggy-looking breaded veal cutlets.

steam, *v.* Add: **I. 7. c.** (Earlier examples.)

1842 C. Fox *Jrnl.* 29 May (1882) viii. 156 Steamed away to London Bridge and saw the Maurices. **1849** *Ibid.* 13 June xv. 244 Steamed to Chelsea, and paid Mrs. Carlyle a humane little visit.

9. d. *to steam open* (later examples). Similarly, *to steam* (a postage stamp, label, etc.) *off*.

1920 M. Webb *House in Dormer Forest* xvi. 214 The kettle having complied, she began to steam the letter open. **1944** R. Lehmann *Ballad & Source* I. vi. 49 She used to send us postcards... We steamed the stamps off. **1979** 'J. le Carré' *Smiley's People* (1980) xix. 242, I can still tap your phone, steam open your mail.

II. 11. Colloq. with *up*. **a.** To stir up or rouse (ardour, etc.). *rare.*

1919 F. Hurst *Humoresque* 97 Ed says he'd never get him to steam up his nerve enough to call at a girl's house after her. **1931** *Daily Express* 21 Sept. 19/2 He was trying to steam up interest in the contest.

b. To rouse or excite (a person), esp. to anger; to agitate, upset.

1922 H. C. Witwer in *Collier's* 17 June 22/4 Are you asking me to go with you so's to steam Rags Dempster up? **1964** Wodehouse *Frozen Assets* iii. 61 She's one of those calm, quiet girls you'd think nothing would steam up. *a* **1974** R. Crossman *Diaries* (1977) III. 366 The Department have steamed me up into the idea that I have got a terrible series of difficulties here.

12. With *up*: Agric., to subject (an animal) to steaming up (see *steaming *vbl. sb.* 5).

1947 V. C. Fishwick *Dairy Farming* II. 164 We steam-up our heifers and cows, and feed a balanced milk-production ration during the lactation period. **1959** *Observer* 15 Nov. 3/1 There are no special hazards in artificial twinning provided that the cow is generously fed—steamed up as it is called in farming language—before calving. **1969** N. W. Pirie *Food Resources* iii. 104 The extreme case is the process known as 'steaming up' or 'flushing' ewes before mating. The extra food given.. increases the probability of conception.

steamboat. Add: **a.** (Earlier *attrib.* examples.)

1821 *Deb. Congress U.S.* 28 Dec. (1855) 44 The jurisdiction had only embraced steamboat navigation. **1847** [see *raft-man* s.v. *raft *sb.*¹ 5].

c. *Comb.*, as **steamboat Gothic** *adj. phr.* (*U.S.*), used to designate an ornamented style of architecture typical of houses built by retired steamboat captains in the mid-nineteenth century.

1962 W. Faulkner *Reivers* viii. 166 The big rambling multigalleried multistoried steamboat-gothic hotel where the overalled aficionados..gathered..each February. **1970** K. Platt *Pushbutton Butterfly* (1971) iv. 43 The beautiful old mansions with their bay windows, ornate Steamboat Gothic cornices and mouldings.

steamboating *vbl. sb.*, (*a*) (earlier examples); also **steamboatman** *U.S.*, one who works on a steamboat, esp. a steamboat owner or captain.

1826 Malthus *Diary* 7 July (1966) 263 Dr Brown said that the introduction of Steam boating had quite altered the habits of the people of Glasgow. **1828** Mrs. B. Hall *Let.* 7 June in *Aristocratic Journey* (1931) xxii. 288 Two nights more and we shall have done with it and have no more steamboating in this country. **1875** 'Mark Twain' in *Atlantic Monthly* Jan. 69/1 When I was a boy, there was but one permanent ambition among my comrades... That was, to be a steamboatman. **1910** D. W. Bone *Brassbounder* 251 Sailormen walk fore and aft; steamboat men, athwart. **1929** G. L. Eskew *Pageant of Packets* ii. 101 All the steamboatmen when in New Orleans did their banking at the Banque des Citoyens.

steamed, *ppl. a.* (In Dict., s.v. Steam *v.*) Add: **1.** In the simple senses of the vb. (Examples of *steamed pudding*.)

1884 N. Lake *Menus made Easy* viii. 149 *Pouding soufflé*—a very light steamed pudding. *Ibid.* viii. 150 *Pouding à la Snowdon*—a steamed pudding of suet, breadcrumbs, brown sugar and marmalade. **1945** *ABC of Cookery* (Ministry of Food) xvii. 61 Most modern steamed puddings are made by one or other of the cake-mixing methods already described. **1978** K. Webber *Bk. Winter Cooking* 93 Boiled or steamed puddings can be stodgy.

2. With *up*. **a.** Excited or roused, esp. to anger; agitated, upset. Freq. in phr. *all steamed up*. Rarely without *up*. *colloq.*

1923 H. C. Witwer in *Cosmopolitan* Sept. 72/2, I was a bit steamed up about her making my popular sex ridiculous by going boy-crazy at fifty. **1935** J. Hargan *Gloss. Prison Lang.* 8 *Steamed*, envious, angry. **1936** M. H. Bradley *Five-Minute Girl* v. 79 If she was all steamed up like this over embroidery silks, he thought, what would she be when he told his news? **1943** P. Cheyney *You can always Duck* vi. 101, I reckon he's a bit steamed up over my allure. **1953** K. Amis *Lucky Jim* xiv. 148 People get themselves all steamed up about whether they're in love or not. **1965** M. Bradbury *Stepping Westward* vii. 330 People live naked in this country [*sc.* the USA]. If you get steamed up you let everyone know. **1979** G. F. Newman *List* ii. 22 They sounded pretty steamed with him. **1980** D. Bogarde *Gentle Occupation* ii. 44 The General insists it is sent to all the Brigades. He's getting very steamed up about the bloody little thing.

b. Drunk, intoxicated. Rarely without *up*. *slang.*

1929 M. A. Gill *Underworld Slang* 10/2 *Steamed up*, drunk. **1950** *Landfall* June 126 Little Spike is six foot two and has a reputation for being a hard case when he is steamed-up. **1971** J. Terrell *Bunkhouse Papers* xii. 156 A cowman sat next to the houseman, and he was steamed with liquor so that he slumped a little to one side.

3. With *up*. Of a glass surface, etc.: covered or bedewed with condensed vapour.

1972 G. Lyall *Blame the Dead* xv. 106, I went along to the buffet car..and stared at the steamed-up window.

steam-engine. b. (Later examples.)

1934 *Railway Wonders of World* I. 309/1 There is one factor, however, where the steam engine will probably always score, and that is in its greater length of useful life and lower maintenance costs. **1970** F. McKenna *Gloss. Railwaymen's Talk* p. iii, The crash of the coal hopper.. and the noise of steam engines preparing for their journeys no longer disturb the peaceful inhabitants of Mortimer Terrace and Highgate Road.

steamer. Add: **5. b.** *steamer rug, trunk* (earlier example).

1886 in *New Canaan Hist. Soc. Ann.* (1959) 19/1 A steamer trunk I believe they call them; something to hold necessary articles on their voyage. **1890** S. Hale *Let.* 22 Dec. (1919) 253 It is..so cold..that we are sitting close up to the grate..and all wound about with the heaviest steamer rug! **1977** H. Fast *Immigrants* vi. 365 They were covered by a big steamer rug.

6. c. A steam locomotive engine or train.

1837 W. Tayler *Diary* 22 Sept. (1962) 51 We passed the Southampton rail road and was just in time to see the steamer go past, with about forty cars fastened to it full of gravel.

1961 *Times* 16 Aug. 9/4 This strong feeling, stimulated in this decade by the departure of the 'steamer', has served to produce something of a golden age in railway literature. **1972** *Times Lit. Suppl.* 10 Nov. 1377/3 All over Britain people are banding together to buy, restore and, they hope, run a steamer. **1975** 'J. Lymington' *Spider in Bath* ix. 156, I meant a real locomotive. A steamer. **1981** *Railway Mag.* Mar. 115/3 No. 765 is the first main-line steamer of the decade to be returned to active duty.

7. a. (Earlier example.)

1870 *Daily News* 15 Oct. 7/4 Alarming Fire at the Gaiety Restaurant... The 'call' for engines was rapidly responded to, no fewer than eight steamers being soon present.

8. b. = *long-neck clam* s.v. *long-neck 2 b, freq. eaten as a delicacy. Also *steamer clam*.

1909 *Rep. Mollusk Fisheries Mass.* (Mass. Comm. Fisheries) 179 Small clams, or 'steamers', are shipped in the shell. **1947** P. A. Morris *Field Guide Shells of our Atlantic Coast* 89 *Mya arenaria*... Known by such names as 'long clam', 'soft-shelled clam', 'steamer clam', and 'long-necked clam', it lives in the muds and gravels between the tides. **1960** J. J. Rowlands *Spindrift* 84 The delicious steamer clam of the North Atlantic is becoming scarce. **1977** [see *quahaug, quahog].

† **10.** The name of a back-stroke in swimming (see quots.). *Obs.*

1861 'R. Harrington' *Swimming* 10 The 'steamer'.. consists in striking the water violently with the foot, raising each leg alternately out of the water to do so. **1879** *Boy's Own Ann.* I. 415/3 The Steamer... Lie on the back, point your feet as much as possible, and then strike them alternately into the water, the knees being kept quite stiff.

11. *Rhyming slang.* [Abbrev. of *steam tug* = 'mug'.] = Mug *sb.*⁵ 1; also *spec.* a male homosexual, esp. one who seeks passive partners.

1932 G. S. Moncrieff *Café Bar* vii. 63 The mug became pleasanter... 'I'm a porter, at some service flats in Victoria...What's your friend do?' the steamer asked genially. **1936** J. Curtis *Gilt Kid* xxxvi 258 If you think I'm going to make a steamer of myself and let you hang about half a dozen more charges on me, you're mistaken. **1958** *Times Lit. Suppl.* 2 May 237/4 Terry..spending his time..among the young homosexuals and their 'steamers'. **1968** G. J. Barrett *Guilty, be Damned!* viii. 95 You might get yourself caught. The Police are a lot sharper than steamers give them credit for being. **1978** M. Puzo *Fools Die* iv. 48 The third player at the table was a 'steamer', a bad gambler who chased losing bets.

steamie (stī·mi). *Sc.* Also **steamy.** [f. Steam *sb.* + -ie.] A public wash-house.

1926 *Glasgow Herald* 19 Oct. 8 The perambulator holds much besides..the baby... In the poorer parts of the city the washing is conveyed to and from the 'steamy' thereon. **1935** *Scottish Educ. Jrnl.* 8 Mar. Suppl. p. vi/2 Modern sanitation has..caused the coining of 'steamie' (the public wash-house). **1958** *3rd Statistical Acct. Scotl.* V. III. xvii. 560 Attendances at the 'steamies' in the year were just short of 1,800,000. **1969** *Dumfries & Galloway Standard* 29 Oct. 1 There was still a need for the 'steamie' as launderettes were not as cheap as the council maintained. **1978** *Times* 18 May 24/1 It was the talk of the steamie, so to speak. It was a topic of conversation at the company.

steaming, *vbl. sb.* Add: **4. a.** (Earlier example.)

1836 *Southern Lit. Messenger* II. 696 Steaming from Washington to Baltimore is an improvement upon that route at least.

b. *Comb.*, as **steaming light,** a white light carried on the masthead of a steamship under way at sea by night.

1909 *Man. Seamanship* (Admiralty) II. i. 29 *Navigation lights*,..oil ones..consist of steaming light, in charge of 2nd captain of forecastle, who is responsible for placing it; after steaming light..; starboard bow light..; and port bow light. **1947** *Sea Breezes* IV. 139/2 A steamer appeared, also carrying steaming lights.

5. With *up*. In *Agric.*, the provision of extra food to farm animals as preparation for reproduction.

1943 R. Boutflour in *Agriculture* L. 306 It is now over twenty years since I coined the expression 'steaming up'; the reason for its choice was to imply that a definite preparation was required. **1947** V. C. Fishwick *Dairy Farming* II. 156, I believe in steaming-up and preparation for calving. This is how you get the milk. **1953** K. Russell *Princ. Dairy Farming* xiii. 153 The amount of steaming-up ration to be fed as concentrates is then decided by two factors—the condition of the cow or heifer and her probable milking capacity. **1960** *Farmer & Stockbreeder* 19 Jan. 122/3 Cows require 'steaming up' prior to calving.

steaming, *ppl. a.* Add: **3.** Used as a substitute for a strong expletive: consummate, 'blithering'. *slang.*

1962 *Listener* 13 Dec. 1024/3 A cautionary tale concerning a real steaming nit of a British civilian. **1965** A. Garner *Elidor* xix. 147 Roland! You great steaming chudd! Come back!

steamless (stī·mlės), *a.* [f. Steam *sb.* + -less.] Without steam; that has run out of steam or is not propelled by steam. Also of a railway: not carrying steam-engines.

1920 *Blackw. Mag.* Apr. 573/2 The N.T.O. insisted on placing the skipper of the steamless tug under arrest. **1967** *Gloss. Sanitation Terms* (*B.S.I.*) 49 *Steamless inlet*, a device to reduce the amount of steam produced in filling a bath with hot water. **1970** *Railway Mag.* Oct. 544/2 My old room facing the now steamless railway.

steam-roller. Add: **b.** (Later examples, esp. of such a force used to a political end.)

Russian steam-roller: with reference to Russian military capacity in the war of 1914–18.

1896 Lloyd George *Let.* 6 Aug. (1973) 106 One of them [*sc.* M.P.s] threatened to pass a steamroller over me yesterday... Killed the Military Lands Bill. Just heard from Balfour. That's their steamroller. **1912** *Chicago Tribune* 3 June 2/2 The Roosevelt adherents..expect through publicity to prevent the operation of the steam-roller. **1916** G. B. Shaw in *N.Y. Times* 9 Apr. VI. 1/3 A combination of the British fleet, the French Army, and the Russian steam roller. **1934** J. Hilton *Goodbye, Mr. Chips* xiii. 89 The Battle of the Marne, the Russian steam-roller, Kitchener. **1952** *Sun* (Baltimore) 7 July 2/4 Meanwhile Taft's men proceed on the lines slammed down by Elihu Root's steamroller. **1976** *Listener* 5 Feb. 132/3 The Soviet military doctrine of the so-called 'steamroller approach'—huge numbers of well-disciplined, fit and adequately trained privates, as distinguished from the élitist concept of an all-volunteer army, like Britain's.

steam-roll *v.*: also **stea·mroll**; (*b*) *fig.*, also, to force or drive in a given direction (cf. *steam-roller *v.* 2); **steam-rolling** *vbl. sb.* (*fig.* examples).

1914 *Times* 29 Aug. 6/2 Our task is stonewalling, and that of the Russians is steamrolling. **1915** F. M. Hueffer *Good Soldier* IV. v. 274 So Edward and Nancy found themselves steam-rolled out and Leonora survives. **1955** *Times* 15 Aug. 5/4 The big screen, Vistavision, the Hollywood technique, and all the rest of it will steamroll the lightness and gaiety of the original idea out of all recognition. **1975** *Times* 21 July 1/8 The ruling party..will steamroll the endorsement through. **1976** *Conservation News* Sept./Oct. 22/1 The main TV companies have made *some* attempt to cover the most excessive speculation and steamrolling of community rights.

stea·m-roller, *v.* [f. the sb.] Also as one word. **1.** *trans.* To crush or level with a steam-roller; to force with a steam-roller.

1913 *New Statesman* 26 July 497/2 To attempt to get through his poems in Classical Prosody is like trying to ride a bicycle over miles of newly-stoned road not yet steam-rollered. **1940** V. Brittain *Testament of Friendship* xix. 361 What had happened..to the mortal remnants of those slaughtered thousands?..Had they been ploughed, exploded and steam-rollered into the soil?

2. *fig.* **a.** To crush or break down, as with a steam-roller; to ride roughshod over; to overwhelm or squash. Freq. in *Pol.* contexts.

1912 *Chicago Daily Tribune* 7 June 1/4 They [*sc.* the Taft men] assent..that they will 'steam roller' the Roosevelt contests with a vengeance. **1918** G. B. Shaw in *Daily Chron.* 12 Jan. 5/2 He hammered poor Mr Walsh with trenchant repetitions of his chivalrous Christian phrase, and steamrollered him amid thunderous plaudits. **1921** *Round Table* June 651 His block majority, with which, if necessary, he could steam-roller opposition. **1930** G. B. Shaw *What I really wrote about War* xi. 283 An Ally [*sc.* Russia] on whom we had depended to steamroller our enemies on their eastern front. **1953** *Manch. Guardian Weekly* 3 Dec. 4/1 It would be a tragedy if the personal intimate side of British elections was steam-rollered into a flat monotony. **1982** *N. & Q.* Apr. 174/2 The book seems..to be an example of that kind of academic system-building where the subtleties of a text are 'steamrollered' in the interests of interpretative 'schemes'.

b. To push (a measure or bill) *through* (a legislative assembly, committee, etc.) by forcibly overriding opposition.

1947 A. W. Grantham in *Hong Kong Hansard* 31 July 257 It is too readily assumed that because there is a majority of Officials the slightest wish of Government is 'steam-rollered' through this Council. **1960** *Times* 1 Mar. 12/3 Certainly, each measure is steam-rollered through. **1964** *Daily Tel.* 27 Feb. 1/1 They accused him..of having 'steam-rollered the Bill through the Cabinet'.

c. To force (someone) *into* (a course of action, situation, etc.).

1959 *Economist* 18 Apr. 212/2 If the Government is steamrollered into granting a flat rate increase for all these pensioners right across the board, this will be the third successive general election which has been immediately preceded by such a step. **1959** P. Bull *I know Face* ii. 42 Luckily Robert and I..were not steam-rollered into a phoney romance to appease the fans and newshawks.

3. *intr.* With *adv.* or *prep.* To proceed (esp. to continue speaking), regardless of opposition or interruption.

1969 D. Francis *Enquiry* I. iii. 38 Gowery steamrollered on. 'You found certain objects.' **1970** J. Porter *Rather a Common Sort of Crime* iii. 31 The Hon. Con steam-rollered happily through the interruption. **1977** *Evening Gaz.* (Middlesbrough) 11 Jan. 13/8 Walker steam-rollered in with a 4–3 finish to win 16–15!

steamship. (Earlier attrib. examples.)

1866 'Mark Twain' *Lett. from Hawaii* (1967) 24 The permanent establishment of a San Francisco and Honolulu steamship line. **1884** *List of Subscribers* (London & Globe Telephone Co.) 11 Culliford & Clark Steam Ship Brokers.

steamy, *a.* Add: **3.** *fig.* Salacious; lustful, sexy, 'torrid'. Cf. Hot *a.* 6 c in Dict. and Suppl.

1970 *Daily Tel.* 19 June 7/4 Making Marilyn Roberts semi-nude curiously lessens the eroticism of one originally steamy scene. **1976** M. Machlin *Pipeline* xii. 139 Once he remembered a steamy necking session out in the middle of a field of oats. **1980** R. McInerny *Second Vespers* (1981) xv. 108 It was a moral outlook, one that had never..been disturbed by the steamy fiction that was her steady diet.

steamy, var. *steamie.

stearate. Add: Also, an ester of stearic acid (later examples).

1899 *Jrnl. Chem. Soc.* LXXV. 358 (*table*) [Amylic] *n*-stearate. **1904** *Ibid.* LXXXVI. I. 284 *iso*Amyl stearate is a neutral, white solid melting at 21°. **1950** Kirk & Othmer *Encycl. Chem. Technol.* V. 845 *n*-Butyl stearate.. is a colorless liquid... It is of value in compounding lubricating oils and as a lubricant for the textile and molding trade, in special lacquers, and as a waterproofing agent. **1976** [see *sorbitan].

steato-. Add: **steatoge·nic** *a.* [*-genic], tending to produce steatosis; **steatorrhœa**: delete and see as main entry below.

1956 *Nature* 14 Jan. 75/1 Experiments on phosphorus poisoning and steathogenic [*sic*] (Handler) diets in which thioctic acid normalizes the fat content of the liver. **1980** *European Jrnl. Cell Biol.* XXII. 567 After 7 days, steatosis developed in all animals on steatogenic diet alone.

steatopyga. Add: **steatopygia,** etc.: also *transf.* in *Archæol.* with reference to figurines that display steatopyga; also **steatopy·gial** *a.*, **steatopy·gism** (both *rare*).

1900 *Jrnl. Anthropol. Inst.* XXX. 253 In..the figure from Adalia..the steatopygia is by no means so pronounced, and is further concealed by the rudeness of the workmanship,..and the abbreviated treatment of the lower limbs. **1901** W. Ridgeway *Early Age of Greece* I. i. 66 In the graves female figurines have been found, some steatopygous, one of slighter type and tattooed. **1910** M. C. Harrison tr. *Mosso's Dawn of Mediterranean Civilization* ix. 150 (*caption*) Neolithic steatopygic figure, Knossos. **1912** Wace & Thompson *Prehist. Thessaly* xii. 225 The..statuettes reported to have been found at Sparta..are clearly of the mainland..type. The legs are short and stumpy, the heads are round, and the steatopygy is most marked. **1923** A. Huxley *Antic Hay* ii. 24 Gumbril's Patent Small-Clothes... A comfort to all travellers, civilization's substitute for steatopygism, indispensable to first-nighters. **1932** J. G. D. Clark *Mesolithic Age in Brit.* iii. 48 Resting on the living rock was found a curious object of shale, thought by the Abbé Breuil to partake of the nature of a phallus or a degenerate steatopygic figurine. **1966** J. Mellaart *Chalcolithic & Early Bronze Ages in Near East & Anatolia* ii. 28 The figurines, both male and female.., show naked or near-naked persons with hands at the waist... The women are slender with small breasts, and exaggerated stomach, enormous navel and a strongly marked pubic triangle, but no steatopygy. **1971** *Ann. Brit. School at Athens* LXI. 66 This figurine..is almost flat, and it is significant that.. about half of those from Crete are steatopygous only when viewed from front or rear; from the side they are seen to be almost flat. **1973** F. Garvie tr. *Buchholz's Prehistoric Greece & Cyprus* 99/2 Thera. Crudely formed, ill-proportioned female clay statuette, with steatopygia, moulded breasts and disc eyes. **1978** J. Updike *Coup* (1979) v. 182 The steatopygial silhouettes of naked mortals.

† **stearrhœa** (stīˌărī·ă). *Med. Obs.* [a. G. *stearrhœa* (J. H. L. Kunzmann (or C. W. Hufeland) 1824, in *Jrnl. d. pract. Heilkunde* LIX. 45), f. Gr. στέαρ (see Steato-) + ῥοία flux.] = next.

1842 W. J. E. Wilson *Pract. & Theoret. Treat. Dis. Skin* xiv. 290 (*heading*) Augmentation of secretion. Stearrhoea. **1913** *Q. Jrnl. Med.* VI. 242 The term stearrhoea, or steatorrhoea, has had an unfortunate history. It was originally employed..to designate passage of liquid fat with the stools, but was applied later, by Erasmus Wilson, to the disease of the skin commonly known as seborrhoea.

steatorrhœa (stīˌătorī·ă). *Med.* Also -rrhea.

[Alteration of prec.: see STEATO-.] **1.** The excretion of fat with the stools.

1859 R. G. MAYNE *Expos. Lex. Med. Sci.* (1860) 1202/2 *Steatorrhœa*, term for a fatty dejection. **1903** E. L. OPIE *Dis. Pancreas* xiii. 315 In some cases of pancreatic disease ..fat is discharged with the fæces as an oily, yellow fluid, and the condition may be designated a true steatorrhœa. **1923** A. E. GARROD *Inborn Errors of Metabolism* (ed. 2) ix. 167 Apart from the steatorrhœa there was nothing to suggest disease of the pancreas in either case. **1956** *Nature* 4 Feb. 237/1 These organisms were obtained from a patient with steatorrhœa.., it would appear that fat synthesis of this type [of bacterium] might make a significant contribution to fæcal fat. **1974** R. M. KIRK et al. *Surgery* vi. 100 The stones may occlude the orifice of the pancreatic duct, preventing exocrine..secretions from reaching the bowel, resulting in steatorrhoea.

2. = SEBORRHŒA. Now *rare* or *Obs.*

1878 W. J. E. WILSON *Lect. Dermatol. 1876–8* 259 Steatorrhœa in its commonest form, namely, steatorrhœa simplex or steatorrhœa oleosa..is illustrated by the model No. 571, which affords a striking example of this affection. **1899** *Allbutt's Syst. Med.* VIII. 759 The name seborrhœa (more correctly steatorrhœa) is not a satisfactory one.

Stechkin (ste·tʃkin). [The name of Igor Yakovlevich *Stechkin*, Soviet engineer, its designer.] A Soviet 9mm automatic or semi-automatic machine pistol. Also *attrib.*

1962 *Guns Rev.* May 23/2 The Soviets..in 1961 produce[d] an article ostensibly similar in function if not in cyclic rate, to the egregious Star of 1934. It is known as the Stechkin Machine Pistol or 'APS' in Russian notation. **1974** 'J. GRAHAM' *Bloody Passage* v. 82 A Stechkin..A true machine pistol. Best I've seen since the Mauser. **1979** K. BONFIGLIOLI *After You with Pistol* xv. 109 The owner of the tan-coloured hand was grasping a large, crude Stechkin automatic pistol... The Stechkin is by no means a lady's handbag-gun. **1981** J. TRENHAILE *Kyril* i. 4 'You have the gun?' In answer Yevchenko pressed a Stechkin into Stanov's hand.

Stedman (ste·dmæn). The name of Fabian *Stedman* (*fl.* 1670), used *attrib.* and in the possessive to designate a method of change-ringing devised by him. Also *absol.*

1731 *Norwich Gaz.* 11 Sept. 4/1 That most noted and harmonious Peal on 7 Bells called Stedman's Triples. **1813** W. SHIPWAY *Campanologia* I. 98 Let a ringer choose what method he will, it must still be on Stedman's principle. **1814** *Ibid.* II. 187 Doubles, or, as commonly called, a Stedman Grandsire, is completed by two singles. **1903** C. D. P. DAVIES *Stedman* 1 Throughout the whole province of Change Ringing, there is no more delightful method than Stedman. **1931** E. MORRIS *Hist. & Art Change Ringing* 78 The Norwich ringers performed a 'touch' of Stedman Cinques. **1957** *Encycl. Brit.* III. 375/2 Stedman's principle, which is *sui generis*, consists in the three front bells ringing their six possible changes, while the remaining pair or pairs of bells dodge. **1975** *Islander* (Victoria, B.C.) 16 Mar. 13/1 They rang a quarter peal of Stedman's triples for Canada's Centennial year.

‖ **steekgras** (stiə·kχras). *S. Afr.* Also † **stick-grass; steek-grass.** [Afrikaans, f. *steek* to prick + *gras* GRASS *sb.*[1]] Any of several grasses of the genus *Aristida* or *Andropogon*, having spiky awns which damage the fleeces of sheep.

1838 J. ALEXANDER *Exped. Discovery Interior Africa* I. ix. 237 The Boschmans have a peculiar mode of fishing in the 'Oup river; they make conical baskets of stick grass, which is as thick and hard as quills. **1893** W. SPILHAUS in J. Noble *Illustr. Official Handbk. Cape & S. Afr.* (rev. ed.) xiv. 314 A year or two ago the Colony was troubled with a particularly obnoxious seed, that of the 'steekgrass' (*Andropogon contortus*, Ness, and *Aristida congesta*, R. and T.). **1896** R. WALLACE *Farming Industries Cape Colony* v. 99 'Steekgrass' is the colonial name applied to a number of species..having long sharp awns attached to their seeds. **1913** C. PETTMAN *Africanderisms* 474 *Andropogon contortus*..is also a steekgras, but it is not the common one. **1954** D. D'EWES *Mydorp* vii. 57 The seeds of the *steekgras* stuck in our stockings and bored relentlessly through to prick the skin. **1975** *Stand. Encycl. S. Afr.* XI. 382/1 Stick-grass or steekgras (*Aristida congesta*) is a biennial pioneer grass.

steel, *sb.*[1] Add: **I. 1. a.** (Later examples.)

With a few exceptions the term is now usu. restricted to iron alloys containing not more than 1·7% carbon.

1866 G. EDE *Managem. Steel* (ed. 4) ii. 15 Steel is a compound of iron and carbon... The carbon rarely exceeds two per cent. **1895** E. L. RHEAD *Metallurgy* xi. 137 Steel proper contains from 0·5 to 1·5 or 1·7 per cent. of carbon. **1946** *Thorpe's Dict. Appl. Chem.* (ed. 4) VII. 47/1 Steel may be roughly defined as an alloy of iron and carbon containing up to 1·7% carbon, all of the carbon being in the combined condition. A second definition, distinguishing it from cast or wrought iron, is that it has been produced in the molten condition, and a third states that steel can be hardened by quenching from a suitably high temperature. There are..certain exceptions to all these definitions. **1967** A. H. COTTRELL *Introd. Metall.* xi. 135 At present, about 80 per cent of steel in Britain is made by the open hearth process. **1976** *Sci. Amer.* July 68/2 For the iron to be made into steel (defined as iron with a carefully controlled carbon content of 1·7 percent or less) the sulfur, the silicon and the excess carbon must be removed. **1983** *Steel Times* Aug. 424/1 Even in the mildest of mild steels, with a carbon content of not more than 0·2% carbon, some other elements are present.

c. The name of a cold shade of grey resembling the colour of steel; steel-grey. Also as *adj.*

1851 E. RUSKIN *Let.* 28 Dec. in M. Lutyens *Effie in Venice* (1965) II. 236 Falkenhayn gave..to Jane a steel glacé silk dress. **1881** [see *BURGUNDY 2 d]. **1895** *Montgomery Ward Catal.* Spring & Summer 9/2 Chambray mixtures in steel or blue with narrow white stripes. **1914** [see *BEAVER[1] 2 c]. **1925** in M. & N. Ward *Home in Twenties & Thirties* (1978) 39 Maids' morning dresses of strong washing gingham..in blue, grey, butcher, or steel.

II. 11. a. *steel-mesh*.

1944 M. LASKI *Love on Supertax* xi. 103 A tall steel-mesh gate. **1976** J. WHEELER-BENNETT *Friends, Enemies & Sovereigns* v. 156 King Peter attributed his father's, King Alexander's, death to the fact that..he had not worn his steel-mesh bullet-proof shirt.

13. *steel man, plant, town.*

1921 *Daily Colonist* (Victoria, B.C.) 1 Oct. 9/3 Mr. Fraser outlined the benefits that would come..from the erection of a steel plant here. **1922** L. MUMFORD in H. E. Stearns *Civilization in U.S.* 10 The steel towns of the Ohio [River]. **1961** *Universe* 27 Jan. 2/4 Steelmen There. Pope John on Monday received members of the council of the European coal and steel authority. **1976** *National Observer* (U.S.) 24 Jan. 1/1 East Chicago, Ind., a smoky Lake Michigan steel town that isn't exactly famous for its esthetic splendor even when the sun shines. **1977** *Times* 19 Dec. 13/3 Sound arguments have been put forward by many respected steelmen for moving away from the large integrated coastal works. **1979** *Steel Times Internat.* Sept. 91/2 The building of a new steelplant.

14. (Earlier example.)

1880 'MARK TWAIN' in *Mark Twain Let. Writer* (1932) iii. 48 The best picture I have had yet is the steel frontispiece to my new book.

15. a. *steel-bright* (later example), *-hard, -sharp, -straight, -strong, -thin*; *steel-blue* (later examples), *-grey* (later examples).

1916 A. HUXLEY *Burning Wheel* 8 The adamant core and the steel-hard chain. **1920** E. SITWELL *Wooden Pegasus* 32 Dusty voice that throbs with heat, Hoping with its steel-thin beat To put stitches in my mind. **1921** J. BUCHAN *Path of King* i. 9 The world put on a new dress, all steel-blue and misty green... Spring had fairly come. **1923** D. H. LAWRENCE *Birds, Beasts & Flowers* 29 Sit beside the steel-straight arms of your fair women. *Ibid.* 177 Steered and propelled by that steel-strong snake of a tail. **1944** BLUNDEN *Shells by Stream* 4 Steel-sharp might Which blows the babe and nurse to atoms in the night. **1944** A. L. ROWSE *English Spirit* xiii. 105 Narrow temples and steel-grey eyes. **1954** L. MACNEICE *Autumn Sequel* 95 She ascends where steelbright rays impinge. **1976** 'Z. STONE' *Modigliani Scandal* II. i. 69 A steel-blue Mercedes coupé. **1977** A. GIDDENS *Stud. in Social & Polit. Theory* 23 His sombre characterization of the 'steel-hard cage' of the modern social order. **1978** 'M. M. KAYE' *Far Pavilions* iv. 73 The steel-grey curtain of the rain.

16. *steel-erector; steel-piercing, -rolling, -using* (examples).

1932 AUDEN *Orators* III. 102 Our steel-piercing bullet, our burglar-proof safe. **1959** *Daily Tel.* 12 Dec. 1/2 Loss of work because of steel shortages in car and other steel-using industries. **1960** *Times* 22 Mar. 12/1 The site of many industries and the country's largest steel-rolling mill. **1974** 'J. ROSS' *Burning of Billy Toober* xi. 100 Almost permanently unemployed but registered as a steel-erector. **1977** *Whitaker's Almanack 1978* 757/1 Some of the country's [*sc.* Nigeria's] more important industrial installations include a steel-rolling mill. **1977** *Time* 19 Sept. 48/3 The relatively brisk pace of the economy is boosting demand in many steel-using industries.

17. *steel-born, -clad* (later example); *steel-barred, -bosomed, -grated, -lined, -nerved, -rimmed, -shafted, -studded, -tipped.*

1909 *Westm. Gaz.* 11 Nov. 5/2 A new steel-studded tyre. **1924** W. J. LOCKE *Coming of Amos* xvi. 211 What kind of steel-nerved wisp of a woman are you? **1926** 'C. BARRY' *Detective's Holiday* vii. 66 A pair of steel-rimmed spectacles. **1929** L. MACNEICE *Blind Fireworks* 8 The steel-bosomed siren calling bitterly. **1930** BLUNDEN *Poems* 139 But steel-born bees, birds, beams invade. **1935** KIPLING in *Times* 17 July 19/4 In the steel-grated prisons where I cast him. **1947** DYLAN THOMAS *Let.* 12 Apr. (1966) 302 In steel-barred rooms, where Mussolini personally had.. interrogated. **1950** *Times* 22 May 4/3 Four golfers and a caddy..were carrying steel-shafted clubs. **1954** L. MACNEICE *Autumn Sequel* 94 The steel-clad troops begin arriving from the rear to rally or harry their humble fellows. **1972** P. BUCKLAND *Irish Unionism* I. viii. 215 Armed raiders who had removed the steel-lined shutters from the windows. **1973** M. RUSSELL *Double Hit* xx. 149, I just don't believe that an accidental swipe with a squash racket, even a steel-shafted one, would have killed a man. **1974** J. AIKEN *Midnight is a Place* x. 290 The duels..which the men..fought, using no weapons but the steel-tipped clogs on their feet. **1978** R. LUDLUM *Holcroft Covenant* xv. 174 The face was strong, the eyes behind the steel-rimmed spectacles alert.

18. steel band, (*a*) *Mus.*, a band composed of musicians who play (chiefly calypso-style) music on steel drums; so **steel bandsman,** a musician in a steel band; (*b*) *Austral.* [BAND *sb.*[2] 12], 'hard thin stratum of ferruginous and siliceous material lying below the sandstone roof and above the opal dirt' (J. S. Gunn *Opal Terminol.* (1971)); **steel bender** (see quot. 1921); **steel-bow(ed)** *a. U.S.*, (of spectacles) having steel frames; **steel driver** *U.S.*, one who makes holes for explosive charges, using a steel stake and a sledgehammer; **steel drum** *Mus.*, a percussion instrument originating in the West Indies, made out of an oil drum with one end beaten down and divided into grooved sections to give different notes; hence **steel drummer; steel-facing,** the process of covering

an engraved metal plate with a film of steel to increase its durability; hence **steel-faced** *a.*; (as back-formation) **steel-face** *v. trans.*; **steel fixer,** a skilled steel worker in the construction industries; **steel frame,** a framework, esp. of a building, made of steel; also *fig.*; freq. *attrib.*; hence **steel-framed** *a.*; also **steel framework; steel guitar** = *Hawaiian guitar* s.v. *HAWAIIAN *a.* and *sb.* II.; **steel-hardened** (later example); **steel orchestra** = *steel band* above; **steel pan** = *PAN *sb.*[1] 1 f; **steel tape,** (*a*) a measuring tape made of steel; † (*b*) tape made of steel for use as a recording medium (*obs.*); **steel trap:** also *fig.* (chiefly in *attrib.* use) and in U.S. phr. *smart as a steel trap* and varr.; **steel wool,** fine strands of steel matted together, used as an abrasive, esp. for scouring.

1949 *Caribbean Quarterly* I. i. 30 The audience was introduced to..Trinidad's own steel band. **1950** *Bull. Austral. Bur. Mineral Resources* No. 17. 27 The first or upper level is indicated by the presence of a very thin and hard band of siliceous sandstone known as the 'Steel Band'. **1960** *Times* 17 Sept. 7/6 The steelband competition of the Trinidad music festival. **1967** *Sunday Mail Mag.* (Brisbane) 8 Jan. 6/7 Then comes eight to twelve feet of quartzite..and often after that, a layer of hard siliceous sandstone known as the 'steelband'. **1974** E. AMBLER *Dr. Frigo* I. 15 He had collapsed while listening to the steel band. **1948** *Trinidad Guardian* 16 June 5/6 (*heading*) Judge advises steel bandsman to mend ways. **1967** *Listener* 31 Aug. 277/2 The steel bandsmen can play anything well: without a conductor and from memory. **1921** *Dict. Occupational Terms* (1927) § 279 *Steel bender*,.. bends steel rods and girders in hand operated or power press, into required shape, to form framework for concrete. **1939** M. SPRING RICE *Working-Class Wives* iii. 53 An unemployed steel-bender in Newcastle. **1963** *Times* 10 June 8/1 The accent on industrialized building and the increasing development over recent years in concreting have created a great new demand for new skilled labour such as scaffolders, concreters, steelbenders and fixers. **1834** in *Proc. Mass. Hist. Soc.* (1924) LVII. 258 Appears to be a pleasant fellow, with frightful whiskers and steel bow spectacles. **1932** Steel-bowed [see *notion-peddler* s.v. *NOTION 9 c]. **1950** W. FAULKNER *Lo* in *Coll. Stories* III. 390 From the pocket of his dressing gown he took a pair of steel-bowed spectacles. **1916** in *Jrnl. Amer. Folk-Lore* (1919) XXXII. 505 He [John Henry] was a steel driver and was famous in the beginning of the building of the C & O Railroad. **1973** A. DUNDES *Mother Wit* 586 The story of John Henry is powerful whether there was an actual steel-driver named John Henry or..not. **1952** *Holiday* Feb. 94/2 Rainbow-uniformed dandies parade and compete in making music on tuned steel drums from oilcans. **1971** *West Indian World* 12 Nov. 7/1 Trinidad.. known..for its calypso singers and steel-drum bands. **1978** *New York* 3 Apr. 31 (Advt.), A cool drink slakes your thirst, steel drums stir your blood. **1960** *Times* 17 Sept. 7/6 This influenced the steeldrummers to discard their tubes and bars and to use tops of oil-drums hung from the neck. **1975** R. L. SIMON *Wild Turkey* (1976) x. 60 The reggae band..had a steel drummer who could go day and night and enough dope to turn on a rock festival. **1884** J. S. HODSON *Guide to Art Illustration* II. iii. 213 The proper thickness of copper having been deposited in the mould, the shell is filed or ground flat on the back, and the face coated with a deposit of iron,—a process commonly called steel facing. **1897** SINGER & STRANG *Etching, Engraving* iii. 61 Steel-faced it may be printed over and over again.., for as soon as the steel face should wear off, the plate can be again immersed in the electrotyper's bath [etc.]. **1937** *Discovery* Mar. 76/2 The burr [in drypoint] being raised, will quickly wear, owing to the rubbing and the pressure it receives in printing. To overcome this a fine film of iron is deposited by an electric process on the plate. This is called steel-facing. An unfaced drypoint will yield only four or five first-class impressions before the burr starts to wear. A steel-faced plate will give as many as fifty. **1961** WEBSTER, Steel-face, v. **1965** ZIGROSSER & GAEHDE *Guide Coll. Orig. Prints* iii. 26 An invention of the mid-nineteenth century..steel-facing... The reason steel-faced prints have acquired a bad name is that they have often been printed in a slipshod manner. *Ibid.* v. 86 It used to be standard practice among professional etchers to steel-face the copperplate. **1936** *Record* Apr. 219/3 There is an awakening of interest amongst another section of building trade workers, namely, the steel benders and fixers. **1949** *Transport & General Workers' Record* June 26/3 (*heading*) Steel fixers and tubular scaffolders. **1974** Steel fixer [see *PRECASTING *vbl. sb.*]. **1898** *Engineering* 8 July 39/3 An architect is made responsible for the general arrangement of the building..while the steel frame or skeleton is the work of a skilful engineer experienced in such matters. **1906** G. A. T. MIDDLETON *Mod. Buildings* IV. xiv. 134/2 Probably the most thorough example of steel-frame construction yet erected in England is that of the Ritz Hotel, Piccadilly. **1922** LLOYD GEORGE in *Hansard Commons* 2 Aug. 1513, I can see no period when they can dispense with the guidance and the assistance of this small nucleus of the British Civil Service, of British officials in India—this 1,200 in a population of 315,000,000... They are the steel frame of the whole structure. **1928** M. MUGGERIDGE in *Young Men of India* XL. 624 There is that amount of truth in the contention of the Die-hard as against that of the sentimental liberal—it must be a steel frame or nothing. **1948** O. BONDY in E. de Maré *New Ways of Building* 70 It was not until the 1890's that the first complete steel-frame buildings were erected in the U.S.A. **1980** J. BOYD-CARPENTER *Way of Life* v. 59 To this day they [*sc.* the *Carabinieri*] are, I believe, the steel frame of the distracted Italian Republic. **1906** G. A. T. MIDDLETON *Mod. Buildings* IV. xiv. 134/2 A steel framework may often be used with considerable economy, as is evidenced by the number of steel-framed structures that are now springing up. **1974** D. SEAMAN *Bomb that could*

Lip-Read iii. 23 His steel-framed grey suitcase. **1906** Steel framework [see *steel-framed* adj. above]. **1940** *Engineering* 1 Nov. 343/3 The steel framework..embedded in the concrete. **1925** *Glasgow Herald* 19 Mar. 8/7 Those two seductive Hawaiian instruments, the ukulele and the steel guitar. **1974** V. GIELGUD *In such a Night* ix. 90 The nerve-battering provided by invisible steel guitars. **1980** *Guardian Weekly* 13 July 1/2 It has all the grisly mod military cons: an isolated water supply, a purified air system, steel-hardened concrete. **1952** S. SELVON *Brighter Sun* xii. 233 Crowds jumped up to the music of steel orchestras. **1971** *News-Advocate* (Barbados) 20 Mar. 7/1 (Advt.), Dancing to the rhythmic beats of the..Elk Owls steel orchestra. **1973** *Nation* (Barbados) 23 Dec. 8 Trinidad's famed steel pans [will] be produced in masse in England. **1983** *Times* 7 Jan. 2/1 [He] was employed for several years by the local education authority to teach steel pan playing in schools. **1900** Steel tape [see TAPE *sb.*[1] 2 a]. **1901** *Electrician* 26 Apr. 7/2 The next [electromagnet] is connected to a microphone circuit to convey the record to the steel tape. **1949** S. J. BEGUN *Magnetic Recording* i. 10 A recording made on steel tape..has been played more than 100,000 times with no measureable deterioration after a slight initial falling off in output level. **1977** J. F. FIXX *Compl. Bk. Running* vii. 90 Measure off a half-mile with a steel tape. **1872** MRS. STOWE *Oldtown Fireside Stories* 57 She was a little thin woman, but tough as Inger rubber, and smart as a steel trap. **1899** A. M. BINSTEAD *Gal's Gossip* 127 He posted sentinel, bright and ready as a new steel-trap. **1921** D. H. LAWRENCE *Tortoises* 32 Little old man, Scuffling beside her..Parting his steel-trap face, so suddenly and seizing her scaly ankle. **1937** E. S. GARDNER *Case of Dangerous Dowager* i. 8 You're going up against a crook who is smart as a steel trap. **1972** *Publishers' Weekly* 17 Apr. 19/1 He's rather amused by what he calls his steeltrap memory. 'I have a tight grip on things in inverse proportion to their importance.' **1896** *Iron Age* 9 Apr. 871/2 The interesting product 'steel wool' is intended for use in all cases where sandpaper, emery paper, pumice stone and materials of a kindred nature are employed. **1947** J. C. RICH *Materials & Methods of Sculpture* vi. 169 'Fire-skin' may be removed by rubbing the work with steel wool, which is available in several grades, varying from very coarse to very fine. **1958** *Listener* 16 Oct. 627/1 Scour round the inside with a steel wool soap-pad. **1977** C. MCCULLOUGH *Thorn Birds* xvii. 444 His thick mane of hair was exactly the color of steel wool.

steel, *v.* **5.** (Later examples.)

1977 *Sci. Amer.* May 61/3 Iron that has been 'steeled' with that much carbon will not deform under stresses of less than 140,000 p.s.i. *Ibid.* Oct. 127/1 It seems evident that by the beginning of the 10th century B.C. blacksmiths were intentionally steeling iron.

steel-head, *sb.* Add: Also **steelhead 1.** (Later example.) Also *attrib.*, as *steelhead trout.*

1911 *Rep. U.S. Comm. Fisheries 1908* 317/2 *Salmo* is represented by..the steelhead trout. **1946** [see *CUT-THROAT 5 b]. **1968** *Times* 22 Oct. 3/3 Another species of trout, the steelhead, has been found to suffer from a vascular disease. **1976** *Vernon* (B.C.) *Daily News* 21 June 20A/5 Sea-going Rainbow are known as *steelhead* trout.

2. (Examples.)

1888 G. TRUMBULL *Names & Portraits of Birds* 112 William Wagner, a well known Washington gunner, tells of hearing it called Water-Partridge and Steel-head. **1917** T. G. PEARSON *Birds Amer.* I. 153/1 They are extremely tough, hardy little birds and gunners know them by such names as Tough-head, Hard-head, Steel-head, etc.

steelheader (stī·lhedəɹ). *N. Amer.* [f. STEEL-HEAD *sb.* + -ER[1].] One who fishes for steelhead trout. Hence **stee·lheading** *vbl. sb.*

1948 *Game Trails in Canada* Feb. 25 The average steelheader that lands one in three seems quite content. **1954** *Daily Progress* (Charlottesville, Va.) 5 Nov. 13/1 They had gone out steelheading that morning. **1964** *Vancouver Province* 14 Feb. 18/1 The Gold River has blanked many a good steelheader. **1971** *Islander* (Victoria, B.C.) 24 Jan. 3/1 She had learned to avoid the cardinal sins of steelheading. **1972** *Daily Colonist* (Victoria, B.C.) 1 Jan. 21/3 April is the awakening month, when..steelheaders shift to trout fishing.

steel he·lmet. [f. STEEL *sb.*[1] + HELMET *sb.*]

1. A helmet made of steel (or other metal), worn as a form of protection in conditions of war.

1916 R. ASQUITH *Let.* 23 May (1980) IV. 263 One fearful addition to the honours of War since I have been away is the steel helmet which we all have to wear now, when in the shell area. **1940** *War Illustr.* 5 Jan. 563/1 It was not until February 1916 that the first steel helmets were issued to the British troops in the last war, but in this war they are an essential part not only of the soldiers' equipment but of that of the police, the A.F.S. and A.R.P. workers. **1978** A. PRICE *'44 Vintage* v. 52 Knitted cap-comforters instead of berets or steel helmets.

2. (With capital initials.) [tr. G. *STAHLHELM.] An organization, founded in 1918 by F. Seldte, of German ex-service men (and others, from 1924) drawn mainly from the Nationalist Party and having a strong conservative bias; a member of this.

The organization was dissolved in 1935, and refounded under the Federal Republic in 1951.

1925 *Spectator* 28 Mar. 487/1 How could he really fail to stand for monarchy when he is championed by such societies as the 'Steel Helmet League' and the 'Front Fighters'? **1926** *Times* 25 May 14/3 The Nationalist 'Steel Helmet' organization. *Ibid.*, Most of the 'Steel Helmets' wore iron crosses. **1932** H. R. KNICKERBOCKER *Germany —Fascist or Soviet?* 137 The four militant organizations

are the Republican Reichsbanner, the National Socialist Storm Troops, the Conservative Steel Helmets, and the Communist Red Front. **1933** E. A. MOWRER *Germany puts Clock Back* 94 The Steel Helmet, or Confederation of Front-line Soldiers, the most respectable of the private armies, was founded on Christmas Day, 1918. **1983** C. MCCARRY *Last Supper* 32 A Steel Helmet, wearing two Iron Crosses on his civilian jacket..seized the weedy young man.

Hence **steel-he·lmeted** *a.*, wearing a steel helmet.

1926 *British Worker* 10 May 1/3 Thus all the display of steel-helmeted troops, all the tearing about of motor-cars filled with special constables..have failed of their object. **1977** N. SAHGAL *Situation in New Delhi* xvi. 159 One of the students informed him it was the police or perhaps the army—many were steel-helmeted and the boy could not tell.

steeling, *vbl. sb.* **3.** (Later example.)

1977 *Sci. Amer.* Oct. 125/1 If bloomery iron is treated in a certain way, it can be transformed into an alloy that is for most purposes far superior to bronze. That treatment is steeling, and its initial discovery was probably accidental.

steel pen. 1. (Later examples.)

1892 'Mark Twain' *Amer. Claimant* vii. 71 Steel pens on his table with the ink-bottle. **1922** JOYCE *Ulysses* 8 He fears the lancet of my art as I fear that of his. The cold steelpen. **1975** *Islander* (Victoria, B.C.) 16 Feb. 10/3 By 1803 producers could no longer keep pace with the consumption of quills and the first steel pen came into use. **1983** *Daily Tel.* 10 Mar. 14/4 The steel-pen writing is brisk and persuasive.

steely, *a.* Add: **5. b.** *steely-eyed.*

1964 D. F. DOWD in I. L. Horowitz *New Sociology* 60 Steely-eyed, if amiable, technicians. **1976** *Saturday Night* Mar. 82/3 A haughty Trudeau is seated at a press conference giving some questioner that familiar steely-eyed look of his.

6. Also in *Comb.*

1922 JOYCE *Ulysses* 252 Bronze By Gold Heard The Hoofirons, Steelyringing.

steenbrass. Now usu. **steenbras.** Substitute for def.: Any of several marine fishes of the family Sparidæ, esp. the red steenbras, *Petrus rupestris*, the silver steenbras, *Sparodon durbanensis*, or the white steenbras, *Lithognathus lithognathus*. (Further examples.)

1893 [see *KINGKLIP]. **1931** *Times Lit. Suppl.* 16 Apr. 301/2 Many of the Cape fish, such as the geelbek,..the steenbras. **1958** *Cape Times* 20 May 2/2 Fishermen have been bringing in some fine, big red-steenbras... Anglers on the rocks have..had good hauls of smaller fish including white stumpnose, galjoen and white steenbras. **1973** [see *KABELJOU].

steening, *vbl. sb.* Add: Also **steyning. 1.** (Later examples.)

1926 T. E. LAWRENCE *Seven Pillars* (1935) I. x. 80 The well was old, and broad, with a good stone steyning, and a strong coping round the top. **1939** *Oxoniensia* IV. 94 They might have been stones used to edge or pave the mouth of the well, as similar fragments of stone occurred in most wells, but they could not be taken as steyning.

steenstrupine (stī·nstrupīn). *Min.* [f. the name of Knud J. V. *Steenstrup* (1842–1913), Danish geologist + -INE[5].] A silicate and phosphate of rare-earth and other elements (chiefly cerium, sodium, calcium, iron, and manganese), found as dark brown or black rhombohedral crystals. Also (*rare*) **stee·nstrupite.**

1882 J. LORENZEN in *Mineral. Mag.* V. 67 The analysis of this mineral has shown it to be a new species, and, as suggested by Prof. Johnstrup, I have therefore given it the name of *Steenstrupine*, after Mr. *Steenstrup*, who.. first found the mineral. **1901** *Meddelelser om Grønland* XXIV. 204 The steenstrupite is found partly in pegmatitic veins and partly embedded in grained albite. **1943** *Mineral. Abstr.* VIII. 369 The unit-cell dimensions of steenstrupine..show a similarity to those of apatite. **1977** *Jrnl. R. Soc. Arts* CXXV. 406/2 The peralkaline nepheline syenites of Ilimaussaq in south Greenland, where the main mineral is steenstrupine (silicate and phosphate of rare earths, sodium, niobium, tantalum, thorium and uranium).

steentjie (stiᵊ·nkʸi). *S. Afr.* Also **steen(t)je.** [Afrikaans, dim. of Du. *steen* stone.] Either of two small marine fishes of the family Sparidæ, *Spondyliosoma emarginatum* or *Sarpa salpa* (= *STREPIE).

1893 H. A. BRYDEN *Gun & Camera S. Afr.* xx. 448 We caught also..steenje, another small fish, which we cut up principally for bait. **1913** W. W. THOMPSON *Sea Fisheries Cape Col.* 154 *Cantharus emarginatus*..Steentje. **1930** C. L. BIDEN *Sea-Angling Fishes of Cape* ix. 166 The men.. caught many steentjies which were scaled and pulped with a baton and baited whole on the big lines. **1957** S. SCHOEMAN *Strike!* iii. 88 There are two species of fish commonly referred to as steentjies, namely the common steentjie..alias strepie,..and the blue or bank steentjie.

steep, *a.*, *sb.*[2], and *adv.* Add: **A.** *adj.* **3. g.** *steepest descent(s)* (Math.), used with reference to a method of finding a minimum of a function of two or more variables by repeatedly

evaluating it at a point displaced from the previous point in the direction that locally involved the greatest drop in its value.

1939 *Proc. R. Soc.* A. CLXIX. 484 In the method of steepest descents the displacement affects all co-ordinates and affects them in the ratio of their residual forces. **1943** *Bull. Amer. Math. Soc.* XLIX. 18 We now choose the line along which the motion proceeds so that the descent is as steep as possible (lines of steepest descent). **1974** ADBY & DEMPSTER *Introd. Optimisation Methods* iii. 57 The steepest descent method uses the Jacobian gradient g to determine a suitable direction of movement.

6. *steep-fronted.*

1936 *Nature* 21 Mar. 491/2 The test piece is flashed over with a steep-fronted impulse in about a microsecond or less.

C. *adv.* **4.** *steep-cut* (earlier example).

1888 KIPLING *Lett. Marque* (1891) xv. 115 Up rough banks..down steep-cut dips.

stee·phead. *U.S.* (orig. *local*). *Physical Geogr.* [f. STEEP *a.* + HEAD *sb.*] A nearly vertical slope, from the base of which springs emerge, at the head of a pocket valley (see *POCKET *sb.* 13).

1918 *10th & 11th Ann. Rep. Florida Geol. Survey* 27 A characteristic feature of this topography is the development of what is [*sic*] known locally as 'steepheads'. *Ibid.*, The depth of the steephead from the plateau is usually from 50 to 60 or more feet. **1942** O. D. VON ENGELN *Geomorphol.* xxii. 569 Pocket valleys of similar origin are also of widespread occurrence in northwestern Florida where they are known as steep-heads. **1971** [see *pocket valley* s.v. *POCKET *sb.* 13].

steeple, *sb.*[1] Add: **I. 4*.** *transf.* A steeple-shaped formation of the two hands, with the palms facing and the extended fingers rising to meet at the tips.

1940 *Detective Fiction Weekly* 26 Oct. 9/1, I..waited for him to begin. He made a steeple of his hands. 'Now it is a very simple matter.' **1972** T. P. MCMAHON *Issue of Bishop's Blood* (1973) iii. 35 He was sighting at me carefully along the top of the steeple made by his manicured fingers. **1978** G. VIDAL *Kalki* ii. 33 When I put out my hand, she made a steeple with her hands, and bowed. This was my first experience with the Hindu pranam, or greeting.

II. 6. steeple clock, (*a*) a clock fixed to a steeple; (*b*) *U.S.*, an antique mantel or shelf clock (see quot. 1959).

1830 [in Dict., sense 5 a]. **1923** W. I. MILHAM *Time & Timekeepers* xx. 368 There are two kinds of clocks, however, which came into vogue shortly after 1850 and are usually classed among 'antiques' and not among the endless varieties of modern clocks. These are the small, spring-driven, brass shelf clock often in a rose-wood veneer case and the steeple clock, sometimes called the 'Sharp Gothic'. **1959** L. GROSS *Housewives' Guide to Antiques* viii. 114 The steeple clock of the mid-nineteenth century was another popular mantel clock. It takes its name from the steeple-like appearance of its case. **1976** R. B. PARKER *Promised Land* (1977) xvi. 97 There was an old steeple clock with brass works on the mantle.

steeple, *v.* Add: **3.** (Later *poet.* example.)

1922 BLUNDEN *Shepherd* 81 The cornel steepling up in white shall know The two friends passing by.

4. *trans.* To place (the fingers or hands) together in the shape of a steeple.

1968 A. MACLEAN *Force 10 from Navarone* iv. 59 The German captain leaned back in his chair and steepled his fingers. **1975** W. SAFIRE *Before Fall* VI. vi. 446 Nixon.. was relaxed when seated, steepling or folding his hands. **1977** G. SCOTT *Hot Pursuit* ii. 18 He steepled his fingers and looked wisely at me across the desk-top.

steepling *ppl. a.*: also *fig.*

1955 *Times* 28 June 3/2 He struck a steepling blow, but he fell into Kenyon's hands standing just inside the ropes beneath the Nursery Clock Tower. **1977** *Guardian Weekly* 17 July 24/4 McCosker hooked at it and sent a steepling catch to Underwood at mid-on. **1982** *Daily Tel.* 20 Aug. 9/4 The steepling rises in standing charges.

steeplechase, *sb.* Add: **1.** (Earlier examples.)

Quot. 1805 in Dict. is in error for 'Steeple-race'.

1793 *Sporting Mag.* Apr. 57/2 The Hon. Mr. O'Hea and Captain Magrath ran a Steeple-chace, near Galloway. **1803** W. TAPLIN *Sporting Dict. & Repository* II. 486 This kind of chase [*sc.* Wild-Goose chase]..was long since changed to a *train scent*, (that is, a drag across the country;) better known by the denomination of a *steeple chase.*

3. (Earlier example.)

1892–3 *T. Eaton & Co. Catal.* Fall & Winter 67/1 Games..errand boy, steeplechase, yacht race [etc.].

steepled, *ppl. a.* Add: **5.** Of the fingers or hands: brought together in the form of a steeple.

1971 P. O'DONNELL *Impossible Virgin* x. 212 Tapping the tips of his steepled fingers together. **1981** 'L. EGAN' *Miser* (1982) ii. 26 'Not much criminal practice,' said Jesse, brooding over his steepled hands.

steepler (stī·pləɹ). *Cricket.* [f. STEEPLE *v.* + -ER[1].] = SKYER, esp. one from which the batsman is caught out.

1959 *Times* 7 Aug. 4/2 He was caught and bowled off a steepler. **1963** *Times* 5 Feb. 3/6 Parfitt at long-on judged a steepler well to end the innings. **1976** *Cricketer Internat.* Sept. 35/2 Crockham Hill caught everything, including two stupendous steeplers in the last over.

steer, *sb.*[1] Add: **2. a.** *Comb.* Designating events or participants in a rodeo, as *steer roper, wrestler; steer bulldogging, roping, wrestling.*

1910 *Oregon Daily Jrnl.* 30 Sept. 18/5 Steer roping contest for championship of the northwest. **1912** *Oregon Sunday Jrnl.* 18 Sept. 2/1 Among the many events to be featured at the Round-Up this year is the world's championship wild steer bull-dogging contest. **1914** *World's Work* (N.Y.) Feb. 445/2 It by no means follows that a good steer roper is a good calf roper. **1922** *N.Y. Times* 12 Nov. I. 5/2 One of the conditions of the steer-wrestling contests is that the contestants will suffer a 'ten-second fine' for 'loosening or knocking off horns'. **1923** *Ibid.* 11 Aug. 16/4 Steer Wrestlers Here. Twenty-three contestants in the 'steer-wrestling' or 'bull-dogging' competitions..arrived in New York yesterday. **1924** *Glasgow Herald* 17 June 9 The 'steer-roping', which at Saturday's display met with some public disapproval, was withdrawn. **1968** R. F. ADAMS *Western Words* (rev. ed.) 305/1 *Steer wrestling*, one of the five standard rodeo events; also called *bulldogging*. The contestant rides alongside a running steer, jumps from his saddle to the steer's head, stops it, and twists it to the ground with its head and all four feet pointing in the same direction. **1976** *Billings* (Montana) *Gaz.* 16 June 1-c/3 University of Wyoming steer wrestler Shawn Madden took an early lead in the second go-round of steer wrestling, throwing his animal in 3.67 seconds. **1979** *Sunset* Apr. 6/3 Horses will compete in 36 classes, including calf and steer-roping events.

b. Special combination. **steerhide** *N. Amer.*, the hide of a steer; *spec.* leather made from this or from the hide of a similar beast.

1921 *Jrnl. Amer. Leather Chemists Assoc.* May 295 (*heading*) On certain characteristics of fresh steer hide. **1925** J. R. ARNOLD *Hides & Skins* ii. 32 The thickest part of a steer hide is over the rump. **1948** H. G. KATES *Luggage & Leather Cards Manual* 184 Steerhide leather is extensively used because of its adaptability to tooling. **1979** *PN Rev.* IX. 39/1 All winter your brute shoulders strained against collars, padding and steerhide over the ash hames.

steer (stiᵊɹ), *sb.*[3] *slang* (orig. *U.S.*). [f. STEER *v.*[1]] A piece of advice or information; a tip, a lead. (See also quot. 1970.)

1899 C. H. HOYT *Texas Steer* (typescript) IV. 21 You're going back to Texas to give the voters of my district a steer. What's that steer to be? **1924**, etc. [see *BUM *a.*]. **1926** *Flynn's* 16 Jan. 638/2 An' divvy with th' crooked barkeep for a steer or some kind of a tip if th' stunt panned out ok. **1935** L. E. LAWES *Cell 202, Sing Sing* iv. 553 You're both on the wrong steer..thinkin' about the devil when all the while it's the man himself deserves your attention. **1959** 'M. M. KAYE' *House of Shade* x. 127 All I've done is to give you a wrong steer, and make bad worse. **1970** D. FRANCIS *Rat Race* vi. 79 I'd have to go round the Luton complex...could probably get a steer home from there, from the twenty-four hour radar. **1982** *Times* 21 Apr. 16/1 Steers from Smiths Industries on its financial performance are obviously worth listening to.

steerer. 3. (Earlier example.)

1873 'J. MORRIS' *Wanderings of Vagabond* xix. 210 Let us now take a peep into the brace room, while the steerer and his victim are on their way to it.

steering, *vbl. sb.* Add: **1. b.** *Meteorol.* The process by which pressure systems, precipitation belts, etc., are moved by temperature gradients or winds.

1919, etc. [see *steering line*, sense 3 b below]. **1944** HEWSON & LONGLEY *Meteorol. Theoret. & Appl.* xxiv. 428 The changes in pressure at the earth's surface are controlled by atmospheric movements at levels of 4 km and higher, the process whereby this occurs being known as 'steering'. **1956** S. PETTERSSEN *Weather Analysis & Forecasting* (ed. 2) I. xiii. 277 From the point of view of theory it is difficult to make any distinction between the steering and the blocking of the movement of sea-level cyclones. **1959** R. E. HUSCHKE *Gloss. Meteorol.* (Amer. Meteorol. Soc.) 541 *Steering*, in meteorology, loosely used for any influence upon the direction of movement of an atmospheric disturbance exerted by another aspect of the state of the atmosphere.

c. *Electronics.* The switching of pulses from one part of a circuit to another.

1956, etc. [see *steering circuit, diode*, sense 3 a below]. **1962** SIMPSON & RICHARDS *Physical Princ. Junction Transistors* xvi. 410 In many bistable circuits,..trigger pulses of opposite polarity must be used alternately..or pulses of one polarity must alternately be switched from one collector or base to another. The latter process, which is usually performed by diodes or auxiliary transistors, is called steering. **1969** J. J. SPARKES *Transistor Switching* v. 122 Two quite distinct methods of controlling the penetration of pulses to the transistor are available, namely pulse steering by applying logic voltages to S_A or R_A, or pulse gating via additional diodes at the pulse inputs.

2. (Later examples.)

1970 K. BALL *Fiat 600, 600D Autobook* ix. 103/2 The first step in dismantling the steering is to remove the steering wheel. **1977** *Western Morning News* 30 Aug. 8/2 (Advt.), Austin 2200. Blue. Power-assisted steering.

3. a. (sense 1 a) *steering-bridge*; (sense *1 b) *steering principle*; (sense *1 c) *steering circuit, diode*; (sense 2) *steering angle, arm, axle, circle, knuckle, lever, linkage, rod.*

1936 *Proc. Inst. Automobile Engineers* XXX. 757 The angle of the front wheels or 'steering angle' is particularly important. **1902** A. C. HARMSWORTH et al. *Motors & Motor-Driving* 216 With a broken steering arm..a car.. may be hurled into a ditch. **1978** D. CLARKE *Car* 84/1 All but the last type of box require a system of linkages to

take the movement created by the drop arm..to the steering arms on the wheels. **1912** *Motor Man.* 87 Details of steering axle with steering arms and connecting bar. **1902** *Chambers's Jrnl.* Oct. 739/1 He left the bridge, roused all hands, and arraigned them on the steering-bridge. **1912** *Motor Man.* 88 Whichever wheel is on the inside of the steering circle turns through a wider angle than the outer wheel does. **1970** K. BALL *Fiat 600, 600D Autobook* ix. 108/2 The steering circle diameter on the Multipla is 28 feet 10 inches. **1956** L. P. HUNTER *Handbk. Semiconductor Electronics* xv. 23 In order to operate the trigger in a binary fashion, it is necessary to provide steering circuits. **1971** J. H. SMITH *Digital Logic* iv. 58 Steering circuits..control the operation of the gates. **1957** R. F. SHEA *Circuit Engin.* x. 337 (*caption*) Bistable multivibrator with steering diodes for high-speed triggering. **1962** SIMPSON & RICHARDS *Physical Princ. Junction Transistors* xvi. 410 Steering diodes may be used in a similar way at other electrodes and for other circuits than the simple bistable one. **1904** *Sci. Amer. Suppl.* 27 Aug. 23953/2 The front axle is provided with ball-bearing steering knuckles. **1970** K. BALL *Fiat 600, 600D Autobook* viii. 98/2 Remove the upper end shaft retaining nut from the steering knuckle and the two washers. **1866** *Eng. Mech.* 6 Apr. 33/1 Velocipede. SL is the steering lever. **1915** *Autocar Handbk.* (ed. 6) xv. 219 This steering lever is mounted upon a short spindle or shaft which is carried in bearings in the steering gear box. **1970** K. BALL *Fiat 600, 600D Autobook* viii. 97/2 The steering linkage also differs from that on the Sedan. **1945** F. A. BERRY et al. *Handbk. Meteorol.* x. 818 Guiding of surface systems by the upper-level flow has been referred to as the steering principle. **1963** *Meteorol. Gloss.* (Met. Office) (ed. 4) 254 Application of the steering principle is most successful in the type of situation..in which almost straight thickness lines intersect a well marked pattern of surface vorticity. **1909** R. W. A. BREWER *Motor Car* xv. 148 The transverse arm on the off side is connected to the steering rod, generally by means of a knuckle joint. **1977** 'J. GASH' *Judas Pair* ix. 106 She twisted something near the steering-rod. The engine muted instantly into a deep, steady thrum.

b. steering box, a housing attached to the body of a motor vehicle that encloses the end of the steering column and the gearing that transmits its motion to the next members; **steering column,** a columnar assembly in a motor vehicle or motor bicycle carrying at its top the steering wheel or handle-bars and transmitting their motion to the rest of the steering gear; **steering line** *Meteorol.* (see quot. 1959); **steering lock,** (*a*) (example); (*b*) also, a similar mechanism fitted to the steering assembly of a motor vehicle, as an anti-theft device; **steering-oar** (earlier and later examples); † **steering pillar** = *steering column* above (*obs.*); **steering post** = prec.; on early motor vehicles (see also quot. 1904); **steering wheel,** (*b*) (examples); (*c*) also, a road wheel of a motor vehicle by which steerage is effected.

1913 W. E. DOMMETT *Motor Car Mechanism* xiii. 118 The method of mounting the steering box and column is clearly shown. **1970** K. BALL *Fiat 600, 600D Autobook* ix. 103/1 The linkage between the steering box and wheels is via a pair of symmetrical track rods and a central link rod connecting the pitman arm to a relay level. **1903** *Motor* 27 May 348/1 The steering column of the motor-bicycle is..a vitally important part. **1931** D. L. SAYERS *Five Red Herrings* xxiii. 262, I was jammed up behind the steering-column. **1976** *Derbyshire Times* (Peak ed.) 3 Sept. 20/5 The single steering column stalk that operates lights, wipers, washers and flasher has too much to do. **1919** J. BJERKNES in *Geofysiske Publ.* I. II. 1 As the line thus gives the momentaneous direction of the cyclone, it may, for practical purposes, be called the steering line. **1923** N. SHAW *Forecasting Weather* v. 155 The dividing line of the cyclone from the centre towards the eastern or advancing side is called the steering line or more recently, warm front. **1959** R. E. HUSCHKE *Gloss. Meteorol.* (Amer. Meteorol. Soc.) 541 *Steering line*, according to Bjerknes' cyclone model, the line of convergence (corresponding to the warm front of a wave cyclone) which tends to be parallel to the direction of motion of the cyclone at the line's point of juncture with the cyclone center. **1955** *Times* 10 May 7/6 The greatest asset in this connexion.. is the admirable steering lock, with which the car can be turned in 37ft. and can be driven into small parking spaces. **1960** O. GREGORY tr. *Spoerl's Living with Car* 212 Various devices.., some of which lock ignition and steering simultaneously... These locks are by no means impossible for a thief to cope with. However, if he has to choose between a car with one of these steering locks and one without he is likely to choose the one without. **1971** A. PRICE *Alamut Ambush* ix. 104 The new Triumph has a steering lock—it would be a verra difficult car to move. **1816** H. KER *Travels* 30 In endeavouring to run the outside of a sawyer, I ran with my stem athwart it, and unshipped my steering oar, which I lost. **1938** B. L. BURMAN *Blow for Landing* 298 The black lifted the steering oar... The rafts began to speed down the water. **1904** A. C. HARMSWORTH et al. *Motors & Motor-Driving* 218 Looseness between steering wheel and end of steering pillar can be found at any time. **1921** W. H. BERRY *Mod. Motor Car Practice* xvi. 324 A long chassis with a large engine involves a steering pillar of considerable length. **1904** A. B. F. YOUNG *Compl. Motorist* iv. 118 The steering post, being situated in the middle of the car, and the steering tiller available on either side, it is possible for the driver to sit either on the right-hand or left-hand side of the car. **1969** J. GORES in *Ellery Queen's Mystery Mag.* Dec. 145/2 He was impaled on the steering post. **1902** A. C. HARMSWORTH et al. *Motors & Motor-Driving* 217 The free or direct gear moves with the impulse or pressure brought against the steering wheels or one of them by any ruts or obstructions on the road. **1907** R. W. WHITMAN

Motor-Car Princ. x. 159 The irreversible type is used for all but the lightest cars, and..it prevents any movement from being transmitted from the wheels to the steering wheel or lever. **1912** *Motor Man.* 231 A weakened [tyre] cover will, as a general rule, give a considerable period of further service mounted on a steering wheel. **1915** *Autocar Handbk.* (ed. 6) xv. 220 It is very important..that the steering road wheels should be easily movable by means of the hand steering wheel... If a car is travelling on a much-rutted road the steering wheels tend to fall into the ruts. **1970** K. BALL *Fiat 600, 600D Autobook* ix. 103/2 Pry off the horn button from the steering wheel with a screwdriver.

steering (stiᵊ·riŋ), *ppl. a.* [f. STEER *v.*[1] + -ING[2].] *steering committee* (orig. *U.S.*): a committee set up to determine the order of business for another body, or to manage the general course of an operation. Also *steering group, sub-committee.*

1887 *Courier-Jrnl.* (Louisville, Kentucky) 6 Feb. 2/2 A steering committee upon the order of business for the remainder of the session was appointed. **1918** H. W. DODDS *Procedure in State Legislatures* iv. 56 Just as the power of standing committees developed when the number of bills introduced had become too large for consideration by the whole house, so the steering committee emerged when measures approved by the standing committees increased until a further selective agency became an irresistible temptation. **1955** *Times* 19 July 6/4 The steering committee making arrangements for the conference of Ministers on further European integration met to-day in Brussels. **1966** N. JOHNSON *Parliament & Administration* i. 22 The Chairman of the Estimates Committees presides over sub-committee A (now called the Steering sub-committee). **1974** 'E. LATHEN' *Sweet & Low* ii. 20 Only three people are really important..the members of the steering committee. **1977** *Wandsworth Borough News* 7 Oct. 14/5 The Steering Group officially recognised by the Council's Recreation Committee at that meeting has met regularly and has discussed such matters as management, development, and use of the building. **1979** 'D. MEIRING' *Foreign Body* v. 57 We have all read the Steering Committee's exhaustive report.

steersman. Add: **1. a.** Also, one who sits at the stern of a canoe and steers. *N. Amer.*

1774 HEARNE & TURNOR *Jrnls.* (1934) 122 The Pataroon or Steersman of each Cannoe has 50£ pr annom. **1791** [in Dict.] **1801** [see *MIDDLEMAN 5 c]. **1897** E. COUES *New Light Early Hist. Greater Northwest* I. 29 The steersman, finding himself within reach of the shore, jumped upon the rock with one of the midmen. **1968** [see *BOWSMAN].

Stefan (ste·fæn). *Physics.* The name of Josef *Stefan* (1835–93), Austrian physicist, used *attrib.* and in the possessive with reference to a law discovered by him (see J. Stefan 1879, in *Sitzungsber. der Österreich. Akad. der Wissensch. in Wien* LXXIX. 391), as **Stefan('s) constant, law** = *Stefan–Boltzmann constant, law* (see next).

1898 *Sci. Abstr.* I. 391 The author's attempt to realise Kirchhoff's conception of an equivalent to an absolutely black body in the shape of a uniformly heated hollow space with a small opening, and test Stefan's law, which maintains that the radiation of such a body is proportional to the 4th power of its absolute temperature. **1923** GLAZEBROOK *Dict. Appl. Physics* III. 711/1 By the Stefan-Boltzmann law.., and a knowledge of Stefan's constant, the whole radiation from the blackened surface is known. *Ibid.* IV. 569/1 Apart from this constant Stefan's law leads straight to Maxwell's law of radiation pressure. **1962** *Newnes Conc. Encycl. Electr. Engin.* 718/1 Here ϵ is the Stefan or total-radiation constant. **1966** [see *Planck('s) equation* s.v. *PLANCK]. **1979** T. B. AKRILL et al. *Physics* xxviii. 377/2 For a non-black or 'grey' body we can apply Stefan's law in the form $P_{tot} = \epsilon\sigma AT^4$, where ϵ is called the total emissivity of the body and is a number always less than 1.

Stefan–Boltzmann (ste·fæn bōᵘ·ltsmæn). *Physics.* The names of Josef *Stefan* (see prec.) and Ludwig Eduard *Boltzmann* (1844–1906), Austrian physicists, used *attrib.* with reference to a law independently discovered by them (L. E. Boltzmann 1884, in *Ann. der Physik u. Chemie* XXII. 291; see also prec.), as **Stefan–Boltzmann constant,** the constant in the Stefan–Boltzmann law, equal to $5{\cdot}67 \times 10^{-8}$ J m^{-2} s^{-1} K^{-4}; **Stefan–Boltzmann law,** the law which states that the total radiation emitted by a black body is proportional to the fourth power of its absolute temperature.

1898 *Sci. Abstr.* I. 391 The Stefan-Boltzmann law is confirmed to within 3 to 8%. **1915** R. A. HOUSTOUN *Treat. Light* xxv. 448 By considering such a cycle Boltzmann proved that the total radiation from a black body was proportional to the fourth power of its absolute temperature. This law had been stated previously but erroneously by Stefan as holding good for all bodies, and as the amended version was due to Boltzmann, it is very often referred to as the Stefan-Boltzmann law. **1954** [see *LANGLEY]. **1958** CONDON & ODISHAW *Handbk. Physics* VI. i. 15/1 The total radiation crossing unit area in unit time in all directions in one hemisphere is usually written σT^4, where σ is called the Stefan-Boltzmann constant. **1977** I. M. CAMPBELL *Energy & Atmosphere* i. 13 This is related to T through the Stefan-Boltzmann law: $E(T) = \sigma T^4$ where σ is known as the Stefan-Boltzmann constant and has value $5{\cdot}672 \times 10^{-8}$ for T in Kelvin and $E(T)$ in Wm^{-2}.

stegomyia (stegomai·ă). [mod. L. (F. V. Theobald 1901, in *Jrnl. Trop. Med. & Hygiene* IV. 235/1), f. STEGO- + Gr. μυῖα fly.] A mosquito of the genus formerly so called, now usually regarded as a sub-division of the genus *Aedes*, which includes tropical and subtropical species responsible for the transmission of yellow fever.

1911 A. ALCOCK *Entomol. for Med. Officers* v. 96 (*heading*) Genera of the Stegomyia type. **1915** G. B. SHAW in *New Statesman* 10 July 326/2 The same result could have been obtained by inoculation with anopheles vaccine and stegomyia emulsion. **1920** *Glasgow Herald* 6 July 6/4 The success of General Gorgas's methods can be gauged..after a campaign extending over eight years against the stegomyia. **1932** *Discovery* Oct. 326/2, I [*sc.* Robert Ross] then examined a small *Stegomyia*. **1971** J. D. GILLETT *Mosquitos* iii. 46 Other *Stegomyia* also at times glide through the water head-first.

steigerite (stai·gĕrait). *Min.* [f. the name of George *Steiger* (1869–1944), U.S. chemist + -ITE[1].] A monoclinic hydrated aluminium vanadate, $AlVO_4.3H_2O$, found as a canary-yellow powdery coating on vanadium-containing concretions.

1935 E. P. HENDERSON in *Amer. Mineralogist* XX. 769 A new yellow, hydrous aluminum vanadate, $Al_2O_3 \cdot V_2O_5 \cdot 6{\cdot}5H_2O$, which has been named steigerite, is here described. **1959** *Amer. Mineralogist* XLIV. 336 Electron micrographs of steigerite from Gypsum Valley, Colorado ..show that the material is composed of thin, poorly developed laths and angular flakes.

stein. (Earlier and later examples.)

1855 *Trans. Hist. Soc. Lancs. & Cheshire* VII. 190 The 'Moss Pottery'..was..confined to the making of common red-clay ware, for domestic use, as jowls, steins, flower-pots, &c. **1897** [see PRETZEL, BRETZEL]. **1915** *Sat. Even. Post* 13 Feb. 52/3 So I sat down and et with 'em and had a few steins of beer. **1976** *Billings* (Montana) *Gaz.* 27 June 1-c/6 Beer was flowing at 50 cents a stein—three for $1. **1981** *West Lancs. Even. Gaz.* 18 July 8/4 A night on the town turned sour when a 22 year-old man hit a glass collector in the face with a beer stein.

Steinberger (ʃtai·nbāɹgəɹ). [Ger., f. the name *Steinberg* (also used) of the vineyard where the wine is produced.] A white wine produced near Hattenheim in the Rheingau.

1833 C. REDDING *Hist. & Descr. Mod. Wines* vii. 205 The Steinberger..takes rank after the Schloss-Johannesberger among these wines. It has the greatest strength, and yet is one of the most delicate, and even sweetly flavoured. **1894** J. L. W. THUDICHUM *Treat. Wines* 201 The *Johannisberg* is the only rival of the Steinberg. **1907** *Hatch, Mansfield Price List* Jan. 18 (Advt.), *Steinberg*, magnificent wine, with great body. **1926** M. SHAND *Bk. Wine* vi. 187 The number of classified districts in the Rheingau, from one or other of which all but three of its wines (Steinberger, Markobrunnen, and Schloss Vollrads) take their first name. **1951** *Good Housek. Home Encycl.* 508/2 The best Hock, which is sold under a number of well-known names, e.g. Johannisberger, Steinberger. **1965** O. A. MENDELSOHN *Dict. Drink & Drinking* 322 Starting at simple Steinberger and ending at the almost hallowed Steinberg Kabinett.

Steiner (ʃtai·nəɹ). *Math.* The name of Jakob *Steiner* (1796–1863), Swiss geometer, used *attrib.* and in the possessive to designate various mathematical concepts suggested by him: **a.** *Steiner triple* or *triplet system* (see quot. 1939); so *Steiner triplet*; also *Steiner system*, a generalization of the triple system to other numbers (see quot. 1974). [Steiner first described such systems in *Jrnl. f. d. reine u. angewandte Math.* (1853) XLV. 181.]

1939 R. C. BOSE in *Annals Eugenics* IX. 354 Steiner (1853) proposed the problem of arranging N things in triplets, such that every pair occurs in just one and only one triplet. Such an arrangement may be called a simple triple system or a Steiner's triplet system. **1963** H. J. RYSER *Combinatorial Math.* viii. 100 The Steiner triple system of order 7 is the same as the projective plane of order 2 in the preceding chapter. **1966** *Annali di Matematica* LXXI. 199 The *Steiner* system $S(5, 8, 24)$ is an arrangement of 24 elements in sets of 8, such that any 5 of the elements belong to exactly one set. **1974** I. ANDERSON *First Course Combinatorial Math.* vii. 102 A Steiner system $S(l, m, n)$ is a collection of m-element subsets of an n element set B such that every l-element subset of B lies in exactly one of the m-element sets. *Ibid.*, A Steiner triple system is an $S(2, 3, n)$ for some n. **1980** *Sci. Amer.* May 14/2 Since Steiner triplets are not ordered, the solution is of course unique.

b. Used with reference to the problem of finding the set of line segments of minimum total length needed to connect a given set of points in a metric space.

1941 COURANT & ROBBINS *What is Math.?* vii. 359 In Steiner's problem three fixed points A, B, C are given. It is natural to generalize this problem to the case of n given points. *Ibid.* 360 To find the really significant extension of Steiner's problem we must abandon the search for a single point P... Given n points..to find a connected system of straight line segments of shortest total length such that any two of the given points can be joined by a polygon consisting of segments of the system. **1961** *Canad. Math. Bull.* IV. 143 Given a triangle T with the vertices a_1, a_2, a_3, to find in the plane of T the point p

which minimizes the sum of the distances $|pa_1| + |pa_2| + |pa_3|$. p, called the Steiner point of T, is unique. **1968** *SIAM Jrnl. Appl. Math.* XVI. 1 A Steiner minimal tree for given points $A_1, .., A_n$ in the plane is a tree which interconnects these points using lines of shortest possible total length. In order to achieve minimum length the Steiner minimal tree may contain other vertices (Steiner points) beside $A_1, .. A_n$. **1979** *Sci. Amer.* Apr. 37/1 First is the Steiner problem of the shortest roads linking many cities.

Steinert (ʃtai·nĕɹt). *Path.* [The name of H. G. W. *Steinert* (b. 1875), German physician, who described the disease in 1909 (*Deutsche Zeitschr. f. Nervenheilkunde* XXXVII. 58).] *Steinert's disease* = *myotonia atrophica* s.v. *MYOTONIA 2 b.

1932 *Index Medicus* XII. 429/1 (*heading*) Cachetic form of myotonia atrophica (Steinert's disease). **1948** F. B. CARLSEN tr. *Thomasen's Myotonia* vii. 94 Steinert's disease invariably set in in later life, generally between the ages of twenty and thirty. **1963** [see *MYOTONIA 2 b].

Steinhäger (ʃtai·nhēgəɹ). Also **-haeger,** (*erron.*) **-hager.** [Ger., f. the name *Steinhagen* of the town, in Westphalia, where it is produced.] A spirit made from juniper-berries; a measure or glassful of this.

1959 M. CROSLAND tr. *Rovan's Germany* 180 Many fine spirits are made in Germany..a white *eau-de-vie* in Westphalia called *Steinhäger*. **1964** L. DEIGHTON *Funeral in Berlin* xl. 239 Coffee, doughnuts and tiny glasses of Steinhager. **1966** R. THOMAS *Cold War Swap* i. 9 Steinhaeger is best when drunk ice cold and washed down with a liter or so of beer. **1968** 'J. LE CARRÉ' *Small Town in Germany* xvii. 273 He drank a Steinhager from the tray. The mat stuck to the stem of the glass. **1976** R. PERRY *One Good Death* xi. 172, I handed him the flask of Steinhager... Muller took a generous swig.

Steinheim (ʃtai·nhaim). The name of a village twelve miles north of Stuttgart, West Germany, used *absol.* or *attrib.* in **Steinheim skull** to designate a Middle Pleistocene fossil hominid known from a skull found there in 1933 by Karl Sigrist and described as *Homo steinheimensis* by F. Berckhemer in 1936 (*Forschungen & Fortschritte* XII. 349/2).

1946 F. E. ZEUNER *Dating Past* ix. 296 If the geological age of the Steinheim Skull were settled..we should at least know whether he, or his ancestor, lived in the middle Pleistocene. **1973** B. J. WILLIAMS *Evolution & Human Origins* x. 170/1 The Steinheim skull also comes from a gravel pit. *Ibid.* 170/2 In facial features Steinheim could indeed be a good ancestor for the later Neandertals.

Stein–Leventhal (ʃtain le·vĕntāl). *Path.* [The names of I. F. *Stein* (1887–1976) and M. L. *Leventhal* (1901–71), U.S. gynæcologists, who described the condition in 1935 (*Amer. Jrnl. Obstetrics & Gynecol.* XXIX. 181–91).] *Stein–Leventhal syndrome*: a hormonal abnormality in women characterized by enlarged polycystic ovaries, infertility, and oligomenorrhœa, often with hirsutism.

1980 F. A. SIMMONS in Meigs & Sturgis *Progr. in Gynecol.* II. 334 These curves are common in the 'Stein-Leventhal syndrome', a condition wherein poor or non-ovulatory cycles are thought to be caused by polycystic ovaries. **1972** *Brit. Med. Jrnl.* 20 May 457/2 A second ovarian cause of amenorrhoea is the Stein–Leventhal Syndrome.

Steinmann (ʃtai·nmæn). *Surg.* The name of Fritz *Steinmann* (1872–1932), Swiss surgeon, who described the device in 1907 (*Zentralbl. Chir.* XXXIV. 939), used in the possessive and *attrib.* to designate a surgical pin that may be passed through one end of a major bone for traction or setting.

1916 ROBERTS & KELLY *Treatm. Fractures* i. 77 (*heading*) Technic of application of Steinmann's nails. **1925** WILSON & COCHRANE *Fractures & Dislocations* xv. 546 Steinmann's pin has now been largely superseded in general use by the calipers, or ice-tongs, which grasp, but do not penetrate the bone. **1933** P. B. MAGNUSON *Fractures* iii. 25 The use of the Steinmann pin is not advisable except in skilled hands. **1974** A. HENRY in R. M. Kirk et al. *Surgery* xvi. 370 A Thomas's splint is fitted, and skeletal traction set up through a Steinmann's pin.. in the proximal tibia.

Steinway (stai·nweᵢ). The name of Henry Engelhard *Steinway* (1797–1871), celebrated German piano-builder, used *attrib.* or *absol.* to designate a piano manufactured by him or by the firm which he founded in New York in 1853.

1875 T. YELVERTON *Teresina in Amer.* ii. 181 In the latter there was one of the best pianos (a Steinway) that I touched in America, where good pianos are by no means rare. **1889** G. B. SHAW in *Star* 13 July 4/4 She was presented with a Steinway Grand. **1905** A. BENNETT *Sacred & Profane Love* III. v. 260 The piano, a Steinway in a hundred Steinways. **1933** E. WHARTON *Human Nature* 172, I always knew fashionable people could barely distinguish a barrel-organ from a Steinway. **1965** J. M. CAIN *Magician's Wife* (1966) ii. 18 At one end of the room was a Steinway baby grand. **1977** *New Yorker* 19 Sept.

64/2, I pictured a sea of Steinways to choose from in Juilliard's basement. **1981** J. JOHNSTON *Christmas Tree* 130 There was a piano, a beautiful Steinway concert grand... The only reality was the room with the Steinway.

Steinwein (ʃtai·nvain). [Ger.] The name of a dry white wine produced in the Steinmantel vineyards, near Würzburg, Bavaria, and sold in special bottles called Bocksbeutel.

1833 C. REDDING *Hist. & Descr. Mod. Wines* vii. 206 The Steinwein of 1748, brought in 1832 seventy pounds the ahm. This may serve to show how much these wines gain by age. **1843** *Penny Cycl.* XXVII. 456/2 The Steinwein must not be confounded with the Steinberger wine of the Rhine. **1920** G. SAINTSBURY *Notes on Cellar-Bk.* xii. 177 The popular form seems to have been..the *bocksbeutel* flasks of Steinwein. **1966** H. YOXALL *Fashion of Life* xiii. 212 At Würzburg..a dinner with salmon and *Steinwein*. **1980** *Times* 9 Aug. 8/7 The Steinburg vineyard, whose fame caused Franconian wines to be referred to as 'Steinwein'.

Stelazine (ste·lăzīn). *Pharm.* Also **stelazine.** [f. *Stel-*, of unkn. origin + *AZINE.] A proprietary name for trifluoperazine.

1958 *Brit. Med. Jrnl.* 12 July 91/1 A short pilot trial of the drug stelazine was undertaken with 25 chronic psychotic patients. **1958** *Trade Marks Jrnl.* 23 July 738/2 *Stelazine* 772,994. Azine chemical compounds for use in medicine and pharmacy. Smith, Kline & French Laboratories Ltd., London. Manufacturing Chemists.—9th January, 1958. **1965** *Nursing Times* 5 Feb. 187/2 In hospital the phenothiazine tranquilizers seem most useful for relieving the fear and tension prominent at this stage. Stelazine (trifluoperazine) seems particularly useful. **1976** H. FERGUSON *Confessions of Long Distance Acid Head* 45 This hospital did not even stock the drug Stelazine which had, apparently, cured me when I had my first nervous breakdown.

stele. Add: **2.** [ad. F. *stèle* (Van Tieghem & Douliot 1886, in *Bull. de la Soc. bot. de France* XXXIII. 216).]

stelk (stelk). *Anglo-Ir.* [Prob. ad. Ir. *stailc* stubbornness, sulkiness, (in Co. Donegal) starch: cf. STALK *sb.*[1]] A cooked vegetable dish made with onions, mashed potatoes, and butter, or other ingredients.

1844 W. CARLETON *Traits & Stories Ir. Peasantry* (new ed.) II. 167 Norah..sent you a crock of her own lard. When you're makin'..*sthilk*.., if you slip in a lump of this, it'll save you the price of butther. *Ibid.*, *Sthilk* is made by bruising a quantity of boiled potatoes and beans together. The potatoes, having first been reduced to a pulpy state, the beans are but partially broken. It is then put into a dish, and a pound of butter or rendered lard thrust into the middle of it. **1890** D. A. SIMMONS *List Words & Phr. S. Donegal* 16 *Stelk*, champ, food made of pounded potatoes. **1949** 'M. INNES' *Journeying Boy* xiii. 162, I greatly fear that for dinner now we shall have to fall back upon stelk. **1971** *Guardian* 10 July 9/6 Stelk.. is alternatively known as champ and consists of spring onions simmered in milk.

stell, *sb.*[4] (Earlier example.)

1801 J. BARROW *Trav. Interior S. Afr.* I. vi. 360 The animal had been shot through the body by a *stell-roar* or trap-gun, set by a Hottentot.

stellacyanin (stelăsai·ănin). *Biochem.* [f. the name *Estelle* (see quot. 1967) after *PLASTOCYANIN.] An intensely blue copper-containing protein found in the latex of the Japanese lac tree, *Rhus vernicifera.*

1966 W. G. LEVINE in J. Peisach et al. *Biochem. of Copper* 377 Omura,..in the course of purifying *Rhus* laccase.., separated a second, blue, copper protein which Peisach later named stellacyanin. **1967** J. PEISACH et al. in *Biochem. Jrnl.* CCXLII. 2857/2 The authors would like to thank..Dr. Tsuneo Omura, whose permission was kindly given to name stellacyanin after Mrs. Estelle Peisach. **1976** *Nature* 27 May 346/1 Stellacyanin..and umecyanin have about 110 and 130 amino acids with a single cysteine residue plus one disulphide bridge.

‖ **Stella Maris** (ste·lă mæ·ris). [L., lit. 'star of the sea'.] A title given to the Virgin Mary (see SEA-STAR 1), used allusively of a protectress or a guiding spirit.

1876 GEO. ELIOT *Dan. Der.* III. v. xxxvii. 113 If a man could paint the woman he loves..as the *Stella Maris* to put courage into the sailors..so much the more honour to her. **1897** O. WILDE *Let.* 31 May (1962) 583 Even for the sheep who has no shepherd there is a Stella Maris to guide it home. **1913** W. J. LOCKE *Stella Maris* i. 1 Stella Maris—Star of the Sea! That was not her real name. No one could have christened an innocent babe so absurdly. *Ibid.* 2 Walter Herold.., one night of storm and dashing spray, seeing the light, burning steadily like a star.., cried: 'Stella Maris! What a name for her!'

stellar, *a.* Add: **1.** (Later example.)

1975 *Physics Bull.* Nov. 484/3 Astrophysicists use the term white dwarfs to describe objects which are stellar, that is to say luminous by themselves, but with a low luminosity and a small radius.

3. Having the quality of a star (STAR *sb.*[1] 5 in Dict. and Suppl.); leading, outstanding. So **ste·llardom,** stardom. orig. and chiefly *U.S.*

1883 *N.Y. Mercury* 3 Nov. 2/2 Effie Ellsler's dramatic stellardom is at an end and the supporting cast will be disbanded. **1883** *Sunday Mercury* (N.Y.) 4 Nov. 7/3 A fine specialty performance will be given by selected stellar artists. **1912** M. B. Leavitt *Fifty Years Theatr. Managem.* xxx. 464 In those days a theatrical star was obliged to work his way up to the rungs of the legitimate ladder until he was found worthy of ranking in stellardom. .. It made good actors,..who have since taken their places as leaders in the stellar ranks. **1932** Kaufman & Ryskud *Of Thee I Sing* I. iv. 75 The two centre chairs are conspicuously empty, obviously waiting for the stellar pair. **1950** J. Dempsey *Championship Fighting* 26 It is only in..'partial' punches that the body-weight does not play a stellar role. **1958** Wodehouse *Cocktail Time* xviii. 156 A man of regular habits, he would normally have shrunk from playing a stellar role in an E. Phillips Oppenheim story. **1964** W. C. Putnam *Geol.* ix. 215/1 Second of the factors is the nature of the ground. San Francisco, 1906, and Long Beach, 1933, both [earthquakes] in California, are stellar examples of the importance of this control. **1976** *Times Lit. Suppl.* 25 June 804/5 The most spectacular book sale held this spring... The stellar attraction was the whole Book of Daniel, twelve leaves, from the Trier copy of the 42-line or Gutenberg Bible, 1455. **1977** *Amer. N. & Q.* XV. 94/1 He has eschewed the glitter of Hollywood which has lured and made stellar personalities out of so many of his fellow novelists.

stellarator (ste·lăreɪ·tǫɹ). *Physics.* [f. Stellar *a.* (see quot. 1951) + L. *-ātor* (see -or 2 and cf. Eng. *generator*).] One kind of toroidal apparatus for producing controlled fusion reactions in a hot plasma, distinguished by the fact that all the controlling magnetic fields inside it are produced by external windings.

1951 L. Spitzer *Proposed Stellarator* (U.S. Atomic Energy Commission NYO-993) 4 Since the proposed system generates power and neutrons by reactions similar to those occurring in the stars, the device analyzed below is called a 'Stellarator'. **1967** Condon & Odishaw *Handbk. Physics* (ed. 2) iv. xi. 209/2 The feature that distinguishes a stellarator from a torus with solenoidal windings alone is the presence of helical windings around the channel which cause the magnetic-field lines to be twisted in such a way that as the particles drift they are constantly turned back toward the channel center. **1980** *Ann. Rep. U.K. Atomic Energy Authority 1979–80* 32/1 Another type of toroidal magnetic trap studied at Culham is the stellarator. **1981** *Nature* 19 Feb. 625/1 The tokamak must create the longitudinal current in the plasma by a transformer effect, and so must be pulsed. The stellarator, on the other hand, can in principle be run statically.

stellate, *a.* Add: **3. b.** *Esp.* in *Anat.*, as *stellate cell*, any of various types of cell with long processes, as a Langerhans cell, a Kupffer cell, or an astrocyte (sense *a); *stellate ganglion*, the lowest of the three cervical ganglia of the sympathetic trunk; *stellate reticulum*, a layer of cells with long processes within the enamel organ of a developing tooth.

1890, etc. [see *Langerhans]. **1895** A. H. Smith *Dental Microsc.* p. xvii, Stellate reticulum of enamel organ. **1899** [in Dict., sense 3 a]. **1901,** etc. [see *Kupffer]. **1918** *Gray's Anat.* (ed. 20) 935 The accelerator fibres of the heart leave mainly through the second and third thoracic nerves and pass to the stellate ganglion. **1921** Tilney & Riley *Form & Function Central Nervous Syst.* xli. 749 Immediately beneath the layers of large and medium-sized pyramidal cells is a stratum containing a number of small monopolar stellate cells belonging exclusively to the Golgi type II. **1945** [see *astrocyte]. **1969** W. A. Beresford *Lect. Notes Histol.* xxiv. 169 'Mesectoderm' cells..induce overlying ectodermal lamina to separate into tooth germs and provide for each an enamel organ with its pulp/stellate reticulum and inner and outer epithelia. **1980** *Gray's Anat.* (ed. 36) 1129/1 The cervicothoracic (stellate) ganglion is..much larger than the middle cervical ganglion, being probably formed by the coalescence of the lower two cervical segmented ganglia with the first thoracic.

stellate, *v.* (Later example.)
1948 [see *stellation 7].

stellated, *a.* Add: **1. b.** *Geom.* Of a polygon, polyhedron, or polytope: capable of being generated from a convex polygon, etc., by extending the edges, etc., until they once more meet at a new set of vertices, etc. [The sense is due to L. Poinsot, who used F. *étoilé* in *Jrnl. de l'École Polytechn.* (1810) IV. 41).]

1859 A. Cayley in *Phil. Mag.* XVII. 123 It is shown by Poinsot..that, besides the regular polyhedrons of ordinary geometry, there are (of course in an extended signification of the term) four new regular polyhedrons, viz. an icosahedron, which I will call the great icosahedron.., and three dodecahedrons, which I will call the great dodecahedron.., the great stellated dodecahedron.., and the small stellated dodecahedron. **1931** *Proc. Cambr. Philos. Soc.* XXVII. 206 Consider..the 'small stellated dodecahedron' {$\frac{5}{2}$, 5}, bounded by pentagrams. **1952** Cundy & Rollett *Math. Models* iii. 83 These four beautiful solids were unknown to the ancient world and were not discovered until modern times. The two with star faces—the two stellated dodecahedra—were found by Kepler (1571–1630); the others with regular faces and star vertices—the great icosahedron and dodecahedron—by Poinsot (1777–1859). **1976** I. Lakatos *Proofs & Refutations* i. 62 Take for instance the 'great stellated dodecahedron' (fig. 15). It consists, like the 'small stellated dodecahedron' of pentagrams, but differently arranged. It has 12 faces, 30 edges and 20 vertices, so that $V-E+F=2$.

stellation. **7.** (Later examples.)
1938 H. S. M. Coxeter et al. *Fifty-Nine Icosahedra* i. 1 We enumerate and describe the polyhedra that can be five Platonic solids by stellation, *i.e.*, by extending or 'producing' the faces until they meet again, always preserving the rotational symmetry of the original solid. **1948** —— *Regular Polytopes* xiv. 264 The first stellation of {5, 3, 3} is constructed by stellating the 720 pentagons into { }'s, and the 120 dodecahedra into {$\frac{5}{2}$, 5}'s. The result is {$\frac{5}{2}$, 5, 3}. **1971** *Sci. Amer.* Dec. 114/2 The five Platonic solids are complete of their symmetrical kind. By mixing regular polygon faces, however, Archimedes made 13 more solids. These can be variously extended by putting cells on faces, called stellation, or by cutting away cells whenever that process can reveal new regular polygon faces within.

stelled, *a.* For † *Obs.* read '*Obs.* exc. *poet.*' and add: **b.** (Later example.)
1949 Blunden *After Bombing* 41 Thence the eye soon travelled over much green plain, Swathed with plats blush-dyed, with blue meres stelled.

Steller (ste·ləɹ). The name of Georg Wilhelm *Steller* (see next), used *attrib.* or in the possessive to designate certain animals associated with his explorations, as *Steller's* (*eider*) (*duck*), a black and white duck with reddish underparts, *Polysticta stelleri*, found in Siberia, Alaska, and Canada; *Steller('s) jay*, a blue jay with a dark crest, *Cyanocitta stelleri*, found in western North America; *Steller's sea-cow*, an extinct sirenian, *Hydrodamalis stelleri*, once found in the Bering Sea; *Steller('s) sea-lion*, a large sea-lion, *Eumetopias jubata*, found in the northern Pacific. Cf. Stellerine.

1814 tr. *G. H. von Langsdorff's Voyages & Travels* II. i. 23 My curiosity was particularly directed to..Steller's sea-cow. **1828** C. L. Bonaparte *Amer. Ornithol.* II. 44 The Steller's Jay is one of those obsolete species alluded to in the preface to this volume. **1884** *Bull. U.S. Nat. Mus.* No. 27. 162 Steller's Duck..Arctic and subarctic coasts. **1902** *Bull. Amer. Mus. Nat. Hist.* XVI. 111 The northern and southern Sea Lions, often known respectively as Steller's Sea Lion and Forster's Sea Lion. **1917** Steller's jay [see *mountain jay* s.v. *mountain 9 c]. **1938** P. A. Taverner *Birds of Canada* 104 Steller's Eider... The smallest and the least eider-like of any of the birds known under that name. **1947** L. G. Ingles *Mammals California* 88 Steller Sea Lion bulls weigh up to 1,800 pounds. **1948** *Pacific Discovery* Jan. 14/2 Steller jays, bald eagles, mountains of wild geraniums,..these are the Teton country. **1957** P. J. Darlington *Zoogeography* vi. 400 Steller's Sea Cow..was hunted to extinction. **1963** *Times* 11 Feb. 14/4 One of these was an account of Steller's Eider Duck. **1968** G. Maxwell *Raven seek thy Brother* viii. 103 No one had ever proved whether or not the..beautiful little duck called Steller's Eider..actually bred in Varanger Fjord. **1972** *Village Voice* (N.Y.) 1 June 75/2 In the dense fir forests you can hear..the scream of the steller's jay. **1972** L. Hancock *There's a Seal in my Sleeping Bag* vii. 145 The rocky shores..are used by the Steller sea-lions as hauling-out grounds. **1975** *Islander* (Victoria, B.C.) 27 Apr. 3/1 He has been studying California and Steller's sea lions in Barkley Sound. **1976** *Ibid.* 27 June 2/2 When the potatoes were small, many Steller jays came to rob the patch.

stellerite (ste·lĕrəit). *Min.* [ad. G. *stellerit* (J. Morozewicz 1909, in *Bull. internat. de l'Acad. des Sci. de Cracovie* (Math.-Nat. Classe) II. 350). f. the name of Georg Wilhelm *Steller* (1709–46), German naturalist and explorer of the Komandorski Islands in the Bering Sea where the first samples were found: see -ite[1].] A zeolite, $CaAl_2Si_7O_{18}.7H_2O$, found as tabular orthorhombic crystals.

1909 *Jrnl. Chem. Soc.* XCVI. II. 1028 A new zeolite, stellerite, has been found at the N.W. Cape of Copper Island, one of the Komandorski islands in the Aleutian Group. **1968** *Amer. Mineralogist* LIII. 511 *Stellerite*, a valid orthorhombic end member of a continuous series with monoclinic stilbite. **1973** *Lithos* VI. 85 The walls of large cavities and fractures are lined with polycrystal aggregates of stellerite.

stelleroid (ste·leroid). *Zool.* [f. mod.L. *Stelleroidea* (J. W. Gregory in E. R. Lankester *Treat. Zool.* (1900) III. xiv. 237), f. F. *stelléride* (J. B. P. A. de Lamarck *Hist. Nat. Animaux sans Vertèbres* (1816) II. 527): see Stellerid and -oid.] A star-fish or a similar invertebrate belonging to the class Stelleroidea.

1900 J. W. Gregory in E. R. Lankester *Treat. Zool.* III. xiv. 237 This list of characters is quite sufficient to mark off the Stelleroids from all other Echinoderms. **1935** Twenhofel & Shrock *Invertebr. Paleontol.* vi. 190 Stelleroidea—..referring to the starlike appearance of the stelleroid. **1962** D. Nichols *Echinoderms* iv. 62 They were burrowers, and, like some recent stelleroids, sat in the substratum with only the arm-tips exposed above the surface. **1970** R. M. Black *Elements Palaeontol.* ix. 105 The stelleroids, too, are vagrant.

Stellite (ste·ləit), *sb.* Also **stellite.** [? f. L. *stell-a* star + -ite[1].] Any of various cobalt-based alloys that usu. contain chromium and small amounts of tungsten and molybdenum and are used for their great hardness and their resistance to heat.

A proprietary name in the U.S.

1913 *Engin. Mag.* XLV. 840/2 It is an alloy of cobalt, chromium, and tungsten and has been called 'Stellite' by its inventor, Mr. Elwood Haynes. **1916** *Official Gaz.* (U.S. Patent Office) 3 Oct. 267/2 The Haynes Stellite Co.... *Stellite*... Metal alloys. **1937** *Jrnl. R. Aeronaut. Soc.* XLI. 330 The mean thermal expansion coefficients of the stellites were obtained between 20°C and a higher temperature ranging from 100°C to 850°C. **1975** Bram & Downs *Manuf. Technol.* iii. 79 Stellite can be purchased in bars of round or square section for manufacturing cutting tools. **1980** *Daily Tel.* 11 Sept. 7 (Advt.), The new 1300cc 'A' Plus unit with stellite faced valves.

Hence **ste·llite** *v. trans.*, to coat with Stellite; **ste·llited** *ppl. a.*, **ste·lliting** *vbl. sb.*

1934 *Jrnl. R. Aeronaut. Soc.* XXXVIII. 329 If only one part is to be 'stellited' it is better to treat the valve. *Ibid.*, The 'stelliting' of the valve stem and neck..would also be beneficial. **1937** *Ibid.* XLI. 330 The cracking of stellited steels is attributed to this difference in expansion. **1959** *Engineering* 9 Jan. 49/2 The journal portions of extension shafts on the vanes are Stellited and ground to provide a wear resistant bearing surface.

Stellwag (ste·lvag). *Med.* [The name of Carl *Stellwag* von Carion (1823–1904), Austrian ophthalmologist.] *Stellwag's sign*: orig., retraction of the upper eyelid in thyrotoxicosis (called also Dalrymple's sign); now often applied to the diminished blinking that normally accompanies it.

1887 *Brit. Med. Jrnl.* 26 Mar. 680/1 A married woman, aged 32, with marked retraction of the left upper eyelid (Stellwag's sign), no proptosis, and no goitre. **1907** *Practitioner* Nov. 733 [In exophthalmic goitre] the margins of the eyelids are unduly separated (Dalrymple's sign); the upper lid tardily follows the eyeball in its downward movements (von Graef[e]'s sign); there is diminished frequency of winking (Stellwag's sign). **1950** R. I. Pritikin *Essent. Ophthalm.* vii. 321 Stellwag's sign—infrequent winking. **1973** R. S. Dillon *Handbk. Endocrinol.* vi. 248/1 'Stellwag's sign' is retraction of the upper eyelids producing apparent widening of the palpebral opening associated with infrequent and incomplete blinking.

stem, *sb.*[1] Add: **4. i.** *pl.* The legs. *slang.*
1860 Hotten *Dict. Slang* (ed. 2) 227 *Stems*, the legs. **1927** *Vanity Fair* XXIX. 67/2 Among some of Conway's more famous expressions are:..'Stems' and 'Gambs' (legs). **1970** C. Major *Dict. Afro-Amer. Slang* 109 *Stems*, the legs.

j. More fully *drill stem*: (*a*) (also *auger stem*) in percussion drilling, a heavy metal rod above the bit in a string of tools, used to provide added weight; (*b*) in rotary drilling, = *grief stem* below; also, the entire drilling column; *grief stem*: in rotary drilling, the rod at the top of the drilling column, having a square cross-section so that it fits in and is turned by the rotary table; = *kelly *sb.*[2] 4.

drill-stem test: a test of the potential of a well in which a sample of the oil or gas is allowed to run into the drill pipe for a short time before the hole is completed, the flow being measured and the fluid recovered; so *drill-stem testing* vbl. sb.

1880 J. F. Carll *Geol. Oil Regions Warren, Venango, Clarion & Butler Counties* III. xxviii. 300 On the down stroke the auger-stem falls 20 inches, while the sinker-bar goes down 24 inches. **1907** *Internat. Libr. Technol.: Rock Boring* 13 The tools consist of a rope socket, sinker bar, jars, stem, and bit. *Ibid.* 15 Auger or Drill Stem.—This part of the string of tools..is made about 30 feet long in some cases. **1922** B. Redwood *Petroleum* (ed. 4) II. 402 The lower end of the drilling-rod or casing with the bit is passed through the rotary table provided with grip rings or square 'grief' stem and clamped tight enough to cause it to revolve. **1938** D. Hager *Practical Oil Geol.* (ed. 5) viii. 252 After drilling into an oil stratum, some idea of productivity may be gained by making a drill-stem test. **1939** —— *Fund. Petroleum Industry* ix. 210 When the kelly is deep enough for a joint of drill stem, the kelly and bit are pulled out. **1951** K. K. Landes *Petroleum Geol.* ii. 51 The string of tools consists of the bit..; the stem, into which the bit fits; the jars; and the socket. **1962** E. J. Lynch *Formation Evaluation* viii. 291 Drill stem testing is the most hazardous of all drilling operations. **1965** E. Lehner et al. in G. J. Williams *Econ. Geol. N.Z.* xix. 350/2 Both the Taramakau and Arahura wells.. encountered faint traces of oil in the Brunner Sandstone at 5,700 ft and 5,030 ft respectively, but drill-stem tests of these zones yielded only salt water. **1973** J. W. Jenner in Hobson & Pohl *Mod. Petroleum Technol.* (ed. 4) iv. 108 A drilling bit..is attached to a heavy drill stem suspended by a cable from a cantilever arm, the Walking Beam, at th surface. **1976** L. St. Clair *Fortune in Death* i. 8 The drill stem had snapped... 'Stem crystalize?' 'Yep. Damned basalt is hard as the drill.'

k. Similarly, a drill used by a burglar or safe-breaker. *U.S. Criminals' slang.*
1914 Jackson & Hellyer *Vocab. Criminal Slang* 81 *Stem, noun.* Current among yeggs. A steel drill. **1926** J. Black *You can't Win* x. 133 Get the 'dan' and 'stems' (drills), and put them safely away. **1935** *Flynn's* 16 Mar. 102/1, I was inserting a 'stem' (drill) in a brace when I heard a most peculiar noise.

5*. a. A principal railway line, from which other tracks may branch; = *main stem.
1832 [see *main stem]. **1869** *Bradshaw's Railway Man.* XXI. 426 *Assets.* Main Stem..Lebanon Branch extension ..Richmond Branch. **1934** in *Amer. Ballads & Folk*

Songs i. 24 The manifest freight Pulled out on the stem behind the mail.

b. A street, esp. one frequented by beggars and tramps (see also quot. 1923); also, = *main street a; *transf.*, an act of begging. *U.S. slang.*

1914 [see *mooch *v.* 6]. **1923** N. Anderson *Hobo* i. 4 Every large city has its district into which these homeless types gravitate. In the parlance of the 'road' such a section is known as the 'stem' or the 'main drag'. **1929** *Amer. Speech* IV. 345 *Stem*, act of begging; also a street where one begs. **1931** 'D. Stiff' *Milk & Honey Route* v. 59 The hobo also damns the hash houses along the stem. **1936** *New Republic* 15 July 289/1 The appearance of the applicant is perhaps not so important as in the case of private residences or on the 'stem'. **1955** D. W. Maurer in *Publ. Amer. Dialect Soc.* xxiv. 133 This is all done on the *stem* or street.

7. (sense 5 b) *stem-form, -formant*; *stem-final, -formative, -initial* adjs. (all also *absol.* as sb.); *stem-forming* adj.

1949 E. A. Nida *Morphology* (ed. 2) ii. 34 All stem-final vowels before vowel suffixes are lost. **1965** *Canad. Jrnl. Linguistics* X. 130 It seems that Chipewyan and Navaho treat it as a stem final, while Mattole treats it as a stem initial. **1973** A. H. Sommerstein *Sound Pattern Anc. Greek* ii. 17 There is a large class of nouns ending, in the nominative singular, in *-eús* [-éws]..which have a stem-final [w] when a consonant follows. **1928** O. Jespersen *Internat. Lang.* II. 97 The bare stem-form of many adjectives would not be euphonious enough. **1966** *English Studies* XLVII. 53 These genetive [*sic*] 'causative' objects disappear in early ME and give place to objects in stem-form or prepositional types. **1935** G. K. Zipf *Psycho-Biol. of Lang.* iv. 144 A stem-formative (or stem-formant) is a morpheme added to the root either at the beginning (prefix) or at the end (suffix) or tucked inside (infix) to make the stem (root plus formant) to which endings are added. **1964** K. L. Pike in D. Abercombie et al. *Daniel Jones* 428 The stem is made up of three monosyllabic roots plus three following stem-formative syllables. **1968** Chomsky & Halle *Sound Pattern Eng.* 130 In short, there are 'stem-forming' vowels / i / and / u / which are deleted in final position..but which remain before certain affixes. **1978** *Language* LIV. 220 Most stem-forming suffixes consist of a single segment, and a certain amount of homophony results. **1949** E. A. Nida *Morphology* (ed. 2) ii. 15 When stem-initial consonants are aspirated, the reduplicated consonant has the same point of articulation, except that it is unaspirated. **1965** [see *stem-final* above]. **1977** *Word 1972* XXVIII. 223 The inappropriate stem-initial consonants.

8. stem analysis *Forestry*, (an) investigation of the past growth of a tree by study of a series of cross-sections of its trunk taken at different heights; **stem borer,** an insect larva that bores into plant stems; **stem cell** *Biol.*, (*a*) a cell in the stem of an organism (*nonce-use*); (*b*) a cell of a multicellular organism which is capable of giving rise to indefinitely more cells of the same type and from which certain other kinds of cell arise by differentiation; **stem-cup,** a Chinese porcelain goblet of a type with a wide shallow bowl mounted on a short base, first made in the Ming dynasty; **stem family** *Sociol.* [tr. F. *famille-souche* (F. le Play *La Réforme Sociale en France* (1866) I. iii. 249)], a family unit in which property descends to a married son who remains within the household, other (esp. married) children achieving independence on receipt of an inheritance; **stemflow** *Forestry*, precipitation which reaches the ground after running down the branches and trunks of trees; **stem ginger,** a superior grade of crystallized or preserved ginger; **stem-glass,** (*a*) a tall narrow glass vase for the display of a single flower or flowers; (*b*) a drinking-glass mounted on a stem; **stemline** *Med.*, the group of cells having a chromosome number that is (one of) the most frequent in a mixed population, esp. of tumour cells; **stem mother** *Ent.* = *fundatrix 2; **stem root,** a root that develops from the stem of a plant, esp. on a lily from just above the top of the bulb; so **stem-rooting** *a.*, producing roots of this kind; **stem rust,** any of various fungus diseases of plants that produce spots on the stems; *esp.* that caused by *Puccinia graminis* on wheat and other cereals, marked by rows of black telia on the stems; **stem stitch:** also used more widely in *Embroidery* (earlier example); **stem succulent,** a plant chiefly native to dry regions, distinguished by a fleshy stem and often very small leaves or spines; **stemware,** stemmed glass drinking-vessels; **stem-winder,** (*c*) also, an enterprising or energetic person; an impassioned talker or public speaker; (*d*) *slang*, a rousing speech; **stem-winding** *a.* (earlier example); also *transf.*

1895 W. Schlich *Man. Forestry* III. I. vi. 83 The investigation of the progress of increment throughout the life of a tree is called a stem analysis. **1974** *Forest Sci.* XX. 75/2 In the comparisons to be discussed, three sets of stem analysis data were used, one for noble fir, one for Douglas-fir, and one for red alder. **1921** H. T. Fernald *Appl. Entom.* xxxiii. 340 Superfamily Tenthredinoidea (The Saw-flies and Stem Borers). **1939** *Geogr. Jrnl.* XCIII. 140 The crop [of rice] would be destroyed by rats and stem-borers. **1972** J. Minifie *Homesteader* xv. 121 The Manitoba maple was subject to a stem-borer which destroyed its growing tips. **1885** Stem cell [in Dict., sense 7]. **1896** E. B. Wilson *Cell* iii. 111 In *Ascaris megalocephala univalens*..each of the first two cells receives two elongated chromosomes... In one of them, which is destined to produce only somatic cells, the thickened ends of each chromosome..degenerate... In the other cell, which may be called the stem-cell.., all the chromatin is preserved and the chromosomes do not segment into smaller pieces. **1959** W. Andrew *Textbk. Compar. Histol.* vi. 234 The cells of this organ, while they include many lymphocytes, apparently serve as stem cells for all of the types of white corpuscles. **1970** *Sci. Jrnl.* June 32/3 The production of a continuous supply of spermatozoa from the testis is ensured by the continued existence of germ cells which form a reservoir of stem cells from which future spermatozoa are derived. **1915** R. L. Hobson *Chinese Pottery & Porcelain* II. xii. 208 In the Bushell collection there are some beautiful reproductions of the Ch'êng Hua 'stem-cups'. **1942** *Burlington Mag.* June 151/2 The part of the base immediately below the stem-cup is a rather squat form of the Venetian bell base. **1980** *Catal. Fine Chinese Ceramics* (Sotheby, Hong Kong) 36 A *Longquan* (Lung Ch'üan) *Celadon stemcup* with plain circular bowl raised on a ribbed columnar foot. **1936** Zimmerman & Frampton *Family & Society* vii. 125 The stem-family..unites one married child to the paternal household, and supplies all the other offspring in a 'state of independence [with a dowry] which the patriarchal family does not give them'. **1947** P. H. Landis *Your Marriage & Family Living* i. 5 The stem family is about halfway between the great family of historic rural societies and the small, individualistic family of urban industrial societies. **1977** P. Laslett *Family Life & Illicit Love in Earlier Generations* v. 211 In certain eighteenth-century areas of Austria and Germany, where a stem family arrangement prevailed..the old were allotted a familial situation which gave to retirement an institutional form. **1941** *Jrnl. Forestry* XXXIX. 521/1 Stemflow, although apparently not related to crown-length density, tree height, basal area, or crown area, does tend to increase with excess or deficit of height of tree as compared with adjacent trees. **1967** M. E. Hale *Biol. Lichens* vii. 96 Stemflow on trees..has been shown to be enriched, relative to throughflow, with potassium and calcium. **1980** Spurr & Barnes *Forest Ecol.* (ed. 3) ix. 230 Airborne dust, ash, and gaseous aerosols may also become attached to or impacted on tree surfaces and carried to the soil as throughfall and stem flow. **1922** A. Ward *Encycl. Food* 224 Crystallized ginger is also made from the young roots. The best grades, selected for uniform size and appearance, are called 'stem ginger'. **1977** *Times* 2 Sept. 10/5 The menu offers..bananas with stem ginger. **1922** Joyce *Ulysses* 224 He took a red carnation from the tall stem-glass. **1974** L. Deighton *Spy Story* xvii. 187 A stem glass from the ice-box, really cold Beefeater and..seven per cent dry vermouth. **1979** B. Hines *Price of Coal* I. 14 He..took down a tinted stem-glass..[and] selected his favourite bloom. **1953** Levan & Hauschka in *Jrnl. Nat. Cancer Inst.* XIV. 5 Chromosome numbers and the concept of 'stemline'.... The chromosome class with the highest number of cells, and the adjacent classes, represent the types mainly responsible for growth and characterize the principal stemline of each tumour. **1962** *Lancet* 27 Jan. 218/2 Several cases of mosaicism in mongolism have been reported... Two of these cases were mosaics with three stemlines, 46, 47, and 48 chromosomes; and two with two stemlines, 46 and 47 chromosomes. **1972** *Science* 23 June 1340/3 The establishment of this pattern was based on the analysis of stemlines, sidelines, and single deviating cells in 80 primary and 20 metastatic tumors. **1878** *Entomologist's Monthly Mag.* XIV. 224 An enormous single egg, from which, undoubtedly, will come the stem-mother. **1907, 1923** [see *fundatrix 2]. **1979** *Vole* Dec. 40/2 Here the mated females [*sc.* grain aphides] hide the eggs that will come through the winter. These hatch in the late spring.., becoming 'stem mothers' that grow up to fly in quest of grain fields. **1901** G. Jekyll *Lilies for Eng. Gardens* iii. 8 It should be planted six to seven inches deep, as it forms stem roots. **1978** B. Mathew *Larger Bulbs* 85 Some species [of *Lilium*] produce a tuft of roots from the stem just as it emerges from the bulb, these stem-roots partly acting as extra support. **1896** T. W. Sanders *Encycl. Gardening* (ed. 2) 19 *Artocarpus*... Propagation: By stem-rooting firm shoots in Feb., March; suckers at any time. **1974** H. G. W. Fogg *Compl. Handbk. Bulbs* vii. 91/1 This stem-rooting lily should be planted at least 4 ins deep. **1899** M. A. Carleton *Cereal Rusts U.S.* 57 The stem rust..is not constant in occurrence, but will occasionally miss one or two years. **1923** *Jrnl. Agric. Res.* XXIV. 979 There are several biologic forms of stem rust of wheat. **1946** K. S. Chester *Nature & Prevention of Cereal Rusts* xiv. 199 Disproportionate emphasis..has been laid on stem rust..in wheat. **1979** Tanous & Rubinstein *Wheat Killing* (1980) ix. 57 There's some stem rust around... The black spores of the rust were clearly visible. **1873** *Young Englishwoman* June 299/1 The embroidery is worked..in satin and stem-stitch, and point-russe. **1897** J. C. Willis *Man. & Dict. Flowering Plants & Ferns* I. iii. 182 In the stem-succulents the leaves are reduced to scales or thorns. **1966** E. Palmer *Plains of Camdeboo* xvi. 258 It is the stem succulents..that are the most typical of Cranemere —Euphorbias, Stapeliads, and other such..most of them leafless, the work of the leaves being done by the stems and so protected from excessive transpiration; or bearing leaves for a short time only. **1929** *Sears Catal.* Fall 898/1 One of the newest creations in stemware. **1966** H. Nielsen *After Midnight* (1967) xv. 194 A small, circular dinner table..had been meticulously set with china, silver and stemware. **1926** in J. F. Dobie *Rainbow in Morning* 85 He's a stemwinder and go-getter. **1942** Berrey & Van den Bark *Amer. Thes. Slang* § 422/5 *Speech-maker*,.. stemwinder, vitalics, *a forceful talker*. **1973** T. H. White *Making of President 1972* (1974) viii. 210 After all the calls to unity,..a stemwinder in the old tradition from Hubert Humphrey,..appearances by Muskie and Kennedy, Sargent Shriver was formally nominated for Vice-President. **1977** *Time* 3 Jan. 55/2 The 1,008 cadres and 24 fraternal foreign delegations..endured no fewer than 55 speeches, including an eight-hour stem-winder by Le Duan. **1867** *Rep. Comm. Patents 1866* (U.S.) 1115 Either side of the case of the stem-winding watch is opened by pressure upon the head of the winding arbor. **1966** *Atlantic Monthly* Sept. 90 A stem-winding sermon by Reverend Cecil Todd..can be obtained by sending one dollar to *Revival Fires* in Joplin, Missouri.

stem, *v.*[2] Add: **6. a.** *Skiing.* [ad. G. *stemmen.*] To decelerate (esp. before a turn) on a traverse descent by weighting the upper ski and angling its outer edge into the snow, causing the ski to turn downhill.

1904 E. C. Richardson *Ski-Running* 41 (*caption*) *The proper way to stem*... A good stemming-spoor is at once known by the broad track of the braker. **1935** *Punch* 6 Feb. 164/3 *Stemming.*—Your ordinary straight running will lack the easy confidence..it should have unless I first show you how to stem, which is the only legitimate way of applying the brake other than using complicated turns like the christiania or the telemark. **1948** H. Innes *Blue Ice* x. 249 Jill stopped then and I stemmed. We were standing at the end of the snowshed. **1970** N. Fleming *Czech Point* i. 22, I sideslipped at first, grew tired of it, stemmed for a while and then had just started to take the slope straight when the shot was fired.

b. The vb. in Comb., as **stem-Christiania,** a turn made by stemming and lifting the lower ski parallel as the manœuvre is completed (less advanced than the full *Christiania); hence as colloq. abbrev. **stem-Christie** (also *v. intr.*, to turn in this manner); **stem turn,** an elementary turn made by stemming and then bringing the unweighted ski parallel in the new line; also *transf.* and as *v. intr.*, to make a stem turn or turns.

1922 V. Caulfeild *Ski-ing Turns* xii. 228 The Stem-Christiania, like the Stem turn, is mainly used for down-hill turning on hard snow... A downhill turn can..be made more sharply by the Stem-Christiania than by any other means except a jump. **1961** *Times* 7 Jan. 7/6 The tried and basic essentials of the snowplough and the stem-christiania in particular. **1936** *Sierra Club Bull.* Feb. 57, I soon began very short linked stem-christies with the aid of the inner pole. **1942** 'N. Shute' *Pied Piper* ii. 20 At each new slope of snow he thought to see John come hurtling over the brow, stem-christie to a traverse, and vanish in a white flurry that sped down into the valley. **1972** 'M. Yorke' *Silent Witness* vi. iv. 141 They stem-christied inexpertly over the wide plateau. **1922** V. Caulfeild *Ski-ing Turns* vii. 123 The Stem turn is impossible in heavy soft snow or breaking crust... There are two forms of the Stem turn... The Pure Stem turn is only possible on gentle slopes... The Lifted Stem Turn can be employed for down-hill turning on moderately steep slopes. **1938** [see *Christie]. **1959** M. Gilbert *Blood & Judgement* ix. 98 Petrella..started straight off down the pavement, did a stem turn at the corner... 'At least he can still walk,' said Borden. **1973** J. Goodfield *Courier to Peking* xiii. 171 They were moving downhill with the competence of an Olympic skier in a slalom race, stem-turning neatly at every snake-like twist of the road.

stem, *v.*[4] Add: **3. a.** (Earlier examples.)
In quot. 1724 the senses of *stem* and *strip* are confused (*D.A.E.*).

1724 H. Jones *Present State of Virginia* 40 It lies till they have Leisure or Occasion to *stem* it (that is pull the leaves from the Stalk) or *strip* it (that is to take out the great Fibres). **1797** G. Imlay *Topogr. Descr. Western Terr. N. Amer.* (ed. 3) 248 This done, you stem the tobacco, or pull out the middle rib of the leaf.

b. (Earlier example.)
1873 *Trans. Illinois Dept. Agric.* X. 61 The grapes were pressed without stemming.

4. To beg or 'panhandle' on the streets. Cf. *stem *sb.*[1] 5* b. *U.S. slang.*
1924 'Digit' *Confessions 20th Century Hobo* 12 *Stemming*, begging, cadging. **1931** 'D. Stiff' *Milk & Honey Route* viii. 84 Panhandling falls into two classes: the domestic type..and a kind carried on in the streets and known as 'stemming'. **1937** *Lit. Digest* 10 Apr. 12/2 *Stemming*, panhandling in cities.

5. a. *fig.* To derive or take origin *from*; to spring from. (The principal modern sense.) orig. *U.S.*

1932 A. H. Quinn *Soul of Amer.* 131 The policy of vigorous intervention in the affairs of the nations bordering on the Caribbean Sea stems from Roosevelt's administration. **1937** R. S. Morton *Woman Surgeon* i. 15, I realize now that my apparent indifference to suitors for marriage stemmed from my determination to study medicine. **1942** W. Faulkner *Go down, Moses* 86 He knew what he had seen in his father's face..something.. not stemming from any difference of race nor because one blood strain ran in them both. **1949** *Here & Now* (N.Z.) Nov. 27/1 From this stemmed a whole line of high-grade thrillers. **1952** B. Smalley *Study of Bible in Middle Ages* 358 Both literal and spiritual exposition stemmed from Origen. **1958** *Times* 11 Feb. 11/7 The whole of this trouble has stemmed entirely from your own behaviour. **1961** I. Fleming *Thunderball* xviii. 194 The source of his wealth was unknown but did not stem from funds held in Italy. **1976** H. Wilson *Governance of Britain* x. 183 Bills normally stem from the legislative arm. **1979** J. Grimond *Memoirs* vi. 98, I believe that much that is wrong with attitudes and organisation in Britain stems from the war.

b. To extend *back to* in origin. Also, to arise *out of*.

1937 *Sun* (Baltimore) 24 Nov. 2/7 Wall Streeters said the controversy stemmed back to the annual report issued by Mr. Gay as president of the exchange last August. **1959** N. Lofts *Heaven in your Hand* 99 The whole thing stemmed back to the beliefs in the African witch-doctors. **1965** *Listener* 11 Nov. 740/1 It is sometimes claimed that race antipathy stems out of the same order of cultural differences. **1974** *Times Lit. Suppl.* 26 Apr. 430/3 There is still doubt and conflict here, stemming back to the ancient world.

stemma. Add: **1. c.** A diagram which represents a reconstruction on stemmatic principles of the position of the surviving witnesses in the tradition of the transmission of a text, esp. in manuscript form. Cf. *stemmatology.

1930 W. P. Shepard in *Mod. Philol.* XXVIII. 130 The claim that the Lachmannian method can deduce..the *ipsissima verba* of the author, takes no account of the fact that an author may change his mind...and the final and most telling objection to this method..is the fact that in practice it..leads to the establishment of a dichotomous 'stemma', a family tree. **1942** J. B. Severs in *English Institute Ann. 1941* 79 And so we deal with each of our families in turn, until the manuscripts in each family have been completely outlined... Thus we build up a complete stemma, or genealogical chart, for our whole body of manuscript. **1949** *Oxf. Classical Dict.* 889/1 Where one witness depends on two or more other witnesses, i.e. where the transmission is 'contaminated' and the stemma 'convergent', it is seldom possible to ascertain the type of interrelationship by stemmatics. **1962** E. J. Dobson in Davis and Wrenn *English & Medieval Studies* 136 We thus arrive at the stemma shown by the unbroken lines in the diagram. **1982** *N. & Q.* Feb. 77/2 The meticulous detail of information is made clear by stemma diagrams. for those with the courage to penetrate the complexity.

stemma·tic, *sb.* and *a.* [a. G. *stemmatik* (P. Maas in *Byzantinische Zeitschrift* (1937) XXXVII. 289), f. L. *stemma(t-)* Stemma + -ic.] **A.** *sb.* *pl.* **stemmatics** = *stemmatology.

1949 [see *stemma 1 c]. **1958** B. Flower tr. *P. Maas' Text. Criticism* 42 Errors arising in the course of transcription are of decisive significance in the study of the interrelationships of manuscripts—I may be allowed to use the term 'stemmatics'. **1968** Reynolds & Wilson *Scribes & Scholars* v. 140 The classic statement of the theory of stemmatics is that of Paul Maas. **1975** *Times Lit. Suppl.* 15 Aug. 928/4 There has been a tendency to see stemmatics as something born in an instant. **1980** *Early Music Gaz.* Jan. 13/3 Source studies and stemmatics play an increasingly important role in musicology.

B. *adj.* Of or pertaining to a textual stemma or stemmata; concerning the reconstruction of the interrelationships between the readings of manuscripts of a text, esp. as *stemmatic theory.*

1958 B. Flower tr. *P. Maas' Text. Criticism* 44 Such errors are so rare..that we cannot rely on being able to find one to establish every stemmatic relationship. **1968** Reynolds & Wilson *Scribes & Scholars* 140 In practice the stemmatic theory has serious limitations. **1980** *Early Music Gaz.* Jan. 13/3 Alejandro Planchart concluded that traditional stemmatic theory (after Paul Maas) best fits the text transmission of plainsong.

stemmatology (stemătǫ·lŏdʒi). [f. L. *stemma(t-)* Stemma: see -ology.] The study which attempts to reconstruct the tradition of the transmission of a text or texts (esp. in manuscript form) on the basis of the relationships between the readings of the various surviving witnesses; this sphere of scholarship.

1942 *Essays & Studies* XXVII. 43 In what may be called stemmatology, the veterans..have done brilliant work. **1981** *N. & Q.* Feb. 1/2 Mr. Hamer..attempts a 'provisional' stemma. The darkness of conflicting and 'contaminated' evidence might have made a less brave man shrink from stemmatology.

stemming, *vbl. sb.*[2] Add: **2.** *Skiing.* The action of *stem *v.*[2] 6 a.
1904, etc. [see *snow-ploughing *vbl. sb.*] **1935** P. Frankau *I find Four People* v. 311 We circled on our awkward attempts at stemming, frozen and undecided.

Sten (sten). Also **sten.** [Acronym f. the initial letters of the surnames of the designers, R. V. *S*hepherd and H. J. *T*urpin + *En*field, Greater London (see quot. 1942[1]), after *Bren.] More fully, *Sten gun.* A type of light, rapid-fire, sub-machine-gun. Also *fig.* and *attrib.*

1942 *Times* 16 July 8/3 It [*sc.* the gun] was known as the Schmeisser gun, and the Sten was merely a slight modification of it... The Ministry of Supply was not justified in taking to itself high praise for the speed of production of the Sten, and in giving the country the false impression that 'Colonel S' and 'Major T' invented and designed the gun. **1942** 'G. Orwell' *Diary* 7 Aug. in *Coll. Ess.* (1968) II. 442 Last night for the first time took a Sten gun to pieces. **1959** 'N. Blake' *Widow's Cruise* 99 He went off into a sten-gun burst of Greek. **1971** B. W. Aldiss *Soldier Erect* 255 It was bloody murder. I had shed the set, and went in firing the sten from the hip. *Ibid.* 261 In my ammo pouch, against the sten magazines, I had stuffed the picture of Hanuman. **1974** S. Gulliver *Vulcan Bulletins* 18 Automatic weapons like Stens and Sterlings.

Hence **Ste·n-gun** *v. trans.*, to shoot at or kill with a Sten gun; **Ste·n-gunner,** one who operates a Sten gun.

1949 Koestler *Promise & Fulfilment* II. v. 280 There is..no conceivable justification..for Sten-gunning the representative of an international body. **1961** *Times* 8 Mar. 13/6 The Tunisian stengunners outside the luxury hotel: the Hiberno-Scandinavian-Afro-Asian chatter in the café below United Nations headquarters.

Stender (ste·ndəɹ, ʃt-). *Biol.* and *Med.* Also **stender.** [The name of Wilhelm P. *Stender*, a manufacturer of Leipzig.] *Stender dish*: a shallow glass dish.

1900 Dorland *Med. Dict.* 637/1 Stender-dish. **1904** *Bot. Gaz.* XXXVII. 12 A simpler..method is to float a quantity of these spores on the surface of water half filling a stender dish. **1978** G. C. Brown *Introd. Histotechnol.* xi. 174 Other items that should be stored at the staining bench in adequate quantities are:..Stender dishes for frozen sections.., in three sizes.

Stendhalian (standā·liăn), *a.* and *sb.* [f. *Stendhal*, the nom-de-plume of the French writer Henri Beyle (1783–1842) + -ian.] **A.** *adj.* Characteristic or suggestive of the writings of Stendhal. **B.** *sb.* A follower or devotee of Stendhal or his works.

1907 G. B. Shaw *Mahor Barbara* Pref. 150 The sensation first came to me from Lever and may have come to him from Beyle, or at least out of the Stendhalian atmosphere. **1928** *Sunday Express* 8 Apr. 7/4 The Stendhalians are invariably cynics who delight in heartlessness and selfishness. **1937** Wyndham Lewis *Blasting & Bombardiering* 53, I was not surprised, luckily, when I became a lion, to find this gilded tamer a tough customer. I began studying her ways with curiosity, this spoilt and cocksure goddess of the ocean wave. I filled a notebook with Stendhalian observations. **1950** *Essays & Studies* III. 86 In this sense the Stendhalian analogy of crystallization may be applied equally to the reasons of the intellect as to the reasons of the heart, of which reason itself comprehends nothing. **1968** E. Hyams *Mischief Makers* vi. 99, I had gone flabby; no Stendhalian hero. **1980** A. Alpers *Life Katherine Mansfield* ix. 163 Murry [was] ..starting a huge Stendhalian novel.

stengah (ste·ŋă). Also **stingah.** [ad. Mal. *setengah* half.] A half measure of whisky with soda (sometimes water), popular amongst the British in Malaysia.

1899 *Munsey's Mag.* XXI. 697/1 The majority of young Englishmen and Germans fared equally well when they limited the numbers of whiskies and sodas (or 'stengahs') which they drank. **1903** W. Del Mar *Around World through Japan* vii. 64 A 'peg' of whiskey and tonic-water, followed by a *stengah* (the Malay word for half, usually pronounced *stinger*) or split drink. **1927** *Blackw. Mag.* June 726/1 At this establishment we learn all about gin-slings and 'Stingahs'. **1948** M. Laski *Tory Heaven* i. 14 Stewards..to bring him..stengahs and pahits whenever he wanted them. **1966** D. Forbes *Heart of Malaya* xiii. 156 Malays could not be seen publicly knocking back *stengahs*. **1975** 'G. Black' *Big Wind* i. 18 When I was a boy in Malaysia a whisky-soda was called a *stengah*. It hasn't been for twenty-five years.

steno (ste·no). Colloq. abbrev. of: **a.** Stenographer (also *attrib.*), or abbrev. Stenographic *a.* orig. and chiefly *U.S.*

1906 J. F. Kelly *Man with Grip* 36 This bright young lady was on the waiting list of a steno. agency. **1928** 'L. North' *Parasites* ii. 34 (*heading*) 'Stalls' and stenos. *Ibid.* vii. 89 That Whispering Slim guy seems to be falling for that steno' of his. **1935** *Motion Picture* Nov. 40/2 Frances Dee..skyrockets to new importance with an amazingly fine performance as a small town steno who wins a five-thousand-dollar lottery. **1958** E. Birney *Turvey* v. 49 There was a good-lookin steno all alone. **1971** D. E. Westlake *I gave at Office* (1972) 19 He led the unsuccessful attempt to ban miniskirts in the steno pool. **1978** P. Boardman *Worlds of Patrick Geddes* ix. 342 A social workers' weekly..sent a stenotypist to take them [*sc.* lectures] down and a department editor to check up on the steno.

b. Stenography. *U.S.*
1946 *N.Y. Times* 1 Apr. 42/3 (Advt.), Girl, genl office work, typ, some steno. **1954** *Los Angeles Times* 21 Mar. 3/2 (Advt.), Clerk—lite steno..$200 start. **1973** *N.Y. Law Jrnl.* 20 July 16/2 (Advt.), *Law secretary*...legal experience and excellent skills required, dictaphone and light steno.

steno-. Add: **stenoba·thic** *a.* *Biol.* [Gr. βάθ-ος depth], (of aquatic life) limited to or found at only a narrow range of depths; **stenohaline** (-hēɪ·ləin) *a.* *Biol.* [Gr. ἅλιν-ος of salt], (of aquatic life) adapted to only a narrow range of salinity; **stenohy·dric** *a.* *Biol.* [Hydro-], adapted to only a narrow range of humidities; **steno·phagous** *a.* *Zool.* [-phagous] (see quot. 1926); **stenopo·dium** *Zool.* [Podium], a narrow, two-branched crustacean limb the flexibility of which is provided by joints; **stenothermal** *a.* (further examples); [ad. G. *stenotherm* (K. Möbius 1871, in *Jahresber. d. Commission z. wissensch. Untersuchung d. deutschen Meere in Kiel* (1873) I. 139)]; hence **ste·notherm, stenothe·rmic** *adjs.*; **stenoto·pic** *a.* *Biol.* Gr. τόπ-ος place] (see quot. 1949).

1902 Stenobathic [see *eurybathic *a.*]. **1975** B. Fell *Introd. Marine Biol.* xi. 92 A deep-water stenobathic species, when brought too rapidly to the surface in a net, suffers disruption of the internal organs. **1930** *Biol. Rev.* V. 350 Most stenohaline marine invertebrates are poikilosmotic: their body fluids have an osmotic pressure which is the same as that of the external medium. **1973** P. A. Colinvaux *Introd. Ecol.* xx. 278 When temperature or salinity may fluctuate widely without seriously affecting individuals, the species are called eurythermal or euryhaline; when slight changes of temperature or salinity are fatal to animals or plants, they are called stenothermal or stenohaline. **1953** E. P. Odum *Fund. Ecol.* iii. 27 Stenohydric—Euryhydric refers to water. **1974** *Ciba Foundation Symp.* XX. 56 In general, relative humidities below 60% (temperatures of 21–28°C) are deleterious for these stenohydric species. **1926** A. S. Pearse *Animal Ecol.* iii. 72 Animals that have a narrow range of foods are called stenophagous and those that eat a wide variety are euryphagous. **1976** *Environmental Entomol.* V. 46/2, 21 (46%) of 46 identified species of phytophagous insects found associated with *A*[*mbrosia*] *dumosa* in southern California were stenophagous. **1932** Borradaile & Potts *Invertebrata* xii. 298 Since..the phyllopodium possesses the same two rami, and bears them, though not as a distal fork, yet in the same way as a great number of limbs of the first type, it is well not to use a name which might imply that there is a constant difference in respect of the rami between the limbs of the two types. We shall therefore call the first type the stenopodium, referring to its usually slender form. **1967** P. A. Meglitsch *Invertebrate Zool.* xviii. 755/1 A good case can be made for thinking of stenopodia as the more primitive form of crustacean appendage. **1888** Stenotherm [see *eurytherm *a.*]. **1964** *Oceanogr. & Marine Biol.* II. 284 Most stenosaline organisms live either in the ocean (polystenosaline forms) or in fresh water (oligostenosaline forms). These terms are analogous to steno- or eurytherm and just as relative in their meaning. **1937** *Brit. Birds* XXX. 247 It should be borne in mind that whereas the adult bird is stenothermal (warm-blooded), in the young the thermotaxic arrangements are undeveloped. **1973** P. A. Colinvaux *Introd. Ecol.* xx. 279 Between about 55°C and 40°C the algal mats are largely made up of filamentous blue-green algae, but these plants are rather stenothermal and will not actively grow at temperatures below 40°C. **1926** A. S. Pearce *Animal Ecol.* ii. 34 Animals are often classified into two groups: stenothermic and eurythermic, the former being restricted to a narrow range of temperature changes and the latter having ability to live through a wide range. **1965** B. E. Freeman tr. *Vandel's Biospeleology* xxiii. 384 Summer cysts containing the adults of stenothermic species [of copepods] are formed during the warm season. **1949** J. H. Kenneth *Henderson's Dict. Sci. Terms* (ed. 4) *Stenotopic*, having a restricted range of geographical distribution. **1967** *Oceanogr. & Marine Biol.* V. 546 This species is also stenotopic; it needs exposed rocky shores, but where the wave-action is not too strong. **1976** *Nature* 24 June 695/1 A major terminal extinction event..will tend selectively to eliminate the larger, more specialised, more stenotopic species.

stenog (stenǫ·g). Also with full point. Colloq. abbrev. of Stenographer. So as *v. intr.*, to write in shorthand or type. orig. and chiefly *U.S.* Cf. *steno.

1905 'O. Henry' in *N.Y. World Mag.* 2 Apr. 3/1 Not being able to stenog, she could not enter that bright galaxy of office talent. **1909** *Fra* (East Aurora, N.Y.) Mar. 82/1 The Stenog wanted a new set of curtains. **1912** Wodehouse *Prince & Betty* ix. 132 If I was good enough for him to marry when I was a stenog., he's good enough for me to marry when I'm a plute. **1941** B. Schulberg *What makes Sammy Run?* vi. 103 'Start typing up that last scene, Ellen,' Kit told the stenog. **1967** *Spectator* 20 Oct. 468/3 Julie Andrews, as tireless Millie the faithful stenog.

stenographer. Add: Now also *spec.* a shorthand typist.
1921 *Dict. Occup. Terms* (1927) § 939 *Stenographer*,.. shorthand typist; wholly engaged in taking down letters, ..etc., in shorthand from dictation of another, and in transcribing them on typewriter. **1978** A. Maling *Lucky Devil* xxx. 159 Taught herself stenotyping, and now she's a legal stenographer.

stenographist. (Examples.)
1839 *Spirit of Times* 20 Apr. 74/2 Paris swarms with scribblers of indifferent merit, authors of well founded pretensions, editors, stenographists, vaudevillists, translators, compilers, and correctors of work. **1845** *Times* 19 Aug. 4/5 England alone has despatched 30 stenographists (short-hand writers). **1905** J. Joyce *Let.* 28 Feb. (1966) II. 83 If I had..a clever stenographist I could *certainly* write any of the novels I have read lately in seven or eight hours.

stenol (stī·nǫl). *Chem.* [f. *st(er)ol, blended with inserted *en* (see -ene); cf. *stanol.] Any sterol having one carbon–carbon double bond in its skeleton.
1949, 1958 [see *stanol]. **1968** *Lipids* III. 397/2 Derivatives of many Δ^{24}-stenols undergo loss of two nuclear hydrogen atoms together with the side chain. **1979** *Nature* 25 Jan. 287/2 Core 16..was analysed for its sterol (stenol + stanol), stanone, and sterene contents.

stenosing (stĭnōᵘ·siŋ), *a.* *Med.* [f. Stenos(is + -ing[1].] Causing or characterized by stenosis.
1903 A. Stengel tr. *Riegel's Dis. Stomach* II. 392 This condition..closely resembles the stenosing gastritis described by Boas. **1971** *Brit. Med. Bull.* XXVII. 26/2 A stenosing ulcer of the small bowel.

stenosis. Delete ‖. For '**1866**...(1880)' read '**1880**...(ed. 4)', and add earlier example.
1872 [see *mitral *a.*].

stenotype. Add: **2.** = Stenotyper. Also *stenotype machine.*
1913 *Chambers's Jrnl.* Feb. 189/2 An English device of this kind appeared upon the market—the stenotype. **1942** *Amer. Cinematographer* Apr. 190/1 A secretary sits over in one corner with a stenotype, faithfully recording, ad verbatim, everything that's said. **1946** A. Huxley *Let.* 27 Oct. (1969) 555 We have just sent her stenotype machine over. **1970** J. Hansen *Fadeout* xix. 161 He left them with a fat young sergeant and a..woman who ran a stenotype machine. **1976** *New Yorker* 5 Apr. 82/2 The stenotype machine, above which Gnusse could see the court reporter's fingers poised.

stenotypy. Add: **stenotyper** (later examples); **stenotypist** (later examples); also **stenoty·pic** *a.*, of, pertaining to, or printed by stenotypy (*rare*); **stenoty·ping.**
1889 *Cent. Dict.*, Stenotypic. **1908** *Chambers's Jrnl.* Oct. 765/1 A simple machine, the 'stenotyper', has been devised. *Ibid.* Oct. 766/1 One stenotypist can be retained.. solely for dictation purposes. **1946** *Nature* 17 Aug. 217/2 A judicious combination of this privilege [*sc.* emendation] with verbatim reporting by the new stenotypic method has resulted in a lively and informal record. **1951** *Catal. Exhibits, South Bank Exhib., Festival of Britain* p. liii, The only British system of stenotyping (machine shorthand). **1978** Stenotyping [see *stenographer]. **1979** 'A. Hailey' *Overload* II. viii. 148 A male stenotypist, who was keeping the official commission record, flipped back through the folded tape of his notes. **1980** 'E. Anthony' *Defector* ix. 223 There was a stenotyper who recorded every word spoken.

Stensen (stī·nsən, ste·nsən). *Anat.* Also (*erron.*) **Stenson.** The name of Niels *Stensen* (1638–86), Danish anatomist, used in the possessive and with *of* to designate structures investigated by him, as: **a.** *Stensen's duct*, the duct of the parotid gland. Cf. Stenonian *a.*

[**1803** C. Bell *Anat. Human Body* III. II. ii. 463 The duct of this gland [*sc.* the parotid] was discovered by Needham, and afterwards by Steno: it is very often called Steno's duct.] **1867** W. Sharpey et al. *Quain's Elem. Anat.* (ed. 7) II. 816 The parotid duct, named also Stenson's duct (d. Stenonianus), appears at the anterior border of the gland. **1977** G. D. Zuidema *Johns Hopkins Atlas Human Functional Anat.* xiii. 56/2 Surrounding the [oral] cavity and exiting into it are the major salivary glands: the parotid (Stensen's duct), the submaxillary (Wharton's duct), and the sublingual glands.

b. Each of two (sometimes four) canals through the bony palate, running from just behind the incisor teeth to each half of the nasal cavity; also the orifices of these canals in the bony palate.

1867 W. Sharpey et al. *Quain's Elem. Anat.* (ed. 7) I. 45 This aperture may be seen to be divided into four smaller foramina, two of which placed laterally are the incisor foramina, called also foramina of Stenson. **1871** Canals of Stenson [see *Jacobson]. **1893** Stenson's canal [see *Jacobsonian *a.*]. **1936** L. B. Arey *Developmental Anat.* (ed. 3) vii. 182 Fusion between the median palatine processes and the palate is incomplete, so that in the midplane there is a gap, the incisive foramen, flanked by the incisive canals (of Stenson). These become covered with mucous membrane. **1969** G. N. C. Crawford tr. *Donath's Anat. Dict.* 480 *Stenon's* (*Stensen's*) *foramen*, foramen incisivum: one of the inferior openings of the incisive canals.

stent (stent), *sb.*[5] *Med.* Also **Stent, stint.** The name of Charles R. *Stent* (d. 1901), English dentist. **a.** Used *attrib.*, *absol.*, and in the possessive to designate a substance invented by him for taking dental impressions; also, an impression or cast of a part or body cavity made of this or a similar substance, and used to maintain pressure on it so as to promote healing, esp. of a skin graft.

The form *Stents* is a proprietary name.

1878 [see *impression material* s.v. *impression *sb.* 9]. **1899** *Trade Marks Jrnl.* 15 Feb. 155 *Stents*... A composition, sold in tablet form, specially intended for taking impressions of the gums and for like dental purposes. Caroline Stent, 5, Coventry Street, London, W.C.; dentist and manufacturer of dental composition. **1920** H. D. Gillies *Plastic Surgery of Face* i. 10 An impression of the Sulcus is taken with warm Stent. *Ibid.*, The dental composition used for this purpose is that put forward by Stent, and a mould composed of it is known as a 'Stent'. **1939** S. Fomon *Surgery of Injury & Plastic Repair* ii. 128 Over irregular areas and where the base lacks resistance, such as on the eyelids and neck, and in inaccessible areas, like the nose and mouth..the use of dental modeling compound, commonly referred to as stent, is invaluable, as it acts in the dual capacity of pressure dressing and splint. *Ibid.* xvii. 1268 All cicatricial tissue beneath the surface is removed to form a pocket into which a stent mold covered with a razor graft, raw side out, is buried and sutured in place. **1961** Webster, *Stent*, also *stint*. **1961** *Brit. Med. Dict.* 1350/1 *Stent's composition*, a proprietary form of composition used in dentistry, and in skin grafting. **1964** R. Battle *Plastic Surgery* x. 234 An impression of the raw surface [of the eyelid] must be taken in Stent's wax.

b. A tube implanted temporarily in a vessel or part.

1964 *Jrnl. Prosthetic Dentistry* XIV. 1168 All stents must be removed daily and cleaned. A pipestem cleaner is effective in cleaning the tube. **1975** *Year Bk. Ear, Nose & Throat* 114 Packing consists of a rayon basket with cellulose sponges in the meatus. Sutures and packing are removed after 7 days. Stents are not used. **1978** *Sci. Amer.* Apr. 67/1 A soft Teflon tube called a stent is placed in the vessel to keep the lumen open and facilitate the suturing. **1980** D. M. Mahoney in R. C. A. Weatherley-White *Plastic Surg. Female Breast* vii. 203/2 At the time of the surgery, the physician lacerates the common bile duct and the liver. Both are successfully repaired but the common bile duct, of course, requires a stint.

stenter, *sb.* Add: **2.** (Later examples.) In mod. use, a machine through which fabric is carried mechanically while under sideways tension.

1911 C. Salter tr. *Potleyn's Dressings & Finishings for Textile Fabrics* xxii. 244 The stenter and drying machine has been greatly improved by the addition of the longitudinal stretching device. **1939** A. J. Hall tr. *Hünlich's Textile Fibres & Materials* iv. 152 The fabric may be dried and stretched simultaneously on a hot air stenter. **1962** J. T. Marsh *Self-Smoothing Fabrics* ii. 9 The alternative use of radiant heat units mounted over the stenter is also quite common. **1973** *Materials & Technol.* VI. vii. 458 By the time the fabric has travelled the 50 or 60 ft.. which represents the length of the stenter, it is practically free from alkali and shows no further inclination to contract.

stenter, *v.* Add: Also, to pass (fabric) through a stenter. (Later examples of the vbl. sb.)

1946 A. J. Hall *Stand. Handbk. Textiles* iv. 228 The modern machine undertakes both the drying and the stentering. **1962** J. T. Marsh *Self-Smoothing Fabrics* xi. 171 The water content of the fabric may be reduced to a figure which prevents any migration during stentering. **1973** *Materials & Technol.* VI. vii. 458 Cotton fabrics are treated with the caustic soda solution while they are in open width on a powerful 'stentering' machine in which the fabric is anchored by its edges as it travels through.

stentorophonic, *a.* **2.** (Later example.)
1914 C. Mackenzie *Sinister St.* II. iv. i. 834 The whirr of the ventilating fans, the stentorophonic orchestra, the red-faced raucous atom on the stage combined to irritate him beyond farther endurance.

† **stentorphone** (ste·ntǭɹfōᵘn). *Obs.* [f. Stentor[2] + *-phone.] An electrical device for reproducing sounds, esp. the human voice, with increased intensity.

1921 *Punch* 2 Feb. 86/1 At Oxford Circus I have known What townmen call the 'stentorphone'. **1927** *Dancing Times* June 357/1 *Al fresco* dancing under cover is provided, with stentorphone music.

step, *sb.* Add: **I. 1. f.** Chiefly *pl.* Any of various children's games (see quots.). Cf. *Grandmother's* (*Foot*)*steps* s.v. *grandmother *sb.* 1 d.

1909 J. H. Bancroft *Games for Playground* 188 Step... The object of the game is for the players who are lined up in the rear to advance forward until they cross the line where the counter is stationed [etc.]. **1940** N. Marsh *Surfeit of Lampreys* (1941) ix. 127 The childish game of Steps in which, whenever the 'he' has his back turned, the players creep nearer. **1969** I. & P. Opie *Children's Games* vi. 189 'May I?' is the usual name, but sometimes the game is known as 'Steps', 'All Sorts', 'Walk to London', 'Variety', or 'Mother, May I?'.

3. a. Also in phr. *a step in the right direction*; *a step up*, a rise in social status; a higher position on a ladder of success.

1877 C. Reade *Woman-Hater* I. i. 190 A little money was given her for a bad purpose. She has used it for a frivolous one. That is 'a step in the right direction'— jargon of the day. **1919** H. S. Walpole *Jeremy* xii. 294 Going to school..was a mixed business; but the balance was now greatly to the good. It was a step in the right direction towards liberty and freedom. **1926** N. Coward *Easy Virtue* II. 86, I don't consider my position in this house a step up... It's been..the most demoralising experience. **1939** L. M. Montgomery *Anne of Ingleside* xxi. 137 'It'll be a step up for a Plummer if you marry a Mitchell,' Ma said. **1954** *Encounter* May 52/1 Eventually she became a model—a further step-up—and she received her first film-part in that capacity. **1974** J. Pope-Hennessy *R. L. Stevenson* i. 32 The Thomas Stevensons ..made..a final move to..Heriot Row. This was in all senses a step up, for Heriot Row..was considered one of the most delectable residential streets in Edinburgh. **1976** *Glasgow Herald* 26 Nov. 6/1 Extensions of the fishing limits around our coasts to 200 miles..are a step in the right direction.

c. *Astronautics.* = *stage *sb.* 11* b.
1932 D. Lasser *Conquest of Space* vi. 103 Each step, as it is called, is a complete rocket motor, containing fuel, combustion chambers, exhausts, etc. **1956** *Spaceflight* I. 5/2 Each extra step multiplies the total weight by a factor of up to ten, so that..rockets of more than five stages are not often contemplated. **1966** H. O. Ruppe *Introd. Astronautics* I. ii. 35 There are cases when a two-step design can do the mission but a one-step rocket cannot.

4. b. Similarly *out of step.*
1961 *Listener* 9 Nov. 768/2 There is also the problem, with direct current lines, of providing what is called the reactive power—power where the current is out of step with the voltage—for the operation of converter equipment.

5. step by step. a. Also, with pauses at regular intervals. (Later examples.)
1893 [in Dict., sense 5 c]. **1966** *McGraw-Hill Encycl. Sci. & Technol.* XIII. 355/1 A shaft which can be driven step-by-step in a vertical direction and subsequently can be moved step-by-step in a rotary direction.

c. (Further examples.) *Esp.* (of mechanisms and the like) moving with pauses at regular intervals; *spec.* in *Teleph.*, with reference to one type of automatic switching, in which successive switches establish contact by a step-by-step movement first in a vertical and then horizontally in a rotary direction.

1845 *Brit. Pat. 10,838* 15 The Invention of causing the two elementary actions..to produce a step by step motion of an indicator in two contrary directions, for the purpose of giving signals. **1879** *Specifications of Patents* (U.S.) 9 Dec. 392/2 An electro-automatic central [station] for telephone exchanges provided with a step-by-step action. **1911** A. B. Smith *Mod. Amer. Telephony* xxvi. 700 Their devices were usually based on some step-by-step ratchet action. **1933** [see *Strowger]. **1938** G. H. Sewell *Amateur Film-Making* iv. 46 The apparatus available to the amateur printer is all of the step-by-step type. Here the films remain stationary for a fraction of time opposite the printing aperture while the exposure is made. **1973** [see *Strowger].

d. *fig.* Involving or comprising a series of distinct stages or operations, often devised to facilitate the accomplishment of something.

1918 C. I. Lewis *Survey of Symbolic Logic* ii. 134 This is a 'step by step' definition. **1937** Michell & Belz *Elem. Math. Analysis* II. x. 608 The elementary fractions are of the first type.., as we shall now demonstrate by the use of a step-by-step process. **1957** K. A. Wittfogel *Oriental Despotism* 284 The step-by-step rise of a new system of landed property. **1968** *Daily Tel.* 8 Nov. 17 Very clear instructions and step-by-step diagrams for making a glove puppet. **1980** 'R. B. Dominic' *Attending Physician* xx. 182 [He] had been subjecting Fournier's narrative to step-by-step dissection.

7. e. The movement through a fixed linear or angular distance made by a stepping device (see *stepping *ppl. a.*) in response to an applied voltage pulse.

1957 *Control Engin.* Jan. 74/1 The simple rugged construction of this new unit leads to high reliability, speeds to 100,000 steps per minute. **1964** *IEEE Trans. Automatic Control* IX. 102/2 Several companies..offer step motors with maximum stepping rates in excess of 3000 steps per second. **1974** T. E. Beliny in B. C. Kuo *Theory & Applications of Step Motors* x. 209 Load torque may actually vary somewhat from step to step.

8. b. A melodic interval of a single degree of the scale (i.e. a tone or semitone) (later examples). Cf. Leap *sb.*[1] 7.

1907 C. E. Kitson *Art of Counterpoint* iv. 50 If the.. third and fourth crotchets are discordant with the C.F. the part must proceed in the same direction by step to the next concord. If the next step will not produce a concord, the passage must be rearranged. **1930** A. M. Richardson *Helps to Fugue Writing* v. 27 If the two missing beats were supplied thus..the result would be impossible cacophony. The only thing to do is to transpose this last group a step lower. **1952** A. O. Warburton *Melody Writing & Analysis* i. 7 When a melodic part moves by step it is said to be 'conjunct'. When it moves by leap it is 'disjunct'. **1971** A. Hopkins *Talking about Sonatas* iv. 58 The Exposition of the Hammerklavier ends with three giant steps.

10. b. *to watch* (or *mind*) *one's step*, to be careful about one's actions, to tread warily. (Chiefly admonitory.) *colloq.*

1934 'G. Orwell' *Burmese Days* viii. 139 You watch your step. Tom Lackersteen may be a drunken sot, but he's not such a bloody fool that he wants a niece hanging round his neck for the rest of his life. **1935** D. L. Sayers *Gaudy Night* vii. 154, I can have a word with her and tell her to mind her step. **1955** M. Gilbert *Sky High* xii. 168 The Inspector... Bit of an awkward mood... I'd mind my step, if I were you. **1977** P. D. James *Death of Expert Witness* i. 23 He seems to be taking quite an interest in you... You'd better watch your step.

II. 12. d. (Earlier example.)
1816 Jane Austen *Emma* I. xiii. 240 They arrived, the carriage turned, the step was let down.

f. *Eton Fives.* The shallow step which divides the court into an inner and outer part.
1890 A. C. Ainger *Fives* 463 The vertical face of the 'step' does not reckon as part of the floor of the court. **1897** [see *hole *sb.* 4 e]. **1975** *Oxf. Compan. Sports & Games* 290/2 Running across the court is a shallow step 10 ft. (3·05 m.) from the front wall, dividing the court into an inner or upper court and an outer or lower court.

g. *to go up the steps*: to be committed or appear for trial at a higher court, esp. the Old Bailey. Also in related phrs. *slang.*
1931 [see *bottle *sb.*[2] 1 g (*a*)]. **1938** F. D. Sharpe *Sharpe of Flying Squad* 334 *Up the Steps*, being committed to the Sessions or Assizes. **1952** 'J. Henry' *Who lie in Gaol* iv. 62 They think it's wonderful 'to go up the steps' —to be sent for trial at the Old Bailey. **1962** *John o' London's* 25 Jan. 82/1 You'll go up the steps.

III. 15. c. *Aeronaut.* An edge built across the float or hull of a seaplane or hydroplane, giving its outline the form of an inverted step and designed to facilitate its separation from the water; *on the step*, with the part of the hull forward of the step out of the water.

1911 *Flight* 25 Nov. 1026/1 Each hydroplane has two steps, the middle step being halfway back from the bow. **1913** *Aeroplane* 24 Apr. 482/1 The [flying] boat got up on its step in a few yards. **1934** W. Nelson *Seaplane Design* vi. 54 Floats without steps tend to cling to the water with a tenacity that requires abnormal power for the take-off. **1935** *Sun* (Baltimore) 10 Oct. 24/4 As the clipper reached Middle River its speed increased until it was flying over the water on the hydroplane step. **1936** J. GRIERSON *High Failure* v. 91 After about half a mile of almost imperceptible acceleration, [the seaplane] Robert Bruce 'got on to the step' and began to hydroplane. **1952** A. Y. BRAMBLE *Air-Plane Flight* xi. 167 The floats are curiously shaped on their under sides, having a sudden discontinuity of surface known as a 'step'. **1983** D. STINTON *Design of Aeroplane* ix. 359 It is necessary to break down the suction by ventilation..and this is done by making a step about half-way along the planing bottom, slightly aft of the aircraft CG.

17*. A change in the value of some quantity, esp. voltage, occurring over a negligibly short interval of time. Freq. *attrib.*

1940 in *Chambers's Techn. Dict.* 806/1. **1958** W. G. HOLZBOCK *Autom. Control* iii. 20 Assume that Figure 3-3c represents the change in level seen in Figure 3-5 after a step change in valve position..closes the valve slightly. **1959** W. I. CALDWELL et al. *Frequency Response or Process Control* ii. 15 If the input to the controller undergoes a step change of 1 psi, then the controller output will be a step equal in magnitude to the setting of proportional gain. **1962** SIMPSON & RICHARDS *Physical Princ. Junction Transistors* xv. 372 $i_0 i$ s the change of output current resulting from the application of a sudden step of input current. **1973** *Nature* 23 Nov. 220/1 Where *C* is membrane capacitance, *i* is membrane current and *V* is the magnitude of the applied voltage step. **1975** G. J. KING *Audio Handbk.* ii. 41 Although a perfect step-wave (i.e. one of zero rise time) cannot, of course, be produced, a good evaluation of amplifier rise time is possible.

18. step-collar, a collar with a V-shaped opening at the junction of the collar and lapel (cf. *step-roll* (*collar*) below); **step-dance** (examples); also as *v. intr.*; **step-dancer** (examples); **step flaking** *Archæol.*, secondary flaking of a flint tool to produce a strong, ridged cutting edge; **step function** *Math.* and *Electronics*, a function that increases or decreases abruptly from one constant value to another; **step-gable** = *corbie-gable* s.v. CORBIE 3 (cf. STEPPED (*ppl.*) *a.*, quot. 1833); hence **step-gabled** *a.*; **step iron,** an iron projection fixed into a wall or the like to serve as a support for the foot when ascending; **step motor,** a stepping motor (see *STEPPING *ppl. a.*); **step pattern** *Art Hist.*, a simple geometric pattern progressing in steps; **step printer** *Cinemat.* (see quots.); hence (as back-formations) **step print** *v. trans.*, **step printing; step print** *sb.* (see quot.); **step response,** the output of a device in response to a step input (*STEP *sb.* 17*); **step rocket,** a rocket of two or more stages; **step-roll (collar),** a rolled step-collar (cf. ROLL-COLLAR); **step saver** *U.S.*, a kitchen designed to reduce the necessity of walking between units, etc.; also *attrib.*, as *step-saver kitchen*; hence **step-saving** *a.*; **step-stool,** a stool which can convert into a short step-ladder; **step wedge** *Photogr.*, a line of contiguous rectangles each of a uniform neutral shade but getting progressively darker from white (or light grey) at one end to black (or dark grey) at the other; also *transf.*; **step-wise** *adv.* (see as main entry below).

1895 J. P. THORNTON *Sectional Syst. Cutting* 104 Step collar vest. **1977** *Summit* (Austin Reed Mag.) Autumn 41 Step collar dress suit..with satin facings £69. **1887** KIPLING *Plain Tales from Hills* (1888) 103 Orth'ris began rowlin' his eyes an' crackin' his fingers an' dancin' a step-dance for to impress the Headman. **1946** D. HAMSON *We fell among Greeks* xix. 204 The Bishop of Kozáni, who was in full regalia on the speaker's platform, executed a step-dance. **1950** A. CLARKE *Coll. Plays* (1963) 297 It was younger than the mayflies That step-danced above it. **1969** in Halpert & Story *Christmas Mumming in Newfoundland* 67 Sometimes janneys 'step-dance'. **1896** STUART & PARK *Variety Stage* iii. 42 The sentimental vocalist, the male impersonator..and the step-dancer were familiar performers. **1969** in Halpert & Story *Christmas Mumming in Newfoundland* 214 True step-dancers in 'Coughlin Cove' have learned their art from their fathers or grandfathers. **1931** R. A. SMITH *Sturge Coll. Flints from Britain* 30 Implement of triangular section... There is some undercutting along both sides, sometimes called resolved flaking or step flaking. **1959** J. D. CLARK *Prehist. Southern Afr.* vi. 146 The Fauresmith tools were made by using what is known as *step flaking*. **1971** *World Archaeol.* III. 161 Macroscopic inspection also revealed woodworking wear in the form of distinctive step-flaking (the result of progressive wear and resharpening of the working edge). **1946** H. CRAMÉR *Math. Methods Statistics* vi. 53 Any non-decreasing function..may be represented..as the sum of a step-function and an everywhere continuous function, both non-decreasing and uniquely determined. **1947** R. LEE *Electronic Transformers & Circuits* iv. 99 It is obtained by applying a step function voltage to the series R_1L_1C circuit. **1967** *Oceanogr. & Marine Biol.* V. 32 Assuming the sea to be at rest at *t* = 0, elevations were found due to northerly stress fields, the stress magnitude varying in time either exponentially, or as a step-function, or as a single half sine wave. **1971** J. H. SMITH *Digital Logic* iv. 74 In the circuit described here the input signal is a step function. **1921** *Glasgow Herald* 8 Jan. 6 It is a whitewashed house, with step-gables. **1937** *Times Lit. Suppl.* 18 Dec. 954/3 The step-gabled houses at Llanedwen. **1978** R. FEDDEN et al. *Hughenden Manor* (1980) 8 Its delightful step-gabled entrance, wood-strutted to the yard. **1912** F. N. TAYLOR *Main Drainage of Towns* vii. 139 Step irons are let into the walls of the shaft.., but sometimes a small wrought-iron ladder is substituted. **1973** R. D. SYMONS *Where Wagon Led* xvi. 260 The wagon was swept down at right angles to the team. My neighbour yelled for the rope, which I threw. He caught the loop and fastened it to the step-iron. **1961** E. M. GRABBE et al. *Handbk. Automation, Computation, & Control* III. xxii. 55 Small step motors have three to six times as much stall torque as the same size a-c servo motor. **1974** B. C. KUO *Theory & Applications of Step Motors* i. 3 High-speed printers of up to 3000 lines per minute can be driven satisfactorily with step motors. **1908** *Encycl. Relig. & Ethics* I. 842/1 'Step' patterns occur in the *cloisonné* settings of Teutonic jewels. **1959** E. A. FISHER *Anglo-Saxon Archit. & Sculpt.* 74 The simple *step pattern* also was common in Celtic art of the pagan period, though it was rare in Celtic Christian art and may have been an independent invention of the Celtic people. **1953** K. REISZ *Technique Film Editing* xiii. 207 Shot 32..was too short for the present film and had to be step-printed to the needed length. **1960** O. SKILBECK *ABC of Film & TV* 125 *Step Print.* Most Positives are made on a continuous process machine in which they run in contact with the Negative; but for some purposes, step printing, Frame by frame, is used. **1930** *Sel. Gloss. Motion Pict. Techn.* (Acad. Techn. Bureau Hollywood), *Step printer*, machine which prints a positive, a frame at a time. **1959** W. S. SHARPS *Dict. Cinematogr.* 120/1 *Step printer*, a printer in which the film to be printed and the raw stock are moved intermittently, and are stationary whilst being exposed one frame at a time. **1959** ZIMMERMAN & MASON *Electronic Circuit Theory* viii. 368 (*caption*) Approximating the step response of a linear *RC* coupling circuit including stray capacitances. **1967** *Electronics* 6 Mar. 9/1 (Advt.), Step response over the full 4½-inch span..is 40 milliseconds. **1932** D. LASSER *Conquest of Space* vi. 104 The step-rocket will ascend to a far greater height than a unit rocket of the same weight. **1946** *Sun* (Baltimore) 23 Dec. 2/4 The 'Tiamat' is a 'step' rocket—that is, it has a rocket booster mounted on its tail. **1966** H. O. RUPPE *Introd. Astronautics* I. ii. 26 Optimization of step rockets poses some very interesting problems. **1881** *Record of Fashion* 27 July 178/2 Step roll is the most suitable style for most of the goods now fashionable. **1901** P. N. HASLUCK *Tailoring* 99 Step-roll collar vest. **1967** *Boston Sunday Globe* 23 Apr. B59/3 The large kitchen..is a stepsaver when the dining room is being used. **1974** *State* (Columbia, S. Carolina) 1 Apr. 9-B/8 (Advt.), Spanish style home includes 3 bedrooms, 2 full baths, cozy den, patio, step saver kitchen with built-ins, enclosed garage and central air. **1978** *Detroit Free Press* 16 Apr. F9/5 (Advt.), 4 bedroom Quad..featuring..step-saving kitchen with all built-ins. **1966** J. POTTS *Footsteps on Stairs* (1967) iii. 38 Hazel had to laugh, just at the sight of him up there on the step-stool. **1931** *Phil. Trans. R. Soc.* A. CCXXX. 91 The intensities were estimated by covering part of the lines with a step-wedge of aluminium foil..and making use of the known absorption-coefficient of aluminium for CuKα rays. **1936** F. R. NEWNES *Technique Colour Photogr.* (ed. 2) iii. 39 The print from the blue filter negative will show less contrast than the others... If the white end of the step wedge is white, then the black end will only be a dark grey. **1962** *Which?* May 135/1 A black and white film's characteristic curve can be obtained by photographing a grey step wedge..and measuring the densities of the grey steps in the picture in relation to their known real densities. **1971** *Jrnl. Oil & Colour Chemists' Assoc.* LIV. 881 A method of achieving this was evolved using a step-wedge produced by gradually increasing the exposure in strips across the film.

step, *sb.*² Colloq. abbrev. of *stepfather, -mother, -son,* etc. Cf. STEP- and the associated main entries.

1895 C. M. YONGE *Long Vacation* ii. 15 Anyone would have thought those poor boys were her steps, not good old Lamb's. **1913** ROWNTREE & KENDALL *How Labourer Lives* iii. 227 There are three 'steps', Mr. Hopwood's children by a former marriage. **1933** G. HEYER *Why shoot Butler?* ii. 23 'You should not encourage your friend to talk disloyally about her brother.'..'He's only a "step".' **1939** A. THIRKELL *Before Lunch* ii. 43 She's an angel. Not a bit like a step. I really think she married father so that she could look after Denis and take me about a bit. **1954** E. EAGER *Half Magic* 155 *Step*,..short for *step-father*. **1971** O. NORTON *Corpse-Bird Cries* iv. 68 They're not her natural parents. They're both steps.

step, *v.* Add: **I. 2. c.** Of an electromechanical device: to move a small, fixed distance in response to an input pulse.

1957 GOODE & MACHOL *System Engin.* iv. 48 The switch steps up through the various banks, taking 0·1 sec to arrive at the first and 0·1 sec to go to each succeeding one. **1958** [see *STEPPING *ppl. a.*]. **1964** *IEEE Trans. Automatic Control* IX. 98/1 The idea of mechanically stepping in angle goes as far back as the clock escapement. **1974** B. C. KUO *Theory & Applications of Step Motors* i. 4 Many solenoid type motors can step only in one direction. **1978** [see *STEPPER 4].

3. e. *step-and-repeat* adj. phr. In photographic printing, etc., involving or pertaining to a procedure in which performance of an operation and progressive movement of something involved in it occur alternately. Also *absol.*

1933 N. MONTAGUE in W. Atkins *Art & Pract. Printing* III. xii. 91 The second method consists of exposing a negative on to a coated plate, moving it a definite distance, exposing again and repeating the process... Thus by means of this 'step and repeat' method..one negative may be used for printing a large number of copies. **1954** J. SOUTHWARD *Mod. Printing* (ed. 7) II. xxxv. 388 The key forme is now made up..for step and repeat. **1967** E. CHAMBERS *Photolitho-Offset* vi. 65 Where multiple repeats are required with great precision step-and-repeat machines are necessary. These are most versatile, and can be used for multi-negative work for postage stamps, labels, cheque backgrounds and the like. **1977** J. HEDGECOE *Photographer's Handbk.* 256 (*heading*) Step-and-repeat images.

II. 12*. To cause to move or progress intermittently; to cause to assume successively larger or smaller values.

1960 *McGraw-Hill Encycl. Sci. & Technol.* XIII. 356/1 Magnets are provided to step the shaft by means of a pawl mechanism. **1971** *Sci. Amer.* June 85/1 If a series of adjacent loops is energized in sequence, a bubble will be stepped along from one loop to the next. **1977** *New Scientist* 7 Apr. 9/2 You can 'step' the laser from one frequency to another in this way, but cannot tune it continuously. **1977** *Offshore Engineer* Aug. 7/1 In the case of the larger Bass Strait fields..price rises are likely to be stepped.

III. 15. step on or upon —. b. *to step on the gas*: see *GAS *sb.*² Also, *to step on it* († *her*). *colloq.* (orig. *U.S.*).

1923 R. CROTHERS *Mary the Third* II. i. 53 This is life! Go on, Lynn! *Step on her!* (Lynn bends lower over the wheel.) **1926** MAINES & GRANT *Wise-Crack Dict.* 13/1 *Step on it*, hurry. **1930** F. L. PACKARD *Jimmie Dale & Blue Envelope Murder* xxii. 316 Then for heaven's sake step on it, old man! **1939** G. GREENE *Confidential Agent* IV. ii. 283 'Step on it, Joe.' They ricocheted down the rough path. **1957** 'N. SHUTE' *On Beach* i. 27 Get up into it, and I'll step on it and show you how she goes. **1974** K. CLARK *Another Part of Wood* vi. 234 His aim was to complain to M. Jean Zay that he was not getting enough drink. 'Tell him to step on it' he repeated. **1981** C. LEOPOLD *Night Fishers of Antibes* lxxv. 201 All he had to do was to put the Citroën into second and step on it.

IV. 21. step down. a. Also *fig.*, to withdraw or retire from office. orig. *U.S.*

1890 *Stock Grower & Farmer* 3 May 3/2 If the bureau cannot do this, let the members of it, the lunkheads, step down and resign. **1945** *Sun* (Baltimore) 22 Sept. 5-0/1 (*heading*) Henry Ford steps down: Grandson becomes president of motor company. **1983** *Times* 30 Aug. 1/2 Mr Menachim Begin has pledged to make a final announcement..about..his intention to step down as Israel's sixth prime minister.

c. Substitute for def.: to reduce (the voltage of a supply); to reduce the voltage of (a supply). (Later examples.)

1938 [see *scanning coil* s.v. *SCANNING *vbl. sb.* 5]. **1978** *Gramophone* Jan. 1340/3 It is also very safe, since it uses only a 12-volt supply, stepped down by a small isolating mains transformer.

25. step off d. *intr.* To die. Cf. *step out*, sense 26 e below. *slang. rare.*

1926 E. WALLACE *Man from Morocco* iii. 21 There will only be the bit of money I have when I—er—step off.

26. step out. c. (Earlier example.)

1806 J. DAVIS *Post-Captain* xii. 74 The sailors were making a run of the tackle-falls, and Mr. Hurricane..was heard to exclaim,..'Step out, men! step out! Walk away with him, cheerly!'

e. To die; to disappear. *U.S. slang.* ? *Obs.*

1844 *Yale Lit. Mag.* IX. 381 Of the other pieces..some will be found in the present number..and the remainder have '*stept out*'. **1851** T. A. BURKE *Polly Peablossom's Wedding* 177 Ay, dead!—stepped out!—d—d—dead as Tecumseh! **1903** A. D. MCFAUL *Ike Glidden* xxx. 277 He is the cause of my ruin. Yes, that is why he stepped out when he did.

f. To appear in company or society; *spec.* to accompany or walk out (*with* a person of the opposite sex); to consort (*with* a lover). *N. Amer. dial.* and *colloq.*

1907 'MARK TWAIN' in *Harper's Mag.* Dec. 44/2, I thought what a figure I should cut stepping out amongst the redeemed in such a rig. **1918** *Dialect Notes* V. 28 *To step* [*out*], vb. i. To go out with a jane. Usually with an unvirtuous intention. General, but especially college communities. **1934** T. E. SULLINGER *Children of Divorce* 9 It affords the father an opportunity to find out how his former wife is spending his alimony, who she is 'stepping out with'. **1936** L. LEFKO *Public Relations* 27 She must be cultured—none of those speak-easy belles you step out with will do. **1940** *Chatelaine* June 59/3 Sally's stepping out again! **1955** D. W. MAURER in *Publ. Amer. Dial. Soc.* xxiv. 190 [Support] will continue as long as she does not have anything to do with men; as soon as she 'steps out' and the fact becomes known, her support stops. **1977** *Detroit Free Press* 11 Dec. 11-B/1 Woodard believes Rae is stepping out with Frank.

g. To parachute *out* of a (disabled) aircraft. *R.A.F. slang.*

1942 B. J. ELLAN *Spitfire* p. x, If you are unlucky enough to get shot down yourself, you..step out. **1953** R. CHISHOLM *Cover of Darkness* I. ii. 24 He climbed to ten thousand and he and his observer stepped out as we used to say.

h. *to step out of line*: see *LINE *sb.*² 28 b.

28. step up. g. (Later examples.) Also (more commonly), to increase (the voltage of a supply).

1909 *Electrician* 2 July 463/1 By means of the three resonance relays..the telephone current was stepped up to 10^{-3} amperes and audible working obtained. **1956** A. H. COMPTON *Atomic Quest* i. 14 Step up the voltages used in our experiments with nuclei, and we should expect to produce interesting nuclear reactions. **1980** J. W. HILL *Intermediate Physics* xxi. 205 The transformer can step up or step down voltages.

h. *fig.* To raise to a higher level or standard, by a stage or stages. More widely, to advance gradually; to increase, intensify.

1920 *Glasgow Herald* 8 July 7 They would suggest that this increase..should be 'stepped up' over a period of years. **1931** *Amer. Speech* VII. verso rear cover (Advt.),

Can you 'step-up' education to meet the new requirements of society? **1938** *Sun* (Baltimore) 5 Sept. 8/8 Soon after they had cleared the Hanover street bridge they stepped up their stroke. **1941** *Punch* 19 Feb. 173/2 People have..stepped their ideas up..about the telephone; I mean, nowadays very few of them actually brush their hair before answering. **1958** *Spectator* 18 July 117/1 The output..could be quickly stepped up. **1967** *Listener* 23 Mar. 390/2 An Aden nationalist leader says terrorist activity will be stepped up when U.N. mission arrives. **1978** K. HUDSON *Jargon of Professions* ii. 50 The war in Vietnam was being stepped up. **1982** *Times* 25 Oct. 6/1 The Solidarity underground..stepped up its pressure this weekend on..the beleaguered Polish leader.

V. 29. The vb.-stem in combination with advbs. and preps. **step-on** *a. U.S.*, that may be operated by pressure of the foot. See also STEP-DOWN, -IN, -OUT, -UP below.

1945 *Richmond* (Va.) *Times-Dispatch* 9 Nov. 24 (Advt.), Step-on pail. **1978** *Detroit Free Press* 5 Mar. A 20 (Advt.), Powerful cleaner has..Convenient step-on switch, easy-change bag holder.

step-. Add: (Later examples of derived and associated forms in somewhat limited currency.) *step-papa* (earlier example).

The concept of orphanage is no longer relevant to the meaning of the *step-* combinations. Consequently, the relationships of *step-brother, -sister,* etc., may be considered to refer reciprocally to children of a later as well as a former marriage: i.e. step-brother = half-brother, etc. A step-parent may be created by marriage to a divorced or a bereaved person.

1893 'MARK TWAIN' in *Century Mag.* Jan. 346/2 Yes; he's my steppapa, and the dearest one that ever was. **1924** G. B. STERN *Tents of Israel* vii. 105 Val,..the eldest step-grandchild..had returned from Vienna especially not to miss the occasion. **1936** M. MITCHELL *Gone with Wind* xlvii. 844 Ah ain' gwine leave Miss Ellen's gran' chillun fer no trashy step-pa ter bring up. **1959** 'E. H. CLEMENTS' *High Tension* v. 83 His step-cousin's [neck] rose..from an open-necked shirt. **1960** M. SPARK *Ballad Peckham Rye* vii. 130 Your step-dad's on about young Leslie. **1962** *Listener* 10 May 828/1 A comic private detective, besides step-mum and callous dad. **1974** D. FRANCIS *Knock Down* xii. 146 My new step-mama will be able to maintain us in the style to which we are accustomed. **1980** M. MCMULLEN *My Cousin Death* (1981) vii. 82 He's some kind of step-relative, and he's on his uppers. **1982** *Listener* 23—30 Dec. 12/1 Christmas for many will either be as desolate as an Oxford Street Santa's heart or so extended—what with the myriads of stepfathers, stepmothers, step-siblings, step-uncles and step-aunts—as to conjure up images of those family groupings which American family therapists love to gather for what they call 'working together'.

ste·p-down, *a.* and *sb.* [f. vbl. phr. *to step down*: see STEP *v.* 21.] **A.** *attrib.* or as *adj.*

1. In sense 21 c of the vb.: causing or pertaining to a reduction in voltage or some other quantity.

1893 [in Dict., s.v. STEP *v.* 21 c]. **1929** *Exper. Wireless* VI. 307/2 This reduces the step down ratio required in the potentiometer. **1947** R. LEE *Electronic Transformers & Circuits* vi. 172 Driver transformers are usually step-down because the grid potentials are relatively low. **1959** *Motor Man.* (ed. 36) 95 It has a stepdown gear which reduces the speed of the propeller shaft in a ratio which is usually between 5 to 1 and 4 to 1. **1961** *Wall St. Jrnl.* (Eastern ed.) 14 Nov. 10 (Advt.), With Con Edison's step-down rates, the more electricity you use, the less it costs per kilowatt-hour. **1981** *Daily Tel.* 22 July 12/6 For really rough going there is a 1·96:1 step-down ratio for four-wheel-drive.

2. In sense 21 a of the vb. Esp. from or in which one steps to a lower level. Chiefly *U.S.*

1949 *Newsweek* 28 Nov. 56 (*caption*) Hudson enters lower-priced field with smaller version of step-down car. **1954** *Sun* (Baltimore) 8 Mar. 20/1 Dr. Clifford M. Witcher ..calls his device a 'step-down detector'. When the user approaches a curb, a flight of steps or the edge of a subway platform, the detector buzzes a warning. **1966** A. R. BELLAMY in *Biochim. & Biophys. Acta* CXXIII. 102 (*title*) RNA synthesis in exponentially growing tobacco cells subjected to a step-down nutritional shift. **1976** *Billings* (Montana) *Gaz.* 27 June 5-D/4 (Advt.), Three bedrooms, step-down living room, and main floor family room with fireplace.

B. *sb.* **1.** A reduction or decrease.

1922 [see *REGENERATIVE *a.* (and *sb.*) 2 b]. **1962** A. NISBETT *Technique Sound Studio* ix. 165 The result is a sudden step-down in volume and quality just before the second signal appears on the tape.

2. The act of stepping down or withdrawing from a position.

1973 *Guardian* 7 June 2/4 In a cleverly disguised step-down from the position that these controversial logs would never be released, Mr Nixon's spokesman said today that..the documents would now be turned over.

Stephanian (stɪfēⁱ·niăn), *a. Geol.* [ad. F. *stéphanien* (A. de Lapparent *Traité de Géol.* (ed. 3, 1893) 819), f. *Stephan-us*, latinized form of Saint-*Etienne*, name of a city in central France where it is represented: see -IAN.] Of, pertaining to, or designating the uppermost division of the Carboniferous in Europe, especially where represented by coal-bearing formations. Also *absol.*

1901 *Jrnl. Geol.* IX. 196 The flora of the Caradons stage contains a mingling of uppermost Westphalian species with Stephanian types. **1903** A. GEIKIE *Text-bk. Geol.* (ed. 4) II. 1051 Stephanian. **1912** A. J. JUKES-BROWN *Student's Handbk. Stratigr. Geol.* (ed. 2) x. 317 A few small tracts of Stephanian measures occur in Brittany and Normandy... The most northern of these is at Littry.., the Coal-measures here..being conformably overlain by Permian Beds, and they belong, therefore, to the latest phase of the Stephanian. **1931** GREGORY & BARRETT *Gen. Stratigr.* vii. 118 The Stephanian is absent from the N. of France and Westphalia, and is present in numerous coalfields preserved in basins in the pre-Palaeozoic rocks of the Central Plateau of France. **1959** [see *DINANTIAN *a.*]. **1963** [see *SAKMARIAN *a.*].

stephanotis. For mod.L. read 'mod.L. (L. M. A. A. Dupetit-Thouars *Genera Nova Madagascariensia* (1806) 11' and add: **1.** (Earlier examples.)

1843 *Curtis's Bot. Mag.* LXX. 4058 (*heading*) Copious-flowering Stephanotis. **1869** S. R. HOLE *Bk. about Roses* iii. 41 The stove, truly, is a gladness and refreshment—gay..with the golden Allamandas,..the bridal Stephanotis.

2. (Earlier and later examples.)

1895 *Army & Navy Co-op. Soc. Price List* 716/1 A new fragrant Toilet Water in Jockey Club, White Rose, Stephanotis, [etc.]. **1980** M. FORSTER *Bride of Lowther Fell* xvi. 244 The perfume was..stephanotis, faint and unbelievably fragrant.

Stephen: see *even Stephen* s.v. *EVEN *a.* 14 d.

ste·p-in, *sb.* and *a.* [f. vbl. phr. *to step in*: see STEP *v.* 24.] **A.** *sb.* A garment or shoe put on by stepping into it; *spec.* in *pl.*, loose drawers (more recently, brief panties) for women (chiefly *U.S.*).

1922 *Woman's Home Compan.* June 70 (*caption*) The children like to wear step-ins. **1928** *Sunday Dispatch* 15 July 16 The same *couturière* is all for 'step ins' for swimmers. **1934** [see *SCANTY *sb.*]. **1939** M. B. PICKEN *Lang. Fashion* 131/1 *Step-in*, shoe with no obvious method of fastening, usually held on snugly by an elastic gore. **1946** WODEHOUSE *Money in Bank* xxvii. 234 A bottle of brandy which she..kept stowed away..in a drawer underneath her step-ins. **1958** S. ELLIN *Eighth Circle* (1959) II. xx. 208 Her brassière and step-ins plastered wetly to her body. **1964** P. WHITE *Burnt Ones* 86 Eileen began to pull. Her step-ins had eaten into her. **1975** J. GORES *Hammett* (1976) xxii. 153 The upended torcher wore no step-ins under her tight red sequins.

B. *attrib.* or as *adj.* Designating a garment or shoe of this type.

1923 *Weekly Dispatch* 18 Feb. 12 Step-in cami-knickers. **1960** *Amer. Speech* XXXV. 79 These dresses were listed variously as..'a step-in sundress of acetate', and 'a step-in charmer'. **1975** *Times* 3 May 8/2 The 'step-in' dress with buttons down the front, which permitted a woman to step into a dress instead of pulling it over her head.

Stepin Fetchit (ste·p'n fe·tʃit). *U.S.* Also **Steppin Fetchit.** [*Stepin Fetchit*, the stage-name of Lincoln Theodore Perry (b. 1902), a popular Black vaudeville actor noted for playing a series of fawning characters in Hollywood films of the 1920s and 1930s. For earlier uses of the sb. phr. *step and fetch it* applied to persons, see *Eng. Dial. Dict.*, *Dialect Notes* (1903) II. 301, (1914) IV. 113.] A type of a shuffling, obsequious, Black servant. Hence, any servile Black man; an Uncle Tom. Also *attrib.*

1940 F. SCOTT FITZGERALD *Let.* 1 Feb. (1964) 597 This way of looking at war gives great scope for comedy without bringing in Stepin Fetchit and Hattie McDaniel as faithful negro slaves. **1951** MCWHINEY & SIMKINS in A. Dundes *Mother Wit* (1973) 588/1 The mere mention of a ghost makes him shake as actively as Step'in Fetch'it under the influence of an Arctic breeze. **1967** P. WELLES *Babyhip* xvii. 121 He shrugged his shoulders in his phoney Steppin Fetchit pose and went home leaving a perfectly groovy [chess] strategy unfinished. **1968** *N.Y. Times* 21 Feb. 56/3 He talks disparagingly of comics and other artists who don't fill the role of rebel. Among these he lists Danny Thomas, Jack Benny, Woody Allen and 'Stepin Fetchit Negroes doing the same thing under a new veneer'.

stepless, *a.* Add: **2.** Continuously variable; capable of being given any value within a certain range.

1969 *Jane's Freight Containers 1968–69* 23 adv., Speed control is through the Wessex Carbon Pile, providing smooth, stepless acceleration. **1971** *Engineering* Apr. 111/2 (Advt.), This equipment provides stepless current adjustment over the full range of the..generator. **1973** *Physics Bull.* Feb. 110/1 The instruments in this range produce stepless DC voltage outputs. **1978** *Amateur Photographer* 2 Aug. 6/2 Stepless shutter speed range of 1/1000th to 4 seconds + B.

Hence **ste·plessly** *adv.*

1958 *Times Rev. Industry* Oct. 67/3 The rate of travel..can be varied steplessly by a knob on the body. **1977** *Design Engin.* July 73/2 Frequency and amplitude can be steplessly varied during operation.

Stepney (ste·pni). Also **stepney.** [f. the name of *Stepney* Street, Llanelli, the place of manufacture.] **1.** A spare wheel for a motor vehicle, comprising a ready-inflated tyre on a spokeless metal rim, which could be clamped temporarily over a punctured wheel. Also *stepney wheel.* Now *Hist.* exc. in Bangladesh, India, and Malta, where = any spare wheel.

1907 *Westm. Gaz.* 3 Dec. 4/3 The popularity of the Stepney Wheel has never been more clearly demonstrated than at the Olympia Show. **1910** G. K. CHESTERTON *Alarms & Discussions* 179 Then he said, 'And I left the Stepney behind.' **1911** *Daily Chron.* 5 Jan. 4/7 Wales claims the origin of the 'stepney', the spare wheel and tyre. **1928** *Evening News* 7 Aug. 9/2 None of your detachable wheels, rims, or Stepneys! **1929** H. NICOLSON *Let.* 22 July in J. Lees-Milne *Harold Nicolson* (1980) xvi. 376 [In Berlin he was like] a stepney wheel of a car that is seldom taken out of the garage. **1937** *Autocar Handbk.* (ed. 13) xi. 196 With the introduction of pneumatic tyres came the puncture, and soon the 'Stepney' appeared: a spare rim and tyre fitted with clamps. **1971** *Listener* 11 Nov. 653/1 After jacking up the car, one of them turned to me and said: 'Have you a Stepney?' 'Yes, in the boot,' I answered... It takes an old Edwardian like me to know that a Stepney was an attachable wheel-rim, which came in about 1907 and went out about ten years later. You wouldn't hear the term in England now, but in Malta it is the *ordinary* word for a spare wheel. **1973** *Opinion* (Bombay) July 31 It helps to have a few holes in the roof of the car and to go about without a Stepney. **1975** J. DAY *Bosch Bk. Motor Car* 178 An early attempt to make puncture mending less troublesome was the Stepney spare wheel of T. M. and W. Davies in 1904. **1977** *Navbharat Times* (Bombay) 2 June (Advt.), Yezdi stepney wheel complete with tyre, tube, hub and bearings. **1980** L. LEWIS *Private Life of Country House* iii. 35 About 1920 we bought a secondhand T model Ford... In case of punctures there was a Stepney wheel to be clamped on to the rim to get you home.

2. *fig.*

1928 E. SUTTON tr. *A. Londres's Road to Buenos Ayres* ii. 18, I told her I had a woman already in Buenos Ayres, that she could only be my little sweetheart, as we say, or my 'stepney', if you like that better. **1929** E. LINKLATER *Poet's Pub* xxvi. 282 Redemption being carried as a kind of stepney on the best of all possible worlds. **1979** P. NIHALANI et al. *Indian & Brit. Eng.* I. 167 Dr X may not be able to give his talk—we'd better arrange for a stepney.

ste·p-out. *Oil Industry.* [f. vbl. phr. *to step out*: see STEP *v.* 26.] In full *step-out well.* A well drilled beyond the established area of an oil or gas field to find out if it extends further.

1948 *Bull. Amer. Assoc. Petroleum Geologists* XXXIII. 1082 Step-out well No. 55, 1 kilometer east of No. 26, is now drilling. **1955** *N.Y. Times* 28 Aug. I. 72 Common practice is to drill 'step outs' at specified distances in all four directions from a discovery well. **1962** F. E. WELLINGS in M. J. Wells *Oil Industry Tomorrow* 32 Step-out wells make [*sic*] the usual division between exploratory wells and development wells. In development wells, if they are intelligently done, the percentage of success is very high,..but less so for the step-out wells. **1977** *Offshore Engineer* Aug. 19/3 Well 3/8-3, south of Ninian, found a hydrocarbon accumulation which has been extended into block 3/7 to the west with a step-out. **1979** *Jrnl. R. Soc. Arts* CXXVII. 407/1 The Gas Council..have indicated a wish to sink step-out wells on the Goathorn Peninsula and on the Studland Peninsula.

stepped, (*ppl.*) *a.* Add: **1.** Also *spec.* of the float or hull of a seaplane or hydroplane. Cf. *STEP *sb.* 15 c.

1911 *Flight* 9 Dec. 1074/2 The float consists of a three-stepped hydroplane. **1951** [see *hard chine* s.v. *HARD *a.* (*sb.*) 22].

2. Carried out or occurring in stages or with pauses, rather than continuously.

1935 *Proc. R. Soc.* A. CLII. 597 The prolific branching of the main part of the first stroke of a series arises solely from downward branching in the stepped leader which precedes it. **1944** *Jrnl. Iron & Steel Inst.* CL. 128A, The causes of the distortion of steel parts during heat treatment are analysed and methods of preventing it, including austempering and other forms of stepped quenching, are discussed. **1977** J. HEDGECOE *Photographer's Handbk.* 125 If you project a transparency using a zoom lens, you get a similar effect to stepped zoom. **1981** *Sci. Amer.* Mar. 28/1 Subsequent leaders..move an order of magnitude faster than the first stepped leader in the freshly ionized gas.

3. With *up.* Raised by degree to a higher standard or level; increased, intensified.

1933 *Sun* (Baltimore) 22 Nov. 20/2 Demands were being made on brewers for a 'stepped-up' beverage, whereas the normally brewed beer runs about four per cent. **1941** *Battle of Britain Aug.–Oct. 1940* (Ministry of Information) 26 Twenty Dornier 215's were encountered over the London Docks flying in a diamond formation escorted by Me 109's 'stepped up' to 22,000 feet. **1955** *Times* 22 Aug. 9/6 Mr. Sinclair Weeks, Secretary of Commerce, to-day announced 'a stepped up programme' to make public as quickly as possible non-classified research reports of industrial significance by the Atomic Energy Commission. **1963** P. FLEMING *Kolchak* xx. 212 They themselves were frightened men, and this combined with the necessarily stepped-up tempo of the interrogation to make them hectoring and exigent. **1976** *National Observer* (U.S.) 23 Oct. 4/4 But he concedes that post officials are 'a little concerned' about handling the stepped-up volume of business over the Christmas season if the strike continues.

stepper. Add: **2. a.** (Earlier and later examples.) Now *Obs.*

1846 *Swell's Night Guide* 59, I does the safe, if they cops me it's nix; six veeks, a fly at the stepper and turn

up. **1891** 'F. W. CAREW' *No. 747* xvi. 188 Toiling under our heavy burdens up that everlasting staircase—as Tony Klism said, it was ever so much worse than 'the stepper'.

4. In full *stepper motor.* A stepping motor (see *STEPPING *ppl. a.*).

1961 *Control Engin.* May 116/1 Applications employing interlocked steppers are ballistic missile prelaunching exercises, reconnaissance drone control, and precise positioning of radioactive fuel elements. *Ibid.* Nov. 103/1A, Stepper motor resembles the conventional ac servomotor except that its winding is excited by a stream of pulses from a multivibrator. **1976** NASAR & BOLDEA *Linear Motion Electric Machines* ix. 255 Some of the advantages of linear stepper motors are ease of control..and locking force. **1978** R. P. HUNTER *Automated Process Control Systems* xiv. 328 The VR stepper requires its windings to be energized in the proper sequence for predictable operation. Also, it can be made to step bidirectionally.

stepping, *vbl. sb.* Add: **1. e.** With *up.* The action of *STEP *v.* 28 h.

1958 *Listener* 24 July 112/1 This would involve a stepping-up of supplies from Persia and from Venezuela. **1965** D. FRANCIS *For Kicks* i. 8 There were trials and prison sentences..and a stepping-up of regular saliva and urine tests.

f. The step-by-step movement of a stepping device (see *STEPPING *ppl. a.*).

1960 *McGraw-Hill Encycl. Sci. & Technol.* XIII. 356/2 These selectors may be arranged for..absorbing the digit pulsed without any stepping of the switch. **1964** [see *STEP *sb.* 7 e]. **1974** B. C. KUO *Theory & Applications of Step Motors* i. 4 There are many different versions of switches and actuators which give stepping motion through the principle of solenoid action. **1977** *Engin. Materials & Design* Aug. 41/1 Medium power switching types provide operating voltage ranges from 6 to 110V dc and 6 to 220V ac, with latching and stepping facilities in selected items.

stepping (ste·piŋ), *ppl. a.* [f. STEP *v.* + -ING².] Of an electric motor or other electromechanical device: designed to make a rapid succession of small, equal movements in response to a pulsed input, each pulse causing one movement.

1957 GOODE & MACHOL *System Engin.* iv. 48 Consider a stepping switch acting as a line finder in a telephone system. **1958** J. G. TRUXAL *Control Engineers' Handbk.* v. 69 Besides stepping relays and the Ledex rotary solenoid, few digital devices are available that can step from one point to another rapidly enough to be useful as a control-system output actuator. **1975** *Physics Bull.* July 319/3 The precision divided tables may be fitted with calibrated hand wheels or driven by either stepping motors or DC gear motors. **1979** *Nature* 12 July 121/1 The electrodes were advanced through the brain with a stepping microdrive..until a cell or process was penetrated.

ste·p-up, *a.* and *sb.* [f. vbl. phr. *to step up*: see STEP *v.* 28.] **A.** *attrib.* or as *adj.* **1.** Causing or pertaining to an increase in voltage.

1893 [in Dict., s.v. STEP *v.* 28 g]. **1903** C. H. SEWALL *Wireless Telegr.* IV. 149 The transmitter of the DeForest system uses..a step-up transformer to increase the voltage. **1947** R. LEE *Electronic Transformers & Circuits* v. 142 Input Transformer,.. Step-up ratio 1:20. *Ibid.* 188 If the transformer is step-up, C_1 = [etc.]. **1977** *Gramophone* Oct. 744/2 A particular criticism levelled towards moving-coil cartridges is that their output is so low as to require a step-up transformer.

2. Furnished with a step to a higher level. Also, of a room with such a feature.

1958 J. MYERSCOUGH *Procession of Lancs. Martyrs & Confessors* xix. 260 A typical example of the 'step-up' chapel provided for Catholics in the days of persecution may be seen..in Alston Lane, Grimsargh. **1979** *Arizona Daily Star* 5 Aug. (Advt. Section) 13/1, 2 bdrms., 2 full baths, step-up dining area.

B. *sb.* **1.** An increase in rate or quantity; an intensification.

1922 GLAZEBROOK *Dict. Appl. Physics* II. 889/1 The amplifying action of a valve is usually more a question of potential step-up than of ratio of power output to power input. **1944** *Times* 26 May 2/1 We should probably finish the war with very nearly the same step-up in engine power over the present war period as occurred between 1914–18. **1972** *Newsweek* 10 Jan. 14/3 Coupled with the increasingly hard line adopted by both negotiating teams in recent weeks, the step-up in the air war might even jeopardize the continuation of the talks themselves.

2. Chiefly *pl.* A step taken on to a platform (such as a bench, etc.) and back again, repeated as a fitness exercise.

1973 *Observer* 7 Oct. 28/1 A middle-aged actor keeping his paunch at bay with weights and interminable step-ups. **1978** *Kingston* (Ontario) *Whig-Standard* 15 July 15/2 Stations 5–9 include such exercises as..chin-ups and step-ups.

stepwise (ste·pwəiz), *adv.* and *a.* Also with hyphen. [f. STEP *sb.* + WISE *sb.*¹ II.] **A.** *adv.*

1. Like or in a series of steps.

1888 [in Dict., s.v. STEP *sb.* 18]. **1902** *Jrnl. R. Inst. Brit. Architects* 20 Dec. 101 The balustrade of its upper flight rising step-wise, and showing at intervals the sockets of its colonnade. **1950** *Jrnl. Neurophysiol.* XIII. 193 The response builds either step-wise or abruptly to a complex series of peaks at 5–8 msec.

2. *Mus.* = *by step* s.v. STEP *sb.* 8 b.

1955 G. ABRAHAM in H. Van Thal *Fanfare for E. Newman* ii. 12 A purely harmonic cadence, the separate parts moving stepwise, even chromatically, with a much less decisive cadential effect rhythmically.

3. In a series of distinct or separate stages; with intermittent pauses, not continuously.

1971 J. Z. YOUNG *Introd. Study Man* iii. 55 It may be that the new mRNA is synthesized stepwise in the nucleus. **1972** *Science* 2 June 1014/2 The details of..whether the velocities increase smoothly or stepwise..cannot be resolved without additional data.

B. *adj.* **1.** *Mus.* Of musical progression, etc.: occurring or arranged regularly in steps (STEP *sb.* 8 b in Dict. and Suppl.).

1920 S. MACPHERSON *Melody & Harmony* i. 4 In the above extracts there is a considerable amount of 'conjunct' (i.e. step-wise) movement. **1930** A. M. RICHARDSON *Helps to Fugue Writing* vii. 37 A progression like the following does not look interesting..but by supplying embellishments it can be made much more effective... The thirds need not be stepwise. **1949** W. PISTON *Counterpoint* (rev. ed.) i. 20 In the Corelli example the upward octave skip is both preceded and followed by a downward stepwise movement. **1952** A. O. WARBURTON *Melody Writing & Analysis* i. 7 After stepwise movement it is usually wise to leap in the opposite direction. **1979** *Early Music* Jan. 138/2 The manuscripts..specify 'clarinet' on the title-page, and include stepwise movement in their lower register.

2. = *step-by-step* attrib. or quasi-adj. s.v. STEP *sb.* 5 c.

1934 *Jrnl. Amer. Chem. Soc.* LVI. 913/1 The development of a complex polymer must require a series of consecutive reactions... It is necessary to assume that the first-formed polymers possess an ordinary double bond and are capable of being isolated. This scheme will be called 'stepwise'. **1954** [see *DEGRADATION¹ 4 c]. **1960** KOESTLER *Lotus & Robot* I. i. 43 The result of this stepwise dismantling of reality is that consciousness alone remains. **1975** N. CHOMSKY *Logical Struct. Linguistic Theory* x. 469 Transformational analysis permits the stepwise formation of complex phrases from already constructed simple phrases. **1980** *Jrnl. R. Soc. Arts* May 348/2 Step-wise mutations in the bacterium..are shown to produce corresponding changes in the synthesized amidase.

sterane (stiᵊ·rēⁱn, ste·rēⁱn). *Min.* and *Chem.* [f. *STER(OID + -ANE.] Any of a class of saturated hydrocarbons with a steroid structure which are found in crude oils and derived from the sterols of ancient organisms; orig., † the compound whose molecule consists of the saturated nucleus only.

1951 *Jrnl. Chem. Soc.* 3527 The substance (II) shall be named gonane (preferred) or sterane (alternative). **1969** *Geochim. & Cosmochim. Acta* XXXIII. 1307 The C_{27}, C_{28} and C_{29} steranes are present in the bitumen from the Green River oil shale and..they were derived from cholesterol, ergosterol, stigmasterol or related sterols. **1973** *Nature* 27 Apr. 603/1 These steranes and terpanes represent the complete reduction of naturally occurring steroids or terpenoids to relatively stable fossil molecules which preserve intact the complete molecular skeleton and stereochemistry of their precursors. Steranes..provided convincing proof for the biological origins of petroleums.

stercolith (stɜ̄·ɪkolɪþ). *Med.* [f. L. *sterc-us* dung + -O + -LITH.] = STERCOROLITH.

1910 *Practitioner* July 106 We came upon another abscess cavity, from which a typical appendicular stercolith was dislodged. **1973** EARNEST & SLEISENGER in Sleisenger & Fordtran *Gastrointestinal Dis.* cxii. 1531/1 A smaller, rounded, smooth hard mass of stool which cannot be expelled is called a stercolith or enterolith.

stercorarian. Restrict † *Obs. rare* to senses in Dict. and add: **B.** *adj. Biol.* [f. mod.L. *Stercoraria*, name of a section of the genus *Trypanosoma* (C. A. Hoare 1964, in *Jrnl. Protozool.* XI. 203/1).] Used to designate those species of *Trypanosoma* which occur in the digestive tract of the secondary host, and are transmitted in its fæces. Cf. *SALIVARIAN *a.*

1964 C. A. HOARE in *Jrnl. Protozool.* XI. 203/2 It must be admitted that not all Stercorarian trypanosomes can be fitted with certainty into the above subgenera,..since the life cycle of many of them is still unknown. **1971, 1977** [see *SALIVARIAN *a.*].

stercorite (stɜ̄·ɪkŏrəit). *Min.* [f. L. *stercorāre* to manure with dung, f. *stercus* dung + -ITE¹.] A hydrated acid phosphate of sodium and ammonium, $Na(NH_4)HPO_4.4H_2O$, occurring as colourless triclinic crystals; microcosmic salt.

1850 T. J. HERAPATH in *Q. Jrnl. Chem. Soc.* II. 73 This being the first instance in which the ammonio-phosphate of soda has been met with as a natural production, I propose to class it amongst our minerals under the name of 'Stercorite'. **1975** *Mineral. Abstr.* XXVI. 125/2 The X-ray powder data of stercorite from Chincha Islands, Peru, prove that the mineral corresponds with [*sic*] the synthetic $Na(NH_4)HPO_4.4H_2O$.

stercorolith. Add: Now superseded by *STERCOLITH.

stereo². Add: **stereo card,** a card on which are mounted a pair of stereoscopic photographs; **stereo pair,** a pair of photographs showing the same scene from slightly different points of view, so that when viewed appropriately a single stereoscopic image is seen.

1975 *New Yorker* 26 May 11/3 (Advt.), Also on view are such items as a glass-plate camera from the period and stereo cards mounted in hand-operated viewers. **1976** *Church Times* 3 Dec. 6/5 Two handsome books contain, respectively, a stereoscopic viewer with facsimiles of old stereocards, and a lively illustrated history of stereographic photography. **1943** H. T. U. SMITH *Aerial Photographs* iii. 79 Ink marks..detract seriously from the effectiveness of depth perception, particularly when on only one photo of a stereo pair. **1966** [see *photo-interpretation* s.v. *PHOTO- 2]. **1977** J. HEDGECOE *Photographer's Handbk.* 299 One reason why stereo cameras have gone out of favour is that you can easily make stereo-pairs with an ordinary camera.

stereo (ste·rio, stiᵊ·rio), *a.* and *sb.*³ *colloq.* [Abbrev. of *STEREOPHONIC *a.*, *STEREOPHONY.] **A.** *adj.* = *STEREOPHONIC *a.*

1954 *Wireless World* Jan. 7/1 The first full-length stereophonic film to be released was Walt Disney's 'Fantasia', ..with stereo sound photographically recorded on four tracks on a separate sound film. **1958** *Times* 20 Jan. 10/4 What stereo discs have achieved is to combine these two channels in one groove traced by one stylus. *Ibid.*, When the B.B.C. establishes stereo broadcasts..it may give a new lease of life to radio manufacturers. **1960** *Practical Wireless* XXXVI. 299/2 A variety of instrumental, choral and solo recordings, both monaural and binaural (stereo). **1963** *Which?* Jan. 3/1 Equipment for reproducing stereo records is more elaborate than for mono, and relatively few stereo record players are sold. **1976** 'A. YORK' *Dark Passage* xii. 144 The stereo cassette recorder was switched on. **1977** G. SCOTT *Hot Pursuit* xv. 147 My heartbeat began to pound as though stereo headphones had been clamped to my head.

B. *sb.* **1.** = *STEREOPHONY.

1956 *Radio & Television News* Aug. 35/1 (*heading*) What should stereo do? **1957** *Audio* Jan. 12/2 Tape is the only logical medium for stereo. **1958** *Observer* 20 Apr. 10/5 In adding another dimension to our listening, stereo represents a most significant advance in the quest for concert-hall realism. **1961** *John o' London's* 5 Oct. 396/4 The new Decca issue..presents both works in first-class stereo. **1970** *Jrnl. Audio Engin. Soc.* XVIII. 624 The transmission of four-channel stereo by means of FM-multipler techniques. **1972** *Daily Tel.* 9 Aug. 13/1 The BBC is to start programmes in stereo on Radios 1 and 2 from the end of September.

2. Any stereophonic apparatus for playing records or tapes. Cf. *STEREOGRAM².

1964 *House & Garden* Nov. 62/2 (*caption*) The wall unit houses the stereo. **1971** *Daily Tel.* 24 Sept. 12, I am unable to understand pop words seated 3ft away from a high quality stereo. **1976** P. CAVE *High Flying Birds* iv. 52 Lorna showed us into the villa and put some low, slow background music on the stereo. **1980** J. GARDNER *Garden of Weapons* I. iii. 37 You got full sound in the Charlton house? And a stereo? I can't live without music.

stereo-. Add: **ste:reo-acu·ity,** the sharpness of the eyes in discerning separation along the line of sight; **ste·reo-camera,** a camera for simultaneously taking two photographs of the same thing from adjacent viewpoints, so that they will form a stereoscopic pair; **stereo-ci·lium** *Anat.*, an immotile cell process of certain epithelial cells of the male reproductive tract and the labyrinth of the ear, similar to a cilium at low magnifications only; hence **stereo-ci·liary** *a.*; **stereo-comparator,** substitute for both defs.: an instrument enabling two different photographs of the same region to be seen simultaneously, one by each eye, either to detect any change (in the case of photographs of the night sky taken at different times) or to make measurements of the area depicted in stereoscopic photographs; (further examples); **ste:reocontro·l** *sb. Chem.*, the control of a synthesis by the choice of reagents and reaction conditions so as to produce a product with a desired stereochemical conformation; also as *v. trans.*, to control thus; **ste:reocontro·lled** *ppl. a.*; **ste·reo-diagram,** a diagram intended to show the three-dimensional structure of something; **ste:reoelectro·nic** *a. Chem.*, pertaining to the relative positions of the electron orbitals in reacting molecules; hence **ste:reoelectro·nically** *adv.*; **ste:reofluoro·scopy** *Med.*, the production of X-ray images which can be interpreted in three dimensions; = *stereoradiography* below; hence **stereoflu·oroscope,** an instrument for producing such images; **ste:reofluorosco·pic** *a.*; **stereogno·sis.** [Gr. γνῶσις means of knowing], the stereognostic sense or faculty; **stereo-isomer** (earlier example); **stereo-isomeride** (later example); now *obs.*; **ste:reoisomeriza·tion** *Chem.*, the conversion of one stereoisomer into another; hence (as back-formations) **ste:reoiso·merize** *v. intr.*, to undergo stereoisomerization; *trans.*, to cause the stereoisomerization of (a

compound); **ste:reoiso·merized** *ppl. a.*, **-iso·merizing** *vbl. sb.*; **stereomi·crograph,** a micrograph that conveys a vivid impression of depth, such as one obtained with a scanning electron microscope; **stereomi·croscope,** a binocular microscope that gives a stereoscopic view of the subject; **ste:reomuta·tion** *Chem.*, the conversion of a *cis-* to a *trans-*isomer or vice versa; **stereo-photo** *a.*: see *stereo-photographic* adj. in Dict.; also as *sb.*, a stereophotograph; **ste:reophotogra·mmetry,** photogrammetry by means of stereophotography; hence **ste:reophotogramme·tric** *a.*; **stereo-photograph** (examples); **stereopla·nigraph** *Cartography* [a. G. *stereo-planigraph*], a machine which plots a map of an area semi-automatically under the guidance of the operator as he views a stereoscopic pair of aerial photographs of it; **ste·reoplotter,** an instrument used for plotting maps of an area from stereoscopic aerial photographs that are projected on to the plotting table; **ste:reo-plo·tting** *a.* that is a stereo-plotter; **stereora·diograph,** a stereoscopic radiograph; so **ste:reoradiogra·phic** *a.*, **-gra·phically** *adv.*; **ste:reoradio·graphy** = *stereofluoroscopy* above; **Ste·reoscan, ste·reoscan,** a proprietary name for a scanning electron microscope; hence (as *stereoscan*), a picture obtained with a scanning electron microscope; **ste·reoviewing** *vbl. sb.*, stereoscopic viewing.

1942 *Summary Progress Rep. Tests of Stereoscopic Vision* (Harvard Univ. Psycho-Educational Clinic, Publ. Bd. No. 55797) 1 The objects of this investigation were: (1) to appraise tests of stereo-acuity in current use. **1974** *Nature* 13 Sept. 141/1 Stereoacuity falls in the region of the resolution of the Calcomp plotter, so it could not be measured. **1959** *Observer* 7 June 3/4, I have had a stereo-camera for five years now and the range of new experiences it can offer is constantly widening. You can photograph people and they are three-dimensional people, frozen in a moment of time. **1961** *New Scientist* 19 Oct. 173/2 The Japanese workers used a stereocamera to record the contours of waves generated by the model. **1977** Stereo-camera [see *stereo pair* s.v. *STEREO²]. **1933** Stereocilium [see *kinocilium* s.v. *KINO-]. **1950** A. W. HAM *Histol.* xxviii. 659/2 The epithelium [of the epididymis] is tall and regular, and tufts of large nonmotile stereocilia..project toward the lumen from the free margins of the cells. **1970** J. BABEL et al. *Ultrastructure Peripheral Nerv. System* 270 Every vestibular sensory cell carries 80–100 stereocilia..and one kinocilium... The stereocilia of the hair cells of the cristae are exceedingly long. **1979** *Nature* 30 Aug. 832/2 The stereociliary array [in the herring utricle] consists of rows of stereocilia which decrease in height the further away they are from the kinocilium. **1908** Stereocomparator [see *stereoplotter* below]. **1939** *Geogr. Jrnl.* XCIII. 240 An improved stereo-comparator for air triangulation. **1950** *Jrnl. R. Aeronaut. Soc.* LIV. 619/2 In cases where co-ordinate measurements are made in a precise stereocomparator, the prints are made on a non-distorting surface such as sensitised aluminium foil. **1975** J. B. HARLEY *O.S. Maps* i. 11 By means of self-recording stereocomparators precise pairs of measurements of co-ordinates are made on pairs of overlapping aerial photographs. **1970** *Jrnl. Macromol. Sci.: Chem.* A. IV. 1014 A very useful technique for the exploration of the stereocontrol of ionic polymerizations. **1979** *Tetrahedron Lett.* Oct. 3805 Unique stereocontrol in aldolization at C6 of penicillanates through modification of solvent and cation has been observed. **1959** Stereocontrol *v.* [see *STEREOREGULATE *v.*]. **1969** *Jrnl. Amer. Chem. Soc.* XCI. 5675 (*heading*) Stereo-controlled synthesis of prostaglandins $F_{1\alpha}$ and E_1 (*dl*). **1975** *Ibid.* XCVII. 5873 Formylation followed by acidic treatment effects cyclobutyl ring cleavage to an enol lactone which constitutes a net stereocontrolled geminal alkylation with introduction of a one-carbon and a three-carbon chain differentially functionalized. **1945** M. F. GLAESSNER *Princ. Micropalaeont.* v. 96 (*caption*) Stereo-diagram of a segment of *Loftusia persica* Brady. **1979** *Nature* 13 Dec. 681/2 (*caption*) Stereodiagram of three molecules in the crystal lattice as they are stacked along the *c* axis in what looks like a continuous double helix. **1956** *Jrnl. Amer. Chem. Soc.* LXXVIII. 6273/1 The tendency of bromine to adopt the axial orientation in the bromination of an enol would seem to indicate that stereoelectronic control is unusually large in this case since the opposing steric effect is certainly quite large. **1972** *Ibid.* XCIV. 3657/1 It is expected that σ-π conjugation would have similar stereoelectronic requirements to p-π conjugation. **1956** *Ibid.* LXXVIII. 6272/2 Addition reactions to the Δ^4-double bond..take place predominantly from the α- rather than the β-direction despite the fact that these are stereoelectronically controlled. **1978** *Further Perspectives Org. Chem.* (CIBA Found. Symp. New Ser. No. 53) 94 The cyclization step is likely to be stereoelectronically impeded. **1932** *Lancet* 2 Jan. 47/2 The perfection of a stereofluoroscope for use in hospitals is reported. **1942** *Radiology* XXXVIII. 392/1 Stereofluoroscopes continue to gather dust, or go to the junk heap. **1928** *Lancet* 3 Mar. 442/2 The latest development in stereofluoroscopic work. *Ibid.*, The law governing stereofluoroscopy has not been fully recognised. The law stated simply is that angles of vision which the X rays make with the body should be identical with those made by the vision of the observer of the body... Early in the days of X rays the tubes were of a size which rendered stereofluoroscopy impossible. **1964** *Radiology* LXXXII. 125 A test of depth perception in 62 subjects strongly suggested that binocular stereofluoroscopy provided an advantage in the perception of depth over the clues available from motion parallax alone. **1900** DORLAND *Med. Dict.* 637/2 Stereognosis. **1905** A. W. CAMPBELL *Histol. Stud. Localisation Cerebral Function* viii. 205 Damage to this part of the brain is attended by disorder of high and combined forms of sensation, such as the muscle sense and that of stereognosis. **1980** D. JENSEN *Human Nerv. System* xiv. 212/2 Faulty stereognosis provides an early indication of cortical damage. **1894** Stereo-isomer [see *HEXONIC *a.*]. **1938** *Biochem. Jrnl.* XXXII. 1627 The pentose phosphoric acid most readily attacked..is not the *d*-arabinose-5-phosphoric acid..but is the stereoisomeride *d*-ribose-5-phosphoric acid. **1943** *Jrnl. Amer. Chem. Soc.* LXV. 1524/2 Oxidation ..was the only reasonable interpretation six years ago when the stereoisomerization of carotenoids was still unexplored. **1977** *Jrnl. Organometallic Chem.* CXXV. 185 This approach has enabled us to determine the lowest energy (threshold) rearrangement mode occurring in the stereoisomerization of these [β-diketonate] complexes. **1952** *Jrnl. Gen. Physiol.* XXXVI. 306 Some of them [*sc.* carotenoids] stereoisomerize even at room temperature. *Ibid.*, A general procedure for stereoisomerizing carotenoids is to heat them in solution. **1962** L. ZECHMEISTER *Cis-Trans Isomeric Carotenoids* v. 56 An attempt to stereoisomerize β-carotene epoxides..did not afford *cis* compounds but furanoid oxides. *Ibid.* 51 The following ratios of unchanged to stereoisomerized starting material were found in the recovered pigment. **1952** *Jrnl. Gen. Physiol.* XXXVI. 306 The possibility that there exists a stereoisomerizing enzyme—a vitamin A or retinene isomerase. **1956** *Nature* 17 Mar. 516/2 Electron micrographs and stereomicrographs showing the surface characteristics and microfibrillar texture of keratin fibres were exhibited. **1975** J. G. EVANS *Environment Early Man Brit. Isles* i. 10 (*caption*) Stereomicrograph of pollen grains of *Fraxinus excelsior*, ash, ×2570. **1962** *Radiology* LXXIX. 31/1 The image on the output phosphor was viewed through a low-power stereomicroscope. **1978** FRIEDMAN & SANDERS *Princ. Sedimentol.* xiii. 417/2 Cuttings from cable-tool drilling are..examined with a stereomicroscope. **1955** *Jrnl. Chem. Soc.* 3446 The *cis*-nitro-acid..on reduction gives about equal quantities of *cis-* and *trans*-amino acid, thus indicating that some stereomutation occurs during reduction. **1975** *Jrnl. Amer. Chem. Soc.* XCVII. 238/2 Pyrolysis of cyclopropane or its substituted derivatives causes..stereomutation (e.g., *trans* → *cis-cyclopropane-1,2-d*$_2$). **1972** *Science* 9 June 1116/2 Turnbull's contribution consists chiefly of figuring the Field Museum's Trinity (Albian Cretaceous) mammal teeth by means of stereophotos. **1930** *Geogr. Jrnl.* LXXV. 159 Stereophotogrammetric methods. **1936** *Ibid.* LXXXVII. 99 To carry out a stereo-photogrammetric examination of the northern aspect and valleys of Mount Everest. **1913** *Engin. News* 27 Mar. 604/2 A method by which the troubles arising in photogrammetric surveys are eliminated is based on the principle of making the necessary measurements on stereoscopic pictures, and is called stereophotogrammetry. **1950** *Engineering* 14 July 28/3 The application of aerial photography and stereophotogrammetry to large-scale railway surveys. **1980** I. NEWTON in K. B. Atkinson *Devel. Close Range Photogrammetry* I. vi. 129 Stereophotogrammetry has made it possible to analyse the size and shape of the palate in far greater detail than hitherto. **1865** H. SIDGWICK *Let.* Apr. in A. & E. M. Sidgwick *Henry Sidgwick* (1906) iii. 129, I got your stereo-photograph (what is the short for it?) at 113 Rue de Sebastopol. **1902** *Year Bk. Photogr.* 1902 162 The effect described will be at once apparent in the stereo-photograph. **1959** *Observer* 7 June 3/4 To look at your first stereo-photograph can be an experience as climactic as seeing the Mediterranean for the first time. **1980** I. NEWTON in K. B. Atkinson *Devel. Close Range Photogrammetry* I. vi. 127 An analysis of surgically corrected abnormal faces has been undertaken in the USA from stereophotographs taken pre- and post-operatively. **1906** J. A. FLEMER *Elem. Treat. Phototopographic Methods* x. 309 Dr. Pulfrich has devised a stereoplanigraph which is being made by the Carl Zeiss firm in Jena. **1974** P. R. WOLF *Elements of Photogrammetry* xiv. 320 Each projector of the C-8 stereoplanigraph has the customary three angular rotations, but translations are introduced as movements of the reference mirrors. **1908** *Geogr. Jrnl.* XXXI. 544 An instrument..which makes the plotting of points and the reading of heights nearly automatic. To distinguish it from the stereo-comparator it has been called a stereo-plotter, as it combines the offices of the stereo-comparator and plotting board. **1979** *Photogrammetric Engin. & Remote Sensing* XLV. 802/1 Systemhouse has developed a universal analytical stereoplotter system with the primary theory of operation of an analytical stereoplotter being applied to its fullest extent. **1927** *Geogr. Jrnl.* LXX. 358 (*heading*) An attempt to describe Mr. Wild's stereo-plotting machine—the Autograph. **1975** J. B. HARLEY *O.S. Maps* i. 11 With the 1:10 000 series control points are plotted on a stable plastic sheet... A stereo-plotting machine is then used to derive map and contours, with the operator plotting the detail..in relation to the control points. **1945** *Light Metals* VIII. 269/2 The stereo radiograph, corresponding to a multiplicity of shots from different angles, is produced on a single film, an important economic advantage. **1965** D. N. & M. O. CHESNEY *Radiographic Photogr.* xiii. 319 In order to have perception of depth, various methods are available for viewing stereoradiographs. **1975** *Radiology* CXV. 455/1 Stereo radiographs usually eliminate the problem of matching sources, since the two views are more similar, but they do not always permit accurate 3-D reconstruction. **1936** *Amer. Rev. Tuberculosis* XXXIV. 517 Stereoradiographic examination of the chest on the same day revealed a widely disseminated infiltrative process throughout the right lung field. **1965** D. N. & M. O. CHESNEY *Radiographic Photogr.* xiii. 319 The anteroposterior projection of the sacro-iliac joints made stereoradiographically. *Ibid.* 317 In stereoradiography a pair of radiographs is taken. **1968** *Official Gaz.* (U.S. Patent Office) 12 Mar. TM 69/2 Cambridge Instrument Company Limited, London... *Stereoscan* for electron microscopes and parts and fittings therefor... First use on or about Mar. 31, 1966. **1970** AMBROSE & EASTY *Cell Biol.* xi. 377 The form of the ruffles is clearly seen in the Stereoscan picture. *Ibid.*, The Stereoscan allows us to see these contacts directly. **1970** *New Scientist* 27 Aug. 419/2 We were able to confirm using the stereoscan microscope that, as skin ages, the amount of extension possible in the fibre network is progressively limited. **1973** *Trade Marks Jrnl.* 11 Apr. 701/2 *Stereoscan*... Stereoscopic scanning electron microscopes and parts and fittings therefore included in Class 9. Cambridge Instrument Company Limited..; manufacturers. **1974** *Physics Bull.* Mar. 103/1 Many types of instrument are discussed, from the earliest use of a single lens to the present day field-ion and stereoscan electron microscopes. **1979** *Nature* 1 Mar. 102/2 Plates, including stereoscans, of pollen grains and spores. **1968** *Times* 1 Nov. 6/8 The idea is to carry automated photographic mapping to the full extent possible and to use stereo-viewing for interpretation. **1973** *Nature* 17 Aug. 413/1 Recently, direct stereoviewing has been developed for use in the scanning microscope.

ste·reoblock. *Chem.* [ad. It. *stereoblocchi* (G. Natta et al. 1957, in *Chim. e Industr.* XXXIX. 276/2): see STEREO- and BLOCK *sb.*] A segment of a polymer chain possessing stereoregularity. Also *attrib.*, as *stereoblock polymer*, a polymer the chains of which contain such segments.

1957 *Chem. Abstr.* LI. 12536 New linear high polymers of α-olefins are described which are distinguished by the presence within the macromol. of chain portions of different steric configurations. Designated 'stereoblock' polymers.., these differ from the block copolymers in that the former are obtained by polymerization of a single type of monomer. **1959** GAYLORD & MARK *Linear & Stereoregular Addition Polymers* 483 As originally defined by Natta, stereoblock polymers of α-olefins are polymers consisting of macromolecules containing crystallizable segments having isotactic structure and noncrystallizable segments. Actually, these macromolecules may consist either of both isotactic and atactic segments or of adjacent isotactic segments having different steric configuration... In practice, amorphous polymers, considered as atactic, may contain very short stereoblocks. **1979** C. H. BAMFORD in R. N. Haward *Developments in Polymerisation* II. 253 The 'hopping' of a propagating chain from an isotactic to a syndiotactic matrix (or vice versa) could give rise to a stereoblock daughter polymer. **1980** *Jrnl. Polymer Sci.: Polymer Physics Ed.* XVIII. 630 The length and the type of junctions between isotactic stereoblocks affects the crystallization of polypropylene.

stereochemistry. Add: Also, the stereochemical properties *of* something; a stereochemical configuration or arrangement. [After G. *stereochemie* (V. Meyer 1890, in *Ber. d. Deut. Chem. Ges.* XXIII. 568), *stereochemisch* adj. (Auwers & Meyer 1888, in *Ibid.* XXI. 789).]

1905 *Rep. Brit. Assoc. Adv. Sci. 1904* 169 The stereochemistry of nitrogen has..attracted considerable attention. **1969** *Jrnl. Leeds Univ. Union Chem. Soc.* XI. 43 The important stereochemistries are octahedral.., tetrahedral.., and square planar. **1972** *Nature* 21 Jan. 180/3 Nyholm's researches in transition metal chemistry..were particularly concerned with metal carbonyl, metal–metal bonded systems, unusual stereochemistries and..the reactivity of coordinated organic groups. **1972** DEPUY & CHAPMAN *Molec. Reactions & Photochem.* vi. 84 It is often observed that related thermal and photochemical processes, while both stereospecific, give products with differing stereochemistry.

Hence **stereoche·mically** *adv.*, as regards the relative spatial positions of atoms; **stereoche·mist,** an expert or specialist in stereochemistry.

1890 *Jrnl. Chem. Soc.* LVIII. II. 970 The authors regard the existence of stereochemically isomeric azo-compounds as doubtful. **1935** TIPSON & STILLER in Harrow & Sherwin *Textbk. Biochem.* ii. 84 Where the two monoses are not identical, four stereochemically different disaccharides are theoretically capable of existence. **1937** *Nature* 10 July 49/2 The stereochemist is catered for in the precise directions given for the synthesis and complete optical resolution of that useful stereochemical agent, α-phenylethylamine. **1963** *Times* 23 Jan. 15/1 From there he went in 1905 to the University of Zurich to work with Alfred Werner, the distinguished stereochemist. **1974** GILL & WILLIS *Pericyclic Reactions* iii. 75 The stereochemically distinguishable modes of addition on a single olefin component.

stereogram² (ste·rio-, stiᵊ·riogræm). [f. *STEREO *a.* + *RADIO)GRAM².] An instrument for stereophonic reproduction of sound combining a radio and gramophone in a single cabinet; a stereo radiogram. Cf. *STEREO *sb.*³ 2.

1958 *Daily Mail* 27 Aug. 3/5 Some stereograms have both amplifiers and speakers inside one cabinet. **1976** *Gramophone* Nov. 914/1 Continuing their tradition of manufacturing single-unit stereograms, Thorn Consumer Electronics have just announced the latest in the line. **1977** M. JANCATH *Seatag* II. xi. 122 The stereogram in the corner stopped.

stereology (steri-, stiᵊri,ǫ·lŏdʒi). [f. STEREO- + -LOGY.] The science of the reconstruction of three-dimensional structures from two-dimensional sections of them.

1963 *1st Internat. Congr. Stereology* ii. 2 With the aid of a Greek pocket dictionary we [*sc.* Hennig & Elias] coined the word Stereology. **1967** *Proc. 2nd Internat. Congr. Stereology* 1 Stereology is three-dimensional interpretation of flat images or extrapolation from two to three-dimensional space. *Ibid.* 2 An announcement in a few journals brought 11 scientists together on the Feldberg Mountain in the Black Forest 11–12 May, 1961. The word stereology was then coined and the International Society for Stereology was founded. **1974** *Nature* 29 Nov. 412/1 Stereology has revealed that there is a continuous increase in mitochondrial volume during the cycle. **1977** N. T. James in Meek & Elder *Analytical & Quantitative Methods Microsc.* 20 An absolute requirement of stereology is that the test samples should be representative of the test tissue.

Hence **stereolo·gical** *a.*, of, pertaining to, or by stereology; **stereolo·gically** *adv.*; **stereo·logist,** one skilled in stereology.

1963 *1st Internat. Congr. Stereology* v. 3 The..arithmetical and geometrical principles which I now assume to be those generally employed by nature and therefore of common interest to stereologists. *Ibid.* vi. 21 The stereological field includes all the investigations in respect to the three-dimensional structures. **1972** *Science* 12 May 655/1 Like all stereology..stereological shape determination is a matter of geometrical probability. *Ibid.*, Simple shapes can be identified stereologically by measuring the length and width of each profile of a feature in section, classifying the quotients length/width of many profiles, and applying mathematical rules elaborated by the reviewer. **1975** *Nature* 10 July 151/2 Articles on high resolution, dark-field microscopy, in-focus phase contrast and stereological techniques. **1981** *McGraw-Hill Yearbk. Sci. & Technol.* 376/2 This relationship has been defined stereologically by Eq. (5).

stereometer[1]. Add: **3.** *Cartography.* Any of various instruments for measuring the parallax of a feature depicted in a stereoscopic pair of aerial photographs.

1911 *Encycl. Brit.* XXV. 900/1 The stereometer may be regarded as a modification of the stereocomparator. **1940** C. A. Hart *Air Photogr.* vi. 178 In order that the stereometer may be moved easily while maintaining a direction parallel to the base-line of the pair of photographs, a parallel-guidance mechanism is often fitted. **1974** P. R. Wolf *Elem. Photogrammetry* viii. 150 Parallaxes of points may be measured stereoscopically. This method employs a stereoscope in conjunction with an instrument called a parallax bar and also..a stereometer.

stereometric, *a.* Add: Hence **stereome·trically** *adv.*

1890 in Webster. **1928** B. J. Leggett *Theory & Practice Radiol.* III. ix. 430 If..we radiograph this pelvis stereometrically and then reconstruct the pelvis in space by the method above, we are in a position to directly and accurately measure this diameter by means of suitable orthographic apparatus. **1972** *Science* 9 June 1136/1 The relative volume of extracellular space was estimated stereometrically.

stereophonic (sterio-, stiəriofǫ·nik), *a.* [f. Stereo- + Phonic *a.* (*sb.*).] Giving the impression of a spatial distribution in reproduced sound; *spec.* employing two or more channels of transmission and reproduction so that the sound may seem to reach the listener from any of a range of directions.

1927 *Wireless World* 26 Jan. 117/2 A marked improvement in quality of reproduction will be noticed, due to the phase-difference introduced by the distance between loud-speaker and phones. This phase-difference also varies with the frequency of the sounds reproduced, and thus a constantly varying difference in phase produces the stereophonic effect so superior to ordinary reproduction. **1940** *Nature* 3 Aug. 174/1 Demonstrations of the stereophonic reproduction of music and speech were given at the Carnegie Hall. **1953** *Sun* (Baltimore) 3 Aug. 16/7 Many picture men believe in the long-range possibilities of theater TV, but they also think that standardization of stereoscopic, stereophonic, wide screen equipment will have to come first. **1957** *Observer* 8 Sept. 10/2 An incredibly lush stereophonic music score exploits all the Hollywood clichés and adds some of its own. **1958** *Times* 14 Jan. 8/3 The B.B.C. last night carried out an experiment in the London area in stereophonic sound transmission. **1958** *Listener* 16 Oct. 605/1 Recently stereophonic disks have been introduced which have brought 'stereo' within reach of a much wider public. **1969** *Islander* (Victoria, B.C.) 6 July 5/2 Travellers enjoy a movie and stereophonic music through individual ear phones. **1977** *Rep. Comm. Future of Broadcasting* 13 in *Parl. Papers 1967–77* (Cmnd. 6753) VI. 1 The BBC.. introduced stereophonic broadcasting in 1966. Stereo was also provided by the IBA's local radio VHF transmissions.

Hence **stereopho·nically** *adv.*, in a stereophonic manner; as regards, or by means of, stereophony.

1940 *Nature* 3 Aug. 174/1 A symphony concert produced in Philadelphia was transmitted over telephone wires to Washington and there reproduced stereophonically. **1958** *Times* 4 Aug. 10/2 After a fly has been buzzing stereophonically round her..a flash-back considerately takes over the task of explanation. **1969** *Amateur Photographer* 16 July 75/1 Testing for stereo should be done with recordings which have been proved to be satisfactory themselves stereophonically. **1972** K. Bonfiglioli *Don't point that Thing at Me* ii. 16 Martland..busied himself coaxing a few more decibels out of the stereo equipment... The Flying Scotsman whooped stereophonically for a level-crossing. **1977** *Gramophone* Sept. 504/2 You cannot distinguish the players stereophonically but quality and playing are excellent.

stereopho·nics, *sb. pl.* [f. prec.: see -ic 2.] Stereophonic techniques; stereophonic sound.

1958 *Punch* 22 Jan. 153/1 As yet there has been no talk of marrying these clever stereophonics with the echo-chamber, but it is bound to come. **1973** D. Francis *Slay-Ride* v. 65 Arne..switched Beethoven on again fortissimo... I sat resignedly..while the stereophonics shook the foundations. **1975** *Listener* 2 Oct. 448/1 Getting your ears in phase with different sound sources, otherwise called stereophonics.

stereophony (steri-, stiəri,ǫ·fŏni). [f. as prec.: see -y[3].] Stereophonic reproduction; stereophonic sound.

1950 *Wireless World* Sept. 327/1 The first broadcast of stereophony, the system in which sources of sound are restored to their relative positions in space, took place in France on June 19th, 1950. **1958** *Manch. Guardian* 30 June 5/3 Even on monaural discs the equipment..gives a remarkable impression of stereophony. **1972** *Daily Tel.* 31 Jan. 7/5 The BBC started experiments with stereophony in 1926.

stereopsis (steri-, stiəri,ǫ·psis). [f. Gr. στερε-ός solid (see Stereo-) + ὄψις power of sight.] The ability to perceive depth and relief by stereoscopic vision.

1911 Stedman *Med. Dict.* 826/1 *Stereopsis*, stereoscopic vision. **1920** *Arch. Opthalmol.* XLIX. 64 When the facts are known it is quite possible that visual tests will not stop with measuring the acuity of each eye and the color sense, but that a certain standard of stereopsis will be required. **1961** *Lancet* 22 July 168/2 No stereopsis has been demonstrated in any of these patients. **1972** *Sci. Amer.* Aug. 86/3 By providing very precise localization of objects in visual space, stereopsis can be regarded as the *raison d'être* of binocular vision.

Hence **stereo·ptic** *a.*; **stereo·ptically** *adv.*, with an appearance of depth; stereoscopically.

1931 *Arch. Ophthalmol.* VI. 139 Stereoptic reactions of a grade above the lowest are observable in ordinary fishes, but the lowest grade is probably not represented in any surviving species. **1972** *Sci. Amer.* Apr. 65/3 One can also feed the three-dimensional coordinates into a computer and obtain simple ball-and-stick drawings that can be viewed stereoptically, enabling one to visualize the folded chain of the protein in three dimensions. *Ibid.* Aug. 120/2 No one believes that the anatomical retina could possibly be given information on any large scale from the mind; there are few, if any, outgoing pathways. Therefore, Julesz conjectures, the eidetic image is formed later than the image on the retina but earlier than the stereoptic image, perhaps much earlier. **1976** *Billings* (Montana) *Gaz.* 24 June 16-A/1 More pictures were taken Wednesday on Viking's fourth pass over the site, Chryse, by Viking's twin cameras. They were lined up to photograph overlapping strips which can be viewed stereoptically for a three-dimensional effect.

stereoregular (sterio-, stiəriore·giŭlăɹ), *a.* *Chem.* Also **stereo-regular.** [f. Stereo- + Regular *a.*] Of a polymer: having each substituent atom or group on the main polymer chain oriented in the same manner on the chain with respect to the neighbouring atoms or groups. Of a reaction: giving rise to such a polymer.

1959 Gaylord & Mark *Linear & Stereoregular Addition Polymers* 473 The most significant new developments have involved the successful preparation of stereoregular polymers from monomers..which previously had only yielded atactic polymers. **1961** *Industr. Chemist* Feb. 74/2 Although stereoregular polymers prepared with Ziegler-type catalysts are now of considerable commercial interest, research is continuing in attempts to obtain stereo-specific polymers by other, and perhaps less complex, catalysts. **1962** *Engineering* 12 Jan. 57/1 The increasing commercial availability of the 'stereo-regular' synthetic rubbers. **1971** *Sci. Amer.* Dec. 50/3 One striking example of the ability of catalysts to perform highly selective molecular alterations is the stereoregular polymerization of olefins pioneered by Karl Ziegler and Giulio Natta. **1979** C. H. Bamford in R. N. Haward *Developments in Polymerisation* II. 251 Stereoregular poly(methyl methacrylate) can function, under suitable conditions, as a template for the polymerisation of the monomer.

Hence **ste:reoregula·rity.**

1959 *Jrnl. Polymer Sci.* XLI. 80 A decrease in polymerization temperature could result in increased regularity as a result of a simultaneous improvement in stereoregularity, linearity, and head-to-tail structure. **1964** *Ibid.* II. A. 4642 Stereoregularity was noticed in acetaldehyde polymerization only when solid boron-trifluoride-etherate was used as the initiator. **1976** *Makromol. Chemie* CLXXVII. 1475 The four catalyst systems employed for the *cis* polymerization..give polymers with a different degree of stereoregularity. This could depend on a different stereoregulating power of the catalysts. **1979** *Jrnl. Polymer Sci.: Polymer Chem. Ed.* XVII. 2022 There was no significant difference between the stereoregularities of the polymers obtained with cesium naphthalene and fluorenylcesium as catalyst. This result indicates that the difference in initiator anions ..did not affect stereoregulation.

stereoregulate (sterio-, stiəriore·giŭlēit), *v.* *Chem.* [f. Stereo- + Regulate *v.*] *trans.* To cause (a polymerization or its product) to be stereoregular. So **ste:reore·gulated** *ppl. a.*, **-re·gulating** *vbl. sb.* and *ppl. a.*; **ste:reoregula·tion.**

1959 Gaylord & Mark *Linear & Stereoregular Addition Polymers* 476 Even the application of free radicals as initiators does not rule out the formation of stereoregulated addition polymers. *Ibid.* 478 The propagation reaction will be stereoregulated or stereocontrolled and the polymer will be tactic, which means that the individual monomers in the chain will overwhelmingly be added in one of the different possible ways and the resulting macromolecule will display stereospecificity, i.e., a high degree of internal orderliness. *Ibid.*, Experience..has shown that stereoregulating effects are strongest in the case of Ziegler-type catalysts. *Ibid.* 479 It appears that a methyl group at the alpha carbon atom is favorable for stereoregulation. **1961** *Industr. Chemist* Feb. 73/1 (*heading*) Nature of the stereoregulating catalysts. **1962** E. L. Eliel *Stereochem. Carbon Compounds* xv. 448 The 'stereoregulated' (i.e. isotactic or syndiotactic) polymers have higher densities, higher melting points, and lower solubility than the atactic polymers. **1976**, **1979** Stereoregulating, -regulation [see *stereoregularity]. **1979** C. H. Bamford in R. N. Haward *Developments in Polymerisation* II. 249 The stereoregulating influence of added prepolymers has been shown to be more pronounced at lower temperatures.

stereoselective (ste:rio-, stiə:riosĭle·ktiv), *a.* *Chem.* [f. Stereo- + Selective *a.*] Of a reaction: producing a particular stereoisomeric form of the product preferentially, irrespective of the configuration of the reactant; = *stereospecific *a.* 1 a.

Orig. (quot. 1957) = *stereospecific *a.* 2.

1957 *Jrnl. Amer. Chem. Soc.* LXXIX. 1595/2 Comparable *trans* eliminations in open-chain systems are stereoselective. **1959** *Ibid.* LXXXI. 110/1 A stereospecific process is one in which the configuration of the product is related to that of the reactant... A stereoselective process is one in which there is no relationship between the configuration of the reactant and that of product but one in which there is a definite driving force for forming one of the possible stereoisomeric products. **1962** E. L. Eliel *Stereochem. Carbon Compounds* xv. 436 All stereospecific processes are stereoselective, but not all stereoselective processes are stereospecific. For example, the low-temperature free-radical addition of hydrogen bromide to the 2-bromo-2-butenes..is both stereospecific and stereoselective in that the cis olefin gives *meso*-2,3-dibromobutane whereas the trans olefin gives the *dl*-dibromide. At higher temperature the reaction is still stereoselective (formation of 75% *dl* isomer to 25% meso isomer) but no longer stereospecific; both olefins now give the same product mixture. **1968** [see *stereospecific *a.* 2]. **1979** *Canad. Jrnl. Chem.* LVII. 646/1 Our synthesis [of sucrose] was predicated upon the stereoselective methoxybromination of the diene 8.

Hence **ste:reosele·ctively** *adv.*; **ste:reosele·ction, ste:reoselecti·vity,** the property or fact of being stereoselective.

1956 *Jrnl. Amer. Chem. Soc.* LXXVIII. 1171/1 The decreased stereoselectivity of ketonization..in the 2-methyl-3-phenylindanone system as compared with the 1-phenyl-2-benzoylcyclohexane situation. **1957** *Ibid.* LXXIX. 1594/2 The fact that eliminations from I and II ..occur stereoselectively rules out the possibility of an intermediate carbanion of any but very short half-life. **1968** S. I. Miller in V. Gold *Adv. in Physical Org. Chem.* VI. 186 We shall then describe a number of causes of stereoselection associated with bonding, steric, thermodynamic [etc.] factors. **1974** Gill & Willis *Pericyclic Reactions* vi. 191 Likewise the *trans*-alkene is formed in the other processes with fairly high stereoselectivity. **1979** *Tetrahedron Lett.* 4867 (*heading*) Stereoselection in the $AlCl_3$-catalysed ene additions of chloral to 1,2-dialkyl ethylenes. **1979** *Canad. Jrnl. Chem.* LVII. 646/1 A new synthesis of sucrose in which all of the steps involved proceeded with complete stereoselectivity. **1980** *Chem. in Brit.* XVI. 518/1 He was able to put one of the carbon-carbon bonds in place stereoselectively creating two of the six contiguous chiral centres on the left hand side of the molecule.

stereospecific (ste:rio-, stiə:riospĕsi·fik), *a.* *Chem.* Also **stereo-specific.** [f. Stereo- + Specific *a.*] **1. a.** Of a reaction: = *stereoselective *a.* Also of a catalyst: causing a reaction to be (more) stereoselective.

1949 *Jrnl. Amer. Chem. Soc.* LXXI. 3866/2 The reaction giving rise to the acetates was at least 90% stereospecific since the ratio of the yield of the acid phthalate of racemate I to the yield of the acid phthalates of other alcohols was ten to one. Stereospecificity of even a higher degree was found in the II series. **1957** [see *atactic *a.* 3]. **1958** *Chem. Abstr.* LII. 1106 (*heading*) Isotactic and other stereospecific polymers. **1959** *Times Rev. Industry* Feb. 57/1 Using the stereospecific catalysts developed in Germany by K. Ziegler, Professor Natta in Milan had succeeded in producing, from propylene gas, useful polymers which he characterized as isotactic. **1960** *Times* 19 July (Royal Soc. Tercentenary Suppl.) p. xiv/3 The past decade has seen the emergence of various stereospecific syntheses. **1966** *Petroleum Handbk.* (ed. 5) 215/2 There are two possibilities in building up the polymer chain [of propylene]; the methyl groups can be situated either at random on both sides of the chain or on one side of the chain only (stereospecific polymerization). **1968** [see sense 2 below]. **1978** J. A. Brydson *Rubber Chem.* i. 8 Stereospecific catalyst systems..led to the availability of..rubbers with a much more controlled molecular architecture than had been obtained before. **1979** *Jrnl. Org. Chem.* XLIV. 3374 The determination of the configuration at C-24 of the revised structure of oogoniol..was accomplished by the stereospecific synthesis of the model compounds..which contain the oogoniol side chain.

b. Of a polymer, esp. rubber: = *stereoregular *a.*

1959 *Times* 27 Apr. (Rubber Industry Suppl.) p. ii/7 Polymer chemists have long yearned to synthetize such stereo-specific polymers found to occur so prevalently in Nature. **1961** [see *stereoregular *a.*]. **1966** *Petroleum Handbk.* (ed. 5) 224/1 In the last few years..a new family of synthetic rubbers—the stereospecific rubbers—have been developed. **1974** J. Ford-Smith tr. *Heinisch's Dict. Rubber* 447/1 Examples of stereospecific polymers are cis-1:4-polyisoprene and cis-1:4-polybutadiene.

2. Of a reaction or process: yielding a product, or having a rate, that depends on the particular stereoisomeric form of the starting material.

1959, **1962** [see *stereoselective *a.*]. **1968** I. L. Finar *Org. Chem.* (ed. 4). II iv. 126 The term stereospecific reaction has been used in the same sense as stereoselective reaction, but now the tendency is to restrict the use of stereospecific to a reaction in which different stereoisomers produce different products or act at different rates. **1970** J. H. Quastel in Ehrenpreis & Solnitsky *Neurosci. Res.* III. 15 Stereospecific effects are also observed in cerebral amino acid exchange reactions; for example, elevated cerebral L-lysine is more effective than elevated cerebral D-lysine in increasing L-lysine exchange. **1978** *Biochem. Pharmacol.* XXVII. 653/1 The finding that the R-enantiomer of gliflumide binds less strongly [than the S-enantiomer] indicates that antidiabetic drugs are bound to plasma by a stereospecific process.

Hence **ste:reospeci·fically** *adv.*, in a stereospecific manner: **ste:reospecifi·city,** the property or state of being stereospecific.

1949 Stereospecificity [see sense 1 a above]. **1955** *Jrnl. Amer. Chem. Soc.* LXXVII. 4567/1 *cis*-Propenyllithium prepared from *cis*-2-bromopropene with lithium in ether.. reacts stereospecifically with benzaldehyde..to give..the alcohol with the same configuration as that of the starting bromide. **1965** Phillips & Williams *Inorg. Chem.* I. x. 383 Other monomers can be stereo-specifically polymerized. **1970** *Jrnl. Polymer Sci.* A. VIII. 988 The stereospecificity of these catalysts is remarkable since they yield polybutadienes with a very high 1,4 content. **1970** J. H. Quastel in Ehrenpreis & Solnitsky *Neurosci. Res.* III. 15 Studies of the stereospecificity of amino acid uptake in vivo have shown..that the L-amino acid usually penetrates the brain to a greater extent than the corresponding D-isomer. **1979** *Jrnl. Chem. Soc.: Chem. Comm.* 918/2 Oxidation of labelled hexa-1,5-dienes with permanganate generates four new chiral centres with complete stereospecificity. *Ibid.* 920/2 The oxidative cycloaddition constitutes a method for the formation, stereospecifically, and in a single step, of four chiral centres from an achiral, acyclic reactant.

stereotactic (sterio-, stiəriotæ·ktik), *a.* [f. Stereo- + Tactic *a.*[2] (*sb.*[2]).] †**1.** *Biol.* = Thigmotactic *a.* *Obs. rare.*

1902 *Q. Jrnl. Microsc. Sci.* XLVI. 171 There appear to me to be yet two possible explanations of the penetration [of sperm into the egg]: (1) it is due to reaction to a stereotactic stimulus; (2) it is purely mechanical.

2. *Biol.* and *Med.* = *stereotaxic *a.*

1954 *Amer. Jrnl. Roentgenol.* LXXI. 441/2 A new type of stereotactic instrument for use in man. **1961** *Lancet* 2 Sept. 552/1 Stereotactic operations for parkinsonism. **1976** *New Scientist* 26 Feb. 427/2 Some of the patients will undergo stereotactic brain surgery, which enables lesions to be placed very precisely in the required parts of the brain.

Hence **stereota·ctically** *adv.*

1934 in Webster. **1949** *Jrnl. Neurophysiol.* XII. 371 Electrolytic lesions were made..at a site determined stereotactically. **1966** *Arch. Neurol.* XIV. 334/2 Cross and Green, utilizing microelectrodes placed stereotactically in the hypothalamus of rabbits, studied the activity of single units, presumably single cells.

stereotaxic (sterio-, stiəriotæ·ksik), *a.* *Biol.* and *Med.* [f. Stereo- + Tax(is + -ic.] Involving or designed for the accurate three-dimensional positioning and movement of objects inside the brain.

1908 Horsley & Clarke in *Brain* XXXI. 63 The foundation of the stereotaxic instrument is a rigid quadrilateral rectangular frame..the ends of which..can be approximated by joints which slide on the lateral bars. **1919** S. Paget *Sir V. Horsley* 189 R. H. Clarke..also devised a stereotaxic apparatus, probably the most complex of all the mathematical instruments of physiology. **1935** *Arch. Neurol. & Psychiatry* XXXIV. 162 With the aid of the Horsley-Clarke stereotaxic instrument, lesions were placed in..the hypothalamus in forty adult cats. **1969** *New Scientist* 30 Jan. 230/1 The stereotaxic.. procedure reaches deep areas of the brain. **1971** *Nature* 8 Jan. 131/1 The brain was exposed from above and a fine knife lowered between the two hemispheres until the blade was at the correct stereotaxic setting for the supraoptic commissure. **1971** 'D. Halliday' *Dolly & Doctor Bird* xvi. 247 Nothing short of stereotaxic surgery will ever obliterate the events of..that night.

Hence **stereota·xically** *adv.*

1964 *Jrnl. Neurophysiol.* XXVII. 754 Electrolytic lesions produced by passing 3-mA anodal d.c. for 10 sec. through a stereotaxically guided stainless steel electrode with 1 mm. bared at the tip. **1979** *Nature* 4 Jan. 52/1 I.c. cannulae (o.d. 0·6 mm) were implanted stereotaxically in male Wistar rats.

stereotaxis (sterio-, stiəriotæ·ksis). *Biol.* [f. Stereo- + Taxis.] †**1.** *Biol.* = Thigmotaxis. *Obs. rare.*

1897 C. B. Davenport *Exper. Morphol.* I. iv. 105 (*heading*) Effect of molar agents in determining the direction of locomotion—thigmotaxis (stereotaxis) and rheotaxis. **1902** *Q. Jrnl. Microsc. Sci.* XLVI. 175 The passage of the spermatozoa through the gelatinous coat..is more or less in a radial direction as regards the egg... The phenomenon is possibly due to stereotaxis, but a purely mechanical explanation seems to the author more probable.

2. *Biol.* and *Med.* Also **ste·reotaxy.** Stereotaxic surgery.

1959 Schaltenbrand & Bailey (*title*) Einführung in die stereotaktischen Operationen. Introduction to stereotaxis. **1959** L. V. Amador et al. in *Ibid.* I. 5/2 Most of the childrens [*sic*] brains were discarded because at the present stage stereotaxy will most likely be applied to adults only. **1974** C. B. T. Adams in R. M. Kirk et al. *Surgery* xiv. 284 This may be done by 'bimedial leucotomy' cutting the white matter under direct vision from above or by undercutting the orbital part of the frontal cortex from below (by open operation or by stereotaxis). **1974** *Lancet* 13 July 106/1 (*heading*) Stereotaxy for obesity.

stereotype, *sb.* Add: **3. b.** A preconceived and oversimplified idea of the characteristics which typify a person, situation, etc.; an attitude based on such a preconception. Also, a person who appears to conform closely to the idea of a type.

1922 W. Lippman *Public Opinion* vi. 93 A stereotype may be so consistently and authoritatively transmitted in each generation from parent to child that it seems almost like a biological fact. **1935** G. W. Allport in C. Murchison *Handbk. Social Psychol.* xvii. 809 Attitudes which result in gross oversimplifications of experience and in prejudgements..are commonly called biases, prejudices, or stereotypes. **1948** Krech & Crutchfield *Theory & Probl. Social Psychol.* II. v. 171 The concept of stereotype ..refers to two different things. (1)..a tendency for a given belief to be widespread in a society... (2)..a tendency for a belief to be oversimplified in content and unresponsive to the objective facts. **1960** T. Hughes *Lupercal* 42 Who lived at the top end of our street Was a Mafeking stereotype, ageing. **1968** W. E. Lambert et al. in J. A. Fishman *Readings Sociol. of Lang.* 487 American students of English-speaking backgrounds who are in the process of studying the French language have a generally negative set of stereotypes about the basic personality characteristics of French-speaking people. **1974** *Howard Jrnl.* XIV. 102 The stereotypes which society has of the offender, are quickly matched by stereotypes which many offenders create of society. **1981** *Church Times* 23 Oct. 9/1 The neatly dressed unmarried lady (never without handbag)..is definitely not the narrow stereotype our media would have us think she is.

c. *Zool.* A stereotyped action or series of actions performed by an animal (see *stereotyped *ppl. a.* c).

1966 R. A. Hinde *Animal Behaviour* xxiii. 389 In captivity animals often develop behaviour stereotypes or tics which are repeated monotonously. **1971** *Sci. Amer.* June 117/1 Although subordinate males had no chance to mate with hens at the display grounds, they did perform mock matings... Mounting a pile of dry cow manure or a log or simply squatting on the ground, they would go through the stereotype of mating actions: treading the object, fluttering their wings, lowering their tail and even in some cases ejaculating.

stereotyped, *ppl. a.* Add: **b.** Also *spec.* in *Psychol.* (see quot. 1934).

1934 H. C. Warren *Dict. Psychol.* 262/2 *Stereotyped* (*responses*), characterizing certain responses which are always performed in substantially the same manner. **1950** T. Wiesengrund-Adorno et al. *Authoritarian Personality* iv. xvi. 627 He..personalizes his own stereotyped hostility. **1957** [see *consummatory *a.* 2]. **1968** A. Storr *Human Aggression* ii. 11 This mechanism is easily set off, and, like other emotional responses, it is stereotyped and, in this sense, 'instinctive'.

c. *Zool.* Of an animal's action or behaviour: repeated though serving no obvious purpose.

1934 E. S. Russell *Behaviour of Animals* i. 8 Their [*sc.* chaffinches'] behaviour is mainly instinctive, independent of previous experience, and to a considerable extent stereotyped and invariable. **1950** G. Sircom tr. *Hediger's Wild Animals in Captivity* viii. 88 M. Holtzapfel ..was able..to remove a long standing stereotyped movement in an armadillo..kept on a slippery and therefore unbiological floor, by providing a layer of earth ten inches deep. **1970** R. A. & B. M. Maier *Compar. Animal Behavior* xvii. 356 Stereotyped movements displayed during courtship can evolve out of general movement patterns.

Hence **ste·reotypedness,** the quality or state of being stereotyped.

1977 *Spare Rib* June 37/1 The visual equivalents tend to have a journalistic stereotypedness about them. **1979** *Trans. Philol. Soc.* 140 We also learn a good deal about.. live Old Persian speech, many features of which, owing to the wretchedly small number of OP inscriptions and the stereotypedness of most of them, have remained hitherto well beyond our reach.

stereotypical (sterio-, stiərioti·pikăl), *a.* [f. Stereotype *sb.* + -ical.] Of, pertaining to, or resembling a stereotype (sense *3 b). Also *spec.* in *Psychol.*, designating behaviour which is repeated without variation irrespective of the particular circumstances. Cf. Stereotypic *a.*

1949 *Commentary* July 41/2 The stereotypical Negro, the unstinting giver. **1950** T. Wiesengrund-Adorno et al. *Authoritarian Personality* iv. xix. 747 Only by identifying stereotypical traits in modern humans..can the pernicious tendency towards all-pervasive classification.. be challenged. **1957** P. Lafitte *Person in Psychol.* v. 50 Co-operation on a ground of stereotypical distrust is not the same as co-operation on a ground of confidence. **1968** S. Stuckey in A. Chapman *New Black Voices* (1972) 439 Small wonder we have been saddled with so many stereotypical treatments of slave thought and behavior. **1975** *Nature* 25 Dec. 750/1 Furthermore, a similar class of stereotypical behaviour is elicited by the DA agonist apomorphine. **1981** *Times Lit. Suppl.* 29 May 602/1 It boasts some fine performances and some stereotypical Allen dialogue.

stereotypically, *adv.* Add: (*c*) so as to constitute a stereotype (sense 3 b); in a stereotypical manner; (*d*) *Zool.*, as a stereotyped action.

1950 T. Wiesengrund-Adorno et al. *Authoritarian Personality* i. iii. 98 The inner conflict is replaced by a new conflict between groups: the stereotypically moral 'we' and the stereotypically immoral 'they'. **1950** G. Sircom tr. *Hediger's Wild Animals in Captivity* viii. 76 The path stereotypically followed by a dingo. **1975** *Daily Tel.* 19 Nov. 14/3 Do all minority faces look stereotypically alike or are they depicted as genuine individuals? **1977** *Time* 26 Dec. 40/1 A throughly Episcopal church in Darien, Conn., an almost stereotypically proper and affluent Northeastern suburb. **1980** *Times* 3 June 11/6 The Nordic countries, which we might stereotypically expect to be sober and/or tormented, are just that.

stereotyping, *vbl. sb.* Add: **3.** *Zool.* = *stereotypy 3.

1950 G. Sircom tr. *Hediger's Wild Animals in Captivity* viii. 76 This stereotyping is not serious, yet it shows clearly enough that the animals's surroundings need enrichment. **1967** *Animal Behaviour* XV. 64 There were no significant differences between any of the groups in spot-picking, route-tracing, or total stereotyping.

stereotypy. Add: **2.** (Further examples.)

1934 H. C. Warren *Dict. Psychol.* 262/2 *Stereotypy*, a pathological phenomenon consisting in the endless repetition of fragmentary or apparently senseless words, apparently useless movements, or of certain postures. **1948** [see *catatonic *a.*]. **1976** M. Hamilton *Fish's Schizophrenia* (ed. 2) iii. 58 A stereotypy is a movement which is not goal-directed and which is carried out in a uniform way, but some mannerisms which are abnormal exaggerations of expressive movements may be confused with stereotypies; however they are not executed in such a rigid way.

3. *Zool.* The frequent repetition by an animal of an action that serves no obvious purpose.

1934 E. S. Russell *Behaviour of Animals* v. 98 Examples of this stereotypy or rigidity of instinctive behaviour. **1967** *Animal Behaviour* XV. 63/2 In this stereotypy, a bird would repeatedly touch the tip or side of the bill to a particular spot. **1981** *Ibid.* XXIX. 4/1 Inappropriate and often perseverant stereotypy can be elicited in many animals by raising them in abnormal and restricted environments.

4. *gen.* The state or quality of being stereotyped (sense b).

1950 T. Wiesengrund-Adorno et al. *Authoritarian Personality* i. iii. 94 One striking characteristic of the imagery in anti-Semitic ideology is its *stereotypy*. **1973** O. Sacks *Awakenings* i. 36 Witty and precise in her speech without significant stereotypy or stickiness of thought. **1976** *Word* 1971 XXVII. 128 This stereotypy of semantic relations of nouns reported by Bowerman decreases as mean sentence length increases.

steric, *a.* Add: Hence **ste·rically** *adv.*, by, or as regards, the three-dimensional arrangement of atoms.

1918 *Jrnl. Chem. Soc.* CXIV. i. 127 If only one group is sterically hindered, then reactions can take place at both tertiary amino-groups. **1959** *New Scientist* 19 Feb. 399/1 Propylene,..using a Natta catalyst, can be turned into straight chains of polypropylene molecules, with all the arms or side groups projecting in the same direction: that is to say, they are sterically oriented. **1974** Gill & Willis *Pericyclic Reactions* v. 130 The face to face interaction of the two π-systems is the only mode that is geometrically and sterically feasible.

stericks. (Later example.)

1859 C. Bray *Let.* 26 June in *Geo. Eliot Lett.* (1954) III. 94 We had Sara in strong stericks all the way home, because she had missed her final chance of explanation and advice from you.

sterilant (ste·rilănt). [f. Steril(ize *v.* + -ant[1].] A sterilizing agent. **a.** An agent used to make something free of plant life or micro-organisms; a herbicide or disinfectant.

1955 *Proc. 12th Ann. North Central Weed Control Conf.* 45/1 The estimated erodibility of the soil treated with various sterilants..is shown in Table 3. **1960** *Farmer & Stockbreeder* 12 Jan. 71/1 In the dairy, a mobile washing-up trough on castors is filled with 30 gallons of hot water with added detergent and sterilant. **1960** *Agronomy Jrnl.* LII. 707/2 The best way to relieve the erosion hazard is to eliminate the weeds with soil sterilants so that crops can be produced that will leave a protective cover. **1973** *Nature* 20 Oct. 455/2 Leaflets were..surface sterilized in 1·5% sodium hypochlorite solution (10 min). The sterilant was removed by 8 successive washes with sterile water. **1979** *Arizona Daily Star* 8 Apr. A1/2 The formula containing 2,4,5-T is used as a sterilant on driveways and brick walks to prevent plant growth.

b. An agent used to render an organism incapable of producing offspring.

1961 *Jrnl. Economic Entomol.* LIV. 688/1 Tests were made to determine the effect of the sterilant on copulation by treated males and on zygote formation in females inseminated by them. **1973** J. J. McKelvey *Man against Tsetse* iii. 197 When the flies are released, they can transmit this sterilant venereally to their mates and again shatter the potential of the fly to increase in numbers. **1978** *Times* 4 Sept. 13/4 There is no problem in producing sterilization substances; the difficulty lies in finding a sterilant that does not have side effects and in persuading wild animals to take it regularly.

sterile, *a.* Add: **8. sterile-male** *Biol.*, used *attrib.* to designate the technique of controlling a natural population by releasing large numbers of sterile males into it, so that females that mate only with these do not reproduce.

[**1955** E. F. Knipling in *Jrnl. Econ. Entomol.* XLVIII. 459/1 The purpose of this paper is to consider the possibility of controlling insects by releasing sexually sterile males among the existing population.] **1959** *Science* 9 Oct. 903/1 The possibility of controlling animal populations by the sterile-male method is not necessarily limited to insects. **1975** *Nigerian Jrnl. Entomol.* I. 181 Because of the ease of preparation, good-keeping quality and re-useability of the [bat's wing] membrane, it..may have a very important role in the control of tsetse flies in Africa by the use of the sterile male technique. **1980** *Adv. Vet. Sci. & Compar. Med.* XXIV. 166 Screwworm populations subjected to autocidal control by the sterile-male technique.

sterilization. Add: Also *fig.*, esp. in *Econ.* Cf. sense 6 of the *vb.*

1938 *Times* 15 Feb. 14/2 'Sterilization' is the word used to describe the Treasury's policy of limiting the expansion of credit which would follow if the imports of gold were used as a basis for the expansion of the note issue or bank credits. **1942** *Sun* (Baltimore) 20 Jan. 10/2 The fall or 'sterilization' of Singapore could contribute to a clearing of the way for Japanese naval debouchment into the Indian Ocean. **1955** *Times* 3 Aug. 12/2 The Swiss authorities would be the first to admit that the sterilization of deposits..cannot be held to promise a solution of the problem posed by the existence of too much money and lack of sufficient opportunities for investing it. **1968** R. A. Mundell *International Econ.* xviii. 256 Sterilization (or neutralization) policy is a specific combination of monetary and exchange policy. When the central bank buys or sells foreign exchange the money supply increases or decreases. The purpose of sterilization policy is to offset this effect. **1974** A. K. Swoboda in Johnson & Nobay *Issues in Monetary Econ.* 66 Keeping the money supply at its initial level involves a lower rate of sterilization operations when non-trade goods are present.

sterilize, *v.* Add: **4.** Also, to render harmless.

1939 *Economist* 8 July 64/2 The Reich Government..is evidently thinking a good deal more of its own commercial ambitions in East Europe and..of ways and means of breaking up the embryonic 'Peace Front' by sterilising Poland as an anti-German military power.

6. *Econ.* To inhibit the use of resources in order to exercise control over the economy, esp. to control the balance of payments by taking offsetting action to hold down the money supply.

1930 *Economist* 3 May 1007/2 The directors conclude a somewhat discursive report with remarks upon the wastefulness of sterilised gold, which they regard as one of the principal causes of the fall in commodity prices. **1935** A. D. Gayer *Monetary Policy & Econ. Stabilization* ii. 32 By the consistent utilization of the devices of 'offsetting' and 'sterilising'..the total media of payments were regulated independently of her [*sc.* America's] resources. **1936** *Sun* (Baltimore) 6 June 16/1 Such action would be aimed primarily at sterilizing the mobile gold received from abroad. **1938** *Times* 15 Feb. 14/2 No more gold will be 'sterilized' by the Treasury. **1942** *Sun* (Baltimore) 11 Feb. 1/3 Such a technique would 'sterilize' the money..; that is, keep it from building up the nation's purchasing power. **1944** *Ibid.* 3 Apr. 9/3 WPB suggested..that legislation be proposed to 'sterilize' such reserves to keep them from affecting the economy. **1968** R. A. Mundell *International Econ.* x. 149 The practice of 'sterilizing' the monetary effects of foreign exchange (or gold) purchases and sales has become widespread as countries look for means of adjusting the balance of payments other than that implicit in price level (or interest rate) adjustments. **1977** C. & D. S. Ammer *Dict. Business & Econ.* 180 Even though the United States is no longer on the gold standard, the term [*sc.* sterilization] continues to be used with the understanding that not gold but other reserves ..are being sterilized.

7. *Town Planning.* To preserve (a piece of land) from building or other development.

1935 A. P. Herbert *What a Word!* i. 18 The Town-planners and Green-Belters, when they propose to 'sterilize' a given area, mean that it shall *not* be sterile, that it shall produce *nothing* but vegetation and natural life, as opposed to buildings. **1937** *Times* 27 July 11/2 The council have also accepted offers to sterilize, free of compensation, the attractive Duncombe Farm estate at Ivinghoe, comprising 64 acres, on the understanding that there will be no building on the adjoining land. **1942** *Rep. Comm. Land Utilisation in Rural Areas* 71 in *Parl. Papers* (Cmd. 6378) IV. 497 The term 'green belt'..has come..to mean a belt of open land..to be 'preserved' from building (or, as is often said, 'sterilised'). **1973** *Town & Country Planning* Nov. 495 Any urban growth entailing the phased development of land would become impossible under site-value rating, except given some action to sterilize the land concerned, which would be precisely contrary to the aim of stimulating development.

sterks, sturks (stɔ̄·ɹks). *Austral. slang. rare.* [? abbrev. f. Stercoraceous *a.*] A fit of depression, irritation, annoyance. Also **ste·rky** *a.*, having loose bowels from fear.

1941 Baker *Dict. Austral. Slang* 71 *Sterks, give one the*, to infuriate, annoy, depress. **1944** J. Devanny *By Tropic, Sea & Jungle* 162 The croc disappears, and there's Ernest, standing up to his waist in the water..scared as hell, but too game to come out... So my dad goes in. He's a bit sterky too. **1959** D. Forrest *Last Blue Sea* 24 He just gives me the sturks.

sterling, *sb.* and *a.* Add: **A.** *sb.* **4. c. sterling area,** the group of countries (chiefly of the British Commonwealth, from 1947 officially known as *scheduled territories*: see *scheduled *ppl. a.* b) that from 1931 to 1972 pegged their exchange rates to sterling, or kept their reserves in sterling and not in gold or dollars, and transferred money freely amongst themselves; also *sterling bloc(k)*, *group*; **sterling balances,** deposits in sterling which are held in British banks by overseas creditors (see also quot. 1948).

1932 B. Blackett in *Times* 23 Jan. 12/4 What I have called the sterling area is sufficiently large and diversified to enable it to be to a very large extent self-contained. **1935** *Economist* 5 Jan. 1/2 The devaluation of the dollar and of the currencies of the sterling group..means that the currency value of the world's existing gold supply has immensely increased. *Ibid.* 26 Jan. 216/2 They might reasonably hope for a moderate increase in trade during the coming year, particularly, between countries within the 'sterling *bloc*'. **1937** A. Huxley *Ends & Means* v. 41 This has already been done in the case of the Sterling Bloc, which is composed of countries whose rulers have decided that it is worth while to co-ordinate their separate national plans so that they shall not interfere with one another. **1948** G. Crowther *Outline of Money* (ed. 2) v. 170 Overseas countries, especially those of the Commonwealth, were content during the war to sell more to Britain than they bought from her, and to take bank deposits in London..in payment of the difference. These were the famous 'sterling balances'. **1949** Koestler *Promise & Fulfilment* xv. 166 On February 22, 1948, Palestine was at short notice expulsed from the Sterling Block. **1956** R. S. Sayers *Financial Policy* viii. 235 The Sterling Area became a legal entity, an area inside which payment in sterling was unrestricted. **1977** *Time* 24 Jan. 14/1 In the past three decades, few remnants of that empire have bedeviled the British more than the 'sterling balances'—deposits from governments and private parties abroad that are kept in British banks and government bonds. **1979** H. Wilson *Final Term* 3 The Sterling Area was dismantled at a stroke.

B. *adj.* **3. c.** *absol.* Sterling silver tableware.

1974 *State* (Columbia, S. Carolina) 3 & 4 Mar. G2/1 Sterling promises to grow both more valuable, and more beautiful, with time. Its luminous beauty..is destined to take on the soft, lustrous patina..prized by so many collectors of antique silver. **1977** 'E. McBain' *Long Time no See* iv. 48 The women cleaned house for other women, soaping fine china and polishing heavy sterling.

Sterling (stɔ̄·ɹliŋ), *sb.*[2] The proprietary name of a sub-machine gun made by the Sterling Armament Company Limited.

1958 J. Boland *League of Gentlemen* v. 100 Orderly piles of automatic weapons of all types—Stens, Brens, and Sterlings—were on the racks. **1969** M. Pugh *Last Place Left* vii. 46 One of them ran his torch up and down me, while a second man held the tip of his Sterling into the light. **1971** J. Wainwright *Last Buccaneer* ii. 237 He hefted the Sterling sub-machine gun. **1974** S. Gulliver *Vulcan Bulletins* 18 Automatic weapons like Stens and Sterlings. **1975** *Trade Marks Jrnl.* 21 May 1034/2 Sterling 976,708. Sub-machine guns; and parts and fittings included in Class 13, sold in kits for modifying the calibre of rifles. Sterling Armament Company Limited, 9 Berkeley St., London. Manufacturers and Merchants.— 22nd June 1971.

stern, *sb.*[3] Add: **2. a.** Also, the rear part of an aircraft.

1931, etc. [see *stern-post]. **1942** *R.A.F. Jrnl.* 16 May 17 There is..a turret in the extreme stern.

8. sterndrive [*drive *sb.* 6] *Naut.* (chiefly *N. Amer.*), an inboard engine connected to an outboard drive unit at the rear of a power-boat; **stern-line** (earlier example); **stern-trawler,** a trawler whose nets are operated from the stern of the vessel.

1968 *N.Y. Times* 9 Feb. 31 When they appeared on the water about eight years ago, they looked like outboards with the power head sawed off... Variously called stern drives, inboard-outwards, [etc.].., they are one of the hottest items in recreational boating. **1976–7** *Sea Spray* (N.Z.) Dec./Jan. 94/1 'Best way to beat the opposition is to join 'em' would seem to be the philosophy behind a decision by C. W. F. Hamilton Marine Ltd to offer OMC, MerCruiser and Volvo sterndrives through its dealers. **1880** 'Mark Twain' *Tramp Abroad* xvii. 157 Lay her in shore and stand by to jump with the stern-line the moment she touches. **1961** *Times* 9 Aug. 5/2 A large stern-trawler..has been ordered by J. Marr and Son, of Hull. **1977** *Grimsby Even. Tel.* 5 May 8/3 A new French stern trawler landed over 1,700 kits of blue ling on Grimsby Fish Docks this week. **1982** *Daily Tel.* 29 July 2/4 The last modern stern trawler fleet in Britain was being forced into an increasingly nomadic existence.

Stern (stɔ̄ɹn), *sb.*[4] The name of Avraham *Stern* (1907–42), used *attrib.* in *Stern gang* or *group* to designate a militant Zionist organization (officially *Loḥame Ḥerut Yisra'el* Fighters for the Freedom of Israel) founded by him in 1940.

1944 *Nation* 2 Dec. 685/1 The so-called Stern gang, with 150 active members. **1947**, etc. [see *Irgun]. **1959** I. Jefferies *Thirteen Days* iv. 51 One of the two terrorist groups was called the Stern Gang. **1963** D. Leitch in Sissons & French *Age of Austerity* iii. 64 The Stern Gang was responsible for the murder of Lord Moyne, Minister resident in Cairo, in November 1944. **1978** L. Heren *Growing up on The Times* iii. 86 The Stern Gang was a savage organisation which even Koestler could not defend despite his theory that ruthlessness was essential for human progress.

Hence **Ste·rnist** *a.* and *sb.*

1944 *Nation* 2 Dec. 685/2 Nathan Friedman-Yellin, the thirty-one-year-old Sternist chief. *Ibid.* 686/1 The Sternists' chief weapon is murder. **1949** Koestler *Promise & Fulfilment* i. viii. 92 The Sternists were believers in unrestricted and indiscriminate terror. *Ibid.* ii. v. 279 Then a Sternist girl came in who once made international news. **1978** L. Heren *Growing up on The Times* iii. 83 Goldschmidt..assumed that the Sternist philosophy, which was never made clear to me, would prevail.

Sterno (stɔ̄·ɹno). *U.S.* A proprietary name for solidified alcohol supplied in containers for use as fuel for cooking stoves, etc.

1915 *Official Gaz.* (U.S. Patent Office) 11 May 672/1 *Sterno.* S. Sternau & Co., New York, N.Y. Filed Mar. 24, 1915... Solid fuel composed mainly of alcohol. Claims use since Jan. 12, 1915. **1935** Z. N. Hurston *Mules & Men* i. ix. 186 Somebody had squeezed the alcohol out of several cans of Sterno and added sugar, water and boiled-off spirits of nitre and called it wine. **1958** *Washington Post* 1 Nov. 1/3 They drink stuff like..Sterno (liquefied and strained 'canned heat'). **1969** *Trade Marks Jrnl.* 30 July 1237/2 *Sterno*... Solid fuels consisting mainly of alcohol. Colgate-Palmolive Company,..New York, State of New York 10022, United States of America; Manufacturers. **1978** R. Doliner *On Edge* (1979) xvii. 223 There was no electricity... He cooked his meals over a can of Sterno. **1979** P. Theroux *Old Patagonian Express* i. 16 Survival techniques at home.. cooking on Sterno stoves and the like.

sterno-. Add: **sternopleu·ron** (also **-pleurum,** † **-pleura;** pl. **-pleura**) *Ent.*, in flies, each of the two hard plates of the body wall to which the middle two legs are attached, protecting parts of the sides and parts of the underside; so **sternopleu·ral** *a.*, of or pertaining to the sternopleuron, or to the sternum and the pleura.

1884 C. R. Osten-Sacken in *Trans. Entomol. Soc. London* 503 *Sternopleura*; it is that portion of the mesosternum which, from its position, forms a part of the pleura... It is convenient to have a separate name for it. *Ibid.*, *Sternopleural suture*, horizontal suture below the dorsopleural and parallel to it; separates the mesopleura from the mesosternum. *Ibid.* 510 Sternopleural bristles. **1925** A. D. Imms *Gen. Textbk. Entomol.* iii. 600 The sternopleuron is situated below suture 2 and above the anterior coxa [of the thorax of Diptera]. *Ibid.*, The mesopleuron is the area in front of the root of the wing between the noto- and sterno-pleural sutures. **1961** J. E. Collin *Brit. Flies* VI. 108 Sternopleura with a bare polished patch. **1975** *Nature* 25 Dec. 666/2 Studies of polygenes affecting sternopleural bristle number in *Drosophila melanogaster*. *Ibid.* 668/1 The distribution of small bristles on the ventral part of the sternopleurum.

stern-post. Also in an aircraft (see quot. 1969).

1931 *Flight* 10 Apr. 324/2 Cases have occurred of the raised flange on plate NA 507, which secures the stern-post to the top longeron, bending and cracking in way of the taper pin. **1939** C. H. L. Needham *Aircraft Design* II. ix. 157 Where tail-plane supporting members attach to the fin stern post, the resultant compressive load should be taken into account. **1969** *Gloss. Aeronaut. & Astronaut. Terms* (*B.S.I.*) iii. 5 *Stern post*, a single member terminating a fuselage, hull or float. **1979** D. B. Thurston *Design for Flying* xv. 233 The afterbody sternpost and deadrise angles required for a stable hull are set forth... These data are based upon actual water handling characteristics of many seaplanes.

sternutator (stɔ̄·ɹniutēⁱtǫɹ). [f. *sternutat-* (in Sternutatory *a.* and *sb.*, etc.) + -or.] A substance that causes nasal irritation; *esp.* a poison gas that causes irritation of the nose and eyes, pain in the chest, and nausea.

1922 *Encycl. Brit.* XXXII. 111/2 The sternutators were originally considered from the point of view of putting a man temporarily out of action by a violent fit of sneezing. **1951** Kirk & Othmer *Encycl. Chem. Technol.* VII. 136 The introduction of sternutators in the form of aerosols as a gas warfare agent was a deliberate attempt by the Germans in World War I to penetrate the gas-mask canister and thus force demasking. **1971** [see *juglone].

stern-wheel, *v.* (Earlier example.)

1807 in *Tennessee Hist. Mag.* (1919) V. 62 Struck a large and stubborn sawyer, two or three feet below the surface of the water in a rapid current—stern wheeled with rapidity.

steroid (stiə·roid, ste·roid). *Biochem.* [f. *ster(ol + -oid.] Any of a large class of naturally occurring or synthetic organic compounds characterized by a nucleus of 17 carbon atoms in the form of four fused rings (three containing six carbon atoms and one containing five) and with varying substituents and degrees of unsaturation, the members of which include sterols, many sex and adrenocortical hormones, insect hormones, bile acids and alcohols, cardiac-active glycones, and some sapogenins and alkaloids,

and many of which have important pharmacological uses. Freq. *attrib.* or as *adj.*, as *steroid chemistry*, *hormone*.

anabolic steroid, a steroid whose anabolic effects predominate over the masculinizing ones.

1936 Callow & Young in *Proc. R. Soc.* A. CLVII. 194 The term 'steroids' is proposed as generic name for the group of compounds comprising the sterols, bile acids, heart poisons, saponins and sex hormones. **1956** *Nature* 28 Jan. 188/2 We have observed the effects on wool growth of treating sheep with adrenocorticotrophic hormone and with cortical steroids. **1959** S. Duke-Elder *Parsons' Dis. Eye* (ed. 13) xiv. 151 In ophthalmology, these steroids may be administered locally or systemically. **1960** [see *progestational *a.*]. **1961** *Nature* 30 Sept. 1368/1 New anabolic steroids with low androgenic activity. **1969** *Times* 16 Jan. 4/8 Steroid hormones, which are excreted in the urine, are one of the many body chemicals that follow a daily fluctuation. **1969** J. A. Vida *Androgens & Anabolic Agents* i. 2 Since anabolic steroids promote protein synthesis in the muscular system, these drugs find important application in clinical medicine to speed up healing of wounds. **1971** *Brit. Med. Bull.* XXVII. 26/1 The amount of oestrogen in steroid contraceptives influences the hazard of thrombo-embolism. **1972** *Nature* 21 Jan. 125/1 Steroids are small and rather simple molecules which nevertheless elicit a complex array of biochemical responses in their target tissues, often leading to profound changes in growth and differentiation. **1974** M. C. Gerald *Pharmacol.* xix. 355 Anabolic steroids..have been used by football players, weight lifters, and athletes. *Ibid.* xxi. 384 Digitalis glycosides consist of an aglycone fraction that has a steroid nucleus. **1977** *Daily News* (Perth, Austral.) 19 Jan. 49 The British National Racehorse Trainers' Federation has officially admitted that anabolic steroids have been in common use throughout the training profession for several years.

Hence **steroi·dal** *a.*, possessing the structure of, or pertaining to, a steroid.

1938 *Jrnl. Amer. Chem. Soc.* LX. 1736/2 Arguments are advanced to indicate that the steroidal hormones and bile acids do not originate from cholesterol. **1957** *Ibid.* LXXIX. 3222/1 Bromination of the steroidal sapogenin side chain. **1963** Katzman & Elliott in Florkin & Stotz *Comprehensive Biochem.* X. iii. 72 A number of methods have been developed for the quantitative estimation of the steroidal estrogens. **1972** *Lancet* 22 Sept. 611/1 Little is known about the growth and steroidal activity of the human ovarian follicle. **1977** *Listener* 18 Aug. 208/2 The wild yam, the source of steroidal compounds used as the starting material for oral contraceptives.

steroidogenesis (stiroi:d-, steroi:dodʒe·nĕsis). *Biochem.* [f. *steroid + -o + -genesis.] The biosynthesis of steroids.

1951 *Recent Progress Hormone Res.* VI. 220 We have investigated the effects of several variables on the ACTH-controlled steroidogenesis of the isolated adrenal. **1970** *Nature* 28 Nov. 885/1 This second monograph.. deals with the pathways of steroidogenesis of the ovary and testis. **1979** *Experientia* XXXV. 159/2 The cyclic AMP-dependent protein kinase may be involved in the processes of spermatogenesis and steroidogenesis as animals reach sexual maturity.

Hence **steroi:doge·nic** *a.*, pertaining to or having the property of steroidogenesis.

1951 *Recent Progress Hormone Res.* VI. 240 Our data suggest the possibility that steroidogenic organs should be visualized as assembly-line processing plants, where specific groups are introduced in a highly specialized way. **1959** *Endocrinology* LXV. 29 Saline extracts of beef diencephalon were assayed in the decerebrate dog for steroidogenic activity. **1974** D. & M. Webster *Compar. Vertebr. Morphol.* xiii. 313 In amphibians the chromaffin tissue is embedded in the ventral portion of the kidney along with steroidogenic tissue.

sterol (stiə·rǫl, ste·rǫl). *Biochem.* [The ending of *cholesterol, *ergosterol, etc., used substantively.] Any of a class of solid, unsaturated steroid alcohols that occur naturally both free and in combination as esters or glycosides and are classified according to their origin as mycosterols, phytosterols, zoosterols, and marine sterols.

1913 *Biochem. Jrnl.* VII. 617 It is now proposed to limit the terms zoo- and phyto-sterol to sterols which are found as tissue constituents of animals and plants respectively. **1939** A. Huxley *After Many a Summer* i. v. 65 Those sterols!..Always linked up with senility. The most obvious case, of course, was cholesterol. **1959** L. F. & M. Fieser *Steroids* xi. 358 The characteristic sterol fraction from freshwater green algae (*Chlorophyceae*) is the common sitosterol mixture of higher plants. **1975** *Lipids* X. 542/1 Sterols have been isolated from both adult and juvenile ivy in free and esterified form. Stigmasterol..is the main component. **1979** *Jrnl. Org. Chem.* XLIV. 3378/1 Algae and fungi produce sterols with 24β configurations, while most higher plants produce sterols with 24α configurations.

-sterol (stĕrǫl), *suffix. Biochem.* [f. *chole)-sterol.] A formative element in the names of many sterols, as in *ergosterol, *phytosterol.

-sterone, formative element in the names of some steroids, as in *androsterone, *progesterone. [ad. G. *-steron*, f. *ster(ol* *sterol + *ket)on* Ketone. App. first used in G. *androsteron* (*Zeitschr. f. phys. Chem.* (1934) CCXXIX. 167) and *luteosteron* (*Ber. d. Deut. Chem. Ges.* (1934) LXVII. 1271).]

sterrettite (ste·rĕtəit). *Min.* [See quot. 1940 and -ite[1].] A hydrous basic phosphate of scandium occurring as transparent, usually colourless, orthorhombic crystals, and now identified with kolbeckite (eggonite).

1940 Larsen & Montgomery in *Amer. Mineralogist* XXV. 513 A few specimens contain orthorhombic crystals distinct from any previously described mineral... The mineral is named sterrettite in honor of Dr. Douglas B. Sterrett who was one of the first investigators of the Utah and Nevada variscite deposits. **1959** *Bull. Geol. Soc. Amer.* LXX. 1648 Powder patterns of sterrettite and kolbeckite showed that their cell sizes are identical with that of synthetic $Sc(PO_4).2H_2O$.., indicating that the major cation might be scandium rather than aluminum. **1968** *Amer. Mineralogist* LIII. 1227 In the course of this work, the supposed Al phosphate sterrettite..was identified..to be scandium phosphate, $Sc(PO_4).2H_2O$, identical with kolbeckite.

stertorousness. (Example.)

1845 Poe *Facts of Mr. Valdemar's Case* in *Amer. Rev.* Dec. 563/1 The stertorous breathing ceased—that is to say, its stertorousness was no longer apparent.

stet. Add: (Earlier and later examples.) The direction occasionally signifies that a non-standard or irregular form should be retained. Cf. Sic *adv.*

1755 J. Smith *Printer's Gram.* xi. 277 Where words are struck out that are afterwards again approved of, they mark dots under such words, and write in the Margin, *Stet.* *a* **1966** 'M. na Gopaleen' *Best of Myles* (1968) 295 A colossal imposition who will..cause your heart to beat like a sludge-hammer (*stet*).

Also as *v. trans.*, to write 'stet' against (an accidental deletion, miscorrection, etc.).

1895 G. B. Shaw *Let.* 16 Dec. (1965) I. 581 If you disagree with my deletions, you can dot them under and 'stet' them. **1968** K. Martin *Editor* i. 19 On the same line I had 'stetted' another word which had been accidentally crossed out. **1975** J. Butcher *Copy-Editing* iii. 26 Stet American and all unusual spellings that are to be retained.

Stetson (ste·tsǫn). Also **stetson.** The name of John Batterson *Stetson* (1830–1906), American hat manufacturer, used *attrib.* and *absol.* to designate hats made by the company founded by him, esp. one with a broad brim and high crown associated with cowboys of the western U.S. Also applied *loosely* to other hats in this style.

A proprietary term in the U.S.

[**1895** *Montgomery Ward Catal.* Spring & Summer 274/1 Cow Boys' sombrero hats... J. B. Stetson's 'Boss of the Plains'. **1897** *Sears, Roebuck Catal.* 234/1 The world famous J. B. Stetson sombrero hat..crown 4½ inch; brim 4 inch.] *c* **1900** in *Amer. Mail Order Fashions* (1961) 27 Every railroad man knows that the Stetson hat is just the right style and shape for his business. **1903** *Everybody's Mag.* Dec. 739/2 'Send for me if you want me again,' says Redruth, and hoists his Stetson and walks off. **1906** *Official Gaz.* (U.S. Patent Office) 6 Mar. 290/1 *Stetson.* Hats and caps. John B. Stetson Company, Philadelphia, Pa. Filed Apr. 14, 1905. Used ten years. The name 'Stetson'. **1924** *Westm. Gaz.* 26 Aug. 5/5 The modern coster wants to wear a Stetson hat. **1940** W. Faulkner *Hamlet* i. ii. 38 Pat Stamper..standing there at the gate to his rope stock pen, with that Stetson cocked and his thumbs still hooked in the top of his pants. **1953** D. Cushman *Stay away, Joe* 19 He had donned for the occasion a new flame-red shirt and a thirty-dollar Stetson hat. **1967** *Telegraph* (Brisbane) 17 Apr. 4/1 He was amused at the surprise that greeted his announcement that the hat he was wearing—a snappy, small-brimmed business model—was also a Stetson... 'Stetson is a brand name. We make all kinds of men's hats.' **1970** *Guardian* 31 Dec. 2/7 His [*sc.* J. B. Stetson's] first hat was a 10-gallon Western, an ancestor of the stetsons worn by President Lyndon Johnson. **1978** D. Bloodworth *Crosstalk* ii. xxi. 164 Idi Amin..went around twice talking to the hijacked passengers..wearing..a big stetson hat.

Hence **Ste·tsoned** *a.*, wearing a Stetson.

1935 A. G. Macdonell *Visit to America* xiii. 235 Fort Worth, overgrown village of spurred and Stetsoned cow-punchers. **1969** *Time* 8 Aug. 55/1 John Wayne as a pistoled and Stetsoned Captain Bligh. **1972** D. A. Pailin in Cox & Dyson *20th-Cent. Mind* III. iv. 120, I remember a stetsoned rancher arriving at a Texas seminary. **1976** *National Observer* (U.S.) 1 May 7/3 They're often joined by Russian trawlermen or Stetsoned Oklahomans.

Steuben (st(i)ū̆be·n, st(i)ū·bĕn). The name of the *Steuben* Glass Works at Corning, N.Y., founded in 1903, used *attrib.* to designate fine glassware made there, esp. the decorative engraved crystal produced since 1933. Also *absol.*

1920 *Official Gaz.* (U.S. Patent Office) 16 Mar. 529/2 *Steuben.* Corning Glass Works, Corning, N.Y. Filed Nov. 19, 1919. No claim being made to the exclusive use of the word 'Steuben' apart from the mark as shown in the drawing. *Particular description of goods.*—Cut Glass and Artistic Glassware. *Claims use* since about Jan. 1, 1904. **1941** G. S. & H. McKearin *American Glass* ix. 422 A small, well formed vase blown from gold-ruby glass..is an opaque pink. It may also be a Steuben piece. **1948** J. S. Plaut *Steuben Glass* ii. 9 Steuben glass is made by the so-called 'off-hand' process. **1958** 'S. Marlowe' *Second Longest Night* i. iii. 19 He was carrying a cocktail glass, expensive Steuben, in his right hand. **1967** 'R. Foley' *Fear of Stranger* (1968) viii. 89 Her Steuben bud vase held an assortment of brushes. **1968** L. J. Braun *Cat who turned on & Off* ii. 23 Oh, what a lovely shop... She's got some old Steuben... Look at this decanter! **1979** *United States 1980/81* (Penguin Travel Guides) 556 Corning Glass Center, Corning, New York: An area has been set aside for visitors where the fine Steuben glass is hand-formed and engraved.

Stevengraph (stī·vĕngraf). [Proprietary term, f. the name of the inventor Thomas *Stevens* (1828–88), a ribbon weaver of Coventry + -graph.] A type of coloured woven silk picture produced during the late 19th century by the firm founded by Stevens.

1879 *Bookseller* 2 June 510/1 Mr. Stevens, of Coventry, has produced two interesting 'Stevengraphs'—one the four-horse Mail Coach..the other the Railway Locomotive. **1928** *Trade Marks Jrnl.* 4 July 1075 Stevengraph ... Woven labels of cotton, or in which cotton predominates. Thomas Stevens (Coventry) Limited, Stevengraph Works,..Coventry. **1957** W. S. Le Van Baker *Silk Pictures of Thomas Stevens* i. i. 15 An amazing third-dimensional effect is achieved in the tiny woven silk pictures known as 'Stevengraphs'. **1964** A. Adburgham *Shops & Shopping* xix. 226 The weaving of dress labels was a development from the Stevengraphs... Many other ribbon firms copied the idea of Stevengraphs to see them through the depression. **1967** *Daily Tel.* 21 Oct. 13/1 A pair of Stevengraphs, illustrating Columbus leaving Spain and arriving in the New World made £105 (Hicks). **1971** G. A. Godden *Stevengraphs* i. 13 The..decorative silk-work pictures known as Stevengraphs were first introduced ..as a gimmick at the York Exhibition which was opened on 7 May 1879. **1980** *Daily Tel.* 8 Dec. 7/1 (Advt.), Wednesday 10 December 12 noon Baxter Prints & Stevengraphs. Viewing: Day Prior and Morning of Sale until 11 am.

stevensite (stī·vĕnzəit). *Min.* [f. the name of Edwin A. *Stevens* (1795–1868), U.S. inventor and businessman and founder of the Stevens Institute of Technology in Hoboken, New Jersey: see -ite[1].] A brown, pink, or white magnesium-containing mineral of the montmorillonite group.

1889 A. E. Foote in *Naturalist's Leisure Hour* XIII. 31/1 Stevensite 5c to 1·50. **1896** A. H. Chester *Dict. Names Minerals* 257 Stevensite... This name was suggested in 1873, at a meeting of the N. Y. Lyceum of Natural History, but not published, though soon after used on labels. **1916** *Amer. Mineralogist* I. 44 In the old Hartshorn quarry, in Springfield Township, Essex County, New Jersey, Mr. Louis Reamer..discovered a single vein of a peculiar mineral, called by the quarrymen 'magnesium' (= talc?)... It proved to be essentially identical with the hitherto imperfectly known stevensite. **1962** W. A. Deer et al. *Rock-Forming Minerals* III. 231 Brindley (1955) described stevensite as a talc-saponite interlayered mineral, whereas Faust and Murata (1953) and Faust *et al.* (1959) regard it as a smectite with a defect structure in which a small proportion of layers with the 'attributes of talc' play only a minor role. **1979** *Sci. Amer.* Apr. 81/2 A good example is the alkaline chemical sedimentation that deposits limestones, cherts, phosphates and magnesium-based clay minerals, including attapulgites, sepiolites and stevensites.

Stevenson (stī·vĕnsən). *Meteorol.* [The name of Thomas *Stevenson* (1818–87), Scottish engineer and meteorologist (and father of R. L. Stevenson), who devised it (*Jrnl. Scottish Meteorol. Soc.* (1864) I. 122).] *Stevenson* (formerly *Stevenson's*) *screen*: a wooden box supported on a stand above the ground and made with doubly louvred sides and a double top with ventilation holes, so that thermometers or other instruments mounted inside it are sheltered from sunlight and precipitation and effectively register properties of the outside air.

1881 W. Marriott *Hints Meteorol. Observers* 10 The thermometers must be mounted in the Stevenson's screen. **1884** *Bull. Philos. Soc. Washington* VI. 24 The Stevenson screen and the double louvre screens in general. **1923** Glazebrook *Dict. Appl. Physics* III. 491/1 Owing to the intense solar radiation experienced in the tropics, it has been held..that the ordinary Stevenson screen is unsuitable for use in low latitudes. **1970** J. P. Glasgow in H. W. Mulligan *African Trypanosomiases* xv. 355 Meanwhile maximum temperatures (in a Stevenson screen four feet above the ground) rose to 41°.

Stevensonian (stīvĕnsōᵘ·niăn), *a.* (and *sb.*). [f. the name of the writer Robert Louis *Stevenson* (1850–94) + -ian.] Of, pertaining to, or characteristic of R. L. Stevenson or his writings. Also as *sb.*, an admirer of R. L. Stevenson or of his writings.

1897 T. Davidson *Let.* 29 Mar. in R. B. Perry *Thought & Char. of W. James* (1935) I. 756 Nothing but a happy-go-lucky Stevensonian adventuresomeness. **1900** *Fortn. Rev.* Jan. 97 He did not underrate *Kidnapped* as a whole. 'By far the most human of my labours hitherto,' is his verdict, anticipating that of all the true Stevensonians. **1913** *Smart Set* No. 3. 45/1 That polished peanut style that passes for Stevensonian English in the 'culture' clubs. **1923** G. K. Chesterton in *Illustr. London News* 22 Sept. 512/2 It was..the essence of the Stevensonian spirit that the melancholy was not incurable even if the misfortune was incurable. **1928** *Observer* 22 Jan. 15/1 Bouvet Island is a Stevensonian treasure island, bare, uninhabited, bleak. **1939** *John o' London's Weekly* 9 June 363/2 A new detective novelist..whose equipment includes..a good homespun prose with Stevensonian echoes. 1974 J. Pope-Hennessy R. L. Stevenson vi. 118 Meredith did..start off on a portrait of Louis Stevenson, in the guise of 'Gower Woodseer' [in]..The Amazing Marriage [but] the Woodseer of the later part..is not consistent with the Stevensonian youth we encounter at the beginning.

stevioside (stī·viosaid). Chem. [ad. F. stévioside (Bridel & Lavieille 1931, in Jrnl. de Pharm. et de Chim. XIV. 103), f. mod.L. Stevi-a, f. the name of P. J. Esteve (d. 1566), Spanish botanist + -ia¹: see *-oside.] A glycoside, $C_{38}H_{60}O_{18}$, present in the leaves of a Paraguayan herb (Stevia rebaudiana) and comparable in sweetness to saccharin.
1931 Chem. Abstr. XXV. 4553 Dieterich named the principle eupatirin, which B[ridel] and R. [Lavieille] changed to stevioside. 1968 Times 3 Dec. 10/8 The weed, known to botanists as Stevia rebaudiana..grows up to 1½ ft. high. Its leaves have a surprisingly sweet taste, the active principle being known as stevioside. 1978 Nature 9 Feb. 495/1 Stevioside, a triterpene glycoside, which is readily extracted from the leaves of S. rebaudiana, is being produced on a commercial scale in Japan... Stevioside, which is approximately 300 times sweeter than sucrose, is not a permitted food additive elsewhere.

stew, sb.² Add: **III. 8.** (sense 5) stewgravy, -jar; **stew-bum** U.S. slang, a tramp, spec. one who is habitually drunk (cf. *bum sb.⁴ 1).
1902 Bookman (N.Y.) Aug. 543/2 The dictum of the ordinary tramp (the 'gay-cat' and stew-bum). 1928 [see *Pegasus 1]. 1952 B. Harwin Home is Upriver xiii. 127 How come you to be a drunk damn' stew-bum when I found you? 1922 Joyce Ulysses 167 Scoffing up stew-gravy with sopping sippets of bread. 1913 D. H. Lawrence Sons & Lovers 79 The stew-jar was in the oven.

stew (stiū, stū), sb.⁴ U.S. colloq. abbrev. of *stewardess c.
1970 D. Harper Hijacked (1971) 23 If a stew flies five years, she'll keep on as long as the company lets her. 1975 B. Mezo Master of Paradise (1976) v. iii. 122 She had been with Pan Am herself as a 'stew'. 1979 S. Barlay Crash Course i. 6 I'm Mara. I used to be a stew myself.

stew, v.² Add: **2. b.** Also (of an infusion of tea, etc.), to 'stand' on the leaves, etc. Also transf., of the pot containing it.
1906 Rep. Brit. Assoc. Adv. Sci. 783 There is found in tea and coffee an astringent substance which gives the well-known bitter taste to the infusions when they are allowed to 'stew'. 1942 R.A.F. Jrnl. 3 Oct. 23 An unpromising enamel teapot stands on top, quietly stewing. 1979 W. H. Canaway Solid Gold Buddha xxiv. 158 The tea stewed for fifteen minutes or so.

c. (Later examples of to leave to (or let) stew in one's own juice.) Also in the senses: To be left to one's own devices, to be kept in a state of uneasy suspense, and ellipt., as to leave (one) to stew, to let (one) stew.
1921 Galsworthy To Let ii. vii. 184 'Please don't let me bother you if you've got people.' 'Not at all...I want to let them stew in their own juice for a bit.' 1928 W. S. Maugham Ashenden vii. 116, I left her to stew in her own juice for a week before I went to see her. She was in a very pretty state of nerves by then. 1934 G. Orwell Burmese Days ii. 38 Office babus are the real rulers of this country now... Best thing we can do is to shut up shop and let 'em stew in their own juice. 1961 B. Ferguson Watery Maze xv. 378 The Japanese in Tenasserim could safely be left to stew in their own juice once we had Rangoon. 1976 W. Greatorex Crossover 181 'It was me,' Calder said. 'I made up the story.' Calder let them stew in the silence. 1980 Church Times 7 Oct. 9/2 After letting us stew for three months, the Lord served up a miracle in the form of a perfect house for us in Beverley.

3. c. To fret; to suffer anxiety or suspense; to be in an agitated state.
1917 S. Lewis Innocents xviii. 208, I was suspicious of these fellows that are always getting and stewing over their wives in public. 1930 E. B. White Let. (1976) 92 White has been stewing around for two days now, a little bit worried. 1932 'A. Bridge' Peking Picnic xi. 32 He seemed to be stewing, so I told him to come over and have a cocktail later on. 1949 E. Pound Pisan Cantos lxxx. 92 Stewing with rage concerning the landlady's doings with a lodger unnamed. 1956 W. H. Whyte Organization Man x. 119 They don't want a man to fret or stew. 1974 H. Harris Double Snare xvii. 133, I wouldn't let them go to life imprisonment... Why shouldn't they stew a little? 1979 Tucson Mag. Mar. 13/1 City planners don't just sit around and stew over traffic congestion.

steward, sb. Add: **1. e.** One employed on a train to serve meals, drinks, etc., to passengers and to attend to their needs. Also one with similar duties on a motor coach or aeroplane.
1906 Railway Epicurean July 2 Harvey's chefs and stewards have the food problems of a continent at their command. 1915 Proc. Amer. Assoc. Dining Car Superintendents xv. 44 Instead of the steward asking, 'All on one check?' we instruct our stewards to, in a quiet way, get around to one of the party and ascertain whether one or more checks are desired. 1928 Lit. Digest 13 Oct. 70/3 The 'Nitecoach' carries a crew of three, driver, steward, and porter. 1931 Sci. Amer. Oct. 236/3 The steward, who now becomes the [airline] passenger's guide on land, is trained to supply any desired information. 1939 [see air hostess s.v. *air sb.¹ B. III. 4]. 1955 F. O'Connor Wise Blood i. 15 There was a steward beckoning people to places and handing out menus. 1975 Economist 21 Jan. 20/2 The £100,000 placed in the plane at Heathrow was recovered and the only damage, to a luckless steward, was one police dog-bite. 1979 P. Theroux Old Patagonian Express iii. 59 It was the steward from the dining car... 'Lunch!' he yelled. 'First call for lunch!'

11. Also occas. = shop steward s.v. *shop sb. 9 c.
1943 Sun (Baltimore) 13 Oct. 8/4 Union local stewards are also pointing out that they are at war with the enemy.

13. steward boy, for (sense 1 d) read: = house-boy s.v. House sb.¹ 23 (later examples).
1962 Sat. Even. Post 5 May 80/3 Her evening spent in helping Fossey's old stewardboy to beat carpets. 1977 Daily Times (Lagos) 25 Feb. 22/4 (Advt.), Steward boy, Houseboy, wanted.

stewardess. Add: **c.** A female attendant on a passenger aircraft who attends to the needs and comfort of the passengers; = air hostess s.v. *air sb.¹ B. III. 4. Also, a similar attendant on other kinds of passenger transport.
1931 United Airlines News Aug. 5/1 (caption) Uniformed stewardesses employed on the Chicago-San Diego divisions of United. The picture shows the original group of stewardesses employed. 1937 Sun (Baltimore) 22 Apr. 6/1 (Advt.), America's first railroad adds another to its long list of firsts... A uniformed Stewardess who is also a graduate nurse! She looks after your comfort en route [etc.]. 1958 'Castle' & 'Hailey' Flight into Danger i. 12 He looked into the aircraft..joined shortly by a stewardess..who smiled at him and made fast the door. 1968 M. Woodhouse Rock Baby v. 124 The stewardess announced our landing. 1978 R. Ludlum Holcroft Covenant xi. 40 In two chairs against the wall sat the captain of the 747 and the stewardess assigned to its first-class lounge. 1982 Christian Sci. Monitor 28 July 6/1 Anna Jones, 33, a British Airways stewardess, was happy to volunteer for a special room overlooking the Strand.

stewardship. Add: **2. b.** Eccl. The responsible use of resources, esp. money, time, and talents, in the service of God; spec. the organized pledging of specific amounts of money etc. to be given regularly to the Church. Also attrib.
1899 C. M. Sheldon (title) His brother's keeper; or, Christian stewardship. 1931 D. W. P. Strang Studies in Christian Stewardship i. 4 The Christian stewardship can claim no less an authorship than that of our Lord Himself. 1938 H. Gerlinghaus Money Raising vi. 98 This lack of development of a social consciousness on the part of students is clearly due to the fact that practically no systematic effort is made to implant the idea of stewardship in undergraduates. 1950 Christian Cent. 22 Nov. 1392/2 There is only one legitimate answer to the financial problem..to..teach our people to practice Christian stewardship. 1957 National Council Outlook May 12/1 The delegates asked for consideration of stewardship and fund-raising efforts in the Council. 1959 Christian Stewardship of Money i. 5 The movement..has broadened out into an attempt to recover and to teach the principles of the right Christian attitude towards the use of money. For want of a less archaic word this attitude is called 'Stewardship'. 1960 Oxf. Diocesan Mag. May 15/3 The book deals with man's attitude to the natural world in general and his Christian stewardship in relation to its resources. attrib. 1931 D. W. P. Strang Studies in Christian Stewardship i. 5 The Stewardship movement has..developed momentum mainly in the New World. 1942 Christian Cent. 1 July 829/1 (heading) United stewardship campaign might go even further. 1962 P. Ferris Church of England ii. 175 Stewardship advisers..were soon appointed in most dioceses. 1979 Oxf. Diocesan Mag. Dec. 2/2 Stewardship campaigns are essentially teaching missions. Ibid. 10/1 Pastoral reorganisation and the stewardship movement have worked together to foster a sense of responsibility for ministry.

stewartite (stiu·əɹtait). Min. [f. the name of the Stewart mine, Pala, San Diego County, California, where it was found + -ite¹.] A hydrous basic phosphate of manganese and ferric iron, $MnFe_2(PO_4)_2(OH)_2.8H_2O$, found as pleochroic orange-yellow to colourless triclinic crystals.
1912 W. T. Schaller in Jrnl. Washington Acad. Sci. II. 143 Stewartite. Probably triclinic. A hydrous manganese phosphate from the Stewart Mine, after which it is named. 1975 Neues Jahrb. für Mineralogie: Abhandl. CXXIII. 146 Laueite, pseudolaueite, and stewartite occur in a similar paragenesis which is confined to the late stage hydrothermal leaching and oxidation of primary triphylite-lithiophilite, $Li[Fe,Mn]PO_4$, in granitic pegmatites.

stewed, ppl. a.¹ Add: **a.** (Examples with reference to tea.)
1906 A. Bennett Old Wives' Tale iv. iii. 509 The lounge tea, which in any case would have been undrinkably stewed. 1924 Kipling Debits & Credits (1926) 309 Drinking stewed tea with your meat four times a day. 1977 M. Beckham One-Way Cemetery i. 3 The old man poured some stewed tea into a couple of mugs.

d. slang (orig. U.S.). Drunk. Also in phr. stewed to the ears (eyebrows, gills, etc.). Cf. *pickled ppl. a.¹ 2.
1737 Pennsylvania Gaz. 6 Jan. 2 The Drinkers Dictionary... Stew'd. 1871 A. A. Wright Diary in J. Wright Generations of Men (1959) v. 63 A very jolly party..we kept it up till daylight. I got pretty well stewed. 1912 Pedagogical Seminary Mar. 97 [expressions denoting] Intoxication..'half stewed'. 1922 S. Lewis Babbitt xxix. 347 He saw you out the other night with a gang of totties, all stewed to the gills. 1925 Wodehouse Sam the Sudden iii. 29 'My opinion is that he was as tight as an owl.' 'Stewed to the eyebrows.' 1930 J. Dos Passos 42nd Parallel 9 They're a bunch o bums and hypocrytes, stewed to the ears most of em already. 1945 B. Macdonald Egg & I (1946) iv. xvi. 176 Vewgene got stewed and ran into a tree. 1958 P. de Vries Mackerel Plaza vi. 82 A casual observer not familiar with him would have thought he was stewed to the gills as he rose and wobbled over to join me. 1978 J. Cashil Mortal Friends v. ii. 522 He wondered if Cushing had noticed himself. He wondered if Cushing was stewed.

stewing, vbl. sb. **b.** (Earlier and later examples.)
1726 J. Gay Let. in Corresp. J. Swift (1963) III. 168 Take a knuckle of Veal... In a Stewing pan put it. 1921 Daily Colonist (Victoria, B.C.) 25 Oct. 6/1 (Advt.), Boneless Stewing Beef, per lb. 14¢. 1948 Good Housek. Cookery Bk. ii. 159 Stewing steak is commonly used, but other meats (kidney and liver) may be made into a brown stew. 1966 M. Kelly When the Bough Breaks xxiii. 104 He took out the shopping list... Stewing steak, mushrooms, onions. 1978 R. Malouf Fried Brings up Rain xiv. 149 She'd got stewing beef in the oven.

stewy, a. Add: Also, resembling stew.
1967 [see gas-stove s.v. *gas sb.¹ 7].

Steyr (ʃtaiɹ). The name of a town in Upper Austria used attrib. and absol. to designate a kind of automatic pistol made there.
1920 H. B. C. Pollard Automatic Pistols iii. 24 The latest product of the Steyr factory, the celebrated army-model of the Mannlicher firm, is the 1912 Steyr Mannlicher pistol. Ibid. 26 Taking the Steyr in comparison with the Parabellum, it is less complicated than the latter, but not so well designed or easy to shoot with. 1934 G. Burrard Identification of Firearms & Forensic Ballistics vii. 144 The rifling found in revolvers and self-loading pistols may be divided conveniently into the following..types: Steyr type, four grooves; right-hand twist; grooves and lands of equal width. Used in all earlier self-loading pistols, such as the Steyr. 1972 R. Gadney Seduction of Tall Man ii. v. 154 This is a Steyr pistol, made in Austria... Fixed barrel, detachable seven shot magazine. 1981 A. Price Other Paths to Glory iv. viii. 211 The battered remains of an Austrian Steyr which he had bought.

sthreal, sthreel, varr. *streel sb.

‖ **stiacciato** (stiatʃā·to). Sculpture. Also schiacciato. [It. schiacciato, (Tuscan) stiacciato, pa. pple. of schiacciare (stiacciare) to flatten.] Very low relief. In full stiacciato-relievo or relievo stiacciato. Also attrib.
1862 J. C. Robinson Ital. Sculpture of Middle Ages 26 To him [sc. Donatello] is..due the invention of that..method of low relief, which is often..called the 'Donatello style'... The ancient Florentine writers on art designated this style 'relievo stiacciato', (flattened relief). Ibid. 110 This beautiful bas-relief..is an example of a very subdued description of 'stiacciato' relievo. 1929 R. Glazier Man. Historic Ornament 55 Donatello also carried the art of low flat relief called 'Stiacciato' to the greatest perfection. 1940 Burlington Mag. Mar. 76/1 The design of the buildings in lowest relief in the background is hardly imaginable without the precedent of Donatello's stiacciato reliefs. 1947 J. C. Rich Materials & Methods of Sculpture ii. 8 Flat relief, stiacciato-rilievo, is the lowest possible true relief. In this form, the effects are achieved by means of contour outlines and finely incised lines. The projection is very slight and there are no undercuts in this type of relief. 1957 H. W. Janson Sculpture of Donatello II. 32/1 A schiacciato panel 'reads' more like a picture than like a conventional relief.

stibadium (stibēᵢ·diəm). rare⁻¹. [a. L. stibadium a. Gr. στιβάδιον, dim. of στιβάς bed of straw.] A semicircular couch.
1840 Browning Sordello v. 174 He that sprawls On aught but a stibadium suffers..goose, Puttest our lustral vase to such an use?

stibio-. Add: **stibiopa·lladinite** Min., a palladium antimonide, approximately Pd_5Sb_2, occurring as white or grey hexagonal crystals; **stibiota·ntalite** Min., an oxide of antimony and tantalum, $SbTaO_4$, occurring as transparent brown or yellowish orthorhombic crystals in which niobium replaces some of the tantalum.
1929 P. A. Wagner Platinum Deposits & Mines S. Afr. i. 12 The other [mineral]..a palladium antimonide, described by H. R. Adam.., from the pegmatitic ore of Tweefontein,..it is proposed to..name stibiopalladinite. 1976 Amer. Mineralogist LXI. 1249 Stibiopalladinite from the type locality, Farm Tweefontein, near Potgietersrust, Transvaal, South Africa, is considered to have a composition between Pd_5Sb_2 and Pd_8Sb_3. 1893 G. A. Goyder in Proc. Chem. Soc. IX. 184 (heading) Stibiotantalite: a new mineral. 1973 Canad. Mineralogist XII. 77/2 Sb, Bi, and Nb... As are characteristic trace elements in many well-differentiated Li-rich pegmatites... Stibiotantalite is presently known from about 10 localities in the world, all of them of this type.

stibocaptate (stibokæ·ptēᵢt). Pharm. [f. Stib- + -o + -capt- in 'antimony dimercaptosuccinate', a chemical name for the substance (see Mercaptan) + -ate⁴.] An antimony-containing drug used in the treatment of schistosomiasis and administered by intramuscular injection.

1962 Trop. Dis. Bull. LIX. 1281/1 (Index), Stibocaptate. 1967 Lancet 21 Jan. 130/1 Monthly intramuscular injections of stibocaptate 10 mg. per kg. body-weight proved to be the best treatment for Zanzibari schoolboys infected with Schistosoma haematobium. 1977 Proc. R. Soc. Med. LXX. 765/2 Stibocaptate (Astiban)..is the only antimonial used for schistosomiasis in the United Kingdom. 1978 Nature 22 June 628/1 Stibocaptate (antimony dimercaptosuccinate, Astiban), one of the more than 10,000 antimonials that have been synthesised, is still on the WHO list as a complementary drug.

stibophen (sti·bəfen). Pharm. [f. Stib- + -o + Phen(-).] The compound pentasodium antimonybis(catechol-3,5-disulphonate) heptahydrate, which is used principally in the treatment of schistosomiasis, and is administered by intramuscular or intravenous injection.
1940 Brit. Pharmacopœia 1932 Add. iii. 22 (heading) Stibophen. 1941 Q. Jrnl. Pharmacy & Pharmacol. XIV. 21 A further study of the effect of pH on the stability of stibophen solutions is being made. 1950 Amer. Jrnl. Trop. Med. XXX. 266 Fouadin, also known as 'stibophen', another trivalent [antimony] compound, which contains 13.6 per cent antimony, was introduced in 1929 for the treatment of Egyptian schistosomiasis. 1977 Brit. Jrnl. Dermatol. XCVII. 308 The patient was treated with 25 injections of stibophen..2 ml intramuscularly on alternate days.

stichtite (sti·ktait). Min. [f. the name of R. C. Sticht (1856–1922), of Australia + -ite¹.] A hydrated carbonate of magnesium and chromium, $Mg_6Cr_2(OH)_{16}CO_3.4H_2O$, occurring as trigonal crystals of a pink, lilac, or purple colour.
1910 W. F. Petterd in Papers & Proc. R. Soc. Tasmania 167 Stichtite... This is beyond doubt an unrecorded mineral species which has hitherto been known under the name of kammererite... The writer has great pleasure in dedicating this new mineral species to Mr. Robert Sticht, the..general manager of the Mt. Lyell Mining and Railway Company. 1977 Austral. Gemmologist XIII. 103/1 Stichtite..occurs as a deep purple to rose pink soft waxy mineral as knots and veins in serpentine.

stick, sb.¹ Add: **I. 4. b.** Proverbial phr. a stick to beat (someone or something) with (perh. with ref. to the proverb: see quot. 1782). Also contrasted with carrot (= reward): see *carrot sb. 2 a.
1653 D. Osborne Let. 24 July (1903) 125 What reason had I to furnish you with a stick to beat myself withal? [1782 F. Hopkinson Misc. Essays I. 266 A proverb..naturally occurs on this occasion: It is easy to find a stick to beat a dog.] 1889 G. B. Shaw in English Illustr. Mag. Oct. 49 A few of us go to Bayreuth because it is a capital stick to beat a dog with. 1928 D. H. Lawrence in Evening News 8 May 8/4 The last stupid stick with which the old can beat the young. 1962 Listener 5 Apr. 597/2 Israel has sometimes been just another stick with which the Arabs beat each other. 1966 Ibid. 9 June 824/2 The Liberals had been glad to use the horror [at Turkish atrocities] felt by people in Britain as a stick with which to beat the Conservatives.

c. Hence (without article) transf., severe physical handling, 'punishment'; fig. unfavourable criticism, censure, reproof. Usu. in phrases to come in for, get, give, or take stick.
1942 E. African Ann. 1941–2 115/1 The Italians nipped across from Diredawa, and, as the troops say, 'gave us stick'. 1956 People 13 May 14/8 As usual the Australians are getting plenty of stick from the armchair critics. 1967 J. Burke Till Death us do Part vii. 116 He went out on the booze... She didn't half give him some stick when she found out. 1977 J. Wainwright Nest of Rats iii. v. 205 We took some stick, and we gave some stick... I belted that face. 1980 Daily Tel. 11 Apr. 19/2, I told him that he could expect trouble from the branches... He will come in for some stick over this.

i. A conductor's baton.
1849 Hamilton's Celebrated Dict. s.v. Battre la mesure, To mark the time by beating with the hand or with a stick etc. 1884 F. Niecks Dict. Mus. Terms s.v. Bâton, A stick used for beating time. Ibid. s.v. Taktstock, a conducting stick. 1920 A. Boult Handbk. Technique Conducting 7 The conductor has, therefore, had to learn to show his ideas on the interpretation of a work by means of his stick and hand. 1931 Times Lit. Suppl. 14 May 394/2 Stickless conducting..may suffice in ordinary class-room teaching; but in the interests of festival work ..it is better to accustom all singers to watch a stick and to train up a generation of conductors who know the technique of using it. 1955 Times 2 Feb. 6/3 There were moments when a loose movement of the stick gave away a little concentration in the quiet entries of the strings. 1978 Gramophone Feb. 1390/1 His mentor was band leader Joe Loss. 'When Loss used a stick the bounce and freedom within a beat was masterly.'

5. c. (Earlier example.)
in front of the sticks: at the wicket, batting.
1840 J. C. W. in Sporting Mag. Aug. 333 (Cricketing Extraordinary) New stumps are wanted to the number of six, So, good Mr. Charon, pray lend us the sticks! 1924 Lawrence & Skinner Boy in Bush ii. 22, I was captain of the first football eleven... And not bad in front of the sticks.

e. Assoc. Football. Phr. between the sticks: between the goal-posts, keeping goal.
1950 Sport 7–11 Apr. 14/3 Good news for Reading fans is that goalkeeper George Marks is expected to be back between the Elm Park sticks at the start of season 1950–1951. 1976 Wymondham & Attleborough Express 19 Nov. 23/4 Wortwell could not produce the form of recent weeks and crashed heavily to their hosts. David Loome took over between the 'sticks'.

7. to up stick(s). For (lit. and fig.) read: also fig. (usu. in form up sticks), to prepare to move, pack up, get going; to pack up and go, remove oneself. (Earlier and further examples.)
1839 Knickerbocker XIV. 141 Why, in the name of common sense, do you not up sticks and off? 1854 [see *up v. 5 b]. 1859 D. Bunce Trav. with Dr. Leichhardt ix. 141 The place was so muddy as to render it necessary to 'up sticks' and start for another..camping ground. 1877 Harper's Mag. Jan. 213/1 If any man tries hard words with me, I knocks him down, up sticks, and makes tracks. 1920 C. A. W. Monckton Some Experiences of New Guinea Resident Magistrate xvi. 183 Up sticks and away for Port Moresby and Sir Francis Winter. 1958 P. Scott Mark of Warrior ii. 168, 0700 we up sticks here and get well under cover a mile into the jungle. 1967 Economist 21 Oct. 306/1 If businesses can up sticks from Quebec.. 1976 G. Green Great Moments in Sport: Soccer i. 28 Neil Franklin..upped sticks..and departed to Bogotá. 1978 Guardian Weekly 26 Mar. ..

II. 8. f. Of bread, esp. as French stick. Cf. *baton sb. 1 c, bread-stick s.v. *bread sb. 10, *grissino.
1909 [see bread stick s.v. *bread sb. 10]. 1943 A. Simon Conc. Encycl. Gastron. IV. 59/2 (heading) Grissini or sticks.. 1980 Sunday Times (Colour Suppl.) 20 Jan. 37/3 French stick, long thin loaf with thick golden crust.

10. Also, in Journalism, a measure of copy, corresponding to about two column-inches.
1898 Scribner's Mag. May 579/1 He talked amiably enough; he said nothing he ought not to have said, but Linton [sc. a reporter] got five sticks out of it (a half column). 1932 G. Greene Stamboul Train ii. i. 44 They've asked me for a quarter of a column, but they'll cut it down to a couple of sticks. 1966 —— in New Statesman 25 Feb. 254/1 The ceremony could not possibly rate more than a couple of sticks in tomorrow's paper.

e. (Later examples in jazz and rock music.) Hence pl., a nickname for a drummer. Naval slang.
1926 Melody Maker Sept. 56 The tambourine is..played with the sticks. 1933 Metronome Oct. 51/2 Playing with the sticks widely separated on the head of the snare drum is a common fault. 1977 Gay News 24 Mar. 32/3 Drummer Rat Scabies (also 19) is very fast with the sticks.
1909 J. R. Ware Passing Eng. 234/1 Sticks,..drummer. 1916 'Taffrail' Carry On 27 A drummer goes by the name of sticks, from the implements with which he beats his drum. 1950 Kerr & James Wavy Navy 263 Sticks, the ship's drummer.

i. = joy-stick s.v. *joy sb. 10. Also occas., a gear lever in a motor vehicle.
1914 H. Rosher Let. 11 Aug. in In R.N.A.S. (1916) 13 Mr. Strutt, our instructor.., controls the engine switch and covers your hand on the stick. 1927 C. A. Lindbergh We v. 76 Pulling the stick back to go up. 1929 Daily Express 7 Nov. 1/1 On two occasions the pilot had to pull his stick back sharply, and once we only just cleared the tree tops. 1948 W. Fortescue Beauty for Ashes xxvii. 210 From their demeanour and that of the pilot, who handed 'the stick' over to a friend while he came to see that I was comfortable, one might almost have thought that this was my own private bomber. 1971 R. Dentry Encounter at Kharmel iii. 42 Pepper threw the gear stick into neutral, applied the handbrake firmly, switched off... She..moved the stick back to first. 1977 R.A.F. Yearbk. 31/1 The Hawk is very docile in the stall and..control is immediately regained once the stick is moved forward.

j. The propeller of an aircraft (rare). dead stick: see *dead a. D. 2.
1917 Editor 21 Apr. 358 The propeller itself is generally known as the 'prop' or 'stick'.

k. = ski stick s.v. *ski sb. 2 b.
1961 Webster, Stick rider, a skier who makes excessive use of ski poles. 1963 Amer. Speech XXXVIII. 207 Sticks,... Slang for skis, and also for ski poles. 1972 'M. Yorke' Silent Witness ii. 14 He stacked his skis and sticks in..the rack. 1977 C. Wood James Bond, Spy who loved Me ii. 15 He stamped hard into his skis.. and stabbed at the snow with his sticks.

l. Surfing. A surfboard.
1967 J. Severson Great Surfing Gloss., Stick, a surfboard. 1967 Internat. Surfing III. iii. 29 Because of the lack of length when changing to a shorter board, different techniques from surfing a longer board are required. Margo Godfrey and Mike Purpus offer some helpful advice on adjusting to the short stick.

11. c. (Earlier and later examples.)
1879 Macmillan's Mag. Oct. 503/1 'What tools will you want?'..He said, 'We shall want some twirls and the stick (crowbar)'. 1934 P. Savage Savage of Scotland Yard xxiii. 252 It's a fair cop. I'll go quiet, and here's my stick (jemmy). 1960 [see *loid].

g. the sticks: a remote, thinly populated, rural area; the backwoods; hence, in extended (freq. depreciatory) use, any area that is off the beaten track or thought to be provincial or unsophisticated; esp. in phr. in the sticks. orig. U.S.
1905 N. Davis Northerner 78 Billy is a cane-brake nigger; he'll take to the sticks like a duck to water when he's scared. 1914 R. Lardner in Sat. Even. Post 7 Mar. 8/1, I will have to slip you back to the sticks [i.e. the minor baseball leagues]. 1921 Daily Colonist (Victoria, B.C.) 22 Oct. 11/3 Judge Landis..has not yet consigned Babe Ruth to oblivion for..playing in the sticks for the..money. 1926 Whitman & McBride Jazz xiii. 154 They had..all the real New Yorker's prejudice against 'the sticks'. 1936 J. Dos Passos Big Money 62 Mighty nice of you to ask me out here in the sticks... It makes you feel good to see folks from the other side... This is the nearest thing to Paree I've seen for some time. 1941 W. C. Handy Father of Blues (1957) ix. 148, I continued..playing for dances, touring on the road and through the sticks and giving concerts. 1958 C. Koch Boys in Island 101 What can y' expect, way out here in the sticks? You would pick on a dame from back of beyond. 1968 J. Lock Lady Policeman ix. 79 Where's that? I know, it's somewhere in the sticks. 1971 N.Z. Listener 27 Sept. 3/2 The 'real' New Zealand..is out there in the sticks, under the open sky, where men have dirt on their elbows. 1974 New Society 7 Feb. 309/1 The idea of 'provincial' journalists working 'out in the sticks' has strong pejorative overtones. 1977 Daily Express 30 Jan. 19/2 More tight-fisted tourists take cars from Swindon rather than Heathrow... 'Most people regard us as a bunch of farmers from out in the sticks,' he said.

h. A cigarette or cigar; spec., a cigarette made with marijuana; also stick of tea, weed (cf. *tea sb. 7 c, *weed sb.¹ 3 c). See also cancer stick s.v. *cancer sb. 5, Thai stick s.v. *Thai a.
1929 W. H. Downing Digger Dialects 27 Consumption stick, a cigarette. 1935 E. Farjeon Nursery in Nineties 348 Papa..smokes all day long; but only affords himself the cheapest..sticks, except when Aunt Mary Albery sends him a hundred Coronas for Christmas. 1938 [see *Mary 1 d]. 1940 R. Chandler Farewell, my Lovely xii. 68 Evidence of what? That a man occasionally smoked a stick of tea, a man who looked as if any touch of the exotic would appeal to him. 1957 C. MacInnes City of Spades i. v. 58 'I'll roll you a stick.'...'Good stuff. And what do they make you pay for a stick here?' 1959 L. Lipton Holy Barbarians iii. 78 Rolling their sticks of 'tea', they looked like a ring of kindergarteners. 1965 W. Soyinka Road 24 Say Tokyo reaches out a stick of weed to him which he accepts behind his back. 1978 T. Williamson Technicians of Death xiv. 121 He got his first 'buzz' with a blend of Thai stick and opium.

12. c. U.S. slang. = *shill sb.
1926 Maines & Grant Wise-Crack Dict. 14/1 Stick, a confederate who wins or loses at dealer's will. 1931 G. Irwin Amer. Tramp & Underworld Slang 182 The cash the 'stick' wins is handed back to the operator of the game ..and the stick never has enough of his employer's money to make it worth his while to decamp. 1966 E. V. Cunningham Helen (1967) ix. 129 A shill is also called a stick, and the role of the shill or stick is to make the customer relax and feel at ease.

13*. Mil. **a.** A number (usu. five or six) of bombs dropped in quick succession from an aircraft.
1940 Illustr. London News 18 May 669 (caption) A 'stick' of five bombs has just burst across her bows. 1940 Times 6 Dec. 4/1 Seeing a convoy in the road, we dropped a stick plumb in the centre of it. 1942 Tee Emm (Air Ministry) II. 100 If you, as bomb aimer..watch a stick of bombs on a town you'll notice that they appear to move outwards... When they hit they form a curved stick. 1956 W. Slim Defeat into Victory vi. xxi. 500 British officers, watching from the ground the fall of each stick of bombs. 1975 T. Allbeury Special Collection iv. 18 There were dull thuds as another stick of bombs was dropped.

b. A group of parachutists jumping in quick succession.
1943 Combined Operations (Min. of Information) ii. 19 They [sc. airborne troops] practise dropping from an aircraft, first in 'slow' then in 'quick' pairs, until they are proficient enough to drop in 'sticks'. 1949 F. Maclean Eastern Approaches iii. ix. 414, I had decided to jump first with the others following in a 'stick'. 1955 J. Thomas No Banners xiv. 127 The despatcher yelled hysterically: 'Now, a nice stick of three!' 1974 C. Ryan Bridge Too Far iii. iii. 156 As the pilot held it steady on course, Mitchell saw the entire stick of paratroopers jump right through the fire. 1982 Times 5 June 4/6 The 15 marines in our 'stick' jumped through the [helicopter] door..with weapons ready.

III. 14. e. (Further examples.) Also with other adjectives.
1846 Swell's Night Guide 49 Which of us had hold of the crappy..end of the stick? 1930 E. Waugh Vile Bodies viii. 143 My private schoolmaster used to say, 'If a thing's worth doing at all, it's worth doing well.'..But these young people have got hold of another end of the stick, and for all we know it may be the right one. They say, 'If a thing's not worth doing well, it's not worth doing at all.' 1934 C. Day Lewis Hope for Poetry vii. 42 Although Lewis's analysis convinces us..as being correct in detail, we are compelled to feel that Lawrence rather than Lewis had got hold of the right end of the stick. 1939 'G. Orwell' Coming up for Air iv. vii. 283 Listen, Hilda. You've got hold of the wrong end of the stick about this business. 1959 'M. Cronin' Dead & done With iv. 61 I've had the rough end of the stick ever since I got here. 1977 P. Scott Staying On (1978) i. 14 Always..I have the mucky end of the stick. But then I am only part of the fixtures and fittings.

i. up the stick: pregnant. slang. Cf. up the pole s.v. *pole sb.¹ 1 b.
1941 Baker Dict. Austral. Slang 71 Stick, up the: (of a girl or woman) to be pregnant. 1958 [see *bun sb.² 1]. 1968 R. Lait Chance to Kill i. 10 Mary up the stick; funny how everyone counts the months. 1976 J. I. M. Stewart Memorial Service ix. 160 Do you know what it's like, Cyril, to be a decent and penniless young man who isn't sure he hasn't got his girl up the stick?

j. to cut one's stick: see Cut v. 43; more than you can shake a stick at: see *shake v. 5 b.

IV. 15. a. stick microphone, mike.
1961 C. Willock Death in Covert xii. 212 The interviewer from ITN..was holding a stick microphone. 1961 Listener 19 Oct. 622/3 Uncle Dimbleby is seated (with stick mike) among a 'representative cross-section' of the British public. 1976 B. Jackson Flameout (1977) iv. xii. 204 He saw him [sc. the pilot] speak into the stick microphone attached to his headset.

16. stick-and-carrot adj. phr. [see *carrot sb. 2 a], characterized by both the threat of punishment and the offer of reward; **stick-back** a., designating a kind of wooden chair having a back formed by upright rods or sticks; **stickball** U.S., a game played with stick and ball; spec. (a) improvised baseball played with a stick and soft ball; (b) an American Indian ball game resembling lacrosse, played by the Indians of the South-eastern U.S.; **stick bean**, a runner bean; **stick-bomb**, a bomb or grenade with a protruding rod or stick for firing or throwing (cf. also *stick v.¹ 35); **stick dance**, any of various kinds of folk-dance in which the dancer holds a stick and (in some dances) beats it against the sticks of other dancers; **stick-dressing**, the art of making shepherd's crooks (cf. stick-dresser, sense 15 b); **stick-fighting** W. Indies, a kind of martial art; hence **stick-fighter**; **stick-figure**, a matchstick figure (see match-stick (c) s.v. *match sb.² 5), a pin-man; **stick fixed** Aeronaut., the control column of an aircraft held in one position; freq. attrib.; **stick force** Aeronaut., the force or effort needed to move the control column of an aircraft or hold it in position; **stick free** Aeronaut., the control column of an aircraft allowed to move freely, unguided by the pilot; freq. attrib.; **stick (hand-)grenade**, a grenade with a handle; **stick-handling** vbl. sb. orig. and chiefly N. Amer., the handling of one's stick in ice hockey (or occas. in other sports) (cf. stick-work (a)); also fig.; hence (as back-formation) **stick-handle** v. intr., to control the puck (in ice hockey) with one's stick; also **stick-handler**; **Stick Indian** Canad. colloq., a member of the North American Indian peoples inhabiting the forests of British Columbia and the Yukon [properly a loan transl. of Chinook Jargon stick siwash forest Indian, a term used by the Coast Indians for those of the interior in this area.]; **stick loaf**: occas. used = French stick, sense 8 f above; **stick-man**, (a) slang, a pick-pocket's accomplice (cf. stick slinger); (b) U.S. colloq., a croupier; (c) W. Indies = stick-fighter above; (d) rare = stick-figure above; **stick-shaker** Aeronaut. colloq., a device which causes the control column to vibrate when the aircraft is close to stalling; **stick shift** N. Amer., a manually operated mechanism for changing gear; a gear lever; **stickwork**, (b) something made from, or fashioned by the use of, sticks.
1963 Times 4 Mar. 11/7 President Ayub has..given himself stick-and-carrot powers to deal with the 'Ebdonians'. 1977 'J. le Carré' Hon. Schoolboy vi. 136 It's a stick and carrot job. If you don't play, the comic will blow the whistle on you... That's the bad news... The good news is five hundred US into your hot little hand. 1783 Narragansett Hist. Reg. (1884) II. 314 Three good large Windsor or Stickback Chairs. 1923 Heal & Son Catal.: Kitchen Furniture 2 Unpolished Stickback Windsor Small Chair..12/6. 1963 Guardian 21 Aug. 6/6 A pale oak gate-leg table is set off by six flame stick-back chairs. 1978 Cornish Guardian 27 Apr. 10/4 (Advt.), Swivel chairs, ..wheel back and stick back chairs, easy chairs [, etc.]. 1824 Nantucket Inquirer 12 Jan. 3/5 No person shall play Foot-ball or Poke, Stick-ball or Swinger, within the compact part of the Town of Nantucket. 1934 E. Newhouse You can't sleep Here xii. 154 Two contending stickball teams left the gutter to see what was up. 1946 Life 11 Nov. 91/1 On the Cherokees' Qualla Indian Reservation in North Carolina..the Wolftown Wolves met the Wolftown Bears in a crucial game of stickball... The game of stickball, which is a primitive version of modern lacrosse, was centuries old when De Soto led a Spanish expedition through Cherokee territory in 1540. 1947 Commentary May 464/1 Sometimes we became so engrossed by a punchball or stickball game that night would fall without anyone's being aware of it. 1953 Sun (Baltimore) 1 Apr. (B. ed.) 12/1 Governor McKeldin can remember romping barefooted in the neighborhood, playing stickball. 1979 United States 1980/81 (Penguin Travel Guides) 533 A recreation of an early-18th-century Cherokee Village, where Cherokees in costume dance, play at stickball, work at crafting baskets, [etc.]. 1981 TV Picture Life Mar. 32/1 A group of young black children playing stickball on the streets of New York. 1906 Dialect Notes III. 158 Stick bean,..pole bean. 1980 J. Gardam Sidmouth Lett. 134 D'you want some beans?.. Stick beans? 1916 in C. F. S. Gamble Story N. Sea Air Station (1928) xiii. 222 The silent firing of projectiles varying in size from the Mills grenade to the 250-lb. stick bomb. 1925 Fraser & Gibbons Soldier & Sailor Words 270 Stick-bomb, a type of trench-mortar bomb attached to a hollow steel rod which passed down the bore of the projectile. 1940 Illustr. London News 10 Feb. 167/1 A charge with 'stick-bombs' —a form of hand grenade. Ibid. 167/2 (caption) The patrol reaches the enemy lines, overwhelming them with the threat of 'stick-bombs'. 1899 Kipling From Sea to Sea II. xxv. 12 A Zanzibar stick-dance, such as you see at Aden on the coal boats. 1907 Sharp & Macilwaine Morris Bk. i. 39 In the Stick and Handkerchief dances, pairs..stand near enough to clap hands or tap sticks with each other. 1950 Blesh & Janis They All played Ragtime (1958) 13 The banjo-ragtime rhythms of dances like the buck and wing, the Virginia Essence, the stick and the sand dances, and the soft-shoe routines. 1982 N. Painting Reluctant Archer vii. 113, I was also roped in..to play the piano for rehearsals of the stick dance which David Raeburn had introduced into his production of The Shoemaker's Holiday. 1968 P. Jennings Living Village 187 Stick-dressing..is the making of shepherds' crooks. A stick is dressed down, a ram's horn is put on top of it and the whole thing is polished. 1956 Caribbean Q. IV. iii & iv. 194 Later this aristocrat's masque was adopted by batonyé or stick fighters. 1968 E. Lovelace Schoolmaster i. 12 'Who say that?' Miguel asked hotly, growing angry, and moving up and down like a stickfighter in a rage. 1956 Caribbean Q. IV. iii. & iv. 192 Antagonisms are relaxed from time to time by fêtes, when the traditional pastimes of dancing, singing and stick-fighting are enjoyed, with liquor and food. 1974 Trinidad Guardian 2 Nov. 5/2 (Advt.), African culture in all forms. Dance, Stick-fighting, Drumming, Calypso, [etc.]. 1965 I. A. Richards in Times Lit. Suppl. 27 May 439/2 A stick-figure man is very different from any man but is a little like his silhouette or his shadow on a screen. 1976 New Yorker 9 Feb. 94/3 Weiss projects no character, he remains a stick figure. 1978 S. Sheldon Bloodline xxiv. 259 He was like a stick figure drawn by a child, with angular arms and legs, and a dry, unfinished face sketched on top of his body. 1945 Jrnl. R. Aeronaut. Soc. XLIX. 617/2 The stick movements the pilot has to make to control the aircraft are related to the stick fixed stability, while the stick forces he has to exert are related to the stick free stability. 1961 A. W. Babister Aircraft Stability & Control iii. 63 The stick fixed static margin is related to the elevator movement (or stick movement) to trim the aircraft. 1937 Jrnl. R. Aeronaut. Soc. XLI. 960 The stick forces required to operate them [sc. flaps] increased too rapidly with speed. 1942 Tee Emm (Air Ministry) II. 85 The stick force needed, say, to take violent avoiding action may be much too great. 1961 A. W. Babister Aircraft Stability & Control iii. 63 We shall now derive the relation between the stick force the pilot has to apply to hold the aircraft in a glide and the stick free static margin. 1983 E. Brown Wings of Weird & Wonderful xvi. 100 The latter aircraft was the less pleasant to fly as the stick continually hunted either side of neutral, and there was no build up in stick force with increase in speed. 1945 Stick free [see stick fixed above]. 1961 Stick free [see stick force above]. 1918 E. S. Farrow Dict. Mil. Terms 586 Stick grenade, a grenade attached to a stick and thrown over short distances like a dart. 1979 O. Sela Petrograd Consignment 53 Boris..took out two stick grenades and a Mills bomb. 1923 Stick hand-grenade [see *hairbrush 2]. 1929 N.Y. Times 10 Mar. xii. 8/1 Trottier staged a prize play when he stickhandled his way through the entire American team. 1969 M. Braithwaite Never sleep Three in Bed (1970) xv. 182 Back in 1926, he really couldn't skate very well, but he could stick-handle like a fiend. 1915 Official Ice Hockey Guide 17 Hill of Cornell.. is very fast, a good stick handler. 1958 Rosetown (Saskatchewan) Eagle 29 May 10/1 Dick..Elliot, plugger type, stick handler, back-bone of the team. 1904 Ice Hockey & Ice Polo Guide 35 Stick-handling, like confidence, coolness, strength and speed, is acquired by practice... The better you can handle your stick the more effective player you will be, because stick-handling is one of the essentials of the game. 1962 Times 28 Feb. 4/4 Cool lacrosse at Cambridge... Cambridge..played calmly, showing glimpses of..skilful stick-handling. 1976 Ottawa Citizen 24 Dec. 2/3 They did some nifty stickhandling through government red tape. [1869 L. Smith Let. 30 Oct. in Rep. Indian Affairs 1969 (U.S.) (1870) 567 Twice a year most of the Indians make a trip up the Stikine River to Talyan, at which place the Stick tribe reside.] 1885 F. Schwatka Rep. Mil. Reconn. Alaska 1883 76 The so-called 'Stick' Indians of the interior are seen in the villages near the trading stores. 1887 G. M. Dawson Notes on Indian Tribes of Yukon 14 They are classed with the 'Stick Indians', by the coast tribes. 1963 R. Symons Many Trails vii. 72 Snowshoes are known only as a strange accoutrement of the 'Stick Indians'. 1980 Times 15 Dec. 1/8 Britons returned home with..wines, stick loaves and under-ripe Camemberts. 1861 J. Binny in H. Mayhew London Labour Extra Vol. (1862) 282/2 While drinking at the bar, one of the women tries to rob him... A man who is called a 'stickman', an accomplice..of hers, comes to the bar... If they have by this time secured the booty, it is passed to the latter, who thereupon slips away. 1931 Amer. Speech VII. 116 Stick-man,..a croupier in a gambling joint. 1952 Evening News (Port of Spain, Trinidad) 28 Jan. 8/2 A Trinidad stickman held his stick at both ends when going into action. 1958 Newark (New Jersey) Star-Ledger 23 Mar. 102/4 Then the stickman rakes in the dice, picks them up, and tosses them back to the shooter. 1966 J. Dos Passos Best Times ii. 47, I sent Arthur a cartoon..of warmongers..hanging from the arc lights on Fifth Avenue, while two stickmen..danced the carmagnole in the foreground. 1975 [see *pogue]. 1980 J. Scott Hunted i. 8 He shoved the counters forward and the stickman flicked them into place. 1962 Flight Internat. LXXXI. 330/1 At 70 kt the stick shaker rattled again, but there was still plenty of aileron control. 1979 Daily Tel. 9 Aug. 7/2 The F.A.A. has proposed that an additional 'stick-shaker' be installed to warn the pilot when the plane loses enough speed and lift to approach a stall. 1960 Wall St. Jrnl. 13 Oct. 26 (Advt.), 'Welcome back, standard transmission.'.. A great majority of those who buy sports cars specify the 'stick shift' for the fun of it. 1976 'E. McBain' Guns (1977) vii. 200 Bucket seats in beige leather, stick shift on the floor. 1923 Kipling Irish Guards in Great War I. 164 They [sc. the trenches] had been French..and [were] riveted with strange French stickwork. 1929 B. Oliver Cottages of England iii. 44 The pricked treatment.., as also the scratched patterns, were equally the common finish to cottage exteriors all over Suffolk and Essex. Nothing looks better than this delightful Essex 'stickwork'. 1933 B. Rackham Guide to Ital. Maiolica i. 5 Lead-glazed earthenware with decoration done..by scratching with a pointed instrument.. through a surface layer of..'slip'... Wares of this kind are..called..sgraffito... The phrase a stecco ('stickwork') is generally used..in referring to this process. 1977 Penguin Dict. Decorative Arts 759/1 Stickwork. Small objects such as chess-men, egg-cups, snuff-boxes, etc. made from sticks of various types of wood assembled by the same technique as in Tunbridge ware and then turned on a lathe.

Stick, sb.⁵: see *Stickie, Sticky.

stick, v.¹ Add: **I. 1. f.** To inoculate, to give a hypodermic injection to; to introduce a hypodermic needle into (a person). U.S. colloq.
1946 Sun (Baltimore) 19 June 10/1 Though he [sc. the traveler] surely has been 'stuck' for every known disease, no telling how often he will be halted by health officers and cast into quarantine. 1963 New Yorker 25 May 42/2 'Fraid I've got to stick you once more. 1969 E. Welty in New Yorker 15 Mar. 43/1 The floor nurse came in to feed Mr. Dalzell, then stick him with a needle.

II. 6. e. Vingt-et-un. To decline the opportunity of adding to one's hand.
1931 W. V. Tilsley Other Ranks 147 A little group in the centre of the room sprawled on their blankets, playing pontoon. 'I'll stick!' 'Twist one!' 'Busted!' 1950 [see *bust v.² f]. 1956 R. Fuller Image of Society vii. 188 'In other words the bank is sticking at sixteen,' said Cawsey, amused at his own turn of phrase. 1976 G. Sims End of Web iii. 22 Hello, young Clive. Still sticking on seventeens? .. I'm coming round for another pontoon lesson shortly.

7. b. (Further example.)
1977 J. F. Fixx Compl. Bk. Running iv. 49 One must stick at one's sport if it is to continue doing any good.

c. Also with a thing as object.
1907 J. M. Synge Lett. to Molly (1971) 172, I cannot 'stick' these plays any more. 1922 A. S. M. Hutchinson This Freedom iv. iii. 279, I couldn't stick the place. 1928 Daily Tel. 27 Mar. 9/1, I resigned..because I could stick the chief's bullying no longer. 1960 D. Storey This Sporting Life i. ii. 29, I couldn't stick the sight of him standing up there against the Batley skyline.

8. c. Also, of a criminal charge: to be substantiated, take effect. Of an order or decision made by a court of justice, legislature, or other authority: to be implemented or complied with. Hence with wider application, esp. in phr. to make (something) stick, to make (that thing) effective; to clinch; to substantiate.
1932 [see *rap sb.¹ 4 e]. 1942 Sun (Baltimore) 12 June 1/7 A..program of cooperation designed (1) to hasten the defeat of Germany and (2) to make that defeat stick. 1944 Ibid. 7 Feb. 1/1 The Department of Labor..would be empowered to hand down decisions 'which will stick and not be vetoed by any other Federal agency'. 1951 E. Paul Springtime in Paris xvi. 324 Every officer at the Commissariat of the 5th had been itching to get Oudin, on any charge at all that could be made to stick. 1963 'S. Woods' Taste of Fears xiv. 148 'They couldn't make it stick,' said his uncle, positively... 'No evidence.' 1971 A. Price Alamut Ambush xii. 147 God knows whether the Americans and the Russians can make the cease-fire stick. 1981 Times Lit. Suppl. 9 Jan. 25/1 Picasso now taught himself how to use a poetic, half-theatrical imagination to make his art 'stick', while at the same time..taking pains to avoid the illustrational.

12. c. to stick in one's throat, (b) of a statement, proposal, notion, belief, etc.: to be difficult to swallow, to be unacceptable.
1843 Dickens Let. 1 Feb. (1973) III. 434 Your dedication to Peel stuck in my throat. 1885 E. W. Hamilton Diary 3 May (1972) II. 855 To luncheon..with F. Rothschild to talk over politics... What sticks in his throat is Chamberlain's programme—his quack remedies for the agricultural labourer, of whose real wants he has no experience. 1924 G. B. Shaw St. Joan Pref. p. li, The truth sticks in our throats with all the sauces it is served with. 1938 W. S. Maugham Summing Up lxxvi. 310 This notion has long stuck in my gizzard. 1953 A. J. Toynbee World & West vi. 98 The new religions which were now being offered..would have stuck in a philosopher's throat if the missionary had not sugared the strange pill for him. 1958 C. P. Snow Conscience of Rich xxxi. 232, I didn't like refusing, but it stuck in my gullet to help that blasted group of reds. 1976 A. Price War Game ii. i. 193 Weston would find the accident..sticking in his throat, a question much too sharp to be swallowed.

18. d. imp. (or in constructions with equivalent force) and followed by up as a coarse expression of contemptuous rejection. Also ellipt. Similarly euphem. phr. stick it in your ear (U.S.). Cf. *shove v. 10 e, *stuff v.¹ 14* a.
1922 S. Lewis Babbitt xix. 240 Bad luck, old dear, and you can stick your job up the sewer! 1939 R. Stout Some Buried Caesar xi. 153 All right. Take your name and stick it up your chimney and go to hell. 1941 Baker

Dict. Austral. Slang 71 Stick it!, a contemptuous ejaculation. 1948 D. Ballantyne Cunninghams xx. 105 He had a good mind to tell Basil Fisher to stick his run. 1960 Wentworth & Flexner Dict. Amer. Slang 520/1 Stick it (something) up your (one's) ass (taboo),..the strongest reply to the question, 'What shall I do with this?' 1971 P. Driscoll White Lie Assignment ii. 20 If you do earn your thousand pounds you can stick it, d'you hear? Stick it right up where it belongs. I don't want a penny of it. 1973 Houston Chron. 21 Oct. 12/7 Members of the House are suggesting to members of the Senate that they take this idea and stick it in their ears. 1977 Daily Tel. 22 June 17/8 After the hearing Mr Jeeves said: 'They can stick their cottage. I shall not move into it.'

21. b. (Earlier example.)
1876 [see *cold a. 12 b].

c. to be stuck for: to be at a loss how to obtain; to be unable to think of. colloq.
1937 in Partridge Dict. Slang. 1963 'J. le Carré' Spy who came in from Cold iv. 32 'Who's Mr. Ironside?'..'I don't think he exists... He's her big gun when she's stuck for an answer.' 1965 A. J. P. Taylor Eng. Hist. 1914–45 viii. 267 The Conservatives were strong in resistance. They were stuck for a positive programme. 1969 Guardian 31 July 6/1 Any time you're stuck for a meal.. come around. 1977 D. Beatty Excellency iv. 53 He might have discovered a snag..got stuck for some spare.

23. a. to be stuck with: to be saddled with, unable to get rid of (an unwanted person or thing). orig. U.S.
1848 [in Dict.]. a 1860 Providence Jrnl. in J. R. Bartlett Dict. Americanisms (ed. 3) 458 We got stuck with a bad lot of paper, and were obliged to stick it on to our readers. 1943 F. J. Bell Condition Red iv. 47 So it was our coal, and we were stuck with it. 1959 W. Kennedy in M. Ross Arts in Canada 136/1 We architects of the mid-twentieth century seem to be stuck with the gods who made us—Gropius, Corbusier. 1962 E. O'Brien Lonely Girl iv. 36 He bought me a grey astrakhan with a red velvet collar, and a flared skirt. 'I'm stuck with you now,' he said.. while he surveyed the coat from behind. 1972 Guardian 22 Mar. 16/2 Westminster..cannot apply a totally British solution to an Irish problem. But being stuck with it, the British Government has to try to muddle through. 1979 R. Jaffe Class Reunion (1980) 14 Emily hated being 'petite', which was a euphemism for getting stuck with all the short boys on blind dates.

d. For (U.S. slang) read (slang, orig. U.S.) and add: also esp., to be fond of, enamoured of, in love with.
1886 Lantern (New Orleans) 20 Oct. 3/2 Poor Charles Ernest is so stuck on a fairy named Emma Brown, that she can make him do anything she wishes. 1897 Kipling Captains Courageous x. 221 I'm not stuck on myself any just now—that's all. 1909 J. Masefield Tragedy of Nan i. 9 'Er father, as she's so stuck on—'E was 'ung. 1938 G. Greene Brighton Rock vii. ii. 295 I'd stake you a fiver she's straight. Why—you told me yourself—she's stuck on you. 1939 A. Huxley After Many a Summer ii. iii. 207 You'd say she was kind of stuck on the fellow. 1967 P. Shaffer Black Comedy 55, I don't mean that's why he popped the question... He's always been stuck on you. 1974 A. Lurie War between Tates (1977) vi. 128 Sandy, who was rather pathetically stuck on her for a while, took her to hear The Magic Flute.

e. to get stuck into: to lay into, make a physical attack on (someone); to make a serious start on, get down to (a task, a meal, etc.). Hence to get stuck in, to pitch in, get down to it. colloq. (orig. Austral.).
1941 Baker Dict. Austral. Slang 31 Get stuck into, to engage a person in a bout of fisticuffs. To tackle a job with a will. 1942 G. Casey It's harder for Girls 228 A bit o' peace..after you an' Winch nearly getting stuck into each other at the pub. 1948 C. Fry Thor with Angels 3 You get stuck Into some work, you whitebellied weasel. 1958 I. Cross God Boy xix. 160 Though arithmetic wasn't my best subject, I was quite glad at the idea of getting stuck into some figures. 1962 Observer 18 Feb. 23/4, I heard a terrific clanging downstairs and went down to see Pancho getting stuck into the gas meter with an iron bar. 1974 A. Morice Killing with Kindness iii. 31 He reached out a hand and promptly got stuck into his beloved evening paper.
1948 S. Matthews Feet First x. 57, I have no time for that 'get stuck in' policy that is sometimes advised in cup-ties or local derbys. Once one side starts tackling with too much vigour there is inevitable retaliation and loss of tempers. 1959 G. Slatter Gun in my Hand v. 51 Gives us a hand sometimes on the mixer..Gets into his old mocker and gets stuck in. 1961 B. Crump Hang on a Minute, Mate 158 Mrs Wagner brought in two plates of food..and told them to get stuck in. 1971 Where Sept. 260/2 He flung out his arms like a Petticoat Lane trader, and got stuck in. In five minutes he had an audience.

IV. 26. stick to —. f. Also to stick to one's last (with allusion to the proverb: see Last sb.¹ 2 c).
1927 Galsworthy Castles in Spain 92 Conrad had always a great regard for..workmen who stuck to their last and did their own jobs well. 1939 A. Powell What's become of Waring? viii. 227, I don't know why he wanted to meddle with writing at all. It wasn't his avocation. He should stick to his last. 1956 A. Wilson Anglo-Saxon Attitudes i. i. 19 You're not a member of the University Press Syndic... Stick to your last.

27. stick with —. e. To adhere to (an account, plan, etc.); to be faithful to, support.
1915 J. London Jacket iv. 28 Stick with it. Don't ever let'm know. 1958 M. L. King Stride toward Freedom vii. 127 We would stick with them through their difficulties. 'We must remain together,' we kept repeating. 1976 M. Machlin Pipeline xlviii. 505 'I've known all along it was a Goddamn fool plan.'..'Then why have you stuck with it so long?'

V. 27*. stick around. intr. To wait, remain in the vicinity, not to go away. colloq. (orig. N. Amer.).
1912 J. Sandilands Western Canad. Dict. & Phrase-bk., Stick around, wait about, hang around, or loaf around. The Canadian sport will stick around in the expectation of meeting the boys and having a good time. 1915 S. Lewis Trail of Hawk iii. 28 Stick around, son, and sit in any time, and I'll learn you some pool. 1919 E. Jordan Girl in Mirror iv. 79 I'm going to 'stick around', and guide them for a few days. 1943 P. Cheyney You can always Duck vi. 99 'I'm givin' no guarantees,' I tell her. 'But maybe I'll stick around. We'll see. So long, honey.' 1959 'S. Ransome' I'll die for You ix. 109 You stick around, Mr. Fisher, where I can find you when I want you? 1979 A. Fox Threat Warning Red xvi. 248 You'll be asked to come over here next week..and you'll have to stick around for a day or two.

32. stick out. b. Also in various colloq. phrases, esp. to stick out a mile. to stick out like a sore thumb: see *sore a. 9 f.
1933 E. Waugh Scoop ii. i. 119 'Have you noticed it?' 'Yes..it sticks out a mile.' 1949 [see road-sign s.v. *road sb. 9 b]. 1952 M. Allingham Tiger in Smoke i. 16 You couldn't miss him. He stuck out like a lighthouse. 1977 'H. Carmichael' Grave for Two iv. 48 'You're saying she's left him.'..'Sticks out a mile.'

c. (Earlier and later examples of to stick it out.)
1882 'Mark Twain' Let. 17 May (1917) I. 419, I have promised Osgood, and must stick it out. 1889 —— Conn. Yank. xxvi. 334 The proprieties required me to stick it out. 1914 G. B. Shaw Misalliance 17, I really couldnt stick it out with Jerry, mother. 1929 J. B. Priestley Good Companions iii. ii. 509 If you went, I'm durned sure I couldn't stick it out another week. 1981 P. P. Read Villa Golitsyn i. vi. 41 He stuck it out for a week and then shinnied down a drainpipe.

e. (Later example.) to stick one out (examples).
1910 A. Bennett Clayhanger i. vii. 49 She would not hear a word about the toast being a little hard... Maggie ..'stuck her out' that the toast was in fact hard. 1915 F. M. Hueffer Good Soldier iv. i. 224 That checked Florence a bit; but she fell back upon her 'heart' and stuck out that she had merely been conversing with Edward. 1916 A. Bennett Lion's Share vii. 53, I knew he was going to be ill when I left him in the cabin, but he stuck me out he wasn't.

i. to stick one's neck out: see *neck sb.¹ 3 e.

34. stick up. k. For Austral. read orig. Austral. and add U.S. examples. Also, (Criminals' slang) of the police: to hold up (a suspect).
1904 'O. Henry' Cabbages & Kings xvii. 302, I couldn't take it with me, not knowing but what the monkeys might stick me up. 1926 J. Black You can't Win vii. 81 Anyway we'll sure be stuck up and frisked at Evanston. 1971 Black Scholar Sept. 32/2 It was the night he, Shotgun and Big Daddy stuck up the policy clearing house. 1978 S. Brill Teamsters iii. 76 They had served time for sticking up a variety store in Akron, Ohio.

† **n.** Cricket. To put a batsman on the defensive. Obs.
1864 Baily's Mag. Sept. 297 Grundy and Wootton.. put every batsman on the defensive, stuck them up, man after man, over after over,..and then sent each back to the Pavilion. 1891 W. G. Grace Cricket ix. 243 There are some bowlers who, by their wonderful accuracy of length, stick up the batsmen and get wickets on the most perfect grounds. 1904 P. F. Warner How we recovered Ashes v. 70 Rhodes stuck up all the batsmen, with the exception of Trumper.

o. In phr. stick 'em up: an (armed) robber's order to his victim to raise his hands above his head; = hands up! s.v. Hand sb. 54. Usu. imp. slang (orig. U.S.). (Cf. put them (or 'em) up (ii) s.v. *put v.¹ 53 a (e).)
1931 [see *reach v.¹ 12 c]. 1938 G. Greene Brighton Rock v. iii. 203 The children were scouting among the rubble with pistols from Woolworth's... Someone said in a high treble: 'Stick 'em up.' 1972 Wodehouse Pearls, Girls, & Monty Bodkin vi. 76 Sequences of spine-chilling drama, with people telling other people to stick 'em up and prodding them in the stomach with pistols.

VI. 35. stick-bomb = sticky bomb s.v. *sticky a.² 5; **stick-jaw**: also spec., toffee; also transf.; **stick-pin** U.S., any (ornamental) pin that is merely stuck in (as distinguished from a safety pin), esp. a tie-pin; also attrib.; hence **stick-pinned** ppl. a.; **stick-slip**, alternate movement and cessation of movement of one surface over another as a result of frictional forces; freq. attrib.; **sticktight**, (b) in full, sticktight flea; a small flea, Echidnophaga gallinacea, which infests poultry; **stickwater**, the liquid that is squeezed out when cooked fish are compressed during the manufacture of fish meal and fish oil.
1943 J. H. Fullarton Troop Target xxiv. 175 Along comes a Jerry tank. Kiwi goes in with stick bombs. Jerry hops out with his shirt-tail alight. Kiwi shoots him. 1971 'A. Hall' Warsaw Document viii. 89 The main doors breached with five stick-bombs. 1894 R. Wells Toffy & Sweets 14 Stick-jaw. 1932 L. Golding Magnolia Street ii. v. 352 Hush, Annie, hush! Here's some stick-jaw! 1938 Auden & Isherwood On Frontier iii. ii. 108 Gone to a demonstration, I suppose, to shout stickjaw slogans with the rest. 1950 'R. Crompton' William—the Bold i. 13 'Look! There's some real stick-jaw toffee,' said Henry. 1895 Montgomery Ward Catal. Spring & Summer 172 Fine solid gold scarf and stick pins. 1906 'O. Henry' in N.Y. World Mag. 18 Dec. 2/2 He wore his tie drawn through a topaz ring instead of fastened with a stick pin. 1928 'S. S. Van Dine' Greene Murder Case v. 64, I was only looking for that old emerald stick-pin you borrowed and never returned. 1973 M. Crowell Greener Pastures 132 A dapper tree sparrow, dark stickpin spot on his soft gray breast. 1939 Proc. R. Soc. A. CLXIX. 378 The friction is fluctuating violently, and the measurement again shows that the motion is proceeding by a process of 'stick-slip'. 1940 Phil. Trans. R. Soc. CCXXXIX. 1 Certain substances..are able to prevent this 'stick-slip' motion and allow continuous sliding to take place. 1958 Engineering 14 Mar. 339/2 The movement of the ram is completely free from 'stick-slip' phenomena because of the small but significant clearance between piston and cylinder. 1959 Times 13 Oct. 4/4 This 'stick-slip' as it is called can make the steering stiff to move initially. 1975 McGraw-Hill Yearbk. Sci. & Technol. 21/2 The jerky motion happens because—under the pressure and temperature conditions of the shallow part of the Earth's lithosphere—rock exhibits a property known as stick-slip. 1915 F. C. Bishopp Fleas as Pests (U.S.D.A. Farmers Bull. 683) 7 The sticktight flea, or southern chicken flea,..the most important of our live-stock infesting species. 1955 W. W. Denlinger Compl. Boston ii. 77 These sticktights abound in neglected..chicken houses. 1962 Gordon & Lavoipierre Entomol. for Students of Med. xxxv. 218 Another species of burrowing flea,..the so-called 'sticktight flea', on rare occasions attacks man. 1915 Rep. U.S. Comm. Fisheries 1914 App. III. 25 An apparatus for evaporating the water which is separated from the oil and known as 'stick water' has recently been installed... The residue or 'stick' will average about 9 per cent ammonia. 1945 Poultry Sci. XXIV. 379/1 The stickwater from fish-meal manufacture added materially to the riboflavin potency of a feed. 1965 G. H. O. Burgess et al. Fish Handling & Processing x. 237 Acidification of the dilute stickwater coagulates some of the fine suspended solids and these are sometimes recovered by further centrifuging.

b. stick-at-it colloq., a plodding conscientious person; **stick-to-it-iveness** colloq. (orig. U.S.), dogged perseverance.
1909 H. G. Wells Tono-Bungay ii. ii. 162 I'm a boiler-over, not a simmering stick-at-it. 1867 in E. B. Custer Tenting on Plains (1889) xvi. 520 With the stick-to-itiveness of a fox-hound when once on a trail. 1908 Daily Express 15 May 1/4 Success..is mostly hard work. It's work and it's stick-to-it-iveness. You've got to keep at it all the time. 1934 J. A. Lee Children of Poor i. 19 With the irresponsibility of my..father and my mother's stick-to-itiveness, I can..fashion an edifice and then.. set the whole show toppling. 1979 N.Y. Rev. Bks. 8 Feb. 10/3 This man who made his million apparently more by stick-to-itiveness than brilliance.

c. stick-on a., that sticks on or can be stuck on; adhesive.
1925 J. W. Bigger Handbk. Bacteriol. 60 Stick-on labels may be used, but these must never be licked in a bacteriological laboratory. 1941 Sun (Baltimore) 13 Aug. 16/6 There also will be speakers on the stick-on spray.. and other matters of interest to apple growers. 1962 L. Deighton Ipcress File i. 13 In crude stick-on letters the film title said Jay. Leeds. 1967 Punch 22 Nov. 780/1 False eyelashes, interchangeable wigs, adhesive eyebrows, stick-on fingernails. 1972 Guardian 11 Aug. 9/6 Another story book you make yourself with stick-on shapes.

stick, v.² Add: **7.** To strike (a person) with a stick.
1937 Partridge Dict. Slang 830/1 Sticked (, be), (to be) caned. 1962 M. Duffy That's how it Was iv. 44 The whole of 3A was sticked..Miss Wilkinson..smacking the outstretched palms. 1982 Sunday Sun-Times (Chicago) 17 Oct. 138/1 Edmonton's Ken 'The Rat' Linseman was suspended..for sticking Toronto's Russ Adam during an exhibition game [of hockey].

stickability (stikăbi·lĭti). colloq. [f. Stick v.¹ + Ability.] Capacity for endurance, persistence, perseverance, staying power.
1888 Voice (N.Y.) 10 May 7/3 Stickability..is the most important ability a farmer can possess. 1905 British Weekly 28 May 193/2 To be able to take rebuffs happily and still go on requires, to use a coined word, 'stickability'. 1922 Chambers's Jrnl. Sept. 634/2 The foreigner has supplanted the middle and lower class Chilian in nearly every branch of industry in which the quality best described as 'stickability' is required. 1962 Daily Tel. 17 Aug. 18/1 All too many lack any degree of 'stickability' and flit from job to job like butterflies. 1976 P. Donovan Relig. Lang. iv. 45 Those engaged in such activities get their stickability..from the quite reasonable assumption that unless someone persists.., there is little chance of any discovery being made.

stickage. (Earlier and later examples.)
1734 J. T. Desaguliers Course of Experimental Philos. I. iii. 89 We are to have regard to..the Quantity of Stickage or Friction; which differ according to the..Nature of the Materials. 1961 New Scientist 27 July 216 Most metals tend to stick to other metals they touch in a deep vacuum. Bearings must be found to prevent such stickage. 1968 Encycl. Polymer Sci. & Technol. IX. 33 Too much waxing can lead to further stickage.

sticker¹. Add: **3. a.** (Later examples in sense 'one who persists in a task'.)
1916 Anzac Bk. 130 He was no 'sticker', and in the third year of his medical course he had side-tracked himself. 1967 C. Fremlin Prisoner's Base xii. 84 Daphne did not believe in dropping things; she was, as she would have told you, a Sticker. 1979 N. Hynd False Flags viii. 71 Bobby wasn't any quitter. He was a sticker.

d. (Earlier and later examples.)
1832 P. Egan Bk. Sports xxii. 344/2 At the bat and the bottle they find him a sticker. 1855 F. Lillywhite Guide

to Cricketers 56 A 'sticker' with 'confidence' was all that was required, to have turned the 'tide' in their favour. 1977 *Times* 12 July 10/1 When Chappell was adding 55 with O'Keeffe, who is well known as a sticker, there were visions of England having to make 175, perhaps 200, today.

f. *U.S. colloq.* A thorn or bur.

1889 H. H. MCCONNELL *Five Years a Cavalryman* iv. 35 The leaves when submitted to the action of fire in order to burn off the sharp stickers, are used as food for cattle. 1898 G. F. ATHERTON *Californians* 231 Trennahan..plucked the 'stickers' from his trousers. 1899 M. GOING *Field, Forest, & Wayside Flowers* 350 When the 'stickers' are at last picked or rubbed off, they fall to the ground. 1945 B. MACDONALD *Egg & I* (1946) III. viii. 94 My hair and shoulders were full of twigs and stickers.

5. a. For *U.S.* read orig. *U.S.* and add later examples. Also *spec.* a small adhesive notice designed to be stuck in a conspicuous place and used to publicize a cause, commodity, or place. Also *attrib.* in **sticker price** *N. Amer.*, the advertised price (of a commodity).

1919 *Nation* (N.Y.) 117/2 Defendants..had printed millions of seditious 'stickers'. 1934 J. M. CAIN *Postman always rings Twice* ii. 13 About three o'clock a guy came along that was all burned up because somebody had pasted a sticker on his wind wing. I had to go in the kitchen to steam it off for him. 1943 K. TENNANT *Ride on Stranger* xvii. 185 Plastering it [*sc.* a vessel] with stickers demanding the guest's release. 1955 W. GADDIS *Recognitions* I. v. 192, I left all my luggage there covered with the most adorable stickers from everywhere, my dears, every chic hotel you ever heard of. 1959 *Listener* 21 May 884/1 An English 'sticker' about Nuclear Disarmament on the door of..the students' canteen. 1962 E. GODFREY *Retail Selling & Organ.* ix. 91 Special delivery instructions..should be written clearly..on a special sticker attached to the despatch docket. 1967 [see *BUMPER *sb.*¹ 5 c]. 1970 *Globe & Mail* (Toronto) 25 Sept. B4/7 The company said the sticker or suggested retail prices, which include federal excise taxes..are up an average of $136. 1976 J. I. M. STEWART *Memorial Service* xvi. 273 His magazine's supposed to be coming out tomorrow. Have you seen the stickers for it?

b. A postage stamp. *Criminals' slang.*

1904 'No. 1500' *Life in Sing Sing* 253/1 *Stickers*, postage stamps. 1926 J. BLACK *You can't Win* ix. 107 You're a cinch to get some coin and a bundle of stickers out of every 'P.O.' You can peddle the stamps anywhere.

Hence **sti·cker** *v. trans.*, to affix a sticker (sense 5 a) to; **sti·ckered** *ppl. a.*; **sti·ckering** *vbl. sb.*

1972 *Daily Colonist* (Victoria, B.C.) 3 Feb. 48/8 The system started in 1963 in Monterey Park, Calif., where 5,000 stickered homes had been broken into only 19 times and about 6,000 non-marked homes suffered more than 2,000 burglaries. 1976 *Publishers Weekly* 29 Mar. 41/1 The titles are produced by Dent in London. Dutton warehouses its inventory in this country and the titles are stickered for the U.S. market here. 1977 *Periodical* XL. 196 Nothing vexes me more than to go into a bookshop and find not just one price sticker on the book jacket, increasing the price, but sometimes two or more... The stickering is a burdensome business.

Stickie, Sticky (sti·ki). *Ir. slang.* [f. STICK *v.*¹: see -Y⁶, -IE.] A member of the official I.R.A. or Sinn Fein. Usu. in *pl.* Also **Stick** *sb.*⁵

1972 *Times* 21 Aug. 10/3 Who'll stop the boys fighting with the stickies (official IRA)? 1978 F. BURTON *Politics of Legitimacy* 188 'Stickies' is the widespread term used to designate the Official IRA/Sinn Féin. I heard two accounts of the origin of the term and both were somewhat apocryphal. One referred to the fact that the Official IRA 'stuck' to the existing organization whereas the Provisionals broke away. The other explained the name by referring to the fact that the Provisional IRA Easter Lilies were pinned to their supporters' clothes whereas the Officials had theirs stuck on. 1978 D. MURPHY *Place Apart* ii. 37 The Officials are also known as the Stickies (or Sticks). *Ibid.* vi. 110 Her son..was 'executed' last year as a punishment for deserting from the Stickies. 1979 *An Phoblacht* 29 Sept. 3/5 In a typical pro-British statement..the Sticks' chairman in South Antrim, Kevin Smyth, accused the IRA of 'gross sectarianism' in bombing the Lisburn premises. 1979 J. B. BELL *Secret Army* 446 The term Stickies or Sticks came from the Official Republican innovation of putting gum on the back of the Easter Lily commemoration badge while the Provos stuck to the conventional pin.

stickily (sti·kili), *adv.* [f. STICKY *a.*² + -LY².] In a sticky manner.

1908 KIPLING *Actions & Reactions* (1909) 102 The Hive was half hidden by smoke... They heard a frame crack stickily. 1937 *Times* 12 July 5/1 The game started rather stickily. 1942 'A. BRIDGE' *Frontier Passage* iii. 39 'Is he dead?'..'Supposed to be—very stickily finished off by the Reds, when they caught him.' 1973 *Times Lit. Suppl.* 9 Mar. 257/2 The book manages, in fact, to make something stickily implausible out of several familiar conventions.

stickiness¹. Add: Also, hesitancy, stubbornness; awkwardness, unpleasantness.

1933 C. MACKENZIE *Water on Brain* viii. 115 Major Hunter-Hunt let his emotion over the stickiness of the Treasury evaporate in a deep sigh. 1947 'N. BLAKE' *Minute for Murder* viii. 167 He had not imagined..that there was anything more in Billson's recalcitrance..than his usual official stickiness. 1948 WODEHOUSE *Spring Fever* xiii. 126 The intense stickiness of the situation. 1962 J. D. MACDONALD *Girl* xii. 186 You do seem to have involved her in some sort of stickiness.

sticking, *ppl. a.* Add: **3.** (Earlier examples of *sticking-out*.) Also † *sticking-off*.

1834 C. M. YONGE *Let.* 4 July in C. Coleridge *Charlotte Mary Yonge* (1903) iv. 123 There were two great sticking-out boxes like pulpits. 1843 C. RIDLEY *Let.* Feb. in U. Ridley *Cecilia* (1958) x. 118 Really it will be tiresome if he grows up with large, sticking-off ears. 1892 C. M. YONGE *That Stick* I. ii. 32 She had such horrid sticking-out ears.

sticking-point. Add: **2.** A point at which one sticks and beyond which one refuses to go; a subject upon which one will not yield or compromise; an obstacle.

1965 *Listener* 23 Sept. 441/1 As a politician he has been mild to the point of compromise. But one sticking point for him has been India's unity. 1970 *Globe & Mail* (Toronto) 25 Sept. 3/2 An early sticking point is expected to come when the Russians raise the question of reopening the Port of Vancouver to the supply ships. 1981 *Church Times* 27 Mar. 12/5 It is not the matter of women priests that is the main sticking point.

stick-in-the-mud. Add: Hence **stick-in-the-muddish** *a.*

1936 M. MITCHELL *Gone with Wind* xxviii. 481 It wasn't hidebound and stick-in-the-muddish like the older towns and it had a brash exuberance that matched her own. 1959 A. SALKEY *Quality of Violence* x. 158 He's slow and easy and a little 'stick-in-the-muddish'.

stick-out, *a.* and *sb.* Add: Also **stickout.**

A. *adj.* **2.** *U.S. slang.* Outstanding.

1948 *Daily Progress* (Charlottesville, Va.) 27 Jan. 9/1 After that, you have to scratch your head to think of another stickout box office attraction. 1958 *Washington Post* 19 June C1/6 Kramer's only hope for a stickout newcomer would be Australia's Mal Anderson against Gonzales.

B. *sb.* **2.** *U.S. slang.* **a.** A horse that seems a certain winner. **b.** An outstanding sportsman.

1937 D. RUNYON in *Collier's* 11 Sept. 70/3 This mare Cara Mia is a stick-out. 1942 BERREY & VAN DEN BARK *Amer. Thes. Slang* § 636/4 Good player,..*stickout.* 1949 *Sun* (Baltimore) 2 July 9/1 A 'stickout' on paper, Nokomis was in front most of the way along the six-furlong route. 1951 *Daily Progress* (Charlottesville, Va.) 26 Sept. 13/8 His former players now coaching send stickouts in his direction. 1958 *Washington Post* 26 June C1/1 As for third base, ball players and fans alike have no range of choice. Frank Malzone of the Red Sox is a stickout.

stickum (sti·kə̆m). *N. Amer. colloq.* [f. STICK *v.*¹ + UM, 'UM.] A sticky or adhesive substance; gum, paste; pomade. Also *transf.* and *fig.*

1909 *Dialect Notes* III. 376 *Stickum*,..mucilage, paste. 1936 *Christian Sci. Monitor* 14 Aug. 1/3 See that he is keeping the quality of his postage stamp stickum up to standard. 1963 C. D. SIMAK *They walked like Men* ix. 48 We used to shave them and give them facials and all of them wanted stickum on their hair. 1969 R. WILLIAMS in D. Knight *100 Yrs. Sci. Fiction* 303 The fact still remains, this machine makes every man self-sufficient, it takes the stickum right out of society. 1978 J. IRVING *World according to Garp* viii. 157 Wet with sweat and sweet with the lush stickum of sex. 1980 *Globe & Mail* (Toronto) 23 Aug. 1/1 (*heading*) Gaudaur strikes at stickum. Football commissioner Jake Gaudaur has instructed referees to rule ineligible players who use excessive amounts of adhesive on their hands.

stick-up, *a.* and *sb.* Add: **A.** *adj.* (Earlier example.)

1854 'C. BEDE' *Further Adventures of Mr. Verdant Green* ii. 5 A modest-looking young gentleman, who appeared to be..ill at ease in his frock-coat and 'stick-up' collars.

B. *sb.* **2.** *slang* (orig. *Austral.*, now chiefly *U.S.*). **a.** = HOLD-UP b.

[1887 W. H. SUTTOR *Austral. Stories Retold* 41 A body of men, mostly armed, met us. We at first thought it was a case of 'stick up'.] 1904 'O. HENRY' in *McClure's Mag.* Apr. 611/1 The first 'stick-up' I was ever in happened in 1890... It will explain how most train robbers start. 1910 H. LAWSON *Skyline Riders & Other Verses* 62 Scott that fired at Brummy Hughson, when the 'stick-ups' used to be. 1930 *Punch* 26 Feb. 236/1 He was reminded..that he had worked on the Babylon stick-up, and consented to come clean. 1944 *Sun* (Baltimore) 18 Mar. 12/1 The bank manager told police that the bandit..drew a gun and said: 'This is a stickup.' 1955 H. KURNITZ *Invasion of Privacy* (1956) vii. 53 Morley..told him they wanted to fake a stick-up. Then he shoots..the holdup man. 1972 [see *PULL *v.* 19 f].

b. One who robs by 'sticking up' his victims; = HOLD-UP a. ? *Obs.*

1905 *N.Y. Times* 2 Jan. 4/1 The 'stick-up' is always a powerful man, whose duties are to intimidate intruders and kill them if necessary, while the others are at work on a safe. 1936 [see *BIMBO²].

c. *attrib.* = *HOLD-UP 2; *esp.* in *stick-up man.*

1905 *N.Y. Times* 2 Jan. 4/1 The man..is declared to be a typical 'yeggman of the stick-up' class. 1909 G. R. CHESTER *Making of Bobby Burnit* xiv. 169 Our local Hicks would rather be robbed by a lot of friendly stick-up artists. 1924 G. C. HENDERSON *Keys to Crookdom* 396 *Assaulter*, rough guy, hard bird..stick-up man, thug. 1930 [see *COOL *v.* 3 e]. 1935 D. RUNYON in *Cosmopolitan* Jan. 63/3 A fast stick-up job without any foolishness about it, maybe leaving any parties we come across tied up good and tight. 1950 *Times* 7 Feb. 8/4 It was the story of a 'stick-up' plot being hatched. 1973 'H. HOWARD' *Highway to Murder* vii. 80 The old man got knocked off by a stick-up guy at the filling station where he worked. 1979 G. F. NEWMAN *List* i. 3 Any moron could get a gun and become a stickup man.

sticky, *a.*² Add: **1. a.** (Later examples.) Also *fig.*

1909 G. STEIN *Three Lives* 27 The horses dragged the carriage slowly over the long road, sticky with brown clay. 1939 *Amer. Speech* XIV. 262 The listener often hears,..if the subject is a thief, 'He has sticky fingers.' 1940 N. MITFORD *Pigeon Pie* v. 95 'I have just labelled a few little things of my own..' she said, putting a sticky one firmly on to the giant radiogram. 1956 H. GOLD *Man who was not with It* (1965) i. 3 They were caught..like the flies caught wriggling in sticky-paper. 1976 A. MILLER *Inside Outside* i. 16 To safeguard the money from the sticky fingers of some of the boys.

c. *sticky wicket* (earlier example). Also *fig.*, esp. in phr. *to bat* (or *be*) *on a sticky wicket*: to contend with great difficulties (*colloq.*).

1882 *Bell's Life in London* 29 July 4/6 The ground..was suffering from the effects of recent rain, and once more the Australians found themselves on a sticky wicket. 1952 *National News-Let.* 24 Jan. 244 It must be clearly understood that Mr. Churchill was batting on a very sticky wicket in Washington. 1957 P. KEMP *Mine were of Trouble* ix. 177 Until substantial reinforcements could arrive we should be batting, in the language of Mr. Naunton Wayne, on a very sticky wicket. 1964 *Language* XL. 239 Enmeshed in these remarks, however, is a sticky wicket. 1971 *Cabinet Maker & Retail Furnisher* 24 Sept. 517 When it comes..to moulded plastics of various kinds, then the timber producer is on a stickier wicket.

d. *fig.* Sickly, mawkish, sentimental, 'soppy'.

1864 [implied at STICKINESS¹]. 1915 R. FROST *Let.* 11 Nov. (1964) 17 He needn't go calling himself sticky names like Gayheart in public. 1925 N. COWARD *Fallen Angels* I. 16, I hope you're not..hurt at our refusing to call you Jasmin?..It's a sticky name, isn't it—for the house? *a* 1961 O. SITWELL in WEBSTER s.v., Invest childhood with a sticky but romantic gloss.

e. *colloq.* (orig. *U.S.*). Of the weather: humid, muggy.

1895 in *Funk's Stand. Dict.* *a* 1961 in WEBSTER, A hot and sticky hour or two on shore. 1977 *Washington Post* 30 June F2/4 Hot, sticky summer weather—the kind of weather that seems to attack the mind as well as the body with its oppressiveness. 1983 *National Trust* Spring 16/1 On one of those stifling, sticky days of this curious summer.., at rehearsal, the Philharmonia Orchestra and Norman del Mar were all in shirt sleeves.

3. a. (Later U.S. examples.)

1937 *Sun* (Baltimore) 28 June 12/1 Several recent offerings have been described as 'sticky' by dealers—meaning only partially sold. 1960 *Wall St. Jrnl.* (Eastern ed.) 5 Dec. 7/2 Underwriters released two 'sticky' corporate debt issues to the free market.

b. *Econ.* Of prices, interest rates, wages, etc.: resistant to change, slow to respond to altered conditions.

1930 *Economist* 6 Sept. 453/1 In many cases the amount of available stock has been limited, and when a fair supply has been in sight prices have proved surprisingly 'sticky'. 1936 J. M. KEYNES *Gen. Theory Employment, Interest & Money* IV. xvii. 232 Wages tend to be sticky in terms of money, the money-wage being more stable than the real wage. 1939 G. MYRDAL *Monetary Equilibrium* vi. 136 When we talk about sticky and flexible prices we are already thinking in terms of indices of different price levels. 1971 D. C. HAGUE *Managerial Economics* iv. 94 The idea that the kinked demand curve is likely to be found where there is oligopoly has led to the widespread feeling that in oligopoly prices will be 'sticky'. 1978 *Daily Tel.* 18 Dec. 14/7 Building society rates tend to follow movements in market rates only rather erratically and usually with a time-lag. As economists say they tend to be 'sticky'.

4. *colloq.* **a.** Of a person: difficult to cope with, awkward, uncooperative; strait-laced, punctilious, particular, tending to make difficulties (*about* or *over* something).

1882 L. TROUBRIDGE *Life amongst Troubridges* (1966) 162 Rather a sticky audience who evidently thought it vulgar to laugh, and only sniggered into their pocket handkerchiefs. 1925 T. E. LAWRENCE *Let.* 3 Nov. (1938) 486 I've got too many subscribers, so am very sticky over these last copies. 1933 C. MACKENZIE *Water on Brain* viii. 115 Personally I've always advocated the spending of money. The only snag is the Treasury. They've been sticky lately. 1935 WODEHOUSE *Luck of the Bodkins* iii. 34 He didn't actually call me a waster..but his manner was sticky. 1937 E. BOWEN in *New Statesman* 6 Nov. 727/2 Be a shade too punctilious and you are sticky; make a little too free and you are a pariah. 1940 GRAVES & HODGE *Long Week-End* ix. 135 Even the stickiest British families seemed ready to abandon their mistrust of the cinema, if the vulgar American scene could only be replaced by a wholesome British one. 1953 J. BINGHAM *Five Roundabouts to Heaven* iii. 26 Sometimes aunt Emily was a bit sticky about paying up. 1960 *Times* 15 Oct. 8/7 We had to be immaculately turned out... Father was very 'sticky' about this. 1972 J. PHILIPS *Vanishing Senator* (1973) III. ii. 127 Bernstein will tell you. If he acts sticky have him call me.

b. Of a situation, issue, period of time, etc.: awkward, presenting great difficulty, disagreeable owing to hardship or danger; of a social function: slow to start, stiff, uncomfortable.

1915 D. O. BARNETT *Lett.* 86 We had a rather sticky time in the trenches..as the enemy's artillery and snipers showed 'a certain liveliness'. 1930 'SAPPER' *Finger of Fate* 17 You have the alternative of a sticky five minutes with three savage Alsatians. 1930 V. SACKVILLE-WEST *Edwardians* i. 17 What was Miriam's party like, Lucy? Sticky, as usual? 1946 WODEHOUSE *Money in Bank* xix. 155 It is a human trait to keep on hoping, however sticky the outlook. 1955 *Times* 22 Aug. 2/7 The play became rather sticky and it looked like one or two fouls before the umpires blew on a B.A.O.R. player. 1958 *Listener* 16 Oct. 621/3 This medley of the fine arts and show business may be accommodated to a pleasant pattern later on: it made a sticky start. 1958 [see *PATCH *sb.*¹ 5 b]. 1960 L. COOPER *Accomplices* I. ii. 17 It was the stickiest do I've ever been in and I thanked God I'd been taught to fight. 1976 *Lancs. Evening Post* 7 Dec. 1/5 Preston South Labour MP Mr Stan Thorne..faced the prospect of a sticky interview with Government whips. 1977 B. PYM *Quartet in Autumn* xv. 127 He was so used to sticky church occasions that a lunch with two former colleagues should have been well within his powers. 1979 *Nature* 7 June 461/2 The sticky issues, however, will be over the appropriate forms of accountability and responsibility.

c. Phr. *to come to a sticky end* (or occas. *finish*): to die or come to grief in violent or exceptionally unpleasant circumstances.

1915 H. ROSHER *In R.N.A.S.* (1916) 40, I wish we could get out to the front... I would much rather come to a sticky end out there than here. 1930 J. COLLIER *His Monkey Wife* xviii. 255 Even if our love affair did come to a horrible sticky end, yet there's so much between us. 1930 J. B. PRIESTLEY *Angel Pavement* xi. 566 Never mind, he'll come to a sticky finish before he's done. 1959 F. MACLEAN *Back to Bokhara* iii. 152 The reformers..have usually come to a sticky end. 1970 'D. HALLIDAY' *Dolly & Cookie Bird* ix. 142 The heroines I've seen come to a sticky end because while the murderer's still running around no one calls in the police. 1980 *Church Times* 19 Dec. 12/3 Some cast away all chances of redemption till they come to a sticky end.

5. a. Special collocations: **sticky-back**, a small photograph or poster with a gummed back; also *attrib.* or as *adj.*; **sticky bomb**, an anti-tank grenade covered with an adhesive substance to make it stick to its target; **sticky dog** *Cricket colloq.*, a sticky wicket; **sticky end** *Genetics*, an end of a DNA double helix at which one strand extends a few nucleotides beyond the other, unpaired; **sticky-fingered** *a.*, apt to steal, light-fingered; **sticky tape** = *adhesive tape* s.v. *ADHESIVE *a.* I.

1913 A. H. DAWSON *Dict. Slang* 184 *Sticky-backs*, photographs about the size of a postage-stamp with gummed backs. 1922 JOYCE *Ulysses* 56 Stamps: stickyback pictures. 1939 'G. ORWELL' *Let.* 4 Jan. in *Coll. Essays* (1968) I. 378 The commonsense thing to do would be to accumulate the things we should need for the production of pamphlets, stickybacks etc. 1940 W. S. CHURCHILL *Let.* 6 June in *Second World War* (1949) II. I. viii. 149 It is of the utmost importance to find some projectile which can be fired from a rifle at a tank... The 'sticky' bomb seems to be useful for..this. 1962 L. DEIGHTON *Ipcress File* vi. 38 It was a sticky bomb about as big as two cans of soup end to end; on impact its very small explosive charge spread a sort of napalm through tank visors. 1925 D. J. KNIGHT in *Country Life* 18 July 95/1 If you..get a chance of bowling on one of these 'sticky dogs', as we call them. 1928 *Daily Express* 9 July 17/1 Should he bat first or should he put Somerset in and hope for a 'sticky-dog' wicket? 1982 P. TINNISWOOD *More Tales from Long Room* vii. 86 That great Groundsman in the sky has secured his covers... And when the sun appears again, as appear it always will, there will be no 'sticky dog' and play will be resumed on time. 1968 *New Scientist* 18 July 142/1 This [enzyme] can be used for linking up small nucleotide sequences by what Professor Khorana calls the 'sticky end' technique. 1976 *Sci. Amer.* Dec. 108 (*caption*) The circle of viral DNA replicates, producing multiple copies that are then cleaved by a specific viral enzyme to give rise to the linear form with 'sticky' ends. 1980 AYALA & KIGER *Mod. Genetics* ix. 327 The purified p184 DNA is mixed with purified DNA from another plasmid..also possessing *Eco* RI sites. The mixture is cleaved with pure *Eco* RI enzyme at a temperature that permits the sticky ends to come apart and reanneal..to form larger hybrid plasmids. 1890 BARRÈRE & LELAND *Dict. Slang* II. 305/2 *Sticky-fingered*.., thievish or covetous. 1932 D. ACLAND *Sticky Fingers* xxv. 314 What a crew we are—sticky-fingered, every one of us. 1982 *Daily Tel.* 23 July 30/3 Mr Steel announced menacingly that a list of sticky fingered policemen had been made available. 1958 *Times* 4 Aug. 9/1 What is the sense of combining the most obdurate kind of fastening (sticky tape, for instance) with the flimsiest of paper bags? 1973 R. PARKES *Guardians* ix. 162 The naked body had been strapped into the armchair with yards of sticky tape.

b. *Comb.* **sticky-out** *a. colloq.*, that protrudes or sticks out.

1928 D. L. SAYERS *Unpleasantness at Bellona Club* x. 118 She has a bad skin and rather sticky-out teeth. 1957 *Woman's Jrnl.* May 51/1 People who had bright red hair and sticky-out teeth and glasses weren't the sort..that any parents wanted to adopt.

sticky (sti·ki), *sb.* [Ellipt. use of STICKY *a.*²] **1.** *slang.* Something that is sticky, *spec.* (*a*) an adhesive material; (*b*) a sticky wicket.

1859 HOTTEN *Dict. Slang* 102 *Sticky*, wax. 1937 PARTRIDGE *Dict. Slang* 831/1 *Sticky*,..sticking-plaster: lower and lower-middle class coll. 1954 A. G. MOYES *Australian Batsmen* 184 Again, the 'sticky' provides plenty of excitement. 1966 I. JEFFERIES *House-Surgeon* viii. 154 Bring me some more sticky and that pint of blood in the fridge. 1967 PARTRIDGE *Dict. Slang* Suppl. 1390/2 *Sticky*,..since late 1940's, usu. cellulose tape (Sellotape, Scotch tape, etc.). 1975 *Daily Mail* 3 June 11/2 As well as cash, the thieves took 'stickies'—the slang term for postage, national insurance and TV licence stamps. 1982 *Private Eye* 13 Aug. 13/2 All you ever get in the Alps is some ghastly sticky made out of rotting Edelweiss.

2. A film of which the action is slow-moving. *nonce-wd.* (*humorously* after *movies*).

1936 E. M. FORSTER *Abinger Harvest* 51 British ladies and gentlemen turn the movies into the stickies for old Elstree's sake.

stickybeak (sti·kibīk). *Austral.* and *N.Z. colloq.* Also **sticky-beak.** [f. STICKY *a.*² + BEAK *sb.*¹] An inquisitive person; one who sticks his nose into others' affairs, a Nosey Parker. Also *attrib.*

1920 B. CRONIN *Timber Wolves* ix. 159 I've told the girls to give out that we've gone fishing, if any stickybeaks get to asking why we ain't visible no more. 1934 *Bulletin* (Sydney) 7 Feb. 10/1 One objection to 'party' telephone lines is the stickybeak subscriber. 1936 M. FRANKLIN *All that Swagger* xlix. 442 Fortunately such 'sticky-beaks' were few. 1948 D. W. BALLANTYNE *Cunninghams* I. v. 24 He wasn't like those other sticky-beak kids, he reckoned. 1965 R. MCKIE *Company of Animals* i. 30 Fire is a strange thing to a jungle elephant. It's foreign. And because it's foreign and he's a sticky-beak, he investigates. 1970 K. BENTON *Sole Agent* vi. 69 She rang me up, she did, the sticky-beak, the prying old cow!

Hence **sti·ckybeak, sticky-beak** *v. intr.*, to pry, to snoop; **sti·ckybeaking** *vbl. sb.*

1934 *Bulletin* (Sydney) 12 Sept. 20/2 Special traps were set outside for any crocs. that might stickybeak. 1945 L. GLASSOP *We were Rats* II. xi. 65 You deny me the right to think as I like... You must prod, and pry, and stickybeak. 1966 P. CARLON *Running Woman* xv. 144, I paused on my way back, looking at those houses. Such a temptation..to sticky-beak. 1969 *Coast to Coast 1967–68* 9 Mind your own business... I don't have to put up with you and your sticky-beaking into my affairs.

stictane (sti·ktēⁱn). *Biochem.* [f. mod.L. *Stict-a* (f. Gr. στικτός spotted), name of a genus of lichens + -ANE.] Any of a class of pentacyclic triterpanes, unsaturated derivatives of which occur in some lichens.

1973 W. J. CHIN et al. in *Jrnl. Chem. Soc.: Perkin Trans.* I. 1437 Ten new triterpenes, derived from a new triterpane system for which the name stictane is proposed, have been isolated from the hexane extractives of the lichens named in the title [sc. *Sticta coronata*, etc.]. 1976 *Ibid.* 857/1 Stictane..and the related flavicane..have an 8α-methyl group and a boat structure for ring B rather than the usual 8β-methyl group and chair ring B, hitherto found in pentacyclic triterpenoids. 1979 *Org. Mass Spectrometry* XIV. 160/1 We present detailed analysis of the mass spectral fragmentations of ten stictane triterpenoids.

stictic (sti·ktik), *a.*² *Chem.* [tr. G. *stictinsäure* stictic acid (Knop & Schnedermann 1846, in *Jrnl. f. prakt. Chemie* XXXIX. 367), f. as prec.: see -IC.] *stictic acid*: a depsidone, $C_{19}H_{14}O_9$, found in many lichens.

1868 WATTS *Dict. Chem.* V. 431 According to Knop and Schneedermann [*sic*]..the acid of *Sticta pulmonacea*, which they call stictic acid, is distinct from cetraric acid, though very much like it in composition and properties. 1935 *Jrnl. Chem. Soc.* 1380 We have now made a detailed comparison of the two acids..and found them to be identical in every way. Accordingly, since stictic acid was discovered and named first.., we propose that the name scopuloric acid should be abandoned. 1963 *New Scientist* 14 Mar. 588/1 The lichen synthesises appreciable amounts of the extra-cellular, water-soluble constituents of lichen acids, atranorin and stictic acid. 1979 *Lichenologist* XI. 321 Chemical examination of two similar specimens..showed stictic acid to be present in the squamulose specimens and absent from the others.

stiction (sti·kʃən). [f. *st*(*atic fr*)*iction*.] = *static friction* s.v. *STATIC *a.* 3 b.

1946 *Jrnl. R. Aeronaut. Soc.* L. 365/1 'Stiction' might be caused by foreign bodies silting up round these pistons. 1953 *Electronic Engin.* XXV. 11/2 Armature stiction due to residual magnetism in the iron circuit..is normally overcome by facing the pole piece or armature, or both, with some non-magnetic coating. 1976 *Gramophone* Nov. 905/2 The bearings use tiny ball-races and the 'stiction' is of a very low order, some five-thousandths of a gram in both planes.

stiddy (sti·di), *adv. E. Anglian dial.* [f. STEAD *sb.*] Instead (*of*).

1946 J. W. DAY *Harvest Adventure* iv. 57 If they'd left the rabbits to the workin' man we'd ha' had 'em in our bellies stiddy of the poor things suffocatin' in their burrows. 1971 *Country Life* 11 Mar. 533/1 Owd Bob, he cut a slice or tew out of it [a stranded sturgeon] with a hatchet an' all the roe runned away. That's where he done wrong. If we'd a-kept that roe in we'd a-got four bob a pound for it... Stiddy o'that we only got fi'pun from the fishmonger in Ely.

stiebel, var. *SHTIBL.

Stiegel (stī·gĕl). The name of Henry William *Stiegel* (1729–85), German-born American manufacturer of iron and glass, used *attrib.* and *absol.* to designate glassware made by him or resembling his work.

[1773 *Pennsylvania Packet* 8 Mar. 3/2 An Elegant Assortment of Henry William Stiegle's Glass is to be Sold.] 1906 *Bull. Pennsylvania Museum* 5 Jan., The Stiegel glassware was of better quality than any produced elsewhere in the United States. 1922 *Country Life* May 49/2 I'd rather have it than any other piece of Stiegel I ever saw. 1949 *Hobbies* June 104/1 Until quite recently the majority of our collectors were in the habit of calling all old blown glass either Stiegel or Wistar. *Ibid.* 104/2 All can not be the proud possessors of Stiegel sugar bowls in sparkling cobalt, creamers in lovely shades of amethyst. 1961 E. M. ELVILLE *Collector's Dict. Glass* 18/2 Other factories followed the Stiegel fashion, especially in Ohio and Pittsburgh districts. 1974 J. GARDNER *Corner Men* viii. 64 Peter was..about to pour vodka into a Stiegel-type glass. 1977 FLEMING & HONOUR *Penguin Dict. Decorative Arts* 759/1 The term 'Stiegel glass' is commonly applied to most c18 American pattern-moulded wares.

Stieltjes integral (stī·ltjəz). *Math.* [Named after Thomas-Jan *Stieltjes* (1856–94), Dutch-born French mathematician, who first considered such integrals in 1894 (*Ann. de la Faculté des Sci. de Toulouse* VIII. J.2).] A definite integral in which the value of a function is summed, not uniformly over an interval, but in accordance with some other function which assigns weightings continuously or discontinuously within the interval.

[1910 H. LEBESGUE in *Compt. Rend.* CL. 86 On désigne sous le nom d'*integrale de Stieltjes*..l'opération fonctionnelle faisant correspondre à $f(x)$ un nombre défini de la façon suivante.] 1914 *Proc. London Math. Soc.* XIII. 131 The definition of the integral of a continuous function with respect to a monotone increasing function given by Stieltjes..is defined to be the Stieltjes integral. 1952 J. C. C. MCKINSEY *Introd. Theory Games* ix. 169 Since the Stieltjes integral is defined by means of a complicated limiting process, it should not be an occasion for surprise that it does not always exist. 1980 A. J. JONES *Game Theory* 276 Technically we are using Stieltjes integrals here.

stifado (stifā·do). Also **stifato, stiffado, stiphado, stuffado.** [a. mod.Gr. στιφάδο, presumably f. It. *stufato* stew; cf. STUFATA.] A Greek dish of meat stewed with onions and other vegetables, esp. tomatoes.

1950 E. DAVID *Bk. Mediterranean Food* 76 Stuffado (a Greek ragoût). Cut 2 lbs. of steak into large pieces. Brown them in oil with 3 lb of small onions and several cloves of garlic. Into the same pan put ½ pint of thick and highly seasoned tomato purée and a glass of red wine. Simmer slowly for 4 or 5 hours. 1952 M. LO PINTO *New York Cookbk.* ix. 145 (*heading*) Stifato (beef stew). 1962 M. SOPER *Encycl. European Cooking* 329 (*heading*) Greek beef stew. *Stiphado.* 1975 I. MURDOCH *Word Child* 210 'What's for din-dins at your place?' 'Smoked salmon. Stifados. Lime soufflé.' 1977 T. HEALD *Just Desserts* i. 16 I'd like the Fritto misto and then the Stiffado.

stiff, *a.*, *sb.*, and *adv.* Add: **A.** *adj.* **I. 2. b.** *stiff one*, *stiff 'un*, (*a*) earlier example; (*b*) *slang*, a racehorse certain to lose or not to run at all (cf. sense 2 f below and *STIFF *sb.* 3 b); (*c*) *slang*, a forged note or cheque (properly sense 1: cf. STIFF *sb.* 2).

1823 P. EGAN *Grose's Dict. Vulgar Tongue* (rev. ed.), *Stiff ones*, of no use, dead men. 1871 'HAWK'S-EYE' *Turf Notes* 11 Most assuredly it is the bookmakers that profit by the 'safe uns', or 'stiff uns', as..horses that have no chance of winning are called. 1890 BARRÈRE & LELAND *Dict. Slang* II. 306/2 There are two bookmakers in Melbourne nicknamed 'the Undertakers', because of their fondness for laying against *stiff 'uns*, which, in this case, means horses that are certain not to run. 1895 A. GRIFFITHS *Criminals I have Known* 228 He had been 'took' with the 'stiff uns' on him, and was sent to the 'boat' (penal servitude). 1897 in Farmer & Henley *Slang & its Analogues* (1903) VI. 365/2 Do not invest money Until you read The Rialto. Never on stiff 'uns, wrong 'uns, or dead 'uns. 1953 *Sat. Even. Post* 4 Apr. 18/3 'I put over a couple of stiff ones,' is the way a paperhanger describes an operation.

e. Intoxicated, drunk. Cf. *STIFF *sb.* 4 c. *U.S. slang. rare.*

1737 *Pennsylvania Gaz.* 6 Jan. 2/2 He's Stiff. 1900 *Dialect Notes* II. 65 *Stiff, adj.*,..very drunk. 1957 N. ALGREN *Walk on Wild Side* I. 6 Getting stiff on the courthouse steps while denouncing the Roman Catholic clergy was a feat which regularly attracted scoffers and true believers alike. 1975 G. V. HIGGINS *City on Hill* i. 9, I always got stiff on the Fourth because it was the only way I could listen to all that crap.

f. *Sport.* Of a horse or athlete: certain (to win). Also of an event: certainly won, 'in one's pocket'. Cf. sense 2 b above. *slang.* ? *Obs.*

1890 BARRÈRE & LELAND *Dict. Slang* II. 306/2 *Stiff for*, (sporting Australian), certain for. The metaphor here is something that cannot be diverted (or averted). After the Melbourne Derby and Cup of 1880, Grand Flaneur was considered *stiff for* every race for which he was entered. 1912 *Punch* 21 Aug. 168/3 He ought to have this event absolutely stiff at the next Olympic Games.

g. Unlucky. *Austral.* and *N.Z. slang.*

1919 W. H. DOWNING *Digger Dialects* 47 *Stiff* (adj.), unlucky. 1922 A. WRIGHT *Colt from Country* 124 'On'y just got cut out of second place,' declared Knocker. 'Ain't a man stiff?' 1930 *Bulletin* (Sydney) 3 Dec. 22/3 ''Struth!' gasped Chips. 'If we're not stiff! Nothing doing for two days, Tommy.' 1958 *N.Z. Listener* 27 June 6/3 Then came the third Test... Maybe they were a bit stiff to lose that, but once again I think it was just that lack of solidity in the middle of the pack.

6. Also *fig.*

1939 F. THOMPSON *Lark Rise* ii. 42 Their talk was stiff with simile. 1977 B. PYM *Quartet in Autumn* iv. 34 That season of the year was stiff with festivals.

8. e. Severe, stern; angry.

1856 THACKERAY *Miscellanies* II. 272 The old gent cut up uncommon stiff. 1930 W. S. MAUGHAM *Cakes & Ale* viii. 104, I wrote a pretty stiff letter to the librarian.

f. *Cricket.* Of a batsman: tending to play stubbornly in a defensive manner.

1869 *Field* 28 Aug. 176/3 This lucky escape seemed to endow the 'stiff batsman' with more than ordinary vigour. 1877 C. BOX *English Game of Cricket* xxvi. 461 *Stiff bat*, usually applied to a batsman who stubbornly defends his wicket. 1885 R. H. LYTTELTON in *J. Lillywhite's Cricketers' Compan.* 16 Midwinter..[was] a stiff and careful bat.

11. a. *stiff as a poker* (earlier and further examples); *to keep a stiff upper lip* (earlier example); hence *stiff-upper-lip* adj.; also as *v. intr.* and *adv.*; also *stiff-upper-lipped* adj., *stiff-upper-lippery*, *stiff-upper-lippish* adj., *stiffupperlippishness*.

1798 G. COLMAN *Heir at Law* (1800) III. ii. 34 The last lord Duberly's father..with a wig as wide as a wash-tub and stuck up as stiff as a poker. 1815 *Massachusetts Spy* 14 June 4/4, I kept a stiff upper lip, and bought [a] license to sell my goods. 1876 E. W. HEAP *Diary* 22 Mar. in *Publ. Amer. Dial. Soc.* (1969) LII. 55 Frozen as stiff as a poker last night. 1934 J. BUCHAN *Free Fishers* xviii. 297 Old Utterson is as stiff as a poker, and would keep us arguing till midnight. 1958 *Listener* 16 Oct. 621/3 She criticizes herself for being too stiff-upper-lipped about the tragedy that she faces. 1961 A. O. J. COCKSHUT *Imagination of Charles Dickens* viii. 116 He oscillated between indignation, self-pity, and reticence of the stiff-upper-lip English school. 1961 *John o' London's* 19 Oct. 447/2 The second film contains a firmly disciplined..undercurrent of Miniverish stiff-upper-lippery. 1963 *Listener* 3 Jan. 42/1 It was all very improbable and too stiff-upper-lippish to have been written by anybody but an anglophile Frenchman. 1973 *New Society* 31 May 483/2 MPs, in praising stiffupperlippishness, used sex as a stalking horse. 1977 *Broadcast* 14 Nov. 10/3 The British are stiff upper-lipping through power cuts. 1978 W. F. BUCKLEY *Stained Glass* xv. 153 As you say, we must be stiff-upper-lip formal. 1979 *Verbatim* Autumn 14/2 Two of the men, an upper-class Indian and an English Colonel, share a British university education, and they have a private joke of using *pip-pip!, I say*, and *stiff upper lip* expressions to each other. 1979 *Evening News* 6 Feb. 7/4 With rakish Parker, Prince Philip found that he could relax from the strains of state business and stiff-upper-lippery. 1983 *Listener* 20 Jan. 33/1 Nigel Anthony is the stiff-upper-lipped adventurer.

b. In predicative use (cf. *RIGID *a.* 1 d): to an extreme degree, as *to bore* (*scare*, etc.) *stiff*.

1905 *McClure's Mag.* May 100/1 He was scared stiff to hear that Morrow was in town. 1918 W. J. LOCKE *Rough Road* ix. 107, I think I ought to tell you that you're boring Durdlebury stiff. 1923 R. D. PAINE *Comr. Rolling Ocean* ix. 150 When that crazy fireman broke loose just now, I was scared perfectly stiff. 1928 F. B. YOUNG *My Brother Jonathan* II. v. 284 She bores everybody she meets stiff with talking about him. 1933 [see *FEED *v.* 6 h]. 1952 [see *FRONT *sb.* (and *a.*) 3 b]. 1956 *English* Summer 46 Billy Temple, who announced in Westminster School Hall that 'the longer poems of Milton bored him stiff'. 1964 I. MURDOCH *Italian Girl* xii. 137, I was scared absolutely stiff of Otto finding out.

11*. *Math.* Of a differential equation: having a solution that shows completely different behaviour over widely different scales of time (or other independent variable).

1952 CURTISS & HIRSCHFELDER in *Proc. Nat. Acad. Sci.* XXXVIII. 235 In the study of chemical kinetics, electrical circuit theory, and problems of missile guidance a type of differential equation arises which is exceedingly difficult to solve by ordinary numerical procedures. A very satisfactory method of solution of these equations is obtained by making use of a forward interpolation process... The differential equations to which this method applies are called 'stiff'. 1973 *Physics Bull.* June 340/2 'Stiff' differential equation problems also have been the principal reason for combined analogue-digital (hybrid computer) simulation. 1979 *Nature* 18 Jan. 201/2 It would be interesting to extend the present work to the case of Bianchi-type models filled with a perfect fluid of equation of state: $p = (\gamma - 1)\rho$, which covers Zel'dovich's stiff equation of state.

II. 16. Delete 'Now only of spirits-and-water.' (Later examples.) Also of tea and *loosely*, denoting a generous measure.

1904 H. JAMES *Golden Bowl* I. xviii. 212 You must awfully want your tea..so let me give you a good stiff cup. 1919 WODEHOUSE *Coming of Bill* (1920) II. xi. 213 Mr. Penway's eyes..fell upon the bottle of Bourbon... He sprang at it and poured himself a stiff dose. 1929 T. WOLFE *Look Homeward, Angel* xxxv. 522 A stiff drink of gin. 1951 'J. WYNDHAM' *Day of Triffids* i. 21, I was shaky again, and..could have done with a stiff drink. 1970 G. F. NEWMAN *Sir, You Bastard* vii. 192, I think you've earned a very stiff scotch. 1975 D. LODGE *Changing Places* ii. 67 'Wow,' said the woman. 'You mix a stiff drink.' 1977 K. O'HARA *Ghost of Thomas Penry* xv. 149 They went off and poured themselves a stiff one each.

19. b. *Finance.* = TIGHT *a.* 10 c. Now *rare* or *Obs.*

1845 *Punch* 11 Oct. 164/2 Money's stiff they say. 1912 *Q. Rev.* July 103 Money is in such keen demand all the world over that the rates tend to become high, whereupon it is called 'stiff' or 'tight'.

20. b. *stiff-brim*, *-leg*.

1896 'MARK TWAIN' in *Harper's Mag.* Aug. 356/1 It was the first season anybody wore that kind—a black stiff-brim stove-pipe. 1930 *Engineering* 10 Oct. 449/2 The material—sand and gravel—was..unloaded by means of a stiff-leg derrick. 1932 W. FAULKNER *Light in August* ii. 27 He wore a tie and a stiffbrim straw hat that was quite new. 1943 *Sun* (Baltimore) 26 Nov. 6/3 The tower was first welded together on the ground. It was then lifted into its foundation by heavy lines attached to two high especially built 'stiff-leg' cranes.

c. **stiff-arm** *v. trans.* (occas. *absol.*), to fend off or push with a stiff arm; hence *fig.*, to rebuff or reject; also as *sb.* (usu. *attrib.*) in Rugby and U.S. Football, the act of tackling or fending off a tackle with a stiff arm; **stiff-neck**, an obstinate, haughty or self-righteous person; cf. STIFF-NECKED *a.*

1927 G. S. LOWMAN *Pract. Football* vii. 85 In all stiff arm and shifting have the body slightly inclined toward the tackler. *Ibid.* xii. 218 When playing in the line, the defensive center should stiff-arm his opponent back into the play. 1934 CRISLER & WIEMAN *Pract. Football* vii. 86 The ball carrier is more easily stopped in the hole than anywhere else, since he cannot dodge, sidestep, reverse or stiff-arm at that point. 1945 *Tee Emm* (Air Ministry) V. 50 Ward off attack by kicking or stiff-arming the shark... Kick or stiff-arm a shark to push him away. 1950 *Sport* 24–30 Mar. 9/1 Vindictive hacking and 'stiff arm' jabs were the rule rather than the exception. 1968 *Listener* 8 Aug. 189/2 There had been the fury of the British backs meeting persistent stiff-arm and late tackling, although these almost automatic fouls were only occasionally penalised. 1973 [see *PLACE *sb.*¹ 9 d]. 1974 A. A. THOMPSON *Swiss Legacy* xviii. 185 She tried to slam the door, but McAllister stiff-armed it violently against the wall. 1921 *Blackw. Mag.* Feb. 251/1 The stiff-necks of Victoria's entourage would have been painfully shocked. 1962 E. SNOW *Red China Today* (1963) i. 18 The young stiff-neck dismissed them angrily and told me I could do as I pleased about disposing of my excess but if I carried it I'd have to pay. 1975 *New Yorker* 28 Apr. 124/2 A repatriated American stiffneck who has been imprisoned by the North Vietnamese finds his particular solace in not budging an inch from the mindless chauvinism he set out with.

d. *stiff-bosomed*, *-elbowed*, *-jointed*, *-legged*, *-limbed*, *-shirted*, *-winged*; **stiff-arsed** *slang* (see quot. 1937).

1937 PARTRIDGE *Dict. Slang* 831/2 *Stiff-arsed*, haughty; supercilious. 1971 B. MALAMUD *Tenants* 70 If you think you..are going to be stiffassed and uptight by what I say, maybe we ought to call it off before we start? 1925 WODEHOUSE *Carry On, Jeeves* ix. 238 Then lay out one of the gents' stiff-bosomed [shirts]. 1942 W. FAULKNER *Go Down, Moses* 74 A white stiff-bosomed collarless shirt beneath a pique vest. 1956 H. GOLD *Man who was not with It* (1965) vi. 50, I..walked imperfectly awake, stiff-elbowed, thick-tongued, and dim-sighted. 1973 M. AMIS *Rachel Papers* 190 In this fashion, with twelve stiff-elbowed tugs, he has wanked into her head. 1743 W. ELLIS *Mod. Husbandman* Oct. i. 126 Its stiff-jointed, high-coloured, long Straw. 1876 'MARK TWAIN' *Tom Sawyer* xvi. 142 They felt rusty, and stiff-jointed, and a little homesick. 1854 DICKENS *Hard Times* III. i. 263 In staggering over the universe with his rusty stiff-legged compasses, he had meant to do great things. 1922 JOYCE *Ulysses* 428 Bloom..blunders stiffiegged, out of the track. 1914 'SAKI' *Beasts & Super-Beasts* 65 Old Shep, the white-nozzled, stiff-limbed collie. 1973 M. AMIS *Rachel Papers* 23 In normal circumstances, with her embarrassment in any kind of pre-coital conversation, her unassumingly pretty face, the stiff-limbed movements: you were a plaything of her unease. 1918 G. FRANKAU *One of Them* iv. 34 The patient lights of brougham or rarer car shine, Waiting stiff-shirted squires and ladies fair. 1933 DYLAN THOMAS *Let.* Nov. (1966) 51 Please don't go all stiff-shirted on me. 1901 *Practitioner* Mar. 241 Why not an 'Epic of Malaria'?..But such a theme would surely spur the stiffest-winged Pegasus to empyrean flights. 1977 *Devon Wetlands* (Devon County Council) vii. 26 Fulmars..have a characteristic stiff-winged, gliding flight.

B. *sb.* **2. b.** Money. *slang.*

1897 'OUIDA' *Massarenes* i. 6 They are astonishing—biggest income in the United States... Made their 'stiff' there, and come home to spend it. 1930 BELLOC *New Cautionary Tales* 58 He wrang his hands, exclaiming, 'If I only had a bit of Stiff How different would be my life!'

3. b. A racehorse which is unlikely to win; *spec.* one which is not intended to win. Cf. *STIFF *a.* 2 b. *U.S. slang.*

a 1890 in Barrère & Leland *Dict. Slang.* (1890) II. 306/1 'What do they mean by a *stiff* in the race?' 'That means generally a horse that on public form should win the race, and that either the jockey, trainer, or horse has been 'fixed' so that he will not win.' 1935 D. RUNYON *Money from Home* 197 There is also a rumor that Follow You is a stiff in the race. 1944 *Sun* (Baltimore) 21 Sept. 17/5 We either get shut out or find we are on a stiff which won't run.

4. a. Also a tramp; a migratory or unskilled worker.

1901, etc. [see *BINDLE² a]. 1915 *Truth* 20 Jan. 90/2 A hopeless shortage of the best labour on the one hand, and an unusual proportion of 'stiffs' on the other. 1927 W. EDGE *Main Stem* iii. 16 No stiff ever got information about a job through a telephone directory. 1963 H. GARNER in R. Weaver *Canad. Short Stories* (1968) 2nd Ser. 41 Who would listen to a harvest stiff in the middle of the tobacco country? 1976 E. WARD *Hanged Man* iii. 13 The driver..reached out to pull Burnett into the dusty cab. Construction stiff. A wandering freemasonry.

b. A mean, disagreeable, or contemptible person (freq. *big stiff*). Also joc. and *loosely*, a man, a fellow; *working stiff*, an ordinary working man. *slang* (orig. *U.S.*).

1882 in *Colorado Quarterly* (1956) Winter 271 Cap. Cline, that lonely old stiff..is now demonotized. 1896 G. ADE *Artie* ii. 17 There I set like a big stiff for five hours. *Ibid.* [illegible] iv. 36, I do n't come in here to give coin to no such stiffs as you. 1914 [see *NANNY-GOAT 3]. 1925 C. CONNOLLY *Let.* 28 Feb. in *Romantic Friendship* (1975) 61 He was described..as 'rather a stiff' which is true at present. 1930 *Punch* 26 Feb. 228/2 He said 'You big stiff!' in a very loud voice and went. 1930 J. DOS PASSOS *42nd Parallel* v. 403 Charley said that working stiffs ought to stick together for decent living conditions. 1936 WODEHOUSE *Laughing Gas* viii. 82 He had told me this man was a pretty good sort of old stiff. 1949 *Daily Ardmoreite* (Ardmore, Oklahoma) 23 Feb. 18/6 A select group of working stiffs in high government circles have run into 20 assorted kinds of complications. 1951 E. PAUL *Springtime in Paris* vii. 139 Hold your trap, you old stiff. 1967 *Sun* 23 Apr. 5/8 A bad customer..a stiff who orders the table d'hôte and nothing to drink. 1975 *N.Y. Times* 8 Jan. 35/1 And if a black man did buy a house, hey, we knocked on his door and said hello. If he was a nice guy, great. If he was a stiff, well, I know lots of white stiffs, too. 1977 *Guardian Weekly* 10 July 15/2 The idea of two young working stiffs [*sc.* Woodward and Bernstein] carrying off the prize is irresistible to youngsters with their careers before them.

c. A drunkard (see also quot. 1969). Cf. *STIFF *a.* 2 e.

1907 J. LONDON *Road* 170 Robbing a drunken man they call 'rolling a stiff'. 1969 *Telegraph* (Brisbane) 3 Oct. 42/1 We get all types of no-hopers here—hoboes, bums, 'alcos', homosexuals and 'stiffs' (methylated spirits drinkers). 1976 N. THORNBURG *Cutter & Bone* vi. 135 It had taken a good part of the day just to locate the poor stiff.

d. *Football.* A member of a reserve team. Freq. in *pl. slang.*

1950 *Sport* 22–28 Sept. 4/1 On Saturday the Forest 'stiffs' romped home to a 5–1 victory over Halifax. 1967 M. PROCTER *Rogue Running* xxv. 164 He became one of the 'stiffs', a second-team man who only played for the first team when a better man was ill, or injured. 1970 *Sun* 5 Sept. 28/6 (*heading*) Gunners sign Metchick for stiffs.

stiff, *v.* Restrict † *Obs.* to senses in Dict. and add: **3.** To cheat; to refuse to pay or tip (a person). *slang* (orig. and chiefly *U.S.*).

1950 *Sat. Even. Post* 15 July 124/3 It was a signal for the waiter to hustle over and put the arm on the customer who was trying to stiff him. 1968 J. M. ULLMAN *Lady on Fire* (1969) vi. 85 He's still haggling over the bill..trying to stiff me for a thousand less than agreed on. 1978 *Detroit Free Press* 5 Mar. B 3/1 Some New York waiters will tell you that the wealthiest men stiff you on a tip. 1979 *Globe & Mail* (Toronto) 31 May 2/5 Agents are still going bankrupt, but now it means they are leaving debts with the wholesalers and airlines. They aren't stiffing the consumer. 1982 *Washington Post* 9 Jan. 21/2 What is McCarthy doing when he refuses to tip a waiter who has given good service?.. He may be cursed by the waiter he stiffs. *Ibid.* 21/3 Instead of stiffing his servers, McCarthy should be stiffing their employers.

4. To kill; to murder. Cf. STIFF *sb.* 3 a. *slang.*

1974 R. L. SIMON *Wild Turkey* (1976) xiv. 104 'The Japanese girl..was found stiffed in an air-conditioning duct.'..'Stiffed?' '*Asesinato.*' 1978 C. EGLETON *Mills Bomb* vii. 73 Did she blow their cover too? Is that how they got stiffed in Prague?

stiffener. Add: **2. a.** Also *spec.* a card, such as a cigarette card, used to stiffen a packet, envelope, etc.

1926 *Chambers's Jrnl.* 10 July 497/2 'Stiffener', the name by which the cigarette-card has always been, and still is, known in the trade. 1967 A. DAVIS *Package & Print* 55 Cigarette cards..were originally intended as stiffeners (and always known in the trade by this name). The first printed stiffeners were produced in America in 1879. 1971 D. POTTER *Brit. Eliz. Stamps* x. 114 As an added attraction they include a stiffener—a card which reduces damage to the envelope in transit—with informative technical and background data.

3. A fortifying or reviving drink, *spec.* an alcoholic one. *slang.*

1928 D. L. SAYERS *Unpleasantness at Bellona Club* ii. 12 Dick Challoner..took the gasping Fentiman away into the deserted library for a stiffener. 1973 G. MITCHELL *Murder of Busy Lizzie* xiv. 161 I'll buy you a stiffener in the bar. 1978 J. PUDNEY *Thank Goodness for Cake* 128 Another visit to the loo for a stiffener... There are always drinks on the way home.

stiffening, *vbl. sb.* Add: **2. c.** *Naut.* Heavy goods taken on board a ship as ballast (see quot. 1894). Cf. *stiffening order*, sense 3 in Dict.

1894 H. PAASCH *From Keel to Truck* (ed. 2) 463/1 *Stiffening*, is the term given to any weighty substances taken on board a vessel for the purpose of making her 'Stiff', i.e. less crank. Stiffening (whether consisting of ballast or a portion of the outward cargo) is put in vessels which do not remain upright without having sufficient weight in the bottom. 1902 B. LUBBOCK *Round the Horn* 29 We are waiting now for our 'stiffening', as we dare not take our last 400 tons of coals out until we get a like weight of grain. 1924 R. CLEMENTS *Gipsy of Horn* vi. 115 We heard one morning that we were chartered, and proceeded to shift ship down to the coal-tips to take in our 'stiffening'—just sufficient coal, that is, to ballast the ship. 1975 *Islander* (Victoria, B.C.) 23 Feb. 6/3 The *Pamir* was taken in tow with a small amount of coal ballast aboard for stiffening.

stiff-necked, *a.* Add: Also *transf.*

1963 *Times* 23 Apr. 13/2 Fortunately the recent exchanges promise slightly less stiff-necked attitudes on both sides.

stiffness. Add: **1. b.** *spec.* (*a*) the force required to produce unit deflection or displacement of an object; (*b*) the maximum deflection of a beam divided by its span or length. (Later examples.)

1893 H. T. BOVEY *Theory of Structures* iii. 190, $dP_1/dl = EA/L$, and EA/L is consequently a measure of the longitudinal stiffness of a bar. *Ibid.* vi. 389 If D is the maximum deflection of a girder of span l under a load W, then W/D, or more usually D/l, is a measure of the stiffness of the girder. 1922 GLAZEBROOK *Dict. Appl. Physics* I. 808/2 The stiffness of a beam is usually measured by the maximum deflection, when loaded, divided by the span. 1925 J. CASE *Strength of Materials* xxiv. 386 The 'stiffness' of a spring is the load required to produce unit deflection. 1969 C. O. RASPOR in W. R. R. Park *Plastics Film Technol.* iv. 88 The tensile modulus is often used as a measure of film stiffness. This quantity is obtained by calculating the ratio of a stress to strain at a certain point on a tensile stress–strain curve. 1978 *Sci. Amer.* Dec. 116 (*caption*) Stiffness of an isolated muscle (the change in the force developed by the muscle when it is stretched, divided by its change in length) increases as the muscle is stretched.

stiffy (sti·fi). *U.S. slang.* [f. STIFF *a.* or *sb.* + -Y⁶.] **1.** A beggar who pretends to be paralysed.

1917 'A.-NO. 1' *Coast to Coast with Jack London* ii. 22 The fellow..was a hobo. He introduced himself as 'Stiffy Brandon'. His moniker indicated that for a beggar craft he had chosen the one which imposed upon the credulous..the awful affliction of the paralytic. 1956 S. HARRIS *Skid Row, U.S.A.* ii. 38 Red Bill was the best stiffy in New York.

2. A naïve or stupid person.

1965 *Liberator* (N.Y.) Aug. 23/1 'You a trick, Dan—a stiffy,' Herman said. 'You so square Little Orphan Annie could put game on you.' *Ibid.*, Stacked broads rushed in on the arms of stiffies straight from the cornfields; you know—them cats with the cowboy hats and ice-cream suits. Some stiffs. 1966 C. KEIL *Urban Blues* v. 118 Negro artists who find their way into white concert halls still find it necessary to 'hip' those 'stiffies' in the audience who insist on clapping their hands in a martial manner.

Stiggins (sti·ginz). The surname of a character in Dickens's *Pickwick Papers* (1836–7), used as the type of the pious humbug.

1916 [see *HOLY *a.* 4 c]. 1931 A. HUXLEY *Let.* 25 Sept. (1969) 355 That horribly snuffling Stiggins tone! 1935 *Times* 5 Jan. 6/3, I do protest strongly at any attempt to revive the activities of the Prudes on the Prowl, the spying of the Stigginses, and the chortling of the Chadbands.

stigmal, *a.* Delete *rare*⁻⁰ and add examples.

1957 *Imms's Gen. Textbk. Entomol.* (ed. 9) 683 It runs along the front margin as the marginal vein and gives off a short stigmal vein. 1962 GORDON & LAVOIPIERRE *Entomol. for Students of Med.* xli. 246 The hard and the soft ticks..possess a pair of stigmata, surrounded by a rounded chitinised plate (known as the stigmal plate) on each side of the body.

stigmasterol (stigmæ·stərǫl). *Biochem.* [ad. G. *stigmasterin* (Windhaus & Hauth 1906, in *Ber. d. Deut. Chem. Ges.* XXXIX. 4380), f. PHYSO)STIGMA: see *-STEROL.] A phytosterol, $C_{29}H_{48}O$, present in Calabar beans (*Physostigma venenosum*) and soya beans.

1907 *Chem. Abstr.* I. 1002 The compound..from the tetrabromide is termed stigmasterol. 1940 *Industr. & Engin. Chem.* Aug. 1138/1 Because of its chemical structure, stigmasterol is an excellent material for the synthesis of the hormone progesterone. 1975 *Lipids* X. 344/2 Sterol patterns of adult and juvenile ivy leaf were essentially identical, with quantities of sterol per 100 g fresh leaf being:..stigmasterol, 9 mg.

stigmergy (sti·gmədʒi). *Ent.* [ad. F. *stigmergie* (P.-P. Grassé 1959, in *Insectes Sociaux* VI. 62), f. Gr. στίγ-μα pricking + ἔργ-ον work: see -Y³.] The process by which the results of an insect's activity act as a stimulus to further activity.

1960 tr. P.-P. Grassé in *Insectes Sociaux* VI. 79 The stimulation of the workers by the very performances they have achieved is a significant one inducing accurate and adapted response, and has been named stigmergy. 1965 *Symp. Zool. Soc. Lond.* XIV. 128 Experimental evidence would seem to be desirable before accepting stigmergy as the explanation of all co-ordinated constructional behaviour. 1981 *Atlantic Monthly* July 49 There is a similar phenomenon in entomology known as stigmergy.

Hence **stigme·rgic** *a.*

1970 G. ORDISH tr. *R. Chauvin's World of Ants* i. 41 At some point there seems to be a brake on the stigmergic process when the stimulation has gone beyond a certain stage. 1971 E. O. WILSON *Insect Societies* xi. 229/2 Stigmergic responses are relatively major elements in nest construction by social insects.

‖ **Stijl** (stail). [Du., = style.] The name of a movement in art and architecture associated with the Dutch periodical *De Stijl* (1917–32), founded by Theo van Doesburg and Piet Mondrian, and devoted to the principles of neo-plasticism. Freq. with article *de*. Also *attrib.* and as *adj.*

1934 [see *NEO-PLASTICISM]. 1936 A. H. BARR *Cubism & Abstract Art* 142 In his *Volume construction* of 1918..the sculptor Vantongerloo applied the *Stijl* love of rectangles to sculpture. *Ibid.* 158 Other of Gropius' buildings at Dessau were even more *Stijl* in character. 1945 R. MOTHERWELL tr. *Mondrian's Plastic Art* 14 In 1915, Theo van Doesburg..was making analogous researches. Together we formed a small group of artists and architects: the De Stijl Group. 1962 H. MYERS tr. *Pingaud's Holland* 168 The *De Stijl* ('The Style') group, founded in 1917, is dedicated to the fight for a geometric art. 1974 *Encycl. Brit. Macropædia* IX. 516/1 Whereas the Bauhaus offered freedom of expression.., de Stijl offered dogmatic strait-jackets of verticals, horizontals, and primary colours.

stilb (stilb). *Physics.* [a. F. *stilb* (A. Blondel in *Recueil des Trav. & Compt. Rend. des Séances* (Internat. Commission on Illumination, 1923) V. 88), f. Gr. στίλβ-ειν to glitter.] A unit of luminance equal to one candela per square centimetre.

1940 in *Chambers's Techn. Dict.* 1942 *Jrnl. Optical Soc. Amer.* XXXII. 355/2 The names, *phot* and *stilb* were..coined by Blondel (1921) and are in general use on the Continent. 1953 [see *NIT *sb.*⁴]. 1963 VAN WIJK & UBING in W. R. van Wijk *Physics of Plant Environment* iii. 90 In photometry a black body at the temperature of melting platinum (2042·16 °K) is used as a standard light source. It has by definition a surface brightness of 60 candelas per cm² or 60 stilb, when viewed in a direction perpendicular to the surface.

stilbamidine (stilbæ·midīn). *Pharm.* [f. STILB(ENE + *amidine* (f. AMID(E + -INE⁵).] A diamidine derivative, $(H_2N)(HN)C{\cdot}C_6H_4{\cdot}CH{:}CH{\cdot}C_6H_4{\cdot}C(NH)(NH_2)$, of stilbene which has antiprotozoal properties and has been used, usu. in the form of its isethionate, in the treatment of trypanosomiasis and leishmaniasis.

1941 [see *PENTAMIDINE]. 1958 J. H. BURN *Lect. Notes Pharmacol.* (ed. 5) 110 Stilbamidine is no longer used [in the treatment of leishmaniasis] because it is toxic to the 5th nerve. 1977 *Jrnl. Pharmacol. & Exper. Therapeutics* CCI. 555/1 Pentamidine, stilbamidine and their close analogs are known to interfere with nucleic acid and protein synthesis.

stilbœstrol (stilbī·strǫl, -bē·strǫl). *Pharm.* Also (*U.S.*) **stilbestrol.** [f. STILB(ENE + ŒSTR(US + -OL.] † **a.** A synthetic derivative, $HOC_6H_4{\cdot}CH{:}CH{\cdot}C_6H_4OH$, of stilbene having slight œstrogenic properties; also, any derivative of this. *Obs.*

1938 E. C. DODDS et al. in *Nature* 5 Feb. 248/1 In view of the fact that 4:4'-dihydroxystilbene is the mother substance of a series of œstrogenic agents, we suggest that it may be termed stilbœstrol. 1938 *Nature* 2 July 34/1 (*heading*) Oestrogenic activity of alkylated stilbœstrols. 1943 *Vitamins & Hormones* III. 233 This substance..is known as diethylstilbestrol, the term stilbestrol being used for the parent substance.

b. A powerful synthetic nonsteroidal œstrogen, $HOC_6H_4C(C_2H_5){:}C(C_2H_5)C_6H_4OH$, used, often in the form of its dipropionate, in hormone therapy, as a postcoital contraceptive, and as a growth-promoting agent esp. in cattle and sheep; = *diethylstilbœstrol* s.v. *DIETHYL 2.

1939 *Lancet* 7 Oct. 788/2 Oral administration of the synthetic œstrogen, stilbœstrol, will inhibit implantation of the ovum in the rat and rabbit. 1943 *Vitamins & Hormones* III. 233 In Great Britain the substance is commonly known as stilbœstrol, and in the present paper, it will be referred to as diethylstilbestrol. 1957 [see *ŒSTROGENIC *a.*]. 1963 *Times* 28 Apr. 12/6 Farmers were warned yesterday that breeding stock should not be supplemented with stilboestrol. 1976 SMYTHIES & CORBETT *Psychiatry* xi. 169 Stilboestrol..is given for cancer of the prostate.

‖ **stile antico** (stī·le antī·ko). *Mus.* [It., lit. 'old style'.] The strict contrapuntal style of the sixteenth century, esp. as exemplified in the works of Palestrina.

1944 W. APEL *Harvard Dict. Mus.* 550/1 *Palestrina style*... As early as the 17th century this style, called *stile antico*, had become classical. 1959 *Collins Mus. Encycl.* 624/2 *Stile antico*, the contrapuntal style of the 16th cent. as practised by Italian composers and formulated by Italian theorists in the 17th and early 18th cent. 1968 *New Oxf. Hist. Music* IV. x. 521 This consciously conservative tendency among Catholic composers..resulted in the ultimate petrifaction of the Palestrinian *stile antico*, which artificially survived well into the eighteenth century. 1974 *Early Music* July 197/1 In the *stile antico* section for 'Domine fili unigenite'. 1976 *Ibid.* July 274 Cavalli's final work, the Requiem, looks backwards also, but to the more sober *stile antico* of Palestrina.

‖ **stile concitato** (stī·le kǫntʃitā·to). *Mus.* [It., lit. 'excited style'.] A baroque style developed by Monteverdi, emphasizing dramatic expression and excitement.

[illegible] *stile concitato.* 1947 M. F. BUKOFZER *Mus. Baroque Era* (1948) ii. 38 Monteverdi enriched the *concertato* style by an important innovation: the *stile concitato* (style of agitation) in which he turned the measured tremolo of the *gorgia* to dramatic effect. 1959 *Listener* 16 Apr. 692/3 He admired the *stile concitato* of Monteverdi. 1967 *Ibid.* 18 May 665/2 These were part of Monteverdi's *stile concitato*, a style he cultivated..to express wrath, anger, indignation.

‖ **stile rappresentativo** (stī·le ra:prezentatī·vo). *Mus.* [It., lit. 'representative style'.] The vocal style of recitative used by Italian musicians of the early seventeenth century (see quot. 1938).

1886 F. PRAEGER tr. *E. Naumann's Hist. Mus.* I. xv. 525 At first the opera was variously styled according to the individuality of the composer... The style itself was generally called *Stile rappresentativo.* 1915 *Strad* Jan. 299/1 The so-called *stile rappresentativo* of Florence was claiming more and more adherents and ministers; Monteverde, likewise, was moved by the ever-expanding ripple. 1938 *Oxf. Compan. Mus.* 896/2 *Stile rappresentativo*... The term was used by the early Roman composer of oratorio, Cavalieri, and the contemporary Florentine composers of opera..to indicate the use of their invention of recitative..which aimed rather at *representing* the sense and the natural inflection of the speaking voice than at providing the enjoyment of pure musical beauty. 1947 M. F. BUKOFZER *Mus. Baroque Era* (1948) ii. 25 The emergence of the *stile rappresentativo* or recitative about the year 1600 has often been regarded as the most important turning point in the entire history of music. 1959 N. C. CARPENTER *Mus. in Medieval & Renaissance Universities* iv. 357 University learning..led such people as Vincenzo Galilei..to experiment in a new musical style based upon his concept of Greek music, the *stile rappresentativo*, with its close affinity between words and music. 1980 *Early Music* July 298/1 But we should not let the polemics of the day in favour of monody obscure the very close, though perhaps not immediately apparent, bond between the expressiveness of the *stile rappresentativo* and the concept of melody that pervades the fabric of many polyphonic works from the second half of the 16th century.

stiletto, *sb.* Add: **2.** Restrict † *Obs. rare*⁻¹ to sense in Dict. and add: **b.** Short for *stiletto heel* (see sense *5).

1959 *New Statesman* 10 Oct. 464/3 She came..smooching forward, her walk made lopsided by the absence of one heel of the stilettos. 1960 *Guardian* 18 Mar. 10/6 Wearing stilettos, you are, of course, tiptoeing. 1967 O. WYND *Walk Softly, Men Praying* ii. 13 She wobbled slightly on worn-over stilettos.

5. **stiletto heel**, a very narrow, high heel on women's shoes, fashionable esp. in the 1950s; a shoe with such a heel; hence **stiletto-heeled** *a.*

1953 *Daily Tel.* 10 Sept. 8/4 One of the models..in the..winter collection..has the new stiletto heel, 3½ in high and just large enough at the base to cover a sixpence. 1981 M. HARDWICK *Chinese Detective* ix. 73 She was in black slip and stockings and four-inch stiletto heels. 1959 *Times* 13 May 10/7 The iniquitous effect of stiletto-heeled shoes on the modern woman's feet. 1973 M. AMIS *Rachel Papers* 124 At this kind of speed it was advisable to place the stiletto-heeled shoe, kept in a side-pocket for this purpose, over the gear-stick to prevent it jiggering like a pump-handle.

still, *a.* and *sb.*³ Add: **A.** *adj.* **1. d.** (Earlier example.) Used also of soft drinks, to distinguish them from the carbonated variety.

1777 P. THICKNESSE *Year's Journey* I. v. 17 The difference between still Champagne, and that which is mousseux, is owing to..the time of year in which it is bottled. [illegible]

3. a. (Allusive examples of *still small voice*: 1 *Kings* xix. 12.)

1769 T. GRAY *Ode for Music* 6 Sweeter yet the still small voice of Gratitude. 1847 C. BRONTË *Jane Eyre* II. iv. 102 Strong wind, earthquake, shock, and fire may pass by; but I shall follow the guiding of that still small voice which interprets the dictates of conscience. 1874 J. G. WHITTIER *Poet. Wks.* 458/1 Speak through the earthquake, wind, and fire, O still, small voice of calm! 1918 L. STRACHEY *Eminent Victorians* 64 In such a situation the voice of self-abnegation must needs grow still and small indeed. 1953 P. C. BERG *Dict. New Words in English* 14 The still small voice of Professor Bryant..has a few good words to say in favour of these 'abominations'. 1983 *Daily Tel.* 25 Mar. 20/3 [illegible] still small voice [illegible] the protagonists or critics of the Budget?

b. **still-air**, *a.* (*a*) *Aeronaut.*, applicable or calculated for a state of no wind; (*b*) not employing forced draught.

1913 *Captain* Sept. 1072/2 A machine with still-air speed of 57 miles per hour was sent up to fight a gale for 400 yds. [illegible] 1951 'N. SHUTE' *Round Bend* iv. 179 She had tankage for twelve hundred gallons, giving her a still-air range of about two thousand miles. [illegible] machines... Reasonably good results, on a small scale,..

can be got from the more old-fashioned still-air machines. 1961 P. W. Brooks *Mod. Airliner* iv. 100 A payload of 6,000 lb. was required for a still-air range of 3,500 miles. 1977 *Shooting Times & Country Mag.* 13–19 Jan. 26/2 Small incubators—those that take 100–200 pheasant eggs—are nearly always the 'still air' type and depend on convection currents for ensuring air movement.

B. *sb.*[2] † **5.** *slang.* A still-born child; a still-birth. *Obs.*

1864 Hotten *Slang Dict.* 247 *Stills*, the undertaker's Slang term for still-born children. 1897 [see *miss *sb.*[4]].

6. a. An ordinary photograph, as distinguished from a motion picture; *spec.* a single shot from a film (or a specially posed photograph of a scene from it) for use in advertising. Freq. with defining word, as *cinema still, film still*, etc.

1916 *Independent* 5 June 86 (*caption*) A striking 'still' from the film 'The Fall of a Nation'. 1922 *Glasgow Herald* 12 Oct. 6 Mr Johnson succeeded in exposing 25,000 feet of film and in taking some 1000 'stills'. 1938 *Archit. Rev.* LXXXIII. *caption facing* p. 72 Two aspects of a turbulent age are seen in 'The Temple of Janus',..wherein the figures of the Furies are arrested as in a cinema 'still'. 1945 H. Read *Coat of Many Colours* xxix. 142 We may select 'stills' for their closed form—for their pictorial composition—but the film itself is essentially open form. 1957 *Times* 25 Nov. 11/3 The story of that enchanting film *The Red Balloon* illustrated with film stills and with a photographic cover in colour. 1962 E. Snow *Red China Today* (1963) xv. 109 McDermott and I used our Canons taking stills. 1972 C. Weston *Poor, Poor Ophelia* (1973) xxvi. 164 The photo was a standard publicity still. 1976 *Oxf. Compan. Film* 66/1 *Frame stills* are reproductions of single frames from the film itself... They convey the true feeling of a film more exactly than can a posed production still. 1978 'A. Garve' *Counterstroke* I. 88 George had brought along a full-face 'still' of Lacy.

b. *attrib.* (as *sing.* or *pl.*).

1922 *Opportunities in Motion Picture Industry* (Photoplay Research Society) 47 Ask the first director you meet where you can find the still man. *Ibid.* 48 Still pictures are made for the publicity department. *Ibid.* 50 Seldom are the many lights placed for the movie camera exactly suited to the still camera. *Ibid.*, The still cameraman is of necessity a versatile flea. 1925 R. Beetham in E. F. Norton *Fight for Everest 1924* 324 His time was so fully taken up with cinema work that most of the still photography had to be done by other members of the party. 1928 [see *news-reel* s.v. *news *sb.* (*pl.*) 6 c]. 1963 *Movie* July/Aug. 27/4 The still pictures have greater effect than the newsreel shots. 1964 C. Willock *Enormous Zoo* ix. 165 Roger spent one whole afternoon trying to get the moment of entry and exit from the sandstone burrow with a stills camera. 1964 M. McLuhan *Understanding Media* II. xx. 193 The physical and psychic *gestalts*, or 'still' shots, with which they [*sc.* Freud and Jung] worked were much owing to the posture world revealed by the photograph. 1974 *Times* 16 Nov. 10/6 It was through stills photographs that the public were first introduced..to the stars. *Ibid.* 10/7 The stills men would retouch the negative. 1974 M. Taylor tr. *Metz's Film Lang.* i. 12 Rudolf Arnheim recognises that..still photography produces an impression of reality much weaker than that of the cinema. 1981 *Gossip* (Holiday Special) 54/3, I met her on a film I did a while back. She was the still photographer.

7. *Naut.* An instruction to cease work and stand to attention conveyed to a ship's crew by the boatswain's pipe.

1933 'L. Luard' *All Hands* 140 Pipe the still, Cox'un. 1963 [see *pipe-down].

still, *adv.* Add: **6. c.** *still and all*: nevertheless, even so; after all. *colloq.*

1829 G. Griffin *Collegians* I. vii. 140 Lord K..gave him a lease o' that farm... Still an' all, Myles do be poor, for he never knew how to keep a hoult o' the money. 1928 F. N. Hart *Bellamy Trial* iv. 104 Still and all, I believe that he was there precisely when he said he was. 1942 G. Marx *Let.* 16 Dec. (1967) 32 Still and all, as Lardner would say, it's a very cozy little place. 1963 A. Lubbock *Austral. Roundabout* 77 'Still-and-all,' they said, 'it's no use worrying over things y' can't help, is it?' 1969 *Guardian* 18 Aug. 9/5 Still and all, it is surely time to desist in good grace. 1978 R. Moore *Big Paddle* (1979) i. 4 Still and all, if you see something I haven't, let me know.

stillage, *sb.*[1] Add: **2.** Now *spec.* a pallet, frame, or similar structure used for storage of goods. Also *collect.*

1963 K. Hudson *Industrial Archaeol.* vii. 111 Old fittings and furnishings which might well remain *in situ* at old factories..include..wooden stillages. 1970 *Cabinet Maker & Retail Furnisher* 25 Sept. 629/1 Each stillage is 36 in × 24 in × 6 ft high, and provides storage for 150 headboards... Customers who take regular bulk deliveries are offered these stillages on loan. 1976 *Listener* 22 July, Stacking of components in wooden stillage instead of metal bins.

stillage (sti·lèdʒ), *sb.*[2] Chiefly *U.S.* [f. Still *sb.*[1] + -age.] The residue remaining in a still after a fermentation, usu. of grain or molasses, and removal of the alcohol by distillation.

1940 *Sun* (Baltimore) 5 Apr. 11/7 The experiments..were designed to determine the value of distillers' dry rye grains and stillage as substitutes for other grains in feeding cattle for the market. 1945 *Industr. & Engin. Chem.* June 534/1 Distilleries processing molasses are frequently confronted with the problem of disposing of stillage. 1963 *Agric. & Biol. Chem.* XXVII. A19/2 (*heading*) Nutritional studies on the utilization of distiller's stillage. 1979 *Nature* 6 Dec. 551/1 The total output of stillage in Brazil is equivalent in biological oxygen demand to the untreated sewage produced by a city of 200 million inhabitants.

still-birth. Add: Also **stillbirth. 1.** (Further examples.) Also, formerly, birth of a child alive or with a beating heart, but not breathing.

1880 *Brit. Med. Jrnl.* 9 Oct. 596/2 Stillbirth—Resuscitation after two hours and five minutes. 1913 R. W. Johnstone *Text-bk. Midwifery* xxxvii. 397 The infant is born in a state of suspended animation—its heart continues to beat, but it makes no effort to breathe or to move. 'Still birth' is therefore not the same thing as the child's being born dead, although death may supervene if prompt treatment is not applied. 1920 O. St. J. Moses *Man. Obstetr.* xviii. 269 In the pale or white form of still-birth the chances of recovery are much less. 1940 Brown & Gilbert *Midwifery* l. 468 In certain circumstances, the fœtus continues to live as a fœtus external to the mother for some time—10 to 20 minutes—and it may die as a fœtus and be counted as a still-birth. 1947 G. I. Strachan *Textbk. Obstetr.* viii. 699 The term 'stillborn' is therefore applied when the fœtus fails to maintain an independent existence, although some authorities..have attempted to make a division into (*a*) dead birth in which the fœtus is obviously dead and exhibits neither heart beat nor respiration, and (*b*) stillbirth in which the heart is beating but respiration is never established. It is simpler, however, to employ the term 'stillbirth' to cover both these states. 1971 *Reader's Digest Family Guide to Law* 197/2 Even a still-birth—where a child is born dead after at least 28 weeks of pregnancy—must be registered as a birth. 1972 S. G. Clayton et al. *Obstetrics by Ten Teachers* (ed. 12) liii. 642 If the heart is beating after delivery, even though there is no sign of respiration, the death of a child born after 28 weeks should not be included as a stillbirth but as a neonatal death.

2. A still-born child.

1963 Butler & Bonham *Perinatal Mortality* 210 Only six of the 'stillbirths' weighed 1,000 grams or less and were born before 28 weeks gestation. 1969 M. M. Garrey et al. *Obstetrics Illustrated* 469/1 A stillbirth is a baby that does not breathe or show any other sign of life after being completely separated from its mother.

still-born, *a.* Add: Also **stillborn. 1.** (Further examples.) Also, formerly, born alive but not breathing.

1896 W. A. N. Dorland *Man. Obstetr.* II. vi. 706 It is a very common occurrence for a child to be born with the respiratory functions in abeyance: such a child is said to be asphyxiated. If efforts at resuscitation prove ineffectual, it is said to have been stillborn. 1899 H. D. Chapin in C. Jewett *Practice of Obstetrics* xxviii. 617 Fœtal death must be distinguished from asphyxia... In the latter the heart is pulsating, reflexes are present, and there may be feeble attempts at respiration... The distinction between a dead born and a still born infant can usually be made by the rapid fall of rectal temperature in the former. 1936 O'D. Browne *Man. Pract. Obstetr.* xix. 170 Technically, if an infant is born alive but never breathes, it is said to be stillborn; if the heart has ceased to beat after birth, deadborn. 1953 *Act* 1 & 2 *Eliz. II* c. 20 § 41 'Still-born child' means a child which has issued forth from its mother after the twenty-eighth week of pregnancy and which did not at any time after being completely expelled from its mother breathe or show any other signs of life, and the expression 'still-birth' shall be construed accordingly. 1955 W. P. D. Logan in Holland & Bourne *Brit. Obstetr. & Gynæcol. Practice: Obstetrics* xxxix. 1140 In certain countries children born alive but dying within a stipulated number of days are registered as stillborn.

B. as *sb.*, a still-born child. Also *fig.*

1913 J. London *Let.* 30 Jan. (1966) 369 For goodness sake, don't let's have a still-born of it. 1977 *Lancet* 30 July 257/1 She gave birth to a pair of male twins, one of which was a stillborn with no malformations. The other boy is normal and healthy.

stilleite (ʃti·lə,əit). *Min.* [ad. G. *stilleit* (P. Ramdohr in *Geotektonisches Symp. zu Ehren von H. Stille* (1956) 482), f. the name of Hans W. *Stille* (1876–1966), Ger. geologist: see -ite[1].] Native zinc selenide, ZnSe, found as grey or black cubic crystals.

1957 *Amer. Mineralogist* XLII. 584 (*heading*) Stilleite. 1970 *Mineral. Abstr.* XXI. 316/1 Mix crystals of ZnSe.., known as the mineral stilleite, and CdSe..were synthesized at 800° to 1200°C.

still-fishing, *vbl. sb. N. Amer.* [f. Still *a.* + Fishing *vbl. sb.*[1]] The practice of fishing from one spot, esp. with a baited line.

1883 *Century Mag.* July 383/1 The Floridians..use a long rod or pole for still-fishing. 1902 J. Turner-Turner *Giant Fish Florida* iii. 49 Still-fishing, the old-fashioned method of angling, is practised under the lea of some islands only on days that do not permit of your getting out into the Pass. 1963 P. A. Parsons *Compl. Bk. Freshwater Fishing* xxiii. 250 Crappies may be caught by drifting, trolling, still-fishing, and casting. 1971 *Islander* (Victoria, B.C.) 7 Nov. 13/2 Which method calls for the most skill—bait fishing (sometimes called still fishing) or artificial lure angling?

Hence (as back-formation) **still-fish** *v. intr.*

1903 *Outing* XLII. 716/1 We could..troll or cast or still-fish with light tackle. 1953 *Sunday Sun Mag.* (Baltimore) 25 Oct. 28/4 Whether you fish on a comfortable party boat..or simply sit on a piling and still-fish,..it's the best kind of relaxation there is. 1974 M. Hoyt *Thirty Miles for Ice Cream* i. 4 We still-fished for perch.

still-house. (Later U.S. examples.)

1834 A. Pike *Prose Sk. & Poems* 24 Our party reached the still house in the valley. 1884 'C. E. Craddock' *In Tennessee Mountains* II. 118 Josiah Tait had put his troubles in to soak at the still-house. 1927 *Greensboro* (N.C.) *Daily News* 24 Apr. II. 7/5 One time I was sent fur to bury a feller who had been found dead by a still house over yon side the Roan. 1976 *Lancet* 18 Dec. 1358/2 Mountaineers' distrust of government for the past half century has had vastly more to do with the antisocial environmental, economic, and developmental policies of half a dozen Federal agencies..than with overturning mash barrels and knocking down stillhouses.

still-hunt, *sb.* Add: **2.** (Earlier and later examples.) Also *transf.* and *fig.*

1828 M. S. Bidwell *Let.* in *Toronto Publ. Libr. MSS* B104. 69 Under the guidance of Mackenzie, who did not conduct himself with the caution and reserve of a new member, the House went on a still hunt for grievances. 1876 *N.Y. Tribune* 28 Aug. 4/4 It will be well for the Republican managers to bear in mind that a 'still hunt' is Gov. Tilden's favorite campaign method. 1916 'B. M. Bower' *Phantom Herd* ii. 32 I'm out on a still hunt for some real boys. 1936 H. Hagedorn *Brookings* ix. 131 He took to riding afternoons in Forest Park along the western edge of the city, on a persistent, still-hunt for a site. 1948 E. N. Dick *Dixie Frontier* 241 Sometimes a candidate..slipped off and went on a 'still hunt'; that is, he visited the people house-to-house and attended small gatherings unheralded.

still-hunt, *v.* **2.** (Earlier and later examples.)

1858 *Harper's Mag.* Oct. 615/2 An old woodsman..had been, without success, still-hunting. 1863 E. H. Walshe *Cedar Creek* 107 You see I'm often away trapping or still-hunting. 1942 W. Faulkner *Go down, Moses* 149 Once, still-hunting with Walter Ewell's rifle, he saw it [*sc.* a bear] cross a long corridor of down timber where a tornado had passed.

Stilling (ʃti·liŋ), *sb.*[3] *Ophthalm.* The name of Jakob *Stilling* (1842–1915), German ophthalmologist, used *attrib.* and in the possessive with reference to a test for colour-blindness based on the use of pseudo-isochromatic plates which he devised (*Arch. Ophthalmol.* (1879) VIII. 164).

1896 J. E. Jennings *Color-Vision & Color-Blindness* vii. 75 Stilling's plates are of practical value, and should be used in conjunction with the worsted test. 1925 M. Collins *Colour-Blindness* iii. 59 These students were examined [for colour blindness] in the course of the ordinary laboratory period by means of Stilling's Tables, and some of them failed completely to pass the tests. 1935 *Discovery* Jan. 15/2 Colour blindness tests (such as the Edridge Green or Stilling tests) are not much use in choosing men for this work. 1954 S. Duke-Elder *Parsons' Dis. Eye* (ed. 12) xx. 345 Stilling's original tests have now been largely replaced by Ishihara's.

still life. Add: **a.** (Earlier *attrib.* examples.) Also *fig.*

a 1784 G. A. Stevens *Let.* in T. Wilkinson *Mem.* (1790) IV. 196 My existence now cannot properly be called living, but what painters term *still life*; having since February 13, been confined in this town gaol for a London debt. 1821 P. Egan *Life in London* II. i. 156 It was not the still-life Beauty of the Sculptor and the Artist. 1831 F. Reynolds *Playwright's Adventures* vii. 112 This still-life personage, devoting the whole of her mind and time to her pianoforte.

b. A painting of such objects. Pl. *still lifes.*

1957 *Encycl. Brit.* XXI. 408/1 The first signed and dated pure still life..was painted in 1504. 1961 R. B. Long *Sentence & its Parts* ix. 206 Sometimes regular plurals replace even firmly established native irregular plurals: for example, in *she does still lifes*. 1970 *Oxf. Compan. Art* 1097/1 The development of the typical still life took place mainly in the Netherlands. 1981 *Daily Tel.* 30 Dec. 10/3 After the war, the artist returned to his delicate, vague, surrealist landscapes and little still-lifes.

Still's disease (stilz). *Path.* [Named after Sir George Frederic *Still* (1868–1941), English physician, who first described the disease in 1896 (*Med.-Chir. Trans.* (1897) LXXX. 47).] A condition affecting children, characterized by arthritis and ankylosis and now believed to be a form of rheumatoid arthritis.

1905 *Lancet* 27 May 1424/2 Though Still's disease was of the nature of a multiple joint affection no changes in the bones or the cartilages were observed. 1936 W. Sheldon *Dis. Infancy & Childhood* xxiii. 596 In some children the changes in and around the joints make up almost the entire clinical picture, but more often there is an accompanying lymphatic reaction, as shown by enlargement of the superficial lymphatic glands and enlargement of the spleen (Still's disease). 1974 A. Henry in R. M. Kirk et al. *Surgery* xv. 303 Rheumatoid arthritis occurring in childhood (Still's disease..) is probably not a separate entity.

Stillson (sti·lsən). Also **Stilson, stilson.** The name of Daniel Chapman *Stillson* (1830–99), used *attrib.* and *absol.* to designate an adjustable pipe wrench invented by him in 1869 and originally manufactured by his employers, the Walworth Company of New York.

1902 *Sears Catal.* 563/3 The Stillson Pipe wrench is too well known to require a lengthy description of same. 1903 *Sci. Amer.* 10 Oct. 266/1 Your kit is not complete unless it includes the famous Stillson wrench which is particularly adapted for turning out the best work without crushing the pipe in the least. 1945 *Walworth 1842–1942* (Walworth Company Inc.) 42 The sales of the Stillsons are highly sensitive to changes in economic conditions. 1960 E. L. Delmar-Morgan *Cruising Yacht Equipment & Navigation* xiv. 162 It used to be said by the old-time mechanics that refractory pipe joints suddenly made themselves loose when they saw you pick up a Stillson. 1961 J. Seymour *Fat of Land* v. 67, I spend half the night

..working in the blizzard or the deep frost with 'stilsons' ..and other esoteric devices. 1971 F. Hamilton *World Encycl. Dogs* 119 The undershot lower jaw [of a boxer].. gives a leverage similar to that of the plumber's Stilson wrench; once contact is made it cannot be broken; a pull just makes it tighter. 1978 J. Gores *Gone, no Forwarding* (1979) x. 62 Balland was out of his car, eight-inch Stillson wrench..in hand.

still-stand. Add: Also **stillstand. 1. b.** *Physical Geogr.* A condition in which there is no crustal movement in a region or no change in sea-level.

1896 *Bull. Geol. Soc. Amer.* VII. 393 The English School denies..the probability or even the possibility of a period of still-stand long enough for essentially complete subaerial denudation close to sea level, but assumes the possibility of a period of still-stand or of slight depression continuous and long enough to allow the sea waves to plane off the sinking lands. 1937 Wooldridge & Morgan *Physical Basis Geogr.* xv. 218 The cycle [of erosion] can only move uninterruptedly to its close if it coincides with a period of still-stand, *i.e.* of unvarying base-level, unaffected by major climatic changes. 1966 J. Wyckoff *Rock, Time, & Landforms* iv. 103 All believe that the reduction of vast, lofty highlands by erosion demands time spans of the order of 25 million years, during which there is virtual stillstand—that is lack of uplift. 1972 R. A. Davis *Princ. Oceanogr.* xx. 322 One of the most prevalent and significant types of evidence is the presence of terraces which were formed during still-stands of sea level.

still water. 2. (Earlier example.)

1832 W. D. Williamson *Hist. Maine* I. 66 The Metawamkeag..has frequent falls and intervening still-waters.

stillwellite (sti·lwelait). *Min.* [f. the name of F. L. *Stillwell* (1888–1963), Austral. geologist + -ite[1].] A borosilicate of lanthanons and calcium, $(Ln,Ca)BOSiO_4$, found as brown rhombohedral crystals.

1955 McAndrew & Scott in *Nature* 10 Sept. 509/2 The presence of stillwellite was discovered in 1954 during a mineralogical investigation of radioactive ore..34 miles east of Mt. Isa in north-western Queensland. 1971 *Doklady Acad. Sci. U.S.S.R.: Earth Sci. Sections* CCI. 182/1 The above data on our synthesis of the lanthanum analogs of stillwellite indicate that this borosilicate may be more widespread in nature than was previously thought.

stilt, *sb.* Add: **1.** Delete ? *Obs.* exc. *dial.* Occas. also with reference to other farm implements. (Later examples.)

1957 E. E. Evans *Irish Folk Ways* x. 129 The Irish were amazed when they first saw a ploughman with a Scots plough both driving the horses and holding the stilts. 1971 *Country Life* 20 May 1203/1, I take the 'stilts' of the big grass cutter and struggle behind it. 1973 *Ibid.* 22 Feb. 474/1 My going to the plough that morning wasn't the first occasion upon which I had set my hands to the stilts.

6. stilt heel, (a shoe with) a high heel; **stilt-heeled** *a.* (later examples); **stilt-root,** an aerial root, arising from the trunk or lower branches of a tree, and acting to provide support; hence **stilt-rooted** *a.*; **stilt-walker** (*a*) (earlier example).

1973 R. Rendell *Some lie & some Die* vi. 49 She was.. dressed..in..full, longish skirt, stilt heels. 1948 'P. Wentworth' *Traveller Returns* xi. 64 The sheer black stockings, and the stilt-heeled shoes. 1980 'L. Egan' *Motive in Shadow* iii. 39 She was wearing..stilt-heeled black patent leather pumps. 1894 F. W. Oliver tr. *Kerner's Nat. Hist. Plants* I. 756 Trees whose erect trunks are supported by tabular roots and those which are provided with stilt-roots may at the same time develop columnar roots from their branches. 1930 *Discovery* Nov. 381/1 No account of jungle vegetation..is complete without some mention of the trees with the curious stilt roots and those with the even stranger buttress roots. 1974 H. MacInnes *Climb to Lost World* xii. 221, I pointed..at a stilt-rooted tree which had grown up with stilts at least fifteen feet clear of the ground... Young stilt roots were growing into the swamp from its base. 1863 A. J. Munby *Diary* 20 May in D. Hudson *Munby* (1972) 162, I saw.. two young female acrobats or stiltwalkers..forlorn and pitiable in their satin shoes & spangles.

Stilton. Add: **a.** *Stilton cheese*: orig. also applied to similar cheeses made elsewhere, but since 1969 restricted to that made in the counties of Leicester, Derby, and Nottingham by members of the Stilton Cheese Makers Association. (Further examples.)

a 1864 J. Clare in *Sel. Poems & Prose* (1967) 25 He.. seldom got astride of a saddle save when he gave old Dobbin a holiday from the plough to carry his Dame to the Fair to sell her Stilton cheese. 1930 L. G. D. Acland *Early Canterbury Runs* 1st Ser. ii. 30 In the 'fifties and early 'sixties [in New Zealand] they got two shillings a pound for their butter, tenpence for cheese, and eighteenpence for stilton cheese. 1969 *Trade Marks Jrnl.* 5 Feb. 234/1 Stilton 831,407 Class 29. Cheese. The Chairman, The Stilton Cheese Makers Association, Melton Mowbray, Leicestershire.—28th Feb. 1962. 1973 *Leicester Mercury* 24 Dec. 8/1 The public in the States is being..misled into purchasing the Purity brand of Stilton cheese in mistaken belief that it is the genuine product produced in England.

b. (Earlier example.)

1826 E. Craven *Mem. Margravine of Anspach* I. v. 190 The Margrave had at his table good cream, and Stilton, or Berkeley hundred, made under my direction.

c. *attrib.*

1966 *Daily Tel.* 15 Nov. 15/2 The Stilton recipe..goes back to the 18th century... In seeking the trade mark, Stilton makers expressed fears that the same fate would befall the Stilton..as happened to the Cheddar. 1969 *Ibid.* 11 Apr. 19/5 Yesterday,..the Stilton Cheesemakers Association..obtained their trademark. 1971 *Sunday Times* (Colour Suppl.) 28 Mar. 36/1 Much the best way of tackling the half cheese on the table is with a knife with a long sharp pointed blade—not a Stilton spoon. 1973 *Leicester Mercury* 24 Dec. 8/1 (*heading*) Stilton men up in arms over 'spurious cheese'.

stilty, *a.* **2.** (Earlier example.)

1845 W. B. S. Taylor *Hist. Univ. Dublin* 339 The *stilty dignity* which is to be found in some other places.

‖ **stilyaga** (stilya·gā). Also **stilyag.** Pl. **-gi.** [*Russ.* (*colloq.*), lit. 'stylish person'.] In the U.S.S.R.: a young person who affects stylish dress as an expression of rebellion, nonconformity, etc.

1955 H. Hodgkinson *Doubletalk* 55 The *stilyag* who goes in for 'style' in his clothes. 1959 *Listener* 5 Feb. 236/2 Soviet Russia has her..'teddy-boys', who are there called *stilyagi*. *Ibid.*, The introspective, thoughtful, humble type has taken the place of the *stilyaga*. 1960 *Guardian* 25 Feb. 4/1 Young people in Russia..wear the 'stilyagi' clothes that mark them as non-conformists. 1965 'A. Burgess' in *Times Lit. Suppl.* 22 Apr. 317/2 A trip to Russia showed me that *stilyagi* behaved much like our own (as they were then) teddy-boys. 1973 T. Allbeury *Choice of Enemies* xxiv. 129, I had difficulty in changing the money because one of the 'stilyagi' recognised me.

‖ **Stimmung** (ʃti·muŋ). Also **stimmung.** [Ger.] Mood, spirit, atmosphere, feeling.

1909 W. M. Urban *Valuation* v. 123 We may cite certain impressionistic or symbolic styles where the general mood or *Stimmung* is almost palpable. 1923 R. Fry *Let.* 13 May (1972) II. 534 Words as you use them..give me more of what the Germans call *stimmung*..than painting can ever do. 1939 E. H. W. Meyerstein *Let.* 4 Apr. (1959) 222 He [*sc.* Eliot] cannot keep the poetic *stimmung* up as long as Bridges can. 1948 Wyndham Lewis *Let.* 18 Oct. 463, I could not help imbibing from my very American father much Stimmung, a certain sentiment, and a lot about the Civil War. 1961 *Times* 23 Mar. 16/6 The Riffelalp had its own *stimmung*. 1972 *Guardian* 12 Aug. 8 The 'Kaiserwalzer'..is..[an] emanation of the Viennese spirit, Stimmung. 1980 R. Adams *Girl in Swing* iv. 59 In memory the whole *Stimmung* changes and our recollections become like a story we have read before.

stimoceiver (sti·mosīvəɹ). [f. Stim(ulate *v.* + -o + Re)ceiver[1].] A radio transmitter and receiver implanted in the head which transmits information about the brain and receives signals which electrically stimulate the brain.

1967 *N.Y. Times* 22 Mar. 49/1 Dr. Delgado's laboratory is developing an instrument, called a 'stimoceiver', that takes advantage of the increasing sophistication of circuit technology and microminiaturization. 1968 J. M. R. Delgado et al. in *Jrnl. Nerv. & Mental Dis.* CXLVII. 331/1 The integration of the three-channel units for radio stimulation and EEG telemetry constitutes the stimoceiver. *Ibid.* 338/2 A new instrument called 'stimoceiver' has been developed for the stimultaneous multichannel recording and stimulation of the brain by FM radio waves in completely unrestrained subjects. 1979 *Amer. Jrnl. Orthopsychiatry* XLIX. 367/2 The grim prospects of a burgeoning psychosurgical technology for widespread telemetric surveillance of behavior via transponders, stimoceivers, and brain pacemakers *à la* Schwitzgebel, Meyer, and Delgado are adumbrated.

stimulable, *a.* Add: (Later example.) Hence **stimulabi·lity.**

1944 *Mind* LIII. 90 Most of the essayists, having a good deal of vitality themselves, are eminently stimulable. 1975 *Year Bk. Ear, Nose & Throat* 36 If nerve stimulability is lost.

stimulant, *sb.* **2. b.** Delete parenthetical comment and add earlier examples.

1848 Mill *Pol. Econ.* II. v. vi. 425 Among luxuries of general consumption, taxation should by preference attach itself to stimulants. 1859 —— *Liberty* v. 180 To tax stimulants for the sole purpose of making them more difficult to be obtained, is a measure differing only in degree from their entire prohibition.

stimulatory, *a.* and *sb.* Restrict *rare* to sense b in Dict. and add: **a.** (Later examples.)

1968 *Times* 21 Oct. 7/1 The buds exert a differential influence during the season, being inhibitory in winter and stimulatory in spring. 1975 *Nature* 7 Aug. 487/2 Either germinating seeds or soil may provide the substrates for growth of the stimulatory micro-organisms. 1981 *Times* 27 Aug. 9/2 Even with the stimulatory measures, however, the deficit does not seem likely to be excessive in the years to come.

stimulus. Add: **2. a.** (Earlier example.)

1791 W. Enfield *Hist. Philos.* II. i. 18 Among the philosophical works of Cicero, we do not now find his *Hortentius*,..which Augustine confesses operated upon his mind, as a powerful *stimulus* to the pursuit of wisdom.

3. c. *Psychol.* Any specific change in physical energy or an event (whether internal or external to the organism) which excites a nerve impulse and gives rise to a reaction.

1894 Creighton & Titchener tr. *Wundt's Human & Animal Psychol.* 16 The processes of motion which, by their operation upon our senses, give rise to sensations, we commonly denominate *stimuli*, or more particularly *sense-stimuli*... Thus we regard the sound-waves of the air or the light-waves set up in surrounding space as stimuli corresponding to our sensations of sound and light. 1919 J. B. Watson *Psychol.* i. 9 The goal of psychological study is the ascertaining of such data and laws that, given the stimulus, psychology can predict what the response will be; or..given the response, it can specify the nature of the effective stimulus. 1957 E. R. Hilgard *Introd. Psychol.* (ed. 2) 596/1 *Stimulus*, some specific physical energy impinging on a receptor sensitive to that kind of energy... Any objectively describable situation or event..that is the occasion for an organism's response. 1980 E. L. Deci in E. Staub *Personality* ii. 43 People do not respond to objective external stimuli; they respond to stimuli *as they perceive them*.

5. *attrib.* and *Comb.*, as (sense *3 c) *stimulus-complex, control, -error, intensity, -object, -pattern, -situation, -threshold, -value, -word*; **stimulus diffusion** (see quot. 1940); **stimulus generalization,** the fact that the response elicited by one stimulus can also be elicited by other stimuli associated with but not identical to the original; **stimulus-response,** abbrev. form of *stimulus-and-response*, used *attrib.* or as *adj.* to denote this process, esp. when considered as the basic element in the study of sense perception, learning or behaviour modification; = S-R s.v. *S 4 a.

1924 R. M. Ogden tr. *Koffka's Growth of Mind* 87 Fine differences in the stimulus-complex may lead to opposite reactions. 1954 W. H. Melching in E. L. Wikes *Secondary Reinforcement* (1966) II. ii. 143 The presence (or absence) of the buzzer during conditioning and extinction was assumed to be an important component of the stimulus complexes. 1956 *Psychol. Monogr.* LXX. v. 2/2 A method which involved a greater degree of stimulus control than has usually been achieved in research on discrimination. 1979 H. K. Rodewald (*title*) Stimulus control of behavior. 1940 A. L. Kroeber in *Amer. Anthrop.* XLII. 1 (*title*) Stimulus diffusion. *Ibid.*, It is the idea of the complex or system which is accepted, but it remains for the receiving culture to develop a new content. This somewhat special process might therefore be called 'idea-diffusion' or 'stimulus diffusion'. 1978 *Language* LIV. 207 If diffusion is to be thought of as operating between the sub-areas, it can only be 'stimulus diffusion'. 1909 E. B. Titchener *Text-bk. Psychol.* I. § 66.218 The observer tends to judge, not in terms of sensation, but in terms of stimulus... This error,..is known technically as the stimulus error. 1949 *Mind* LVIII. 452 The Stimulus-error and the Constancy Hypothesis are particular forms of this fallacy. 1943 C. L. Hull *Princ. Behavior* xii. 183 The reaction involved in the original conditioning becomes connected with a considerable zone of stimuli other than, but adjacent to, the stimulus conventionally involved in the original conditioning; this is called *stimulus generalization*. 1977 in Honig & Staddon *Handbk. Operant Behavior* xi. 316/2 Another possibility is that the mechanism underlying conditioned reinforcement is stimulus generalization. 1909 *Amer. Jrnl. Psychol.* XX. 4 A progression of stimulus intensities such that the differences of corresponding sensation between any consecutive pairs are equal to one another. 1933 *Psychol. Abstr.* VII. 538/1 The relationship between stimulus intensity and duration in the motor nerve of the frog. 1921 *Psychol. Rev.* XXVIII. 398 The dependence of a stimulus-object upon its setting is especially familiar in the case of contrasting colors or objects. 1970 *Jrnl. Gen. Psychol.* Apr. 151 The strength with which an unfamiliar stimulus-object elicits a particular mediational process. 1924 R. M. Ogden tr. *Koffka's Growth of Mind* iii. 137 The phenomenon corresponding to a given stimulus-pattern. 1950 *Mind* LIX. 187 A red shape presents a stimulus pattern that I react to immediately. 1921 *Psychol. Rev.* XXVIII. 390 The response member of a stimulus-response couple may consist of a group of reactions. 1927 L. L. Bernard *Introd. Social Psychol.* viii. 109 Tropism is not a stimulus-response process in the same sense that reflexes and instincts are. It makes use of stimulus-response mechanisms. 1957 E. R. Hilgard *Introd. Psychol.* (ed. 2) i. 21/1 Stimulus-response theory (or S-R theory, as it is commonly called) asserts that all behavior is in response to stimuli. 1964 E. A. Nida *Toward Sci. Transl.* iii. 40 In most actual instances of communication, verbal symbols enter into a chain of stimulus-response situations. 1980 *Dædalus* Spring 23 This evidence of central control over receptors ..affected the picture of the stimulus-response relationship that had dominated psychology for decades. 1923 Ogden & Richards *Meaning of Meaning* iii. 139 The excitation of part of an engram complex, which is called up by a stimulus..similar to a part only of the original stimulus-situation. 1977 A. Giddens *Stud. in Social & Polit. Theory* i. 76 All descriptive predicates, however 'theoretical', are learned in conjunction with definite stimulus-situations. 1897 C. H. Judd tr. *Wundt's Outl. Psychol.* 254 The stimulus from which the resulting psychical process, for example, a sensation, can be just apperceived, is called the *stimulus-threshold*. 1935 L. Bloomfield in C. Hockett *Bloomfield Anthol.* (1970) 310 His audience will respond only to the exact stimulus-value of his words. 1962 *Science Survey* XV. 251 Now the 'stimulus value' of a moving object depends not only on the actual capacity of the eye to detect and evaluate movement, but [etc.]. 1905 *Psychol. Bull.* II. 249 The influence of the grammatical form of the stimulus-word on the reaction is rather striking. 1971 *Jrnl. Gen. Psychol.* Apr. 281 Stimulus words were carefully selected in order to control for associative response frequencies.

sting, *sb.*[1] Add: **7*.** *Aeronaut.* A rod-like support used in wind-tunnel testing (see quot. 1933).

1933 *Gloss. Aeronaut. Terms* (*B.S.I.*) III. 20 *Sting*, a

light rod attached to and extending backwards from a body for convenience of mounting for test in a wind tunnel. 1948 *Jrnl. R. Aeronaut. Soc.* LII. 240/1 In this work the model was supported from the rear by means of a sting. 1959 *Engineering* 6 Feb. 188/1 The model support consists of a quadrant and sting.

sting, *sb.*[2] Add: **5. a.** Also in phr. *a sting in the tail* and varr.

1657 [in Dict.]. *c* 1820 Blake *On Homer's Poetry* in *Compl. Writings* (1972) 778 Those who will have Unity exclusively in Homer come out with a Moral like a sting in the tail. 1926 *Times* 7 Sept. 17/5 The sting of this book is in its tail. 1952 A. Christie *They do it with Mirrors* 192 Don't say it. I'm suspicious of these village parallels. They've always got a sting in the tail. 1979 A. Williamson *Funeral March for Siegfried* xxxii. 165 He..added a sting in the tail. 'Of course, if the murderer were one of you, an interloper would not be necessary.'

b. Freq. in phr. *to take the sting out of* (something). (Earlier and later examples.)

1860 *Bailey's Mag.* Oct. 42 Hayward's..rare defence completely took the sting out of the Surrey bowling. 1942 *R.A.F. Jrnl.* 3 Oct. 24 It has taken the sting out of the Adjutant. 1956 B. Holiday *Lady sings Blues* (1973) xix. 157, I was so happy I cried. People like Lena took the sting out of other little people. 1977 *Jrnl. R. Soc. Arts* CXXV. 464/1 This defence consists in establishing..that the derogatory words—or at least their sting—were true.

c. *Austral. slang.* (*a*) Strong drink, 'stingo'; (*b*) a drug, *spec.* one administered to a racehorse in the form of an injection.

1929 K. S. Prichard *Coonardoo* 60 'Misses his three square meals a day and sting,' Bob explained. 1949 L. Glassop *Lucky Palmer* 36 They're going to give it the sting. They'll hit it with enough dope to win a Melbourne Cup. 1958 F. Hardy *Four-Legged Lottery* 173 The 'smarties' soon found stings that didn't show on a swab. 1972 J. de Hoog *Skid Row Dossier* 4 You can share a bottle of sting (methylated spirits) down a lane.

d. *slang* (chiefly *U.S.*, orig. *Criminals'*). (*a*) A burglary or other act of theft, fraud, etc., esp. one that is carefully planned in advance and swiftly executed; (*b*) a police undercover operation designed to ensnare criminals.

1930 *Liberty* 20 Sept. 77/1 The hustlers would sit around planning their stings and I guess about half of the jobs pulled in southern Ohio that year started in my parlor. 1955 *Publ. Amer. Dial. Soc.* xxiv. 76 The *sting* we described involved a wallet obviously *on its feet*. 1975 *Courier-Mail* (Brisbane) 11 Mar. 6/8 A transaction between a jewellery salesman and a professed buyer with $230,000 in his pocket was intercepted yesterday by a cab driver who made off with the cash. Investigators believe the theft was a set-up 'sting'. 1976 *National Observer* (U.S.) 13 Mar. 6/3 The Sting also produced leads to three murders and several other unsolved major crimes. 1977 *Tel.* (Brisbane) 24 Aug. 25/2 'Sting' officers operated in old warehouses and run-down storefronts, developed close contacts with loose-lipped thugs who believe they dealt with fellow criminals. 1982 *Sunday Times* 14 Nov. 15/2 The Miami 'sting' was so well set up that it survived a remarkable breach of security. 1983 *Observer* 30 Jan. 10/2 His second reaction was to inform the American authorities and get their approval for an elaborate and costly 'sting'.

sting, *v.*[1] Add: **2. e.** *to sting* (someone) *for* (something): to induce (someone) to give (money, etc.) by begging or borrowing in an exploitative manner. *slang* (orig. *U.S.*).

1903 *Kansas City Daily Star* 21 Apr. 6/7 An undergraduate is no longer 'stuck' for a dinner, a seat at a play, a railroad ticket; he is 'stung'. 1940 N. Marsh *Surfeit of Lampreys* (1941) ii. 34 We hope to sting Uncle G. for two thousand [pounds]. 1973 Wodehouse *Bachelors Anonymous* iii. 26 He wants to make a touch... He even stung me for a bit the other day. 1976 P. Cave *High Flying Birds* iv. 46 You still letting that bum sting you for drinks?

f. To swindle or overcharge (someone); to involve (someone) in financial loss. Freq. in *pass.* Cf. sense 2 d in Dict. *slang* (orig. *U.S.*).

1905 [see *bundle *sb.* 2 h]. 1922 S. Lewis *Babbitt* iii. 27 Guess I'll have to get down to the office now and sting a few clients. 1923 *Daily Mail* 22 Jan. 8 [He] told me.. he stood to lose some enormous number of millions of marks if Germany went *phut*... He seemed hurt when I said I was very glad if he got stung for trading with the enemy. 1927 Wodehouse *Small Bachelor* vii. 121 'How much did you pay?' 'Three hundred dollars.' 'You were stung... The stock is so much waste paper.' 1943 K. Tennant *Ride on Stranger* vii. 62 In this world you've got to sting or get stung. 1955 M. Allingham *Beckoning Lady* v. 82 You sting 'em when the time comes. 1974 'E. Lathen' *Sweet & Low* xi. 108 Big names do not like getting stung. 1981 *London Mag.* July 15 I've no idea how much her son pays her... I like to think she's really stinging her son.

stingaree. (Earlier example.)

1838 *Papers Mirabeau Buonaparte Lamar* (1922) II. 87 To crown the whole Sergeant Bryant was cut on the foot with an oyster shell, and Mr. Edington was stung by a Stingaree.

stinge (sti·ndʒ). [Back-formation from Stingy *a.*] A stingy person. Hence as *v. intr.* (*rare*), to behave in a stingy manner.

1914 'Bartimeus' *Naval Occasions* xxi. 185 Accustomed to tribute tendered with a lavish hand, Arabella decided that this must be a 'proper stinge'. 1937 *Boy's Own Paper* 3 Dec. 143/1 The barber can't stand stinges. 1937 V. Sackville-West *Pepita* II. iii. 205, I couldn't see why a person ready to spend hundreds of pounds should be equally ready to stinge over a stamp or a ball of string. 1977 C. Boylan in D. Marcus *Best Irish Short Stories* II. 113 Elizabeth had got herself a job. 'Well, I had to... My, Morgan has become a stinge.'

stinger[1]. Add: **3.** Also *Austral.*, an exceptionally hot or cold period of time.

1899 'S. Rudd' in Murdoch & Drake-Brockman *Austral. Short Stories* (1951) 103 My! it'll be a stinger to-night. 1942 E. Langley *Pea Pickers* (1958) II. vii. 167 The next day was a stinger... It dawned sultry red.

4. A long structure attached to the stern of a pipe-laying barge which supports the pipe as it enters the water and prevents it from buckling.

1958 *Offshore Drilling* Oct. 11/2 The 'stinger' is final cradle [*sic*], submerged 85 feet off the stern of the lay barge, which holds the pipe to a 2500 ft. radius to prevent any undue strain during the process of lowering it into the underwater trench. 1966 M. J. Lamb in *Exploiting the Ocean* (Marine Technol. Soc.) 296 A 'stinger' is used in deep water to limit the sag in the pipe. 1969 [see *overbend *sb.*]. 1976 *Offshore Platforms & Pipelining* 6/3 Key changes include..the addition of 160-ft truss-type stinger in place of the usual pontoon stinger.

stinger[3] (sti·ŋəɹ). Corruption of *stengah. Also used as the name of various other mixed drinks or cocktails (see quots. 1973, 1976).

1901 *Scribner's Mag.* Jan. 106 Two 'stingers' were brought... A 'stinger'..is a noggin of Scotch whiskey, enlivened by much or little, according to individual taste, of the local buzz-water. 1903 [see *stengah]. 1916 H. L. Wilson *Somewhere in Red Gap* ix. 376, I found 'em in the palm grill, or whatever it's called, drinking stingers. 1928 [see *round *sb.*[1] 20 a]. 1942 D. Powell *Time to be Born* (1943) x. 242 Have another daiquiri... Or change to a stinger. 1961 I. Fleming *Thunderball* xv. 156 After they had had coffee and a stinger at the bar they separated and went to the [gaming] tables. 1973 *Sat. Rev. Society* (U.S.) May 45/1 B & B Stinger. 3 parts B & B Liqueur, 1 part White Creme de Menthe. 1976 *Scotsman* 24 Dec. (Weekend Suppl.) 3/7 A Stinger..is a better drink, being creme de menthe well laced with brandy, and stronger.

stinging, *ppl. a.* Add: **1. a.** *stinging lizard*, one of several North American lizards, esp. a spiny lizard of the genus *Sceloporus*, also called a scorpion.

1870 J. C. Duval *Adventures Big-Foot Wallace* xlv. 294 They chaw tobacco and drink whiskey even in the wintertime, when the 'cow-killers' and stinging-lizards are all frozen up. 1889 H. H. McConnell *Five Years a Cavalryman* 77 The 'scorpion' or 'stinging lizard' abounds. 1926 J. K. Strecker in J. F. Dobie *Rainbow in Morning* (1965) 61 The true scorpion is popularly called a 'stinging lizard', this misnomer being in common use throughout the state of Texas.

stingo. Add: **b.** *fig.* Vigour, energy, vim; *to give* (a person) *hot stingo* = *to give it hot* s.v. Hot *a.* 11 d. ? *Obs.*

1885 *Punch* 22 Aug. 86/3 It's rare fun, by Jingo! I give 'em hot stingo. 1927 *Daily Tel.* 19 July 15 To keep in good trim and add stingo to your efforts in sport. 1927 Galsworthy in A. A. Horn *Ivory Coast* Foreword 5 A gorgeous book..full of sheer stingo. 1928 *Observer* 18 Mar. 23/3 Some shanties, sung by Raymond Newell and a chorus, are full of stingo.

stingy, *a.* Add: **2.** (Earlier example.)

1781 J. Woodforde *Diary* 22 Feb. (1924) I. 302, I was very stingy this morning alias in a bad humour and made Nancy uneasy by my talking.

4. d. Meagre, spare, circumscribed.

1927 E. M. Forster *Aspects of Novel* viii. 205 [James's] characters..are constructed on very stingy lines. They are incapable of fun, of rapid motion, of carnality.

6. Of the brim of a hat: narrow. Also as *sb.*, a narrow-brimmed hat. *U.S.* (*Black English*).

1965 *Liberator* Aug. 23/2 He was neat from toe to stingy. 1969 N. Cohn *AWopBopaLooBop* (1970) ix. 86 He wore Stingy Brim straw hats, tight pants, lurid shirts.

stink, *sb.* Add: **1. c.** A contemptible person, a stinkard. *slang.*

1916 Joyce *Portrait of Artist as a Young Man* 8 Rody Kickham was a decent fellow but Nasty Roche was a stink. 1918 E. Pound *Let.* 4 June (1971) 137 Meredith is, to me, chiefly a stink. I should never write on him as I detest him too much ever to trust myself as critic of him. 1950 R. Moore *Candlemas Bay* v. 281 And the rest of you little stinks, shut up, too! 1972 D. Devine *Three Green Bottles* 102 That stink, Celia Armitage, had somehow found out.

3. a. Also, a row or fuss; a furore (later examples). Now chiefly in phrs. *to raise* (*kick up, make*) *a stink.*

1907 [see *jim-hickey]. 1913 Kipling *Diversity of Creatures* (1917) 293 We *mustn't* be tried! It'll make an infernal international stink. 1942 *Tee Emm* (Air Ministry) II. 81 Do you do it merely because there's a stink if you don't? 1948 'N. Shute' *No Highway* ii. 31, I remember the Russians kicking up a stink. 1959 'M. Cronin' *Dead & done With* iv. 56 The first thing he'd do when he got back was see his M.P. and kick up a stink. 1976 L. Sanders *Hamlet Warning* (1977) iii. 31 [She] commanded a world press. She could raise a tremendous stink if she chose to do so.

b. *like stink*, furiously, intensely. Cf. Like *adv.* 1 b. *colloq.*

1929 R. C. Sherriff *Journey's End* I. 40 If you see a Minnie coming..you have to judge it and run like stink sometimes. 1938 M. Allingham *Fashion in Shrouds* xv. 240 It's raining like stink. 1945 'P. Woodruff' *Call Next Witness* II. v. 114 He clapped in his heels and rode like stink. 1955 M. Allingham *Beckoning Lady* iii. 40 The telephone's here..and when it rings you have to run like stink before the caller gives up. 1972 D. Devine *Three Green Bottles* 11 She wasn't really clever, she just worked like stink.

c. In other colloq. phrases (parallel to the use of *hell* and similar words).

1942 *R.A.F. Jrnl.* 16 May 26, I bet they've been giving old Jerry stink this afternoon. 1977 I. Shaw *Beggarman, Thief* I. ii. 21 We'd've been in a stink of a mess without him.

4. Now used principally of Chemistry.

1914 'I. Hay' *Lighter Side School Life* iv. 116 Master Nixon..had pointed out that it would be a good thing to enrol as a member some one who understood 'Chemistry and Stinks generally'. 1928 R. Cullum *Myst. Barren Lands* xi. 107 You can't afford to use up the source that gives you a living so you can carry on with your stinks. 1945 'R. Crompton' *William & Brains Trust* vii. 129 'English isn't bad, 'cause ole Sarky can't see what you're doin' at the back, an' Stinks isn't bad, 'cause you can get some jolly good bangs if you mix the wrong things together.' 1955 [see *lab *sb.*[1]]. 1961 A. Wilson *Old Men at Zoo* i. 37 Eventually..the laboratory work will be on a scale that will make this place look like a school stinks room.

5. stink beetle = *stink bug*; **stink bomb,** a small hand-missile which emits a nauseating smell when broken, typically thrown by schoolboys; also *transf.*; **stink-bug,** substitute for def.: a shield bug of the family Pentatomidæ, which includes many species that feed on plants and eject a strong-smelling liquid if attacked; (earlier example); **stink-finger**: in coarse slang phr. *to play* (*at*) *stink-finger* (see quot. 1903) (now *rare* or *obs.*); **stink-fish,** (*a*) *S. Afr.* = *bamboo-fish* s.v. *bamboo *sb.* 2; (*b*) *Ghana* = *stinking fish* (*b*) s.v. *stinking *ppl. a.* 2 a.

1889 H. Vaughan-Williams *Visit to Lobengula* (1947) xxv. 162, I must mention the stink beetles, as they are called. They all emit the most horrible stench when killed or even touched. You get them all over South Africa. 1979 *Jrnl. Arid Environments* II. 101 When stink-beetles of the genus *Eleodes* are placed in a bottle.., their fumes will kill other insects placed inside the bottle. 1915 D. O. Barnett *Let.* 26 June 192 It seems that the tobacco stores had amalgamated with a stink-bomb dépôt. 1922 A. Haddon *Green Room Gossip* iv. 117 At this juncture there was a good deal of sneezing and coughing in the auditorium... Both stink bombs and 'electric snuff' were thrown from the gallery. 1958 'J. Byrom' *Or be he Dead* xiii. 167 He lobbed this [conversational] grenade..as innocently as a child with a stinkbomb. 1974 D. Ramsay *No Cause to Kill* I. 38, I used to sit in the library wishing I had a stink bomb to set off under her nose. 1877 Bartlett *Dict. Americanisms* (ed. 4) 647 Squash-bug... A small yellow bug, injurious to the vines of squashes, melons, and cucumbers... In Connecticut, called a stink-bug. 1903 Farmer & Henley *Slang* VI. III. 369/1 *To play at stinkfinger*,..to grope a woman. 1934 H. Miller *Tropic of Cancer* 282, I had no Odette Champsdivers with whom to play stinkfinger. [1902 *Marine Investigation in S. Afr.* I. 116 Bamboesvisch. Stinkvisch. Scarce in Cape Town market, but common in Saldanha Bay.] 1913 Stink-fish [see *bamboo-fish* s.v. *bamboo *sb.* 2]. 1962 C. Baeta *Prophetism in Ghana* ii. 17 Adherents of the Twelve Apostles Church..are not allowed to eat pork, stink-fish, shark's meat or snails.

stink, *v.* Add: **2. b.** Also *to stink of* (or *with*) *money*: to be 'offensively' rich. *slang.*

1877 E. Peacock *N.W. Lincs. Gloss.* 239/2 A very proud man is said to '*stink wi' pride*', a very rich one to '*stink o' brass*'. 1922 Joyce *Ulysses* 9 Touch him for a guinea. He's stinking with money. 1932 I. Brown *Marine Parade* xii. 152 We must do our best. He stinks of money. Will you fix up about rooms and for God's sake let's have a decent dinner.

c. To exhibit or savour of moral (artistic, etc.) decay. Of persons: also, to be despicable or completely incompetent. Of actions, phenomena, etc.: also *spec.* in phr. *to stink to* (*high*) *heaven*. *colloq.*

1934 J. T. Farrell *Young Manhood of Studs Lonigan* viii. 121, I watched you guys go through signal practice. You stunk! 1956 H. Kurnitz *Invasion of Privacy* xv. 99 She regarded Zorn bleakly. 'You stink, Michael Zorn,' she said. 1963 'D. Cory' *Hammerhead* iv. 61 Sofia was the actress of the family. I stink.

1936 *Metronome* Feb. 61/2 *Stinks*, what one pub thinks of another's tunes. 1940 'N. Blake' *Malice in Wonderland* I. v. 61 Big Business does rather stink, doesn't it? 1959 H. Hobson *Mission House Murder* iii. 21 Rock an' Roll came in..now that's out—that stinks too. 1963 C. D. Simak *They walked like Men* ix. 53 'How did you know that?' 'Just a guess,' I said. 'This whole thing stinks to heaven.' 1973 *Times* 23 May 16/4 The affairs of Lonrho stunk to high heaven. 1979 R. Jaffe *Class Reunion* (1980) II. ii. 196 Chris would make it be like the old days. But the old days had stunk too.

5. b. With *up*. To cause (a place) to stink. Also *fig. colloq.* (orig. and chiefly *U.S.*).

1956 B. Holiday *Lady sings Blues* (1973) vii. 65 The manager got panicky and began to holler at me that I was stinking up his Grand Terrace. 1967 C. Drummond *Death at Furlong Post* xi. 140 Harassed fat women cooking *sauerbraten* and stinking up the place with the smell of vinegar. 1977 D. MacKenzie *Raven & Ratcatcher* iii. 44

A rumour..that..I was allowed to resign rather than stink-up the fair name of the Serious Crimes Squad. 1979 R. Jaffe *Class Reunion* (1980) III. iii. 319 He was sweet even though he did stink up her bedroom because he always forgot to open the window.

stinkaroo, var. *stinkeroo.

stinker. Add: **1.** (Later examples as a term of abuse.) Also in weakened uses, esp. banteringly and in mock-contempt. Now *slang.*

1922 [see *house *sb.*[1] 4 c]. 1936 M. Allingham *Death of Ghost* xx. 237, I will show that stinker! 1949 A. Huxley *Let.* 6 Mar. (1969) 593 Saying what a stinker he is, both in bed and out. *c* 1951 T. Roethke *Selected Lett.* (1968) 170 After your generous words, I feel a terrible stinker questioning everything. 1962 Wodehouse *Service with Smile* viii. 120 Is that you, Stinker? 1975 *Daily Mail* 30 May 3/6 A gang of 'real stinkers' have raided a top wartime air ace and stolen his most prized souvenir—a 6ft. German propellor.

3. Also formerly, a rank cigar or cigarette. (Earlier and later examples.)

1834 *Proc. Geol. Soc.* II. 21 The greater part of the workings are only shallow pits, touching merely the sulphureous beds, locally called 'stinkers'. 1924 N. Coward *Rat Trap* I. 11 There are cigarettes in that silver box, Keld; stinkers on one side, and opulent Turkish on the other. 1935 Wodehouse *Luck of Bodkins* iv. 42 Have you such a thing as a stinker?..And a match? 1961 J. W. Anderson *Fur Trader's Story* iii. 23 At Moose Factory, I saw the last of the sulphur matches ('stinkers' we used to call them) which were in their day considered a great advance on the striking of flint on steel to make fire. 1970 B. Cartland *We danced All Night* vii. 198 Virginian cigarettes were first called 'stinkers', then 'gaspers', and were considered a little vulgar.

6. *fig.* **a.** A strongly-worded letter; a disagreeable review or other communication. *slang.*

1912 *World's Work* Apr. 509/1 The principal content of this mail proved to be, as usual, a very long and wordy 'stinker'. 1936 Wodehouse *Young Men in Spats* vi. 157 For weeks..Stiffy had been yearning to write an absolute stinker to old Wivelscombe, telling him exactly what he thought of him. 1945 L. Durrell *Spirit of Place* (1969) 81, I was afraid..that you would write me a stinker calling me a peach fed sod. 1953 E. M. Forster *Hill of Devi* 228, I composed a stinker... H. H. supervised it and decried any attempt at moderation.

b. More widely, something repugnant because of its difficulty or unendurable nature. *colloq.*

1917 Kipling *Diversity of Creatures* 241 The second stanza..of that Ode is what is technically called a 'stinker'. 1941 S. J. Baker *Dict. Austral. Slang* 72 *Stinker*, a disagreeable, highly unpleasant and often humid day. 1947 *Penguin New Writing* XXIX. 100 During the war the standard [of films] was undeniably high, with the exception of a few frank stinkers such as *Half-Way House*. 1959 D. Hewett *Bobbin Up* 170 Already the sky was pale and smoky with the promise of 'another stinker'. 1967 *Listener* 9 Feb. 196/2 Stylistically, the Royal Victoria Hospital is indeed a stinker. 1979 H. R. F. Keating *Inspector Ghote draws Line* xv. 149 The headache..has become a real stinker now.

c. In phr. *to come a stinker* = *to come a cropper* s.v. Cropper[3] (chiefly in the work of P. G. Wodehouse). *slang.*

1923 J. Manchon *Le Slang* 293 *To come a stinker*, ramasser une pelle. 1936 Wodehouse *Laughing Gas* iv. 41 And then..the engagement went and busted itself up. One moment, it was buzzing along like a two-year-old... The next, it had come a stinker. *a* 1975 —— *Sunset at Blandings* (1977) iii. 26 Lack of the stuff [*sc.* wealth] is always the rock on which the frail craft of love comes a stinker where Blandings is concerned.

stinkeroo (stiŋkĕrū·). *slang* (orig. *U.S.*). Also **stinkaroo.** [f. Stink *sb.* or *v.*: see *-eroo.] Something of a very low standard; a very bad performance. Also, a furore or 'stink'.

1934 D. Runyon in *Collier's* 24 Nov. 8/3 The contest.. turns out to be something of a disappointment, and, in fact, it is a stinkeroo, because there is little skill and no science whatever in it. 1946 *Sun* (Baltimore) 21 June 16/4 The fight last night was a 'stinkeroo' and should be investigated. 1951 J. B. Priestley *Festival at Farbridge* ii. 54 They've sunk two-and-a-half million dollars in this new stinkaroo that opens tonight. 1958 B. Nichols *Sweet & Twenties* ix. 120 It caused an absolute stinkaroo in the women's colleges. 1973 *Daily Tel.* 24 Nov. 11/7 Many of the critics' notices [of the play] were unfavourable, referring to it as 'an antique', 'a puerile affair' and 'a stinkaroo'.

stinking, *ppl. a.* Add: **1. e.** As an intensifier: 'offensively', in *stinking drunk, rich* (somewhat *derog.*); also *absol.* and const. *with*, having too much (money, etc.). Cf. sense *2 b of the vb. *colloq.*

1887 *Lantern* (New Orleans) 12 Feb. 3/1 Dey had four bottles er booze and got stinkin' 'fore two o'clock. 1926 L. H. Nason *Chevrons* iii. 96 He went off and got stinking drunk. 1934 E. Waugh *Handful of Dust* iii. 115 'Tight that night.' 'Stinking.' 1940 E. Pound *Cantos* lvi. 62 Sangko stinking with graft. 1945 E. S. Gardner *Case of Golddigger's Purse* ii. 9 Not only is he rich but he's *stinking* rich. 1956 A. Christie *Dead Man's Folly* ii. 27 Stinking with money—absolutely stinking! 1965 *New Statesman* 19 Mar. 462/2 The father meets old cronies and gets stinking drunk. 1978 N. Marsh *Grave Mistake* iv. 111 She was in affluent circumstances, stinking rich in fact. 1980 D. Bogarde *Gentle Occupation* i. 22 The only thing to do is to get absolutely stinking... It was the best thing in the world for despair: a good skin-full.

2. a. stinking fish, (*a*) in allusion to the phr. *to cry stinking fish* (sense 1 a): something worthless or rotten; (*b*) *Ghana* (see quot. 1973).

1935 E. R. Eddison *Mistress* vi. 98 Have I not proof of 's loyal mind within reason: his refusing on't when Lessingham did offer it? Nay, 'twas but stinking fish then: 'twas under suzerainty. 1968 J. Abruquah *Torrent* ii. 29 You are treated nice when you are a stranger instead of having little bits of dried stinking fish and kenkey thrown at you. 1973 K. A. Sey *Ghanaian Eng.* vii. 90 *Stinking fish*, fish preserved with salt and used for seasoning 'soups' and 'stews'. 1981 N. J. Crisp *Festival* xi. 260 Vincent Consel said to his editor, 'I detect the faint aroma of stinking fish.'

stinkingly, *adv.* Add: **2.** Excessively, extraordinarily. Cf. sense *1 e of the ppl. a. *colloq.*

1906 R. Fry *Let.* 14 Jan. (1972) I. 248 The Raphael.. [is] stinkingly pretty and gives me no aesthetic pleasure. 1951 M. Kennedy *Lucy Carmichael* II. i. 65 He is.. frightfully good-looking..and stinkingly rich. 1957 J. Braine *Room at Top* xxvi. 207 We got really stinkingly sozzled... I don't believe that I've ever drunk so much before. 1979 N. Freeling *Widow* xxvi. 158 A bungalow, very large and super and stinkingly rich.

stinko (sti·ŋko), *predic. a. slang* (orig. *U.S.*). [f. Stink *v.* + *-o[2].] **a.** Of a very low standard; very bad. Cf. *stinkeroo. *rare.*

1924 F. Scott Fitzgerald *Let.* 20 Dec. (1964) 174, I thought *The White Monkey* was stinko.

b. Intoxicated; blind drunk. Also as quasi-advb., *stinko drunk*, (joc.) *paralytico*.

1927 *New Republic* 9 Mar. 72/1 The following is a partial list of words denoting drunkenness now in common use in the United States..stinko. 1936 P. G. Wodehouse *Laughing Gas* ix. 92 Are you really such a poor judge of form as to imagine that I am stinko? 1942 E. Waugh *Put out More Flags* iii. 182 'Darling, she was plastered.' 'Are you sure?' 'My dear, stinko paralytico.' 1954 G. Smith *Flaw in Crystal* 55 They generally come every night for a week and get stinko every night. 1960 S. H. Courtier *Gently dust Corpse* ii. 21 We ought to be getting stinko together today. An anniversary, see. 1974 D. Ramsay *No Cause to Kill* II. 132 Jessie's a lush. Stinko most of her waking time. 1976 *National Observer* (U.S.) 6 Nov. 12/6 Poor unfortunates who get stinko drunk every day on skid row.

c. Over-supplied (*with* money, etc.). *rare.*

1960 *Sunday Express* 28 Feb. 3/3, I should be a stockbroker or someone stinko with money.

† **Stinkomalee** (sti:ŋkŏmălī·). *Obs.* Also **Stinkomiles.** [Fanciful combination f. Stink *sb.* or *v.* + Trincomalee.] A disrespectful sobriquet of London University. Hence **Sti:ncomale·an** *a.*

1825 T. Hook in *John Bull* 25 Dec. 413/1 In consequence of the nature of the property [beside a stagnant pond] the first act of the Council has been to give a new and distinguishing name to the Institution—instead of the London College, or Carmarthen-street University, as heretofore, it is in future to be called—*Stinkomalee*! 1827 T. Creevey *Let.* 14 Apr. in J. Gore *Creevey Papers* (1963) xiii. 223 The enlightened moderns who are now founding Stinkomiles College at the end of Gower Street. 1844 A. H. Clough *Let.* 24 Nov. (1957) I. 141 It would be far better to be at Stinkomalee (the London University acknowledges that agnomen,..does it not?). 1849 —— *Let.* 15 Feb. in *Ibid.* I. 242, I have accepted the Stincomalean position. 1851 T. Arnold *Let.* 14 Sept. (1966) 205, I saw..that you had been made Professor of English at University College, London. Do you still call that institution by the irreverent name of Stinkomalee?

stink-pot. Add: **5. a.** A term of abuse for a person or (*rarely*) a thing. *slang.*

1854 T. Woolner *Let.* 4 Oct. in *Geo. Eliot Lett.* (1954) II. 176, I will not..display the filthy contaminations of these hideous satyrs and smirking moralists—..stink pots of humanity. 1916 Joyce *Portrait of Artist* v. 230 Go away from here, he said rudely. Go away, you stinkpot. 1928 D. H. Lawrence *Let.* Aug. (1932) 744 Whether I shall have the strength to put my nose into that stinkpot of an island [*sc.* England], I don't know. 1948 D. Ballantyne *Cunninghams* xx. 267 They can call me miserable old stinkpot. 1959 I. & P. Opie *Lore & Lang. Schoolch.* ix. 155 These syllables are used..to turn a verb or adjective into a descriptive noun, as:..fuss-pot, stinkpot, [etc.]. 1973 R. Ludlum *Matlock Paper* xii. 112 *Nowhere* does your signing this little stinkpot say you *agree* to retire from the scene.

b. A machine which emits foul exhaust fumes: esp. a truck or motor-boat. *slang.*

1972 B. Garfield *Line of Succession* (1974) II. 155 Mario had..a hatred of stinkpot powerboats. 1977 H. Fast *Immigrants* I. 55 They're gone now, all of them [*sc.* fishing-boats with sails]. Nothing but stinkpots—I'm sorry—oil burners. 1978 D. Bagley *Flyaway* xxi. 179 The truck broke through..and it killed them... Lousy stinkpots! Never have liked them except when I'm in a hurry.

stinkweed. Add: **c.** Any of several other plants with an unpleasant smell.

1932 [see *khaki bos, bush* s.v. *khaki D.]. 1954 *Sun* (Baltimore) 4 June 21/1 Due to the fact that its leaves, when crushed, are very disagreeable to get next to, the *ailanthus* is also known as 'stinkweed'. 1970 J. H. Gray *Boy from Winnipeg* 153 My job was to graze the cattle slowly..keeping them out of the stinkweed which, I was warned, spoiled the butter.

stinky (sti·ŋki), *a.* (*sb.*). [f. Stink *sb.* + -y[1].] = Stinking *ppl. a.* Also as *sb.*

1888 Kipling *Plain Tales* 238 Now we've got the [joss-]sticks mixed with a lot of glue, and they..smell stinky. 1949 N. R. Nash *Young & Fair* I. ii. 17 What if you had a stinky summer? 1958 S. A. Grau *Hard Blue Sky* v. 317 'And he does smell,' she said, 'phew.'..'Go on, stinky,' Annie told him. 1966 L. Kirstein *Rhymes & More Rhymes of PFC* 163 Stinky shorts, dead shoes, sodden shirt. 1972 D. Ramsay *Little Murder Music* 190, I think human nature is pretty stinky. 1982 *Observer* 25 Apr. 7/6 To get where he is and stay there he must have had to do some pretty stinky things. His is a pretty stinky country, after all.

stint, *sb.*[1] Add: **II. 6. a.** Also, a portion of land allotted for pasturing a limited number of sheep or cattle.

1849 *Gloss. Provincial Words Teesdale, Co. Durham* 125 *Stint*, a limited number of cattle gaits. 1904 in *Eng. Dial. Dict.* V. 768/1 The marshes of Skinburness, &c.,..were not enclosed in 1811 as were the commons, but were divided into stints, 400 being made out of 1,008 acres. 1954 M. Beresford *Lost Villages* vi. 204 Abandoned cornfields tumble first to grass and weed... The flocks which had been stinted could now have their stint enlarged. 1975 *Country Life* 11 Dec. 1676/2 As winter sets in, the salt marshes on the English side of the Solway will be grazed by..hill sheep from the Lake District. The 'stints' (pasturage for sheep and cattle) are owned by the Solway-side farmers.

7. a. Also, a period of time spent on a particular job; a turn (at doing something).

1955 S. Wilson *Man in Gray Flannel Suit* (1956) xxii. 170 After college had come a brief stint in the Army. 1957 *Economist* 21 Dec. 1073/1 No really outstanding executives for private business have ever been ready to take on a stint in the hardest jobs that industry in this country has to offer. 1965 *Listener* 24 June 933/2 This is the end of my stint for *The Listener*. 1976 H. Wilson *Governance of Britain* vii. 130 His three or four weeks' compulsory stint as 'Minister in Attendance' at Balmoral. 1978 S. Brill *Teamsters* viii. 307 He..then served a stint as a railroad brakeman.

stint, var. *stent *sb.*[5]

stipe[2] (stəip). *slang.* Also *rarely* **stip.** [Abbrev. of Stipendiary *a.* and *sb.*] **1.** A stipendiary magistrate.

1860 Hotten *Dict. Slang* (ed. 2), *Stipe*, a stipendiary magistrate. 1956 S. Hope *Diggers' Paradise* xvi. 141 The 'stipe'..had failed to find a permanent job to his liking in Britain. 1966 F. Shaw et al. *Lern Yerself Scouse* 65 *De Stipe*, the Liverpool Stipendiary Magistrate. 1978 *New Society* 28 Sept. 710/1 Roberts devoted the remainder of his..speech to remembering odd little incidents in the early career of the senior 'stip'.

2. A stipendiary racing steward. Chiefly *Austral.*

1922 *Daily Mail* 13 Nov. 11 The *Daily Mail* scheme for invigorating turf supervision... The plea for 'stipes' is a newspaper stunt. 1930 *Bulletin* (Sydney) 8 Jan. 35/1 And all the stipes and vets. and docs. galore..Wouldn't change that. 1963 *Truth* (Wellington, N.Z.) 9 July, Stipe should have acted [at Trentham races]. 1969 D. Francis *Enquiry* i. 13 Stipendiary Stewards, officials paid by the Jockey Club... The Stipe who had been acting at Oxford was notoriously the most difficult. 1977 *Australian* 15 Jan. 20 The racing page screamed Stipes Probe Jockey.

stiphado, var. *stifado.

stipple, *sb.* Add: **3.** *stipple-engraved* adj.

1936 [see *line-engraved* s.v. *line *sb.*[2] 32]. 1961 E. M. Elville *Collector's Dict. Glass* 77/1 A stipple engraved wineglass with the standing figure of a small boy with a bird on his wrist. 1973 *Country Life* 6 Dec. 1931 (Advt.), Stipple engraved goblet..circa 1785.

stipple, *v.* **1. c.** (Earlier example.)

1765 T. H. Croker et al. *Compl. Dict. Arts & Sciences* II. s.v. *Miniature*, The whole appears as if strippled [*sic*] or wrought with points.

stipulation. Add: **2. c.** *U.S. Law.* An agreement between opposing parties (or their counsels) relative to the course of a judicial proceeding; a requirement or condition of such an agreement.

1802 *South Carolina Rep.* (1817) II. 162 The Court opened a stipulation entered into between the insurers and the insured, by which it was agreed that one case should decide all others. 1828 *U.S. Supreme Court Rep.* XXVI. 448 The want of possession, if consistent with the stipulations of the parties,..has never been held to be, *per se*, a badge of fraud. 1876 *Ibid.* (1877) XCIV. 278 Stipulations between counsel relative to the course of proceeding in a cause pending in this court cannot be withdrawn by one party without the consent of the other. 1909 *Northeastern Reporter* LXXXVIII. 786/1 By that stipulation the defendants had the right of inspection before final acceptance of the goods. 1948 *Pacific Reporter* (1949) XCIX. 956/1 The contents of the proposed stipulation with reference to the repairs of the damage as to sidewalks, are merely instructions to workmen. 1977 *National Observer* (U.S.) 8 Jan. 18/1 Agreements to something he [*sc.* your counsel] calls 'stipulations' pour from his eager lips so thick and fast that soon a miasma of confusion covers you. Eventually you learn that a stipulation is a point or condition agreed upon between the litigants.

stipulative (sti·piulătiv), *a.* [f. Stipulate *v.* + -ive.] That stipulates or specifies as an essential condition. *spec.* in *stipulative definition* (Logic), the act of stipulating what a word, phrase, etc. shall be used to mean; a definition so stipulated.

1950 R. Robinson *Definition* ii. 19 By 'stipulative definition' I mean establishing or announcing or choosing one's own meaning for a word. 1956 J. Hospers *Introd. Philos. Anal.* i. 51 We are stating what *we* are going to mean by it... We are stipulating a meaning, and we have a stipulative definition. 1965 E. J. Lemmon *Beginning Logic* 33 A definition such as Df ↔ may be called a *stipulative* definition. 1976 A. R. Lacey *Dict. Philos.* 48 Stipulative definitions are *prescriptive* in that they prescribe how a word is to be used. 1981 P. A. Angeles *Dict. Philos.* 59 When Norbert Wiener coined the word 'cybernetics' he gave it the stipulative definition: 'the science of communication and control systems'.

stir, *sb.*³ Add: **1.** Also without article, esp. in phr. *in stir*. (Later examples.)

1926 [see *jolt sb. 4 b]. 1939 J. Steinbeck *Grapes of Wrath* ii. 19 When you been in stir a little while, you can smell a question comin' from hell to breakfast. 1970 G. F. Newman *Sir, You Bastard* ii. 79 Tasting stir, Goldfoy suddenly realized he was on the wrong side of thirty for acquiring the habit. 1977 'E. Crispin' *Glimpses of Moon* xii. 250 You get better conditions than that in stir.

2. *Comb.* (Designating) a person deranged, etc., by long imprisonment, esp. as *stir-crazy*. Also *fig.* *Criminals' slang* (chiefly *U.S.*).

1908 J. M. Sullivan *Criminal Slang* 24 *Stir crazy*, prison crazy, a man whose mind has become affected by serving long sentences. [illegible]

sti·r-fry, *v.* orig. *U.S.* [Stir *v.*] **a.** *trans.* Chiefly in Chinese cookery: to fry (meat, vegetables, etc.) rapidly on a high heat, while stirring and tossing them in the pan. Also *absol.*

1959 C. B. T. Lee *Chinese Cooking for Amer. Kitchens* 82 Add all ingredients except egg roll skin and beaten egg. Stir-fry for 3 minutes. *Ibid.* 102 Stir-fry onions for 2 minutes. 1969 *Guardian* 16 July 16/4 Stir-fry the vegetables. 1972 K. Lo *Chinese Food* i. 11 After the vegetables have been stir-fried for a minute or two the meat is returned to the pan and stir-fried together with the vegetables. 1976 'M. Delving' *China Expert* viii. 93 'You were in trouble with police,' she said, stir-frying thin strips of beef and vegetables. 1982 *Daily Tel.* 14 Jan. 15/4 Draw in prepared beef and stir/fry for 10 mins.

b. The vbl. phr. used *attrib.*

1959 *House Beautiful* June 146/1 [illegible] 1976 *Publishers Weekly* 19 Apr. 76/1 Soufflés, stir fry cookery, casseroles. 1981 J. Mann *Funeral Sites* xii. 77 Ian..had spent some time making a Chinese stir-fry dinner.

Also **sti·r-fried** *ppl. a.*; **stir-fry·ing** *vbl. sb.*

1959 C. B. T. Lee *Chinese Cooking for Amer. Kitchens* 64 *Chow* (stir-frying). This is the most popular method of Chinese cooking and is called either stir-frying or sautéing at high heat. *Ibid.* 65 Do not try to attempt more than two stir-fried dishes at one meal. 1972 K. Lo *Chinese Food* i. 45 All that quick-fried dishes require is a very short period of stir-frying over high heat with a couple of tablespoons of oil. 1980 *Redbook* Oct. 136/2 To save time and money in the kitchen, Denise does a lot of stir-frying and uses a slow cooker. 1981 *Sunday Tel.* 6 Dec. 13/5 Stir-fried shredded beef with bean sprouts and celery.

stirk. 1. (Later examples.)

[illegible] cattle,..64 accredited calves, 49 non-accredited calves and 81 store cattle and stirks... Charolais heifer stirks sold to £148,..Hereford heifer stirks sold to £107. 1978 *Morecambe Guardian* 14 Mar. 22 (Advt.), Rearing Calves and Stirks, Fat Cattle and Slaughter Cows. 1979 L. Derwent *Border Bairn* i. 15 Her brother, the shepherd..accepted me more or less as one of his flock. A yowe or a gimmer, a stirk or a stot.

Stirling¹ (stɜ̄·rlɪŋ). [The name of the Revd. Robert *Stirling* (1790–1878), Scottish minister and engineer.] *Stirling* (or † *Stirling's*) *cycle*, the thermodynamic cycle on which an ideal Stirling engine would operate, consisting of an isothermal expansion, a drop in temperature at constant volume by giving up heat to a regenerator, an isothermal compression, and an increase in temperature at constant volume by gaining heat from the regenerator; *Stirling* (or † *Stirling's*) *engine*, orig., an external-combustion air engine invented by Stirling (*Brit. Pat. 4081* (1816)); more widely, a mechanical device used to provide either power or refrigeration and operating on a closed regenerative cycle, the working fluid being cyclically compressed and expanded at different temperatures; also *ellipt.* as *Stirling*.

1845 *Minutes Proc. Inst. Civil Engineers* IV. 348 (*heading*) Description of Stirling's improved air engine. *Ibid.* 359 In Mr. Stirling's engine the intense heat of the fire did not come into actual contact with the pistons. 1887 *Encycl. Brit.* XXII. 523/1 Stirling's cycle is theoretically perfect whatever the density of the working air. 1889 C. H. Peabody *Thermodynamics of Steam-Engine* xi. 174 A recent hot-air engine made on the same principle as Stirling's hot-air engine. 1943 E. H. Lewitt *Thermodynamics Applied to Heat Engines* (ed. 3) iii. 57 The Stirling cycle is thermodynamically reversible owing to the action of the regenerator. 1963 *Engineer* CCXIV. 1063/1 A Stirling cycle machine operates on a closed regenerative thermodynamic cycle. 1973 *Sci. Amer.* Aug. 81/2 In practice Stirling engines do not work on the Stirling cycle. It is not possible to have isothermal (constant temperature) compression and expansion processes. 1980 *Times* 16 Oct. (Internat. Motor Show Suppl.) p. xiv/8 Most of the technology of the Stirling has been established since the Second World War,..but mainly for vehicle and industrial duties rather than aircraft.

Stirling². The name of Allan *Stirling* (1844–1927), Scottish-born American engineer, used *attrib.* to designate a water-tube boiler invented and patented by him (*U.S. Pat. 381,595* (1888)), usu. consisting of three interconnected upper steam and water drums and one or two lower water drums, connected by banks of inclined water-tubes which are heated by combustion gases and bent to enter the drums radially.

1889 *Amer. Machinist* 23 May 12/1 (Advt.), The Stirling Water Tube Boilers have unusually large steam and water spaces and well-defined circulation. 1924 F. J. Drover *Coal & Oil Fired Boilers* II. v. 143 For from 1,000 to 10,000 sq. ft. of heating surface the standard Stirling boiler consists of three steam drums and two mud drums. 1940 H. M. Spring *Boiler Operator's Guide* iv. 117 The Stirling boiler..is one of the first types of bent-tube boiler to come into common use.

Stirling³. *Math.* The name of James *Stirling* (1692–1770), Scottish mathematician, used *attrib.* and in the possessive to designate concepts in the theory of numbers, as **Stirling('s) approximation** or **formula**, either of two functions of an integer *n* which are approximations for factorial *n* when *n* is large, viz. $n! \sim n^n e^{-n}$ and (more accurately) $n! \sim \sqrt{2\pi n}\, n^n e^{-n}$; **Stirling('s) number**, a member of either of two arrays used in combinatorics, first described by him (*Methodus Differentialis* (1730)), *spec.* (*a*) the number of ways of arranging the integers 1 to *m* in *n* disjoint non-empty ordered sets, the first element of each ordered set being the least; (*b*) the number of ways of partitioning the integers 1 to *m* into *n* disjoint non-empty sets.

[illegible] $(e/N)^N$ if *N* is large. 1962 W. J. Moore *Physical Chem.* (ed. 4) vii. 233 This expression is evaluated by means of the Stirling formula, $\log N! = (N + \tfrac{1}{2}) \log N - N + \tfrac{1}{2} \log 2\pi$. 1928 *Amer. Math. Monthly* XXXV. 77 The Stirling Numbers are characterized by many very beautiful properties. 1933 *Tôhoku Math. Jrnl.* XXXVII. 255 (*caption*) Table of Stirling's numbers of the first kind. *Ibid.* 277 The Stirling number of the second kind can be obtained by aid of a problem of probability. 1966 F. N. David et al. *Symmetric Functions & Allied Tables* v. 226 Stirling's Numbers of the first kind...1 1 2 6 24 120 720...1 3 11 50 274 1764...1 6 35 225 1624...1 10 85 735...1 15 175...1 21...[etc.]. *Ibid.* 223 Stirling's Numbers of the second kind...1 1 1 1 1 1 1...1 3 7 15 31 63...1 6 25 90 301...1 10 65 350...1 15 140...1 21...[etc.].

stirrer. Add: **1. d.** One who stirs up trouble or discontent; an agitator, a trouble-maker. *colloq.* (chiefly *Austral.*).

1963 T. & P. Morris *Pentonville* xi. 247 Other prisoners described him as a *stirrer* rather than a leader in that he got others to do the dirty work for him. 1970 *Sunday Truth* (Brisbane) 28 June 4/3 He was no stirrer. He came quietly to Brisbane made the scene for six months or so then quietly drifted back a few weeks ago to his favorite haunts around Kings Cross. 1973 C. Mason *Hostage* vii. 104 'You're a born boat-rocker.' 'D'you think so? A stirrer, they'd call it at home.' 1977 *Sounds* 9 July 15/2 He's an absolute stirrer with very little concept of what our job is. 1982 *Observer* 13 June 16/1 Jessica Mitford is what Australians call a stirrer, meaning a person with a talent for causing trouble.

stirrup, *sb.* Add: **1. c.** *up in the stirrups* (earlier example).

1812 J. H. Vaux *Vocab. Flash Lang.* in *Mem.* (1964) 277 A man who is in swell street, that is, having plenty of money, is said to be up in the stirrups.

2. c. Delete † *Obs.* and add: Now *spec.* the strap itself. Also, a similar strap attached to women's stretch trousers or slacks. orig. *U.S.* in recent use.

1955 *Sun* (Baltimore) 7 Jan. (B ed.) 15/1 [Baseball] The stockings, modeled after those of the Boston Red Sox. The stirrup—that cutaway portion which extends down into the shoe—will be orange. 1963 *Women's Wear Daily* 23 Sept. 15/1 Rayon/nylon twill stretch pants..hi rise, back zip, no stirrups. 1967 *Boston Sunday Herald* 26 Mar. (Advt. Section), Comfy elastic waist, neat stitched pleat, self-fabric stirrups. 1980 *Times* 19 Feb. 8/4 The major buyers have bought..denims, dungarees, stirrup trousers (they used to be called ski pants) in the now obligatory stretch fabrics.

e. Also, each of a pair of supports for holding the legs of a female patient raised and apart, as during childbirth.

1936 H. J. Stander *Williams' Obstetrics* (ed. 7) xv. 406 In the hospital she is..placed on a suitable delivery table,..the legs held in position by adjustable stirrups. 1977 M. French *Women's Room* (1978) i. 69 The humiliation of being in stirrups and having people peer at her exposed genitals whenever they chose.

8. stirrup pump, a portable hand pump held steady by a stirrup-like foot-plate and used, esp. in the war of 1939–45, for extinguishing small fires and incendiary bombs with water drawn from a bucket and directed by a hose.

[1902 *Shand, Mason & Co.'s Portable Fire Appliances* 3b, This is another adaptation of the 'London Brigade' Hand Pump. The pump is..provided with a stirrup, by means of which it can be used with an ordinary house pail.] 1939 C. C. Ramsay *Fire-Fighting in Peace & War* vi. 41 The Home Office specification insists on a stirrup [illegible]

stishovite (stɪ·ʃəvait). *Min.* [f. the name of S. M. *Stishov*, Russian geochemist, who first synthesized it in 1961: see -ite¹.] A dense, tetragonal polymorph of silica, formed at very high pressure and found in meteorite craters. Cf. *coesite.

1962 E. C. T. Chao et al. in *Jrnl. Geophysical Res.* LXVII. 419/1 Stishov and Popova (1961) recently synthesized a high-density polymorph of SiO_2 [illegible] we propose to name the new mineral stishovite. [illegible]

stitch, *sb.*¹ Add: **I. 2. d.** *pl.* Fits of laughter; esp. in phr. (*to have*, etc., *someone*) *in stitches*. Occas. *sing.* = *laugh sb.* 4 b. (See also sense 2 a, quot. 1601.)

1935 *Motion Picture* Nov. 41/1 A laugh festival that will have you in stitches from its opening scene to its ridiculous but uproarious climax. [illegible] 1980 O. Blakeston *For crying out Shroud* vi. 56 I've got some new gear that will give you stitches. 1981 D. M. Thomas *White Hotel* IV. i. 139 She had them in stitches with her absurd—but true—anecdotes. 1983 *Listener* 20 Jan. 38/4 The sardonic puppets, C4s, had my anglophone family in stitches.

II. 7. (Earlier *fig.* example.)

1837 F. D. Maurice *Let.* Feb. in F. Maurice *F. D. Maurice* (1884) I. xiv. 224, I consider..whether we ought to take up our stitches (not intentionally dropped) at the age of twenty-four [*i.e.* go to a university].

III. 13. stitch welding, a form of spot welding in which a series of overlapping spot welds is produced by a machine which makes each weld and advances the work automatically; hence (as back-formation) **stitch weld** *sb.* and *v. trans.*; **stitch-welded** *ppl. a.*; **stitch welder,** a machine that performs stitch welding.

1934 *Welding Industry* Aug. 223/1 A continuous spot welding machine is shown... This is often called stitch welding. *Ibid.* Dec. 348/1 A development which is the logical consequence of the attempt to speed up the spot welding process is the so-called continuous spot or stitch welder. 1946 *Philips Resistance Welding Handbk.* i. 18 Stitch welders, which have been described as the sewing machines of the resistance welding industry, are either pneumatically or mechanically driven to produce a very large number of spots in rapid succession. 1951 *Trans. Inst. Welding* June 90/1 A seam weld is considered better than a stitch weld, because of its more regular formation. 1958 *Times Rev. Industry* July 26/2 The components [of the gas turbine] are stitch-welded around the circumference. 1961 J. A. Oakes *Welding Engineer's Handbk.* xxiii. 243 (*caption*) Set-up for stitch welding a steel door. 1972 *Automobile Engineer* Jan. 12/1 The fuel tanks are stitch-welded to the sides of the chassis. 1978 D. R. Andrews *Soldering, Brazing, Welding & Adhesives* iii. 65 For stitch welding the electrodes are automatically opened and closed between the making of consecutive welds and the work is moved while the electrodes are parted.

stitch, *v.*¹ Add: **II. 3.** (Later *fig.* examples.)

c 1862 E. Dickinson in *Poems* (1955) I. 300, I saw no Way—The Heavens were stitched—I felt the Columns close. 1936 L. MacNeice tr. *Aeschylus' Agamemnon* 68, I stitched this murder together; it was my title. 1961 *Daily Tel.* 16 Nov. 21/3 The precast concrete sections are 'stitched' together with 33 miles of 1¼ in diameter high tensile steel strand. 1973 M. Amis *Rachel Papers* 220 My father..crossed his little legs and stitched his fingers.

9. stitch up. f. Of a criminal, etc.: to cause (a person) to be convicted, esp. by informing or manufacturing evidence. Also *gen.*, to swindle, to overcharge exorbitantly.

1970 G. F. Newman *Sir, You Bastard* v. 142 Your confederate has just about stitched you up. 1977 *New Society* 7 July 6/2 Both Sheila and Gary have many stories of being 'stitched up' by the police or fleeced. Gary says the Dip Squad—the special police patrol looking for pickpockets—are 'a bunch of wankers'. 1977 *Woman* 3 Sept. 30/3 After shelling out £1.50 for a fold-up version [of an umbrella] she found that she'd been stitched up... Two spokes were broken. 1978 F. Branston *Sergeant Ritchie's Conscience* i. v. 69 Those [rivals] who wouldn't be frightened he stitched up, his favourite method being to sell an opponent some drugs, then inform on him to the police.

sti·tchdown. [f. Stitch *v.*¹ + Down *adv.*] A shoe or boot on which the lower edge of the upper is turned outward and stitched on to the sole; a veldt-shoe. Also *stitchdown shoe*.

1840 E. Stiff *Texan Emigrant* vii. 95 During the intolerable hot weather in the summer of 1838, the same Kentucky jeans pants, the same pair of stitchdowns,..adorned the tall and disproportioned outward man. 1916 F. Plucknett *Introd. Boot & Shoe Manuf.* xxx. 258 The 'veldtschoen', or 'stitchdown', has passed through so many stages that no description of it would correctly describe its production in all factories. The principal idea is that the edge of the upper (except just around the heel) shall be turned outward instead of being folded over the edge of an insole, and that this projecting flap shall be attached to the sole with a vertical thread seam. 1940 *Chambers's Techn. Dict.* 888/2 *Veldtschoen*,..a sandal-like form of shoe in which the upper is attached directly to the sole by a row of stitches near the edge. Also called *stitchdown shoes*. 1969 T. C. Thorstensen *Pract. Leather Technol.* xv. 248 The stitchdown shoe differs from the Littleway in that the insole is fastened to the upper by flairing the upper outward rather than inward.

sti·tchless, *a.* [f. Stitch *sb.*¹ + -less.] Without stitches; *spec.* (formerly) of a tennis ball, put together without stitches or a stitched seam. Also, 'without a stitch', unclothed.

1927 *Daily Tel.* 22 Mar. 15/6 W. H. Powell..was beaten. ..It was due..to the fact that Powell is not yet used to the stitchless ball. 1939 Joyce *Finnegans Wake* 451 I'd plant you..on the electric ottoman in the lap of lechery, simpringly stitchless with admiracion. 1953 M. Peake *Mr. Pye* xvi. 128 She had left her clothes on a high rock... 'My child..' said Mr. Pye '..my stitchless child.'

stiver, *v.* Add: **2.** *trans.* To ruffle (the hair); to make it bristle or stand on end. Also with *up*.

1886, etc. in *Eng. Dial. Dict.* 1924 Galsworthy *White Monkey* I. viii. 61 Michael stivered his hair. 1926 —— *Silver Spoon* I. v. 34 Michael stivered up his hair.

stivered *ppl. a.* (later examples); **stivery** *a.* (examples).

1892 S. Hewett *Peasant Speech* (E.D.D.), Didee iver zee sich a stivery head as 'er 'th agot? 'Er luketh 's-of 'er'd been drawed drii a brimbly 'ädge back'ards. 1918 Galsworthy *Five Tales* 127 He looked like a stuffed man..sitting there, with..his stivered hair. 1928 —— *Swan Song* I. iii. 19 Dabbing at his hair, bright and stivery, he straightened his tie and ran down. 1939 N. Marsh *Overture to Death* v. 58 The stivered grass was washed with colour, and before him his own attenuated shadow appeared.

stoat, *sb.* Add: *fig.* Also, a treacherous fellow; a sexually aggressive man, a lecher.

a 1960 E. M. Forster *Maurice* (1971) xxx. 138 His feeling for Dickie required a very primitive name... What a stoat he had been! 1978 C. Egleton *Mills Bomb* xxii. 208 'Would it surprise you to learn that he was a fag?' 'You've got it wrong; everyone knew he was a stoat.'

stochastic, *a.* Restrict Now *rare* or *Obs.* to sense in Dict. and add: **2. a.** Randomly determined; that follows some random probability distribution or pattern, so that its behaviour may be analysed statistically but not predicted precisely; *stochastic process* = *random process* s.v. *Random *a.* 1 b.

[1917 L. von Bortkiewicz *Die Iterationen* 3 Die an der Wahrscheinlichkeitstheorie orientierte, somit auf 'das Gesetz der Grossen Zahlen' sich gründende Betrachtung empirischer Vielheiten möge als Stochastik..bezeichnet werden.] 1923 A. A. Tschuprow in *Metron* II. 461 Every stochastical (1) theory of statistics sees in the empirical statistical numbers images of certain really significant quantities—reflected confuse[d] images blurred..by the Chance. [*Note*] (1) I use the word 'stochastical' as synonymous to 'based on the theory of probability'—cf. J. Bernoulli, *Ars Conjectandi*, Basileae, 1713, p. 213 'Ars Conjectandi sive Stochastice nobis definitur ars metiendi quam fieri potest exactissimi probabilitates rerum' and L. v. Bortkiewicz, *Die Iterationen*. 1934 *Proc. Nat. Acad. Sci.* XX. 376 A stochastic process is defined by Khintchine to be a one parameter set of chance variables: $x(t)$, $-\infty < t < \infty$. 1943 *Rev. Mod. Physics* XV. 32 That we should be able to idealize Brownian motion as a Markoff process appears very reasonable. But we should be careful not to conclude too hastily that every stochastic process is necessarily of the Markoff type. 1957 *New Scientist* 20 June 17/3 A new approach to population dynamics was needed, and quite recently this has been provided by J. G. Skellam in the form of a stochastic model which allows the experimentalist to regard his population as a random variable at each instant in time, and is much more flexible than the earlier deterministic equations. 1968 P. A. P. Moran *Introd. Probability Theory* iii. 108 We have already dealt with some simple cases of successive trials in which the probabilities at each stage depend on what has happened before... As the successive stages can be regarded as successive instants of time the sequence of events may be regarded as a 'random' or 'stochastic'..process. 1971 Kimura & Ohta *Theoret. Aspects Population Genetics* iii. 33 In any finite population, gene frequencies are subject to stochastic change due to random sampling of gametes. 1979 *Sci. Amer.* Mar. 64/2 (Advt.), The key was recognizing that the star formation process was 'stochastic'. That is, new massive stars are not necessarily created adjacent to a supernova; rather, a probability exists for their formation.

b. *Mus.* Applied (orig. by Iannis Xenakis (b. 1922), Romanian-born Greek composer) to music in which the overall sound structure is determined, but internal details are left to chance or are established mathematically by composer or computer (by the laws of probability or otherwise).

1958 Y. Xenakis in *Gravesaner Blätter* IV. 112 (*title*) In search of a Stochastic music. *Ibid.* 121 This *glissando* passage has been taken as an example, for it comprises every problem of this Stochastic music controlled by arithmetic. 1963 T. Pynchon *V.* x. 292 He got around to talking stochastic music and digital computers with one technician. 1969 *Sat. Rev.* (U.S.) 28 June 56/2 Though much has been written, especially by Xenakis himself, on the technique of 'stochastic' music, most of it is utterly unintelligible to the layman. 1975 *New Yorker* 19 May 90/1, I heard a Balinese gamelan one night and the Strasbourg Battery in Yannis Xenakis's latest 'stochastic' composition the next. 1978 P. Griffiths *Conc. Hist. Mod. Music* xi. 169 Iannis Xenakis..has also used computers as calculating aids in the composition of his 'stochastic' music, where the musical form is made analogous to a stochastic process (i.e. one ruled by laws of probability, such as a sequence of dice throws).

stochastically *adv.*, delete † and add later examples; **stochasti·city,** the property of being stochastic.

1947 *Biometrika* XXXIV. 228 The efficiency of any two tests would be identical, in the conditions stated, if the coefficient of correlation between them was ±1 because then, of course, they would be functionally, and not stochastically, related. 1968 P. A. P. Moran *Introd. Probability Theory* v. 247 *X* is said to be 'stochastically larger' than *Y* if $F(x) \leqslant G(x)$ for all *x*. 1971 *Jrnl. Statistical Computation & Simulation* I. 42 Refinement in modelling eventuates a requirement for stochasticity. 1979 *Nature* 9 Aug. 459/2 The explanation..is to the contrary of the conventional explanation of these non-seasonal cycles in terms of demographic stochasticity.

stocious (stōᵘ·ʃəs). *slang* (chiefly *Anglo-Ir.*). Also **stotious.** [Of uncertain origin.] Drunk, intoxicated.

1937 *News Chron.* 20 Feb. 8/6 Slang also appeals to our elementary sense of humour, as when we say of a man who is drunk that he is..stotious. 1949 C. Graves *Ireland Revisited* i. 20 Words to discriminate the various degrees of intoxication... You have..spifflu, langers, and stocious. The last word rhymes with atrocious and means thickly speaking drunk... We were unable to find anybody who had ever seen it in print. 1952 *Caribbean Quarterly* II. iv. 27 Since when you become so damn stocious? a 1966 'M. na Gopaleen' *Best of Myles* (1968) 119 A young man charged with delivering them was found stotious in a doorway. 1970 M. Kenyon *100,000 Welcomes* xx. 174 'She's stotious,' Rafferty said. 1980 J. O'Faolain *No Country for Young Men* v. 115 'Coming home stocious five nights a week,' said Doris.

stock, *sb.*¹ Add: **I. 3. d.** (Earlier example in sense of a language group.)

1730 W. Wotton *Discourse Confusion Babel* 15 So that though this will invincibly prove the Gradation and Derivation of different Dialects from a common Stock, yet it will not prove the actual Formation of some essentially different Tongues which I here contend for.

IV. 28. b. (Earlier example.)

1817 W. Scott *Let.* 29 Oct. (1933) V. 4 Like the Highlandman's gun, she wants stock, lock, and barrel, to put her into repair.

V. 41. (Later examples.)

1877 A. I. Root *ABC Bee Culture* 158/1 Our pure Italian stocks could have been opened, and their queens removed, scarcely disturbing the cluster. 1930 W. Herrod-Hempsall *Bee-Keeping* I. vi. 315 A 'Swarm' is a cluster of bees and their queen only; a 'Colony' consists of the bees and queen living on combs containing brood..and food; a 'Stock' includes the latter together with the home in which the bees are residing. 1980 R. J. & W. E. Howe *Practical Beekeeping* vi. 49 When a stock of ten frames is broken up into a number of nuclei, the flying bees from these nuclei will return to their old stand.

VI. 52. b. (Earlier example.)

1870 'Mark Twain' in *Galaxy* Oct. 575/1 The 'chance' theory..is..calculated to inflict..pecuniary loss upon any community that takes stock in it.

c. *fig.* Reputation, esteem, credit.

1930 *Times Lit. Suppl.* 17 Apr. 334/4 He found British stock very high in North Germany. 1942 *R.A.F. Jrnl.* 27 June (recto rear cover), The stock of the R.A.F. is high in the Soviet. 1955 *Times* 24 June 10/2 General Perón's stock still seemed to be rising to-day as the country gradually returned to normal conditions. 1979 A. Boyle *Climate of Treason* viii. 237 This minor triumph sent up the personal stock of Philby.

53. c. = *Stock-car 2. *U.S.*

1951 *Sun* (Baltimore) 11 Oct. B24/1 The Philadelphia district will be well represented when the 100-mile National Championship, for sportsman stocks, gets the green flag at the Langhorne Speedway, Sunday. 1979 *Arizona Daily Star* 1 Apr. c 12/5 Tucson Dragway will run its weekly racing program today, with the junior pro stocks..topping the racing.

56. d. *Theatr.* A stock company; repertory. Chiefly *U.S.*

1916 *Variety* 27 Oct. 12/1 The Alcazar stock is enjoying satisfactory business. 1933 M. Lincoln *Oh! Definitely* vii. 73 'He had been getting three pounds a week in stock' but would 'take two-ten for town'. 1937 *Daily Tel.* 14 Aug. 9/1 No money will induce them [*sc.* good actors] to bury themselves in Stratford..under 'stock' conditions. 1962 *Listener* 16 Aug. 242/2 Between her junior and senior years in college..she played summer stock.

57. a. (Earlier examples.)

1730 C. Carter *Compl. Pract. Cook* 1 A good Stock of strong Broth Well made, and good Gravies well drawn off, are very principal Ingredients in the Composing of all Made-Dishes of boil'd Meats. 1747 H. Glasse *Art of Cookery* ix. 78 An Oyster Soop. Your stock must be made of any Sort of Fish the Place affords.

b. (Later examples without prefix in Paper-making.)

1924 S. Leicester *Pract. Stud. for Paper Manufacturers* v. 116 The mistakes in sizing are some of the most difficult to elucidate... The stock used may be the cause. 1963 R. R. A. Higham *Handbk. Papermaking* ii. 45 Distinct variations occur between one batch of stock and another with regard to treatment, colour, temperature, consistency, retention of additives, etc.

c. Cinematographic film.

1897 C. F. Jenkins *Picture Ribbons* 27 The film is of transparent celluloid, one side of which is coated with a sensitive emulsion, that for the negative being much more rapid than the positive stock. 1909 *Moving Picture World* 3 July 11/2 The non-inflammable film stock is now being issued by so many manufacturers. 1938 *Times* 15 Mar. 12/3 A twelve-minute film on 16 mm. stock, shown privately in Liverpool. *Ibid.*, The technical quality of the film is excellent, super-panchromatic stock giving rich quality to shots which are themselves carefully composed. 1974 C. Priest *Your Bk. of Film-Making* i. 23 The film is twice as wide as 8 mm film stock.

VIII. 62. stock split *U.S.*, the division of a stock into an increased number of shares; hence **stock splitting**; cf. *split-up* s.v. *Split-.

1955 *Times* 6 July 9/3 According to the Associated Press, the directors of General Motors Corporation have to-day recommended a three-for-one stock split to be voted on by stockholders at a special meeting on September 23. 1959 *Economist* 28 Feb. 788/1 Stock splitting (the American equivalent of the British scrip or bonus issue). 1967 *N.Y. Times* (Internat. ed.) 11–12 Feb. 9/6 Your first bonus report will be our list of 30 stock split candidates. 1977 *Dædalus* Fall 85 Tests indicate that stock prices quickly adjust to changes in public information (announcements of stock splits, dividend increases, etc.).

63. a. *stock-agent*, *-auction*, *-breeding*, *-carrying*, *-driver*, *-driving*, *-farm* (later examples), *-farmer* (later example), *-feed* (later examples), *-inspector*, *-rearing*, *-sale*, *-station* (earlier examples), *-theft*, *-thief*, *-trader*, *-train* (examples), *-yard* (later examples); *stock-proof* adj.; **stock and station** *Austral.* and *N.Z.*, used *attrib.* to designate firms or their employees dealing with farm products and supplies; **stock horse** (earlier example); **stock-rider** (later examples); **stock-route** (later

examples); **stock-whip** (earlier and later examples); also as *v. trans.*, to beat with a stock-whip.

1933 *Press* (Christchurch, N.Z.) 9 Sept. 15/7 Dealers and stock-agents use various terms..to make failing mouthed sheep sound younger. 1977 *Weekly Times* (Melbourne) 19 Jan. 11/3 Barney, the stock agent, was looking him straight in the eye when he said: 'If I were you, Clarence, I'd sell the lot and run some sheep.' 1881 *Adelaide Observer* 22 Oct. 44/1 He was suspicious of all stock and station salesmen. 1908 in D. J. Gordon *Handbk. S. Austral.* 327 (Advt.), Bagot, Shakes, & Lewis, Limited. Stock and station agents. 1930 L. G. D. Acland *Early Canterbury Runs* 1st Ser. ii. 13 Ford and Newton, who were the leading Christchurch Stock and Station Agents. 1965 G. McInnes *Road to Gundagai* vii. 113 Here were the big mortgage and stock-and-station houses where wool was finally baled and cleaned for export. 1948 W. Faulkner *Intruder in Dust* (1949) vi. 134 Monday was stock-auction day at the sales barns behind the Square. 1937 R. H. Lowie *Hist. Ethnological Theory* viii. 114 This yields..the sequence of (a) hunting-gathering; (b) hoe-culture; (c) hoe-culture with stock-breeding; (d) 'agriculture'. 1957 *Times Lit. Suppl.* 8 Nov. 678/3 Professor Nichols makes a comprehensive review of the genetic basis of modern stockbreeding. 1866 J. Murray *Descr. Province Southland* 9 The stock-carrying capacity of the natural herbage is of course variable. 1960 *Farmer & Stockbreeder* 12 Jan. 78/1 He brought back the idea of loose-housing..and introduced it at Langhill to cater for the additional stock-carrying capacity of the next-door, buildingless farm then being acquired. 1851 *Lyttelton* (N.Z.) *Times* 19 Apr., A settlement of whale-fishers and stock-drivers. 1871 *Republican Rev.* (Albuquerque, New Mexico) 27 May 1/3 M. Maloney..arrived here on Thursday, being sent ahead to employ stock drivers. 1867 H. Phillips *Jrnl. Rockwood* 29 Sept. 88 (typescript), T.A.P. & I.I. stock driving. 1874 J. C. McCoy *Hist. Sk. Cattle Trade* 92 [The farm] is allowed to lay awaste, whilst its owner has turned to stock-driving. 1848 *Senate Rep. 30th U.S. Congr. 1 Sess.* No. 75. 29 Some five hundred head of beef cattle were taken from the government stock farm. 1912 M. Nicholson *Hoosier Chron.* 27, I own a stock farm near Lexington. 1894 *Harper's Mag.* Apr. 676/2 'Crit' Marston, the young blue-grass stock-farmer, is a favorite throughout all that section. 1960 *Farmer & Stockbreeder* 15 Mar. 113/1, I am growing ten acres each of stockfeed peas and beans. 1970 *Oxford Times* 30 Oct. 14 Demand for stockfeed potatoes would be far greater than usual. 1846 H. Weekes in Rutherford & Skinner *New Plymouth Settlement* (1940) i. vi. 124 'Peter' was an excellent stock-horse, would follow cattle like a dog. 1888 *Century Mag.* Feb. 507/1 At every shipping point..stock inspectors..jealously examine all the brands on the live animals or on the hides of the slaughtered ones. 1930 L. G. D. Acland *Early Canterbury Runs* 1st Ser. iii. 47 He then became Stock Inspector in the North Island, but quarrelled with his superiors. 1948 V. Palmer *Golconda* ii. 15 He might have been a country teacher or a stock-inspector. 1915 *N.Z. Jrnl. Agric.* 20 Feb. 190 If the long shoots of this plant [sc. *Eleagnus*] are interlaced while the hedge is growing it makes a close and excellent stock-proof fence. 1960 *Farmer & Stockbreeder* 15 Mar. 123/2 The Monmouthshire style of hedging..gives a real stock-proof fence. 1915 *Chambers's Jrnl.* Jan. 47/2 A son of his anticipated this kind of stock-rearing many years ago in Manitoba. 1960 *Farmer & Stockbreeder* 29 Mar. 12/3 An 81-acre Northants stock-rearing and feeding farm has been sold for £9,800. 1908 E. J. Banfield *Confessions of Beachcomber* II. iii. 314 A stockrider..in..flash riding-boots. 1973 *Parade* (Melbourne) Sept. 30/2 The authorities were able to choose exactly the sort of men they wanted from the hundreds of adventurers, prospectors, settlers and stock-riders who offered their services. 1901 M. Franklin *My Brilliant Career* xxxiv. 286 An overgrown old orchard, skirting one of the great stock-routes. 1977 *Meanjin* (Austral.) XXXVI. I. 69 Cattle cross on the stockroutes. 1948 W. Faulkner *Intruder in Dust* (1949) vi. 134 Stock-sale day unlike Sunday was a man's time. 1843 J. Backhouse *Narr. Visit Austral. Colonies* xxiii. 264 Accompanied by the Agricultural Superintendent, we walked to a stock-station..where three men are placed in charge of some cattle. 1847 A. Harris *Settlers & Convicts* xiii. 252 It was ..not till noon..that we succeeded in finding the nearest stock-station. 1904 *Transvaal Agric. Jrnl.* July 573 Stock theft has always been a great source of worry and trouble to the farming community of this country. 1955 L. G. Green *Karoo* xvii. 199 Crime in the karoo usually means stock-theft. *Ibid.* ix. 112 The hunt for a stock-thief who fled into the poort. 1958 *Johannesburg Sunday Times* 28 Sept. 14/9 A quiet-spoken, slightly-built man has become the terror of stock-thieves in the Evaton and Losberg areas. 1942 W. Faulkner *Go Down, Moses* 248 A back-street stock-trader's boarding house. 1948 —— *Intruder in Dust* (1949) vi. 134 The men with their stock-trader walking-sticks not even stopping. ? 1906 in J. V. Allen *Cowboy Lore* (1971) I. 19 Another train run in to my stock train. 1961 R. P. Hobson *Rancher takes Wife* xiv. 171 By the time the stock train pulled in..we had a count on the herd. 1852 *Harper's New Monthly Mag.* Dec. 25/1 The Australian 'stockman' is a sort of Europeanized Tartar... His food is beef and 'damper'... In his 'run' the stockman is king; his cattle are his subjects; his saddle is his throne; his sceptre is the stock-whip. 1853 J. Rochfort *Adventures of Surveyor in N.Z.* iv. 42 If the natives had not lent her [*sc.* the mare] to me he would have gone over and stock-whipped them. 1901 M. Franklin *My Brilliant Career* i. 4 Father came to my rescue, despatching the reptile with his stock-whip. 1936 *shaw's Railway Man* xxi. 428 *Expended*... Union stock whip and blanket, And bury me deep. 1955 J. Cleary *Justin Bayard* viii. 111 He tried to hit me, and old Thaddeus stockwhipped him. 1958 R. Srow *To Islands* i. 17 What's the use of holding it against Mr. Heriot that he used to be a handy man with a stockwhip? 1869 *Bradshaw's Railway Man* xxi. 428 *Expended*... Union stock yards Chicago.. $100,000. 1911 C. E. W. Bean *'Dreadnought' of Darling* xv. 145 The wind..piled it uselessly over every fence and stockyard. 1929 K. S. Prichard *Coonardoo* iv. 51 They wandered from the stock-yards to the shade-miah. 1958 L. Durrell *Mountolive* viii. 162 The mauve-veiled evening voices of Alexandrians uttering stockyard quotations. 1963 *Times* 16 Jan. 6/6 Born in the stockyards district (where he still lives) and where as a 'stockyards cowboy' he once herded cattle from pen to pen. 1978 D. Greig *Daisy* v. 54 In Chicago we stayed at the Hotel on Lake Michigan, near the famous Stockyards where, as they used to say, the unfortunate animals went in whole at one end and came out the other processed into fifty different products.

b. *stock carp*.

1785 J. Woodforde *Diary* 20 Oct. (1926) II. 211 Mr. Townshend..sent me 20 brace and ½ of stock Carp.

65. stock book (earlier examples); also *spec.* a book in which a record is kept of the animals which make up the stock of a farm; **stock-boy,** (*a*) *Austral.*, an Aboriginal employed to look after cattle or other stock; (*b*) *U.S.*, a boy employed by a business firm to look after stock; **stock-building** = *stock piling vbl. sb.*; **stock control** (see quot. 1943); **stock cube,** a cube of concentrated, dehydrated meat stock sold for use in making soups, stews, etc.; **stock culture,** an uncontaminated culture of a micro-organism maintained continuously and available as a source of experimental material; **stock knife,** (*a*) delete †, add later examples, and substitute for def.: a knife for cutting wood, *esp.* one used by a clogger for shaping the soles of clogs; (*c*) a stockman's knife; **stock-pot** (earlier and later examples); **stock rail** *Railways*, each of the outer fixed rails at a set of points.

1835 J. F. Cooper *Monikins* I. ii. 32 Love was a sentiment much too pure and elevated for one whose imagination dwelt habitually on the beauties of the stock-books. 1847 A. Harris *Settlers & Convicts* xiii. 260 Outside the yard..is..set a table with the stock book, pens, and ink, and in that the cattle are registered. 1882 W. D. Hay *Brighter Britain!* I. viii. 202 We keep a stock-book, in which every beast is entered. 1937 E. Hill *Great Austral. Loneliness* xli. 305 In Kimberley and the Territory lubras are even to-day recognised as the best 'stock-boys'. 1955 J. Cleary *Justin Bayard* iv. 58 The stockboys had roped the piebald now and thrown a saddle on him. 1972 R. Milner in W. King *Black Short Story Anthol.* 376 This receptionist thought I had come about a stock-boy job, you dig. 1979 D. Anthony *Long Hard Cure* x. 86 He owns a chain of department stores..one of those self-made men, who went to work at fourteen as a stockboy. 1967 A. Battersby *Network Analysis* (ed. 2) xiii. 221 They will be high during the first few months because of retail stock-building. 1977 *Financial Results of Oil Majors 1976* (Shell Internat. Petroleum Co.) 8 The increase in demand, combined with some stock-building at the end of the year in anticipation of a significant rise in oil prices, raised oil production outside the USSR, Eastern Europe and China to 47 million barrels daily in 1976. 1943 *Princ. Production Control (B.S.I.)* 7 *Stock control*, the means by which the correct quantity and quality of material and components are made available according to the production plan, and excessive stocks avoided. 1962 A. Battersby *Guide to Stock Control* v. 48 The calculations can conveniently be summarized on the Stock Control Form..and a specimen set of figures is shown. 1976 J. Lund *Ultimate* iii. 29 They talked..on the economics of warehousing and stock control. 1965 *Listener* 26 Aug. 317/2 Add enough water to almost cover the meat, and the stock cubes. 1979 *Times* 29 Sept. 15/5 Do you keep a stockpot..and boil it daily?..Cookery books..have a sneaky way of implying that stock cubes will never do. 1903 *Jrnl. Hygiene* III. 2 Gelatin plates were then made from the broth culture; if only a single species developed, agar tube-cultures were prepared and used as the stock-cultures of the organism. 1979 *Jrnl. Appl. Bacteriol.* XLVII. 381 The maintenance of stock cultures of lactic acid bacteria in small microbiological laboratories may present a technical problem. 1955 R. P. Hobson *Nothing too Good for Cowboy* i. 16 He cut the moosehide wrapping with his stock knife. 1968 J. Arnold *Shell Bk. Country Crafts* 105 The shaping of clogs from these clefts is done with a stock knife... It consists of a stout blade with a long projecting handle. 1978 *Lancashire Life* Apr. 49/2 The tools a sole-cutter used were three in number—stock-knife, hollower and gripper. 1845 E. Acton *Mod. Cookery* i. 3 Never..set the soup by in it, but strain it off..and fill the stock-pot immediately with water. 1917 *Harrods Gen. Catal.* 964/1 Extra heavy bellied Stockpot, enamelled Pearl Grey outside and in. 1928 'O. Douglas' *Eliza for Common* x. 128 Some quite dull books read like that—as if the author had simply thrown everything in, a sort of stock-pot of a book. 1931 R. Campbell *Georgiad* i. 18 His melancholy recipes for 'happiness'.. How to 'rechauffe' the stock-pot of desire. 1960 E. David *French Provincial Cooking* 158 The pot.. is usually a tall straight-sided or slightly bulbous stock-pot made of earthenware, copper, enamelled iron, or heavy aluminium. 1982 *Daily Tel.* 14 Jan. 15/5 Put..chine bone into stockpot, cover with water,..and simmer. 1850 *Civil Engin. & Archit. Jrnl.* XIII. 270/1 The top of the switch not being mitred into the underside of the bearing surface of the stock rail, it is not liable to be locked by the barbing over of the stock from the pressure of the wheels. 1890 W. H. Cole *Notes on Permanent-Way Material* i. 20 The points or switches are..so adjusted to their respective stock-rails that when one switch is pressed against its stock-rail the other is drawn away, and thus one line of metals or the other is made continuous. 1935 E. Beal *Railway Modelling in Miniature* ii. 36 Then solder the stock-rail for the other track.

B. *adj.* For '(in attributive use only)' read '(usu. in attributive use)'.

1. a. *stock model*; *stock-type* adj.; **stock shot** *Cinemat.* = *library shot* s.v. *Library¹ 3; **stock size** (later examples).

1926 *Daily Colonist* (Victoria, B.C.) 4 July 3/1 (Advt.), The car was a stock model in every respect except for a 48-gallon gasoline tank and changes in the top, back seat and tire carrier. 1941 B. Schulberg *What makes Sammy Run?* vi. 93 A shoe-string producer told him he had bought the stock shots from *Hell's Angels*. 1974 *Radio Times* 14 Mar. 11/4 Processed chases up and down stock-shot ski slopes. 1952 M. Laski *Village* xix. 262 Margaret was lucky, she was stock-size, not like Wendy herself who had always had to have everything made for her. 1980 *Country Life* 3 July 78/2 Our model girl was stock size and everything was too big for her. 1958 *Spectator* 1 Aug. 170/1 A stock-type Vauxhall Velox.

2. *stock author, burlesque, comedy* (earlier example), *star*; **stock actor** (earlier example); also *stock actress*; **stock character,** a dramatic character representing a type in a conventional manner and recurring in many works; (cf. sense 3 a below); **stock piece** (earlier and later examples); **stock play** (earlier example).

1839 Marryat *Diary Amer.* 2nd Ser. II. xiii. 121 The American *stock* actors, as they term those who are not considered as *stars*. 1921 E. O'Neill *Diff'rent* II. 243 She resembles some passé stock actress of fifty made up for a heroine of twenty. 1824 J. Decastro *Mem.* 154 T. Dibdin, esq., succeeded him..as the stock author of that theatre. 1864 'P. Paterson' *Glimpses of Real Life* xxv. 240 Jones keeps a stock author, and does not rely on outsiders. 1916 *Variety* 27 Oct. 12/1 Stock burlesque at the Lyric is moderately successful. 1864 H. Morley *Jrnl.* 16 Jan. (1866) 325 The *gracioso* was a popular addition made by Lope de Vega to the stock characters of a Spanish play. 1893 [see *Cloak sb. 6]. 1976 *Country Life* 12 Feb. 346/1 In *Albert Herring*, Britten took stock characters—pompous mayor, stolid policeman,..overbearing lady of the manor. 1812 *Dramatic Censor 1811* 27/1 Among all the stock comedies which the Theatres are in the habit of representing, this is, in our idea, one of the very best. 1804 W. Cooke *Mem. C. Macklin* 408 It was always one of the stock pieces which he engaged himself to perform. 1843 *Ainsworth Mag.* IV. 135 His..acting contributed greatly to the success of the drama, though it had not sufficient stamina to become 'a stock piece'. 1708 L. Downes *Roscius Anglicanus* 8 Note, That these being their Principal Old Stock Plays. 1856 A. C. Ritchie *Mimic Life* I. ii. 44 As the 'stock star' of a popular theatre, in Boston, she had shone several years in the dramatic firmament.

3. a. Also with reference to fictional characters of a standardized or conventional type (cf. *stock character*, sense 2 above); also *transf.*

1940 W. S. Maugham *Books & You* p. xii, The characters..are not very interesting, and most of them are the stock figures of Victorian fiction. 1951 M. McLuhan *Mech. Bride* (1967) 118/2 The 'good girl' is the nineteenth-century stock model which has long been merged with the mother image. 1960 [see *Exurbanite a.]. 1963 [see *Cliché 3]. a 1963 L. MacNeice *Astrol.* (1964) vi. 200 Catering for stock-type 'individuals' (all humanity being divided into 12 groups). 1980 J. Gardner *Garden of Weapons* II. vii. 185 A man full of bounce, like the stock uncle known to all large families.

b. Special collocations: **stock bowler** *Cricket*, a reliable but unspectacular bowler; hence **stock bowling; stock response,** an automatic and superficial reaction to a literary device (see quot. 1939); also *transf.* and *fig.*

1968 *Listener* 11 July 61/2 Connolly, in 1964 a strenuous but pedestrian fast bowler, has reduced his pace, developed swing and cut, and become an admirably steady stock-bowler. 1976 J. Snow *Cricket Rebel* 37, I could no more be regarded as a stock bowler relying on line and length to keep the scoring in check. *Ibid.* 77 Only occasionally did he call upon me to do a stock bowling job with the intention of closing the game up. 1925 I. A. Richards *Princ. Lit. Crit.* xxv. 203 Against these stock responses the artist's internal and external conflicts are fought, and with them the popular writer's triumphs are made. 1939 Brooks & Warren *Understanding Poetry* 639 *Stock response*, the general uncritical response made on conventional or habitual grounds to a situation, subject, phrase, or word in literature. Advertisers frequently attempt to appeal to stock responses. 1957 A. Thwaite *Home Truths* 40 Or will it seem Merely the self-duped mind's harangue at Death, The stock-response still raging in the shroud? 1961 K. Tynan *Curtains* i. 8 The stock response of terror in the face of matricide has vanished. 1966 'K. A. Saddler' *Gill Edge* ix. 128 'Well,' he said continuing, and just in time as I was running out of stock responses. 1975 *Times* 20 Sept. 6/3 If Agatha Christie works almost entirely with what the critics call 'stock responses', she knows..how to take advantage of our responding in a stock way to..stock situations.

4. In non-*attrib.* use.

1966 *Listener* 25 Aug. 288/3 The authors gave us sharply observed characters—stock, but none the worse for it. 1977 *Hot Car* Oct. 88/2 The diesel comes stock with a servo. 1979 *Jrnl. R. Soc. Arts* Nov. 776/2 The stereotypes are used in a relatively straightforward way, as stock as the London brick.

stock, *sb.*⁴ Add: **b.** A discordant intrusion of igneous rock which has a roughly oval cross-section and steep sides, and is smaller than a batholith.

1898 *Jrnl. Geol.* VI. 706 It will be found advantageous to discriminate between bysmalith and stock by limiting the term stock to such bodies as occupy nearly vertical tubes or funnels of indefinite depth in rocks of any and all kinds..and which maintain such a relation to them as to appear to belong to the category of dikes. 1916 *Yukon Territory* (Canada Dept. Interior) iii. 35 Occasional pebbles derived from the various dikes and stocks outcropping along the valleys. 1944 [see *Laccolith]. 1955, 1957 [see *Intrude v. 5]. 1977 A. Hallam *Planet Earth* 69/1 The upper surface [of a batholith] is generally irregular, with upwardly projecting stocks and dikes that may be the only surface clue to the much larger body at depth.

stockade, *sb.* Add: **2. c.** A prison, esp. a military one.

1865 *Atlantic Monthly* Mar. 286/2 'Is it a pen?'..'Yes, yours,' retorted one of the guard, with a grin,—'the Stockade Prison.' 1882 W. D. Hay *Brighter Britain!* I. 23 A man..on a subsequent conviction, might be sent to the Stockade (prison) without the option of a fine. 1905 [in Dict., sense 2 a]. 1906 *Daily Colonist* (Victoria, B.C.) 16 Jan. 2/5 [Two men] will be sent to the convict stockade on the islands in the tropics off the coast. 1945 *Richmond* (Va.) *News-Leader* 10 Oct. 8/1 'Stockade' in army language is synonymous with jail. 1979 P. Gosling *Zero Trap* xix. 191 'What's a glasshouse?'..'A prison,' Skinner explained... 'The stockade,' Laura amplified. 'That's what they call it in the [U.S.] Army.'

stock-broker, stockbroker. Add: **b. stockbroker belt,** any prosperous residential area in the Home Counties favoured by stockbrokers or other affluent businessmen; also *transf.* of similar areas elsewhere; usu. with *the*; similarly **Stockbroker('s Tudor,** a facetious term for a style of mock-Tudor architecture supposed to be favoured by such people.

1960 [see *belt sb.¹ 5 a]. 1961 *Spectator* 14 Apr. 523 They live in the Sussex stockbroker-belt. 1968 *Listener* 22 Aug. 248/2 A moderately successful novelist..bored with his stockbroker-belt home. 1976 J. I. M. Stewart *Young Pattullo* ii. 24 She had been brought up in a stockbroker belt in the Home Counties. 1981 M. Jon *Wallington Case* iv. 24 It's a large house in..the stockbroker-belt, not far from Wilmslow. 1939 O. Lancaster *Homes Sweet Homes* 70 (*heading*) Stockbrokers Tudor. 1940 Graves & Hodge *Long Week-End* xi. 180 In 'Stockbroker's Tudor' houses..ingenuity was displayed in olde-worlde disguise. 1958 J. Cannan *And be a Villain* i. 6 Next best were these Stockbrokers' Tudor jobs built immediately before the first world war. 1959 *Good Food Guide* 98 Unlike Victorian Gothic, Stockbroker Tudor and suchlike varieties, this medieval mansion is perfect in every detail, except date. 1980 D. Clark *Poacher's Bag* iv. 81 It had been an old inn. Now it had sprouted wings in stockbroker Tudor.

stockbrokerage (later examples).

1952 S. Kauffmann *Philanderer* (1953) ii. 26 It reminded her of her father (a minor officer of a stockbrokerage firm). 1972 *Publishers Weekly* 6 Mar. 24/1 Her husband..has his own stockbrokerage firm on Wall Street.

sto·ck-car. [f. Stock *sb.*¹ + Car *sb.*¹] **1.** A truck or wagon for transporting cattle or other livestock by rail. *U.S.*

1858 *Pennsylvania Rail Road Ann. Rep.* 14 The rolling stock [included]..188 Eight-wheeled Stock Cars. 1875, 1898 [in Dict. s.v. Stock *sb.*¹ 63 a]. 1920 *Proc. 3rd Nat. Country Life Conf.* 12 They shot me across the country in a stock car. 1949 *Exciting Western* May 55/1 Another group was bringing the cattle up the Texas trail to ship east in the stock cars.

2. A racing car which has the basic chassis of an ordinary commercially produced vehicle but is extensively modified for use in racing. orig. *U.S.*

1914 *Automobile* 9 Apr. 792/2 It was a more strenuous test than ever staged in America in the days of stock cars. 1927 *Sat. Even. Post* 21 May 71 America's fastest stock car. 1935 Eyston & Lyndon *Motor Racing & Record Breaking* viii. 73 They have been rivalled by reconstructed 'stock' cars. 1960 *Daily Tel.* 26 Sept. 1/8 (*heading*) Stock car ploughs into 10-deep crowd. 1982 *Times* 16 June 10/3 Andretti raced cars, stock cars, midgets, whatever presented itself.

3. *attrib.*, as *stock-car driver, racer, racing,* etc.

1914 *Automobile* 9 Apr. 800/1 Chairman Richard Kennerdell and his Contest Board are at work on the stock car rules. 1955 *Times* 17 May 10/1 A woman stock car driver..died in a Coventry hospital to-night from injuries received in an accident while she was competing in a stock car meeting. 1960 *News Chron.* 26 Sept. 1/5 He organises stock-car and jalopy racing. 1969 A. Lurie *Real People* 114 Nick Donato proposed that we should all go and watch what he called 'my kind of racing'—at the stock-car track over in Dryden. 1973 'D. Jordan' *Nile Green* vi. 29 Ramshackle, blaring taxis, weaving round the squares with the recklessness of stock-car drivers. 1976 B. Bova *Multiple Man* (1977) viii. 87 Become a stock car racer. It's a helluva lot safer. 1976 A. Davis *Television* 67 The programmes close with classified results..including American and Gaelic football, stock car racing, canoe slalom and hurling.

stocker. Add: **6.** A warehouseman or stock-keeper. Also (*U.S.*), an assistant engaged to look after stock held for sale by a business firm.

1921 *Dict. Occup. Terms* (1927) § 940 *Stock keeper, stocker*, a warehouseman..who keeps stock book showing amount of stock (as distinguished from stores). 1976 *Billings* (Montana) *Gaz.* 27 June 5-E/4 At age 14 or 15, cashiers, salesmen, stockers, baggers, gas pumpers, car washers,..can work. 1979 *Arizona Daily Star* 5 Aug. (Advt. Section) 5/10 Full time person to work in our yard as salesperson and stocker.

7. *U.S. colloq.* A stock-car; a stock-car racer.

1976 *Harper's Mag.* Jan. 20 You simply can't believe the noise of these engines. Stockers, motorcycles, needle-nosed dragsters..tear the night apart for hours. 1976 *Time* 27 Sept. 82/3 Stock cars... Richard Petty, king of the stockers, won $378,865 last year. 1978 *Time* 25 Sept. 88 Members of the National Association for Stock Car Auto Racing..rolled up to the 'diplomatic entrance' in their Day-Glo colored 'stockers'.

stock exchange. Add: **b.** *attrib.*

1849 J. Francis *Chronicles & Characters Stock Exchange* xv. 288 It was proved that one million had been wasted in commissions and military preparations; in Stock Exchange transactions and Stock Exchange jobbing. 1877 R. Giffen (*title*) Stock Exchange securities. 1922 Joyce *Ulysses* 521, I shall..suck my thumping good Stock Exchange cigar. 1940 T. S. Eliot *East Coker* III. 11 The Stock Exchange Gazette, the Directory of Directors. 1957 *Encycl. Brit.* XXI. 420/1 After 1954 all members of the council (from 30 to 36 in number) were to be elected by stock exchange members generally.

stock-fish, stockfish. Add: **2.** [ad. Afrikaans, f. Du. *stokvis* stockfish, hake.] Also † **stok-fish.** The South African hake, *Merluccius capensis*, of the family Gadidæ, a large marine food fish. *S. Afr.*

1823 W. W. Bird *State of Cape of Good Hope in 1822* viii. 159 The hottentot, jacob evert, elft, hake or stockfish, the king klipfish, the steen brazen, and the stompneus are all of excellent quality. 1853 L. Pappe *Edible Fishes Cape of Good Hope* 31 The cured or dried Cape Stok-fish is an excellent dish. 1913 W. W. Thompson *Sea Fisheries Cape Colony* ii. 48 The larger fish, such as..steenbras, stockfish,..are caught with fish bait. 1930 [see *kingklip]. 1947 K. H. Barnard *Pict. Guide S. Afr. Fishes* iii. 83 The stockfish has become one of the mainstays of the South African fishing industry.

stockholding, *vbl. sb.* or *ppl. a.* Add: (Further examples.) Also, (of or pertaining to) the practice of holding material in stock.

1961 *Wall St. Jrnl.* 26 Apr. 3 Ralph Gish, manager of the funds' investment departments, said common stock holdings have been increased in these industries. 1962 A. Battersby *Guide to Stock Control* ii. 18 If the cost of the material goes up, the amount of capital invested in stock also increases, and so does the stockholding cost. 1965 *Mod. Law Rev.* XXVIII. v. 555 Loss to other outlets of fast-selling titles..can have a devastating effect on stockholding booksellers. 1971 *Engineering* Apr. 123 (Advt.), This van is just one of many which provide a service from stock-holding depot to the customer... With 29 stock-holding depots strategically placed throughout the British Isles, you are assured personal service. 1981 *Times* 26 Jan. 15/2 A group of smaller stockholding concerns is said to have formed a consortium through which it can buy and distribute cheaper foreign produced steel. 1981 J. Sutherland *Bestsellers* i. 18 The stockholding bookshop would be sacrificed to the bookstand in the supermarket.

Stockholm (stǫ·khōᵘm). The name of the capital city of Sweden, used *attrib.* in **Stockholm syndrome** (see quots.); **Stockholm tar,** a kind of tar prepared from resinous pinewood and used in shipbuilding, skin ointments, etc.

1978 *Practitioner* Feb. 297/1 Mr Vaders had a mild case of 'Stockholm syndrome'... Named after the dramatic and unexpected realignment of affections in the Sveriges Kreditbank robbery, this syndrome consists of a positive bond between hostage and captor, and feelings of distrust or hostility on the part of the victim towards the authorities. 1980 C. Moorehead *Fortune's Hostages* ix. 183 The phrase 'Stockholm syndrome' was coined, probably in America, to explain this strange affection the victims of kidnappings and sieges come to feel for the men who hold them prisoner. 1867 *Chambers's Encycl.* IX. 296/1 The Stockholm tar, which is so widely used in shipbuilding. 1929 R. Hughes *High Wind in Jamaica* ii. 50 He did envy the chap whose job it was to dip his hand in a great pot of aromatic Stockholm tar, and work it into the dead-eyes. 1976 *Country Life* 5 Feb. 298/3 Sail-making..was very hard on the hand, and..open cracks in the skin, were a frequent source of trouble... For this 'Stockholm' tar was provided by the ship owners.

Stockholmer (stǫ·khōᵘməɹ). [f. prec. + -er¹.] A native or inhabitant of Stockholm.

1938 *Daily Tel.* 22 Jan. 12/3 Modern Stockholmers..are accustomed to fairly substantial midday meals in restaurants. 1968 P. B. Austin *On Being Swedish* iii. 21 To be *jäktad* (literally 'chased'), to be the victim of a mad whirl of engagements such as can only lead to stomach ulcers, is the contemporary Stockholmer's life-style. 1983 *Times* 30 June 13/4, I live in Stockholm and, like the vast majority of Stockholmers, buy a monthly card that gives me unlimited access to the underground and buses.

stocking, *sb.* Add: **7. a.** *stocking-heel* (earlier example).

1888 Kipling *Story of Gadsbys* 1 (*stage direction*), Spreads stocking-heel on open hand for inspection.

c. *stocking knitter* (later example), *-maker* (later examples).

1921 *Dict. Occup. Terms* (1927) § 374 *Stocking knitter*,..a frame hand..who attends a power-driven frame adjusted to knit elastic hosiery. 1779 in J. R. Anderson *Burgesses of Glasgow* (1935) 117 Wilson, Gabriel, stocking-maker. 1812 J. Melish *Trav. in U.S.A.* II. 55 Professions exercised in Pittsburg:..stocking-makers, taylors, printers, book-binders.

d. stocking bar, a counter or bar in a shop at which stockings are sold; **stocking cap,** a knitted woollen hat with a long tapered end which hangs down from the crown; **stocking filler,** a small present suitable for putting in a Christmas stocking; also *fig.*; **stocking mask,** a thin nylon stocking pulled over the face to disguise the features, used esp. by criminals; hence **stocking-masked** *a.*; **stocking stuffer** *N. Amer.* = *stocking filler* above; **stocking tights** = *tights sb. pl.* c; **stocking top** (later examples).

1962 *Guardian* 23 Feb. 8/3 The idea of a stocking bar came from America eight years ago. 1965 Stocking bar [see *bar sb.¹ 28 b]. 1902 Stocking cap [in Dict., sense 7 b]. 1978 *Times* 4 Mar. 22/5 The fishermen still wear their shirts and trousers of Portuguese tartan and long black stocking caps. 1959 *Listener* 10 Dec. 1054/1 A useful stocking-filler at 2s. is a gardener's measuring beaker, graduated for almost all fertilizers, insecticides, etc. 1973 *Radio Times* 20–27 Dec. 3 How's this for a stocking-filler? *Radio Times* has a complete Christmas package for you. 1979 M. Babson *Twelve Deaths of Christmas* xx. 109 Just tiny bits and pieces. Stocking fillers, small tokens. 1966 *Times* 16 May 10 Three men in stocking masks raided Martins Bank in South Audley Street. 1978 G. Greene *Human Factor* I. ii. 25 He might object to a stocking mask all the same. 1971 *Daily Tel.* 24 Sept. 2/8 Six stocking-masked bandits..ambushed a lorry at Bethnal Green yesterday. 1977 N. Adam *Triplehip Cracksman* xviii. 186 Corny stocking-masked villains. 1976 *Globe & Mail* (Toronto) 7 Dec. 24/7 Rockefellers don't stress paperbacks this time of year—the market for stocking stuffers is apparently limited. 1977 *Time* 17 Jan. 28/2 Around holiday season, stocking-stuffer items like *The Slipper and the Rose* usually show up, all covered in glitter and good will. 1967 *Economist* 5 Aug. 517/1 Manufacturers report orders..up by 23 per cent, mainly thanks to the invention of stocking tights..to go under mini skirts. 1977 J. Wainwright *Nest of Rats* I. i. 9 One leg of her stocking-tights badly torn. 1859 Geo. Eliot *Adam Bede* III. vi. liii. 305, I can count a stocking-top [in knitting] while a man's getting 's tongue ready. 1935 N. Mitchison *We have been Warned* I. ii. 19 To wear the sgian dhu in his stocking top, as he was allowed to with the kilt. 1978 R. H. Lewis *Antiquarian Bks.* ii. 47 Nina Hamnet..remembered..for her disconcerting habit of keeping her money in her stocking tops.

stocking-foot. Add: **d.** The loose or pendent part of a stocking-cap. *rare*⁻¹.

1921 D. H. Lawrence *Sea & Sardinia* v. 161 The old boy brings his stocking-foot over the left ear.

Hence **stocking-footed, (-feeted)** *a.*, having stocking-feet; in stocking-feet.

1926 J. F. Dobie *Rainbow in Morning* (1965) 99, I had a bay, white-faced, and stocking-footed horse called Buck. 1973 Stocking-feeted [see *leather-jacketed* s.v. *Leather *sb.* 5 d].

stock-in-trade. Add: **b.** Also *attrib.*

1931 L. H. Myers *Prince Jali* ii. 26 The point was that his parents and the world each presented the other with something intelligible, they presented stock-in-trade figures between whom a stock-in-trade intercourse was possible. 1970 P. Oliver *Savannah Syncopators* 100 These repertoires of traditional songs, stock-in-trade lines and phrases and sudden original words and verses. 1977 *Word 1972* XXVIII. 190 Bloomfield discussed such stock-in-trade instances as *adder* (for *nadder*), *newt* (for *eft*), and *apron* (for *napron*), to illustrate one type.

stockist (stǫ·kist). [f. Stock *sb.*¹ + -ist.] One who stocks (certain) goods for sale.

1922 *Autocar* 10 Nov. 52 We are Stockists of High grade Cars. 1941 *Picture Post* 3 May 32/1 Write for style folder and name of nearest stockist to Swallow Raincoats Ltd. 1956 *People* 13 May 14/1 (Advt.), See the many other advantages at your local stockist. 1977 D. J. Elliott in D. Marcus *Best Irish Short Stories* II. 150 Our range of Sheer Elegance shampoos and lipsticks. Ask for samples at our nearest stockists.

stock-jobber. b. (Examples.)

1833 *Niles' Reg.* XLIV. 570/1 The 'black-leg' in the gambling houses..more fairly takes the chances of the play, than the stock-jobber on 'change. 1911 H. S. Harrison *Queed* 107 If a man became the greatest stock-jobber in the world, who would remember him after he was gone.

stock-keeping, *vbl. sb.* (Earlier example.)

1844 W. Wakefield in *N.Z. Company Rep.* (1845) xvii. 139 The sport of hunting them [*sc.* deer]..would afford a manly amusement to the young Colonists, fitting them for the more serious occupations of stock keeping and wool growing.

stockman. Add: **1.** (Further Austral. examples.)

1862 R. Henning *Let.* 2 Nov. (1966) 113 Mr Palmer was not at home, but we were received by an amiable stockman. 1911 C. E. W. Bean *'Dreadnought' of Darling* xi. 100 The subordinate hands on the station—..the groom, stockman, rouseabouts—live in the men's huts, close behind the homestead. 1929 K. S. Prichard *Coonardoo* 5 Nowadays..aboriginal stockmen usually receive a small wage as well as payment in kind by rations. 1944 *Living off Land* iv. 94 Before the sun rises..the stockman's 'boys' are back with the horses. 1962 A. Upfield *Will of Tribe* iii. 27 'You have two white stockmen.'..'Just the two,' replied the cattleman. 1978 O.

WHITE *Silent Reach* xviii. 191 'Where are the stockmen?' ..'Out looking for the Dalziel girl.'
Hence **sto·ckmanship**, the art of raising or looking after livestock.
1959 *Farmer & Stockbreeder* 22 Dec. (Suppl.) 7/2 As science develops..the art of good farming will remain—the art of knowing when and how to work the soil, the art of stockmanship. **1969** J. G. S. & F. DONALDSON *Farming in Britain Today* II. xviii. 170 Efficient planning of the battery can almost eliminate the need for skilled stockmanship. **1979** *Daily Tel.* 1 May 11/2 Stockmanship, record-keeping, observation, simple disease control..will be the necessary tools of his [*sc.* the pigman's] trade.

stock-market. Add: **1. c.** *attrib.*
1930 W. C. BROOKS *How Stock Market Really Works* viii. 121 There followed closely upon the Hatry crash the break of the stock market boom in New York. **1951** M. MCLUHAN *Mech. Bride* (1967) 7 Speed of communication and movement makes possible at the same time such diverse facts as stock-market operations, international armies, [etc.]. **1964** —— *Understanding Media* xxi. 207 The classified ads (and stock-market quotations) are the bedrock of the press. **1977** *Listener* 17 Apr. 204/1 The great stock market crash of 1929.

stock out. *Business.* [f. STOCK *sb.*[1] + OUT *adv.*] An occurrence of being out of stock of an item wanted by customers.
1957 CLARK & GOTTFRIED *Dict. Business & Finance* 339/1 Stock-out. **1967** E. DUCKWORTH in Wills & Yearsley *Handbk. Management Technol.* 107 Because delivery from the wholesaler was so short the consequences of any 'stock out' would not be serious. **1969** J. ARGENTI *Managem. Techniques* v. 26 Symptoms: frequent stock-outs or frequent disposals of surpluses. **1979** *Washington Post* 22 June A 12/3 Its allocation rules 'do not prevent long lines, stockouts, and early closings at retail outlets'.

stockpile (stǫ·kpəil), *sb.* orig. *U.S.* Also **stock-pile, stock pile.** [f. STOCK *sb.*[1] + PILE *sb.*[3]] **1.** A pile of coal or ore accumulated at the surface after having been mined.
1872 *Trans. Amer. Soc. Civil Engineers* II. 30 This covers the cost from miners' hands to cars or stock pile. **1912** C. E. VAN BARNEVELD *Iron Mining in Minnesota* 140 For lighter stripping work and stock-pile loading, the 70-C Bucyrus is quite largely used. **1958** *Engineering* 7 Feb. 188/1 Limestone is fed into a swing hammer-mill either direct from tipping lorries or by bulldozing from a stock pile.
2. a. A reserve or store of goods or commodities, esp. one accumulated in anticipation of shortage or market fluctuation.
1942 *Sun* (Baltimore) 15 Jan. 1/2 The facilities of new car dealers will be used to store for a year or more an estimated 130,000 new passenger automobiles under a 'stock pile' plan. **1943** *Times* 15 Dec. 5/6 The complete success achieved was due to..the statesmanship of all nations represented, especially those with resources outside their occupied lands and therefore able to make stockpiles of supplies at the expense of other nations less fortunate. **1957** *Economist* 30 Nov. 791/1 In the postwar years the cartel has not been restrictive. Helped by stockpile buying of industrial diamonds and by a demand for gems as a hedge against inflation, its policy has been to hold prices down rather than force them up. **1958** *Manch. Guardian* 25 Feb. 16/6 No wool from the British stockpile is included in this week's catalogue. **1962** *Economist* 20 Jan. 249/3 The United States will not now authorise sales of stockpile tin below £965 a ton. **1970** *Listener* 23 July 107/3 Most European countries keep a 2 months' stockpile of oil by government command. **1972** D. HASTON *In High Places* xii. 150 The supplies were flowing well through the icefall;..there was a great stockpile at Camp I.
b. *spec.* An accumulation of nuclear weapons.
1946 *Rep. Internat. Control Atomic Energy* (Dept. of State, Washington) III. i. 31 How can a strategic balance be maintained between nations so that stockpiles of fissionable materials will not become unduly large in one nation and small in another? **1947** *Nature* 11 Jan. 48/1 A.D.A. should take over..the right of ensuring that any dangerous products were consumed in [atomic] power plants and that no excessive stockpile be produced. **1955** [see *ATOMIC *a.* 2 e]. **1957** *Times* 6 Nov. 9/6 Mr. Dulles said to-night that the United States was considering the problem of establishing stockpiles of nuclear weapons in Europe for N.A.T.O. forces' use in case of emergency. **1969** *Daily Tel.* 16 Sept. 22/7 The threat of nuclear war was increasing every day with the mounting nuclear stockpiles. **1976** *Survey* Summer–Autumn 193 The total explosive energy that could be released by the strategic stockpile is a measure frequently used to compare US and Soviet forces.
c. *fig.*
1945 J. STEINBECK *Cannery Row* xvii. 104 The sea rocks and the beaches were his stock pile. **1957** *Listener* 21 Nov. 826/1 Imperialist behaviour built stockpiles of national resentment. **1966** *Electronics* 31 Oct. 23 Stockpiles of good technical men in some of the aerospace companies. **1982** R. LUDLUM *Parsifal Mosaic* viii. 111 What he learned—what he *thinks* he learned—has turned him into a stockpile of nitro.

sto·ckpile, *v.* orig. *U.S.* [f. the sb.] **1.** *trans. Mining.* To heap up (ore, coal, etc.) in piles at the surface.
1921 E. W. DAVIS *Magnetic Concentration of Iron* 136 It may be necessary to mine, crush, and roast perhaps three tons of ore, cob, fine grind, and concentrate two tons of ore, dewater and agglomerate one ton of ore, stockpile one-half ton of ore, and dispose of two tons of tailings. **1937** —— *Magnetic Roasting of Iron Ore* 3 The ore being treated is a coarse tailing product rejected from existing concentration plants. It is in ideal physical condition for this process and is of no value at the present time, altho it has been mined, crushed, and stock-piled.
2. a. To accumulate a stock of (something); *spec.* to build up a stock of (nuclear weapons). Also *absol.*
1943 *Sun* (Baltimore) 28 Apr. 7/4 The Government at last began to 'stockpile' 100 octane gasoline. **1947** *Ibid.* 1 Jan. 6/3 It can be assumed that similar weapons..are now being perfected and stock-piled for future use. **1957** *Times* 18 Nov. p. xxxii/1 Decisions to buy and stockpile or to hold off and release stocks are, no doubt, dictated by consumer demand. **1959** *Listener* 16 July 88/2 President De Gaulle's refusal to allow American nuclear warheads to be stockpiled on French territory. **1974** G. MARKSTEIN *Cooler* lxxvii. 254 She..had stock-piled the pills the medical officer gave her until she had collected a fatal dose. **1976** *Country Life* 11 Mar. 638/2 It looked as if inflation would mean ever-advancing prices so wine merchants..started to stockpile.
b. *fig.*
1959 *News Chron.* 9 July 4/5 To stockpile acting talent of splendid calibre. **1959** *Daily Tel.* 2 Sept. 16/1 Employers urged to 'stockpile' labour. **1966** [see *SACK *v.*[1] 8 a]. **1975** *Language for Life* (Dept. Educ. & Sci.) xxiv. 347 Pre-service education is not a phase in which the intending teacher must stockpile resources for a lifetime.
Hence **sto·ckpiled** *ppl. a.*; **sto·ckpiler.**
1951 *Business Week* 24 Nov. 26 (*heading*) Stockpilers are dipping in now and then to keep both civilian and military industry going. **1972** *Sci. Amer.* Jan. 22/1 There could be uncertainties about the performance of stockpiled weapons. **1979** *Guardian* 23 May 15/1 Agonised consumers deciding whether to take the stock-piled bread out of the freezer and fill it up with petrol.

sto·ckpiling, *vbl. sb.* orig. *U.S.* [f. *STOCKPILE *v.* + -ING[1].] **1.** The action of making a stockpile of goods, raw materials, or nuclear weapons.
1943 *Sun* (Baltimore) 20 July 9/4 Suspension of purchases of Australian wool for stockpiling. **1946** *Rep. Internat. Control Atomic Energy* (Dept. of State, Washington) II. v. 25 It is, furthermore, clear that the stockpiling of appreciable quantities of fissionable material suitably denatured, must precede the development of these safe power reactors. **1953** E. HYAMS *Prophecy of Famine* 41 But to do so would upset our stock-piling and other contingency supplies. **1959** *Listener* 18 June 1066/2 France's refusal to allow stockpiling of American atomic weapons on French soil for the Nato fighter-bomber force.
2. *attrib.* and *Comb.*, as *stockpiling act, policy, programme, purchase, purpose.*
1946 *Times* 26 July 7/1 This week President Truman has signed the Strategic and Critical Materials Stockpiling Act. **1945** *Sun* (Baltimore) 30 Mar. 7/5 The committee will review the food procurement and stockpiling policies of war agencies. **1943** *Ibid.* 20 July 9/4 Officials of agencies concerned with the stockpiling program. **1944** *Times* 14 Feb. 7/1 Certainly it would hardly be wise to start a 'stockpiling' programme. **1949** *Times* 10 Sept. 5/7 The article provides safeguards against..injuring British interests in their commercial export as a result of stockpiling purchases. **1947** *Sun* (Baltimore) 20 Dec. 2/4 To facilitate the procurement of such raw materials by the United States for stockpiling purposes.

sto·ck-take, stocktake, *sb.* [f. the vb.] An instance of stock-taking.
1972 J. BROWN *Chancer* (1974) xii. 139 Three bad stock-takes and you're out. First stock-take I had it wasn't bad... The next big stock-take, the area manager, he came down. **1973** *Daily Tel.* 13 Apr. 18 The introductory work of stocktaking and separating VAT stock took two hours above that required by an efficient business for a normal necessary periodical stocktake. **1979** *Ibid.* 29 June 3/1 The committee decided to lock the door to ensure the 'privacy' of the club bars and stock after a stock-take before handing over to a new steward.

stock-work. (Earlier and later examples.)
1808 R. JAMESON *Syst. Mineral.* III. xi. 255 A Stockwerk, is a mountain-mass of greater or less extent, traversed in all directions by a very great number of small veins. **1957** *Mineral. Mag.* XXXI. 588 In these the tin ores occur as stockworks of topaz-bearing greisen.

stød (stœd). *Linguistics.* [Da., lit. 'push, jolt'.] A glottal stop or catch (see quot. 1973).
1954 PEI & GAYNOR *Dict. Linguistics* 204 *Stød*, the Danish term for *glottal stop*.., often used by phoneticians for other languages, too. **1964** J. C. CATFORD in D. Abercrombie et al. *Daniel Jones* 36 *Voiced creak* may be one form of the 'stød' in Danish. **1973** J. D. O'CONNOR *Phonetics* vii. 237 Danish has the same stress system as English with the addition that each stressed syllable may or may not have the 'stød' or glottal stop added to it (in fact, it is rarely a complete stop but rather a short period of creaky voice). **1977** C. F. & F. M. VOEGELIN *Classification & Index World's Lang.* 140 Subdialectal division into Northern and Southern on the basis of the so-called 'stød' feature, usually a 'glottal creak' in Southern corresponding to pitch-stress phenomena in Northern Danish. **1980** *Amer. Speech* LV. 61 The Danish words that have the glottal catch or stød are generally those with accent 1 elsewhere in Scandinavia.

stodge, *sb.* Add: **1. b.** Also *spec.* heavy and usu. fattening food (often with little nutritional value). *colloq.*
1963 R. I. MCDAVID *Mencken's Amer. Lang.* 296 *Pudding* implies what we normally call *stodge.* **1963** *Times* 13 Feb. 5/2 If the prisoner could not tolerate all the 'stodge' he became undernourished. **1976** *Milton Keynes Express* 25 June 4/5 Remember that no exercise programme will work if not backed by sensible eating patterns and cut out stodge from today. **1980** *Times* 28 Nov. 3/2 The writers complain of surviving on stodge like potatoes and rice.
2. b. (Earlier example.)
1903 FARMER & HENLEY *Slang* VI. 373/2 *Stodge*,..a heavy meal.
c. Food of any kind. *slang.*
1890 BARRÈRE & LELAND *Dict. Slang* II. 307/2 *Stodge*,.. (popular and thieves), food. **1917** 'TAFFRAIL' *Sub* ii. 72 Cream, jam, mineral waters and all other sorts of 'stodge'. **1929** F. C. BOWEN *Sea Slang* 133 *Stodge*, food, generally used in the gunroom only. **1940** M. MARPLES *Public School Slang* 167 *Stodge* (Rugby),..= food—e.g. 'I've got a box of stodge.'
4. A hard effort; an unfulfilling occupation.
1846 J. C. PATTESON *Let.* in C. M. Yonge *Life J. C. Patteson* (1874) I. iii. 58 Reading books for the second or third time is light work compared to the first stodge at them. **1873** C. M. YONGE *Pillars of House* II. xxiv. 34 To let him go on here in the stodge is a bit of short-sightedness I can't understand.
5. = *STODGER.
1922 E. V. LUCAS *Genevra's Money* xxiii. 152 How silly of us to think he was going to be a stodge.

stodge, *v.* Add: **4.** (Examples.)
1912 F. M. HUEFFER *Panel* I. iii. 93, I tell you, I'm tired! Used up! I must have comfort, quiet! I can't stodge away any more. *Ibid.* 98, I plodded and stodged for just that, and nothing else. **1928** —— *Last Post* II. iii. 259 They ought no longer to go stodging along in penury. **1939** D. JONES *Let.* 17 Jan. in R. Hague *Dai Greatcoat* (1980) ii. 89 Writing is odder than painting... One seems to stodge on and scratch out for hours and days and then sometimes..something breaks through. **1959** *Listener* 29 Oct. 748/2 Poor Dr. Bronowski seems fated to the *pas seul*... His fellows stodge around, looking severe and sagacious and sound and sensible.
5. To walk or trudge through mud or slush; to walk with short heavy steps. Occas. *trans.*, to trudge through (mud). *dial.* or *colloq.*
1854 A. E. BAKER *Northampt. Gloss.* II. 306 *Studging*, walking with short heavy steps; always used with the adjunct along. 'He goes *studging* along.' **1902** *Aberdeen Weekly Free Press* 7 June 3/6 A polissman wha was comin' stodgin' doon the street. **1920** W. DEEPING *Second Youth* xxiii. 195 The 'Old Man' and his orderly stodged back again up a waterlogged communication trench. **1929** —— *Roper's Row* viii. 83 She had seen the feet of cattle stodging the mire in Melfont. *Ibid.* xiii. 138 A very stout woman..stodged round the grave after the service was over.

stodger (stǫ·dʒəɹ). *colloq.* [f. STODG(Y *a.* + -ER[1].] A stodgy person; one who is lacking in spirit or liveliness. Hence **sto·dgery**, behaviour characteristic of such a person.
1905 *Punch* 25 Jan. 62/1 The other regular old stodgers who go to all the parties within a radius of six miles. **1907** *Punch* 9 Jan. 20/2 Well, father's quite right, they are the most awful stodgers. You know they are. **1920** W. DEEPING *Second Youth* iv. 31 If you were starving, Miles, I suppose you would walk down Oxford Street and say nothing. What stodgery! We middle-class people are hopeless!

stodgy, *a.* Add: **2. a, b.** (Earlier and later examples.) Also applied to other objects, activities, etc.
1874 L. TROUBRIDGE *Life amongst Troubridges* (1966) 89 We had meant to play Rats and Ferrets, but we had to begin a stodgy game of Old Maid. **1885** C. M. YONGE *Nuttie's Father* I. x. 111 One of the stodgey [*sic*] old clergymen in books. **1976** J. I. M. STEWART *Memorial Service* ii. 24 The stodgy lime-streaked effigy of Provost Harbage..is really more congruous with the spirit of the place. **1977** *National Observer* (U.S.) 1 Jan. 5/4 It was a stodgy old company when he came to it as president of the international division. **1977** *Time* 31 Jan. 13/2 Leidigkeit, 38, has brought scandal and notoriety to Bonn's Ermekeil Strasse, formerly a quiet, slightly stodgy row of shops, middle-class town houses and student flats.
d. Applied *loosely* to music, its performance, interpretation, etc.
1934 C. LAMBERT *Music Ho!* v. 294 The stodgy and academic imagination of *Verklarte Nacht.* **1959** *Times* 12 Jan. 12/3 It was surprising that Miss Puppulo was so stodgy in some early miniatures at the start of the programme. **1974** *Early Music* Apr. 81 It is so easy..for four viols to be too stodgy. **1978** R. DONINGTON in J. M. Thomson *Future of Early Music in Britain* 14 The dodge.. is to get that massive resonance without sounding in the least thick and opaque and stodgy and Straussian.

stoep. Delete ‖ and add: Also † **stoop** (cf. STOOP *sb.*[3]). **a.** (Earlier and later examples.)
1797 A. BARNARD *Let.* 10 July in *S.Afr. a Century Ago* (1901) 57 As for the young Dutchmen, I saw hardly any; the young ones prefer smoking their pipes on the *stoep*. **1798** —— *Jrnl.* 21 May in *Lives of Lindsays* (1849) III. 457 Their *stoop* was covered with a set of large idle boors in their blue jackets, sons of the family—men who do hardly anything beside eating and smoking. **1804** J. BARROW *Trav. Interior S. Afr.* II. ii. 104 He..parades the *stoop*, or raised platform before the door. **1805** *Gleanings in Afr.* ii. 17 A stone terrace, extending the whole length of the house, and elevated a few feet above the level of the street, is the grand promenade of the family; this is called the *Stoop.* **1939** tr. *E. N. Marais's My Friends the Baboons* ix. 101 We could follow the whole tragedy step by step from the stoep of Mr. van Heerden's house. **1966** E. PALMER *Plains of Camdeboo* ii. 26 A wide

stoep running round three sides in the manner of Karoo houses. **1980** *Listener* 17 July 66/3 A farmer's wife in Natal..sitting..on the stoep.
b. *stoep lantern, plant*; **stoepsitter**, one who habitually sits idly on the stoep of his house.
1971 *Het Suid Western* 14 May 9 Wrought iron porch or stoep lanterns, wired in new condition—R5 each. **1961** *Argosy* Mar. 20 She attended to her stoep plants. **1934** C. R. SWART *Africanderisms* (typescript), *Stoepsitters*,..a sluggard or lazy person; sometimes humorously applied by townsmen to farmers, who used to spend much of their time on the stoep, drinking their favourite beverage, coffee. **1948** O. WALKER *Kaffirs are Lively* 92 They don't work. They're stoep-sitters, coffee-tipplers and pipe-spitters. **1972** *Sunday Times* 24 Sept. 19 Topical remarks, especially by the three 'stoepsitters', are made through the play.

stog, *v.*[2] Add: **1.** Also *fig.*
1928 J. Y. T. GREIG *Breaking Priscian's Head* 60 Old pedantic grammarians stogged to the neck in Latin, have done their work too well.

stogy, *a.* and *sb.* Add: Now freq. *stogie.* **A.** *adj.* **b.** (Examples.)
1903 FARMER & HENLEY *Slang* VI. 373/2 *Stogy-cigar*, a rough coarse cigar. **1930** J. OMWAKE *Conestoga Six-Horse Bell Teams* 118 The Conestoga wagon gives its name to the Stogie cigar, a great thin coarse one, supposed to have been originally a foot long and made for the delectation of the wagoner.
B. *sb.* **b.** (Earlier and later examples.)
1873 J. O'CONNOR *Wanderings of Vagabond* 52 After the lunch liquors and cigars (red-eye and stogies), the best the place afforded, were introduced by the host. **1916** C. SANDBURG *Chicago Poems* 47 He lighted a three-for-a-nickel stogie. **1930** J. DOS PASSOS *42nd Parallel* 19 He was smoking a thin black stogy of a sort Fainy had never seen before. **1957** V. PACKARD *Hidden Persuaders* ix. 103 The man who puffs on his cigar is sucking his thumb while the man who chews vigorously on his stogie is a nail biter.

Stoic, *a.* Add: **3.** *Comb.*, as *Stoic-Christian, -Epicurean, -Megaric* adjs.
1933 A. N. WHITEHEAD *Adventures of Ideas* iii. 43 These doctrines have all weakened the Stoic-Christian ideal of democratic brotherhood. **1948** L. SPITZER *Linguistics & Lit Hist.* 15 *Pantagruélisme*, the name given by Rabelais to his stoic-epicurean philosophy. **1966** *Philos. Rev.* LXXV. 246 Rescher takes these..as evidence that Arabic logic was directly influenced by the Stoic-Megaric tradition.

stoicheiometry. Add: Now usu. thus (in the U.K.) or as **stoichio-**. In mod. use, the quantitative relationship between the substances in a reaction or compound. (Further examples.)
1971 W. F. PICKERING *Mod. Analytical Chem.* iv. 153 It can also be used to determine the stoichiometry of reactions occurring in solution. **1975** R. F. BROWN *Organic Chem.* viii. 192 Stoichiometry, the weight relations in chemical reactions, must be kept in mind in the study of rates of reaction.
stoichiometric *a.* (further examples).
1921 *Jrnl. Geol.* XXIX. 533 The stoechiometric relation between MgO and FeO in olivine and bronzite. **1962** J. H. WHITE *Inorganic Chem.* ix. 108 The formation of solid solution does not appear to be very different [from that of metallic compounds] except that..no stoichiometric relationship between the constituent metals exists. **1965** PHILLIPS & WILLIAMS *Inorganic Chem.* I. viii. 296 Metal fluorides but not hydrides are stoicheiometric.

stoicize, *v.* (Earlier example.)
1718 C. HAYES tr. *Addison's Dissertation upon Roman Poets* 42 Pompey..ought to have been very much stoiciz'd indeed, who, despoil'd of all the Goods of Fortune, could place the Sum of his Felicity in meer naked Virtue.

Stoico- (stōᵘ·iko), combining form of L. *stōicus* or Gr. στωϊκός STOIC, as in *Stoico-Platonic, -sybaritical* adjs.
1979 M. A. SCREECH *Rabelais* iii. 38 The inerrant Stoico-Platonic Christian sage. **1822** M. EDGEWORTH *Let.* 12 June (1971) 406 He and Harriet and Fanny too declare it is too much trouble to hold a parasol. I believe you too are of the same *Stoico-sybaritical* sect so I will waive the subject.

Štokavian (ʃtǫkā·viăn), *sb.* (and *a.*). Also **Shtokavian, Stokavian, stokavian.** [f. Serbo-Croat *štokavščina* (*štokavski* adj.): see -IAN.] A widely spoken dialect of Serbo-Croat on which the literary language is based. Also *attrib.* or as *adj.*
[**1911** *Encycl. Brit.* XXIV. 695/2 Servian is sometimes called *shtokavski* because the Servian word for 'what' is *shto*, whereas the Croats say *cha* for *shto*, and therefore their language is called *chakavski*.] **1925** P. RADIN tr. *Vendryès's Lang.* 291 In Italy, in the province of Campobasso, there is a Serbo-Croatian colony..which..speaks a dialect of the Stokavian type. **1939** L. H. GRAY *Foundation of Lang.* 355 *Serbo-Croatian*, with three dialects conventionally named according to the way in which they form the word for 'what': *Štokavian* (the basis of the literary language), *Čakavian*, and *Kaykavian.* **1949** R. JAKOBSON *Slavic Languages* 4 Serbocroatian from East to West presents three basic groups: Štokavian, Čakavian and Kajkavian. **1964** M. PARTRIDGE *Serbo-Croatian* 13 Three distinct basic dialects exist in spoken Serbo-Croatian. They are referred to as *čakavian, kajkavian* and *štokavian* according to whether the word 'ča?', 'kaj?' or 'što?' is used respectively as the interrogative pronoun meaning *what*? **1974** *Encycl. Brit. Macropædia* XVI. 867/1 The literary Serbo-Croatian language was formed in the first half of the 19th century on the basis of the Shtokavian dialects that extend over the greater part of the Serbo-Croatian territory in Yugoslavia. **1976** *Language* LII. 375 The dat. sg. + *i* here..represents a morphological change rather than a phonological difference between kajkavian and štokavian. **1977** *Archivum Linguisticum* VIII. 91 In štokavian Serbo-Croat (on which the standard language is based), the rising accent is a disyllabic one.

stoke(s (stōᵘk(s)), *sb.*[4] *Physics.* [f. *STOKES[1].] Proposed in Ger. by M. Jakob 1928, in *Zeitschr. f. techn. Physik* IX. 22/1.] The unit of kinematic viscosity in the C.G.S. system, equal to 1 cm.^{2} sec.$^{-1}$
1931 G. BARR *Monogr. Viscometry* i. 4 Jacob [*sic*] has proposed that the C.G.S. unit be called the 'stokes', but the suggestion has not yet had time to bear fruit. **1934** [see *CENTISTOKE(S)]. **1961** V. L. STREETER *Handbk. Fluid Dynamics* I. 14 The unit of one square centimeter per second is called a stoke. The centistoke (= 0·01 stoke) is often a more convenient unit. **1964** SABERSKY & ACOSTA *Fluid Flow* i. 10 In the c.g.s. system,..the unit of absolute viscosity..is called a poise, and the unit of kinematic viscosity, 1 cm.^{2}/sec., is called a stoke.

stoke, *v.*[2] Add: **1. a.** Also, to feed or build up (a fire).
1942 E. LANGLEY *Pea-Pickers* x. 148 The hut was warmed by a little red fire which the fair-haired comrade stoked. **1971** G. JONES in Jones & Elis *Twenty-Five Welsh Short Stories* 106 That night, when he went into the house, he saw that the big iron double bed had been moved down into the middle of the kitchen and a great furnace of a fire stoked up in the chimney.
c. To excite, thrill, elate. *slang* (chiefly *Surfing*).
1963 *Pix* 28 Sept. 63 A good stomping movement that 'stokes' the tourists is worth two extra points. **1965** *S. Afr. Surfer* I. 3/3 Your magazine stoked me out of my mind. *Ibid.* 7/1 We will let him stoke you on some of the modern variations of body riding.
2. Also *absol.* with *up.*
1901 'R. ANDOM' *Troddles & Us & Others* iv. 47 Troddles stoked-up on bread-and-butter pudding to such an extent that I wondered how on earth he could..drag himself about. **1946** R. LEHMANN *Gipsy's Baby* 29, I have often noticed how much less greedy children of the proletariat are than others. One would imagine that they would be more absorbed in the problem of stoking up. **1975** J. SYMONS *Three Pipe Problem* xvi. 155 They sat in one of the high-backed compartments where the punters came to stoke up after their losses.
3. stoke-up *slang*, a large or sustaining meal.
1955 J. THOMAS *No Banners* xv. 133 Later..it would be possible to go to the black-market eating-houses for an occasional 'stoke-up'.
stoked *ppl. a.* (later examples in sense *1 c of the vb.); also, keen or 'hooked' *on. slang.*
1963 [see *BOARD *sb.* 1 b]. **1968** *Surfer Mag.* Jan. 47/3, I realized they're really stoked. **1969** *Sunday Mail* (Brisbane) 2 Feb. 20/3 I'm stoked on Chinese food. **1970** *Studies in English* (Univ. Cape Town) I. 33 People bitten by the *surf bug*..are really *stoked* on surfing. **1976** *N.Y. Times Mag.* 12 Sept. 40/1 Something like 10 million Americans..are stoked on floating about three inches over the paved surfaces of planet earth. Their flotation device is the new, Nasworthy-improved skateboard. **1977** *Skateboard Special* Sept. 2/1 The guy was really stoked but he fell off a nose wheelie and ended up taking a trip to McDonalds.

Stokes[1] (stōᵘks). *Physics.* The name of Sir George *Stokes* (1819–1903), Irish-born physicist and mathematician, used in the possessive and *attrib.* to designate concepts and phenomena discovered by him or arising out of his work: **a.** *Stokes' theorem*: the theorem that the line integral of a vector function round a closed path is equal to the surface integral of the curl of the function over any surface bounded by the path.
1893 J. J. THOMSON *Notes Recent Res. Electr. & Magnetism* i. 10 Now by Stokes' theorem $\int (Xdx + Ydy + Zdz)$ taken round a closed circuit is equal to [etc.]. **1940** E. T. BELL *Development Math.* xviii. 364 Stokes' theorem, its proof, and its generalizations have developed into a thriving industry of modern analysis. **1975** R. L. FERRARI *Introd. Electromagnetic Fields* vi. 109 We have postulated Maxwell's equations in their integral form... Using the vector calculus rule, Stokes' theorem, these can be transformed into differential relationships required to hold everywhere in space.
b. *Stokes' law*: the statement (not always true) that in fluorescence the wavelength of the emitted radiation is longer than that of the radiation causing it. Also *Stokes' line, shift*, etc., with reference to spectral emission lines at a lower frequency than the stimulating or incident radiation.
[**1902** *Encycl. Brit.* XXXII. 124/1 According to the experimental law of Stokes, the wave-lengths of the fluorescent radiation are longer than those of the radiation which excites it.] **1926** R. W. LAWSON tr. *Hevesy & Paneth's Man. Radioactivity* v. 57 It is the same condition as that with which we meet in optics in connection with the occurrence of fluorescence according to Stokes' law. **1949** P. PRINGSHEIM *Fluorescence & Phosphorescence* ii. 163 The fourth row of the same table shows..the distances of the first Stokes line from the exciting line. *Ibid.* vii. 556 Even if the absorption and emission correspond to the same electronic transition,..relatively large Stokes shifts have a great probability [in crystal phosphors]. **1973** *McGraw-Hill Yearbk. Sci. & Technol.* 307/1 Other consequences of the transient nature of the scattering are that the Raman-shifted (Stokes) pulse is both narrowed in time and delayed with respect to the exciting pulse. **1975** D. H. BURRIN in Williams & Wilson *Biologist's Guide to Princ. & Techniques Pract. Biochem.* v. 146 The energy emitted from these molecules in regaining the ground state within a period of less than 10^{-8} s gives rise to a fluorescent peak, showing the Stokes' shift. **1978** P. W. ATKINS *Physical Chem.* xvii. 562 (*caption*) Stokes and anti-Stokes rotational Raman lines.
c. *Stokes' law* (or *formula*): the statement that the resisting force on a spherical particle moving through a fluid is $6\pi\eta Vr$ (where η is the viscosity of the fluid, V the speed of the particle, and r its radius), so that its limiting rate of fall is $2gr^2\rho/9\eta$ (where g is the acceleration due to gravity and ρ the difference in density between the particle and the fluid).
1910 *Rep. Brit. Assoc. Adv. Sci. 1909* 407 To test Stokes's formula for air, the size, density, and terminal velocity of fall of some spherical spores were determined. **1936** *Discovery* Nov. 349/2 This period [of settlement of dust particles] is dependent on the application of Stokes' law, and hangs on particle size, density, etc. **1968** P. A. P. MORAN *Introd. Probability Theory* ix. 431 Assume that the particle, besides being spherical, is large enough for Stokes law to give a good estimate of the resisting force. **1974** G. S. ORMSBY in P. L. Moore et al. *Drilling Practices Manual* vi. 158 Since Stokes Law applies in a sand trap, large quantities of barites..may be settled from weighted drilling fluids. **1983** *Sci. Amer.* Apr. 128/2 Many of the grains in Middle Eastern coffee are too large to fall according to Stokes's law.

Stokes[2] (stōᵘks). The name of Sir Wilfrid *Stokes* (1860–1927), English engineer, used *attrib.* and *absol.* to designate a type of trench mortar invented by him.
1915 W. S. CHURCHILL *Let.* 7 Sept. in M. Gilbert *Winston S. Churchill* (1972) III. Compan. II. 1167 In the early part of June, Lloyd George and I were shown the Stokes gun in action. **1919** [see *LEWIS[3]]. **1919** *Athenæum* 25 July 664/1 'Stokes', the name of the inventor of the T.M., has, by metonymy, come to mean the trench mortar gun itself (so 'Nissen' = hut; 'Armstrong' = hut). **1923** KIPLING *Irish Guards in Great War* II. 146 Our own two-inch Stokes in the front line strove to cover the noise by separate rapid fire. **1930** G. B. SHAW *What I really wrote about War* (1931) 241 The thermit shower was produced by firing from Stokes guns a cloud of shells packed with it. **1974** A. PRICE *Other Paths to Glory* I. i. 16 Their dead hanging on the unbroken barbed wire among the dud shells and unexploded Stokes mortar bombs.

Stokes–Adams (stōᵘks,æ·dămz). *Med.* The names of William *Stokes* (1804–78) and Robert *Adams* (1791–1875), Irish physicians, who described the condition in 1846 (*Dublin Q. Jrnl. Med. Sci.* II. 73) and 1827 (*Dublin Hosp. Rep.* IV. 414) respectively, used *attrib.* to designate occasional transient cessation or extreme slowness of the pulse, esp. when caused by heart-block.
1903 R. H. BABCOCK *Dis. Heart* xxiv. 634 The diagnosis of Stokes-Adams disease presents difficulty when the paroxysms are characterized only by vertigo and increase of an already existing bradycardia. **1922** *Lancet* 13 May 933/2 In a woman of 49 with complete heart-block, who was under observation for seven days and nights, severe Stokes-Adams attacks were recurring almost every minute. **1947** SCHERF & BOYD *Cardiovascular Dis.* xxii. 318/1 In this book the term Stokes-Adams attacks will embrace all types [of circulatory standstill] resulting from a change of cardiac activity irrespective of whether they are due to cardiac standstill or tachycardia... We deal with a syndrome and not a disease entity. **1974** *Ciba Symposium* New Ser. XX. 133 AV blocks of all kinds, from first-degree block to complete block, including the Stokes-Adams syndrome.

stokesite (stōᵘ·ksəit). *Min.* [f. *STOKES[1] + -ITE[1].] A hydrated silicate of calcium and tin, $CaSnSi_3O_9.2H_2O$, found as colourless, transparent orthorhombic crystals.
1899 A. HUTCHINSON in *Phil. Mag.* XLVIII. 480 Among the specimens..recently acquired for the Cambridge Mineralogical Museum, has been found a crystal whose characters prove it to belong to a new mineral species. This mineral I propose to call Stokesite in honour of Sir George Gabriel Stokes, Bart., whose jubilee as Lucasian Professor was this year celebrated by the University. **1977** *Mineral. Mag.* XLI. 413 Stokesite in very small amounts has been reported at two localities in Czechoslovakia..while at Corrégo do Urucum, Brazil, it has been found..as spherical clusters of crystals up to 3 cm diameter.

Stokowskian (stǫkǫ·vskiăn), *a.* and *sb.* [f. the name of Leopold *Stokowski* (1882–1977), English-born American conductor + -AN.] **A.** *adj.* Of, pertaining to, or characteristic of Stokowski.

1961 *Times* 19 June 9/2 Inappropriate infusions of Stokowskian sensuousness marred..his interpretation of Orff's *Carmina Burana.* **1977** *Gramophone* Apr. 1556/2 The Stokowskian concentration and persuasiveness will be hard for anyone to resist. **1978** *Ibid.* Jan. 1263/3 The Schubert is by Stokowskian standards given an unmagical performance.
B. *sb.* An admirer of Stokowski.
1975 *Gramophone* Aug. 322/2 It is very much a performance for Elgarians as well as for the Stokowskians. **1978** *Ibid.* Jan. 1308/2 Still, Stokowskians will want this, although I am sure they will lament with me that CBS scheduled this record before the *Pastoral* Symphony, which Stokowski was due to record when he died.

stole, *ppl. a.* Add: (Later examples.) Now *colloq.*
1923 [see *HISN, HIS'N]. **1976** *Billings* (Montana) *Gaz.* 20 June 5-D/1 (Advt.), *Found* in Missoula: Male Great Dane Cross. Approx. 1½ yrs. old. Stole in Blgs. last Fall or Winter.

stolewise (stōᵘ·lwəiz), *adv.* [f. STOLE *sb.*[1] + WISE *sb.*[1]] Draped like a stole.
1922 JOYCE *Ulysses* 19 Buck Mulligan slung his towel stolewise round his neck.

Stolichnaya (stalī·tʃnaiä). [Russ., lit. 'of the capital, metropolitan'.] The proprietary name of a variety of Russian vodka.
1966 L. DEIGHTON *Billion-Dollar Brain* xi. 98 Stolichnaya..the only vodka I will drink. **1969** *Official Gaz.* (U.S. Patent Office) 25 Feb. TM 145 Stolichnaya. V/O Sojuzplodoimport. **1973** *Radio Times* 20 Dec. 110/4, 1 bottle Russian vodka (Nureyev recommends Stolichnaya). **1975** *Trade Marks Jrnl.* 4 June 1163 Stolichnaya Vodka. ..998,200. Vodka. Vsesojuznoje Objedinenie Sojuzplodoimport..32/34 Smolenskaja Square, Moscow, U.S.S.R.; Manufacturers and Merchants. —11th Sept. 1972. **1977** J. WAMBAUGH *Black Marble* (1978) i. 3 He..stealthily withdrew the bottle of Stolichnaya from the pocket of his raincoat.

‖ **stolkjærre** (stu·lkyerə). Also **stolkjaerre.** [Norw. (Bokmål), f. *stol* seat, STOOL *sb.* + *kjærre* cart.] A two-wheeled cart with seats for two persons. Hence as *v. intr.*, to ride in such a cart.
1885 *One & a Half in Norway* 125 The court-yard of the station was quite busy with carrioles and stolkjærres. **1924** *Public Opinion* 9 May 454/2 The traveller in stolkjærre or automobile ascends to snow-mantled plateaux. **1932** *New Yorker* 9 Apr. 59/2 Yachting on the Trollfjord, stolkjærreing through the Naeroedal, automobiling in the Baltic Capitals.

‖ **Stollen** (ʃtǫ·lən). Also **stollen, Stolle.** [Ger.] A rich fruit loaf, often made with nuts added.
1906 E. OSWALD *German Cookery for English Kitchen* 200 Stolle... Prepare the yeast as in recipe 'Napfkuchen'. **1927** *Daily Express* 20 Dec. 5/3 Germany has a cake similar to the Polish Strutzel, called Stollen. The dough should be made in the same way with quarter pound each chopped mixed peel, melon, angelica, prunes, and raisins, ..for a filling, and the icing made with lemon juice. **1959** H. SLESAR *Grey Flannel Shroud* (1960) i. 15 Fine European baked goods, the stollen and strudel and delicate little kuchen. **1975** *Woman* 17 May 15/3 There are more than 200 types of bread made in Germany, like..Stollen, a famous dryish bread filled with glacé fruit, almonds, raisins and currants. **1977** [see *PFEFFERKUCHEN]. **1977** *Sunday Tel.* (Colour Suppl.) 5 June 13/2 Church remains an integral part of the day, the marzipan in the stollen bread.

stolon. Add: Hence **stolo·nial** *a.*, of or pertaining to stolons.
1911 [see *EPICARDIAC *a.*]. **1964** *Oceanogr. & Marine Biol.* II. 317 High salinities..caused..reduction of stolonial material and fusion of hydranths.

‖ **stolovaya** (stalōᵃ·vaiä). [Russ.] A canteen, a cafeteria.
1943 E. M. ALMEDINGEN *Frossia* ix. 329 It is dinner time. Come on, let us eat at the station *stolovaya.* **1976** 'S. HARVESTER' *Siberian Road* vii. 75 The swingdoors of the *stolovaya* shut behind the Russian. **1982** *Spectator* 27 Mar. 21/3 The food in a Russian *stolovaya* (or 'diner').

STOLport (stǫ·lpǭɹt). orig. *U.S.* Also **STOL-port, stolport,** etc. [f. *STOL* (see *S 4 a), after *AIRPORT.] An airport for aircraft which need only a short runway for take-off and landing.
1968 *N.Y. Times* 14 Jan. 1/1 A stolport would serve planes that make a 'short take-off and landing'. They use runways much shorter than those required by commercial jets. **1968** *Science News* 7 Sept. 230 (*caption*) Frenetic ground travel to and from New York's main airports may be replaced by STOL-ports along the Hudson river. **1975** *Sunday Sun* (Toronto) 12 Oct. 17/1 As the plan suggests, the airport and the STOLport would take over all of Toronto's air travel eventually. **1976** *Globe & Mail* (Toronto) 16 Feb. 5/5 Nordair [is] getting permanent access to the Toronto island Airport and the Victoria STOLport in Montreal. **1980** *Times* 3 June 19/5 The company has drawn up plans for a 2,000ft 'stolport' (short takeoff and landing airport) in the east Shetlands basin. **1982** *Times* 24 June 3/4 A group of companies is proposing to build a small airport, to bring a different kind of transport interchange to the docklands known as a Stolport (Stol stands for short-take-off-and-landing).

Stolypin (stalī·pin). [The name of Pyotr Arkadyevich *Stolypin* (1862–1911), Russian conservative statesman.] **1.** *Stolypin's necktie*, the noose. *colloq.*
1909 J. R. WARE *Passing Eng.* 234/2 *Stolypin's necktie* (*Europ. Politics, 1897*), the final halter. This term was brought into fashion in 1907 (Nov.–Dec.), at a Duma then recently assembled in St Petersburg. One Rodicheff, an extreme Radical, brought in the term on 30th November 1907. **1974** *Encycl. Brit. Micropædia* IX. 583/1 Stolypin ..instituted a network of courts-martial... Within the few months of their existence they used 'Stolypin's necktie' (the noose) to execute more than 1,000 defendants.
2. Used *attrib.* and *absol.* to designate a type of railway carriage made for the transport of prisoners.
1970 HARARI & HAYWARD tr. *Amalrik's Involuntary Journey to Siberia* xi. 127 This was a so-called 'Stolypin' car, specially constructed for the transport of prisoners... They are named after the Tsarist Prime Minister and Minister of the Interior who introduced them after the first Russian Revolution of 1905. **1974** T. P. WHITNEY tr. *Solzhenitsyn's Gulag Archipelago* I. II. i. 491 The prisoners got used to calling this kind of railroad car a *Stolypin* car, or, more simply, just a *Stolypin.*

stoma. Add: **3.** *Surg.* A permanent opening made into a hollow organ; *spec.* one made from outside the body. Freq. *attrib.*, as *stoma patient, therapy.*
1937 R. SCHINDLER *Gastroscopy* xiv. 269 In cases of gastric resection the stoma of resection is generally easily seen if the 85° instrument is used. **1943** H. L. BOCKUS *Gastro-Enterol.* I. xxxi. 633 The stoma which is too large allows food to enter the jejunum immediately after eating. **1952** *Jrnl. Amer. Med. Assoc.* 25 Oct. 812/2 A special clinic devoted to intestinal stomas of all types has been established recently at the hospital. **1977** *Lancet* 15 Oct. 806/1 How often in a lifetime does a patient with an ileostomy have to return to hospital or seek advice from the Ileostomy Association or a stoma therapist? **1978** *Jrnl. R. Soc. Med.* LXXI. 519 All of these patients have become well rehabilitated, helped particularly by the excellent stoma-therapy services at these two hospitals. **1978** K. P. KRETSCHMER *Intestinal Stoma* 115 A stoma patient is well advised when entering an unfamiliar locality to inform himself first of bathroom facilities. **1980** *Recent Adv. Surg.* X. 281 Courses for the training of stomatherapists. **1981** *West Lancs. Evening Gaz.* 14 Jan. 14 (Advt.), Victoria Hospital Gastro Enterology Services Unit Stoma Therapy Nursing Service..are looking for a person who has the ability to communicate and assess the patient's psychological and social needs.
Hence **sto·mal** *a.*
1952 W. M. CRAPPER in F. A. Jones *Mod. Trends Gastro-Enterol.* xviii. 464 A stomal ulcer may appear, as we have seen recently, when a woman aged 73 developed ulceration 9 months after gastro-jejunostomy for a duodenal ulcer. **1979** J. P. DELANEY et al. in Najarian & Delaney *Gastrointestinal Surg.* 191 The only way symptoms of stomal gastritis can be relieved is by diversion of the upper gastrointestinal juices away from the stomach.

stomach, *sb.* **6. a.** For † *Obs.* read Now *rare* and add later examples.
1859 EARL GRANVILLE *Let.* in E. Fitzmaurice *Life Granville* (1905) I. xii. 344, I ought to..tell you of..the enormous weight off my stomach when I failed [to form a government]. **1965** E. B. WHITE *Let.* July (1976) 533 The city is very strange this summer—alternately deserted and packed, and the nearness of Harlem always in everybody's stomach.
10. a. *stomach muscle, ulcer, upset*; good for the stomach, *stomach powder.*
1965 P. O'DONNELL *Modesty Blaise* xviii. 199 Instinct tensed her stomach-muscles an instant before the woman dropped on her with both knees. **1911** E. WHARTON *Ethan Frome* vii. 135 I've a good mind to go and hunt up those stomach powders I got last year... Maybe they'll help the heart-burn. **1972** V. CANNING *Rainbird Pattern* vi. 115 He was restless himself from a substantial dinner and lay awake for hours wishing he had brought some stomach powder. **1945** A. HUXLEY *Let.* 27 May (1969) 527 He interferes with the normal functioning of his own body and worries or strains himself into stomach ulcers. **1961** L. MUMFORD *City in History* xv. 473 Definite ailments, like stomach ulcers and high blood pressure, seem to be aggravated by the strain of living, say, within sound of a busy motorway or airport. **1960** L. COOPER *Certain Compass* 23 Adrian said that he had a stomach upset, and went back. **1976** D. CLARK *Dread & Water* v. 102 Mugs..if used communally..can serve to pass germs among the party, causing stomach upsets.

stomachy, *a.* Restrict *dial.* to senses in Dict. and add: **4.** Of the voice or vocal sounds: deeply resonant, as if produced in the stomach. *colloq.*
1936 E. M. FORSTER *Abinger Harvest* II. 101 The soloist ..invites his 'friends' in a stomachy voice to rise in their shirt-fronts and shout. **1975** 'D. RUTHERFORD' *Mystery Tour* iv. 71 His rich stomachy laugh.

stomatal, stomatitis, stomato-. Add: Also with pronunc. (stōᵘ-).

stomiatoid (stōᵘ-, stǫ·miătoid), *sb.* and *a.* *Zool.* [f. mod.L. name of suborder *Stomiatoidei*, f. generic name *Stomias* (H. R. Schinz in G. L. C. F. D. Cuvier *Thierreich* (1822) II. 310), f. Gr. στόμ-α mouth + -IA[2]: see -OID.] **A.** *sb.* A deep-sea fish of the suborder Stomiatoidei, distinguished by a large mouth and rows of photophores on its sides. **B.** *adj.* Of or pertaining to a fish of this kind or the suborder as a whole.
1957 E. LE DANOIS *Fishes of World* vi. 178 The smaller stomiatoids with formidable teeth prey on other mid-water, deep-sea fishes. **1974** *Nature* 8 Nov. 98/1 They [*sc.* hatchetfishes] form one of the major groups of the stomiatoid fishes. **1976** *Jrnl. R. Soc. Arts* Apr. 251/2 Stomiatoid and lantern fishes..have large, highly sensitive eyes and marvellous arrays of light organs.

stomion (stǫ·miǫn). *Gr. Archæol.* Pl. **stomia.** [a. Gr. στόμιον, dim. of στόμα mouth.] The entrance to an ancient tomb.
1934 E. GJERSTAD et al. *Swedish Cyprus Expedition* I. 35 Three chambers..roughly circular, or oval in shape with horizontal floors, and steeply sloping tunnel-shaped stomia. *Ibid.* 47 The proper dromos of Chamber B was never excavated. The stomion is very long and slightly curved eastwards. **1946** *Ann. Brit. School at Athens* XLI. 79 The tombs were entered by a narrow passage (*dromos*), sometimes wider below than above, with a rectangular doorway (*stomion* in *SEC* descriptions) closed by a limestone slab. **1969** V. KARAGEORGHIS *Cyprus* iii. 152 Towards the middle of the 7th century the same tomb was re-used by cutting a passage through the filling of the dromos of the first burial to provide access to the *stomion* of the chamber. **1970** *Mariner's Mirror* LVI. 390 Anchors ..were broken in antiquity before being used as fill in the stomion of a tomb last closed in the 11th century B.C.

stomium (stōᵘ·miŏm). *Bot.* [mod.L., coined in Ger. (K. Goebel 1901, in *Organogr. d. Pflanzen* II. 753), f. Gr. στόμιον (see prec.).] In ferns, a part of the wall of the sporangium which ruptures to release the spores.
1905 I. B. BALFOUR tr. *Goebel's Organogr. of Plants* II. 575 A point of opening which we may designate the stomium occurs in all sporangia which discharge their spores into the air. **1936** A. J. EAMES *Morphol. Vascular Plants* xii. 263 The annulus [of the Cyatheaceæ] is complete and oblique (nearly vertical) with a poorly defined stomium. **1969** F. E. ROUND *Introd. Lower Plants* xiii. 156 The sporangial head has a cluster or ring of thickened cells (annulus) which acts as the dehiscence mechanism working on a weaker region of unthickened cells (stomium).

stomp (stǫmp), *v.*[2] Chiefly *U.S.* (orig. *dial.*). [Var. STAMP *v.* in senses of branch II.] **1. a.** *intr.* = STAMP *v.* 2 a. Also *fig.*
1803 J. DAVIS *Trav. U.S.A.* x. 382 He began to *stomp* upon me, and ax if I had yet got enough. **1936** C. CARMER *Listen for Lonesome Drum* 74, I stomped on his hand. **1961** C. MCCULLERS *Clock without Hands* vii. 148 He took down his records of German lieder..and stomped on them, stomping with such despair and fury that not a groove of the records remained unshattered. **1971** B. MALAMUD *Tenants* 65 In the last piece Harry..is painted white by three brothers after they had considered stomping on him..for what he did. **1973** *Observer* 25 Nov. 28/4 They stomped all over Newport County.., winning 3–1.
b. = STAMP *v.* 2 b.
1914 *Dialect Notes* IV. 156 *Stomp, v.i.*, var. of *stamp*, to strike the foot forcibly and noisily downward. **1917** *Ibid.* 400 *Stomp*.., tread heavily or noisily with one or more feet... 'He *stomped* on the floor as hard as he could.' **1928** J. PETERKIN *Scarlet Sister Mary* 147 Tell Doll not to stomp so hard. **1940** *Time* 29 July 40 They banged, rattled, beat, blew, stomped. **1969** *New Scientist* 17 July 119/2 The astronaut will be asked to stomp on the surface several times to produce observable seismic signals. **1982** B. CHATWIN *On Black Hill* xvi. 77 The bull bellowed; horses stomped in their stalls.
c. = STAMP *v.* 2 e.
1919 E. POUND *Quia Pauper Amavi* 18 He stomped into my bedroom. **1941** *Time* 13 Oct. 15/3 Mr. Ford stomped out, grinding his teeth. **1953** W. BURROUGHS *Junkie* (1972) ii. 23 Whitey was stomping up and down the length of the bar trying to promote some free drinks. **1956** J. MASTERS *Bugles & Tiger* 87 Biniram unpacked my suitcase, threw my pyjamas on the bed, and stomped out. **1967** G. STEINER *Lang. & Silence* 138 He stomps like a boisterous giant through a literature often marked by slim volumes of whispered lyricism. **1971** B. W. ALDISS *Soldier Erect* 89 He came stomping along the edge—for a moment I thought he was going to dive in after me, boots and all! **1981** 'J. GASH' *Vatican Rip* vii. 61 I'd never seen people move so fast... Everybody simply stomped hurriedly past.
2. a. *trans.* = STAMP *v.* 3 a. Also *fig.*
1916 in H. Wentworth *Amer. Dial. Dict.* (1944) 593/1 Before you stomp all that snow off. **1941** H. SKIDMORE *Hawk's Nest* 2 [She] stomped the red clay from her feet. **1954** *Ladies' Home Jrnl.* Oct. 116/3 Any passerby could look through the glass and see if the teacher inside were perhaps being pinned to the wall or stomped into the floor. **1971** B. PATTEN *Irrelevant Song* 40, I will make all that is possible step out of time To a land of giant hurrays! where the happy monsters dance And stomp darkness down. **1981** M. C. SMITH *Gorky Park* I. iii. 45 The host's carload entered, stomping snow off their boots.
b. *to stomp one's feet.* Cf. STAMP *v.* 3 c.
c **1927–34** J. TOOMER in *Black Scholar* (1971) Jan. 8, I teased the girls. I sent notes. I stomped my feet and made strange noises. **1941** *Sat. Even. Post* 10 May 113/2 Fern stomped feet against the floor. **1955** *Birmingham* (Alabama) *News* 14 July 55/2 She stomped her feet in the manner in which she testified she saw Colin walking toward the car to take the battery. **1972** *Jazz & Blues* Nov. 5/1 They jitterbugged to 'One O'Clock Jump' and stomped their feet to 'Maple Leaf Rag'.
c. To stamp or trample on (a person, etc.). Also *transf.*

1934 C. CARMER *Stars fell on Alabama* 165, I fixes to stomp him to death... There I was stompin' jest like I'd stomped a thousand coons. **1942** *R.A.F. Jrnl.* 16 May 28 The Indians then began a victory dance. Before the dance was over..the..British fliers were stomping the ground in customary style. **1959** N. MAILER *Advts. for Myself* (1961) 201, I never could stomach the relish with which soldiers would describe how they had stomped some faggot in a bar. **1967** *Daily Progress* (Charlottesville, Va.) 1 May c2/6 Jerry got into an argument with his mother and his father shouted: 'I'm gonna stomp you!' **1975** D. LODGE *Changing Places* ii. 86 He saw Carol jumping up and down on the mountainous figure of the black wrestler, 'Stomp me baby, stomp me,' he moaned.
d. With *out.* = STAMP *v.* 3 d. Chiefly *transf.*
1936 *Sun* (Baltimore) 4 Dec. 12/3 We are against crime. Crime should be 'stomped' out. **1940** W. FAULKNER *Hamlet* I. iii. 82 That first Snopes will turn around and stomp the fire out. **1941** *Time* 29 Dec. 22/1 Castillo would use his new powers..to stomp out Nazi propaganda agents. **1976** *Science* 10 Sept. 982/1 Despite government efforts to stomp it out..the banned anticancer drug Laetrile has a steady..market..in the United States.
e. To beat *out* (a rhythm) with one's foot.
1973 *Black World* Mar. 61/2 Arms open wide, he stomped out a savage drum beat: 'Kill! kill! kill! kill!'
f. To tramp or trudge between (a series of places).
1977 'J. LE CARRÉ' *Hon. Schoolboy* xv. 331 He was reduced to stomping the air-freight agencies, asking about a firm called Indocharter.
3. Chiefly *Jazz.* **a.** *trans.* To perform (a dance) to a lively, stamping rhythm.
1926 B. KRENZ (*song-title*) Stomp your stuff. **1926** in R. S. Gold *Jazz Lexicon* (1964) 297 When they start dancin'—Stompin and prancin'—the dance called the sugar foot stomp. **1978** *Amer. Poetry Rev.* July/Aug. 45/2 'Stomping the blues' is also dancing with the get down style of dance-beat-oriented people.
b. With *off.* To beat (a tempo) with one's foot as a signal to a jazz band to start to play; also, to signal to (a band) in this way. Also *absol.* or *intr.*
1925 in R. S. Gold *Jazz Lexicon* (1964) 298 (*tune-title*) Stomp off, let's go. **1960** H. O. BRUNN *Story Orig. Dixieland Jazz Band* vi. 68 For this reason LaRocca was not allowed to 'stomp off' his band in the usual fashion. **1961** *Artesian* Winter 33 They stomped off the solid beat. **1970** W. APEL *Harvard Dict. Mus.* (ed. 2) 441/2 In the earliest forms of jazz the leader 'stomped off' the tempo (gave it by tapping his foot).
c. *intr.* To dance or play a stomp. Cf. *STOMP *sb.* 1.
1925 (*tune-title*) Everybody stomp. **1929** (*tune-title*) I'm gonna stomp, Mr. Henry Lee. **1937** C. CONNOLLY in L. Russell *Press Gang!* 80 And then dancing, while.. David stomps on the piano. **1957** D. HAGUE in S. Traill *Concerning Jazz* 112 A resurgence of swing in evidence.. and Dixieland still stomping here and there. **1968** *Daily Mail* 16 Mar. 6/1 'Ullo, darlin', can you stomp?' my rocker friend Jonny asked gaily. I looked puzzled. 'It's the new Rocker dance,' he explained. **1974** *Ibid.* 16 Oct. 6/4 He does not stomp quite so energetically these days.
Hence **stomped** (stǫmpt) *ppl. a.*
1946 R. BLESH *Shining Trumpets* iv. 95 A stomped and hand-clapped rhythmic base. **1950** —— & JANIS *They all played Ragtime* ix. 176 Nor is erudite musical analysis needed to differentiate ragtime from jazz when one has heard him play the *Maple Leaf* in the authentic St. Louis manner and then follow with his own complex stomped version 'along the lines of jazz creation'.

stomp (stǫmp), *sb.* orig. and chiefly *U.S.* [f. prec.] **1. a.** Chiefly *Jazz.* A lively dance, usu. involving heavy stamping; also, a tune or song suitable for such a dance; stomping rhythm. Also *attrib.*
1912 (*tune-title*) Stomp dance. **1923** (*tune-title*) House rent stomp. **1926** *Amer. Mercury* Apr. 388/1 Hot jazz (which the Charleston and the Stomp—ye gods, what a name!—are bringing back, worse luck!). **1929** WODEHOUSE *Summer Lightning* iv. 108 Leopold's justly famous band, its cheeks puffed out and its eyeballs rolling, was playing a popular melody with lots of stomp in it. **1933** *Fortune* Aug. 90/3 Gene Gifford has composed and arranged some of the neatest exercises in *stomp* (very fast) time. **1940** *Swing* June 24/2 Fundamentally, there are two types of jazz—blues and stomps... Stomp tunes are gay; blues are mournful. **1952** *Mademoiselle* Dec. 120/3 The great era of the stomp was the twenties. **1956** H. KURNITZ *Invasion of Privacy* xiii. 85 She opened..with '*Vissi d'arte*' from Tosca..and to close, a hot and authentic stomp. **1968** *Daily Mail* 16 Mar. 6/5 Others did the stomp, an accelerated calypso, one of the most energetic and difficult dances I have ever tried to learn. **1977** J. WAINWRIGHT *Do Nothin' till you hear from Me* vii. 116 An outfit, straight from a 'viper session' could take a stomp, play it at..a nice, bouncy pace, and it came out faster than seemed mortally possible.
b. A heavy stamping step to the beat of such a dance.
1927 *Observer* 6 Feb. 15/7 Once you get the stomp—the peculiar beat of the foot—and you both hit the floor and not a neighbour's ankle, it is quite suitable as a ballroom dance. **1940** *Time* 25 Nov. 41/1 Dancer Massine [pieces out] simple footwork with deft body movements, well-timed claps and stomps. **1942** *Sat. Even. Post* 14 Feb. 20/2 A fast double shuffle that should have climaxed in a stomp. **1971** B. MALAMUD *Tenants* 217 Some of the youths try to imitate the newly married couples shaking their hips and shoulders but give it up and break into a stomp, shake, and whirl.
2. A party characterized by lively dancing to popular music; *spec.* a rent party.
1926 WHITEMAN & MCBRIDE *Jazz* viii. 177 The 'stomp' consisted of a barbecue with music afterwards, during and before. The guests raised a purse to save their host's home and also composed a new blues for the occasion. **1940** [see *G.I. 2 a]. **1967** E. A. GOLLSCHEWSKY in *Coast to Coast 1965–66* 86 The stomp crowd breaking up down at the Junior Citizens' Hall. **1977** P. DICKINSON *Walking Dead* III. iii. 230 The villagers met..for dances—those noisy nights half-way between revivalist meetings and beer-hall stomp.
3. A heavy, tramping gait or walk; *on the stomp*: tramping or trudging from place to place.
1971 B. W. ALDISS *Soldier Erect* 205 The parade-ground stomp was out in Dimapur, where it raised too much dust; the fashion was for a sort of brisk stroll, a gun-fighter's walk. **1977** 'J. LE CARRÉ' *Hon. Schoolboy* xvi. 392 He went on the stomp for refugee and orphan stories. **1982** *Times* 6 Sept. 7/1 A stomp along the cliff path, talking all the way.

stomper (stǫ·mpəɹ), *sb.* [f. *STOMP *v.*[2] + -ER[1].] **1.** *pl.* Shoes or boots; *spec.* large, heavy shoes. Cf. STAMPER *sb.* 4; *waffle stomper* s.v. *WAFFLE *sb.*[1] *U.S. slang.*
1899 B. W. GREEN *Word-Bk. Virginia Folk-Speech* 366 *Stompers*, large heavy shoes. **1945** L. SHELLY *Jive Talk Dict.* 19/1 *Stompers*, pair of shoes. **1974** K. MILLETT *Flying* (1975) II. 233 The Left wears its jeans and stompers. **1979** B. MALAMUD *Dubin's Lives* vii. 257 Dubin wore two scarves..waffle stompers, earmuffs.
2. *Jazz.* A person who performs a stomp.
1925 in B. Rust *Jazz Records 1897–1942* (1972) I. 758 (*recording artists*) The Dixie Stompers. **1927** *Music* 5 Jan. 50/2 (*heading*) Red and Miff's Stompers. **1944** *Amer. Speech* XIX. 268 In sharp contrast to the power-aggressiveness group [of names of boys' clubs] is a small but well defined group that might be called the rakish: Top Hats..Ramblers, Stompers, Hepcats. **1959** *Encounter* Sept. 51/1 Richard Waring..turns Oberon into a vocal, melodramatic, and *bravura* stomper. **1968** *Blues Unlimited* Nov. 17 A host of harpists, guitarists and stompers.

stompie (stǫ·mpi). *S. Afr. slang.* [a. Afrikaans, dim. of *stomp* STUMP *sb.*[1]] A cigarette butt; also, a partially-smoked cigarette, esp. one stubbed out and kept for relighting later.
1947 L. ABRAHAMS in B. Sachs *H. C. Bosman: S. Afr. Opinion—Trek Anthol.* (1971) 235 He stubbed out the stompie on the kerb. Pushing it into his pocket, he came over. **1959** J. MEIRING *Candle in Wind* ii. 32 She pulled a stompie out of her pocket and lighted it. **1969** A. FUGARD *Boesman & Lena* 37 The whiteman stopped the bulldozer and smoked a cigarette... He threw me the stompie. **1981** A. PATON *Towards Mountain* v. 34 The smell [of tobacco smoke] was made worse by his habit of keeping stompies in his pockets, a stompie being a cigarette not fully smoked, then stubbed out, and stored away for future use.

stomping (stǫ·mpiŋ), *vbl. sb.* Chiefly *U.S.* [f. *STOMP *v.*[2] + -ING[1].] **1.** The action of stamping or treading heavily.
1819 M. EDGEWORTH *Let.* 28 Jan. (1971) 164 Made such a stomping about the room that Mr. Sneyd could not think what was the matter. **1950** BLESH & JANIS *They all played Ragtime* ix. 166 The term 'stomp', used to designate a hot number of dynamic rhythm, was derived in New Orleans from the stomping of bare feet in the Bamboula and the Congo. **1976** *National Observer* (U.S.) 27 Mar. 2/5 The debate..was marked by howls, foot-stomping, angry exchanges.
2. *Jazz.* The action of dancing or playing a stomp.
1930 R. WRASKOFF (*song-title*) Stomping! Hot stomp. **1936** in R. S. Gold *Jazz Lexicon* (1964) 297 (*song-title*) Stomping at the Savoy. **1941** W. C. HANDY *Father of Blues* i. 6 That was real stomping. **1963** *Pix* 28 Sept. 62 Rubber soled sneakers are fine for stomping but you can get an extra two points for owning a pair of huaraches.
3. *transf.* An attack in which the victim is trampled upon. More generally, a beating. Cf. *STOMP *v.*[2] 2 c. *U.S. colloq.*
1958 *Washington Post* 13 Sept. D3/5 A coroner's jury yesterday found Robert C. Gerald..responsible for the death by stomping of Hazel R. White. **1967** *Daily Progress* (Charlottesville, Va.) 1 May c2 (*heading*) Threatened with 'stomping' boy kills parents, granddad. **1971** J. MANDELKAU *Buttons* xiv. 155, I grabbed him by the hair when he didn't answer and started to swing him around and punch him in the face... You know what he cried? 'I know where it's at. I can take a stomping!' **1977** L. O'DONNELL *Aftershock* xv. 217 The beating and the stomping weren't necessary.
4. *Comb.*, as **stomping ground** = *stamping ground* s.v. STAMPING *vbl. sb.* 3.
1854 in *Amer. Speech* (1940) XV. 397/1 Crossing the top of said ridge to a white oak & 2 chesnut saplings by the edge of a stomping-ground. **1937** *Dialect Notes* VI. 617 This is the stomping ground of Pecos Bill. **1950** A. LOMAX *Mister Jelly Roll* 42 New Orleans was the stomping grounds for all the greatest pianists in the country. **1977** 'J. LE CARRÉ' *Hon. Schoolboy* xiii. 307 The East was his natural stomping ground.

stomping (stǫ·mpiŋ), *ppl. a.* Chiefly *U.S.* [f. *STOMP *v.*[2] + -ING[2].] **1.** *Jazz.* That plays, or is played, in the manner of a stomp; exciting with a heavy, 'swinging' rhythm; also, that dances the stomp.
1927 H. FORD et al. (*song-title*) Stompin' fool. **1950** BLESH & JANIS *They all played Ragtime* viii. 160 New Orleans, not St. Louis, made the real impact on Chicago music,..as it developed into stomping jazz with much ragtime retained. **1956** M. STEARNS *Story of Jazz* (1957) xvi. 187 A series of stomping bands swung along this circuit. **1968** P. OLIVER *Screening Blues* v. 151 Joe Pullum, a Houston singer who was accompanied on *Joe Louis Is the Man* by Andy Boy's stomping piano. **1972** *Jazz & Blues* Sept. 12/2 The number seems to be based on 'Way Down Upon The Swanee River' but is played by Fats as a fast boogie-woogie piece with a stomping left hand.
2. = STAMPING *ppl. a.* *U.S.*
1942 *Time* 9 Feb. 28/2 Brazil's Aranha..announced to stomping, cheering crowds that Brazil..had..'broken her diplomatic and commercial relations with Germany'.

stompneus. Substitute for def.: Either of two edible fishes found off the coast of southern Africa, the red and silver *Chrysoblephus gibbiceps* or the silvery *Rhabdosargus globiceps*, both of the family Sparidæ. (Later examples.)
1945 *Cape Argus Mag.* 20 Oct. 1 Often we hooked two together—silverfish, panga, stompneus, elft. **1953** *Cape Times* 4 Mar. 2/4 Kabeljou, yellowtail, white stompneus and stockfish were all caught in Hermanus.

-stomy (stŏmi), f. Gr. στόμ-α mouth, opening + -Y[3], used in *Surg.* to form the names of operations in which (*a*) an opening is made into the internal organ denoted by the preceding element, as in *COLOSTOMY, GASTROSTOMY; or (*b*) a permanent connection is made between the internal organs indicated, as in *gastro-duodenostomy* and *gastro-gastrostomy* s.v. *GASTRO-.

stone, *sb.* Add: **1. c.** A meteorite; now *esp.* one containing a high proportion of silicates or other non-metals.
1628 J. HOSKINS *Let.* in N. Wallington *Hist. Notices* (1869) I. i. 14 As it is reported, there fell divers stones, but two is certain, in our knowledge. **1796** *Gentleman's Mag.* LXVI. 845/1 Various instances are alleged of such falling stones, or, as they may be denominated, extinguished meteors. **1802** *Phil. Trans. R. Soc.* XCII. 212 Have not all fallen stones, and what are called native irons, the same origin?..Are all, or any, the produce or the bodies of meteors? **1809** *Jrnl. Nat. Philos.* XXIII. 233 Account of a meteoric stone..that fell in the circle of Ichnow. *Ibid.*, Several persons..got out the stone, which was above two feet beneath the surface of the snow... A professor of natural philosophy..considered it..as ferruginous. **1826**, etc. [see *IRON *sb.*[1] 1 d]. **1977** A. HALLAM *Planet Earth* 24 Freshly fallen stones are usually quite cool to the touch.
d. A fashion shade of yellowish or brownish grey; stone-colour. Also *attrib.* or as *adj.* Cf. sense 19 in Dict.
1848 E. RUSKIN *Let.* 10 May in W. James *Order of Release* (1947) v. 107 A stone silk dress with two broad flounces. **1865** M. EYRE *Lady's Walks in South of France* i. 10 The colours most in vogue are some shade of grey, stone, or buff. **1890** [see *box-cloth* s.v. *BOX *sb.*[2] 24]. **1907** *Yesterday's Shopping* (1969) 157/2 Paints mixed ready for use... White, light stone, dark stone, middle stone, black. **1923** *Daily Mail* 2 June 1 In delightful shades of Fawn,..Dove Grey, Stone, Beaver. **1977** *Times* 18 Aug. 23/6 *Rover* 3·5 *litre*..blue with stone leather interior.
2. g. *artificial stone* (see quot. 1967).
1722 *Brit. Pat.* 447 Thomas Ripley..and Richard Holt..have been at much labour..for the finding out and inventing 'A certain compound liquid metall never before known and used by the Antients or Moderns, by which artificiall stone and marble is made.' *c* **1778** [see *LITHODIPYRA]. **1868** *Building News* 10 Apr. 248/2 (*heading*) Ransome's artificial stone. *Ibid.* 3 July 448/2 A method of manufacturing artificial granite..has just been patented by Mr. P. M. Parsons. **1935** *Economist* 9 Feb. 321/1 The two trades..which represent the largest consumers of cement are 'public works contracting, etc.', and 'artificial stone and concrete manufacturing'. **1935** [see *RECONSTRUCTED *ppl. a.* a]. **1967** *Gloss. Highway Engin. Terms* (*B.S.I.*) 37 *Artificial stone*, a form of precast concrete in which the finished surface resembles that of natural stone.
j. (Earlier example in *Curling*.)
1824 [see *HOG *sb.*[1] 10].
7. a. (Earlier and later S. Afr. examples in sense 'diamond'.)
1884 MRS. CAREY-HOBSON *At Home in Transvaal* 184 He had placed no stones in the bank since Graham had been on the Fields. **1891** E. GLANVILLE *Fossicker* xxix. 292 The cooling mud has closed around the 'stones', taking the impress of every angle and facet. **1946** S. CLOETE *Afr. Portraits* 109 His favourite stone was his blink klippie—his shining stone—the first diamond to be found in Africa. **1972** *Panorama* Dec. 27 'Stones' are usually over one carat (a carat being 200 milligrams). Anything smaller falls in the 'melee' category.
b. *spec.* in *Criminals' slang*, a diamond (see also quot. 1955).
1904 'No. 1500' *Life in Sing Sing* 252/2 *Stone*, diamond. **1936** J. CURTIS *Gilt Kid* xxiv. 240 Ten nicker for a little stone like that. **1955** D. W. MAURER in *Publ. Amer. Dial. Soc.* XXIV. 122 A man's tie-pin, seldom worn nowadays, was a prop. If it had a diamond setting, it was referred to as a stone.
c. *Austral.* Opal or opal-bearing material; an opal. Also *to be on stone*, to have struck opal stone. Also *N.Z.* (see quot. 1965).
1895 *Rep. N.S.W. Dept. Mines* 68 A patch of stone was taken about the end of the year which brought £1,200. **1921** K. S. PRICHARD *Black Opal* iv. 33 You don't suppose Jun'd try to take the stones off of him, do you? **1924**

T. C. WOLLASTON *Opal* iv. 61 The men were not 'on stone', it seemed, but perhaps I could change the luck? 1965 G. J. WILLIAMS *Econ. Geol. N.Z.* iii. 20/2 *Stone*, a miner's term for payable [sc. auriferous] quartz. 1967 A. KALOKERINOS *In Search of Opal* 18 Stones that are worth $2,000 or more on the field are found at a rate that would not exceed one per week.

14. c. In phr. *to give a stone and a beating to* (Racing slang): to outrun easily, despite carrying a heavier weight. Also *transf.*, to surpass. Now *rare.*
1885 *Daily News* 4 Feb. 5/2 *Canis vulpis* is, as a rule, able to give, intellectually speaking, and in language germane to the matter, 'a stone and a beating' to the majority of his pursuers. 1906 *Punch* 18 Apr. 286/3 Their Smokeroom is deliciously comfy, and can give a stone and a beating to ours at the Camellia.

18. a. attrib. *stone-cliff*, *-heap* (later examples). **b.** Objective, etc. *stone-haunting* adj. **c.** Instrumental, etc. *stone-bearded*, *-faced* (later examples), *-flagged*, *-headed* (earlier example), *-horned*, *-ribbed*, *-strewn* adjs.
1922 JOYCE *Ulysses* 141 A man supple in combat: stonehorned, stonebearded, heart of stone. 1912 E. POUND *Ripostes* 26 Storms, on the stone-cliffs beaten. 1932 W. FAULKNER *Light in August* xvi. 355 This time he indicates the stonefaced woman; she may or may not be listening to what he is saying. 1973 M. WOODHOUSE *Blue Bone* ix. 82 The Eisenwald Volksklinik was..a huge stonefaced structure. 1904 E. WHARTON *Italian Villas* i. 53 The house is built about three sides of a raised stone-flagged terrace. 1978 J. WAINWRIGHT *Jury People* v. 16 The room had a stone-flagged floor. 1933 AUDEN *Poems* (ed. 2) 43 By pot-holed becks A bird stone-haunting, an unquiet bird. 1829 G. GRIFFIN *Collegians* I. viii. 170 The difference which existed between..an English halberd and a stone-headed gai-bulg. 1868 *N. & Q.* 15 Aug. 165/2 The game Set-a-Foot is still played by the rising generation who frequent Park Square, Regent's Park, under the name of Stone Heaps. 1941 F. THOMPSON *Over to Candleford* 356 They ran..and wrestled the whole way, or pushed each other over stone-heaps or into ditches. 1977 *New Yorker* 17 Oct. 37/3 Her stone heap... My mother spreads out soapy white laundry on these stones, so that the hot sun will bleach them even whiter. 1922 Stone-horned [see *stone-bearded* above]. 1936 L. B. LYON *Bright Feather Fading* 19 The bone-bare garden steep, the stone-ribbed land. 1853 M. ARNOLD *Poems* 179 The climbing gourd-plant's leaves Muffled its walls, and on the stone-strewn roof Lay the warm golden gourds. 1974 R. ADAMS *Shardik* x. 71 The bear's trail led on through the bushes to emerge in open, stone-strewn woodland.

19. a. *stone-bright*, *-cold* (examples in quasi-advb. attrib. use, esp. in phr. *stone cold sober* = utterly sober), *-comfortless* adjs.
1916 E. POUND *Lustra* 26, I have known the stone-bright place, The hall of clear colours. 1913 F. H. BURNETT *T. Tembarom* xxxiv. 435 It'd be stone-cold safe to rush things. 1937 T. RATTIGAN *French without Tears* III. i. 65 Are you stone-cold sober? 1958 A. SILLITOE *Saturday Night & Sunday Morning* vii. 111 We've been stone-cold sober since Canning Circus. 1969 C. ARMSTRONG *Seven Seats to Moon* v. 59, I could have been stone-cold-dead in Chicago! 1979 O. SELA *Petrograd Consignment* 144 Unlike the other revolutionaries, the Bolsheviks..were resolutely stone-cold sober. 1924 D. H. LAWRENCE in M. Magnus *Mem. Foreign Legion* 13 There I had a big and lonely, stone-comfortless room. 1769 J. WEDGWOOD *Let.* 1 Dec. (1965) 85 We have nobody making white ware here, only stone white ware. 1949 E. POUND *Pisan Cantos* lxxxiv. 129 Carrara Snow on the marble Snow-white against stone-white.

b. Intensively with adjs. in non-similative use (after *stone-broke* adj., sense 20 a below): completely, utterly, 'plumb', as *stone crazy*, *drunk*, *mad*, etc. Also in adj. relation to sb., complete, utter, 'dead'; excellent. Cf. *stone ginger* (b), sense 20 a below. *slang.*
1928 *Lawn Tennis & Badminton* 23 June 255/2 Few could have foreseen that the two doubles would have been the 'stone certainties' for Britain that they proved to be. 1933 PARTRIDGE *Words, Words, Words!* 214 India gives us..*piache*, mad... On the analogy of *stone mad*, *stone piache* was employed for a change. 1935 Z. N. HURSTON *Mules & Men* (1970) I. iii. 66 You must be stone crazy! Why, dis hide is worth five thousand dollars. 1947 K. TENNANT *Lost Haven* ix. 126 Oh, don't let him think of the punt again—that was the stone finish! 1959 *Esquire* Nov. 70 *Stone*, adjective meaning complete. Example: He's a stone musician. 1960 *Observer* 25 Dec. 7/7 If..he were stone rich and lived in a big drum in the country. 1968 *Blues Unlimited* Dec. 12 First things developed was the set of four reissue albums labelled 'Legendary Masters'; three being stone blues albums. 1970 D. M. DAVIN *Not here, not Now* III. vi. 202 This was the finish, the stone end of it. 1978 *N.Y. Times* 30 Mar. A21/1 A little later another patrol..declared him stone drunk, and confiscated his documents and his car keys.

c. As *adj.*, excited; intoxicated with drink or drugs, 'stoned'. *U.S. slang. rare.*
1945 L. SHELLY *Jive Talk Dict.* 19/1 *Stone* (adj), excited or intoxicated. 1960 R. G. REISNER *Jazz Titans* 165 *Stone*, drunk or high.

20. a. stone-breaker (earlier U.S. example); **stone-broke** *a.* (earlier and later examples); **stone bruise** chiefly *N. Amer.*, an injury to the feet caused by walking on stony ground; hence **stone-bruised** *a.*; **stone cell** (earlier example); **stone cist** *Archæol.*: see CIST 1 a; **stone-craft,** the art or skill of working in stone; sculpture; **stone cream,** a traditional blancmange-like sweet served cold on a base of jam; **stone-dust,** dust or powder made of particles of broken stone; hence **stone-dusting,** the introduction of stone-dust to the air in a mine to render the coal-dust less combustible; **stone-dust** *v. trans.*, with place as obj.; **stone face** *U.S. colloq.*, a person whose features reveal no emotions; a poker-faced person; esp. in phr. *great stone face* in playful allusion to Hawthorne's tale (see bracketed quot. 1850); **stone-field,** an expanse of ground covered with large stones; *spec.* = *FELSENMEER; **stone frigate** *Naut. slang*, a Naval shore establishment or barracks (see quot. 1948); formerly *spec.* a naval prison; also *transf.*; **stone garland** *Geomorphol.*, a low bank or terrace of large stones occurring on a steep slope and curved downwards so as to resemble a garland or necklace; **stone ginger,** (*a*) see sense 17 b in Dict.); (*b*) *slang*, a certainty, a 'sure thing' (cf. sense 19 b above); also as adj., certain; **stone guard,** an attachment serving to prevent stones entering the air-intake system of a motor vehicle or aeroplane; a similar device protecting another part of a vehicle; **stone-hand** (later examples); **stone kist,** var. *stone cist* above; **stone line** *Geomorphol.*, a layer of isolated stones between subsoil and underlying rock; also, the line of stones that this appears as in a section through the soil; **stone net** *Geomorphol.*, a network of stone rings or polygons; **stone pavement** *Geomorphol.*, an area of ground covered with large flattish stones; **stone polygon** *Geomorphol.*, a naturally occurring arrangement of stones in the approximate form of a polygon; **stone ring,** (*a*) *Geomorphol.*, a natural circle of stones on the ground, similar to a stone polygon; (*b*) *Archæol.* = *stone circle*; **stone river,** a dense, stream-like accumulation of rocks and large stones occurring along a valley bottom or down a slope; *esp.* one of those in the Falkland Islands; **stone run** = *stone river* above; **stone stripe** *Geomorphol.*, one of the evenly spaced bands of coarse rock debris separated by finer material that occur on slopes in cold environments.
1827 S. RODMAN in B. Swan *New Bedford in 1827* (1935) 8 Occupied most of the day at my house lot. Made a further trial of my stone breaker, the weight raised by a horse. 1886 H. BAUMANN *Londinismen* 196/2 Stone-broke. 1933 *Bulletin* (Sydney) 20 Dec. 10/3 There was a hardy war-time story of a stonebroke Digger. 1981 O. BERNIER *Pleasure & Privilege* xii. 197 Naples wasn't exactly short of mobility... Some were stone broke. 1805 LEWIS & CLARK *Orig. Jrnls. Lewis & Clark Expedition* (1904) II. 290 We have a lame crew just now,..one with a bad stone bruise. 1885 *Cent. Mag.* Nov. 29/1 Angy, who was complaining of a stone-bruise, got up. 1976 T. WALKER *Spatsizi* xi. 122 The continuous descent over rough ground lamed one saddle horse with a stone bruise. 1909 'O. HENRY' *Roads of Destiny* xxi. 354 Five of my best staff-officers fell, suffering extremely with stone-bruised heels. 1875 BENNETT & DYER tr. *Sachs's Textbk. Bot.* ii. 106 The polyhedral stone-cells (sclerenchyma) in the flesh of pears are arranged in groups. 1888, 1934 Stone cist [see *passage grave* s.v. *PASSAGE sb. 16 b]. 1903 J. R. HARRIS *Dioscuri in Christian Legends* 37 We recognized stonecraft amongst the arts of the Dioscuri. 1931 *Catholic Bull.* (Dublin) June 578 Metal-work, stone-craft, and architecture. 1861 MRS. BEETON *Bk. Househ. Managem.* 747 Stone cream of tous les mois... ½ lb. of preserve, 1 pint of milk, 2 oz. of lump sugar, 1 heaped tablespoonful of tous les mois, 3 drops of essence of cloves, 3 drops of almond-flavouring... When rather cool, but before turning solid, pour the cream over the jam. 1973 E. SPRIGGE *Life of Ivy Compton-Burnett* v. 78 They liked fish, too, and junket, and that old favourite among puddings, stone cream. 1896 M. E. WILKINS *Madelon* xxix. 330 Damned foolishness, that does more harm to the world than the shattering of all the commandments into stone-dust. 1920 *Chambers's Jrnl.* Mar. 266/2 This fact is taken advantage of to localise explosions in some American mines by mixing the first rush of air with stone-dust. 1930 *Engineering* 28 Feb. 295/3 Stone-dusting in coal mines was not considered to be injurious in Poland. 1975 *Telegraph* (Brisbane) 13 Nov. 16/2 No agreement was reached between union and management on stone dusting Kianga No. 1. [1850 HAWTHORNE *Great Stone Face* in *Nat. Era* 16 Jan. 16/1 The Great Stone Face, then, was a work of Nature..formed on the perpendicular side of a mountain by some immense rocks, which had been thrown together in such a position, as, when viewed at a proper distance, precisely to resemble the features of a human countenance.] 1949 *Life* 5 Sept. 82/2 (*heading*) The great stone face [of Buster Keaton]. 1960 *Newsweek* 25 Jan. 90/2 Here is the Great Stone Face on the most famous element of this vanishing art, his dead pan. 1972 J. MOSEDALE *Football* iv. 47 Even in high school his classmates called him the 'Great Stone Face'. 1977 *Rolling Stone* 21 Apr. 88/3 Only a stoneface could resist smiling. 1906 *Jrnl. Geol.* XIV. 103, I feel sure that these immense block-fields of Bear Island are formed in quite the same manner as the Falkland stone-runs... The only differences between the two occurrences are differences of topography and age: in Bear Island a great plain forming a stone-field, in the Falkland Islands valleys filled at the bottom by stone-rivers. 1959 A. H. MCLINTOCK *Descr. Atlas N.Z.* p. xv, At the timber line there is a locally heavy scrub belt..passing into snow-tussock grasslands, stone-fields, and herb moor. 1978 O. WHITE *Silent Reach* viii. 87 It gets harder when you hit the..stonefields. 1917 M. T. HAINSSELIN *Grand Fleet Days* iv. 15 Where I met her was in a Stone Frigate—that is to say, a Naval Shore Establishment. 1929 F. C. BOWEN *Sea Slang* 134 *Stone frigate*, a naval gaol or, more recently, any shore establishment. 1948 PARTRIDGE *Dict. Forces' Slang* 182 *Stone frigates*, Royal Naval Barracks or Shore Establishments; they are usually named after the old frigates. 1955 'N. SHUTE' *Requiem for Wren* iii. 81 She found that *H.M.S. Mastodon* was a stone frigate. It was Exbury Hall, about three miles up the Beaulieu River from the Solent. 1979 *Mariner's Mirror* LXV. 51 H.M.S. *Thunderer* (our title as a 'stone frigate') has since prospered... It is planned amongst other things to produce a book on the history of the college. 1932 E. ANTEVS *Alpine Zone Mt. Washington Range* iv. 62 A balsam fir forest..grows normally up to the stone garland. 1977 R. J. SMALL *Study of Landforms* x. 326 If the slope becomes a little steeper, the polygons give way to 'stone garlands'. 1936 J. CURTIS *Gilt Kid* iv. 41 It was stone-ginger, you thought, that you'd get a smashing job up here. 1943 J. A. W. BENNETT in *Amer. Speech* XVIII. 90 'That's a stone ginger' (a dead certainty) conceals the name of a famous and unbeatable horse, Stone Ginger [in New Zealand]. 1972 G. F. NEWMAN *You Nice Bastard* 348 *Stone ginger*, a million; certainty. 1936 *Times* 19 Oct. 8/4 The chromium-plated radiator has an integral stoneguard. 1947 *Jrnl. R. Aeronaut. Soc.* LI. 287/2 The best solution of the problem of the stone guard would be to abolish the guard, and eliminate stones and other refuse by momentum-separation. 1958 *Times* 22 Sept. 12/6 Superficially it [sc. a motor car] had many attractive qualities..a detachable silver stone-guard before the radiator. 1981 *Buses* Dec. 535/1 This ex-Liverpool Atlantean..has acquired a stone guard in front of the windscreens. 1921 Stone hand [see *IMPOSER b]. 1978 L. DAVIDSON *Chelsea Murders* xxv. 156 He..was rapidly rewriting lines for the stone-hand. 1926 Stone kist [see *round barrow* s.v. *ROUND *a.* 15 a]. 1980 D. K. CAMERON *Willie Gavin* vi. 54 There was hardly a year when the winter ploughs did not turn up an old hunter..crouched still in his cold stone-kist. 1938 C. F. S. SHARPE *Landslides & Related Phenomena* iii. 24 This layer [of rock fragments] outcrops in natural and artificial cuts and marks the approximate boundary between the base of..the 'B' horizon of the soil and the 'C' horizon or parent rock material. Where well developed it appears as a broken line of stones suggesting the name *stone-line* here used. 1969 C. OLLIER *Weathering* iv. 46 The profiles in many tropical countries have rock..overlain by a stone line, overlain in turn by fairly uniform, fine grained 'soil'. 1975 R. V. RUHE *Geomorphology* vii. 127/3 A stone-line surface usually differs topographically from the present land surface. 1949 *Jrnl. Geol.* LVII. 143 Stone nets, stone stripes, and soil stripes have formed on high, flat erosion surfaces..in the Wind River Mountains, Wyoming. 1977 D. & V. WEYMAN *Landscape Processes* iv. 69/2 Stripes are found on slopes above 4° and seem to be stone nets elongated by downslope movements of slope debris. 1969 E. WATSON tr. *Tricart's Geomorphol. of Cold Environments* II. ii. 109 Moist Climates with Severe Winters: Mountain Type... Stone polygons and stone stripes (as well as stone pavements, which are typical), are fairly frequent. 1977 D. & V. WEYMAN *Landscape Processes* iii. 52/1 In general, desert surfaces show only a shallow weathering layer. Bare rock outcrops are common and many other areas have a stone pavement of coarse material. 1924 HUXLEY & ODELL in *Geogr. Jrnl.* LXIII. 208 We propose here to style the two forms 'stone-polygons' and 'fissure-polygons' respectively. The stone-polygons are represented at one extreme by isolated stone-circles, while at the other they may become drawn out into a series of elongated mud-strips separated by strips of stone. 1950 [see *PATTERNED *ppl. a.* b]. 1970 R. J. SMALL *Study of Landforms* x. 327 The reason why stone polygons as a whole vary so much in size (their diameters range from 0·5 to 15 metres) is not understood. 1924 *Geol. Mag.* LXI. 509 (*heading*) Formation of 'stone rings' in rocks which are being shattered by frost action. 1954 J. R. R. TOLKIEN *Fellowship of Ring* I. vii. 141 The stone-rings upon the hills. 1957 J. K. CHARLESWORTH *Quaternary Era* I. xxvii. 572 The severer frost in the lower, sodden layers produces the finer material and brings it to the surface, pushing the coarser to the sides. The stone-rings so produced grow outwards from their centres to build a polygonal network. 1980 *Sci. Amer.* July 67/1 As a result the term now coming into favour as a description of these megalithic enclosures is stone ring. 1877 C. W. THOMSON *Atlantic* II. iv. 246 At the mouth of the valley the section of the 'stone river' exposed by the sea is like that of a stone drain on a huge scale. 1894 J. GEIKIE *Gt. Ice Age* (ed. 3) xl. 723, I do not think there can be much doubt that the 'stone-rivers' of the Falkland Islands are of the same nature and origin as the rubble-drifts already described in connection with the glacial phenomena of Europe. 1956 W. EDWARDS in D. L. Linton *Sheffield* 20 Newer Drift... This is well developed on the hillsides in the Millstone Grit country—for example, on Burbage Moor..—its content of large gritstone blocks betraying its presence, especially where these are concentrated in 'stone-rivers'. 1969 C. OLLIER *Weathering* xii. 214 Block-streams (stone rivers) also have sharp edged and angular blocks, and occur in the same areas as blockfields. 1906 *Jrnl. Geol.* XIV. 101 The large old stone-runs of the Falkland Islands evidently were formed in a period of the past with a climate more severe than the present. 1950 *Geol. Mag.* LXXXVII. 106 The stone runs of the Falklands extend over a greater area than is at present exposed, since they are masked by vegetation. 1934 *Proc. Geologists' Assoc.* XLV. 174 Fig. 24 ..shows the stone-stripes in cross section one to two inches thick lying in shallow depressions in the clay-loam. 1978 A. L. BLOOM *Geomorphology* xv. 363 Like sorted polygons, stone stripes require active freeze-thaw processes but are not restricted to regions of permafrost.

b. stone-fish: esp. the highly venomous *Synanceja verrucosa*, of the family Scorpænidæ, a bottom-dwelling fish resembling a small rock, found in tropical seas and bearing venom glands at the base of the dorsal fin spines; (later examples); **stone roller** (earlier example).
1908 E. J. BANFIELD *Confessions of Beachcomber* I. iv. 143 Beware of the stone fish.., the death adder of the sea. 1947 I. L. IDRIESS *Isles of Despair* xxxv. 234 The lancet of the hideous little stone fish in his salamander coat. 1971 *Islander* (Victoria, B.C.) 20 June 5/2 The ugliest fish in the sea (and one of the most dangerous) is the stonefish. 1878 C. HALLOCK *Sportsman's Gaz.* 386 The 'stone toter', or 'stone roller', is a far better variety.

stone, *v.* Add: **1. b.** In colloq. phr. *stone the crows*: see *CROW sb.[1] 3 d. Similarly *stone me*: an exclamation of astonishment.
1961 SIMPSON & GALTON *Hancock* 38 *Tony*: Any room for a littl'un? (*Laughs*). *They stare at him frostily. Tony (laugh dries)*: Cor, stone me. 1967 *Listener* 21 Dec. 815/2 Mrs Dale speaks. 'Why hello, Jim—Cor, stone me, what a booze up we had last night up the BMA.' 1979 J. WAINWRIGHT *Tension* 183 Stone me!—next thing I know I have a..hand-grenade here in my pocket.

3. a. Also, to cover or shut *up* with stones. Also *fig.*
1889 V. MCNABB *Let.* 24 Apr. in F. Valentine *Father Vincent McNabb* (1955) I. ii. 62 Every little fountain of grief seems stoned up. 1953 A. BRYANT *Story of England* iii. 68 His [sc. Jesus'] body had vanished from the tomb in which it had been stoned up.

9. a. To become intoxicated with drink or drugs (with *out*, to the point of unconsciousness). **b.** *trans.* To render intoxicated or (*fig.*) ecstatic. Also *refl.* Chiefly as (*ppl.*) *a.*: see *STONED *ppl. a.* and *a.* 7. *slang* (orig. *U.S.*).
1952 G. MANDEL *Flee Angry Strangers* 139 I'd rather stay with the tea. It's great pod. I don't want to stone out. 1959 *Jazz* Fall 290, I heard Phineas Newborn play 'I'll Remember April' two Mondays ago at The Five Spot and he completely stoned me. *a* 1961 T. CAPOTE in WEBSTER (1961) s.v., Planned to stone himself with vodka. 1972 J. BROWN *Chancer* iii. 38 You smoke Egyptian Black, that will stone you out of your head.

stone age. Add: **b.** *fig.*, esp. as the type of an outmoded or unsophisticated era. Also *attrib.* and as *adj.*
1927 KIPLING in *Maclean's Mag.* 15 Sept. 52/4 The old lady..was primitive Stone-Age—bless her! She looked on us as a couple of magicians. 1937 F. SCOTT FITZGERALD *Let.* 5 July (1964) 15 The girls who were what we called 'speeds' (in our stone-age slang) at sixteen were reduced to anything they could get at the marrying time. 1959 M. LASKI *Offshore Island* III. 84 I've enjoyed civilization too much to be happy in a new stone age. 1973 R. THOMAS *If you can't be Good* (1974) xvi. 147 Back in the mid-fifties was back in the stone age. It was way before the Pill. 1981 *Quarto* May 4/2 In the age of computerised type-setting, the technology of the book trade seems more and more stone age.

stone-blue. Add: **2.** (Earlier and later examples.)
1860 C. BRONTË *Emma* in *Cornh. Mag.* Apr. 494 Bright stone-blue is a colour they like in dress. 1962 *L. L. Bean Catal.* Spring 11 Ladies Knee Sox..Colors: Stone Blue, Stone Green.

stone-boat. **2.** For *U.S.* read *U.S.* (chiefly *North.*) and *Canad.* (Earlier and further examples.)
1859 N. P. WILLIS *Convalescent* 75 A *stone-boat* would run glibly over such a shallow snow! 1901 'R. CONNOR' *Man from Glengarry* 189 In the afternoon the colt was put through her morning experience, with the variation that the stone-boat was piled up with a fairly heavy load of earth and stone. 1962 J. ONSLOW *Bowler-hatted Cowboy* viii. 79 A stone-boat is best described as a heavy wooden sled, on which can be hauled rocks and stones..dead cows, sick cows, or other heavy objects.

stoned, *ppl. a.* and *a.* Add: **7.** *slang.* **a.** Drunk, extremely intoxicated (see also quot. 1952). Freq. const. *on*. Chiefly *predic.*, esp. in phr. *to get stoned*. Cf. *STONE *sb.* 19 c. orig. *U.S.*
1952 *Life* 29 Sept. 67/2 Like boiled snails, bop jokes certainly are not everybody's dish, but those who acquire the taste for them feel cool, gone, crazy and stoned. *Ibid.* 67/3 *Stoned*, drunk, captivated, ecstatic, sent out of this world. 1955 *Amer. Speech* XXX. 305 *Stoned out of his skull*, intoxicated to an intense degree. 1957 J. KEROUAC *On Road* (1958) I. xiii. 90, I had finished the wine..and I was proper stoned. 1968 *Listener* 28 Nov. 735/2 He would only be taken in charge if he was drunk: were he to spend his ten shillings on getting stoned out of his mind the police would happily accommodate him. 1972 R. REID *Canadian Style* (1973) iv. 144 Then they all laugh and get stoned. 1976 P. CAVE *High Flying Birds* ii. 18 We drive off the ferry at Roscoff late in the afternoon, both well and truly stoned on cut-price booze.

b. In a state of drug-induced euphoria, 'high'; also, incapacitated or stimulated by drugs, drugged. orig. *U.S.*
1953 ANSLINGER & TOMPKINS *Traffic in Narcotics* 315 *Stoned*, under the influence of drugs. 1956 'E. MCBAIN' *Cop Hater* (1958) ix. 85 You're an H-man..and we know you copped three decks a little while back. Are you stoned now, or can you read me? 1967 M. M. GLATT et al. *Drug Scene* viii. 97 Addicts know these dangers, one for example describing graphically how in a 'stoned' state he had stepped out in front of a car. 1971 'D. HALLIDAY' *Dolly & Doctor Bird* x. 129 They're all lying around in there wearing beads and stoned out of their skulls on French Blues. 1981 M. LEITCH *Silver's City* viii. 65 If he'd been pissed, he reflected, instead of stoned, he might still be in khaki, but, as it was, the old man had a down on drugs, and so it was a dishonourable discharge or nothing.

c. *fig.*
1952 [see sense 7 a above]. 1963 R. I. MCDAVID *Mencken's Amer. Lang.* 742 A cool cat..is..much of the time stoned on wine, pot.., heroin or an overdose of Zen Buddhism. 1969 *Listener* 17 July 88/3 We are, by any definition, stoned on liberty, smashed by self-fulfilment; the real need now is for silence and what used to be called classical restraint—and irony. 1980 *Times Lit. Suppl.* 31 Oct. 1220/5 He [sc. Tom Robbins] is also a moralist, and although superficially he belongs to the 'stoned' school of American fiction, along with Brautigan, Kotzwinkle et al, there is a more interesting comparison to be made with the work of Aldous Huxley.

d. With *out*.
1968 A. DIMENT *Great Spy Race* iii. 39 He..[was] chortling in his stoned out way. Tim was really blocked. 1972 R. K. SMITH *Ransom* i. 23 Joyboy had been a stoned-out junkie. 1977 *Rolling Stone* 13 Jan. 51/1 We even have a comedy collection—the Firesign Theatre's *Forward into the Past*, a double album's worth of puns, alliterations, slapstick and stoned-out mayhem.

stone jug. Add: **3.** *Rhyming slang.* = MUG *sb.*[5] Cf. *STEAMER 11.
1923 J. MANCHON *Le Slang* 296. 1974 P. WRIGHT *Lang. Brit. Industry* xiv. 128 When the Duke of Edinburgh visited one of the Astley collieries in the 1950s, he so satisfied the miners..that one of them was moved to say to the local press, 'He's no stone jug, yon mon.' This was the highest possible praise from a South Lancashire miner.

stonemason. (Earlier Amer. example.)
1758 C. SMITH *Let.* 23 Feb. in *Lett. to Washington* (1899) II. 269 Our Stone Masons has been Sick, Ever Since you have been Away, and our Stone Work is much Behind hand.

stone-wall, *sb.* Add: **2. b.** As an epithet for one who seeks to confound by dogged resistance. Chiefly applied to Thomas Jonathan ('*Stonewall*') Jackson (1824–63), Confederate general during the American Civil War. (The earlier *fig.* use.)
1862 *Texas Almanac Extra* 18 Sept. 1/1 Stonewall Jackson was marching on Baltimore with 40,000 men. 1863 G. MEREDITH *Let.* 7 Jan. (1970) I. 185 Busy my good sir, so as to drive the pen as fast as Stonewall Jackson is driving the federals. 1867 *John Lillywhite's Cricketers' Compan.* 46 'Young Stonewall'—as he [sc. H. Jupp] has been called—was in immense form. 1902 E. B. V. CHRISTIAN in Alverstone & Alcock *Surrey Cricket* iii. 82 The eleven..received very valuable additions in Tom Humphrey and Jupp, or 'young Stonewall', the 'old Stonewall' being Mortlock, who still played. 1970 R. LOWELL *Notebook* 120 Above your fire the blood-crossed flag of the States, A Stonewall Jackson, a Twenty-two at half-cock.

3. (Examples in *Cricket* and *Lawn Tennis.*)
1885 F. GALE *Hon. Robert Grimston* iii. 25 The well-known stone-wall cricketer, Mr. A. Haygarth. 1895 J. N. PENTELOW *England v. Australia* 37 Lucas played the stonewall game. 1932 E. BOWEN *To North* vi. 57 He played..a stonewall game and beat Emmeline.

stonewaller (later examples).
1895 J. N. PENTELOW *England v. Australia* 20 Boyle gave the young stonewaller still better assistance. 1958 *Economist* 2 Aug. 361 He indicated to the Austrians that he thought Mr Dulles the chief western no-man and habitual stonewaller. 1971 *Jamaican Weekly Gleaner* 17 Nov. 27/1 Mrs. A. D. Scott, regarded as one of the toughest stonewallers ever to play locally, was not very happy on the very fast surface.

stonewall, *v.* Add: **b.** For 'orig. and chiefly *Australian*' read orig. *Austral.* and add later examples; *trans.* (later examples); also, to block (an enquiry, request, etc.); to obstruct (a person or organization). Now chiefly *N. Amer.*
1964 M. GOWING *Britain & Atomic Energy 1939–1945* xiii. 344 The Combined Policy Committee discussed the matter but the Americans stonewalled. 1972 *Accountant* 23 Mar. 373/2 Often in the past, the Budget speech has been preceded by the unreality of questions to the Chancellor which his junior Ministers have had to stonewall with the traditional: 'I cannot anticipate my right hon. Friend's Budget statement.' 1974 *Newsweek* 11 May 23/2 The President himself..served notice that he would stonewall any further demands for tapes in the Watergate scandal. 1974 *Globe & Mail* (Toronto) 10 Dec. 6/2 What the Government does not seem to realize is that in the process of stonewalling the Opposition, it has, itself, compromised the independence of Mr. Munro. 1976 D. HIRO *Inside India Today* 260 The Congress administration stonewalled again, when Mishra died..and the opposition demanded a 'high power' inquiry. 1982 *Daily Tel.* 25 Jan. 12/7 The Nixon administration..also gave the world 'stonewall' as a verb and then got out in the attempted practice thereof.

stoney (stǭᵘ·ni). *dial.* Also *Sc.* **staney, stanie; stonie.** [Var. STONY *a.*] A child's coloured marble made of stone or a stone-like material.
1856 *N. & Q.* 2nd Ser. I. 283/2 Stone marbles are called *stoneys*, and clay ones *commoneys*, though *Dutch alleys* are only *stoneys* enamelled. 1868 *Little Corporal* May 67/3 Chinies, Stonies, and Agates, some large and some small. 1885 'J. STRATHESK' *More Bits from Blinkbonny* (ed. 2) ii. 33 Those played with were called 'taas', and consisted of 'marbles, stanies, frenchies, moral-leggers', etc. 1919 W. WINGATE *Poems* 74 Reddies and stanies for 'moshie' or 'ring'. 1956 G. E. EVANS *Ask Fellows who cut Hay* xxiv. 215 Single marbles were placed in a long line, as many marbles as there were players... The player stood at the end of the line, an agreed distance from the first marble, or *stoney* as it was called. 1965 *Press & Jrnl.* (Aberdeen) 13 Apr. 6/4 A good 'staney', a hard stone boolie which could be hurled against the school wall without breaking.

stonk (stǫŋk). [?Echoic.] **1.** *dial.* Also **stunk.** (See quot. 1841). Also, a game of marbles; a coloured marble.
1841 JAMIESON *Dict. Sc. Lang.* II. 503/1 *Stunk*, the stake put in by boys in a game, especially in that of marbles. 1896 *Manch. City News* 10 Oct., The game is called 'stonks' oftener than marbles. *Ibid.*, A brown or other coloured marble is a 'stonk' and counts one.

2. *Mil. slang.* A concentrated artillery bombardment. Also *fig.*
1944 W. ROBSON *Let.* 8 May (1960) 94 Our gunners were in readiness for the great stonk we requested at nightfall. 1947 D. M. DAVIN *Gorse blooms Pale* 197, I wasn't so crackers I wasn't still listening for that bloody stonk to come screaming down on us. 1961 *Times* 27 Nov. 6/5 The 'stonks' that Mr. Brown and his regional organizers are now going to bring down. 1975 D. CLARK *Premedicated Murder* iv. 65 'We were AGRA.' 'Army Group Royal Artillery..thickening up on other people's stonks and barrages.' 1981 LD. HAREWOOD *Tongs & Bones* ii. 45 You could never tell..if your arrival would bring down an artillery 'stonk' on your head.
Hence as *v. trans. Mil. slang*, to bombard with concentrated artillery fire.
1944 *Daily Tel.* 15 May 6 Here was one more message before we left—that British troops on a captured ridge were being 'stonked' heavily. 1946 R. ALLEN *Home made Banners* xi. 136 Moaning Minnie..was the name they gave to the German multiple mortars that stonked their positions, wherever they were, a minimum of twice and a maximum of several dozen times in each twenty-four hours.

stonker (stǫ·ŋkəɹ), *v. slang* (chiefly *Austral.* and *N.Z.*). [f. *STONK + -ER[5].] *trans.* To render useless; to put out of action, thwart. Also, to kill, destroy; to defeat or outwit. Now chiefly as *pa. pple.*
1919 W. H. DOWNING *Digger Dial.* 48 *Stonker*, exterminate; kill; strike out. 1928 *Bulletin* (Sydney) 15 Feb. 26/4 Then one [shell from a gun] comes in and stonkers 'Iggins and the Company Sergeant-Major. 1941 BAKER *Dict. Austral. Slang* 72 *Stonker*, to defeat, outwit, put out of action. 1945 R. L. SEDDON *Whims of W.A.A.F.* 4 Benzine restrictions have stonkered my car. 1959 G. SLATTER *Gun in my Hand* xiv. 201 He went and stepped on a bloody mine. Stonkered the poor bastard properly.
Hence **sto·nkered** *ppl. a.* (also *spec.* drunk; excessively intoxicated; extremely tired).
1924 *Truth* (Sydney) 27 Apr. 6 *Stonkered*, to be very drunk. 1925 FRASER & GIBBONS *Soldier & Sailor Words* 271 *Stonkered*, put out of action. 1932 A. W. UPFIELD *Royal Abduction* 250 'Why don't they shut off the confounded thing?' 'Too stonkered with surprise, I'll bet.' 1940 F. SARGESON *Man & Wife* (1944) 76 Once they were a bit stonkered the boys would want to have a bo-peep at the bird while he was asleep. 1948 V. PALMER *Golconda* xix. 161 There were one or two old chaps on the executive who were glad enough to see me stonkered. 1963 A. LUBBOCK *Austral. Roundabout* 44 Two on 'em there was, lyin' stonkered in the road. 1967 K. GILES *Death & Mr. Prettyman* ii. 59 It won't help. I know when I'm stonkered. 1970 *Private Eye* 22 May 16 I'm pretty stonkered. Where can we get a snooze round here?

stony, *a.* For '7–9 stoney' read '7– stoney' and add: **2. d.** Of a meteorite or meteoritic material consisting mostly of silicates and other non-metals.
1802 *Phil. Mag.* XIII. 23 (*heading*) Experiments and observations on certain stony and metalline substances which at different times are said to have fallen upon the earth. 1866 *Catal. of Meteorites* (Geol. Survey of India) 8 Two classes of meteorites or solid bodies which have been known to fall to the earth's surface, namely, 1st, stony masses, or aërolites, (often with particles of iron)—and 2nd, masses chiefly iron, or aërosiderites. 1898 *Amer. Jrnl. Sci.* CLV. 63 It seems probable that certain of the stony meteorites that have been found are really the matrices in which some of the iron nodules, formed perhaps many miles distant, were embedded at the moment they entered our atmosphere. 1926 E. A. FATH *Elem. Astron.* xiv. 196 The three classes have the following general composition:..Iron Meteorites... Stony-iron Meteorites... Stony Meteorites.—These consist essentially of silicate minerals with minor amounts of metallic alloys and sulphides. 1981 *Times* 23 Jan. 14/5 The fossil falls into the class known as 'H-chondrite', stony meteorites containing a large amount of iron.

5. c. (Examples of phr. *stony silence.*)
1911 M. BEERBOHM *Zuleika Dobson* vii. 90 The Duke did not try to break the stony silence in which Zuleika walked. 1979 A. BRINK *Dry White Season* iv. iv. 301 They were still staring at me in stony silence, their young dark faces expressionless.

d. (Earlier example.)
1886 [see *BROKE *ppl. a.* 3].

6. b. *stony-faced*; **c. stony-broke** *a.* (earlier example); **stony-iron** *sb.* and *a.*, used to designate meteorites which contain appreciable quantities of both stony material and iron.

1890 Stony-broke [see *pebble-beached* adj. s.v. *PEBBLE *sb.* 5 b]. 1933 M. ARLEN *Man's Mortality* xv. 315 Manteuffel, staring stony-faced towards the darkness..appeared not to have heard his question. 1975 F. BRESLER *You & Law* 81 Even in this stony-faced sector of the law, fairness prevails. 1918, 1962 Stony-iron [see *MESOSIDERITE]. 1969 *Times* 9 Apr. 7/3 Stony-iron meteorites have been found in several Hopewell burial mounds. 1978 D. W. SEARS *Nature & Origin of Meteorites* iii. 73 Stony-iron meteorites are traditionally defined as having approximately equal proportions of stony material and iron. At various times, four groups of stony-iron meteorites have been defined, but since two of these contain only one meteorite each we need here consider only two in any detail: the pallasites and the mesosiderites.

stooge (stūdʒ), *sb. slang* (orig. *U.S.*). [Origin unknown; the possibility that it represents an altered form of *student* has been suggested (students having frequently been employed as stage assistants).] **1. a.** A stage hand. **b.** A stage assistant, esp. one who acts as the butt or foil for a leading character; a feed, straight man. **c.** The assistant of a conjuror or similar performer. **d.** *transf.*
1913 *Sat. Even. Post* 1 Nov. 64/4 Ben, I want you to plant one of your stooges in that coop with a couple of smoke-pots, so that we'll get the effect of Jack coming through the thickest of it. 1929 *Variety* 24 July 1/1 Stuges perform on the floor with dead-pan faces and unconscious feet beating out the time-step. 1936 R. E. SHERWOOD *Idiot's Delight* II. iii. 129, I was a stooge for Zuleika, the Mind Reader. 1936 WODEHOUSE *Laughing Gas* xxii. 238 You expect me, do you, not only to act as a stooge for you in front of the camera, but to sit smiling in the background while you horn in and swipe my interview? 1940 —— *Eggs, Beans & Crumpets* 166 She's a conjuror's stooge... A conjuror's assistant, don't you know. 1941 *Punch* 14 May 468/1 A lament for the absence of that long-suffering, superbly resilient feminine stooge, *Margaret Dumont* [as partner of the Marx brothers]. 1955 W. R. MATTHEWS *Brit. Philosopher as Writer* 12 In it [*sc.* one kind of philosophical dialogue] the author sets up one or more 'stooges', who..can be made to ask just the questions which he, in the person of another character, can answer. 1967 M. ARGYLE *Psychol. Interpersonal Behaviour* vi. 112 Subjects took part in three three-minute discussions with stooges trained to stare, at distances of two, six and ten feet. 1977 *Spare Rib* June 28/2 Our humour is inter-reactive, there's no stooge. 1979 *Sci. Amer.* May 22/1 A common method of cheating is to rely on what magicians call a 'stooge': someone who is watching behind a screen and sending secret signals to the psychic by any one of scores of little-known techniques. 1982 *Jewish Chron.* 9 July 10/3 Their roles—ventriloquist and dummy, reciter and interrupter, smart alec and stooge.

2. A newcomer, a novice (in certain *spec.* contexts: see quots.).
1930 J. LAIT *Big House* i. 6 A first-timer [in prison] is a 'stooge'. 1935 J. HARGAN *Gloss. Prison Lang.* 8 *Stooge*, first offender. 1942 FORBES & ALLEN *Ten Fighter Boys* 55 As the squadron was on readiness most of the day, training was difficult to do, but we did some. However, within a fortnight the squadron moved south to 'K': we stooges were left behind with one pilot to finish off training.

3. A person whose function is merely to carry out another's directions; an unquestioningly loyal or obsequious subordinate, a lackey; a person used as an instrument by someone behind the scenes, a cat's paw. Also *fig.*
1937 H. G. WELLS *Brynhild* vi. 85, I have to..proclaim you. Be your Aaron. Your John the Baptist. Your—Stooge! 1937 —— *Star Begotten* vii. 128 He assembled by wire and telephone all his most trusted henchmen, tools, stooges, subordinates, intimates. 1944 *Times* 6 Jan. 8/3 'If the Beveridge plan is adopted, does it mean that we shall all become State stooges?'.. 'No... I have been a Civil servant myself, and I rather resent the suggestion that a Civil servant is a stooge.' 1948 *Observer* 7 Mar. 4/2 Communists have no use for democratic Socialists except as stooges, and the end they are working for is not Socialism but the totalitarian police-State. 1951 *Negro Hist. Bull.* (U.S.) Feb. 111/1 Black stooges mouthing the sentiments of the white politicians. 1952 R. A. KNOX *Hidden Stream* xvii. 157 If I see a rather nice picture in a shop and..buy it, is that because..I can't resist buying it? But if so, surely my will is not really free: it is just a sort of stooge. 1957 M. SPARK *Comforters* iii. 69 At first I thought she was running a gang, but now, all things considered, I think she may be their stooge. 1960 *Washington Post* 28 Apr. A 22/1 But..his habit of surrounding himself with stooges and sycophants, inevitably led to the debacle after an egregiously fraudulent election. 1978 *Detroit Free Press* 5 Mar. A 11/1 Joshua Nkomo and Robert Mugabe..branded the moderate African leaders as 'sworn stooges of Premier (Ian) Smith'.

4. *R.A.F.* In war-time: a flight during which one does not expect to encounter the enemy. Also *attrib.* and in *stooge-around*.
1942 T. RATTIGAN *Flare Path* I. 37 It's a raid, I suppose, Teddy. It's not exactly a practice stooge-around. 1945 C. H. WARD-JACKSON *It's a Piece of Cake* 57 *Stooge patrol*, a patrol on which the pilot does not see, or expect to see, the enemy. 1952 M. TRIPP *Faith is Windsock* v. 86 At one stage we saw a Fortress orbiting slowly, presumably on a stooge with a team of W/Ops jamming enemy frequencies.

5. *attrib.* and *Comb.*
1940 *Punch* 1 May 482/3 His Lordship..was taken on as a stooge fiance in order to camouflage Felicity's unprincipled intentions. 1948 PARTRIDGE *Dict. Forces' Slang* 183 *Stooge pilot*, a pilot employed in flying-training aircraft carrying untrained navigators or gunners. 1957 R. N. CAREW HUNT *Guide to Communist Jargon* xxxiii. 115 The Communists..introduced one-party government, and the various stooge-parties the existence of which they permitted had no influence in determining policy. 1958 *Observer* 16 Mar. 17/6 The sententious, almost stooge-like, quality associated with the chorus in Greek tragedy.

stooge (stūdʒ), *v.* [f. the sb.] **1.** *intr.* To act as a stooge (senses 1 and 3) (*for* someone). *slang.*
1939 R. CHANDLER *Big Sleep* xviii. 135 We're glad to stooge for a shamus of his standing. 1955 J. THOMAS *No Banners* xxii. 214 One of the *Milice* degenerates who stooged for the Gestapo. 1973 'G. BLACK' *Bitter Tea* viii. 122 For a time Jeremy stooged in espionage, but that must have hurt his feet. 1979 *Sci. Amer.* May 22/3 That Strang often stooged for Geller is well established.

2. *slang* (orig. *R.A.F.*). Of an aircraft: to cruise (*about*, *around*, etc.). Hence *gen.*, to drift, wander, move randomly.
1941 *Illustr. London News* 9 Aug. 165/2 (*caption*) We just stooged about watching the bombers drop their loads. 1942 T. RATTIGAN *Flare Path* I. 19 We were stooging along over the Dutch coast. 1942 [see *BEAT *v.*[1] 40 i]. 1953 G. HEYER *Detection Unlimited* xiii. 208 Stooging round with me, and thinking how much better you could do the job yourself. 1956 'J. WYNDHAM' *Seeds of Time* 136 The streets became..full of crowds stooging around. 1958 M. K. JOSEPH *I'll soldier no More* xiii. 238 Been in 691 Squadron, stooging around the Channel ports all winter. 1973 'N. GRAHAM' *Murder in Dark Room* xviii. 128, I noticed the Austin in the mirror... That made me think I'd seen another black Austin stooging around fairly recently.
Hence **stoo·ging** *vbl. sb.*
1944 *Times* 26 Jan. 4/2 Anti-submarine patrols form another vitally important part of the Coastal Air Force's work; it is often dull, monotonous 'stooging', but is also often well rewarded. 1960 D. STOREY *This Sporting Life* I. ii. 62, I don't want the thought of your stooging always lying over me. 1978 J. GARDNER *Dancing Dodo* xviii. 134, I did some stooging—keeping cave, you know.

† **stook** (stuk), *sb.*[2] *Obs. slang.* Also **stoock.** [Possibly ad. G. *stück* piece.] A pocket-handkerchief. Also *Comb.*, as **stook-buzzer, -hauler,** one who steals pocket-handkerchiefs.
1859 HOTTEN *Dict. Slang* 103 *Stook*, a pocket-handkerchief. *Stook hauler*, or *buzzer*, a thief who takes pocket-handkerchiefs. 1862 H. MAYHEW *London Labour* Extra vol. 25 *Stook-buzzers*, those who steal handerchiefs. 1889 E. SAMPSON *Tales of Fancy* 18 A dirty face, and a still more dirty 'stook'. 1893 P. H. EMERSON *Signor Lippo* xiv. 48 All I get is my kip and a clean mill tog, a pair of pollies and a stoock.

stookie (stu·ki), *Sc.* and *north. dial.* Also **steuke, stoukie.** [dial. var. STUCCO *sb.*] **1.** Plaster of Paris; any plaster-like substance.
1796 *Edin. Mag.* May 385 The carved wood an' polish'd stoukie. 1948 *Proc. Sc. Anthrop. & Folklore Soc.* III. iii. 83 When the doorstep had been washed, the careful housewife could draw designs and patterns with white 'stookie'. 1968 in *Sc. Nat. Dict.* (1974) IX. 60/3 My stooky halfed in two and I had to go back into hospital.

2. A plaster statue, a wax figure or dummy. Also *transf.*, a slow-witted person, a block-head.
a 1828 T. BEWICK *Howdy & Upgetting* (1850) 13 Dinna sit there leyke steuke, and sit and say nowse. 1895 W. C. FRASER *Whaups of Durley* xv. 219 Jamie sat like a stookey wi' a face as red as a partan's tae. 1903 J. LUMSDEN *Toorle* 193 Because, ye stupid stookie, I step aside for none. 1931 A. J. CRONIN *Hatter's Castle* II. iii. 256 Did ye notice the stookies in the window? 1934 J. BUCHAN *Free Fishers* xix. 314 Give her your arm..and don't stand glowering like a stookie. 1948 *Aberdeen Press & Jrnl.* 27 May 4/1 The civic representatives all standing like 'stookies' as they had not got the words of the Psalm they were singing.

stool, *sb.* Add: **2. b.** (Earlier example.)
1836 DICKENS *Let.* ?27 July (1965) I. 157 If you write me word that you will give him a stool, he shall sit himself upon it forthwith.

18. a. (Earlier example.) Also *transf.*, a person employed as a decoy by criminals.
1825 *Huntington* (N.Y.) *Town Rec.* (1889) III. 322 No person [shall] be permitted to gun with macheanes or stools in sd. Town. 1847 J. ROACH *Let.* 20 May in T. Coleman *Passage to America* (1972) xi. 183 There is three hundred emigrants in the Rochester tonight... The head man is a 'Stool'—make him jump.

b. A police informer. Cf. *stool-pigeon*, sense 19 b in Dict. and Suppl.
1906 G. E. STEVENS *Wicked City* 233 Under others were inscribed: 'He is a "stool".'..'He was croaked by the cops.' 1915 J. LONDON *Jacket* ii. 10 They laughed at him and turned him away..for the stool that he was. 1932 E. WALLACE *When Gangs came to London* xv. 129 I'm not so sure that I want to tell you anything—I never was a stool. 1939 J. STEINBECK *Grapes of Wrath* xx. 338 I'll come for ya tonight. Maybe I'm wrong. There's stools aroun' all a time. 1962 B. COBB *Murder: Men Only* i. 12 He said he wasn't a stool, he wasn't giving anybody away.

19. b. stool-pigeon (*a*) (earlier examples); (*b*) a police informer.
1830 *Workingman's Gaz.* (Woodstock, Vermont) 1 Dec. 79/2 A wag who keeps an oyster cellar in Newark advertises, among other things, 'wildbirds domesticated and stool pigeons trained to catch voters for the next Presidency—warranted to suit either party.' 1836 W. IRVING *Astoria* I. 137 One man..was used like a 'stool pigeon', to decoy the others. 1844 [see *ROPER 5]. 1845 *Yankee* (Boston) 9 Aug. 2/6 If this business is so profitable to thieves, how much do those [*sc.* police officers] make out of it who encourage the stool pigeon business? 1849 *Bankers' Mag.* Aug. 89 The senior high constable of Philadelphia..recollected that Harry White..who he had been lately using as a 'stool pigeon', or secret informer, had informed him..that 'a big thing' was coming off shortly. 1850 *Congress. Globe* 18 July 1403/1 Sheltering this aggression, on the part of the United States, behind 'poor New Mexico', who is only a stool-pigeon. 1910 E. A. WALCOTT *Open Door* 134 Rafferty..assured the chief that he would pass word to certain stool-pigeons to keep their eyes and ears open for trace of the missing canvas. 1930 *Times Lit. Suppl.* 4 Dec. 1047/3 Occasionally a masterful rogue arose who shot a few people as 'stool-pigeons', even though they had never imparted any information to the police. 1974 J. THOMSON *Long Revenge* ii. 23 A stool pigeon planted in a local Gestapo prison to eavesdrop on the detainees.

stool, *v.* Add: **4. a.** (Earlier example.)
1842 W. P. HAWES *Sporting Scenes* I. 55 I'll tell you all about that..the next time we're stooling snipe together.

b. *intr.* To act as a stool or stool-pigeon; to inform *on* (someone). Chiefly *U.S. slang.*
1911 [see *BOOB *sb.* 1]. 1938 *Amer. Speech* XIII. 191/2 *To stool*, to act as a stool-pigeon. 1950 PATTERSON & CONRAD *Scottsboro Boy* III. v. 224 There were little mice in Kilby. They ran in the cells. They weren't the trouble that the big rats were, though, them that stooled on you. 1960 'E. MCBAIN' *See them Die* v. 48 You'd stool on Pepe for that rotten cop?..A stoolie is a stoolie. 1973 'B. MATHER' *Snowline* xi. 133 'I stand in a sort of special relationship with these bums. If they thought I was stooling on them—well, you see what I mean?' 'No,..I don't see that putting me in touch..could possibly be construed as stooling. I'm not a policeman.'

stoolie (stū·li). *U.S. slang.* [f. *STOOL *sb.* 18 b or *stool(-pigeon)* (s.v. STOOL *sb.* 19 b) + -IE.] A police informer, a stool-pigeon.
1924 G. C. HENDERSON *Keys to Crookdom* 419 *Stool pigeon*, a spy, stoolie, squealer. One who betrays his fellow crooks. 1930 'E. QUEEN' *French Powder Mystery* p. xiv, Without the stool-pigeon a huge percentage of felonies would remain unsolved... Our problem is to find a 'stoolie' who will part with the tip. 1947 *Sun* (Baltimore) 13 Sept. 8/1 Victor Mature..succeeds in winning sympathy for Nick against odds, overcoming even the stigma which attaches to the stoolie. 1958 'E. MCBAIN' *Killer's Payoff* (1960) viii. 79 The policeman trusted the stoolie's information... The stoolie trusted the policeman... Cops were averse to working with pigeons they did not know and trust. 1960 [see *STOOL *v.* 4 b]. 1974 T. P. WHITNEY tr. *Solzhenitsyn's Gulag Archipelago* I. iii. 128 'There is no way out! You have to confess to everything!' whisper the stoolies who have been planted in the cell. 1978 LAROSA & TANENBAUM *Random Factor* (1979) 60 When Parker put the pressure on, some stoolie somewhere would turn the man he was looking for.

stoop, *sb.*[3] Add: **b. stoop ball** *N. Amer.*, a ball game resembling baseball, but in which the ball is thrown against a stoop or building rather than to a batter.
1941 B. SCHULBERG *What makes Sammy Run?* ix. 166 Kids yelling at each other in a stoop-ball game. 1947 *Commentary* May 463/2 As one of a large family of games such as stoopball..it demands an ability to maneuver freely. 1959 J. D. SALINGER in *New Yorker* 6 June 102/2 Stoopball, for the information of rural readers, is a ball game played with the support of a flight of brownstone steps or the front of a apartment building. 1978 G. A. SHEEHAN *Running & Being* vii. 90 We knew our block... Knew which steps to get pointers in stoop ball.

stoop, *v.*[1] Add: **III. 14. stoop crop** *N. Amer.*, a crop whose cultivation demands stoop labour; **stoop labour** *N. Amer.*, agricultural labour performed in a stooping (or squatting) position; **stoop tag** *N. Amer.* = *SQUAT TAG.
1928 *Sat. Even. Post* 10 Mar. 170/2 He does heavy field work—particularly in the so-called 'stoop crops' and 'knee crops' of vegetable and cantaloupe production. 1939 J. STEINBECK *Grapes of Wrath* xix. 316 Lettuce, cauliflower, artichokes, potatoes—stoop crops. 1967 *PTA Mag.* (U.S.) June 5 He was one of the migrant workers who follow the course of stoop crops through California fields and valleys. 1943 *Sun* (Baltimore) 3 Aug. 11/1 Asparagus cutting ordinarily is a specialized stoop-labor job. 1959 *Economist* 7 Mar. 876/1 Some harvesting can be done mechanically, but most crops still require back-breaking 'stoop' labour. 1972 *Islander* (Victoria, B.C.) 23 Apr. 13/2 Most moved down to southern Ontario to do back-breaking stoop labor in its sugar-beet fields. 1979 G. SWARTHOUT *Skeletons* 89 There are eight million illegals in the country... It would be one thing if they were all agricultural—stoop labor, fruit- and produce-pickers, so on. 1898 F. P. DUNNE *Mr. Dooley in Peace & War* (1899) 179 Little Flora an' little Fauna playin' stoop-tag aroun' a whale..or engagin' in some other spoort iv childhood! 1955 W. GADDIS *Recognitions* I. iii. 146 The critics!..They're like a bunch of old maids playing stoop-tag in an asparagus patch.

stoopid (stū·pid), *a.* and *sb.* Non-standard (often joc. or playful) repr. of STUPID *a.* and *sb.*
1848 THACKERAY *Vanity Fair* xl. 362 Hold your tongue, you stoopid old fool. *Ibid.* lv. 499 Shut your mouth, you old stoopid. 1854 MRS. GASKELL *Let.* 2 Sept. (1966) 303 Many happy returns of your birthday..I hope it won't be a 'stoopid' day. 1902 F. G. ELLERTON *Let.* 11 Aug. in

S. Bailey *John Bailey* (1935) 83 Glad you had a little slap at that stoopid old Swinburne. 1913 R. BROOKE *Let.* 24? Nov. (1968) 537 I'm going to spend 5 days a week there, and three in London (that's 8, stoopid). 1956 N. MARSH *Off with his Head* (1957) x. 211 Lor', Dulcie, what a stoopid gel you are. 1978 C. STORR *Winter's End* xi. 131 'I didn't like your being angry with me.' He began with a denial..then checked himself. 'I was stoopid.'

stop, *sb.*[2] Add: **I. 1. g.** An act of stopping and questioning a suspected person.
1968 J. LOCK *Lady Policeman* iii. 21 These stops winkled out the juveniles who had absconded. 1970 P. LAURIE *Scotland Yard* ii. 47 He appreciates encouragement, and has been advised..about stops on the street.

II. 12. Also in *Photogr.*, a diaphragm or (orig.) a perforated plate for reducing the effective diameter of the lens of a camera or enlarger (now usu. built into the apparatus); hence used as a unit of change of relative aperture (or exposure or film speed), a reduction of one stop being equivalent to a halving of any of these.
1858 SUTTON & WORDEN *Dict. Photogr.* 255 The principle of this form of lens will be best understood by discussing, in the first place, the case of a single plano-convex lens, with a stop in front. 1883 *Photogr. Simplified* 23 Always focus with the largest stop, so as to get as much light as possible, and afterwards insert a stop which gives the necessary sharpness. 1902 G. B. SHAW *Let.* 11 Aug. (1972) II. 282 A slow plate and a suitably small stop will prolong the exposure sufficiently to make it manageable by hand with a cap. 1955 MORGAN & LESTER *Leica Man.* (ed. 13) ii. 75 With lenses of short focal length, the addition of 62·5mm extension results in the true aperture being nearly 2 stops smaller than the marked f-numbers. 1961 G. MILLERSON *Telev. Production* iii. 36 Lens diaphragms are graduated in units called stops or f-numbers. 1977 H. INNES *Big Footprints* II. iii. 179 The light's going to be tricky... It'll soon be dusk. If I were you I'd open up a stop. 1979 *Amat. Photographer* 10 Jan. 90/1 The extra two stops of film speed obtained by raising 400ASA to 1600 are invaluable under such conditions.

III. 16. (Later examples.) Now freq. in phrase *to pull out all the stops*, to make every possible effort.
1927 *Oxford Mag.* 20 Oct. 3/2 He may be said to have 'pulled out all his stops'. He gave the University a speech which for ease, eloquence and felicity could not readily be surpassed or indeed equalled. 1955 A. L. ROWSE *Expansion of Elizabethan England* 123 As his rebellion progressed Tyrone had to pull out the Catholic stop. 1957 *Economist* 5 Oct. 20/2 A Russian admiral on a good will naval visit to the Syrian port of Lattakia was serenading nationalism with all the right stops out. 1965 MRS. L. B. JOHNSON *White House Diary* 20 Dec. (1970) 341 This evening we gave a State Dinner... We opened up all the stops and the Christmas carols rang forth. 1974 A. PRICE *Other Paths to Glory* I. vi. 77 'But they have no idea who did it?' 'Not from what I heard... I know they're pulling out all the stops, though.' 1978 P. MCCUTCHAN *Blackmail North* ii. 20 We'll be doing our best, all stops out.

IV. 17. d. *Cryptography.* A character representing a punctuation mark.
1915 [see *NULL *sb.*[1] b]. 1939 F. PRATT *Secret & Urgent* 18 *Stops* are punctuation marks, usually sentence endings, for which special characters are provided, sometimes placed after each word.

e. Short for *full stop*: (*a*) as used, spelt out, in a telegram; (*b*) = *PERIOD *sb.* 11 b.
1936 [see *LITERARY *a.* 3 b]. 1964 F. CHICHESTER *Lonely Sea & Sky* xxxii. 333 Another exciting telegram.. which read, 'Delighted to see that you have achieved your ambition to beat your own record Stop.' 1977 *Times* 7 Oct. 15/5 Sir, Almost all who write you on this subject assume that high productivity is desirable, stop.

18. b. (Later examples.) Also = *PERIOD *sb.* 11 b.
1861 GEO. ELIOT *Let.* 6 Oct. (1954) III. 456 There is a point of disgust..which one feels must make a full stop, and call for a *Finis* in friendship. 1923 P. SELVER tr. *Čapek's R.U.R.* I. 10 It was in the year 1920 that old Rossum the great physiologist, who was then quite a young scientist, betook himself to this distant island for the purpose of studying the ocean fauna, full stop. 1962 *Observer* 1 July 8/5 The controversy has been between those who say yes, full stop, and those who say yes, but... 1971 R. AMBERLEY *Ordinary Accident* x. 92 Once he sends for a lawyer then that will be full stop.

V. 28. a. (Earlier example in sense 'the game of Newmarket'.)
1886 W. B. DICK *Mod. Pocket Hoyle* (ed. 11) 343 Newmarket, or Stops. This game is played in a similar manner to the game of 'Boodle'.

† **28*.** *Cricket.* A fielder standing three or four yards behind the wicket, a little on the off side. *Obs.* Cf. *BACK-STOP b, *long-stop* s.v. LONG *a.*[1] 18 d.
1773 in H. T. Waghorn *Cricket Scores, Notes, &c.* (1899) 90 All England. May, Lumpey, bowlers; Minshul, Miller, Parmore (stop). Hampshire. Bret, Nyren, bowlers; Small, Sutton, Lear (stop). 1851 W. CLARKE in W. Bolland *Cricket Notes* 129 In laying out your field, you should be careful in selecting good men for your principal places, such as wicket keeper, point, stop, short slip.

V. 29. stop band *Electronics*, a band of frequencies which are highly attenuated by a filter; **stop bath** *Photogr.*, a bath for arresting the process of the preceding bath, esp. development, by neutralizing any of its chemical that may still be present; **stop bead** [BEAD *sb.* 6 b] (see quot. 1964); **stop-butt,** a slope or bank constructed behind the targets at a rifle-range to stop bullets; **stop button,** a button or switch which is pressed or pulled to stop the action of a machine; **stop chords** *Jazz*, chords played on the first beat of every bar or every other bar, as the only accompaniment to a solo; **stop chorus** *Jazz*, a solo accompanied by stop chords; **stop consonant** = sense 19 b; **stop-cylinder,** a printing press in which the cylinder is stopped to permit the return of the reciprocating carriage; **stop lamp,** a light on the rear of a motor vehicle, which is automatically illuminated when the brakes are applied; **stop light,** (*a*) = *stop lamp* above; (*b*) *N. Amer.* = *stop sign* below, also *fig.*; **stop list,** (*a*) a list of persons, etc., deprived of particular rights, privileges, or services; *spec.* a list of persons with whom members of an association are forbidden to do business; (*b*) a list of prohibited books; (*c*) a list of words to be omitted from a concordance or index; hence **stop-list** *v. trans.*, to include in a stop list; **stop log,** a log or plank, or a beam or plate of concrete or steel, fitting between vertical grooves in walls or piers to close a water channel; **stop-netting** = *stop-net* (*b*); **stop sign,** a sign indicating that traffic should stop; *N. Amer.* spec., a red traffic-light; **stop signal,** a signal indicating whether a train should stop; **stop time** *Jazz*, a stop chorus or a series of stop chords; **stop volley** *Lawn Tennis* (see quot. 1928); **stopway,** an area at the end of an airfield runway in which an aircraft can be stopped after an interrupted take-off; **stop word,** a word (usu. one of a set of the words most frequently occurring in a language or text) that is automatically omitted from or treated less fully in a computer-generated concordance or index.
1922 Stop band [see *passband* s.v. *PASS *sb.*[2] 17 b]. 1959 KUH & PEDERSON *Princ. Circuit Synthesis* xiii. 200 The frequency where the pass- and stopbands coalesce is called the cutoff frequency. 1978 *Internat. Jrnl. Electronics* XLV. 247 Filters are designed with various frequency characteristics for the pass band and the stop band and also with predefined zeros in the stop band. 1898 H. MACLEAN *Pop. Photographic Printing Processes* iv. 42 To counteract some more or less over-toning..what is termed a 'stop' bath is used. 1967 E. CHAMBERS *Photo-litho-Offset* v. 58 The required number of contacts are made on lith type plates, developed, passed into the stop-bath and etch-bleached in the usual manner. 1980 D. FRANCIS *Reflex* xiii. 151, I set out the trays of developer and stop bath and fixer. 1876 *Encycl. Brit.* IV. 496/1 An inner or stop bead is mitred round on the inside to complete the groove or channel for the lower sash. 1964 J. S. SCOTT *Dict. Building* 155 *Stop bead*, a bead mitred round the inner edge of a sash window to prevent the inner sash from swinging into the room. 1976 R. DAY *All about House Repair & Maintenance* 62 Broken sash cords are easily removed... First remove the stop bead on the inside. 1864 A. WALKER *Rifle* (ed. 2) 114 If at a smaller angle it would, instead of acting as a stop-butt [etc.]. 1923 KIPLING *Land & Sea Tales* 177 The long shed of the Village Rifle Club reeked with the oniony smell of smokeless powder, machine-oil, and creosote from the stop-butt. 1963 W. H. FULLER *Small-Bore Target Shooting* i. 25 Stop butt or bullet catchers. 1940 N. MARSH *Surfeit of Lampreys* (1941) xiv. 209 When we'd got about half-way d-down she started screaming... I shoved down the stop button. So we stopped... Just below the first floor. 1977 J. WAINWRIGHT *Day of Peppercorn Kill* 120 He glanced at the tape-recorder, pressed the 'stop' button and said, 'We need a new reel.' 1941 *Musical Q.* Jan. 52 The second chorus is played with great feeling by the clarinet to a background of 'stop chords', a very effective device of the New Orleans style. 1958 R. HARRIS in P. Gammond *Decca Bk. of Jazz* iii. 45 *Dipper Mouth* with a hot theme stated by the ensemble; a superb Dodds solo against stop chords; [etc.]. 1942 BERREY & VAN DEN BARK *Amer. Thes. Slang* § 578/10 *Stop chorus*, a chorus in which the orchestra plays only one note in every one or two measures as a background for a tap dancer or other soloist. 1968 P. OLIVER *Screening Blues* ii. 67 The singer 'reading on down' to each new chapter [in sermon-like recitations] against the stop chorus of the pianist or a full jazz band. 1975 Stop consonant [see *PRESSURIZATION]. 1978 *Maledicta* II. 111 Rising tones in Thai do not co-occur in syllables ending in a stop consonant. *a* 1877 KNIGHT *Dict. Mech.* I. 671/2 The stop-cylinder press, designed for woodcut printing. 1980 B. CRUTCHLEY *To be Printer* ii. 21 'And what do you know about printing?' I was about to reply..: 'Well, I can tell the difference between a two-rev and stop-cylinder,' (those were basic types of printing machine). 1959 *Motor Man.* (ed. 36) viii. 217 Sidelamps, headlamps, rear lamps and stop lamps should also be examined for bulb failure. 1979 *Southern Star* (Eire) 29 Sept. 2/6 Defendant was fined £3 for having no stop lamp, £3 for having no rear lamp and £3 for having no number plate lighting. 1930 D. MACKAIL *How Amusing!* 190 His stop-light flickered almost ceaselessly as he crawled round the square. 1931 O. NASH *Hard Lines* 45 But there is no stoplight For a talkative cosmopolite. 1950 *How to drive Car* (ed. 18) xi. 88 Most cars are now fitted with direction indicators and 'stop' lights. The latter are automatic if properly maintained and come on when the brakes are applied. 1978 *Verbatim* Sept. 7/2 The tremendous role that traffic signals play in our national consciousness (it can be argued that the entire Interstate system was built in order to get around—and under and over—stoplights). 1920 *Daily Tel.* 18 May 16/5 The association published his name on their stop list, the object of which was to prevent all members of the association having any trade relations with the offending agent. 1949 *Rep. Committee on Resale Price Maintenance* 54 in *Parl. Papers 1948–49* (Cmd. 7696) XX. 383 Only the 'open price-cutter' who advertised that he was committing a breach of the conditions of sale was immediately stop-listed. 1958 *Times* 13 Mar. 11/3 A nation with what is reported to be over 1,000 books on the stop-list had got enough censorship. 1963 *Times* 28 Jan. 9/2 Merseyside was on the original list but within a few months, because of the motor industry's plans and other new projects, was put on the stop-list. Last week it was restored to the active list. 1963 [see *re-list* vb. s.v. *RE- 5 a]. 1970 *Computers & Humanities* IV. 167 This program segment provides for two stoplists. 1974 *Times* 5 Feb. (Europa Suppl.) p. xiv/5 With credit cards the vendor has to check against the stop list (to check on people who have had their accounts stopped, or cards that have been reported missing). 1975 O. SELA *Bengali Inheritance* xxi. 185, I intend to hold your passport... I'm also putting you on a stop list at Kai Tak [airport]... I don't want you disappearing. 1979 J. E. ROWLEY *Mechanised In-House Information Syst.* I. 74 Words in a title are compared with a stop-list, to suppress the generation of useless index entries. 1930 *Engineering* 6 June 725/3 Each weir is divided into six bays by piers, between which stop logs can be placed, while for emergency regulation..low level Stoney sluices are provided. 1973 *Detroit Legal News* 30 Aug. 13/2 Two feet of the south stop-log chamber wall at the Fairview Pumping Station was removed and had to be replaced with new reinforced concrete. 1927 Stop-netting [see *RUN-BACK 2]. 1981 *Sunday Tel.* 4 Oct. 16/3 A badly-flighted lob cleared the stop-netting. 1934 *Amer. Speech* IX. 114/2 Those who drive have to make allowances for *stop streets* and *stop signs*. 1951, 1972 [see *RUN *v.* 40 d]. 1976 *S. Wales Echo* 27 Nov. 9/3 Pleaded guilty by letter to failing to conform to a stop sign while on his motor-cycle. 1923, 1963 Stop signal [see *HOME B. 2 d]. 1929 *Musical Q.* XV. 611 As to what possibilities free-will tricks as the jazz 'break', stop-time, the harmony chorus, an exaggerated syncopation, etc., hold for the development of musical form beyond jazz itself, he would be bold who would predict. 1966 *New Yorker* 11 June 135 The Onward was playing 'Victory Walk', an engaging stop-time number. 1978 *N.Y. Times* 30 Mar. C16/2 Even in strongly swinging situations, jumping brightly through crisply muted breaks and stop time, the singing flow of Mr. Vaché's playing is never lost. 1915 M. E. MCLOUGHLIN *Tennis as I play It* 56 That is when a 'stop-volley' is employed to drop a ball just over the net. 1928 B. NUTHALL *Learning Lawn Tennis* 106 One of the most useful strokes in the game..is what is called the 'stop volley'... It is necessary to be quite close to the net to play it. The racket is just put in the way of the ball, which drops dead on the other side of the net. 1978 *Times* 4 July 19/2 Ground strokes were spiced with many a delicate angled cross-court chip, the stop volley, the lob and smash. 1960 in *Guide to Civil Land Aerodrome Lighting* (*B.S.I.*) 7 *Stopway*, a defined rectangular area at the end of a runway which has been selected or prepared as a suitable area in which aircraft can be stopped after an interrupted take off but which is not suitable for use as part of the runway. 1980 *Observer* 2 Nov. 6/8 A.. DC-10 from Delhi ran from the runway on to the stopway, the hard section on either side, which is meant to be firm enough to take the weight of aircraft. 1969 *Computers & Humanities* III. 135 If stop words are desired, the user can either specify his own or request a standard list which is encoded within BIBCON. 1979 J. E. ROWLEY *Mechanised In-House Information Syst.* I. 74 The stop-list or stopword list contains words under which entries are not required, such as the, he, is, a and in some circumstances, machine, processing, plant, etc. 1982 *N. & Q.* Oct. 385/1, I understand that a microfiche concordance of the stop words will soon be available.

stop, *v.* Add: **II. 15. c.** (Earlier example.) Also, to hit (game).
1845 *Punch* 25 Jan. 46/2 Out they [*sc.* the hares] rushed from every quarter—so many—that it was often impossible to 'stop' more than one out of half-a-dozen.

e. *colloq.* (orig. *Mil.*). To be hit by (a bullet). Phrases *to stop one*: to be hit or killed; *to stop a packet*: see *PACKET *sb.* 1 f.
1901 *Boy's Own Paper* 5 Oct. 14/2 After the battle of Spion Kop, one man, who was hit in seven places, said that he had stopped a whole volley himself. 1915 *Sphere* 6 Nov. 144/1 A man's troubles begin rather than end when he 'stops' a German or Turkish bullet. 1916 E. V. LUCAS *Vermilion Box* clxxxiv. 213 Poor boy, I do so hope he manages not to 'stop one'—which is what being hit is called here. 1929 J. BUCHAN *Courts of Morning* I. xiii. 152 If I hadn't thought of that head-crashing dodge, I think I might have stopped a bullet. 1933 H. S. WALPOLE *Vanessa* IV. i. 682 Maurice stood there wishing that he might 'stop one' before he had to go over the top. *a* 1976 A. CHRISTIE *Autobiogr.* (1977) v. ii. 234 You stop one, you've had it, and you've left behind a young widow.

f. To drink; usu. in phr. *to stop one*, to take a drink. *Austral. slang.*
1924 *Truth* (Sydney) 27 Apr. 6 *Stop one*, to take a drink. 1929 K. S. PRICHARD *Coonardoo* xxix. 279 Geary poured himself a drink. 'Hi, Dick,' he called, 'could you stop one?' 1936 A. RUSSELL *Gone Nomad* x. 78 Then, jerking his finger knowingly, 'I s'pose yer could stop one?' I could. I needed that rum. 1937 PARTRIDGE *Dict. Slang* 835/2 *Stop a pot*, 'to quaff ale,' C. J. Dennis.

21. a. (Later examples in conventional phrases indicating that there is no obstacle to an action.)
1951 M. KENNEDY *Lucy Carmichael* II. iii. 94 You.. make an entrance if you like. I'm not stopping you. 1970 *Globe & Mail* (Toronto) 28 Sept. 23/2 (Advt.), Make the

break with Tradition. What's to stop you? Certainly not the price. **1973** R. THOMAS *If you can't be Good* (1974) xx. 180 'I wanta see Connie Mizelle,' he said. 'What's stopping you?' 'Not a damn thing,' he said. 'Let's go.'

c. *to stop the show* (orig. *U.S.*); to cause an interruption of a performance by provoking prolonged applause or laughter, or requests for encores. Cf. *show-stopper, -stopping* adj. s.v. *SHOW *sb.*[1] 22.

1926 *Amer. Speech* I. 437/1 When an act proves to be such a wow that it is forced to respond to encore after encore and the remainder of the acts on the program must wait until the audience will allow them to go on, it is said to 'stop the show cold'. **1933** *Fortune* Aug. 92/1 Jim Europe had stopped the show with *St. Louis Blues*. **1957** R. HART-DAVIS *Let.* 19 May in *Lyttelton–Hart-Davis Lett.* (1979) II. 103 The Gibbon quotation stopped the show long enough for me to consult my scrappy notes. **1966** 'M. RENAULT' *Mask of Apollo* vi. 107 This line, as I had feared it might, stopped the show. **1977** *Times* 13 June 15/4 *The Merchant of Venice* [was] performed by the Ibadan Boys' Grammar School... A British widower['s].. son..was cast as the Prince of Morocco. His opening line stopped the show: 'Mislike me not for my complexion—.'

d. To give a still picture of (a moving object).

1937 *Star* (Kansas City) 8 Aug. 3 The camera 'stops' the action of a chorus in training. **1937** *Discovery* Nov. 353/1 Anyone can find a gannet, and any shutter working to 1/500 sec. will 'stop' it. **1950** A. HUXLEY *Themes & Variations* 161 On Alexander's [tomb] the monster has been 'stopped', as the photographers say, in the act of shooting up from the doorway leading into the vault.

33. a. stop down. Also *absol.* and *intr.* for *pass.*

1971 P. PURSER *Holy Father's Navy* xiv. 75 Can you stop down to make it look like dusk? **1978** *SLR Camera* Aug. 46/1 As the lens is focused through these various degrees of magnification the lens automatically stops down.

b*. stop off = senses 21 a, 23 b, and 24 a. Now *rare* or *obs.*

1891 W. B. YEATS *Let.* 21 Jan. (1954) I. 162 Ellis..may do some of my chapters himself... Providence has stopped off his terrible activity for the present with twelve lectures for the University Extension. **1892** —— *Lett.* (1954) II. 201, I helped to stop off another man of learning the other day who came trying to get a book from Unwin to do. **1902** H. JAMES *Wings of Dove* VII. xxv. 382 Having suffered him to insist almost convicted her of indelicacy. Why hadn't she stopped him off? **1904** G. B. LANCASTER *Sons o' Men* 47 Stop that row, Tommy... Stop it off. **1929** T. E. LAWRENCE *Let.* 22 July (1938) 666 Dirty Dogs, they *have* stopped off poor Trotsky.

III. 35. b. For *U.S.* read orig. *U.S. to stop over* (earlier and later examples); *to stop off* (earlier and later examples).

1855 *Knickerbocker* XLVI. 604 He had 'stopped off', he said, to see a friend. **1857** M. J. HOLMES *Meadow-Brook* xvi. 182 Wishing to see a friend of his who lives here, we have stopped over one train. **1873** 'MARK TWAIN' & WARNER *Gilded Age* xxiv. 218 Once when you renewed your ticket after stopping over in Baltimore. **1892** *Harper's Mag.* Feb. 437/2, I stopped off overnight to see about something for a friend. **1925** D. H. LAWRENCE *Let.* ?7 Nov. (1962) 864 It is great fun stopping off in Switzerland to see you. **1952** M. LASKI *Village* xii. 173 'Shall we stop off soon and eat our lunch?' asked Roy and at the next field-gate they dismounted. **1970** G. F. NEWMAN *Sir, You Bastard* viii. 244 Stopping off after court for an early liquid lunch. **1971** *Daily Tel.* 29 Dec. 10 Many people suffer from jet fatigue and on long-distance routes often go to the additional expense of stopping over somewhere on the way to recuperate.

c. *to stop in*: to pay a brief visit, 'drop in'. *U.S.*

1904 *Dialect Notes* II. 421 *Stop in, vb. i.*, to call. 'I *stopped in* at his house one day.' **1925** T. DREISER *Amer. Tragedy* (1926) I. II. xxxvi. 402 He stopped in, not at all sure that on this first occasion he would be able to broach the dangerous subject. **1953** J. CHEEVER in *New Yorker* 22 Aug. 23/2 He was rude to his friends when they stopped in for a drink. **1963** *Jrnl. Amer. Med. Assoc.* 26 Oct. 459/1 He was found dead in his crib by a family friend who had stopped in at the home. **1979** *Yale Alumni Mag.* Apr. (Suppl.) cn17/3 Classmates are eagerly invited to stop in!

d. With *by*: (*a*) as *adv.*, = sense 35 c above; (*b*) as *prep.*, to call at, visit (a place). orig. *U.S.*

1905 *Dialect Notes* III. 96 *Stop by, v. phr.*, to call, to visit. 'I believe I'll *stop by* and see Bud.' **1923** *Ibid.* V. 244 *Stop by*, v. phr., to visit. '*Stop by* my house.' **1928** F. N. HART *Bellamy Trail* v. 172 They were going to stop by for her. **1943** T. PRATT *Barefoot Mailman* i. 11, I picked him up when I stopped by at St. Augustine. **1953** N. GORDIMER *Lying Days* II. v. 48 It was Ludi, he would stop by at the old Plaskett's on the way to say hullo—. **1957** *New Yorker* 2 Nov. 89/3 Don't wait..stop by your favorite shop and try one today. **1964** MRS. L. B. JOHNSON *White House Diary* 8 Apr. (1970) 103, I had asked Mrs. MacArthur and her son..to stop by the White House to warm up and have a cup of tea. **1973** M. AMIS *Rachel Papers* 20, I mentioned that Gloria would probably be stopping by later on.

36. b. *to stop out* (examples); *spec. N. Amer.*, to interrupt one's higher education for a time in order to pursue some other activity.

1926 I. S. COBB *Some United States* xi. 257 I'm a Virginian—at present stopping out in Kentucky. **1942** BERREY & VAN DEN BARK *Amer. Thes. Slang* § 214/7 *Stop out*, to stay away from home all night. **1971** *Less Time, More Options* (Rep. Carnegie Commission on Higher Educ.) vii. 28 Colleges and universities can assist by.. encouraging students to have work or service experience before entering college, to stop out while in college to obtain it, or both. **1977** *N.Y. Times* 16 Jan. IV. 9/1 Paul Marantz is stopping out. He's one of the estimated two million college undergraduates..who last year left school to spend some time in the outside world, or to try out some other form of education, but who do plan to return eventually and earn their degrees.

37. e. *Bridge.* To refrain from increasing one's bid beyond a specified level. Const. *in*.

1959 *Listener* 5 Feb. 265/1 The British pair stopped in Five Hearts. **1964** FREY & TRUSCOTT *Official Encycl. Bridge* 533/1 *Stopping below game*, the decision to 'stop on a dime' in two no trump or three of a major may be influenced by a variety of factors.

IV. 43. a. stop-tap, the time at which drinks cease to be served in a public house.

1938 F. D. SHARPE *Sharpe of Flying Squad* xxiii. 240 Bob said that they hadn't *passed by* any public houses and that it was after 'stop tap' that they were passing the shop. **1940** DYLAN THOMAS *Portr. Artist as Young Dog* 228 If you go for a constitutional after stop-tap along the sands you might as well be in Sodom and Gomorrah. **1960** V. JENKINS *Lions down Under* 103 The 'five o'clock, to six o'clock swill' in the bars of New Zealand cities—for 'stop-tap' is at six—is also a phenomenon to be avoided. **1975** R. LEWIS *Part of Virtue* vi. 147 Next evening, after stop tap, he was putting some crates out behind the pub.

b. *Cinematogr.* Combinations of the verb with a sb., with reference to the technique of stopping the camera between frames in order to produce special effects, esp. animation; as *stop-action, -frame, -motion, -shot*, etc.

1912 F. A. TALBOT *Moving Pictures* 201 When the 'stop' call was given the witch disappeared from the stage... The strange effects produced in the witch's cave were obtained both by double printing and the 'stop-motion'. **1915** J. B. RATHBUN *Motion Picture Making* 73 Trick street scenes, commonly known as 'stop' pictures. **1933** G. H. SEWELL *Commercial Cinematography* x. 155 Stop-motion is..the method of cine-photography in which one, two, or three frames..are taken at one time, the camera being stopped and the subject re-arranged after each shot or group of shots. **1959** HALAS & MANVELL *Technique Film Animation* xxii. 274 The technique for stop-action puppet work must be worked out in terms of single motion-picture frames. **1966** *Listener* 14 July 67/1 The stop-shots neatly made each point. **1968** *Guardian* 22 Mar. 10/4 The stop-frame technique in which the puppets are photographed separately for each movement. **1976** R. B. PARKER *Promised Land* (1977) xx. 122 Powers was quiet. We all were. It was like a stop frame in instant replay. **1980** *Sci. Amer.* Apr. 84/1 A glass is a solid that can be regarded as a stop-action photograph of a liquid.

44. stop and frisk *a.*, of or pertaining to the stopping and searching of suspects by the police; so **stop-and-search, stop-search-question; stop-and-start** *a.*, alternately stopping and starting; **stop-me-and-buy-one,** a travelling vendor of refreshments, usu. ice-creams [from the slogan on the refrigerated box at the front of Wall's Ice-Cream tricycles]; also *attrib.*; **stop-off,** (*b*) the act of stopping off (see 35 b); a place where one stops off; also *attrib.*; **stop-out,** (*a*) *colloq.*, one who stays out late; (*b*) *N. Amer.*, a student who interrupts his or her studies for a time in order to pursue some other activity; an interruption of studies for this purpose; also *attrib.*; **stop-over,** (also **stopover**), (*a*) (earlier and later examples); (*b*) permission given to a passenger to break his journey (now *rare* or *obs.*); (*c*) a place where a journey is broken; also *fig.*

1967 *Economist* 21 Oct. 286/1 The cases to be heard this year are a mixed bag. Those involving the criminal law and the police—particularly 'stop and frisk' laws—may be the most controversial. **1975** *New Yorker* 2 June 101/1 A Terry stop is what civil libertarians sometimes refer to as stop-and-frisk. **1974** *Spartanburg* (S. Carolina) *Herald* 25 Apr. A 10/1 A federal court judge began hearing arguments Wednesday on whether to halt the hotly debated police 'Operation Zebra' stop-and-search dragnet for the black killer or killers of 12 whites. **1950** J. G. DAVIS *Dict. Dairying* 62 The feed to the dies is done with a variable stop-and-start motion, allowing the strip time to stop while the die punches out the shape. **1961** *Times* 4 May 13/6 The stop-and-start tendencies of our economy. **1976** *Woman's Day* (U.S.) Nov. 50/2 If you do a lot of stop-and-start driving,..change every three months or 3,000 miles. [**1935** *Automobile & Carriage Builders' Jrnl.* LXXV. 4/1 'Stop me and buy one.' The latest type of cycle carrier for ice-cream vending.] **1935** *Food* Oct. 3/1 A holiday spent in a number of South Coast towns suggests that England is..becoming as ice-cream-minded as North America. The last three hot summers have provided a golden harvest for the familiar tricycle. But even more recent..has been the appearance..of the 'ice-cream parlour'... 'Come in and have one' is evidently proving as alluring a slogan as the more familiar 'Stop me and buy one'. **1936** N. COWARD *To-Night at 8.30* 49 Asked if I'd got an ice-cream wafer... What did she think I was, a 'Stop me and buy one'? **1939** N. MONSARRAT *This is Schoolroom* xi. 228 To..buy an ice from the stop-me-and-buy-one man. **1947** DYLAN THOMAS *Let.* 11 Apr. (1966) 300 There were stop-me-&-buy-one bicycle boys selling, not ice-cream, but bottles of Chianti. **1979** D. ROBINSON *Eldorado Network* III. xliii. 288 Ice-cream. Stop-me-and-buy-one, the Eldorado man on a tricycle. **1912** J. SANDILANDS *Western Canad. Dict. & Phrase-bk.*, *Stop-off* or *stop-over privileges*, an arrangement made with the ticket agent to break a railway journey at some place where the passenger wishes to make a short halt. **1931** C. BEATON *Jrnl.* Feb. in *Wandering Years* (1961) x. 226 En route for home there was a three-week stop-off in Paris. **1947** *Sun* (Baltimore) 18 Jan. 4/2 Is the police court merely a stop-off between one back room and the next? **1958** *Times Lit. Suppl.* 7 Mar. 125/3 London, New York, Paris, Rome come to life not as tourist centres, holiday stop-offs, but as places of work. **1977** *Horse & Hound* 14 Jan. 25/2 Cost of the trip is £530 return (excursion, 21 days–6 months, no stop-offs). **1906** E. DYSON *Fact'ry 'Ands* ii. 24 'See,' cried Annie—'See, you dirty stop-out!' She placed the hat on the floor and danced wildly amongst the feathers. **1941** BAKER *Dict. Austral. Slang* 72 *Stop-out*, an inveterate gadabout, esp. a woman. **1966** F. SHAW et al. *Lern Yerself Scouse* 27 *Yer a derty stopout*, you are a nocturnal reveller. **1971** *Less Time, More Options* (Rep. Carnegie Commission on Higher Educ.) vi. 13 That service and other employment opportunities be created for students between high school and at stop-out points in college. *Ibid.* 21 Those who plan to continue with academic study either directly or after a stop out. **1971** *Time* 27 Sept. 79/3 Still, many stop-outs do better academically than their less-seasoned classmates, if only because they are a year older. **1974** *Globe & Mail* (Toronto) 31 Oct. 1/5 The so-called stopout students, those who postponed entering university immediately after high school graduation, now are starting to go back to school. **1881** *Harper's Mag.* Apr. 767/2 They are allowed stop-over tickets which give them the privilege of turning their stock out at any place for the winter, and then sending them on in the spring to market. **1885** *Outing* (U.S.) Nov. 150/2 There I took advantage of what, in railroad parlance, is called a 'stop-over'. **1905** *Chambers's Jrnl.* Jan. 87/1 At Vancouver I stepped on board a Canadian Pacific Railway steamer bound for Hong-kong, with a stop-over on my second-class ticket. **1928** *Blue Peter* July p. iv (Advt.), A Convenient Stopover. Honolulu is a regular port of call for passenger steamers crossing to or from the Orient. **1953** I. LEVIN *Kiss before Dying* I. ii. 11 College would only be an unnecessary stopover on the road to..success. **1959** *Economist* 20 June 1106/1 Mr Khrushchev will round off his Scandinavian tour in August by a two or three day stop-over in Helsinki. **1976** *National Observer* (U.S.) 22 May 18/6 From New York the round-trip economy fare is about $1,600, with stopover privileges in London and Nairobi. Because it's a very long flight, you can use the stopovers. **1973** *Time Out* 2–8 Mar. 10/3 More recently they have been performing stop-search-question late night patrols in Hornsey which have resulted in such serious crime detection as arresting people with a quid's worth of dope.

stop-and-go (stɒp ənd gəᵘ), *a.* Also **stop and go.** [f. STOP *v.* + AND *conj.* + GO *v.*] **1.** = *STOP-GO *adj.* 1.

1926 G. FRANKAU *My Unsentimental Journey* xi. 145 St. Louis; where the men..have the most elaborate 'stop and go' signs for their traffic. **1935** D. L. SAYERS *Gaudy Night* viii. 177 Its modern bustle of cars and complication of stop-and-go lights. **1942** *Policy on Rotary Intersections* (Amer. Assoc. State Highway Officials) 1 There is discussion of design speeds, interweaving lengths, curves, geometric shapes, sight distance, roadway widths, cross slope, curbs, islands, lighting, stop-and-go control and special adaptations of rotary intersections. **1962** N. FREELING *Love in Amsterdam* I. 61 Don't use your brain. Be a farmer holding a stop-and-go sign.

2. = *STOP-GO *adj.* 2.

1943 F. L. WRIGHT *Autobiogr.* (rev. ed.) IV. 329 No need to get tangled up in spasmodic stop-and-go traffic. **1972** *Jazz & Blues* Nov. 32/1 He also evinces real invention over the stop-and-go rhythms of *Departure*. **1977** 'E. MCBAIN' *Long Time no See* vii. 94 His stop-and-go typing irritated Carella.

3. = *STOP-GO *adj.* 3.

1961 *Times* 5 Jan. 11/5 The short-term 'stop-and-go' remedies of recent years..deal with the symptoms rather than the disease. **1961** *Hansard Commons* 7 Nov. 850 Before the right hon. Member for Monmouth we had the present Prime Minister as Chancellor. He was a 'stop and go' man. **1962** *Listener* 10 May 817/2 The new-found interest in planning in Tothill Street was a direct reaction to the stop-and-go policies of 1959 and 1960.

stopbank (stɒ·pbæŋk). *Austral.* and *N.Z.* Also **stop-bank.** [f. STOP *sb.*[2] + BANK *sb.*[1]] A levee, an embankment.

1950 G. WILSON *Brave Company* xiii. 196 The clean lines of the river and the stopbank. **1958** S. ASHTON-WARNER *Spinster* 54 All them kids on the stopbank. **1965** S. T. OLLIVIER *Petticoat Farm* xv. 198 Stop-banks were raised against rivers. **1977** *N.Z. Herald* 5 Jan. 1-6/4 The Auckland Harbour Board controls to the tidal boundary at Bonds Rd, Matatoki, a vast acreage that has built up outside the stopbanks.

stope, *v.* Add: **2.** *Geol.* Of magma or a magmatic body: to make *its way* by stoping; also, to subject to stoping.

1908 *Amer. Jrnl. Sci.* CLXXVI. 19 The latter are regarded as then stoping their way up into the overlying shell. **1932** F. F. GROUT *Petrogr. & Petrol.* III. 202 No batholith is known to have stoped its way to the actual surface. *Ibid.* 203 Some rocks are stoped and assimilated more readily than others. **1962** W. T. HUANG *Petrology* iv. 104 If the specific gravity is lower than the corresponding solid rock, a magma could stope its way into rocks of similar chemical composition.

Hence **stoped** *ppl. a.*

1932 F. F. GROUT *Petrogr. & Petrol.* III. 203 The stoped blocks may dissolve before moving far. **1970** K. C. JACKSON *Textbk. Lithol.* ii. 38 The margin of the magma body becomes cluttered with scattered stoped blocks of wall rock.

stop-go (stɒp gəᵘ), *a.* and *sb.* [f. STOP *v.* + GO *v.*] **A.** *adj.* **1.** Of signs or lights: indicating alternately to traffic that it should stop or that it should go.

1918 *Wells Fargo Messenger* Feb. 94/3 The copper flashed us a smile as he gave his stop-go apparatus another twist. **1952** M. STEEN *Phoenix Rising* ii. 50 They were.. held up..by 'Stop'-'Go' signs. **1965** *Motor* 17 July 1/2 The long queues of cars waiting at the wrongly timed stop-go lights.

2. Alternately stopping and going, or acting and not acting.

1960 *Times* 10 Oct. 16/1 In their new 'stop-go' style they were infinitely the more dangerous. **1973** 'M. INNES' *Appleby's Answer* iii. 32 Their taxi made only a tedious stop-go progress. **1980** *Times Lit. Suppl.* 2 May 503/4 Would that English historical journals discussed, for example, the implications for future research of the stop-go policy of recruitment of graduates to history departments.

3. *Econ.* Of, pertaining to, or designating a policy of alternately restricting demand, in order to contain inflation, and expanding credit, in order to reduce unemployment.

The earlier designation of the policy was *stop-and-go* (see *STOP-AND-GO *a.* 3).

1962 *Daily Tel.* 21 Feb. 10 It is precisely these 'stop-go' policies of successive Chancellors which have been a major cause of our export troubles. **1965** *Listener* 3 June 817/2 The British Government then in office found its negotiating position undermined by gossip and arguments at home about the imminence of devaluation of the pound sterling as the only way out of the old stop-go circle. **1971** *Business Week* 13 Nov. 146/3 Yet Ulman and Flanagan conclude that governments have a strong tendency to choose stop-go policies. **1975** J. DE BRES tr. *Mandel's Late Capitalism* xiv. 455 The 'Stop-Go' pattern of the British economy in the first post-war Tory era is the classical example of such a relatively autonomous credit cycle. **1979** *Dædalus* Spring 47 In Sweden, special factors reduced the country's vulnerability to uncontrollable money wage increases, hence to disruptive policies of the stop-go variety.

B. *sb.* *Econ.* A stop-go policy; the economic cycle resulting from this.

1964 S. BRITTAN *Treasury under Tories* vii. 208 This was the event which turned the business community violently against 'stop-go' and made it look with a less jaundiced eye on national planning. **1966** *Listener* 2 June 808/2 Does more inflation mean more difficulties with the balance of payments and more 'stop-go'? **1972** *Accountant* 23 Mar. 365 What evidence is there to convince management and industry that the new phase of expansion which the Chancellor's proposals should generate will not, as on so many previous occasions, culminate within some 18–24 months in a revival of 'stop-go' and balance of payments difficulties? **1976** K. JOSEPH *Monetarism is not Enough* 10 We refused to believe that it was the drug which had caused the need for a stop, hence we still say 'stop-go', but it is the go which causes the stop, not vice-versa.

stoping, *vbl. sb.* Add: **b.** *Geol.* The process by which intruding magma detaches blocks of the surrounding rock.

1903 R. A. DALY in *Amer. Jrnl. Sci.* CLXV. 272 (*heading*) The hypothesis of overhead stoping† by deep-seated magmas. [*Note*] †A technical mining term meaning to excavate upwards or sideways to remove ore. **1903** —— in *Bull. U.S. Geol. Surv.* No. 209. 102 Magmatic stoping would tend to weaken the earth's crust immediately above the intruding body. **1939** BAILEY & WEIR *Introd. Geol.* xlii. 237 On the whole, stoping is probably the most important method of emplacement. **1977** A. HALLAM *Planet Earth* 69/1 Intrusions are emplaced in various ways: for example,..by stoping, a process of gradual movement involving the dislodging of the blocks of country rock and their incorporation—perhaps even assimilation—into the magma.

stoppable (stɒ·păb'l), *a.* [f. STOP *v.* + -ABLE; cf. STOPPABILITY.] That can be stopped.

1934 in WEBSTER. **1977** *Time* 3 Jan. 20/1 Stop most anyone you see—they're generally stoppable—and he or she will soon be spinning you a web of recollection to entertain you both. **1982** *Economist* 16 Oct. 18/1 The dangers threatening Lebanon's Palestinians are not academic; they are both foreseeable and stoppable.

stoppage. Add: **7. b.** A cessation of work owing to disagreement between employer and employees; a strike or a lock-out.

1902 *Encycl. Brit.* XXV. 554/1 The adjustment of differences that might otherwise lead to stoppage. *Ibid.* XXXIII. 14/1 To distinguish stoppages as strikes or lock-outs according to the source of the original demand for a change of conditions would lead to a very arbitrary and misleading classification. **1926** *Publishers' Circular* 29 Dec. 895/3, 1926... The year of the General Stoppage. **1966** *Listener* 1 Sept. 302/2 An American-owned engineering works in north-west London is to close down because of stoppages and the economic squeeze. **1976** *West Lancs. Evening Gaz.* 8 Dec. 1/2 More than 350 Blackpool Corporation busmen agreed unanimously to give authority to the TGWU's North Lancashire District Committee to call them out on a one-day stoppage in support of their Fylde colleagues.

stopped, *ppl. a.* Add: **2. d.** *Bridge.* (See quots.) Cf. *STOPPER *sb.* 7 e.

1901 R. F. FOSTER *Foster's Bridge* 35 A Suit is Stopped when you can make one trick in it, or can compel the adversary to quit it and lead something else. **1929** M. C. WORK *Compl. Contract Bridge* iv. 20 A suit is stopped when the bidder holds such cards in it that he can be sure of taking at least one trick in that suit.

9. *Carpentry.* Of a chamfer, housing, etc.: closed, not running the whole length of a member. Cf. STOP *v.* 30.

1918 *Woodwork Joints* 167 The sketch to the right shows 'stopped housing', the groove coming to within ½in. of the front edge of the shelf. **1934** P. A. WELLS *Design in Woodwork* ii. 15 The number of joints can be trebled by variations such as 'through', 'stopped' or 'secret' dovetails. **1949** W. J. WEST *Woodwork* ix. 78 To cut stopped housings start by using a mallet and chisel to chop out a slot of the required depth. **1979** A. B. EMARY *Woodworking* iii. 18 At (a) is seen the stopped housing joint where the recess has been terminated a short distance from one edge and the piece which fits into the recess has been cut to fit round the stopped end.

stopper, *sb.* Add: **1. g.** *Assoc. Football.* A player whose function is to block attacks on goal from the middle of the field. Also *attrib.* as *stopper centre-half.*

1934 D. JACK *Soccer* 124 The defensive pivot..is essentially a 'stopper', a destructive player if you like. **1941** *Daily Mail* 10 Feb. 4/2 Though occasionally out-witted by Lawton, Dykes made himself a nuisance as a stopper. **1951** *Sport* 16–22 Mar. 9/1 He was an admirable foil for two clever attacking halves, for he was a stopper pure and simple. **1961** *Times* 10 Feb. 19/6 A fundamental change in tactics with the arrival of the 'stopper' centre-half. **1978** *Time* 3 July 53/2 He spent the first half, at his own behest, in the unlikely role of a stopper on Paolo Rossi—and very nearly gave away a goal.

7. d. *Rowing.* The after part of a rowlock.

1897 J. JEFFERY *Rowing* 8 That part against which the oar is pressed in rowing is called the 'thowl', and the opposite, or after-thowl, is called the 'stop', or 'stopper'. **1904** G. RIXON *Rowing & Sculling* 2 In some stock gigs it will be found that there is not sufficient room between the thowl and stopper, causing the oar to stick or 'lock' on a full reach forward.

e. *Bridge.* A card of such value that it can reasonably be counted on, in conjunction with other cards in the same suit, to take a trick in that suit. Cf. *STOPPED *ppl. a.* 2 d.

1901 R. F. FOSTER *Foster's Bridge* 112 When the make is original, a guarded king is very likely a stopper in the dealer's suit. **1913** F. IRWIN *Auction High-Lights* 101 To bid 'a no-trump' declares nothing actually, except general help. The bid is often made on three stoppers. *Ibid.* 105 Four diamonds to the jack might not prove a stopper if the card next to the jack did not happen to be the ten-spot, but it is, and a sequence-stopper is always safe. **1933** *Times* 24 Jan. 13/4 This is a conventional and artificial response. It does not guarantee a 'stopper' in Spades. **1959** T. REESE *Bridge Player's Dict.* 222 Some players bid 1 NT on a fair balanced hand even when they have no stopper. **1978** *Detroit Free Press* 2 Apr. 19c/2 If North has a diamond stopper, he bids no trump.

f. Something which attracts and holds attention; something striking or impressive. *colloq.* Cf. *show-stopper* s.v. *SHOW *sb.*[1] 22.

1968 *Punch* 21 Feb. 269/1 'What's your snap reaction, Jack?' broke in Gringeworth. 'It's um, well, certainly a stopper,' said Tubstraw. 'My God! It's the stopper of the century!' exclaimed Gringeworth. **1973** *Times* 21 Feb. 13/8 'A memorable image.' 'It's a stopper.'

stopping, *vbl. sb.* Add: **I. 1. b.** *stopping off*: also *attrib.*; *stopping out*; *stopping-over.*

1942 BERREY & VAN DEN BARK *Amer. Thes. Slang* § 45/2 Small town,..*stopping-off place.* **1966** 'A. HALL' *9th Directive* iv. 41 I'll need up-to-the minute information ..final itinerary..stopping-off points, so forth.

1971 *Time* 27 Sept. 79/1 The trend of stopping out is growing..partly because the draft law now gives young men with high lottery numbers a new freedom. **1977** *N.Y. Times* 16 Jan. IV. 9/1 Stopping out..has become so popular on some campuses that the notion of graduating in four years seems almost quaint.

1932 *New Yorker* 4 June 38 You leave Seattle July 9, and do a bit of stopping-over at Yokohama, Tokio, and other Japanese ports.

III. 7. *stopping distance*; **stopping-house** *Canad.*, a house offering accommodation to travellers; a boarding-house or rooming-house; **stopping-place,** (*a*) (earlier and later examples); (*b*) *Canad. Hist.*, a stopping house, or a settlement where groups of travellers customarily stop for food and lodging; **stopping rule** *Statistics*, any rule in sequential testing or sampling for deciding when an investigation should be terminated, dependent on the cumulative trends in the results obtained.

1947 *Highway Code* 10 The good driver knows how stopping distances increase with speed, and drives accordingly. **1883** *Prince Albert* (Saskatchewan) *Times* 18 Apr. 1/5 The road from Carrot River crosses the South Saskatchewan at this point where there is now a first class ferry and stopping house. **1912** H. FOOTNER *New Rivers of North* 235 None of the stopping-houses along this trail have progressed beyond the most primitive stage. They provide a floor for you to sleep on and a fire-place, in some cases a stove for you to cook your food on; that is all. **1970** R. & J. PATERSON *Cranberry Portage* i. 4, I got a stoppinghouse here... My rooms is all full up. **1827** A. SHERWOOD *Gazetteer Georgia* 37 *Camp c.* in the N.W. part of the Warren..and well known as a stopping place. **1836** C. FOX *Jrnl.* 31 Aug. in *Memories of Old Friends* (1882) ii. 5 Dr. Buckland was an outside *compagnon de voyage*, but often came at stopping places for a little chat. **1878** J. M. LEMOINE *Chronicles of St. Lawrence* 21 When being jolted in a two-wheeled post stage, without springs, over these villainous roads, the traveller will do well to fix beforehand the stopping places (for meals), as hostelries are few and far between. **1909** A. D. CAMERON *New North* 28 We 'make tea' at Sturgeon Creek (the Namao Sepee of the Indians), the first of the 'stopping-places' or Waldorf-Astorias of the wilderness. **1950** Stopping place [see *LAY-BY *sb.* 1 c]. **1953** *Jrnl. R. Statistical Soc.* B. XV. 9 Thus if a history of the population is available over some period of time, $\lambda+\mu$ can be estimated from the observed number of incidents and the U_i, and $\lambda/(\lambda+\mu)$ from the proportion of the incidents that are births; the details will depend on the stopping rule. **1960** P. ARMITAGE *Sequential Med. Trials* ii. 17 The design of the trial is determined entirely by the stopping-rule. **1978** *Brit. Jrnl. Cancer* XXXVIII. 760/1 Investigators were also asked whether they used any formal or informal stopping rules for the early termination of trial if treatment differences should develop.

stopple, *sb.*[1] Add: **1. d.** = *ear-plug* (*b*) s.v. *EAR *sb.*[1] 16. *U.S.*

a **1961** in WEBSTER, s.v., A stopple must be fitted into the ear canal so that noise does not leak around the edges. **1965** J. M. CAIN *Magician's Wife* (1966) xvi. 121 He must put stopples in his ears to account for his failure to answer, in case his phone had rung. **1977** *New Yorker* 12 Sept. 161/3 (Advt.), Noise? No problem. Flents Ear Stopples seal it out.

stoppo (stɒ·pəᵘ). [f. STOP *sb.*[2] or *v.* + -O[2].] **1.** *slang.* A rest from work.

1938 J. PHELAN *Lifer* xii. 120 What's the chances for a 'stoppo', Reg?

2. *Criminals' slang.* An escape, a get-away, esp. in phrase *to take stoppo*, to make a rapid escape in order to avoid detection. Freq. *attrib.*, with reference to rapid escape by car, in *stoppo car, driver, man.*

1935 G. INGRAM *Cockney Cavalcade* v. 78 'I took stop-o.' In other words, ran away from the police. **1949** PARTRIDGE *Dict. Underworld* 712/1 *Take stoppo*, to be obliged to run away... To take heed when the look-out cries 'Stop!' **1974** *Listener* 7 Nov. 595/3 Boys who are going to be stoppo drivers—driving stolen escape cars. **1975** M. KENYON *Mr Big* xx. 192 Walk, then, to the stoppo car... And wait... Till Slicker comes. **1978** J. GARDNER *Dancing Dodo* xxxv. 279 Dobson..held the clutch down, took off the hand brake and let his toe press gently on the accelerator... 'It's a stoppo man's take-off,' the instructor said.

stop-watch (stɒ·pwɒtʃ), *v.* [f. the sb.] *trans.* To time with a stop-watch.

1973 J. WAINWRIGHT *Devil you Don't* 19 We being timed?..Stop-watched? **1977** B. FREEMANTLE *Charlie Muffin* xvii. 166 Cuthbertson..had insisted on final rehearsals..stop-watching the journey and testing the surveillance.

stop-work (stɒ·pwɜːk), *a.* Also as one word. [f. STOP *v.* + WORK *sb.*] **1.** *Austral.* and *N.Z.* Designating a meeting that requires employees to stop working in order to attend. Also *ellipt.* as *sb.*

c **1926** 'MIXER' *Transport Workers' Song Bk.* 25 With their silly bluff and twaddle, And their stop-work meetings, too, By which I'm not allowed to work Till their business is through. **1941** *Argus* (Melbourne) 15 Nov. 3/6 A stopwork meeting of builders' workers. *Ibid.*, Mr. Howitt said congress regretted the stopwork meeting. **1946** K. TENNANT *Lost Haven* (1947) x. 138 Jack Starbrace called a little stop-work meeting and addressed it. **1957** *Landfall* 11 Apr. 278 But it was a good day for a stop-work. **1977** *N.Z. Listener* 15 Jan. 6/4 A great many immigrants, probably the majority, were never involved in any kind of trade unionism in Britain and would not have recognised a 'stop-work' meeting if they had actually fallen over one!

2. *N. Amer.* Designating an order requiring work to stop.

1972 *Even. Telegram* (St. John's, Newfoundland) 28 June 1/1 The federal cabinet has authorized a stop-work order. **1973** *N.Y. Law Jrnl.* 26 July 2/1 This is an article 78 proceeding to rescind a 'stop work' order issued by the New York City Department of Buildings.

storage. Add: **1.** (Examples relating to *Computers.*)

1946 GOLDSTINE & VON NEUMANN in J. von Neumann *Coll. Wks.* (1961) V. 24 It may be seen from the fact that each binary digit requires essentially one relay or one pair of vacuum tubes..that this form of storage rapidly becomes quite expensive. **1964** T. W. MCRAE *Impact of Computers on Accounting* ii. 38 These early machines.. were fitted with magnetic tape storage. **1973** C. W. GEAR *Introd. Computer Sci.* vii. 309 We must develop a scheme for allocating storage as it is needed and 'recovering' it when structures using it are deleted.

2. b. *cold storage* (earlier and later examples).

1877 *Illustr. London News* 3 Mar. 203/1 A..company called Cold Storage Wharf..undertakes to provide cold storage accommodation for fresh meat..and all produce of a perishable nature, from all parts of the world. **1926** *Daily Express* 11 May 1/3 No difficulty was experienced at any of the cold storage centres. **1933** *Discovery* Apr. 126/1 Canning and freezing both eliminate the risk of spoilage by moulds and bacteria which are a constant danger in cold storage. **1946** *Nature* 21 Dec. 920/2 Fruit cold-storage research has continued.

fig. **1897** *Outing* XXX. 367/1 Still that stony stare and his reiterated, impertinent queries... A cold-storage air arose between us. **1920** W. PERRETT *Peetickay* 3 It seemed ..that the thing to do would be to apply the stenographic principle to the consonants... That plan is now in cold storage. **1951** A. HUXLEY *Let.* 22 July (1969) 637 It seems to be rather a shame that this anthology-with-comments, which cost me a lot of work..should remain indefinitely in cold storage. **1979** A. BOYLE *Climate of Treason* xi. 420 If he did idly question why the Soviet Union had kept their guests in cold storage for so long, Philby soon found a ready answer.

d. *Computers.* The placing or keeping of data and instructions in a device from which they can be retrieved as needed.

1909 *Sci. Proc. R. Dublin Soc.* XII. 80 The present design of the machine provides for the storage of 192 Variables of twenty figures each. **1945** J. VON NEUMANN in B. Randell *Origins Digital Computers* (1973) 358 R has also the properties of a memory. Indeed, it is the natural medium for long time storage of all the information obtained by the automatic device on various problems. **1958** *Listener* 11 Dec. 983/2 For high-speed storage, the magnetic cores of the present machines will give way to films of special iron alloys deposited *in vacuo* upon ceramic plates. **1978** J. P. HAYES *Computer Archit. & Organization* v. 325 The physical processes involved in storage are sometimes inherently unstable, so that stored information may be lost over a period of time.

5. *storage device, medium, register, space.*

1946 *Electr. Engin.* LXV. 389 Each is a complete adding and subtracting machine, and functions as a storage or memory device. **1955** *IRE Trans. Electronic Computers* IV. 16/1 This report deals with a storage device utilizing magnetic cores to achieve fairly large amounts of information storage with a relatively moderate amount of circuitry. **1972** D. LEWIN *Theory & Design Digital Computers* vi. 185 Storage devices may have either destructive or non-destructive read-out of the stored data. **1947** A. W. BURKS et al. in J. von Neumann *Coll. Wks.* (1963) V. 40 There still remains the problem of automatic integration of this storage medium with the machine. **1966** *Sci. Amer.* Sept. 80/1 Magnetic materials ..supply the principal storage medium in computers. **1946** *Ann. Computation Lab. Harvard Univ.* I. 14 Each storage register consists of twenty-four electro-magnetic counter wheels. **1965** HOLLINGDALE & TOOTILL *Electronic Computers* 114 In most computers an individual storage register is not a separate entity, either physically or conceptually, and the term *storage location* is more appropriate. **1936** *Discovery* May 158/1 It may well revolutionise library methods and solve the eternal question of storage-space. **1979** D. MALLETT *Greatest Collector* ix. 84 Richard Wallace..bought no furniture, not wishing, perhaps, to add to the already acute problems of storage space.

6. storage heater, (*b*) = *night storage heater* s.v. *NIGHT *sb.* 14; hence **storage heating,** heating by means of storage heaters; **storage life** (see quot. 1971[1]); **storage location** *Computers*, a place in a store capable of storing one unit of data and usu. specifiable by an address; **storage protection** *Computers*, the protection of storage locations against unauthorized or accidental reading or writing; **storage ring** *Physics*, an approximately circular accelerator in which particles can be effectively stored by being made to circulate continuously at a high energy; **storage tube** *Electronics*, any of various kinds of electron tube which store the information or image applied to them so that it can be retrieved at a later time; **storage unit,** (*a*) one of a set of domestic cupboards; (*b*) a unit that serves as storage for a computer; **storage wall,** a partition wall consisting of cupboards, often designed to be opened from either side.

1961 *Listener* 19 Oct. 629/1 For an existing house the storage-heater system works out very favourably. A good-looking storage-heater with thermostatic control.. costs £22 10s. 6d. **1977** B. PYM *Quartet in Autumn* x. 88 Mrs Pope's sister apparently being too mean to switch on the storage heaters before January. **1961** *Listener* 19 Oct. 629/1 A storage-heating installation. *Ibid.* 629/2 The chief disadvantage of storage-heating is the comparatively slow reaction to sudden big outside-temperature changes. **1971** *Gloss. Terms Quality Assurance* (*B.S.I.*) 8 *Storage life*, the specified length of time prior to use for which items which are inherently subject to deterioration are deemed to remain fit for use under prescribed conditions. **1971** *Country Life* 9 Sept. 643/1 How can the storage life of vacuum-packed bacon be extended? **1949** D. R. HARTREE *Calculating Instruments & Machines* 96 Different means have to be used for identification of a storage location on a wire. **1964** T. W. MCRAE *Impact of Computers on Accounting* i. 10 If we have 100 storage locations, each one of which can store one character, we will divide this up into ten words, each word containing ten storage locations. **1965** G. B. DAVIS *Introd. Electronic Computers* 223 A storage protection feature is an optional feature not usually found on small computers. **1970** O. DOPPING *Computers & Data Processing* ix. 128 In a computer with storage protection, a program can modify the contents of storage only within a prescribed area assigned to the program. **1956** G. K. O'NEILL in *Physical Rev.* CII. 1418/2 (*heading*) Storage-ring synchrotron: device for high-energy physics research. *Ibid.*, Two 'storage rings', focusing magnets containing straight sections one of which is common to both rings, are built near the accelerator. **1965** *New Scientist* 18 Mar. 692/2 They accelerated two comparatively low energy beams of electrons to 300 MeV in a linear accelerator; they steered these beams into two 'storage rings' built adjacent to one another in the form of a figure of eight, and finally made them collide head-on at the point where the two rings touch. **1978** *Nature* 20 July 202/1 PETRA, Europe's 19 GeV on 19 GeV electron-positron storage ring under construction..near Hamburg, has circulated its first beam around the ring and stored it 'for several minutes'. **1946** *Radar: Summary Rep. & Harp Project* (U.S. Nat. Defense Res. Comm., Div. 14) 144/1 Storage tube. **1947** *Electronics* Sept. 80/1 Specific applications of this storage tube include use in simultaneous multicolor and three-dimensional presentation of radar or sonar data. **1975** D. G. FINK *Electronics Engineers' Handbk.* VII. 30 The correlation is a storage tube that receives a visual input to a photoemissive film..and later compares the original image with a similar image. **1951** *Catal. of Exhibits, South Bank Exhib., Festival of Britain* 125/2 Storage unit. **1964** T. W. MCRAE *Impact of Computers on Accounting* vii. 199 A magnetic tape reel is a remarkably compact storage unit. **1978** J. MCNEIL *Consultant* xi. 117 Two lines of storage units ran at right angles to the wall. **1978** J. KELLOCK *Elements of Accounting* xii. 215 Hardware includes the storage unit, arithmetic unit, control unit, and all the input and output devices used with the computer. **1945** NELSON & WRIGHT *Tomorrow's House* xi. 132 (*caption*) It is only a step to the use of such equipment [*sc.* storage cabinets] to form the walls themselves—a device which we have named the 'storagewall'. **1959** *House & Garden* July 43/2 The dining-room, linked to the kitchen by a storage wall. **1970** *Observer* 18 Oct. 35/1 Ideally, every room in every house needs a storage wall.

store, *sb.* Add: **11. b.** *Computers.* = *MEMORY *sb.* 2 d.

1837 C. BABBAGE in B. Randell *Origins of Digital Computers* (1973) 21 The Store may be considered as the place of deposit in which the numbers and quantities given by the conditions of the problem are originally placed, in which all the intermediate results are provisionally preserved, and in which at the termination all the required results are found. **1919** A. MACFARLANE *Lect. on Ten Brit. Physicists* 80 Directive cards to transfer numbers from the store to the mill and from the mill to the store. **1948** *Nature* 8 May 712/1 The general ideas for a large automatic calculating machine are to be found in the designs of Charles Babbage for an 'analytical engine'... It was to work by means of plungers passing through punched cards, and was to contain a 'store' for numbers... The main components of any digital computing machine were then described... There must be a store to hold numbers and instructions. **1948** *Proc. R. Soc.* A. CXCV. 283 Stores have been operated with thirty-two lines and with sixty-four lines, each line containing thirty-two digits; 12 in. diameter cathode-ray tubes were used. **1964** F. L. WESTWATER *Electronic Computers* iv. 59 To read a word out of the store we have to open a gate at the end, and this permits pulses to escape. **1968** *Brit. Med. Bull.* XXIV. 191/1 The basic configuration of any computer consists of a store, a suitable input and output device, and a control mechanism. **1977** *Sci. Amer.* Sept. 130/1 In the context of electronics 'memory' (or, in British usage, 'store') usually refers to a device for storing digital information.

12. a. (Further examples.)

The use of the word in this sense has not become common in the U.K. except in *Comb.*, as *chain store, department store* (see under the first elements), *store detective* (see sense 13 d below), in which it still refers to a large shop.

1839 W. WAKEFIELD in *N.Z. Jrnl.* (1840) No. 9. 112 It partly belongs to Captain Mayhew, an American, who has a store on it. **1862** *Times* 1 Sept. 5/1 At one corner of the street was a little provision and drapery store kept by an old woman. **1956** H. G. DE LISSER *Cup & Lip* ii. 22 The shops—or stores, as they are invariably called in the West Indies—were open. **1975** *Encounter* Jan. 41/2 But for chrissake—that's 'Christ's sake' in American, chaps—has anybody ever gone away from a shop—meaning 'store', youse guys—empty-handed through ignorance of some one of these local variants?

13. b. *store barn, cupboard* (earlier example), *-shop*; *store-boat* (earlier and later examples).

1926 D. H. LAWRENCE *David* xii. 87 And she shall have her handmaidens about her, and her store-barns of wool. **1797** *State Papers & Publick Documents U.S.* (1815) II. 436 On the 21st of January, the ice began to give way, and their store-boat arriving on the 28th, they proceeded on the 31st for the Natchez. **1944** T. D. CLARK *Pills, Petticoats, & Plows* 25 In the Louisiana sugar belt, barge store boats eased along the back ways of sugar plantations receiving stolen goods. **1841** C. FOX *Jrnl.* 5 May in *Mem. Old Friends* (1882) vii. 123 We went all over his comfortable house..choosing papers, positions of store cupboards, and other important arrangements. **1888** C. M. YONGE *Our New Mistress* xii. 109, I went into one of those great store shops where they sell all sorts of things. **1972** E. WHITE in W. King *Black Short Story Anthol.* 366 Jill.. passes the store-shops of the Jews.

d. For *U.S.* and *colonial* read chiefly *N. Amer. store buyer, detective, porch*; *store cheese, clothes* (examples), *pants, tea* (earlier example), *teeth* (examples); **store church**: see quot. 1948; = **storefront church*; **store pay** (examples).

1965 *Harper's Bazaar* Feb. 21/3 The entirely new role of the store buyer. **1980** *Times* 12 Aug. 8/4 Store buyers.. still come to Paris, but..to see the ready-to-wear. **1982** *Times* 3 Aug. 6/1 This week, there weren't any store buyers. **1863** P. S. DAVIS *Young Parson* 61 One plate of 'store cheese', and half a bread-basket of ginger crackers. **1894** *Rep. Vermont Board Agric.* XIV. 25 A full cream store cheese is run through a grinder. **1948** H. L. MENCKEN *Amer. Lang.* Suppl. II x. 591 A *store-church* is one set up in a vacant store or in the front room of a dwelling house. **1961** C. HIMES *Black on Black* (1973) 60 She hid in Rev'end Sinner's store church when she run away. **1840** *Knickerbocker* XVI. 262, I felt an awe of young ladies in 'store clothes'. **1872** [see *boiled shirt* s.v. *BOILED *ppl. a.* 2]. **1944** B. JOHNSON *As much as I Dare* 294 These young men did not want to give up their store clothes. **1907** *St. Nicholas* Oct. 1106/2 He wondered how the store detectives worked to find a man who might be picking pockets in a great crowd. **1968** J. LOCK *Lady Policeman* xix. 157 They had been detained by the store detective. **1979** R. RENDELL *Make Death love Me* vi. 59 A woman had grabbed him and he'd only just escaped the store detective. **1891** 'O. THANET' *Otto the Knight* 4 Thar, store pants an' gallowses! Make haste an' putt 'em on! **1932** *Atlantic Monthly* Apr. 475/1 Steve-john..was..a bronze perfection—Celini's 'Perseus' in store pants. **1942** W. STEGNER *Mormon Country* 126 Smart alecs had money to jingle in their store pants. **1842** R. H. BONNYCASTLE *Canada & Canadians* II. 180 A quintal of fish..is worth 12s. 6d. in hard cash, or 14s. 6d. store pay. **1891** S. M. WELCH *Recoll. Buffalo 1830–40* 353 The workmen were to receive..only half cash, the remainder in trade—store pay, *i.e.*: in orders on the employers or other stores for such goods as they needed. **1905** J. S. CARTER *Story of Dundas* 51 The store-keeper bought the settlers' produce but would give them only trade in return, or what was known as 'store pay'. **1934** C. M. WILSON *Backwoods Amer.* ii. 16 Hired boys are among the most cherished perpetrators of store-porch mirth. **1949** B. A. BOTKIN *Treas. S. Folklore* p. xix, The rural south is a land of the out-of-doors come up to the door and even indoors, where the 'gallery', the store-porch, the kitchen..are made for story-telling. **1843** 'R. CARLTON' *New Purchase* I. ix. 64 'Tisn't nun of your spice-wood or yarb stuff, but the rale, gineine *store* tea. **1878** *Brooklyn Monthly* June 185/1 It occurred to me that a brief description of the sensations experienced might be of interest to any of my readers who are contemplating a new set of 'store teeth'. **1951** C. LYNCH-ROBINSON *Last of Irish R.M.s* vi. 113 When I first got my new 'store teeth' in, they worried me. **1975** *Budget* (Sugarcreek, Ohio) 20 Mar. 8/6 Mrs. Gintz is a sister to O.K. Brown, the dentist that pulled my last teeth, and made me some store teeth.

14. In *Comb.* with adjs. or ppl. adjs., as **store-bought,** bought (often ready-made) from a store; also *fig.*; also **store-boughten** *U.S.*; **store-wide,** operating or applying throughout the whole of a store.

1952 J. STEINBECK *East of Eden* xvi. 181 Would you say they were made clothes or store bought? **1953** *Manch. Guardian Weekly* 1 Oct. 2 She swayed like a riven oak over her failure to compete with 'powder and store-bought hair'. **1962** *Times* 4 May 9/6 It has become 'common sense' to substitute a store-bought, ready-made universe for the disquieting uniqueness of actuality. **1970** *Islander* (Victoria, B.C.) 22 Nov. 5/1 Those home-garden farmers aimed for near total independence from store-bought produce. **1981** *Farmstead Mag.* Winter 63/2 Pickled mushrooms..bring an outrageous price, in delicatessen shops, if you can find them and the storebought ones don't taste nearly as good as those pickled at home. **1883** ZEIGLER & GROSSCUP *Alleghanies* 91 Two good-natured-looking young men dressed in..'store-boughten' coats, and homespun pantaloons. **1933** L. I. WILDER *Farmer Boy* viii. 54 Clothes..made of store-boughten cloth, woven by machines. **1974** M. LAURENCE *Diviners* ii. 29 Storeboughten cookies are looked down on. **1938** *Sun* (Baltimore) 8 Sept. 3/1 A union demand for..a store-wide seniority plan. **1979** *Tucson* (Arizona) *Citizen* 3 Oct. 10/A (Advt.), Tremendous storewide savings.

store, *v.* Add: **4. d.** *Computers.* To retain a physical representation of (data or instructions) that enables them to be subsequently retrieved.

1909 *Sci. Proc. R. Dublin Soc.* XII. 78 An Analytical Machine must have some means of storing the numerical data of the problem to be solved. **1937** H. H. AIKEN in *IEEE Spectrum* (1964) Aug. 69 It is necessary that numbers may be removed from the calculating units and temporarily stored in storage positions. **1945** J. VON NEUMANN in B. Randell *Origins Digital Computers* (1973) 356 A distinction must be made between the specific instructions given for and defining a particular problem, and the general control organs which see to it that these instructions..are carried out. The former must be stored in some way. **1948** *Nature* 8 May 712/2 In all these machines there is provision for storing numbers, say in the scale of 2, in certain places. **1964** F. L. WESTWATER *Electronic Computers* ix. 144 Inside a computer, alphabetical characters and numerals are both stored as numbers. **1972** D. LEWIN *Theory & Design Digital Computers* vi. 184 The speed of computers is limited by the time required to store and retrieve information.

e. *Computers.* To transfer *into* a store or storage location.

1964 *Ann. N.Y. Acad. Sci.* CXV. 654 The speed of the computer is fixed by the length of time required to read information from or store information into one of the 1,024..12-bit memory locations. **1973** C. W. GEAR *Introd. Computer Sci.* ii. 37 The CPU can be told to load a number into its accumulator from a specific cell in the memory..or to store a number from the accumulator into memory.

6. *Comb.*: **store-and-forward** *Telecommunications*, used *attrib.* with reference to a data network in which messages are routed to one or more intermediate stations where they may be stored before being forwarded to their destinations.

1963 *On Line Data Processing* (Inst. Electr. & Electronics Engineers) 63 The store and forward switching system must interconnect with line switching facilities. **1980** R. L. FREEMAN *Telecommunication Syst. Engin.* ix. 429 The ARPANet connects dispersed computers of various manufacture and varying design. The subnet providing that connection is a form of store and forward system and must deal with such problems as routing, buffering, synchronization, [etc.].

stored, *ppl. a.* Add: **1. c.** *stored program*, a program that is stored in a computer in the same way as data, *esp.* one that can be automatically manipulated like data; freq. *attrib.*

1957 D. D. MCCRACKEN *Digital Computer Programming* i. 4 We speak of modern computers as being stored program machines. **1964** *Ann. N.Y. Acad. Sci.* CXV. 654 The LINC is a small stored-program digital computer which uses transistor circuitry and a random-access ferrite-core memory. **1970** O. DOPPING *Computers & Data Processing* vi. 93 The use of a stored program, which can be modified by the machine itself, is one of the basic ideas in the design of a computer. **1983** *Sci. Amer.* Jan. 92/3 The stops and combination pistons [of the organ].. are under the control of microprocessors having stored-program capabilities.

storefront, *sb.* (and *a.*). orig. and chiefly *U.S.* Also **store-front, store front.** [f. STORE *sb.* + FRONT *sb.*] **1. a.** The side of a shop facing the street; (a building with) a shop window.

1880 G. W. CABLE *Grandissimes* 376 A large porte-cochère..[opened] upon the banquette immediately beside and abreast of the store-front. **1922** F. FARRINGTON *Meeting Chain Store Competition* iii. 37 You can make your store front as conspicuous as a red front Atlantic and Pacific Tea Co. store front and still make it infinitely more attractive. **1945** *Planning Store of Tomorrow* (National Retail Merchants Assoc.) 6/1 Some are specialists on store fronts, but can do or supervise an intelligent job on interior layout. **1962** E. SNOW *Red China Today* (1963) lxx. 538 Some of the old foreign store fronts (such as Whiteaways) now exhibited shining lathes and other machines. **1974** R. L. SIMON *Wild Turkey* iii. 19 [A] shocking pink storefront temporarily labeled 'The Institute of Oral Love'. **1977** *Guardian Weekly* 4 Dec. 12/5 This indifference..turns storefronts not into show-cases for the articles sold by the shops but into museums of all the goods that were once sold there or will be sold at some future date. **1979** *United States 1980/81* (Penguin Travel Guides) 73 This little deli is just a storefront on a shopping-center strip, so look carefully.

b. A room or rooms at the front of shop premises, esp. as used for some other purpose.

1972 *N.Y. Times* 3 Nov. 22/2 The two fully staffed and fully equipped Nixon storefronts in the area contain busy volunteers. **1973** *Houston Chron.* (Texas Mag.) 14 Oct. 2/1 The Hare Krishna sect..is big business. It began in an East Village store front in New York. **1976** *National Observer* (U.S.) 7 Feb. 20/3 Performances first held in storefronts and lofts and later in streets, parks, and other public places. **1982** S. PARETSKY *Indemnity Only* vii. 83 She ran a clinic in a shabby storefront down the street.

2. *attrib.* and *Comb.*, as *storefront cinema, headquarters, industry, location, restaurant, theatre, window*; **storefront church,** a shop building used as a church or meeting place, esp. by small evangelical groups; also *storefront mission, synagogue, temple.*

1938 C. HIMES *Black on Black* (1973) 167 For one hundred and fifty dollars he leased a storefront church in the heart of the slums and poverty. **1957** *Economist* 28 Sept. 1031/2 The Negro newcomer from the South..may attend a 'storefront church', of uncertain denomination, in a rented shop with the display window painted over. **1968** P. OLIVER *Screening Blues* ii. 62 Infinitely smaller, but important because of their numbers and their devout followings of tiny congregations, are the multitude of store-front churches which line the streets in the Negro areas of the main urban centres. **1973** A. DUNDES *Mother Wit* 175 Professor Dillard..examines the names of store-front churches... The individual words in store-front church names may be SE (Standard English). **1967** *Daily Tel.* 16 Feb. 19/6 The Storefront Cinema..is a rented shop that has been converted into a miniature cinema. **1976** *National Observer* (U.S.) 3 Apr. 5/1 He has more than 25 store-front headquarters in every one of the state's nine congressional districts. **1967** *Economist* 6 May 563/2 Everywhere in Bangkok there are store-front industries where people are beating metal, mending things by hand, getting used to machines. **1970** *Globe & Mail* (Toronto) 25 Sept. 11/4 Government information centres in 'store-front locations'. **1978** *Sunday Mail TV Suppl.* (Brisbane) 23 July 2/2 The pair operate a store-front mission in a lower class area. **1978** *Chicago* June 208/1 We've always liked to brag about Chicago's Mom-and-Pop storefront restaurants. **1975** A. BERGMAN *Hollywood & Le Vine* (1976) vii. 80 A store-front synagogue in Brooklyn. **1965** D. HENDERSON in S. Henderson *Understanding New Black Poetry* (1973) III. 269 The ritual is black The ritual is in the storefront temple On the corner. **1973** E. BULLINS *Theme is Blackness* 7 Sweating out long, hot summers in store-front theaters. **1976** *National Observer* (U.S.) 2 Oct. 15/1 A Teletype machine, set in the store-front window of the bureau.

3. *attrib.* passing into *adj.* **a.** Of, pertaining to, or designating legal aid or citizen's advice organizations which operate from shop premises in order to be easily accessible.

1971 *Sunday Sun* (Brisbane) 31 Oct. 20/2 Why..can't we have legal offices in poor areas offering cheap legal aid along the lines of America's 'store-front' lawyers. **1973** *Courier-Mail* (Brisbane) 9 June 1/3 A 'store front' legal and general advice system will be available cheaply to Brisbane people in a few weeks. **1973** *Globe & Mail* (Toronto) 2 Aug. 8/7 (Advt.), Store-front Citizenship Office now open..to answer your questions about how you can become a Canadian citizen. **1974** *Index-Jrnl.* (Greenwood, S. Carolina) 18 Apr. 6/1 A group called Resource One operates a storefront 'people's computer', a cross between a hip encyclopedia and a community bulletin board. **1974** *Courier-Mail* (Brisbane) 25 Apr. 2/4 Australia's first 'store front' legal aid office will open in Ipswich tomorrow. **1975** *Weekend Mag.* (Montreal) 1 Nov. 8/2 Robert Cooper..is a lawyer with an extraordinary string of..achievements in his chosen field, including the opening of Canada's first storefront law office. **1979** N. HARTLEY *Quicksilver* i. 19 Storefront law offices, giving free legal advice to minority groups.

b. Of, pertaining to, or characteristic of a storefront church.

1972 J. L. DILLARD *Black English* v. 217 We see evidence of formal styles of Negro Non-Standard in the speech of the storefront preachers. **1973** *Black World* Sept. 31 'Tambourines' has the impelling, fervent, and nervous rhythms of the 'store-front' spiritual or gospel hymn. **1976** *National Observer* (U.S.) 1 May B 2/3 Store-front religion offered solace and precious respectability to mothers whose families were collapsing around them. **1978** J. UPDIKE *Coup* (1979) v. 173, I hope that stuff hasn't taken you in; it's just our usual native storefront I'm-comin'-home-Jesus routine.

stork (stɔːk), *v. U.S. slang.* [f. the sb., with reference to the nursery fiction that babies are brought by the stork: see sense 1 c.] *trans.* To make pregnant.

1936 A. HUXLEY *Eyeless in Gaza* xxv. 353 What would you do if the fever frau had the misfortune to be storked? **1968** R. STOUT *Father Hunt* (1969) xiii. 157 'Didn't she stop because she was pregnant?'.. 'Yes,' he said. 'She was storked.' **1977** *Amer. Speech* 1975 L. 67 *Stork vt*, make pregnant. 'Jim storked her; that's why she's not back up here this year.'

sto·rkbird. *nonce-wd.* [f. STORK *sb.* + BIRD *sb.*] = STORK *sb.*

1922 JOYCE *Ulysses* 416 Madam, when comes the storkbird for thee?

storm, *sb.* Add: **I. 1. f.** *any port in a storm*: see *PORT *sb.*[1] 1 c.

g. In pl. *ellipt.* for storm windows. *N. Amer.*

1952 *Home Building in Canada* Oct.–Nov. 22/2 If you are wondering which windows and when to protect with storms, the answer is simple—all of them, from October to April. **1968** *Globe & Mail* (Toronto) 17 Feb. 45 (Advt.), Complete with drapes, aluminum, storms and screens. **1973** *N.Y. Law Jrnl.* 1 Aug. 3/2 Alwin J. Dovale, installer of storms and screens. **1977** *Chicago Tribune* 2 Oct. II. 7/1 Maybe next year I can afford real storms with the money I'll save on utility bills.

3. a. Freq. in phr. *to weather the storm.*

1671 [see WEATHER *v.* 4 b]. **1802** [in Dict.]. **1849, 1853** [see WEATHER *v.* 4 b]. **1868** [in Dict.]. **1924** *Nation & Athenæum* 26 Jan. 603/1 His plight was serious; but he weathered the storm. **1934** F. W. CROFTS *12.30 from Croydon* viii. 95 He had come to an arrangement with his uncle whereby he hoped to weather the storm.

e. *up a storm* adv. phr., vehemently, violently, with enthusiasm or energy. *U.S. colloq.*

1953 J. STREET *Civil War* iv. 55 The editors just r'ared back in the omnipotence of Jove and pontificated up a storm. **1956** B. HOLIDAY *Lady sings Blues* (1973) xviii. 149 After Marietta taught me, I knitted up a storm and got real fancy. I made cable-knit sweaters for Bobby Tucker and his little boy. **1965** *Charlottesville* (Va.) *Daily Progress* 29 Apr. 6/1 When I ask him to go to the store for me he starts to wheeze up a storm and tells me he is a sick man. **1967** *Boston Sunday Herald Mag.* 30 Apr. 19/3 Right now she's cooking up a storm in preparation for the rash of friends who will be stopping by on their way to Expo. **1972** *TV Guide* (U.S.) 15 Jan. A54/1 Aretha Franklin sings up a storm and impersonates top female vocalists. **1974** K. MILLETT *Flying* (1975) v. 518, I will console myself with material goods. I will shop up a storm. **1983** *Oxford Times* 29 Apr. 3/7 Youngsters from the First Yarnton Brownies have been knitting up a storm to make a blanket for Mother Theresa in India.

III. 6. a. *storm flake, -month, -rack, -song, -spirit.*

1876 Storm flake [see *scroll-leaved* adj. s.v. *SCROLL *sb.* 6 a]. **1894** *Stonyhurst Mag.* Feb. 233 And like the storm-months smote the earth. **1878** O. WILDE *Ravenna* 14 As from the storm-rack comes a perfect star! **1926** J. N. CAMERON in *Oxford Poetry* 14 The haggard storm-rack of disastrous days. **1925** BLUNDEN *English Poems* 40 While on her soul the stormsong bursts, and groanings Knell through roof and flue. **1929** —— *Near & Far* 41 Storm-spirit, coil your lightnings round mad towers.

b. *storm-bitten, -damaged, -driven, -rent* (later example), *-threatened, -tormented, -torn, -troubled* adjs.

1939 W. B. YEATS *Last Poems* 6 A small forgotten house that's set On a storm-bitten green. **1980** *New Age* (U.S.) Oct. 26/1 Eight acres of storm-damaged apricots. **1841** J. G. WHITTIER *Poet. Wks.* (1898) 190/2 Loose rock and frozen slide, Hung on the mountain-side, Waiting their hour to glide Downward, storm-driven! **1900** W. S. CHURCHILL in *Morning Post* 1 Jan. 6/2 These tall figures, full of animated movement, clad in dark flapping clothes, with slouch, storm-driven hats. **1850** E. B. BROWNING *Poet. Wks.* (1904) 141/1, I lack your daring, up this storm-rent chasm To fix with violent hands a kindred god. **1977** Storm-threatened [see *QUANTUM 5 d]. **1844** POE in *Columbian Mag.* Dec. 275/2 Storm-tormented ocean of his thoughts. **1876** J. G. WHITTIER *Poet. Wks.* (1898) 247/2 The storm-torn plumes Of old pine-forest kings. *c* **1958** E. M. FORSTER *Life to Come* (1975) 199 They flew round and round the basilica.., they shot through its roof into the storm-torn night. **1850** E. BRONTË *Wuthering Heights* 489 No coward soul is mine, No trembler in the world's storm-troubled sphere.

d. storm apron *U.S.*, a waterproof sheet used to cover the front of an open carriage in wet weather; **storm boat** *Mil.*, a light but powerful boat used for conveying attacking troops across rivers; **storm cellar** orig. and chiefly *U.S.*, a cellar or dugout made to be a place of refuge from a storm; also *transf.* and *fig.*; **storm centre** (later *fig.* examples); **storm choke,** a safety valve installed in an oil-well pipe below the ocean surface, designed to stop the oil flow should it exceed a predetermined rate as a result of damage at the wellhead; **storm coat** orig. and chiefly *U.S.* (earlier and later examples); **storm collar,** a coat-collar which may be turned up and fastened close round the neck; **storm door**: for *U.S.* read orig. *U.S.* and add later examples; **storm drain,** a drain built to carry away excess water in times of heavy rain; **storm flag,** (*a*) (example); **storm-flap,** a piece of material designed to protect an opening or fastening from the effects of rain, as on a tent, coat, etc.; **storm-house** (example); also, a shelter from the weather on a boat; **storm-jib** (earlier example); **storm lantern** orig. *U.S.* = *hurricane-lamp* s.v. HURRICANE 3 a; **storm-proof** *a.*, (*a*) (later example); also, protected from or affording protection from stormy weather; **storm rubber** *N. Amer.*, a rubber overshoe; **storm sewer** *U.S.* = *storm drain* above; **storm shutter** (earlier example); **storm surge** *Oceanogr.*, an abnormal raising of the sea level in a region as a result of the wind and atmospheric pressure changes associated with a storm; **storm-trysail** (earlier and later examples); **storm wind**: *spec.* a wind having a speed within certain limits (see quots. and cf. STORM *sb.* 1 b); **storm window,** (*c*) *N. Amer.*, a detachable window put up in winter to form an insulating double window.

1895 *Montgomery Ward Catal.* Spring & Summer 591/2 *Storm Aprons.* These aprons are held firmly in position on the dash..forming an unbroken water-shed over front of dash. No mud, snow, or rain can settle inside of carriage. **1943** L. I. WILDER *These Happy Golden Years* xxix. 260 Back in his [buggy] seat, he unrolled the rubber storm apron. **1945** *Sun* (Baltimore) 27 Feb. 3/1 The sergeant..took them back to the road to carry the stormboat down to the river and launch it. **1945** *Finito! Po Valley Campaign* (15th Army Group) 12 Each 20-foot, powered, plywood storm boat. **1920** G. ADE *Hand-Made Fables* 30 The Money-lender beat it to a Storm-Cellar. **1929** J. F. DOBIE *Vaquero of Brush Country* 151 Storm cellars in north Texas, Oklahoma, and Kansas still preserve its architecture. **1962** F. I. ORDWAY et al. *Basic Astronautics* xii. 503 One..suggestion is that a special 'storm cellar' be constructed within the spacecraft, a well-shielded area into which the crew could retreat. **1971** J. H. GRAY *Red Lights on Prairies* ii. 36 When the first oratorical thunder clapped, the chief, the mayor..took to the storm cellars to wait for the storm to blow over. **1977** J. CLEARY *Vortex* i. 8 People build storm cellars to retreat to. **1965** *Listener* 30 Sept. 481/2 Europe is no longer the storm centre in world affairs. The clouds have shifted to Asia. **1978** M. PUZO *Fools Die* xxix. 335 She was having a good time standing outside the party storm center. **1966** P. HINDE *Fortune in North Sea* viii. 154 The first safety valve is installed and left at the bottom of each production well at sea, and is known as the Storm Choke. **1975** *North Sea Background Notes* (Brit. Petroleum Co.) 40 Precautions are taken to shut down production automatically on any failure of the wellhead or flow-line by installing suitable safety valves. These are the 'storm choke' in the well bore,..and the surface safety valve. **1830** J. F. WATSON *Annals of Philadelphia* 179 In the year 1749, I met with the incidental mention of a singular over-coat, worn by captain James as a storm coat, made entirely of beaver fur. **1849** THOREAU *Week Concord Riv.* 250 He ran along over the wet stones like a wrecker in his storm coat. **1953** 'S. RANSOME' *Drag Dark* (1954) i. 16 The corpse..wore..a tan gabardine storm-coat, and big galoshes. **1974** 'J. ROSS' *Burning of Billy Toober* i. 7 His stiff-fabric stormcoat. **1981** *Daily Tel.* 30 Mar. 18/5 Snug, high-collared storm coats are ready to roam Tibetan mountains. **1898** *T. Eaton & Co. Catal.* Spring & Summer 124/1 Men's Klondike mining coats,.. with 6-inch storm collar and capot to pull over the head. **1899** [see *EMPIECEMENT]. **1931** *Daily Mail* 26 May 1/4 (Advt.), West Riding suiting coats... Smart Storm Collar and pull-in Belt. **1939** H. M. MINER *St. Denis* ii. 25 Storm doors or built-on entries are put on the houses in winter. **1977** *Grimsby Even. Tel.* 27 May 17/7 (Advt.), Freehold semi-detached house... Porch with storm door. Entrance Hall. **1960** C. ACHEBE *No Longer at Ease* ii. 16 His car was parked close to a wide-open storm drain from which came a very strong smell of rotting flesh. **1974** N. GORDIMER *Conservationist* 218 The English-language evening paper published a picture of a pet dog being rescued from a flooded storm-drain by the fire brigade. **1896** *Weather Bureau Bull.* (U.S.) No. 80. 7 Two storm flags (red with black centers), displayed one above the other,..announce the expected approach of tropical hurricanes. **1929** *T. Eaton & Co. Catal.* Spring & Summer 373/1 Palmetto Tent... Insect-proof mosquito door and rear window with storm flap operated from inside. **1968** J. IRONSIDE *Fashion Alphabet* 41 *Trench-coat*... This short-cape effect is often called a 'storm cape' or 'storm flaps'. **1972** *Village Voice* (N.Y.) 1 June 13/2 (Advt.), Nylon Mountain Tent... Rear screen window with storm flap. **1973** *Shooting Times & Country Mag.* 7 July 37/2 (Advt.), Zip full length from neck to hem, covered by storm flap. **1836** T. POWER *Impressions Amer.* I. 31 She.. had stump-royal masts, and a storm-house abaft. **1839** *Southern Lit. Messenger* V. 8/2 The James Cropper..was fitted with..a storm house over the wheel. **1887** *Harper's Mag.* Dec. 119/1 Two men..were bending down at the storm-house in front of her parlor-door. *c* **1810** W. HICKEY *Mem.* (1960) xiii. 207 It blew so hard we could scarcely carry a close-reefed mainsail and storm-jib. **1895** *Montgomery Ward Catal.* Spring & Summer 553/2 Cold Blast or Storm Lantern: is made on the same principle as street lamps, with wind break. **1923** W. DEEPING *Secret Sanctuary* xx. 207 He..lit the storm-lantern he used at night, and extinguished the lamp. **1964** D. VARADAY *Gara-Yaka* vi. 51, I hurried to the hut with a storm lantern. **1976** *Norwich Mercury* 17 Dec. 6/7 If the light fails, you use a storm lantern. **1901** Storm-proof [see *COVER-ALL, COVERALL *sb.*]. **1909** *Chambers's Jrnl.* May 335/2 The lamp is stormproof, and is unaffected by cold weather, while it constitutes the safest form of street-

1917 *Punch* 27 June 409/3 Special 'storm troops'—men picked for their youth, vigour and daring, to carry out counter-attacks—are now a feature of the German Armies. 1922 C. E. Montague *Disenchantment* ix. 125 Canadians and Australians..were the 'storm troops', the men who had to be sent for to do the tough jobs. 1924 J. Ross *Years of My Pilgrimage* xxx. 283, I had heard from my friend, Sir Henry Wilson, F.M., that the use of storm troops was an extravagant way of utilizing men. 1933 J. Buchan *Prince of Captivity* iii. i. 274 They were violent German nationalists..true storm-troops, ready for any forlorn hope and prepared to use any means however devilish. 1943 C. Day Lewis *Word over All* 42 Spent as storm-troops after defeat or triumph, Deeply indifferent. 1973 *Black Panther* 11 Aug. 8/3 Tommassi said he was offered $5,000 to use his storm troops as registrars.

b. *spec.* The troops of the Nazi *Sturmabteilung.

1923 *Times* 15 Jan. 10/4 Bands of 'storm troops' paraded the streets, singing the Fascist war songs. 1933 *Granta* 26 Apr. 370/1 We believe that only now has come the time to see these issues in themselves, out of the context of cruelty and outrage which Herr Hitler's storm-troops created for them. 1954 B. & R. North tr. *M. Duverger's Pol. Parties* i. i. 36 In the case of the National Socialist Storm Troops the initial element was the squad (*schar*). 1982 T. Keneally *Schindler's Ark* xxxiii. 344 An influential officer in the S.A. (the Sturmabteilung, or Storm Troops).

2. *sing.* **a.** A branch or detachment of storm troops.

1935 [see *Nazi *a.*]. 1938 J. Cary *Castle Corner* 483 A soldier picked by lot for the storm troop.

b. *attrib.*

1939 S. Spender tr. *E. Toller's Pastor Hall* i. 14 Go and ask the Stormtroop Leader to come in. 1958 *Times* 3 Sept. 11/3 The outbreaks will have served a useful purpose if they oblige the public to understand that the Storm Troop mentality exists in England, too. 1981 *Listener* 2 July 7/1 Röhm..was jailed on 29 June [1934] by a worried Heines, and other storm-troop leaders..were due to arrive the next day.

Hence **sto·rm-troop** *v. intr.*, to behave in an aggressive manner like storm-troops; so **sto·rm-trooping** *ppl. a.*

1960 *News Chron.* 5 Jan. 4/5 The storm-trooping bird-brains of Notting Hill. 1974 *Times* 10 May 22/6 The National Union of Students goes storm-trooping about the country's universities suppressing..freedom of speech. 1977 *Sounds* 9 July 31/3 This is the nucleus of the band, who charge in with a stormtroopin' instrumental romp through the Byrds 'Eight Miles High'.

Storthing. (Later examples with spelling *Storting.*)

1955 *Times* 23 June 11/6 King Haakon VII..was formally elected by the Storting, the Norwegian Parliament, on November 18, 1905. 1964 *Ann. Reg.* 1963 267 The Storting on 28 March passed a Bill concerning Norway's participation in a common defence system for the European section of NATO. 1977 *Time* 26 Sept. 17/3 From the beginning, the race for control of the Storting (Parliament) had been regarded by most Norwegians as a 'destiny election' for the Labor Party's brand of socialism.

story, *sb.*¹ Add: **I. 4. e.** Also in catch-phrase *that's the story of my life*, etc., used of something that supposedly epitomizes one's life or experience.

1964 *Punch* 11 Mar. 385/3 It's the story of my life—looking for small watch-straps. 1969 *Time* 30 May 22/3 In 13 years, he's been a hard-liner in criminal cases. That's the story of his life.

6. d. (*that is*) *another story* (earlier and later examples); (*to be*) *the same story*: a repetition of some occurrence; similarly *a different story*; *the old story*: see *old *a.* 7 e.

1818 Scott *Heart Midl.* II. xiii. 308 But if she's gaun to look after the kye at St. Leonard's, that's another story. 1905 *American Mag.* May 107/1 It has been the same story in every strike the man has undertaken, though it has been a longer job in most cases. 1940 A. Christie *Sad Cypress* ii. iii. 128 As a matter of fact, it was Nurse O'Brien who set me on the track; but that's another story. 1958 'Castle' & 'Hailey' *Flight into Danger* i. 20 The met report was reasonable... In a month or so's time it'll be a very different story. 1966 'H. Calvin' *Italian Gadget* v. 64 'Brains?..I haven't shown much evidence of them here... But that's another story,' he added hastily. 1979 J. Crosby *Party of Year* xxiii. 146 Let's look at the back stairs.—Same story there... The door was of steel.

e. For *U.S.* read orig. *U.S.* and add later examples. Now chiefly = *news story* s.v. *news *sb.* (*pl.*) 6 b.

1905 E. Wallace *Four Just Men* v. 86 'A very good story,' said the chief complacently, reading the proofs. 1942 *Sphere* 27 June 409/1 Each regional editor acquires stories from his own Embassy or exiled Government as well as sending out his own reporters for stories of special interest to his country. 1961 C. Willock *Death in Covert* xii. 203 One headline said: *Regency Bucks Ride Again*, and the sub-head to the same story complained: *Last time a man was blown up.* 1976 *Task of Broadcasting News* (B.B.C.) ii. 17/1 'Story' is only a journalist's professional jargon for an item of news. The proper place for it is a news bulletin.

III. 9. *story-ballet*, *-film*; **storyboard,** a large surface on which is displayed a series of rough drawings representing a shot-by-shot breakdown of a planned film (*spec.* used of advertising commercials); **story-book:** also *attrib.*; freq. *fig.* with allusion to the conventionally

nominally equal to the franc, with its submultiple the *stotinka* (pl. *-ki*), or centime. 1933 *Whitaker's Almanack* 546/1 Bulgaria—Lev of 100 *Stotinki*. 1976 A. Grey *Bulgarian Exclusive* vii. 48 I'll bet you fifty *stotinki* that he'll start telling us anti-communist jokes.

stotious, var. *stocious.

stotty (stǫ·ti). *north. dial.* [Origin unknown.] In full, *stotty cake*. A soft roll split and filled with meat or cheese.

1971 *Guardian* 5 July 4/6 The traditional local [Tyneside] foods such as leek and suet puddings, stotty cake. 1975 *Times* 20 Sept. 11/5 The bar snacks..include..stotties, which are a local version of a bap, split and filled with meat or cheese. 1982 *Times* 12 Aug. 3/4 Most of the gunners come from north-east England, and Councillor Joseph Hall, Mayor of Sunderland, welcomed them as they left their aircraft. They were given..large portions of locally made pease pudding and stotty cake.

stoush (staʊʃ), *v.* *Austral.* and *N.Z. slang.* Also † **stouch.** [Orig. uncertain: perh. rel. to *stashie* uproar, quarrel (*E.D.D.*, *S.N.D.*).] To thrash or beat (a person); to punch or strike; to fight.

1893 J. A. Barry *Steve Brown's Bunyip* 66 I'll get stoushed over this job yet. Brombee's got it in for me. 1894 *Bulletin* (Sydney) 5 May 13/3 'Then 'e biffed me.' 'And did yer stouch him back?' 1900 H. Lawson *On Track* 148 'If you don't,' said Steelman, 'I'll stoush you.' 1924 Kipling *Debits & Credits* (1926) 309 'What *your* crowd down under are suffering from is growing pains. You'll get over em in three hundred years or so—if you're allowed to last so long.' 'Who's going to stoush us?' Orton asked fiercely. 1941 K. Tennant *Battlers* xxvi. 281 What with not being allowed to stouch any of the coves in charge of this turnout. 1945 [see *quilt *v.*² a]. 1965 E. Lambert *Long White Night* 79 Get out of that bloody car while I stoush yer!

stoush (staʊʃ), *sb.* *Austral.* and *N.Z. slang.* Also † **stouch.** [f. the vb.] Fighting; also, *to take stoush*, to receive a beating. A brawl or fight; a scrap, 'punch-up'.

1908 H. Fletcher *Dads & Dan between Smokes* 32 He looked as though he liked bein' hit an' took stoush fer breakfast every mornin'. 1914 C. J. Dennis in *Bulletin* (Sydney) 16 July 47/1 Wot's jist plain stoush wif us..is 'valler' [*sc.* valour] if yer far enough away. 1924 *Truth* (Sydney) 27 Apr. 6 *Stouch*, a fight; to assault. 1945 R. S. Close *Love me Sailor* 149 It was like the old days when I got Ernie into some stoush ashore just for the hell of fighting him out of it. 1952 J. Cleary *Sundowners* (1960) iii. 129 The warmonger. You start any more stoushes..and..it'll be the finish of you. 1966 *Weekly News* (N.Z.) 22 June 59/4 The final folly was that it was the Lions and not Otago who were principally responsible for the 'stoush' of the first half of Saturday's game. 1970 D. M. Davin *Not here, not Now* iv. i. 229 I've played football against him. He's a good man in the stoush, no doubt about that.

stout, *a.* Add: **I. 3. a.** Also, *stout fellow*: see *fellow *sb.* 9 a.

II. 6. a. For ?*Obs.* read 'Now only *U.S. dial.*' and add later examples.

1882 'Mark Twain' *Stolen White Elephant* 269 Your word 'stout' means 'fleshy'; our word 'stout' usually means 'strong'. 1913 *Dialect Notes* IV. 54 That calf's terrible stout; he pretty near pulled me all over the field. 1962 W. Faulkner *Reivers* iv. 82 Let Lucius get out... He's younger than me and stouter too for his size.

12. c. *stout party*, a fat person (*humorously*). Esp. in catch-phrase *collapse of stout party* (see quot. 1975).

1855 *Punch* 25 Aug. 80/1 (*caption*) *Stout Party*: Well, I'm sure! What can possess those skinny creatures to wear round hats, I can't think,—making themselves so conspicuous! 1949 M. Allingham *More Work for Undertaker* vii. 102 'Do you will the stout party to give you the sixpences?' he ventured. 1957 R. G. C. Price *Hist. Punch* iii. 96 The florescence of the 'collapse of stout party' type of caption comes later [than the 1860s]. 1959 [see *pearly *sb.* b]. 1975 R. Pearsall *Collapse of Stout Party* 4 To many people Victorian wit and humour is summed up by *Punch*, when every joke is supposed to end with 'Collapse of Stout Party', though this phrase tends to be as elusive as 'Elementary, my dear Watson' in the Sherlock Holmes sagas.

stouter. Add: (Later example.) Cf. *strouter.

1819 L. A. Anspach *Hist. Island Newfoundland* 430 The place where the operation of curing the cod-fish is performed, is a *stage* or covered platform erected on the shore, with one end projecting over the water, which is called the *stage-head*, and which is fortified with stouters.

stove, *sb.*¹ Add: **6.** (sense 3) *stove-plant* (earlier example); (sense 5) *stove-lid* (U.S.), *-oven*, *-tile*, *-wood* (U.S.); **stove enamel,** a vitreous enamel that is sufficiently heat-resistant to be used on stoves; hence **stove-enamelled** *a.*, **stove-enamelling**; **stove lifter** *N. Amer.*: see Lifter 2 a.

1907 *Yesterday's Shopping* (1969) 11/1 Stove enamel polishing paste..tin, 0/2. 1949 Kirk & Othmer *Encycl. Chem. Technol.* IV. 165 White baking finishes may be classified in four general types: (1) kitchen-cabinet enamels, (2) refrigerator finishes, (3) washing-machine finishes, and (4) stove enamels... Stove enamels are intended for use on the trimmings for stoves... A maximum temperature of 200°F is encountered. 1958 *Observer* 13 Apr. 10/2 A durable finish chromium rather than aluminium, and vitreous enamel..rather than stove enamel. 1912 C. H. B. Quennell in L. Weaver *House & its Equipment* 103 A few years ago, and in the case of the cheaper ones to-day, baths were 'stove-enamelled'. 1977 *Custom Car* Nov. 85/1 (Advt.), 100E Jaguar IRS complete with crossmember, stove enamelled, all new parts. 1939 *Jrnl. R. Aeronaut. Soc.* XLIII. 607 It is desirable therefore, that in cases where stove-enamelling treatments or other processes involving re-heating at elevated temperature have to be applied to duralium, [etc.]. 1876 'Mark Twain' *Tom Sawyer* i. 17 She could have seen through a pair of stove lids just as well. 1929 W. Faulkner *Sound & Fury* 318 'I came in here to burn them up' ..I says, looking at him and opening the stove lid. 1886 *Harper's Mag.* Nov. 835/1 We'll have a real egg and cinder flip with the hot stove-lifter in it when we get back. 1927 M. de la Roche *Jalna* xxv. 306 She up and shied the stove lifter at my 'ead. 1855 E. Acton *Mod. Cookery* (rev. ed.) ii. 70 Set the dish into a gentle oven... A stove-oven, if the heat be properly moderated, will answer for the baking. 1968 *Globe & Mail* (Toronto) 13 Feb. 30/4 (Advt.), 4 Bedrooms, large family room. Fireplace! Walkout! Stove Oven! 1778 W. Cowper *Let.* 3 Dec. (1904) I. 151, I made Mr. Wrighte's gardener a present of fifty sorts of stove plant seeds. 1860 *Inventory Objects Mus. Ornamental Art, S. Kensington* 51/1 German enamelled stove tile; allegorical figure under an arcade.—Dated 1567. 1936 *Burlington Mag.* Sept. 111/1 The occurrence..on the jug of a relief corresponding to one on a green-glazed stove-tile. 1960 R. G. Haggar *Conc. Encycl. Cont. Pott. & Porc.* 269/2 Stoves and stove-tiles were made from the sixteenth until the eighteenth centuries, the earlier stove-tiles having flat surfaces with relief decorations and concave cylindrical backs. 1867 D. R. Locke *Swingin' round Cirkle* 159, I held a stick of stove wood suspended over his head. 1929 W. Faulkner *Sound & Fury* 332 Then she..stacked stovewood into her crooked arm. 1972 *News & Observer* (Raleigh, N.C.) 30 Dec. 4/2 We don't hear much about stove wood [nowadays].

stove, *v.*¹ Add: **5. b.** To heat so as to fuse a coating to the object being coated. Also *absol.*, and *intr.* (of the coating) for *pass.*

1951 *Industrial Finishing* IV. 184/1 Unless the article is suspended approximately equidistant from the emitting surfaces there will be a risk of it being unevenly stoved. 1954 *Archit. Rev.* CXVI. 132 The undersides of most metal deckings are ribbed, and the steel ones are usually finished with red oxide, 'stoved' on. 1962 D. W. Hislop in H. W. Chatfield *Sci. Surface Coatings* xviii. 531 A finish which stoves in half an hour at 150° C on sheet metal may require three times as long..at the same temperature when applied to a heavy casting. *Ibid.* 532 To be sure that a finish has been stoved adequately, a recording instrument is used with a thermocouple in contact with the painted metal. *Ibid.* 537 These lamps may be arranged in banks to give a high heating intensity. They enable paint films on suitable objects..to be stoved in times of a few seconds. 1977 *Hot Car* Oct. 73/1 It first etches and then stoves so that the finished coating (they say) is a really corrosion resistant lacquer around five times the thickness of factory wheel lacquer. 1979 J. D. Sandars et al. *Man. Colour Matching* 129 Apply 25–30 micron dry films to burnished degreased mild steel. Stove for appropriate times.

stove, *ppl. a.* Add: **1.** (Earlier example.) Also **stove-in.**

1850 H. Melville *White Jacket* I. iv. 20 Eternally talking of line-tubs, Nantucket, spermoil, stove boats, and Japan. 1897 Kipling *Captains Courageous* iv. 98 They found..a gin-bottle, and a stove-in dory, but nothing more. 1979 'A. Hall' *Scorpion Signal* xix. 223 A stove-in radiator with rusty water blowing out of it.

2. stove-up. Run-down, exhausted; worn out. Chiefly *predic.* of persons. *N. Amer. slang.*

1901 A. C. Hegan *Mrs Wiggs of Cabbage Patch* ix. 127 If I was n't so stove up, an' nobody was n't lookin', I'd jes' skitter 'round this here yard like a colt! 1942 Berrey & Van den Bark *Amer. Thes. Slang* § 129/12 *Physically run-down*,..stove-up. 1955 R. Hobson *Nothing too Good for Cowboy* xvi. 175 You look stove-up, boy, what's the trouble with that hind leg of yours? 1960 H. Lee *To kill Mockingbird* viii. 81 Mr. Avery'll be in bed for a week—he's right stove up. He's too old to do things like that. 1974 D. Sears *Lark in Clear Air* i. 18 An elderly man in levis and stove-up range-boots was..in the lower bunk.

stoved, *ppl. a.*¹ Add: **5.** *stoved enamel* = *stove enamel* s.v. *stove *sb.*¹ 6.

1926–7 *Army & Navy Stores Catal.* 311 The patent 'Peveril' grate..in best bright black stoved enamel, which does not require blackleading. 1967 *Times Rev. Industry* May 53/3 A good deal of paint is exported as the cellulose finish of a car, the stoved enamel surface of a washing machine or the paintwork of a jet airliner.

stoven, *ppl. a.* (Earlier example.)

1851 H. Melville *Moby Dick* II. xiv. 122 The terrific wreck of the stoven planks.

stove-pipe. Add: **3. b.** More recently, = *drain-pipe* (fig.) s.v. *drain *sb.* 5. Also *attrib.*, as *stove-pipe trousers*.

1955 T. H. Pear *Eng. Social Differences* vii. 176 Narrow stove-pipe trousers. 1970 *Globe Mag.* (Toronto) 26 Sept. 5/2 The greasers..wear stove pipes, grey with pin stripes. 1978 *Sunday Times Mag.* 18 June 43/2 The names are Mazurca and Maier, cheap pink shirts against blond hair, stovepipe trousers above bare ankles and feet.

c. A portable trench mortar. *U.S. Mil.*

1920 H. H. Bissell *Hist. Sixty-Third U.S. Infantry* 37 It didn't prevent their finding ranges, or getting the maximum of performance out of the old 'Stove-pipes'. 1957 R. Leckie *Helmet for my Pillow* vi. 236, I remained..spared the ordeal of carrying mortar shells to the 'stove-pipe' crews.

stovies (stō·viz), *sb. pl. Sc.* and *north. dial.* [f. Stove *v.*¹ 6.] A dish of potatoes stewed in a pot; a potato stew. Also *attrib.*

1893–4 R. O. Heslop *Northumb. Words* II. 687 Hey! lass, is the stavies [*sic*] no ready yit? 1894 J. Inglis *Oor Ain Folk* iv. 40 One day there was a fine dish of 'stoved taties' for dinner... The lads would..have 'the stovies' finished before he had a chance to start. 1907 in *Ochtertyre Ho. Bk. Accomps* 258 *Stovies*, potatoes stewed fine with dripping or fat bacon, onions and spice, and served hot. 1939 M. Spring Rice *Working-Class Wives* vi. 144 Her main dish is 'stovies' made with onions, potatoes and water; she never puts either dripping or meat into them. 1971 S. Walker *Highland Cookbk.* 77 Plain stovies are excellent with beef stew or mince..Put as many peeled and sliced potatoes as you need in a pot [etc.]. 1973 *Courier & Advertiser* (Dundee) 26 Feb. 7/1 A liquid gas portable stove..would have barely supported our stovie pot!

stoving, *vbl. sb.* **b.** (Later examples.)

1952 *Industrial Finishing* V. 201/2 How efficiency of drying..can be increased with existing stoving equipment. 1962 D. W. Hislop in H. W. Chatfield *Sci. Surface Coatings* xviii. 531 All paint stoving ovens can be a hazard as the solvent vapour/air mixture is a potential fire risk. 1972 *Materials & Technol.* V. viii. 215 The use of these oils in water-thinnable stoving paints has been more successful. *Ibid.* xi. 363 Anti-corrosive stoving primers for cars. 1982 W. M. Morgans *Outl. Paint Technol.* (ed. 2) I. xiii. 221 The combination of urea resin with non-drying alkyd is used very widely for stoving finishes of almost all types.

stow, *v.*¹ Add: **6. stow away. a.** (Earlier example of jocular use.)

1858 M. Tuckett *Diary* 24 Nov. (*c* 1975) 21 We stowed away a good breakfast.

b. (Later example.) Also, to conceal oneself on board a train or aeroplane.

1929 R. Hughes *High Wind in Jamaica* vii. 168 Otto was a Viennese by birth, but had stowed away in a Danube barge when he was ten years old, had taken to the sea, and thereafter generally served in English ships. 1973 *Times* 8 Aug. 7/3 A Brazilian youth was sent home ..after stowing away on board a South African Airways aircraft by mistake.

stowaway. Add: **1.** (Earlier and later examples.) Also, one who steals a passage by aeroplane.

1850 *Morning Chron.* 22 July 6/1 All the passengers were summoned on deck that their names might be read over, their tickets produced, and a search made in the steerage, and in every hole and corner of the ship, for 'stow-aways'. 1922 [see *scratch *sb.*¹ 5 b]. 1973 *Times* 8 Aug. 7/3 (*heading*) Stowaway takes wrong flight.

2. *gen.* Something stowed away. Also, a place where things may be stowed.

1913 E. F. Benson *Thorley Weir* iii, For all these weeks Charles had never touched the cupboard except to insert some further stowaway. 1915 W. J. Locke *Jaffery* xx, Of all the stowaway places under my control..only one is locked. 1928 *Daily Express* 21 Mar. 5/3 The window-seat top lifts up, and this makes another good stowaway for toys.

STP (es tī pī). orig. *U.S.* Also **S.T.P.** [Prob. f. the initial letters of *S*cientifically *T*reated *P*etroleum, the commercial name of a motor-oil additive which was being extensively advertised when the drug first appeared.] A synthetic hallucinogenic drug chemically related to amphetamine but having effects similar to those of LSD; 2,5-dimethoxy-4-methylamphetamine.

1967 *Village Voice* (N.Y.) 13 Apr. 7/2 STP is a new psychedelic drug which..has become the most sought-after high on the psychedelic scene. 1967 *Guardian* 4 July 7/1 A new drug being used by the 'hippies' who inhabit San Francisco... For want of a name the hippies call it STP—a trade mark for an oil additive. 1969 *Oz* May 21/2 Alexander 'Sashe' Shulgin, the brilliant chemist who is best known to the public for his synthesis of STP (which was later illegally distributed in a dose form twice that 'recommended'). 1971 *It* 2–16 June 8/5 Those 80,000 doses of S.T.P. in the peanut butter jar that we were saving for an emergency. 1974 M. C. Gerald *Pharmacol.* xvii. 328 DOM is 2,5-dimethoxy-4-methylamphetamine to the chemist and STP (serenity, tranquillity, and peace) to the drug user.

strabismic, *a.* Add: (Earlier and later examples.) Also *fig.*

1855 *Lancet* 12 May 479/1 If the strabismic eye be brought into play, it assumes a normal condition, and moves in obedience to the will. 1915 H. De Sélincourt *Realms of Day* ii. 8 The weak and dim-eyed, who blame life for their own strabismic will-lessness. 1955 W. Gaddis *Recognitions* ii. viii. 695 A small figure clutching a filthy dollar bill fixed him with a strabismic stare. 1975 *Nature* 20 Nov. 199/1 If a kitten is made strabismic by sectioning an eye muscle..their neurones are found to receive input from one eye or the other, but very few receive input from both.

‖ **stracciatella** (stratʃiate·lā). [It.] A soup made with stock, eggs, and cheese.

actualities of American life as vividly and independently as any 'straight' novelist.

10. c. straight A('s) *U.S.*, uniform top grades; **straight-armed** *a. Cricket*, with the arm unflexed; *spec.* † designating a style of round-arm bowling with a straight arm, or an exponent of this style (*obs.*); **straight arrow** *N. Amer. slang*, an honest or genuine person; also as *adj.* and *adv.*; **straight-backed** *a.*, (*a*) (earlier example of a person); **straight chain** *Chem.*, a chain of atoms that is neither branched nor closed in on itself to form a ring; usu. *attrib.* (with hyphen); **straight cut** *Cinemat.*, a complete cut between sequences (as opposed to a fade or a dissolve); **straight drive** *Cricket*, a drive in which the ball is struck back down the pitch towards or past the bowler; also as *v. trans.*; hence **straight driver, straight driving** *vbl. sb.*; **straight-edge(d) razor** = *straight razor* below; also *ellipt.*; **straight eight** *Mech.*, (a motor vehicle having) an internal combustion engine with eight cylinders arranged in a straight line; freq. *attrib.*; similarly *straight four, straight six*; (cf. *IN-LINE *a.* 1 a); **straight-faced** *a.*, solemn, serious (cf. sense 8 a in Dict. and Suppl.); hence **straightfacedly** *adv.*, **straightfacedness; straight fight,** an election in which there are only two candidates; **straight goods** *U.S. slang*, the truth; an honest person; **straight-grain(ed)** *a.* (see quot. 1929); **straight job** *U.S. slang*, a single-unit truck, one with its body built directly on to its chassis; **straight leg** *U.S. Mil. slang*, a member of the ground staff as opposed to one of the flying personnel (see also quot. 1967); **straight mute,** a simple cone-shaped mute for a trumpet or trombone; **straight pein** *a.*, designating a type of hammer which has the pein in line with the handle; freq. *absol.* as *sb.*; **straight razor,** a razor with a long blade that folds into its handle for storage, a cut-throat razor (see *CUT-THROAT 1 d); **straight-run** *a. Chem.*, (of a petroleum fraction) produced by distillation without cracking or other chemical alteration of the original hydrocarbons; **straight stitch,** in *Embroidery*, a single, short, detached stitch; also as *adj.*, designating a simple type of sewing-machine; hence **straight-stitching** *a.*; **straight time** *a.* orig. and chiefly *U.S.*, of or relating to remuneration received for work performed within normal or regular hours; also *absol.* (cf. OVERTIME *sb.* a in Dict. and Suppl.).

1926 *Amer. Oxonian* July 98 It isn't merely four years of football, four years of straight A, and ten thousand activities that make a winner [of a Rhodes Scholarship]. **1948** *Chicago Daily News* 20 Sept. 18/2 In pre-medical college Jim S. was a brilliant student—straight A's. [illegible] knows o' t'time.' 1937 P. K. Devine Devine's Folk Lore of Newfoundland 33 To make strange, to be afraid or timid. 'Don't make strange,' said to a guest sitting down to eat. 1966 Amer. Speech XLI. 295 [Newfoundland] Don't make strange. Said to make a guest feel at home. 1974 P. Gzowski Bk. about this Country 173/1 The luxury of a babysitter is rare—besides, the baby makes strange, and no babysitter with knowledge aforehand would come near!

III. 16. c. For † Obs. read Now rare and add strange-moulded.

1917 D. H. Lawrence Look! We have come Through! 135 Also she who is the other has strange-moulded breasts.

Stra·ngelove. Also strangelove. The name of the character Dr. Strangelove from the film of that title (1963) directed by Stanley Kubrick, used transf. to designate one who ruthlessly considers or plans nuclear warfare. Freq. attrib. and with Dr.

1968 Listener 16 May 638/1 Dr Strangelove is still at it... He has realised that the hydrogen bomb may yet prove to be a limited..means of annihilating the human race. His current concern is with finding ways of doing it more cheaply..and with greater guts. 1972 Village Voice (N.Y.) 1 June 10/4 In the strangelove language of the ABC, the accident exceeded the 'maximum credible accident' established as a possibility for the installation in the AEC's Hazards Summary. 1973 Guardian 22 Feb. 11/3 The Strangelove school, of which Dr Kissinger is a charter member, sees the world as a series of problems that can be manipulated by US money, technology, and bombs. Ibid. 28 Feb. 10/6 Professor William Shockley the exponent of sterilisation for low IQ subjects..went into a Dr Strangelove act. 1975 University (Princeton Univ.) Winter 5/1 The Physicists makes a Strangelove even out of Einstein. 1976 'R. B. Dominic' Murder out of Commission xv. 137 Dean Kennison was no Dr Strangelove, yearning to set off bigger and better bangs. 1980 Times 12 Aug. 10/7 Nervousness about a latterday Dr Strangelove getting his itchy finger on the button.

Hence (Dr.) Stra·ngelovean, -ian a.; Stra·ngelovism, a word or expression characteristic of one who toys with the concept of nuclear war.

1967 Newsweek 27 Mar. 47 Words like deterrent, credibility, overkill and doomsday machine became familiar, and were even kidded in such movies as 'Dr. Strangelove'. Now development of anti-ballistic missiles has produced a second generation of Strangelovisms. 1969 Washington Post 23 Apr. A16/4 Mendel Rivers..suggested that nuclear weapons be used if necessary to 'bring this crowd to its knees'... Such 'strangelovisms' from that source are of course not new. 1971 P. Dickson Think Tanks 4 Outside Washington, D.C.,..a group of analysts is fighting the wars of the 1990's in a $50,000 Strangelovean game room to see who wins, why, and with what weapons. 1977 Time 11 Apr. 13/1 The concepts are often Strangelovian. 1978 Chatelaine (Canada) Dec. 41/3 There was something Dr. Strangelovian about these top-level intellectuals who discussed top-secret scenarios.

strangeness. Add: **3.** Particle Physics. A quantized property of hadrons, now attributed to the s quark, that is conserved in strong and electromagnetic interactions but not in weak ones and is represented by a quantum number S equal to the hypercharge of a particle minus its baryon numbers.

1956 M. Gell-Mann in Nuovo Cimento IV. Suppl. 852 Since we have S=0 for ordinary particles and S≠0 for 'strange' ones we refer to S as 'strangeness'. 1960 New Scientist 5 May 1126/2 Like electric charge, the total magnitude of strangeness remains constant in a nuclear process. Not so, however, for the decay phenomena... Decay forces violate strangeness-conservation. 1963 S. Tolansky Introd. Atomic Physics (ed. 5) xxiii. 397 Whilst the strangeness number seems to play a basic part in the baryon reactions it does not operate in the case of the leptons... The concept of isospin is hardly appropriate to the leptons and with this falls away the significance of strangeness too. 1965 H. Muirhead Physics of Elementary Particles ix. 396 The classification of particles using the hypercharge quantum number is more economical in numbers than one involving strangeness. 1972 [see *hypercharge]. 1981 Sci. Amer. June 57/1 Strangeness conservation is now understood to be not a fundamental principle like energy conservation..but a consequence of the detailed theory of the strong interactions.

stranger, sb. (and a.). Add: **3. b.** (Earlier example.)

1798 Coleridge Frost at Midnight 20 Only that film, which flutter'd on the grate, Still flutters there... Ah me!..How often in my early school-boy days, With most believing superstitious wish Presageful have I gaz'd upon the bars, To watch the stranger there!

4. b. (Earlier examples.)

a 1674 T. Traherne Centuries of Meditations (1927) iii. ii. 151, I was a little stranger, which at my entrance into the world was saluted and surrounded with innumerable joys. 1787 J. Woodforde Diary 6 May (1926) II. 320 Mrs. Custance was brought to bed of a Boy about 11 o'clock this Morn'. She with the little stranger as well as can be expected.

c. Now in gen. colloq. use, to address one who has not been seen for some time.

1934 E. O'Neill Days without End ii. 59 Hello, Stranger. 1969 New Yorker 3 May 34/3 'Well, stranger, where've you been?' she greeted me. 'Why didn't you come back like you said?' 1973 Weekly News (Glasgow) 11 Aug. 5/1 (caption) Hello, there, stranger! 1977 F. Parrish Fire in Barley iii. 31 'Mornin', stranger,' said the landlord. 'How's the old lady keepin'?' 'Fairish,' said Dan.

d. Esp. in phr. to be (quite) a stranger and varr., said of an infrequent visitor.

c 1807 Jane Austen Watsons in J. Leigh Mem. Jane Austen (1871) 349 'So Emma,' said he, 'you are quite a stranger at home.' 1860 C. M. Yonge Friarswood Post-Office vii. 115 Ha! Harold King! Well, to be sure, you are a stranger! 1910 A. Bennett Clayhanger iii. vii. 378 'Well, Mr. Clayhanger,' said the steward... 'You're quite a stranger.' 1916 Joyce Portrait of Artist (1969) 219 You are a great stranger now. 1937 A. Uffield Mr. Jelly's Business xx. 211 Hello, Mr. Muir! You're quite a stranger. 1962 G. Avery Greatest Gresham ix. 162 Well, if it isn't the kiddies from next door. Why, you are strangers these days.

e. Austral. and N.Z. An animal which has strayed from a neighbouring flock or herd.

1852 J. R. Clough Jrnl. 11 Feb. in Deans Lett. 1840–54 (1937) 290 Branded 57 calves..counted all the other cattle; 201 of them strangers. 1933 L. G. D. Acland in Press (Christchurch, N.Z.) 16 Dec. 21/8 Stranger, a sheep of a neighbour's on your own run. 1965 J. S. Gunn Terminol. Shearing Industry ii. 28 Stranger, a strange sheep, probably from an adjoining property, which has joined the flock being shorn. 1972 P. Newton Sheep Thief xvi. 137 There was nothing unusual in..having a few 'strangers' (neighbour's sheep) on the place.

5. b. (Earlier examples.)

1705 House of Commons Jrnl. 31 Oct. 6/2 Ordered, That the Serjeant at Arms, attending this House, do from time to time, take into his Custody any Stranger or Strangers, that he shall see, or be informed of to be, in the House, or Gallery, while the House, or any Committee of the Whole House, is sitting. 1795 tr. C. P. Moritz's Trav. 58 The members call aloud to the gallery, withdraw! withdraw! On this the strangers withdraw.

strangle, sb. Add: **4.** strangle-hold (non-attrib. examples of fig. use).

1930 G. B. Shaw Apple Cart p. x, This purely inhibitive check on tyranny has become a stranglehold on genuine democracy. 1939 Daily Tel. 18 Dec. 6/4 Hitler knows and fears the stranglehold of the British and Allied blockade. 1980 I. Colegate Shooting Party (1982) 7 The stranglehold of the rich on the life-blood of the working man.

strangle, v. Add: **3. b.** Also with off.

1918 D. H. Lawrence New Poems 38 The frost has..ruthlessly strangled off the fantasies Of leaves.

strangler. Add: **1. a.** spec. in Bot., an epiphytic plant which eventually sends its roots to the ground and smothers its host.

1895 J. Rodway In Guiana Forest 91 The strangler is now ready for its deadly work. The forest giant..is bound by cords which are stronger than iron bands. 1952 P. W. Richards Tropical Rain Forest ii. 21 The third section of dependent plants, here termed stranglers..begin life as epiphytes and later send roots to the ground. 1960 N. Polunin Introd. Plant Geogr. xiv. 435 Stranglers..begin life as epiphytes but later send down roots to the soil. 1976 Hortus Third (L. H. Bailey Hortorium) 288/1 Clusia..dioecious trees and shrubs, occasionally more or less epiphytic or stranglers.

b. strangler fig, vine.

1955 Sci. Amer. Apr. 74/2 The strangler fig in the tropical jungle, which kills other trees to reach the light, is a rare type. 1962 Times 6 Apr. 7/2 Strangler figs..envelop and kill other trees. 1976 Publishers Weekly 12 Jan. 50/3 'Nanny' grows upon the family like a strangler vine upon a tree.

2. = *choke sb.[1] 7.

1925 E. W. Knott Carburettor Handbk. i. 29 Easy starting devices... First, stranglers or air chokes which reduce the main air supply by means of a suitable shutter or similar device, the use of which increases the suction on the main fuel orifice or jet far beyond the normal state of affairs. 1976 J. Watson Understanding your Car v. 27 A second butterfly valve, mounted above the spray tube..is known as a strangler, and by cutting off most of the air it greatly increases the suction on the jet to give a very rich mixture for starting.

stranskiite (stræ·nski,əit). Min. [ad. G. stranskiit (H. Strunz 1960, in Naturwissenschaften XLVII. 376/1), f. the name of I. N. Stranski (b. 1897), Bulgarian-born physical chemist: see -ite[1].] An arsenate of zinc and copper, $Zn_2Cu(AsO_4)_2$, found as blue triclinic crystals.

1960 Amer. Mineralogist XLV. 1315 (heading) Stranskiite. 1978 Ibid. LXIII. 213 Inclusions of intergrown stranskiite, $Zn_2Cu(AsO_4)_2$, and schultenite, $PbHAsO_4$, in massive tennantite from Tsumeb, Southwest Africa, have been investigated by X-ray diffraction and X-ray fluorescence.

strap, sb. Add: **15*.** Typogr. Short for strap-line, sense 17 below.

1960 A. Hutt Newspaper Design vii. 128 Essentially the strap is a single-line affair. 1981 A. Graham-Yooll Forgotten Colony xviii. 238 A photograph of the man..was splashed over the front page of the Buenos Aires evening newspaper..with a strap that read: 'This is how our English friends see us.'

16. a. With the meaning 'that has a strap', as (sense 3) strap watch.

1926 Daily Colonist (Victoria, B.C.) 11 July 9/4 (Advt.), Strap Watch. Guaranteed accurate and dependable. Handsome case. Leather strap. 1962 K. Orvis Damned & Destroyed xxiv. 181, I dropped my eyes to my strap-watch.

17. strap-end Archæol., the metal fastening on a strap (sense 3); strap-game (examples); strap handle Ceramics, a handle on a vessel such as a jug or ewer which is in the form of a loop and flattened like a narrow strap; hence strap-handled a.; straphanger, (later examples); also fig., one who commutes to work by public transport; strap-line Typogr., a subsidiary heading printed above a headline; strap-rail (examples).

1973 Oxf. Univ. Gaz. CIII. Suppl. v. 18 Mr A. R. Lake: Presented a 12th-century bronze strap-end from Bicester, Oxon. 1977 Antiquaries Jrnl. LXII. 420 Belt-buckles and strap-ends of the later Roman Empire. 1847 Knickerbocker XXIX. 281 He was accused of having 'come the strap-game' over a native. 1873 J. H. Beadle Undevel. West vii. 140 A score of 'smart Alecks' relieved of their surplus cash by betting on the 'strap game'. 1939 J. D. S. Pendlebury Archaeol. Crete iii. 134 The size of the strap handles is also an indication of date. 1972 Trans. Oriental Ceramics Soc. XXXVIII. 65 A stoneware ewer, ovoid with a short spreading neck and strap-handle. 1957 V. G. Childe Dawn Europ. Civilization (ed. 6) vii. 131 Strap-handled jugs. 1915 W. Cather Song of Lark ii. v. 195 In the street-car..she sat at the waistcoat buttons of a fat strap-hanger. 1950 A. J. Deutsch in D. Knight 100 Yrs. Sci. Fiction (1969) 169 The other seats were filled, and there were a dozen strap-hangers. 1981 Times 5 Aug. 10/4 Washington commuters..are not strap-hangers like New Yorkers, Londoners and Parisians. 1960 A. Hutt Newspaper Design vii. 128 The use of subsidiary lines—strap-lines over headings, tag-lines following them—has been a feature of headline practice. Ibid., Strap-lines are not suitable over double-column headlines. 1979 Guardian 9 Oct. 10/7 Lord Beaverbrook..sometimes put a strap line over the story saying that the piece didn't reflect editorial policy. 1874 B. F. Taylor World on Wheels xiii. 105 Years ago, he rode on a train of the old Utica & Adrian Railway—strap-rail at that. 1948 Exhibit Finder (Museum of Sci. & Industry, Chicago) 5 The story of the early days of railroading is further traced by samples showing the progress of rail manufacture from strap rail, flat as a pancake, to the heavy crowned rail of today.

strap, v.[1] Add: **1. e.** to strap (oneself) in: in an aircraft, to fasten one's safety belt. Also absol. (occas. with up or without adv.).

1913 Flight 20 Sept. 1040/2 Neither the pilot nor passenger was strapped in. 1919 J. Buchan Mr. Standfast ix. 173 He signalled to me to strap myself in..and proceeded to practise 'stunts'—the loop, the spinning nose-dive, and others. 1958 'Castle' & 'Hailey' Flight into Danger v. 72 Better strap yourself in... You must have watched the pilot quite a lot. 1962 L. Deighton Ipcress File v. 31 The steward helped him strap in. 1970 'R. Llewellyn' But we didn't get Fox vii. 69 She waited for me to strap, started a jet..and taxied down to the... 1977 R. Air Force Yearbk. 29 The excellent level of the straps is a noteworthy point and enables the pilot to strap in tightly and securely. 1977 'O. Jacks' Autumn Heroes v. 69 Gerry Steinberg was strapping up beside his pilot.

f. intr. for pass. To admit of being fastened by means of a strap.

1924 A. D. Sedgwick Little French Girl i. v. 37 Grey shoes strapping across the instep with a buckle.

7. Comb.: strap-down a. Astronautics, applied to an inertial guidance system in which the gyroscopes are fixed to the vehicle rather than mounted in gimbals; strap-on a. that can be attached by a strap or straps; in Astronautics, applied to a booster rocket mounted on the outside of the main rocket so as to be jettisonable; also as sb., such a booster.

[1962 Fernandez & Macomber Inertial Guidance Engin. viii. 308 The strapped-down gyro reference package..has become widely used as a guidance aid in ballistic missiles where high accuracy is not required.] 1963 Slater & Ausman in C. T. Leondes Guidance & Control Aerospace Vehicles iii. 82 A system of this sort..is sometimes inelegantly called 'strapdown'. 1983 Times 8 June 2/8 The IMU system uses specially designed and positioned gyros attached to the body of the missile, called strapdown gyros. 1966 Sci. News 13 Aug. 104/1 Solid propellant strap-ons could be used to raise the Saturn V's orbital payload..to as much as 427,000 pounds. 1968 New Scientist 31 Oct. 231 The vehicle..appeared to have a two-stage core with four strap-on boosters. 1975 Aviation Week 12 May 21/1 Viewed from below a cluster booster, the procedure would appear like the petals of a flower opening if all four strap-ons separated at the same moment. 1981 J. Sutherland Bestsellers x. 111 Such 'novelties' as strap-on shark fins. 1982 Aviation Week 14 June 18 The U.S. vehicle..uses strap-on solid boosters and integral liquid propulsion to launch itself.

S-strap: see *S I. 2 c.

stra·p-hang, v. colloq. Also straphang. [Back-formation from straphanger s.v. Strap sb. 17.] intr. To be a straphanger in a railway carriage, etc. Also fig.

1908 O. Jespersen in Englische Studien LXX. 119 You strap-hang on the Subterranean. 1917 Daily Mail 28 Aug. 2/5, I think those weary girls look like tired little flowers as they strap-hang for half an hour or more. 1931 Galsworthy Maid in Waiting vii. 55 The only difference..between Parties is that one Party sits in the National 'Bus, and the other Party strap-hangs. 1937 W. H. Auden in Auden & MacNeice Lett. from Iceland

v. 55 The bowler hat who straphangs in the tube. 1959 Times Lit. Suppl. 24 Apr. 237/4 Miss Charles straphangs from fashion on a journey whose destination does not interest her. 1974 K. Royce Trap Spider viii. 141, I still wonder what happened to her; it's not usually a happy ending with girls who strap-hang with rats like Laurie Yates. 1982 Barr & York Official Sloane Ranger Handbk. 100/1 In the Europe Supermarket in Old Brompton Road, or strap-hanging in the tube from Gloucester Road, astonishingly you meet more people you know.

Hence stra·p-hanging vbl. sb. and ppl. a.

1919 Electrician LXXXII. 497/2 The somewhat elaborate provision made for 'strap hanging'. 1920 Cycling 5 Feb. p. i (Advt.), The strap-hanging problem is easily solved by the satisfied owner of a Rudge-Whitworth. 1928 Daily Express 22 Dec. 7/2 (heading) Straphanging rule dispute. 1945 [see *keynote v.]. 1957 L. Durrell Justine i. 53 Here, where the general impression of British culture suggested parsimony, indigence, intellectual strap-hanging—here I would pass the evening alone. 1972 C. Fremlin Appointment with Yesterday i. 8 Every strap-hanging commuter in London.

strapless, a. Add: spec. of women's dress: without shoulder straps. Also absol.

1935 Mademoiselle Feb. 91/3 Strapless and backless brassiere. 1946 Vogue Aug. 90/2 In another seventy-five years he'll probably be all for Picasso and your strapless frock. 1955 N. Fitzgerald House is Falling i. 13 Her strapless, white swim-suit. 1969 A. Lurie Real People 92 Anna May came out, in a cerise strapless satin evening gown. 1973 Country Life 8 Mar. 633/2 Slinky dresses that have the finest of straps or are completely strapless. 1980 Daily Tel. 13 Oct. 19/3 Strongest revival of all—the straight-across strapless.

‖ **strapontin** (strapontæ̃). [Fr.] A tip-up seat, usu. additional to the ordinary seating in a theatre, taxi, etc., esp. in France.

1926 W. J. Locke Old Bridge v. xviii. 270 Perella insisted on sitting on the little seat, so that Silvester should be at the back with Beatrice. 'He loves it—hates strapontins.' 1927 Observer 29 May 12/3 As for the strapontins, which, at every performance of a successful play, block up all the gangways, actors and managers agree that they are dangerous. 1934 H. Miller Tropic of Cancer 179 Carl was sitting opposite us, on the strapontin. 1965 P. H. Newby One of Founders iv. 113 Hedges..climbed in behind Prudence, seated himself on a well-upholstered strapontin, and allowed himself to be driven off.

strapped, ppl. a. **2.** For U.S. slang read slang (orig. U.S.) and add: (Further examples.) Now freq. const. for. Also in extended use. Also cash-strapped adj.

1935 Sun (Baltimore) 13 Mar. 2/6 PWA is not yet 'strapped' for funds. 1936 L. C. Douglas White Banners ix. 193 If he had been strapped, the chances were he would have bought a hat to-day. 1936 M. Franklin All that Swagger xlviii. 437 Also she was strapped for ready money. 1952 Wodehouse Pigs Have Wings i. 23 A bit strapped for the ready, eh? 1958 L. Wolff Low Level Mission iii. 76 The Axis powers had always been strapped for oil; the specter of a shortage..haunted the two dictators. 1960 J. Lodwick Asparagus Trench 7 Fortunately, neither was strapped for children. 1968 Sunday Mail Mag. (Brisbane) 8 Dec. 13/3 Strapped for cash he and the dancers tossed their luggage by night from fire escape. 1973 Times 7 Dec. 18/7 (caption) If only we could be sure the Stringalongs are as strapped for petrol as we are, we could risk asking them to dinner. 1976 Time 5 Apr. 37/1 Cash-strapped Chrysler is stepping up preparations to bring out its first domestic subcompact late next year. 1977 Time 25 July 48/3 By the spring of 1974, the whipsaw effect of recession and rising costs—particularly for oil which fuels 80% of Con Ed's generating capacity—left the company strapped. 1979 Church Times 2 Nov. 10 The Roman Church is almost everywhere strapped for ready cash. 1982 Times 18 Dec. 9/1 Cash-strapped countries like Iran and Libya are producing flat out.

strapper[1]. Add: **2.** (Later Austral. examples.)

1963 M. L. West Gallows on Sand i. 3 The strappers who stood round the tracks in the misty mornings trying to pick Saturday's winners. 1970 Sunday Truth (Brisbane) 19 July 30/6 He checked to see if he was registered as a strapper with any turf club in the country.

5. A worker who furnishes or secures a thing with straps.

1881 Instructions to Census Clerks (1885) 79 Leather Goods..Maker. Strapper. 1921 Dict. Occup. Terms (1927) § 047 Strapper, puts up straps of wood or steel in support of roof in machine cut coal face. Ibid. § 345 Strapper, attaches straps to bags, trunks, etc.

strappy (stræ·pi), a. [f. Strap sb. + -y[1].] Of footwear or clothes: having straps.

1977 Observer (Colour Suppl.) 31 July 18/3 Bags, handluggage, strappy evening sandals. 1978 Detroit Free Press 16 Apr. (Detroit Suppl.) 20 (Advt.), Strippy, strappy stripes of canvas, wedged high, to put spring snap in your steps. 1980 Daily Tel. 21 July 13/1 There are strappy Fifties-style sundresses.

Strasburg. Add: Now usu. with spelling Strasbourg. **a.** Strasbourg goose, a goose fattened in such a way as to enlarge the liver for use in pâté de foie gras (see Pâté 1 a in Dict. and Suppl.); S. pâté, formerly = S. pie in Dict. but now usu. a goose pâté not enclosed in pastry.

1857 C. M. Yonge Dynevor Terrace I. xv. 240 A Strasburg goose nailed down and crammed before a fire becomes a Strasburg pie. 1877 E. S. Dallas Kettner's Bk. of Table 277 The liver of the Toulouse duck..is by most good judges preferred even to that of the Strasbourg goose. 1969 J. Fraser Clap Hands v. 76 Byron's passion for Liquorice Allsorts was rapidly diminishing... Every week..he was fed a large box of them, like a Strasbourg goose. 1860 Dickens Uncommercial Traveller (1861) xiv. 204 An unopened Strasbourg pâté fresh from Fortnum and Mason's. 1980 D. Bloodworth Trapdoor x. 56 They sat perched on high stools..with a pot of taramasalata,..an opened tin of Strasbourg pâté, a loaf of brown bread.

b. Used to designate the European Parliament, established in 1958, which has its seat in the premises of the Council of Europe at Strasbourg.

1972 Guardian 14 July 10/2 There is now some recognition at Strasbourg that with the coming of new members it must change its ways. Ibid. 23 Oct. 12/2 No doubt progress towards an effective European Parliament must take time. There are nevertheless measures which can be taken at once to lift Strasbourg out of its futile and legalistic rut. 1976 Times 8 Mar. 13/1 (heading) Who governs, Strasbourg or Westminster?

strass[2]. Delete rare⁻⁰ and add: **2.** A kind of waxed straw with a silky appearance, used for dress trimmings, etc.

1926 Westm. Gaz. 10 Mar. 7/3 Raspberry red strasse (a sort of waxed straw) was made into rosettes for a trimming on one black frock. 1927 Daily News 8 Apr. 2/2 Beneath the large strass-trimmed finish to the belt in front fell a full panel of white georgette trimmed with strass.

stratal, a. (Later examples.)

1966 S. M. Lamb Outl. Stratif. Gram. 1 These several systems may be called stratal systems, and each may be said to be associated with a stratum of linguistic structure. 1967 D. G. Hays Introd. Computational Linguistics viii. 162 A stratal conversion is a mapping or transduction of the representation of an utterance on one stratum into the representation of the same utterance on an adjacent stratum.

strategic, a. Add: **2.** Of, pertaining to, or designating nuclear weapons intended to destroy an enemy's capacity to make war. Cf. *tactical a. 1 c.

1957 Listener 26 Dec. 1056/1 Nobody has managed..to draw an effective distinction between 'strategic' and 'tactical' atomic weapons. 1957 Times 18 Nov. p. x/2 Between 1946..and 1950, strategic stockpiling had hardly begun. 1958 Spectator 21 Feb. 219/1 If Russia were to launch a major attack on the West, even with conventional forces only, the West would have to hit back with strategic nuclear weapons. 1961 Listener 14 Dec. 1012/1 The idea was not to create a complete system of strategic deterrence under Nato, but one large enough to meet certain specific problems. 1965 H. Kahn On Escalation v. 92 A high degree of escalation could easily be involved..if it were a strategic-weapons submarine. 1969 Times 27 Oct. 9/1 The discussions which will open between the United States and Russia in Helsinki next month are...the long awaited Salt discussions—the strategic arms limitation talks. 1978 Orbis XXII. 319 A desire to strengthen the linkage between theater- and strategic-nuclear forces. 1979 Financial Rev. 28 Sept. 10/5 It is not that the doctrine of strategic deterrence is being discarded, but that it is being constantly adapted to new technologies as each side seeks to prevent the other from gaining a decisive advantage.

3. Of, pertaining to, or designating materials essential to a country for fighting a war.

1958 Economist 26 July 283 A relaxation of the embargo on strategic exports to communist countries has been in the wind for some time. 1959 Listener 10 Dec. 1023/1 On the British side there is the question of the so-called strategic controls. This is the agreement between Great Britain and certain other Western countries not to export to the U.S.S.R. certain goods which might be used for military purposes. 1969 Plano & Olton Internat. Relations Dict. iii. 76 Some of the most critical strategic materials include foodstuffs, aluminum, cadmium, copper, magnesium, [etc.]. 1981 Financial Rev. (Sydney) 24 Apr. 54/2 Germanium is one of about two dozen metals called 'strategic' because they are vital to defence and industry, but available in large supplies only from foreign sources.

4. Special collocations: strategic bombing, the bombing of an enemy's territory with the aim of disrupting its economy and destroying morale; hence strategic bomber, an aircraft used for this purpose; cf. *tactical a. 1 b; strategic hamlet, a settlement or reservation for accommodating potential terrorists or their supporters under surveillance (esp. with reference to the Vietnam war); strategic studies, the analysis of conflict in international relations in all its aspects.

1961 Listener 14 Dec. 1012/1 If it becomes unwise..to consider basing MRBMs or strategic bombers in Europe. 1977 Sci. Amer. Aug. 26/3 The Backfire's capability as a strategic bomber—defined as a bomber that can reach the other country's territory—is certainly less significant than that of U.S. bombers based in Europe or on aircraft carriers. 1941 Nineteenth Cent. Sept. 163 Bombing of cities..is a true example of strategic bombing. 1966 Listener 27 Oct. 616/2 The very foundation stone of the independent service—strategic bombing—becomes a stumbling block. 1963 Times 21 Jan. 9/7 A vast campaign to build 'strategic hamlets' has been launched in every province... President Ngo Dinh Diem has described 1962 as the 'year of the strategic hamlets'. 1973 Black Panther 30 June 11/3 Smith's program involves the concepts of the 'strategic hamlet' and 'pacification'; to round up peasants from areas where guerilla forces show signs of gathering support (they will then be relocated to centers where they are surrounded by police barricades). 1975 New Yorker 5 May 131/1 What were we doing to the South Vietnamese, with our 'strategic hamlets' and 'free-fire zones'? 1959 Times 18 Feb. 12/1 The chairman and council of the Institute for Strategic Studies held a reception yesterday evening. 1981 Listener 5 Nov. 530/1, I..started a postgraduate seminar in strategic studies.

strategize (stræ·tĭdʒəiz), v. U.S. [f. Strategy + -ize.] intr. To formulate a strategy or strategies; to plan a course of action. Hence stra·tegizing vbl. sb.

1943 Sun (Baltimore) 8 Nov. 6/3 The delay in bringing the bill to the House floor for action developed because both sides were 'strategizing'. 1977 Dædalus Fall 134 Four competing hypotheses can be posed for the explanation of kinship rules: detailed genetic control, rational strategizing, complete cultural determinism, and coupled cultural and genetic control. 1978 New Scientist 21 Sept. 873 Men in dark suits and homburg hats will be commissioning think tanks to strategize. 1983 Washington Post 3 June A3 Back in those days [sc. the 1960s]..you didn't have to strategize and study and do the kind of homework on your cases that you have to do now.

strategy. Add: **2. d.** In (theoretical) circumstances of competition or conflict, as in the theory of games, decision theory, business administration, etc., a plan for successful action based on the rationality and interdependence of the moves of the opposing participants; also transf. (see quot. 1979).

1944 von Neumann & Morgenstern Theory of Games i. 44 The same arguments which forced us to consider sets of imputations instead of single imputations necessitate the abandonment of that narrow concept of 'standard of behavior'. Actually we shall call these sets of rules the 'strategies' of the game. 1954 Psychol. Bull. LI. 406/2 A strategy is a set of personal rules for playing the game. For each possible first move.., your opponent will have a possible set of responses. 1965 H. I. Ansoff Corporate Strategy vi. 118 A grand or mixed strategy is a statistical decision rule for deciding which particular pure strategy the firm should select in a particular situation. 1969 R. Farquharson Theory of Voting iv. 20 Any procedure can be represented as a game by assuming that each voter makes a plan in advance regarding the course of action he will take in every division which can arise. Any such plan may be called a 'strategy'—the voter's set of strategies constitutes the complete range of such possible plans. 1979 Science 25 May 795/2 Gideon Louw..laments the widespread biological use of the word 'strategy' because of the implication of rational choice..but..there is no simpler way to label possible evolutionary designs.

Stratfordian (strætfɔə·ɹdiăn), sb. (a.) [f. the name of the town Stratford-upon-Avon, War., birthplace of William Shakespeare + -ian.] **1.** One who lives in or was born in Stratford-upon-Avon.

1821 J. Saunders Let. 8 June in A. Mathews Mem. Charles Mathews (1839) III. ix. 204 Intreating a line when you have anything desirable to impart to the Stratfordians. 1909 'Mark Twain' Is Shakespeare Dead? 58 Stratfordians who were not Stratfordians of Shakespeare's day, but later comers. 1963 Times 12 Feb. 11/4 It is likely that the Stratfordians thus deprived of some edification from the pulpit were less put out than those who now find the harmonies of a concert-platform sadly broken.

2. A supporter of the view that Shakespeare was the author of the plays generally attributed to him. Also as adj. Cf. *Shakespearian a. (and sb.) b.

1908 G. Greenwood Shakesp. Probl. Restated 172 Really, really, there must be some limits even to Stratfordian demands on our credulity! Ibid. 226 The futilities which are gravely trotted out by enthusiastic Stratfordians as valuable evidence to illustrate the life of Shakespere. 1912 [see over-prove s.v. *over- 27 a]. 1930 P. Allen Case for E. de Vere as Shakespeare 6, I remained an orthodox Stratfordian until 1923. Ibid. 26 All these discoveries and inferences..were fast and firmly establishing the case for Oxford, at the same time that they were destroying utterly the Stratfordian arguments. 1962 Economist 28 July 364/2 His work..made him a 'convinced Stratfordian'.

stratification. Add: **1. b.** The placing of seeds close together in layers between layers of moist sand, peat, or the like in order to preserve them or promote germination; also extended to the placing of seeds in such a medium other than in layers.

1914 Moon & Brown Elem. of Forestry vi. 103 Commercial houses rarely practice stratification, because they have storehouses where moisture conditions are kept uniform. 1928 Jrnl. Forestry XXVI. 775/2 Stratification of these seeds for one to four months previous to planting has been found to hasten greatly their germination. 1976 H. L. Edlin Nat. Hist. Trees xiv. 181 The seeds of ash trees..and many other common genera..require stratification for sixteen months.

2. e. spec. in Sociology, the formation and establishment of social or cultural levels resulting from differences in occupation and political, ethnic, or economic influence.

1927 P. Sorokin Soc. Mobility ii. 13 Unstratified society..is a myth which has never been realized in the history of mankind... The forms and proportions of stratification

vary, but its essence is permanent. 1944 S. Putnam tr. da Cunha's Rebellion in Backlands ii. 117 It was natural that the deep-lying layers of our ethnic stratification should have cast up so extraordinary an anticlinal as Antonio Conselheiro. 1962 Guardian 22 June 20/2 The so-called gulf between science and literature..led to no social stratification. 1981 R. Fletcher Sociol. vii. 200 Specific interest groups and a changing ethnic composition are the elements most likely to be troublesome in..problems of social stratification.

f. The existence in a lake or other body of water of two or more distinct layers differing in temperature, density, or the like.

1898 Amer. Naturalist XXXII. 26 It is in a condition of 'inverse stratification', as Forel calls it, when the colder water is above the warmer. 1935 P. S. Welch Limnology iv. 51 Exceptional meteorological conditions may..prevent stratification completely. 1952 Phil. Trans. R. Soc. B. CCXXXVI. 355 (heading) Water movements in lakes during summer stratification. 1972 M. G. Gross Oceanography vii. 191 Well-developed density stratification of the open ocean inhibits strong vertical currents. 1973 P. A. Colinvaux Introd. Ecol. xviii. 249 If a lake is highly productive, thermal stratification has some interesting consequences for the bottom water.

g. Variation in the richness of the fuel-air mixture during the period of its introduction into the cylinder of an internal-combustion engine.

1922 H. R. Ricardo Internal-Combustion Engine I. v. 75 In order to increase the range of power as far as possible, every effort is made to encourage stratification. 1981 Sci. Amer. May 45/1 The stratification also makes it possible to burn fuel-air mixtures so lean in fuel that they would not burn if the fuel were uniformly mixed with the air.

2*. Statistics. The (usu. notional) division of a population into distinct groups from each of which a proportion of an overall sample may be taken.

1920 A. L. Bowley Elements of Statistics (ed. 4) ii. iv. 336 The stratification of a universe of measurable objects is also treated by Mr. Yule. 1934 Jrnl. R. Statistical Soc. XCVII. 608 This method of stratification..gives an improvement in precision. 1957 Kendall & Buckland Dict. Statistical Terms 282 The process of stratification may be undertaken on a geographical basis, e.g. by dividing up the sampled area into sub-areas on a map. 1966 Rep. Comm. Inquiry Univ. Oxf. II. 420 In drawing the sample stratification by college and subject was used to the greatest possible extent. 1967 G. Wills in Wills & Yearsley Handbk. Management Technol. 187 This process of stratification..can only be carried out if you have details of the relevant population of shops by region, type, etc. 1977 Lancet 28 May 1142/2 With stratification for hospital, age, and year of admission, the maximum-likelihood estimate of uniform risk ratio was 3·3.

stratificational (strætifikē[1]·ʃənăl), a. [f. Stratification + -al.] **1.** Linguistics. Of or pertaining to the concept of language as a series of strata or structural layers, esp. stratificational grammar, linguistics, whereby language is envisaged and analysed in terms of a number of different strata, each with its own rules of formation and related to each other.

1962 S. M. Lamb (title) Outline of stratificational grammar. Ibid. 3 The code relating each pair of neighboring strata is a set of stratificational rules. Ibid. 6 Stratificational analysis may be described as a process of emicization followed by the description of the results. 1966 Georgetown Univ. Monogr. Ser. Lang. & Linguistics xvii. 87 The picture of the organization of language in terms of four strata can conveniently be called stratificational. 1968 P. M. Postal Aspects Phonol. Theory iv. 89 By abandoning any vestige of a natural relation between phonetic and phonemic representations, stratificational phonemics has completely lost contact with these early motivations. 1970 Canad. Jrnl. Linguistics XV. 97 Since this hypothesis is independent of the concept of strata, it seems sensible to call the grammar to be developed here a relational network grammar, rather than a stratificational grammar. 1972 D. G. Lockwood Introd. Stratificational Linguistics i. 5 Stratificational theory may eventually be able to provide evidence on the relation of the neural networks to the storage of knowledge. 1977 P. Strevens New Orientations Teaching of English vi. 79 Very little..stratificational theory..could be thoroughly taught and learned during, say, a two-year training college course.

2. Of or pertaining to social or cultural strata.

1963 New Society 3 Oct. 30/3 The evolution of American jazz has been correctly recognised..as a musically expressed protest movement against the existing stratificational order. 1968 Canad. Jrnl. Linguistics XIII. 126 The variables studied..showed stratificational patterns clearly identifiable with different social levels of the community.

Hence stratifica·tionalism, adherence to the theory that language comprises several structural layers; stratifica·tionalist, one who holds this theory.

1968 South Atlantic Bull. Mar. 1/3 Dashing across the empty plains from a distant Danish horizon comes a new band, the troop of Stratificationalism. 1969 R. I. McDavid in 2nd & 3rd Lincolnland Conf. on Dialectology (1972) 1 There seems to be diffidence on the part of Lamb's stratificationalists. 1973 Amer. Speech 1969 XLIV. 287 Pike or Lamb might charge that James D. McCawley's 'Prelexical Syntax' is arcane from the point of view of a tagmemicist or stratificationalist. 1978 Language LIV. 170 If it can be said to have a dominant philosophy, it would be stratificationalism.

stratified, ppl. a. Add: **1. a.** (Earlier example.)

1799 [see Polygenous a. 1].

e. Of a lake or other body of water: exhibiting stratification. [tr. F. stratifié (F. A. Forel 1880, in Bibliothèque Univ.: Arch. des Sci. physiques et nat. IV. 94).]

1881 Proc. Boston Soc. Nat. Hist. XXI. 66 In summer the return current takes place at no great depth, because then the water is stratified according to its temperature. 1910 E. M. Wedderburn in Murray & Pullar Bathymetrical Surv. Sc. Fresh-Water Lochs I. 104 As winter draws on the lake becomes thermally stratified. 1957 G. E. Hutchinson Treat. Limnol. I. v. 334 Not only can a lake oscillate as a whole, but if it is stratified, the various layers of different density can oscillate relative to one another. 1972 M. G. Gross Oceanography xi. 301 Tidal currents [in an estuary] cause mixing between layers, so that the waters become only moderately stratified. 1980 G. E. Fogg in Barnes & Mann Fund. Aquatic Ecosystem ii. 37 Sometimes lakes are permanently stratified, as for example in tropical regions.

f. Sociol. (Cf. *stratification 2 e.)

1927 P. Sorokin Social Mobility ii. 11 If the economic status of the members of a society is unequal, if among them there are both wealthy and poor, the society is economically stratified. 1960 V. Packard Status Seekers xxii. 325 Life is said to be more stable and serene in clearly stratified societies. 1969 C. D. Darlington Evol. of Man & Soc. xviii. 422 The urban society was still stratified and even more highly stratified, but the strata and their boundaries had changed.

g. stratified charge: in an internal-combustion engine, a rich mixture for ignition in each cycle followed by a lean one for combustion, usu. achieved by having a side chamber in each cylinder into which the mixture for ignition is introduced; freq. attrib.

1931 D. R. Pye Internal Combustion Engine vi. 148 (caption) Thermal efficiencies obtainable at weak petrol-air mixtures with 'stratified charge' operation. 1953 H. R. Ricardo High-Speed Internal Combustion Engine (ed. 4) xviii. 366 Two alternative schemes were tried:..(2) The use of a stratified charge in which the mixture immediately in the zone of the sparking-plug is very much richer than that in the main body of the combustion space. 1976 National Observer (U.S.) 22 May 9/4 The Japanese car is..powered with a larger version of the stratified-charge engine. 1981 Sci. Amer. May 45/1 Direct-injection stratified-charge engines have efficiency advantages comparable to those of the diesel.

2. Statistics. Employing, or obtained by means of, the technique of stratification.

1920 A. L. Bowley Elements of Statistics (ed. 4) ii. iv. 333 In a non-stratified selection we should have had $\sigma = \cdot 0141$. 1934 Jrnl. R. Statistical Soc. XCVII. 567 This method has been called by Professor Bowley the method of stratified sampling. 1956 B.B.C. Handbk. 1957 104 This Survey employs part-time interviewers..who each day question some four thousand people—a sample or cross-section of the adult public... (Technically speaking, the sample used is a stratified quota sample.) 1980 Nature 1 May 5/3 One [technique] is stratified sampling, in which the site was first divided into 20 metre by 20 metre squares, and then four randomly selected one metre by one metre trenches excavated within each of the larger squares.

stratiform, a.[2] Delete rare⁻¹ and add later examples.

1885 [see Cumuliform a.]. 1944 H. C. Willett Descriptive Meteorol. iv. 92 These high stratiform clouds are spread over large areas where convection is extensive in the lower atmosphere. 1978 Bull. Amer. Meteorol. Soc. LIX. 518 (heading) An analytical model of snowflake growth in stratiform clouds.

stratify, v. Add: **1. c.** To preserve or promote the germination of (seeds) by stratification.

1905 Terms Forestry & Logging (U.S. Dept. Agric., Bureau of Forestry) 23 Stratify, to preserve tree seeds by spreading them in layers alternating with layers of earth or sand. 1916 J. W. Toumey Seeding & Planting vii. 104 Some of the pines and junipers germinate so slowly that the seed is usually stratified for a year before sowing. 1949 Q. Jrnl. Forestry XLIII. 169 The seed is stratified in wet sand for six weeks before sowing at a temperature of 34° F. 1960 New Scientist 12 May 1210/3 When seeds of certain shrubs or trees are stratified—stored in moist sand at 41° F. to break their dormancy—the amount of moisture in the sand may markedly influence the percentage germination subsequently achieved.

2. c. Statistics. To subdivide (a population) into groups in order to take a stratified sample. Also absol.

1949 F. Yates Sampling Methods for Censuses & Surveys iii. 25 A population may be stratified for two or more different characteristics... Thus we may stratify farms according to size and according to geographical regions. 1966 Rep. Comm. Inquiry Univ. Oxf. II. 420 It was possible to stratify by college and proposed subject. Ibid. 424 The names under Social Studies were further stratified according to college thus: Nuffield, St. Antony's, and others. 1967 G. Wills in Wills & Yearsley Handbk. Management Technol. 187 In most cases..the statistician can..stratify the list of 27,000 names [of shops] in terms of the region of the country they are in, [etc.]. 1970 J. E. Freund Statistics xi. 285 Stratified sampling..can be very effective provided one stratifies with respect to truly relevant characteristics of the population.

3. Also, to become stratified.

1935 P. S. Welch Limnology iii. 15 Criteria which would make a lake include only those bodies of standing water which are of considerable expanse and which are deep enough to stratify thermally. 1980 R. S. K. Barnes in Barnes & Mann Fund. Aquatic Ecosystems i. 14 Water bodies stratify when stable density differences are generated, often as a result of surface heating.

stratigraphic, a. Add: (Later transf. examples.) Also fig.

1972 Computers & Humanities VII. 39 One of the most characteristically archaeological of these uses is the generation and printing of stratigraphic backplots of artifacts and other items in a site. 1976 Jrnl. Asian Studies XXXV. 636 Another stratigraphic approach to Indian society.

strato-. Add: **1.** Also used in forming other terms, as stra·totype Geol., a particular group of strata chosen as defining a named stratigraphic unit or boundary; freq. attrib.; stra·tovolca:no Geol. [ad. G. strato-vulkan (K. von Seebach 1866, in Zeitschr. d. deutsch. geol. Ges. XVIII. 644)], a volcano built up of alternate layers of lava and ash.

1965 Bull. Amer. Assoc. Petroleum Geologists XLIX. 1701/1 The most effective means of providing a chronostratigraphic unit with fixed and uniform limits..appears to be by the designation of a specifically bounded section of rock strata as the stratotype of this unit. 1969 Proc. Geol. Soc. Lond. Aug. 142 A standard stratigraphical scale expresses the relative ages of rock in agreed terms defined by boundary points in stratotype sections. 1972 J. A. Couvering in Bishop & Miller Calibration of Hominoid Evolution 247 Stage-age names..are put in quotation marks when used in a sense which differs from the strict definition of the stratotype. 1978 Nature 16 Nov. 258/2 The rich foraminiferal and molluscan faunas at these localities establish the correlation of the beds..with the Jemmys Point Formation in Gippsland, the stratotype of the Kalimnan Stage (early Pliocene). [1885 A. Geikie Text-bk. Geol. (ed. 2) 227 Von Seebach..distinguished two volcanic types. 1st, Bedded Volcanoes (Strato-Vulkane), composed of successive sheets of lava and tuffs, and embracing the great majority of volcanoes.] 1957 G. E. Hutchinson Treat. Limnol. I. i. 25 The typical large volcano with a well-developed crater is an intermediate or composite structure, usually built of alternating layers of lava and cinders, and therefore called a stratovolcano. 1973 Sci. Amer. Aug. 67/1 These later lavas form the great stratovolcanoes, some still active, that dominate the Andean chain. 1977 Whitaker's Almanack 1978 1037/2 Merapi, a strato-volcano on the island of Java, has been active for several centuries.

2. [f. *strato(sphere.)] **a.** Used to form the names of various kinds of high-altitude aircraft, as stratocruiser, -fortress, -freighter, -jet, -liner, -plane, -tanker.

All but stratoplane are the names of aircraft built by the Boeing Airplane Company, but only Stratoliner (formerly also Stratocruiser) is a proprietary name (in the U.S.).

1944 Sun (Baltimore) 15 Nov. 8/2 An announcement by the Boeing Aircraft Company of a 'Stratocruiser' for post-war production... A military prototype of the Stratocruiser..is undergoing tests. 1966 D. Francis Flying Finish iv. 48 The pressurized stratocruiser which took us [to New York]. 1953 Britannica Bk. of Year 28/1 The Boeing YB-52, an eight-engine jet swept-wing Stratofortress designed gradually to replace the intercontinental B-36, made its first flight. 1981 Nature 17 Dec. 606/2 The massive wide-area saturation bombing from B-52 stratofortresses alone ultimately added up to 800,000 tonnes of bombs. 1947 Sun (Baltimore) 10 July 2 (caption) The Boeing Aircraft Company's 'Stratocruiser', passenger-carrying counterpart of the 'Stratofreighter', is shown as it takes off on its initial test flight at Seattle. 1947 Daily Progress (Charlottesville, Va.) 18 Dec. 10/1 The XB4 Stratojet, described as potentially the world's most powerful plane, made its initial test hop yesterday. 1955 Times 20 June 3/5 One 'Stratojet' bomber succeeds another as the latest weapon of defence. 1938 Sun (Baltimore) 8 June 3/2 Two other new transports, the Boeing 'Stratoliner' and the Douglas DC-4. 1939 Official Gaz. (U.S. Patent Office) 26 Dec. 805/1 Boeing Aircraft Company... Stratoliner. For airplanes and structural parts thereof. 1955 M. McCarthy Sights & Spectacles (1959) 162 Stratoliners from Kansas City ferry-in patrons for Guys and Dolls. 1933 Daily Progress (Charlottesville, Va.) 16 Nov. 6/4 Clues to the tail-winds that may push future strato-planes on high-speed flights through the stratosphere were sought..here today by astronomers watching the Leonid meteor shower. 1936 Discovery Apr. 125/2 The efficiency of the reaction-motor beginning where that of the propeller leaves off, at approximately the 10-mile level,..raises the question as to whether certain makers of hush-hush 'stratoplanes' are not working along the wrong lines. 1955 Ann. Rep. 1954 (Boeing Airplane Co.) 15/2 The new military airplane, to be known as the KC-135 Jet Stratotanker, will be an advanced version of the prototype. 1980 Times 1 Feb. 3/3 (caption) A woman stratotanker pilot in the United States Air Force at RAF Fairford, Gloucestershire, yesterday, after a refuelling mission.

b. Used in the sense 'travelling in, suitable for travel in, the stratosphere', in stratonaut [f. -naut after Aeronaut, *astronaut], stratosuit.

1934 Amer. Speech IX. 236/2 According to C. E. Mason, the New York Times coined the word stratonaut to describe venturers in the stratosphere. 1936 Nature 27 June 1053/1 A general introduction which is packed with references to Magellan and Copernicus on one hand, and the stratosphere explorers (stratonauts) and theoretical investigators of the expanding universe on the other. 1937 C. G. Philp Conquest of Stratosphere 6 Seventy-four years ago..those gallant pioneers, Glaisher and Coxwell,..claimed to have reached a height of 7 miles, into the lower stratosphere..without a single essential of the modern stratonaut, for they had no oxygen apparatus. 1945 Sun (Baltimore) 23 June 3/2 (caption) An airman wears a new

flexible pressurized 'Strato-suit' of rubberized fabric..for use in high-altitude flying. 1949 Jrnl. Brit. Interplanetary Soc. VIII. 40 Doubtless considerable improvements have been effected in stratosuits since the Haldane-Davis original.

stratopause (stræ·topǭz). Meteorol. [f. *strato(sphere + Pause sb.] The upper limit of the stratosphere, separating it from the mesosphere. (In current use at a greater height than formerly: cf. *stratosphere 2.)

1950 S. Chapman in Jrnl. Atmospheric & Terrestrial Physics I. 121 Taking stratosphere to denote the nearly isothermal region above the troposphere, its upper boundary, where the temperature first begins to increase upwards more rapidly than is common in the lower stratosphere, would be the stratopause. 1963 Q. Jrnl. R. Meteorol. Soc. LXXXIX. 156 At its meeting in June 1962, the executive committee of the World Meteorological Organization passed a resolution on the terminology..for the high atmosphere. It is as follows:..(a) Stratosphere: Region (situated between the tropopause and stratopause) in which the temperature generally increases with height. (b) Stratopause: The top of the inversion layer in the upper stratosphere (usually around 50 to 55 km). 1979 Jrnl. Atmospheric Sci. XXXVI. 1616/1 Large [meridional wind] oscillations are seen near the stratopause (~50 km) with a period of about 2 days and an amplitude as large as 30 m s⁻¹.

stratosphere (stræ·tŏsfiə̆ɹ). [f. Stratum + -o + Sphere sb.] † **1.** Geol. [ad. G. stratosphäre (E. Suess Das Antlitz der Erde (1901) III. i. i. 4).] (See quots.) Obs.

1908 H. B. C. Sollas tr. Suess's Face of Earth III. i. 2 So great is the part played by stratified deposits in the structure of the earth's crust that we might be tempted to speak of the stratosphere of the earth in contradistinction to the scoriosphere of the moon. 1909 Ibid. IV. xv. 546 The stratosphere, or younger sedimentary envelope has been formed almost entirely at the expense of the Sal envelope.

2. Meteorol. The region of the atmosphere extending from the top of the troposphere up to a height of about 50 km. (the stratopause), in the lower part of which there is little temperature variation with height in temperate latitudes and in the higher part the temperature increases with height; formerly, the lower part of this region only (up to a height of about 20 km.).

1909 Sci. Abstr. A. XII. 208 (heading) Variation in height of the stratosphere (isothermal layer). 1909 W. N. Shaw Free Atmosphere in Region of Brit. Isles 47 M. Teisserenc de Bort has introduced the words 'troposphere' and 'stratosphere' to denote these two layers. Ibid. 48 Such evidence as we have goes to show that the stratosphere is a region of comparative calm. 1923 Daily Mail 26 Feb. 5/4 In this stratosphere it has been ascertained from balloon soundings that the temperature ceases to fall with an increase in height,..up to a level of 13½ miles, the highest attained so far by any instrument of man. 1934 Discovery Mar. 57/2 Professor Piccard, the Belgian physicist, was the first to make a successful flight [by balloon] into the stratosphere. 1937 Jrnl. R. Aeronaut. Soc. XLI. 414 It is likely to be many years before stratosphere flight, flights at heights of 40,000 feet and over, will become the commercially paying proposition which its enthusiasts believe it ultimately will be. 1951 'J. Wyndham' Day of Triffids ii. 39 Somewhere high up in the stratosphere, he and Comrade Baltinoff found themselves attacked by the planes. 1951 [see exosphere s.v. *exo-]. 1960 M. Nicolet in J. A. Ratcliffe Physics of Upper Atmosphere ii. 19 The stratosphere is essentially that region where the temperature increases, or at least does not decrease, with altitude. 1963 [see *stratopause]. 1980 Nature 27 Nov. 347/1 Air samples, collected cryogenically at different heights of the stratosphere, were analysed for carbon dioxide.

3. Oceanogr. The bottom layer of the ocean, in which (by analogy with the original meaning of sense *2) there is little temperature variation with depth.

1937 Nature 26 June 1085/1 The oceanic troposphere, like the corresponding section of the atmosphere, is a relatively shallow layer marked by steep temperature gradients which contrast strongly with the more even conditions of the stratosphere. 1942 H. U. Sverdrup et al. Oceans iv. 141 From analogy with the atmosphere, Defant (1928) has applied the terms troposphere and stratosphere to two different parts of the ocean. Troposphere is applied to the upper layer of relatively high temperature that is found in middle and lower latitudes and within which strong currents are present, and stratosphere to the nearly uniform masses of cold deep and bottom water. Ibid., Within the oceanic stratosphere the salinity is very uniform. 1966 R. W. Fairbridge Encycl. Oceanogr. 940/2 The oceanic stratosphere refers to the nearly uniform masses of cold deep water and bottom water.

4. transf. and fig. An upper region, esp. in a hierarchy; a high, or the highest, plane, level, or rank.

1951 M. McLuhan Mech. Bride (1967) 62/1 The bathroom has been elevated to the very stratosphere of industrial folklore. 1952 Observer 3 Feb. 5/8 Their father was a big Paris dealer who moved in the stratosphere with Anatole France and de Goncourt. 1958 Listener 9 Oct. 578/2 To come down from the stratosphere of critical intentions to the rough terrain of poetry itself. 1965 B. Sweet-Escott Baker Street Irregular vii. 188 Since August 1943 the stratosphere had begun to take our work in the Balkans seriously. a 1974 R. Crossman Diaries (1975) I. 609 They were astonished that Harold had upped me into the stratosphere. 1975 Country Life 16 Jan. 138 With the fall-front secretaire..we are well-up in the stratosphere in terms of furniture... It was sold for 45,000 gn. 1980 Daily Tel. 3 Nov. 16 At various times she has advanced three different reasons for keeping the [minimum lending] rate in the stratosphere.

stratospheric (strætŏsfe·rik), a. [f. prec. + -ic.] **1.** Of or pertaining to the stratosphere; occurring or performed in the stratosphere.

1920 W. J. Humphreys Physics of Air x. 191 Let a stratospheric column be dropped bodily a distance dh, and let the surrounding air come in until equilibrium is again established. 1935 Jrnl. R. Aeronaut. Soc. XXXIX. 194 This plea is often put forward by the protagonists of so-called stratospheric aviation. 1945 Washington Post 10 Oct. 1/3 Development of even greater bombers capable of operating at stratospheric altitudes..is a certainty. 1959 Daily Tel. 30 Nov. 1/6 A balloon descending from a stratospheric study of the planet Venus landed safely in a meadow 20 miles north of Manhattan, Kansas, to-day. 1980 Nature 27 Nov. 347/2 The stratospheric CO_2 mixing ratio is not constant with height but rather decreases with increasing height.

2. fig. On a scale, from a point of view, or of a pitch, so high as to be suggestive of the stratosphere; esp. of a cost: = *astronomical a. 1 b.

1935 Sun (Baltimore) 16 Feb. 2/1 This bill..requires Congress to abdicate. It requires the country to lean on a dubious dream... Changes made by the Senate committee..in no sense change the stratospheric realities of the bill itself. 1958 K. Goodwin in P. Gammond Decca Bk. of Jazz xiii. 149 His ability to produce stratospheric screamers with apparent ease was utilized to add bite and drive to the brass sections. 1966 Economist 19 Nov. 778/1 It might at least consider supporting stratospheric French suggestions for so-called 'solving of the key currency problem' by the admitted archaism of raising the world price of gold. 1973 Daily Tel. 17 Nov. 29/3 To prevent gilt-edged prices from falling even further and yields from escalating to even more stratospheric heights. 1980 Jewish Chron. 29 Feb. 17/3 Stratospheric admission prices for a suburban commercial venture.

stratosphe·rically, adv. colloq. [f. prec. + -al[1] + -ly[2].] To a stratospheric degree, 'astronomically'.

1946 J. W. Day Harvest Adventure xx. 331 This type of oratory was something very rich and rare, obviously from an intellect stratospherically above the plain, blunt, and fairly honest opinions held by the rank and file of Newark Labour. 1961 B. Fergusson Watery Maze x. 237 The stratospherically high priority afforded him for the journey did not really do him much good.

Stratovision (stræ·tovi̇ʒən). U.S. Television. Also stratovision. [f. *strato(sphere + *tele)vision.] A proprietary name for a system whereby television programmes are broadcast to a wide area by retransmission from a circling aircraft.

1945 Cleveland (Ohio) Plain Dealer 10 Aug. 1/5 The stratovision system simply puts the antenna and transmitter in an airplane flying in lazy circles 30,000 feet above the earth. 1946 Official Gaz. (U.S. Patent Office) 14 May 197/1 Stratovision for radio receiving and/or transmitting sets, television receiving and sending apparatus, [etc.]. 1948 Sun (Baltimore) 6 Oct. 19/4 If the World Series goes beyond five games, the East Coast and mid-West can see the remaining contests. A linking of the two networks by stratovision plane would provide the largest audience in television broadcast history. 1959 Washington Post 24 Dec. A1/7 Four transmitters on the Purdue campus will send two simultaneous programs to the plane, which will travel in a 10-mile circle 23,000 feet above Montpelier... The system, known as 'stratovision', is designed to explore the value of the plane in reaching hundreds of thousands of students at one time. 1964 B. Grob Basic Television (ed. 3) v. 85 The UHF channels 66 to 83 are available for experimentation with this airborne television system called stratovision.

stratum. Add: Now usu. with pronunc. (strā·tə̆m, pl. strā·tă). **6. a.** (Earlier and later examples of social stratum.)

1886 T. H. S. Escott in P. Bailey Leisure & Class in Victorian England (1978) iii. 58 A social movement..is now steadily progressing on a lower social stratum. 1927 P. Sorokin Social Mobility 141 There has never existed a society in which vertical social mobility has been absolutely free and the transition from one social stratum to another has had no resistance. 1973 E. Berckman Victorian Album 50 This murder..must have been in too drab a social stratum..to attract even contemporary attention.

b. Earlier and later examples of social strata: cf. sense 6 a in Dict. and Suppl.

1890 W. James Princ. Psychol. I. iv. 121 Habit..keeps different social strata from mixing. 1937 R. H. Lowie Hist. Ethnological Theory vi. 57 It is hard to understand how Morgan could have missed the social strata of the caste-ridden Oceanians. 1968 G. P. Mitchell Dict. Sociol. 182 In theory social strata are made up of individuals and families.

c. Statistics. Each of the groups into which a population is divided in the technique of stratified sampling.

1920 A. L. Bowley Elements of Statistics (ed. 4) ii. iv. 332 It may happen..that the universe consists of different regions or strata in which the chances are different, and the question arises whether we should proceed at random.., or..partially arrange the choice so as to take the same proportion out of each region or stratum. 1952 A. Hald Statistical Theory with Engin. Applications xvii. 495 In sampling investigations of industrial products stratified sampling is often useful. For example, when a lot is being loaded a random sample of items may be taken from every truckload, the truckloads being the strata. 1960 Jrnl. Amer. Statistical Assoc. LV. 105 If the sample is allocated to the strata in proportion to the number of elements in the strata, it is virtually certain that the stratified sample estimate will have a smaller variance than a random sample of the same size. 1980 Hawkins & Weber Statistical Analysis xi. 295 Stratified sampling is appropriate only if the variable of interest is relatively homogeneous within strata and heterogeneous among strata.

¶ **d.** The form strata used as sing., with pl. stratas.

1937 Times Lit. Suppl. 27 Nov. 910/2 He marries a penniless girl of a lower social strata than himself. 1946 M. Peake Titus Groan 159 The mixture of cunning and honesty which he did not yet perceive to be a still deeper strata of Steerpike's cleverness. Ibid. 177 The abstract language with which they communicated their dizzy stratas of conjecture. 1971 Timber Trades Jrnl. 14 Aug. 20/2 The era of the family business is now a complex structure of top, middle and lower stratas. 1980 Good Housekeeping Nov. 181/4 After this comes a strata of accessories.

7. Delete ? Obs. **a.** (Later examples.)

1955 T. H. Pear Eng. Social Differences i. 23 Within the wholesale trade there are interesting strata-differences. 1962 Conveyancing (Strata Titles) Act in Statutes of New S. Wales 1961 129 'Strata plan' means a plan which..shows the whole or any part of the land comprised therein as being divided into two or more strata, whether or not any such stratum is divided into two or more lots. 1977 [see b below]. 1982 Polit. Sci. Q. XCVII. 482 The manipulation of ethnicity or of strata-local forces.

b. (in sing. form.)

1955 T. H. Pear Eng. Social Differences vi. 144 Neighbouring families for whom class-consciousness..was usually submerged below stratum-consciousness. 1977 Dædalus Fall 73 Country-wide strata consciousness and a less pronounced stratum formation of ethnic, religious, or regional elements.

c. Special Combs.: strata-bound a., confined to a single stratum or group of strata; strata title Austral. and N.Z., the freehold or leasehold of or title to a stratum or storey (or more than one) of a building.

1962 Econ. Geol. LVII. 272 It is..reasonable to expect that a majority of strata-bound ore fields will be readily explainable on explicit grounds of tectonic history. 1979 Nature 15 Nov. 247/1 Uranium mineralisation, for the most part, is strata-bound and occurs in breccia matrix and vugs. 1962 Statutes of New S. Wales 1961 128 (title) Conveyancing (strata titles) act. 1977 Courier-Mail (Brisbane) 7 Apr. 17/2 Many strata title (home units) property owners risk serious financial loss because of inadequate legal insurance on their units. 1977 N.Z. Herald 5 Jan. 2-16/2 (Advt.), Home unit, Avondale, as new, strata title, 2 brms, internal garage.

Straussian (strau·siăn), a. and sb. [f. the name of (1) the German composer Richard Strauss (1864–1949), or (2) the Viennese family of composers of whom Johann Strauss II (1825–99) was the foremost member + -ian.] **A.** adj. **1.** Of, pertaining to, or characteristic of the music of Richard Strauss.

1910 G. B. Shaw in Nation 19 Mar. 969/2 To those of us who are neither deaf nor blind nor anti-Straussian critics (which is the same thing), she was a superb Elektra. 1949 A. Huxley Ape & Essence ii. 24 Flawlessly pure of all Wagnerian lubricity and bumptiousness, all Straussian vulgarity! 1954 Grove's Dict. Mus. (ed. 5) VIII. 129/1 He wrote..some fine songs for voice and orchestra,..of which 'Hymnus' and 'Gesang der Apollopriesterin' (Op. 33) are the most characteristic of Straussian rapture. 1963 Times 8 Mar. 15/3 Earlier in the work he had not shown the same ability to sustain a true Straussian intensity of feeling. 1979 Guardian 26 Mar. 14/5 It was no disappointment to find Italian verismo rather than Straussian opulence.

2. Of, pertaining to, or characteristic of the music of the Strauss family.

1935 Punch 8 May 561/2 The scheme and time of the play..has prevented him from suggesting or—except in a brief snatch of the Blue Danube—borrowing from the glories of the Straussian epoch. 1958 Listener 21 Aug. 285/1 The commonplace of a Straussian waltz.

B. sb. An admirer of Richard Strauss; an exponent of his music.

1959 Times 17 Nov. 16/4 With that renowned Straussian, Mr. Rudolf Kempe, now..available to take charge, the London Symphony Orchestra assembled all its forces on Sunday..so that the second half of the programme could be devoted to Also sprach Zarathustra. 1967 Guardian 13 June 7/3 Straussians all think it [sc. Die Frau ohne Schatten] the masterpiece. 1977 Gramophone June 64/1 A lack of sensuousness and ardour in the first half of the work is something which many Straussians won't readily forgive.

stravaig, v. For Sc. and north. read Sc., north, and Ir. and add further examples. Also in general literary use, and occas. trans. with road as obj.

1831 S. Lover Legends & Stories of Ireland p. xxiv, Stravaig, to ramble. 1876 C. M. Yonge Three Brides I. xvi. 279 Her own children, which it is a shame to see stravaging about the place! 1929 E. Bowen Last September i. viii. 93 They do be stravaging about always and not contented at all. 1934 T. E. Lawrence Let. 23 Nov.

(1938) 830 Visit it, sometime, if you still stravage the roads of England in a great car. **1958** S. Beckett *From Abandoned Work* 14, I might be sprawling in the sun now sucking my pipe..wondering what there was for dinner, instead of stravaging the same old road in all weathers.

Stravinskian (străvi·nskiăn), *a.* (and *sb.*). Also **Stravinskyan.** [f. the name of *Stravinsky* (see below) + -AN.] Of, pertaining to, or characteristic of the Russian-born composer Igor Fyodorovich Stravinsky (1882–1971) or his music. Also as *sb.* = *STRAVINSKYITE.

1925 F. Toye *Well-Tempered Musician* iii. 74 Stravinsky himself was put up to defend Tchaikowsky, at whom all the Stravinskians..had been constantly sneering for years. **1947** D. Milhaud in *Stravinsky's Poetics of Music* p. xi, The *Poetics of Music* brings to light the indissoluble relationship between the two aspects of the Stravinskyan temperament: that is, his music and his philosophy. **1958** *Times* 22 Feb. 3/7 Wishart had absorbed his Stravinskian predilection. **1962** *Times* 29 Mar. 8/5 Sir William, whose attitude to critics is Stravinskyan..is ready to forecast critical reactions to his work. **1968** *Listener* 1 Aug. 153/2 Britten's exuberant cantata..is informed by a Stravinskian economy of gesture and dramatic style. **1978** *Gramophone* July 174/1 The second [movement is] a sonata—very Stravinskian yet it could not be by Stravinsky but only by Malcolm Williamson.

So **Stravi·nskyite,** a devotee of Stravinsky's music.

1924 C. Gray *Survey Contemp. Music* 132 The devotees of the Russian ballet, the Stravinskyites, seek the satisfaction of normal human activities in art. **1949** G. F. Knight in *Penguin Music Mag.* IX. 82 The majority of the musical world divided into two irreconcilable camps: the Stravinskyites and the Anti-Stravinskyites. **1961** *Times* 12 Apr. 6/5 Elsewhere all praise—and a rap on the knuckles for all those Stravinskyites who stayed at home.

straw, *sb.*¹ Add: **I. 1. d.** The colour of straw, a pale brownish-yellow.

1799 in M. Edgeworth *Parent's Assistant* (1800) (ed. 3) VI. 119 Mr. Davis, slate-color and straw. **1897** *Sears, Roebuck Catal.* 231/2 Silk Mitts..in the following colors: ..sky blue, lemon, straw, cardinal. **1923** *Daily Mail* 19 Feb. 5 A full range of new colourings, including Peach, Lemon, Straw, Rose. **1942** *R.A.F. Jrnl.* 3 Oct. 15 A heavy, oily liquid, from straw to black in colour. **1978** A. S. Byatt *Virgin in Garden* xi. 111 Red was defiance, gold avarice, straw plenty. Green was hope, but sea-green was inconstancy.

3. *poppy straw*: see *POPPY *sb.* 8.

II. 5. g. (Earlier and later examples.) Now usu. made from paper or plastic.

1851 *London at Table* iii. 52 *Mississippi Punch.* Let them use a glass tube or straw to sip the nectar through. **1860** Bartlett *Dict. Americanisms* (ed. 3) 90 *Cobbler*,..a drink made of wine, sugar, lemon, and pounded ice, and imbibed through a straw or other tube. **1888** Ruskin *Præterita* III. ii. 57, I..saw the Bishop of Oxford taught by Sir Robert Inglis to drink sherry-cobbler through a straw. **1926** 'O. Douglas' *Proper Place* xxxi. 286 She.. soon had Alistair supremely happy drinking lemonade through a straw. **1926** [see *soda straw* s.v. *SODA¹ 7]. **1953** Dylan Thomas in *Listener* 17 Sept. 459/2 They gave him a bottle with a straw. **1967** R. A. Waldron *Sense & Sense Devel.* vi. 116 A drinking-*straw* is nowadays usually made of plastic. **1982** H. Engel *Ransom Game* viii. 45, I settled for a vanilla shake... The straw stood up unaided in..the froth.

h. Used as a means of deciding something by chance (lit. by choosing the shortest (or longest) from among several straws held so as to conceal one end); phr. *to draw a straw* or *straws*, to draw a lot or lots.

1832 [see Draw *v.* 34]. **1939** Wodehouse *Uncle Fred in Springtime* i. 13 It was the person on whom life had thrust the..task who must be considered to have drawn the short straw. **1959** R. Bradbury *Day it rained Forever* 47 Sundays we draw straws for who wears the suit the extra night.

8. c. (Earlier and later examples.) Now esp. *a straw in the wind* (usu. in *pl.*).

a **1654** J. Selden *Table-Talk* (1689) 31 Take a straw and throw it up into the Air, you shall see by that which way the wind is. **1799** W. Cobbett *Porcupine's Works* (1801) X. 161 'Straws' (to make use of Callender's old hackneyed proverb) 'Straws serve to show which way the wind blows.' **1823** Byron *Don Juan* xiv. viii, You know or don't know that great Bacon saith 'Fling up a straw, 'twill show the way the wind blows.' **1927** A. Adams *Ranch on Beaver* vii. 99 'As straws tell which way the wind blows,' remarked Sargent, 'this day's work gives us a clean line on these company cattle.' **1939** Madge & Harrisson *Britain, by Mass-Observation* ii. 107 Yet through agents in the constituencies, and straws in the wind like West Leicester, came a slightly better indication of popular sentiment. **1960** C. P. Snow *Affair* xxv. 334 There have been other things, straws in the wind, maybe, which give reason to think that contemporary standards among a new scientific generation are in a process of decline. **1975** *Language for Life* (Dept. Educ. & Sci.) xii. 189 These are straws in the wind. What they indicate is the degree to which learning and the acquisition of language are interlocked. **1983** *Listener* 27 Jan. 3/1 As MPs have already pointed out in the debate, Captain Nick Barker of HMS *Endurance* had detected straws in the wind.

e. *to have straws in one's hair* (and varr.): to be insane, eccentric, or distracted.

1890 'L. Carroll' *Nursery 'Alice'* x. 39 That's the March Hare, with the long ears, and straws mixed up with his hair. The straws showed he was mad—I don't know why. Never twist up straws among *your* hair, for fear people should think you're mad!] **1923** Wodehouse *Inimitable Jeeves* vii. 72 When your uncle the Duke begins to feel the strain a bit and you find him in the blue drawing-room sticking straws in his hair, old Glossop is the first person you send for. **1925** —— *Carry On, Jeeves!* vi. 142 His [*sc.* a psychiatrist's] outlook on life has become so jaundiced through constant association with coves who are picking straws out of their hair. **1937** D. L. Sayers *Busman's Honeymoon* xviii. 346 (*heading*) Straws in the hair. **1962** 'S. Woods' *Bloody Instructions* ix. 100 Dennis Dowling..brought with him an atmosphere of mingled drama and insanity. Antony thought: 'definitely straws in the hair' as soon as he opened the door.

10. e. *potato straw*: see *POTATO *sb.* 7.

f. A plastic phial in which bull semen is stored for artificial insemination.

1966 *Canad. Jrnl. Compar. Med. & Vet. Sci.* XXX. 109 The use of plastic straws would..encourage volume storage of high quality semen from young sires. *Ibid.* 111/1 Better fertility results can be anticipated with straw packaged semen as compared with that packaged in glass ampoules. **1982** *Sunday Times* 12 Sept. 45/2 The firm..specialises in artificial insemination..in cattle, and expects Pickles [*sc.* a bull] eventually to produce 40,000 'straws', or phials, of semen a year. These straws will be frozen, and sold to cattle breeders all over the world at about £50 a time.

III. 11. (Earlier examples.)

1829 P. Egan *Boxiana* 2nd Ser. II. 681 *Hall*..went briskly into the ring, and tossed up his Dunstable straw. **1849** *Theatr. Programme* No. 5. 45/2 (Advt.), Charles Vyse, manufacturer of Leghorns and Straws to the British and Foreign Courts.—30 Ludgate-street.

IV. 13. a. *straw pulp.*

1888 Cross & Bevan *Text-bk. Paper-Making* vi. 101 The presse-pâte system is largely adopted for straw pulp. **1937** E. J. Labarre *Dict. Paper & Paper-Making Terms* 238/2 Straw pulp is prepared by cooking straw with soda.

b. *straw-clutching.*

1962 L. Davidson *Rose of Tibet* iii. 65 Every bit of straw-clutching, every bit of hope..was followed instantly by a reaction of dismay.

c. *straw-bottomed* ppl. adj.

1749 Smollett tr. *Le Sage's Gil Blas* (1750) II. iv. xi. 137 We quitted the hermitage, leaving..two old straw-bottomed chairs.

14. straw ballot = *straw vote*; **straw basher** *slang*, a straw hat or boater; **straw-blond(e** *a.*, applied to hair of a pale, yellowish blond colour; also *absol.*, this colour; **straw-board** (earlier and later examples); also, a piece of this material; **straw boss** orig. *U.S.*, a subordinate or assistant foreman; **strawboy** *Ir.* (see quots.); **straw braid** (earlier example); **straw-bug** *slang*, a strawberry; **straw-dry** *a.*, as dry as straw, very dry; **straw-foot**: see *HAY-FOOT; **straw-gold,** the colour of straw; = sense 1 d above; **strawline,** a light rope used to pull a heavier one into position, esp. in Logging; **straw-man,** (*b*): delete *U.S.* and add examples; **straw-pale** *a. rare*⁻¹, as pale as straw; **straw poll** orig. *U.S.* = *straw vote*; **straw potatoes,** very thinly cut potato chips; **straw tick** *U.S.* [Tick *sb.*²], a straw-filled mattress; **straw vote**: for *U.S.* read orig. *U.S.* (earlier and later examples).

1932 Straw ballot [see *straw poll* below]. **1967** *Canad. Ann. Rev. 1966* 63 RIN..polled 27·7 per cent of the vote in a Université de Montréal straw ballot. **1901** Straw basher [see *BASHER²]. **1931** A. J. Cronin *Hatter's Castle* ii. xii. 421 A stiff, board-like straw-basher. **1928** E. O'Neill *Strange Interlude* i. 25 Her straw-blond hair, framing her sunburned face, is bobbed. **1973** A. Hunter *Gently French* v. 47 Her hair was a warm straw blonde. **1850** *Rep. Commissioner Patents 1849* (U.S. Patent Office) 305 [The] said process is peculiar to the use of strawboard. **1862** *Harper's Mag.* June 135/1 He was making a personal examination of straw-board shoes provided for those who have gone to be soldiers. **1885** G. F. Green in Rattray & Mill *Forestry & Forest Products* xviii. 474 Wood-pulp boards, straw-boards, and mill-boards are sometimes referred to as 'paste-boards'. **1956** H. Williamson *Methods Bk. Design* xix. 321 Millboards are harder and more solid than strawboards. **1894** W. H. Carwardine *Pullman Strike* ix. 117 These employees..had been so ground between the upper millstone of 'low wages' and the nether millstone of 'high rents', the continued oppression of the 'straw bosses', [etc.]. **1915** S. Lewis *Trail of Hawk* ii. xiii. 132 He had laughed away the straw boss who tried to make him go ask for a left-handed monkey-wrench. **1945** 'N. Shute' *Most Secret* viii. 172 Them Frenchies won't work right without they have a straw-boss. **1976** L. St. Clair *Fortune in Death* x. 98 Dime-stores, cafeterias, moving to a new job..every time some greasy straw boss ran his hand up my skirt. **1894** C. R. Browne in *Proc. R. Irish Acad.* 3rd Ser. III. 352 Mr. Michael Lavelle..informs me that he has heard that sometimes, on the occasion of a wedding, 'straw-boys' go round with long straw masks on, and if they do not get either money or liquor will threaten to break the windows and furniture of the house. **1937** C. M. Arensberg *Irish Countryman* iii. 106 The 'strawboys'—privileged masqueraded figures whose mock-dangerous invasion of the wedding feast has been dignified to represent a last remnant of a primeval bride-capture. **1968** A. Gailey in *Folk Life* VI. 90 In parts of Fermanagh there survive even to the present day traces of an old ceremony performed by groups of..young men, disguised latterly in..straw masks.., but in former times..wearing complete suits of straw. They interrupted the festivities following the solemnisation of marriages in the country districts, and were known simply as the Strawboys. **1864** *Harper's Mag.* Oct. 578/2 He laid all kinds of evil results at the door of straw braid. **1908** A. Huxley *Let.* 29 June (1969) 28 Latest News Stop Press Strawbugs for tea. **1959** I. & P. Opie *Lore & Lang. Schoolch.* ix. 155 These syllables [*sc.* -*bug*, -*gog*, etc.] are used..to replace the second half of a word, as: newbug, rasbug, strawbug, goosegog, and wellygogs. **1951** W. de la Mare *Winged Chariot* 47 Unlike the plant called 'everlasting', this [*sc.* poetry] Never straw-dry, sapless, or sterile is. **1963** *Glamour* Nov. 23 Even hair that's straw-dry turns silky. **1963** A. Lubbock *Austral. Roundabout* 3 Here are fine expenses of pasture, turning to straw-gold in summer. **1977** J. Aiken *Last Movement* i. 20 Her hair had been..a pale Scottish straw-gold. **1956** *Amer. Speech* XXXI. 152 *Strawline*,..a small-size wire rope which hauls the heavy logging cables into position. **1975** *Islander* (Victoria, B.C.) 22 June 7/4 A strawline was taken across the river by boat, then each cable was pulled to the other side by the horses. **1896** L. T. Hobhouse *Theory of Knowl.* 59 The straw man was easily enough knocked over by the critic who set him up. **1934** A. Woollcott *While Rome Burns* 76, I have often challenged one of these straw-man authorities. **1946** Koestler *Thieves in Night* 328 The authorities..only got the Rumanian captain and his crew, who couldn't give away much as all their dealings had been with straw men under assumed names. **1981** 'M. Hebden' *Pel is Puzzled* xviii. 180 He seemed active enough, but there seemed an awful lot lacking in him... Was he really just a straw man? **1922** W. B. Yeats *Seven Poems* 13 Under the shadow of stupid straw-pale locks. **1932** C. E. Robinson *Straw Votes* iv. 52 The newspaper or magazine conducting a straw poll by the ballot-in-the-paper method prints a straw ballot in the publication for a certain period of time before an election. **1944** *Chicago Tribune* 26 Oct. 12/2 (*heading*). New deal area lifts F.D.R. in N.Y. straw poll. **1958** *Spectator* 6 June 722/1 In my own straw poll I found two electors who were going to vote Liberal for the first time. **1978** *Nature* 6 Apr. 484/3 A straw poll taken three weeks ago at a meeting of faculty professors..voted 23 to 3 against approving the proposal. **1904** C. H. Senn *New Cent. Cookery Bk.* (rev. ed.) 596 Pommes Pailles (Straw Potatoes). **1959** *Times* 6 Apr. 13/5 Serve with sweet corn and straw potatoes. **1931** *Amer. Speech* VII. 169 Most of these [mattresses and ticks] were filled with corn husks, straw or hay, and were called 'husk ticks', 'hay mattresses', and 'straw ticks'. **1949** L. I. Wilder *Long Winter* viii. 68 They must fill the straw ticks with hay, because there was no straw in this new country. **1954** W. Faulkner *Fable* 195 He was sleeping on a straw tick in the lodge room over the store. **1866** *Cleveland* (Ohio) *Leader* 6 Oct. 4/2 A straw vote taken on a Toledo train yesterday resulted as follows; A. Johnson 12; Congress, 47. **1887** *San Francisco Thunderbolt* 4 Nov. 1 The straw vote taken at the 'Report' office is unreliable. **1977** R. Holland *Self & Social Context* v. 175 A special session on legal registration produced a straw vote which revealed an even balance of viewpoint.

strawberry. Add: **I. 1. b.** A sea anemone, probably a variety of *Actinia equina*, the body of which resembles a strawberry.

1856 G. Tugwell *Man. Sea-Anemones* ii. 33 Here is 'the strawberry', whose body is mottled with red and green, after the fashion of that pleasant fruit. **1856** Geo. Eliot *Jrnl.* 8 May–26 June in *Lett.* (1954) II. 243 It was a crescendo of delight when we found a 'Strawberry'..in a low tide pool. **1971** *Oxf. Bk. Invertebrates* 14/1 Particularly common is the 'strawberry' variant [of the beadlet anemone] in which the body is crimson with green spots.

c. The fruit of certain seaweeds resembling a strawberry. (Perhaps a misunderstanding by Kipling.)

1897 Kipling *Capt. Cour.* iii. 53 The hook had fouled among a bunch of strawberries, red on one side and white on the other.

4. (Later examples.)

1897 *Sears, Roebuck Catal.* 222/1 They [*sc.* scarves] consist mostly of combination colors, just a few of which are blue, lavender, light green, cherry, strawberry, [etc.]. **1922** Joyce *Ulysses* 492 A blond feeble goosefat whore in a tatterdemalion gown of mildewed strawberry lolls spread-eagle in the sofacorner. **1954** [see *ALIZARIN]. **1974** *Harrod's Xmas Catal.* 19 Luggage..in colourful Vinyl: light tan, orange, blue, or strawberry.

5*. Applied to things resembling a strawberry in shape or colour. **a.** An emery bag in the shape of a strawberry.

1903 K. D. Wiggin *Rebecca of Sunnybrook Farm* vi. 66 She polished her needles to nothing, pushing them in and out of the emery strawberry. **1937** [see *emery cushion* s.v. *EMERY *sb.* 3]. **1976** P. Clabburn *Needleworker's Dict.* 99/1 *Emery cushion* (emery bag, emery ball, strawberry), small pincushion, often in the shape of a strawberry, which is filled with emery powder... If needles become damp and rusty they are run through the cushion to make them shiny and smooth again.

b. A sore or bruise, esp. one caused by friction with the ground. *N. Amer. colloq.*

1921 *Daily Colonist* (Victoria, B.C.) 13 Oct. 11/4 'Strawberry', or open sore, on his hip, caused by sliding bases and constantly reopened. **1937** *Pittsburgh Press* 11 Jan. 27/5 Here are some expressions commonly used..: *Strawberry*, a bruise from sliding [etc.]. **1981** *Washington Star* 19 Mar. D 1 'Look at that,' he said, hitching up his knickerbockers to reveal matching strawberries just above both knees, red and angry-looking. At least, they used to call them 'strawberries'. He still does.

c. A nose having the colour of a strawberry, esp. as the result of heavy drinking.

1949 Partridge *Dict. Slang.* (ed. 3) Addenda 1188/1 *Strawberry*,..a red nose: Cockney's. **1980** C. Smith *Cut-Out* ix. 62 His nose..had turned..to the characteristic boozer's strawberry.

II. 6. Also *strawberry pink, red.*

1939 Joyce *Finnegans Wake* 207 The lellipos cream to her lippeleens and the pick of the paintbox for her pommettes, from strawbirry reds to extra violates. **1952** A. G. L. Hellyer *Sanders' Encycl. Gardening* (ed. 22) 113 [*Cirrhopetalum*] *Amesianum*, old gold, single haired, lower sepals strawberry-red. **1956** G. Durrell *My Family & Other Animals* ii. 28 (*heading*) The strawberry-pink villa.

7. a. *strawberry bed* (later examples), *box* (also *fig.*).

1787 J. Woodforde *Diary* 15 Oct. (1926) II. 352, I was very busy this morning in my Garden making some new Strawberry Beds. **1892** W. B. Yeats *Countless Kathleen* ii. 34 My asparagus and strawberry beds Are trampled into clauber. **1936** 'R. Hyde' *Passport to Hell* vi. 93 Life just one strawberry-box after another. **1951** *Dict. Gardening* (R. Hort. Soc.) IV. 2042/1 An infected Strawberry bed should be cleared by burning all the plants. **1965** G. McInnes *Road to Gundagai* iii. 35 All about us they [*sc.* passengers] were vomiting into 'strawberry boxes'.

b. *strawberry cream* (earlier example), *shortcake.*

1792 J. Woodforde *Diary* 26 June (1927) III. 359 Dinner..a very fine Leveret rosted, Strawberry Cream, Jelly. **1841** L. B. Swan *Jrnl.* 20 June (1904) 28 We had a new dish, 'Strawberry Short Cake' very fine indeed. **1953** G. W. Brace *Spire* xi. 93, I had strawberry shortcake... Hot biscuits, yellow cream, and quite often wild berries.

c. *strawberry-breasted.*

1878 G. M. Hopkins *Poems* (1967) 77 Star-eyed strawberry-breasted Throstle.

8. strawberry bass (earlier and later examples); **strawberry blite,** substitute for def.: a herb, *Chenopodium capitatum*, with triangular leaves and heads of small flowers followed by fruit resembling a strawberry; (later examples); **strawberry blond(e)** *a.*, applied to hair of a light reddish blond colour; as *sb.*, the colour itself; a person with hair of this colour; **strawberry dish** *Silver-work* (see quot. 1977); **Strawberry Fields** *slang* [prob. f. *Strawberry Fields Forever*, title of a song (1967) by John Lennon and Paul McCartney] = *LSD²; **strawberry guava,** a shrub or small tree, *Psidium cattleianum*, of the family Myrtaceæ, native to tropical America and bearing white flowers and large edible berries; also, the red or yellow fruit of this tree; **Strawberry Hill** *Archit.*, the name of the house in Twickenham bought in 1747 and rebuilt by Horace Walpole after the Gothic style, used *attrib.* to designate the style of early Gothic Revivalist architecture inspired and epitomized by this house; **strawberry perch** (earlier example); **strawberry pot,** a large garden pot with pockets in its sides, designed to contain growing strawberry plants; **strawberry tomato,** substitute for def.: *U.S.*, a ground-cherry of the genus *Physalis* or its edible fruit; (earlier and later examples); **strawberry tree,** (*b*) add: also = Madroño; (later examples); (*c*) (earlier and later examples); **strawberry weevil,** a small black and white beetle, *Anthonomus signatus*, found in eastern North America, where it lays its eggs in strawberry buds, so that no fruit is formed.

1867 T. F. De Voe *Market Assistant* 294 Calico bass, speckled bass, or partridge-tailed bass.—This fish is also known among our fishermen as the 'strawberry bass'. **1947** B. W. Dalrymple *Panfish* 81 You'd think there be Strawberry Bass..in there. **1900** L. H. Bailey *Cycl. Amer. Hort.* I. 290/2 The common Strawberry Blite..has been introduced to the trade as a pot-herb. **1943** Fernald & Kinsey *Edible Wild Plants* iii. 180 The Strawberry-Blite, one of the most striking plants of Canadian clearings, on account of its masses of brilliant red pulpy fruits, may be used as a potherb like spinach. **1970** *Beaver* Winter 23 Strawberry spinach, also known as ..strawberry blite, is similar to its close cousin, lamb's quarters. **1884** E. W. Nye *Baled Hay* 98 That is what is ..sprinkling my strawberry blonde hair with gray. **1887** *Courier-Jrnl.* (Louisville, Kentucky) 6 Feb. 12/2 Seventeen young women, with hair ranging from strawberry blonde to deep crimson, are seated..on a long platform. **1895** Palmer & Ward *Band played On* (song) 4 Casey would waltz with a strawberry blonde, And the Band played on. **1958** *Daily Express* 17 Mar. 1/4 An unassuming strawberry blonde. **1977** B. Bainbridge *Injury Time* ii. 19 Alma's hair, rinsed to an unusual shade of strawberry blonde. **1941** *Burlington Mag.* Aug. 68/1 A set of four strawberry dishes—also silver-gilt, by Paul Crispin, 1734. **1977** Fleming & Honour *Penguin Dict. Decorative Arts* 766/1 *Strawberry dish*, a type of late C17–C18 English silver dish, deeper but no larger than a plate, rather thin with punched decorations in the C17, more substantial and usually with a scalloped rim in the C18. It is improbable that such dishes were used originally only for strawberries. **1971** *Tel.* (Brisbane) 27 Oct. 3/1 A youth had seven tablets of LSD, known as 'Strawberry Fields', when picked up in a city hotel by detectives. **1976** H. Ferguson *Confessions Long Distance Acid Head* 22 Then came the fatal trip which led..to my fleeing to India to forget. I was on Blue Cheer, I think, though it could have been Strawberry Fields. **1901** L. H. Bailey *Cycl. Amer. Hort.* III. 1460/2 Strawberry Guava. Shrub or small tree, 10–20 ft. high... Pulp fleshy, soft and juicy, purplish red next the skin,..sweet and acid, with a strawberry-like fragrance and flavor. **1976** *Monitor* (McAllen, Texas) 7 Nov. 1c/1 (*caption*) The strawberry guava, a shrublike tree, produces a fig-like fruit almost cherry-sized that can be eaten raw and can also be made into jellies. **1836** R. Griffin *Hist. Audley End* v. 127 The chapel..was newly fitted up..according to the fashion of the day, with..clustered pilasters, and a groined ceiling, in the style called after its patron, *Strawberry Hill Gothic*. **1891** T. G. Bonney in *Hist. Houses of United Kingdom* 90 A room..now serves as a chapel... It is..a specimen of Strawberry Hill Gothic. **1928** A. Huxley *Point Counter Point* xix. 344 The fantastic towers and pinnacles of Gattenden Castle, built..in the most extravagant style of Strawberry Hill Gothic. **1977** *Times* 6 Aug. 3/3 The eleventh duke [of Norfolk] rebuilt it [*sc.* Arundel Castle] in the Strawberry Hill baronial fantasy style. **1877** C. Hallock *Sportsman's Gazetteer* 378 Strawberry Perch. ..*Pomoxys hexacanthus*. **1946** M. Free *All about House Plants* ix. 67 'Strawberry pots' made of earthenware.. are much used, suitably planted, for patio decoration in California. **1977** Jellicoe & Allen *Town Gardens to live In* xi. 125/2 (*heading*) A handmade strawberry pot. It will take twenty-eight plants. **1862** M. D. Colt *Went to Kansas* ix. 133 The strawberry tomatoes..are indigenous to the soil. **1919** E. L. Sturtevant *Notes on Edible Plants* 432 *P. lanceolata*..was among the strawberry tomatoes grown at the New York Agricultural Experiment Station in 1886. **1969** *Oxf. Bk. Food Plants* 126/2 The ground cherry, which is also called 'Strawberry Tomato'.., is an annual, native in parts of eastern and central North America. **1792** A. Menzies *Jrnl.* 2 May in *Menzies' Jrnl. of Vancouver's Voy.* (1923) 20 The Oriental Strawberry Tree..at this time a peculiar ornament to the Forest by its large clusters of whitish flowers & ever green leaves, but its peculiar smooth bark of a reddish brown colour will at all times attract the Notice of the most superficial observer. **1813** H. Muhlenberg *Catalogus Plantarum Americæ Septentrionalis* 25 *Euonymus Americanus*, (burning bush, strawberry tree). **1838** J. C. Loudon *Arboretum & Fruticetum Britannicum* II. 1117 The Arbutus, or Strawberry Tree...Robust evergreen shrubs, or low trees. **1866** *Trans. Illinois Agric. Soc.* VI. 391 The Strawberry Tree, with its delicate foliage, green wood and beautiful berries. **1975** D. McClintock *Wild Flowers of Guernsey* 155 The Strawberry Tree..has established itself in Jersey. **1884** *Ann. Rep. Michigan State Hort. Soc. 1883* 155 This strawberry weevil..was described by Thomas Say. **1929** Auchter & Knapp *Orchard & Small Fruit Culture* xiv. 486 Strawberry Weevil..lays its eggs in the flower buds and then girdles the stem so as to prevent further development. **1976** *Islander* (Victoria, B.C.) 30 May 10/2 Another expense was combating the strawberry weevil.

strawflower (strǭ·flauᵊɹ). *N. Amer.* [f. Straw *sb.*¹ + Flower *sb.*] = Everlasting *sb.* 4; esp. *Helichrysum bracteatum*, a perennial herb of the family Compositæ, native to Australia, whose variously coloured flowers are often dried and preserved.

1924 L. H. Bailey *Man. Cultivated Plants* 786 Strawflower..frequently grown as an 'everlasting'. **1933** L. Bloomfield *Language* x. 160 The *straw-* in *strawberry* is phonetically the same as the *straw-* in *strawflower*. **1972** F. Mason *Roads to Liberty* 10 She had tucked a pair of blue-dyed strawflowers into the dark hair above her ear.

straw hat. Add: Also **strawhat.** **2.** *U.S. Theatr.* A summer theatre; a theatre operating during the summer only and presenting various productions or companies. Freq. *attrib.*

1935 *Variety* 12 June 62/4 (*heading*) Nearly 100 straw hat troupes will dot eastern landscape. **1936** A. Green in *Esquire* Sept. 160/3 A summer stock (legit) is a 'straw hat'. **1946** *Life* 5 Aug. 81/1 More than 125 straw-hat theaters now adorn the eastern sea-coast. **1948** *Sun* (Baltimore) 24 June 16/1 They appeared in the same roles last year on the strawhat circuit. **1952** *Ibid.* 10 June (B* ed.) 10/1 The play is not the thing this week at the Maryland strawhat. **1968** *N.Y. Times Bk. Rev.* 23 June vii. 5/4 They finagle their way into the strawhat dramatic workshop of a famous theater guru.

Hence **straw-hatter,** a straw-hat theatre, or play presented in one; **straw-hatting** *vbl. sb.*, work in straw-hat theatres.

1949 *N.Y. Times Mag.* 21 Aug. 24/4 The trick of operating a successful straw-hatter is to build up a steady clientele—a sizable number of people who get the habit of regular attendance because they have learned that the general average of production is good. **1954** *Wall St. Jrnl.* 4 Aug. 9 The Howard Lindsay-Russell Crouse straw hatter ('Life with Father') subsequently made Broadway. **1950** *Richmond* (Va.) *Times-Dispatch* 9 Dec. 5/1 Straw-hatting is arduous.

strawish, *a.* For †*Obs.* read *rare* and add later examples.

1598 J. Florio *Worlde of Wordes* 253/1 *Pagliaresco*, made of strawe, strawish. **1978** B. Freemantle *Clap Hands* viii. 50 Thinning strawish hair... A very ordinary sort of man.

strawy, *a.* **4.** (Later example.)

1957 T. Hughes *Hawk in Rain* 20 With love so like fire they dared not Let it out into strawy small talk.

straying, *vbl. sb.* **b.** For †*Obs. rare*⁻¹ read Now only *arch.* and add later examples.

1884 *Eng. Illustr. Mag.* Dec. 152/2 Thoughts that had gone astraying half across the globe. **1936** Auden *Look, Stranger!* 56 And Garbo's and Cleopatra's wits to go astraying.

stray-running, *ppl. a. rare*⁻¹. [f. Stray *a.* or *v.*² + Running *ppl. a.*] That runs astray.

1914 D. H. Lawrence *Prussian Officer & Other Stories* 251 What am I frightened of him for? Why, for you, you stray-running little bitch.

streak, *sb.*¹ Add: **2. g.** (Examples.)

1892 G. M. Sternberg *Man. Bacteriol.* i. viii. 75 We commonly make a streak upon the surface of cooked potato or solidified blood serum in studying the development of various bacteria on these culture media. **1939** A. J. Salle *Fund. Princ. Bacteriol.* ix. 133 The last streaks should thin out the culture sufficiently to give isolated colonies. **1969** Sirockin & Cullimore *Practical Microbiol.* ii. 17 It will be observed that away from the initial streak, the growth is less dense and discrete colonies are present.

h. *Hairdressing.* A strand or strands of hair fashionably tinted, esp. in a light colour. Cf. *HIGH LIGHT 1 b.

1949 *Queen* 21 Dec. 39 Light streaks are again fashionable. These can be tinted to match an evening dress—the colour can be washed out next day. **1956** Ashley & Stevenson *Hair Design & Control* xi. 127 When introducing blonde streaks, or tipping, it assists to segregate the streak or section of hair concerned by means of a piece of cloth or Cellophane. **1966** J. S. Cox *Illustr. Dict. Hairdressing* 144/1 *Streaks*, light strands of hair deliberately contrived to enhance the appearance. **1979** R. Rendell *Make Death love Me* v. 40 She and Pam argued as to whether it was possible to put blonde streaks in one's hair at home.

3. c. *like a streak* (earlier examples). Also *quick as a streak* and *transf.*

1839 *Knickerbocker* XIII. 298, I see him yesterday afternoon..starting off like a streak, to go to Norridgewock. **1849** C. Lanman *Lett. from Alleghany Mts.* xi. 89 The water wheeled my head round to the hole, and in I went quick as a streak. **1887** M. E. Wilkins *Humble Romance* 376 He went past me like a streak when I was coming up the road. **1920** C. Sandburg *Smoke & Steel* 138 Maybe I will light out like a streak of wind.

d. *slang* (orig. *U.S.*). A rapid move; (a journey undertaken at) a fast rate. Also *fig.*

a **1861** T. Winthrop *John Brent* (1862) xxii. 243 She's got the old man to take care of and follow off on his next streak. **1865** A. D. Whitney *Gayworthys* 141 She's going a good streak, ain't she? **1875** J. G. Holland *Sevenoaks* iv. 60 We'll wopse 'im up in some blankits, an' make a clean streak for the woods. **1909** R. A. Wason *Happy Hawkins* 280 She was in the habit of estimatin' just how little nourishment it would take to run her to the next feed, gettin' it into her in the shortest possible time, an' then makin' a streak for it. **1960** *Twentieth Cent.* Dec. 556 His streak to stardom.

e. *to talk a streak*, to talk fast or constantly; *to talk a blue streak*: see *blue streak* (*b*) s.v. *BLUE *a.* 13.

1915 J. London *Jacket* v. 37 He sleeps most of the watch, and we can talk a streak. **1968** T. Wolfe *Electric Kool-Aid Acid Test* xxvii. 373 Robertson's talking a streak. It's a grand speech.

4. b. *colloq.* (orig. *Austral.*). A tall, thin person.

1941 Baker *Dict. Austral. Slang* 73 *Streak*, a tall, lean person. **1947** K. Tennant *Lost Haven* iii. 57 Thank goodness he hadn't told that long, gabbling streak about Cherry. **1959** I. & P. Opie *Lore & Lang. Schoolch.* ix. 169 There is a fusion of terms between those for the thin and lanky lad and those for the overgrown... Epithets include:..streak or streaker, Tower of London, walking barge pole, [etc.]. **1966** *Listener* 3 Mar. 317/1 That long streak of misery in a blue shirt.

6. b. (Earlier examples.) Also in phr. (*on*) *a losing* (or *winning*) *streak*, (experiencing) a series of losses (or wins). Hence, a series (of games, etc.) of a specified kind.

1843 *Knickerbocker* XXI. 303, I had 'struck a streak of bad luck'. **1865** 'Mark Twain' in Harte & 'Twain' *Sk. Sixties* (1926) 205 There never was a bad James in the Sunday-school books that had such a streak of luck as this sinful Jim. **1871** B. Harte *Luck of Roaring Camp* 34 We've had a streak of bad luck since we left Poker Flat. **1912** C. Mathewson *Pitching in a Pinch* xi. 233 But what's a new hat against a losing streak or a batting slump? **1950** *Daily Ardmoreite* (Ardmore, Okla.) 30 Apr. D. 6/2 Last year's edition of the Indians set one of the hottest paces in the league before folding with a long losing streak. **1963** A. Baron *Lowlife* xi. 107 The old song inside my head, *don't be mad, don't walk out of a winning streak*. **1967** *Boston Globe* 5 Apr. 51/5 It is also fair enough to figure the 76ers will end the Celtics' streak of eight straight National Basketball Assn. titles. **1968** *Globe & Mail* (Toronto) 3 Feb. 35/6 Toronto Marlboros snapped Montreal Junior Canadiens' unbeaten streak at 10 games. **1972** 'H. Carmichael' *Naked to Grave* xvi. 183 Mrs Davey won quite a lot of money... She said something to him about Mrs Davey's winning streak. **1973** G. Moffat *Deviant Death* ix. 125 The police were on a losing streak and they knew it. The questions were just form. **1980** *Times* 19 Feb. 11/5 Their winning streak includes the Boat Race record.

6*. Any of various virus diseases of plants which cause discoloured stripes to appear on their leaves; = *streak disease*, sense *7 below.

1930 *Discovery* June 196/1 Other important virus diseases of plants include..streak of maize,..and many others. **1936** J. Johnson in *Phytopathology* XXVI. 285 The writer has repeatedly observed a disease of tobacco in the field that is characterized by a necrosis of, or along, the veins of the leaf... The term 'tobacco streak' is.. proposed as perhaps the simplest and most suggestive for this malady. **1939** *N.Z. Jrnl. Sci. & Technol.* A. XX. 365 In New Zealand, pea-streak has not been observed other than at Palmerston North. **1952** Gram & Weber *Plant Dis.* iii. 387/2 Heavy dressings of stable manure make the symptoms of tomato streak worse. **1963** L.

Bos *Symptoms Virus Dis. in Plants* 39 The term 'streak', such as in 'cocksfoot streak' or 'pea streak', is confusive and insufficient. The name does not indicate whether the streak is chlorotic or yellow, such as in 'cocksfoot streak', or is necrotic, such as in 'pea streak'. **1977** J. Kranz et al. *Diseases, Pests & Weeds in Tropical Crops* 21 Maize streak has not been reported from Europe or the Americas.

7. streak camera, a camera which uses the principles of streak photography; also, an electron-optical analogue of this allowing the resolution of events of the order of a picosecond duration and used esp. in high-speed spectroscopy (see quots. 1973, 1977); **streak culture,** substitute for def.: a bacterial culture made by drawing the point of an infected needle or the like over the surface of a solid culture medium; (earlier and later examples); **streak disease** = sense 6* above; **streak photography,** a form of photography in which film is automatically and rapidly moved past the open shutter of a camera, allowing a one-dimensional record of high-speed events to be made which can be reconstituted optically; so **streak photograph; streak plate,** (*a*) *Bacteriology*, (a vessel containing) a streak culture; (*b*) *Min.*, a small tablet of unglazed porcelain on which minerals may be rubbed to ascertain the colour of the streak (sense 2 d); **streak virus,** a virus causing a streak disease in plants.

1962 *Sci. Amer.* May 102/2 For this purpose we use a streak camera. **1973** *Ibid.* June 60/2 In the streak camera, which has an electronic circuit fast enough to measure picosecond events, light from a slit is focused onto a cathode where electrons are released and accelerated towards a phosphorus substance, which emits light. A voltage increasing with time..streaks the electrons across the phosphor so that electrons released at earlier times appear at a different position on the phosphor than electrons released later. **1977** *Jrnl. R. Soc. Arts* CXXV. 772/2 Direct linear measurement of pulse durations by electron-optical streak-cameras, in which the time-into-space transformation is brought about by deflecting an electron-optical slit image across the output phosphor of an image-tube. **1892** G. M. Sternberg *Man. Bacteriol.* i. viii. 75 Koch made 'streak cultures' by drawing the point of a platinum needle, charged with bacteria, over the surface of a gelatin or agar plate. **1926** A. Bulleid *Text-bk. Bacteriol.* vii. 62 An examination of the culture tube with the naked eye will reveal the presence of 'colonies' on or in the medium, according to whether the culture was made on the surface (streak culture) or into the medium (stab or shake culture). **1923** W. F. Bewley *Dis. Glasshouse Plants* vi. 132 The organism from the tomato can cause a number of 'stripe' or 'streak' diseases of other plants. **1925** *Rep. Proc. Imperial Bot. Conf.* 132 (*heading*) Streak disease, an infectious chlorosis of sugar-cane. *Ibid.* 133 Streak disease in maize has been known in Natal for many years. **1938** *Jrnl. Agric. Res.* LVI. 747 A virus disease of peas.., manifested by a streaking of the stems and leaves and a spotting of the pods, was observed under greenhouse conditions..in the fall and winter of 1934. The disease.. resembles the streak disease described by Linford, in 1929, as occurring in pea fields throughout the United States. **1970** Liebscher & Koehler tr. *Fröhlich & Rodewald's Pests & Dis. Tropical Crops* 240 Leaves [of sugar cane] infected with streak disease exhibit light-coloured, short and long streaks along the veins. **1950** *Jrnl. Appl. Physics* XXI. 448/2 (*caption*) Streak photograph illustrating the motion of the platform of the instrument shock testing machine. *Ibid.* 445/1 Methods of streak photography are easy to perform and can often be done with ordinary laboratory equipment. **1952** G. A. Jones *High Speed Photogr.* ix. 181 Streak photography is mainly of value in the case of luminous objects. **1980** *Sci. Amer.* May 102/2 The course of growth over time also appears to be continuous, as can be seen in time-lapse motion pictures, multiple-exposure photographs and streak photographs. (A streak photograph is made with a camera in which the lens is left open and the film moves at a constant speed.) **1895** *Buck's Handbk. Med. Sci.* Suppl. 83/1 'Streak-plates' are made on gelatine and agar, after the medium has been poured in the plates and become solidified, by drawing an infected needle across them in four or five parallel courses. **1898** Brush & Penfield *Man. Determ. Mineral.* (ed. 15) v. 228 The streak of a mineral..may be quickly determined by rubbing it on a piece of white, unglazed porcelain... Pieces of unglazed porcelain, called streak-plates, are made especially for this purpose. **1964** J. Sinkankas *Mineral. for Amateurs* viii. 202 Hematite crystals appear quite black, but when rubbed across a porcelain streak plate, the characteristic deep red trace shows the true color. **1966** *McGraw-Hill Encycl. Sci. & Technol.* III. 616/2 Streak plates are incubated in a closed vessel in which the air is replaced by an inert oxygen-free gas. **1930** *Ann. Appl. Biol.* XVII. 623 Our maize streak virus was taken from a naturally streaked maize plant. **1948** *Phytopathology* XXXVIII. 421 To determine the incidence of the streak virus in wild and crop plants, collections were made in areas near infected tobacco fields.

streak (strīk), *sb.*² *slang* (orig. *U.S.*). [f. *STREAK *v.*² 6 c.] An act of running naked in a public place; = *STREAKING *vbl. sb.* 4. Also *attrib.* Cf. Streek, streak *v.* 5 in Dict. and Suppl.

1974 *Newsweek* 4 Feb. 63/3 A student who participated with 125 others in a co-educational streak has been suspended from school. **1974** *Daily Tel.* 11 Mar. 16 High spirits may account for some streaks, and sheer frustration or a desire to insult society for others. **1974** *Newsweek* 18 Mar. 42/3 Any number of other streak-watchers didn't react at all. **1980** *Times* 5 Jan. 3/5 [He] ran down Dean Street, Soho, on New Year's Eve, wearing only shoes while taking part in a 'streak' for charity.

streak, *v.*² Add: **2. b.** *Bacteriology.* To draw an infected needle or the like lightly over the surface of a solid culture medium in order to initiate a culture in which there is a varied density of growth: used with either the needle or the medium as obj. Also, to transfer (a bacterial specimen) in this way.

1910 [implied at *STREAKING *vbl. sb.* 1]. **1927** R. A. Kelser *Man. Veter. Bacteriol.* vi. 70 By going to the end of the streak with a sterile needle and streaking that portion down, the end of such down streak will contain but very few bacteria. **1934** A. T. Henrici *Biol. of Bacteria* xii. 203 It is often advisable to streak a second or even a third plate without recharging the wire loop. **1949** Kelly & Hite *Microbiology* xi. 147 The specimen is streaked out with a sterile inoculating needle. **1969** Sirockin & Cullimore *Practical Microbiol.* ii. 17 Streak out a loopful of the broth culture using the aseptic techniques described. **1976** Williams & Shaw *Micro-Organisms* x. 124 (*caption*) Streaking a plate with pure cultures. **1977** *Lancet* 29 Oct. 906/1 A bacteriological loop is used to sweep across the surface of a young culture of the isolate and is then streaked across one end of the strip. **1980** *Nature* 21 Feb. 793/1 Faecal specimens were collected on sterile 'Culturettes'..and streaked onto MacConkey plates.

c. *Hairdressing.* To tint (the hair) with streaks (*STREAK *sb.*¹ 2 h). Cf. *STREAKING *vbl. sb.*¹ 3.

1965 R. Corson *Fashions in Hair* xiv. 625 Women developed an experimental urge and began streaking their hair.

6. a. (See Streek, streak *v.* 5. The spelling *streak* is now standard, probably through assoc. of the vb. with Streak *sb.* 3 c.) (Further examples in gen. use.)

1768 A. Ross *Fortunate Shepherdess* 51 O'er hill an' dale she forcefully did dreel; A' road to her was bad an' gueed alike, Nane o't she wyl'd, but forret still did streak. **1834** *New Monthly Mag.* XLI. 465 Away we 'streaked' at the rate of twelve miles an hour against the current. **1888** P. Gillmore *Days & Nights by Desert* xx. 170 True, it was wounded; but as it 'streaked' across the plain, from the pace it was going, no one would have thought so. **1915** H. Rosher *In Royal Naval Air Service* (1916) 50, I climbed to 2,000 feet and streaked off over the Channel. **1931** *Punch* 28 Oct. 456/1 The village kids..used to spend most of their leisure in pushing the door ajar in order to set it [*sc.* an electric bell] going and then streak for home. **1955** 'A. Gilbert' *Is she Dead Too?* vii. 128 'So you opened the door—' 'And Tom [*sc.* a cat] streaked out.' **1973** E. Lemarchand *Let or Hindrance* viii. 94, I got out at last, and streaked up to the bungalow.

b. *trans.* To cause to move fast or like lightning.

1912 W. Deeping *Sincerity* xviii. 137 She shot well, very few of her arrows streaking their way through the sunlight to stand slantingly in the grass. **1928** *Daily Express* 19 June 17/2 He hit only three 4's, and..streaked one from Staples dangerously through the slips when 28. *Ibid.* 25 June 17 Chapman..'streaked' several shots through the slips. **1970** J. Howard *Please Touch* 2 When supersonic transports streak 330 passengers to their destinations.

c. *intr.* To run naked in a public place as a stunt. Cf. *STREAK *sb.*² *slang* (orig. *U.S.*).

1973 [implied at *STREAKING *vbl. sb.*¹ 4]. **1974** *Runner's World Mag.* Feb. 9/1 During the winter of 1958–9 a group of us 'streaked' all over Berkeley. **1974** *Daily Tel.* 5 Mar. 3/6 At Memphis State University, the dean issued a warning that students caught 'streaking' would be suspended. **1974** *Globe & Mail* (Toronto) 18 Mar. S1/1 Phil Esposito, stripped as though to streak, held court in the cluttered quarters, tall, dark, unquiet. **1979** *Daily Tel.* 12 Jan. 9/3 The girls..had danced on the lawns in the nightdresses, 'streaked' to chapel and enjoyed midnight parties.

Hence **strea·ker²,** one who runs naked in a public place; also *attrib.*; **streak-in** [*-IN³], a communal act of running naked in a public place.

1973 *Time* 10 Dec. 14/2 Streakers generally race nude between two unpredictable points. **1974** *Newsweek* 4 Feb. 63/3 One Los Angeles radio station broadcast 'streaker alerts' to warn the populace that naked youths were on the loose. **1978** D. Bloodworth *Crosstalk* xiii. 104 The streaker had invaded the Brasserie Lipp in Paris at lunch-time. **1978** J. Irving *World according to Garp* xi. 210 A young woman had reported that she was approached by an exhibitionist—at least, by a streaker. **1974** *Kingston* (Ontario) *Whig-Standard* 8 Mar. 3/2 The mass streak-in started near Victoria and Leonard Halls. **1974** *Times* 9 Mar. 4/8 Some of the students were arrested when a 'streak-in' turned into a riot.

streaking, *vbl. sb.* (In Dict. s.v. Streak *v.*²). Add:

1. (Examples in *Bacteriology*: cf. *STREAK *v.*² 2 b.)

1910 Hiss & Zinsser *Text-bk. Bacteriol.* viii. 148 (*heading*) Separation of bacteria by surface streaking. **1949** Kelly & Hite *Microbiology* xi. 147 Though the streaking must be done carefully and according to some plan, it must be performed speedily. **1973** R. G. Krueger et al. *Introd. Microbiol.* iv. 169/1 Aerobic organisms..are inoculated onto the surface of the solidified medium by a procedure called streaking.

2. *Television.* A picture condition in which the trailing edges of areas of a particular colour are extended by streaks of the complementary colour.

1956 Amos & Birkinshaw *Television Engin.* II. (*caption facing* p. 27) An image of test card C as reproduced by a video-frequency amplifier giving short-term streaking. **1975** B. Grob *Basic Television* (ed. 4) xix. 411 The streaking is especially evident trailing to the right after the edges of numbers or letters in the picture... The cause is phase distortion with time delay for low video frequencies up to about 200 kHz.

3. *Hairdressing.* (See quot. 1966.)

1966 J. S. Cox *Illustr. Dict. Hairdressing & Wigmaking* 144/1 *Streaking*, the bleaching of a few strands of hair in the coiffure. **1975** *Time Out* 30 May 63/4 (Advt.), Streaking including cut & blow £5.50. **1976** *Southern Even. Echo* (Southampton) 3 Nov. 2/3 (Advt.), Fashion cutting, blow drying, shampoo and set and streaking.

4. The act of running naked in a public place. Cf. *STREAK *v.*² 6 c, *STREAK *sb.*²

1973 *Time* 10 Dec. 14/2 Another statistic in a growing Los Angeles-area fad: streaking. **1974** *Washington Post* 6 Mar. B3 Lady Godiva established the political importance of streaking. **1977** D. Morris *Manwatching* 210 The phenomenon of 'streaking'..is a strange example of an act that *only* has value as a deliberate Overexposed Signal.

streaky, *a.* Add: **2. b.** Also *absol.* as *sb.*

1969 *Listener* 17 Apr. 535/3 An angst-ridden fly on the ceiling: 'What is my life? Hanging about dustbins, crawling up drains, promiscuous sex on a rasher of *streaky*. There must be more to life than just pleasure.' **1973** *Tel.* (Brisbane) 5 Apr. 29/3 If she [*sc.* a housewife] wants half-a-pound of streaky she is likely to be called 'dear' by Britain's grocers. **1979** A. Parker *Country Recipe Notebk.* vi. 82 Pot-roasted pork streaky (belly of pork) is better cold than hot.

stream, *sb.* Add: **2. f.** (*a*) (Later example.)

1937 W. R. Inge *Rustic Moralist* iv. i. 234 What ought the helpless *intelligentsia* to do? Not to float with the stream, a feat which any dead dog can accomplish.

4. f. See *ON STREAM *adv. phr.* and *a.*

6. a. Examples of time considered as a continuous flow.

1597 Shaks. 2 *Hen. IV* iv. i. 70 Wee see which way the streame of Time doth runne. **1739** Hume *Treat. Hum. Nature* II. iii. 276 When we turn our thought to a future object, our fancy flows along the stream of time. **1846** J. S. Mill in *Edin. Rev.* Oct. 356 Authentic history, as we ascend the stream of time, grows thinner and scantier. **1896** L. T. Hobhouse *Theory of Knowl.* ii. xix. 465 Think of the whole page as the stream of time advancing from the top downwards. **1914** B. Russell *Our Knowl. External World* vi. 167 A truer image of the world..is obtained by picturing things as entering into the stream of time from an eternal world outside.

d. *stream of thought* = *STREAM OF CONSCIOUSNESS.

1890 W. James *Princ. Psychol.* II. xix. 79 This consciousness must have the unity which every 'section' of our stream of thought retains so long as its objective content does not sensibly change. **1921** J. Varendonck *Psychol. of Day-Dreams* iv. 293 Only such terminations of fore-conscious streams of thought as are in relation with acknowledged desires..of our conscious life can come to the surface. **1938** W. S. Maugham *Summing Up* 223 Of the other experiments that have been made [by the novelist] the most important is the use of the stream of thought... It was tempting to explore greater depths of character by an imaginative picture of the subconscious of the persons of his invention. **1961** *John o' London's* 2 Feb. 109/3 The author includes many.. remarks..on the Ulysses characters, providing fascinating insights into what Joyce was getting at with his stream-of-thought vignettes.

e. *Educ.* A division in a school according to ability or to subjects studied; a group of pupils selected in this way.

1938 [see *MULTILATERAL *a.* 4]. **1946** M. L. Jacks *Total Education* iv. 59 The Headmaster of a Senior School tells me that his main difficulty lies not with the A stream nor with the C stream,..but with that solid mass in the middle. **1953** *Manch. Guardian* 11 May 6/10 On entry at eleven, each child is given intelligence tests and group tests in arithmetic and English, and on the result is placed in the appropriate stream (both the Douglas schools have a six-stream, Ramsey a four, and Castle Rushen a three-stream entry). **1959** *Observer* 20 Dec. 22/2 Perhaps we can totter along somehow, and for a time, by prolonging the 'stream' system into university education. **1966** J. Partridge *Middle School* iii. 41 It is clearly recognized that the 'A' stream contains the most intelligent boys and the 'D' stream the least. **1977** J. Aiken *Last Movement* i. 31, I never actually taught her..because she was in the science stream and I was tutoring in business methods.

9. a. *stream-name.*

1922 E. Ekwall *Place-Names of Lancashire* 25 It *might* be an old stream-name..and might have given name to the forest. **1960** P. H. Reaney *Orig. Eng. Place-Names* v. 81 It is also possible that this (and other names) may contain a stream-name *ec(c)les*, as in Ecchinswell.

b. *stream-cut* adj., *-cutting* vbl. sb.

1970 R. J. Small *Study of Landforms* ix. 314 Pediments form not at the bases of fault-scarps but beneath the retreating walls of stream-cut valleys. **1973** *Nature* 2 Mar. 40/1 In a streamcut channel, the Ganurgarh shales intercalated with limestone have been exposed. **1957** G. E. Hutchinson *Treat. Limnol.* I. i. 47 Since there is little or no stream-cutting below such lakelets, they may persist longer than do the lakes dammed by the main mass of the slide.

c. stream-flow orig. *U.S.*, flow of water in streams and rivers; the rate or amount of this in any one stream or from any particular area; **stream function** *Physics*, a mathematical function of position defined so that

lines along which it has a constant value are the streamlines of a flow or the lines of force of a field.

1902 W. P. Mason *Water-Supply* (ed. 3) vi. 269 The forest acts as a 'governor' of stream flow, rather than as a means of increasing precipitation. **1922** Glazebrook *Dict. Appl. Physics* I. 498/2 It has been possible..to deduce the total run-off or stream flow from a drainage area from the difference of rainfall and the computed evaporation over that area. **1971** *Sci. Amer.* Sept. 142/3 The cooling water required by power plants already constitutes 10 percent of the total U.S. streamflow. **1979** *Bull. Amer. Meteorol. Soc.* LX. 560 (*heading*) A method for assessment of effects on streamflow by orographic cloud seeding in the Colorado Rocky Mountains. **1879** H. Lamb *Treat. Math. Theory Motion of Fluids* iv. 67 If *P* move about in such a manner that the value of ψ does not alter, it will trace out a curve such that no fluid anywhere crosses it, *i.e.* a stream-line. Hence the curves ψ = const. are the stream-lines, and ψ is called the 'stream-function'. **1937** S. L. Green *Hydro- & Aero-Dynamics* ii. 19 When $w = \phi + i\psi$ is a function of $z = x + iy$ the conditions $\phi_x = \psi_y$, $\phi_y = -\psi_x$ are satisfied, and these conditions are exactly the same as those satisfied by the velocity potential and the stream function for an irrotational motion in two dimensions. **1979** Bertin & Smith *Aerodynamics for Engineers* ii. 44 The existence of a stream function is a necessary condition for a physically possible flow.

stream, *v.* Add: **II. 14.** *Educ.* In a school, to divide (pupils) into streams (sense *6 e); to place (a pupil) in a stream.

1957 *Listener* 12 Dec. 997/1 The Russians..neither stream nor select their children... All are expected to reach a common standard. **1966** J. Partridge *Middle School* iii. 41 The boys are thus streamed according to recognized ability. **1973** *People's Jrnl.* (Inverness & Northern Counties ed.) 4 Aug. 9/2, I suppose they had their problems trying to get me 'streamed'. **1980** *Times* 7 May 15/3, I got myself streamed at Manchester Grammar towards the sciences.

streamed, *a.* Restrict †*Obs.* to sense in Dict. and add: **2.** *Educ.* Of a school, esp. a comprehensive school: organized in classes of pupils of like rather than mixed ability. Also of pupils: grouped according to ability.

1962 *Guardian* 30 Mar. 10/5 There are some very good streamed schools. **1968** *Listener* 4 July 8/3 In streamed comprehensive schools..all advantages lie with the children of middle-class parents. *a* **1974** R. Crossman *Diaries* (1977) III. 391 She has got four different streamed groups and a fifth group of backward children. **1978** *Times* 28 July 4/8 The bottom classes of streamed schools..often.. get the worst teachers.

streamer, *sb.* Add: **2. f.** A long narrow strip of coloured paper used as a festive decoration or rolled up to unwind when thrown (at a celebration, etc.).

1857 C. M. Yonge *Dynevor Terrace* I. xvi. 262 The [Christmas] tree became more laden, and the streamers and glass balls produced a more brilliant effect. **1918** A. Bennett *Jrnl.* 14 Nov. (1932) II. 242 The feature of last night was girls with bunches of streamers which they flicked in your face as you passed. **1930** E. Rice *Voyage to Purilia* xii. 155 The room was gay, with streamers of coloured paper and with large gas balloons. **1959** M. Shadbolt *New Zealanders* 239 She..found herself waving farewell to her tearful parents from the deck of a ship. Only then, as the tangling streamers snapped across the widening water, did she remember to be surprised at herself. **1980** *Daily Tel.* 25 Nov. 15/6 This useful company also supplies carnival and party novelties including balloons.., dance streamers,..lucky dip prizes and stocking fillers.

g. = *BANNER *sb.*¹ 2 c. Also *transf.*

1909 G. R. Chester *Making of Bobby Burnet* xix. 230 Use two-inch streamers clear across the page. **1922** U. Sinclair *They call me Carpenter* 88 The headlines flamed before my mind's eye—streamer heads, all the way across the sheet. **1938** F. D. Sharpe *Sharpe of Flying Squad* xxv. 254 Then a very good crime reporter called a portico thief a cat-burglar and the story got a streamer headline. **1957** [see *BANNER *sb.*¹ 2 c]. **1963** *Times* 19 Apr. 8/5 Invitations to browse in these places, moreover, occasionally decorate the discreet vans of the Stationery Office. But this form of streamer on the van is technically known as a 'filler', which means that it is used only when the space is not required for boosting some new publication.

h. *Angling.* A fly with feathers attached, which simulates a small fish; also, the feathers so employed. Freq. *attrib.*, chiefly as *streamer fly.* orig. and chiefly *U.S.*

1919 D. Carroll *Fishing, Tackle & Kits* 265 The white bucktail with the red feather streamers in the shape of a tail made the trout strike. **1929** *Field & Stream* June 65/3 (Advt.), Dixie Wiggler..different colored streamers. **1930** *Forest & Stream* Mar. 202/2 A lure which is justly famous in a few restricted localities..is the Streamer Fly... This fly is tied in more or less conventional hackle style with a two- or three-inch streamer feather as a tail. **1952** J. Veniard *Fly Dressers' Guide* xiii. 140 *Streamer Flies.* This type of fly is immensely popular in America, but its possibilities in the United Kingdom have never been seriously considered. The nearest approach we have to it is the 'lure', which consists usually of two or three hooks in tandem with a long wing. The idea is very much the same..the main difference being that many of the 'streamers' are tied to represent the small fish on which game fish feed, whereas most of the lures are just brightly coloured 'attractors'. **1971** K. Draper *Trout Flies in N.Z.* ii. 41 One of the most popular types of tie is the Parson's Glory style of streamer. This fly is closely allied to the American streamer fly which is tied in much the same manner except that the long neck or saddle hackles 'stream' free. **1975** D. J. Collyer *Fly-Dressing* xi. 167 Seven years ago I received..two very interesting flies, both streamers..designed by Lew Oatman, an American, to imitate small fish.

3. b. Restrict †*Obs.* to sense in Dict. and add: In mod. use, a long, thin component or appendage of the tail of some comets.

1909 G. F. Chambers *Story of Comets* iii. 23 Although comets usually have but one tail, 2 are not uncommon, whilst even that number is often increased by the presence of slender streamers, which are virtually independent tails. **1931** [see *JET *sb.*² 4 c (i)]. **1972** D. C. Knight *Comets* 13 The tail of Donati's comet of 1858 was some 50 million miles long..and was split into two or three streamers.

f. A filamentary luminosity sometimes seen to extend from an electrode in a gas when the potential difference is not great enough to produce a spark or arc; a similar feature that extends from a cloud or something on the ground prior to a stroke of lightning along the same path.

1910 *Encycl. Brit.* V. 883/1 Bright curved streamers starting from the negative terminal. **1934** *Physical Rev.* XLVI. 101/2 The breakdown in argon is similar to that in helium except that no glow discharge is observed and no anode streamer forms. **1935** *Proc. R. Soc.* A. CLII. 597 The leader to the first stroke consists of a series of streamers moving downwards in a step-by-step manner. **1953** Meek & Craggs *Electr. Breakdown of Gases* iii. 159 In a 6-cm. gap with a 0·05-mm. point, the streamers gradually lengthen, as the voltage is raised, to 1·1 cm. and then breakdown occurs. **1972** *Jrnl. Physics D* V. 2179 Photomultiplier investigations..have shown that the streamers are weakly luminous channels, and that almost all the ionization phenomena are concentrated at their tip. **1979** J. G. Navarra *Atmosphere, Weather & Climate* ix. 298 When the leader stroke approaches the ground, a discharge streamer is extended from the ground up to the leader and completes the channel.

6*. [f. the sb.] *Educ.* With a categorizing letter prefixed: a child belonging to that stream in a school, or one whose abilities are adapted to such a stream; esp. *C-streamer*, a child of little academic ability.

1966 *Guardian* 4 Apr. 6/1 C-stream children in a competitive society have the makings of failure before they start... Let us take it that eventually the C-streamer has been taught to read. **1967** *Punch* 20 Dec. 952/3 These.. riddles will hold ten-year-olds enthralled until *next* Christmas—or even, in the case of certain C-streamers, the one after. **1970** *Sunday Times* (Colour Suppl.) 18 Jan. 8/1 They are C-streamers, from poor homes.

7. b. *Physical Chem.* Applied to a type of molecular orbital possessing a single nodal plane and formed out of identical atomic orbitals of the same phase from each of the atoms in the ring or chain backbone of the molecule.

1966 Phillips & Williams *Inorg. Chem.* II. xxvii. 348 The lowest benzene π (aromatic) orbitals of the 'streamer' kind, one from each benzene, have nodes only in the planes of the benzene rings, and taken in combination are therefore either of the A_{1g} or A_{2u} type symmetry. **1974** Gill & Willis *Pericyclic Reactions* iv. 117 The lowest hexatriene level (ψ_1) is the typical 'streamer' orbital (i.e. $\pm\pm\pm\pm\pm\pm$) which is of *A* symmetry with respect to C_{2v} and of *S* symmetry with respect to the mirror plane σ_{yz}.

streaming, *vbl. sb.* Add: **f.** *Educ.* The practice of dividing pupils in a school into streams (sense *6 e). Also, the allocation of pupils to different schools according to ability.

1954 *Brit. Jrnl. Psychol.* XLV. 147 Present methods of streaming in primary and secondary schools do not give the more backward as good a chance to make up. **1961** *New Left Rev.* Jan.–Feb. 6/2 A streaming system..with secondary modern, technical and grammar-type schools. **1966** D. Jenkins *Educated Society* v. 200 An argument for turning all Secondary schools into Comprehensive schools with a strong streaming system. **1971** *Where* Sept. 262/2 The practice of streaming—classifying children according to their abilities in reading, writing and arithmetic—is gradually becoming suspect. **1980** *Daily Tel.* 10 Dec. 16/6 The proposals include phasing out of church schools, single sex schools, and streaming.

streaming, *ppl. a.* Add: **1. b.** Of a cold: accompanied by copious running of the eyes and nose.

1923 W. R. Inge *Lay Thoughts* (1926) iii. vii. 229 Persons..suffering from streaming colds, which are quite as infectious as..mumps or chicken-pox. **1982** *Guardian Weekly* 26 Dec. 22/1 One is sorry for foreigners, both as a Christian duty and because they tend to have streaming colds.

stream-line, *sb.* Add: Now usu. **streamline.** **1. a.** Also *Aerodynamics.* In mod. use, a line such that, at any instant, the direction of the tangent at any point is the direction of the flow of fluid at that point. (This definition is equivalent to that in Dict. for the special case of steady flow.) (Earlier and later examples.)

1868 W. J. Rankine in *Engineer* 16 Oct. 285/1 A stream-line is the line, whether straight or curved, that is traced by a particle in a current of fluid. **1945** R. von Mises *Theory of Flight* ii. 23 In the case of an unsteady flow the streamlines, *i.e.*, the curves whose tangents have the velocity direction are, in general, not the pathways of the particles. **1971** G. M. Hidy *Waves* iv. 47 Streamlines generally bear no relation to particle paths because different fluid elements form different streamlines at given times. **1980** Bober & Kenyon *Fluid Mech.* iv. 124 If the flow is steady then a fluid particle will move along a streamline.

b. *attrib.* (*a*) Designating motion of a fluid that is free from turbulence, so that it can be represented by a pattern of streamlines that either is constant or changes steadily with time.

1898 [in Dict.]. **1916** H. Barber *Aeroplane Speaks* 62 To secure a stream-line motion of the air free from eddies. **1957** G. E. Hutchinson *Treat. Limnol.* I. v. 251 The only simple way of observing regular streamline or laminar flow is adjacent to some smooth surface over which a slow current is passing. **1979** A. L. Lydersen *Fluid Flow & Heat Transfer* i. 2 This type of motion, where the velocity at a certain point is constant and independent of time, is termed laminar flow. It is sometimes called streamline flow or viscous flow.

(*b*) Having or being a shape such that the flow of a fluid round it is smooth, and there is no separation of streamlines from the surface; more widely, shaped so as to reduce air or water resistance; *streamline wire*, a wire of elongated cross-section.

1907 F. W. Lanchester *Aerodynamics* i. 20 In an actual fluid, bodies of other than streamline form experience resistance apart from that directly due to viscosity. *Ibid.* iii. 102 In the hydrodynamic theory of an inviscid fluid, every conceivable body is of stream-line form. **1909** [in Dict.]. **1914** *Automobile Topics* 30 May (Advt., back cover), That beautiful stream-line Car. **1918** W. L. Cowley *Aeronautics* iv. 93 The following table gives the resistance coefficients of aeroplane stream line wires, of fineness ratio 3:1. **1919** G. Whale *Brit. Airships* 160 The remaining two engines are carried in a small streamline car situated amidships. **1928** E. Cadbury in C. F. S. Gamble *Story North Sea Air Station* xxii. 408, I..seized a scarf, goggles and helmet, tore off my streamline coat, and, semi-clothed..took a running jump into the pilot's seat. **1929** *Jrnl. R. Aeronaut. Soc.* XXXIII. 360 In aeronautical nomenclature a 'streamline body' is one about which the flow of a real fluid..approximates very closely to a steady flow of the hypothetical inviscid fluid, except in a very thin layer called the 'boundary layer', surrounding the exposed surfaces. *Ibid.* 361 The ideally streamline aeroplane cannot exist. **1936** J. C. Corlett *Rigging & Airframes* v. 108 On the final inspection of streamline wires they must be checked for 'safety'. **1953** M. Rauscher *Introd. Aeronaut. Dynamics* vi. 259 The theory of the ideal fluid comes very close to representing the actual flow conditions about a streamline body in a fluid of low viscosity.

(*c*) *fig.*

1933 S. Spender *Poems* 44 Where only a low streamline brightness Of phosphorus on the tossing hills is white. **1942** *R.A.F. Jrnl.* 2 May 21 The camp is the last word in streamline modernity. **1967** B. J. Banfill *Pioneer Nurse* xi. 129 Snow, in long streamline drifts, covered the now familiar objects.

c. Used predicatively as *adj.* in preceding senses.

1907 F. W. Lanchester *Aerodynamics* i. 27 If..we assume continuity as hypothesis, then all bodies must be streamline. **1922** *Daily Mail* 3 Nov. 2 The body of the car is streamline according to British practice. **1929** *Jrnl. R. Aeronaut. Soc.* XXXIII. 366 The b.h.p. required.. should not be seriously influenced by interference, provided that the interference does not cause the flow to cease being streamline. **1936** B. Jones *Elem. Pract. Aerodynamics* vii. 115 Wherever possible the airplane parts that would cause parasite drag are made streamline in shape. **1971** J. W. Ireland *Mech. of Fluids* viii. 234 Oil of specific gravity 0·9 and viscosity 1 poise is pumped through a 5 cm diameter pipe at the rate of 280 litres/min. Show that the flow is streamline.

d. A contour of a body that is coincident with a streamline of flow round it; *loosely*, a smooth, flowing outline.

1917 D'A. W. Thompson *Growth & Form* xvi. 673 The naval architect learns a great part of his lesson from the investigation of the stream-lines of a fish; and the mathematical study of the stream-lines of a bird..has helped to lay the very foundations of the modern science of aeronautics. **1936** B. Jones *Elem. Pract. Aerodynamics* vii. 117 Any non-streamlined body can have its resistance or drag greatly reduced merely by the addition of a blunt nose and a tapered tail. If the contour is also a continuous curve the shape approaches the ideal streamline. **1943** Koestler *Arrival & Departure* ii. 41 He was able to re-draw in his mind the curve of Odette's knees.., the streamlines of her jumper. **1944** W. Fortescue *Mountain Madness* xxiv. 162 If my curls were grey, at least I had kept my stream-line.

strea·mline, *v.* Also **stream-line.** [f. the sb.] **1.** *trans.* To give a streamline form to. Chiefly as *STREAMLINED *ppl. a.*, *STREAMLINING *vbl. sb.*

1918 H. Barber *Aeroplane Speaks* (ed. 6) 61 From the designer's point of view it always pays to stream-line detrimental surface. **1927** Haldane & Huxley *Anim. Biol.* xiii. 316 The air-sacs..are used to stream-line the body. **1927** *Daily Tel.* 27 Sept. 9/5 It appears that Flight-Lieut. Kinkead came down because the spinner, a metal fitting streamlining the propeller boss, came off.

2. *fig.* **a.** To slim; to remodel on smooth, uncluttered lines. Also *absol.*

1935 P. B. Hawk (*title*) Streamline for health. **1937** *Denver Post* 27 Jan. (Mag. Section) 9 (*heading*) Streamline your dance frock.

b. To simplify, esp. in order to make more efficient or better organized.

1936 *Sun* (Baltimore) 2 Nov. 13/5 Those who watch financial fashion observe a tendency to streamline capital set-ups for tax purposes. **1947** *Hansard Commons* Written Answers 2 Dec. 46, I am anxious to do all I can to streamline controls. **1950** A. Huxley *Let.* 16 Mar. (1969) 620 Chapters might be 'streamlined'. **1957** *New Yorker* 26 Oct. 60/2 The producers got panicky and decided to 'streamline' the production, stressing professional finish. **1958** *Post Office Mag.* Apr. 117/2 The Ministry are co-operating with us all..on streamlining general building methods. **1974** *Whitaker's Almanack 1975* 812/1 Through its advisory services, it helps its member governments to ..streamline health services.

streamlined (strī·mləind), *ppl. a.* Also **stream-lined.** [f. prec. + -ed[1] (early uses derived from the sb.).] **1.** Having a streamline form; designed so as to reduce air or water resistance.

1913 *Aeroplane* 30 Oct. 480/1 Its [*sc.* an aeroplane's] small span and carefully streamlined body. **1916** [see *clean *a.* 13 c]. **1928** C. F. S. Gamble *Story North Sea Air Station* 10 Her two stream-lined gondolas..were designed to be capable of sustaining severe impact loads when alighting on water. **1930** *Observer* 16 Feb. 17/5 She has oval, instead of stream-lined funnels. **1948** M. Laski *Tory Heaven* iv. 58 I'd thought of..one of the old Lagondas..but I believe everything's streamlined these days. **1968** O. S. Nock *Railway Enthusiast's Encycl.* 279 From 1923 onwards, the 'Cheltenham Flyer' of the G.W.R. claimed the honours until the advent of the L.N.E.R. streamlined trains. **1977** C. McCullough *Thorn Birds* xvii. 439 Nowadays even the ocean liners were streamlined, hulls shaped like destroyers to get there faster.

2. *fig.* **a.** Having smooth, flowing, or elongated lines; slender.

1934 H. Read *Art & Industry* 3 'Streamlined' is popularly, if inaccurately, used as a term of approval for the design of any object in daily use. **1935** *Amer. Speech* X. 194/1 Terminology from other fields aids the fashion editors... The *streamlined* silhouette came in with the new automobile. **1937** *Nation* 15 May 559/2 The cows came from their stables down a runway into a streamlined building. **1944** A. Holmes *Princ. Physical Geol.* v. 63 The effects of shearing or flowage give the rocks a new structure, due to the stream-lined arrangement of the platy and elongated minerals. **1951** M. McLuhan *Mech. Bride* (1967) 96/1 Streamlined, synthetic blondes—these are at once abstract and exciting. **1966** Wodehouse *Plum Pie* 75 Dieting continues to be all the go..and the number of those who hope to become streamlined by pushing their plates away untasted increases daily. **1976** *Country Life* 1 Apr. 805/2 The two aspects of the ubiquitous Deco that have caused such confusion in recent terminology, the one still curved and flowing (Streamlined Moderne), the other angular and machine-style (Zig-Zag Moderne).

b. Efficient; simplified, having inessentials removed.

1937 *Words* May 100/1 Our streamlined professors. **1942** E. Paul *Narrow St.* xvii. 128 Clever women..manipulated high statesmen, financiers and stream-lined executives as if the men were marionettes. **1947** A. Huxley *Let.* 9 Mar. (1969) 568 We might think in greater detail about..a stream-lined construction for the revised Goldsmith. **1957** L. F. R. Williams *State of Israel* ix. 156 This..end is secured by an ingenious, stream-lined, procedure which makes the guillotine unnecessary. **1971** *Daily Tel.* 15 Jan. 15/5 A streamlined, yet cosy, kitchen overlooks a tiny paved garden. **1979** *Tucson* (Arizona) *Citizen* 20 Sept. 3 c/2 The majority's reasons for dissolving the former trustees included a desire..to make 'more streamlined' decisions regarding control of health care costs and treatment of indigents.

strea·mliner. [f. as prec. + -er[1].] **a.** A stream-lined train.

1938 *Times* 14 Oct. 15/6 One of the latest streamliners to be put into operation is that of the Reading Line between New York and Philadelphia. **1946** D. C. Peattie *Road of Naturalist* v. 55 To read the country while you run is possible even from a streamliner. **1958** J. Kerouac *On Road* ix. 229 A hundred and ten miles an hour straight through, an arrow road, sleeping towns, no traffic, and the Union Pacific streamliner falling behind us in the moonlight. **1967** O. Wynd *Walk Softly* v. 72 The Fuji is a great train, a streamliner, everything shiny and new. **1982** Whitehouse & Allen *E. Treacy—Railway Photographer* 85 A streamliner takes the 5.25 to Euston past Wavertree.

b. One who streamlines.

1943 F. L. Wright *Autobiogr.* iv. 321 Invention.. looking for salvation to the engineer, the streamliner, and the elevator, has been trying to..hold the profits of superconcentration.

c. *gen.* Something streamlined.

1968 *Church Times* 13 Sept. 18/3, I hear you have sold your old house and are looking for a streamliner.

strea·mlining, *vbl. sb.* [f. as prec. + -ing[1].] **1. a.** Streamlined shape or design. **b.** The action of giving something a streamlined shape.

1918 H. Barber *Aeroplane Speaks* (ed. 6) 61 The weight of the stream-lining is always paid for many times over by the greater velocity and consequent increase of lift due to the decreased drift. **1921** *Discovery* Apr. 98/1 This probably is partly due..to the careful streamlining of the machine [*sc.* an aircraft] and the sensible distribution and installation of the engines. **1932** D. L. Sayers *Have his Carcase* vi. 76 The car glided away amid the reverent murmurings..of persons gathered..to admire its streamlining. **1936** *Discovery* Feb. 40/1 To diminish air-resistance by the streamlining of both engine and train. **1950** *Times* 5 May 5/6 A large engine does not necessarily mean a large wash. Bad streamlining and excessive speed create the wash. **1955** *Sci. Amer.* Mar. 90/3 The streamlining of birds of course is the envy of all aircraft designers. **1965** H. Hood in R. Weaver *Canad. Short Stories* (1968) 2nd Ser. 219 'That's quite a car, Mister.' 'Yes, that's what they call streamlining.' **1980** *Daily Tel.* 24 Sept 12/2 The drag factor of a car is becoming an increasingly significant item... It is, loosely, what used to be referred to as the streamlining effect.

2. *transf.* and *fig.* **a.** The action or result of giving smooth, flowing lines to something.

1934 *Punch* 19 Dec. 691/1 This 'streamlining' bunkum is spreading from the sordid realms of mere mechanics to those of everyday life, even to human physiognomy. **1937** G. Frankau *More of Us* vi. 70 'I find these modern figures far too arty,' Continued Sophie, obviously not repining Tho' hers had long since faltered from streamlining. **1940** Graves & Hodge *Long Week-End* xi. 181 The use of streamlining as a modern style in domestic objects.

b. Simplification of procedures, an organization, or the like to improve efficiency.

1959 *Daily Tel.* 24 July 5/7 (*heading*) Railway streamlining to be speeded. **1965** *Listener* 3 June 822/2 The streamlining of the administrative and financial planning structure was as urgent as was the physical planning in 1961. **1970** *Daily Tel.* 18 Mar. 2/5 As a result of a streamlining operation the weekly expenses had been cut from £1,200 to £700. **1978** R. Lewis *Inevitable Fatality* ii. 42 If there's to be a *streamlining*..some heads will have to roll.

stream of consciousness. [f. Stream *sb.*] **1.** *Psychol.* An individual's thoughts and conscious reaction to external events experienced subjectively as a continuous flow. Also *loosely* (influenced by sense *2), an uncontrolled train of thought or association.

1855 A. Bain *Senses & Intellect* 359 The concurrence of Sensations in one common stream of consciousness,—in the same cerebral highway. **1890** W. James *Princ. Psychol.* I. ix. 239 Consciousness..does not appear to itself chopped up in bits... A 'river' or a 'stream' are the metaphors by which it is most naturally described. In talking of it hereafter, let us call it the stream of thought, of consciousness, or of subjective life. **1908** W. McDougall *Introd. Social Psychol.* i. 15 Psychology must not regard the introspective description of the stream of consciousness as its whole task. **1928** D. H. Lawrence *Lady Chatterley's Lover* xiv. 253 The quiver was going through the man's body, as the stream of consciousness again changed its direction, turning down-wards. **1942** M. McCarthy *Company She Keeps* vi. 251 Damn my stream of consciousness, her mind said. **1959** Penfield & Roberts *Speech & Brain Mechanisms* iii. 47 Ganglionic patterns that preserve the record of the stream of consciousness. **1975** C. Fremlin *Long Shadow* xx. 142 Cynthia's stream-of-consciousness soon meandered obediently back to the matter in hand. **1979** K. R. Popper in Popper & Eccles *Self & its Brain* iii. 157 When we—actively—try to be passive, there may be something like a stream of consciousness; but normally we are active, and then there is.., rather, organized procedures of problem-solving.

2. *Lit. Criticism.* A method of narration which depicts events through this flow in the mind of a character; an instance of this.

[**1918** M. Sinclair in *Little Rev.* Apr. 6 In identifying herself with this life which is Miriam's stream of consciousness Miss Richardson produces her effect of..getting closer to reality than any of our novelists.] **1939** S. O'Casey *Let.* Apr. (1975) I. 792, I differ from you.. in your contention that my 'dream fantasies & streams of consciousness' are 'foreign to my best style'. **1961** W. C. Booth *Rhetoric of Fiction* iii. xi. 324 The deep plunges of modern inside views, the various streams-of-consciousness that attempt to give the reader an effect of living thought and sensation, are capable of blinding us to the possibility of our making judgments not shared by the narrator or reflector himself. **1964** M. McLuhan *Understanding Media* xxix. 296 Here [*sc.* in *David Copperfield*] was the stream of consciousness, perhaps, in its original form before it was adopted by Proust and Joyce and Eliot. **1971** B. Malamud *Tenants* 162 Bill took on a sort of stream-of-consciousness and heavily overworked association. **1978** I. B. Singer *Shosha* xiii. 233, I had also read in a literary magazine about the kind of literature called the 'stream of consciousness'.

3. *attrib.* (freq. with hyphens).

1931 *N. & Q.* 1 Aug. 74/1 This is in part a development from the 'stream of consciousness' method. **1942** *Q. Jrnl. Speech* Feb. 4/2 *This Lonely Heart*, a stream-of-consciousness play. **1955** L. P. Hartley *Perfect Woman* iii. 24 Do you think the stream-of-consciousness method has come to stay, or have Joyce and Virginia Woolf exhausted it? **1958** *Listener* 16 Oct. 603/2 The late Dorothy M. Richardson was one of the earliest exponents of the 'stream-of-consciousness' novel. **1975** B. Garfield *Death Sentence* (1976) iii. 22 He darted from topic to topic... He wasn't a stream-of-consciousness talker. **1982** *Times* 7 Apr. 9/8 A stream-of-consciousness chess match.

streek, streak, *v.* **5.** This sense is now usu. spelt *streak* and regarded as part of Streak *v.*[2] in Dict. and Suppl. (q.v., sense 6).

streel, *v.* Add: (Earlier and later examples.) Hence as *sb.*[1], a straggling, untidy procession of persons; **stree·ler,** a disreputable, idle person; **stree·ling** *vbl. sb.*

1805 E. Cavanagh *Let.* 20 Aug. in Londonderry & Hyde *Russian Jrnls.* (1934) ii. 182 In walk'd a Grenadier of a Man..& after him streal'd in at his heels a Girl. *c* **1874** D. Boucicault in M. R. Booth *Eng. Plays of 19th Cent.* (1969) II. 202, I was thrying to get away from him..but he was at my heels all the way, and Tatthers at his heels. A nice sthreel we made along the road. **1907** J. M. Synge *Playboy* ii. 48 An ugly young streeler with a murderous gob on him. **1927** E. Bowen *Hotel* xvi. 184 Miss Fitzgerald's party going forward in the leisurely and spread-out manner called in Ireland 'strealing'. **1937** G. Frankau *More of Us* ii. 30 Remembered she her own pre-ducal streelings: Some boy she had filched from her own mother dear; When all the world that either knew was Ealing's? **1943** J. Stuart *Taps for Private Tussie* iv. 57 It was after four o'clock when Aunt Vittie and Grandma came strealin up the Turnpike from town. **1971** T. Kilroy *Big Chapel* ix. 177 She never went anywhere without a streel of children.

streel (strīl), *sb.*[2] Chiefly *Anglo-Ir.* Also **sthreal, sthreel.** [ad. Ir. *s(t)raoill(e)* untidy or awkward person; cf. *straille* wench or untidy girl and prec.] A disreputable, untidy woman; a slut.

1842 S. Lover *Handy Andy* xliii. 322 To marry a thrampin' sthreel like that—a great red-headed Jack. **1909** G. B. Shaw *Press Cuttings* 36 Not out o bed yet! Go and pull her out be the heels, the lazy sthreel. **1919** —— *O'Flaherty V.C.* in *Heartbreak House* 185, I thought that covetious sthreal in there was a walking angel; and now if ever I marry at all I'll marry a Frenchwoman. **1922** Joyce *Ulysses* 354 She did look a streel tugging the two kids along with the flimsy blouse..like a rag on her back and a bit of her petticoat hanging like a caricature. **1936** M. Franklin *All that Swagger* xx. 188 That streel must have gone off with James Fullwood. **1961** 'F. O'Brien' *Hard Life* i. 11 A streel of a girl with long lank fair hair arrived to look after myself and the brother. **1970** D. M. Davin *Not Here, not Now* i. i. 8 She certainly kept the house in trim, even if she always looked a bit of a streel. **1978** D. Murphy *Place Apart* iv. 78 Jimmy wouldn't like a streel.

Hence **stree·lish** *a.*; **stree·lishness.**

1936 M. Franklin *All that Swagger* xxii. 218 Belike she has picked up some of me brogue as well as some civilised habits, if she doesn't fall back into her streelishness. **1974** E. O'Brien in *New Review* Apr. 35/2 We saw this wild creature coming... Her costume was streelish.

streepje, var. *strepie.

street, *sb.* Add: **2. f.** *the street*: also = *Fleet Street* s.v. *fleet *sb.*[2] 2 b.

1932 *News Chron.* 11 Feb. 6/3 A year ago he was coming back as Editor to the Street. **1963** L. Meynell *Virgin Luck* v. 101 The Street isn't the best place to come looking for a job at the moment. **1976** 'J. Welcome' *Grand National* viii. 123 Things were bad on the street... Two dailies were..expected to be unable to survive.

g. *Physics.* More fully *vortex street* [tr. G. *wirbelstrasse*]. An arrangement of vortices in which they form two parallel lines with clockwise rotation in one and anticlockwise rotation in the other; similarly *cloud street.*

1927 *Proc. R. Soc.* A. CXVI. 170 These vortex bands.. roll up and form what is commonly known as a vortex street. **1929** *Aircraft Engin.* I. 124/3 Vortices in a 'street' of two rows. **1936** *Proc. R. Soc.* A. CLIV. 68 For any given street the distance between consecutive vortices remains remarkably constant. **1954** [see *cloud street* s.v. *cloud *sb.* 12]. **1956** A. A. Townsend *Struct. Turbulent Shear Flow* vii. 144 For higher Reynolds numbers, either the vortex street forms very close to the cylinder or the circulation is itself unstable. **1973** *Times* 29 Jan. 14/4 As a fish swims it produces a turbulent 'vortex street' of whirling water behind it. **1978** R. S. Scorer *Environmental Aerodynamics* ix. 340 Cloud streets occur over land temporarily in the morning, and occasionally in the evening. *Ibid.* 341 Streets are common over the sea, particularly where the air stream is being slowly warmed.

h. *the street*: the streets regarded *loosely* as the realm of the common people, and esp. as the source of popular political support.

1931 [see *Nazi *sb.*]. **1954** B. & R. North tr. *M. Duverger's Pol. Parties* i. i. 38 The Storm troops wrested from the Communist and Socialist crowds their dominance of the street. **1969** *Listener* 24 Apr. 555/3 This was the street taking over a modern state in a way which hasn't happened, I think, at any other time in our history. It was as if this country had been taken over by the Black and Tans.

i. *the street* (U.S. slang): the world outside prison or other confinement, freedom.

[**1931** G. Irwin *Amer. Tramp & Underworld Slang* 185 *Streets*, freedom, and so called by prisoners in confinement.] **1935** J. Hargan *Gloss. Prison Lang.* 8 *Street, hit the*, to be released. **1956** B. Holiday *Lady sings Blues* (1973) xviii. 144, I was too busy thinking about 'the street' all the time and the life I'd left. **1966** J. Mills *Panic in Needle Park* xix. 184 It is no accident that our patients refer to the world outside [the hospital] as 'the street'; they cherish their mobility, the opportunity to escape difficult relationships, very highly. **1977** E. Leonard *Unknown Man No. 89* iv. 35 The jury believed Robert Leary and he was allowed to return to the street.

3. a*. *on the street*: (*a*) *U.S. slang*, outside prison, at liberty; (*b*) *slang*, by illicit trafficking (with reference to the acquisition of drugs); (*c*) *colloq.*, out of work, unemployed.

1935 N. Ersine *Underworld & Prison Slang* 73 *Street*, *n.*, figuratively, freedom. 'Another year will see him on the street.' **1951** W. Faulkner *Requiem for Nun* iii. 251 They worked their fines out on the street. **1977** *New Yorker* 24 Oct. 64/3 A number of men who are heterosexual on the street practice homosexuality in prison. **1979** *Guardian* 30 Oct. 8/5 We have either an extremely successful therapeutic service, or people are obtaining the drugs which they want 'on the street'. **1980** J. Wainwright *Venus Fly-Trap* 12 It's *my* living, too... If I upset that crowd, I could be on the street.

g. Also (with hyphens) *attrib.* Also (*U.S.*) *the man on the street.* Similarly *the woman in the street.*

1926 Galsworthy *Silver Spoon* iii. xi. 305 She had the political cynicism of the woman in the street. **1928** *Amer. Speech* IV. 134 The American newspaper man.. speaks a patois bewildering to the man on the street. **1942** G. Greene *British Dramatists* 20 We notice the quality which reached its height in the great comedies ..a kind of man-in-the-street poetry. **1962** A. Nisbett *Technique Sound Studio* vii. 130 A sort of convention has arisen whereby 'man-in-the-street' interviews are cut together by simple editing. **1964** R. K. Goldsen in I. L. Horowitz *New Sociol.* 89 We must endeavor to make sociological knowledge as inescapable for men-on-the-street as are..the virtues of the latest detergent. **1973** *Observer* (Colour Suppl.) 4 Feb. 15/4 He really wanted to please the man on the street and the man on the street knew it. **1977** E. W. Hildick *Vandals* i. ii. 17 Tape-recorded man- and woman-in-the-street interviews.

h. *by a street*: by a wide margin (esp. of a sporting victory).

1962 *Times* 5 Nov. 4/1 Oxford..could have won by a street before half time. **1971** *Daily Tel.* 28 Sept. 30/1 He already knew what most of us had already calculated—that Bodell had won by a street. **1977** *Time Out* 17–23 June 65/5 The Scots should win the drinking by a street. **1982** *Age Monthly Rev.* (Melbourne) Mar. 11/3 Any label embracing such a wide range of usage is too wide by a street.

i. *to be up* (*down*, † *in*) *one's street*: to be suited to someone's taste or ability.

1903 Farmer & Henley *Slang* VII. 10/1 *Street*.., a capacity, a method; a line: *e.g.* 'That's not in my street' = 'I am not concerned' or 'That's not my way of doing,' etc. **1929** *Publishers' Weekly* 21 Dec. 2813/2 A great many of the books published today are, as the saying is, right up her street. **1937** *Forward* 13 Nov. 1/2 We Labour people can the more easily say these things because some of his activities were 'up our street'. **1945** E. Waugh *Brideshead Revisited* ii. iv. 259 She is a jolly attractive girl, the sort of girl any chap would be glad to have—artistic, too, just down your street. **1955** Knight & George *Advice to Student of French* 67 The historical line of enquiry is outside your scope, but the analysis of the book or books is right down your street. **1960** L. Cooper *Accomplices* i. vi. 55 John Pollard got me the job and...I loved it... It was right up my street. **1977** *It* May 28/1 If you like Miles Davis's 'In a Silent Way' then Don Cherry has a new release which is just up your street.

j. *to play* or *work both sides of the street* (orig. and chiefly *U.S.*): to ally oneself with both sides, to behave inconsistently and opportunistically.

1938 *Sun* (Baltimore) 8 Sept. 1/2 Our friends of the New Deal have the devil's own nerve when it comes to working both sides of the street... Mr. James A. Farley..can reel off a speech as pious as the heart could wish and he can play the part of Jobmaster General with all the ruthlessness of the Tammany school of politics. **1951** E. Kefauver *Crime in Amer.* xvii. 202 He played both sides of the street and made contributions to candidates of both major parties. **1969** *Listener* 13 Feb. 196/3 Amnesty International has to play both sides of the political street in seeking to obtain the release of political prisoners... Information usually comes either from the press or through prisoners' friends, but known domestic or foreign opponents of a regime are not..necessarily the best channel for bringing influence to bear.

4. a. *street accident, band, battle, beggar* (later example), *bookie, -bookmaker, -bookmaking, clothes, cry* (earlier example), *decoration, fair, -fight* (further examples), *-fighter, game, gang, market, meeting, music* (earlier example), *organ, party, patrol, photographer, piano, preacher, riot, -rioter, -singer* (earlier example), *-singing* (later example), *song, theatre, -trader, -trading, vendor, violence, warfare.*

1892 Kipling *Many Inventions* (1893) 164, I heard Keller saying, as though he were watching a street accident, 'Give him air. For God's sake, give him air.' **1980** J. Hone *Flowers of Forest* i. i. 14 An essential witness in a street accident. **1838** Dickens *Nicholas Nickleby* (1839) ii. 6 Street bands are on their mettle in Golden Square. **1977** *New Statesman* 2 Sept. 292/1 The traditional Trinidadian street bands and dancers. **1936** *New Yorker* 7 Mar. 29/1 In 1923..the Commissioner..in a street battle routed the brownshirts. **1978** 'A. Stuart' *Vicious Circles* 21 Last night's riots in Milan where fascists had fought communists in a running street battle. **1976** *Birmingham Post* 16 Dec. 5/5 Patrick Haplin, a street beggar, will celebrate Christmas in prison—for the tenth consecutive year. **1939** *John o' London's Weekly* 2 June 320/2 He gets himself..into minor social difficulties, finding himself one day..in the police cells in Kennington, accused of being a street-bookie. **1980** G. M. Fraser *Mr. American* xxiv. 479 Unless ultimately he could break them, he might as well go back to catching street bookies. **1939** H. Hodge *Cab, Sir?* xvi. 238 We [*sc.* cabmen] have come to look on the police-court dock as a normal trade risk—just as street-bookmakers and prostitutes do. **1981** R. Samuel *East End Underworld* xiv. 179 They kept a big cat's meat shop, but his real money came from street bookmaking. **1908** M. Morgan *How to dress Doll* vii. 63 (*heading*) Dolly's street clothes. Here is Dolly dressed for a walk. **1981** M. C. Smith *Gorky Park* i. 10 In his street clothes Arkady was slovenly. **1858** *Punch* XXIV. 103 The value of the houses ..is daily diminishing by reason of the Street Cries, which render the place uninhabitable. **1911** *Encycl. Brit.* XXIX. 759/2 Street decoration. **1969** *Guardian* 18 Dec. 9/1 There are fewer street decorations..store displays are less lavish. **1872** B. Jerrold *London* xix. 158 These street-fairs are held chiefly on Saturday nights and Sunday mornings. **1982** 'E. McBain' *Beauty & Beast* viii. 130 Calusa's street fairs during..March and April, when the tourists were thickest. **1930** E. Pound *XXX Cantos* ix. 35 And he fought in Fano, in a street fight. **1976** *Sunday Mail* (Glasgow) 26 Dec. 1/2 Late-night revellers were terrified when several running street fights broke out. **1970** K. Platt *Pushbutton Butterfly* (1971) ix. 102, I promised my mother I would only marry a street fighter. **1890** *Public Ledger* (Philadelphia) 9 Dec. 6/5 Mr. Stewart Culin..recently delivered a lecture in Brooklyn, on children's street games. **1969** I. & P. Opie *Children's Games* p. vi, There is no town or city known to us where street games do not flourish. **1942** E. Waugh *Put out More Flags* i. 32 A Glasgow millionaire..who had started life in a street gang. **1979** *Amer. Speech* 1976 LI. 61 Their speech is closer to standard than is that of the adolescent street-gang members. **1870** D. J. Kirwan *Palace & Hovel* xxxiv. 507 The roughest audience..wandered right and left..to..choke the thoroughfare to buy in the street market, which was now—eleven o'clock—at the height of commercial prosperity. **1922** V. Woolf *Jacob's Room* viii. 157 The street market in Soho is fierce with light. **1982** *Listener* 16 Dec. 34/3 Beware street-market tapes at silly prices, even if they have well-known brand names. [**1923** R. Macaulay *Told by Idiot* i. xii. 46 She attended street labour meetings in the east [end].] **1933** 'G. Orwell' *Down & out in Paris & London* xxv. 183 There were street meetings... In the East India Dock Road the Salvation Army were holding a service. **1982** R. Manheim tr. *Grasse's Headbirths* vi. 85 Street meeting in the pedestrian zone. **1829** *Harlequin* 20 June 48 Street music is on the march... At Ascot, some glee-singers received two sovereigns for singing before the Royal Stand. **1849** E. Ruskin *Let.* 28 Oct. in M. Lutyens *Effie in Venice* (1965) I. 54 This Milan is a most wonderful place for street organs. **1964** G. Mitchell *Death of Delft Blue* i. 15 If you go there [*sc.* Amsterdam], be sure to look out for the street organs, the barrel-organs, you know. **1953** *Times* 3 June 8/1 The most popular events were the street parties. In some 30 or 40 streets the inhabitants had clubbed together to hold parties, starting, as a rule, with tea for the children. **1977** *New Yorker* 27 June 52/1 London's celebrations of the Queen's Silver Jubilee seemed like one mammoth street party. **1976** *Guardian* 12 Apr. 20/7 All good police officers know that the street patrol on foot..is the classic champion over the scourge of street crime. **1945** E. Waugh *Brideshead Revisited* i. vi. 132 Here..is a group taken by a street photographer. **1981** 'S. Caudwell' *Thus was Adonis Murdered* xviii. 230 The street photographers and sellers of souvenirs continued about their business. **1857** *Punch* XXXII. 40 All music sounds alike to him, whether it be the Handel of the organ-loft or the handle of the street piano. **1903** [see *plunk *v.* 1 b]. **1978** L. Deighton *SS-GB* xix. 166 Douglas stopped to give a penny to an old man at the handle of a street piano. **1878** *Golden Hours* X. 85/2 Moxy looked up quickly into the face of an old black 'mammy' who..had paused for a moment to listen to the words of the street preacher. **1916** G. B. Shaw *Androcles & Lion* p. xii, The horror of the High Priest was perfectly natural: he was a Primate confronted with a heterodox street preacher uttering what seemed to him an appalling and impudent blasphemy. **1977** J. Gillis *Killers of Starfish* (1979) v. 35 He pretended he was a street preacher once and people put pennies and dimes in his hat to save the sinners. **1980** L. St. Clair *Obsessions* i.16 Perhaps there would be no more street riots and shooting. **1900** Kipling *Let.* 24 July in C. Carrington *Rudyard Kipling* (1955) xiii. 314 We advanced against 'em [*sc.* the Boers] as if they were street-rioters that we didn't want to hurt. **1789** C. Burney *Hist. Mus.* III. 64 It seems to have been the wish of illiterate and furious reformers, that all religious offices should be performed by field-preachers and street-singers. **1958** E. Routley *Eng. Carol* 228 Television probably accounts in part for the decline of street-singing. **1891** R. Fry *Let.* 4 Mar. (1972) I. 129 There is a good deal of spontaneous music in the Italians... Their street songs are perfect of their kind. **1959** W. R. Bird *These are Maritimes* x. 274 There were street songs brought out over the years from the old country. **1959** G. Wickham *Early Eng. Stages* I. iii. 51 Specially erected platforms..in market squares or other open spaces..are usually known by such names as 'booth theatres', *théâtres de la foire*, or simply 'street theatres'. **1977** *Spare Rib* May 16/1 I'd like to do street theatre, but it's not that easy in a place like Sheffield. **1870** D. J. Kirwan *Palace & Hovel* xxvi. 395 These dog-sellers are the keenest street-traders to be found in London. **1979** S. Brett *Comedian Dies* i. 17 [He] spoke with the brash confidence of an East End street-trader. **1903** *Act 3 Edw. VII* c. 45 § 2 Any local authority may make bye-laws with respect to street trading by persons under the age of sixteen. **1977** J. Thomson *Case Closed* viii. 99 They'd kept their street-trading licence and..they'd go round the local markets selling clothing. **1872** B. Jerrold *London* ii. 23 Stopped by street vendors of all descriptions. **1978** N. Longmate *Hungry Mills* vii. 100 His heart went out to the inexperienced street-vendors he encountered. **1977** *Times* 22 Jan. 4/4 The two days of street violence [in Cairo] which took more than 60 lives. **1830** in *Times* (1982) 7 June 14/7 Locomotive Carriages might be used with great advantage in cases of Riot and Street Warfare. **1938** 'G. Orwell' *Homage to Catalonia* x. 174 Few experiences could be..more nerve-racking than those evil days of street warfare.

b. *street architecture, -corner* (earlier and further examples), *-crossing, -end* (earlier example), *island, lamp* (earlier example), *map, -name* (later example), *plan, -side* (later examples).

1933 J. Betjeman *Ghastly Good Taste* vi. 99 The true eighteenth century tradition, which lavished adornment on the interior and did not worry as much about street architecture. **1978** *Architectural Design* 5 June 314/2 Shaw's Albert Hall Mansions of 1879–81..are excellent as street architecture. **1836** Dickens *Sk. Boz* 2nd Ser. ii. 22 The policeman at the street corner. **1944** [see *feel *v.* 9 e]. **1978** J. Wainwright *Jury People* xxxvi. 118 Yobbo types, tearaways, bully-boys, street-corner louts. **1875** 'Mark Twain' in *Atlantic Monthly* May 571/1 Go on until you know every street-crossing, the character, size, and position of the crossing stones, [etc.]. **1956** D. Gascoyne *Night Thoughts* 26 Street-crossing islands stand becalmed. **1977** R. L. Wolff *Gains & Losses* ii. iii. 243 The virtuous street urchin who has never heard of Christ..raises himself..to the proprietorship of a muddy street-crossing as a sweeper. **1890** Kipling *Life's Handicap* (1891) 79 A lamp at a street-end. **1919** J. Buchan *Mr. Standfast* x. 187 A hundred yards away a bomb fell on a street island. **1934** *Sun* (Baltimore) 31 May 5/3 A hard-driving taxi driver ignored a red signal, threatened the traffic policeman's knees, missed the street island by a hair [etc.]. **1799** C. B. Brown *Arthur Mervyn* I. iv. 33 [The room's] height and spaciousness were imperfectly discernible..by gleams from a street lamp. **1964** L. Deighton *Funeral in Berlin* vii. 52 Spectacles produced a street map and..began marking circles here and there. **1978** T. Allbeury *Lantern Network* x. 141 The street map showed it as a small road off the Brighton Road. **1970** J. McN. Dodgson *Place-Names Cheshire* I. p. xliv, As a rule, street-names not recorded before 1700 are excluded. **1929** *Woolley's Ludlow Guide* (ed. 18) 44 (*caption*) Street plan of Ludlow. **1978** W. J. Burley *Wycliffe & Scapegoat* iii. 55 Wycliffe studied the street plan. 'Here we are. Albert Terrace.' **1974** *New Yorker* 29 Apr. 47/1 The odd oarsman..looks forlorn seen from the bridges or the streetsides. **1977** *Antiquaries Jrnl.* LVII. 251 The building extended only for the width of the streetside room..from the street frontage.

c. *street lighting* (examples).

1916 G. B. Shaw *Androcles & Lion* p. lxxii, The sportsmen, the musicians, the physicists, the biologists will get their apparatus for the asking as easily as their bread or, as at present, their paving, street lighting, and bridges. **1979** *Time* 8 Jan. 23/2 In 1975 Gacy became a trustee of the Norwood Park Township Street Lighting District.

d. *attrib.* passing into *adj.*, with reference to the streets as the focus of modern urban life, esp. among the poor and contrasted with polite society. Often with the implication of illegal dealings (esp. drug-trafficking), or the sharp-wittedness needed to survive 'on the streets'. orig. *U.S.*

1967 'T. Wells' *Dead by Light of Moon* xiii. 126 A street merchant is a con artist who pretends to sell stolen goods. **1967** *Trans-Action* Apr. 5/1 Street culture exists in every low income ghetto. It is shared by the hustling elements of the poor, whatever their nationality or color. **1972** *N. Y. Times* 24 Dec. iv. 6 Murphy called a news conference to announce that 57 pounds of heroin with an estimated street value of more than $10-million had been stolen. **1973** D. Barnes *See the Woman* (1974) 73 His name is Frederick L. Pepper... He's got a street name of 'Red Pepper'. **1976** R. Condon *Whisper of Axe* i. xv. 87 The street price for one kilogram of heroin is one million two hundred thousand dollars. **1979** W. J. Fishman *Streets of East London* 74/1 A street culture based on the pub. **1980** *Brit. Med. Jrnl.* 6 Dec. 1511/1 Phencyclidine is now a class 2 controlled substance in the United States—and after marijuana has become the most widely abused 'street drug' in North America. **1982** R. Leigh *Girl with Bright Head* ii. 12 She wasn't street tough but neither was she a runaway kid up from the provinces.

e. street-Arab (earlier and later examples); **street-boy** (earlier example); **street-child,** a homeless or neglected child who lives chiefly in the streets; **street crime** *U.S.*, a crime such as robbery, assault, etc., committed on the streets; **street fighting,** fighting conducted in the streets, esp. on a large scale for political or revolutionary ends; **street floor** *U.S.* = Ground-floor; **street furniture,** objects such as post-boxes, road-signs, litter bins, etc., placed in the street for public use or assistance (orig. a planners' term); **street girl,** a homeless or neglected girl who lives chiefly in the streets; a prostitute; **street-grid,** an arrangement of streets crossing at right angles to each other; **street hockey** *N. Amer.*, a variety of ice hockey played on the street; **street jewellery,** painted enamel advertising plates considered as collectors' items; **street kid** = *street child* above; **street-legal** *a.*, applied to a motor vehicle which satisfies the legal requirements for roadworthiness; **street level,** (*a*) ground-floor level; (*b*) *fig.*, the level of direct contact with the public or of operation on the streets; **street-light,** (*b*) (later examples); **Street name** *U.S.* [after *Wall Street], the name of a stock-brokerage firm, bank, or dealer in which stock is held on behalf of the purchaser; **street people** orig. and chiefly *U.S.*, (*a*) homeless or vagrant people who live on the streets, esp. as a protest against the conventional values of society; (*b*) people involved in petty crime in the urban underworld; (*c*) *spec.* people dealing in the illicit supply of drugs 'on the street'; **street rod** orig. *U.S.* (see quot. 1954); hence **street rodding** *vbl. sb.*; **street-scape,** a view or prospect provided by the design of a (city) street or streets; **street scene,** the spectacle of life in the streets; **street-smart** *a. U.S. slang* = *street-wise* adj. (*b*) below; also **street-smarts,** the ability to live by one's wits in an urban environment;

street-to-street *a.*, of fighting: taking place in the streets; **street tree,** a tree planted at the side of a street to enhance the view; **street urchin** (later examples); **street village,** a long, narrow village formed of buildings along either side of a main street; **street warden,** (*a*) an air-raid warden assigned to a particular street or streets; (*b*) a warden selected to look out for certain social problems in a particular street or streets; **street-wise** *a. slang* (orig. and chiefly *U.S.*), (*a*) familiar with the outlook of ordinary people in an urban environment; (*b*) cunning in the ways of modern urban life; **street worker** orig. *N. Amer.*, a social worker whose concern is with juvenile delinquents.

1859 G. A. Sala *Twice round Clock* 388 Street Arabs, threw 'cart-wheels' into the midst of the throng. **1924** Lawrence & Skinner *Boy in Bush* 49 The children.. were singing..with a sort of street-arab abandon. **1854** Dickens *Hard Times* i. xvi. 127, I was a ragged street-boy. **1863** —— *Mrs Lirriper's Lodgings* i, in *All Year Round* Extra Christmas No. 3 Dec. 8/2 You must allow me to inform you..that my grandson is *not* a street-child. **1959** I. & P. Opie *Lore & Lang. Schoolch.* xii. 232 The street-child today with his soot-blackened face and red-daubed nose, rattling a tin, is a much more demure creature than his predecessors. **1982** G. Wagner *Children of Empire* vii. 123 Three-quarters of his young life had been spent in the workhouse, yet technically he was a 'street child'. **1973** *Listener* 20 Sept. 364/1 You'd expect.. New York to take a highly sophisticated view of the drug problem, for it is more subtle in its operations here,..and responsible for more street crimes—robberies and rapes—than in any other State. **1978** *Chicago* June 162/2 The state's attorney's office..must prosecute virtually all local street crime, leaving meager resources for long, complex investigations. **1832** F. Maceroni (*title*) Defensive instructions..on..street and house fighting. **1900** W. S. Churchill in *Morning Post* 12 July 7/7 The cavalry halted on the hills for a while, the general being desirous of obtaining the formal surrender of Heilbron, and so preventing street-fighting or bombardment. **1981** *Times* 6 July 13/1 Within a few months there have been three major eruptions of street fighting, all of which have included an ethnic element. **1927** Doubman & Whitaker *Organization & Operation of Department Stores* vii. 162 The first or street floor of a store is the most desirable for selling. **1972** H. Kemelman *Monday the Rabbi took Off* xxvi. 170 Why would he take an apartment on a street floor here? **1944** J. C. Riddell *Rep. Post War Housing* 11 In all future planning there should..be the closest co-operation between all departments and services responsible for the erection of street furniture and small buildings. **1976** *Cumberland & Westmorland Herald* 27 Nov., We don't want to waste officers' time..on more street furniture which costs a fortune today. **1907** G. B. Shaw *Major Barbara* i. 206 You have had the education of a lady... Don't talk like a street girl. *a* **1911** [see *most *adv.* 4]. **1979** *Maledicta* III. 11 For a budding sexologist I must have been uncommonly naive, but I swear that I didn't catch on that these were street girls' joints till it dawned on me, while sorting out my notes, that I seemed to have a disproportionate number of sexual idioms. **1948** *Antiquity* XXII. 173 The wide road leading to its main entrance from the east is plainly out of alignment with the street grid on the west side of the Forum block. **1964** *Listener* 27 Aug. 300/2 The avenue slices diagonally across the basic street-grid. **1964** *Globe & Mail* (Toronto) 15 Dec. 41/8 He brought a fresh approach to street hockey when he began trying to teach Frank how to shoot a puck. **1976** *New Yorker* 26 Apr. 90/3 Meynell and I played marbles, mumblety-peg, running games, street hockey, primitive baseball, stoop ball, games of imagination. **1978** Baglee & Morley *Street Jewellery* 9 Street jewellery, flashing in the winter sunlight, gleaming in gaslight..—the enamel sign. **1982** *Arts North* June 9 (*caption*) Street jewellery—one of the saucier enamel signs from the fascinating exhibition at the Dorman Museum. **1929** E. Wilson in *New Republic* 17 Apr. 256/2 The money with which the street-kids have been playing craps. **1977** *Rolling Stone* 5 May 55/1 You could go to New Delhi or Calcutta, there are thousands of street kids there. **1976** *Casper* (Wyoming) *Star-Tribune* 29 June 19/1 (Advt.), Yamaha 125. Street legal. Good condition. **1980** *Dirt Bike* Oct. 58/3 The rear fender is plastic, which is a rarity on street legal bikes. **1934** *Archit. Rev.* LXXV. 214/3 The storey built above the street-level floor was called a solar. **1963** 'J. le Carré' *Spy who came in from Cold* xv. 140 Branch Secretaries with..a good record of stimulating mass action at street level. **1974** 'E. Lathen' *Sweet & Low* xi. 114 He reached street level. **1976** *Times* 6 Sept. 2/3 Plans are advanced to open..an office in Belfast..accessible to both communities. An effort will be made to concentrate organization at street level. **1982** G. F. Newman *Men with Guns* ix. 69 Kohn avoided contact with street-level hoods... Now he was a respectable businessman. **1955** E. Blishen *Roaring Boys* iii. 128 The street lights..were on early. **1969** L. Michaels *Going Places* 27 Streetlights glowed in a receding sweep. **1930** C. F. Hodges *Wall Street* 383 When a security is registered in the name of a recognized brokerage firm, usually members of the Exchange, it is said to be in a 'Street Name'. **1933** *North Western Reporter* CCXLVI. 660/1 The court properly instructed as to the 'street name' custom of the exchange..submitted to the jury the question of fact [etc.]. *Ibid.* 664/2 Evidence sustained finding that customer's repudiation of broker's purchase of stock in 'street name', instead of customer's name, was made within reasonable time. **1976** E. Stewart *Launch!* (1977) 35 'And I want to know how much the brokerage houses are holding in street names.' All the brokerage houses had street-name accounts, stock bought and traded for clients in the broker's name. **1967** *Trans-Action* Apr. 5/1 In Los Angeles, members of.. street groups sometimes call themselves 'street people', 'cool people', or simply 'regulars'. **1969** *Guardian* 24 May 1/3 The precincts inhabited by Berkeley's hippies and 'street people'. **1972** *National Observer* (U.S.) 27 May 7/2 There's evidence that methadone has become almost as popular as heroin among addicts. Street people say so. **1976** *Billings* (Montana) *Gaz.* 4 July 2-B/3 'At first, we got mostly street people,' said Nyberg. 'Lately our patients have begun to be from the higher social levels.' **1954** *Amer. Speech* XXIX. 103 *Street rod* (*job, roadster*, etc.), *n.* A hot rod suitable for street use, one of the competition types. The 'street rod' is distinguished from the 'track car', which is intended primarily for drag-strip or lakes racing. **1972** *World of Wild Wheels* (Custom Car) 58/1 Street rods don't have to be American based—just tinged with American thought. **1977** *New Society* 3 Mar. 436/2 Custom-car cruisers, in their glistening, overpowered improvisations... The monthly influx of around 250 'street rods' causes a solid traffic jam. **1976** *Panorama* (Austral.) Dec. 4 One of Australia's fastest-growing sports is street rodding—turning pre-1948 cars into sparkling, high-performance vehicles which belie their age. **1924** *Glasgow Herald* 8 Mar. 9 Where aerial invaders left ugly.. scars in the streetscape noble new buildings have already appeared. **1979** *Jrnl. R. Soc. Arts* Nov. 770/2 Giorgio Grassi and others have designed housing blocks as long arcaded streetscapes. **1870** D. J. Kirwan *Palace & Hovel* xxvi. 407 The great..attraction among the multifarious street scenes of London, is the Punch and Judy show. **1979** *N.Y. Times Mag.* 30 Sept. 37/2 The girls' mother, Ada, is down on the stoop, watching the street scene as if it were television. **1976** *National Observer* (U.S.) 1 May 5/1 Rizzo is tough, street-smart, charming in his own special way. **1976** *N.Y. Times* 9 Aug. 30 To be free, however, requires street-smarts, the cunning of the survivor. **1978** *Time* 3 Apr. 61 Norris also sought out local black leaders and followed their street-smart advice. **1978** *New Yorker* 20 Nov. 113 They thought always about winning, and, one way or another, they almost always did win. Like the A's, these Yankees have street-smarts. **1983** *Underground Grammarian* VII. iii. 7/2 Frank will be demoted to the lowliest rank in education, teacher, so that those adaptable street-smart kids can go and apply their skills in *her* classroom. **1945** *Finito! Po Valley Campaign* (15th Army Group) 33 Street-to-street battles. **1976** *Southern Even. Echo* (Southampton) 12 Nov. 10/3 She recalls the street-to-street fighting that became an everyday feature. **1911** W. Solotaroff *Shade-Trees in Towns & Cities* p. ix, This book treats particularly of the planting and care of street-trees. **1981** *Garden* CVI. 443/2 They [*sc.* local authorities] might be persuaded to plant street trees. **1977** Street urchin [see *street-crossing*, sense *4 b]. **1978** J. Krantz *Scruples* vi. 168 Jake..had a droll and artful street-urchin look to him and typically Black Irish coloring. **1949** *Ann. Assoc. Amer. Geographers* XXXIX. 261 *Street Villages* (*Strassendörfer*), the name being used only for those villages which were founded on an existing route. **1974** C. Taylor *Fieldwork in Medieval Archaeol.* vi. 142 Caxton..is now a long street village on either side of the Old North Road. **1940** N. Last *Diary* Oct. *Nella Last's War* (1983) 78 Our street warden called tonight... He wanted to know if we had buckets, stirrup-pumps, blankets, bandages, [etc.]. **1973** *Daily Tel.* 8 Jan. 2/2 Mrs Green said her organisation wanted street wardens whose job 'for perhaps £1 a week' would be to call on old people each morning. **1980** *Church Times* 11 Apr. 6/4 Much is said about almost every activity that laypeople can undertake—forming a 'Jesus gang', personal witness, acting as street wardens, helping the bereaved, befriending the elderly. **1965** *New Yorker* 27 Mar. 78 A [social] worker therefore had to be wary as well as trustful, be security minded as well as loving, and be 'street-wise' as well as compassionate. **1971** *N.Y. Times* 18 June 37 Take a dirt-poor Sicilian peasant kid fresh out of steerage. Make him scrappy and street-wise. **1977** H. Fast *Immigrants* v. 321 Al Smith, street-wise Catholic from New York. **1980** *Times Lit. Suppl.* 1 Aug. 867/5 The learned men on the council of the SPR were not, as we would say now, 'street-wise'. **1981** *Daily Tel.* 27 Nov. 16/2 Their [*sc.* young blacks'] values place a premium on being 'street-wise',..that is, being able to survive in the rough and tough world of the streets. **1964** *Maclean's Mag.* 25 Jan. 23 Almost all of them have quit school. The street-worker has become so friendly with them that he can sometimes return stolen goods before the police are even aware of the theft. **1973** 'J. Patrick' *Glasgow Gang Observed* xxi. 219 Adolescents..did not know how to react to the non-evaluative, non-judgmental approach of the street worker.

street-car. **1.** For *U.S.* read *N. Amer.* and add further examples.

1887 *Grip* (Toronto) 5 Feb. 6/2 Toronto law is plain and hard—no streetcars out on Sunday. **1915** D. R. Campbell *Proving of Virginia* 108 So I shall bid you good-by and take a street car home. **1929** M. de la Roche *Whiteoaks* iv. 58 They boarded a street car and stood together, swaying, hanging by the straps,..oblivious of the other passengers. **1931** W. G. McAdoo *Crowded Years* iv. 44 The street cars went like sleepy tortoises; they were pulled by mules. **1947** *Partisan Rev.* XIV. 366 They returned in a streetcar, although Jasper wanted to take a taxi. **1947** T. Williams (*title*) A streetcar named Desire. **1968** *Globe & Mail* (Toronto) 15 Jan. 17/1 Bus companies reported delays and power failure contributed to the problems of city street cars. **1974** *Plain Dealer* (Cleveland, Ohio) 26 Oct. 5-D/1 The people who came to the Barons-Rangers game that night long ago came by streetcar and bus and by shank's mare as well as by auto. **1978** *Detroit Free Press* 16 Apr. (Record) 9/4 The Warsaw street cars stop sixty or seventy meters from the gates of the ghetto.

2. A shell. *Mil. slang.*

1920 C. R. Herr *Company F Hist.: 319th Infantry* 22 The air was filled with the sounds of the shells as they lazily went on their way towards the back lines of both sides 'Street cars'..the boys called them. **1950** R. Chandler *Let.* 18 May (1966) 78 Doesn't he [*sc.* Partridge] overlook some of the most commonly used words of soldier-slang? E.g...'street cars' or 'tram cars' for heavy long range shells.

streetman. Add: **2. b.** A petty criminal who works 'on the street', esp. as a pickpocket or drug pedlar. *U.S. slang.*

1908 'O. Henry' *Gentle Grafter* 161 I'd like to shake hands with Parley-voo Pickens, the greatest street man in the West. **1974** *Publishers Weekly* 21 Jan. 80/3 He is playing partner to the pusher whose street man is keeping the girl hooked. **1981** W. H. Hallahan *Trade* ii. 41 They were the perfect team, the tough street man and the elegant boardroom manipulator.

Strega (strē̆·gä). The proprietary name of a kind of Italian liqueur flavoured with orange; a drink or glassful of this.

1910 *Trade Marks Jrnl.* 19 Jan. 104 Liquore Strega. Liqueurs. The firm trading as Ditta Giuseppe Alberti. **1920** D. H. Lawrence *Lost Girl* xvi. 358 Pancrazio took her to the place where she could drink coffee and a strega. **1922** —— *Aaron's Rod* xiv. 203 The waiter rattled off a list, beginning with Strega and ending with cherry brandy. **1938** E. Ambler *Cause for Alarm* iv. 63, I found..an empty Strega bottle. **1948** W. S. Maugham *Here & There* 25, I ordered coffee and strega, which is the best liqueur they make in Italy. **1974** K. Clark *Another Part of Wood* v. 179 The only penalty was that the meal traditionally ended with quantities of Strega, which gave one a sore head the next day.

strength, *sb.* Add: **1. d.** Freq. in phr. *strength of character.*

1836 J. H. Newman *Parochial Sermons* III. i. 3 Of course men who make such sacrifices, often evidence much strength of character in making them. **1863** N. Hawthorne *Our Old Home* i. 74, I have not found reason to suppose that the English dowager of fifty has actually greater courage, fortitude, and strength of character than our woman of similar age. **1919** G. B. Shaw *Heartbreak House* ii. 52 You know, Ellie has remarkable strength of character. I think it is because I taught her to like Shakespear when she was very young. **1957** *Oxf. Dict. Chr. Ch.* 583/2 It was owing to Gregory, in whom firmness and strength of character were tempered by gentleness and charity, that many of these evils were conquered. **1975** *Economist* 15 Mar. 38/3 Mike Denness had the job sewn up after his batting performances against India last summer... His batting efforts..show his strength of character.

j. (Later examples.) In *Telecommunications* also with following numeral, indicating signal strength as shown on a meter.

1914 *Rep. Brit. Assoc. Adv. Sci. 1913* 132 These instructions would include directions for simultaneous observations of..the strength of the time-signals..and the average strength and frequency of strays. **1923** *Radio Times* 28 Sept. 2/1 We can take the following as useful ranges from one of the main broadcasting stations for good strength on the head telephones. **1968** J. Sangster *Touchfeather* xvi. 188 He finally got the message, strength five. 'What do you want?' **1979** P. Niesewand *Member of Club* i. 6 'I've got them, but they're only hearing me strength two.'..'How do you hear them?' 'Strength five.'

2. e. *the strength of*: the point or meaning of, the essential facts about (*ellipt.* in quot. 1958). Esp. in phr. *that's about the strength of it*: that is what it amounts to (cf. *that's, (about) the size of it* s.v. *size *sb.*[1] 10 f); *to get the strength of*: to understand. *colloq.* (orig. and chiefly *Austral.* and *N.Z.*).

1908 H. Fletcher *Dads & Dan: between Smokes* 112 'So yous thinks I'se wore out,..an' past patchin' an' mendin'?' 'That's about ther strength of it.' **1916** C. J. Dennis *Moods of Ginger Mick* 63 Then, bit be bit, Mick gits the strength uv it. **1926** K. S. Prichard *Working Bullocks* xv. 136 Now..I'll just give you the strength of Red Burke... They say there never was a good Burke. **1943** N. Marsh *Colour Scheme* v. 93, I don't get the strength of it myself. He wouldn't say much. **1946** K. Tennant *Lost Haven* (1947) ix. 129 If it hadn't been for her engine..you might just as well have left her on the sandbar to go to pieces... That's about the strength of it. **1958** F. Norman *Bang to Rights* i. 10 The strength was that he'd got nicked for ponceing off his old woman. **1965** A. Prior *Interrogators* x. 188 'Just passing and you saw the door was open?' He laughed. 'Well, yes, that's just about the strength of it.' **1969** *Advertiser* (Adelaide) 12 May 5/4 Get the strength of this: You [*sc.* Australians] talk about bankos and trunks—is that English? **1974** J. Cleary *Peter's Pence* vi. 178 'What's the strength of all this?' 'Strength?' Kessler's English didn't run to Australian colloquialisms. 'What's the point, the meaning?'

f. *give me strength*: used as an expression of exasperation.

1967 'S. Woods' *And shame Devil* 251 'Give me strength,' said O'Brien helplessly. 'I'll try to explain.' **1970** K. Benton *Sole Agent* xviii. 194 'You make all my plans sound so drab and sordid.' 'Oh give me strength!'

16*. *strength through joy* [tr. G. *kraft durch freude*]: a movement founded in Germany by the National Socialist Party in 1933 to promote physical and cultural recreational activities among working people. Also *transf.* and *fig.* (Freq. with capitals and hyphens.)

1935 G. Dimitrov *Working Class against Fascism* in *Rep. 7th World Congress* (1936) 43 In the Hitler Youth Leagues, in the sports organizations, in the *Kraft durch Freude* organisations (*Strength Through Joy*). **1939** *Ann. Reg. 1938* iii. 201 During the first days after the 'Anschluss', 10,000 [Austrian] workers..inclined to Communism and Socialism, had been invited for a fourteen-day trip to Germany where they found a hearty welcome and were shown the institutions of the *Kraft durch Freude*

(strength through joy) movement. **1943** *Tribune* 4 June 19/1 The strength-through-joy brigades you will have met Whose mouths are baggy and whose hair is scented. **1962** L. R. Banks *End to Running* ii. iv. 177 Full of an awful sort of phoney strength-through-joy. **1967** T. Stolper tr. *G. Stolper's German Economy* v. 152 Annual paid vacations, inexpensive theaters and concerts, and all the other activities of the party's leisure time organization—'Strength through Joy' (*Kraft durch Freude*). **1973** 'G. Black' *Bitter Tea* x. 156 The girl..looked as if she had graduated from one of Lee Kuan Yew's strength-through-joy courses. **1975** *Listener* 16 Jan. 71/3 Physical fitness was a Nazi fetish..bronzed young Germans cultivating 'strength through joy'. **1979** J. Gardner *Nostradamus Traitor* xv. 86 'They sent me up to Scotland.' It was a toughening-up course at a Strength Through Joy Camp... Survival. Living off the land.

17. *strength-(to-)weight ratio*; *strength-giving* (earlier example), *-sapping, -showing* adjs.

1845 J. R. Lowell in *Amer. Rev.* Aug. 137, I saw them in all higher moods, and durst Face their strength-giving eyes. **1961** *Times* 6 Dec. 3/4 Both boxers kept up a strength-sapping pace. **1939** N. de V. Hart *Bridge Players' Bedside Bk.* 133 North's Two Spades is a true strength-showing reverse, because South has to raise the bidding level to three in order to put North back to clubs. **1967** *Bridge Players' Encycl.* 490/1 *Strength-Showing Bids*, in some special situations a suit bid can be used to show strength rather than length or control. **1978** *Jrnl. R. Soc. Arts* CXXVI. 682/2 The primary incentive for the development of titanium was without doubt its strength-to-weight ratio and the potential of this property in aircraft construction. **1945** F. S. Stewart *Airframe Materials* i. 2 The strength-weight ratio of materials used in airframes is of such great importance.

strep (strep), colloq. abbrev. of (*a*) Streptococcus (freq. *attrib.*); (*b*) *streptomycin.

(*a*) **1927** *Amer. Speech* II. 313/2 A streptococcus infection of the throat becomes a 'strep throat'. **1941** R. Chandler *Let.* 1 Feb. (1981) 19 Awfully sorry to hear you had been sick. I know what the streps can do to a person. **1956** [see *staph]. **1962** A. Lurie *Love & Friendship* iv. 68 There's germs flying round in the air.., pneumonia and bronchitis and strep. **1966** H. Kemelman *Saturday the Rabbi went Hungry* iv. 25 It's a strep infection, the doctor says. **1974** [see *shet, shed]. **1980** *Daily Tel.* 6 Oct. 14/6 Strep infections can cause breathing disorders, shock, bleeding and meningitis in newborn babies.

(*b*) **1959** J. Braine *Vodi* vi. 85 They'd tried strep. and P.A.S. **1961** C. Cockburn *View from West* i. 6 Hallucinated myself by the effects of streptomycin—we soon all got very matey with this potent drug and called it strep.

strephosymbolia (stre:fosimbō̆ᵘ·liä). *Psychol.* [mod.L., f. Gr. στρέφειν to turn, twist + σύμβολ-ον (see Symbol *sb.*[1]) + -ia[1].] (See quot. 1937.)

1925 S. T. Orton in *Arch. Neurol. & Psychiatry* (Chicago) XIV. 610 The term 'congenital word-blindness' because of its association with the acquired condition and the implications therefrom, does not seem to be properly descriptive of this disability, and I would therefore like to offer the term 'strephosymbolia'..as a descriptive name for the whole group of children who show unusual difficulty in learning to read. **1937** —— *Reading, Writing & Speech Problems in Children* 214 *Strephosymbolia*, a delay or difficulty in learning to read..characterized by confusion between similarly formed but oppositely oriented letters, and a tendency to a changing order of direction in reading. **1968** E. J. Kahn *Harvard* xiv. 223 There are students who should be exempt from language requirements because they suffer from strephosymbolia—an affliction that makes it difficult..to see the ends of written words in unfamiliar languages.

strepie (strī·pi). *S. Afr.* Also **streepie, streepje.** [Afrikaans, f. *streep* Stripe *sb.*[3] + -ie.] A small fish, the striped karanteen, *Sarpa salpa*. Cf. *karanteen.

1913 [see *bamboo-fish* s.v. *bamboo *sb.* 2]. **1945** *Cape Argus* 27 Jan. 4/8 Half an inch of her wool..tied to a 'streepie' hook, caught about a dozen small fish. **1957** S. Schoeman *Strike!* iii. 89 The strepie, with its beautiful gold and silver colouring, is the favourite prey of the elf. **1974** *Stand. Encycl. S. Afr.* X. 320/1 Strepie..is one of the most beautiful and best-known bait fishes in southern Africa.

‖ **strepitoso** (strepitǭ·so), *a.* (*adv.*) *Mus.* [It., lit. 'noisy, loud'.] A direction indicating that a composition be played in a spirited or boisterous manner. Also as *sb.*, a piece designed to be played in this manner.

1801 T. Busby *Compl. Dict. Mus.*, *Strepitoso*.., a word signifying that the movement to which it is prefixed is to be performed in an impetuous, boisterous style. **1876** Stainer & Barrett *Dict. Mus. Terms* 410/1 *Strepitoso* (*It.*), noisy, impetuous. **1946** E. Blom *Everyman's Dict. Mus.* 596/2 *Strepitoso* (It. = noisy), a direction suggesting a forceful and spirited perf., but more often used in the sense of a climax growing in force and speed. **1966** *Listener* 17 Feb. 256/3 The first act finale of both operas ends with a canonic *andantino* in A flat leading to a prolonged *strepitoso* in C major.

† **strepsinema.** *Cytology. Obs.* [f. Gr. στρεψι-, comb. form of στρέφειν to twist + νῆμα yarn.] A condition of the nucleus during cell division, characterized by pairs of chromosomes twisted around one another or in the form of twisted rings; in most cases applied to diplotene nuclei.

1900 H. H. Dixon in *Proc. R. Irish Acad.* VI. 2 In the next stage (fig. 2) the nucleus is in what I would suggest to call the 'strepsinema' condition. The chromatin appears in much the same condition as in the preceding stage, except that in many places it may be seen that the two portions of the thread are more or less loosely twisted together. **1911** *Q. Jrnl. Microsc. Sci.* LVII. 15 The onset of synizesis coincides with the entrance of the nucleus into strepsinema. **1925** E. B. Wilson *Cell* (ed. 3) ii. 126 A noteworthy peculiarity of the spireme sometimes seen is a twisting of the longitudinal halves about each other to form a strepsinema; but this is much less common in the somatic mitoses than in meiosis.

† **strepsitene.** *Cytology. Obs.* [ad. F. *strepsitène* (V. Grégoire 1907, in *La Cellule* XXIV. 372), f. prec.: see *-tene.] = prec.

1911 *Q. Jrnl. Microsc. Sci.* LVII. 14 Following immediately after the bouquet is the strepsitene or diplotene stage, in which the conjugants which were temporarily united in the pachytene loops separate again. **1925** E. B. Wilson *Cell* (ed. 3) vi. 544 Sooner..or later.., the threads are plainly longitudinally double (diplonema); and the two threads, especially in the later stages, are often spirally twisted about each other to form the strepsinema or strepsitene. These various conditions cannot as yet be very logically separated as distinct stages.

strepto-. Add: **streptostylic** *a.*, (further examples); now used with reference to the free articulation of the quadrate bone with the squamosal rather than to any taxonomic group; [ad. G. *Streptostylica*, name of a group (H. Stannius 1856, in von Siebold & Stannius *Handb. der Zootomie* (ed. 2) II. 45)]; so **stre·ptostyly,** streptostylic condition. Also used as comb. form of Streptococcus, Streptococcal *a.*, as in *streptodornase, and of *Streptomyces* (see *streptomycete), as in *streptovaricin.

1933 *Univ. Calif. Publ. Zool.* XXXVII. 524 A comparison of the streptostylic condition in these salamanders with that in certain reptiles..shows great differences in both skeleton and muscles, such that no homology is possible. **1980** *Nature* 21 Feb. 779/1 Streptostylic quadrates are found in fossil lizards before the appearance of mesokinesis and occur in animals with limited mesokinetic potential such as *Uromastix* and *Ctenosaura*. **1925** J. S. Kingsley *Vertebrate Skeleton* (ed. 2) 335/2 (Index), Streptostyly. **1933** *Univ. Calif. Publ. Zool.* XXXVII. 521 In a study of some Pacific coast salamanders, it was found that at least three species..possess a pivoting squamosal... Streptostyly, so far as the writer has been able to learn, is otherwise unknown in Amphibia. **1973** *Nature* 11 May 72/2 The articulation of the maxillae with the nasals and premaxillae reveals how streptostyly was associated with an akinetic skull in *Hesperornis*.

streptocarpus. (Earlier example.)

1828 J. Lindley in *Bot. Reg.* XIV. 1173 (*heading*) Cape Streptocarpus.

streptodornase (streptodǭ·ɹnēⁱz). *Pharm.* and *Biochem.* [f. Strepto(coccal *a.* + *d(e)o(xy)r(ibo)n(ucle)ase.] A deoxyribonuclease, or a group of such enzymes, produced by some streptococci and used in conjunction with streptokinase to bring about the dissolution of purulent and fibrinous exudates.

1949 S. Sherry et al. in *Jrnl. Clinical Investigation* XXVIII. 1094/1 The word 'Streptodornase' has been employed in this article as an abbreviation of streptococcal desoxyribose nuclease. **1977** *Proc. R. Soc. Med.* LXX. 571/1 Skin testing to streptokinase-streptodornase, candida and phytohaemagglutinin proved negative.

streptokinase (streptokəi·nēⁱz). *Pharm.* [f. as prec. + *kinase.] An enzyme produced by hæmolytic streptococci which activates plasminogen to form plasmin and is given intravenously to dissolve intravascular blood clots.

1944 Christensen & Macleod in *Jrnl. Gen. Physiol.* XXVIII. 581 Streptococcal fibrinolysin, also a misnomer in the light of present knowledge, may be termed 'streptokinase', analogous to 'enterokinase' or 'mold kinase'. **1969** *Brit. Med. Jrnl.* 29 Mar. 812/1 Nine patients with arteriographically proved pulmonary embolism have been treated by a 36-hour infusion of streptokinase. **1978** F. X. Hasselberger *Uses of Enzymes* x. 122 A controlled experiment..confirmed the superiority of streptokinase over heparin (a chemical anticoagulant) in reducing deaths due to acute myocardial infarction.

streptomycete (streptomǝi·sīt). *Bacteriology.* [f. mod.L. *Streptomycetes*, pl. of *Streptomyces*, generic name (Waksman & Henrici 1943, in *Jrnl. Bacteriol.* XLVI. 339), f. Strepto- + Gr. μύκης fungus.] A bacterium of the genus *Streptomyces*, which belongs in the family Streptomycetaceæ of spore-forming, mould-like bacteria and comprises aerobic forms that grow as branching filaments, form spores in chains, and occur chiefly as saprophytes in soil.

1956 Carter & Smith *Microbiol. & Pathol.* (ed. 6) xxxii. 494 The streptomycetes are of medical importance because they are the source of the antibiotics, streptomycin, [etc.]. **1961** *Lancet* 12 Aug. 377/2 A large variety of mould products have been reported effective in cancer in man, from myxomycetes to streptomycetes. **1977** R. W. Thoma *Industr. Microbiol.* 51 Rigorous proof for genetic recombination via transduction or transformation among streptomycetes is still lacking. **1978** *Nature* 31 Aug. 844/2 C. Stuttard..reported the isolation of virulent transducing phage in a chloramphenicol-producing streptomycete.

streptomycin (streptomǝi·sin). *Pharm.* [f. prec.: see *-mycin.] An antibiotic, $C_{21}H_{39}N_7O_{12}$, produced by the soil bacterium *Streptomyces griseus*, which was the first drug to be successful against tuberculosis but is now used chiefly in conjunction with other drugs because of its toxic effects. Also *Comb.*

1944 A. Schatz et al. in *Proc. Soc. Exper. Biol. & Med.* LV. 67/1 Because of its similarity to streptothricin, this substance may be designated as streptomycin, derived from the generic name that has recently been given to the aerial-mycelium producing and sporulating group of actinomycetes, namely *Streptomyces*. **1948** 'G. Orwell' *Let.* 4 Feb. in *Coll. Ess.* (1968) IV. 404 We are now sending for some new American drug called streptomycin which they say will speed up the cure. **1961** *Lancet* 29 July 247/2 Streptomycin-resistant strains [of gonococci] remained.. susceptible to penicillin. **1973** *Sci. Amer.* Sept. 130/2 Streptomycin and isoniazid have diminished the need for state tuberculosis sanatoriums. **1974** M. C. Gerald *Pharmacol.* xxvi. 457 Streptomycin binds to the ribosome in such a manner that incorrect amino acids are laid down, thus resulting in the formation of a nonfunctional protein. **1978** *Antimicrobial Agents & Chemotherapy* XIII. 430 Combinations of penicillin and gentamycin have been shown to be synergistic against all strains of enterococci, including those resistant to penicillin and streptomycin.

streptosolen (streptosǭᵘ·lĕn). [mod.L. (J. Miers 1850, in *Ann. Mag. Nat. Hist.* 2nd Ser. V. 208), f. Strepto- + Gr. σωλήν pipe.] A climbing evergreen shrub, *Streptosolen jamesonii*, of the family Solanaceæ, which is native to Colombia and Ecuador and bears clusters of orange flowers.

1938 J. S. Dakers *Mod. Greenhouse* vii. 105 Streptosolen..has orange scarlet flowers produced in bunches. **1952** R. Genders *Greenhouse* xv. 150 Streptosolen..is a grand evergreen plant for covering a wall of a lean-to house. **1976** *Homes & Gardens* Aug. 75/2 At the minimum heat mentioned, little more than that needed to keep out winter frost, climbers one can grow include..blue plumbago, orange streptosolen, [etc.].

streptothricin (streptoþri·sin, -þrəi·sin). *Biochem.* [f. mod.L. *streptothric-*, Streptothrix + -in[1].] Each of a group of related antibiotic but toxic compounds, $C_{13}H_{21}N_6O_7$-$(C_6H_{12}N_2O)_n$, produced by the soil bacterium *Actinomyces lavendulæ*.

1942 Waksman & Woodruff in *Proc. Soc. Exper. Biol. & Med.* XLIX. 207 A new substance can now be added to this list of antibiotic agents. This substance has been obtained from a soil *Actinomyces*, and is designated as *streptothricin*, derived from the early generic designation *Streptothrix*, given to this group of organisms. **1955** *Sci. News Let.* 21 May 326/3 Two other antibiotics, streptothricin and noformicin,..are effective against Newcastle disease virus in the test tube. **1963** Barber & Garrod *Antibiotic & Chemotherapy* vi. 89 This first antibiotic of any value resulting from this investigation was streptothricin, which..was subsequently found to be too toxic for clinical use. **1972** *Jrnl. Antibiotics* XXV. 501 All the available preparations of antibiotics of this type were mixtures of six streptothricins differing in the number of L-β-lysine residues (1 in streptothricin F, 2 in streptothricin E and so on to 6 in streptothricin A). **1978** *Res. in Vet. Sci.* XXV. 110 The number of worms remaining in the infected dogs after one or more treatments with this streptothricin complex.

streptothricosis (stre:ptoþrikǭᵘ·sis). *Vet. Sci.* Also **-trichosis.** [f. as prec. + -osis.] A usu. chronic, sometimes fatal disease caused by actinomycetes and producing scabs on the skin of cows and other farm animals, esp. in the wet season in tropical regions.

1927 *Ann. Rep. Veter. Dept., Northern Provinces, 1926* (Nigeria) 10 Contagious Impetigo or Streptothricosis... This disease made its appearance in the Government herd [of cattle] at Vom. **1971** D. L. Doxey *Veter. Clinical Path.* xiv. 267 Although formerly classed as of differing aetiology, cutaneous streptothricosis in cattle, mycotic dermatitis in cattle, sheep, goats and horses, and strawberry foot rot of sheep, are all caused by the same organism, *Dermatophilus congolensis*. **1973** Ainsworth & Austwick *Fungal Dis. Animals* (ed. 2) 135 Dermatophilosis, especially in cattle (streptotrichosis) and sheep (mycotic dermatitis, strawberry foot-rot), occurs in many tropical and temperate countries as acute local infections of the skin following damage by prolonged soaking, insolation, tick bites or other agencies.

streptovaricin (streptovæ·risin). *Pharm.* [f. mod.L. *Strepto-myces* (see *streptomycete) + Vari(ous *a.* + *-my)cin.] Each of a group of related antibiotics produced by the bacterium *Streptomyces spectabilis*.

1957 P. Siminoff et al. in *Amer. Rev. Tuberculosis* LXXV. 582 A new antimicrobial complex, streptovaricin, is produced by *Streptomyces spectabilis*, n. sp. The complex consists of at least five microbiologically active, closely related components which have been named streptovaricins *A*, *B*, *C*, *D*, and *E*, respectively. **1972** *Accounts Chem. Res.* V. 60/1 The streptovaricins all contain an identical carbon skeleton and..they differ from one another in the degree of oxygenation and degree of acetylation.

streptozotocin (streptozǫ·tǒsin). *Pharm.* [f. as prec. + *-zoto-*, of unkn. origin + *-MY)-CIN.] An antibiotic substance obtained from *Streptomyces achromogenes* that damages insulin-producing cells and is used to induce diabetes in laboratory animals.

1960 J. J. VAURA et al. in *Antibiotics Ann. 1959–60* 234 Streptozotocin is a new antibiotic produced by a streptomycete isolated from the soil. **1972** *Jrnl. Pharm. Sci.* LXI. 491/1 Chemically streptozotocin is a *N*-methyl-*N*-nitrosourea derivative..of D-glucosamine. **1976** A. MARBLE et al. in G. S. Avery *Drug Treatment* xv. 421/1 Streptozotocin, another hyperglycaemic drug, has been similarly used in cases of malignant insulin-secreting tumour. **1980** *Nature* 3 Jan. 100/1 Proinsulin mRNA was purified from rat B-cell tumours induced with streptozotocin and nicotinamide.

Strepyan (stre·piăn), *a.* *Archæol.* Also **Strepyian.** [ad. F. *Strépyien*, f. *Strépy*, name of a town (the type site) in Belgium: see -AN.] Of or belonging to a palæolithic culture of Europe supposed to have existed before the Chellean. Freq. *absol.*

[**1904** A. RUTOT in *Bulletin Société d'Anthropologie de Bruxelles* XXIII. Mém. I. 15 Les industries éolithiques quaternaires et des pièces qui se rapportent absolument à notre transition de l'Éolithique au Paléolithique ou du Mesvinien au Chelléen, c'est-à-dire au Strépyien.] **1910** J. MCCABE *Prehistoric Man* i. 12 It is usual to admit three stages in the earlier Paleolithic, the names of which are taken from the French sites where we find them best exhibited... Advanced students, like M. Rutot, add an earlier stage (the Strepyian). **1911** W. J. SOLLAS *Ancient Hunters* v. 112 The distinctive character of the Strepyan industry, according to M. Rutot, is that all the implements retain a considerable part of the original crust of the flint nodule. **1914** J. GEIKIE *Antiquity of Man in Europe* ii. 43 The 'Strepyan', on the other hand, is marked by the presence not only of simple flakes but of primitive forms of the Chellean *coup de poing*. **1927** PEAKE & FLEURE *Apes & Man* vi. 90 The Strépyan, more often termed by others pre-Chellean, are accepted under the latter name by most archaeologists as being merely a very early type of Chellean. **1948** A. L. KROEBER *Anthropology* xvi. 631 Rutot's Mesvinian stage of the Eolithic is recognized as probably a Belgian facies of the oldest Levalloisian or Pre-Mousterian, his Strepyan as being Chellean—all of them Palaeolithic and Pleistocene. **1961** L. D. STAMP *Gloss. Geogr. Terms* 532 *Acheulian*,..a cultural stage.. characterized by a certain type of chipped stone implements. The more usually accepted stages are:..Eolithic, Strepyan, pre-Chellean.

stress, *sb.* Add: **I. 3. g.** *Psychol.* and *Biol.* An adverse circumstance that disturbs, or is likely to disturb, the normal physiological or psychological functioning of an individual; such circumstances collectively. Also, the disturbed state that results.

1942 *Endocrinology* XXXI. 420 When the normal animal is subjected to stress the adrenal cortices show hypertrophy. **1953** FRUTON & SIMMONDS *Gen. Biochem.* xxxvii. 843 Similar reduction in the adrenal ascorbic acid and cholesterol is observed when normal animals are subjected to a variety of stress [*sic*] (injury, cold, heat, drugs, toxins, lack of oxygen, etc.). **1955** H. BASOWITZ et al. *Anxiety & Stress* i. 7 Anxiety has been defined in terms of an affective response; stress is the stimulus condition likely to arouse such response. **1959** *New Scientist* 12 Nov. 927/1 Some examples of the diseases thought to result from stress are high blood pressure, peptic ulceration and coronary thrombosis. **1968** PASSMORE & ROBSON *Compan. Med. Stud.* II. xxxvi. 8/1 Parenthood itself can be a stress for the immature adult. **1973** R. M. MAY *Model Ecosystems* iii. 60 Equation (3.21) tends to require that each species encounters greater competitive stress from its own, rather than from the other, species. **1976** *Sci. Amer.* July 55/1 The familiar human experience described as stress (caused by many factors, including fear, physical trauma, severe heat or cold or even extreme joy) has as a common denominator an increased secretion of adrenal steroids. **1978** S. LEVINE et al. in H. Ursin et al. *Psychobiol. of Stress* i. 4 When the psychologically threatening or arousing aspects of a situation were altered, classical stresses such as fasting and heat no longer activated the pituitary-adrenal system. **1979** *McGraw-Hill Yearbk. Sci. & Technol.* 374/1 Cacti suffering from water stress become fully rehydrated within 24 hr following a heavy rain.

5. c. Substitute for def.: A force acting on or within a body or structure and tending to deform it; now usu. the intensity of this, the force per unit area. (Further examples.)

As orig. defined by Rankine the stress was the equal and opposite reaction of the body to the force, rather than the force itself (see quots. 1855, 1856 in Dict.).

1876 *Encycl. Brit.* IV. 285/2 There are three kinds of stress, due to tension, compression, and shearing. *Ibid.* The ultimate strength of the material is measured by the maximum intensity of stress which it can bear, or in other words, by the stress which the unit area of cross section can bear. **1925** J. CASE *Strength of Materials* i. 2 When we wish to give the stress a numerical value it is desirable.. to refer to the stress in relation to the area of the cross section... The total force acting on a section, divided by the area of that section, is called the stress intensity or, more often, simply the stress. [*Note*] In future when we use the word 'stress' without qualification it must be understood to mean 'intensity of stress'. **1938** LAURSON & COX *Mech. of Materials* i. 2 Total stress is a force... Intensity of stress, however, is expressed in units of force divided by units of area. **1960** H. K. PRESTON *Practical Prestressed Concrete* i. 3 The same beam..is prestressed by a force of 54,000 lb... This force creates a uniform compressive stress of +1,000 psi over the entire cross section of the beam. **1979** *Nature* 23 Aug. 670/1 Arctic sea-ice breaks under wind stress throughout the year, exposing leads of open water.

5*. Phr. *stress(es) and strain(s)* (with reference to senses 3 and 5; cf. *strain and stress* s.v. *STRAIN *sb.*^{2} 10*).

1854 C. PATMORE *Angel in House* I. VIII. viii. 118 Puzzled and fagg'd by stress and strain. **1856** *Phil. Trans. R. Soc.* CXLVI. 481 (*heading*) Elements of a mathematical theory of elasticity... Part I. On stresses and strains. **1935** *Discovery* Sept. 259/1 The interdependent mechanical stresses and strains. **1952** *Sat. Rev.* (U.S.) 20 Sept. 38/2 There never are stresses in government, but *stresses* and *strains*. **1959** M. STEEN *Tower* I. vi. 85, I realised what the last few years, with their stresses and strains, had done to us both. **1960** *Times* 13 Jan. 15/2 Wrestling once again with unknown aeronautical quantities and resolving new propositions in stress and strain. **1979** *Jrnl. R. Soc. Arts* CXXVII. 363/2 New stresses and strains in the relationships between..Britain and the remaining territories.

III. 10. (sense 3 d) *stress area*; (sense 3 g) *stress reaction, situation, symptom*; (senses 5 c, 8) *stress-pattern*; (sense 8) *stress-difference, -point, -shift*; **stress analysis** *Engin.*, the theoretical or experimental study of the stresses within a mechanical structure in relation to its function; hence **stress analyst**; **stress-breaker** *Dentistry*, a device attached to or incorporated in a partial denture to reduce the occlusive forces that have to be borne by the underlying tissue and the teeth to which the denture is attached; so **stress-breaking** *vbl. sb.* and *ppl. a.*; **stress-broken** *ppl. a.*; **stress concentration** *Engin.*, a local increase in the stress inside an object; also, a stress raiser; **stress contour** *Phonetics*, a sequence of varying levels of stress within an utterance; **stress corrosion** *Metallurgy*, the development of cracks as a result of the combined effects of stress and corrosion; freq. *attrib.*; **stress diagram** *Mech.*, a diagram that represents graphically the stresses within a framed structure; **stress-dilatancy** *Physics*, dilatancy that occurs as a result of applied stress; **stress disease,** a disease that occurs as a result of continual exposure to stress; **stress fracture** *Med.*, a fracture of a bone caused by the repeated application of a high load; **stress-free** *a.*, pertaining to or possessing freedom from mechanical or biological stress; **stress grading** *vbl. sb.*, the grading of timber according to its strength, as estimated from the number and distribution of knots and other visible defects; so **stress grade** *sb.* and (with hyphen) *v. trans.*; **stress-graded** *ppl. a.*; **stress-group** *Phonetics*, a group of syllables forming a rhythmic unit with one primary stress; **stress incontinence** *Med.*, a condition found chiefly in women in which a (usu. small) escape of urine occurs when the intra-abdominal pressure increases suddenly, as in coughing or lifting; **stress interview,** an interview in which there is a deliberate attempt to subject a candidate to stress by the nature of the questioning; **stress mark,** (*a*) *Phonetics*, a symbol or a diacritical mark indicating that a syllable carries stress; (*b*) *Photogr.*, a mark on a photographic print caused by friction or pressure on the film surface; hence **stress-marked** *a.*; **stress maximum** *Phonetics*, the tonic accent; **stress mineral** *Petrol.*, a mineral whose formation in metamorphic rocks is believed to be dependent on shearing stress; **stress-neutral** *a.* *Linguistics*, designating a derivational or inflectional suffix which plays no part in the placing of stress within a word; hence **stress-neutrality**; **stress phoneme** *Linguistics*, a phoneme whose contrastiveness consists in a distinctive degree of stress; **stress raiser** *Engin.*, a feature in the shape or composition of an object that gives rise to a local increase in stress; **stress relaxation** *Engin.*, a decrease of stress occurring in a material when the associated deformation remains constant; **stress relief** *Metallurgy*, the reduction of residual stress in a material by thermal treatment; also **stress-relieve** *v. trans.*, **-relieved** *ppl. a.*, **-relieving** *vbl. sb.* (freq. *attrib.*); **stress-strain** *adj. phr.* *Engin.*, pertaining to or depicting the relation between mechanical stress and the strain it produces; **stress-timed** *a.* *Phonetics*, designating or pertaining to a language in which primary stresses occur at approximately equal intervals, irrespective of the number of unstressed syllables in between; hence **stress-timing**.

1926 PIPPARD & BARROW in *Building Res. Board Techn. Paper* No. 1. 1 The bow girder..presents an interesting problem in stress analysis. **1980** *Strain* XVI. 132/2 The stress analysis of turbine components for the new hydro-electric pumped storage system. **1950** M. HETÉNYI *Handbk. Exper. Stress Analysis* p. v, Several principal methods and literally hundreds of individual tools and artifices constitute the 'arsenal' of the experimental stress analyst. **1976** B. JACKSON *Flameout* (1977) ii. 32 His career as stress analyst with Lockheed Aircraft. **1973** *Times* 17 Apr. 1/2 His brief will be to review the functions and relationship of the two bodies to enable them to make the most effective contribution in strengthening the voluntary housing movement in stress areas. **1930** H. P. BOOS in I. G. Nichols *Prosthetic Dentistry* xxxvii. 600 Stress-breakers can be used in conjunction with the tube successfully. **1955** J. OSBORNE *Dental Mech.* (ed. 4) x. 215 (*caption*) Split casting type of stress breaker. **1930** L. M. FARNUM in I. G. Nichols *Prosthetic Dentistry* xxxvi. 593 Stress-breaking construction is indicated where there are no posterior abutments on one or both sides of the mouth. **1963** C. R. COWELL et al. *Inlays, Crowns, & Bridges* xi. 118 This form of bridge incorporates a stress-breaking device, which allows limited movement at one of the joints between pontic and retainer. **1973** D. H. ROBERTS *Fixed Bridge Prostheses* ix. 152 The dovetail and slot introduces a certain degree of 'stress-breaking' between the two parts of the bridge, and because of this the retainers..are far less likely to fail. **1955** J. OSBORNE *Dental Mech.* (ed. 4) ix. 150 In cases when the teeth are periodontally affected, stress-broken designs may be employed. **1925** TIMOSHENKO & LESSELLS *Appl. Elasticity* i. 10 A semi-circular groove in a strip subjected to tension..also produces very high stress-concentration. **1936** [see *stress raiser* below]. **1977** E. J. HEARN *Mech. of Materials* xviii. 477 If..stress concentrations such as notches, keyways, holes, etc., are present in the bar, these will result in local stress increases. **1958** A. A. HILL *Introd. Linguistic Struct.* 28 Stress contours differ from pitch contours in that two phrases are never united into a single stress contour. **1971** *Language* XLVII. 269 It appears that the stress contours of English sentences are determined in a simple and regular way by their underlying syntactic structures. **1931** *Jrnl. Iron & Steel Inst.* CXXIV. 723 Stress corrosion of metals. **1967** A. H. COTTRELL *Introd. Metallurgy* xxiii. 467 In stress-corrosion cracking there is usually very little overall corrosion. **1973** A. PARRISH *Mech. Engineers' Ref. Bk.* (ed. 11) v. 77 The higher Mo bearing steels offer more resistance to stress corrosion cracking than 18/8; stress relief treatment (two hours at 870°C) after fabrication considerably reduces the risk of cracking. **1873** J. G. MEDLEY *Roorkee Treat. Civil Engin. in India* (ed. 3) I. xxv. 550 Loads on Roofs naturally divide themselves into two sets... Hence two distinct Stress-diagrams must be drawn, one for each system of load. **1919** PIPPARD & PRITCHARD *Aeroplane Structures* viii. 72 Probably the most satisfactory method of determing the forces in the individual members of a structure is by means of the stress diagram. **1965** G. M. MILLS *Theory of Structures* ix. 168 The variation of stress along a given axis may be shown graphically by means of a stress diagram. **1924** O. JESPERSEN *Philos. Gram.* xvii. 231 The old compound *mankind* (now stressed on the second syllable) comprises all human beings, but the younger *mankind* (stressed on the first syllable) is opposed to *womankind*. (The stress-difference, as made in N.E.D., is not, however, recognized by everybody.) **1971** *Language* XLVII. 261 The analysis given..correctly predicts the existence of a stress difference associated with the two readings of sentences like *The parable shows what suffering men can create*. **1944** G. W. S. BLAIR *Survey Gen. & Appl. Rheol.* iii. 31 The exceptions [to this rule] are..(*c*) Materials whose consistency is increased by increasing the stress (as distinct from the strain) applied to them. This phenomenon has been little studied, but may be referred to as 'stress-dilatancy'. **1962** *Proc. R. Soc.* A. CCLXIX. 500 (*heading*) The stress-dilatancy relation for static equilibrium of an assembly of particles in contact. **1979** *Geotechnique* XXIX. 341 Rowe's (1962) stress-dilatancy relation..allows indirect measurement of this angle [*sc.* of interparticle friction] based on triaxial compression tests on dense..samples. **1948** *Observer* 13 June 5/5 Absenteeism which arises..from those once..despised causes which passed under names such as neurasthenia and described to-day as stress diseases. **1966** G. E. EVANS *Pattern under Plough* viii. 96 It is more enlightened and scientific in psychosomatic and stress diseases for medicine to address itself as much to the man as to the actual disease. **1952** R. WATSON-JONES *Fractures* (ed. 4) I. xv. 343 (*heading*) Fatigue or stress fractures. **1983** *Brit. Med. Jrnl.* 12 Nov. 14491/1 Stress fractures are widely recognised in running. **1946** *Nature* 5 Oct. 475/1 After discussing the effect of swelling on the sorption isotherm he proceeded to derive a stress-free isotherm. **1961** *Economist* 21 Oct. 249/1 A stress-free cruising speed of 70-75 mph. **1978** D. BLOODWORTH *Crosstalk* xxiv. 188 Stress-free mice are far better performers. **1941** *Grading Rules for Structural Timber* (*B.S.I.*) 2 A further standard for the compressive stress grading of these species for use in compression and tension members..is being prepared. **1944** (*title*) Grading rules for stress-graded timber. (B.S.I.) *Ibid.* 4 The present revision has been undertaken..to provide for stress grades higher than 800 lb. f. **1971** *Timber Trades Jrnl.* 21 Aug. 26/2 Typical yields of sawn timber have been..graded by the stress-grading machine installed by Timber-lab at Princes Risborough. *Ibid.* 23/3 The timber for all the main structural components was visually stress-graded to a minimum of 50 grade before use. **1973** *Materials & Technol.* VI. i. 27 In Britain, four basic stress grades are specified for sawn softwood, and three for laminated timber. *Ibid.*, Visual stress grading is not a difficult operation, but requires

considerable experience. **1876** H. SWEET in *Trans. Philol. Soc. 1875–6* 473 We find..that every sentence can be analyzed into smaller groups characterized by one predominant stress-syllable, round which the others group themselves... In our first sentence there are two such stress groups... A word is, phonetically speaking, a stress-group. **1959** J. T. PRING *Colloq. Eng. Pronunc.* 56 A stress-group is formed by a strongly stressed, prominent syllable, together with any unstressed, non-prominent syllables which cluster about it. **1935** A. W. BOURNE *Midwifery for Nurses* ii. 24 Stress incontinence is due to a weakening of the supports of the bladder. **1972** LAW & FRIEDMAN *Midwifery* xiv. 334 The patient is then asked to strain down and any tendency to prolapse of the vaginal walls is noted. She is then asked to cough to determine whether any stress incontinence is present. **1955** *Explorations* Feb. 7, I examined stress interviews as well as non-directive ones. **1978** *Jrnl. R. Soc. Arts* CXXVI. 270/1 Whilst stress interviews in which the interviewer sets out to be provocative or rude may have been appropriate for the selection of American Special Services personnel during the war, I would not recommend them for civilian use. **1888** H. SWEET *Hist. Eng. Sounds* 8 The stress marks are put before the element on which the stress begins. **1918** *Photo Miniature* Mar. 41 *Stress marks*, scummy appearance or black lines on a bromide or D.O.P. print, caused by..the sensitive paper being rubbed against the negative or other sheets of paper, or any sharp pressure. **1919** *Brit. Jrnl. Photogr. Alm.* 251 Free from tendency to give rise to stress or abrasion marks. **1961** *Amer. Speech* XXXVI. 221 Tone patterns illustrated by Kingdon's tonetic stress marks in ascending order of complexity. Indicates the force of each pattern on basic grammatical constructions. **1968** L. A. MANNHEIM tr. *Fritsche's Faults in Photography* III. 331 Most enlarging papers are largely protected against stress marks by an emulsion supercoating. **1964** R. H. ROBINS *Gen. Linguistics* iv. 138 In a word stressed on a non-initial syllable, in stress-marked languages, the stressed articulation usually begins on the consonant. **1969** *Computers & Humanities* III. 136 The next step in this project is to replace orthographic entries with phonetic entries... Then, stress maxima will be determined and various features of meter tabulated and analyzed. **1971** *Language* XLVII. 588 Thus in *the mangy dog*, the stress on the first syllable of *mangy* is a stress maximum. **1918** A. HARKER in *Q. Jrnl. Geol. Soc.* LXXIV. p. lxxvii, Shearing stress manifestly favours the production of sericite and the chlorites, of albite [etc.]... These may conveniently be styled stress-minerals. **1952** H. RAMBERG *Origin Metamorphic & Metasomatic Rocks* 119 It has yet to be proved..that any of the suggested stress minerals really are such. **1965** G. J. WILLIAMS *Econ. Geol. N.Z.* x. 158/1 The chief constituent of all specimens he examined is a variety optically identical with the stress-mineral antigorite. **1971** *Language* XLVIII. 269 If word stress is assigned prior to syntactic transformations, then it follows automatically that transformationally attached affixes are stress-neutral. **1972** *Ibid.* XLVIII. 336 He [*sc.* Lakoff] suggests that the NSR [*sc.* Nuclear Stress Rule] might precede the assignment of word-stress; this destroys the principle of the phonological cycle, and again fails to explain the stress-neutrality of transformationally placed affixes. **1954** S. ROBERTSON *Devel. Mod. Eng.* (rev. ed.) iv 77 In English, any word of two or more syllables has its own stress-pattern **1968** R. A. LYTTLETON *Myst. Solar Syst.* vi. 193 Tektites reveal series of dark and light bands associated with the internal stress-pattern. **1980** *Early Music* July 403/2 Freed from the obligations of setting a poetic text, from the need to conform to text stress-patterns and changes of poetic mood, a composer might indulge in the exploration of thematic material to the full. **1933** L. BLOOMFIELD *Language* xvii. 295 In modern English verse..the author shapes his wording so that stress-phonemes at certain intervals. **1968** CHOMSKY & HALLE *Sound Pattern Eng.* 26 He [*sc.* the speaker] need not make a choice among various 'stress phonemes'. **1932** D. JONES *Outl. Eng. Phonetics* (ed. 3) xxviii. 223 The lengths separating the 'stress-points' or 'peaks of prominence' of the syllables. **1956** *Kenyon Rev.* XVIII. 466 Mr. Chatman has shown the metrical stress-points in each line, but in my judgment he has misplaced them in lines 2, 12 and 14. **1936** *Trans. Amer. Inst. Mining & Metall. Engineers* CXX. 32 Yet another field in which correlation of metallographic and mechanical methods is needed is the study of the relative seriousness of 'inherent' and 'imposed' sources of stress concentration —'stress raisers', as Gillett has aptly called them. **1978** R. J. GRAY in McCall & French *Metallogr. in Failure Analysis* 240 The surfaces must be free of machining marks that could serve as stress raisers where a fissure and subsequent fracture could occur. **1966** LAZARUS & OPTON in C. D. Spielberger *Anxiety & Behavior* x. 227 The second phase involved the plan to manipulate 'ego-defense' processes so as to reduce stress reactions while subjects watched a stressful film. **1979** D. A. BAKAL *Psychol. & Med.* iii. 86 The capacity of any situation to produce stress reactions depends on the characteristics of individuals. **1943** *Jrnl. Chem. Physics* XI. 127/1 As a result of their studies of stress relaxation..of polyvinyl acetate held at constant elongation these authors conclude that the polymer is composed of a netted system of chains through which interpenetrates a system of relatively free chains. **1959** *Jrnl. Iron & Steel Inst.* CXCII. 198/3 Stress relaxation tests at a constant total strain of 0·15% for times exceeding 20 000 h on three low-alloy steels. **1979** R. P. BROWN *Physical Testing of Rubbers* xi. 200 Stress relaxation is the measurement of change of stress with time under constant strain. **1935** *Symp. Welding Iron & Steel* (Iron & Steel Inst.) II. 42 Stress relief by heat treatment reduced the stresses to approximately 10 per cents of those existing in the plates as welded. **1973** [see *stress corrosion* above]. **1980** *Metallography* XIII. 69 A stress-relief treatment at 600–650°C results in the transformation of ferrite primarily to $M_{23}C_6$ carbide. **1935** *Symp. Welding Iron & Steel* (Iron & Steel Inst.) II. 47 The whole member was now stress-relieved in the furnace and delivered to the machine shop for completion of the work. **1980** *Metallography* XIII. 59 Large components are invariably stress relieved to reduce the residual stresses generated in welding. **1935** *Symp. Welding Iron & Steel* (Iron & Steel Inst.) II. 46 The stress-relieved grid was next planed on a planer and planed top and bottom. **1980** *Metallography* XIII. 59 (*heading*) Microstructural transformations in stress relieved type 316 stainless steel weld metal. **1938** D. K. BULLENS et al. *Steel & its Heat Treatment* (ed. 4) I. v. 140 Finish machining is then done and a final stress-relieving draw given..at 1050°, holding 48 hr. at temperature and furnace cooling. **1956** *Jrnl. Iron & Steel Inst.* CLXXXIII. 99/2 Stress relieving of 11-ft. dia. electrically welded steel duct. **1980** *Jrnl. Nucl. Materials* XCI. 189 The stress-relieving treatments [of Zircalloy-4] were made at..773, 793 and 813 K, during 1 and 2 hr. **1888** H. SWEET *Hist. Eng. Sounds* 124 This law of stress-shift in weak diphthongs explains the INorth. *am*=*eom*: weak *eom* became first *eam*.., then *eám*, and finally, by dropping the almost inaudible *e*, *am*. **1930** T. SASAKI *On Lang. R. Bridges' Poetry* 91 It has been the rule in the English blank verse since Chaucer not to tolerate stress-shift (or inversion of accent) in the fifth foot. **1972** M. L. SAMUELS *Linguistic Evol.* iii. 36 Later in Germanic stressing on the root-syllable was generalised, and because of this stress-shift the voiced allophones..were no longer in complementary distribution, [etc.]. **1959** *New Scientist* 12 Nov. 927/2 Much was still to be understood of the intermediate steps between the stress situation and the decrease in circulating eosinophils. **1972** 'T. COE' *Don't lie to Me* xviii. 149 That premonition of something being wrong that sometimes strikes people in stress situations. **1886** K. PEARSON in I. Todhunter *Hist. Theory Elasticity* I. 503 There exist certain materials for which even in a state of ease the stress-strain relation is not linear; that is to say the stress-strain curve..is not a straight line even for very small elastic strains. **1923** GLAZEBROOK *Dict. Appl. Physics* V. 56/2 This will be the most convenient place in which to treat of the stress-strain relations of a doped fabric. **1956** *Nature* 24 Mar. 561/1 Papers..dealt with the measurement of residual stresses in cold-drawn tubes;..and stress-strain characteristics of metal at high rates of strain. **1973** J. G. TWEEDDALE *Materials Technol.* I. iv. 79 Having derived the respective stress and strain values it is possible to study the tensile characteristics of a material from a graph comparing these values—a stress–strain diagram. **1958** *Times Lit. Suppl.* 17 Oct. 596/4 A cold in the head is more often than not a stress-symptom with which one must learn to live. **1977** P. DICKINSON *Walking Dead* I. i. 24 At a certain point of over-crowding..[rats] manifest stress symptoms. **1946** K. L. PIKE *Intonation Amer. Eng.* III. 35 Many non-English languages..tend to use a rhythm which is more closely related to the syllable than the regular stress-timed type of English. **1956** [see *ISOCHRONIC *a.* 3]. **1980** *English World-Wide* I. 1. 108 RP is stress-timed, with primary stress recurring at roughly even intervals through a sentence. **1964** M. A. K. HALLIDAY et al. *Linguistic Science* iii. 72 The English type of rhythm is known as 'stress-timing'.

stress, *v.*^{1} Add: **3. d.** In contexts of *Biol.* and *Psychol.*: cf. *STRESS *sb.* 3 g.

1973 *Country Life* 7 June 1859/2 The transfer to a new environment stresses the calves, and it is now that latent infection will show itself. **1975** *Verbatim* Sept. 5/2 An analysis of the tapes will show exactly how stressed he was, stated the author of a book on certain intelligence methods. **1979** *Sci. Amer.* Nov. 65/1 When the reovirus-infected mice were stressed by injection with a large dose of glucose, however, it became quite clear that their ability to metabolize glucose had become impaired.

stressable (stre·săb'l) *a.* *Linguistics.* [f. STRESS *v.*^{1} + -ABLE.] Capable of being stressed. Also **stressabi·lity.**

1964 W. S. ALLEN in D. Abercrombie et al. *Daniel Jones* 6 In verse, as in the language, not every heavy syllable is stressed; it is only *stressable*. **1964** *English Studies* XLV. 495 He investigates its relation to Latin phonology, which in the realm of accent is characterized by the category of stressability. **1972** *Language* XLVIII. 295 The statement 'B is stressable' is to rule out cases like anaphoric noun phrases, which are not stressable. **1977** *Ibid.* LIII. 28 We agree that the stressability of prepositions is to be handled by the rules of the grammar rather than poetry, but we differ substantially on the specific analysis.

stressed, *ppl. a.* Add: **3.** *Engin.* Subjected to mechanical stress; *spec.* = *PRESTRESSED *ppl. a.*; *stressed skin*, an outer covering of an aircraft or other structure that bears a significant part of the stresses and contributes to the overall strength and stiffness; usu. *attrib.*

1930 *Flight* 11 Apr. 411/2 Recently a monoplane of orthodox aerodynamic design was completed by the company, employing the stressed-skin type of wing construction. **1951** *Archit. Rev.* CX. 342/1 The roof is of stressed-skin plywood construction formed by two skins of exterior ply panel grade on an internal timber framework. **1954** [see *DIAGRID]. **1966** *Daily Tel.* 10 Aug. 18/2 American steel was being bought for the 'highly stressed' parts of future British-built nuclear submarines. **1968** *Punch* 13 Nov. 688/3 British Railways 'revolutionary gas-turbine advanced passenger train with stressed-skin, aircraft-type construction'. **1979** *Jrnl. Magnetism & Magn. Materials* XI. 76 Uniaxially stressed semi-conductors.

stressful, *a.* Add: Also, causing or inclined to cause stress.

1952 *Psychosomatic Med.* XIV. 311/2 Characteristically in the patient with chronic fatigue, the stressful activity is implicit rather than explicit. **1966** O. NORTON *School of Liars* vi. 104 'How do you protect a man like him from stress?'..'By not being stressful yourself.' **1972** C. M. PARKES *Bereavement* iii. 32 Situations that tend to produce alarm are regarded as *stressful*. **1978** *Detroit Free Press* 5 Mar. D4/4 Are you always changing things in your life, changing jobs, changing residences?..If so, subtract two years. Too much change is stressful.

stressman (stre·smæn). *Engin.* Pl. **stressmen.** [f. STRESS *sb.* + MAN *sb.*^{1}] = *stress analyst* s.v. *STRESS *sb.* 10.

1935 C. G. BURGE *Compl. Bk. Aviation* 260/2 The 'stressman' may be able to suggest practical methods of lightening the weight. **1954** VEALE & RADFORD *Aircraft for All* iv. 62 Stressmen..check the strength of every major part. **1960** *Times* 1 Apr. 2/3 (Advt.), Hunting Aircraft Limited have vacancies for..aerodynamicists, stressmen, design draughtsmen. The work programme embraces both high and low speed military and civil aircraft. **1975** *Offshore Engineer* Sept. 134/1 There is provision for override so that when needed the system can operate continuously—say during a period of bad weather or, as Desmond Thurgood, Seatek's chief stressman, points out, during installation.

stressor (stre·səɹ). *Psychol.* and *Biol.* [f. STRESS *sb.*, *v.*^{1} + -OR.] A single condition or agent that constitutes a stress for an organism (see *STRESS *sb.* 3 g).

1950 H. SELYE *Physiol. & Path. of Exposure to Stress* 9 The expression systemic stress is used here to denote a condition in which..extensive regions of the body deviate from their normal resting state. In accordance with the common usage of the word 'stress', the term 'systemic stress' is sometimes loosely employed also to denote the stimuli which cause systemic stress. In this sense, it is preferable however, to speak of alarming stimuli or 'stressors'. **1958** *Proc. 10th Internat. Congr. Entomol.* IV. 727 (*heading*) Crowding as a stressor [in insects]. **1962** *Lancet* 27 Jan. 200/2 It has long been known that the response of schizophrenic patients to stressor agents is dulled. **1969** *Nature* 4 Oct. 18/2 Stress results from a threat, real or apparent, to the biological integrity of the animal. More simply—a 'stressor' can be defined as any stimulus or situation which causes maladaptive behaviour. **1972** C. M. PARKES *Bereavement* i. 4 Loss of a close relative is normally a major stressor.

stressy (stre·si), *a.* *rare.* [f. STRESS *sb.* or *v.*^{1} + -Y^1.] Characterized by stress, *spec.* in the prosody of G. M. Hopkins; in which stress is conspicuous.

1880 G. M. HOPKINS *Lett. to R. Bridges* (1955) 107, I think you have missed the clue. You take the rhythm for three triple time, iambs and anapaests say, and four feet to a line (except the refrain). But to get this you have to skip..a whole foot as marked and stressy as any other foot. **1961** *Times Lit. Suppl.* 18 Aug. 549/4 Neither of these versions reveals the bold, thoroughgoing 'stressy' flexibility of genuine sprung rhythm.

stretch, *sb.* Add: **1. i.** *Baseball.* An action used in pitching (see quots.).

1939 E. J. NICHOLS *Hist. Dict. Baseball Terminol.* 75 *Stretch*,..a pitcher's straightening of his arms above his head preliminary to delivering the ball. **1951** H. TURKIN *Official Encycl. Baseball* 572 The pitching delivery can be broken down and analyzed to reveal six distinct actions: windup, stretch, leg lift, stride, body pivot and follow through... The stretch brings the pitching arm behind the head. **1976** *Webster's Sports Dict.* 428/2 *Stretch*,..a movement a pitcher uses instead of a windup when there are runners on base. *Ibid.* 429/1 The stretch, with its integral pause, allows the pitcher to throw to the base to try to pick off the runner or to keep him close to prevent his stealing without interrupting the pitching motion and making a balk.

j. *Aeronaut.* Modification of an existing aircraft design to increase its capabilities, esp. by lengthening the fuselage; capacity for this allowed for in a design.

1954 *Economist* 11 Sept. 2/2 However much 'stretch' may have been designed into the two machines—and the evidence suggests it was not too great—these changes in elements outside the designer's control mean modifications..delays. **1960** *New Scientist* 30 June 1640/1 The modifications involved in stretch are chiefly concerned with stress and control parameters. **1976** *Farnborough Internat. Exhibition* (Official Programme) 46/2 The Lynx design is capable of considerable 'stretch', says Westland.

2. d. Freq. in phr. *by any* (or *no*) *stretch of the imagination*.

1942 T. BAILEY *Pink Camellia* xv. 122 Peter could not, by any stretch of the imagination, be compared to that Satan of the Scriptures who came so inopportunely to the Garden. **1957** *Pract. Wireless* XXXIII. 573/2 Most neutral leads..have registered between 5 and 25 volts R.M.S., voltages which cannot, by any stretch of the imagination, be called lethal. **1977** A. ECCLESTONE *Staircase for Silence* iv. 77 The church they belong to seems hopelessly stuck fast in a way of life that by no stretch of the imagination can be described in terms of leaven or salt or light.

4. c. Now chiefly *at full stretch*: to capacity; working fully or as hard as possible.

1934 G. B. SHAW *On Rocks* I. 221, I am an overworked.. man,..having to keep my mind at full stretch all the time struggling with problems. **1955** *Times* 1 June 10/1 United States tire manufacturers are still working at full stretch. **1977** *Evening Post* (Nottingham) 27 Jan. 4/5 Wilford power station, Nottingham, has been at full stretch to meet heavy demand.

6. g. *Linguistics.* A definable extent (of text or speech).

1961 M. A. K. HALLIDAY in *Word* XVII. 250 Language is patterned activity. At the formal level, the patterns are patterns of meaningful organization: certain regularities are exhibited over certain stretches of language activity. **1964** *English Studies* XLV (Suppl.). 56 It certainly seems worth while trying to bring more system into stretch-of-speech analysis in general by studying the potentialities and realities of absence of sound, sound

zero. **1967** D. G. HAYS *Introd. Computational Linguistics* x. 171 As the text is being prepared, each stretch between unit boundaries is compared with the contents of the exclusion list. **1972** J. MCH. SINCLAIR *Course in Spoken Eng.: Grammar* 3 Strict grammatical relationships need only be made across stretches of language a few words long. **1973** A. H. SOMMERSTEIN *Sound Pattern Anc. Greek* i. 1 The final chapter..contains a summary..of the rules of the part of Greek phonology that I have investigated, a stretch of text written in the form it would have before the application of these rules, [etc.].

7. b. Also *loosely*, a prison sentence (freq. with preceding numeral signifying the number of years). Also *transf.*

1949 'M. INNES' *Journeying Boy* ix. 109 If we were getting him a stretch, we could go to bed feeling we had done something useful. **1951** P. BRANCH *Lion in Cellar* xx. 222 He's in Joe Gurr again. He got nicked in Cardiff on a snout gaff... It's only a two stretch and a lot of the Boys had their collars felt. **1957** G. THOMAS *Gazooka* 42, I owed it to him now to see that he would not be saddled with another stretch under the probation officer or in a reform school for an offence which was not..immoral. **1960** S. CLAYDON *Lesson in Murder* v. 75, I was going to serve my stretch, come out, and get a job. **1967** J. MORGAN *Involved* 40 What do you think he'll get for this lot, guv'nor, ten? I suppose a ten stretch is the least he can expect. **1976** K. BONFIGLIOLI *Something Nasty in Woodshed* ix. 106 'Porridge'..means penal servitude. There is a legend..that if..on the last morning of your 'stretch', you do not eat up all your nice porridge, you will be back in durance vile within the year.

8. Chiefly *U.S.* **a.** Also *attrib.* (esp. = 'home-stretch') as *stretch run, turn*.

1934 D. RUNYON in *Collier's* 3 Mar. 42/3 Gallant Godfrey comes to the conclusion that Westrope is working on him in a stretch run. **1944** *Sun* (Baltimore) 14 Apr. 14/1 Backers of Sollure had no worries all through the stretch run. **1949** *Time* 10 Oct. 42/2 They had less reason to thank their own bats than the batty stretch-run performance of the Cardinals. **1972** *N.Y. Times* 4 June v. 1/4 Shortly after the field hit the stretch turn, Run the Gantlet moved into the lead position. **1978** *Detroit Free Press* 2 Apr. 2E/1 Craig's Corner, a 7-year-old gelding, won the Carolina Cup steeplechase at Camden, S.C., with a strong stretch run before a race record crowd of 30,000-plus. **1979** *Internat. Herald Tribune* 31 Oct. 23/3 It would be the clubhouse turn in the United States, but this was the Vaal Racing Club in the Orange Free State of South Africa, where the horses run clockwise, so it was really the stretch turn.

b. *home-stretch*: see *HOME-STRETCH. *back stretch* = *back-straight* s.v. *BACK- B.

1839 *Picayune* (New Orleans) 2 Apr. 2/2 He went to work himself, soon passed the old black, made all sorts of a brush while rounding the last turn and commencing the back stretch. **1868** H. W. WOODRUFF *Trotting Horse Amer.* xii. 122 On the back-stretch..Mr. Duffy asked me if I could ride it out without tiring. **1903** A. ADAMS *Log of Cowboy* xv. 237 He was speeding her on the back stretch. **1931** *Daily Express* 21 Sept. 11/5 Box stalls were built around the walls, while in other parts of the ballroom were reproduced in miniature the back stretch and the home stretch of the track. **1933** *Boys' Mag.* XLVII. 119/2 He shot by a little group of runners, and in the back-stretch was hard upon the heels of the four leaders. **1948** *Life* 21 June 32 (*caption*) Dewey, Taft and Stassen will get away fast, but watch out for Dark Horse Vandenberg on the backstretch. **1970** *Toronto Daily Star* 24 Sept. 21/4 The jock had to check her twice on the backstretch.

c. *transf.* and *fig.*

1949 *Sun* (Baltimore) 9 Aug. 14/3 Baseball's 1949 stretch is now only a few furlongs away. Oddly enough, conditions in the two leagues have been somewhat reversed since last April. **1957** *Baseball Digest* Jan.–Feb. 15 Despite Joss' great work in the stretch, the season ended with the Indians in second place. **1972** *Publishers' Weekly* 12 June 9 (Advt.), Put your money on *Miss Elizabeth Arden* to win in the stretch. **1976** *Billings* (Montana) *Gaz.* 20 June 1E/3 Friday when he led the second round by one shot, Mahaffey expressed disappointment that shoddy putting down the stretch prevented him from running away from the field. **1976** *National Observer* (U.S.) 6 Nov. 5/1 Ford, at the very last, was doing something right. Most of the polls showed he was closing fast on Jimmy Carter, who was limping badly down the stretch.

stretch, *v.* Add: **I. 1. b.** Also (*slang*), to kill (a person). Cf. LAY *v.*^{1} 56 b in Dict. and Suppl.

1902 KIPLING *Traffics & Discoveries* (1904) 11 He said if you stretched a man at his prayers you'd have to hump his bad luck before the Throne as well as your own. **1953** M. GILBERT *Fear to Tread* viii. 107 Once..Annie had a husband. She got tired of him, so she 'stretched him with a bottle'.

V. 16. a. Also, to straighten, to remove the curl from (hair).

1963 W. SOYINKA *Lion & Jewel* 9 Her hair is stretched Like a magazine photo. **1971** C. ACHEBE *Girls at War* (1972) 58 Now Abigail was a lady; she could sew and bake ..put on powder and perfumes and stretch her hair.

VI. 21. c. *colloq.* To eke out (food), esp. to serve a greater number of people than originally intended.

1923 *Chambers's Jrnl.* Christmas 858/2 The problem of how to stretch a supper made for two to fit three. **1951** H. MACINNES *Neither Five nor Three* xiii. 184 She began worrying how far she could stretch the beef stew now simmering on the stove. **1974** N. FREELING *Dressing of Diamond* 80 She had made stuffing for the trout, to stretch them a bit. **1977** C. MCCARRY *Secret Lovers* viii. 98 She wondered if I was free to have dinner at her house... They were having something that the cook could stretch for three.

d. *Cinemat.* To adapt (a silent film) for projection on sound equipment by duplicating alternate frames so that the speed of action is not distorted.

1953 L. J. WHEELER *Princ. Cinematogr.* v. 145 Many occasions arrive when it is desirable to 'stretch' an original negative, that is, assuming an old, silent negative, is required to be printed into a sound film. **1965** *Listener* 11 Feb. 231/3 If a silent film (sixteen frames/sec.) is projected on a sound projector (twenty-four frames/sec.), the action appears too fast. To rectify this, silent films are sometimes 'stretched' by making a special print in which every other frame is printed twice. **1969** *Observer* 26 Jan. 7/4 The technique of stretching them [*sc.* silent films] to run at 24 or 25 frames a second..has been known and used for very many years. **1976** *Oxf. Compan. Film* 664/2 Makers of compilation films often incorporate old footage without stretching it, with the result that modern audiences have come to regard all silent film as comic.

e. *Engin.* To increase the capability or power of (an aircraft, power plant, etc.).

1960 *New Scientist* 30 June 1640/1 Engine power again was an essential factor in stretching the Viscount. **1967** *Economist* 8 July p. xxvi/3 The newest Gardener engine now runs at 1,800 rpm to give 180 bhp—still with remarkable fuel economy. This is near the far limit to which this engine can be stretched without supercharging. **1979** *Nature* 19 July 187/3 Why is it necessary to scale up to 1,300 MW? Doesn't the ability to replicate and perhaps stretch the 250 MW Dounreay prototype fast reactor give sufficient insurance against the risk of long-term uranium scarcity?

22. d. Also in colloq. phr. *to stretch it* (or *things*): to go too far, to go beyond the limits of credibility; to exaggerate.

1965 M. ALLINGHAM *Mind Readers* xx. 212 Rightie ho. If you say so. Any friend of yours is a friend of mine but that's stretching it. **1974** M. HASTINGS *Dragon Island* xi. 96 'I can't believe that by some chance Jones found himself with *two* people in whom he had special interest.' 'That would be stretching it.' **1975** R. STOUT *Family Affair* ix. 83 Everyone in Washington is connected..with Watergate. That's stretching it, but not much. **1980** R. HILL *Killing Kindness* v. 46 It's stretching things a bit... Still, it's worth checking.

e. *colloq.* To cause (someone) to exert himself to the utmost of his talents or abilities, esp. with regard to learning or employment.

1951 C. MORGAN *Breeze of Morning* I. v. 25, I had found that he always 'stretched' me in the way I liked, and gave me confidence by having such unswerving confidence in himself. **1960** C. DAY LEWIS *Buried Day* i. 23 Under such conditions, an only child may become precocious, stretching himself unnaturally to meet the adult world on its own terms. **1968** *Guardian* 21 Nov. 10/5 He is satisfied by being a bishop; he says he felt a need to be stretched by some equally big job. **1978** D. DEVINE *Sunk without Trace* iv. 36, I hear good reports of your work... I fear, however, we're not stretching you enough. **1983** *Times* 8 Jan. 3/8 Local education authorities..could ensure..that the curriculum suited and stretched all children.

23. b. *Jazz.* To play without restraint, esp. in a solo. Const. *out*.

1961 *N.Y. Times Mag.* 25 June 39/3 When a cat stretches out, he can make the moon on his own thrust or horn. **1962** *Down Beat* 5 July 35/3, I heard this group in person, at the Village Gate, and they stretched out. **1968** *Crescendo* June 12/2 Everybody really had a chance to stretch out and play what and how they wanted to.

VII. 24. a. **stretch forming** *vbl. sb.* *Mech.*, a process in which sheet metal under tension is shaped by the pressure of a punch to the required contour; hence **stretch-form** *v. trans.*; **stretch mark,** a linear mark on the skin (esp. of the stomach or thighs) when it has been distended by pregnancy or obesity; = *STRIA 2 e; **stretch receptor** *Physiol.*, a sensory receptor that responds to the stretching of tissue; **stretch reflex** *Physiol.*, a reflex contraction of a muscle resulting from the stretching of the same muscle; **stretch spinning** (see quot. 1957).

1951 G. SACHS *Princ. & Methods Sheet-Metal Fabricating* v. iii. 470 In most instances a part is stretch-formed from a rectangular blank. **1973** J. G. TWEEDDALE *Materials Technol.* II. iv. 85 (*caption*) Stretch forming a curved shape. **1942** *Iron Age* 4 June 49 (*heading*) Stretch-forming contoured sheet metal aircraft parts. **1951** *Archit. Rev.* CIX. 166/1 It can be spun into a bell-like shape, or it can be shaped on a stretch-forming machine, which pulls it into shape over a former. **1973** J. G. TWEEDDALE *Materials Technol.* II. iv. 85 Stretch forming uses a principle that involves applying a uniaxial tension to a thin plate,..and then pushing a controlled shape progressively into the surface of the tensioned sheet. **1960** F. W. GOODRICH *Maternity* iii. 59 The development of 'stretch marks' in the skin of the abdomen and thighs is not unusual. **1970** D. MARLOWE *Echoes of Celandine* viii. 143 'The marks you see..are not stretch-marks even though I have had three children.'..She then turned around..revealing scars..across her thighs. **1972** G. BOURNE *Pregnancy* vi. 110 Stretch marks may occur at puberty on the buttocks and also on the breasts, especially if a girl is overweight. **1980** *Family Med. Guide* (R. Soc. Medicine) ix. 210/1 Cushing's syndrome is characterized by..thinning of the skin which may lead to large stretch marks, and a tendency to bruise easily. **1936** C. L. EVANS *Starling's Princ. Human Physiol.* (ed. 7) xxxvii. 906 The lung evidently possesses stretch receptors similar in their responses to those of muscle. **1961** *Listener* 23 Nov. 858/1 If the animal [*sc.* an octopus] cannot use information from internal stretch receptors in its muscles, it will not be able to define the relative positions of the sense organs on the suckers that it uses to pick up an object. **1969** J. H. GREEN *Basic Clin. Physiol.* ix. 52/2 As one breathes in, the lungs expand, and stretch receptors in the lungs send sensory information up the vagus nerve to the respiratory centre cutting short inspiration. **1916** C. ASAYAMA in *Q. Jrnl. Exper. Physiol.* IX. 278 Quick stretching of tibialis anticus by a sharp pull on its tendon ..elicits a reflex contraction of the muscle... It may be termed a stretch-reflex. **1978** B. W. PAYTON in G. Ross *Essentials Human Physiol.* ix. 478 Stretch reflexes occur in all muscles but are particularly obvious in those involved in maintaining posture. **1925** *U.S. Patent* 1,528,219 This extract..prevents adhesion of the filaments in multiple spinning, and thereby greatly promotes 'stretch spinning'. **1957** *Textile Terms & Definitions* (Textile Inst.) (ed. 3) 96 *Stretch spinning*, a process of spinning whereby the filaments are substantially stretched at some stage between extrusion and collection.

b. Used *attrib.* or as *adj.* to designate various (usu. synthetic) fibres or fabrics which are elastic or capable of stretching, and garments, etc. (which may stretch to provide close fitting) made from them. Occas. *absol.* as *sb.*

1956 *Jrnl. Textile Inst.* XL. 280 A stretch yarn in which the deformation is produced by suitable combinations of heat-setting and twisting. **1957** *Times* 18 Nov. 11/4 Ties..by Jacques Fath with matching nylon stretch socks. **1959** *Times* 12 Jan. 11/3 Courtauld's process to obtain resilient stretch-nylon yarn. **1961** *Listener* 16 Nov. 825/2 My favourite example comes from a very modish American magazine... 'As contemporary..as C. P. Snow and stretch-pants.' **1962** *Guardian* 23 Feb. 8/5 Until recently I had never found a fine stretch which did not ladder quickly. **1963** *Harper's Bazaar* Jan. 50 Stretch slacks and cardigan in gold and silver. **1963** *Daily Mail* 24 Aug. 5/1 (Advt.), 100% Nylon Stretch Tights. **1963** *Economist* 7 Sept. 840/2 The most promising..growth area ..is..in 'stretch' fabrics. **1964** *Woman* 18 Jan. 13 Keep your stretch pants slender. **1968** *Vogue* 15 Apr. 60 Snug-topped bubble dress..of sun red stretch. **1972** *Times* 22 May 10/2 Too-tight stretch nylon socks cause many foot troubles. **1977** 'J. FRASER' *Hearts Ease in Death* xiv. 166 The chair..was fitted with one of those cheap stretch covers with a large floral pattern. **1978** J. GORES *Gone, no Forwarding* xiii. 76 A wide-hipped woman in red stretch slacks. **1980** *Times* 19 Feb. 8 The now obligatory stretch fabrics.

stretchabi·lity. [f. STRETCHABLE *a.*: see *-BILITY.] Capability of being stretched.

1940 H. R. MAUERSBERGER *Matthews's Textile Fibers* (ed. 5) xx. 863 On cold-drawing, nylon becomes exceedingly strong and elastic. But elasticity means more than 'stretchability'. The degree to which a stretched material recovers its original length is a measure of true elasticity. **1959** POTTER & CORBMAN *Fiber to Fabric* (ed. 3) iv. 60 Stretch yarns..when subjected to certain methods that use heat to set crimp in thermoplastic filament yarns.. increase their stretchability. **1962** *Listener* 22 Mar. 510/2 What he does..is to measure the 'strength' of a flour.. by measuring the 'stretchability' of the dough it makes. **1967** E. CHAMBERS *Photolitho-Offset* i. 9 It [*sc.* zinc] also stretches more before breaking—this stretchability being sometimes useful to correct mis-register. **1978** *Detroit Free Press* 5 Mar. 30 (Advt.), We tested it for launderability, durability, stretchability and shape retention.

stretchable, *a.* (Later examples.)

1889 A. JAMES *Diary* 16 Nov. (1965) 56 She seems as large a joke as ever, an embodiment of the stretchable, a purely transatlantic and modern possibility. **1975** L. BLUE *To Heaven with Scribes & Pharisees* iv. 39 Jewish cooking..inclines to casseroles and stews, which are infinitely stretchable.

stretched, *ppl. a.* Add: **5.** Of an aircraft, engine, etc.: increased in size or operating capacity; based on a smaller or less powerful design. Cf. *STRETCH *v.* 21 e.

1960 *New Scientist* 30 June 1639/2 The Super-VC 10, which BOAC has just ordered off the drawing-board, is an example of a 'stretched' aeroplane. **1966** *Wall St. Jrnl.* 13 Jan. 1 Their orders of 'stretched' jets—conventional models expanded to carry more passengers—will also have to be carefully considered. **1967** *Economist* 8 July p. xxvi/3 Operators are suspicious of 'stretched' engines. **1972** *Daily Tel.* 26 Sept. 6/7 The car is a stretched version of the latest Daimler Double-Six, itself developed from the Jaguar XJ12. **1978** *Ibid.* 23 Feb. 6/8 Improvements in technology mean that the British Hovercraft Corporation's 'stretched' SRN-4 is likely to be used on some of the established 100–150 mile European routes.

stretcher, *sb.* Add: **II. 6. a.** (Earlier example.)

1833 *Reg. Deb. Congr. U.S.* 22nd Congr. 1 Sess. App. p. xli, [Duty] on square wire used for the manufacture of stretchers for umbrellas..twelve per centum ad valorem.

9. a. Also, a camp-bed used as a spare bed in a house, etc. Chiefly *Austral.* and *N.Z.*

1943 *Amer. Speech* XVIII. 86 A common article of furniture [in New Zealand] is a *stretcher*—a folding camp bed or cot, often used to provide temporary sleeping accommodation in a house. **1974** *Weekend Mag.* (Montreal) 18 May 21/1 All summer cottages in those days had two or three camp cots or 'stretchers', with flat wire springs and small mattresses, which could be folded up and stuffed under beds for use when unexpected or surplus guests arrived. **1980** B. MASON *Solo* 30 Tim, I got the stretcher out. It's quite sound. Needs a dust, that's all. I'm giving you three blankets. That should be enough.

11. (Later examples.)

1885 *Outing* (U.S.) Oct. 77/1 The trout..were lusty, vigorous fellows, and with a 'Silver Doctor' as a stretcher, I managed to forget myself..completely. **1938** W. C. PLATTS *Mod. Trout Fishing* vii. 59 Two flies—a stretcher, or tail fly, and one dropper—is rather risky. **1963** A. N. MARSTON *Newnes Encycl. Angling* 249/1 *Stretcher*, the

bottom fly on a wet-fly cast made up of two or more flies. Usually called a tail fly.

III. 12. **stretcher case,** an injured or sick person needing conveyance on a stretcher; **stretcher strain** *Metallurgy*, a furrowed marking on the surface of a metal produced by local deformation.

1917 'CONTACT' *Airman's Outings* v. 118 On this occasion there was good reason for the delay, as we ceded the right of way to a hospital ship and waited while a procession of ambulance cars drove along the quay and unloaded their stretcher cases. **1978** R. V. JONES *Most Secret War* xxxvi. 310 The Navy would not take him because as a stretcher case he would occupy as much space on board ship as four men standing up. **1931** *Metal Progress* Sept. 90/1 Stretcher strains (or more appropriately 'worms') are the shop names for the phenomenon known as the 'Lines of Lüder', after Lüder of Magdeburg, who first described them in 1860. **1971** *Steel in U.S.S.R.* I. 899 (*heading*) Causes of the formation of strain lines (stretcher strains) when drawing stainless-steel tubes.

stretcher (stre·tʃəɹ), *v.* [f. the sb.] *trans.* To carry *off* or convey (an injured or sick person) on a stretcher.

1976 *Daily Mirror* 15 Mar. 30/6 The sickening blow of seeing Gary Locke stretchered off in only the seventh minute. **1978** J. UPDIKE *Coup* (1979) i. 7 The beer-crazed mob of American boobs cheers..the crunched leg of the unhome-team left tackle as he is stretchered off the field. **1980** K. ROYCE *Third Arm* v. 52 He did not himself feel shock until after Adams had been stretchered from the car. **1982** *Times* 11 June 6/4 Casualties..were stretchered to a field hospital.

stre·tchiness. [f. STRETCHY *a.* + -NESS.] The quality of being stretchy; elasticity.

1963 *Economist* 7 Sept. 840/2 A slight stretchiness can be woven into two directions. **1976** A. FARRER *Interpretation & Belief* 208 In this way I stretch my terms and I do not know how much they ought to be stretched. At this point the stretchiness of terms in my descriptions to myself of my own experience comes into play.

stretching, *vbl. sb.* Add: **c.** **stretching-board,** (*c*) *Leather-manuf.* = *stretch-bench* s.v. STRETCH *v.* 24; **stretching-course** (later example).

1976 T. WALKER *Spatsizi* v. 47 He taught me how to skin, showing me how to turn the hides fur side out before they were quite dry and then pull them back on the stretching board. **1973** L. RUSSELL *Everyday Life Colonial Canada* v. 60 Each layer of bricks was called a course; the parallel arrangement was a stretching course and the transverse a heading course.

stre·tch-out. Chiefly *N. Amer.* Also unhyphened. [f. STRETCH *v.* + OUT *adv.*] **1.** A practice of requiring workers, esp. in textile industries, to do extra work or operate extra machines for little or no additional pay.

1933 *Sun* (Baltimore) 30 June 12/1 The 'stretch-out' is a scheme for getting more work done in the textile mills with less labor. **1933** E. CALDWELL *God's Little Acre* v. 84 The mill can't get us back unless they shorten the hours, or cut out the stretchout, or go back to the old pay. **1934** *Sun* (Baltimore) 17 Aug. 1/3 Wage increases, shorter hours, differentials in the higher wage brackets, and an end to the 'stretch-out' are objectives to be sought in the strike. **1943** *Ibid.* 14 June 10/7 A managerial stretch-out which prostrates war workers is intolerable.

2. A practice of slackening production schedules as an economy, so that a set quantity will be produced over a longer period; a postponement of the date of fulfilment of orders or contracts, etc.

1946 [see sense 3 below]. **1952** *N.Y. Times* 16 Aug. (Late city ed.) 14/1 The North Atlantic defense program, already handicapped by a 'stretch-out'. **1959** *Wall St. Jrnl.* 14 Jan. (Eastern ed.) 2/4 The stretchout is understood to apply to the date at which the two aircraft were to go into flight and become 'operational'. **1960** *Times* 21 Nov. (Canada Suppl.) p. xiii/3 Stretch-out of deliveries has been broadly accepted by the industry. **1969** *Look* 29 Apr. 57/3 Support on the part of so many in the diocese made this patient waiting over weeks of time much easier. But the long stretch-out freed us. **1979** *Aviation Week & Space Technol.* 13 Aug. 9 Neither the abandonment of the B-1 by the current Administration nor its stretchout of the MX missile..speeded up SALT.

3. *attrib.*

1934 *Sun* (Baltimore) 1 Sept. 2/4 It has failed to do anything about solving the 'stretch-out' problem. **1946** *Ibid.* 10 May 15/3 A 'stretch-out' plan, under which those employed will work less hours weekly, will be adopted soon. **1960** *Times* 21 Nov. (Canada Suppl.) p. xiii/2 The viability of a satisfactory stretch-out arrangement depends on the transfer of contracts from the weaker to the stronger mining operations. **1967** *Canad. Ann. Rev. 1966* 75 The Prime Minister announced a major increase in the amounts allocated to vocational and technical training in the stretch-out period.

stretto, *adv.* and *sb.* Add: **A.** *adv.* (Earlier example.)

1740 J. GRASSINEAU *Mus. Dict.* 240 *Stretto*, shortened, is often used to signify that the measure is to be short and concise, therefore quick.

B. *sb.* **a.** Also *transf.*

1962–3 *Sight & Sound* Winter 19/1 Finally, there are the flashbacks and then the stretto of flashbacks, as if, at the end, Colin Smith were still attempting to make up his mind. **1963** J. WIESENFARTH *H. James* v. 104 The *coda* begins in Chapter XII and ends with Chapter XIV in a *stretto*. **1979** *UCT Studies in English* (Univ. Cape Town) Sept. 38 Pope mimics the convention: *The Rape of the Lock* is threaded with premonitory phrases which he gathers into a *stretto* as the climax draws near.

b. *stretto maestrale* [cf. *MAESTRALE] (see quot. 1946).

1876 STAINER & BARRETT *Dict. Mus. Terms* s.v. *Maestrale*, *Stretto maestrale*, a term sometimes applied to the stretto of a fugue when in canon. **1910** E. PROUT *Anal. J. S. Bach's Forty-Eight Fugues* 13 As the subject appears in a complete form in all the groups of the entries now under notice,..we have here an example of a stretto maestrale. **1946** E. BLOM *Everyman's Dict. Mus.* 672/1 *Stretto maestrale*,..a S[tretto] in which the fugal subject not only appears in close, overlapping entries, but is carried through from beginning to end at each entry. **1948** G. OLDROYD *Technique & Spirit of Fugue* ix. 143 It is a specimen of '*stretto maestrale*' in which a phrase in its full length is repeated in canon throughout all the strands. **1959** J. V. COCKSHOOT *Fugue in Beethoven's Piano Mus.* v. 68 This four-fold entry foreshadows the final section, with an effect of *stretto maestrale*.

streusel (strǫi·zĕl, strū·zĕl, ‖ʃtroi·zĕl). Chiefly *U.S.* [Ger., f. *streuen* to sprinkle.] A crumb-like topping for cakes and pastries made from fat, flour, cinnamon, and sugar; a confection with such a topping. Freq. *attrib.*, esp. as *streusel cake, kuchen*.

1909 L. MEIER *Art of German Cooking & Baking* xvii. 335/2 (*heading*) Streusel Coffee Cake. Preparation of the Streusel. **1910** M. MALZBENDER *Pract. Man. for Confectioners, Pastrycooks & Bakers* (rev. ed.) iii. 66/2 Streusel Kuchen. Proceed same as No. 6 only put streusel on top before baking. **1952** L. J. MITCHELL *Lüchow's German Cookbk.* xi. 183 (*heading*) Apple crumb cake (Apfel Streuselkuchen). *Ibid.*, For the Streusel, coarsely mix ⅜ cup butter, the flour, sugar and cinnamon with a pastry blender. **1957** M. E. SHOWALTER *Mennonite Community Cookbk.* i. 27 (*heading*) 'Streusel Kuchen' *Raised Coffee Cake*... Let rise 1½ hours. Sprinkle with streusel crumbs. **1960** 'A. KNOX' *Cooking Austrian Way* (rev. ed.) 202 (*heading*) Streusel Cake *Streuselkuchen*... This is a cake with a yeast dough and a streusel mixture on top. **1966** W. S. RAMSON *Austral. Eng.* viii. 161 Some unrecorded borrowings from German may be in local use in parts of South Australia. Price noted..*streuselkuchen*, 'cake', [etc.]. **1976** *Woman's Day* (U.S.) Nov. 104, no. 14 Mashed Potatoes/Peach Streusel/Gingerbread. **1977** C. MCFADDEN *Serial* (1978) xx. 46/1 Sam wanted to buy a Sara Lee streusel cake.

streuth, var. 'STREWTH in Dict. and Suppl.

strewage (strū·ĕdʒ). [f. STREW *v.* + -AGE.] = STREWING *vbl. sb.* b.

1902 J. H. SKEINE *Pastor Agnorum* 266 The waking of that Syrian refugee..after his sleep among the great strewage of rocks on a down of Canaan. **1929** R. BRIDGES *Test. Beauty* III. 894 Vestiges of his stony asceticism imbue all time, thick as the strewage of his flinty tools, disseminate wheresoe'er he hath dwelt. **1940** C. F. C. HAWKES *Prehist. Found. Europe* ii. 13 The basement-beds below it, formed of the strewage of older land-surfaces.

strewn field. *Geol.* Also **strewnfield.** [f. STREWN *ppl. a.* + FIELD *sb.*] A part of the earth's surface over which tektites of a similar age are found.

1937 *Proc. R. Soc. Victoria* XLIX. 167 There are no records of the discovery of australites in towns within the australite 'strewn field'. **1961** *Sci. Amer.* Nov. 63 (*caption*) Estimates of ages of tektites in strewn fields are based on ages of rocks with which they are found. **1973** *Nature* 16 Feb. 431/1 Tektites..are apparently limited to four large areas (strewnfields) and four corresponding ages.

'strewth. Add: Also **streuth, strewth, 'strooth, 'struth, struth.**

1913 A. J. REES *Merry Marauders* ix. 149 'Strooth! he's looked up all our lines. **1915** [see *KING-PIN 2]. **1925** [see *GARN *int.*]. **1933** M. LOWRY *Ultramarine* ii. 75 Gawd strewth, you're some fellow, you are! **1938** P. LAWLOR *House of Templemore* xvii. 186, I have made a string bookshelf just like you had. Streuth! So you have. **1954** A. SETON *Katherine* xiv. 235 "Struth,' said Edmund. ..'High time I got me *some* wife.' **1975** P. G. WINSLOW *Death of Angel* iv. 86 Strewth, they've made a mess of this office. **1977** *Sunday Sun* (Brisbane) 30 Jan. 29/1 Struth! What next? says Sam.

stria. Add: **2. d.** (Earlier and further examples.) Also in *sing.*

1859 *Phil. Trans. R. Soc.* CXLVIII. 3 The discharge did not exhibit the uniform white light of the Torricellian vacuum, but striæ in confused or irregular forms. **1883** [see *STRIATED *ppl. a.* 1 e].

e. *Med.* Also *stria atrophica* [mod. L. *atrophica* (see ATROPHIC *a.*)]. A stretch mark; *stria gravidarum* [L., gen. pl. of *gravida* pregnant woman], one on a pregnant woman, usu. darker than the surrounding skin; *stria albicans* (pl. *albicantes*) [L., pres. pple. of *albicāre* to make or be white], a former stria gravidarum that has become light-coloured following delivery.

1867 *Jrnl. Cutaneous Med.* I. 142 None of the early writers have alluded to any other cause of the striae atrophicae than over-distension. **1880** *Trans. Amer. Gynecol. Soc.* IV. 141 These cicatrices are usually described as red or white shining striæ, marking the skin..of pregnant and multiparous women. **1884** R. & F. BARNES *Syst. Obstetric Med. & Surg.* I. viii. 294 Striæ gravidarum. **1906** T. W. EDEN *Man. Midwifery* I. 51 *Striæ gravidarum* appear on the abdominal wall... They are pearly..when recent, but afterwards they become pale and silvery (*striæ albicantes*). **1968** D. C. BETHEA *Introd. Maternity Nursing* iii. 25 Thin, red streaks, called striae gravidarum, occur on the breasts of some pregnant women. **1970** C. LERCH *Maternity Nursing* v. 61/2 In subsequent pregnancies new striae appear in addition to the silvery white markings of the previous pregnancy.

striatal: see *STRIATUM.

striate, *a.* Add: **2.** *Anat.* Epithet of the striatum (esp. in sense *a) and the blood vessels supplying it.

1890 BILLINGS *Med. Dict.* 599/2 *Striate arteries*, small twigs from anterior and middle cerebral arteries that.. supply corpus striatum. **1902** D. J. CUNNINGHAM *Text-bk. Anat.* 837 An inferior striate vein descends on each side from the substance of the corpus striatum. **1907** *Arch. Neurol.* III. 42 Homonymous visual defects are due to disease and destruction of the striate cortex. **1921** TILNEY & BASSETT *Form & Functions Central Nerv. Syst.* xliv. 805 In its process of evolution from the lower vertebrates to mammals, the primordial portion of the striate body corresponds to the globus pallidus. **1948** A. BRODAL *Neurol. Anat.* iv. 73 The striate body, the corpus striatum, consists of large grey nuclear masses.. subdivided by fibre strands into different portions. **1968** PASSMORE & ROBSON *Compan. Med. Stud.* I. xxiv. 52/2 The striate area..contains extra fibres which widen the outer lines of Baillarger into a broad white band. *Ibid.* 72/1 The striate artery..supplies the medial part of the head of the caudate nucleus and putamen and the anterior part of the internal capsule. **1978** *Nature* 3 Aug. 423/1 Posteriorly, the striate cortex is so distinct and uniform that its borders are visible, in histological sections, even to the naked eye.

striate, *v.* Add: Also *fig.*

1979 D. HOFFMANN in *Harvard Guide Contemp. Amer. Writing* xii. 579 Clarity and strength striate the poems of Charles Edward Eaton.

striated, *ppl. a.* Add: **1. a.** Also *fig.*

1923 H. CRANE *Let.* 2 Mar. (1965) 129 Striated with nuances, nervosities, that we are heir to.

d. Of muscle: = STRIPED *ppl. a.* 1 c.

1846 W. B. CARPENTER *Man. Physiol.* iii. 199 When the striated Muscular Fibre is examined still more closely, it is found to contain an assemblage of very minute elements. **1851, 1866** [in Dict., sense 1 a]. **1959** W. ANDREW *Textbk. Compar. Histol.* ii. 48 The muscular tissue in man and other vertebrates is of three chief types: (1) smooth, (2) cardiac, and (3) skeletal. The last two types frequently are referred to as striated because of the cross-striations on the fibers. **1982** [see *SKELETAL *a.* b].

e. Of an electric discharge: exhibiting striæ (sense 2 d).

1852 *Phil. Trans. R. Soc.* CXLII. 100 In a well-exhausted receiver containing a small piece of phosphorus, the discharge is throughout its course striated by transverse non-luminous bands. **1883** *Rep. Brit. Assoc. Adv. Sci. 1882* 31 A stria, with its attendant dark space, forms a physical unit of a striated discharge. **1942** J. D. STRANATHAN *'Particles' Mod. Physics* iii. 71 Small amounts of impurity in the gas affect the striated positive column greatly. **1973** J. YARWOOD *Electricity & Magnetism* xiv. 503 For a striated positive column, the number of ions per unit volume is least at the bright edge of the striation.

striation. Add: **2. b.** *Electr.* = STRIA 2 d (in *sing.*).

1902 *Encycl. Brit.* XXVIII. 47/2 The bright parts of the striations are slightly concave to the positive electrode. **1942** J. D. STRANATHAN *'Particles' Mod. Physics* iii. 67 As the pressure is still further reduced..the striations become more coarse and indistinct, and the Crookes dark space lengthens. **1973** [see *STRIATED *ppl. a.* 1 e].

striatum (strəiˌēɪ·tə̆m). *Anat.* Pl. **striata.** [mod.L., neut. of *striātus* STRIATE *a.*]. **a.** The corpus striatum, a body of grey matter within each cerebral hemisphere comprising the lentiform nucleus (i.e. the globus pallidus and the putamen) and the caudate nucleus; some writers also include the claustrum. **b.** (Now the more usual use.) = *NEOSTRIATUM, i.e. the putamen and the caudate nucleus only.

[**1803** C. BELL *Anat. Human Body* III. 1. iii. 87 The corpora striata are smooth, cineritious convexities... These bodies are called striata, from the intermixture of the medullary matter, which gives the appearance of striæ when they are cut.] **1889** J. LEIDY *Elem. Treat. Human Anat.* (ed. 2) xv. 751 The striatum, or corpus striatum, appears in the body of the lateral ventricle as a smooth, convex gray eminence projecting from its outer wall. **1948** A. BRODAL *Neurol. Anat.* iv. 73 The putamen and the caudate nucleus together constitute the neostriatum (often collectively called on account of their similarities the striatum). **1971** N. G. SUTTON *Anat. Brain & Spinal Medulla* vii. 102 The claustrum is often considered part of the striatum, that is a detached part of the putamen, but it has also been thought to be an included portion of the cortex of the insula. **1978** *Nature* 23 Feb. 767/1 The rats were decapitated and the corpus striatum ipsilateral to the lesion, the contralateral unlesioned striatum and striata from unlesioned animals were assayed.

Hence **stria·tal** *a.*, of or pertaining to the striatum.

1926 *Jrnl. Nerv. & Mental Dis.* LXIV. 9 All kinds of striatal tremors..are action-tremors. **1937** *Arch. Neurol. & Psychiatry* (Chicago) XXXVIII. 737 One can infer that hippocampal activity..through the striatal connections regulates attitudes of emotional expression. **1974** D. & M. Webster *Compar. Vertebr. Morphol.* xi. 268 It is not definitely established whether some or all of the bird's striatal structures are related to the mammalian striatum.

striature. Add: Also *fig.*
1918 F. Hackett *Ireland* xiii. 362 This striature of Catholics and Protestants, nationalists and anti-nationalists, Irish and Scotch-Ulstermen, is by no means so insufferable as the tenor of argument may indicate.

strich, var. *stritch.

strickle, *v.* (Later examples.)
1927 *Jrnl. Inst. Metals* XXXVII. 25 Red-hot sand was used to fill this space and strickled off level with the top of the mould. **1934** *Proc. Inst. Brit. Foundrymen 1932–1933* XXVI. 548 It [*sc.* stucco] is a splendid material and can be swept or strickled to very fine limits.

strict, *a.* Add: **I. 3. a.** (Later example.)
1897 F. Thompson *New Poems* 68, I, the boundless strict savannah Which God's leaping feet go through.
8. e. Logic. *strict implication*: a relationship holding between propositions in which it is impossible for the antecedent to be true and the consequent false. Cf. *material implication* s.v. *material *a.* 2 b.
1912 C. I. Lewis in *Mind* XXI. 526 Intensional disjunction bears the same relation to inferential or 'strict' implication that extensional disjunction bears to the algebraic or 'material' implication. **1933** C. A. Mace *Princ. Logic* iv. 68 This is clearly a different sense of implies, and is sometimes called strict implication. **1947** H. Reichenbach *Elem. Symbolic Logic* viii. 379 The calculus of strict implication..constitutes a system of this kind. **1977** *Fontana Dict. Mod. Thought* 206/1 The systems of strict implication are the basis of contemporary modal logic.
II. 11. d. *strict liability*: a liability which does not depend upon intent to commit an offence.
1896 *Rep. Cases N.Y. Court of Appeals* CLI. 142 The weight of the argument..is in favor of the rule of strict liability which requires a public official to assume all risks of loss. **1926** *Law Q. Rev.* XLII. 51 The description of the rule in *Rylands* v. *Fletcher* as an example of absolute liability in tort is unhappy in view of some half dozen exceptions which are admitted as qualifications of it. 'Strict liability' seems to be a better term. **1935** *California Law Rev.* May 431 Liability on the ground of nuisance should not be confused with the doctrine of absolute or strict liability for certain classes of lawful acts. **1945** W. T. S. Stallybrass *Salmond's Law of Torts* (ed. 10) ii. 20 A period of strict liability, an 'unmoral' period, is succeeded by a period of fault liability, a 'moral' period. **1953** *N.Y. Univ. Law Rev.* XXVIII. 1076 The courts have made no inroads upon strict liability for damage done by animals ferae naturae, nor upon scienter liability. **1979** *Internat. Jrnl. Sociol. of Law* Feb. 54 In short, by substantially removing the issue of intention from the crimes of employers, the 1844 Act took a substantial step towards the doctrine of strict liability.
12. b. *strict tempo*: in Music, a strict and regular rhythm; freq. used *attrib.* with reference to a kind of ballroom dancing to music with such a rhythm.
1936 F. G. Hawkes *Studies in Time & Tempo* vi. 35 If the proper rhythmical effect..is to be secured, the observance of accurate and strict *tempo* becomes an absolute necessity. **1958** P. Gammond *Decca Bk. Jazz* xxv. 320 The vacuities of 'strict tempo' and the morbid sex neurosis of the modern 'sob' song. **1959** F. Newton *Jazz Scene* xiii. 230 'Strict tempo' dancing, the foundation of the mass ballroom vogue among the British working class,..grew in a direction diametrically opposed to jazz. **1961** *Listener* 23 Nov. 887/2 A champion strict-tempo dancer. **1978** F. Mullally *Deadly Payoff* vi. 81 Strict-tempo ballroom dancing: the slow foxtrot, the quickstep, the waltz. **1982** Warner & Sandilands *Women beyond Wire* ii. 19 The strict-tempo orchestra which reeled off foxtrots and quicksteps.

strictarian (striktɛ̄ə·riăn), *sb.* and *a. rare.* [f. Strict *a.*: see *-arian.] **A.** *sb.* One holding rigidly conformist views. **B.** *adj.* Characteristic of a strictarian.
1867 [see *-arian]. **1926** *Chambers's Jrnl.* Mar. 153/1, I was not churlish enough or strictarian enough to decry that acceptance. **1931** A. L. Rowse *Politics & Younger Generation* ix. 251 These are points in Marxism, particularly in the orthodox, strictarian interpretation of it, that are open to criticism.

‖ **stricti juris** (stri·ktəi dʒūə·ris, stri·kti yuə·ris), *adv. phr. Law.* Also **stricti iuris.** [L., lit. 'of strict law'.] Strictly according to the law; according to law as opposed to equity. Also as quasi-*sb.*, the practice of strict interpretation of the law.
1684 G. Mackenzie *Institutions Law of Scotland* IV. i. 342 Some actions are stricti juris; in which the Judge is to follow the strict prescript of the Contract upon which the Action is raised. **1704** T. Wood *New Inst. Imperial or Civil Law* iv. iii. 327 Some *stricti juris*, where the Judge cannot depart from the strict terms of the Contract. **1845** H. Broom *Sel. Legal Maxims* v. 242 Defences which were not admissible or valid *stricti iuris*. **1971** R. D. Baker *Judicial Rev. in Mexico* vii. 185 The rule of *stricti iuris* (*estricto derecho*) requires the courts to confine their attention to and base their decisions exclusively on those conclusions of law. **1977** A. Watson *Society & Legal Change* ii. 12 *Stipulatio*..was a unilateral formal contract of strict law (*stricti iuris*) but the formalities were far from complicated.

strictly, *adv.* Add: **9.** *colloq.* (chiefly *U.S.*). Definitely; exclusively. Phr. *strictly for the birds*: see *bird *sb.* 5 d.
1938, 1945 [see *Mickey Mouse 2]. **1947** [see *model *sb.* 7 e]. **1951** [see *man-hungry* s.v. *man *sb.*[1] 20]. **1977** *Amer. Speech* 1975 L. 67 *Strictly*.., absolutely, honestly, sincerely, definitely. 'My teacher is handsome, strictly!'

‖ **stricto sensu** (stri·kto se·nsu). Also *erron.* **strictu sensu.** = *sensu stricto.
1931 [see *Folsom]. **1972** *Mod. Law Rev.* XXXV. i. 55 In the case of custom *stricto sensu* lack of consensus enhances the importance of the consolidating effect of time. **1976** *Times Lit. Suppl.* 17 Sept. 1176/3 A philosophical analysis of the feasibility of the structuralist enterprise *stricto sensu*. **1979** *Nature* 22 Feb. 599/1 Thus 'mosaic' RNAs produced by splicing are *strictu sensu* recombinant molecules in that they contain data drawn from different parts of the genome.

stricture, *sb.*[1] Add: **I. 1. b.** *Phonetics.* Partial or complete closure of the air-passage in the articulation of speech sounds.
1943 K. L. Pike *Phonetics* vii. 120 At the time in the production of some sound when any moveable part of the vocal apparatus causes any *stricture* (the partial or complete closure of an air passage) it becomes an *articulator*. **1962** B. M. H. Strang *Mod. Eng. Structure* 31 Articulated sounds may further be differentiated by the variable shape of the articulators and strictures involved in their production. **1964** J. C. Catford in D. Abercrombie et al. *Daniel Jones* 26 The articulatory stricture generates turbulent airflow. *Ibid.* 32 These five major phonatory stricture types.

strictu sensu, ¶ var. *stricto sensu.

stride, *sb.* Add: **1. b.** Esp. in phr. *to take* or *make strides*: to make progress.
1926 J. S. Huxley *Essays in Pop. Sci.* 21 Great strides have been taken in this field too during the last twenty years. **1934** *Discovery* Dec. 362/2 Photography for all purposes has made immense strides latterly. **1956** J. B. Wilson *Lang. & Pursuit of Truth* i. 14 The development of a good system of notation made it possible to take great strides in our mathematical knowledge. **1976** *Field* 18 Nov. 979/1 Great strides have been made in short term forecasting in the last five years.
c. Phr. *to lengthen* (or *shorten*) *one's stride.* Also (U.S.) *to lengthen* (or *shorten*) *stride.*
1925 E. F. Norton *Fight for Everest: 1924* 32 On April 8 we lengthened our stride and covered 12 miles. **1978** *Washington Post* 24 Mar. b6/2 Her many backers had some anxious moments as she shortened stride after a clear lead at the head of the stretch. **1980** H. D. Westacott *Walker's Handbk.* (ed. 2) iv. 60 On level ground use your natural stride and resist any temptation to lengthen it. On a gradient the stride should be shortened.
3. c. (Earlier and later *fig.* examples.) Freq. in phrases, as *to get into one's stride, to hit one's stride, to put* or *throw* (someone) *out of* (or *off*) *his stride.*
1890 S. Webb *Let.* in J. MacKenzie *Victorian Courtship* (1979) viii. 104, I had a bad week... But I have 'got into my stride again' now. **1919** *Punch* 12 Mar. 210/1 The operator won the first game before I could get into my stride. **1933** A. Powell *From View to Death* i. 19 Conversationally, Zouch was getting back into his stride and he knew that by the evening he would be in good form. **1941** E. S. Gardner *Case of Haunted Husband* viii. 53 He threw me out of my stride for a whole half day. **1946** K. Tennant *Lost Haven* (1947) xiv. 226 'Why, I had a talk with him only the other night.' 'Did you?' Dipper asked, put out of his stride. **1955** H. Kurnitz *Invasion of Privacy* (1956) xii. 77 It was late in the working day..but ..Louis Stradling was just hitting his full stride. **1967** N. Freeling *Strike out where not Applicable* 85 'Who told you that?'..The young man was thrown out of his stride. **1978** D. Devine *Sunk without Trace* xxii. 202 He was disappointed. He didn't think I'd let a man like Max Sapiro put me off my stride. **1983** *Listener* 27 Jan. 21/1 But when Mr. Maccoby gets into his stride of explaining the mechanisms of what has so often been the loathsome behaviour of so-called Christians to Jews, his book becomes of potent interest.
d. Also (chiefly *U.S.*) without possessive adj. (Later examples.)
1941 B. Schulberg *What makes Sammy Run?* iii. 48 It was funny to see him taking the Vendome [*sc.* an expensive restaurant] in stride too. **1974** *Publishers Weekly* 7 Jan. 49/2 Coach John Wooden..has taken it all in stride. **1976** *New Society* 3 June 521/1 Everyone understood what it meant to be photographed and took the request in stride.
4. c. Delete ? *Obs.* and add later examples.
1922 Joyce *Ulysses* 344 A navy threequarter skirt cut to the stride showed off her slim graceful figure to perfection. **1939** *Country Life* 11 Feb. p. xxxiii/2 (Advt.), There is plenty of stride; the knees are well formed; the lines and run of seams are perfectly executed.
d. *pl.* Trousers. Also occas. breeches; jeans. *slang.*
1889 A. G. Murdoch *Scotch Readings* (Ser. 3) 26 His two legs, which were encased in a pair of all but skin-tight 'strides'. **1889** Clarkson & Richardson *Police!* xxv. 346 If the 'Peter' (cash-box) can be found, that is at once appropriated, as also are a man's 'strides' (trousers). **1914** Jackson & Hellyer *Vocab. Criminal Slang* 81 *Strides*,..a pair of trousers. **1924** *Truth* (Sydney) 27 Apr. 6 *Strides*, trousers. **1932** L. Mann *Flesh in Armour* 291 His tunic and light coat were of the ultra fashionable style, and his strides would not have disgraced an officer of the Guards. **1947** D. M. Davin *For Rest of our Lives* xxxviii. 196 Trying to get his strides up. **1950** 'N. Shute' *Town like Alice* ix. 261 Could you get into a pair of my strides? **1960** 'A. Burgess' *Doctor is Sick* xxvi. 211 He handed a crumpled bundle to Edwin, saying: 'You'll 'ave to take my strides.'..The trousers, Edwin found, were too short. **1973** M. Amis *Rachel Papers* 186 The Oxford University candidate was to be seen in T-shirt and khaki strides. **1980** B. Mason *Solo* 91, I wiped damp hands on my serge strides.
6*. Ellipt. for *stride piano* (see sense 7 b below).
1956 Panassié & Gautier *Guide to Jazz* 260/2 *Stride*, a piano style much in use by soloists about 1930. **1969** *Listener* 6 Feb. 186/3 Peterson stands at the end of a long and honourable tradition of jazz piano playing originally known as 'Harlem stride'. The stride refers to a left-hand vamping method using alternating tenths and note clusters at least an octave apart. **1975** *New Yorker* 19 May 6/2 Jaki Byard, who displays a confident feel for ragtime, stride, and more modern piano styles, has fun with bassist Major Holley.
7. b. *Jazz.* Used *attrib.* to designate a style of piano-playing in which the left hand alternately plays a single note and a chord that is an octave (or more) higher; *esp.* in **stride piano** (hence **stride pianist**); also *stride accent, bass, tempo,* etc.
c **1938** N. E. Williams *His Hi de Highness of Ho de Ho* 35/2 'Gut tempo' and 'stride tempo' usually are intelligible only to our own musicians. **1950** Blesh & Janis *They all played Ragtime* x. 192 He could play the ragtime stride bass, but it bothered him because his stomach got in the way of his arm, so he used a walking bass instead. **1952** B. Ulanov *Hist. Jazz in Amer.* iv. 29 'Stride piano', the particular pride and joy of Fats Waller and, before him, of innumerable ragtime pianists, comes from the blues. *Ibid.* 30 The blues is usually played in unaccented four/four time or with stride accents. **1955** L. Feather *Encycl. Jazz* 289 His lacy, charming melodies sometimes contrast with 'stride' passages of great intensity. **1959** *Jazz Rev.* June 14/1 He [*sc.* James P. Johnson]..developed the New York style of 'stride' piano from the rags of Scott Joplin and the southern Negro cotillion and set dances. **1978** *Listener* 29 June 841/2 An exhilarating two hours of Fats Waller numbers..accompanied on stage by the celebrated stride pianist, Luther Henderson. **1983** *Listener* 20 Jan. 10/2 Its earliest landmark was James P. Johnson's stride-piano showpiece 'Carolina Shout'.

stride, *v.* Add: Pa. pple. also **strode** (*colloq.*).
1936 'R. West' *Thinking Reed* iii. 89 From youth he had strode through the twenty-four hours at the pace of a Marathon race. **1951** R. Lynd *Essays on Life & Lit.* i. 27 But a *gauche* big farmer's son in a white coat..had strode past her roughly. **1972** *Observer* (Colour Suppl.) 24 Sept. 23/2 The clear mental picture of the battlefront with which he had so boldly strode into Samsonov's headquarters. **1980** *New Yorker* 24 Mar. 127/1 (Advt.), No consumer advocate has yet strode forth to defend and protect the interests of those who can afford a $30,000 sport coupe.
7. *Jazz.* To play stride piano (see *stride *sb.* 7 b).
Found only in the gerund or participial form *striding.*
1944 *Metronome* Nov. 17/3 Alberta Simmons, from down in the Jungles, could beat the average man 'striding'. **1958** P. Gammond *Decca Bk. of Jazz* xv. 187 Nobody else has compounded so many pianistic devices—the delayed note, the tremolo, the dazzling run, the striding bass—into such a homogeneous quiddity.

stridency. Add: (Examples in sense 1 b of *strident *a.*)
1968 Chomsky & Halle *Sound Pattern Eng.* 329 Stridency is a feature restricted to obstruent continuants and affricates. **1976** *Word 1971* XXVII. 220 This appearance of the Stridency distinction among Stops and Continuants evidently does not generalise immediately to the affricates. **1979** *Canad. Jrnl. Linguistics* XXIV. i. 20 It incorporates both articulatory (anteriority-coronality) and acoustic (stridency) features.

strident, *a.* Add: **1. b.** *Phonetics.* Of the articulation of a consonantal sound: characterized by friction that is comparatively turbulent. Also as *sb.*, a consonant articulated in this way.
1956 Jakobson & Halle *Fundamentals of Lang.* 31 Strident/mellow: acoustically—higher intensity noise *vs.* lower intensity noise; genetically—rough-edged *vs.* smooth-edged. *Ibid.* 42 Mellow constrictives, opposed to strident constrictives, or strident plosives (affricates) opposed to mellow plosives (stops proper) do not appear in child language before the emergence of the first liquid. **1965** *Amer. Speech* XL. 9 *T* cannot follow a dental stop or *S* follow a strident (sibilant). **1968** Chomsky & Halle *Sound Pattern Eng.* 329 Strident sounds are marked acoustically by greater noisiness than their nonstrident counterparts. **1976** *Word 1971* XXVII. 220, /s/..[and] /ʃ/..also embody the Strident vs. Mellow distinction and are both + Strident.

strider. Add: **2.** *U.S.* = *pond-skater* s.v. *pond *sb.* 4.
1974 A. Dillard *Pilgrim at Tinker Creek* xi. 189, I read that striders are attracted to any light. **1978** *Sci. Amer.* Apr. 134/1 When the strider stands on the surface film, it is supported by its long, slender middle and hind legs.

strife. Add: **1. f.** *Austral. colloq.* Trouble, disgrace, difficulties. Freq. in phr. *in strife.*
1963 A. Lubbock *Austral. Roundabout* 45 'By cripes!' said the landlord, 'I bet you was in strife after that.' **1966** *Sunday Mail Mag.* (Brisbane) 9 Jan. 2/2 He's having trouble with his irrigation. His sudax is coming along all right but he's often in strife priming his pump. **1966** P. Mathers *Trap* 15, I reckon she needs..a bloody flogging... She'll get us all in strife, he finished. **1969** 'A. Garve' *Boomerang* iv. 147 'Keep close on my tail,' he called. 'If you get in any strife, bang on your horn.'
5. *strife-weary* adj.; *strife-torn* ppl. adj.
1972 R. D. Walshe in G. W. Turner *Good Austral. Eng.* xi. 227 The ego ceases to be a shifting strife-torn no-man's-land between the armies of the id and the superego. **1983** *Times* 30 Mar. 7/2 She flies there today on her third tour of the strife-torn Brahmaputra valley state in the past 10 weeks. **1949** Koestler *Promise & Fulfilment* iii. i. 302 Millions of war-worn, strife-weary people longing to find peace.

striggle (stri·g'l), *colloq.* [perh. f. Str(aggle *v.*[1] + W)iggle *sb.*] A wavy line.
1906 W. De Morgan *Joseph Vance* xxx. 286 I've got him [*sc.* a fly out of the milk]! But he's brought a long striggle of cream out with him. **1963** V. Nabokov *Gift* iv. 220 This old diary, which was written in an even hand with little striggles and was in a home-made code.

strikable (strəi·kab'l), *a.* [f. Strike *v.* + -able.] **a.** That may be struck. **b.** Of an issue: that may provoke an industrial strike.
1904 J. P. Mannock *Billiards Expounded* I. i. 23 The various 'strikable faces' of a ball. **1977** *Washington Post* 18 May c2 The idea of 5-day mail service is a 'strikeable issue'.

strike, *sb.* Add: **6. d.** Infestation of a sheep or cow with flies whose larvæ burrow into the skin; an occurrence of this. Freq. with preceding sb.
1933, etc. [see *blow-fly strike* s.v. *blow-fly b]. **1933**, etc. [see *Mules]. **1934** *Bulletin* (Sydney) 26 Sept. 22/1 The C.S.I.R. regards dipping in the light of only 'perhaps rendering the sheep less favorable for strike, but a measure not to be relied upon'. **1937**, etc. [see *fly-strike* s.v. *fly *sb.*[1] 11]. **1952** I. E. Newsom *Sheep Dis.* vi. 140 In South Africa..*L*[*ucilia*] *cuprina* is thought to be responsible for 90 per cent of the strikes either alone or in combination with other flies. **1972** *TV Vet Sheep Bk.* xlviii. 143/1 In Britain strike usually starts when the lambs start scouring. **1975** *N.Z. Jrnl. Agric.* Sept. 65/1 One measure for blowflies' resistance to insecticides is the time that it takes, after a spray or dip, for implanted larvae to establish a strike. **1977** *Bulletin* (Sydney) 22 Jan. 16/3 The blowfly costs rural industry $70 million a year in sheep and cattle strike.
e. A sudden military attack concentrated on selected targets; also occas. *concr.*, the force used in such an attack. Also (chiefly with reference to the use of nuclear weapons) preceded by a qualifying word, as *first-strike, pre-emptive strike, second strike*: see under the first elements.
1942 [see *strike patrol*, sense 20 below]. **1943** T. Dudley-Gordon *Coastal Command at War* ii. 16 When the Admiralty desires a special reconnaissance or strike to be 'laid on'. **1943** *Yank* 19 Nov. 3 But, when the last strike returned, there were no bullet holes, no torn fabric and the pilots climbed out unhurt. **1945** [see *air strike* s.v. *air B. III. 2]. **1963** *Ann. Reg. 1962* 520 The purpose of these bases can be none other than to provide a nuclear strike capability against the Western hemisphere. **1972** *Newsweek* 10 Jan. 1/1 Described by the Pentagon as 'protective-reaction' strikes, the bombings in fact signaled to the world the continuing U.S. interest in Southeast Asia. **1979** H. Kissinger *White House Years* xxiii. 983 The Son Tay raid was accompanied by a two-day strike by 200 airplanes against North Vietnamese supply installations.
f. *bird-strike*: see *bird *sb.* 9.
9. a. Freq. with preceding qualifying word, as *general, outlaw, selective, sit-down, stay-away* (*-down, -in*), *sympathetic, wild-cat strike*: see under the first elements. Also *fig.*
1907 R. Dunn *Shameless Diary of Explorer* xv. 201 Miller's stomach went on strike after we washed in the glacier stream.
b. *transf.* A concerted abstention from a particular economic, physical, or social activity on the part of persons who are attempting to obtain a concession from an authority or to register a protest; esp. in *hunger strike, rent strike* (see *hunger *sb.* 4, *rent *sb.*[1] 4 c).
1889, etc. [see *hunger strike* s.v. *hunger *sb.* 4]. **1911** G. B. Shaw *Getting Married* 220 Ive told our last four Prime Ministers that if they didnt make our marriage laws reasonable there would be a strike against marriage. **1934** *Sun* (Baltimore) 8 Nov. 10/4 People with fixed incomes necessarily buy less. There are indignation meetings and 'buyers' strikes'. **1937** *Ibid.* 30 Aug. 8/1 The falling birth rate indicates that 'mankind cannot be forced or bribed to produce children'... The present 'birth strike' will continue until necessary social readjustments are effected. **1938** *Ibid.* 28 Jan. 22/1 Forty-eight tenants of an apartment building..started a 'strike' January 1, demanding rent reductions. **1965** B. Pearce tr. *Preobrazhensky's New Economics* 167 A consumers' strike is the limit which arises to state planning whenever the state's prices exceed the level acceptable to the private market. **1970**, etc. [see *rent strike* s.v. *rent *sb.*[1] 4 c]. **1976** *Gramophone* Dec. 1052/1 The Lysistrata plot about the women stopping a war by going on sexual strike.
11. (Earlier examples.) Also, the sudden discovery of an accumulation of natural gas. orig. *U.S.* (For 67 d read 68 d.)
1852 L. Clappe *Shirley Lett. Calif. Mines* (1922) 131 They are always longing for big strikes [of gold]. **1855** H. Helper *Land of Gold* 296, I may make a 'strike', but that is mere speculation. **1864** [see *oil strike* s.v. *oil *sb.*[1] 6 e].
12. Restrict *U.S.* to senses in Dict. and add: **a.** Also *Ninepins.* (Earlier and later examples.) Also *fig.*
1859 *Atlantic Monthly* Nov. 641 Strike: terms of the game of nine-pins. **1939** H. Miller *Cosmological Eye* 219 *Of Human Bondage* was a great book, he thought. I thought so too and I scored another strike for the constable on my mental blackboard. **1958** [see *frame *sb.* 11 g]. **1974** *Cleveland* (Ohio) *Plain Dealer* 13 Oct. c 8/3 Marge opened her third game with nine strikes in a row, but left the 5-9 pins on her first ball in the 10th frame on the way to her 275 game and 614 series.
b. (*a*) Earlier examples.) (*b*) Also, a pitched ball recorded against the batter. (Earlier and later examples.) Esp. as one of three counts against the batter.
1841 *Picayune* (New Orleans) 25 May 2/2 If 'Edith' wishes to see 'a great strike'.., let her walk down Water street..and see *the* 'bachelors' make the ball fly. **1845** in *Appletons' Ann. Cycl. 1885* (1886) X. 77/2 Players must take their strike in regular turn. **1856** *Spirit of Times* 22 Nov. 197/2 The striker should also be compelled to run on such occasions, strike or no strike. **1867** *Ball Players' Chron.* 4 July 6/2 Their batting was of a superior character, two of their players..each making some powerful strikes. **1868** H. Chadwick *Base Ball Player's Bk. Reference* 75 Mills called 'one strike' on him. **1912** C. Mathewson *Pitching* 12 It put me in the hole with the count two balls and one strike. **1942** *Sun* (Baltimore) 3 Apr. 18/7 The machine will throw 75 per cent more strikes in a given number of pitches than a human. **1974** *Anderson* (S. Carolina) *Independent* 19 Apr. 4B/7 The Citadel scored on a missed third strike and two errors.
(*c*) *fig.* Usu. preceded by a numeral or enumerative adj. and const. *against* or † *on.* Something to one's discredit, a black mark.
1938 *New Republic* 26 Jan. 336/1 All movements for social good will..have two strikes on them before they start. *a* **1939** in E. J. Nichols *Hist. Dict. Baseball Terminol.* (Unpublished Ph.D. thesis, Pennsylvania State College) Appendix I, p. iv, You therefore are starting with two strikes against him. It's up to you to hit one into the bleachers and send yourself home. **1943** *Official Rep. Deb. House of Commons Canada* 31 May 3196/2, I am a little afraid that a man who approaches that board claiming exemption as a conscientious objector goes to bat with three strikes against him. **1956** B. Holiday *Lady sings Blues* (1973) xxi. 169 The only evidence they've got is on me. I've got one strike against me. **1962** J. Glenn in *Into Orbit* 16, I knew that I might have a couple of small strikes against me... I was not a college graduate. ..Also..I was probably a little older than most of the men NASA was considering. **1968** *Globe & Mail Mag.* (Toronto) 13 Jan. 3/2 The student council also did nothing. Strike one for student power. **1975** *Listener* 13 Feb. 204/1 One of the main strikes against Ted Heath was that he did not 'come over' on the box. **1979** 'S. Woods' *This Fatal Writ* 129 The discovery of your man, injured, would have been an additional strike against him.
c. *Cricket.* The right of the batsman to receive the next ball. Also without article.
1886 *Cricket* 20 May 137/1 Seeing over sixty runs scored, he, strange to relate, did not succeed in getting a strike. **1955** [see *farm *v.*[2] 6]. **1963** A. Ross *Australia 63* iii. 83 He played McKenzie fine of Harvey at cover, called euphorically, in an effort to keep the strike, for a second, and was run out. **1976** J. Snow *Cricket Rebel* 57 Geoff Boycott took first strike leaving 'Ollie' [Milburn] at the non-striker's end.
d. *U.S. Football.* A forward pass, straight into the hands of the receiver.
1947 *Richmond* (Va.) *Times-Dispatch* 9 Nov. B7/7 Brown threw a perfect 'strike' to Elliott on the 10-yard ribbon, but the lanky freshman end dropped the ball. **1972** J. Mosedale *Football* v. 72 To this strike-tossing forward passer..went the plaudits of the nation.
20. (sense 3 c) *strike bushel* (earlier example); (sense 6 e) *strike aeroplane, aircraft, carrier, Command, patrol, power, trainer, wing*; also *strike-attack, -reconnaissance* attrib. phrases; (sense 8) *strike-faulting* vbl. sb.; (sense 9) *strike action, benefit, call, committee, leader* (example), *meeting, money, movement, notice, record, wave, weapon*; *strike-free, -happy, -prone, -ridden, -torn* adjs.; **strike-bound** *a.*, immobilized by a strike; **strike-breaker** (earlier example); **strike-breaking** *vbl. sb.*, the action of a strike-breaker; also as *ppl. a.*; hence (as back formation) **strike-break** *v. intr.*; **strike force,** (*a*) a military force equipped to deliver a (nuclear) strike; (*b*) a police unit organized for rapid and effective action against crime; **strike-slip** *Geol.* (orig. *U.S.*), the component of the slip of a fault in a horizontal direction, parallel to the strike; also as *adv.*; freq. *attrib.*, esp. in *strike-slip fault*, a fault in which motion was predominantly parallel to the strike; **strike zone** *Baseball*, an imaginary rectangle 17 inches wide, stretching from the height of the batter's armpits to that of his knees, within which the pitcher must throw the ball for the pitch to be called a strike.
1949 *Britannica Bk. of Year* 606/2 Minority groups..threatening and, from time to time, taking strike action by way of protest. **1977** M. Edelman *Polit. Lang.* vii. 131 Wage demands their fellow workers would otherwise be free to back with strike action if necessary. **1965** *New Scientist* 22 Apr. 217/1 The Hawker Siddeley Buccaneer naval strike aeroplane might be modified to suit the RAF. **1957** *Times* 22 Aug. 6/6 The supersonic strike aircraft which Hawker Aircraft are developing as a private venture. *Ibid.*, The fact that it is described as a strike aircraft indicates that it can be used as a bomber as well as a fighter. **1980** *Daily Tel.* 24 Sept. 4/8 Iraqi transport aircraft have been withdrawn to the safety of bases in Jordan, beyond the reach of Iran's strike aircraft. **1977** *R.A.F. News* 11–24 May 1/5 The two squadrons operate in the same maritime strike attack role. **1896** *Rep. Proc. Internat. Typogr. Union N. Amer.* 22/1, $48,087.18..[were] paid during the two years in strike and lockout benefits. **1949** *Britannica Bk. of Year* 687/1 *Strike-bound*, prevented from moving, travelling, sailing etc. by a strike or strikes. **1956** *B.B.C. Handbk. 1957* 121 The editors of strike-bound national dailies and periodicals. **1982** *Daily Tel.* 3 Aug. 22/4 Strike-bound Sealink ships have..moored at the two Holyhead berths. **1961** *Economist* 6 May 525/1 Individual exporters and importers should be allowed to send their own staff in to get their own goods on and off the ships, if their staff will agree to strike-break in this way. **1904** *N.Y. Even. Post* 4 Aug. 2 [Half of] the strike breakers are men who, having been idle for a time, simply wanted a chance to make a little ready money. **1905** *Amer. Mag.* May 107/2 It is quite a new profession, this strike breaking, a curious evolution of modern industrial methods. **1920** *Manch. Guardian News Bull.* 10 Sept. 2/1 A direct incitement to strike-breaking. **1978** S. Brill *Teamsters* x. 362 The police sent an armed convoy to escort a strikebreaking truck. **1858** Trollope *Three Clerks* I. i. 12 Young Tudor had produced a very smart paper on the merits—or demerits—of the strike bushel. **1976** Strike call [see *stay-away* s.v. *stay *v.*[1] 31]. **1966** *Daily Tel.* 18 Aug. 1/4 'Straight Laced' is the first multinational strike-carrier exercise for some years. **1968** *Ann. Reg. 1967* 27 The White Paper envisaged..the merging of the RAF's Fighter and Bomber Command into a new Strike Command. **1949** *Britannica Bk. of Year* 210/2 The men's allegiance to the strike committee outweighed their loyalty to the union. **1925** N. E. Odell in E. F. Norton *Fight for Everest: 1924* 299 There is considerable evidence of strike-faulting which would explain this. **1965** G. J. Williams *Econ. Geol. N.Z.* iii. 30/2 Some ore was won from a 390 ft shaft, but both strike- and cross-faulting seem to have disrupted the lode. **1961** *Listener* 14 Dec. 1011/2 The development of nuclear strike forces by Britain and France. **1973** *Black Panther* 1 Sept. 11/2 Federal and local strike forces smashed into homes and offices in a series of pre-dawn raids. **1947** *Sun* (Baltimore) 2 Jan. 17/1 If operations are strike-free, enough steel can be turned out to restore within a few months a balance between supply and demand. **1982** *Times* 23 Mar. 8/7 The reductions in strike-free days. **1955** *Times* 26 May 11/5 We are being placed on a par with other 'strike-happy' industries. **1913** W. Owen *Let.* 19 Oct. (1967) 201 You should set up as Suffragette, Dublin-Strike-Leader, or Schoolmistress, so that you would be *obliged* to speak for seven hours a day. **1978** P. Boardman *Worlds of Patrick Geddes* vii. 246 The strike leader had the reputation of being a dangerous man. **1926** *Brit. Gaz.* 12 May 1/7 Large crowds of them congregated in the streets, while some abortive strike meetings were held in the squares. **1913** D. H. Lawrence in *Westm. Gaz.* 13 Sept. 2/2 Strike-money is paid in the Primitive Methodist Chapel. **1932** *Sun* (Baltimore) 13 Sept. 8/3 The certain futility of the 'strike' movement. **1926** *Brit. Gaz.* 12 May 2/2 The *Weston Mercury*, Weston-super-Mare, reports that after strike notices had been received the local branch of the Typographical Association decided to return to work. **1942** Strike patrol [see *rover[1] 3 e]. **1959** *Time* 23 Feb. 22/3 U.S. strike power is clearly supreme now. **1961** *Daily Tel.* 22 Apr. 9/2 The strike-prone motor industry. **1963** *Times* 2 Feb. 9/2 Adopting American nuclear warheads for its strike-reconnaissance aircraft missiles defending the North American continent. **1938** *Encycl. Brit. Bk. of Year* 614/1 In 1936 there were but 156 strikes..; a rather typical strike record for Canada. **1967** *Spectator* 8 Dec. 706/2 Two of our favourite illusions are that we are among the most strike-ridden nations on earth, and that every strike brings chaos in its wake. **1913** W. Lindgren *Mineral Deposits* ix. 121 The strike-slip is the component of the slip parallel with the fault slip. *Ibid.* 126 The expressions 'normal' and 'reverse' may be used in connection with oblique and dip faults, even when these are strike-slip or oblique slip faults. **1932** C. R. Longwell et al. *Textbk. Geol.* I. xii. 315 (*caption*) Broken lines show the displacement (slip), and its three components—throw, heave, and strike-slip—measured along axes at right angles to each other. **1964** W. C. Putnam *Geology* vi. 146/2 Ordinarily, in order to establish whether or not movement has been strike-slip or dip-slip, it is necessary to have layered rocks with strongly differing dips cut by the fault. *Ibid.* 147/1 The actual movement as demonstrated by the outcrop was strike-slip. **1971** I. G. Gass et al. *Understanding Earth* xxiii. 327/2 The San Andreas is called a strike-slip fault. **1977** *Sci. Amer.* Apr. 36/2 In Mongolia most earthquakes are associated with strike-slip faulting. **1977** *Belfast Tel.* 28 Feb. 7/9 The crisis at strike-torn Leyland deepened. **1967** *Observer* 26 Nov. 2/6 The Jaguar strike-trainer which was born..in the Anglo-French agreement of 1965. **1957** *Encycl. Brit.* XXI. 469/1 The strike waves that accompanied the Russian revolutions of 1905 and 1917. **1955** *Times* 26 May 11/5 The apparently indiscriminate use of the strike weapon. **1944** *Hansard Commons* 7 Mar. 1910 In conjunction with the strike wings of Coastal Command and R.A.F. fighters our Light Forces have constantly attacked enemy convoys in the Channel. **1948** *Sporting News Dope Bk.* 119 The umpire shall rule it a ball even though it passed over the heart of the plate within the strike zone. **1950**

Official Baseball Rules ii. 17 The strike zone is that space over home plate which is between the batter's armpits and the top of his knees when he assumes his natural stance.

strike, *v.* Add: The pa. pple. *stricken* remains common in U.S. (esp. legal) use. (Examples in various senses.)
See also sense 13 a below.
1906 *Federal Reporter* (1907) CXLVII. 451 All of the testimony given by the witness..is withdrawn and stricken out of this case. **1938** *Congress Rec.* 24 May 7405/2 That the Committee do..report the bill back to the House with the recommendation that the enacting clause be stricken out. **1976** *National Observer* (U.S.) 9 Oct. 7/4 A new trend in comics has stricken down many of the old taboos.
III. 13. a. Also (*U.S.*) const. *from.* The *absol.* use has now been revived in the U.S., esp. in legal contexts and *colloq.*, in the *imp.*, annulling or reversing what the speaker has just said.
The pa. pple. form *stricken* is common in the legal examples of this sense.
1829 *Rep. Supreme Court Tennessee* (1832) IX. 229 That an attorney may be stricken from the roll for good cause, none can doubt. **1915** *Southwestern Reporter* CLXXV. 661/1 No further steps..were taken in the case until the February term, 1904, of the Magoffin circuit court, when it was stricken from the docket. **1973** *N.Y. Law Jrnl.* 19 July 4/2 The Convention..voted 132 to 49, to strike that section from the Constitution.
1957 *Reports Supreme Court Kansas* (1958) CLXXXI. 623 In our opinion the reply was erroneously stricken. **1965** *Pacific Reporter* CCCCIV. 230/2 Where..a second clause appears which expresses a different intent and declares a life estate plus a remainder which is void under the rule, the qualifying clause will be stricken. **1973** *N.Y. Law Jrnl.* 31 Aug. 19/2 Motion to strike the statement of readiness is granted. **1978** *N.Y. Times* 29 Mar. B3/4 Over strong objections from the prosecutor, Sybil R. Moses, Judge William J. Arnold ordered the question stricken.
1963 R. I. McDavid *Mencken's Amer. Lang.* xi. 754 In television we might note *mark it* and *strike it*, directions to stage hands to chalk out the position for scenery and then rub out the mark for the next set. **1976** R. M. Stern *Will* I. ii. 17 Do you..believe that the crash was not an accident? Strike that. We will look into it with an open mind. **1977** H. Greene *FSO-1* ii. 16, I don't give a damn what the congressman says. Strike that: I *do* give a damn.
b. Phr. *to strike* (a medical practitioner, etc.) *off the register*: to remove (that person's name) from the register of qualified practitioners and thereby forbid him or her to practise. Usu. *pass.*
1911 G. B. Shaw *Doctor's Dilemma* p. xciii, Execute the doctor, if necessary, *as* a doctor, by striking him off the register. **1936** A. Christie *Cards on Table* xvi. 157, I heard him say he'd get Dr. Roberts struck off the—Medical Register, would it be? **1951** 'E. Crispin' *Long Divorce* xvi. 199 We can and shall get him struck off the register.
IV. 24. b. Also with particular kind of work as obj.
1878 Trollope *Is he Popenjoy?* III. xix. 251 She had on one occasion threatened to strike lecturing.
c. (Later examples.) Now only *N. Amer.* Also in wider but analogous contexts.
1930 J. Dos Passos *42nd Parallel* I. 117 She'd worry Mac about striking his boss for more pay. **1941** *Sun* (Baltimore) 23 Sept. 12/2 Now the affected union, the Seafarers' International Union of the AFL, serves formal notice that it will strike every ship on which it has contracts. **1946** *Ibid.* 16 Jan. 4–0/1 They [*sc.* students] held a mass meeting, staged a snake dance, struck their classes, and otherwise asserted themselves, in protest over the resignation of..a football coach at the university. **1950** Patterson & Conrad *Scottsboro Boy* II. vii. 137 Right here we struck the whole squad. No one would do any work till the question of slowing down the work was settled. **1968** *Globe & Mail* (Toronto) 1 Feb. B10/3 The union will strike company plants in five U.S. cities. **1978** *N.Y. Times* 29 Mar. A20/4 Photoengravers voted 177 to 0 yesterday to strike The New York Times and The Daily News.
V. 28. f. *Cinemat.* To make (another print) *from* a motion picture film.
1970 A. Fowles *Dupe Negative* xiv. 192 I've got four hundred feet of 35 mm. ECO original here... How long will it take to strike a master positive? *Ibid.* 196 The piece of film that actually runs through the camera is called the original..from which all subsequent prints are struck.
30. a. (*b*) (Later examples with a match as subject.)
1957 'R. West' *Fountain Overflows* i. 14 And I think the matches are wet, they won't strike. **1962** J. Braine *Life at Top* ii. 31, I heard a match strike and smelled cigar smoke.
(*c*) Also *imp.* as a mild imprecation (cf. sense 46 c in Dict. and Suppl.). Chiefly *Austral.* and *N.Z.*
1936 A. Russell *Gone Nomad* vi. 44 'Strike a light!' he broke in suddenly. 'See them?' **1960** I. Cross *Backward Sex* ii. 39 'Strike a light,' he hissed... 'Get over here, quick,' he said. 'Have a bloody look, man.'
44. b. (Later examples.)
1904 *Sun* (N.Y.) 23 Aug. 1 The storm twister struck Willow Lakes about 9 o'clock. **1976** *Daily Mirror* 16 July 13/3 Earthquakes killed 275 people and injured 2,000 early yesterday as they struck Indonesia's tropical holiday island of Bali.
46. c. *strike me pink*: see *pink *a.* 8. Also (*Austral.* and *N.Z.*) *ellipt.* as *strike!*
1915 C. J. Dennis *Songs of Sentimental Bloke* (1916) 43 'Ah, strike!' she sez. 'I wish that I could die!' **1960** B. Crump *Good Keen Man* 116 Strike, he went crook! Who the hell was responsible? Had we been blasting fish?
VII. 66. e. Also, in pass. constr., to be favourably impressed by (an idea, suggestion, etc.). Now *colloq.*
1889 J. K. Jerome *Three Men in Boat* i. 17 The only one who was not struck with the suggestion was Montmorency. **1938** C. P. Conigrave *Walk-About* ix. 50, I don't think he's too struck on my going back to Rosewood. **1940** 'N. Shute' *Old Captivity* iv. 110, I don't know that I'm so struck on this, sir.
68. d. (Earlier examples.) *to strike it rich* (earlier examples). Also in similar *fig.* phrases.
1835 C. F. Hoffman *Winter in West* II. 47, I hear that he has lately struck a lead. **1852** L. Clappe *Shirley Lett. Calif. Mines* 216 When a company wish to reach the bed-rock as quickly as possible, they sink a shaft..until they 'strike it'. **1854** *California Daily Chron.* 19 May 3/7 Messrs. Emory & Bacon, just above the claim of Messrs. Meredith & Co., have also struck it rich. **1862** 'Mark Twain' *Let.* (1917) I. iii. 76 Well, if you haven't 'struck it rich—' that is, if the piece of rock you sent me came from a *bona fide* ledge—and it looks as if it did. *c* **1863** T. Taylor *Ticket-of-Leave Man* III. 47 He..had to bolt to Australia—struck an awfully full pocket at the diggings, and is paying off his old ticks like an emperor.
fig. **1884** 'Mark Twain' *Huck. Finn* xxi. 208 We struck it mighty lucky. **1953** [see *pay-dirt* s.v. *pay- 2]. **1975** *Sydney Morning Herald* 15 Nov. 55 West Indies batsmen struck pay dirt aplenty in the SCG yesterday. **1977** A. C. H. Smith *Jericho Gun* vi. 85 He didn't mind a penny. It was what he had always thought he would do when he struck it.
e. *intr.* Colloq. phr. *to strike lucky*, to hit a vein of good fortune.
1951 *Sport* 6–12 Apr. 17/1 Birmingham struck lucky because several London clubs refused to give Graham Warren a trial. **1984** *Financial Times* 31 Jan. 17/7 The Bush strikes lucky more often than any fringe theatre has a right to.
VIII. 71. Also, to reach (a figure, loss, or profit) by balancing an account.
1880 *Tax Cases* I. 500 In striking their annual profits so as to fix the sum divisible as dividend, the Railway Company have gone upon actual expenditure, and not upon a mere estimate of probable wear and tear. **1932** *Economist* 16 Jan. 127/2 For years past the banks have been building up contingency reserves by appropriations made before and after striking their net profits. **1955** [see *clearing-bank* s.v. *clearing *vbl. sb.* 8]. **1980** *Daily Tel.* 30 July 1/4 Last year's loss..was struck after allowing for depreciation of £87 million and interest payments of £188 million.
75. e. For (See quot.) read: 'To induce (a person) to pay money on the promise of getting him votes, legislative favors, etc.' (*D.A.E.*), and add earlier examples.
1859 G. W. Matsell *Vocabulum* 87 *Strike*, to get money from candidates before an election, under the pretense of getting votes for them. **1883** M. de L. Landon *Wit & Humor of Age* 345 He had a way of striking the politicians who wanted a favor out of the Governor.
76. a. (Earlier and later examples.) Also, † of an electric charge, to pass as a spark (cf. sense 43 a); of an electric discharge, to come into being; also *transf.* of the tube containing it.
1777 T. Cavallo *Compl. Treat. Electr.* III. iii. 163 When the jar is charging, and the charge is become so high as to strike through half an inch of air. **1827** *Phil. Mag.* I. 344 If the distance be greater than that over which the charge can *strike* in the form of a spark, or with explosion. **1929** *Ibid.* VIII. 1100 The uncertain delay which occurs between the instant at which the requisite voltage is applied to the lamp and that at which the discharge strikes. **1962** J. H. & P. J. Reyner *Radio Communication* v. 237 Once the tube has struck, however, the current can be maintained with a somewhat lower anode potential.
c. *trans.* To bring (an arc) into being. Cf. sense 30 a.
1891 [in Dict., sense 76 a]. **1930** *Engineering* 7 Feb. 173/2 Oil..played an important part in quenching the arc which was struck when those contacts were separated. **1950** Gill & Simons *Mod. Welding Technique* xi. 129 On occasion it may be found difficult to strike an arc. **1976** C. Bradshaw *Metall. for Schools* xi. 143/1 An arc is struck between the electrode and the workpiece.
77. b. *U.S. Naut.* (See quot. 1952[1].)
1952 J. V. Noel *Naval Terms Dict.* 212 *Strike*..to work for, as in..'he is striking for chief'. *Strike for*..to learn the trade of. **1952** *MSTS Bull.* May 9/1 Few and far between are those who don't 'strike' for a rating during their short or long Navy career.
IX. 79. strike down. e. *trans.* To hold invalid (chiefly in legal contexts). *U.S.*
1894 *Congress. Rec.* 12 Dec. 267/1, I do not care who strikes down class legislation in this country. **1951** *Federal Reporter* (1952) CXCIII. 250/2 The court's opinions make abundantly clear its intention to strike down the entire arrangement. **1964** *Mod. Law Rev.* XXVIII. III. 343 Their main agreement had been struck down by the Restrictive Practices Court. **1979** *Tucson* (Arizona) *Citizen* 3 Oct. 4C/2 The decision..struck down a..Superior Court ruling.
82. strike off. a. *spec.* in *pass.*, of a medical practitioner, solicitor, etc.: to be struck off the register (see sense 13 b above).
1937 A. J. Cronin *Citadel* IV. xxi. 424 You remember the case of Jarvis, the manipulator, several years ago, when he got some cad of a doctor to anaesthetise for him. *He* was struck off, instanter. **1958** J. Cannan *And be Villain* i. 20 He'd be struck off if he was the least bit naughty. **1965** A. Christie *At Bertram's Hotel* xvi. 153 We still call him Dr. Stokes although he's been struck off. **1983** *Times* 12 Oct. 3/4 Mr Parsons is asking Mr Justice Vinelott to order that Mr Davies be struck off.
83. strike out. i. (Earlier and later examples in Baseball.) Also *fig.*
1853 *Oregonian* (Portland) 2 July 1/5 No doubt they will find that strikers have struck out. **1866** *N.Y. Herald* 28 Aug. 8/2 Pennington was third man at the bat, and struck out. **1937** *New Yorker* 19 June 30 The senator had his hopes, but he struck out on three wide 'ha's'. **1974** *Los Angeles Times* 13 Oct. III. 10/2 Garvey grounded to short. Ferguson struck out.
k. *trans.* Of a pitcher in Baseball, to put (a batter) out by pitching three strikes to a batter. *U.S.*
1939 E. J. Nicholls *Hist. Dict. Baseball Terminol.* (Unpublished Ph.D. thesis, Pennsylvania College) 75 *Strike-out king*, a pitcher who is noted for the large number of times he strikes out opposing batters. **1968** *Washington Post* 4 July C1/8 It was the third time in the game that he struck out the side. **1975** *New Yorker* 14 Apr. 92/2 He struck out two of the first three Yankee batters, without really trying his fastball.
X. 88. strike-back, used *attrib.* to designate the capacity of making a retaliatory nuclear strike; **strike me blind** *slang* (see quots.); **strike-out,** an out in baseball, called when a batter has made three strikes; also *attrib.* and *fig.*; **strike-over** *U.S.*, the typing of a character on a spot occupied by a character typed previously; **strike through** *Printing* (see quots.).
1962 *Listener* 29 Mar. 539/2 It was clear that we would soon..have a sufficiently invulnerable strike-back nuclear capacity. **1966** Schwarz & Hadik *Strategic Terminology* 44 *Strike-back capability*, nuclear forces which could survive an enemy first strike and then be used against him in a second strike. **1901** S. H. King *Dog-Watches at Sea* 146 Rice was known as 'strike me blind'. **1936** B. M. Adams *Ships & Women* viii. 180 The dish..called 'strike-me-blind'. Boiled rice, with black-strap molasses. **1911** J. B. Foster *How to Pitch* 72 It happens to be a pitcher..of the strike-out kind. **1922** E. J. Lanigan *Baseball Cycl.* II. 39/1 Another top-notcher joined them in the person of Thomas Ramsey, eminent strike-out king. **1937** *Philadelphia Rec.* 23 Mar. 15/1 Mr. Roosevelt has..grown into the stature of a strike-out king. **1967** *Boston Sunday Herald* 14 May II. 3/3 Six of his strikeouts came in those innings. **1978** M. Puzo *Fools Die* xvi. 170 After Pfc. Hiller was recalled, his case would be evaluated by a Regular Army board. Another strikeout. **1950** *Richmond* (Virginia) *Times-Dispatch* 3 Oct. 1/8 A patent on typewriter type designed to permit strikeovers on letters in about 11 per cent of common typing errors. **1978** W. White *W. Whitman's Daybooks & Notebooks* I. p. xxii, Corrections, strike-overs, inserted words..I have transcribed them exactly the way Whitman has left them. **1958** T. Landau *Encycl. Librarianship* 290/2 *Strike through*, penetration of the type impression from the verso to the recto of a page due to improper pressure or faulty make-ready. **1979** G. A. Glaister *Gloss. Bk.* (ed. 2) 464/1 *Strike through*, a fault caused when the oily medium in printing ink soaks into and through the paper, making it translucent.

striker. Add: **I. 2. e.** For † *Obs.*[−0] read *rare*[−1] and add example.
1970 J. H. Gray *Boy from Winnipeg* 69, I also first got to know horses that were strikers.
3. f. *Tanning.* One who smooths and stretches skins either by hand or by means of a machine. Also *striker-out.*
1921 *Dict. Occup. Terms* (1927) §338 *Striker, striker-out*, (i) lays wet hide or skin on a slate or marble slab or table, and rubs it with a hard 'slicker' tool, of stone or steel, to stretch it, drive out excess of moisture, smooth it, and to close grain; (ii) sets rollers of a machine..in motion..and passes skin or hide between revolving rollers. **1972** *Classification of Occupations* (Dept. Employment) III. 24/2 *Finishing machine operator*... Other titles include..Striker.
4. (Earlier example in Baseball and later examples in Cricket.) In Assoc. Football and Hockey, a forward whose main function is to seek to score goals. In *Rugby Football* = *hooker[1] 6.
1816 [see *play *v.* 17 f]. **1862** *Sunday Mercury* (N.Y.) 13 July 6/2 The Excelsiors led off, Young being their first striker, and he sent the ball flying to left field. **1891** W. G. Grace *Cricket* viii. 235 It is the striker's duty to call [for a run] if the ball is hit in front of the wicket. **1963** J. Greaves *Soccer* vii. 73 If John White or another Spurs' player is bringing the ball up..I move into a position ready to race through and be first to the ball when he pushes it forward. It is the ball goal-strikers dream of. *Ibid.* 74 Remember, the striker never takes it for granted ..he goes after the ball on every occasion. **1973** *Daily Mail* 24 July 27/1 John White, Bristol's 30-year-old reserve hooker,..replaces ex-Coventry striker John Gray. **1974** M. Weir *Women's Hockey for Seventies* 96 Before the ball is hit the right striker is sprinting out to the right wing and the right wing is cutting into the space she has made. *Ibid.* 8 It is confusing for a defence to have to cater for elusive strikers. **1974** *Encycl. Brit. Macropædia* 257/2 The striker does not have to run after he has hit the ball. **1980** *Daily Tel.* 20 Mar. 34/3 Wales, without Chester striker Ian Rush, could not break down the Irish defence.
6. b. (Later examples.)

1898 *Harper's Mag.* Apr. 700/2 My 'striker' had just left me, with instructions to have my horse fed. **1929** B. Davis *Truth about Geronimo* 107 Geronimo's son demanded the post of striker (servant) to me. **1948** *Time* 14 June 9/3 He takes the same attitude toward Congress as he would to a striker who fails to put the proper polish on his boots.
c. (Earlier and later examples.)
1836 *Spirit of Times* 9 July 162/2 An awkward looking *striker* of old Thompson's holding her by the cheek of the bridle. **1853** 'P. Paxton' *Stray Yankee in Texas* 335 To a few he [*sc.* John Murel] confided the extent of his design, and to each of these gave the authority to enlist all the minor villains of their acquaintance. The latter were termed Strikers and used but as tools. **1873** J. H. Beadle *Undevel. West* xi. 184, I had published a severe criticism of this Judge Smith. His 'strikers' now had me at Court as defendant. **1883** 'Mark Twain' *Life Mississippi* xxix. 315 [Murel's gang of robbers] was composed of two classes: the Heads or Council..[and] the active agents..termed strikers.
d. An engineer's apprentice on a steamboat. Also in extended use: see quots. 1944, etc.
Cf. also quot. 1891 at sense 3 b in Dict.
1872 [see *mud-clerk* s.v. *mud *sb.*[1] 5]. **1875** 'Mark Twain' in *Atlantic Monthly* XXXV. 70/2 He turned up as apprentice engineer or 'striker' on a steamboat. **1944** K. D. McCracken *Baby Flat-Top* 53 In the Navy a striker is a seaman or fireman who is working particularly hard in order to convince his superiors that he ought to become a petty officer of some kind. **1955** C. S. Forester *Good Shepherd* 104 An electrician's mate and his striker stood behind him. **1963** *Amer. Speech* XXXVIII. 45 *Striker*, a [truck] driver's helper. **1970** *National Fisherman* Aug. 21-A/1 Emery Brown as rigman or 'striker' [on a shrimp boat]..testified in person during this trial.
III. 18. striker-boat (examples); hence *striker boatsman.*
1891 Striker boat [in Dict., sense 3 b]. **1950** *Richmond* (Va.) *Times-Dispatch* 23 July (Mag. Section) 5/1 When a bunch [of fish] is spotted, a striker boat, manned by a striker boatsman, is sent out to indicate the direction in which the fish are moving.

striking, *vbl. sb.* Add: **I. e.** *Tanning.* The process of smoothing and stretching skins. Also *striking-out.* Freq. *attrib.*
a **1877** Knight *Dict. Mech.* III. 2429/1 Striking-machine. **1882** *Encycl. Brit.* XIV. 385/2 For striking or pinning by hand the hide is dampened with water, thrown over a beam, and worked all over the grain side with a striking pin. *Ibid.*, Striking machines are now very generally used for the operation. **1897** C. T. Davis *Manuf. Leather* (ed. 2) 378 The 'striking out' was performed on mahogany tables. **1920** *Conquest* Nov. 38/3 Stretching and smoothing [hides] with a striking-pin (a two-handled tool triangular in sections). **1942, 1953** [see *setting *vbl. sb.*[1] 13 a].
3. striking-circle *Hockey* (see quots.); **striking distance** (earlier example); **striking force,** (*a*) the force with which a projectile strikes; (*b*) a military force held in readiness for sudden attack; **striking-plate** (*b*) (see quot. *a* 1877); **striking platform** *Archæol.*, a flat area on a core of flint or stone on which a blow is struck to detach a flake; **striking price** *Stock Exch.* (see quots. 1973, 1982).
1890 F. S. Creswell *Hockey* 11 No goal can be scored unless the ball be hit by one of the attacking side from within the striking circle. **1906** *Official Handbk. Hockey Assoc.* 120 In front of each goal shall be drawn a white line 4 yds. long, parallel to, and 15 yds. from, the goal line. This line shall be continued each way to meet the goal line by quarter-circles having the goal posts as centres. The space enclosed by these lines and the goal lines, including the lines themselves, shall be called the striking circle. **1961** F. C. Avis *Sportsman's Gloss.* 219/1 *Striking circle*, in Hockey the space immediately in front of goal, really a rough *semi*-circle, 15 yards from the goal line. **1751** B. Franklin *Exper. & Observations Electricity* 62 A needle..will draw the fire from the scale silently at a much greater than the striking distance. **1881** Striking force [in Dict.]. **1917** T. E. Lawrence *Lett.* (1938) 230 Force 3 is our striking force (of perhaps 6,000 not bad men) and may be able to rush Deraat, or at least should cut off the garrison there. **1944** [see *air strike* s.v. *air *sb.*[1] B. III 2)]. **1965** J. A. Michener *Source* (1966) 793 The well-trained Jews of the Palmach—an abbreviation for the Plugat Machatz, 'striking force', organized in 1941 to resist the threatened German invasion. *a* **1877** Knight *Dict. Mech.* III. 2429/2 *Striking-plate*, the device by which the wooden centering of an arch is lowered when the arch is completed. **1876** *Encycl. Brit.* IV. 311/2 Figure 55 shows the *striking plates* and *wedges* by which the centre is lowered after the completion of the arch. **1913** *Proc. Prehistoric Soc. E. Anglia* I. III. 311 The flaking..is of a very high order, dexterous vertical blows, with well-masked cones of percussion, and striking platforms being supplemented by the most regular and fine edge-work. **1949** K. P. Oakley *Man Tool-Maker* 25 Each blow is delivered obliquely downwards near the edge of some conveniently placed flattish area (the striking platform), usually the scar of a flake previously struck off. **1977** L. L. Johnson in Hill & Gunn *Individual in Prehistory* x. 218 Collapsed platforms were noted only where there was no retouch on the striking platform. **1961** K. S. Most *How to make Money on Stock Exch.* iii. 35 Suppose I have a..well-founded belief that Woolworths' shares are going to rise in price..I shall have to pay out £335 plus purchase costs for every 100 shares. I may not have this money available..so I arrange to give the price of a call option, say, 5s. per share, for the right to buy 100 Woolworths' shares at any time during the next twelve weeks at a price of, say, 66s., being the 'striking price' at the end of the previous Account. **1973** N. Searle *Successful Investments* 85 *Striking price*, the price at which the holder of an option has the right to effect a purchase or sale. **1982** *Times* 9 Nov. 19 With a tender offer for sale, investors tender at the price they are prepared to pay. The issuing house works down the list to the lowest price at which the issue is totally subscribed. This becomes the 'striking price'.

Strindbergian (strindbɜ̄·ɹgiăn), *a.* [f. the name of the Swedish dramatist Johan August *Strindberg* (1849–1912) + -ian.] Of, pertaining to, or characteristic of Strindberg or his writings.
1913 G. B. Shaw *Quintessence of Ibsenism* (Completed ed.) Pref. p. xiii, An eminent bacteriologist filled three columns of The Times with a wild Strindbergian letter. **1934** C. Lambert *Music Ho!* iv. 265 It [*sc.* Soupault's *Death of Nick Carter*] reads..like an elaborate stage direction from some super-Strindbergian play. **1954** *Encounter* May 51/1 [The Swedes] are inclined to examine themselves with a touch of Strindbergian introspection. **1966** J. Fowles *Magus* xliv. 281 Brought up, like bacilli in a test-tube, on a culture of such pure Strindbergian melancholia. **1974** R. Rendell *Face of Trespass* viii. 70 An old Swedish film full of pale Strindbergian people.

Strine (strəin), *a.* and *sb. joc.* (orig. *Austral.*). Also **'strine.** [imit. of alleged Austral. pronunc. of *Australian*, coined by Alistair Morrison in 1964 under the pseudonym 'Afferbeck Lauder' (Strine pronunc. of 'alphabetical order').] **A.** *adj.* Australian. **B.** *sb.* **a.** An Australian. **b.** The English language as (allegedly) spoken by Australians.
1964 A. Morrison in *Sydney Morning Herald* (Sat. Mag.) 19 Dec., (*heading*) New light on the Strine language, by Afferbeck Lauder, Professor of Strine Studies, University of Sinny. *Ibid.*, Selected translations of everyday words will be of interest..also to overseas vistas and to the many New Strines in our mist. **1965** 'A. Lauder' (*title*) Let stalk Strine. **1965** *Listener* 2 Sept. 340/1 While I was there they discovered a new dialect or speech pattern called Strine. Strine is simply the way the word 'Australian' sounds if you slur and twist it enough. **1967** *Daily Express* 6 May 13/6 He said in a broad Strine accent: [etc.]. **1973** E. McGirr *Bardel's Murder* iv. 93 Iced beer stops up the nose which is why you Yanks and also the Strines talk so funny. **1974** *Times* 21 Dec. 10/6 'The *legs*, Ealing, go for the *legs*!' she exhorted in a strong 'strine accent. **1980** [see *roomette].

string, *sb.* Add: **I. 1. f.** Also *fig.*, a limitation, condition, or restriction attached to something. Freq. in phr. *no strings attached* (cf. *no strings* s.v. *no *a.* 5 d). Also (with hyphen) as adj. phr. Hence *strings-attached a.* (*rare*). orig. *U.S.*
1888 in *Dict. Amer.* (1951) II. 1665/1 Bob Ingersoll says there is a string to it. **1930** *Randolph Enterprise* (Elkins, W. Va.) 19 Dec. 4/2 All the propositions with a string to them remind us of the..First of April joke. **1948** G. E. Kirk *Short Hist. Middle East* viii. 242 The masses are accustomed to poverty and will listen to their own political leaders rather than to foreigners who offer them opulence with a political 'string' attached. **1951** in M. McLuhan *Mech. Bride* (1967) 90/1 It has for its elements ..imagination with no strings attached. **1953** S. Plath *Johnny Panic & Bible of Dreams* (1977) II. 151 Would he ask her out..just for herself, no strings attached? **1960** *Washington Post* 16 Nov. A16/2 Much has been said about the desirability of aid without strings, and a strong case can be made for this in some areas where the need is economic. Certainly any strings ought to be obvious. **1969** *Daily Tel.* 12 Dec. 1/1 The Government is to give a new £7 million loan to Upper Clyde Shipbuilders... The new loan would not carry any 'strings'. **1971** *Nature* 16 Apr. 420/2 A 'substantial' effort will be made in the category called *l'aide au développement*, a strings-attached arrangement whereby state loans proffered for industrial development must be repaid if the project proves successful and profitable. **1976** *Women's Rep.* Sept./Oct. 2/1 The feminist-run clinics in Australia..who persuaded the government to fund them (no strings attached). **1980** *Forest Products News* (Wellington, N.Z.) XVII. 1. 2/2 As a gesture of goodwill, NZFP has given 'no-strings-attached' aid to an experimental forestry venture in Northland. **1981** J. B. Hilton *Playground of Death* v. 58, I could aspire to be his assistant editor... He was very proud of the *Examiner*'s freedom from strings.
i. Also *fig.* in *to pull strings*, to exert influence privately. Cf. *string-pulling* vbl. sb., sense 32 below.
1924 M. Kennedy *Constant Nymph* III. xvi. 213 With half a dozen strings within her reach, she had not made up her mind which to pull. **1938** M. Allingham *Fashion in Shrouds* xxii. 404 I've been trying to pull a few strings myself..but there's an ominous frigidity on all sides. **1955** G. Greene *Loser takes All* i. v. 26 Rice is still short, but I'm certain Aunt Marion can pull strings with the grocer. **1960** *News Chron.* 30 Jan. 3/8 She admits she will pull any strings to get things done. **1979** R. Jaffe *Class Reunion* (1980) II. viii. 265 He couldn't be dumb or they wouldn't have accepted him at Le Rosay. On the other hand, his father had strings to pull everywhere.
m. Also in phrases (freq. *attrib.*) with *sealing-wax*, used to denote the unpretentious apparatus with which great discoveries may be made.
1962 *Daily Tel.* 5 Mar. 20/5 The traditional British method of scientific research with 'string and sealing wax' will pay rich dividends. **1969** *New Scientist* 28 Aug. 422/2 Systems which are..still in the string and sealing-wax stage of development. **1972** *Physics Bull.* July 393/1 The individual with his sealing wax and string has been replaced by the battalion with a multimillion pound particle accelerator. **1975** *Nature* 2 Oct. 349/1, I have been told that it is impossible to 'put the clock back'. The assumption is that the age of string, sealing wax and enthusiasm has gone for ever. **1976** *Sci. Amer.* Oct. 138/2 Blackett's world was no longer Rutherford's string-and-sealing-wax one.
n. A hoax or trick. Cf. String *v.* 15 in Dict. and Suppl., *stringer 9. *U.S. slang.*
1851 T. A. Burke *Polly Peablossom's Wedding* 92 Of course Mabe was innocent of the 'string'. **1937** E. H. Sutherland *Professional Thief* iii. 69 Many other short-con games have been played, including the gold-brick,.. the strap, the string (a variation of the string). [*sic*: ? *read* strap].
o. A fashion shade of the natural colour of string, a light greyish-brown. Also *attrib.* or as *adj.* Cf. *string-colour, -coloured*, sense 31 b in Dict.
1914 *Queen* 24 Oct. 2 (Advt.), Colours—champagne, silver, Wedgwood, sky, string. **1923** *Daily Mail* 7 June 6 In Ivory, String, Beige, Light Grey. **1949** *Dict. Colours Interior Decoration* (Brit. Colour Council) III. 26/1 *String*, a colour standardised by B.C.C. in 1934. A similar colour is here shown under the name of String Beige. **1963** *Harper's Bazaar* May 17 (Advt.), In navy, string, cedar, nut brown or black calf... In cardinal, white or string calf. **1972** *Vogue* June 13/1 A kind of warm biscuit shade that some paint-makers bluntly term 'string'.
p. (See quot. 1964.) Usu. *attrib.* (see *string underwear, vest*, sense 32 below) or as *adj.*
1964 *Which?* Apr. 123/1 There are four main types of knit for men's underwear—*plain*; *interlock*; *cellular, mesh* or *eyelet*; and *string*... String fabrics are mesh fabrics, but of a very open structure—the holes may be nearly one inch across, and the fabric is usually in the form of thick strands joined together. This type originated in Norway, where the fishermen used to cut up their old fishing nets and wrap them round their bodies to keep warm when fishing in icy weather. **1966** 'A. York' *Eliminator* viii. 156 His underwear was Norwegian string. His coat was a Burberry.
4. c. (Later examples.) Freq. (with hyphen) *attrib.*
1943 J. B. Priestley *Daylight on Saturday* xxxi. 245 He was one of that select..group of second-string personages for whom the party..had always to provide. **1958** *People* 4 May 19/1 The man who may take over as second string to Tony Lock is Mike Allen, of Northampton. **1965** *Times Lit. Suppl.* 25 Nov. 1058/3 Moore was a kind of second-string Clarence Darrow. **1977** C. McCullough *Thorn Birds* iv. 80 The big Queensland blue brute that led the dog pack took a slavish fancy to the priest and followed him without question, meaning Frank was very definitely the second-string man.
d. (Later *attrib.* examples.) Also, of a team.
1934 *Times* 14 Feb. 6/3 In the first string match P. Q. Reiss (R.A.F. Club) just beat S. N. Capel-Cure by three games to two. *Ibid.* 3 Mar. 6/4 The match was decided on the last fight, that between the first-string welter-weights. **1951** *Sport* 27 Jan.–2 Feb. 3/1 On Saturday, 'Archie' kept goal for the Rochdale second string. **1972** J. Mosedale *Football* iii. 32 Walter Camp named him a second-string All-American. **1976** *Norwich Mercury* 19 Nov. 10 Terry Medwin's finest moment so far as Norwich City Reserves' coach was in defeat. The second string went down 3–2 in September at Tottenham.
6. c. A very scanty bikini (see quot. 1974[1]).
1974 *W* 14 June 17/2 The latest—The String—looks like a winner on the beaches of other countries too... Held by thin strings, it's just two tiny triangles—front and back—worn with a mini-bra. **1974** *Times* 13 Aug. 5/6 The String, a sort of cache-sexe sized bathing suit from Brazil which is now sweeping America. **1977** *Courier-Mail* (Brisbane) 5 Nov. 1/4 They were what we call 'strings'—just a string holding them up.
II. 12. c. (Examples.)
1848 B. A. Baker *Glance at New York* 11, I have beat Miss Wilson one string. **1855** J. Holbrook *Ten Years among Mail Bags* 60 Just allow me twenty on a 'string'. **1871** G. W. Peck *Adventures of Terence McGrant* iii. 22 I'd do it to him half a string. **1924** *Billiards Mag.* June 46/1 Kreshel beat the coast's amateur three-cushion titlist, 80–44. The score of the first block was 40–14, with the string completed in 110 innings.
14. d. *transf.* = *stable *sb.*[1] 2 b. *U.S. slang.*
1913 G. J. Kneeland *Commercialized Prostitution in N.Y. City* iv. 77 A single girl, at times a 'string' of girls, 'working' for them [*sc.* pimps] on the street or in houses. **1946** *Amer. Mercury* Sept. 272/2 Promoters of commercialized prostitution look to two main sources for replenishing their 'stables' or 'strings' of girls. **1982** L. Block *Eight Million Ways to Die* (1983) x. 87 She wants out of my string of girls.
15. b. Orig. (more fully *string of tools*), the drilling bit and weights that occupy the hole in drilling for oil, etc.; in mod. use, the entire drilling assembly in the hole (so *drilling string*); also, the coupled lengths of drill pipe or of casing in the hole.
1895 W. T. Brannt *Petroleum* vii. 182 The string of tools—the bit, the auger-stem and jars, with the sinker-bar—are [*sic*] more than sixty feet long. **1929** Babbitt & Doland *Water Supply Engin.* vii. 160 The only tools on the string in spudding are usually the auger stem and the spudding drill. **1939** D. Hager *Fund. Petroleum Industry* viii. 181 A string of cable tools consists of the bit, stem, jars, sinker bar or sub, and rope socket. The parts of the string are all joined by tool joints and fastened to the drilling cable or line by means of the rope socket. **1947** *Richmond* (Va.) *Times-Dispatch* 12 Mar. 11/2 Pacific Western's well contains the longest 'string' of casing ever run into a well—16,406 feet. **1963** G. Sell *Petroleum Industry* iii. 53 The swivel is so designed as to allow the drilling string to rotate freely on roller bearings. **1976**

M. Machlin *Pipeline* xxvii. 318 Can you imagine old Wilbur all touted out in greasy coveralls, working the string on some well up in the slope? **1979** R. Piper *Story of Oil* vi. 23 When boring for oil, a separate engine, apart from the one that raises and lowers the drill string, is needed to turn the drill stem.

c. *Math.*, etc. A sequence of symbols or linguistic elements in a definite order.

1932 Lewis & Langford *Symbolic Logic* iii. 49 Propositions are not strings of marks, or series of sounds, except incidentally. **1940** W. V. O. Quine *Math. Logic* vii. 284 Now *x* is a string of accents, symbolically 'Ac *x*,' if every initial segment of *x* ends in an accent. **1954** *Jrnl. Assoc. Computing Machinery* I. 120/2 A finite, possibly null, sequence of members of the alphabet is called a string. **1955** N. Chomsky *Logical Struct. Linguistic Theory* (microfilm, Mass. Inst. Technol.) viii. 356 There are cases where similar strings have intuitively quite different interpretations, but where we can account for the difference by assigning different markers to them. **1958** [see *identifier 2 c]. **1970** J. Lyons *Chomsky* 58 The ambiguity of such strings as *old men and women*. **1977** *Word* 1972 XXVIII. 51 The surface string of such sentences indeed looks perfectly straightforward—an adjective with comparative inflection and a comparative marker. **1979** *Sci. Amer.* Oct. 138/3 It was hoped that by transforming the statements of mathematics into strings of meaningless symbols to be combined according to the rules of logic, whatever unavowed principles of reasoning had given rise to the paradoxes would be revealed.

d. *Computers.* A linear sequence of records or data.

1956 *Jrnl. Assoc. Computing Machinery* III. 147 Areas are set aside for shuttling strings of control fields back and forth until a completely sorted sequence is obtained. **1964** C. Dent *Quantity Surveying by Computer* iii. 34 After the second pass tapes A and B contain the data in strings of four items. **1979** Page & Wilson *Introd. Computational Combinatorics* iii. 49 Two strings of *r*, *s* items respectively are each in ascending order in the main store of a computer.

16. e. A continuous series of successes or of failures. orig. and chiefly *U.S.*

1890 Barrère & Leland *Dict. Slang* II. 313/2 A common expression in America is 'to get in a string', applied to any kind of fortunate series. **1898** H. H. Blossom *Checkers* 170 Well, I've had my hard luck, and 'played out the string'. **1967** *Boston Herald* 8 May 16/5 Womack preserved the victory that ended a four-game losing string for New York. **1968** *Globe & Mail* (Toronto) 15 Jan. 21/1 The victory stretched the Canadiens' unbeaten string to nine games. **1973** *Times* 17 Apr. 14/6, I try to take it in my stride and relax, and not get too nervous about continuing a string. **1976** *Billings* (Montana) *Gaz.* 27 June 1-F/2 The Mustangs stretched their scoreless string to 12 innings before finally connecting in the fourth inning.

f. *Sport.* (See quot. 1961.) Also *spec.* in Bowling, a succession of strikes. *N. Amer.*

1961 Webster, *String*, a fixed or standard number of turns at play in a game or competition. **1970** *Globe & Mail* (Toronto) 28 Sept. 21/4 Fred Harrison failed to win any..prize money in the Ace Bowling Centre's men's open five pin tournament but he..included a perfect 450 game in his 10-game string. **1979** Ritger & Allen *Compl. Guide Bowling Spares* 228/1 *String*, a number of continuous strikes. Also, in some areas, one game of bowling.

17. a. (Earlier examples.)

1875 *Chicago Tribune* 23 Nov. 7/3 [She] always had a full string at measuring-time. **1889** *Current Lit.* Apr. 314/1 Presently his week's 'string' averaged twelve thousand a day.

b. (See quots.)

1892 *Dialect Notes* I. 207 When he [*sc.* a correspondent] comes to make up his bill, he takes all the articles he has written for a given period and pastes them together, end to end. This he calls his string. **1913** W. G. Bleyer *Newspaper Writing & Editing* iii. 55 On some papers the correspondents clip out all of their news stories and paste them together in a 'string' which they send in once a month, so that the telegraph editor may pay them according to the length of the 'string'.

III. 25. (Later examples.)

1854 *Trans. Mich. Agric. Soc.* VI. 177 The strings of fence will average eight and three-quarter rails high. **1903** A. Adams *Log of Cowboy* 17 On the Mexican side there was a single string of high brush fence.

28. (Earlier example.)

1809 T. D. W. Dearn *Bricklayer's Guide* 101 This projection frequently occurs, and in many instances serves as an agreeable relief to the eye, if of no other use; it is sometimes called a string.

30*. *Billiards.* A string-line, a baulk-line. *U.S.*

1857 *Spirit of Times* 30 May 200/1 The player in hand can play at any ball, the largest half of which lies outside the string. **1872** 'Mark Twain' *Roughing It* 336 Cheese it, pard; you've banked your ball clean outside the string. **1964** Sullivan & Crane *Young Sportsman's Guide to Pocket Billiards* viii. 77 Through the head spot is drawn the 'head string'. This is a line that passes through the head spot and the two center diamonds on the opposing side rails (near the head end of the table). There are comparable designations—'foot spot' and 'foot string'—at the opposite end of the table. **1974** *Rules of Game* 80/1 Each player takes a cue ball, and plays it against the foot cushion from behind the head string.

31. a. string box (earlier and later examples); **string man** (later *Hist.* example), **player**; **c.** instrumental, as *string-soled*, *-tied* adjs.

1839 Dickens *Nicholas Nickleby* xxxvii. 354 Paper, pens, ink, ruler, sealing-wax, wafers, pounce-box, string-box, fire-box..all had their accustomed inches of space. **1926–7** *Army & Navy Stores Catal.* 120/2 Household string box..containing a ball of fine, medium, and coarse brown string. **1980** R. Adams *Girl in Swing* xix. 255 She came back with the other two cassettes, the lemon-squeezer, the string-box and two brown-paper parcels. **1971** *Country Life* 23 Dec. 1776/3 The peacock for the most distinguished person at the high table was carried into the dining-hall with pompous ceremony on a gold or silver-gilt charger by the most elegant lady of the assembled company, attended by trumpeters, pipers and string-men. **1923** *Daily Mail* 6 Feb. 7 All the string-players pulled their weight. **1979** *Jrnl. R. Soc. Arts* CXXVII. 385/2 The Council has approved this year's awards of scholarships to enable young string players and singers to undertake advanced studies. **1924** *Blackw. Mag.* Oct. 556/2 We steal softly on our string-soled shoes past the latrine. **1925** J. Gregory *Bab of Backwoods* xiii. 285 There was a string-tied canvas bag, as long as her open palm. **1980** *Farmer & Stockbreeder* 2 Feb. 5/3 Hay from £10 to £16 (£10, loose in stack; in bales, string-tied, £10 to £10 10s.).

32. string analysis *Linguistics*, a method of analysing sentences as linear strings; **string art** *U.S.*, the art of making decorative pictures by winding yarn round nails driven into a flat surface; **string bass** *Jazz*, a double-bass; also *transf.*, the player of a double-bass; **string-bean**, (*a*) (earlier examples); (*b*) *U.S. colloq.*, a tall thin person; also *transf.*; **string bikini** = *string c above; **string bog** *Physical Geogr.*, a boggy area containing long, high banks of silty material; **string correspondent** = *stringer 11; **string cot** = *string bed* (cf. Cot *sb.*[4] 1); **string drum**, a musical instrument, consisting of a rectangular box over which strings are stretched, and played by striking the strings with a stick; **string figure**, a figure made by passing a length of string round the fingers of both hands (cf. Cat's-cradle); so **string game**; **string glove**, a glove knitted or crocheted of coarse mercerized cotton yarn; **string man** = *stringer 11 (see also sense 31 a in Dict. and Suppl.); **string point, proof**: in sugar manufacture, a degree of concentration at which the boiled sugar may be drawn out in the form of a thread; **string-pulling** *vbl. sb.*, the act of exerting influence, esp. behind the scenes; cf. Wire-pulling *vbl. sb.*; hence **string-puller**; **string puppet**, a puppet actuated by means of strings, a marionette; also *fig.*; **string slum** *U.S.*, a row of unsightly buildings along the side of a road (see quot. 1963); **string tie** orig. *U.S.*, a very narrow necktie worn as a bow; **string-tone** *Mus.*, the sound of bowed stringed instruments; hence **string-toned** *a.*; **string vest**, a man's vest or singlet made from an open-knit fabric (cf. sense 1 p above); also *string underwear*.

1960 *Language* XXXVI. 63 Positively, it leads to the development of a string constituent analysis in which grammatical strings..are discussed. **1962** Z. S. Harris (*title*) String analysis of sentence structure. **1972** Hartmann & Stork *Dict. Lang. & Linguistics* 221/2 A string analysis of the sentence *Today we heard three shots in the park* would be as follows: *We heard shots* is the elementary sentence; *today* is an adjunct to the left of the elementary sentence; *in the park* is an adjunct to the right of the elementary sentence; *three* is an adjunct to the left of the word *shots*. **1972** *Creative Crafts* Aug. 21/1 Our ship..is an excellent example of free-standing string art, the foolproof craft which makes curves from straight lines. **1975** *String Art Encycl.* 41 (*caption*) A traditional fruit display looks different..when you stitch it using the string-art technique. **1947** *Melody Maker* Aug. 7/1 Their instrumentation..is a combination of piano, flute, cello, violin, string bass and trumps. **1930** *Ibid.* Jan. 2/1 The pianist and string bass must be particularly complimented on the steadiness of their playing. **1956** M. Stearns *Story of Jazz* (1957) xvii. 205 The string-bass began to 'walk', or play melodic figures instead of pounding away at one or two notes. **1975** R. H. Rimmer *Premar Experiments* (1976) i. 70 Ellen, I know you can't help it, but you remind me of a starving, stringbean kitten that wandered into our house when I was a kid. **1977** *New Yorker* 3 Oct. 80/2 'Did Germany need living space?' Hellmann asked, translating the stringbean's German word. **1974** *McCall's* Nov. 10/1 Winter vacation time is coming and the string bikini is still with us—better, if not bigger, than ever. **1976** 'E. McBain' *Guns* (1977) vii. 74 The..white string bikini. **1956** *Contrib. Gray Herbarium Harvard Univ.* CLXXVIII. 62 These bog ridges are the strings of the Strangmoor of European authors. **1959** *Geogr. Jrnl.* CXXV. 145 A particularly well defined form [of patterned ground feature] are the string bogs, or *strängmoore*, which occur particularly in eastern Canada. **1973** A. L. Washburn *Periglacial Processes & Environment* iv. 151 Although string bogs or closely similar features have also been observed far north of tree line and well within the zone of continuous permafrost.., most investigators agree they are not necessarily indicative of permafrost. **1960** *Spectator* 24 June 920 Later he became a 'string correspondent' sending items to all the local papers, and he also sold jokes at a dollar apiece. **1969** B. Moore *Workers in World News* i. 6 To return to our Paris correspondent, as well as the news coming to him through the newspaper in whose office he worked, he would probably have his own 'stringers'—or string correspondents—in the different provincial centres. **1895** Kipling *Day's Work* (1898) 178 Scott..laid himself down to rest on a string cot. **1960** R. P. Jhabvala *Householder* ii. 83 A string-cot had been put up for her in the living-room. **1940** *Amer. Speech* XV. 125 'Ionisation', written for percussion instruments, requires the use of bongo, sirens,..guiro, claves, maracas, lareole, and string-drum. **1976** D. Munrow *Instruments Middle Ages & Renaissance* 8 Various names have been used for the string drum... The thick gut strings are stretched over an oblong sound box and tuned to the key-note and fifth of the pipe so as to provide a drone accompaniment. All the strings are struck at the same time with a small stick held in the right hand. **1902** *Man* II. 146 Many travellers have stated that various peoples, more or less primitive, amuse themselves by making string figures to which the general term of 'cat's cradle' is usually applied. **1963** K. Vonnegut *Cat's Cradle* v. 20 His fingers made the string figure called a 'cat's cradle'. **1879** *Jrnl. Anthropol. Inst.* IX. 26 Now as to the origin of the string games among these Malays (Dayaks) and Polynesians, it is evident that they did not learn them from Europeans. **1929** *Encycl. Brit.* X. 601/2 In particular it is found that the string game called 'cat's cradle' in various forms is of very wide diffusion, being found even in Australia. **1949** J. Tey *Brat Farrar* xxiv. 217 Did I put my string gloves in the locker? **1978** A. Morice *Murder by Proxy* iii. 32 His coat, cap and string gloves..were neatly arranged on a chair. **1943** C. Hollingworth *Germans just behind Me* ix. 250 By means of bribing his assistant I got a telephone call to my own 'string man' in Belgrade in order that my paper should know I was alive. **1968** M. Allingham *Cargo of Eagles* viii. 98 I'm the string man in these parts.. I write for the Gazette at Nine Ash and keep a watching brief for the Globe in town. It's called stringing. **1909** Jones & Scard *Manuf. Cane Sugar* vii. 198 The highly concentrated juice is boiled to 'string' or crystallizing point. The admission and subsequent discharge of the juice are so regulated, that by the time the latter has reached the point of withdrawal, it has been concentrated to 'string' point. **1909** H. C. P. Geerligs *Cane Sugar* 214 The consistency of the liquid being such that a sample can be drawn out in threads, the liquid is said to be boiled to 'string proof'. **1961** *Guardian* 27 Sept. 10/4 International string-pullers still try to make the Congo dance to their tunes. **1977** D. Ramsay *You can't call it Murder* v. 51 Judith contrived, with the aid of a string puller..to gain admittance. **1948** *Amer. Rec.* 330 The same political manoeuvres, corruption, and string-pulling by moneyed interests..were discernible. **1970** R. E. Johnson *God Keepers* (1971) xiv. 146 The choice between public-opinion pressure and interpersonal string-pulling was probably exaggerated. **1937** W. S. Lancaster (*title*) Hand puppets and string puppets. **1970** G. F. Newman *Sir, You Bastard* vi. 126 The visit was nothing more than a test to see just how much the firm's man he was, to see how he would interpret the string-puppet role. **1965** R. Brett *Dead Side of Mike* iv. 39 Two Italian string-puppets in silver armour. **1939** *Sun* (Baltimore) 25 Mar. 6/2 A bill designed to halt the growth of string slums along the public highways by conservative zoning has been reported below the Judiciary Committee of the State Senate for weeks. *Ibid.* 24 Oct. 12/ The string slums walling in sections of the highways are composed of hot-dog stands, ramshackle overnight cabins, automobile graveyards, cheap dance halls, poolrooms and any sort of other hideous business places. **1950** *Ibid.* 28 Apr. 18/3 Once string slums come into the picture, they stay. **1895** Montgomery Ward *Catal.* 304/1 Men's..string ties. **1916** *Daily Colonist* (Victoria, B.C.) 2 July 12/6 (Advt.), Red, white and blue string ties. **1942** L. D. Cash *Seat of Scornful* xi. 152 He wore..a shiny black alpaca suit and a string tie. **1976** L. Henderson *Major Enquiry* xvi. 106 He was dressed in..a string tie. **1928** J. P. Dunn *Student's Guide to Orchestration* xi. 14 String tone permeates every orchestral movement of any length. **1968** A. Nilsen *Introd. Organ* ii. 90 Undulating stops..are usually of string-tone. **1930** *Oxf. Compan. Mus.* 665/1 *Gamba*..a species of slightly string-toned diapason. **1967** D. Penner *Ritual* vii. 70 He..wore a string vest. **1972** *Guardian* ..string underwear.. **1979** ..string vests..hell out of a rowing machine.

string, *v.* Add: **1. c.** To fit (a thing) with the necessary strings or ties to keep it firm or in place.

[**1805** *Edin. Bk. Prices* 63 Stringing or banding.] **1931** *Henley's ABC Gliding & Sailflying* 252 Having sewn up all the edges neatly, the next operation is 'stringing' the wing to keep the fabric tight to the ribs.

5. c. *to string on*: for *vulgar* read *Austral.* and *N.Z. slang*. (Earlier example.)

1882 A. Bathgate *Waitaruna* 142 A barmaid in one of its hotels..is popularly known as 'Goodall's stringer'... She makes herself agreeable to those who frequent the house, and so she 'strings them on' and induces them to spend their money there.

10. a. Also *spec.* to place (pipes) end to end along the line of a trench, in preparation for welding them together.

1949 *Our Industry* (Anglo-Iranian Oil Co. Ltd.) (ed. 2) v. 163 The pipes are strung out along the line of the trench and placed into position alongside or over it on temporary supports, and the lengths are then connected by electric welding. **1957** *Oil & Gas Reporter* VI. 1141 The device of 'stringing pipe' for oil and gas pipe lines does not, within and of itself, constitute a transportation of property. **1966** *Petroleum Handbk.* (Shell Internat. Petroleum Co.) (ed. 5) 266/2 The construction phases consist of: clearing and grading the right of way, ditching, 'stringing' and 'welding' the pipe, [etc.]. **1968** *Sunday Mail* (Brisbane) 29 Sept. 12/2 The first pipes will be 'strung out' this week.

11. b. *fig.* To stretch (something) *out* in order to make it last.

1867 'Mark Twain' *Sk. New & Old* (1875) 73 What is the use of stringing out your lives to a lean and withered old age? **1894** — in *North Amer. Rev.* Apr. 447 It [*sc.* the story] is not strung out as I have strung it out, but it is all there. **1977** P. Hill *Fanatics* 123 They're just stringing it out, putting off the evil hour.

12. d. To extend or continue. Const. *along*, *out*.

1869 'Mark Twain' *Lett. to Publishers* (1967) 21 So much of the 400 or 500 pages still left are reprint, and so will string out a heap. **1877** — in *Atlantic Monthly* Nov. 591/1 Isaac knelt down and began to pray; he strung along, and strung along, till everybody had got tired. **1896** — in *Harper's Mag.* Aug. 351/1 Well, the time strung along and along, and that fellow never come!

d. *to string along with*: to accompany, to agree with, to support or go along with (usu. without undue enthusiasm). Occas. without const. *colloq.* (orig. *U.S.*).

1927 *Vanity Fair* (N.Y.) Nov. 67/2 To this day the Mayor's..'The Family Time' but the players still string along with the theatrical paper [*sc.* Variety]. **1937** J. Steinbeck *Of Mice & Men* iii. 74 ..how you an' him string along together. **1946** *Sun* (Baltimore) 31 May 15/1 The majority of the bettors decided to string along with Blind Path, a well bred youngster. **1950** 'S. Ransome' *Deadly Miss Ashley* ix. 103 String along with you? Don't let me down. **1955** M. Allingham *Beckoning Lady* iii. 39 She..had thought it best to string along with them. **1960** Wodehouse *Jeeves in Offing* vii. 74 I string along with that school of thought. **1972** L. P. Davies *What did I do Tomorrow?* iii. 114, I wasn't going to be taken in. I'll string along, I thought. **1978** A. Gilchrist *Cod Wars* xi. 109 If at some particular moment, they were stringing along with three other departments and accepting..a continued tough line of policy, then my warning telegrams might seem tactless, tiresome, inept.

e. *to string out*: to be under the influence of a drug. Cf. *strung *ppl. a.* 4 c. *U.S. slang.*

1967 Wentworth & Flexner *Dict. Amer. Slang* Suppl. 708/1 *String out*, to be or be addicted to narcotics; to 'hang' or 'be strung out'. **1970** *Sunday Tel.* 20 Dec. 6/6 How long did you string out for?

15. a. For *U.S. slang* read *slang* (now chiefly *U.S.*). (Earlier and later examples.)

1812 J. H. Vaux *Vocab. Flash Lang.* in *Mem.* (1964) 251 To banter or jest with a man by amusing him with false assurances or professions, is also termed *stringing him*, or *getting him in tow*. **1846** *Swell's Night Guide* 133/1 *Strung*, to, to impose on a person's belief by some joke or lie. **1908** A. M. Binstead *Pink 'Un & Pelican* v. 115 She'd string him for fifty bob on an old tea-chest an' a jar o' pickled inyuns! **1931** P. MacDonald *Crime Conductor* i. 5 'It isn't!' said the Assistant Editor incredulously. 'You're stringing me!' **1959** 'R. Macdonald' *Galton Case* xviii. 147 They were stringing you. They just don't want a woman in the way. **1982** H. Engel *Ransom Game* i. 5, I guess I don't have any reason to believe they'd string me.

g. *to string* (someone) *along*: to fool or deceive (someone); *spec.* to encourage (someone) to remain in a state of misplaced confidence. Cf. sense 15 a in Dict. *colloq.* (orig. *U.S.*).

1902 G. H. Lorimer *Lett. Merchant* xvii. 270 Clytie.. **1943** K. Tennant *Ride on Stranger*.. **1962** A. ..

16. *intr.* To work as a stringer (sense *11).

1960 G. Edinger *Twain shall Meet* xv. 187 European journalists, stringing for papers in America or Britain. **1966** E. West *Night is Time for Listening* ii. 49 'It's not an assignment,' Davison said. 'I've been stringing.' **1972** *Maclean's Mag.* June 82/1 Fred Cleverly is a CBC news reporter in Winnipeg. He also strings for the Toronto *Star*. **1977** J. Shaw *Bagpiper*.. iii. 202 An old newspaperman..who occasionally strings for us when there's anything of interest happening in that area.

‖ **stringendo** (strindʒe·ndo), *adv.* (*a.*) and *sb.* *Mus.* [It., gerund of *stringere*, to press, squeeze, bind together.] **A.** *adv.* (*a.*). A direction indicating that a composition be played with increasing speed and excitement. Also *transf.* and *fig.*

1853 Geo. Eliot *Let.* 2 Dec. (1954) II. 129 Mrs. Pitt scolds the servants, *stringendo & fortissimo*, while I am dressing. **1894** G. B. Shaw in *World* 25 Apr. 24/2 Wagner thought it sufficient to indicate the necessary changes of tempo by such hints as 'ritenuto', 'stringendo', and the like. **1922** Joyce *Ulysses* 207 *Stephen* (*Stringendo.*) He has hidden his own name, a fair name, William, in the plays, a super here, a clown there, as a painter of old Italy set his face in a dark corner of his canvas. **1959** *Times* 30 Oct. 4/7 A tendency to exaggerate the stringendo passages. **1977** *Gramophone* Nov. 837/1 His tempo for the funeral march is barely faster than Boult's, but it seems much more so thanks in part to his *stringendo* manner.

B. *sb.* (Pl. **stringendi.**) A passage played in this manner.

1937 R. Jaques tr. *A. Cortot's Stud. Mus. Interpretation* 154 From the bars before the *stringendo* the upper C sharps of the left-hand part may be played by the right hand. **1978** *Gramophone* Feb. 1405/2 One finds the latter [*sc.* Karajan]..readier to indulge in *stringendi*, urging things on, where Giulini's combination of concentration and steadiness is compelling in quite a different way.

stringent, *a.* **5.** (Example.)

1870 J. K. Medbery *Men & Mysteries Wall St.* v. 69 Money is 'very active', and the loan market 'stringent'.

stringer. Add: **5. a.** Also, a long horizontal member serving to support or tie together a bridge.

1940 *Sun* (Baltimore) 24 May 19/5 The last span between piers 36 and 37 was closed today, engineers said, and work is being rushed on floor beams, stringers and decking. **1960** [see *cream-truck* s.v. *cream *sb.*[2] 7]. **1976** *Columbus* (Montana) *News* 24 June 6/6 The best method to improve the bridges is to install more stringers.

d. (Earlier example.)

1848 *Rep. Comm. Patents 1847* (U.S.) 72 One patent has been granted for improvements in the rail, and the manner of fastening it to the stringers.

f. *Aeronaut.* A spanwise member of a wing, parallel to the spars, used to give lateral stiffness to the ribs; also, a longitudinal member of a fuselage, serving to reinforce and stiffen the skin and assisting it to carry direct load.

1918 *Flight* 4 July 740/2 The main [wing] ribs consist of ply wood webs socketted into grooved spruce flanges, which are tapered off..except where they are met by a longitudinal stringer. **1920** *Ibid.* 12 Aug. 879/2 To this main [fuselage] structure is added stringers which bring the outside form up to a streamline shape. **1928** Chatfield & Taylor *Airplane & its Engine* xii. 212 As even the main ribs are very light, they must often be supported against tipping over sidewise. This is done by means of light wood stringers, which run parallel to the spars. **1932** M. Langley *Metal Aircraft Construction* v. 113 The slightly curved contours of stringers may be achieved by rolling them to template immediately they come off the draw bench. **1945** *Aeroplane* 17 Aug. 185/2 The multiplicity of stringers forms an impressive skeleton upon which a preformed skin is laid in large panels of over 30 ft. in length. **1961** B. Fergusson *Watery Maze* vii. 160 This was *Bachequero's* first venture in anger since her conversion, and she was no doubt straining every plate and stringer to do herself credit. **1973** 'A. Hall' *Tango Briefing* ix. 111 The whole of the airframe began shivering as the stringers took the strain.

g. *Surfing.* (See quot. 1962.)

1962 T. Masters *Surfing made Easy* 65 *Stringers*, pieces of wood laminated into the surfboard foam for decoration and rigidity. **1965** J. Pollard *Surfrider* ii. 21 Those strips of wood used in foam boards to add lateral strength are called 'stringers'. **1968** W. Warwick *Surfriding in N.Z.* 16/3 Today about 50% of all surfboards are built with a centre stringer of either wood or fibreglass, whereas up to about 1966 nearly all boards had some kind of stringer.

9. = *string *sb.* 1 n. *U.S. slang. rare.*

1851 T. A. Burke *Polly Peablossom's Wedding* 89 He never lacked assistance from his acquaintances whenever he had concocted a 'stringer'.

10. *Metallurgy.* A microstructural feature consisting of a narrow vein of inclusion or alloy constituent oriented parallel to the direction of metal working.

1942 C. G. Johnson *Metallurgy* (ed. 2) x. 187 Fig. 103 illustrates a slag stringer in steel that caused failure in the hardening operation. **1959** *Jrnl. Iron & Steel Inst.* CXCI. 353/1 Forging draws out the carbides into long stringers running in the direction of hot working. **1976** *Sci. Amer.* Nov. 106/3 The most significant difference between bending and stretching is the role of microscopic impurities or inclusions that are not metallic... During hot-rolling they become elongated into 'stringers'.

11. A newspaper correspondent paid in proportion to the quantity of his published work (cf. *string *sb.* 17 b). Hence, a correspondent employed part-time; *spec.* one employed to report on events in a particular place. Also *transf.* Freq. *attrib.* orig. *U.S.*

1952 *Time* 21 Jan. 7/1 Saporiti was in Portugal when he first started as a stringer (part-time correspondent) for *Time* in the spring of 1946. **1952** *Iowa Quest* 31 Jan. 3 (*heading*) 11 students hold 'stringer' jobs. **1956** *Sun* (Baltimore) 28 Aug. 4/6 Even the wire services used mainly stringer correspondents to cover trial. **1958** *Spectator* 31 Oct. 570/2 A free-lance reporter (formerly a stringer for *Confidential*). **1962** *Rep. Comm. Broadcasting 1960* 315/2 in *Parl. Papers 1961–2* (Cmnd. 1753) IX. 259 Organisation of independent television news: camera crews and 'stringer' cameramen. **1970** *Radio Times* 30 Apr. 10/4 Every weekday BBC radio puts out 5½ hours of news and current affairs programmes to Britain. It employs 17 full-time foreign correspondents and nearly 100 stringers, mostly newspapermen. **1973** H. Trevelyan *Diplomatic Channels* vii. 116 They [*sc.* intelligence services] employ stringers to get caught and occasionally exchanged; but these are regarded by proper spies as an inferior form of life and are not admitted to the international spy confraternity. **1979** E. Koch *Good Night Little Spy* viii. 65 He was a so-called 'stringer'; he was not attached to any one newspaper and freelanced for several of them.

stringily (stri·ŋili), *adv.* [f. Stringy *a.* + -ly[2].] So as to be stringy, so as to resemble string.

1940 G. Frankau *Self-Portrait* i. 19 My childhood acquired..a horror of macaroni, with which—cooked stringily à l'Anglais—we were compulsorily fed when we stayed for 'afternoon class'. **1976** Scott & Koski *Walk-In* (1977) ii. 17 One lonely and sullen girl in cut-offs and sweat shirt, hair hanging stringily to her waist.

stringing, *vbl. sb.* **1. b.** (Later examples.)

1952 *Iowa Quest* 31 Jan. 3/5 'Stringing' is interesting and has provided many aspiring journalists with valuable on-the-job training. **1970** A. Fowles *Dupe Negative* iii. 23 I'd shot an interview with him..on a stringing job for the BBC. **1973** *Times* 3 July 18/7 Lyall then did more years in the editor's chair..before devoting himself to stringing.

† **stringlet** (stri·ŋlet). *U.S. Obs.* [f. String *sb.* + -let, after *ringlet*.] A long wisp of hair.

a **1852** F. M. Whitcher *Widow Bedott Papers* (1856) xv. 154 Them great long stringlets a danglin' down her cheeks. **1874** *Rep. Vermont Board Agric.* II. 600 Faded-out hair upon either side, with stringlets hanging half-way to the ground from hip and shoulder.

stringy, *a.* Add: **1. c.** Designating defective cotton or wool, esp. cotton which has been imperfectly scutched.

1902 W. I. Hannan *Textile Fibres Commerce* 115 The cotton which is struck off by the beater blades of the scutcher should be removed away from the beater's course immediately; any delay at this stage may cause the fibres to become contorted into very curious shapes, and such cotton is then termed *stringy*. **1932** E. Midgeley *Technical Terms Textile Trade* II. 215 *Stringy*, wool partially matted in fibre and drawn into a slightly ropy form. The stringing of wool is usually due to inefficient scouring. **1950** *Mercury Dict. Textile Terms* 481/1 *Stringy cotton*. This is a defective cotton produced by ginning wet or unripe seed cotton, or sometimes by a wrong adjustment of the brushes that take the lint from the ginsaws.

2. Also of hair: thin, tending to hang in strands. Also *Comb.*

1956 J. Cheever in *New Yorker* 14 Jan. 26/1 Her light hair was long and stringy. **1981** P. Theroux *Mosquito Coast* xviii. 234 The stringy-haired man.

stringy-bark. Add: **a.** (Earlier example.)

1801 [see *blackbutt].

b. (Earlier example.) Also, the wood of one of these trees.

1848 W. Westgarth *Australia Felix* vi. 73 These natives appear to like also the fruit of the pandanus, of which large quantities were found in their camps, soaking in water contained in vessels formed of stringy-bark. **1901** M. Franklin *My Brilliant Career* i. 3 The stringy bark roof of the salt-shed..protected the troughs from rain. **1928** 'Brent of Bin Bin' *Up Country* vi. 94 On that early journey when it rained they hove to under the drays, well-covered by tarpaulins supplemented by stringy-bark lean-tos. **1977** *Weekly Times* (Melbourne) 19 Jan. 39/2 The basic materials used are local gum and stringy bark.

c. Also as *sb.*, an inhabitant of the outback, an uncouth person.

1836 J. F. O'Connell *Residence Eleven Yrs. New Holland* 49 Let us suppose the suitor an old 'stringy-bark', such being the soubriquet in which inland settlers rejoice. **1861** H. Earle *Ups & Downs* 59 She would never have had the bad taste to prefer a stringy bark like me to such a fine-looking, first-class fellow as yourself. **1892** H. Nisbet *Bushranger's Sweetheart* iv. 30 He was a larikin of the larikins, this tiny Stringy Bark, who haunted my thoughts.

strio-. For 'striæ' read 'striatum' and add: **strioni·gral** *a.*, epithet of nerve fibres running from the corpus striatum to the substantia nigra; **striopa·llidal** *a.*, epithet of nerve fibres running from the neostriatum to the globus pallidus (the palæostriatum).

1920 S. W. Ranson *Anat. Nerv. Syst.* xi. 164 The function of the substantia nigra is equally obscure... There terminates within it a bundle, consisting of both direct and crossed fibers from the corpus striatum, the strionigral tract. **1970** *Brain Res.* XVII. 125 As part of an analysis of the intrinsic and extrinsic connections of the caudate nucleus some information has been obtained on the termination of the strio-pallidal and strio-nigral fibres. **1937** J. H. Globus *Pract. Neuroanat.* i. 156 Through the red nucleus, the rubrospinal tract is brought in continuity with part of the striopallidal system. **1970** Strio-pallidal [see *strio-nigral* adj. above].

strioscopy (strəi,ǫ·skǒpi). *Physics.* [f. Strio- (here of unkn. significance) + -scopy.] A form of electron microscopy in which the beam is focused on to the specimen as a hollow cone of particles, giving a bright image on a dark field.

1967 *Jrnl. Electron Microsc.* XVI. 11 The contrasts obtained by the contrast-stop method are better than those of strioscopy. **1974** *Physics Bull.* Sept. 397/2 Castaing contributes a chapter on the prism-mirror analyser, Fert on strioscopy. **1979** J. R. Fryer *Chem. Applications Transmission Electron Microsc.* ii. 62 Normal dark field technique either tilts the incident beam, or displaces an objective aperture, so that the primary beam does not contribute to the final image, whilst strioscopy physically stops the axial primary beam passing.

So **striosco·pic** *a.*

1972 P. W. Hawkes *Electron Optics* iii. 89 In a commercial instrument..that makes provision for strioscopic illumination, as this type of illumination is called, a third condenser..is provided below the annular aperture to match the latter exactly to the objective aperture. **1973** *Nature* 17 Aug. 412/1 Strioscopic methods are also of potential value for high resolution work.

strip, *sb.*[2] Add: **1. f.** A sequence of small drawings telling a comic or serial story in a newspaper, etc. Freq. as *comic strip*. Also *transf.* orig. *U.S.*

1920 L. N. Flint *Editorial* x. 229 In the paper..the week-day issues contain a preponderance of syndicate features—'comics', 'strips'..and continued stories. **1920** C. Sandburg *Smoke & Steel* 47 The comic strips in the papers. **1928** *Daily Sketch* 7 Aug. 4/2, I keenly appreciate the qualities that make Pop the greatest comic strip in the world. No comic strip artist..has the same facile and generic lines the creator of Pop possesses. **1939** Joyce *Finnegans Wake* 537 Such wear a frillick for my comic strip, Mons Meg's Monthly, comes out aich Fanagan's Week. **1943** D. Powell *Time to be Born* iv. 95 She had a curious impression of being in a Buck Rogers strip..and gazing into another planet. **1955** Auden *Shield of Achilles* ii. 38 Mild-looking middle class boys Who read the comic strips. **1967** *Listener* 21 Dec. 821/3 This feedback from strip to pop and back into strip again is very noticeable. **1979** *Tucson* (Arizona) *Citizen* 20 Sept. 7n/6 Why, you might wonder, would Universal and NBC risk an expensive space comic strip like 'Buck Rogers'?

g. = *air-strip* s.v. *air *sb.*[1] B. III. 7. Also *fig.* See also *fighter strip* s.v. *fighter 4, *landing-strip* s.v. *landing *vbl. sb.* 8.

1936 W. H. McCormick *Mod. Bk. Aeroplanes* xi. 106 The strip extends across the landing-ground. **1944** *Yank* 14 Jan. 10 As a draftsman working for the Australian government, he helped plan both strips. **1958** 'N. Shute' *Rainbow & Rose* i. 9 This is the only strip in the vicinity? **1962** M. McLuhan *Gutenberg Galaxy* 64 Greek celature as a take-off strip for the medieval manuscript culture. **1977** *Whitaker's Almanack 1978* 756/2 Several flying strips are also in use by light aircraft [in Nigeria].

h. A street noted for its night-clubs, bars, gambling houses, etc. Freq. with def. article and capital initial (orig. with reference to *Sunset Strip* in Hollywood: see quot. 1974). *slang* (chiefly *N. Amer.*).

1939 *California* (Federal Writers' Project, Calif.) 193 Further west on Sunset Boulevard..is a section popularly known as 'the Strip'. **1941** B. Schulberg *What makes Sammy Run?* vii. 124 The wind sweeping down the Strip from the sea. **1957** *MacLean's Mag.* 6 July 33/1 The many-tongued enclave known as the Strip is cut off..by a near-Gothic stone pile that straddles the Avenue [*sc.* Spadina Avenue in Toronto] just north of College Street. **1967** W. Murray *Sweet Ride* vi. 89 The Place is located in the heart of the Strip. It..had once catered to a touristy clientele. **1968** *Globe & Mail* (Toronto) 3 Feb. 23/1 Visit the main floor bar of the Brown Derby at Dundas and Yonge, the crossroads of the Yonge Street 'strip' which includes seven bars within a block-and-a-half. **1971** *Guardian* 8 July 3/1 Bangkok has its own strip, the new Petchburi Road extension: miles of girlie bars, short time hotels, and soul food snack bars. **1974** *Encycl. Brit. Macropædia* XI. 109/2 Sunset Boulevard meanders 21 miles west from the state park..to the sea. A one-mile section of the boulevard becomes the 'Sunset Strip', or simply the 'Strip'. **1976** *Publishers Weekly* 21 June 62/2 Rush Street was a nightlife strip, virtually deserted during the day. **1978** S. Brill *Teamsters* iv. 124 Just over to the right is the strip—a row of flickering neons wrapped around bold signs that advertise 'go-go girls' and 'live dancers'.

i. *to tear* (someone) *off a strip*, *to tear a strip off* (someone) and varr.: to upbraid or reprimand (someone); *to lose a strip*, *to have a strip torn off*, to be reprimanded or receive a dressing-down. *colloq.* (orig. *R.A.F. slang*).

1940 N. Monks *Squadrons Up* ii. 56 For any breaches of discipline..he would 'tear a strip' off the luckless pilots. **1940** 'N. Shute' *Landfall* i. 25 Dickens tore me off a strip just now. **1942** T. Rattigan *Flare Path* i. 30, I didn't particularly like doing it, and I had the hell of a strip torn off about it afterwards. **1952** E. F. Davies *Illyrian Venture* iv. 71 Nicholls used to tear tremendous strips off Trayhorn. **1957** L. P. Hartley *Hireling* 42 If my wife saw me wearing one, she would tear me off a strip. *a* **1963** J. Lusby in B. James *Austral. Short Stories* (1963) 225 It's all right... I've just lost a strip, too. **1967** *Listener* 31 Aug. 264/1 Mr Kosygin..tore great strips off almost every major industry for inefficiency, or shoddy work, or both. **1979** 'M. Hebden' *Death set to Music* ix. 99 He'd clearly suspected it might have been Nosjean's [idea] and had been hoping to be able to tear a strip off him.

j. A track used for motor-racing. See also *drag strip* s.v. *drag *sb.* 1 f. *U.S.*

1941 *Sun* (Baltimore) 30 Aug. 13/1 Dick Pending has the racing strip in good condition and unless more rain comes tomorrow, the track will not be too bad. **1946** *Ibid.* 2 Oct. 16/5 The racing strip has been brought around slowly to peak form. **1977** *Custom Car* Nov. 5/4 The drivers of these American cars are also quite happy with the racing. They enjoy chasing the slower cars down the strip.

k. = *strip light* (*b*), sense 6 a below.

1970 R. Crawford *Kiss Boss Goodbye* ii. iii. 68 The basement was..warmed by wall-heaters and lit by soft-pearl strips. **1981** I. McEwan *Comfort of Strangers* x. 133 The room was small, windowless and heavily perfumed. It was lit by a fluorescent strip.

l. *Cricket.* The narrow band of ground lying between the wickets.

1976 J. Snow *Cricket Rebel* 30 None of England's fast bowlers had been particularly menacing during the first Test on a typically sluggish Edgbaston strip. **1977** *Sunday Times* 9 Jan. 28/6 MCC's other team in Bengal found a better wicket for batting at Dacca than the mutilated strip at Calcutta.

5*. *colloq.* The clothing worn by and distinguishing a football team.

If the original sense is 'clothing to which a player strips down', this sense should properly be placed under Strip *sb.*[3]

1974 *Evening News* (Edinburgh) 8 Oct. 16/3 Postal United, the East of Scotland League club, had their strip stolen from a car in the Hailesland Park area. **1977** *Shoot* 18 June 4/4 The national strip of Zambia is green jerseys, orange shorts, and black stockings. **1981** 'G. Gaunt' *Incomer* xiv. 87 The [football] team were..passing flagon bottles around. Frank & Bob were..in a corner, having changed into strip early, and managed to grab a bottle between them.

6. (sense *1 f), *strip advertisement*, *form*, *heroine*; (sense *1 g) *strip landing ground*; **strip architecture** *U.S.*, the types of building or other features characteristic of strip development; **strip-built** *a.* *rare*[−1], that has been subjected to strip or ribbon development; **strip cartoon**, a sequence of cartoons (sense *2) telling a (comic) story; freq. *attrib.*; hence **strip cartoonist**; **strip chart, stripchart**, a long roll of (usu. graduated) paper on which the pen of an automatic recording device can trace changes of a measured quantity with time by moving the paper past the pen at a constant rate; usu. *attrib.*, designating recorders using such rolls; **strip-cropping**, (*a*) *U.S.*, a system of land cultivation in which crops of different types and habits of growth are sown alternatively in strips along the contours of a hill, etc., to prevent soil erosion (cf. *contour cropping* s.v. *contour *sb.* 4); (*b*) the practice of growing crops in strips (cf. *strip farming* below); **strip-cultivation**, (*a*) = *strip farming* below; (*b*) *Archæol.*, the practice of using strip lynchets in farming; **strip development** *U.S.* = *ribbon development* s.v. *ribbon *sb.* 10 a; **strip-farm** *v. trans.* *U.S.*, to cultivate (land) in strips along the contours of a hill, etc., to prevent soil erosion (cf. *strip-cropping* (*a*) above); **strip farming** *Hist.*, a system of land cultivation in which the land was divided up into long narrow strips and allocated to different peasant-farmers; **strip-grazing** *Agric.*, a system of farm management in which strips of land are alternately grazed and kept empty; rotational grazing; hence **strip-graze** *v. trans.*, to graze (land or livestock) in this way; **strip-grazed** *ppl. a.*; **strip light**, (*a*) *Theatr.*, any device to provide diffused stage lighting by mounting several lamps in a row, as on a batten, in a trough, etc.; (*b*) a lighting device, now usu. in the form of a tubular fluorescent lamp, for providing a continuous line of light; also as *v. trans.*; hence **strip-lighted** *ppl. a.*; **strip-lighting** *vbl. sb.*; **strip line, stripline** *Electr.*, a *microstrip; **strip-lit** *a.* = *strip-lighted* ppl. adj. above; **strip-lynchet** *Archæol.*, a horizontal terrace used for cultivation; a long, narrow lynchet (sense *2 b); **strip mill** *Metallurgy*, a rolling mill specially designed for the production of metal strip; **strip mine** *U.S.*, a mine worked by strip-mining; also as *v. trans.*, to obtain or exploit by strip-mining; **strip-mined** *ppl. a.*, **strip-miner**; **strip mining** *vbl. sb. U.S.*, a method of mining in which surface material is removed in successive parallel strips to expose the mineral, the spoil from each new strip being placed in the previously excavated one; **strip packaging**, a method of packaging small items, liquids, etc., in which individual sachets are formed (from plastic or metal foil), filled and heat-sealed in a single process; **strip park** orig. *U.S.*, a long, narrow park developed alongside a road, canal, etc. (cf. *strip development* above); **strip printer**, a photocomposing device which prints characters on a strip of paper or film; also, any device which prints on a narrow roll of paper; **strip steak** *U.S.* (see quot. 1962) (cf. *strip-loin*); **strip system** = *strip farming* above; **strip-wound** *a.*, wound with strips, esp. of metal.

1938 Strip advertisement [see *before A. 5 c]. **1976** *New Yorker* 15 Mar. 27/3 'Strip' architecture—the endless miles of trailer parks, gas stations, used-car lots, Taco Bells, etc.,..that fan out from every American metropolis—has its own validity. **1936** C. Day Lewis *Noah & Waters* 15 Strip-built roads that stray Out like suckers to drain the country. **1936** *Discovery* Dec. 384/1 Shopkeeper's bill of the early 18th century. Note the smokers conversing about their tobacco, quite in the modern 'strip-cartoon' style. **1950** *Times* 2 Mar. 6/5 Separate or detachable sections or supplements comprised wholly or mainly of strip cartoons. **1967** E. Short *Embroidery & Fabric Collage* iii. 60 The coverlet tells the story of Tristan, in a series of scenes showing different incidents, in the manner of a strip cartoon. **1974** *Listener* 24 Jan. 118/1 Under the strip-cartoon image lies a message that is often puritanical. **1953** *New Internat. Yearbk. 1952* 47/1 J. C. Bancks, Australia's most popular strip-cartoonist, was creator of Ginger Meggs. **1950** *Instruments* XXIII. 260/3 (Advt.), New 'Pneumatic Capacilog' air-operated strip-chart recorder is completely self-contained. **1966** *N.Y. Times* 3 Feb. 33 The computer recorded wave variations that often are undetectable to the eye of a physician using the traditional strip chart. **1978** *Nature* 12 Oct. 520/2 The outputs are recorded on stripchart recorders, allowing a maximum resolution of 100 ms. **1936** *Sun* (Baltimore) 18 Aug. 3/5 This would be effected through..a blending of 'soil depleting' crops with grasses by a system of 'strip-cropping'—a strip of crops and a strip of grass. **1949** Martin & Leonard *Princ. Field Crop Production* v. 125 Strip cropping, now widely advocated, has been practiced for generations in sections of Pennsylvania. **1976** *S. Wales Echo* 26 Nov. 8/6 If you have plenty of cloches you could accomplish what is called strip cropping. This means that sowings are arranged in alternate strips so that cloches can be moved sideways from one strip to the next and back as required. **1932** Kendrick & Hawkes *Archaeol. in England & Wales 1914–31* x. 173 The Celtic system..lasted to reach its height in Roman times, and makes a striking contrast to the strip cultivation of the Saxon and medieval open fields. **1974** C. Taylor *Fieldwork in Medieval Archaeol.* iii. 28 These terrace-like features [*sc.* strip lynchets] on hillsides are the remains of medieval strip cultivation. **1955** *Sun* (Baltimore) 7 Jan. 19/5 Shops in long-established business districts are predominantly in 'strip' developments; that is, strung out along principal highways that bisect the neighborhood. **1980** *Blair & Ketchum's Country Jrnl.* Oct. 68/1 The arrival of new kinds of people or a new industry or housing developments and strip developments loomed all the larger in many small towns. **1943** *Sun* (Baltimore) 8 Sept. 18/1 The corn rows follow the lay of the land on the contour and the land is strip-farmed..with the corn rows acting as dams to check losses of soil and moisture. **1913** A. D. Hall *Pilgrimage Brit. Farming* xiv. 103 The strip farming..prevails over all the land [of the Isle of Axholme] which we may suppose to have been dry in medieval times. **1962** H. R. Loyn *Anglo-Saxon England* i. 20 Pre-Saxon strip farming has been recognized at sites in Wessex..and in Cumberland, Northumberland and South Scotland. **1949** *Radio Times* 15 July 17/2 Those inter-planetary adventures we find, in strip form, in almost every comic. **1960** *Farmer & Stockbreeder* 12 Jan. 78/1 One part [of a herd] is housed and milked in a modern and double row cowshed and is strip-grazed in summer. *Ibid.* 79/1 The kale is no longer strip-grazed. **1971** *Power Farming* Mar. 29/1 It was particularly useful for direct-drilling kale, which could then be strip grazed. **1976** *Burnham-on-Sea Gaz.* 20 Apr. 22/3 Although being ..strip grazed on a paddock system..the herd has shown that it can milk well. **1960** *Farmer & Stockbreeder* 15 Mar. 133/1, I shall be going in for milk production. Please suggest a ration based on strip-grazed beet tops, swedes, kale, hay, oats, barley and beet pulp. **1955** *Times* 6 June 4/5 Some farmers are such convinced believers in strip grazing that in the larger fields they use two electric fences, one at the feeding face and one as a back fence to keep the animals off the grass that should be starting to grow again. **1975** *Country Life* 26 June 1702/3 Strip-grazing..involves using two swards, one solely for grazing..and used for a succession of years..and the other sward used more often as a shorter ley for conservation. **1967** *Listener* 21 Dec. 822/1 The idea of a strip-heroine for middle-aged onanists is surely a gloomy one. **1938** *Flight* 21 July 60/1 It is in fact a strip landing ground with natural wind buffers. **1920** S. Lewis *Main Street* xviii. 221 Sending to Minneapolis for..a strip light. **1927** *Proc. Inst. Civil Engin.* CCXXIV. 160 The manometer is illuminated by a 'strip light'. **1934** S. Gold *Neon* xxii. 61 The question of strip-lighting the building over the entire front is usually entertained only by cinemas. **1963** Parker & Smith *Scene Design & Stage Lighting* xvi. 292 One form of stage-lighting instrument that predates the invention of the incandescent lamp is the striplight, which produces the effect of a line of light by means of a number of sources. **1972** P. Lively *Driftway* i. 1 Big strip lights on the ceiling reached away almost as far as you could see. **1981** 'J. Ross' *Dark Blue & Dangerous* xx. 109 A corridor flanked with strip-lighted offices. **1926** *Gloss. Terms Electr. Engin.* (Brit. Engin. Stand. Assoc.) 146 *Strip lighting*, a system of lighting in which a number of lamps, usually of tubular form, and installed in line with one another, so as to give the impression of a more or less continuous strip of light. **1934** S. Gold *Neon* xxii. 61 A combination of colours in strip-lighting gives a charming effect to an otherwise straightforward display. **1976** L. Deighton *Twinkle, twinkle Little Spy* iv. 39 The entrance hall..was brightly lit by indirect strip-lighting set into the ceiling. **1952** *Proc. IRE* XL. 1658/2 In the case of strip lines, the line conductor is a thin narrow ribbon of metal either cut from sheet or deposited. **1967** *Electronics* 6 Mar. 58/2 The military believes recent advances in stripline versions of Butler matrixes..can produce faster memory units. **1974** *Physics Bull.* Apr. 153/3 The copper conductors..are suitable for high resolution stripline and ground plane applications. **1960** *Guardian* 14 Apr. 9/3 Illuminated..by ..strip-lit shelves. **1973** M. Amis *Rachel Papers* 131 Cat's crap on the strip-lit kitchen floor, musty wine-shop smells from the dining-room, objects tingled to flayed senses. **1928** *Antiquity* II. 172 Their..observations..of the ..long-strip lynchets of Saxon and medieval times. **1929** *Ibid.* III. 174 The strip lynchets..on sloping ground, are made stable..by the facing of masonry. **1975** J. G. Evans *Environment Early Man Brit. Isles* vii. 168 Today, where ridge and furrow and strip lynchets are preserved they are generally under permanent pasture. **1983** *Out of Town* June 26/2 Bands of quite difficult ground are often stepped and striped by patterns of 'strip lynchets'... The strip lynchets..were gradually bitten into the hillslopes by ploughs that were hauled (approximately) along the contours. **1910** H. P. Tiemann *Iron & Steel* 286 Bar mills, also called merchant mills or, on account of the special product which they make..rod mill, hoop mill or strip mill. **1945** *Times* 26 Feb. 5/7 The strip mill for light sheet and tin plate, the continuous billet mill..[etc.] all belong to this type. **1980** *Times* 19 Feb. 2/5 Wide sheet steel from the BSC's strip mills..is widely used in the manufacture of domestic 'white goods'. **1934** *Coal Age* Oct. 376/3 The spread of trailer operation at Southwestern strip mines reflects a number of advantages. **1970** *New Scientist* 21 May 364/2 Nuclear explosions are also planned for strip-mining large deposits of non-ferrous metals in the northern territories. **1976** *Billings* (Montana) *Gaz.* 1 July 2-A/4 The high court has dramatized the need for a national strip mine law so that everybody plays according to the same rules in extractable resource development. **1978** *Peace News* 6 Oct. 7/2 Stewart Udall, Secretary of the Interior, gave approval to WEST to strip-mine vast areas of Indian land for coal. **1936** *Coal Age* Oct. 415/1 Strip-mined coal, under early production conditions,.. generally sold at prices substantially under those for deep-mined coal. **1980** *Sci. Amer.* Oct. 160/3 The reclamation of strip-mined land involves the relatively simple processes of flattening the piles of overburden, replacing the top-soil and replanting it. **1946** *Sun* (Baltimore) 19 Nov. 4/5 Mark McCauley, Davis (W. Va.) strip miner convicted of first degree murder..was sentenced to be hanged. **1977** *Economist* 23 Apr. 52/3 Despite the small proportion of stripminers in the United Mineworkers Union, the union as a whole has withdrawn its support for federal legislation. **1935** *Coal Age* Feb. 91/1 This first-and-second method of strip mining cannot be employed economically with shovel equipment which must operate down in the cut. **1949** *Hansard Commons* 19 May 706 Strip-mining, as it is known in America, and opencast work in regard to gypsum..or any..base metal is essentially a mining problem. *Ibid.* There were..technicians in the country.. familiar with American strip-mining methods. **1977** *Economist* 23 Apr. 52/2 The technique of stripmining——clearing the topsoil above a coal seam to scoop out the coal with bulldozers——once seemed an answer to low productivity. **1969** L. S. Mounts in W. R. R. Park *Plastics Film Technol.* v. 140 Many products can be packaged in water soluble films with advantage. These include..industrial and agricultural products like sprays, chemical additives and strip packaging of seeds. **1975** C. F. Ross *Packaging of Pharmaceuticals* i. 4 Sachets, filled automatically on suitable strip-packaging machines. **1938** *Sun* (Baltimore) 24 June 12/3 The financial operations..seem likely to leave the Eastern United States with their third 'strip park', the others being the Shenandoah Skyway and the park at Natchez, Miss. **1972** *Times* 7 June 4/3 A number of smaller strip-parks, which people could walk to..would be..useful. **1962** *Amer. Lithographer* Apr. 90/3 An automatically-timed exposure light has been added to the Strip Printer Photo Composing Machine Model 299. **1965** R. R. Karch *Graphic Arts Procedures* (ed. 3) xiii. 331 The Strip Printer is used to produce lines of type in various sizes on paper or 35 mm. film. **1976** *Times* 8 June 10/5 By adding a simple keyboard and strip printer to a standard telephone, the telephone terminal could..interact with a computer. **1962** J. N. Winburne *Dict. Agric. & Allied Terminol.* 769/2 *Strip steak*, the steak cut from the loin strip of a beef carcass. **1977** *Rolling Stone* 30 June 111/3 His guitar style is taut and as lean as a strip steak. **1954** J. Keith *Fifty Yrs. Farming* xi. 125 Over a great part of Britain there developed the common-field and strip system. **1965** R. Whitlock *Short Hist. Farming in Britain* i. 20 In conjunction with the pattern determined by the type of plough, arose the Saxon strip system of fields. **1907** Hobart & Ellis *Armature Construction* xi. 265 Windings for a strip-wound barrel type of armature. **1962** *Times* 26 Feb. (Canada Suppl.) p. vii/2 (Advt.), Hydraulic pressure cylinders in steel, stripwound, etc.

strip, *sb.*[3] Add: **2.** *colloq.* (orig. *U.S.*). An act or the practice of removing one's clothes or of striptease. See also *strip *v.*[1] 24 b.

1928 *Variety* 12 Dec. 46/3 Why do women principals try to do strip numbers against the competition of experienced runway specialists?..Columbia, by the way, seems to be leery of the limit in strip at this telling. **1956** B. Holiday *Lady sings Blues* (1973) v. 54 He kept on doing this slow elaborate strip. **1966** *Guardian* 9 July 8/3 Perhaps ten [clubs] provide regular striptease. Up to ten provide strip occasionally. **1971** R. Petrie *Thorne in Flesh* vi. 86 Dahlia does a strip... I auction the things she takes off.

strip, *v.*[1] Add: **I. 1. a.** Also const. *down*, *off*, in *intr.* for *refl.* use.

1947 'A. P. Gaskell' *Big Game* 22 You're not supposed to strip off but I had. **1962** D. Francis *Dead Cert* xi. 124 I'm glad it's you that's got to strip off and get soaked, and not me. **1976** D. Barnes *Yesterday is Dead* ii. 272 After taking a leak he'd strip down and jack off.

e. *to strip* (*well*, etc.): to have a good body, to have a pleasing appearance when stripped.

1815 T. Belcher *Art of Boxing* ix. 33 James Belcher.. stripped remarkably well, and displayed much muscle. **1932** D. L. Sayers *Have his Carcase* ix. 106 He strips better than I should have expected... Better shoulders than I realised, and, thank Heaven, calves to his legs. **1955** T. H. Pear *English Social Differences* ix. 201 Such boys, to use the drill-instructor's expression, 'strip better'.

h. (Later *intr.* examples.)

1897 *National Police Gaz.* 26 May 14/2 It is fully expected that he will not only strip in much better fettle at Epsom than he did for the Guineas, but run a remarkably different horse altogether. **1973** *Times* 26 Feb. 12/8 Skymas runs in the Wills Premier Chase at Haydock this week, and will certainly strip fit.

i. *intr.* To perform a strip-tease act. *colloq.* (orig. *U.S.*).

1929 *Variety* 25 Sept. 53/3 She has the unadornment stuff to herself, since the other gals never strip beyond regulation soub garb. **1939** Joyce *Finnegans Wake* 68 She stripped teasily for binocular man. **1962** J. D. MacDonald *Girl, Gold Watch & Everything* vii. 86 I'm working a place, Rio's, up North Miami, singing and sort of stripping some, but not down to raw. **1976** 'E. McBain' *Guns* (1977) ii. 49 'Jocko said you used to be a stripper.' 'Yeah, but..I haven't been stripping for seven, eight years now.'

3. b. Delete 'now *rare*'. (Later examples.)

1881 'Mark Twain' *Prince & Pauper* xxxiii. 385 The pickpockets had stripped him of his last farthing. **1919** G. B. Shaw *Heartbreak House* ii. 81 Are you one of those who are so sufficient to themselves that they are only happy when they are stripped of everything, even of hope? **1936** J. Buchan *Island of Sheep* i. vii. 130 They had only to get hold of Haraldsen..to strip him bit by bit of his possessions. **1950** C. S. Forester *Midshipman Hornblower* 262 The last visit of Spanish ships of war had stripped the place of almost all its stores, and many of the dockyard hands had been pressed as seamen at the same time.

4. b. To pull off the winter growth of hair from (a dog); to pluck. Cf. *stripping *vbl. sb.*[1] 1 a.

1930 E. C. Ash *Pract. Dog Bk.* xi. 197 Stripping a coat is in the varieties of Terriers most important. Powdered chalk is well rubbed into it. The long hair is then plucked out. **1931** *Daily Tel.* 21 May 1/3 Dogs stripped.

7. e. *slang* (orig. *U.S.*). To unpack or unload (a load, container, lorry, etc.).

[**1950** *Western Folklore* IX. 119 *Pulled, stripped*, or *gutted a load*, lost a load of logs.] **1963** *Amer. Speech* XXXVIII. 45 *Strip a load, v. phr.*, to unload a truck. **1968** *Wall St. Jrnl.* 27 Sept. 34/2 Management agreed to allow the dockworkers to strip and stuff containers in which mixed types of cargo had been packed. **1970** *Times* 16 Sept. (Road Haulage Suppl.) p. vii/9 The [overladen] container..should be devanned (or stripped, to use container parlance) and delivered on two vehicles. **1972** *Guardian* 8 May 20/2 According to the dockers' leader..those terms..are for..the same guarantees over 'stuffing' (packing) and 'stripping' (unpacking) the containers. **1973** *Amer. Speech 1969* XLIV. 208 *Strip her*, unload a trailer.

8. (Later examples.) Now freq. in contexts of the inspection or repair of motor vehicles, engines, etc. Also with *down*. Cf. *stripped *ppl. a.* c.

1937 *Discovery* May 164/1 Part of the necessary machinery could not be stripped down to parts small enough to be carried by mules over the narrow and difficult trails. **1958** *Listener* 13 Nov. 778/1 He drives the thing [*sc.* a car] straight into the repair shop at the back and has it stripped down. **1972** *Daily Tel.* (Colour Suppl.) 20 Oct. 10/4 At the end of 36,000 miles the engines were stripped and every component measured and meticulously examined. **1981** B. Hines *Looks & Smiles* 26 His bike..had also been stripped down to the frame.

II. 11. a. Also *spec.*, to remove paint or varnish from woodwork, etc. Cf. *stripped *ppl. a.* b.

1908 P. N. Hasluck *Cassell's House Decoration* 171/2 All the washing and stripping should be done first. The wallpaper must be removed, and the paint stripped. **1956** *Pract. Householder* July 596/1 A number of preparations.. are intended to strip off only one coat at a time... I prefer the type which strips several coats. **1981** *New Homemaker* Apr. 90/1 (Advt.), Stripping isn't the soul-destroying job it used to be... Powerful Ronseal strips without scraping.

14. Also, to harvest (a crop).

1891 R. Wallace *Rural Econ. & Agric. Austral. & N.Z.* i. 6 Twenty acres of grain can be stripped per day. **1938** *Sun* (Baltimore) 6 Sept. 2/7 The corn almost ready to strip. **1979** *Verbatim* Summer 8/1 In Queensland a wheat crop is *headed*, in Victoria *stripped*.

III. 17. c. (Later example.)

1978 D. Bagley *Flyaway* xxv. 230 I've got a spare differential... The bastards are always stripping so I've made it a habit to keep a spare.

23*. *Physics.* **a.** To deprive (an atom or ion) *of* an electron, or (a molecule) *of* an atom. Also *absol.*

1933 O. H. Blackwood et al. *Outl. Atomic Physics* xiv. 305 Throughout the interior of a star, atoms do not exist in what we consider their ordinary conditions... Near the center of the star they are assumed to be stripped of nearly all their planetary electrons. **1936** *Trans. Faraday Soc.* XXXII. 350 One empirical molar weight of $C_2H_4S_4$ in suspension in water is first 'stripped' of two sulphur atoms with sodium hydroxide. **1954** H. E. Huntley *Nuclear Species* i. 4 By stripping the atoms which lay in its path of one or more of its orbital electrons the swiftly moving particle produced large numbers of positively charged ions and free electrons. **1969** *Times* 22 Apr. 6/3 Large amounts of energy are needed to strip the calcium atoms of their electrons before accelerating them into the target of plutonium atoms. **1970** *Sci. Amer.* Aug. 32/3 The ions are stripped not only in the terminal but also halfway down the positive acceleration column. **1978** L. Vályi *Atom & Ion Sources* i. 35 At low impact energies only the outer shell can be stripped of its electron.

b. To remove (an electron or other particle) *from* an atom, ion, nucleus, etc. Also const. *off*.

1935 B. Jaffe *Outposts of Science* ix. 349 This recoiling nucleus spends its energy of motion in stripping electrons from other atoms near it. **1947** *Physical Rev.* LXXII. 1003 A simple theory of neutron production, according to which the proton is 'stripped' from the deuteron by striking a target nucleus. **1958** *Ann. Physics* III. 275 In a deuteron stripping reaction, the rôle of the incident deuteron is to present at the target nucleus surface a neutron or proton ready to be captured (or 'stripped' off). **1979** *Sci. Amer.* Aug. 122/2 The star..becomes a white dwarf: a star with a core consisting of a highly compressed gas of atomic nuclei (mostly helium nuclei) and the electrons stripped from them.

23. a.** *Oil Industry.* To separate (crude oil or gas) into fractions, to fractionate; to extract or recover (a light fraction) *from* a mixture.

1922 D. T. Day *Handbk. Petroleum Industry* II. 324 The great bulk of crude handled was still stripped in batch stills. **1931** Hoffert & Claxton *Motor Benzole* viii. 226 It is essential that the benzole should be stripped from the wash oil as completely as possible. **1938** A. E. Dunstan et al. *Sci. of Petroleum* II. ii. xxv. 1559/2 The latent heat of vaporization of the components stripped from the oil is supplied by the sensible heat of the oil. **1979** *Liquefied Petroleum Gas* (Shell Internat. Petroleum Co.) 3 The aim of this scheme is to 'strip' the large amounts of associated gas which were previously flared and then separate LPG and other heavier gas liquids for export.

b. *Chem.* To extract or recover (a solute) *from* a solvent previously used in its extraction.

1962 Cotton & Wilkinson *Adv. Inorg. Chem.* xxxii. 906 The protactinium can be stripped from the solvent by aqueous acid fluoride solutions. **1980** *Sci. Amer.* Jan. 59/3 Both are recovered by using a complexing agent to dissolve the metal selectively into a dilute solution and then stripping the metal from the solution.

23*.** *Printing.* To mount (copy) in the correct position on a sheet for use in making a printing plate; freq. const. *in*. Also, to make (a flat) thus.

1937 R. W. Polk *Pract. of Printing* (ed. 2) xli. 291 If a number of pictures are to be used together, and the sizes of the originals are not in proper proportion to each other, separate exposures are made for each size, the necessary reductions made, and the resulting negatives are 'stripped' together in proper position on the composite negative plate... If the sizes are not in proper proportion, separate negatives must be made and stripped in on the plate. **1948** R. R. Karch *Graphic Arts Procedures* viii. 232 After the negatives or positives of illustrations and type matter have been prepared.., the job must be stripped on a layout so that press plates may be made. *Ibid.* 233 Positives used for stripping flats for the deep-etch process are stripped on a piece of transparent acetate. **1964** R. W. & E. W. Polk *Pract. of Printing* (ed. 6) xl. 304 (*caption*) A masking sheet with negatives stripped in. *Ibid.* 306 Positioning and attaching films on a masking sheet is called stripping a flat. **1967** Karch & Buber *Offset Processes* v. 142 After the film is developed, the negative is washed and dried. It is then 'stripped' or placed in a predetermined position on a special type of paper for the plate maker who will expose the image onto the offset plate. **1975** J. Butcher *Copy-Editing* iv. 39 Combined line and half-tone may be used for a photograph that needs some lettering or a scale; this is usually done by stripping a line negative into a half-tone negative.

IV. 24. a. strip cell, a cell in which a prisoner is subjected to sensory or physical deprivation; **strip-down**, (*a*) *U.S. colloq.*, a car which has been stripped down and reassembled so as to improve performance; (*b*) the dismantling or disassembly of an engine, etc. (cf. sense 8 above); **strip-poker** orig. *U.S.*, a game of poker in which a losing player sheds a garment as a penalty or forfeit; **strip-search**, a search of a prisoner during the course of which he is stripped naked; also as *v. trans.*; = *skin-search* s.v. *skin *sb.* 13; **strip the willow**, a Scottish country dance performed by couples in longways sets.

1971 *New Society* 1 July 15/1 A strip-cell. This contains only a mattress on a bare floor. **1973** *Black Panther* 7 July 9/2 If the guards wanted to they could turn on a light in the ceiling, but I was always kept in the dark, and nude. That is part of the deprivation, why the soul breaker is called a strip cell. **1950** *Sun* (Baltimore) 6 Oct. (B ed.) 7/4 Juveniles who have been racing the highways in stepped-up strip-downs. **1963** *Times* 24 May (London Underground Suppl.) p. xii/5 The trains were withdrawn from service and sent to Acton, where they undergo a comprehensive strip-down. **1969** *Daily Tel.* 18 Feb. 5/5 Repair is in progress. A strip-down survey of the refinery is being undertaken. **1929** M. Lief *Hangover* i. 9 'How about a fast game of strip-poker?' she suggested. **1935** G. Greene *England made Me* ii. 86 Two girls playing strip poker. **1961** *Times* 14 June 16/3 There had been a 'strip-poker' party that night. **1978** D. Williams *Treasure up in Smoke* xix. 169 'He..suggested some kind of poker.'..'Strip poker?' **1947** Strip-search [see *run-over 1]. **1970** G. F. Newman *Sir, You Bastard* vii. 189 The arrested men were strip-searched and made to await the DI's pleasure. **1979** *Tucson* (Arizona) *Citizen* 3 Oct. 10c/5 A woman sentenced for drunken driving should not be strip-searched. *Ibid.*, Subjected to extreme trauma by a strip-search. **1924** *Scottish Country Dance Bk.* I. 16 Strip the Willow or Drops of Brandy... Running step is used all through this dance. **1980** L. Lewis *Private Life of Country House* xii. 166 'Strip the Willow', a country dance in which couples in turns came from the ends of two rows to perform some steps in the middle.

b. [Sometimes, f. *strip *sb.*[3] 2.] In various

slang or *colloq.* Combs. in sense *1 i, as *strip act, bar, dancer, girl, party, show, song*; **strip club**, an establishment providing entertainment in the form of strip-tease; **strip joint** [JOINT *sb.* 14 a in Dict. and Suppl.] *slang* = prec.

1950 A. COOKE in *Manch. Guardian Weekly* 13 July 13/2 The all-American cult of the 'strip act'. **1963** R. I. McDAVID *Mencken's Amer. Lang.* 728 Today the higher-priced girls are often connected with burlesque or work in strip-bars. **1975** D. LODGE *Changing Places* ii. 96 One of the South Strand strip bars. **1960** *Spectator* 12 Aug. 236 That strip clubs have been multiplying in London recently is generally known. **1962** *Ibid.* 7 Dec. 883 The strip-club owner who intends to fight..on a campaign against entertainments tax. **1973** J. M. WHITE *Garden Game* 36 The neon lights of the strip-clubs and restaurants. **1946** D. RUNYON *Short Takes* 236 There were cut-outs of guys with their arms around hula dancers and around strip dancers. **1961** *Times* 21 Sept. 15/2 The strip-girl loved by a foolish, tiresome but engaging missionary. **1951** *Sun* (Baltimore) 27 June 30/3 Prince Georges County Sheriff Carlton Beall began a crackdown on what he called 'strip joints'. **1959** *Times Lit. Suppl.* 13 Mar. 148/4 Gambling rooms, saloons and strip joints. **1975** D. LODGE *Changing Places* ii. 95 He now stands gawping incredulously at the strip-joints that jostle each other all along Cortez Avenue. **1959** *Times* 19 June (Queen in Canada Suppl.) p. iv/5 Police are always cracking down on private strip parties. **1972** J. BROWN *Chancer* viii. 110 These strip parties—young business blokes—you know. **1967** *Listener* 5 Oct. 437/3 One of the old Windmill strip shows. **1971** *New Scientist* 10 June 641/2 Their dirty raincoats..have been snapped up by strip-show patrons. **1937** HART & KAUFMAN *You can't take it with You* III. 171 She kept singing a strip song while Mrs. Kirby undressed.

strip, *v.*[2] Add: **1. b.** *strip cup* (see quot. 1962).

1941 ROADHOUSE & HENDERSON *Market-Milk Industry* iv. 67 (*caption*) Strip cup used for detecting abnormal milk. The first stream of milk from each teat is milked into the strip cup through the fine mesh screen. **1950** *N.Z. Jrnl. Agric.* Mar. 265/1 Every cow should be tried, using a strip cup, before putting on the machines. **1955** J. G. DAVIS *Dict. Dairying* (ed. 2) 37 The routine use of strip-cups in the cowsheds will assist in the prevention of mastitis spreading. **1962** J. N. WINBURNE *Dict. Agric. & Allied Terminol.* 769/1 *Strip cup*, a small metal cup or vessel with a fine wire strainer or inner liner into which the first streams of milk from each teat are milked from the cow for examination to detect any indication of mastitis infection or any other abnormal condition of the milk or udder. **1975** CAMPBELL & MARSHALL *Sci. providing Milk for Man* xiv. 337 The California Mastitis Test (CMT) is much more sensitive in detecting inflamed quarters than is the strip cup.

stripe, *sb.*[3] Add: **1. d.** *pl.* A prison uniform (with reference to the stripes with which it is patterned). *U.S. slang.*

1887 *Courier-Jrnl.* (Louisville, Kentucky) 29 Jan. 3/2 He changed his stripes for a suit of citizens' clothes. **1905** B. TARKINGTON *In Arena* 22 I'm going to clear this town of fraud, and if Gorgett don't wear the stripes for this my name's not Farwell Knowles! **1940** W. FAULKNER *Hamlet* III. ii. 212 He had never seen convicts' stripes before either. **1943** P. STURGES in Gassner & Nichols *Best Film Plays 1943–44* 279/1 He's going to be in jail, Trudy, for a long time. He can't do you any good in stripes, honey.

e. A narrow strip of magnetic material along the edge of a cine film on which the sound may be recorded.

1954 R. H. CRICKS tr. *Bau's How to make 8 mm. Films as Amateur* 169 The magnetic stripe is coated between the perforations and the edge of the film. **1972** *Amateur Photographer* 12 Jan. 65/3 Fujicascope SH1... Sound unit: Magnetic sound stripe, 6w amplifier. **1973** *Sci. Amer.* Dec. 49/1 Sights and sounds the camera records stay together on the super 8 film in synch during processing. Spoken comment can be added to the magnetic stripe during projection.

f. *U.S.* A line which forms part of the marking on a sports pitch or court. Cf. LINE *sb.*[2] 7 f.

1967 *Boston Herald* 1 Apr. 17/1 Kennedy led the visitors with 17 points, 11 from the foul stripe. **1974** *State* (Columbia, S. Carolina) 3 Mar. 1-D/7 We wanted to keep him off the foul line (Stewart made one of two from the stripe).

2. *to pull stripes*: see *PULL *v.* 19 h.

8. a. For *U.S.* read orig. *U.S.* and add later examples.

1943 L. ADAMIC *My Native Land* 137 Trubar scored a great cultural victory and set a national-linguistic precedent for men of his stripe. **1968** *Guardian* 9 Apr. 9/3 Negro organisers of all stripes, urging their footloose young 'to keep your cool'. **1979** *Daily Tel.* 6 Sept. 4/2 Guyana, led by a Socialist of another stripe.

stripe, *v.*[2] Add: **2. d.** To apply a magnetic stripe to (a cine film). Cf. *STRIPE *sb.*[3] 1 e.

1954 R. H. CRICKS tr. *Bau's How to make 8 mm. Films as Amateur* 169 You then send your film to a suitable firm which 'stripes' it—i.e., coats a narrow strip of magnetic material along its whole length. Two methods of striping have been proposed. **1960** R. BATEMAN *Movie-Making as Pastime* ix. 58 A 'magnetic stripe' system is becoming more widely used... Experiments in 'striping' 8 mm film have been made.

striped, *ppl. a.* Add: **1. b.** *striped bass*, a large North American fresh-water or marine bass of the genus *Roccus*, esp. *R. saxatilis* (earlier and later examples); *striped gopher*, a ground squirrel, *Citellus decemlineatus*, found in North America; *striped mouse*, a mouse with one or more stripes along its back, found in Africa and belonging to the genus *Rhabdomys* or *Lemniscomys*; *striped squirrel*, one of several small rodents with striped markings, esp. the North American chipmunk; (earlier examples); *striped tuna* = *skipjack tuna* s.v. *SKIPJACK 4.

1796 J. MORSE *Amer. Universal Geogr.* I. 203 The Striped Squirrel is still less than the [red squirrel]. **1818** *Amer. Monthly Mag.* II. 295 The striped bass..is another excellent salt-water fish. **1854** THOREAU *Walden* 323, I am on the alert for the first signs of spring, to hear..the striped squirrel's chirp. [**1900** H. A. BRYDEN *Animals Afr.* v. 48 The pretty little Striped Barbary Mouse..is a very different kind of animal.] **1922** *Pacific Fisherman* Feb. 12/2 Striped tuna..is required to be designated with the qualifying adjective 'striped'. **1932** *Discovery* Nov. 364/2 Pythons are so sluggish that they have been nibbled to death by striped mice. **1941** E. T. SETON *Trail of Artist-Naturalist* 299 Gone..also the striped gopher, whose labyrinths are only four inches down. **1951** TRESSLER & LEMON *Marine Products of Commerce* (ed. 2) xx. 445 Skipjack..is also known as the striped tuna from the markings on the body. **1956** W. R. BIRD *Off-Trail in Nova Scotia* ii. 48 I've been after striped bass in many places up and down the coast of America but this is the best of them all. **1963** G. H. THOMSON *Crocus Country* xx. 133 At this time [*sc.* 1905] there were quite a few striped gophers left, though later they entirely disappeared, driven out, it was said, by their bigger grey cousins. **1973** *Stand. Encycl. S. Afr.* IX. 248/1 Some, like the striped mouse.., are diurnal. **1974** *Calhoun Times* (St. Matthews, S. Carolina) 18 Apr. 3/3 Some 200 striped bass are swimming through Lakes Marion and Moultrie with special tags attached.

c. (Earlier example.)

1850 *Phil. Trans. R. Soc.* CXL. 515 Muscles are now named according to their function, voluntary and unvoluntary; or according to their structure, striped and unstriped.

f. *striped trousers*, (typically worn by civil servants, businessmen, etc.), used allusively to indicate the wearer's status, and, by extension, bureaucracy, formality, etc.; by metonymy, a civil servant, etc.; so *striped-trouser(ed)* adj.; similarly (chiefly *U.S.*) *striped pants.* Cf. *PIN-STRIPE, *PIN-STRIPED *a.*

1933 DYLAN THOMAS *Let.* Oct. (1966) 37 Oh to look..different from the striped trouser lads. **1945** 'G. ORWELL' in *New Saxon Pamphlets* III. 38 The striped-trousered ones will rule, but so long as they are forced to maintain an intelligentsia, the intelligentsia will have a certain amount of autonomy. **1946** KOESTLER *Thieves in Night* 220 At the end emerged the striped-trousered finished product. **1958** S. HYLAND *Who goes Hang?* xxxiii. 140 He was almost in tears. Black tears with striped trousers. **1968** W. SAFIRE *New Lang. Politics* 398/2 The diplomatic niceties of 'striped-pants' diplomacy. **1972** M. GILBERT *Body of Girl* xxiv. 206 A crook in striped trousers turns my stomach. **1974** 'D. KYLE' *Raft of Swords* x. 97 They may be useful. So may the striped-trousers in the Foreign Secretary's entourage. **1976** G. MARKSTEIN *Man from Yesterday* xi. 59 'Who's had to apologize?' 'The gentlemen in striped pants.' **1977** *Time* 9 May 22/3 His youthful diplomatic appointee's aversion to striped-pants airs. **1977** 'J. LE CARRÉ' *Honourable Schoolboy* i. 24 Ring every damned striped-pants in the Colony! **1981** *New Standard* 1 Sept. 14/6 An old-fashioned striped-trouser diplomat.

g. Of cine film: having a magnetic stripe (*STRIPE *sb.*[3] 1 e).

[**1956** J. J. ROSE *Amer. Cinematographer Hand Bk. & Ref. Guide* (ed. 9) 7 (Advt.), Your pre-striped film with magnetic sound lip-synchronized to your picture.] **1972** E. & D. SCHULTZ *How to make Exciting Home Movies* xi. 131 Your splicer can get magnetized when it's used on a film already striped.

2. (Earlier example.)

1839 *Morning Post* (Boston) 4 July 2/2 She made a remark, which, if reported in full, would make one of O.F.M. [*sc.* Our First Men] feel decidedly striped.

striper (strəi·pəɹ). *colloq.* [f. STRIPE *sb.*[3] + -ER[1].]

1. Usu. as *two* (*two and a half, three, four*)-*striper*: an officer in the Royal Navy or U.S. Navy (from the stripes worn to denote rank). In later use, in the army, a lance-corporal (*one-striper*), corporal (*two-striper*), sergeant (*three-striper*).

1917 M. T. HAINSSELIN *Grand Fleet Days* xv. 118 But nowadays you find them lolling and sprawling in all the most comfortable armchairs, while the three-striper has to take a high chair or else go to his cabin! **1918** L. E. RUGGLES *Navy Explained* 146 *Two striper.* Instead of saying 'he was a lieutenant', many men say 'he was a two-striper', meaning that he wore two stripes on his cuff. If the officer was a lieutenant commander he would wear two-and-a-half stripes, hence 'he was a two-and-a-half striper'. **1920** *Blackw. Mag.* Mar. 320/2 A dapper two-and-a-half-striper, R.N., dashed alongside in an obviously Navy gig, and scrambled aboard. **1936** *Nat. Geogr. Mag.* LXIX. 799/2 The three-striper looked me up and down. **1950** G. HACKFORTH-JONES *Worst Enemy* i. 20 It made me remember how I felt when some pompous four-striper came slumming or snooping on board my submarine. **1954** *Sun* (Baltimore) 11 Dec. 13/4 Some 250 other 'stripers' were named brigade officers for the period ending March 17. **1977** [see *SPEC *sb.*[4]]. **1978** A. PRICE *'44 Vintage* iii. 41 A two-striper like himself. **1978** H. WOUK *War & Remembrance* i. 10 Bill, isn't that a three-striper slot?

2. = *striped bass* s.v. *STRIPED *ppl. a.* 1 b.

1945 *Richmond* (Va.) *Times-Dispatch* 29 Aug. 22/1 It may be stated that in the West as in the North, the fish is called the striped bass or simply the striper. **1955** *Field & Stream* June 51/1 So far as is known, this is the first fresh-water lake in the entire country to be stocked with adult stripers. **1961** J. STEINBECK *Winter of our Discontent* xv. 292 Stripers come in sometimes. **1974** *Spartanburg* (S. Carolina) *Herald-Jrnl.* 21 Apr. B6/5 A hybrid is a cross between a striper and a white bass.

stripey (strəi·pi). [f. STRIPE *sb.*[3] + -Y[6] (in sense 2, ad. Afrikaans *strepie*): cf. STRIPY *a.*] **1.** *Navy slang.* A long-service able seaman; one with good-conduct stripes.

1942 *Gen* 1 May 42/1 'What's the buzz, Stripey?'.. 'Dunno,' curtly responded the Bosun's Mate. **1945** 'TACKLINE' *Holiday Sailor* i. 8 Stripey was a small, middle-aged A.B. **1977** [see *RATE *sb.*[1] 9 c].

2. *Afr.* Also **stripie.** = *STREPIE.

1964 A. TREW *Smoke Island* viii. 133 Ezekiel was fishing in a deep pool; at his side were four or five mullet and as many 'stripeys'. **1969** *Guardian* 8 Mar. 7/6, I glimpsed his outstretched pectoral fins for a moment and saw the flash of purple. He was a stripie.

stripiness (strəi·pinės). [f. STRIPY *a.* + -NESS.] The condition of being stripy.

1958 tr. *Herberts' Artists' Techniques* III. 273 The tendency to stripiness can be avoided by stippling the paint on with a sponge. **1960** *Textile Terms & Definitions* (Textile Inst.) (ed. 4) 144 *Stripiness* (warp knitting), longitudinal defects caused by yarn variation or structural distortion in warp-knitted fabric.

strippable (stri·pāb'l), *a.* [f. STRIP *v.*[1] + -ABLE.] **1.** Of a coating: capable of being stripped off or removed.

1950 *Effects Atomic Weapons* (U.S. Sci. Lab., Los Alamos) x. 333 Another possibility in connection with protection against radioactive contamination is to use strippable coatings. **1980** *Daily Tel.* 16 Dec. 11/4 The wallpapers are trimmed, washable and strippable.

2. *U.S.* Of a mineral deposit: capable of being strip-mined.

1975 *N.Y. Times* 24 Mar. 20/2 Their entrepreneurial thrust to exploit what industry spokesmen call the West's 'Persian Gulf' of strippable coal. **1978** *Time* 17 Apr. 74/3 When they do [legislate], Western strip-mine owners fear, up to 80% of the region's strippable tonnage will be ruled off limits.

stripped, *ppl. a.* Add: **a.** (Further examples. Cf. *STRIP *v.*[1] 23*, 23**.)

1931 HOFFERT & CLAXTON *Motor Benzole* iv. 61 After the removal of the benzole in the scrubbers, the gas is usually referred to as stripped gas. **1933** O. H. BLACKWOOD et al. *Outl. Atomic Physics* xiv. 305 The electrons and the more or less stripped nuclei together are assumed to form a gas, which acts like a perfect gas even at huge densities. **1947** *Physical Rev.* LXXII. 1008/2 It is just the narrowness of the cone which distinguishes the stripped neutrons from those produced by direct nuclear encounters. **1978** *Nature* 7 Sept. 41/2 Stripped iron nuclei in a hydrogen plasma under central solar conditions, according to the classical Debye—Hückel model, would undergo phase separation for concentrations well below the cosmic abundance value.

b. *spec.* Of wood (esp. pine) used for furniture or domestic woodwork, etc.: that has had the accretions of paint or varnish removed, so as to reveal the natural grain and colour.

1934 M. ALLINGHAM *Death of Ghost* ix. 105 The high narrow room with its top lights and stripped pine panelling. **1966** A. CHRISTIE *Third Girl* iv. 31 Long Basing.. had two antique shops, one mostly consisting of stripped pine chimney pieces. **1976** *Listener* 15 July 49/2 He likes corner cabinets and stripped pine. **1976** *Lancs. Even. Post* 7 Dec. 14/5 (Advt.), Chairs include set eight stripped beech. **1981** 'M. YORKE' *Hand of Death* xx. 189 A customer called. Lynn sold her a stripped-pine chair.

c. *stripped-down* adj., that has had all superfluous or extraneous parts removed; also *fig.* Esp. (orig. *U.S.*) applied to a motor vehicle so adapted in order to improve engine performance. Also of a machine: disassembled, dismantled. Cf. STRIP *v.*[1] 8 in Dict. and Suppl.

1946 [see *HOPPED *a.* 3]. **1958** *Times* 24 Nov. (Canada Suppl.) p. iv/4 In 1959, General Motors will introduce a small, stripped-down Chevrolet, which will be billed as a 'new' small car. **1961** R. B. LONG *Sentence & its Parts* 494 'Kernels.' This term is applied to stripped-down nucleuses [of sentences]. **1973** A. MACVICAR *Painted Doll Affair* vi. 66 A pimpled hairy youth in overalls wriggled out from underneath a stripped-down car. **1975** *New Yorker* 7 July 78/3 'Blue Lou' is short, stripped-down, and full of business. **1978** *Archit. Design* 5 June 310/2 The stripped-down classicism promoted by men [*sc.* architects] like Burnet and Richardson. **1979** J. GARDNER *Nostradamus Traitor* xlix. 237 They travelled in a stripped-down Heinkel 111. **1980** J. CARTWRIGHT *Horse of Darius* v. 66 He laid out the stripped-down Kalashnikov, the plasticine, the detonators.

stripper[1]. Add: **1. b.** *colloq.* (orig. *U.S.*). A performer of strip-tease.

1930 *Variety* 3 Dec. 54/5 (*heading*) Detroit censor pinches four stock strippers. **1945** P. CHEYNEY *I'll say she Does!* ii. 42, I was a stripper one time... I had a feature spot in the programme. **1950** *Manch. Guardian Weekly* 13 July 13/2 A couple of famous strippers and a beloved Broadway comedian appeared in something called 'Wine, Women, and Song'. **1960** *New Left Rev.* Mar.–Apr. 45/2 The calculated obscenity of the stripper's act. **1972** C. WESTON *Poor, Poor Ophelia* v. 27 She unzipped her.. mini-dress, and with the grace of a professional stripper, stepped out of it. **1980** *Times* 21 Oct. 12/6 (*caption*) Why should it only be male strippers who are subsidized?

2. b. *Oil Industry.* A still for the fractionation of oil or oil products.

1930 H. S. BELL *Amer. Petroleum Refining* (ed. 2) xii. 222 Reboiling coils are sometimes used as strippers. **1938** A. E. DUNSTAN et al. *Sci. of Petroleum* II. II. xxv. 1559/2 Reheaters, when used, usually have been placed on the oil stream flowing from one of the lower plates of the stripper. **1961** D. PETRIE *Petroleum* xi. 64 In order to make the fractionation process even more exact, each fraction when it leaves the tower is led into another smaller fractionating tower called a 'side stripper'. **1975** W. G. ROBERTS *Quest for Oil* (rev. ed.) viii. 85 The purpose of the stripper is to remove small amounts of vapour from the liquid entering it.

c. *Physics.* In a particle accelerator, a section containing metal foil or gas at a high positive potential, which removes electrons from the ions passing through it. Also *attrib.*

1959 *Nuclear Instruments & Methods* IV. 123/2 The image size and beam divergence required at the object point for the accelerator are limited by the dimensions of the stripper tube. **1974** J. B. A. ENGLAND *Techniques in Nuclear Structure Physics* I. iii. 268 There is usually little to choose in final beam intensity between foil strippers and gas strippers for hydrogen and helium ions. **1979** *Sci. Amer.* Apr. 47/2 The aiming mechanism would first direct the negative-hydrogen beam at the target and then pass it through a gas 'stripper' for neutralization.

3. (Earlier example.)

1843 J. H. GREENE *Exposure Arts & Miseries Gambling* (1845) 121 When cards are prepared as I have above described, they are called *strippers*.

4. a. A bleaching agent or solvent used to remove colour from fabrics before re-dyeing. Cf. STRIP *v.*[1] 21.

1909 A. MORRIS in Rothery & Edmunds *Mod. Laundry* II. xl. 150 Permanganic acid in the form of its potash or soda salt is a powerful oxidizing stripper. **1957** *Woman* 16 Nov. 25/4 Light over dark won't go, but if you want to try it you'll need a colour 'stripper'.

b. A chemical preparation used to remove paint, varnish, etc., from a surface; *paint stripper*: see *PAINT *sb.* 6.

1937 A. JONES *Cellulose Lacquers, Finishes & Cements* xii. 241 Wax has been objected to as a thickening agent as it settles out and the paint remover or stripper requires frequent and repeated shaking. **1949** C. H. HAYWARD *Woodworker's Pocket Bk.* 4 Many proprietary strippers are now available. **1979** C. CURZON *Leaven of Malice* ii. 22, I had bought..cleaning agents, strippers, filler for the cracks..in the plaster.

stripper[2]. (Later example.)

1917 L. A. KLEIN *Princ. & Practice Milk Hygiene* i. 5 Cows in which lactation is about to cease are called 'strippers'.

stripper[3] (stri·pəɹ). *Oil Industry.* [? f. STRIPPER[2], by analogy with a low-yielding milk cow.] More fully, *stripper well.* An oil well in which production has dwindled to only a few barrels a day.

1930 *Oil & Gas Jrnl.* 18 Dec. 48/2 The proration committee is working out a plan safeguarding the interest of operators owning small and shallow producing wells... Nothing will be done that might cause the abandonment of the stripper wells. **1931** *Ibid.* 8 Jan. 26/1 There would apparently be nothing to do but to abandon these old strippers as having exceeded their useful life. **1976** *Watertown* (S. Dakota) *Public Opinion* 6 July 7/5 Throughout the United States marginally productive oil wells known as 'strippers' account for more than 13 percent of domestic production. **1980** *Fortune* 24 Mar. 64/1 Independents will also get a break on oil from 'stripper' wells, which produce ten or fewer barrels a day.

strippeuse (stripǝ̄·z). *joc.* [An alteration of *STRIPPER[1] 1 b, perh. after *danseuse.*] A (female) performer of strip-tease.

1939 *Life* 2 Jan. 15/2 Last year blonde Della Carroll was première strippeuse at New York's Paradise restaurant. **1960** [see *IN *a.* 2 a]. **1962** *Times Lit. Suppl.* 26 Oct. 827/5 An eminent 'strippeuse' at Las Vegas.

stripping, *vbl. sb.*[1] Add: **1. a.** (Later examples.)

1952 W. H. KIRK *Sewell's Dogs & their Management* (rev. ed.) iv. 76 Proper trimming or stripping is a long, tedious, and continuous work. **1974** R. RENDELL *Face of Trespass* xiv. 134 Mr Greenberg doesn't have a surgery on Saturday afternoons... We're only open for clipping and stripping.

b. (Further examples. Cf. *STRIP *v.*[1] 23*, 23**, 23***.)

1922 D. T. DAY *Handbk. Petroleum Industry* II. 324 Stripping may be held to mean a complete removal of all light fractions down to those of lubricating value, and represents the initial refining process. **1943** *Ann. Reg. 1942* 364 Saha..suggested that both the stripping of the atoms and their high velocities are due to a nuclear reaction analogous to fission. **1949** MELCHER & LARRICK *Printing & Promotion Handbk.* 283/2 When two or more photographic negatives are used together to make one printing plate, the process of combining them is known as 'stripping'. **1955** [see *DESORPTION]. **1972** *Physics Bull.* Mar. 145/1 After stripping the nitrogen ions..will behave like deuterons or alpha particles in the injection channel into the synchrotron. **1977** L. VÁLYI *Atom & Ion Sources* iv. 251 An alternative method for the production of multiply charged ions is the stripping of electrons in the interaction of singly charged positive ions. **1980** J. R. WALKER *Graphic Arts Fund.* vi. 116/1 (*caption*) Light table makes good work surface for stripping.

3. *stripping agent, bench, cement, column, desk, foil, operation, still*; **stripping-bill(e)**, a bladed implement or bill used in besom-making; **stripping-knife**, (*b*) a knife used in the stripping of wallpaper or paint from surfaces.

1958 M. G. LARIAN *Fund. Chem. Engin. Operations* xiii. 495 In desorption dissolved gases are removed from a liquid by contacting the liquid with a suitable gas (the desorbing or stripping agent), or a volatile liquid is separated from a relatively nonvolatile solvent. **1967** E. CHAMBERS *Photolitho-Offset* v. 51 A stripping bench, layout and lining tables are essential... The former provides the illuminated working surface on which the various images can be positioned. **1968** J. ARNOLD *Shell Bk. Country Crafts* v. 97 The cuttings are prepared for use with chopping and stripping-billes for the coarse and finer work. **1974** P. W. BLANDFORD *Country Craft Tools* ii. 38 The besom broom maker called his general-purpose tool a 'chopping bill' and had a lighter one with more curve to the point called a 'stripping bill'. **1967** E. CHAMBERS *Photolitho-Offset* v. 50 With face-up stripping a stripping cement is first applied to the base support. **1930** H. S. BELL *Amer. Petroleum Refining* (ed. 2) xii. 222 If withdrawn as a finished product, the light ends must be removed and this is usually accomplished in a separate small stripping column. **1967** E. CHAMBERS *Photolitho-Offset* v. 55 (*caption*) Bench-type illuminated stripping desk. **1972** *Physics Bull.* Mar. 144/3 An aluminium stripping foil of density 40 μg cm[-2] was therefore used at the high energy end of the Linac to convert the particles to [14]N[7+]. **1927** W. DEEPING *Kitty* xx. 254, I could lend you a plank and a couple of step-ladders, and a stripping-knife. **1951** *Good Housek. Home Encycl.* 210/1 A paper-hanger's stripping knife, which is a flat, fairly flexible, steel-bladed knife. **1948** R. R. KARCH *Graphic Arts Procedures* viii. 232 (*heading*) The stripping operation. **1980** J. R. WALKER *Graphic Arts Fund.* vi. 116/2 (*caption*) Stripping operations require sharp knives, scissors and brushes for opaquing pinholes. **1931** HOFFERT & CLAXTON *Motor Benzole* viii. 225 The function of a modern stripping still is to remove the remaining benzole from the hot oil leaving the preheater.

strippy, *a.* Delete *rare* and add later examples.

1963 *Times* 20 Apr. 11/6 Veneered with carefully matched figured or burr walnut,..with drawer fronts, fall flaps and door framing straight or herringbone cross-banded with strippy walnut. **1978** *Detroit Free Press* 16 Apr. (Detroit Suppl.) 20 (Advt.), Strippy, strappy stripes of canvas [sandals], wedged high, to put spring snap in your steps.

strip-tease. *colloq.* (orig. *U.S.*). Also **strip tease, striptease.** [Back-formation from next.] **1.** A kind of entertainment in which a female (occas. a male) performer undresses gradually in a tantalizingly erotic fashion before an audience, usu. to music; an instance of this.

[**1930** *Variety* 1 Oct. 49/4 Girls have the strip and tease down to a science.] **1936** *Variety* 2 Dec. 70/5 An undersea ballet, veil waving number and a mild strip tease by the entire chorus, which required little feeling, were nicely executed. **1937** *Daily Tel.* 29 Apr. 22/2 Can anything be said in defence of the present public interest in 'strip-tease' and nudist or semi-nudist displays on stage? **1943** *Scrutiny* XI. 286 The business-cum-Riviera set of which he is the representative in fiction (on the stage—strip-tease) were very pally with Goering and Co. **1960** *News Chron.* 23 Sept. 10/2 Strip-tease..can be banal. **1978** G. GREENE *Human Factor* II. iii. 75, I thought dinner in the Café Grill and afterwards a spot of strip-tease.

2. In *transf.* and *fig.* use.

1937 *Hansard Commons* 20 Apr. 1623 We had a display of what I believe is now known as 'strip-tease', in which we were kept in tantalising expectation of what was to come. **1956** E. LINKLATER *Dark of Summer* iv. 62 The whole female art of novel-writing—is an exquisitely prolonged strip-tease. **1969** I. & P. OPIE *Children's Games* 13 They snatch the girls' ties or hair ribbons and call it 'Strip Tease'. **1982** J. O'FAOLAIN *Obedient Wife* i. 26 'Do you feel I owe you a confidence?' 'No..if I come here to do a strip-tease, it doesn't mean *you* have to.'

3. In *attrib.* use. **strip-tease artist**, a performer of strip-tease.

1936 *N.Y. Post* 15 Sept. 13/1 Gypsy Rose Lee is at once the Bernhardt, the Duse and the Joan Crawford of Strip-Tease girls. **1939** A. HUXLEY *After Many a Summer* I. vi. 71, A strip-tease dancer in a Western mining-camp. **1944** 'G. ORWELL' in *Horizon* X. 237 A strip-tease act. **1947** H. A. SMITH *Low Man on Totem Pole* viii. 68 Miss Lee turned to a paragraph in the magazine in which Henry L. Mencken was represented as having coined a word to describe a strip-tease artist. **1953** C. DAY LEWIS *Italian Visit* i. 25 Whoever would master the truth by which your provocative, charming Strip-tease universe lives. **1958** N. MARSH *Singing in Shrouds* v. 101 That damn' spiritual striptease session. **1968** P. OLIVER *Screening Blues* vi. 251 As the strip-tease artist compares with the artist's model, so the seductive effects of slow unveiling are more stimulating erotically than the starkly naked. **1979** C. MACLEOD *Family Vault* xi. 71 Does it disgust you..that your..husband once made a fool of himself over a striptease dancer?

Hence as *v. intr.*, to perform a strip-tease act; **strip-teaseuse**, joc. alteration of *STRIP-TEASER (cf. *CHANTEUSE, *STRIPPEUSE, etc.); **strip-teasing** *vbl. sb.* and *ppl. a.*; **stripteuse** = *strip-teaseuse* above.

1937 G. FRANKAU *More of Us* 185 Dalliest thou, strip-teasing and beachcombing, On some far southern beach of Gallic joy. **1937** *Variety* 31 Mar. 69/1 Kraus is accused by John S. Sumner, head of the vice society, of permitting strip-teasing in his show. **1941** *Sun* (Baltimore) 8 Mar. 20/2 (*caption*) Strip-teaseuses Betty Coette..and Winnie Garrett. **1942** *Time* 28 Sept. 40/3 Gipsy Rose Lee, strip-teuse turned woman of letters. **1951** *Sun* (Baltimore) 27 June 30/3 A blond stripteuse was arrested at a Silver Hill night spot. **1957** *Times Lit. Suppl.* 11 Oct. 611/1 Little Rose Louise was..new to burlesque, and able to gasp at strip-teasing Flossie. **1958** *Listener* 18 Sept. 418/2, I.. drove to a night club, where a girl stripteased while lashing a whip. **1960** *News Chron.* 22 Sept. 3/1, I have given up strip-teasing to be with my husband. **1962** *Guardian* 23 Feb. 9/4 A strip-teasing woman. **1977** J. MITFORD *Fine Old Conflict* vii. 118, I was temporarily in despair, hoping against hope that something would turn up. It did, in the shape of a former stripteaseuse whom I had met at a PW party.

strip-teaser. orig. *U.S.* [f. STRIP *v.*[1] + *TEASER[1] 2 g.] A performer of strip-tease; an ecdysiast or stripper.

1930 *Variety* 26 Nov. 40/2 The main b[ox] o[ffice] lure is the girls, those known as 'strip teasers'. **1935** E. E. CUMMINGS *Let.* 29 Jan. (1969) 135, I recommend the Irving Place Burlesk (stripteasers in excelsis). **1952** *Chambers's Jrnl.* Feb. 71/1 There was not a pinpoint of light anywhere, not a glimmer from the long line of caravans parked round the square, where the Parisian Strip-teasers slept beside their watchful mothers. **1960** *News Chron.* 25 Feb. 5/8 They got Trixie Kent, our strip-teaser, to autograph one or two scraps of clothing. **1982** *Washington Post* 22 Jan. (Weekend section) 47/1 Despite its new soigné image, the essential pulse of Baltimore remains the rhythmic tic-toc of a stripteaser's tassels.

stritch (stritʃ). Also **strich.** [Origin uncertain.] A musical instrument resembling a straightened alto saxophone. (Chiefly associated with the American jazz musician Roland Kirk, b. 1936.)

1960 *Downbeat* 4 Aug. 13/1 Kirk haunted music stores, examined all kinds of antique instruments, many of them remnants of the 19th century... When he found what he was looking for..they weren't saxophones at all... One was a stritch, the other a manzello. **1962**, etc. [see *MANZELLO]. **1969** *Punch* 12 Mar. 393/2 The manzello and strich..sound fine as solo instruments or as sudden quasi-orchestral interludes in a tenor solo. **1977** *Time* 19 Dec. 53/1 Kirk played the manzello (a quasi-saxophone), the stritch (a horn resembling a dented blunderbuss) and the tenor sax together.

strivingly, *adv.* (Later example.)

1890 W. JAMES *Princ. Psychol.* II. xxi. 315 The impulse to take life strivingly is indestructible in the race.

strobe (strōᵘb), *a.* and *sb.* [f. first syllable of STROBOSCOPIC *a.* and related words.] **A.** *adj.* **1.** = STROBOSCOPIC *a.*

1942 *Amer. Cinematographer* Sept. 422/3 Adjust the speed of the projector until the bars on the 'strobe' band corresponding to the number of blades in the projector shutter appear to stand still. **1949** H. LURAY *Strobe* viii. 110 (*caption*) A classic strobe shot. Drop of milk splashing on a plate covered by a thin layer of milk. **1962** *Amer. Jrnl. Physics* XXX. 925/2 Strobe photography is an area that has not been saturated at science fairs. **1966** *McGraw-Hill Encycl. Sci & Technol.* XIII. 187/1 Stroboscopic or 'strobe' photography is generally understood to refer to pictures of both single and multiple exposure taken by flashes of light from electrical discharges. **1978** *Oceans* May–June 39/1 Flash bulbs or strobe (electronic flash) lighting is often used to restore lost colors and provide light for an exposure.

2. Special collocations: *strobe disc*, a disc with alternate light and dark sectors of equal size for checking the speed of rotation of something, its appearance being steady only when this speed is related in a definite way to the periodicity of the illumination; *strobe lamp, light*, an electric light that can be made to flash on and off rapidly and automatically; also (*U.S.*), an electronic flash for a camera; so *strobe-lighted, -lit* adjs.; *strobe pulse* (Electronics) (see quot. 1971).

1942 *Amer. Cinematographer* Sept. 423/1 Check the strobe disc for synchronization. **1967** *Nature* 23 Dec. 1173/1 The correct speed of a gramophone turn-table may be checked by viewing a strobe disk illuminated by electric light supplied by the a.c. mains. **1974** *Sci. Amer.* Aug. 108/3 The solution of the problem of stopping fast action lies in an electronic flash lamp. The duration of most 'strobe' lamps owned by amateurs is about 5 × 10[-4] second, which is sufficiently brief for many events of interest. **1975** *Gramophone* Jan. 1297 (Advt.), Dots on the outer rim of the SR717 are illuminated by the built-in strobe lamp. **1962** N. MAXWELL *Witch-Doctor's Apprentice* vi. 68 Augusto helped a lot, lugging the Strobe light and holding his flashlight so that I could check my camera settings. **1971** R. BUSBY *Deadlock* vii. 98 Pulsating strobe lights left Leric with a fragmented picture of girls. **1978** *Sci. Amer.* June 128/1 My best clue came from examining the floating drops under a strobe light set near the frequency of oscillation, so that the flashing light effectively slowed the vibration of the drops. **1979** *Listener* 18 Oct. 517/3 A strobe-lighted production number that sends images flashing around the vast auditorium. **1967** P. WELLES *Babyhip* ii. 36 It's not everyone who can have a strob-lit [*sic*] dream of love. **1972** *Listener* 23 June 845/1 Weird strobe-lit collages. **1946** *Jrnl. Inst. Electr. Engineers* XCIII. IIIA. 318/2 The operator had under his control a variable-range strobe pulse which he could bring into coincidence with the target echo. **1967** *Electronics* 6 Mar. 117/1 (*caption*) The level detector senses the moment when C_{int} reaches zero volts and generates a strobe pulse. **1971** *Gloss. Electrotechnical, Power Terms* (*B.S.I.*) III. vi. 18 *Strobe pulse*, a pulse, of duration less than the period of a recurrent phenomenon, used for scrutinizing a particular epoch of that phenomenon... In radar, a strobe pulse is sometimes made to follow automatically the echo from a moving object.

B. *sb.* **1.** = *strobe disc*, sense *A. 2 above.

1942 *Amer. Cinematographer* Sept. 423/1 Watch the strobe for any changes in projector speed.

2. *Electronics.* = *strobe pulse*, sense *A. 2 above.

1946 *Jrnl. Inst. Electr. Engineers* XCIII. IIIA. 319/1 In the A.I. Mark IV [radar] equipment, the search was a progressive outward movement of the strobe from zero to maximum range, followed by a rapid fly-back to zero. **1953** *Electronic Engin.* XXV. 191/1 Photographing a C.R.O. trace using a reference waveform as a strobe. **1959** *Ibid.* XXXI. 136/2 The *X* displacement may be dissociated from the phase of the strobe (time scale). **1980** D. G. GREEN *Digital Techniques & Systems* iv. 49 The strobe determines the times at which the *S* and *R* input signals should be effective.

3. a. = *strobe light*, sense *A. 2 above. **b.** Stroboscopic photography.

1949 H. LURAY *Strobe* v. 73 You will know which size flash bulb your strobe approximates. *Ibid.* vii. 101 Strobe is used outdoors very much like flash. **1962** *Amer. Jrnl. Physics* XXX. 925/1 The strobe was flashing at three times the frequency of the wave. **1968** J. D. MACDONALD *Pale Grey for Guilt* (1969) xii. 153 Maybe the music got too loud... Maybe it was the strobes. **1975** J. RATHBONE *Kill Cure* II. vi. 48 For one brief moment she saw the man full face, caught in the jerky light of the strobe.

strobe (strōᵘb), *v.* [f. prec.] **1.** *trans. Electronics.* To gate (*GATE *v.*[1] 2 a) by means of a strobe pulse.

1947 *Wireless World* Aug. 290/1 Work was also carried out on 'strobing' a portion of the time base. **1950** [see *GATE *v.*[1] 2]. **1981** NASHELSKY & BOYLESTAD *Devices* xi. 391 The output remains high unless strobed.

2. *intr. Cinemat.* and *Television.* To exhibit or give rise to strobing.

1959 HALAS & MANVELL *Technique Film Animation* xix. 232 Fairly light colours do not strobe so much as white. **1965** *Punch* 12 May 684/1 Before my very first TV appearance the studio manager tried to get me to change clothes with him... He then explained that the black and white check I was wearing would 'strobe'—an optical illusion whereby the wearer of any pattern containing vertical or horizontal stripes appears to vibrate. **1982** *Observer* 12 Dec. 26/3 Don't blink, don't sniff, don't stick your chin up, don't slouch, don't wave your hands about, don't wear stripes (they 'strobe'), look at the person you're talking to, *smile.*

3. *fig.* (*intr.* and *trans.*) To flash.

1977 R. E. HARRINGTON *Quintain* vi. 53 Possible explanations flashed through his mind..thoughts strobing across his mind. **1980** J. MCNEIL *Spy Game* ix. 103 The fire strobed rosy light onto the burnished yew furniture.

Hence **strobed** *ppl. a.*

1980 D. C. GREEN *Digital Techniques & Systems* iv. 49 A clocked or strobed flip-flop will change only when a clock pulse is received.

strobilanthes (strobilæ·nþīz). [mod.L. (K. L. Blume *Bijdragen tot de Flora van Nederlandsch Indië* (1826) 781), f. STROBILUS + Gr. ἄνθος flower, in reference to the shape of the young inflorescence.] A herb or sub-shrub of the genus *Strobilanthes*, belonging to the family Acanthaceæ, native to tropical Asia, and bearing clusters of blue or white tubular flowers.

1836 *Curtis's Bot. Mag.* LXIII. 3517 (*heading*) Mr. Sabine's strobilanthes. **1918** R. N. PARKER *Forest Flora Punjab* 386 In most floras all the Strobilanthes are described as shrubs but I have not found the Punjab species shrubby. **1944** J. CORBETT *Man-Eaters of Kumaon* (1946) 17 A bed of strobilanthes, the bent stalks of which were slowly regaining their upright position, showed where.. the tigress had passed. **1979** A. J. HUXLEY *Success with House Plants* 375/1 A strobilanthes cannot tolerate temperatures below about 55°F.

strobing (strōᵘ·biŋ), *vbl. sb.* [f. *STROBE *v.* + -ING[1].] **1.** The action of *STROBE *v.* 1.

1959 *Electronic Engin.* XXXI. 130/2 Strobing consists of sampling a repetitive waveform at what to all intents and purposes is an instant in the repetition period. **1981** NASHELSKY & BOYLESTAD *Devices* xi. 390 Figure 11.5 shows how a 311 comparator can be used with strobing.

2. a. *Cinemat.* Jerkiness in what should be a smooth movement on the screen. **b.** *Television.* An irregular movement and loss of continuity sometimes seen in lines and stripes in a television picture.

1959 HALAS & MANVELL *Technique Film Animation* xix. 231 When objects are panned through the screen area, strobing always appears to be worst (on standard 5 inch field) when the distance of movement is between ·125 and ·500 inch per frame. **1961** G. MILLERSON *Technique Television Production* iii. 52 Where..tiling, venetian blinds, etc., are seen at such a distance that their surface appears as close horizontal lines, line-beating or strobing will occur. **1973** R. DOUGALL *In & Out of Box* xxiii. 280 Ties..could be noisy or even loud, so long as they had no horizontal or diagonal patterns which caused 'strobing'. **1980** *Radio Times* 25 Nov.–5 Dec. 102/2 There are a few rules about television clothes... Some patterns can make the picture go all funny—what's known as strobing.

strobo- (strōᵘ·bo), formative element f. the first syllable of STROBOSCOPE, etc., as in **stro·botorch**, a light source designed to give very brief flashes of light at a known rate; **stro·botron [*-TRON]**, a gas-filled cold-cathode discharge tube used as a strobotorch, the flashing rate being determined by the frequency of the voltage applied to a control grid.

1951 *Electronic Engin.* XXIII. 187/3 It is..possible to link the flashing rate of the strobotorch to an electrical contractor..fitted on the mechanism. **1970** *Nature* 15 Aug. 731/1 A sheet of white translucent plastic was placed between the geometrical figure and the strobotorch. **1937** GERMESHAUSEN & EDGERTON in *Electronics* Feb. 12/1 In this paper a tube, named the Strobotron, is described which has been developed primarily for producing stroboscopic light. **1949** *Jrnl. R. Aeronaut. Soc.* LIII. 460/1 The cathode ray tubes, amplifiers, and strobotron lamps, together with their power supplies, are built into one case. **1966** *McGraw-Hill Encycl. Sci & Technol.* XIII. 190/1 A thyratron or a strobotron can be used as a trigger tube in place of the switch *S*.

stroboscope. Add: Also with pronunc. (strōᵘ·bǒ-).

stroboscopic, *a.* Add: Also with pronunc. (strōᵘbǒ-). Also, involving or pertaining to rapid flashes of light. (Further examples.)

1949 H. LURAY *Strobe* i. 22 Real stroboscopic lighting equipment with ample light output can do wonderful things. **1959** L. A. MANNHEIM *Successful Flash Photogr.* 130 (*heading*) Stroboscopic shots. *Ibid.*, A stroboscopic flash outfit. **1971** *Nature* 11 June 397/2 Lately there has been a fad in this area for dances to be held under stroboscopic lighting. **1980** J. W. HILL *Intermediate Physics* iv. 35 Stroboscopic photography (where photographs are taken by a camera at intervals of fractions of seconds on the same film) shows this.

Hence **strobosco·pically** *adv.*, by means of a stroboscope or stroboscopic illumination.

1919 *Proc. Nat. Acad. Sci.* V. 174 All measurements were made by the Michelson interferometer viewed stroboscopically. **1932** *Phil. Mag.* XIII. 163 The nature of the instability induced by sound in smoke jets has been examined stroboscopically and by means of photography. **1951** *Electronic Engin.* XXIII. 430/1 The valve structure may be observed stroboscopically while subjected to vibration and the offending portion detected visually. **1977** D. GOLDSTEIN et al. *Test System for Evaluations of Armors using Duplicate Fragments* (U.S. Patent 4,044,599), By using stroboscopically-controlled illumination, the high-speed flight of the fragments can be captured on one photograph.

stroboscopy (strob-, strǒbǫ·skǒpi). [f. STROBOSCOPE + -Y[3].] **a.** The use of stroboscopic techniques or apparatus. **b.** Stroboscopic effects.

1932 *Sun* (Baltimore) 17 Sept. 16/4 Such effects are produced through the principle of stroboscopy, applications of which already make possible the photography of rapidly whirling objects. **1971** CHIN-WU KIM in W. O. Dingwall *Survey of Linguistic Sci.* 22 A slowed-down or stationary appearance in stroboscopy is due to an electro-mechanical apparatus of the stroboscope. **1980** S. L. LYONS *Exterior Lighting* ii. 31 A special effect of flicker is that termed stroboscopy in which rotating or rhythmically moving objects may give the illusion of being stationary.

stroganoff (strǫ·gānǫf). Also **stroganov, strogonoff** and with capital initial. [a. Fr., f. the name of the 19th-cent. Russian diplomat Count Paul *Stroganov.*] A dish of strips of beef cooked in a sauce containing sour cream. In full, *beef stroganoff, bœuf stroganoff.*

1932 A. HEATH *Good Food* 30 Beef *Strogonoff*..slices from a fillet of beef..cut..into shortish, thin strips.. onions and mushrooms..cook..in butter..add..thick sour cream. **1937** 'COUNTESS MORPHY' *Kitchen Library* IV. 50 (*heading*) Bœuf Strogonov (Russian). **1944** E. M. ALMEDINGEN *Dasha* vi. 235 All of it sounds as tidy as pieces of beef in the Strogonov dish. **1955** *Good Food Guide 1955–56* 341 Steak maison, tournedos Rossini and bœuf Stroganoff. **1961** *Listener* 28 Dec. 1107/2 In Leningrad's smartest restaurant I ordered beef strogonoff. **1964** 'D. SHANNON' *Death-Bringers* (1966) vi. 72 It's Beef Stroganov, and it's hot. **1969** D. LAMBERT *Angels in Snow* ii. 48 'What's for dinner?' Harry asked. 'Strogonoff,' said his wife. **1980** A. N. WILSON *Healing Art* xiv. 157 Did you get Gale to fix you..her strogonoff, followed by bilberry strudel?

Stroh (strōᵘ). The name of the inventor Charles *Stroh* (fl. 1901) used *attrib.* to designate stringed instruments (chiefly of the violin class) having an aluminium plate and horn attachment in place of a wooden body, formerly used for recording purposes.

1902 *Encycl. Brit.* XXXI. 766/1 The recent invention of the Stroh violin. **1923** [see *PHONOFIDDLE]. **1934** C. LAMBERT *Music Ho!* iv. 257 The old pre-electric horn recording, with its euphoniums instead of 'cellos, and its handful of Stroh violins. **1947** F. W. GAISBERG *Music on Record* vi. 79 Stringed instruments we recorded by a subterfuge. We substituted the Stroh violin for violins and violas. **1979** *Oxf. Jun. Compan. Mus.* (ed. 2) 314/1 *Stroh violin*, a violin that has an aluminium plate and a trumpet bell instead of the normal wooden body. Invented in 1901 by Charles Stroh for use in the early recording studios, where primitive microphones could only pick up sound aimed directly at them (which..the trumpet bell could do). There are also Stroh violas, cellos, mandolins, and guitars.

stroke, *sb.*[1] Add: **13. f.** *to put* (someone) *off* (his) *stroke*, to distract (someone) from his course of activity; to disconcert or disturb. *colloq.*

a **1914** JOYCE *Stephen Hero* (1944) xx. 103 Besides girls praying put me off my stroke. **1922** — *Ulysses* 285 Put you off your stroke. **1965** J. GALE *Clean Young Englishman* IV. 167 The note put me right off my stroke. I was trying to tell the audience what the war in Algeria was really like... But somehow I never finished what I wanted to say. **1977** R. PERRY *Dead End* iii. 41 She must have..seen the bodies..but it didn't put her off her stroke at all.

14. d. *to pull a stroke*, to play a dirty trick. Cf. *PULL *v.* 19 e. *slang.*

1970 P. LAURIE *Scotland Yard* 293 *Pull a stroke, to*, to play a dirty trick. **1974** J. MCVICAR *McVicar* I. II. i. 109 It would be wrong to let Charlie go... He's pulled too many strokes.

17. b. For ? *nonce-use* read *colloq. rare* and add later example.

1914 JOYCE *Dubliners* 153, I don't say Hynes.—No, damn it, I think he's a stroke above that.

d. In *Telegraphy*, the name of the signal for an oblique stroke. Now usu. *colloq.*, a spoken representation of a solidus. Freq. used as *conj.* to indicate or stress alternatives: or else, alternatively.

1884 W. LYND *Pract. Telegraphist* i. 27 The oblique stroke is to be signalled 'stroke', thus—'FI three stroke five FF', meaning 3/5 (three shillings and fivepence). **1965** M. ALLINGHAM *Mind Readers* xv. 153, I have my own feel, of course, which would be 'glad stroke laughing-at' in his case. **1971** J. YARDLEY *Kiss a Day keeps Corpses Away* ii. 39 The Truman stroke Eisenhower regime. **1974** G. MARKSTEIN *Cooler* xlvii. 171 ABPQ stroke 113 stroke 1. Ah yes. Is that your national registration number? **1977** N. J. CRISP *Odd Job Man* iii. 28 One dozen cardigans, stroke thirty-three, blue, for knitwear.

e. *spec.* in *Logic* = *Sheffer('s) stroke* s.v. *SHEFFER.

1925 WHITEHEAD & RUSSELL *Principia Math.* (ed. 2) I. p. xvi, The symbol '*p*|*q*' is pronounced: '*p* stroke *q*'... All the usual truth-functions can be constructed by means of the stroke. **1952** R. L. WILDER *Introd. Found. Math.* ix. 220 Since P[rincipia] M[athematica] was first published, with its two undefined symbols (or 'logical constants')..it has been shown that *one* undefined symbol..[is] sufficient. The symbol referred to is | and is called 'stroke'. **1975** P. K. BASTABLE *Logic* 180 Later he [*sc.* Russell] preferred to become acquainted with implication through defining it as 'Either not *p* or *q*' or, like Sheffer and Nicod, in terms of the stroke functor.

24*. *Basket-making.* A single movement analogous to a stitch in sewing or knitting; the result of this.

1912 T. OKEY *Art of Basket-Making* vii. 59 Where an even number of pairs of leagues or sticks is used, the centre strokes lie alternate, and not side by side. *Ibid.* 154 *Stroke*, any complete movement in basket-work: analogous to a stitch in needlework. **1960** E. LEGG *Country Baskets* iii. 29 And now for the strokes, which are comparable with 'stitches' in knitting... The strokes are indeed simple and few.

25. c. stroke-maker *Cricket*, a batsman who plays attractive, attacking strokes; hence **stroke-making; stroke-ornamented** *a. Archæol.* (see quot. 1970); **strokeplay** *Cricket*, the playing of attractive, attacking strokes; hence **stroke-player.**

1927 *Observer* 5 June 21/5 A beautiful stroke-maker, he [*sc.* H. W. Austin] is pleasant to watch. **1976** J. SNOW *Cricket Rebel* 84 Nurse could be a brilliant and savage stroke-maker on his day, but could graft when necessary. **1956** R. ALSTON *Test Commentary* iii. 23 One of the features of the morning's play was the stroke-making of.. Van Geloven. **1977** *World of Cricket Monthly* June 87/1 He impressed the Lord's gathering with his crisp stroke-making. **1925** V. G. CHILDE *Dawn Europ. Civilization* xii. 172 The second ware may..be called stroke-ornamented pottery. Its forms are rather more angular than those of spiral-meander pottery. *Ibid.* xviii. 272 Hut foundations..yielded sherds with curvilinear decoration and others recalling the Danubian stroke-ornamented ware. **1970** BRAY & TRUMP *Dict. Archaeol.* 222/1 *Stroke-ornamented ware*, pottery with zigzag patterns made by a series of distinct jabs rather than continuous lines. It was current during the centuries after 4000 BC in Bohemia, west Poland, Bavaria and central Germany. **1930** C. G. MACARTNEY *My Cricketing Days* ii. 14 Perhaps this sort of cricket was an aid to stroke play, perhaps not, but as far as I can see, it never did me any harm. **1979** *Daily Tel.* 19 May 29/1 Any doubts about his form or fitness were violently dismissed in a morning of rich strokeplay. **1935** *Times* 20 July 13/5 Some of the English cricketers now getting past their prime are still stroke-players. **1963** A. ROSS *Australia 63* x. 183 There were, on the England side, three stroke-players capable of enhancing any Test, on the Australian side two.

stroke, *sb.*[2] Delete † *Obs.* and add: Also, an act of stroking, esp. by way of caress.

1953 H. E. BATES *Nature of Love* iv. 36 She gave her hair a long deep casual stroke with the brush.

2. An act of copulation. *slang. rare.*

1785 GROSE *Classical Dict. Vulgar Tongue* s.v., *Stroke. To take a stroke*, to take a bout with a woman. **1976** P. CAVE *High Flying Birds* ii. 19, I happened to be engaged upon a variation of the sexual act known as the 'Birmingham Stroke' at the time our little love-nest started rolling.

3. A comforting gesture of approval or congratulation (see also quot. 1964). Hence, a flattering or friendly remark, etc., esp. one made in order to help or manipulate another. Cf. *STROKE *v.*[1] 1 e. Now chiefly *U.S. colloq.*

1964 E. BERNE *Games People Play* 15 By an extension of meaning, 'stroking' may be employed colloquially to denote any act implying recognition of another's presence. Hence a *stroke* may be used as the fundamental unit of social action. An exchange of strokes constitutes a *transaction*, which is the unit of social intercourse. **1969** T. A. HARRIS *I'm OK, You're OK* iii. 45 The Adult has something to work on: what must I do to gain their strokes, or their approval? **1973** *Houston* (Texas) *Chron. Texas Mag.* 14 Oct. 4/1 The popular saying around PDAP [*sc.* The Palmer Drug Abuse Program] is 'different strokes for different folks', and that's the basis of the program. **1978** M. PUZO *Fools Die* xi. 122 He started off dishing out some nice strokes. With an admiring smile he told me how smart I was, how honest, so absolutely reliable. **1981** *TV Picture Life* Mar. 12/3 Let's face it, everybody needs their strokes and that would be very ego-satisfying.

4. Special combinations: **stroke book**, a pornographic book; **stroke house** *U.S.*, a cinema where pornographic films are shown.

1972 *Pussycat* XXXIII. LIX. 10/1 For a stroke book, the quality of writing is astonishingly good. **1978** T. GIFFORD *Glendower Legacy* (1979) 73 I'm just going to pig out at home, look at a stroke book... As a matter of fact, I've taken to writing for stroke books. **1971** *Atlantic Monthly* July 52 He would camp in the 42nd Street stroke houses and come back with tales of what they were getting away with now.

stroke, *v.*[1] Add: **1. e.** In recent use, to reassure (a child, etc.) by approval or congratulation (see also quot. 1964). Hence, to manipulate (another) by means of flattery, persuasion, etc.; to compliment. Cf. *STROKE *sb.*[2] 3. Chiefly *U.S. colloq.*, esp. in political contexts.

1964 [see *STROKE *sb.*[2] 3]. **1969** T. A. HARRIS *I'm OK, You're OK* iii. 48 If a two-year-old concludes *I'm OK*, does this mean his *OK* is the product of 'self-stroking' and, if so, how does a small child stroke himself? **1973** T. C. HUSTON in L. Chester et al. *Watergate* iv. 43 Mr. Hoover should be called in privately for a stroking session at which the President [*sc.* Nixon] explains the decision he had made. **1975** *Atlantic Monthly* Mar. 44 It's Show Biz, man—a bunch a' egomaniacal people using a captive audience to stroke themselves. **1977** *Time* 17 Oct. 20/1 Carter also stroked the Jerusalem government by promising that the U.S. would never attempt to impose a Middle East settlement. **1978** *New Yorker* 9 Jan. 41 He tells his client, 'It's looking pretty good. We'll stay on top of it.' This is what is known as 'stroking' the client. **1981** *Observer* 11 Jan. 6/5, I think he's still a little kid from Hoboken, who likes to be stroked by Presidents.

stroke, *v.*[2] Add: **I. 4*.** Of a bell: to chime the strokes of (the hour, etc.). *poet. rare*[-1].

1902 HARDY *Poems of Past & Present* 132 As the hope-hour stroked its sum, You did not come.

III. 6. b. Of an oarsman or crew: to row at (a certain number of strokes per minute).

1928 *Times* 11 Aug. 5 The winner stroked an average of 28 to Gunther's 30. **1976** C. FREUD in *Webster's Sports Dict.* 431/1 With 500 yards to go, the Cairo Police, stroking 38, edged past Oxford.

7. *Sport.* To hit or kick (the ball) smoothly and elegantly; to score in this manner.

1960 J. FINGLETON *Four Chukkas to Australia* xvi. 136 He..raved of the manner in which Cowdrey stroked the ball. **1962** [see *CROSS *sb.* 22 e]. **1972** *Even. Telegram* (St. John's, Newfoundland) 24 June 1/1 Bernie Allen stroked his first home run of the season. **1976** *Wymondham & Attleborough Express* 3 Dec. 27/4 Wortwell fought back and were awarded a spot kick only for Webb to stroke the ball straight to the keeper.

stroll, *v.* **3.** Delete † *Obs.* and add later examples. Chiefly *U.S.* in recent use.

1956 H. GOLD *Man who was not with It* (1965) vi. 50 Her laughter rang out as we strolled a business street of the suburb. **1974** *New Yorker* 3 June 76/3 (Advt.), Hike forest trails, stroll lovely gardens. **1977** *Gay News* 24 Mar. 23/1 They taxi to the Toilet and stroll the dock strip at 3 am.

stroller. Add: **6.** A child's push-chair, esp. a collapsible buggy.

1920 *Sears, Roebuck & Co. Catal.* Fall 1049/2 Stroller Style Carriage... Body made of selected reeds finished in the natural shellac color. **1922** *Ibid.* Spring 590/2 Select your stroller or collapsible sulky from this page and you will be satisfied with service given. **1932** *Babies, just Babies* Dec. 63/3 (Advt.), Baby walker, stroller and go-cart combined. **1954** W. MCCULLOUGH *Illustr. Handbk. Child Care* vii. 144 We saw the Smithsonian Institute and art galleries today. So glad we brought along his stroller. We just fold it up and pack it in the car. **1962** J. R. BERNARD in *Southerly* XXII. II. 97 We keep stroller as meaning saunterer but we add the meaning of baby's push-cart. **1977** N. SAHGAL *Situation in New Delhi* i. 5 Young women in bright coats briskly pushing strollers carrying rosy bundled babies.

7. A casual shoe. Normally in *pl.*

1948 *Woman & Beauty* Oct. 118/2 For shoes I picked a pair of black suède strollers. **1953** R. CHANDLER *Long Good-Bye* xi. 64 On his feet were black moccasin type ties, the kind..that are almost as comfortable as strollers. **1958** J. CANNAN *And be Villain* viii. 177 She stepped into her 'strollers'. **1970** *Washington Post* 30 Sept. B5/3 (Advt.), Softy suede stroller... Soft-walk into fall.

strom, strum. For *dial.* read *dial.* and *Naut.* and add: **2.** Chiefly in form **strum. b.** *Naut.* (See quots.) Also *strum-box, -plate.*

1894 H. PAASCH *From Keel to Truck* (ed. 2) 172/1 *Strainer; strum; strum-box*, terms applied to perforated plates, wire-clothes or any other objects fitted to allow the entry or exit of water or other fluids, but preventing the passing of any refuse matter. **1948** R. DE KERCHOVE *Internat. Maritime Dict.* 739/2 *Strum plate*, a plate fitted in pump suctions, deck scuppers, sea cocks, having a number of small holes in it to allow water to pass, but designed to stop foreign matter that would clog the piping. **1962** A. G. COURSE *Dict. Naut. Terms* 192 *Strum box*, a square metal box with perforated sides fitted round the bottom of a suction pipe in a ship's bilges. **1975** *B.S.I. News* July 21/2 Strum boxes for ships.

stroma. Add: **2. c.** *Bot.* The colourless fluid surrounding the grana inside a chloroplast.

1914 M. DRUMMOND tr. *Haberlandt's Physiol. Plant Anat.* i. 37 In some cases the characteristic pigment [of chromoplasts] is suspended in a colourless protoplasmic matrix (or stroma)..in the shape of minute globules or vesicles (grana). **1979** KRAMER & SCOTT *Cell Concept* iv. 80 The chemical changes in photosynthesis go on in the chlorophyll in the granules on the lamellae and are completed in the stroma.

stromatolite (strōᵘ·mătolǝit). *Geol.* [ad. G. *stromatolith* (see next): see -LITE.] A laminated calcareous sedimentary structure built up by algae or bacteria. Cf. *oncolite* s.v. *ONCO-.

1930 PEACH & HOME *Chapters on Geol. Scotl.* vi. 214 The structures referred to above as resembling the *Stromatolites* of Kallowsky [*sic*] are considered to be of the nature of a chemical deposit brought about by the action of organisms such as algae. **1957** *Prof. Papers U.S. Geol. Surv.* No. 294-D. 129/1 The present study discriminates between fossil algae and stromatolites. **1969** DUNBAR & WAAGE *Hist. Geol.* (ed. 3) viii. 182/2 Since these deposits preserve only the gross form of the colony, the algae cannot be identified biologically and the deposits are simply called stromatolites. **1979** D. ATTENBOROUGH *Life on Earth* i. 22 The blue-green pillars of Hamelin Pool are living stromatolites.

Hence **stromatoli·tic** *a.*, of the nature of or pertaining to stromatolites.

1933 *Trans. Geol. Soc. S. Afr.* XXXV. 35 The main interest of the rocks is centred in the abundance amongst them of stromatolitic limestones. **1955** *Jrnl. Paleont.* XXIX. 723/2 (*heading*) Recent stromatolitic sediments from south Florida. **1978** *Sci. Amer.* Sept. 86/3 Microfossils have been identified in some 45 stromatolitic deposits.

stromatolith (strōᵘ·mătoliþ). *Geol.* [f. mod.L. *stromat-* STROMA + -LITH.] †**a.** A laminated rock structure with a complex interleaving of igneous and sedimentary components. *Obs.* **b.** [a. G. *stromatolith* (E. Kalkowsky 1908, in *Zeitschr. der Deutsch. geol. Ges.* LX. 68).] = *STROMATOLITE.

1916 W. G. FOYE in *Jrnl. Geol.* XXIV. 791 The noun 'Stromatolith' may be defined as a rock mass consisting of many alternating layers of igneous and sedimentary rocks in sill relationship. **1918** *Jrnl. Geol.* XXVI. 608 The stromatoliths are the sedimentary equivalent of the calcareous and siliceous 'sinter' of the hot springs. **1933** *Q. Jrnl. Geol. Soc.* LXXXIX. 415 These bodies—stromatoliths—are composed of chlorite or chlorite with zones of a brown or green-brown isotropic mineral. **1963** D. W. & E. E. HUMPHRIES tr. *Termier's Erosion & Sedimentation* xi. 223 The Cyanophyceae can give rise to carbonaceous substances (bogheads) in the form of 'water-blooms' and can also precipitate calcareous 'biscuits', stromatoliths and calcareous muds.

Hence **stromatoli·thic** *a.*, of the nature of or pertaining to a stromatolith.

1916 W. G. FOYE in *Jrnl. Geol.* XXIV. 791 The term 'batholithic' does not describe the true character of these areas and the term 'stromatolithic' is suggested in its place. **1936** *Geogr. Jrnl.* LXXXVII. 534, I recognized..a palaeozoic series which extends from the stromatolithic Cambrian to a lower carboniferous formation. **1963** D. W. & E. E. HUMPHRIES tr. *Termier's Erosion & Sedimentation* xvii. 343 The stromatolithic limestones of the Precambrian.

Strombolian, *a.* Add: Also **strombolian.** Applied to (the stage of) a volcanic eruption in which there are repeated or continuous explosions of moderate force with the ejection of gases and bombs of lava. (Earlier and later examples.)

1897 I. C. RUSSELL *Volcanoes N. Amer.* i. 9 These two volcanoes belong to the explosive type,..but illustrate two quite well marked phases of that type, which..are termed the Strombolian stage and the Vesuvian stage,—the former characterized by long-continued but mild activity, the second by periods of rest broken by explosions of extreme violence. **1974** *Nature* 2 Aug. 385/1 Between these eruptions, activity on the volcano had been restricted to strombolian explosions and the extrusion of lava at the bottom of..the summit crater. **1976** P. FRANCIS *Volcanoes* iii. 108 Strombolian activity..is a bit noisier than Hawaiian, but it's still not particularly dangerous.

strong, *a.* **1. a.** Delete note and add: *strong man*: see as main entry below. Also *strong woman*, a designation for a woman who publicly exhibits feats of strength, as in a circus.

[**1936** J. S. CLARKE *Circus Parade* v. 46 Sandwina..was, until recently, the strongest woman in the world.] **1952** R. MANNING-SANDERS *Eng. Circus* xx. 284 Coming now to the *strong man* act, let us say at once that 'strong' men and 'strong' women, *are* strong men and women. **1953** K. TENNANT *Joyful Condemned* xx. 193 Stretched on the sofa lay Rene McGarty..looking like something between a lady wrestler and a circus strong woman. **1975** *Listener* 28 Aug. 275/3 Mildly humiliating experiences, such as being..manhandled by a circus strongwoman.

f. *strong silent* (*type*, etc.): see *SILENT *a.* 1 a.

6. b. (Earlier examples.)

1840 J. S. MILL in *Westm. Rev.* XXXIII. 260 The strong points of each [*sc.* Bentham and Coleridge] correspond to the weak points of the other. *c* **1869** TAYLOR & DUBOURG *New Men & Old Acres* I. 22 Perhaps you didn't know Bunter was such a way... Humour is his strong point.

9. d. Also *fig.* in phr. *strong meat* (alluding to Heb. v. 12: see quot. 1526 in Dict.), applied to something acceptable only to strong or instructed minds.

1837 [see MEAT *sb.* 1 b]. **1909** H. G. WELLS *Tono-Bungay* I. i. 26 Gulliver was there unexpurgated, strong meat for a boy perhaps. **1965** *Listener* 21 Oct. 640/2 *Nineteen Eighty-Four* was prefaced by a warning that it was not for nervous listeners. There was no such warning before Shirley Jenkins's *The Child*... Yet this too was.. strong meat, dealing..with the thought stream of a woman on the point of giving birth.

10. d. Of a field of force.

1903 J. J. THOMSON *Conduction of Electricity through Gases* ii. 21 If we..apply a strong electric field between the plates. **1930** PAULING & GOUDSMIT *Structure of Line Spectra* ix. 159 The resultant magnetic moment of the atom in a very strong field, such that the spins and orbital moments of the individual electrons are quantized relative to the field. **1978** PASACHOFF & KUTNER *University Astron.* xxvii. 699 A related prediction of general relativity is that 'clocks' run slower in a stronger gravity field than they do in a weaker.

e. *Physics.* Applied to the strongest of the four known kinds of force between particles, which acts between nucleons and other hadrons when closer than about 10^{-13} cm. (so that protons in an atomic nucleus remain bound together despite the repulsive force due to their electric charge), and which conserves strangeness, parity, and isospin.

1947 *Nature* 4 Oct. 453/2 We refer to any particle with a mass intermediate between that of a proton and an electron as a meson... In using this term, we do not imply that the corresponding particle necessarily has a strong interaction with nucleons. **1953** *Physical Rev.* XCII. 833/2 Let us suppose that both..have interactions of three kinds: (i) Interactions that rigorously conserve isotopic spin. (We assume these to be strong.) (ii) Electromagnetic interactions...(iii) Other charge-dependent reactions, which we take to be very weak. **1954** *Progress Theoret. Physics* XII. 107/2 Contrary to the case of electric charge, *v*-charge is defined only for such particles that have strong nuclear interactions and its conservation is violated by the weak interactions responsible for decays. **1973** *Sci. Amer.* Aug. 34/2 All the particles discovered to date participate in strong interactions except the photon and the four weakly interacting leptons. **1975** *Nature* 5 June 453/1 The electromagnetic interaction is responsible for the force between charged particles, its strength is 1/137 that of the strong force. **1978** PASACHOFF & KUTNER *University Astron.* ix. 261 For nuclear fusion to begin, atomic nuclei must get close enough to each other so that the nuclear force, technically called the strong force, can play its part.

12. b. Restrict † Obs. to sense in Dict. and add: Also in more recent Canad. use, *strong wood(s)* [tr. Canad. Fr. *bois fort(s)*], a region of thick afforestation. Freq. *attrib.* (see also quot. 1921).

1794 D. M'GILLIVRAY 12 Oct. *Jrnl.* (1929) 34 Soon after their departure 2 tribes of Assinoboines arrived..called Strong Wood &..Grand River Assiniboine [*sic*]. **1800** A. HENRY *Jrnl.* 5 Sept. in E. Coues *New Light Early Hist. Greater Northwest* (1897) I. iii. 83 We had a quarter of a mile of strong wood to pass through. **1861** *Canad. Naturalist* Dec. 438 The Strong-wood Reindeer inhabit the thickly wooded parts of the District. **1921** A. HEMING *Drama of Forests* 15 The several zones of the Canadian wilderness are locally known as the Coast Country—the shores of the Arctic Ocean and Hudson Bay; the Barren Grounds—the treeless country between Hudson Bay and the Mackenzie River; the Strong Woods Country—the whole of that enormous belt of heavy timber that spans Canada from east to west; the Border Lands—the tracts of small, scattered timber that lie between the prairies and the northern forests; the Prairie Country; the Mountains; and the Big Lakes. **1969** E. W. MORSE *Fur Trade Canoe Routes* II. iv. 45 The North Saskatchewan formed roughly the boundary of the 'strong woods' region where the furs were harvested.

14. d. (Earlier *fig.* example.)
1745 Young *Night-Thoughts* viii. 15 Men, who think nought so strong of the romance, So rank Knight-errant, as a Real Friend.

19. c. *Math.* Of a mathematical entity or concept: implying more than others of its kind; defined by more conditions.
1950 W. Feller *Introd. Probability Theory* I. viii. 156 We shall prove a much stronger statement. 1955 M. Loève *Probability Theory* 18 Now we can..use the supplementary requirement that the additive property of P remains valid for denumerable sums... This is the celebrated Borel strong law of large numbers. 1964 A. P. & W. Robertson *Topological Vector Spaces* iii. 47 This topology is denoted by β(E′,E) and is sometimes called the strong topology on E′. 1971 G. Glauberman in Powell & Higman *Finite Simple Groups* i. 44 Here the condition of p-stability is too strong to be useful. 1979 *Proc. London Math. Soc.* XXXVIII. 338 We say that a linear operator L on 𝒜 is a strong Feller operator if Lf∈𝒞 whenever f∈𝒜.

22. b. (Earlier example.) Occas. *transf.* with reference to non-Teutonic languages.
1833 *Philol. Museum* II. 385 No *weak* verb ever in process of time became *strong*, while strong verbs do become weak. 1946 Binchy & Bergin tr. *Thurneysen's Gram. Old Irish* 335 According to the way in which these stems are formed, two main classes of verbs can be distinguished, for which the terms 'strong' and 'weak' verbs are borrowed from the grammar of the Germanic languages. Strong verbs are without exception primary... Weak verbs are for the most part denominative. 1962 C. Watkins *Indo-European Origins of Celtic Verb* ii. 116 In the strong (non-derivative) verb..the present still exhibits numerous divergent formations.

24. b. Hence *fig.*, as (one's) *strong suit*: something at which one excels. Also *strong card*, a particular advantage or forte. *colloq.*
1865 'Mark Twain' *Sketches New & Old* (1875) 33 Jumping on a dead level was his strong suit. 1884, 1898 [see Suit *sb.* 20 a]. 1899 Ade *Fables in Slang* 138 Marie was a Strong Card. The Male Patrons of the Establishment hovered around the Desk long after paying their Checks. Within a Month the Receipts of the Place had doubled. 1936 E. M. Forster *Abinger Harvest* i. 16 As my husband points out, that is one of our strong cards. 1940 G. Frankau *Self-Portrait* lxiii. 388 Adaptability has always been one of my strong suits. 1970 R. Lowell *Notebook* 140 Dating children with trash was your strong suit.

25. *strong-armed* (example), *-blooded*, *-bodied* (later examples), *-charactered*, *-elbowed* (later example), *-gutted*, *-jawed*, *-membered*, *-muscled*, *-thewed*.
1785 Strong-bodied [see *baas]. 1850 J. G. Whittier *Elliott* in *National Era* 10 Jan. 6/4 Strong-armed as Thor! 1907 *Munsey's Mag.* Dec. 309/1 The piccola looked up at the dark..strong-jawed face. 1922 Joyce *Ulysses* 525 He wrote pencilled messages offering his nuptial partner to all strongmembered males. 1922 D. H. Lawrence *England, my England* (1924) 82 He looked so strong-blooded and healthy. 1926 —— *Plumed Serpent* vii. 130 White men sitting there would have been strong-muscled and frank. a 1930 —— *Etruscan Places* (1932) iii. 74 This sense of vigorous, strong-bodied liveliness is characteristic of the Etruscans. 1930 S. Spender *Twenty Poems* 3 Weapons men use, stone, sling, and strong-thewed bow He will not know. 1931 R. Graves *To Whom Else?* 11 With their strong-gutted and capacious bellies Digested stones and glass like ostriches. 1944 Blunden *Shells by Stream* 31 Strong-elbowed and with wondrous beard, Whose statue's this? 1960 C. Day Lewis *Buried Day* ii. 44 My father..[had] the expression of an actor playing the part of a strong-charactered, resolute, if moody, man. 1978 W. F. Buckley *Stained Glass* xvii. 169 He was strong-jawed, with a splotchy face that showed the ravages of frostbite.

26. strong-back, (*a*) substitute for def.: any of several plants used in the West Indies to make medicinal infusions; (later examples); (*b*) (later examples); also in extended uses, esp. a beam placed across an access cover to secure it in position; **strong-eyed** *a.* chiefly *N.Z.*, of a sheep-dog: possessed of good powers of controlling sheep; hence **strong eye,** (*N.Z.*, a sheep-dog with) this ability; **strong joint** *U.S. slang* (see quots. 1935, 1938); **strong stress** *Prosody*, accentuation which falls on syllables separated by a varying number of unstressed syllables, characteristic of certain poetic traditions, as Old English alliterative verse.
1927 M. W. Beckwith *Notes Jamaican Ethnobotany* 28 Strong-back..For a weak back drink a little as tea each day. 1927 G. Bradford *Gloss. Sea Terms* 176/2 *Strongback*, a steel (or wood) beam placed across a hatch to support the sections of the hatch covers... A spar lashed to and running between the old style davits to steady them and to aid in controlling and securing the boat. 1953 C. S. Forester *Hornblower & Atropos* xvi. 253 The next morning Hornblower watched launch and longboat start off with strongbacks erected in their sterns, and blocks and tackles rigged on them. 1953 *Caribbean Q.* III. i. 10 There are two or three kinds of strong-back (to make strengthening tea). 1970 M. Slater *Caribbean Cooking for Pleasure* 21/2 In Jamaica, fish soup is 'Fish Tea' and sometimes called 'Strong Back', but this term is applied to anything nourishing 'make strong back'. 1977 *Austral. Sailing* Jan. 38/3 The strongback on which the mast is stepped is made as a complete trussed girder beam which is placed as a unit inside the shell. 1934 J. Lilico *Sheep Dog Mem.* 4 My father..taught me how to prevent the several faults that the strong-eyed young dog will assuredly acquire if not taken in time. 1949 G. W. C. Hartley *Shepherd's Dogs* ii. 3 Excessively strong-eyed dogs. *Ibid.* 5 The pup is from a 'strong-eye' strain. 1952 *Arena* (N.Z.) xxxi. 2 Shepherding the stragglers would be Charlie's strong-eyes, Beau and Belle. 1977 *Field* 13 Jan. 55/2 A dog which can 'will' his sheep into submission requires what shepherds term a strong (or dominant) eye. 1926 Maines & Grant *Wise-Crack Dict.* 14/1 *Strong joint*, unfair or cheater's game. 1935 N. Ersine *Underworld & Prison Slang* 73 *Strong joint*, a crooked gambling house. 1938 F. Chester *Shot Full* xi. 98 A 'pick-out'. This is another form of 'strong-joint', or never-win game. 1963 Strong-joint [see *flat joint* s.v. *flat *a.* 15]. 1959 *PMLA* LXXIV. i. 588 The two main alternative principles of English meter..are actually two kinds of stress—strong stress (the Old English, the *Piers Plowman* tradition) and syllable stress (the Chaucer-Tennyson tradition). 1973 *Studies in Eng. Lit.: Eng. Number* (Tokyo) 22 There was no mention of Anglo-Saxon verse as a forerunner of sprung rhythm... There is..little reason to suppose that Hopkins derived his sprung rhythm from strong-stress verse.

strong, *adv.* Add: **1. c.** *to come out strong* (earlier example); *to come on strong* (orig. U.S.), to adopt or exhibit aggressive behaviour; to perform or contest successfully.
1860 *Players* I. 147 Many a fair lady must have lost her heart when she saw Mr. J. M. on his twentieth birthday, when he came out 'strong' at a fancy dress ball. 1970 H. E. Roberts *Third Ear* 5/2 *Come on strong*, to do something to a superlative degree, to an extreme; to 'pour it on'. 1976 *Honolulu Star-Bull.* 21 Dec. E-1/5 Young guys who've never had experience come on too strong and that's how a fight with a host starts. 1979 *Tucson* (Arizona) *Citizen* 28 Apr. B. 1/1 The Kings came on strong at the onset of the third period, outscoring the Suns, 15–6.

strong, *v.* Restrict † *Obs.* to sense a and add: **b.** (Later examples.) Also, to cause (one) to smell strong. Now only *U.S. local*.
1913 H. Kephart *Our Southern Highlanders* xiii. 283 A verb will be coined from an adverb... Or from an adjective... 'Baby, that onion 'll strong ye!' 1941 E. P. O'Donnell *Great Big Doorstep* xviii. 250 The coffee gunna strong you soon, darling.
c. *to strong it*, to behave excessively, to exaggerate. *slang.*
1964 *New Statesman* 10 Apr. 555/2 'To strong it' means to overdo something, like taking more than 30 purple hearts in one night. 1970 G. F. Newman *Sir, You Bastard* iii. 108 Don't you think that's stronging it?

strong, *sb. Austral. slang.* [f. the adj.] In phr. *the strong of* (a person or thing) = *the strength of* s.v. *strength *sb.* 2 e.
1916 A. Wright *Under Cloud* 31 Don't yer want to own up? Some reason for wantin' to preserve yer incog, I suppose. What's th' strong of it? 1938 X. Herbert *Capricornia* 566 What's the strong of you? What's the questioning for? I've done nuthin'. 1959 E. Lambert *Glory thrown In* 161 'What's the strong of this joint?' demanded Doc brusquely. 'Not an undertaker's is it?'

strong arm, *sb.* (and *a.*) orig. *U.S.* **A.** *sb.* Used *transf.* and *fig.* with reference to power: see sense 1 b of the adj. Cf. Long *a.* 1 c.
2. With *the*: physical force or violence considered as a means of action, *spec.* in the course of robbery. Cf. Strong hand.
1836 M. Holley *Texas* xiv. 322 This military council.. distributed lots to the inhabitants, contrary to all law, but that of the *strong arm*. 1903 A. H. Lewis *Boss* 316 He was all for th' strong-arm, an' th' knock-about! It's a bad system. Nothin's lost by bein' smooth, Gov'nor. 1907 J. London *Road* 169 Into the man's back goes his knee; around the man's neck..passes his right hand, the bone of the wrist pressing against the jugular vein. Barber Kid throws his whole weight backward... It is the strong arm. 1948 *Daily Express* 2 Oct. 1/5 Modern youngsters..get panicky and use the strong arm.
3. A criminal who resorts to violence; one who is employed or hired to use force against persons.
1907 J. London *Road* 159 A world of rods and gunnels, blind baggages and 'side-door Pullmans'..'strong arms' and 'bindle-stiffs'. 1932 'Spindrift' *Yankee Slang* 60 *Strong arm*, bouncer or 'chucker out' for speakeasies and Honkytonks. 1978 N. J. Crisp *London Deal* ii. 21 He was a strong arm for an ambitious East End team which.. incurred the displeasure of the incumbent mob.
B. *attrib.* or as *adj.* (stress on first syllable). **1.** Of a person: having or showing strength of arm; physically powerful; *spec.* of a criminal: resorting to violence, esp. for hire or in the case of robbery. Freq. in phr. *strong-arm man*.
1897 Eldridge & Watts *Our Rival, the Rascal* ix. 281 The ordinary robber of to-day is frequently and familiarly styled by his 'pals', a 'strong arm man'. 1904 'No. 1500' *Life in Sing Sing* 257/2 *Strongarm guy*, highway robber. 1931 *Times* 24 Sept. 11/2 A large force of gunmen and 'strongarm men'..was soon effective in breaking down the resistance of the strikers. 1947 J. Mulgan *Report on Experience* x. 125 It comes down to cases with two strongarm men who call round and offer a beating-up. 1953 J. Phelan *Underworld* i. 21 A strong-arm chap is, roughly speaking, a bruiser, one who, for a pound or two, will punch some one on the nose, or start a fight in a pub, or even sometimes assault the police. 1973 D. Westheimer *Going Public* ii. 27 Your strongarm boy won't always be around to save your skin.
2. Of an action: involving the use of physical violence. Also of policies, etc.: characterized by a display of (excessive) force; heavy-handed, oppressive, bullying.
1901 'J. Flynt' *World of Graft* ii. 18, I had been inclined to believe that the 'strong-arm' crimes were committed by men who were transients in the city. 1930 P. MacDonald *Link* vii. 109 The law was in a pair, and with..a little strong-arm stuff of their own, got the two apart. 1951 E. Paul *Springtime in Paris* xi. 202 The Communists began strong-arm tactics to discourage the timid from seeing things at night. 1973 Wodehouse *Bachelors Anonymous* xii. 160 The strong-arm methods favoured by both counsellors might, of course, be resented, for he had no official knowledge that his love was returned. 1978 S. Brill *Teamsters* vii. 288 The Vice-President has a record of strong-arm robbery.
Hence **stro·ng-arm** *v. trans.*, to treat roughly or manhandle (a person); to rob with violence, to coerce; to seize (something) by force; also *intr.*, to proceed in an aggressive, bullying manner; **stro·ng-arming** *vbl. sb.*
1903 *Monthly Maroon* (Chicago) June 444 If he refused, Phil..was to strong-arm him while Tommy took away the badge. 1937 E. Hemingway *To have & have Not* iii. vii. 130 Don't try to strong arm it away from me. 1941 Baker *Dict. Austral. Slang* 73 *Strongarm, to*, to act in a bullying fashion. 1948 Lait & Mortimer *New York* xi. 114 Mugging..in old Chicago days called 'strong-arming'. 1954 D. Dodge *Lights of Skaro* vi. 215, I strong-armed my way out like a hero. 1965 D. Francis *Odds Against* xiii. 179 He was strong-arming Brinton. 1969 *Daily Colonist* (Victoria, B.C.) 27 Feb. 35/1 A new trial was ordered by the B.C. Court of Appeals for a man jailed..in connection with a Prince George strongarming. 1977 *Observer* (Colour Suppl.) 5 June 42/3 The OAS had financed themselves initially by strong-arming contributions from rich settlers, who usually shared their sympathies.

strongers (strǫ·ŋəɹz). *slang.* [f. Strong *a.* + *-er⁶.] **1.** = *soogee-moogee *a.* *Naut.*
1929 F. Bowen *Sea Slang* 135 *Strongers*, any strong cleaning preparation such as spirits of salt. 1961 F. H. Burgess *Dict. Sailing* 202 *Strongers*, a compound cleaning mixture, caustic, etc., made up for use; also spirits of salts, acids, etc., undiluted.
2. = Strong drink. *slang. rare.*
1939 Joyce *Finnegans Wake* 58 Swiping rums and beaunes and sherries and ciders and negus and citronnades too. The strongers.

stro·ng-ha·nded, *a. rare.* [f. Strong hand + -ed².] † **1.** = Strong *a.* 5 e. *Obs.*
1818 H. B. Fearon *Sk. Amer.* 224 The wealthy or 'strong-handed' farmer..owns five to twelve hundred acres.
2. Of a ship: well-manned. Also by synecdoche, of a ship's captain: in charge of a well-manned ship.
1827 J. F. Cooper *Red Rover* I. ii. 45 It is plain enough, by the manner in which his [a captain's] sails are furled, that he is strong-handed. 1844 —— *Afloat & Ashore* I. v. 129 He took us on board purely out of a national feeling, for his ship was strong-handed without us.
3. Forceful, imperious.
1949 I. Deutscher *Stalin* vi. 178 Stalin..secured a more impressive abode for his Commissariat through a strong-handed intervention in a somewhat comic scramble between the commissars for accommodation.

strongly, *adv.* Add: **1. i.** *Physics.* By means of the strong interaction (see *strong *a.* 10 e).
1960 P. Roman *Theory Elementary Particles* v. 460 The K⁻, but not the K⁺, can react strongly with nuclear matter. 1977 *Sci. Amer.* Oct. 58/2 A strongly decaying hadron exists for only 10^{-23} second before it breaks up into less massive hadrons.
2. b. (Examples in *Math.*: cf. *strong *a.* 19 c.)
1955 M. Loève *Probability Theory* ix. 442 Since..T is bounded and linear, every Bλ is..a strongly closed linear subspace of B. 1966 E. H. Spanier *Algebraic Topology* ix. 510 Let X be a strongly simple space. 1979 *Proc. London Math. Soc.* XXXVIII. 507, K is said to be a spectral compact convex set if A and V are in spectral duality, and K is said to be strongly spectral if in addition [etc.].

strong man. 1. A man of great physical strength; *spec.*, one who displays his strength professionally, as in a circus.
1699, etc. [see Strong *a.* 1]. 1841 Thackeray *Loose Sk.* ii, in *Britannia* 15 May 315/3 A vast number of booths and exhibitions..no less than four companies of strongmen..'the Indian strong men'; 'the strong men with the fairy pony', &c. 1872 Browning *Fifine at Fair* 23 The Strong Man, whom..by-and-by You shall behold do feats. 1908 *Variety* 16 May 15 Blocksom and Burns, the 'strong' men,..have bookings at hand which, if accepted, will keep them busily playing in the west all summer. 1954 [see *beefcake* s.v. *beef *sb.* 5]. 1977 C. Wood *James Bond* xii. 99 He was..the product of a union between the strong man of a travelling circus and the Chief Wardress at the Women's Prison.
2. A dominating man; the man who exercises effective control of an organization; *spec.*, one who exercises absolute political power.
1859 Mill *Liberty* iii. 119, I am not countenancing the sort of 'hero-worship' which applauds the strong man of genius for forcibly seizing on the government of the world and making it do his bidding in spite of itself. 1879 [see Strong *a.* 3 d]. 1886 Kipling *Plain Tales from Hills* (1888) 89 He called on the biggest and strongest man that the Government owned, and explained that he wanted an appointment at Simla on a good salary. The compound insolence of this amused the Strong Man. 1924 A. D. Sedgwick *Little French Girl* ii. ix. 165 They were both agreed on the necessity of a strong man for France and on many lopped heads. 1943 E. A. Peers *Spain in Eclipse* i. ii. 39 The erstwhile 'strong man', President Azaña, emerged out of his silence. 1961 P. Kemp *Alms for Oblivion* ii. 29 Pluto had been nominated Minister Without Portfolio; Marshal Phibul, the 'Strong Man' of the country,..was under arrest. 1969 D. Acheson *Present at Creation* (1970) xxi. 187 Colonel Juan Perón, the President and 'strong man' of Argentina. 1977 *Westindian World* 3–9 June 12/1 The other is Major General Ziaur Rahman, who became President of Bangladesh only a few weeks ago (though he had been the country's strongman for some time).
3. *attrib.*
1962 C. Walsh *From Utopia to Nightmare* ix. 122 A strong-man government. 1977 *Times* 12 July 14/3 Strongman rule has provided the region with comparative stability..and, in the case of South Korea and Singapore, phenomenal economic growth.

strong-minded, *a.* Add: **b.** (Earlier example.)
1843 Dickens *Mart. Chuz.* (1844) iv. 42 Then there was the widow of a deceased brother.., who being almost supernaturally disagreeable, and having a dreary face and a bony figure and a masculine voice, was..what is commonly called a strong-minded woman.
strong-mindedness (earlier example.)
1849 C. Brontë *Shirley* II. xii. 294 With all her strictness, with all her 'strongmindedness', she could gain no command over them.

strong point. 1. *one's strong point*: see Strong *a.* 6 b.
2. *Mil.* [tr. G. *feste stellung*: see Strong *a.* 8 and Point *sb.*¹ A. 19.] A specially fortified position in a defence system. Also *transf.* and *fig.*
1915 E. Dane *Battles in Flanders* x. 183 An orchard, triangular in shape and bounded along each face by a road, which the Germans had fortified. This, one of the strong points of the German second line, the Devons carried by storm. 1931 *Times Lit. Suppl.* 30 Apr. 350/3 Billets, water-supply, roads, wiring, strongpoints, fighting. 1946 D. L. Sayers *Unpopular Opinions* 100 Nobody was quite ready to coerce Britain into giving away her colonies, dependencies, and scattered strong-points. 1957 *Times* 16 Feb. 7/5 Its simple existence is a strong-point in the struggle to maintain our standards against the spread of the 'new illiteracy'. 1978 L. Heren *Growing up on The Times* iii. 66 The Jewish Agency..had been determined to establish as many Jewish strongpoints as possible in an effort to extend its territorial claims. 1980 *Sci. Amer.* Mar. 56/1 Preserved under the city is a record of its development since the year of its founding in A.D. 71 as a strongpoint for the Roman Ninth Legion.
3. *R.A.F.* (See quots.)
1946 D. Hamson *We fell among Greeks* i. 18 The strongpoint is normally a powerful bolt and shackle fitted into the framework of the aircraft at the point of exit of the parachutist. To it are attached the static-lines from each man's parachute. 1951 *Gloss. Aeronaut. Terms* (B.S.I.) iii. 15 *Strong point*, a fitting in an aircraft which is capable of transmitting a shock load and to which the static line or strop is attached. 1958 P. Kemp *No Colours or Crest* iii. 40 The sergeant clipped the static lines of our parachutes to the 'strong-point', a stout wire running along the fuselage beneath the roof.

Strongyloides (strǫndʒiloi·dīz). *Med.* and *Vet. Sci.* Also **strongyl-**. [mod.L. (B. Grassi 1879, in *Rendiconti R. Ist. Lombardo di Sci. e Lett.* XII. 233), f. *Strongyl-us* (see Strongyle¹) + *-oides* (see -oid).] Nematode worms of the genus so named; also, = *strongyloidiasis.
1911 *Jrnl. Exper. Med.* XIV. 6 Most of the patients were not diarrheic, so that the strongyloides were incidentally found as a result of the routine stool examinations. 1929 A. S. Chandler *Hookworm Dis.* vii. 287 [illegible] 1958 *Jrnl. Amer. Med. Assoc.* 22 Nov. 1651/2 Careful examination of serial sections of large bowel failed to reveal any Strongyloides in this tissue. 1974 Passmore & Robson *Compan. Med. Stud.* III. xii. 56/1 In many British soldiers who acquired the infection when prisoners of war in the Far East, strongyloides has persisted for more than 25 years.

strongyloidiasis (strǫndʒiloidai·ăsis). *Med.* and *Vet. Sci.* [f. *Strongyloides + *-iasis.] Infection with or a disease caused by nematode worms of the genus *Strongyloides* (family Rhabditidæ), esp. *S. stercoralis*, a threadworm infesting the human and the canine gut in tropical and subtropical regions, causing diarrhœa.
1930 E. C. Faust *Human Helminthol.* xxxiii. 538 (*heading*) Strongyloidiasis. 1935 *Arch Path.* XIX. 782 In certain cases in which strongyloidiasis had been active for several months the presence of filariform, postfilariform and preadolescent worms in the respiratory passages suggested hyperinfection. 1977 *Lancet* 28 May 1121/2, 1 patient with strongyloidiasis had an increased total number of IgG-labelled cells in the lamina propria.

strongyloidosis (strǫ:ndʒiloidǝ̄ᵘ·sis). *Med.* and *Vet. Sci.* [f. as prec. + -osis.] = prec.
1911 *Jrnl. Exper. Med.* XIV. 7 Of 54 cases of strongyloidosis..none had been subject to diarrhea. 1957 Smith & Jones *Vet. Path.* xiii. 494 It is generally believed that severe and fatal strongyloidosis is likely to occur only in man or animals debilitated from faulty nutrition or other factors. 1969 J. R. Georgi *Parasitol. for Veterinarians* viii. 134 Canine strongyloidosis has been much feared in the past and demands respect in the future as a menace to human health. 1980 *Acta Parasitol. Polonica* XXVII. 89 Anatomical and pathological changes in animals that have died of the strongyloidosis consist mostly of changes in the respiratory and alimentary systems.

strongylosis (strǫndʒilǝ̄ᵘ·sis). *Vet. Sci.* and *Med.* [f. mod.L. *Strongyl-us* (see Strongyle¹) + -osis.] Infection with or a disease caused by nematode worms (strongyles) of the genus *Strongylus* or the family Strongylidae containing it, which are parasites of many domestic animals.
1892 O. Fleming tr. *Neumann's Treat. Parasites* iv. ii. 580 (*heading*) Bronchial and pulmonary strongyloses. 1928 *Daily Tel.* 7 Aug. 9/4 Strongylosis, or grouse disease, is an infection produced by parasitic worms in the intestines, which are found to be present in almost all grouse. 1969 J. R. Georgi *Parasitol. for Veterinarians* ix. 164 Although *Nematodirus* spp. infections are ordinarily not associated with clinical disease, serious outbreaks have been attributed to *N. battus*. This strongylosis is characterized by..very severe and debilitating diarrhea. 1981 *Equine Vet. Jrnl.* XIII. 35 (*heading*) Ischaemic myocardial fibrosis and aortic strongylosis in the horse.

strontian (strǫ·ntiăn), *a.* *Min.* [f. Stront(ium + *-ian 2.] Of a mineral: having a (small) proportion of a constituent element replaced by strontium.
1930 W. T. Schaller in *Amer. Mineralogist* XV. 573 Strontium—strontian. 1959 [see *ellestadite].

strontium. Add: Now usu. with pronunc. (strǫ·ntiǝ̆m). *strontium 90*, a radioactive isotope of strontium which is one of the chief products of the fission of uranium 235, can pass from fall-out into plants and animals and hence into human tissue (where it is concentrated in bones and teeth), and has been used in radiotherapy.
1955 *Sci. News Let.* 28 May 345/1 Strontium 90 is of particular interest because, being chemically similar to calcium, it may be deposited in human bone. 1961 *Lancet* 12 Aug. 366/1 Measurements of strontium 90 in human bone in the United Kingdom. 1978 *Jrnl. R. Soc. Arts* CXXVI. 257/1 Two fission products, Strontium 90 and Caesium 137 have particularly awkward half-lives of about 30 years.

strool (strul). *Sc.* Also **strule**. [? ad. Sc. Gaelic **srùil* stream (*srùlach* abounding in streams).] A stream (of water or other liquid). Also *fig.*
1867 G. W. Donald *Poems* 17 The water comes doon in perfect strools upon's. 1920 D. H. Edwards *Men & Manners* 236 There's juist a dreeble o' a strule at the burn spoot. 1922 Joyce *Ulysses* 756 Sending me that long strool of a song out of the Huguenots to sing in French.

'strooth, var. 'Strewth in Dict. and Suppl.

strop, *v.*¹ Add: **1.** Also *transf.* and *fig.*
1841 [in Dict.]. 1944 'Palinurus' *Unquiet Grave* iii. 72 The parrot stropping its beak on the bars of the cage. 1957 C. Day Lewis *Pegasus* 25 The river endlessly stropping its tides against the embankment. 1974 'J. Herriot' *Vet in Harness* vii. 51 'I think a beef sandwich would go down rather nicely, Jim,' he murmured, as he stropped his carving knife on a steel.

strophanthidin (strofæ·nþidin). *Pharm.* [f. next + *-idin.] A poisonous steroidal aglycone, $C_{23}H_{32}O_6$, which is prepared by hydrolysis of strophanthin-K and is a stimulant of heart muscle.
1888 T. R. Fraser in *Proc. R. Soc. Edin.* XIV. 373 All mineral acids..resolve strophanthin, even in the cold, into glucose and a substance which I have named strophanthidin. 1962 H. L. Kern et al. in A. Pirie *Lens Metabolism Rel. Cataract* 394 The aglycone, strophanthidin, abolished the active transport of sugars..by intestine of hamsters. 1978 *Nature* 1 June 389/1 The aglycone strophanthidin was preferred to cardiac glycosides such as ouabain because its effects are more readily reversible.

strophanthin (strofæ·nþin). *Pharm.* [f. Strophanth(us + -in¹.] Any or all of several polycyclic glucosides obtained from certain varieties of plants of the genera *Strophanthus* and *Acokanthera* and used as cardioactive drugs; *G-* or *g-strophanthin* (also *strophanthin-G, -g*) [L. *g(rātus* agreeable], the primary active component, $C_{29}H_{44}O_{12}$, of the mixture obtained from *S. gratus*, *A. schimperi*, or *A. ouabaio*; also called *ouabane*; *H-, h-strophanthin*, etc., [L. *h(ispidus* hairy], a mixture of the glucosides obtained from the seeds of *S. hispidus*; *K-, k-strophanthin*, etc., [*K(OMBÉ)], $C_{36}H_{54}O_{14}$, the principle extract from *S. kombe*, also called *kombé.
1873 *Pharm. Jrnl.* III. 524/2 From this extract [of seeds of *Strophanthus Hispidus*, D.C.] Dr. Fraser succeeded in separating a very powerful active principle, which he proposes should be named strophanthin. 1927 *Blackw. Mag.* May 582/1 Strophanthin, its [sc. strophanthus'] laboratory derivative is a drug commonly used by the medical profession. 1940 *Elsevier's Encycl. Org. Chem.* Ser. III. XIV. 270, *h-Strophanthin* is a mixt. of glycosides from the seeds of *Strophanthus hispidus*. 1960 J. Lewis *Introd. Pharmacol.* xii. 356 Ouabain (G-strophanthin) is a crystalline glycoside obtained from *S. gratus*. 1962 H. L. Kern et al. in A. Pirie *Lens Metabolism Rel. Cataract* 389 The cardiac glycosides, gitalin and strophanthin K, produced effects [on the transport of potassium and anaerobic glycolysis in calf lens] comparable to ouabain. 1978 *Acta Physiol. Acad. Sci. Hungaricae* LI. 166 The therapeutic and toxic effects of strophanthin and digoxin on the heart were studied in the 6–7-week old human embryo.

stroppy (strǫ·pi), *a. colloq.* [? abbrev. of Obstreperous *a.* with altered stem-vowel.] Bad-tempered, rebellious, awkward, obstreperous, unruly. Hence **stro·ppiness**.
1951 H. Hastings *Seagulls over Sorrento* ii. i, in *Plays of Year 1950* IV. 76 There ain't nothing clever about answering him back and being stroppy. 1964 J. Burke *Hard Day's Night* iv. 80 'Have you got a licence for it?' 'Oh, don't be so stroppy.' 1968 [see *shit *sb.* 1 g]. 1969 *New Statesman* 31 Oct. 633/3 It anyhow seems to me (perhaps out of stroppiness) that the good toys aren't necessarily the ones that teach how to count or measure. 1973 J. Wainwright *Devil you Don't* 116 All that balls about 'with respect'...Honesty. Stroppiness. 1977 *Guardian Weekly* 27 Feb. 5/4 Those who didn't know him well..missed the stroppy radicalism of his spirit. 1983 *Listener* 14 Apr. 37/1 Susan, on the other hand, is streetwise and stroppy.

stroud. Add: **1.** (Later U.S. examples.)
1846 T. L. McKenney *Mem.* I. i. 21 It was not so much a competition in blankets, and strouds, and calicoes..as in whiskey. 1872 *Rep. Indian Affairs 1871* (U.S.) 459 Blankets, leggings, strouds, paints.
3. (Later examples.)
1751 G. Croghan in *Pennsylvania Colonial Rec.* (1851) V. 531, I gave him a Strowd Shirt, Match Coat [etc.]. 1934 P. H. Godsell *Arctic Trader* 39 A dozen other drivers, all clad in..blue stroud leggings and fur caps.

Strouhal (strau·ăl, strū·ăl, ‖ strǝ̄ᵘ·hal) *Mech.* [The name of Čeněk (or Vincent) *Strouhal* (1850–1922), Czech scientist.] *Strouhal number*: a dimensionless number used in the study of the vibrations produced in a body by a fluid flowing past it, defined as νd/v (or v/νd) where v is the fluid velocity, ν the frequency of the vibration, and d the effective diameter of the body.
1949 *Proc. R. Soc.* A. CXCVIII. 175 The results are expressed in the form of Strouhal number: S(R) = fd/U, where R = Uod/ν. 1975 *Offshore Engineer* Dec. 42/3 For an isolated stationary cylinder the Strouhal number is fairly constant for a wide range of Reynolds numbers.

strouter (strau·təɹ). *Newfoundland.* [Perh. f. dial. form of Strut *sb.*²: cf. also Stouter.] A heavy post used to support and strengthen the end of a fishing stage or wharf; = Stouter.
1895 *Jrnl. Amer. Folklore* VIII. 31 *Strouters*, the outside piles of a wharf, which are larger and stronger than the inner ones. 1937 P. K. Devine *Folk Lore Newfoundland* 50 *Strouters*, the perpendicular posts at the front end of a fishing stage, jammed firmly into the sea bottom, and having rails nailed across to make the ladder for getting into and out of boats. 1973 *Even. Telegram* (St. John's, Newfoundland) 25 Oct. 3 You could hear a tin can bonking against the strouters or the rocks down there in the landwash. 1975 V. Butler *Little Nord Easter* 61 For the shores [of a wharf] there'd be strouters. They'd be called strouters and they'd be a little larger, probably seven or eight inches in diameter. 1975 *Canad. Antiques Collector* Mar.–Apr. 23/1 In the fishery we have such terms as: strouters (perpendicular posts which support the front of a fishing-stage).

Strowger (strau·gəɹ). *Teleph.* The name of Almon B. *Strowger*, U.S. telephone engineer, used *attrib.* with reference to an exchange switching system proposed by him in 1891 (U.S. Patent 447,918), involving successive step-by-step switches.
1900 K. B. Miller *Amer. Telephone Pract.* (ed. 4) xxxiv. 459 The most important idea..in all of the Strowger automatic work, was that of simplifying the contacts for the different line wires. 1901 J. E. Homans *ABC of Telephone* xx. 268 A Strowger exchange in Atlanta, Ga., has as many as 500 subscribers. 1933 K. B. Miller *Telephone Theory & Pract.: Automatic Switching* iii. 22 Of the three general types of automatic or machine switching systems, 'step-by-step', 'power-driven', and 'all-relay', the former, also called the 'Strowger system' will be treated in this chapter. 1967 *Times Rev. Industry* July 19/1 There are three types of telephone exchange in use; one is the Strowger which has been the standard British equipment ever since we introduced the automatic exchange. 1973 *Nature* 16 Feb. 416/1 The antiquated Strowger step-by-step switching on which Britain's telephones deplorably now depend.

struck, *ppl. a.* Add: **4. b.** *struck-off*: of persons in certain professions, debarred from practising by having one's name deleted from the register of those qualified. Cf. *strike *v.* 13 b, 82.
1963 *Sunday Express* 27 Jan. 23/3 A struck-off doctor—now a dope addict. 1972 E. Routley *Puritan Pleasures of Detective Story* ii. vii. 74 An unexpected meeting with a struck-off solicitor turned private detective. 1976 E. Ward *Hanged Man* xxix. 189 Are *you* really a doctor?.. You sound like a struck-off vet.
7. e. *struck joint* (Building): a joint in which the mortar between two courses of bricks is sloped inwards so as to be flush with the surface of one but below that of the other.
1876 *Notes Building Construction* II. xiii 219 Struck Joints should be formed by pressing back the upper portion of the joint while the mortar is moist, so as to form a sloping surface which throws off the wet. 1948 *Archit. Rev.* CIV. 20 (*caption*) Brick external walls are finished in buff face brick with struck joints. 1978 S. Martin *Build your own House* (ed. 7) v. 75 The flush joint and the struck weathered joint are formed with a trowel.
7*. Of, pertaining to, or affected by an industrial strike. Chiefly *U.S.*
1894 S. & B. Webb *Hist. Trade Unionism* ii. 80 Finding that the yarn was for a 'struck shop'. 1937 *N.Y. Times* 16 June 1/6 The Republic Steel Corporation..asked for a mandamus writ to compel Postmaster General Farley..to deliver to people in its struck plants 'all matter properly mailable'. 1939 *Sun* (Baltimore) 21 July 2/6, 60,000 butchers in retail shops would refuse to handle 'struck products'. 1947 *Ibid.* 14 Oct. 24/1 The league said it would withdraw its charges.., provided the union would delete its 'closed shop' and 'struck work' clauses. 1962 *Aeroplane* 12 Apr. 16/1 Since 1958 the same eight airlines..have had a mutual aid agreement whereby a 'struck' company receives from the others any excess revenues attributable to the strike, less additional expenses. 1968 *Globe & Mail* (Toronto) 3 Feb. B5/3 Men and women who make a living by working on struck newspapers. 1977 *Time* 27 June 35/2 Not since World War II, when President Roosevelt threatened to call out the armed forces to reopen struck mines, has the union played such an important role in the nation's well-being.

struck (strʌk), *sb.* [Subst. use of Struck *pa. pple.* and *ppl. a.*] A bacterial disease of sheep causing sudden convulsive death after few symptoms; orig. more loosely in *dial.* use.
[1784: see Strike *v.* 45 d]. 1903 *Jrnl. S.-E. Agric. College* xii. 86 First and most prominent is the disease commonly known as 'struck', or 'struck in the blood'. These terms appear to be somewhat loosely applied to any cases of sudden death in sheep... There is, however, especially in the marsh lands of Kent,..a disease in which few symptoms of a definite character are ever seen, owing to the rapid approach of death, and to which the term 'struck' is intended to be applied. [1904 *Eng. Dial. Dict.* V. 827/1 Ken[t]. A sheep which dies suddenly of a disease akin to apoplexy is said to die 'struck'.] 1929 *Jrnl. Compar. Path. & Therapeutics* XLIII. 1 The term 'struck' is applied by the farming community [of Romney Marsh] to a rapid and fatal disease where post-mortem examination reveals an acute inflammatory condition in one or more of the following parts of the body: areas of muscular tissue, organs in the abdominal cavity, and organs in the thoracic cavity. 1966 T. Dalling *Internat. Encycl. Vet. Med.* I. 640 The classical type [of *Clostridium perfringens* Type C] was shown to be the cause of 'Struck' (Romney Marsh disease). 1972 *TV Vet Sheep Bk.* lxii. 169/1 All the clostridial diseases—enterotoxaemia, pulpy kidney, lamb dysentery, struck (blackleg),..can and should be prevented by vaccinating the ewe flock.

structural, *a.* Add: **2.** *structural engineering*, the branch of civil engineering concerned with large modern buildings and other structures; so *structural engineer*.
1896 *Engineering Index 1892–5* II. 412 Structural engineering. Courses in —. See Engineering Education. 1912 *Register of Former Students* (Mass. Inst. Technol.) 65 Burleigh, Cha(rle)s R(andall)..structural engineer and chief draughtsman. 1924 *Times Trade & Engin. Suppl.* 29 Nov. 248/2 Structural engineers are irregularly employed. *Ibid.*, Some of the finest structural engineering in the world was done in the Black Country. 1943 A. Rand *Fountainhead* i. i. 15 Of course, no one denies the importance of structural engineering to a future architect. 1977 *Modern Railways* Dec. 488/1 The APT project placed a complex set of inter-related demands on the structural engineer.
4. d. *Chem.* Of or pertaining to the arrangement of atoms in a molecule; *structural formula*, a plane schematic method of representing the structure of a molecule by using punctuation (as $CH_3{\cdot}CH{:}CH_2$) or lines (as $CH_3{-}CH{=}CH_2$) to indicate the position and nature of the bonds between constituent atoms; *structural isomerism*, a form of isomerism in which molecules having the same constituent atoms may have different structures, the atoms being joined in different sequences; so *structural isomer*.
1872 *Phil. Mag.* XLIII. 241 (*heading*) On the relations between the atomic hypothesis and the condensed symbolic expressions of chemical facts and changes known as dissected (structural) formulæ. 1876 *Ibid.* II. 162 To so-called normal butylic alcohol is generally assigned the structural formula $CH_3(C_2H_4)OH$. 1926 J. Read *Textbk. Org. Chem.* xii. 215 The various kinds of isomerism encountered up to the present point are all included under the general title of structural isomerism. Structural isomers are substances possessing the same molecular formula but different structural formulæ. 1951 I. L. Finar *Org. Chem.* I. i. 6 It is always desirable to show the arrangement (if known) of the atoms in the molecule, and this is done by means of structural formulæ or bond-diagrams. 1980 C. W. Spangler *Org. Chem.* i. i. 17 Such branching produces structural isomers, compounds having identical molecular formulas, but whose carbon backbones are arranged differently in three-dimensional space. *Ibid.* 18 If another atom (or group) is bonded to the carbon system (say, chlorine), positional isomerism as well as structural isomerism becomes possible. *Ibid.* 19 There are many different methods of writing structural formulas for organic molecules.
e. *Biol.* Of a gene: that specifies the amino-acid sequence of a polypeptide.
1959 Pardee, Jacob, & Monod in *Jrnl. Molecular Biol.* I. 177 The situation revealed with the present system, namely a genetic 'complex' comprising, besides the 'structural' genes (z, y) a repressor-making gene (i) whose function is to block or regulate the expression of the neighbouring genes, is, so far, unique for enzyme systems. 1966 E. A. Carlson *Gene* xxiv. 229 The repressor was assumed to be incomplete. Thus the structural genes would be transcribed and their enzymes would synthesize a metabolite. 1976 F. J. Ayala *Molecular Evolution* ii. 12 Substitutions in the DNA nucleotide sequence of a structural gene may result in changes in the amino acid sequence of the polypeptide encoded by the gene, although this is not always the case because of the degeneracy of the genetic code.
5. In the Social Sciences, Psychology, and other disciplines, such as Linguistics, connected with the analysis of social, mental, or linguistic organization. **a.** Of, pertaining to, involving, or resulting from those aspects of a system concerned with the formal laws and relations of its structure, as distinguished from function or phenomenon; also, relating to or connected with the 'deep' structures that are considered to generate 'surface' structures. Cf. *deep structure* s.v. *deep *a.* IV. c; *functional *a.* 2 c.
1884 W. James in *Mind* IX. 19 The contrast is really between two *aspects*, in which all mental facts without exception may be taken; their structural aspect, as being subjective, and their functional aspect, as being cognitions. 1890 O. T. Mason in *Ann. Rep. Smithsonian Inst.* (1891) 527 A complete syllabus of anthropology would include—first, what man is, and second, what man does. What man is may be denominated *structural anthropology*; what man does, *functional anthropology*. 1908 *Philos. Rev.* XVII. 651 The book is a very pronounced example of the structural type of psychology. 1917 *Internat. Jrnl. Amer. Linguistics* I. 177 The existence of phonetic shifts, and the presence of structural similarities are too numerous. 1932 M. Fortes tr. *Petermann's Gestalt Theory* ii. 33 The concept leads to a theory of reactions..which is characterized by Köhler's key-word 'structural reaction'. 1940 C. C. Fries in *Language* XVI. 199 (*heading*) On the development of the structural use of word-order in modern English. 1944 *Social Res.* XI. 99 A Beethoven symphony where from a part of the whole we could grasp something of the inner structure of the whole itself. The fundamental laws, then, would not be piecemeal laws but structural characteristics of the whole. 1952 C. C. Fries *Structure of English* iv. 58 One of the basic assumptions of our approach here to the grammatical analysis of sentences is that all the structural signals in English are strictly formal matters that can be described in physical terms of forms, correlations of these forms, and arrangements of order. 1952 A. R. Radcliffe-Brown *Structure & Function in Primitive Society* 11 When we are dealing with a structural system we are concerned with a system of social positions. 1958 *Eng. Jrnl.* XLVII. 479 Structural ambiguity, on the other hand, results from the arrangement of the words, that is, from the structure of the utterance. It is sometimes known as syntactic ambiguity and, in older logic books, as amphiboly. 1962 R. Jakobson *Sel. Writings* I. 654 An analysis of the structural laws which underlie language and its evolution necessarily leads us to ascertain a limited set of actually given structural types. 1963 Jacobson & Schoepf tr. *Lévi-Strauss's Structural Anthropology* p. viii, Two papers..are published here for the first time in conjunction with fifteen others that seem to me to elucidate the structural method in anthropology. 1964 H. Hartmann *Ego Psychol.* xiv. 289 The genetic viewpoint had to be supplemented by a structural approach, though Freud never quite explicitly stated this. 1974 *Howard Jrnl.* XIV. 37 Worsening structural inequalities... (We shall use the term 'structural' to refer to housing, education, employment, income and race.) 1974 I. Rossi *Unconscious in Culture* 70 The same structural code is at work in mind, society, and physical reality. 1975 G. Steiner *After Babel* ii. 77 Hamann's opening statement..and his..dictum that theories of language and of economics will prove mutually explanatory..set out *in nuce* much of Lévi-Strauss's structural anthropology. 1976 G. S. Klein *Psychoanal. Theory* 10 Basic tendencies that are often implied in the 'structural point of view' of contemporary psychoanalysis.
b. Special collocations: *structural analysis*, analysis of a system in terms of its general characteristics or structure; hence *structural analyst*; *structural change* (see quot. 1972); *structural description* = *structural analysis* above; *structural-functional* adj., that takes account of both structure and function; so *structural-functionalism*; *-functionalist* adj. and sb.; *structural grammar* (see quot. 1975); *structural integration*, a technique of deep massage developed by Ida P. Rolf (see *Rolf); *structural linguistics*, the study of a language viewed as a system made up of interrelated elements without regard to their historical development (cf. *descriptive linguistics* s.v. *descriptive *a.* 3 b); hence *structural linguist*; *structural linguistic* adj.; *structural psychology*, an approach to the study of consciousness which relies on the introspective analysis of simple experience into elements; *structural semantics*, the study of the sense relations that may be established between words or groups of words; hence *structural semanticist*; *structural-semantic* adj.; *structural unemployment*, unemployment resulting from reorganization in the structure of industry due to technological change, etc., rather than from fluctuations in supply and demand; *structural word* = *empty word* s.v. *empty *a.* and *sb.* C (cf. *grammatical word* s.v. *grammatical *a.* 2; *structure word* s.v. *structure *sb.* 8).
1898 E. B. Titchener in *Philos. Rev.* VII. 465, I believe..that the best hope for psychology lies today in a continuance of structural analysis. 1901 H. Oertel *Lectures on Study of Lang.* i. 45 We are here concerned with the method only of Humboldt's structural analysis. 1940 *Language* XVI. 216 This is the neglect of the method of structural analysis, i.e. of organized synchronic description. 1974 tr. *Wertheim's Evol. & Revol.* i. 91 Weber's views are more than once mentioned as having paved the way for structural analysis. 1979 F. Kermode *Genesis of Secrecy* iv. 80 When the structural analysts have done their work, interpretation may take over. 1964 Structural change [see SC s.v. *S 4 a]. 1972 R. A. Palmatier *Gloss. Eng. Transformational Gram.* 168 *Structural change*.., the generalization of the operation which a particular transformation performs..; the right-hand side of a transformational rule. 1964 Structural description [see SD s.v. *S 4 a]. 1977 *Language* LIII. 15 The ö of *hösli* was not affected because, after all, phonetically it didn't meet the structural description of Lowering. 1947 T. Parsons in Parsons & Henderson tr. *Weber's Theory Social & Econ. Organization* 20 A second type [of conceptual scheme]..may be called a generalized structural-functional system. 1977 J. D. Douglas in Douglas & Johnson *Existential Sociol.* i. 6 The classical structural-functional paradigm of social theory grew out of and progressively diverged from this mechanistic model. 1958 *Listener* 28 Aug. 308/1 The first great proponent of structural-functionalism, Radcliffe Brown, failed to get beyond metaphor and analogy when he sought to explain his method. 1977 *Scott. Jrnl. Sociol.* I. 186 A sense that differs both from American structural-functionalism, in which structure is merely descriptive, and from French structuralism, in which it is 'reductive'. 1976 E. Leach *Culture & Communication* i. 3 Others..offer structural-functionalist explanations. 1977 *Dædalus* Summer 64 Like many structural-functionalists, he regards as unproblematical the processes by which corporations and other 'surface structures' come into existence. 1949 W. O. Birk (*title*) Structural grammar for building sentences. 1975 *Language for Life* (Dept. Educ. & Sci.) 594 *Structural grammar*, a grammar intended to explain the working of language in terms of the functions of its components and their relationships to each other without reference to meaning. 1963 *Systematics* I. 66 (*heading*) Structural integration. 1975 *Sat. Rev.* (U.S.) 22 Feb. 15/3 The body reformation methods called structural integration (commonly known as rolfing, after its developer Ida Rolf). 1949 *Archivum Linguisticum* I. i. 89 The functional interpretations of language given by structural linguists are justified. 1951 *Language* XXVII. 8 We must remember the particular attention which structural linguists in the United States have bestowed upon non-culture languages. 1958 *New Statesman* 6 Sept. 288/3 The new advance guard, the Structural Linguists, round on the New Critics as amiable old pipe-smoking fuddy-duddies. 1954 U. Weinreich in *Word* X. 389 Structural linguistic theory now needs procedures for constructing systems of a higher level out of the discrete and homogeneous systems that are derived from description. 1962 *New Yorker* 10 Mar. 158/2 For the scientific study of language the Structural Linguistic approach is superior to that of the old grammarians. 1940 *Amer. Speech* XV. 438/2 Harris, Zellig S. Rev. of L. H. Gray, Foundations of Language... Some useful remarks on 'structural' linguistics. 1941 A. W. de Groot in *Archives Néerlandaises de Phonétique Expérimentale* XVII. 71 (*title*) Structural linguistics and phonetic law. 1945 *Word* I. 58 With structural linguistics as the chief interest of some workers, area may be expected to play a more central role,—to be, in effect, a specialized point of departure. 1948 *Lingua* I. i. 23 'Structural linguistics' is a method, it is not a science and 'structuralism' is nothing but a collective name for general linguistic and explanatory grammatical examination of certain phenomena. 1959 J. C. Catford in Quirk & Smith *Teaching of English* vi. 168 Professor Quirk..is talking about the approach to the scientific study of language which is known as 'structural linguistics'. 1972 D. Lodge *20th Cent. Lit. Crit.* 545 Structural linguistics goes beyond the description of any particular language to pursue the 'deep structures' that are common to all languages. 1898 E. B. Titchener in *Philos. Rev.* VII. 449 (*title*) The postulates of a structural psychology. 1933 J. C. Flugel *Hundred Yrs. Psychol.* iv. ii. 229 (*heading*) 'Structural' and 'functional' psychology. 1980 J. M. Carroll (*title*) Toward a structural psychology of cinema. 1973 *Archivum Linguisticum* IV. 67 The structural-semantic category system of Soskin and John (1963). 1977 J. Lyons *Semantics* I. iv. 102 One of the points that Saussure and other structural semanticists have insisted upon is that each language has, not only its own stock of forms, but also its own system of meanings or concepts. 1962 D. H. Hymes in J. A. Fishman *Readings Sociol. of Lang.* (1968) 103 An ethnographic semantics..should be a structural analysis, achieving the economies of the rules of a grammar in relation to a series of analyses of texts. In the past generation Jakobson and his associates have done most to develop such a structural semantics. 1977 J. Lyons *Semantics* ix. 270 From its very beginnings structural semantics..has emphasized the importance of relations of paradigmatic opposition. 1932 A. H. Hansen *Econ. Stabilization* ix. 148 By structural unemployment..we mean unemployment caused by changes in the structure of industry that are of a non-recurring type. 1966 *Economist* 29 Jan. 407/2 The boom of the past years has brought jobs for Negroes and for the unskilled, in the teeth of cries that 'automation' and 'structural unemployment' meant that such people had no hope of finding work. 1979 *Guardian* 31 May 735/3 The spectre of 'structural unemployment'—the likelihood that some of today's young people will never have a job. 1940 Bryant & Aiken *Psychol. Eng.* iv. 38 A distinction is made between 'full' and 'empty' words, the latter being what are known as 'structural' words. 1966 J. Derrick *Teaching Eng. to Immigrants* i. 8 If we take our first example again, 'The man is hitting the horse' we find we can substitute any one of a hundred or more different items for the words *man*, *hit* and *horse*... But if we try to do the same thing at one of the other places in the sentence, we find we can only substitute a limited set of words... The words in these places, where the choice is limited, are what writers have called the structural or grammatical words of the language, while the other words have been called content words or lexical items.
c. = *structuralist 2.
1953 *College Composition & Communication* IV. iv. 124 The structural objection to the traditional use of *language* is that the account of the mechanisms becomes distorted beyond reason. 1959 *Word* XV. 176 The newly established chair of General Linguistics and Phonetics (which was, incidentally, the first structural academic position to be established anywhere).

structuralism (strǫ·ktiŭrăliz'm). [f. Structural *a.* + -ism.] **1.** *Psychol.* A method, connected esp. with the American psychologist E. B. Titchener (1867–1927), of investigating the structure of consciousness through the introspective analysis of simple forms of sensation, thought, images, etc.
1907 J. R. Angell in *Psychol. Rev.* XIV. 64 If you adopt as your material for psychological analysis the isolated 'moment of consciousness', it is very easy to become so absorbed in determining its constitution as to be rendered..oblivious to its artificial character. The most essential quarrel which the functionalist has with structuralism..arises from this fact. 1927 M. Bentley in C. Murchison *Psychologies of 1925* 390 However important or trivial we shall find the accomplishments of structuralism to be, we must recognize the gain in clear thinking which accrued to Titchener's sharply drawn distinction between the analytical psychology of structure and the descriptive psychology of mental operation and functional performance. 1930 *Times Lit. Suppl.* 19 Jan. 508/3 Modern schools of psychology, Structuralism, and Functionalism. 1968 *Internat. Encycl. Social Sci.* XV. 610 The movement called 'structuralism' which was founded in Germany by Wilhelm Wundt and transplanted to the United States by Edward B. Titchener of Cornell University.
2. Any theory or method in which a discipline or field of study is envisaged as comprising elements interrelated in systems and structures at various levels, the structures and the interrelations of their elements being regarded as more significant than the elements considered in isolation; also, more recently, theories concerned with analysing the surface structures of a system in terms of its underlying structure. **a.** *gen.*
1951 *Mind* L. 270 Braithwaite evidently believes that the whole philosophy of structuralism breaks down over the question of a combining relation. 1968 *Sunday Times* 10 Mar. 52 Structuralism is a technique for analysing any kind of symbolic system. Its break with the past consists in refusing to take note of the appropriateness of symbols for the things they symbolise. 1969 P. Anderson in Cockburn & Blackburn *Student Power* 246 Namier's legacy to English historiography was thus inevitably equivocal. His structuralism was rapidly suppressed from memory. 1970 M. Lane *Structuralism* 31 Structuralism, then, is a method whose primary intention is to permit the investigator to go beyond a pure description of what he perceives or experiences..in the direction of the quality of rationality which underlies the social phenomena in which he is concerned. 1971 C. Maschler tr. *Piaget's Structuralism* i. 4 We come upon at least two aspects that are common to all varieties of structuralism: first, an ideal..of intrinsic intelligibility supported by the postulate that structures are self-sufficient. 1972 *Sci. Amer.* Sept. 50/3 Structuralism recognizes that information about the world enters the mind not as raw data but as highly abstract structures that are the result of a preconscious set of step-by-step transformations of the sensory input. 1973 *Film Comment* May/June 52/1 In recent years, structuralism and semiology have received much attention as methods for analyzing and interpreting film... Structuralism..attempts to analyze comparatively the deep structures, thus locating those distinctive features common to all of man's cultural and social expressions. 1975 *New Rev.* II. xiv. 55/1 (*title*) Is your structuralism really necessary? *Ibid.* 57/2 Is not the case for temporalism overwhelmingly stronger than the case for structuralism? 1978 *History Workshop* Autumn 3 British Marxist structuralism exalts theoretical practice to the point where it seems to become an end in itself. 1980 *London Rev. Bks.* 15 May 3/2 Structuralism is the philosophy of those in the universities and thereabouts who are not philosophers.
b. *Linguistics.* Applied to theories in which language is considered as a system or structure comprising elements at various phonological, grammatical, and semantic levels, esp. after the work of F. de Saussure (1857–1913).
1945 E. A. Cassirer in *Word* I. 99 (*title*) Structuralism in modern linguistics. *Ibid.* 104 If the adherents and defenders of the program of linguistic structuralism are right, then we must say that in the realm of language there is no opposition between what is 'formal' and what is merely 'factual'. 1953 A. Martinet in A. L. Kroeber *Anthropol. Today* 577/1 It would seem that the teaching of Ferdinand de Saussure has, directly or indirectly, influenced most of linguistic structuralism. 1964 *English Studies* XLV (Suppl.). 33 They intend..to stress the importance of semantic studies..(as a necessary counterpart to formal structuralism). 1968 J. Lyons *Introd. Theoret. Linguistics* x. 443 It is one of the cardinal principles of 'structuralism', as developed by de Saussure and his followers, that every linguistic item has its 'place' in a system and its function, or value, derives from the relations which it contracts with other units in the system. 1972 *Language* XLVIII. 419 Structuralism proper in linguistics began with phonology. 1976 *Archivum Linguisticum* VII. 152 With the rise of structuralism, linguistics turned back upon itself, so to say, and tended to abstract away from the social matrix of language.
c. *Anthrop.* and *Sociol.* The theories or methods of analysis concerned with the structure or form of human society or social life; also, following the work of the French anthropologist Claude Lévi-Strauss (b. 1908), theories concerned with the deeper structures of communication from which the surface structures or 'models' evolve.
1955 R. Firth in *Jrnl. R. Anthropol. Inst.* LXXXV. 1 All British social anthropologists are structuralists in their use of the analytical principles developed by this method. But the rigidity and limitations of a simple structuralism alone have come to be more widely perceived. 1969 A. G. Frank *Latin Amer.* (1970) ii. 68 The pioneering service..of those latter students of economic development and cultural change is precisely that they drop all pretense and practice of social scientific structuralism. 1973 J. Rex *Discovering Sociol.* ix. 118 French structuralism has to be sharply distinguished from the structuralism of Radcliffe-Brown with which it compares itself, and the structuralism of Simmel and Weber, of which it remains largely ignorant. 1978 J. Z. Young *Programs of Brain* 299/1 *Structuralism*, a movement in social science originated by Claude Lévi-Strauss, which supposes that social structures depend upon certain basic characteristics of human brain programs.

structuralist (strǫ·ktiŭrălist), *sb.* (and *a.*). *Social Sci.* and *Humanities.* [f. Structural *a.* + -ist.] **1.** An advocate or adherent of a structural approach (see *structural *a.* 5 a) or of a theory of structuralism.
1907 J. R. Angell in *Psychol. Rev.* XIV. 67 Dwelling as the structuralist is supposed to do upon the problem of determining the irreducible elements of consciousness and their characteristic modes of combination. 1922 W. McDougal in *Amer. Jrnl. Psychiatry* Jan. 347 Mott and the other structuralists..would..dismiss the second alternative with contempt. 1949 *Archivum Linguisticum* I. 184 A survey of post-neogrammarian views on language, viz. those of Saussure..and the Danish 'structuralists'. 1965 *Economist* 25 Sept. p. xxvi/2 Some extreme structuralists argue that demand deflation is actually likely to increase cost inflation. 1975 G. Steiner *After Babel* ii. 83 More than a century before the modern structuralists, Humboldt notes the distinctive binary character of the linguistic process. 1982 *Listener* 18 Nov. 17/2 Structuralists and post-Structuralists maintain that the notion of the author as the creator of his works is merely a modern consolation prize.
2. *attrib.* or as *adj.*
1907 J. R. Angell in *Psychol. Rev.* XIV. 62 The most lucid exposition of the structuralist position still remains, so far as I know, Titchener's paper, 'The Postulates of a Structural Psychology'. 1929 W. B. Pillsbury *Hist. Psychol.* xvi. 271 The structuralist school..holds that consciousness is directly observable and is composed of simple, definitely describable elements. 1955 *Jrnl. R. Anthropol. Inst.* LXXXV. 1 The air of enchantment which for the last two decades has surrounded the 'structuralist' point of view. 1955 *Times* 11 June 5/5 In contrast to much of to-day's painting..which is considered finished when it has reached the stage which, in any other epochs, would have been called the sketch, Structuralist painting is carried to completion in strict obedience to the scientific laws of colour-form structure. 1963 *Indian Econ. Rev.* Feb. 67 The monetarist-structuralist controversy is an argument over the remedy for inflation, the monetarist taking supply as given and recommending a contraction of demand, the structuralist..taking demand as given [etc.]. 1965 N. Chomsky *Aspects of Theory of Syntax* ii. 67 Such a system is apparently what is implicit in modern taxonomic ('structuralist') grammars. 1970 J. Lyons *Chomsky* 29 The 'structuralist' approach was by no means confined to Boas and his successors in America. 1970 *Sunday Times* 15 Nov. 32/2 Today when we go to Paris, we can read structuralist novels, look at structuralist paintings and hear structuralist music. Even newspaper cartoons and gourmet meals are subjected to structuralist interpretations. 1976 *Archivum Linguisticum* VII. 152 Even present-day transformational grammar, though quite remote from classical structuralist thought, singles out a 'fluent speaker' set apart from the disturbing influences of social variation. 1979 A. R. Peacocke *Creation & World of Science* i. 29 This more narrowly functionalist account of myth has given way, following the lead of Lévi-Strauss, to a 'structuralist' account.

structuralistic (strǫktiŭrăli·stik), *a.* [f. *structuralist *sb.* + -ic]. Characteristic of a structuralist or of structuralism; structuralist.
1957 H. Whitehall in N. Frye *Sound & Poetry* 134 Both criticism and linguistics, in their modern phases, are structuralistic in attitude. 1972 *Jrnl. Social Psychol.* LXXXVII. 37 Mental imagery and image formation have recently become a topic of concern again, after avoidance of such structuralistic phenomena for 50 years or more. 1978 *Studies in Eng. Lit.: Eng. Number* (Tokyo) 134 Her idea of 'structure' is structuralistic, and this is no coincidence.

structuralize (strǫ·ktiŭrăləiz), *v.* [f. Structural *a.* + -ize.] *trans.* = Structure *v.*; also, to apply structural theories or analysis to (something). So **stru:cturaliza·tion**; **stru·cturalized** *ppl. a.*; **stru·cturalizing** *vbl. sb.*
1936 K. Mannheim *Ideology & Utopia* iii. 128 All that has become intelligible, understandable, rationalized, organized, structuralized, artistically, and otherwise formed, and consequently everything historical seems in fact to lie between these two extreme poles. 1953 C. E. Bazell *Linguistic Form* iii. 29 Distributional criteria seldom provide grounds here for a narrower structuralization. 1964 H. Hartmann *Ego Psychol.* iii. 64 Once structuralization has taken place, they [sc. ego interests of the id] become partly independent in the service of the ego. 1965 *Canad. Jrnl. Linguistics* Spring 142 Once Kiowa was structuralized, however, it turned out that it made the same distinctions in vowel type that the Tanoan languages made. 1975 *Times Lit. Suppl.* 21 Nov. 1396 (*heading*) Structuralizing in Sumatra. 1977 *Dædalus* Fall 70 Far-reaching changes in the political regimes could affect the functioning of economic institutions and the structuralization of social hierarchies.

structurate (strǫ·ktiŭreⁱt), *v. rare.* [f. Structure *sb.* + -ate³.] *trans.* = prec. So **stru·cturated** *ppl. a.*
1937 *Mind* XLVI. 512 The physical sciences are in a happy position, because their numerical world is mathematically structurated. 1978 *Eng. Jrnl.* Dec. 20/1 You say you've been grammaticized, transformalized, and structurated.

structuration (strǫktiŭrēⁱ·ʃən). [f. Structure *v.* + -ation.] The condition or process of organization in a structural form.
1927 *Brit. Jrnl. Psychol.* XVIII. 6 After many such changes an image may represent a vast structure of meanings, and so 'structuration' may be attributed to it. 1934 *Mind* XLIII. 110 In man the main feature, because the *dynamic* feature, of his development consists in the structuration of his propensities into sentiments. 1952 W. Sprott *Social Psychol.* iii. 52 All groups..tend towards a structuration which is evidenced by..a role system, an ethos, a certain formalization of action-patterns. 1973 A. Giddens *Class Struct. Advanced Societies* xi. 201 The influence of paratechnical relations upon working-class structuration. 1973 *Screen* Spring/Summer 11 Semiology opens out that process in its disengagement of the heterogeneity of codes at work in the structuration of film. 1980 *Times Lit. Suppl.* 26 Sept. 1072/1 It might be more appropriate to speak of the differential structuration of life chances as constituting classes.

structure, *sb.* Add: **3. d.** Also *spec.* in *Linguistics*, *deep structure*: see *deep *a.* IV. c; *surface structure*: see *surface *sb.* 6 d.
1961 Y. Olsson *On Syntax Eng. Verb* ii. 27 Both collocation and colligation operate syntagmatically... They are examples of *structures* constituted by *elements*. 1965 *Language* XLI. 73 The nested structures are phrase types which are in clear structural contrast in the language.
8. *structure-dependence*, *dependency*, *sensitivity*; *structure-borne*, *-dependent*, *-independent*, *-sensitive* adjs.; **structure-function** *attrib. phr.*, pertaining to both structure and function; **structure plan** *Local Government*, a plan drawn up by a local planning authority for the development, use, conservation, etc., of a prescribed area of land; hence **structure planning**, the preparation of such a plan; **structure word** = *structural word* s.v. *structural *a.* 5 b.
1962 A. Nisbett *Technique Sound Studio* ii. 38 Structure-borne noises are almost impossible to eliminate and rebuilding work is complicated by the need to avoid all noisy work when nearby studios are recording. 1972 *Lebende Sprachen* XVII. 37/2 *Structure-borne noise*, a condition when the sound waves are being carried by a solid material. 1976 N. Chomsky *Reflections on Lang.* i. i. 33 The principle of structure-dependence is not learned, but forms part of the conditions for language learning. 1976 *Times Lit. Suppl.* 17 Dec. 1590/4 By structure dependency he [sc. Chomsky] means, and I mean, that the significance of any feature is determined by its position in a structure. 1965 Structure-dependent [see *analysability]. 1978 *Logophile* VIII. 5/2 All syntactic operations in language are structure-dependent. 1963 *Canad. Jrnl. Linguistics* VIII. 59 (*heading*) A structure-function description of Terena phrases. 1977 *Dædalus* Fall 115 For the moment I want to draw the kind of structure-function distinction we might adopt, say, in discussing walking: what it is versus where it might take you. 1964 E. A. Power *Introd. Quantum Electrodynamics* i. 5 Also even the structure-independent radiation damping term depends on the arrow sign of time—unlike the fundamental

equations themselves—because in calculating the radiation field boundary conditions at infinite time are involved. **1976** N. Chomsky *Reflections on Lang.* I. i. 33 Construct a structure-dependent rule, ignoring all structure-independent rules. **1971** *Act* 19 & 20 *Eliz. II* c. 78 §7 The local planning authority shall..prepare and send the Secretary of State..a structure plan for their area. **1980** *Oxford Times* 12 Dec. 10/4 Policy E1 of the Structure Plan says that the Council will restrain growth of employment in the county as a whole. **1973** *Times* 8 Sept. 14/6 He made it plain that he did not expect an end to the structure planning process, the framework for regulating development which every local planning authority is expected to prepare. **1976** *Alyn & Deeside Observer* 10 Dec. 16/1 The strange spectacle of two Conservative councils locked in a fierce exchange over structure planning. **1936** *Jrnl. R. Aeronaut. Soc.* XL. 593 The strength of a crystal across a crystal plane is a structure-sensitive property. **1970** *Language* XLVI. 261 We demonstrate the use of transformational rules applied to the output of structure-free grammars as a means of generating symmetrical strings which are not only structure-sensitive but context-sensitive as well. **1976** *Times Lit. Suppl.* 17 Dec. 1590/4 My emphasis on structure sensitivity in 'natural' situations leads me to look for leads in linguistics. **1956** *Publ. Amer. Dial. Soc.* xxvi. 59 A listing by parts of speech brings out the numerical superiority of nouns, and the relatively small number of structure words borrowed. **1965** Structure-word [see *functor 2].

structure, *v.* Delete *rare* and add: **a.** (Later examples.) Also, to establish a hierarchy of relationships or a pattern in (something). Also *absol.* and *loosely*, to construct, form, or organize. So **stru·cturing** *ppl. a.*

1933 *Mind* XLII. 176 It is the architectonic (supreme and structuring) principle or spirit regulating in their mutual relations all other structuring activities. **1949** M. Mead *Male & Female* xv. 296 This has been an accurate picture of the way in which we have structured our society, with women as keepers of the house.., and men as keepers of women in the house. **1955** Auden *Shield of Achilles* i. 16 When, on some windless day Of dejection, unable To name or to structure. **1960** L. Pincus *Marriage* III. 221 An inhibited response structured by..these unconscious dilemmas. **1962** *Amer. Speech* XXXVII. 107 Bloomfield..calls /r/ a 'vocaloid' and says it may be syllabic..but he has it function as a consonant when structuring the final consonant clusters of English. **1971** *Sci. Amer.* Aug. 91/1 Input/Output tables provide management..with a powerful new tool for forecasting and measuring the indirect as well as the direct inter-industry relationships that structure our industrial economy. **1974** R. Helms *Tolkien's World* v. 98 The parallel plot-lines of Book III..are both structured by the laws of providential control and of the cause-and-effect morality. **1978** *Guardian Weekly* 10 Sept. 17/4 A secret study memorandum..which helped structure Carter's decision last month to use trade restrictions. **1979** *Nature* 23 Aug. 652/1 The chemical exchanges among uplands, marshes and coastal waters are important in structuring these ecosystems.

b. To give (someone or something) a place in a structure; to absorb or integrate into a pattern or system.

1954 J. A. C. Brown *Social Psychol. Industry* vi. 174 Aggressiveness as a response to this has become structured into his basic personality. **1955** C. Cherry in B. L. Evans *Stud. in Communication* iii. 60 The individuals were to be structured into social coherence. **1959** M. Dolinsky *There is no Silence* iii. 44 You don't easily structure the unusual or the exotic. **1977** A. Giddens *Stud. in Social & Polit. Theory* i. 70 Criticism thus cannot be terminated within the sphere of science itself, but must concern itself with the standards or values which structure science as one mode of activity among others.

c. To present or manipulate (a situation, etc.) in such a way as to elicit desired responses or effects.

1966 *Listener* 10 Mar. 342/2 Parents who are trying to train their children make use of verbal explanations to structure the situation. **1971** J. B. Carroll et al. *Word Freq. Bk.* p. ix, His assistance was also essential in..structuring the survey to reflect curriculum divisions and publications diversity. **1973** *Howard Jrnl.* XIII. 282 Treatment..is a matter of so structuring the environment..that aversive consequences..follow unwanted or undesirable behaviour. **1975** *Language for Life* (Dept. Educ. & Sci.) x. 145 He must structure the learning so that the child becomes positively aware of the need for a complicated utterance.

structured, *ppl. a.* Add: **1.** (Later examples.)

1929 A. N. Whitehead *Process & Reality* 138 The enduring object..may be conceived as independent of the structured society. **1940** C. S. Sherrington *Man & his Nature* xi. 353 There is the difficulty that the outward process on analysis proves always to be 'granular', quantal, 'structured'; the inner process to be structureless, non-quantal. **1962** H. A. Gleason in Householder & Saporta *Problems in Lexicography* 91 Occasionally sets of language phenomena are observed which, though apparently structured, seem not to be so rigorously structured as this. **1968** *Guardian* 12 June 8/4 The taking of hallucinogenic drugs, where the objective is..to erase the influence of structured experience and organisation of perceptual material in order to explore some kind of world beyond the structures. **1980** P. Hill *Savages* iii. 46 The boys are not criminals... They need a structured group, a sense of belonging.

2. a. Organized or arranged so as to produce a desired result. Also *loosely*, formal, organized, not haphazard.

1959 *Listener* 17 Dec. 1079/1 The use of measured or structured arguments to get at the causes of such interesting phenomena as millionaires or strikes. **1968** *Brit. Med. Bull.* XXIV. 189/2 The computer can accept data only in a highly structured (digital) form. **1975** *Language for Life* (Dept. Educ. & Sci.) v. 59 In certain cases the object was to equip the mother with the ability to work through a structured programme with her child. **1976** *Ann. Rep. Manpower Services Comm. 1975–76* iv. 30/1 The review is intended to provide a framework within which choices can be made and resources allocated in a more structured way. **1977** *Cornish Times* 19 Aug. 8/7 It will provide all learning opportunities through the medium of formal teaching processes, planned excursions, structured and unstructured play and recreational experiences.

b. Of a computer program: organized in a logical way to facilitate debugging and modification; *spec.* composed of linked but distinct modules each having one entry point and one exit point, so that the program may be read straight through; so *structured programming* vbl. sb.

1966 *Information Processing 1965* II. 448/1 (*heading*) On solving large structured programs. **1972** *Bit* XII. I. 38 (*heading*) An experiment in structured programming. **1974** D. D. McCracken *Simplified Guide to Fortran Programming* vi. 110 This makes it possible to write the program without a go to... This..is one aspect of structured programming, which holds great promise of converting programming from a hit-or-miss craft into an engineering science. **1978** L. A. Leventhal *Introd. Microprocessors* vi. 250 In structured programming only single logic structures are used. Bohm and Jacopini showed that any program could be written by using only three structures. 1. A sequential structure... 2. A conditional structure of the if-then-else type... 3. A loop structure of the do-while type... The computer executes P repeatedly as long as A is true. **1983** *Personal Computer World* Sept. 216/2 If a program is badly structured..altering one part of the program may have unforeseen effects in a completely different part.

structuredness (strʌ·ktiŭɹdnės). [f. Structured *ppl. a.* + -ness.] The quality of being structured.

1966 S. Beer *Decision & Control* xiv. 350 He refers to the gain in structured-ness that accompanies the entropy change in a biological system as a gain in the *maturity* of that system. **1970** *Jrnl. Gen. Psychol.* July 69 The tasks of the Wechsler..are not composed of stimuli all of the same degree of structuredness. **1974** *Lang. Sciences* Aug. 27/2, I am personally convinced that the focus on structuredness has been and will be as heuristic..in semantics as it has been in phonology and grammar. **1977** *Lancet* 22 Jan. 172/1 The data are too few at present to indicate whether the use of H.C.G. in addition to myelin or cancer basic protein will significantly improve the specificity of the M.E.M. test and the related test for the 'structuredness of cytoplasmic matrix' as means for detecting the presence of malignant disease.

stru·cturing, *vbl. sb.* [f. Structure *v.* + -ing[1].] The action of the verb.

1951 M. McLuhan *Mech. Bride* (1967) 77/1 The emotional structuring which results..from affection that is tendered or withdrawn. **1958** *New Scientist* 25 Sept. 887/2 While apes and infants struggle towards some sort of structure-formation, the human artist of today struggles to free himself from all structuring or imagery. **1962** H. A. Gleason in Householder & Saporta *Problems in Lexicography* 92 Crossing grammatical structure in various ways are..other types of structurings, some of them, at least, so different in form as to be unrecognizable as structure by the methods of descriptive linguistics. **1965** H. I. Ansoff *Corporate Strategy* (1968) i. 19 One part of the administrative problem is concerned with organization: structuring of authority and responsibility relationships, work flows, information flows, [etc.]. **1970** T. Lupton *Managem. & Social Sci.* (ed. 2) ii. 35 The factors which influence the structuring of social relationships. **1977** M. Edelman *Polit. Lang.* v. 92 This kind of cognitive structuring exemplifies the selective perception of information to reinforce established beliefs. **1980** *Amer. Speech* LV. 42 A toponymic dialect would be shaped..also by other important factors:..by the structuring of onomastic fields.., by the sequence of naming, [etc.].

structurism (strʌ·ktiŭriz'm). [f. Structure *sb.* + -ism.] The artistic theory or practice of a structurist (sense *2).

1963 *New Republic* 16 Feb. 26/3 Biederman..has written several books expounding Structurism as a future imperative. **1965** *Minneapolis Tribune* 21 Mar. (Arts Sect.) 1/4 His work evolved into the painted aluminium reliefs that represent his latest effort. These carried him through an attitude called constructionism to structurism. **1970** *Time* 26 Jan. 37 Biederman himself, having grandly declared that both painting and sculpture were obsolete, arrived at what he has come to call 'structurism'—reliefs that have the dimension of sculpture and the color of painting. **1973** *Phaidon Dict. Twentieth Cent. Art* 35/1 Biederman rejected Constructivism and termed his 'new art' Structurism—three-dimensional constructions in which small single-coloured rectangles are placed on a solid-coloured background, at right angles to the background and to each other.

structurist. Restrict *rare* to sense in Dict. and add: **2.** An artist whose work emphasizes underlying structural forms and processes in nature; applied esp. to the U.S. artist Charles Biederman (b. 1906) and by him to earlier artists in whose work he perceived this tendency. Also *attrib.* or as *adj.*

1958 C. Biederman *New Cézanne* viii. 61 Monet alone is the great Structurist innovator for the 19th century. It is Cézanne who is the great Structurist innovator for the 20th century! *Ibid.* 77 The genuine Structurist is indifferent to the slick materials of industry. **1959** —— in *Structure* 2nd Ser. I. 16/1 Artists no longer found their art but only their method of art, not in the creations but only in the creative method of nature's structural process. I call this the Structurist direction of art. **1960** *Guardian* 2 Feb. 7/5 The post-cubist collage-makers and other pioneer abstract 'structurists'. **1967** *Times* 17 Mar. 12/5 The trustees [of the Tate Gallery] have also recently acquired 'Structurist Relief. Red Wing No. 20' 1954–65, by Charles Biederman. *Ibid.* Biederman's 'structurist' work developed from a study of Mondrian. **1970** *Sat. Rev.* (U.S.) 17 Oct. 54 He and others 'in curriculum development and teacher training have been guilty for years of structuring canvases and of choosing the brushes and paints'. And yet how many structurists are willing to abandon that on which they depend? **1976** J. van der Marck in *Charles Biederman* (Minneapolis Inst. Arts) 64 That gradual development, from the simple to the complex, which is a tenet of the Structurist philosophy.

structurization (strʌ:ktiŭrəizē¹·ʃən). [f. Structure *sb.* + -ization.] The process of giving a structure to something or of arranging material into an organized pattern.

1950 *Brit. Jrnl. Psychol.* Dec. 129 Do [rumours]..take their origin almost wholly from 'within' the mind... Or do they arise as 'constructions' or structurizations of the stimulus situation, produced by the peculiar attitude, which itself was formed under the influence of the special situation of a crisis? **1955** H. Read *Icon & Idea* iii. 60 Then, on the basis of this symbolic activity and of intuitive thought generally, a representational structurization of space becomes possible. **1958** C. Biederman *New Cézanne* viii. 62 Having failed, out of fear of the consequences of Cézanne's geometrical structurizations, they [*sc.* the Cubists] now switched their attention to the surface aspects of Cézanne. **1969** *New Scientist* 6 Nov. 304/3 For him [*sc.* Grèco],..the concept of structure and the process of structurization..are essential for the understanding of human intelligence.

structurize (strʌ·ktiŭrəiz), *v.* [f. Structure *sb.* + -ize.] *trans.* To give a structure to (something), to organize structurally.

1958 C. Biederman *New Cézanne* iv. 37 The brush stroke structurizes color. *Ibid.* 48 Art is now directed towards a 'constant reality', structurized by 'constant relationships'. **1968** *Times* 18 Dec. 9 Research capacities are being structurized to the optimum in every single economic branch in order to meet the country's requirements of the scientific and technological revolution.

strudel (strū·d'l, ‖ʃtrū·d'l). [a. Ger., lit. 'eddy, whirlpool'.] A baked sweet of Austrian origin, made of very thin layers of pastry with a filling, usu. of fruit. Also used *attrib.* to denote the kind of dough or pastry used in such confections. *apfelstrudel, apple strudel*: see *apple *sb.* B. II.

1893 *Encycl. Pract. Cookery* II. 525/2 *Strudels*, a kind of pancake or fritter made in Germany. **1903** *Jewish Encycl.* IV. 257/2 The *strudel*, or single-layered jelly or fruit cake, takes the place of the pie for dessert. **1915** L. Kander *'Settlement' Cook Book* (ed. 7) xxi. 251 Prepare Strudel dough,..fill with kraut and other above ingredients. **1923**, etc. [see *apple strudel* s.v. *apple *sb.* B. II]. **1932** [see *lekach]. **1941** B. Schulberg *What Makes Sammy Run?* ix. 225 *Strudel*,..still hot... Sammele used to say I made the best *strudel* in the whole world. **1950** *Here & Now* (N.Z.) Dec. 26/1 Whether visitors to Vienna went to the big expensive hotels..or ordered a quick meal in a coffee-house..they always kept haunting memories of *schnitzels, strudels* and cakes. **1960** [see *schnitzel]. **1972** F. B. Maynard *Raisins & Almonds* 31 She was baking while I watched..the miracle of the strudel. **1981** *Times* 19 Mar. 8/6 Feta cheese and phyllo or strudel pastry are..sold in shops specializing in Greek or Cypriot foods.

'strue (strū), schoolboy's abbrev. of Construe *v.* and *sb.*

1903 Farmer & Henley *Slang* VII. 14/2 *Strue*, verb, (schools'), 'construe'. **1906** D. Coke *Bending of Twig* viii. 122 But mind, you've promised to sap out the 'strues, and 'strue them to me. *Ibid.* 125 Russell..made Lycidas extend his 'struing services to six more mornings' work.

struggle, *sb.* Add: **3.** *Comb.*, as **struggle-buggy** *U.S. slang*, a motor vehicle; *spec.* an old and battered one; **struggle meeting,** [tr. Chinese *dòuzhēng huì*], in Communist China: a meeting at which those who have aroused official or public disfavour are criticized or denounced.

1925 *College Humor* Sept. 20/2 I'll say you can park in my struggle buggy. **1946** Struggle-buggy [see *rinky-dink *a.*]. **1966** F. Schurmann *Ideology & Organization in Communist China* v. 318 'Struggle meetings' were held throughout China in which offending cadres were attacked, and by mass demand removed from office. **1973** *Times* 21 Mar. (China Trade Suppl.) p. viii/5 Officials who have been through 'struggle meetings', because they were considered to be bureaucratic..are likely to be sufficiently shaken by the experience to avoid arousing such resentments in the future.

struma. Add: **1. c.** *struma lymphomatosa* [mod.L., coined in Ger. (H. Hashimoto 1912, in *Arch. f. klin. Chir.* XCVII. 219)], = *Hashimoto's disease* s.v. *Hashimoto; *Riedel's struma* [*Riedel], a rare condition of uncertain status in which the thyroid becomes hard, nodular, and fibrotic.

1931 *Arch. Surg.* XXII. 548 Recent writers have stated that struma lymphomatosa is the early stage of Riedel's struma, despite the fact that Hashimoto..definitely

rejected such a relationship. **1947** H. Selye *Textbk. Endocrinol.* 714/2 The etiology of struma lymphomatosa is unknown but its histologic characteristics are not typical of inflammatory lesions. **1966** [see *Riedel]. **1974** S. L. Robbins *Path. Basis Dis.* xxix. 1328/1 In years past, Riedel's struma was thought to be the fibrotic end-stage of struma lymphomatosa.

strumble (strʌ·mb'l), *sb.* *rare*[−1]. [f. Strumble *v.*[1]] A rumble.

1938 Belloc *Sonnets & Verse* 180 Beneath His feet are implacable fate, and panic or night, and the strumble of the hungry river of death.

strumitis (strumǝi·tis). *Path.* [f. Strum(a + -itis.] Inflammation of a goitrous thyroid gland.

1889 *Buck's Handbk. Med. Sci.* VII. 96/1 Inflammation of the thyroid gland (thyroiditis, strumitis) is most commonly seen as the accidental or intentional result of remedial measures employed in the treatment of goitre. **1955** S. C. Werner *Thyroid* v. 745 Thyroiditis appears in acute, subacute, and chronic forms... If the acute disorders have their onset in pre-existing goiter, the disease is often termed strumitis.

strumming (strʌ·miŋ), *ppl. a.* *rare.* [f. Strum *v.* + -ing[2].] Sounding like a strummed instrument.

1887 Hardy *Woodlanders* III. vii. 140 She fancied that she could hear, above the sound of her strumming pulse, the vehicle. *a* **1911** D. G. Phillips *Susan Lenox* (1917) II. xxv. 553 As a menace, as a prophecy, the old women and the hunchback and the strumming piano had gone forever.

strung, *ppl. a.* Add: **3.** Also, extended, continuing in a long series.

1978 *Language* LIV. 91 This falls short of accounting for strung-out MI correspondences where relative orderings remain the same.

4. For 'String *v.* 4' read 'String *v.* 3'. **a.** *strung-up* (later examples). Also (*N. Amer. slang*) *strung out* (overlaps with sense c below).

1933 M. de la Roche *Master of Jalna* xv. 198 She worries greatly over the child and that keeps her in a strung-up state. **1967** *N.Y. Times* 18 Aug. 22 'These are very strung-out kids with individual hang-ups,' said Jim Fouratt..describing the modern runaway. **1974** M. Haskell *From Reverence to Rape* 208 Martha Vickers' spoiled, strung-out younger sister. **1979** K. M. Peyton *Marion's Angels* ix. 137 You don't think she might have—? Oh, Christ! She was strung up when she left me. We both were. **1980** *Globe & Mail* (Toronto) 20 Mar. 11/5 She takes to the streets daily in response to calls from tenants to investigate nuisance neighbours who might be strung out emotionally.

c. *strung out*: weak or ill, esp. as a result of drug addiction; hence, addicted to, using, or 'high' *on* drugs. *slang* (orig. and chiefly *U.S.*).

1959 *Esquire* Nov. 70 *Strung Out*, in bad physical condition. **1960** *Jazz Rev.* Nov. 8/2 Unfortunately it was at this period he acquired the 'monkey' and frequently was strung out. **1965** *N.Y. Post* 3 Dec. 45/1 If one spends time talking to marijuana users, one can only conclude that the entire college population is 'strung out' thrice weekly. **1966** *San Francisco Chron.* 29 June 3 Acid does nothing for me... I love to get strung out on pot. **1973** *Black World* Aug. 59/1 The horns..lef' by some strungout junky musician. **1977** *Guardian Weekly* 30 Oct. 15/4 Young people get strung out on heroin.

strunt, *sb.*[1] Add: Also, the whole tail.

a **1930** D. H. Lawrence *Phoenix* (1936) 16 Wag thy [*sc.* a puppy's] strunt, then!

strunzite (strʌ·nzəit). *Min.* [ad. G. *strunzite* (C. Frondel 1957, in *Neues Jahrb. f. Min.: Monatshefte* 222), f. the name of K. Hugo *Strunz* (b. 1910), German mineralogist: see -ite[1].] A hydrated basic phosphate of manganese and iron, $MnFe_2(PO_4)_2(OH)_2.8H_2O$, found as tiny yellow hair- or lath-like monoclinic crystals and produced by the weathering of manganese and iron phosphates.

1958 C. Frondel in *Naturwissenschaften* XLV. 38/1 Among the outstanding problems in the mineralogy of the hydrated phosphates of iron and manganese has been the identity of a straw yellow mineral first recognized in 1947... The writer takes pleasure in naming this mineral after Dr. Hugo Strunz, Professor of Mineralogy at Berlin. **1975** *Mineral. Record* VI. 71/2 Stewartite, strunzite, and cacoxenite, which in our specimens are confined to oxidized assemblage[s], are the products of weathering.

strut, *sb.*[3] Add: **c.** A type of slow and complicated dance or dance-step.

[**1917**: see *shimmy *sb.*[2] 1]. **1937** [see *Big Apple* s.v. *big *a.* B. 2]. **1970** C. Major *Dict. Afro-Amer. Slang* 111 *Strut*.., a fancy-step slow dance. **1979** R. B. Gillespie *Crossword Mystery* i. 17 He..executed a few soft-shoe steps which merged into a strut.

strut, *v.*[1] Add: **7. f.** *to strut one's stuff*: to display one's ability. *U.S. slang.*

1926 C. Van Vechten *Nigger Heaven* II. vi. 242 Some one cried, Strut your stuff, Lasca! **1935** [see *roller *sb.*[1] 18 d]. **1941** *Sun* (Baltimore) 30 Aug. 13/1 Rain today made the prospect for off-going for the first card, thus giving the 'mudders' an opportunity to strut their stuff. **1972** *N.Y. Times* 3 Nov. 28/1 The company is going to strut its stuff, with nothing more in mind than to entertain, in Washington Square, near the Arch. There will be singing and dancing and acting and acrobatics. **1977** *Rolling Stone* 30 June 121/2 (Advt.), Each run is equipped with a super, custom-designed sound system, so you can 'strut your stuff' or 'space walk' to your favorite tunes.

g. *intr.* To dance the strut. Cf. *strut *sb.*[3] c.

1975 *Time Out* 8 Aug. 67/1 D'you wanna shake, strut, shimmy, jive, twist, waltz, mash potato, tango, tap or conga?

'struth, var. 'Strewth in Dict. and Suppl.

struthonian (strūþō^u·nian), *a.*, (*sb.*) *joc.* [Irreg. f. L. *struth-io*: see -ian and Struthious *a.*] Tending to 'hide one's head in the sand', like the ostrich (see *sand *sb.*[2] 2 d), and so to ignore unwelcome facts (Koestler's word). Also as *sb.*, one who does this. Hence **strutho·nianism.**

1963 A. Koestler in *Encounter* July 7/1 One need not be an economist to find these figures disconcerting. But there is a struthonian answer to them (from struthio the Latin for ostrich). Your Old Struthonian will start with some disparaging remarks about statistics in general, and ..then explain that Britain started the Industrial Revolution, and that accordingly it is only natural..that other nations, which started later, should grow at a faster rate. **1966** *Listener* 18 Aug. 223/2 Housing, with doubtful justification, lags behind needs, and continues to take a lion-hearted and struthonian view of the climate. *Ibid.* 224/1 A fine piece of struthonianism is the failure to accept the fact that teenagers are now in practice sexually active. *Ibid.* 25 Aug. 279/1 It is unfortunate that Professor Wisdom should perpetuate Arthur Koestler's barbaric derivative struthonian... If *leonine*, then *struthionine*, or nothing.

strüverite. Add: **2.** *Min.* An oxide of titanium containing tantalum and ferric iron, found as black tetragonal crystals having a metallic lustre. [a. It. *strüverite* (F. Zambonini 1907, in *Rend. dell'Accad. d. Sci. fis. e matem.* (Napoli) XIII. 46).]

1907 *Mineral. Mag.* XIV. 411 Strüverite. **1908** *Ibid.* XV. 78 The mineral strüverite occurs as a rare accessory constituent of the pegmatite which is found in large detrital masses in the neighbourhood of Craveggia (Val Vigezzo, northern Piedmont). **1917** [see *mossite]. **1974** *Mineral. Abstr.* XXV. 236/1 Strüverite, tanteuxenite and tantalanatase from the pegmatites near the town of Chepelare, Central Rhodopes [Bulgaria].

Struwwelpeter (strū·əlpītəɹ, ‖ʃtru·vəlpē:tər). Also *erron.* **Struwelpeter.** The name of a character in a children's book of the same name by Heinrich Hoffmann (1809–94), used *attrib.* to designate (a person with) long, thick, and unkempt hair. Also *transf.* Also *Comb.*, as *Struwwelpeter-haired*. Hence **Struwwelpeterdom,** (of hair) the condition of being thick and untidy. Cf. *shockheaded Peter* s.v. *shock-headed *a.*

1909 W. J. Locke *Septimus* iii. 37 He passed his hand through his Struwelpeter hair. *Ibid.* xxii. 351 His hair..reached the climax of Struwelpeterdom. **1920** E. Sitwell *Wooden Pegasus* 78 As he rides on his rocking-horse All Struwwelpeter-haired. **1927** 'G. Daviot' *Man in Queue* vii. 82 The artist's eyebrows disappeared in the Struwwelpeter hair. **1958** P. Kemp *No Colours or Crest* x. 228 The Commissar was a small, squat man with very bright eyes and a *Struwwelpeter* shock of black hair. **1959** C. Spry *Favourite Flowers* xxv. 188 The flowers [of *Anemone alpina*] were over and I saw for the first time what Mr. Drake calls the 'Struwwelpeter' seed heads. **1973** R. W. Clark *Einstein* xi. 269 William Rothenstein, making notes for his remarkable portrait of Einstein who is presented as a Struwelpeter character, smiling from an aureole of almost electrified hair.

strychninization (stri:kninəizē¹·ʃən). [f. as next + -ization.] The act of applying strychnine.

1898 in *Syd. Soc. Lex.* **1933** *Ann. Rep. London Co. Council* IV. III. 143 After strychninisation of the gyrus uncinatus and lobus pyriformis, fits could be induced by olfactory stimulation. **1951** J. F. Fulton *Frontal Lobotomy* ii. 47 Strychninization of the posterior hypothalamus activates the dorsomedial nuclei. **1974** *Proc. Soc. Exper. Biol. & Med.* CXLV. 979 (*heading*) Subcortical multiple unit activity..during strychninization of the cerebral cortex in the female rat.

strychninize (stri·kninəiz), *v.* [f. Strychnine + -ize.] *trans.* To apply strychnine to.

1934 in Webster. **1938** *Jrnl. Neurophysiol.* I. 69 The distribution of the spikes within the cytoarchitectonic area, of which a part has been strychninized. **1951** P. D. MacLean in J. F. Fulton *Frontal Lobotomy* ii. 57 Then we strychninized the nucleus. **1980** *Pavlovian Jrnl. Biol. Sci.* XV. 58/1 The response was made single by strychninizing a larger area of cortex.

Hence **stry·chninized** *ppl. a.*

1938 *Jrnl. Neurophysiol.* I. 72 Farther away from the strychninized locus spikes are absent. **1978** H. Takeuchi in Chalazonitis & Boisson *Abnormal Neuronal Discharges* 161 The strychninized somatic abnormal biopotential was similarly subjected to temperature changes.

strychnization (striknəizē¹·ʃən). [f. Strychn(ia + -ization.] = *strychninization.

1916 *Q. Jrnl. Exper. Physiol.* IX. 356 Slight strychnisation of the dorsal mechanisms..gives rise to a typical complex of symptoms. **1924** *Proc. R. Soc.* B. XCVI. 272 Local strychnization of the central nervous system..yields good results in investigations on the sensory mechanisms of the spinal cord.

stryddag (strē¹·tdaχ). *S. Afr.* Pl. **stryddae.** [Afrikaans, lit. 'struggle day, day of battle'.] An Afrikaner political party rally. Also *transf.*

1950 *Cape Times* 26 July 1/3 Mr. C. R. Swart, Minister of Justice, told a Nationalist Party stryddag here yesterday that he was not prepared to reintroduce public hangings. **1961** *Ibid.* 11 Jan. 1/6 In Britain he will be able to give the moderately-phrased versions of *apartheid* far more expertly and convincingly than in heated debates in Parliament or during *stryddae* in the platteland. **1972** *Daily Dispatch* 14 Apr. 10 The day of the big stryddag is over. Those large crowds will never again drive across the veld to hear emotional appeals to the blood. **1978** *Pace* Dec. 51 On Sunday, Chief Buthelezi is going to..address an all-black *stryddag* in Soweto.

stschi, var. Stchi in Dict. and Suppl.

Stuart (stiū·ăɹt). Also formerly **Stewart.** The name of the British royal family from 1603 to 1688, used *attrib.* to designate that period of history and applied esp. to artefacts, buildings, etc., of that date or style.

1873 C. M. Yonge *Pillars of House* IV. xxxvi. 50 Here's the dining-room—This is the middle period, the Stewart style part. **1880** E. Glaister *Needlework* viii. 87 Some oak chairs, probably of early Stuart date. **1922** S. Weyman *Ovington's Bank* ix. 104 Long-backed Stuart chairs. **1937** *Discovery* Aug. 231/1 The penny of Tudor and Stuart times. **1973** *Country Life* 5 Apr. 909/2 He has treated the wainscot of the Stuart bedroom..in a similar manner to the Stuart staircase. **1981** V. Glendinning *Edith Sitwell* i. 12 Renishaw Hall..retained within its walls..most of its Stuart and Regency past.

stub, *sb.* Add: **7. e.** A short length of wire used in flower-arranging. Cf. *stub wire*, sense 11 below.

1951 R. A. Birch et al. *Mod. Florist* ix. 83 On the bench is fixed the wire tidy..a set of upright metal cylinders or holders into which the wires, or 'stubs', are placed. *Ibid.* 88 Next we come to 'invisible wiring' with ordinary stubs. **1960** V. Stevenson in T. A. Price et al. *Retail Florist's Handbk.* iii. 66 Stem wires, often called stub wires (one assumes because they are stubbed into the design), vary from 3½ to 18 inches in length... Twenty gauge..is the most widely used wreath stub. **1963** M. Smith *Arranging Flowers* viii. 83 With a very fragile stem it is best to lay a stub-wire against it..and to twist fuse-wire round both the stub and the stem to bind them together.

9. a. Also *spec.* the butt or stump of a cigar or cigarette.

1855 'Q. K. P. Doesticks' *Doesticks, what he Says* xvi. 133 Perhaps they expect us to smoke 'stubs', like the newsboys. **1869** [in Dict.]. **1873** J. H. Beadle *Undevel. West* 787 Even little darkeys watch for the 'old stubs' as they are thrown away. **1914** B. M. Bower *Flying U Ranch* 187 He spat upon the burnt end of his cigarette stub from force of the habit that fear of range fires had built. **1973** T. Pynchon *Gravity's Rainbow* (1975) III. 309 The two of them sit there, passing a cigarette back and forth, till it's smoked down to a very small stub.

11. stub-end, (*b*) *U.S.*, the unconnected end of a stub track; (*c*) a cigarette stub (in quot. *fig.*); **stub station** *U.S.*, a railway station at which the tracks terminate; **stub-switch** (examples); **stub-tail,** (*b*) also used of maize (later example); (*c*) a short and thick or broad tail; also *fig.*; **stub-toed** *a.*, of a shoe: having a broad toe; **stub track** *U.S.*, a railway track, usu. at a terminus, connected to another at one end only; also *stub-end track* (cf. *stub-end* (*b*) above); **stub wing** *Aeronaut.* (see quot. 1956); hence **stub-winged** *a.*; **stub wire** = sense 7 e above.

1896 *Engineering News* XXXVI. 27/1 When a long stub-end track gets full of empties, the cars at the stub end are likely to remain for weeks and months. **1900** *Ibid.* XLIV. 377/2 Stub-end tracks should generally be in pairs, with crossovers near the ends, so that the engine of one incoming train can be got out without waiting for its train. **1903** W. M. Camp *Notes Track Constr.* I. vi. 466 An arrangement that is sometimes provided where inbound, outbound and transfer houses are consolidated at one point is to have parallel stub tracks, with the inbound house on one side, the outbound house on the opposite side and the office between them, at the stub ends of the tracks. **1932** Auden *Orators* III. 85 Stub-end of year that smoulders to ash of winter. **1916** J. A. Droege *Passenger Terminals* v. 104 The head or stub station is an end-of-the-line station. *Ibid.*, Practically all the terminals in New York City are built on the stub station plan. **1929** *Amer. Railway Engineering Assoc. Man.* xiv. 960 The through and loop types of station are superior to the stub station from the standpoint of train operation. **1885** G. Mordecai *Rep. Terminal Facilities* 6 The tracks are of good material, laid with stub switches and railfrogs. **1903** W. M. Camp *Notes Track Constr.* I. 292 The stub switch, with its open joint in winter and tight joint in summer, with a loose head block to be tampered with every few days. **1873** *Spider & Fly* IV. i. 19 This corn is the Illinois growth of 1857, and is called 'stub-tail' because about one-third of it is rotten. **1938** L. MacNeice *Zoo* 234 They [*sc.* bears] showed their stub-tails. **1973** *Times* 17 May 35/1 Compared with the 1100 it has a more streamlined bonnet curving down to a low radiator grille, and a stub tail somewhat reminiscent of the Hillman Avenger. **1930** J. Dos Passos *42nd Parallel* 134 He'd..clatter up and down stairs making a tremendous racket with his stubtoed ironplated shoes. **1896** *Engineering News* XXXVI. 27/1 The empty car storage tracks on Mr. Derr's diagram are very long stub tracks, which are objectionable. **1921** *Railway Engin. & Maintenance of Way Cycl.* 348/1 A freight terminal is commonly considered an important freight station served by stub tracks. **1956** *Railway Track & Structures Cycl.* (ed. 8) 425/2 Bumping posts are obstructions placed at the end of stub tracks. **1931** *Flight* 2 Jan. 16/2 The lower stub wings form part of the landing gear structure. **1956** W. A. Heflin *U.S. Air Force Dict.* 496/2 *Stub wing*,.. 1. A short wing, esp. as used on certain autogiros. 2. That part of a wing on certain airplanes that lies next to the fuselage, to which the rest of the wing, separately built, is attached... 3. Short for 'stubwing stabilizer.'..*Stubwing stabilizer*, a hydro-stabilizer on a flying boat. **1958** *Times Rev. Industry* Aug. 39/2 The [Rotodyne] fixed stubwing..takes over the task of supporting the aircraft. **1957** *Times Survey Brit. Aviation* Sept. 2/6 Bristol are sending the stub-winged twin-rotor Type 173. **1960, 1963** Stub wire [see sense 7 e above]. **1976** *Eastern Even. News* (Norwich) 22 Dec. 4/8 (Advt.), Oasis (dry and wet). Dried flowers. Stub wires and a large selection of containers.

stub, *v.*[1] Add: **9. a.** Also *fig.*

1957 *Economist* 19 Oct. 194/1 At a time when the Middle East has become more of a happy hunting ground for Russians seeking friends and influence than ever before, it is on Turkey that they are always stubbing their toe. **1967** [see *goods *sb. pl.* 2 b]. **1976** 'J. Ross' *I know what it's like to Die* xix. 126 It was a reputable organisation. At least, insofar as it hadn't stubbed its corporate toes on, or interfered with, anything under the supervision of the superintendent's own bailiwick.

b. Also *transf.* and *fig.*

1875 C. B. Lewis *Quad's Odds* 480 The writer will stub along through life with a heart full of joyfulness. **1878** B. F. Taylor *Between Gates* 241 An old whaler stubbing about estimated him [*sc.* a whale] at sixty barrels.

12. *trans.* To extinguish (a cigarette) by pressing the lighted end of the stub against a hard object. Freq. with *out*. Also *fig.*

1927 *Daily Express* 28 Oct. 5 A new glass ash tray with cigarette rests has a glass stopper fitting in at the back which is used for stubbing one's smokes... On the stubber a Greek girl dancing, scarf in hand [is represented]. **1930** J. Cannan *No Walls of Jasper* 116 He stubbed out his cigarette and smiled at her. **1955** P. Larkin in *Listener* 8 Sept. 373/1, I lie Where Mr. Bleaney lay, and stub my fags On the same saucer-souvenir. **1962** J. Braine *Life at Top* xxvi. 277, I looked round for an ashtray and for the fifth time since nine o'clock stubbed out my cigarette on the floor. **1970** R. Lowell *Notebk.* 214 A hand prepared to stub out liberty. **1974** 'E. Ferrars' *Hanged Man's House* xv. 149 There was always something that you could do with a cigarette, light it, draw on it, tip ash off it, stub it out. **1978** S. Radley *Death & Maiden* xi. 109 She stubbed out her cigarette with sudden vigour.

stubbed, *ppl. a.* Add: **6.** Chiefly, of a toe: injured by being struck against something. Cf. Stub *v.*[1] 9 a in Dict. and Suppl. orig. *U.S.*

1890 *Brighton* (Colorado) *Reg.* 11 Jan. 4/1 Montana is the 'stubbed-toe State'. **1958** J. G. MacGregor *North-West of 16* viii. 111, I don't suppose a city boy, or even a modern farm boy, knows what a stubbed toe is. Maybe that's just as well, because they were most painful. **1977** *Evening Gaz.* (Middlesbrough) 11 Jan. 4/5 None of the hair-raising stunts he performs for his TV series Some Mothers Do 'Ave 'Em led to any injury greater than a stubbed finger. **1978** B. Bainbridge *Young Adolf* xviii. 101 Testily he kicked at the..wood... Recoiling, he curled his stubbed toes within his boot.

7. *stubbed-out*: of a cigarette, extinguished by being pressed against a hard object.

1975 O. Sela *Bengali Inheritance* xxv. 218 A single stubbed out Stuyvesant in the ashtray. **1979** *N.Y. Rev. Bks.* 8 Feb. 13/4 No other writer since Noël Coward can have so constantly punctuated dialogue and action with a ritual pattern of lighting-up, inhalation, smoke-rings, and stubbed-out butts.

stubber. Add: **2.** A contrivance against which a cigarette is stubbed out.

1927 [see *stub *v.*[1] 12].

stubble, *sb.* Add: **5.** objective, *stubble-burner, -burning*; instrumental, as *stubble-covered* adj.; **stubble-fed** *a.*, of poultry: fed on the stubble left in a reaped field; hence **stubble-feeding; stubble-grown** *a.* = *stubble-covered* adj. above; **stubble-jumper** *slang* (chiefly *Canad.*), a prairie farmer (see also quot. 1946); **stubble-quail,** a brown, black, and white quail, *Coturnix pectoralis*, native to southern Australia.

1980 *Sunday Times* 24 Aug. 1/7 Can stubble-burners be controlled? **1973** *Times* 24 Aug. 2/5 The National Union of Agricultural and Allied Workers yesterday called for regulations to control stubble burning. **1976** A. Price *War Game* I. ii. 48 There hadn't been so much stubble-burning this year, he noted approvingly. **1916** Joyce *Portrait of Artist* (1969) ii. 61 Stephen often glanced with mistrust at his trainer's flabby stubblecovered face. **1882** W. D. Hay *Brighter Britain!* I. 224 Cricket-fed turkey would shame any stubble-fed bird altogether. **1928** *Daily Express* 6 Oct. 4/6 These stubble-fed geese are the best of all for eating. **1960** G. E. Evans *Horse in Furrow* xii. 172 There were two kinds of shacking; *Lammas shack* ..and *Michaelmas shack*, stubble-feeding after the corn harvest. **1916** Joyce *Portrait of Artist* (1969) v. 227 Good evening, gentlemen, said the stubblegrown monkeyish face. **1946** *California Folklore Q.* Apr. 164 'Top hands', 'sodbusters', 'hay stopers', 'stubble jumpers',..denote farmers who have turned to mining, and these terms are always opprobrious. **1961** *Sun* (Vancouver) 4 July 1/1 The prairie farmer, to those of us who don't know him well, is a stock comic character. Clod-hopper, we call him, and stubble-jumper. **1973** *Islander* (Victoria, B.C.) 19 Aug. 12/1 An authentic stubble-jumper from the prairies was looked upon as being at the very bottom rung of the social and employment ladder. **1848** J. Gould *Birds Austral.* V. plate 88 The name of Stubble Quail has been given to it by the colonists of Van Diemen's Land, from the great numbers that visit the fields after the harvest is over. **1921** Matthews & Iredale *Man. Birds Austral.* I. 224 Stubble-quail... Head, neck, entire back and scapulars rufous-brown and black streaked with white. **1965** *Austral. Encycl.* VII. 316/1 The stubble-quail, which is closely allied to the quail of Europe, is confined to southern Australia and Tasmania.

stubborn, *a.* Add: **4.** *stubborn-hearted* (later example).

1906 W. B. Yeats *Poems, 1899–1905* 55 Women are hard and proud and stubborn-hearted.

Stubbsian (stʌ·bziăn), *a.* [f. the surname *Stubbs* (see below) + -ian.] **1.** Of, pertaining to, or characteristic of the English painter George *Stubbs* (1724–1806), or his work.

1960 *Times* 4 Nov. 16/7 Ferneley's brand of aristocratic elegance is splendidly represented, especially in the 'Five Hunters in the Park at Deene' (a Stubbsian subject with a better understanding of space than Stubbs). **1963** *Times* 1 May 5/7 Stubbs, of course, ranges over the whole field of conversation pieces, portraiture, wild life and landscape as well as horse painting;..and there is a picture of almost Stubbsian quality by George Garrard.

2. Of or pertaining to William *Stubbs* (1825–1901), historian and Bishop of Oxford, or his historical opinions.

1979 *English Hist. Rev.* XCIV. 487 By 1934 Kenneth Pickthorn's study of *Early Tudor Government*..still framed its argument in Stubbsian terms by denying that the period had much of a constitution at all aside from the medieval idea of the supremacy of law.

stubby (stʌ·bi), *sb.* *Austral. slang.* Also **stubbie.** [f. Stubby *a.*] **1.** A short, squat beer-bottle with a capacity of 375 ml. Also *Comb.*, as *stubby beer bottle*.

1957 *Encycl. Brit.* IV. 106/2 A variety of standardized forms and sizes of bottles are in use, including the so-called Stubby, Steinie, Packie, Export and Single Trip bottles. **1966** *Sunday Truth* (Brisbane) 8 May 33/7 Well-known transport man generally likes to have a 'stubbie' of beer at home each evening. **1968** F. Hardy *Unlucky Australians* 49 He threw an empty stubby into the box and went to the refrigerator for a full one. **1969** *Sunday Truth* (Brisbane) 23 Nov. 37/7 Both as a safety measure and a fight against litter, I advocate the absolute banning of the stubby beer bottle. **1972** G. Morley *Jockey rides Honest Race* 165 Phil opened the freezer and pulled out four stubbies. **1977** *Mod. Boating* (Austral.) Jan. 88/1 Buy another stubby at the bar to gather strength to try again in half an hour.

2. Usu. *pl.* Shorts (see Short *sb.* 6 d in Dict. and Suppl.).

1977 *Australian* 7 Apr. 3 Stubbies—the football shorts with pockets—have become an international fashion... Although the Stubby is a very Australian name—thought of in the context of short shorts to go with short bottles of beer—Mr Phillips is confident they will become as American as apple pie. **1982** *Guardian* 18 Dec. 11 'Stubbies', slang for both the shorts they wear..and the bottles from which they drink. *Ibid.* 11/4 (*caption*) 'Stubbies' (shorts) and pot bellies predominate on Australia's popular side.

‖**Stube** (ʃtū·bə). Also **stube.** [Ger., lit. 'room, parlour'.] = *bierstube.

1946 S. Spender *European Witness* 138 He invited me to join the other officers in going to a drinking Stube. **1956** E. Berckman *Strange Bedfellow* vii. 66 The dimness and coolness of the village *Stube*. **1967** J. Eastwood *Little Dragon from Peking* xvii. 153 Hilde, the maid of all work, had already mopped the wooden floor of the *Stube*. **1969** A. Glyn *Dragon Variation* viii. 232 In a gasthof in the Maria-Josephastrasse, Jeff Falkner sat at a window, staring out at the snow. The others were downstairs playing chess in the stube. **1980** J. Cartwright *Horse of Darius* x. 149 A *stube* full of hearty people drinking beer.

stuc, stuck. Restrict † *Obs.* to sense in Dict. and add: **2.** = Stucco *sb.* 1 b.

1932 T. Corkhill *Conc. Building Encycl.* 207 *Stuc*, plasterwork to imitate stone. **1971** *Country Life* 14 Oct. 969/1 Its walls are of *stuc*, by M. Germain, an imitation of dressed Caen stone.

stuccador(e (stʌkădō°ɹ). Also **stuccodore.** [irreg. ad. It. *stuccatore*; cf. Sp. *estucador*.] A worker in stucco. Cf. Stuccoer.

1952 *Archit. Rev.* CXI. 201/3 Its Tapestry Room with Floral Zephyrs by James 'Athenian' Stuart and a Saloon with walls marvellously decorated by Italian stuccadores. **1956** *Essays in Crit.* VI. 324 The iconography of other eighteenth-century arts (Tiepolo's painting, the interior plasterwork of Dublin stuccodores), is still that of the High Renaissance. **1972** *Canad. Antiques Collector* Mar.–Apr. 15/1 The interiors..are as exuberant and varied as the stuccodores through seventy years of changing taste (1730–1800) could make them. **1973** *Country Life* 10 May 1306/2 The Schmuzer family..whose activity as mason-stuccadors persisted right into the Rococo period. **1978** A. Laing in A. Blunt et al. *Baroque & Rococo* IV. 241/1 In Bavaria it was a Wessobrunner stuccador, Joseph Schmuzer (1683–1752), who was most successful in creating a local practice as the architect of parish churches, extending his competence from stucco to masonry.

stucco, *sb.* Add: **2. b.** A house plastered with stucco.

1976 C. Weston *Rouse Demon* (1977) xi. 50 The Simmons house turned out to be a two-story Monterey-style stucco. **1981** P. Mallory *Killing Matter* xvi. 167 A blue stucco at the corner of Delgado and Harding.

stuccoist (stʌ·kou,ist). = Stuccoer. Cf. *stuccador(e.

1945 J. Lees-Milne *Jrnl.* 26 Sept. in *Prophesying Peace* (1977) 238 Lord Leigh assured me..that the plasterwork was by Cipriani, which I find hard to believe unless there was an earlier stuccoist of the same name as the well-known painter. **1969** M. Whinney *Home House* 19 In November 1775 Joseph Rose, Adam's principal stuccoist, was paid £603.10.0.

stuck, *ppl. a.* Add: **5.** With advs. forming adjs. with reference to attachment or sealing by adhesive, etc., as *stuck-down, -on*.

1908 Kipling *Actions & Reactions* (1909) 101 The Hive shook beneath the shattering thunder of a stuck-down quilt being torn back. **1940** *Chambers's Techn. Dict.* 815/2 *Stuck-on soles* (*Shoes*), shoe soles in which the upper inner sole and the outer sole are attached together by means of strong cement; used for women's and children's shoes. **1960** *Farmer & Stockbreeder* 15 Mar. Suppl. 7 A whitewood bedside cabinet is given a stuck-on veneer finish. **1978** D. Francis *Trial Run* xviii. 228 One of them handed me a stuck-down envelope.

stuckness (stʌ·knės). [f. Stuck *ppl. a.* + -ness.] The condition of being stuck or unable to move or progress.

1969 J. S. Bruner in Elkind & Flavell *Stud. Cognitive Devel.* 225 Let me characterize that growth briefly as moving from a diffuse distractibility in the weeks immediately following birth to a stage of stuckness where attention has an 'obligatory character', to use Stechler's phrase. **1974** R. M. Pirsig *Zen & Art of Motorcycle Maintenance* xxiv. 278 The first [problem] is stuckness, a mental stuckness that accompanies the physical stuckness of whatever it is you're working on.

stud, *sb.*[1] Add: **II. 5. d.** One of a series of small devices protruding slightly above the surface of a road and used to demarcate traffic lanes; *spec.* = *cat's eye 5.

1935, etc. [see *road stud* s.v. *road *sb.* 9 b]. **1939** L. MacNeice *Autumn Jrnl.* viii. 33 The metal studs in the sleek macadam. **1943** *Ann. Reg. 1942* 386 A traffic stud insecurely fixed..flew up and injured a cyclist. **1958** *Spectator* 8 Aug. 190/2 The road studs known as cat's eyes. **1975** R. Hoban *Turtle Diary* xxxii. 151 There were reflecting studs in the road. **1978** *Highway Code* 18 Coloured reflecting road studs may be used with white lines—white studs mark the lanes or centre of the road, while the edge of the carriageway may have red studs on the left-hand side and amber by the central reservation of dual carriageways. Green studs may be used across lay-bys and side roads.

e. = *ear-stud* s.v. *ear *sb.*[1] 16.

[**1873**: see *ear-stud* s.v. *ear *sb.*[1] 16.] **1968** J. Ironside *Fashion Alphabet* 172 *Stud*, a plain 'knob' on the ear. **1979** N. Freeling *Widow* xxiii. 143 Garnet studs in the pale fleshy lobes of the ears.

6. See also *press-stud.

7. g. One of a number of metal pieces set into the tyre of a motor vehicle to improve roadholding in slippery conditions. *N. Amer.*

[**1909** *Westm. Gaz.* 11 Nov. 5/2 Messrs. B. F. Goodrich will exhibit their all-rubber non-skids, in which the rubber studs form an integral part of the tread, as well as a new steel-studded tyre.] **1963** *Pop. Mechanics* Sept. 24 PM tried one version of Season Safety Stud tires on a glassy smooth skating rink and found performance impressive. **1976** *National Observer* (U.S.) 13 Nov. 10/4 In Illinois, Minnesota, and Wisconsin studded tires have been barred this year for the first time. In Michigan a new law permits studs only if they meet certain rigid—and some say impossible—specifications.

IV. 9. stud-ear-ring = *ear-stud* s.v. *ear *sb.*[1] 16; cf. sense 5 e above; **stud welding,** a method of welding in which an arc is struck between a stud and the base metal, producing a pool of molten metal into which the stud is driven to form a weld.

1919–20 *T. Eaton & Co. Catal.* Fall & Winter 394/3, 14k stud earrings, set with diamond. **1966** [see *ear-stud* s.v. *ear *sb.*[1] 16]. **1973** E. Lininton *Crime by Chance* i. 10 Sue..put on makeup and small stud earrings. **1941** A. C. Davies *Sci. & Pract. Welding* iv. 298 (*heading*) Stud welding or studding. **1962** *Engineering* 20 Apr. 527/1 Stud welding has the advantage that there is no distortion of the flanges of the steel beams. **1975** Bram & Downs *Manuf. Technol.* ii. 57 Stud welding is a form of electric-arc welding.

stud, *sb.*[2] Add: **4. c.** = Stud-horse 2. Chiefly *U.S.*

[**1890** J. P. Quinn *Fools of Fortune* 188 Next to the banking games in the estimation of the betters comes poker, both 'draw' and 'stud'.] **1933** D. Runyon in *Collier's* 28 Jan. 8/1 A proposition may be only a problem in cards, such as..how often a pair of deuces will win a hand in stud. **1942** W. Faulkner *Go down, Moses* 26 'Stud,' Mr Hubert said. 'One hand. You to shuffle, me to cut, this boy to deal.' **1979** Reese & Flint *Trick 13* 64 In seven-card stud a player who stays in the pot till the finish receives two cards face down, four face up, and one face down.

d. A man of (reputedly) great sexual potency or accomplishments; a womanizer, a habitual seducer of women. In weakened uses: as a familiar term of address among men; a boy-friend or escort.

Particularly common since *c* 1960.

1895 W. Rye *Gloss. E. Anglia* 217 *Stud*, a nickname given to a man from his love of venery (Wilton, 1877). **1909** *Dialect Notes* III. 377 *Stud* (*-horse*), *n.*, a stallion. Also used as a term of familiar address among men. 'Hello, old stud, how are you?' **1955** *Amer. Speech* XXX. 305 [Wayne University slang] *Stud*, a ladies' man. **1959** C. MacInnes *Absolute Beginners* 165 The chick sits on cushions in the front part, with a brolly, and her stud heaves the thing along with a hop pole, just like gondolas. **1970** B. Moore *Luck of Ginger Coffey* iv. 88 Throwing her across this bed yesterday, pleased with yourself for being the great stud. **1964** S. Bellow *Herzog* 154 Still in fleeting moments the young and glossy stud—such as he really had never been. **1974** M. Haskell *From Reverence to Rape* 250 In the sixties we came to realize that the figure of the stud (the gamekeeper, the 'macho' Latin, the gigolo) is, like the sex-starved woman, largely a figment of male homosexual fantasy. **1981** S. Rushdie *Midnight's Children* iii. 395 A notorious seducer; a ladies'-man; a cuckolder of the rich; in short, a stud.

e. Hence (without explicit sexual significance): a man, a fellow, esp one who is well-informed; a youth. *U.S. slang* (chiefly *Blacks'*).

1929 M. A. Gill *Underworld Slang* 10/2 *Stud*, man. **1944** D. Burley *Orig. Handbk. Harlem Jive* 70 If you're a hipped stud, you'll latch on. **1946** Mezzrow & Wolfe *Really Blues* (1957) 379 *Stud*, guy, man. **1963** E. J. Gaines in *Sewanee Rev.* Autumn 550, I mean a stud's going to drink eggnog, and he isn't going to put whiskey in it. **1967** W. Murray *Sweet Ride* x. 169 We're looking for a couple of studs..Jimmy the Head and Jawbone. **1970** R. D. Abrahams *Positively Black* ii. 46 But who's this stud they call Billy?

5. (sense 1) *stud-fee, -master*; **stud-book:** also *transf.* and *fig.*, a catalogue of aristocratic pedigree, esp. = *Burke *sb.*, *Debrett *sb.*; **stud-poker** (now the commonest form): (earlier and later examples).

1888 Kipling *Story of Gadsbys* (1891) viii. 122 'Fraid you won't be entered in the Stud Book correctly unless you go Home? **1933** D. L. Sayers *Murder must Advertise* xi. 198, I think I know the stud-book pretty well. I was not aware that you had a cousin Bredon. **1982** R. Barnard *Death & Princess* xii. 122 Lady Dorothy..can drone on endlessly about her family tree... She's a sort of walking stud-book. **1922** Joyce *Ulysses* 540 What's our studfee?..You fee men dancers on the Riviera, I read. **1953** X. Fielding *Stronghold* 48 My stud-fee's as low as you like—I'd do it for nothing. **1937** E. Rickman *On & Off Racecourse* i. 3 Their best horses are placed at stud..or are sold to other stud-masters. **1975** *N.Z. Jrnl. Agric.* Sept. 60/3 New Zealand studmasters have proved their ability through the years. **1864** W. B. Dick *Amer. Hoyle* 167 *Stud poker*.., in all essential particulars, is like the other Poker games. **1922** S. Lewis *Babbitt* xxv. 295 All the way north he pictured the Maine guides: simple and strong and daring, jolly as they played stud-poker in their unceiled shack. **1959** 'B. Mather' *Achilles Affair* I. i. 13 He and Finnessy were playing murderous stud poker. **1978** E. Tidyman *Table Stakes* II. iii. 191 The best stud poker player who ever sat down among them.

6. *attrib.* or *adj.* use of sense 4 d above: manly, displaying a masculine sexual character. Hence *loosely* in commendation: fine, excellent. Chiefly *U.S. slang*.

1944 Auden *Sea & Mirror* in *For Time Being* iii. 56 The stud contralto gargling through her maternal grief. **1969** *Sat. Rev.* (U.S.) 5 July 28 There's nothing like a head-shrinker..for putting you where you're at. That's stud. **1971** *Black Scholar* Sept. 45/1 He had learned the stories about stud broads..but he knew Christine 'used' to be a stud broad. **1977** *Amer. Speech* **1975** L. 67 *Stud*. 1: *adj*, outstanding, having all the attributes approved of by the group.

studded, *ppl. a.* Add: **1. c.** Of a tyre: provided with studs (*stud *sb.*[1] 7 g).

1966 *Better Homes & Gardens* Jan. 14/2 Studded snow tires made their debut in many of the snow belt states last winter. **1970** *Toronto Daily Star* 24 Sept. 39/1 Studded tires were causing costly damage to city streets.

student[1]. Add: **2. b.** A scholar at an institute of primary or secondary education. orig. *U.S.*

1900 E. E. Brown in N. M. Butler *Monographs on Educ. in U.S.* 183 In these laboratories [high school] students perform representative experiments in the science they are pursuing. **1924** *Junior High School Clearing House* Mar. 6/1 It was felt..that to single out the student who excelled in scholarship was usually to recognize native abilities but not necessarily serious effort. **1936** *Evening Citizen* (Glasgow) 29 Aug. 4/6 [In the United States] even schoolboys and schoolgirls are students. **1962** *Dict. Canad. English: Beginning Dict.* 644/2 *Student*... 2. a person who is studying in a school, college, or university: *That high school has 3,000 students.* **1976** *Times* 5 Aug. 14/7 We have primary school students, presumably working for BAs in Plasticine; the National Union of School Students; and graduation day for high school students... Formerly people were schoolboys or schoolgirls until they became undergraduates.

c. An inexperienced user of illegal drugs; *spec.* one who takes small or occasional doses. *U.S. drug-users' slang.*

1936 [see *joy-popper* s.v. *joy *sb.* 10]. **1949** [see *junkie]. **1951** [see *Saturday night 1].

5. a. *student activism, activist, body, counselling, demonstrator, duel, exchange, grant, hostel, leader, politics, protest, revolt, revolution, riot, unrest, violence*; appositive, as *student nurse*.

1977 *Hongkong Standard* 14 Apr. 4/4 The University of the Philippines, a hotbed of student activism before Mr Marcos declared martial law in September 1972. **1969** 'E. Lathen' *When in Greece* x. 114 He had almost forgotten his role as a student activist. **1906** W. James *Mem. & Stud.* (1911) xiv. 362 Above all things, offer the opportunity of higher personal contacts. A university provides these anyhow within the student body. **1979** A. Price *Tomorrow's Ghost* ii. 26 We're not infiltrating the delectable student body. **1959** *Listener* 19 Mar. 515/1 This sort of thing must surely have been studied by university psychology departments before now—particularly in America, where 'student counselling' is respectable. **1968** 'J. le Carré' *Small Town in Germany* ii. 13 Student demonstrators..overturned the American Ambassador's car. **1911** Student-duel [see *mensur]. **1979** J. Leasor *Love & Land Beyond* iii. 48 Men..take pride in bearing scars of student duals on their cheeks. **1971** K. Dick *Ivy & Stevie* 42 I've been to Potsdam, Amsterdam, Königsberg... Sort of student exchange holidays. **1965** *Students' Handbk. 1965* (Univ. Coll. London Union) 43 For years the Councils of the N.U.S...have reiterated a call for the abolition of the Means Test on parental incomes used in the assessment of student grants. **1974** Student grant [see *inflation-proofing* s.v. *inflation 8]. **1960** *N.U.S. Year Bk.* 35 This year marks the eleventh in which the Union has maintained a student hostel in the Bloomsbury area. **1977** R. Barnard *Blood Brotherhood* xvi. 182 It was the address of a student hostel. **1962** E. Snow *Red China Today* (1963) i. 20 Huang Hua, whom I knew as a student leader when I taught briefly at the American-supported Yenching University. **1932** *Lancet Commission on Nursing* v. 46 Each 'student nurse', as they are called, is attached to a nursery class with part-time practical work with little children. **1956** K. Hulme *Nun's Story* xvii. 287 She listened deliberately now to the talk of her student nurses. **1954** P. Toynbee *Friends Apart* ii. 35, I used my freedom to become..violently caught up in the excitement of student politics. **1965** *Granta* Summer 9/1 The worst tactical mistake the SRC could make is not to dissociate itself from the old idea of 'student protest'. **1976** D. Clark *Dread & Water* vi. 133 We'd had a student protest at the gate. **1969** *Listener* 8 May 630/2, I said that western civilisation today was being challenged from within... The most obvious symptom is the outbreak of what is commonly called 'student unrest', or 'student revolt'. **1978** P. Boardman *Worlds of Patrick Geddes* xi. 403 In 1930, an event occurred which took the old and hardy critic of universities himself by surprise: a student revolt. **1968** *Punch* 31 July 168/1 When the student revolution intervened the group moved to England. **1968** *Listener* 4 July 1/1 'Student riots', 'student violence', 'the movement for student power'—these are phrases that have been used to cover a wide variety of actions and attitudes and opinions. **1966** J. Mitford in *Vogue* (U.S.) 15 Mar. 93/1 Campuses throughout the country, undergoing the 'wave of student unrest', are producing their share of women individualists. **1968** Student violence [see *student riot* above].

b. student card, a card or ticket issued to members of a student body, and usu. entitling the holder to certain privileges; **student power,** the exercise of authority within a school, college, or larger sphere by students (cf. *power *sb.*[1] 4 f); **student(s') union** = Union *sb.*[1] 10 c; also applied to a similar association or building at other centres of higher education, and *loosely*, to a national association of students, formed to promote the welfare and views of its members (cf. *N.U.S.* s.v. *N II. 1); **student teacher,** a student of a university or training college who teaches in a school for a certain period, as part of the qualification for a teaching certificate (cf. Pupil Teacher); hence **student-teachership; student-teaching; student-teacher** *a.*, designating the relation between students at a school or college and their teacher or teachers (cf. *pupil-teacher* adj. s.v. *pupil *sb.*[1] 3 b).

1973 *Sat. Rev. Society* (U.S.) May 53/2 Cost of the 16-hour program: $50 ($30 for anyone with a student card). **1975** P. Theroux *Gt. Railway Bazaar* iii. 51 She was a student and..had a student card... A card got each one a 50 percent reduction on the ticket. **1968** *Listener* 21 Mar. 365/3 He will go to any lengths to placate and excuse the brutally vocal ginger group of Student Power. **1973** *Times* 15 Oct. 17/1 Student power alone would not have toppled the government unless circumstances were very much in their favour. **1891** Students' Union [see Union *sb.*[1] 10 c]. **1916** J. Buchan *Power-House* v. 134 It had something to do with the Slav States of Austria and an Italian Students' Union, and it threatened..to be dangerous. **1967** M. Kenyon *Whole Hog* i. 12 The student union cafeteria always had..fried, boiled or scrambled eggs. **1977** P. Johnson *Enemies of Society* xii. 171 It was Margaret Thatcher..who, in the winter of 1970–1, changed the wording of the official regulations to allow public money to be handed over to the Student Unions. **1982** A. Taylor *Caroline Minuscule* iii. 30 The forthcoming motion the Students' Union were planning..deploring violence. **1909** *Rep. Board Educ. 1907–8* 57 During 1907 the new method of providing for the preliminary education of Elementary School Teachers, which is known as the 'Bursar System', has been brought into operation... The prospective Teacher either goes direct into a Training College..or..obtains an appointment as a Student-Teacher. **1953** W. Moore *Bring Jubilee* (1955) xiii. 122 That personal, face-to-face, student-teacher relationship. **1973** E. McGirr *Bardel's Murder* ii. 47 When I started..there was no difficulty in getting student teachers because they could use the rod. **1910** *Rep. Board Educ. 1908–9* 56 Upon..the passing of the necessary leaving examination, two alternative courses are immediately open to him—to proceed to a Training College or to proceed to Student-Teachership. **1970** *Daily Tel.* 23 Oct. 13/8 In the 'bad old days' of student-teachership, despite our huge classes we knew how to deal with 'emotional and learning disorders' without the aid of drugs! **1929** Myers & Harshman *Training Secondary Sch. Teachers* 17 Regulations relative to prerequisites for student-teaching.

studentess (earlier U.S. example); **stu·denting** *nonce-wd.* [-ing[1]], studying; an object of study.

1834 *Knickerbocker* IV. 120 The Collegiate Institute..was originally designed to afford its fair studentesses all the advantages usually obtained by the best educated of the other sex. **1922** Joyce *Ulysses* 739 That delicate looking student..nearly caught me washing through the window only for snapped up the towel to my face that was his studenting.

Student[2] (stiū·dĕnt). *Statistics.* The pseudonym of William Sealy Gossett (1876–1937), English brewery employee, used *attrib.* and in the possessive to designate statistical concepts devised by him, as *Student('s) (t-)distribution*, a statistical distribution which is that of a fraction whose numerator is drawn from a normal distribution with a mean of zero and whose denominator is the root mean square of *k* terms drawn from the same normal distribution (where *k* is the number of degrees of freedom); *Student's (t) test*, a test for statistical significance that uses tables of this distribution.

1929 *Nature* 17 Aug. 267/1 This is not to say that the deviation from 'Student's' *t*-distribution..may not have a real application. **1935** R. A. Fisher *Design of Experiments* iii. 38 (*heading*) Student's *t* test. **1937** Yule & Kendall *Introd. Theory Statistics* (ed. 11) xxiii. 440 We proceed to give one or two examples of the way in which the 'Student' distribution is generally used to test the significance of various results obtained from small samples. **1938** *Biometrika* XXX. 223 As the theory of mathematical statistics has developed, the significance of 'Student's' test has been elaborated from many angles and deeper meanings associated with it than its author had ever dreamed of. **1968** P. A. P. Moran *Introd. Probability Theory* vii. 326 Since *t* is scale-invariant its distribution is independent of σ, and is known as 'Student's *t*-distribution with $n-1$ degrees of freedom'. **1980** *Brit. Med. Jrnl.* 29 Mar. 890/2 The serial changes in mean daily volume, time taken, and flow were pooled and assessed with Student's two-tailed *t* test.

stu·dentish, *a.* [f. Student + -ish[1].] Pertaining to, characteristic of, or resembling an (undergraduate) student; student-like, esp. in dress or opinion.

1934 *Times Lit. Suppl.* 9 Aug. 552/3 Her spectacles, her studentish clothes, her pride in her degrees..combine to make her an almost lovable character. **1979** *Church Times* 7 Dec. 7/3 Their very appearance was not studentish: 'puritans predominated, all with clean short hair, inexpensive shoes, mass-produced clothes well cared for'.

studentize (stiū·dĕntəiz), *v.* *Statistics.* Also **Studentize.** [f. *Student[2] + -ize.] *trans.* To subject (data) to studentization; chiefly as **stu·dentized** *ppl. a.*, applied esp. to data that have been standardized by division throughout by their estimated standard deviation, and to quantities derived therefrom; **stu:dentiza·tion** (see quot. 1957).

1938 *Suppl. Jrnl. R. Statistical Soc.* V. 80 This theory has been developed further, and quite a number of large-sample results have been 'studentized'. By this we mean that a statistic whose sampling distribution involves the unknown standard deviation of the population is modified so that its distribution involves only quantities calculated from the sample. **1939** *Biometrika* XXXI. 21 In a recent paper H. O. Hartley..has suggested a systematic method of obtaining probability levels for 'studentized' functions. **1957** Kendall & Buckland *Dict. Statistical Terms* 284 *Studentisation*, the process of removing complications due to the existence of an unknown parent scale-parameter by constructing a statistic whose sampling distribution is independent of it; especially by dividing a statistic which is of a certain degree in the observations by another statistic of the same degree. **1959** H. Scheffé *Analysis of Variance* ii. 28 In the next chapter we will apply the distribution of the Studentized range. **1972** *Biometrika* LIX. 165 Duncan's modification also starts with the studentized range test.

stud-horse. Add: **2.** (Earlier and *absol.* examples.) (The usual form is now *stud *sb.*[2] 4 c or *stud-poker* s.v. Stud *sb.*[2] 5.)

1879 *Rep. Cases at Law Supreme Court of State of Arkansas* (1881) XXXIV. 442 The proof..showed that he did bet chips, or checks, at a game played with cards, called 'stud', or 'stud-horse', poker; being somewhat different from other games called 'straight poker' and 'draw poker'. **1882** in *Colorado Q.* (1956) IV. III. 271 Brownie Lea is leading stud horse at Del Norte. **1920** C. E. Mulford *Johnny Nelson* ii. 19 He's a travelin' edicator in th' innercent game of draw—or was it studious, Nelson?

studiable, *a.* Delete *nonce-wd.* and add: Also, capable of being studied or observed.
1971 *Jrnl. Gen. Psychol.* Jan. 172 The activities of single cells studiable by microelectrode techniques. **1982** *Times Lit. Suppl.* 23 Apr. 465/3 He does give a studiable picture of how Los Angeles may have sleep-walked into this unassimilable truth.

studio. Add: **2. c.** *Cinematogr.* A room in which a cinematographic film is shot. Hence, a film-making complex including film studios and attendant offices and premises (also in *pl.*); the company which runs this. Cf. *film studio* s.v. *FILM sb. 7 c.
1911 C. N. Bennett et al. *Handbk. Kinematogr.* xiii. 102 Covered-in studios provided with expansive glass roofs for daylight work..are hardly among the first flights of commercial Kinematographic enterprise. **1923** *Variety* 15 Nov. 18 (heading) Hollywood studios moving to less costly locations. **1928** *Morning Post* 20 Oct. 4/3 The 'dedication' of Fox's new 8,000,000-dollar Movietone Studio at Los Angeles. **1937** A. Huxley *Let.* 12 July (1969) 425 With regard to the handling of my work, I am prepared to authorize you to take up the matter with film studios. **1956** H. Kurnitz *Invasion of Privacy* iii. 26 The actor was undoubtedly a handsome youngster... Five studios were squabbling over him now. **1971** *Guardian* 2 Dec. 11/2 The studio, Paramount, was 'not happy with it' [sc. a film] and failed to promote it.
d. *Radio* and *Television.* In a broadcasting station, etc.: a room from which items are broadcast live or in which they are recorded for subsequent transmission; the premises housing such a studio or studios. Also *pl.*
1922 J. Reith *Diary* 29 Dec. (1975) ii. 119 Newcastle... Here I really began my BBC responsibility. Saw transmitting station and studio place. **1923** *Radio Times* 5 Oct. 38/3 'Romeo and Juliet' is being..broadcast from our own studio. **1938** *Encycl. Brit. Bk. of Year* 633/2 The apparatus will operate satisfactorily either in natural or artificial light, the technique used in the television studio being somewhat similar to that in a film studio. **1958** M. Bragg *Without City Wall* xxviii. 255 The television studios..were on the edge of town. **1973** *Daily Tel.* 6 Jan. 1/2 Mr Wilson, Leader of the Opposition, visited a television studio yesterday morning to pre-record his contribution to the programme.
e. A room used for recording and editing music, etc. to be reproduced on a gramophone record or similar medium.
1928 [see *recording studio* s.v. *RECORDING vbl. sb. 5]. **1932** *Daily Mirror* 21 Oct. 10/3 It is good watching a record being made in a gramophone studio. **1955** L. Feather *Encycl. Jazz* 341 The bands of [illegible], Count Basie, Hot Lips Page and Eddie Durham, the last two being studio-assembled combinations. **1977** *Rolling Stone* 24 Mar. 55/4 Ten months devoted to Fleetwood Mac's album has left Buckingham spindly and studio wan.
f. = *studio flat*, sense 3 b below, orig. *U.S.*
1941 A. Powell *Time to be Born* (1943) v. 100 Amanda's furnished 'studio'..was..a one-room apartment. **1962** P. Moyes *Death on Agenda* vi. 94 [The flat] was what is known on the Continent as a studio—that is to say, a one-room bachelor apartment with its own tiny hallway, off which led a box-sized kitchen and a dwarf bathroom. **1977** I. Shaw *Beggarman, Thief* i. 361 The room was a nasty little one-room studio near the university.
3. a. (Later examples in broadcasting and recording: cf. sense 2 above.)
1922 [see sense 2 d above]. **1944** L. MacNeice *Christopher Columbus* 16 The radio dramatist must..be studio-conscious, remembering what results can be obtained. **1956** B. Holiday *Lady sings Blues* (1973) iii. 35 Benny [Goodman] was a radio-studio musician who talked a lot then about having his own band one day. **1972** P. Black *Biggest Aspidistra* ii. 28 The studio announcer introduced the band and the place, and wished the nation good night.
b. Special Combs. (senses *2 c, d) *studio audience, manager*; **studio apartment** *U.S.* = *studio flat* below; **studio bed, couch,** a couch which converts into a bed; **studio flat,** a flat containing a spacious room with large windows, which is or resembles an artist's studio; more recently, a small one-roomed flat; **studio party,** an informal party held in an artist's studio; also, a social gathering at a film studio; **studio portrait,** a posed photograph, as taken in a photographer's studio; **studio potter,** a potter (freq. one of a small group) who works in a studio producing hand-thrown pottery; hence **studio pottery;** **studio theatre,** (an) experimental theatre.
1903 *Archit. Rec.* July 240 The most economical way of combining a good high studio with an economical disposition of space would be to make the studio apartment two-storied in the service and living portions, and only one-storied in the space devoted to the studio. **1929** *Washington Post* 1 Sept. 6/5 Studio Apartment $100. **1949** *Archit. Rec.* Nov. 125 At the foot of the San Jacinto mountains, this small group of studio apartments is keyed to the needs..of a specific class of tenant—artists who spend part of the year in Palm Springs. **1978** J. Irving *World according to Garp* xix. 428 Roberta went to ..Duncan's live-in studio... Duncan's studio-apartment. **1932** *B.B.C. Year Bk. 1933* 20 Comedians probably most need the stimulus of the crowded hall, for which the 'studio audience' is a poor substitute. **1977** B. Pym *Quartet in Autumn* x. 90 The radio offered a choice of comedy, with a braying studio audience. **1963** M. McCarthy *Group* vi. 118 Across the room, a big lumpy studio bed was covered with a black velveteen spread. **1931** Sears, Roebuck *Catal.* Spring 663 New 'Sun Bed' Latest Style Couch... This new type davenport-bed or studio couch. **1977** A. Schofield *Venom* i. 3 His bedroom..had been designed..as a bed-sitter..and the bed itself was a studio couch. **1934** A. Huxley *Let.* 13 Oct. (1969) 385 We are in London for the winter—having found a studio flat, miraculously large. **1970** K. Giles *Death in Church* ii. 31 He had what they call a studio flat—bed, gas fire and tiny kitchenette. **1937** R. Chandler in *Black Mask* Jan. 19/1 [illegible] 'We get one in the radio column about every second month. We're a small station still.' **1980** S. Brett *Dead Side of Mike* i. 13 'I'm sorry, I don't speak BBC. What's an SM?' Studio Manager. Knob-twiddler, tape-machine starter and what you will.' **1909** E. Nesbit *Daphne in Fitzroy St.* xvi. 275 'Meals in studios are always rather like picnics.' 'Oh, yes,' she said. 'I've been to lots of studio parties. They're great fun, aren't they?' **1938** R. H. Heaton *Perfect Hostess* 110 (heading) Miss Eighteen borrows the Attic for her first Studio Party for her young friends. [illegible] **1938** M. Allingham *Fashion in Shrouds* [illegible] studio portraits of Miss Adamson. [illegible] **1940** D. Leach *Potter's Bk.* iii. 43 A potter's prime need is good clay. Whether he be industrial, peasant or studio potter the raw material of which pots are made is of fundamental importance. [illegible] We saw the sort of simple country pots that we as studio potters (people actively engaged in making pots ourselves—by hand in a machine age) so much admire. [illegible] A studio potter (generally a small pottery with perhaps half a dozen partners or employees)..machinery plays the least possible part. [illegible] Sotheby's Belgravia studio pottery sale was also making high prices. [illegible] studio or theatre exists..to prevent dramatic art from being wiped out by the commercially minded. [illegible] the type of experimental studio-theatre that flourished in Stanislavsky's lifetime. **1972** J. Elsom *Theatre outside London* x. 172 There will be a high fly-tower, a studio theatre, fine foyers, a restaurant and good backstage facilities.

‖**studiolo** (stūdiǫ·lo). [It., lit. 'small study'.] A private study hung with paintings.
1926 J. E. Jeffery tr. *R. de la Sizeranne's Celebrities Ital. Renaissance* 180 She [sc. Isabella d'Este] heard that the decorations ordered for her *studiolo* at Mantua were not progressing, for Lorenzo Leonbruno, the painter, was a dawdler. **1958** *Times* 14 Nov. 16/4 Later developments, such as the 'Studiolo' group, Pietro Candido, Salviati, and Tibaldi, are not represented. **1968** N.Y. City (Michelin Tire Corp.) 27 The major attraction of the Italian section is the 'studiolo', or private study, from the Ducal Palace at Gubbio (Umbria). **1978** M. Girouard *Life in Eng. Country House* vi. 166 The books..in the *franche* [illegible] paintings in the *studiolo* of Federigo Montefeltro at Urbino.

‖**studium** (stiū·diŏm). *Hist.* [late L. (4th-cent.) use of L. *studium* Study *sb.*] = Study *sb.* 9. Also, = next.
1610 [see Study *sb.* 9]. **1673** J. Ray *Observations Journey Low-Countries* 342 The *Studium*, called the *Sapienza*, where are the public Schools [in Siena]. **1834** [see next]. **1922** *Encycl. Brit.* XXXIII. 602/1 Immediately after 1168 admissions to Oxford as a *studium* and a *studium generale* begin to multiply. **1936** *Times Lit. Suppl.* 2 May 361/2 Out of such enterprises developed the medieval universities [illegible] schools were not made. Later, of course, there was deliberate creation: there was the *studium* at Naples, brought into being by Frederick II. **1981** P. Kibre *Scholarly Privileges Middle Ages* ii. 19 The measure requiring the rectors..to take the oath that they would not seek the removal of the *studium* from the city [of Bologna] was retained.

‖**studium generale** (stiū·diŏm dʒenerē¹·li, -ā·lī). *Hist.* Pl. **studia generalia** (-ē¹·liă, -ā·liă). [med. L., f. prec. + *generāle*, neut. sing. of *generālis* General *a.*] A medieval university which did not only receive scholars from its own locality (an earlier equivalent of the universitas[?] University); = *general study* s.v. Study *sb.* 9.
1834 *Edin. Rev.* Oct. 225 The oldest word for an university institution of higher education, was *studium*, and *studium generale*..a term employed in the twelfth and thirteenth centuries, and retained in those which followed. **1895** H. Rashdall *Univ. Europe Middle Ages* I. 9 A Studium Generale meant a School of general resort, but in its origin the expression was a wholly popular and extra-legal one. *Ibid.* 10 In the latter half of the Thirteenth Century this unrestricted liberty of founding Studia Generalia gradually ceased. **1928** *Daily Tel.* 10 July 12/4 Where the liberal Arts and Sciences of a complete Studium Generale may be cultivated and practised. **1966** H. Wieruszowski *Medieval Univ.* I. i. 16 The term *studium generale* was used for schools of higher learning until late in the fifteenth century... Gradually..the term 'university' replaced the word *studium*.

study, *sb.* Add: **6. b.** *a quick study* (also *transf.* in general contexts).
1900 J. K. Jerome *Three Men on Bummel* i. 13 Muriel is master of six pieces already, as perhaps you know; and all the other children are quick studies. **1954** M. Ewer *Heart Untouched* ix. 164 She had learnt something in these last few days. She was a quick study. **1974** P. De Vries *Glory of Hummingbird* xiii. 197 We'll brush him up. He's a quick study. He's not a Neanderthal.
7. b. Hence applied to the face registering an expression of incredulity, etc. (*colloq.*).
1886 C. M. Yonge *Chantry House* II. xiv. 136 Emily's countenance was a study. **1964** C. Chaplin *My Autobiogr.* x. 156 You should have seen his face watching you, it was a study! **1973** *Press & Jrnl.* (Aberdeen) 3 Aug. 7/3 We stopped for lunch at a little rustic inn. Specialité de la maison—chicken and chips with frozen peas on the side. Archie's face was a study.
12. (sense 5) *study-leave, tour*; (sense 8) *study-bedroom*; **study circle,** a group that meets regularly to discuss a particular topic of study; **study group** = prec. freq. an investigative committee formed by a political, industrial, or other body for this purpose.
1930 *Times Educ. Suppl.* 26 July 332/4 Study-bedrooms for 108 inmates. **1978** J. I. M. Stewart *Full Term* i. 15 Schools that are a bit lavish with study-bedrooms for senior boys. **1931** M. MacNeice *Friends' Wreath* vi. 84 They've no team spirit, they won't take part In our study circles and community art. **1973** R. D. Skinner *Robert Emm* iv. 31 Study circles, Chapter meetings. **1926** *Scribner's Mag.* Apr. 672 The Foundation is accumulating a body of literature on Positive Health, for the use of individuals and organized study groups. **1948** *Ann. Reg. 1947* 223 Before the conference came to an end thirteen of the participating countries decided to create a study group to examine the possibility of creating a general European Customs union. **1974** R. Crossman *Diaries* (1975) I. 79 The only thing I need really mention is the dinner held at Arnold Goodman's house for my Rent Bill study group. **1981** *Times* 10 Oct. 1/37 The huge expansion of universities has..made it more difficult for them [sc. students] to get study-leave. **1982** M. Duke *Flashpoint* vi. 44 A spell of study-leave before you take on the new job. **1937** *John o' London's Weekly* 7 May 209/2 A tin to keep the stumps of the study tour away from the studious and lose the bristles of my toothbrush during study-tours. **1977** *Jrnl. R. Soc. Arts* CXXV. 519/2 Approximately 90 former Bursary winners made study tours abroad.

study, *v.* Add: **I. 1. a.** Now only *U.S. colloq.*, to make a close study of (a subject), to 'bone' *up* (*on, in*), esp. in preparation for some display of knowledge (*intr.* use of sense 7 b).
1946 *Chicago Daily News* 25 June 13/3 A'll git a li'l closer, an' study up on him!' **1976** K. Robinson *Landscape with Dead Dons* xii. 114, I am sure that if you once studied up a little in psychology you would be as struck as I was. **1970** N. Armstrong et al. *First on Moon* vi. 132 He had studied up on everyone so he could tell wonderful stories about them. **1980** J. Ball *Then came Violence* ii. 16 'We know quite a lot about financial things.' 'Yes, he knew you'd been studying up?' [illegible] the chief asked.
2. a. (Later examples with *about*.)
1844 *Yorks. Comet* No. 1. 1 Moore Ah studied about it an' war it puttered me. **1895** *Dialect Notes* I. 374 Study, talk, discuss, consider... 'I studied about her to my man when I got home.' **1940** W. Faulkner *Hamlet* iv. i. 'We, I was about minded one night when I was studying them out. Studying about something else and forget how long the time was. **1969** 'Malcolm X' *Autobiogr.* (1966) xiii. 306, I studied about if I just *should* happen to say something to her—what would her position be?'
II. 7. b. (Examples.)
1880 'Mark Twain' *Tramp Abroad* 412 Studying up the subject of Alpine climbing. **1902** Joyce *Ulysses* 728 He knows a lot..about the body and the insides..often wanted to study up that myself.

stuff, *sb.*¹ Add: **II. 3. c.** Esp. in phr. *bit of stuff*. Now chiefly in slang use, with or without epithet, of a woman or girl. Cf. *BIT sb.² 4 i.
1828 *Sukalton's Log Bk.* III. 164, I entered the house in great spirits, fancying myself, he made use of a slang phrase, a very good bit of stuff. **1858**, etc. [in Dict.]. **1909** in J. M. Ware *Passing Eng.* 31/1 He waited for a girl..to the scene to get the Comedy Theatre. He was an elderly cove and he had got a decent bit of stuff. **1920** M. Aldiss *Soldier Erect* 10 The infantry myth that one spent one's whole leave yanking it up some willing bit of stuff in a pub yard?
4. d. (Earlier example.)
1851 S. Rutter *Hints to Gold Hunters* 12 The principal use of the washpan is in rewashing the partially washed stuff taken from the rocker.
III. 6. g. Narcotics, 'dope'. Phr. *on the stuff*, addicted to drugs, on drugs. *slang* (orig. *U.S.*).
1929 *Amer. Speech* IV. 345 Stuff, dope. **1934** [see *MAIN-LINE *v.*]. **1947** J. Pollock *Underworld Speaks* 84/1 *On the stuff*, addicted to dope. **1952** *Sunday Times* 3 Feb. 5/4 There has lately been a lot of research into the sale of narcotics (or 'junk' or 'stuff') and their effects on addicts. **1959** 'F. Newton' *Jazz Scene* 292 Jive talk.. contains all the fancy-dress devices of private languages.. the never-ending substitution of new passwords into the group for new codes..the use of neutral and general words for highly specific things (e.g. *on the stuff*, or simply *on* for drug addiction). **1965** *New Statesman* 20 Aug. 248/3 Addicts have a secret language, which changes like a code. The commonest current name for heroin is 'stuff'. **1973** L. Hellman *Pentimento* 290 'His room-mate's on the stuff.' This then new way of saying dope..was no surprise. Years before she had told me her son was on the stuff. **1976** H. Ferguson *Confessions Long Distance Acid Head* 65 'Yes. You were the bloke who got done for someone else's stuff..weren't you?' It was a junkie whom I had met in Ashford.
7. c. See also *HOT STUFF a.

e. *to do one's stuff*: to do what is required or expected of one; to perform one's role. *colloq.*
Quot. 1663 may belong to another sense.
1663 G. Fox *Jrnl.* (1694) I. 266 A while after, when the priest had done his stuff, they came to the friends again. **1922** *Radio News* (U.S.) IV. 854/1 (*caption*) Take a look at S. M. Brown, Chief on the Mauretania, 'doing his stuff' in the saloon. **1930** *Pep* T. E. Lawrence *Let.* (1938) 677 That portable was good at Miranshah. I hope yours is doing its stuff. **1933** *Bulletin* (Sydney) 19 Apr. 29/4 Australia will be represented in this event by Alan Bruce, who has been doing his stuff in London. **1946** F. Sargeson *That Summer* 14 If you know how to do your stuff you never could tell but what it mightn't end up in a date. **1959** [see *ANTI-AIRCRAFT a.]. **1967** G. F. Fiennes *I tried to run a Railway* vi. 70, I go when I can for the fun of hearing Richard doing his stuff. **1972** Wodehouse *Pearls, Girls, & Monty Bodkin* ii. 27 The Bishop and assistant clergy and the bridesmaids should be encouraged to line up and do their stuff. **1976** *Daily Times* (Lagos) 27 Aug. 30/1 Ghana's Johnny Francois and a few others did their stuff abroad and, gradually, the panel took root.
f. *that's the stuff* (*to give them* or *to give the troops*): that is what is particularly appropriate to the situation; that is what is required.
1923 'Bartimeus' *Seaways* vii. 98 George Grayson and his Flock of fascinating Flappers presents a screaming farce: The Giddy Governess!! That's the stuff to give the troops! **1927** *Daily Express* 13 Oct. 12 That, if one may be pardoned the colloquialism, was to say the stuff to give them. **1942** H. C. Bailey *Dead Man's Shoes* xv. 63 'A new married man's health and well-being,' he said. 'that's the stuff,' said Mr. Ogmore. **1977** P. D. James *Death of Expert Witness* II. i. 53 Inspector Blakelock.. 'we always glad to have suggestions from anybody.' 'That's the stuff,' said Mr. Ogmore. **1977** P. D. James *Death of Expert Witness* II. i. 53 Inspector Blakelock.. that's the stuff to give the troops, he would invariably say.
g. *to know one's stuff*: to be experienced or knowledgeable in one's subject, profession, etc. *colloq.*
1927 *Amer. Speech* II. 277 Know your onions or know your stuff, have grasp of your subjects. **1935** *Swing Music* June 11/1 The Little Man of the Rhythm Clubs did himself proud in this last topic. He knows his stuff. **1938** G. Greene *Brighton Rock* iii. i. 108 'The coroner's a man who knows his stuff.' **1945** K. A. Knox *God & Atom* x. 132 All I have written could have been written by much more better by someone who, in an expressive modern phrase, knew his stuff. **1952** J. Steinbeck *East of Eden* xxiii. 46 It's a fair, Kate sure knows her stuff. **1967** M. Argyle *Psychol. Interpersonal Behaviour* ix. 166 His [sc. the supervisor's] influence will be accepted more readily if it is believed that he really knows his stuff. **1973** A. Christie *Postern of Fate* iii. vii. 174 'He gave me a lot of knowledge about planting things.' 'Yes, he knew his stuff, as you might say.'
h. *gen.* Used *loosely* to denote any collection of things about which one is not able or willing to particularize (a weakened application of senses 6, 7); material, matter, business. *colloq.*
1922 [see sense 7 e above]. **1949** 'G. Orwell' *1984* ii. ii. 123 You thought I was a good Party member. Pure in word and deed. Banners, processions, slogans, games, community hikes—all that stuff. **1967** K. Brattigan *Trout Fishing in America* (1970) 83 One spring day she had me ascend to the attic and clean up some boxes of stuff and throw out some stuff and put some stuff back into its imaginary proper place. **1977** J. D. MacDonald *Condominium* xxxvi. 370 Once they left we were going to start having fun with this stuff.
f. Hence, with preceding epithet.
1920 etc. [see *to do one's stuff* e above]. **1932** S. Gibbons *Cold Comfort Farm* xiii. 191 She had best not pull any Cinderella stuff on me. **1939** *Punch* 5 July 9/1 'Sam,' they said to him, 'what's wrong? You can bowl much better stuff than that.' **1948** *Sporting Mirror* 21 May 7/3 Jack Martin may also be suitable for fast stuff on occasions. **1964** *Times Lit. Suppl.* 26 Apr. 420/3 The principal message [of Hochhuth's comedy *Lysistrata und die NATO*] is largely straightforward feminist stuff. **1976** *National Observer* (U.S.) 9 Oct. 24 The threat of another oil embargo is always serious stuff here. **1978** D. Williams *Treasure up in Smoke* viii. 110 What he said was pretty strong stuff... He fairly laid into Mr O'Hara.
9. b. *N. Amer.* In various sports, the spin or 'work' imparted to a ball in order to make it vary its course; the type of control which effects this. Also *fig.*
1905 *Sporting Life* (U.S.) 9 Sept. 1/1 If I tried some of the stuff that certain pitchers use and escape bumping, I have an idea that the fielders would never stop..hitting. **1933** *Harper's Weekly* 13 Sept. 21/2 Wellman, the giant Brown, is another (pitcher) who has the 'stuff'. **1927** *Daily Tel.* 21 Feb. 13/6 T. A. Workman, their captain, was in wonderfully good form against Commander S. W. Beadle, finding an almost perfect length for an American service which had plenty of 'stuff' on it. Beadle could not do anything with it, and was kept on the defensive throughout. **1936** J. T. Farrell *World I Never Made* v. 68 The O'Neills are proud of their name, and they got as much stuff on the ball in the game of life as old Three-fingered Brown has when he toes the mound. **1947** *Sun* (Baltimore) 3 Apr. 20/1 He is only 20 years old, has a good arm and has much of the well known stuff on the ball. **1967** Varner & Harrison *Table Tennis* v. 51 These spinners are often one-ball hitters: they vary their 'stuff' until you yield a loose return, which they efficiently kill. **1970** J. H. Gray *Boy from Winnipeg* 152 That got us seats behind home plate where we could watch the stuff, mainly curves, that the pitchers were putting on the ball. **1981** *Washington Star* 30 Apr. c 4/1 'I really had good stuff tonight,' the lefthander said in a post-game radio interview. 'My slider wasn't great at the beginning, but my fastball really was good.'
10*. Phr. *not to give a stuff*, etc. = *not to give a fuck* s.v. *FUCK sb. 2. Cf. *STUFF v. 14* a. Chiefly *Austral.* and *N.Z. slang*.
1976–7 *Sea Spray* (N.Z.) Dec./Jan. 62/1 Well, deep down inside I don't really give a stuff. **1974** *Bookseller* 19 Jan. 333/1 A word or two of criticism: I don't give a stuff for your great managing director. **1977** *Bulletin* (Sydney) 22 Jan. 100/3 The list goes on and on and on and as it grows so does the feeling amongst the blokes in the bush that no one gives a stuff. **1979** R. Gormmer *Burgess's Daughter* i. 42 In the end no one cares a stuff who's in jail or what war's on, so long as it's far away. **1980** B. Mason *Solo* 207, I don't give a stuff if it's so or not. That spoke to me. Opened up my life, things I'd forgotten.
11. c. stuff-over *a.*, applied to chairs, etc., which are upholstered by having the material drawn over the frame of a fixed seat and secured beneath; also *absol.* as *sb.*, a stuff-over seat.
1925 R. S. Bowers et al. *Furniture Making* xxxi. 353 Stuff-over chair and settee. **1963** *Times* 2 Feb. 11 The slip-in seat is almost universal and the stuff-over almost unknown in Portuguese Chippendale chairs. **1972** *Country Life* 1 June 1414/1 Regency mahogany dining chairs..with the stuff-over seats. **1976** *Liverpool Daily Post* 11 Dec. (Advt.), For sale, stuff-over roll back (Chesterfield Settee in silk damask.

stuff, *v.*¹ Add: **8. c.** To pack or load (a freight container). *slang.*
1967 Kerr & R. Gran *Cargo Handling* vi. 115 Containers can come into..a Consolidation Depot where they are stuffed with miscellaneous general cargo. **1968** [see *STRIP v.¹ 7 c]. **1972** *Timber Trades Jrnl.* 13 May 44/1 The dockers threaten to continue the ban until their demands are met which include the exclusive right to stuff (pack) and unstuff (unpack) containers. **1972** *Nature* 11 Aug. 301/2 British dockers are..asking that members ..of the union should stuff and strip all containers at the centres at which containers are stuffed with goods.
14*. a. Used in coarse expressions of contempt or defiance. Cf. *FUCK v. 2; *STUFFED *ppl. a.* 6.
1955 P. Larkin *Less Deceived* 30 Ah, were I courageous enough To shout *Stuff your pension!* **1958** F. Norman *Bang to Rights* 168 The greeter just got up and told him to stuff his job. **1963** J. Wain *Strike Father Dead* iv. 205 Very well, they could keep their stuff. And still it, I wasn't even going to stay in the same miserable country. **1965** 'T. Hinde' *Games of Chance* I. iii. 99 'Stuff you,' I said. **1973** J. Porter *It's Murder with Dover* i. 3 'He should have taken a stronger line.. Told old Creech to stuff it.' **1976** W. Trevor *Children of Dynmouth* xi. 204 He goes up to him and tells him to stuff himself and in a half half-minute he's belting the old lorry up the London road. **1977** *Time* 28 Mar. 11/1 Stuff the country. We need a safe thing we're going to do. He won the election and now he's doing it.
b. *vulg. slang.* (With male subject) to copulate with (someone). Occas. *intr.*
1960 B. Moore *Luck of Ginger Coffey* iv. 35 Trying to stuff another man's wife, is that your idea of being a friend? **1977** F. Raphael *Cracks in Ice* (1979) 133 Nature..can also be applied, since it was originally adjectival, to a pregnant woman and to a sausage, both of which, in vulgar parlance, can clearly claim to have been stuffed. **1982** J. Scott *Clynch of Mandarin* vi. 63 You come all the way from the city..to stuff your own woman and bring her. **1983** *Sunday Times* 16 Jan. 35/3 He was sacked later from a Danish ship..for [illegible].
7. Special collocations: **stuffed monkey,** a type of biscuit or cake made with almonds; **stuffed olive,** a stoned (usu. green) olive filled with pimento; **stuffed owl** [from the title of an anthology, ult. derived from Wordsworth's *Misc. Sonnets* III. xiii], used *attrib.* with reference to poetry which treats trivial or inconsequential subjects in a grandiose manner; hence **stuffed-owlish** *a.*; **stuffed pepper,** a cooked dish of green or red pepper (capsicum) de-seeded and filled with tomatoes, rice, meat, etc.; **stuffed shirt** *colloq.* (orig. *U.S.*), one who is pompous and conservative, but usu. ineffectual; hence **stuffed-shirted** *a.*; **stuffed-shirtedness**; **stuffed vine leaves,** an eastern (esp. Greek or Turkish) dish consisting of vine leaves wrapped round a savoury mixture of rice, onion, etc.
1892 I. Zangwill *Children of Ghetto* I. 14 The confectioners' shops, crammed with 'stuffed monkeys' and 'bolas'. **1943** A. L. Simon *Conc. Encycl. Gastron.* IV. 127/2 *Stuffed Monkey* (S. Africa... Place the one half on a flat baking-sheet, cover with the filling or stuffing... Put the other half of the pastry over the filling; press the edges firmly together. **1962** *Listener* 11 Jan. 107/3 'Stuffed monkey' is a rich cake popular in the Jewish cuisine. **1928** Wodehouse *Money for Nothing* ix. 262 Anchovy sandwiches, stuffed olives. **1967** K. Giles *Death in Diamonds* iv. 78 Another cupper and a plate of stuffed monkeys. **1897** Kipling *Captains Courageous* ix. 198 Try a stuffed olive. **1920** [see *angel's-food-cake*]. **1976** N. Bagnold? J. Powers *Fifth Defector* i. 18 Here is a thing what'll stuffed olives..or stuffed-owl nerves. **1930** Wyndham Lewis & Lee *(title)* The stuffed owl: an anthology of bad verse.] **1941** Blunden *Thomas Hardy* xii. 264 Hardy's poems.. have their share of stuffed-owl expressions. **1957** R. A. Knox *On Eng. Translation* 12, I will not entertain you.. with choice specimens of really stuffed-owl renderings in this field, such as that famous translation from the Italian, about the medieval story of a woman who was turned into a horse. **1960** *Guardian* 23 Mar. 6/7 Of the longer pieces, 'The Cruel Place' is the most stuffed-owlish. **1864** V. Davies *Let.* 8 Oct. in C. W. Woodward *Mary Chesnut's Civil War* (1981) xxxvi. 669 Colonel Lubbock was there, stuffed peppers... **1971** I. Blundell *Deinot Loss & Stuffed peppers.* **1960** 'E. McBain' *Killer's Payoff* ix. 89, I was going to call you for that stuffed-pepper recipe.. you used for the last buffet. **1978** H. Kaplan *Damascus Cover* vii. 69 Boys..carried trays laden with soup and stuffed peppers. **1966** C. Nicholas ii. 34 They're all stuffed shirts. Another stuffed shirt, Sir? Shubert Brothers... **1930** G. N. Horn? *Dead Snort* 29 'He's an inveterate stuffed shirt.' **1946** *Nature* 8 Mar. vii. 143 His, a bit lonely, and a bit stuffed-shirted. **1960** *Spectator* 2 Dec. 888 [illegible]. **1960** 'D. T. Lavin' *Fire* 3 They'll have to learn that.. **1961** Muriel Spark... the [illegible] stuffed-shirtedness; there was nothing in [illegible] of a stuffed-shirted ... **1978** D. [illegible] the stuffed shirt of pop... **1939** A. Heath *Open Sesame* 120 Vine leaves, Stuffed. **1978** H. Kaplan *Damascus Cover* vii. 69 A man sitting on a straw stool eating stuffed vine leaves sunk in goat's milk.

stuffer. Add: **3.** An advertising leaflet or similar material enclosed with other literature, esp. when sent by post.
1942 [see *FILLER¹ 2]. **1971** *Oxf. Univ. Gaz.* (Ann. Rep. Delegates Univ. Press) 5 The Promotion Department had to prepare, produce, and distribute 850,000 stuffers, 550,000 prospectuses. **1972** *Publishers Weekly* 31 Jan. 93/1 The prices they wish printed on the mailing piece, circular, stuffer, etc. **1976** *New Yorker* 12 Apr. 120/1 The new stuffer is a word-and-picture collage printed on one side and a full chronology of Tharp choreographies on the other.

stuffiness. Add: **4.** A formal or strait-laced attitude.
1926 Galsworthy *Silver Spoon* iii. vi. 262 The book breaks through the British 'stuffiness' which condemns any frank work of art. **1933** *Times Lit. Suppl.* 2 Nov. 776/4 We see Angrove's clerical stuffiness gradually dispelled by the Greek sun and sea. **1978** J. H. L. Anderson *Death in North Sea* vi. 112 Mr Wilson more than made up for his initial stuffiness... I took him into a small bar.

stuffado, var. *STIFADO.

stuffage. Delete † *Obs.* and add: **1.** (Later examples.)
1943 *Aeronautics* Jan. 34/2 An aeroplane should be stripped for action and not a stuffage of gadgets. **1962** C. Roberts *Buried Books* 13 The discarded papyrus MS served as stuffage for the binding cannot conceivably be later.

stuffed, *ppl. a.* Add: **6.** Phr. *get stuffed*: used as a coarse imprecation. Cf. *STUFF *v.*¹ 14* a, b.
1952 M. Tripp *Faith is Windsock* x. 155 'Get stuffed,' he said savagely. **1968** R. Kirkby? R. Weaver? ... **1977** R. Kirkman? 'Why don't you tell Leopold to go get stuffed?' 'Because we need the foreign currency.' **1975** *Weekend Mag.* 2 Nov. 16/2 [He] told Tony Gibson to go and get stuffed; that he had no intention of telling them anything important, he proposed to deal only with the government. **1979** R. Rendell *Make Death love Me* vii. 69 Who're you giving orders to? You can get stuffed.
7. Special collocations: **stuffed monkey,** [see above]

stuffing, *vbl. sb.* Add: **1. a.** The putting of fraudulent votes into a ballot-box. Also *ballot stuffing*. Cf. Stuff *v.*¹ 8 d.
1976 *Birmingham Post* 16 Dec. 2/9 Special see-through ballot boxes were used to show that two pre-voter 'stuffing' had been stopped. **1977** *Time* 25 Apr. 48/1 Drazen had gathered evidence of 'ballot stuffing' in [illegible] was even sought to prove ballot stuffing by the boss of Quitman County. **1979** *Internat. Jrnl. Sociol. of Law* Feb. 71 In fact larger towns, during the civilian rule, they occasionally employed illegal means..and the stuffing or disappearance of ballot boxes was not uncommon.
2. c. *to put stuffing into*: to add strength or substance to. *colloq.*
1938 [see *KICK sb.¹ c]. **1977** P. Harcourt *At High Risk* iii. iv. 179 The whisky and wine I had drunk were making me rash..but they were putting some needed stuffing into me. **1979** J. Sherwood *House of Hyenas* iv. 54 She really puts the stuffing back into these women. Some of them even learn to cope with their dreadful husbands.

stuffless, *a.* Delete *nonce-wd.* and add earlier and later examples.
1896 Belloc *Verses & Sonnets* 23 The tiny stuffless voices of the dark. **1932** W. De La Mare *Early Novels of Wilkie Collins* in J. Drinkwater *Eighteen-Sixties* 90 The best of them [sc. Collins' characters] are triumphantly in the round;..those in the background are somewhat stuffless in effect.

stuffy, *a.* Add: **2. b.** (Earlier example.)
1813 Jane Austen *Pride & Prejudice* I. xvi. 173 They were superior to the broad-faced stuffy uncle Philips, breathing port wine, who followed them into the room.
5. Prim, formal, strait-laced, pompous; boring, conventional.
1895 Kipling *Day's Work* (1898) 181 'You might have come to me to begin with,' said Scott stiffly... 'Well, you need't be stuffy about it.' **1904** E. Nesbit *Phoenix & Carpet* i. 11 It's awfully stuffy for a chap not to be allowed out in the evenings. **1922** J. Cannan *Misty Valley* xv. 245 Cousin Innes was very stuffy about it when you got engaged. But all we wanted was for you to be happy, Claire. **1948** E. Waugh *Let.* 24 Nov. (1980) 293, I went to one stuffy upper class dinner party on my first night & I go to another tonight. **1965** *Listener* 16 Sept. 418/2 Some of them have grasped the BBC's main dilemma, which is that it ought..to be in the widest, not in any 'stuffy' sense, an instrument of culture and education. **1979** R. Jaffe *Class Reunion* (1980) i. ii. 34 Every man I meet wants to go into business. They're all so stuffy.

stuggy, *a.* Add: Also *transf.*
1945 P. Woodruff *Call Next Witness* II. 59 Quite as heavy a weight as his stuggy little brown mare was meant to carry. **1960** L. Fielden *Natural Bent* ix. 250 Ships with high noses and short noses, ships with long lines and stuggy lines.

Stuka (stū·ka, ʃ-). Also **stuka.** [Abbrev. of G. *sturzkampfflugzeug* dive-bomber.] A dive-bomber of the German air force, esp. as used in the war of 1939–45.
1940 *Sun* (Baltimore) 23 May 1/5 The awful destruction which German *Stukas* (dive bombers) wreak. **1942** *Ann. Reg. 1941* 12 On January 10 German Stuka dive-bombers ..made an attack..on a British convoy. **1942** *Daily Tel.* 23 Mar. 1/5 Fighters and stukas operate from the more forward flying ground at Tmimi. **1946** A. Lee *German Air Force* iii. 37 The Stuka pilot went straight to a specialist dive-bomber school. **1956** D. M. Davin *Sullen Bell* vii. 151 Sounds worse than a lot of screaming stukas. **1962** *Times* 7 June 16/5 After the drone of Stukas has died away. **1969** G. Macbeth *War Quartet* 21 So we would die..Picked into pulp by stukas. **1973** S. B. Jackman *Guns covered with Flowers* viii. 127 This was how the Stuka pilots must have seen it [sc. Warsaw], open and defenceless in the bomb-sights. **1980** *Daily Tel.* 16 Oct. 16/5 Sunset and sand. Stukas and sand. Food, flies, hospital, home—and sand.
Hence **stu·ka** *v. trans.*, (*pass.*) to be attacked by Stukas.
1946 *Amer. Speech* XXI. 208, I served with the British Eighth Army in the desert.., but I never heard *sand happy. Bomb happy*, yes, when a bloke had been Stuka'd too often. **1977** N. Faulks *No Mitigating Circumstances* vii. 80 On one lovely afternoon, I was sitting there with the G3 when we were Stuka'ed. **1980** *Daily Tel.* 16 Oct. 16/6 One moves from page to page, one moment 'brewing up', another being 'stuka-ed', to the cameo of a desert sunset.

stultificatory (stʊ·ltifikē¹·tŏri), *a.* [f. Stultification: see -ory².] = Stultifying *ppl. a.*
1931 *Times Lit. Suppl.* 21 May 400/2 Further penetration into the processes of the universe is self-stultificatory. **1972** D. Bell in Cox & Dyson *20th-Cent. Mind* I. vi. 180 A transcendental and intuitionist philosophy which Mill saw as internally inconsistent, obscurantist, and stultificatory of the progress of both science and philosophy.

stuma: see *STUMER 3.

stumblebum (stʊ·mb'lbʊm). Also **stumble bum, stumble-bum.** *slang* (orig. and chiefly *U.S.*). [f. Stumble *v.* + *BUM *sb.*⁴] **a.** A worthless, clumsy, or inept person; a 'down and out', a drunkard.
1932 E. Hemingway *Death in Afternoon* 297 American word would be awkward bum, stumble-bum, flat-footed tramp. **1935** *Punch* 11 Dec. 652/3 An American gangster is stated to have begun his career by starting as a stumble-bum. **1936** [see *PALOOKA]. **1940** G. Frankau *Self-Portrait* xxxi. 178 The good old English word for a posterior, which possesses a different meaning (cf. 'stumblebum'—a guy who comes home drunk) in America. **1954** *Sun* (Baltimore) 11 Feb. 1/4 He became a ragged stumble-bum, raging drunkenly through the village and selling mediocre poetry. **1955** D. Keene *Who has Wilma Lathrop?* x. 93 He had a stubble of beard that made him look like a stumblebum. **1966** A. La Bern *Goodbye Piccadilly* xii. 113 These stumble-bums may have stumbled across the real culprit. **1970** *Guardian* 25 Sept. 10/5 Iago is a red-necked farm boy... He's a stumblebum. **1981** *Times Lit. Suppl.* 11 Sept. 1042/1 The Eisenhower of the war years has lost much of his lustre, and the successful organizer of Montgomery, Bradley and Patton has been reduced by some writers to the level of a strategic stumble-bum.
b. *attrib.* or as *adj.* Also *fig.*
1940 *Topeka* (Kansas) *Daily Capital* 16 Jan. 4/4 Russia's stumblebum campaign in Finland. **1952** B. Wolfe *Limbo '90* (1953) vi. xxii. 372 It made its slapstick stumblebum way back and forth. **1975** N. Freeling *What are Bugles blowing For?* xii. 75 Airs and graces with plain stumblebum cops. **1981** *Washington Post* 23 July c9/4 Joe Benjamin has a nihilistic, stumblebum-drunk son (splendidly played by Michael Rothhaar).

stumer. Add: **1. a.** Also *attrib.*, as *stumer cheque.*
1926 F. M. Ford *Man could stand Up* II. iii. 140 Two [were] awaiting court-martial for giving stumer cheques. **1944** P. Cheyney *They never say When* i. 18 Tell him to get in touch with Effie and get that stumer cheque from her and issue a writ against Swayle. **1962** *Listener* 11 Jan. 98/1 People who cash stumer cheques. **1972** L. Lamb *Picture Frame* xvii. 149 Nice old Mr Murgatroyd got your picture back from the man who gave you a stumer cheque.
b. *Austral.* Also **stoomer** (-ū-). (See quots.)
1898 *Bulletin* (Sydney) 17 Dec. (Red Page), A *stoomer* or *stumer* is a man without money. **1900–10** O'Brien & Stephens *Materials Dict. Austral. Slang* (MS.), *Come a stoomer*, stake a bet and lose everything. **1941** S. J. Baker *Dict. Austral. Slang* 73 *Stumer*, (in gambling or racing) a bankrupt, a defaulter... *Come a stumer*, to crash financially, esp. in a racing bet.
2. *gen.* Something which is worthless; a failure, a 'flop', a 'dud'. Also used of persons.
1886–96 A. R. Marshall in Farmer & Henley *Slang* (1903) VII. 18/1 The merry stumer. **1902** *Sporting Times* 1 Feb. 3/1 He..had given her as security a 'stumer' in the shape of an unfinished history of Corsica. **1923** Wodehouse *Inimitable Jeeves* xii. 132 The agony of having put his little all on a stumer that hadn't finished in the first six. **1925** Fraser & Gibbons *Soldier & Sailor Words* 273 *Stumer*, an expression used commonly to denote a shell that had failed to explode. **1928** Galsworthy *Swan Song* II. iv. 140 There is no good in me... You've pitched on a stumer. **1934** *Punch* 10 Jan. 50/1 *Myself*... No, drama. Young couple sight-seeing in mine. Old Miner guide. Roof falls, water rises. *Daphne.* Sounds like a stumer. Are you going to give it a run? **1970** *Daily Tel.* 9 Feb. 15/1 While in the course of a year countless shares will establish new lows only half a dozen will turn out to be real stumers and eventually worthless. **1976** *Times* 21 Feb. 15/3 Eclecticism guarantees that in a period like this the [Tate] collection will come to include a fair proportion of stumers. **1980** R. Hill *Spy's Wife* iv. 25 Don't be such a stumer!.. Fetch them.
3. Also **stuma**. A state of agitation; a sweat or 'stew'.
1932 Auden in *Rev. Eng. Studies* (1978) Aug. 284 Poor old Ma in a perfect stuma. **1936** —— *Look, Stranger!* 36 Behind your simple sense of humour You hide the boss's simple stuma. **1941** S. J. Baker *Dict. Austral. Slang* 73 *Stumer, in a*, in a 'stew', worried, angry.

stumm, var. *SHTOOM *a.*
1974 L. Deighton *Spy Story* iv. 41 So far, both sides have kept stumm about these operations. **1981** G. Markstein *Ultimate Issue* 209 Keep stumm about how you heard... You can always say you picked up a rumour.

stummick, repr. dial. and pop. pronunc. of Stomach *sb.*
1888 Kipling *Soldiers Three* (1890) 38 He was pegged out..on his stummick, a peg to each arm an' leg. **1936** M. Mitchell *Gone with Wind* xxii. 372 De Yankees is comin'!.. Dey'll run dey baynits in our stummicks! **1947** K. Tennant *Lost Haven* x. 157 If I was ever *you*, Mollie, or if I ever developed scales on me stummick, which amounts to the same thing..I'd be able to wriggle under rocks. **1977** J. Aiken *Five-Minute Marriage* v. 88 You didn't ought to go out giving lessons on an empty stummick.

stump, *sb.*¹ Add: **2. c.** *up a stump*: perplexed, in difficulties (see also quot. 1834). Cf. *up a tree* s.v. Tree *sb.* 7. *slang* (orig. and chiefly *U.S.*).
1829 S. Kirkham *Eng. Gram.* 206 Hele [= he will] soon be up a stump. **1834** W. G. Simms *Guy Rivers* II. 241 Brooks..in backwood parlance, was 'considerably up a stump'—that is to say, half drunk. **1880** 'Mark Twain' *Tramp Abroad* xxxvi. 402 The public reciter..would find himself 'up a stump' when he got to the church bell. **1924** Galsworthy *White Monkey* I. xii. 100 Look here, Uncle Soames, I'm up a stump. **1944** Duncan & Nickols *Mentor Graham* 147 For once in his life, work had him so up a stump that he could not snatch a moment for study or reading.
8. Restrict † *Obs.* to sense in Dict. and add: **b.** To leave one's home, job, or settled way of life, to move. Also without possessive pronoun. Cf. *to pull up stakes* s.v. Stake *sb.*¹ 1 e.
1955 'A. Gilbert' *Is she Dead Too?* xiii. 227 Seems to have pulled up his stumps now he's married again. Wonder if they left an address. **1974** M. Butterworth *Man in Sopwith Camel* I. i. 19 I've been trying to bully him into pulling up stumps and doing something with the rest of his life.
9. c. An act of stumping a batsman out. Also *stump-out*. Cf. Stump *v.*¹ 8.
1859 *All Year Round* 23 July 305/2 All clever catches, and clever stumps too. **1871** 'Thomsonby' *Cricketers in Council* 38 A stump-out may send the batsman back to his friends. **1912** A. A. Lilley *Twenty-Four Years in Cricket* v. 61 Stover's wicket-keeping was remarkable... He..was always able to gather the ball with ease, and thus create for himself the maximum of certainty in.. effecting a possible stump.
d. *pl.* Close of play, when stumps are drawn. Chiefly *Austral.*
1954 J. H. Fingleton *Ashes crown Year* xxv. 268 England carried on to stumps. **1962** *Times* 3 Dec. 3/2 He looked to be coasting through to 'stumps' when Benaud bowled him. **1977** *World of Cricket Monthly* June 30/2 Bold tactics by Intikhab..carried the Pakistani score to 6-249 at stumps.
†**16*.** A stringed instrument of the lute family (see quots.). *Obs.* exc. *Hist.*
a **1623** in *R. Johnson's Compl. Works for Solo Lute* (1972) 22 (*music title*) Alman To the Stumpe. **1947** E. Blom *Everyman's Dict. Mus.* 674/1 *Stump*, an obs. string instrument of the Cittern type invented *c.* 1600 by Daniel Farrant. **1961** A. Birch in A. Baines *Mus. Instr. through Ages* vii. 166 There were other instruments too, for accompanying the voice or for solo playing, 'stump', 'poliphant', 'penorcon', but little more than their names has survived. **1976** D. Munrow *Instr. Middle Ages & Renaissance* ix. 83/4 Of the stump there are no surviving examples or descriptions though the name does suggest a small instrument. One piece of stump music is extant, however, entitled *Alman R. Johnson to the stump by F.P.*..giving the impression that the stump was a wire-strung equivalent of the theorbo.
17. (sense 2) *stumpwood*; (sense 14) *stump oration, oratory* (earlier example), *speech* (earlier example).
1831 *Constellation* (N.Y.) 12 Feb. 98/2 You see, sir, I want an office, for, as I told 'em in my stump horation twict, the man..is the very one that ought to be awarded. **1811** E. Fletcher *Let.* 11 Jan. (1965) 26 For you must know that the people in these parts get into office by 'Stump oratory' or praising and electioneering for themselves. **1820** J. Flint *Lett. from Amer.* (1822) 251 The harangues are called stump-speeches. **1953** *Forestry Abstr.* XIV. 296/2 Miscellaneous notes are appended on: splitting stumpwood with explosives.., and Spruce and Pine stumpwood for the manufacture of fibreboards. **1977** *Ibid.* XXXVIII. 44/2 (*heading*) Determining the volume of stumpwood and rootwood in *Picea abies*.
18. stump-grubber, a machine designed to excavate the stumps of trees after the trees have been felled (cf. *stump-machine*); **stump-grubbing,** the excavation of tree-stumps by manual or mechanical means; **stump-jump** *a.*: also *absol.* as *sb.*; **stump jumper** *U.S.*, a countryman or hillbilly (cf. *stubble-jumper* s.v. *STUBBLE *sb.* 5); **stump plant,** a cutting consisting of a short cut-back stem and roots which may or may not be pruned; **stump topgallant mast** (earlier example); **stump water** *U.S.*, the rain-water which collects in the stumps of hollow-trees, associated esp. with folk remedies and charms; **stump word,** a word formed by abbreviating a single longer one, esp. by reducing it to a single syllable (freq. the first) or the minimum necessary for understanding; cf. *CLIPPING *vbl. sb.*² 2 c; **stump-work** (later *attrib.* examples).
1971 *Sylwan* CXV. x. 19 Investigation of the technical and economic efficiency of the Odynice stump-grubber. **1977** *Forestry Abstr.* XXXVIII. 580/1 The specifications of the stump-grubbers are tabulated, and data are given on stump-grubbing performance. **1938** M. Richardson in B. A. Botkin *Treas. S. Folklore* (1949) III. i. 442 He went on the stump-grubbing gang, soon as he got to the Farm. **1961** *Forestry Abstr.* XXII. 582/1 (*heading*) Using a vibratory shock method in stump-grubbing. **1977** Stump-grubbing [see *stump-grubber* above]. **1911** E. M. Clowes *On Wallaby* xi. 297 Those people who..were once in undisputed possession of these mountains and forests—before the days of the axe and saw, the 'stump-jump', and the 'mallee roller'. **1936** J. H. Street *Look Away* xiii. 87 That's the home of the hillbillies. Some folks call 'em 'stump-jumpers'. **1944** in H. Wentworth *Amer. Dial. Dict.* 606/2 She musta been one o' these West Virginia stump-jumpers. **1953** *Brit. Commonw. Forest Terminol.* I. 36 *Cutting, root and shoot*, one consisting of a pruned tap-root and cut-back stem. Syn...stump plant. **1960** *Forestry Abstr.* XXI. I. 31/1 Stump plants made from 1-year seedlings were planted in pots. **1804** in *Naval Docs. U.S. Wars with Barbary Powers* (U.S. Office Naval Rec.) (1942) IV. 346 You will know my ship by her having stump top Gl Masts. **1892** J. C. Harris *Uncle Remus & Friends* 290 De way ter git rid er ha'nts wuz ter git some prickly-pear root en bile it in stump-water en sprinkle it 'roun' de yard. **1972** J. S. Hall *Sayings from Old Smoky* 133 'His head is full of stump water.' That is, 'He don't use his brain.'..Possibly 'stump water in the head' meant originally that the person had been affected by magic, that is, was 'teched in the head', or dazed. **1922** O. Jespersen *Language* II. ix. 169 We come to those changes which result in what one may call 'stump-words'. ..Words may undergo violent shortenings both by children and adults. **1963** *Amer. Speech* XXXVIII. 156 Other stump words or clipped forms such as *info, auto.* **1971** Stump word [see *MELO]. **1938** *Burlington Mag.* Oct. 172/2 Stuart 'stump-work' embroidery. **1958** *Times* 25 Nov. 18/6 A highly important Venetian glass mirror with stumpwork panels. **1971** *Country Life* 10 June 1426/2 A mirror framed in stumpwork embroidery made in England three centuries ago.

stump, *a.* **4. stump-nose,** substitute for def.: = Stompneus in Dict. and Suppl.; (earlier example).
1838 [see *KLIPFISH 1].

stump, *v.*¹ Add: **2. b.** (Earlier example.)
1803 G. Colman *John Bull* IV. i. 41 Now, Sir, you and I'll stump it.
13. (Examples.)
1857 A. Lincoln in H. Binns *Life A. Lincoln* (1927) 181 Like the boy that stumped his toe..it hurt too bad to laugh. **1891** *Harper's Mag.* Feb. 364/2 Mus' be powerful sorrowful ter set at home an' shed tears lest he mought her stumped his toe on the road.
17. b. (*c*) In extended use, const. *with*.
1956 'C. Blackstock' *Dewey Death* ii. 35, I hope the department will stump up with a decent wreath. **1958** *Listener* 9 Oct. 569/2 The Americans stumped up with *The Old Man and the Sea*.
c. (Earlier example, with person as obj.)
1853 J. Palliser *Solitary Rambles* v. 126, I..reminded him how completely he had stumped me up that afternoon.

stumpage. 1. (Earlier example.)
1835 *Knickerbocker* V. 423 Such rough words as tariff jobbing, cuts, stumpage.

stumper. 3. (Further example.)
1833 in P. Norman *Scores & Annals W. Kent Cricket Club* (1897) 73 Herbert..is such a wonderous stumper.

stumping, *vbl. sb.*¹ Add: **1. b.** (Earlier example.)

1844 W. Lillywhite *Hand-bk. Cricket* 15 No byes, overthrows, stumping, or catching out behind wicket allowed.
3. *Comb.*, as **stumping machine** *N. Amer.* = *stump-machine* s.v. Stump *sb.*¹ 18; **stumping powder,** an explosive used for clearing land of tree stumps.
1871 G. Easton *Trav. Amer.* xiii. 129 Until time has so destroyed the roots that, with the aid of a pair of oxen and a 'stumping machine', the stumps can be removed with comparative ease. **1906** *Daily Colonist* (Victoria, B.C.) 30 Jan. 5/5 An experiment in clearing Island lands by the aid of a steam stumping machine is shortly to be made in the vicinity of Duncan. **1973** L. Russell *Everyday Life Colonial Canada* ii. 31 Later a 'stumping machine' was used. It consisted of a very heavy wooden tripod, with a stout, vertical iron screw mounted at the apex. **1921** *Daily Colonist* (Victoria, B.C.) 6 Apr. 5/4 The regulations under which the Provincial Department of Agriculture will issue funds for the purchasing of cheap stumping powder to farmers have been approved. **1955** R. P. Hobson *Nothing too Good for Cowboy* i. 11 The cattle company..has exploded like a can of stumping powder.

stun, *v.* Add: **6.** *Comb.* **stun gas,** a gas that incapacitates by causing temporary confusion and disorientation; **stun grenade,** a grenade that only stuns through its sound and flash; **stun gun,** a gun that fires shot which stuns without causing serious injury.
1968 *Punch* 21 Feb. 253 Here is a run-down of anti-crowd devices which other nations, notably America, are developing:..stun gas; gas which temporarily blinds. **1977** *Times* 19 Oct. 1/4 The 'stun' grenades which played such a vital part in enabling the West German commando unit to overcome the terrorists..were supplied by Britain. **1981** A. Winch *Blood Money* xxi. 236 The stun grenades..looked like unmarked beer cans and provided a deafening explosion, a blinding flash. **1971** *Sunday Times* 30 May 5 The stun gun has already been used effectively by the Alameda County Sheriff's Department who are called in whenever student riots at Berkeley become too much for the local police. **1975** *Nature* 13 Feb. 495/3 One hopes that the [polar bear tagging] expedition will be equipped with a supply of stun-guns.

stung, *ppl. a.* Add: **2.** *Austral. slang.* Drunk. Cf. *STUNNED *ppl. a.* 2.
1919 W. H. Downing *Digger Dialects* 48 *Stung*,.. drunk. **1952** T. A. G. Hungerford *Ridge & River* 62 The old bloke's stung already, and the pubs aren't even opened yet! **1965** W. Dick *Bunch of Ratbags* 219 We had arrived at Doreen's sister's wedding-reception about an hour ago and by now we were all half stung. **1970** K. Slessor *Bread & Wine* 154 To a total abstainer, the line 'I am stung, stung to the heart of me' would convey, I suppose, nothing so coarse as the associations which I find in it.

stunned, *ppl. a.* (In Dict. s.v. Stun *v.*) Add: **2.** *Austral.* and *N.Z. slang.* Drunk. Cf. prec.
1919 W. H. Downing *Digger Dialects* 48 *Stunned* (adj.), drunk. **1933** 'P. Cadey' *Broken Pattern* xii. 129 I'm afraid I got a bit stunned... I had one over the odd.
3. Phr. *like a stunned mullet* [Mullet¹]: dull, stupefied. *Austral. slang.*
1953 Baker *Australia Speaks* 267 [Similes] Dullness: (looking) like a stunned mullet. **1963** J. O'Grady *Things they do to You* 147, I returned and lay on the bed like a stunned mullet. **1977** *Australian* 16 May 1 Mr Hawke said yesterday the Federal Government had responded like a stunned mullet to his acceptance of the proposed Industrial Relations Bureau.

stunner. 1. (Earlier example.)
1829 P. Egan *Boxiana* 2nd Ser. II. 417 The blow was a *stunner*, and visible on his forehead.

stunt, *sb.*² Add: **1. a.** (Earlier example.) Also *spec.* in aerobatics.
1892 R. H. Davis *West from Car Window* 13 They went about it as gleefully as schoolboys at recess doing 'stunts'. **1909** *Flight* 11 Sept. 552/2 He made the machine dart down as though it were going to pitch to earth head foremost, but when within about 20 feet of the ground, without effort he brought it horizontal again. These are what Americans style 'stunts'. **1915** W. E. Dommett *Submarine Vessels* viii. 88 Of course, nothing in the nature of 'stunts', such as are performed by airmen, are tried. **1927** C. A. Lindbergh *We* i. 13 We did a few stunts over the fair-grounds to get everyone's attention.
b. (Earlier and later examples.) Also *spec.* in Advertising, Journalism, etc., a 'gimmick' or device for attracting attention.
1878 S. Butler *Let.* 7 Feb. (1955) 174 It was a stunt for advertising the books, so I sent them. **1919** 'R. N. Etienne' *Strange Tales from Fleet* 27 The 'stunt' was over, and two brief hours had prevented the twelfth Cruiser Squadron from cutting off the enemy light forces. **1922** *Daily Mail* 13 Nov. 11 The plea for 'stipes' is a newspaper stunt. **1930** *Lancet* 7 June 1264/1 It has even been whispered, Sir, that there are too many 'stunts' (if this word may appear in your columns), too little science, but this is just malicious gossip. **1942** *Sun* (Baltimore) 29 Dec. 13/2 It would be a good stunt for us fellows to learn all the tricks of the Coast Guard and then we could turn rum runners! **1968** J. R. Ackerley *My Father & Myself* vii. 61 My brother's assignment was what we called a 'stunt', a common affair, in this case important if only because the Brigadier had set his heart on it. **1975** *Sunday Times* (Colour Suppl.) 20 July 12/2 His bisexuality..was attracting sensational publicity. How much of this was simply a stunt?
c. A stint, a task, an exercise. orig. *U.S.*
1880 *Hermean* (N.Y.) 256 Ye have heard it said by those of old time, 'A rolling stone gathers no moss,' but by reason perhaps of its having been used as a 'stunt' for our childhood in the copy books, we seldom realize how beautiful and full of wisdom is the adage. **1904** G. H. Lorimer *Old Gorgon Graham* 85 And you set the other at a twelve-hour stunt of making all the beds you've mussed. **1921** G. B. Shaw in G. C. Williamson *John Keats Memorial Vol.* 176 Milton can do a stunt of geniality, as in *L'Allegro.*
d. In wider use, a piece of business, an act, enterprise, or exploit.
1904 *Sun* (N.Y.) 8 Aug. 5 He took lessons in holding the life net..and the other stunts firemen are taught. **1913** R. Brooke *Coll. Poems* (1918) p. lxxxiv, Then I do my pet boyish-modesty stunt and go pink all over. **1920** C. Bax *Square Pegs* 21 *Hilda.* You'll drive me frantic If you're not just the teeniest bit romantic. *Gioconda.* It isn't done. You're absolutely wrong In asking me to do that stunt. So long! **1928** [see *PEEP *sb.*¹ 2 b]. **1964** C. Hassall *Rupert Brooke* vii. 277 'So of course you were frank and boyish?' asked Mrs. Cornford, on hearing he [sc. Rupert Brooke] had just met Henry James. 'Oh yes,' he said, 'Of course I did the fresh, boyish stunt, and it was a great success.'
2. a. *attrib.*, as *stunt artist, flying, pilot*, etc.
1904 W. H. Smith *Promoters* iii. 75 He might have made a successful actor, of the modern 'stunt' sort. **1916** C. Winchester *Flying Men* 112 Trick flying, or 'stunt' flying as it is colloquially called, can only be attempted with impunity by those aviators who have had some experience. **1922** H. L. Wilson *Merton of Movies* 174 Ain't I a good stunt actress? **1931** *Morning Post* 18 Feb. 6/4 (*heading*) 'Stunt' pilot's escape. **1931** *Everyman* 23 Apr. 388/2 We are on the eve of a reaction from the 'stunt Press', he believes—the Press of competitions and coupons and catchpenny sensations. **1938** M. McCarthy in *Partisan Rev.* Feb. 35 In the actual production of Gielgud's *Hamlet* and Welles's *Caesar*, the exploiter, that is, the stunt artist, wears a more successful disguise. **1971** *Flying* Apr. 46/3 He is in the stunt-flying business. **1976** M. Maguire *Scratchproof* iv. 53 'Remember to stretch the line tight,' the stunt arranger emphasized. **1977** D. Anthony *Stud Game* vi. 37 After the war Dusty became a stunt pilot for the movies. **1981** *Times Lit. Suppl.* 13 Feb. 177/5 The black-and-white plates which illustrate the book..reveal completely new aspects of the works of art reproduced, without ever verging on the stunt photography which so often distorts Baroque sculpture in books written by less scrupulous authors.
b. Special Comb.: **stunt-drive** *v. intr.*, to drive a car for stunts (sense 1 a), esp. for making dangerous film sequences; hence **stunt-driving** *vbl. sb.*; **stunt man,** one who performs dangerous feats, esp. as a stand-in for a film actor.
1966 J. Cleary *High Commissioner* v. 96, I used to stunt-drive in the old Ealing comedies. **1975** *New Yorker* 21 Apr. 92/3 One can disregard obvious high-risk occupations, such as stunt-driving. **1930** *Aberdeen Press & Jrnl.* 23 Jan. 2/6 To those who wish to get plenty of excitement out of life our advice is—Be a movie stunt man. **1953** C. A. Lindbergh *Spirit of St. Louis* II. vi. 275 We walked over to a group of pilots, mechanics, and stunt men. **1953** Dylan Thomas *Let.* (1966) 416, I cry to myself as I kick clear of the cling of my stuntman's sacking. **1968** P. Geddes *High Game* viii. 101 He'd worked, off and on, as a stunt-man in movies. **1977** *New Yorker* 27 June 84/3 The successful stunt man explained that his plan had been to scale the tower a day earlier.

stunt (stʊnt), *v.*³ [f. Stunt *sb.*²] **a.** *intr.* To perform stunts (in quots. with reference to aerobatics). **b.** *trans.* To use (an aeroplane) for the performance of stunts.
1917 'Contact' *Airman's Outings* p. xxiii, They could turn, climb, and stunt quicker than any two-seater. **1928** *Daily Mail* 9 Aug. 7/1 The aeroplane was apparently 'stunting' at a height of between 2,000 and 3,000 feet. **1928** *Daily Tel.* 18 Sept. 11/4 There is no reason why the autogiro should not be stunted. **1953** C. A. Lindbergh *Spirit of St. Louis* II. vi. 421 DHs aren't built like Jennies. ..And you can't stunt 'em like a Jenny either—no rolls or loops. **1970** L. Deighton *Bomber* iv. 64 When the Luftwaffe was officially born in 1935 Peter Redenbacher was stunting a Bücker Jungmann biplane above the heads of Hitler, Göring, the foreign Press and a deliriously happy German crowd.
Hence **stu·nter, stu·ntist,** one who performs or organizes stu·nts; **stu·nting** *vbl. sb.* and *ppl. a.*
1914 G. Hamel *Flying* 212 It [sc. the Royal Flying Corps] talks shop..and indulges in rude health and proper pride, discouraging 'stunting' and heroics. **1922** *Daily Mail* 2 Nov. 5 Some of the members of the Committee.. went on the Council as economy 'stunters'. **1923** *Glasgow Herald* 14 July 7 Close on five o' clock the 'stunting' planes came to earth. **1925** *Public Opinion* 31 July 107/3 When the political stuntists saw fit to mobilise. **1927** C. A. Lindbergh *We* vi. 94 The De Havilands were not considered safe for hard stunting. **1928** *Observer* 18 Mar. 17/2 Two officers..who are considered to be the best 'stunters' in the force. **1940** G. Frankau *Self-Portrait* xlvi. 285 The stunt merchant, for all his stunting, had a creed. **1970** R. Blake *Conservative Party* viii. 253 The Conservatives, in spite of having a real point, incurred the charge of 'stunting' when they used such expressions as 'Gauleiter Laski'. **1978** *Detroit Free Press* 16 Apr. (Cartoon Suppl.) 1/2 (Advt.), Boys' deluxe MX motocross style bike... Not for stunting or off-road use.

stunty, *a.* Add: **3.** Having the character of a stunt, extravagant, 'gimmicky'.
1981 *Daily Tel.* 27 Jan. 13/4 Cardin's minis do not look stunty or contrived.

stupe, *sb.*² Add: (Later U.S. examples.) Also as *adj.*
1967 'T. Wells' *Dead by Light of Moon* (1968) i. 15 His assistant, a big stupe called Jersey Eng. **1967** E. McGirr *Here lies my Wife* v. 151 He carries on..all the time. He's stupe. **1977** *Time* 25 Apr. 46/3 She tells him that true love has washed away her sins and the pure and simple stupe embraces her.

stupend (stiūpe·nd), *v. rare.* [Back-formation from Stupendous *a.*] *trans.* To amaze, dumbfound. (G. B. Shaw's word.)
1904 G. B. Shaw *Let.* 6 Dec. (1972) II. 470 You will be stupended at my meanness in this obvious & cheap retort. **1927** —— *Let.* 2 Feb. in *To a Young Actress* (1960) 113 The discovery that you actually wanted me to shew that daub to Charlotte has perfectly stupended me.

stupendous, *a.* Add: Now freq. in trivial use.
1892 E. Dowson *Let.* in *N. & Q.* (1962) Mar. 102/1, I have become the victim of the most stupendous cold which has ever occurred to me, & fear that I shall spend tomorrow in bed. **1939** *Airman's Gazette* Dec., All aircraft off duty being allowed to..view the stupendous, side-splitting entertainment. **1942** E. Paul *Narrow St.* ii. 15 The *marc* was undiluted—stupendous, in fact. **1959** C. L. Wrenn *Word & Symbol* (1967) 35 That stupendous scholar, Max Förster.

‖**stupor mundi** (stiū·pǫɪ mu·ndī, stiū·pǫɪ mʊ·ndəi). L. phr.: the marvel of the world; an object of admiring bewilderment and wonder. Cf. Stupor 2 b.
The phrase was originally used by the thirteenth-century historian Matthew Paris to describe the Emperor Frederick II of Germany.
1879 *Encycl. Brit.* IX. 732/2 The general contemporary opinion regarding Frederick II. is expressed in the words *stupor mundi*:..wonder and perplexity are the predominant sentiments which..[the contemplation of his career] even yet awakens. **1946** R. Lowell *Lord Weary's Castle* 5 Over the drum-beat of St. Stephen's choir I hear him, *Stupor Mundi.* **1980** J. A. T. Robinson *Roots of Radical* ii. 22 What is easier is to preen ourselves on our 'comprehensiveness' without realising that it is in every sense a *stupor mundi*, an object of incredulous disbelief to the world and even our fellow Christians.

Sturge–Weber (stɜ̄·ɪdʒ we·bəɪ). *Path.* [The names of W. A. *Sturge* (1850–1919) and F. P. *Weber* (1863–1962), English physicians, who described the syndrome in 1879 and 1922 respectively.] *Sturge-Weber syndrome* or *disease*: a congenital syndrome in which a diffuse malformation of blood-vessels on one side of the head produces port-wine nævus on the face and lesions of the brain, usually resulting in fits and mental retardation.
1935 H. Bergstrand in *Abstr. 2nd Internat. Neurol. Congr., London* 124 Sturge–Weber's disease is a syndrome comparable with von Recklinghausen's..and Bourneville's diseases. **1974** J. H. Menkes *Textbk. Child Neurol.* x. 412/2 The coincidence of a facial vascular disease and seizures suggests Sturge–Weber disease.

sturine (stiūᵊ·rīn, -in). *Biochem.* Also **-in.** [a. G. *sturin* (A. Kossel 1896, in *Zeitschr. f. physiol. Chem.* XXII. 180) f. pop.L. *sturiōnem* Sturgeon: see -ine⁵.] A protamine extracted from the testicles of fish of the genus *Accipenser*, esp. *A. sturio* and *A. guldenstadtii*, sturgeons of the Baltic and Caspian Seas.
1896 [see *SALMINE]. **1960** *Adv. Protein Chem.* XV. 34 In the case of sturine, both procedures..yield two NH_2-terminal amino acids, alanine and glutamic acid. **1977** *Bull. Exper. Biol. & Med.* LXXXIII. 710 The proportion of added DNA to be bound with spheroblasts untreated with sturine was 18%.

‖**Sturmabteilung** (ʃtu·ɹmaptai:luŋ). [Ger., lit. 'storm detachment'.] A paramilitary force forming part of the German National Socialist Party, founded in 1921 and deprived of power in 1934. Abbrev. S.A. (s.v. *S 4 a).
1923 *Times* 22 May 11/1 At the headquarters of the [Nazi] organization near the Gärtnerplatz Theatre, the visitor is confronted in the main hall by the notice 'Sturm-Abteilung' (Storm-troop section). Newspaper correspondents..are not very welcome. **1932** E. Lengyel *Hitler* x. 145 Germany is divided into five Sturm Abteilung inspections. **1977** M. Walker *National Front* i. 19 They [sc. Mosley's army] had neither the numbers nor the organization of Hitler's *Sturmabteilung* or Mussolini's *Squadristi.* **1977** [see *S.A.* s.v. *S 4 a].

‖**Sturmbannführer** (ʃtu·ɹmbanfǖ:rəɹ). Ger., = 'battalion leader'.] An officer in the *Schutzstaffel* (formerly part of the *Sturmabteilung*).
1955 J. Thomas *No Banners* xxv. 246 The *sturmbannführer* literally foamed at the mouth. **1965** M. Spark *Mandelbaum Gate* i. 13 Freddy..felt an urge to explain that he was not a mass-butcher and that he had never desired to become a *sturmbannführer*. **1978** T. Allbeury *Lantern Network* viii. 99 A fresh-faced blond German with the insignia of a Sturmbannführer.

Sturmer (stɜ̄·ɹməɹ). Also **sturmer.** The name of a village near Haverhill, on the Suffolk–Essex border, used *attrib.* or *absol.* in

Sturmer apple, Pippin to designate a late-ripening dessert apple belonging to a variety developed there in the 1830s by S. and J. Dillistone (fl. 1827–50) and distinguished by yellowish-green skin, sometimes slightly russetted, and crisp, creamy-white flesh.

1831 *Catal. Fruits cult. in Garden* (Hort. Soc.) (ed. 2) 37 Apples... Sturmer Pippin. **1847** *Gardeners' Chron.* 27 Feb. 135/2 When the Nonpareil is getting over, the Sturmer Pippin, then in perfection, comes in to supply its place. **1851** R. Hogg *Brit. Pomology* 190 The Sturmer Pippin was raised by Mr. Dillistone, a nurseryman at Sturmer, near Haverhill, in Suffolk, and was obtained by impregnating the Ribston Pippin with the pollen of the Nonpareil. **1877** E. S. Dallas *Kettner's Bk. of Table* 35 Dessert Apples... February—Sturmer Pippin. **1950** *N.Z. Jrnl. Agric.* Aug. 191/1 Good sources of vitamin C include..sturmer apples. **1959** [see *DELICIOUS *a.* 2 b]. **1971** R. Petrie *Thorne in Flesh* x. 133 A flow of customers beyond the carefully stacked pyramids of Sturmers and Jonathans swelled and dwindled.

‖ **Sturm und Drang** (ʃtuɹm unt draŋ): see *storm and stress* s.v. Storm *sb.* 3 d.

[**1844** F. L. J. Thimm *Lit. of Germany* 85 This period, so styled by Goethe, after the title of one of the dramas of Klinger, 'Sturm und Drang'.] **1857** C. Kingsley *Two Yrs. Ago* III. i. 29 One of the Sturm-und-drang party, of course; the express locomotive school, scream-and-go-ahead. **1873** Gostwick & Harrison *Outl. German Lit.* xvi. 228 That time of 'Sturm und Drang', when writing wild poetry was regarded as the object of life. **1925** L. P. Smith *Words & Idioms* iii. 105 In that wild period, which was called at the time *Genieperiode*, but has since acquired the name of *Sturm und Drang*, the great watchwords *Genius, Originality*, and *Creative* acquired a resonance.. which they had certainly never possessed in England. **1950** M. J. C. Hodgart *Ballads* ix. 152 The philosophical and political obsessions which attracted the *Sturm und Drang* writers to folksong. **1973** *Listener* 28 June 862/3 The Romantic poets..loved the word 'storm' as a synonym for energy, in phrases like *Sturm und Drang*, 'storm and thrust'.

transf. **1885** H. Conway *Family Affair* III. v. 85 His *sturm und drang*, his emotional days, were well over... He had not bowed his knee to the intense, nor sacrificed on the altar of the incomprehensible. **1909** A. Nicolson *Let.* 24 Mar. in H. Nicolson *Sir Arthur Nicolson* (1930) xi. 305 When we have passed through the present 'Sturm und Drang' period, I should not be surprised if we were to find both France and Russia gravitating rapidly towards and Central Powers. **1938** E. Bowen *Death of Heart* I. v. 86 Eddie..wrote some pamphlets, which were printed by a girl who had a press in a loft. Arts and crafts had succeeded *Sturm und Drang*. **1962** J. Heller *Catch-22* iii. 29 Only when all the *Sturm und Drang* had been left far behind would he tip his flak helmet back wearily on his sweating head and stop barking directions to McWatt at the controls. **1978** H. Frank *Single* vi. 137 She wanted..mortification, punishment, Sturm und Drang.

S-turn: see *S 2 c.

Sturt (stɒ̄ɹt), *sb.*[4] The name of Charles *Sturt* (1795–1869), Australian explorer, used *attrib.* or in the possessive in **Sturt('s) (desert) pea** to designate a plant collected by him in 1844, *Clianthus formosus*, a trailing perennial herb of the family Leguminosæ, native to desert regions of Western Australia and bearing racemes of red or white flowers blotched with black at the base.

1865 J. E. Tenison-Woods *Hist. Discovery & Exploration Austral.* I. i. 29 Some of the species are engraved in one account of his [*sc.* Dampier's] voyage, amongst which appears the beautiful clianthus, known to the colonists as Sturt's Desert Pea. **1911** W. R. Guilfoyle *Austral. Plants* 114 'Sturt's Desert Pea'..or 'Australian Glory Pea' (biennial), flowers scarlet. **1933** [see *CLIANTHUS]. **1936** I. L. Idriess *Cattle King* iv. 27 The bluebush a sombre carpet splotched by scarlet patches of Sturt's desert pea. **1949** D. Walker *We went to Australia* xix. 183 The sturt pea is the most sensational [wild flower of West Australia]. **1966** *Times* 11 Nov. (W. Australia Suppl.) p. iv/2 There is the spectacular Sturt pea, sprawling crimson and black-hearted, on the red earth. **1977** *Caravan World* (Austral.) Jan. 37/3 The scarlet splash of Sturt's Desert Pea..mellowing the stark red earth. **1979** D. Bellamy in *Radio Times* 1–7 Dec. 5/1 Many Australian plants..have adapted to withstand bush fires... Sturt's Pea, the first plant to bloom after the ravages of fire.

stushie (stɒ̄·ʃi, sti·ʃi).). *Sc.* Also **stashie, stishie**. [Origin unknown.] A disturbance, uproar, row, fracas.

1824 G. Smith *Douglas Travestie* 12 Mony an aukward stashie was he in. **1840** G. Webster *Ingliston* xxviii. 289 The haill toun's been in a stushie about it. **1926** 'H. MacDiarmid' *Penny Wheep* 8, I lo'e the stishie o' Earth in space. **1959** M. Pugh *Chancer* 68 Big trouble from the film company... They're in a stushy. Some of the survivors started creating hell about the film. **1973** *People's Jrnl.* (Inverness & Northern Counties ed.) 24 Nov. 17/4 (*heading*) Army 'Samaritans' caused a stushie on the road to Durness. **1980** G. Hammond *Reward Game* vii. 96 Do you still want Briesland House, or did this morning's stishie put you off it for life?

stuss (stɒs, stɒʃ). *U.S.* [ad. Yiddish *shtos*, perh. a. G. *stoss* push, stack.] A form of faro.

1912 A. H. Lewis *Apaches of N.Y.* 35 Jigger owned a stuss-house in Forsyth Street. *Ibid.*, Between them they would divide the harvest of the stuss. **1913** [see *pipe-fiend* s.v. *PIPE *sb.*[1] 11 b]. **1930** D. Runyon in *Sat. Even. Post* 5 Apr. 48/3, I do not wish to play stuss. *Ibid.* 72/1 It is what is called a stuss house, and many prominent citizens of the neighbourhood are present playing stuss. **1975** *Way to Play* 207/3 Stuss is also known as Jewish faro. It is a simplified form of faro, with a larger percentage in the house's favor.

stutter, *sb.*[2] Add: Also *transf.*

1974 C. Ryan *Bridge Too Far* iv. vi. 263 As he neared the ground the stutter of machine guns and the dull thud of mortar bursts seemed to engulf him. **1981** M. E. Atkins *Palimpsest* ii. 12 The car engine—with..the merest apology of stutter—cut out completely.

stutter, *v.* Add: **1. b.** (Later examples.) Also *fig.*

1931 Nordhoff & Hall *Falcons in France* 236 Guns were stuttering faintly on every side. **1935** J. Steinbeck *Tortilla Flat* ii. 26 The lawyer..climbed into his Ford and stuttered down the hill. **1963** *Times* 11 Feb. 4/1 Meanwhile, the F.A. Cup stutters forward, step by step. **1976** *Scottish Rev.* Summer 8 Tractors stutter in and out of fields.

2. Also *fig.*

1929 *Oxford Poetry* 12 And the map stutters inarticulate lines.

stuttering, *ppl. a.* (Later *transf.* and *fig.* examples.)

a **1918** W. Owen *Poems* (1963) 44 Only the stuttering rifles' rapid rattle Can patter out their hasty orisons. **1946** D. C. Peattie *Road of Naturalist* iv. 44 Turning out editorials of a rare charm and delicate fancy while the stuttering telegraph battered on his eardrums. **1980** L. Cody *Dupe* iii. 23 Even Selwyn's stuttering typewriter was silent.

S-twist: see *S 8.

style, *sb.* Add: **II. 13. c.** Proverbial phr. *the style is the man.*

[**1624** R. Burton *Anat. Melancholy* (ed. 2) 7 It is most true, *stylus virum arguit*, our stile bewrayes vs. **1753** G. Buffon *Histoire Naturelle* VII. p. xvii, Le style est l'homme même.] **1901** G. B. Shaw *Caesar & Cleopatra* 208 Going to Caesar's books, and concluding that the style is the man. **1901** A. Whyte *Bible Characters: Stephen to Timothy* civ. 72 If the style is the man in Holy Scripture also,..we feel a very great liking for Luke. **1942** H. F. Heard *Reply Paid* (1943) ix. 141 Usually I don't like to have my style modified. 'The style is the man.' **1978** *Language* LIV. 284 In describing Achilles' speech, we have also been describing his character, since 'style is the man'.

III. 23. (Later examples.) Also *gen.*, one's characteristic manner of acting or reacting. Phr. *to cramp one's style*: see *CRAMP *v.* 5 c.

1937 C. Odets *Golden Boy* i. iii. 47 Joe knows his own needs, as he says. Don't ask him to change his style. **1970** G. F. Newman *Sir, You Bastard* viii. 247 It hadn't been his original intention to make her suffer, but he couldn't alter his style. **1978** S. Brill *Teamsters* vi. 260 Lying low just isn't his style.

24. b. Also, more generally, attractive or impressive quality; originality.

1967 *Trans-Action* Apr. 11/1 Style is difficult to define as it has so many referents. It means to carry one's self well, dress well, to show class... A person with style must also show respect..for another's superior power. **1968** *Listener* 13 June 761/3 Jack, the prime Kennedy,.. had a quality which practically everybody recognised as 'style'. **1979** R. Jaffe *Class Reunion* (1980) ii. xi. 288 How much more dignified it would have been if Rusty could have trusted her... But Rusty had never had style.

V. 28. (in sense 21 d) *style manual, sheet*; **style analysis**, analysis of the characteristic style of an artist, writer, composer, etc., or of a school or period, on the basis of which attribution of a particular work can be made; **style-book** (*b*): for *U.S.* read orig. *U.S.* and add examples; **style critic**, an expert in style analysis; **style-setter**, someone who or something which sets the fashion; so **style-setting** *ppl. a.*

1927 E. Rickert *New Methods for Study of Lit.* 274/2 (Index), Subjectivity in style analysis. **1953** M. Schapiro in A. Kroeber *Anthropol. Today* 290/1 The refinement of style analysis has come about..through problems in which small differences had to be..described precisely. **1955** H. Read *Icon & Idea* vi. 112 Then are many other factors which can be used in style analysis..but though all these stylistic traits build up to an index of the painter's personality, they do not..indicate the painter's awareness of a self. **1969–70** *Computers & Humanities* IV. 41 These results, minor as they are, are of a nature that has not been achieved in any other use of the computer for style analysis in music. **1973** *Black World* Nov. 5/1 Conventional histories of music and style-analysis texts generally ignored the subject of the Black man's contribution to music. **1898** (*title*) Stylebook of the Chicago Society of Proofreaders. **1930** [see *NEGRO 1 d]. **1981** K. Waterhouse *Daily Mirror Style* 5 In most newspaper offices there is to be found a manual known as the style-book which lays down..the paper's rules on the usage of words and punctuation. **1959** *Times* 26 May 13/5 On that hypothesis David stays as the author of the 75 pictures,..finally reinstated by modern style-critics. **1978** *Jrnl. R. Soc. Arts* CXXVI. 718/2 He may be the less inclined to indulge in wild or subjective speculation than the style critic. **1922** (*title*) Style manual of the Government printing office (U.S. Govt. Printing Office). **1964** E. D. Seeber (*title*) A style manual for students. **1959** *News Chron.* 1 July 3/1, I have a feeling it is going to be a style-setter. **1960** *Ibid.* 29 Feb. 6/7 Princess Margaret has always been a style-setter. **1955** Keepnews & Grauer *Pictorial Hist. Jazz* iii. 44 Earl Hines..quickly developed into an outstanding and style-setting pianist. **1960** *Farmer & Stockbreeder* 1 Mar. 44/2 The Fordson Dexta with its proved style-setting three cylinder Diesel engine. **1924** H. L. Mencken *Let.* 7 Dec. (1961) 272 Have you such a thing as a Style Sheet for The Atlantic? **1982** R. Quirk *Style & Communication in Eng. Lang.* i. 16 This is not, of course, to say that the existence of the 'style sheet' mentality is always advantageous.

style, *v.* Add: **7.** To design, arrange, make, etc., in a particular (esp. fashionable) style.

1934 J. Rorty *Our Master's Voice Advertising* 11 'Styling' clothes, kitchens, automobiles—everything, in the interest of more rapid obsolescence and replacement. **1936** H. L. Mencken *Amer. Lang.* (ed. 4) 194 The American liking for short cuts in speech, *e.g.*..*to style* for *to cut in accord with the style*. **1958** J. Cannan *And be a Villain* i. 25 The fashions of the day, styled to suit *gamines*. **1958** *Observer* 25 May 17/2 Mgr. Knox seems to have styled his translation to fit in with the least sufferable conceptions of this, really, stout-hearted young girl. **1976** 'Z. Stone' *Modigliani Scandal* iii. iv. 137 Her hair had been styled by Sassoon.

-style, *suffix*, forming adjs. and advbs. **1.** Appended to adjs. **a.** Forming adjs. having the sense 'resembling or characteristic of something that is—'.

1934 *Discovery* June 174/1 He will have to put up at a Japanese style hotel. **1958** *Listener* 11 Sept. 368/1 He was dressed in European-style clothes. **1960** W. Thorp *Amer. Writing in 20th Cent.* 296 A series of events.. convinced [them] that revolution, Russian-style, would never take place in America. **1966** [see *SEDAN 1 c]. **1975** *Country Life* 20 Mar. 747/2 Elegant regency-style restaurant.

b. Forming advbs. with the sense 'in a manner that is—'. *colloq.*

1967 G. Kelly in *Coast to Coast 1965–6* 95 Bill Beatty leant across the table, his cigarette dangling casual style from his fingers. **1976** T. Sharpe *Wilt* (1978) vi. 61 We're here idyllicstyle, cruising down the river in the good old summertime.

2. Appended to sbs., forming advbs. and adjs. with the general sense '(in a manner) characteristic of or befitting—'.

The advb. use is highly colloq.

1944 W. Saroyan *Human Comedy* xxiv. 166 He whistled, newsboy style. **1949** *Amer.-German Rev.* Apr. 10 Meals were served family-style. **1973** *Guardian* 31 Jan. 5/2 An election-style budget.

styled, *ppl. a.* Restrict † *Obs. rare* to sense in Dict. and add: **2.** Of a person's hair: professionally arranged, cut, or set.

1958 *Spectator* 27 June 833/2 Another of the old gang in hacking jacket and styled hair mincing over to a seat. **1977** E. Leonard *Unknown Man No. 89* ii. 13 His styled hair..glistened.

3. With prefixed sb., adj., etc., forming combs. indicating the origin or provenance of the style.

1958 *Spectator* 11 July 56/2 Manhattan-styled clerical tuxedos. **1966** *Melody Maker* 15 Oct. 19 After an Alpert-styled 'Taste of Honey'.., he introduced a couple of new numbers. **1976** B. Bova *Multiple Man* (1977) v. 64 Their Bavarian-styled paneled dining room.

styler (stəi·ləɹ). [f. Style *v.* + -er[1].] **1.** = *STYLIST 2 a, b. *rare*.

1960 *Lebende Sprachen* V. 35/2 [Clothing industry vocabulary] *Designer, styler, stylist*, Modellmacher, Modelleur (Bekleidung). **1972** *Classification of Occupations* (Dept. Employment) III. 257/2 Dolls' hair styler.

2. A device for styling hair.

1971 *Wall St. Jrnl.* (Eastern ed.) 1 Apr. 1/5 Bercy Industries..has doubled its own forecast of men's hair styler sales... The response to stylers cheers retailers. **1976** *Honolulu Star-Bull.* 21 Dec. B-3/1 (Advt.), Save $5! women's 1200-watt styler-dryer.

styling. Add: **2.** The action, process, or result of *STYLE *v.* 7.

1928 *Publishers' Weekly* 9 June 2370 Recourse to art for investing conventional merchandise with fresh or added appeal has been the chief reliance of post-war sales strategy. It has been termed 'Styling' in some quarters. **1939** *Archit. Rev.* LXXXVI. 62/1 It is scarcely conceivable that any but an American designer would have designed such a building: it is 'styling' applied to architecture. One can only hope that this same 'styling' will not later be applied to Town Halls. **1958** *Observer* 25 May 11/3 The styling of clothes for small children has now passed out of the hands of British makers. **1959** *Daily Tel.* 27 Aug. 11/3 This same firm which now also makes Pilot [television] models with slightly different styling. **1966** *Listener* 17 Nov. 712/2 A new automobile... Engineering and 'styling' costs were $9,000,000. **1975** G. Howell *In Vogue* 206/2 Teddy Boys would go to a barber for 'styling'. **1980** *Times* 29 Feb. 23/4 Smooth, wedge-shaped styling makes for a distinctive appearance.

stylism (stəi·liz'm). [f. Style *sb.* + -ism.] A stylistic device or effect; emphasis on style.

1928 H. Crane *Let.* 5 Feb. (1965) 317, I am not as original in some of my stylisms as I had thought I was. **1942** [see *ABSTRACTIONAL]. **1958** *Times* 30 Sept. 3/1 The German company..proves to be a finely cast body of actors with an inevitably stronger feeling for classical stylism. **1971** R. apRoberts *Trollope* i. 15 Both symbolic explanation and *stylism*, then, are overworked for fiction.

stylist. Add: **1. b.** *transf.* In sport or music, one who plays with style.

1897 K. S. Ranjitsinhji *Jubilee Bk. Cricket* iii. 118 A young player of much promise—essentially a stylist, with brilliant strokes all round the wicket when set. **1898** —— *With Stoddart's Team* (ed. 3) i. 34 Hayward is essentially a stylist. **1969** *Listener* 3 Apr. 470/3 He's also perhaps the most original clarinet stylist in the British Isles. **1981** *Best of Karate '81* Spring 35/2 (*caption*) A Japanese stylist from Indiana, he has won virtually every major title available.

2. orig. *N. Amer.* **a.** In industry, esp. the retail clothing and car industries, one whose job is to create, co-ordinate, or promote the latest fashions or designs for a firm's commodities.

1928 *Daily Express* 18 July 12 The post of 'stylist' has been created during the last year in some of the better Canadian and American stores. The duties consist of linking departments,..informing the bag department of the trend of fashion in the shoe section, and the glove department what is happening in the costume department. *Ibid.* 24 Sept. 7 Harrods Fashion Buyers and Stylists. **1956** *Stanford Law Rev.* July 628 The General Motors stylist Harley Earl cannot be split into five pieces. **1958** *Spectator* 20 June 817/3 The quality and style of 'St. Michael' garments are the product of close collaboration between the Buying Executives, Stylists and Technologists. **1978** *Jrnl. R. Soc. Arts* CXXVI. 538/1 Mr. Nader caused a greater fundamental change to the appearance of cars than did a hundred stylists or engineers.

b. One who styles hair; = *hair-stylist* s.v. *HAIR *sb.* 10.

1937 R. Stout *Crime on her Hands* (1939) ii. 23, I could have got a job as a stylist, or..started a hat shop. **1962** *Times* 3 Feb. 9/4 Where are the barbers or even the hairdressers now? They have become stylists to a man or woman. **1968** [see *CRIMPER[1] 1 b]. **1979** *West Lancs. Evening Gaz.* 10 Nov. 11 (Advt.), Wanted—all-round Stylist—male/female.

stylistician (stəilisti·ʃən). [f. Stylistic *a.* and *sb.* + -ian.] One who studies stylistics.

1939 *Language* XV. 257 One misses the subtler approach of the 'stylisticians' of today. **1948** L. Spitzer *Linguistics & Lit. Hist.* iv. 136 In general, stylisticians have rather shied away from Diderot. **1957** *Archivum Linguisticum* IX. 148 He shares the weakness of other stylisticians, in claiming to have found the Philosopher's Stone. **1969** Crystal & Davy *Investigating English Style* 12 The stylistician is on precisely the same footing as anyone else. **1980** *Times Lit. Suppl.* 11 Jan. 28/1 He makes extensive use of the rhetorical device known to stylisticians as 'free indirect speech'.

stylometric (stəilome·trik), *a.* [f. Style *sb.* + -o + *-metric.] Of or pertaining to stylometry.

1935 H. W. B. Joseph *Essays in Anc. & Mod. Philos.* i. 1 The first book of Plato's *Republic*..may be earlier in date than those which follow. This view about it has strong support from 'stylometric' investigations. **1964** *Guardian* 17 June 9/3 Mr Thomson..determined..to use stylometric methods to investigate the Baconian controversy. **1971** *Computers & Humanities* V. 302 The goals are:..to establish stylometric criteria based on preferred sounds, to see if significant differences in phonemic and phonetic frequencies exist between prose and verse. **1982** *Times Lit. Suppl.* 26 Nov. 1322/2 Most unsettling is his claim, on stylometric grounds long invaluable in the study of Plato, that the so-called 'common books'..are Eudemian, not Nicomachean.

stylometry (stəilǫ·mėtri). [f. as prec. + -metry.] The technique of making statistical analyses of the features of a literary style, esp. by means of a computer. Hence **stylo·metrist**, one who practises this technique.

1945 R. G. Collingwood *Idea of Nature* 58 Scholars.. have analysed the language of the Platonic dialogues statistically... Whatever view is taken of Platonic 'stylometry' in its more detailed development. **1953** *Classical Q.* XLVII. 80 This admission that only equal amounts of text should have been compared..had the effect of largely invalidating his own and most earlier attempts to order the dialogues by relative affinities of style. Stylometrists ignored the warning... The effect was..to discredit mechanical stylometry. **1972** *Ibid.* May 89 (*heading*) The new stylometry. *Ibid.*, Stylometry can be defined as the use of numerical methods for the solution of literary problems. **1979** *Sci. Amer.* Nov. 34/2 The 50,000 words of the 14 epistles attributed to St. Paul are a challenge to the stylometrist. *Ibid.* 34/3 Forensic stylometry is at least an inhibition to the temptation on the part of police officers to adjust the admissions they present in court.

stylopization (stəi:lǫpəizēi·ʃən). *Ent.* [f. as Stylopized *pa. pple.* and *ppl. a.* + -ization.] The state of being or becoming stylopized.

1882 *Trans. Entomol. Soc. London* 228 The effects of stylopisation vary very much in different specimens. **1956** G. Lapage *Veterinary Parasitol.* xiv. 416 The Strepsitera are commonly known as 'stylops'. The bees and wasps in which they are parasitic undergo, as a result of their effects, changes known as 'stylopisation'. **1980** *Ann. Entomol. Soc. Amer.* LXXIII. 448/1 Morphological characters often modified as a result of stylopization include clypeal pigmentation, pollen-carrying hairs, and wing venation.

stylostatistics (stəi:lostăti·stiks). [f. Style *sb.* + -o + Statistics.] The application of statistical methods to the analysis of features of literary style.

1956 G. Herdan *Language as Choice & Chance* i. 8 The main division [of the book] is according to the four main branches of quantitative linguistics: Stylostatistics, Statistical Linguistics, Information Theory, and Linguistic Duality. *Ibid.* ii. 14 That the 'mood' of a text of considerable length should be reflected in the relative frequency of a linguistic form..,—this is precisely what stylostatistics claims to achieve. **1962** A. Ellegård *Who was Junius?* vi. 97 (*heading*) Identification by stylostatistics. **1964** G. Herdan in H. G. Lunt *Proc. 9th Internat. Congr. Linguists* 323 Stylostatistics leads to establishing the general laws for the use of language as a necessary preliminary of the determination of divergencies from these laws in individual style.

stylus. Add: Pl. **styluses, styli. 2. a.** Substitute for def.: A tracing-point used to produce the written record in a chart recorder, telegraph receiver, or the like. (Further examples.)

Quot. 1892 in Dict. belongs to the next sense.

1966 *McGraw-Hill Encycl. Sci. & Technol.* XI. 382/1 By the use of a coated chart or an inked ribbon between the stylus and the chart, an intermittent record is made. **1971** Magrab & Blomquist *Measurement Time-Varying Phenomena* v. 182 A voltage will be applied to the coils of the writing system which will cause the slider and writing stylus to move in proportion to the voltage, thus obtaining a recording.

b. = *NEEDLE *sb.* 3 e.

1879 *Year-Bk. of Facts* 88/2 In that patent he [*sc.* Edison] describes a means of recording ordinary telegraph signals by a chisel-shaped stylus indenting a sheet of paper. **1892** W. Gillett *Phonograph* 12 Instead of one stylus serving for the two purposes, there are now two; one, the recorder, having a very keen edge,..and the other, the reproducer, having a tiny knob highly polished. **1904** S. R. Bottone *Talking Machines & Records* 48 The recording stylus is now seldom made of steel, except only in the very commonest forms of recorders; but is usually made of..sapphire. **1943** *Gramophone* July 29/3 The stylus is a small sapphire carried on the toe of an L-shaped holder. **1960** *Practical Wireless* XXXVI. 370 (Advt.), Turnover sapphire styli. **1962** A. Nisbett *Technique Sound Studio* iv. 87 The hot-stylus method.. helps to smooth the wall as the cutter moves on. **1975** *Times* 17 Sept. 16/3 Nowadays, you have a 'cartridge' with a 'stylus' in it, and the [gramophone] arm is called a 'pick-up'.

fig. **1936** Dylan Thomas *Coll. Poems* (1952) 37 The grooved land rotating, that the stylus of lightning Dazzle this face of voices on the moon-turned table.

stymie, *sb.*[2] (Earlier example.)

1834 *Rules of Musselburgh Golf Club* in C. B. Clapcott *Rules of Golf of 10 Oldest Golf Clubs* (1935) 66 With regard to Stimies the ball nearest the hole if within six inches shall be lifted.

stymie, *v.* Add: **2.** *fig.* To impede, obstruct, frustrate, thwart (a person, an activity, or a project).

1902 G. Ade *Girl Proposition* 70 In about 8 minutes he had the Regular Fellow stymied and Hazel was leaning against him. **1922** A. Haddon *Green Room Gossip* vii. 154, I looked like being stymied because I couldn't sing in tune. **1933** Wodehouse *Mulliner Nights* v. 183 There came the shrill cry of a Hunting Bishop stymied by a hat-stand. **1938** *East Liberty* (Pittsburgh) *Tribune* 26 Aug. 1/5 (*heading*) New recreation center at Mellon Field 'stymied'. **1946** J. W. Day *Harvest Adventure* xvi. 274 Mr John Loverseed..raised this whole question..in spite of repeated attempts by the Ministry of Agriculture to stymie him. **1957** *Economist* 2 Nov. 389/1 It is hard to visualise Afghans and Albanians, Salvadoreans and Sudanese quickly triumphing over the very real difficulties that have stymied the five-power group. **1966** L. Durrell in *Sat. Even. Post* 4 June 68/3 Coco was for selling her to a local clinic, but once more we were stymied by this public holiday. The clinic was shut. **1980** *Daily Tel.* 28 July 24/3 France's participation in the military force had merely been to stymie an intended intervention by troops from Pacific area countries.

stymied, *ppl. a.* Add: Also *fig.*

1937 Partridge *Dict. Slang* 844/1 *Stymied*,..awkwardly placed; nonplussed. **1974** P. De Vries *Glory of Hummingbird* vii. 97, I was worried about a stymied affair with a girl with whose parents I was getting on swimmingly.

styptic, *a.* Add: **1. b.** *styptic pencil*, a stick of styptic substance used to stem the bleeding of small cuts.

1908 *Sears, Roebuck Catal.* 799/2 Styptic Pencils. Used when shaving. Will instantly stop bleeding. **1936** G. Greene *Gun for Sale* vii. 211 Plying his styptic pencil, sticking the cotton-wool on the longer wounds. **1961** I. Fleming *Thunderball* i. 9 Bond dabbed with the blood-stained styptic pencil at the cut on his chin. **1978** *Observer* 26 Mar. 29/7 He had dropped his styptic pencil down the washbasin.

Stypven (sti·pvĕn). *Med.* Also **stypven.** [f. Styp(tic *a.* and *sb.* + Ven(om *sb.*] The dried and purified venom of Russell's viper for use in solution as a local hæmostatic and a blood coagulant; **Stypven time**, coagulation time measured when Stypven is added as a coagulant.

Stypven is a proprietary name in the U.S.

1940 *Lancet* 17 Aug. 195/1 This venom..is very efficient in bringing about the rapid formation of a strong clot when applied to a bleeding surface. It is supplied under the name of Stypven by Messrs. Burroughs Wellcome and Co. **1948** *Official Gaz.* (U.S. Patent office) 20 Jan. 409/2 Burroughs Wellcome & Co. (U.S.A.) Inc., New York.. *Stypven*. For dry viper venom... Claims use since 1937. **1955** *Lancet* 1 Oct. 692/1 'Stypven' times were estimated on all the samples as follows: To 0·1 ml. of warmed plasma was added 0·1 ml. of stypven... Immediately after, 0·1 ml. of calcium chloride was added, and the clotting-time was recorded. **1961** *Ibid.* 2 Sept. 503/1 In these men, the whole-blood clotting-time, silicone clotting-time, calcium clotting-time.., prothrombin-time, and 'Stypven' time were estimated in duplicate weekly for eight weeks. **1976** *Ibid.* 6 Nov. 995/2 The stypven time (Russell viper venom) of P.P.P. reflects the amount of available active phospholipids which may be dietary in origin.

styrene (stəiə·rīn). *Chem.* [f. Styr(ax + -ene.] **1.** A colourless, toxic, aromatic liquid, $C_6H_5{\cdot}CH{:}CH_2$, orig. obtained from the storax tree (hence called Styrol or Styrolene) and now recovered as a by-product of petroleum. Also called *vinylbenzene* and *phenylethylene*.

1885 I. Remsen *Introd. Study Compounds of Carbon* 343 Styrene... This hydrocarbon is contained in liquid storax... It is formed by distilling cinnamic acid with lime. **1926** H. G. Rule tr. *J. Schmidt's Text-bk. Org. Chem.* ii. iii. 331 Styrene, phenyl-ethylene, vinyl-benzene, $C_6H_5{\cdot}CH{:}CH_2$, is the simplest representative of the olefine derivatives [of benzene]. **1947** *Sun* (Baltimore) 17 Apr. 2/5 Styrene was not explosive, Thomas said, but would burn as rapidly as gasoline. **1956** *Sci. News* XLII. 41 Hydrocarbons and low boiling organic compounds..are becoming increasingly important industrially owing to the development of petrochemicals,..which include polythene, styrene and synthetic rubber. **1978** J. R. Holum *Org. & Biol. Chem.* viii. 163 Polystyrene, made from styrene, has phenyl groups on alternate carbons of the main chain.

2. = *POLYSTYRENE. Also *attrib.*

1938 *Encycl. Brit. Bk. of Year* 147/1 To the growing family of synthetic resins and plastics, 1937 saw added Styrene, characterized by clarity, low initial colour, and thermoplastic properties. **1969** *Islander* (Victoria, B.C.) 6 July 8/2 Keeping food hot or cold en route is important —a styrene foam chest has exceptional insulating ability. **1972** J. Potter *Going West* 45 Every can was found to have a double bottom. Nuggets of gold lay embedded in white styrene between the two sheets of tin.

3. Special Combs.: **styrene-acrylonitrile,** the combination of styrene and acrylonitrile, esp. as copolymers in a rubber; usu. *attrib.*; **styrene-butadiene,** the combination of styrene and butadiene, esp. as copolymers in a rubber; usu. *attrib.*; **styrene monomer,** the monomeric form of styrene; = sense *1; **styrene oxide,** the toxic epoxide, $C_6H_5{\cdot}\underset{O}{\underbrace{CH{-}CH_2}}$; **styrene plastic,** any or all of the plastic materials that may be made using styrene; **styrene resin,** any compound formed by the polymerization of styrene.

1957 H. R. Simonds *Conc. Guide to Plastics* ii. 71 A styrene-acrylonitrile copolymer is available from Bakelite Co. **1977** R. A. Dickie in M. O. W. Richardson *Polymer Engin. Composites* iii. 177 Results are presented..on several rubber-modified polystyrene and styrene-acrylonitrile copolymers. **1958** *Times Rev. Industry* Apr. 51/1 The synthetic S.B.R. (styrene-butadiene rubber) being tough and hard wearing, is good for tire treads. **1978** D. R. Paul in Paul & Newman *Polymer Blends* II. xii. 59 A patent..describes bonding an ethylene-propylene-diene terpolymer rubber..to styrene-butadiene rubber.. by a graft. **1947** I. Thomas *Injection Moulding Plastics* iii. 168 The base material for polystyrene is styrene monomer. **1971** R. D. Deanin in Tobolsky & Mark *Polymer Sci. & Materials* xiv. 335 These [polyesters] are mixed with about one-half their weight of styrene monomer. **1939** *Jrnl. Amer. Chem. Soc.* LXI. 997/2 The reaction of styrene oxide and hydriodic acid results in a primary alcohol. **1979** *Experientia* XXXV. 241/2 The observed differences in the metabolism of styrene oxide can be assigned theoretically to 2 possible factors. **1943** D. W. Brown *Handbk. Engin. Plastics* ii. 27 (*heading*) Styrene plastics. **1947** *Brit. Catal. Plastics* 51/2 (*heading*) Methyl methacrylate and styrene plastics. **1937** R. S. Morrell *Synthetic Resins* i. 16 The vinyl and styrene resins are used as lacquers and in nitro-cellulose finishes. **1959** E. C. Bernhardt *Processing of Thermoplastic Materials* iii. 628 (*heading*) Styrene resins.

Hence **sty·renated** *ppl. a.*, containing styrene in chemical combination; **styrena·tion**, the process of chemically adding styrene.

1954 Kirk & Othmer *Encycl. Chem. Technol.* XIII. 146 Styrenated drying oils, and styrenated alkyds as a base for many protective coatings..illustrate still further uses for styrene. *Ibid.* 177 The rate of the styrenation reaction and the clarity of the product are influenced by the kind and amount of unsaturation present in the oil. **1972** H. Warson *Applic. Synthetic Resin Emulsions* v. 252 It is not readily possible to form a styrenated oil or alkyd in emulsion form directly. *Ibid.* 253 A maleinized oil is used as the basis for styrenation in the ammoniacal form.

styria, styriate(d, varr. Stiria, Stiriate(d.

Styrofoam (stəiə·rǒfōᵘm). Chiefly *U.S.* Also **styro-**. [f. *Poly)styr(ene + -o + Foam *sb.*] A proprietary name for a variety of foam plastic.

1950 *Official Gaz.* (U.S. Patent Office) 11 July 403/1 Dow Chemical Co...*Styrofoam*... For irregular solid masses of multicellular expanded synthetic resinous material and granular masses of the same material comminuted. **1962** *Punch* 19 Dec. 881/1 How to make.. lambs from glass fibre, angels from styrofoam. **1969** *Jane's Freight Containers 1968–69* 438/3 Insulated container. Rectangular, with styrofoam insulation between aluminium panels. **1971** *Islander* (Victoria, B.C.) 11 July 2/1 The styrofoam in the deckhead was wonderful in keeping it relatively cool below. **1973** R. Hayes *Hungarian Game* ii. 20 Each item was cradled in plush-lined styrofoam... Nothing could rattle. **1978** G. Vidal *Kalki* vi. 135 Geraldine was referring to the top part of a huge Styrofoam statue of Vishnu. **1981** *Times Lit. Suppl.* 6 Mar. 254/2 A styrofoam beaker of instant coffee.

suabe (swā·bə, swēⁱb). *Mus.* [It., ad. G. *Schwabe* Swabian.] *suabe flute*: an organ flute-stop.

1855 E. J. Hopkins *Organ* 119 *Suabe-flute*,..a tenor c Manual Stop of 4 feet, formed of wood pipes, with inverted mouths. It's tone is liquid and clear, and not so loud as the Wald-flute. **1907** *Musical Times* 1 Aug. 514/2 Swell Organ..Voix celestes..Suabe flute 4 ft. **1954** *Grove's Dict. Mus.* (ed. 5) VI. 358/2 *Suabe flute*, a 4-ft open flute stop of medium scale, said to have been invented by William Hill. The tone is a soft variety of that of the Clarabella.

suave, *a.* Add: The pronunc. (swāv) is now standard. **3.** (Earlier example.)

1831 F. Reynolds *Playwright's Adventures* iv. 63 St Alm was anything but *suave*.

suavify, *v.* For *rare*[−0] read *rare*[−1] and add example.

1825 *Spirit of Public Jrnls. for 1823* (ed. 2) 444 Eating much tends to suavify the mood.

sub, *sb.* Add: **4.** (Earlier examples in Sporting and Printing contexts.)

1864 *Field* 9 July 22/1 Lillywhite was caught by Yescombe, a 'sub'. **1876** *Scribner's Monthly* Apr. 838/1 He consented finally to allow another printer to take his place in the 'Clarion' office—temporarily, and as his 'sub' only. **1887** *Irish Times* 24 May 7/7 D. Carbery c. sub. b. W. G. Downey 1.

6. (Earlier examples of sense 'subscription'.)

1805 M. L. Weems *Let.* 9 Jan. (1929) II. 310 In 18 hours subscriptioneering I obtain'd from the Legislature 100 subs. to Sydney. **1833** J. Romilly *Diary* 12 Mar. (1967) 30 Fairly bullied Waud & Jones into subscribing to my Blencowe cause:—got 4 others subs today. **1898** W. S. Chirchill *Let.* 5 Aug. in R. S. Churchill *Winston S. Churchill* (1967) I. Compan. ii. 956, I have to pay £40 for one charger, £35 for the other & £20 subs to the mess.

8. a. = Submarine *sb.* 3. Also Comb., as *subchaser* = *submarine chaser* s.v. *SUBMARINE *sb.* 3 b.

1917 J. M. Grider *Diary* 29 Sept. in *War Birds* (1927) 21 We were supposed to look out for gulls which they say usually follow in the wake of a sub. **1918** L. E. Ruggles *Navy Explained* 124 *Sub-chaser*, a small, swift, light draft boat used to hunt submarines. **1931** 'Taffrail' *Endless Story* xxi. 333 'Sub-chaser' 28, manned by the French, broke down in the Atlantic 700 miles from the Azores and was given up for lost. **1936** *Nat. Geogr. Mag.* LXIX. 799/1 Seamanship..includes instruction on how to.. maneuver..such craft as subchasers and motor launches. **1968** A. Diment *Bang Bang Birds* ii. 16 Boris snooping round Holy Loch and the nuclear subs. **1977** *New Yorker* 29 Aug. 20/1 A subchaser lurches forward on the calm water and comes to a stop as a black sub surfaces at its side.

b. = *SUBMARINE *sb.* 4 b. *U.S. colloq.*

1955 *Sat. Even. Post* 1 Jan. 16 'I tell you,' a sandwich-shop operator said, 'Subs are taking over.' **1976** R. B. Parker *Promised Land* ii. 5, I was ready to settle for Ugi's steak and onion subs.

sub, *v.* Add: **2.** (Earlier examples.) In *gen.* use, to act as a substitute. Also *trans.*, to substitute (something). Chiefly *U.S.*

1853 'Mark Twain' *Let.* 26 Oct. (1917) I. i. 26, I am subbing at the Inquirer office. *Ibid.*, If I want it, I can get subbing *every night* of the week. **1926** *Amer. Mercury* Dec. 465/2 When a new act was placed last on a programme, *Variety* put it: 'Fred and Daisy Rial subbed in the walk-out assignment.' **1943** *Sun* (Baltimore) 17 Sept. 8/2 (*heading*) Subbing camera for gun, corporal 'shoots' zeros. **1950** A. Lomax *Mister Jelly Roll* (1952) 218 The lord of New Orleans piano was scratching hard for a living.., subbing for other piano players who showed up drunk on their jobs. **1974** *Globe & Mail* (Toronto) 24 July 10/2 Toronto Executive Alderman Arthur C. Eggleton subbing for Mayor David Crombie. **1981** B. Granger *Schism* (1982) x. 88 Father Malachy is subbing for the pastor at St. Mary's... The pastor broke his leg, jogging.

3. (Earlier examples.) Also *to sub up*: to pay up or subscribe.

1874 C. Holloway *Jrnl. Visit to N.Z.* 22 Apr. (*typescript*) I. 57 In some instances the dissipated individual had to sub a few shillings of the Landlord to help him on the road. **1874** Hotten *Slang Dict.* 314 *Sub*, to draw money in advance. **1942** O. Jespersen *Mod. Eng. Gram.* VI. 546 *Sub* = subsidy or subsistence.., also subscription ..and as a vb., esp. *sub up* 'subscribe'. **1958** G. Mitchell *Spotted Hemlock* vii. 75 'Wasn't that rather expensive?'.. 'I believe Tony Bianchi subbed up.'

4. Also, *to sub the purple*: see *PURPLE *sb.* 7 b.

5. [Substratum 4.] In the manufacture of photographic film: to coat with a substratum (see quot. 1965). Chiefly as *vbl. sb.*, the process of applying a substratum; the substratum itself.

1941 T. T. Baker *Photographic Emulsion Technique* x. 179 The film base may be wiped or cleaned prior to subbing... The cleaned and substratumed film base is coated at a fairly rapid rate. **1958** H. Baines *Sci. Photogr.* vi. 83 The rear side of roll film and sheet film is subbed (substratum coated). **1965** M. J. Langford *Basic Photogr.* ix. 161 The manufacturer first 'keys' both sides of the film base or coats them with a foundation layer of gelatin and cellulose ester known as the 'subbing' layer. Next, the emulsion is coated over the subbing on the face of the film. **1977** J. Hedgecoe *Photographer's Handbk.* 263/1 Other non-porous surfaces should be pre-coated with the subbing which is normally supplied with the emulsion.

sub, *Latin prep.* Add: **14*. sub specie æternitatis,** 'under the aspect of eternity', i.e. viewed in relation to the eternal; in a universal perspective. [Cf. Spinoza *Ethices* (*a* 1677), in *Opera Posthuma*, 1677, v. xxix. 254.] Hence **sub specie temporis**, viewed in relation to time rather than eternity.

1896 W. Caldwell *Schopenhauer's System* v. 268 Art enables us somehow to see things *sub specie aeternitatis*. **1911** *Encycl. Brit.* XXI. 441/2 The nature of any fact is not fully known unless we know it in all its relations to the system of the universe, or, in Spinoza's phrase, *sub specie aeternitatis*. **1925** A. Huxley *Let.* 21 Apr. (1969) 247 There, on the other side of the water, are one hundred and five million beings whose sole function—if you look at their lives sub specie aeternitatis—is to provide people like us with money. **1935** E. R. Eddison *Mistress* 20 This man, as I have long observed him, looked on all things *sub specie æternitatis*; his actions all moved..to slow perfection. **1952** V. A. Demant *Relig. & Decline of Capitalism* iii. 70 Hence what was true *sub specie aeternitatis* in the liberal aim is being lost. **1973** G. M. Brown *Magnus* vii. 139 If..we could look with the eye of an angel on the whole history of men, *sub specie aeternitatis*, it would have the brevity and beauty of this dance at the altar.

1928 L. Hodgson in A. E. J. Rawlinson *Essays on Trinity & Incarnation* viii. 378 Perhaps the best one can do is to speak of God as ἀπαθὴς *sub specie aeternitatis* but παθητικός *sub specie temporis*. **1944** W. Temple *Let.* 12 Jan. (1963) 142, I have treated the Son and the Spirit as God *sub specie temporis* and the Father as God *sub specie eternitatis*. **1960** *Encounter* XV. 77 *Sub specie temporis* his Combination Rooms say more to us than Beckett's wet and windy plains.

14. sub specie mortis,** in the face of death.

1955 *Times* 26 May 3/4 The ninth symphony, we are told, is poignant in that it was his last and written *sub specie mortis*. **1964** *Listener* 21 May 849/3 Written *sub specie mortis*, they are his [*sc.* Mahler's] most 'existentialist' works.

14*. sub verbo** = *sub voce*, sense 15 in Dict.; abbreviated *s.v.* (see *S 4 a).

1902 J. M. Baldwin *Dict. Philos. & Psychol.* II. 358/2 Many citations in Eisler, Wörterb. d. philos. Begriffe, sub verbo.

15. sub voce (examples). Cf. Voce.

1859 *N. & Q.* 23 Apr. 341/1 Skinner, *Gloss.*, *sub voce*, evidently understands the word in this sense. **1871** *Ibid.* 9 Dec. 487/1 See Halliwell's *Dict.*, sub voce 'Braid'.

sub-, *prefix.* Add: **I. 1. a. subti·dal** *Ecol.*, situated or occurring below the low tide mark.

1939 Clements & Shelford *Bio-Ecol.* x. 313 The subtidal community reaches up into the tidal area. **1979** R. Brewer *Princ. Ecol.* v. 231 It is usual to recognize three zones on both sandy and rocky shores. These are the intertidal zone itself, a supratidal zone above it.., and the subtidal zone.

(*b*) In derived advbs., as **subgla·cially** *adv.*, under an ice sheet or glacier.

1909 Webster, Subglacially. **1978** *Nature* 8 June 456/2 Lava flows which were erupted subglacially in southwestern Iceland.

b. subglo·ttal, -lary·ngeal, -te·ctal (the tectum of the skull); **sub-ge·nual** *a.*, situated below the knee; **sub-gingi·val** *a.*, situated or occurring beneath the gums, esp. between the gum-margins and the teeth.

1934 Webster, Subgenual. **1935** R. E. Snodgrass *Princ. Insect Morphol.* xvii. 527 Proximally, below the 'knee', is a large fan-shaped subgenual organ. **1978** H. V. Daly *Introd. Insect Biol. & Diversity* vi. 109/1 Subgenual organs are found in many insects but are lacking in Archeognatha, Coleoptera, and Diptera. **1898** H. H. Burchard *Text-bk. Dental Path. & Therapeutics* xxiv. 456 By subgingival deposits are meant calculi which are first deposited in the annular depression between the gum-margin and a tooth. **1979** Williams & Elliott *Basic & Appl. Dental Biochem.* xii. 224 Dental plaque covers the tooth (supragingival plaque) and extends over the tooth surface of the gingival pocket (subgingival plaque). **1932** W. L. Graff *Language & Languages* i. 33 Even if we assume that the subglottal force of expiration is the same.., on leaving the larynx its strength is.. weakened. **1970** *Language* XLVI. 313 It would seem difficult to ascertain whether a change in fundamental frequency is due to a change in the tension of the laryngeal muscles or to a change in the subglottal air pressure. **1901** *Proc. Zool. Soc. London* I. 281 The sub-laryngeal pouch is essentially a cæcal diverticulum of the ventral wall of the larynx, between the thyroid and cricoid cartilages. **1949** Koestler *Insight & Outlook* 391 A continuous flow of physiological processes, involving.. sublaryngeal movements (inner speech). **1940** *Chambers's Techn. Dict.* 818/2 *Subtectal*, lying beneath the roof, as the roof of the skull. **1975** *Nature* 30 Oct. 738/1 In the vertebrates below mammals, the tectal and subtectal areas are the main centres of termination of sensory pathways.

(*b*) *subconjuncti·vally, -glo·ttically, -pi·ally.*

1907 *Ophthalmoscope* V. 383 The conjunctival cicatrice had been divided subconjunctivally. **1974** *Nature* 11 Oct. 553/2 Grafts..placed on Fischer hosts that had been inoculated subconjunctivally..were rejected abruptly. **1975** *Year Bk. Ear, Nose & Throat* 317 Direct laryngoscopy showed adducted cords with an absent lumen subglottically. **1950** *Jrnl. Neurophysiol.* XIII. 192 A fine steel needle electrode thrust subpially into the substance of the acoustic tubercle.

e. *su·b-ice*; **subso·lidus** *Geol.*, existing or occurring in conditions corresponding to a point in a phase diagram below a solidus, i.e. when the system is wholly solid.

1959 *Times* 9 Jan. 11/6 Apart from the exposed mountains near the coast, sub-ice hill and dale occur. **1973** *Nature* 20 Apr. 539/3 New chapters in the history of the continent will be based on the results of continued palaeomagnetic studies, much deep sea drilling and to a lesser extent sub-ice drilling. **1952** B. Mason *Princ. Geochem.* x. 232 We must..consider the phase changes that may take place in solid solutions in the subsolidus region. **1979** *Nature* 15 Mar. 220/1 This mineral assemblage could be produced by hot-pressing the above composition in subsolidus conditions.

2. a. *subadjacent* (later example); *sub-irrigate*: see also *SUB-IRRIGATION.

1922 Joyce *Ulysses* 653 He..gained retarded access to the kitchen through the subadjacent scullery.

3. a. su·b-floor, a floor serving as a base for another floor; **su·b-frame,** a secondary frame; *spec.* (*a*) in carpentry and building, the frame for the attachment or support of a window or door-frame, or of panelling; (*b*) in a vehicle, the frame on which the coachwork is built, as distinct from the chassis; **substra·tosphere,** the upper part of the troposphere, immediately below the stratosphere.

1893 J. P. Allen *Pract. Building Construction* viii. 122 Double floors may sometimes mean that the flooring or floor-boards are laid on the joists in two thicknesses.. the bottom thickness being straight-jointed..; while the other..is laid in the ordinary way above the sub-floor, as the bottom one is called. **1929** W. C. Huntington *Building Construction* vi. 264 This type of construction is greatly superior to that which rests the studs on a sole plate placed on top of the sub-floor. **1973** *Building Materials* (MTP Construction) 186 Any sub-floor can be levelled and smoothed. **1929** *Motor* 1 Oct. 33 (Advt.), A sub-frame, on which the body is built, isolates the coachwork from the chassis and prevents distortion; the sub-frame being mounted on the chassis by supports having a certain degree of flexibility. **1944** N. W. Kay *Pract. Carpenter & Joiner* vi. 138 Care must be taken to prevent any damage by shrinkage. One method..is to form a sub-frame within the main frame. D shows a sub-frame, to carry the glazing, tongued into the frame of the door. **1968** D. Braithwaite *Fairground Archit.* v. 88 At least two further wagons were required for 'loose stuff'— sub-frames, gates and shutters, roundings and so on. **1979** *Daily Tel.* 1 Dec. 18 (Advt.), For Sale. Mini 850, 1969, 47,000 miles... New sub-frame, reconditioned gear-box. **1937** *Popular Sci. Monthly* Nov. 68/2 Aerial Battles miles above the earth..are foreshadowed by a substratosphere plane placed under test by the U.S. Army Air Corps. **1952** *Chambers's Jrnl.* Feb. 84/2 Another day Byrd flew over a never-ending succession of mountain-ranges, which looked as if they were suspended in the substratosphere.

II. 5. b. *sub-leader* [Leader[1] 12], *-network*; **submuni·tion** chiefly *U.S.*, (usu. *pl.*) small, short-range guided missiles; also *sing.*; **su·bstorm** *Meteorol.*, a disturbance of the earth's magnetic field restricted to certain, usu. polar, latitudes and typically manifested as an aurora and other upper atmospheric phenomena.

1913 S. O'Casey *Let.* 8 Mar. (1975) I. 23 The Editor recently, in a sub-leader, advised all his readers to go and see pictures exhibited in the Central Branch of the Gaelic League. **1975** *Aviation Week & Space Technol.* 6 Oct. 15/2 Improvement of the BLU-63 submunition bomblet with two basic sizes of fragments. **1983** *Financial Times* 2 Dec. 2/3 Clusters of anti-armour submunitions which would be fired from stand-off positions. **1956** J. Klein *Study of Groups* iv. 50 This will be useful when we wish to analyse sub-networks. **1978** *Sci. Amer.* June 95/1 Clos's design is based on the idea of building a large network out of smaller networks called subnetworks. **1961** Akasofu & Chapman in *Jrnl. Geophysical Res.* LXVI. 1339/2 Each such event, which Birkeland.. called a polar elementary storm, is here called a *DP* substorm. **1969** *New Scientist* 25 Sept. 669/3 These are the 'substorms' which manifest themselves as aurorae and associated magnetic disturbances. **1979** *Nature* 22 Feb. 649/1 Measurements were made during a magnetic substorm which appeared together with the polar light on 25–26 February, 1978.

c. (*a*) *sub-code* [*CODE *sb.*[1] 3 d], *-cycle, -entry, -function, -genre, -hierarchy, -item, -literature, -part* (later examples), *-plot* [Plot *sb.* 6], *-problem, -sense, -society, -substantiality, -system* (later examples), *-theme, -theory, -topic, -war, -world*; **sub-goal** *Psychol.*, something that must be achieved on the path to the main objective.

1960 R. Jakobson in T. A. Sebeok *Style in Lang.* 352 For any speech community..there exists a unity of language, but this over-all code represents a system of interconnected subcodes. **1967** *Language* XLIII. 752 Variants of one and the same code, realized by means of different information channels, are called 'subcodes'. **1973** S. Heath in *Screen* Spring/Summer 215 The distinction between cinematic codes and sub-codes is initially made in *Langage et Cinéma* as that between general and particular cinematic codes. **1953** A. K. C. Ottaway *Educ. & Society* 45 A sub-cycle was here set up within the whole economic setting. **1974** *Sci. Amer.* June 78/2, I have omitted the important subcycles of nitrogen and phosphorus, two elements that are strongly related to the origin of life and to biological processes in general. **1876** C. A. Cutter *Rules for Dictionary Catalog* 13 Class entry with specific or class subentry. **1979** G. N. Knight *Indexing* vi. 106 If it is a sub-entry itself that errs with excessive references, then the remedy is to turn it into a cross-reference to a separate heading having its own sub-entries. **1949** Koestler *Insight & Outlook* x. 153 Like all means towards an end (or subfunctions in non-teleological language) [etc.]. **1969** H. R. F. Keating *Inspector Ghote plays Joker* iii. 35, I regard it as a sub-function of my post to make person-to-person contact with as many people in your department as possible. **1976** *N.Y. Times* 13 Jan. 40/5 This was because she belonged to, if in fact she hadn't pretty much created a subgenre of the mystery novel. **1980** *Times Lit. Suppl.* 7 Nov. 1262/2 One of the major subgenres of science fiction (works dealing with robots and computers). **1932** E. C. Tolman *Purposive Behav.* 459/1 (*Index*) Sign-objects.. become sub-goal-objects. **1967** M. Argyle *Psychol. Interpersonal Behaviour* v. 91 Driving a car from A to B involves the sub-goals of getting the engine started, getting the car moving in top gear, and getting to the intermediate points X, Y and Z. **1977** *Dædalus* Fall 121 The theory includes the representation of the task environment, the definition of goals and subgoals,..and the order in which alternative courses of action will be explored. **1962** H. C. Conklin in Householder & Saporta *Probl. Lexicogr.* 128 Subhierarchies of varying 'depths' are often discernible within larger hierarchic structures. **1928** in W. K. Hancock *Australia* (1930) v. 90 In the existing Customs tariff there are 259 items or sub-items which provide *ad valorem* duties of 40 per cent. or over. **1961** Webster, *Subliterature*,..inferior literature that does not survive the test of time. **1974** *Times Lit. Suppl.* 1 Feb. 105/1 Though it is important to acquaint oneself with best-sellers and sub-literature.., the chief stress should be on literature of enduring aesthetic worth. **1949** Koestler *Insight & Outlook* x. 135 A sub-whole composed of sub-parts, which in themselves are sub-wholes, and so on. **1976** Lieberman & Rhodes *Compl. CB Handbk.* xi. 233 Each applicant..must follow the procedure prescribed by Subpart 1 of Part 1 of this chapter. **1916** C. Hugon tr. *Creizenach's Eng. Drama in Age of Shakespeare* v. 255 In those cases where a comic sub-plot runs side by side with the main plot. **1962** G. K. Hunter *John Lyly* iv. 237 The sub-plot episode of the pages and Grim the Collier of Croydon in Edwardes' Damon and Pithias. **1907** W. James *Mem. & Stud.* (1911) x. 236 So the great problem splits into two sub-problems. **1980** *Sci. Amer.* Feb. 30/3, I shall treat such issues as special subproblems of allocation. **1947** E. Partridge *Usage & Abusage* 134/2 *Titanic*..is frequently employed with the sub-sense of 'extraordinarily powerful'. **1981** *Dictionaries* II.–III. 168 Brockhaus lists it as an example illustrating a sub-sense of *Blut*. **1951** E. E. Evans-Pritchard *Soc. Anthropol.* i. 13 Its boundaries include..peoples of near and further Asia, north Africa, and parts of Europe—an almost limitless number of..societies and sub-societies. **1922** Joyce *Ulysses* 385 *Entweder* transsubstantiality *oder* consubstantiality but in no case subsubstantiality. **1957** V. W. Turner *Schism & Continuity in African Society* i. 1 This book is..an attempt to analyse..the form and functioning of a sub-system, the village, within a wider system, the totality of Ndembu society. **1972** W. Labov *Language in Inner City* ii. 64 BEV..is best seen as a distinct subsystem within the larger grammar of English. **1949** M. Mead *Male & Female* xvii. 345 In this story there is a sub-theme of the girl's mother's flirtation with a younger man. **1981** A. Paton *Towards Mountain* xxxiv. 307 The birth and rise of Afrikaner nationalism is one of the most powerful subthemes of my life story. **1951** Parsons & Shils *Toward Gen. Theory of Action* i. 28 Economic theory..only becomes a distinctive subtheory of the general theory. **1897** O. J. Nave *Topical Bible* 4 Under the subtopic, *Instances of*, are grouped all the illustrative facts that occur in the Scriptures relating to each subject. **1923** *Notes from Ireland* Nov. 120/2 All this time a sub-war rages. **1949** Koestler *Promise & Fulfilment* i. x. 114 The year..ended with the first encounters in the sub-war between the British Foreign Office and the future State of Israel. **1890** W. James *Princ. Psychol.* ii. xxi. 291 The popular mind conceives of all these sub-worlds more or less disconnectedly. **1983** *Times* 10 Jan. 22/7 A sort of sub-world of smacked bottoms and scrawny mothers-in-law.

(*b*) With derived adjs., as *sub-intentional, -intentioned, -systemic.*

1957 P. Lafitte *Person in Psychol.* ix. 121 A projective test..is designed to elicit what might be called sub-intentional behaviour, including the whole range of covert behaviour that the person cannot report directly. **1968** *Internat. Encycl. Soc. Sci.* XV. 387/2 Subintentioned deaths are those in which the deceased played an important indirect, covert,..or unconscious role in his own demise. **1961** Webster, Subsystemic. **1966** S. Beer *Decision & Control* xvi. 428 The prefrontal lobe of the cortex..has no specific sub-systemic control responsibilities. **1977** *Dædalus* Summer 81 The functionings of their subsystemic parts..and of the whole are to be understood with the aid of general systems theory.

d. *sub-smile* (earlier example); **sub-optimiz·a·tion** (see quot. 1967).

1964 T. W. McRae *Impact Computers on Accounting* iii. 93 Cyberneticists are for ever emphasizing the dangers of sub-optimization. **1967** E. Duckworth in Wills & Yearsley *Handbk. Management Technol.* 119 *Sub-optimization*—the achievement of optimum working of, say, one department of a company without regard to the effect this may have on the rest of the organization. **1852** C. M. Yonge *Two Guardians* viii. 140 A certain sub-smile about the corners of his mouth.

6. a. *sub-conductor.*

1947 H. G. Farmer *Royal Artillery Concerts* v. 15 For many years Zavertal would rehearse the orchestra for months without giving the slightest heed to the military band, which was left to the sub-conductor (the Serjeant Major) and his subordinates. **1976** D. Steele in H. Procter-Gregg *Beecham Remembered* i. 109 He came across the orchestra bridge to start untangling the confusion caused by Weingartner's being quite out of touch..with his sub-conductors.

7. a. Also of immaterial or abstract entities. *substring* (*STRING *sb.* 15 c), *-tree* (Tree *sb.* 6 b (*e*)), *-unit*; **su·b-channel** *Radio*, a distinct division of a channel or frequency band; **su·bgrain**, a small grain contained within another grain in a metal; **sub-hori·zon**, a layer within an existing archæological or soil horizon; **su·blattice** *Physics*, a coextensive part of a fuller lattice, obtained by considering all the members having some property not possessed by the other members; **su·b-shell** *Physics*, in an electron shell, the complete set of orbitals capable of being occupied by electrons of identical azimuthal quantum number *l*.

1959 *Wall St. Jrnl.* 30 June 1/4 Multiplexing..is a technique by which a radio station can divide its regular channels into subchannels and transmit two or more sound signals at the same time. **1970** J. Earl *Tuners & Amplifiers* i. 24 Other features found in tuner-amplifiers.. are automatic mono/stereo switching.., a low-pass filter for reducing the stereo sub-channel noise when the aerial signal is not quite strong enough for noise-free stereo reception, [etc.]. **1955** *Phil. Mag.* XLVI. 1343 Recent work..has shown that many of the dislocations left inside a metal after deformation are arranged along surfaces forming low angle boundaries between neighbouring regions of crystal, these latter being called subgrains, cells, or particles. **1975** *Nature* 10 Apr. 489/1 Granular xenoliths..show various strain effects, including undulose extinction.., slip-planes, and subgrain development. **1928** *Bull. Amer. Soil Survey Assoc.* IX. 36 Other sub-horizons are designated as A_1, A_2, etc. **1973** P. A. Colinvaux *Introd. Ecol.* iii. 45 (*caption*) There may be several subhorizons in each of the main horizons and roots may penetrate them all. **1959** W. F. de Jong *Gen. Crystallogr.* ii. 101 Either the direct lattice..is congruent (similar) with the elementary Bravais lattice, or one is a sub-lattice of the other. **1973** H. D. Megaw *Crystal Structures* viii. 174 Physicists who are less used to describing any but very simple periodic structures,..use the very misleading term 'sublattice' for a Bravais array. **1976** *Physics Bull.* July 294/2 In a crystal the atomic lattice can divide into two interpenetrating sublattices so that most, if not all, of the neighbours of an atom on one sublattice belong to the other. **1930** Ruark & Urey *Atoms, Molecules & Quanta* ix. 272 The *n*, *l*, and *s* quantum numbers can still be assigned to individual electrons, and, therefore, the shells and subshells can be designated in terms of these numbers. **1959** G. Troup *Masers* 161 We take as an example Cr^{+++} which has 3 electrons in the unfilled outermost sub-shell. **1980** H. H. Sisler et al. *Chem.* viii. 209 For electrons in *s* subshells, we find that the probability distribution..is independent of direction in space and varies only with distance from the nucleus. **1955** N. Chomsky *Logical Struct. Linguistic Theory* (microfilm, Mass. Inst. Technol.) vi. 202b *Z'* differs from *Z* only in that it contains a substring *Y* replacing the substring *X* of *Z*. **1966** D. G. Hass in *Automatic Transl. of Lang.* (NATO Summer School, Venice, 1962) 145 In natural languages, texts can be segmented into recurrent substrings. **1972** *Computer Jrnl.* XV. 232/2 Each coordinate of *T* specifies a set to which a substring of *A* belongs. **1947** *Proc. Cambr. Philos. Soc.* XLIII. 26 We call *S* a subtree of *L* if $p_0(S) = 1$ and $p_1(S) = 0$. **1972** R. J. Wilson *Introd. to Graph Theory* iv. 51 Let $T_1, \ldots, T_k$ be the subtrees obtained from *B* by removing the vertex *v* and every edge incident to *v*. **1976** J. S. Gruber *Lexical Structures in Syntax & Semantics* ii. i. 219 Instead of writing the lexical attachment rules as transformations, we will write them as terminal subtrees. **1936** *Economist* 7 Mar. 530/1 The promoters buy specified blocks of securities and deposit them with named trustees, who issue an agreed number of 'sub-unit certificates' against them. **1950** *Cold Spring Harbor Symp. Quantitative Biol.* XIV. 69/1 The phenomenon that the asymmetric unit in the crystal is a submultiple..of the molecular weights found in the ultracentrifuge, and..that excelsin..splits into 3n subunits. **1980** *Times* 15 Jan. 14 DNA is a long chain-like molecule composed of four different chemical subunits.

b. *sub-caste, -clan, -clone* (also as vb. trans.), *-flight* [*FLIGHT *sb.*[1] 1 h], *-nation* (also *-nationalism*), *-unit* (later example).

1892 H. H. Risley *Tribes & Castes Bengal* I. 78 The Bauris are divided into the following nine sub-castes. **1974** tr. *Wertheim's Evolution & Revolution* iii. 240 Even if a sub-group within a caste..has constituted itself as a separate sub-caste, such a collective step will generally imply a disruption of recognized family ties. **1954** Subclan [see *PARAMOUNT *a.* 1 c]. **1961** *Virology* XIII. 160/2 Subclones showing the morphology characteristic of the superinfecting virus were plated for virus release. *Ibid.*, Clones showing the morphology characteristic of the original virus were subcloned. **1977** *Jrnl. Protozool.* XXIV. 28/1 The plausible inference can be drawn that doublet and singlet subclones from a single source do not differ in the kinds of genes or of other molecules they contain. **1978** *Nature* 7 Dec. 579/2 We subcloned the mixed progeny of the cross. **1941** *Hutchinson's Pictorial Hist. War* 14 May–8 July 216/2 Dive-bombers usually approach at about eight thousand feet or so, and on arriving over the target break up into sub-flights of three. **1935** A. M. Carr-Saunders in Huxley & Haddon *We Europeans* viii. 256 We might..contrast America with Europe, regarding the European nations as sub-nations, that is as communities with a lower degree of distinction. **1967** M. Ayub Khan *Friends not Masters* x. 183 We will remain 'sub-nations' if we do not join together to offer united resistance to power pressures. **1957** *Economist* 7 Sept. 739/2 This theoretically sensible policy ran up against the sub-nationalism of the local peoples. **1974** G. W. Choudhury *Last Days United Pakistan* i. 1 Emerging Bengali regionalism or sub-nationalism. *a* **1944** K. Douglas *Alamein to Zem Zem* (1946) 13 He allotted me two tanks, as a troop, there not being enough on the squadron strength to make sub-units of more than two tanks.

(*b*) *subnational.*

1977 *Jrnl. Commonwealth & Compar. Pol.* XV. 236 In 1968 the 81 successful UNIP candidates included 24 politicians at subnational levels.

c. *sub-area, -centre, -epoch, -interval* (later examples), *-phase, -zone.*

1926 *British Gaz.* 12 May 2/3 A service of corporation 'buses has been started in the sub-area of Keighley. **1980** *Amer. Speech 1976* LI. 235 The second map..shows the boundaries of twelve dialect areas and subareas of Scottish English. **1953** L. Kuper *Living in Towns* 304 Library books had been issued from sub-centres in local schools. **1977** *Lancet* 5 Nov. 946/1 The health centre and subcentres provided additional support. **1910** *Geol. Fören. i Stockholm Förhandl.* XXXII. 1146 (*heading*) Gothi- and Finiglacial sub-epochs. **1940** A. H. Sutton in *Bull. Geol. Soc. Amer.* LI. 1402 Subepoch—subseries. These terms as herein proposed are applicable to the first subdivisions of epochs and series respectively. **1962** D. R. Cox *Renewal Theory* ii. 30 To obtain (1) from first principles, divide the time interval (O, *t*) into a large number *k* of small subintervals of length Δt, where $k\Delta t = t$. **1980** A. J. Jones *Game Theory* ii. 109 Then divide the interval [0, 1] into three equal subintervals. **1936** *Discovery* Oct. 329/1 Measles and certain skin affections..are the result of these sub-phase mutations. **1977** *Antiquaries Jrnl.* LVII. 392 It is simply a sub-phase of one phase..of a conservative..coinage. **1903** Subzone [see *SECULE]. **1969** Bennison & Wright *Geol. Hist. Brit. Isles* ix. 213 The 6 goniatite stages called after goniatite genera..were formally called zones but they are stages further divided into 16 zones (and many subzones).

d. *sub-depot; subdiscipline, -field, -speciality, -specialty, -task.*

1938 *Times* 13 Sept. 17/6 The opening of a further sub-depot for recruits at the air station at Dishforth, Yorks. **1976** *Daily Mail* (Hull) 30 Sept. 5/1 Councillors at Selby, concerned at the possibility of an Army Ordnance sub-depot being purchased by the Central Electricity Generating Board. **1958** W. Stark *Sociol. Knowl.* i. 31 The relation of the two sub-disciplines to each other. **1982** *Sci. Amer.* Nov. 52/1 Answers to these questions call for close collaboration among earth scientists from many subdisciplines. **1902** W. James *Var. Relig. Exper.* ix. 195 For them the soul is only a succession of fields of consciousness: yet there is found in each field a part, or sub-field, which figures as focal and..from which..the aim seems to be taken. **1964** Gould & Kolb *Dict. Soc. Sci.* 510/2 Political behaviour has come to be regarded as a sub-field, within the social sciences. **1968** N. Chomsky *Lang. & Mind* ii. 24 Linguistics..is simply the subfield of psychology that deals with these aspects of mind. **1977** *Dædalus* Fall 59 These subfields have been codified and systematized in an attempt to bring them into closer relation with theoretical frameworks. **1971** *Optometry Today* 15 Vision care needs of the aging patient have virtually produced a subspeciality within the opto-metric profession: vision care of the aging. **1980** *Jrnl. R. Soc. Med.* LXXIII. 758/1 Between four and seven reviews within six subspecialities of medicine. **1963** *Lancet* 5 Jan. 42/2 Some of the subspecialties such as skins and eyes. **1961** *Communications Assoc. Computing Machinery* IV. 438/2 A request, by a task, to call in and execute a sub-task causes..a new task (the subtask called) to be added to the task list, with the appropriate precedence and priority. **1971** *New Society* 26 Aug. 373/1 My first subtask is..to move the ramp... This sets up the subsubtask of computing the coordinates. **1982** *Sci. Amer.* Jan. 123/2 The several concurrent processes can be different subtasks of a single program.

e. *Math.* Prefixed to sbs. to denote an entity which is contained in some similar entity, in that each of its elements is also an element of the latter and that it shares the characterizing properties of the latter, as *subalgebra, -field, -formation, -graph, -manifold, -matrix, -module, -object, -ring*. Also Subgroup b, *SUBSEQUENCE[2], Subset *sb.*[2]

1933 Subalgebra [see *LATTICE *sb.* 3**]. **1979** *Proc. London Math. Soc.* XXXVIII. 315 Let *Ax* be the closed subalgebra of *L(Hx)* generated by the operators $T_1, \ldots, T_N$ and set *H*. **1940** E. T. Bell *Devel. Math.* xi. 239 The final outcome may be roughly described as an analysis of the structure of fields with respect to their possible subfields and superfields. **1971** G. Higman in Powell & Higman *Finite Simple Groups* vi. 209 *Q(a)* is the real subfield of the field of the 5-th roots of unity. **1966** tr. *Gericke's Lattice Theory* iv. 71 We shall show that the set of sub-formations of a formation..relative to a given axiom system..that satisfies a condition yet to be formulated forms a complete lattice. **1931** *Proc. Nat. Acad. Sci.* XVII. 125 A subgraph *H* of a graph *G* is a graph formed by dropping out arcs from *G*. **1979** Page & Wilson *Introd. Computational Combinatorics* iv. 76 Given a graph G = (P, L) then G' = (P', L') is a subgraph if P' is a subset of P and L' is a subset of L. **1963** H. Flanders *Differential Forms* v. 52 A manifold **M** is called a submanifold of a manifold **N** provided there is a one-to-one smooth mapping *f*: **M**→**N** which has this..property. **1970** G. K. Woodgate *Elem. Atomic Struct.* viii. 152 The matrix of *H'* which has to be diagonalized breaks up into submatrices of given *M*. **1980** A. J. Jones *Game Theory* iii. 149 There are nine 2 × 2 submatrices obtained by deleting the *i*th row and *j*th column from *A*. **1965** J. J. Rotman *Theory of Groups* iv. 68 A subset *W* of the *R*-module *V* is a submodule of *V* in case it is a subgroup of

V which is closed under scalar multiplication. **1981** *Amer. Math. Monthly* LXXXVIII. 53 Submodules of finitely generated free modules [illegible]

8. *sub-entitle* vb. (earlier U.S. example), *-functional* adj.

[illegible]

9. *sub-classify* vb. (earlier example); *sub-classification* (earlier example), *-component, -kind*; **sub-carrier** *Telecommunication*, a carrier wave used to modulate another carrier; **sub-fraction**: see also as main entry in Suppl.; **sub-level** *Physics*, each of a group of energy levels of an atom or nucleus which coincide under a coarse approximation or when some factor (as a magnetic field) is removed; **sub-line** *Genetics*, a variant arising in an inbred line and distinguished by a trait us. inherited from a genetically impure ancestor; **sub-passage** *sb. Biol.* and *Med.*, the passage of a strain of micro-organisms cultivated in one animal through another, esp. to increase the virulence; also as *v. trans.*; hence **sub-passaging** *vbl. sb.*; **subsa·tellite** *Astronautics*, a satellite of a satellite; *spec.* a small artificial satellite released from another satellite or spacecraft; **sub-u·nderwriter** *Econ.*, one who underwrites part of a liability (esp. a share issue) underwritten by another; so **sub-u·nderwrite** *v. trans.*, **sub-u·nderwriting** *vbl. sb.*

[illegible]

III. 12. a. *sub-Himalayan* (earlier example). **1847** [see *SHERPA 1].

b. *sub-equatorial* (also *fig.*).

1909 WEBSTER, Subequatorial. **1935** H. H. BASHFORD *Lodgings for Twelve* 108 Apart from the excitements incident to the relief of Ladysmith and Mafeking, the Boer War—at any rate to the average undergraduate—was a sub-equatorial and not very important affair. **1977** *Sci. Amer.* Apr. 106/2 This is the earliest-known evidence of metallurgy in the entire subequatorial region.

14. a. Also with the sense 'numerically less than', and in derived advbs. *sub-literary, -molecular, -morphemic, -optimal* (hence *-optimally*), *-optimum, -phonemic* (hence *-phonemically*); **sub-bitu·minous** *a. Geol.*, (of coal) of inferior quality to bituminous; intermediate in rank between bituminous coal and lignite; **subce·llular** *a. Biol.*, smaller than a cell; occurring inside a cell; **sub-compa·ct** *a. U.S.*, designating a car which is smaller than a compact one (see *COMPACT *ppl. a.*[1] II. 1 b); also *absol.* as *sb.*; **subfree·zing** *a.*, designating or characterized by a temperature lower than the freezing-point of water; **sublu·minal** *a.* [L. *lūmen, lūmin-* light], having or being a speed less than that of light.

1908 *Econ. Geol.* III. 136 The term 'sub-bituminous' was recommended and formally adopted by the [U.S. Geological] survey for all official publications. **1949** F. J. PETTIJOHN *Sedimentary Rocks* xii. 366 Subbituminous, semibituminous, and semianthracite coals are transitional coal types. **1979** B. L. C. JOHNSON *Pakistan* xi. 161/1 The estimates of reserves of Lower Tertiary lignitic to sub-bituminous coal range between 449 and 478 million tonnes. **1953** *New Biol.* XV. 120 There are too many examples in which patterns arise..within single cells (and thus demand a theory dealing in sub-cellular units). **1964** G. H. HAGGIS et al. *Introd. Molecular Biol.* ii. 20 Some enzymes are localized in certain sub-cellular structures. **1978** *Sci. Amer.* Dec. 68/2 Myoglobin combines with the oxygen released by red cells, stores it and transports it to the subcellular organelles called mitochondria. **1967** *Wall St. Jrnl.* 24 Feb. 1/1 AMC [*sc.* the American Motors Corporation] also is thinking of building a 'subcompact' car that would compete directly in size and price with Volkswagen. **1971** *Flying* Apr. 68/2 (Advt.), A different-looking subcompact with the spirit of a sporty car. **1980** *Times* 12 Dec. 24/3 Chrysler extended the close-down of its Belvidere, Illinois, assembly plant, which makes subcompact cars. **1958** *N.Y. Times* 15 Dec. 2/6 The Weather Bureau warned that subfreezing temperatures would continue today. **1979** P. THEROUX *Old Patagonian Express* iii. 51 Two feet of snow in Boston. Chaos and death. Power cuts in sub-freezing weather. **1952** *New World Writing* Apr. 234 Even *aficionados* of murder fiction will concede..that except in the hands of a few writers it has been a sub-literary product—characters unreal, dialogue artificial, plots highly improbable. **1980** *Jrnl. R. Soc. Arts* Apr. 302/1 He illuminates..this curious sub-literary genre in such a way that even serious students gain a profitable exposure to materials not often considered in length. **1969** *Physics Today* May 45/3 Particles traveling at subluminal and those traveling at luminal velocities are two entirely distinct kinds of objects. **1980** *Sci. Amer.* Aug. 76/2 The expansion that appears from the earth to be superluminal would be relativistic but still subluminal when measured by the slower clocks of the source itself. **1935** *Discovery* Dec. 353/1 Raindrops may form on sub-molecular electrically-charged units, or ions. **1964** G. H. HAGGIS et al. *Introd. Molecular Biol.* 338 Study of the mechanism of enzyme action is in a sense submolecular biology. **1947** C. F. HOCKETT in *Language* XXIII. 321 A scholar deciphering a dead language written in a non- [illegible]

b. Similarly combined with sbs. (forming words used chiefly *attrib.*), as *sub-cabinet* (chiefly *U.S.*), *-microgram, -proletariat, -threshold, -zero*; **submi·llimetre** *a.*, less than a millimetre in size or length; pertaining to or employing electromagnetic waves of such a length; also **su:bmillime·tric** *a.*

1956 R. J. DONOVAN *Eisenhower* v. 66 Almost as soon as the Cabinet rises each week, Rabb meets with a group of sub-Cabinet officers. **1974** P. GORE-BOOTH *With Great Truth & Respect* 389 The process of an important and difficult decision is remarkable... It passes through a hierarchy or a sort of sub-cabinet or both. **1981** *Economist* 24 Jan. 24/3 Lower appointments to sub-cabinet jobs are still being made and will be for some weeks. **1965** PHILLIPS & WILLIAMS *Inorg. Chem.* I. xvi. 576 Work on natural polonium has been limited to the sub-microgram scale. **1976** *Nature* 10 June 454/1 Gibberellins are amongst the most potent of the naturally occurring plant growth regulators and exert maximal activity in most tissues when present in sub-microgram quantities. **1955** *Jrnl. Appl. Physics* XXVI. 1384/1 Submillimeter radiation was produced when a pulsed, bunched high-energy electron beam was passed through a simple rectangular wave guide. **1973** *Physics Bull.* May 305/3 Submillimetre spectroscopy, or far infrared spectroscopy as it is more frequently termed, is a field in which there has been much activity during the last 15 years. **1976** *Sci. Amer.* June 127/1 All one had to do was to drop his view to the submillimeter level, where little creatures abound that are still new to science. **1975** *Nature* 6 Mar. 39/2 The mystery of the submillimetric limb brightening [of the sun]. **1958** *Listener* 30 Jan. 186/1 Today the population of the delta belongs to the class which, I believe, sociologists call the rural sub-proletariat. They are landless or almost landless peasants who seek casual employment in agriculture or anything that comes along. **1974** M. B. BROWN *Economics of Imperialism* iv. 87 Nor can we overlook..the role of migrant labour as a 'sub-proletariat' in Europe today. **1979** *Dædalus* Spring 105 The institutionalization of a subproletariat, and the creation of ethnic ghettos in the large urban areas are..examples of the changes taking place. **1937** BEST & TAYLOR *Physiol. Basis Med. Practice* lxiii. 1225 If a second stimulus also of subthreshold strength..be sent into the nerve an impulse is set up. **1976** *Ann. Rev. Microbiol.* XXX. 234 Negative chemotaxis appears to be largely an all-or-none response to a threshold concentration, but weaker effects caused by prolonged exposure to subthreshold concentrations probably also occur. **1942** O. NASH *Face is Familiar* 137 And nobody is prompter In the face of hell, high water, and sub-zero thermomter. **1980** R. MCCRUM *In Secret Place* xii. 113 He's having a sub-zero feud with Hayter.

c. Compounded with a further prefix, as **sub-mi·cro-** *Chem.*, involving amounts less than those typical of microanalysis; also used as an independent word.

1945 *Jrnl. Biol. Chem.* CLXI. 589 In order to estimate the P content of these solutions, a submicroprocedure..having a range of 0·2 to 3 γ of P was employed. **1964** N. G. CLARK *Mod. Org. Chem.* xxiv. 496 The isolation of minute quantities of material from biological sources has necessitated even greater refinements, so that submicro techniques (requiring 30–50 μg) have been developed during recent years. **1974** [see *MICRO- 8 b].

18. *sub-neolithic* (also *fig.*), *-Roman* adjs.

1905 A. J. EVANS in *Ann. Brit. School at Athens* X. 22 This stratum, to which the name 'Early Minoan I.' may be conveniently applied, shows naturally a greater survival of Neolithic elements... In its general complexion indeed it may be described as 'Sub-Neolithic'. **1956** E. E. CUMMINGS *Let.* 11 Mar. (1969) 248 Good Freudians were quick to suggest that my superego suffers from subneolithic trends. **1962** H. R. LOYN *Anglo-Saxon England* i. 39 Wheel-made pottery of sub-Roman character. **1977** *History* LXII. 175 We cannot, however, expect that these works will ever provide information about the sub-Roman centuries.

IV. 19. *sub-historical, -literate, -mature, -moral, -solid* adjs.; **sub-econo·mic** *a.*, not justifiable on purely economic grounds; **subinhi·bitory** *a.*, (of a dose of a drug, chemical, etc.) enough to hinder but not prevent microbial growth; **sublu·minous** *a.*, dim; *spec.* in *Astr.*, of less luminosity than the normal; **subse·xual** *a. Genetics*, characterized by or being a form of parthenogenetic reproduction in which the first division of meiosis occurs, with crossing-over, but not the second (reduction) division; **subso·cial** *a. Biol.*, applied to species of spiders or insects that live gregariously but without a fixed social organization; **subvo·cal** *a.*, designating an unarticulated level of speech comparable to thought; hence **subvo·cally** *adv.*

1948 *Rep. Native Laws Commission 1946–48* (Dept. Native Affairs, S. Afr.) 4/2 Government assistance..in respect of sub-economic schemes has all along been linked with the condition that the municipality should bear a share of the loss. **1971** *Leader* (Durban) 7 May 1/5 The Verulam Town Board has announced its intention to erect..100 sub-economic houses. **1980** *Sci. Amer.* Jan. 50/2 He created a two-way grid of categories based on the degree of geological knowledge (known deposits, inferred deposits and probable deposits) and on current economics (economic, subeconomic and uneconomic). **1940** K. MANNHEIM *Ideology & Utopia* 128 Besides this subhistorical biological element a spiritual, transcendental element is also to be found in this sphere. **1973** R. J. W. EVANS *Rudolf II* ii. 45 Such a view..has survived in subhistorical writing and belles-lettres. **1956** M. HYNES *Med. Bacteriol.* (ed. 6) ix. 123 Serial culture of susceptible organisms in the presence of sub-inhibitory concentrations of an antibiotic results in the emergence of bacteria that can flourish in the presence of enormous concentrations of the antibiotic. **1976** *Ann. Rev. Microbiol.* XXX. 64 Some strains of *M. osloensis* produce such a preponderance of coccal cells that their true nature can only be ascertained in films from media with subinhibitory concentrations of penicillin. **1958** J. BERRY in J. A. Fishman *Readings Sociol. of Lang.* (1968) 743 In subliterate societies (in most of Tropical Africa for example) where book-production is at the best financially hazardous, the need for exotic type can have a deterrent effect on book production. **1973** R. A. CRAMPSEY *Puerto Rico* 13 In 1940 the bulk of the people were subliterate or illiterate. **1864** Subluminous [in Dict.]. **1959** *Encounter* July 53/2 The

photography is that chocolate-marshmallow kind of subluminous chiaroscuro. **1969** O. J. EGGEN in S. S. Kumar *Low-Luminosity Stars* I. i. 22 Photometric parallaxes have been derived for the 27 stars in Table III which are very probably subluminous. **1976** *Progress in Sci. Culture* (E. Majorana Centre) Spring 52 Extragalactic sources can be classified in order of luminosity as follows (1) subluminous galaxies, such as M 31, (2) normal galaxies, [etc.]. **1899** Submature [see *PENEPLANATION]. **1922** C. A. COTTON *Geomorphol. N.Z.* I. xxviii. 415 The coast has passed through the stage of youth and has become sub-mature. **1951** *Jrnl. Sedimentary Petrol.* XXI. 128 Definition of the four stages of textural maturity... I. Immature stage... II. Submature stage. Sediment contains very little or no clay, but the non-clay portion..is still itself poorly sorted... III. Mature stage... IV. Supermature stage. **1965** G. J. WILLIAMS *Econ. Geol. N.Z.* xiii. 192/1 A submature valley formed on this surface, later to be dammed by a basalt flow. **1946** *Mind* LV. 115 A will-less saint would be a sub-moral being, a fine creature perhaps, but not a responsible moral agent. **1937** C. D. DARLINGTON in *Nature* 30 Oct. 761/2 Other mechanisms occur in the dog roses and with certain kinds of parthenogenesis whereby, as in *Œnothera*, a large part of the genes are prevented from recombining. With such systems stability has been achieved at the expense of variability, and we have arrived at what we may call a sub-sexual method of reproduction. **1947** —— & MATHER *Elements of Genetics* xii. 266 As compared with sexual species variation is much reduced but it still occurs. The new apomictic species is thus often subsexual. **1928** W. M. WHEELER *Social Insects* i. 13 The insects included in categories (1) to (5) may be designated as 'infrasocial'; those of (6), which are more interesting for our purposes may be called 'quasisocial' or 'subsocial'. **1958** *Science* 2 May 1046/1 Social organization in the main groups of social bees..did not arise..through subsocial family groups. **1976** *Sci. Amer.* Mar. 101/3 Michener's second evolutionary route he calls subsocial. On this route only one level of behavior precedes eusociality; it is characterized by solitary rather than communal nest building. The solitary female [spider] remains at the nest, however, and cares for her young. **1922** JOYCE *Ulysses* 674 The decocted beverages, allowing for subsolid residual sediment of a mechanical mixture, water plus sugar plus cream plus cocoa, having been consumed. **1934** M. TEN HOOR in *Jrnl. Philos.* XXXI. 534 The responsibility for descriptive analysis and psychological definition..has been accepted..by..the behaviorists... This theory..in its extreme form..contends that thought is nothing but subvocal speech. **1980** A. KENNY *Aquinas* iii. 78 Aquinas has a clear grasp of the relationship between the intellect and the imagination when thought takes place in mental images or in subvocal speech. **1961** E. J. FURLONG *Imagination* vii. 77 The words 'the Chapel' are subvocally present to me along with the visual object.

b. In derived advbs., as *submaturely*.

1900 *Proc. Boston Soc. Nat. Hist.* XXIX. 309 In central France..the initial form was an uplifted and submaturely dissected peneplain, in which valleys with incised meanders have been..developed. **1913** *Bull. Geol. Soc. Amer.* XXIV. 201 The submaturely dissected scarp.

19*. a. With adjs. derived from the names of persons, used to designate something in the manner of but inferior to their work, characteristic style, etc. (Chiefly in nonce-formations.)

1934 E. SITWELL *Aspects Mod. Poetry* i. 18 Mr. Housman was followed by a school of poets, rather loosely held together by their sub-Wordsworthian ideals. **1959** *Listener* 5 Feb. 258/2 A laboured sub-Wodehousian straining after slapstick instead of farce. **1962** *John o' London's* 10 May 459/2 The opening has a sub-Chaplinesque quality. **1967** J. PHILIP et al. *Best of Granta* I. 16 Following the editorial come five sub-Miltonic stanzas. **1977** P. JOHNSON *Enemies of Society* xi. 154 One prominent sub-Marxist 'scientist' who constantly uses the crisis-mechanism, to justify, among other things, the use of positive censorship, when possible, is Herbert Marcuse.

b. Hence, prefixed simply to the names of persons.

1963 *Times Lit. Suppl.* 5 Apr. 235/2 Here is the eternal sex-life of the American college girl told in the eternal sub-Salinger..style. **1968** J. BINGHAM *I love, I Kill* xi. 132 It was called *Deeper in the South*..kind of sub-Tennessee Williams. **1977** *Listener* 28 July 122/3 A pregnant older lady who paints sub-Ernst surrealities.

20. b. *subtranslucent* adj.

1868 J. D. DANA *Syst. Mineral.* (ed. 5) v. 194 Plasma... Rather bright-green to leek-green, also sometimes nearly emerald-green, and subtranslucent or feebly translucent. **1955** BROWN & DEY *India's Mineral Wealth* (ed. 3) 623 The bloodstones are subtranslucent, dark green chalcedony speckled with red; the moss agates..perfectly translucent stones.

c. su:bacroce·ntric *a. Cytology* = *subtelocentric* adj. below; **subhe·dral** *a.*, applied to crystals having partially developed faces, or incompletely bounded planes; **su:bmetace·ntric** *a. Cytology*, applied to a chromosome with the centromere almost in the middle, so that the two chromosome arms differ slightly in length; also *ellipt.* as *sb.*; **su:bteloce·ntric** *a. Cytology*, applied to a chromosome with a centromere near one end, but not as near as in an acrocentric chromosome; also *ellipt.* as *sb.*

1906 W. CROSS et al. in *Jrnl. Geol.* XIV. 698 *Subhedral*, partly bounded by crystal faces, hypantomorphic, hypidiomorphic. **1961** M. J. D. WHITE *Chromosomes* (ed. 5) ii. 23 Intermediate types exist..so that we may describe particular chromosomes 'subacrocentric' or 'metacentric'. **1963** *Austral. Jrnl. Zool.* XI. 8 Four pairs of subacrocentrics, whose short arms are in most cases large enough to be distinctly visible in the preparations. **1963** *Jrnl. Nat. Cancer Inst.* XXXI. 642 The diploid complement of 22 pairs of chromosomes consists of 5 median metacentrics, 3 subtelocentrics, [etc.]. **1964** *Ibid.* XXXII. 858 This system was arrived at by the arrangement of metacentric chromosomes in descending size order followed by submetacentric and subtelocentric chromosomes arranged in a similar manner. **1964** *Hereditas* LII. 211 Chromosomes..with arm ratios of 3·0 or higher are classed as ST ('subtelocentric'). **1973** *Nature* 5 Oct. 262/1 The diploid karyotype of *U. limi* consisted of twenty-two chromosomes, eighteen metacentrics and four submetacentrics. **1975** G. ANDERSON *Coring* ii. 35 (*caption*) Porous network of medium-grained euhedral to subhedral dolomite rhombs. **1976** *Jrnl. Cellular Physiol.* LXXXVII. 104 The four groups (I, metacentric; II, submetacentric; III, subacrocentric; and IV, acrocentric) were defined for this purpose by the ratio, short/long arm length. **1980** *Canad. Jrnl. Genetics & Cytol.* XXII. 421 There are three SAT-chromosomes (arm ratio (r) = 1·74–1·92), four metacentric chromosomes (r = 1·07–1·14), seven submetacentric chromosomes (r = 1·22–1·68) and seven subtelocentric chromosomes (r = 1·75–2·42).

e. *subcircular* (later examples), *-rectangular* adjs.

1940 *Antiquity* XIV. 16 The hopelessly decayed traces of a large wooden object, apparently subrectangular in plan. **1957** G. E. HUTCHINSON *Treat. Limnol.* I. ii. 171 Subcircular, a less perfect approach to circular form. **1970** R. J. SMALL *Study of Landforms* iv. 121 King..has postulated that in many parts of Africa stream incision along joints has given rise to a subrectangular drainage pattern. **1979** *Geogr. Mag.* July 668/3 Sub-circular pans on the Essex marshes.

g. *subcurative, -fertile*; **subcli·nical** *a.*, not giving rise to any observable symptoms; **subpa·tent** *a.*, (of a parasite or parasitic infection) present but not detectable; of or pertaining to such an infection; **subte·rtian** *a. Med.*, applied to a severe form of malaria caused by the sporozoan *Plasmodium falciparum* and to the sporozoan itself; also *ellipt.* as *sb.*, subtertian malaria.

1919 R. ROSS *Suggestions for Care of Malaria Patients* 7 A severe type in which the paroxysms..are often found to recur every day and at irregular times..is caused by the malignant tertian parasite, sometimes called the 'subtertian' parasite (*Plasmodium falciparum*). **1926** *Q. Rev. Biol.* I. 399/2 In many infections the patent period is followed by a Subpatent Period of indefinite length. **1930** M. F. BOYD *Introd. Malariol.* ii. 32 It would appear that resistance may be established earliest to subtertian, and more slowly in tertian and quartan. **1946** *Nature* 17 Aug. 243/2 With the addition of subcurative doses of 'Mapharsen', the amount of penicillin required to cure rabbit syphilis is reduced to a fraction of that required when penicillin is used alone. *Ibid.* 5 Oct. 487/2 L.V. is responsible for a certain number of cases of epididymal inflammations, many of them of a subclinical type. **1947** *Ann. Rev. Microbiol.* I. 49 Subpatent infections persisted in some individual ducks for as long as eight months after they had been inoculated with sporozoites. **1954** MARTIN & HYNES *Clin. Endocrinol.* (ed. 2) viii. 187 It is difficult to assess the efficiency of therapy in a subfertile male. **1954** *Brit. Med. Jrnl.* 6 Feb. 293/2 It has been found that in indigenous East Africans the sickle-cell trait affords a considerable degree of protection against subtertian malaria. **1971** *Observer* (Colour Suppl.) 31 Oct. 10/1 Like many men her husband seemed to be subfertile but was by no means infertile. **1974** J. R. BAKER in *Trypanosomiasis & Leishmaniasis* (Ciba Foundation Symposium No. 20) 32 Parasites may often be subpatent, that is, too scanty to be detected by microscopic examination. **1978** *Jrnl. R. Soc. Med.* LXXI. 507 The spectrum of illness is wide, from severe and prostrating to mild and, probably, subclinical. **1979** *Tropenmedizin u. Parasitologie* XXX. 239/1 Infected cows treated with subcurative doses of trypanocidal drugs. **1979** E. NNOCHIRI *Textbk. Imported Diseases* iv. 59 *P[lasmodium] falciparum* infections, in contrast, are insidious in onset with irregular fever which subsequently becomes subtertian (i.e. between 36 and 48 hours) in periodicity.

h. *subclinically, -terminally*.

1954 *Amer. Jrnl. Public Health* XLIV. 575/2 It was ascertained that previous infection of a child with Type 2 or Type 3 virus failed to prevent his becoming infected subclinically with Type 1 virus. **1963** R. P. DALES *Annelids* ix. 182 The rectum opening subterminally at a dorsal anus.

22. *subfertility*; **sub-song,** the part of a bird's song that is softer and less well defined than its characteristic series of notes and is believed to have no territorial significance.

1948 MARTIN & HYNES *Clin. Endocrinol.* viii. 157 Several examinations are advised before diagnosing subfertility and they should be performed as soon as possible after ejaculation. **1962** H. LOURIE *Question of Abortion* xxiii. 201 The sub-fertility clinics, the clinics dealing with sterile marriages. **1971** *Daily Colonist* (Victoria, B.C.) 21 Aug. 22/2 Male subfertility is the main factor in 45 per cent of childless marriages. **1925** E. M. NICHOLSON in *Field* Dec. 31/3 Even the chaffinch..has a very low rambling, warbling sub-song with no fire or decision about it. **1948** *Brit. Birds* XLI. 51 The sub-song was occasionally replaced by the typical loud burst of song characteristic of this species [*sc.* the redstart]. **1979** *New Scientist* 17 May 537/1 Subsong is a rather soft and rambling type of singing..in which the bird seems to try out various sounds for itself.

23. *subsulphide*.

1868 Subsulphide [see *PLUMBOUS *a.* 2]. **1976** *Nature* 15 Jan. 109/3 Vanadium subsulphide β-V_3S is known to have a tetragonal unit cell.

‖ **suba** (ʃɒ·bā). [Hungarian.] A type of long sheepskin cloak worn by Hungarian shepherds.

1925 G. A. BIRMINGHAM *Wayfarer in Hungary* xiv. 130 The *suba* is a long cloak of sheepskin reaching to the ankles. It has no sleeves... Only when the weather is bad is the *suba* fastened in front. **1939** *Times Lit. Suppl.* 4 Nov. 636/4 The Hungarian shepherd's 'suba', a magnificently voluminous sheepskin cloak. **1971** *Nat. Geogr. Mag.* Apr. 481 (*caption*) Shepherds wear huge suba to escape autumn's chill. **1979** J. SNOWDEN *Folk Dress of Europe* 25 In the last century, the *suba*, a sheepskin cloak, was developed from primitive peasant wear to a garment of considerable elaboration and costliness... Circular, or nearly so,..the *suba* was cut with radial sections joined to a shoulder yoke. The long seams..were decorated..on the skin side.

subacid, *sb.* **1.** (Earlier example.)

1785 A. SEWARD *Let.* 7 June (1811) I. 75 That tetchy unprovoked spleen..clouding and staining the lustre of fine talents, and many excellent qualities... Let us all take warning, and correct our acids and sub-acids of every sort.

subacute. Add: **c.** *subacute sclerosing panencephalitis*, a frequently fatal degenerative disease of the central nervous system, caused by reactivation of a measles virus some years after the original infection.

1950 J. G. GREENFIELD in *Brain* LXXIII. 150 The name *subacute sclerosing encephalitis* therefore appears fully justified. Dr. van Bogaert's term 'leuco-encephalitis' emphasizes the characteristic damage to the white matter, but leaves out of account the cortical changes which are also important. Perhaps the term 'Panencephalitis' already adopted by Pette (1942) for forms which attack both grey and white matter could be usefully employed here, i.e. 'Subacute sclerosing panencephalitis'. **1967** *Brit. Med. Jrnl.* 5 Aug. 352/2 Measles complement-fixing and haemagglutination-inhibiting antibodies have been found in the serum of 22 patients with subacute sclerosing panencephalitis in significantly higher titre than in controls.

subadole·scent, *sb.* and *a.* [SUB- 19.] **A.** *sb.* = *PREADOLESCENT *sb.* **B.** *adj.* = *PREADOLESCENT *a.* a.

1957 R. A. HEINLEIN *Door into Summer* ii. 32 Did you ever try to discuss with a subadolescent something the child does not want to talk about? **1977** W. M. SPACKMAN *Armful of Warm Girl* 33 An enlarged snapshot of two baby boys and two sub-adolescent girls.

subadult (sɒbæ·dɒlt, sɒbădɒ·lt), *a.* and *sb.* [SUB- 19.] **A.** *adj.* Not fully adult. **B.** *sb.* A subadult individual.

Applied chiefly to animals (cf. *preadult* s.v. *PRE- B. 1).

1903 *Nature* 3 Dec. 112/1 A subadult Australian barn-owl in which large bunches of the nestling down are retained on the legs. **1934** WEBSTER, Subadult *n.* **1946** *Nature* 28 Dec. 927/1 The author distinguishes between juvenile, young, sub-adult and adult, in that order of ascending age [of mammals]. **1962** B. HARRISSON *Orang-Utan* iii. 108 Sub-adults, who have left their mothers, cuddle in pairs to get additional warmth. **1975** *Sci. Amer.* May 56/3 Lions are termed cubs until they are two years old and subadults between the ages of two and four. **1976** *Ibid.* Apr. 118/3 Subadult crocodiles often form a semicircle where a channel enters a pan, facing the inrushing water and snapping up the fish that emerge from the river.

‖ **subak** (sū·bāk). Also **soebak.** [Balinese.] A Balinese rice-growers' co-operative, organized to ensure equitable distribution of water for irrigation.

1921 *Man. Netherlands India* (Admiralty) xi. 373 In the..construction of aqueducts..the Balinese excels. Moreover, the natives have their own irrigation associations or *subaks*. **1926** H. NORDEN *Byways of Tropic Seas* ix. 159 Balinese economic life is governed by *Soebak*, a communistic organization which came into being for the joint irrigation of rice fields. **1937** M. COVARRUBIAS *Island of Bali* (1972) iii. 67 Disputes concerning ricefields or irrigation water are settled by the council of a special agricultural society, the *subak*. **1957** K. G. WITTFOGEL *Oriental Despotism* 25 In Bali the peasants are obliged to render labor service for the hydraulic regional unit, the *subak*, to which they belong. **1972** *Times* 11 Nov. 13/4 The brilliant green *sawahs*..are tended by village co-operatives, the *subaks*.

subaltern, *sb.* **3.** (Earlier examples.)

1685 tr. *Arnauld & Nicole's Logic* II. ii. 169 If they differ in Quantity only, and agree in Quality, as A.I. and E.O. they are call'd Subalterns. **1816** *Elements of Logic* II. iii. 47 Propositions which differ only in quantity are called subalterns.

sub-aqua (sɒ:bæ·kwă), *a.* [Adj. use of L. phr. *sub aqua* under water: cf. SUB- I.] Of or pertaining to underwater swimming or diving performed (esp. as a sport) with the assistance of an aqualung. Also *ellipt.* as *sb.*, sub-aqua swimming.

1955 *Neptune* Aug. 31/1, I was..surprised to learn that..few members of the British Sub-Aqua Club were interested in..spearfishing. **1959** *Listener* 8 Jan. 67/3 The 'sub-aqua-jet' which provides motive power for an under-water swimmer. **1962** *Underwater Swimming* ('Know the Game' Ser.) 3/1 The formation of sub-aqua clubs throughout the country. **1976** *Milton Keynes Express* 1 June 11/2 Saturday's attractions include demonstrations of..swimming, life-saving, sub aqua and

weight-lifting. **1978** *Times* 14 July 26/3 Sub-aqua diving is one of the country's leading growth sports.

subaqueous, *a.* Add: **4.** In *fig.* use, lacking real substance or strength; wishy-washy.

1960 C. DAY LEWIS *Buried Day* i. 15 The whole picture, clear yet elusive, is bathed in a brooding, sub-aqueous light. **1970** H. BRAUN *Parish Churches* xix. 228 During the last fifty years an inevitable reaction has introduced pallid sub-aqueous treatments [of stained-glass windows], less obstructive to light but lacking all the ancient warmth and liveliness. **1977** *Listener* 28 July 122/3 A sort of subaqueous, loopy, transcendental speculation about female identity.

subarctic, *a.* Add: **2.** [SUB- 18.] Also **Subarctic.** Applied to a European climatic period that followed the Arctic and preceded the Preboreal.

1876 [see *ARCTIC *a.* 3]. **1935** *Discovery* July 198/2 Relics from Arctic and Subarctic times during and soon after the last glaciation are still to be found in Scotland. **1973** P. A. COLINVAUX *Introd. Ecol.* vii. 93 There is peat between the Dryas-bearing bottom mud and the first line of stumps, a gradation probably, but one that could be used as a stratigraphic unit. It represented the subarctic period.

Subarian (siŭbā̆rĭ·ăn), *a.* and *sb.* Also **Subaraean.** [f. Akkadian *Subar(tu* 'Assyria' + -IAN.] **A.** *adj.* Of or pertaining to the Subarian people (see below) or their language. **B.** *sb.* **a.** (A member of) an ancient people of northern Mesopotamia in the 3rd and 2nd millennia B.C., sometimes identified with the Hurrians. **b.** The language (written in cuneiform) of this people. Cf. *HURRIAN *sb.* and *a.*, *MITANNI.

1923 C. J. GADD *Fall of Nineveh* 20 In his own building records Nabopolassar says, 'I slew the Subaraean, and turned the enemy's land into mounds and ruins.' *Ibid.*, Throughout these references, it is most probable that the enemy is the same, though described indifferently as Subaraean and Assyrian. **1926** —— in *Revue d'Assyriologie* XXIII. 63 The suffix *-ia* which is in constant use to form shortened names has long been established as a characteristic of Subaraean. **1939** [see *PAPUAN *sb.* 2]. **1964** G. ROUX *Anc. Iraq* xi. 166 Babylon was attacked by a coalition of Elamites, Guti, 'Subarians' (Assyrians) and people from Eshnunna. **1974** *Encycl. Brit. Micropædia* V. 222/3 The Hurrian language, once improperly called Mitannian or Subarian, exists chiefly in four varieties of cuneiform.

sub-asse·mbly. [SUB- 7 a.] A unit assembled separately but designed to be incorporated, with other such units, into a larger manufactured product; also, the production of sub-assemblies. Cf. *ASSEMBLY 1 c.

1919 *Brit. Manufacturer* Nov. 29/1 Parallel with the Finished Part Stores is the sub-assembly stores. **1924** W. J. HISCOX *Factory Lay-Out, Planning & Progress* ii. 29 The term 'sub-assembling' is used when two or more component parts are assembled together to form one part... Sometimes the 'sub-assembly' merely covers the fitting of a pin to a certain part. **1941** *Sun* (Baltimore) 21 Feb. 30/5 The Omaha plant will be used to assemble bombers from sub-assemblies and parts supplied by the automotive and other non-aeronautical enterprises. **1952** F. ALLEN *Big Change* II. vii. 111 The Ford assembly line, with its subassemblies, was unique. **1967** *Electronics* 6 Mar. 171/1 (*caption*) Three subassemblies constitute IBM's new mass production computer for varied aerospace applications. **1973** J. G. TWEEDDALE *Materials Technol.* II. iv. 73 Such mechanically-wrought products can form the primary material for further mechanical manipulation..into fairly complex sub-assembly components.

Hence (as a back-formation) **sub-asse·mble** *v. trans.*; **sub-asse·mbling** *vbl. sb.*

1924 Sub-assembling [see *sub-assembly* above]. **1940** *Sun* (Baltimore) 28 Nov. 1/2 Parts are manufactured and sometimes 'sub-assembled' at plants where idle machinery and man power are available.

subatla·ntic, *a.* Also **Subatlantic, sub-Atlantic. 1.** (In Dict. s.v. SUB- 1 e.)

2. [SUB- 18.] Pertaining to or designating a European climatic period that followed the Sub-boreal and which is still current. Also *absol.*

1876 [see *ARCTIC *a.* 3]. **1935** *Discovery* July 198/2 A fair number of these species has even survived the last climatic phase, the humid Subatlantic time (after 500 B.C.). **1957** E. E. EVANS *Irish Folk Ways* xiv. 186 The onset of the sub-Atlantic climatic deterioration in the last millennium B.C. seems to have accelerated the peat-forming processes. **1975** J. G. EVANS *Environment Early Man Brit. Isles* vi. 149 The growing of cereals [in Yorkshire]..would have been difficult..at any time since the onset of the Sub-atlantic.

subau·dible, *a.* [SUB- 14, 19.] **1.** Not loud enough to be audible. Also *fig.*

1839 [in Dict. s.v. SUB- 19]. **1928** E. BLUNDEN *Undertones of War* vii. 109 A good joke: but with this sub-audible meaning.

2. Of a frequency: lower than the lowest audible frequency. Of a sound: too low-pitched to be audible.

1922 *Proc. IRE* X. 253 It may be at sub-audible, or super-audible frequencies. **1978** *Gramophone* June 128/2 Superimposed on this are subaudible tones successively at 4, 5, 6, 8, 10 and 12 Hertz.

Sub-bo·real, *a.* Also **sub-Boreal.** [SUB- 18.] Pertaining to or designating a European climatic period that followed the Atlantic and preceded the Subatlantic. Also *absol.*

1876 [see *ARCTIC *a.* 3]. **1935** *Discovery* July 198/2 The Atlantic period was followed by another drier, continental phase, the Subboreal, lasting from 2500 till 500 B.C. **1963** H. N. SAVORY in Foster & Alcock *Culture & Environment* iii. 27 Primitive man, at least under sub-Boreal climatic conditions, had a choice of two main routes which gave him relatively easy access to the Glamorgan uplands from the coastal area of settlement. **1975** J. G. EVANS *Environment Early Man Brit. Isles* iv. 84 The sequence is overlain by peat of Atlantic and Sub-boreal age.

‖ **subbotnik** (subɒ·tnik). Pl. **-niki,** (anglicized) **-niks.** [a. Russ. *subbótnik*, f. *subbóta* Saturday: cf. SABBATH.] In the Soviet Union, the practice or an act of working voluntarily on a Saturday, for the benefit of the collective; = *SATURDAYING *vbl. sb.*

The practice originated with workers on the Moscow-Kazan railway in Moscow on 10 May 1919. The meaning given in quot. 1920 is imprecise.

1920 *19th Cent.* Sept. 399 This mutilation was due to an accident which had happened to him while he was a *subbotnik*. *Subbotniki*..are workmen who work on Saturday (*Subbota*) for the benefit of the Government: there is quite a large *subbotnik* movement in Russia. **1921** L. TROTSKY *Defence of Terrorism* viii. 136 The flourishing, unprecedented in the history of humanity, of labor voluntarism in the form of *subbotniks* (Communist Saturdays). **1959** C. LANDAUER *Europ. Socialism* I. xxvii. 772 The response to the call for *subbotnik* work left much to be desired. **1975** T. P. WHITNEY tr. *Solzhenitsyn's Gulag Archipelago* II. III. i. 14 Soon after that there began the Communist 'subbotniki'—'voluntary Saturdays'. **1979** *Nature* 16 Aug. 532/3 The Vietnamese economy is in such an urgent state that 75% of the proceeds of this year's *Subbotnik*, the Saturday in April when Soviet citizens contribute a day's work for the good of the economy, are to be devoted to Vietnam.

sub-bo·ttom, *a.* [SUB- 1 e.] Of or pertaining to what is underneath the sea-bed.

1949 *Trans. Amer. Geophysical Union* XXX. 7 The most clear-cut sub-bottom echoes are found in the region which shows least topographic relief. **1968** [see *PROFILING *vbl. sb.* 3]. **1975** *Petroleum Rev.* XXIX. 103/1 Sub-bottom profiler.

sub-ca·libre, *a.* Also (*U.S.*) **sub-caliber.** [SUB- 5 c.] Of a projectile: smaller in calibre than the gun from which it is fired, and discharged from a secondary tube set inside the main barrel. Also, of, pertaining to, or employed in the firing of sub-calibre projectiles.

1876 E. H. KNIGHT *Amer. Mech. Dict.* III. 2434/1 *Subcaliber projectile*,..a projectile for cannon or small-arms, of smaller diameter than the bore of the gun from which it is fired. **1909** *Teachers' Assembly Herald* 13 Apr. 19/1 Arms. Double-barreled shot-gun, pocket rifles, sub-calibre or auxiliary barrels. **1917** W. S. CHURCHILL in M. Gilbert *Winston S. Churchill* (1977) IV. Compan. I. 131 This after all is only applying the sub-calibre principle to actual service. **1946** T. C. OHART *Elements Ammunition* vii. 179 The subcaliber gun is mounted on a large gun tube for practical purposes. **1967** F. W. HACKLEY et al. *Hist. Mod. U.S. Mil. Small Arms Ammunition* I. II. vii. 93 *Subcaliber cartridges*. These..were at first called 'ball cartridges for artillery drill cartridges'. They were designed to be fired from subcaliber barrels mounted within..the main artillery gun tube. **1973** J. QUICK *Dict. Weapons & Mil. Terms* 426 Subcaliber ammunition is adapted for firing in weapons of larger caliber by subcaliber tubes.

sub-ca·tegory. [SUB- *7 e, 9.] A subsidiary category; a subsection of a category; *spec.* in *Maths.* and *Linguistics.*

1909 WEBSTER, Sub-category *sb.* **1931** A. POPE *Introd. Lang. Drawing & Painting* II. iv. 63 It is well to stop in the joyous enthusiasm of thinking out new categories and subcategories. **1949** KOESTLER *Insight & Outlook* iv. 37 First, bisociation is not the same thing as ambiguity; ambiguity is merely a subcategory of it. **1956** R. REDFIELD *Peasant Soc. & Culture* 24 Typologies of Latin-American cultures or of the peasant subcategory of such cultures. **1972** A. G. HOWSON *Handbk. Terms Algebra & Anal.* xix. 98 Hence, *Ab* is a *functor* from G to the *subcategory* of Abelian groups and homomorphisms. **1982** *Sci. Amer.* Mar. 91/1 He divided the galaxies into two main classes, ellipticals and spirals, with several subcategories.

Hence **sub-ca·tegorize** *v. trans.*, to place in or divide into sub-categories; **su:b-categoriza·tion.**

1965 N. CHOMSKY *Aspects of Theory of Syntax* ii. 95 Rules..which analyze a symbol in terms of its categorial context, I shall henceforth call *strict subcategorization rules*. **1965** Subcategorize [see *pre-adjectival* s.v. *PRE- B. 1 d]. **1971** J. ANDERSON in A. J. Aitken et al. *Edin. Stud. Eng. & Scots* 69, I would like to consider some of the set of phenomena that we could reasonably require a subcategorisation of the modal verb in English to provide an explanation for. **1979** *Dictionaries* I. 14 The initial capitals are subcategorized and provided with lower-case letters and figures.

subce·ption. *Psychol.* A blend of *subliminal perception* (see quot. 1949).

1949 MCLEARY & LAZARUS in *Jrnl. Personality* XVIII. 171 An experiment which confirms the notion that subjects give discriminatory galvanic skin responses to visual stimuli presented at tachistoscopic speeds too brief for correct verbal report. The implied perceptual process is termed [illegible]

sub-cheese (sɒbtʃī·z). *Mil. slang* (orig. *Anglo-Indian*). Also **sub-cheeze, -chiz,** [ad. Hind. *sab* all + *chiz* thing (see CHEESE *sb.*[3]).] The lot; everything; all that there is. Also in phr. *the whole sub-cheese*.

[illegible]

su·b-clause. [SUB- 5 c, 9.] **1.** *Law.* A subsidiary section of a clause. Also *transf.*

[illegible]

2. *Gram.* A clause that is subordinate to a main clause.

[illegible]

subcli·max. *Ecol.* [f. SUB- 22 + *CLIMAX 4 b.] A point in an ecological succession at which a plant community is prevented by climatic or other factors from reaching its natural climax.

[illegible]

subconscious, *a.* Add: **1. a, b.** *absol.* (Later examples.)

[illegible]

subconsciousness. 1. (Earlier example.)

1874 G. H. LEWES *Probl. Life & Mind* (Ser. 1) I. 141

Neural processes which formerly were accompanied by Consciousness sink into Sub-Consciousness.

subcontinent. Add: Now also with main stress on second syllable. (Later examples.) More recently applied commonly to India, Pakistan, Bangladesh, and Sri Lanka.

[illegible]

subcontinental, *a.* Add: **3.** [SUB- 5 b.] Of or pertaining to a subcontinent, *spec.* the Indian subcontinent (see prec.).

[illegible]

subcontract, *v.* Add: **3.** (Later examples.) Formerly, of the sub-contractor; now often with contractor or work as subj. Also with *out*.

[illegible]

sub-crea·tion. [SUB- 5.] J. R. R. Tolkien's word for the process of inventing an imaginary or secondary world, different from the primary world but internally consistent.

[illegible]

Also **sub-crea·tor,** one who engages in sub-creation.

[illegible]

sub-crea·tive, *a.* [SUB- *5 c (*b*). 8.] Of or pertaining to sub-creation or secondary creation. Hence **sub-creatively** *adv.*

[illegible]

subcri·tical, *a.* Also **sub-critical.** [SUB- 14.] **1.** *Metallurgy.* Less than the critical temperature above which ferrite changes into austenite; *subcritical annealing*, annealing in which the temperature is not raised above this.

[illegible]

2. Of a flow of fluid: slower than the speed at which waves travel in the fluid.

[illegible] transition between what might be called supercritical and subcritical flow. **1981** *Sci. Amer.* Apr. 138/2 A ridge across a streambed provides an example of how a barrier can create supercritical flow, a standing wave and a hydraulic jump in an initially subcritical flow.

3. *Nucl. Physics.* Containing or being less than the critical mass (see *CRITICAL *a.* 7 b).

[illegible]

subcu·ltural, *a.* Also **sub-cultural.** [SUB- 5, 7.] Of or pertaining to a subculture. Also, that is inferior to or below the general cultural level.

[illegible]

su·bculture, *sb.* Also **sub-culture. 1.** *Biol.* and *Med.* [SUB- 9.] A culture (of bacteria or the like) started from another culture; the process of starting a culture in this way.

[illegible]

2. [SUB- 7.] A group or class of lesser importance or size sharing specific beliefs, interests, or values which may be at variance with those of the general culture of which it forms part.

[illegible]

sub-culture, *v. Biol.* and *Med.* [f. prec., sense 1.] *trans.* To produce a subculture of. Hence (with variable stressing) **subcultured** *ppl. a.*, **subculturing** *vbl. sb.*

[illegible]

sub-deb. *slang* (chiefly *U.S.*). Now *rare*. [*SUB- 14 b.] A girl who will soon 'come out' as a social debutante. Hence, less specifically, a girl in her mid-teens. Also *fig.* and *attrib.*

[illegible] Margaret, two years older than Alva..and there was Joyce, the subdeb. **1939** [see *DATING *vbl. sb.* c]. **1944** W. S. MAUGHAM *Razor's Edge* vii. 278 The living-room would do very well for the sub-deb dances which it would be her pleasant duty to give. [illegible]

Hence **sub-debu·tante** *U.S.*

[illegible]

subdialect. Add: Hence **su:bdiale·ctal** *a.*, of or pertaining to a subdialect.

[illegible]

subdivision. Add: **2. c.** *N. Amer.* An area of land subdivided into plots for the erection of houses; a housing estate.

[illegible]

subdominant, *a.* Add: **2.** *Ecol.* Designating a species which is prevalent in a community, but below the dominant in precedence. Also as *sb.* Cf. *DOMINANT *a.* 8.

[illegible]

subduct, *v.* Restrict 'Now *rare*' to senses in Dict. and add: **5.** *trans.* and *intr. Geol.* To move sideways and downwards underneath a neighbouring lithospheric plate.

[illegible]

Hence **subducted, subducting** *ppl. adjs.*

[illegible]

subduction. Restrict 'Now *rare*' to senses in Dict. and add: **5.** *Geol.* [a. F. *subduction* (A. Amstutz 1951, in *Arch. des Sci.* IV. 326).] The sideways and downward movement of the edge of a lithospheric plate into the mantle beneath a neighbouring plate; **subduction zone,** a strip along which this is occurring.

[illegible]

subdue, *v.* Add: **2. f.** In phr. *to be subdued to what one works in*: to become reduced in capacity to the standard of one's material (in allusion to SHAKES. *Sonnets* cxi.).

1907 W. RALEIGH *Shakespeare* iv. 107 Shakespeare accepted the facts, and subdued his hand to what it worked in. **1912** L. STRACHEY *Landmarks in French Lit.*

iv. 92 Their [*sc.* the Elizabethans'] work has vanished from the stage, and is today familiar to but a few of the lovers of English literature. Shakespeare alone was not subdued to what he worked in. **1926** G. M. Trevelyan *Hist. Eng.* v. iii. 559 When a man, in defending his country from foreign conquest, has to rely on certain forces, he ceases to be capable of criticizing them. He becomes subdued to the material in which he works.

su·bdwarf, *sb.* and *a.* *Astr.* [Sub- 22.] **A.** *sb.* A star which when plotted on the Hertzsprung–Russell diagram lies just below the main sequence, being less luminous than dwarf stars of the same temperature. Cf. *subgiant.

1939 G. P. Kuiper in *Astrophysical Jrnl.* LXXXIX. 548 Three classes of objects of special interest are expected to be found..: (1) white dwarfs; (2) intermediate white dwarfs or, more generally, stars not over 2 or 3 mag. below the main sequence... The second group extends almost along the whole main sequence. Since these stars merge into the main sequence and are much more similar to main-sequence stars than to white dwarfs.., the name 'subdwarfs' is suggested for this class of stars, in analogy with 'subgiants'. **1962** *New Scientist* 3 May 218/2 Some hot subdwarfs are found from their spectra to have helium but virtually no hydrogen. **1979** *Nature* 24 May 305/1 The observations of CH Cygni reported here were made to determine whether a symbiotic star is a binary system composed of an M6 giant and a hot subdwarf, or whether it is a cool star surrounded by a thick corona.

B. *adj.* Designating such a star.

1981 *Nature* 8 Oct. 432/2 The most likely explanation.. is that the atmospheres are untypical of the subdwarf stars as a whole.

sub-edit, *v.* Add: (Earlier example.) Also *absol.*

1855 D. G. Rossetti *Let.* 23 Jan. (1965) I. 241 He sub-edits the *Leader*. **1915** Wodehouse *Psmith, Journalist* xx. 145, I am Psmith. I sub-edit.

sub-editor. (Earlier example.)

1834 [see *city-editor* s.v. *city 9].

subfamily. Add: **1. b.** A subdivision of a human family, *spec.* one living within a primary family group (see quot. 1964).

1964 *Census of Population 1960* (U.S. Dept. Commerce) I. 1. p. lviii/2, A subfamily is a married couple with or without own children, or one parent with one or more own children under 18 years old, living in a housing unit and related to the head of the household or his wife. **1970** S. L. Barraclough in I. L. Horowitz *Masses in Lat. Amer.* iv. 129 Some such units have incomes close to those of sub-family producers.

2. *transf.*, esp. in *Linguistics*.

1856 W. D. Whitney in *Jrnl. Amer. Oriental Soc.* V. 195 The various sub-families and even closer kindred dialects had deviated too widely from their original and from one another. **1972** R. J. Wilson *Introd. Graph Theory* viii. 119 We call a transversal of a subfamily of *S* a partial transversal of *S*. **1978** *Language* LIV. 181 The Southern sub-family merged *i* with *ī* and *u* with *ū*.

sub-fra·ction. [Sub- 9.] **1.** *Math.* (In Dict. s.v. Sub- 9.)

2. *Biochem.* Any one of the portions into which a fraction may be further divided. Cf. *fraction *sb.* 7.

1946 *Nature* 5 Oct. 474/1 The division of the combined cystine in wool into four sub-fractions of different chemical reactivity. **1962** V. N. Orekhovich et al. in A. Pirie *Lens Metabolism Rel. Cataract* 324 We succeeded in dividing β-crystallin of cattle lens into $β_1$- and $β_2$-crystallin..and γ-crystallin was divided into three sub-fractions. **1978** *Jrnl. Neurochem.* XXX. 563 A subfraction, derived from the microsomal fraction of rat cerebral cortex,.. appears to be enriched in receptor sites for a number of potential neurotransmitters.

Hence **su:bfractiona·tion** *Biochem.*, the process of separating a fraction into further components.

1955 *Biochem. Jrnl.* LX. 615/1 This subfractionation has not brought to light any enzymic heterogeneity in the granules. **1978** *Jrnl. Neurochem.* XXX. 783 The possibility that these findings might reflect merely contamination of myelin with other membranes was tested by subfractionation.

subfusc, -fusk, *a.* and *sb.* Add: Now usu. in form **subfusc**. **a.** Also *spec.* of clothing: dark, as prescribed by the regulations of the Universities of Oxford and Cambridge for examinations and other formal occasions; (later examples).

1930 W. J. Locke *Town of Tombarel* v. 163 Cousin Hortense in some sort of unremarkable subfusc raiment. **1973** *New Society* 1 Nov. 259/3 His clothes very subfusc—grey suit, polished black shoes, the only brightness a purple and red bowtie. **1978** G. Greene *Human Factor* v. i. 232 Two women who might have been sisters in their similar subfusc clothing waited by what he guessed was a confessional box.

fig. (Later examples.)

1927 C. Parsons in *Oxford Poetry* 24 Lost in what corner of this maze, With mind already dyed subfusc. **1949** C. P. Snow *Time of Hope* v. xxxiii. 280 Allen..made subfusc, malicious, aunt-like jokes at Getliffe's expense. **1958** L. Durrell *Balthazar* x. 210 The frail subfusc moonlight glancing along the waves. **1970** N. Marsh *When in Rome* iv. 106 Mailer seemed to me to be, in a subfusc sort of way, cocksure.

b. (*b*). (Later examples with reference to university examination dress.)

1944 A. L. Rowse *Eng. Spirit* xxxvii. 260 Black-gowned young men and women, all dutifully clad in *sub fusc*. **1961** E. Williams *George* xx. 320, I was able to keep up the illusion of study by twice donning sub-fusc and walking down to the Examination Schools in white tie and mortar-board.

su·bgiant. *Astr.* [Sub- 22.] A star which when plotted on the Hertzsprung–Russell diagram lies between the main sequence and the giants; a star similar to a giant of the same spectral type but less luminous. Cf. *subdwarf *sb.* and *a.*

1937 *Astrophysical Jrnl.* LXXXV. 383 Three bright 'subgiants' having well-determined trigonometric parallaxes. **1943** W. W. Morgan et al. *Atlas Stellar Structure* 6 For the stars of types F–K, class IV represents the subgiants and class III the normal giants. **1978** H. L. Shipman *Introd. Astron.* xi. 296 Star *A* becomes a rather unusual type of star, a subgiant—a low-mass, very dim, small red giant.

su·bgrade. **1.** *Road* and *Railway Engineering* [Sub- 3 a.] The layer, either natural or constructed, lying immediately beneath the foundations of a road or railway line. Cf. Substratum 4 b.

1893 G. A. Perkins et al. in *Rep. Mass. State Highway Comm.* v. 78 The subgrade, or the ground on which the large stones rest, should be thoroughly compacted by rolling. **1906** *Engin. Rec.* 14 Apr. 478/3 The reduced quantity of broken stone required, when it is laid on a firm sub-grade. **1930** *Engineering* 1 Aug. 139/3 Maintenance of a mile of gravel road, including the sub-grade. **1962** *Ibid.* 30 Mar. 439 The stresses transmitted to the subgrade by a high quality structural are so small. **1979** *Railway Gaz. Internat.* Jan. 52/2 Protection of the sub-grade against frost.

2. [Sub- 9 (*a*).] A subsidiary grade; one within a grade.

1919 *Sociol. Rev.* XI. 90 We might perhaps recognise.. an intermediate group, concerned chiefly with relations between sub-grades. **1931** J. S. Huxley *What dare I Think?* vi. 218 In this stage of thought there are, of course, many sub-grades.

subgroup. Add: **a.** No longer 'Chiefly *Nat. Hist.*'. (Later *gen.* examples.)

1935 *Planning* II. xli. 6 Utility Services..has been working in three or four sub-groups, one of which dealing with transport has produced the survey contained in broadsheet No. 21. **1961** Y. Olsson *Syntax Eng. Verb.* ii. 30 One sub-group is independent, for its extra-lingual correlation, of the speech-situation in which its terms are employed. **1978** K. Hudson *Jargon of Professions* 10 Changes in the technology of communication have made it far easier for one cultural sub-group to hear the members of another group talking.

b. More widely, any group all of whose elements are elements of a larger group. (Earlier and later examples.)

1887 *Amer. Jrnl. Math.* IX. 51, I use 'self-conjugate sub-group' in translating Klein's 'ausgezeichnete Untergruppe' and Jordan's 'groupe permutable.' **1937**, etc. [see *proper *a.* 5 c]. **1975** I. Stewart *Concepts Mod. Math.* vii. 104 If I gave you a group of order 615 you would know, without any information about the multiplication table, that its subgroups cannot have any orders other than 1, 3, 5, 15, 41, 123, 205, and 615.

Hence **su·bgroup** *v. trans.*, to divide or classify into subgroups; **sub-grouping,** a subsidiary grouping or subgroup; the action of dividing or classifying into these.

1922 E. Wallace *Flying Fifty-Five* xi. 67 They also were grouped and sub-grouped and indexed. **1956** J. Klein *Study of Groups* 168 It must be established that these sub-groupings are not thrown up by chance. **1960** *Amer. Speech* XXXV. 216 The two final chapters [of H. M. Hoenigswald, Language Change & Linguistic Reconstruction] deal with..the procedures for the sub-grouping of language families. **1977** *Lancet* 1 Jan. 9/1 The I.Q. data from this study have been subgrouped according to whether the immersion accident occurred in the hot or the cold months of the year. **1978** *Language* LIV. 468 Dyen concerns himself with three topics: the AN homeland, the subgrouping and external relationships of the AN language, and reconstruction.

subharmo·nic, *sb.* and *a.* Also **sub-harmonic.** [Sub- 9.] **A.** *sb.* An oscillation with a frequency equal to an integral submultiple of another frequency. Freq. *attrib.*

1924 W. N. Bond in *Nature* 8 Mar. 355/2 The production of the half frequency easily, of the third frequency with care, and of the fourth frequency faintly..leaves little doubt that the frequencies obtainable are all submultiples of the fundamental applied frequency... These forced vibrations might be described as sub-harmonics. **1940** H. F. Olson *Elem. Acoustical Engin.* vii. 137 It has been analytically shown..that subharmonics are possible in certain vibrating systems. **1952** [see *frequency divider* s.v. *frequency 5]. **1961** M. L. Gayford *Acoustical Techniques & Transducers* iii. 67 A curvature of the sides of the [loudspeaker] cone assists the suppression of subharmonics. **1976** *Gloss. Terms Mech. Vibration & Shock* (B.S.I.) 12 *Subharmonic response*, a response of a mechanical system exhibiting some of the characteristics of resonance at a frequency having a period that is an integral multiple of the period of the periodic excitation. **1978** A. B. Pippard *Physics of Vibration* I. ix. 253 Each subharmonic is stable over a limited range of excitation.

B. *adj.* Involving or being a subharmonic.

1940 H. F. Olson *Elem. Acoustical Engin.* vii. 137 Another feature of subharmonic phenomena is the relatively long time required for 'build up'. **1962** A. Nisbett *Technique Sound Studio* 245 Paper cones are often corrugated to reduce any tendency to 'break up' radially and produce sub-harmonic oscillations. **1978** A. B. Pippard *Physics of Vibration* I. ix. 253 The pin is shown making contact with the cone every other cycle of the latter, and therefore responding in the octave subharmonic mode.

sub-head. Add: **3.** (Examples from newspaper journalism.)

1889 T. Campbell-Copeland *Ladder of Journalism* vi. 42 The first line..should consist of from twelve to fifteen letters, presenting in the briefest form..the subject of the article; beneath which, the sub-head of twelve words or thereabouts, making a line and a half, should be placed. **1927** *Amer. Speech* II. 239/2 For a very long story, 'subheads' are usually provided, brief crosslines in bold face type the same size as the body type. **1961** C. Willock *Death in Covert* xii. 203 One headline said: *Regency rakes ride again*, and the sub-head to the same story complained: *Last time a man was blown up*. **1979** D. Anthony *Long Hard Cure* vii. 64 The news story..was on the front page, under the subhead: *Maniac claims fourth victim*.

Hence **su·bhead** *v. trans.*, to furnish with a sub-heading; also *fig.*

1877 *Harper's Mag.* Dec. 45/2 One of them was heading and sub-heading cable dispatches from the seat of war. **1949** *Scrutiny* XVI. 52 He [*sc.* C. E. M. Joad] contrasts the present period, which he subheads as 'foreheads defiantly low', with the happy time of his youth. **1978** W. White in W. Whitman *Daybks. & Notebks.* II. 415 The account, which totals more than 26 column inches in the *Times*, ends with a section subheaded 'The Poet Greets His Friends'.

subheading. (Earlier example.)

1874 *Catal. Apprentices' Libr.* (N.Y.) p. v, Headings containing a large number of titles are subdivided into sub-headings to facilitate reference.

sub-human, *a.* Add: **1. b.** as *sb.* One who is less than human; a person of sub-human instincts.

1957 R. Campbell *Portugal* iv. 62 This..moray..was fed chiefly on recalcitrant slaves..devoured before the gloating eyes of the subhuman who was its owner. **1970** G. Jackson *Let.* in *Soledad Brother* (1971) 247 Would you like to know a subhuman... I'm not a very nice person...more kin to the cat than anything else, the big black one.

Hence **sub-huma·nity**, the quality of being sub-human, less than human existence; (*rarely*) a level of creation below the human race; **sub-hu·manly** *adv.*, in a sub-human manner, bestially.

1909 G. B. Shaw in *Nation* 28 Aug. 787/2 Mr. Chesterton..finally excogitates, as a proof of my superhumanity or sub-humanity, exactly the reason that would have been given by one of Wellington's private soldiers. **1929** A. Huxley *Do what You Will* 75 They live..subhumanly..they sink..towards a repulsive subhumanity. **1939** J. Cary *Mister Johnson* 157 They have become a new kind of creature, a sort of subhumanity which can smile and eat and live at a level of corruption and misery which would kill a real human being. **1966** 'H. MacDiarmid' *Company I've Kept* ii. 50 As soon as you make allowances..then you are opening the floodgate for mediocrity and..you are submerged under a tide of subhumanity. **1970** G. Greer *Female Eunuch* 262 Is it too much to ask that women be spared the daily struggle for superhuman beauty in order to offer it to the caresses of a subhumanly ugly mate?

su:b-irriga·tion. [Sub- 2 a.] The irrigation of land from beneath the surface, esp. by means of underground channels or pipes.

1880 *News & Press* (Cimarron, New Mexico) 19 Aug. 1/6 Sub irrigation for wheat would be too expensive. **1904** *Bull. Bur. Census, U.S. Dept. Comm. & Labor* No. 16. 25/1 In one of the systems of subirrigation the water is carried through pipes 14 inches below the surface. **1930** *Amer. Speech* VI. 11 Subirrigation is accomplished by allowing water to stand in deep ditches from which it soaks laterally. **1979** R. Adams et al. *Dry Lands* viii. 116/1 Sub-irrigation..is normally achieved either by creating an artificial water-table just below the ground surface or by burying a drip irrigation system or a network of perforated pipes.

Hence **subi·rrigate** *v. trans.* (also *absol.*), **subi·rrigated** *ppl. a.*

1903 [in Dict. s.v. Sub- 2 a]. **1933** *Nat. Geogr. Mag.* Feb. 189 (*caption*) The soil is remarkably fertile and is subirrigated by springs in the surrounding hills. **1950** H. B. Roe *Moisture Requirements in Agric.* viii. 229 On Elgin Bench..the subirrigated district includes an area of about 60,000 acres. **1976** D. Goldberg et al. *Drip Irrigation* i. 10 Drainage systems were utilized to function as a double action system, draining surplus water yet sub-irrigating through the same medium in a reverse operation.

subitaneous, *a.* Add: **b.** *Biol.* Of the egg of a small aquatic invertebrate: hatching soon after it is laid. Cf. *resting egg* s.v. *resting *ppl. a.* 1 b.

1950 *Adv. Genetics* III. 240 From the fertilized eggs (the ephippial eggs) there emerge, after a shorter or longer resting period, exclusively females, the eggs of which (the subitaneous eggs) develop parthenogenetically in the brood chamber of the female. **1979** *Nature* 30 Aug. 722/1 Both quick-hatching (subitaneous) and resting eggs are produced but neither kind was believed to be fertilised, though in *Chaetonotus* two types of subitaneous egg have been reported.

subitize (sʊ·bitaiz), *v.* *Psychol.* [f. L. *subitus* Subite *a.* + -ize.] *intr.* and *trans.* To apprehend immediately (the number contained in a small sample). Hence **su·bitizing** *vbl. sb.*

1949 E. L. Kaufman et al. in *Amer. Jrnl. Psychol.* LXII. 520 A new term is needed for the discrimination of stimulus-numbers of 6 and below... The term proposed is *subitize*... We are indebted to Dr. Cornelia C. Coulter, the Department of Classical Languages and Literatures, Mount Holyoke College, for suggesting this term. *Ibid.*, If no discontinuities had appeared in the results, no distinction between subitizing and estimating could have been drawn. **1971** *Jrnl. Gen. Psychol.* Jan. 121 The number of items in an array capable of being subitized. **1981** *Nature* 15 Oct. 569/2 Judgements of 'small' numerosities..are ordinarily attributed to subitizing.

subject, *sb.* Add: **III. 16.** (Earlier example.) Also, *first* (*second*) *subject*, the primary (or subsidiary) theme of a composition, esp. in sonata-form; (earlier and later examples).

1752 C. Avison *Ess. Mus. Expression* i. ii. 28 In the greater Kinds of musical Composition, there is a principal or leading *Subject* or Succession of Notes, which ought to prevail, and be heard throughout the whole Composition. **1771** C. Burney *Present State Mus. in France & Italy* 49 The first subject is judiciously returned to while it still vibrates on the ear. **1876** Stainer & Barrett *Dict. Mus. Terms* 411/1 In sonata form there should be two chief subjects, called first and second. **1883** Grove *Dict. Mus.* III. 752/1 The Father of the Symphony [*sc.* Haydn] enriched his new Art-form with a Second Subject, so constructed as to enhance the beauty of the Primary Theme by the introduction of some form of expression distinctly opposed to it. **1955** J. F. Russell in H. Van Thal *Fanfare for Ernest Newman* 148 It is difficult to remember a Mozart first subject, for example, in which the common chord is not melodically employed. **1977** *Gramophone* June 90/2 Walton brilliantly exploits every conceivable kind of antiphony..the Worcester Cathedral choristers angelically distanced in the gentle second-subject at 'The glorious company of the Apostles'.

IV. 18. a. (sense 8) *subject clause, complement*, (also 13) *-oriented* adj.; (sense 13) *subject-changer*; (sense 14) *subject card, catalogue* (earlier example), *cataloguing, entry* (earlier example), *heading, index* (earlier example), *list* (earlier example), *reference* (earlier example); **subject-term** *Logic* = sense 7 b.

1869 C. A. Cutter in F. L. Miksa *Charles Ammi Cutter* (1977) ii. xxiv. 168 The subject-cards would..be copied from these author-cards. **1982** D. L. Foster *Managing Catalog Department* (ed. 2) iii. 72 The best way to divide the catalog..is to divide the subject cards from the others within each drawer. **1873** M. Dewey in G. Dawe *Melvil Dewey* (1932) 323 By this plan any books may be found without a catalogue since the library is in itself a full classed subject catalogue. **1900** E. W. Hulme in *Libr. Assoc. Rec.* 5 Nov. 571 (*heading*) Principles of dictionary subject-cataloguing in scientific and technical libraries. **1748** Richardson *Clarissa* (1785) II. 160 But I asked him, If he had any news by his last letters from London: A question which he always understands to be a *subject-changer*; for otherwise I never put it. **1978** I. Murdoch *Sea* 164 'Been to Ireland lately?' This always set Perry off and was a guaranteed subject-changer. **1957** R. W. Zandvoort *Handbk. Eng. Gram.* iii. vi. 165 *What*..may introduce a subject clause, an object clause,..a predicative clause, or a clause preceded by a preposition. **1939** H. E. Palmer *Gram. Spoken Eng.* (ed. 2) ii. 80 A certain number of adverbs may be used as subject-complements, i.e. as complements to..verbs of incomplete predication. **1869** C. A. Cutter in *N. Amer. Rev.* CVIII. 115 In the New Catalogue, on the contrary, the subject entry is the fullest. **1874** *Catal. Libr. Mercantile Libr. Assoc. San Francisco* p. vi, Subject-headings, when there are two or more titles, are denoted by a separate line in the same [fount]. **1973** M. Amis *Rachel Papers* 60, I indent subject-headings, co-ordinate footnotes, mark cross-references in red and blue biros. **1861** *Catal. N.-Y. State Libr.: Gen. Libr., 1st Suppl.* p. xii, *Subject-Index.*—In the Index following the catalogue, the subjects of the books are arranged alphabetically. **1875** C. A. Cutter in *Nation* 14 Oct. 252/1 'Analysis'—that is, reference under subjects to topics discussed in certain books incidentally but not at sufficient length to justify the insertion of the book in the subject-list. **1964** *Language* XL. 77 Middle voice embraces at least five subtypes: (1) subject-oriented action, [etc.]. **1975** *Language for Life* (Dept. Educ. & Sci.) xii. 189 The primary school teacher is likely to conceive of his task in terms of integrated rather than subject-oriented work. **1876** *Public Libr. in U.S.A.* (U.S. Bureau Educ.) i. xxvii. 542 A dictionary catalogue (author- and anonymous-title entries with imprints,.. subject-references to the classed part). **1880** W. H. S. Monck *Introd. Logic* v. 39 A particular proposition is not limited to some *only* of the objects denoted by the subject-term. **1980** A. Kenny *Aquinas* ii. 34 The word 'substance' can be used to refer to the thing that sentences such as the above are about: the object for which the subject-term of the sentences stands.

b. Used appositively in senses 7–9, as *subject-object, -predicate, -verb* adjs.

1933 *Jrnl. Philos.* XXX. 65 We have described those features of subject-object situations whereby the narrowness of a point of view is escaped. **1936** J. R. Kantor *Objective Psychol. Gram.* xvi. 222 The logical essense of these cases can be clearly discerned in the fact that the nominative and accusative are subject-object cases. **1977** Douglas & Johnson *Existential Sociol.* p. xi, Experimental or other methodological protocol can produce the so-called subject-object dualism. **1900** B. Russell *Leibnitz* ii. 12 The question whether all propositions are reducible to the subject-predicate form is one of fundamental importance to all philosophy. **1980** A. Kenny *Aquinas* ii. 51 Sentences which are of subject-predicate form. **1935** G. K. Zipf *Psycho-Biol. of Lang.* v. 234 One cannot determine *a priori* what actual proportion of spoken English consists of simple subject-verb sentences. **1979** *Amer. Speech 1976* LI. 134 Of the nine problems covered, subject-verb agreement receives a thorough treatment.

su·bjecthood. [-hood.] **1.** [Subject *sb.* I.] The state or condition of being a subject; = Subjection 2.

1927 C. C. Martindale *Christ is King* v. 93 The vast duty of our subjecthood almost narrows itself to this. **1968** *Economist* 17 Feb. 16/1 In the Kenya independence settlement devised by Mr Duncan Sandys in 1963,..they were offered the chance either to acquire Kenyan citizenship, or to have full rights of British subjecthood, including passports.

2. *Gram.* [Subject *sb.* II.] The state of being a subject (of a sentence, etc.).

1976 *Classical Q.* XXVI. 38 The 'with' idiom expresses not the idea of means, but rather that of subjecthood or even agency. **1979** *Trans. Philol. Soc.* 223 The insertion of this reciprocal marker is shown to be sensitive to the transitive subjecthood, either actual or at some initial stage of derivation, of the affix it replaces.

subjective, *a.* **4. d.** (Earlier example.)

1853 J. S. Le Fanu in *Dublin Univ. Mag.* Dec. 723/1 Was this singular apparition..the invention of my poor stomach? Was it, in short, *subjective* (to borrow the technical slang of the day) and not the palpable aggression and intrusion of an external agent?

subjectivistic, *a.* (Earlier example.)

1884 W. James in *Unitarian Rev.* Sept. 210, I will.. speak rather of *subjectivism*, and the *subjectivistic* point of view.

subjectivity. **4.** (Earlier example.)

1854 A. G. Henderson tr. *V. Cousin's Philos. Kant* viii. 177 The subjectivity of human reason; this it is that troubles Kant.

subjee. Add: Also **sabji, sabzi, subzee.** (Earlier and later examples.)

1826 W. Ainslie *Materia Indica* II. ii. i. 39 Banghie.. (Tam[ool])... Subjah..(Duk[hanie] and Hind[oostanie]). **1883** W. Dymock *Vegetable Materia Medica of Western India* 603 Cannabis sativa, Linn., Var. Indica... Leaves, Bhâng, Siddhi, Sabzi (Hind. Beng. and Bomb.). **1893** —— et al. *Pharmacographia Indica* III. 320 *Sabzi* or *Sabji*, an infusion of Bhang with black pepper, anise and sugar. In Bengal milk, and cucumber and melon seeds are added. **1938** R. P. Walton *Marihuana* xi. 190 Bhang. Also known as Subjee—is the larger leaves and capsules of the cannabis compressed in balls and sticky layers with here and there some flowers between. *Ibid.* 195 *Subzee*, an infusion of bang.

subjunct (sʊ·bdʒʊŋkt). *Gram.* [f. L. *subjunctus*, pa. pple. of *subjungĕre* Subjoin *v.*: cf. Adjunct *ppl. a.* and *sb.*] In Jespersen's terminology, a word or group of words of the third rank of importance in a phrase or sentence. Cf. *primary *sb.* 9, *adjunct *sb.* 5 b.

1914 O. Jespersen *Mod. Eng. Gram.* II. xii. 283 The adjunct in *perfect simplicity* is a shifted subjunct of the adjective contained in the substantive *simplicity*, cf. *perfectly simple*. We may call these *shifted subjunct-adjuncts*. **1924** —— *Philos. Gram.* vii. 97 For tertiary we may use the term *subjunct*, and quaternary words.. may be termed *sub-subjuncts*. **1935** [see *adjunct *sb.* 5 b].

subjunctive, *sb.* Add: **1. b.** *Comb.*, as **subjunctive-equivalent**, an expression which conveys the subjunctive mood by a construction involving an auxiliary verb and an infinitive.

1927 E. A. Sonnenschein *Soul of Grammar* ii. 87 Modern English makes a large use of 'subjunctive-equivalents,' e.g. expressions formed by combining a tense..of the verbs 'shall', 'will', 'may', 'let', with an infinitive. **1965** F. Behre in *English Studies* Apr. 89 But now is perhaps the right moment to question the fitness of using the term 'subjunctive-equivalent' in contemporary English.

su·blanguage. [Sub- 5 c.] A specialized language or system of notation that occurs only in certain contexts or is used only by certain people among those who speak the same ordinary language.

1934 Webster, *Sublanguage*, a subordinate language; a dialect. **1951** J. Holloway *Lang. & Intell.* x. 182 These sub-languages include arithmetic and geometry..; chess notation; musical notation [etc.]. **1966** M. Gross in *Automatic Transl. of Lang.* (NATO Summer School, Venice, 1962) 134 Of course a translation from L_1 to L_2 need not be an exact mapping between L_1 and L_2, but there may be a large sublanguage of L_1. **1972** *Science* 23 June 1304/3 In a sub-language..such as the jargon of surgeons, the information is carried mainly by the kernels. **1973** G. W. Turner *Stylistics* i. 26 Such sub-languages as the language of telegrams, newspaper headlines, advertisements or knitting patterns.

sub-le·thal, *a.* [Sub- 19, 20.] **a.** *Med.* Of a drug, treatment, etc.: having an effect (only just) less than lethal.

1896 [in Dict. s.v. Sub- 20 g]. **1910** Hanson & Zinsser *Textbk. Bacteriol.* xii. 195 (*heading*) Active immunization with sublethal doses of fully virulent bacteria. **1937** *Ann. Reg. 1936* 59 Experimental epidemiologists showed the importance of latent and sub-lethal infection. **1947** *Radiology* XLIX. 303/1 At sublethal doses, the minimum granulocyte count occurs at about the same time as in non-survivors. **1977** J. L. Harper *Population Biol. Plants* xvi. 493 These are the pathogens that kill young seedlings,..that convert sub-lethal damage done by other causes into lethal damage.

b. *Genetics.* Of an allele or a chromosomal abnormality: = *semi-lethal *a.*

1935 *Jrnl. Heredity* XXVI. 357/2 Hadley reported the inheritance of a sub-lethal, hairless defect in Holsteins [*sc.* a breed of cattle]. **1946** *Nature* 16 Nov. 722/2 When a gene is sublethal, as are those for hæmophilia and achondroplasic dwarfism, its elimination by natural selection is in approximate equilibrium with its appearance by mutation. **1961** R. D. Baker *Essent. Path.* xi. 274 'Sublethal genes' are those which produce malformations compatible with life in the uterus but responsible for death soon after birth.

Hence **suble·thally** *adv.*

1958 *Science* 4 July 32 (*heading*) Delayed deaths in sublethally X-rayed F_1 hybrid mice injected with parental strain spleen cells. **1978** *Nature* 13 Apr. 625/2 Sublethally irradiated adult BALB/c mice.

sublimate, *v.* Add: **5. c.** *trans.* in *Psychoanal.* To refine or direct (instinctual energy), esp. that of the sexual impulse, so that it is manifested in more socially acceptable ways. Also *absol.* and *intr.*

1910 J. J. Putnam in A. A. Brill tr. *Freud's Three Contrib. Sexual Theory* p. vii, The instincts with which every child is born..may be refined ('sublimated')..into energies of other sorts. **1916** C. E. Long tr. *Jung's Coll. Papers Anal. Psychol.* 141 Here we are confronted by an energetic effort to sublimate the fear into an eager desire for knowledge. **1921** R. Macaulay *Dangerous Ages* vi. 112 You have some bad complexes, which must be sublimated. **1953** J. Strachey et al. tr. *Freud's Compl. Psychol. Wks.* VII. 50 The perversions..—by being 'sublimated'—are destined to provide the energy for a great number of our cultural achievements. **1967** M. L. King *Trumpet of Conscience* iv. 69 This rare opportunity for bloodletting was sublimated into arson. **1974** 'S. Woods' *Done to Death* 195 If she had guilt feelings..she might have sublimated them this way.

absol. and *intr.* **1933** J. Jastrow *House that Freud Built* vi. 136 We sublimate as we grow in psychic stature. **1955** H. Hartmann in A. Freud *Psychoanal. Study of Child* X. 12 Melanie Klein..equates the capacity to cathect ego activities with libido with the capacity to sublimate. **1973** H. McLeave *Question of Negligence* xxiii. 183 Some boy jilted her..thirty years ago. Now she sublimates like mad and expends all her pent-up emotion on her patients.

sublimating *vbl. sb.* (examples in *Psychoanal.*).

1913 E. Jones *Papers on Psycho-Anal.* xx. 416 (*heading*) The value of sublimating processes for education and re-education. **1923** J. S. Huxley *Ess. Biologist* vii. 276 Dominant ideas at work in the sublimating process.

sublimated, *ppl. a.* Add: **4.** *Psychoanal.* Of a (sexual) instinct, feeling, etc.: that has been refined and made more socially acceptable.

1911 *Amer. Jrnl. Psychol.* XXII. 436 If the transference is successful, be it a purely erotic feeling, or a sublimated one of respect.., there springs up the feeling of sympathy. **1923** J. S. Huxley *Ess. Biologist* vii. 271 A sublimated instinct has more and higher values attached to its satisfaction than one unsublimated. **1951** E. Jones *Ess. in Applied Psychol.* II. xiii. 320 A given sublimated interest ..may represent one of the described stages. **1966** G. Onn tr. *Wyss's Depth Psychol.* i. ii. 194 Sublimated ideas may also temporarily sink back into the unconscious, regress and become symbols of complexes.

sublimation. Add: **5. c.** *Psychoanal.* The refining of instinctual energy, esp. that of the sexual impulse, and its manifestation in ways that are socially more acceptable.

1910 A. A. Brill tr. *Freud's Three Contrib. Sexual Theory* 58 It must be through these roads that the attraction of the sexual motive powers to other than sexual aims, the sublimation of sexuality, is accomplished. **1920** B. Low *Outl. Psycho-Anal.* (ed. 2) iii. 81 If the sublimation-process can afford an adequate outlet for the psychic energy accompanying the primitive desires, we achieve a fairly satisfactory adjustment. **1925** I. A. Richards *Princ. Lit. Crit.* xxxi. 232 If we do not extend the 'sublimation' theory too far..it may be granted that in some cases the explanation is in place. **1943** H. Read *Educ. through Art* vi. 177 Sublimation is thus the transformation of instinctive egoistic drives, wishes and desires onto socially useful or socially approved thoughts, ideals and activities. **1957** G. Faber *Jowett* v. 84 [His] extraordinary energy..may, perhaps, have been derived..from a perpetual 'sublimation' of the energy which most men release in acts of sex. **1977** R. L. Wolff *Gains & Losses* vii. 404 *Zoe*..is the first novel to sound the notes which novelists were so often to repeat. Scepticism of Christian evidences, sublimation of doubt in sex, [etc.].

6. b. *Psychoanal.* The result of the refinement or transmutation of sexual or instinctual energy.

1926 *Internat. Jrnl. Psycho-Anal.* VII. 44 Thus Leonardo's genital activity..was wholly merged in his sublimations. **1955** H. Hartmann in A. Freud *Psychoanal. Study of Child* X. 13 We know much more about the origin of specific contents of sublimations. **1973** *Jrnl. Genetic Psychol.* Mar. 153 It is out of the basic societal repression/inhibition of drives that sublimations are born.

Hence **sublima·tional** *a.*

1934 in Webster. **1935** *Mind* XLIV. 348 Sublimational, substitutional or Changeling psychology may be Freudian, but it surely is not the only 'scientific' psychology. **1943** A. Huxley *Let.* 4 Mar. (1969) 487 A revival of cerebrotonic philosophy in some..form, with a practical system of sublimational outlets, seems to be the only hope.

sublimatory, *a.* Restrict † *Obs.* to senses in Dict. and add: **3.** *Psychoanal.* Pertaining to sublimation of instinctual energy or of the sexual drive.

1943 A. Strachey *New Ger.-Eng. Psycho-Anal. Vocab.* 66 Relating to sublimation; sublimatory. E.g...sublimatory processes. **1955** H. Hartmann in A. Freud *Psychoanal. Study of Child* X. 16 We will tend to see in sublimation..a continuous process which..does not exclude temporary increases or decreases in sublimatory activities. **1968** *Psychoanal. Rev.* LV. 10 This concrete orientation occurred along with a reduced capacity for fantasy release or other sublimatory behavior. **1981** *Internat. Jrnl. Psychoanalytic Psychotherapy* VIII. 461 The newly liberated creative capacity permitted an important sublimatory release.

subliminal, *a.* Add: **c.** In collocations which denote exploitation of the idea that people can be unconsciously influenced by messages or other stimuli projected just below the threshold of awareness, as *subliminal advertising, propaganda*, etc.

1957 *Times* 18 Sept. 9/5 The report in your columns to-day from your New York Correspondent on subliminal advertising must be taken as a timely warning of an encroachment, if not upon the physical freedom, certainly upon the free will of the cinema and television audiences of the near future. **1957** *Technology* Nov. 328/4 The process—christened 'subliminal projection' because the message is transmitted at sub-threshold intensities—is ready for commercial exploitation. *Ibid.* 334/4 'Subliminal' propaganda—briefly flashing a suggestion on a cinema or television screen for subconscious observation. **1958** *Times* 5 July 7/2 A committee of the Institute of Practitioners in Advertising has reported on the subject of 'subliminal communication'. **1968** *Punch* 23 Oct. 563/1 Won't it [*sc.* the Government] use every trick in the book—including subliminal TV appeals and pressures—to make us buy more and more? **1975** *Perceptual & Motor Skills* XLI. 847 (*title*) Effect of subliminal stimuli on consumer behavior: negative evidence. **1981** J. E. Alcock *Parapsychol.* iv. 72 It is even unclear from the reports whether the increase in popcorn and cola sales occurred only after the exposure to subliminal advertising.

Hence **subli·minally** *adv.*, in a manner which is subliminal or below the threshold of sensation or consciousness.

1892 F. W. H. Myers in *Proc. Soc. Psychical Res.* VIII. 438 Similar subliminal activity is going on also along the *red to violet* spectrum of which we are supraliminally as well as subliminally cognisant. **1902** W. James *Var. Relig. Exper.* x. 237 Subjects who are in possession of a large region in which mental work can go on subliminally. **1963** *Observer* 7 Apr. 22/2 The way it [*sc.* a book] is presented, with a cover more than subliminally reminiscent of the Leopard. **1977** J. Gardner *Werewolf Trace* xxiv. 195 They also had a tape on the video link..which Harvester wanted to play through subliminally while Joseph watched the news.

sublingual, *a.* (*sb.*) **A.** *adj.* **1.** Delete † *Obs.* and add later examples.

1958 *Martindale's Extra Pharmacopœia* (ed. 24) I. 67 *Aleudrin*... Isoprenaline sulphate, available as Solution containing 1% for inhalation, and as Sublingual Tablets of 20 mg. **1980** *Amer. Speech* LV. 52 Nitroglycerine sublingual tablets given to heart patients for angina.

Hence **subli·ngually** *adv.*, under the tongue.

1945 *Proc. Soc. Exper. Biol. & Med.* LVIII. 185/1, 4000 units of penicillin in 0·05 gm of zephiran was administered sublingually to a man. **1961** *Lancet* 9 Sept. 587/2 Testosterone was given sublingually. **1980** *Monthly Index Med. Specialities* Feb. 76/3 Lingraine... Ergotamine tartrate 2mg; green tab. Migraine [etc.]... 1 sublingually at onset of attack.

su:b-lingui·stic, *a.* **1.** [Sub- 19.] Not fully linguistic; expressed in a level below that of language.

1933 L. Bloomfield *Language* ix. 148 These shortened forms occur in various languages; their relation to normal speech is obscure, but evidently they represent a kind of *sub-linguistic* communication, in which the ordinary meaning of the forms plays no part. **1956** J. Whatmough *Language* i. 7 May 'thought' be not merely sub-linguistic, but also non-linguistic, or both? **1977** 'A. Burgess' *Beard's Roman Women* v. 108 I'm American, so there won't be any language problem. Not that there'll be any need for language. Sublinguistic activity, let's call it.

2. [*sub- 5 c (*b*).] Of or pertaining to a sub-language.

1976 *Amer. Speech 1974* XLIX. 266 The changes involved in these expressions are not sublinguistic.

sublittoral, *a.* Add: (Further examples.) *spec.* in *Ecol.*, applied to the inshore biogeographic zone normally taken as extending from mean low tide to the edge of the continental shelf.

1909 E. Warming *Œcol. Plants* iv. xli. 172 Sub-littoral 'region':—Ranges from below low-tide mark down to a depth of twenty fathoms (40 metres); here algae of all colours are represented. **1931** [see *profundal *a.* and *sb.*]. **1937** T. A. & A. Stephenson in *Trans. R. Soc. S. Afr.* XXIV. 360 It [*sc.* a particular zone of a beach] is occasionally exposed to a considerable extent, when maximal spring tides coincide with calm weather. This region we propose to call the Sublittoral Fringe. **1971** *Nature* 9 Apr. 402/2 In the Mediterranean, *Dardanus* inhabits sandy bottoms with some exposed rocks..from sublittoral to moderate depths down to 100 m.

B. *sb.* The sublittoral zone.

1961 in Webster. **1964** V. J. Chapman *Coastal Vegetation* i. 1 So far as algae are concerned, the sublittoral will extend downwards to the point where algae cease to grow. **1980** Hiscock & Mitchell in J. H. Price et al. *Shore Environment* II. i. 333 The real downward extent of the sublittoral..is often ignored by field workers.

Hence **subli·ttorally** *adv.*, in sublittoral regions.

1964 *Oceanogr. & Marine Biol.* II. 260 Sublittorally, the distribution of the radioactivity on the sea bottom is not uniform. **1971** *Nature* 9 Apr. 402/2 All are essentially warm temperate to tropical species occurring sublittorally in the Mediterranean, the Eastern and South Atlantic.

sublunar, *a.* and *sb.* Add: **A.** *adj.* **2.** *Navigation* and *Astr.* Applied to a point on the surface of the earth which lies on a line joining the centre of the moon and the centre of the earth, i.e. a point at which the moon is vertically overhead.

1938 P. V. H. Weems *Air Navigation* (ed. 2) xvi. 263 The geographical position of a heavenly body is the point on the earth's surface that has the body in its zenith; in other words the substellar, subsolar, or sublunar point. **1971** *Nature* 31 Dec. 537/2 Most large earthquakes occur after the epicentral region has passed the sublunar or subsolar point.

su:b-machi·ne-gun. [Sub- 5 b.] A light portable machine-gun firing ammunition of the same type and calibre as a pistol. Cf. *Thompson.

1926 [see *Thompson]. **1934** *Sun* (Baltimore) 4 Dec. 3/5 None of these companies manufacture the deadly submachine gun in popular use by criminals. **1942** J. Steinbeck *Moon is Down* 2 Grey-helmeted men who carried sub-machine-guns in their arms. **1951** 'J. Wyndham' *Day of Triffids* vi. 112, I dropped down, pulling Jocella with me as the clatter of a sub-machine gun began. **1965** J. A. Michener *Source* (1966) 816 Little Vered in her boxlike hat came darting in with her submachine gun spurting. **1973** G. Greene *Honorary Consul* iii. iii. 132 A stranger stood there waving a sub-machine gun at him.

su·b-man. [*sub- 14 b.] A man of markedly inferior development or capacities.

1921 R. A. Freeman *Social Decay & Regeneration* 248 As we are accustomed to speak of a man whose bodily and mental qualities are such as to lift him far above the common level, as a super-man, so we may conveniently refer to one who is to a like degree below the average as a sub-man. **1939** Dylan Thomas *Let.* 14 Sept. (1966) 237 Submen from the islands of crabs. **1951** N. M. Gunn *Well at World's End* xxix. 277 The Cromagnons, those sub-men who painted their bisons on the walls of far interior caves. **1964** *Punch* 28 Oct. 655/2 Peter Cook with his sub-man monologues. **1981** P. Audemars *Gone to her Death* vii. 123 What do they do now? Send these apes—these sub-men—to a reform school.

submarginal, *a.* (*sb.*) Add: **2.** [Sub- 19.] Of land: not capable of being farmed profitably.

1930 *Economist* 9 Aug. 272/1 It permits the survival of sub-marginal farms which plainly ought to be driven out of cultivation by the operation of economic forces. **1938** *Encycl. Brit. Bk. of Year* 24/2 In the autumn of 1937 a long-term programme for agriculture was announced [in the U.S.]. It included:..crop insurance, retirement of submarginal land, and price adjustment payments. **1970** E. Flores in I. L. Horowitz *Masses in Lat. Amer.* ix. 336 Public lands could not be given to anybody simply because they were submarginal.

submarine, *a.* and *sb.* Add: The noun may also be stressed on the final syllable. **A.** *adj.* **1.** Also *fig.*

1917 T. S. Eliot *Prufrock & Other Observations* 35 His laughter was submarine and profound Like the old man of the sea's. **1925** A. Huxley *Those Barren Leaves* v. i. 348 He found himself adding, with a kind of submarine laughter below the surface of his voice: 'Do you think you *can* make an end?'

2. (Further examples.) *submarine boat* (earlier examples.)

Later examples tend to merge with the recent *attrib.* use of the *sb.* (sense 3 b below).

1807 T. Jefferson *Let.* 16 Aug. in *Writings* (1853) V. 165, I have ever looked to the submarine boat as most to be depended on for attaching them [*sc.* torpedoes to the cable of a ship]. **1818** *Monthly Mag.* Feb. 46/2 His boat at this time he called the submarine boat, or the plunging boat. **1897** *Knowledge* 1 Jan. 20/1 All the great naval Powers are busily engaged in bringing submarine warfare to a perfect system of attack. **1919** *Daily Mail Year Bk.* 76/2 One and a half year's unrestricted submarine war. **1940** *Times* 11 June 7/4 Someone has blundered—Dr. Goebbels, or the German submarine command, or both. **1979** O. Sela *Petrograd Consignment* 17 The chiefs of the [German] civil, military and naval cabinets..had decided to end the blockade by declaring unrestricted submarine warfare.

B. *sb.* **3. b.** *attrib.* and *Comb.* in many obvious uses. **submarine chaser,** a small patrol boat equipped for military operations against submarines; **submarine pen**: see *pen *sb.*¹ 2 e; **submarine scout** = *blimp 1.

1908 C. Field *Story of Submarine* 183 The rise of the Russian submarine flotilla. **1914** C. M. Domville-Fife *Submarines* 9 The submarine fleets of England, France, Russia, [etc.]. **1915** W. E. Dommett *Aeroplanes & Airships* vi. 75 In place of an enemy camp or railway junction, we get the submarine base or dockyard. **1917** *Daily Mail* 5 Mar. 5/4 Expert officials of the [U.S.] Navy department devised a scheme for placing on board merchant ships.. two or three small boats as submarine-chasers. **1917** *Jane's All the World's Aircraft* 78c The 'Blimp' or Submarine Scout, evolved by the British Naval Air Service. **1931** W. G. Carr *By Guess & by God* 261 Lieutenant Johnson was 'submarine-minded'. He loved them. **1941** *Hutchinson's Pict. Hist. of War* 14 May–8 July 193 Fast revenue cutters of the U.S. coastguard service..are now in service with the Royal Navy as submarine chasers. **1942** *Sun* (Baltimore) 17 Mar. 1/5 That the submarine-killer is not instantly available is due to the same old cause—we did not start early enough. **1954** P. K. Kemp *Fleet Air Arm* 35 The first of the famous S.S., or Submarine Scout, airships. **1959** *Encounter* Jan. 13/1 The Polaris submarine-launched missile. **1975** B. Meyrick *Behind Light* xiv. 190 They had to raise and lower the submarine nets. **1979** J. Sherwood *Hour of Hyenas* xiv. 163, I know Georgiades' boat..former Nazi navy submarine chaser.

4. *slang.* † **a.** A doughnut. *U.S. Obs.*

1916 *Independent* 9 Oct. 77 'Two submarines and a mug of murk—no cow!' orders the waiter. **1942** Berrey & Van den Bark *Amer. Thes. Slang* § 91/25 *Doughnuts*,.. sinkers, submarines.

b. A type of sandwich; = *hoagie; *poor boy* (*sandwich*) s.v. *poor *a.* (*sb.*) 8. Freq. *attrib.* as *submarine roll, sandwich*. Cf. *sub *sb.* 8 b. Chiefly *U.S.*

1955 *Sat. Even. Post* 1 Jan. 16/2 The submarine is a noble edifice built of meats, cheeses, fish—preserved and pickled—and fresh vegetables and greens, all stuffed into a whole long loaf of bread and laved generously with oil, herb-flecked vinegar and other delicious lubricants. **1961** Webster, Submarine sandwich. **1967** *Amer. Speech* XLII. 279 (*title*) The submarine sandwich: lexical variations in a cultural context. **1973** Submarine roll [see *hoagie]. **1973** *Kingston* (Ontario) *Whig-Standard* 11 July 7/2 Who are the biggest fans of the Jaycee beer garden where beer and submarine sandwiches are sold? **1979** *Tucson* (Ariz.) *Mag.* Sept. 68/2 Real bargains in pizzas, submarines and dinner platters.

submarine *v.*: (*b*) *intr.*, to act or move like a submarine (*fig.*); **submariner** (-mæ·rinəɹ), a member of the crew of a submarine; **su·bmarining** (also -ī·ning) *vbl. sb.*, the use of or activity with submarines; also *attrib.* and as *ppl. a.*; **submarinism** *disused* = *submarining* vbl. sb. above.

1911 *Chambers's Jrnl.* Feb. 170/1 'Ought to prove a tidy job for us, though,' he muttered with some anxiety, 's'long as she don't take to submarinin' first.' **1915** *Times* 1 Feb. 9/3 All is fish which comes into the net of the submariner. **1915** *Glasgow Herald* 30 Mar. 8 The commander of the U16, which sank the British steamer Dulwich and the French vessels Ville de Lille and Dinorah, discussed 'submarinism' from the standpoint of one who had experience. **1915** *Spectator* 13 Nov. 645/2 America's last word as to submarining in the North Sea. **1917** R. Lord *Captain Boyd's Battery, A.E.F.* (1919) 24 *Submarine*,.. let's submarine, etc.—to submerge, to make onself scarce in the presence of impending duty. **1918** G. Frankau *One of Them* xxii. 170 That Hand before whose Thumb the Cave-men bow, Whose oiled Palm guides the submarining mermans. **1927** *Westm. Gaz.* 23 May 6/2 The war advanced the study of submarinism and aviation. **1946** G. Millar *Horned Pigeon* viii. 96 A submariner named Mike Caplatt produced an original musical comedy. **1966** M. R. D. Foot *SOE in France* iv. 62 The most active submarining spell. **1971** *Wall St. Jrnl.* 12 Mar. 1/4 Occupants [of a crashing car] could 'submarine' under an inflating bag and thus not be protected by it. **1972** J. Broome *Convoy is to Scatter* i. 24 Submarining then was a human rather than a technological way of life, full of individuals who stamped one's memory with their character. **1976** J. Lee *Ninth Man* i. 5 If they stumbled into trouble on the beach, they could claim they were submariners. **1981** *Sunday Times* 26 Apr. 3/5 The 10-year-old dummy [in a test of safety belts] 'submarined'—that is, he slid forward under the belt, which tightened across his stomach and around his neck.

submerge, *v.* **3.** Delete 'Now *rare*' and add later examples.

1903 A. H. Burgoyne *Submarine Navigation* II. 162 Having reached the 'limit of visibility' it becomes necessary to submerge. **1915** *Glasgow Herald* 30 Mar. 8 In the vicinity of the enemy or when weather conditions make it necessary we submerge. **1930** W. Faulkner *As I lay Dying* 146 We submerge in turn, holding to the rope, being clutched by one another. **1958** J. Lewis in C. S. Lewis *Lett. to Amer. Lady* (1969) 72 He comes up for air now and then, blows a few pathetic bubbles, then submerges again. **1974** P. Lovesey *Invitation to Dynamite Party* xiv. 172 Put the boat in diving trim... To submerge, push down the ballast-levers.

submerged, *ppl. a.* Add: **a.** Also *Naut.*, operating or being under water (esp. of or relating to a submarine).

1902 H. C. Fyfe et al. *Submarine Warfare* 258 When running submerged the submarine is lighter than her displacement. **1914** C. W. Domville-Fife *Submarines* 10 Each of the 1,500 surface warships..carries the means for delivering submarine attacks in its torpedoes and submarines and submerged discharging tubes. **1928** C. F. S. Gamble *N. Sea Air Station* 311 Their maximum submerged speed was, for a limited period, as much as 9 knots.

c. Engin. *submerged-arc welding*, a method of arc welding with a bare metal electrode in which both arc and electrode tip are entirely covered by a loose flux powder fed to the welding area.

1945 *Industry & Welding* Apr. 78 (*heading*) Submerged arc welding steps up production of invasion boat assemblies. **1952** [see *burden *sb.* 6]. **1975** Bram & Downs *Manuf. Technol.* ii. 57 The submerged-arc welding process must be automatic.

submergence. **b.** (Earlier example.)

1871 Geo. Eliot *Middlemarch* (1872) I. i. iii. 33 The secondary importance of ecclesiastical forms and articles of belief compared with that spiritual religion, that submergence of self in communion with Divine perfection.

submergible, *a.* Delete *rare* and add later examples.

1936 *World Petroleum* VII. 246 (*heading*) Submergible barges for Gulf Coast drilling. **1977** *Offshore Engineer* May 98/3 A submergible, electrically driven tide recorder is available from Benthos.

submersible, *a.* and *sb.* Add: **B.** *sb. spec.*, a small submersible vessel designed for use in underwater exploration or drilling or recovery operations, etc.

1959 *Time* (Atlantic ed.) 23 Mar. 56/2 The *Skipjack* is the consummation of a long program to give the U.S. its first true submersible designed primarily for underwater work. **1967** *Listener* 18 May 657/2 Three of the bombs were soon recovered. The fourth—in the sea—nearly eluded an armada of warships, 'submersibles', and 'submarines'. **1973** D. Kyle *Raft of Swords* (1974) vi. 50 An oil company..had..been considering the purchase of a German submersible..small submarines for underwater industrial and defence use. **1980** *Daily Tel.* 14 Nov. 17 The film never gets round to it, being so preoccupied with salvage vessels, submersibles and diving bells, trying to locate, miles down, the dear old Titanic.

su:b-microsco·pic, *a.* [Sub- 14.] Too small to be seen even with the aid of a microscope; also *absol.*

1912 *Chem. Abstr.* VI. 1014 (*heading*) Methods for the recognition of submicroscopic structures. **1938** S. Chase *Tyranny of Words* iii. 20 The submicroscopic, which we do not consciously see or feel. **1954** *Sat. Rev.* (U.S.) 19 June 48/1 'Reality' is apperceived on three levels: macroscopic, microscopic, sub-microscopic. **1978** H. McLeave *Borderline Case* (1979) xiii. 133 A submicroscopic particle that struck and then went to earth.

Also **su:b-microsco·pical** *a.*; **su:b-microsco·pically** *adv.*

1934 *Amer. Jrnl. Sci.* CCXXVII. 284 A regularly repeated twinning of submicroscopically small units can give rise to an apparently homogeneous crystal. **1949** *Jrnl. Iron & Steel Inst.* CLXIII. 270/2 This heat might melt a sub-microscopically thin layer of metal. **1953** C. Wakeley *Med. Dict.*, Sub-microscopical. **1954** *Ann. Reg. 1953* 373 The determination of the structure of molecules in the sub-microscopical genes that regulate heredity. **1961** *Lancet* 2 Sept. 546/1 We have to continue the search for the basis of this specificity at the sub-microscopical level. **1976** *Dermatologica* CLIII. 209 No submicroscopical alterations of the cytoplasm were found in the interacting cells. **1981** *Acta Crystallogr.* A. XXXVII. 754/1 The diffraction of such submicroscopically intergrown twins is calculated for lamellae and blocks.

submi·niature, *a.* [Sub- 14.] Even smaller than what is described as 'miniature'; very much reduced in size. Chiefly used in *Electronics* and *Photogr.*

1947 *Electronics* June 160/2 (caption) Oscillator circuits of the two units..are printed on the outer surface of a steatite cylinder housing the subminiature tube. **1956** *Spaceflight* I. 28/1 There are two alternative designs, one using subminiature valves and the other transistors. **1968** *Amateur Photographer* 1 May 13/2 Sizes [of camera] below 35mm are usually called sub-miniature. **1977** 'J. le Carré' *Hon. Schoolboy* vi. 123 Four lozenges of sub-miniature film..and a battered subminiature camera.

Hence **submi:niaturiza·tion**, the development or use of subminiature devices, esp. in electronics.

1949 *Aviation Week* 11 Apr. 18/1 Not content with 'miniaturization' of electronic equipment for airborne installations, engineers are now utilizing 'subminiaturization' of this material to reduce its size and weight. **1957** *Circulation* XVI. 764/1 The transition from antisubmarine warfare to phonocardiography involved mainly subminiaturization of the transducer and adaptation of the amplifiers to the recording instruments used in routine clinical phonocardiography. **1960** *Analog Science Fact/Fiction* Nov. 108/2, I don't know what genius indulged his yen for subminiaturization,..but he carried it too far.

submission. Add: **I. c.** In legal use, a theory of a case put forward by an advocate. Cf. Submit *v.* 7.

1922 *Westm. Gaz.* 20 Dec. 7/1 In my submission..this woman was called by the police as the only corroboration which they produced. **1923** *Ibid.* 4 Jan. 3/7 It was put to the Court that there should be no difference in the rates of wages for similar work in different localities,..but the Court could not uphold this submission. **1976** *Daily Tel.* 20 July 3/2 In my submission it is nonsense.

submucosal, *a.* Add: (Examples.) Also, = Submucous *a.* 2 b.

1951 Whitby & Hynes *Med. Bacteriol.* (ed. 5) v. 55 The sub-mucosal lymphatic plexus is much nearer to the surface than that of the skin. **1975** *New Yorker* 22 Dec. 75/1 There was no evidence of any intradermal, submucosal, or subcutaneous hemorrhaging. **1978** *Jrnl. R. Soc. Med.* LXXI. 359 The ileal spout was excised and showed a significant submucosal haematoma.

Hence **submuco·sally** *adv.*

1977 *Lancet* 8 Oct. 771/2, 0·5–1·0 ml was injected submucosally.

subna·sal, *a.* **1.** [Sub- 1 b.] *Anat.* (See quots.)

1882 A. Thomson et al. *Quain's Elem. Anat.* (ed. 9) I. 80 Subnasal or spinal point, the middle of the inferior border of the anterior nasal aperture at the base of the nasal spine. **1951** B. Z. Seligman *N. & Q. Anthrop.* (R. Anthrop. Inst.) (ed. 6) i. 11 *Nasal height*,..the distance between the nasion, which is the point of junction of the nasal and frontal bones in the midline, and the subnasal point, which is the point at which the lower end of the nasal septum meets the upper lip.

2. [Sub- 19.] Not quite or somewhat nasal in tone. *rare*⁻¹.

1936 R. Lehmann *Weather in Streets* i. ii. 39 Cool voice, with an edge of sub-nasal gentility.

subnormal, *sb.* Add: **2.** [Sub- 14: see sense b of the adj., below.] *Educ.* and *Psychol.* One who is below normal in academic or general ability.

1916 L. M. Terman *Measurement of Intelligence* vi. 78 Conversely, we may say regarding the subnormals that:—the child testing at (about) 90 is equaled or excelled by 80 out of 100. **1956** J. F. Horner *Summary of Scientology* 15 The children who get extra attention are the subnormals. **1975** N. O'Connor in Kirman & Bicknell *Mental Handicap* iv. 102 The generally slowed reaction time to both visual and auditory, simple and complex stimulus-response situations which characterizes all subnormals.

subnormal, *a.* Add: **b.** *Educ.* and *Psychol.* Of a level of intelligence and general ability which is below a predetermined standard of normality. See also *educationally subnormal* adj. phr. s.v. *educationally *adv.*

1919 H. Woodrow *Brightness & Dullness in Children* ii. 22 In 1904, the French Minister of Public Instruction made him [*sc.* Binet] a member of a commission appointed for the purpose of organizing classes for subnormal children... How were subnormal children to be positively distinguished? **1935** C. L. Burt (*title*) The subnormal mind. *Ibid.* ii. 77 In another 12 per cent. the parent, though not intellectually subnormal, was more or less unstable. **1940** A. O. Heck *Educ. Exceptional Children* xxiii. 342 Frequently, children are referred to as bright, average and subnormal. **1958** K. Lovell *Educ. Psychol.* xii. 149 Those whose IQ's are within the range from about 80–55 will usually be classified as educationally subnormal. **1975** N. O'Connor in Kirman & Bicknell *Mental Handicap* iv. 102 Within the subnormal group, abnormal EEG was a poor indicator for prognosis.

subnormality. Add: **b.** *spec.* The condition of being mentally subnormal; *severe subnormality* (see quot. 1959). Also *Comb.*, as **subnormality hospital**, a hospital for patients who are severely subnormal.

1935 C. L. Burt *Subnormal Mind* i. 8, I shall restrict myself to those forms of subnormality which seem to be chiefly mental in their origin. **1959** *Mental Health Act* 7 & 8 *Eliz. II* lxxii § 4 (2) In this Act 'severe subnormality' means a state of arrested or incomplete development of mind which includes subnormality of intelligence and is of such a nature or degree that the patient is incapable of living an independent life. **1965** *Mod. Law Rev.* XXVIII. v. 580 Persons who..receive local authority services for sub-normality. **1968** *Economist* 21 Dec. 37/2 Had she been of below average intelligence she could have been sent to a subnormality hospital. **a 1974** R. Crossman *Diaries* (1977) III. 664 No, out on tour studying the sub-normality hospitals, but I thought I would come down and put in an appearance.

subnu·clear, *a.* *Physics.* [Sub- 14.] Smaller than or occurring in an atomic nucleus; relating to such particles or to phenomena involving them.

1964 *New Scientist* 20 Feb. 458/1 The frontier of physics has lain, since the war, in the field of sub-nuclear particles. **1964** M. Gowing *Britain & Atomic Energy 1939–45* 16 These two theories [of Schroedinger and Bohr]..form the basis of all modern atomic, nuclear, and sub-nuclear physics. **1969** *Times* 5 Feb. 13/7 The classification is part of the conceptual order which physicists are trying to impose on the sub-nuclear particles. **1981** C. H. L. Smith in J. H. Mulvey *Nature of Matter* iii. 61 In the 1930s.. protons and neutrons were the only sub-nuclear particles known.

suborbital, *a.* and *sb.* Add: **A.** *adj.* **2.** [Sub- 19.] Being or having a trajectory that does not make a complete orbit of a planet.

1959 *N.Y. Times Mag.* 11 Oct. 18/1 The moment has come, after months of training, testing and short, sub-orbital flights, when one of seven carefully chosen men climbs into a space capsule perched high on the nose of an Atlas rocket. **1967** *New Scientist* 16 Nov. 424/1 The Soviet Union seems to have developed a sub-orbital missile, and the implications of the new weapon have been quickly realized. **1977** A. Hallam *Planet Earth* 28/2 This hypothesis demands that the impact is sufficiently catastrophic to vaporize large amounts of surface and subsurface rock, the gases being ejected into suborbital trajectories.

subordinate, *a.* Add: **2. b.** *subordinate legislation* Law (see quots.).

[**1841** G. C. Lewis *On Govt. of Dependencies* 52 Legislation is subordinate when the sovereign person or body delegates the legislative power to an inferior authority, which issues or makes the law.] *Ibid.*, A power of subordinate legislation is sometimes *direct*;..the laws made in virtue of it are issued avowedly..by the subordinate legislature. **1901** C. Ilbert *Legislative Methods & Forms* p. v, Chapter III deals with what I have called subordinate legislation, that is to say, that part of the law which is enacted, not directly by the supreme legislature, but under delegated powers. **1917** *Erskine May's Treat. Laws Parl.* (ed. 12) xxiii. 567 Something must be said here of those administrative orders, rules and regulations which constitute what is sometimes called delegated or subordinate legislation. **1975** J. P. Morgan *House of Lords & Labour Government* ii. 63 An equally important function is their work on subordinate legislation, a subject that requires a section to itself. **1980** *Oxf. Compan. Law* 758/1 Subordinate or delegated legislation takes many forms, rules, regulations, and orders made by Ministers of the Crown, frequently in the form of statutory instruments; [etc.].

subo·rdinator. [f. Subordinate *v.* + -or.] Something which subordinates; *spec.* in *Gram.*, a subordinating conjunction.

1962 C. L. Barber in F. Behre *Contrib. Eng. Syntax* 25 Clauses of time.., most of them introduced by *when*..; the next commonest subordinators in these clauses are *until*..and *while*. **1965** *Language* XLI. 242 Connectors are divided into subordinators (subordinating conjunctions)..and coordinators. *Ibid.*, Subordinators link the sentence they introduce with either a preceding or a following superordinate structure. **1978** *Ibid.* LIV. 140 In only one case in our data did we observe Level I possibly in use as a status subordinator between adults.

subpœna, *sb.* Add: **2. c.** In Lat. phrases: *subpœna ad testificandum* [L., in order to testify], *subpœna duces tecum* [Duces tecum] (see quot. 1980).

1768 Blackstone *Comm.* III. xxiii. 382 In the hands of third persons they [*sc.* books and papers belonging to the parties] can generally be obtained by rule of court, or by adding a clause of requisition to the writ of *subpoena*, which is then called a *subpoena duces tecum*. **1808** E. H. East *Rep.* IX. 476 The precedents of the common subpœna ad testificandum are scarcely more ancient than that of the subpœna duces tecum. **1891** *Weekly Notes* 12 Dec. 195/1 This was a motion..to set aside a writ of *subpœna duces tecum*. **1909** *Law Jrnl. Rep. King's Bench Div.* LXXVIII. 120/2 This case must not be taken as a precedent for any supposed rule that a person summoned on *subpœna ad testificandum* may get it set aside by swearing that he can give no relevant evidence. **1944** *All England Law Reports* (1945) I. 274 The party on whose behalf the motion is made has been required by a *subpœna ad testificandum* and *duces tecum* to appear before the district auditor. **1965** *Annual Practice* I. 881 Any party in any cause or matter may by *subpœna ad testificandum* or *duces tecum* require the attendance of any witness before an officer of the Court. **1980** *Oxf. Compan. Law* 1195/1 It [*sc.* a subpœna] takes two forms, *subpoena ad testificandum*, when the recipient is called to give evidence, and *subpoena duces tecum*, when he is required to bring documents or papers relevant to the controversy for examination by the court.

su·bpopulation. **1.** (In Dict. s.v. Sub- 3.)

2. a. [Sub- 7 a.] A population forming part of a larger population. **b.** [Sub- 9.] One derived or originating from some other population.

1959 *Heredity* XIII. 217 This gene flow..prevented the divergence of the two sub-populations. **1961** *Lancet* 9 Sept. 586/2 Serious outbreaks have occurred mainly because substantial poorly vaccinated subpopulation groups have remained—for example, lower socioeconomic groups in the U.S.A. **1964** S. Lieberson in J. A. Fishman *Readings Sociol. of Lang.* (1968) 553 Such subpopulations as social classes, juvenile delinquents, racial, ethnic, and tribal populations, age groups, regions, and occupational groups. **1971** J. Z. Young *Introd. Study Man* xxvii. 385 Like all populations, ours is composed of many sub-populations, differing from each other not only outwardly but in their gene structure. **1974** J. W. Drake in Carlile & Skehel *Evolution in Microbial World* 53 Microbial subpopulations frequently become extinct. **1979** *Sci. Amer.* Mar. 53/1 Could it be that they represent a sub-population of tumor cells endowed with the particular characteristics making for successful metastasis?

su·bprogram. *Computers.* [Sub- 5 c.] = *subroutine.

1947 *Math. Tables & Other Aids to Computation* II. 358 Nor can it [*sc.* a computer] be directed to repeat automatically sub-programs within the same total program. **1965** *Math. in Biol. & Med.* (Med. Res. Council) iv. 205 All the sub-programs that enter into the fidac system are listed in a manual, which specifies for each what user-input parameters are required and what values they may take on. **1979** *Sci. Amer.* Dec. 87/1 The most important technique for limiting the complexity of computer programs is the use of subprograms: self-contained pieces of programming that are named, stored in a library and called on to perform their particular computation as part of the execution of other programs.

su·b-range. **1.** [Sub- 5 b.] A subsidiary range (of mountains).

1859 [in Dict. s.v. Sub- 5 b].

2. [Sub- 7 c.] A range of values or conditions within a larger range.

1956 A. A. Townsend *Struct. Turbulent Shear Flow* iii. 45 In this subrange [of Reynolds numbers], the motion is independent of the viscosity. 1968 Fox & Mayers *Computing Methods for Scientists & Engineers* iii. 50 The computed value at r_e, and the given $y_0 = a$,..provide boundary values for solution by linear equations in this first sub-range.

subregion. Add: (Later examples, chiefly in non-geographical use.)
1898 A. N. Whitehead *Treat. Universal Algebra* I. i. 125 A region defined by any ρ independent letters lying in a region of ν−1 dimensions, where ρ is less than ν, is called a subregion of the original region. 1959 G. & R. C. James *Math. Dict.* 374/1 *Subregion*, a region within a region. 1974 *Nature* 11 Oct. 531/1 The periventricular areas of the hypothalamus were further dissected into four subregions and assayed for adrenaline. 1977 *Verbatim* Dec. 7/2 Cultural maps, such as those provided by Odum and Vance, would have been more useful than his reprinted essay from PMLA in identifying the subregions of the South.
sub-regional *a.* (later examples).
1946 *Richmond* (Va.) *News Leader* 7 Feb. 3/3 The Richmond sub-regional office of the Veterans Administration will be open..until 5 p.m. 1966 [see *leisure *sb.* 6 a and c]. 1977 *Lancet* 14 May 1054/1 We were surprised to read..that the treatment of leukæmia should no longer be regarded as regional or subregional.

su·broutine. *Computers.* [Sub- 5 c.] A routine designed to be stored in a computer's memory so that longer, self-contained programs can make use of it any number of times without its being written into the program each time.
c 1946 Goldstine & von Neumann in J. von Neumann *Coll. Wks.* (1961) V. 25 Both..machines are controlled by instructions punched into several tapes and they can be ordered to switch from one to the other as desired. They are usually referred to as 'master routine' and 'subroutine' tapes. 1948, 1951 [see *routine *sb.* (*a.*) 1 d]. 1956 G. A. Montgomerie *Digital Calculating Machines* xii. 248 We then transfer control to this subroutine whenever it is required, and arrange for its last action to be the transfer of control back to the main programme. 1959 M. H. Wrubel *Primer of Programming for Digital Computers* iv. 100 Use the subroutine for sinh x and cosh x to construct a program for calculating f = sinh (x+y)/cosh x cosh y. 1973 C. W. Gear *Introd. Computer Sci.* iv. 156 Any built-in subroutines and functions required by the program are loaded with it. 1980 R. L. Duncan *Brimstone* xi. 272 'We..can't handle the master program... So we'll attack the subroutines.'.. The computer responded.

sub-Saha·ran, *a.* [Sub- 1.] Situated or originating in regions bordering on the Sahara desert.
1955 *Ann. Amer. Acad. Political & Social Sci.* Mar. 13/1 The fact that sub-Saharan Africa has so large a number of distinguishable languages makes impressive documentation. 1969 *Times* 22 Oct. (Ghana Suppl.) p. i/2 Ghana was the pacesetter for modern Africa when it became the first sub-Saharan black country to move from colonial status to independence. 1978 J. Updike *Coup* (1979) iii. 121 This French villa spun of sub-Saharan materials.

su·bsample. [Sub- 9(*b*).] A sample drawn from a sample.
1909 Webster, Sub-sample, *n.* & *v.t.* 1913 *Econ. Geol.* VIII. 134 Each sample has thus been divided into 10 subsamples which may be used to estimate roughly the probable error. 1939 *Brit. Jrnl. Psychol.* XXX. 76 Burt chose his subsample of persons to be not only equal in average to one another, but equal to the average of all. 1959 H. Barnes *Oceanogr. & Marine Biol.* i. 32 (*caption*) Stempel (Suction) pipette. Used for taking an aliquot from a plankton sample. The sub-sample is contained between the curved part of the plunger and the barrel of the pipette. 1972 H. J. Eysenck *Psychology is about People* ii. 92 The actual mean scores for P, E and N in the general population, and in various sub-samples graded by sex, age and class are known.
Hence **su·bsample** *v. trans.*, **su·bsampling** *vbl. sb.*
1909 Subsample *v.* [see the sb. above]. 1959 H. Barnes *Oceanogr. & Marine Biol.* i. 32 If larger nets are employed then it [*sc.* the catch] may have to be sub-sampled and only a fraction counted... There are various ways of such sub-sampling. 1969 R. Lange *Chem. Oceanogr.* v. 79 It is..useful to organize the numbers of the bottles for subsampling in such an order that [etc.]. 1971 *Nature* 4 June 290/2 They were subsampled for metal analysis and placed 0·8 m above ground in three locations downwind of Swansea.

subscriber. Add: **2. d.** One who pays a regular sum for the hire of a telephone line.
1878 (*title*) List of subscribers (Bell Telephone Co. of N.Y.). 1922 [see *dialling tone* s.v. *dialling *vbl. sb.* 4]. 1934 Haldane & Huxley *Animal Biol.* xii. 272 The telephone bells of all the subscribers would start ringing. 1978 *Broadcast* 6 Mar. 10/2 Viewdata is transmitted along normal telephone lines... Identifying the caller, the computer responds with a personalized greeting to the subscriber.
3. Special Comb.: **subscriber trunk dialling,** a telephone service by which subscribers can make trunk calls without the assistance of an operator, by dialling the exchange code and the number required; abbrev. *STD* (see *S 4 a).
[1950 *Post Office Electr. Engineers' Jrnl.* XLIII. 170/2 If, at some future date, subscriber-to-subscriber trunk dialling is introduced, a national numbering scheme and a translator trunk dialling system may be introduced.] 1952 *Prof. Papers Inst. P.O. Electr. Engineers* No. 203. 1 Subscriber Trunk Dialling in the United Kingdom... The possibility of extending the range over which subscribers can dial their own calls has received increasing attention by many telephone administrations. 1979 M. Underwood *Victim of Circumstance* III. iii. 183 With subscriber trunk dialling and unitemised telephone accounts, it's very difficult to trace calls.

subscript, *sb.* Add: **2. b.** *Computers.* A symbol (notionally written as a subscript but in practice usually not) used in a program, alone or with others, to specify one of the elements of an array.
1957 *Proc. Western Joint Computer Conf.* Feb. 190/1 The programmer may also employ subscripted variables having three independent subscripts. 1966 R. V. Jamison *FORTRAN Programming* vi. 83 We cannot write these subscripts in the usual lowercase manner with an actual lowering of the subscript. Instead we write, for example, X(4), X(7) for x_4, x_7. 1973 C. W. Gear *Introd. Computer Sci.* iii. 145 A three-dimensional array..is an array of two-dimensional arrays. It is accessed by specifying three subscripts. Thus, if the array is B, we can refer to B[I, J, K]. 1982 R. S. Forsyth *Pascal at Work & Play* xii. 172 Execution errors occur when a program attempts something illogical or impossible... Typical examples are division by zero and trying to use a subscript outside the bounds set for the array. 1983 [see *subscripted *a.*].
So **su·bscripting** *vbl. sb.*, the action of providing with a subscript or subscripts; the use of subscripts.
1959 *Communications Assoc. Computing Machinery* Feb. 4/1 Unfortunately, many algebraic languages now in use..do not allow subscripting of subscripts. 1972 Bergmann & Bruckner *Introd. Computers & Computer Sci.* x. 342 In FORTRAN the subscripting is different; the subscripts run from 1 through *n*. 1981 A. R. Miller *BASIC Programs* iii. 39 A matrix is referenced by its name, which can be a single alphabetic character, or a string of characters. The indices are given as subscripts except in computer programs, where subscripting is not possible.

subscripted (sɒ·bskriptid), *a.* [f. Subscript *sb.* + -ed².] Having a subscript, provided *with* a subscript; *spec.* in *Computing*, specified out of an array by means of a subscript or subscripts.
1957 [see *subscript *sb.* 2 b]. 1972 W. Labov *Language in Inner City* iv. 153 The subscripted parentheses indicate optional elements that co-occur. 1983 *Daily Tel.* 19 Sept. 12/6 Subscripted variables take the form A(x), x being the value (or address) of the subscript... Subscripted string variables, needed to store strings of text, include a dollar sign, thus: 40 LET A$(6) = 'BUG'.

subscription. Add: **10. a.** **subscription book,** (*b*) (earlier example); **subscription television** (also **T.V.**) *N. Amer.*, a television service which provides programmes for subscribers.
1870 'Mark Twain' *Lett. to Publishers* (1967) 31 You will make the finest success of it that has ever been made with a subscription book. 1955 *How to unscramble Subscription T.V.* (Zenith Radio Corporation, U.S.), Subscription T.V. can provide you and your family with the best of entertainment..major sports events..education..carefully prepared programs for your children..at a *nominal* price..when *you* want it..*if* you want it..and without ever having to leave the family room of your home. 1962 *Rep. Comm. Broadcasting 1960* 271 in *Parl. Papers 1961–2* (Cmnd. 1753) IX. 259 We now recommend that no service of subscription television be authorised. 1973 C. Sagan *Cosmic Connection* viii. 62 An unmanned roving vehicle on Mars could probably be supported by subscription television.

subsea·, *a.* and *adv.* Chiefly *Oil Industry.* [Sub- 1 e.] **A.** *adj.* Situated or occurring beneath the surface of the sea.
1909 [see *air-borne *a.*]. 1962 *Offshore* July 19 Equipment manufacturers have accelerated their research and development of sub-sea wellheads and auxiliary equipment. 1977 *Financial Times* 1 Apr. 11/1 Figures..suggest that over the life of North Sea oil development at least $20bn (at current prices) will be spent to make sure that platforms, pipelines, and sub-sea well systems are safe and in good working order. 1980 F. C. F. Earney *Petroleum & Hard Minerals from Sea* iv. 129 The oil companies' interest in subsea completion systems.
B. *adv.* Below the surface of the sea.
1971 *Bull. Amer. Assoc. Petroleum Geologists* LV. 1694/2 The gas-oil contact is at 6,636 ft subsea. 1977 *Offshore Engineer* May 49/1 The top of the reservoir formation is 1,800m subsea.

sub-sequence² (sɒ·bsīkwĕns). [Sub- 7 a, *e.] A sequence contained in or forming part of another sequence; *spec.* in *Math.*
1908 [see *oscillatory *a.* 3]. 1958 R. C. Moore *Introd. Hist. Geol.* (ed. 2) iv. 80 The second division of the Huronian Sequence, named the Cobalt Sub-sequence, has an aggregate thickness of more than 12,000 feet. 1972 A. G. Howson *Handbk. Terms Algebra & Anal.* xxii. 109 This definition formalises the notion of a subsequence as a sequence derived from the original sequence by the omission of a number (not necessarily finite) of terms. 1975 N. Chomsky *Logical Struct. Linguistic Theory* ix. 329 The term arrangement of *t** may not be the same as the term arrangement of the transformation *T* defined in terms of *t*, although the latter term arrangement must be a subsequence of the former.

subsequent, *a.* and *sb.* Add: **A.** *adj.* **2. d.** Applied to a stream or valley that has developed its course so as to follow rock that is more easily eroded, and consequently in most cases following the strike of the rock. (Further examples.)
1954 W. D. Thornbury *Princ. Geomorphol.* v. 113 Because of the coincidence of subsequent valleys with belts of weak rock it is usually concluded that any valley which follows such a course is a subsequent valley. This may not be true, for the valley may have been on the weak rock from the beginning. 1970 R. J. Small *Study of Landforms* vii. 233 The most significant feature..will be the appearance and growth of 'subsequent' streams, which by the process of headward erosion will extend along lines of geological weakness.
B. *sb.* **2.** *Physical Geogr.* A subsequent stream (see sense A. 2 d in Dict. and Suppl.).
1895 W. M. Davis in *Geogr. Jrnl.* V. 144 The Welland, Gwash, Chater, and Eye being parts of consequent streams that have been captured by subsequents. 1956 D. L. Linton *Sheffield* 42 Its headstream the Doe Lea has the aspect of a true subsequent, but from Staveley northwards though it maintains the direction of the Doe Lea this is no longer that of the strike of the rocks. 1970 R. J. Small *Study of Landforms* vii. 234 In an area where the structure comprises a series of anticlines and synclines, one is tempted..to regard all synclinal streams as longitudinal consequents and all anticlinal streams as subsequents.

su·bsere. *Ecol.* [Sub- 5.] A secondary sere (see quot. 1926). Cf. *prisere.
1916 [see *prisere]. 1926 Tansley & Chipp *Aims & Methods in Study of Vegetation* ii. 19 A new sere beginning after the succession has been stopped by the destruction of some later phase..we distinguish as a secondary sere or subsere. 1938 Weaver & Clements *Plant Ecol.* (ed. 2) iii. 78 Seres..on secondary areas, such as lumbered, burned, flooded, or otherwise denuded ones, are termed subseres. 1964 V. J. Chapman *Coastal Veg.* i. 3 Should an area of dune that has developed to forest become destroyed by burning a new succession would arise, but this would be known as a subsere.

subserve, *v.* **3. a.** Delete † *Obs.* and add later *poet.* examples.
1840 Browning *Sordello* iii. 533 Old engagements out he blots For aye: Taurello shall no more subserve. 1968 T. Kinsella *Nightwalkers* 60 We dwell together in urgency; Dominate, entering middle age; subserve, Aborting vague tendencies with buttery smiles.

subset, *sb.*² Add: A set all the elements of which are contained in another set. (Further examples.)
1911 *Trans. Amer. Math. Soc.* XII. 285 Among such [compact] assemblages one very important class are those which have the property that the first derived set (*E'*) of every subset (*E*) of (*D*) is closed. 1928 *Amer. Jrnl. Math.* L. 521 In studying these difficulties we are led to the introduction of a new notion of a subset of a point set being connected through the complement of the point set. 1961 M. A. K. Halliday in *Word* XVII. 276 Subsets progressively differentiated as the degree of collocational likeness set as defining criterion increases. 1968 E. T. Copson *Metric Spaces* v. 67 Let *a* and *b* be two points of a subset *A* of a metric space *M*. 1977 J. L. Harper *Population Biol. Plants* xiii. 416 There are 45 species of *Heliconius*, each specializing on its own sub-set of *Passiflora* species. 1980 *Amer. Speech 1976* LI. 165 We cannot always extrapolate from a corpus to the language in general or even to some subset of the language.

subshrub. (Later examples.)
1886 G. Nicholson *Illustr. Dict. Gardening* II. 168/1 Hypericum... An extensive genus..of greenhouse or hardy, evergreen or deciduous, herbs, shrubs, or subshrubs. 1957 R. E. Woodson et al. *Rauwolfia* i. 11 The smallest species appears to be..a truly rhizomatous subshrub only six inches tall. 1980 *Country Life* 13 Mar. 790/1 Among sub-shrubs profitably treated as herbaceous and cut back to the ground now is mint-scented, sun-basking *Elsholtzia stauntonii*.

subsidiarity (sǝbsidiæ·riti). [tr. G. *subsidiarität* (1931, paraphrasing Pope Pius XI in *Rundschreiben über die gesellschaftliche Ordnung* (*Quadragesimo Anno* § 80); cf. F. *subsidiarité* and Subsidiary *a.*] The quality of being subsidiary; *spec.* the principle that a central authority should have a subsidiary function, performing only those tasks which cannot be performed effectively at a more immediate or local level.
1936 B. W. Dempsey tr. *O. von Nell-Breuning's Reorganization of Social Econ.* x. 206 The Pope repeats the same statement..: 'Of its very nature, the true aim of all social activity should be to help individual members of the social body, but never to destroy or absorb them.' This is the frequently mentioned and famous principle of *Subsidiarity of Social Activities*, also called the principle of *Subsidiarity of Associations*, a fundamental principle of Christian social doctrine. 1964 S. Attanasio tr. H. Küng's *Structures of Church* vii. 215 Is there a *criterion* with respect to the exercise of the papal pastoral office in the individual dioceses?.. Post-Vatican theology developed a criterion: This is the *principle of subsidiarity* which, according to Pius XII, 'is valid for social life in all its organizations, and also for the life of the *Church* without prejudice to her hierarchical structure'. 1967 *New Catholic Encycl.* XIII. 762/1 The principle of subsidiarity is broadly concerned with the limits of the right and duty of the public authority to intervene in social and economic

affairs. 1976 J. P. Wogaman *Christian Method of Moral Judgment* v. 142 According to the doctrine of subsidiarity, as developed in various papal encyclicals, social problems should be dealt with at the most immediate (or local) level consistent with their solution. 1980 J. H. Whyte *Church & State in Mod. Ireland* (ed. 2) vi. 163 Concepts such as vocationalism, the principle of subsidiarity, and the danger of excessive State control. 1982 *Times* 18 Sept. 7/5 The 'principle of subsidiarity'—a meaningless or even misleading phrase in English—is being discussed in the European Parliament in connection with eventual revision of the Treaty of Rome. It is defined to mean that the European Community's activities should be limited to those which are better performed in common than by member states individually.

subsidiary, *a.* Add: **1. c.** *subsidiary company*, a company controlled by a holding company. Cf. Subsidiary *sb.* 2 c(*b*).
1916 F. G. Underhay *Income Tax* 272 (*Index*), Subsidiary company. 1928 *Daily Mail* 25 July 18/6 The net dividends received from the Subsidiary Companies amounted to £24,808 1s. 11d. 1970 M. Greener *Penguin Dict. Commerce* 170 A subsidiary company must state in its accounts the name of its ultimate holding company and the country where this is incorporated.

subsistence. Add: **11.** Also, with reference to the process of farming, esp. on poor agricultural land or with simple technology, merely to maintain a bare living, and without producing a significant surplus for trade, as *subsistence agriculture, crop, economy, farming; subsistence farm, farmer*; (sense 10) **subsistence level,** the economic level at which only the bare necessities of life can be provided; **subsistence wage,** the amount of money a person must earn in order to achieve a minimal standard of living.
1937 Subsistence agriculture [see *cash-crop* s.v. *cash *sb.*¹ 3]. 1940 Subsistence crop [see *subsistence farm* below]. 1940 E. Huntingdon *Princ. Econ. Geogr.* 711 (Index), Subsistence economy. 1962 R. S. Thoman *Geogr.* III. viii. 150/1 The crudest of subsistence economies need not involve markets. 1971 *World Archaeol.* III. 171 Hunting.. is considerably less important in the overall subsistence economy than foraging. 1940 Whitbeck & Williams *Econ. Geogr. S. Amer.* (ed. 3) ii. 57 Subsistence crops are found both on tiny subsistence farms and on the commercial plantations. 1951 A. L. Rowse *England of Elizabeth* vi. 231 He was no mere subsistence farmer. 1980 *Jrnl. R. Soc. Arts* Mar. 175/2 Let us take as an example..a subsistence farmer debating whether to produce a surplus for sale. 1949 W. Smith *Econ. Geogr. Gt. Brit.* I. i. 44 As long as subsistence farming was practised..price fluctuations had only a limited significance. 1962 *Listener* 22 Mar. 496/1 They are undergoing the transition from subsistence farming to being a landless proletariat. 1978 A. J. Huxley *Illustr. Hist. Gardening* i. 13 The final stage of collapse back into subsistence farming after conquest and pillage is..all too frequent. 1923 H. W. B. Joseph *Labour Theory of Value in Karl Marx* ii. 44 Marx believed in the so-called 'iron law of wages', in accordance with which there is a constant tendency under capitalism for wages to sink to the bare subsistence-level. 1978 A. J. Huxley *Illustr. Hist. Gardening* i. 11 Communities have to raise themselves above subsistence level before they can really afford to grow..plants not strictly utilitarian. 1926 *British Worker* 10 May 3 Mr. Herbert Smith told to running bursts of sympathy the story of the miners' patient struggle to retain a subsistence wage. 1982 *Washington Post* 17 July A4/4 Many..have also, it is alleged, worked long hours for subsistence wages.

subsolar, *a.* Add: **2.** Also *Astr.* (Further examples.)
1910 G. L. Hosmer *Textbk. Pract. Astron.* xiv. 175 If an observer measures an altitude of the sun he locates himself on the circumference of a circle whose centre is the sub-solar point. 1938 [see *sublunar *a.* 2]. 1970 N. Armstrong et al. *First on Moon* ix. 195 As you get closer to the subsolar point you can definitely see browns and tans on the ground. 1978 Pasachoff & Kutner *University Astron.* xv. 409 Because of Mercury's slow rotation, the subsolar point is not always at the same place on the surface and so is not eternally heated.

subso·nic, *a.* (and *sb.*). [Sub- 14.] **A.** *adj.* **1. a.** Pertaining to, involving, capable of, or designating speeds less than the speed of sound. Cf. *supersonic *a.* (and *sb.*) 2 a.
1937 *Jrnl. R. Aeronaut. Soc.* XLI. 1099 The drag coefficient rises..as the velocity of sound is reached, and.. remains very much higher than the sub-sonic figure. 1946 *Ibid.* L. 907/2 The National Physical Laboratory has examined the possibility of using flexible walls in high-speed subsonic tunnels. 1958 *Times* 9 Jan. 5/5 It is..to carry 95–100 passengers, cruising at high subsonic speeds of around 600 m.p.h. 1973 *Times* 5 Feb. 13/5 Capacity now exceeds demand for subsonic flight in the mass market. 1978 *Jrnl. R. Soc. Arts* CXXVI. 685/1 The breakdown of the direct operating costs (DOC) per passenger mile for a typical subsonic aircraft is indicated in Table III.
b. *ellipt.* as *sb.* An aircraft not made to travel faster than sound. Cf. *supersonic *a.* (and *sb.*) 2 b.
1970 *New Scientist* 10 Dec. 445/2 This aircraft will cruise at a higher altitude than the subsonics. 1975 *Nature* 31 Jan. 299/1 Ozone reduction could then be kept near the current reduction due to aircraft alone..for fleets up to 4,000 747-class subsonics at 11 km..or 1,000 at 13 km. 1977 *Jrnl. R. Soc. Arts* CXXV. 361/1, I originally planned to cover supersonics as well as subsonics.

2. = *infrasonic *a.* 1. Cf. *supersonic *a.* 1. *rare.*
1961 in Webster. 1976 *Gramophone* Sept. 513/2 Arm mass problems are reduced on record warps, causing less intermodulation and unwanted sub-sonic cone movement of the loudspeakers.
Hence **subso·nically** *adv.*
1962 *Economist* 8 Sept. 944/1 The aircraft have to fly mainly supersonically but in part subsonically. 1977 *Daily Tel.* 12 Dec. 2/7 The Indonesian decision to allow Concorde to fly subsonically through its air space.

su·bspace. [*sub- 7 e.] **1.** *Math.* A space (*space *sb.*¹ 16*) that is wholly contained in another space, or whose points or elements are all in another space.
1931 H. P. Robertson tr. *Weyl's Theory of Groups & Quantum Mech.* i. 20 ℜ is decomposed into mutually perpendicular sub-spaces. 1946 *Nature* 12 Oct. 513/1 He has studied problems of deformation of sub-spaces, of 'imbedding', of automorphism, and of the variation of multiple integrals. 1968 P. A. P. Moran *Introd. Probability Theory* v. 226 Particular subspaces of the space of all distributions. 1979 *Proc. London Math. Soc.* XXXVIII. 221 Let *X* and *Y* both be subspaces of the Hilbert space of all square summable sequences.
2. *Sci. Fiction.* A physical space in which motion and communication are supposed to occur free of their usual limitations, or subject to different laws.
1955 *Mag. Fantasy & Sci. Fiction* Aug. 106/1 The subspace radio had announced his coming, and preparations consonant with his exalted rank had been made. 1976 L. Niven in R. Bretnor *Craft Sci. Fiction* 180 *FTL, hyperdrive, hyperspace, subspace*, all refer to means of traveling faster than light in an otherwise relativistic universe.

subspecies. Hence **su:bspecia·tion** [cf. *speciation], the evolutionary development of a subspecies or subspecies.
1942 E. Mayr *Systematics & Origin of Species* vii. 169 Subspeciation, that is geographic variation, has actually brought about the formation of unquestionably new species of birds. 1952 E. O. Dodson *Textbk. Evolution* xvi. 314 Subspeciation is the ordinary prerequisite to speciation in the neo-Darwinian scheme. 1956 Peterson & Fisher *Wild America* iii. 43 Roger nearly always starts a difficult subject, like..subspeciation, when I'm threading my way through the stickiest London traffic. 1978 *Nature* 22 June 603/1 Four species [of *Plasmodium*]..are stipulated, three of which exhibit subspeciation.

substance. Add: **8. b.** *substance P* (Biochem.): an undecapeptide thought to be involved in the synaptic transmission of nerve impulses, esp. pain impulses.
[1931 Euler & Gaddum in *Jrnl. Physiol.* LXXII. 80 This standard preparation, which we call *P*, dissolved easily in water to form a practically clear solution.] 1934 Gaddum & Schild in *Ibid.* LXXXIII. 1 This unidentified substance has been known in the laboratory for some time as substance P, and will be referred to under this name. 1964 W. G. Smith *Allergy & Tissue Metabolism* vi. 71 Substance P, which is a pharmacologically active polypeptide..stimulates guinea pig ileum. 1979 *Sci. Amer.* July 69/3 Enkephalin and another peptide, substance P, have been implicated in pain perception, substance P with the transmission of pain-related impulses and enkephalin with their suppression.

sub-sta·ndard, *a.* [Sub- 14.] **1.** Of a quality or size less than that which is normally or officially regarded as standard.
1909 in *Cent. Dict.* Suppl. 1930 *Daily Express* 6 Oct. 13/1 The famous 'Marshella'..sub-standard [stockings]... Special lot..slightly sub-standard. 1947 *Mind* LVI. 313 A race of philosophers from whom it is the fashion to expect somewhat sub-standard ratiocination. 1952 C. P. Blacker *Eugenics: Galton & After* 312 The mother is frequently sub-standard mentally. 1964 D. MacArthur *Reminiscences* vi. 157 He took a substandard force and welded it into a weapon so deadly as to take command of the air whenever it engaged the enemy. 1975 M. Sullivan *Watch how you Go* i. 22 They lived all their long lives under sub-standard conditions, and accepted them.
2. Of speech: not conforming to standard usage; *spec.* employing forms which are widely used but are considered incorrect.
1933 L. Bloomfield *Language* iii. 50 In such communities the non-standard language can be divided, roughly, to be sure, and without a sharp demarcation, into *sub-standard* speech, intelligible at least, though not uniform, throughout the country, and *local dialect*. 1951 Trager & Smith *Outl. Eng. Struct.* 84, *I have knowed* can be called substandard or 'incorrect'. 1964 *English Studies* XLV. (Suppl.) 149 St. Mary's Lane, Lewes, is called 'Simmery Lane' in local sub-standard speech. 1977 *Word 1972* XXVIII. 264 It is much more prevalent in the north, especially in substandard and rural speech.
3. *Cinemat.* Of film: less than 35 mm. wide; *spec.* 16 mm. wide.
1934 *Discovery* Feb. 47/2 [Films] shot on standard 35 mm film-stock before being reduced to the sub-standard non-flam 16 mm normally required for class-room use. 1940 *Chambers's Techn. Dict.* 818/1 Sub-standard sizes in use are 17·5, 16, 9·5 and 8 mm., as contrasted with the standard 35 mm. 1953, 1959 [see *narrow gauge 2].

substantial, *a.* Add: **9.** (Later examples.) More recently also in a somewhat weakened sense, esp. 'fairly large'.
1957 W. S. Churchill *Hist. Eng.-Speaking Peoples* III. 218 A substantial section of the population, which included the most prominent if not always the most powerful of French citizens, were largely exempt from taxation. 1976 *Sunday Times* 30 May 24/4 (Advt.), It is mandatory that candidates have experience of..the control and motivation of a substantial work force.

substantia nigra (sɒbstæ·nʃⁱă nəi·gră). *Anat.* [mod.L., = black substance.] A curved layer of grey matter in the brain that extends from the pons to the subthalamic region on each side, separating the tegmentum of the midbrain from the crus cerebri, and forming part of the extra-pyramidal system.
1882 *Quain's Elements Anat.* (ed. 9) II. 315 A section into the crus cerebri shows the two parts of which it is composed to be separated from one another by a tract of dark coloured grey substance known as the substantia nigra. 1923 [see *Parkinsonism]. 1961 *Lancet* 26 Aug. 446/1 A tiny old hæmorrhage in one substantia nigra, and a small periaqueductal glial scar, were the only focal lesions found. 1976 Smythies & Corbett *Psychiatry* iv. 41 The dopamine system is divided into several portions. The main one has its cell bodies in the substantia nigra.

substantive, *a.* **1. e.** (Earlier example.)
1854 T. Troubridge *Let.* 30 Dec. (MS.), I daresay they will make my Brevet rank *substantive* which is the new word they have coined for a *real* Lt. Colonelcy.

su·bstation. [Sub- 7 d.] **1.** A building or establishment subordinate to a principal station or office.
1881 *Rep. Indian Affairs* (U.S.) 45 The temporary establishing of two substations for the police..resulted in the locking up of the offenders. 1891 [see Sub- 7 d]. 1933 *Jrnl. R. Hort. Soc.* LVIII. 162 Only two varieties have been recommended for extended trial at the sub-stations. 1982 'M. Hebden' *Pel & Bombers* ii. 12 The sous-brigadier who ran the substation at St. Blaize.
2. A station at which electrical current is switched, transformed, or converted, intermediate in rank between a generating station and a low-tension distribution network.
1901 [see Sub- 7 d]. 1933 *Archit. Rev.* LXXIII. 2 The scheme of building included..a university sub-station for electric power which supplies the laboratories. 1968 [see *feeder 10 a]. 1976 P. R. White *Planning for Public Transport* viii. 162 Substation costs were greatly reduced, and more powerful locomotives permitted.
3. *Telephony.* (See quots.)
1922 W. Aitken *Automatic Telephone Systems* I. 3 We have, therefore, a *sub-station* at the subscriber's office, which may consist of one, or a plurality of direct lines, or a small switchboard. 1940 *Chambers's Techn. Dict.* 818/1 *Sub-station* or *subscriber's station* (*Teleph.*), a subscriber's telephone located on his premises.

subste·llar, *a.* **1.** *Navigation* and *Astr.* [Sub- 1 a.] Applied to a point on the surface of the earth which lies on a line joining some particular star and the centre of the earth, i.e. a point at which the star is vertically overhead.
1910 C. L. Poor *Nautical Sci.* vii. 154 To find the substellar point..it is necessary for the observer to know the sidereal time at which the observation is made. 1938 [see *sublunar *a.* 2]. 1967 P. van de Kamp *Princ. Astrometry* ii. 19 The terrestrial longitude and latitude of a substellar point equal the Greenwich Hour Angle (GHA) and declination of the star.
2. *Astr.* [Sub- 14.] Much smaller than a typical star.
1973 *Physics Bull.* Nov. 648/1 The shock wave preceding a black hole of substellar mass could explain both the scale of damage caused and the visual effects observed. 1978 *Nature* 24 Aug. 781/1 The resolution and sensitivity of the spinning infrared interferometer suggest other applications, for example, to cool features other than planets (such as dust rings or substellar companions).

substituend (sɒbsti·tiuĕnd). [ad. L. *substituend-us*, gerundive of *substituĕre* Substitute *v.*] A thing that can be put in the place of another, *spec.* in *Linguistics* or *Logic.*
1955 N. Chomsky *Logical Struct. Linguistic Theory* (microfilm, Mass. Inst. Technol.) viii. 432 We would naturally expect the substituend Y_1 to play the same role in the overall structure of the sentence as did the term Y_2 for which it is substituted. 1965 *Philos. Rev.* LXXIV. 473 We must have..an appropriate nominal *substituend* of *x*. 1976 *Analysis* XXXVI. 81 The anaphoric substituend of a proform, if there is one, is (possibly a grammatical variant of) its antecedent.

substitutable, *a.* Delete *rare* and add later examples.
1905 W. James *Meaning of Truth* (1909) v. 132 Reality ..is always defined as a terminus within the general possibilities of experience; and what knows it is defined as an experience *that 'represents' it, in the sense of being substitutable* for it in our thinking. 1941 *Mind* L. 167 Both the original word and all signs substitutable for it stand independently for universals. 1961 R. Brain *Speech Disorders* v. 57 Even a phoneme is abstract, since it stands for and represents an indefinite variety of mutally substitutable phones. 1980 A. Kenny *Aquinas* ii. 58 It is tantamount to a predicate variable for which no predicate is substitutable.
Hence **substitutabi·lity.**

1907 W. James *Meaning of Truth* (1909) vii. 175 The relations..which we epistemologists study, relations of adaptation, of substitutability, of instrumentality, of reference and of truth. 1922 tr. *Wittgenstein's Tractatus Logico-Philosophicus* 171 Equations express the substitutability of two expressions. 1957 A. C. L. Day *Outl. Monetary Econ.* xxxi. 395 There may also be a high or low degree of substitutability in consumption in each country. 1979 *Nat. Westminster Bank Q. Rev.* Nov. 33 Each occupation should be compared with a group of workers where there is some close, short-run substitutability i.e. where there is a fair degree of movement between the two labour markets.

substitute, *sb.* Add: **I. 3.** (Earlier examples.)
1777 *Jrnls. Continental Congress U.S.* (1907) IX. 1002 The laws which have been enacted in the State of Pennsylvania, permitting the furnishing of substitutes to perform militia duty. 1779 J. Woodforde *Diary* 30 Sept. (1924) I. 266, I let my man Ben have my little Mare to go to Norwich this morning to try to get a Substitute to serve for him in the Militia.
4. b. *spec.* in *Sport*, a player who replaces another after a match has begun. Abbrev. *sub* (see Sub *sb.* 4 in Dict. and Suppl.).
1849 [in Dict., sense 4 a]. 1916 [see *bench *sb.* 1 c]. 1951 *Sport* 30 Mar.–5 Apr. 6/1 Of course, the idea of substitutes in both Association and Rugby football is not new. 1976 *Southern Even. Echo* (Southampton) 17 Nov. 23/4 When Saints beat Wolves 6–2 in the Second Division last month, the appearance..of Martin Patching as substitute was lost into obscurity because of the emphatic margin of victory.
II. 5. b. A person or thing that becomes the object of love (or another emotion) deprived of its natural outlet. Formerly only with qualifying noun, as *father, mother substitute*: see the first elements in Suppl. Cf. *surrogate *sb.* (*a.*) 2.
1956 L. Durrell *Justine* i. 78 For her we, her lovers, had become only mental substitutes for this first childish act—so that love, as a sort of masturbation, took on all the colours of neurasthenia. 1964 C. Isherwood *Single Man* 23 Jim is the substitute I found for a real son. 1973 E. Caldwell *Annette* (1974) ii. iv. 50 I'd say that enormous teddy bear is a substitute till some boy comes along with the real thing she's after.
6. e. *Philol.* A word that can stand in the place of another, e.g. a pronoun.
1807 Webster *Philos. & Pract. Gram. Eng. Lang.* 15 *Substitutes*, words which are used in the place of other words or of sentences. 1933 L. Bloomfield *Language* ix. 146 In every language we find certain forms, *substitutes*... In English, the pronouns are the largest group of substitutes. 1958 C. F. Hockett *Course in Mod. Linguistics* xxx. 253 The substitutes in this sentence are the morpheme *he* (in the word *his*) and the morpheme *do* (in the word *did*). *He* refers to *John*: it is John's hat which John puts on.

substitute, *v.* Add: **4.** (*c*) More recently, used incorrectly for Replace *v.* 3 a.
1974 *Daily Tel.* 25 July 6/7 The tribunal concludes that British Rail's proposal to compensate..at rates of four, five and six per cent. are inadequate and substitutes them with levels of five, 7½ and 10 per cent. 1978 *Maledicta* ii. 176 Most commonly they are typically formed by substituting *diavolo* with other terms. 1980 *Coal: Energy for Future* (Shell Internat. Petroleum Co.) 3 OECD coal demand is likely to..grow much more rapidly as national actions to substitute oil by coal begin to take effect.
5. Freq. with *for.* (Later examples.)
1953 [see *furan]. 1962 *Listener* 17 May 883/1 But how could it satisfactorily substitute for the complexity and psychological depth abandoned in hacking the novel down to size? 1965 *Language* XLI. 239 A construction.. which may substitute for a word..is a phrase. 1975 *Sci. Amer.* Feb. 36/1 An ion of ferrous iron..can easily substitute for a magnesium ion.

substitution. Add: **7.** Also, the replacement of one atom or group of atoms in a molecule by another. (Further example.)
1964 N. G. Clark *Mod. Org. Chem.* xix. 381 The typical reactions of aromatic hydrocarbons are those of nuclear substitution, whereby one or more of the available hydrogen atoms attached to the nucleus are replaced by substituents derived from the reagent.
11. *gen.* Replacement (*of* one thing) *by* another. See sense 4 of the vb. in Dict. and Suppl.
1888 [see Gravitation 1]. 1938 R. D. Charques *Footnotes to Theatre* ii. 89 Some of the advocates of the.. pictorial theatre went even so far as to recommend the substitution of the living actors by two-dimensional puppets. 1969 G. Stedman Jones in Cockburn & Blackburn *Student Power* 28 The political upheavals within the university in the last years are not the product of an imaginary substitution of workers by students. 1978 *Financial Results Oil Majors, 1977* (Shell Internat. Petroleum Co.) 7 In Europe, oil products demand declined by 1 per cent due to warmer weather, a generally weak economic performance and substitution by other energy sources, particularly hydro-electricity.
12. Special Comb. **substitution group** *Math.*, a group all the elements of which are substitutions (Substitution 5 b); now usu. called *permutation group*.
1889 [see *intransitive *a.* 4]. 1916 G. A. Miller et al. *Theory & Applications Finite Groups* i. 10 It will be proved..that every finite group can be represented as a substitution group.

substitutional, *a.* Add: **2. c.** *Metallurgy.* Of an alloy: involving the substitution at certain lattice sites of atoms of the minor component for those of the major component; *substitutional site*, a lattice site in an alloy at which atomic substitution occurs.
1940 [see *interstice 1 b]. 1966 C. R. Tottle *Sci. Engin. Materials* viii. 186 Where complete or partial solid solubility exists and the solute atoms are too large to occupy interstitial positions, they replace solvent atoms at random in the lattice, and are accordingly referred to as substitutional. 1969 *Physics Bull.* May 167/2 To have the desired effects the implanted impurity atoms must eventually occupy isolated substitutional sites within the lattice. 1980 Chomel & Cottu in P. Haasen *Strength of Metals & Alloys* II. 1017 Substitutional alloy softening only results from a thermal component reduction of the flow stress.

substituti·vity. *Logic.* [f. Substitutive *a.* 2 b + -ity.] The capacity of terms to function as logically equivalent substitutes for one another (see quot. 1965).
1940 W. V. Quine *Math. Logic* 96 This restriction gives rise..to the following..substitutivity principle. 1943 [see *identical *sb.* 1]. 1945 *Mind* LIV. 358 Subject to certain important restrictions pointed out recently by Prof. W. V. Quine, we may accept what he calls the principle of substitutivity as applied to class-identity. 1959 K. R. Popper *Logic of Sci. Discovery* 343 One of our axioms would become redundant, *i.e.* our axiom of substitutivity. 1965 *Jrnl. Philos.* LXII. 139 Quine, Frege, and Russell approach problems connected with *oratio obliqua* constructions determined to defend Leibniz's Law, the principle of substitutivity, sometimes referred to as 'the indiscernibility of identicals'. 1976 *Language* LII. 3 The sense of 'referential' represented by a pair of heavy parentheses is different from the standard one in terms of substitutivity of identicals.

substrate, *sb.* Add: **2.** *Biochem.* The substance upon which an enzyme acts, i.e. whose reaction it brings about.
1907 *Bio-chem. Jrnl.* II. 143 When the relative amount of substrate is large, catalase is rapidly changed into an inactive form. 1938 [see *adenosine b]. 1962 van Heyningen & Waley in A. Pirie *Lens Metabolism Rel. Cataract* 336 Proteolytic enzymes are often characterized by their action on substrates which are not known to be the natural substrates of the enzyme *in vivo*. 1978 J. R. Holum *Org. & Biol. Chem.* xiii. 270 Many enzymes are named by attaching the suffix -ase to the name of the compound, called the substrate, whose reaction the enzyme catalyzes.
3. *Biol.* The surface or material on which any particular organism occurs or grows.
1908 W. M. Bayliss *Nature of Enzyme Action* ii. 7 A name is frequently needed for the substances on which enzymes exert their activity... On the whole, 'substrate,' already used by many writers, seems to answer the purpose best. 1949 W. C. Allee et al. *Princ. Animal Ecol.* x. 158 The surface of water is an important substrate for life, though not nearly so important as the surface of land. 1967 M. E. Hale *Biol. Lichens* iv. 61 A large group of lichens..have high fidelity for limestone and other basic substrates. 1976 *Nature* 15 July p. xiii (Advt.), An order of Crustacea common on sandy or muddy substrates all round the European coasts. 1977 J. L. Harper *Population Biol. Plants* xxiv. 765 The variations in substrate that undoubtedly occurred..in the field were eliminated by using a standard potting compost throughout the experiment.
4. Any underlying bulk phase, layer, etc., on which something is deposited. Cf. Substratum 4.
1937 *Nature* 24 July 158/1 In recent years, the practice has grown up among workers in surface chemistry of using the word 'substrate' to denote the bulk phase underlying a surface film, regardless of the fact that this word has been in general use for a much longer time to denote the substance upon which an enzyme acts. 1954 *Electronic Engin.* XXVI. 296 The applied metal film..adheres well to the substrate. 1960 [see *epitaxially *adv.*]. 1967 *Times Rev. Industry* May 76/1 The use of polythene in coating paper and other substrates for packaging a variety of goods. 1974 *Physics Bull.* June 225/3 Spiller and Segmuller's x ray waveguide consists of a 30–50 nm thick layer of boron nitride between a substrate and cover layer of sapphire.

substratum. Delete ‖ and add: **5. a.** *Linguistics.* Elements or features of a language which are identified by linguists as being relics of, or due to the influence of, an earlier extinct language, usually of the same region. Cf. *superstratum 2 a.
1922 O. Jespersen *Language* xi. 192 Many scholars have recently attached great importance to the..influence exerted by one language on another in those cases in which a population abandons its original language and adopts that of another race... There is thus created what is now generally termed a *substratum* underlying the new language. 1933 L. Bloomfield *Language* xxi. 386 There is no sense in the mystical version of the substratum theory, which attributes changes, say, in modern Germanic languages, to a 'Celtic substratum'. 1956 J. Whatmough *Language* iv. 51 We have superimposed or adjacent languages (superstratum, substratum, and adstratum). 1972 H. Kurath *Studies in Area Linguistics* 120 The phonemic system of Gullah shows some clear influence of the African substratum.
b. *attrib.* and *Comb.*, as *substratum influence, language*, etc.; **substratum theory,** a theory that attributes linguistic change to the influence of a substratum language.
1933 L. Bloomfield *Language* xxi. 386 The substratum theory attributes sound-change to transference of language: a community which adopts a new language will speak it..with the phonetics of its mother-tongue. 1937 J. Orr tr. *Iordan's Introd. to Romance Linguistics* i. 12 An historical summary of the substratum problem. 1952 R. Hall in *Lingua* III. 144 The basic prerequisite for the possibility of substratum influence is a language transfer which takes place through a stage of bilingualism. 1954 *Word* X. 395 Diachronic dialectology deals ..with convergence, i.e. it studies partial similarities increasing at the expense of differences (traditionally, substratum and adstratum studies..and the like). 1962 Burrill & Bonsack in Householder & Saporta *Probl. Lexicogr.* 189 Words which had the force of generic terms in substratum languages may not be understood as generic terms by the present-day populace. 1973 *Archivum Linguisticum* IV. 110 In regard to the so-called Black English, William A. Stewart was the first to advocate a creole substratum theory. 1980 *English World-Wide* I. i. 150 It is not legitimate to compare a static description of creole with a static description of a substratum language.

substructure. Add: **su·bstructured** *a.* [-ed²], having a substructure.
1952 Gerth & Martindale tr. *M. Weber's Anc. Judaism* iii. x. 254 Babylonia and Egypt knew no unified, religiously substructured ethic. 1971 *Black Scholar* June 52/2 The substructured prison movements are gaining momentum.

subsumptive, *a.* (Earlier example.)
1807 R. Kirwan *Logick* II. ix. 521 Those *subsumptive* and illative words, *but, now, therefore.*

subta·bulate, *v. Math.* [Sub- 9.] *trans.* To expand (a mathematical table) by systematic interpolation; to evaluate (a tabulated function) for a set of values of the argument in between the tabulated ones. Hence **subta·bulated** *ppl. a.*; also **su:btabula·tion.**
1924 Whittaker & Robinson *Calculus of Observations* iv. 57 We may obtain without difficulty formulae for subtabulation based on central-difference formulae. 1936 *Suppl. Jrnl. R. Statistical Soc.* III. 87 The differences of subtabulated values. *Ibid.* 93 Then..the original table is subtabulated over the range required to tenths, hundredths or thousandths. 1947 *Math. Tables & Other Aids Computation* II. 286 He had made independent subtabulations in each interval. 1952 D. R. Hartree *Numerical Analysis* v. 78 A set of subtabulated values. 1956 F. B. Hildebrand *Introd. Numerical Analysis* v. 146 This problem would occur..if a function were initially tabulated for increments of 0·1 in *x* and it were required to subtabulate the function for increments of 0·01. 1975 *Nature* 16 Oct. 541/1 The principle adopted in this project was to compute accurate values at rather widely spaced values of the argument, and to produce values at the required tabular interval by systematic interpolation or subtabulation.

subtee·n, *sb.* (and *a.*) orig. *U.S.* [*sub- 14 b.] **a.** A child belonging to the age-group next below teenage. Freq. *pl.*
1952 *Amer. Speech* XXVII. 73 Bonds department store, 16 Feb.: Adv. announcing opening of 'Infants to Subteens Shop'. 1960 *Twentieth Cent.* Nov. 389 The sub-teens, the pocket-money market,..have never been considered anything but children. 1964 *Discovery* Oct. 31/2 Even the sub-teens of the 1960's interpret all too readily the Freudian symbolism of Peter Pan. 1976 *Publishers Weekly* 7 June 75/1 Subteens and teens with an ambition to get into journalism.
b. *attrib.* passing into *adj.*, esp. with reference to clothing designed for the older preteenage girl.
1953 *Sun* (Baltimore) 22 July 5/5 (Advt.), Very special purchase! Regular 5.98 and 7.98 sub-teen cottons. 1962 *Sunday Express* 4 Feb. 14/6 Dating is now an accepted thing in the age group known as 'subteen'. 1968 P. Welles *Babyhip* xxii. 151 Armed with enormous shopping bags, they descended on the subteen department. 1978 M. Farren *Feelies* 146 The sub-teen girls in the crowd went even wilder.
Also **subtee·nage** *a.*; **subtee·nager.**
1959 *New Statesman* 7 Nov. 631/3 He is a pervert.. interested emotionally only in pubescent girls, subteenagers as they are known in American advertising. 1960 P. Goodman *Growing up Absurd* v. 117 This is the meaning, surely, of the publicity that has been trumped up for the Little League, the baseball teams of subteenagers sponsored and underwritten by various business firms. 1963 *Times Lit. Suppl.* 8 Feb. 91/4 (Advt.), He lives in Switzerland and is married with two sub-teenage daughters. 1968 *Punch* 3 July 3/2 A friend is worried at Wimbledon's chauvinistic effect on his sub-teenage kids. 1977 *Washington Post* 20 Feb. D4/2 He keeps horses and a pony for two subteenage daughters. 1980 *Ibid.* 5 Oct. A22/3 There are a 'tremendous' number of subteenagers who abuse a variety of drugs and alcohol.

subtense, *sb.* Add: **a.** Also, the angle subtended by a line at a point. (Further examples.)
1958 *Engineering* 21 Feb. 231/3 Fig. 4 shows the composite picture for the Horseshoe Falls, the angles signifying the angular subtense of any part of the Falls at the floodlights. 1974 *Nature* 3 May 86/2 The test strips.. had a subtense of between 1° and 2°. *Ibid.* 13 Dec. 535/2 Stereo blending does not work well when the angular subtense of the line joining the speakers exceeds about 60°.

su·btest. [Sub- 5 c, 7 a.] A test which is subsidiary to or forms part of a main test, esp. (*Psychol.*) in aptitude assessment.
1939 *Brit. Jrnl. Psychol.* July 21 In normal mental test theory there is a test variable (consisting, if need be, of several sub-tests which are used additively). 1961 *Lancet* 26 Aug. 487/1, I also heartily endorse his call for a refinement in the constituent subtests of psychological batteries, so that intellectual functions may be more precisely identified. 1968 W. E. Lambert in J. A. Fishman *Readings Sociol. of Lang.* (1968) 480 It is clear from these analyses that the subtests of the Modern Language Aptitude Test..are generally highly correlated with intelligence. 1976 *Word 1971* XXVII. 320 It includes subtests of vocabulary, oral comprehension, sentence completion, spelling, and grammar.

su·btext. † **1.** (See Sub- 3 a in Dict.)
2. [Sub- 5 c.] An underlying theme in a piece of writing (esp. in a novel or play). Also *transf.*
1950 E. R. Hapgood tr. *C. Stanislavski's Building Character* viii. 113 What do we mean by subtext? What is it that lies behind and beneath the actual words of a part?..It is the manifest, the inwardly felt expression of a human being in a part, which flows uninterruptedly beneath the words of the text, giving them life and a basis for existing. 1960 S. Moore *Actor's Training: Stanislavski Method* iii. 27 An artistic, rich imagination will also contribute a great deal when an actor interprets the lines and fills them with the meaning that lies behind, the 'subtext'. 1964 *Evergreen Rev.* Dec. 78/1 The modern style of interpretation..digs 'behind' the text, to find a sub-text which is the true one. 1973 *Times* 2 Jan. 7/8 Also admirable was the manner in which Prince underlined the subtext of naturalism that lies beneath the very obvious symbolic superstructure. 1978 G. Vidal *Kalki* i. 16 Whenever I got the chance I gave my pitch, which, basically, is the subtext of *Beyond Motherhood.*

subtha·lamic, *a. Anat.* [Sub- 1 b.] Situated below the thalamus.
1882 *Quain's Elements Anat.* (ed. 9) II. 326 The fibres [of the crura]..are seen diverging at the side of the subthalamic tegmental region into the inner capsule. 1962 *Gray's Anat.* (ed. 33) 1025 The floor of the diencephalon.. forms the subthalamic tegmental region. This, together with the anterior part of the floor and the immediately adjacent parts of the side wall, comprise the hypothalamus. From the functional point of view, however, the subthalamic tegmental region is usually excluded from the hypothalamus. 1973 *Brit. Med. Jrnl.* 15 Dec. 666/1 A vascular lesion of the subthalamic nucleus results in sudden onset of violent choreic movements in the contralateral half of the body.

subtha·lamus. *Anat.* [Sub- 1 f.] A region of grey and white matter in the brain at the base of the diencephalon, below the thalamus and adjacent to the substantia nigra and the red nucleus.
1920 S. W. Ranson *Anat. Nervous Syst.* xiv. 222 The hypothalamus consists of three parts: (1) the pars optica hypothalami,..(2) the pars mamillaris hypothalami, and (3) the subthalamus. 1946 F. W. Jones *Buchanan's Man. Anat.* (ed. 7) 1378 The inferior surface of the thalamus is related to..the subthalamus, which intervenes between the thalamus and the tegmental part of the mid-brain. 1974 D. & M. Webster *Compar. Vertebr. Morphol.* xi. 261 Just ventral to the dorsal thalamus is the ventral thalamus (sometimes called the subthalamus), which is primarily concerned with somatic motor functions.

Subtiaba (subtiā·bā). [The name of a village, (San Juan Bautista de) *Subtiaba*, earlier *Sutiaba*, (see quot. 1891): perh. of Nahuatl origin.] **a.** (A member of) an Indian people of western Nicaragua. **b.** The Tlapanec language of this people (no longer spoken), formerly considered to have Hokan affinities but now regarded as Otomanguean. Formerly also **Subtia·ban.** Also *Comb.*, as **Subtiaba-Tlapanec,** a group of related central American Indian languages, including Subtiaba.
[1878 S. Habel in *Smithsonian Contrib. Knowl.* No. 269. 24, I proceeded to Leon. Here I collected as many words and sentences as I could of the *Raburochi* language, spoken in the neighboring village of Sutiaba.] 1891 D. G. Brinton *Amer. Race* 159 The Subtiabas are inhabitants of the valley of that name near the modern city of Leon in Nicaragua. 1911 Thomas & Swanton *Indian Languages of Mexico & Central Amer.* 77 Mangue..was the most northwesterly tribe of the series, the area occupied extending..northwards from the territory of the Subtiaba (Squier's Nagrandans) 'along the Gulf of Fonseca'... *Subtiaban* (*Synonyms*: Nagrandan, Maribi). This language..forms a dictinct family. 1925 E. Sapir in *Amer. Anthropol.* XXVII. 402 Subtiaba, a language now spoken by only a small number of Indians in a village near Léon, on the Pacific slope of Nicaragua... For a long time the language was believed to be an isolated one.. But it appeared later that it is very closely related to Tlappanec or Yopi, a language spoken in the state of Guerrero in Southern Mexico. 1935 P. Radin in *Internat. Jrnl. Amer. Linguistics* VIII. 45/1 Lehmann succeeded in demonstrating quite clearly that Tlappanec was closely related to the Subtiaba language of Nicaragua. 1965 *Canad. Jrnl. Linguistics* Spring 100 The third constituent of Hokan-Coahuiltecan, Subtiaba-Tlappanec. 1978 *Language* LIV. 507 Both papers are crucially concerned with a particular language known in two dialectal forms, Subtiaba (extinct, of Nicaragua) and Tlapanec (still spoken in Guerrero, Mexico).

subtilin (sǝ̄·btilin). *Pharm.* [f. L. *subtil-is* slender + -in¹.] Any of a group of polypeptides of differing antibiotic activity (*subtilin A, B, C*) derived by culture from *Bacillus amyloliquefaciens* (orig. identified as *B. subtilis*), the most potent of which are used against Gram-positive bacteria and certain pathogenic fungi.
1944 Jansen & Hirschmann in *Arch. Biochem.* IV. 298 The substance thus differing from tyrothricin has been named *subtilin* by the authors. 1948 *Sun* (Baltimore) 9 Jan. 5/1 Recently publicized 'wonder drugs' include.. subtilin. 1948 C. H. Hassal in *Nature* 28 Feb. 318/1 By use of the latter procedure, followed by dilution of the alcoholic extract with adjustment of the *p*H to 2·3, an active concentrate, which we will term subtilin C, was obtained. 1966 *McGraw-Hill Encycl. Sci. & Technol.* XIII. 230/2 Subtilin A, the major component of the subtilin family, is separated by partition chromatography. 1976 J. S. Glasby *Encycl. Antibiotics* 334/1 A further antibiotic isolated from cultures of *Bacillus subtilis*, subtilin may be produced by both surface and submerged growth on various media. The associated antibiotic subtilin C..appears to be identical..in all respects except that it gives no colour reaction with $FeCl_3$ solution.

subtilisin (sɒbti·lisin). *Biochem.* [f. L. *subtīlis* slender + -in¹.] Any of a group extracellular proteinases derived from strains of *Bacillus amyloliquefaciens* (orig. identified as *B. subtilis*).
1953 Güntelberg & Ottesen in *Compt. Rend. des Travaux du Laboratoire Carlsberg: Ser. Chim.* XXIX. 47 Since..the *B. subtilis* proteinase appears to be a rather well defined enzyme..we feel that it will be appropriate to give it a name, and we propose to call it 'subtilisin'. 1968 A. White et al. *Princ. Biochem.* (ed. 4) xii. 255 This is indicated by the finding that the subtilisins, proteolytic enzymes of *Bacillus subtilis* of different genetic origin, possess entirely different amino acid sequences. 1980 *Developmental Biol.* LXXVIII. 383/2 Protease activity of subtilisin and trypsin was confirmed with Azocoll..as substrate at pH 7·0.

sub-title, *sb.* Add: **1.** (Earlier examples.)
1825 T. H. Horne *Outlines for Classification of Library* 86 To each Volume should be prefixed..an Alphabetical Table of the several Titles and sub-titles. 1865 Geo. Eliot *Let.* 16 Sept. (1956) IV. 203 Mr. Lewes..thinks my suggestion as to the sub-title acceptable.
3. *Cinemat.* and *Television.* A caption which appears on a cinema or television screen, esp. to translate the dialogue or to explain the action. Freq. in *pl.*
1909 *Moving Picture World* 27 Feb. 235/1 If the audience is not given time to read the sub-titles or if they are indistinct..the spectators lose the thread. 1924 Wodehouse *Leave it to Psmith* i. 30 What he did not know about erring wives and licentious clubmen could have been written in a sub-title. 1931 B. Brown *Talking Pictures* xi. 287 Another [camera]..photographs a sub-title tablet about a foot across and illuminated by a couple of arc lamps. 1944 [see *dub *v.*⁵]. 1957 M. Summerton *Sunset Hour* x. 140 The French film was mediocre. I ignored the sub-titles, testing my ear on the dialogue. 1975 G. Howell *In Vogue* 5/1 The subtitles to films brought American slang to Britain... 'Beatrix Esmond goes nix on the love-stuff.'

sub-title, *v.* Add: **b.** *Cinemat.* and *Television.* To furnish (a film or programme) with subtitles. Also **su·btitled** *ppl. a.*; **su·btitler**; **su·btitling** *vbl. sb.*
1930 E. V. Knox in *Living Age* 1 Apr. 188 It is a *lingua franca*, or a *lingua californica*... The subtitlers have created a wilderness and called it prose. 1948 *Brit. Film Rev.* Apr. 10 The Cinemas of Great Britain are now showing the sub-titled films so well known to readers in other parts of the world. *Ibid.*, It cannot be said that subtitling in England is uniformly good. 1950 *Jrnl. Soc. Motion Picture & Television Engin.* Nov. 536 Several operations are necessary in order to subtitle pictures. 1968 *Punch* 31 Jan. 154/3 The sub-titler..can sum up a passage of flashy philosophy in one profound-seeming sentence. 1979 K. Conlon *Move in Game* vi. 71 'Tell me some more about academic life.'.. 'Well... There were subtitled foreign films.' 1982 *English World-Wide* III. i. 53 Films are virtually all subtitled.

subtopia (sɒbtōᵘ·piă). Also **Subtopia.** [Blend of Suburb and Utopia: cf. Suburbia.] A disparaging term for: Suburbia regarded as an ideal place. Applied more generally to areas of undifferentiated, ill-planned, and ugly suburban development; unsightly suburbs which encroach on the countryside.
1955 I. Nairn in *Archit. Rev.* CXVII. 365 There will be no real distinction between town and country. Both will consist of a limbo of shacks, bogus rusticities, wire and aerodromes, set in some fir-poled fields... Upon this new Britain the *Review* bestows a name in the hope that it will stick—*Subtopia.* 1960 Koestler *Lotus & Robot* ii. 277, I loathe crooners and swooners,..neon and subtopia. 1963 A. Ross *Australia 63* iv. 102 The descent from Utopia to Sub-topia is steep and short. 1971 *Country Life* 2 Sept. 566/1 Will there still be English villages as we know them, or will they have merged into an unending subtopia in which town and country have become indistinguishable? 1976 W. J. Burley *Wycliffe & Schoolgirls* vii. 123 The killer was a man of the suburbs..at home in a neatly patterned subtopia.

subtopian, *a.* and *sb.* Also **Subtopian.** [f. prec. + -an.] **A.** *adj.* Of, pertaining to, or characteristic of subtopia.
1955 I. Nairn in *Archit. Rev.* CXVII. 372 The other is the panic reflex to the spread of Subtopia, which attempts improvements using standards which are themselves Subtopian. 1963 *Times Lit. Suppl.* 3 May 321/2 For a man with such a harrowing tale to tell Mr. Camp ought not to be so reassuringly readable..He will be avidly read by the subtopian commuters and their desperate wives. 1973 J. Leasor *Host of Extras* i. 24 This subtopian hinterland of back-to-back houses and outside privies.
B. *sb.* A resident of subtopia.
1958 N. Mackenzie *Conviction* 11 Those parts of it [*sc.* Britain] that remain unspoiled are falling into the hands of the subtopians. 1972 I. Broat (*title*) The Subtopians.
Hence **subto·pianism,** the characteristics or ideals of subtopia; **subto·pianize** *v. trans.*, to render subtopian.
1959 *Cambr. Rev.* 25 Apr. 447/2 One can imagine some of them..trying to show that this eclipse was a bad thing, for which broadcasting, subtopianism, Trade Unions and the Welfare State were jointly to blame. a 1963 C. S. Lewis *Poems* (1964) 62 One huge celestial charabanc, will stink and roll Through patient heaven, subtopianized from pole to pole. 1970 *New Scientist* 13 Aug. 342/2 It needed the motor-car to..subtopianize suburbia.

subtotal, *sb.*, *a.* (and *v.*) [f. Sub- + Total *a.* and *sb.*, *v.*] **A.** *sb.* (stressed *su·btotal*) [Sub- 9.] An intermediate total; a total of part of a group of numbers to be added.
1906 *U.S. Patent 823,474*, Fig. 4, showing means for printing marks or characters indicating both totals and subtotals. 1921 J. A. V. Turck *Origin Mod. Calculating Machines* 168 A feature common to recording of added columns of numerical items is the distinguishing characters for clear, sub-totals and totals by the use of letters, stars and other marks. 1952 D. R. Hartree *Numerical Analysis* ii. 20 After each contribution is added, a subtotal is taken, then the next contribution is set and added. 1977 *New Yorker* 29 Aug. 54/2, I kept the new totals in conformity with their figures but changed the supporting details and some subtotals.
B. *adj.* (stressed *subto·tal*) *Surg.* [Sub- 20 g.] Involving the removal of only part of an organ or tissue.
1908 *Practitioner* Dec. 788 Surgeons adopted what has been called hysterectomy with intraperitoneal treatment of the stump, or subtotal hysterectomy. 1977 *Lancet* 29 Oct. 899/2 The natural history of the disease may be interrupted by ablative therapy (subtotal thyroidectomy or the use of radioiodine).
Hence **su·btotal** *v. trans.*, (*a*) to add (numbers) so as to obtain a subtotal; (*b*) to obtain a subtotal from the contents of (a register, etc.).
1936 *Suppl. Jrnl. R. Statistical Soc.* III. 95 The contents of any register may be totalled,..or sub-totalled, i.e. printed without clearing the register. *Ibid.* 99 Several prints of the function may be obtained by inserting more non-add steps after position 8, and sub-totalling register 5 on each of these. 1956 G. A. Montgomerie *Digital Calculating Machines* xii. 250 This causes the accumulator to be sub-totalled into register 117.

subtracter. Add: **3.** *Electronics.* = *subtractor.
1950 W. W. Stifler *High-Speed Computing Devices* xiii. 284 The subtracter which is subtracting a large number from a smaller generates an extra carry pulse at the end of the arithmetic operation. 1970 *IEEE Trans. Computers* XIX. 720/1 A cascade of these subtracters, controlled by a multiplier recorder, provides multiplication.

subtracting, *vbl. sb.* Add: Also as *ppl. a.*
1956 J. L. Stewart *Circuit Theory & Design* ix. 289 (*caption*) A two-tube subtracting circuit.

subtractive, *a.* Add: **a.** Also in *Linguistics*, of a morph or morpheme. Cf. *replacive *a.*
1948, etc. [see *replacive *a.*]. 1953 [see *portmanteau *sb.* 4 d]. 1968 *Amer. Speech* XLIII. 203 Primary graphemic shortenings..may be divided into the subtractive and the replacive.
c. *Photogr.* Of or pertaining to the production of a coloured photographic image by passing white light through a series of filters which absorb or subtract different parts of the spectrum. Cf. *additive *a.* c.
1906 E. J. Wall tr *Konig's Natural-Color Photogr.* i. 23 (*heading*) Three-color printing, or the subtractive method of three-color photography. 1916 G. L. Johnson *Photogr. in Colours* ix. 141 Processes..which depend on the 'three-colour' principle are daily growing in favour... There are two forms of this process, the 'subtractive' one..and the 'additive' method. 1935 [see *additive *a.* c]. 1957 V. J. Kehoe *Technique Film & Television Make-Up* 219 The dye images form the composite color pictures by subtractive synthesis. 1978 *SLR Camera* Dec. 61/1 This subtractive method is the most commonly used in modern colour printing.
B. *sb.* Something that is subtracted or deducted from another quantity; *spec.* in *Linguistics*, a subtractive morph or morpheme.
1949 E. A. Nida *Morphology* (ed. 2) iv. 103 Such bound forms are either (1) nonclitics—additives, replacives, subtractives. 1954 *Word* X. 224 The same comment applies to 'subtractives'. 1979 *Daily Tel.* 21 Nov. 18 Apart from the purchase of a stamp..the ½p is no more than an additive to or subtractive from some other price.

subtractor (sʊbtræ·ktəɹ). *Electronics.* [f. SUBTRACT *v.* + -OR.] A circuit or device that produces an output dependent on the difference of two inputs or of multiples of them. Cf. *SUBTRACTER.

1950 W. W. STIFLER *High-Speed Computing Devices* 450/1 (Index), Subtractor [*in text as* subtracter]. **1953** A. D. & K. H. V. BOOTH *Automatic Digital Calculators* vi. 36 An adder or subtractor requires the provision of some form of register in which the sum is to be stored. **1970** J. EARL *Tuners & Amplifiers* v. 118 The signals from these [microphones] are fed into an 'adder/subtractor' network, giving two outputs, one L+R and the other L−R. **1977** J. G. GRAEME *Designing with Operational Amplifiers* vii. 177 To combine addition and subtraction with integration, the summing and differencing techniques of adders and subtractors are applied to integrators.

subtribe. (Later examples of peoples.)

1857, etc. [see *HAPU]. **1958** G. LIENHARDT in Middleton & Tait *Tribes without Rulers* 103 A tribe is divided into subtribes, its largest political segments. **1977** *Time* 19 Dec. 21/3 Its population of 2·5 million citizens includes members of 76 ethnic groups, mostly subtribes of the Tswana.

subtype. Add: *spec.* a subdivision of a type of micro-organism.

1951 WHITBY & HYNES *Med. Bacteriol.* (ed. 5) xii. 203 By preparing specific Vi phages more than 20 types and sub-types of the typhoid bacillus have been recognized. **1963** *Lancet* 12 Jan. 92/2 Three serotypes are known, but subtypes of type 2 have recently been demonstrated in some animal species. **1979** *Sci. Amer.* Jan. 66/1 That particular subtype of the influenza virus had been the agent of the pandemic of 1918, which killed 20 million people worldwide.

Hence as *v. trans.*, to assign to a subtype; to classify in terms of subtypes; **su·btyping** *vbl. sb.*

1973 *Lancet* 20 Oct. 867/1 Relatives of 9 blood-donors were also subtyped; all had the same subtype as the index case to which they were related. *Ibid.* 869/1 The value of subtyping as an epidemiological tool. **1977** *Ibid.* 15 Oct. 803/2 A multiply resistant strain of type-19 (not yet subtyped) *Streptococcus pneumoniæ* was isolated. **1980** *Brit. Jrnl. Psychiatry* CXXXVII. 502/1 Subtyping of schizophrenia into paranoid and non-paranoid subtypes.

Subud (subu·d). [Contraction of Skr. *suśīla* good disposition, *budh* to awake, learn, *dharma* custom (see quot. 1968).] A system of exercises by which the individual seeks to approach a state of perfection through the agency of the divine power; hence, a movement (founded in 1947 and led by the Javanese mystic Pak Muhammad Subuh, b. 1901) based on this system.

1958 J. G. BENNETT *Concerning Subud* vi. 111 Subud.. the perfect harmony of the inner life (Budhi) and outer life (Sušila) that is attained when our entire being is submitted to the Will of God. **1959** A. HUXLEY *Let.* 12 Aug. (1969) 874 Subud is simply a technique for reproducing the quaking of the early Quakers—a release via the muscles. **1962** *Lancet* 26 May 1125/2 As Subud has taken some hard knocks in your columns, I feel that someone ought to speak up for the 5000–6000 members of the Subud movement in this country. **1968** E. VAN HIEN *What is Subud?* ii. 25 Subud is a contraction of three Sanskrit words: Susila Budhi Dharmi. In Subud terminology, these have been interpreted as follows: Susila means 'right living'. Budhi refers to 'the higher powers and capacities latent in man himself'. Dharma means 'submission to the Will of God'. Taken together, they mean 'Right living according to the highest that is possible for man in submission to God's Will'. **1969** M. SUBUH *Basis & Aim of Subud* 5 It is also necessary to explain that Subud is neither a kind of religion nor a teaching, but it is a spiritual experience awakened by the Power of God. **1972** N. SAUNDERS *Alternative London* xviii. 176 Subud forms a link between psychotherapy and mysticism as roads to self-realisation.

suburban, *a.* and *sb.* Add: **A.** *adj.* **4.** Special collocations: *suburban line*, a railway line which runs between the centre of a city and its suburbs; *suburban neurosis*, a form of neurosis said to occur esp. among suburban housewives which is associated with feelings of boredom, loneliness, and lack of personal fulfilment; *suburban sprawl*, the straggling and often ill-planned expansion of the suburbs of a city over a large area of adjacent countryside; an instance of this.

1869 *Bradshaw's Railway Man.* XXI. 379 The Suburban line, from the Salt River station to Wynberg, is now open. **1926** *Times* 6 May 3/1 Skeleton services were run on main and suburban lines, and more trains are promised to-day. **1972** C. FREMLIN *Appointment with Yesterday* i. 10 South Coast, this [ticket] office... Suburban line, opposite Platform Six. **1938** S. J. L. TAYLOR in *Lancet* 26 Mar. 759/1, I hope to show that environment plays no less a part in the production of what I venture to call 'the suburban neurosis' than it does in the production of physical disease. **1962** *Listener* 6 Dec. 948/2 The so-called 'suburban neurosis' is due to society's having failed to provide a constructive role for these mothers. **1983** *Jrnl. Amer. Acad. Child Psychiatry* XXII. 172 (*heading*) The nuclear family, suburban neurosis, and iatrogenesis in Auckland mothers of young children. **1949** H. BLUMENFELD in *Social Forces* Oct. 59/1 The Association poses the alternative of 'self-contained towns' versus 'suburban sprawl'. **1958** *Listener* 19 June 1022/3 The transformation of most of the country into a gigantic suburban sprawl. **1972** *Country Life* 6 Jan. 18/1 The suburban sprawl that characterises much of the eastern seaboard of the northern United States.

B. *sb.* **2. b.** (Earlier and later examples.)

1841 S. BAMFORD *Passages in Life of Radical* (ed. 2) I. xxxiv. 203 He passed on, leaving those warm-hearted suburbans capering and whooping like mad. **1926** R. MACAULAY *Crewe Train* II. vi. 129 Don't waste time arguing about the accepted premises of life, of which one is that suburbans are dull. **1977** *Transatlantic Rev.* LX. 197 She laughed..being confused by Mr and Mrs Superb the Semi-Detached Suburbans strolling their Sealyhams, for woodpeckers.

Also **suburbaniza·tion**, the act of suburbanizing or the condition of being suburbanized; an instance of this; **subu·rbanized** *ppl. a.*, rendered suburban; **surbu·rbanly** *adv.*

1921 *Edin. Rev.* Jan. 111 The local feeling of the less suburbanised Home Counties continues to object. **1926** *Daily Tel.* 3 Aug., In the urbanisation or suburbanisation of the country motor transport is destined to be even more effective than railways. **1938** *Archit. Rev.* LXXXIII. 216/3 It is gratifying to find *Country Life* adding its own opposition to a tendency which, if not soon halted, will result in literally nation-wide suburbanization. **1951** N. PEVSNER *Middlesex* 55 Finchley Parish had only 1,500 inhabitants in 1801 and still only 7,000 in 1871. Thereafter suburbanization set in. **1963** S. S. IKRAMULLAH *Purdah to Parliament* ii. 17 The mentality and attitude of those who lived in these parts were also suburbanly correct. **1977** *Time* 25 Apr. 35/2 We are going to go on with suburbanized homes. **1978** H. CARPENTER *Inklings* iv. 64 They still went on walking tours, until the increasing suburbanisation of the countryside and the outbreak of war brought that annual event finally to a halt.

Suburbia. Add: Now often **suburbia.** (Earlier and later examples.) Freq. rather disparagingly. Also in N. Amer. and general contexts, and (*poet. nonce-use*) as quasi-*adj.*

1895 E. PUGH (*title*) A street in Suburbia. **1922** L. MUMFORD in H. E. Stearns *Civilization in U.S.* 13 'Suburbia' is used here in both the accepted and in a more literal sense. On one hand I refer to the fact that the growth of the metropolis throws vast numbers of people into distant dormitories where..life is carried on without the discipline of rural occupations and without the cultural resources that the Central District of the city still retains. **1925** WODEHOUSE *Sam the Sudden* xiv. 99 The early morning patois of Suburbia, which is the English language filtered through toast and marmalade. **1936** T. SHARP *Eng. Panorama* vi. 94 H. G. Wells..anticipated with extraordinary accuracy (and, as it seems to-day, with a maddening optimism) the universal suburbia which is already upon us. **1947** AUDEN *Age of Anxiety* III. 76 A married tribe commutes, mild from suburbia. **1967** MCLUHAN & FIORE *Medium is Massage* 72 It gave us darkest suburbia and its lasting symbol: the lawnmower. **1970** G. F. NEWMAN *Sir, You Bastard* iii. 92 The deposit on their admission to suburbia was managed jointly.

subvent, *v.* Restrict † *Obs.* rare⁻¹ to sense in Dict. and add: **2.** = SUBVENTION *v.*

1921 *Discovery* Nov. 293/1 The excavations authorised and subvented by the French Government began in 1880 and have been continued to the present time. **1965** *New Statesman* 23 Apr. 646/1 The only question is, should the taxpayer continue to subvent [the Catholic education system]? **1976** *Times Lit. Suppl.* 25 June 793/2 A..fear of subventing the profits of commercial firms.

subve·rsive, *sb.* [f. the adj.] A subversive person; one who wishes to overthrow a political regime. Also *transf.* and *fig.*

1887 G. MEREDITH *Let.* Feb. (1970) II. 853 Londoners, ..ladies, dandies, mild revolutionists, total subversives, would mob together. **1927** *Weekly Dispatch* 23 Oct. 1/2 The fight against subversive anti-Fascism ceased. The subversives were crushed. **1951** *Manch. Guardian* 30 June 5/5 They [*sc.* the Chinese] have had quite a lesson in Korea and would hesitate before moving into Burma, except as subversives. **1954** I. DEUTSCHER *Age of Permanent Revolution* 14 He [*sc.* Trotsky] stakes everything on the change and upheaval that Time, the great subversive, must bring about. **1977** F. ORMSBY *Store of Candles* 49 At high tide the sea is under the city, A natural subversive. **1978** 'J. HIGGINS' *Day of Judgment* v. 75 You specialized in handling subversives, revolutionary movements generally and so on.

subvi·tal, *a.* (*sb.*) [SUB- 19 a.] **a.** *Genetics.* Of a gene: causing the death of a significant proportion of the individuals carrying it, but not as many as a semi-lethal gene. Also as *sb.* Cf. *LETHAL *a.* 1 d, *SEMI-LETHAL *a.* and *sb.*

1948 E. HADORN in *Symp. Soc. Exper. Biol.* II. 181 We may even come across mutants, or organ systems of lethals, that behave during a first sensitive period as subvital factors, during a second period as semi-lethals, and during a third period as true lethals. **1951** T. DOBZHANSKY *Genetics & Origin of Species* (ed. 3) iii. 67 About 57 percent of the second and 49 percent of the third chromosomes which were free of lethals or semilethals were 'subvital' in homozygotes. A subvital is a deleterious gene or gene complex which causes..the death of less than half of the homozygotes. **1962** [see *LETHAL *a.* 1 d]. **1978** *Acta Embryol. Exper.* I. 101, 80·9% of them survive the ill effects of the mutation; it is, therefore, considered a subvital mutation.

b. *Biol.* Not fully alive; having only some of the characteristics of living systems.

1954 *New Biol.* XVI. 10 Amongst the energy-providing materials are needed also 'sub-vital' systems which, using the energy provided, can grow and split and thus reproduce themselves.

su:bvocaliza·tion. [SUB- 22; cf. *subvocal* s.v. *SUB- 19.] The act or process of articulation with the lips or other speech organs silently or with barely audible sound, esp. while reading.

1947 G. T. BUSWELL in *Scientific Monthly* June 542/2 Completely silent reading, or 'nonoral' reading..is carried on without subvocalization. **1960** A. W. EDFELD *Silent Speech* II. vi. 88 The quotient between the rate of oral reading and the rate of silent reading was used as the measure of subvocalization. **1966** *New Scientist* 29 Dec. 738/1 'Subvocalization' in its most familiar form..consists of audible whispering while reading to oneself. **1974** *Nature* 8 Nov. 121/1 They were instructed to hold their tongues firmly between their teeth and lips while listening and were told to minimise subvocalisation.

Hence [as a back-formation] **subvo·calize** *v. trans.* and *intr.*, to utter or form (words) by subvocalization; **subvo·calizer**; **subvo·calizing** *ppl. a.*

1947 G. T. BUSWELL in *Scientific Monthly* June 542/2 Few persons listening to a lecture follow the speaker by subvocalizing after him the words he speaks. **1947** —— in *Elem. School Jrnl.* Dec. 193/2 Persons who subvocalize in silent reading have a much slower rate than those who suppress all tendencies to deal with words separately. *Ibid.* 194/2 They were subvocalizers, the victims of a method of teaching reading that fixed oral-reading habits first. **1964** *Jrnl. Educ. Psychol.* LV. 339 Subvocalizers exhibited a higher mean lip movement and a slower mean breathing rate than did nonsubvocalizers. **1966** *Science* 16 Dec. 1467/2 An individual who subvocalizes to any great extent is limited to a top reading speed of approximately 150 words per minute—a maximum attainable while reading aloud. **1966** *New Scientist* 29 Dec. 738/3 Of 17 subvocalizing college students out of 50..nearly all managed to reduce their involuntary vocal activity to nil within five minutes. **1978** K. AMIS *Jake's Thing* xv. 153 Jake had subvocalized an oath.

subway, *sb.* Add: **a.** (Earlier and further examples.) Also, for the passage of vehicles.

1825 HOOD & REYNOLDS *Odes & Addresses* 7 Speak up —or hath he hid his name To crawl thro' 'subways' unto fame Like Williams of Cornhill? **1869** *Bradshaw's Railway Man.* XXI. 454 Thames Subway. Incorporated..for making and maintaining a subway, under the river Thames, from Deptford to the Isle of Dogs. Length, 582 yards, with various roads and approaches. **1954** *Gloss. Highway Engin. Terms* (*B.S.I.*) 25 *Subway*, an underground passageway or tunnel to permit traffic movement or to accommodate pipes and cables underneath a structure, road or railway.

b. For *U.S.* read 'Chiefly *N. Amer.* (orig. *U.S.*)' and add earlier and further examples. Freq. *attrib.*

1893 *Massachusetts Acts & Resolves* 1420 The mayor of the city of Boston shall appoint..three commissioners.. to be known as the board of subway commissioners. **1906** *Daily Colonist* (Victoria, B.C.) 27 Jan. 2/5 An aged woman was killed by a subway train today while trying to go into a car in which the door had been closed. **1919**, etc. [see *METRO²]. **1941** B. SCHULBERG *What makes Sammy Run?* i. 22 They were walking down the steps to the subway arm in arm. **1951** E. PAUL *Springtime in Paris* v. 106 They stood there staring at the subway map. **1968** *Globe & Mail* (Toronto) 17 Feb. 48/3 (Advt.), An apartment.. minutes from the Davisville subway station. **1971** *New Society* 18 Aug. 322/2 The [Glasgow] underground (never called the tube)... Glaswegians persist in calling it the subway. **1979** R. JAFFE *Class Reunion* (1980) I. viii. 117 The subway kiosk in Harvard Square.

2. Special Comb.: **subway alumni** *sb. pl.* (U.S. slang), city-dwelling supporters of a college football team who, though not graduates of the college, attend games or follow the results through the news media (also *transf.*).

1947 *Sun* (Baltimore) 3 Nov. 15/8 Many letters have come in attacking Army for dropping Notre Dame... These letters came from Notre Dame's subway alumni, not from Notre Dame. **1960** *Washington Post* 7 Apr. D7 Silky Sullivan, the hero of the 'subway alumni', went into the 1958 Derby as the sentimental favorite. **1982** *Chicago Sun-Times* 26 Oct. 91 Faust would be having a devil of a time staying in the good graces of Notre Dame alumni (real and subway).

Hence as *v. intr.* (N. Amer. colloq.), to travel by subway or underground railway.

1929 M. LIEF *Hangover* 307 He subwayed up to Times Square. **1945** *PM* (N.Y.) 15 Apr. M4/2 We subwayed to Brooklyn. **1968** *Globe & Mail* (Toronto) 17 Feb. 46 (Advt.), You drive a Mercedes, but want to subway to the office.

subzee, var. SUBJEE in Dict. and Suppl.

‖**succah** (su·kä, suka·). Also **sukkah,** [Heb. *sukkāh*, lit. 'hut'.] One of the booths in which a practising Jew spends part of the Feast of the Tabernacles. Cf. *SUCCOTH.

[**1819** *Christian Spectator* I. 126/2 They made booths, in Hebrew *succoth*, that is, sheds or hovels of thorn bushes.] **1875** J. PICCIOTTO *Sk. Anglo-Jewish Hist.* xvi. 140 The Succoths [*sic*] or tabernacles were then, as at present, decorated with fruits and flowers. **1905** *Jewish Encycl.* XI. 660/2 The sukkah or booth was to be a structure especially built for the festival. **1925** *Public Opinion* 4 Sept. 220/3 A sukkah for use during the Feast of Tabernacles. **1970** *New Yorker* 20 June 32/2 The Rabbi's disciples escorted her personally into the *sukkah*. **1979** *Jewish Chron.* 7 Dec. 39/1 Tradition has it that the minute Yom Kippur ends, you dash out and erect the first plank of the Succah. Tradition in our house has it that as soon as Succot is over, you dash out looking for the first Chanucah presents. **1981** C. POTOK *Bk. of Lights* (1982) vi. 191 'What do you want built, chaplain?' 'It's called a succah. It's a kind of booth or hut with wooden sides and an open roof covered with leaves and branches.'

succeed, *v.* Add: **13.** Also in proverbial phr. (see quots.).

1840 T. H. PALMER *Teacher's Man.* 223 'T is a lesson you should heed, Try, try again; If at first you do n't succeed, Try, try again. **1857** W. E. HICKSON *Try Again* in *Moral Songs* 8 'Tis a lesson you should heed, Try, try, try again. If at first you don't succeed, Try, try, try again. **1915** E. B. HOLT *Freudian Wish & its Place in Ethics* iii. 103 The child is frustrated, but not instructed; and it is in the situation where, later on in life, we say to ourselves, 'If at first you don't succeed, try, try again!' **1960** I. JEFFERIES *Dignity & Purity* v. 91 Not to worry... If at first you don't succeed, try, try, try again. *a* **1976** A. CHRISTIE *Miss Marple's Final Cases* (1979) 39 You musn't give up, Mr. Rossiter. 'If at first you don't succeed, try, try, try again.'

‖**succès** (sükse). [Fr., = SUCCESS *sb.*] Used in phrases with reference to types of artistic success or acclaim, as *succès de scandale* (də skandal), success due to notoriety or scandalous character; *succès d'estime* (destim), a critical rather than a popular or commercial success; *succès fou* (fū), a success marked by wild enthusiasm. Also *transf.* and *fig.*

[**1826** *New Monthly Mag.* Dec. 578 Merely that lukewarm approbation, which in Paris is termed *un succès d'estime*.] **1859** *Once a Week* 13 Aug. 136/1 My second attempt..will be something more substantial than a mere *succès d'estime*. **1878** J. A. C. MORISON *Gibbon* vi. 86 The book was..a *succès fou*. **1887** R. CHURCHILL *Let.* 2 Mar. in W. S. Churchill *Lord Randolph Churchill* (1906) II. xvii. 291, I think the Government are earning a rather second-rate kind of *succès d'estime*. **1896** G. B. SHAW *Our Theatres in Nineties* (1932) II. 35 Mr Cartwright..enjoyed ..a sort of *succès de scandale*. **1908** MRS. H. WARD *Diana Mallory* III. xvi. 331 She would find herself a *succès fou*—people tumbling over each other to invite her, and make a show of her. **1919** 'C. DANE' *Legend* 56 The first two books were a *succès d'estime*. **1928** *Observer* 1 Jan. 8 The success which Victor Margueritte's novel 'La Garçonne' made all over Europe..was chiefly a succès de scandale. **1948** W. FORTESCUE *Beauty for Ashes* xix. 142 It was hard work creating something from nothing, but very great fun, and had the usual *succès fou* of all dramatic performances coached by the founder. **1965** A. J. AYER in *Listener* 4 Nov. 700/2 The result was *Language, Truth and Logic*... Though it had an almost immediate *succès de scandale*, its tenets..had a respectable philosophical ancestry. **1977** *Daily Tel.* 17 Mar. 14/3 It has already won a *succès d'estime* in the United States, and seems likely also to become a cult work here. **1978** *Christian* V. 86 What caught on with a succès fou and drew in the spiritual élite of the generation in tens of thousands was a Benedictine reform movement. **1979** M. HILEY *Victorian Working Women* I. iv. 48 *The Pictorial World*..was obviously hoping for some *succès de scandale* by splashing women in trousers across its front page.

success, *sb.* Add: **3. b.** *success of esteem*, *success of scandal*, tr. *succès d'estime*, *succès de scandale* s.v. *SUCCÈS.

1880 Success of esteem [in Dict.]. **1916** G. SAINTSBURY *Peace of Augustans* iii. 144 The extraordinary power of the close of *Vathek* has secured it..a success of esteem. **1926** C. E. MONTAGUE *Rough Justice* III. vii. 99 Notorious novels, successes of scandal, that lived as hard and about as long as super-impudent ball-dresses. **1939** D. CECIL *Young Melbourne* vii. 191 *Glenarvon* had a success of scandal; three editions were called for within a few weeks. But it dealt the death blow to..Caroline's social position. **1958** W. PLOMER *At Home* xii. 175 They [*sc.* publishers' readers] are liable to advise the rejection of typescripts that might have popular success and be moneymakers, or success of esteem followed perhaps by durability and influence.

6. a. *attrib.*, as *success ethic, hunter, rate, value*, etc.

1923 W. STEVENS *Let.* 11 Feb. (1967) 236 Aside from this absurd hero-worship, or success-worship, the town is purely a business place. **1946** *Nature* 17 Aug. 242/2 A success-rate of syphilis prevention of more than 97 per cent was claimed to be unequalled by any other mode of treatment. **1949** *Success-goal* [see *open-class* s.v. *OPEN *a.* 22 a]. **1951** M. MCLUHAN *Mech. Bride* (1967) 35/2 They remain avid customers for the success manuals and beauty treatments which by themselves constitute a large line of merchandise. **1955** KOESTLER *Trail of Dinosaur* 93 The same is true of obsessional success-hunters in every field. **1957** R. K. MERTON *Social Theory* (rev. ed.) v. 170 The distribution of success-values among economic and social strata. **1965** H. HENDIN in A. Giddens *Stud. in Social & Polit. Theory* ix. 311 His legal ambitions were excessive and he found it impossible to compromise with his grandiose success fantasies. **1977** *Time* 13 June 44/2 They are an uncommonly interesting lot, whose lives and habits illuminate what achievement means today in the society that invented the success ethic.

b. *Comb.*, as **success story** *colloq.* (orig. *U.S.*), (*a*) an account of a success; (*b*) an instance of a successful venture, an achievement, etc.

1925 *Ladies' Home Jrnl.* Feb. 28/2 (*heading*) A great success story. **1938** *Time* 14 Nov. 84/2 Last year, when the U.S. Circuit Court of Appeals ruled that physicians might send contraceptives by mail, her career became a 'success story'. **1954** W. K. HANCOCK *Country & Calling* vii. 203 Departments would naturally prefer to get 'a good press', whereas we were bound by our instructions to write critical history, not 'a success story'. **1973** *Nature* 9 Nov. 58/1 The study of X-ray sources is one of the great success stories of present-day astrophysics. **1978** *Jrnl. R. Soc. Arts* CXXVI. 755/2 A comparative success story is the inundation of the ancient city of Nagarjunakonda to make way for a great hydro-electric project.

succession. Add: **IV. 14. f.** *Ecol.* The sequence of ecological changes in which one group of plant or animal species is replaced by another.

1860 H. D. THOREAU in *N.Y. Weekly Tribune* 6 Oct. 6/6 (*heading*) The succession of forest trees. **1899** *Bot. Gaz.* XXVII. 95 The ecologist..must study the order of succession of the plant societies in the development of a region. **1904** *Univ. Nebraska Stud.* IV. 332 Such succession herbaria are the natural outgrowth of formational ones. **1926** TANSLEY & CHIPP *Aims & Methods in Study of Vegetation* ii. 7 Vegetation, when left to itself, tends to change in a definite direction..and this change we call succession. **1957** G. E. HUTCHINSON *Treat. Limnol.* I. xv. 834 It is not impossible that the element plays some part in regulating phytoplankton succession. **1975** *Sci. Amer.* May 90/1 Forest succession proceeds too slowly for it to be observed directly.

g. *Geol.* A group of strata whose order represents a single chronological sequence.

1940 *Bull. Amer. Assoc. Petroleum Geologists* XXIV. 309 Near Las Vegas an apparently conformable succession of marine beds, mostly limestone, is designated as the Bird Spring formation. **1976** *Jrnl. Geol. Soc.* CXXXII. 121 The study area covers..the eastern half of the flysch succession. **1979** D. ATTENBOROUGH *Life on Earth* ii. 36 The limestones at the top of the Moroccan succession are about 560 million years old.

V. 15. succession powder (earlier example); **succession state**, a state which comes into existence after the overthrow or division of a previous state (used orig. of those states which succeeded the dismembered Austro-Hungarian Empire in 1919).

a **1821** MRS. PIOZZI in A. Hayward *Autobiogr., Lett. & Lit. Remains Mrs. Piozzi* (1861) I. 356 In Italy it was supposed to have been the succession powder mingled with chocolate whilst in the cake, not in the liquid we drink. Acqua Toffana, and succession powder (polvere per successione) were administered, as I have heard, with certain although ill-understood effects. **1924** Succession state [see *NATIONALISTICALLY *adv.*]. **1943** C. HOLLINGWORTH *German just behind Me* ii. 14 Like Romania it [*sc.* Yugoslavia] is a 'Succession State'. **1973** *Times Lit. Suppl.* 23 Mar. 318/2 Now that the breakaway of Bangladesh has effected a second partition of the Indo-Pakistani subcontinent, there has been renewed interest in all three succession states in the long-standing controversy over whether the first partition was either inevitable or necessary.

successional, *a.* Add: **2. c.** *Ecol.* Of or pertaining to ecological succession. Cf. *SUCCESSION 14 f.

1922 R. H. YAPP in *Jrnl. Ecol.* X. 13 The Successional Habitat practically agrees with Clements' developmental concept of habitat. **1967** M. E. HALE *Biol. Lichens* vii. 99 Successional stages leading to forested stands. **1979** *Sci. Amer.* Feb. 73/2 The Swiss have come increasingly to rely on natural tree types and natural successional trends as a basis for their silviculture.

successor. Add: **c.** *attrib.*, as *successor-designate*; **successor state** = *succession state* s.v. *SUCCESSION 15.

1958 D. TAIT in Middleton & Tait *Tribes without Rulers* 197 His companion is generally his successor-designate in the office. **1974** P. GORE-BOOTH *With Great Truth & Respect* 388, I set up a committee of three, consisting of Colin Crowe, Dennis Greenhill, my successor-designate, representing the Foreign Office and Jack Johnston representing the Commonwealth Office, to meet daily. **1930** *Economist* 9 Aug. 274/1 A century ago the present 'successor States' of the Hapsburg and Ottoman Empires might have been economically self-sufficient. **1971** H. MACMILLAN *Riding Storm* xvi. 537 The complicated intrigues and rivalries among the successor states of the old Turkish Empire.

succinct, *a.* **2.** (Later example.)

1958 S. J. PERELMAN *Most of S. J. Perelman* 491 Now, Messieurs, exposition is wearisome, so I will be succinct.

succinylcholine (sʊ:ksinəilkōᵘ·līn). *Pharm.* Also **succinyl choline.** [f. SUCCINYL + *CHOLINE.] The ion $[—CH_2COO·(CH_2)_2 \overset{+}{N}(CH_3)_3]_2$ formed by esterification of succinic acid with choline; also, a halogen salt of this, given intravenously as a short-acting muscle relaxant and local anæsthetic; = *SUXAMETHONIUM.

1950 *Chem. Abstr.* XLIV. 2124 It is shown that succinyl choline produces strong curarizing effects. **1952** [see *SCOLINE]. **1965** J. POLLITT *Depression & its Treatment* vi. 80 Occasionally recovery from succinyl choline is delayed and artificial respiration required for an extended period. **1974** M. C. GERALD *Pharmacol.* iii. 52 Alcohol.. and the skeletal muscle relaxant succinylcholine are broken down by the enzymes alcohol dehydrogenase and pseudocholinesterase, respectively.

succinylsulphathiazole (sʊ:ksinəilsʊlfaþəi·āzōᵘl). *Pharm.* Also **-sulf-.** [f. SUCCINYL + *SULPHATHIAZOLE.] A poorly absorbed sulphonamide derivative which is used in the treatment of gastrointestinal infections and is inactive until hydrolysed to sulphathiazole in the body; 4′-(thiazol-2-ylsulphamoyl)-succinanilic acid, $C_9H_8NS·NH·SO_2·C_6H_4·NH·CO·(CH_2)_2COOH$.

1941 *Proc. Soc. Exper. Biol. & Med.* XLVIII. 129 (*heading*) Succinyl sulfathiazole, a new bacteriostatic agent locally active in the gastrointestinal tract. **1981** H. J. ROGERS et al. *Textbk. Clinical Pharmacol.* xix. 647 Only 5–10% of these drugs are absorbed... Examples are phthalylsulphathiazole and succinylsulphathiazole, both of which hydrolyse to sulphathiazole.

succorance (sʊ·kərǣns). *Psychol.* Also **succourance.** [f. SUCCOUR *v.* + -ANCE.] A term used in some forms of personality assessment to describe the need for help, sympathy, and affection as a psychogenic force. Hence **su·ccorant** *a.*

1938 H. A. MURRAY *Explorations in Personality* ii. 83 *Succorance* (Succorant attitude), to seek aid, protection or sympathy. *Ibid.* iii. 181 The Succorance drive seeks a nurturant O. *Ibid.* 182 The Succorant need is always a sub-need. **1944** L. MUMFORD *Condition of Man* ii. 75 He builds his life around the themes of rejection and succorance. **1953** *Brit. Jrnl. Psychol.* Nov. 333 *Succourance*: some heroes show their great need of support, encouragement, care and protection. **1973** *Jrnl. Genetic Psychol.* June 185 Femininity involved being more..succorant. **1977** H. G. BURGER in B. Bernardi *Concept & Dynamics of Culture* 421 The nine behavioral systems of man were stated on unclear empirical grounds by Beatrice Whiting (1963:7): succorance, or asking help from others, [etc.].

succotash. (Earlier example.)

1751 J. MACSPARRAN *Diary* 4 Aug. (1899) 47 Mʳdined with us upon Suckatash and Ham.

‖**Succoth** (suko·t). Also † **Souccoth; Succot, Sukkot(h.** [a. Heb. *sukkōṯ*, pl. of *sukkah*: cf. *SUCCAH.] = *Feast of Tabernacles* s.v. TABERNACLE *sb.* 1 b.

1882 tr. *L. Kompert's Scenes from Ghetto* 102, I hope you will be back in time for the *Souccoth*. **1888** H. POLANO *Talmud* III. 244 Making a tabernacle for thyself during Succoth. **1905** *Jewish Encycl.* IX. 583/1 Number of days on which the several ceremonies of Sukkot are observed. **1907** I. ZANGWILL *Model of Sorrows* ii, in *Ghetto Comedies* 20 When *Succoth* (Tabernacles) came, again no money, no bread. **1921** *Daily Colonist* (Victoria, B.C.) 16 Oct. 15/7 The Jewish folk of Victoria will observe Succoth, the original Thanksgiving Day of ancient Israel, better known as the Feast of the Tabernacle—on October 16. **1944** M. SAMUEL in M. W. Weisgal *Chaim Weizmann* I. 88 The Jews were more transfigured by *their* celebration of Shavuoth and Sukkoth than the Russian peasants by *their* thanksgiving celebrations. **1973** *Synagogue Light* Sept., Passover and Succoth are of seven days duration. **1974** *Times* 8 Oct. 10/1 The Soviet authorities allowed.. about 90 Jews to hold a picnic..to mark the *Sukkot*, a religious festival. **1979** [see *SUCCAH].

succubus. 1. (Later examples.)

1950 A. CLARKE *Coll. Plays* (1963) 315 Branduv is sleeping with a succubus. **1958** L. DURRELL *Balthazar* vii. 167 Thirst *can* be quenched like this, by inviting a succubus to one's bed. **1969** J. UPTON tr. *R. Diaz Sánchez's Cumboto* 261 The dream reoccurred many times, it was the work of a clever succubus who came to my cot regularly to conduct her oneiric concert. **1977** A. CARTER *Passion of New Eve* ii. 27, I would..remember the myth of the succubus, the devils in female form who come by night to seduce the saints.

succuss, *v.* Add: **b.** *Homœopathy.* To shake (a preparation of a drug) vigorously.

1910 *Encycl. Brit.* XIII. 646/1 To make the 2× potency, 10 drops or 10 grains of this first dilution or trituration are mixed with 90 drops of pure alcohol..and are succussed or triturated. **1938** D. SHEPHERD *Magic of Minimum Dose* 264 The 12th potency..is prepared..by diluting one drop in a hundred and shaking or succussing violently. **1974** *Homoeopathy* June/July 86 Between each dilution he [*sc.* Hahnemann] succussed (shook vigorously) the medicine.

succussion. Add: **c.** *Homœopathy.* The vigorous shaking of a preparation of a drug.

1848 HEMPEL & QUIN tr. *Jahr's New Manual* II. 1059 *Succussion*, shaking. **1910** *Encycl. Brit.* XIII. 646/1 The continuation of the dynamization of trituration or succussion develops a spiritual acurative agency. **1938** D. SHEPHERD *Magic of Minimum Dose* 21 This trituration or succussion is a most important part of the preparation of the drug. **1974** *Homoeopathy* June/July 86 This process of dilution and succussion seemed to give the remedy more energy and a greater healing effect.

such, *dem. adj.* and *pron.* Add: **A. 4. β.** sich (further examples).

1846 DICKENS *O. Twist* (rev. ed.) xv. 81 Oh, you naughty boy, to make me suffer sich distress. **1863** *Southern Confederacy* (Atlanta) 13 May 2/1 The buryal squad organized *fust* and foremost, and begun to inter their money and spoons and 4 pronged forks and sich like. **1890** KIPLING in *Scots Observer* 28 June 149 Don't call your Martini a cross-eyed old bitch; She's human as you are— you treat her as sich. **1938** M. K. RAWLINGS *Yearling* xvi. 185 'Is that true, Buck?' Buck whittled busily. 'Now if you was to tell me a tale,' he said, 'I'd not ask you no sich of a question.' **1953** E. SIMON *Past Masters* III. 191 'But I did it in my own time,' said Monro... 'There ain't no sich thing, old son.' **1981** P. MACDONALD *One Way Street* i. 9 These bloody English..have put their imprint on this place in sich a way as to make yew want to heave.

B. V. 24. Also in phr. *or some such* (also *somesuch*): or some such thing.

1967 D. FRANCIS *Blood Sport* iii. 35 He was in France on business wasn't he, or somesuch. **1972** *Daily Tel.* 11 Apr. 22/5 Plan will be to approach them with proposals for short and sharp bursts of selling with stamps—stamp weeks or somesuch. **1973** R. PARKES *Guardians* vii. 122 The doctor believed it might aid expiation or abreaction or some such.

suck, *sb.*¹ Add: **1. c.** An act of fellatio. *coarse slang.*

1941 G. W. HENRY *Sex Variants* II. 1177 A real suck seems to be one in which orgasm and ejaculation are induced. **1972** *Screw* 12 June 21/2 They start their separate ways through a variety of fucks and sucks and lesbian encounters.

† **9.** A breast-pocket. *Criminals' slang. Obs.*

1821 D. HAGGART *Life* 26 He returned the screaves to his lil, and placed it in his suck. **1923** *Chambers's Jrnl.* 6 Oct. 716/1, I..pulled the dub of the outer jigger from his suck.

10. *slang.* A sycophant; esp. a schoolboy who curries favour with teachers. Cf. SUCK *v.* 25 e in Dict. and Suppl.; *sucker-up* s.v. *SUCKER *sb.* 14.

1900 FARMER *Public School Word-Bk.* 197 *Suck, subs.* (University), a parasite, a toady. **1907** B. M. CROKER *Company's Servant* xx. 213 He was just a suck—that's all. **1916** JOYCE *Portrait of Artist* (1969) i. 11 We all know why you speak. You are McGlade's suck. **1955** W. GADDIS *Recognitions* II. ii. 373 The shade of the boy whom he had not seen since they were boys together (Martin was Father Joseph's 'suck') lived on the air as though they had parted only minutes before.

11. *pl.* as *int.* Used as an expression of contempt, chiefly by children. Also in phr. *sucks to you* and varr. *slang.*

1913 C. MACKENZIE *Sinister Street* I. I. vii. 98 This kid's in our army, so sucks! **1922** F. HAMILTON *P.J.: Secret Service Boy* iv. 178 'S', he announced, '*u,c,k,s,t,o,y,o,u*.' **1935** N. MITCHISON *We have been Warned* I. 28 Brian is a baby. Oh sucks, oh sucks on Brian. **1945** E. WAUGH *Brideshead Revisited* II. v. 287 It's great sucks to Bridey. **1952** 'C. BRAND' *London Particular* xv. 191 A most regretable air of sucks to you. **1968** *Melody Maker* 30 Nov. 24/5 This is a rotten record—yah boo and sucks. **1974** *Times* 4 Mar. 9/5 Sucks boo, then, with acting like this, to that new National Theatre down the road. **1978** 'J. LYMINGTON' *Waking of Stone* ii. 45 'Sucks to you!' she said..tossing her head so her pigtails swung. **1983** *Listener* 19 May 11/1 The council treated the urbane Mr Cook to the politician's equivalent of 'Yah, boo, sucks'.

12. *Canad. slang.* A worthless or contemptible person. Cf. *SUCK *v.* 15 f; *suck-hole* s.v. *SUCK-.

1974 *Globe & Mail* (Toronto) 8 Mar. 1/6 The teachers are copping out. They're now saying, if we can't have our way, then we're going to be sucks and refuse to work. **1975** *Citizen* (Ottawa) 28 Oct. 1/1 A neighbor described Rob as 'a quiet guy who was always getting put down a lot. Lots of people used to call him a suck... He didn't do much socially or in the way of sports.'

suck, *v.* Add: **I. 1. f.** (See quot. 1960.) With person or part as obj. Cf. sense 23* below. *coarse slang.*

1928 in A. W. Read *Lexical Evidence from Folk Epigraphy Western N. Amer.* (1935) 78, I suck cocks for fun. **1960** WENTWORTH & FLEXNER *Dict. Amer. Slang* 527/2 *Suck v.i., v.t.* 1 [taboo] to perform cunnilingus or, esp., fellatio. **1972** *Screw* 12 June 21/2 Characters fuck and suck each other like *real* people do. **1973** E. BULLINS *Theme is Blackness* 79 You heard what I said, bitch.. take me to dinner and suck mah dick and et cetera fa dessert.

II. 9. c. *to suck the hind tit* or *teat*: to be inferior or have no priority. Also *intr.* with *on. slang* (orig. *U.S.*).

1940 W. V. T. CLARK *Ox-Bow Incident* iv. 244 'Well,' he said, 'if you like to suck the hind tit.' **1951** N. MONSARRAT *Cruel Sea* III. vi. 179 You have n't a hope... As far as radar is concerned, corvettes are sucking on the hind tit. **1963** *Time* 8 Nov. 47, I don't want these kids around here to suck on a hind tit when it comes to getting a good education. **1975** *Weekend Mag.* (Montreal) 31 May 20/2 Radio, no matter what you've read about the Radio Revolution, still sucks the hind teat at the CBC.

10. a. *suck it and see* (see quot. 1951). Now used *attrib.* and *absol.* (also with hyphens) to denote experimental methods.

1951 PARTRIDGE *Dict. Slang* (ed. 4) Add. 1189/2 *Suck it and see!* A derisive c[atch-]p[hrase] retort current in the 1890's. **1968** *New Scientist* 3 Oct. 10/1 Biologists..prefer to employ the 'suck it and see' approach adopted by Harold Wilson to politics rather than the impractical (?) idealism of Michael Foot. **1973** *Nature* 2 Mar. 16/2 In the best tradition of 'suck it and see' Fowlis has attempted to use such a velocimeter to measure the flow of both mercury and the liquid alloy NaK. **1976** *New Scientist* 16 Dec. 636/1 Types of experiment that could be usefully or uniquely performed in space:..'suck-it-and-see' experiments to explore a new environment (such as the plant growth and spider-web-spinning variety). **1979** *SLR Camera* June 42/3 It's difficult to lay down any hard and fast recommendations for using fill-in lighting; it's really a suck-it-and-see situation.

III. 15. e. To practise fellatio (or cunnilingus). *coarse slang.*

1928 in A. W. Read *Lexical Evidence from Folk Epigraphy Western N. Amer.* (1935) 78 My cock is only 10 ins long so if any one would like to suck meet me here 9 pm. **1960** [see sense 1 f above]. **1975** E. HANNON *Doors* 123 White chicks dig suckin, that's a fact. That's cause suckin's sophisticated. **1977** M. T. BLOOM *13th Man* (1978) viii. 148 The pimp said: 'She wouldn't suck so she couldn't make a living. I had to send her back.'

f. To be contemptible or disgusting. *slang.* Cf. *SUCK *sb.*¹ 12.

1971 *It* 2–16 June 3/2 Polaroid sucks! For some time the Polaroid Corporation has been supplying the South African government with large photo systems..to use for photographing blacks for the passbooks..every black must carry. **1976** G. V. HIGGINS *Judgment of Deke Hunter* vi. 59, I had a lousy summer... I thought it sucked, and I bet next summer'll suck too. **1978** M. GORDON *Final Payments* xi. 193 All the hotels have the same pictures. The last one, the food sucked.

VI. 22. b. suck (a)round. *intr.* To go about behaving sycophantically. Occas. *ellipt.* Cf. sense 25 e in Dict. and Suppl. *slang* (orig. and chiefly *U.S.*).

1931 *Princeton Alumni Weekly* 22 May 798/1 If 'drag' or 'hot dope' is necessary one usually 'sucks around' for it. **1934** G. ADE *Let.* 27 June (1973) 186 As for the Landis party on July 10th I have had no invitation but maybe I could suck around and get one. **1940** M. MARPLES *Public School Slang* 169 Thus a boy is said to *suck round*, if he tries to ingratiate himself. **1941** B. SCHULBERG *What makes Sammy Run?* xi. 209 The tycoon who spends the first part of his life sucking and crushing, and the last part giving away dimes. **1979** 'A. HAILEY' *Overload* III. xiv. 273 Logically, she should go to the city editor. She might have done it, too, if the son-of-a-bitch hadn't handed her that coach-and-team crap earlier today. Now it would look as if she was sucking around him because of it.

23*. suck off. *trans.* To cause (someone) to experience an orgasm by fellatio or cunnilingus. *coarse slang.* Cf. sense 1 f above.

1928 in A. W. Read *Lexical Evidence from Folk Epigraphy Western N. Amer.* (1935) 79 When will you meet me to suck me off? **1941** G. W. HENRY *Sex Variants* II. 1176 The object of *suck* can be either the organ or the person; but the object of *suck off* is usually the person, who is mentioned within the idiom, e.g. 'to suck him off'. **1959** W. BURROUGHS *Naked Lunch* 76 Equilibrists suck each other off deftly. **1969** FABIAN & BYRNE *Groupie* (1970) vii. 50 He listened superciliously..and, spreading his legs, asked me to 'suck him off' to make him less uptight. **1971** *Guardian* 27 Sept. 14/5 One American GI is forcing a Vietnamese woman to suck him off. **1976** J. CROSBY *Snake* (1977) XXXV. 222 Elf has had a busy night... Sucking me off till all hours.

25. e. For '*Schoolboy slang*' read '*slang* (orig. *Schoolboys*)'. (Later examples.) Cf. *sucker-up* s.v. *SUCKER *sb.* 14.

1936 M. MITCHELL *Gone with Wind* xl. 719 We hear how you suck up to the Yankees..to get money out of them. **1945** E. WAUGH *Brideshead Revisited* II. iv. 261, I imagine she's been used to bossing things rather in naval circles, with flag-lieutenants trotting round and young officers on-the-make sucking up to her. **1957** R. K. MERTON *Social Theory* (rev. ed.) viii. 270 Data in *The American Soldier* on what was variously called brown-nosing, bucking for promotion, and sucking up. **1963** D. OGILVY *Confess. Advert. Man* (1964) i. 15, I despise toadies who suck up to their bosses; they are generally the same people who bully their subordinates. **1966** [see *CRAWL *v.*¹ 3 c]. **1979** J. COOPER *Class* (1980) vi. 131 Harry Stow-Crat also has to suck up to neighbouring farmers in case he should want to hunt over their land.

suck-. **suck-hole**: restrict † to sense in Dict. and add: (*b*) *U.S.*, a whirlpool, a pond; (*c*) *Canad.* and *Austral. slang*, a term of abuse (cf. *SUCK *sb.*¹ 12); hence as *v. intr. slang* (orig. and chiefly *Canad.*), to curry favour.

1909 *Dialect Notes* III. 377 *Suck-hole, n.*, a whirlpool. Common [in East Alabama]. **1961** PARTRIDGE *Dict. Slang* Suppl. 1302/2 *Suck-hole, v.*, to toady, as in 'He won't suck-hole to anyone'; hence, to cringe; low Canadian; C. 20. **1964** F. O'ROURKE *Mule for Marquesa* 200 They rode on toward the small water hole... Dolworth led them off a plateau down the rocky trail to the suckhole under the rock ledge. **1966** P. MATHERS *Trap* 12 Our progressive mayor..and his pack of scabby suckhole mates. **1968** J. WAINWRIGHT *Edge of Extinction* 48 He can roast to hell—then go suckholing to Old Nick. **1970** *Globe Mag.* (Toronto) 31 Oct. 4/2 No matter how strong I could become there was still someone in this city of 470,000 who thought I was a suckhole. **1972** J. METCALF *Going down Slow* vii. 128 Can't even fix yourself a sandwich without suckholing round that man.

suck-egg. Add: **c.** Used *attrib.* to designate a dog regarded as the type of viciousness or worthlessness. Also *transf.* *U.S. dial* (chiefly *South* and *Midland*).

1892 *Dialect Notes* I. 232 He is as mean as a suck-egg dog. **1927** P. GREEN *Unto Such Glory* in *One-Act Plays for Stage & Study* 3rd Ser. 104 He's a dirty low-down suck-egg dog. **1931** *Virginia Q. Rev.* Jan. 102 Hayes got up and slunk off like a suck-egg dog caught in the hen-house. **1958** 'W. HENRY' *Seven Men at Mimbres Springs* ix. 107 But I will be a suck-egg son of a bitch if I can't tie my good arm behind me, stand on my bad leg only, and still whup me the living daylights out of any skinny little Alabama bast—.

sucker, *sb.* Add: **I. 1. b.** For *U.S.* read orig. *N. Amer.* (Earlier and later examples.)

1838 *Patriot* (Toronto) 29 May 1/2 It's true that pigs has their troubles like humans—constables catches 'em, dogs bites 'em, and pigs is sometimes as done-over suckers as men. **1927** A. CONAN DOYLE *Case-Bk. Sherlock Holmes* 92 I'll see this sucker and fill him up with a bogus confession. **1941** [see *PLAY *v.* 24 a]. **1957** *Essays in Crit.* VII. 47, I confess to being a sucker myself, if not for Malory, for Welsh legend. **1960** P. GOODMAN *Growing up Absurd* iii. 65 Our present poor are absolute sheep and suckers for the popular culture which they cannot afford, the movies, sharp clothes, and up to Cadillacs. **1973** L. MEYNELL *Thirteen Trumpeters* iv. 57 He got..a tiny percentage out of the total takings of the Casino. The more suckers who turned up and the more each sucker spent the better pleased he was. **1979** *Financial Rev. Survey* (Sydney) 22 Oct. 11/2 Look at the advertising man himself. He's the biggest sucker in town. From rotary engines to studded blue jeans—you'll find 'em at the agency. **1981** M. GEE *Dying, in Other Words* 58 Elsie laughed when she told about Pelham and called her a sucker, and said that she ought to ask him for money, men often liked giving you money, it was part of the game.

9. g. *Golf.* (See quot. 1931.) orig. *U.S.*

1931 *Daily Express* 2 Sept. 1/5 The United States Golf Association passed a special rule permitting 'suckers'—that is, balls embedded in the mud—to be lifted and cleaned without penalty. **1963** *Times* 9 Jan. 4/3 There do not seem to have been any 'suckers', although some of Ray's towering drives were repeatedly expected to produce them.

III. 13. For *local* read *colloq.* (orig. *local*) and add later examples. Also *spec.* (chiefly *N. Amer.*), a lollipop; *all-day sucker*: see *ALL *a.* IV. b.

1907 *Dialect Notes* III. 250 *Sucker, n.*, a kind of hard candy held by a small wooden stick and sucked. 'Let's buy suckers.' **1938** *Times* 13 Jan. 14/5 One of them said: 'I'll buy some suckers.' **1956** J. SYMONS *Paper Chase* xii. 91 A window in which gobstoppers, liquorice bootlaces and sherbet suckers nestle. **1962** J. LUDWIG in R. Weaver *Canad. Short Stories* (1968) 2nd Ser. 242 'I got no money for suckers,' the woman said nastily. **1971** *Islander* (Victoria, B.C.) 19 Sept. 4/3 The small children eagerly hunted suckers that had been hidden in a large hay wagon. **1977** E. JONG *Loveroot* 45 Little sugar suckers with sour centers.

IV. 14. (sense 1 b) *sucker bait, bet, list, punch, trap*; **sucker-bashing** *Austral. slang* (see quots. 1945, 1953); **sucker-disk** = SUCKER *sb.* 10; **sucker-rod** (earlier example); **sucker-up** = *SUCK *sb.*¹ 10 (cf. SUCK *v.*¹ 25 e in Dict. and Suppl.).

1939 *Amer. Speech* XIV. 80/2 *Mootch* is a derisive term applied to a careful customer... Retailers lose money on the 'mootch', because he buys only those things offered as 'sucker bait' or 'specials'. **1976** 'TREVANIAN' *Main* (1977) xiii. 249 'Have you any reason to think you might be in trouble?' he asks. But she is not taking sucker bait like that. She smiles. **1945** J. A. ALLAN *Men & Manners in Austral.* 89 Before that the settlers had cut the scrub a foot above ground, piled the refuse round the stumps, and fired it as the new shoots appeared. Even after that, 'sucker bashing'—which had raised the cost of clearing to 15/– an acre—had still been needed. **1953** BAKER *Australia Speaks* iii. 80 *Sucker bashing*, work at cutting down saplings. **1962** *Australasian Post* 25 Oct. 40 Whilst sucker-bashing at Mirambigo Station. **1920** *Collier's* 26 Mar. 22/3 You actually intend *makin'* a sucker bet like that? **1979** *Tucson* (Arizona) *Citizen* (Weekender Mag.) 28 Apr. 9/3 Don't buy much insurance. Cover your potential catastrophic losses with insurance, but not your minor setbacks. Remember that the way insurance companies make money is by taking as many sucker bets as possible. **1964** *Oceanogr. & Marine Biol.* II. 412 The functional histology of the sucker-disk of two British regular echinoids..has been described. **1977** *Playgirl* May 76/2 The sucker-disc mouth [of a lamprey] was stuck solidly to the smooth skin on J. T.'s right side. **1910** *Collier's* 17 Dec. 25/1 'Sucker lists', as the promoters call the roster of victims..are traded and passed on. **1966** T. PYNCHON *Crying of Lot 49* v. 114 After a week of anxiously watching the mailbox..getting nothing but sucker-list stuff through the regular deliveries. **1981** E. AMBLER *Care of Time* v. 65 If they're pulling names on the sucker list, they can forget mine. I'm not available. **1947** *Amer. Speech* XXII. 122/2 *Sucker punch*, a hit or punch delivered without warning. **1950** J. DEMPSEY *Championship Fighting* 50 The right lead is called a sucker punch. **1979** N. HYND *False Flags* xxii. 201 It was a sucker punch... The fist landed, breaking his nose. **1865** *Harper's Mag.* Apr. 571/1 Small engines are used in most cases, with hardly sufficient power to raise the sucker-rod out of a deep well. **1953** POHL & KORNBLUTH *Space Merchants* xvi. 156 Warren Astron had never returned to his sucker-trap on Shopping One. **1973** *Sunday Advocate-News* (Barbados) 16 Dec. 3/5 So this Christmas, shop wisely, avoid the sucker traps. **1911** F. SWINNERTON *Casement* ii. 66 'Suckers-up' (those who sought by illegitimate means to ingratiate themselves with the manager). **1976** P. LIVELY *Stitch in Time* i. 10 Toady, said Maria to it [*sc.* a cat] silently, sucker-up.

sucker, *v.* Add: **4.** *trans.* To cheat, to trick. *slang* (orig. and chiefly *U.S.*).

1939 *Sat. Even. Post* 14 Oct. 78/1 It was a little deal I got suckered on. **1948** *Chicago Tribune* 27 Mar. 1. 1/4 Apparently we are again going to be suckered into approval of a glorified world WPA. **1958** J. & W. HAWKINS *Death Watch* (1959) 87 We're going to sucker the killer out in the open. **1971** L. GRIBBLE *Alias the Victim* xii. 184 He had been suckered badly. What had to be done was to get away. **1978** J. GORES *Gone, no Forwarding* (1979) xv. 90 Delaney suckered us into making a payment which he now claims is an admission of guilt because we made it.

sucket. For † *Obs.* read 'Now rare exc. *arch.* and *Hist.*' and add: **a, b, d.** (Later examples.)

a. 1929 E. LINKLATER *Poet's Pub* xii. 144 The table already gleamed with..jumbals and marchpane and suckets of one kind and another. **1959** P. VANSITTART *Tournament* xiv. 115 Suckets shaped as unicorns, swans, frogs.

b. 1917 A. WAUGH *Loom of Youth* 10 'Those who can, do, while those who can't, teach.' This choice sucket.. comes consolingly to the ears of one whom the chances and caprices of life may have thrown casually on the preceptorial beach.

d. 1938 CURRIER & BUHLER *Marks Early Amer. Silversmiths* 165 Forks were apparently unknown except for serving—to which use were doubtless put the small sucket-forks..for sweetmeats. **1956** G. TAYLOR *Silver* v. 112 The three prongs were curved, unlike the two prongs of the sucket fork. **1977** FLEMING & HONOUR *Penguin Dict. Decorative Arts* 768/2 *Sucket fork*, an implement with a spoon at one end and a two-pronged fork at the other, intended for eating fruit, especially *succade*.

sucking, *vbl. sb.* Add: **3. a. sucking-pot** (later example); **sucking reflex** *Biol.*, the instinct to suck as possessed by the young of all mammals; **sucking response** *Biol.*, the action of sucking as a response to some stimulus or influence; **sucking-up** *slang*, sycophancy.

1843 C. A. F. PARKE *Let.* 19 Aug. in U. Ridley *Cecilia* (1958) xi. 125 She uses a sucking pot, but the Old Crab thinks that she sucks in wind. **1923** T. P. NUNN *Education* 167 An infant is born in vigorous possession of the sucking reflex. **1974** *Biol. Abstr.* LIX. 2593/2 An otherwise normally developed female rabbit without ears may have lost them when still in the nest due to a 'sucking-reflex' among its siblings, such as that which occurs among young mice. **1938** *Jrnl. Genetic Psychol.* LIII. 369, 49 per cent of sleeping infants gave sucking responses to stimulation of the lips. **1975** *Jrnl. Compar. Physiol. & Psychol.* LXXXVIII. 796 Monitoring sucking responses to a rubber teat revealed that..the vigorous oral activity continued largely unabated. **1946** B. MARSHALL *George Brown's Schooldays* ii. 7 'Thank heaven my people sent me here with a decent grub box.' 'But what has a grub box to do with being caned..?' Brown asked. 'The gentle art of sucking-up, of course... Not to the beaks.' **1978** 'M. INNES' *Ampersand Papers* I. v. 44 He wasn't doing any sucking-up act on Archie.

sucking, *ppl. a.* Add: **6. sucking louse,** a blood-sucking ectoparasite of mammals belonging to the order Siphunculata (or Anoplura); **sucking stomach** *Zool.*, a stomach in certain invertebrates that expands so as to provide a food reservoir (formerly interpreted as the means by which the animal imbibed fluid).

1910 R. DOANE *Insects & Disease* iv. 54 The sucking lice..are suspected of carrying some of these same diseases. **1950** *N.Z. Jrnl. Agric.* Jan. 68/1 Sucking louse: This parasite [of pigs] is very common in New Zealand. **1962** GORDON & LAVOIPIERRE *Entomol. for Students of Med.* xxxvi. 223 Members of the order Anoplura, all of which are known as 'sucking lice' possess 'sucking' mouthparts borne on an elongated head. **1886** F. R. CHESHIRE *Bees & Bee-Keeping* I. vii. 94 Cook calls the honey-sac the 'sucking stomach', using an old, but extremely misleading, title. **1925** A. D. IMMS *Gen. Textbk. Entomol.* 98 The organ is then known as the food-reservoir or 'sucking stomach', but the latter expression is misleading and incorrect.

suckle, *v.* **3.** *intr.* Delete *rare* and add later examples.

1966 P. SCOTT *Jewel in Crown* I. 28 Their children, three girls and two boys to date (apart from the one still suckling..) sat on the front benches. **1977** *Sci. Amer.* Aug. 80/3 Since the evicted joey may continue to suckle for another four months, the female red kangaroo may have three offspring in the 'pipeline' at any one time: a dormant blastocyst, a small joey nursing and developing in the pouch and a larger young-at-foot still suckling.

sucrase (sⁱū·krēⁱz). *Biochem.* [f. F. *sucre* SUGAR *sb.* + *-ASE.] An enzyme that catalyses the hydrolysis of disaccharides to monosaccharides; *spec.* that which catalyses the hydrolysis of sucrose to glucose and fructose; = INVERTIN, *INVERTASE, *SACCHARASE.

1900 in B. D. JACKSON *Gloss. Bot. Terms*. **1901** *Jrnl. Chem. Soc.* LXXX. I. 180 The isolation of 'sucrase' the actual enzyme of cane sugar inversion from yeast in a pure form appears..to be hopeless. **1954** A. WHITE et al. *Princ. Biochem.* xvii. 397 Specific disaccharases for sucrose and lactose, named sucrase and lactase, respectively, are supposed to occur also in the intestinal juice. **1974** *Encycl. Brit. Micropædia* IX. 640/1 Sucrase is produced by the mucous membrane cells lining the walls of the small intestine. **1981** M. TOPOREK *Basic Chem. Life* xix. 271 Evidence at present indicates that..maltase, sucrase, and lactase are not actually secreted into the intestinal lumen.

‖**sucrier** (sükrie). [Fr.] A sugar-bowl, usu. made of porcelain and with a cover.

1869 C. SCHREIBER *Jrnl.* 9 Oct. (1911) I. 50 A Bow (sprigged) sucrier with cover and acorn top. **1904** E. DILLON *Porcelain* p. xxi, Sèvres porcelain. Two small *sucriers*..Gros bleu and green ground, with birds on branches painted in white reserves. **1960** *Times* 18 June 11/2 But odd pieces—*sucriers*, cup and saucers, teapots—can be obtained at moderate cost. **1975** *Country Life* 4 Dec. (Suppl.) 43/1 Chelsea-Derby sucrier, c. 1770.

sucrose. Mark sense in Dict. † *Obs.* and add: **1. b.** *spec.* a white crystalline sugar, $C_{12}H_{22}O_{11}$, which can be derived from sugar-cane, sugar beet, and in lesser quantities from most other plants, and is used as a sweetener; = *SACCHAROSE.

In chemical terms, sucrose is an optically active disaccharide composed of D-fructose and D-glucose and having a structure described by the systematic name α-D-glucopyranosyl-(1,2)-β-D-fructofuranoside.

1857 W. A. MILLER *Elem. Chem.* III. ii. 54 Cane sugar or Sucrose ($C_{12}H_{11}O_{11}$).—This variety of sugar is chiefly otained from the sugar cane. **1888** BLOXAM *Chemistry* (ed. 6) 644 Sucrose fuses at 160°C. (320°F.), and does not crystallize on cooling. **1903** A. J. WALKER tr. *Holleman's Textbk. Org. Chem.* I. 274 On hydrolysis sucrose yields *d*-glucose and *d*-fructose in equal proportions. **1964** N. G. CLARK *Mod. Org. Chem.* viii. 138 Molasses is the dark syrup remaining after the removal of crystallized sugar from evaporated sugar-cane juice or the aqueous extract of sugar beet: it contains between 40 and 50 per cent of sucrose (table sugar). **1980** C. W. SPANGLER *Org. Chem.* I. xii. 248 Lactose and sucrose are two of the more common disaccharides.

2. *attrib.* and *Comb.*, as **sucrose (density) gradient** *Biochem.*, a gradient of sucrose concentration used in the centrifugation of biological media to prevent convection currents; freq. *attrib.*; **sucrose phosphate,** any of the esters that can be formed between sucrose and phosphoric acid; **sucrose phosphorylase,** a bacterial enzyme which catalyses the breakdown of sucrose, ultimately producing glucose-1-phosphate and fructose.

1944 *Jrnl. Exper. Med.* LXXIX. 304 Concurrent experiments..performed without the protective action of a sucrose gradient showed no indication of a sedimentation boundary. **1947** *Ann. Rev. Microbiol.* I. 362 Friedewald & Pickels.., by centrifugation in a sucrose density gradient so as to reduce convection, noted differences between PR8 and Lee strains. **1968** H. HARRIS *Nucleus & Cytoplasm* iii. 43 (*caption*) Sucrose-density-gradient sedimentation pattern of a crude extract of *Escherichia coli* cells exposed to [¹⁴C] uracil for 20 seconds. **1979** *Biochim. & Biophysica Acta* DLXIV. 191 Sucrose density gradient analysis of the postribosomal fraction of muscle and liver revealed that the sedimentation profiles of the synthetases of the two tissues were similar. **1938** *Chem. Abstr.* XXXII. 5920 The rabbit paw was injected with 10 cc. of 2% aq. solns. of..Ca sucrosephosphate. **1960** *Plant Physiol.* XXXV. 269/2 Any sucrose-phosphate which is formed is ultimately dephosphorylated by enzymes in sugar beet tissue at some stage prior to storage in the root. **1979** *Infection & Immunity* XXIV. 868/1 Hydrolysis of sucrose phosphate would be expected to yield glucose 6-phosphate and fructose. **1943** *Jrnl. Biol. Chem.* CLI. 360 It is possible to obtain active preparations of sucrose phosphorylase relatively free of invertase and phosphatase. **1977** *Jrnl. Molecular Catalysis* II. 453 The interest in sucrose phosphorylase lies in the fact that a stable and re-usable insoluble preparation can be useful for both preparative and analytical purposes.

suction. Add: **4. b. suction dredge** *Engin.*, a type of dredge employing a suction pump, used in the dredging of soft material from sea-beds and river bottoms; hence **suction dredger,** a vessel which carries a suction dredge; **suction dredging** *vbl. sb.*; **suction gas,** the town gas produced by a suction plant; **suction lift** *Mech.*, the height to which a liquid can be drawn up a pipe by suction; **suction plant,** a form of gas producer (see PRODUCER 3) in which the blast is induced by suction; **suction pressure** *Bot.* [tr. G. *saugkraft* suction force (Ursprung & Blum 1916, in *Ber. d. Deutsch. bot. Ges.* XXXIV. 539)], the force with which a cell can imbibe water, being the difference between the pressure exerted by the cell walls on the cell contents and the osmotic pressure of the contents; **suction stroke,** in an internal-combustion engine, a piston stroke in which fresh mixture is drawn into the cylinder; also, *suction sweeper*.

1901 *Daily Colonist* (Victoria, B.C.) 27 Oct. 3/2 Next Monday..the first suction dredge ever operated in the western part of the Dominion will be given a trial. **1940** *Sun* (Baltimore) 4 Dec. 6/3 Excavations by huge dipper and suction dredges already are under way at both ends of the canal. **1977** *New Yorker* 20 June 68/2 Suction dredges are portable, cheap, irresistible to a certain class of lone, adventuring miner. **1911** *Daily Colonist* (Victoria, B.C.) 6 Apr. 14/5 Plans are being prepared for a new suction dredger of the type of the King Edward for use in British Columbia coast waters. **1930** *Engineering* 13 June 760/1 The sand backing was filled in over the bank by suction dredgers. **1974** H. R. COOPER *Pract. Dredging* (ed. 2) i. 10 (*caption*) A powerful pump, a floating platform, a pipe and disposal system...that is the simple anatomy of the Suction Dredger. **1965** G. V. WILLIAMS *Econ. Geol. N.Z.* vii. 69/2 These sands were washed beyond the narrow confines of the Ohinemuri River..where they were worked by suction-dredging some years ago. **1974** H. R. COOPER *Pract. Dredging* (ed. 2) p. x, During the 12 years since the first edition of *Practical Dredging* was published, trailing suction dredging methods have become increasingly important. **1907** *Daily Mail Year Bk.* 75/2 Suction-gas has been adapted to marine purposes. **1936** BONE & HIMUS *Coal* xxiv. 417 By the year 1901 'Suction-Gas Plants' were established on the market. *Ibid.* 418 A typical 'suction gas', generated from gas-coke, with air saturated with steam at 51·7°C, contains CO_2=5·15, CO=25·45, H_2=13·10, CH_4=0·30, and N_2=56·00 per cent. **1909** N. HAWKINS *Mech. Dict.* 559/2 Suction lift. **1940** KRISTAL & ANNETT *Pumps* ii. 103 It is a generally accepted rule that 15-ft. suction lift is a safe operating condition. **1976** C. P. KITTREDGE in I. J. Karassik et al. *Pump Handbk.* II. 148 A positive value of h_s is called a suction head while a negative value of h_s is called a suction lift. **1909** *Rep. Brit. Assoc. Adv. Sci. 1908* 826 A suction plant costs less and occupies less ground space, but the gas made in it is not so strong as in the older form of pressure plant. **1920** H. C. GREENWOOD *Industr. Gases* III. 344 Suction plants have an advantage in the reduction of risk of carbon monoxide poisoning owing to the prevailing negative pressure. **1922** W. STYLES in *Biochem. Jrnl.* XVI. 728, I propose for this quantity, already described as a force and a power, but which is in reality a pressure, the term 'suction pressure'. **1958** *New Biol.* XXV. 38 Water moves from the soil to the leaves along a gradient which most European workers call a gradient of suction pressure or suction force and most Americans, a gradient of diffusion pressure deficit. **1978** *Physiol. Plant Path.* XIII. 275 Infection of tomato plants by *Meloidogyne javanica* resulted in increased suction pressure in the root system. **1904** R. T. MECREDY *Dict. Motoring* 169 The Suction Stroke... The descent of the piston naturally causes a vacuum in the combustion chamber, which at first was air and gas tight. **1933** V. L. MALEEV *Internal-Combustion Engines* v. 59 Temperature t_a of the gases in the cylinder at the end of the suction stroke is higher than the outside temperature t_a. **1941** NEWTON & SEEDS *Motor Vehicle* (ed. 3) xi. 172 The displacement of the piston on the suction stroke represents potential ability for forming a vacuum in the cylinder. **1920** *Chambers's Jrnl.* Nov. 830/1 A suction-sweeper that we have examined recently runs the electric type very close indeed. **1926–7** *Army & Navy Stores Catal.* 114/1 Whirlwind Suction Sweeper. Its revolving Brush sweeps the carpet... Its powerful suction sucks the dust into the dustproof container.

c. *spec.* in *Aeronaut.*, used *attrib.* to designate various devices concerned with controlling flow conditions in the boundary layer, as *suction aerofoil, control, slot*, etc.

1933 *Gloss. Aeronaut. Terms* (*B.S.I.*) VII. 58 *Suction face*, the side of an airscrew blade formed by the upper surfaces of its aerofoil elements. **1946** *Jrnl. R. Aeronaut. Soc.* L. 431/1 The suction aerofoil exhibits a large discontinuous fall of velocity followed by a gentle rising velocity from the position of the suction slot to the trailing edge. **1950** *Ibid.* LIV. 159/2 The suction wing principle must be associated with the flying wing layout for it to be truly advantageous. **1960** *Aeroplane* XCIX. 268/2 In spite of official reluctance to admit the potentialities of suction control of the boundary layer, the enthusiasts persist in their efforts. **1977** *Jrnl. R. Soc. Arts* CXXV. 350/1 The US..flew a modified twin-jet reconnaissance aircraft..in 1966 with suction slots which also achieved a high degree of wing laminar flow.

suctorian. Add: In mod. use *spec.* a protozoan of the class or subclass Suctoria, the adult form of which is usually sessile, lacking cilia and feeding by the use of suctorial tentacles. Also as *adj.* = SUCTORIAL *a.*

1931 R. R. KUDO *Handbk. Protozool.* xxxiii. 399 The body of a suctorian may be spherical, elliptical, dendritic, etc. **1939** *Jrnl. Cellular & Compar. Physiol.* XIV. 410 The tentacles of the suctorian protozoan Ephelota coronata.. are very long and thin. **1975** *Nature* 7 Aug. 467/2 Microtubules have also independently evolved into many other organelles of motility, such as..suctorian tentacles and haptonemata. **1980** J. N. FARMER *Protozoa* xvii. 678/1 The tentacles of suctorians included in this family are of one type, the feeding tentacles.

sucupira (su:kūpīᵊ·rä). [a. Pg., f. Tupi *sucupira*.] A dark brown hardwood obtained from trees of the genus *Bowdichia* or *Diplotropis*, both native to South America, esp. Brazil, and belonging to the family Leguminosæ; also, a tree of either of these genera.

1924 RECORD & MELL *Timbers Trop. Amer.* II. 270 The woods commonly known as 'sucupira' are of a deep chocolate-brown color. **1950** *Archit. Rev.* CVII. 124 The photograph..shows..an office partition in 'sucupira', a rich purple hardwood. **1977** *Transatlantic Rev.* LX. 86 The colossal Ceibas, para nuts and sucupiras with their blue flowers high in the sun.

sudak. (Later example.)

1973 *Nat. Geogr. Mag.* May 612/1 All the strange but delicious bounty of the Volga, handsome, fat fish with names like sazan, sudak.

Sudan. Add: **2.** *Chem.* Used *attrib.* to designate various azo and diazo dyes mostly derived from 2-hydroxynaphthalene and anthraquinone, used as industrial dyes and biological stains: as *Sudan I* (also *1*), the orange-yellow azo dye, C_6H_5ROH (where R= —N:N·$C_{10}H_6$—); *Sudan II* (also *2*), the brown azo dye, $(CH_3)_2·C_6H_3·N:N·C_{10}H_6OH$; *Sudan III* (also *3*), the red diazo dye, $C_6H_5·N:N·C_6H_4·ROH$; *Sudan IV* (also *4*), the scarlet diazo dye, $CH_3·C_6H_4·N:N·C_6H_4(CH_3)·ROH$; *Sudan black* (*B*), the black diazo dye,

$$C_6H_5·R·R \quad \overset{NH}{\underset{NH}{\diamond}} \quad C(CH_3)_2.$$

1894 A. G. Green tr. *Schultz' & Julius' Syst. Survey Org. Colouring Matters* 66 (*table*) Sudan I..Benzene-azo-β-naphthol. $C_{16}H_{12}N_2O$. *Ibid.* 70 (*table*) Sudan II.. Xylene-azo-β-napthol. $C_{18}H_{16}N_2O$. *Ibid.* 86 (*table*) Sudan III..Benzene-azo-benzene-azo-β-naphthol. $C_{22}H_{16}N_4O$. **1907** *Practitioner* Nov. 635 Fresh sections stained with Sudan III. **1956** [see *polybase]. **1961** R. D. Baker *Essent. Path.* iv. 40 The lipid is bound in the organ and does not have the physicochemical form necessary to absorb Sudan dye. **1966** T. S. & C. R. Leeson *Histology* i. 16/1 Fat can be detected in sections which have not been exposed to fat solvents by stains such as Sudan III, Sudan IV, and Sudan black B. **1974** Passmore & Robson *Compan. Med. Stud.* III. i. xxi. 14/2 Sudan black B stains the cytoplasm of the myeloid series, the intensity of the staining increasing with maturation.

3. *Sudan grass* (*U.S.*), a tall annual grass, *Sorghum sudanense*, which is cultivated for hay in dry regions of the United States. Also *ellipt.*

1912 *Yearbk. U.S. Dept. Agric. 1911* 72 Sudan grass.. is another example of a new forage crop that has become popular almost in one season. **1929** C. C. Deam *Grasses of Indiana* 325 Sudan grass has only recently been introduced into Indiana and its use as a hay crop is on the increase. **1949** *Hoard's Dairyman* 25 Oct. 756/3 Frost-nipped cane, sudan, pig weeds, Johnson grass, and flax are poisonous to cattle. **1964** Mrs. L. B. Johnson *White House Diary* 6 July (1970) 176 You can look down on the church spire in the valley below and the fields in between, with Sudan grass waving in the wind. **1978** J. Updike *Coup* (1979) i. 28 In the wide belt of transition between withered sudan and stark desert, there were islands of what had been, before the drought, pasture land.

Hence **su:danophi·lia** *Med.* [*-philia], the condition in which cells containing particular fatty or lipid structures can be stained with a Sudan dye; hence **su:danophi·lic** *a.*, capable of taking up Sudan stains.

1911 Stedman *Med. Dict.* 840/2 *Sudanophilia*,..a condition in which the leucocytes contain minute fat droplets which take a brilliant red stain when treated with 0·2 per cent Sudan III. **1954** E. W. Dempsey in R. O. Greep *Histology* xxvii. 745 (*caption*) The two sections are from two phases of secretion and illustrate the increased sudanophilia of the rodlike mitochondria during the phase of extrusion of fat from the cells. **1956** *Nature* 7 Jan. 48/1, I observed certain sudanophilic corpuscles which do not appear to have been previously described. **1961** R. D. Baker *Essent. Path.* iv. 40 Fat occurring normally in adipose tissue, adrenal cortex and corpus luteum absorbs Sudan dyes and is called sudanophilic. **1979** *Atherosclerosis* XXXIII. 486 Sudanophilia is evident in the upper thoracic portion and in the area of the renal arteries. **1980** *Ibid.* XXXV. 103 Polar coordinate mapping was used to determine the rate of progression of spontaneous sudanophilic coeliac lesions on the aortic wall in White Carneau pigeons.

Sudanese, *sb.* Add: Also in *Comb.*, as **Sudanese-Guinean** (see quots.).

1954 Pei & Gaynor *Dict. Linguistics* 207 *Sudanese-Guinean*, a family of African Negro languages, spoken by an estimated total of 50,000,000 persons... Some linguists consider Sudanese and Guinean as two independent families. **1967** M. Schlauch *Language* ii. 39 In a wide belt stretching across Northern Africa, bounded on the South by a line extending Eastwards from the shores of the Gulf of Guinea and then dipping still farther to the South, we find a chain of languages grouped together and known as Sudanese-Guinean.

Sudanic (sudæ·nik), *a.* and *sb.* [f. Sudan + -ic.] **A.** *adj.* = Sudanese *a.*; *spec.* of or pertaining to the Sudan or an extensive group of African languages spoken there and elsewhere in central, northern, and eastern Africa. **B.** *sb.* (One of) the Sudanic group of languages.

1912 D. Westermann *Shilluk People* i. 32 Hamitic languages..differ from the Sudanic languages chiefly in the grammatical gender. *Ibid.*, Numerous Shilluk-words, which most probably are Sudanic, are found in languages generally counted as Hamitic. **1913** N. W. Thomas *Anthrop. Rep. Ibo-Speaking Peoples of Nigeria* i. 141 The languages of West Africa, commonly called Sudanic, and spoken by the true negro, have been classified into four main groups—Eastern Sudanic, Central Sudanic, Middle and Western Sudanic. **1931** C. K. Meek *Sudanese Kingdom* iv. 184 Mlle. Homburge has recently written a paper attempting to prove a close connection between Ancient Egyptian, Fulani, Sudanic, and Bantu. **1936** *Discovery* June 171/1 The Nilotes of the Nile Valley, speaking Negro (Sudanic) languages and extending from the Anglo-Egyptian Sudan some 200 miles south of Khartum into Uganda. **1956** E. E. Evans-Pritchard *Nuer Relig.* iii. 104 They think easily in terms of Spirit but not in terms of medicines, the idea of which as it obtains among their Sudanic neighbours they seem scarcely able to grasp. **1956** A. W. Southall *Alur Society* ii. 24 The Bendi are also Sudanic speakers. **1957** Ld. Hailey *African Survey 1956* iii. 84 Negro (including Sudanic, Bantu, and Nilotic), and Hamito-Semitic. **1972** J. Biggs-Davison *Africa—Hope Deferred* iii. 24 The Sudanic economy was mainly rural and pastoral. **1977** *Sci. Amer.* Apr. 110/3 Ehret suggests that the names applied to cattle and sheep by many modern Bantu-speakers were probably derived from the non-Bantu languages known collectively as Central Sudanic. **1980** *Cambr. Encycl. Archaeol.* 342/1 The formative processes of the Early Iron Age complex took place in the country to the north-west, in the 'sudanic' belt of open grassland savanna on the northern fringes of the equatorial forest.

Sudanization (sudănəizēi·ʃən). [f. as prec. + -ization.] The action or process of making Sudanese in character, *spec.* with reference to the independence of the Sudan from Great Britain in 1956.

1951 *Britannica Bk. of Year* 44/1 The year was one of progress towards the government's declared object—the Sudanization and independence of the Sudan. **1955** *Times* 2 Aug. 5/2 The first stage in the 'Sudanization' of important posts held by foreigners, had been completed. **1970** H. Trevelyan *Middle East in Revolution* 19 Commissions were to be established to guide the Governor-General, to supervise Sudanisation of the Civil Service,..and to supervise the formation and work of the Constituent Assembly. **1978** S. Lloyd *Suez 1956* i. 12 A Sudanisation Committee to deal with the administration and defence forces.

Sudano- (sudā·no), used as comb. form of Sudan and its derivatives, as in *Sudano-Sahelian* adj.; **Sudano-Guinean** = *Sudanese-Guinean* s.v. *Sudanese *sb.* Cf. *sudanophilia (s.v. *Sudan).

1939 [see *Nilo-]. **1954** Pei & Gaynor *Dict. Linguistics* 207 Some linguists consider Sudanese and Guinean as two independent families; others, notably Delafosse, consider Sudano-Guinean and Bantu to be members of a larger linguistic group. **1979** *Nature* 18 Jan. 167/3 UNCOD therefore proposed giant transnational projects like..a joint livestock management programme in the Sudano-Sahelian countries.

sudden, *a.* **3. b.** Delete 'see quots.' and add: (*a*) a single toss used to decide an issue; hence in *Lawn Tennis*, a game played to break a tie; also in general sporting use (usu. *attrib.*), designating an additional competition or period of extra time in which the first to concede a game or score is immediately eliminated; (further examples); (*b*) *U.S.*, a potent alcoholic drink; (*c*) (see quot. 1886 in Dict.).

1863 C. Reade *Hard Cash* I. vii. 205 America is fertile in mixtures: what do we not owe to her? Sherry-cobbler, gin-sling..sudden death. **1865** 'Mark Twain' in *Californian* 18 Mar. 8/3 Our reserve (whom we had..kept out of sight and full of chain-lightning, sudden death and scorpion-bile all day..) came filing down the street as drunk as loons. **1927** W. E. Collinson *Contemp. Eng.* 36 *Sudden death* [is used] for a game played to bring a set to a sudden, decisive conclusion without playing out the full number. **1939** *Sun* (Baltimore) 21 July 15/8 Skipper Bill Barrow, of the Rochester Yacht Club, sailed his Thisbe II to victory today in a sudden-death race against defending champion Aphrodite. **1945** *Ibid.* 3 Mar. 7/2 Tech meets the winner tonight, and got this break by having its name picked out of a hat when the 'sudden death' playoff plan was decided on. **1946** L. P. Hartley *Sixth Heaven* viii. 162 'Game-ball all,' was called... 'Shall we play it out?' said Dick, 'or shall we have sudden death?' **1961** *Times* 29 Aug. 3/4 Player and..J. Herbert tied for the lead.. and then had a sudden-death play-off. **1972** 'E. Lathen' *Murder without Icing* xxvi. 224 'I hear that it wasn't a bad game.'..'Not bad! When it went into sudden death overtime?' **1974** *Times* 22 Jan. 10/7 The WCT circuit as a whole contains a controversial innovation: a 13-point tie-break with a 'sudden death' finish. This means that the first player to score seven points wins the tie-break whether he leads by two points or not. **1977** *Evening Gaz.* (Middlesbrough) 11 Jan. 14/6 These matches are 'sudden death' affairs, a single match in each round either home or away depending on the luck of each draw.

Sudeten (sudē^i·tən), *a.* and *sb.* [Ger., the name of the *Sudeten* mountains in north-eastern Czechoslovakia.] **A.** *adj.* Of, pertaining to or designating the predominantly German-speaking area of Czechoslovakia in the vicinity of the Sudeten mountains (the Sudetenland) which was annexed by Germany from 1938 to 1945. Freq. as *Sudeten German*.

1937 *Times* 20 Oct. 13/2 (*heading*) Czechoslovakia and the Sudeten Germans. *Ibid.* 6 Dec. 11/5 (*heading*) Sudeten German quarrels. *Ibid.*, Dissensions within the Sudetendeutsch Party. **1939** *Encycl. Brit. Bk. of Year* 526 At the time of the annexation by Germany of the Sudeten areas of Czechoslovakia there were in the country some 5,000 refugees from the old Reich and from Austria. **1946** W. S. Churchill *Victory* 131 Henlein, Sudeten-German leader, committed suicide. **1959** W. F. Leopold in J. A. Fishman *Readings Sociol. of Lang.* (1968) 355 Sudeten Germans with Bavarian dialect adapt themselves slowly to Swabian. **1966** S. Mann *Collecting Playing Cards* iv. 84 (*heading*) The Franconian or Sudeten pattern (Sudeten-deutsch). **1968** [see the *sb.* below]. **1974** *Listener* 25 Apr. 530/2 The Sudeten 'problem' was being manipulated both by appeasers here and..by Hitler. **1982** S. G. Duff *Parting of Ways* xv. 135 Gradually, up to 1933, the Sudeten Germans had become reconciled to the [Czechoslovak] Republic.

B. *sb.* An inhabitant of the Sudetenland; a Sudeten German.

1938 H. Nicolson *Diary* 13 May (1966) 341 The Sudetens could not approve of a pro-Russian and anti-German policy. **1943** *Amer. Speech* XVIII. 200 The term *Sudetens*, extremely frequent in the news columns of 1938, did not exist before that year. **1968** K. Martin *Editor* xii. 252 The Sudetens had some real grievances, even though they were the best-treated minority in Europe... The Czech government knew that their real problem had nothing to do with Sudeten grievances.

Sudetic (sudē^i·tik), *a.* Now *rare.* [f. *Sudet(en *sb.* + -ic: cf. G. *sudetisch*.] Of or pertaining to the Sudeten region of Czechoslovakia.

1907 *Muret-Sanders Encyclopaedic Eng.-German & German-Eng. Dict.* II. 710/3 *Sudeten..Gebirge*,..Sudetic Mountains. **1928** C. Dawson *Age of Gods* xii. 270 A movement of population was certainly taking place at this period, for the skulls of the Lengyel people belong not to the old 'Sudetic' type of the Danube region, but are distinctively Nordic. **1928** P. Selver tr. *Benes' My War Memoirs* xix. 481 The Austrian Minister..sent..the.. Allied Governments a protest against the attempt to retain the Sudetic Germans within Czechoslovakia. **1934** Priebsch & Collinson *German Lang.* i. ii. 37 Of less moment..are..the fair broad-heads of East Baltic type on the eastern periphery and a very primitive strain, called by Günther Inner Asiatic or Sudetic (from the Sudetes). **1938** *Manch. Guardian* 12 May 6/3 It is not clear what is meant by the 'extreme limit' to which the Czecho-Slovak Government is asked to go in its 'concessions' to the Sudetic German minority.

sudoite (sū·do,əit). *Min.* [ad. G. *sudoit* (G. Müller 1962, in *Naturwissenschaften* XLIX. 205/2), f. the name of Toshio *Sudo* (b. 1911), Japanese mineralogist and crystallographer: see -ite[1].] (See quot. 1963.)

1963 *Amer. Mineralogist* XLVIII. 214 G. Müller (1962) proposes 'sudoite' as a name for this dioctahedral series of phyllosilicates, as chlorite is the name of the analogous trioctahedral series. **1977** *Mineral. Abstr.* XXVIII. 16/1 An essentially regular interstratification of mica (sericite) and chlorite (sudoite) was found in an alteration area of the Matsumine Kuroko deposit of the Hanaoka mine.

suds, *sb. pl.* Add: **4. c.** For *U.S. slang* read '*slang* (orig. and chiefly *U.S.*)' and substitute for def.: Beer. (Earlier and later examples.)

1904 G. V. Hobart *I'm from Missouri* iii. 52 Who.. hoists a few dippers of suds?.. Dad! **1924** *Truth* (Sydney) 27 Apr. 6 *Suds*, beer. **1925** Fraser & Gibbons *Soldier & Sailor Words* 273 Suds, ale. **1926** *Flynn's* 16 Jan. 638/2 The boozeclerk give us th' high sign he had doped th' suds or skat. **1931** 'D. Stiff' *Milk & Honey Route* 177 Fill up on 'suds' for a dime. **1943** C. L. Sonnichsen *Roy Bean* 171 The bear..was still consuming his free bottle of suds. **1962** *Radio Times* 17 May 43 Let's split to your pad for some suds. **1975** *Globe & Mail* (Toronto) 8 Feb. 1/2 Before then, Labatt had only a marginal share of the suds market in Quebec. **1977** *Mod. Boating* (Austral.) Jan. 30/1 The figure propped half-standing on a bar stool, with his face in a glass of suds. **1979** *Tucson* (Arizona) *Mag.* Sept. 60/3 Sip suds out of glass jars while you wait.

5. in the suds a. (Later U.S. examples.)

1816 U. Brown *Jrnl.* 28 Sept. in *Maryland Hist. Mag.* (1916) XI. 234 We both in the sudds pretty much. *Ibid.* 29 Sept. 235 Thinking that I was not out of the sudds yet. **1887** R. T. Cooke *Happy Dodd* xxvii. 295, I shan't leave Mis' Payson in the suds.

suds (sɒdz), *v.* [f. the sb.] **1.** *trans.* To lather; to cover with soap-suds, or wash in soapy water.

1834 'C. Packard' *Recoll. Housekeeper* 12 Ma'am Bridge was *sudsing* the clothes in a tub before her. **1939** N. S. Colby *Remembering* ii. 62 She dipped my hair in a basin of hot water, sudsed it, rinsed it, and dried it with a towel. **1976** *S. Wales Echo* 27 Nov. 6/3 (Advt.), Rub-a-Dub Doll. Soap her and suds her. See how much fun a bath can be. **1981** P. Theroux *Mosquito Coast* xv. 185 The..splash of our foot-operated wheel sounded like a washing machine sudsing clothes.

2. *intr.* To form suds. *U.S.*

1893 M. A. Owen *Voodoo Tales* 5 An impertinent housewife had dared to affirm that her soap wouldn't 'suds'. **1972** *Fortune* Jan. 73/1 Detergent foam first became a matter of national concern in the early 1960's, when Representative Henry S. Reuss of Wisconsin, among others, pointed out that detergents were persisting, and sometimes sudsing, in the environment.

So **su·dsing** *vbl. sb.* and *ppl. a.*

1844 'J. Slick' *High Life N.Y.* II. 20 I'd gin myself a good sudsing in the wash hand basin. **1879** *Scribner's Monthly* Oct. 940/2 As soon as they begin to boil, remove them to the 'sudsing'-water. **1881** S. P. McLean *Cape Cod Folks* 167 A good poundin', and boilin', and sudzin', you need. **1957** T. Sturgeon in D. Knight *100 Yrs. Sci. Fiction* (1969) 134 Slim heard more water running and sudsing noises, and, by ear, followed the operation through a soaping and two rinses. **1971** *New Yorker* 6 Nov. 5 (Advt.), This rich, sudsing, mentholated cleanser was developed by dermatologists. **1978** *Nature* 6 Apr. p. xxvii/2 The concentrated detergent powder dissolves quickly to provide fast action, minimal sudsing, and free rinsing.

sudsable (sɒ·dzăb'l), *a.* [f. *suds *v.* + -able.] Capable of forming soap-suds; also of garments: washable in soapy water. Hence **sudsabi·lity.**

1951 *Sun* (Baltimore) 15 Dec. 10 (Advt.), She never has too many blouses..so lovable..so wearable..so sudsable. **1959** *Wall St. Jrnl.* 16 Dec. 9/2 More folks are becoming more conscious of the sudsability of their tap water. **1970** *Globe & Mail* (Toronto) 25 Sept. 16/2 (Advt.), Tam-and-scarf set in thick suds-able hand-crocheted acrylic.

sudser (sɒ·dzəɹ). *U.S. slang.* [f. Suds *sb. pl.* + -er[1].] A soap opera.

1968 *New Yorker* 30 Mar. 114/2 It has the suggestions of sadness and 'depth' that make it a kind of high-class sudser for women. **1975** *Ibid.* 5 May 31/1 This NBC half-hour TV sudser expired after fifteen months. **1982** *Washington Post* 8 Dec. c10 Clooney's autobiography.. has been turned into another drably shabby TV sudser.

sudsy, *a.* Add: Also *transf.* and *fig.* (Earlier and later examples.)

1866 *Harper's Mag.* Sept. 544/2 He's gone! across the sudzy sea. **1980** *Times Lit. Suppl.* 17 Oct. 1160/1 Thanks to Arianna Stassinopoulos's votive ministrations, Maria Callas has graduated from opera to the sudsier, sublimer realm of soap opera.

suède. *Delete* ‖ and add: Now usu. **suede. 1.** Now applied to other kinds of leather finished to resemble undressed kid-skin. Also an article, usu. a shoe, made of suede.

1923 [see *sand *sb.*[2] 1 i]. **1957** M. B. Picken *Fashion Dict.* 211/1 *Suede*.., leather, usually calf, finished by special process, with flesh side buffed on emery wheel to produce napped, velvety surface. **1968** V. Canning *Melting Man* viii. 237 The only spare shoes were a pair of ginger suèdes. **1970** *Daily Tel.* 2 Mar. 14 Ankle-length, shiny, wet-look coats, suèdes and leathers were often trimmed with fur. **1975** C. Calasibetta *Fairchild's Dict. Fashion* 324/2 *Suede*, leather, usually lambskin, doeskin, or splits of cowhide..that has been buffed on the flesh side to raise a slight nap. **1982** T. Heald *Masterstroke* v. 103 A heavy dew underfoot..soaked through Bognor's suedes, moistening his socks.

2. *attrib.* and *Comb.*, as *suede-coloured*, *-gloved*, *-like*, adjs.; **suede brush**, a brush with which to brush suede; **suede cloth** = *suedette; **suede-footed** *a.* = *suede-shoed* adj. below; **suedehead** *slang* (see quot. 1970); **suede shoe**, a shoe made with a suede upper; chiefly used *attrib.* to denote: (*a*) resemblance to the rough texture of suede; (*b*) *fig.*, something which displays a spurious smartness (*U.S. colloq.*); **suede-shoed** *a.*, wearing suede shoes.

1951 *Catal. of Exhibits, South Bank Exhib., Festival of Britain* 30/1 Suede brush; Federation of British Rubber Manufacturers Association. **1967** 'K. O'Hara' *Unknown Man* ix. 81 A rubber suede-brush she used to buff the key-case. **1930** *Daily Express* 30 July 5/4 Suede cloth, which made its real appearance in furnishing last year. **1979** *Arizona Daily Star* 5 Aug. J 5/2 (Advt.), Soft supple suedecloth is in several styles. **1897** Suède-coloured [in Dict.]. **1938** J. W. Day *Dog in Sport* iv. 64 It will take many generations of stupid women in Bayswater and *suède*-footed young men in Kensington to ruin the character of this eminently sensible working dog. **1979** —— in *East Anglian Mag.* Aug. 531/2 None of your suede-footed, whey-faced, sniffling little intellectuals. **1981** J. Johnston *Christmas Tree* 121 Her suede-gloved hands clasped on her knee. **1970** *Time* 8 June 37 The skinheads are lineal descendants of the rockers—with an added touch of mindless savagery. When their hair grows a trifle longer, they refer to themselves as suedeheads. Skins or suedes, they specialize in terrorising such menacing types as hippies and homosexuals, Pakistani immigrants and little old ladies. **1974** P. Cave *Mama* (new ed.) iv. 25 The suedehead kids weren't expecting any 'bovver'. **1971** *Country Life* 28 Oct. 1107/1 When some browsing animal blunders against them bursting their [*sc.* the puffballs'] suede-like skin. **1952** *News* (San Francisco) 27 Feb. 10/1 (*heading*) 'Suede-shoe boys' renew racket here. Home-owners warned on repair work. **1973** M. Amis *Rachel Papers* 29 Chronic bronco was reserved for nicotined oldsters with suede-shoe lungs. **1979** *Tucson* (Arizona) *Citizen* 20 Sept. 1B/6 There are also a lot more 'pseudo-high rollers' in Phoenix, too, which is Mano's polite description of a phony. 'Suede shoe types,' he calls them. **1980** D. Marlowe *Rich Boy from Chicago* iv. 52 He edited the college magazine (pre-Beat poetry, suede-shoe satire). **1938** *New Statesman* 21 May 863/2 The abusive semi-illiterate or the sleek, shinily tailored, down-at-heel, suède-shoed play-boy, who hawks inferior goods on their doorstep.

sueded (swē^i·dėd), *a.* [f. Suède + -ed[2].] Of leather: buffed on the flesh side to raise a slight nap. Also of fabrics, etc.: provided with a nap.

1956 *Gloss. Leather Terms* (*B.S.I.*) 5 A fine soft leather ..sueded on the flesh side. **1962** *L. L. Bean Catal.* Spring 12 Ladies' bush coat and pant..styled from sueded cotton poplin. **1971** *Leader* (Durban) 7 May 5/5 (Advt.), Men's bri-nylon sueded warm winter shirts. **1976** *National Observer* (U.S.) 30 Oct. 9/3 (Advt.), Made of strong and supple full grain steerhide with the rough side out. Rich, sueded finish. **1978** *Textiles* (Manchester) VII. 46/2 Patterned and sueded fabrics.

suedette (swe^ide·t). Also **suèdette.** [f. Suède + -ette.] A material designed to imitate the texture of suede, esp. a type of cotton or rayon fabric with a suede-like nap.

1915 *Chambers's Jrnl.* May 413/1 A cover of water-proofed suedette. **1930** *Daily Express* 30 July 4/5 To make a smart..tea cosy, cut out four pieces of material.. in suede, velvet, or suedette. **1960** *Pract. Wireless* XXXVI. 350/2 The cabinet is finished in cream rexine with a royal blue suedette surround and a blue and gold scale. **1962** *Punch* 23 May 785/3 Massive Mums in tartan trews and suèdette jackets. **1963** *Punch* 10 July 54/2 Apple-green suedette wallpaper. **1971** *Sunday Times* 6 June 33 Swimming in suede is the new thing; swimming in cotton suedette the next best. **1977** *Cosmopolitan* Feb. 19/1 Wore brown suedette shoes with thin black suits and thick regional accents.

‖**suerte** (suē^ə·ɹte). [Sp., lit. 'chance, fate, luck': cf. Sort *sb.*[1]] An action or pass performed in bull-fighting; one of the three stages of a bull-fight; = *tercio, tertio 2 a.

1838 *Q. Rev.* LXI. 418 'Suertes' or manners of killing the bull. **1893** Chapman & Buck *Wild Spain* v. 58 It is in this phase of the fight that we trace the origin of several of the *suertes* which are practised in the modern Corrida de Toros. **1910** *Encycl. Brit.* IV. 790 The fight is divided into three divisions (*suertes*). *Ibid.*, Then begins the *suerte de picar*, or division of lancing. **1932** R. Campbell *Taurine Provence* 61 The estocada is the climax, to hasten ..which, all the other suertes (actions, passes, and feats) must be devoted. **1957** A. MacNab *Bulls of Iberia* v. 53 The *banderilla* act is a 'decorative' *suerte* rather than one of 'punishment'. **1967** McCormick & Mascareñas *Compl. Aficionado* i. 24 'The suerte of the varas' means the picador's work.

Suess (sūs). The name of Hans E. *Suess* (b. 1909), Austrian-born U.S. chemist, used *attrib.* to designate certain phenomena in radio-carbon dating, as **Suess effect**, the reduction in the proportion of carbon 14 in the atmosphere and plant life during the twentieth century as a result of the increased burning of fossil fuels, which lack that isotope; **Suess wiggle**, each of a series of relatively short-term irregularities, of disputed existence and origin, in the calibration curve obtained by dendrochronology for radio-carbon dating.

1957 *Proc. R. Soc.* A. CCXLIII. 562 An accurate assessment of the Suess effect can yield valuable data on the carbon cycle. **1976** *Nature* 8 July 128/1 There have probably also been periods of irregular fluctuation spanning a few hundred years (the so-called Suess 'wiggles'). **1977** *Sci. Amer.* May 86/3 There is uncertainty in interpreting the present era of solar activity from carbon-14 evidence because of the Suess effect. **1979** *Nature* 5 July 48/1 (*heading*) Confirmation of the Suess wiggles: 3200–3700 BC.

suet. Add: **2. suet-brained** *a.*, stupid; **suet crust**, a form of heavy pastry made with suet, esp. used for meat or fruit puddings; **suet-faced** *a.*, having a face with an unhealthy or colourless complexion; **suet-headed** *a.*, stupid.

1921 *Public Opinion* 26 Aug. 199/2 Even among the most suet-brained readers of the Morning Post there are some [etc.]. **1845** E. Acton *Mod. Cookery* xvi. 406 (*heading*) Common suet-crust for pies. **1906** *Mrs. Beeton's Bk. Househ. Managem.* xxxi. 889 Suet crust..flour..suet.. baking-powder..salt..water. **1951** *Good Housek. Home Encycl.* 671/1 Make 6–8 oz. suet-crust pastry. **1977** 'E. Crispin' *Glimpses of Moon* xii. 231 Mrs Clotworthy is making a steak-and-kidney pudding with a thick suet crust. **1922** Joyce *Ulysses* 166 A pallid suetfaced young man polished his tumbler knife fork and spoon with his napkin. **1937** E. Pound *Let.* 10 Mar. (1971) 291 Make it clear..that 200 words per subject is all that wildcat editing can get over on the suet-headed Brits.

suevite (swē^i·vəit). *Petrogr.* [ad. G. *suevit*, f. L. *Suēvia*, *Suēbia*, name of a region in W. Germany (see Swabian *a.* 1 a): see -ite[1].] A type of welded braccia found associated with impact craters, similar to a tuff but showing signs of impact metamorphism; orig. such a rock from the Ries crater near Nördlingen in W. Germany.

1938 *Mineral. Abstr.* VII. 74 The tuffs (suevite) of the Nördlinger Ries are supposed to be rocks fused by the impact of the meteorite. **1970** *New Scientist* 23 July 174/3 The so-called 'suevite' rocks of the Ries are almost identical to some of the surface samples from the fragmented lunar 'regolith'.

suey pow (sū·i pau). *U.S. slang.* Also **suey-pow, sui pow.** [Orig. unknown.] (See quot. 1914.)

1914 Jackson & Hellyer *Vocab. Criminal Slang* 82 *Suey pow, noun*, current amongst opium smokers. A sponge or rag used to cool and cleanse the face of an opium bowl. **1926** *Variety* 29 Dec. 7/4 The dopes and hop heads, with their 'stem',..'sui pow', [etc.]. **1939** [see *joy-pop* s.v. *joy *sb.* 10].

Suez (sū·iz, sū·ēz). The name of an Egyptian port [Arab. *al-Suways*] at the head of the Red Sea, used *attrib.* and *absol.* to denote the military and political crisis which resulted from the nationalization of the Suez Canal in 1956; *Suez group* (now *Hist.*), a group of Conservative MPs who opposed the withdrawal of British troops from the Suez Canal Zone in 1954; hence applied to other groups advocating the presence of British troops in the Middle or Far East.

1955 *Ann. Reg. 1954* 34 Anglo-Egyptian talks have been recently renewed in Cairo..and the so-called 'Suez group' in the Conservative Party, about 40 in number and led by Captain Waterhouse, had consequently become restive. **1958** H. Nicolson *Diary* 18 June (1968) 350, I am very worried about the Lebanon situation, fearing it may prove a repetition of Suez. **1961** *Guardian* 6 Dec. 18/1 Captain Charles Waterhouse, one of the original 'Suez rebels' in the winter of 1956–7. **1962** *Hansard Commons* 13 Nov. 281/1 The hon. Member for Leeds, East spoke of my hon. Friend the Member for Inverness as being a member of the Suez Group. **1966** *New Statesman* 3 June 804/1 The cabinet's Suez Group (Wilson, Healey, Stewart and Bottomley) are prepared to bring back a good many servicemen following the end of Confrontation but want to maintain the bases till the late 1970s. **1968** M. Jones *Survivor* iii. 55 She could not remember events like Suez and Hungary. **1972** R. R. James *Ambitions & Realities* ii. 104 What became known as 'the Suez Group' constituted the first organized element in the Conservative Party that viewed Heath with hostility. **1981** A. Price *Soldier no More* ix. 122 Ever since Suez the Americans had been bad friends with the Israelis.

suffer, *v.* **3. b.** (Earlier example const. *under*.)

1836 Dickens *Let.* 15 Nov. (1965) I. 195, I..am still suffering under..a head-ache.

suffering, *ppl. a.* Add: **3. d.** *suffering cat*(*s*)*!* an exclamation expressing surprise or annoyance. Also *the suffering Moses* (cf. Moses 1 c), etc.

1869 'Mark Twain' *Innocents Abroad* v. 52 The suffering Moses!—there ain't money enough in the ship to pay that bill! **1897** Kipling *Captains Courageous* vi. 134 'Sufferin' Christianity!' sez Counahan (he always said that whin..he was not feelin' good). **1907** S. E. White *Arizona Nights* xv. 217 Suffering cats, think how that fellow sized us up for a lot of pattern-made fools. **1931** S. Lewis *Sel. Short Stories* (1935) 162 Suffering cats! You might have been one of your uncles still puttering around with dirty pitchforks back on the farm! **1948** G. H. Johnston *Death takes Small Bites* v. 122 She doesn't think I've got any guts.' 'Well, sufferin' cat! What does she want? Alexander the Great?' **1977** J. Porter *Who the Heck is Sylvia?* vi. 54 Oh, suffering cats, with that bunch of lecherous thugs it could have been *anybody!*

sufficient, *a.* Add: **1. c.** For *rare* read '*rare* exc. in allusion to or imitation of Matt. vi. 34'.

1766 A. Adams *Let.* 13 Oct. in L. H. Butterfield et al. *Adams Family Corr.* (1963) I. 56 Sufficient to the Day is the Evil thereof. **1917** H. B. Twyford *Purchasing & Storing* 323 A 'sufficient unto the day' policy has brought some rude jolts to many manufacturing establishments. **1921** Galsworthy *To Let* i. xii. 114 He never looks happy—not really happy. I don't want to make him worse, but of course I shall have to, when Jon comes back. Oh! well, sufficient unto the night! **1928** D. H. Lawrence *Lady Chatterley's Lover* ii. 18 Sufficient unto the day is the evil thereof. Sufficient unto the moment is the *appearance* of reality. **1960** C. Day Lewis *Buried Day* ii. 34 They watch the spring rise inexhaustibly—a breathing thread out of the eddied sand, sufficient to their day. **1967** S. Beckett *Stories & Texts for Nothing* v. 93, I haven't been damned for what seems an eternity, yes, but sufficient unto the day, this evening I'm the scribe. **1983** E. Rossiter *Lemon Garden* v. 72 'What about this hospital business?' Sufficient, I thought, unto another day.

2. c. *sufficient condition* (see quot. 1930). Cf. *necessary condition* s.v. *necessary *a.* 1 d.

1914 B. Russell *Our Knowledge of External Wrld.* iv. 109 In the hypothetical sense, continuity may be allowed to be a *necessary* condition if two appearances are to be classed as appearances of the same thing. But it is not a *sufficient* condition, as appears from the instance of the drops in the sea. **1923** C. D. Broad *Sci. Thought* xiii. 499 Certain brain-events are the necessary and sufficient conditions of the occurrence of all our different sensations. **1930** L. S. Stebbing *Mod. Introd. Logic* xv. 271 A condition X is a *sufficient condition* of an occurrence A provided that whenever X is present A occurs. But if A may occur when X is absent, then X, though a sufficient is not a *necessary* condition of A. **1948** Ambrose & Lazerowitz *Fund. Symbolic Logic* v. 83 The sufficient condition for *q*'s truth is given by '$p \supset q$'. **1949** [see *necessary *a.* 1 d]. **1965** E. J. Lemmon *Beginning Logic* i. 28 Hence we shall say that, whenever it is the case that if *P* then *Q*, *P* is sufficient condition for *Q*, and, whenever it is the case that only if *P* then *Q*, *P* is a *necessary condition* for *Q*.

d. *sufficient statistic*, a statistic that contains all the information in the observations it is based on that is relevant to the estimate being made.

[**1922** R. A. Fisher in *Phil. Trans. R. Soc.* A. CCXXII. 316 The statistic chosen should summarise the whole of the relevant information supplied by the sample. This may be called the Criterion of Sufficiency.] *Ibid.* 359 In the case of the normal curve of distribution it is evident that the second moment is a sufficient statistic for estimating the standard deviation. **1972** A. W. F. Edwards *Likelihood* ii. 18 If we were certain that no other model would ever be contemplated, then the sufficient statistic could replace the original data as raw material for inductive inference.

suffisance. Restrict † *Obs.* to senses 1–6 and add: **7.** Now only with Fr. pronunc. (sufizaṅs). (Later *Lit.* examples.)

1917 D. H. Lawrence *Phoenix II* (1968) 82 The police-officer turned, saluted politely, and said, with the polite, intolerable *suffisance* of officialdom: 'Good evening! Trouble here!' **1925** —— *St. Mawr* 25 At the same time he was free of the Englishman's water-tight *suffisance*. **1957** S. Smith *Coll. Poems* (1975) 344 Ah me the *suffisance* I drew therefrom What strength, what glory from that fattening fluid.

suffix, *sb.* Add: **3. suffix ablaut**, variation in the vowel of a suffix; **suffix language** (earlier example).

1900 E. Björkman *Scand. Loan-Words in M.E.* I. 112 Here *aȝ* might depend on suffix-ablaut as in O.E. *fæȝen*, *fæȝ(e)nian*. **1977** *Archivum Linguisticum* VIII. 80 We must now examine cases where SF apparently fails, despite a following [l] rather than [ł]. Several cases, such as

falaed, *alaer* may be explicable on grounds of 'suffix ablaut'. **1874** H. Bendall tr. *Schleicher's Compar. Gram.* 3 The Indo-European is therefore a suffix-language, together with the neighbouring languages of the Finnish stem.

Hence **su-ffixual** *a.* = Suffixal *a.*

1901 J. Huguenin *Secondary Stress in Anglo-Saxon* 13 The inflected cases in which the suffixual syllable is lengthened by position are, the genitive and dative singular feminine, the accusative singular masculine, and the genitive plural. **1964** [see *mora[1] 3 b].

suffling (sɒ·f'liŋ), *vbl. sb.* [f. Suffl(e *v.* + -ing[1].] A sound as of blowing or heavy breathing.

1904 H. F. Day *Kin O'Ktaadn* i. 11 The..whummle of horses and..sufflings of..cattle hint that 'fodder-time' is at hand. **1933** W. de la Mare *Lord Fish* 37 It was so full of the suffling and sighing, the music and murmuration of water.

Suffolk. Add: **a.** *Suffolk ham*; **Suffolk latch** (see quot. 1972); **Suffolk sheep**, a black-faced hornless sheep of a breed first developed in East Anglia, distinguished by a short fleece, large size, and the production of lean meat.

1855 E. Acton *Mod. Cookery* (rev. ed.) xiii. 256 The receipt for the Suffolk ham. **1966** *Times* 28 Dec. 9/6 A genuine Suffolk ham man is Mr. R. Stiff of Kersey. He uses his own pigs and douses the hams in black treacle, spices and stout. They are then floated in tubs of sweet pickle. **1940** *Chambers's Techn. Dict.* 819/2 *Suffolk Latch* (*Join.*), a variant of the *Norfolk Latch*. **1972** *Country Life* 13 Jan. 98/1 The great variety of country door-latches would reward a study in depth: the well-known type, the 'Suffolk latch', is usually operated by a pivoted blade, which passing through the door and depressed by thumb pressure, lifts the latch. **1981** 'G. Gaunt' *Incomer* xviii. 117 Les Taunton thumbed down the Suffolk latch of the Queen's Head taproom and entered. [**1794** A. Young *Gen. View Agric. Suffolk* xiii. 33 The Norfolk breed of sheep spread over almost every part of the county; and as the most famous flocks are about Bury..it has been observed, that they ought rather to be called the Suffolk breed.] **1893** J. Wrightson *Sheep* viii. 75 The original Suffolk sheep existed in famous flocks during Arthur Young's time.

c. Also, = Suffolk punch, sheep.

1831 W. Youatt *Horse* 39 The immense power of the Suffolk is accounted for by the low position of the shoulder. **1902** *Encycl. Brit.* XXV. 193/2 The Suffolk..probably took its origin in the crossing of improved Southdown rams with the old horned Norfolk ewes. **1928** [see *actioned *ppl. a.*]. **1960** G. E. Evans *Horse in Furrow* xiii. 181 He knew the Suffolks so well..that he could pick out a horse's breed by studying him. **1979** C. MacLeod *Luck runs Out* (1981) ii. 21 Those gorgeous Clydesdales and Percherons and Belgians and Suffolks..with their brasses polished like gold. **1980** 'D. Shannon' *Felony File* vii. 173 He's found the sheep... They're sixty dollars each. They're Suffolks.

Suffolker (sɒ·fəkəɹ). [f. Suffolk + -er[1].] A native or inhabitant of Suffolk.

1849 Dickens *Dav. Copp.* (1850) xi. 117 The men generally spoke of me as..'the young Suffolker'. **1910** H. M. Doughty *Chron. Theberton* iii. 40 We Suffolkers never misplace H's. **1952** M. Allingham *Tiger in Smoke* vi. 100 We come from the same part of the country, sir. We're all Suffolkers. **1978** *East Anglian Daily Times* 7 Dec. 8/6 First was depicted the hitherto unseen paintings of a dead Suffolker, the late Cecil Howard Lay.

suffrage, *sb.* Add: **11. b.** Also *adult suffrage.*

1906 W. R. Cremer in *Hansard Commons* 25 Apr. 1572 If they once opened the door and enfranchised ever so small a number of females..it ultimately meant adult suffrage... Hon. Members had not really thought out what adult suffrage must lead to. **1910** *Hansard Commons* 11 July 55 The Member for Clitheroe explained with great explicitness what his object is. The hon. Gentleman's object is adult suffrage. That adult suffrage, of course, includes the vote for all adult women... The result of this adult suffrage, when it does come, will be a total electorate of 23,000,000 instead of 7,000,000. In that total electorate there will be a considerable majority of women. **1939** G. B. Shaw *Geneva* i. 16 The president and parliament are elected by adult suffrage every two years.

suffragette. Add: Also as *v. intr.* (in quot. *fig.*). **suffragettish, -ism** (examples). Also **suffragetty** *a.*

1909 H. G. Wells *Ann Veronica* vii. 134 And her straight hair was out demonstrating and suffragetting upon some independent notions of its own. **1912** C. S. Churchill *Let.* 7 Feb. in M. Soames *Clementine Churchill* (1979) vi. 76 Amy is kind, but more Suffragetty, Christian Sciency and Yankee Doodle than ever. **1913** G. B. Shaw *Let.* 4 Feb. in *B. Shaw & Mrs. Campbell* (1952) 79 That is the sort of thing that you vaguely lump into a cloud of abomination as Suffragettism. **1957** E. Hyams *Speaking Garden* 75 What, in their time, were more ludicrous than suffragettism or antisepsis or anti-slavery? **1970** G. Greer *Female Eunuch* 295 The history of suffragettism.. is beyond the scope of this book. **1974** V. Nabokov *Look at Harlequins* (1975) iv. iv. 173 Mrs. Noteboke, a stout dark lady in suffragettish tweeds.

suffragi (sufrāgī). Also **suffraggi.** [a. *sufragī*, repr. Egyptian Arab. pronunc. Turk. *sofrajı*, f. Arab. *sufra* food, dining-table + Turk. agent-suffix *-jı*.] A waiter, butler or steward.

1924 *Blackw. Mag.* Feb. 246/2 Our suffragi and cook led a sort of troglodyte life for days. **1959** W. Thesiger *Arabian Sands* xiii. 247 In the new 'hotel'..there was electric light, fans, and tinned food served by a Sudanese suffragi. **1972** R. Maugham *Escape from Shadows* iii. 136 Suffragis sprang up from nowhere with drinks, and I was offered one. **1979** *Stand* XX. iv. 34/2 The suffraggi puts a whisky before me.

suffragist. Add: Hence **suffragi·stic** *a.*, **suffragi·stically** *adv.*

1907 M. Beerbohm in *Sat. Rev.* 13 Apr. 457/2 The shrill suffragistic cheers which punctuated the first performance. **1909** *Daily Chron.* 13 Nov. 6/4 The pageant had been preceded by an excellent entertainment, including several 'suffragistic' and other playlets. **1923** K. D. Wiggin *My Garden of Memory* (1924) xiii. 121 Ella intrusively and suffragistically fluttered into the nest,..sadly complicating the family arrangements.

suffusive, *a.* (Earlier example.)

1872 Geo. Eliot *Middlem.* I. ii. xvi. 295 That agreeable after-glow of excitement when thought lapses from examination of a specific object into a suffusive sense of its connections with all the rest of our existence.

sugar, *sb.* Add: **1. e.** *colloq.* A lump or tea-spoonful of sugar.

1962 L. Deighton *Ipcress File* xxiii. 150 He poured coffee into a black wedgwood cup and put four sugars in. 'Raise the sugar count,' he said. **1978** C. MacLeod *Rest you Merry* (1979) ii. 18 'Why don't I make us a cup of coffee?' 'Great idea. Three sugars in mine.' **1982** *Sunday Tel.* 18 Apr. 8/6 How many sugars they were allowed in their tea.

2. d. *slang* (orig. *U.S.*). A narcotic drug: spec. (*a*) heroin; *brown sugar* (see quot. 1974); (*b*) LSD (taken on a lump of sugar).

1935 A. J. Pollock *Underworld Speaks* 116/2 *Sugar and salt*, poisonous habit forming drugs; any of the white narcotics. **1951** *Evening Sun* (Baltimore) 27 Mar. 4/1 Dope in general was 'cement'..'sugar', etc. **1956** H. Gold *Man who was not with It* (1965) iii. 27 You'll dream about the sugar yet. You'll wake up hot for it. No joy-popping, hear? Stay off, kid. **1967** M. M. Glatt et al. *Drug Scene* Gloss. 116 *Sugar*, dose of LSD on sugar lump. **1973** K. Royce *Spider Underground* viii. 118 We sat in a corner of this dark, smoke-infested hole that smelled of.. third-rate pot... 'What a place to pick,' I complained. 'It's the sort of dump the fuzz raid three times a week.'.. 'Relax, man. They hit us last night... That makes it safe, man. I'm not carrying sugar or anything. I don't touch the stuff.' **1974** *Indonesian Observer* 26 July 3/2 French police said this year they have seized 50·6 pounds (23 kilograms) of 'brown sugar' in the suitcases of 13 Chinese arriving at Orly airport enroute to Amsterdam. The brown sugar is 33 per cent heroin diluted with 60 per cent caffein and strychnine. **1978** D. MacKenzie *Raven settles Score* (1979) 32 No more Hong Kong brown sugar. We'll be out of business. **1979** *Observer* 25 Nov. 4/1 Detectives call them the 'sugar people' and they are young, rich and blue-blooded. They are also heroin addicts. It is in an ironic double reference to the 'sugar daddy' parents and to the expensive white powder they inject or sniff.

e. *colloq.* A term of endearment. Also in Comb., as *sugar-babe*, *-baby*, *-pie*, etc.

1930 *Dialect Notes* VI. 85 *Sugar-pie*,..common term of endearment. **1930** J. H. Combs in B. A. Botkin *Folk-Say* v. 245 A-settin' on the ice till my feet got cold, sugar-babe. **1936** M. Mitchell *Gone with Wind* xxvi. 455 Scarlett said gratefully: 'Thank you, Sugarbaby.' **1936** J. Curtis *Gilt Kid* vi. 68 When am I going to see you again, sugar? **1944** L. A. G. Strong *Director* xvii. 135 See here, sugar. I'll take care of you. **1951** S. Spender *World within World* i. 26 No, you don't, sugar, you don't go out with your cold. **1962** J. D. MacDonald *Girl, Gold Watch & Everything* vii. 87 What you do for a living, sugar? **1976** P. Flower *Crisscross* i. 10 'What's funny, sugar?' Sibyl said... Would he ever get Sibyl to stop calling him sugar? **1980** D. Brierley *Blood Group O* 76 Okay, sugar, what are you looking for?

3. a. *sugar of lead*, *sugar of milk* (later examples).

1847 C. J. Hempel tr. *Rau's Organon of Specific Healing Art* lxii. 128 If triturated with sugar of milk, it [*sc.* phosphorus] changes to phosphoric acid in a very few hours. **1864** P. Squire *Compan. Brit. Pharmacopœia* 161 *Sugar of Milk*... Crystallized Sugar obtained from the Whey of Cow's Milk by evaporation. **1895** *Montgomery Ward Catal.* Spring & Summer 252/3 Artists Tube Oil Colors..Silver White, Sugar of Lead, Terre Verte. **1975** *Nature* 23 Oct. 632/2 Something needed to be done to stop the watering of milk..and even so flagrant a malpractice as the use of 'sugar of lead', as lead acetate was called, to sweeten beer.

4. a. (*a*) *sugar basin* (earlier examples), *basket*, *bin*, *cube*, *dish* (earlier examples), *factory*, *icing* (later *comb.* and *fig.* examples), *industry*, *kettle* (earlier example), *knife* (later example), *lump*, *mill* (later examples), *mule*, *ration*, *refinery* (earlier example; also *fig.*), *scoop*, *thermometer*, *trade* (later example), *worker*. (*b*) *sugar field*, *grove*, *island* (later examples), *land*, *plantation*.

1785 *Daily Universal Reg.* 1 Jan. 3/2 (Advt.), Oval pierced sugar and cream basons, 10 oz. to 15 oz. a pair. *a* **1828** D. Wordsworth *Jrnl.* (1941) II. 81 A sugar-basin made of cocoa-nut. **1917** F. H. Bigelow *Historic Silver of Colonies* 472/1 (Index), Sugar baskets. **1981** *Sunday Tel.* 18 Jan. 13/2 Garrads have augmented the exhibition with antique castors.., as well as sugar baskets, boxes, tongs and nippers. **1922** Joyce *Ulysses* 58 There he is,..leaning against the sugar-bin in his shirt-sleeves. **1897** Sugar cube [see *cube *sb.*[1] 1 b]. **1978** T. Allbeury *Lantern Network* xi. 169 She was screwing up the paper from the sugar cubes. **1742** W. Ellis *Timber-Tree Improved* II. 151, I was told..that this Wood makes fine Sugar-dishes, and other Turners-ware. **1765** J. Wedgwood *Let.* 17 June (1965) 34 The articles are..a slop basin, sugar dish with cover, [etc.]. **1908** Kipling *Actions & Reactions* (1909) 96 They [*sc.* bees] took to cadging round sugar-factories and breweries. **1958** O. Caroe *Pathans* xxvi. 429 Peshawar, always famous for its sugar-cane, has been enriched with finer varieties which have turned the old village industry of *gur* into the great sugar-factories which now sustain the life of Pakistan. **1930** W. K. Hancock *Australia* iv. 81 Polynesians in their wild state never clamoured for admission to the Queensland sugar-fields. **1792** G. Imlay *Topogr. Descr. Western Terr. N. Amer.* 136 Luxuriant sugar groves. **1847** *Ex. Doc. 31st U.S. Congress 1 Sess. House* (1849) No. 5. iii. 629 A ridge covered with sugar maples, formerly an Indian sugar grove. **1948** E. N. Dick *Dixie Frontier* 247 A clump numbering from one hundred to three hundred trees was chosen for the operation. Such a clump came to be called a sugar grove. **1930** E. Waugh *Labels* vii. 180 Gaudi has again introduced his 'sugar-icing' motive, translating it from tile and mosaic into carved stone. **1979** 'M. Hebden' *Pel & Faceless Corpse* xii. 123 The pink shirt had suddenly become sugar icing-coloured and hideously wrong. **1887** *Encycl. Brit.* XXII. 628/1 There are numerous modified and subsidiary processes connected with refining, as well as with all branches of the sugar industry. **1779** Sugar island [see *scuttle *v.*[2] 1 a]. **1980** *Jrnl. R. Soc. Arts* Apr. 271/1 The UK has traditionally bought 50 per cent of the sugar consumed here on the world market, principally from the Sugar Islands of the Caribbean. **1834** J. Kemper in *Wisconsin Hist. Coll.* (1898) XIV. 444 If ardor leads some of the [Sioux] hunters beyond the boundary stake, they can be punished by the soldiers by having their sugar kettles broken or their lodges torn down. **1949** *Caribbean Q.* I. 8 It was..the stalwart, armed with hoes and..sugar knives.., whose work would 'make or break' the proprietor. **1692** *Calendar Virginia State Papers* (1875) I. 44 We marcht to the Suggar Land. **1883** Sweet & Knox *On Mexican Mustang through Texas* vii. 82 A great deal of the finest sugar-lands in the world. **1974** *Guardian* 23 Jan. 12/6 As far as sugar lands are concerned,..the Government is now the largest landowner. Tate and Lyle sold their lands to the last government. **1901** Kipling *Kim* xii. 307 She chuckled like a contented parrot above the sugar lump. **1964** D. Francis *Nerve* ix. 122 The dope has been given to the horses on sugar lumps. **1882** W. D. Hay *Brighter Britain!* I. viii. 221 If all the farmers in the district were to combine to grow beet-root on every acre they could plough,..even then it would hardly pay the sugar-mills, or possibly the farmers either. **1971** *Advocate-News* (Barbados) 24 Apr. 10/1 (Advt.), ½ acre house plots and/or cottage with sugarmill and swimming pool. **1908** *U.S. Dept. Agric. Farmers' Bull.* No. 334. 24 Sugar mules are those shipped south to use on the sugar farms of Georgia, Louisiana, and other Southern States. **1960** V. Williams *Walk Egypt* 71 A sugar mule, now, was a big fellow. He ate big, but he pulled big, and he would look big before the wagon. **1883** 'Mark Twain' *Life on Miss.* xl. 419 The great sugar plantations border both sides of the [Mississippi] river. **1978** 'A. York' *Tallant for Disaster* ii. 28 The burnt earth roadway which led to the sugar plantation. **1917** Sugar ration [see *ration 3 c]. **1978** L. Deighton *SS-GB* xxv. 237 Drink up your tea, that's a good boy. It's the last of the sugar ration. **1794** A. Young *Trav. France* (ed. 2) II. xix. 539 The sugar refinery is a considerable business, there are 10 large and 17 smaller houses engaged in it. **1896** G. Meredith *Let.* 17 June (1970) III. 1236, I..can own her sweet to the ear, wondering what it is in her that extracts her deadly bitter from a sugar-refinery. **1916** *Daily Colonist* (Victoria, B.C.) 1 July 8/2 Mothers had been remembered by most of the workers, for there were bread boards, and sleeve holders, sugar scoops and wooden spoons. **1960** R. A. Parker *Family of Friends* 89 The old days of the Quaker garb and the sugar-scoop bonnet were gone forever. **1977** *Time* 14 Nov. 21/1 The Concordski whistled down the runway for 33 seconds, sucking in air through four 'sugar scoop' intakes slung beneath its body. **1913** M. H. Neil *Candies & Bonbons & how to make Them* 24 A sugar thermometer is generally used for testing the boiling syrup. **1887** *Encycl. Brit.* XXII. 625/1 Within the first twenty years of the 16th century the sugar trade of San Domingo expanded with great rapidity. **1973** *Sunday Express* (Trinidad) 1 Apr. 12/5 A delegation of sugar workers is to ..protest what they call the 'abandonment of the cane-growing industry'.

b. *sugar-broker* (example), *-growing* (earlier example), *-maker* (later examples), *-making* (later examples), *-planter* (later examples), *-producer*, *-producing* (earlier example), *rationing*, *-refiner* (later examples).

1841 *Picayune* (New Orleans) 10 June 2/3 Several dealers in sugar and sugar brokers were yesterday summoned before Recorder Bertus. **1816** *Niles' Reg.* 6 Apr. 81/1 The representatives of the sugar-growing states insist on a certain duty upon that article. **1835** J. J. Audubon *Ornith. Biogr.* III. 439 With large ladles the sugar-makers stirred the thickening juice of the maple. **1899** W. A. MacKay *Pioneer Life in Zorra* 171 Not infrequently would the sugar-makers remain in the woods most of the night boiling down the sap. **1828** M. O'Brien *Jrnls.* (1968) I. iii. 27 During sugar-making time it will contain a furnace and other vessels. **1953** R. F. V. Heuston *Salmond's Law of Torts* (ed. 11) xiv. 566 In *Indermaur* v. *Dames* itself the hole in the floor was a defect but a necessary incident of sugar-making. **1842** *Niles' Reg.* 14 May 176/3 (*caption*) Sugar planters of Louisiana. **1926** J. Masefield *Odtaa* i. 4 In the seventies others, from all parts of England, settled as sugar-planters along the northern sea coast in the Pituba country. **1983** A. Brookner *Look at Me* iv. 56 The wealthy sugar planter's daughter. **1881** *Harper's Mag.* Apr. 646 We met one of the largest sugar producers. **1974** *Guardian* 23 Jan. 12/4 Jamaica is the biggest sugar producer in the Commonwealth Caribbean. **1866** 'Mark Twain' *Lett. from Hawaii* (1967) 135 Maui.. that deservedly famous sugar-producing region. **1918** *Times* 29 Jan. 3/1 When sugar rationing actually came

into operation, the workers..had to face considerable pressure. **1976** J. Lee *Ninth Man* 77 Talking about sugar rationing. **1879** G. W. Bagby *Canal Reminiscences* 10 What was their petty thieving compared to the enormous pillage of the modern sugar refiner and the crooked-whiskey distiller? **1979** *Dædalus* Summer 113 Sugar refiners, soap boilers, glass blowers, and brewers..depended on continuously fired furnaces.

c. *sugar-cured* (earlier U.S. examples), *-free*, *-pink*.

1848 A. Prentice *Let.* 20 June in *Tour in U.S.* vi. 56, I tasted some excellent sugar-cured ham. **1889** *Judge* (U.S.) 12 Jan. 222/2 Beautiful red, sugar-cured ham. **1924** *Amer. Jrnl. Physiol.* LXVII. 635 Three other totally depancreatized dogs had been used for studying the administration of insulin..for several weeks, during which time their urine was never sugar-free for a period of more than 6 or 7 hours at a time. **1978** *N.Y. Times Mag.* 23 July 22/3 The absence of what had formerly been desirable is now proudly advertised: not only lead-free gas, but salt-free diets and sugar-free soft drinks. **1961** *House & Garden* Feb. 48 A..sofa covered in sugar-pink taffeta. **1978** 'M. M. Kaye' *Far Pavilions* xxi. 299 Rajastham.. where..men..painted their houses blinding white or sugar-pink.

5. a. sugar-almond (later examples); **sugar aquatint**, a method of etching in which the artist draws his dark areas on a copper plate with a solution of black water-colour and sugar; **sugar-bag,** (*a*) a bag or sack for containing sugar, *esp.* a bag made of coarse thick paper specially coloured or (*Austral.* and *N.Z.*) of fine sacking; also used as a measure of quantity; (*b*) (in Austral. Aborigines' speech) a wild bees' honeycomb; **sugar-cake** (later examples); **sugar-camp** (earlier and later examples); **sugar card,** a ration card entitling the holder to a ration of sugar; **sugar-coated** *ppl. a.* (earlier example); also *fig.*; **sugar-crusher,** (*a*) a machine for crushing sugar-cane; (*b*) an implement for crushing sugar for use at table; **sugar daddy** [cf. *daddy 3] *slang* (orig. *U.S.*), an elderly man who lavishes gifts on a young woman; also *transf.*; **sugar-house molasses** (earlier examples); **sugar mouse,** a sweet made of sugar in the shape of a mouse; **sugar nippers,** (*a*) an implement for cutting loaf sugar into lumps; (*b*) a pair of sugar tongs; **sugar-on-snow** *U.S.*, a delicacy made by pouring hot maple syrup on snow (Snow *sb.*[1] 4 a); **sugar-paper,** coarse paper such as that used for making sugar-bags; **sugar puff,** (*a*) a puff (see Puff *sb.* 5) made with sugar; (*b*) *pl.*, the proprietary name of a breakfast cereal; **sugar rag** *U.S.* = *sugar-teat*; **sugar sack,** a bag made of fine sacking for containing sugar; the sacking itself; **sugar sand** *U.S.*, a fine sand raised by the sap of the maple tree which results in a gritty sediment in maple syrup unless removed; **sugar shell** *N. Amer.*, a spoon with a shell-shaped bowl for serving sugar; **sugar sifter,** (*a*) see quot. 1875; (*b*) = *sugar caster*; **sugar snow,** (*b*) *N. Amer.*, a snowfall in the maple sugar season (see quot. 1932); **sugar soap,** an alkaline abrasive used to remove paint, and in solution for cleaning paintwork; **sugar stick,** also *fig.*; **sugar-teat** (later example); also **sugar-tit; sugar trough** *U.S.*, a wooden trough used for collecting maple sap; **sugar vase,** a tall sugar-container for use at table; **sugar-weather** *Canad.*, spring weather, characterized by cold nights and warm days, that starts the sap running in maple trees.

1935 *Amer. Speech* X. 193/2 The 'bonbon [fashion] shades' included *icing blue* and *sugar almond pink*. **1973** G. Greene *Honorary Consul* iii. ii. 124 It [*sc.* a missal] might have been a first Communion present, for it closely resembled the sugar almonds..distributed on such occasions. **1962** D. Bland *Illustration of Bks.* (ed. 3) viii. 155 Picasso used sugar aquatints in his Buffon, making two plates, one to print grey and the other black. **1764** *New Hampsh. Hist. Soc. Coll.* (1889) IX. 156, [I sent] also lb 14½ Sugar bag with it. **1830** R. Dawson *Present State of Australia* 136 The strange native pointed with his tomahawk to the tree and..repeated the words, 'Choogar-bag, choogar-bag, choogar-bag!' (sugar-bag) their English expression for honey, or anything sweet. **1864** R. Henning *Let.* 27 Nov. (1966) 185 The other [aboriginal] has been..climbing gum-trees after 'sugar-bags', or wild honeycombs. **1882** *Cassell's Family Mag.* Nov. 756/2 The crowns..have two square corners like the bottom of a sugar-bag. **1913** D. H. Lawrence *Sons & Lovers* vii. 164 There's something very blue; is it a bit of sugar-bag? **1927** M. Terry *Through Land of Promise* 104 We found the others clustered round a bauhinia tree...'We've got a sugar bag.' **1928** V. Palmer *Passage* i. v. 44 It was Uncle Tony standing with a sugar-bag over his shoulders. **1948** F. A. Iremonger *William Temple* v. 81 A nine-year-old boy in a Bethnal Green school, who handed to his teacher one morning an untidy piece of blue paper torn from a sugar-bag. **1963** *N.Z. Listener* 6 Sept. 9/2 Reference to the price of a 'sugar' bag full of oysters. It drew my attention to the frequency with which we in New Zealand refer to a 'sugar bag' as a basic unit of quality. **1967** A. & D. Reid *Paddle Wheels on Wanganui* 71 On another trip the same cabin boy acquired a sugar-bag of apples. **1923** Sugar cake [see *sauerbraten]. **1977** A. Wilson *Strange Ride of Rudyard Kipling* ii. 110 The Durbar Room at the Queen's beloved Osborne House—not a very happy sugar-cake Moghul decoration. **1779** M. Patten *Diary* (1903) 400, I went to our shugar Camp and covered some fire steads with brush where we had Cabbage and french Turnip seed sowed to preserve them from Cattle. **1805** R. Sutcliff *Trav. N. Amer.* (1811) 184, I saw several sugar camps..where the sap is collected in small wooden troughs. **1959** R. Campbell *I would do it Again* ii. 7 The neighbours gathered at the sugar camps. **1966** *Publ. Amer. Dial. Soc.* xxxviii. 66 *Sugar camp.* This characteristically Midland [Illinois] term appears only once in the field interviews but with much more frequency in the checklists. **1917** H. H. Henson *Jrnl.* 11 Dec. in *Retrospect* (1942) I. vi. 217, I started the day by filling up the new sugar cards for the household. **1875** 'Mark Twain' in *Atlantic Monthly* Aug. 195/1 Stephen sweetened him up and put him off a week. He called then..and came away sugar-coated again. **1935** *Motion Picture* Nov. 81/1 That keen humor, barbed sometimes, pointed always, but never other than good-natured and sugar-coated, has passed beyond our ken. **1977** R. L. Wolff *Gains & Losses* ii. 197 The earliest [High Church] novelists..whose fiction amounted to little more than sugar-coated tracts. **1870** A. S. Stephens *Married in Haste* 366 He held a sugar-crusher in one hand. **1901** Kipling *Kim* xv. 403 He felt..that his soul was out of gear with its surroundings—a cog-wheel unconnected with any machinery, just like the idle cog-wheel of a cheap Beheea sugar-crusher laid by in a corner. **1962** J. B. Priestley *Margin Released* i. i. 11 In winter, toddy, for which we had those silver sugar-crushers. **1926** G. Frankau *My Unsentimental Journey* ii. 32 There came another woman to the sofa: and spoke to me of 'sugar-daddies'. **1935** Wodehouse *Luck of Bodkins* xxi. 266 The morning papers had come aboard, reassuring citizens ..that sugar daddies were still being surprised in love-nests. **1959** [see *door-mat b]. **1973** *Times* 13 July (Motor Racing Suppl.) p. iii/2 The oil and petrol companies, for a long time the sugar-daddies of top class motor racing. *Ibid.* 20 Sept. 3/7 Norma Levy, a prostitute, had a 'sugar daddy' called Bunny who paid her rent and gave her a Mercedes car. **1848** W. E. Burton *Waggeries* 35 Encomiums on the sweets of married life were drowned in sugar-house molasses. **1886** B. P. Poore *Perley's Reminisc.* I. 39 Many of the passengers visited the bar to imbibe Holland gin and sugar-house molasses—a popular morning beverage. **1931** A. Uttley *Country Child* xii. 115 She pinched the stocking from the toe to the top... There was a tin ball..filled with comfits, and an orange, and a sugar mouse. **1965** 'M. A. Gibbs' *Sugar Mouse* xv. 155 A sugar mouse, its chocolate eyes run to smudges, its paper ears flattened,..and its sugar hardened into rock. **1790** *Pennsylvania Packet* 1 Mar. 1/1 This Day ..will commence the Sale of a Large and General Assortment of..screw drivers, iron holders, sugar nippers. **1840** Barham *Ingol. Leg.* 1st Ser. 240 With those great sugar nippers they nipp'd off his 'flippers'. **1858** P. L. Simmonds *Dict. Trade Products*, *Sugar-nippers*, tools for cutting loaf-sugar into lumps. **1921** *Glasgow Herald* 14 July 5 A pair of George II. silver sugar nippers. **1981** *Sugar nippers* [see *sugar basket*, sense 4 a above]. **1947** *Publ. Amer. Dial. Soc.* viii. 9 *Sugar on snow*,..'waxed' maple sugar served on snow. **1948** *Richmond* (Va.) *Times-Dispatch* 9 Jan. 16/1 As serious a breach of etiquette as eating 'sugar-on-snow' with a knife or beating one's grandmother in public. **1973** M. Crowell *Greener Pastures* 173 It never fails to remind me..of our introduction to sugar-on-snow. **1926** *Paper Terminol.* (Spalding & Hodge, Ltd.) 24 *Sugar paper*, a common quality of wrapping paper made principally from paper waste. Used.. for sugar bags. **1972** *Guardian* 5 Dec. 16/7 Drawing paper ..Grey or off-white, good quality sugar paper. **1711** Sugar puff [see *ratafia 1]. **1736** [In Dict., sense 4 a]. **1957** *Trade Marks Jrnl.* 1 May 460 Sugar Puffs... Cereal preparations coated with sugar and flavoured with honey... Quaker Oats Limited. **1959** *Elizabethan* Apr. 10/1 You've taken all the Sugar Puffs which are sweet already and left me with one mouldy old bit of Shredded Wheat. **1962** J. Braine *Life at Top* xiii. 173, I want Sugar Puffs, Daddy, I do. And yoggy. And cheese. **1855** J. E. Cooke *Ellie* 203 Are you going..to make a sugar-rag for that baby up there? **1895** 'Mark Twain' in *Harper's Mag.* Dec. 136/1 Somebody fetch this sick doll a sugar-rag. **1938** *Daily Progress* (Charlottesville, Va.) 15 Feb. 1/6 Mayor J. Fulmer Bright..dubbed the concessions offered by the State a 'sugar-rag dipped in paregoric'. **1891** Kipling *Light that Failed* ii. 18 Has any man here a needle? I've got a piece of sugar-sack. **1929** B. L. Burman *Mississippi* 78 Two beds, one made of automobile cushions nailed together and covered with a few folded sugar-sacks. **1965** S. T. Ollivier *Petticoat Farm* x. 140 The thin tired figure with the..sugar-sack apron and dishevelled hair. **1882** *Vermont Agric. Rep.* VII. 64 In the process of sugar making there was a point where it would combine with the lime, making 'sugar sand' or the malate of lime. **1949** [see *nitre *sb.* 1 d]. **1975** *Islander* (Victoria, B.C.) 25 May 5/3 The strained [maple] syrup should sit to allow sugar sand to settle to the bottom of the mixture. **1895** Sugar shell [see *flat-ware* s.v. *flat A. *adj.* 15]. **1916** *Daily Colonist* (Victoria, B.C.) 12 July 7/1 Sale Goes Merrily On!..Sugar Shells, fine silver plate, plain, for 50c. **1875** Sugar sifter [in Dict., sense 4 b]. **1906** Galsworthy *Man of Property* i. vi. 88 Now, what did you give for that sugar-sifter? **1976** *Deeside Advertiser* 9 Dec. 9/6 She presented a cut glass sugar sifter to Mrs. Brockley, past president. **1826** A. Anderson *Diary* 20 Mar. in G. Sellar *Narrative* (1916) viii. 124 Gordon awakened us by shouting 'A sugar snow.' There had been a light shower of it during the night, and the air was soft. Holes were rebored, and there was a fine run of sap. **1932** L. I. Wilder *Little House in Big Woods* 92 It's called sugar snow, because a snow this time of year means that men can make more sugar... The snow will hold back the leafing of the trees, and that makes a longer run of sap. **1973** M. Crowell *Greener Pastures* 149 Sugar snow is falling in those distinctive great feathery flakes that foretell the beginning of a maple sap run. **1930** C. H. Eaton *Painting & Decorating* IV. xiii. 843 Sugar soap has a softening action on the water, and is not so liable [as soda]..to cause undue softening of the paint film. **1958** *Woman* 22 Feb. 14/3 Walls must be washed, brushed... Paintwork washed with sugar-soap, rinsed and allowed to dry. **1963** W. Tee *Painting & Decorating* viii. 67 When you have removed all traces of the sugar soap, mop up surplus moisture. **1892** *Irish Daily Independent* 4 July 5/5 We are not sugarsticks. *Ibid*, Sugarsticks..men whose steadfastness would melt away before a passing cloud. **1936** W. B. Yeats *Let.* 21 Dec. (1940) 124 He [*sc.* Wilfred Owen] is all blood, dirt & sucked sugar stick. **1938** M. K. Rawlings *Yearling* v. 51 The 'coon nibbled at his flesh and cried again. 'He wants his sugar-teat,' Fodder-wing said maternally. **1892** *Dialect Notes* I. 232 Sugar-tit. **1936** M. Mitchell *Gone with Wind* viii. 145 Prissy produced the sugar-tit..and the baby's wails subsided. **1958** S. A. Grau *Hard Blue Sky* 118 So she went into the bedroom and picked up the sugar tit and tucked it into his mouth. **1779** in *Proc. Mass. Hist. Soc.* (1886) 2nd Ser. II. 453 Made Sugar Troughs and Katch.[d] some Sap. **1837** R. Bird *Nick of Woods* II. iv. 90 What should I do but see the old sugar-trough floating in the bushes. **1946** C. Richter *Fields* 17 She lifted the long bundle from out of the sugar trough. **1848** H. R. Forster *Stowe Catal.* 144 A pierced sugar-vase—with goats' heads. **1956** G. Taylor *Silver* ix. 202 *Sugar Vases.* Among the many varieties of vases is one based on the Greek volute-krater. **1981** *Sunday Tel.* 18 Jan. 13/1 Tate and Lyle's own collection..includes silver gilt sugar vases with tops, and the pierced ladles used with them. **1826** A. Anderson *Diary* 18 Mar. in G. Sellar *Narrative* (1916) viii. 124 Have had no sugar-weather this week; frosty with strong winds, and some snow. **1942** G. Campbell *Thorn-Apple Tree* 97 When the March sun began to honeycomb the snow, and the sun was warm on the south side of the house, then came sugar weather.

b. sugar glider, a flying phalanger, *Petaurus breviceps*, found in Australia and New Guinea; **sugar squirrel** (later example); = *sugar glider* above.

1937 *Discovery* Dec. 365/1 Only fifteen inches in total length, with a lovely ash-grey coat.., the Sugar Glider is usually a gregarious creature. **1941** E. Troughton *Furred Animals of Australia* 95 'Sugar Glider' is now adopted as being brief and suitable for popular use. **1972** *Sci. Amer.* Sept. 56/1 Males of the sugar glider..go even further. **1932** *Victorian Naturalist* XLIX. 97 When one has kept the 'Sugar Squirrel' in captivity and suffered keen bites from its long piercing teeth, one is able to appreciate the spitfire temper concealed in these beautiful little creatures.

c. sugar-berry (earlier and later examples); also, one of several other North American species of *Celtis*; **sugar (snap) pea,** add: = *mange-tout; (later examples); **sugar-tree,** (*a*) (earlier and later examples).

1818 W. P. C. Barton *Compendium Floræ Philadelphicæ* I. 151 Celts occidentalis... Sugar-berry Tree. American Nettle Tree. **1896** *Chicago Rec.* 17 Feb. 4/6 He laid the groundwork..by cutting a sugarberry sprout. **1948** *Florida Anthropologist* May 19 This vegetation includes sugarberry, banyan, mulberry, papaya, saw palmetto and small plants. **1969** T. H. Everett *Living Trees of World* xiv. 129/1 The closely related sugarberry (*C. laevigata*), native from Indiana and Illinois southward, ..has a maximum height of 90 feet. **1907** A. French *Bk. Veg.* 198 Pea, edible-podded or sugar, is a type of pea with tender pods, which are eaten. **1951, 1972** Sugar pea [see *mange-tout]. **1980** *Ecology Center* (Berkeley, Calif.) *Newslet.* Oct. 6/2 A great crop of Sugar Snap Peas. **1705** R. Beverley *Hist. & Present State Virginia* ii. 21 The Honey and Sugar-Trees are likewise spontaneous, near the Heads of Rivers. **1949** *Chicago Tribune* 13 Mar. i. 6/4 The Crane Naval depot encroached upon some fine old sugar trees in Martin county.

sugar, *v.* Add: **2. a.** (Later examples.) *to sugar the pill* = *to gild the pill* s.v. Gild *v.*[1] 1 b.

1794 Ld. St. Helens *Let.* 14 Oct. in A. Paget *Paget Papers* (1896) I. 66 They [*sc.* the Prussian Cabinet] have no right to complain, as I observe that you continued to gild and sugar over the pill which you were directed to administer. **1936** V. W. Brooks *Flowering of New England* xv. 287 He liked to administer doses of moral quinine, and he never thought of sugaring his pills. **1954** N. Mitford *Madame de Pompadour* xviii. 237 To sugar the pill of what was, in fact, his dismissal, a Cardinal's hat was procured for Bernis by Stainville. **1955** E. Pound *Section: Rock-Drill* lxxxix. 55 Louis Philippe suggested that Jackson stand firm And not sugar his language. **1978** J. Carroll *Mortal Friends* v. ii. 521 The bishop sugared the request with his smile. **1978** [see *pill *sb.*[1] 1 b].

c. To flatter. Also const. *up*.

1923 J. Manchon *Le Slang* 300 *To sugar a person up*, flatter quelqu'un. **1939** R. Chandler *Big Sleep* ii. 25 It won't get you anything. Sugaring them never does. **1958** R. Stout *And four to Go* iii. 172 There was no point in trying to sugar him. The damage..had been done the second he saw me. **1962** W. Faulkner *Reivers* x. 219 When I sugars up a woman, it aint just empty talk.

4. Also *transf.*

1906 G. B. Shaw *Let.* 4 Apr. in *Florence Farr, Shaw, Yeats* (1946) 26 Your standard of work [*sc.* in acting]..is far too low... You sugar disgracefully except where you see your way to an effect. **1882** 'F. Anstey' *Vice Versa* viii. 166 Although (to use a boating expression) he 'sugared' with some adroitness, he was promptly found out, for his son had been a dashing and plucky player.

5. Used in imprecations, esp. as pa. pple.: = Blow *v.*[1] 29. *euphem.*

1886 Mrs. H. Wood in *Argosy* XLI. 270 'Stephenson says he had blue eyes. Now Dick's are brown.' 'Eyes be sugared,' retorted the lawyer. **1903** [see *amateur 3 b]. **1903** Kipling *Traffics & Discov.* (1904) 107 War's declared at midnight. *Pedantics* be sugared! **1942** *Tee Emm*

(Air Ministry) II. 78 Real pilot be sugared. Real little show-off, more like! 1962 B. Glanville *Diamond* xxi. 339 'They wouldn't talk to me.'..'Sugar them; you're too good for them.'

6. *trans.* To 'cook' or 'doctor'; *spec.* to give a specious impression of the amount of trade done by (a place of business, etc.). *colloq.*

1892 Stevenson & Osbourne *Wrecker* xv. 239 Out of six thousand mats [*sc.* bags of rice], only twenty were found to have been sugared; in each we found..about twelve pounds of drug. 1894 *Daily News* 26 Dec. 5/3 'Sugaring a house'..in Birmingham..denoting a system of creating a fictitious appearance of business by privately giving away money to be spent at its bars.

sugarallie (ʃugərа·li). *Sc. colloq.* Also **sugarellie** (-e·li), **-olly** (-ǫ·li), etc. [A shortened form of *sugar alicreesh*, 16th-cent. Sc. *sukker lagrace, succour alacreische*, f. Sugar *sb.* + Du. *lakk(e)ris* Liquorice, licorice.] **a.** Liquorice.

1812 P. Forbes *Poems* 21 Sulphur, salt fish, sugar allie. 1842 *Children in Mines Rep.* App. 465 in *Parl. Papers 1842* XVI. 1 Mother gives me 3d., which I spend in sugar-alleys and sweeties. 1876 S. R. Whitehead *Daft Davie* iii. 53 The stock of candy and liquorice (known in that countryside by the name of 'sugar-ally'). 1915 A. S. Neill *Dominie's Log* xviii. 206 To-night I have a great craving for a stick of twisted sugarelly—the polite call it liquorice. 1921 —— *Carroty Broon* xvii. 229 Long tubes of sugarella. 1947 L. Derwent *Clashmaclavers* 87 Whiles a lucky-bag I'd try, Or sticky sugar-alla buy.

b. *Comb.* **sugarallie button**, a round sweet made of liquorice; **sugarallie hat**, a tall cylindrical hat, esp. as formerly worn by policemen; **sugarallie water**, a drink made by vigorously shaking a container in which water and a stick of liquorice have been placed.

1887 A. D. Willock *Rosetty Ends* xx. 148 Havin' discovered that Flossie had a weakness for sugarellie buttons, Simpson has made it a habit to hae ane or twa o' thae sweetmeats in his pouch on courtin' nichts. 1887 J. McBain *Arbroath* 104 Their bonnets were replaced by 'sugarelle hats'. 1904 'H. Foulis' *Erchie* 65 The sugar-aully hats the polis used to hae. 1953 J. J. Lavin *Compass of Youth* I. vi. 54 Yelling derisively: 'Sugarawlly Hat,' to the might and majesty of the law. 1889 J. M. Barrie *Window in Thrums* xxii. 207 Bairns..shook their bottles of sugarelly water into a froth. 1923 W. D. Lyell *Justice-Clerk* I. iv. 26 What say ye to a sma' bottle o' sugaralie water? 1947 J. F. Hendry *Fernie Brae* 10 Shall we make some sugarolly water?

sugar-bird. Add: **3.** Also, an African honey-eater of the genus *Promerops*.

1798 Lady A. Barnard *Jrnl.* Apr. in *Lives of Lindsays* (1849) III. 408 The sugar-bird's tail..is long and elegantly formed. 1913 D. Fairbridge *That which hath Been* 30 The emerald-throated sugar-birds..darted from one pink protea to another. 1973 S. Cloete *Company with Heart of Gold* 155 A sugar bird returned to its infinitesimal nest in the grey bush.

sugar-bush. Add: **1.** (Later examples.)

1842 [see *eighty 2 c]. 1896 *Vermont Agric. Rep.* XV. 38 Does the location of a sugar bush determine the quality of the sugar? 1950 *N.Y. Times Mag.* 23 Apr. 46/2 'Sugar bush' is rarely heard in New Jersey, but in Pennsylvania it is the normal term. 1973 L. Russell *Everyday Life Colonial Canada* xi. 144 The settler who would exploit his grove of maple trees or 'sugar-bush' cleared narrow roads through the woods.

2. Also, one of several other species of *Protea* rich in nectar. Cf. Protea in Dict. and Suppl.

a 1823 J. Ewart *Jrnl.* (1970) ii. 14 The Protea of Linn[æus] called by the colonists the sugar bush, from the quantity of sweet juices the large and beautiful flowers contain. 1931 V. Sampson *Kom Binne* 28 The wild arums, the sugar-bush goblets of pink or cream. 1970 M. Muller *Cloud across Moon* 239 The masses of white and pink sugar bushes were covered with nearly opened sticky, stiff flowers.

3. *U.S.* An evergreen shrub, *Rhus ovata*, native to southwestern North America and bearing yellow flowers followed by dark red berries.

1900 *West Amer. Sci.* X. 61 The Sugar-bush is a handsome evergreen shrub. 1931 G. H. Vansell *Nectar & Pollen Plants California* 49 Sugar bush..of coastal southern California blossoms in winter. 1949 *Nature Mag.* Nov. 424/1 There is the gray of some manzanitas, the silver of white sage, the dark green of sugar bush. 1982 M. Millar *Mermaid* vi. 57 Drought-resistant native plants like ceanothus and sugar-bush.

sugar-candy. Add: Hence **sugar-ca·ndyish** *a.*, resembling sugar-candy.

1874 Disraeli *Let.* Aug. in *Lett. to Lady Bradford* (1929) I. vii. 135 Her manners not only sugary but sugar-candyish. 1927 J. Masefield *Midnight Folk* 172 A bowl of raspberries and cream with blobs of sugar-candyish brown sugar.

sugarellie, var. *sugarallie.

sugarer (ʃu·gərəɹ). *slang.* [f. Sugar *v.* + -er[1].] One who shirks, *spec.* at rowing. Cf. Sugar *v.* 4.

1904 W. G. East *Rowing & Sculling* 20 A sugarer, a man who, whilst rowing correctly, avoids putting in a full share of work. 1925 W. Deeping *Sorrel & Son* xviii. 171 It was necessary to be neither a funk nor a sugarer.

sugaring, *vbl. sb.* Add: **4.** Bribery.

1891 J. P. Quinn *Fools of Fortune* 285 This payment is what the 'fakirs' call 'sugaring', and I have never known one of these officials for whom the dose could be made too sweet. 1902 S. E. White *Blazed Trail* xvi. 117 The old-time logger found these two individuals susceptible to the gentle art of 'sugaring'.

sugar-loaf. Add: **3.** *sugar-loaf hill* (earlier and later examples).

1799 Malthus *Diary* 9 July (1966) 131 We..saw Doverfield..with his sugar loaf hills covered with snow. 1859 D. Bunce *Travels with Dr. Leichhardt* iv. 29 There are two lofty sugar-loaf hills..which may be seen from Hobart Town. 1969 [see *favela].

sugar-maple. Add: (Earlier example.) Also, the light-coloured wood of this tree.

1731 P. Miller *Gardeners Dict.* s.v. Acer, There is another Sort of Maple, which is very common in Virginia, and is known by the name of the Sugar Maple. 1936 *Wood Products* Mar. 11/1 During recent years the U.S. Forest Products Laboratory..has investigated the weight and hardness of sugar maple. 1980 *Family Handyman* Sept. 63/2 A cubic foot of poplar weighs 26 lbs. compared to 39 lbs. for a cubic foot of sugar maple.

sugarolly, var. *sugarallie. **sugarro**, var. *saguaro.

suggan. a. Restrict *Anglo-Irish* to first two meanings and add later examples of both. Also *attrib.* in *suggan chair*.

1888 Yeats *Fairy & Folk Tales* 133 She lulls them to rest in the low *suggaun* chair. 1922 Joyce *Ulysses* 489 Bloom..leading a black bogoak pig by a sugaun. 1957 E. E. Evans *Irish Folk Ways* xv. 207 When..the load is placed directly on the ass a simple 'sugan' is sometimes made—a ring of straw some fifteen inches in diameter bound with a fine straw rope. 1977 C. Rocks in *Winter's Tales 23* 128 My da goes to his own sugan chair inside the hearth.

b. For 'a coverlet' read: *N. Amer.* (Usu. in form *soogan, sugan* (sū·gǎn).) A thick blanket or padded quilt suitable for camping out. (Earlier and later examples.)

For evidence of earlier but limited currency of this sense in Scotland see *S.N.D.*

1907 S. E. White *Arizona Nights* 72 Sitting cross-legged on his 'so-gun' in the middle of the floor. 1915 *Dialect Notes* IV. 245 *Soogan*,..sheep herder's blanket. 'When they move, they just roll up the soogan and are off.' 1926 *Amer. Speech* I. 653/1 [Hobo lingo.] *Sugan*, a bed comforter. 1955 R. Hobson *Nothing too Good for Cowboy* vii. 66 [I] saw Jimmy John shove him back into the soogans. 1974 D. Sears *Lark in Clear Air* ii. 32 No matter how quick I turned out of my soogans I would see him up prowling around.

suggest, *v.* **1. e.** (Earlier example.)

1751 Fielding *Amelia* I. iii. iii. 187 The thought of going back at first suggested itself.

suggestibility. Add: **1.** Also in contexts where hypnosis is not involved.

1890 W. James *Princ. Psychol.* II. xix. 97 This suggestibility is greater in the lower senses than in the higher. 1908 W. McDougall *Introd. Social Psychol.* iv. 97 The measure of the suggestibility of any subject is, then, the readiness with which he thus accepts propositions. Of course, the proposition is not necessarily communicated in formal language, it may be implied by a mere gesture or interjection. 1924 W. B. Selbie *Psychol. Religion* iv. 87 Such cases are generally those of persons in a high condition of suggestibility, and it often happens that suggestions do not become active..until they have..incubated in the unconscious realm. 1962 L. Deighton *Ipcress File* 224 Tricyandamino-propene..can change brain's nerve cells and cells of membrane that sheath the cells... From this change the suggestibility of the subject is increased. 1972 *Jrnl. Social Psychol.* LXXXVI. 11 He demonstrated the importance of motivational and experiential factors in determining suggestibility.

suggestible, *a.* Add: **1.** Also in contexts where hypnosis is not involved.

1903 W. James in *Harvard Monthly* Mar. 6 There is no test..by which, if a title or decoration, a public badge or mark, were to be won by it, some weakly suggestible or hauntable persons would not feel challenged. 1908 W. McDougall *Introd. Social Psychol.* iv. 100 Children are..inevitably suggestible..because of their lack of knowledge and lack of systematic organisation of such knowledge as they have. 1921 *Discovery* Nov. 294/1 Children are more suggestible than grown persons, and women are more suggestible than men. 1955 *Times* 17 May 3/3 Orchestras being suggestible by professional training cannot wholly eliminate from their playing the implications of a conductor's gestures. 1981 F. Hoyle *Ice* ii. 39 He is very suggestible; if he is told he is stupid, he thinks he is stupid and behaves accordingly.

suggestio falsi (sʊ̆dʒe·stio fæ·lsəi). Pl. **suggestiones falsi.** [mod.L., = suggestion of what is false.] A misrepresentation of the truth whereby something incorrect is implied to be true; an indirect lie. Often in contexts with *suppressio veri.

1815 H. Maddock *Princ. & Pract. Chancery* I. 208 Whenever *Suppressio veri* or *Suggestio falsi* occur..they afford a sufficient ground for setting aside any Release or Conveyance. 1855 *Newspaper & Gen. Reader's Pocket Compan.* I. 4 He was bound to say that the *suppressio veri* on that occasion approached very nearly to a positive *suggestio falsi*. 1898 Kipling *Stalky & Co.* (1899) 36 It seems..that they had held back material facts; that they were guilty both of *suppressio veri* and *suggestio falsi*. 1907 W. de Morgan *Alice-for-Short* xxxvi. 389 That's suppressio veri and suggestio falsi! Besides, it's fibs! 1962 J. Wilson *Public Schools & Private Practice* i. 19 It is rare to find a positively verifiable untruth in a school brochure: but it is equally rare not to find a great many *suggestiones falsi*, particularly as regards the material comfort and facilities available. 1980 D. Newsome *On Edge of Paradise* 7 There are undoubted cases of *suppressio veri*; on the other hand, he appears to eschew *suggestio falsi*.

suggestion. Add: **8. suggestion-book** (later examples).

1931 W. Holtby *Poor Caroline* iv. 113, I put it down in the suggestion-book six weeks ago. 1967 V. Gielgud *Conduct of Member* i. 11 The Suggestion Book of the Fonthill Club was much like others of its kind. Its contents.. dealt with the apparent shortcomings of the House Committee.

suggestive, *a.* Add: **2. c.** (Earlier example in sense 'likely to make suggestions'.)

1846 Dickens *Let.* 5 Oct. (1977) IV. 629, I shall have the greatest satisfaction..in putting you in communication with two or three gentlemen who I am sure will be most valuable, willing, and suggestive advisers.

d. (Earlier and later examples.)

1888 [implied at *suggestiveness]. 1895 C. D. Warner *Golden House* iii. 24 Her judges were cosmopolitans who had seen the most suggestive dancing in all parts of the world. 1924 R. Macaulay *Orphan Island* xiv. 182 There were complaints, too, of fashions in dress, which, on the part of the younger females, were becoming immodest and suggestive. 1926 T. Dreiser *Amer. Tragedy* I. xviii. 132 They..lay there laughing and yet in a most suggestive position. 1978 L. Meynell *Papersnake* iii. 51 Mabel, archpriestess of the art of suggestive repartee.

suggestiveness. Add: (Later examples in sense 2 d of the adj.)

1888 M. S. van de Velde *Random Recollections of Courts & Society* x. 252 Some foreigners..wondered at the excessive licence she permitted herself on the English stage, and the marked suggestiveness of her looks and gestures. 1913 Mrs. Belloc Lowndes *Diary* 7 Jan. (1971) 42 We talked of English prudery, and..of *The Rosary* and its wonderful success, and I said I felt sure this was owing to the suggestiveness of certain scenes. 1963 L. Deighton *Horse under Water* xxiii. 100, I noticed Singleton's lip curl..at H.K.'s suggestiveness.

suggestology (sʊdʒestǫ·lŏdʒi). [f. Suggest *v.* + -ology.] The study of suggestion, a branch of parapsychology originated by a Bulgarian, Dr. Georgi Lozanov. Similarly **suggestopae·dia, suggesto·pedy** [Gr. παιδεία education], the application of suggestology to education, teaching by suggestion.

1970 Ostrander & Schroeder *Psychic Discoveries behind Iron Curtain* xxii. 293 With suggestopaedia the Bulgarians have expanded time in a very real sense, teaching you in a minute what usually takes many weeks to learn. 1970 *New Society* 31 Dec. 1155/1 This conference is called an 'International Symposium on the Problems of Suggestology'... Among the listed attractions is a visit to the research centre of suggestology in Sofia. 1973 Ostrander & Schroeder *Psi* xxii. 293 Suggestology is *not* hypnosis. With suggestology you are always in the waking state and aware of everything around you. 1978 Hall-Pozharlieva & Pashmakova tr. *Lozanov's Suggestology & Outl. of Suggestopedy* i. 1 Suggestology, the science of suggestion, and its concomitant penetration into pedagogy, suggestopedy, is a newly developing science. 1980 *San Francisco Bay Guardian* 16–23 Oct. 30 (Advt.), Hypnosis/Self-Hypnosis with Dr. Leonard Elkind. Weight. Smoking. Autogenics. Suggestology. Self-Improvement.

sugi (sū·gi). Also 8 ssugi, suggi. [Jap.] = *cryptomeria.

1727 [see *hinoki]. 1795 tr. *C. P. Thunberg's Trav.* (ed. 2) III. 123 Ssugi signifies Cedar wood. 1876 *Trans. Asiatic Soc. Japan* IV. 53 One piece of *sugi* of 6 by 3 by 0·4 ft. 1916 E. H. Wilson *Conifers & Taxads of Japan* 69 The Cryptomeria, or Sugi as it is called in Japan, is the noblest of the Japanese conifers. 1954 [see *Japanese cedar* s.v. *Japanese *a.* b]. 1970 J. Kirkup *Japan behind Fan* iv. 137 A bus took me on the long winding road, through groves of immense *sugi* or Japanese cedar.

suh (sʊ̆), chiefly *U.S. Southern* and *Black* pronunc. of Sir *sb.* Cf. *sah.

1894 Kipling *Day's Work* (1898) 50 Most of your prominent siahs, suh, are impo'ted from Kentucky. 1901 W. Churchill *Crisis* I. iv. 38 'T'ank you, Mistah Cantah,' wailed the poor woman, 't'ank you, suh... De Lawd Jesus'll rewa'd you, suh.' 1911 [see *hide *v.*[1] 2 c]. 1929 W. Faulkner *Sartoris* II. 117 'Come on here and get in,' he commanded. 'Naw, suh. I'll walk.' 1940 H. G. Wells *Babes in Darkling Wood* III. ii. 252 Cutting facts dead unless they wear the old school tie! 'Don't know you, suh.' 1971 *Black Scholar* Sept. 38/2 'Yes suh, officer,' she replied.

Sui (swē[i]). Also Suy, Swi. [Chinese *sui.*] The name of a dynasty which ruled in China from 581 to 618 A.D. and re-unified the country after the divisions of the Northern and Southern Dynasties period. Freq. *attrib.*

1738 J. B. Du Halde *Descript. China & Chinese Tartary* I. 194 Thus ended the Dynasty nam'd Swi, the last of five petty ones. 1797 *Encycl. Brit.* IV. 653/1 The

whole of their [*sc.* China's] emperors..are comprehended in 22 dynasties, mentioned in the following table..Chin.. Swi..Twang [etc.]. 1845 *Encycl. Metrop.* XVI. 550/1 The tower of Kao-ming-chi, with its gardens, temple and pavilions, erected by Kao-tsu of the Swi dynasty. 1897 J. MacGowan *Hist. China* xix. 261 Yang-Kien was forty-eight years old when he became Emperor... He gave his dynasty the name of Sui. 1910 *Encycl. Brit.* VI. 195/2 This period of disorder was brought to a close by the establishment of the Suy dynasty. 1958 W. Willetts *Chinese Art* I. v. 310 We are concerned only with the Sui and the first four reigns of the T'ang... The Sui Emperors were great patrons of Buddhism. 1969 *Guardian* 2 July 3/2 A 15-inch terra cotta horse of the Sui period (seventh century AD), is offered for £6,000. 1972 *Trans. Oriental Ceramics Soc.* XXXVIII. 29 Much that is ascribed to Sui ..should be placed in the opening generations of T'ang. 1979 A. Henning tr. *Myrdal's Silk Road* (1980) i. 11 Tashkent was known during the Sui dynasty.

‖**suiboku** (su·ibǫku). [Jap., lit. 'liquefied ink', f. *sui* water + *boku* ink stick.] A style of Japanese painting in black ink on a white surface characterized by bold brush-work and subtle gradations of tone (see quot. 1970).

1912 E. F. Fenollosa *Epochs Chinese & Jap. Art* II. xi. 43 Kakei..made a decided change in Chinese landscape style: the 'In' style..in that he introduced the utmost decorative splendour of *notan*, or dark and light beauty. He made the strong shapes of his touches of glowing ink 'look as if they were falling in drops'. This is 'suiboku', or wet ink. 1959 R. Saito *Jap. Ink Painting* 13 Because of the value of the light and dark color of the sumi, and the taste and interest which come from each variation of the brush, suiboku painting really reveals the true spirit of the Oriental people. 1959 *Times Lit. Suppl.* 17 Apr. 218/4 The distinction is most apparent, and most interesting, in the field of *suiboku*—the monochrome painting practised by the *bunjin-ga*. 1970 *Oxf. Compan. Art* 1114/1 The essentials of *suiboku* were bold composition in the Chinese style, strength of brush-work, and nuance in the tone of the ink.

suicide, *sb.*[2] Add: **c.** (Further examples, esp. as *suicide letter, note, pact.*) Also *spec.* in *Mil.* use, designating highly dangerous or deliberately suicidal operations and persons, etc., involved in them, as *suicide aircraft, mission, squad*, etc.

1897 'Mark Twain' *Following Equator* lvii. 546 In India, the annual man-killings by snakes are..as forecastable as are the tiger-average and the suicide-average. 1911 *Daily Colonist* (Victoria, B.C.) 19 Apr. 4/3 Before he shot himself..he shot Miss Bovee three times, they having previously entered into a suicide pact. 1916 'Boyd Cable' *Doing their Bit* iii. 47 You bombers of the 'Suicide Clubs' might note this. 1923 Kipling *Irish Guards in Gt. War* I. 67 There seemed no meaning or reason in the affair, unless it was a suicide-party of Germans who had run from the attack of the day before and had been ordered thus to die. 1928 A. C. Havlin *Hist. Company A, 102nd Machine Gun Battalion* 3 We were to serve as 'suicide squads' in the..26th Division. 1929 D. Hammett *Dain Curse* vii. 65 Your husband's letter sounded enough like a suicide letter..so you murdered him. 1938 'E. Queen' *Four of Hearts* xxii. 293 Park left a suicide note to efface his trail and vanished. 1942 *R.A.F. Jrnl.* 13 June 15 If a 'suicide squad' job came along..it would be assumed that every man was ready for that sort of thing. 1945 *News Chron.* 1 June 4/5 Conferences..are believed to have included plans to counter the..suicide plane. According to a Tokio statement, these suicide attacks..are being developed by the..Japanese Naval Command. 1946 *Jrnl. R. Aeronaut. Soc.* L. 293 As with the pages devoted to German aircraft, so with those given to Japanese. They are full and informative, and end with brief interesting notes on suicide aircraft and the Baka flying bomb. 1954, 1956 [see *kamikaze *a.* 1]. 1963 'D. Cory' *Hammerhead* x. 123 Fedora brushed what was left of the suicide pill on to the palm of his hand. 1969 R. Rendell *Best Man to Die* xv. 147 It's the seat on the driver's left that's called the suicide seat. 1971 *New Scientist* 11 Mar. 531/2 No civil defence measure would improve this state of affairs except perhaps the issue of suicide pills. 1974 'S. Woods' *Done to Death* 127, I haven't told him about the suicide note. *Ibid.* 129 Mr Maitland's theory about the suicide letter is right. 1976 A. White *Long Silence* xviii. 158 'And the third objective?'..'Obviously to flee..and eventually return to England. None of you strikes me as the kind of fool who would accept a suicide mission.' 1977 A. Giddens *Stud. in Social & Polit. Theory* ix. 308 Her husband left her seven months later, and this precipitated her suicide attempt. 1978 *Times* 21 Nov. 6/4 Mr Jones had forced his followers to make a suicide pact with him. He predicted..all 1,200 members of the sect would die. 1979 T. Sharpe *Wilt Alternative* xvi. 149 Could have left a suicide squad to cover their retreat.

d. *Comb.*, as **suicide blonde** *slang*, a woman with hair dyed blonde (esp. rather amateurishly), a peroxide blonde; **suicide clause**, a clause in a life insurance policy which releases the insurer from liability if the insured commits suicide within a specified period; **suicide squeeze** *Baseball*, the action of a runner on third base in running for home as the ball is pitched (cf. *squeeze play).

1942 Berrey & Van den Bark *Amer. Thes. Slang* § 430/4 Bottle baby,..peroxide, suicide blonde, an artificial blonde. 1959 J. Braine *Vodi* vii. 104 'You don't have to whitter on about one little suicide blonde.' 'She's a real blonde,' Tom said. 1973 A. Sillitoe *Men, Women & Children* 174 The snow-white hair of a suicide-blonde flashed around: 'Hey up, Margaret!' 1902 C. L. Greene *Medical Examination for Life Insurance* 357 There can be little doubt that in the case of persons insured under policies containing a suicide clause, such deaths are very generally reported as accidental. 1976 'L. Black' *Healthy Way to Die* x. 112 Eddie asks her if there is a suicide clause in the life policy. 1955 P. Richards *Mod. Baseball Strategy* xi. 129 The 'suicide-squeeze', which has the runner going home on the pitch, is absolutely certain to work—*if* the batter bunts the ball on the ground. 1974 *Los Angeles Times* 13 Oct. III. 9/5 It is properly called a 'suicide squeeze' because it calls for the runner to arrive at home plate at the same time as the ball.

suicidology (s[i]ūisəidǫ·lŏdʒi). [f. Suicide *sb.*[2] + -ology.] The study of suicide and its prevention. Hence **suicido·logist.**

[1929 W. A. Bonger in *Psychiatrisch-Juridisch Gezelschap* 9 Feb. 3 De wetenschap van de zelfmoord, the *suicidologie* (cursivering van mij) zou men haar kunnen noemen, is ruim een eeuw oud.] 1964 E. S. Shneidman in *Contemp. Psychol.* IX. 371/2, I thank Louis Dublin, the Grand Old Man of Suicidology, for this book because in it he..has given us all new clues to suicide. 1967 —— *Bull. Suicidology* July 7/2 The 10-point program for suicide prevention here outlined is a mutual enterprise whose successful development depends on the active interest, support, and activities of 'suicidologists'. 1969 *Nature* 4 Oct. 12/2 The Johns Hopkins University in collaboration with the National Institutes of Mental Health has established a course in 'suicidology'. 1970 L. Lasagna in O. G. Brim et al. *Dying Patient* 96 The 'suicidology' program concept. 1976 E. S. Shneidman *Suicidology* 7 Suicidology is defined as the scientific study of suicidal phenomena. 1976 R. K. McGee in *Ibid.* 482 The volunteer suicidologist has become a vital component of the suicide prevention scene.

suid (s[i]u·[i]d), *sb.* and *a.* *Zool.* [f. mod. L. *Suidæ*: see Suidian *a.* and *sb.*] = Suidian *a.* and *sb.*

1957 P. J. Darlington *Zoogeogr.* vi. 403 Fossil suids.. are known only from the main part of the Old World. 1969 [see *bunodont *a.* and *sb.*]. 1970 B. G. F. Weitz in H. W. Mulligan *African Trypanosomiases* xviii. 419 Table 18.1 shows the results obtained by the Haemagglutination-Inhibition Test on ten suid feeds. 1976 D. Pilbeam in C. J. Jolly *Early Hominids of Africa* 509 Most workers can..sort..bovids from suids. 1980 *E. Afr. Med. Jrnl.* LVII. 333 *G. brevipalpis* fed mostly on suids.

sui generis. Add: Also used *attrib.*

1944 S. Putnam tr. *E. da Cunha's Rebellion in Backlands* ii. 60 Such a climate tends to create a *sui generis* pathology throughout the whole of the northern coastal strip. 1963 J. Lyons *Structural Semantics* i. 2 In this theory meaning is defined as a *sui generis* 'reciprocal relation between name and sense, which enables them to call up one another'. 1977 *Time* 4 Apr. 41/3 The superlative interpretations by the *sui generis* Budapest Quartet come from tapes of live performances at the Library of Congress in 1959 and 1961.

suikerbos (sōi·kəɹbǫs). *S. Afr.* Also **suikerbosch, -bossie, zuikerbosch.** [Afrikaans, f. *suiker* Sugar *sb.* + *bos* Bush *sb.*[1]] = Sugar-bush 2 in Dict. and Suppl.

[1793 tr. *C. P. Thunberg's Trav. Europe, Africa, & Asia* I. 292 The *Protea mellifera* (*Tulp-boom* and *zuyker-boom*) contains in its calyx a sweet juice.] 1822 W. J. Burchell *Trav. S. Afr.* I. ii. 18 The delicate Humming birds..are, in Southern Africa,..called by the Dutch colonists *Suiker-vogels* (sugar birds), from having been observed.. to feed principally on the honey of the flowers of the *Suiker-bosch* (sugar-bush). 1852 C. Barter *Dorp & Veld* vii. 74 We came upon knolls covered with the evergreen *Suiker bos* a graceful shrub. 1887 A. A. Anderson *Twenty-Five Years in Waggon* I. 210 The fine flat-topped Kameel doorn is very common, palms, baobab,..zuikerbosch, acacia. 1937 S. Cloete *Turning Wheels* 104 As he rode past a clump of soikerbos [*sic*] a duiker sprang out. 1950 M. M. Kidd *Flowering Plants Cape Peninsula* Pl. 34 June–Aug. Sugar Bush, Suikerbossie. 1952 *Cape Times* 4 Sept. 5/4 The five dozen selected proteas include..two varieties of the furry suikerbos type. 1971 *Cape Argus* 10 July 5 Now is the time to see all those lovely Proteas and suikerbossies in full bloom.

sui-mate (s[i]ū·əi,mē[i]t, s[i]ū·i-). *Chess.* [f. L. *suī* of oneself + Mate *sb.*[1]] = Self-mate *sb.* Also *ellipt.* (sui) and as *v. intr.*

1870 *Dubuque Chess Jrnl.* 1 Nov. 7 White sui-mates in ten moves. 1890 [see *retractive *a.* 3]. 1907 S. S. Blackburne *Terms & Themes Chess Problems* 21 Sui-mate Problem. One in which one player..compels the other to mate him. 1965 *New Statesman* 16 Apr. 625 This one is a 'sui in 5', and..what matters is for White to commit suicide..by forcing Black to mate him in five moves. 1966 *Ibid.* 11 Nov. 718/3 Many readers have let me know their gradual addiction to sui-mates.

suine (s[i]ū·əin), *a. nonce-wd.* [f. L. *sūs, su-* pig + -ine[1]: cf. L. *suīnus* (see Swine).] Piglike, porcine.

1922 Joyce *Ulysses* 554 The suine scions of the house of Lambert.

suing, *vbl. sb.* Add: **4. a.** (Later examples.)

1946 *Law Rep.* (King's Bench Division) 18 Sept. (1947) 93 The plaintiff was incapable of suing. 1983 *Weekly Law Rep.* 22 July 884 By the grant of a stay the plaintiffs would be deprived of the juridical advantage of suing as plaintiffs in the Admiralty Court.

suiseki (su,isē·ki). [Jap., *sui* water + *seki* stone(s).] The Japanese art of arranging stones on a tray, often one containing shallow water.

1929 *Encycl. Brit.* III. 855/2 Some stones are placed on a tray with low-growing grass or bamboo... Another way of enjoying them, which has been for centuries and is still popular among the Japanese, is known as *sui-seki*... A natural stone of desirable shape is placed in a porcelain or bronze tray or dish with sand and water. 1972 *Islander* (Victoria, B.C.) 2 Dec. 7/2 Very short sections [of a yew log], an inch or two in thickness, could be used as bases for viewing stones after the Japanese suiseki fashion... The art of suiseki is popular in Japan. 1976 *N.Y. Times* 8 Aug. 24 As with all suiseki, the stones must be as found in nature, though they may be cleaned with a soft cloth or brush. However, they are never polished or sculpted.

suisse. Add: **2.** A soft French white cheese resembling *Neufchâtel. Usu. in the form *petit suisse*: see *petit *a.* (*sb.*) 5.

1892 Stevenson & Osbourne *Wreckers* v. 70 When I called for a *suisse*..I was bluntly told there were no more.

suit, *sb.* Add: **V. 18.** † **h.** A gold watch, usu. with seals, case, etc. *Criminals' slang. Obs.*

1718 C. Hitching *Regulator* 13 They [*sc.* pickpockets] greatly benefit; either by a Suit, *alias* Gold-watch,..or by a Wedge Lobb, *alias* Gold or Silver Snuff-Box. 1839 H. Ainsworth *Jack Sheppard* II. xiv. 40 A fence, or receiver,..bargaining with a..pickpocket, for a *suit*,—or to speak in more intelligible language, a watch and seals.

19. b. Now usually, a jacket and trousers of the same material, sometimes with matching waistcoat, and esp. for formal or office use.

1932 G. Greene *Stamboul Train* I. i. 7 He..required no longer..his suit from Savile Row..to hearten him. *a* 1953 E. O'Neill *Hughie* (1959) 8 He wears an ill-fitting blue serge suit. 1960 C. Day Lewis *Buried Day* ii. 43, I am standing..in a white suit and holding my broad-brimmed round straw hat.

c. Cf. **trouser suit*.

e. *birthday suit*: see also s.v. *birthday 3.

g. = *bathing-suit* s.v. *bathing *vbl. sb.* 2, *swim-suit* s.v. *swim *sb.* 10 b.

1883 L. Troubridge *Life amongst Troubridges* (1966) 165 Walked along... meaning to bathe... Ran down in our suits. 1949 D. Smith *I capture Castle* x. 162 We didn't bathe because none of us had brought suits. 1977 *Times* 16 June 13/6 The suit in our picture..is the first suit for ages..to cover up the spare tyre.

20. a. *long suit*: see Long *a.*[1] 5 b, *5 c.

b. (Earlier *fig.* example.)

1851 H. Melville *Moby Dick* I. v. 47, I quickly followed suit, and descending into the bar-room accosted the grinning landlord.

VII. 24. a. (senses 19 b, c) *suit coat, -jacket*; **suit bag**, (*a*) a protective covering for a suit which is not being worn; (*b*) a travelling bag designed to contain a suit of clothes; **suit length**, a piece of material of the right size for making into a suit; also *fig.*; **suit-weight**, used *attrib.* of fabrics of an appropriate thickness for making up into suits.

1966 *Olney Amsden & Sons Ltd. Price List* 5 Suit bag zipped 51/9 doz. 1978 W. Stovall *Presidential Emergency* i. 1 He set down his suit bag, underseater and attaché case. 1971 D. E. Westlake *I gave at the Office* (1972) 15 A guy..whose suitcoat collar was turned up indoors. 1972 *National Observer* (U.S.) 27 May 1/4 Wallace removed his suit coat, handed it to an aide, and moved forward to greet well-wishers. 1965 M. Shadbolt *Among Cinders* x. 79 The crumpled..suit-jacket with sleeves too short. 1977 *Transatlantic Rev.* LX. 69 She had taken off her suitjacket. 1924 J. Joyce *Let.* 30 Sept. (1957) II. 221 There is now a special cheap edition..about 1/11½ per normal novel suitlength real continental. 1971 D. Lees *Rainbow Conspiracy* ii. 24 The foreman weaver in most mills is allowed to take any end pieces as part of his perks. .. More often than not he finds himself with a suit length. 1955 *Archit. Rev.* CXVII. 351 (*caption*) Light suit-weight Cheviot tweed by Michal Illan. 1963 *Guardian* 10 May 8/4 Tweed, flannel or other suit-weight woollen.

b. In *Bridge*; freq. as opp. to *no trump(s) *phr.*, as *suit-bid, -break, call, contract, declaration, double, game, -jump*; **suit preference signal**, a play of a card of a certain rank to indicate which suit one wishes one's partner to return.

1917 E. Bergholt *Royal Auction Bridge* I. 90 In some circles, the practice of raising partner's suit-bid, when no other bid has intervened, is considerably overdone. 1962 *Times* 24 Oct. 3/7 Why be forced into a higher contract which may be in jeopardy through unlucky suit-breaks? 1907 Suit call [in Dict., sense 24 a]. 1977 *Homes & Gardens* Feb. 14 Presumably he also appreciates the point I made above about playing unbalanced hands in a suit contract. 1910 W. Dalton *Saturday Bridge* iv. 65 (*heading*) Defensive suit declarations by the dealer. 1927 *Observer* 13 Mar. 27 The suit double..has several interesting aspects. 1910 W. Dalton *Saturday Bridge* vii. 89 There are two distinct games at Bridge, the No Trump game and the suit game. 1929 M. C. Work *Compl. Contract Bridge* 52 When determining whether to make a suit-jump of two or three..do not be influenced..by Queens or Jacks of other suits. 1934 H. Lavinthal in *Bridge World* June 5/1, I am offering a new convention for the defense. I call this convention the High-Low Suit Preference Signal. 1981 *Times* 14 Nov. 17/6 Where there is any risk of confusion, suit preference signals should not be applied to the first trick.

suit, *v.* Add: **9. d.** To fit (someone) *up* with a specific type of clothing, as for sport, protection, etc. Cf. *kit *v.*[1] 2. *U.S.*

1945 M. H. Allee *Smoke Jumper* iii. 24 A man suited up for smoke jumping would almost as soon fall into the fire itself as into deep water. 1970 *New Yorker* 24 Oct. 140/3 Yale suited up sixty men, including four quarterbacks. 1976 *Daily Tel.* 1 Sept. 3/3 Only when everyone [*sc.* U.S. policemen] is suited up is the order given to tackle a disorderly crowd. 1979 *Tucson Mag.* Apr. 66 (Advt.), Dave Bloom & Sons will suit you up for all your active sport needs.

13. b. *suit yourself*: do (or think) as you please, please yourself.

1897 Kipling *Captains Courageous* i. 21 'You stole it.' 'Suit yourself. We stole it ef it's any comfort to you.' 1932 W. Faulkner *Light in August* xxi. 478 'I reckon I'll ride back here,' she says... 'Suit yourself,' I says. And we drove off. 1953 K. Tennant *Joyful Condemned* xiii. 120 'Just suit yourself.' Miss Pilcher shrugged her broad shoulders. 1977 'M. Underwood' *Murder with Malice* xiii. 118 'I'll probably call back later.' 'Suit yourself,' the woman said, indifferently.

16. (Earlier and later examples.) Also, to match or be in accord.

1816 Jane Austen *Emma* III. ii. 20 Frank Churchill is a capital dancer, I understand—We shall see if our styles suit. *a* 1817 —— *Persuasion* (1818) IV. v. 91 Mr. Elliot is an exceedingly agreeable man..but we should not suit. 1971 'D. Halliday' *Dolly & Doctor Bird* xiii. 193 I've done an Eysenck personality inventory on you both... You wouldn't suit.

17. e. To dress oneself *up* in clothing designed for a specific task or purpose.

1959 J. Blish *Clash of Cymbals* viii. 191 We should suit up at the half-hour. 1967 *Boston Sunday Herald* 30 Apr. I. 22/2 Jim Lyle headed for the flight line to suit up for a routine mission aboard one of the giant radar picket planes. 1975 'A. Hall' *Mandarin Cypher* xi. 170 'Time to suit up, isn't it?'..I got into the wet-suit. 1978 G. A. Sheehan *Running & Being* xv. 206 He will suit up and get out on the roads.

sui·tcase. (In Dict. s.v. Suit *sb.* 24.) Add: **1.** Now more generally, a piece of luggage in the form of an oblong case, usu. with a hinged side and a handle, for carrying clothes and other belongings.

1902 [in Dict. s.v. *suit *sb.* 24]. 1942 W. Faulkner *Go Down, Moses* 235 The boy waked him at last and got him and the suitcase off the train. 1981 D. M. Thomas *White Hotel* iv. i. 133 She realized they were travellers, for they were weighed down by rucksacks and suitcases.

2. Phr. *to live out of* (or *from*) *a suitcase* (or *suitcases*): to move between temporary accommodation, esp. hotels and boarding houses; to be a wanderer, to have no fixed abode.

1946 L. Durrell *Let.* 25 Sept. in Durrell & Miller *Private Corr.* (1963) 229, I can't tell you what wonderful peace and quiet it is, having a house of your own after so many years living from suitcases in hotels. 1960 J. Weightman tr. *H. de Montherlant's Sel. Essays* 181 To live for years on end out of a small suitcase..seemed so much part and parcel of my everyday life. 1969 *Photoplay* Jan. 69/1 'It never occurred to me it would take ten years to settle down,' Audrey said recently, after ten years of living out of suitcases. 1975 C. Egleton *Skirmish* xiii. 132 He had spent the greater part of his life living out of a suitcase.

3. a. *attrib.* Designating devices small or compact enough to be fitted into a suitcase, usu. in connection with secret or criminal activities, as *suitcase bomb, radio*, etc.

1954 *Richmond* (Va.) *Times-Dispatch* 25 Mar. 16 (*heading*) Now the 'suitcase A-bomb'. *Ibid.* 16/2 All of which means that a 'suitcase atom bomb' is no longer a figment of the imagination. 1972 T. Ardies *This Suitcase is going to Explode* xiii. 134 Suitcase bombs have been discussed..in public. *Ibid.* xvii. 188 Very damning stuff—such as the plans for constructing a suitcase nuclear bomb. 1974 L. Deighton *Spy Story* xix. 207 Our boy with the suitcase radio set came in five by five. A powerful signal.

b. *Comb.*, as **suitcase farmer** *N. Amer.*, a farmer who is resident on his farm for only a small part of the year (see quots.).

1941 R. Diller *Farm Ownership, Tenancy, & Land Use* 2 'Suitcase farmer' is a term used of farmers on the Great Plains who put in a crop of wheat in the fall and come back to harvest it the next summer, after having spent the winter in their permanent homes elsewhere. 1956 *Saturday Night* (Toronto) 13 Oct. 15/1 The wheat-marketing problem means the end of the 'suitcase farmer', who has been accustomed to spend only a few weeks on his land each spring and summer for seeding and harvesting. 1970 Duckham & Mansfield *Farming Syst. World* II. ii. 114 A 'suit-case' farmer moves seasonally between his several farms.

sui·tcaseful. [f. prec. + -ful.] As much as a suitcase will hold.

1928 *Daily Express* 11 Aug. 5/1 Sand tray with which the little ones can play by an open window. If it is possible to persuade a friend to bring back a suitcaseful of shore sand.., so much the better. 1965 J. Fleming *Nothing is Number* II. v. 83 He..brought back a suitcaseful of these books. 1979 J. Sherwood *Hour of Hyenas* xvi. 189 If you'll let me have that suitcaseful of guns that madam lent you.

suite. Add: **2. c.** (Earlier example.) Now freq. with reference to a three-piece suite of two armchairs and a sofa.

1805 *Times* 7 Nov. 4/2 An elegant drawing-room suite of 5 curtains, chairs, &c...suites of chairs, sofas. 1920 'O. Douglas' *Penny Plain* v. 47 Can you imagine it furnished with a 'suite'..and a grand piano? 1974 I. Murdoch *Sacred & Profane Love Machine* 244 She had chosen..the maroon armchairs of corded velvet (they could not afford a 'suite').

d. (*a*) (Earlier example); (*b*) substitute for def.: a set of instrumental compositions (orig. of movements in dance style) to be played in succession; also, an assemblage of movements from opera or ballet scores; (later examples).

1760 J. Mainwaring *Mem. Life G. F. Handel* 68 The two first movements of Handel's seventh suite in the 1st Vol. of his Lessons formerly stood for the Overture in his famous opera of Agrippina. 1893 G. B. Shaw *Music in London 1890–94* (1932) III. 1 The usual two or three concertos..selection of overtures, suite from the latest 'incidental music' composed for the theatres. 1902 *Encycl. Brit.* XXXI. 42/1 Edward Alexander MacDowell..has written..symphonic poems, overtures, and suites for orchestra. 1928 *Grove's Dict. Mus.* (ed. 3) V. 184/2 The composer..accepts the term 'suite' as one which allows him a freer hand than symphony and one which indicates more definiteness of design than symphonic poem. 1977 *Zigzag* June 39/4 Only in the lengthy 'Rangers At Midnight' suite do the band really go over the top into arty cleverness.

e. *Geol.* A group of related minerals, rocks, or the like, esp. ones from the same place.

1845 [in Dict., sense 2 a]. 1882 A. Geikie *Text-bk. Geol.* 648 The earliest system or connected suite of deposits in the Palæozoic series has received the name of Cambrian. 1934 *Bull. Amer. Paleont.* XXI. No. 71. 18 It is sometimes..convenient to bracket several intimately related members together into what are here called formational suites or in common speech 'suites'. 1937 Hatch & Wells *Petrol. Igneous Rocks* (ed. 9) v. ii. 273 As it is difficult to speak of a suite of associated rocks of one age and derived from a common magmatic source as a 'province', the term comagmatic assemblage is preferred. 1951 [see *nováčekite]. 1963, etc. [see *ophiolite suite* s.v. *ophiolite b]. 1972 *Sci. Amer.* Mar. 34/2 Delta and river deposits sweep back across the continent, covering the miogeocline with a suite of continental shales and conglomerates. 1974 *Nature* 15 Nov. 219/2 X-ray diffraction analysis of residue in specimens from the eastern United States showed the mineral suite: chlorite, gibbsite, illite, kaolinite, [etc.]. 1978 *Sci. Amer.* Sept. 85/2 Boundaries between such layers, where one characteristic suite of fossils gives way to another, provide the basis for dividing geologic time into eras, periods and epochs. 1980 *Encounter* May 16/1 Those rocks (second sample though they might be from a suite already sampled by Shackleton) were..among the most precious data which could possibly have been obtained for geology.

f. A set of jewellery, esp. one containing matching pieces.

1869 S. R. Hole *Bk. about Roses* viii. 109 Let him display..casket after casket of lustrous gems. Then invite her to select her *suite*. 1888 *J. Simmons Illustr. Trade Catal.* 25 (*heading*) Coloured bright gold brooches and earrings. Separately, or in suites to match. 1936 *Watchmaker & Jeweller* Feb. 187/2 (*caption*) A costume suite of gilt jewellery. 1981 M. Babson *Bejewelled Death* i. 14 The Orpington Bequest..was the last of the monumental suites..to have survived.

g. The set of components which forms a lavatory; hence also, (matching) bathroom furniture or fittings.

1926–7 *Army & Navy Stores Catal.* 325 The 'Colonial' Closet Suite, comprising strong Vulcanware closet-pan and trap..polished mahogany seat..cistern, cover and brackets, brass chain pull and pottery handle. 1951 *Catal. of Exhibits, South Bank Exhibition, Festival of Britain* 125/1 W.C. suite. 1966 *Guardian* 5 July 8/5 A tiled bathroom, with or without a coloured 'suite'. 1973 J. Thomson *Death Cap* i. 15 Pale green bathroom suite, fitted carpets, the lot.

h. *Computers.* A collection of related programs which can be run one after the other without interruption.

1967 *Oxford Computer Explained* 9 This Suite is run twice a day. 1980 R. McCrum *In Secret State* xi. 98 It was just a suite of programmes that wasn't in Lister's index.

suiting, *vbl. sb.* Add: **5.** Now in wider use, and freq. *sing.* Also applied to the finished garments.

1923 A. Huxley *Antic Hay* iii. 34 A very small man.. popped out from a canyon..between two stratified precipices of mid-season suitings. 1930 *Daily Express* 8 Sept. 11/5 This cardigan type of tailored suit is made of a soft suiting tweed. 1957 L. Durrell *Bitter Lemons* 44 His rusty, moth-bedevilled business suiting and wrinkled dicky suggested extremes of dreadful indigence. 1980 J. B. Hilton *Anathema Stone* iii. 28 [He] left the farm with a sample of his suiting in the jaws of a bull-mastiff.

Suk (sūk), *sb.* and *a.* **A.** *sb.* **a.** An East African people who inhabit an area on the Uganda-Kenya border; a member of this people. **b.** The Nilotic language spoken by the Suk. **B.** *adj.* Of or pertaining to this people or their language.

1902 H. Johnston *Uganda Protectorate* II. xix. 847 The Sūk, like the Turkana, pierce the lower lip... The Sūk women sometimes shave the head. 1902 *Encycl. Brit.* XXXIII. 541/2 The languages spoken in the Uganda Protectorate belong to the following stocks..*Masai* (Bari, Masai, Elgumi, Turkana, Sūk, &c.). 1930 [see *Nandi *sb.*[2] and *a.*]. 1936 *Discovery* June 171/2 The Nandi, Masai, Turkana, and Sūk are perhaps the best-known tribes of this group. 1947 [see *Kipsigis]. 1953, etc. [see *Sebei]. 1963 *Times* 6 May 19/6 Mr. Jomo Kenyatta, president of the Kenya African National Union, today reminded a crowd already nursed to a happy pitch of electoral enthusiasm by the chanting and dancing of Suk and Kalenjin tribesmen. 1974 *Encycl. Brit. Micropædia* VII. 347/2 Linguists often divide the Nilotic languages into a western group..and a southern group, including Nandi and Suk.

suk, var. *souk.

sukebind (s[i]ū·kbəind). [Arbitrary formation: cf. Bind *sb.* 2, 3.] Name given by Stella Gibbons (see quot. 1932) to an imaginary plant associated with superstition and fertility, hence used allusively with reference to intense rustic passions.

1932 S. Gibbons *Cold Comfort Farm* v. 75 In the fulness of summer, when the sukebind hangs heavy from the wains..'tes the same. 1968 *Listener* 19 Sept. 379/3 The sukebind twines lushly over the grave of Mary Webb, another esteemed pre-war novelist, and since Stella Gibbons planted the fatal seed, nobody has bothered to hack away the undergrowth to discover what lies beneath. 1975 *Times* 15 July 14/3 A Country Sports Fair conjures images of..young couples competing among the sukebind. 1982 W. Golding *Moving Target* 106 Mr Trevelyan's fascinating book... He climbed one or two family trees where the Sukebind was a-blowing.

sukey (sū·ki). *dial* and *colloq.* Also with capital initial and **suckey, sukie, suky.** [Dim. of *Susan, Susanna*, fem. name.] A tea-kettle.

The nursery rhyme 'Polly put the kettle on' (see quot. 1981) is known from 1841: *Oxf. Dict. Nursery Rhymes*.

1823 'J. Bee' *Slang* 167 Sukey, a tea kettle. 1875 E. Tweddell *Rhymes Cleveland Dial.* 40 Suckey was bolin' a gud un when we gat there. An' Ah..helpt to fettle t'tea. 1877 E. Peacock *N.W. Lincs. Gloss.* 244/2 *Suky*, a child's name for a tea-kettle. 1898 J. D. Brayshaw *Slum Silhouettes* 182 Now, Sukey boils; fill the teapot, Dick. 1919 'W. N. P. Barbellion' *Diary* 13 Feb. (1920) 95 Our sukie is an old copper one, and sings sometimes in splendid imitation of an orchestra tuning up. 1952 M. Laski *Village* i. 13 I'll just get the sukey going, and then we'll have a nice cup of tea. 1981 *Jrnl. Lancs. Dial. Soc.* Jan. 46 Many people remember *sukey* as a name for the kettle. 'Sukey's boiling,' they would say. This must originate in the nursery rhyme 'Polly put the kettle on'.

sukh, var. *souk.

sukiyaki (sukiya·ki, ‖skī·yaki). Also **suki-yaki.** [Jap.] A Japanese dish, consisting of very thin slices of beef fried with vegetables in sugar and soy sauce, and often served with rice.

1920 *Japan Advertiser* 22 Aug. 5/1 Another name by which this dish [*sc.* nabe] is usually known outside of Tokyo, is suki-yaki. This is derived from suki, which means a spade, and yaki, to cook. 1932 H. A. Phillips *Meet Japanese* xvii. 185 Beef *sukiyaki* tasted good after a long day's jaunt. 1935 B. Woon *San Francisco & Golden Empire* v. 62 The best *suki-yaki* restaurant is not in the Japanese quarter but in a Japanese hotel near the corner of California Street and Grant Avenue. Here tasteful *suki-yaki* dishes are cooked in chafing-dishes, Japanese style. 1943 H. Mears *Year of Wild Boar* iii. 51 The Japanese who patronized this place..did so only to sample American culture, as in New York the American might dine in a Japanese *sukiyaki* restaurant. 1952 R. Cutforth *Korean Reporter* xvi. 147 There are other famous meals—*Sukiyaki*—a fry of chicken or beef with vegetables and soya. 1964 I. Fleming *You only live Twice* xxii. 253 A highly spiced dish of *sukiyaki*, the national dish of beef stew. 1970 P. & J. Martin *Jap. Cooking* 72 Put a sukiyaki pan or a large, heavy frying pan over a portable cooking stove. 1977 *Time* 19 Dec. 43/1 (*caption*) Drama Coach Lee Strasberg cooks sukiyaki in Manhattan.

sukkah, var. *succah. **Sukkot(h**, var. Succoth.

sulfa-, altered and U.S. form of *sulpha-, used to form the name of certain drugs (in British English *sulpha-* also occurs): **sulfame·razine** (also -izine)[*mer + *azine], the readily absorbed sulphonamide $CH_3 \cdot C_4H_2N_2 \cdot NH \cdot SO_2 \cdot C_6H_4 \cdot NH_2$, now rarely used except in Sulphatriad; N^1-(4-methylpyrimidin-2-yl)sulphanilamide; **su:lfaquino·xaline** [*quinoxaline], the sulphonamide $C_8H_5N_2 \cdot NH \cdot SO_2 \cdot C_6H_4 \cdot NH_2$, used as a coccidiostat in the treatment of cæcal coccidiosis in poultry; N^1-quinoxalin-2-ylsulphanilamide.

1943 A. D. Welch et al. in *Jrnl. Pharmacol. & Exper. Therapeutics* LXXVII. 357 The chemistry of this compound, which will be referred to as sulfamerizine, has been described in the publications of several groups. 1945 *Brit. Med. Jrnl.* 3 Feb. 155/1 One new compound—sulphamerazine (or sulphamerizine)—has been the subject of much work in America. 1961 A. Goth *Med. Pharmacol.* xli. 444 Sulfamerazine and sulfamethazine resemble sulfadiazine in most respects, except for the fact that they are excreted more slowly by the kidney. 1962 H. Burn *Drugs, Med. & Man* xx. 200 Various new substances were prepared in this way, among them sulphathiazole, sulphadiazine and sulphamerazine. These compounds were not only different from sulphanilamide, but were much more potent. 1977 *Approved Names 1977* (Brit. Pharmacopœia Commission) 81 Sulfamerazine. 1944 *Jrnl. Biol. Chem.* CLVI. 343 Recently, a new sulfonamide, sulfaquinoxaline, was introduced. 1961 *New Scientist* 21 Dec.

742/1 Sulfaquinoxaline and sulfadimidine were administered to turkeys either in their food or by intramuscular injection. 1976 *Nature* 17 June 621/2 The number of antagonists is now large, some of the best-known being.., sulphaquinoxaline and actinomycin D. 1977 *Approved Names 1977* (Brit. Pharmacopoeia Commission) 81 Sulfaquinoxaline.

Sulfasuxidine (sʊlfăsʊ·ksidīn). *Pharm.* [f. *sulfa- + *sux- + *-idine.] A proprietary name for the drug succinylsulphathiazole.

1942 *Official Gaz.* (U.S. Patent Office) 6 Jan. 9/2 Sharp & Dohme, Incorporated, Philadelphia...*Sulfasuxidine* for pharmaceutical preparations useful as bactericides and as antiseptics. 1943 *Trade Marks Jrnl.* 26 May 223/1 *Sulfasuxidine*...Pharmaceutical substances for human use and for veterinary use..consisting wholly of sulphur compounds...Sharp & Dohms Ltd...1st April, 1942. 1976 A. I. Braude *Antimicrobial Drug Therapy* i. 11 Succinylsulfathiazole (Sulfasuxidine) is a good example of how the para-NH_2 group becomes free after slow hydrolysis from its inactive form to the active sulfathiazole.

sulham (su·lhām). Also † **silham, sulam; selham.** [a. Arab. *zulḥam.*] A large Arab hooded cloak (properly distinguished from the burnous).

1791 W. Lempriere *Tour from Gibraltar to Morocco* ix. 229 They then were obliged to uncover their cap or turban..and to wear instead of the *haick* the *sulam*, which is a cloak made of white or blue woollen cloth. 1809 J. G. Jackson *Acct. Empire of Marocco* 138 The Berebbers wear drawers, and a cloak of dark blue cloth, called a Silham. 1817 J. Riley *Narr. Loss Amer. Brig 'Commerce'* 198 The cloak, or sulam, is made of coarse black cloth. 1891 Hall Caine *Scapegoat* (ed. 2) viii. 172 He drew forth from the folds of his selham a long knife. 1903 *Westm. Gaz.* 15 Jan. 5/1 The ladies..all wearing Moorish sulhams. 1907 F. Campbell *Shepherd of Stars* vi. 72 Strange faces look out from the jellab and sulham hoods. 1921 *Chambers's Jrnl.* Jan. 22/2 A 'warm man' of merchandise, with robes of a surpassing whiteness, rich cloth selham or burnous, and brilliant lemon-coloured slippers. 1951 W. Blunt *Black Sunrise* xvii. 201 After two or three courses Ismail took off his *selham* and *haik*. 1975 C. Calasibetta *Fairchild's Dict. Fashion* 64/1 *Burnoose*.., travelling cape..worn by Moors and Arabs in northern Africa. Also called a *selham*.

Suliote (s[i]ū·liŏ[u]t) *sb.* (and *a*). Also **Souliot(e), Suliot.** [ad. Gr. Σουλιώτης: see -ote.] An inhabitant of the Suli mountains in Epirus, of mixed Greek and Albanian origin. Also *attrib.* or as *adj.*

1812 Byron *Ch. Har.* II. lxxii, Oh! who is more brave than a dark Suliote? 1827 F. G. Halleck *Alnwick Castle & Other Poems* 10 Bozzaris ranged his Suliote band. 1832 T. Gordon *Hist. Greek Revol.* I. I. i. 76 The Souliotes, the flower of Albanian warriors, were driven from Epirus. 1852 G. F. Bowen *Mount Athos, Thesaly, & Epirus* viii. 214 The Suliote hamlet of Kiafa. 1897 W. A. Phillips *War of Greek Independence* vii. 127 Marko Botzares, the Suliot hero. 1900 'Odysseus' *Turkey in Europe* ix. 404 The Suliots have somehow acquired in popular estimation the reputation of being Greeks. As a matter of fact they were a tribe of Christian Albanians. 1910 *Encycl. Brit.* XI. 494/2 On the night of the 21st of August [1823] occurred the celebrated exploit of Marko Botzaris and his Suliotes: a successful surprise attack on the camp of the Ottoman vanguard, in which the Suliote leader fell. 1914 D. J. Cassavetti *Hellas & Balkan Wars* xv. 188 Old people would shake their heads sadly and ask if these empty-headed dolls could possibly belong to the same country as the grand Souliot women. 1952 C. M. Woodhouse *Greek War of Independence* i. 13 Parts of Crete remained..independent of all foreign rule. So to some extent did Souli, a wild precipitous district above the River Akheron in Epirus (Southern Albania); but the Souliotes were not strictly Greeks. 1973 D. Dakin *Greek Struggle for Independence* i. 31 The wily Souliots, hoping to find out what Ali was really up to, sent a token force only. *Ibid.*, Back in Souli, Tzavellas and the chief of Souliot chiefs, Georgios Botsaris, decided to defy Ali.

sulk, *sb.*[2] **1. b.** (Earlier examples.)

1792 W. B. Stevens *Jrnl.* 8 Oct. (1965) I. 48 The strange Sulk of a Day and a half, during our Northern Tour. 1836 J. Romilly *Diary* 21 Oct. (1967) 104 Much discussion (in wch the V.Ch. never joined, he being in a grand sulk).

Sulka (su·lkā). The name of *Sulka & Co.*, shirtmakers and hosiers (est. 1895), of London and New York, used *attrib.* to designate exclusive fabrics (esp. silk) and garments made, designed, or sold by them. Freq. as *Sulka tie*.

1925 *Trade Marks Jrnl.* 21 Oct. 2307 *Sulka*... Shirtings and handkerchiefs, being linen piece goods. A. Sulka & Company.., New York City, New York, United States of America; merchants and manufacturers. 1944 A. Huxley *Time must have Stop* (1945) vi. 69 The beautifully fitting pearl-grey suit, the Sulka tie. 1963 E. Linington *Death of Busybody* x. 126 Mendoza's custom-tailored gray Italian silk and Sulka tie. 1981 'E. Lathen' *Going for Gold* ix. 99 Brad, wrapped in yards of Sulka silk, was pacing the floor.

sulky, *sb.* Add. **4.** (*b*) *sulky cultivator, plow* (earlier examples).

1867 *Trans. Ill. Agric. Soc.* VI. 49 Driving a sulky plow, and plowing his one-fourth acre. 1868 *Rep. Iowa Agric. Soc. 1867* 154 It is then plowed with double-shovel, or sulky cultivators.

sull (sʊ̆l), *v.* *U.S.* [Back-formation from Sullen *a., adv.* and *sb.*] *intr.* Of an animal, to balk; of a person, to become sullen or to sulk.

1869 *Overland Monthly* III. 127 A mustang..will both 'sull', (have the sulks) and 'buck'. 1891 'O. Thanet' *Otto the Knight* 29 The ox, he sulled,..an' Jim jes' guv 'im one on the head. 1902 *Dialect Notes* II. 246 *Sull*,..to hold a position with imperturbable obstinacy and a total disregard of surroundings, as a possum, or a hog in a corner. 1903 *Ibid.* 332 *Sull*,..to sulk; to balk. 'My oxens sull whenever they get hot.' 'She is a quare child and sulls whenever she is contrairied.' 1929 W. Faulkner *Sound & Fury* 87 'She sulling again, is she,' Roskus said. 1938 M. K. Rawlings *Yearling* xxv. 327 Do he ever come here drunk, remember he ain't human when he gits to sullin'. 1949 *10 Story Western* May 11/2 Tell them slow motion sons to keep them cattle comin' before this drive balks and sulls. 1959 W. Faulkner *Mansion* 10 All Frenchman's Bend knew Houston: sulking and sulling in his house all alone by himself since the stallion killed his wife four years ago.

Hence (rarely) as *sb.*, a sulky fit, a 'sulk'.

1972 E. Welty *Optimist's Daughter* II. iv. 97 He's been in a sull ever since you married Judge McKelva and didn't send him a special engraved invitation to the wedding.

Sullan (sʊ·lăn), *a.* (and *sb.*). [f. the name of Roman general and dictator Lucius Cornelius *Sulla* (*c* 138–78 B.C.) + -an.] Of or pertaining to Sulla or his party, or the laws and political reforms instituted by him. Also as *sb.*, a supporter of Sulla.

1866 W. P. Dickson tr. *Mommsen's Hist. Rome* IV. v. iii. 90 The young Gaius Caesar..brought to trial..another Sullan officer Gaius Antonius. *Ibid.* v. 167 Antonius, originally a Sullan like Catilina. 1892 W. W. Fowler *Julius Caesar* ii. 29 Caesar..fled in disguise into the mountains of Samnium. Here he was pursued and captured by the Sullan bloodhounds, who were everywhere. 1905 G. S. Gordon *Let.* 15 Nov. (1943) 10, I *cannot* get out of my head even now the Ontological proofs of the existence of God, the Sullan Constitution, Pericles and Athenian finance, and why Pleasure cannot be an ultimate Ethical end. 1923 T. R. Holmes *Roman Republic* I. i. 59 The Sullan reign of terror was never forgotten by the Romans. 1949 L. R. Taylor *Party Politics in Age of Caesar* i. 21 The *optimates*, who were determined to save what they could of their Sullan prerogatives. 1974 A. Watson *Law Making in Later Roman Republic* vi. 95 It is not proved that all these *leges* are Sullan. 1976 *Classical Q.* XXVI. 105 The Sullan reforms tell us a little about this.

sullen, *a.* Add: **6.** *sullen-eyed, -faced; sullen-blooming, -looking, -smiling.*

1879 O. Wilde in *Time* July 402 No sullen-blooming poppies stain thy hair. 1961 R. S. Thomas *Tares* 47 And given to watching, sullen-eyed, Love still-born, as it was then. 1914 Joyce *Dubliners* 117 A very sullen-faced man. 1919 J. Masefield *Reynard the Fox* I. 29 Surly, Tall, shifty, sullen-smiling.

sulph-, sulpha- (sʊ·lfă). *Pharm.* Also (chiefly *U.S.*) **sulfa-.** [f. *sulpha(nilamide.)] Formative element in the names of drugs derived from sulphanilamide, as **sulphace·tamide** [Acetamide], the sulphonamide $CH_3CO \cdot NH \cdot SO_2 \cdot C_6H_4 \cdot NH_2$, which is used in the form of the sodium salt in the treatment of eye infections; N^1-acetylsulphanilamide; **sulphadi·azine** [*azine b], the readily absorbed sulphonamide $C_4H_3N_2 \cdot NH \cdot SO_2 \cdot C_6H_4 \cdot NH_2$, which is used in the treatment of meningococcal meningitis; N^1-pyrimidin-2-ylsulphanilamide; **sulphadi·midine** [Di- + *pyri)midine], the readily absorbed sulphonamide $(CH_3)_2C_4H_2N_2 \cdot NH \cdot SO_2 \cdot C_6H_4 \cdot NH_2$, used in the treatment of a wide range of systemic and urinary tract infections; N^1-(4,6-dimethylpyrimidin-2-yl)-sulphanilamide; **sulphafu·razole** [*fur(an + *pyr)azole], the readily absorbed sulphonamide $(CH_3)_2C_3NO \cdot NH \cdot SO_2 \cdot C_6H_4 \cdot NH_2$, which is used in the treatment of infections of the urinary tract; N^1-(3,4-dimethyliosoxazol-5-yl)sulphanilamide = *sulphisoxazole* below; **sulphagua·nidine** [Guanidine], the poorly absorbed sulphonamide $HN{:}C(NH_2) \cdot NH \cdot SO_2 \cdot C_6H_4 \cdot NH_2$, formerly used in the treatment of intestinal infections; N^1-guanidinylsulphanilamide; **sulphame·thazine** [Meth(yl + *azine] = *sulphadimidine* above; **sulphame·thizole** [Me(thyl + Thi(o- + *dia*)*zole* (f. Diazo- + *-ole)], the readily absorbed sulphonamide $CH_3 \cdot C_2N_2S_2 \cdot NH \cdot SO_2 \cdot C_6H_4 \cdot NH_2$, which is used in treating coliform infections of the urinary tract; N^1-(5-methyl-1,3,4-thiadiazol-2-yl)sulphanilamide; **su:lphametho·xazole** [Meth(yl + *is*)*oxazole* s.v. *iso- b], the sulphonamide $CH_3 \cdot C_3HNO \cdot NH \cdot SO_2 \cdot C_6H_4 \cdot NH_2$, used in the treatment of respiratory and urinary tract infections, and as a component of the preparation co-trimoxazole; N^1-(5-methylisoxazol-3-yl)sulphanilamide; **su:lphametho:xypyri·dazine** [*methoxy- + *pyridazine], the long-acting sulphonamide $CH_3O \cdot C_4N_2H_2 \cdot NH \cdot SO_2 \cdot C_6H_4 \cdot NH_2$, used in the treatment of systemic and urinary tract infections; N^1-(6-methoxypyridazin-3-yl)sulphanilamide; **Sulphame·zathine**, a proprietary name for sulphadimidine (sulphamethazine); **sulphapy·ridine** [Pyridine], the readily absorbed sulphonamide $C_5H_4N \cdot NH \cdot SO_2 \cdot C_6H_4 \cdot NH_2$, which is used chiefly in the treatment of dermatitis herpetiformis; N^1-pyridin-2-ylsulphanilamide; **sulpha·rsenite**, any compound containing the elements sulphur, arsenic, and oxygen; **sulphasa·lazine** [Sal(icyl + *azine], the sulphonamide $C_5H_4N \cdot NH \cdot SO_2 \cdot C_6H_4 \cdot N{:}N \cdot C_6H(OH)COOH$, which is used in conjunction with corticosteroids in the treatment of ulcerative colitis; 4-hydroxy-4′-(2-pyridylsulphamoyl)azobenzene-3-carboxylic acid; **Sulphatri·ad**, a proprietary name for a mixed sulphonamide drug containing sulphadiazine, sulphamerazine, and sulphathiazole, used in the treatment of acute infections; **sulphiso·xazole** *Pharm.* [*isoxazole* s.v. *iso- b] = *sulphafurazole* above.

1941 *Pharm. Jrnl.* 29 Nov. 188/3 Sulphacetamide is a name which it has been proposed should be adopted for the preparation now known under the trade mark 'Albucid'. 1975 *Prescribers' Jrnl.* XV. 139 Sulphacetamide in the form of eye drops and ointment can produce sensitivity reactions on the skin around the eyes. 1940 R. O. Roblin et al. in *Jrnl. Amer. Chem. Soc.* LXII. 2002/2 In order to avoid possible confusion between sulfapyridine and sulfapyrimidine, the term sulfadiazines is suggested for these [pyrimidine] compounds. 1943 [see *sulphathiazole]. 1956 I. L. Finar *Org. Chem.* II. xviii. 668 Sulphadiazine..is less toxic than Sulphathiazole; it is the most widely used of the 'sulpha' drugs, its main use being for mild infections. 1980 *Jrnl. Med. Microbiol.* XIII. 131 At therapeutic levels in blood, trimethoprim and sulphadiazine singly produced mainly a bactericidal action on pathogens responsible for urinary-tract infections. 1950 *Brit. Med. Jrnl.* 12 Aug. 409/1 Sulphadimidine ('sulphamezathine') and Sulphamerazine (U.S.P.) have almost identical qualities. 1961 [see *sulfaquinoxaline* s.v. *sulfa-]. 1977 *Martindale's Extra Pharmacopoeia* (ed. 27) 1479/2 Sulphadimidine penetrates into cerebrospinal fluid less readily than sulphadiazine and is usually less effective than sulphadiazine in meningeal infections. 1961 *Lancet* 22 July 178/1 Rebollo..claimed a cure [for meningitis due to *Pseudomonas pyocyanea*] with sulphafurazole, given orally. 1976 *Ibid.* 11 Dec. 1276/1, 91 men with nongonococcal urethritis (N.G.U.) were randomly treated with..sulphafurazole (sulfisoxazole). 1941 *Jrnl. Amer. Med. Assoc.* 3 May 2019/2 For 2-sulfanilamidopyrimidine it [*sc.* the Council on Pharmacy and Chemistry of the American Medical Association] adopted the term sulfadiazine and for sulfanilylguanidine the term sulfaguanidine. 1943 *Listener* 16 Sept. 321/2 A quite new drug, sulphaguanidine, shows promise for the treatment of bacillary dysentery; it succeeds here..because much of it is not absorbed, and it therefore remains to act in the intestine. 1958 E. Newby *Short Walk in Hindu Kush* xviii. 220 Everyone..was now suffering from dysentery. We all munched sulphaguanadine tablets but even these failed. 1977 *Martindale's Extra Pharmacopoeia* (ed. 27) 1483/1 Sulphaguanidine has been employed for the treatment of local intestinal infections..though it has now been largely superseded by the less toxic sulphonamides, phthalylsulphathiazole and succinylsulphathiazole. 1942 *Lancet* 30 May 639/1 In the summer of 1941 our attention was drawn by Drs. Martin and Rose of the research laboratories of Imperial Chemical (Pharmaceuticals) Ltd. to a near relation of sulphadiazine..to which the name sulphamethazine has been given. 1951 A. Grollman *Pharmacol. & Therapeutics* xxi. 434 Sulfamethazine.., the methyl derivative of sulfamerazine resembles the latter in action and is used for the same purposes as sulfamerazine and sulfadiazine. 1965 *Pharmacopeia U.S.A.* (ed. 17) 785 U.S.P. XVII Title..Sulfamethazine. Other Designation(s)...Sulphadimidine (*BP*). 1978 Spinelli & Enos *Drugs in Vet. Pract.* x. 135/1 Foot rot... The following can be used: 1. Sulfamethazine, starting with one intravenous injection [etc.]. 1952 *Chem. Abstr.* XLVI. 686 Sulfathiazole, sulfadiazine, and sulfamethizole have been detd. by this method in various pharm. prepns. 1977 *Lancet* 16 Apr. 863/2 Sulphamethizole and alkali were prescribed pending a culture report. 1960 *Antibiotics & Chemotherapy* (N.Y.) X. 572 A new sulfonamide compound, sulfamethoxazole, is identified chemically as 5-methyl-3-sulfanilamidoisoxazole. 1977 *Lancet* 2 July 4/1 Much of the shigellosis could be successfully treated with ampicillin trihydrate and closely related antibiotics, or with co-trimoxazole (trimethoprim and sulphamethoxazole). 1981 H. J. Rogers et al. *Textbk. Clin. Pharmacol.* xix. 653 Sulphamethoxazole is about 50% metabolised so that much reaches the urine in an inactive form. 1956 *Antibiotic Med. & Clin. Therapy* III. 386 A new antibacterial sulfonamide, sulfamethoxypyridazine, has been studied in 67 patients. 1980 *Biochem. Pharmacol.* XXIX. 984/1 Kidney weight/body weight ratio, DNA and protein concentrations of kidney cortex were determined in 55-day-old rats repeatedly pretreated with saline, PAH, sulfamethoxypyridazine, cyclopenthiazide and phenobarbital, respectively. 1943 *Trade Marks Jrnl.* 6 Oct. 421/2 *Sulphamezathine*... Pharmaceutical organic substances being sulphanilamido compounds for veterinary use. Imperial Chemical (Pharmaceuticals) Ltd. 1944 *Pharmaceutical Jrnl.* 8 Apr. 154A/3 Imperial Chemicals (Pharmaceuticals), Ltd., now offer their 'Sulphamezathine' brand of sulphadimethylpyrimidine in the form of a stable solution of the sodium salt. 1970 *Country Life* 26 Feb. 491/1 The drinking water should be dosed with sulphamezathine to prevent coccidiosis. 1939 *Jrnl. Amer. Med. Assoc.* 7 Jan. 49/2 Recent reports from investigators indicate that a pyridine derivative of sulfanilamide..is apparently more promising in the treatment of certain types of pneumonia than sulfanilamide itself... The Council has therefore adopted the term 'sulfapyridine'. **1942** *Times* 30 Nov. 2/3 In 1939, by skilled use of the new drug sulphapyridine, it [*sc.* the fatality rate in spotted fever] was brought down to 33 per cent. **1957** [see *M AND B]. **1967** [see *SULPHANILAMIDE]. **1981** H. J. Rogers et al. *Textbk. Clin. Pharmacol.* xix. 649 Sulphapyridine in low doses over prolonged periods may control dermatitis herpetiformis. **1859** Sulpharsenite [in Dict.]. **1868** J. D. Dana *Syst. Min.* (ed. 5) ii. 84 (*heading*) Sulpharsenites, sulphantimonites, sulphobismuthites. **1905** *Nature* 6 Apr. 534/1 To these minerals a third must now be added in hutchinsonite, a new sulpharsenite from the Binnenthal, which also contains thallium as an important constituent. **1954** Sulpharsenite [see *HUTCHINSONITE]. **1961** *Brit. Med. Jrnl.* 21 Jan. 5220/2 Sulphasalazine ('salazopyrin', 'asulfidine') has been used extensively in Sweden and America since 1941 in the treatment of colitis. **1977** *Lancet* 29 Oct. 931/1 Inflammatory bowel disease was diagnosed and the patient was put on intramuscular corticotrophin, sulphasalazine, and codeine phosphate. **1948** *Trade Marks Jrnl.* 12 May 348/2 Sulphatriad... Pharmaceutical preparations of sulphonamides... May & Baker Limited... Manufacturing chemists. **1950** 'N. Shute' *Town like Alice* vi. 184 The flies would probably result in dysentery but she knew what to do about that; she had plenty of sulphatriad. **1968** J. H. Burn *Lect. Notes Pharmacol.* (ed. 9) 101 If three sulphonamides are used together, as in Sulphatriad, only one-third of the amount of each need be used. **1965** *Pharmacopeia U.S.A.* (ed. 17) 785 U.S.P. XVII Title... Sulfisoxazole. Other Designation(s)... Sulphafurazole (*BP*). **1976** [see *sulphafurazole* above]. **1977** *Lancet* 2 July 4/1 Most of the *S. flexneri* strains were resistant to tetracycline, streptomycin, and sulphafurazole diethanolamine (sulfisoxazole dialomine). **1952** H. Beckman *Pharmacol. in Clin. Pract.* 648 The principal ones [*sc.* sulphonamides] in current use are sulfadiazine, sulfamerazine, Sulfamethazine..and sulfisoxazole.

sulpha (sʌ·lfă). *Pharm.* Also (chiefly *U.S.*) **sulfa.** [f. *SULPHA(NILAMIDE.] Any of the drugs derived from sulphanilamide. Usu. *attrib.*, as *sulpha drug*.

1942 B. Bliven *Men who make Future* i. 11 Within the past year or two, the magical effects of sulfanilamide in curing a whole series of diseases have been supplemented by..the other 'sulfa' drugs. **1951** E. A. McCourt *Home is Stranger* xiv. 211 She had asked Weary to get a prescription of some kind from Dr Harrington—sulpha, perhaps. **1956** *Sci. News* XLI. 18 Modifications of sulphanilamide have led to the extensive sulphanamide or sulpha group of drugs. **1967** *New Scientist* 16 Feb. 384/2 A fatal type of malaria..is now being effectively controlled by a new sulpha drug. **1973** E. Arnold *Proving Ground* (1974) xiv. 169 The nurses..applied sulfa powder and repacked and rebandaged the wound. **1975** B. Wood *Killing Gift* (1976) III. ii. 94 I've given her some sulfa and a shot of codeine so she'll sleep.

sulphæmoglobin (sʌ:lfhīmo,glōᵘ·bin). *Biochem.* and *Med.* Also **sulph-hæmoglobin, -hemoglobin,** (*U.S.*) **sulfhemo-.** [f. Sulp(h- + Hæmoglobin.] A sulphur-containing derivative of hæmoglobin, produced by its reaction with soluble sulphides or sulphides absorbed from the alimentary tract, and giving rise to the greenish discoloration found in putrefying cadavers.

1896 A. Bruce tr. *Thoma's Text-bk. Gen. Path.* I. iii. 43 In poisoning by charcoal fumes, carbonic-oxide-hæmoglobin is formed, and in poisoning by sulphuretted hydrogen, sulph-hæmoglobin or sulph-hæmatin. **1908** Hall & Defren tr. *Abderhalden's Text-bk. Physiol. Chem.* xxiv. 561 This green shade is due to the formation of sulphhemoglobin, which, however, has never been prepared in a pure state. **1947** K. Simpson *Forensic Med.* xxviii. 314 Sulphaemoglobin forms naturally as post-mortem decomposition sets in. **1980** *Amer. Jrnl. Physiol.* CCXXXVIII. H745/2 This report concerns a method for 'labeling' red blood cells..by the formation of sulfhemoglobin.

Hence **su:lphæmoglobinæ·mia** [Gr. αἷμα blood], the presence of sulphæmoglobin in the blood, caused by drug-potentiated absorption of hydrogen sulphide from the alimentary tract, or direct assimilation of the sulpha group from any sulphonamide.

1910 *Jrnl. Amer. Med. Assoc.* 17 Dec. 2181/2 The patient was observed for some weeks, and as the ordinary blood examination failed to reveal any polycythemia or other abnormality to account for the condition, a tentative diagnosis of sulphemoglobinemia was made. **1961** A. S. MacNalty *Brit. Med. Dict.* 1373/2 It [*sc.* sulphæmoglobin] is produced under the influence of many substances such as nitrates, chlorates, nitrites..etc., causing enterogenous cyanosis or sulphaemoglobinaemia. **1980** *Amer. Jrnl. Clin. Path.* LXXIII. 245/1 In one specimen, marked methemoglobinemia and sulfhemoglobinemia were also demonstrated.

sulphane (sʌ·lfēⁱn). *Chem.* Also (*U.S.*) **sulf-.** [a. G. *sulfane* (Fehér & Laue 1953, in *Zeitschr. für Naturforschung* VIIIb. 687/1): see -ANE.] Any of the hydrides of sulphur, H_2S_x.

1955 *Chem. Abstr.* XLIX. 15590 The results reported in this series, on compds. of the form H_2S_n, M_2S_n, and X_2S_n (M = alkali metal, X = halogen), indicate that they have a paraffin-like chain structure. The generic name *sulfanes* is suggested. **1968** Burton & Machmer in G. Nickless *Inorg. Sulphur Chem.* x. 340 The sulphanes are extremely sensitive compounds and hence are very difficult to prepare in the pure state. **1979** *Geophysical Research Lett.* VI. 807/1 We discuss the possible importance of gaseous elemental sulfur (particularly S_2, S_3..and S_4) and sulfanes (H_2S_n) in the lower atmosphere of Venus.

sulphanilamide (sʌlfăni·lămaid). *Pharm.* Also (*U.S.*) **sulf-.** [f. *sulphanilic* s.v. Sulph- (f. Anil(ine + -ic) + Amide.] **a.** The amide of sulphanilic acid, which has wide bacteriostatic activity, has been used, esp. topically, in the treatment of infections due to hæmolytic streptococci, and is the parent compound of the sulphonamides; *p*-aminobenzenesulphonamide, $H_2N{\cdot}C_6H_4{\cdot}SO_2{\cdot}NH_2$.

1937 *Jrnl. Amer. Med. Assoc.* 17 Apr. 1340/2 The Council [on Pharmacy and Chemistry] has therefore formally adopted the nonproprietary name 'Sulfanilamide' for para-aminobenzenesulfonamide. **1942** *Times* 9 Oct. 2/4 A further long list of requirements was sent back from Moscow... The articles dispatched..have included:—530,000 blankets, 10,000 kilos sulphanilamide. **1953** M. Lowry *Let.* 25 Oct. (1967) 345 The chief engineer has an ulcerated throat, and the ship itself is running on sulphanilamide. **1962** J. Heller *Catch-22* xli. 428 Snowden..shifted the position of his hips a bit so that Yossarian could begin salting the wound with sulfanilamide. **1967** *Martindale's Extra Pharmacopoeia* (ed. 25) 1376/1 Sulphanilamide..together with its earlier derivatives, sulphapyridine and sulphathiazole, has been largely superseded by more effective and less toxic compounds. **1974** R. M. Kirk et al. *Surgery* ii. 27 Sulphanilamide powder is sometimes used (wrongly) on raw surfaces, in abscess cavities and body spaces such as the peritoneal and pleural cavities.

b. Any substituted derivative of this compound.

1961 in Webster. **1962** H. A. Krebs in A. Pirie *Lens Metabolism Rel. Cataract* 331 The effects..of sulphanilamides on the metabolism of *p*-aminobenzoic acid.

sulphatase (sʌ·lfătēⁱz). *Biochem.* Also (*U.S.*) **sulf-.** [a. G. *sulfatase* (C. Neuberg 1924, in *Naturwissenschaften* XII. 799/2), f. *sulfat* Sulphate *sb.*: see -ASE.] Any of a group of enzymes found chiefly in mammalian tissues which catalyse the hydrolysis of sulphuric acid esters.

1924 *Chem. Abstr.* XVIII. 3600 The new enzyme sulfatase. **1953** *Biochem. Jrnl.* LI. 585/1 The sulphatases differ with respect to the type of sulphuric acid ester upon which they act. **1964** A. White et al. *Princ. Biochem.* (ed. 3) xi. 175 No sulfatase capable of effecting hydrolysis of sulfate esters of carbohydrates is known to be present in animal tissues. **1980** *Jrnl. Path.* CXXX. 243 The giant lysosomes contained both acid phosphatase and aryl sulphatase.

sulphate, *sb.* Add: **3. sulphate ion,** the ion SO_4^{2-}; **sulphate process** *Paper-making*, a method of manufacturing a tough brown paper involving the digestion of wood chips by sodium hydroxide and sodium sulphate to form the pulp; so **sulphate pulp**; cf. *KRAFT; **sulphate-reducing** *a. Biol.*, (of a process or micro-organism) bringing about the reduction of sulphate ions to sulphide; *spec.* applied to bacteria of the genera *Desulphovibrio* and *Desulphatomaculum*, which do this as part of their respiratory metabolism.

1897 G. S. Newth *Text-bk. Inorg. Chem.* (ed. 9) xi. 105 SO_4'', in the same way, stands for the sulphate ion, with its two negative charges. **1978** J. R. Holum *Org. & Biol. Chem.* xix. 411 The sulfur dioxide is then oxidized to sulfate ion, which is excreted by the kidneys. **1894** G. Clapperton *Pract. Paper-Making* iv. 32 The former [process]..employs a solution of sodium compounds containing a large percentage of sulphate of soda, and is known as the sulphate process. **1963** R. R. A. Higham *Handbk. Papermaking* v. 98 There are three basic alkaline processes, which are: Soda process, Sulphate (Kraft) process, [illegible]. **1934** *Sci. Amer.* Apr. 55/1 Most soda pulp mills changed over to the kraft process (which is also often called the sulfate process). **1907** G. Clapperton *Pract. Paper-Making* (ed. 2) iv. 36 During recent years the demand for 'sulphate' pulp has increased largely, owing to the demand for strong brown paper. **1962** F. T. Day *Introd. Paper* iii. 19 The wood chips are cooked in digesters in a solution of caustic soda to produce soda pulp, or with a mixture of caustic soda and sulphate of soda to produce sulphate pulp. **1966** F. E. Dean *Paper* ii. 36 Unbleached sulphate or 'kraft' pulp..is used mainly for wrapping paper. **1926** *Science* 1 Jan. 24/1 The oil-field waters in which the sulphate-reducing bacteria occur are similar in general composition to seawater. **1954** *New Biol.* XVII. 67 Vastly greater amounts of sulphide are formed in nature by a single group of micro-organisms called the sulphate-reducing bacteria... Their sulphate-reducing process corresponds to the respiration of more normal organisms. **1979** *Arch. Microbiol.* CXXI. 261/1 Some sulfate-reducing bacteria are able to utilize colloidal sulfur as respiratory substrate.

sulphate, *v.* (Earlier example.)

1888 D. Salomons *Managem. Accumulators* (ed. 3) v. 58 If the positives sulphate the surface becomes very hard.

sulphathiazole (sʌlfăþai·ăzōᵘl). *Pharm.* Also (*U.S.*) **sulf-.** [f. *SULPHA- + *thiazole* s.v. Thio- 1.] The readily absorbed sulphonamide $C_9H_9NS{\cdot}NH{\cdot}SO_2{\cdot}C_6H_4{\cdot}NH_2$, now rarely used; thiazol-2-ylsulphanilamide.

1939 *Jrnl. Amer. Chem. Soc.* LXI. 3593/2 The potentiometric titration curves for the acidification of 2% solution of the sodium salts of 2(*p*-aminobenzenesulfonamido)thiazole (Sulfathiazole) and sulfapyridine are submitted in Fig. 1. **1943** *Endeavour* Apr. 42/1 Sulphathiazole and sulphadiazine are now among the physician's sheet anchors in the treatment of meningitis, gonorrhoea and pneumonia. **1967** [see *SULPHANILAMIDE]. **1978** J. Irving *World according to Garp* i. 7 Sulfathiazole was for the clap—with lots of water recommended.

sulphatide (sʌ·lfătaid). *Biochem.* Also (*U.S.*) **sulf-.** [Sulphat(e *sb.* + -IDE.] Any of the group of lipids consisting of the sulphuric acid ester of a cerebroside.

1884 J. L. W. Thudichum *Treat. Chem. Constitution of Brain* i. 22 The albuminous substances of the brain may be considered as nitrogenised sulphatides, inasmuch as sulphur is an essential constituent. **1954** A. White et al. *Princ. Biochem.* xxxii. 801 A number of less clearly defined lipids have also been recognized [in the brain], such as sulfatides... Only one sulfatide has been isolated and studied, cerebron sulfuric acid. **1966** *Lancet* 24 Dec. 1421/2 The predominance of sulphatide in the white matter of the brain in this disease has been noted by Lees. **1978** *Nature* 7 Dec. 625/1 The negative charge imparted by the sulphate group of the sulphatide does not seem to be the direct cause of its strong adhesion capacity, as other negatively charged lipids, such as ganglioside or phosphatidyl inositol, did not adhere strongly.

sulphation. Add: Conversion into a sulphate; incorporation of a sulphate ion, SO_4^{2-}, into a molecule. (Further examples.)

1957 *New Biol.* XXIV. 52 There is an informed opinion which asserts that sulphation is not necessary for metachromasia and that mucopolysaccharides without sulphur are also metachromatic. **1971, 1972** [see *SOMATOMEDIN]. **1975** *Nature* 24 Jan. 269/2 Somatomedin, previously known as 'sulphation factor',..stimulates the incorporation of $^{35}SO_4$ into costal cartilage. **1977** *Lancet* 22 Jan. 168/1 The mammary gland itself may be the site of sulphation [of vitamin D].

sulphazin (sʌ·lfăzin). *Pharm.* Also (chiefly *U.S.*) **sulfazin.** [Russ.] A drug consisting of a suspension of one per cent purified sulphur in peach oil, given intramuscularly to induce fever.

1970 *Time* 29 June 30/3 A Soviet drug called Sulfazin, which induces fever and temperature, is administered as a punishment. **1977** *Lancet* 23 July 185/1 There are reports of the use of sulphazin to produce a painful fever. **1979** H. Fireside *Soviet Psychoprisons* iv. 82 If a patient speaks out against such brutality, he is subjected to punishment by overdoses of drugs or injections of sulfazin that make it 'painful for him even to stir'. **1981** M. C. Smith *Gorky Park* I. xii. 185 Sulfazin was one of the favorite narcotics of the KGB.

sulphetrone (sʌ·lfĕtrōᵘn). *Pharm.* [f. Sulphone with insertion of *t)etr(asodium* (f. Tetra- + Sodium).] A trade name for the drug *SOLAPSONE.

1947 *Lancet* 20 Dec. 897/2 Two studies were made—(1) to assess the efficacy of streptomycin; and (2) to evaluate the possible synergic action of streptomycin and sulphetrone. **1959, 1974** [see *SOLAPSONE].

sulph-hæmoglobin, var. *SULPHÆMOGLOBIN.

sulphide, *v.* Add: (Example.) Also **su·lphiding** *vbl. sb.*

1950 R. W. Moncrieff *Artificial Fibres* viii. 105 The yarn was wound into skeins, and these were washed, sulphided, bleached and washed. **1955** E. E. Loening et al. in R. S. Schultze *Sci. & Applic. Photogr.* 62 In the case of sulphiding, the left part of the curves is..absent. **1982** *Photogr. Sci. & Engin.* XXVI. 223/1 Sulfiding of the cathode can be minimized by poising the potential of the cathode more positive than −0·55 v.

sulphidic (sʌlfi·dik), *a. Chem.* [f. Sulphid(e *sb.* + -IC.] Of or containing sulphides.

1929 H. Schneiderhöhn in P. A. Wagner *Platinum Deposits & Mines S. Afr.* xvii. 208 (*heading*) The sulphidic ore minerals of the felspathic Harzburgite and Merensky 'reef'. **1959** *Times* 23 Sept. 19/6 Sulphidic minerals, mainly pyrites. **1978** *Metals* (Shell Internat. Petroleum Co.) 4 Rich sulphidic ores, such as those in the massive copper belt of Zambia/Zaire.., have been recovered by underground mining for many years.

sulphinpyrazone (sʌlfinpi·răzōᵘn). *Pharm.* Also (*U.S.*) **sulf-.** [f. Sulphin(ic *a.* + *PYRAZ(OLE + -ONE.] The uricosuric drug $C_6H_5{\cdot}SO{\cdot}(CH_2)_2CH{\cdot}C_3HN_2O_2(C_6H_5)_2$, which promotes excretion of urates by inhibiting their reabsorption by kidneys; 1,2-diphenyl-4-(2-phenylsulphinylethyl)pyrazolidine-3,5-dione.

1958 *Arthritis & Rheumatism* I. 532 The present communication is concerned with G-28315 (4-[phenylsulfoxyethyl]-1,2-diphenyl-3,5-pyrazolidinedione). [*Note*] The generic name of this compound is sulfinpyrazone. **1961** *Lancet* 30 Sept. 763/2 Aspirin..could within a few days overcome the effects for which probenecid, sulfinpyrazone, or zoxazolamine had been prescribed. **1978** *Times* 2 Mar. 16/6 The American research project was set up to test the theory that the risk of this further thrombosis might be reduced by treatment with a drug, sulfinpyrazone, which acts on the blood platelets, the small cells that start the process of thrombosis. **1981** H. J. Rogers et al. *Textbk. Clin. Pharmacol.* xvii. 569 Sulphinpyrazone (Anturan)..is a uricosuric drug related to phenylbutazone which prolongs platelet survival..without prolonging the bleeding time.

sulpho-. Add: In mod. use often repr. *SULPHONYL, as in *sulphochlorination, -lipid* below. **su:lphobromophtha·lein** *Pharm.* [Bromo- + Phthalein] = *bromosulphthalein* s.v. *BROMO-; **su:lphochlorina·tion** *Chem.*, the introduction of the chlorosulphonyl group, $ClSO_2$—, into a molecule; **su·lpholipid** *Biochem.*, any of a class of lipids whose structures terminate with the sulphonate group, $—SO_3^-$.

1945 *Jrnl. Amer. Med. Assoc.* 4 Aug. 1001/2 A positive Hanger test or a strong sulfobromophthalein dye retention also provides valuable evidence [of hepatitis]. **1974** M. C. Gerald *Pharmacol.* ii. 34 Frequently employed diagnostic agents include..sulfobromophthalein for liver function tests. **1931** *Jrnl. Amer. Chem. Soc.* LIII. 2648 Phenols can be converted in phenolpolysulfonylchlorides by the action of chlorosulfonic acid. It also was noticed that the reagent caused four distinct types of reaction: namely, sulfonation, sulfochlorination, chlorination, oxidation. **1980** *Chem. in Brit.* XVI. 466/1 The sulphochlorination of paraffins to alkane sulphonyl chlorides is conducted in the presence of uv light. **1930** M. Bodansky *Introd. Physiol. Chem.* (ed. 2) iii. 70 Aminolipids, sulfolipids, etc.—groups which are at present not sufficiently well characterized for classification. **1977** D. E. Metzler *Biochem.* ii. 112/2 Chloroplasts contain a large amount of a special sulfolipid.

sulphonamide (sʌlfǫ·nămaid). *Chem.* and *Pharm.* Also (*U.S.*) **sulf-.** [f. Sulphone + Amide.] **a.** Any organic compound that is an amide of a sulphonic acid, characterized by the group $—SO_2N=$; *spec.* any of the drugs derived from sulphanilamide (and so containing this group). Freq. *attrib.*

1881 *Jrnl. Chem. Soc.* XL. 602 The [1:4:3] acid..is converted by ammonia into a sulphonamide crystallising in needles. **1947** *Sci. News* IV. 60 When new drugs like the sulphonamides or D.D.T. are developed, their chemical properties have been reported for thirty years or so, and it is their biological effects which are the true modern discoveries. **1959** *Times* 7 Dec. (Agric. Suppl.) p. vii/4 To prevent coccidiosis in chickens, nitrophenol, a sulfonamide..is added to the feed. **1964** N. G. Clark *Mod. Org. Chem.* xx. 414 In most cases, the sulphonamides have convenient melting-points, and are admirably suitable for characterizing both sulphonic acids and amines. **1974** [see *POTENTIATED *ppl. a.*]. **1977** *Martindale's Extra Pharmacopoeia* (ed. 27) 1468/1 Because they are similar in chemical structure to *p*-aminobenzoic acid, sulphonamides interfere with the synthesis by micro-organisms of folic acid from *p*-aminobenzoic acid... The sulphonamides have been largely replaced by antibiotics in the treatment of infections.

b. *attrib.* and *Comb.*, as *sulphonamide drug, group* (of atoms or of drugs); *sulphonamide-resistant* adj.

1943 *Times* 16 June 5/7 Recent American figures suggest that one death occurs from the Sulphonamide drugs in every 2,571 deaths from all causes. **1959** *Sci. News* LI. 96 Antithyroid activity was first observed in some of the sulphonamide drugs, but the first compound used clinically, in 1943 by Astwood in America, was thiourea. **1979** Davies & Littlewood *Elementary Biochem.* iv. 83 Sulfonamide drugs are not effective in open, suppurating wounds; such wounds contain pus and other materials that are a source of *p*-aminobenzoic acid, which antagonizes the action of the sulfonamide drugs. **1939** *Brit. Med. Jrnl.* 5 Aug. 269/2 Sulphanilamide consists of a benzene ring to opposite ends of which are attached an amino group and a sulphonamide group. **1942** *Times* 21 Sept. 5/7 Another most important factor in saving life has been the series of new drugs, of which the sulphonamide group is the most important. **1942** *Proc. Soc. Exper. Biol. & Med.* L. 336 The present report is concerned with the *in vitro* and *in vivo* production of sulfonamide resistant strains of staphylococci. **1968** *Times* 12 Oct. 18/8 One of the organisms sometimes responsible for travellers' diarrhoea is now sulphonamide-resistant. **1981** H. J. Rogers et al. *Textbk. Clin. Pharmacol.* xix. 649 Sulphadiazine is now only rarely used (with benzylpenicillin) in the treatment of meningococcal meningitis since sulphonamide-resistant meningococci are common.

sulphonate, *v.* (Earlier example.) Hence **su·lphonated** *ppl. a.*, **su·lphonating** *vbl. sb.*

1882 *Jrnl. Chem. Soc.* XLII. 196 The author could not obtain the salt 'A'..by sulphonating pure cymene. **1902** Sulphonating [in Dict.]. **1936, 1966** [see *SOAPLESS *a.* b]. **1972** *Materials & Technol.* V. 302 By the use of energetic sulphonating agents such as sulphur trioxide.., fatty acids can be sulphonated at the alpha carbon atom. The sulphonated acids have useful surface-active properties.

sulphonium (sʌlfōᵘ·niʌm). *Chem.* Also (*U.S.*) **sulf-.** [f. Sulph(ur *sb.* + *-ONIUM.] A hypothetical monovalent complex cation having a central sulphur atom bonded to three hydrogen atoms; also, any derivative of this in which one or more of the hydrogen atoms is replaced by organic radicals. Usu. *attrib.*

1894 [see *IODONIUM]. **1942** *Jrnl. Amer. Chem. Soc.* LXIV. 1165/1 The ability of dialkyl sulfides to react with ω-halogenated ketones with the subsequent formation of sulfonium halides has been known for some time. **1975** R. F. Brown *Org. Chem.* xxix. 945 The sulfonium ions (R_3S^+) are much more stable than are the analogous oxonium ions (R_3O^+).

sulphonyl (sʌ·lfǒnail). *Chem.* Also (*U.S.*) **sulf-.** [f. Sulphone + -YL.] The divalent radical $—SO_2—$, derived from a sulphonic acid group by removal of the —OH group. Usu. *attrib.*

1920 *Chem. Abstr.* XIV. 1947 Place 3 g. of pulverized sulfonyl chloride in a round-bottomed flask. **1953** *Chem. & Engin. News* 5 Jan. 91/3 The inorganic name of the radical SO_2 is sulfuryl, while its organic name is sulfonyl. **1975** R. F. Brown *Org. Chem.* xxix. 956 Some of the sulfonyl chlorides and esters have been used so often that trivial names have been coined.

Hence **su:lphonyla·tion,** conversion into a sulphonyl compound; (as a back-formation) **sulpho·nylate** *v. trans.*

1956 *Chem. Abstr.* L. 10677/1 (*heading*) Friedel-Crafts acylation and sulfonylation reactions. **1979** *Tetrahedron Lett.* Sept. 3790 The mild conditions used in this sulphonylation provide some advantages over the more usual preparative methods for unsymmetrical sulphones. **1980** *Chem. Abstr.* XCIII. 843/2 Thiazole hydrobromide was sulfonylated with..arsenesulfonyl chlorides to give the corresponding 7-sulfonylthiazolium chlorides.

sulphonylurea (sʌ:lfǒnail,yūᵊrī·ă). *Pharm.* Also (*U.S.*) **sulf-.** [f. *SULPHONYL + Urea.] Any of the group of hypoglycæmic drugs containing the active grouping $—SO_2{\cdot}NH{\cdot}CO{\cdot}NH—$, which are used orally in the treatment of diabetes.

1956 *Science* 6 Apr. 583/2 A statistically highly significant hypoglycemic response occurred in 34 of the patients with diabetes who were given the sulfonylurea. **1966** *New Scientist* 24 Nov. 433/1 The longing of diabetics for a hypoglycaemic drug which could be taken orally..was realized ten years ago when the sulphonylureas and diguanides were introduced. **1974** M. C. Gerald *Pharmacol.* xxv. 442 The oral hypoglycemic agents, the sulfonylureas such as tolbutamide and the biguanide phenformin, are useful agents for the treatment of the stable maturity-onset diabetes.

sulphur, *sb.* Add: **1. f.** The colour of sulphur, a greenish-yellow.

1924 R. Campbell *Flaming Terrapin* ii. 32 Panthers' eyes..Flashed their pale sulphur on the sunless air. **1963** *Listener* 10 Jan. 84/2, I don't like the colours, especially the Ribena, pillarbox, scrofula, and sulphur.

5. c. (Earlier example.)

1801 M. Edgeworth *Belinda* I. viii. 240 Helena and her young companions now came into the room, bringing with them the sulphurs at which they had been looking.

8. sulphur bacterium *Biol.*, any of the bacteria which derive their energy from the oxidation of sulphur or inorganic compounds of sulphur; **sulphur cycle** *Ecol.*, the cycle of changes whereby sulphur compounds are interconverted between sulphates and hydrogen sulphide in the air and sulphates, sulphides, and sulphur in organisms and the soil; **sulphur print** *Metallurgy*, a print on photographic bromide paper showing the distribution of sulphur as sulphides in a steel surface with which it has been placed in contact; **sulphur shower** (earlier example); **sulphur soap,** a medicinal soap containing elemental sulphur for use in treating skin complaints; **sulphur-spring** (earlier examples).

1891 A. B. Griffiths *Res. on Micro-Organisms* viii. 179 All belong to the class of 'sulphur-bacteria'..—that is, bacteria which in the presence of free hydrogen-sulphide oxidize sulphur, forming sulphuric acid. **1939** Clements & Shelford *Bio-Ecol.* iii. 101 Hydrogen sulphide is also acted upon by a remarkable group of sulphur bacteria. **1962** W. W. Umbreit *Mod. Microbiol.* xv. 276/2 There are three major types of photosynthetic bacteria. The first two of them, the thiorhodaceae and the chlorobacteriaceae, are sulfur bacteria. **1979** Arms & Camp *Biology* x. 165 Purple and green sulfur bacteria (Thiorhodaceae) use hydrogen gas and hydrogen sulfide (H_2S) as hydrogen donors. **1967** *New Scientist* 9 Nov. 333/1 The continuation and evolution of life depends upon a stable eco-system. A characteristic of such a system is that cyclical transformations of the major biological elements take place within it—the carbon cycle, the nitrogen cycle, the sulphur cycle and so on. **1973** R. G. Krueger et al. *Introd. Microbiol.* xxx. 745/2 Bacteria can carry out each of the processes in the sulfur cycle; bacteria and certain colorless blue-green algae are the only known living agents of the steps involving interconversions of inorganic forms of sulfur. **1977** I. M. Campbell *Energy & Atmosphere* viii. 289 More efficient dispersal of sulphur dioxide at source cannot be regarded as an acceptable long term solution, since that merely transfers the problem to another region or country, the problem intensified by the fact that the anthropogenic term in the sulphur cycle is of the same order of magnitude as the natural terms. **1912** *Jrnl. Iron & Steel Inst.* LXXXV. 380, I have adopted a method which is virtually a modification of the well known method of obtaining a 'sulphur print'. **1977** R. B. Ross *Handbk. Metal Treatments & Testing* 373 Chemical analysis and micro-examination..require laboratory equipment and skilled personnel, whereas the Sulphur print may be used in relatively unsophisticated conditions. **1854** Thoreau *Walden* 340 The sulphur-like pollen of the pitch pine soon covered the pond and the stones and rotten wood... This is the 'sulphur showers' we hear of. **1894** A. Watt *Art of Soap-Making* xxi. 172 Sir H. Marsh's Sulphur Soap... A few drops of otto of roses are added to give the soap an agreeable fragrance. **1925** G. Martin *Mod. Soap & Detergent Industr.* II. II. iv. 34 Sulphur soaps, when dissolved in water, slowly evolve sulphuretted hydrogen, which gives them an unpleasant smell. **1953** J. Davidsohn et al. *Soap Manuf.* I. xxii. 505 Sulfur soaps are frequently prepared in combination with beta-naphthol, tar, glycerine and camphor. **1785** T. Jefferson *Notes on Virginia* vi. 59 We are told of a Sulphur-spring on Howard's creek of Greenbriar. **1811** W. J. Hooker *Jrnl. Tour in Iceland in 1809* 195 We could not resist the present temptation of alighting from our horses, to visit one of the sulphur-springs that lay in our route.

9. sulphur-bottom (whale), substitute for def.: Sibbald's rorqual (see *SIBBALD); (further examples); **sulphur (-crested) cockatoo,** a white cockatoo, *Kakatoe galerita*, with a yellow crest, native to Australia; **sulphur tuft,** a toadstool, *Hypholoma fasciculare*, with a yellow cap tinged with brown; **sulphur whale** (earlier example).

1851 H. Melville *Moby Dick* I. xxxi. 221 Adieu, Sulphur-Bottom! **1934** R. Campbell *Broken Record* iv. 94 These blue whales are the great sulphur-bottoms. **1959** A. C. Hardy *Open Sea* II. xv. 280 It [*sc.* Sibbald's rorqual] has also been called the sulphur-bottom whale on account of a yellowish scum of diatoms which these whales usually carry when they first return to polar waters again after visiting warmer latitudes for breeding. **1811, 1893** Sulphur (-crested) cockatoo [in Dict.]. **1908** E. J. Banfield *Confessions of Beachcomber* I. i. 17 Sulphur-crested cockatoos sail down upon the red raiment of the tree. **1963** *Times* 8 June 14/3 Probably the most talkative..is one of the four sulphur-crested cockatoos. **1909** E. W. Swanton *Fungi* 115 'Sulphur-tuft'... Taste intensely bitter. Poisonous. **1979** *Country Life* 25 Oct. 1423/1 The fruiting bodies of sulphur tuft..also grow on tree stumps. **1829** T. C. Haliburton *Hist. & Statistical Acct. Nova Scotia* II. ix. 404 Fish-Whale Species. Sulphur Whale.

sulphur-. In words in the Dict. beginning thus the second 'u' when unstressed is marked with the pronunc. (-iŭ-); this is now often pronounced (-ə-) or (-ŭ-).

sulphuretum (sʌlfərī·tʌm). *Ecol.* Also (*U.S.*) **sulf-.** Pl. **sulphureta, -tums.** [mod.L., f. Sulphur *sb.* + *-ETUM.] An ecological community of organisms, mainly consisting of sulphur bacteria, which metabolizes sulphur compounds in a closed subcycle of the larger environmental sulphur cycle.

1925 L. G. M. Baas-Becking in *Ann. Bot.* XXXIX. 615 The natural ecological community of these [sulphur] bacteria is a miniature cycle in itself, and will be called a sulphuretum. **1967** *New Scientist* 9 Nov. 333/2 One such ecosystem is known as the sulfuretum, based primarily on the sulphur bacteria and essentially anaerobic. *Ibid.*, Palaeochemical evidence, based on fractionation of the sulphur isotopes, has shown that sulfureta were active at least 2×10^9 years ago.

sulphydryl. Add: Also (*U.S.*) **sulfhydryl.** (Examples.) = *MERCAPTO(-) b, *THIOL b.

1924 *Biochem. Jrnl.* XVIII. 1020 The sulphydryl compounds are apparently incapable of combining directly with molecular oxygen. **1946** *Nature* 3 Aug. 155/2 Manganese dioxide is reduced with great ease to form divalent manganese ion by sulphydryl compounds, for example, thioglycollic acid. **1978** *Bull. Amer. Acad. Arts & Sci.* Feb. 10 Elwood Jensen had already made important contributions to..our understanding of the role of sulfhydryl groups in protein structure.

sulpiride (sʌ·lpiraid). *Pharm.* [a. F. *sulpiride*, prob. f. *sul(f-* Sulph- + *pir-*, alteration of *pyr-* Pyr(o-: see -IDE.] An anti-emetic and neuroleptic drug used in the treatment of gastro-intestinal disorders, vertigo, and psychiatric conditions; *N*-(1-ethylpyrrolidin-2-ylmethyl)-2-methoxy-5-sulphamoylbenzamide, $(C_2H_5)C_4H_8N(CH_2){\cdot}NH{\cdot}SO_2{\cdot}C_6H_3(OCH_3){\cdot}CO{\cdot}NH_2$.

1970 *Jrnl. Amer. Med. Assoc.* 10 Aug. 1076/1 The new drug sulpiride was tested to determine its effectiveness in the treatment of ulcerative colitis. **1976** *Lancet* 18 Dec. 1358/1 We concluded that sulpiride should be prescribed with care in hypertensive patients. **1979** *Nature* 11 Jan. 94/2 The antipsychotic drugs, molindone and sulpiride, and the antiemetic drug, metoclopramide, are dopamine antagonists when tested in the anterior pituitary or the brain.

sul ponticello: see *PONTICELLO b.

sultana. Add: **5.** (Later *Hist.* example.)

1935 P. P. Argenti *Occupation of Chios by Venetians (1694)* p. xxxix, The enemy fleet..consisted of twenty great sultanas, and thirty galleys and galliots, all under the command of the *Capoudân Pasha*.

7. Also produced in other parts of Turkey, Greece, and Australia. (Further examples.)

1855 E. Acton *Mod. Cookery* (rev. ed.) xxi. 442 Sultana raisins are well adapted to these puddings, as they contain no pips. **1920** C. L. T. Beeching *Mod. Grocer & Provision Dealer* III. viii. 163 The sultana raisin may be said to share in the good qualities of both the currant and the Valencia. **1938** C. J. Elliott *Retail Grocery Trade* xii. 108 The Australian sultana is a little larger than the Turkey and Smyrna variety. **1966** A. Uttley *Recipes from Old Farmhouse* 58 Add one ounce of sugar and one ounce of sultanas.

9*. = *busy Lizzie* s.v. *BUSY *a.* 11; *patient Lucy* s.v. *PATIENT *a.* 5.

1938 M. K. Rawlings *Yearling* xxvi. 360 The church was decorated with..donations of house plants; sultanas and geraniums, aspidistras and coleas [*sic*]. **1977** [see *PATIENT *a.* 5].

10. sultana grape, the white seedless grape from which sultanas are made.

1861 Mrs. Beeton *Bk. Househ. Managem.* 666 Sultana Grape... The white or yellow grape..produces the Sultana raisin. **1931** C. L. T. Beeching *Law's Grocer's Man.* (ed. 3) 513/2 The vine which grows the sultana grape is vigorous and upright. **1979** *Illustr. London News* Jan. 66/3 The sultana grape vineyards start a few kilometres to the east of Ayios Nikolaos.

sultanate. 1. (Earlier example.)

1822 tr. *Malte-Brun's Universal Geogr.* I. xxii. 590 It would be rather interesting to enumerate the various denominations which designate the different states. The use of the terms empire, kingdom, sultanat, khonet, and others, will be learnt in the descriptive part of this work.

sultanize, *v.* **2.** (Earlier examples.)

1804 J. Mackintosh *Let.* 14 Aug. in *Mem.* (1835) I. v. 212 The Governor..is..an..intelligent man; but every Englishman who resides here very long, has..his mind either emasculated by submission, or corrupted by despotic power. Mr. Duncan may represent one genus, the Braminised Englishman; Lord W— is indisputably at the head of the other, the Sultanised Englishman. **1876** *Hansard Commons* 16 Mar. 103 It was not a wise thing to endeavour even in India to Sultanize the Crown.

sul tasto: see *TASTO b.

sultry, *a.* Add: **2. b.** (*c*) (Earlier example.)

1880 'Mark Twain' *Tramp Abroad* xxv. 250 It was getting pretty sultry for me. I said to myself, 'Is it possible she is going to stop there, and wait for me to speak? If she does, the conversation is blocked.'

(*d*) Of a woman: lascivious or sensual, arousing sexual desire; also *transf.* and in *Comb.* orig. *U.S.*

1940 *Time* 7 Oct. 63/2 He watches..another become a sultry, sirenic dancer. **1946** *Sun* (Baltimore) 25 Apr. 12/1 There is also a ballet touch to Miss Horne's sultry song number, 'Love'. **1949** R. Harvey *Curtain Time* xvi. 160 Miss Nethersole specialized in sultry rôles and her performance in Daudet's *Sapho* was considered scandalous. **1956** *People* 13 May 4/4 Certainly none of the sultry Continental sirens stood a chance when Diana strolled on to the beach. **1977** C. Storr *Tales Psychiatrist's Couch* i. 6 She exuded an air of unsatisfied sexuality... She was what I'd call sultry. **1978** *Times* 30 Nov. 16/8 A trip to Rio to see the real thing—*real* sultry-eyed temptresses.

Sulu[1] (sū·lu). [Prob. ad. Sama-Bajaw dial. f. Tau Sug *sulúg* current.] = *TAU SUG.

1816 [see *MACASSAR 2]. **1898** D. C. Worcester *Philippine Islands & their People* viii. 201, I had made numerous attempts in Mindanao, Basilan, and Sulu to get an explanation of the Moro aversion to pork. **1908** N. M. Saleeby *Hist. Sulu* i. 133 Jolo is the Spanish representation..of the word Sulu... The complete form of the word is Sulug... The Sulus pronounce it and write it Sūg. *Sūg* means a sea current. *Ibid.* iii. 155 The ancient Sulus..had many myths relating to the marriages and heroic deeds of their gods. **1923** S. Y. Orosa *Sulu Archipelago & its People* v. 67 The people of the Sulu Province number over 170,000, roughly grouped as Sulus and Samals. The dominating and most advanced people are the Sulus or *Tao-Sug*, 'people of the current'. *Ibid.* vi. 72 The Sulu is of the brown or Malay race. **1936** G. A. Malcolm *Commonwealth of Philippines* iii. 39 The Sulus of whom I would speak..are Moros living in the Sulu Archipelago in the Philippine Islands. **1977** C. F. & F. M. Voegelin *Classification & Index World's Lang.* 41 Taw Sug = Tausug = Sulu = Joloano Sulu. Palawan, Philippines, northeast coast of Borneo. Closely related to Maranao.

‖ **sulu**[2] (sū·lu). [Fijian.] In Fiji: a length of cotton cloth wrapped about the body to form a sarong; hence, a type of sarong worn by both sexes (typically from the waist to the knee by men, and to the ankle by women). Also, a similar fashion garment worn by women.

1850 D. Hazlewood *Feejeean & English Dict.* 129/1 *Sulu-ma*, v. to put on a sulu, or dress... The difference between malo and sulu seems to be in the way in which it is worn: malo is sulu when put round the body and not between the legs. **1897** 'Sundowner' *Rambles in Polynesia* 7 For many years yet..the Polynesian islander will continue to wear his *sulu* or lava-lava, as the case may be. **1921** W. A. Chapple *Fiji—its Problems & Resources* ii. 22 His [*sc.* the Fijian's] sulu is his only garment,..a rectangular piece of cotton cloth that he folds round his loins and tucks in upon itself. **1926** *Glasgow Herald* 25 Sept. 4/5 Clad only in their sulus (or kilts). **1944** W. E. Harney *Taboo* (ed. 2) 135, I had only a loincloth—a sulu, as it is called. **1970** *Honey* June 86 Vivid multi-coloured patchwork slit sulu 11 gns, and tie top, 84s. **1977** *Times* 20 July 1/7 The staff of the Fijian High Commission had turned out in pinstripe *sulu* skirts and morning jackets.

sulvanite (sʌ·lvănait). *Min.* [f. Sul(phur *sb.* + Van(adium + -ITE[1].] A bronze-coloured sulphide of copper and vanadium, Cu_3VS_4, that usu. occurs massive, rarely as crystals having cubic symmetry, and is often chemically altered.

1900 G. A. Goyder in *Jrnl. Chem. Soc.* LXXVII. 1094 (*heading*) Sulvanite, a new mineral. **1974** *Amer. Mineralogist* LIX. 307/2 In all occurrences, sulvanite is coated with alteration minerals consisting of malachite, volborthite, and azurite.

sum, *sb.*[1] Add: **5.** (Later examples.)

1946 *R.A.F. Jrnl.* May 152 The *Bulletin* slowly built up a sum of good will among contributors. **1967** G. Steiner *Lang. & Silence* 31 Literature, philosophy, theology, law, the arts of history, are endeavours to enclose within the bounds of rational discourse the sum of human experience.

6. d. = *logical sum* s.v. *LOGICAL *a.* 7.

1918 C. I. Lewis *Survey of Symbolic Logic* iii. 185 The 'sum', $a + b$, denotes the class of those things which are either members of a or members of b (or members of both). **1934** W. V. Quine *System of Logistic* xvii. 171, ε'α may be called the sum of the class of classes α. **1968** P. A. P. Moran *Introd. Probability Theory* iv. 185 The advantage of using half-open intervals is that if two of them abut, their sum is again a half-open interval. **1981** W. Marciszewski *Dict. Logic* 53 The union (sum) of sets: $x \in X \cup Y \equiv (x \in X) \vee (x \in Y)$.

14. *Comb.*, as **sum check** *Computers*, a check on the accuracy of a group of digits in which they are added together and the result compared with a previously computed sum (which may accompany the group as a check digit); also applied to similar checks in which a quantity other than a sum is employed; = *summation check* s.v. *SUMMATION[2] 6; so **sum-checked** *a.*

1962 R. V. Oakford *Introd. Electronic Data Processing Equipment* ii. 31 If a single *R* check bit is changed, the sum check will fail..in that row, but not in the four columns. **1972** *Computer Jrnl.* XV. 196/2 A similar routine deals with sum-checked binary input.

sum, *v.*[1] **1. a.** Delete 'Now *rare*' and add further examples.

1935 *Lancet* 11 May 1123/1 For the pig,..the combination of virus plus hæmophilic organism is much more potent than was to be expected from summing the mild diseases caused by the two agents acting separately. **1947** *Electronic Engin.* June 179/1 Suppose..that it is desired to sum the voltages from *n* sources. **1962** F. I. Ordway et al. *Basic Astronautics* vii. 325 The average lifetime..may be easily determined by summing all lifetimes and dividing by 100. **1971** *Nature* 24 Dec. 485/2 He summed data of six previous studies..and demonstrated in the total sample an over-representation of the last-but-one position. **1977** J. G. Graeme *Designing with Operational Amplifiers* vii. 177 The number of signals that can be summed is limited only by increasing circuit errors.

b. Restrict † to *pass.* and add later example of *intr.* use.

1966 G. C. Hemmens *Structure of Urban Activity Linkages* i. 6 The matrix of linkage coefficients is a stochastic matrix where each row sums to one.

Sumac, var. *SOUMAK.

sumach, sumac. 2. a. (Later examples.)

1936 W. Faulkner *Absalom, Absalom!* vi. 214 The old street of the slave quarters—a jungle of sumach and persimmon. **1965** A. Lurie *Nowhere City* (1966) xv. 162 The sumac held them back with its woolly, awkward stems. **1980** *Hunting Ann. 1981* 42/2 Here, the berries of black haw, sumac, bittersweet and greenbrier are important even in winter.

Sumatra. Add: **c.** A variety of tobacco yielding a light-coloured leaf.

1911 B. Miall tr. *Cabaton's Java, Sumatra, & other Islands of Dutch East Indies* xi. 229 The Manilla variety [of tobacco]..has been..less largely used than the Deli (Sumatra) tobacco. **1912** A. E. Tanner *Tobacco* xii. 85 Leaf used for making British cigars consists of Sumatra, Borneo, and Havana. **1969** *Times* 24 Nov. (Congo Suppl.) p. vi/6 The main varieties [of tobacco] grown are Kentucky, White Burley, Sumatra and heavy Dark Western. **1975** N. Freeling *What are Bugles blowing For?* xiv. 84 He sat..smoking a small Dutch cigar as a treat: rather heavy—they had mixed some dark Brazilian leaf in with the light Sumatra tobacco.

Sumatran, *a.* Add: **b. Sumatran tiger,** a small tiger belonging to the subspecies *Panthera tigris sumatræ*.

1908 *Proc. Zool. Soc.* 890 Mr. R. I. Pocock..exhibited photographs of a Sumatran Tiger, recently purchased by the Society. **1945** F. Harper *Extinct & Vanishing Mammals of Old World* 310 The Sumatran Tiger.., although less common than formerly, is still numerous in various districts. **1976** *Guardian* 19 Apr. 7/6 Six Sumatran tiger cubs were born at Whipsnade last year.

sumbitch (sʌ·mbitʃ). *U.S. slang.* Contraction of *SON OF A BITCH.

1975 O. Sela *Bengali Inheritance* xiii. 108 That sumbitch Winston would go far in Russia. **1976** M. Machlin *Pipeline* liii. 533 Play that 'I'm Dreaming of a Fat Paycheck'. That's a sumbitch! **1977** *Time* 7 Mar. 34/2 Strauss, you are a rich sumbitch. **1981** P. Mallory *Killing Matter* iv. 53 The sumbitch has sure got him a way with the womenfolk.

Sumero-. Add: **Sume·rogram,** a character or group of characters representing a Sumerian word, used in written Hittite (Akkadian, etc.) as a substitute for the equivalent (longer) word in that language.

1952 O. R. Gurney *Hittites* vi. 121 Hittite texts are liberally interspersed with purely Akkadian and Sumerian words, the latter usually written by single signs, the use of which as 'ideograms' (or better, 'Sumerograms') can often be recognized only by means of the context, for they may be the same signs that are normally used for mere syllables. **1965** J. Puhvel in W. Winter *Evidence for Laryngeals* 83 The Sumerogram used for *Hattusa-* reveals its meaning of 'Silver City'. **1983** *Trans. Philol. Soc.* 102 The increasing tendency to make use of Sumerograms and Akkadograms in place of syllabically written Hittite words.

‖ **sumi** (sū·mi). [Jap., = ink, blacking.] (See quot. 1958); = INDIAN INK. Cf. next.

1911 [see *dry brush* s.v. *DRY *a.* C. 3]. **1958** M. L. Wolf *Dict. Painting* 285 *Sumi*, Japanese ink or blacking, composed of a mixture of carbon and glue molded into sticks or cakes. When rubbed into water on an inkstone, it becomes the common medium of the painter and writer. **1970** *Globe & Mail* (Toronto) 26 Sept. 26/4 There's another show close to sellout in Kazuo Hamasaki's Japanese sumi paintings.

‖ **sumi-e** (sū·mie). Also **sumi-ee, sumiye, sumi-ye.** [Jap.] Japanese ink painting; also *collect.*, sumi pictures. Cf. prec.

1938 D. T. Suzuki *Zen Buddhism & its Influence on Jap. Culture* I. ii. 24 Calligraphy in the Far East is an art just as much as *sumiye* painting. **1960** H. Hayward *Antique Coll.* 271/1 *Sumi-ye*, Japanese ink-pictures, i.e. painted in black only. **1965** W. Swaan *Jap. Lantern* xiii. 146 Lessons in *sumi-e* (Japanese style, ink painting). **1977** *Time* 17 Jan. 34/3 Every cut of the chisel seems to possess the final, unlaboured rightness of a brush stroke by a master of *sumi-e* (ink painting). **1981** G. MacBeth *Kind of Treason* xx. 196 On the wall he'd hung the *kakemono*.., a thin scroll in *sumi-ee* with a house under a mountain.

summa. 5. b. (Later example.)

1941 Swenson & Lowrie tr. *Kierkegaard's Concluding Unscientific Postscript* II. ii. v. 528 If it is postulated and granted that it is easy to understand that God becomes a particular man, so that the difficulty first emerges in the next fact, that He becomes a lowly and despised man—then in *summa summarum* Christianity is humor.

summability (sʌmăbi·liti). [f. Summable *a.*: see *-BILITY.] The property of being summable.

1904 *Q. Jrnl. Pure & Appl. Math.* XXXV. 43 In the treatment of certain questions fundamental in the theory we gained nothing by the introduction of the idea of absolute summability. **1968** P. A. P. Moran *Introd. Probability Theory* viii. 351 The study of the relative strengths of various methods of summability of series.

‖ **summa cum laude** (sʌ·ma kʌm lǭ·di, su·ma kum lau·di, -e), *adv.* (*adj.*, *sb.*) *phr.* Chiefly *U.S.* [L., 'with highest praise'.] With highest distinction: designating a degree, diploma, etc., of the highest standard. Also *transf.* and *fig.* Occas. *ellipt.* as *summa*. Cf. *MAGNA CUM LAUDE.

1900 [see *MAGNA CUM LAUDE]. **1951** S. F. Nadel *Found. Social Anthropol.* i. 18 You may be said to have accomplished assimilation *summa cum laude*. **1962** *Listener* 16 Aug. 242/1 He was psychoanalyzed..presumably that Buddy calls one of those '*summa-cum-laude* thinkers and intellectual men's-room attendants'. **1970** G. Greer *Female Eunuch* 295 Mrs Friedan is a *summa cum laude* graduate of Smith College. **1976** *N.Y. Rev. Bks.* 24 June 4/2 When Commencement Day arrived, Mrs. Plath..[came] to watch her daughter being awarded her *summa cum laude* certificate. **1976** *Time* 20 Dec. 18/2 James R. Schlesinger..*summa cum laude* and Ph.D in economics at Harvard. **1977** *New Yorker* 19 Sept. 47/3 She had just graduated summa cum laude from the University of Washington, in Seattle. **1978** F. Maclean *Take Nine Spies* iv. 132 At Hamburg Sorge had taken his doctor's degree in political science *summa cum laude*. **1980** M. Babson *Dangerous to Know* vi. 40 An interview with one of the graduates, *summa cum laude*, of the Fat Farm. **1980** *Sci. Amer.* Aug. 35/1 He left Harvard three years later with a most irregular but steadily brilliant record, graduating in three years with an A.B. summa in chemistry.

summand. Delete *rare* and add further examples.

1943 *Mind* LII. 243 Even in logic, 'all' means something more... It means..completeness of the logical summands. **1964** E. A. Power *Introd. Quantum Electrodynamics* iv. 53 Suppose then $n_j \geq 1$, and then in each of the *N* summands on the right one gets a contribution of $\lambda_k = j$ for that particular summand. **1979** *Proc. London Math. Soc.* XXXVIII. 213 If (G,H) is a countably generated pair with $cd_R G \leq 1$, and $I_R G$ a direct summand of I_G, then $H_f G$.

summarizable (sʌ·măraizăb'l), *a.* [f. Summariz(e *v.* + -ABLE.] Capable of being summarized.

1970 *Nature* 23 May 774/2 In the last 18 pages chairmen attempt to summarize their sessions, but this is disappointing; it is not summarizable material. **1977** M. Cohen *Sensible Words* 139 Conventional intellectual historians who read merely for summarizable ideas.

summary, *sb.* Add: **4.** Special Comb.:

summary punch, a card punch that automatically punches the results obtained by a tabulator from a number of other cards; hence as *v. intr.*; **summary-punched** *a.*, **summary punching** *vbl. sb.*

1935 *Astron. Jrnl.* XLIV. 180/1 The wiring for the tabulator and summary punch is changed very little during the cycle. **1949** E. C. Berkeley *Giant Brains* iv. 50 The reproducer..can..summary punch, or copy totals or summaries obtained in the tabulator into blank cards in the reproducer. **1956** G. A. Montgomerie *Digital Calculating Machines* viii. 154 Automatic punches can also be connected to the tabulator to act as summary punches. **1957** N. Chaplin *Introd. Automatic Computers* xv. 341 Summary punching produces, by machine, cards that may contain variable and modified information derived from other cards. *Ibid.* 342 A summary punch machine..usually does not produce more than one hundred summary punched cards per minute. **1970** O. Dopping *Computers & Data Processing* iv. 75 The summary punch can punch information coming from the registers of the tabulator.

summate, *v.* Delete *rare* and substitute for def.: To add together or combine; *spec.* in *Physiol.*, with reference to nerve impulses, etc. Also *intr.* and *fig.*

1922 *Jrnl. Optical Soc. Amer.* VI. 550 When quite differently weighted, in terms of the relative powers of the three elementary processes to generate brilliance, the three chromatic curves should summate to yield the visibility curve. **1932** P. Bloomfield *Imaginary Worlds* xiv. 246 Happiness does not summate. The happiness of ten million individuals is not a millionfold the happiness of ten. **1935** *Discovery* May 140/1 In order to see more clearly in a bad light, we instinctively keep on blinking and peering so that the recurring slight pressures by the eyelids are, when summated, capable of evoking phosphenes. **1935** Winton & Bayliss *Human Physiol.* (ed. 2) ix. 349 Responses which are partially or completely super-imposed are said to summate. **1951** G. Humphrey *Thinking* i. 17 The implication that stimuli may be linearly summated is accepted by representative objective psychologists. **1957** *Encycl. Brit.* III. 866/1 Similar documents may be assembled and summated before they are journalized. **1962** W. Nowottny *Lang. Poets Use* iv. 78 The particulars which inhabit these schemes, though extraordinarily difficult to summate, permit themselves to be assimilated to a common ideogram of decline. **1970** *Jrnl. Gen. Psychol.* LXXXIII. 144 According to the second principle, two responses having the same form summate. **1971** A. C. Guyton *Basic Human Physiol.* vi. 63/2 Not only can discharges from separate presynaptic terminals summate with each other, but rapidly successive discharges from the same presynaptic terminal can also summate.

2. *trans.* To summarize.

1955 G. Gorer *Exploring Eng. Character* xiv. 269 If the 25 per cent of the population who say that they are influenced either regularly or occasionally by the advice of horoscopes are summated, one finds that there are very few categories where there is a variation of more than 3 per cent from the national norm. **1976** J. Bayley *Uses of Division* I. i. 24 It remained for Proust to summate the retrospective social novel.

Hence **summa·ted** *ppl. a.*

1938 J. Newton *Introd. Metallurgy* xiii. 406 In slag calculations use is sometimes made of 'summated' percentages by means of which oxides of similar chemical properties are grouped together and treated as a single constituent.

summation[2]. Add: **3. a.** (Further examples.) Now *esp.* such addition in an electronic device.

1962 M. G. Hartley *Introd. Electronic Analogue Computers* iii. 23 An arrangement for the summation of three voltages. **1977** J. G. Graeme *Designing with Operational Amplifiers* vii. 175 This characteristic makes possible signal summation and subtraction through the simple connection of summing or differencing resistors to the amplifier inputs. **1981** F. W. Hughes *Op Amp Handbk.* viii. 206 The output signal may be a direct mathematical summation of the input signals or may include a determined amount of gain.

b. Substitute for def.: The process or effect by which repeated or multiple nerve impulses can produce a response that each impulse alone would fail to produce. (Further examples.)

1956 A. C. Guyton *Textbk. Med. Physiol.* v. 45/1 If impulses occur too far apart in time..temporal summation will not occur. **1979** Spence & Mason *Human Anat. & Physiol.* xi. 293 During spatial summation, nerve impulses in many different stimulatory presynaptic cells travelling to a single postsynaptic cell may all arrive at the postsynaptic cell very close together in time.

c. *Psychol.* Cumulative action or effect (see quots.).

1921 E. J. Kempf *Psychopathol.* i. 62 The tendency to suppress our affections may accumulate; that is, a summation of the repressing or suppressing egoistic wishes may occur. **1924** J. Rivière et al. tr. *Freud's Coll. Papers* I. 95 An assumption which is not improbable in itself—namely, that a *noxia* such as coitus interruptus attains its effect by summation. According to the disposition of the person..a longer or shorter time will be required before the effect of this summation becomes evident. **1955** J. Strachey et al. tr. *Freud's Compl. Psychol. Wks.* II. ii. 174 Even a hysteric can retain a certain amount of affect that has not been dealt with; if, owing to the occurrence of similar provoking causes, that amount is increased by summation to a point beyond the subject's tolerance, the impetus to conversion is given.

6. *summation network*; **summation check** *Computers* = *sum check* s.v. *SUM *sb.*[1] 14.

1954 *Computers & Automation* Dec. 22/1 Summation check. **1969** Jordain & Breslau *Condensed Computer Encycl.* 498 One weakness of the summation check is its inability to detect transposed digits. **1968** D. Eadie *Introd. Basic Computer* xv. 347 In most analog computers the summation network is combined with an operational amplifier.

summative, *a.* Delete *rare* and add: (Later examples.) Also, cumulative, pertaining to accumulation.

?**1930** W. C. Williams *Sel. Essays* (1954) 103 We've got to experiment with technique long before the final summative artist arrives. **1931** *Brit. Jrnl. Psychol.* July 25 All such views of perception may be distinguished from summative or integrative theories by being called 'response' theories of perception. **1936** *Jrnl. Psychol.* II. 80 (*caption*) The summative efficiency of the samples. **1938** W. Benary in W. D. Ellis *Sourcebk. Gestalt Psychol.* viii. 105 In these examples brightness differences are the *reverse* of what a summative theory would have demanded. **1968** W. A. Scott in Lindzey & Aronson *Handbk. Social Psychol.* (ed. 2) II. xi. 218 We shall use the term *summative* to designate a scale that is scored by adding the response scores on its component items.

summatively (sʌmē̆ⁱ·tivli), *adv.* [f. Summative *a.* + -LY[2].] Additively, cumulatively.

1936 *Mind* XLV. 270 Everything that can be described 'organically' can also be described 'summatively'. It is simply a question of convenience. **1951** G. Humphrey *Thinking* iii. 103 Watt professes to hold..a contributory theory of mental energetics, one which derives motive power in the kind of experiment which he performed,..summatively from task and reproductive tendency. **1976** *Nature* 4 Mar. 59/1 Baylor *et al.* showed that the cones of the red-eared turtle, *Pseudemys scripta elegans*, are summatively and reciprocally coupled over distances up to 50 μm.

summator (sʌmē̆ⁱ·tǫɹ). [f. Summate *v.* + -OR.] **1.** *Electr. Engin.* That which sums; *spec.* a device which sums the analogue or digital information it receives. Cf. Integrator.

1930 *Engineering* 11 Apr. 482/1 The summator proper consists of two parts, a series of small dials giving the total kilowatt hours recorded by all the individual meters and larger dials, on which the maximum demand in kilowatts is aggregated. **1953** *Proc. Inst. Electr. Engineers* C. I. 44/1 The summator operates on the same principle of current balance as the telemeter and its error term is the same. **1974** *Jrnl. Appl. Physiol.* XXXVII. 748/1 A problem..is the inherently slow response time of the continuous discharge integrators (usually called analog summators, or merely integrators) used to supply this running average.

2. *Psychol.* In full, *verbal summator*: (see quots.).

1936 B. F. Skinner in *Jrnl. Psychol.* II. 71 The verbal summator is a device for repeating arbitrary samples of speech obtained by permuting and combining certain elemental speech-sounds. *Ibid.*, Apart from its use as a test, the summator is valuable in the study of other aspects of verbal behavior. *Ibid.* 73 The verbal summator..evokes latent verbal responses through summation with imitative responses to skeletal samples of speech. **1957** C. E. Osgood et al. in Saporta & Bastian *Psycholinguistics* (1961) 293/1 Skinner (1936) has devised a 'verbal summator' technique for studying language behavior... Samples of meaningless speech sounds are repeated until the subject perceives some meaningful form—a kind of verbal inkblot. **1970** *Jrnl. Gen. Psychol.* Oct. 143 Skinner hoped to measure the strength and relative importance of verbal responses and intended that the verbal summator, or Tautophone, as it was subsequently named, should become the instrument for doing so.

summer, *sb.*[1] Add: **4. a.** (*e*) (Further examples.)

1798 J. Woodforde *Diary* 11 June (1931) V. 121 Master Neville Custance called on us..being very lately come home from School for the Summer Vacation. **1875** Trollope *Prime Minister* (1876) I. xv. 237 The lawyer's regular summer vacation had not yet commenced. **1942** O. Nash *Good Intentions* 179 A summer cold Is to have and to hold. **1970** J. Creasey *Part for Policeman* vi. 53 What's the matter with him? Summer 'flu? **1975** *Times* 19 Apr. 9/2 Kathy had been in bed with a so-called summer cold..sniffing and sneezing. **1980** P. Harcourt *Tomorrow's Treason* I. i. 23 What with leave and summer flu, we're already short of staff. **1982** R. Timperley *Face in Leaves* i. 11 The long summer vacation was stretching out ahead of me.

(*g*) (Later examples.)

1873 H. James *Let.* 24 Mar. (1974) I. 355, I walk abroad in my summerest clothes and am warm. **1979** *Times of India* 17 Aug. 3/4 A wag remarks that half the city's population migrates to cooler climes during the 'summerest' month of May.

c. *summer apple* (example), *pearmain* (example), *pippin* (earlier example).

1795 J. Jay *Let.* 12 Dec. in *Columbia Lit. Columns* (1970) XIX. III. 43 Ten are Summer Pippins, a very large fair Yellow apple. **1870** J. W. McClung *Minnesota* xi. 154 Among the varieties [of apples]..are..Summer Pairmain, [etc.]. **1930** J. Dos Passos *42nd Parallel* II. 145 They ate sweet summerapples.

f. *U.S.* Designating tourists or those who visit a place for a summer holiday. Cf. *summer boarder*, sense 6 a below.

1886 *Leslie's Monthly* Feb. 203/1 Old Sampson don't like the Summer gentry. **1889** W. D. Howells *Hazard of New Fortunes* I. 135 She frankly gave up her house to the summer-folks (as they call them in the country). **1892** *Rep. Vermont Board Agric.* XII. 139 To these more prominent places may be added a multitude of..attractive homes to the summer guest. **1898** E. N. Westcott *David Harum* 286 Our friend had met quite a number of the 'summer people'. **1938** *Sun* (Baltimore) 24 Mar. 10/2 New England has been declining. Her rural areas are given over to a sort of subsistence farming or to the entertainment of 'summer people'. **1971** H. T. Walden *Anchorage Northeast* 19 So few 'summer people' are here that the term has little or no usage. **1977** *New Yorker* 10 Oct. 112/3 He is the native by the side of the road who, having been called stupid by the summer person exasperated at his inability to provide directions to Portland, says, 'Mebbee, but at least I ain't lost.' **1980** J. Coates *Sentimental Education* 124 She belonged to the town—she was not one of the summer people.

5. Indirect objective, *summer-going* adj.; instrumental, *summer-painted, -soothed, -tranced* adjs.; similative, *summer-happy, -kind, -merry, -sweet* adjs.; 'in or during summer', *summer-basking, -born, -green, -idle, -opened, -running, -shaded, -still, white* adjs.

1931 R. Graves *Poems 1926–30* 69 You are no more than weather, The year's unsteadfastness To which, now summer-basking,.. The mind pays no honour. **1975** *Language for Life* (Dept. Educ. & Sci.) xviii. 267 Many children..are likely to continue to need special help in the junior school, particularly those summer-born children who may have had only two years of early schooling. **1954** J. Betjeman *Few Late Chrysanthemums* 43 Oh sun upon the summer-going by-pass Where ev'rything is speeding to the sea. **1930** J. Dos Passos *42nd Parallel* 137 There was a blue haze at the end of every street of brick houses and dark summergreen trees. **1917** D. H. Lawrence *Look! We have come Through!* 104 And we're going to be summer-happy And summer-kind. **1955** E. Bowen *World of Love* iv. 67 The summer-idle water dawdled in shallows. **1917** Summer-kind [see *summer-happy* above]. **1957** E. Blunden *Poems of Many Years* 279 By the arched grey bridge of summer-merry streams. **1887** J. R. Lowell in *Atlantic Monthly* Feb. 250 And listen while Old Hundred pours Forth through summer-opened doors. **1937** E. Muir *Coll. Poems* (1960) 80 The lint-white stubble plain From which the summer-painted birds have flown A year's life on. **1972** *Trout & Salmon* Feb. 10/2 Clearly the nets are taking an excessive proportion of summer-running salmon. **1850** J. G. Whittier *Poet. Wks.* (1898) 340/1 Down the summer-shaded street A wasted female figure..Came rushing. **1883** R. Bridges *Prometheus the Firegiver* 37 Piloting over the wind-dappled blue Of the summer-soothed Aegean. **1925** A. Huxley *Sel. Poems* 38, I am a pool of waters, summer-still. **1945** W. de la Mare *Burning-Glass* 42 Summer-sweet as that wild rose. **1881** O. Wilde *Poems* 66 We too might waste the summer-tranced day. **1918** D. H. Lawrence *New Poems* 9 The flagged, clean pavement summer-white.

6. a. summer boarder *U.S.*, one who lives at a boarding-house in the country in summer; hence **summer-board** *v. trans.*, to take (someone) as a summer boarder; **summer-boarding; summer camp** orig. and chiefly *U.S.*, a camp providing recreational and sporting facilities during the summer holiday period, usu. for children; **summer cottage** *N. Amer.*, a cottage, usu. at a holiday resort or in the country, occupied during the summer; hence **summer cottager,** one who occupies a summer cottage; **summer country** *N.Z.* (see quot. 1898); **summer eggs** (examples); **summer kitchen** *N. Amer.*, an extra kitchen, adjoining a house or separate from it, used for cooking in hot weather; **summer-long** *adv.* and *adj.*, (lasting) throughout the summer; **summer master** *Canad. Hist.*, a person in charge of a trading post for the summer only; **summer mastitis,** a severe inflammation of the udder of cows usu. associated with the bacteria *Corynebacterium pyogenes* or *Peptococcus indolicus*; **summer pruning,** the selective cutting back of branches of trees or shrubs during the growing season; hence **summer prune** *v.*; **summer-pruned** *ppl. a.*; **summer pudding,** a pudding made of stewed fruit (freq. raspberries and red currants) and bread; **summer resort,** a popular place of resort in the summer, esp. a summer holiday resort; also, the act of visiting such a place; **summer resorter** *U.S.*, one who frequents summer resorts; **summer road** *Canad.*, a road suitable for use all year round, as opp. to one used in winter only by sleighs; **summer sale,** a sale of merchandise at reduced prices in the summer, esp. by shops wishing to clear their seasonal stock; **summer sausage** *U.S.*, a type of dried or smoked sausage which can be made in winter and kept until summer; **summer school,** a school or course of education conducted by a university, etc., in the summer, esp. during the long vacation; **summer stock** *U.S.*, theatrical productions by a repertory company organized for the summer season, esp. at holiday resorts, freq. *attrib.*; **summer term,** that term of an academic year or of legal sessions which occurs before the summer vacation; **summer theatre,** a theatre operating only in summer; **summer-tilth** *dial.*, fallow land; the cultivation of such land; **summer-weight** *a.*, of clothes: light, suitable for wear in summer; also *transf.*; **summer wood** = *late wood* s.v. *LATE *a.*[1] 4.

1903 K. D. Wiggin *Rebecca* x. 107 Mother has summer-boarded a lot o' the school-marms. 1847 H. N. Moore *Fitzgerald & Hopkins* 73 And stated also that there were several summer boarders from the city present. 1879 *Harper's Mag.* July 164 A few quiet summer boarders took shelter for a season's rest. 1897 [in Dict., sense 4 a (*f*)]. 1949 *Sat. Even. Post* 25 June 47/2 At the end of one unusually arduous summer he put an ad in a Portland paper for summer boarders. 1880 *Harper's Mag.* Sept. 536/1 Summer boarding here can be had for one dollar per week. 1893 *McClure's Mag.* I. 242/2 The camp was founded by Mr. Ernest Berkeley Balch as a summer camp for boys. 1948 *Sat. Even. Post* 23 Oct. 87/2 He wants to send every youngster in Lawrence to summer camp for at least two weeks. 1958 R. Liddell *Morea* III. ii. 238 There [Cerigo] monasteries are, regrettably, regarded merely as summer camps for visitors. 1979 *Country Life* 24 May 1640/1 At the age of 14..I was packed off to a summer camp in the Welsh hills. 1840 *Montreal Transcript* 22 Dec. 402/2 Some owners of lots also propose putting up summer cottages. 1902 W. D. Howells *Literature & Life* 49 A few houses of the past remain, but the type of the summer cottage has impressed itself upon all the later building, and the native is passing architecturally, if not personally, into abeyance. 1958 *Edmonton Jrnl.* 28 June 25/1 Schools and universities are closing their doors for the next few months and many Canadian households will begin the annual exodus to summer cottage or camp. 1948 *Chicago Tribune* 20 June VII. 12/5 Many summer cottagers will be happy to know that the same house makes a similar type of cream that repels chiggers. 1971 *Islander* (Victoria, B.C.) 2 May 6/1 In this strange fantasyland live 300 permanent residents and another 3,200 summer cottagers. 1898 Morris *Austral Eng.* 444/2 *Summer country, n.*, in New Zealand (South Island), country which can be used in summer only; mountain land in Otago and Canterbury, above a certain level. 1922 W. Perry et al. *Sheep Farming in N.Z.* vii. 88 The higher country..which is likely to hold snow to some depth in the winter months, is termed 'summer country'. 1947 P. Newton *Wayleggo* (1949) 14 A large proportion of the country [in the South Island]—the shady and hindermost areas—is suitable for summer grazing [of sheep] only... Such country is known as 'summer country'. 1884 A. Sedgwick tr. *Claus's Elem. Text-bk. Zool.* x. 418 The so-called summer eggs..produce generations containing no males. 1952 J. Clegg *Freshwater Life Brit. Isles* xii. 169 These so-called 'summer eggs' are laid, perhaps twenty or more at a time. 1874 *Southern Mag.* XIV. 124 There was Charley's wife..flitting about from house to summer-kitchen. 1939 H. M. Miner *St. Denis* ii. 25 Airy summer kitchens, which do not retain the heat of the stove, are built onto the sides of the houses. Too exposed to be warm, these annex kitchens are evacuated in winter. 1924 E. Sitwell *Sleeping Beauty* xxvi. 95 When the thickest gold will thrive Summer-long in the combs of the honey-hive. 1960 C. Day Lewis *Buried Day* ii. 31 On and on droned the voices, blending slumbrously with..the summer-long hum of insects. 1980 *Beautiful Brit. Columbia* Summer 39 In the summer, you may examine thousands of items at the summer-long Crafts Centre. 1913 I. Cowie *Company of Adventurers* 228 Many of these journals were kept by a 'summer master', who was quite often a very illiterate laborer, who could barely scrawl phonetics in the book during the real master's absence on the annual voyage to and from headquarters with the furs and for the outfit. 1967 A. M. Johnson in *Saskatchewan Jrnls.* (Hudson's Bay Rec. Soc.) p. xxviii, He sent Bird to Buckingham House with instructions to leave the summer master in charge there. 1934 R. G. Linton *Vet. Hygiene* (ed. 2) vi. 446 The well-known suppurative form of mastitis..is especially prone to attack dry cows and virgin heifers during the summer months... This form is often referred to as epidemic mastitis or summer mastitis. 1970 W. H. Parker *Health & Dis. Farm Animals* xv. 212 Infection of a dry cow or unbred heifer with..summer mastitis, is as common in beef as in dairy breeds. 1786 Summer prune *v.* [in Dict., sense 5 a]. 1980 V. Canning *Fall from Grace* vii. 118 They summer pruned the wistaria. 1960 *News Chron.* 6 Aug. 6/4 The summer-pruned laterals are further shortened. 1707 J. Mortimer *Whole Art of Husbandry* xvii. 396 To the Boughs that put out in Spring, give a Summer pruning a little after Midsummer. 1725 [in Dict., sense 4 a (*e*)]. 1806 W. Pontey *Forest Pruner* 235 As a general rule, we think summer is preferable to winter-pruning. 1895 *Meehan's Monthly* May 87/1 Summer pruning is especially effective with coniferous trees... One who understands this business of summer pruning an evergreen can so manage that the tree forms an absolutely perfect specimen. 1972 G. E. Brown *Pruning Trees, Shrubs & Conifers* iii. 50 Summer pruning..promotes spur formation. 1933 E. C. Carver *Pract. Catering* vi. 114 Summer pudding. Thin slices of stale bread, stewed fruit... Serve with cream or custard. 1974 P. Haines *Tea at Gunter's* xx. 206 Heaping my plate with summer pudding..I looked at the bread on my plate, oozing deep crimson juice. 1832 *Louisville* (Kentucky) *Public Advertiser* 12 July 3/5 He has prepared his House and Garden at the lower end of Jefferson Street, for the purpose of making it a general *Summer Resort*. 1846 *Chambers's Miscellany* XIV. cxxi. 32 Musselburgh,..another pleasing summer resort, is situated two miles eastward. 1853 E. T. Turnerelli *Kazan* II. i. 4 This village is a favourite place of summer-resort for the inhabitants. 1873 J. H. Beadle *Undevel. West* xv. 257 For a summer resort one can spend weeks very pleasantly there. 1882 G. W. Peck *Peck's Sunshine* (1883) 125 He said he should at once begin..by boarding at a summer resort hotel. 1974 *Times* 12 Nov. 14/1 Mr and Mrs Ronald Heywood own a 56-bedroom two star hotel in a summer resort on the east coast. 1889 *Advance* (Chicago) 19 Sept. 673/3 At Astoria the summer resorters distribute themselves to the various beaches. 1907 'Mark Twain' in *N. Amer. Rev.* Nov. 327 They respected these elegant summer-resorters. 1820 S. H. Wilcocke in L. F. R. Masson *Les Bourgeois de la Compagnie du Nord-Ouest* (1890) II. 224 With the summer road they were acquainted and that, therefore, they followed. 1909 *Gow Ganda* (Ontario) *Tribune* 17 Apr. 6/2 What will be the cry on the summer roads when we reach those points where the dense forest and rocks obstructs the view ahead? 1974 E. C. Stacey *Peace Country Heritage* i. 7 A few farmers used the..summer road. 1899 J. F. Fraser *Round World on Bicycle* xxvi. 324 All the millinery shops in Oxford Street begin their early summer sales or spring-clearance sales. 1923 A. Huxley *Antic Hay* xvi. 223 If I wait till the summer sale, the crêpe de Chine will be reduced by at least two shillings. 1976 *Times* 2 Aug. 16/3 The usual summer sales hiatus. 1893 F. E. Rhorer *Meat Man's Friend* 33 By making summer sausage the same as above, but allowing the meat to be very coarse, it is called Salami. 1965 *House & Garden* Jan. 60 Summer sausage or Thüringer. These terms are interchangeable with dried cervelas. In fact, all dried sausages of this type are called summer sausage. 1976 T. Gifford *Cavanaugh Quest* (1977) x. 181 She sliced thick chunks of summer sausage. 1860 J. C. Patteson *Jrnl.* Sept. in C. M. Yonge *Life J. C. Patteson* (1874) I. ix. 473 In taking away natives to the summer school, it must be understood that some..are taken..merely to teach us their languages. 1871 E. Eggleston *Hoosier Schoolmaster* i You might teach a summer school. 1919 M. Beer *Hist. Brit. Socialism* II. iv. xiv. 294 In 1906 a Fabian Summer School was established. 1967 B. Jefferis *One Black Summer* (1968) i. 1 The grounds and buildings would be full of summer school students: doctors who longed to pot; dressmakers who yearned to try their hands at sculpture. 1971 *Daily Tel.* (Colour Suppl.) 3 Dec. 9/2 The lecturer..led his summer school audience down the howling avenues of Joycean puns. 1981 V. Glendinning *Edith Sitwell* xvi. 205 In August Edith had lectured..at a summer school in Cambridge. 1942 Berrey & Van den Bark *Amer. Thes. Slang* § 587/4 *Straw hat*, a summer stock theater, in which plays are tried out. 1955 J. P. Donleavy *Ginger Man* vii. 64, I was once approached by a talent scout in summer stock. 1965 *New Statesman* 2 July 20/1 There is a very funny story about Maury Stein, a Summer Stock actor at Indian Lake. 1977 I. Shaw *Beggarman, Thief* III. vi. 262 'Where've you acted before?'..'Well..noplace.'..'Not even summer stock?' 1853 Root & Lombard *Songs of Yale* 4 Presentation Day is the sixth Wednesday of the Summer Term, when the graduating Class..are presented to the President as qualified for the first degree, or the A.B. 1859 J. A. Symonds *Let.* Feb. (1967) I. 181, I always connect it in my mind with that interminable Harrow Summer Term. 1922 *Times* 11 Oct. 11/5 During the last weeks of the Summer Term, at the request of the Lord Chancellor, I undertook the trial of undefended suits for divorce, and heard about four hundred cases. 1980 C. Fremlin *With no Crying* ii. 8 It looked like being the best summer term ever... O-levels were still a full year away. 1801 *Monthly Mirror* June 414 'Make hay while the sun shines,' has been found a most salutary maxim at the summer theatres. 1938 L. Bemelmans *Life Class* II. vii. 189 They were.. Bavaria's greatest peasant actors... Their theater, part of the inn, was not the usual..summer theater, a converted old barn, but a real theater. 1981 N. Crisp *Festival* i. 15 Who in their right mind..would have dreamed of a summer theatre at..a somewhat shabby would-be genteel spa. 1818 in Thirsk & Imray *Suffolk Farming 19th Cent.* (1958) 104 To leave all the muck, dung and compost made the last year and all hay, clover hay and summertilths. 1903 in G. E. Evans *Farm & Village* (1969) 160 Beans and Peas to be twice clean hoed or a clean summertilth. 1970 in —— *Where Beards wag All* viii. 89 Ploughing a long fallow or summer-tilth was a very hard and slow job for the man and his horses. 1883 *Graphic* 14 Apr. (Advt., rear cover), Youth's overcoat, summer weight. 1931 *Daily Tel.* 22 May 9/6 Summer-weight weaves in hopsack, tweed, and knitted mixtures. 1968 A. Diment *Bang Bang Birds* v. 66 It's hell trying to keep a crease in bottle green, summer-weight cavalry twill. 1977 *Time* 27 June 46/2 The story also has some pretty serious problems, or perhaps more accurately, some puzzling aspects for what is intended as summer-weight entertainment. 1896 W. R. Fisher in W. Schlich *Man. Forestry* V. i. 6 It [*sc.* spring-wood] contains less woody substance than the summer- or autumn-wood of the same annual zone. 1930 *Forestry* IV. 10 The greater length of the summer wood tracheids of the Sitka spruce is in accordance with the observations of Lee and Smith. 1982 *Sci. Amer.* July 35/2 These make the directly visible springwood ring, followed once the tree is great with leaf by a wider, denser, darker ring of mixed fibrous growth and small summerwood vessels.

b. summer crookneck, a small yellow or orange summer squash with a curved neck; **summer grape** (later examples); **summer squash,** substitute for def.: any of several varieties of the gourd *Cucurbita pepo* whose fruits are eaten young; (examples).

1890 *Amer. Naturalist* XXIV. 731 Summer crooknecks appeared in our garden catalogues in 1828. 1969 *Oxf. Bk. Food Plants* 122/1 'Summer Crookneck'..has bright yellow or orange, warty fruits, shaped like a crooked club. 1834 J. J. Audubon *Ornith. Biogr.* II. 92 The Summer Grape..occurs in all the barren lands of the Western Country. 1949 *Amer. Photography* Apr. 244/3 The summer grape is somewhat similar to the blue grape. 1815 W. Bentley *Jrnl.* 14 Aug. (1914) IV. 346 A more free use has been made of the summer squash than ever before known. 1902 *Harper's Bazaar* Sept. 766 There was nothing in her larder except a summer-squash pie. 1981 *Farmstead Mag.* Winter 37/1 Winter squash, of course, shares space in seed catalogs with its sister vegetable—the summer squash.

summer, *sb.*[3] Add: **1.** (Later example of *summer-up*.)

1960 J. Bayley *Characters of Love* iii. 130 Here the confident summer-up of Othello might become a little uneasy.

2. *Electronics.* A circuit or device that produces an output dependent on the sum of two or more inputs or of multiples of them.

1958 W. J. Karplus *Analog Simulation* ix. 234 Since the output voltage is proportional to the sum of the input voltages, this circuit is termed 'summer'. 1968 Passmore & Robson *Compan. Med. Stud.* I. ii. 5 The summer would have many input voltages, each one representing the factors for heat gain..or the heat loss. 1981 R. G. Irvine *Operational Amplifier Characteristics* vii. 176 The gain of this circuit may be changed from unity by modifying the value of the feedback resistor on the inverting summer.

summer-game. 2. (Earlier example.)

1859 G. W. Matsell *Vocabulum* 117 *Summer game*, playing merely for amusement. *Summer game*, playing a game for the benefit of another person with his money.

summerize, *v.* Restrict *nonce-wd.* to sense in Dict. and add: **1.** (Later example.)

1891 H. James *Let.* 7 June (1981) III. 342 The Curtises go to India—or believe they do—in the autumn; so I suppose they summerize at the Barbaro.

2. *trans.* To prepare (something) for summer. Also *intr.* for *refl.* *U.S. colloq.*

1935 *Evening Sun* (Baltimore) 3 Apr. 9 Let Hutzler's summerize your home. 1949 *Sun* (Baltimore) 2 June 4 Come to K. Katz and 'summerize'! 1962 *Harper Motors, Inc.* (Charlottesville, Va.) *Advertising Let.*, This time of year you're probably deluged with offers from service stations to 'summerize' your car—that is, drain the anti-freeze, check the motor, and get ready for summer driving.

Hence **summeriza·tion,** the act or process of preparing a thing for summer.

1974 *Old Times* (Upper Canada College) Autumn 8 With..the summerization of the Patrick Johnson rink.. the quality of extracurricular programs..is greatly enhanced.

summer-land, summerland, *sb.* **2.** (Earlier examples.)

1861 *Herald of Progress* (N.Y.) 13 July 3/4 (*heading*) Tidings from the summer land. A plea in behalf of little children. 1869 *Spiritualist* 17 Dec. 19/3 He found that he could pass through the upper air with ease, and at last they reached what the Spiritualists call the 'Summer Land', but in reality the compound essence of seventeen summers distilled into one, would not equal it in loveliness. 1890 W. James *Princ. Psychol.* I. x. 394 The odd thing is that persons unexposed to spiritualist traditions will so often act in the same way when they become entranced, speak in the name of the departed,..send messages about their happy home in the summer-land, and describe the ailments of those present.

summers (sʌ·məɹz), *adv.* *U.S.* [f. Summer *sb.*[1] + -s.] During the summer; each summer (for a number of years).

1907 'Mark Twain' *Christian Sci.* II. viii. 235 It [*sc.* a local Christian Science church] can appoint its own fan-distributors, summers. 1936 H. W. Horwill *S.P.E. Tract* XLV. 192 A peculiar use of the plural form is illustrated in..'A niece of theirs had earned her way through College by waiting on the Atkins-Smythes' table, summers'. This usage is said to be a relic of the old adverbial genitive. 1976 *National Observer* (U.S.) 1 May 8/2 The limits have riled Pevsner since college days, when he worked summers in Europe.

summer season. Restrict *rare* to sense in Dict. and add: **2.** A period in summer for which people are employed in connection with seasonal or holiday entertainment, trade, etc.

1952 W. Granville *Dict. Theatrical Terms* 159 *Seasonal shop*, an engagement for the summer season in, say, a concert party or in a touring company that visits theatres which open only in the summer months. 1973 *Melody Maker* 4 Aug. 50/6 The Teign Valley Stompers from Teignmouth, Devon, who are at present enjoying a summer season on Pontin's Holiday Camps' Devon circuit, have engaged a new trombonist.

summer-time. 2. Add: (Later examples.) Now adopted in the U.K. for daylight saving from March to October (see quot. 1982). Cf. *British Summer Time* s.v. *British *a.* 5; *double summer-time* s.v. *double *a.* 6.

1916 *Times* 26 Aug. 7/2 Of the changes which have already proved themselves to be changes for the better, that which immediately affects the greatest number of people is the introduction of 'summer time'. 1937 D. L. Sayers *Busman's Honeymoon* vii. 158 October 2nd—sun would be setting about half-past five. No, it was Summer Time. Say half-past six. 1967 [see *British Standard Time* s.v. *British *a.* 5]. 1982 *Whitaker's Almanack 1983* 142 In the United Kingdom, Summer Time, one hour in advance of G.M.T. will be kept between 01h G.M.T. on the last Sunday in March and 01h G.M.T. on the day following the fourth Saturday in October. Thus, in 1983, Summer Time will be in force between March 27 and October 23.

summing, *ppl. a.* Add: **2.** *Electronics.* That performs summation; producing an output dependent on the sum of the inputs.

1948 *Electronics* Apr. 124/3 The summing amplifier..is widely used in d-c and a-c servomechanisms. 1960 Rogers & Connolly *Analog Computation in Engin. Design* ii. 13 When used in this manner, the operational amplifier is known as a summing amplifier. 1963 B. Fozard *Instrumentation Nucl. Reactors* ix. 107 The point *b* corresponds to the summing junction in a computing

amplifier. 1967 *Electronics* 6 Mar. 120/2 A network between the output of a_1 and the summing junction of a_2 can be set to bring the total loop phase shift to 360°. 1981 F. W. Hughes *Op Amp Handbk.* viii. 208 The inverting AC summing amplifier is similar to the inverting DC summing amplifier, except for the input capacitors.

summit, *sb.* Add: **3. b.** The highest level, *spec.* with reference to politics and international relations; also *ellipt.* for *summit conference, meeting*, etc., sense 4 below.

1950 W. S. Churchill in *Times* 15 Feb. 4/2 It is not easy to see how things could be worsened by a parley at the summit, if such a thing were possible. 1955 *Newsweek* 11 Apr. 44/1 Only if the Big Four Foreign Ministers reached 'a substantial measure of agreement' would a further conference be convened—at the summit. 1957 P. Frank *Seven Days to Never* i. 33 We haven't knuckled under, not at the Summit or anywhere else, and..the alliance stands. 1958 *Listener* 14 Aug. 220/2 Then came the Czestochowa raid; the decision for this must have been taken at the summit. 1959 *Economist* 11 July 92/2 While an agreement not to mention time limits may be enough to get from the foreign ministers' level to the summit, to reach a settlement there will require something more. 1967 *Spectator* 30 June 757/1 The most certain result of the Glassboro summit, in fact, is no more than that Mr. Johnson's standing at home is now rather higher. 1978 R. Ludlum *Holcroft Covenant* xxxi. 365 Since I left Brazil, I've not owned a weapon... I should like to have one now. Only for the duration of the summit.

4. summit meeting, a meeting between heads of government, etc., to discuss matters of international significance (cf. sense 3 b above); also *transf.*; similarly *summit conference, talks*.

1955 *Times* 23 June 8/3 The senator's resolution demanding that the United States should refuse to attend the 'summit' conference. 1959 *Encounter* Aug. 33/2, I was running out of pennies, and had to have a summit conference with the [telephone] operator. 1977 *Whitaker's Almanack 1978* 590/1 A unified political command for Egypt, Syria, and the Sudan was agreed at the end of the two-day tripartite summit conference in Khartoum. 1955 *N.Y. Times* 5 May 2/5, I say at this moment I see no reason for that summit meeting. 1963 *Ann. Reg. 1962* 208 Krushchev suggested a Summit meeting. 1977 Summit meeting [see SSRC s.v. *S 4 a]. 1955 *Times* 7 May 8/7 There are certainly no indications that Washington has modified its resistance to 'summit talks' with Russia in advance of preparatory soundings.

summit, *v.*[2] [f. *summit *sb.* 3 b.] *intr.* To take part in summit meetings.

1972 *Time* 5 June 40 Prime Minister Indira Gandhi is willing to summit with the chap (probably at the end of the month). 1973 *New Scientist* 5 July 30 Nixon the President, summiting and clowning with the visiting Brezhnev, and Nixon, the suspect, seeking to elude the Watergate noose. 1979 *Daily Mail* 23 Jan. 5/2 When he is not summiting in the sunshine there is apparently nothing like All Creatures Great and Small..to help him forget crumbling pay norms and secondary picketing.

summitee·r. [f. *summit *sb.* 3 b + -eer.] One who takes part in summit meetings. Hence **summitee·ring.**

1957 *Time* 16 Dec. 21/2 The man who..must lead NATO along the course the summiteers lay down. 1958 *Daily Mail* 31 July 4/3 Macmillan has emerged as a worthy Summiteer. 1962 *Observer* 18 Feb. 11/7 No one knows either what the country thinks about the pay pause or Mr. Macmillan's summiteering. 1978 *Guardian Weekly* 23 July 6 Hot from their discussions on how to save the world from an energy crisis, the six summiteers repaired to the 14th century Gymnich Castle, near Bonn, for dinner. 1982 *Economist* 20 Mar. 12/1 This month's summiteers are likely to turn a blind eye to the subject.

su·mmitry. [f. *summit *sb.* 3 b + -ry.] The practice of convening or holding summit meetings, or of using them as a diplomatic device.

1958 *Economist* 8 Feb. 479/1 The Western dislike of time-wasting summitry is due..to a feeling that even an inconclusive get-together would fill the democracies with a false sense of security. 1967 *Spectator* 28 July 97/1 Mr. Macmillan likened the preliminaries to the summitry so dear to his heart to a stately minuet. 1972 Ld. Gladwyn *Mem.* xvi. 276 The so-called science of 'Summitry' was now being pursued with zeal and intelligence by the Foreign Office. 1979 *Time* 2 Apr. 26/1 Jimmy Carter initiated his most stimulating success as President six months ago: his summitry that broke 30 years of bloodshed and stalemate to make possible a peace treaty between Egypt and Israel.

summoned, *ppl. a.* Add: Also in comb. with adv., as *summoned-up*.

1977 *New Yorker* 19 Sept. 108/2 Calling beautiful coaches for the harshly summoned-up emergency.

Sumner[2] (sʌ·mnəɹ). The name of Thomas H. *Sumner* (1807–76), U.S. shipmaster, used *attrib.*, in the possessive, and *absol.*, with reference to a method devised by him in 1837 of finding one's position on the surface of the earth, employing an approximate value of latitude or longitude based on dead reckoning, in conjunction with an astronomical observation, to calculate a number of positions that define a line that must contain the true position; so *Sumner line* = *position line* s.v. *position *sb.* 7 b.

1849 H. Raper *Pract. of Navigation & Naut. Astron.* (ed. 3) 345 (*heading*) Position on a line of bearing. [*Note*] Or 'Sumner's Method'. 1881 S. T. S. Lecky *'Wrinkles' in Pract. Navig.* II. viii. 201 Unless the error of the latitude is greater than that assumed, the ship must be somewhere on this 'Line of position', which, for convenience, will henceforth in these pages be termed a 'Sumner line', after the American seaman who first brought this useful problem prominently to the notice of the profession. 1901 J. R. Walker *Explanation of 'New Navigation'* 8 The straight line is called the Sumner Line, or Line of Position. 1919 [see *position *sb.* 7 b]. 1924 R. Clements *Gipsy of Horn* xii. 228, I worked a Sumner, or position by double altitude. 1976 *Oxf. Compan. Ships & Sea* 845 *Sumner's position line*, a systemized method of finding a ship's position by means of a sight.

sumo (sū·mo). [Jap.] In Japan, a form of wrestling in which a wrestler wins a bout by forcing his opponent outside a circle or making him touch the ground with any part of his body except the soles of his feet. Freq. *attrib.*, esp. as *sumo wrestler, wrestling*; also *absol.*, a sumo wrestler.

1880 W. E. L. Keeling *Tourists' Guide Yokohama* 23 The wrestlers (*sumô*)..will not fail to interest him. 1893 *Jap. Soc. Trans. & Proc.* I. 19, I have seen English wrestling, and found it similar to Japanese *wrestling* (*Sumô*, not *Ju-jitsu*). 1923 J. Street *Mysterious Japan* ix. 103 The kind of wrestling known as *sumo* still maintains its ancient prestige as the national sport. 1934 [see *all-in 2]. 1936 K. Nohara *True Face of Japan* v. 220 Our taste for prodigies..is gratified by the corpulence of the *Sumo*. 1938 Bush & Kagami *Japanalia* 156/1 *Sumô* wrestlers are huge fellows. 1958 *Times* 27 Dec. 7/6 Such are the advance preliminaries of a bout of *sumo*, the national sport of Japan. 1964 I. Fleming *You only live Twice* i. 17 It is only the *sumo* wrestlers who drink *saké* in these quantities without showing it. 1966 *New Scientist* 28 July 182/3 Sumo wrestling, in which two monstrous men charge one another, clinch briefly and separate, with one the winner, usually in the space of a few seconds. 1974 *Daily Tel.* (Colour Suppl.) 22 Feb. 39/3 If one is to understand Japan in any depth, a realisation of the significance and the enjoyment of Sumo is as important as it is to appreciate the influence of soccer if one is to understand Britain. 1977 *Time* 4 July 52/2 Surpassing even such traditional Japanese sports as sumo wrestling, *bēsubōru* has become Japan's favorite sport. 1978 M. Kenyon *Deep Pocket* x. 125 This character was a sumo-wrestler, hewn from a cliff-face.

sumotori (sumotō·ri). [Jap., f. prec. + *tori* active partner in the performance of techniques.] A sumo wrestler.

1973 *Newsweek* 13 Aug. 92 Anyone who [tries] socking a sumotori in the stomach will gladly go back to brick walls. 1974 *Daily Tel.* (Colour Suppl.) 22 Feb. 39/3 The mature *sumotori* (as Sumo wrestlers are called) is about six feet tall and will weigh anything from 16 to 24 stone.

sump, *sb.* Add: **2. a.** (Further examples.) Also *fig.*

1963 T. & P. Morris *Pentonville* iii. 69 Pentonville represents one of the sumps of the English prison system; a receptacle into which the sludge is continuously drained. 1969 *Gloss. Terms Water Cooling Towers* (*B.S.I.*) 6 *Sump*, a lowered portion of the cold water basin floor for draining down purposes. 1975 *Sci. Amer.* Oct. 23/3 As fast as the heavy water leaked out it was collected in a sump and pumped directly back into the reactor.

c. A depression in the bottom of the crank-case of an internal-combustion engine, which serves as a reservoir of lubricating oil.

1907 *Westm. Gaz.* 9 Nov. 14/3 The oil is forced by a gear-driven pump from a sump in the crank-chamber. 1929 [see *skew gear* s.v. *skew *a.* 2]. 1950 *Brit. Repair Man.: Cars* 38/2 The oil filling orifice is housed in the valve top cover, and the sump, which is a steel pressing, has a capacity of 7½ pints. 1980 J. McClure *Blood of Englishman* i. 9 Droopy was removing the sump... He..extended a hand for a No. 8 ring spanner.

5. sump guard, a cowling for protecting the sump of a motor vehicle from perforation on poor roads.

1968 *Guardian* 6 May 5/5 The 1800 I was driving was standard except for a sump guard, essential on East African roads. 1980 J. Barnett *Palmprint* vii. 62 A heavy stone clanged against the sumpguard.

sump'n (sʌmᵖn). Also **somepin, sumpin,** etc. Repr. colloq. (chiefly U.S., esp. Blacks') pronunc. of Something *sb.*, (*adj.*,) and *adv.*

1880 [see *kin var. Can *v.*[1]]. 1882 *Indianapolis Jrnl.* 5 Aug. 4/6 They's somepin kindo' hearty-like about the atmosphere. 1929 *Amer. Mercury* Sept. 50/2 Done sumpin' to fine captain one time didn't aim to do. 1938 C. Himes *Black on Black* (1973) 167 Ef'n yo' is God, den gimme somp'n tuh eat. 1951 X. Herbert in Murdoch & Drake-Brockman *Austral. Short Stories* 298 Git to jiggery out of it, you stinkin' rottin' black sumpen. 1961 Wodehouse *Ice in Bedroom* v. 41 Why not Heels Incorporated or Doublecrossers Limited or sump'n? 1976 *National Observer* (U.S.) 17 Apr. 17/1, I just want to tell him sump'n!

sumpter, *sb.* **3.** For † *Obs.* read 'Now *rare*' and add later example.

1958 L. Durrell *Balthazar* ii. 32 A liquid-eyed camel.. humped down the narrow street threatening to knock us down with its bulging sumpters of *bercim*.

sun, *sb.* Add: **I. 1. e.** (*i*) *the sun is over the foreyard* (*Naut.*): it is noon (the time at which the first drink of the day is taken).

1844 [see Fore-yard[2] 1]. 1862 'Vanderdecken' *Yacht Sailor* ix. 123 It will be a favourable time to 'make the sun over the foreyard', and serve out grog in moderation to all hands. 1903 H. Holmes *Life & Adventures* 11 The sun's over the fore yard; no doubt they have spliced the main brace. 1962 W. Granville *Dict. Sailors' Slang* 115/2 *Sun over the foreyard*, time for drinking in the wardroom. Eight bells in the forenoon watch: mid-day. It is a traditional Naval convention never to drink before the sun clears the foreyard.

2. a. *rising sun* (later examples). Also, (*a*) as a decorative motif; (*b*) as the emblem of Japan (with ref. to the literal meaning of the country's name in Japanese: see *Nippon).

1840 J. Madison *Papers* III. 1624 [At the Constitutional Convention, 1787] Doctor Franklin, looking towards the President's chair, at the back of which a rising sun happened to be painted, observed..that painters had found it difficult to distinguish in their art, a rising, from a setting, sun. 1841 J. F. Cooper *Deerslayer* II. xii. 206 You are a man whose fathers came from beyond the rising sun; we are children of the setting sun. 1863 *Chambers's Encycl.* V. 683/1 *Japan* (native name, *Nipon*.. i.e., the Land of the Rising Sun). 1895 'C. E. Craddock' *Mystery Witch-Face Mountain* 185 Some [quilts] were of the 'log cabin' and 'rising sun' variety. 1897 *Far East* 20 Mar. 83/2 The children of the Rising Sun. 1935 J. C. Lincoln *Cape Cod Yesterdays* 109, I ducked my tousled head under the..'rising-sun comforter' and fell asleep in spite of the racket. 1942 *R.A.F. Jrnl.* 2 May 1 It is difficult to form an exact picture of the air strength of the land of the Rising Sun. 1983 *Jewish Chron.* 27 May 15/3 The cupped-hand emblem replaces the now familiar rising sun logo [of the Jewish Welfare Board].

4. b. (*d*) For quot. 1727 substitute:

1688 J. Walker tr. *Pascal's Thoughts* xxxi. 246 This Dog is mine, said those poor Children; That's my place in the Sun: This is the beginning and Image of the Usurpation of all the Earth.

(Further examples.)

Quot. 1897 comes from a speech by Bernhard von Bülow, Chancellor of Germany; quot. 1901 is an example from Emperor Wilhelm II earlier than that in Dict.

[1897 *Times* 7 Dec. 5/5 We desire to throw no one into the shade, but we also demand our own place in the sun-light.] 1901 *Times* 20 June 5/4 We have..fought for our place in the sun and have won it. It will be my business to see that we retain this place in the sun unchallenged, so that the rays of that sun may exert a fructifying influence upon our foreign trade and traffic. 1926 Galsworthy *Silver Spoon* I. iii. 22 Five million pounds spent on the organised travel of a hundred thousand working men..would infect the working class with a feverish desire for a place in the sun. The world is their children's for the taking. 1928 C. R. Longwell in *Theory Continental Drift* (Amer. Assoc. Petroleum Geologists) 145 Perhaps the very completeness of this iconoclasm, this rebellion against the established order, has served to gain for the new hypothesis a place in the sun. 1939 L. MacNeice *Autumn Jrnl.* viii. 35 Sun shines easy, but I no longer Docket a place in the sun. 1951 'J. Tey' *Daughter of Time* vii. 96, I sure would hate a brother who took my credit and my women and my place in the sun. 1967 V. Lincoln *Private Disgrace* (1968) iii. 37 Lizzie longed for a place in the sun. But..her longing for popularity was self-defeating.

8. b. (Earlier example.)

1749 [see *cascade *sb.* 2 b].

II. 11. a. *sun-flush, -glaze, -glint* (earlier example), *-glory, -mote*; *sun-hero, -man*.

1924 G. B. Shaw *St. Joan* ii. 27 Joan (rising, with a sunflush of reckless happiness irradiating her face). 1958 C. Tomlinson *Seeing is Believing* (1960) 1 A quick gold, dyeing the uncovering beach With sunglaze. 1880 J. E. Watt *Poet. Sk.* 85 Oor sun-glints o' glory are followed by gloom. 1929 D. H. Lawrence *Pansies* 117 Men should group themselves into a new order Of sun-men..walking each in his own sun-glory. 1911 F. H. Woods in *Encycl. Relig. & Ethics* IV. 355/1 Cúchulainn as a sun-hero..was directly connected with Lug, the sun-god. 1929 Sun-man [see *sun-glory* above]. 1933 W. de la Mare *Fleeting* 96 The sun-motes where the mosses drowse.

b. *sun-film, -scorch.* See also *suntan *sb.* (and *a.*).

1930 E. Pound *XXX Cantos* ii. 11 Snipe come for their bath, bend out their wing-joints, Spread wet wings to the sun-film. 1907 W. De Morgan *Alice-for-Short* xix. 208 With her hair shaken out and only the least little shade of sun-scorch from long exposure on the inexhaustible sands.

c. *sun-canopy* (later example), *-filter, -lotion, -shield, -umbrella* (earlier examples).

1923 *Heal & Son Catal.: Kitchen Furnit. & Garden Furnit.* 10 Hammock..with sun canopy and fittings complete. 1970 *Cape Times* 28 Oct. 18/4 You can select your material from our large range of fabrics in tweed, sun-filter, satin, taffeta, shantung and parchment. 1979 P. Niesewand *Member of Club* vi. 40 The sun filter curtains were..green, yellow and orange stripes. 1967 H. Pinter *Tea Party* 49 You're off to Spain... What sun lotion do you use, Lois? 1974 W. Garner *Big enough Wreath* ix. 118 One of the two guards stepped out of the gate-house, sun-shields hiding his eyes. 1977 G. Scott *Hot Pursuit* iii. 29 The driver had my wrist... His other hand had to stay on the wheel and his knife was behind the sunshield. 1831 *Boston* (Mass.) *Transcript* 31 May 3/2 Light Sun Umbrellas..are offered at low prices. 1867 A. D. Whitney *Summer in Leslie Goldthwaite's Life* viii. 173 Miss Craydocke appeared..under her great brown sun-umbrella.

d. = affording maximum access to the sun; used, worn, etc., for sun-bathing; as *sun balcony, loggia, parlour, porch, room; sun-dress, -suit, -top; sun-chair.*

1971 'D. Halliday' *Dolly & Doctor Bird* vii. 90 A sun balcony..ran round the..side of the villa. 1976 'W. Trevor' *Children of Dynmouth* i. 19 Mrs Dass was reclining on a sun-chair in the bow-window. 1942 R. Godden *Breakfast with Nikolides* vi. 138 Her spotted sun-dress, her sun hat and sandals. 1976 I. Murdoch *Henry & Cato* II. 319 Gerda was wearing a sun dress with shoulder straps. 1965 Sun loggia [see *pram-park* s.v. *pram[2] 3]. 1911 Sun-parlour [see *solarium 2 a]. 1940 Auden *Another Time* 92 The poor old fat banker in the sun-parlour car. 1918 M. B. Cooke *Threshold* 53 Joan went in search of Mr. Farwell and found him reading in the sun porch. 1955 Sun-porch [see *make *sb.*[2] 13]. 1977 *Stornoway Gaz.* 27 Aug. 7/5 (Advt.), For Sale. Detached stone-built house..containing living room, bedroom, kitchen, bathroom, sun porch. 1907 E. Wharton *Fruit of Tree* III. xxiv. 349 A glazed 'sun-room', mosaic pavements, a marble fountain. 1935 *Archit. Rev.* LXXVIII. 167 It contains nine different types of flat, each with an open balcony and a glass-enclosed sun-room that can be thrown open in fine weather. 1977 *Age* (Melbourne) 18 Jan. 9/6 (Advt.), Comp. an imposing ent. hall, a large and charming sittingroom,..mod. kitchen opens to an excellent sunroom. 1929 *Punch* 17 July p. xxxv/2 (Advt.), If preparing for a sun-bath, a swim, or both, slip into the Jantzen Sun-suit! 1971 'D. Halliday' *Dolly & Doctor Bird* xi. 147, I got out into the garden in my sunsuit. 1937 *Night & Day* 29 July 22/2 Deeply to be deplored are such things as..sun-top dresses. 1972 W. Ellis *Knife Edge* vi. 114 Emma..innocently seductive in her shorts and sun-top.

12. a. *sun-worshipper* (*transf.* example); *sun-worship* (earlier example); *sun-clouding, -creating, -defying, -disdaining, -enticing, -screening* adjs.

1930 R. Campbell *Adamastor* 91 Stripped are the great sun-clouding planes. 1847 Emerson *Poems* 84 None so backward in the troop,..But knows the sun-creating sound. 1879 Longfellow *Poet. Wks.* (1910) 137 There is a mountain in the distant West That, sun-defying, in its deep ravines Displays a cross of snow upon its side. 1904 W. de la Mare *Henry Brocken* xiii. 150 The sun-disdaining eagle. *Ibid.* vii. 79 His sun-enticing thatch of hair. 1958 *Which?* I. iv. 17/2 CR had a number of the preparations tested for their sunscreening quality. 1813 *Monthly Rev.* LXXI. 477 The sun-worship of the Persians, and the manicheism of the Zend-Avesta..are classed with the monotheism of the Jews. 1966 B. H. Deal *Fancy's Knell* v. 77 Her red bathing suit [was] brilliant against her white skin. Evidently she wasn't the sun worshiper the others were.

b. *sun-alight, -ambered, -bedazzled, -begotten* (later example), *-bemused, -bitten, -black, -blazoned, -bleached, -brown* (earlier example), *-bruised, -caught, -charged, -coloured, -compelled, -dappled, -dark, -darkened, -dazed, -delighted, -desired, -detested, -dimmed, -dozed, -driven, -eaten, -faded, -fed, -flaked, -flecked* (also *fig.*), *-flooded, -flushed, -fondled, -forgotten, -freckled, -gilded, -glazed, -kissed* (later examples, *spec.* of fruit, freq. with commercial spelling *-kist*), *-lashed, -licked, -mellowed* (*fig.*), *-parched, -ripened, -scarred, -scrubbed, -sculptured, -shafted, -shot, -shy, -soaked* (also *fig.*), *-stained, -strewn, -struck, -swung, -whitened* adjs.

1904 Hardy *Dynasts* I. I. v. 32 Till we sight Famed Milan's aisles of marble, sun-alight. 1951 W. de la Mare *Winged Chariot* 23 Sun-ambered, weathered, sweet as new-mown hay. 1946 —— *Traveller* 18 A dwindling, sun-bedazzled moon. 1912 —— *Listeners* 24 A sea Of sun-begotten grain. 1957 L. Durrell *Bitter Lemons* 118 We'll all subside into sun-bemused tranquillity. 1920 H. G. Wells *Mr. Britling sees it Through* I. iii. 72 It was a tall, lean, sun-bitten youngish man of forty perhaps. 1923 D. H. Lawrence *Birds, Beasts & Flowers* 46 Columns dark and soft, Sunblack men, Soft shafts, sunbreathing mouths. 1947 Dylan Thomas *Let.* 3 Aug. (1966) 318 Sunblack webfooted waterboys..bleed from the heat. 1919 V. Woolf *Night & Day* xx. 275 The sun-blazoned windows. 1835 J. E. Alexander *Sketches in Portugal* xi. 267 Peasants with long and sun-bleached hair floating about their shoulders..stood behind fruit and vegetable baskets. 1979 *Arizona Daily Star* 5 Aug. A10/3 The approaching slick has not hurt business at the long strip of sun-bleached sand. 1861 A. J. Munby *Diary* 19 May in D. Hudson *Munby* (1972) 93 His frank intelligent face.. has a pure rich sunbrown tint. 1957 L. Durrell *Bitter Lemons* 138 How could such a sun-bruised world be transformed? 1932 D. Gascoyne *Roman Balcony* 9 A rusty and serrated leaf, Alive with sun-caught moisture. 1942 E. Bowen *Seven Winters* 32 A pinkish suncharged gauze. 1926 D. H. Lawrence *Sun* iv. 13 She stood a few steps, erect, in front of the sun-coloured woman. 1922 Joyce *Ulysses* 712 He would hear and somehow reluctantly, suncompelled, obey the summons of recall. 1924 R. Campbell *Flaming Terrapin* v. 80 The sun-dappled herds a-skipping to the song. 1983 A. Price *Gunner Kelly* II. 77 The sun-dappled pools where the stream idled between the trees. 1924 Galsworthy *Forest* iv. i, (*stage direction*) Franks comes in. Very sun-dark and thin. 1926 D. H. Lawrence *Sun* iii. 11 The child too was another creature, with a peculiar quiet, sun-darkened absorption. *Ibid.* i. 6 She went home, only half-seeing, sun-blinded and sun-dazed. 1942 J. Masefield *Generation Risen* 70 Sun-delighted earth. 1925 Blunden *English Poems* 55 It glittered mist and fire amain, Sun-desired, desiring. 1931 R. Campbell *Georgiad* i. 25 Shame to show your sun-detested sight Among the sons of valour and delight. 1917 D. H. Lawrence *Look! We have come Through!* 101 The stars, in their sun-dimmed closes. *a* 1918 W. Owen *Poems* (1920) 18 So we drowse, sun-dozed. 1909 E. Pound *Personae* 48 The stars of heaven sheathe their glory And sun-driven forth-goeth Settentrion. 1926 D. H. Lawrence *Plumed Serpent* xx. 343 She stepped across the sun-eaten plaza. 1887 Kipling *From Sea to Sea* (1899) I. 34 The maroon cloth..is..neither strained nor meagre nor sunfaded. 1926 D. H. Lawrence *Sun* iv. 13 Her sun-faded fair hair in a little cloud. 1917 E. Pound *Lustra* 184 The air is solid sunlight, *apricus*. Sun-fed we dwell there. 1934 S. Spender *Vienna* iii. 30 The once sun-flaked walls. 1844 J. R. Lowell *Poems* 17 Dim vistas, sprinkled o'er with sun-flecked green, Wound through the thickset trunks. *a* 1950 J. Cleary in Murdoch & Drake-Brockman *Austral. Short Stories* (1951) 438 Her laugh is a warm, tumbling sound, sun-flecked and musical. 1904 M. A. von Arnim *Adventures of Elizabeth in Rügen* 156 Up there in the sun-flooded space among the shimmering bracken. 1862 G. M. Hopkins *Poems* (1967) 10 So those Mermaidens crowded to my rock, And thicken'd, like that drifted bloom, the flock Sun-flushed. *a* 1960 M. Trist in 'B. James' *Austral. Short Stories* (1963) 258 He was a nice baby, blue-eyed, fair-haired and with sun-flushed skin. 1935 L. MacNeice *Poems* 42 The light on the sun-fondled trees. 1881 O. Wilde *Poems* 219 Bare to sun-forgotten fields the fire of the sun! 1925 S. O'Casey *Let.* 7 Feb. (1975) I. 131 One can hardly look for the blossoming of roses in these sun-forgotten places. 1916 W. B. Yeats *Eight Poems*, Imagining a man, And his sun-freckled face. 1892 Stevenson & Osbourne *Wrecker* xii. 190 Day after day, in the sun-gilded cabin, the whiskey-dealer's thermometer stood at 84°. 1960 J. Betjeman *Summoned by Bells* iii. 26 Only one harbinger of future woe Came to me in those far, sun-gilded days. 1915 W. J. Locke *Jaffery* iii. 36 A fair-bearded..giant..ran up and laid a couple of great sun-glazed hands on my shoulders. 1920 Sunkist [see *Sunshine State* s.v. *sunshine *sb.* 6]. 1979 N. & I. Lyons *Champagne Blues* 172 I'll have a nice glass of tomato juice with a quarter of a Sunkist lemon. 1891 O. Wilde *Picture of Dorian Gray* ix. 161 The green, flickering, sun-lashed garden. 1926 D. H. Lawrence *David* ii. 18 He beats himself against the sun-licked pebbles. 1849 C. Brontë *Shirley* xxxvi. 625 My intention was then formed, but not mature for communication; now it is ripe, sun-mellowed, perfect. 1848 J. R. Lowell *Poems* 2nd Ser. 65 The next heart-beat, the wind-hurled pile,..Bursts rattling over the sun-parched roof. 1915 G. Frankau *Tid'apa* i. 7 Do you know our churchyard at Aden; lone tombs on a sun-parched plain. 1935 *Discovery* June 162/2 The fruit is fully sun-ripened and canned immediately after gathering. 1897 J. L. Allen *Choir Invisible* xv. 159 Frenzied fightings and awful deaths had left but the sun-scarred dust. 1955 P. Larkin *Less Deceived* 19 And how remote that bare and sunscrubbed room. 1955 S. Spender *Coll. Poems 1928–53* 159 Already you are beginning to become Fallen tree-trunk with sun-sculptured limbs. 1910 W. de la Mare *Three Mulla-Mulgars* xiv. 193 Nod lifted his face and saw..the vast sun-shafted precipices. 1890 R. Bridges *Shorter Poems* IV. xiv. 75 I'll sit with my love in the scented hay: And watch the sunshot palaces high. 1936 C. Day Lewis *Noah & Waters* 52 Then plunge out of heaven upon his prey, Slanting and swiftsure as a sun-shot ray. *a* 1973 J. R. R. Tolkien *Silmarillion* (1977) xvi. 135 What errand have you, Dark Elf, in my lands? An urgent matter, perhaps, that keeps one so sun-shy abroad by day. 1910 H. G. Wells *Hist. Mr. Polly* vii. 212 He..dreamt.. of the East and West Indies until his heart ached to see those sun-soaked lands before he died. 1960 *Times* 29 Feb. 15/1 Falla's four sun-soaked dances of Spain. 1916 D. H. Lawrence *Twilight in Italy* 36 Her hands and her face were all sun-bleached and sun-stained. 1916 Blunden *Harbingers* 3, I still can watch the purple-slumbrous main Fretting the sun-strewn air. 1794 T. Dwight *Greenfield Hill* VII. 154 Idolatry fans off the vernal breeze, And sun-struck Labour, phrenzied, sinks to peace. 1896 A. E. Housman *Shropshire Lad* xliii. 62 By blowing realms of woodland With sunstruck vanes afield..Content at heart I followed With my delightful guide. 1963 A. Lubbock *Austral. Roundabout* 14 Grey salt-bush, tufts of coarse brownish grass, and stony soil merge into the sun-struck distance. 1874 J. R. Lowell in *Atlantic Monthly* May 588 Indifferent as the figures on a slate Are to the planet's sun-swung curve Whose bright returns they calculate. 1957 T. Hughes *Hawk in Rain* 39 He smiles in a mirror, shrinking the whole Sun-swung zodiac of light to a trinket shape On the rise of his eye. 1848 J. R. Lowell *Poems* 2nd Ser. 64 A great cloud edged with sun-whitened spray.

c. *sun-clear* (later example), *sun-gloved, -gold, -haired, -heavy, -leaved, -round, -sweet* adjs.

1945 W. de la Mare *Burning-Glass* 36 The grass takes on a shade Of paradisal green, sun-clear. 1939 Dylan Thomas *Map of Love* 6 Comes love's anatomist with sun-gloved hand Who picks the live heart on a diamond. 1911 E. Pound *Canzoni* 5 Guerdoned by thy sun-gold traces. 1938 S. Spender *Trial of Judge* I. 18 Let the nordic Sunhaired head be matched against cloud drifts. 1918 D. H. Lawrence *New Poems* 9 The glimmer of the limes, sun-heavy sleeping, Goes trembling past me up the College wall. 1939 Dylan Thomas in *Poetry* Feb. 26 The sun-leaved holy candlewoods. 1918 E. Sitwell *Clown's Houses* 14 Like wooden bumpkins' sun-round stare. 1937 W. de la Mare *This Year, Next Year* 50 Came the woodman with his axe into the sun-sweet glade.

d. *sun-drunk, -fast, -flashed, -glittering, -glowing, -honeyed, -peering* adjs.; *sunbask* vb. intr.

1967 C. B. Christesen in *Coast to Coast 1965–6* 29 When..taxed on this subject while sun-basking by herself on the top deck. 1925 A. Huxley *Selected Poems* 16 The sun-drunk petals. 1962 *Economist* 21 Apr. 250/1 The French have produced a [plastic] geranium which is guaranteed to be sunfast. 1905 Sun-flashed [see *iridesce *v.*]. 1916 Blunden *Harbingers* 11 Odysseus came..And called without my strong sun-glittering gates. 1926 D. H. Lawrence *Sun* iv. 14 Like a blot of ink on the pale, sun-glowing slope. 1953 Dylan Thomas *Under Milk Wood* (1954) 44 There's the clip clop of horses on the sun-honeyed cobbles of the humming streets. 1923 D. H. Lawrence *Birds, Beasts & Flowers* 84 John, oh John, Thou honourable bird Sun-peering eagle.

13. a. sun arc *Cinematogr.*, an arc lamp used to simulate sunlight in film production; **sun-back,** a low-cut back of a garment; also *attrib.*; **sunbaking** *vbl. sb. Austral.*, sunbathing; **sun-bath** (earlier example); **sun-bather,** one who takes a sun-bath; hence (as back-formation) **sun-bathe** *v. intr.*; **sun bed,** (*a*) a lightweight bed or couch for sun-bathing; (*b*) a bed designed for artificial sun-bathing in ultraviolet light; **Sunbelt** *U.S.* (also as two words) [*belt *sb.*[1] 5 a], a zone consisting of the most southerly states of the U.S., extending from California in the west to the Carolinas in the east; **sun-bonnet** (earlier and later examples); hence **sun-bonneted** *a.*; **sun-break,** (*c*) = *brise-soleil; **sun club,** a club for sun-bathers or naturists (*naturist 2); **sun compass,** a navigational device for finding true north from the observed direction of the sun, allowing for the time of day; also *fig.* and *attrib.*; **sun-cream,** a creamy preparation rubbed on the skin to protect it from sunburn or to promote sun-tanning; **sun-cure** (earlier example); **sun-dance** (earlier example); **sun-deck,** (*a*) (earlier example); (*b*) *N. Amer.*, a terrace or balcony situated so as to catch the sun; **sun-drenched** *a.*, soaked with sunshine; having (typically) very sunny weather; **sun-dry** *a.* = Sun-dried *a.*; **sun-dust,** the motes in a sunbeam; hence **sun-dusted** *a.*; **sun-extinct** *a. poet. nonce-wd.*, inwardly dead; **sun-eye** *poet.*, the sun; **sun flash,** a flash of sunlight; a device or pattern resembling this (see quots.); **sun furnace,** an apparatus constructed of mirrors designed to concentrate solar energy for use in high-temperature experiments and research; **sun gear** *Mech.* = *sun wheel* (*a*); **sun-glade** (earlier example); **sun-glass,** (*a*) (earlier examples); (*c*) *pl.*, spectacles with tinted lenses for protecting the eyes from sunlight; = *dark glasses* (*b*) s.v. *dark *a.* 13 c; hence **sun-glassed** *a.*, wearing sun-glasses; **sun-glow,** (*a*) also, the colour of this; **sun-grazer** *Astr.* (see quot. 1982); so **sun-grazing** *a.*; **Sun Gun** *Cinemat.*, a proprietary term for a portable incandescent lamp; **sun-helmet** (earlier example); **sun hot** (see quot. 1961); **Sun King** [see *roi soleil], a sobriquet of Louis XIV of France; also *transf.* and *attrib.*; **sun-lamp,** an electric lamp designed to emit radiation of a similar type to that of sunlight; now esp. one that produces ultraviolet light for therapeutic purposes or to produce an artificial sun-tan; hence **sun-lamped** *a.*; **sun lounge,** (*a*) a room built largely of glass to admit the maximum amount of sunlight; (*b*) *U.S.* = *sun-bed* (*a*) above; **sun-lounger** = *sun-bed* (*a*) above; **sun-motor,** a machine which converts solar energy to another form of energy, such as electrical or mechanical energy; **sun-oil,** (*a*) oil rubbed on the skin to prevent sunburn or promote tanning; (*b*) = *sunflower oil* s.v. Sunflower 4; **sun-painting** = *sun-printing* below; **sun-print** *Photogr.*, a print made from a negative by means of sunlight; a daylight print; so **sun-print** *v.*, **-printing** *vbl. sb.*; **sun-roof,** (*a*) = *sunshine roof* s.v. *sunshine *sb.* 6; (*b*) a part of the roof of a house which is suitable for sunbathing; **sun-scald,** (*a*) esp. damage to trees caused by the bark being dried by excessive heat and wind; (earlier and later examples); **sun-scorch,** the burning of leaves by sunlight when a plant lacks sufficient water; also = *sun-scald*; **sunscreen,** (*a*) (see sense 11 c in Dict.); (*b*) a preparation intended to screen the skin from ultraviolet rays and thereby prevent sunburn; **sun-seeker,** (*a*) *Astronautics*, a photoelectric device used in satellites and spacecraft which maintains its orientation with respect to the sun and can be used to direct instruments and provide navigational information; (*b*) one seeking a sunny place for a holiday or to live in; **sun-shaft:** for *U.S.* read orig. *U.S.* and add later and *fig.* examples; **sunship** [-ship], a mock title for the sun; **sunspecs** *colloq.* = *sun-glass (*c*); **sun-thickened oil,** a polymerized oil of a honey-like consistency, produced from linseed oil by action of the sun and used as a base in oil-painting; **sun valve,** a mechanical device which used the heat of the sun as it appeared or disappeared to turn a

lighthouse light off or on; **sun visor,** a projecting shield on a cap, or a hinged screen mounted inside (formerly also outside) a motor vehicle, to shade the eyes from bright sunshine; **sun-wheel,** (*a*) (examples); (*b*) (example).

1928 *Amer. Speech* III. 366 'Back-spot'.., 'baby-spots', 'sunarcs', 'twins', 'floods' and others. 1930 *Sel. Gloss. Motion Pict. Technician* (Acad. Motion Pict. Arts & Sci., Hollywood), *Sun lamps*, a large lamp (*Sun Arc* or *Sun Spot*) reflecting its light by means of a parabolic mirror. 1933 *Sun* (Baltimore) 11 Aug. 8/7 Her sunback suit she casts aside, She is a nudist—off go things. 1934 *Times* 22 June 17/6 Many swimming and bathing suits now have a 'sun-back' and a high throat line. 1955 J. Potts *Death of Stray Cat* ii. 18 Summer people..in..their sunback dresses. 1935 E. Dark *Return to Coolami* xxiv. 262 He had wondered..if Susan liked surfing, if she liked sunbaking. 1977 *Best of Austral. Angler* 63/1 The middle of the day is mostly for sunbaking and dreaming. 1866 *Galaxy* 15 July 544 What you want..is a sun-bath daily. 1941 A. Christie *Evil under Sun* vi. 107, I oiled myself and sunbathed. 1978 'A. York' *Tallant for Disaster* i. 17 Supposing the *Gazette* did learn that Mistress Castanos does sunbathe in the altogether? 1929 *Daily Express* 14 Jan. 19/3 The groups of Lido sun-bathers. 1973 H. Nielsen *Severed Key* xviii. 189 Sunny walked on the beach... The sun-bathers and surfers were far behind her. 1967 *Punch* 11 Jan. p. viii/2 The optimistic can snap up adjustable folding sun beds. 1979 *Sunday Express* 28 Jan. 16/6 Ever sat down on a foreign beach for a bit of blissful solitude only to find your local pub bore a couple of sunbeds away? 1980 *West Lancs. Even. Gaz.* 9 July 14 (Advt.), 'Mermaid Sontegra' Canopy Sun Beds £2 per half-hour session or course of 6 £10. 1983 *Daily Tel.* 31 Jan. 15/8 Sunbed lamps are designed to cut down ultra-violet B light, which burns before it tans. 1969 K. P. Phillips *Emerging Republican Majority* v. 438 Chart 134 illustrates how the electoral votes of the Sun Belt will have almost tripled in the half-century between 1920 and 1970. 1976 *National Observer* (U.S.) 24 Apr. 12/1 The movement is away from the noise, dirt, crime, and congestion of the oldest urban centers and to the so-called Sunbelt. 1980 *Christian Sci. Monitor* (Midwestern ed.) 4 Dec. 2/2 In some of the wooded parts of this bustling Sunbelt city, white-tailed deer have been spotted. 1837 *Southern Lit. Messenger* III. 332 She had on a deep sun-bonnet. 1941 J. Masefield *In Mill* 130 All the horses were wearing sun-bonnets and ear-flappers. 1981 M. Byrd *California Thriller* (1984) x. 82 She wore a sun-bonnet..and carried a clipboard. 1839 *Southern Lit. Messenger* V. 113/2 The bevy of sun-bonnetted lasses, who gave us of their pies and apples. 1947 *Archit. Rev.* CII. 148/1 Covering one-third of it is a key pattern of loggia-like sun-breaks, the scale of which is exactly double that of the rest of the elevation. 1969 J. Elliot *Duel* III. iv. 275 The other creatures on the beach..sat under sun-breaks, walked in and out of the water, tinkered with boats. 1936 *Sun Bathing Rev.* June–July 43/2 (*heading*) Non-nudist sun clubs. 1950 *Sun Bather* Spring 23/1 That's my ideal sun-club. In the National Trust Land, on reserved beaches. In our own gardens, where suitable. 1978 *Lancashire Life* July 31/3 Although Lancashire has four Sun Clubs (naturist terminology for nudist camps), none is on the coast. 1925 *Nat. Geogr. Mag.* Nov. 523/2 In clear weather the sun compass enabled us to do accurate navigation... Mr. Albert H. Bumstead..invented it for our trip and I consider it a great contribution to science. 1947 *New Biol.* III. 14 The sun may be either to the left or to the right of a marching hopper and it appears that the hopper while marching keeps its direction with reference to the sun. That such 'sun-compass orientation' exists was proved by ingenious experiments in the field. 1967 J. Grierson *Heroes of Polar Skies* iv. 65 Byrd..expected to maintain his heading by the sun compass. 1966 L. Cohen *Beautiful Losers* I. 91 There is a tube of sun cream in the glove compartment. 1863 L. M. Alcott *Hospital Sketches* v. 70 Very soon after leaving the care of my ward, I discovered that I had no appetite, and cut the bread and butter interests almost entirely, trying the exercise and sun cure instead. 1849 M. H. Eastman *Dahcotah* p. xxii, The Sioux worship the sun. The *sun dance* is performed by young warriors who dance, at intervals of five minutes, for several days. 1897 M. Kingsley *Trav. W. Afr.* 130 The captain is on top of the sun deck most of the time. 1950 J. D. MacDonald *Brass Cupcake* (1955) ii. 15 The apartment has a big bedroom, sun deck, living room. 1970 R. Lowell *Notebk.* 111 Thirty raspberry bushes stacked on my sundeck. 1924 A. J. Small *Frozen Gold* ii. 47 He was no longer in the sun-drenched Spring. He had flung himself back..into the winter. 1929 *Radio Times* 8 Nov. 421/3 Honey..from sun-drenched meadows. 1979 R. Gillespie *Crossword Mystery* i. 27 The sun-drenched sidewalk. 1885 W. B. Yeats in *Dublin Univ. Rev.* June 110/2 And with a sun-dry weed He wrote it on the sands. 1909 H. G. Wells *Tono-Bungay* I. ii. 51 The seaports of the sun-dry Levant. 1849 Thoreau *Week Concord Riv.* 373 The particles of golden light, the sun-dust, have..fallen like seeds on the earth. 1964 W. Golding *Spire* i. 10 Those two men posed so centrally in the sundust. 1946 R. S. Thomas *Stones of Field* 42 The breeze could bring..songs to his ear from the sun-dusted moor. 1929 D. H. Lawrence *Pansies* 120 It is only immoral to be dead-alive, Sun-extinct And busy putting out the sun in other people. 1931 C. Day Lewis *From Feathers to Iron* 23 That golden seed extends Beneath the sun-eye, the father, To ear at the earth's end. 1960 H. Hayward *Antique Coll.* 272/1 *Sun flash horse brass*, a face piece, extremely popular in Kent,..originally a disc of latten..with its centre hand-raised into a high dome or boss and encircled with a wide, flat rim. 1971 J. S. Gunn *Opal Terminol.* 46 *Sunflash*, pattern exhibiting flashes of colour, usually weak, in a dark potch background. 1949 *Sun* (Baltimore) 25 Feb. 17/3 A sun furnace which..can concentrate the temperature of the sun's surface on a space about three inches in diameter. 1955 *Sci. News Let.* 21 May 328/2 French scientists are using the sun furnace to produce and study some minerals which are made at temperatures too high for ordinary furnaces. 1935 R. Trautschold *Standard Gear Bk.* xi. 173 The relative speed of the driven internal gear in terms of the speed of the driving sun gears. 1975 *Sci. Amer.* Dec. 120/2 When the motor drives the cylinder, the idler gear and its companion rotate as a planetary system around the sun gear that is fixed to the base. 1849 H. Melville *Mardi* I. xxxix. 152 He would not be able to perceive us, owing to our being in what mariners denominate the sun-glade, or that part of the ocean upon which the sun's rays flash with peculiar intensity. 1804 M. Lewis *Jrnl.* 19 Aug. in *Orig. Jrnls. Lewis & Clark Expedition* (1904) I. ii. 112 The main chief Brack fast with us & beged for a Sun glass. 1806 W. Clark *Jrnl.* 2 Apr. in *Ibid.* (1905) IV. xxiv. 236 An Indian whome I hired for a Sun glass. 1927 A. Conan Doyle *Case-Bk. of Sherlock Holmes* xii. 306 He *had* grey-tinted sun-glasses. 1976 'B. Shelby' *Great Pebble Affair* 119 My sunglasses fell from my hand, cracking one of the hand-ground lenses. 1961 *John o' London's* 6 July 25/1 The sun-glassed eyes of the vacation-bound. 1972 K. Bonfiglioli *Don't point that Thing at Me* x. 86 Hatted and sun-glassed to the point of anonymity. 1977 *Western Morning News* 30 Aug. 4 (Advt.), 1976 Vauxhall Chevette 4-door Saloon. Sunglow. Low mileage, family saloon. 1965 *Observer* 17 Oct. 13/4 The comet may herald the return of a family of 'sun-grazing' comets..which produced some spectacular effects in the last century. *Ibid.*, One theory has suggested that this group of 'sun-grazers' may have been formed in the wake of the sun after it passed through a cosmic dust cloud. 1982 *New Scientist* 21 Oct. 158/3 Comets that pass near the Sun are called 'sungrazers'. 1961 *Official Gaz.* (U.S. Patent Office) 14 Nov. TM 49/1 Sylvania Electrical Products, Inc... *Sun Gun*, for motion picture camera lamps and reflectors. 1969 J. Whale *Half-Shut Eye* iii. 30 The battery-powered hand-lights which cameramen call sun-guns. 1976 *Listener* 12 Feb. 171/2 By shooting a gun numerous times and flashing a sun-gun, we persuaded hordes of bats to fly round the cave. 1879 *Cornh. Mag.* XXXIX. 516 Saint-Luc wore a sun-helmet. 1873 C. I. G. Rampini *Lett. from Jamaica* 179 Rockatone (stone) at ribber-bottom (bottom of the river) no know sun hot. 1961 F. G. Cassidy *Jamaica Talk* vi. 109 The oldest, and still current expression [for *noon*] is *sun hot*. 1939 O. Lancaster *Homes Sweet Homes* 26 Few of his fellow-sovereigns enjoyed the robust health of the Sun King. 1976 N. Thornburg *Cutter & Bone* xi. 258 The women were heavy and overdressed, with elaborate Sun King coiffeurs [sic]. 1977 *Time* 8 Aug. 37/1 Yves Saint Laurent, the Sun King of fashion. 1885 *List of Subscribers, Classified* (United Telephone Co.) (ed. 6) 87 Electric Sun Lamp & Power Co., Limited. 1934 L. Mumford *Metropolitan Milieu in City Devel.* (1946) ii. 34 Finally the sun lamps.. overcame the lack of real sunlight in these misplanned domestic quarters. 1957 C. MacInnes *City of Spades* II. v. 141 You're getting so pale... You must have some sun-lamp treatment. 1980 J. Hone *Flowers of Forest* I. 61 An unreal tan... Something assumed..with lotions or sun-lamps. 1976 'Trevanian' *Main* (1977) xi. 219 There is a lighter tone to his sun-lamped bronze around the ears, indicating that his haircut is fresh. 1910 *Bradshaw's Railway Guide* Apr. 1020 Linden Hall Hydro... Splendid winter garden and sun-lounge. 1971 [see *double-glazing* s.v. *double *a.* 6]. 1979 M. Babson *So soon done For* xiv. 103 The sun lounges, the chairs, the cushions..all belonged to the Norrises... She was reclining in one of the sunlounges. 1972 D. Lees *Zodiac* 191 We found ourselves side by side on one of the sun loungers. 1980 *Daily Tel.* 10 Dec. 3/7 A morning on your sun-lounger on one of Tobago's deserted beaches. 1884 *Cassell's Family Mag.* Mar. 252/1 The Sun-Motor. Our illustration gives a general view of the machine constructed by Captain J. Ericsson of New York, for utilising the sun's heat in producing mechanical power. *Ibid.* 252/2 The sun-motor may be very useful in some hot parts of the globe. 1952 'J. Wyndham' in 'E. Crispin' *Best SF* (1955) The main batteries charged by the sun-motor. 1945 'L. Lewis' *Birthday Murder* (1951) x. 151 Her face bare of lipstick and shining with sun oil. 1981 *Sci. Amer.* Feb. 62/3 The production of 'sunoil' amounted to 5.6 million tons in 1979–80. 1876 C. M. Yonge *Three Brides* I. ix. 142 The likeness of a young man..where the hard verities of sun-painting had refused to veil the haggard trace of early dissipation. 1971 *Country Life* 8 July 104/1 In the 1840s, before artists reacted violently against the threat posed by the new so-called sun paintings. 1858 *Lake Price Man. Photogr. Manip.* 218 Such a negative would suffer considerably by being sun-printed. *Ibid.*, Injured by sun-printing. 1928 Blunden *Undertones of War* viii. 78 A large sunprint on view at headquarters suspected many enemy mine-shafts. 1966 T. Pynchon *Crying of Lot 49* ii. 38 A small automobile with a sun roof. 1972 *Country Life* 15 June (Suppl.) 22/2, 5 bedrooms, bathroom, 2 w.c.'s, flat sun-roof. 1980 *Daily Tel.* 23 Jan. 14/4 Electrically operated sunroof and windows and central locking system are included as standard. 1855 *Trans. Mich. Agric. Soc.* VI. 158 The tree has received a sun scald, and the sap soured in consequence. 1932 Felt & Rankin *Insects & Diseases of Ornamental Trees & Shrubs* iv. 116 Beech, spruce and pines are subject to sun-scald. 1967 *New Scientist* 30 Nov. 546/2 The temperature of the fruit [*sc.* tomatoes] directly exposed to the sun is at least 5 to 10 deg C higher than the surrounding air, and this high temperature frequently causes sunscald. 1928 B. D. Jackson *Gloss. Bot. Terms* (ed. 4) 469/1 Sun scorch. The burning of foliage when the soil is parched. 1932 Felt & Rankin *Insects & Diseases of Ornamental Trees & Shrubs* iv. 115 The leaves may transpire more water than the roots can take up in a given length of time. This condition will cause sun-scorch of the leaves. 1950 *N.Z. Jrnl. Agric.* Jan. 5/2 Do not place them [*sc.* unripe tomatoes] at a window exposed to strong sunlight, as this will induce sun scorch and render the skin tough. 1969 *Gloss. Landscape Work* (*B.S.I.*) v. 28 Sun scorch (sunscald), damage caused to bark by unaccustomed exposure to the sun, for example, following the sudden removal of shade. 1958 *Which?* I. iv. 17/2 Some sunscreens are lotions, some oils, some creams, others aerosols. 1980 *Daily Tel.* 22 Feb. 15/2 Any exposed area of skin should always be protected, either by a moisturiser or—in hot sun—by a sunscreen. 1956 *Nature* 7 Apr. 645/1 The Royal Aircraft Establishment..is also studying the design of a sun-seeker for carriage in the rocket. The sun-seeker would be used for measurements of solar radiations and for obtaining ultra-violet pictures of the sun at high altitude. 1963 M. Caidin *Man-in-Space Dict.* 198 As used in manned spacecraft or robot satellites, the sunseeker 'seeks out' the sun by its brightness. An automatic pilot notes the position and angle of the sunseeker, and fires reaction jets to keep the spacecraft oriented on the basis of the position of the sun. 1970 *Times* 31 Dec. (Rev. of Year) p. vii./5 Sunseekers are beginning to look farther afield than the popular Spanish mainland. 1975 D. Francis *High Stakes* ix. 141 Selling dream retirement homes to elderly sun-seekers. *a* 1918 W. Owen *Poems* (1963) 64 Who's prejudiced Against a grimed hand when his own's quite dust, Less live than specks that in the sun-shafts turn. 1941 Blunden *Thomas Hardy* iv. 67 The secret of that apparent indifference was his lifelong purpose..of striking for truth under the intense sunshafts of philosophic poetry. 1974 F. Warner *Meeting Ends* I. 1 A sunshaft strikes the steeple by my room. 1836 Poe *Four Beasts in One* II. 206 You need not look up at the heavens; his Sunship is not there—at least not the Sunship adored by the Syrians. *That* deity..is worshipped under the figure of a large stone pillar. 1975 *Observer* (Colour Suppl.) 20 June 13/4 The Reactolite 90/20 lenses..capable of withstanding the impact of a 1⅛ oz. steel ball dropped from a height of 50 in., more than double the requirement of the United States' stringent sunspec regulations. 1976 *Punch* 11 Aug. 234/1 Choose a chair and pull up a glass, push up the sunspecs and just drink in this room. 1935 E. Neuhaus tr. *Doerner's Materials of Artist* iii. 105 Sun-thickened oil is to be preferred to boiled oils, as also to the resin-oil varnishes. 1975 U. Dix tr. *Wehlte's Materials & Techniques of Painting* 389 Sun-thickened oil darkens rapidly when stored in tin canisters. 1910 *Chambers's Jrnl.* Sept. 620/2 One great feature of this beacon is the sun-valve, whereby the light is ignited and extinguished automatically at varying periods, according to the time of year. 1936 W. H. McCormick *Mod. Bk. Lighthouses* xi. 92 The light is automatically turned on and off..by placing the light in charge of an 'AGA' Sunvalve. 1975 Hague & Christie *Lighthouses* v. 159 Early in the present century the.. operation of unattended lighthouses..was revolutionised by the invention of the sun- or light-valve. This..consists of an arrangement of reflective gold-plated bars supporting a suspended absorbent black rod; when lit by the sun this rod absorbs the direct heat and that reflected from the other bars and expands downwards thereby cutting off the supply of gas. The first sun-valve was put into operation..near Stockholm in 1907. 1926 *Daily Colonist* (Victoria, B.C.) 8 July 2/1 Bright Sunshine is fine—enjoy it all the more by wearing a sun visor... A necessity to campers, sportsmen, etc. 1936 *Times* 19 Oct. 8/2 The inside fittings include..sun visors, footrests, etc. 1978 L. Heren *Growing up on The Times* ii. 30, I..had a large American Ford V8 car fitted with a sun visor projecting over the windscreen. 1827 Sun-wheel [see *planet-wheel* s.v. *planet *sb.*[1] 4]. 1965 *Daily Mail* 28 Oct. 7/3 If we convert the reverse wheel into a sun wheel (stationary wheel) by introducing a brake band, we'll get an intermediate gear ratio and three speeds. 1973 T. Pynchon *Gravity's Rainbow* I. 100 The symbol used is a rude mandala, a red circle with a thick black cross inside, recognizable as the ancient sun-wheel from which tradition says the swastika was broken.

b. sun-bear (later examples); **sun gem,** for *cornutus* substitute *cornuta* and add: distinguished by tufts of feathers on either side of the head; (examples); **sun-perch** (earlier and later examples); **sun plant,** (*a*) a small, half-hardy, annual herb belonging to one of several varieties of *Portulaca grandiflora*, native to Brazil and bearing single or clustered terminal flowers which open in sun; (*b*) a plant that grows best in full sunlight; **sun-rose:** cf. *helianthemum; (earlier example); **sun-spider** = Solpugid; **sun-trout** (earlier example).

1894 N. B. Denys in W. W. Skeat *Malay Magic* (1900) v. 183 The Malayan Sun-bear, the only animal of the bear species in the Peninsula... It is black in colour, with the exception of a semi-lunar-shaped patch of white on the breast, and a yellowish-white patch on the snout and upper jaw. 1931 *Times Educ. Suppl.* 19 Sept. (Home & Classroom Suppl.) p. iv/3 Mr. Charles Tonge has presented a young Malay sun-bear. 1965 R. McKie *Company of Animals* ix. 146 Sun bears can become dangerous as their power increases with age. 1861 J. Gould *Monogr. Trochilidæ* IV. 212 (*heading*) Sun Gem. 1912 Brabourne & Chubb *Birds S. Amer.* 144 *Heliactin..bilophum*.. Sun-Gem. 1804 Lewis & Clark *Orig. Jrnls. Lewis & Clark Exped.* (1905) VI. 174 In this lake there is also.. Sunperch. 1876 'Mark Twain' *Tom Sawyer* xiv. 123 They were back again with some handsome bass, a couple of sun-perch and a small catfish. 1902 W. S. Gordon *Recoll. Old Quarter* 177 How full were the holes of crawfish, turtles, sun-perch, grindles, and of daring, voracious pike. 1887 G. Nicholson *Illustr. Dict. Gardening* III. 202/2 Sun-plant. Fl[owers] yellow, purple,..terminal. 1900 B. D. Jackson *Gloss. Bot. Terms* 260/2 Sun-plants, plants which prefer full sun-light: their stems are often short, the leaves have the palisade cells well developed. 1963 *Oxf. Bk. Garden Flowers* 140/2 Sun Plant. The yellow, pink, scarlet or purple cup-shaped flowers of this little plant from Brazil open in direct sunshine and close in shadow. 1979 W. M. M. Baron *Organization in Plants* (ed. 3) iii. 42 Shade plants can utilize low light intensities more efficiently than sun plants. 1822 Sun-rose [see *helianthemum]. 1959 *Southwest Rev.* XLIV. 137/1 An arachnid frequently, and naturally, confused with the true vinegarone is the solpugid—or wind-scorpion, wind-spider, or sun-spider. 1974 *Stand. Encycl. S. Afr.* X. 217/1 The sun-spider can easily be distinguished from all other arachnids by the two immense jaws at the front of the head. 1884 G. B. Goode *Fisheries U.S.: Nat. Hist. Aquatic Animals* I. 362 In the Southern Atlantic States it [*sc.* the squeteague] is called 'Grey Trout', 'Sun Trout', and 'Shad Trout'.

‖ **sun** (sun), *sb.*[2] Pl. **sun.** [Jap.] A Japanese

unit of length, equivalent to approximately 1·19 inches (3·03 centimetres).

1727 [see *SHAKU 1]. 1888 *Encycl. Brit.* XXIV. 490/2 Japan... Sun, 10 = shaku (11·948 inches = 10/33 metre), 6 = ken, 60 = cho. 1956 K. Tomiki *Judo* i. 22 Regulations require that the surrounding mats be all 5 *sun* (about 6 inches) lower than the contest area.

sun, *v.* Add: **2. b.** Now esp. = *sun-bathe* vb. s.v. *SUN *sb.* 13 a. (Later examples.)

1933 V. Woolf *Jrnl.* 13 Apr. in *Writer's Diary* (1953) 197 But we go today and I shall sun, with only a few books. 1968 *Sat. Rev.* (U.S.) 23 Nov. 48/1 Three beaches where you can swim and sun stark naked. 1976 E. Dewhurst *After Bail* vii. 90 The sun never does anything to my lily-white skin... Alan doesn't let me sun for too long.

sunbeam. Add: **1. d.** Someone, esp. a woman or girl, who enlivens or cheers another. Cf. (*little*) *ray of sunshine* s.v. *RAY sb.[1] 1 a.

1886 C. M. Yonge *Chantry House* II. xxi. 190 She was always a sunbeam, with her ever ready attention. 1900 C. H. Chambers *Tyranny of Tears* iv. 128 We're all very sorry you're going—particularly cook. Cook's very strong in her attachments... Cook's words was, 'This'll be a dull 'ouse when the little sunbeam's gone.' 1943 F. Thompson *Candleford Green* viii. 133 Girls..of the type then called 'sunbeams in the home': good, affectionate, home-loving girls. 1970 G. Heyer *Charity Girl* x. 150 She couldn't conceive how she had ever contrived to exist without 'our sweet little sunbeam'.

sunburn, *sb.* Add: **2.** The name of a fashion colour.

1923 *Daily Mail* 11 Sept. 11 Nude, Sunburn, Mulatto, and all shades. 1932 *Barker's Sales Catal.* 27 Poplin tennis shirts... Guaranteed fast self colours of blue, champagne, helio, ivory, sunburn, white and light grey.

sunburn, *v.* **2.** (Later examples.)

1873 J. H. Beadle *Undevel. West* xxix. 640 An Indian will 'sunburn' as much or even more than a white man. 1928 *Daily Mail* 6 Aug. 12/6 One girl tells me she 'doesn't sunburn easily'. 1962 L. Deighton *Ipcress File* i. 14 He had a clear complexion that sunburnt easily.

sunburst. Add: **2. b.** *attrib.* of things designed or arranged as conventional or stylized representations of the sun and its rays; esp. *sunburst clock*, a clock framed by radiating arms; *sunburst pleat* = *sun-ray pleat* s.v. *SUN-RAY 2 b.

1908 *Sears, Roebuck Catal.* 362/2 Salt and pepper shakers. In beautiful sunburst pattern. 1920 *Glasgow Herald* 29 Apr. 6 Her bouquet was of 'sunburst' roses. 1927 A. E. W. Mason *No Other Tiger* xxiii. 260 They sold the lot—the emerald ring, the diamond sunburst ear-rings and all. 1939 M. B. Picken *Lang. of Fashion* 113/2 *Sunburst p[laits]*, accordion-like plaits that are narrow at top and wider at the bottom, thus producing a flare. Fabric plaited on bias so that plaits radiate from a center. 1949 M. Steen *Twilight on Floods* III. iv. 426 The gilt sunburst clock over the fireplace. 1962 M. Kelly *Due to a Death* ix. 152 There was a sunburst window over the hairdresser's door. 1969 M. Tripp *Malice & Maternal Instinct* i. 6 A sunburst clock on one of the blue walls in the main room. 1980 *News & Observer* (Raleigh, N. Carolina) 28 Oct. 17/7 The sun set Sunday on the familiar sunburst insignia.

sundae (sʌ·ndeⁱ). orig. *U.S.* Also (*rarely*) **sundi.** [Origin uncertain. There exist a number of differing accounts both of the invention of the dish and of the coinage of its name.

The name is generally explained as an alteration of *Sunday*, either because the dish originally included left-over ice-cream sold cheaply on Monday, or because it was at first sold only on Sunday, having, according to some accounts, been devised to circumvent Sunday legislation. The alteration of the spelling is sometimes said to be out of deference to religious people's feelings about the word *Sunday*. For several accounts see H. L. Mencken, *The American Language* Suppl. I. (1945), pp. 376–7.]

A confection of ice-cream topped or mixed with crushed fruit, nuts, syrup, whipped cream, etc. locally also called *college ice*.

1897 W. A. Bonham *Mod. Guide for Soda Dispensers* 126 Peach Sundae. Ice cream, vanilla or peach..5 ounces. Crushed or sliced peaches..2 ounces. Serve with a spoon. Pear, orange, raspberry and other fruit sundaes are made by adding the syrup or fruit to the ice cream. 1904 *N.Y. Evening Post* 21 May (Sat. Suppl.) 4/7 The Sundi, so popular at the confectioner's, can be prepared at home. Make a rich vanilla ice cream and over it pour the juice of your preserved fruits. 1904 *Minneapolis Times* 15 June 6 In one of the Jersey City churches fans and lemonade are distributed. Some brands of 'sundae' might be added with propriety. 1910 *Chambers's Jrnl.* July 431/1 A sundae—a mixture of ice-cream, soda-water, and raspberry juice. 1927 A. P. Herbert *Plain Jane* 88 I'm fizzy and fiery and fruity and tense, So let's have a sundae and hang the expense! 1951 T. Sterling *House without Door* ii. 22 Year after year..Schrafft's had been serving lamb and mint jelly and hot fudge sundaes to others. 1970 *Kay & Co.* (Worcester) *Catal.* 1970/71, 896 Six Bohemian sundae glasses in the Zorka design... Perfect for all sweets.

Sundanese, *sb.* (Earlier example.)

1878 [see *MADURESE *a.* and *sb.*].

‖ **sundang** (sūndā·ŋ). [Malay.] A heavy two-edged sword used in Malaysia.

1902 *Encycl. Brit.* XXX. 497/1 Malays use...short broad swords called *sūndang*. 1936 G. B. Gardner *Keris* i. 39 The *sundang* is two edged and may be straight or sinuous. 1947 R. Winstedt *Malays* 165 The type..is closer to the Bugis *sundang* or short sword. 1972 M. Sheppard *Taman Indera* 133 The largest member of the kris family is the *Sundang*, the sword kris. It originated in the Celebes... The Sundang is a cutting and slashing weapon and is not intended for thrusting.

Sunday, *sb.* Add: **1. c.** *a month of Sundays* (earlier example); *Sunday out* (earlier example); hence *Sunday outer*; *Sunday-go-to-meeting clothes* (earlier examples); also *ellipt.* as *Sunday-go-to-meetings*.

1831 Sunday-go-to-meetings [see *GO-TO-MEETING *sb.*]. 1831 J. R. Motte *Diary* 28 Aug. in A. H. Cole *Charleston goes to Harvard* (1940) 100 Rose at 7, and having shaved and dressed myself,—in Sunday-go-to-meeting clothes, started for a walk to Boston. 1841 *Punch* 21 Aug. 65/1 A veritable footman,..upon the occasion of his 'Sunday out'. 1846 D. Corcoran *Pickings from Picayune* 49 The hoosier asked him if he thought him 'darn'd fool enough to dirty his Sunday-go-to-meetin' clothes'? 1847 J. Codman *Sailor's Life & Sailor's Yarns* 25 He..dressed himself in his 'Sunday-go-to-meetings'..and bade adieu to home. 1849 G. E. Jewsbury *Let.* 29 Mar. (1892) 286 If I don't get a better letter from you, or at least a letter with something in it, you may pass 'a month of Sundays' at breakfast without any letter from me. 1879 Dickens *Life Charles James Mathews* I. i. 30 A couple advanced who evidently did not belong to the usual class of 'Sunday outers'.

d. *pl. ellipt.* for: (*a*) Sunday clothes or best; (*b*) Sunday newspapers.

(*a*) 1901 'Mark Twain' in *Century Mag.* Nov. 26/2 Tommy was..not in his Sundays, but in his dreadful work-clothes. 1933 *Punch* 14 June 663/1 Tom was busy brushing up his Sundays to go a-calling at the station-master's house. 1944 E. Carr *House of All Sorts* 89 Neither of them noticed the dust on his 'Sundays' as they smiled off down the street.

(*b*) 1949 E. Benn *Happier Days* x. 116 The Sundays and Weeklies were outside the squabbles of the Dailies. 1963 *Listener* 24 Jan. 175/3 An English reviewer, writing in one of the 'posh Sundays'..recently claimed that only Dubliners are now writing outstanding prose. 1976 T. Stoppard *Dirty Linen* 9 They each carry several newspapers, a whole crop of the day's papers and the Sundays. 1983 *Listener* 27 Jan. 18/3 There are the smart Sundays, the *Guardian's* Agenda page on Monday mornings, and pieces such as this in the literate weeklies.

3. *Sunday('s) suit* (earlier example not in the possessive); carrying out an activity only on Sundays or for pleasure (on the analogy of **Sunday driver*, **Sunday painter*), as *Sunday architect, artist, golfer, novelist, poet, sailor*; **Sunday best** (earlier examples); also **Sunday's best** and *transf.* and *attrib.*; **Sunday child** (earlier example); **Sunday closing,** the closing on Sundays of shops, except for the sale of certain commodities, or of public houses, etc.; **Sunday driver,** one who drives chiefly at week-ends, freq. an unpractised, slow, or unskilful driver; **Sunday face:** for (*Sc.*) read (orig. *Sc.*); (later example); also **Sunday's face;** hence **Sunday-faced** *a.*; **Sunday joint,** a roasted joint of meat traditionally served for Sunday lunch; **Sunday lunch,** the traditional large meal served at midday on Sunday; **Sunday observance,** the keeping of Sunday as a day of rest and worship; **Sunday painter,** an amateur painter, one who paints purely for pleasure; often applied to a naïve painter (*NAÏVE *a.* 1 c), esp. Henri Rousseau; **Sunday punch** *U.S. slang*, a knock-out blow (of the fist); also *transf.*; **Sunday supplement,** an illustrated section issued with a Sunday newspaper, sometimes characterized by the portrayal of voguish living.

1978 *Listener* 6 Apr. 439/1 A small temple of individualism..by a Sunday architect. 1978 *Times* 12 Apr. 16/5 Those who think the Berlin Wall was built..for Sunday artists to exhibit their wares on. 1794 Sunday's best [see Best *a.* 8 d]. [1844 G. E. Jewsbury *Let.* 17 Sept. (1892) 143 So, on the whole, you may set it down as one of the best good deeds you ever did—quite a 'Sunday best.'] 1846 *Amulet* 12 Some urchins, dressed out 'in their Sunday's best', all neatly clean. 1846 *Godey's Mag.* July 8/2 Like most of the nobility he dresses with the utmost plainness, hardly above the substantial Yankee 'squire' in his Sunday best. 1849 N. P. Willis *Rural Lett.* iii. 325 It was that kind of Sabbath weather in which Nature seems dressed and resting—every tree looking its 'Sunday best'. 1859 [see Best *a.* 8 d]. 1969 R. Blythe *Akenfield* ii. 59 Sunday-best suits. 1886 C. M. Yonge *Chantry House* I. i. 8 He was punished for 'telling fibs', though the housemaid used to speak..of his being a 'Sunday child'. 1850 *Punch* 31 Aug. 92/2 The Sunday closing of the country Post was considered no other than an unmeaning rant of a party. 1863 *Punch* 28 Mar. 130 (*caption*) Probable effect of Mr. Somes's Sunday Closing Bill. 1881 *Act* 44 & 45 *Vict.* c. 61 s. 5 This Act may be cited as the Sunday Closing (Wales) Act, 1881. 1932 U. Sinclair *Candid Remin.* II. ix. 60 He would join the church, sign pledges, vote for Sunday closing. 1971 *Reader's Digest Family Guide to Law* 660/2 Some areas—parts of Wales and Monmouthshire—have Sunday closing [of public houses] by law. 1925 *New Yorker* 11 July 11/1 The Sunday painter is to the art-artist what the Sunday driver is to the owner of the Hispano or Rolls-Royce. 1942 *Sun* (Baltimore) 26 Jan. 18/3 Sunday drivers and sightseers accounted for more than seventy per cent of the total number of cars passing along the Eastern avenue road. 1975 L. Deighton *Yesterday's Spy* xx. 161 The Sunday drivers creeping along the promenade. *a* 1779 D. Graham *Writings* (1883) II. 51 Put on a Sunday's face, and sigh as ye were a saint. 1786 Burns *What ails ye Now* in *Poems ascribed to R. Burns* (1801) 29 Wi' pinch I put a Sunday's face on, An' snoov'd awa' before the Session. 1906 E. Dyson *Fact'ry 'Ands* xiii. 165 His Trowsis had er slitherin' chin, 'n' ther Sunday face iv er sick sheep. 1910 T. S. Eliot in *Harvard Advocate* 26 Jan. 114 Sunday: this satisfied procession Of definite Sunday faces. 1934 Dylan Thomas 18 *Poems* 25 For, sunday faced, with dusters in my glove, Chaste and the chaser, man with the cockshut eye. 1928 J. Buchan *Runagates Club* xii. 319 His clothes ..were workman-like, and looked as if they belonged to him—no more the uneasy knickerbockers of the Sunday golfer. *c* 1921 D. H. Lawrence *Mr. Noon* in *Mod. Lover* (1934) 172 They were socialists and vegetarians... None of the horrors of Sunday joints. 1980 'M. Hebden' *Pel under Pressure* v. 47 He was lying on the floor, trussed up like a Sunday joint. 1932 E. M. Delafield *Thank Heaven Fasting* III. ii. 263 Mr. Pelham was sleeping, after Sunday lunch. 1973 'M. Underwood' *Reward for Defector* viii. 63 They sat down to roast lamb, roast potatoes, cauliflower with a cheese sauce and brussel sprouts... 'Mrs Tidmarsh enjoys cooking a proper Sunday lunch.' 1960 *News Chron.* 9 Mar. 6 Mr. Bratby may be a professional painter, but he is a Sunday novelist. [1785: see Observance 1 a.] 1857 *Punch* 4 July 4/2 Having put down the Sabbatarians and secured rational liberty to the millions in respect to Sunday observance. 1973 J. Wainwright *High-Class Kill* 209 Pornographic literature—and blue films—and illegal gambling—and anything else the Sunday Observance crowd can think up. 1925 Sunday painter [see *Sunday driver* above]. 1948 R. O. Dunlop *Understanding Pictures* iv. 26 Chief of these 'Sunday' painters was the *Douanier* Rousseau—so-called because he was for long a customs official. 1961 M. Leake tr. *Bouret's Henri Rousseau* 170 After the publication of this text [*sc.* R. Grey's *Henri Rousseau*] in 1922, the label 'Sunday-painters' became attached to the naïf and primitive painters and to the popular realist masters, and still survives. 1980 B. Bainbridge *Winter Garden* xii. 88 He gathered there were few actual artists in the room. A General was pointed out to him and an Admiral, both retired. He supposed they were Sunday painters, rather like Churchill and Roosevelt. 1979 M. McCarthy *Cannibals & Missionaries* iii. 73 The Senator..calls himself a 'Sunday poet', so he doesn't publish. 1929 D. Runyon in *Cosmopolitan* Oct. 64/1 If you argue with Dave the Dude too much he is apt to reach over and lay his Sunday punch on your snoot. 1944 W. W. Elton et al. *Guide Naval Aviation* iv. 71 The real 'Sunday punch' of naval aviation is the torpedo bomber. 1979 E. Newman (*title*) Sunday punch. 1973 H. Nielsen *Severed Key* iii. 27 As the day cleared, a few hardy Sunday sailors took out their boats. 1830 in M. R. Mitford *Stories Amer. Life* I. 280 Sampson stood, in his Sunday suit, showing with his teeth an air of joyous satisfaction. 1905 E. Wharton *House of Mirth* II. ix. 429 The photographer whose portraits of her formed the recurring ornament of 'Sunday Supplements'. 1913 [see *RINKY-DINK a.]. 1958 J. Blish *Case of Conscience* I. iii. 36 Stop sounding like a Sunday supplement. You underestimate your own intelligence. 1979 M. Tabor *Baker's Daughter* i. 13 A basement in a Sunday supplement conversion.

Sundayfied *a.* (earlier example).

1870 *Bazar Bk. Decorum* 164 We are apt to be, as the French say, *endimanchés*, which we may translate by the coined word *Sundayfied*.

Sunday-school. Add: **1. a.** (Earlier *attrib.* example).

1792 *Looker-On* 24 Mar. 36, I really once detected her knitting stockings, for prizes to the Sunday-school girls.

b. *transf.* A school in which instruction in Socialist principles is given on a Sunday.

1901 *Young Socialist* Apr. 2 We ought to..muster as large an army as possible of young soldiers of our cause... This is already being done in our Socialist Sunday Schools. 1922 J. Buchan *Huntingtower* x. 198 Wee Jaikie went to a Socialist Sunday School last winter. 1930 A. P. Herbert *Water Gipsies* xv. 217 Ernest assumed that it would be a treat for Jane to spend her Sunday afternoon at a proletarian Sunday School. 1978 *Times* 5 May 15/5 As long ago as 1918 to 1925 I attended a William Morris Sunday School in an English industrial city.

2. Used *attrib.* or as *adj.* with allusion to the sanctimoniousness, sentimentality, or strict morals held to be inculcated by Sunday-schools: primly moral.

1843 Dickens *Mart. Chuz.* (1844) xxvii. 333 'Not the truth?' cried Tigg... 'Don't use that Sunday-school expression, please!' 1894 G. B. Shaw *Let.* 4 July (1965) I. 448 Ober Ammergau was a miserable, genteelified, Sir Noel Patonesque Sunday School piece of illustrated Bibleism: Bayreuth is very different. 1931 *Amer. Mercury* Nov. 352/2 *Gone Sunday-School*, said of a circus that has abolished the grift. *Ibid.* 354/2 *Sunday-school show*, a show on which gambling games for the public have been prohibited. 1952 S. Kauffmann *Philanderer* (1953) iii. 54 No, it doesn't matter how good he *is*; how good he tries to be, human good, not Sunday-school good. That's what matters. 1973 *Time* 25 June 6/2 Like the circus before it, the carnival is today largely a 'Sunday school' operation.

Su·nderland. The name of a town in Tyne and Wear, England, used *attrib.* to designate (*a*) a type of coarse cream-coloured ware, usu. decorated with a pink lustre and transfers, made there in the late eighteenth and nineteenth centuries; also similar ware made elsewhere; (*b*) less frequently, a type of coarse

brown earthenware made in Sunderland in the nineteenth century.

1870 W. Chaffers *Marks Pott. & Porc.* (ed. 3) 587 The ware made here was..decorated with the pink metallic lustre so usual on the Sunderland jugs. 1874 [see *Newcastle pottery* s.v. *NEWCASTLE[1] 2]. 1911 J. F. Blacker *19th-Cent. Eng. Ceramic Art* xvi. 396 The purple and pink lustre..decorated the white ware, which must be distinguished from 'Sunderland ware', the brown earthenware, resembling what is known in the trade as 'rockingham'. 1920 [see *PRATT]. 1937 J. R. Hodgdon *Collecting Old Eng. Lustre* v. 34 There are many large bowls, mostly of the late Sunderland ware... Mrs Harpur..is an authority on Sunderland lustre. 1951 John & Baker *Old Eng. Lustre Pott.* xvi. 97 *Sunderland Ware* has always been reserved for a robust glazed brown earthenware lined with a white glaze and suitable for cooking. 1975 P. D. James *Black Tower* iii. 87 A splendid Sunderland lustreware jug commemorating Trafalgar. 1979 'J. Gash' *Grail Tree* xvi. 165 Dull pink lustres, universally known as 'Sunderland' ware, don't always come from Sunderland.

sundown, sun-down. Add: **2.** (Earlier example.)

1873 *Kansas Mag.* Sept. 207/2 A flaring sun-down dangled by its strings.

3. *U.S. colloq.* Used *attrib.* to designate one who practises as a doctor or lawyer, etc., outside normal working hours or in addition to his principal occupation. Cf. *SUNDOWNER 3.

1897 *Boston Transcript* 5 Aug. 5/1 There are sundown doctors, sundown lawyers and sundown ministers. 1904 L. Derville *Other Side of Story* 42 A sundown doctor.. [is] a doctor who practices his vocation after four o'clock, when he can leave his desk in some Government office. 1949 *Sun* (Baltimore) 29 June 8/3 Attorneys practising in the county without maintaining offices there have come to be known as 'sundown' lawyers.

sundowner. Add: **1.** (Earlier examples.)

[1846 C. P. Hodgson *Reminisc. Austral.* 302 A 'Sundowner' (? a task requiring no great exertion which lasts until sundown).] 1868 *Sydney Punch* 14 Nov. 198 (*heading*) The song of the sundowner.

2. orig. *Colonial* (esp. *S. Afr.*). **a.** An alcoholic drink taken at sunset. Also *transf.* and *attrib.*

1909 *Daily Chron.* 20 Oct. 6/7 The 'sundowner' refreshment of the West African late afternoon. 1921 *Chambers's Jrnl.* Jan. 43/1 Surely you are not going to refuse a solitary sundowner. 1932 E. Waugh *Black Mischief* viii. 298, I said we'd drop into the Brethertons for a sundowner. 1947 W. S. Maugham *Creatures of Circumstance* 37 See you again at six for a sun-downer. 1966 D. Varaday *Gara-Yaka's Domain* x. 113 We had a good notion the herd would return for a 'sundowner'. 1978 G. Greene *Human Factor* II. iv. 86 He sits there on a hot evening swilling his sundowners without a care in the world.

b. An evening drinks party.

1962 *Pretoria News* 9 Nov. 9/3 The war raged on, much to the delight of our sundowner guests. 1971 D. Creed *Trial of Lobo Icheka* xv. 149 I'll throw a Sundowner at your place tonight... Invite anyone you feel I might like to meet. 1973 J. J. McKelvey *Man against Tsetse* ii. 73 At the 'sundowners', or cocktail hours, heavy white socks seem safer and more comfortable than black ones.

3. *U.S. colloq.* One who practises as a doctor, etc., outside normal working hours or in addition to his principal occupation. Cf. *SUNDOWN, SUN-DOWN 3 and *MOONLIGHTER 3.

1886 *Turf, Field & Farm* 10 May 399/3 The night doctors gather the 'stiffs' and the 'sundowners' flourish around them with scalping knives. 1904 *Sun* (N.Y.) 14 Aug. 17 The Washington sundowner is so called because he practises a profession, usually medicine or dentistry, after the close of Government office hours, or after sundown.

sundries, *sb. pl.* Add: **a.** (Earlier examples.) Also (chiefly *Austral.*), in Cricket: the extras, or runs scored otherwise than off the bat.

1755 Fielding *Voy. to Lisbon* 182 The whole pitiful 30 l. came pure and neat into the captain's pocket,.. attended with the value of 10 l. more in sundries, into the bargain. 1794 A. Young *Trav. France* II. xix. 421 Live Stock..Corn..Tobacco..Sundries. 1867 *Australasian* 16 Mar. 332/1 With sundries forty-five, the innings closed for the very long score of 211. 1976 *0–10 Cricket Scene* (Austral.) 13/1 That 104 was seven runs more than the 11 New Zealanders could muster between them..excluding the 15 sundries.

b. sundriesman (earlier example).

1885 *List of Subscribers, Classified* (United Telephone Co.) (ed. 6) 84 (*heading*) Druggists' sundriesmen.

sundriness. (Later example.)

1878 W. Barnes *Outl. Eng. Speech-Craft* 86 The goodness of a speech should be sought in..its fulness of words for all things and time-takings which come, with all their sundrinesses, under the minds of men of the speech, in their common life.

sunfish, *sb.* Add: **3.** *U.S. colloq.* A manner of bucking. Cf. Sunfish *v.*

1903 *Wide World Mag.* Mar. 548 A broncho named 'E.A.'..used a combination of 'sunfish' and 'twister'. 1939 P. A. Rollins *Gone Haywire* 260 One prodigious forward jump, then a 'sunfish', and the beast raced into a 'circle buck'.

sunfish, *v.* Add: (Later example.) Also *transf.*

1923 *Century Mag.* CVI. 831/2 Down across Texas it went sunfishin', back-flippin', side-windin'... The Staked Plains used to be heavily timbered until that big wind swiped the trees off. 1971 A. P. McInnes *Dunlevy* 86 Sometimes the mare sunfished, but the girl stuck solidly. 1979 D. Anthony *Long Hard Cure* xxv. 195 He'd ducked ..and gone to one knee, sunfishing a little. His right arm moved..and his pistol boomed loud.

Hence **su·nfisher,** a horse that sunfishes; **su·nfishing** *ppl. a.* and *vbl. sb.*

1913 in *Dialect Notes* IV. 28 Sunfisher. 1923 *N.Y. Times* 16 Aug. 10/1 As Yak went by the sun-fishing longhorn sort o' sheered. 1924 W. M. Raine *Troubled Waters* v. 47 Rocking chair [an outlaw horse]..was a noted fence rower, weaver, and sunfisher. *Ibid.* 52 Neither side-bucking nor pitching, sunfishing nor weaving could shake the lean-loined, broad-shouldered figure from his seat. 1961 R. F. Adams *Old-time Cowhand* 298 A 'sunfisher' was a hoss that twisted his body into a crescent, or, in other words, when he seemed to try to touch the ground with first one shoulder and then the other, lettin' the sunlight hit his belly. 1967 *Sunday Mail Mag.* (Brisbane) 12 Feb. 3/1 In the flash of a second, he'd switched from 'sunfishing' to 'corkscrewing'.

sunflower. Add: **4. sunflower oil** (earlier example); **Sunflower State** *U.S.*, a nickname for Kansas.

1768 *Pennsylvania Gaz.* 6 Oct. 2/3 The sun-flower oil may prove equally valuable with the best Florence oil, for diet or medicine. 1888 *Harper's Mag.* June 39/1 Her citizens affectionately speak of Kansas as the 'Sunflower State'. 1904 *Minneapolis Times* 7 June 6 The floods in Kansas are subsiding. There was danger for a time that the Sunflower state would grow a crop of pond lilies. 1965 Mrs. L. B. Johnson *White House Diary* 2 Sept. (1970) 315 It was a pleasant journey back to the Ranch, flying over the flat, rich lands of Kansas..sunflowers everywhere, as big as salad plates. You can see why it's called the Sunflower State.

Sung (suŋ), *sb.* Also 7 **Sunga,** 8 and Pinyin **Song.** [Chinese *sòng*.] **a.** The name of a dynasty which ruled in China from 960 to 1279; a member of this dynasty. Also *attrib.*

1673 J. Ogilby *Nieuhoff's Embassy from East-India Co.* I. 249 The Tartars, after a long and tedious destructive War with this Family Sunga for seventy three years, conquer'd the whole Empire, extirpating the whole Family, and set up a new one call'd Ivena. 1738 J. B. Du Halde *Descr. Empire of China & Chinese Tartary* I. 206 (*heading*) The Nineteenth Dynasty, call'd Song. 1745 tr. J. F. Gemelli Careri in A. & J. Churchill *Coll. Voy. & Trav.* (ed. 3) IV. II. iv. 313/1 When China was rul'd by the family of Sung. 1831 *Canton Miscellany* I. 28 Hwuytsung, an Emperor of the Sung Dynasty. 1893 D. C. Boulger *Short Hist. China* v. 57 The folly of the Sungs had completed the discomfiture of the Kins. 1925 B. Rackham in R. Fry et al. *Chinese Art* 16 A further wide expansion of craftsmanship is shown by the manifold variety of wares of the Sung period. 1958 W. Willetts *Chinese Art* I. iii. 133 The Sung Emperors wanted to surround themselves with examples of ancient high art. 1977 *N.Y. Rev. Bks.* 26 May 21/2 (Advt.), To the Sung, poetry was a part of everyday life. 1979 *China Now* Mar./Apr. 9/2 For many practising potters the heights of Chinese ceramics were achieved during the Song Dynasty.

b. Used *attrib.* and *absol.* of the arts, design, and porcelain of the Sung period.

1885 *Trans. Asiatic Soc. Japan* XII. 171 The angular forms..of the Sung dynasty, usually called the *Sung-pan* ..or '*Sung-block*' printing. 1906 R. Fry *Let.* 2 Dec. (1972) I. 275 He's got..some first-rate early Chinese Sung pieces. 1933 *Burlington Mag.* Nov. 204/1 It is obviously a copy of a Sung or Yüan celadon dish. 1937 E. Linklater *Juan in China* xii. 222 I've plenty of things to show you, Ming, Sung, pictures, anything you like. 1943 D. Welch *Maiden Voyage* xviii. 149 This is another type of Sung porcelain called Ying-Ching or 'shadow blue'. 1961 *Guardian* 19 May 9/7 The European eye cannot fail to respond to what it would call the romanticism of Sung landscapes. 1976 M. Delving *China Expert* ii. 36, I shall hope that whoever stole the Sung vase will make his little slip.

sungar, sangar, *sb.* Delete ‖ and add: Now chiefly *Mil.* Also **sanger.** The spelling *sangar*, with pronunc. (sæ·ŋəɪ), is now usual. (Later examples.) Also, a strong point or fortified look-out post.

1938 D. Forbes *My Life in S. Afr.* v. 68 Small sangers were put up at other commanding positions to hold the enemy back until the fort was finished. 1944 *N.Y. Times* 25 Apr. 5/5 It was about noon..when he climbed out of his own 'sanger',—a type of rocky foxhole characteristic of this section [of Italy]. 1951 G. Wilson *Brave Company* vii. 135 Our sangars are much more elaborate. We have erected roofs reinforced by sandbags and screw pickets. 1962 *Times* 2 May 14/7 The pickets settled down in their *sangars* [in a wadi, W. Aden]. 1974 *Sunday Times* 17 Feb. (Colour Suppl.) 27/3 High look-out platforms called *sangars* [in Northern Ireland]. 1979 *Observer* 4 Mar. 11/1 The man on sangar duty must look out through a narrow slit and observe movements for up to four hours at a time. 1982 *Times* 12 June 5/4 To the commando crouching in his damp 'sanger', a slit trench built up with rocks, there has seemed little reason for the wait.

sungar, *v.* (Earlier example.)

1900 W. S. Churchill in *Morning Post* 25 July 5/7 Both infantry and guns are strongly sangared among the rocks and stones of the kopjes.

sungates, *adv.* For †*Obs.* read 'Now only *arch.*' and add: Also **singates, sungaets.** (Later examples.)

1879 *Shetland Times* 20 Sept. 3/5 They paired and proceeded to the house, walking once round it 'singates', *i.e.* sun ways, or from left to right; that was to secure luck to the pair. 1890 J. Service *Notandums* 100 She was tell't to tak her withershins nine times through a heap o' unwatered yarn, to tak the cat through't sungates aboot as many times again. 1916 A. Huxley *Burning Wheel* 49 Though they turn sungates to its widdershins. 1931 [see *NOUST].

sunk, *ppl. a.* **2.** Restrict Now *rare* or *Obs.* to senses in Dict. and add: **d.** *colloq.* Of a person: in a hopeless position, in trouble, in a mess. Freq. *hyperbolical.*

1922 [see *SPURLOS VERSENKT]. 1934 A. P. Herbert *Holy Deadlock* 103 'Hell!' thought Mr. Ransom, 'we're sunk!' 1941 M. Allingham *Traitor's Purse* xx. 231 You can't say you're afraid we're sunk... Everyone's relying on you. 1951 J. Frame *Lagoon* 56 If visitors come tonight I am sunk. 1960 G. Sanders *Mem. Professional Cad* II. v. 136 If you go to a party with the impedimenta of a date, an overcoat or a hat, you are sunk.

4. b. *sunk band, cord*, a strip of cloth or string on which a binding is constructed, fitted in to furrows across the spine of the book; *sunk-enamel*, champlevé; *sunk garden*, a (portion of a) garden created below the natural level of the surrounding land, a sunken garden (Sunken *ppl. a.* 4).

1889 W. Matthews *Mod. Bookbinding* 27 [Raised-band sewing] is three or four times the cost of the ordinary, or sunk-band, sewing... Sunk-band is the ordinary style of the book sewing of our time. Here the sheets are sawed with three or five furrows to admit the bands of twine. 1959 L. M. Harrod *Librarians' Gloss.* (ed. 2) 268 Sunk Bands (Cords)... Cords or bands..placed in grooves sawn into the backs of sections of a book. 1965 L. S. Darley *Introd. Bookbinding* 61 (*caption*) Sawing for sunk cords. 1929 *Times* 2 Nov. 10/4 The coral inlay, and red sunk-enamel (champlevé) on the handle. 1922 J. Buchan *Huntingtower* v. 93 A path which wound down to the sunk garden. 1973 *Country Life* 15 Nov. 1591/1, I bask on a stone seat in the sunk garden.

sunken, *ppl. a.* Add: **2.** *Comb.*, as *sunken-eyed*.

1851 H. Melville *Moby Dick* I. xxxiv. 253 Your whales must be seen before they are killed; and this sunken-eyed Platonist will tow you ten wakes round the world, and never make you one pint of sperm the richer. 1971 S. Hill *Strange Meeting* 191 The men glanced up apprehensively as they passed along, and their faces had the sunken-eyed look of suppressed fear.

4. *sunken bath, living-room.*

1925 F. Scott Fitzgerald *Great Gatsby* (1926) v. 110 Through dressing-rooms and poolrooms, and bathrooms with sunken baths. 1979 J. Melville *Wages of Zen* xi. 110 The tiled sunken bath was drained and clean. 1970 *Globe & Mail* (Toronto) 26 Sept. 42/3 (Advt.), 3 Bedrooms, 3 Washrooms, Electric Light Fixtures, Sunken Living Rooms. 1976 M. Machlin *Pipeline* xxxvii. 408 The entrance hall, which gave onto a white-carpeted, sunken living-room, looked as though a regiment of cavalry had galloped through it.

sunker (sʌ·ŋkəɪ). *Newfoundland.* [f. Sunk *ppl. a.* + -er[1].] A submerged rock. Also *fig.*

c 1880 in G. S. Doyle *Old Time Songs & Poetry of Newfoundland* (1927) 29 We'll rant and we'll roar on deck and below, Until we see bottom inside the two sunkers. 1896 *Jrnl. Amer. Folklore* IX. 33 Among the peculiar words connected with the fishing I note the following:..*sunker*, a breaker. 1951 *Newfoundland & Labrador Pilot* I. 134 Duck Road shoal, about 4 cables north-eastward of Eastern head and Anchor Cove Sunkers. 1966 A. R. Scammell *My Newfoundland* 63 The words don't seem to have a clear channel from me brain to me lips. Too many sunkers for 'em to ground on, I spose. 1973 *Maclean's Mag.* Jan. 16/1 At dusk, the Nordfjeld slammed onto a 'sunker', a rock that's awash at high tide, one mile off Flowers Cove.

sunket (sʌ·ŋkėt), *sb.*[2] *dial.* [Origin obscure.] A simpleton, a silly fellow.

1823 E. Moor *Suffolk Words & Phrases* 409 Sunket. A child sickly and unpromising is so called—'Ah! 'tis a poor sunketing thing.' 1895 W. Rye *Gloss. Words E. Anglia* 219 *Sunket*,..a contemptuous appellation of a silly fellow. 1940 C. P. Snow *Strangers & Brothers* 21 'You can't take sides with those sunkets against me,' said George. His voice had risen. We were used to the odd Suffolk words as his temper got up.

sunlight, *sb.* Add: **1. c.** *artificial sunlight*: see *ARTIFICIAL *a.* 5.

sun-lighted, sunlighted, *ppl. a.* (Earlier U.S. example.)

1787 M. Tilghman *Let.* 6 Oct. in *Maryland Hist. Mag.* (1926) XXI. 220 The bright, Sun-lighted Wedding proceedings.

su·nlighting, *vbl. sb.* [f. Sunlight *sb.* + -ing[1].] **1.** The process, degree, etc., of the illumination of buildings by sunlight.

1961 [see *DAYLIGHTING *vbl. sb.* 2]. 1977 *Washington Post* 10 Apr. B5/1 Maximum sunlighting and, therefore, heat capture are assured in the winter when the sun is low.

2. (See quot. 1977.)

1977 D. N. Baron in *Brit. Med. Jrnl.* 22 Oct. 1080/1 If taking on paid outside work during night-time hours is moonlighting then let us call doing outside work (though unpaid) during daytime hours sunlighting. 1978 *Lancet* 14 Jan. 89/2 Then follow 3–4 years of internship or resi-

dency training..with opportunities of moonlighting and sunlighting. 1983 *N.Y. Times* 7 Feb. A2/3 Bureaucrats [in Madrid] practiced moonlighting to such an extent it turned into sunlighting.

Sunnism (sʌ·niz'm). Also **Sunniism.** [f. Sunna or Sunni + -ism.] The doctrines or principles of the Sunnites.

1892 *Chambers's Encycl.* IX. 398/2 The moderate Shiism that has been the national religion of Persia since the native royal line of Safides ascended the throne in 1499 is more Koranic than Sunnism. 1911 D. S. Margoliouth *Mohammedanism* v. 174 Nadir Shah attempted to substitute Sunnism. 1953 O. Caroe *Soviet Empire* iv. 50 The Samanids came from Baltch and enforced a rigorous form of Hanafi Sunniism. 1962 G. E. von Grunebaum *Mod. Islam* 11 The catholicity of Sunnism. 1983 S. Akhavi in N. R. Keddie *Relig. & Pol. in Iran* vii. 129 His [*sc.* Shariati's] thought has more affinities to Shi'ism than to Sunnism.

sunny, *a.* **2. b.** Restrict † *Obs.* to phrases in Dict. and add: *sunny South*: the southern states of the U.S.

1846 *Spirit of Times* (N.Y.) 18 Apr. 96/2 The wish of his heart should always be, peace and prosperity to the 'Sunny South'. 1950 *Chicago Tribune* 11 Mar. 8/3 Eric, the redbird..flew by, fat 'n' sassy from a sojourn in the sunny South.

5. b. *sunny side* (*fig.* or in fig. contexts), (*a*) in phrases expressive of cheerfulness or optimism, esp. *on the sunny side of the wall*; (*b*) *on the sunny side of* (an age): on the right side of, i.e. less than (cf. Shady *a.* 2 b); (*c*) *sunny side up*: of an egg, fried on one side only; hence *sunnyside egg*.

(*a*) 1831 E. Trelawny *Adventures Younger Son* II. viii. 61 Then, only looking at the sunny side of things, all was bright. 1837 [in Dict., sense 5]. 1858 Trollope *Doctor Thorne* I. vi. 141 Mary..was..of the same age as Frank; but, as I..have so often said before, 'Women grow on the sunny side of the wall.' 1890 W. S. Gilbert *Gondoliers* II. 119 Live to love and love to live—You will ripen at your ease, Growing on the sunny side—Fate has nothing more to give. 1970 C. Major *Dict. Afro-Amer. Slang* 111 *Sunny side* (*of the street*), the 'good life'; luxury, leisure and comfort.

(*b*) 1865 *Atlantic Monthly* XV. 711 How many of us,.. on the sunny side of thirty, have gone through the 'Paradise Lost'? 1967 *Boston Sunday Globe* 23 Apr. (TV Week) 2/1 That's mighty high flying for a young Negro actor still on the sunny side of 30.

(*c*) 1901 *Dialect Notes* II. 149 *Sunny side up*,..of eggs, to fry [*sic*] on only one side. 1948 *Royal Air Force Rev.* Jan. 20/2 It's whizzo when you get a fried egg sunny-side-up for tea. 1953 A. Christie *After Funeral* xii. 101 Worried, bad-tempered and irritable in the office. *But* since his uncle's death that's all changed. He's like the breakfast eggs (if we had 'em). Sunny side up! 1967 [see *ONCE *adv.* B. 8 e]. 1971 H. Howard *Murder One* vii. 92 If he's made a funeral in the family he'll fry like a sunny-side egg. 1979 R. Fiennes *Hell on Ice* iii. 32 Ginnie ladled her sunny-side-up eggs' with semolina.

c. *Sunny Jim*, the name of an energetic character employed as the proprietary name for a brand of breakfast cereal; also used allusively, as a term of address, and as a nickname. Also *Sonny Jim* (influenced by Sonny).

'Sunny Jim was the creation of an American schoolgirl called Ficken (not Fincken) and the various jingles which accompanied him were written by Miss Minnie Hanff. It is believed that Sunny Jim was the winning entry in a competition run by the Force Food Company to find a suitable advertising character to promote "Force".' —C. Fincken (A. C. Fincken & Co. Ltd., manufacturers of Force), private let. to ed., 24 June 1983.

1903 *Poster*, High o'er the fence leaps Sunny Jim 'Force' is the food that raises him. 1904 *Trade Marks Jrnl.* 30 Mar. 381 'Sunny Jim'... Cereal Food Products. The firm trading as The 'Force' Food Company, 6, Holborn Viaduct, London, E.C.; Manufacturers. *c* 1904 *Story of Sunny Jim* (Force Food Co.), Jim Dumps was a most unfriendly man, Who lived his life on a hermit plan. He'd never stop for a friendly smile, But trudged along in his moody style. Till 'Force' one day was served to him, Since then they call him Sunny Jim. 1911 Chesterton *Innocence of Father Brown* xii. 315 Sir Aaron Armstrong was..comic... It was like hearing that Sunny Jim had hanged himself. 1911 'I. Hay' *Safety Match* xii. 187 Mr. Blunt..cleared the topmost rail... 'Now then, Sunny Jim!' remarked a reproving voice. 1916 *Punch* 5 Apr. 229/1 [He] says he's quite a Sunny Jim, That buoyant health and youthful vim Are sticking out all over him. 1943 *Current Biogr.* (1944) 779/2 Vandegrift—so cheerful that he has earned the nickname 'Sunny Jim'—is..the toughest Marine Corps leader who ever charged at the head of his troops. 1960 D. Storey *This Sporting Life* I. v. 127 'All right, all right! Don't preach, sonny. Hey!' she calls to the M.P. 'Hey sonnyjim! What kinda car has Arthur Machin got?' 1962 S. L. Goldberg *Joyce* i. 3 What his [*sc.* James Joyce's] early works do not portray.. is the aspect of his character that earned him the family nickname, 'Sunny Jim'. 1967 A. Wilson *No Laughing Matter* II. 127 Does your Mother know you're out, Sonny Jim? 1976 *Times* 6 Apr. 16/1 The new Prime Minister [*sc.* James Callaghan]..enjoys life... He is not called Sunny Jim for nothing.

6. a. *sunny golden, -winking.*

1922 Joyce *Ulysses* 406 She dare not bear the sunnygolden babe of day. *Ibid.* 216 He walked by the treeshade of sunnywinking leaves.

sunnyasee, sunnyasi. Add: Also **sannyasi** (now the most usual form), **sanyas(s)i, san(n)yas(s)in.**

1812 J. Malcolm in *Asiatick Researches* XI. 267 That crowd of holy mendicants, Sanyásís and Fakirs, with whom India swarms. 1854 Geo. Eliot tr. *Feuerbach's Essence Christianity* xvii. 167 No regenerate man could assume the rank of a Sanyassi,..if he had not previously paid three debts. 1876 Sannyasin [see *BHIKSHU]. 1891 Monier Williams *Brahmanism & Hinduism* 55 He was a Sannyāsī and an unmarried Smārta Brāhman. 1938 W. S. Maugham *Writer's Notebk.* (1949) 280 At his birth his horoscope was taken, and the astrologer said that he would either become a very rich, successful man, a king among men, or a sanyasin. 1957 *Contributions to Indian Sociol.* I. 17 Now, what one is in the habit of calling Indian Thought is for the very great part the thought of the sanyasi. 1960 E. R. Leach *Aspects of Caste in S. India, Ceylon & N.W. Pakistan* 6 It is open to every man to become a *sannyasi* and receive the adulations of his society. 1978 *Times* 5 Aug. 7/8 Dom Bede Griffiths, a Benedictine monk who has spent the past 16 years of his life as a *sannyasi*—a kind of hermit—at a Hindu ashram at Kerala in India. 1980 *Daily Tel.* 13 Feb. 15/4 She joined a group of sanyassins, and became fascinated with the guru's writings.

sun-ray. Add: **2. b.** *sun-ray pleat, pleated* adj., *pleating*.

1903 *Young Woman* XI. 318/2 For evening dresses accordion pleated—or sun-ray pleated—point d'esprit net is an ideal material. 1922 Joyce *Ulysses* 731 The orange petticoat I had on with sunray pleats. 1935 *Times* 4 Nov. 9/3 Sunray pleating gives fullness from the knee. 1959 *Spectator* 2 Jan. 10/2 Short office skirts and dropping Spanish skirts and brisk housewifely sun-ray pleats. 1972 *Country Life* 25 May 1354/2 Sunray pleating is again in fashion. 1978 *Lancashire Life* Apr. 79/2 A tailored jacket which can be teamed with a sunray pleated skirt.

3. An (artificial) ultraviolet ray used for medical or cosmetic treatment. Chiefly *attrib.*, esp. in *sun-ray lamp*; formerly also *ellipt.* for *sun-ray treatment*.

1928 *Daily Express* 27 June 3/6 The speedy development of sun-ray clinics all over the country. *Ibid.*, The treatment of disease by artificial sun-rays. 1930 M. Kennedy *Fool of Family* x. 89, I wonder if sunray treatment would do her good... Sir Ivor knew nothing of sunray, and he had no faith in doctors. 1954 'N. Blake' *Whisper in Gloom* II. xii. 164 Does this young lady..own a sun-ray lamp? 1977 C. Fremlin *Spider Orchid* xxii. 149 He looked from the bedside table to the sunray lamp.

sunrise. Add: **b. sunrise industry,** a new and expanding industry; cf. *sunset industry* s.v. *SUNSET 3.

1980 L. C. Thurow *Zero-Sum Society* (1981) iv. 95 We do need the national equivalent of a corporate investment committee to redirect investment flows from our 'sunset' industries to our 'sunrise' industries. 1980 *Economist* 23 Aug. 16/2 Those who try to shelter dying jobs in sunset industries, and thereby blight the prospects of growth of good jobs in sunrise ones. 1983 *Times* 20 Apr. 21/7 The traditional 'sunset' industries are a pain in the neck for the Industry Secretary. However much he tries to brush them under the carpet in favour of the glamorous 'sunrise' sector of high technology, they persist in creeping back into the public consciousness.

sunset. Add: **1. b.** *to ride* (*go, sail*, etc.) *off into the sunset*, phr. derived from a conventional closing scene of many films used, freq. ironically, to denote a happy ending.

1967 H. Harrison *Technicolor Time Machine* (1968) iii. 28 He takes the girl with him and together they sail into the sunset to a new life. 1976 W. Goldman *Magic* III. xii. 207, I didn't even bother getting mad at your crack about me going off into the sunset. 1977 *Times* 17 Feb. 6/4 Our black hero..rides off to freedom in the sunset.

3. *sunset-red* (examples; also as *sb.*), *-tinted* adjs.; **sunset-gun** (earlier U.S. example); **sunset home,** a home (Home *sb.* 8) for the elderly, a 'twilight' home; **sunset industry,** an old and declining industry.

1840 Thoreau *Jrnl.* 16 June in *Writings* (1906) VII. 141 To hear..the bittern begin to boom from his concealed fort like a sunset gun! 1978 *Dædalus* Spring 220 A society that increasingly emphasizes..singles bars for the young and..sunset homes for the elderly. 1980, 1983 Sunset industry [see *sunrise industry* s.v. *SUNRISE b]. 1837 E. B. Browning *Epistle to Canary* (1913) 11 A spark of light from highest dawn, Which glows and opens..till sunset reds are likest to them. 1934 Webster, Sunset-red adj. 1964 *New Yorker* 5 Sept. 86 (Advt.), Slip into the run-about shift...mirage aqua, sunset-red, cactus-green, burnt-clay. 1979 *Arizona Daily Star* 5 Aug. (Parade Suppl.) 24/2 (Advt.), These are handsome..books, smartly bound in sunset red, desert tan and cavalry blue. 1876 J. G. Whittier *Mabel Martin* (new ed.) 40 And sad the uncompanioned eves, And sadder sunset-tinted leaves. 1954 L. MacNeice *Autumn Sequel* 33 The sunset-tinted Balloons were down.

b. *N. Amer. Pol.* Applied to legislation whereby a government agency or programme is automatically terminated at the end of a fixed period unless formally renewed.

1976 *National Observer* (U.S.) 5 June 1/1 The hottest political idea of the year is something called the sunset bill. *Ibid.* 1/5 Colorado recently became the first state to adopt sunset legislation. 1976 *Wall St. Jrnl.* 25 June 1/1 Colorado's new 'Sunset Law'. The experimental measure would terminate in six years the state's regulatory agencies..unless they justify their existence. 1978 *Globe & Mail* (Toronto) 15 Feb. 1/3 The Ontario Government is on the verge of embracing sunset law. 1982 *Times* 25 Feb. 7/8 It is not a wilderness protection Bill.. but a wilderness sunset Bill, that would end wilderness protection.

Hence **su·nset** *v.*, (*a*) *intr.*, to decline, sink (*rare*); (*b*) *trans.*, to subject to, or terminate by means of, sunset legislation (see sense 3 b above). (*N. Amer.*)

1933 V. McNabb *Nazareth or Social Chaos* 30 The prodigal's prosperity which sunsets in beggary. 1978 *Canadian Broadcasting Corporation News* 1 June (*oral quot.*), [The] act would be sunsetted out of existence. 1979 *National Jrnl.* 17 Mar. 438/1 In the debate on the sunset bill..when it was passed by the Senate, we tried to figure out what exactly they thought they were sunsetting. 1982 *N.Y. Times* 28 Sept. B.10/4 His impatience is also displayed in his plans to 'sunset' the aeronautics board well before it is scheduled to expire.

sunshade. Add: **2.** (Earlier U.S. example.)

1842 *N.Y. Times* 22 Mar. (Advt.), Umbrellas, parasols and sun-shades manufactured at Newark, N.J.

5. *pl.* = *sun-glass* (*c*) s.v. *SUN *sb.* 13 a. Cf. *SHADE *sb.* 11 e.

1965 'Lauchmonen' *Old Thom's Harvest* ii. 24 The minister took off his sunshades and his naked eyes followed the girl. 1967 *Sunday Times* 28 May 18, I give a lot of speeches with my sunshades on.

sunshine, *sb.* Add: **2. a.** (Later example.) Now freq. as a colloq. form of address to any person. Cf. also (*little*) *ray of sunshine* s.v. *RAY *sb.*[1] 1 a.

1942 Berrey & Van den Bark *Amer. Thes. Slang* § 184/1 Hap, Happy, Sunshine, a cheerful person. 1972 M. Gilbert *Body of Girl* iv. 43 Hullo, Sunshine. What can we do for you? 1976 P. Cave *High Flying Birds* ii. 17, I turned back to the ticket man. 'OK now, sunshine?' 1976 *Daily Mirror* 16 Mar. 12/1 Mike Reid's cheery Cockney greeting seems to switch on every girl in the place. 'Hello sunshine,' he says.

6. sunshine law *U.S.*, a law making the official meetings and records of certain government agencies accessible to the public; **sunshine roof,** on a motor vehicle, a roof that can be slid open; = *sun-roof* (*a*) s.v. *SUN *sb.* 13 a; **Sunshine State,** (*a*) *U.S.*, any of several states (see quots.); (*b*) *Austral.*, Queensland; **sunshine-yellow** *a.* and *sb.*, (of) a bright shade of yellow.

1972 *Atlantic Monthly* Sept. 22 The state of Florida has adopted a 'sunshine law' which requires that all official meetings in which public business is transacted be open to the public. 1977 *National Observer* (U.S.) 22 Jan. 16/5 We need an all-inclusive 'sunshine law' in Washington so that special interests will not retain their exclusive access behind closed doors. 1982 *Times Lit. Suppl.* 26 Feb. 225/5 'Sunshine laws' have opened committee hearings to public scrutiny. 1929 *Daily Express* 12 Jan. 4/7 The royal coupé, with a sunshine roof. 1954 J. Trench *Dishonoured Bones* II. v. 72 He got out the old Austin..unfastening the sunshine roof. 1977 *Horse & Hound* 14 Jan. 44/3 (Advt.), Land-Rover..sunshine roof. 1893 L. Wagner *Significance of Names* 36 New Mexico is The Sunshine State. 1918 S. S. Visher *Geogr. S. Dakota* 60 South Dakota is known as 'the Sunshine State', not because it surpasses in this respect..states..in the south-west, but because of the contrast between South Dakota and the Eastern States and northern European countries from whence most of the persons not born in South Dakota came. 1920 *Monthly Weather Rev.* (U.S. Weather Bureau) Mar. 154/2 In this 'Sunshine State' [*sc.* California] we have 'Sunkist Orange'. 1947 *Time* 17 Mar. 42/2 Employees..are happier in The Sunshine State [*sc.* Florida] where living is so pleasant and healthful. 1962 C. Rohan *Delinquents* 128 'If you ask me, all Brisbane's full of coppers and all of them bastards,' she said, expressing in one concise sentence the full theory of central government of the sunshine state. 1976 *Daily Record* (Glasgow) 30 Nov. 28/6 And although the Sunshine State has seen a few fancy sights the citizens of Palm Springs are in for an extra special treat next week. 1971 'A. Gilbert' *Tenant for Tomb* i. 22 The plumber..wanted to install a bright blue bath..and had to be coaxed into substituting a sunshine-yellow one. 1975 A. Fraser *Whistler's Lane* x. 161, I bought myself a blouse in sunshine yellow.

sun-spot. Add: Now usu. **sunspot. 2. b. sunspot cycle,** the recurring increase and decrease in the number of sunspots, with a period averaging just over 11 years.

1922 H. S. Jones *Gen. Astron.* v. 126 There is a remarkable connection between the Sun-spot cycle and the occurrence of magnetic storms on the Earth. 1977 *Jrnl. R. Soc. Arts* CXXV. 157/1 Since the solar output appears to change by a good deal less than 1 per cent, even during solar flares, it is not surprising that weather events show very little correlation with the sunspot cycle.

3. *Cinemat.* A powerful arc lamp used to imitate the light from the sun in colour cinematography; = *sun arc* s.v. *SUN *sb.* 13 a.

1930 [see *sun arc* s.v. *SUN *sb.* 13 a]. 1976 H. R. F. Keating *Filmi, Filmi, Inspector Ghote* iii. 29 We are using a great number of different lights for different purposes] in filming, Five-Ks, Two-Ks, Sunspots, Solars, Babies.

4. A place that affords plentiful sunshine.

1976 *U.S. News & World Rep.* 2 Feb. 26/1 Most travelers favor sun spots. St. Maarten, in the Netherlands Antilles, leads the parade. 1983 *Listener* 6 Jan. 36/3 (Advt.), First choose your holiday sun-spot, then choose where to stay.

sunstone, sun-stone. Add: **5.** [tr. ON. *sólarsteinn*.] A stone whose exact properties

are uncertain, mentioned in several medieval Icelandic sources.

'A semi-precious stone capable of being used as a burning-glass': P. G. Foote in *ARV: Jrnl. Scandinavian Folklore* (1956), XII. 26–40.

1874 Cleasby & Vigfusson *Icelandic-Eng. Dict.* 579/2 *Sólar-steinn*, m. a sun-stone or loadstone, = leiðarsteinn, used by sailors to find the place of the sun on a cloudy day. 1947 J. E. Turville-Petre tr. *Story of Rauð & his Sons* 24 The King..sent a man out to observe the weather, and there was not a patch of clear sky to be seen. The King then asked Sigurd to determine how far the sun had travelled. He gave a precise answer. So the King had the sun-stone held aloft, and observed where it cast out a beam; the altitude it showed was exactly as Sigurd had said. 1968 *Carnegie Mag.* May 152/1 In overcast weather, a 'sunstone' determined the position of the sun. 1970 B. E. Gelsinger in *Mariner's Mirror* LVI. 222 Thorkild Ramskou..suggested that the sunstone was a crystal such as Iceland spar which polarizes light... The sunstone could thus indicate the position of the sun even though the sky was completely overcast. This description..harmonizes with non-Icelandic references to the sunstone... Pliny the Elder..described the sunstone or *solis gemma* as a white stone which casts rays of the sun. 1980 M. Magnusson *Vikings!* vii. 191 Unfortunately, today's scholars do not rate the so-called sun-stone as a Viking Age navigational aid..; nothing is sacrosanct in the severe world of scholarship.

sunt. Add: Also **sont.** For *Acacia arabica* substitute *Acacia nilotica*. (Later example.)

1901 *Knowledge* June 138/2 The timber forming a raft is generally of the 'sont' tree.

sun-tan, *sb.* (and *a.*). (In Dict. s.v. Sun *sb.* 11 b.) Add: **1. a.** (Later examples.)

1958 M. K. Joseph *I'll soldier no More* xiii. 237 They're just out for a bit of suntan. 1980 *West Lancs. Evening Gaz.* 11 Aug. 10 (Advt.), A guaranteed suntan without sunburn.

b. In *Comb.* designating cosmetics which provide protection against sunburn and promote suntanning, as *sun-tan lotion, oil*, etc.

1934 *Beautycraft* July 19/1 To acquire a brown, healthy skin..it must be anointed plentifully with one of the good Sun-tan oils now on the market. 1938 E. Ambler *Cause for Alarm* vi. 90 The points of his dress collar..were.. smeared with grease and sun-tan powder. 1951 Koestler *Age of Longing* ii. 36 She felt herself go slightly pale under the sun-tan make-up. 1962 'E. McBain' *Like Love* ix. 132 Contents medicine cabinet...one tube suntan lotion, one bottle Seconal, one toothbrush. 1976 P. Parish *Medicines* II. xli. 242 The effectiveness of suntan applications is..related to their ability to cut out the burning effects of the sun's rays.

2. *pl.* **a.** Lightweight, tan-coloured summer uniform worn by military personnel. **b.** Trousers forming part of this uniform or similar slacks for casual wear. *U.S.*

1937 *Amer. Speech* XII. 75/1 *Suntans*, summer uniform, made of lightweight material with sheen. 1945 E. Newhome in *New Yorker* 10 Feb. 22/1 He had removed only his tie and was lying..in his suntans. 1947 J. Bertram *Shadow of War* VIII. v. 279 We stared at the Commodore's drab suntans. 1958 'E. Dundy' *Dud Avocado* I. i. 7 The Left Bank uniform of the day, dark wool shirt and a pair of old Army suntans. 1960 J. Updike *Rabbit, Run* 98 He takes clean Jockey pants, T-shirts,..a pair of laundered suntans..and a sports shirt from the closet. 1966 *Times* 28 Mar. (Austral. Suppl.) p. viii/4 The streets are full of people in shorts and suntans. 1972 W. McGivern *Caprifoil* (1973) xiii. 217 Admiral Burkholder..wore suntans, and the collar of his shirt was open.

3. A light-brown fashion colour. Also as *adj.*

1937 [see *MIST *sb.*[1] 1e]. 1976 *Horse & Hound* 3 Dec. 17 (Advt.), Deep pile Borg washable numnah, foam filled. Pony or F.S. general purpose. Cream or suntan.

Hence as *v. trans.* and *intr.*, to expose (oneself) to the sun in order to acquire a tan; **su·n-tanned** *ppl. a.*; **su·n-tanning** *vbl. sb.* and *ppl. a.*

1821 Sun-tanned [in Dict. s.v. Sun *sb.* 12 b]. 1876 'Mark Twain' *Tom Sawyer* xviii. 185 That swarthy, suntanned skin of his. 1932 *Sun* (Baltimore) 5 Sept. 6/2 The millions busy today suntanning themselves, picnicking in the country. 1938 W. de la Mare *Memory* 16 The suntanned soldiers. 1959 *Chambers's Encycl.* III. 762/2 Genetically or environmentally induced melanization (sun-tanning) of the skin may serve a useful function in screening out injurious short-waved fractions of the sunlight from the sensitive underlying tissues. 1961 *Times* 29 Nov. 13/6 Just the right amount of suntanning. 1976 B. Shelby *Great Pebble Affair* 117 Donnely and I were suntanning on the roof. 1977 N. Faulks *No Mitigating Circumstances* vii. 99, I had a little lawn tennis at Monfalcone as well as at Trieste, and had an idle, suntanning time.

Suntory (sʌntō·ri). Also **Suntori.** The proprietary name of a Japanese whisky.

1959 R. Kirkbride *Tamiko* ii. 11 Here he was..without even a drink in his hand. 'A double Suntory,' he said to the baaten. 1960 *Trade Marks Jrnl.* 21 Dec. 1659/2 Suntory 809,445. All goods included in Class 33 [*sc.* alcoholic beverages]. Kabushiki Kaisha Kotobukiya (a Corporation duly organised and existing under the laws of Japan)..Osaka, Japan; Manufacturers.—15th Aug. 1960. 1967 'J. H. Roberts' *February Plan* I. i. 17 He..remembered enough of his long unused Japanese to order a bottle of Suntori. 1975 R. L. Duncan *Dragons at Gate* (1976) iii. 99, I have ordered Kobe steaks... I have also requested a bottle of Suntory.

sun-up, sunup. For *local*, chiefly *U.S.* read: *local U.S.* (chiefly *Midland*), *Caribbean*, and formerly (perh. rendering Afrikaans *sonop*) *S. Afr.* Add: freq. in phr. *from sun-up to sun-down*. (Earlier and further examples.)

1712 T. Banister *Let.* 12 Nov. in *Coll. Connecticut Hist. Soc.* (1924) XXI. 377 Wee Set out by or before Sun up, for Wyndham. 1826 J. F. Cooper *Last of Mohicans* I. iv. 69 One would think such a horse as that might get over a good deal of ground atwixt sun-up and sun-down. 1887 Rider Haggard *Jess* xxxii. 305 Will you consent to marry me to-morrow morning at sun up, or am I to be forced to carry the sentence on your old uncle into effect? 1903 K. D. Wiggin *Rebecca of Sunnybrook Farm* x. 102, I could teach school from sun-up to sun-down if scholars was all like Rebecca Randall. 1920 [see *KLOMP]. 1949 *Caribbean Q.* I. III. 45 Your face turned to sun-up. *a* 1963 S. Plath *Crossing Water* (1971) 47 The blue hour before sunup. 1965 'Lauchmonen' *Old Thom's Harvest* viii. 99 Another hour and it was sun-up. 1976 A. Haley *Roots* (1977) cxiii. 646 Twenty-eight wagons were packed and ready to roll on the following sunup.

sunwards, *adv.* **2.** (Earlier example.)

1851 H. Melville *Moby Dick* III. xxx. 189 Here, too, life dies sunwards, full of faith.

sunyasi, var. Sunnyasee, sunnyasi in Dict. and Suppl.

‖ **sunyata** (sūnya-tā, ʃ-). *Buddhism.* Also **çûnyatâ.** [Skr. *śūnyátā* emptiness, non-existence, f. *śūnyá* empty, void.] The concept of the essential emptiness of all things and of ultimate reality as a void beyond worldly phenomena.

1907 D. T. Suzuki *Outl. Mahâyâna Buddhism* vii. 173 The emptiness of things (*çûnyatâ*) does not mean nothingness..but..conditionality or transitoriness of all phenomenal existences. 1916 A. Coomaraswamy *Buddha & Gospel of Buddhism* v. 318 The *Prajñāpāramitās* are filled with..texts upon the Emptiness (*Sunyata*) of things. 1938 B. L. Suzuki *Mahayana Buddhism* i. 15 Sunyata is what is left behind after an endless series of negations, and is therefore the most positive and fundamental of ideas. 1951 E. Conze *Buddhism* v. 130 We must now make an effort to understand this all-important idea of *Emptiness*... What we call *emptiness* in English is *sūnyatā* in sanskrit. 1978 C. Humphreys *Both Sides Circle* v. 57 What I call the mystical metaphysics of the *Madhyamika* (Middle Way) School, founded by Nagarjuna and expanded through several centuries into the ultimate concept of *sunyata*, 'no-thing-ness'.

‖ **Sun Yat-sen** (sʊn yæt sen). Also **Sun Yatsen.** The Cantonese form of the personal name Sun I-xian, adopted by Sun Wen (1866–1925), founder in 1911 of the Republic of China, used *attrib.* to designate a modern style of clothing in China.

1946 O. Lang *Chinese Family & Society* ix. 77 Those who wear long Chinese gowns are usually old-fashioned men... Modern-minded officials wear black coats with high collars, the so-called 'Sun Yat-sen jackets', and tight trousers tucked into black or khaki puttees—a Western garment common in China. 1965 'Han Suyin' *Crippled Tree* I. xvi. 222 Most of us had come to wear the Japanese students' uniform, which later was termed the Sun Yatsen suit, and is now spoken of as the Communist garb. 1977 'S. Leys' *Chinese Shadows* (1978) ii. 75 Impeccably cut Sun Yat-sen jackets. (*Note*) Chung-shan chuang, which a silly vogue in the West persists in calling a 'Mao jacket'—as if the present regime had invented it.

Sun Yat-senism (sʊn yætse·niz'm). Also as one word. [f. prec. + -ism.] The political principles of Sun Yat-sen, which included Chinese nationalism, democracy, and the people's livelihood (the 'three principles of the people').

1927 *Observer* 17 July 20/2 The understanding between Chiang Kai-shek and Feng Yu-hsiang is precariously maintained by intermediaries, not by Sun Yat-senism. 1931 tr. *P. M. D'Elia's Triple Demism of Sun Yat-Sen* 41 Some authors have not hesitated to believe that they could make a certain distinction between 'Sunyatsenism' and 'Sunwenism'. 1957 Chiang Kai-shek *Soviet Russia in China* I. i. 36 Officers and cadets at the Academy.. formed a rival group named Society for the Study of Sun Yat-senism. 1979 *World Today* June 244 During his long career, Ho Chi Minh made a point of reassuring both Chiang Kai-shek and Mao Tse-tung that he was dedicated, first, to Sun Yat-senism and, later, to Marxism-Leninism-Mao Tse-tung thought.

sup (sʌp), *sb.*[2] *Math.* [f. *SUP(REMUM).] Supremum (of).

1940, 1949 [see *INFIMUM]. 1968 E. T. Copson *Metric Spaces* i. 14 The supremum or least upper bound of *A*..is denoted by sup *A*.

sup, *v.*[1] Add: **2.** (Later examples.) Also const. *up*.

1869 [in Dict., sense 1 b]. 1898 J. MacManus *Bend of Road* 94 There wasn't a man ever supped from a noggin in Corradooey he couldn't sweep the floor with! 1952 M. Tripp *Faith is Windsock* v. 86 Now sup up, as Arthur [a north-countryman] says, and have one on me. 1971 D. Lees *Rainbow Conspiracy* ix. 152 Sup up first—that's let the beer get cold. 1977 Scollins & Titford *Ey up, mi Duck!* III. 30 Men were content just to gossip and 'sup'.

supe. Add: (Later examples.) Also as *v. intr.*, to act as a supernumerary in a theatre. Cf. *SUPER *v.* 1.

1893 W. K. Post *Harvard Stories* 107 Do you remember the time..that we saw the old man suping in that spectacular play? 1915 J. B. Rathbun *Motion Picture Making* 72 'Westerns' are invariably taken in their proper locale, the 'supes' usually being drawn from the ranches and towns surrounding the studio. 1977 [see *SCOFF *v.*[1] 2 c].

super, *sb.* Add: **II. 3. a.** (Earlier examples.) Also in the context of films. Cf. *EXTRA *sb.* b.

1838 *Actors by Daylight* I. 112/1 Many of the old supers of course remained. 1844 Dickens *Let.* 3 Jan. (1977) IV. 9 That extraordinary compound of odd scents peculiar to a theatre..accompanies me, as I meet perspiring supers in the narrow passage. 1924 Galsworthy *White Monkey* II. ii. 125 The lurid professions—film-super, or mannequin. 1930 E. Waugh *Vile Bodies* ix. 155 The Colonel's somewhere in that little crowd singing the hymn... He was crazy to be allowed to come on as a super.

5. (Earlier and later examples.)

1857 F. Cooper *Wild Adventures in Austral.* 59 'Scotchy'..introduced me as a particular friend to Wilder the owner of that run, under the impression that a 'super' was required. 1864 C. R. Thatcher *Songs of War* 12 The Super issued Quite a rabid Proclamation. 1916 A. Bennett *These Twain* iii. 33 The sole lecture of his répertoire, but it had served to raise him ever so slightly out of the ruck of 'Supers'. 1939 'F. O'Brien' *At-Swim-Two-Birds* 80 The policemen were rounded up and marched across the prairies to the Circle N, as fine a body of men as you'd hope to see, myself and the super as proud as be damned at the head of them. 1953 'N. Blake' *Dreadful Hollow* xv. 197 My dear old Super, Pop was always threatening to cut us off with a shilling. 1977 T. Berger *Who is Teddy Villanova?* i. 1 My apartment, on the door of which the churlish super had posted a notice that tended to humiliate.

8. *Bookbinding.* [Origin unknown.] = *MULL *sb.*[7] (see quots. 1914, 1970). *U.S.*

1914 J. J. Pleger *Bookbinding* III. 15 *Super*, a thin, loosely-woven, starched cloth glued on the back of books. 1916 *Rebacking Bks.* (Worcester County, Mass., Law Library) 7 The back had been hand sewed.., a strip of super between the bands pasted on the boards. 1940 Perry & Baab *Binding of Bks.* iii. 48 Super cloth (a sized cheese cloth) is too light for reinforcing purposes. 1967 [see *MULL *sb.*[7]]. 1970 R. K. Kent *Lang. Journalism* 130 *Super*,..a loose, open-weave cotton material used as reinforcement for the binding of a book: also called *crash*, or in England, *mull*.

9. [Short for *SUPERDUPER *a.* or *SUPER-BOMB.] A colloquial name for the fusion or hydrogen bomb. Now chiefly *Hist.*

1951 W. L. Laurence *Hell Bomb* i. 3, I first heard about the hydrogen bomb in the spring of 1945 in Los Alamos... They were already considering preliminary designs for a hydrogen-fusion bomb, which in their lighter moments they called the 'Super-duper' or just the 'Super'. 1958 J. Cleugh tr. *Jungk's Brighter than Thousand Suns* xvi. 259 In the discussions among those 'in the know' that arose in consequence of the news of the explosion in the Soviet Union one word was continually repeated which outsiders, at that time, would scarcely have understood. It was 'Super'. 1982 *New Scientist* 2 Sept. 642/2 It seems that the realisation of the Super, with its implications of mega-deaths, is not technically out of reach of any power which can produce a fission bomb.

III. [Absol. use of the adj.] **10.** *colloq.* High-octane or top-grade petrol.

[1965: see *PREMIUM 7 b.] 1967 'G. Douglas' *Death went Hunting* vi. 47 He wanted petrol—four of super, it was. 1978 *Country Life* 25 May 1502/2 French petrol has always been expensive... A gallon of super..costs about £1.37.

super (sū·pəɪ), *v.* **I. 1.** [f. Super *sb.* 3 a.] *intr.* To appear in a play or film as an extra or supernumerary. *Theatr. slang.*

1889 [implied in *supering* s.v. Super *sb.*]. 1920 J. Ferguson in *Northern Numbers* 98 In Town she 'supered' and would just 'walk on'. 1938 G. B. Shaw *Let.* 20 Sept. in *B. Shaw & Mrs. Campbell* (1952) 323 All the élite of the profession over forty rushed down to Pinewood to super in it [*sc.* a film]. 1976 *New Yorker* 16 Feb. 26/1 Chance for man to super in new Met production of Aida.

2. [f. *SUPER *sb.* 8.] *trans.* To back (a book) with super. *U.S.*

1914 J. J. Pleger *Bookbinding* III. 125 After enough books have been head-banded and supered.., put the backs together.

II. 3. [Short for Superannuate *v.*] *trans.* To remove (a pupil) from a school or form on account of age. Chiefly *pass.* Cf. Superannuate *v.* 3 a. *School slang.*

1902 'C. Turley' *Godfrey Marten, Schoolboy* xi. 135 'I have been in the Lower Fourth exactly four terms,' he went on, 'and my people are getting sick, and Sandy says I shall be 'supered' in a term or two.' 1923 E. W. Hornung *Old Offenders & Few Old Scores* 239 He was in our house, and super'd, poor beast! 1937 R. A. Knox *Double Cross Purposes* v. 84 He always was a scug, till the day he was super'd. 1945 T. Rattigan *Love in Idleness* I. 8 He was super'd from Eton... Removed for not being in a high enough form.

4. [Short for Superimpose *v.*] *trans.* To superimpose (a caption, etc.) on a film. *Television slang.*

1964 [see *CUE *v.*[2] a]. 1964 [see *CAPTION *sb.* 4]. 1966 G. N. Leech *Eng. in Advertising* vi. 60 In the transcription [of a television commercial], italicised portions are 'supered' and capitalised portions are in the spoken commentary. 1975 *Listener* 9 Jan. 38/2 Why not super the characters' names towards the end of a play?

super, *a.* Restrict *Trade colloq.* to senses in Dict. and add: **1.** Now also preceding the sb. as *super foot.* Cf. *superficial foot* s.v. SUPERFICIAL *a.* 2 b.

1949 *Gloss Terms Timber (B.S.I.)* 23 Super. foot, a square foot of timber of the thickness stated, e.g. 'per super. foot of 1½ in'. 1953 *Brit. Commonw. Forest Terminol.* I. 57 Foot, superficial... A unit of volume equal to 1/12th of a cubic foot. In Australia and New Zealand, applied to standing trees, logs and sawn timber of any thickness, usually under the synonym *Super foot.* 1971 *Sunday Australian* 8 Aug. 4/3 A brick veneer home of this size contains 11,000 superfeet of timber.

3. *slang.* Very good or pleasant, first-rate, excellent, 'smashing'. Also as *int.*

This sense has developed from and overlaps with sense 2 in Dict.

[1837 DICKENS *Pickw.* xli. 445 I'll be upon the wery best extra-super behaviour!] 1895 *Army & Navy Co-op. Soc. Price List* 15 Sept. 1079 White or black super opera bodice. 1923 [see *CABARET 2b]. 1932 *Daily Express* 25 June 7/4 We have race meetings here, and super bathing. 1946 J. B. PRIESTLEY *Bright Day* xi. 327 This is jolly good, though. Super. 1954 'R. CROMPTON' *William & Moon Rocket* i. 27 'Wizard,' said William. 'Super,' said Ginger. 1968 *Sunday Times* 25 Aug. 4/6 Doesn't almost everyone describe something mildly nice as 'super'. 1976 *Evening Post* (Nottingham) 13 Dec. 7/2 His wife Lee, said: 'Isn't it super? We can't get over it.'

super-, *prefix.* Add: This prefix, particularly in senses of branches II and III, has continued to be an important formative element in English. The sections in this article together with the following main entries contain a selection of those formations found more frequently during the last hundred years or so.

I. 2. a. (*b*) **su:perexcha·nge** *Physics* [ad. F. *superéchange* (H. A. Kramers 1940, in *Magnétisme* (Centre Nat. de la Recherche Sci. de France) III. 49)], an exchange force that acts between the electrons of two cations through those of an intervening anion, as in some antiferromagnetic materials; **superinvest** *v.*: delete † and add later example; **superprona·tion** = SUPINATION.

1950 *Physical Rev.* LXXIX. 354/1 This indicates that the superexchange directly through the O ion may be more powerful than that between locations making angles of 90° with the O ions. 1967 J. S. KOUVEL in J. H. Westbrook *Intermetallic Compounds* xxvii. 529/2 A remarkable feature of these superexchange interactions in ionic materials is that they almost always give rise to an antiparallel rather than parallel (i.e., ferromagnetic) alignment of moments. 1980 A. S. CHAKRAVARTY *Introd. Magnetic Properties Matter* xiv. 458 The superexchange mechanism..becomes the dominant factor if the separations between the magnetic ions are too large for the direct exchange mechanism to be operative. 1922 *19th Cent.* Oct. 594 Even sordidness itself has put off all its vileness, and is seen superinvested in beauty. 1907 *Practitioner* Apr. 486 The arms are fixed in a position of rigid extension, and frequently superpronation, so that the palms looked outward.

II. 4. a. (*a*) *super-legal, -moral, -muscan* [L. *musca* fly], *-physical, -sensational* adjs.; **su:peradiaba·tic** *a. Meteorol.*, being or involving a lapse rate greater than that of dry air when it rises and expands adiabatically (viz. approximately one degree centigrade per 100 metres), or a temperature gradient in any other fluid greater than that of an adiabatic expansion of the fluid during upward motion; **superlu·minal** *a.* [L. *lūmen, lūmin-* light], having or being a speed greater than that of light; **superna·tional** *a.* = *SUPRANATIONAL *a.*; hence **superna·tionalist** *a.*; **super-re·al** *a.* = *SURREAL *a.*; also as quasi-*sb.*; **super-reali·stic** *a.* = *SURREALIST, *SURREALISTIC *adjs.*

1925 *Nature* 28 Feb. 301/2 The frequent superadiabatic lapse-rates which occur in the bottom layer of the atmosphere. 1975 *Ibid.* 30 Oct. 748/1 Within the continental tectosphere..the thermal gradients are superadiabatic, and the dominant mechanism of heat transport is conduction, not advection. 1978 *Ibid.* 26 Oct. 726/2 The theory of corona formation [on the sun] is not well developed, and the computation of acoustic fluxes is critically dependent on the theory of superadiabatic convection. 1920 H. G. WELLS *Outl. Hist.* 478/2 He was a real monarch, super-legal. 1959 K. R. POPPER *Logic Sci. Discovery* ix. 236 Saying that they are 'spread with super-luminal velocity' is about as helpful as saying that twice two turns with super-luminal velocity into four. 1975 *Physics Bull.* Jan 13/1 The prospect of discovering superluminal particles is so appealing that the search is started afresh whenever there is an improvement or extension of experimental technique. 1980 Superluminal [see *subluminal* adj. s.v. *SUB- 14 a]. 1922 W. R. INGE *Outspoken Ess.* II. 131 It would be too absurd to suppose that our own State is the only specimen of these superhuman and supermoral individualities. 1960 K. AMIS *New Maps of Hell* (1961) iii. 84 Religious or quasi-religious feelings..attach themselves to the super-intelligent or super-moral alien power. *a* 1902 S. BUTLER *Way of All Flesh* (1903) lxxx. 370, I..saw a fly alight on..hot coffee on which the milk had formed a thin skin... I noted with what..almost supermuscan effort he..made for the edge of the cup. 1929 S. LESLIE *Anglo-Catholic* xv. 208 The Bees which buzzed..throughout the Papal City, resembling supermuscan flies perched on the walls. 1898 F. W. MAITLAND *Roman Canon Law in Church of England* i. 8 The cosmopolitan, the 'extra-national', or 'super-national' tone of the work of these two English canonists. 1928 G. B. SHAW *Intelligent Woman's Guide Socialism* lxxxiii. 450 Substitute supernational morality, law, and action, for the present international anarchism. 1977 *Irish Democrat* Mar. 3/1 The process of merging the national Governments of western Europe into one supernational administration. 1979 *Dædalus* Winter 190 Andreas Papandreou and his Panhellenic Socialist Movement (PASOK) have been able to capitalize on a supernationalist stand. 1889 *Cent. Dict.*, Superphysical. 1924 W. B. SELBIE *Psychol. Relig.* 277 Various forms of super-physical life. 1935 S. BECKETT *Echo's Bones*, The sphincter..Potwalloping now through the promenaders This trusty all-steel this super-real Bound for home like a good boy. 1942 *Horizon* July 41 The best Winchester pictures..possess the quality of super-real mystery. 1952 R. CAMPBELL *Lorca* iv. 63 There are also curious nonsensical excursions into the super-real. 1926 A. HUXLEY *Essays New & Old* 185 The adventures of Felix the Cat are super-realistic in the highest degree. 1955 S. SPENDER *Making of Poem* vi. 103 The golden Romanticism has..claims to set up a super-realistic reality. 1888 Supersensational [see *non-spatial* s.v. *NON- 3]. 1943 H. READ *Educ. through Art* ii. 28 As for the mental activity called intuition, by which we do not mean any super-sensational faculty of the mind, but the apprehension of abstract quantities and relations..it is the basis of a fourth type of art.

(*b*) *superluminally*.

1979 *Nature* 18 Jan. 182/1 (*heading*) Superluminally expanding radio sources and the radio-quiet QSOs.

(*c*) In related sbs., as *supernationalism, -nationalist, -nationality*; **super-re·alism** = *SURREALISM; **super-re·alist** = *SURREALIST *sb.*; **super-rea·lity** = *SURREALITY.

1917 G. B. SHAW *Platform & Pulpit* (1962) 106 Supernationalism will be limited by general psychological homogeneity. 1980 *Encounter* May 94/1 It thus seems natural to expect an author such as Fleming..to endow his heroes with patriotism and his villains with a super-nationalism of the nastier kind. 1941 L. B. NAMIER in *Time & Tide* 5 July 558/1 The outlook and ideas of the modern supernationalists..are very largely of German origin. 1916 E. HOLMES *Nemesis of Docility* i. 16 Those nations cling tenaciously to their respective nationalities, as against the supernationality of Germany. 1933 *Bull. Mus. Mod. Art* Oct. 2/1 Superrealism is the most conspicuous movement. 1952 R. CAMPBELL *Lorca* iv. 65 The dream region of 'super-realism'. 1931 'WYNDHAM LEWIS' *Diabolical Principle* 64 The cultural message of *Transition* is still further defined by the incorporation of the *dream-aesthetic* of the Super-realists into a body already reeking with 'romance'—indeed putrid with the excessive decomposition of that condition. *Ibid.* 65 The *infantile* is the link between the Super-realists and Miss Stein, as it is between Miss Stein and Miss Loos. 1934 H. READ in *Cinema Q.* III. I. 17 Some painters call it a super-reality (*surréalité*). 1935 D. GASCOYNE *Short Survey Surrealism* v. 109 The more recent ideas of surrealism, which conceive super-reality as existing in the material world, objectively, as well as subjectively in the automatic thought of the unconscious. 1945 H. READ *Coat of Many Colours* xxxix. 198, I believe that in general the plastic arts will tend towards rationality and the poetic arts towards superreality.

5. c. *superfamily* (later examples), *superspecies* (later examples); **su·pergalaxy** *Astr.*, a supercluster; *spec.* = *local supercluster* s.v. *LOCAL *a.* 2 d; hence **supergala·ctic** *a.*

1953 E. MAYR et al. *Methods & Princ. Syst. Zool.* iii. 52 The age of specialization has resulted in a general pushing upward of the categories, subfamilies becoming families, and families becoming superfamilies. 1978 *Nature* 16 Nov. 264/2 All three families of humans and apes are included in the superfamily Hominoidea. 1971 *New Scientist* 29 July 245/1 His final choice of supergalactic equator gave highly positive concentration indices for galaxies in the northern galactic hemisphere. 1982 *Nature* 2 Dec. 409/1 At high supergalactic *z* coordinates the galaxy density is much lower than near the supergalactic plane *z* = 0. 1946 G. GAMOW in *Nature* 19 Oct. 549/1 The realm of galaxies as seen through Mt. Wilson telescope represents only a small part of a much larger system (a 'super-galaxy' in the super-Shapley sense) rotating round a distant centre. 1955 *Sci. Amer.* Mar. 42/1 One source of radio emission, extending over a long path across the sky, coincides with the plane of the supergalaxy of which the Milky Way is a part. 1971 *New Scientist* 29 July 245/1 The Supergalaxy is, in turn, composed of smaller clusters of galaxies, including the local cluster of about a dozen members, our Galaxy being one of them. 1931 E. MAYR in *Amer. Mus. Novitates* No. 469. 2, I propose for *Artenkreis* the more convenient term, Superspecies. I define superspecies as a systematic unit containing geographically representative species that have developed characters too distinct to permit the birds to be regarded as subspecies of one species. 1976 E. DELSON in C. J. Jolly *Early Hominids of Africa* 535 This reliance on superspecies is perhaps..a recognition of our own uncertainties.

f. *Biol.* **superfe·male**, a female with a higher ratio of X chromosomes to autosomes than normal females; **su·permale**, a male in which this ratio is lower than in normal males, or the ratio of Y chromosomes to autosomes is higher.

1922 *Amer. Naturalist* LVI. 63 If the intersexes result from an intermediate ratio of X [chromosomes] to autosomes because the X has a net female tendency, then it might be expected that by increasing the ratio of X to autosomes a superfemale would be produced, and conversely, a supermale by increasing the relative number of autosomes. 1955 [see *INTERSEX]. 1959 *Lancet* 12 Dec. 1088/1 The inappropriateness of the term 'superfemale' is emphasised by the discovery of a human 3X 2A individual whose primary and secondary sex characters are underdeveloped. 1969 *Guardian* 8 Mar. 3/1 The theory that so-called supermales—men with an extra male chromosome—might be born criminals has lost its first test in a United States court.

6. a. *super-minister; super-ministry.*

1946 *Nature* 24 Aug. 247/2 What the Haldane Report recommended, however, was, not placing responsible ministers under a super-minister, but the consolidation or grouping of departments into a small number of super-ministries with one responsible minister for each. *a* 1974 R. CROSSMAN *Diaries* (1977) III. 666 Benn and Crosland are now super-Ministers of industry and planning, the same level as Barbara and me. 1937 L. HART *Europe in Arms* xv. 191 Ideas of possible organization range from a super-Ministry which should..absorb the..existing departments to a small Ministry, superimposed, which should guide the Government. 1975 *Globe & Mail* (Toronto) 25 Nov. 5/3 Liberal leader Robert Nixon also pledged to end the three super-ministries if his party were elected.

b. *super-organism* (later examples), *-priority, -quality, -system*; **supergra·vity** *Physics*, (a theory of) gravity as described or predicted by a supersymmetric quantum field theory; **supersy·mmetry** *Physics*, a very general type of mathematical symmetry which relates fermions and bosons; hence **su:persymme·tric** *a.*

1976 *Physics Lett.* B. LXII. 335/1 The first order formulation with torsion is closely related to the description of supergravity in superspace. 1980 *Nature* 21 Feb. 717/3 Known as supergravity, the new theory attempts to treat the familiar gravitational field as only a component part of a more elaborate network of forces and fields. 1971 E. O. WILSON *Insect Societies* (1972) i. 1/2 The giant of all such 'superorganisms' is a colony of the African driver ant *Anomma wilverthi*. 1973 P. A. COLINVAUX *Introd. Ecol.* xl. 551 The ideas of the superorganism and the social entity no doubt acquired much of their plausibility from the prevalence of the phenomenon of ecological dominance in plant communities of the temperate zones. 1917 W. S. CHURCHILL *Let.* 19 Aug. in M. Gilbert *W. S. Churchill* (1977) IV. Compan. I. 141 At present the Admiralty claim a super priority upon all supplies. 1952 *Times* 30 June 6/7 Lord De L'Isle and Dudley, V.C...announced on June 7 that the R.A.F. was to be equipped with the GA5 as an all-weather fighter and that it would have 'super-priority'. 1922 JOYCE *Ulysses* 312 His superb highclass vocalism, which by its superquality greatly enhanced his already international reputation. 1960 *Times* 3 Oct. (John Harvey Advt. Suppl.) p. ii/3 Super-quality surgical steel. 1974 *Proc. 17th Internat. Conf. High Energy Physics* I. 254/1 A supersymmetric theory. 1982 *Nature* 26 Aug. 801/1 During the past year there has also emerged a growing interest in particle physics theories that not only unify the description of the three basic interactions (strong, weak and electromagnetic) but which are also supersymmetric. 1974 B. ZUMINO in *Proc. 17th Internat. Conf. High Energy Physics* I. 254/1 Fermi-Bose supersymmetry was introduced by Wess and the author. It connects Bosons with Fermions. Its existence was suggested by dual models (when formulated as two-dimensional field theories) and the name supergauge symmetry in four dimensions seemed a natural choice. The supergauge algebra having only a finite number of generators in four dimensions, it seems now reasonable to avoid the word gauge and adopt the expression Fermi-Bose supersymmetry, or simply supersymmetry, suggested recently by Salam and Strathdee. 1977 *Physics Today* Apr. 49/3 As far as I know, the only natural way to keep a scalar boson massless is to have a 'supersymmetry',..which puts scalar fields in the same multiplet as massless fermion fields. 1978 *McGraw-Hill Yearbk. Sci. & Technol.* 356/1 Gauge supersymmetry treats all fundamental particles..on the same basis, accomplishes a fusion of space-time symmetries and internal symmetries.., and promises new types of renormalizable field theories as possible models for unified interactions, perhaps even including gravitation. 1934 WEBSTER, Supersystem. 1940 BRYANT & AIKEN *Psychol. Eng.* ii. 8 'Universal grammar', a norm or super-system which will comprehend all the various local systems. 1975 *Bio Systems* VII. 15/2 Having pure chemical systems for each of the three subsystems, we connect them to form a chemical supersystem.

c. *super-being, -block* [BLOCK *sb.* 14], *-boss* [BOSS *sb.*[6]], *-brain, -car, -carrier* [*CARRIER 1 m], *-cinema, -city, -computer, -crook* [CROOK *sb.* 13], *-grid* [*GRID 8 a], *-gun, -hero, -heroine, -journalist, -liner* [LINER[2] 8 a], *-magic, -male* (see also sense 5 f), *-nation, -patriot, -port* [PORT *sb.*[1] 1], *-profit, -race* (examples), *-rich* [RICH *sb.* 11], *-salesman, -salesmanship, -ship, -sleuth, -speed, -spy, -stud* [*STUD *sb.*[2] 4 d], *-tanker* [*TANKER[1] 1 a]; **su·peralloy** *Metallurgy*, an alloy capable of withstanding high temperatures, high stresses, and often highly oxidizing atmospheres; **su·perbike**, a motor cycle with a nominal engine capacity of 750 cc. or more; **Super Bowl** *U.S. Football* [after *ROSE BOWL 2], the final of the National Football League championship, contested annually since 1967 (from 1970, a play-off between the winners of the two sections of the League, the National and the American conferences); **su·perchurch**, (*a*) a church formed by the amalgamation of separate churches; (*b*) a very large church; **su·percrat** *N. Amer.*, a powerful

bureaucrat; **Super Glue**, the proprietary name of a strong adhesive; also **superglue**; **su·pergrass** [*GRASS *sb.*[1] 11*] (see quot. 1979); **super-hi·ghway** *N. Amer.*, a road designed for high-speed traffic, a motorway; also *fig.*; **su·perjet** a very large or fast jet aeroplane; also *attrib.* and *fig.*; **su·perloo** *colloq.*, a public convenience on certain British railway stations which offers a range of washing facilities, including showers; **su·perrat**, a rat that is resistant to the action of the usual rat poisons; **superset** *Math., Linguistics*, etc., a set (*SET *sb.*[2] 10 c) which includes another set or sets; **su·persound** sound which is too intense to be endured, or of too high a frequency to be perceived (cf. *ULTRASOUND); **su·perstate**, a dominant political community, esp. one formed from an alliance or union of several nations; *spec.* = *SUPERPOWER 3; **su·perstore**, a large store selling a variety of goods and typically situated away from a town's main shopping area; a small hypermarket; **su·perwoman** (later examples); in recent use, a woman who fills successfully concurrent roles as career-woman, wife, and mother.

Many of these formations are not 'chiefly nonce'.

1953 C. L. CLARK *High-Temperature Alloys* xvi. 269 Up to the time of the introduction of these superalloys it was generally agreed that any alloy intended for high-temperature service should be processed and heat treated. 1981 *McGraw-Hill Yearbk. Sci. & Technol.* 325/1 Superalloys can operate for extended periods of time at temperatures about 1200°F (650°C), and provide resistance to hot corrosion and erosion. 1930 *Daily Express* 8 Sept. 2/4 A patient imagines—quite seriously—that he is a kind of super-being. 1980 I. WATSON *Gardens of Delight* iv. 28 Who's this 'God' you were telling my people about? An alien superbeing—is that it? 1970 *Cycle World* Oct. 34 Suzuki's entry in the 'Superbike' field just happens to be an excellent touring bike. 1976 *Good Motoring* Nov. 5/2 A strong quota of 'superbikes' in the 750cc-plus category. 1928 *Survey* (N.Y.) 1 Mar. 696/1 It is necessary to examine more closely the structure and use of these superblocks. 1975 *New Society* 14 Aug. 375/3 A whole superblock in front of the museum, including many publishers offices and landmarks.., would have gone. 1916 *Blackw. Mag.* June 813/2 The German super-bosses. 1977 *Listener* 12 May 608/2 The man..he most admires in Italian public life today, the Fiat superboss. 1966 *Los Angeles Times* 22 Oct. III. 1/5 The capacity of the Coliseum for the Super Bowl will be 90,000. 1979 *Arizona Daily Star* 1 Apr. H 4/1 One would think the question of desirability of rising profits was of the same genre as who should win the Super Bowl. 1928 G. CAMPBELL *My Mystery Ships* xi. 208 To find out what its [*sc.* the name's] origin is or what it means, we shall have to wait till the person with the super-brain who thought of it appears before the Invention Board. 1975 *Pix* (Austral.) 13 Nov. 42/5 Futurologist and 'super brain' Dr. Kahn..on a recent visit said the world would be better off if Australians worked harder. 1920 *Motor* 3 Nov. 113 (Advt.), The Supreme development of the British super-car. 1977 *Belfast Tel.* 17 Jan. 9/4 There is everything from the family saloon to the specialised super-cars. 1969 D. ACHESON *Present at Creation* xxii. 195 We urged sending a powerful naval force, including the newly commissioned supercarrier *Franklin D. Roosevelt*. 1978 *Times* 28 Jan. 4/8 Mr Brown was seeking to make optimum use of the Navy's power in face of the strengthened Soviet Navy, which is making supercarriers increasingly vulnerable. 1970 *Time* 25 May 76 Episcopalians are potential participants in the proposed multichurch Protestant merger, the Church of Christ Uniting. Should the Episcopal Church join the new super-church, [etc.]. 1977 *Time* 26 Dec. 41/2 The faithful throng to gaudy superchurches with 5,000 to 10,000 seats, green shag wall-to-wall carpeting, pit orchestras and Jesus rock bands. 1923 *Gramophone* Apr. 7/1 A terrace opposite Holland House (now I believe about to be pulled down for a super-cinema). 1931 *Ann. Reg.* 1930 48 Of the existing cinemas 85 per cent are now wired, and there has been a great advance in the building of new super-cinemas. 1955 *Times* 9 May 3/1 One of Madrid's super-cinemas, the Coliseum. 1958 A. TOYNBEE *East to West* 103 You find yourself interned in one of the standardized super-cities of the modern world. 1971 *Americana Ann.* 103 Paolo Soleri..planned..gigantic supercities towering high in the air or floating on water. 1968 N. WALFORD tr. *O. Johannesson's Great Computer* iv. 108 The generating of controlling computers must be entirely computer-controlled. This task..was performed..by linking together about a hundred computers..and combining them..to form a unit known as the supercomputer. Such a unit had sufficient capacity to breed new computers of its own type. 1982 *Times* 30 Apr. 17/2 The market for supercomputers, as they are generally known, is also set for rapid growth. 1972 *Newsweek* 11 Dec. 25 A slimmed-down, tidied-up..Nixon Inc. with..four or so White House-level supercrats. 1978 *Globe & Mail* (Toronto) 17 July 2/4 [The Ottawa] system has fewer than a dozen super-powerful bureaucrats who dominate the policy-making process. They are the supercrats, and Mr. Gotlieb is one of the more successful ones. 1934 S. G. HEDGES *Plague Panic* xxvi. 211 The organized police systems of the world had failed so utterly to bring this super-crook to book. 1979 *Daily Mail* 29 Jan. 6/3 The other gel was a super-crook. 1977 *Drive* May–June 91/3 One of the latest superglue products..is a two-part, metal-to-glass adhesive. 1978 R. MARK *Office of Constable* xiii. 163 The age of the supergrass had arrived. 1979 *Observer* 8 Apr. 1/3 Twomey and Carpenter claim they were framed by a 'supergrass'—a police informer hoping for lenient treatment in return for turning in other villains. 1983 *Listener* 19 May 7/1 Following information from a supergrass, dozens of people alleged to be members of it had been arrested. 1950 *Times Rev. Industry* May 32/1 The projected 275/300kV British super-grid transmission system. 1979 *Nature* 8 Nov. 123/2 The extension of the Union-wide 'supergrid' of 1500 V DC transmission lines should, theoretically, allow power stations to be sited anywhere. 1915 *Chambers's Jrnl.* Oct. 661/2 The journalistic words 'superguns' [etc.]. 1929 *Encycl. Brit.* III. 367/1 As fast as the Liège forts fell to the super-guns. 1972 *Village Voice* (N.Y.) 1 June 53/2 Lee and Harry are inclined to put him down for killing the hawk with his supergun. 1917 'CONTACT' *Airman's Outings* 211 The super-heroes of the war. 1980 *Dædalus* Spring 119 The only people foolish enough to believe in fairy tales and superheroes (the last survivals in the mythology of atheism). 1970 *Times* 22 Dec. 8 Miss Comic Strip will be..selected for her..desirability as an imaginary superheroine. 1925 *Amer. City Mag.* Apr. 373/1 The Super-Highway is unique... It will furnish an express motor traffic highway. 1949 *Word Study* May 1/2 A super-highway toward..success. 1978 J. A. MICHENER *Chesapeake* xiii. 799 She found Route 2, which took her to Route 695, the superhighway circumnavigating Baltimore. 1958 *Daily Herald* 3 Mar. 1/1 They [*sc.* aircraft firms] are talking in terms of a super jet liner capable of crossing the Atlantic in four hours with 150 passengers. 1964 S. BELLOW *Herzog* (1965) 241 The superjet carried him to Chicago in ninety minutes. 1978 *Detroit Free Press* 16 Apr. (Parade Suppl.) 20/3 Except for her, however, not a single person arrested fits the image usually associated with the superjet, 'fast lane' set. 1916 L. CURTIS *Let.* 13 Nov. in *Let. People India* (1917) 27 As a sort of super-journalist much of my information has been derived from pumping people with first-hand knowledge. 1976 *Listener* 6 May 554/1 The superjournalists have evidently conquered the supermarket, for the rise of Mr Bob Woodward and Mr Carl Bernstein is now being presented as almost as spectacular a saga as the fall of President Nixon. 1928 *Manch. Guardian Weekly* 31 Aug. 180/3 To enlarge their docks for the building of a super-liner. 1963 *Economist* 27 July 322 Cunard is still chasing its ambition of a new super-liner. 1969 *Daily Tel.* 17 Dec. 1/4 The charge for using the 'superloo' at Euston and Victoria will go up from 6d to 1s. 1972 *Travelling* Autumn 43/3 Edinburgh's Waverley Station..will provide superloos, catering facilities. 1921 T. R. GLOVER *Jesus in Exper. Men* i. 8 The early Christian..really used the Gospel as a sort of super-magic. 1972 D. KENNEDY *Recoll. Assiniboine Chief* 156 The old witch saw that she was thwarted by super-magic. 1970 G. GREER *Female Eunuch* 194 Her sister, killed because of the maleficent supermale. 1914 E. BARKER *Nietzsche & Treitschke* 25 Treitschke looks to war as the expression of an exclusively national supernation. 1977 P. JOHNSON *Enemies of Society* xix. 248 Is there to be a huge African super-nation, based only on colour, but with immense racial differences within it? 1917 *N.Y. World* 7 Mar. 10/1 At a Carnegie Hall meeting of..super-patriots, Irving T. Bush was hissed because he defended the President of the United States. 1945 [see *SALISBURY STEAK]. 1977 *Private Eye* 13 May 16/1 That super-patriot the late Lord Beaverbrook. 1969 *Sunday Times* 16 Feb. 30/4 In addition to the container ship, another ship of the future is the LASH vessel (lighter-aboard-ship) that will cruise at high speeds, pausing only briefly at super-port gathering points to pick up or discharge its fleet. 1970 *Daily Colonist* (Victoria, B.C.) 9 Aug. 5/3 A second superport, providing modern bulk handling facilities. 1946 *Sun* (Baltimore) 9 July 2/2 He declares that Mr. Norton is trying to..swell employers' 'super-profits'. 1974 B. PEARCE tr. *Amin's Accumulation on World Scale* II. ii. 392 The origin and dynamics of the superprofits of monopolies. 1912 C. SAROLEA *Anglo-German Problem* i. 59 The German is convinced that he belongs to a super-race. 1979 *Jrnl. R. Soc. Arts* CXXVII. 324/2 A single pathogen genotype able to attack all of the components, the so called 'super-race'. 1974 *N.Y. Times* 13 Nov. 47/8 Armed with a new toxic rodent killer, the City Health Department opened a campaign yesterday to destroy a strain of 'super rats' breeding in the South Bronx. 1977 *New Scientist* 28 Apr. 200 This company has developed an anti-coagulant which is particularly effective against super rats..which are already prevalent in the United States and are now being reported in Europe and Asia. 1981 *Oxford Jrnl.* 27 Feb. 6 Pest controllers are battling against a breed of 'super-rats' which are immune to normal poisons. 1969 *Times* 5 May (Wall St. Suppl.) p. xii/2 Many of the super-rich of the United States live in Texas. 1982 *Country Life* 11 Mar. 666/2 This brilliant..novel about the super-rich in France. 1934 WEBSTER, Supersalesman. 1936 O. NASH *Primrose Path* 127 And a bright super-salesman Has sold you a pup. 1978 M. PUZO *Fools Die* xvi. 172 A very soft-selling supersalesman. 1933 *Sat. Even. Post* 7 Jan. 21 Super-salesmanship—1932 model. 1968 Supersalesmanship [see *SELL *v.* 3 j]. 1970 *Psychonomic Sci.* XXI (4) 235/3 There is no predetermined hierarchy of supersets and subsets. 1976 J. S. GRUBER *Lexical Structures in Syntax & Semantics* II. ii. 278 In this case we cannot allow the derived tree to be a subset of that in the lexical environment. The only alternative is to require that the derived tree be a superset of that in the lexical environment. 1937 *Sun* (Baltimore) 16 Nov. 12/1 The reasons that prompted the commission to advise against construction by this country of superships to rival the Queen Mary. 1974 *National Rev.* (U.S.) 1 Mar. 261 New sources of competitive coal have opened up in Australia, Canada, and South Africa, and again the specter of superships rises to plague us. 1974 *Aiken* (S. Carolina) *Standard* 24 Apr. 4-A/5 It's doubtful if the FBI will long retain, or ever again seek, the super-sleuth. 1942 *Pop. Sci. Monthly* Feb. 49/2 Sound waves too powerful for the human system to bear, and others too high in pitch for the human ear to hear, are new miracle-working tools in science and industry. In dozens of laboratories, scientists are perfecting sound-generating devices, and discovering new uses for supersound and ultrasound. 1952 *Chambers's Jrnl.* 1 June 363 Scientists have been having the time of their lives exploring super sound. 1927 *Glasgow Herald* 1 June 15 If we cleaved our way above the ocean at the superspeeds now contemplated. 1961 *Times* 17 Nov. 17/5 (Advt.), Current Ferranti activities include..Atlas (most advanced super-speed computer in the world). 1937 KOESTLER *Spanish Testament* i. 24 If one had taken them seriously, one might have imagined that half Esturil consisted of super-spies. 1980 R. HILL *Spy's Wife* viii. 53 Is she another Kremlin super-spy? 1918 O. GREGORY *Meccania* iv. 91 The Super-State must borrow from the Socialists the conception of an all-embracing power and activity. 1929 B. RUSSELL *Marriage & Morals* xv. 173 The control of the super-State over education would be a positive safeguard against war. *Ibid.*, Loyalty to the international super-State should everywhere be fought. 1935 J. E. C. WELLDON *Forty Years On* ii. 76, I have felt that the Darwinian theory of the survival of the fittest..was responsible for the German doctrine of the super-state, which, as the Germans conceived it, could only be Germany. 1941 A. HUXLEY *Let.* 27 Nov. (1969) 471 The super-states based on the three centres of heavy industry and advanced technology—Europe, North America and East Asia. 1959 N. MAILER *Advts. for Myself* (1961) 272 The iron commissars of the Soviet superstate. 1974 M. B. BROWN *Econ. Imperialism* ix. 225 The nation states, apart from the super-states—USA, EEC, Japan and the USSR—are forced into a client relationship with the giant companies. 1978 *New York* 3 Apr. 45/3 Loyalty to the superstate as a substitute for the supernatural. 1965 *Punch* 7 July 1/2 Why bother with exports when the superstore will take anything you care to make? 1980 *Times* 13 Feb. 3/2 Comparing superstores with local supermarkets is like comparing apples with oranges. 1975 *Time Out* 19 Sept. 25/1 Petersen, the latest in the 'Alvin Purple' brigade of Australian superstuds. 1921 *Mex Fuel Oil* (Anglo-Mexican Petroleum Co.) 7 These losses are being made good by the building of several supertankers, commencing with the *San Florentino*..18,000 tons. 1953 *Wall St. Jrnl.* 1 July 4/2 The S.S. New Jersey Sun, second of four supertankers being built for Sun Oil Co. 1977 *Whitaker's Almanack 1978* 1035/2 A serious problem would be the shipping and supertankers passing round Cape Horn. 1958 *Listener* 18 Dec. 1040/1 A picture of a girl maybe, fantastically beautiful, a blonde superwoman. 1975 S. CONRAN (*title*) Superwoman. 1976 *National Observer* (U.S.) 11 Sept. 20/5 The superwoman image ignores the reality of the average working woman or housewife.

III. 9. a. (*a*) *superdense, -fast, -luminous, -sumptuous*; **su·percolo:ssal** *a. U.S. colloq.*, very large, very good, stupendous; **super-coo·l** *a. slang* (orig. and chiefly *U.S.*), very cool (*COOL *a.* 4 e), relaxed, fine, etc.; also *absol.* as *sb.*; **superfatted** *a.*, (*b*) *slang*, of persons: overweight, fat; **superio·nic** *a. Physics*, having a high ionic electrical conductivity; also as *sb.*, a superionic substance; **superma·ssive** *a. Astr.*, having a mass many (i.e. typically between 10^6 and 10^9) times that of the sun; **superwea·k** *a. Particle Physics*, pertaining to or being a proposed interaction several orders of magnitude weaker than the weak interaction which would not be invariant under charge conjugation and space inversion jointly.

1934 WEBSTER, Supercolossal. 1937 *Amer. Speech* XII. 241/1 Supercolossal is an adjective heard several times orally in Colorado. 1938 WODEHOUSE *Code of Woosters* v. 130 Big is right, though perhaps 'super-colossal' would be more the *mot juste*. 1947 —— *Full Moon* v. 92 'Her profile. Lovely, don't you think?' 'Yup.' 'And her eyes. Super-colossal.' 1976 *National Observer* (U.S.) 17 July 6/3 The brand-new National Air and Space Museum here is a supercolossal mixture of show biz and science. 1970 T. WOLFE *Radical Chic & Mau-Mauing Flak Catchers* 131 The pimp style was a supercool style that was much admired or envied. 1975 *Radio Times* 23–29 Aug. 11/4 James Coburn was the nicest of all those Bond-type supercools. 1978 *Hot Car* July 91/3 They were super-cool amongst the sixties surfing set in the USA. 1981 *Times* 22 July 11/2 That style had itself been borrowed from younger Jamaicans, and the super cool they affected. 1967 *Listener* 27 Apr. 545/1 These observations..imply an origin [of the Universe] from a superdense state 10,000 million years ago. 1977 *Time* 19 Sept. 50/1 The whole principle of diesel ignition is to raise the temperature of the fuel mixture by compressing it into a superdense mass in the cylinder. 1980 *Lok Sabha Deb.* (Delhi) 5 Aug. 264 A superfast train like the K.K. Express runs late by 5 to 6 hours. 1982 *Economist* 3 Apr. 120/3 It is at the frontiers of R and D now being done into superfast computers. 1927 WODEHOUSE *Small Bachelor* iii. 48 'Important people!' Mr. Waddington snorted sternly, 'A bunch of super-fatted bits of bad news.' 1947 L. HASTINGS *Dragons are Extra* ix. 212 A bald, double-chinned type who looked very like a super-fatted edition of ex-President Hoover. 1972 W. L. ROTH in *Jrnl. Solid State Chem.* IV. 60/1 Such solids, which may be called super ionic conductors, exhibit ionic conductivities that can be as large as inverse ohm-centimeters at temperatures ranging from near room temperature to 1200°C. 1972 *New Scientist* 11 May 321 A small and hardly known group of compounds called 'superionics' reveal exceptional electrical conductivity in the solid state. 1980 *Jrnl. Physics & Chem. Solids* XLI. 1323/1 The superionic conductors are characterized by their high ionic but very low electronic conductivity at room temperature. 1968 D. MOORE tr. *Schatzman's Struct. Universe* i. 15 If our present interpretations are correct, quasars are very remote and super-luminous. 1977 *Jrnl. R. Soc. Arts* CXXV. 215/1 The enigmatical 'quasars'..are thought to be immensely remote and super-luminous. 1967 Supermassive [see *RELATIVISTIC *a.* 2b]. 1981 *Economist* 24 Jan. 97/1 Still higher output would result from the collision of supermassive black holes containing, say, a mass equivalent to a million of earth's suns. 1922 JOYCE *Ulysses* 497 It is immense, supersumptuous. 1970 *Physical Rev.* D. II. 257 (*heading*) Unitarity and the phase of the mixing parameter in superweak theories. 1979 CHENG & O'NEILL *Elem. Particle Physics* ix. 206 If the Wolfenstein model is correct, there exists a 'superweak' fifth force in nature.

(*b*) *supercolossally*.

1966 *New Yorker* 1 Oct. 184 Both supercolossally ambitious and energetic men.

b. *super-civilized* adj. (also *absol.* as *sb.*); **superallo·wed** *a. Nucl. Physics*, (of a beta decay) having an exceptionally high probability of occurrence as measured by the product of the half-life of the initial state and a function of the energy and momentum of the emitted electron.

1950 *Rev. Mod. Physics* XXII. 397/2 The very lowest $\log_{10} ft \sim 3$ to 4 are allowed transitions between nuclei having similar nuclear wave functions. These transitions are called the superallowed transitions, while allowed transitions between nuclei not having very similar wave functions have log *ft* ranging from 4 to 6. 1964 *Physical Rev. Lett.* XII. 301/1 The transition from the ground state of ^{33}Cl to the 5·1-MeV excited state of ^{33}Ar is superallowed and has a large matrix element for neutrino absorption. 1975 *Nature* 18 Sept. 179/2 They start..from basic input experimental data—the lifetimes and energy release of certain particularly simple nuclear β decays known as superallowed transitions. 1929 'R. CROMPTON' *William* i. 12 The Outlaws never made the pretence affected by the super-civilised, of indifference to their neighbours' affairs.

10. b. su·pervoltage *Physics* and *Med.*, a higher than usual voltage; *spec.* a voltage in excess of 200 kilovolts; usu. *attrib.* with reference to the use of X-rays generated using such voltages.

1934 *Illinois Med. Jrnl.* LXVI. 286/2 Much of the improvement claimed for super-voltage is available at 200 kilovolts. 1956 A. H. COMPTON *Atomic Quest* i. 14 The new physics laboratory..would include space for super-voltage equipment. 1976 *Lancet* 6 Nov. 992/2 Patients who received T.N.I. were all treated with supervoltage or megacurie equipment.

12. b. superoxide: in mod. use distinguished from peroxide (q.v. in Suppl.) and restricted to the anion O_2^-; (further examples).

1950 *Chambers's Jrnl.* Apr. 255/2 It [*sc.* a new life-saving apparatus] has depended upon finding a method for fairly large-scale production of the chemical used, namely potassium superoxide. 1965 PHILLIPS & WILLIAMS *Inorg. Chem.* I. xiii. 491 The metal oxides so far discussed contain the anion O^{2-} only. There are also two other series of oxides, the peroxides and superoxides which contain the anions O_2^{2-} and O_2^- respectively. 1979 *Experientia* XXXV. 245/2 The explanation for such action is the hypothesis that vitamin C can act as an antoxidant as well as oxidant, by generating superoxide.

IV. 13. supercalender *sb.* (earlier example); **super-tax**: abolished as an official term in the U.K. in 1929, but still in common (esp. *attrib.*) use; cf. *SURTAX *sb.*

1894 Super-calender [see *SHEAVE *sb.*[2] 3]. 1931 *Times Lit. Suppl.* 16 July 556/4 The incidence of income-tax and supertax on business profits. 1972 *Daily Tel.* 14 Jan. 13 Our friends, on hearing that we own two houses, put us in the super-tax class. 1978 F. OLBRICH *Desouza pays Price* v. 21 The Taj Mahal Hotel['s]..clientele consisted exclusively of those in the super-tax bracket.

VI. With reduplication of the prefix.

17. Used as an intensifier Cf. branch III.

1871 Super-superabundance [in Dict., sense 10a]. 1871 Super-superabundant [in Dict., sense 9a(*a*)]. 1871 Super-superabundantly [in Dict., sense 9a(*b*)]. 1934 'J. SPENSER' *Limey breaks In* ix. 158 This warder was another of the variety known amongst prison populations as super-super bastards. 1937 A. CALDER-MARSHALL in C. Day Lewis *Mind in Chains* 60 The second stage was reached, where super- and super-super-films were made. 1963 *Supermarket & Self Service* (Johannesburg) Aug. 16/1 A further 'super-supermarket' of American-style dimensions, is under contemplation. 1974 T. P. WHITNEY tr. *Solzhenitsyn's Gulag Archipelago* I. II. iv. 590 Things were neat and clean, they said, and it was always warm, and the only work was mental work—and all of it super-supersecret.

18. Used to denote a further increase in rank or degree. Cf. branch II.

1971 *Nature* 26 Nov. 182/1 The observations would therefore seem to exclude super-clusters or the still greater hierarchy of super-super-clusters as the source of X-ray background. 1979 *Ibid.* 12 Apr. 615/1 Since nuclear shell theory predicts a further island of stability at Z = 164–184..we must consider whether the three peaks originate from decay of super-superheavy elements. 1980 *Sci. Amer.* July 112/1 In this way super-supercoiled molecules can be created, molecules that have many more super-helical twists than are usually present.

super-acid, *a.* Add: Also **super acid. 3.** Of, pertaining to, or designating a non-aqueous solution having very great protonating power.

1927 CONANT & HALL in *Jrnl. Amer. Chem. Soc.* XLIX. 3047 We investigate certain of the properties of these solutions in which salt formation is exceptionally complete, and we propose to call such solutions 'superacid solutions'. 1940 GLASSTONE *Textbk. Physical Chem.* xii. 959 On account of the very marked tendency of the $CH_3 \cdot CO_2H_2^+$ ions to lose the proton they have taken up from the acid, the solutions exhibit acidic properties, e.g., in catalysis, of an exceptionally strong nature and so they have been called super-acid. 1968 *Accts. Chem. Res.* I. 202/1 The acidity of fluorosulfuric acid can be considerably increased by the addition of SbF_5 and SbF_5—SO_3, and the resulting solutions are the most highly acidic media known... These systems may therefore be justifiably called superacid media. 1979 *Science* 5 Oct. 14/3 Two superacid systems used very frequently are HSO_3F-SbF_5 (Magic Acid)..and HF-SbF_5 (fluoroantimonic acid).

B. *sb.* A solution of a strong acid in a very acidic solution, which is an extremely effective protonating agent; any acid stronger than some standard acid in either Brönsted acid (i.e., proton-donor) or Lewis acid (electron-acceptor) systems.

1968 *Jrnl. Amer. Chem. Soc.* XC. 2726/1 (*caption*) Chemistry in super acids. 1979 *Science* 5 Oct. 13/3 All protic acids stronger than 100 percent sulfuric acid should be classified as superacids. *Ibid.* 14/2 It is suggested that those [Lewis acids] stronger than anhydrous aluminum chloride..should be categorized as superacids. *Ibid.* 16/3 The astonishing acidity of Magic Acid and related superacids allows protonation of exceedingly weak bases.

Hence **superaci·dic** *a.* = *SUPER-ACID *a.* 3.

1979 *Science* 5 Oct. 14/3 These superacidic systems can be 10^{16} times stronger than 100 per cent sulfuric acid.

su:peraci·dity. [f. SUPER-ACID *a.*, after ACIDITY.] **a.** *Med.* (See quot.)

1900 DORLAND *Med. Dict.* 653/2 *Superacidity*, increase of the normal acidity of the gastric secretion.

b. *Chem.* The quality or state of being a superacid.

1927 *Jrnl. Amer. Chem. Soc.* XLIX. 3061 Our work indicates that the proximate cause of superacidity in a solution is an abnormally high value of the hydrogen-ion activity. 1979 *Science* 5 Oct. 14/2 It should always be kept in mind that superacidity encompasses both Bronsted and Lewis acid systems.

superadiabatic: see *SUPER- 4 a (*a*).

su:peraerodyna·mics, *sb. pl.* [f. SUPER-, with reference to the *upper* atmosphere.] The study of motion of and in a gas so rarefied that it has to be treated as a collection of individual particles rather than a continuous fluid. Hence **su:peraerodyna·mic** *a.*

1934 A. F. ZAHM in *Jrnl. Franklin Inst.* CCXVII. 153 (*heading*) Superaerodynamics. 1952 W. F. HILTON *High-Speed Aerodynamics* xiii. 351 The Mach number should be greater than the Reynolds number for superaerodynamic conditions to prevail. *Ibid.* 353 There is no reason why *M* should not be less than unity; i.e., a subsonic superaerodynamic flow could exist. 1957 *Jrnl. Aeronaut. Sci.* XXIV. 527/1 The mechanics of the kinetic theory of gases is employed to describe the drag force on the nose of a missile moving in the superaerodynamic region of the atmosphere. 1960 *McGraw-Hill Encycl. Sci. & Technol.* XIII. 293/2 It is convenient to divide superaerodynamics into three flow regimes.

superallowed, -alloy: see *SUPER- 9 b, 6 c.

supera·nnuable, *a.* [f. SUPERANNU(ATE *v.* + -ABLE.] (Of a post or salary) that entitles the holder to superannuation (sense 2).

1950 *Times* 27 Apr. 1/4 (Advt.), The above appointments are superannuable under the British Electricity Authority and Area Boards Scheme. 1960 *Times Lit. Suppl.* 18 Nov. 748/2 The appointment is superannuable. 1971 *Daily Tel.* 13 Apr. 6/6 (Advt.), A permanent superannuable post paying up to £2,691 per annum. 1978 *Nature* 18 May p. xxvii/1 (Advt.), Annual salary (superannuable) will be within the professional range which has a minimum of HK$131,640.

superannuate, *sb.* (Earlier example.)

1816 *Hist. Colleges Winchester, Eton & Westminster* 46 Boys between eighteen and nineteen years old, called superannuates.

superau·dible, *a.* **1.** [SUPER- 9 a.] Very loud. *nonce-use.*

1921 D. H. LAWRENCE *Tortoises* 47 That fragile yell, that scream, Super-audible.

2. [SUPER- 4 a.] Of a frequency: greater than the highest audible frequency; ultrasonic. Of a sound: too high-pitched to be audible. Now *rare.*

1922 [see *SUBAUDIBLE *a.* 2]. 1926 R. W. HUTCHINSON *Wireless* 237 The resultant wave is anything from 2,000 metres to, say, 12,000 metres in length (super-audible frequency). 1944 *Proc. IRE* XXXII. 735/2 Superaudible noise impulses which may be present.

super-bike: see *SUPER- 6 c.

superbity. Delete *rare* and add later examples.

1924 G. B. SHAW *St. Joan* p. ix, She went to the stake without a stain on her character except the overweening presumption, the superbity as they called it, that led her thither. *a* 1945 E. R. EDDISON *Mezentian Gate* (1958) xxix. 141 In her nose, a critical outward-regarding superbity that judged without appeal. 1979 C. P. SNOW *Coat of Varnish* xxxviii. 297 In spite of his superbity, he might possess a kind of self-protective cunning.

su·perbomb. [SUPER- 6 c.] **a.** Also **super bomb.** A fission bomb. *Obs.* exc. *Hist.*

1940 O. R. FRISCH in *Ann. Rep. Progr. Chem.* XXXVI. 16 Since the energy release in this reaction would be about 10^8 times larger than in ordinary chemical reactions..it has been feared that it might form the basis for the construction of a super-bomb exceeding the action of ordinary bombs by a factor of 10^6 or more. 1941 in H. D. Smyth *Gen. Acct. Devel. Atomic Energy Mil. Purposes* (1945) iv. 38 In such a reaction the energy would be released at an explosive rate which might be described as a 'super bomb'. 1964 M. GOWING *Britain & Atomic Energy 1939–45* i. 34 A chain fission reaction was a possibility, the inevitable question had arisen. Could this energy be harnessed for making a super bomb?

b. A fusion or hydrogen bomb. Cf. *SUPER *sb.* 9.

1951 W. L. LAURENCE *Hell Bomb* i. 3 'Is it true about the superbomb?' I asked him. 'Will it really be as much as fifty times as powerful as the uranium or plutonium bomb?' 1961 *New Statesman* 15 Sept. 330/1 For this interpretation not least among the exhibits in evidence is Krushchev's grim emphasis on his 100-megaton superbomb, which American experts regard as militarily poor but incredibly dirty. 1975 *Sci. Amer.* Oct. 106/2 It called for the fastest possible development of the hydrogen bomb, which was widely referred to at the time as the superbomb (or simply the Super).

Super-Bowl: see *SUPER- 6 c. **supercalender**: see SUPER- 13.

su:percalifra:gilistice:xpialido·cious, *a.* Also **supercalifragilistic**; formerly also other varr. [Fanciful: cf. *SUPER *a.* 3.] A nonsense-word used esp. by children, now chiefly expressing excited approbation: fantastic, fabulous.

Made popular by the Walt Disney film 'Mary Poppins' in 1964. The song containing the word was the subject of a copyright infringement suit brought in 1965 against the makers of the film by Life Music Co. and two song-writers: cf. quots. 1949, 1951. In view of earlier oral uses of the word sworn to in affidavits and dissimilarity between the songs the judge ruled against the plaintiffs.

1949 PARKER & YOUNG (*unpublished song-title*) Supercalafajalistickespialadojus. 1951 —— (*song-title*) Supercalafajalistickespeealadojus; or, The super song. 1964 R. M. & R. B. SHERMAN (*song-title*) Supercalifragilisticexpialidocious! 1967 *Decisions U.S. Courts involving Copyright 1965–66* 488 The complaint alleges copyright infringement of plaintiff's song '*Supercalafajalistickespeealadojus*' by defendants' song '*Supercalifragilisticexpialidocious*'. (All variants of this tongue twister will hereinafter be referred to collectively as 'the word'.) 1971 *Daily Tel.* 6 Nov 13/5 If you can stand more than a day of Supercalifragilisticexpialidocious entertainment you can settle in at the concrete Contemporary Resort Hotel. 1972 *Atlanta Constitution* 9 Apr. 20A/1 Disney World, the new supercalifragilisticexpialidocious tourist attraction created by the folks who brought you Mickey Mouse, is packing them into Orlando. 1980 *Amer. Speech* LV. 266 Whatever the ancestry of *supercalifragilisticexpialidocious*, it entered the general public consciousness as a result of the wonderful world of Disney. 1982 *N.Y. Post* 29 July 64/3 His eyes are willing unspoken words to life as though they were part of one of those supercalafragilistic electronic scoreboards.

supercharge, *sb.* Restrict *rare* to senses in Dict. and add: **3.** *Engin.* An explosive charge of higher than usual pressure in the cylinders of an internal-combustion engine; increased pressure of the charge.

1912 E. BUTLER *Evol. Internal Combustion Engine* vi. 69 It has been proposed by Clerk and others to employ super-compression of the charge.., the combustion cylinder thereby receiving a super-charge of air. 1948 *Petroleum Handbk.* (Shell Petroleum Co. Ltd.) (ed. 3) xxv. 356 The extra fuel has a certain cooling effect, and this permits an increase in boost (degree of supercharge). 1971 L. J. K. SETRIGHT *Power to Fly* v. 120 The immediate effect of the new fuel was to allow a greater degree of supercharge.

su·percharge, *v. Engin.* [SUPER- 9 b.] *trans.* To increase the pressure of the fuel-air mixture in (an internal-combustion engine).

1919 W. J. WALKER tr. *Devillers's Automobile & Aero Engines* xxiv. 384 Each compressor, having two compression strokes per revolution, can supercharge two cylinders. 1924 *Glasgow Herald* 6 May 4 The proposition of super-charging engines for sea-going and for road as well as for air service. 1966 *McGraw-Hill Encycl. Sci. & Technol.* VII. 208/2 Supercharging a two-cycle diesel engine requires some means of restricting or throttling the exhaust in order to build up cylinder pressure.

supercharged, *pa. pple.* and *ppl. a.* Add: **2.** *Engin.* **a.** Of the fuel-air mixture in an internal-combustion engine: increased in pressure by mechanical means. **b.** Of a vehicle or its engine: equipped with a supercharger.

1919 W. J. WALKER tr. *Devillers's Automobile & Aero Engines* xxiv. 387 The volume of the supercharged mixture remains sensibly constant during injection. 1943 J. B. PRIESTLEY *Daylight on Saturday* vii. 41 Where did Germany get her first super-charged aero engines from? 1955 *Times* 5 May 16/5 The newest of our engines to be publicly announced is a supercharged turboprop of 4,000 h.p. 1980 *Daily Tel.* 21 May 14/3 The preview of supercharged prototype models in Turin.

su·percharger. *Engin.* [f. *SUPERCHARGE *v.* + -ER[1].] A compressor that increases the pressure of the fuel-air mixture supplied to the cylinders of an internal-combustion engine. Cf. *BLOWER[1] 3 d.

1921 A. W. JUDGE *Automobile & Aircraft Engines* x. 452 The power required to drive the supercharger is about 6 per cent. 1946 *Happy Landings* July 3/1 The supercharger control..should be operated smartly. 1960 G. MAXWELL *Ring of Bright Water* II. x. 144 The supercharger screamed, dial needles moved with incredible rapidity towards red zones; I had a glimpse of the speedometer hovering at 145 m.p.h. 1980 *Daily Tel.* 21 May 14/3 Manufacturers have been increasingly attracted to the turbocharger because it is driven by waste gases from the exhaust, while

the supercharger is driven by a power take-off from the engine.

su·percharging, *vbl. sb. Engin.* [f. *SUPERCHARGE *v.* + -ING[1].] The action or use of a supercharger.

1922 *Encycl. Brit.* XXX. 41/1 Supercharging for high flying. 1937 *Discovery* Dec. 386 (caption) The McClelland two-stroke petrol turbine engine. The unfinned cylinders are pumps providing a supercharging effect. 1980 *Daily Tel.* 21 May 14/3 Fiat officials indicated that the ultimate intention was to apply supercharging to its small and medium volume production models.

superchurch: see *SUPER- 6 c.

su·perclu:ster. *Astr.* [SUPER- 5 c.] A cluster of objects that are themselves clusters (in quot. 1930, of stars, but now only of galaxies).

1930 R. J. TRUMPLER in *Lick Observatory Bull.* XIV. 187/2 It seems worth while to examine the hypothesis that our Milky Way system..together with the two Magellanic Clouds and about a hundred globular clusters form a cluster of extra-galactic objects which we may call the 'supercluster'. 1958 *Astron. Jrnl.* LXIII. 260/2 The local super-cluster of galaxies is an irregular assembly of groups, clouds and clusters dominated by the Virgo cluster in its center. 1970 *Nature* 19 Dec. 1137/1 Wolfe and Burbidge argue from these observations that super-clusters of galaxies..must definitely be ruled out. 1978 PASACHOFF & KUTNER *University Astron.* i. 6 Before we could enlarge our field of view another 100 times we might see a supercluster—a cluster of clusters of galaxies. 1982 *Daily Tel.* 1 Dec. 1/6 The largest-known object in the universe, a supercluster of millions of galaxies, has been discovered.

Hence **su·perclu:stering** *vbl. sb.*

1960 *Soviet Astron.* III. 910 We conclude that not only clustering, but also superclustering, is a general characteristic of the distribution of galaxies. 1971 D. W. SCIAMA *Mod. Cosmol.* vii. 96 The suggested scale of their clustering would be much the same in relation to their separation as is the case for the clustering or super-clustering of galaxies.

su·percoil, *sb. Biochem.* [SUPER- 5 c.] A coiled coil; *spec.* a structure sometimes assumed by DNA in which the double helix itself is coiled or looped. Cf. *SUPERHELIX.

1965 PEACOCKE & DRYSDALE *Molecular Basis Heredity* iv. 168 In the fibre, these molecules take the form of 'super-coils' in which the axis of the DNA helix is itself coiled with a pitch of 120Å and a diameter of 100Å. 1976 *Nature* 10 June 516/1 Supercoils could be introduced into progeny molecules either before sealing of the closed circular form, or subsequent to an initial closing of the molecule. 1980 *Sci. Amer.* July 100/2 Because the forces that stabilize the double helix are strong the closed circular molecules resist such underwinding, and..they compensate by forming supercoils.

su·percoil, *v. Biochem.* [f. prec. sb.] **a.** *trans.* To make (a molecule) into a supercoil. **b.** *intr.* To become a supercoil.

1967 *Jrnl. Molecular Biol.* XXV. 28 Only molecules in which both strands are intact can be supercoiled in either sense. 1971 *Nature* 29 Oct. 591/2 Segrest and Cunningham start with the tropocollagen molecule, a three-strand rope formed by supercoiling two identical helices termed α_1 and α_2. 1982 *Sci. Amer.* July 87/1 When the ethidium is removed, the ring [of DNA] supercoils.

Hence **su·percoiled** *ppl. a.*, having the structure of a coiled helix; **su·percoiling** *vbl. sb.*, the action or result of forming a supercoil; also *attrib.*

1967 *Jrnl. Molecular Biol.* XXV. 23 It would be desirable to have independent evidence on the direction of supercoiling and to obtain an estimate of the number of supercoiling turns. *Ibid.* 27 The supercoiled form of polyoma DNA can be converted to the unsupercoiled form by one single-strand scission. 1980 *Sci. Amer.* July 100/1 Affecting DNA's in a wide range of sizes and shapes (including some that are not organized into a double helix), supercoiling takes a variety of forms. 1980 *Proc. Nat. Acad. Sciences* LXXVII. 2445/1 A massive tangle of catenated supercoiled molecules.

supercolossal: see *SUPER- 9 a (*a*).

supercondu·cting, *ppl. a. Physics.* [f. SUPER- 9 b, tr. Du. *suprageleidend* (H. K. Onnes 1913, in *Versl. van de gewone Vergad. d. Wis- en Natuurk. Afdeeling, K. Akad. v. Wetensch. te Amsterdam* XXI. 1390).] Possessing no electrical resistivity; employing a substance in this state.

1913 H. K. ONNES in *Proc. Sect. Sci. K. Akad. Wetensch. Amsterdam* XV. 1429 A thread of super-conducting mercury, if an ordinary conducting particle were present anywhere in the current path, could show resistance at that spot. 1935 *Discovery* July 213/2 The fascinating transition of metals into the super-conducting state when near to absolute zero. 1958 *Listener* 11 Dec. 984/1 A current once established in a superconducting lead ring will continue to flow indefinitely. 1966 C. R. TOTTLE *Sci. Engin. Materials* vi. 127 The superconducting state can be destroyed by the application of an external magnetic field. 1978 *Jrnl. R. Soc. Arts* CXXVI. 608/2 The savings from introducing superconducting generators are relatively small.

Hence (as a back-formation) **supercondu·ct** *v. intr.*, to conduct electricity without any resistance.

1964 *New Scientist* 20 Aug. 441 (*heading*) USSR field windings that superconduct. 1976 G. K. HAINES *Supercold, Superhot* v. 30 More than twenty-five elements have now been found that will superconduct.

supercondu·ction. *Physics.* [SUPER- 10 a.] = *SUPERCONDUCTIVITY; conduction of electricity without resistance.

1940 E. F. BURTON et al. *Phenomena Temperature Liquid Helium* x. 322 It might be possible to explain the discrepancy..in the apparent number of superconduction electrons. 1959 *Electronic Engin.* XXXI. 589/2 Domains of ferromagnetism alternating with domains of superconduction, the latter having a magnetization antiparallel to the applied field.

su:perconducti·vity. *Physics.* [SUPER- 10 a.] The property of having zero electrical resistivity exhibited by some substances at temperatures close to absolute zero.

1913 H. K. ONNES in *Proc. Sect. Sci. K. Akad. Wetensch. Amsterdam* XV. 1428 (*heading*) Experiments on the possible influence of contact with an ordinary conductor upon the superconductivity of mercury. 1934 *Times Lit. Suppl.* 26 July 531/3 A similar property of superconductivity, acquired at specific low temperatures, has since been found also in a number of other metals and in certain alloys. 1959 *Sci. News* LI. 21 The idea of loss-free electromagnetic devices was immediately suggested when superconductivity was first discovered, and a number of efforts have since been made to find metals or alloys that would exhibit superconductivity at higher than [liquid] helium temperatures. 1978 *Jrnl. R. Soc. Arts* CXXVI. 608/2 Superconductivity..offers the prospect of lower capital costs and higher efficiencies than conventional generators for large units.

So **supercondu·ctive** *a.* = *SUPERCONDUCTING *ppl. a.*

1913 H. K. ONNES in *Proc. Sect. Sci. K. Akad. Wetensch. Amsterdam* XVI. 116 The actual degree of conductivity of the superconductive mercury. 1955 H. B. G. CASIMIR in W. Pauli *Niels Bohr* 119 There exist many metals that do not become superconductive in the temperature range in which they have been investigated. 1966 K. MENDELSSOHN *Quest for Absolute Zero* ix. 199 In 1930 de Haas and Voogd had found that wires of lead–bismuth alloy remained superconductive in magnetic fields as high as 20,000 oersted. 1972 *Physics Bull.* Oct. 615/1 The highest transition temperature for superconductive materials reported so far is 21 K in a three element compound, niobium–aluminium–germanium.

supercondu·ctor. *Physics.* [f. SUPER- 6 c, tr. Du. *suprageleider* (H. K. Onnes 1913, in *Versl. van de gewone Vergad. d. Wis- en Natuurk. Afdeeling, K. Akad. v. Wetensch. te Amsterdam* XXI. 1390).] A substance that becomes superconducting at sufficiently low temperatures; also, such a substance in the superconducting state.

1913 H. K. ONNES in *Proc. Sect. Sci. K. Akad. Wetensch. Amsterdam* XV. 1429 A pushing forward of the electrons in the galvanic current through a super-conductor without performance of work. 1931 *Ann. Reg.* 1930 61 Meissner discovered that copper sulphide is a superconductor. *Ibid.*, Niobium carbide becomes a superconductor at 10° A. 1955 H. B. G. CASIMIR in W. Pauli *Niels Bohr* 119 At temperatures above the transition point..superconductors behave in all respects like these normal metals. 1959 *Sci. News* LI. 21 The real difficulty in the use of superconductors in magnetic devices is the fact that superconductivity is destroyed by fields of only a few hundred oersted. 1969 ROSE-INNES & RHODERICK *Introd. Superconductivity* i. 5 About half the metallic elements are known to be superconductors and..a large number of alloys are superconductors. 1983 *New Scientist* 24 Mar. 802/1 Like most of the organic superconductors already known, the new material loses its resistance only under high pressure.

su·perconscious, *a. Psychol.* [SUPER- 4 a.] Transcending human or normal consciousness. Also *absol.*

1884 F. W. H. MYERS in *Proc. Soc. Psychical Res.* II. 219 We shall come, perhaps, to find *super-conscious* as necessary a term as *sub-conscious*. 1904 HARDY *Dynasts* I. v. iv. 166 In that immense unweeting Mind is shown One far above forethinking; purposive, Yet superconscious. 1921 *Public Opinion* 28 Jan. 90/2 What the world needs is a man whose genius will come from the superconscious, the divine. 1940 A. HUXLEY *Let.* 12 Jan. (1969) 449 It looks as though there were a kind of spiral development, from unconscious animal, through conscious human up to what for lack of better words may be called super-conscious spiritual. 1953 R. F. C. HULL tr. *Jung's Psychol. & Alchemy* in *Coll. Wks.* XII. ii. 268 There are people who can never understand the unconscious as anything but a *sub*-conscious, and who therefore feel impelled to put a superconscious alongside or possibly above it.

Hence **superco·nsciousness.**

1898 *Advance* (Chicago) 27 Jan. 107/2 [Jesus] had now reached the state of mind known to Hindus as samahdo or super-consciousness, and there was no pain for him on the cross. 1901 *Harper's Mag.* CII. 788/1 Since this unpleasant whirl of superconsciousness had swept over him. 1940 *Mind* XLIX. 130 The condition of 'super-consciousness' which most Indian philosophers and mystics agree to be attainable in this life. 1962 M. SADHU *Samadhi* i. 13 The question of the higher aspects of consciousness in man, often called simply the Superconsciousness, is becoming more and more urgent.

su·percontinent. *Geol.* [SUPER- 6 c.] One of the large land masses that are thought to have existed in the geological past and from which two or more of the present continents are thought to be derived; = *PROTOCONTINENT. Cf. *GONDWANALAND, *LAURASIA, *PANGÆA.

1963 *Sci. Amer.* Apr. 89/1 According to Wegener all the continents had been joined in a single supercontinent about 200 million years ago. 1969 *Times* 25 Apr. 13/8 Australia and Antarctica are also thought to have been joined together, the land mass they formed being part of the supercontinent of Gondwanaland. 1970 [see *PANTHALASSA]. 1977 A. HALLAM *Planet Earth* 210 In the late Permian we thus encounter a single sea-less supercontinent, a phenomenon probably unique in the history of the Earth.

su:percontra·ction. [SUPER- 13.] The contraction of a hair or fibre to less than its original length after treatment with heat or chemicals.

1933 ASTBURY & WOODS in *Phil. Trans. R. Soc.* A. CCXXXII. 337 The appearance of the phenomenon of 'super-contraction' just described is perhaps the most striking manifestation of this change. *Ibid.* 359 A general effect to which we have ventured to attach the term 'super-contraction', to distinguish it from the property of a normal stretched hair, when wetted, of returning exactly to its original unstretched length. 1954 ALEXANDER & HUDSON *Wool* iii. 75 Supercontraction of unstrained fibres was first observed by Speakman. 1977 F. KIDD in R. S. Asquith *Chem. Natural Protein Fibers* ix. 394 Woods..recorded 50% supercontraction of Cotswold wool.

So **su:percontra·ct** *v. intr.* and *trans.*, to (cause to) undergo supercontraction; **su:percontra·cted, -contra·cting** *ppl. adjs.*

1933 *Phil. Trans. R. Soc.* A. CCXXXII. 365 With respect to the *minimum* (super-contracted) length,..permanent set develops even in hot water. 1953 R. W. MONCRIEFF *Wool Shrinkage* xxv. 352 Fibres..which had not been treated with diepoxybutane in the phosphate buffer did not subsequently supercontract in metabisulphite. 1954 ALEXANDER & HUDSON *Wool* iii. 76 Phenol is a much more effective supercontracting agent than formamide. 1962 W. J. ONIONS *Wool* ii. 38 Steam set fibres are also resistant to boiling 5 per cent sodium bisulphite, which supercontracts untreated fibres. 1977 F. KIDD in R. S. Asquith *Chem. Natural Protein Fibers* ix. 394 White skunk hairs..supercontract more in sodium metabisulfite than do human hairs.

supercoo·l, *v.* [SUPER- 9 b.] **a.** *trans.* To subject to supercooling.

1932 F. F. GROUT *Petrogr. & Petrol.* III. 232 In the absence of seeding some magma intrusions may be greatly supercooled. 1959 B. CHALMERS *Physical Metallurgy* vi. 265 The liquid of composition *O* is supercooled until a nucleus in one of the phases forms. 1976 *Nature* 22 July 323/2 The cell contents are readily supercooled.

b. *intr.* To undergo supercooling.

1940 *Lancet* 17 Feb. 303/2 The capacity to supercool may be abolished by soaking the skin in water, and skin which does not supercool may be induced to do so by leaving it unwashed for a week. 1959 B. CHALMERS *Physical Metallurgy* vi. 244 When a metal sample is sufficiently subdivided into isolated drops, some supercool to the extent of about $0{\cdot}2T_E$ before nucleation occurs. 1975 *Rev. Mod. Physics* XLVII. 454/1 The $A \to B$ transition [in liquid ^3He] strongly supercooled the first time the transition was made from temperatures considerably above T_0.

super-cool, *a.*: see *SUPER- 9 a (*a*).

supercoo·led, *ppl. a.* [SUPER- 9 b.] **a.** Liquid though below the freezing point. **b.** Apparently solid, but formed from a liquid without a definite change of phase and having (on the atomic scale) the disorder of a liquid.

1898 *Chem. News* 29 July 59/1 (*heading*) The variation with the temperature of the number of nuclei which are formed in different supercooled liquids. 1937 *Times* 14 Oct. 11/4 Local conditions may cause the moisture in a cloud to become, in the meteorologist's phrase, 'supercooled'. 1947 J. C. RICH *Materials & Methods Sculpture* xi. 329 Glass is sometimes referred to as a supercooled liquid, because the material has no definite melting point. 1966 C. R. TOTTLE *Sci. Engin. Materials* iii. 81 Any material possessing the properties of a supercooled liquid, having failed to crystallize and therefore being amorphous in structure, can be called a glass. 1980 *Jrnl. R. Soc. Arts* May 369/2 Silver iodide smoke particles can also provide 'kernels' on which ice crystals can grow in supercooled cloud.

supercoo·ling, *vbl. sb.* [SUPER- 9 b.] The cooling of a liquid to below its freezing point without solidification or without crystallization occurring; more widely, cooling to below the temperature of a phase transition without the change of phase occurring; reduction in the temperature of a phase transition.

1898 *Chem. News* 29 July 59/1 The number of points increase with strong super-cooling. 1922 GLAZEBROOK *Dict. Appl. Physics* I. 947/2 The supercooling of a vapour without condensation is analogous to the supercooling of a liquid without crystallization. 1963 R. A. FLINN *Fund. Metal Casting* ii. 21 Some supercooling usually ensues before initial crystallization. 1975 *Rev. Mod. Physics* XLVII. 463/2 This would explain..experiments [on ^3He] at melting pressure..where *no* superheating of the $B \to A$ transition is observed while supercooling of the $A \to B$ transition is common. 1980 S. A. MORSE *Basalts &*

Phase Diagrams iii. 28 Supercooling results from the fact that the nucleation of a crystal is a random process, requiring the accidental arrangement..of atoms in the pattern of the crystal structure, after which a crystal nucleus may grow at a great rate.

supercrat: see *SUPER- 6 c.

supercri·tical, *a.* [SUPER- 9 a.] **1.** Highly critical.
1610, 1661 [in Dict. s.v. SUPER- 9a]. **1937** *Even. News* 12 Feb. 11/1 Only the super-critical, rather cosmopolitan-minded audience of the West End still look coolly on their simple robust brand of humour.

2. Of, pertaining to, or designating a fluid at a temperature and pressure greater than its critical temperature and pressure.
1934 *Econ. Geol.* XXIX. 457 The first work on critical and supercritical phenomena of solutions was done by Hannay and Hogarth in 1879. **1968** *New Scientist* 27 June 697/1 Supercritical boilers probably present more problems. **1972** *Physics Bull.* Apr. 236 The use of supercritical helium for cooling superconducting and other low temperature equipment. **1981** E. CORLETT *Revolution Merchant Shipping* 40/1 Natural gas..at normal temperatures ..is supercritical... It cannot be liquefied by pressure unless below −82° C.

3. Of a flow of fluid: faster than the speed at which waves travel in the fluid. Of an aerofoil: giving rise to such a flow over much of its surface when its speed relative to the bulk fluid is subcritical, but in such a way that flow separation is largely avoided.
1941, etc. [see *SUBCRITICAL *a.* 2]. **1967** *Aviation Week* 24 July 25/1 Whitcomb describes the shape as a 'supercritical airfoil' because it enables aircraft in subsonic flight to achieve these higher critical Mach numbers before the normal drag rise is experienced in transonic flight. The shape of the upper surface of the transonic wing permits predominantly supersonic air flow to be maintained across the wing. **1969** *Ibid.* 17 Feb. 22/1 Improved version of the National Aeronautics and Space Administration's 'supercritical wing'..will begin test flights in mid-1970. **1977** *McGraw-Hill Yearbk. Sci. & Technol.* 422/1 Supercritical compressor blades. **1979** *Time* 2 Apr. 21/2 With only two engines and a 'supercritical' wing that cuts aerodynamic drag, it is the most fuel-efficient commercial jet flying today.

4. *Nucl. Physics.* Containing or being more than the critical mass (see *CRITICAL *a.* 7 b).
1950 GLASSTONE *Sourcebk. Atomic Energy* xiv. 394/2 The presence of stray neutrons in the atmosphere makes it impossible to prevent a chain reaction in a supercritical mass. **1958** J. CLEUGH tr. *Jungk's Brighter than Thousand Suns* xii. 191 If he passed the point or was not quite quick enough in breaking contact, the mass might become 'super-critical' and produce a nuclear explosion. **1973** *Nature* 23 Mar. 251/1 Assuming a relative ^{235}U abundance throughout the [hypothetical] planet equal to the terrestrial crustal abundance, there is just enough energy to disperse the planet if the ^{235}U could be assembled into a super-critical mass.

Hence **su:percritica·lity,** supercritical state.
1959 *Times* 5 Mar. 2/4 (Advt.), Join a section concerned with the assessment of criticality hazards in the handling and processing of fissile materials and to carry out theoretical research into the factors governing criticality and supercriticality. **1976** L. SANDERS *Hamlet Warning* (1977) I. xi. 95 The blast..would be well into the kiloton range. The trick was..to bring the mass to supercriticality uniformly and at the same instant.

su·percurrent. *Physics.* [f. *super-* in *SUPERCONDUCTOR, etc.] An electric current flowing without dissipating energy, as in a superconductor.
[**1936** *Nature* 16 May 824/1 The assumption for the supra-conductor is, that $I = I_c + I_s$... That is to say, there is a third sort of current I_s, call it the supra-current.] **1940** E. F. BURTON et al. *Phenomena Temperature Liquid Helium* x. 302 We must now recognize three entirely different kinds of electric current..: (1) the displacement current in an insulator..; (2) the ordinary conduction current..; (3) the super-current. **1962** *New Scientist* 22 Nov. 454/3 Superconductivity can also be destroyed by increasing the supercurrent itself beyond a critical value. **1966** K. MENDELSSOHN *Quest for Absolute Zero* x. 242 Superfluid flow, just as a persistent supercurrent, is distinguished by zero entropy. **1980** *Sci. Amer.* May 43/1 In the absence of a magnetic field, and with a current below a critical value, a supercurrent flows through the Josephson junction just as if the superconductor were not interrupted by the insulating barrier.

su:per-du·per, *a. colloq.* (orig. *U.S.*). Also **sooper-dooper, super-dooper, super duper.** [A reduplicated extension of SUPER *a.*] Especially splendid, powerful, etc.; exceptional, particularly good.
1940 *N.Y. Times* 27 Sept. 27/7 (Advt.), After seeing this new M-G-M sooper dooper musical smash, our little voice went pattering all over the house. **1942** *Fortune* Feb. 104/3 Joining the Hearst organization he rapidly became what he calls a super-dooper circulator in Milwaukee, Detroit, Boston. **1942** *Sun* (Baltimore) 5 Mar. 8/8 A week ago officers knew before the curtain rose that something super-duper was about to be staged. **1949** *Hansard Commons* 26 May 1490 They are now subsidising the building of a 'super-duper' liner to compete with the Cunard flagships. **1951** *People* 3 June 2/2 British Railways put on a super-duper Pullman car special from Victoria to Epsom. **1958** D. WALLACE *Forty Years On* i. 15 Super-duper rockets with fancy-nancy warheads. **1971** *News-Advocate* (Barbados) 20 Mar. 8/1 Now this promises to be a super duper weekend. **1975** *New Yorker* 24 Mar. 33/1 The Colony range from the modest Villa Torino unit..to the superduper Viking Executive units. **1978** M. PUZO *Fools Die* xvi. 181 There was one super-duper bicycle that I was sorry she had bought.

super-e·go. *Psychoanal.* [SUPER- 5.] A Freudian term for that aspect of the psyche which has internalized parental and social prohibitions or ideals early in life and imposes them as a censor on the wishes of the ego; the agent of self-criticism or self-observation. Also *transf.*
1924, etc. [see *ID²]. **1938** *Times Lit. Suppl.* 26 Feb. 132/4 When the moral superego takes charge and the ego is no longer coercive but submissive. **1945** AUDEN *Coll. Poetry* 242 The fatal ease with which Conscience, i.e., the voice of God, is replaced by 'my conscience', i.e., the Super-Ego. **1958** M. ARGYLE *Relig. Behaviour* xii. 157 If religious behaviour is derived from the super-ego in some way, religion should have an irrational 'super-ego' quality about it. **1958** W. J. H. SPROTT *Human Groups* x. 173 The importance of the primary group can be expressed by saying that the group acts as the 'super-ego' of its members. **1965** *Listener* 2 Sept. 337/2 Lord Reith—one of the few twentieth-century Britons who have been willing to cast themselves as the community's superego. **1972** *Jrnl. Social Psychol.* LXXXVI. 157 The less dramatic resolution of the Oedipal complex experienced by females causes them to have a weaker superego than males. **1981** W. EBERSOHN *Divide Night* xi. 153 Freud..had decreed that the human psyche had three parts, the Id, the Ego, and the Super-ego.

superelevation. Add: **2.** Also *transf.* (see quot.)
1930 *Engineering* 14 Feb. 193/3 The spillway channels [of the dam] are curved, requiring large superelevation of the floor toward the outside.

supere·ssive, *a.* (and *sb.*) *Gram.* [f. L. *superesse* to be higher than, survive, remain + -IVE.] Designating a case or grammatical relation which expresses position above or on top of. Also *absol.* as *sb.*
1903 [see *INTROESSIVE *a.*] **1951** W. K. MATTHEWS *Languages U.S.S.R.* vi. 99 Marr and M. Brière..recognise secondary cases—a locative, an inessive..a superessive, a disjunctive, [etc.]. **1954** PEI & GAYNOR *Dict. Linguistics* 207 *Superessive,* in certain languages (notably, languages of the Finno-Ugric family) a declensional case, having the same denotation as the English preposition *on* or *upon.* **1971** D. I. SLOBIN in W. O. Dingwall *Survey Linguistic Sci.* 310 A variety of Hungarian case endings on nouns indicating such locative relations as illative, elative,..and superessive—that is, in plain English,..the directional notions of *into, out of,*..and the positional notion of *on top of.*

supere·tte. orig. and chiefly *U.S.* [f. *SUPER(MARKET + *-ETTE.] A small supermarket.
1938 *Sat. Even. Post* 17 Sept. 85/3 It also developed a store called the 'Super-ette', which is a compact, limited-stock, self-service store. **1956** *Sun* (Baltimore) 10 Feb. 23/1 There were an estimated 67,500 'superettes' in 1953. **1963** *Listener* 10 Jan. 75/1 Supermarkets and superettes (the latter still large by British standards) together took over four-fifths of all American retail food trade in 1958. **1976** *Daily Times* (Lagos) 3 Nov. 12/1 (Advt.), A spacious van for traders, commercial houses,..supermarkets and superettes.

super-exchange, -fatted: see *SUPER- 2 a (*b*), 9 a (*a*).

superfe·cta. *U.S.* [f. SUPER- 6 c, after *perfecta.*] A method of betting in horse-racing whereby the bettor must pick the first four finishers of a race in the correct order.
1972 *Compton Yearbk. 1971* 532/2 *Superfecta,* a system of betting on races in which the bettor must pick the first, second, third, and fourth horses in this sequence in a specified race in order to win. **1972** *N.Y. Post* 1 Mar. 63/5 Besides superfectas, seats, TV..the track also plans to improve the lighting. **1973** *Sunday Mirror* 9 Sept. 20/2 The gang went to work on trotting races in New York, fixing 'superfecta' races. **1977** *Time* 21 Nov. 46/3 The growth of exotic betting devices—superfectas and the like—with their huge pay-offs represents an additional impetus to crooked horsemen.

superfemale: see *SUPER- 5 f.

su·perfix. *Phonetics.* [f. SUPER- 2, after *prefix, suffix,* etc.] A sequence of stress or other suprasegmental phonemes which is treated as part of the grammatical structure of words and phrases.
1951 TRAGER & SMITH *Outl. Eng. Structure* II. 56 Suprasegmental morphemes consisting of patterns of stress, with the possibility of including plus junctures, are called superfixes. Those consisting of pitches and a terminal juncture are called intonation patterns. *Ibid.,* A morphemic phrase consists of two or more bases, with their suffixes, and a superfix. **1954** [see *SIMULFIX]. **1956** H. WHITEHALL in *Kenyon Rev.* XVIII. 415 In their *Outline,* Trager and Smith have..assembled some..linguistic tools. Their exposition of the superfix..throws..light on the distinction between the verse of Wyatt, Donne,..Hopkins, and Eliot in which superfix patterns function freely..as part of the rhythm, and the verse of Surrey, Pope, Tennyson, in which the superfixes are frozen..in the rhythm. **1965** [see *high-tone* s.v. *HIGH *a.* 22a]. **1972** HARTMANN & STORK *Dict. Lang. & Linguistics* 226/1 *Superfix,* term used to describe the suprasegmental or prosodic features of a particular word or utterance, e.g. the stress pattern which distinguishes modifier + noun *green house* from the compound noun *greenhouse.*

su·perflow. *Physics.* [f. *super-* in next.] Flow of a superfluid.
1939 *Canad. Jrnl. Res.* XVII A. 163 An ordinary flow of the whole fluid is superposed on the super-flow. **1966** [see *LAMBDA 4]. **1975** *Nature* 10 Jan. 93/3 It is superflow in the film which is responsible for the well known and dramatic phenomenon of the beaker of helium which empties itself while remaining upright.

superfluid (stress variable), *sb.* and *a.* *Physics.* [SUPER- 6 c, 9 a.] **A.** *sb.* (*su·perfluid*) A fluid that exhibits superfluidity.
1938 P. KAPITZA in *Nature* 8 Jan. 74/2 The helium below the λ-point enters a special state which might be called a 'superfluid'. **1950** [see *quantum liquid* s.v. *QUANTUM 7 a]. **1965** *Economist* 22 May 935/2 Helium itself behaves oddly. If cooled below 2·2 degrees K it becomes a 'superfluid'. **1974** *Nature* 15 Mar. 195/2 With the discovery of the new phases of ^{3}He, the number of known superfluids in nature has been doubled. **1982** *McGraw-Hill Yearbk. Sci. & Technol.* 227/1 The neutrons in a neutron star..are expected to form a *p*-wave-paired superfluid whose properties should be related to those of ^{3}He-A and B.

B. *adj.* (*superflu·id*) Exhibiting or pertaining to superfluidity.
1941 *Physical Rev.* LX. 357/2 We might regard liquid helium as if it consisted of a 'mixture' of two liquids—one is 'superfluid' without viscosity..and the other is 'normal'. **1947,** etc. [see *NORMAL *a.* 2 k.] **1968** C. G. KUPER *Introd. Theory Superconductivity* ii. 20 At any temperature $T < T_c$ only a fraction..of the electrons are in the condensate ('superfluid' electrons) and the remainder are 'normal' electrons. **1974** *Nature* 6 Dec. 441/2 Liquid ^{3}He.. under its own saturated vapour pressure..was found to undergo a superfluid transition at 0·93 mK.

superflui·dity. *Physics.* [SUPER- 10.] The property of flowing without viscosity or friction which, with other exceptional properties, is exhibited by the isotopes of liquid helium below certain temperatures; an analogous property of other collections of particles (as the electrons in a superconductor) that exhibit quantum effects on a macroscopic scale.
1938 *Physical Rev.* LIV. 952/2 The transport properties ..of liquid helium, when passing the λ-point actually change in a very conspicuous manner; thus one speaks of a 'superfluidity' and of a 'super-heat-conductivity'. **1955** H. B. G. CASIMIR in W. Pauli *Niels Bohr* 131 The Bose-Einstein gas is the only model known at present that leads to macroscopic wave functions at finite temperatures and the super-fluidity of liquid helium has almost certainly something to do with this model. **1967** *New Scientist* 25 May 454/1 In superconductivity, as in its counterpart superfluidity, we are faced with phenomena that are quite outside our ordinary physical experience. **1975** *Rev. Mod. Physics* XLVII. 430/1 Observation of fourth sound [in ^{3}He] proves the existence of superfluidity. **1976** *Sci. Amer.* Dec. 56/1 Under the name of superconductivity, superfluidity is also displayed by the conduction electrons in a great many metals and alloys.

su:perfluore·scence. *Physics.* [SUPER- 6 c.] The co-operative emission of radiation by a system of atoms as a result of fluorescence and the spontaneous correlation of excited atomic states; also, superradiance.
1966 *Jrnl. Appl. Physics* XXXVII. 682 (*heading*) Studies of ruby superfluorescence and population inversion. **1974** *Sci. Amer.* June 31/1 Since short-pulse laser systems must store large amounts of energy prior to pulse amplification, high gain coefficients in large-aperture amplifiers present two difficult problems. The first is termed superfluorescence. This is simply the normal fluorescence emitted spontaneously by the excited laser material, amplified by the gain of the material itself. **1975** BONIFACIO & LUGIATO in *Physical Rev.* A. XI. 1507/2 The system spontaneously creates correlations, i.e., a macroscopic dipole which gives rise to a pulse whose maximum intensity is proportional to N^2 and whose time duration is proportional to N^{-1}. We call this phenomenon superfluorescence. **1980** *Nature* 8 May 70/1 Superfluorescence produces radiation pulses which have much larger amplitudes than those which one would obtain in normal incoherent atomic radiation processes.

Hence **su:perfluore·scent** *a.*
1973 *Appl. Physics Lett.* XXII. 79/2 Figure 1 illustrates the repetitive superfluorescent pulses observed at 3370 Å in N_2. **1977** R. L. BYER in Harper & Wherrett *Nonlinear Optics* ii. 89 For efficient superfluorescent operation the input noise field must be amplified by approximately 10^{14}.

superfluous, *a.* (*sb.*) Add: **5.** Special collocations: *superfluous hair,* bodily hair considered to be unattractive in women, esp. on the face; *superfluous woman,* a woman unlikely to marry, because of a surplus of women over men in the population; also *superfluous girl.*
1876 GEO. ELIOT *Dan. Der.* II. III. xxi. 49 The sad faces of the four superfluous girls, each, poor thing.. having her peculiar world which was of no importance to any one else. **1800** in C. W. Cunnington *Feminine Attitudes* (1935) ii. 44, I shall sell a compound to take off all superfluous hair. **1873** *Young Englishwoman* Aug. 414/1

Will you kindly tell us..whether you know of any depilatory that may be safely used for the removal of superfluous hair? **1933** D. L. SAYERS *Murder must Advertise* iv. 69 Do you suffer from superfluous hair? **1976** CADOGAN & CRAIG *You're a Brick, Angela!* v. 74 Superfluous hair, poor complexions and excessive perspiration preoccupied many readers. **1886** L. M. ALCOTT *Jo's Boys* i. 21 There is plenty for the 'superfluous women' to do... I am very glad..that my profession will make me a useful..spinster. **1911** G. B. SHAW *Getting Married* Pref. 140 In our population there are about a million monogamically superfluous women, yet it is quite impossible to say of any given unmarried woman that she is one of the superfluous. **1978** CADOGAN & CRAIG *Women & Children First* vii. 133 The 1921 census showed a 1,700,000 surplus of women over men as a result of the slaughter of the war years..the so-called superfluous woman.

su·perfly, *a.* and *sb. U.S. slang.* [SUPER- 9 a: cf. FLY *a.* 1.] **A.** *adj.* **a.** Very good, excellent, the best (esp. in the context of drugs). **b.** *spec.* Typical of the film character Super Fly (see quot. 1975¹). Also with capital initial.
1971 R. WOODLEY in *Esquire* Apr. 79/1 'That,' he said in crisp, sure tones, 'is top-shelf coke. Super-fly.' **1973** ... *New York* 30 Aug. 29/1 ... **1974** *Florida FL Reporter* XIII. ... A kind of Swahili-speaking superfly image. **1975** WENTWORTH & FLEXNER *Dict. Amer. Slang* Suppl. 747/2 *Superfly,*..very wonderful, desirable, or attractive... Became popular after the 1972 motion picture *Super Fly,* about a cocaine dealer in Harlem. **1975** *Los Angeles Times* 14 July ii. 5/3 Last year a ninth-grader impressed the entire student body on several occasions by wearing flashy 'super fly' suits. **1976** *National Observer* (U.S.) 8 May 16/3 ... No Superfly suits, no platform shoes. **1977** E. LEONARD *Unknown Man No. 89* vii. 87 ... Get his superfly hair fixed up.

B. *sb.* [From the title of the film: see sense b of the adj. above.] One who sells illegal drugs, a 'pusher'.
1973 *Black Panther* 7 July 7/3 The high level dope pushers, the 'Super Flys', were the target. **1974** *Black World* Sept. 25/2 *Long Black Song* tells us, here in the 1970's, that the days of darky entertainers, superflies, sweetbacks, and Melindas, if not over, are numbered.

superfœtation, var. SUPERFETATION.

superfusate (s'ūpəɹfiū·zėɪt). *Med.* [f. *SUPERFUSE *v.* + -ATE, after *filtrate, precipitate.*] Any solution which has been used in the process of superfusion.
1970 *Proc. Soc. Exper. Biol. & Med.* CXXXIII. 1373/2 The presence of a constant concentration of LH in the superfusate in this system furnishes an appropriate control to evaluate the release obtained in response to hypothalamic extracts. **1979** *Experientia* XXXV. 225/2, 5-min fractions of superfusate were collected serially in glass vials.

superfuse, *v.* Add: **1. b.** *Med.* To subject (tissue) to, or employ (fluid) in, the technique of superfusion. Also, of a liquid, to flow over the surface of (tissue) in a thin layer. Cf. *PERIFUSE *v.*
1953 *Brit. Jrnl. Pharmacol. & Chemotherapy* VIII. 322/1 Two tissues were suspended one above the other and the same fluid was superfused over them both. **1964** *Ibid.* XXIII. 360 The blood superfused the second tissue and was then returned to the jugular vein by gravity. **1975** *Nature* 25 Dec. 754/1 The exposed subesophageal ganglia were superfused with continuously flowing snail Ringer. **1978** *Ibid.* 29 June 765/2 Each stream of blood superfused a separate collagen strip which was excised from the Achilles tendon of a rabbit.

Hence **superfu·sed** *ppl. a.,* subjected to superfusion; **superfu·sing** *ppl. a.,* that superfuses.
1953 *Brit. Jrnl. Pharmacol. & Chemotherapy* VIII. 322/2 ... superfused muscle. **1977** *Nature* 6 Jan. 65/2 Test solutions were assayed..by their effects on isolated superfused smooth-muscle organs. **1980** *Ibid.* 3 Jan. 93/1 (*caption*) Potassium chloride was added..to the superfusing fluid for 2-min periods at intervals of 16 min.

superfusion. Add: **1. b.** *Med.* The technique of causing a stream of liquid to run over the surface of a piece of suspended tissue, keeping it viable and allowing the interchange of substances between it and the fluid to be observed.
1953 *Brit. Jrnl. Pharmacol. & Chemotherapy* VIII. 321/1 A piece of intestine may be suspended in air and kept in good condition by a stream of a suitable solution running over its surface... This technique may be called superfusion, since the fluid runs over the tissue, by analogy with perfusion, in which the fluid runs through the tissue. **1970** *Proc. Soc. Exper. Biol. & Med.* CXXXIII. 1373/2 Continuous superfusion of a single pituitary gland might permit new approaches to the study of mechanisms and dynamics of LH release. **1980** *Nature* 3 Jan. 92/2 Superfusion of these slices for 2 min with Krebs' solution containing added KCl..increased the tritium overflow.

supergalactic, -galaxy: see *SUPER- 5 c.

supergene (s'ū·pədʒīn), *a. Min.* [f. SUPER- 1 a: see *-GEN 3.] Of an ore or mineral: enriched or deposited by a downward-moving solution; involving deposition by a downward-moving solution.
1914 F. L. RANSOME in *Bull. U.S. Geol. Survey* No. 540. 153 The suggestion is offered that minerals deposited by generally downward-moving and initially cold solutions may be termed supergene minerals. **1944** [see *paramorphic* s.v. *PARA-¹ 2 c]. **1977** A. HALLAM *Planet Earth* 112 Where leaching of sulfide ore deposits occurs, residual red and brown iron hydroxyoxide cappings (gossans) are left, and other elements can be carried down and precipitated in a zone of 'supergene' enrichment near the water table.

su·pergene, *sb. Genetics.* [f. SUPER- 6 c + *GENE.] A group of closely linked genes, freq. having related functions.
1949 DARLINGTON & MATHER *Elem. Genetics* ii. 46 Thus the cross between male and female is a back-cross for the X-Y pair of chromosomes or, if you like, the X-Y super-gene, and the offspring are of each sex. *Ibid.* v. 118 The differences could be interpreted as two differences so closely linked as never to recombine (two-gene system). Or one of them could be regarded as associated with an inversion inhibiting recombination (one super-gene system). **1978** *Nature* 13 July 164/1 The Z complex seems to provide an example of a 'supergene'—a large chromosomal segment with multiple genes involved in similar, or closely related functions.

su·pergiant, *sb.* and *a.* [SUPER- 6 c.] **A.** *sb.* **a.** A very large star that is even brighter than a giant, in many cases despite being relatively cool. (The dominant sense.)
1927 H. N. RUSSELL et al. *Astronomy* II. xxi. 725 Certain very bright stars, much more brilliant than the ordinary giants, are sometimes called super-giants. **1962** *Listener* 15 Feb. 370/2 A very brilliant white super-giant such as Rigel. **1978** PASACHOFF & KUTNER *University Astron.* xi. 294 The sun..is only one-millionth as luminous as the most brilliant of the red supergiants.

b. A supergiant galaxy (see sense b of the adj. below).
1975 S. VAN DEN BERGH in A. Sandage et al. *Stars & Stellar Systems* IX. xii. 531 Assuming the brightest stars in the Sc giant galaxy M33 to be similar to those in the Sc supergiant M100.

B. *adj.* **1.** *Astr.* **a.** Designating a star that is a supergiant.
1930 R. H. BAKER *Astronomy* xx. 372 Super-giant stars are extraordinarily luminous giants. **1973** *Daily Tel.* 29 Sept. 13/1 Under the best conditions it might be possible with glasses to glimpse the red supergiant star Antares in Scorpius. **1981** *Nature* 15 Oct. 513/1 Red giant and supergiant stars have long been favourites of professional and amateur astronomers.

b. Of a galaxy: in the brightest of five luminosity classes.
1960 S. VAN DEN BERGH in *Astrophysical Jrnl.* CXXXI. 216 The nomenclature for the luminosity classes has been chosen to agree with that used in the Yerkes system of stellar luminosity classification: (I) supergiant galaxy, (II) bright giant galaxy, (III) normal giant galaxy, (IV) subgiant galaxy, and (V) dwarf galaxy. **1978** *Sci. Amer.* Nov. 103/1 If [the cluster]..harbors a centrally located supergiant elliptical galaxy, which is a strong radio source and is surrounded by an X-ray-emitting cloud and a massive halo of stars.

2. *gen.* Extremely large.
1977 *Time* 2 Aug. 37/1 In June a high-pressure air mass began building up just east of the Rockies. It stayed there, with some up-and-down movement of air, and slowly turned into a supergiant mass. **1981** *Sci. Amer.* Nov. 66/2 Several supergiant natural-gas fields were found north of the Arctic Circle.

su:pergranula·tion. *Astr.* [SUPER- 6 b.] A pattern of large convective cells, each thousands of miles across, covering the surface of the sun. So **supergra·nular** *a.,* of or pertaining to supergranulation; **supergra·nule,** an individual cell of this kind.
1962 R. B. LEIGHTON et al. in *Astrophysical Jrnl.* CXXXV. 474 Some of the properties of the larger cells suggest that they may be a giant system of convective cells—a 'supergranulation'—associated with the granulation but originating in deeper layers where the scale height is relatively great. **1964** *Astrophysical Jrnl.* CXL. 1120 The velocity cells (called 'supergranules')..have an average diameter of 32000 km. **1967** K. O. KIEPENHEUER in J. N. Xanthakis *Solar Physics* xiii. 385 Even large spots, with their strong magnetic fields and complex structures, seem to be fed into the patterns of the supergranular network. **1973** *Nature* 14 Dec. 421/1 Spicules..cluster favourably in regions of enhanced magnetic fields along supergranular boundaries within the chromosphere. **1976** D. F. GRAY *Observation & Analysis Stellar Photospheres* xviii. 442 The gas flow in a supergranule mimics that of the granule, but the size of the convective cell is about 20 times as large..and there is no brightness variation across a supergranule. **1977** *New Scientist* 13 Jan. 77/1 The supergranulation 'cells', unlike the small convective granulations visible on the Sun's surface, are of the order of 15 000 to 30 000 km across.

supergrass, -gravity: see *SUPER- 6 c, b.

su·pergroup. a. [SUPER- 5 c.] A group composed of a number of other groups.
1943 M. SCHLAUCH *Gift of Tongues* 63 Finno-Ugric and Nenets..together form a super-group. **1969** *Proc. Geol. Soc.* Aug. 145 The following formal lithostratigraphical divisions are recognized: *Supergroup, Group, Formation, Member, Bed. Ibid.,* A supergroup consists of two or more adjacent and naturally related or associated groups. **1972** *Sci. Amer.* Sept. 133/2 In the next step most of these groups are combined, five at a time, to form 'super-groups' of 80 conversations each.

b. [SUPER- 6 c.] In rock music: a group formed by star musicians from different bands. Also *loosely,* an exceptionally talented or successful group.
1970 *Times* 7 Jan. 7/1 What the pop world calls a super-group is a group formed by star musicians from existing groups. **1976** *Sounds* 11 Dec. 7 I can only hope and pray that..two supergroups will emerge, but they have one hell of a name to live up to. **1976** *New Musical Express* 17 Apr. 23/3 They're what's already been described as a pub-rock supergroup. **1980** *Washington Post* 4 Dec. 9 Last night, The Police filled a packed Warner Theater with a lithe, sensual and utterly danceable brand of rock that is going to propel them to supergroup status in the next few years.

superheat, *v.* Delete 'in order to increase its pressure' and add: More widely, to heat (a substance) above the temperature of a phase transition without the change of phase occurring.
1869 *Amer. Jrnl. Sci.* XCVII. 22 To subject the oils to a temperature above their boiling points, or in other words, to super-heat their vapors. **1939** CARPENTER & ROBERTSON *Metals* II. xv. 1194 Marked changes in the structure of cast iron could be produced by superheating the melt, i.e. heating to a temperature considerably higher than that required to melt the metal.

superheated, *ppl. a.* Add: **1. a.** More widely, heated above the temperature of a phase transition without the change of phase occurring.
1931 G. W. TYRRELL *Volcanoes* vi. 161 Some of the material was melted up by the ascent of a highly super-heated lava.

superheating, *vbl. sb.* Add: (*a*) Also more widely (cf. *SUPERHEAT *v.*).
1980 S. A. MORSE *Basalts & Phase Diagrams* iii. 28 Superheating of crystals above their melting temperature is a rare phenomenon.

superhea·vy, *a.* (and *sb.*). [SUPER- 9 a.] **a.** *gen.* Extremely heavy, heavier than the normal. Occas. as *sb.*
1952 *Sci. Amer.* May 44/1 It is a job for accurate balancing and gyroscopic controls, and therefore an ideal spot for Hevimet, super-heavy Carboloy created-metal. **1974** *Physics Bull.* Dec. 578/3 The quarks give way to..superheavy mesonic matter and, ultimately, neutrons, protons and the lighter mesons. **1979** *Daily Times* (Lagos) 22 Sept. 30/1 That's how wrestling 'superheavy' Ray Apallon begins the open challenge to 'any of your Nigerian heavyweight wrestlers'.

b. *Nucl. Physics.* Of, pertaining to, or designating an element with an atomic mass or atomic number greater than those of the naturally occurring elements; *spec.* having an atomic number of 110 or more and belonging to a group having a limited range of proton/neutron ratios which confer enhanced stability against radioactive decay. Also as *sb.,* such an element.
1955 J. A. WHEELER in W. Pauli *Niels Bohr* 183 The superheavy nuclei that are neutron stable. **1966** L. DEIGHTON *Ipcress File* xviii. 107 Tritium is also called super-heavy hydrogen. **1970** *Physics Bull.* Dec. 534/2 The success of this view..has led to the suggestion that there exists a further 'island' of stability around mass number 300: superheavy nuclei which may have lifetimes from a fraction of a second, up to many years. **1976** *New Scientist* 18 Feb. 344/3 The radioactive counts from the mercury region showed the most promise for a superheavy. **1979** *Nature* 16 Aug. 549/2 There is no convincing evidence that superheavy elements have been discovered. **1980** *Physical Rev.* C. XXI. 1664/2 The recent theoretical estimates of low barriers..are suggesting the hope to detect superheavies in the ^{48}Ca + ^{248}Cm reaction.

superhe·lical, *a. Biochem.* [f. *SUPERHELIX, after HELICAL *a.*] Belonging to or consisting of a superhelix.
1966 *Jrnl. Gen. Physiol.* XXXXIX. 125, I wonder whether you would explain again how you calculated the number of superhelical turns. **1974** *Nature* 5 Apr. 476/3 The superhelical structure imposes upon the DNA molecule a topological restraint. **1980** *Sci. Amer.* July 108/1 It is possible to gain a general understanding of how a left-handed superhelical coil..is transformed into a right-handed interwound superhelix by considering the twisting of a telephone cord.

Hence **superhe·lically** *adv.*; also **su:perheli·city,** the state of being superhelical.
1974 *Nature* 20 Sept. 248/2 The affinity of the repressor for the operator..increases with increasing negative superhelicity up to a factor of approximately 14 for the DNA with −160 superhelical turns. **1978** *Ibid.* 12 Jan. 118/2 Superhelically wound oligonucleosome fibres. **1980** A. KORNBERG *DNA Replication* i. 25 Superhelicity, supercoiling and superhelix are terms for the twisting upon itself of the double helix.

su·perhelix. *Biochem.* Pl. **-helices.** [SUPER- 5 c.] A helix formed from a helix; *spec.* a three-dimensional structure sometimes assumed by polypeptides, in which double protein or DNA helices are themselves coiled into a higher-order helix. Cf. *SUPERCOIL *sb.*

1964 G. H. HAGGIS *Introd. Molecular Biol.* iv. 80, α-Helices probably twist together like the strands of a rope, in keratin and myosin, to form super-helices. **1971** *Nature* 5 Nov. 27 (*caption*) Since the normal DNA double helix is right handed, the superhelix is more likely to be left handed. **1980** *Sci. Amer.* July 100/1 In the chromatin ..of higher organisms the DNA is wound around a core of protein to form a left-handed solenoidal superhelix.

su·perhet, colloq. abbrev. of next. Also *fig.*
1926 *Glasgow Herald* 12 Jan. 10 The real heart of a superhet set is the first detector. **1926** R. W. HUTCHINSON *Wireless* 236 The multivalve..'Super-Het' is scarcely a receiving set..for a beginner. **1937** [see *direct vision* s.v. *DIRECT *a.* 1 b]. **1951** R. HOGGART *Auden* vi. 195 So the scene for a work such as this [sc. *The Age of Anxiety*] must be a time-ridden, newspaper-headline-obsessed, 'superhet' city. **1960** *Practical Wireless* XXXVI. 342/2 The output is fed by a jack into the I.F. portion of a six transistor superhet. **1976** *CB Mag.* June 1/2 (Advt.), This handsomely styled 23-channel solid-state CB two-way radio features a..dual-conversion superhet receiver with RF stage.

superhe·terodyne, *a.* and *sb. Radio.* [f. *SUPER(SONIC *a.* (and *sb.*) + *HETERODYNE *a.*] **A.** *adj.* Employing or involving a method of radio reception (also used in television) in which a signal from a tunable local oscillator is combined with the incoming carrier wave to produce an ultrasonic intermediate frequency whose value is fixed and predetermined, so that it is unnecessary to vary the tuning of the subsequent amplifier and detector and increased selectivity and amplification are possible.
1922 *Wireless World* 1 Apr. 11/1 The Armstrong superheterodyne principle, in which the incoming signals are heterodyned before the first detector valve. **1934** *Times Rev. Year 1933* 1 Jan p. ix/4 Superheterodyne receivers were especially popular. **1966** *McGraw-Hill Encycl. Sci. & Technol.* XI. 257/1 Frequency-modulation (FM) receivers are almost always superheterodyne. **1976** *Gramophone* July 232/2 The superheterodyne circuit made modern radio possible. **1977** W. TUTE *Cairo Sleeper* vii. 128 'Hafiz the barman has a wireless set.'.. It was a superheterodyne job with valves.

B. *sb.* A superheterodyne receiver.
[**1921** *Q.S.T.* May 16/1 If a good U.S. amateur with such a set and an Armstrong Super could be sent to England, reception of U.S. amateurs would straightway become commonplace.] **1922** *Ibid.* July 7/1 Super-regeneration is..the method that makes two tubes do all the work that ten used to do in the super-heterodyne. **1933** K. HENNEY *Radio Engin. Handbk.* xvii. 449 The h-f superheterodyne seldom has high sensitivity, unless the first or h-f tube is regenerative. **1940** [see *CHASSIS 5]. **1965** *Wireless World* July 336/2 The various oscillators in superheterodynes have all set their own problems.

super-highway: see *SUPER- 6 c.

superimpo·sable, *a.* [f. SUPERIMPOSE *v.* + -ABLE.] Capable of being superimposed.
1920 in WEBSTER. **1925** *Rep. Proc. Imperial Bot. Conf. 1924* 41 Repeated examination in the same and different years of families derived from the same matings or selfings..gave rise to curves so similar as to be in many cases superimposable. **1933** *Jrnl. Theol. Stud.* XXXIV. 97 Private objects, such, e.g., as the different elliptical shapes seen by individuals at different distances, &c., from a circular plate, are often not compatible, congruent, or superimposable. **1971** *Nature* 9 July 106/2 The system is symmetric, for it is superimposable on its mirror image. **1979** *Dædalus* Summer 90 They are both interchangeable and superimposable.

superimpose, *v.* Add: **4.** *intr.* Of two figures or the like: to be capable of being brought into coincidence; to occupy the same positions in relation to their contexts.
1971 *Nature* 2 July 12/2 Counts from the highest polysome fraction superimpose on the zones corresponding to the two major light chains. **1972** *Sci. Amer.* Aug. 95/1 The degree to which the fields do not superimpose can be measured, and in this case there was a range of six degrees of horizontal disparity and two degrees of vertical disparity. **1975** *Nature* 10 Jan. 127/1 The sheets could then superimpose, forming the observed structures. **1978** *Ibid.* 27 July 389/2 (*caption*) Note that the peaks do not superimpose.

superimposed, *ppl. a.* **1. b.** (Earlier and later examples.)
1875 J. W. POWELL *Explor. Colorado Valley* 166 The beds in which the streams had their origin..have been swept away. I propose to call such superimposed valleys. **1977** A. HALLAM *Planet Earth* 76/2 Alternatively, in superimposed drainage, the river courses may have initially developed on a cover of rocks whose structure was different to that of the rocks beneath, the upper beds having since disappeared through erosion.

su:perindivi·dual, *a.* (and *sb.*). [SUPER- 4 a.] Of or pertaining to that which is above or greater than the individual. Also as *sb.*
1916 F. VON HÜGEL *German Soul* 92 The moral relation between the individual and the super-individual unity. **1924** W. B. SELBIE *Psychol. Relig.* 150 The crowd will then act and feel, and express itself together. In doing so it becomes an entity or super-individual. **1936** J. KANTOR *Objective Psychol. Gram.* iv. 49 Those who would reject psychology from linguistics because linguistic phenomena are superindividual. **1943** *Mind* LII. 342 His [*sc.* Kant's] phrase *Bewusstsein überhaupt,* 'consciousness in general', though sometimes supposed to stand for a super-individual consciousness, might equally well..be taken to mean a character which all individual consciousnesses have in common.

Hence **su:perindivi·dualist, -individuali·stic** = *SUPERINDIVIDUAL *a.*; also [SUPER- 9 a], that favours a high degree of individualism.
1934 WYNDHAM LEWIS *Let.* 2 Nov. (1963) 223 Handicapped as we are under a super-individualist legislation—which allows the utmost licence in criticism of the State, in contradistinction to the Individual. **1943** *Mind* LII. 342 The majority of Kant's followers have, I suppose, adopted what I may call the 'super-individualist' interpretation. **1958** W. STARK *Sociol. Knowledge* i. 19 American society was also super-individualistic.

superinfe·ct, *v. Med.* [Back-formation from next.] *trans.* **a.** To cause (an infected cell) to be further infected *with* an organism of a similar kind. **b.** Of a bacterium or virus: to infect (a cell that already contains organisms of a similar kind).
1954 *Jrnl. Bacteriol.* LXVII. 696/1 Lysogenic cells were superinfected with phages. **1971** *Nature* 23 Apr. 496/3 By superinfecting it with a mixture of cat leukaemia and sarcoma virus the defective human virus might be helped. **1980** *Internat. Jrnl. Radiation Biol.* XXXVII. 120 When cells of *E. coli* are superinfected by phage λ, the phage DNA can appear in three distinct forms.

Hence **superinfe·cted, superinfe·cting** *ppl. adjs.*
1954 *Jrnl. Bacteriol.* LXVII. 698/2 The superinfected culture produces both the carried type and the superinfecting type of phage. **1961** *Virology* XIV. 220 The genetic incorporation of the superinfecting P2 was studied by examining the progeny of the superinfected cells. **1976** *Path. Ann.* XI. 259 Bacterial and, to a lesser extent, viral infections were also encountered in patients with multiple superinfecting organisms. **1981** *Virology* CIX. 74/1 The same concentration of PAA was also applied to the superinfected Raji cells.

su:perinfe·ction. *Med.* [SUPER- 15.] **a.** An infection occurring after or on top of an earlier infection, *esp.* as a consequence of treatment of the latter by broad-spectrum antibiotic or other therapy. **b.** The further infection of cells that are already infected with a similar organism, *esp.* as a technique in virology and immunology.
1922 *Stedman's Med. Dict.* (ed. 7) 972/2 *Superinfection,* ..a fresh infection added to one of the same nature already present. **1954** *Jrnl. Bacteriol.* LXVII. 702/2 The type of phage produced after superinfection was studied both in mass culture and in single burst experiments. **1961** *Lancet* 12 Aug. 352/2 Bacteræmia developed as a superinfection during antibiotic therapy. **1974** M. C. GERALD *Pharmacol.* xxvi. 459 While the incidence of superinfection is low with narrow-spectrum antibiotics such as penicillin, it is a common occurrence with tetracycline and other broad-spectrum antibiotics. **1981** *Virology* CIX. 72/1 Superinfection of the nonproducer cells with EBV..is known to induce synthesis of EA and MA.

superinte·lligent, *a.* **1.** [SUPER- 4 a.] Above or beyond the range of intelligence.
1960 [see *super-moral* s.v. *SUPER- 4 a].

2. [SUPER- 9 a.] Very highly intelligent.
1971 B. DE FERRANTI *Living with Computer* ix. 80 Recent work..demonstrates that by transplanting brain tissues super-intelligent animals can be produced. **1977** *Time Out* 28 Jan.–3 Feb. 60/4 (Advt.), Our labour force.. is super-intelligent.

superintendent, *sb.* and *a.* Add: **A.** *sb.* **1. e.** A police officer next above the rank of inspector.
1832 *Observer* 24 June 4/4 Mr. Thomas, the Superintendent of Police, F Division, came before..the presiding Magistrate. **1836** [in Dict., sense 1 a]. **1885** *Encycl. Brit.* XIX. 337/1 All promotions in the [police] service up to the rank of superintendent are made from the next rank below. **1907** G. B. SHAW *Major Barbara* Pref. 171 Those who pester our police superintendents with confessions of murder might very wisely be taken at their word and executed. **1936** G. HEYER *Behold, here's Poison* iv. 71 Things are more serious than I had supposed. This is Superintendent Hannasyde, of Scotland Yard. **1977** 'E. CRISPIN' *Glimpses of Moon* viii. 150 Not at all, Superintendent. Partial deafness must be quite a handicap in your profession.

superionic: see *SUPER- 9 a (*a*).

superiority. Add: **5.** Special Comb.: **superiority complex,** (*a*) *Psychol.,* an attitude of superiority which conceals actual feelings of inferiority and failure; (*b*) *gen.,* an exaggerated sense of personal superiority; (cf. *inferiority complex* s.v. *INFERIORITY 2).
1929 A. ADLER *Probl. Neurosis* vi. 87 In his business we find the man with a 'superiority complex': but if he were to lose his position..he would promptly go back to the expression of inferiority and make capital out of it. **1936** H. PREECE in *Crisis* Dec. 364/2 Each expression of this interest in the Negro is the manifestation of a definite superiority complex. **1945** A. L. ROWSE *English Spirit* xxxiii. 232 The English have a singular faculty for depreciating their great men. (Is it perhaps a form of superiority-complex?) **1979** *Nature* 11 Oct. 424/2 The Egyptian scientific community does not like or agree with the Israeli superiority complex.

superius (s'ūpī°·riǒs). *Mus.* [a. L. *superius,* neut. (used as *sb.*) of *superior* SUPERIOR *a.*]. (See quots. 1801, 1876.) Also *transf.,* a person who sings this part.
[**1519** O. PETRUCCI *Motetti de la Corona* (heading) Libro secundo. (Superius.)] **1776** J. HAWKINS *Gen. Hist. Sci. & Pract. Music* II. I. vii. 86 *Quinible*..may rather mean a high part.., which in general lies above the tenor..and at others between the contretenor and the superius or treble. **1801** T. BUSBY *Dict. Mus., Superius,*..the name by which the contrapuntists of the fifteenth and sixteenth centuries distinguished the *upper part* of any composition. **1876** STAINER & BARRETT *Dict. Mus. Terms* 412/1 *Superius,*..a name given to the upper part in a composition by the writers of the sixteenth century. **1907** *Grove's Dict. Mus.* (ed. 2) III. 631/1 'Le Parangon des Chansons', printed by 'Jaques Moderne'..in nine volumes..and..so arranged that the Superius and Tenor sit facing each other, on opposite sides of the table—the Superius reading from the lower half of the left-hand page, and the Tenor from the upper half. **1954** G. REESE *Music in Renaissance* i. 16 The *texture* in which a vocal top line (or, as this part is interchangeably called, treble, superius, cantus, or discantus) is supported by a subordinate, instrumental tenor and contratenor was much in vogue. **1970** *Proc. R. Mus. Assoc. 1969–70* 95 The fourth and final line of the text..telescopes the beginning and end of Sandrin's superius. **1977** *Early Music* Apr. 243/3 Surely..the superius should cadence on C with the other voices. *Ibid.* July 419/2 The superius of his motet *Sufficiebat*..is closely related to the tenor of Hayne's chanson *Mon souvenir.*

superjet: see *SUPER- 6 c.

superlative, *a.* Add: **2. c.** *superlative surprise,* the name given to an especially complicated method of change-ringing. Cf. SURPRISE *sb.* 5 b.
1788 W. JONES et al. *Key to Art of Ringing* xi. 179 Superlative Surprise... The above is an original composition of our own on purpose for this work, and has never yet been rung: the principle upon which it is founded, will..give it credit among amateurs of the art, for..it will plainly appear the most even treble bob peal..discovered. **1845** *Cambr. Chron.* 6 Dec. 4/3 The company of change-ringers, of Saffron Walden, performing..upwards of 1,200 changes of Superlative Surprise. **1874** W. BANISTER *Art & Science Change Ringing* 33 (*heading*) Superlative Surprise. **1931** E. MORRIS *Hist. Change Ringing* 458 Harry Withers..once..conducted a peal of Superlative Surprise at Selly Oak, Birmingham. **1965** W. G. WILSON *Change Ringing* 237/2 Superlative Surprise Major, 114.

su·perlattice. [SUPER- 5.] **1.** *Metallurgy.* An ordered arrangement of some of the atoms in a solid solution extending through large parts of it and coexisting with the disorder of the remaining atoms; also, a solid solution possessing this; = *SUPERSTRUCTURE 3.
1932 *Proc. R. Soc.* A. CXXXVI. 211 A structure of this kind is usually termed a superlattice (überstruktur). **1951** N. F. M. HENRY et al. *Interpret. X-Ray Diffraction Photographs* xv. 211/1 The use of the word 'superlattice' is unfortunate. The original German word *Uberstruktur* conveyed perfectly the concept of a new type of ordered structure being imposed on the existing disordered phase. This has nothing to do with the type of lattice. **1966** C. R. TOTTLE *Sci. Engin. Materials* viii. 187 Domains form and grow during normal superlattice formation, according to composition, temperature, and time of annealing. **1967** A. H. COTTRELL *Introd. Metallurgy* xiv. 190 When formed from true metals the structure is often an ordered solid solution or superlattice, in which the two species are arranged in some regular alternating pattern. **1978** *Nature* 9 Nov. 168/2 Evidence..that an FeNi L1₀ superlattice exists in the taenite of the Cape York and Toluca meteorites. **1979** [see *SUPERSTRUCTURE 3].

2. *Physics.* A small-scale periodicity in the composition of a semiconductor.
1970 *IBM Jrnl. Res. & Development* XIV. 61 We consider a one-dimensional periodic potential, or 'superlattice', in monocrystalline semiconductors formed by a periodic variation of alloy composition or of impurity density during epitaxial growth. **1977** *McGraw-Hill Yearbk. Sci. & Technol.* 380/1 Photocurrent measurements in the superlattice structures have made it possible to observe simultaneously quantum states and associated anomalous conductance.

superloo, -luminal, -male: see *SUPER- 6 c, 4 a (*a*), 5 f (and 6 c).

superman. Add: **1.** Also *loosely,* a man of extraordinary power or ability; a superior being.
1925 H. V. MORTON *Heart of London* 110 Above the kneeling priests is the Pharaoh, that ancient superman. **1942** *R.A.F. Jrnl.* 13 June 15 They are neither freaks nor super-men. *Ibid.* 18 The stories told in the newspapers to emphasise the superman qualities which the Commandos deny possessing. **1959** KOESTLER *Sleepwalkers* v. ii. 471 Both considered themselves supermen and started on a basis of mutual adulation. **1969** G. JACKSON *Let.* 28 Dec. in *Soledad Brother* (1971) 179 How could there be a *benevolent* superman controlling a world like this.

2. (With capital initial.) The name of an invincible hero with superhuman powers, including that of flight, introduced in an American comic strip (1938). Also *transf.* and *allusively.*
1938 *Action Comics* June 1 So was created..Superman! champion of the oppressed, the physical marvel who had sworn to devote his existence to helping those in need! **1940** *Time* 26 Feb. 44/3 Last week Superman took to the air in earnest, as a three-a-week serial. **1942** H. HAYCRAFT *Murder for Pleasure* ix. 191 Converting Strangeway's wife Georgia into what Miriam Allen de Ford calls 'a sort of female Superman'. **1958** *Times Lit. Suppl.* 1 Apr. p. xx/4 The impression remains of a sense of values associated with 'Superman' and American comics. **1968** S. ELLIN *Valentine Estate* III. iv. 142 'How the hell did he come to miss me?'..'You're Superman,' the first man answered. 'Bullets bounce off you.' **1976** *Survey* Winter 1 The..*New York Times*..in the past presented Henry Kissinger as a species of diplomatic superman. **1977** *New Scientist* 14 Apr. 59/3 Schlesinger..is riding high at present—pictured in a superman suit on a recent *Time* magazine cover. **1980** F. WELDON *Puffball* 77 'Now it's our turn.' 'I don't want it to be,' she said, as if he, like Superman, could turn the world the other way.

Hence **su·permanhood.**
1910 *Dublin Rev.* Oct. 344 Human nature..is likely to remain still exactly the same. Those who believe it to be travelling towards moral supermanhood have obviously not studied it. **1924** W. J. LOCKE *Coming of Amos* xix. 254 My vanity was pricked by what seemed to be her lack of confidence in my supermanhood. **1964** *Punch* 1 Apr. 507/3 Convincingly exciting, despite hero's slight leaning towards supermanhood.

su·permarket. orig. *U.S.* Also **super market.** [SUPER- 6 c.] **1. a.** A large self-service shop, selling a wide range of groceries and household goods, and freq. one of a chain of stores.
1933 *N.Y. Times* 25 Feb. 28/1 In a move interpreted by the trade as an effort to help both corporate chains and independent wholesale grocers fight the competition of 'super-markets' which have sprung up in the last two years, the Associated Grocery Manufacturers of America, Inc., yesterday drew up a proposed model law for States which may seek to prevent the sale of standard grocery products at or below purchase price. *Ibid.* 5 Mar. 11/4 For three months now a large supermarket in New Jersey has been doing a business reputed to average $100,000 a week. **1933** *Chain Store Age* (Gen. Merchandise ed.) June 95/1 The 'One-stop-drive-in super market' provides free parking, and every kind of food under one roof. **1949** R. GRAVES *Seven Days in New Crete* 121 We buyers..drifted round with our baskets, silently helping ourselves to whatever we wanted... The procedure recalled that of an American super-market. **1959** *Spectator* 25 Sept. 409/1 This applies particularly to supermarkets, whose whole economy depends on people going in to buy a can of beans and coming out with a dazed expression and three pounds' worth of groceries. **1969** *Islander* (Victoria, B.C.) 6 July 8/2 Your supplies..come from a good delicatessen and..your super market. **1978** *Oxford Consumer* Mar. 4/1 The change from counter to self-service stores in Britain only started in the late 1950's/early 1960's, when a supermarket was defined as having a minimum sales area of 2,000 sq. ft. **1979** M. BOYCE *I was There!* 70 The pithead baths is a supermarket now.

b. *transf.* and *fig.*
1962 *Listener* 26 July 127/2 The Marshall Plan and this vast new Supermarket have destroyed the roots of that sickness. **1973** J. W. POLIER in A. E. Wilkerson *Rights of Children* p. xiv, The 1970 White House Conference on Children announced that prepared reports would offer 'a supermarket of proposals'.

2. a. *attrib.,* as *supermarket chain, company, shopping,* etc.
1933 *N.Y. Times* 26 Feb. 15/8 The independent and corporate chain stores are standing together against the alleged menace of the 'super-market' competition. **1934** *Archit. Rec.* LXXVI. 206 Markets are now built up to the street line and are of the 'super-market' type. **1951** C. W. MILLS *White Collar* I. ii. 25 As supermarkets mushroomed..the chains began to imitate their supermarket competitors. **1963** *Times Rev. Industry* Aug. 65/2 Premier..is one of the few supermarket companies proper with a highly developed and well integrated scheme. **1967** G. WILLS in Wills & Yearsley *Handbk. Management Technol.* 192 A product which cannot gain distribution in the major supermarket chains may have a very high direct percentage distribution yet a low percentage sterling distribution. **1975** D. LODGE *Changing Places* iii. 115 We seem to have fixed on the same day for supermarket shopping. **1977** F. PARRISH *Fire in Barley* ii. 24 A lifetime of deep-frozen scampi, supermarket Riesling, [etc.].

b. supermarket trolley, a wire basket on wheels which a supermarket customer can push around the shop collecting goods for purchase; also (U.S.) *supermarket cart.*
1972 *Even. Telegram* (St. John's, Newfoundland) 28 June 29/6 (Advt.), 5 supermarket carts. **1970** *Cape Times* 28 Oct. 21/3 (Advt.), Lawn mowers, kitchenware, glassware, 22 supermarket trollies. **1977** *Irish Times* 8 June 8/8, I wandered in to sit with the farmers, and saw a heifer that would fit into a supermarket trolley sell for £60. **1982** BARR & YORK *Official Sloane Ranger Handbk.* 95/1 He and his chums stage a mixed doubles wheelbarrow/piggy-back/supermarket trolley race down the High/the Broad/the Cornmarket.

Hence **su:permarketee·r** [-EER], a person or company involved in a supermarket business.
1960 *Spectator* 13 May 712 Nor have many of the supermarketeers realised that self-service in itself gets you nowhere. **1964** *Punch* 29 Jan. 174/3 Other supermarketeers include London Grocers. **1973** *Guardian* 17 Feb. 13/6 To ask them to cut their profits still further would, in the view of the supermarketeer, be both unjust and ineffective.

su·permart. [SUPER- 6 c.] = prec.
1954 *Archit. Rev.* CXVI. 234 Beneath the two major blocks..the lower floors will house shops, a supermart, a highways terminal, an art museum, [etc.]. **1961** *Guardian* 17 Feb. 8/6 The baby sitting up in the supermart trolley. **1978** P. VAN GREENAWAY *Man called Scavener* xiv. 193 Take any name familiar to our age: a supermart..a refrigerator.

supermassive: see *SUPER- 9 a (*a*).

su·permind. 1. [SUPER- 6 c.] A mind of exceptional capacity or ability; a person possessing such a mind.
1918 *Daily News* (Chicago) 5 Jan. 10/6 (*heading*) Four super-minds ruling in Loop. **1962** 'S. RANSOME' *Without Trace* v. 50 Loot that wasn't dropping soon enough because..that high-geared supermind of Lynch's kept hanging fire. **1975** J. TAYLOR *Superminds* (1976) ix. 161 The view of a group of defence workers..who advised their government to liquidate all superminds as being a menace to security.

2. [SUPER- 5 a or 6 b.] An extended or superior mind that is a composite of many individual minds.
1941 L. MACNEICE *Poetry of W. B. Yeats* vi. 126 This mind, for Yeats, is not so much the mind of God as the super-mind of humanity. **1965** *Listener* 1 July 26/1. This civilization exists in a corporate form, the individual thinking units of which have long ago been synthesized into one super-mind.

su·permu:ltiplet. *Physics.* [SUPER- 6 b.] **a.** A group of transitions in an atom between spectral terms of different multiplicity, all the transitions involving the same change in the orbital quantum number l of an electron from the same initial value.
1927 RUSSELL & MEGGERS in *Sci. Papers Bureau of Standards* (U.S.) XX. 331 The five terms first mentioned are evidently closely related, and their combinations with the triad ^{3}P, ^{3}D′, ^{3}F can be arranged in the form of a 'supermultiplet'. **1935** CONDON & SHORTLEY *Theory Atomic Spectra* ix. 245 The set of all lines arising in transitions between two polyads..having the same parent configuration is known as a supermultiplet. **1972** I. I. SOBEL'MAN *Introd. Theory Atomic Spectra* xxxi. 313 By summing eqn. (31.48) over all transitions $J \to J'$ within the given multiplet we obtain the line strength of this multiplet... Summing over LL'..gives the line strength of the supermultiplet.

b. In particle physics, a multiplet (sense *b) in the broader sense, comprising particles of different hypercharge as well those of different charge.
1952 *Ann. Rev. Nucl. Sci.* I. 44 The two states belong to the same supermultiplet (in this case a single charge multiplet with $S = \frac{1}{2}$ and $T = \frac{1}{2}$). **1964** *New Scientist* 27 Feb. 523/3 The discovery of the omega-minus confirms that other groupings of particles into families or 'supermultiplets' are valid. **1972** G. L. WICK *Elementary Particles* v. 82 They thought that all these particles might belong to a larger multiplet, or supermultiplet, which connects both different isotopic spin and different strangeness.

supernacular, *a.* Add: (Later examples.) Also *transf.*
1920 G. SAINTSBURY *Notes on Cellar-bk.* 18 Some of the finer kinds [of sherry] are really supernacular—the best 'Tio Pepe', for instance. **1958** [see *DEVADASI].

supernatant, *a.* Add: **B.** *sb. Biol.* and *Med.* A supernatant substance.
1922 *Brit. Med. Jrnl.* 19 Aug. 297/2 To this high refinement..Otto, Munter, and Winkler attribute the potency of their products as compared with supernatants obtained by the centrifuge only. **1955** *New Biol.* XIX. 91 The supernatant is decanted and again spun usually at about 10,000 to 20,000 *g* for twenty minutes. **1977** *Proc. R. Soc. Med.* LXX. 192/2 The supernatants were decanted into plastic counting vials and mixed with 10ml Instagel.

su·pernate, *sb. Biol.* and *Med.* [f. prec., after *filtrate, precipitate.*] = *SUPERNATANT *sb.*
1943 *Jrnl. Immunol.* XLVI. 326 After centrifuging these mixtures the supernates were tested for the presence of a sheep-cell hemolysin. **1979** *Experientia* XXXV. 193/2 After decantation of the supernate into counting vials.

supernation, etc.: see *SUPER- 4 a (*a*) and (*c*), 6 c.

supernaturalize, *v.* Add: Hence **su:pernaturaliza·tion.**
1933 *Downside Rev.* LI. 729 St Thomas gives no support ..to the original views of Rousselot..as to the need of a kind of supernaturalization of the (natural) intellectual powers in order to accept and assent to revelation. **1961** E. L. MASCALL *Grace & Glory* i. 15 The sanctification and the supernaturalization of our whole being, body and soul alike—this is the purpose for which the Catholic Church and its whole sacramental equipment exist.

su:pernorma·lity. [f. SUPERNORMAL *a.* + -ITY.] **a.** The quality of exceeding what is normal; an instance of this.
1909 O. J. LODGE *Survival of Man* I. i. 2 Assertions concerning psychological supernormalities have not only excited attention, but have rather notably roused the interest of careful and responsible students. **1922** *19th Cent.* Oct. 600 One of the first indications of potential genius in school children is unusual precocity or mental supernormality. **1955** W. NAYLOR *Silver Birch Anthol.* 8 The faculty of being able to deliver, week after week, words of wisdom..in this spontaneous fashion, is in itself evidence of supernormality. **1977** D. MORRIS *Manwatching* 277 Having exhausted one line of supernormality, we switch to another, selecting a new element for improvement and dwelling on that until it too has become stale.

b. [SUPER- 9 a.] The quality of being exceedingly normal. *rare.*
1945 *New Yorker* 7 Apr. 74/2 Now that the end [of the war] is practically at hand, the supernormality of the English is surprising.

superno·va. *Astr.* Pl. **-novae, -novas.** [SUPER- 6 c.] **1.** A star that undergoes a sudden and temporary increase in brightness like a nova but to a very much greater degree, as a result of an explosion that disperses most of the stellar material.
1934 BAADE & ZWICKY in *Physical Rev.* XLV. 138/1 Supernovae flare up in every stellar system (nebula) once in several centuries. **1934** —— in *Proc. Nat. Acad. Sci.* 15 May 254 The extensive investigations of extragalactic systems..have brought to light the..fact that there exist two well-defined types of..novae which might be distinguished as common novae and super-novae. **1939** [see *NOVA 2]. **1965** *Listener* 20 May 741/1 Only three supernovae, the stars of 1054, 1572, and 1604, have appeared in our Galaxy since records began. **1976** *Sci. Amer.* Dec. 89/1 Some supernovas may leave behind cosmic ashes in the form of a neutron star or black hole. **1977** *Whitaker's Almanack 1978* 155/1 One important source of radio noise is the Crab Nebula, which is known to be the remains of the supernova of A.D. 1054. **1978** PASACHOFF & KUTNER *University Astron.* xi. 296 On the average, Type I supernovae reach an absolute magnitude of −19 at peak brightness, while that of Type II supernovae is about 2 magnitudes fainter.

2. *fig.*
1965 *Listener* 14 Jan. 84/3 Look at the soft gleam of D major near the start of the development (bar 170)—the distant glow of fiery *supernova,* into the heart of which we are at length to be flung. **1974** *State* (Columbia, S. Carolina) 1 Apr. 10A/1 As far as the networks were concerned, the satiric supernova had burned out. **1979** *Tucson* (Arizona) *Citizen* 20 Sept. 7B/6 (*heading*) 'Buck Rogers' no supernova, but it won't wink out, either.

3. *attrib.,* as *supernova explosion, remnant.*
1960 *McGraw-Hill Encycl. Sci. & Technol.* XIII. 303/2 The remains of an old supernova explosion. **1975** *Sci. Amer.* Mar. 29/2 The Crab Nebula is the remnant of a supernova explosion. **1960** *McGraw-Hill Encycl. Sci. & Technol.* XIII. 304/1 Unexplained radio sources also may be from supernova remnants. **1978** PASACHOFF & KUTNER *University Astron.* xi. 296 Optical astronomers have photographed two dozen of these stellar shreds, which are known as supernova remnants.

supernumerary, *a.* and *sb.* Add: **A.** *adj.* **1. d.** *Genetics.* Of a chromosome: additional to the normal complement of autosomes and sex chromosomes. Cf. sense f of the *sb.* below.
1907 E. B. WILSON in *Biol. Bull.* XII. 304 The unpaired chromosome may be either present or absent in either the male or female, and hence is without significance in sex-production. It is in fact a kind of supernumerary chromosome, which I shall designate as the '*s*-chromosome' in order to distinguish it from the odd sex-chromosome of the usual type. **1927** *Jrnl. Agric. Res.* XXXV. 782 The distribution of a supernumerary chromosome to the four daughter cells has been studied and found to be erratic. **1969** BROWN & BERTKE *Textbk. Cytol.* xviii. 380/1 Supernumerary chromosomes are usually much smaller than the autosomes and for that reason have often been termed 'fragment' chromosomes.

B. *sb.* **1. e.** (Earlier example.)
1755 C. CHARKE *Life* 115 A poor, beggarly Fellow, who had been sometimes Supernumerary in Drury-Lane Theatre.

f. *Genetics.* A chromosome which may be absent from normal individuals of either sex, having little or no effect on phenotype and occurring irregularly.
1909 E. B. WILSON in *Jrnl. Exper. Zool.* VI. 150 The chromosomes in question are the ones which in earlier papers I have called the 'supernumeraries'. **1917** E. E. CAROTHERS in *Jrnl. Morphol.* XXVIII. 469 Another variation which has attracted attention in our collection of Acridian material..is the presence in certain individuals of one, or sometimes two, entities which I shall designate as supernumeraries. They possess the staining capacity of chromatin. **1969** BROWN & BERTKE *Textbk. Cytol.* xviii. 381/1 A..peculiarity of supernumeraries which seems to indicate that they have genes for their own survival is preferential fertilization in corn.

superordinate, *a.* (*sb.*) Add: **a.** Delete 'Now only in *Logic*'. (Later examples.) Also *spec.* in *Gram.*
1934 PRIEBSCH & COLLINSON *German Lang.* vi. 318 The older construction is characterized by the precedence of the superordinate infinite, e.g. *lassen tragen.* **1949** C. E. BAZELL in *Travaux du Cercle Linguistique de Copenhague* V. 77 If..the functions of one member (e.g. the substantive) alone are similar to the functions of the whole group, this member is said to be superordinate. **1954** *Theology* LVII. 326 It would not necessarily be apparent which was the subordinate and which the superordinate sex. **1970** T. LUPTON *Managem. & Social Sci.* (ed. 2) iii. 80 The superordinate manager must work hard to create conditions for the subordinate such that the latter will feel that he is

being supported and encouraged. **1974** W. Rees-Mogg *Reigning Error* i. 21 Only those who appreciate my superordinate quality are fit to live in our new world of Nazism/Fascism/Soviet Communism. **1979** *Trans. Philol. Soc.* 37 In Italian promotion of the subordinate *si* into the superordinate clause..results in the unacceptable sequence *si si*.

b. *sb.* Also, that which is of a superior order or category.

1934 Priebsch & Collinson *German Lang.* vi. 318 Extensions of the infinitives and participles..sometimes follow their superordinates. **1957** R. W. Brown in Saporta & Bastian *Psycholinguistics* (1961) 505/1 The concrete noun..is likely to be more picturable than its superordinate. **1969** Greenfield & Bruner in J. S. Bruner *Beyond Information Given* (1974) xxi. 388 An itemized superordinate in labeling form. **1972** *Jrnl. Social Psychol.* LXXXVI. 302 The use of a superordinate ..functionally eliminated one element from the set.

superordination. Add: **2. b.** The condition of belonging to a higher or more powerful category or class; opp. *subordination*.

1952 V. O. Key *Politics, Parties, & Pressure Groups* (ed. 3) i. 4 Politics deals with human relationships of superordination and subordination,..of the governors and the governed. **1959** G. D. Mitchell *Sociol.* iii. 41 The relationship between father and son is one of super- and sub-ordination. **1975** A. Ryle *Frames & Cages* xiii. 121 The concept of superordination recurs through the book.

su:perorga·nic, *a.* (and *sb.*). *Sociol.* [Super- 4 a.] Applied to the social and cultural aspects of life which evolve from and transcend the individuals in society. Also *absol.* as *sb.*; occas. *transf.*

1862, 1876 [In Dict. s.v. Super- 4 a]. **1917** A. L. Kroeber in *Amer. Anthropol.* XIX. 163 (*title*) The superorganic. **1932** A. Raven (*title*) Civilization as divine superman. A superorganic philosophy of history. **1962** E. E. Evans-Pritchard *Ess. Social Anthropol.* ii. 34 Spencer clearly formulated the theory of gradual modification of species or structure..and..extended his evolutionary ideas to include the social or super-organic. **1968** *Internat. Encycl. Social Sci.* XV. 124/2 It was Spencer who coined the term 'superorganic', which, following its use by Kroeber in 1917 in his article 'The Superorganic', has been accepted as designating the unique and distinct elements in human behavior, and therefore as synonymous with 'culture'. **1973** P. A. Colinvaux *Introd. Ecol.* xl. 550 A first grand generalization was the persuasive philosophy of Clements and his disciples..who likened the climax stage to some superorganic being.

supero·vulate, *v. Physiol.* [Back-formation from *superovulation.] **a.** *intr.* To produce abnormally large numbers of ova at a single ovulation. **b.** *trans.* To cause (an animal) to do this. So **supero·vulated** *ppl. a.*

1956 *Nature* 3 Mar. 429/1 Fallopian tubes from a superovulated female, killed approximately 12 hr. after mating to one of these males, were fixed in Bouin. **1961** M. C. Chang in C. A. Villee *Control of Ovulation* 185 It seems that ovulation can be easily induced in the pregnant rabbits and that about half of the pregnant animals superovulate; that is ovulate a larger number of eggs than expected. **1970** *Sci. Jrnl.* May 50/2 Hormonal treatment can be used to get the female to superovulate. **1971** *Nature* 10 Sept. 125/2 Random bred female albino mice.. were superovulated with gonadotrophins. **1979** *Ibid.* 25 Jan. 298/2 Two-celled embryos were collected from super-ovulated donor ewes early on day 2 of their oestrous cycle. **1979** *New Scientist* 26 Apr. 269/2 [They] superovulated 14 heifer cows up to 10 times in rapid succession—and the cows responded with crops of up to 19 eggs at a single ovulation.

So **su:perovula·tion** [Super- 10 b].

1927 *Amer. Jrnl. Anat.* XL. 213 Superovulation, or the liberation of an unusual number of ova invariably occurs. **1981** *Sci. Digest* Aug. 89/1 They were beginning to suspect that superovulation..might be disrupting the internal reproductive environment.

superpe·rsonal, *a.* [Super- 4 a.] Transcending the limits of what is personal. Also *absol.* Hence **su·perperson, superpe·rsonalism, su:perpersona·lity.**

1860, 1899 [in Dict. s.v. Super- 4 a]. **1924** G. B. Shaw *Saint Joan* p. xvii, An appetite for evolution..[is] a superpersonal need. **1926** W. D. Lighthall *Superpersonalism* 26 We can proceed a stage further, and call it the Person of Evolution. But its vast and complex personality requires a differentiating term. May we not apply to it the term 'Superperson'? and call its personality a 'super-personality', its point of view the 'superpersonal', the system of its study, 'Superpersonalism'? **1927** J. S. Huxley *Relig. without Revelation* i. 52 Metamorphosed from a divine person into a super-person. **1939** Auden in *I Believe* (1940) 20 But this does not warrant ascribing to a culture a super-personality, conscious of its parts as I can be conscious of my hand or liver. **1963** *Observer* 31 Mar. 11/3 The God hypothesis asserts the existence of some kind of supernatural personal or superpersonal being. **1977** *Rolling Stone* 21 Apr. 68/2 A chair is made by a person, so the world has to be made by a superperson.

superpla·stic, *a.* and *sb. Metallurgy.* [Super- 9 a, 10: see *superplasticity.] **A.** *adj.* Of, pertaining to, or designating a metal capable of extreme plastic extension under load; involving or characteristic of such materials.

1947 *Chem. Abstr.* XLI. 2375 The superplastic alloy cannot be a mixt. of solid solns. of Al and Zn. **1970** *New Scientist* 12 Mar. 504/2 One of the most suitable titanium alloys—Ti-318—is superplastic at 930°C if correctly worked, and can be made to stretch and flow like plastic. **1978** *Jrnl. R. Soc. Arts* CXXVI. 689/1 It is possible that the single heat cycle combination of superplastic forming and diffusion bonding will revolutionize the fabrication of titanium sheet structures for aircraft applications. **1978** *Nature* 16 Nov. 209/2 The consolidated product can have very fine grain sizes which in turn leads to great ductility at ambient temperature—even to superplastic behaviour.

B. *sb.* A superplastic metal.

1969 *New Scientist* 2 Jan. 22/2 This metal, after treatment which reduces the grain size to about a micrometre, behaves like a superplastic at room temperature—it can be stretched by a factor of about 10 in one direction without breaking. **1971** *Britannica Yearbk. Sci. & Future 1972* 406 While the superplastics are only starting to shed their image as laboratory curiosities, the fiber composites have almost arrived.

Hence **superpla·stically** *adv.*; **su:perplasti·city** [tr. Russ. *sverkhplastichnost'* (Bochvar & Sviderskaya 1945, in *Izvestiya Akad. Nauk SSSR: Otdelenie tekhnicheskikh Nauk* IX. 824)], the state or quality of being superplastic.

1947 *Chem. Abstr.* XLI. 2375 (*heading*) Superplasticity in zinc-aluminum alloys. **1969** *Sci. Jrnl.* June 75/2 At one time, it was considered that superplastic metallic alloys were amorphous-like and for this reason behaved superplastically. **1977** *Jrnl. R. Soc. Arts* CXXV. 348/1 The exploitation of superplasticity in titanium sheet opens the way for shaped components of reduced cost. **1978** *Ibid.* CXXVI. 688/1 When the temperature of the blank reaches 950°C the argon pressure is increased at a programmed rate to expand the blank into the tool superplastically.

superposable, *a.* Add: Hence **su:perposabi·lity**, the property of being superposable.

1913 *Jrnl. Chem. Soc.* CIII. 839 There are evidently two vertical planes of symmetry cutting diagonally through the ring carbon atoms, and it might be objected that the presence of these two planes of symmetry are in reality the cause of the superposability. **1963** R. Ballabh *Hydrodynamic Superposability* i. 1 The idea of superposability as regards fluid motions does not seem to have engaged the attention of mathematicians in a formal way until the year 1940. **1973** *Engin. Fracture Mechanics* V. 555 The same is practically true for elastomers also but, in addition, their ultimate properties are well described by a failure envelope which approximates time-temperature superposability.

su·perpower. Also super power. [Super- 6.] **1.** [Super- 6 c.] orig. and chiefly *U.S.* Electrical power produced by the co-ordination and interconnection of existing power plants for greater economy and efficiency. Freq. *attrib.* Now *Hist.*

1921 W. S. Murray in *Prof. Papers U.S. Geol. Survey* No. 123. 11 On first presenting the subject considered in this report to the late Secretary of the Interior..I used the word 'superpower' to describe a system that would furnish power to the railroads and the industries within the territory between Boston and Washington that has now become more familiarly known as the superpower zone. **1921** *Independent* CVII. 316/2 The problems certain to arise within the circuit of the super-power zone are like those connected with the interstate and intrastate services of the railroads. *Ibid.* 317/2 The probable issues of super-power seem to make such a body [*sc.* a Federal control agency] inevitable. **1926** *Encycl. Brit.* (ed. 13) 681/2 The basis for a super-power system lies in the economy effected by the interconnection of electric power systems whose peak loads are reached at different times. **1983** T. P. Hughes *Networks of Power* xi. 297 Only one month after World War I ended, William S. Murray, a consulting electrical engineer, urged the secretary of the interior to prepare the ground for this superpower system.

2. [Super- 6 b.] Power of a greater kind or degree than the ordinary.

1922 D. H. Lawrence *Aaron's Rod* xviii. 269 Newly flushed with his own male super-power, he was going to have his reward. **1928** G. B. Shaw *Intelligent Woman's Guide* lxxxiii. 454 The more power the people are given the more urgent becomes the need for some rational and well-informed superpower to dominate them. **1931** *Times Lit. Suppl.* 26 Sept. (Home & Classroom Suppl.) p. iv/2 A super-power amplifier with an undistorted output of 125 watts. **1970** *New Society* 5 Feb. 231/3 This is an entity which has recently escaped from the dominance of Europe: and is now shadowed by the dominance of American superpower. **1975** *Microwave Jrnl.* XVIII. 50/3 The power of conventional microwave generating devices can be increased appreciably only through a matched increase of the..electron beam intensity. It is therefore tempting to use intense Relativistic Electron Beams to generate 'super power' microwaves. **1977** *N.Y. Rev. Bks.* 23 June 39/2 We do not, and cannot hope to, have a monopoly of 'superpower' in our world, as Rome had in hers.

3. [Super- 6 c.] A nation or state having a dominant position in world politics; one which has the power to act decisively in pursuit of interests which embrace the whole world; *spec.* the United States of America and the Union of Soviet Socialist Republics.

[**1930** *Economist* 12 July 63/2 The most important contributions to the Conference..were descriptions of what has actually taken place in the..control of super-power zones.] **1944** W. T. R. Fox *Super-Powers* ii. 20 There will be 'world powers' and 'regional powers'. These world powers we shall call 'super-powers', in order to distinguish them from the other powers..whose interests are great in only a single theater of power conflict. **1957** *Foreign Affairs* XXXV. 177 Britain is no longer a Super-Power. **1967** *Spectator* 30 June 758/2 Almost by definition superpowers do not have to care about face... America will continue to be a superpower whatever it does about Vietnam. **1971** *Guardian* 9 Dec. 12/1 China has..been drawn into a Super Power type of defence of Pakistan. **1977** E. Heath *Travels* viii. 166 What I saw when I arrived was a laudable achievement, possible only to a military commander with the resources of a superpower. **1978** J. Updike *Coup* (1979) ii. 54 Capital investments cleverly pried from the rivalry between the two superpowers (and that shadowy third, China, that has the size but not as it were the mass, the substance, to be called super).

superpronation: see *super- 2 a (*b*).

superra·diant, *a. Physics.* [Super- 9 a.] Involving or exhibiting superradiance.

1954 R. H. Dicke in *Physical Rev.* XCIII. 102/2 For want of a better term, a gas which is radiating strongly because of coherence will be called 'super-radiant'. **1974** *Sci. Amer.* June 126/1 The optical gain of the rapid discharge is so large that emission becomes superradiant, which means that the unit will lase without an optical cavity. **1978** *Nature* 20 Apr. 742/2 The word superradiant has often been misused..but is now taken to mean either the coherent radiation generated by a system of atoms possessing an externally created macroscopic polarisation rather akin to a phased array of dipoles, or alternatively the incoherent fluctuating radiation produced by a system of excited atoms possessing no initial polarisation but instead developing a cooperative decay behaviour through quantum correlations.

Hence **superra·diantly** *adv.*; also **superra·diance**, the spontaneous emission of coherent radiation by a system of atoms, esp. when the coherence is due to the initial correlation of the atoms by an external macroscopic polarization.

1965 *Physical Rev. Lett.* XIV. 589 (*heading*) Nuclear superradiance in solids. **1974** *Sci. Amer.* June 126/2 If the ultraviolet pulses are focused by a cylindrical lens to a line on the surface of the dye.., the dye will often lase superradiantly in visible light along the direction of the line. **1979** *New Scientist* 8 Mar. 763/2 The wave analogue of this process was called 'super-radiance': incident waves in certain modes would be amplified (rather than absorbed) by a rotating black hole, and would carry away some of the black hole's rotational energy.

super-rat: see *super- 6 c. **super-real,** etc.: *super- 4 a (*a*) and (*c*).

su:perregenera·tion. *Electronics.* [Super- 6 b.] Regenerative amplification in which self-oscillation is prevented by repeated quenching of the signal at an ultrasonic frequency.

1922 *Q.S.T.* July 7/1 At a meeting of the Institute of Radio Engineers, Edwin Howard Armstrong on June 7th gave his new invention of super-regeneration to a tense and expectant audience. **1922** E. H. Armstrong in *Proc. IRE* X. 244 This new result is obtained by the extension of regeneration into a field which lies beyond that hitherto considered its theoretical limit, and the process of amplification is therefore termed super-regeneration. **1943** F. E. Terman *Radio Engineers' Handbk.* ix. 664 Receivers employing superregeneration find their chief usefulness in the wavelength range 0·5 to 10 meters. **1975** R. L. Shrader *Electronic Communication* (ed. 3) xviii. 435/2 This type of superregeneration is often produced in RF amplifiers and is characterized by a wide band of spurious signals that it generates.

Hence **superrege·nerative** *a.*, employing or characterized by superregeneration; **superrege·neratively** *adv.*; **superrege·nerator**, a superregenerative device.

1922 *Q.S.T.* July 9/2 The super-regenerative amplifier. *Ibid.* 11/1 In the super-regenerator there is periodically sufficient positive resistance to wipe out this oscillation and hence it is not heard. **1934** *Wireless Engineer* XI. 35/1 The receiver may be made to function super-regeneratively by increasing the s.g. voltage to the point of 'squegging'. **1948** Slurzberg & Osterheld *Essent. Radio* v. 232 Superregenerative detector circuits are used in light, compact, portable code receivers. **1959** R. L. Shrader *Electronic Communication* xvii. 542 When coupled to an antenna, the superregenerator radiates a very broad signal. **1965** *Guardian* 18 Jan. 16/3 The popular super-regenerative receiver..is the simplest and cheapest on the market. **1975** R. L. Shrader *Electronic Communication* (ed. 3) xviii. 434/2 A demodulator used in the past in the VHF range is the superregenerative detector.

supersaturation. Add: Also, the state of being supersaturated.

1876 *Phil. Mag.* II. 216 This is a consequence..of the condition of supersaturation being maintained even in solutions from which crystals of sugar are being deposited. **1941** [see *kern *sb.*² 3]. **1983** *Sci. Amer.* June 108/3 The concentration of dissolved nitrogen in the dolphin's muscle tissue was indicative of a degree of supersaturation that in a human diver would have been dangerous.

superscript, *sb.* Add: **2.** A superscript character.

1901 [see Subscript *sb.* 2]. **1927** [see *Kronecker delta]. **1945** F. A. Ficken in F. A. Berry et al. *Handbk. Meteorol.* ii. 144 In work with tensors, indices occur as superscripts as well as subscripts. **1970** G. K. Woodgate *Elem. Atomic Struct.* iv. 67 The notation for a level specifies the value of *j* as a subscript to the letter code for *l*. The superscript gives the..multiplicity of the term.

supersensible, *a.* (*sb.*) **a.** (Earlier example.)

1798 A. F. M. Willich *Elem. Crit. Philos.* 180 The *supersensible substratum* of nature is that object, of which we can determine nothing in an affirmative sense.

supersensitive, *a.* Add: (Further examples in *Physiol.* Cf. *supersensitivity 2.)

1949 Cannon & Rosenblueth *Supersensitivity of Denervated Structures* ii. 11 They are often quite easily rendered supersensitive to some agent, e.g., adrenaline, by preliminary treatment with one or another chemical substance, e.g., cocaine or thyroxine. **1962** *Nature* 3 Feb. 487/1 The increased secretory activity seen in the supersensitive submaxillary glands. **1978** *Life Sciences* XXIII. 1283 Rats were supersensitive to norepinephrine as well as to dopamine.

su:persensiti·vity. [Super- 10.] **1.** Great or excessive sensitivity.

1934 [see *frayed *ppl. a.*²]

2. *Physiol.* The state or fact of a tissue or organ having an increased sensitivity to stimuli, as manifested by a longer or increased response, a reduced threshold, or increased susceptibility.

1949 Cannon & Rosenblueth *Supersensitivity of Denervated Structures* ii. 11 The term supersensitivity covers several possibly different phenomena. **1959** *Jrnl. Physiol.* CXXXXVII. 178 Organs chronically deprived of their motor nerves develop an increased sensitivity to the neurohumoral transmitter and to other chemical agents. This phenomenon is observed in several types of tissue, e.g. striated muscle, smooth muscle, ganglia and glands, and is known as denervation supersensitivity. **1963** *Pharmacol. Rev.* XV. 226 The changes and modifications of dose-response curves of sympathomimetic amines by various drugs or procedures known to cause super- and subsensitivity to this group of substances. **1974** *Sci. Amer.* Jan. 48/3 This partial denervation could conceivably lead to chemical supersensitivity, accessory sprouting of collateral nerve fibers or the formation of new synaptic contacts.

superse·xual, *a.* **1.** [Super- 4 a.] Beyond or outside the sphere of sexuality.

1895 *World* 20 Mar. 15/2 She had resolved from the very outset to maintain her companionship with Lucas on a supersexual basis. **1976** *Encounter* June 51 A pop-star pseudo-Christ has to be at once sexy, bisexual and super-sexual, so as to cater simultaneously for all possible needs.

2. [Super- 9 a.] Having strong sexual appetites, highly sexed.

1970 R. D. Abrahams *Positively Black* v. 112 This is done commonly by picturing men as supersexual animals who cannot control themselves when they see another woman.

su·persign. **1.** [Super- 3 a.] A diacritical mark written or printed above a letter.

1947 H. Jacob *Planned Auxiliary Lang.* 40 Supersigns were introduced to make Esperanto fully phonetic. **1958** J. Berry in J. A. Fishman *Readings Sociol. of Lang.* (1968) 745 The supersigns interfere with spacing and involve 'kerning'.

2. [Super- 6 b] A combination of letters, figures, etc., forming a unit.

1976 J. J. White in *Visible Lang.* X. 81 It is possible to view them as iconic 'supersigns'; i.e., as *collections* or *configurations* of symbolic signs (viz. words). *Ibid.*, Whether iconicity occurs at the sign or supersign level would be something which semiotic analyses would have to consider.

superso·nic, *a.* (and *sb.*) [f. Super- 4 a + L. *son-us* sound + -ic, as tr. F. *ultra-sonore*.] **1.** Of, pertaining to, or designating sound waves or vibrations with frequencies greater than those audible to the human ear or greater than 20,000 Hz.

This use of *supersonic* is now deprecated in scientific contexts, *ultrasonic* being the preferred term.

1919 *Electrician* 25 Apr. 494/2 The French have experimented with a system in which a continuous wave signal in heterodyned to a supersonic frequency. **1930** *Daily Express* 22 May 6/5 The wireless enthusiast, whether.. crystal-set owner or disciple of the supersonic heterodyne, will still have his moments of doubt. **1957** I. Murdoch *Sandcastle* xiv. 224 The next act was to blow a long blast upon the supersonic whistle. **1975** *Gramophone* Jan. 1412/3 CD-4 employs a carrier tone at supersonic frequency (30,000 Hz). **1980** *Daily Tel.* 5 Mar. 17/4 The Fisheries Agency team is using supersonic waves of 24 kilohertz which..have proved effective in repelling the dolphins without affecting ordinary fish.

2. a. Involving, pertaining to, capable of, or designating speeds greater than (*spec.* up to five times) the speed of sound. Cf. *hypersonic *a.*, *sonic *a.*

1934 *Jrnl. R. Aeronaut. Soc.* XXXVIII. 866 (*heading*) Supersonic wind channel for model tests. **1936** *Aircraft Engineering* Sept. 260/2 The wing shows what the Germans call a 'supersonic profile', because the aeroplane is supposed to fly the greater part of its route at supersonic speeds. **1948** 'N. Shute' *No Highway* xii. 313 It [*sc.* an aircraft] was at the speed of sound... He said he'd been through to the supersonic zone several times. **1953** *Hansard Commons* 11 May 881 As for the problem of the supersonic bang, I must tell the hon. Gentleman that there is absolutely no solution in sight. In fact, we are probably in for some rather noisy times. **1972** *Nature* 18 Aug. 379/2 A low density tunnel for simulating supersonic and hypersonic flight at altitudes of 20 to 70 miles. **1978** *Jrnl. R. Soc. Arts* Dec. 37/1 The Concorde is to-day the precursor of much larger supersonic transport aircraft in years to come.

b. *ellipt.* as *sb.* An aeroplane capable of flying at speeds greater than the speed of sound.

1947 *Times* 1 Jan. 3/3 It is generally assumed here that cooperation in research will cover tests now being carried out by both countries in the field of supersonics, guided missiles, and the development of jet aircraft engines. **1962** *Listener* 5 July 36/2 The demand for supersonics came from the aircraft manufacturers. **1968** *Economist* 11 May 71/1 With jumbo jets and the supersonics on the horizon, aviation insurers had enough to worry about already. **1973** *Times* 24 May (Aviation Suppl.) p. vi/5 When supersonics have, by public demand, taken over all long-haul air services..we may well wonder what all the argument was about.

3. *colloq.* **a.** Very fast. **b.** Excellent, wonderful, exciting, etc. Also as *int.*

1947 *Argus Week-End Mag.* 25 Oct., Isn't he simply supersonic! **1954** C. Davies *Let.* 24 Dec. in B. Russell *Autobiogr.* (1969) III. ii. 96 My thoughts were speeding along with yours at a super-sonic rate. **1955** G. Dorman *Swooping Vengeance* iv. 31 Ginge's eyes gleamed and he said eagerly,..'Gee, Wing! This is supersonic! Can I be in it too?' *Ibid.* v. 44 'Absolutely supersonic!' was Ginge's enthusiastic comment. **1963** *Listener* 17 Jan. 109/2 The unique problems with which the continent [*sc.* Africa] is struggling in an age of supersonic speed of change. **1972** R. Godden *Diddakoi* iv. 83 Miss Brooke made girdle scones..'Supersonic!' said Clem. **1980** *Daily Mail* 23 July 33 (Advt.), Young men required for supersonic battle of wits.

Hence **superso·nically** *adv.*

1952 *Chambers's Jrnl.* June 363/2 They [*sc.* rats] are being lured forth by their mating-note, produced supersonically, and trapped as they emerge. **1970** *Daily Tel.* 2 Sept. 26/6 The St David's Civic Society..has launched a national campaign to stop the Concorde flying supersonically over land.

superso·nics, *sb. pl.* (const. as *sing.*). [f. prec.: see -ic 2.] The science of sound waves or vibrations with frequencies greater than those audible to the human ear or greater than 20,000 Hz.

1925 *Nature* 9 May 690/1 The method adopted to measure the vibrational energy in the water at the high frequencies used in supersonics is interesting. **1952** *Chambers's Jrnl.* June 363/1 The urge to explore supersonics afresh arose from the infestation of reservoirs by seagulls... Scientists have been..studying the artificial application of supersonics, and..have evolved a method of producing a high-intensity beam of sound of any desired frequency. **1977** *Jrnl. R. Soc. Arts* CXXV. 361/1 Supersonics has a place by the end of the century.

supersound: see *super- 6 c.

su·perspace. *Physics.* [Super- 5 d.] A concept of space-time arising out of the attempt to quantize the gravitational field, in which points are defined by more than the usual four co-ordinates; also, a space of infinitely many dimensions postulated to contain actual space-time and all possible spaces.

1971 *Daily Tel.* (Colour Suppl.) 7 May 35 Inside superspace there is neither space nor time. **1972** *Nature* 15 Dec. 382/2 This point of view, which leads to the notions of 'superspace' and quantum fluctuation of three geometries, was described by Professors K. Kuchař and J. A. Wheeler. **1974** A. Berry *Next 10,000 Yrs.* 111 All the stars and galaxies that we can see are on the curved, solid part of the doughnut, while the hole in the middle represents the mysterious region of Superspace. **1975** *Physics Lett.* B. LVI. 178/2 The usual supersymmetry transformation is then a linear transformation in the 8 dimensional 'superspace' of z^A. **1976** B. Zumino in Arnowitt & Nath *Gauge Theories & Mod. Field Theory* 262 We call superspace a space whose points are labeled by four commuting coordinates x^α and by a number of additional totally anticommuting coordinates θ^a, $\theta^a\theta^b + \theta^b\theta^a = 0$, which also commute with the x^α. **1978** *Sci. Amer.* Feb. 138/3 The superspace theories are elegant but technically complicated. **1980** P. Davies *Other Worlds* v. 104 We may construct a different world for each shape of space. Stitching them altogether [*sic*] gives us an infinite-dimensional superspace. Contained in superspace are all the possible spaces... Each space of superspace will contain its own superworld of all possible particle arrangements.

su·perstar. [Super- 6 c.] **1.** An outstanding performer in the theatre, music, sport, etc.; something exceptionally successful, advanced, etc. Freq. *attrib.*

1925 W. Deeping *Sorrell & Son* 130 You wouldn't expect a couple of cinema super-stars to be running away from publicity. **1936** 'Riff' & 'Raff' *They're Off!* v. 40 A..relation of my own was running a horse with no less a person than our super-star jockey in the saddle. **1969** N. Cohn *AWopBopaLooBop* (1970) xx. 188 He became a superstar but he wasn't happy. **1972** *Guardian* 24 Feb. 14/4 [David] Frost's importance is not the super-star status. **1976** *Nature* 8 Apr. 471/1 Superstar technology is the name given by a working party of the Council for Science and Society to highly innovatory, large scale technical projects. **1978** *Detroit Free Press* 5 Mar. c 14 (Advt.), Hertz, the superstar in the Rent-A-Car industry has an immediate vacancy. **1980** M. Fonteyn *Magic of Dance* 32 From the Charleston it was an easy step to the emergence of a male superstar dancer. **1982** *London Rev. Bks.* IV. xxiv. 20/2 Star quality, however, was not at all what was looked for in those who played opposite a superstar like Kean.

2. *Astron.* A very important or powerful heavenly body. *rare.*

1929 S. Leslie *Anglo-Catholic* iv. 58 He..meditated..upon sun and moon, whose counter-changes and performances in the sky he followed with deep amaze. These super-stars of heaven never ceased to vary in effects. **1964** [see QSO s.v. *Q II. 2 b].

Hence **su·perstardom.**

1973 *Harper's Mag.* Oct. 111 Apparently, in the eyes of one dazzled by his own celebrity,..superstardom puts a man above the law. **1977** *Times Lit. Suppl.* 22 Apr. 479/2 Hollywood exacts its price for superstardom.

super-state, -store: see *super- 6 c.

superstratum. Add: **2. a.** *Linguistics.* A language responsible for linguistic change (esp. in vocabulary) in another upon which it is imposed and over which it is temporarily dominant. Cf. *substratum 5 a.

[**1932** E. Tappolet in *Archiv für das Studium der neueren Sprachen* CXI. 234 Von ähnlichen Erwägungen geleitet, sprach von Wartburg (Leipzig) über die Wirkung des *Superstratums*.] **1953** J. B. Trend *Lang. & Hist. Spain* xii. 167 Spanish-speaking America offers an exceptionally favourable field for examining the linguistic concepts of substratum and superstratum. **1976** W. F. H. Nicolaisen *Scottish Place-Names* vi. 84 The place-names created by a certain language form an adstratum to English names in one place, a superstratum in another, and a substratum in most places.

b. *attrib.* and *Comb.*, as *superstratum influence, language*.

1957 *Publ. Amer. Dial. Soc.* 1956 xxvi. 100 A substratum influence is one derived from a dominated language, a superstratum from a dominant language. **1960** *Amer. Speech* XXXV. 234 Substratum languages can affect all features of grammar, whereas superstratum languages tend to affect vocabulary only. **1978** *Canad. Jrnl. Ling.* 1977 XXII. 206 After surveying the distribution, function, and status of each Romance language..it discusses the thorny problem of the influence exerted by the so-called substratum and superstratum languages.

superstructure. Add: **1. c.** *Geol.* [tr. G. *oberbau* (C. E. Wegmann 1935, in *Geol. Rundschau* XXVI. 332).] A relatively shallow overlying layer of an orogenic belt that is unaffected by plutonic activity or metamorphism.

1944 *Proc. Geologists' Assoc.* LV. 69 A distinction must be drawn, he [*sc.* Wegmann] maintains, between happenings in the non-migmatitic superstructure (Oberbau) and the migmatitic infrastructure (Unterbau) lying below. **1972** J. G. Dennis *Structural Geol.* xvii. 394 In many orogenic belts, the superstructure has not been preserved in place: most of it has been eroded or transported to the external zone as allochthons.

d. *Biochem.* The higher-order structure of a protein or enzyme molecule which is superimposed upon the sequence of amino-acids or nucleotide bases.

1962 A. Spector in A. Pirie *Lens Metabolism Rel. Cataract* 334 The *N*-terminal residues of native proteins are probably not readily accessible to the enzyme since such groups are masked or buried in the superstructure of the molecule. **1973** *Nature* 7 Sept. 23/1 The term superstructure will be used to include the secondary and higher order structures that might be super imposed upon the primary base sequence of a nucleic acid. **1981** *Sci. Amer.* Feb. 60/2 (*caption*) Helical superstructures might be formed with increasing salt concentration.

2. b. *Pol.* and *Econ.* In Marxist theory, the institutions and culture which are considered to result from or reflect the economic system on which a society is based.

[**1903** *Social Democrat* VII. 274 The Greeks attained to a high pitch of civilization, with a slave class as the basis of its economic superstructure.] **1904** N. I. Stone tr. *Marx's Contributions Critique Pol. Econ.* 11 The sum total of these relations..constitutes the economic structure of society—the real foundation, on which rise legal and political superstructures. **1926** M. Eastman *Marx, Lenin & Sci. of Revolution* iv. 50 It is obvious that if the material basis positively *determined* the superstructure, we should not have to *disregard* the superstructure and examine the basis, for the one could be directly inferred from the other. **1943** J. A. Schumpeter *Capitalism, Socialism, & Democracy* xi. 121 We now turn to the cultural complement of the capitalist economy—to its socio-psychological superstructure, if we wish to speak the Marxian language—and to the mentality that is characteristic of capitalist society. **1960** E. R. Goodman in J. A. Fishman *Readings Sociol. of Lang.* (1968) 729 Man considered language as an element in the Marxist superstructure dependent upon the economic base of society... Just as this base might be changed by force, so Man thought, the linguistic superstructure should be impelled to develop toward its ultimate goal. **1975** *Chinese Econ. Stud.* VIII. iv. 10 The superstructure refers to the national government, army, law, and other political systems and their corresponding ideological forms, such as philosophy, literature, and fine arts. **1977** R. Williams *Marxism & Lit.* ii. vi. 111 Cultural work and activity are not now, in any ordinary sense, a superstructure.

3. *Metallurgy.* = *superlattice 1.

1932 *Proc. R. Soc.* A. CXXXVI. 216 The type of superstructure represented by Fe_3Al gives other lines [in an X-ray diffraction photograph]. **1979** *Nature* 11 Oct. 469/2 Ordered solid solutions (superstructures or superlattices), in which atoms of one kind segregate into a particular set of lattice positions, are usually obtained by slow cooling at the critical ordering temperature.

super-super-: see *super- 17, 18. **super-symmetric, -symmetry:** see *super- 6 b.

su:pertechnolo·gical, *a.* **1.** [Super- 9 a.] Involving or employing highly advanced technology.

1968 *Economist* 14 Dec. p. xxv/3 Some sort of middle course will be found, between the super-technological and the purely sociological. **1977** P. Johnson *Enemies of Society* xii. 164 The galaxy of star universities which formed the vertebrae of the super-technological East Coast.

2. [Super- 4 a.] Beyond or superseding the technological.

1973 *Nature* 6 Apr. 382/1 A branch of this water-drawing and wood-hewing supertechnological fraternity would be preserving health.

supervision. Add: **1. b.** Special Comb.: **supervision order**, a court order placing a child or young person under the supervision of a local authority or a probation officer in cases of delinquency, petty crime, etc.

[**1933** *Act* 23 & 24 *Geo. V* c. 12 § 62 An order placing him..under the supervision of a probation officer, or of some other person appointed for the purpose by the court.] **1938** *Act* 1 & 2 *Geo. VI* c. 40 Supervision order in place of order committing to care of fit person. **1968** J. Lock *Lady Policeman* xv. 126 The juvenile court placed her under a Supervision Order and she returned home. **1980** *Times Lit. Suppl.* 28 Nov. 1347/3 The usual treatment [in dealing with schoolboy truancy] is to place the child under a supervision order: he is then seen by a social worker or probation officer from time to time.

supervoltage, -weak, -woman: see *super- 10 b, 9 a (*a*), 6 c.

suphrosyne: see *sophrosyne.

supje, var. Sopie in Dict. and Suppl.

supp (sʌp). Colloq. abbrev. of Supplement *sb.*¹ 1 b. Usu. *colour supp.*

1968 *Punch* 6 Nov. 646/2, I don't want to..get myself interviewed in a colour-supp series. **1974** *Listener* 17 Jan. 95/1 To decorate a Mini outrageously for a colour supp. **1975** J. Symons *Three Pipe Problem* xv. 136, I read this *Observer* colour supp. piece, you see.

supper, *sb.*¹ Add: **1. c.** (Earlier example.)

1818 H. B. Fearon *Sk. Amer.* 44 A mechanic..has 3 meals a-day, coffee with fish or meat for breakfast; a hot dinner; and tea (called supper) in the evening.

4. *supper-bell* (earlier and later examples), *-board* (example), *-hour* (earlier example), *-table* (earlier example), *-tray* (earlier example); *supper-eater* (earlier example); **supper club**, a restaurant serving suppers and usu. providing entertainment; **supper dance**, (*a*) a dance after which the man escorts his partner into supper; (*b*) a dancing party at which supper is served; **supper house** (earlier example); **supper-party** (earlier example).

1770 P. V. Fithian *Let.* 30 Nov. in *Jrnl. & Lett.* (1900) I. 9 About seven the supper Bell rings. **1940** W. Faulkner *Hamlet* i. iii. 77 Walking on toward the brazen sound of Mrs. Littlejohn's supper-bell. **1918** G. Frankau *One of Them* xx. 154 Gay with a thousand supper-boards, whose drink Was poured to rag-time tunes by Herman Finck. **1927** H. Crane *Let.* 19 Dec. (1965) 313, I don't think I'll dare attend that supper club again. **1977** *New Yorker* 8 Aug. 66/3 Only after a time of heading a hotel and supper-club trio did he turn his hand to composing and arranging. **1885** F. C. Baylor *On Both Sides* 64 'A supper-dance, then,' he stupidly insisted. **1928** 'Brent of Bin Bin' *Up Country* xvi. 284 Thank heaven, here was the supper dance for which he was committed to good old Mrs Mac. **1948** M. Laski *Tory Heaven* viii. 110 'May I have the supper-dance?' he had said to Penelope. **1979** S. Smith *Survivor* xxvii. 253 The Christmas Eve supper dance was one of the biggest nights of the year in [hospital] sp9. **1799** Malthus *Diary* 29 May (1966) 37 We told him that we were no supper eaters. **1814** Jane Austen *Mansfield Park* II. x. 227 Previous inquiries.. about the supper-hour. **1855** Trollope *Warden* xvi. 265 It was a London supper-house. **1808** *Monthly Pantheon* I. 684/2 The more gay and fashionable may go to balls or supper parties. **1781** R. F. Greville *Diary* 6 Aug. (1930) 12 At this time the distribution of the Supper Tables was mentioned to Him. **1847** C. Brontë *Jane Eyre* I. v. 74 Monitors, fetch the supper-trays!

supper, *v.* **1. b.** (Earlier U.S. example.)

1805 R. Parkinson *Tour Amer.* 68 Going to look at the horses after what is called suppering them up at night.

supperward. Add: Also as *adv.* (*U.S. rare*) = Supperwards *adv.*

1932 W. Faulkner *Light in August* xv. 330 The townspeople began to move supperward.

supplantal. (Example.)

1891 *Harper's Mag.* June 69/1 The excitements of the day had..withdrawn his mind from..his fear of supplantal.

supple, *a.* Add: **8.** *supple-faced, -tempered, -thewed.*

1931 V. Woolf *Waves* 100 The little men at the next table... Supple-faced, with rippling skins. **1865** J. R. Lowell *Ode Recited at Commemoration* vi. 27 They could not choose but trust In that sure-footed mind's unfaltering skill, And supple-tempered will. **1959** R. Graves *Coll. Poems* 317 Free from the cramps of yesterday, Clear-eyed and supple-thewed.

supple-jack. Add: **3.** *U.S.* A toy representing the human figure, the limbs of which are manipulated by a string. Also *fig.* Cf. Supple *a.* 2 c. ?*Disused.*

1776 M. Cutler *Jrnl.* 17 June in *Life & Correspondence* (1888) I. 55 They made us several presents of the small affairs in the cabins, such as sweetmeats, cayenne-pepper, supple-jacks, cassada or bread..trinkets etc. **1791** W. Maclay *Jrnl.* 25 Feb. (1927) xiv. 390 Schuyler is the supple-jack of his son-in-law Hamilton. **1835** A. B. Longstreet *Georgia Scenes* 13 Bob Simons danced..like a 'Suple-Jack'..when the string is pulled with varied force, at intervals of seconds. **1853** P. Kennedy *Blackwater Chron.* x. 147 His body spread out as usual in his favorite position of a supple-jack distorted to the utmost. **1871** W. Whitman *Democr. Vistas* 30 How the millions of sturdy farmers and mechanics are thus the helpless supple-jacks of comparatively few politicians. **1904** *N.Y. Times* 8 July 5 Those political supplejacks who go about with sanctimonious moan, saying: 'The President is wrong, but we must support the President.'

supplement, *v.* Add: Freq. const. *by* and (more recently) *with.* (Later examples.)

1890 G. Gissing *Emancipated* I. i. v. 151 Then he strolled away and supplemented his meal with a fine bunch of grapes. **1940** H. G. Wells *Babes in Darkling Wood* iii. iii. 270 She realised she had forgotten her lunch, and she supplemented her tea with two boiled eggs. **1946** Bible (Rev. Stand. Version) 2 Pet. i. 5 Make every effort to supplement your faith with virtue. **1952** S. Kauffmann *Philanderer* (1953) vi. 97 When everyone else was making money, Robert was forced to supplement his insufficient income with his savings. **1977** K. M. E. Murray *Caught in Web of Words* vi. 105 He had spare time in which to supplement his income by literary work.

supplemental, *a.* Add: **a.** (Later examples.) Now chiefly *U.S.*

1952 *Sun* (Baltimore) 28 Aug. 1/3 The supplemental defenses of anti-aircraft artillery and infantry posts. **1958** *Yearbk. Agric.* 1957 (U.S. Dept. Agric.) 769/1 *Supplemental irrigation*,..irrigation during dry periods in regions where normal precipitation supplies most of the moisture for crops. **1966** K. Amis in *New Statesman* 14 Jan. 52/1, I have never before met *supplemental* for *supplementary* except as the term for the second-bite-at-the-cherry examination for pass students at the University College of Swansea. **1969** D. Acheson *Present at Creation* (1970) xviii. 159 A supplemental budget estimate was prepared. **1976** *Amer. N. & Q.* XV. 2/1 Supplemental to this special number of *AN&Q*. **1980** *Amer. Speech* 1976 LI. 202 All of those partial records will be included in the atlas as supplemental evidence.

supplementarity (sʌplɪmentæ·rɪti). *rare.* [f. Supplementary *a.* + -ity, after F. *supplémentarité* (J. Derrida).] The condition or quality of being supplementary.

1976 G. C. Spivak tr. *Derrida's Of Grammatology* ii. iv. 314 In as much as we designate the impossibility of formulating the movement of supplementarity within the classical logos. **1979** C. Norris in *PN Rev.* X. 38/1 Writing is the example *par excellence* of a supplementarity which enters into the heart of all intelligible discourse.

supplementary, *a.* Add: **b.** (Further examples.)

Esp. of welfare payments. *Supplementary Benefits* replaced National Assistance in 1966.

1841 *Hansard Commons* 5 Mar. 1361 He had..felt it his duty late in the year, to bring forward a supplementary estimate of 22,000*l*, to carry into effect the recommendation of the naval and military commission, which increased the estimates for the whole year to 6,185,000 *l.* **1920** W. S. Churchill *Let.* 3 Mar. in R. S. Churchill *Winston S. Churchill* (1969) II. Compan. ii. xiii. 985 The House is now discussing peacefully the supplementary Estimates of the Navy. **1940** *Economist* 27 Jan. 141/2 The direct cost involved in the change over to a new central organisation for the payment of supplementary pensions is..nil. **1966** *Hansard Commons* 7 Mar. 1730/2 Will the right hon. Lady confirm whether supplementary benefits will now be the subject of Parliamentary Questions to her? **1966** *Whitaker's Almanack 1967* 1109/1 The supplementary pension may be claimed by persons over pension age and the supplementary allowance by persons aged 16 or over but under pension age, who are not in full-time work. **1973** B. Mather *Snowline* vii. 85 He was getting nothing—unemployment pay, supplementary allowance—not a sausage. **1977** in R. Crossman *Diaries* III. 124 The House of Commons must annually debate and pass three Consolidated Fund Bills, authorizing the issue of the supply of funds to cover the civil and defence estimates and supplementary estimates. **1977** *Proc. R. Soc. Med.* LXX. 602/1 That there are five million people in receipt of supplementary benefit is as much a sign of affluence as poverty for of households receiving benefit 96% have television, 68% washing machines, 62% a refrigerator and 23% a car.

suppletion. Restrict † *Obs. rare* to sense in Dict. and add: **2.** *Linguistics.* The replacement of a form which is missing from a grammatical paradigm by one derived from a different root. Also *attrib.*

1933 L. H. Gray in *Language* IX. 84 Athematic verbs, as well as their semantic equivalents elsewhere, seem peculiarly liable to suppletion by other verbs to furnish their aorists. **1942** Bloch & Trager *Outl. Linguistic Anal.* iv. 58 Suppletion may be regarded as an extreme kind of internal change, in which the entire base..is replaced by another form. **1951** [see *morpholexical *a.*]. **1959** F. W. Householder in Saporta & Bastian *Psycholinguistics* (1961) 21/1 It is always possible (with due allowance for irregularity and suppletion) to derive a related expression of structure B. **1978** *Language* LIV. 21 Cases of feminine suppletion like *mon amie* are not included.

suppletive, *a.* Restrict *rare* to sense in Dict. and add: **2.** *Linguistics.* Displaying suppletion. Also as *sb.*, a suppletive form.

1926 L. Bloomfield in *Language* II. 161 If in a construction all the component forms are irregular, the whole form is *suppletive*. If *go* be taken as the stem of the verb, then the past *went* is suppletive. **1933** *Language* IX. 83 'Buy'..appears only in the aorist, serving as suppletive to the present and imperfect of ὠνέομαι. **1934** Priebsch & Collinson *German Language* ii. ii. 140 For the suppletive comparatives and superlatives cf. the Germanic section. **1957** S. Potter *Modern Linguistics* iv. 101 *Aller* is suppletive (*aller, vais, irai*). **1964** R. H. Robins *Gen. Linguistics* v. 207 Such roots involving total variation among their allomorphs are sometimes called suppletive. **1976** J. S. Gruber *Lexical Structures in Syntax & Semantics* ii. ii. 303 The past tense is not used in those cases in which we have a suppletive form.

supply, *sb.* Add: **I. 4. b.** *in short supply*: see *short *a.* 15 a.

II. 6. b. Also, a supply teacher.

1957 A. Wilson *Bit off Map & Other Stories* 152 'Why can't they get a Supply in?' 'Supply teachers need notification.' **1974** M. Higgins *Changeling* i. 7 Your replacement is only a supply, and..the Head'd be only too happy to have you back.

11. (Earlier and later examples of phr. *supply and demand*.)

1843 Carlyle *Past & Present* iv. v. 368 The sixpence a day and supply-and-demand principle. **1919** M. Beer *Hist. Brit. Socialism* I. ii. iv. 152 We have been dealing with pure theory, leaving out of account such factors as supply and demand. **1936** J. M. Keynes *Gen. Theory Employment, Interest & Money* v. xxi. 292 Prices are governed by the conditions of supply and demand. **1976** J. Snow *Cricket Rebel* 19 Most boys wanted to bat and because I could do both I usually found myself caught up with a ball in my hand due to the law of supply and demand.

III. 12. a. *supply train* (earlier example), *wagon* (earlier example); *supply-boat* (earlier example); also (partly with ref. to the supplies of an army and partly *gen.*) *supply base, depot, line, ship, station, store*; **supply day**, a day on which the House of Commons debates an Opposition motion criticizing the Government's proposed expenditure (cf. sense 10 a in Dict.); **supply drop**, the dropping of supplies by parachute; **supply house**, (*a*) *U.S.*, a commercial establishment selling supplies; (*b*) *Canad.*, a hut, tent, lean-to, or other structure, used as a storehouse; **supply-side** *a. Econ.* (orig. *U.S.*), pertaining to the supply side of the economy; hence, designating a policy designed to increase the incentives to produce and invest, by means of tax cuts; hence **supply-sider**, an advocate of this policy; **supply teacher**, a teacher supplied by the education authority to fill a (temporary) vacancy; hence, one who is regularly employed to do this; hence (as a back-formation) **supply-teach** *v. intr.*, to work as a supply teacher; **supply teaching** *vbl. sb.*

1958 L. Uris *Exodus* i. xviii. 101 It was a fenced-in area containing several acres of trucks and other rolling stock and a dozen enormous warehouses. During the war the depot had been a major supply base for the Allies in the Middle East. **1840** J. F. Cooper *Pathfinder* II. ix. 73 We shall lie in wait..to intercept their supply-boats. **1946** *May's Treat. Parliament* (ed. 14) xxv. 686 The House had attempted to counter this tendency [*sc.* the government's postponement of the discussion of estimates] by making one day each week a compulsory supply day. **1959** *Listener* 12 Mar. 441/1 The time allotted to the opposition for the criticism of policy and administration in supply days and so forth would not need to be curtailed. **1976** H. Wilson *Governance of Britain* i. 19 Defeated on a snap vote on a Supply Day debate on the stocks of cordite in government depots, he threw in his hand. **1918** E. S. Farrow *Dict. Military Terms* 596 Main supply depots are established at advanced bases or at convenient positions on the railway. **1921** *Daily Colonist* (Victoria) 18 Mar. 3/2 Two all-metal monoplanes have made an initial flight..carrying 1,000 pounds of gasoline each for the supply depot at Hay River. **1947** 'N. Shute' *Chequer Board* iv. 86 Last job was a Dakota squadron in South-East Asia Command. Supply drops, I suppose. **1978** T. Allbeury *Lantern Network* iii. 33 You have been given details of..suitable areas for supply drops. **1897** *Sears, Roebuck Catal.* 1 Sears, Roebuck & Co., (Incorporated), Cheapest Supply House on Earth, Chicago. **1905** L. Mott *Jules of Great Heart* 161 A voyageur showed him to the supply-house, and he got some pemmican, tea and bread, and a blanket. **1957** V. J. Kehoe *Technique Film & Television Make-Up* xiii. 194 Get a small sized balsa wood head form from a hat supply house. **1975** *New Yorker* 7 July 73/1 While discarded manufactured objects found in the street or in junk shops may be richly charged with poetic and psychological associations..this is not the case with new supply-house items. **1942** *R.A.F. Jrnl.* 3 Oct. 11 We were detailed to attack Jerry's supply lines. **1956** D. L. Linton *Sheffield* p. xxiii, It can serve the other towns of the region with wholesale and retail goods, professional and social services only over 'supply lines' that are relatively costly in maintenance, operation, or time.

1915 J. M. de Robeck in M. Gilbert *Winston S. Churchill* (1972) III. Compan. i. 753 The passage of supply ships for the Fleet through the Dardanelles with the forts still intact is a problem to which I can see no practical solution. **1975** *BP Shield Internat.* May 6 To enable it to continue operations for several days when supply ships cannot come alongside. **1976** *Wall St. Jrnl.* 9 Apr. 8/1 (*heading*) Supply-side fiscalism. *Ibid.* 15 Nov. 26/4 Supply-side fiscalists..agree that tax changes do not affect total demand, but they emphasize the effects on supply. **1980** *N.Y. Times* 22 June iv. 20 They recommend capital formation and other supply-side policies that have recently become fashionable. **1980** *Wall St. Jrnl.* 28 Feb. 24/3 Reception to 'supply-siders' was still hostile..when they criticized the economic models being used by the congressional budget committees for assuming that higher government spending was better for the economy than lower tax rates. **1981** *Christian Sci. Monitor* (Weekly Internat. ed.) 7 Sept. 20/3 The supply-siders who persuaded President Reagan to seek a balanced budget by cutting taxes. **1909** F. Ash *Trip to Mars* xxxiv. 262 Airships are of no use without a supply-station. **1885** *List of Subscribers, Classified* (United Telephone Co.) (ed. 6) 204 Supply Stores... Army & Navy Auxiliary Co-operative Supply, Limited... Civil Service Supply Associations, Limited. **1946** W. Faulkner *Portable Faulkner* 752 A pair of offices up a flight of stairs above the supplystore. **1968** *New Statesman* 22 Mar. 376/1, I am now supply teaching in London. **1980** J. Barnes *Metroland* iii. i. 138, I was supply teaching in Wandsworth at the time: twenty-five quid a week for the privilege of having your bicycle tyres let down each week by different kids at different schools. **1902** Supply teacher [in Dict.]. **1963** S. Marshall *Experiment in Education* iii. 115, I happened to be ill..and the only supply teacher the L.E.A. could find at short notice was an Indian teacher. **1969** R. Godden *In this House of Brede* xv. 341 Father Gervase has gone as supply teacher for a fortnight to Bishop Palin's Grammar School for Boys. **1976** *Rhyl Jrnl. & Advertiser* 9 Dec. 20/5 (Advt.), Applications are invited from Qualified Teachers..who wish to be included on the Authority's list of Supply Teachers for Primary and/or Secondary Schools. **1957** *Kingston* (Ontario) *Whig-Standard* 24 Jan. 17/6 She told the students something of the practice and supply teaching possible while at college. **1976** *Times Lit. Suppl.* 13 Aug. 1006/5 He was a student at the London School of Economics, wanted to be a writer and did supply teaching for a living. **1860** H. Greeley *Overland Journey* 55 Our route..was no longer encumbered with great army supply-trains. **1866** A. D. Richardson *Secret Service* xix. 241 Their retreat was a stampede, leaving behind great quantities of ammunition..supply-wagons and ambulances.

support, *sb.* Add: **II. 7. d.** The solid substance or material on which a painting is executed.

1892 J. G. Vibert *Sci. of Painting* viii. 96 A picture is composed of three altogether distinct elements:— 1. The support, or the material substance painted on, as wood, canvas, stone, paper, etc. **1926** A. P. Laurie *Painter's Methods & Materials* iv. 53 Well-seasoned panels of wood form an excellent support for pictures. **1958** M. L. Wolf *Dict. Painting* 286 The *support* is covered with the *ground*..for evenness, and is then ready to receive the actual painting.

8. *Math.* The smallest closed set of elements outside which a given function or mapping is zero.

1964 A. P. & W. Robertson *Topological Vector Spaces* i. 18 Let *K*(S) be the set of real (or complex) valued functions continuous and of compact support on the separated locally compact space S. **1967** MacLane & Birkhoff *Algebra* iv. 143 Show that the set of all functions *f* with finite support constitute a ring..under pointwise sum and convolution product. **1980** D. L. Cohn *Measure Theory* vii. 200 Define functions g_1 and g_2 by $g_1 = h_1$ and $g_2 = h_2 - (h_1 \wedge h_2)$. Then g_1 and g_2 are nonnegative, their supports are included in U_1 and U_2 respectively.

III. 9. *attrib.* **a.** That provides support or acts as a support.

1953 F. P. Magoun in *Speculum* XXVIII. 460 At least some of the language of the *Riddles* is traditional, since verses from these appear in the support-evidence. **1962** *Listener* 29 Mar. 549/1 The British..could be fairly allowed to bring back a division [from Germany] to this country and let that division act in a support role. **1964** *Language* XL. 26 Old French developed a 'support vowel'..only where there would have been an unwieldy cluster otherwise. **1967** *Times Rev. Industry* June 20/2 The ratio of 'support staff' to salesmen. **1972** *Guardian* 2 Feb. 1 The dollar sank close to its new 'support floor'. **1975** *Offshore* Sept. 9/1 Now converted into a support ship for the North Sea. **1976** M. Machlin *Pipeline* liii. 526 By one p.m. Simon Orloff had climbed the twelve foot support member. **1976** *Spare Rib* Dec. 8/2 Up till now I've always done support gigs. **1977** C. McFadden *Serial* (1978) ix. 24/1 He was uptight about the support money he gave her. **1980** *New Age* (U.S.) Oct. 21/1 How important it is to use your personal support network, and your head, when dealing with illness.

b. *spec.* Designating stockings or tights reinforced with elastic yarn to support the muscles and veins of the legs.

1970 *Vogue* May 64/2 Ours is the *original* support stocking. **1971** *Ibid.* Nov. 26/1 Support tights don't have to be thick and ugly. *a* **1975** R. Crossman *Diaries* (1977) III. 640 These surgeons come and go but they know nothing about support stockings. **1975** *Guardian* 25 Mar. 13/5 Support hose or support tights are supplied through the Hospital Supply Service. **1976** *Times* 26 Mar. 10/3 The support tights market in America has been booming.

10. Special Combs., as **support barge**, a barge providing assistance for offshore oil-drilling; **support buying**, the purchase of a commodity, a currency, or stocks and shares, in order to encourage a price rise; **support cost**, the cost of supporting something; *spec.* the cost of supporting the armed services; **support group**, (*a*) a group of musicians taking a subordinate part in a concert; (*b*) a group of people giving support to a charitable or political organization; **support line** *Mil.*, the second line of troops in a battle; a trench occupied by such troops; **support price**, a minimum price for agricultural produce, maintained by support buying or deficiency payments; **support trench** *Mil.*, a trench forming part of a line of strong points in the rear of the strong points of the firing line.

1976 *Offshore Engineer* Apr. 5/1 Field operator Occidental has given a letter of intent for a long-term charter of the *Bredford*, the first purpose-built semi-submersible support barge. **1932** *Economist* 9 Jan. 69/2 The time to support prices was opportune. Support buying of high-grade bonds, particularly in the railroad list, has resulted in improvement of prices and a stronger tone in all security markets. **1969** *Times* 13 Jan. 11/1 There will.. be support buying to support the levies in some cases. **1958** *Spectator* 14 Feb. 192/2 To talk about the foreign policy of a small Power would be a little ridiculous if international affairs consisted purely of rocket sites and support costs. *a* **1974** R. Crossman *Diaries* (1976) II. 208, I went across to the Party meeting on German support costs. **1969** *Listener* 10 July 59/1 After all this and the excellent support groups, the Rolling Stones were a musical disappointment. **1976** *Spare Rib* Dec. 13/1 Our support group is small so we have very few problems with disagreements on tactics, etc., but have to work harder. **1977** *Lancashire Life* Dec. 75/1 CARE's Wigan support group invited the organisation to inspect the building. **1917** W. Owen *Let.* 4 Feb. (1967) 430 We worked back through the reserve, & support lines to the crazy village where the Battalion takes breath. **1918** *Aussie* Aug. 10/1 In that sector of the front..there was a small town... Subsequently, as our Support Line settled down in front of it, it became the object of minute attention on the part of Fritz. **1971** S. Hill *Strange Meeting* ii. 110 Garrett had asked Hilliard to write to the men's relatives, as soon as they got into the support line the following day. **1943** *Sun* (Baltimore) 30 Jan. 18/4 So-called support prices, the minimum which canners may pay this year to growers who participate in the subsidy program,..are as follows. *Ibid.* 2 Oct. 6/2 The food planners tell us solemnly that this is not a subsidy program; it is a support-price program. **1949** A. McLintock *Descr. Atlas N.Z.* 44 Since the war 'support schemes' have also been established for wool and export meat, based on the reserve funds built up during the war and post-war periods of stabilisation.. [but] in the long run 'support prices' cannot diverge very far from market realisations. **1974** *Times* 15 Jan. 1/7 The Council..had rejected..an immediate 10 per cent rise in the support price for beef paid to farmers. **1915** Support trench [in Dict., sense 1 e]. **1923** Kipling *Irish Guards in Great War* I. 40 The line of support-trenches was held.

support, *v.* Add: **12. c.** *Sport.* To be a supporter or follower of (a team, etc.). Cf. *supporter 5 b.

1952 J. Arlott *Concerning Soccer* viii. 122 The spectator has the loudest word; for the good of football he should support good football. **1962** K. Wolstenholme *Book of World Soccer* 55/2 When you think of all that, what other football club is there worth supporting? **1979** E. John in K. Keegan *Against the World* x. 071 I've always supported England, I've stood on the terraces at Wembley, so what I relish now is the sheer luck of having the privilege to travel with the team.

supportative (sǝpǭ°·ɹtǝtiv), *a. rare.* [f. support *v.* + -ative.] = Supportive *a.*

An unnecessary formation, since the shorter *supportive* is completely established.—Ed.

1972 *Nature* 24 Mar. 154/2 These two basically different types of control are complementary and supportative. **1976** *Times* 22 July 14/2 (Advt.), A perceptive, sensitive and intelligent individual capable of organising our president in a supportative manner. **1981** *Spectator* 7 Feb. 19/2 Then follow the supportative words, the substructuration of Belonging.

supporter. Add: **3. g.** A jock-strap.

1895 *Montgomery Ward Catal.* Spring & Summer 488/3 The best fitting, most comfortable and effective supporter yet devised. Used by ball players, athletes and the theatrical profession generally. **1978** R. Doliner *On the Edge* (1979) iv. 66 The Senator pulled on his supporter, made a cup of his hand and laid himself gently to rest in the elastic sling.

5. b. One who supports a particular form of sport or who makes a practice of following the fortunes of a particular team, by attending matches, etc.

1922 *Glasgow Herald* 30 June 8 An enthusiastic supporter of baseball. **1928** *Daily Mail Year-bk.* 84/2 The supporters of the Chelsea F.C. **1972** T. Stoppard *Jumpers* i. 40 That he [*sc.* God] should have been taken up by a glorified supporters' club is only a matter of psychological interest. **1973** *Times* 24 Apr. 8/4 You have been wonderful supporters. I think I am very fortunate to play my last match in front of such wonderful people. **1976** *Milton Keynes Express* 2 July 42/6 There was a unanimous vote that a supporters' club be formed to further promote the need for a track in the area as soon as possible. **1980** *Daily Tel.* 19 Sept. 3/3 More than 30 supporters were ejected during the match, eight were still in police custody.

supporting, *ppl. a.* Add: **5.** Of actors or their roles, or of items in a programme of entertainment, usu. at a cinema: subordinate, less important.

1933 P. Godfrey *Back-Stage* v. 62 The fake star can be made to twinkle brightly by absorbing the surrounding light of the 'supporting cast'. **1939** *Chatelaine* Nov. 24/4 Some will continue for a while in minor productions and supporting roles. **1947** M. Gilbert *Close Quarters* vii. 108, I reached the cinema in time for the beginning of the supporting picture. **1953** [see *big stuff* s.v. *big *a.* B 2]. **1966** *Listener* 23 June 918/1 The supporting performances ..are enthusiastically full-blooded. **1977** *Rolling Stone* 21 Apr. 31/3 Blondie begins to seek out that untapped audience with a supporting slot on Iggy Pop's American tour.

supportive, *a.* Delete *rare* and add later examples.

1954 H. C. Shands in *Amer. Jrnl. Orthopsychiatry* XXIV. 84 It is necessary that the anxious individual have available a supportive pattern of relationship to depend upon through the learning period. **1962** Henderson & Gillespie *Text-bk. Psychiatry* xi. 286 This sort of supportive psychotherapy is relevant in every case. **1965** *Listener* 30 Sept. 501/3 Supportive material, notes by the editor, and letters by other hands are made to fill in the picture. **1972** *Science* 20 Oct. 229/3 She appears to interpret her findings as supportive of a smaller proportion of genetic variance among blacks than among whites. **1973** *Black Panther* 23 June 6/2 She has vein trouble in her legs for which she wears supportive stockings. **1978** G. Vidal *Kalki* vi. 149 'Senator White says that..he's going to call Mr. Kalki, as a witness... Will Mr. Kalki be supportive?' 'Hopefully, Kalki is supportive of all of us all of the time and for all time.' **1980** *Daily Tel.* 6 Dec. 12/4 Most American psychotherapists now advertise themselves as 'supportive'.

Hence **suppo·rtiveness.**

1968 A. J. Tannenbaum in H. L. Foster *Ribbin'* (1974) i. 4 They have worked with the children..in order to win their confidence and provide supportiveness. **1978** *Nature* 17 Aug. 698/1 We have also tried to determine whether there was a dorso-ventral pattern of host supportiveness of tumour growth.

supposably, *adv.* (Earlier example.)

a **1866** J. Grote *Exam. Util. Philos.* (1870) vi. 107 The happiness of any supposably actual being.

suppose, *v.* Add: **8. b.** Also *I suppose*, ellipt. for *I suppose so*, as a hesitant or reluctant affirmative.

1959 'E. McBain' *'Til Death* (1961) v. 67 'I think that's wise, don't you?' 'I suppose.' **1973** S. Cohen *Diane Game* (1974) xii. 103 'Look how much good information is published by..guys at universities.' 'Yes, I suppose.' **1976** 'Trevanian' *Main* (1977) x. 206 'Is she a viable?'..'I suppose. She had reason and opportunity.'

d. *pass.* Now esp., to be expected, intended, or meant; to have as a duty, to be obliged.

The pronunc. of the pa. t. is often colloquially modified from (sǝpǭu·zd) to (sǝpǭu·st). The negative is, idiomatically, freq. used to mean 'to have a duty or obligation not (to do something)'.

1859 Dickens *Tale of Two Cities* iii. x. 223 We saw the man, who was supposed to be at the gate, standing silent behind him. **1863** *Proc. Linnean Soc.* VII. p. xxvii, Still less does it seem consistent with that impartiality which every reviewer is supposed to possess. **1894** J. Pope *Mem. Rt. Hon. Sir John A. Macdonald* I. ii. 24 It appears that Mr. Baldwin considered this notice as sufficient to relieve him of the ordinary obligations which are supposed to govern the actions of Cabinet Ministers. **1902** 'R. Connor' *Glengarry Days* ii. 43 Girls are not supposed to be soldiers, are they, Margaret? **1914** G. B. Shaw *Misalliance* 86 Look here, Mr Percival: youre not supposed to insult my sister. **1931** *Morning Post* 31 Jan. 6 Officers..were not 'supposed' to keep a scrap log. **1949** E. Caldwell *This Very Earth* xi. 112 What's a girl supposed to do on Tuesday nights..? Bring her diary up to date? **1953** *N.Y. Herald-Tribune* 29 Apr. 4 They were supposed to address the lecturer as comrade. **1963** E. Albee *Who's Afraid of Virginia Woolf?* (1964) 84 When Daddy retired, he'd take over the college... That's the way it was supposed to be. **1969** A. Christie *Hallowe'en Party* xiii. 141, I brought her in lots of things that she was not supposed to eat. **1976** M. Machlin *Pipeline* xi. 134 That's more than the whole job was supposed to cost in the first place. **1976** P. & W. Proctor *Women in Pulpit* vi. 106 If that outburst was supposed to shock me because I'm a woman—forget it, brother!

supposite, *sb.* For † *Obs.* read *rare* and add: **1, 2.** (Later examples in form *supposit*.)

c **1882** G. M. Hopkins *Sermons & Devotional Writings* (1959) ii. iii. 146 A person is defined a rational (that is/ intellectual) supposit. **1929** tr. *St. Thomas Aquinas's Summa contra Gentiles* IV. iv. xxxiv. 144 Relative terms, nouns or pronouns, relate to the same supposit.

‖ **suppositio materialis** (sʊpǫzi·ʃio mǟtiᵊri-ǟ·lis). *Logic.* [med. L.] Reference to a word or phrase used simply as an example within a statement, and devoid of its normal semantic function.

1843 Mill *Logic* I. ii. ii. 29 This employment of a word to denote the mere letters and syllables of which it is composed, was termed by the schoolmen the *suppositio materialis* of the word. **1921** W. E. Johnson *Logic* I. x. 169 The scholastic logicians introduced the phrase 'suppositio materialis'..but modern logicians have interpreted this phrase as equivalent to what they call the 'universe of discourse'. **1935** H. Straumann *Newspaper Headlines* ii. 67 The two sentences: 1. 'The first line of Gray's *Elegy* states a proposition, and 2. 'The first line of Gray's *Elegy*' does not state a proposition. Both utterances are true, but in the second case the validity entirely depends on the inverted commas. This phenomenon used to be well known in mediaeval scholasticism under the name of *suppositio materialis*, and it still plays an essential part in semantics. **1961** [see *HYPOSTASIS 8].

suppositious, *a.* Restrict Now *rare* or *Obs.* to senses 1 and 2 and add: **3. a.** (Later examples.)

1905 JOYCE *Let.* 12 July (1966) II. 97 We might take a small cottage outside Dublin... Not that I imagine that the atmosphere of our suppositious cottage could.. become more unpleasant to you than the atmosphere you are at present breathing. **1957** G. E. HUTCHINSON *Treat. Limnol.* I. iv. 231 Most of the evidence is purely suppositious. **1978** P. W. J. RILEY *Union of England & Scotland* 4 Although the island comprised more than one kingdom the term 'Great Britain' was already respectable usage... James [VI/I] now contemplated for this suppositious entity not only one king but one kingdom. **1982** *Christian Sci. Monitor* 26 Aug. 9/2 Being at the center of population has some commercial advantage for local inhabitants. The last stop of the suppositious point was at Mascoutah, Ill., in 1970—a town now dethroned by the new centre.

supposititious, *a.* **1. b.** (Later *fig.* example.)

1934 H. G. WELLS *Exper. Autobiogr.* I. v. 265 Russia.. is now no longer a Communism nor a democratic Socialism... It is a novel experimental state capitalism... It is the supposititious child of necessity in the household of theory.

suppositous (sʌpǫ·zitəs), *a. nonce-wd.* [f. L. *suppositus*, pa. ppl. of *suppōnĕre*: see SUPPOSITUM.] Supposed, assumed.

1922 JOYCE *Ulysses* 686 An infinity rendered equally finite by the suppositous probable apposition of one or more bodies equally of the same and of different magnitudes.

suppress, *v.* Add: **8. a.** To prevent or inhibit (an action or phenomenon); *esp.* to eliminate, partly or wholly (electrical interference or unwanted frequencies).

1929 T. E. SHEA *Transmission Networks & Wave Filters* I. i. 20 For demodulation and recognition of signals the carrier frequency and one sideband may be suppressed. **1933** *Popular Sci.* Jan. 57/2 Interference troubles are present in abundance... To suppress these oscillations, 25,000-ohm resistors are placed in each spark plug lead. **1964** R. F. FICCHI *Electr. Interference* iv. 29 Shielding is the only practical method of suppressing interference which is radiated directly from a source. **1969** J. H. GREEN *Basic Clin. Physiol.* xvi. 91/2 The hormones of the adrenal cortex have an action in suppressing allergic responses. **1977** *Lancet* 5 Nov. 954/2 A 6-day course of oral dexamethasone at a dosage which would completely suppress A.C.T.H. in a person with adrenal insufficiency. **1980** PIERCE & POSNER *Introd. Communication Sci. & Systems* x. 224 A data signal with dc suppressed is sent through..single-sideband telephone links.

b. To fit with a suppressor.

1948 *Electronic Engin.* XX. 95 Garages and service stations are asked to co-operate in 'suppressing' cars already on the road. **1955** *Times* 31 Aug. 5/1 Everyone, he said, should beware of people who told them that all appliances must now be suppressed, especially if they were trying to sell suppressors. **1970** *AA Bk. Car.* 332 (*heading*) Suppressing the coil and the dynamo.

suppressant (sǒpre·sănt). [f. SUPPRESS *v.* + -ANT[1].] An agent that suppresses or restrains; *spec.* (more fully *appetite suppressant*) a drug which suppresses appetite.

1958 *Jrnl. Amer. Med. Assoc.* 24 May 437/1 We were particularly impressed with the marked reduction in side-effects with the new drug, as compared with our previous experience with other appetite suppressants. **1968** *McGraw-Hill Yearbk. Sci. & Technol.* 189/1 material that is applied directly to the burning fuel to reduce the intensity or rate of burning is termed a suppressant. **1974** M. C. GERALD *Pharmacol.* ii. 28 One group of appetite suppressants is composed of indigestible gums that swell when they come into contact with the stomach's fluids. **1977** P. THEROUX *Consul's File* 73 Ayer Hitam was malarial, and the tablets we took..were only suppressants.

suppressed, *ppl. a.* Add: **c.** *Forestry.* Of a tree: growing in the lower levels of a forest.

1893 [see *DOMINANT *a.* 6]. **1938** J. S. BOYCE *Forest Pathol.* xvi. 388 In Germany suppressed trees of artificially infected red beech were found to be more susceptible to decay. **1976** [see *OVERTOPPED *ppl. a.*].

d. Fitted with an interference suppressor.

1959 *Which?* Aug. 91/1 All [clothes driers]..were stable.., and suppressed for TV, though none was completely suppressed for radio. **1970** *AA Bk. Car.* 332/3 (*caption*) On cars not fitted with suppressed high-tension leads or caps, insert individual line suppressors in each HT lead.

e. suppressed-carrier *Telecommunications*, usu. *attrib.* (see quot. 1924).

1924 S. R. ROGET *Dict. Electr. Terms* 251/1 *Suppressed carrier wave telephony*, a system of Carrier Current Telephony in which the excess of unmodulated carrier wave is filtered out and not transmitted, but re-introduced in the receiving apparatus in sufficient quantity to prevent distortion. **1935** [see *PILOT *sb.* 5*]. **1959** [see *PHASING *vbl. sb.* 1]. **1974** HARVEY & BOHLMAN *Stereo F.M. Radio Handbk.* ii. 12 In the Zenith-G.E. system..a suppressed-carrier or balanced modulator is employed.

suppressible, *a.* Add: Hence **suppressibi·lity**, capacity for being suppressed.

1973 *Clin. Endocrinol.* II. 369 There was a good correlation between a normal TRH response and normal thyroid suppressibility by T_3. **1977** *Lancet* 5 Nov. 954/2 Inadequate A.C.T.H. suppressibility in patients with Addison's disease while on treatment may be due to the maintenance of a secondary pituitary hyperplasia by inadequate replacement therapy.

suppression. Add: **7. a.** *Psychol.* The restraint or repression of an idea, an activity, or a reaction by something more powerful.

1880 W. JAMES *Coll. Ess. & Rev.* (1920) 197 What is this volition?..the permanent suppression of an idea although it may be immediately and urgently pleasant. *Ibid.*, What do we mean by 'suppression'? Either complete oblivescence, or such presence as to evoke the steady sentiment of aversion or negation. **1894** CREIGHTON & TITCHENER tr. *Wundt's Human & Animal Psychol.* 206 It may sometimes be observed that these phenomena of suppression do not extend to the entire image. **1951** T. C. RUCH in S. S. Stevens *Handbk. Exper. Psychol.* v. 172/2 Suppression exists in two forms. The first is termed 'suppression of motor activity'..and the second form is 'suppression of electrical activity'. **1971** K. H. PRIBRAM *Languages of Brain* vii. 138 Thus some sort of suppression of responsiveness must occur when an imbalance in the ordinary mode of excitation is produced, and their suppression exceeds the malfunction produced solely by disuse. **1974** ATKINSON & BIRCH in Atkinson & Raynor *Motivation & Achievement* xv. 274 The impact of some feature of the immediate environment is not so much an instigation to activity as the opposite, suppression of an activity.

b. *Psychoanal.* The action or result of (consciously) inhibiting an unacceptable feeling, desire, or memory. Cf. *REPRESSION 2 c.

1913 A. A. BRILL tr. *Freud's Interpretation of Dreams* v. 199 The theory of repression..asserts that such repressed wishes still exist, contemporaneously with an inhibition weighing them down. Language has hit upon the truth when it speaks of the 'suppression' of such impulses. **1926** J. S. HUXLEY *Ess. Pop. Sci.* vii. 72 Repression, suppression, sublimation, and the rest are [psychological] realities; and we are finding out how our minds do work. **1955** E. MOSBACHER tr. *Ferenczi's Final Contrib. Psycho-anal.* iv. 265 With suppression one does not feel the pain, only the *effort* which is necessary to 'alienate over' the pain. With repression one does not even feel this any longer. **1969** H. NAGERA et al. *Basic Psychoanal. Concepts* II. ix. 43 The possibilities range from complete suppression..to an intensity which is greater than the ideational content would lead one to expect. **1977** A. SHERIDAN tr. *Lacan's Four Fundamental Concepts Psycho-anal.* ii. 27 Is it not possible to see emerging from the text itself..the reality of the disappearance, of the suppression, of the *Unterdrückung*, the passing underneath?

8. *Electr.* Prevention of electrical interference.

1933 *Jrnl. Inst. Electr. Engineers* LXXIII. 543/2 Devices for the suppression of interference from many items of electrical plant, particularly domestic appliances, could be incorporated in future designs. **1964** R. F. FICCHI *Electr. Interference* vii. 110 Capacitors are used when suppression is required on two commutator motors. **1970** *R.A.C. Guide & Handbk.* 56 Radio Interference Suppression. Regulations made by the Postmaster General.

9. *Phonetics.* The lowering of normal stress levels in verse; an instance of this.

1956, 1973 [see *PROMOTION 1 f].

‖**suppressio veri** (sǒpre·ʃio vīᵊ·rai). [mod. L., = suppression of what is true.] Misrepresentation of the truth by concealing facts which ought to be made known. Cf. *SUGGESTIO FALSI.

1755 CHESTERFIELD in *World* No. 105. 632 Here is not only the *suppressio veri*, which is highly penal, but the *crimen falsi* too. **1815, 1855** [see *SUGGESTIO FALSI]. **1889** *Athenæum* 20 Apr. 500/3 There is an unintentional *suppressio veri* in his assertion. **1905** *Spectator* 25 Feb. 286/2 The English Church Union could hardly subscribe *ex animo* to an interpretation containing an important *suppressio veri*. **1950** M. HAY *Foot of Pride* v. 135 It would not be easy to find a more flagrant case of *suppressio veri* than this omission..of any reference to the notorious Rohling scandal. **1979** J. MELVILLE *Wages of Zen* iii. 28, I told him exactly who I am and, with a touch of *suppressio veri*, what I'm about.

suppressor. Add: **1. b.** *spec.* A device for stopping a machine or part to which it is fitted from causing electrical interference.

1930 *Engineering* 14 Nov. 626/1 A diagram..is given in Fig. 16, the transmitting and receiving suppressors being marked TS_1 and TS_2. **1948** *Electronic Engin.* XX. 95 An ignition suppressor for fitting in the H.T. supply lead from the coil. **1955** [see *SUPPRESS *v.* 8 b]. **1970** [see *SUPPRESSED *ppl. a.* d].

2. *Genetics.* A gene in whose presence the effects of some other gene are not expressed. Also *suppressor gene*.

1928 *Zeitschr. f. Induktive Abstammungs- und Vererbungslehre* XLVI. 85 (*heading*) The genetics of 'black suppressor' in Drosophila melanogaster. **1932** *Amer. Naturalist* LXVI. 323 That the suppressor is a translocated wild-type allelomorph rather than a mutation in another gene has been proved in certain cases. **1960** *Heredity* XV. 91 The phenotypic manifestation of the suppressor gene is hidden by the mutation *en*. **1966** *Ann. Rev. Microbiol.* XX. 409 The best characterized and most intensively studied of the suppressor mutations which affect the translation process are the suppressors of the two classes of mutants, 'amber' and 'ochre'.

3. *Electronics.* = *suppressor grid*, sense 4 below.

1937 F. E. TERMAN *Radio Engin.* (ed. 2) iv. 128 The virtual cathode in conjunction with the plate and suppressor grid forms the equivalent of a triode tube in which the suppressor is the grid. **1959** [see *GRID 5 a]. **1968** ROMANOWITZ & PUCKETT *Introd. Electronics* vi. 237 The suppressor is usually connected to the cathode and is thus at full negative potential with respect to the plate.

4. Special Combs.: **suppressor (T) cell** *Immunol.*, a thymus-dependent lymphocyte which can suppress the stimulation of antibody production in lymphocytes in the presence of antigen; **suppressor grid** *Electronics*, in a thermionic valve, a coarse grid situated between electrodes (usu. the screen grid and the anode) so as to stop secondary electrons emitted by the latter from reaching the former.

1972 *Jrnl. Immunol.* CVIII. 590/1 It is possible that there are separate populations of activator (*x*) and suppressor (*y*) T cells. **1979** *Jrnl. Exper. Med.* CXLIX. 1018 The suppressor T cells regulated the DH [*sc.* delayed hypersensitivity] in the induction stage. **1981** *Nature* 23 July 357/2 Nonspecific suppressor cells may be one explanation for the severe immunodeficiency and the recurrent infectious complications characteristic of patients with chronic GvHD [*sc.* graft-versus-host disease]. **1931** *Electronics* Nov. 176/2 In order to avoid the effects of secondary emission from the plate..an auxiliary electrode was inserted to suppress this secondary current. The advantages of this same sort of suppressor grid are utilized in the..power pentode design. **1944** *Electronic Engin.* XVII. 163 The suppressor grids are generally operated at cathode potential. **1974** HARVEY & BOHLMANN *Stereo F.M. Radio Handbk.* ii. 13 If a modulating signal is now applied, the bias on the suppressor grid is alternately raised and lowered in sympathy with the modulating signal.

supra-. Add: **I. 1. a. supracru·stal** *a.* and *sb.* *Geol.*, (a stratum, formation, etc.) lying above the basement rocks of the crust; **suprafa·cial** *a. Chem.*, (of a concerted reaction undergone by a molecule) involving the formation of two new bonds on the same face of the molecule; **suprali·ttoral** *a. Ecol.*, applied to a biogeographic zone normally taken as extending from mean high tide to the limit of influential sea spray or land vegetation; also *ellipt.* as *sb.*

1946 *Amer. Jrnl. Sci.* CCXLIV. 851 The supracrustal formations are subjected to folding, plastic deformation, fractures and thrusts in which the blocks of the basement also take part. **1973** *Nature* 21 Sept. 138/1 The Isua supracrustals may represent a shallow-water shelf facies. **1965** WOODWARD & HOFFMANN in *Jrnl. Amer. Chem. Soc.* LXXXVII. 2512/1 In the first process, here designated suprafacial, the hydrogen atom is associated at all times with the same face of the π-system. **1980** E. N. MARVELL *Thermal Electrocyclic Reactions* i. 5 Thus for a π bond a suprafacial process forms new connections on the same side of the nodal plane. **1909** WARMING & VAHL *Œcol. Plants* IV. xli. 173 Thus arises a kind of supra-littoral 'region'. **1949** T. A. & A. STEPHENSON in *Jrnl. Ecol.* XXXVII. 298 We therefore formally propose that the three main zones of the shore be called: Supralittoral Fringe, Midlittoral Zone, Infralittoral Fringe. **1967** *Oceanogr. & Marine Biol.* V. 464 The plants and animals living in the supralittoral zone can either tolerate or need a permanent or almost permanent emersion, but with moistening by sprays and waves. **1974** *Nature* 22 Feb. 520/2 The environmental significance of each of these.. seems to reflect various aspects of turbulence, turbidity, and/or desiccation in the littoral or supralittoral.

b. *supranuclear*: delete 6 b (further examples); *suprapatellar*, *-umbilical*, *-ventricular* (examples referring to VENTRICLE 1); **supra-meatal** (-mi,ēᵊ·tăl) *a.*, situated above the acoustic meatus; **suprao·ptic** *a.* situated above the optic chiasma; **suprase·llar** *a.*, situated or occurring above the sella turcica.

1893 W. MACEWEN *Pyogenic Infective Dis. of Brain & Spinal Cord* i. 9 The apex of this triangular depressed area points forward. The author proposes to name this area the supra-meatal triangle. **1922** *Brit. Med. Jrnl.* 29 July 164/2 Later experience in using this suprameatal angle as a guide has been extensive. **1980** *Gray's Anat.* (ed. 36) iii. 302/2 Immediately above and behind the meatus there is frequently a small depression with a bony spicule (suprameatal spine) in its anterior margin. This lies within the area of the suprameatal triangle. **1977** *Lancet* 12 Nov. 1029/2 One [patient] had periodic vertigo due to a supranuclear vestibular lesion. **1979** *Jrnl. Compar. Path.* LXXXIX. 503 The supranuclear cytoplasm of the superficial epithelial cells. **1921** TILNEY & RILEY *Form & Function Central Nerv. Syst.* xxxi. 550 Situated above the optic chiasm and in communication with the chamber of the third ventricle is a small canal which projects outward over the optic nerve. This is the supra-optic canal. **1980** K. E. MOYER *Neuroanatomy* xxviii. 70/2 The supraoptic nucleus is located directly over the lateral portion of the optic chiasm. **1902** D. J. CUNNINGHAM *Text-bk. Anat.* 293 The joint-cavity may communicate with bursæ situated in relation to the inner head of the gastrocnemius muscle and the tendon of the semi-membranosus muscle, besides the large supra-patellar bursa already described. **1975** L. M. ELSON *It's your Body* v. 284 (*caption*) Suprapatellar bursa. **1934** J. H. GLOBUS *Neuroanatomy* (ed. 6) i. 15 The diamond shaped somewhat depressed space outlined at the base of the brain is called the interpeduncular space..; and because it overlies the sella turcica in the base of the skull, it is often termed the suprasellar space. **1977** *Lancet* 9 Apr. 780/1 Air encephalograms were performed on 7 of the 12 patients with prolactin levels between 15 and 30 μg/l and in all 7 suprasellar extension of the tumour was found. **1906** *Practitioner* Dec. 781 Rolleston thinks that, for practical purposes, it is advisable to adopt Oppenheim's division of the reflex into a supra-umbilical and infra-umbilical zone. **1978** *Acta Path. Japonica* XXVIII. 288 Just after birth.., this female patient showed deep cyanosis and supraumbilical abdominal hernia. **1974** *Ciba Symposium* XX. 133 Localization of the infiltrate in the AV node blocks the supraventricular impulses when the atrial rate increases. **1979** *Brit. Med. Jrnl.* 15 Dec. 1553/2 Disopyramide..has been used to treat supraventricular and ventricular arrhythmias.

d. Phonology. *supradental* (also as *sb.*), *supraglottal* (also as *sb.*), *supralaryngeal*.

1926 B. KARLGREN *Philology & Ancient China* iv. 80 No pronouncement is here made about whether these sounds were hard, supradental ('cerebrals'), *tṣ-*, *tṣ'-*, *ṣ-*, etc., somewhat resembling English heartshaped. **1969** *Language* XLV. 125 To say that the supradentals are merely occurrence phenomena of [r] plus dental leads to undesirable results. **1935** *Amer. Speech* X. 311/2 The quality of vowels depends not only on supra-glottal resonances, but..on the complex and variable sound emitted by the larynx. **1964** CRYSTAL & QUIRK *Prosodic & Paralinguistic Features in Eng.* iii. 39 Supraglottals may be tense or lax. **1979** *Amer. Speech 1978* LIII. 290 The supraglottal articulatory motions and positions used in human language..are not used by other animals at all. **1964** J. C. CATFORD in D. Abercrombie et al. *Daniel Jones* 29 Phoneticians should be able to classify 'voice qualities' and other phonatory activities in as systematic a way as they classify supralaryngeal articulation. **1978** *Amer. Speech* LIII. 292 They carefully measured chimpanzee supralaryngeal cavities.

II. 4. a. *supraconscious* (later example), *-decent*, *-dialectal*, *-individual*, *-literal*, *-logical*, *-normal* (later examples), *-racial*, *-regional*, *-sentential*; **su:pramole·cular**, composed of many molecules; higher in organization than a molecule; **su:prarela·tional**, of a postulated being or power that transcends or includes all that is relational.

1953 J. STRACHEY tr. *Freud's Interpret. of Dreams* in *Compl. Wks* V. vii. 615 We must avoid, too, the distinction between 'supraconscious' and 'subconscious', which has become so popular in the more recent literature of the psychoneuroses, for such a distinction seems precisely calculated to stress the equivalence of what is psychical to what is conscious. **1938** S. BECKETT *Murphy* v. 90 As different..as a *voyeur's* from a *voyant's*, though Wylie was no more the one in the indecent sense than Murphy was the other in the supradecent sense. **1960** P. DORF tr. M. M. Guxman in J. A. Fishman *Readings Sociol. of Lang.* (1968) 768 Even in the formation process of the new written Bashkir national language, the problem of working out a unified, supradialectal literary norm,..was in the twentieth century just as real as it was in eighteenth- and nineteenth-century Germany or Italy. **1936** *Mind* XLV. 293, I would fain hold..that the highest values are, not only supra-individual, but supra-national. **1949** *Archivum Linguisticum* I. 164 A mark of palatalization in the shape of a supraliteral meniscus. **1896** W. CALDWELL *Schopenhauer's System* i. 42 This fondness of Schopenhauer for the supra-logical character of intuition and genius has its dangerous side. **1936** *Essays & Studies* XXI. 136 In the case of words like 'spell' and 'Host' the supra-logical connotation and accidental associations reinforce the literal meaning. **1909** in *Cent. Dict. Suppl.* **1961** *Nature* 8 July 145/1 A supramolecular organization of the enzyme systems. **1976** *Sci. Amer.* July 65/1 One of the major challenges in cell biology today is the mapping of supramolecular structures such as membranes and ribosomes. **1902** W. JAMES *Var. Relig. Exper.* 484 Our supra-normal cognitions, if such there be, and if we are telepathic subjects. **1959** B. WOOTTON *Social Science & Social Pathology* ii. 59 Exceptionally difficult problems, failure to cope adequately with which might well be a sign,..of their lack of the supra-normal qualities which the situation demands. **1977** J. L. HARPER *Population Biol. of Plants* x. 328 Partial closure of stomata..usually occurs when supra-normal concentrations of CO_2 are applied to leaves. **1922** JOYCE *Ulysses* 717 With what antagonistic sentiments were his subsequent reflections affected?.. Abnegation? In virtue of..extraracial attraction, intraracial inhibition, supraracial prerogative. **1973** R. C. VAN CAENEGEM *Birth of Eng. Common Law* i. 14 The justices in eyre or itinerant royal judges..were called *justitiarii totius Angliae*, to indicate that their commission was supraregional. **1910** W. JAMES *Coll. Ess. & Rev.* (1920) 497 Mr. Bradley tumbles to philosophy's call. Down he slides, to the dry valley of 'absolute' mare's nests and abstractions, the habitation of the fictitious suprarelational being which his will prefers. **1936** *Mind* XLV. 538 His initially hopeless attempt to make sense..of freedom and individuality in terms of his supra-relational whole. **1961** Y. OLSSON *Syntax Eng. Verb* ii. 34 This constitutes a new kind of intersectional concord, often exceeding the limits of the sentence (*supra-sentential concord*).

b. *supra-clan*, *-class*, *-Elder*, *-language*, *-party*, *-village*.

1979 *Social Sci. & Med.* XIII. D. 209/2 Within these dispersed people, there was no supraclan political system to organize for mutual defense. **1952** C. BARDSLEY *Bishop's Move* viii. 96 A supra-class Church founded on Life and Love. **1977** M. WALKER *National Front* iv. 84 The NF's supra-class, supra-party appeal. **1958** D. TAIT *Tribes without Rulers* 193 There is no supra-Elder authority to impose a solution on recalcitrants of a district. **1975** *Amer. Speech 1972* XLVII. 253 Such supralanguage phenomena as rhyme and assonance in verse. **1974** tr. *Wertheim's Evolution & Revolution* iii. 245 The creation of supra-village organizations.

6. *supra-body*, *-burgher*, *-consciousness* (earlier example), *-council*, *-language*, *-organism*, *-parliament*, *-system*.

1967 M. AYUB KHAN *Friends not Masters* xi. 199 There was obviously no place for a supra-body of religious experts exercising a power of veto over the Legislature and the Judiciary. **1905** JOYCE *Let.* 19 July (1966) II. 99 Also desirables are..a sizeable beefsteak.., and..an intelligent supra-burgher like yourself to share the meal. **1911** A. MITCHELL tr. *Bergson's Creative Evolution* 275 It is consciousness, or rather supra-consciousness, that is at the origin of life. **1974** *Physics Bull.* Mar. 86/1 A formalized supra-council of scientific institutions. **1980** *Encounter* July 50/1 *Finnegan's Wake*..the project of a supra-language distinct to the given text. **1949** KOESTLER *Insight & Outlook* II. xi. 167 We must expect the ultimate achievement of a proportionate superiority in the mature human supra-organism. **1971** Supra-parliament [see *MULTINATIONAL *a.* and *sb.*]. **1964** *Language* XL. 274 Modern Greek has two independent and partial phonological suprasystems.

7. suprale·thal *a.*, exceeding what is lethal; so **suprale·thally** *adv.*; **suprama·ximal** *a. Physiol.*, greater than what is required to produce the maximum response; hence **suprama·ximally** *adv.*; **supraṭhe·rmal** *a. Physics*, having greater energy than that associated with thermal excitations; **supra-thre·shold** *a. Physiol.*, exceeding the threshold value required for the perception of a stimulus.

1957 *Jrnl. Exper. Zool.* CXXXVII. 426 A total continuous dose of 24 kr, referred to as a supralethal dose, was given to interphase cells at each exposure. **1979** *Nature* 11 Oct. 490/2 Supralethal doses of chemoradiotherapy followed by allogeneic bone marrow transplantation are being used to treat patients with acute leukaemia. **1955** *Jrnl. Exper. Zool.* CXXX. 190 (*heading*) Survival and cell division in supralethally x-irradiated giant amoebae following injection of nonirradiated protoplasm. **1975** *Nature* 20 Nov. 233/2 (*caption*) Supralethally irradiated rats. **1925** LIDDELL & SHERRINGTON in *Proc. R. Soc.* B. XCVII. 497 The reflex mode of employing the motor units is to subject them to an incitement which is 'supramaximal' in the sense that it is in excess..of that which is necessary to evoke in them individually their 'maximal' response. **1977** *Lancet* 30 Apr. 942/1 Desmedt and Borenstein have modified this test..by first applying a train of supramaximal stimuli at 3 Hz for 4 minutes. **1973** *Nature* 26 Oct. 465/1 The sciatic nerve was stimulated supramaximally at a rate of 6.4 Hz with square waves of 0.5 ms duration. **1969** *New Yorker* 12 Apr. 104/3 The three pots are the passive seismic experiment, the solar-wind experiment, and the suprathermal-ion-detector. **1980** *Nature* 29 May 285/1 Of perhaps even greater interest was an image of the suprathermal X-rays, or bremsstrahlung, derived from interactions of the hot electrons with ions from the exploding pusher. **1946** *Ibid.* 27 July 131/2 The strength of the stimulating current was gradually increased to threshold and supra-threshold values. **1980** VAN BOMMEL & DE BOER *Road Lighting* ii. 35 The supra-threshold level of visibility can be expressed in terms of the visibility level attainable.

III. 9. supracondu·cting *ppl. a.* [tr. Du. *suprageleidend*: see *SUPERCONDUCTING *ppl. a.*] = *SUPERCONDUCTING *ppl. a.*; so **supracondu·ction**, **-conductive** *a.*, **su:praconducti·vity**, **supraconductor**; all now *rare*.

1932 *Nature* 10 Dec. 880/2 The application of mechanical stresses..raises the transition temperature of a supraconducting metal. **1937** M. & B. RUHEMANN *Low Temperature Physics* IV. ii. 269 It is by no means clear whether at sufficiently low temperatures all metals become supraconducting. **1932** *Nature* 10 Dec. 879 (*heading*) Electric supra-conduction in metals. *Ibid.*, Currents of electricity started in a ring of metal in the supra-conductive state will continue apparently undiminished in intensity. **1941** *Ibid.* 13 Sept. 317/1 It..appears that, in sufficiently pure and homogeneous samples of tantalum, the changes of electrical resistance..and specific heat accompanying the establishment of the supra-conductive state occur at one and the same temperature. **1930** *Engineering* 16 May 640/3 Some of the discoveries made by the late Professor Kamerlingh Onnes,..as, for instance, supraconductivity. **1962** P. J. & B. DURRANT *Introd. Adv. Inorg. Chem.* xviii. 591 Graphite does not show supraconductivity. **1976** *Progress in Sci. Culture* (E. Majorana Centre) Spring 90 The most important theoretical items were probably the nuclear analogy of supraconductivity..and the accounting of wave functions for deformed nuclei. **1933** *Nature* 14 Oct. 602/1 The use of a supra-conductor (therefore completely free from Joule heating) has been more than once suggested for the production of magnetic fields at low temperatures.

suprachoroid (s[1]ūprăkōᵊ·roid), *sb.* and *a.* *Ophthalm.* Also **-chorioid**, and in L. form **-chor(i)oidea**. [ad. mod. L.: see SUPRA- 1 b and CHOROID *a.* (*sb.*).] **A.** *sb.* A layer of loose cellular tissue lying between the choroid and the sclera. **B.** *adj.* Epithet of this layer.

1892 A. DUANE tr. *Fuchs's Text-bk. Ophthalm.* iv. 245 The suprachoroid..consists of numerous fine non-vascular but richly pigmented lamellæ lying between the chorioid proper and the sclera. **1896** W. A. FROST *Fundus Oculi* i. 12 Between the sclerotic and the choroid is some loose cellular tissue—the suprachoroidea. **1959** S. DUKE-ELDER *Parsons' Dis. of Eye* (ed. 13) i. 6 The greater part of the muscle is composed of meridional fibres running antero-posteriorly on the inner aspect of the sclera to find a diffuse insertion into the suprachoroid. **1962** *Gray's Anat.* (ed. 33) 1258 It [*sc.* the sclera] is separated from the outer surface of the chorioid by an extensive perichorioidal space, which is traversed by an exceedingly delicate cellular tissue, termed the suprachorioid lamina. **1971** M. J. HOGAN et al. *Histol. Human Eye* viii. 386 The suprachoroidea lies between the choroid and the sclera and appears to be derived partly from each tissue. **1978** F. W. NEWELL *Ophthalmology* (ed. 4) i. 15/1 The outermost layer, the suprachoroid (*lamina fusca*), is made up of delicate lamellae composed of elastic and collagenous fibers to form a syncytium.

So **su:prachoroi·dal** *a.*, situated above the choroid.

1887 F. FERGUS tr. *Meyer's Pract. Treat. Dis. Eye* iv. 148 (*caption*) Supra choroidal space. **1918** J. H. PARSONS *Dis. Eye* (ed. 3) xvii. 334 Fuchs..attributes it to slight separation of the ciliary body, so that the aqueous percolates from the anterior chamber into the suprachoroidal space. **1975** *Symposium on Glaucoma: Trans. New Orleans Acad. Ophthalm.* xix. 304 The surgeon should then confirm that the probe may be introduced with equal facility.. into the space between the ciliary body and sclera which anatomists call the suprachoroidal space.

supracleithrum (s[1]ūpră̄klai·þrǒm). *Zool.* Pl. **-cleithra.** [SUPRA- 1 c.] A dermal bone dorsal to the cleithrum in the pectoral arch of some fishes and amphibians.

1905 A. SEDGWICK *Student's Text-bk. Zool.* II. vii. 162 These bones are now often called supracleithrum, cleithrum and clavicle respectively. **1949** A. S. ROMER *Vertebr. Body* vii. 179 Above each cleithrum there are usually additional elements—typically a supracleithrum and post-temporal, and sometimes other bones as well—which curve upward and forward above the gill chamber and anchor the dermal girdle to the skull. **1981** PEARSON & BALL *Lect. Notes Vertebr. Zool.* iv. 52/1 Large clavicles occur in paleoniscoids and Polypterus, with large cleithra and smaller postcleithra, supracleithra, and post-temporals above them.

supracoracoid (s[1]ūprăkǫ·răkoid), *sb.* and *a.* *Zool.* Also in L. form **-coracoideus.** [f. SUPRA- 1 c + CORACOID *a.* and *sb.*] **A.** *sb.* A muscle in some birds, amphibians, and reptiles which passes over the coracoid bone and is attached to the head of the humerus and to part of the sternum. **B.** *adj.* Of, pertaining to, or designating the supracoracoid.

1933 L. A. ADAMS *Introd. Vertebrates* iv. 86 (*in figure*) Supracoracoideus. **1949** SAUNDERS & MANTON *Man. Pract. Vertebr. Morphol.* (ed. 2) ix. 88 The supracoracoideus muscle elevates the wing. **1956** A. S. ROMER *Osteol. of Reptiles* vii. 308 Beneath the anterior end of the glenoid is a supracoracoid foramen (or coracoid foramen), carrying..the supracoracoid nerve. **1974** ANDREW & HICKMAN *Histol. Vertebr.* vi. 107/2 The powerful muscles of flight (pectoral and supracoracoid) are centrally located on the sternum. **1979** *Nature* 15 Mar. 247/2 It has been argued that the structure of the coracoid of *Archaeopteryx* would not have permitted the supracoracoideus muscle to function as a wing elevator.

supraliminal, *a.* (Later examples.)

1903 F. W. H. MYERS *Human Personality* I. i. 14 Sensations, thoughts, emotions, which..by the original constitution of our being, seldom emerge into that *supraliminal* current of consciousness which we habitually identify with ourselves. **1918** [see *OVERLEARN *v.*]. **1931** *Brit. Jrnl. Psychol.* Jan. 305 Another series [of observations] was made in which a supraliminal admixture of spectral light was reduced until the field appeared pure white. **1971** *Jrnl. Gen. Psychol.* Jan. 122 Manipulating intensity..from subliminal to supraliminal luminance results in..emergence of linear detail.

suprana·tional, *a.* [SUPRA- 4 a.] Having power, authority, or influence that overrides or transcends national boundaries, governments, or institutions.

1908 [in Dict. s.v. SUPRA- 4 a]. **1924** J. C. W. REITH in *Radio Times* 29 Feb. 361/2, I like to think that wireless, as with music, is supra-national, a word coined, I believe, by Lord Cecil to indicate that which is above not only nationality, but something more even than international. **1941** *Burlington Mag.* Feb. 38/1 Mediaeval artists shared the common task of glorifying God under the guidance of a supra-national church. **1950** W. S. CHURCHILL in *Hansard Commons* 27 June 2147, I would add, to make my answer quite clear to the right hon. and learned Gentleman, that if he asked me, 'Would you agree to a supranational authority which has the power to tell Great Britain not to cut any more coal or make any more steel, but to grow tomatoes instead?' I should say, without hesitation, the answer is 'No'. **1958** A. J. ZURCHER *Struggle to unite Europe 1940–1958* vii. 80 This first European supranational community set up its administrative offices in Luxembourg on August 10, 1952. **1962** A. SAMPSON *Anat. of Britain* xxvii. 429 The very biggest firms in Britain..belong more to an international, than a national economy. And in the Common Market they are likely to become much more supra-national. **1973** *Observer* (Colour Suppl.) 12 Aug. 19/3 They were 'intergovernmental' rather than 'supranational'—that is to say based on negotiation between sovereign Governments, not on the principle that the institution itself, operating as a unit, could overrule member Governments. **1977** M. WALKER *National Front* ii. 25 His [*sc.* Mosley's] book *The Alternative*, which advocated a European nationalism, a new supra-national state of Europe.

Hence **suprana·tionalism**, **su:pranationa·lity.**

1921 *Glasgow Herald* 14 Oct. 10 It was only a developed sense of supra-nationalism that would in the future make war unthinkable. **1930** *Tablet* 16 Aug. 206/2 The Popes held out..against every threat..which aimed at lowering the supra-nationality of the Papacy. **1955** A. L. ROWSE *Expansion of Elizabethan England* vii. 241 One sees, as against the supra-nationalism of the Habsburgs.., the nationalist assumption..that the Low Countries should govern themselves. **1959** *Times Lit. Suppl.* 3 Apr. 187/3 The actual degree of supranationality in these Communities. **1971** *Mod. Law Rev.* XXXI. VI. 607 It is clear that with the elements of supranationality are commingled features of a more typical international organisation. **1980** *Times Lit. Suppl.* 10 Oct. 1134/1 Supranationalism has lost what appeal it had in the 1950s.

supraopticohypophysial (s[1]ūprăˌǫ:ptikohai-poti·ziăl), *a.* *Anat.* Also **-eal.** [f. *supraoptic* adj. s.v. *SUPRA- 1 b + -O + *HYPOPHYSIAL *a.*] Applied to a tract of nerve fibres in the brain running from the supraoptic nucleus to the hypophysis.

[**1937** *Jrnl. Path. & Bacteriol.* XLIV. 310 The fact.. seems to indicate that at this time the lesion was confined to the supraoptic-hypophyseal system.] **1943** STRONG & ELWYN *Human Neuroanat.* xvii. 317/2 They [*sc.* the connections of the hypothalamus with the posterior lobe of the hypophysis] are unmyelinated fibers which arise principally from the supraoptic and paraventricular nuclei and form a well defined bundle, the supraopticohypophysial tract. **1961** *Lancet* 2 Sept. 525/1 A lesion in the tuberal nuclei, with or without damage to the supraopticohypophyseal system, resulted in diabetes insipidus. **1980** K. E. MOYER *Neuroanatomy* xxxiii. 80 The supraopticohypophysial tract..comes principally from the supraoptic nucleus.

supra-pe·rsonal, *a.* [SUPRA- 4 a.] = *SUPERPERSONAL *a.*

1918 J. H. LECKIE *World to Come* II. 322 The notion.. of attaining some supra-personal state of being is not an idea that can appear reasonable. **1934** M. BODKIN *Archetypal Patterns in Poetry* 276 The sense of..a supra-personal life present within the group, which is made explicit in the writings of St. Paul. **1949** H. READ *Conc. Hist. Mod. Painting* vii. 249 The 'vibrations of the spirit' that then take place are..perhaps supra-personal, in that they assume the archetypal patterns into which mankind projects an explanation of its destiny. **1955** J. BURNABY *Christian Words & Christian Meanings* iii. 48 It may sometimes be wholesome for us to tell ourselves that God is not 'a person', but 'supra-personal', or even 'Absolute Being'. **1958** *Times Lit. Suppl.* 23 May 278/3 Its [*sc.* The Warburg Institute's] character is also suprapersonal, and it is never likely to lose its identity and be submerged in some more general apparatus of historical research. **1972** M. KIRKHAM in *Focus on Robert Graves* No. 3 (1973) 40 The feeling might be called impersonal, in the sense that it reflects..everybody's response to a general condition; a better word would be 'suprapersonal', to signify a going-beyond personality.

suprasegmental (s[1]ū:prăsegme·ntăl), *a.* and *sb.* *Linguistics.* [f. SUPRA- 4 a + *SEGMENTAL *a.* 2 c.] **A.** *adj.* Designating a feature or features of a sound or sequence of sound other than those constituting the consonantal and vocalic segments, as stress, pitch, and intonation in English.

1941 TRAGER & BLOCH in *Language* XVII. 224 These two kinds of phonemes [*sc.* juncture and prosodic phonemes] are usually recognizable only as modifications of other sound-types; they are suprasegmental. **1942** C. F. HOCKETT in *Language* XVIII. 8 Features..which clearly extend over a series of several segmental groupings are *suprasegmental*. **1942, 1952** [see *PROSODIC *a.* 2]. **1953** [see *intrasegmental* s.v. *INTRA- 1]. **1968** P. KRATOCHVÍL *Chinese Lang. Today* ii. 35 The most striking suprasegmental feature of MSC syllables is the characteristic contour known as the tone. **1971** [see *PROSODEME]. **1975** N. CHOMSKY *Logical Struct. Linguistic Theory* iii. 111 In this study, suprasegmental features (pitch, stress, juncture) have not been seriously considered.

B. *sb.* A suprasegmental feature.

1955 N. CHOMSKY *Logical Struct. Linguistic Theory* (microfilm, Mass. Inst. Technol.) vii. 278 It has often been suggested that constituent structure be determined by considerations involving suprasegmentals. **1965** *Word Study* Feb. 2/2 Structural and contextual meanings are signaled largely by intonational clues, by such suprasegmentals as pitch, stress, and juncture. **1975** *Language* LI. 737 How we perceive duration, pitch, and intensity is an area that seldom receives serious attention in the literature on suprasegmentals. **1981** *Amer. Speech* LVI. 306 A summary of paralanguage, including suprasegmentals, hesitation phenomena, and nonlinguistic sounds.

Hence **su:prasegme·ntally** *adv.*, in terms of suprasegmental features.

1957 S. POTTER *Modern Linguistics* v. 105 Sentences may be described *suprasegmentally* in respect of the prosodemes of *length*, *stress* and *pitch*. **1970** J. W. GAIR *Colloq. Sinhalese Clause Structures* vi. 133 The focus may be marked only suprasegmentally, even if the form of the predicator indicates that the focus is elsewhere in the clause.

suprasensuous, *a.* (In Dict. s.v. SUPRASENSUAL *a.*) Add: (Later examples.) Also *absol.* with *the*.

1902 *Pop. Sci. Monthly* Apr. 519 The scientist often has recourse to the suprasensuous. **1947** A. EINSTEIN *Music in Romantic Era* III. xviii. 340 An idea which then led to the assertion that true, ideal music is not heard at all, but is non-sensuous and suprasensuous. **1957** J. I. M. STEWART *Use of Riches* 41 He knew why this picture was neither better nor worse than the Maremma or the La Verna... Brilliantly sensuous, it was yet suprasensuous.

supraspecies (s[1]ū·prăspī:ʃīz, -sīz). [f. SUPRA- 6 + SPECIES *sb.*] (See quot. 1940.) So **supraspeci·fic** *a.*, above the rank of a species.

1940 J. S. HUXLEY *New Systematics* 10 We may substitute the term 'species-group', reserving the term 'supraspecies' for groups of an intermediate nature, in which it is dubious whether the constituent groups are best called subspecies or species. **1942** E. MAYR *Systematics & Origin of Species* vii. 169 The term supraspecies..seems to me to be an unfortunate combination. **1961** Supraspecific [see *NOMENCLATURAL *a.*]. **1975** *Nature* 9 Oct. 516/1 Because supraspecific taxa have different numbers of species, their observed linearity is evidence for the ecological reality of supraspecific taxa.

suprasterol (s[1]ūprăste·rǫl). *Biochem.* [ad. G. *suprasterin* (A. Windaus et al. 1930, in *Ann. d. Chem.* CCCCLXXIII. 20): see SUPRA- + *-STEROL.] Either of two optically active polycyclic isomers (*suprasterol I, II*) of $C_{28}H_{44}O$ produced by prolonged irradiation of vitamin D.

1931 *Chem. Abstr.* XXV. 301 Ergosterol in EtOH, subjected to the action of Hg light at about 75° for 50 hrs., gives a mixt. of suprasterol I..and II. **1943** *Endeavour* Apr. 73/2 Calciferol itself [vitamin D] was also liable to be broken down further to inactive substances—toxisterol and suprasterols I and II. **1976** H. CAMPION et al. in B. E. C. Nordin *Calcium, Phosphate & Mineral Metabolism* xii. 452 Ergocalciferol itself can, under prolonged irradiation, undergo irreversible photoisomerization to compounds known as suprasterol$_2$ I.

supravital (s[1]ūprăvai·tăl), *a.* *Histology.* [SUPRA-.] Of a stain or the process of staining: involving living tissue, esp. blood, outside the body. Hence **supravi·tally** *adv.*

1921 *Arch. Internal Med.* XXVIII. 513 Janus green B .., used as a supravital stain in dilute solutions..stains mitochondria an intense green. *Ibid.* 515 The reticulum present in certain erythrocytes is seen as a delicate network and is best demonstrated by staining supravitally with brilliant cresyl blue or azur II. **1930** *Edin. Med. Jrnl.* XXXVII. 429 Supravital staining.—The process consists of the application of basic dyes to portions of tissues removed during life or immediately after somatic death. **1972** C. GURNEY in C. E. Mengel et al. *Hematology* i. 6 When stained supravitally with a number of special stains, these young cells show small dark granules. **1974** *Nature* 22 Feb. 551/2 The supernatant was then shaken from the wells and supravital stain was added.

supremacist (s[1]upre·măsist), *sb.* and *a.* [f. SUPREMACY + -IST, orig. in *white supremacist*.] **A.** *sb.* One who believes in the supremacy of one of the races or of either of the sexes or of any other social group. **B.** *adj.* That is a supremacist. Orig. and freq. preceded by defining word: see also *male supremacist* s.v. *MALE *sb.* 4, *white supremacist* s.v. *WHITE *a.* 11 e.

1959, etc. [see *white supremacist* s.v. *WHITE *a.* 11 e]. **1961** WEBSTER, *Supremacist*.., an advocate or adherent of some concept of group supremacy; *esp*: *white supremacist*. **1968** in B. & T. Roszak *Masculine/Feminine* (1970) 256 Men..maintain a dominant position for themselves, and as supremacists, try to perpetuate that position of dominance. **1969** *Manifesto for N.Y. Radical Feminists* in J. Hole & E. Levine *Rebirth of Feminism* (1971) 443 The purpose of the male power group is to fulfill a need. That need is psychological, and derives from the supremacist assumption of the male identity. **1975** *Economist* 1 Feb. 24/1 Weaning the more sensible loyalists away from their Protestant supremacist partners. **1976** P. DRISCOLL *Barboza Credentials* v. iii. 217 An ultra-white brotherhood of supremacist bitter-enders. **1982** *Washington Post* 4 Mar. D 1/1 None of this football nonsense of airy polls full of supremacist blather.

So **supre·macism**: see *white supremacism* s.v. *WHITE *a.* 11 e.

Suprematism (s[1]upre·mătiz'm). Also **suprematism.** [ad. Russ. *suprematizm*.] An artistic movement initiated by the Russian painter Kazimir Malevich in 1913; the abstract, geometrical style of art produced by this movement. Hence **Supre·matist**[1] (*a*) *sb.*, an adherent of Suprematism; (*b*) *adj.*, of, pertaining to, or characteristic of Suprematism.

[**1915** K. MALEVICH (*title*) Ot Kubizma do Suprematizma.] **1933** *Times Lit. Suppl.* 2 Feb. 76/3 The various channels in which the Futurist movement has run.. orphism..suprematism. **1936** *Bull. Museum of Mod. Art* Nov.–Dec. 6 Malevich, the Suprematist, passed through a proto-Dada phase in 1914. **1948** H. READ *Art Now* (ed. 3) iv. 104 Malevich and Tatlin revolted against the naturalistic tradition and established a completely geometrical style which they called *Suprematism*. **1955** *Archit. Rev.* CXVII. 226/1 Malevitsch, in *Bauhausbuch* No. 11, hopefully says of his own filleted and rectilinear aesthetic 'thus one may also call Suprematism an aeronautical art'. **1958** *Spectator* 14 Feb. 203/1 His Suprematist work exploiting a simple vocabulary of colours and shapes and rhythms. **1958** *Listener* 31 July 168/3 Malevich and the Suprematists reflected it, in a form so extreme and absolute that it led to the painting of a picture consisting of a white square on a white ground. **1972** [see *RAYONISM, RAYONNISM]. **1972** *Times* 13 Apr. 4/8 A Suprematist construction of about 1916..by Ivan Puni made £3,200. **1980** I. MURDOCH *Nuns & Soldiers* i. 80 He became a cubist, then a surrealist, then a *fauve*: a futurist, a constructivist, a suprematist.

suprematist[2]: see *white supremacist* s.v. *WHITE *a.* 11 e.

supreme, *a.* and *sb.* Add: **A.** *adj.* **2. a.** *Supreme Soviet*: the national legislature of the U.S.S.R.; also, the national legislature of any of its constituent republics.

1936 *Times* 15 June 11/4 The legislative assemblies will consist of one All-Union Parliament called the 'Supreme Council (or Supreme Soviet) of the U.S.S.R.'. **1947** *Ann. Reg. 1946* 218 M. Kalinin, chairman of the Presidium of the Supreme Soviet, resigned for reasons of health. **1957** *Whitaker's Almanack 1958* 950/1 The Union Republics and Autonomous Republics have Supreme Soviets..of their own..although their jurisdiction is severely circumscribed in favour of the central Government. **1974** tr. *Sniečkus's Soviet Lithuania* 67 The Supreme Soviet of the Lithuanian SSR has approved the Five-Year Plan for the economic development of the Republic for 1971–5. **1978** *Ann. Reg. 1977* 490 Article 90 [of the Constitution of the USSR 1977]. The term of the Supreme Soviet of the USSR, the Supreme Soviets of Union Republics, and the Supreme Soviets of Autonomous Republics shall be five years.

3. a. *the supreme sacrifice*: the laying down of one's life for one's country in battle. Also *transf.*

1916 W. M. CLOW *Evangel of Strait Gate* xv. 173 These young men..have gone down not only to the horror of the battlefield but to the gates of death as they made the supreme sacrifice. **1935** J. E. C. WELLDON *Forty Years On* i. 46 Citizenship demands at times the supreme sacrifice—as it was called during the Great War of 1914–18. **1955** J. BURNABY *Christian Words & Christian Meanings* vi. 104 The 'supreme sacrifice', in the cliché of the wartime newspaper, does consist in the carrying of disregard of self to the limit. **1965** J. A. MICHENER *Source* (1966) 491 Because He [*sc.* Christ] offered Himself as the supreme sacrifice two things happened. We were saved and He ascended to Godhood. **1981** HINCHLIFF & YOUNG *Human Potential* v. 98 When one speaks of the dead of two world wars as having made the supreme sacrifice, one means that the sacrifice was made for..one's country.

B. *sb.* **5.** *supreme of chicken* = *suprême de volaille* s.v. *SUPRÊME *sb.* (*a.*) 2 a.

1939 *Vogue's Cookery Bk.* 81 Supreme of Chicken. 1 chicken 4 eggs 1½ cups cream. **1959** A. CHRISTIE *Cat among Pigeons* xiv. 154 Ann Shapland..was sitting at a table..eating Supreme of chicken. **1983** *Out of Town* Dec. 72/2 The pastry case on the Supreme of Chicken.. was a little too generous.

‖**suprême** (süprẹm), *sb.* (*a.*) [F., f. L. *suprēmus*: see SUPREME *a.* and *sb.*] **1.** A kind of sauce (see quot. 1906). Also ‖ *sauce suprême*, *suprême sauce*.

1813 L. E. UDE *French Cook* viii. 191 (*heading*) Filets of fowls sautés au suprême. **1846** A. SOYER *Gastronomic Regenerator* 342 Fillet three fowls.., sauté the same,.. sauce over with a sauce suprême. **1906** *Mrs. Beeton's Bk. Househ. Managem.* lxii. 1671 *Suprême*, a rich, delicately flavoured cream sauce, made from chicken stock, etc. **1936** LUCAS & HUME *Au Petit Cordon Bleu* 73 Pour over the following *suprême* sauce. **1948** *Good Housek. Cookery Bk.* 303 Suprême Sauce. Make as for Velouté sauce, but add up to ¼ pint of cream. **1961** *Harper's Bazaar* Feb. 72/2 There are three kinds of *roux*... Pale—for making *veloutés*, *suprême* sauces and *allemande* sauce.

2. a. In full, *suprême de volaille*: a dish consisting of breast of chicken or other poultry usu. served with a white sauce. **b.** The part of the bird used in making *suprême de volaille*.

1850 THACKERAY *Pendennis* II. i. 6 The supreme de volaille was very good. **1864** M. B. CHESNUT *Diary* 31 Jan. in C. V. Woodward *Mary Chesnut's Civil War* (1981) xxii. 551 Gumbo, ducks and olives, suprême de volaille. **1907** [see *JARDINIÈRE 2]. **1944** A. SIMON *Conc. Encycl. Gastron.* VI. Birds 111/2 The suprêmes are constituted by the meat on each side of the breast, from the point where the wing originates to the extremity of the stomach. **1975** *Times* 22 Feb. 7/2 Chicken Neptune—a suprême stuffed with prawns and butter and served with a shellfish sauce (£2.20). **1979** J. TOVEY *Entertaining with Tovey* 61 For cream soups I use a chicken..stock. Use..the bones of a bird from which you have cut the suprêmes. **1983** *Sunday Tel.* 17 Apr. 18/5 While they ladle out the mulligatawny or dish out the *suprême de volaille*.

supremo (sūprī·mo, sūprē̄ᵊ·mo), *sb.* [f. Sp. (*generalissimo*) *supremo* supreme general.] **a.** A supreme leader or ruler; one holding the highest military or political authority.

The reference in quot. 1944 is to Earl Mountbatten of Burma, whose nickname this was during his period as Supreme Allied Commander, South-East Asia (cf. quot. 1966).

1937 C. S. FORESTER *Happy Return* iv. 43 No expostulation on his part would override the orders given by el Supremo. *Ibid.* 46 'Supremo,' sighed Hernandez.. 'The captain came instantly on hearing your summons.' **1944** *Daily Express* 6 July 2/7 Why the Supremo?.. A handsome, romantic figure. Hence the Latin-sounding nickname. **1958** *Ibid.* 11 July 1/1 Now their advice and complaints can reach the Cabinet or the Prime Minister only through their 'supremo'—the chairman of the staff chiefs. **1966** E. H. COOKRIDGE *From Battenberg to Mountbatten* ix. 188 In June 1946 Lord Mountbatten's post as Supremo in South-East Asia came to an end and he returned to England. **1979** A. FOX *Threat Warning Red* ii. 21 Pat Cleary, a two-star British admiral, was the representative in Brussels of the American NATO supremo in Norfolk, Virginia.

b. *transf.* One who has overall charge of some department of government or sphere of activity.

1963 *Daily Express* 21 Oct. 1/4 Some, particularly in the Research Department, may follow the supremos into resignation. **1972** *Observer* 10 Dec. 2/7 The appointment of a Land Release Supremo with regional teams to unclog the machinery which is holding up the release of land. **1976** H. WILSON *Governance of Britain* iv. 97 The successful attack by other ministers to prevent him [*sc.* Herbert

Morrison] from becoming an economic supremo. **1983** *Private Eye* 17 June 7/1 A short list of possible replacements..included..the ruthless supremo of the Royal Philharmonic Orchestra.

supremum (s[1]uprī·mǒm). *Math.* [L., = highest, neut. of *suprēmus* (see SUPREME *a.* and *sb.*).] The smallest number that is greater than or equal to each of a given set of real numbers; an analogous quantity for a subset of any other ordered set.

1940, 1949 [see *INFIMUM]. **1968** E. T. COPSON *Metric Spaces* i. 13 An ordered field *S* is said to have the supremum property if and only if every non-empty subset of *S*..has a supremum in *S*. **1971** HADLEY & KEMP *Variational Methods in Economics* ii. 53 We now define U^* as the supremum of levels of utility which can be maintained indefinitely.

suq, var. *SOUK.

sur-, *prefix.* Add: **sura·nal** *a. Zool.* = *supra-anal* adj. s.v. SUPRA- 1 b; also as *sb.*, a suranal plate; **surhu·man** *a. Lit.* = SUPERHUMAN *a.* (cf. F. *surhumain*).

1906 J. B. SMITH *Explan. Terms Entomol.* 135 *Suranal*, supra-anal. **1925** A. D. IMMS *Gen. Textbk. Entomol.* 41 The tergum of the last segment, whatever its numerical designation may be, is frequently referred to as the suranal plate or pygidium. **1962** D. NICHOLS *Echinoderms* v. 66 In the urchin immediately after metamorphosis the whole of the aboral surface is covered by an apical disk of plates, consisting of a central suranal, through which the anus opens, a ring of five basals, [etc.]. **1933** T. E. LAWRENCE *Let.* 1 Aug. (1938) 773 He takes figures of to-day and projects their shadows on to clouds, till they grow surhuman and grotesque. **1952** E. POUND *Personae* 56 Beauty That seems to be some quivering splendour cast By the immortal nature on this quicksand And by surhuman fates.

surah. (Earlier example.)

1873 *Young Englishwoman* May 234/1 Surah is a kind of twilled Indian silk tissue, extremely soft and brilliant.

suramin (sǖ·rămin). *Pharm.* Also **Suramin.** [Etym. unknown: perh. f. SURRA.] A complex symmetric urea used in the treatment of trypanosomiasis and filariasis. Also *suramin sodium*.

1941 *Brit. Pharmacopœia 1932* Add. IV. 33 Suramin is the symmetrical urea of the sodium salt of *m*-benzoyl-*m*-amino-*p*-methylbenzoyl-1-aminonaphthalene-4:6:8-trisulphonic acid. **1951** A. GROLLMAN *Pharmacol. & Therapeutics* xx. 416 Atoxyl..was the first drug used successfully in trypanosomiasis. Tryparsamide..was an improvement and with Suramin..is effective in the treatment of the early stage of the disease before the organisms appear in the spinal fluid. **1974** *Encycl. Brit. Micropædia* IX. 687/3 Suramin sodium, a white or pinkish powder soluble in water, is administered in an aqueous or saline solution. **1978** *Nature* 22 June 627/1 Three anti-trypanosomal drugs are listed by the WHO as essential for the treatment of human sleeping sickness caused by African trypanosomiasis. They are melarsoprol (Mel B), pentamidine (Lomidine) and suramin (Antrypol, Germanin).

surbahar (sā·ɹbahāɹ). [Bengali *surbāhār*.] A mellow-toned Indian stringed instrument or esraj, larger than a sitar.

1896 S. M. TAGORE *Universal Hist. Mus.* 88 A distinguished musician Babu Kally Prosonno Banerji.. plays skilfully on the *Víná*, *Sur-báhár* and *Setár*. **1914** A. H. F. STRANGWAYS *Music of Hindostan* iii. 88 Next to the expressive *vinā* comes the dignified *surbahar*. **1927** *Grove's Dict. Mus.* (ed. 3) II. 706/1 The *Surbahar* (Calcutta) has sympathetic strings, a mellow tone, is fatiguing to play and expensive to buy. **1969** R. SHANKAR *My Music* i. 37/1 A cousin of the sitar is the large, deep-toned *surbahar*. **1979** *Radio Times* 6–12 Jan. 54/6 (*heading*) Imrat Khan plays surbahar and sitar.

surcharge, *sb.*[2] Add: **7.** *Civil Engin.* **a.** The part of a load that is above the horizontal plane containing the top of a retaining wall. **b.** A load placed upon uncompacted material to compress it.

1881 *Van Nostrand's Mag.* XXV. 336/2 The author found a wall of slag blocks having a batter of ¼ of the height, and an effective thickness of 1 foot sustained a bank of broken slag 10 feet high, with a surcharge of some 5 feet more. **1930** *Engineering* 30 May 689/3 The heavy 24-in. steel beam..was intended for applying a surcharge to the filling in the bin. **1967** C. A. O'FLAHERTY *Highways* xii. 597 A surcharge of uncompacted material is added on top [of the embankment] to accelerate the outflow of water and the compaction of the underlying compressible material.

† **surculus.** *Bot. Obs.* Pl. **surculi.** [L., = young twig, branch, shoot.] (See quots. 1775, 1849.)

1775 ASH, *Surculus*, a shoot, a sucker, a slip; a middle branch between the larger and smaller ribs of a leaf. **1826** KIRBY & SPENCE *Introd. Entomol.* III. 227 The cocoon..is fastened by one side to the roots or surculi of Typha latifolia. **1849** J. H. BALFOUR *Man. Bot.* 639 *Surculus*, a sucker proceeding from the neck of a plant, and afterwards rooting, as in the Rose.

surd, *sb.* Add: **1.** (See also quots.) (Further examples.)

1908 G. H. HARDY *Course Pure Math.* i. 7 If *a* is a rational number, the two numbers $\pm\sqrt{a}$ are either rational or irrational, and..*generally* the latter. Numbers of this kind, when irrational, are called pure quadratic surds. A number $a\pm\sqrt{b}$..is sometimes called a mixed quadratic surd. **1959** G. & R. C. JAMES *Math. Dict.* 379/2 *Surd*, a sum of one or more irrational indicated roots of numbers. Sometimes used for irrational number. **1962** H. COHN *Second Course in Number Theory* iii. 40 If *a*, *b*, *c* are integers..we define the conjugate surds $\lambda=(a+b\sqrt{D})/c$, $\lambda'=(a-b\sqrt{D})/c$.

sure, *a.* and *adv.* Add: **III. 8. f.** Colloq. phr. *don't (you) be too sure*, do not depend too confidently (upon something).

1866 MAYNE REID *Headless Horseman* iii. 16 'Don't be too sure, all of ye,' said the surly nephew. **1916** G. B. SHAW *Pygmalion* v. 189 But dont you be too sure that you have me under your feet to be trampled on and talked down. **1942** T. BAILEY *Pink Camellia* xviii. 98 Don't be too sure. You're the girl I want, and I'm going to have you.

IV. 9. c. Now often in phr. *that's for sure*, placed at the end of the sentence.

1971 C. BONINGTON *Annapurna South Face* xiii. 156 We can't do it in the next two days..—that's for sure. **1981** C. ROSS *Scaffold* 106 Well, who's telling? Not me, that's for sure.

d. *sure thing*, a certainty, a secure prospect; freq. as asseverative affirmation: Yes, indeed! Also as *attrib. phr. colloq.* (orig. *U.S.*).

1836 J. HILDRETH *Dragoon Campaigns Rocky Mts.* 24, I say, stranger, didn't I say that old 'Slow and Easy' was a sure thing, in the end? **1848** *Sporting Life* 22 Jan. 269/1 Teetotum had the call for the July in the betting, and it was booked a sure thing for her. **1896** ADE *Artie* ii. 9 You never see such a sure-thing crowd in your life. *Ibid.* xvi. 147 'Sure thing,' says he. **1908** J. M. SULLIVAN *Criminal Slang* 2 *Sure thing gambler*, character who bets with suckers at race tracks. **1933** D. L. SAYERS *Murder must Advertise* v. 91 'Should you care to make one in our next dope-raid?' 'Sure thing. When do you expect it?' **1943** *Sun* (Baltimore) 22 Apr. 18/1 Ralph Root apparently thought Overlin was a sure thing. Roberts drew his whip. **1953** W. BURROUGHS *Junkie* iii. 38, I had one of his sure-thing croakers reach for a telephone on me. **1962** P. GREGORY *Like Tigress at Bay* v. 63 'Sure thing, boss,' she said lightly. **1963** N. MARSH *Dead Water* (1964) vi. 158, I appreciate your reluctance to form a theory too soon... But..it looks a sure thing to me. **1979** C. MACLEOD *Family Vault* xxiii. 150 'Would you mind getting this box for us?' 'Sure thing, Mr Verplanck.'

10. c. *sure-fire* adj. phr., certain to succeed or attain the desired end (occas. in predicative use). *colloq.* (orig. *U.S.*). Less frequently, *sure-shot* (chiefly *U.S.*).

[**1901** 'H. MCHUGH' *Down Line* 93 Swift often told himself that he could give Marshall P. Wilder six sure-fires and beat him down to the wire.] **1909** P. G. WILLIAMS in *Sat. Even. Post* 5 June 17/2 *Sure fire*, certain of success. **1912** *Variety* 18 May 8/2 The Rev. William Sunday (Billy) the evangelist or sure-fire evangelist, has done one of the worst 'financial flops' in the history of his travels. **1914** [see *FIXER 1]. **1926** WHITEMAN & MCBRIDE *Jazz* viii. 171 In the old days, it took six months to spread even the most sure-fire song over the United States. **1933** D. L. SAYERS *Murder must Advertise* iii. 41 He thought it was a sure-fire mascot. **1941** B. SCHULBERG *What makes Sammy Run?* ii. 28 The most surefire story sale that's come to Hollywood in years. **1952** J. STEINBECK *East of Eden* 79 The preacher turned over his hole-card, the sure-fire card. **1960** G. E. EVANS *Horse in Furrow* vi. 82 You could get a *sure-shot* cigar for tuppence in those days. **1967** N. FREELING *Strike out if not Applicable* 14 He had certain surefire jokes that were repeated all over Holland. **1974** G. F. NEWMAN *Price* vii. 238 Buy into Nu-Schoenberg... They're sure-fire, I promise you. **1983** *Listener* 16 June 8/1 The search for a sure-fire hit in American network television engages thousands of minds and millions of dollars every year.

B. *adv.* **3. a.** Now also *N. Amer. colloq.* (freq. introduced between subj. and vb., as a mere intensive).

1861 *Trans. Illinois Agric. Soc.* IV. 460 Once successfully transplanted it will live sure. **1876** 'MARK TWAIN' *Tom Sawyer* iv. 83 They're coming, sure. **1908** 'YESLAH' *Tenderfoot S. Calif.* i. 14 It sure was a cold night. **1933** J. COZZENS *Cure of Flesh* i. 20 Sure, the truck came. Is anything wrong? **1953** *Manch. Guardian Weekly* 22 Jan. 7 You sure left an awful mess in Washington. **1969** A. LURIE *Real People* 106 Parts of it were pretty, sure. In a phony way, like this place. **1975** R. DAVIES *World of Wonders* (1977) I. vii. 83 You didn't need feet to fly a plane, but you sure needed brains.

c. For *dial.* read '*colloq.* and *dial.*' and add earlier example. Now chiefly *N. Amer.* in independent affirmative use.

1803 G. COLMAN *John Bull* I. 4 *Den.* Troth, and myself, Mr. Dennis Brulgruddery, was brought up to the church. *Dan.* Why, zure! **1914** WODEHOUSE *Man Upstairs* 133 'Is that a fact?' 'Sure,' murmured Archibald. **1963** MRS. L. B. JOHNSON *White House Diary* 26 Nov. (1970) 11 If it had been a request to chop off one's right hand one would have said, 'Sure'. **1975** R. STOUT *Family Affair* xi. 130 I'm under arrest. I asked if you could finish your lunch, and they said sure, no hurry.

4. a. (Further examples.) *sure as hell* (U.S. slang), most certainly; (*as*) *sure as God made little apples*, etc.: see *APPLE *sb.* 1 b.

1824 SCOTT in *Edin. Weekly Jrnl.* 9 June 181/3 As sure as ever ye sit there, She'll tell the Bailie. **1828** J. NEAL *Rachel Dyer* xix. 238 He will get away if you turn your head... That he will! if you don't look sharp, as sure as my name is Peter P. **1856** C. M. YONGE *Daisy Chain* I. ii. 13 Madam, said I, you'll have to answer for your mother's death, as sure as my name's Dick May. **1944** E. S. GARDNER *Case of Black-Eyed Blonde* xx. 194 I'm telling you just as sure as you're sitting there, that if you don't get men out to Jason Bartsler's place, a murder is going to be committed. **1976** *Listener* 6 May 562/3 Wayne ..introduces me to Commemorativo Tequilla. 'It doesn't hurt your head, but it may hurt your back, as you sure as hell fall over a lot.'

C. Comb. *sure-handed* adj.

1930 M. MEAD *Growing up in New Guinea* iii. 23 The decisive, angry gesture..has taught him to be alert and sure-handed. **1962** *Times* 26 Feb. 4/1 [The French rugby team] were able to start attacks..and, surehanded, to develop them.

surefootedly, *adv.* (Examples.)

1936 *Discovery* Aug. 242/1 Clambering sure-footedly about the larger..trunks. **1977** *Times Lit. Suppl.* 15 Apr. 449/2 Walks surefootedly through the minefield that separates fulsome idolatry from condescending anecdotal chit-chat.

surely, *adv.* **II. 4. a.** (*b*) (Later examples.)

1876 C. M. YONGE *Three Brides* II. viii. 152 'I must go. Can I?' 'Surely, as soon as there is a train.' **1922** E. RAYMOND *Tell England* II. i. 166 'Surely,' answered my companion, which was a new way he had acquired of saying 'yes'. **1975** M. RUSSELL *Murder by Mile* iii. 19 'Like to follow me along?' 'Surely.'

‖**Sûreté** (sürte). [Fr., = SURETY *sb.*, security.] In full, *Sûreté nationale*, the French police department of criminal investigation, controlled by the Ministry of the Interior. Also *transf.* of similar forces elsewhere.

Since 1966 amalgamated with the Prefecture of Police of Paris in the *Police Nationale*. The department was previously known as the *Sûreté générale* and the *Service de la Sûreté*, and was latterly also responsible for policing the provincial towns of France.

1871 *Observer* 9 Apr. 6/4 M. Ranc..was the chief of the *Sûreté Generale* under Gambetta. **1885** *Encycl. Brit.* XIX. 343/2 The *service de sûreté*, or detective department (out of uniform)..comprises a commissary, principal inspectors, brigadiers, and 211 inspectors. **1917** J. F. MACDONALD *Two Towns—One City* III. i. 192 This foreign gentleman..represents the *Sûreté* (or Criminal Investigation Department) of Paris. **1926** D. L. SAYERS *Clouds of Witness* iii. 94, I have written to the Sûreté and the Crédit Lyonnais to produce his papers. **1935** A. CHRISTIE *Death in Clouds* xi. 115 At the Sûreté Poirot renewed acquaintance with the Chief of the Detective Force. **1955** *Times* 9 May 8/3 The Binh Xuyen garrison which has been occupying the headquarters of the Sûreté in Saigon. **1963** A. ORLOV *Handbk. Intelligence & Guerrilla Warfare* xii. 143 An old Soviet informant who was an officer of the French Sûreté Générale (Secret Police). **1973** 'M. INNES' *Appleby's Answer* x. 97 The great Vidocq, who transformed the efficiency of the Paris Sûreté by..insisting that his detectives should never forget a face. **1980** R. GRAYSON *Monterant Affair* xi. 94 The hotel was..within easy walking distance of Sûreté headquarters.

surf, *sb.* Add: **3.** Simple attrib., as *surf-beach*, *-beat*, *line*; locative, as *surf-bathe* vb., *-fish* vb., *-fisherman*, *-fishing* (examples), *lifesaver*, *lifesaving*, *-rider*, *-riding* (earlier example), *-swimming* (example); **surfboard** (later examples); hence as *v. intr.*, to ride on a surfboard (also *fig.*); **surfboarder, surf-boarding** *vbl. sb.*; **surf-bum** *slang*, a surfing enthusiast who frequents beaches suitable for surf-riding; cf. *ski bum* s.v. *SKI *sb.* 2 b; **surf-casting** *vbl. sb.*, fishing by casting a line into the sea from the shore; so (as a back-formation) **surf-cast** *v. intr.*, **surf-caster**; **surf-clam** (examples); **surf day**, a day marked by rough surf along the shore (see quot. 1854); **surf-grass**, any of several species of marine grass of the genus *Phyllospadix* (family Zosteraceæ), having thickened rootstocks and slender stems and growing underwater on rocky shores in temperate regions; **surf music**, a variety of rock music which celebrates the sport of surf-riding; **surf-ride** *v. intr.* [back-formation from *surf-riding* above] = *surfboard* vb. above; also *fig.* and as *v. trans.* and *sb.*; **surf safari** = *SURFARI.

1940 V. BRITTAIN *Testament of Friendship* xii. 192 You'll look at the Rhodes Memorial and the Union Buildings..; you'll..surf-bathe at Durban.., and then you'll begin to think you know everything. **1932** N. PALMER *Talking it Over* 137 Surf-beaches of any size are rare in the world. **1966** *Weekly News* (N.Z.) 19 Jan. 11/1 Mt. Maunganui is probably one of New Zealand's best-known surf beaches. **1977** *Herald* (Melbourne) 17 Jan. 14/4 Within 16 km of Wollongong are 17 superb surf beaches. **1873** 'MARK TWAIN' *Gilded Age* lx. 543 A receding of tides, a quieting of the storm-wash to a murmurous surf-beat. **1974** *Encycl. Brit. Micropædia* IX. 689/2 Surf beat, ocean waves of uncertain origin, with the relatively long periods of 1 to 5 minutes. These low-frequency waves appear to be related to the interaction of normal wind waves and swell. Surf beat is believed responsible for the generation of seiches in bays. **1931** T. E. LAWRENCE *Let.* 14 July (1938) 729 Here is a final report..on the little surf-board target. **1934** WEBSTER, *Surfboard*, *v.i.*, to ride the surf on a surfboard.—*surfboarding*, *n.* **1938** E. HEMINGWAY *Fifth Column* (1939) III. iv. 103 Or what about Malindi where you can surfboard on the beach. **1962** *Coast to Coast 1961–62* 63 He wished he could stand up and walk away from Pammie and go out with the surfboard riders. **1962** M. McLuhan *Gutenberg Galaxy* 248 (*heading*) Heidegger surf-boards along on the electronic wave as triumphantly as Descartes rode the mechanical wave. **1953** *Pop. Mechanics* July 157 Hitching a ride on a beach-bound ocean wave with a featherlight surf-board is rated tops in water sports by practiced surf-boarders. **1969** *Britannica Bk. of Year 1968* 801/1 *Surfari*, a group of surfboarders who travel together in search of good surfing areas. **1964** *Sunday Mail Mag.* (Brisbane) 17 May 1 Surfboarding was virtually forgotten until the late 1930's. **1958** Surf-bum [see *petite *a.* 2]. **1971** Surf-bum [see *pipe-line *sb.* c]. **1975** *Country Life* 16 Jan. 131/2 Surf-cast for corvina..on a California beach and you will probably have to show your California fishing licence. **1968** 'S. Jay' *Sleepers can Kill* xxiv. 248 When you've walked through to the beach, you'll see a surfcaster, fishing by himself. **1928** *N.Y. Times* 8 Oct. 21/4 Charles Vollum of Philadelphia became surf casting champion of the United States today at the annual tournament of the Dover Fishing Club of Philadelphia. **1963** *Weekly News* (Auckland, N.Z.) 8 May 56/6 Pukehina surfcasting beach. Near Te Puke, Bay of Plenty, one of the best surfcasting beaches in New Zealand. **1979** *Angling* July 45/3 As a contemporary guide to the basics of general shorefishing it gives excellent surfcasting instruction. **1884** *Bull. U.S. Nat. Museum* No. 27. 260 Hen Clam, Surf Clam, or Sea Clam. Florida and Gulf of Mexico to Labrador. **1949** [see *skimmer *sb.* 1c]. **1978** *Times* 29 July 3/6 More than two weeks after the wreck..we saw millions of dead molluscs, urchins, razor and surf clams. **1854** G. W. Peck *Melbourne & Chincha Islands* 187 Often when the mornings are still, and the surface of the sea undisturbed by a ripple, the surf will be rolling tremendously on the narrow beaches... These are called 'surf-days', and special allowance is made for them in the charter parties of vessels loading at the islands. **1950** J. S. Learmont *Master in Sail* 60 Surf days did not count as working days. These surf days are peculiar to the northern part of the coast of Chile. **1940** O. H. P. Rodman *Handbk. Salt-Water Fishing* iii. 99 We will make a definite statement in regard to wetting down your cutty hunk line before you really start surf fishing. **1979** 'A. Blaisdell' *No Villain need Be* vii. 120 They like to surf-fish, and they claim rain..drives 'em in toward the beach. **1920** Heilner & Stick *Call of Surf* i. 5 Those great and goodly fish which so frequently take into their capacious jaws the bait of the surf fisherman. **1967** O. E. Middleton in *Coast to Coast 1965–66* 123 The surf-fishermen leaned out over the shallows. **1920** Heilner & Stick *Call of Surf* i. 4 Surf fishing is by no means a new development of the angler's art..but only of late years has it begun to achieve real popularity. **1949** S. K. Farrington *Fishing the Atlantic* iv. 82 Surf fishing at Narragansett should be a revelation. **1923** L. Abrams *Illustr. Flora Pacific States* I. 94 *Phyllospadix torreyi* S. Wats. Torrey's Surf-grass. **1981** *Sci. Amer.* Mar. 92/1 The sea grasses number 12 genera and..about 50 species... Eelgrass and surfgrass are familiar examples in temperate regions. **1977** *N.Z. Herald* 8 Jan. 1–6/4, I have been talking to the wife of a surf lifesaver, and she spoke of the apparent indifference people show on being saved from drowning. **1968** W. Warwick *Surfriding in N.Z.* 1 It is difficult to imagine how closely it was once associated with the surf lifesaving movement. **1887** O. J. Humphrey *Wreck of Rainier* 33 When the surf line was hauled tight the boat would run on the line and be kept head to the sea. **1923** H. Belloc *Sonnets & Verse* 1. 28 Above the surf-line, into the night-breeze; Eastward above the ever-whispering trees. **1965** P. L. Dixon *Compl. Bk. Surfing* 142 If the dory broaches in the surf line and turns over, bail out and get clear of oars and falling boat. **1965** *N.Z. Listener* 17 Dec. 4/1 The million-dollar industry of surf music, surf movies and surf-wear. **1977** *Sounds* 9 July 28/2 The Turtles started out playing surf music at High School hops in LA. **1953** 'S. Rattray' *Bishop in Check* 101 One-half per cent of them play tennis—or swim, or surf-ride. **1958** *Listener* 2 Oct. 494/1 This motorization wave is not something on which the rich alone can surf-ride. **1973** *Times* 1 June (Australia Suppl.) p. i/3 The Whitlam Government has been in office just six months tomorrow. For the Prime Minister it has been 'a surf ride so far'. **1976** *National Observer* (U.S.) 19 June 1/3 Now, surf-riding his victory in the California Presidential primary last week, he's racing to a showdown with Gerald Ford. **1882** *Hawaiian Almanac* 52 At one time they sent their champion surf-rider to compete with chiefs in the sport at Hawaii. **1981** L. Leamer *Assignment* iii. 43 They take this drug. They have learned this from these hippie surf-riders. **1882** *Hawaiian Almanac* 52 Among the various sports and pastimes of the ancient Hawaiians..the principal one..is that of surf-bathing, or more properly speaking, surf-riding. **1962** *Austral. Women's Weekly* 24 Oct. (Suppl.) 3/4 *Surf safari*, a trip around different beaches to find a good surf. **1858** R. M. Ballantyne *Coral Island* xxv. 305 'What sort of amusement is this surf swimming?'..'Each man..has got a short board or plank, with which he swims a mile or more to Sea, and then, gettin' on the top o' yon thunderin' breaker, they come to shore on the top of it.'

surf, *v.* Restrict *rare* to sense in Dict. and add: **2.** *intr.* To go surf-riding; to surf-ride. Also *transf.* and *fig.*

1917 *Chambers's Jrnl.* Apr. 280/2 The depth of the lagoon is trifling.., and this it is which makes surfing there so safe and enjoyable. **1932** *Ibid.* Aug. 462/2, I had snaps, too, of the children, riding or surfing, and of the whole family in their ocean-going yacht. *a* **1957** R. Campbell *Coll. Poems* (1960) III. 83 Over its surge in red tornadoes rolling My heart goes surfing on the waves of fire. **1965** *N.Z. Listener* 17 Dec. 4/5 Once a person is bitten by this surfing bug he seems to become insatiable. He surfs every day he can, the whole year round. **1970** *Motor Boat & Yachting* 16 Oct. 29/1 *La Russhe* surfed handsomely down the backs of the heavy swell and buried herself into the short steep seas on the way. **1976** M. Birmingham *Heat of Sun* iv. 51 Biriwa has..a comparatively safe beach..where you can surf when the tide is right.

3. *trans.* **a.** To ride (a boat) on the surf. **b.** To surf-ride at (a specified place).

1965 P. L. Dixon *Compl. Bk. Surfing* 18 Dories, canoes, sailing catamarans, and a few special motorboats can be surfed by experts. Where waves break far from shore and spill gradually forward. **1967** W. Murray *Sweet Ride* vi. 85 Ten years ago..no one surfed this place but him. **1968** *Surfer Mag.* Jan. 56/1 Paulo surfed a beach break off the famous Rio Copacabana called Posto Six Pier.

Hence **su·rfing** *vbl. sb.*, surf-riding, surf-boarding.

1955 A. Ross *Australia 55* xv. 214 The essential art of surfing is timing. **1959** H. Hobson *Mission House Murder* xviii. 119 When they'd had enough surfing, they brought the boards back up the beach. **1963** *Wall St. Jrnl.* 22 July 1 Surfin' music is characterized by a heavy echo guitar sound, supposed to simulate the roar of the surf. **1971** 'D. Halliday' *Dolly & Doctor Bird* i. 2 Skin diving, rum punches, calypso night-clubs, surfing, dancing, gambling.

surface, *sb.* Add: **1. b.** Also, *to scratch the surface (of)*: see *scratch *v.* 3 a.

3. f. *Aeronautics.* An aerofoil, considered as something whose intended effects arise superficially.

1843 *Mechanics' Mag.* 8 Apr. 277/2 The main surfaces..are here placed one above the other, and each pair are connected together by strong shafts. **1912** W. Wright in C. C. Turner *Romance of Aeronautics* xvii. 178 A smaller surface set at a negative angle in front of the main bearing surfaces or wings will largely counteract the effect of the fore-and-aft travel of the centre of pressure. **1930** P. H. Sumner *Marine Aircraft* ii. 104 The larger the aeroplane the larger the control surfaces become and the loads necessary to move elevators, rudder and ailerons may become too heavy for the pilot to operate their surfaces. **1974** *Encycl. Brit. Macropædia* I. 377/1 The essential components of an airplane are a wing system to sustain it in flight, tail surfaces to stabilize the wing, movable surfaces (ailerons, elevators, and rudders) to control the attitude of the machine in flight, [etc.].

g. *surface-to-air, surface-to-surface* adj. phrs., of, pertaining to, or designating a guided missile designed to be launched from the ground or at sea, and directed respectively at a target either in the air, or elsewhere on the earth's surface. Cf. *air-to-air* adj. s.v. *air *sb.*[1] B. III. 1; *ground-to-air, ground-to-ground* s.v. *ground *sb.* 17 d; SAM s.v. S. 4 a.

1950 *Sun* (Baltimore) 7 Feb. 1/2 Research continues on these surface-to-air missiles. **1951** D. C. Cooke *Jets, Rockets & Guided Missiles* 146 This is a Surface-to-Surface Missile, Air Force, Third Model, Second Modification. **1954** *Jrnl. Brit. Interplanetary Soc.* XIII. 164 One of the first G.E.C. missiles, the Hermes A-1 is a development of the German Wasserfall in a surface-to-surface rôle. **1959** *Listener* 4 June 984/3 A British Thunderbird surface-to-air guided missile. **1962** *Times* 11 Aug. 6/1 The Government yesterday made their expected announcement that Blue Water, the surface-to-surface guided missile..is to be cancelled. **1978** P. McCutchan *Blackmail North* vi. 69 Russia's been supplying Libya with a big range of surface-to-surface and surface-to-air missiles.

6. a. *attrib.* in lit. sense, as *surface film* (also *spec.* in sense (*b*)); (*a*) in reference to the surface of the ground, as *surface exploration, find* (both *Archæol.*), *worker*.

1949 W. F. Albright *Archaeol. of Palestine* iii. 49 The great increase of surface exploration in Palestine in the middle decades of the nineteenth century. **1903** *Proc. R. Soc.* LXXII. 222 The influence of the stroking is therefore limited to a very thin surface film. **1981** O. N. Bishop *Physics* xvii. 161/2 The liquid closest to the surface of the object may show adhesion with it; there is often a surface film of liquid which is carried along with the object. **1917** Surface find [see *Iroquoian *a.* and *sb.*]. **1977** *Antiquaries Jrnl.* LVII. 324 The seal-matrices..derive either from excavations or from surface-finds at known sites. **1963** *Times* 2 Mar. 8/5 The miners' demands include pensions at 50 for underground workers and at 55 for surface workers.

(*d*) *Naut.* Designating ships which move on the surface of the water as opp. to submarine vessels, as *surface craft, ship, vessel, warship*, etc.; also *Comb.*, as *surface-borne, -sailing* adjs.

1905 *Trans. Inst. Naval Archit.* XLVII. 407 Misconceptions exist..as to the relative chances of accidents happening to boats compared with surface craft. **1910** C. W. Domville-Fife *Submarines of World's Navies* ii. 101 This is if the surface warship was steaming in an erratic course. **1915** W. E. Dommett *Submarine Vessels* 5 The term 'submersible vessels' should..be reserved for those which, whilst mainly surface vessels, can be brought to an awash or submerged condition. **1928** C. F. S. Gamble *Story North Sea Air Station* xiii. 224 A pilot might sight..a submarine and a surface-borne craft like a cruiser or destroyer. **1939** *Sun* (Baltimore) 17 Apr. 9/1 The North Haven surface ship to be used in transporting supplies, will carry 314,000 separate items. **1945** *Army & Navy Jrnl.* 18 Aug. 1534 *ASV, Airborne Surface Vessel Detection*, airborne radar devices used to locate surface vessels and surfacing submarines. **1954** *Ann. Reg. 1953* 337 *The Tirpitz*..a well-documented account of the career and sinking of the war-time surface raider. **1975** *Listener* 17 June 77/2 The 'Bismarck'..was later sunk by surface vessels. **1982** A. Melville-Ross *Trigger* ii. 33 You don't have enough surface ships left for you to hoist your admiral's flag.

(*e*) In reference to (chiefly, public) transportation at ground- or sea-level, as opp. to underground or air carriage (orig. *U.S.*); cf. *surface car*, sense 6 d in Dict. Also, *spec.* applied to mail or post, as *surface letter, parcel*, etc.; cf. *surface mail*, sense 6 d below.

1906 'Mark Twain' *Let.* 5 May in C. Clemens *Mark Twain* (1932) 156 My daughters are frequently robbed by conductors on the surface lines. **1909** *N.Y. Even. Post* (Semi-weekly ed.) 4 Mar. 1 On streets leading to these ferries surface travel was blocked by heavily laden vehicles stalled. **1927** *New Republic* 12 Oct. 208/2 Chicago, alas! despite the fact that it could undoubtedly solve its transportation difficulties by surface carriage,..has decided to go in for subways. **1933** *Jrnl. R. Central Asian Soc.* Jan. 81 Surface transport conditions for the necessary stores and spares are bad. **1934** *Air Mail Service* (G.P.O. Green Paper 1), The actual cost incurred for handling, surface transmission, and air conveyance. **1951** *Overseas Air Mails* (G.P.O.) Feb. 1/2 The general regulations applicable to ordinary surface parcels..apply to air parcels. **1956** L. Zilliacus *From Pillar to Post* xiii. 163 An ordinary surface letter..takes a week or more... By air it takes two or three days. **1977** *National Observer* (U.S.) 1 Jan. 2 Adams is also a critic of several leading schemes for deregulation of airlines and surface carriers.

(*f*) *Linguistics.* Of or pertaining to the level of language at which normal communication exists, as opposed to the underlying level revealed by 'deep' semantic and syntactic analysis, esp. as *surface grammar*. See also *surface structure*, sense 6 d below.

1953 G. E. M. Anscombe tr. *L. Wittgenstein's Philos. Investigations* I. 168e In the use of words one might distinguish 'surface grammar' from 'depth grammar'. What immediately impresses itself upon us about the use of a word is the way it is used in the construction of the sentence, the part of its use..that can be taken in by the ear.—And now compare the depth grammar, say of the word 'to mean', with what its surface grammar would lead us to suspect. **1958** C. F. Hockett *Course in Mod. Linguistics* xxix. 249 This most apparent layer constitutes, we shall say, *surface grammar*. Beneath it lie various layers of *deep grammar*, which have much to do with how we speak and understand but which are still largely unexplored, in any systematic way, by grammarians. **1965** N. Chomsky *Aspects of Theory of Syntax* 199 In place of the terms 'deep structure' and 'surface structure', one might use the corresponding Humboldtian notions 'inner form' of a sentence and 'outer form' of a sentence... The terms 'depth grammar' and 'surface grammar' are familiar in modern philosophy in something roughly like the sense here intended. **1967** D. G. Hays *Introd. Computational Linguistics* viii. 155 Their system begins with a surface parser. **1972** *Language* XLVIII. 678 His general discussion of what syntax is all about deals exclusively with surface phenomena, chiefly the order of elements in a sentence. **1977** *Word 1972* XXVIII. 92 Yet the simplest solution superficially is not necessarily the best, and a surface-oriented approach to *td* predicates is faced with problems, too. **1981** A. C. Thiselton in *Believing in Church* iii. 51 Language is said to determine the scope and limits of thought on the basis of vocabulary-stock or even surface-grammar.

c. locative, as *surface-sow* vb.; *surface-sowing, -sown* (all chiefly *N.Z.*), *-swimming* adjs. and sbs.; objective, as *surface skimmer*; *surface sterilization* (hence *surface-sterilize* vb., *-sterilized* ppl. adj.).

1748 Richardson *Clarissa* III. 145, I love to plague thee, who are..a surface-skimmer in learning, with out-of-the-way words and phrases. *a* **1911** D. G. Phillips *Susan Lenox* (1917) I. xiii. 213 We shallow surface-skimmers make such a..fuss. **1921** H. Guthrie-Smith *Tutira* xix. 163 The land is surface-sown with grass and clover seed. **1882** W. D. Hay *Brighter Britain!* I. viii. 197 In spite of..the rough ground, and the mere surface-sowing, our grass will carry four sheep per acre. **1950** *N.Z. Jrnl. Agric.* Apr. 309/2 The uncertain establishment of plants from the surface sowing of clover seeds. *Ibid.* Feb. 121/1 The more fertile surface-sown hill country of the North Island. **1954** Kirk & Othmer *Encycl. Chem. Technol.* XII. 914 Ultraviolet radiation exhibits extremely low penetration. Because of this the major applications have been in air sanitation and surface sterilization of food products, packaging materials.., and working spaces. **1978** *Canad. Jrnl. Bot.* LVI. 226/1 Close the open end of the syringe with a syringe nose cap to keep the seeds and sterilizing solution from being ejected during surface sterilization. *Ibid.*, After the seed is surface sterilized, remove the nose cap and affix a sterile 18-gauge needle to the syringe. **1956** *Nature* 17 Mar. 534/2 The adult female mosquito was surface-sterilized by immersion for 2 min. in a 0·5 per cent solution of mercuric chloride in 50 per cent ethyl alcohol. The surface-sterilized insect was transferred to insect Ringer solution. **1967** K. M. Smith *Insect Virol.* ix. 165 Pupae were surface-sterilized in 70% ethanol for 5 minutes. **1978** *Canad. Jrnl. Bot.* LVI. 225 (*heading*) Rapid, contamination-free sowing of surface-sterilized seeds and spores. **1925** J. T. Jenkins *Fishes Brit. Isles* 73 The coryphænidæ are tropical and sub-tropical fish of pelagic or surface-swimming habits. **1970** *Commercial Fisheries Rev.* Apr. 4/1 The government of American Samoa seeks to broaden the islands' economic base by harvesting surface-swimming tunas.

d. **surface-active** *a.* *Physical Chem.*, (of a substance) able to affect the wetting or surface tension properties of a liquid; hence **surface-activity**; **surface blow** *Engin.*, a device by which the surface water and scum in a steam boiler may be blown off; hence **surface blow-off**, the act of discharging this scum; **surface casing** *Oil Industry*, the length of casing in a bore-hole which is nearest the surface; **surface chemistry**, the study of the chemical processes occurring at the boundaries between different phases; **surface couching** *Embroidery*, a form of couching (Couch *v.*[1] 4 b) in which the couched thread

is held flat on the surface of the fabric by stitches looped over it (cf. *underside couching* s.v. *underside b); so **surface couched** *pa. pple.*; **surface effect,** any effect associated with, or only encountered near, a surface; also *attrib.*, esp. designating an air-cushion vehicle in which the cushion is sealed by rigid sidewalls and flexible seals fore and aft (cf. *sidewall 3 b); **surface mail,** a postal service for conveying mail by land or sea, contrasted with *airmail; the mail conveyed; **surface noise,** a background hiss heard on reproduction of a gramophone record owing to irregularities in the surface of the groove walls; **surface-road** (examples); **surface shelter**: in the war of 1939–45, an air-raid shelter at ground-level; **surface speed,** the speed of which a submarine is capable when moving on the surface; **surface structure** *Linguistics* (esp. in *Generative Grammar*), the syntactic elements forming an utterance or sentence, contrasted with the 'hidden' or not immediately recognizable logical form underlying such elements (the *deep structure*: see *deep *a.* IV. c); a string of such elements arranged with labels and brackets to show the relationship of the constituent parts.

1920 *Chem. Abstr.* XIV. 3256 The changes in the surface tension brought about by acid and alkali are so slight the titration with surface active substances as indicators cannot be significantly disturbing. **1978** P. W. Atkins *Physical Chem.* viii. 240 The material described in this section is put to use in the study of surface-active agents (or surfactants). These agents include long chain molecules, such as soaps and detergents, which accumulate at the water–air interface and lower the surface tension. **1925** *Chem. Abstr.* XIX. 3094 It was proved that in certain diseases the surface activity of the urine not only deviates from its normal value in the quant. sense, but that in conditions such as *morbus maculosis* it suffers qual. changes. **1972** *Materials & Technol.* V. x. 273 (*heading*) Principles of surface-activity. **1859** W. J. M. Rankine *Man. Steam Engine* iii. iv. 453 Another blow-off cock is sometimes so placed as to discharge occasionally the scum, consisting of crystals of salt, which collects on the surface of the water: this is called the 'surface blow'. **1888** R. H. Thurston *Man. Steam Boilers* xii. 446 When using sea-water in the boilers, frequently blowing off from the bottom or a continuous discharge from the 'surface-blow' or 'scum pipes' is essential to keeping the water so fresh as not to produce deposits or incrustation. **1888** *Lockwood's Dict. Mech. Engin.* 361 *Surface blow-off*, the blowing off of the scum which collects on the top of the water in a boiler. **1977** Woodruff & Lammers *Steam-Plant Operation* (ed. 4) v. 254 Surface blowoff is advantageous in skimming or removing oil from the boiler water. **1946** L. C. Uren *Petroleum Production Engineering* (ed. 3) I. xi. 388 Some varieties of surface casing are made of galvanized sheet steel. **1977** *Offshore Engineer* Aug. 28/2 A widely used drilling programme..using 30" conductor pipe, 20" surface casing. **1926** E. K. Rideal (*title*) An introduction to surface chemistry. **1951** A. E. Alexander *Surface Chem.* p. v, The study of surface chemistry gives an unusually clear insight into the real existence and behaviour of molecules. **1975** *McGraw-Hill Yearbk. Sci. & Technol.* 174/2 Conductance monitoring of thin-film electrodes constitutes a powerful new approach to the study of surface chemistry and physics. **1938** A. G. I. Christie *Eng. Medieval Embroidery* 25 Surface Couching. The method of couching familiar to modern workers, that of securing one or more threads by passing another across them.., although well known in the Middle Ages, does not appear to have been extensively used. *Ibid.*, The medieval English embroidery, preserved in the Musée de Cluny,..is surface-couched throughout. **1963** *Opus Anglicanum* (V.&A. Mus. Exhib. Catal.) 14/1 Silver-gilt and silk thread in underside and surface couching and stem stitch. *Ibid.* 44/2 In the band with butterflies the metal threads are surface couched. **1905** R. C. H. Heck *Steam-Engine & Other Steam-Motors* I. iv. 109 The surface-effect.—Of the total interior surface of the cylinder, that part which may be called the clearance-surface—including the cylinder head and the piston-face, with the steam-passages. **1945** [see *mass effect* s.v. *mass *sb.*[2] 10 d]. **1962** *Marine Engin./Log* Oct. 72/2 A surface-effect ship is being developed under a $370,000 MarAd contract. **1979** *Canad. Jrnl. Biochem.* LVII. 106/1 This preference [for phosphatidylcholine] is manifested in the ethereal system, in which surface effects are absent. **1935** *Post Office Mag.* Jan. 2/2 1928, new services introduced and direct air or combined air and surface mails to half the countries of the world. **1946** R. Allen *Home Made Banners* xii. 156 Pop's reply was so long that it came by surface mail. **1956** *B.B.C. Handbk. 1957* 247 It is published in a surface mail edition at an annual rate of 25*s.* **1977** P. Moyes *To Kill Coconut* vi. 82 Look at the date. Three weeks old. Just arrived by surface mail. **1921** *Daily Colonist* (Victoria) 17 Mar. 7/7 The Sonora plays with a total absence of that 'surface noise' or record scratching which you had believed could not be eliminated. **1981** *Hi-Fi Answers* Sept. 87/1 The general idea was to boost high frequencies on recording, so that when an equal and opposite act was applied on play-back it cut out a lot of the surface noise. **1889** *Cent. Dict.*, Surface-road. **1903** *N.Y. Evening Post* 3 Sept. 6/4 The short-haul business is well provided for by the existing surface roads. **1940** *New Statesman* 19 Oct. 375 He is getting worried about his wife and children in their surface shelters. **1902** *Encycl. Brit.* XXXII. 576/2 With her original machinery the *Plunger* was to have had a surface speed of 15 knots. **1976** G. Cook *Silent Marauder* i. 58 The K-class steam-driven submarine..could produce a surface speed of twenty-four knots. **1964** N. Chomsky *Current Issues in Linguistic Theory* i. 10 Thus the syntactic component must provide for each sentence (actually, for each interpretation of each sentence) a semantically interpretable *deep structure* and a phonetically interpretable *surface structure*, and, in the event that these are distinct, a statement of the relation between these structures. **1969** *Neuphilologische Mitteilungen* LXX. 203 The distinction between 'surface' and 'deep' structures should be given up, since no such contrast exists: there are only structures and their meanings. **1971** *Archivum Linguisticum* II. 131, I shall use the traditional term *Article* to refer to *the, a, this*, etc. when I am characterizing them as surface-structure elements. **1975** *Ibid.* VI. 23 Even mutations which are determined by a lexical environment are not all triggered by the surface structure. **1977** E. von Glasersfeld in D. M. Rumbaugh *Language Learning by Chimpanzee* v. 103 In this context it must be said that Chomsky's introduction of the terms 'surface structure' and 'deep structure'..seemed a step in the right direction.

surface, *v.* Add: **3. b.** *fig.* To bring to public notice; *spec.* to produce or expose (a defector, spy, etc.). *U.S. colloq.*

1955 *N.Y. Times* 6 Mar. iv. 2/6 In Moscow last week the authorities 'surfaced' a brilliant British atomic scientist who had disappeared behind the Iron Curtain five years ago. **1963** J. Joesten *They call it Intelligence* i. iv. 45 Now and then secret agents are purposely 'surfaced'. **1973** *N.Y. Times* 20 May i. 64/1 Martin Tolchin, another Times reporter, surfaced one of the stories last October. **1974** *Anderson* (S. Carolina) *Independent* 23 Apr. 1B/2 Rep. Dan Marrett..surfaced the controversial issue.

4. a. (Examples of submarines and divers.) Also *fig.*

1935 *Jrnl. R. United Services Inst.* LXXX. 126 Diving and surfacing were carried out by filling or employing a number of goatskins. **1955** *Times* 18 Aug. 8/3 [The officer]..had an under-water swimming suit with breathing equipment. He failed to surface. **1959** *Listener* 9 Apr. 635/2 The *Skate* surfaced ten times during the voyage. **1965** M. Shadbolt *Among Cinders* xxvi. 276, I swam down a gloomy passage..and surfaced in a gently lighted room. **1974** L. Deighton *Spy Story* xviii. 192 Nuclear subs go faster submerged... When we surfaced they did the usual tests.

b. *fig.* Of persons: to become fully conscious or alert, esp. after sleep. Also, to come to general notice (after a period of seclusion), to appear in public view.

1959 H. Hamilton *Answer in Negative* ii. 33 He was rather silent over the meal... It was only when they had returned to the drawing-room that he really surfaced and returned to the case. **1963** *Times* 11 Jan. 3/7 He went to bed early last night and did not feel well enough to surface today. **1968** 'R. Simons' *Death on Display* xii. 180 'Has there been any sign of that damned Tebaugh woman yet?' 'Afraid not... She still hasn't surfaced.' **1971** 'A. Gilbert' *Tenant for Tomb* v. 73 If there wasn't a reason he'd have—what's the word?—surfaced before this. **1975** *New Yorker* 21 Apr. 133/1 Members of revolutionary committees that were created by the Communists over the past several years in all South Vietnamese provinces have now surfaced.

c. Similarly, of something newly presented to public attention, esp. after being concealed.

1971 *Nature* 26 Feb. 590/1 The proposal surfaced last December with the report of a panel of consultants commissioned by Senator Ralph Yarborough. **1973** *Time* 5 Feb. 51/1 The dispute soon surfaced in the press. **1978** R. Ludlum *Holcroft Covenant* xxvii. 318 She wanted me to be prepared if it ever surfaced, if anyone for any reason ever remembered and tried to use the information.

surfaced, *a.* Add: **2.** Provided with a (special) surface or surfaces. Esp. of paper treated on one side to receive a sharp printed impression. (Usu. without qualifying word.)

1888 *Paper & Printing Trades Jrnl.* Mar. 29/1 The use of highly surfaced super-calendered paper..is extending to this country. **1967** M. A. Khan *Friends not Masters* iii. 23 The only good stretch of surfaced road that existed, to my knowledge, was somewhere in Pabna district. **1971** D. Potter *Brit. Eliz. Stamps* vi. 68 All these stamps were issued on surfaced paper.

3. Of a submarine: that is afloat but not submerged.

1943 *Times* 6 Dec. 4/5 On the sixth day a Liberator attacked a surfaced U-boat near this convoy. **1974** L. Deighton *Spy Story* xviii. 193 If London's reception is poor, surfaced subs in transit monitor for them.

surfacer. Add: **1. b.** *spec.* A woodworking machine for cutting and planing the surface of wooden boards.

1884 J. Kane *Shavings & Sawdust* xviii. 81 Small surfacers..used by cabinet and piano factories..should have all the four rolls driven. **1937** H. Hjorth *Machine Woodworking* v. 139 A single planer or surfacer planes only one side of a board at a time, while a double surfacer planes both sides at the same time.

3. A paint used to smooth any slight unevennesses of a surface before another coat is applied.

1927 *Automotive Manufacturer* July 13/2 For the first coat, use primer and surfacer, half and half, spraying on a medium coat... No sanding will be needed prior to spraying the coat of surfacer. **1954** A. St. J. Masters *Do your own Spray Painting* viii. 75 For amateur use there is no doubt that the best plan is to continue with the cellulose surfacer as previously described. **1979** *Guardian* 30 May 5/3 (Advt.), The body is then given a surfacer, sprayed with a sealer and oven-hardened.

surface wave. Surface *sb.*] **a.** A wave of displacements propagated along the surface of a solid or a liquid.

1887 Ld. Rayleigh in *Proc. Lond. Math. Soc.* XVII. 11 It is not improbable that the surface waves here investigated play an important part in earthquakes, and in the collision of elastic solids. **1900** *Nature* 4 Oct. 562/1 The earthquake wave takes about 110 minutes to travel from its origin to the opposite end of the earth's diameter, but whether it is propagated through the centre of the earth or as a surface wave cannot at present be decided. **1953** H. Kolsky *Stress Waves in Solids* ii. 16 Where there is a bounding surface..elastic surface waves may also occur. These waves..are similar to gravitational surface waves in liquids. **1973** *IEEE Trans. Microwave Theory & Techniques* XXI. 176/1 Their size, design flexibility, and reproducibility make acoustic surface-wave devices excellent candidates for many important applications in radar and communication systems.

b. *Radio.* [tr. G. *oberflächenwelle* (A. Sommerfeld 1911, in *Jahrb. d. drahtl. Telegr.* IV. 166).] A radio wave propagated along the surface of the earth.

1913 [see *space wave* s.v. *space *sb.*[1] 19]. **1943** [see *ground wave* s.v. *ground *sb.* 18]. **1971** [see *sky wave* s.v. *sky *sb.*[1] 9].

surfacing, *vbl. sb.* Add: **1.** (Examples *spec.* of roads.)

1908 C. E. Morrison *Highway Engin.* iii. 46 Gravel roads seem to occupy an intermediate place between those of earth and broken stone, in the tractive force required, character of surfacing, and cost of construction. **1937** *Times* 13 Apr. p. iii/2 Much that can make for safety lies in the expert designing, surfacing and lighting of the highways. **1954** *Gloss. Highway Engin. Terms* (*B.S.I.*) 31 *Surfacing*, the top layer or layers, comprising the wearing course and/or base course but not the base.

2. (Earlier example.)

1853 E. Clacy *Lady's Visit to Gold Diggings Austral.* vi. 85 The riches of Peg Leg Gully were brought to light through the surfacing of three men with wooden legs, who were unable to sink a hole in the regular way.

3. Of a submarine: rising to the surface of the water. Also *fig.* Cf. Surface *v.* 4 a in Dict. and Suppl.

1922 *Glasgow Herald* 27 Apr. 7 Submarine H 42 rose to the surface some 30 yards right ahead of the Versatile... The reason for her surfacing was not known. **1958** *Times* 18 Dec. 11/4 Surfacing is an effective word to indicate those friends who suddenly appear on the doorstep after touch has been lost with them for months. **1970** *Wall St. Jrnl.* 6 Nov. 1/1 The surfacing of high-school students' demands is relatively recent in its own right.

surfactant (sɒ̄ɹfæ·ktănt). *Chem.* [f. initial elements of *surf*ace-*act*ive adj. s.v. *surface *sb.* 6 d + -ant[1].] A surface-active agent. Also *attrib.*

1950 *American Dyestuff Reporter* XXXIX. 379/3 A new word, Surfactants, has been coined by Antara Products, General Aniline & Film Corporation, and has been presented to the chemical industry to cover all materials that have surface activity, including wetting agents, dispersants, emulsifiers, detergents and foaming agents. **1959** *Times* 7 Dec. (Agric. Suppl.) p. vii/1 Since the war, various growth supplements have been recommended, including arsenic acid supplements, surfactants..hormones and antibiotics. **1967** *New Scientist* 28 Sept. 686/2 They propose that a process of solubilization takes place under the influence of a surfactant—a substance which lowers the surface tension of water. **1968** *Gloss. Formwork Terms* (*B.S.I.*) 25 *Surfactant* (surface-active agent, activating agent, deprecated), a chemical which lowers the surface tension of water. *Note.* Surfactants are used in mould oils to reduce the occurrence of blowholes in the concrete face. **1972** *Daily Colonist* (Victoria, B.C.) 27 Feb. 24/7 Doctors believe hyaline membrane..is an unwanted substitute for a totally or partially missing 'surfactant' membrane that in normal babies keeps the lungs from collapsing. **1979** *Enhanced Oil Recovery* (Shell Internat. Petroleum Co.) 5 Injection of water containing surfactants (soap-like chemicals) into the reservoir can lower the oil/water interfacial tension very substantially..and therefore mobilize the oil held by capillary forces.

surfacy (sɒ̄·ɹfĕsi), *a.* Also **surfacey.** [f. Surface *sb.* 1 b + -y[1].] 'On the surface', without depth; superficial.

1887 W. Armstrong in *Art Jrnl.* June 167/1 Titian fails to give the substance of flesh. His flesh is surfacy and without the variety of truth; it is, in fact, without texture. **1957** *Psychol. Rev.* LXIV. 139/1 When we are fixated upon the vase in the Rubin reversible figure, the background recedes, is less surfacy, and..seems to provide a less centrally adequate form of sensory input. **1975** *Citizen* (Ottawa) 3 Nov. 57/3 Neither he nor Dame Peggy wish to deliver glib, surfacy imitations. **1979** *Rydge's* (Sydney) Apr. 105/2, I would still be trying to select the one that had the fine edge, but I wouldn't be as surfacey as to select it on that one aspect.

surfari (sɒ̄ɹfā·ri). [Blend of Surf *sb.* and *safari *sb.*: cf. *surf safari* s.v. *surf *sb.* 3.] A journey made by surfers in search of good conditions for surfboarding; a group of surfers travelling to or around suitable beaches. Phr. *on surfari* (S. Afr.): cf. *on safari* s.v. *safari *sb.* 1 a.

1963 *Pix* 28 Sept. 63 Driving in a surfari to the beach is just one point better than travelling 'Robinson Crusoe' (alone). **1965** J. Pollard *Surfrider* ii. 21 Soon it will be too crowded, other 'surfaris', groups of surfers looking for a surf, are already on the beach. **1965** *S. Afr. Surfer* I. iii. 27 Takkies has made it to the beach at Zinkwazi

where he meets three stokies who are on surfari. **1968** *Surfer* IX. iv. 69 Ever since Endless Summer, surfers visiting Cape St. Francis have been disappointed, day after day, by fast unrideable, rock-riddled waves where Bruce found the best surf of his world surfari.

surfer (sɒ̄·ɹfəɹ). [f. Surf *sb.* or *v.* + -er[1].] One who rides a surfboard; a surfboarder. Also *fig.*

1955 A. Ross *Australia 55* xv. 214 The heads of the surfers bob over several ignored undulations. **1962** M. McLuhan *Gutenberg Galaxy* 144 (*heading*) Peter Ramus and John Dewey were the two educational *surfers* or wave-riders of antithetic period. **1966** T. Pynchon *Crying of Lot 49* vi. 147 What chance has a lonely surfer boy For the love of a surfer chick? **1970** A. Toffler *Future Shock* xiii. 255 Surfers display sores and nodules on their knees and feet as proud proof of their involvement. **1978** G. A. Sheehan *Running & Being* vi. 75 'Surfing is a spiritual experience,' says Michael Hynson, one of the world's top surfers.

surficial (sɒ̄ɹfi·ʃăl), *a.* *Geol.* [f. Surface *sb.*, after *superficial*.] Of or pertaining to the surface of the earth. Cf. Superficial *a.* 1 a.

1892 J. D. Dana in *Amer. Jrnl. Sci.* CXLIV. 166 The outflow retains a thickness of 250 feet quite to its extreme western limit, which it could not have done if it had been a subaerial, or, using a much needed new word, a surficial flow. **1926** [see *hypabyssal *a.*]. **1981** Costa & Baker *Surficial Geol.* ii. 25/2 Both surficial and bedrock geologic maps can be used to identify and classify materials, hazards, and resources.

Hence **surfi·cially** *adv.*, on the surface (esp. of the earth).

1895 J. D. Dana *Man. Geol.* (ed. 4) iv. 806 The trap was poured out surficially from fissures along the eastern margin of the area. **1918** [see *creep *v.* 10]. **1944** C. Palache et al. *Dana's Syst. Min.* (ed. 7) I. 799 The crystals..are often surficially bounded by a yellow or brown alteration shell. **1971** *Nature* 2 July 41/1 In the theory of plate tectonics convergent plate junctures are the loci of orogeny, marked surficially by arc-trench systems.

surfie (sɒ̄·ɹfi). *slang* (chiefly *Austral.*). [f. Surf *sb.* + -ie.] A surfer or surfboarding enthusiast, *spec.* characterized as one of a set of long-haired, sun-tanned young people on a beach. Also *attrib.*

1962 *Austral. Women's Weekly* 24 Oct. (Suppl.) 3/4 *Surfie*, a fond term for a good and keen surfer. **1963** *Sunday Mail* (Brisbane) 10 Nov. 23 He talk surfie talk..'cowabunga, wipe-out, I'm get stoked..yay gremmies'. **1967** *Coast to Coast 1965–66* 254 In one coffee-bar doorway stood the hoodlums..glowering at the pink and orange and green of the surfies going by. **1972** *Sunday Mail* (Brisbane) 26 Mar. 8/2 They have peddled the mushrooms to all-night trippers, located through surfie contacts. **1981** *Times Lit. Suppl.* 30 Jan. 110/5 He agrees to deliver a deal for this scruffy surfie and the plot is primed.

surfy, *a.* (Earlier example.)

1738 A. Hill *Let.* 11 May in G. Sherburn *Corr. Alexander Pope* (1956) IV. 98 The rushing of a watery sound—a kind of hollow, washy murmur, like the workings of a surfy tide.

surge, *sb.* Add: **2. c.** Also, a rapid increase in price, activity, etc., esp. over a short period.

1964 *Ann. Reg. 1963* 191 The final deficit for the fiscal year which ended on 30 June 1963 was $6,200 million, largely because of the surge in business spending and improved tax collection. **1976** *Yellowstone Explorer* July 7/2 The surge in the use of back-country areas is certainly as true here in Yellowstone as it is in other wild places across the country. **1980** *N.Y. Times* 18 Nov. B7/3 The population surge in the Sun Belt has been even greater than expected.

d. (Further examples.)

1911 W. N. Shaw *Forecasting Weather* iii. 72 The last of the charts to represent the classification of isobars..are selected to show what Abercromby calls 'surge'—that is to say, a general alteration of pressure that seems to be superposed upon the changes related to a low pressure centre. **1936** *Discovery* Sept. 289/2 It is thus possible to study the passage of 'surges' travelling along the mission line at 186,000 miles a second. **1973** *Physics Bull.* Mar. 148/3 The high voltage cathode-ray oscillograph..was used to detect lightning discharges and other electrical surges in high voltage transmission lines. **1979** *Time* 8 Jan. 80/3 That includes..keeping a weather-eye on cold surges (masses of low-temperature air moving rapidly down from Siberia).

4*. *Naut.* A rhythmic motion forward and aft that is in addition to any steady speed of the vessel.

1949 K. C. Barnaby *Basic Naval Archit.* xvii. 255 A very uneven drive, such as that given by a single-cylinder paddle-wheel engine, will cause a perceptible surge. **1968** F. N. Spiess in J. F. Brahtz *Ocean Engin.* xv. 566 Stability against horizontal oscillatory motion (surge and sway) and against roll and pitch can chiefly be achieved by providing horizontal extent comparable to or greater than a wavelength. **1977** *Offshore Engineer* May 44/3 During these tests, the data acquisition system recorded..surge, sway and yaw of the lay barge.

5. **surge chamber, tank** *Civil Engin.*, a chamber (often open to the air) connected by a T-junction to a water pipe so as to absorb surges of pressure by filling and drops in pressure by emptying; **surge voltage** *Electr.*, the peak voltage produced in a transmission line by an electrical surge.

1928 *Daily Express* 10 Oct. 12 The pent-up waters sweep through a narrow tunnel to the surge chamber of a newly built power-house, driving the turbo-generators. **1974** *Encycl. Brit. Macropædia* XVIII. 770/2 To assist regulation with long pipelines, a surge chamber is often connected to the pipeline as near as possible to the turbine, thus enabling part of the water in the pipeline to pass into the surge chamber as the turbine is closed. **1909** *Trans. Amer. Soc. Mech. Engineers* XXX. 443 'Surge tank' is a term applied to a stand pipe or storage reservoir placed at the down-stream end of a closed aqueduct to prevent undue rise of pressure in case of a sudden diminution of draft, and to furnish water quickly when the gates are opened, without having to wait for the velocity in the long feeder to pick up. **1930** *Engineering* 3 Jan. 19/2 Each divided tunnel is provided with a separate surge tank. **1975** *North Sea Background Notes* (Brit. Petroleum Co.) 27 Injection water surge tanks, filters and pumps are located on this deck. **1904** E. B. Raymond *Alternating Current Engin.* ii. 76 The surge voltage is an entirely separate phenomenon from that of resonance. **1979** C. A. Gross *Power System Analysis* iv. 118 Surge voltages provide the most stringent test and supply the rationale for the standard impulse voltage waveform.

surgent, *a.* (*sb.*) Add: **A.** *adj.* **1. c.** *Psychol.* A term used by the psychologist R. B. Cattell (b. 1905), in his factorial analysis of personality, to designate a type characterized by resourcefulness and responsiveness considered as a distinct source trait.

1933 R. B. Cattell in *Brit. Jrnl. Psychol.* Jan. 326 The essence of the temperament is expressible by some term conveying the idea of 'leaping' or 'rising up' with facility. 'Repressed' and 'Unrepressed' convey more than we are entitled to infer at present... The word 'Surgent', from the Latin *surgo*, seems most aptly to express the quality which the tests reveal. **1940** J. Bowlby *Personality & Mental Illness* v. 71 This division corresponds roughly to the division of surgent personalities into good- and bad-tempered. **1968** *Psychol. Abstr.* XLII. 68/2 Findings..gave more significance to the covariances between 'surgent' character traits and inventive factors.

Hence **su·rgency**, the attribute possessed by the surgent personality (see sense 1 c of the adj.).

1933 R. B. Cattell in *Brit. Jrnl. Psychol.* Jan. 327 The amount of [trait] 'c' possessed by any individual could be referred to as the degree of Surgency. **1940** J. Bowlby *Personality & Mental Illness* v. 70 The hypomanic personality has much surgency. **1952** *Brit. Jrnl. Psychol.* May 153 The range covered is very comprehensive, including such recent words as..narcoanalysis, surgency and tele. **1973** R. B. Cattell *Personality & Mood* i. 10 Intelligence, ego strength, surgency, and characterological anxiety..are source traits.

surgeon, *sb.* **1. b.** For 'army or the navy' read 'army, navy, or air force' and add: *surgeon-general* (earlier and later examples); also (*U.S.*), the senior medical officer of the Bureau of Public Health or similar state authority.

1706 G. Farquhar *Recruiting Officer* (ed. 2) iv. ii. 49 In short, the Operation will be perform'd with so much Dexterity, that with general Applause you will be made Surgeon General of the whole Army. **1777** *Jrnls. Continental Congress U.S.* (1907) VII. 162 There [shall] be a physician and Surgeon General with the main army. **1869** *Boyd's Business Directory* 111 Governor's Staff [N.Y. State]—..Surgeon-General, Jacob S. Mosher, of Albany. **1917** *Rep. Surgeon General, U.S. Navy* 16 The Surgeon General, as a member of the General Medicine Board, has participated in the work [for the Council of National Defense]. **1973** *Philadelphia Inquirer* (Today Suppl.) 7 Oct. 4 (Advt.), Warning: the Surgeon General has determined that cigarette smoking is dangerous to your health.

c. (Examples of fig. phr. *surgeon's knife*.)

1940 L. MacNeice *Last Ditch* 22 Here she stands who was twenty and is thirty. The same but different and he found the difference A surgeon's knife without an anaesthetic. **1962** *Daily Tel.* 13 Sept. 1/1 Sir Alexander Bustamante said that the Treaty of Rome was a 'surgeon's knife thrust into the Commonwealth body'.

3. b. **surgeon-fish,** substitute for def.: a herbivorous, tropical, marine fish of the family Acanthuridæ, distinguished by sharp spines on either side of the tail; (examples); **surgeon's knot** (see quot. 1968).

1871 *Harper's Mag.* July 191/2 The terror of all, the surgeon-fish,..boldly swims in every quarter, opening and shutting his lancet. **1931** J. R. Norman *Hist. Fishes* v. 97 The Surgeon-fishes..of tropical seas derive their name from the presence of a lancet-like spine on either side of the fleshy part of the tail. **1974** *Environmental Conservation* I. 72 (*caption*) A Surgeon-fish..is prominent on right below. **1733** *Med. Ess. & Obs. Soc. in Edin.* I. 108 By the help of a needle, or a flexible eye'd probe, the surgeons knot is made with the thread. **1945** *Ann. Surg.* CXXI. 440 The artery was secured to the tube..by a No. 3 Deknatel ligature tied tightly behind the holding ridge, using a surgeon's knot. **1968** E. Franklin *Dict. Knots* 27 *Surgeon's knot.* This is a variation of the reef knot in which an extra turn is taken at the start to help prevent the knot from tending to loosen while being completed. Used by surgeons for tying a ligature and by us [*sc.* Scouts] for parcels, etc.

surgeoncy (earlier example); **surgeoness** (earlier example).

1729 *Indenture of Apprenticeship* (Hammersmith Archives: PAF/1/272, Ref. 70), Mary Webb, daughter of John Webb, a poor child of the said parish, apprentice to Anne Saint of St. Leonard's Shoreditch in the County of Middlesex—surgeoness. **1792** *Dublin Even. Post* 18 Feb. 1/2 (Advt.), Wanted, a surgeoncy in a regiment of infantry.

surgery. Add: **1. d.** *Math.* The topological alteration of manifolds by conceptually removing a neighbourhood and replacing it by another having the same boundary; an instance of this.

1961 J. Milnor in *Proc. Symp. Pure Math.* III. 39 Given any imbedding of $S^p \times D^{q+1}$ in a manifold W of dimension $n = p+q+1$, a new manifold W' can be formed by removing the interior of $S^p \times D^{q+1}$ and replacing it by the interior of $D^{p+1} \times S^q$. This procedure will be called surgery. *Ibid.* 40 A surgery of type $(0, n+1)$ replaces W by the disjoint sum $W + S^n$. **1974** *Encycl. Brit. Macropædia* XVIII. 503/1 If M is an oriented manifold of dimension $n \geq 4$, one can, by a succession of surgeries of index 1, kill the whole fundamental group π_1 of M. **1979** M. A. Armstrong *Basic Topology* vii. 162 The result is a surface homeomorphic to the torus. A further surgery will give us the sphere.

2. a. Also, the regular session at which a doctor receives patients for consultation in his surgery.

1938 F. B. Young *Dr. Bradley Remembers* i. 1 Between six and eight..Dr. Bradley 'took' his evening surgery as usual. **1944** J. D. Carr *Till Death do Us Part* xi. 113 I've got to be back..for surgery at half-past ten. **1964** D. Francis *Nerve* v. 73 I'm late for surgery... Those pills ought to keep him quiet. **1975** 'J. Bell' *Victim* ii. 23 Dr. Swallow was dealing with his morning surgery.

b. Hence, a session at which a Member of Parliament, local councillor, etc., is available to be consulted locally by his constituents, usu. on regular occasions. Also, the room or office at which this occurs.

1951 *Hansard* (Commons) 19 Feb. 966 It is a practice of mine..to call personally upon as many of my constituents as I can, and I find that by doing this a different set of problems is presented to me from those which my post-bag or even my weekly 'surgery' bring. **1957** *Times* 22 Apr. 7/7 On the question of surgeries, they are largely a self-imposed task about which MP's cannot complain since they are so often the chosen method of getting votes at the next election. **1964** G. E. Noel *Harold Wilson & 'New Britain'* xiv. 111 As Prime Minister he intends, whenever humanly possible, to retain the system initiated in Ormskirk of personally visiting constituents who have reported problems instead of obliging them to attend 'surgeries'. **1968** *Times* 7 Nov. 11/5, I was at my 'surgery' near the hall when constituents called to complain that they could not gain admittance to the meeting. *a* **1974** R. Crossman *Diaries* (1975) I. 258, I am going to have three successive days sitting on the front bench, followed on Friday by a full day of official visitations and a surgery in Coventry. **1982** P. Turnbull *Dead Knock* vii. 126 Councillor Floyd..was holding a surgery in the Council Chambers.

c. A similar occasion when free advice is provided by lawyers, accountants, or others.

1973 *Observer* (Colour Suppl.) 18 Nov. 39/1 The law surgery..run by the Sheffield Free Legal Information Service. **1980** *Daily Tel.* 7 June 19/3 An increasing number of Citizens' Advice Bureaux have regular 'surgeries' chaired by volunteer local accountants. **1981** *Times* 4 Apr. 2/5 The Asian community is..providing census 'surgeries' for householders.

surgical, *a.* Add: **1. c.** Of garments: worn to cure, correct, or relieve an illness or deformity.

1896 *Woman's Life* 10 Oct. 200/2 (Advt.), Surgical hosiery, belts, etc. **1910** *Bradshaw's Railway Guide* May (Advt. facing p. xv), Bailey's surgical hose. **1955** W. Gaddis *Recognitions* i. i. 24 Her mother..done in by a surgical belt salesman from New York. **1974** D. Ramsay *No Cause to Kill* i. 7 Painfully swollen legs encased in surgical stockings.

d. *fig.* or in fig. contexts.

1939 C. Isherwood *Goodbye to Berlin* 68 The afternoon he came to say good-bye there was a positively surgical atmosphere in the flat. **1962** *Listener* 8 Mar. 400/2 Purchase tax—deliberately uneven and at times deliberately surgical in its effect.

e. Designating swift and precise military attack, esp. from the air. orig. *U.S.*

1965 T. C. Sorensen *Kennedy* xxiv. 684 The idea of..a so-called 'surgical' strike..had appeal to almost everyone first considering the matter, including President Kennedy. **1971** *Harper's Mag.* Nov. 55 Even the language of the bureaucracy—the diminutive 'nukes' for instruments that kill and mutilate millions of human beings, the 'surgical strike' for chasing and mowing down peasants from the air by spraying them with 8,000 bullets a minute—takes the mystery, awe, and pain out of violence. **1974** E. Newman *Strictly Speaking* ii. 63 The war in Indochina produced a host of terms that media folks accepted at their peril: protective reaction strike, surgical bombing, free-fire zone. **1978** *Guardian Weekly* 5 Mar. 9/3 Moscow might be ready to undertake a surgical strike to take out China's nuclear installations.

2. *ellipt.* as *sb.* A surgical case or ward; † a surgical operation. *colloq.*

1828 W. Sewall *Diary* 1 July (1930) 121/2 Sat off for home, accompanied by Reed's son, for the purpose of having a surgical on his foot. **1961** [see *knife *sb.* 1 f]. **1976** C. Storr *Unnatural Fathers* i. 11 I'm awfully muddled, the way surgicals and medicals are mixed up here.

surgically *adv.*: also *fig.* (earlier and later examples).

1805 J. Taylor *Let.* 25 Jan. in *Minutes of Evidence* 200 in *Parl. Papers 1809* II. 1, The lad was brought to Dublin..and was surgically rejected and dismissed before I received the letter. **1965** *Economist* 18 Sept. 1074/2 More surgically still, General de Gaulle calls for an 'interpretation' of the common market treaty which in fact violates it.

Surgicenter (sɒ̄·ɹdʒisɛ:ntəɹ). *U.S.* Also **surgicenter.** [f. Surgi(cal *a.* + Centre *sb.*] The proprietary name for a surgical unit where minor operations are performed on out-patients.

1969 Ford & Reed in *Arizona Medicine* Oct. 801/2 The building to house the Surgicenter is under construction at 1040 East McDowell Road, Phoenix. *Ibid.* 804/2 The Surgicenter..is designed to provide quality surgical care to the patient whose operation is too demanding for the doctor's office, yet not of such proportion as to require hospitalization. **1971** *Official Gaz.* (U.S. Patent Office) 15 June TM183 Surgicenter, Inc., Phoenix, Ariz... *Surgicenter.* For providing facilities for doctors to perform surgical operations on patients... First use Feb. 12, 1970. **1973** *Americana Annual* 450 A trend toward development of more outpatient or ambulatory care services by hospitals was also evident in 1972. A noteworthy development was the emergence of 'surgicenters' where minor surgery can be performed on an out-patient basis. **1977** *Washington Post* 7 Nov. C7/4 The new facility will have 16 operation rooms, plus two 'surgicenters' for patients needing minor surgery that does not require hospitalization. **1981** *National Jrnl.* (U.S. Govt. Research Corp.) 20 June 1113/1 Surgicenters and other new services and programs that will be cultivated in a pro-competitive environment.

surging, *vbl. sb.* Add: **3.** *Electr.* The occurrence of surges in a current; also, a surge.

1904 E. B. Raymond *Alternating Current Engin.* ii. 75 On underground cables, where the ratio of *l* to *c* is much lower than in overhead wires, the tendency to puncture, due to surging, is much less. **1926** R. W. Hutchinson *First Course Wireless* vii. 105 The discharge consists, not of a steady flow, but of a number of rapid oscillations or surgings of electricity to and fro. **1966** *McGraw-Hill Encycl. Sci. & Technol.* XIII. 323/2 Surging in electric circuits corresponds to overshooting.

4. *Mech.* An increased action in a valve spring of an internal-combustion engine owing to its natural frequency of oscillation coinciding with the frequency of operation of the valve.

1931 H. R. Ricardo *High-Speed Internal-Combustion Engine* (ed. 2) viii. 227 Periodic vibrations in the spring itself ('surging'). **1975** M. J. Nunney *Automotive Engine* ii. 80 To lessen any tendency towards surging within the operating speed range of the engine, the valve springs are designed to have a high natural frequency of vibration.

‖ **surimono** (surimō·no). [Jap.] Pl. unchanged; (anglicized) **-s.** A print; *spec.* a small-sized Japanese colour print used to convey greetings or to mark a special occasion.

1899 C. J. Holmes *Hokusai* 9 He..designed many *surimono*—the dainty cards used for festive occasions. *Ibid.* 15 The celebrated designer of *surimonos*. **1910** *Daily News* 16 May 4/5 It is worth while knowing what a surimono is. **1961** *Times* 7 Mar. 22/6 Two fine surimono by Kunisada. **1977** *Times* 18 July 10/3 The Japanese *surimono* was a wood-block print.

Surinam. Add: Also, in names of pidgin or creole languages spoken in Surinam, as *Surinam Negro-English, Taki-Taki*; cf. *Sranan; **Surinam cherry,** (*b*) substitute for def.: an evergreen shrub or small tree, *Eugenia uniflora*, native to tropical America; also, its edible red fruit; (examples).

1895 'F. Franceschi' *Santa Barbara Exotic Flora* 33 The Surinam Cherry..[is] growing too in Montecito. **1920** Britton & Millspaugh *Bahama Flora* 304 *Eugenia uniflora*... Native of South America. Surinam Cherry. **1972** C. D. Adams *Flowering Plants Jamaica* 522 Surinam Cherry. Shrub to 2·5 m. high;..berries red, edible. **1934** *Amer. Speech* IX. 181 (*heading*) Surinam Negro-English. **1964** Surinam Negro-English [see *Negro 1 d]. **1967** R. I. McDavid in G. V. Bobrinskoy *Lang. & Areas* 86 A viable language in its own right—like Surinam Taki-taki.

Hence **Surina·mer** [-er[1]], a native or inhabitant of Surinam.

?**1943** *Holland carries On* (Netherlands Information Bureau, N.Y.) 27/1 The Surinamers are..far from being moulded..into a real national community. **1963** H. Mitchell *Europe in Caribbean* xii. 119 Instruction is modelled on that of the Netherlands, where many Surinamers complete their studies. **1969** *Atlantic Monthly* Nov. 48/3 The border confrontation gave the Surinamers something new to talk about. **1976** *Daily Times* (Lagos) 26 Aug. 24/3 The black Surinamers (the former Dutch Guyanese in South America) could live with, and tolerate, the Indians, Japanese, Lebanese Jews, Ameri-Indians, Caribs, Arowaks and a salad of cultures over the century.

Surinamese (sʲūᵊrinæmī·z), *sb.* and *a.* [f. Surinam + -ese.] **A.** *adj.* Of or pertaining to Surinam or its people. **B.** *sb.* A native or inhabitant of Surinam; *pl.*, the people of Surinam.

?**1964** *Final Rep. Surinam–American Technical Co-operative Service* 37/2 Its objective was to supplement..training facilities provided for Surinamese technicians. **1972** *Guardian* 25 Mar. 12/6 The West Indians and Surinamese who have the luck to find a modern flat, cram it with friends and relations. **1979** *Dictionaries* I. 147 There are native speakers of Surinamese Dutch, and there is indigenous transmission of the language form from generation to generation. **1980** *Times* 18 Mar. 7/4 When it was announced that Mr Bruma would form the new Cabinet, many Surinamese were astonished.

surjection (sɒ̄ɹdʒe·kʃən). *Math.* [f. Sur-, after *injection 4*.] An onto mapping.

1964 W. J. Pervin *Found. Algebraic Topology* i. 11, *f* is a surjection or epimorphism. **1979** *Q. Jrnl. Math.* XXX. 358 The well-known surjection from tensor powers of P_{ab} to lower central factors of P.

Hence **surje·ctive** *a.*, that is a surjection.

1964 S.-T. Hu *Elem. Gen. Topology* i. 7, $f: X \to Y$ is surjective if, for every point y in Y, there exists at least one point x in X such that $f(x) = y$. **1968** D. L. Clarke *Analytical Archaeol.* ix. 360 The taxa of one aspect are related to the taxa of the other aspects as elaborate injective and surjective mappings. **1979** *Proc. London Math. Soc.* XXXVIII. 209 We recall some facts..about the abelian group $A(G)$... If F is a finite subgroup of G, then $A(G) \to A(F)$ is surjective.

surly, *a.* Add: **5.** *surly-looking* adj.

1904 W. H. Hudson *Green Mansions* vii. 97 Two dogs..They were surly-looking brutes. **1954** W. Faulkner *Fable* 141 Followed by a thin wiry surly-looking private.

surmising, *ppl. a.* (Later example.)

c **1862** E. Dickinson *Poems* (1955) I. 348 Sweeter—the Surmising Robins—Never gladdened Tree—Than a Solid Dawn.

‖ **sur place** (sür plas), *adv.* [Fr.] **1.** At the place in question; 'on the spot'.

1915 Lady R. Churchill *Let.* 21 Nov. in M. Gilbert *Winston S. Churchill* (1972) III. Compan. ii. 1284, I can understand that you want to study *sur place* this new phase of warfare. **1939** 'A. Bridge' *Four-Part Setting* xii. 157 You didn't stay and face the situation and the pain *sur place*, did you? **1976** *Listener* 28 Oct. 533/2 It is..exciting to read a regional novel *sur place*.

2. *Ballet.* Without leaving the place where one has been standing.

1930 Craske & Beaumont *Theory & Pract. Allegro in Classical Ballet* 87 *Relevez* sharply *sur place* on the *left pointe.* **1947** *Ballet Annual* I. 28 Her *fouettés* were *sur place*, and one never doubted her ability to complete the thirty-two. **1950** French & Demery *Advanced Steps in Ballet* 37 *Petits battements sautés.* These should be *sur place.*

surplus, *sb.* and *a.* Add: **A.** *sb.* **1. b.** *Polit.* In some systems of election by transferable vote: the votes which are transferred from a candidate who has attained the quota necessary for election to one who has not.

1926 Hoag & Hallett *Proportional Representation* 345 The particular ballots of a candidate to be transferred as his surplus shall be those which have received certain serial numbers. **1950** Theimer & Campbell *Encycl. World Politics* 353/1 In successive counts by the electoral officials the candidates with most preferences are elected and their surpluses over the minimum quota necessary for election transferred according to the voters' preferences until all the seats are filled. **1973** *Irish Times* 2 Mar. 1/1 This was also the first striking example of Fine Gael votes transferring to Labour: Mr. Kyne was elected on the surplus of Mr. Eddie Collins.

B. *adj.* **2.** *surplus value* (Econ., esp. in Marxism), that part of the value of the results of human labour which accrues beyond the amount needed to reproduce the initial labour power.

1816 S. T. Coleridge in D. P. Calleo *Coleridge & Idea of Modern State* (1966) i. 12 The nearest approach to the realization of such a state is a colony, composed of 100 wealthy Planters, and a 100,000 Slaves, the surplus value of whose labor above the price of the scanty food and cloathing centers in the 100. **1887** [see sense B.1 in Dict.]. **1904** W. T. Mills *Struggle for Existence* xxv. 325 Labor produces more than the cost of its own reproduction. This product of labor in excess of the labor cost of producing labor is the 'surplus value' of Karl Marx. **1933** H. G. Wells *Shape of Things to Come* i. §4.51 The entrepreneur, the capitalist, became the villain of his [*sc.* Marx's] piece, using the prior advantage of his capital to appropriate the 'surplus value' of production. **1944** G. B. Shaw *Everybody's Political What's What?* i. 1 He [*sc.* Marx] proved up to the hilt that capital in its pursuit of what he called Mehrwerth, which we translate as Surplus Value (it includes rent, interest, and commercial profit), is ruthless. **1966** T. Pynchon *Crying of Lot 49* iv. 89 How can you be against a corporation that wants a worker to waive his patent rights. That sounds like the surplus value theory to me, fella, and you sound like a Marxist. **1975** *Chinese Econ. Stud.* VIII. iv. 60 Capitalist production is commodity production aimed at reaping surplus value.

3. Of a shop: that sells goods which are surplus to (chiefly, military) requirements.

1951 R. Senhouse tr. *Colette's Last of Chéri* 208 Jean de Touzac—is in the surplus store racket. What a set! **1970** A. Fowles *Dupe Negative* xi. 140, I found a surplus store and bought a duffel bag. **1978** S. Wilson *Dealer's Move* iii. 40, I..drove down to a surplus shop in Hampstead Road, and bought a down-filled sleeping-bag.

Hence as *v. trans.* (U.S. Mil. colloq.), to dispose of (property which is surplus to requirements); also with *out*. Chiefly in *pass.*

1963 D. Broun *Egypt's Choice* (1964) i. 12 The helicopter..used to belong to the United States Marine Corps. It was surplussed out a year ago. **1968** R. West *Sk. from Vietnam* i. 18 Many were 'surplused' during the following month.

surprise, *sb.* Add: **4. b.** Also, in phr. *the surprise of one's life(time)*. Cf. *of one's life* s.v. *life *sb.* 8 a.

1927 W. E. Collinson *Contemp. Eng.* 117 The surprise of his lifetime. **1931** *Daily Express* 15 Oct. 19/3 You will have the surprise of your life.

c. As *int.*: *surprise, surprise*: an exclamation indicating surprise. Sometimes parenthetically. Freq. in irony or sarcasm.

1953 B. Glemser *Dove on his Shoulder* vi. 111 'Roger!' Miss Marsh laughed. 'Surprise! Surprise!' **1962** *Times* 24 Nov. 4/6 The plum Monday spot finally went—surprise, surprise—to our old friend *Naked City*. **1970** A. Price *Labyrinth Makers* xiv. 178 Surprise, surprise! I didn't expect to see you. **1978** I. Murdoch *Sea* 106, I gather you didn't even know Lizzie was living with Gilbert. Surprise, surprise. Everybody knew that. **1982** N. Painting *Reluctant Archer* vii. 124 At the end of the programme the identity of the 'mystery accompanist' was divulged. It was of course (surprise, surprise!) me!

5. *surprise weapon*; *surprise-free* adj.; **surprise-party** (earlier and later examples); (*b*) also, the celebration or function itself.

1968 *Listener* 20 June 791/1 Kahn and Wiener flatly deny that they're making 'predictions': they are merely sketching 'possible scenarios' for the future, based on what they call 'surprise-free projections'. **1858** H. D. Thoreau *Jrnl.* 9 Aug. (1906) XI. 86 There are also regattas and fireworks and 'surprise parties' and horse-shows. **1909** E. Nesbit *Daphne in Fitzroy St.* xvii. 272, I thought you'd like the surprise party. Was I wrong? **1969** N. W. Parsons *Sagebrush Harp* xxi. 118 A vogue for surprise parties began among the English families in our community. **1946** *Rep. Internat. Control Atomic Energy* (Dept. of State, Washington) I. 4 This danger is accentuated by the unusual characteristics of atomic bombs, namely their devastating effect as a surprise weapon, that is, a weapon secretly developed and used without warning.

surprise, *v.* Add: **5. a.** Also *colloq.* as a retort: *you'd be surprised*, the facts are not as you would think.

1926 Maines & Grant *Wise-Crack Dict.* 16/2 *You'd be surprised*, admitting entire satisfaction with results. **1948** 'J. Tey' *Franchise Affair* x. 102 'What else could she have been doing?'..Robert bit back a 'You'd be surprised!' **1964** 'E. McBain' *Ax* v. 82 'I don't think the boys would know without *your* knowing too.'..'Sometimes... You'd be surprised.' **1971** 'D. Halliday' *Dolly & Doctor Bird* ii. 19 'The emergency situation is perhaps more frequent in medicine than in portrait-painting.' 'You'd be surprised,' said the man Johnson gently.

b. *intr.* for *pass.*

1943 *Mod. Lang. Notes* LVIII. 14 They wanted to surprise me, but I don't surprise so easy. **1978** *Guardian Weekly* 1 Jan. 20/4 You don't drive cabs in Harlem if you surprise easily.

surprisingness. (Later example.)

1962 N. Streatfeild *Apple Bough* vi. 83 What never wore off was the surprisingness of Grandfather.

surra. Substitute for def. after 'countries,': caused by the flagellate *Trypanosoma evansi* and characterized by periods of increasingly severe fever and loss of weight, usually leading to death. (Earlier and later examples.)

1883 W. Robertson *Textbk. Practice Equine Med.* xi. 235 Surra may be conveniently defined as a specific blood disease of the horse. **1932** Riley & Johannsen *Med. Entomol.* xviii. 300 They [*sc.* tabanids] transfer by direct inoculation certain trypanosomes of animals, such as *Trypanosoma evansi*, which causes the highly fatal surra of horses. **1962** Gordon & Lavoipierre *Entomol. for Students of Med.* xxiv. 155 There are a number of species [of tabanid] which are vectors of diseases to domestic animals, such as surra.

surreal (sɒ̄rī·ăl), *a.* [Back-formation from *surrealism, *surrealist *a.* and *sb.* Poss. coined (as *surréel*) in Fr. Cf. *super-real* adj. s.v. *super- 4 a (*a*).] Having the qualities of surrealist art; bizarre, dreamlike. So **surrea·lity, surre·ally** *adv.*

1936 D. Gascoyne tr. *Breton's What is Surrealism?* vi. 66 As I said in the *Manifesto*: 'I believe in the future transmutation of those two seemingly contradictory states, dream and reality, into a sort of absolute reality, of surreality, so to speak.' **1937** *Burlington Mag.* Jan. p. xiv/1 Some 'surreal' influence haunts the regions of the Black Forest. **1952** *N.Y. Times Bk. Rev.* 4 May 26/3 I'll agree with that; however I didn't select the surrealism, the distortion, the intensity, as an experimental technique but because reality is surreal. **1956** *Time* 18 June 109/1 Author Gascar's power to evoke disgust, which he does by combining familiar objects in unfamiliar ways, until they become surreal and emetic. **1968** *New Yorker* 25 May 87/1 A surreally funny hour of film..which is goonish, rude, and altogether relieving. **1968** P. Oliver *Screening Blues* vi. 199 A startling flight of sexual fantasy, it [*sc.* 'Coffee Blues']..extends to surreal associations which imply the sexual virtuosity of the singer. **1974** *Encycl. Brit. Micropædia* IX. 693/2 The world of dream and fantasy would be joined to the everyday rational world in 'an absolute reality, a surreality'. **1976** S. Hynes *Auden Generation* vii. 227 As the 'thirties moved on toward the end, there was only the surreal... Even the agents of order were surreal and terrifying. **1980** J. O'Faolain *No Country for Young Men* xv. 319 Scale impresses him. He calls it 'art'. 'Surreality', if you please. **1982** *Times Lit. Suppl.* 21 May 549/1 Surreally hard-edged, the world *Child's Play* projects is one where details have a hallucinatory vividness.

surrealism (sǝri·ăliz'm). Also † in F. form ‖ **surréalisme,** and with capital initial. [ad. F. *surréalisme*, f. *sur-* super- + *réalisme* realism; the precise English equivalent would be *super-realism* (see *SUPER- 4 a (*b*)).] A movement in art and literature seeking to express the subconscious mind by any of a number of different techniques, including the irrational juxtaposition of realistic images, the creation of mysterious symbols, and automatism (q.v., sense *5); art or literature produced by or reminiscent of this movement.

The term *surréalisme*, coined by Guillaume Apollinaire (see quot. 1917), was taken over by the poet André Breton as the name of the movement, which he launched with his *Manifeste du Surréalisme* in 1924; his statement there of the term's meaning is given in quot. 1935.

[**1917** 'G. Apollinaire' *Notes to 'Parade'* in *Table Ronde* (1952) Sept. 45 De cette alliance nouvelle, car jusqu'ici les décors et les costumes d'une part, la chorégraphie d'autre part, n'avaient entre eux qu'un lien factice, il este résulté, dans 'Parade', une sorte de surréalisme.] **1927** C. Connolly *Let.* 21 Apr. in *Romantic Friendship* (1975) 294 His [sc. Brueghel's] realism with people, 'surrealisme' with places, is like Crabbe. **1931** [see *POPULISM b]. **1934** C. Lambert *Music Ho!* II. 78 Surrealism may conveniently be defined as the free grouping together of incongruous and non-associated images. **1935** D. Gascoyne tr. A. Breton in *Short Survey Surrealism* iv. 61 *Surrealism*, pure psychic automatism, by which it is intended to express, verbally, in writing, or by other means, the real process of thought. **1952** R. Bryden in *Granta* 29 Nov. 8/1 Sometimes we find that neither subject suffers from juxtaposition, but that together they form a new kind of experience to Surrealism, which we rather admire. **1970** *Oxf. Compan. Art* 1115/1 Surrealism sought to explore the frontiers of experience and to broaden the logical and matter-of-fact view of reality by fusing it with instinctual, subconscious, and dream experience in order to achieve an absolute or 'super' reality. **1978** *Amer. Scholar* Summer 357 It is clear, from what people say about contemporary surrealism.., that such poetry is supposed to be terribly mysterious, profound stuff.

surrealist (sǝri·ălist), *a.* and *sb.* Also † in F. form ‖ **surréaliste** and with capital initial. [ad. F. *surréaliste*, f. *sur-* super- + *réaliste* realist.] **A.** *adj.* Of, pertaining to, or characteristic of, surrealism. **B.** *sb.* An adherent of surrealism. Also *transf.*

The adjective was coined by Guillaume Apollinaire (see quot. 1918), perhaps (according to Robert) in the sense of F. *surnaturaliste*, and was taken over by the movement founded by André Breton (see prec.).

[**1918** 'G. Apollinaire' *Les Mamelles de Tirésias* (1946) 9 Pour caractériser mon drame, je me suis servi d'un néologisme qu'on me pardonnera car cela m'arrive rarement et j'ai forgé l'adjectif surréaliste qui..définit..une tendance de l'art.] **1918** *Egoist* Apr. 56/1 Surréaliste is the denomination M. Guillaume Apollinaire..has attached to his play, *Les Mamelles de Tirésias*... Thus he must be credited with the foundation of a successor to the *Unanimiste* and *Simultanéiste* schools. **1925** R. Fry *Let.* 1 May (1972) II. 567, I went yesterday..to see the works of the two great Sure-realist [*sic*] painters Miro and Masson. **1925** —— *Let.* 11 Nov. (1972) II. 584 That beastly young Surrealist Masson. **1929** A. Huxley *Do what you Will* i. 167 The Surréalistes..have presented us with the dream-like incoherencies which creative thought uses as its raw material. **1934** *Sun* (Baltimore) 25 Oct. 12/2 The Senator is the surrealist of politics—for surely he is above reality—or below it or to the right or left of it. **1936** D. Gascoyne *Man's Life is this Meat* (verso title-page), With the exception of Nos. 1–6, the poems in this collection are Surrealist poems. **1940** L. Trilling in *Kenyon Rev.* Spring 157 The Surrealists have, with a certain inconsistency, taken from Freud a kind of scientific sanction for their program. **1942** E. Waugh *Put out More Flags* i. 39, I should have thought an air raid was just the thing for a surrealiste..limbs and things flying about in odd places. **1958** *Sunday Times* 26 Jan. 13/4 Behind the Empress, entirely dominating her..was the surrealist figure of Rasputin. **1964** M. McLuhan *Understanding Media* (1967) ii. xvii. 180 The elders of the tribe..had never noticed that the ordinary newspaper was as frantic as a surrealist art exhibition. **1978** K. J. Dover *Greek Homosexuality* iii. 133 'Surrealist' elements are very rare in Greek art, but an exception is the 'phallos-bird' which has the legs, body and wings of a bird but a neck and head in the form of a curved penis.

Hence **surreali·stic** *a.*, characteristic or suggestive of surrealism; **surreali·stically** *adv.*

1930 *Nation* 6 Dec. 326/1 The sheer absurdity of the characters' behaviour produced a sort of surrealistic poetry. **1934** Webster, Surrealistically. **1958** *Spectator* 20 June 813/1 He produced Hauptmann's *Hannele* surrealistically as early as 1895. **1959** M. Pugh *Chancer* 36 Your eyes are so bloody bloodshot that they look like surrealistic marbles. **1979** *United States 1980/81* (Penguin Travel Guides) 129 You might hear along the way that Cleveland isn't the town it used to be, but surrealistically speaking, what place is? **1980** *N. & Q.* Dec. 505/2 As usual in De Quincey's surrealistic dream prose, several ideas coalesce in a single image.

Surrey² (sʌ·ri). The name of a county in southern England, used *attrib.* in **Surrey capon, chicken, fowl,** to designate a fowl specially fattened before being killed and prepared for cooking.

1874 L. Wright *Bk. Poultry* xxii. 319 We have often been asked to describe the large 'Surrey' or 'Sussex' fowls which are so largely reared for the London market. **1910** J. T. Brown *Encycl. Poultry* II. 459/1 'Surrey Fowls'. A trade description for the best produce of the Sussex fattening coops. **1938** [see *Light Sussex* s.v. *LIGHT *a.*¹ 3]. **1971** *Selfridge Xmas Food Catal.* 6/1 Surrey capons.. Plump full-breasted succulent birds. **1971** *Guardian* 27 Nov. 3/8 You tuck the quail into a poulet de bresse (a Surrey fowl would do).

surrogate, *sb.* (*a*). Add: **2. a.** Also as the second element of a Comb. Chiefly in *father-surrogate* s.v. *FATHER *sb.* 12, *mother-surrogate* s.v. *MOTHER *sb.*¹ 16 a.

1950 A. Huxley *Themes & Variations* 46 Not a trace of the divine or the eternal remains, and the notions of State, Nation and Party are therefore free to expand into vast and monstrous caricatures of God. In the service of this God-surrogate and of his prophet, Efficiency, totalitarian dictators find it right and proper to behave with systematic savagery. **1970** Masters & Johnson *Human Sexual Inadequacy* v. 147 *Partner surrogate* has been reserved to indicate the partner provided by the cotherapists for an unmarried man referred for treatment who has no one to provide psychological and physiological support during the acute phase of the therapy. **1979** [see *sex surrogate* s.v. *SEX *sb.* 5].

c. *spec.* A surrogate partner in sex therapy.

1975 M. Cole in S. Jacobson *Sexual Problems* 103 The use of male surrogates for the treatment of vaginismus and frigidity of various types has proved to be even more successful. **1976** T. Sharpe *Wilt* xii. 119 'I was a surrogate,' said Sally. 'A surrogate?' 'Like a sex counsellor.'

d. A woman whose pregnancy arises from the implantation in her womb of a fertilized egg or embryo from another woman.

1978 *Time* 5 June 59 The demand for surrogates remained strong... Despite potential legal problems, some have already opted for surrogate mothers. **1982** *New Scientist* 7 Oct. 16 This slippery slope would begin to steepen if the same technical procedure were to be applied to a non-donor who was not sterile but who acted as a surrogate ('foster mother') for the donor.

B. Now esp. in contexts where the substitute is intended to fulfil the emotional needs of a person. Also used in sense 2 d above.

1955 *Times Lit. Suppl.* 25 Feb. p. ix/1 Poe lived riotously in Dupin... Dupin was to him what Jim Hawkins was to Stevenson or Hadrian VII to Frederick Rolfe, a surrogate self living the life denied to the writer. **1977** *Time* 15 Aug. 50/3 In the life of the mind, Saville lives a surrogate boyhood. **1977** C. McFadden *Serial* (1978) xliii. 92/2 His Surrogate Parent for the session made him drink a lot of lemon-grass tea. **1978** Surrogate mother [see sense 2 d above]. **1979** *Sci. Amer.* June 36/3 Will this research lead..to the use of 'surrogate parents', where, for example, rich women might pay poor women to carry their children? **1979** W. Styron *Sophie's Choice* xi. 316 Sophie found herself acting as a kind of surrogate kin, a younger sister or daughter.

surround, *sb.* Add: **1.** (Earlier example.)

1825 in *N. Dakota Hist. Q.* (1929) IV. 35 The Mandans went out to kill Buffalo, by making whats called a surround, at 8 miles distant from fence.

3. The area or substance surrounding something; the vicinity, surroundings, or environment (*of* something).

1922 *Daily Mail* 11 Nov. 15/4 The inflammation often extends to the surround of the eye and to the wattles and throat (of poultry). **1937** *Nature* 3 July 11/2 Large thermal inertia in the optical parts and small and slow changes in the temperature of the surround are necessary. **1939** *Country Life* 11 Feb. p. xxii (Advt.), All types of Fencing and Tennis Court Surrounds are described in Catalogue 495. **1943** H. J. Massingham *Men of Earth* iii. 10 A country building..in relation to its matrix or surround. **1959** *Listener* 1 Jan. 13/1 It was the country, the flat agricultural surround, that so ravished me. **1962** *Which? Car Suppl.* Oct. 129/1 [There was] creaking noise from steering column surround. **1969** *Daily Tel.* 16 Sept. 14/3 Bipolar cell responses to illumination of the surround have been thought to be mediated by horizontal cells.

surround, *v.* **I. 1.** For † *Obs.* read '*Obs.* exc. *dial.*' and add later example.

1875 [illegible] ii. 17 In winter nearly all the peat-land was drowned, or as the old fen-men say 'surrounded'.

III. 6. The verb-stem in Comb., as **surround sound, surround-sound,** any of various systems of stereophony involving three or more speakers surrounding the listener so as to give a more realistic effect; esp. a four-, five- or six-speaker system employing signal matrixing, with the aim of reproducing the original front-to-back, floor-to-ceiling, and side-to-side sound distribution. Also *attrib.*

1969 *High Fidelity Mag.* Sept. 63/1 Vanguard's initial offering in what it has termed 'Surround Sound' will include the Berlioz Requiem, which calls for four brass bands to be spread around the cardinal points of the hall. **1974** *Nature* 13 Dec. 535/2 The present upsurge of interest in surround-sound was in some measure triggered by engineers and producers playing back such four-track material directly into four amplifiers and loudspeakers distributed approximately in a square near to the corners of the monitor room. **1978** *Broadcast* 6 Mar. 18/3 In radio, engineers are experimenting with surround sound systems as the next step forward from stereophonic sound. **1981** *Hi-Fi Answers* May 58/1 Efforts were made in the mid seventies to market a system of surround sound which went by the name of quadraphonics. **1983** *Listener* 19 May 34/1 It can create a remarkable surround-sound effect.

sursassite (sɜ·rsăsait). *Min.* [ad. G. *sursassit* (J. Jakob 1926, in *Schweiz. Min. und Petrogr. Mitt.* VI. 376), f. *Sursass*, name of the Oberhalbstein region in the Rhaeto-Romance dialect: see -ITE¹.] A hydrated silicate of manganese and aluminium, found as tufts and radial aggregates of reddish brown or yellow monoclinic crystals.

1928 *Chem. Abstr.* XXII. 45 (*heading*) Sursassite, a manganese silicate from Val d'Err. **1964** *Amer. Mineralogist* XLIX. 168 Various chemical formulae have been proposed for sursassite. **1973** *Mineral. Rec.* IV. 290/1 Recently braunite, and sursassite..have been found in Palos Verdes Hills, Los Angeles County, California... The only other occurrence of sursassite known in North America is in New Brunswick..Maine.

sursum- (sɜ·rsəm-), formative element [f. L. *sursum* from below, up] used in terms in *Ophthalm.*, as **sursumdu·ction** [a. F. *sursumduction* (G. T. Stevens 1886, in *Arch. d'Ophtalm.* VI. 545): see DUCTION], vertical movement upwards of one eye alone; the degree to which this action occurs; **sursumve·rgence** [L. *vergentia* (f. *vergere* to bend, turn): see -ENCE], the simultaneous movement of one eye upwards and the other downwards, classified as *left* or *right* according to which eye moves upward; the degree to which this motion occurs; **sursumve·rsion** [L. *version-em*, f. *vertere* to turn: see -ION], the parallel upward movement of both eyes.

1893 G. E. de Schweinitz *Dis. Eye* ii. 76 Sursumduction, or the power of uniting the image of the candle flame, seen through a prism placed with its base downward before one eye, with the image of the same object as seen by the other eye, is ascertained by beginning the trial with a weak prism..and gradually increasing its strength. **1949** W. S. Duke-Elder *Textbk. Ophthalm.* IV. xlv. 3814 Depending on whether the [eye] movement is in, out, up or down, the terms adduction, abduction, supraduction (sursumduction) and infraduction (deorsumduction) are employed. **1975** M. M. Parks *Ocular Motility & Strabismus* xviii. 129/1 Dissociated double hyperdeviation is synonymous with alternating sursumduction which describes the spontaneous elevation of each eye, as the cover-uncover test is performed. **1897** A. Duane *New Classification of Motor Anomalies of Eye* 38 The sursumvergence, i.e., the amount by which the eyes can diverge in a vertical plane, is determined by the strength of prism placed up or down before the eyes, which the latter can overcome when looking at a distant object. **1962** H. W. Brown in G. M. Haik *Strabismus* (Symposium N. Orleans Acad. Ophthalm.) xii. 243 The normal limits of sursumvergence are not well known. **1974** Burian & von Noorden *Binocular Vision & Ocular Motility* xii. 207/2 In some tests the normal limits..are given as..3Δ to 4Δ for sursumvergence and deorsumvergence. **1897** A. Duane *New Classification of Motor Anomalies of Eye* 68 Paralysis..may be had by assuming a weakness of deorsumversion in the former case and of sursumversion in the latter. **1975** M. M. Parks *Ocular Motility & Strabismus* ii. 14/2 Vertical versions are supraversion (sursumversion) and infraversion (deorsumversion).

Sursum corda (sɜ·rsəm kǭ·rdă). [L. *sursum* upwards + *corda*, pl. of *cor* heart.] In Latin Eucharistic liturgies, the words addressed by the celebrant to the congregation at the beginning of the Eucharistic Prayer; in English rites, the corresponding versicle, 'Lift up your hearts'. Also *transf.* and *attrib.*

1559 T. Becon *Displaying of Popish Masse* in *Works* (1563) III. 41b Before it was *Sursum Corda*, Lift vp your hearts vnto the Lord, but now is *Sursum Cauda*, come on, lift vp your heeles. **1744** [see *ANAPHORA 3]. **1837** J. Romilly *Diary* 2 Nov. (1967) 136 Cricket made a long dull oration ending with 'Sursum corda'. **1889** H. M. Luckock *Div. Liturgy* xii. 176 The Gallican was almost alone among the ancient Liturgies in placing the prayers for the Church before the *Sursum Corda* ['Lift up your hearts'], which commenced the more sacred part, the Anaphora in the East, the Canon in the West. **1917** *Daily Chron.* 4 July 2/6 A fine speech ended finely on the sursum corda note. **1934** S. Beckett *More Pricks than Kicks* 31 That..is where I have sursum corda. **1955** W. Gaddis *Recognitions* II. i. 332 This called upon, he took courage: the sursum corda of an extravagant belch straightened him upright. **1971** N. Freeling *Over High Side* i. 6 Oranges..smelt, like everything else, of plastic... Sursum corda, thought Van der Valk; get up off the floor.

surtax, *sb.* Add: *spec.*, an additional income tax at higher rates charged on personal incomes above a certain value; = *super-tax* sb. s.v. SUPER- 13 in Dict. and Suppl. Also *attrib.*

A surtax on personal income was introduced in the U.S. in 1913. In the U.K. the designation *surtax* officially replaced *super-tax* in 1929; this tax was abolished in 1973 when a new graduated system of income tax was established.

1916 *Yale Law Jrnl.* Apr. 427 The Tariff Act of 1913.. provides for levying, assessing and collecting an additional income tax. This additional tax is commonly known as a 'surtax'. **1927** *Rep. Comm. Nat. Debt & Taxation* 416 in *Parl. Papers* (Cmd. 2800) XI. 371 In view..of the already complicated character of the present Income Tax and Super-tax,..we think it might be found convenient to raise the additional revenue..by the introduction of a special graduated Sur-tax applicable to investment income alone. **1949** *Economist* 20 Apr. 718/2 The average surtax-payer will have about £2,650 left. **1954** *U.S. News & World Rep.* 19 Mar. 102/3 Other changes..are provided in the massive tax bill. The normal tax and the surtax on personal income are combined. **1970** *Money Which?* Mar. 4/1 The Surtax Office will work out how much surtax you have to pay, on the basis of the information you give your Tax Inspector for income tax. **1978** *Daily Tel.* 12 May 2/4 The amendment..raised the point at which 'surtax' starts from £7,000 to £8,000 of taxable income.

surtax, *v.* (Later examples in spec. sense of the sb.)

1934 G. B. Shaw *On Rocks* ii. 237, I shall get three and a half per cent..and on that..I shall be income-taxed and surtaxed. **1950** —— *Farfetched Fables* 96 To substitute cost-of-production prices..for prices loaded with enormous rents for the proprietors of London and [illegible] mines, not equivalently surtaxed.

surucucu (surucu·ku). Also **sirocucu, surukuku.** [a. Tupi *surucucú*.] A large, venomous pit viper, *Lachesis muta*, native to tropical America and distinguished by black bands and blotches on a reddish-yellow skin; = *bush-master* s.v. BUSH *sb.*¹ 11.

1845 *Encycl. Metrop.* XXV. 775/2 Surukuku..probably the Boschmeester, or Couesmaster, of the inhabitants of Surinam. **1910** R. L. Ditmars *Reptiles of World* iv. 339 This terrible creature is known under several titles—the Surucucu, the Mapepire and the Bushmaster. **1967** *Times* 23 Nov. 10/7 The Indian girl..heard a surucucu coming through the undergrowth. It's the biggest poisonous snake in Brazil, and really very dangerous because it's aggressive.

surveil (sɜrvei·l). Also **surveille.** [Back-formation from SURVEILLANCE.] *trans.* To exercise surveillance over (someone), subject (someone) to surveillance. Also with a place or area as obj., and *absol.* Hence **survei·lled** *ppl. a.*, **survei·lling** *vbl. sb.*

1960 *Federal Suppl.* (U.S.) CLXXXII. 750/1 The plaintiff also stresses that the store as a whole, and the customer exits especially, were closely surveilled. **1966** *Harper's Mag.* Oct. 37/1 [illegible] **1966** *Guardian* 6 Aug. 4/1 At this same time before I was being surveilled. **1969** *New Scientist* 10 July 101/1 Night surveilling systems for railway marshalling yards. **1975** R. F. Conners *Don't embarrass Bureau* ii. 123 'You'll have to conduct the surveillance'. 'I'm supposed to surveil her?' 'That's right'. **1975** [illegible] 'Where the hell is everybody?'.. 'Surveilling. Big emergency.' **1980** N. Freeling *Castang's City* xvi. 111 A few hints are conveyed by the word 'light'. Not around twenty-four hours; that's 'intense' and needs three separate shifts.. Light means not leaning on people: the surveilled aren't supposed to notice.

surveillance. Add: With pronunc. (sɜrvei·(y)ăns) (cf. the notes s.v. SURVEILLANCE). **b.** *attrib.*, esp. of devices, vessels, etc., used in military or police surveillance.

1947 *Aviation Week* [illegible] It recommended that surveillance radar be developed as an adjunct to airport traffic control... Surveillance radar could be used to control traffic [illegible]. **1958** [illegible] BMEWS will have a long-range surveillance radar system [illegible]. **1962** *Times* 24 July 10/6 New methods of detection by surveillance drones, airborne and ground radar, [etc.]. **1966** Woodhouse *Free Fire* viii. [illegible] **1975** D. Pitts *Target Manhattan* (1976) ii. 26 He switched on the surveillance cameras and looked at the street. **1976** *Honolulu Star-Bull.* 21 Dec. A-6/1 He was on a surveillance team which saw Scanlan and Maiava meet with the informer on different occasions. **1980** *Globe & Laurel* July/Aug. 199/2 Its initial appearance took E Coy—on surveillance duty—completely by surprise.

survey, *sb.* Add: **5. b.** A systematic collection and analysis of data relating to the attitudes, living conditions, opinions, etc., of a population, usu. taken from a representative sample of the latter; freq. = *POLL *sb.*¹ 7 d. Also preceded by a defining word, as (*public*) *opinion survey*, *social survey*: see under the first element in Suppl.

1927 [see *social survey* s.v. *SOCIAL *a.* 12]. **1935** *Fortune* July 65 (*heading*) Fortune applies to factual journalism the technique of the commercial survey. *Ibid.* 66/2 Fortune will present the results of independent surveys of national scope scientifically conducted. *Ibid.* 66/1 It seems obvious that the survey technique is not only as well adapted to journalistic use as to other uses but considerably better adapted. **1959** J. W. Krutch *Human Nature & Human Condition* vii. 127 One survey made by the Gallup Poll may reveal that 61 per cent of all adults could not remember having read one book during the year just passed. **1965** M. Frayn *Tin Men* xiii. 69 The crash survey showed that people were not interested in reading about road crashes unless there were at least ten dead. **1969** *Times* 7 Jan. 8/6 Both science and arts students believe in magic to an equal extent, according to a survey carried out at Ghana University. **1979** [see *SAMPLE *sb.* 7 d].

6. (Later examples.) Also in senses 4 and *5 b. **survey course** *U.S.*, an introductory academic course in which the significant features of a wide subject area are studied.

1911 *Daily Colonist* (Victoria, B.C.) 23 Apr. 14/2 An advertisement was published yesterday..calling for tenders for the purchase of the old survey ship, which formerly served in the war fleets. **1930** L. G. D. Acland *Early Canterbury Runs* 1st Ser. ix. 224 This was in 1852 when a few wooden buildings..were all there was of Christchurch, except survey pegs. **1941** C. Fadiman *Reading I've Liked* (1946) p. xxii, My brother, five years my senior and a student at Columbia College, was at the time taking a conventional survey course that used a sound standard anthology. **1951** M. McLuhan *Mech. Bride* (1967) 47/1 Survey techniques inevitably throw up images of normalcy. **1964** P. Meadows in I. L. Horowitz *New Sociol.* 450 Others exploited survey-questionnaire methods. Indeed, industrial society became in the 'thirties the land of the Gallup Poll—'Galluputia'. **1967** M. Argyle *Psychol. Interpersonal Behaviour* ix. 151 In fact the reliability of survey interviews is not very high. **1978** *N. & Q.* Feb. 82/1 Brief or survey treatment of major authors.

survey, *v.* Add: **1.** *spec.*, to examine the condition of a property on behalf of its prospective buyer.

1866 Geo. Eliot *Let.* 5 Sept. (1954) III. 341 It is a better house than we have..; moreover, the place must be surveyed by a builder before we can come to a final decision.

5. Also (sɜ·rvei). To carry out a survey (sense *5 b) (of a group of people, or its beliefs, living conditions, etc.).

1953 Pohl & Kornbluth *Space Merchants* (1955) iii. 34 Survey the book-buyers, the repeat-viewers. **1958** M. Argyle *Relig. Behaviour* iv. 51 Beliefs have also been repeatedly surveyed by one or two investigators. *Ibid.* vi. 63 Kinsey and Associates [illegible] surveyed over 500 children grouped around the ages of 12, 15 and 18.

survey line. Also with hyphen. [f. SURVEY *sb.* + LINE *sb.*²] **a.** A line along which the measurements and observations are made in a survey.

1889 G. W. Usill *Pract. Surveying* v. 139 The accuracy of a survey..will be assured by arranging the survey-lines so that the offsets shall be as short as possible. **1930** S. W. Perrott *Surveying for Schools* i. 3 It frequently happens that the principal lines of a survey..do not form triangles. **1981** J. Pettit *Site Surveying & Levelling* 9 Measurements can then be carried out between these points, or from the survey lines joining them, to complete the survey.

b. *Dentistry.* A line scribed on a cast of a tooth marking the place of greatest diameter with respect to the chosen line of insertion of the denture.

1949 V. R. Trapozzano *Comprehensive Rev. Dentistry* xx. 539 Draw the labial of a mandibular molar and indicate a typical survey line. **1954** Osborne & Lammie *Partial Dentures* vii. 86 If a carbon marking rod is substituted for the vertical plane and a tooth is rotated about the vertical plane, then an actual line will be produced at the level of the maximum tooth bulge. This is known as the survey line. **1980** R. W. Blakeslee et al. *Dental Technol.* xi. 271 The resultant survey line shows those hard and soft tissues over which the removable partial denture must pass when it is placed and withdrawn by the patient. The survey line also shows the height of contour of each tooth.

surveyor. Add: **6.** *Dentistry.* An instrument used to survey the casts of teeth, esp. to determine parallelism between surfaces on different teeth.

1928 W. E. Cummer in Turner & Anthony *Amer. Textb. Prosthetic Dentistry* (ed. 5) xx. 326 [illegible] **1980** R. W. Blakeslee et al. *Dental Technol.* xi. 271 The surveyor is a platform to which an adjustable vertical tool holder is attached so that it is perpendicular to the platform.

surview, *sb.* **3.** Delete *arch.* and add later examples.

1958 *Medical World* LXXXIX. 9 (*heading*) Surview of the National Health Service 1948–58. **1961** K. Tynan *Curtains* i. 118 Mr. Dallas' play, a scathing surview of the Trojan War, is acted with notable assurance. **1977** *Times Lit. Suppl.* 23 Dec. 1508/1 Dickens's imaginative achievement in creating a whole surview of the chaos that can be unleashed in a riot is masterly.

survivability. (In Dict. s.v. SURVIVABLE *a.*) Add: (Later examples.) Now esp., ability to survive military attack.

1964 *Financial Times* (Defence Survey) 23 Mar. 21/4 The solid fuel missile..which, when widely dispersed in underground silos.., offers reasonable survivability against any first strike. **1972** *Sci. Amer.* July 14/2 Methods of anti-submarine warfare that might eventually threaten the survivability of missile-launching submarines. **1976** *Ibid.* July 64/1 (Advt.), Computer-aided design is used to model..helicopter 'survivability' under the most turbulent conditions. **1980** D. Bloodworth *Trapdoor* xi. 61 This..Airborne Command Post is designed to improve communications and to increase survivability in case of sudden nuclear attack. **1981** *Times* 28 Feb. 15/3, I suggest that..the overriding problem is the flammability of aircraft fuel. If this can be reduced, the survivability problems will diminish.

survivable, *a.* Add: **1.** (Later examples.)

1973 *Washington Post* 13 Jan. A23/3 The only survivable..nuclear force. **1982** *Daily Tel.* 12 Nov. 5/1 The sinking of the destroyer Sheffield, 3,500 tons, and of the Atlantic Conveyor, 14,946 tons, by Exocet missiles have given us some evidence 'that large surface ships are not survivable, or at least not in a cost effective manner'.

2. Capable of being survived (esp. of an accident); not fatal.

1961 in Webster. **1967** *Times Rev. Industry* Feb. 38/3 The attitude to safety in survivable accidents..while officially condoned, is indefensible. **1981** *Brit. Med. Jrnl.* 10 Oct. 963/1 The suggestion that a nuclear war may be survivable. **1982** *Observer* 14 Mar. 5/1 The report..published by the United States Transportation Safety Board in Washington..defines a 'survivable' accident as one in which the forces exerted on passengers do not exceed the limits of human tolerance and in which the aircraft structure remains largely intact.

survival. Add: **3.** Also, used *spec.* in *Anthrop.* with ref. to a theory that from such surviving customs and observances the earlier stages in the evolution of a culture can be reconstructed.

1871 E. B. Tylor in *Proc. R. Inst.* VI. 91 [illegible] **1873** —— *Primitive Culture* (ed. 2) I. i. 16 Among evidences aiding us to trace the course which the civilization of the world has actually followed, is that great class of facts to denote which I have found it convenient to introduce the term 'survivals'. **1920** R. R. Marett *Psychol. & Folk-lore* v. 99 Folk-lore, usually defined as the study of survivals, needs to conceive its object in a dynamic, not a static way. **1937** R. H. Lowie *Hist. Ethnol. Theory* v. 41 Applying the principle of survivals, the author interprets mythological references to matriarchy..as relics of a one-time gynaecocracy. **1944** B. Malinowski *Sci. Theory of Culture* iii. 29 The real harm done by the concept of survivals in anthropology consists in that it functions on the one hand as a spurious methodological device in the reconstruction of evolutionary series; and, worse than that, it is an effective means of short-circuiting observation in field-work. **1969** L. Mair *Introd. Social Anthrop.* ii. 26 Rivers was the last British field anthropologist to interpret usages that he found puzzling as survivals of an earlier stage of society.

4. *attrib.* and *Comb.*, as *survival capsule, car, course, kit, machine, pack, rate, skill, suit, training*; **survival bag,** a large plastic bag used by climbers as a protection against exposure; **survival curve,** a curve showing how the number of survivors varies with the size of a radiation dose or with the length of time after a dose; **survival time** *Biol.*, the time for which a biological system survives after a given dose of a chemical or ionizing radiation; **survival value** *Biol.*, the property of any heritable or other character that renders the individuals possessing it more likely to survive and reproduce; also *transf.*; also, the ability to survive.

1971 *Guardian* 22 Feb. 10/2 Cheap, light plastic or plasticised 'survival bags' can be bought for a few shillings. **1977** *Navy News* Aug. 21/3 The party spent the night practically underwater in polythene survival bags. **1960** *Britannica Bk. of Year* 557/2 The three *survival capsule* was used to mean the pilot's detachable compartment on a rocket. **1962** *Amer. Speech* XXXVII. 273 *Survival car*, a traffic patrol car equipped with all sorts of strapping and cushioning devices to insure survival of the driver in case of a high-speed collision or roll-over. **1961** D. Hoyt *Score* (1962) i. 2 Sometimes I wonder why my high school didn't give me a course in how to take tests. They said it would be a survival course. **1936** E. C. Smith in B. M. Duggar *Biol. Effects of Radiation* II. xxxvi. 893 It has already been mentioned that deviations from the logarithmic type in the S-shaped survival curves have been attributed by many to other varying factors. **1947** *Radiology* XLIX. 322/2 Since the survival curves for the two radiations [illegible]. **1980** *Genetics* XCV. 281 After UV treatment, [mutant] pso1-1 in stationary phase is very sensitive and demonstrates an exponential survival curve. **1944** *Yank* 21 July 21/1 A plastic-boxed survival kit (fishhooks, dextrose tablets, first-aid materials and other stuff). **1962** D. Slayton in *Into Orbit* 24 He would have a survival kit attached to his seat which included a mirror he could use to signal airplanes overhead, some packages of shark repellant and a line for catching fish. **1973** *Times* 17 May 12/6 Compiled a kind of survival-kit beginning with instructions on how to write out a cheque. **1976** R. Dawkins *Selfish Gene* ii. 21 The replicators which survived were the ones which built survival machines for themselves to live in. **1970** 'B. Mather' *Break in Line* ix. 116 Compressed rations that had probably been stolen from American Air Force survival packs. **1953** E. Smith *Guide to Eng. Traditions & Public Life* 240 The increase of population was largely due not so much to a higher birth-rate as to a higher survival-rate. **1976** *National Observer* (U.S.) 6 Nov. 17/2 Such 'survival skills' as filling out a job application and using a telephone book. **1980** *Christian Sci. Monitor* (Midwestern ed.) 4 Dec. B32/1 Survival suits and inflatable life rafts must now be provided by the shipping companies. **1947** *Radiology* XLIX. 359/1 Survival time, which was one of the most sensitive responses, showed effects following daily exposures in the range of 0·1n of fast neutrons and 1r of gamma rays. **1980** *Amer. Jrnl. Hematol.* VIII. 290 The gamma model is so far the best among the nine recommended methods for calculating the mean survival time in ⁵¹Cr-labeled platelet survival study. **1972** *National Observer* (U.S.) 27 May 1/1 Bondurant's school is one of a handful that offer this 'survival' training. **1912** J. S. Huxley *Individual in Animal Kingdom* i. 16 This..view of the individual, as a whole whose diverse parts all work together in such a way as to ensure the whole's continuance, or, as the evolutionist would say, whose structure and working have 'survival-value', cannot stand without some qualification. **1924** J. A. Thomson *Sci. Old & New* xlvii. 280 The notable musical talent of birds..has its survival-value in connection with mating and as an expression of very vital emotion. **1944** A. L. Rowse *Eng. Spirit* xvii. 142 The survival-value of the College must be rated extraordinarily high. **1965** J. D. Chambers in Glass & Eversley *Population in Hist.* xiii. 313 The survival-value of the small man under the impact of enclosures should not be under-estimated. **1966** *Listener* 17 Mar. 385/2 This behaviour has great survival value. So long as the troop sticks together, the prospects of a predator getting a meal are slim.

survivalism (sǝrvəi·văliz'm). *rare.* [f. SURVIVAL + -ISM.] **1.** A theory of survival (see *SURVIVAL 3).

1892 F. W. Maitland *Let.* 4 Sept. (1965) 104, I am putting into the L.Q.R. a protest against Mr. Gomme's 'survivalism'.

2. A policy of trying to ensure one's own survival or that of one's social or national group.

1952 *Round Table* Dec. 26 The persistence of the unceasing attack on 'survivalism'..argues that the Soviet critics recognize..the existence of a national pride. **1953** O. Caroe *Soviet Empire* xiii. 223 The minstrels sang of heroes of the resistance, and their original work is therefore banned as tainted with 'survivalism'. **1982** *New Musical Express* 30 Oct. 19/1 The survivalism of Jamaica's sufferers.

survivalist. For *nonce-wd.* read *rare* and add: **1.** (Later examples with ref. to *SURVIVAL 3.)

1893 F. W. Maitland in *Law Q. Rev.* IX. 44 Had the manner in which Coton Field was occupied in 1835 been brought to the notice of some of our 'survivalists', they would have pronounced it to be an interesting relic of archaic times. **1968** *Encycl. Brit.* IX. 519/2 The controversy between the diffusionists, who believed culture contacts to be the main explanation of peasant beliefs and customs, and the survivalists, who attributed them to the processes of folk memory and to oral tradition handed down through the ages.

2. One who succeeds in surviving; one who makes a policy of aiming to survive. Also *attrib.*

1922 *Glasgow Herald* 6 May 9/2 They cannot deprive the London Scot of his reputation as a dancer and his fame as the survivalist of an institution so noteworthy as the Royal Caledonian Ball. **1953** O. Caroe *Soviet Empire* xi. 177 That, however, is what Communists would nowadays call a feudal survivalist deviation. **1978** *Time* 17 Apr. 2/2, I admire politicians... They're the best of the survivalists. **1980** *Times Lit. Suppl.* 25 Apr. 476/2 As historians come to appreciate the strength of 'survivalist' Catholicism, so Protestantism seems to be less and less a pre-ordained and natural consummation, England's manifest destiny.

survivant, *a.* For † *Obs.* read *rare* and add later example.

1934 F. Scott Fitzgerald *Tender is Night* I. xix. 107 But they were frightened at his survivant will.

survive, *v.* Add: **3.** *intr.* and *trans.* In trivial use. Freq. in phr. *I'll survive.*

1902 Kipling *Traffics & Discoveries* (1904) 30 'But it'll bore you to death,' he says... 'I'll survive,' I says, 'I ain't British. I can think,' I says. **1928** M. Arlen *Lily Christine* xiii. 240 'All this trouble your silly husband has brought on you!' 'Oh, we'll survive that,' she said lightly. **1949** 'J. Tey' *Brat Farrar* xxxi. 278 The fact that we are making him part of the family..will take a lot of the fun out of it for the scandal-mongers. We'll survive, Nell. And so will he. **1958** C. S. Forester *Hornblower in W. Indies* 184 'I don't envy you, frankly.' 'No doubt I'll survive, sir.' **1971** 'F. Clifford' *Blind Side* IV. ii. 157 'It's nice... Cosy.' 'No *fados*, I'm afraid.' 'I'll survive.'

survivor. Add: **1. c.** Special *Comb.*: **survivor syndrome,** the (freq. delayed) symptoms, such as disintegration of personality, nightmares, tension, and guilt, which are classed as a syndrome and can afflict someone who has survived a dehumanizing and degrading experience of terror.

1968 W. G. Niederland in H. Krystal *Massive Psychic Trauma* iv. 63 Only in this way can we understand, in our appraisal of these people, the mental condition from which they suffer today: this survivor syndrome which I have described as a clinical entity. **1979** B. Bettelheim *Surviving* 29 Unable to embark on the strenuous and hazardous task of integrating their personalities, such survivors suffer from a psychiatric disorder which has been named the concentration camp survivor syndrome.

3. *colloq.* One who has the knack of surviving afflictions unscathed.

1971 P. D. James *Shroud for Nightingale* ix. 295 She would be earning a good living somewhere... The Mary Taylors of the world were natural survivors. **1978** J. Anderson *Angel of Death* xiv. 167 You're a survivor, Paul. People like you always come through.

survivorship. Add: **2. b.** The probability of surviving to a given age; the proportion of a population that does this.

1949 L. I. Dublin et al. *Length of Life* (ed. 2) ix. 178 With information available regarding the actual mortality and survivorship of the cohort born in 1890, it becomes possible to compute..the average years of life lived after any attained age. **1954** *Q. Rev. Biol.* XXIX. 105/1 These quantities are nicely summed up by the familiar life-table function, survivorship (l_x)..and by the age-specific birth rate. **1978** *Nature* 5 Oct. 466/1 In higher forms life span and survivorship can be expressed in terms of allometric and Gompertz equations.

4. Special Comb.: **survivorship curve,** a curve showing the proportion of a population surviving at different ages.

1953 E. P. Odum *Fund. Ecol.* vi. 108 The resulting curve is called a survivorship curve. **1976** *Nature* 1–8 Jan. 12/2 Van Valen..has made a notable contribution in this respect by applying the survivorship curve technique of population biologists to the study of extinction rates for numerous fossil taxa.

sus, suss (sʌs), *sb. slang.* **1.** [Abbrev. of SUSPICION or SUSPICIOUS *a.*] Suspicion of having committed a crime; suspicious behaviour, esp. loitering; the sus law. Freq. in phr. *on sus.*

1936 'J. Curtis' *Gilt Kid* xxv. 248 What you nick me for? Sus? **1963** T. & P. Morris *Pentonville* xv. 312 Men who are, in the prison idiom, 'done for sus', that is to say, prosecuted as 'suspected persons or reputed thieves loitering with intent to commit a felony'. **1970** G. F. Newman *Sir, You Bastard* ii. 74 Chance nickings in the street, from anything on sus, to indecent exposure. **1978** G. Williams *Textbk. Criminal Law* xxxvii. 817 Another provision of the Vagrancy Act s.4 (as amended) allows the punishment on summary conviction of 'suspected persons' and 'reputed thieves' who 'frequent and loiter' in certain public places with intent to commit an arrestable offence. Persons 'found' committing the offence can be arrested. In police jargon, the man is 'picked up on sus'. **1981** *Times* 24 Aug. 3/8 The delight at the passing of 'sus' is, however, mitigated by a degree of apprehension about its replacement, the newly created offence of 'interference with vehicles'.

2. [Abbrev. of SUSPECT *sb.*² or SUSPECTED *ppl. a.*] A suspected person, a police suspect.

1936 'J. Curtis' *Gilt Kid* xxix. 281 Yes, there was a bit of a coring match when they claimed me. Picked me up as a sus and then hung a screwing rap on me. **1967** K. Giles *Death in Diamonds* vi. 110 Sorry, old man, they found your chief sus. with his neck broken. **1970** R. Busby *Frighteners* viii. 80 He's going to go running to the law, because if he don't, he's the number one suss. **1977** *Evening Standard* 8 Mar. 8/2 'Sus' is an ugly word whose meaning is now known to nearly every young West Indian living in London. It is short for 'suspected person'. Its widespread and growing use by the police against black youngsters is coming to be regarded by many lawyers.. as a major scandal.

3. *attrib.* and *Comb.*, as *sus book, case, charge, offence*; **sus law:** until 1981, the law by which a person could be arrested on suspicion of committing a crime; effective since the Vagrancy Act (5 *Geo. IV* c. 83) of 1824.

1970 J. Boland *Big Job* xv. 124 The Sus book..was where lists of Suspected Persons were kept. **1977** *Morning Star* 19 Jan. 2/4 These limitations have serious impact in 'sus' (being a suspected person) and 'enclosed premises' charges. **1977** *Evening Standard* 8 Mar. 8/3 A study of a number of 'sus' cases shows that they all conform to a remarkably similar pattern. **1981** *New Statesman* 13 Feb. 3/1 The government is proposing to keep the 'Sus' laws in Scotland, even though they are being repealed in England and Wales. *Ibid.* The 'useful and necessary' provisions of the Vagrancy Act 1824 and the Burgh Police (Scotland) Act 1892 which define sus offences in Scotland. **1981** *Times* 24 Aug. 3/8 The controversial 'sus' law, under which people can be arrested on suspicion that an offence is likely to be committed, is no more. The Criminal Attempts Act, which comes into force today, abolishes section 4 of the Vagrancy Act 1824.

sus, suss (sʌs), *v. slang.* [Abbrev. of SUSPECT *v.*; cf. prec.] **1. a.** *trans.* To suspect (a person) of a crime (cf. *SUSS *sb.* 1). Also in general use.

Participles of the verb are usu. formed with a double final consonant in the stem. The form with final double *s* has now spread to the infinitive. The substantive, however, is still most commonly encountered with a single final consonant (*sus*).

1953 D. Webb *Crime is my Business* x. 202 He turned to Hodge and said, 'Who's sussed for this job?' **1959** *Observer* 11 Oct. 21/4 Commercial artist..pursued by beat blonde he has never seen... Later heavily sus-ed of her murder at the beach house. **1960** [see *LOT *sb.* 2d]. **1966** C. Rougvie *Gredos Reckoning* iii. 49, I sussed a weirdie and asked: 'You queer or something?' **1970** R. Busby *Frighteners* ii. 25 You'll get sussed right off. The club boys'll mark you down for a copper the minute you walk through the door.

b. With obj. clause: to suspect, to imagine or fancy (something) as likely; hence, to feel or surmise.

1958 [see *GET *v.* 27d]. **1960** *Punch* 24 Feb. 284/2, I sussed that all the dodgy bookshops would soon be skint. **1969** *It* 4–17 July 14/1 It wasn't a situation too conducive to free, relaxed chat and one could suss that Mick was a bit fed up with having to reel out witty and intelligent quips for the voracious appetites of the human media. **1977** *Transatlantic Rev.* LX. 192 Mercurially sussing that the largest ingredient of the briefcase was dollar bills, [he] added: 'Were you aware..that the largest ingredient of bank-note paper was Indian Hemp?'

2. To work or figure *out*; to investigate, to discover the truth about (a person or thing). Also with obj. clause and without const.

1966 *Queen* 28 Sept. 28/3 Youth susses things out on its own. **1969** Fabian & Byrne *Groupie* xxix. 207 When chicks came round I enjoyed sussing them out, and trying to guess which one would last and which one would be dropped. **1971** *It* 2–16 June 18/2 Everybody seems to have at least two nicknames plus their birth-signs so every little chickie can think they've got it sussed. **1971** N. Saunders *Alternative London* xxvii. 256 Talk to him to sus him out—if you're not sure of him, don't leave him out of your sight. **1975** *Daily Tel.* 20 Jan. 7/1 'If ever my members sussed out that I can't read, I'd be a gonner,' he said. **1976** P. Cave *High Flying Birds* x. 105 Stay there a minute. I'll go and suss it out. **1977** *Daily Mirror* 10 May 17/1 It took me about half a day to suss out the industry and realise how easy it would be to move in. **1977** *Sounds* 9 July 30/5 Here we have a stylish axe/singer who's sussed the factors that made Benson such a universally popular guitarist. **1980** *Times Lit. Suppl.* 26 Sept. 1064/3 A morning's browsing in a book shop will suffice for you to suss out the market.

susceptance (sǝse·ptăns). *Electr.* [f. SUSCEPT(IBLE *a.* + -ANCE.] In an alternating current circuit, the imaginary part of the admittance, as opposed to the real part or conductance.

1894 Steinmetz & Bedell in *Trans. Amer. Inst. Electr. Engineers* XI. 648 Admittance, conductance and susceptance are thus used as the inverse correspondents of impedance, resistance and reactance, and may be added as vector quantities. **1960** H. W. Jackson *Introd. Electric Circuits* xiv. 307 Susceptance is the ability of an inductance or capacitance to pass alternating current. *Ibid.*, Capacitive susceptance is a $+j$ quantity. **1966** *McGraw-Hill Encycl. Sci. & Technol.* XIII. 330/1 Susceptance is a function involving both resistance and reactance. If resistance is negligible, the $B =$ ..the reciprocal of the reactance.

susceptibility. 2. a. (*b*) (Earlier example.)

1754 Richardson *Sir Charles Grandison* IV. xxxiii. 228 Emily is a good girl; but she has susceptibilities already.

susceptible, *a.* Add: **B.** *sb. Med.* An individual capable of getting a disease because not immune.

1923 *Jrnl. Exper. Med.* XXXVII. 255 The massive lethal dose of a 1:200 dilution or less selects a relatively constant number of susceptibles. **1944** L. E. H. Whitby *Med. Bacteriol.* (ed. 4) iii. 30 When the proportion of susceptibles is high the disease becomes epidemic until the endemic level of susceptibles is again reached. **1980** *Sci. Amer.* July 26/3 A graph shows the cases reported from 1950 on; there are peaks every four to seven years, time to accumulate a pool of new susceptibles 'following the high birth rate' in densely populated areas.

‖ **sushi** (su·ʃi). [Jap.] A Japanese dish consisting of small balls of cold boiled rice flavoured with vinegar and commonly garnished with slices of fish or cooked egg. Also *attrib.* Hence **sushiya** (suʃi·ya), in Japan, a shop which serves *sushi*.

1893 A. M. Bacon *Jap. Interior* xi. 180 Domestics served us with tea and sushi or rice sandwiches. **1910** J. Inouye *Home Life in Tokyo* vi. 77 The most common food taken on such an occasion is *sushi*, which is a lump of rice which has been pressed with the hand into a roundish form with a slight mixture of vinegar and covered on the top with a slice of fish or lobster, or a strip of fried egg, or rolled in a piece of laver. **1928** K. Yamato *Shoji* vi. 77 His *sushi*, to afford the acme of succulence, had to be eaten at the stall. **1936** K. Tezuka *Jap. Food* 74 *Sushi* has been made in many ways since olden times and is prized by rich and poor alike. **1967** D. & E. T. Riesman *Conversations in Japan* 282 We were standing at the *sushi* buffet of the train. **1968** P. S. Buck *People of Japan* xiii. 158 Since sushi is nothing more than the equivalent of a sandwich, or fishy snack, the sushi bar can hardly be described as a den of iniquity. **1970** P. & J. Martin *Jap. Cooking* 53 The *sushiya*, or sushi shop, plays in Japan a role curiously similar to that of the pub in England.

Susian (sū·ziăn), *a.* and *sb.* [ad. L. *Sūsiānus*, Gr. Σούσιος Σουσιανή Susian, f. the name (τά) Σοῦσα Susa (cf. OPers. *Shush*): see -IAN.] **A.** *adj.* Of or pertaining to (Susa, the ancient capital of) Susiana (modern Khuzistan in Iran), its natives or inhabitants, or the language spoken by them. **B.** *sb.* **a.** A native or inhabitant of Susiana or its capital, Susa. **b.** The language of the Susians, known from inscriptions of the third millenium B.C., also known as Elamite (see *ELAMITE *sb.* and *a.*). Also **Susia·nian** *a.* and *sb.*

?**1552** W. Barkar tr. *Xenophon's Cyropædia* IV. sig. Qiᵛ, They had selected a moste goodly tente for Cyrus, and a Seusian woman. **1601** P. Holland tr. *Pliny's Nat. Hist.* VI. xxvii. 138 It receiveth..the river Hedypnus..and one more out of the Susianes countrey. **1857** W. K. Loftus *Trav. Chaldæa & Susiana* xxviii. 372 The details of the Susian and Persepolitan structures. *Ibid.* xxx. 408 A much-defaced and weathered inscription, written in a language which M. Oppert terms 'late Susanian'. *Ibid.* 426 Pliny, referring to Susa, says that 'the Eulæus surrounded the citadel of the Susians'. **1874** A. H. Sayce in *Trans. Soc. Bibl. Archaeol.* III. 466 The Susians, or Susianians proper, who had their seat at Shushan. *Ibid.* 474 This plural in *-ib* (or *-be* after a consonant) meets us again in Susian. *Ibid.* 476 Susian or southern Susianian. *Ibid.* 484 The Susian and Accadian genitive follows the substantive which governs it. **1877** G. Rawlinson *Orig. of Nations* II. iv. 213 The primitive Babylonians and their neighbors and kinsmen, the Susianians. *Ibid.*, Babylonian and Susianian royal names. **1915** P. M. Sykes *Hist. Persia* I. iv. 57 In Elam there are found..proper names.. which belong to a language..known among scholars as Anzanite, Susian, or simply Elamite. *Ibid.* 58 The chief deity..was referred to as..the 'Susian'. *Ibid.* 54 There was a very ancient occupation of the Susian plain. **1948** W. W. Tarn *Alexander the Great* II. II. 311 It remains to consider the Susian satrapy mentioned above. **1965** W. Culican *Medes & Persians* v. 98 Besides Medes, Persians and Susians, R. D. Barnett has listed..Haraiva [*etc.*]. *Ibid.* 102 The tablets..are file-copies kept by the Susian scribes.

Susie-Q (sū·zi kiū). Also **Suzie-Q, Suzi-Q,** and without hyphen. [Origin unknown.] A modern dance of Negro origin; the step characteristic of this dance (see quots.).

1936 Davis & Coots (*song-title*) Doin' the Suzi-Q. *Ibid.* 4 A new dance hit the town, It's really gettin' 'round, It's lots of fun, I found, Doin' the Suzi-Q. **1937** L. Shomer *How to Dance* 37 The Suzi-Q is the latest and most intricate of Fox Trot Steps. To begin with, it combines the features of the tap-dance with the nimble Off-Beat Syncopated Running Steps and Turns. **1938** A. Murray *How to become Good Dancer* 188 Neither truckin' nor the Suzie-Q is a complete dance in itself. Both are skylarking steps that add variety. *Ibid.* 190 The Susie-Q is a solo dance. It is not danced with a partner. **1946** Mezzrow & Wolfe *Really Blues* xiii. 235 And from the old folks' shuffle to the Suzie Q and Sand, wasn't none of them steps new to grandpa—just the names were different. **1956** G. P. Kurath in A. Dundes *Mother Wit* (1973) 106/2 The Susie-Q and Truckin' are said to have developed in New York's Negro quarter, Harlem. **1963** *N.Y. Times Mag.* 27 Oct. 104/2 [The Negroes'] body rhythm and frank sensuality turned the formal European waltz into the closely clutched two-step and one-step,..the Susie Q. and the big Apple.

suspected, *ppl. a.* Add: **3.** *The Suspected*, a moth, *Parastichtis suspecta*, which has reddish-brown fore-wings and is found in Europe and northern Asia.

1908 R. South *Moths Brit. Isles* II. 7 The Suspected... Of this species there are two groups of forms—plain and variegated. **1948** W. J. Stokoe *Caterpillars Brit. Moths* I. 323 The Suspected... The chief British quarters of this species appear to be in Yorkshire. **1973** *Times* 5 May 12/8 The men who christen moths must be poets. Consider some of the enchanting names of those recorded in the garden..: Heart and Dart, Flame Shoulder, Nutmeg, Common Quaker, The Suspected, [etc.].

suspend, *v.* Add: **4. d.** *to suspend disbelief*, to refrain from being sceptical, or from doubting the truth of something. Cf. *SUSPENSION 3 b.

1963 *Listener* 28 Feb. 393/2 By the time he arrived at the cliff-hanging conclusion there was nothing for it but to suspend disbelief. **1979** *Amer. N. & Q.* Feb. 97/1 In suspending disbelief, poets could construct a fictitious transitional zone.

5. b. For *Obs.* (or *dial.*) read Now *rare* and add later example.

1962 *Listener* 27 Sept. 483/1 Dostoyevsky harrows and suspends his reader.

suspended, *ppl. a.* Add: **I. 3.** *suspended animation* (earlier example).

1795 *British Critic* VI. 533 The author having examined the causes of suspended animation in animals that are hanged, drowned, suffocated, or killed by noxious vapours, concludes that it is occasioned solely by the exclusion of vital air from the lungs.

4. b. *suspended sentence* (Law), a sentence which is imposed but remains in suspense provided that the offender commits no further offence within a stipulated period.

The suspended sentence was first introduced in Europe in the late nineteenth cent. Before this the phr. 'to suspend sentence' was used, esp. in the U.S., to denote the remission or commutation of a capital sentence (see quots. 1828, 1860). In Great Britain the suspended sentence became legal only in 1967 (see quot. 1967), and is commonly used in conjunction with the system of probation (see PROBATION 3).

[**1828** De W. Clinton in E. Cowen *N.Y. State Supreme Court Rep.* (1859) IX. 730 If the judiciary be exposed to sudden..attempts on its humanity..to suspend the sentence of the law, what must be the effect on the executive, when it comes before him, backed by judicial authority; a prevalent sentiment against the punishment of death. **1860** N. Howard *Practice Rep. Supreme Court State N.Y.* XX. 119, I have learned by newspapers that the recorder of this city occasionally suspended sentence upon verdicts or pleas of guilty. *Ibid.*, The court does not possess the power to suspend sentence indefinitely. The judge should recommend the prisoner to a pardon and not suspend sentence, in case he thinks no punishment ought to be inflicted.] **1884** *Chicago Legal News* XVI. 392/1 The same ruling might be held to apply as to the enforcement of suspended sentences..if the power of suspension existed. **1912** *Atlantic Reporter* LXXXII. 424/1 The term 'suspended sentence', as used in criminal law, refers to the suspension of the execution of a sentence already imposed, and not correctly to the suspending of a sentence. **1923** *Texas Law Rev.* I. 191 If anyone is to be given a suspended sentence and another chance to 'make good', surely it is the young man who has committed his first misdemeanor. **1947** *Survey* LXXXIII. 219/1 In 1940, 33 percent of our adult offenders were put on probation or granted suspended sentence. **1950** [see *BOX *v.*³ 2 d]. **1950** *Times* 21 Oct. 3/3 Sir Leo Page had suggested to him that the probation system might be strengthened by the suspended sentence as used in France and other countries. **1957** *Alternatives to Short Terms of Imprisonment* (Home Office) 9 We understand from the Association of Chief Police Officers that there is strong support among the police for the courts being given power to impose a suspended sentence. **1967** *Act Eliz. II* c. 80 §39 A court which passes a sentence of imprisonment for a term of not more than two years for an offence may order that the sentence shall not take effect unless, during a period specified in the order..the offender commits in Great Britain another offence punishable with imprisonment.. and in this Part of this Act 'operational period', in relation to a suspended sentence, means the period so specified. **1971** L. Radzinowicz in M. Ancel *Suspended Sentence* p. vi., The suspended sentence is essentially a continental system. It began its meteoric career over seventy years ago, with the Belgian and French laws of 1888 and 1891... From there it made a *tour du monde*... It eventually reached Israel..in 1954, before entering the United Kingdom, as a very late immigrant, in 1967. **1972** J. Wilson *Hide & Seek* viii. 151, I got six months suspended sentence last time and fined twenty rotten quid. **1973** F. Rinaldi *Suspended Sentences in Australia* vi. 85 To every suspended sentence there should be added a supervision order. **1979** T. Skyrme *Changing Image Magistracy* x. 125 After the introduction of suspended sentences other forms of penalty, financial as well as custodial, diminished steadily.

c. *suspended participle* (Gram.), a participle in an absolute clause or phrase whose subject is omitted, resulting in ambiguity; a dangling participle.

1942 E. Partridge *Usage & Abusage* 93/1 *Confused participles*... Here will be treated what are variously known as disconnected or misrelated or suspended participles... Dr Onions cites the following additional examples:– *Calling upon him last summer*, he kindly offered to give me his copy. (Say: *When I called*.) **1972** R. D. Walshe in G. W. Turner *Good Austral. Eng.* 256 This lapse..has variously been called the..*isolated, suspended*, or *dangling participle* (or *phrase*).

d. *suspended disbelief*: see *SUSPEND *v.* 4 d.

1965 *New Statesman* 20 Aug. 262/1 For a moment you forgot these were actors and participated..in the panic of..the St Valentine's day massacre... A moment later the curtain came down, the lights went up. The theatre has its own short way with suspended disbelief. **1977** *N.Y. Rev. Bks.* 26 May 13/4 If in the end I remain in a state of suspended disbelief, it is..because I find it hard to believe that there can be a single explanation for so complex a phenomenon.

II. 6. *suspended ceiling*, a ceiling fixed so as to alter the proportions of the room or to give sufficient space above it to accommodate services.

1933 *Archit. Rev.* LXXIV. 54/3 The suspended ceilings are built of steel, wire hangers, steel bars and expanded metal, and plaster. **1955** [see *INSULATION 3 a]. **1978** *Cornish Guardian* 27 Apr. 17/8 (Advt.), The County Council invite offers to submit fixed price tenders for.. the provision of a suspended ceiling.

B. *sb.* or quasi-*sb. ellipt.* for *suspended sentence*, sense 4 b above. *slang.*

1970 G. F. Newman *Sir, You Bastard* i. 34 The bird'll get a suspended, I don't doubt her old man's had a word somewhere. **1979** M. Page *Pilate Plot* ix. 130 If you co-operate, I can probably get you off with a £20 fine and a month's suspended—and no press publicity.

suspender. Add: **II. 4. b.** (Earlier examples.)

1878 *Queen* 13 July (Advt.), The new stocking suspender (Patent) worn by the leaders of fashion and strongly recommended by the medical profession. **1881** *Queen* 18 June 6/2 (*heading*) The New Stocking Suspender. .. A handsome pair of..suspenders, in a fancy box.

c. suspender belt, an undergarment used for holding up stockings, consisting of a belt and suspenders to which the tops of the stockings are clipped; a garter belt; **suspender clip, end,** the clip on a suspender belt.

1926–7 *Army & Navy Stores Catal.* 667/2 Suspender belts. White only..each 2/6. **1930** A. Huxley *Brief Candles* 303 And then that further humiliation of having to ask him to help her look for her suspender belt! **1976** *Vogue* 15 Mar. 79 Blue satin suspender belt, £6.90. **1973** T. Pynchon *Gravity's Rainbow* (1975) I. 127 Concentrating on gartering her nylons,..suspender-clips glittering silver under or behind her lacquered red fingernails. **1966** *Olney Amsden Price List* 36 *Suspender ends*..Nylon Fitting 6/8 dozen cards.

suspense, *sb.* Add: **3. e.** (Earlier example.)

1869 *Bradshaw's Railway Man.* XXI. 383 From this was deducted 31,383 *l.* transferred from suspense account.

II. Attributive uses and combinations. **6.** *attrib.* Of popular literature, etc.: characterized by the capacity to arouse suspense, excitement, or apprehension, as *suspense novel, story*, etc.

1952 *Spectator* 3 Oct. 452/2 Many of their 'suspense'—as opposed to 'detective'—novels are first-class. **1957** S. Beach in *This Week's Stories of Mystery & Suspense* 327 In the suspense story the focus..is fixed on the effort of a single individual to overcome danger. **1962** A. Lurie *Love & Friendship* xi. 220 It gave him a dissolute, suspense-movie look which Miranda rather liked. **1963** *Listener* 24 Jan. 158/2 Before compressionism could be taken seriously, suspense drama and the literature of confinement had to be brought together. **1972** J. Philips *Vanishing Senator* III. iii. 150 You said yesterday I should be writing suspense novels. Well, maybe I've read too many of them. **1977** *Amer. N. & Q.* XV. 76/2 Norman Donaldson, an authority on suspense fiction, has written a new introduction for this edition. **1980** D. Bloodworth *Trapdoor* xxix. 175 Enemy agents in suspense thrillers who were programmed by post-hypnotic suggestion.

7. *Comb.*, as *suspense-laden* adj.

1963 *Times Lit. Suppl.* 24 May 374/4 *Les Gommes*.. was conspicuous for..sharp characterization and suspense-laden plotting. **1964** *English Studies* XLV. 375 My chief reason for favoring four beats is therefore that the atmosphere seems more mysterious, suspense-laden, and, as it were, inhuman with four beats than with three.

suspension. Add: **I. 3. b.** (*willing*) *suspension of disbelief*: Coleridge's phrase for the voluntary withholding of scepticism on the part of the reader with regard to incredible characters and events. Now freq. in allusive or extended use.

1817 Coleridge *Biog. Lit.* II. xiv. 2 A semblance of truth sufficient to procure for these shadows of imagination that willing suspension of disbelief for the moment, which constitutes poetic faith. **1930** I. A. Richards *Practical Criticism* vii. 277 Coleridge, when he remarked that 'a willing suspension of disbelief' accompanied much poetry, was noting an important fact. **1962** N. Coghill in Davis & Wrenn *Eng. & Medieval Studies* 210 Here indeed is a call upon us for the suspension of our disbelief. **1962** *Listener* 6 Sept. 366/1 Willing suspension of disbelief doesn't exist for television. **1976** T. Sharpe *Wilt* (1978) ix. 98 Wilt looked desperately round the caravan and met the eyes of the police stenographer. There was a look in them that didn't inspire confidence. Talk about lack of suspension of disbelief.

12. suspension dot, one of a series of dots used to indicate an omission or an interval in a printed text; **suspension-feeder,** a bottom-dwelling aquatic animal which feeds on plankton, etc. found in suspension in the surrounding water; so **suspension-feeding** *ppl. a.* and *vbl. sb.*; **suspension period, point** = *suspension dot* above; **suspension polymerization,** polymerization in which the polymer separates out from a dispersion of the monomer in a liquid.

1949 G. Summey *Amer. Punctuation* viii. 109 Suspension dots or 'French dots' (*points de suspension*) occur in groups, usually of three, usually spaced but sometimes closed up. They mark preceding material as unfinished, or left dangling an instant for attention. They are used within sentences or as terminal points—sometimes in place of the usual sentence point, sometimes in addition. (Ellipsis dots, in the same form, are discussed in Chapter 9.) **1925** O. D. Hunt in *Jrnl. Marine Biol. Assoc.* XIII. 567 Those which feed by selecting from the surrounding water the suspended micro-organisms and detritus,..for want of a better term, may be termed Suspension-feeders. **1959** A. C. Hardy *Open Sea* II. v. 106 Most animals on rocks or stones will be suspension-feeders,..because little detritus can remain there. **1975** *Nature* 7 Aug. 521/1 As those samples included some typical 'suspension-feeders' (Porifera, Ectoprocta, Sabellida and so on), a water current able to transport the food items evidently exists and may explain the development of a rich bottom fauna under the Ross Ice Shelf. **1925** *Jrnl. Marine Biol. Assoc.* XIII. 575 The contents of their stomachs resembles closely that of the suspension-feeding molluscs. **1963** R. P. Dales *Annelids* ii. 53 They [*sc.* sabellids] are not the only ones that have adopted suspension-feeding. **1963** H. Shaw *Punctuate it Right!* xvi. 91 When ellipsis periods come at the end of a statement requiring a period, then four of these 'suspension periods' or 'suspension points'.. are occasionally used. **1969** G. Smith in *Lett. Aldous Huxley* 4 He often used suspension points (...) in place of commas or final stops when typewriting. **1972** *Computers & Humanities* VI. 152 Omit some parenthetic clauses of no importance to the context, and..replace them by suspension points. **1944** *India Rubber World* CXI. 173/1 More details on suspension polymerization will be given in another article on this same subject. **1973** *Materials & Technol.* VI. viii. 504 Suspension polymerization is more suited to batch operation and it is difficult to convert it into a continuous process.

suspensoid (sǝspe·nsoid). *Physical Chem.* [a. G. *suspensoid* (P. P. von Weimarn 1908, in *Zeitschr. f. Chem. u. Industr. d. Kolloide* III. 27/2), f. *suspens-ion* SUSPENSION: see -OID.] A lyophobic colloid from which the dispersed phase is readily (and often irreversibly) precipitated by the addition of an electrolyte.

1909, etc. [see *EMULSOID]. **1927** [see *LYOPHOBE *a.*]. **1936** *Jrnl. Faraday Soc.* XXXII. 1166 The variation of sulphur dioxide and black suspensoids during the fog of 23rd December, 1935, in London, is shown. **1954** [see *KERN *sb.*³ 3]. **1975** *Jrnl. Faculty Fisheries & Animal Husbandry* (Hiroshima Univ.) XIV. 24 Suspensoids were collected from the surface and bottom water samples at stations 3 and 13.

suspicion, *v.* For Now *dial.* (chiefly *north.*), *U.S.*, or *rare arch.* read *dial.* and *colloq.* (orig. *U.S.* and add: **a.** (Later examples.)

Quot. *a* **1637** in Dict. appears to be a fortuitous occurrence unrelated to later uses.

1916 H. L. WILSON *Somewhere in Red Gap* ii. 68 Wilbur says I'm too good, not suspicioning. I'm just being wily, so he says he'll write up and fix it. **1919** J. BUCHAN *Mr. Standfast* xxi. 386 If the Boche once suspicions how little he's got before him the game's up. **1937** C. S. FORESTER *Happy Return* xxii. 259 He is in need of distraction, I suspicion. **1938** S. BECKETT *Murphy* iii. 32 Intense Love nature prominent, rarely suspicioning the Nasty. **1946** S. J. PERELMAN in *New Yorker* 5 Jan. 21/3 Our nineteen-year-old son, which he's home from Yale on his midyears and don't suspicion that his folks are rifting. **1959** *Observer* 22 Mar. 23/6 The major is no fool, and he suspicions as quickly as the audience that the presumed Englishman is a wounded Hungarian on the run. **1961** R. P. HOBSON *Rancher Takes Wife* (1962) viii. 114, I quite often suspicioned this trait of Gloria's but when I found it out for sure it was almost too late. **1973** 'D. SHANNON' *No Holiday for Crime* (1974) vi. 88, I suspicioned what she was, but I didn't have no proof.

b. *absol.* or *intr.*

1905 KIPLING *Actions & Reactions* (1909) 40 An' d'you mean to tell me you never suspicioned? **1946** C. MCCULLERS *Member of Wedding* III. 173 In those bridge games..nobody ever drew a good hand, the cards were all sorry, and no high bids made—until finally Berenice suspicioned, saying: 'Less us get busy and count these old cards.'

Susquehannock (sɒskwəhæ·nɒ̆k). Now only *Hist.* Forms: 7 **Sasquehanno, Sasquesahanock, -hanough, Sesquesahamock,** 8 **Susquehannah,** 9 **Susquehanno, Susquehanough; Susquehanna,** etc. [a. the name of this people in a neighbouring Algonquian language, lit. 'person (or people) of the Susquehanna River': the river flows from N.Y. State into Chesapeake Bay.] = *CONESTOGA I. Also *attrib.*

1612 J. SMITH *Map of Virginia* 8 To proceed, 60 of those Sasquesahanocks, came to the discouerers with skins, bowes..and tobacco pipes for presents. *Ibid.*, The description of a Sasquesahanough. *Ibid.* 19 The people differ very much in stature,..some being very great as the *Sesquesahamocks*, others very little, as the Wighcocomocoes. **1676** *Rec. Court of New Castle on Delaware* (1904) 39 If the Sasquehannos should aply to you for anything, you are to use them kindly. **1751** in *New Jersey Archives* (1883) 1st Ser. VII. 598 The Susquahannah Indians only want leave from the Mohawks whom they call their Fathers in order to their accepting of a missionary. **1833** S. KERCHEVAL *Hist. Valley of Virginia* p. xxiv, He ran amongst his men, crying out..these are our friends the Susquehanoughs. **1845** *Encycl. Metrop.* XXV. 937/2 Thus Maryland was inhabited by the Susquehannoes, who were afterwards destroyed by the Iroquois assisted by four nations. **1898** *Contrib. Indian Hist. Lower Susquehanna Valley* (Hist. Soc. Dauphin County, Pa.) 39 Prior to 1600, but how long before is not known, the Susquehannocks were seated upon the river from which they have derived their name. **1910** *Encycl. Brit.* VI. 897/2 *Conestoga*, a tribe of North American Indians of Iroquoian stock... They were sometimes known as Susquehannas... The tribe suffered final extinction in the Indian wars of 1763. **1915** J. BUCHAN *Salute to Adventurers* v. 79, I was with Bacon in '76, in the fray with the Susquehannocks. I speak the Indian tongues. **1940** T. W. CLARKE *Bloody Mohawk* 36 The Iroquois, about 1660, turned their attention to the Andastes, or Susquehannocks of Pennsylvania and southern New York. **1957** *Encycl. Brit.* XII. 683/2 The Iroquoian family occupied three territories, a northern, southern, and southeastern. In the northern area there lived, besides the Iroquois proper, the Conestoga or Susquehanna in Pennsylvania. **1978** *Handbk. N. Amer. Indians* XV. 363/1 The term Sasquesahanough (Susquehannock) was first recited to Capt. John Smith by his Algonquian-speaking interpreter when he was visited by 60 Susquehannocks in 1608. **1978** J. A. MICHENER *Chesapeake* i. 4 The common warriors..felt that for a Susquehannock to pass more than a year in peace would be disgraceful. *Ibid.* 6 He would have to pass two Susquehannock villages to the south.

suss: see *SUS, SUSS *sb.*, *v.*

Sussex. Add: **a.** (Later examples of a breed of cattle; also *ellipt.*)

1886 J. MACDONALD *Pringle's Cattle* (ed. 3) vi. 117 The Sussex breed of cattle possesses several of the characteristics of the Devon, but is larger in frame. *Ibid.*, Some fine specimens of Sussex oxen are shown annually. **1919** K. J. J. MACKENZIE *Cattle* x. 144 Today the Sussex is essentially a beef-breed. *Ibid.*, The Sussex inherits some of the faults of the draught-cattle from which he springs. *Ibid.*, The Sussex bullock has to be thoroughly fattened before he is a really good butcher's animal. **1974** *Country Life* 7 Nov. 1396/1 Today the beef animal is supreme—Welsh Black, Sussex, Galloway.

c. Sussex spaniel, a long-coated, stocky, golden-brown spaniel belonging to a breed developed in Sussex and neighbouring counties; also *ellipt.*

1856 Sussex spaniel [in Dict., sense a]. **1859** [see *Norfolk spaniel* s.v. *NORFOLK b]. **1904** H. COMPTON *Twentieth Century Dog* II. 237 The Sussex spaniel is a smaller dog than the Clumber. **1981** C. I. A. RITCHIE *Brit. Dog* vi. 164 In spite of the popularity of land spaniels, such as the beautiful Sussex, the water spaniel was perhaps the favourite.

sussexite (sɒ·seksəit). *Min.* [f. *Sussex*, the name of a county in New Jersey + -ITE[1].] A basic borate of manganese and magnesium, $(Mn,Mg)BO_2OH$, found as white or yellowish orthorhombic crystals, isomorphous with szaibelyite.

1868 G. J. BRUSH in *Amer. Jrnl. Sci.* XCVI. 140 (*heading*) New borate from Mine Hill, Franklin, Sussex Co., New Jersey—sussexite. **1951** C. PALACHE et al. *Dana's Syst. Min.* (ed. 7) II. 375 The names sussexite and szaibelyite are applied to the halves of the series with Mn > Mg and Mg > Mn, respectively. **1954** [see *HULSITE]. **1974** *Encycl. Brit. Micropædia* IX. 699/1 Sussexite occurs as hydrothermal fibrous veinlets in the U.S. at Franklin, N.J., and Iron County, Michigan.

susso (sɒ·so). *Austral. slang. Obsolescent.* Also **Susso.** [f. SUS(TENANCE + *-O[2].] **a.** State government relief paid to the unemployed, *spec.* during the Depression. Also in phr. *on the susso.*

1941 BAKER *Dict. Austral. Slang* 51 *On the susso*, in receipt of unemployed sustenance. **1942** L. MANN *Go-Getter* 10 Five shillings were five shillings and a handsome help to the sustenance. 'We're on the Susso now.' That was the song they knew and did not sing. **1974** *Times Lit. Suppl.* 15 Feb. 155/4 We're on the Susso now. In the 1930s Melbourne schoolchildren grew up chanting this (to them) cheerful folk song—'Susso' being the state government sustenance available to the unemployed throughout Australia under varying conditions during the Depression.

b. One who draws this relief.

1947 V. PALMER *Cyclone* 8 He thinks it puts hair on his chest knocking about with the sussos. **1963** F. HARDY *Legends from Benson's Valley* 166 The very thought..of the contempt the respectable held for the sussos changed his mood to defiance.

sussy (sɒ·si), *a. slang.* [Shortened f. SUSPICIOUS *a.* or SUSPECTED *ppl. a.* + -Y[1]; cf. *SUS, SUSS *sb.*] Suspicious, suspect, suspected.

1965 L. J. CUNLIFFE *Having it Away* xiv. 97 It seemed a bit sussy to me. **1974** G. F. NEWMAN *Price* iii. 97 Sneed's questions were becoming more accusing; there was something sussy about Roger Dawes. **1978** N. MARSH *Grave Mistake* iii. 95 He's done porridge for attempted blackmail and he's sussy for bringing the hard stuff ashore.

sustain, *v.* **9. d.** Delete † *Obs.* and add later examples.

1884 *Encycl. Brit.* XVII. 88/1 It was in that very opera, *The Siege of Rhodes*, that Mrs. Colman, daughter-in-law of one of the composers, sustained the character of Ianthe. **1939** JOYCE *Finnegans Wake* 49 He may have been the utility man of the troupe capable of sustaining long parts at short notice. **1975** *U.S. News & World Rep.* 3 Mar. 39/2 Students of geopolitics assert that the U.S. has a near-perfect combination to sustain such a role. **1980** M. FONTEYN *Magic of Dance* 312 These ballets seem essential to theatre dance as a whole because they stretch the artist's interpretive powers to the limit in sustaining long roles. **1983** *Financial Times* 16 Feb. 13/4 The solid-voiced baritone Roland Herrmann sustained the killing role of Creon with burly resilience.

sustainable, *a.* Add: **3.** Capable of being maintained at a certain rate or level.

1965 *McGraw-Hill Dict. Mod. Econ.* 501 *Sustainable growth*, a rise in per-capita real income or per capita real gross national product that is capable of continuing for a long time. A condition of sustainable economic growth means that economic stagnation will not set in. **1971** *Nature* 9 July 80/2 The blue whale could have supplied indefinitely a sustainable yield of 6,000 individuals a year. **1976** *Times* 4 Aug. 3/8 The achievement of a sustainable, stationary population.

Hence **sustainabi·lity.**

1972 T. SOWELL *Say's Law* iii. 100 An increase beyond limits of sustainability existing at any given time would lead only to reduced earnings and subsequent contraction of the quantity supplied. **1980** *Jrnl. R. Soc. Arts* July 495/2 Sustainability in the management of both individual wild species and ecosystems..is critical to human welfare.

sustained, *ppl. a.* Add: **1. b.** *sustained yield* (orig. *Forestry*): the quantity that can be periodically harvested from a crop or population without depleting it in the long term; also *attrib.*

1919 RECKNAGEL & BENTLEY *Forest Management* xii. 124 By sustained yield is understood the yield or cut of timber from a forest which is managed in such a way as to permit the removal of an approximately equal volume of timber, annually or periodically, equal to the increment. **1980** PURDOM & ANDERSON *Environmental Sci.* ix. 219/1 Foresters are finding the sustained yield method, which produces a modest annual timber crop, increasingly more desirable. *Ibid.* x. 245/2 The goal of the fishing industry should be to establish a sustained yield. Closed seasons, catch quotas, nets with larger mesh size, and minimum fish size can help achieve a sustained yield.

c. *sustained-release* adj. (Pharm.): applied to a preparation that releases a substance slowly or intermittently into the bloodstream over a period so as to maintain a steady concentration of it, esp. by means of numerous tiny pellets with different coatings contained in and administered orally as a single capsule. Cf. *slow-release* adj. (*b*) s.v. *SLOW *a.* 16 d, *SPANSULE.

1956 *Jrnl. Pharmacy & Pharmacol.* VIII. 975 It was thought that these resins might provide suitable chemical carriers for drugs in sustained release preparations. **1974** SHOTTON & RIDGWAY *Physical Pharmaceutics* xii. 340 Sustained release products can be made by embedding the drug in a hydrophobic matrix from which it is leached out over a period of time. **1979** *Arizona Daily Star* 8 Apr. C10/1 (Advt.), Most products provide short-duration nutritional burst. *Heritage* sustained-release tablets work all day long, up to 12 hours, to release nutrients when you need them.

sustaining, *ppl. a.* Add: **c.** *sustaining pedal*, (*a*) (see PEDAL *sb.* 1 b); chiefly *U.S.*; (*b*) = *damper-pedal* s.v. DAMPER 7 a.

1889 in *Cent. Dict.* s.v. *Pedal*. **1911** H. E. KREHBIEL *Pianoforte & its Music* iii. 47 On some pianofortes there is a third pedal between the other two, called the Tone Sustaining Pedal, the action of which is to withhold the damper from the string or strings struck just before the depression of the pedal. **1922** A. H. LINDO *Pedalling in Pianoforte Music* I. 14 Students..are frequently told that it [*sc.* the right pedal] should be called, not the 'loud', but the 'sustaining' pedal. **1923** [see *ACCENTUATOR]. **1931** G. JACOB *Orchestral Technique* i. 2 In transcribing pianoforte music the effect of the sustaining pedal is often not taken into account. **1976** *Gramophone* Dec. 1016/2 The Gieseking is..an object-lesson..in how to do without the sustaining-pedal as a prop.

d. *sustaining programme*, a radio or television programme which is paid for by the broadcasting station. *U.S.*

1931 F. A. ARNOLD *Broadcast Advertising* 31 Sustaining programs are those which are prepared and paid for exclusively by the broadcasting station and in which the advertiser has no participation whatever. **1952** H. L. EWBANK *Broadcasting* viii. 128 A *sustaining* program is neither paid for by a sponsor nor interrupted by spot commercials. **1961** S. P. LAWTON *Mod. Broadcaster* 85 The network contracts themselves,..and the agreements for carrying sustaining programs, all play an important part in the make-up of schedules of affiliate stations. **1973** J. R. GRIMES *Mod. Radio Programming* xii. 173 *Sustaining*, non-sponsored.

sustentation. Add: **7. b.** *Aeronaut.* The action or condition of being aerodynamically supported either by the lift afforded from the motion of an aerofoil or by means of an air-cushion.

1907 [see *AEROFOIL]. **1939** *Nature* 18 Feb. 272/1 Most modern air transport is by means of the aeroplane, a body heavier than air, depending upon forward movement for sustentation. **1966** *McGraw-Hill Encycl. Sci. & Technol.* I. 197/2 Another form [of air-cushion vehicle] creates high air pressure beneath its structure for sustentation. **1977** T. K. S. MURTHY in *Proc. 2nd Internat. Waterborne Transportation Conf.* (1978) 308 The sustentation of the vehicle above the water surface is therefore partly due to the pressure of the air in the cushion and partly due to the hydrostatic buoyancy of the submerged hulls.

Susu[2] (sʊ·sʊ). Also † **Soosoo, Suzee.** [Native name.] (A member of) a Mande people inhabiting the north-west of Sierra Leone and the southern coastal regions of the Republic of Guinea in West Africa; also, the language of this people. Also *attrib.* or as *adj.*

[**1670** J. OGILBY *Africa* II. 368 The Kingdom of Bena and Sousos, deriving its Name from the inhabitants of its principal Town, which is named Sousos, stands situate about nine days Journey from..the Kingdom of..Serre-Lions.] **1786** J. MATTHEWS *Let.* 20 Feb. in *Voy. River Sierra-Leone* (1791) 13 The river Riopongeos..is..one of the principal rivers for trade... The natives are originally Suzeės. *Ibid.* 20 Nov. 95 The Suzeė language seems to be the root from which the Bagoe..sprung. **1803** T. WINTERBOTTOM *Acct. Native Africans Sierra Leone* I. i. 5 The Bulloms..possessed the whole of the river Kissee, from which they were driven by a nation called Soosoos or Suzees. **1845** *Encycl. Metrop.* XXIV. 579/2 The Sûsûs were well known to the learned and philosophical Arab historian of Africa. **1846** R. G. LATHAM in *Proc. Philol. Soc.* II. 221 The Susu, of which we have a grammar, is allied to the Mandingo. **1911** [see *MALINKE]. **1957** M. BANTON *W. Afr. City* vii. 127 The Susu and Yalunka (or Dyalonke) appear to be two branches of the same people. .. There are four Susu and three Yalunka chiefdoms. **1977** *Whitaker's Almanack 1978* 758 The southern half of Sierra Leone is inhabited by peoples whose languages fall into the Mende group; the northern half by the Temne, and smaller groups such as the..Susu.

susuhunan (sʊsūhūnā·n). [ad. Javanese *sesuhunan.*] The title of the monarchs of Surakarta (also called Solo) and of Mataram in Java.

1817 T. S. RAFFLES *Hist. Java* II. x. 157 The company and the *susuhúnan* should assist each other. **1831** *Canton Misc.* No. 2. 77 Solo, is the residence of the Susuhunan commonly called the Emperor, and whose ancestors in the 13th and 14th centuries reigned over the greater part of Java. **1915** D. M. CAMPBELL *Java* II. xix. 997 The sultan and susuhunan on state occasions frequently adorn themselves in the Dutch general's uniform. **1973** G. M. D. HOWAT *Dict. World Hist.* 1526/1 The Crown Prince of Mataram, later Susuhunan Amangkurat II (*reg.* 1677–1703).

susumber (sŭsɒ·mbəɪ). Also **sosuma, soushumber.** [perh. f. Ewe *sŭsume* or Twi *nsúsúaa* an edible plant + Twi *mbá* young plants.] A prickly shrub, *Solanum torvum*, of the family Solanaceæ, native to the tropics, esp. America and the West Indies, and bearing clusters of white flowers followed by edible berries; also *attrib.*; = *macaw-bush* s.v. MACAW[2].

1814 J. LUNAN *Hortus Jamaicensis* II. 245 There are two varieties, both very common in Jamaica, the berries

about the size of small cherries... They are..known by the names soushumber, cat-nail, Port-Morant tobacco, and macaw bush. **1839** B. M'MAHON *Jamaica Plantership* 27 He then ran after them, flogging, knocking them down, and tumbling them into the susumber bushes, full of thorns. **1913** W. HARRIS *Notes on Fruits & Veg. in Jamaica* 42 The soushumber is used mainly by the natives who..consider it a wholesome vegetable. **1929** M. W. BECKWITH *Black Roadways* 14 Salt cod cooked with..the sosuma berry, is a favourite breakfast dish even upon the tables of the whites. **1953** *Caribbean Q.* III. 1. 10 Susumber berries..grow wild everywhere. **1972** C. D. ADAMS *Flowering Plants Jamaica* 656 S[*olanum*] *torvum*... Gully Bean, Susumber, Turkey Berry... Shrub 1–4 m high... General in the tropics.

susurrate, *v.* For † *Obs. rare*[-0] read *rare* (chiefly *Lit.*) and add: (s[1]ū·sɒrē[1]t, s[1]ūsɒ·rē[1]t). *intr.* (Examples.)

1957 H. WILLIAMSON *Golden Virgin* III. xxvi. 395 While feet susurrated on the parquet floor made smooth.. by scatterings of french chalk. **1968** M. JONES *Day They put Humpty together Again* 45 The lining of her coat susurrated noisily. **1972** *New Yorker* 30 Sept. 6/3 In the Palm Court, violin music susurrates from five to seven.

susurrous, *a.* Add: Also, characterized by, or full of, whispering.

1886 in WEBSTER. **1946** M. PEAKE *Titus Groan* lxiv. 388 The long corridors were susurrous with rumour.

susurrus. (Earlier example.)

1826 *Blackw. Mag.* XX. 146/2 Through the..range of laughter, from faint susurrus to indomitable guffaw.

Sutherland (sɒ·ðəɪlænd). The title of Harriet Elizabeth Leveson-Gower, Duchess of *Sutherland* (1806–68), used *attrib.* in *Sutherland table*, a gate-leg table with rectangular leaves.

1879 *Designs Cabinet Furnit.* (Blyth & Sons) (Index) 2 Sutherland Tables. **1926** R. B. WHIFFEN *Pocket Compend. Furnit.* 99 *Sutherland Table*—A small table with a narrow top (useless when shut) and two large folding leaves, when these are let down it occupies but little space. **1952** F. G. ROE *Victorian Furnit.* xi. 93 The 'Sutherland Table', a Victorian recension of the old flap table with pull-out supports, its name a reminder of that Duchess of Sutherland who had been Queen Victoria's Mistress of the Robes, ..and was..Her Majesty's personal friend. **1979** 'J. GASH' *Grail Tree* xvii. 170 Just telling Jimmo here you're wanting a Sutherland table.

Suthu, Suto, varr. *SOTHO.

sut(t)ringee, var. SITRINGEE in Dict. and Suppl.

suture, *sb.* Add: **4.** *Geol.* In plate tectonics, the junction or line of junction formed by the collision of two lithospheric plates.

1971 *Nature* 18 June 418/2 Within the present continents there are several linear belts of distinctive oceanic and geosynclinal deposits which apparently mark the boundaries (sutures) between once separated continents. **1977** *Sci. Amer.* Apr. 32/1 Most of the sutures in Eurasia appear to be older than 200 million years.

suture *v. trans.*, (*b*) *Geol.*, to join (lithospheric plates) by means of a suture; often const. *together*; **suturing** *vbl. sb.* (further examples in *Geol.*).

1970 *Nature* 14 Nov. 659/1 If..continents are being joined, their suturing prevents further relative motion between the plates on which they ride. **1976** B. E. HOBBS et al. *Outl. Structural Geol.* x. 468 Depositional sites that are subsequently 'sutured' together by convergent plate motion. **1977** *Sci. Amer.* Apr. 32/1 When two continents collide, they suture themselves together to form a larger continent. **1979** *Nature* 6 Dec. 608/2 The act of complete suturing could trap basaltic crust and supracrustals between the two masses.

sux- (sɒks) (before a consonant also **suxa-**), formative element [repr. the sound of *succ-* (sɒks-) in SUCCINYL] in the names of drugs, as in *SULFASUXIDINE and *SUXAMETHONIUM.

suxamethonium (sɒ:ksămeþō·niəm). *Pharm.* [f. *SUX- + *METHONIUM.] = *SUCCINYLCHOLINE. Also *suxamethonium bromide, chloride, iodide.*

1953 J. H. GADDUM *Pharmacology* (ed. 4) xi. 230 Suxamethonium iodide..also causes brief neuromuscular block. **1963** [see *DECAMETHONIUM]. **1977** *Lancet* 18 June 1305/2 Prolonged suxamethonium apnœa during a general anæsthetic occurred in a patient with Goodpasture's syndrome who had recently had plasmapheresis.

Suze (sūz, ‖ süz). [See quot. 1961.] The proprietary name of a yellow, gentian-based aperitif; also, a drink or glassful of this.

1950 D. AMES *Corpse Diplomatique* iii. 22 It..enables one to have another drink... I thought a Suze and a Cinzano? **1961** *Trade Marks Jrnl.* 5 Apr. 483 *Suze*... Aperitif wines having a gentian base. Distellerie de la Suze.., 11, Avenue de Général Leclerc, Maisons-Alfort (Seine), France. **1964** L. DEIGHTON *Funeral in Berlin* xxii. 124, I poured two Suzes into my face. **1974** N. FREELING *Dressing of Diamond* 72 He'd like a big Suze with lots of ice.

Suzie-Q, var. *SUSIE-Q.

Suzie Wong (sʊ·zi wɒŋ). *slang.* Also **Susie Wong.** The name of the leading character in *The World of Suzie Wong* (1957), a novel by R. L. Mason, applied *transf.* to a woman, esp. a prostitute, in Hong Kong who consorts with visiting servicemen, etc.; also *sing.* used generically, and *attrib.*

1962 E. SNOW *Other Side of River* xxxvii. 274 What did happen to all these Suzie Wongs? **1965** *Guardian* 24 July 8/5 A teenage English blonde would be far safer in the Suzy Wong quarter of Hongkong than in a side street in Soho. **1971** *Nat. Geographic* Oct. 547/2 Sailors come to sport with Suzie Wong. *Ibid.* 571/2 The fleets of the world have indeed found this superlative anchorage, and the sailors have found Wan Chai, that traditional world of all the Suzie Wongs. **1977** 'J. LE CARRÉ' *Hon. Schoolboy* I. vii. 151 What's happened to Susie Wong since war-weary GIs..have ceased to flock in for rest and recreation? **1978** P. HARCOURT *Agents of Influence* iii. 60 What are you doing here..enjoying the delights of Suzie Wong land?

‖**suzuribako** (suzuribā·ko). [Jap.] In Japan: a box (often, of finely-wrought lacquer-work) in which an inkstone, ink-stick, several brushes, and a small water container are kept; equivalent to an inkstand.

1967 *Times* 7 Mar. 21/6 A suzuribako by Shiomi Masanari. **1974** *Country Life* 6 June p. xii/2 Detail of a suzuribako decorated with a figure of Kajiwara Kagesuye ..Japanese 19th century. **1981** *Jrnl. R. Asiatic Soc.* I. 120 Eight *suzuribako* appear, one (No. 18) with a concealed *waka* poem.

svabite (svā·bəit). *Min.* [ad. Sw. *svabit* (H. Sjögren 1892, in *Geol. Föreningens i Stockholm Förhandl.* XIII. 789), f. the name of A. *Svab* (1703–68), Swedish mining official: see -ITE[1].] A fluoride and arsenate of calcium found as colourless or light-coloured prismatic crystals of the hexagonal system.

1893 *Jrnl. Chem. Soc.* LXIV. II. 420 Svabite is a new mineral of the apatite group from the Harstig mine. **1966** *Doklady Acad. Sci. USSR: Earth Sci. Sect.* CLXVI. 134/1 Svabite $Ca_5[AsO_4]_3(OH, F, Cl)$, the arsenical counterpart of apatite, is an extremely rare mineral.

Svan (svăn). Also † (Pl.) **Ssuanes.** [Russ., cf. L. *Suani* (also used).] (A member of) a southern Caucasian people living in Svanetiya in western Georgia; also, the language of this people. Also **Sva·nian, Swa·nian.** Also *attrib.*

1601 P. HOLLAND tr. *Pliny's Nat. Hist.* I. VI. iv. 117 You meet with another river called Charien: upon which bordereth the nation of the Salæ, named in old time Phthirophagi and Suani... The river Cobus,..issueth out of Caucasus, and runneth through the country of the Suani abovesaid. **1814** F. SCHOBERL tr. *von Klaproth's Trav. in Caucasus & Georgia* xxiv. 298 The village of Chulam is inhabited by families of Ssuanes. *Ibid.* xxiv. 292 About six German miles to the south-west of the village of Ckaratschai lies the mountain Dshuman-taw, where commence the settlements of the Ssuanes. **1869** D. W. FRESHFIELD *Trav. in Central Caucasus* x. 292 Suanetia is the general name bestowed..on the upper valley of the Ingur, and is derived from the inhabitants, who from very ancient times have been called the Suani, or Suanetians. **1910** [see *SVANETIAN *a.* and *sb.*]. **1939, 1948** [see *LAZ]. **1959** B. GEIGER et al. *Peoples & Lang. Caucasus* iv. 15 Svan... English variants: Svan, Svanetians... The Svan language is a member of the S. Caucasian..language-family, to which belong..Mingrelo-Laz and Georgian, the latter languages forming one group as against Svan. **1962** D. M. LANG *Mod. Hist. Georgia* i. 10 The Svans were cut off for centuries from the main stream of Georgian civilization. *Ibid.* 18 Svanian and Mingrelo-Laz..are separate languages.

svanbergite (svæ·nbɔ̄ɪgəit). *Min.* [f. the name of Lars F. *Svanberg* (1805–78), Swedish chemist: see -ITE[1].] † **a.** = PLATINIRIDIUM in Dict. and Suppl. *Obs.* **b.** [ad. Sw. *svanbergit* (L. J. Igelström 1854, in *K. Vetenskaps-Akad. Förhandlingar* XI. 156).] A basic phosphate and sulphate of aluminium and strontium, $SrAl_3PO_4SO_4(OH)_6$, found as translucent rhombohedral crystals.

1857 C. U. SHEPARD *Treat. Mineral.* (ed. 3) 303 *Svanbergite* (S.), Platiniridium, Svanberg... In small grains and rarely..cubes, with truncated angles. **1866** BRANDE & COX *Dict. Sci., Lit. & Art* II. 532/3 Pissophane. Svanbergite. Amblygonite. **1900** *Mineral. Mag.* XII. 252 Svanbergite is crystallographically very similar both to beudantite and also to hamlinite and florencite. **1979** *Mineral Abstr.* XXX. 422/2 The source of the Sr and P in the svanbergite was probably the basaltic lavas which covered northern Syria in the Quaternary.

Svanetian (svanī·ʃăn), *a.* and *sb.* Also † **Suanetian.** [f. *Svanet(iya* (see *SVAN) + -IAN.] **A.** *adj.* Of or pertaining to the Svans. **B.** *sb.* = *SVAN.

[**1788** G. ELLIS *Mem. Map Countries between Black Sea & Caspian* 77 (*heading*) Georgian language. Carduel dialect. Imretian. Suaneti dialect.] **1854** A. VON HAXTHAUSEN *Transcaucasia* v. 159 His wife was the daughter of a Suanetian prince. *Ibid.* 162 The Suanetians have generally blue eyes and blond hair. **1896** D. W. FRESHFIELD *Exploration Caucasus* I. x. 221 The Suanetian language resembles Old Georgian. **1902** *Encycl. Brit.* XXVI. 619/2 The high valleys of the Caucasus are populated by..Svanetians, Ossets, Pshaves, and Khevzurs in the middle. **1910** *Ibid.* XI. 760/1 The Svanetians, Shvans, or Swanians, on the Upper Ingur. *Ibid.* 761/1 Both the Laz..and the Svanetian present..structural and verbal differences. **1951** W. K. MATTHEWS *Languages U.S.S.R.* v. 87 The rather more divergent Svanetian (Svan). **1959** [see *MINGRELIAN *sb.* and *a.*].

‖**svara** (swa·ra). Also 8 **swara.** [Skr., lit. 'sound, voice'.] In Indian music, a note of a musical scale.

1792 W. JONES in *Asiatick Researches* III. 68 The first of these [notes] is emphatically named *swara*, or *the sound*, from the important office which it bears in the scale. **1891** [see *MURCHANA]. **1927** *Grove's Dict. Mus.* (ed. 3) II. 705/2 The second subject..with only two variations, and after these a *Svarā*, or sol-fa'ed passage, by way of a cadenza. **1968** *Indian Mus. Jrnl.* V. 28 He used to play the svara exercises on a single string. **1972** P. HOLROYDE *Indian Music* vi. 221 The svaras or notes are still used for vocal gymnastic exercises.

svarabhakti (swarab(h)a·kti, svara-). *Philol.* [Skr., vowel-separation, f. *svára* vowel + *bhakti* separation.] The process by which a parasitic vowel is inserted between two consonants. Usu. *attrib.*, esp. as *svarabhakti vowel.*

1880 A. H. SAYCE *Introd. Sci. of Lang.* I. 317 The insertion..of vowels..goes under the technical name of *Swarabhakti*. This name was imported from the Hindu grammarians by Johannes Schmidt. *Ibid.* 318 Prosthesis, or prothesis..is another illustration of *Swarabhakti*. **1888** [see *INDETERMINATE *a.* (*sb.*) 2 e]. **1908** *Indogerm. Forsch.* XXIII. 254 The *-i-* of *pulisa-* and the second *-u-* of *puruṣa* are svarabhakti-vowels. **1942** *Amer. Speech* XVII. 100 A short vowel in E before *r* + a velar is lowered to [ɑ] in the dialect and a svarabhakti [ɪ] develops between the *r* and the velar. **1977** F. COLLINSON in Campbell & Collinson *Hebridean Folksongs* II. 257 The variation of rhythm or melody arising from the presence of a svarabhakti vowel is of constant occurrence in most of the songs.

Hence **svarabha·ktic** *a.*

1894 W. M. LINDSAY *Latin Lang.* 145 The inserted or 'parasitic' vowel (sometimes styled in the terminology of the Sanskrit grammarians 'svarabhaktic' vowel..) is often seen in the older Latin loanwords from Greek. **1965** *English Studies* XLVI. 174 We may here just possibly be ..taking..the *e* to be svarabhaktic.

‖**svarita** (swa·rita). Also **Svarita.** [Skr. *svarita.*] A falling glide used in the recitation of Vedic texts (see quots.). Also in extended use.

1916 A. A. MACDONELL *Vedic Gram. for Students* 448 The Svarita is a falling accent representing the descent from the Udātta pitch to tonelessness. **1955** T. BURROW *Sanskrit Lang.* iii. 113 The accent of the syllable immediately following the udātta is termed *svarita-* and it is described by Pāṇini as a combination (*samāhāra-*) of udātta and anudātta. **1957** *New Oxf. Hist. Music* I. iv. 200 The way of chanting the Rigvedic hymns has definite musical importance, as the three accents employed, the *udātta*, the *anudatta*, and the *svarita*, denote a distinct difference in pitch. **1971** *Canadian Jrnl. Linguistics* XVII. 73 The sandhi-organization..is sufficient peculiar origin for Sanskrit svarita. **1973** A. H. SOMMERSTEIN *Sound Pattern Anc. Greek* v. 122 On the vowel following an acute-accented vowel, as also on the latter part of a circumflex-accented vowel, there was a falling glide. [*Note*] Hereinafter often referred to as *svarita*.

Svedberg (sve·dbɔ̄ɪg). *Biochem.* [The name of Theodor S. *Svedberg* (1884–1971), Swedish chemist.] Also *Svedberg unit.* A unit of time equal to 10^{-13} second used in expressing sedimentation coefficients. Symbol S (*S 4 d).

1942 *Ann. N.Y. Acad. Sci.* XLIII. 176 The members of the conference..indicated a desire to honor Professor The Svedberg... It was unanimously decided:..to adopt, as a convenient practical unit for sedimentation constants, the *Svedberg*, to be denoted by the letter *S* and equal to 10^{-13} times the absolute units, which are in seconds. **1944** *Jrnl. Biol. Chem.* CLII. 682 The average sedimentation constant of the iron hydroxide micelle was 150 Svedberg units. **1970** *Nature* 5 Sept. 1068/2 One can comprehend but not condone the biologist's affection for units such as Å, the Svedberg and mmHg. **1976** *Sci. Amer.* Aug. 63/1 It was the expected size (nine Svedberg units).

svelte. Delete ‖ and add: **b.** *transf.* Elegant, smooth, graceful.

1909 E. POUND *Personae* 43 And first the cities of north Italy I did behold, Each as a woman wonder-fair, And svelte Verona first I met at eve. **1967** *Listener* 30 Mar. 434/1 His earlier work, technically less accomplished, rougher, coarser in execution, left a way open—one felt one could break out of the paint. But now his handling is so skilful, svelte, that all other possibilities are closed. **1974** N. MARSH *Black as he's Painted* xi. 78 Is our svelte hired limousine at the door? **1977** *Gramophone* July 202/1 His rhythmic pungency..in the third piece and the svelte charm of the central waltz..suggest that he might be equally at home with Roussel's symphonic music.

Svengali (svеŋgā·li). The name of *Svengali*, musician and hypnotist, a character in the novel *Trilby* (1894) by George Du Maurier, used *transf.* and allusively to designate one who exercises a controlling or mesmeric in-

fluence on another, freq. for some sinister purpose. Also *attrib.* and *Comb.*

1914 KIPLING *Divers. Creatures* (1917) 145 I'm glad Zvengali's back where he belongs [referring to a dog with a mesmeric stare]. **1919** C. MACKENZIE *Sylvia & Michael* iv. 92 The juggler..passed into the category of the Svengalis, and became one of a long line of romantic impossibilities. **1934** B. DARWIN *Playing the Like* 121 He believes himself a new Svengali with a second Trilby. **1942** *Amer. Speech* XVII. 90/1 The word 'Svengali' shows the player's ability to keep his opponent so 'hypnotized' that he will not be aware of his trickery. **1962** N. FREELING *Love in Amsterdam* I. 40 He fascinated her. Svengali stuff. **1963** AUDEN *Dyer's Hand* 457 It is impossible to represent Christ on the stage. If he is made dramatically interesting, he ceases to be Christ and turns into a Hercules or a Svengali. **1966** N. MARSH *Black Beech & Honeydew* x. 231 A hideous Svengali-like face. **1972** *Maclean's Mag.* Mar. 41/2 He had a strange hypnotic power—not that he was a Svengali, but when he spoke people listened. **1978** M. DICKENS *Open Book* vi. 59 Charles Pick..already showed the infectious Svengali enthusiasm to which many writers beside me owe the fact that they have had the courage to go on writing.

Sverdrup (svɔ̄·ɪdrup). Also **sverdrup.** [Name of H. U. *Sverdrup* (1888–1957), Norwegian oceanographer and meteorologist.] Also *Sverdrup unit.* A unit of flow equal to one million cubic metres per second.

1963 G. L. PICKARD *Descr. Physical Oceanogr.* vii. 117 The most commonly used unit for volume transport is 'one million m^3/sec'..referred to as 'one sverdrup'. **1970** *Sci. Jrnl.* Mar. 58/2 Fifty Sverdrup units of flow approach the east coast of Mindanao in the Philippines and half of this volume turns north into the Kuroshio. **1977** J. D. MACDONALD *Condominium* xxxiv. 341 The..total flow of all the rivers of the world combined..is two sverdrups.

swab, *sb.*[1] (*a.*) Add: **1. c.** (Earlier U.S. example.)

1863 'MARK TWAIN' *Celebr. Jumping Frog* (1867) 73 A sheet was wound around me until I resembled a swab for a Columbiad [cannon].

d. Also † *transf.*, a naval officer. *Obs.*

1793 C. DIBDIN in *Britannic Mag.* I. 25/2 And there's never a swab but the captain knows the stem from the stern of the ship. **1850** H. MELVILLE *White Jacket* II. xliii. 289 Touch your tile whenever a swob (officer) speaks to you.

f. *Oil Industry.* A device in the form of a plunger with a valve, used to raise fluid in a well and induce a flow.

1904 *Dialect Notes* II. 391 *Swab*, n., a tool used in drilling... When water comes in faster than it can be got out by the sand-pump, the swab is run down. The fluid passes through it, and by it several hundred feet of fluid can be raised out of the hole at one run. **1916** A. B. THOMPSON *Oil-Field Devel.* x. 482 The early swab consisted of a hollow steel barrel, around which was wrapped sufficient hemp..to tightly fit the well casing when inserted. **1930** W. H. OSGOOD *Increasing Recovery of Petroleum* I. x. 169 Swabbing..may result in the forming of emulsions when the swab is run too low in the fluid and water is present. **1974** P. L. MOORE et al. *Drilling Practices Manual* ix. 241 Swab pressures are associated with fluid flow, caused by pulling equipment out of a liquid filled bore-hole.

swab, *v.*[1] Add: **II. 7.** *Oil Industry.* To introduce a swab (*SWAB *sb.*[1] (*a.*) 1 f) into (an oil-well) in order to induce a flow.

1916 A. B. THOMPSON *Oil-Field Devel.* x. 482 It was the local custom to swab wells at intervals. **1974** P. L. MOORE et al. *Drilling Practices Manual* ix. 245 This deceleration pressure indicates a well can be swabbed when running pipe into the hole.

swabber[1]. **1. b.** (Later example.)

1931 [see *roach-powder* s.v. *ROACH *sb.*[4] 4].

swabbing, *vbl. sb.* (Examples in the *Oil Industry.*)

1921 W. H. JEFFERY *Deep Well Drilling* xii. 338 Swabbing and agitating are sometimes effective in causing wells to resume flowing. **1930** [see *SWAB *sb.*[1] 1 f]. **1974** P. L. MOORE et al. *Drilling Practices Manual* xii. 302 When a viscous mud is being used, additional mud weight may be required because of swabbing.

Swabian, *a.* and *sb.* Add: **b.** *sb.* (Earlier example.)

1840 BROWNING *Sordello* I. 12 They laughed as they enrolled That name at Milan on the page of gold For Godego.., Loria, and every sheep-cote on the Swabian's fief.

c. The dialect of Swabia. Also *Comb.*

1866 J. MACGREGOR *Thousand Miles in Rob Roy Canoe* (ed. 2) v. 76 They were much delighted..and went back prattling their purest Suabian in a highly satisfied frame of mind. **1886** STRONG & MEYER *Outl. Hist. German Lang.* v. 74 Swabian-Alemanic, spoken in Bavaria as far as the Lech, and in Würtemberg. **1937** D. P. INSKIP tr. *E. Tonnelat's Hist. German Lang.* x. 210 It is difficult to draw a boundary between Swabian and Alemanic proper. **1961** R. E. KELLER *German Dialects* 10 Speaking of a certain dialect, e.g. Swabian, implies that such a dialect has an identity which distinguishes it more or less clearly from other dialects. **1981** R. MANNHEIM tr. *G. Grass's Meeting at Telgte* iv. 22 After thirty years of residence in London, the diplomat Weckherlin still spoke an unvarnished Swabian.

swacked (swækt), *ppl. a. U.S. slang.* [f. SWACK *v.*[1] in Sc. dial. sense 'to gulp, swill' + -ED[1].] Drunk, intoxicated.

1932 *Amer. Speech* VII. 436 A man drunk is 'limp', 'tight', 'swacked'. **1936** WODEHOUSE *Laughing Gas* ix. 93 My father used to drink till he saw the light, and he prided himself on being able to say anything at any time of the day or night, no matter how swacked he might be, without tripping over a syllable. **1965** H. KANE *Devil to Pay* (1966) iii. 17 I'm slightly swacked on champagne. **1977** J. WAMBAUGH *Black Marble* (1978) vi. 71 They said he was bombed, swacked, bagged. By noon? She wasn't sure if it was booze.

swaddy: now generally superseded by *SQUADDIE.

Swadeshi. Substitute for def.: Used chiefly *attrib.* to designate an Indian nationalist movement originating in Bengal, which advocated principally the support of indigenous industries using home-produced materials (esp. cotton), and the boycott of foreign goods. Now (since the partition of 1947) *Hist.* (Later examples.)

1925 S. BANERJEA *Nation in Making* 198 [Jogesh Chunder Chaudhuri] it was who first started an Industrial Exhibition of *Swadeshi* articles as an annexe to the Indian National Congress. That was in 1896. **1936** J. NEHRU *Autobiogr.* xxxv. 266 So far the Congress had thought along purely nationalist lines, and had avoided facing economic issues, except in so far as it encouraged cottage industries and *swadeshi* generally. **1941** L. S. S. O'MALLEY *Mod. India & West* xvi. 762 *Swadeshi* goods..are goods manufactured in India by Indian labour from Indian raw and basic materials under the guidance of concerns whose capital and management are predominantly Indian, with the proviso that foreign raw or basic materials may be used in cases where India cannot supply them. **1970** 'B. MATHER' *Break in Line* xii. 156 Big coloured Swadeshi towels warming..by the stove. **1975** E. SHILS in H. M. Patel et al. *Say not the Struggle Nought Availeth* 68 The Indian political movement..did not cavil at the European substance of higher education. The Swadeshi movement made an issue of it, but it was more concerned with the intentions which were said to underlie it rather than with the substance.

swag, *sb.* Add: **6. b.** *Theatr.* A festooned stage-curtain or drapery, fastened similarly. Also *transf.* and *attrib.*

1959 RAE & SOUTHERN *Internat. Vocab. Techn. Theatre Terms* 58 74 Swag border. **1961** J. OSBORNE *Entertainer* 11 Different swags can be lowered for various scenes to break up the acting areas. **1982** BARR & YORK *Official Sloane Ranger Handbk.* 136/3 Lots of pretty pelmets and a few swags, variations on the theme of stage curtains—not like those dreadful draped net affairs one sees from the bypass.

7. (Earlier U.S. example.)

1848 *Holden's Dollar Mag.* Aug. 475/2 A 'Swag' is often met with in the Western country. It is a concave spot, sunk in below the level by nature.

9. (Earlier example.)

1794 *Sessions Papers Central Criminal Court* Jan. 341/1 There are very few gentlemen here on the jury but what know what a *swag* is; the meaning is, a bundle of clothes that are stolen from any place.

10. Also *N.Z.* Freq. in colloq. phrases *to hump the swag*: see HUMP *v.* 2 in Dict. and Suppl.; *on the swag*: on one's travels.

1853 J. ROCHFORT *Adventurers of Surveyer* vi. 49 Disregarding the state of the roads,..we strapped on our 'swags', consisting of a pair of blankets and a spare pair of trousers, and started for the diggings. **1935** J. GUTHRIE *Little Country* xxi. 312 You shouldered your swag and left to seek the foot of another rainbow. **1941** BAKER *N.Z. Slang* v. 41 Such expressions as *to swag it* and *go on the swag* need no elaboration. **1947** D. M. DAVIN *Gorse blooms Pale* 76 Jack went off on the swag for a few years. **1966** J. K. BAXTER *Pig Island Lett.* 16 No books, no bread Are left in my swag. **1971** *N.Z. Listener* 19 Apr. 56/5 He had a compass in his swag but it was pukero'd.

11. (*Austral.* and *N.Z.* examples in former sense.)

1929 K. S. PRICHARD *Coonardoo* 49 A boy with a swag of ideals, Hughie was still, Mrs. Bessie realized. **1949** F. SARGESON *I saw in my Dream* ix. 75, I suppose you blokes get told a lot of yarns about a crook missis and a swag of kids. **1963** *Weekly News* (Auckland) 5 June 37/2 There was a big swag of fowls on the station running semiwild. **1973** *New Journalist* (Australia) July–Aug. 4/1 It is cheaper to buy a swag of aged situation comedies.. than to produce even the simplest studio-bound program in Australia.

12. a. swag lamp, light *N. Amer.*, an overhead light externally wired so that the flex hangs in a loop across the ceiling towards the power socket; **swagsman** (earlier examples); also *N.Z.*; now *Obs.*

1970 *Toronto Daily Star* 24 Sept. 28/6 (Advt.), Swag lamps; chromed chairs, easy chairs. **1966** M. M. PEGLER *Dict. Interior Design* (1967) 436 *Swag light*, a lamp or light fixture which is hooked into the ceiling with the electric cord..swagged from the hanging point to the nearest wall, and then down to the floor outlet where it is plugged in. **1869** in W. M. HUGO *Hist. First Bushmen's Club Austral. Colonies* (1872) 30 A swagsman, and not ashamed to own it. I have done the 'wallaby' for years past in search of a billet. **1874** A. BATHGATE *Colonial Experiences* xv. 212 One source of annoyance to the squatters is the 'swagsmen'..or men who travel about the country, professedly in search of work, but who do not in reality want it.

b. (Earlier example.)

1829 P. EGAN *Boxiana* 2nd Ser. II. 74 It is impossible to describe the applause bestowed upon Delay by the boys of the Blue Anchor, the Cock and Cross, and the Ship and Gun, near the *great swag* shop in the east.

swag, *v.* Add: **4.** Chiefly *Austral.* and *N.Z.* **a.** Also in extended use, to travel as a swagman (*up* a region). **b.** Also, to carry in a 'swag'; to wander about (the land) as a swagman.

1861 J. HAAST *Rep. Topographical Exploration Western Districts Nelson Province* i. 16 We again started, on the 11th of February, swagging part of the provisions, &c., down the Buller. **1875** J. JENKINS *Diary Welsh Swagman* (1975) 52 It is better than swagging the country..searching for work. **1883** W. S. GREEN *High Alps N.Z.* 247 We would be obliged..to obtain a sheep and 'swag' it up the glacier again. **1901** *Bulletin Reciter* (Sydney) 5 And swagging up the long divide that leads to Daybreak Range We came. **1914** A. A. GRACE *Tale of Timber Town* 116 You'll get the tucker..and you'll help swag it. **1939** J. D. PASCOE *Unclimbed N.Z.* 42 We left the hut in auspicious weather to swag up the Mingha riverbed. **1960** 'A. CARSON' *Rose by any Other Name* ix. 50, I was swagging my way up to the Northern Territory.

5. *Criminals' slang.* † **a.** To steal; to make away with (stolen property). *Obs.* **b.** To push (a person) forcefully, to 'shove'; to take or snatch away roughly.

1846 *Swell's Night Guide* 113/2 Bag, to take away, see pinch and swag. **1886** H. BAUMANN *Londinismen* 200/2 *Swag*..*v.*, plündern, rauben. **1958** F. NORMAN *Bang to Rights* I. 10 So when we got swaged into the meatwagon I asked another geezer the strength of him, and the strength was that he'd got nicked for ponceing. **1978** J. BARNETT *Head of Force* iii. 21 The object is to see if the Commissioner was swagged away by anyone during the demo.

swager[2] (swē[1]·dʒəɪ). Now *rare.* [f. SWAGE *v.*[3] + -ER[1].] One who swages metal.

1881 in *Instructions to Census Clerks* (1885) 91. **1921** *Dict. Occup. Terms* (1927) § 190 s.v., Agricultural machine knife swager. **1954** *Times* 9 Apr. 9/4 In my grandfather's lifetime the swager was a familiar figure in the West Country.

swagged, *a.* [f. SWAG *sb.* 6 + -ED[2].] Draped in swags; decorated with swags.

1959 *House & Garden* July 13 For curtains, we suggest ..swagged muslin. **1970** *Daily Tel.* 21 Oct. 15 Two Moroccan woollen belts, about 54in long,..bound and swagged with other bright wools.

swagger, *sb.*[1] Add: **2.** Also, short for *swagger bag, coat*, etc.: see *SWAGGER-.

1929 *Papers Mich. Acad. Sci., Arts & Lett.* X. 327/2 *Swagger* (hospital slang), a tunic for promenade occasions; 'square-push' tunic. **1939** [see *beaver lamb* s.v. *BEAVER[1] 6]. **1968** J. IRONSIDE *Fashion Alphabet* 38 *Swagger*, a jacket with a very full back, hanging loose in front. **1979** *Arizona Daily Star* 1 Apr. (Suppl.), In-fashion spring bags... Swaggers, shoulder-straps, double handles.

swagger-. Add: *swagger-bag*; **swagger coat,** a three-quarter-length ladies' coat cut with a loose flare from the shoulders (particularly fashionable in the 1930s).

1933 *Bulletin* (Glasgow) 14 Oct. 15/1 A swagger-coat with collarette and gauntlets of black astrakhan. **1938** 'J. BELL' *Port of London Murders* ii. 24 Her hands were pushed into the pockets of an old swagger coat. **1953** 'P. WENTWORTH' *Watersplash* ii. 8 The glove and its fellow had been thrust into the pocket of a blue swagger coat. **1974** *Index-Jrnl.* (Greenwood, S. Carolina) 23 Apr. 3/2 (Advt.), Special selection of baskets, swagger bags, totes, envelopes. **1980** B. BAINBRIDGE *Winter Garden* x. 74 She stood in the gutter in her swagger coat and allowed her teeth to chatter piteously.

swagging, *vbl. sb.* Add: **3.** *Austral.* and *N.Z.* Travelling as a swagman; carrying one's 'swag', back-packing.

1883 W. S. GREEN *High Alps N.Z.* xvi. 268 Descending to the lower camp..and doing the hard swagging work all over again. **1892** *N.Z. Alpine Jrnl.* I. 100 All our dirty work and heavy swagging will be done for us. **1940** W. S. GILKISON *Peaks, Packs & Mountain Tracks* xiii. 102 Swagging—or, if you prefer it, back-packing—is more or less an essential part of every climbing trip. **1960** 'A. CARSON' *Rose by any Other Name* ix. 50 Swagging is an honourable profession in Australia.

Swahili. (Earlier example of the language.)

1847 W. W. GREENOUGH in *Jrnl. Amer. Oriental Soc.* I. 263 The Sooahelee has been called a lingua franca.

Swahilian, *a.* (Example.)

1846 J. R. BROWNE *Etchings Whaling Cruise* xvi. 335 The Sowhelian language is the most generally spoken.

Swainson (swē[1]·nsən). The name of William *Swainson* (1789–1855), English naturalist, used in the possessive to designate birds named in his honour, as **Swainson's † buzzard, hawk,** a dark-coloured buzzard hawk, *Buteo swainsoni*, found in western North America; **Swainson's thrush,** an olive-backed thrush, *Hylocichla ustulata*, found in western North America; **Swainson's warbler,** a brown and white warbler, *Limnothlypis swainsonii*, found

in swamp regions of south-eastern North America.

1858 S. F. BAIRD *Birds Pacific Railway Routes* 19 Swainson's Buzzard,.. more nearly related to a generic form of the Old World. *Ibid.* 252 Swainson's Warbler.. South Atlantic States. **1869** *Amer. Naturalist* III. 31 Swainson's Thrush... Common at Cœur d'Aleñe Mission. **1895** *U.S. Dept. Agric. Yearbk. 1894* 222 The food of Swainson's hawk..is of much the same character as that of the two preceding species. **1912** C. A. REED *Birds Eastern N. Amer.* 359 Swainson's Warbler is a comparatively rare species found in the Southeastern States. **1939** F. C. LINCOLN *Migration Amer. Birds* 79 Observers in the Great Plains saw large flocks of Red-tailed Hawks, Swainson's Hawks and Rough-legged Hawks wheeling majestically. **1972** *Islander* (Victoria, B.C.) 21 May 6/2 In the evening the sound of the Swainson's thrush. **1976** *National Observer* (U.S.) 31 July 5/2 Southern Swainson's warblers would no doubt agree. **1980** *Country Life* 3 July 46/2, I watched huge flocks of broad-winged hawks..and a few Swainson's hawks.

Swakara (swa·kara). [f. the initials of *S*outh *W*est *A*frica + *KARA(KUL.] The coat of a karakul lamb, bred in Namibia, valued as a fur. Chiefly *attrib.*

1966 *Fur Rev.* May 13/1 Selective breeding has improved the..qualities of S.W.A. Persian Lambskins and brought about a changeover..to a flat glossy pelt... In order to spread..the extended range of S.W. African merchandise, a publicity agent has..[come] up with the clever catchword 'Swakara'. **1973** *Country Life* 1 Feb. 302/3 Slim little jackets in Swakara broadtail, dyed delectable shades of peach. **1978** *Lancashire Life* Sept. 110/1 Natural grey Swakara Persian lamb has a unique charm.

swale, *sb.*[3] Add: Also *U.S.*, a hollow between adjacent sand-ridges.

1894 *Dialect Notes* I. 334 *Swale*, low land between sand ridges on the coast beaches [of New Jersey]. **1945, 1976** [see *point bar* (*b*) s.v. *POINT *sb.*[1] D. 14].

swale (swē[1]l), *sb.*[4] *South. dial. local.* [Origin uncertain: cf. SWEAL, SWALE *sb.*, and *swill* sb.[1] 10 in *Eng. Dial. Dict.*] A small broom or brush without a stick for a handle.

1949 K. S. WOODS *Rural Crafts of England* III. vii. 123 Some besoms are made without sticks. These are known as swales, an interesting word meaning 'a small bright fire enough to boil a kettle'. Swales are used to brush the flakes from steel-plate. **1968** J. ARNOLD *Shell Bk. Country Crafts* 100 Like the besom, it has a head of birch..but is without a handle and is called a swale.

Swaledale (swē[1]·ldē[1]l). The name of a region of North Yorkshire used *absol.* or *attrib.* to designate a long-woolled sheep of the hardy hill breed first developed in the area; also, the breed itself or the long coarse wool produced by a sheep of this kind.

1916 W. J. MALDEN *Brit. Sheep & Shepherding* vi. 58 Among the remaining breeds of the northern hills may be mentioned the Swaledale, a very hardy breed. **1944** G. HENDERSON *Farming Ladder* i. 24 The stock consisted of 650 Blackfaced mountain ewes, and thirty pedigree Swaledales. **1961** J. GUNSTON *Profit from Sheep* ii. 22 Swaledales..do very well in a wide range of hard-grazing and cold districts. **1971** *Farmers Weekly* 19 Mar. 43/3 Certain wool types such as Blackface, Devon and Swaledale have met a better market demand than others. **1980** *Times* 3 Mar. 16/2 A Pennine farmer..was saving 50 bales of hay a day among his Swaledale ewes.

swallow, *sb.*[1] Add: **1. d.** *ellipt.* for *swallow dive* below.

1902 *Encycl. Brit.* XXXIII. 121/2 The 'swallow' is one of the most thrilling dives. **1971** 'D. HALLIDAY' *Dolly & Doctor Bird* xi. 148 Sergeant Trotter himself nipped up the diving-board and executed a swallow and somersault.

e. *transf.* A woman employed by the Soviet intelligence service, who seduces men for the purposes of espionage. *slang.*

1972 D. BLOODWORTH *Any Number can Play* ix. 69 You have doubtless read about the..'swallows' of the KGB, the young ladies trained..to bed down intelligence targets, so that they can be comfortably and conveniently bugged and photographed in compromising..positions? **1976** 'M. BARAK' *Secret List H. Roehm* xii. 130, I need a swallow in America. One..who is sexually skilled and expert in obtaining information. **1979** P. WAY *Sunrise* i. 15 Had she been working for the KGB, Joanna would have been..called a 'swallow'. In the CIA she would have been a 'honeypot'.

4. swallow dive, a forward dive in which the arms are extended sideways, to simulate the outline of a swallow, until just before entry into the water; also *fig.*; so **swallow-diving;** hence **swallow-dive** *v. intr.*; **swallow fork** orig. *Amer.*, a forked cut used in marking cattle or sheep on the ear (see quot. 1966); hence **swallow-fork** *v. trans.*, to cut a swallow fork in (the ear); **swallow-forked** *ppl. a.*, shaped so as to cut a swallow fork; **swallow's nest:** also in *swallow's nest soup*, an oriental dish (see BIRD'S-NEST, BIRD-NEST *sb.* 1).

1898 *Swimming Mag.* Oct. 46/1 To Englishmen the term 'swallow' dive, not 'swan', would best convey the notion of this idealistic manner of reaching the water. **1971** 'D. HALLIDAY' *Dolly & Doctor Bird* xi. 148 Sergeant Trotter, reappearing at the top of the diving-board, swallow-dived efficiently. **1976** 'A. HALL' *Kobra Manifesto* xvi. 215 Sassine had come off his high in a swallow dive. **1897** *Encycl. Sport* II. 425/1 The most graceful is that termed 'swallow-diving', the body being shot out from the board [etc.]. **1636** *Plymouth* (Mass.) *Rec.* (1889) I. 1 Every mans marke of his Cattle... Christopher Wadesworth a swallow forke. **1869** *Overland Monthly* III. 126 An over-slope and a slit in the right, and a swallow-fork in the left. **1934** *Amer. Ballads & Folk Songs* xvi. 409 They cropped and swallow-forked his ears. **1966** *Publ. Amer. Dial. Soc.* 1964 XLII. 16 *Swallow fork*, two slits run together to form a *W* or an *M*. **1972** P. NEWTON *Sheep Thief* xvi. 134 It was a pair of swallow-forked earmarkers. **1920** E. & P. SYKES *Through Deserts & Oases Central Asia* iv. 78 Swallows' nest soup is almost unprocurable nowadays. **1976** *Times* 14 Feb. 10/4 A real Thai Chinese restaurant..three colours swallow's nest soup..or even plain shark's fin soup.

swallow, *v.* Add: **1. b.** Also, *to swallow the anchor*, to retire from a sea-faring life; also *transf.*; *to have swallowed the dictionary*: see *DICTIONARY 1 c.

1907 J. MASEFIELD *Tarpaulin Muster* xii. 129 An old sailor..had 'swallowed the anchor' in Colon. **1931** A. R. L. GARDNER *Art of Crime* 253 We are glad to be able to quote these..words to..our readers who may entertain..fears lest the crook proper should one day 'swallow the anchor' and retire permanently from the stage. **1977** *Islander* (Victoria, B.C.) 22 May 6/1 But, now he had 'swallowed the anchor', he was a hard-headed business man.

swallower. 3. Delete † and add later example with *up*.

1855 MRS. GASKELL *Let.* Feb. (1966) 332 Meta's atelier is such a swallower-up of time.

swallow-tailed, *a.* **II. 6.** (Earlier example.)

1824 J. MORIER *Adventures of Hajji Baba* I. p. xl, I sighed for shaven chins and swallow-tailed coats.

swamp, *sb.* Add: **3. a.** *swamp-dweller* (earlier example), *forest, -jungle, land* (earlier examples), *meadow*; **swamp buggy** *N. Amer.*, a vehicle used in swampy regions; *spec.* a tracked vehicle which can pull a heavily loaded trailer; **swamp cooler** *U.S.* (see quot. 1950); **swamp fever,** (*b*) a contagious virus disease of horses, causing anæmia, emaciation, and usually death; **swamp fire** *Canad.*, methane burning in a swampy area; a will-o'-the-wisp (also used in metaphorical comparisons); **swamp plough** *N.Z.*, a type of plough with a large mould-board, for use on heavy soils; **swamp rock,** a type of rock music associated with the Southern U.S.; **swamp Yankee** *U.S. dial.* (see quot. 1963).

1941 *Nat. Geogr. Mag.* June 706 Their 'swamp-buggy' is a seagoing amphibious-looking vehicle. Its 10-foot high wheels are equipped with fat, fin-studded oversized tires which act as propellers; when the odd vehicle leaves the land and takes to water, it begins to swim. **1966** *North* July–Aug. 14/2 When the usually dependable swamp buggy breaks down, it's back to the dog team. **1973** *Globe & Mail* (Toronto) 3 Feb. 7/5 While the public sleeps, a Pandora's Box is opening to release a flood of hovercraft, dune buggies, swamp buggies, trail bikes, air sleds and airboats on the long-suffering landscape of crowded Southern Ontario. **1950** *Newsweek* 14 Aug. 51 In dry climates it is possible to rig up a primitive but highly effective cooling system, called a 'swamp cooler'. It consists simply of a fan blowing over an excelsior mat which is drenched with dripping water. **1979** *Tucson* (Arizona) *Citizen* (Weekender Mag.) 28 Apr. 9/1 A swamp cooler has maybe five moving parts; if it quits, you go up on the roof, look to see which part has quit moving, and replace it. **1890** Swamp-dweller [see *PINE LAND]. **1903** *Rep. Min. Agric. Canada 1902* 85 There has been known in the Red River Valley a peculiar and very fatal disease of horses... It is a disease of low lying and swampy country and it is therefore popularly known as swamp fever. **1975** *Daily Colonist* (Victoria, B.C.) 6 July 2/3 Swamp fever..equine infectious anemia—has ravaged almost half of..the little ponies. **1903** S. E. WHITE *Forest* 122 Like swamp-fire, it lured the imagination always on and on and on through the secret waterways of the uninhabited North. **1954** V. LYSENKO *Yellow Boots* 146 Behind them the swamp fire, like a gigantic Jack o'lantern, bumped and danced and ran around the sky, then finally, as it reached the ground, evaporated into the night air. **1982** H. LIEBERMAN *Night Call* iii. 12 Daughtry's reputation..traveled like swampfire. **1909** GROOM & BALFOUR tr. *Warming's Oecol. Plants* lx. 234 (*heading*) Littoral swamp-forest. Mangrove. **1955** P. A. BUXTON *Nat. Hist. Tsetse Flies* ix. 269 In places there are 'swamp forests'..the trees growing in a few feet of water at all seasons. **1964** G. B. SCHALLER *Year of Gorilla* (1965) viii. 215 The swamp forest that grows in the low country bordering the South China Sea. **1902** D. G. HOGARTH *Nearer East* 108 Torrential floods, which..support a dense swamp-jungle. **1662** in *Connecticut Hist. Soc. Coll.* (1912) XIV. 433 One Parcel of land.. being Swamp land. **1701** *Early Rec. Providence, Rhode Island* (1894) V. 125 A Certaine ffarme or tract of land consisting of upland swampe land & Meadow land. **1697** *Cambridge* (Mass.) *Proprietors' Rec.* (1896) 344 Four Rods of fence, Lyeing att the head of Samuel Hastings Swamp-meadow. **1880** *Harper's Mag.* June 80 Out in the swamp meadow the tall clumps of boneset show their dull white crests. **1951** R. P. HOBSON *Grass beyond Mountains* 41 We sat around..talking of range cows, and tough trails, slough grass and swamp meadows. **1930** L. G. D. ACLAND *Early Canterbury Runs* 1st Ser. iii. 42 They..spent a lot of money in cutting the scrub, crushing it down with rollers, and ploughing it in with swamp ploughs. **1973** *Massey Ferguson Rev.* (N.Z.) Mar.–Apr. 5/1 He..leaves it for two years before getting to work with a 19-inch swamp plough. **1970** *Guardian* 17 Apr. 10/2 Then Creedence. The band['s]..music, like that of Delaney and Bonnie, is called 'swamp rock', and identified with the Southern States of America. **1941** H. KURATH *Linguistic Atlas of New England* II. II. Map 450 The map shows a great variety of terms, largely derogatory and jocular, applied to a person who lives in the country—specifically to an old farmer who seldom visits the village or city. The following terms were recorded in more than one community: rustic,.. swamp Yankee, hayback, hayseed or hayseeder. **1963** *Amer. Speech* XXXVIII. 121 The term *swamp Yankee* may be defined as 'a rural New England dweller who abides today as a steadfast rustic and who is of Yankee stock that has endured in the New England area since the colonial days.' **1975** G. V. HIGGINS *City on Hill* iv. 104 That back country's full of swamp Yankees, guys..that impregnate their own daughters.

b. swamp deer: see also *BARASINGHA; (earlier example); **swamp rabbit,** either of two dark brown rabbits of the south-eastern United States, the cane-cutter, *Sylvilagus aquaticus*, or the marsh rabbit, *S. palustris*; cf. *swamp hare*; **swamp robin** (earlier and later examples); also, = TOWHEE; **swamp wallaby,** a large wallaby, *Wallabia bicolor*, which has reddish or greyish fur with darker markings; **swamp warbler** (earlier example).

1874 T. C. JERDON *Mammals of India* 254 The Swamp Deer... Horns very large and moderately stout. **1845** C. LYELL *Second Visit U.S.* I. 228, I had heard much of the swamp-rabbit, which they hunt near the coast in South Carolina and Georgia. **1875** *Fur, Fin & Feather* (ed. 3) 136/1 The 'swamp rabbit' inhabits the heavy timbered woodlands and river bottoms. **1938** M. K. RAWLINGS *Yearling* v. 51 The pair of black swamp rabbits was not new. **1964** W. H. BURT *Field Guide Mammals* (ed. 2) 223 Swamp Rabbit... This is a rich brownish-gray rabbit with coarse hair; feet rusty. **1769** R. SMITH *Jrnl.* 18 May in *Tour of Four Great Rivers* (1906) 41 The lively Note of the Swamp Robin, the Red Bird and other Birds from the earliest Dawn is entertaining. **1955** Swamp robin [see *JOREE]. **1896** Swamp wallaby [see *brush wallaby* s.v. *BRUSH *sb.*[1] 4]. **1970** W. D. L. RIDE *Guide Native Mammals Austral.* v. 47 The Swamp Wallaby..is usually placed in a separate genus *Wallabia*. **1859** THOREAU *Jrnl.* 30 Apr. in *Writings* (1906) XVIII. 167 This first *off-coat* warmth just preceding the advent of the swamp warblers (parti-colored, red-start, etc.) brings them out.

c. swamp ash (earlier example); **swamp azalea** (later example); **swamp blackberry,** a low-growing, semi-evergreen dewberry, *Rubus hispidus*, found near water and marshy ground in parts of Canada and northern and central U.S.A.; **swamp blueberry,** the high-bush blueberry, *Vaccinium corymbosum*, or its fruit; **swamp-cabbage** (later examples); also, the cabbage palmetto, *Sabal palmetto*; **swamp dewberry** = *swamp blackberry* above; **swamp hickory,** the water hickory, *Carya aquatica*, or the bitternut hickory, *C. cordiformis* (earlier and later examples); **swamp honeysuckle,** (*b*) a honeysuckle of eastern North America, *Lonicera oblongifolia*, with yellowish flowers and red berries; **swamp laurel,** substitute for def.: the sweetbay magnolia, *M. virginiana*; formerly, also the pale American laurel, *Kalmia polifolia*; (earlier and later examples); **swamp lily** (earlier example); (*d*) *Crinum americanum*, which bears white flowers and is native to the south-eastern United States; **swamp mahogany,** a gum tree, *Eucalyptus robusta*, native to coastal regions of eastern Australia; (earlier example); **swamp maple** (examples); **swamp pine** (later examples); **swamp rose** (examples); also, another wild N. Amer. rose, *Rosa palustris*; **swamp sumach,** for *venenata* read *vernix* (later examples); **swamp willow** (earlier example).

1794 W. CLARK *Jrnl.* 15 Sept. in *Mississippi Valley Hist. Rev.* (1914) I. 437 The face [of the land] is nearly covered with a thick growth of Shrubbery, Brush, some Beech, Swamp Ash. **1958** G. A. PETRIDES *Field Guide to Trees & Shrubs* 365 Swamp Azalea... A medium-sized to tall shrub with leaves glossy above. **1854** THOREAU *Jrnl.* 4 Aug. in *Writings* (1906) XII. 419 The swamp blackberry on high land, ripe a day or two. **1903** H. L. KEELER *Our Northern Shrubs* 161 Few trailing plants combine a better effect of flower and foliage than our Swamp blackberry. **1975** E. WIGGINTON *Foxfire 3* 285 Swamp blackberry is found in thickets in low, wet places. **1860** THOREAU *Jrnl.* 30 Dec. in *Writings* (1906) XX. 299 Some ten days later comes the high blueberry, or swamp blueberry, the commonest stout shrub of our swamps. **1917** E. S. BAILEY *Sand Dunes Indiana* 154 There is a chance to study all the sides of a small pond, with the shrub zone of plants in perfect type, such as swamp blueberry, cranberry [etc.]. **1949** *Pacific Spectator* Spring 223 You had to cross the river..to find the low swamp blueberries, lighter blue and sweeter than any other kind. **1880** *Harper's Mag.* June 66 The swamp-cabbage flower..peers above the ground beneath his purple spotted hood. **1938** M. K. RAWLINGS *Yearling* xx. 250 He pulled away layer after layer of the white cores and came at last to the hearts [of palms], crisp and sweet. He said, 'Now I want that fryin' pan, Mr. Penny, please, for my swamp cabbage.' **1942** S. KENNEDY *Palmetto Country* 3 Folks outside the region usually think of the palmetto as the tall palm which is locally called the swamp cabbage or cabbage palm. **1976**

National Observer (U.S.) 22 May 16-A/1 They were forced to subsist on a diet of unpolished rice, swamp cabbage, and tiny fish. **1924** C. Deam *Shrubs Indiana* 109 *Rubus hispidus* Linnaeus. Swamp Dewberry. **1942** L. R. Tehon *Fieldbk. Native Illinois Shrubs* 116 The Swamp Dewberry grows near lakes and marshes, especially at the base of wooded slopes. **1976** *Hortus Third* (L. H. Bailey Hortorium) 985/2 *Swamp dewberry, running blackberry, swamp b.*, slender, hispid, often glandular trailer, laying close to the ground, without prickles. **1806** in *Message from President of U.S., communicating Discoveries made in exploring the Missouri by Captains Lewis & Clark* 65 The growth, on the highest [places is] handsome oaks, swamp hickory, ash, grape vines, &c. **1912** I. S. Cobb *Back Home* 306 He was tough as swamp hickory. **1938** C. H. Matschat *Suwannee River* 161 They alus stuck togither tightern the bark on a swamp hickory. **1958** G. A. Petrides *Field Guide Trees & Shrubs* 47 Swamp Honeysuckle *Lonicera oblongifolia*... A more or less hairless honeysuckle. **1743** J. Clayton *Flora Virginica* 83 *Magnolia Laurifolia*,..Swamp-Laurel. **1869** J. G. Fuller *Uncle John's Flower-Gatherers* 138 The farmers around here call it [*sc. Kalmia*] 'Swamp-Laurel'. **1884** C. S. Sargent *Rep. Forests N. Amer.* 20 Sweet Bay..Swamp Laurel... A tree 15 to 22 meters in height. **1737** J. Brickell *Nat. Hist. N. Carolina* 21 Another Weed, vulgarly called the Swamp-Lillie..grows in the Marshes and low Grounds, and is something like our Dock in its Leaves. **1884** A. Nilson *Timber Trees New South Wales* 71 Swamp Mahogany.—A large tree..with a rough furrowed bark. **1810** Swamp maple [see *maple tree]. **1869** Mrs. Stowe *Oldtown Folks* xiv. 153 Here and there, a swamp-maple seemed all one crimson flame. **1936** E. B. White *Let.* 3 Sept. (1976) 141 Joe and I have gathered boughs of red swamp maple, to decorate the back porch. **1969** T. H. Everett *Living Trees of World* xxii. 221/1 The most important American soft maples are the red or swamp maple..and the silver maple. **1743** M. Catesby *Nat. Hist. Carolina* II. p. xxii, The Swamp Pine grows on barren wet land. **1851** J. S. Springer *Forest Life* 41 This difference is accounted for by..the tardiness with which the swamp Pine matures. **1958** G. A. Petrides *Field Guide Trees & Shrubs* 15 Swamp Pine..similar to Pitch Pine. **1785** H. Marshall *Arbustrum Americanum* 135 Swamp Pennsylvanian Rose..[rises] to a height of four or five feet. **1814** J. Bigelow *Florula Bostoniensis* 121 Swamp rose..grows in swamps and wet grounds. **1902** *Outing* June 272/2 The Carolina or swamp rose..is well known to us all. **1814** Swamp sumach [see *poison dogwood* s.v. *poison *sb.* 5 b]. **1945** H. T. Darlington *Higher Plants of Michigan* 25 Red maple and swamp sumac..may add to the brilliant effect. **1765** J. Bartram *Jrnl.* 31 July in *Trans. Amer. Philos. Soc.* (1942) XXXIII. 17/1 They have yᵉ upland willow oak with A hoary leafe, & yᵉ swamp willow with A narrow leafe.

swamp, *v.* Add: **5.** (Earlier and later examples.) Also with *out*. Also *Canad.*

1784 M. Patten *Diary* 18 Mar. (1903) 480, I swampt out 4 small oak logs the boys saved in cuting wood Ready for hauling out. **1851** J. S. Springer *Forest Life* 84 This is done by an experienced hand, who 'spots' the trees where he wishes the road to be 'swamped'. **1871** R. L. Dashwood *Chiploquorgan* viii. 104 A crew of lumberers have different occupations assigned to them;..the 'swampers', who 'swamp'—cut roads—to the felled trees, to enable the 'teamster' and his assistants to haul them on a 'Bob sled'. **1937** P. K. Devine *Devine's Folk Lore of Newfoundland* 50 To swamp a road or path is to build one with a bedding of boughs to be used in hauling slide loads of wood in winter. **1954** C. Bruce *Channel Shore* 27 [He] had swamped a hauling-road into the middle of the stretch that lay south of the shore road. **1974** D. Sears *Lark in Clear Air* iii. 40 Where the logs came from and who cut them and the names of the horses that swamped them out.

6. *intr.* To work as a bullock-driver's assistant (also casually, in return for having one's 'swag' carried); to make (*one's way*) by obtaining a lift from a traveller. Cf. *swamper 1 c, d. *Austral. slang.*

1926 K. S. Prichard *Working Bullocks* 101 Billy Williams the bullocky, and Ern Collins who was swamping for him, turned their team into the yards on the following Monday. **1937** E. Hill *Ports of Sunset* 96 In they came, across the jagged Leopolds, or up from the desert, 'swamping' with a bullocky, staggering behind a pack donkey, or on Shanks' pony. **1944** M. J. O'Reilly *Bowyangs & Boomerangs* 6 My duties were to help to load and unload, bring the horses in the morning, to harness up, help to corduroy bad patches on the track, [etc.]... Fortunately the chap I 'swamped' for was an exceptionally good sort. **1964** T. Ronan *Packhorse & Pearling Boat* 170 If I broke it for a tenner, I'd roll my swag and swamp my way back to Queensland.

swamper. Add: **1. b.** (Examples.) Also, an assistant to a cook.

1907 *Oregonian* (Portland) 13 Oct. 8/1 He was a swamper in a saloon. **1929** *Collier's* 5 Jan. 33/1 As a result it became pay dirt, and in later years the swamper actually had to pay for his job. **1939** P. A. Rollins *Gone Haywire* 65 Until the call was given, the average cook permitted nobody to approach the fire except the helper whom he rarely had, and who was known as the flunky, roustabout, swamper, or cook's louse. **1962** E. Lucia *Klondike Kate* iii. 81 The [theatrical] company had its own bartenders and swampers. **1979** D. Anthony *Long Hard Cure* ii. 20 He'd returned promptly to his apartment over the tavern. His Negro swamper bore him out.

c. An assistant to a driver of horses, mules, or bullocks. *slang* (orig. *U.S.*).

1870 *Daily Territorial Enterprise* (Virginia City, Nevada) 21 Apr. 3/1 A 'swamper' is a man who goes with the driver of a 10, 12, or 14-mule team as his assistant—the driver being chief engineer and the swamper first-assistant. **1926** K. S. Prichard *Working Bullocks* i. 6 Red Burke shouted to the bullocks... His swamper yelled and danced. **1960** A. Downs *Wagon Road North* 43 Many drivers were accompanied by a 'swamper', who was usually a young fellow apprenticed to the teaming business. The swamper looked after the horses, including rounding them up in the morning, usually about four o'clock, and in general assisted the teamster with the over-all duties of freighting.

d. One who travels on foot but has his swag carried on a wagon; hence, one who obtains a lift. Cf. *swamp *v.* 6. *Austral. slang.*

1901 M. Vivienne *Travels in W. Australia* 284 A 'swamper' is a man tramping without his swag, which he entrusts to a teamster to bring on his waggon... While on foot the swamper will generally leave the track, and prospect. **1929** J. Raeside *Golden Days* 380 With many a swamper's swag on And many a billy black. **1966** T. Ronan *Once there was Bagman* i. 15 My..fellow swamper tossed his swag off [the mailman's truck] here; he was home.

e. An assistant to the driver of a lorry. *N. Amer. slang.*

1929 *Amer. Speech* IV. 345 *Swamper*, a helper on an auto truck. **1953** C. Armstrong *Catch-as-catch-Can* xiv. 114 The driver of this linen service truck, told his swamper ..to stay with it. **1963** *Sun* (Vancouver) 28 Feb. 1/5 A wood truck swamper was charged $25 each for two stolen kisses Wednesday. **1975** E. Iglauer *Denison's Ice Road* viii. 194 We don't have swampers, a second man on the truck, the way the oil-field men have.

2. (Earlier examples.)

[**1735** J. Belcher in *New Hampshire Provincial Papers* (1870) IV. 878 The B B's Pr—st is a jolly Fellow. I hear he stood Kick and Cuff upon the Road with some Swampeers.] **1775** *N. Carolina Gaz.* (New Berne) 24 Mar. 3/3 Fellow Dismalites and Swampers, are we not the Men whom God hath appointed to curb the Insolence of Britain. **1857** J. D. Long *Pictures of Slavery* xvii. 323, I made an appointment to deliver a temperance address to the 'swampers'.

swan, *sb.* Add: **2. c.** (Later examples with allusion to Shakespeare.)

1895 G. B. Shaw *Our Theatres in Nineties* (1932) I. 197 Everyone concerned..is full of earnest belief that the splendor of the Swan will be revealed at last, like the Holy Grail. **1922** Joyce *Ulysses* 186 Shakespeare..does not stay to feed the pen chivying her game of cygnets towards the rushes. The swan of Avon has other thoughts.

e. [f. *swan *v.*[1] 2.] An apparently aimless journey; an excursion made for reconnaissance or for pleasure. *slang* (orig. *Mil.*).

1946 Visct. Montgomery *El Alamein* 45 A recurrence of what was then becoming known in the Eighth Army as the 'annual swan' between Egypt and El Agheila. **1958** *Spectator* 23 May 665/2 The General.., yielding to a very natural temptation to go for a 'swan' early in the battle, was away from his headquarters for over thirty-six hours. **1960** C. Achebe *No Longer at Ease* xvii. 153 But for an African like you, who has too many privileges as it is, to ask for two weeks to go on a swan, it makes me want to cry. **1968** *Listener* 22 Feb. 238/1 It [*sc.* a festival] has become an accepted 'swan' for the British correspondents. **1974** D. Hart-Davis *Peter Fleming* iv. 75 The trip as a whole was designed to be what he later called a 'swan'—a general look round. **1979** D. Clark *Heberden's Seat* vii. 150 'Reed and I may have to go to London for the day.'.. 'It's not just a swan is it?'

4. a. (Further chiefly *poet.* examples.) *swancomb* (fig.), *-flight*, *-meat*, *-plumage*; objective, *swan-delighting* adj.; instrumental, *swan-instructed* adj.; parasynthetic and similative, *swan-breasted*, *-bright*, *-feathered*, *-fledged*, *-soft* adjs.

1930 R. Campbell *Adamastor* 73 The great swan-breasted seraphs soar and sing. **1923** E. Sitwell *Bucolic Comedies* 35 The swan-bright fountains. **1922** Swancomb [see *high-reared* s.v. *high *adv.* 10 a]. **1936** Auden *Look, Stranger!* 41 The swan-delighting river. **1953** R. Graves *Poems* 17 Past either cheek Swan-feathered arrows whistle. **1862** G. M. Hopkins *Vision of Mermaids* (1929), And shake From wings swan-fledged a wheel of watery light. **1959** E. Pound *Thrones* xcviii. 38 The King's job, vast as the swan-flight. **1942** S. Smith *Magic Morning* in *Coll. Poems* (1975) 206 'Charley, Charley, Charley' cry the swan-instructed curlews. **1922** Joyce *Ulysses* 151 Wonder what kind is swanmeat. **1953** E. Sitwell *Gardeners & Astronomers* 37 And Cygnus who gave you all his bright swan-plumage. **1925** —— *Troy Park* 12 In the thick swan-soft fields.

b. **swan dive** *U.S.*, a swallow dive (see *swallow *sb.*[1] 4); hence **swan-dive** *v. intr.*; **swan-drop,** (*b*) (earlier example); **swanproof** *a. nonce-wd.*, not susceptible to the influence of Shakespeare (cf. sense 2 c in Dict. and Suppl.); **swan-shot,** also used in angling as a weight; **swan-song:** hence, any final performance, action, or effort; **Swan Vesta,** the proprietary name of a make of match; cf. Vesta 4.

1898 *Swimming Mag.* Oct. 45/2 The diving..included forward headers,..somersaults and the 'Swan' dive from twenty, thirty, and forty feet. **1912** J. London *Son of Sun* ii. 53, I used to swan-dive a hundred and ten feet in the clear. **1932** E. Hemingway *Death in Afternoon* i. 21 As though a diver could control..[the] speed..of a swan dive. **1972** B. F. Conners *Don't embarrass Bureau* (1973) i. 7 Mrs. Green..executed her swan dive, flopping onto the water with the poise of a stricken bird. **1853** J. Palliser *Solitary Rambles* ii. 55 My own saddle-bags contained..powder and shot, and, by great good luck, some swan-drops. **1905** G. B. Shaw in *Shaw on Theatre* (1958) 103 Since Shakespear's words are still the basis of the dialogue, there are moments when the bard enjoys his own again; for all the players are not as completely swanproof as Mr Tree. **1856** 'Stonehenge' *Man. Brit. Rural Sports* 255/2 Swan-shot or lead, in some form, is required to sink the bait. **1971** *Angling Times* 10 June 6/2 Any float will do that a swan shot can't quite take under. **1976** *Monitor* (McAllen, Texas) 28 Nov. 11A/4 Rockefeller fairly bubbled with optimism during a recent swan song interview. **1978** G. Greene *Human Factor* vi. ii. 319 Ivan made his swan song as an interpreter in a building not far from the Lubianka prison. [**1907** *Yesterday's Shopping* (1969) 24/2 Swan White Pine Vestas. (Bryant & May's.)..Doz... 0/3½.] **1908** *Trade Marks Jrnl.* 12 Aug. 1340 *Swan Vestas*... Matches. Bryant & May, Limited, Fairfield Works, Bow, London..; match manufacturers. **1958** J. Townsend *Young Devils* vii. 59, I collected..a number of loose Swan Vestas from the class. **1977** 'E. Crispin' *Glimpses of Moon* vii. 109 Ling gave his Swan Vesta box an experimental shake.

d. *swan-maiden* (earlier example).

1859 G. W. Dasent *Pop. Tales from Norse* p. lxi, Brynhildr and the Valkyries..became swan-maidens.

swan, *v.*[1] Add: **1.** Also without *it*. Also *transf.*

1938 H. G. Wells *Apropos of Dolores* vi. 304 He began as an Osteopath but afterwards he became a Mind Healer—with Physical Exercises... He taught them to swan (!?) Swan, you know—like swans. Swanning exercises. Some of them swan now quite beautifully. **1962** *Listener* 13 Sept. 386/2 In his painting Andrea can be seen swanning through the water.

2. To move about freely or in an (apparently) aimless way (formerly, *spec.* of armoured vehicles); hence, to travel idly or for pleasure. Freq. with *about*, *around*, or *off*. *slang* (orig. *Mil.*).

1942 *Daily Tel.* 3 Sept. 6/6 Breaking up his armour into comparatively small groups of..tanks, he began 'swanning about', feeling north, north-west and east for them [*sc.* British tanks]. **a 1944** K. Douglas *Alamein to Zem Zem* (1946) 24 It seemed crazy to go swanning off into the mist. **1945** *Times* 17 Mar. 4/2 [General Patton's armour]..is 'swanning' more or less unchallenged amid the open moors of the Hunsrück plateau. **1947** C. Day Lewis *Poetic Image* 111 A few bold or bomb-happy types still swanning around outside. **1961** G. Egmont *Art of Egmontese* i. 15 Another excellent way of making contacts is, of course, 'swanning' on the Continent. **1971** *Petticoat* 17 July 28/1 You can't do that if you're swanning around making films all the time. **1980** D. Bogarde *Gentle Occupation* viii. 200 She swanned about at the party like the Queen Mother.

Swanee (swǫ·ni). Also **Swannee.** [Var. of *Suwannee*, the name of a river in Georgia and Florida.] **1.** *Swanee whistle*, a small woodwind instrument with a slide-plunger to vary the pitch, chiefly used as a toy. Also *Swanee flute*.

1926 S. T. Warner *Lolly Willowes* ii. 114 She bought a Swanee flute. **1930** R. Paget *Human Speech* 239 Various forms of mute for converting the whistle sound of a Swannee Whistle into a breathed sound. **1961** A. Baines *Mus. Instruments through Ages* ix. 235 The Swanee whistle, scored for by Ravel in *Les Enfants et les sortilèges*, is also made by Indian children as a bamboo bird-pipe. **1962** A. Nisbett *Technique of Sound Studio* x. 172 An object (or person) being thrown high into the air might be indicated by the use of a glide up and down on a swannee whistle (this is a whistle which has a slide piston to govern the pitch). **1978** *Times* 15 July 2/4 (*caption*) Pupils..playing a Swanee whistle..and a Melodica, a wind keyboard. **1983** *Daily Tel.* 23 June 18/4 The piece, to be recited by the composer himself, is performed on toy clarinets, saxophones, rattles, swanee whistles, plastic hosepipes and paper-bags.

2. *to go down the Swanee* = *to go down the drain* s.v. *Drain *sb.* 1e; to become ruined or bankrupt. Cf. *River *sb.*[1] 3* c. *slang*.

1977 *Observer* 21 Aug. 1/3 A senior Leyland convener..called on the Government to give Leyland 'latitude' in settling its pay problems. Without that, he said, the company 'would go down the Swanee'.

swank, *sb.*[2] Add: **2.** = Swanker[2].

1913 V. Sackville-West *Let.* 15 Feb. in V. Glendinning *Vita* (1983) v. 54 [He is] a swank, more swank than you could ever dream of. **1923** 'R. Crompton' *William Again* v. 91 He was a pariah, outside the pale, one of the 'swanks' who lived in big houses and talked soft. **1949** W. C. Williams *Autobiogr.* xxxii. 190 We were not concerned with the moving-picture colony or the swanks.

swank (swæŋk), *a.*[2] *colloq.* (chiefly *U.S.*). [f. Swank *sb.*[2] or *v.*] Stylish; 'posh', 'classy'. (Freq. applied to shops, hotels, or apartments.)

1913 [see *swank *sb.*[2] 2]. **1919** W. Deeping *Second Youth* xvii. 145 Look here, come for a ride. Had this new swank machine just a week. **1928** *Publishers' Weekly* 30 June 2578 From honor and riches to poverty and shame—from the swankest hunting set of England to a garret in the Latin Quarter of Paris. **1947** D. Riesman in *University Observer* Winter 20/1 John..refuses to angle for the mastership of..one of the swank Harvard Houses. **1957** L. Stern *Midas Touch* i. xii. 98 These were the women..who patronized the swank Michigan Avenue specialty shops. **1972** 'E. Lathen' *Murder without Icing* (1973) xx. 179 He was thrilled at having a swank apartment. **1981** R. Barnard *Mother's Boys* i. 12 Have you got a big box of chocks? Something real swank?

swank, *v.* Add: **1. b.** To boast.

1874 Hotten *Slang Dict.* 316 *Swank*, to boast or 'gas' unduly. **1914** G. B. Shaw *Fanny's First Play* iii. 211, I used to boast about what a good boy Bobby was. Now I swank about what a dog he is; and it pleases people just as well. **1950** *Sport* 7–13 Apr. 9/2 Lest I may appear to be swanking, let me hasten to add that all of the credit went to someone else. **1960** J. Rae *Custard Boys* i. vii. 80 'You think that I am swanking too much, John?' With his accent the slang word sounded very strange. **1980** *London Rev. Bks.* 17 Apr. 6/2 Anonymity..is no guarantee against a tendency on the part of informants to swank about their supposed religious experiences.

swankily (swæ·ŋkili), *adv. slang.* [f. Swanky *a.*[2] + -ly[2].] In a swanking or ostentatious manner; boastfully.

1924 D. Moore *Fen's First Term* viii. 87 Angela did it first, and did it swankily. **1940** E. F. Benson *Final Edition* xiii. 284, I swankily told my friend..that I had decided not to go to the Coronation but to give my place to someone else. **1951** *Sport* 6–12 Apr. 11/1 You are unfortunate in not..being able to play swankily to the gallery, not having the peculiar knack some players have of catching the eye.

swankiness (swæ·ŋkinės). *slang.* [f. Swanky *a.*[2] + -ness.] The quality of being swanky; swagger.

1920 *Christian World* 2 Sept. 4/2 The average American is free from swankiness. **1965** *Listener* 22 July 125/1 The 'swankiness' inside the school was matched in the streets outside. 'Grammar grubs', the secondary schoolboys shouted at us, and we passed by, noses lifted... We thought them the bottom.

swa·nking, *vbl. sb. slang.* [f. Swank *v.* + -ing[1].] = Swank *sb.*[2]

1900 [in Dict. s.v. Swank *v.* 1]. **1916** *Captain* June 231/1 (*heading*) The perils of swanking. **1918** *Daily Express* 2 Oct. 2/2 History will declare that by swanking the Hohenzollerns fell.

swa·nking, *ppl. a. slang.* [f. Swank *v.* + -ing[2].] That swanks; boastful, ostentatious, pretentious.

1918 *Daily Express* 2 Oct. 2/2 The swanking dustman is a nuisance. So is the swanking duke.

swankpot (swæ·ŋkpǫt). *slang.* [f. Swank *sb.*[2] + Pot *sb.*[1]] An ostentatious or boastful person; one who is full of swank.

1914 *Picture Fun* 26 Dec. 1/6 Brimstone..and Billy kept the old swankpot nicely on the trot. **1927** H. Walpole *Jeremy at Crale* xii. 212 He's an awful swankpot. **1936** J. B. Priestley *They walk in City* v. 115 Silly swank-pot! **1959** I. & P. Opie *Lore & Lang. Schoolch.* xiii. 302 If a boy is under the necessity of coming to school in a new suit his fellows greet him with,..'Swank pot', 'Posh guy'.

swanky, *a.*[2] Add: Also, boastful. Of things: imposing, stylish, 'posh'.

1929 'R. Crompton' *William* i. 9 'I read that too,' interrupted Ginger, 'so you needn't be so swanky.' **1940** C. Day Lewis tr. *Virgil's Georgics* ii. 49 No mansion tall with a swanky gate. **1959** *Spectator* 25 Sept. 406/3 An English producer and a London critic..in the swanky bar of the Excelsior. **1974** *Sunday Tel.* 8 Dec. 8/6 Swanky Christmas presents, beautifully wrapped in red and gold.

swan-neck. **2.** (Later examples.)

1923 G. Sturt *Wheelwright's Shop* 223 *Swan-necks*, curved hooks fastened to the shafts of a dung-cart, for attaching the shafts to the body. **1935** *Discovery* Jan. 9/1 The adjustment of these beams was generally effected by bending the swan-necks in or out so as to alter the arm lengths. **1967** *Gloss. Sanitation Terms* (*B.S.I.*) 51 *Swan-neck*, a short bent delivery pipe attached to the outlet of a tap.

swan-necked, *a.* Add: **2.** (Earlier example.)

1745 W. Ellis *Mod. Husbandman* Aug. vii. 62 Their five-toothed, long, Swan-neck'd, wooden..Rake.

swanning (swǫ·niŋ), *vbl. sb. slang* (orig. *Mil.*). [f. Swan *v.*[1] + -ing[1].] The action of the verb (sense *2).

1951 E. Linklater *Campaign in Italy* v. 257 Some..were indulging in a favourite pastime of the army, known as swanning. The swan..has the habit of taking short flights that create appreciable commotion but have no serious purpose. Officers who spent their spare time in swanning had in a like manner no graver reason than a desire to watch some particular fragment of a battle, or to visit friends. **1960** *Times Lit. Suppl.* 16 Sept. 587/2 The 22nd Armoured Brigade was continually exercised in a swanning role, of the kind which had so often led to defeat in the past. **1975** *Bookseller* 12 Apr. 2095/1 Harold Latham, the Macmillan editor,..was on a casual swanning tour round Georgia.

swanny, *a.* **2.** (Later example.)

1871 G. M. Hopkins *Jrnls & Papers* (1959) 207 Clouds..in burly-shouldered ridges swanny and lustrous.

swan-pan. (Examples with spelling *suan pan*, now the usual form.)

1836 J. F. Davis *Chinese* II. xviii. 296 A little apparatus called a *Suān-pān*, or 'calculating dish'. **1917** S. Couling *Encycl. Sinica* 1/1 *Suan p'an*, reckoning plate, the counting-board used by the Chinese. **1946** G. Stimpson *Bk. about Thousand Things* 207 Virtually all calculations were performed on the abacus, an apparatus resembling the Chinese *suan pan* or the bead-and-frame affairs now used in kindergarten work. **1973** T. R. Tregear *Chinese* vi. 128 A further six hours a week is devoted to arithmetic, when calculating with the abacus or *suan p'an* is learnt.

Swan River. The name of a river in Western Australia, used *attrib.* in **Swan River daisy,** an annual herb of the genus *Brachycome*, esp. *B. iberidifolia*, belonging to the family Compositæ, native to Western and South Australia, and bearing pinnate leaves and blue, violet, or white flowers resembling daisies.

[**1841** J. Lindley in *Edward's Bot. Reg.* XXVII. 9 Mr. Lowe, of Clapton, has also raised the Large Swan Daisy.] **1873** W. B. Hemsley *Handbk. Hardy Trees, Shrubs, & Herbacious Plants* 235 Swan River Daisy.—An erect glabrous annual about a foot high. **1915** W. Stevens *Let.* 25 July (1967) 184 Another new thing was what is called swan-river daisies from Australia. **1957** J. S. Dakers *Annuals* xiii. 92 Swan River Daisy..one of the most beautiful of all our annuals. **1962** R. Page *Educ. Gardener* xi. 302 The cypresses are underplanted with sheets of..the blue Swan River daisy.

Swanscombe (swǫ·nzkǒm). The name of a village in north-west Kent, used *attrib.* to designate a Middle Pleistocene fossil hominid, an early type of *Homo sapiens*, known from parts of a skull found in a gravel pit near Swanscombe in 1935 and subsequent years. Also *Swanscombe skull*.

1938 W. LeG. Clark in *Jrnl. R. Anthropol. Inst.* LXVIII. 58 (*title*) General features of the Swanscombe skull bones. **1940** *Nature* 13 July 51/2 Swanscombe man appears..in gravels heralding the third glacial stage. **1946** F. E. Zeuner *Dating Past* viii. 279 The view..is beginning to be held generally, and especially on the strength of the Swanscombe skull, that *H*[*omo*] *sapiens* evolved during the Penultimate Interglacial. *Ibid.* ix. 298 Swanscombe Man..is a member of the *sapiens* group. **1962** *Listener* 22 Nov. 878/2 For those who think Pleistocene is a substance, Mary Cathcart Boxer's *Mankind in the Making*..will prove a model of clarity that..sorts the jumble of prehistory in a manner that even those as thick of skull as Swanscombe Woman can grasp. **1973** B. J. Williams *Evolution & Human Origins* x. 169/2 The bone of the Swanscombe skull is thinner than that of Peking Man but thicker than in modern man's. **1975** J. G. Evans *Environment Early Man Brit. Isles* i. 1 Many dramatic environmental changes separate Swanscombe Man by more than 150,000 years from the development and eventual spread into Britain of farming communities.

Swansea (swǫ·nzi). The name of a city in South Wales, used *attrib.* and *absol.* to designate pottery and porcelain made at the Cambrian Pottery there from 1764 to 1870.

1863 W. Chaffers *Marks & Monograms on Pott. & Porc.* 151 *Swansea.* This china was introduced about 1800, and was remarkable for the beautiful delineation of birds, butterflies, and shells. **1879** M. E. Braddon *Vixen* II. ii. 19 Old Worcester teacups..or flowered Swansea. *Ibid.* vii. 107 The Swansea tea-set. **1895** *Wales* Aug. 372/2 The best Swansea china is exquisitely beautiful. **1904** [see *duck's egg d]. **1957** Mankowitz & Haggar *Encycl. Eng. Pott. & Porc.* 216/2 These [fakes], however, are mostly distinguishable from true Swansea as the style of decoration and often the forms and shapes are quite dissimilar. **1967** W. H. Boore *Cry on Wind* x. 95 An old, long-settled place..has its own surprising treasures..a bardic chair..some priceless Swansea porcelain [etc.]. **1976** *Western Mail* (Cardiff) 27 Nov. 16/3 (Advt.), Collector wishes to purchase Swansea Pottery and Porcelain.

swanskin. Add: **3. b.** *fig.* Soft and delicate, smooth like swanskin. (Only found in the work of E. Sitwell.)

1925 E. Sitwell *Troy Park* 38 Once, plumaged like the sea, his swanskin head Had wintry white quills. **1936** —— *Victoria of England* xvi. 197 Wild violets beneath their swan-skin leaves.

SWANU (swā·nu). [Acronym f. the initial letters of *S*outh *W*est *A*frica(n) *N*ational *U*nion.] An African nationalist organization in Namibia. Cf. *SWAPO.

1962 *Rep. U.N. Spec. Comm. S.W. Afr.* 14 Sept. 3 Mr. Kozonguizi explained that he represented the South West Africa Union (SWANU)... The aims of SWANU were to achieve independence for South West Africa. **1963** R. First *South West Afr.* v. iv. 200 The following month the South West African National Union, known as the 'First S.W.A.N.U.', was established. **1970** J. Woronoff *Organizing African Unity* iii. 265 Several nationalist groups were formed as of 1959. First was the South West African National Union (SWANU). **1973** *Black World* Oct. 35/2 Free the land FROLIZI. Swing in there SWANU.

swap, swop, *sb.* Add: The spelling *swap* is recommended. **II. 3.** *Finance.* In foreign exchange operations: an exchange of an amount of money at different rates (i.e. a 'spot' sale for a 'forward' purchase). More generally, an arrangement between the central banks of two countries for stand-by credit to facilitate the exchange of each other's currency. Chiefly *attrib.*

1963 *Economist* 14 Dec. 70/1 A permanent system of automatic swap-lines as opposed to the existing three-monthly swaps is favoured together with easier facilities for medium term credit. **1968** *Times* 9 Sept. 1/2 *Swap arrangements.* The 12 members of the Basle central bankers' club have made reciprocal arrangements to make short-term loans to each other in the event of any currency coming under severe pressure. **1970** Sloan & Zurcher *Dict. Econ.* (ed. 5) 425 Swap credits are used especially in periods of emergency when a particular country's currency..comes under pressure because speculators are selling it on the world markets. **1975** *Financial Times* 29 Oct. 7/1 A classic swap is a transaction in which a spot purchase of a given currency, is covered by a forward sale of the same amount. **1979** *Bank of England Q. Bull.* June 131 The Federal Reserve and the US Treasury again repaid some swap debt to other central banks.

4. Special combinations. **swap fund** *U.S. Stock Exchange*, a fund which investors enter by exchanging securities directly for shares in the fund, obtaining a diversified portfolio without selling stock, and thereby avoiding liability for capital gains tax on the sale of these securities; **swap meet** chiefly *U.S.*, a gathering at which enthusiasts discuss, exchange, or trade items of common interest; **swap shop,** an agency for putting people with articles to exchange or trade in touch with one another; also *fig.*

1966 *Economist* 23 July 380/1 The Revenue Service..will no longer permit investors to defer capital gains tax on the appreciation of stocks exchanged for shares of the special swap funds. **1973** *Daily Tel.* 25 Aug. 16/1 A market has been established in them [*sc.* bottles] and regular 'swop-meets' are arranged so that enthusiasts can buy and sell among themselves. **1976** *Billings* (Montana) *Gaz.* 11 July 3-B/1 The swap meet has become an annual event that attracts visitors from Canada and other states to exchange information about antique cars and parts, he said. **1976** *Milton Keynes Express* 18 June 27/6 (Advt.), Dishot Swop Shop. **1976** *Sunday Post* (Glasgow) 26 Dec., Just before half-time some fans not involved in the beer can 'swop-shop' took refuge on the park. **1977** *Skateboard Special* Sept. 2/1 If you want to take up our super Swap-Shop offer now's your chance. **1979** *Guardian* 5 July 4/4 Instead of handing down golden tablets..the Schools Council will become more of a swap shop for ideas.

swap, swop, *v.* Add: **II. 8. a.** Also *colloq.*, to give (something) to (a person) by way of exchange; *to swap horses in midstream*: see *Horse *sb.* 17.

1934 D. Hammett *Thin Man* iii. 14 Right now I'd swap you all the interviews with Mayor-elect O'Brien ever printed..for a slug of whis——. **1940** W. Faulkner *Hamlet* i. ii. 38 The team Stamper had swapped him stopped now with their heads down. **1948** —— *Intruder in Dust* (1949) ix. 192, I swapped Crawford Gowrie a German pistol.

c. Also with indirect obj., to make an exchange of some specified item with (someone). *colloq.*

1976 *Evening Chron.* (Newcastle) 26 Nov., As Coun. Collins says this council work will suit a pensioner, if he will answer this letter and tell me how many council meetings he has in a week, I will swap him.

SWAPO (swā·po). Also **Swapo.** [Acronym f. the initial letters of *S*outh *W*est *A*frica(n) *P*eople's *O*rganization.] An African nationalist organization in Namibia. Cf. *SWANU.

1962 A. K. Lowenstein *Brutal Mandate* vi. 117 SWAPO grew out of the Ovamboland People's Organization. **1970** C. P. Potholm *Four African Polit. Systems* iv. 100 In South West Africa, the South West Africa People's Organization (SWAPO) likewise undertook a modest policy of selective sabotage. **1973** *Times* 25 Aug. 5/1 The Swapo youth wing's statement..says that its protests at 'Boer' injustices in Namibia have been met with imprisonment, torture, brutality and other forms of oppression. **1976** *Plain Truth* Dec. 6/2 The South West Africa People's Organization (SWAPO), despite its terrorist activities and opposition to the conference, has been invited to be the proposed new government's political opposition.

swapping, swopping, *vbl. sb.* Add: **3.** *Finance.* The action or process of making a swap (sense *3).

1957 *Times* 19 Dec. 15/1 There was rather more outright buying of Dominion and Colonial stocks.., as well as a fair amount of swapping among Crown Colony loans. **1971** *Guardian* 8 Sept. 1/8 Of this inflow, £500 millions was used, indirectly, to support the dollar by 'swapping forward'—Britain actually claimed only a small proportion of the foreign currency due to her, and took the rest in foreign IOUs.

Swaraj (swarā·dʒ). *Indian Hist.* Also **swaraj.** [ad. Skr. *svarāj* self-ruling (*svarājya* own dominion), f. *sva* one's own + *rāj* to reign, rule.] Self-government (for India); the agitation in favour of this.

[**1845** *Encycl. Metrop.* XXI. 679/2 The Swa-ráj, or 'Own Sovereignty', secured to him all the territory possessed by Sivá-ji.] **1907** *Westm. Gaz.* 18 Dec. 1/3 The movement known as Swaraj. **1908** *Times* 27 Oct. 8/3 There is a good deal of talk going on in these days about 'swaraj', or the making of India a self-governing country. **1920** M. K. Gandhi *Non-Co-operation* 12 Aug. (1921) 2 Mr. Tilak lived for his country. The inspiration of his life was freedom for his country which he called Swaraj. **1945** R. Hargreaves *Enemy at Gate* 182 It was a deadlock..which forced the sponsors of *Swaraj* to try and 'save face' by endeavouring to shift the whole matter on to a basis of pacifism and 'appeasement'. **1965** J. K. Mittal in *University of Allahabad Studies: Law Section* 39 In 1927, the Swaraj Constitution, based on a declaration of rights,

was framed to give momentum to the fight for *Swaraj* (i.e. Self-Government). **1977** C. Allen *Raj* x. 129/2 All Anglo-India knew that one day *swaraj* (home rule) must inevitably come.

Hence **Swara·jist,** one who advocated self-rule for India; also *attrib.*

1908 *Westm. Gaz.* 24 June 5/1 The family lawyer..introduced him to two men..who were ardent Swarajists. **1923** *Glasgow Herald* 12 Dec. 8 Failing unconditional assent, the Swarajist intention is to obstruct every official measure coming before the Assembly. **1953** Earl Winterton *Orders of Day* x. 133 The Swarajists were very active..in India.

sward, *sb.* Add: **4.** *sward-land* (earlier example); **sward-cutter** (earlier example).

1744 W. Ellis *Mod. Husbandman* Jan. i. 12 (*heading*) The Gloucestershire way of preparing and sowing sward-Land with corn. **1786** R. Sandilands (*title*) A description of the patent instrument called a sward-cutter.

swarf, *sb.*[2] Add: **a.** Also, any fine waste produced by a machining operation, esp. when in the form of strips or ribbons. (Further examples.)

1917 *Yorkshire Post* 3 Jan. 4/6 Rough copper, copper ore, and copper scrap and swarf in the possession of or due under existing contract to a manufacturer. **1953** *Times* 23 Oct. 5/3 There's swarf—chips of wood, metal, etc.—grinding around in your expensive machinery and shortening its life. **1970** P. Dickinson *Seals* ii. 41 Down the inside rim of the second key-hole there was..a thin curl of swarf still attached to the main brass. **1973** J. G. Tweeddale *Materials Technol.* II. vi. 142 In more ductile materials chips may remain partially bonded to each other to form continuous severely-work-hardened ribbons sometimes called swarf.

b. *spec.* The material cut out of a gramophone record as the groove is made.

1935 H. C. Bryson *Gramophone Record* x. 275 When metal is recorded upon..it is often necessary to arrange for the removal of the swarf either by blowing..or by means of a small brush. **1947** *Jrnl. Inst. Electr. Engineers* XCIV. iii. 288/2 By using a suction system to remove swarf continuously while recording, these troubles are avoided. **1977** *Times* 18 Apr. (Gramophone Suppl.) p. iv/7 For a long-playing record, this swarf, a strip narrower than a human hair, might be half a mile long.

Hence (*rarely*) as *v. trans.* with *up*, to make dirty with swarf; **swarfed** *ppl. a.*, dirtied with swarf, mucky. *colloq.*

1914 D. H. Lawrence *Widowing of Mrs. Holroyd* i. i. 4 A man in blue overalls, swarfed and greased. *Ibid.* 5 *Mrs. Holroyd*: Here, take hold, and help me fold it. *Blackmore*: I shall swarf it up.

swarm, *sb.* Add: **2.** (*c*) *spec.* (i) of asteroids or meteors (cf. *meteor-swarm* s.v. Meteor 6 d); (ii) of earthquakes; cf. also *dike-swarm* s.v. *dike, dyke *sb.*[1] 10.

1929 J. Jeans *Universe around Us* iv. 242 The asteroids occur as a single swarm. **1958** C. F. Richter *Elem. Seismol.* i. vi. 71 Certain localities are..visited by earthquake swarms, long series of large and small shocks with no one outstanding principal event. Such swarms are common in volcanic regions. **1959** *Listener* 30 July 172/2 The Trojans, whose mean distances from the Sun are the same as that of Jupiter, so that they lie far beyond the main swarm [of asteroids]. **1962** F. I. Ordway et al. *Basic Astronautics* iii. 105 Many swarms of meteors orbit the Sun and some periodically intersect the orbit of the Earth causing meteor showers. **1979** *Nature* 25 Oct. 661/1 Earthquake swarms, consisting of many earthquakes of nearly equal magnitude within a small area, often occur in areas of recent or current volcanic or tectonic activity. **1981** I. Ridpath *Young Astronomer's Handbk.* 197/1 At various times of the year, the Earth crosses the orbits of certain comets, encountering whole swarms of meteors.

(*e*) *Ecol.* = *hybrid swarm* s.v. *hybrid *a.* 3.

1926 *Nature* 30 Oct. 624/1 Where a specific name has been given to a smaller group within the swarm..we may adopt this name for the minor group. **1963** Davis & Heywood *Princ. Angiosperm Taxon.* xiv. 483 Hybrids..may even become established and form large swarms many miles from either parent.

3. *swarm-formation.*

1946 *Nature* 21 Sept. 423/1 The most important biological attribute of an outbreak centre is to provide conditions for survival and multiplication of locusts at those times when their range of dispersal is at a minimum, and also to provide conditions necessary for an increase in that range of dispersal (by swarm-formation). **1953** J. S. Huxley *Evolution in Action* iii. 72 At least six species [of malaria-carrying mosquitoes] must be distinguished..some mating without swarm-formation, others requiring the stimulus of swarming.

swarmed (swǭ·ɹmėd, swǭɹmd), *ppl. a. poet. rare.* [f. Swarm *v.*[1] + -ed[1].] Of a place: crowded, thronged. Of people: assembled in a crowd, congregated, massed.

1885 G. M. Hopkins *Poems* (1967) 98 How then should Gregory, a father, have gleaned else from swarm-ed Rome? **1951** R. Graves *Poems & Satires* 37 Tormented by his progress he displays An open flank to the swarmed enemy.

swarmer[1]. Add: **1.** Also, a swarmer cell.

1964 *Bacteriol. Rev.* XXVIII. 242/2 If the swarmer is to become a recognizable caulobacter cell, it must develop a stalk after cell division.

3. swarmer cell *Bacteriol.*, a flagellated motile cell produced by the stalked cell of certain species of stalked bacteria.

1950 *Biochimica & Biophysica Acta* V. 41 A study was made of the flagellation of swarmer cells of *Proteus vulgaris*. **1976** *Jrnl. Bacteriol.* CXXVIII. 456/1 The stalked cell [of *Caulobacter crescentus*] divides repeatedly to produce new swarmer cells, whereas the swarmer cell, which cannot divide, loses motility and develops into a stalked cell.

‖ **swart gevaar** (swart χəfā·r). *S. Afr.* [Afrikaans, lit. 'black peril', f. Du. *zwart* Swart, black + *gevaar* danger, peril.] The name given in South Africa to the threat to the Western way of life and white supremacy believed to be posed by the black races. Cf. *yellow peril* s.v. Yellow *a.* 1 d.

[**1939** R. F. S. Hoernlé *S. Afr. Native Policy* 1 To protect White South Africa against 'the Native Danger'—*die donker gevaar* or *die swart gevaar*—is..the simple pole towards which the needle of Native Policy steadily points.] **1948** *Hansard S. Afr.* 20 Jan. 111 In a pathetic attempt to get into power they have dropped Republicanism and adopted the Swart Gevaar. **1970** *Cape Times* 28 Oct. 9/2 They introduce *swart gevaar* where it suits them. **1979** *Economist* 8 Sept. 59/1 Afrikaner Nationalists, brought up for more than half a century on the politics of *swart gevaar* (black danger).

swarth (swǭɹð), *v. poet. rare*[−1]. [f. Swarthy *a.*[1]] *trans.* To make swarthy, to darken.

a 1889 G. M. Hopkins *Poems* (1967) 180 His cheeks the forth-and-flaunting sun Had swarthed about with lion-brown.

swarthily, *adv.* Delete *rare*[−0] and add examples.

1955 J. Thomas *No Banners* v. 40 De Laurière was a tall man, swarthily handsome. **1981** *Times Lit. Suppl.* 20 Feb. 198/4 A swarthily soulful young boy sitting all alone on a chair in a predominantly bare room.

‖ **swartwitpens** (swartvi·tpens). *S. Afr.* Also **swart witpenz, zwart wit pens.** [Afrikaans, f. *swart* black + *wit* white + *pens* belly.] = *sable antelope* s.v. Sable *sb.*[2] 5.

1869 T. Baines *Diary* 31 Aug. (1946) I. 137 We saw a fine troop of *Zwart-wit-pens*..or Harris bucks. **1880** E. F. Sandeman *Eight Months in Ox Waggon* 254 We rode along..hoping to find..a swartwitpense [*sic*] feeding on the luxuriant grass. **1889** H. A. Bryden *Kloof & Karroo* xvi. 284 The Sable antelope, 'zwart wit pens'—i.e. black with white belly. **1939** S. Cloete *Watch for Dawn* xx. 265 Game was abundant: eland, giraffe, wildebeeste, zebra, swart witpenz,..and innumerable rhinoceros. **1951** *Cape Argus* 8 Dec. 4/3 He and Conroy spotted a fine buck—a swartwitpens.

swarve, *v.*[1] (Example of *active* form.)

1906 Kipling *Puck of Pook's Hill* 250 Next floods the brook'll swarve up.

swash, *int.* or *adv.* and *sb.*[1] Add: **A.** *int.* or *adv.* (Later example of spelling *swosh.*)

1927 J. Masefield *Midnight Folk* 92 He swung Black-malkin [*sc.* a cat] round his head and pitched him swosh into the mud.

B. *sb.* **I. 1.** † **b.** *transf.* Nonsense; worthless stuff. Cf. *hog-wash b. *slang. obs.*

1895 W. C. Gore in *Inlander* Nov. 65 *Swosh*,..nonsense; inferior work. **1924** Galsworthy *White Monkey* ii. v. 162 Anyway sentiment was swosh! Cut it out!

5. b. *Physical. Geogr.* The rush of sea water up the beach after the breaking of a wave.

1919 D. W. Johnson *Shore Processes & Shoreline Development* x. 514 Since there are a variety of marks left on the sand by wave action, and the present feature is peculiarly a product of the swash, I have given it the name of 'swash mark'. **1934** *Geogr. Jrnl.* LXXXIII. 485 When the swash dies out the backwash of the wave returns directly down the steepest slope to the sea. **1976** P. D. Komar *Beach Processes & Sedimentation* ii. 14 The return flow of the swash collides with the incoming surf bores.

III. 9. (sense *5 b) *swash mark*, *-slope*, *-zone*; **swash-plate** *Engin.*, a disc mounted obliquely on the end of a revolving shaft, which can impart to a rod in contact with the edge of the disc a reciprocating motion parallel to the axis of the shaft.

1919 Swash mark [see sense 5 b above]. **1982** *Sci. Amer.* Aug. 130/2 Seaward of a swash mark on some beaches one is likely to find smaller diamond-shaped markings left by the backwash. **a 1877** Knight *Dict. Mech.* III. 2467/2 Swash-plate. **1913** W. E. Dommett *Motor Car Mech.* 158 The plungers are driven by a swash-plate mechanism. **1977** *Design Engin.* July 92/1 (Advt.), To obtain precise control in many fields, e.g. valves, or engine throttle position, or a pump swashplate, you need a remote position actuator. **1931** *Geogr. Jrnl.* LXXVIII. 134 They [*sc.* waves]..on nearing high-water mark were busily pushing forward shingle to the top of their swash slope. **1976** P. D. Komar *Beach Processes & Sedimentation* ii. 14 Schiffman ..defines a transition zone between the surf and swash zones.

swashbuckle, *v.* (Later examples.)

1939 W. Fortescue *There's Rosemary, There's Rue* vi. 41 One proud day I was promoted to study the part of Rosalind in 'As You Like It', and I swashbuckled round that flat in imaginary doublet and hose. **1979** R. Blythe *View in Winter* ix. 312, I knew a remittance man in Kenya ..swashbuckling about with a revolver in his belt.

swashbuckler. Add: **b.** A book, film, or other work portraying swashbuckling characters.

1975 *Daily Colonist* (Victoria, B.C.) 27 July 20/3 Clavell's most ambitious novel—an oldfashioned swashbuckler complete with all the popular ingredients. **1977** *Time* 30 May 42/2 *Star Wars* is a combination of *Flash Gordon*, *The Wizard of Oz*, the Errol Flynn swashbucklers of the '30s and '40s and almost every western ever screened.

swastika. Delete ‖ and add: Now usu. pronounced (swǫ·stikă). **2. a.** This symbol (with clockwise projecting limbs) used as the emblem of the German (and other) Nazi parties; = *Hakenkreuz, hakenkreuz. Also, a flag bearing this emblem.

1932 'Nordicus' *Hitlerism* ii. 17 Thousands flocked to his standard—the 'Hakenkreuz'—(swastika), the ancient anti-semitic cross in a color scheme of red-white-black in memory of the colors of the old army. **1933** [see *Aryan *a.* 2]. **1941** G. Ziemer *Educ. for Death* i. 4 A squad of Nazi youngsters in..brown shirts decorated with swastikas. *Ibid.* ii. 30 A luxury hotel managed by a Jew... The swastika over it fluttered gaily. **1951** L. Hagen *Follow my Leader* i. 6 Most of the men in my Sturm wore at least part of a uniform, and all I could do was wear a swastika armlet. *Ibid.* vii. 266 Our compatriots..clung to their German ways and..flew the swastika on our national holidays. **1967** T. Gunn *Touch* 15 A silk tent of swastikas. **1977** E. Heath *Travels* iv. 113 Along this street had stretched the Nazi columns... Gone, now, were the crowds and the bright-red banners flaunting their swastikas over the streets. **1979** J. Burmeister *Glory Hunters* i. 5 In addition to her national flag she [*sc.* a ship] also flew the Swastika.

b. *attrib.* and *Comb.*

1934 *Ann. Reg. 1933* i. 179 Minor acts of defiance towards the Austrian Government..such as..the lighting of Swastika fires and the daring hoisting of forbidden Swastika banners under the eyes of the police..and the hoisting of Swastika flags. **1940** H. G. Wells *All Aboard for Ararat* iv. 101 As regards the olive branch incident, it is to be noted that the leaves were blood-stained and tied with a swastika ribbon. **1946** J. Flanner in *New Yorker* 5 Jan. 46/1 Ten years ago, he [*sc.* Goering] was baying 'Heil' as he strutted the swastika-hung streets. **1957** T. Gunn *Sense of Movement* 36 The swastika-draped bed. **1960** *Jewish Chronicle* 8 Apr. 14/3 The recent swastika-daubings in this country.

Hence **swa·stika'd** *a.*, decorated with or wearing a swastika, esp. as a badge of Nazism.

1965 *New Statesman* 15 Oct. 552/3 Buckley has..described the American Nazi Party as 'two dozen swastika-ed cretins who go about plying their pathology in the fever-swamps of the crazy-Right'. **1969** *Listener* 14 Aug. 225/3 Where do those swastika'd Hell's Angels types fit in?

Swat (swǫt), *sb.*[3] [The name of a district in the Malakand Division of North-west Frontier Province, Pakistan.] = (and superseded by) *Swati. Also *attrib.* or as *adj.*

1897 *Westm. Gaz.* 8 Sept. 2/2 The Afridi rising..was all a matter of wire-pulling on the part of..the Swat Fakirs. **1911** G. P. Gooch *Hist. our Times* vii. 170 A rising began in 1897 among the Swats, Mohmands, and Afridis.

swat, *v.*[1] Add: **2.** Restrict Chiefly *U.S.* to sense in Dict. and add: Also, to crush (a fly, etc.) with a blow. (Later examples.)

1916 A. Huxley *Let.* 29 Sept. (1969) 114 A poem..which..is destined to become a cause of rupture in the world, dividing it up into..Monts and Caps, Mouldiwarpians and Swat-that-Moleites. **1942** *R.A.F. Jrnl.* 18 Apr. 9 The familiar white butterflies should be 'swatted' wholesale. **1958** R. K. Narayan *Guide* vii. 139 He repelled me with a back-stroke of his left hand as if swatting a fly. **1962** K. Kesey *One Flew over Cuckoo's Nest* (1973) 9 One swats the backs of my legs with a broom handle to hurry me past. **1976** *Times Lit. Suppl.* 12 Nov. 1414/2 Identical communities to Tolmers Square have been swatted from the urban map.

Swatantra (swatā·ntrā). [Hindi, (one who is) self-determined or self-motivated.] In full, *Swatantra party*. A liberal conservative political party (the Freedom Party) in the Republic of India from 1959 to 1972. Also *attrib.*

1959 *Hindu* 8 June 1/5 Addressing a..public meeting in Royapettah last evening, Mr. C. Rajagopalachari explained the aims and policies that would be pursued by the new Opposition party, which he said, would be called the Swatantra Party. **1963** H. Tinker *Democratic Ideal in Asia* 23 C. R. Rajagopalachari, the veteran Swatantra statesman. **1966** *Economist* 24 Dec. 1319/3 The rajahs who joined Swatantra in droves were angry with Congress for having reduced them to commoners. **1979** V. L. Pandit *Scope of Happiness* ii. 13 A merger of Congress (O), Swatantra party, Jan Sangh, and Bharatiya Lok Dal.

swatch, *sb.*[1] For '*Sc.* and *north.*' read 'orig. *Sc.* and *north.*' and restrict *north.* to sense 1.

2. Hence, of other materials (see also *S.N.D.*). Also, a collection of samples bound together, a swatch-book. (Later examples.)

1953 *Times* 23 July 1/4 (Advt.), Duffle jackets and duffle coats... Swatches sent on request. **1973** *Sci. Amer.* June 119/2 Continue..until a piece of filter paper or swatch of cotton held close to the exit by means of long metal forceps begins to burn. **1982** *Daily Tel.* 2 Aug. 9/2 He wears swatches of the hats he is currently working on, hat-pinned to his tie for inspiration.

transf. (Later examples.)

1928 P. Grey *Making of King* 6 Ye'll mind an' bring a swatch o' yer wallpaper wi' ye. **1957** *Brit. Commonwealth Forest Terminol.* ii. 192 *Swatch*, a sample sheet of veneer, usually 3 ft. long and the full width of the flitch. **1965** G. McInnes *Road to Gundagai* iii. 38 One's wants were provided for by a swatch of neatly cut squares from the Hobart *Mercury* struck on a nail in the wall. **1973** *Sci. Amer.* Apr. 41/1 A swatch of inks as rendered by Kodak color slide films that maximize consumer satisfaction with the greenness of grass, the blueness of sky, and the healthy glow of complexions. **1981** N. Gordimer *July's People* 54 She knew it was impossible that he could have made free of the still-thick swatch of notes, lying swollen as the leaves of a book that has got wet and dried again.

3. (Later examples.) Also extendedly (esp. without the notion of a sample), a portion, a clump.

1930 *Aberdeen Press & Jrnl.* 19 May 5/2 Swatches from Shakespeare... The miscellany consisted of excerpts from 'Henry IV',..the ghost scene in 'Hamlet', [etc.]. **1950** A. Lomax *Mister Jelly Roll* 30 The hollows of his cheeks and temples showing dark against silvery skin, and up towards the ceiling a swatch of silvery hair. **1961** J. Steinbeck *Winter of our Discontent* 358 A swatch from Lincoln's Second Inaugural. **1963** *Punch* 31 July 165/1, I..consumed unbelievable swatches of it [*sc.* electricity]. **1972** J. Mosedale *Football* ii. 23 A swatch of astroturf in the Hall leads to the present. **1975** *Times Lit. Suppl.* 24 Oct. 1254/1 Mr Boston gives a fair selection: the life and death of the Admirable Crichtoun from *The Jewel*, a reasonable swatch of *Logopandecteision*.

4. Comb. **swatch-book,** a book of samples.

1956 *Archit. Rev.* CXIX. 286/1 One of the first firms to pin their colours to this mast is T. & W. Farmiloe Ltd., the manufacturers of Nine Elms Paints, who present the full range in the form of a truly magnificent swatch-book. **1978** *Times* 26 Jan. 13/5 Some of the collections were just dull... Why not just send for the swatch-book?

swatchel (swǫ·tʃəl). *slang.* Also 9 **schwassle.** [Perh. f. G. *schwätzeln*, frequentative form of *schwatzen* to chatter, tattle.] An older form of *swazzle; also interpreted as the name for Mr. Punch in a Punch and Judy show. Freq. *attrib.*, as *swatchel box*, *cove* (see quots.).

1854 *Housek. Words* 24 Sept. 76/1 A Punch's show [is] a schwassle-box. **1864** Hotten *Slang Dict.*, *Swatchel-cove*, the master of a Punch-and-Judy exhibition who..does the necessary squeak for the amusement of the bystanders. **1887** W. E. Henley *Villon's Good-Night* in J. S. Farmer *Musa Pedestris* (1896) 174 You swatchel-coves that pitch and slam. **1900** *Sat. Rev.* 19 May 613/1 Students of Romany..will find some interest in a list furnished to a friend who handed it on to me by a 'swatchel-cove' or peregrinating Punch-exhibitor. **1921** *Glasgow Herald* 24 June 7 'Swatchel' is Mr Punch, hence 'Swatchel-box' the show, and 'Swatchel cove' the patterer. *c* **1938** A. Hambling *Punch & Judy* 3 Wet the swatchel, and having fixed the thread, put it on the tongue crosswise. **1983** *Listener* 22 Sept. 14/3 The word 'swatchel' is Punch and Judy showmen's slang for the figure of Punch.

swath[1]**, swathe.** Add: **3. c.** Now freq. *to cut a wide swath*.

1902 H. L. Wilson *Spenders* 348 You folks been cuttin' a pretty wide swath here in New York. **1929** *Amer. Speech* V. 119 [Maine] Someone conceited..'feels his oats', 'cuts a wide swath', 'is one of the big bugs'. **1960** I. Wallach *Absence of Cello* (1961) 241 He was determined to cut a wide swath with the girls—no easy trick in Philadelphia.

5. *swath-width* (later example); **swath-board,** a slanting board attached to the cutter-bar of a mowing machine, designed to force the cut grass, etc., into a narrower swath; **swath-turner** (examples).

1952 J. W. Day *New Yeomen of England* vii. 87 After mowing, the lucerne is tedded to remove the wad, left by the swathe board, and is then swept to the tripods and cocked. **1963** *Listener* 28 Mar. 552/1 The swathe-board..of a grass-mower. **1922** Joyce *Ulysses* 699 Grindstone, clodcrusher, swatheturner, carriagesack. **1958** *Times* 27 Oct. 15/4 A swath-turner was used to invert the swath and move it onto dry ground. **1970** G. F. Burnett in H. W. Mulligan *African Trypanosomiases* xxiv. 506 When treating an area of woodland, the aircraft must pass over it on parallel runs at regularly spaced intervals, each of which is referred to as a 'swath width'.

swather[2]**.** (Later examples.)

1929 *Kansas City* (Missouri) *Times* 26 June, The swather, or windrowing machine, is proving almost as popular as the older combine, which it complements. **1958** *Times* 24 Nov. 15/4 There is still a great deal of room for improvement in the design of combine harvesters and swathers. **1976** *Billings* (Montana) *Gaz.* 17 June 6-F/2 (Advt.), Swather, with conditioner. 14' auger head, industrial gas engine.

Swati (swǫ·ti). Also †**Swa(u)tee, Swathi.** [f. *Swat *sb.*[3] + *-i.] A member of a people inhabiting the district of Swat in Pakistan. Also *attrib.* or as *adj.*

1815 M. Elphinstone *Acct. Kingdom of Caubul* ii. xii. 319 The Swautees..appear to be of Indian origin... Swaut and Boonair, their last seats, were reduced by the Eusofzyes in the end of the fifteenth century. **1866** T. Seaton *From Cadet to Colonel* II. 202 Afreedees and Swatees, Afghans and Maguls. **1897** W. S. Churchill in *Daily Tel.* 7 Oct. 11/1 The Swatis, Bonerwals, Mohmands and other frontier tribes with whom the Malakand Field Force is at present engaged are brave and warlike. **1927** *Rep. Admin. Border N.W.F.P. 1925–26* (Calcutta) 7 Extensive smuggling of *charas* into Peshawar which was known to be carried on by Swathi traders. **1955** *Times* 25 June 7/7 Swati politics are quite straight-forward. **1977** D. Murphy *Where Indus is Young* 10 That battered bus, full of Swatis on their way home.

Swatow (swā·tau). The name of a port (now Shantou) in the province of Guandong, China, used to designate a type of porcelain produced in the Ming dynasty (A.D. 1368–1644) (see quots.).

1925 R. L. Hobson *Later Ceramic Wares China* xii. 111 A type of coarse porcelain, distinguished by an iron-red biscuit and accretions of..grit in the base..known among Chinese dealers as Swatow ware. **1945** W. B. Honey *Ceramic Art China* i. 21 The red-and-green and green-and-blue wares made for export in Southern China (the 'Swatow wares' of the English collector). *Ibid.* ii. 127 Plates and dishes of the Swatow class. **1953** S. Jenyns *Ming Pottery & Porcelain* 147 The so-called 'Swatow' plates, which we now believe..to have been made at, or near, Shih-ma in Fukien. **1970** *Oxf. Compan. Art* 235/1 The still unidentified makers of 'Swatow' porcelains, which are chiefly large dishes coarsely but attractively painted in red and green.

swatter (swǫ·təɹ), *sb.* [f. Swat *v.*[1] + -er[1].] An instrument for swatting flies. Also occasionally, one who swats flies (with a swatter).

1917 [see *fly-swatter* s.v. *fly *sb.*[1] 11]. **1923** *Dundee Tel.* 21 July 3/3 We have tried fly-papers, swatters, formaline solution, and nets. **1926** *Glasgow Herald* 4 Oct. 8 Poultry food is made from the Mexican bluebottle, professional 'swatters' making a good living by catching them. **1947** J. Steinbeck *Wayward Bus* i. 8 The death of a fly by swatter, or slowly smothered in the goo of fly paper. **1967** O. Wynd *Walk Softly, Men Praying* iii. 35 The Principal turned back to pick up a swatter on his desk, then lashed out.

S wave: see *S 6.

sway, *sb.* Add: **I. 9. b.** *Naut.* A rhythmic linear motion of a vessel from side to side (as distinguished from the rotatory motion of a roll).

1957 *Trans. Inst. Naval Architects* XCIX. 121/1 Sway accelerations were actually measured to a good approximation in the form of the displacement of the apparent vertical. **1968** Rawson & Tupper *Basic Ship Theory* xii. 427 Disturbances in the yaw, surge and sway modes will not lead to such an oscillatory motion..when the ship is in a seaway. **1977** *Offshore Engineer* May 44/3 During these tests, the data acquisition system recorded waves,..sway and yaw of the lay barge, pull and length of mooring cables, and anchor positions.

12. (Later examples in *Thatching*.)

1949 K. S. Woods *Rural Crafts of Eng.* iv. xiii. 203 The light timbers that support the thatch are 'flues' laid upward from eaves to ridge like rather thin rafters; 'sways' or long laths are laid horizontally across them at frequent intervals. **1966** *Punch* 10 Aug. (Advt. following p. 216), Hazel rods or 'sways' are used in conjunction with iron hooks to fasten the thatch to the roof timbers.

sway-. Add: **sway-bar,** (*b*) chiefly *N. Amer.*, a bar joining the suspension assemblies of corresponding wheels at either side of a motor vehicle so as to reduce rolling when cornering; an anti-roll bar; **sway-brace** *v.* (earlier example).

1973 *Hot Rod* Oct. 108/1 Also known as stabilizer bars, antiroll bars, or just plain sway bars, these little goodies ..can make so much difference you wouldn't believe it. **1979** *Tucson* (Arizona) *Citizen* 3 Oct. c. 14/5 (Advt.), 69 MGB... High performance sway bars, Monza exhaust system. **1894** W. H. Warren *Engin. Construction* xix. 304 High trestle piers of timber present great varieties in design. They should be thoroughly sway-braced.

swayable, *a.* Add: (Later examples.) No longer *rare*.

1978 C. Tomlinson *Shaft* 42 A wind is having its way with all swayable things, Combing through flag and steam, streaming-out hair. **1982** *Washington Post* 4 May c5/4 As hostess, you should generally give the impression of someone who has rules but is swayable.

‖ **swayamvara** (swa:yamvā·rā). *India.* Also **swayambara, swayamvar.** [Skr., lit. 'self-choice'.] A Hindu ceremony in which a woman chooses her husband from amongst several contenders; a symbolic representation of this, preceding an arranged marriage.

1831 H. H. Milman in *Q. Rev.* XLV. 17 A solemn assemblage, called the *Swayambara*, or self-election, where the princess is to designate the favoured suitor by throwing a wreathe of flowers round his neck. **1863** M. Williams *Indian Epic Poetry* 100 Draupadi was about to hold her swayamvara. **1932** J. Nehru *Let.* 23 June in *Glimpses of World Hist.* (1939) 210 There is a long poem about one of these Chalukyan kings, and in this it is stated that he was chosen by his wife at a public swayamvar. **1970** *Times* 28 Feb. 6/1 The wedding began on Monday with the Swayamvara ceremony, in which the Hindu girl accepts her parents' choice of husband by garlanding him.

swayback, *sb.* Add: **1. a.** (Example of the condition in human beings.) Also, an instance of this.

1939 J. Cary *Mister Johnson* 156 She is a huge, lumbering woman... She has a sway back..and long heels like a hen. **1946** *Richmond* (Va.) *News-Leader* 14 Nov. 30 (*heading*) Swayback is figure fault. Exercises quickly correct.

b. A sway-backed horse or lamb. Also *transf.*, of a person.

1874 *Rep. Vermont Board Agric.* II. 402 The buckskin McClellan was a regular hollow or sway back. **1921** S. Kaye-Smith *Joanna Godden* i. 35 'He'd three sway-backed lambs at Rye market on Thursday.' 'Sway-backs!' 'Three. 'Twas a shame.' **1934** S. Beckett *More Pricks than Kicks* 68 A woman..is either: a short-below-the-waist, a big-hip, a sway-back, a big-abdomen or an average. **1974** M. Laurence *Diviners* iv. 80 Make pemmican out of the swayback which dropped dead of exhaustion on the Back Forty.

2. A copper deficiency disease affecting the nervous system of young lambs, causing paralysis. Cf. *renguerra.

1938 *Nature* 5 Mar. 400/1 Swayback..accounts from time to time for many lambs. **1947** *Sci. News* V. 100 Research has been going on into a disease of newborn lambs called swayback. **1960** *Farmer & Stockbreeder* 1 Mar. 105/3 Injection of a copper preparation..into a ewe during pregnancy can prevent swayback in its lambs. **1970** 'J. Herriot' *If only they could Talk* xxv. 149 The diseases which beset the lambs themselves—swayback, pulpy kidney, dysentery. **1980** *Daily Tel.* 16 Feb. 12/6 Thousands of lambs may die from a nervous disorder called swayback.

sway-backed, *a.* Add: (Further *transf.* examples.)

1919 T. K. Holmes *Man from Tall Timber* ix. 101 'Does seem a pretty springtime, after all,' Aunt Tabby ruminated, as she rocked in a swaybacked chair. **1950** *Audio Engin.* Sept. 30/2 If the *lows* and *lower highs* are both present to excess, the system is *sway-backed*. **1965** J. A. Michener *Source* (1966) 730 Shmuel Hacohen, a sway-backed Jew from Russia, sought an opposing judgment. **1976** H. MacInnes *Agent in Place* xviii. 197 It was a steep pull, the stone steps made sway-backed by centuries of..feet. **1977** D. Harsent *Dreams of Dead* 49 On knees and palms, swaybacked like a stricken runner..she begged, 'Pleasure me, pleasure me.'

swayless (swē[i]·lės), *a. poet. rare.* [f. Sway *sb.* + -less.] Not swayed or swaying; unmoved, immovable.

1856 *Tait's Mag.* XXIII. 548/1 A gnarled tree, which..free and swayless in the fresh air grew. **1897** F. Thompson *New Poems* 12 And with her magic singing kept she..That garden of enchanting In visionary May; Swayless for my spirit's haunting.

Swazi (swā·zi), *sb.* and *a.* [ad. Nguni *Mswati*, the name of a former king of the Swazi.] **A.** *sb.* **a.** A (member of a) people of mixed stock, predominantly Nguni, inhabiting the kingdom of Swaziland (independent since 1968) and parts of eastern Transvaal in the Republic of South Africa. **b.** A dialect of Nguni spoken by the Swazi. **B.** *adj.* Of, pertaining to, or characteristic of this people.

[**a 1857** J. Shooter *Kafirs of Natal & Zulu Country* 391 The Amaswazi partially shave their head [*sic*].] **1872** C. A. Payton *Diamond Diggings of S. Afr.* 142 A book of very great interest on new African sport and travel, entitled, I believe, 'Swazi Kafirs and Swazi Game'. **1878** A. Aylward *Transvaal of To-day* 182 The Swazis transferred the fidelity and love they bore him. **1884** K. Johnston *Africa* xxvi. 461 The most..numerous..are the Bechuanas..cut off from the Zulus and Swazi by the Quathlamba range on the east. **1902** *Encycl. Brit.* XXXIII. 111/2 The Swazies are a branch of the Bantu family... Swazieland was first constituted a petty native state in 1843, when the Barabuza people under their chief, Swaze, rose against their Zulu oppressors, and according to custom took their name from the founder of their chieftaincy. **1910** J. Buchan *Prester John* xi. 193 There were tall Zulus and Swazis with *ringkops* and feather head-dresses. **1919** H. H. Johnston *Compar. Study of Bantu & Semi-Bantu Languages* 298 The Zulu-Kafir Languages..Swazi..Tekele. **1937** N. J. van Warmelo in I. Schapera *Bantu-Speaking Tribes* iii. 51 Commencing with the increasing power of the Ngwane Chief Sobhuza (*ca* 1820), the 'Swazi' people gradually began to come into being. **1947** J. Stevenson-Hamilton *Wild Life S. Afr.* xxiii. 188 The man (a Swazi) entered, and almost immediately was attacked by the leopard. **1956** H. Bloom *Episode* v. 72 She wagged her head and huffed her shoulders and muttered in Swazi. **1961** W. Vaughan-Thomas *Anzio* ix. 197 The cheerful Swazi Pioneer..dashed around the football field in an enormous beret. **1973** 'S. Harvester' *Corner of Playground* iii. ii. 181 A Swazi king in eighteen-fiftyfour raided the Tsonga to find boys and girls to sell to the Boers. **1982** *Times* 1 June 8/1 We [*sc.* the Zulus] are their countrymen, and yet they are prepared to sell us out to the Swazis in a clandestine deal.

swazzle (swǫ·z'l). Also **swozzle.** [Var. *swatchel.] In a Punch and Judy show: an instrument consisting of two convex metal pieces bound together with a length of tape stretched from side to side between them, which is held in the mouth of the puppeteer and is used to produce the characteristic squeaking voice of Mr. Punch.

Mayhew's *call* (Call *sb.* 6 e), described in *London Labour* (1861) III. 45/2.

1942 S. de Hempsey *How to do Punch & Judy* 86 Amongst the professional Punch and Judy performers the gadget for the Punch voice is popularly known as a 'Swazzle'. **1951** G. SPEAIGHT in *Oxf. Compan. Theatre* 644/2 [Punch] spoke in a high squeak, formed by inserting a 'swazzle' or squeaker into the mouth of the speaker. **1959** *Times* 13 Aug. 10/6 My friend, removing his top and bottom dentures, put my swazzle in his mouth and gave an expert demonstration of the Punch voice. **1962** *Guardian* 31 Mar. 6/5 The swozzle—the hand-made reeded 'call' which the operator keeps in his mouth to reproduce the Punch squawk. **1973** G. SIMS *Hunters Point* ii. 13 Mr. Punch's high-pitched buzzing voice was.. achieved by using a 'swozzle', a piece of linen stretched between two flat pieces of silver, bound together with more linen, and placed at the back of Mr. Jackman's throat. **1983** *Daily Tel.* 11 Apr. 12/4 The swazzle, the flat metal instrument bound with black thread which the Punchinello keeps in his mouth (at some risk of swallowing) to make Mr Punch's nasal squeak.

swear, *v.* Add: **III. 17*. swear with—.** = sense 12 b. *rare.*

1789 H. WALPOLE *Let.* 2 July (1961) XXXI. 306, I do not propose putting your name.., as I think it would swear with the air of ancientry you have adopted in the signature and notes. **1976** C. OMAN *Oxf. Childhood* 101 It was decreed that she must wear a rose-pink robe which swore most horribly with her greatest asset.

IV. 21. swear off. c. (Earlier examples.) More recently, with obj. expressed. Chiefly *U.S.*

1839 *Spirit of Times* 16 Nov. 434/1 Like swearing off from liquor and going into a grog-shop. **1853** MRS. STOWE *Key to Uncle Tom's Cabin* 91/2 Well, after all, I suppose, Mr. Legree, you wouldn't have any objections to swarin' off? **1922** 'MARK TWAIN' in *Harper's Mag.* Mar. 457/1, I..swore off my taxes like the most conscienceless of the lot. **1960** R. ST. JOHN *Foreign Correspondent* v. 88 He became a newspaper reporter and swore off personal involvement in politics, at least for the time.

swearing, *vbl. sb.* **4.** (Later examples.)

1842 D. G. ROSSETTI *Let.* 1 Sept. (1965) I. 7 Uncle Henry's Swearing-book combines both Bible and Prayer-Book. **1899** Swearing-habit [see *drinking-habit* s.v. *DRINKING *vbl. sb.* 4 c]. **1939** JOYCE *Finnegans Wake* 524 Mr. Cockshott, as he had his assignation with, present holder by deedpoll and indenture of the swearing belt.

swearing, *ppl. a.* **2.** (Earlier example.)

1796 J. WOODFORDE *Diary* 10 Oct. (1929) IV. 312 My Boy, John Brand, left my Service to day, as he had proper Notice so to do, being the most saucy swearing Lad that ever we had.

sweat, *sb.* Add: **II. 3. a.** *cold sweat*: freq. in phr. *in a cold sweat* (also *fig.*) (later examples). Cf. sense 10 in Dict.

1840 LYTTON *Money* (ed. 2) III. vi. 94 'Poor fellow! He'll be ruined in a month.'..'I'm in a cold sweat.' **1941** C. MACKENZIE *Red Tapeworm* xii. 153 He would..have broken out in a cold sweat at the thought of what might have happened. **1966** C. AIRD *Relig. Body* xvii. 158 Cousin Harold must have been in a cold sweat in case his father died before he got to Cullingoak.

4. c. A long training run for schoolboys. *Public Schools' slang.*

1916 E. F. BENSON *David Blaize* xiv. 274 You brutes have been having an innocent happy sweat along the road. **1924** KIPLING *Debits & Credits* (1926) 93 For the juniors, a shortish course..while Packman lunged Big Side across the inland and upland ploughs, for proper sweats. **1983** W. BLUNT *Married to Single Life* iv. 62 Long melancholy 'sweats' (runs) over the downs [at Marlborough].

III. 9. b. *old sweat*: see *OLD *a.* D. 4.

10. b. *no sweat*: see *NO *a.* 5 d.

IV. 11. *sweat labour* (later example), *-stain*; *sweat-absorber*; *sweat-marked*, *-shining*, *-soaked*, *-stained* (examples), *-wet* adjs.; **sweat-band,** (*b*) in *Sport*, a strip of material worn around the (fore)head or wrist to absorb perspiration; **sweat-bath,** a steam-bath or hot-air bath, esp. among N. American Indians; cf. SWEAT-HOUSE 1; **sweat-box,** (*a*) (earlier and later examples); also *U.S.*, a room in which a prisoner undergoes intensive questioning (see quot. 1931); (*d*) *transf.* and *fig.*, *spec.* a heated compartment in which perspiration is induced, to encourage weight loss, etc.; **sweat cooling** *Engin.*, a form of cooling in which the coolant is passed through a porous wall and evenly distributed over the surface, which is cooled by its evaporation; hence **sweat-cooled** *ppl. a.*; **sweat equity** *U.S.*, an interest in a property earned by a tenant who contributes his labour to its upkeep or renovation; **sweat-hog** *U.S. slang*, a difficult student singled out in school or college for special instruction; **sweat-lodge** (later examples); **sweat pants** chiefly *U.S.*, trousers of thick cotton cloth worn by athletes, esp. before or after strenuous exercise; tracksuit trousers; **sweat-rag,** delete *Australian* and substitute for def.: any cloth used for wiping off sweat, or worn round the head to keep sweat out of the eyes; **sweat rug** a rug put on a horse after exercise; **sweat-shirt** orig. *U.S.*, a loose shirt; *spec.* a long-sleeved, high-necked pullover shirt of thick cotton cloth (usu. with a fleecy lining), worn by athletes to avoid taking cold before or after exercise (cf. SWEATER 7 b); hence **sweat-shirted** *a.*; **sweat-shop**: for *U.S.* read orig. *U.S.* and add earlier and further examples; also *fig.*; **sweat-suit** orig. *U.S.*, an athlete's suit consisting of a sweat-shirt and sweat-pants.

1956 S. BECKETT *Malone Dies* 93 A sweat-absorber for the armpit. **1956** R. H. APPLEWHAITE *Lawn Tennis* i. 12 Sweatbands..are worn round the wrist to prevent perspiration running down the arms into the hands. **1977** J. F. FIXX *Compl. Bk. Running* xii. 134 When I started running, I saw a lot of runners wearing sweatbands, so after sweat had dripped into my eyes a few times I went out and bought one. **1877** S. POWERS *Tribes of California* xxvi. 244 [The Shasta Indians] have no assembly chamber..; nothing but a kind of oven large enough that one person may stretch himself therein and enjoy a sweat-bath. **1921** J. HASTINGS *Encycl. Relig. & Ethics* XII. 128/2 When we turn to the Old World, we find a striking resemblance to the American customs in Herodotus's description of the use of the sweat-bath among the Scythians as a means of purification, after mourning. **1963** E. WAUGH *Let.* Sept. in C. Sykes *Evelyn Waugh* (1975) xxvi. 439, I have sat in a 'sweat-bath' and been severely massaged. **1965** S. G. LAWRENCE *40 Yrs. on Yukon Telegraph* xiv. 75 They [sc. some Indians] stayed over a day and all the old men took sweat baths. **1870** *U.S. Navy Gen. Orders & Circulars* (1887) 97 He was.. gagged and confined in a sweat-box of such dimensions that it was impossible to sit down. **1897** *Chicago Tribune* 10 July 1/4 The upper gallery commonly known as the 'sweat box' in regular theaters. **1901** 'J. FLYNT' *World of Graft* 102 He was copped out on suspicion. They put him in the sweat-box, made him cough, an' you know the rest. **1931** Z. CHAFEE et al. in *Rep. Nat. Comm. Law Observance & Enforcement* (U.S.) ii. 38 The original 'sweat box' used during the period following the Civil War..was a cell in close proximity to a stove, in which a scorching fire was built and fed with old bones, pieces of rubber shoes, etc., all to make great heat and offensive smells, until the sickened and perspiring inmate of the cell confessed in order to get released. **1973** 'H. HOWARD' *Highway to Murder* ii. 28, I ought to stick you in the sweat box until you told me the name of your client. **1974** J. ENGELHARD *Horsemen* vi. 38, I never go in a sweatbox... I lose all the weight I want playing tennis. **1948** *Technical Publ. Amer. Inst. Mining & Metall. Engineers* No. 2343. Class E. 1 In designing a sweat cooled part it is imperative to assure a given rate of flow of coolant. *Ibid.*, A less orthodox method consists of making the part to be cooled of a porous material, so that the cooling fluid can be forced through the pores... This method, referred to as 'sweat cooling', was proposed at the Jet Propulsion Laboratory in September 1944. **1969** E. C. ROBERTSON *Now Bks. Rocket Motors* iv. 29 Many devices have been tried to keep the walls of the chamber cool and techniques have ranged from sweat cooling..to the one that is most common today. **1973** *Time* 16 July 43 A group of poor, racially mixed tenants took over a nearby city-owned tenement, stripped the shabby interiors and are building modern apartments to replace the narrow, cold-water flats... In return for their 'sweat equity', the builder-residents will make payments as low as $80 per month and ultimately own the building as a cooperative. **1980** B. VILA *This Old House* v. 83/1 The calculations you make in a sweat equity job are different from those in a project in which you are employing professionals. **1976** *Senior Scholastic* 4 May 41 John Travolta..[is] back in the classroom..as the leader of the sweathogs in ABC's *Welcome Back, Kotter.* **1979** BROOKS & MARSH *Compl. Directory Prime Time Network TV Shows, 1946–Present* 673/1 Gabe's 'sweathogs' were the outcasts of the academic system, streetwise but unable or unwilling to make it in normal classes. **1970** *Islander* (Victoria, B.C.) 22 Nov. 5/1 Little by little they cleared each acre with axe and cross-cut saw. It was slow, sweat-labor. **1973** *New Society* 19 July 137/2 A 'sweat lodge', or hut fashioned from rocks, branches and a sacred blanket. The sauna-like action of a fire inside the hut helps purify his soul along with his body. **1977** *Rolling Stone* 7 Apr. 55/3 She learned of the sweat lodge and the sacred pipe ceremony and the Sun Dance while researching her Indian history book, and then began to understand them as part of the present. **1914** D. H. LAWRENCE *Prussian Officer* 20 His sweat-marked horse swishing its tail. **1957** H. ROOSENBERG *Walls came tumbling Down* v. 127 They had noticed that Nell's green skirt was badly worn—would she try on these sweat pants and see if they fitted? **1978** R. B. PARKER *Judas Goat* vi. 33 My blue sweat pants worn stylishly with the ankle zippers open. **1843** 'R. CARLTON' *New Purchase* I. xi. 73 This luxury..was used only as a 'sweat rag', and not 'as a nose-cloth'. **1930** *Aberdeen Press & Jrnl.* 28 Mar. 7/5 Making a swab with a sweat-rag, he attempted to stop the flow of blood. **1953** X. FIELDING *Stronghold* 256 The dirty old sweat-rag which he had worn round his head for the last three months. **1974** D. STUART *Prince of My Country* v. 32 Father puts down his knife and wipes his face with the sweatrag at his neck. **1971** M. BRANDER *Horseman's Vade Mecum* 439 *Sweat-rug*, a string rug put on under a reversed top rug when a horse has been sweating. **1978** 'F. PARRISH' *Sting of Honeybee* i. 11 She had taken off his saddle and put on a sweat-rug. **1923** D. H. LAWRENCE *Birds, Beasts & Flowers* 172 And dance, and dance, forever dance, with breath half sobbing in dark, sweat-shining breasts. **1929** *Sears, Roebuck Catal.* Spring/Summer 394 Every Man and Boy Wants A Sweat Shirt. **1938** E. HEMINGWAY *Fifth Column* (1939) 291 He'd pull on a rubber shirt over a couple of jerseys and a big sweat shirt over that. **1948** *Daily Express* 4 Sept. 2/5 (*caption*) The fluffy blonde in pale lemon sweat shirt. **1958** J. & W. HAWKINS *Death Watch* (1959) i. 16 She was wearing jeans, moccasins and a white sweat shirt. **1978** L. HEREN *Growing up on The Times* ix. 307 Another [young lad] exchanged his jeans and sweatshirt for a white dinner jacket and plum-coloured trousers. **1977** R. BARNARD *Blood Brotherhood* i. 14 The be-jeaned and sweat-shirted figure. **1892** *Charities Rev.* Jan. 115 What relaxation or excitement can a car-driver or a sweat-shop tailor get except by drinking? **1903** *Bond of Brotherhood* (Calgary, Alberta) 12 June 4/1 Healthy niggers sound in wind and limb well broke to handcuffs, two pair of genuine sweat shop overalls given with each piece of ebony. **1938** *Times Lit. Suppl.* 3 Dec. 767/2 The story of two Jews who, in youth, work in the same tailoring sweat-shop. **1959** *Daily Tel.* 17 Apr. 13/8, I cannot really think that he should want my job. Whitehall, and certainly Downing Street, is nothing but a sweatshop. **1972** *Bookseller* 4 Mar. 1476/1 If 28 jobs were costing only £6,000 a year..then the N.B.L. were running a sweat shop. **1944** K. LEVIS in Murdoch & Drake-Brockman *Austral. Short Stories* (1951) 429 Our shirts sweat-soaked under the midday sun. **1973** 'R. MACLEOD' *Burial in Portugal* i. 29 His sweat-soaked shirt was sticking to his back. **1973** R. BUSBY *Pattern of Violence* vi. There was a dark sweat stain down the back of his shirt. **1932** W. FAULKNER *Light in August* ii. 28 Byron watched him standing there and looking at the men in sweat-stained overalls. **1975** H. R. F. KEATING *Remarkable Case* i. 3 His jacket and trousers were..worn and sweat-stained. **1930** L. W. OLDS *Track Athletics & Cross Country* i. 4 Sweat suits should be fleece-lined, washable and worn for warmth rather than a flashy appearance. **1951** I. SHAW *Troubled Air* x. 158 Archer lay on the mat in his sweatsuit. **1979** J. P. R. WILLIAMS *JPR* iv. 91 An Adidas sweat-suit keeping out the elements. *a* **1963** S. PLATH *Crossing Water* (1971) 58 Tangled in the sweat-wet sheets I remember the bloodied chicks.

sweat, *v.* Add: **I. 2. b.** In slang phrases *to sweat one's guts out* (later examples); also *to sweat blood*, (*a*) to exert oneself to the utmost; (*b*) to be terrified.

1911 G. S. PORTER *Harvester* xvii. 405 He just sweat blood to pacify her, but he couldn't make it. **1924** D. H. LAWRENCE in M. Magnus *Mem. Foreign Legion* 53, I sweat blood every time anybody comes through the door. **1937** 'G. ORWELL' *Road to Wigan Pier* xii. 228 It makes one sick to see half a dozen men sweating their guts out to dig a trench.., when some easily devised machine would scoop the earth out in a couple of minutes. **1953** 'J. TEY' *To love & be Wise* xiii. 163, I expect he sweats blood over his writing. He has no imagination. **1966** R. JEFFRIES *Evidence of Accused* v. 45 You sweated your guts out for months and finished your book, then the public looked the other way. **1973** W. M. DUNCAN *Big Timer* xxi. 138, I was sitting there sweating blood when those damned cops arrived.

d. With *off*. To (cause to) lose (weight, etc.) through strenuous exercise; *spec.* in *Boxing* (see quot. 1955).

1895 KIPLING *Day's Work* (1898) 347, I sweated the beef off 'em, and then I sweated some muscle on to 'em. **1899** —— *Stalky & Co.* 129 We've sweated a stone and a half off him since we began. **1955** F. C. AVIS *Boxing Dict.* 110 *Sweat off*, to lose weight through perspiration caused by vapour baths, etc., in an effort to bring the body to the poundage required for a given championship grade. **1976** *Southern Even. Echo* (Southampton) 12 Nov. 27/5 The..finalist outboxed his opponent, who was weakened after sweating off six pounds during the week.

4. c. *slang.* To subject (a prisoner, etc.) to close interrogation † or torture; to give the 'third degree' to (someone). Cf. *sweat-box* s.v. *SWEAT *sb.* 11.

1764 *Select Trials* I. 285, I..had heard him say, that Capt. Clark was a very great Rascal; and at Admiral Knowles's Trial, he would sweat Capt. Clark if he was examined, and if he could not sweat him there, he would sweat him another way. **1892** 'MARK TWAIN' *Amer. Claim.* xix. 194 It seems a piteous thing to sweat this poor ancient devil for a burglary he hadn't the least hand in. **1926** J. BLACK *You can't Win* xviii. 260, I wasn't taken out of my cell and 'sweated' or third-degreed, or beaten up. **1979** 'J. LE CARRÉ' *Smiley's People* (1980) xix. 237 Probably Mikhel intercepted and read it... We could sweat him, but I doubt if it would help.

II. 9. b. *trans.* With *out*, to await or endure anxiously or with unease. Esp. in phr. *to sweat it out. colloq.*

1876 'MARK TWAIN' *Tom Sawyer* xx. 200 Well, it's a kind of a tight place for Becky Thatcher... Just..let her sweat it out! **1942** E. COLBY *Army Talk* 229 *Sweat*..is a synonym for wait. You sweat a man out when you are waiting for him. You 'sweat out' a chow line while waiting for your turn for the sergeant to put your food in the mess kit. **1945** 'L. LEWIS' *Birthday Murder* (1951) xiii. 191, I haven't much time..but I'll sweat it out awhile. **1960** *News Chron.* 29 Sept. 1 Mr. Khruschev is just sweating it out in New York for an announcement of a manned flight in orbit. **1976** 'D. FLETCHER' *Don't whistle 'Macbeth'* 148, I had no intention of telling Hugo... Let him sweat that one out.

c. *intr.* With *on*, to await anxiously (an event or person); *spec.* in the game of lotto. Also *transf.*, to be close to attaining, as in phr. *to sweat on the top-line. slang* (chiefly *Austral.*) orig. *Mil.*

1917 A. G. EMPEY *From Fire Step* xix. 127 Sometimes you have fourteen numbers on your card covered and you are waiting for the fifteenth to be called. In an imploring voice you call out, 'Come on, Watkins, chum, I'm sweating on "Kelly's Eye".' *Ibid.* 252 *Sweating on leave*, Impatiently waiting for your name to appear in orders for leave. **1919** *Athenæum* 1 Aug. 695/2 'Sweating on the top line' is to be within an ace of obtaining what you want. **1959** S. J. BAKER *Drum* 150 *Sweat on*, to wait, usually to wait anxiously (for something to happen). **1968** S. L. ELLIOTT *Rusty Bugles* in E. Hanger *3 Austral. Plays* I. iv. 62 Wimpy sweats on me see..waits his chance..puts on a hut raid the other night and finds me mosquito net's not down and I lose my stripes.

d. *intr.* To experience discomfort through anxiety or unease (*colloq.*). In phr. *don't sweat it* (U.S. slang), don't worry.

1963 *Amer. Speech* XXXVIII. 271 *Don't sweat it* means 'don't worry about it'. **1973** R. HAYES *Hungarian Game* xxxix. 234 'Hold off for a moment. I want to watch him sweat.' 'The guy's about to faint from pain.' **1976** N. THORNBURG *Cutter & Bone* x. 238 Cutter reached over and covered her hand with his own, patted it. 'Don't sweat it, kid,' he said. 'It's nothing.' **1978** D. DEVINE *Sunk without Trace* ix. 92 No point in being early. Let him sweat.

III. 13. b. *Cookery.* To heat (meat or vegetables, etc.) in a pan with fat or water, in order to extract the juices.

1877 E. S. DALLAS *Kettner's Bk. of Table* 452 *Sweat, to*, is not a pretty phrase, but it expresses clearly..the act of making meat yield its juices by being heated in a pan with little or no water... The heat applied must be low and slow. **1942** [implied at *SWEATING *vbl. sb.* 3 c]. **1953** N. HEATON *Cassell's Cooking Dict.* 171 *Sweat*, to heat gently to extract flavour. **1972** *Guardian* 18 Aug. 11/3 Finely chop one large onion and two cloves garlic. Sweat these in a little oil in a thick saucepan.

Hence **swea·table** *a. rare*, capable of becoming sweated labour or a sweated labourer.

1922 G. B. SHAW in S. & B. Webb *Eng. Prisons* p. xlvi, The supply of sweatable labor. **1928** —— *Intelligent Woman's Guide* xli. 158 Our capitalist traders..were the enemies of every country, including their own, where there was a sweatable laborer to make dividends for them.

sweater. Add: **2. b.** A servant. *Winchester College slang.*

1900 J. S. FARMER *Public School Word-Bk.* 198 *Sweater* ..(Winchester), a servant. **1973** *Country Life* 19 July 147/1 This is a souvenir plaque..showing the famous painting of the 'Trusty Servant' at Winchester College.. or 'Sweater' as he is sometimes called.

5. (Earlier example.)

1846 *Manch. Guardian* 21 Mar. 7/4 A sort of middlemen, called 'sweaters', who get it [sc. tailoring work] by men and women at starvation prices.

6. (Earlier example.)

1845 *Currency Theory Reviewed* 69 It being obvious that the coinage, in the very nature of things, must be for ever, unit by unit, falling under depreciation by the mere action of ordinary and unavoidable abrasion—(to say nothing of the inducement which every restoration of the coinage holds out to the whole legion of 'pluggers' and 'sweaters').

7. b. Now also a similar garment in general informal use; a jumper or pullover.

1895 *Century Mag.* May 25/2 His brawny, muscular chest, which was covered only by a dark, close-fitting 'sweater', was that of an athlete. **1912** J. SANDILANDS *Western Canad. Dict. & Phrase-Bk.*, *Sweater*, a woollen jacket, much worn in Canada during the winter both indoors and outdoors, and sometimes a somewhat gaudy article of wear. **1957** *Times Lit. Suppl.* 25 Oct. 640/1 A tall, bespectacled young man in turtle-necked sweater. **1981** G. SWIFT *Shuttlecock* i. 13 Martin has a red polo-neck sweater and Peter a brown one and they both wear identical child's blue jeans.

9. *attrib.* and *Comb.*, as (sense 7 b) *sweater blouse, coat, dress, -suit*; **sweater girl** *U.S.*, a girl, esp. a model or actress, who wears tight-fitting sweaters; orig. a name applied to the American actress Lana Turner (b. 1921) who wore such a sweater in the film *They won't Forget* (1937), and in subsequent publicity photographs; **sweater-shirt,** (*a*) *U.S.*, a knitted garment that may be worn as a sweater or a shirt; (*b*) = *sweat-shirt* s.v. *SWEAT *sb.* 11.

1925 *Vogue* Early Mar. 60 (*caption*) This straight-line sweater blouse from Molyneux..is fashioned of fine dark-blue tricot covered with an all-over woven pattern in gold thread. **1954** *New Yorker* 27 Nov. 141/1 A wool jersey sweaterblouse, lavender or white, has cap sleeves and a scoop neck ornamented with gold thread and tiny pink felt buds. **1911** Sweater coat [see *LOVAT]. **1963** *Vogue* Dec. 190 Sweater-coat hand-knitted in Italy. **1965** *Harper's Bazaar* May 6 (Advt.), An enchanting cashmere sweater-dress. **1940** *Movie Mirror* June 9/1 (*caption*) Sweet and sophisticated sixteen: Lana Turner, at the time her face hit a thousand papers as the 'Sweater Girl'. **1941** *Life* 14 Apr. 33/2 Mr. Breen's letter left movie-makers wondering..what to do with their up-and-coming sweater girls. **1956** S. ERTZ *Charmed Circle* 71 Among all the 'sweater girls' she looked, in her unrevealing black dress, as if she had strayed in by mistake. **1971** D. MACKENZIE *Sleep is for Rich* iii. 66 Crying Eddie was getting plenty of attention from the sweater girls. **1964** *New Yorker* 12 Oct. 15 Sweater-shirt of pink cashmere. **1977** *Private Eye* 4 Mar. 20/2 (Advt.), American styled printed sweatershirts and T-shirts. **1929** M. LIEF *Hangover* 232 'You're looking fine,' said Whippet, admiring her..slim figure in a neat-fitting sweater-suit. **1964** *Glamour* Sept. 160 Town sweater-suits [are] booted for the summer.

Hence **swea·tered** *a.*, wearing a sweater; clothed in a sweater.

1901 S. E. WHITE *Claim Jumpers* i. 11 Two sweatered and white-ducked individuals. **1936** R. CHANDLER in *Trouble is my Business* (1954) 205 The sweatered man snatched the gun up. **1971** C. MCCULLERS *Mortgaged Heart* (1977) 74 His blue sweatered shoulders were shaking.

sweat-house. Add: **3.** (See quot.) Cf. *sweat-box* (*c*) s.v. SWEAT *sb.* 11. *rare.*

1882 *Harper's Mag.* Nov. 872/2 The grapes for raisin-making..are removed to an airy building known as a 'sweat-house', where they remain possibly a month, till the last vestiges of moisture are extracted.

sweating, *vbl. sb.* Add: **3. a.** Also with *out*. Cf. SWEAT *v.* 12.

1969 BENNISON & WRIGHT *Geol. Hist. Brit. Isles* iii. 43 The last major effect of the metamorphism was the 'sweating out' of synorogenic pegmatites which cross-cut some of the Inverian structures. **1971** I. G. GASS et al. *Understanding Earth* iii. 66/2 The present oceans and atmosphere of the Earth are secondary features due to the subsequent dewatering or 'sweating out' of the Earth's interior.

c. *Cookery.* The action or process of *SWEAT *v.* 13 b.

1942 C. SPRY *Come into Garden, Cook* xi. 137 Cook the sliced vegetables first in a little fat... This preliminary sweating of the vegetables draws out the flavour.

5. b. Extortion of a confession (from a prisoner, etc.) by close interrogation † or torture. Cf. *SWEAT *v.* 4 c.

1824 J. DODDRIDGE *Notes Settlement Indian Wars* II. xii. 122 The torture of sweating..that is of suspension by the arms pinioned behind the backs, brought a confession. **1904** *Cincinnati* (Ohio) *Enquirer* 21 Oct. 4 He confessed, under sweating, that he broke into several offices. **1949** *Amer. Speech* XXIV. 262 The device of *sweating* consisted of suspending the offender from the limb of a tree by his arms, and laying lashes on him.

6. sweating pen *Austral.*, a pen in which sheep are kept (formerly, to sweat so as to soften the wool) before shearing; = *holding pen* s.v. *HOLDING *vbl. sb.* 6 b.

1882 ARMSTRONG & CAMPBELL *Austral. Sheep Husbandry* xv. 176 On each side of the board are built the sheep pens, which are filled from a race on each side.. which is in its turn filled from the sweating pen. *c* **1929** H. B. SMITH *Sheep & Wool Industry in Austral. & N.Z.* (ed. 3) x. 73 After drafting, the sheep to be shorn are run up a ramp into the sweating pens of the shed. **1965** J. S. GUNN *Terminol. Shearing Industry* II. 29 *Sweating pen*, sometimes used in the same sense as 'holding pen', although there is no longer any suggestion of deliberately 'sweating' the sheep.

sweating, *ppl. a.* **2.** (Later examples.)

1976 K. BONFIGLIOLI in *Winter's Crimes 8* 44 The sweating heel of some nameless cheese. **1981** J. B. HILTON *Surrender Value* vi. 47 A sweating expresso machine.

sweaty, *a.* Add: **1. d.** Severe, demanding. *colloq.*

1919 A. LUNN *Loose Ends* iii. 27 'It's a sweaty house for new men.' Cluff shook his head sadly. 'Yes, it's a hard life for new men.' *Ibid.* xiii. 118 These Blues are sometimes rather sweaty. They think it lip if you cut your work for a man who's been a Blue. **1973** M. AMIS *Rachel Papers* 85 I'm not trying to be sweaty or anything, but, um—just out of interest—how long have you known De Forest?

sweatily *adv.*: also *fig.*, anxiously, feverishly.

1975 *Times* 4 Sept. 12/8 Men talking sweatily about the upcoming upcurve in house prices. **1978** W. F. BUCKLEY *Stained Glass* iii. 21 On the occasions when they found themselves trapped in his company at a dinner party, they would sweatily engage other members of the party.. in concentrated, often nonsensical, discussions.

Swede. Add: **4. swede-basher** *slang*, a farm worker; hence, a rustic (cf. *BASHING *vbl. sb.* 3); so **swede-bashing** *a.*

1936 J. CURTIS *Gilt Kid* iv. 40, I know you're not a swede-bashing judy. **1943** HUNT & PRINGLE *Service Slang* 63 *Swede-basher*, agricultural worker; country bumpkin. **1948** A. BARON *From City, from Plough* 84 There's a lot o' these swedebashers go down the farms every night. **1966** *New Statesman* 18 Mar. 363/1 Sir Gerald Nabarro is said to have remarked on TV that the fate of the nation depended on a few swede-bashers. **1976** J. GRENFELL *Joyce Grenfell requests Pleasure* (1977) xiii. 190, I tried to sing a song appropriate for the swede-bashers from Lincolnshire, the Cockneys, Scots.., and so on.

Swede (swīd), *a.* Chiefly *Canad.* [f. the sb.] = SWEDISH *a.*; *spec.* in *Swede saw*, a type of saw having a bow-like tubular frame and a sharp blade with many cutting teeth.

1934 G. BETTANY *Valley of Lost Gold* 29 I've been to the Swede settlement west of the hills. **1950** J. HAMBLETON *Abitibi Adventure* 120 It looked just like any other 'Swede saw', with its tubular metal frame, painted blue, and a thin, keen blade which was kept taut by a clamp. **1971** D. C. BROWN *Yukon Trophy Trails* ii. 39 Louis packed the axe and the Swede saw, and I carried the gun and lunch. **1981** *Nordic Skiing* Jan. 30/2 Lee saws firewood with the Swede saw and I get the honor of splitting it into burning size.

Swedenborgian, *a.* and *sb.* Add: **a.** *adj.* (Earlier and later examples.) **b.** *sb.* (Earlier and later examples.) **Swedenborgianism** (earlier examples).

1791 J. LACKINGTON *Mem.* xxv. 195 The *Swedenborgians*, or *New Jerusalemists*, are gaining ground very fast. **1807** SOUTHEY *Lett. from Eng.* III. lxii. 144 (*heading*) Account of Swedenborgianism. *Ibid.*, The New Jerusalem, or Swedenborgian chapel. **1842** C. FOX *Jrnl.* 6 June in *Memories of Old Friends* (1882) viii. 160 Thomas Carlyle came in..and we presently got, I know not how, to Swedenborgianism. **1856** *Spiritual Herald* June 147 The Lord may be better represented by an animated and intelligent orb..than by a Whewellite or Swedenborgian star. **1914** C. MACKENZIE *Sinister Street* II. III. ix. 682, I should love to be a sort of Swedenborgian with all sorts of fanciful private beliefs. **1920** M. BEER *Hist. Brit. Socialism* II. III. i. 12 William Hill, a Swedenborgian preacher and a grammarian. **1976** *Gramophone* June 32/1 Pitcairn wished to sponsor the promotion and recording of his fellow Swedenborgian's music.

swedge, *v.* Add: Also *intr.* To go *off* or depart without paying. *U.S. Naut. slang. rare.*

1897 KIPLING *Capt. Cour.* v. 124 'Seems kinder unneighbourly to let 'em swedge off like this,' Salters suggested, feeling in his pockets.

Swedish, *a.* and *sb.* Add: **A.** *adj.* *Swedish exercises* = *Swedish drill*; also *fig.*; *Swedish massage*, a system of massage combined with manipulation of the joints and muscles, first devised in Sweden; hence *Swedish masseur*, (fem.) *-euse*, one trained in the practice of Swedish massage; *Swedish modern* = *Danish modern* s.v. *DANISH *a.* (cf. *SCANDINAVIAN *a.* 2). Also with reference to persons living in Sweden, or of Swedish descent.

1799 MALTHUS *Diary* 27 July (1966) 191 The son-in-law told us that a Norway lap must not go into Sweden, nor a Swedish lap come into Norway. **1911** *Daily Colonist* (Victoria, B.C.) 8 Apr. 5/2 (Advt.), Swedish Massage, Medical Sick-Gymnastic, Electric Vibrations... Above treatments highly recommended by leading physicians. **1911** Swedish masseur [see *OPSONIST]. **1912** 'SAKI' *Unbearable Bassington* vii. 124 A sporting cat..watching the Swedish exercises of a well-spent..mouse. **1923** WODEHOUSE *Inimit. Jeeves* xi. 123 If she had knocked off starchy foods and done Swedish exercises for a bit, she might have been quite tolerable. **1948** A. H. RUTT *Home Furnishing* (ed. 2) xiii. 217 Swedish Modern is a favorite style which successfully combines native, Classical, Empire, and Modern ideas. **1958** 'S. MARLOWE' *Second Longest Night* iii. 26 The living-room was Swedish modern ..with black lacquered pieces and high-grained white ash. **1959** A. GLYN *I can take it All* ii. 30 'But you live in—' I was going to say Helsinki, but I remembered..that she was a Swedish Finn '—in Helsingfors?' **1966** J. MITFORD in *McCall's* Mar. 190/2 A splendid Swedish masseuse.. rubs you all over with cream. **1970** D. BAGLEY *Running Blind* vii. 156 The room was decorated in that generalized style known as Swedish Modern. **1975** C. YOUNG *Massage* ii. 16 In the beginning of the nineteenth century a Swedish fencing master, Peter Henrik Ling, introduced a system of movement that consisted of massage and exercises... This method..became the basis for..Swedish massage. **1979** J. TATE tr. *Martenson's Death calls on Witches* v. 33 Her parents were Swedish-Americans.

† sweedle (swī·d'l), *v. Obs. slang.* [Blend of SWINDLE *v.*[2] and WHEEDLE *v.*] *trans.* To swindle by wheedling. Hence **swee·dling** *vbl. sb.* and *ppl. a.*

1908 H. A. JONES *Dolly reforming Herself* IV. 94 I'm not going to be sweedled!—*Matt.* What is sweedled? *Harry.* Sweedling is sweedling! It's part swindling and part wheedling! **1908** *Westm. Gaz.* 4 Nov. 5/2 Dolly is a 'sweedling' extravagant little vixen. **1909** *Daily Chron.* 19 Jan. 4/4 He circumvented Isengrim the Wolf and 'sweedled' King Noble the Lion. **1914** *Angl. Forsch.* XLII. 20 The mind, hesitating between *swindle* and *wheedle*, compromises on *sweedle*. When the result pleases the coiners, it sometimes continues in family use, as *sweedle* in the case of a Nebraska family.

Sweeney (swī·ni). *slang.* Also **sweeney, Sweeny.** [f. the name of *Sweeney* Todd, a barber who murdered his customers, the central character of a play by George Dibdin Pitt (1799–1855), and of later plays.] **1.** In full *Sweeney Todd*. Rhyming slang for 'Flying Squad'. So, a member of the Flying Squad.

1936 J. CURTIS *Gilt Kid* xxii. 223 The slops had been turning up at the block of flats just as he was making his getaway. Yes, and coming along in a jam jar too. That made them look like Sweenies. **1938** F. D. SHARPE *Sharpe of Flying Squad* 333 *The Sweeney Todd*, the Flying Squad. **1956** J. D. CARR *Patrick Butler for Defence* xiii. 140 The Flying-Squad people are called sweenies, from Sweeny Todd. **1967** N. LUCAS *C.I.D.* xiii. 195 By the way, don't bother to call the Sweeny (Sweeny Todd—Flying Squad). **1971** R. BUSBY *Deadlock* I. v. 56, I was with the sweeney before this firm... The Sweeney Todd—crime squad... If you're an old sweeney man too, we'll get along all right. **1977** *Guardian Weekly* 17 July 10/1 Was designed—as they say in The Sweeney—to put the frighteners on Labour knockers.

2. A (nickname for a) barber.

1966 'L. LANE' *ABZ of Scouse* 104 I'm goin' ter Sweeny ter 'ave me hur cut. **1980** *Globe & Laurel* July/Aug. 217/2 It has been noted that a significant number of the ship's company now have a 'Sweeney' Barber special.

sweenied (swī·nid), *a. U.S.* Also **sweeneyed, swinneyed, swyneyed.** [f. next + -ED[2].] Suffering from sweeny. Hence *sweenied-looking* adj.

1861 *Harper's Mag.* Aug. 421/2 The people have been fed on buncombe, while a lot of spavined, ring-boned,.. swyneyed, split-headed..polleviled politicians have had their noses in the public crib. **1872** *Rep. Indian Affairs 1871* (U.S.) 554 The three mules were thin, and one of them lame in the right shoulder, 'sweenied'. **1872** *Borderer* (Las Cruces, New Mexico) 5 Oct. 2/4 God Almighty only knows the age of 'em!—three footed, one-eyed, sweeneyed, spavined, broken-down ex-livery stable stock, 'political hacks', and sway-backed horses. **1960** V. WILLIAMS *Walk Egypt* 188 Mule and tree had grown old and swinneyed together. *Ibid.* 240 He picked up a mule... It was a swinneyed-looking thing.

sweeny. (Earlier examples.)

1813 E. GERRY *Diary* 23 June (1927) 131, I answered he [sc. a horse] was foundered, but was informed that it was another complaint called the sweeney. **1832** J. P.

KENNEDY *Swallow Barn* II. i. 22 He professed to cure the colt's distemper, sweeny, and other maladies.

sweep, *sb.* Add: **I. 1. b.** Also *spec.* with reference to aircraft patrols, usu. offensive, but occas. also for reconnaissance purposes.

1940 *Sun* (Baltimore) 21 Feb. 1/6 In a daylight sweep over the Channel..British fighters set another Nazi E-boat afire. **1942** *Ann. Reg. 1941* 52 About the middle of June the Royal Air Force began to make what were called 'offensive sweeps'..seeking for enemy machines. **1959** R. COLLIER *City that wouldn't Die* x. 167 Some fifty day-fighters and thirty night-fighters had taken part in this spectacular sweep. **1973** 'R. LEWIS' *Blood Money* viii. 106 The helicopter seemed to have completed its sweeps... The shadows..had made spotting difficult.

d. *Sport.* Victory in all the games in a contest, tournament, etc., by one team or one competitor, or the winning of all the places in an event or competition. orig. and chiefly *U.S.*

1960 WENTWORTH & FLEXNER *Dict. Amer. Slang* 531/1 *Sweep n.*, the act or an instance of one athlete or team winning a tournament without losing an individual game or contest. **1974** *State* (Columbia, S. Carolina) 31 Mar. 5-D/4 The Gamecocks claimed all three places for a sweep of the 880 as John Brown rolled home with a time of 1:56.6 to best teammates Mike Sheley and Don Brown. **1977** *Hongkong Standard* 12 Apr. 12/7 John Mayberry also drove in two runs to help Royals complete a season-opening sweep of the three-game series against the Tigers. **1979** *Arizona Daily Star* 1 Apr. C1/1 James Frazier led an Arizona sweep in the high jump with a winning leap of 7-3½, followed by Roger Curtis' 7-1½.

5. b. *Cricket.* An attacking stroke made on the front foot, in which the batsman brings the bat across his body to hit the ball square or backward of square on the leg side.

1888 R. H. LYTTELTON in Steel & Lyttelton *Cricket* ii. 65 George Parr's leg hit..was the sweep to long leg off a shortish ball. **1920** D. J. KNIGHT in P. F. Warner *Cricket* 35 The sweep to leg is a very paying and useful stroke, although not elegant. It is effected by sinking almost down on the right knee and sweeping the ball right round in the direction of long leg. **1955** *Times* 9 May 15/1 He had played some good drives and sweeps. **1970** *Times* 19 Aug. 6/5 Most of the Yorkshire batsmen were obsessed by that ugly and risky stroke, the sweep.

6. d. *Electronics.* A steady movement across the screen of a cathode-ray tube of the spot produced by the electron beam; the moving spot itself, or the line it generates.

1924 *Wireless World* 5 Mar. 705/2 The approximate form of transient phenomena may also be indicated, if the frequency is low enough to enable a single sweep of the ray across the screen to be seen. **1946** *Radar: Summary Rep. & Harp Project* (U.S. Nat. Defense Res. Comm., Div. 14) 144/1 By making this motion rapid and continuous, the point of light becomes a line of light, and is called a sweep. **1958** *New Scientist* 10 Apr. 17/2 A ray of greenish-blue light—the sweep—pivots on the centre of the tube like the spokes of a wheel. **1966** M. WOODHOUSE *Tree Frog* xxi. 154 The bright scanning sweep swung around the orange tube face of the monitor like the seconds' hand of a stop-watch. **1975** G. J. KING *Audio Handbk.* v. 114 The oscilloscope's time-base is switched off and the horizontal sweep provided by high-level signal from the audio oscillator.

e. *Electronics.* A steady, usu. repeated, change in the magnitude or frequency of a voltage or other quantity between definite limits.

1930 *Proc. IRE* XVIII. 590 A single sweep, exposing each tone about 1/150th of a second was found sufficient to give a useful record. **1966** *McGraw-Hill Encycl. Sci. & Technol.* XIII. 336/1 Hyperbolic sweeps may be generated as a modification of the type of circuitry used in the generation of saw-tooth sweep waveforms. **1975** D. G. FINK *Electronics Engineers' Handbk.* xvi. 29 Circuits delivering a linear voltage sweep fall into two categories, the Miller time base and bootstrap time base.

8. b. *fig.* A comprehensive search, esp. in relation to crime investigation; *spec.* a search for electronic listening devices. *colloq.* (orig. *U.S.*).

1966 *Wall St. Jrnl.* 17 Feb. 1/4 In Burns' 'sweeps'.. specialists check furniture, light switches, air vents, drapes, rugs, telephones, pictures and walls with..detection gear. **1973** *Times* 18 June 2/7 One of the largest British-based international companies recently employed a security firm to conduct 30 anti-bugging 'sweeps' on its premises every month. **1974** *Union* (S. Carolina) *Daily Times* 20 Apr. 1/7 Police mounted a room-by-room sweep of hotels..in search of Dantzler. **1978** J. GARDNER *Dancing Dodo* xiii. 93 '*Overshoot?*' Dobson queried reflectively.. 'Will you do a sweep of the files?'

II. 14*. *Aeronaut.* = *sweepback* s.v. *SWEEP- 3.

1914 *Aeroplane* 26 Mar. 358/2 (*caption*) Plan view of the Grahame-White biplane, showing sweep of wings. **1947** *Aircraft Engin.* June 180/2 As can be seen..the sweep is 38°..for the main plane and rather less for the tail plane. **1976** *Farnborough Internat. Exhibition* (Official Programme) 41 Studies indicate that, by adjusting the angle of sweep, fuel consumption..can be materially reduced.

III. 15. f. *Forestry.* The natural curve of a tree or log of wood.

1932 CHAPMAN & DEMERITT *Elem. Forest Mensuration* xi. 179 The extent of the actual loss of boards by reason of crook or sweep depends on the minimum length of a merchantable board. **1946** *Q. Jrnl. Forestry* XL. 52 Many of the trees had a severe 'sweep' which resulted in the very poor output of suitable telegraph pole material. **1957** *Brit. Commonw. Forest Terminol.* II. 192 *Sweep*, the natural bend of a log, generally applied to long gentle bends.

V. 26. (Later examples.)

1923 H. BELLOC *Sonnets & Verse* III. 119 The sweeps have fallen from Ha'nacker Mill. **1968** J. ARNOLD *Shell Bk. Country Crafts* 170 The original form of sweep consisted of a light framework mounted on each stock, or sail-arm, over which a canvas sail was set or furled according to the wind.

29. Also in mine-sweeping.

1915 *Chambers's Jrnl.* June 387/2 Those six small gray ships will return with..a fearsome tale of many mines caught in their sweeps and destroyed. **1923** *Man. Seamanship* (Admiralty) II. 172 The vessel..puts the end of the sweep on a slip somewhere on her quarter-deck. **1943** *His Majesty's Minesweepers* (Min. of Information) 8/1 The thud of the explosion as a mine, caught in a sweep, detonated under a trawler's counter.

33. b. (Examples.) Chiefly at Yale University. ? *Obs.*

1900 *Dialect Notes* II. 65 *Sweep, n.*, a care-taker of college rooms at Yale, where negro boys are employed. **1950** *Harvard Alumni Bull.* 22 Apr. 590/3 In early times, sweeper was in use instead of goody, and even now at Yale College the word sweep is retained.

VII. 34. (sense *6 d, e) *sweep amplifier, generator, oscillator, voltage*; (sense 19) *sweep-ticket*; **sweep-swinger** *U.S.*, an oarsman in a racing boat.

1947 R. LEE *Electronic Transformers & Circuits* i. 4 Make efficient transformers for the non-sinusoidal wave shapes such as are encountered in pulse, video, and sweep amplifiers. **1946** *Radar: Summary Rep. & Harp Project* (U.S. Nat. Defense Res. Comm., Div. 14) 144/1 *Sweep circuit* or *generator*, a circuit which produces at regular intervals an approximately linear or circular, or other form of movement (sweep) of the beam of the cathode-ray tube. **1975** D. G. FINK *Electronics Engineers' Handbk.* xvi. 29 Sweep generators may also be looked upon as integrators with a constant-amplitude input signal. **1939** H. J. REICH *Theory & Application Electron Tubes* xv. 596 Practical sweep oscillators do not furnish a voltage that satisfies the requirements for a perfect sweep voltage. **1967** *Electronics* 6 Mar. 2 (Advt.), All solid-state Hewlett-Packard 3211A sweep oscillators..meet virtually all of your swept frequency testing requirements. **1949** *N.Y. Times* 12 June 48/4 Hundreds of sweepswingers are sweating it out..on Connecticut's Thames River. **1971** L. KOPPETT *N.Y. Times Guide Spectator Sports* xviii. 234 A crewman is a 'sweepswinger'. **1930** *Daily Express* 23 May 3/4 Who sent out the Mayfair Luncheon Club's £20,000 sweep tickets? **1934** J. H. REYNER *Television* vii. 78 The spot can be shifted horizontally or vertically, as required, irrespective of the sweep or work voltages. **1962** SIMPSON & RICHARDS *Physical Princ. Junction Transistors* xvii. 443 The simplest sweep voltage is obtained by suddenly applying a d.c. voltage *V* to a resistor *R* and a capacitor *C* in series and taking the voltage across the capacitor as the output.

sweep, *v.* Add: **2. b.** *Cricket.* To hit (the ball) with a sweep (sense *5 b). Also *absol.* or *intr.*, to play a sweep.

1920 [see *SWEEP *sb.* 5 b]. **1958** D. BRADMAN *Art of Cricket* 80 An inviting half-volley comes along... The greater scoring medium would be to sweep it fine. **1963** *Times* 19 Feb. 4/2 He is a fine cutter and an enthusiastic sweeper. Today he swept only twice, lest the shot should get him into trouble, as it sometimes does. **1965** D. SILK *Attacking Cricket* iv. 60 The batsman must always try to sweep the ball along the ground. **1976** *Star* (Sheffield) 30 Nov., Fletcher eventually fell lbw sweeping at Eknath Solkar.

6. c. *to sweep* (a person) *off his feet*: to affect with overwhelming enthusiasm, to infatuate. Also *transf.* Cf. *to carry* (a person) *off his feet* s.v. FOOT *sb.* 27.

1913 F. L. BARCLAY *Broken Halo* xiv. 151, I remember being swept completely off my feet when I first met Jim. **1937** W. R. INGE *Rustic Moralist* I. ii. 46, I do not approve of concentration camps, or of Jew-baiting, or of sabre-rattling. I only want to understand a movement which has swept a great nation off its feet. **1977** *Daily Mirror* 16 Mar. 13/5 Mr. Lipscombe's daughter Gillian was swept off her feet by De Roth.

8. b. *U.S.* To win every event in (a series of sporting events, etc.), or to take each of the main places in (a contest or event).

[**1942** BERREY & VAN DEN BARK *Amer. Thes. Slang* § 650/7 *Phlanx, sweep the event*, to win all of the main events in all three first places in a meet.] **1960** WENTWORTH & FLEXNER *Dict. Amer. Slang* 531/1 *Sweep...v.t.*, to win a tournament without losing a game or contest. **1974** *Greenville* (S. Carolina) *News* 22 Apr. 15/1, I didn't think either team would sweep this series. **1979** *Tucson* (Arizona) *Citizen* 20 Sept. 8D/3 Montreal swept a doubleheader from New York, 3–1 and 4–1.

13. d. *fig.* To examine (premises, telephone lines, etc.) for electronic listening or recording devices. *colloq.* (orig. *U.S.*).

1966 *Wall Street Jrnl.* 17 Feb. 1/4 The companies also are having their offices regularly 'swept'—checked by professional sleuths to find any hidden transmitters. **1968** [see *SWEEPER 5 b]. **1970** K. BENTON *Sole Agent* xx. 210 This room's all right. It was 'swept' only a few weeks ago. **1979** J. BARNETT *Backfire is Hostile!* iii. 37 'How safely can we speak on this line?'..'The line is swept every fifteen minutes and it is very clean.'

14. b. *Ent.* To drag a net over the surface of (herbage, etc.) in order to catch insects. Cf. SWEEP-NET 2.

1826 KIRBY & SPENCE *Introd. Entomol.* IV. l. 517 For this last operation—sweeping the grass, &c.—..you will find a net invented by Mr. Paul..a very useful implement. **1926** A. H. HAMM in J. J. Walker *Nat. Hist. Oxf. District* 263 *Hemerodromia precatoria* Fln. and *H. raptoria* Mg. have been captured by sweeping water plants in 'Mesopotamia'. **1977** RICHARDS & DAVIES *Imms's Gen. Textbk. Entomol.* (ed. 10) II. III. 1205 The adults are most often obtained by sweeping or shaking the vegetation.

18. a. Also *fig.*

1957 W. S. CHURCHILL *Hist. English-Speaking Peoples* III. vii. 272 Although his generals and Ministers were reluctant and apprehensive a kind of delirium swept the martial classes of the Empire. **1958** P. H. GIBBS *Curtains of Yesterday* xix. 156 That was a gruesome sight! The whole country is swept by typhus. I guess some of us may be unlucky. It may be difficult to dodge.

b. To achieve widespread popularity throughout (a town, country, etc.). Also *spec.* in Politics, to gain control of by an overwhelming margin.

1892 *Times* 9 July 11/1 Mr Gladstone is not likely to 'sweep' the counties any more than he has 'swept' the boroughs. **1931** W. HOLTBY *Poor Caroline* vii. 278 Tell her that C.C.C. is going to *sweep England*. **1950** *Times* 27 Apr. 4/3 Any party which, at the next election, pledged itself to forming a coalition Government no matter how big a majority it obtained would sweep the country. **1960** *Sunday Express* 14 Aug. 12/3 The short cut is sweeping the town. **1970** *Morning Star* 29 May 1 Ceylon's Left wing United Front led by Mrs Sirimavo Bandaranaike swept the polls here today. **1974** *News & Courier* (Charleston, S. Carolina) 10 Mar. 9-A/2 Sweep the Negro vote..and pick up enough whites to come out of the primary with something more than 50 per cent of the ballots.

19. Also *spec.* with an aircraft as subject.

1941 E. SHEPHERD *Mil. Aeroplane* 26 These aeroplanes have to sweep the seas and watch enemy harbours. **1959** R. COLLIER *City that wouldn't Die* iv. 56 At 9.35 p.m. the usual dusk patrol, a few day and night fighters, sweeping the raiders' normal routes. **1976** A. WHITE *Long Silence* vii. 53 We had picked up our fighter escort... Every so often, one or the other would peel off and sweep an observation circuit.

sweep-. Add: **1. sweep hand** = next; **sweep second(s (hand)** orig. *U.S.* = *centre-second(s* s.v. CENTRE, CENTER *sb.* and *a.* 19; hence *sweep-seconds watch*.

1948 *Wrist Watches, Pocket Watches & Clocks* iii. 185 Watch stops... This may be caused by..sweep hand rubbing on dial. **1967** R. MEYERS *Dolphin Rider* (1968) i. 22 Henries..listened while he watched the sweep hand of his watch. **1977** *Times Lit. Suppl.* 24 June 779/1 The sweep hand of my watch is there in order to make seconds easier to read. **1948** *Wrist Watches, Pocket Watches & Clocks* ii. 154 Remove sweep seconds hand. **1953** W. J. GAZELEY *Watch & Clock Making* iii. 48 Nowadays..we have what are termed sweep-seconds watches. At one time these were referred to as centre-seconds. **1962** E. BRUTON *Dict. Clocks & Watches* 170 *Sweep seconds*, American name for centre seconds. **1962** J. D. MACDONALD *Girl, Gold Watch, & Everything* viii. 101 Uncle Omar's gold watch..had an hour hand, a minute hand and a sweep second hand. **1969** *Guardian* 20 Aug. 7/1 All the clocks..have a sweep second hand. **1979** *Sci. Amer.* May 145/1 A sweep second hand on a wristwatch will also serve.

3. With advbs.: **sweepback** *Aeronaut.*, the form of an aircraft wing that is angled backwards, so that the part farther from the fuselage is aft of the nearer part; the angle made by such a wing with a line at right angles to the fuselage; **sweep-forward** *Aeronaut.*, the form of an aircraft wing that is angled forwards, so that the part further from the fuselage is forward of the nearer part; **sweep-out,** an act of sweeping out; (*U.S. colloq.*) a clearance or purge.

1914 *Aeroplane* 19 Mar. 308/1 Owing to the sweep back on the wings the side area of these struts may be regarded as taking the place of tail fins. **1918** H. J. STEPHENS *Gloss. Aeronaut. Words* (ed. 2) 36 *Sweepback*, the angle at which the planes slope backwards each side of the fuselage. **1939** *Aircraft Engin.* Apr. 159/3 The basic characteristic of sweepback on a rectangular wing would appear to be an early stalling of the tips which may or may not produce greater lateral stability. **1968** MILLER & SAWERS *Technical Devel. Mod. Aviation* vi. 204 The slightly greater sweep-back of the 707's wing means that it takes off at a greater angle of attack. **1977** *Jrnl. R. Soc. Arts* CXXV. 349/2 We can use the improved methods..to reduce wing sweepback (thus improving take-off and landing). **1932** *Technical Rep. Aeronaut. Res. Committee 1930–1* I. 39 Sweep-forward increases the maximum lift and considerably delays the stall. **1953** M. RAUSCHER *Introd. Aeronaut. Dynamics* ix. 378 A wing without pronounced sweep-back or sweep-forward. **1975** L. J. CLANCY *Aerodynamics* xvi. 532 Sweep forward would have a de-stabilizing effect. **1947** *Sun* (Baltimore) 16 Aug. 12/8 Governor Lane has ordered a sweep-out at the Board of Supervisors of Elections... About 30 places on the pay roll..are slated to be pulled out from under employés who are on the wrong side of the Democratic factional fence. **1978** *Detroit Free Press* 6 Apr. K 5/3 A total of 47 rookies won jobs this season, reflecting a lot of dead wood on the rosters. 'Next year', predicts our source, 'will see an even bigger sweep-out.'

sweeper. Add: **1. c.** (Later examples.) Now usu. short for *mine-sweeper*.

1915 S. H. CARDEN in M. Gilbert *Winston S. Churchill* (1972) III. Compan. I. 405 Battleships preceded by sweepers making way up towards Narrows. **1941** S. O'CASEY *Let.* 28 Apr. (1975) I. 886 Delighted to hear John [Allen]'s allright on a sweeper. **1979** D. GURR *Troika* i. 5 Losses of submariners trying to run the Baltic minefields without benefit of sweepers were appalling.

d. *Cricket.* A batsman who sweeps (sense *2 b).

1961 *Times* 21 Aug. 3/3 There can be few more effective sweepers. **1963** [see *SWEEP *v.* 2 b]. **1965** D. SILK *Attacking Cricket* iv. 60 The best sweepers bring the bat down on the ball from above as well as across.

e. *Assoc. Football.* One who plays as the last line of defence except the goalkeeper, across the width of the field (i.e. as opposed to a right or left back, etc., in other systems).

1964 *Times* 13 Apr. 4/1 Moore..played a giant part in his role as 'sweeper' of the rear. **1971** *Times* 15 Feb. 9/2 Of the other younger England [hockey] players Perry had a solid game as sweeper. **1973** *Daily Pennsylvanian* 9 Oct. 6 We knew they were using a sweeper, so we had to run to the corners to draw him out. **1976** *Denbighshire Free Press* 8 Dec. 24/2 Even with Bernie Welsh operating as sweeper behind a defensive line of four, Courtaulds were far from impressive at the back when the ball was in the air.

2. a. (Earlier example in *spec.* sense.) Also *attrib.* as *sweeper caste*.

1844 W. H. SLEEMAN *Rambles & Recollections Indian Official* I. viii. 64 The right of sweeping within a certain range is recognised by the caste to belong to a certain member... If any house-keeper..happens to offend the sweeper..none of his filth will be removed. **1859** MRS. R. M. COOPLAND *Lady's Escape from Gwalior* iii. 58 Matrané, a woman of the sweeper caste. **1909** J. HASTINGS *Encycl. Relig. & Ethics* II. 551/2 The sweeper or scavenger caste of Hindustan.

5. b. *colloq.* An electronic device for detecting listening or recording apparatus. Also, a person operating such a device.

1968 *Observer* 16 June 7/2 To help in the job of debugging, Mr. Johnson is having electronic 'sweepers' manufactured... Small ones can be operated by firms' security officers to keep a boardroom 'clean' after it has been 'swept'. It was with one of these 'sweepers' that an attempt to smuggle a bug into a boardroom in the bottom of a coffee pot was recently foiled. **1972** K. BENTON *Spy in Chancery* xi. 116 They sent a 'sweeper' team..and they went through the rooms in Chancery. **1979** F. FORSYTH *Devil's Alternative* x. 232 A secure room regularly checked by the 'sweepers' who are..looking for..listening devices.

6. (Examples.) Also, a partially fallen or drifting tree. *N. Amer.*

1888 S. M. ST. MAUR *Jrnl.* 19 July in *Impressions of Tenderfoot* (1890) vii. 95 We glided through the water at about ten miles an hour, sometimes rushing..within a few inches of a sweeper, as they call the trees which hang across the river. **1929** L. JOHNSTON *Beyond Rockies* 191 'Where are them sweepers, Charlie?' (sweepers being the river term for drifting trees, which may in a moment upset the craft of an unwary skipper). **1977** *New Yorker* 9 May 120/2 Over the cut bank a sweeper had recently fallen, a spruce whose trunk reached into the river... Sweepers tend to trap boats.

7. *Electronics.* A sweep generator or oscillator. *colloq.*

1967 *Electronics* 6 Mar. 2 (Advt.), The main frame of the 3211A contains everything you could hope to find in a sweeper. **1976** *Physics Bull.* Sept. 411/1 The hired equipment included spectrum analysers, signal generators, sweepers, oscilloscopes, [etc.].

sweeping, *vbl. sb.* Add: **3.** *sweeping-brush.*

a **1828** D. WORDSWORTH *Jrnl.* (1941) II. 329 Presently a Man enters with a sweeping brush, to 'arrange'. **1922** JOYCE *Ulysses* 327 Gob, he'd adorn a sweepingbrush,..if he only had a nurse's apron on him.

sweet, *sb.* Add: **1. d.** (Earlier and later examples.) Now freq. *sing.*

1832 F. TROLLOPE *Domestic Manners Americans* II. xxviii. 131 They are 'extravagantly fond'..of puddings, pies, and all kinds of 'sweets'. **1954** J. BETJEMAN *Few Late Chrysanthemums* 95, I know what I wanted to ask you—Is trifle sufficient for sweet? **1968** [see *PUDDING *sb.* 6 a]. **1979** J. COOPER *Class* xii. 202 Everything from lemon water ice to jam roly-poly pudding, Caroline would call 'pudding'. She would never say 'sweet' or 'dessert'.

f. *pl.* Drugs, *esp.* amphetamines. *U.S. slang.*

1961 [see *HOLD *v.* 15 f]. **1979** S. SMITH *Survivor* xxi. 221 A whole load of minor drugs, mostly amphetamines—known as 'sweets', 'blues' and 'black bombers'.

5. b. *pl.* A woman's breasts. *poet.*

1817 KEATS *Poems* 49 Ah! who can e'er forget so fair a being? Who can forget her half retiring sweets? **1870** D. G. ROSSETTI *Poems* (ed. 2) 111 Your silk ungirdled and unlac'd And warm sweets open to the waist.

8. (sense 1 e) *sweet coupon, paper, ration, rationing*; (sense 1 d) *sweet course*; **sweet trolley,** a dining trolley from which a choice of cold sweet dishes may be offered in a restaurant.

1943 N. LAST *Diary* 25 Dec. in *Nella Last's War* (1983) 270 Not a flower, a card—or a sweet, although you had the sweet coupons in your pocket. **1974** G. MARKSTEIN *Cooler* xlvi. 164 Grace spent all the sweet coupons he had left on buying a bar of chocolate. **1892** *Girl's Own Paper* 23 Apr. 476/2 The sweet course can also be arranged for by having some stewed fruit..with a mould of rice or cornflour. **1981** P. VAN GREENAWAY *'Cassandra' Bell* vii. 83 The evening meal..lasted ten minutes... Cherry stabbed a fork at his once or twice, derided the sweet course, and went. **1964** *Guardian* 1 Feb. 8/3 An occasional sweet-paper flutters striped among the bushes. **1979** M. INGATE *Tomb of Flowers* xxi. 153 A few sweet papers, and one or two bottles. **1944** Sweet ration [see *RATION 3 c]. **1978** E. MALPASS *Wind brings up Rain* i. 11 She tried to take back the toffee—she *needed* her sweet ration. **1942** *Times* 24 July 2/6 As a prelude to the introduction of chocolate and sweet rationing..there is heavy selling at some retail shops. **1963** *P.M.L.A.* Dec. p. vii/2 [U.K.] sweet trolley: [U.S.] dessert cart. **1964** L. DEIGHTON *Funeral in Berlin* xv. 93 The steak was O.K. and I was strong-willed enough not to hit the sweet-trolley too hard. **1981** *Radio Times* 19–25 Sept. 21/1 It's irritating being pointed at in a restaurant, like a sweet trolley.

sweet, *a.* and *adv.* **A.** *adj.* **3. b.** Delete † *Obs.* and add later N. Amer. examples of *sweet butter*. See also SWEET WATER.

1925 *N.Y. Produce Rev.* 27 May 95 (Advt.), Specializing in sweet butter. **1952** M. SMALL *Special Diet Cook Bk.* 201 Grocers..catering to the Jewish trade usually carry sweet butter. **1971** S. WALKER *Highland Cookbook* 8 Scones are delicious with sweet butter, in Scotland called fresh butter.

e. (Later examples.) In mod. use also in the *Oil Industry*, of petroleum or natural gas: free from sulphur compounds, *esp.* hydrogen sulphide or alkyl mercaptans.

1863 *Edin. Rev.* Apr. 411 The 'sweetest' kinds of coal (the freest from sulphur) are reserved for the smelting furnace. **1911** *Rep. Brit. Assoc. Adv. Sci. 1910* 612 The Coal Measures include 'sweet', *i.e.*, non-sulphurous, coals at several horizons. **1919** E. W. DEAN *Motor Gasoline Properties* (U.S. Bur. Mines. Techn. Paper No. 214) 25 If the liquid remains unchanged in color and if the sulphur film is bright yellow or only slightly discolored.., the test shall be reported negative and the gasoline considered 'sweet'. **1950** [see *HYDRODESULPHURIZATION]. **1975** *Offshore Engineer* Sept. 44/3 The sweet gas is extracted through wells drilled by a Saipem rig. **1980** *Blair & Ketchum's Country Jrnl.* Oct. 6/3 Light, so-called 'sweet', crude yields a high percentage of automotive gasoline.

4. b. Hence, applied to music, esp. jazz, played at a steady tempo without improvisation, or to this style of playing and its exponents. Cf. *HOT *a.* 8 g. orig. and chiefly *U.S.*

1924 [see *HOT *a.* 8 g]. **1927** *Melody Maker* May 477/1 A really good saxophonist..must be able to render a sweet melody correctly phrased and as though his soul were in it, without a trend to exaggerate sloppy sentiment. **1933** *Fortune* Aug. 47/1 He is decidedly not a *sweet* trombonist—he doesn't play sentimentally with lots of *vibrato*. **1934** S. R. NELSON *All about Jazz* iii. 66 If it is of the melody type, and without much syncopation, the number is treated in the 'sweet' manner. **1956** A. HODEIR *Jazz: its Evolution & Essence* viii. 129 Both 'straight' jazz and 'sweet' music..make use of a sonority and a melodic and harmonic language that are exaggeratedly sugar-coated. **1981** *Oxford Times* 6 Feb. 13/1 The Dorseys' orchestra at this time was sweet rather than swinging, which will disappoint those like myself who prefer the jazzier side of Jimmy and Tommy Dorsey.

5. f. Used as an intensifier in certain slang phrases (often of a coarse nature) meaning 'nothing at all'. See also F.A. s.v. *F III. 3, *FANNY ADAMS 2, S.F.A. s.v. *S 4 a. Also *sweet nothing*.

1958 F. NORMAN *Bang to Rights* I. 28 You can do sweet B.A. about it. **1959** I. & P. OPIE *Lore & Lang. Schoolch.* xvii. 365 They stand on the field and they rave and they shout On subjects they know sweet nothing about. **1973** B. BROADFOOT *Ten Lost Years* ix. 95 The government provided sweet bugger all. Absolutely sweet bugger all. **1973** B. TURNER *Hot-Foot* vi. 43 What had I gained for my trouble? Sweet nothing, that's what.

6. c. (Further examples, with reference to mechanical or technical operations.)

1937 *Times* 11 Dec. 4/7 The engine is, in my opinion, more responsive and sweet than its predecessor. **1955** *Times* 10 May 7/6 The clutch is exceptionally sweet in operation, a point which helps to make the car easily manoeuvrable. **1975** *Washington Post* 25 Jan. A19/1 As J. Robert Oppenheimer said of the hydrogen bomb: 'It was so technically sweet, we had to do it.'

8. d. Also in phrs. *to bet one's sweet life*, *to take one's own sweet time*, *to go one's own sweet way*, and varr.

1889 KIPLING *From Sea to Sea* II. xxxii. 110 The younger ones [sc. Mormons]..*will* mix with the Gentile..and you bet your sweet life there's a holy influence working toward conversion in the kiss of an average Gentile. **1942** BERREY & VAN DEN BARK *Amer. Thes. Slang* § 54/3 Not hurry..take one's (own) sweet time. **1945** A. KOBER *Parm Me* 85 You betcha sweet life I'll give you a buzz. **1946** *Civil & Mil. Gaz.* (Lahore) 19 July 6/4 The station authorities..took their own sweet time in handing the driver the token for him to proceed on his journey. **1968** M. ALLINGHAM *Cargo of Eagles* iv. 52, I let him pass, making sure he'd turn off, but not on your sweet life. He was right with me all the way. **1970** 'D. HALLIDAY' *Dolly & Cookie Bird* vii. 105 You go your own sweet way, or so the evidence tells me. **1975** D. DELMAN *One Man's Murder* ii. 49 So you're finally here. You took your own sweet time about it. **1976** H. MACINNES *Agent in Place* xi. 120 Katie has complicated everything in her own sweet way. **1978** 'G. VAUGHAN' *Belgrade Drop* v. 33 If one single person's seen you get on this lorry..you can bet your sweet life they'll turn it inside out.

9. d. *to keep* (someone) *sweet*: to keep (someone) well-disposed towards oneself, *esp.* by complaisance or bribery.

1939 C. DAY LEWIS *Child of Misfortune* II. vi. 241 It was necessary to keep the wealthier parishioners sweet. **1944** 'N. SHUTE' *Pastoral* viii. 202 Mine won't worry, but I'd like to keep them sweet. **1965** N. GULBENKIAN *Pantaraxia* xi. 228 Mr. Sheets..had what he described as 'a wonderful idea' to keep the Russians sweet politically. **1972** G. BROMLEY *In Absence of Body* vi. 69 Joe Retford..helps to keep him sweet—wines him and dines him and all that. **1978** N. FREELING *Night Lords* iii. 17 The cops were capable of leaking the most dreadful nonsense if one didn't take pains to keep them sweet.

11. *Austral. slang.* Fine, in order, ready.

1898 *Bulletin* (Sydney) 17 Dec. (Red Page), *Sweet, roujig* and *not too stinkin'* are good. **1939** K. TENNANT *Foveaux* 312, 'I brassed a mug yesterday,' he told her, 'and everything's sweet again.' He flashed a roll of notes. **1949** L. GLASSOP *Lucky Palmer* 242 'Everything jake?' he asked. 'She's sweet,' said Max. **1962** S. GORE *Down Golden Mile* 120 Might as well be in it. We'll be sweet for getting back. **1975** X. HERBERT *Poor Fellow my Country* 353 Mossie came in..to say cheerfully, 'She's sweet.'

C. 1. a. sweetback *U.S. slang*, a woman's lover, a ladies' man; a pimp; also *sweetback man* (cf. *sweet man* below); **sweet band** orig. and chiefly *U.S.*, a band which plays sweet music; **sweet biscuit,** a biscuit flavoured with sugar; **sweet-bone(s** (later example); **sweet dreams** *int.*, a farewell to someone going to bed; **sweet Jesus** *int.*, used as an oath or exclamation (cf. *JESUS 1 b); **sweet life** = *DOLCE VITA; hence **sweet-lifer,** one who leads the sweet life; **sweet mama** *U.S. slang* (see quots.); **sweet man** *U.S. slang* = *sweetback* above; **sweetmouth** *v. trans. slang*, to flatter; **sweet music,** light instrumental music of a popular or conventional character (cf. *SWEET *a.* 4 b); also *fig.*, esp. in allusion to love-making; **sweet nothings** *colloq.*, sentimental trivia, endearments; **sweet papa** *U.S. slang* (see quot. 1970); **sweet spot,** the point on a bat, club, racket, etc., at which it makes most effective contact with the ball; cf. *MEAT *sb.* 3 f; **sweet-stuff**: now freq. in *pl.*; † also *euphem.*, gin (*obs.*); **sweet tooth**: also *transf.* and *fig.*

1929 in P. Oliver *Screening Blues* (1968) vi. 206 Had a man, good old sweetback. **1935** A. J. POLLOCK *Underworld Speaks* 117/2 *Sweet back*, a pimp. **1950** BLESH & JANIS *They all played Ragtime* ii. 39 The dapper, foppish 'macks' or 'sweet-back men'..got their gambling stakes from the girls. **1974** Sweetback [see *SUPERFLY *sb.*]. **1935** *Vanity Fair* (N.Y.) Nov. 71/2 Hot musicians look down on sweet bands, which faithfully follow the composer's arrangements. **1938** *Sat. Even. Post* 7 May 23/1 Art Hickman and the first wave of big sweet bands [were] calling the country's dance tunes. **1974** *Listener* 24 Oct. 532/1 Would Albert McCarthy..say that Glenn Miller's was the best dance/swing/'sweet' band? [**1926–7** *Army & Navy Stores Catal.* 6/2 Assorted biscuits. A choice selection of Plain, Sweet, and Fancy kinds.] **1929** W. FAULKNER *Sartoris* 168 Negroes lounged, skinning bananas or small florid cartons of sweet biscuits. **1941** *Ration Craft* 9 The present shortage of sweet biscuits is well known. **1977** *Lancashire Life* Feb. 19/1 Sweet biscuits were unknown until about sixty years ago. Before that the only biscuits made were ship's biscuits. **1969** C. DRUMMOND *Odds on Death* vii. 130 Sister has some Wiltshire sweetbones done under crisp suet crust. **1908** *Sears Roebuck & Co. Catal.* 198/1 Tenor Solos..Good Bye, Sweet Dreams, Good Bye. **1970** *New Yorker* 28 Feb. 70/2 Good night, sleep tight, sweet dreams. **1981** P. NIESEWAND *Word of Gentleman* xvii. 109 'I need some sleep.'..'Sweet dreams, then.' **1932** W. FAULKNER *Light in August* viii. 182 'Come on out,' the blonde woman said. 'For sweet Jesus,' Max said. **1955** F. O'CONNOR *Wise Blood* v. 95 Oh sweet Jesus, come on! **1973** 'D. JORDAN' *Nile Green* xxxiii. 157 Her voice so still, so soft, and I believed her, sweet Jesus, I believed her. **1962** *Sunday Express* 18 Feb. 13/5 Klaus was tired of being respectable and hungered for 'the sweet life'. **1974** M. CECIL *Heroines in Love* ix. 218 The sweet life was turning sour on heroines in the late 1960s. **1967** D. SKIRROW *I was following this Girl* iii. 16 I've been tailing that toffee-nosed sweet-lifer. **1950** A. LOMAX *Mister Jelly Roll* 19 Now these boys used to all have a sweet mama..they was what I would call, maybe a fifth-class whore. **1970** C. MAJOR *Dict. Afro-Amer. Slang* 111 *Sweet mama*, black female lover. **1942** BERREY & VAN DEN BARK *Amer. Thes. Slang* § 443/5 *Beau*,..sweet man. *Ibid.* § 508/3 *Pimp*..sweetman. **1952** S. SELVON *Brighter Sun* ii. 21 Look how Ah take up meself and leave sweetman life in town. **1959** [see *saga boy* s.v. *SAGA[1] 3]. **1972** J. MARYLAND in T. Kochman *Rappin' & Stylin' Out* 211 Damn, Rev., that's some real cruel shit, suggesting a sweet man [pimp] be iced. **1948** *Publ. Amer. Dial. Soc.* ix. 81 Employment [by the Gullahs] of groups of words for..verbs ..or other parts of speech (such as..*to sweet mouth* 'to flatter'). **1950** *Language* XXVI. 330 Not recorded in the Atlas but commonly considered to be of Negro origin are such metaphors as sweet-mouth 'to flatter' and bad-mouth 'to curse'. **1973** J. JONES *Touch of Danger* xli. 238 He went on sweetmouthing me, with his slippery mean eyes. **1967** *Guardian* 28 Sept. 4/5 If pop music should be a fad that passes he sees Radio One as becoming a 'sweet music' station. **1970** *Ibid.* 10 Mar. 1/3 A..choice between ..pop music on Radio 1 and 'sweet' music on Radio 2. **1971** R. GADNEY *Somewhere in England* xxi. 180 A small black girl..offered him 'some sweet music'. **1977** J. WAINWRIGHT *Day of Peppercorn Kill* 99 [They] should be making sweet music, every night of the week. **1981** H. R. F. KEATING *Go West, Inspector Ghote* iii. 29 Rock music, country music, sweet music, pop music—all or any of these..at the touch of a button. **1900** FAZL-I-HUSAIN *Diary* 20 May in A. Husain *Fazl-i-Husain* (1946) ii. 35 The sweet nothings so often talked of in the romantic descriptions. **1934** C. LAMBERT *Music Ho!* III. 212 The blues have a certain austerity that places them far above the sweet nothings of George Gershwin. **1973** M. AMIS *Rachel Papers* 119 Half the guests, including DeForest (after a minute of sweet-nuthins with Rachel), had wisely got the hell out as soon as dinner was over. *c* **1923** in *W. C. Handy's Coll. Blues* (?1925) 28 Ashes in my sweet pa-pa's bed So that he can't slip out. **1941** W. C. HANDY *Father of Blues* x. 141 The sweet papa who happened to be shining around the absentee prisoner's gal at the moment. **1970** C. Major *Dict. Afro-Amer. Slang* 111 *Sweet papa*, a sugar-daddy and sweet man. **1976** *National Observer* (U.S.) 1 May 10/4 The sweet spot—the precise point of contact on the racket face where all the force of a swing goes into the ball without jarring the arm—was considerably farther from the center than anyone had ever suspected. **1976** *Golf International* 13–29 May 21/1 Because we use investment casting, the head weight is distributed over a wider area, increasing the sweet spot. We call this Perimeter Weighting. **1980** *Esquire* Mar. 78 Tennis players, of course, are accustomed to a long racquet, but they're also accustomed to a nice fluffy projectile and the luxury of a forgiving 'sweet spot'. **1835** Dickens in *Even. Chron.* 7 Feb. 3/3 Wretched houses with..'sweet-stuff' manufacturers in the cellars. **1908** *Chambers's Jrnl.* Feb. 204/1 The scent for sweetstuffs is very strongly developed in the Customs officer, and he has found sugar in such an unlikely article as blacking. **1963** *Times* 18 May 9/4 We teach our students the harmful effects of the consumption of sweetstuffs between meals. **1899** J. London *Let.* 29 July (1966) 45 If you're a sweet tooth you will not receive accommodation here except in the fruit line and the candy stores. **1946** Dylan Thomas *Deaths & Entrances* 14 Till the sweet tooth of my love bit dry. **1960** *Times* 5 July 16/5 A symphony for sweet-tooths.

b. *sweet locust* (earlier example); *orange* (earlier example); *sweet pepper-bush* (earlier and later examples); **sweet Alice,** sweet alyssum, *Lobularia maritima* (cf. Alyssum 2) or *Arabis alpina*, another small cruciferous herb with white flowers; **sweet bay,** (*b*) for *glauca* substitute *virginiana*; (later examples); **sweet-bough** *U.S.*, an early variety of apple or the tree that bears it; **sweet buckeye,** a yellow-flowered horse chestnut, *Aesculus octandra*, found in eastern North America; **sweet cane:** = *sweet flag*; (examples); **sweet chestnut:** for *vesca* substitute *sativa*; also, the fruit or timber of this tree; (later examples); **sweet corn** (earlier and later examples); **sweet flag** (earlier example); **sweet gum**(-tree) = Liquidambar 2; **sweet melon** = *spanspek; **sweet olive,** an evergreen shrub, *Osmanthus fragrans*, of the family Oleaceæ, native to eastern Asia and bearing clusters of small fragrant white flowers; **sweet pepper,** (*a*) = *pepper *sb.* 2 b; (*b*) = *sweet pepper-bush*; **sweet potato,** the edible tuber of a perennial vine, *Ipomœa batatas*, native to South America and widely cultivated elsewhere; (examples); **sweet scabious,** (earlier and later examples); also *E. philadelphicus*; **sweet willow,** (*a*) (examples); (*b*) (later examples).

1886 Britten & Holland *Dict. Eng. Plant-Names* 459 Sweet Alice. *Arabis alpina*, L... A corruption of Sweet Alison, which name belongs more properly to *Alyssum maritimum*, L. **1927** V. Woolf *To Lighthouse* I. iv. 38 She was picking Sweet Alice on the bank. **1850, 1903** Sweet bay [see *laurel magnolia* s.v. *laurel *sb.*[1] 6]. **1938** M. K. Rawlings *Yearling* xviii. 217 The sweet bay was still in bloom, filling the sink-hole with its fragrance. **1958** G. A. Petrides *Field Guide to Trees & Shrubs* 303 Sweet-bay Magnolia... A large shrub or small tree with thick, rather leathery, elliptic leaves that are evergreen. **1850** *Rep. Comm. Patents: Agric. 1849* (U.S.) 281 Of summer apples, the best..are the early-harvest and early sweet-bough. **1906** *Harper's Mag.* Apr. 667 He halted under the sweet-bough and gave one branch a shake. **1815** D. Drake *Cincinnati* ii. 77 Sweet buckeye. **1943** R. Peattie *Great Smokies* 155 The sweet buckeye or horse chestnut is found here up to 125 feet in height. **1969** T. H. Everett *Living Trees of World* xxii. 224/2 The largest of the Americans is the sweet or yellow buckeye. **1611** Sweet cane [see Cane *sb.*[1] 2]. **1718** J. Quincy *Compleat Eng. Dispensatory* II. i. 85 Sweet-Cane..is a spicy bitterish Root. **1822** J. Campbell *Trav. S. Afr.: 2nd Journey* I. xx. 226 A constant succession of fresh visitants arrived, several of whom brought us presents of sweet cane. **1838** J. C. Loudon *Arboretum & Fruticetum Britannicum* III. 1983 The term Sweet Chestnut is applied with reference to the fruit. **1909** Elwes & Henry *Trees Gt. Brit. & Ireland* IV. 844 The Sweet or Spanish Chestnut..is..one of the largest trees in England. **1956** *Handbk. Hardwoods* (Forest Prod. Res. Lab.) 72 Sweet chestnut bears a close resemblance to oak but is more easily worked. **1977** *New Yorker* 4 July 22/2 If he could, he would supplement local bounty only with sweet chestnuts and Korean pears. **1981** G. Keynes *Gates of Memory* xxix. 351 Nearer to us were glorious stands of trees,..sweet chestnuts hundreds of years old with twisted trunks. **1646** E. Hopkins *Let.* 20 Mar. in *Coll. Mass. Hist. Soc.* (1863) 4th Ser. VI. 334 Wequash Cooks brother tooke from him..2 bushell of sweet corne. **1810** T. Jefferson *Garden Bk.* (1944) 424 [Sowed].. Sweet or shriveled corn in the N.W. corner. **1909** 'O. Henry' *Roads of Destiny* 364 Cigarettes rolled with sweet corn husk were as honey to Buck's palate. **1917** Will & Hyde *Corn among Indians* 118 The Upper Missouri tribes prepared this 'sweet corn' for winter use in two ways: by boiling it in kettles, and by roasting it in fires. **1974** A. Price *Other Paths to Glory* II. iii. 139 To the north..of the house there had been..a single tiny field of sweet corn. [**1640** J. Parkinson *Theatrum Botanicum* xlviii. 139 This sweet smelling Flagge hath many flaggy long and narrow fresh greene leaves. **1728** R. Bradley *Dictionarium Botanicum* I. s.v., Calamus aromaticus Off. is also call'd Acorus, and in English, The sweet smelling Flag.] **1790** L. Castiglioni *Viaggio negli Stati Uniti* II. 185 Acorus verus..Sweet-flagg. **1700** *Baltimore Rent Rolls* in *Maryland Hist. Mag.* (1924) XIX. 367, 127 acre Sur[veyed]..begun at a bounded sweet gum. **1709** J. Lawson *New Voy. Carolina* 95 The sweet Gum-Tree, so call'd, because of the fragrant Gum it yields in the Spring-time, upon Incision of the Bark, or Wood. **1884** [see Copalm]. **1981** A. Mitchell *Gardener's Bk. Trees* 101/1 For summer foliage and autumn colours the Sweet gum has few equals. **1819** Sweet locust [see *honey-locust* s.v. *honey *sb.* (*a.*) 7 b]. **1883** J. Roth *Man. S. Afr. Gardening* 78 The Water Melons must not be ripped or cut, as required by Sweet Melons. **1970** *Rand Daily Mail* 28 Feb. 7/4 South Africans also speak of..'sweet melons'. [**1789** W. Aiton *Hortus Kewensis* I. 14 Sweet-scented Olive. Nat[ive] of Cochinchina, China, and Japan.] **1861** S. K. Holmes *Jrnl.* 15 Oct. in *Brokenburn* (1955) 61 Mrs. Carson gave Mamma plants of sweet olive..and purple magnolia. **1899** [see *citronelle]. **1958** S. A. Grau *Hard Blue Sky* iii. 122 There was..the winey odor of the sweet olive. **1785** J. Woodforde *Diary* 19 Apr. (1926) II. 185 To a Dozen of sweet Oranges to carry home p[d] 0. 1. 6. **1923** Sweet pepper [see *pepper *sb.* 2 b]. **1944** E. A. Holton *Yankees were like This* 84 The perfume of bush honeysuckle and sweet pepper from the swamps. **1969** *Oxf. Bk. Food Plants* 128/1 The larger-fruited kinds [of *Capsicum annuum*] are quite mild in taste and are known as 'sweet peppers'. **1972** *Country Life* 16 Mar. 625/3 Every garden that can provide lime-free soil ought to contain a bush of the Sweet Pepper, *Clethra alnifolia*. **1814** O. O. Rich *Synopsis Genera N. Amer. Plants* 50 Clethra. Sweet Pepper-Bush. **1901** C. T. Mohr *Plant Life Alabama* 652 Sweet Pepper Bush... Common in the coast plain on swampy banks of pine-barren streams. **1976** *Hortus Third* (L. H. Bailey Hortorium) 286/2 Sweet pepperbush... Summer to autumn. **1750** J. Birket *Some Cursory Remarks* 9 They have..abundance of..the Sweet Potatoe. **1775**, etc. [see Potato 3 a]. **1832** [see Batata]. **1972** Y. Lovelock *Veg. Bk.* I. 233 Sweet potato is now grown throughout the tropics. **1976** M. H. Kingston *Woman Warrior* (1977) 79 My mother liked to look at the ducks and plan how she would dig a pond for them near the sweet potato field. **1789** W. Aiton *Hortus Kewensis* I. 137 Sweet Scabious. Nat[ive]. **1828** C. Rafinesque *Med. Flora* I. 162 *Erigeron Philadelphicum*... Vulgar Names—Skevish, Scabish, Sweet Scabious [etc.]. **1937** *Range Plant Handbk.* (U.S. Dept. Agric. Forest Service) W67 Annual wild-daisy (*E. annuus*) and Philadelphia wild-daisy, misnamed sweet scabious.. are other wild-daisies with similar properties. **1976** *Hortus Third* (L. H. Bailey Hortorium) 1014/1 Sweet scabious..naturalized in California. **1731** Sweet willow [see Willow *sb.* 2 a]. **1776** W. Withering *Bot. Arrangement Veg. Gt. Brit.* 610 Gale..Sweet Willow. Dutch Myrtle. In marshy barren ground. **1839** J. J. Audubon *Ornith. Biogr.* V. 288 A heavy growth of cotton-wood, ash, and sweet-willow. **1855** A. Pratt *Flowering Plants & Ferns Gt. Brit.* V. 56 Sweet Gale, or Dutch Myrtle..is called Sweet Willow.

c. *sweet-breathed* (later examples), *-faced* (later example), *-flavoured* (example), *-fleshed*, *-mannered*, *-souled* (later example), *-tasted* (later example), *-voiced* (later example).

1881 O. Wilde *Poems* 209 Most bounteous Spring! That cans't give increase to the sweet-breath'd kine. **1949** M. Mead *Male & Female* xiv. 283 Life is a race that boys and girls must run clear-eyed, sweet-breathed, well bathed. **1981** M. Warner *Joan of Arc* xiii. 267 The young Joan of Arc, the sweet-faced child of hagiography. **1952** A. G. L. Hellyer *Sanders' Encycl. Gardening* (ed. 22) 9 [*Actinidia*] *purpurea*, sweet-flavoured purple berries. **1923** D. H. Lawrence *Birds, Beasts & Flowers* 41 A rock-living, sweet-fleshed sea-anemone. **1887** G. M. Hopkins *Let.* 25 Dec. (1956) 183 The youngest boy Leo is a remarkably winning sweetmannered young fellow. **1932** D. H. Lawrence *Etruscan Places* i. 12 Those pure, clean-living, sweet-souled Romans, who smashed nation after nation. **1913** J. Masefield *Daffodil Fields* 31 Cropping sweet-tasted pasture. **1919** —— *Reynard the Fox* 11 John Pym..Gross and blunt-headed like a shrike. Yet sweet-voiced as a piping flute.

d. sweet-lip, any of several marine fishes with prominent mouths, esp. an Australian food fish, *Lethrinus chrysostomus*, or a brightly coloured tropical fish of the family Plectorhynchidæ.

1934 T. Wood *Cobbers* xvii. 223 Sweet-lip, and barracouta, a slim silver sword. **1951** T. C. Roughley *Fish & Fisheries Austral.* (rev. ed.) 75 The best-known of the emperor breams is the sweet-lip or red-mouthed emperor. **1974** J. M. Thomson *Fish of Ocean & Shore* xiii. 142 The sweetlip emperor, or simply sweetlip..is highly regarded for the table.

2. a. *sweet-sounding* (later example).

1910 W. de la Mare *Three Mulla-Mulgars* v. 71 When you hear my sweet-sounding..song.

b. *sweet-familiar*, *-sad* (examples).

1865 G. M. Hopkins *Poems* (1967) 21 New-dated from the terms that reappear, More sweet-familiar grows my love to thee. **1946** A. Hutchings in A. L. Bacharach *Brit. Music* xvi. 200 Parts were Arthur Blissy, and none the worse for that; parts were sweet-sad and Englysshe. **1962** R. Prawer Jhabvala *Get Ready for Battle* ii. 97 There was music blaring out of various radios, sweet-sad music played at top volume.

3. sweet seventeen: see Seventeen *a.* 2; now more usually, **sweet sixteen** (cf. Sixteen *sb.* 4); **sweet-throated,** sweet-voiced; **sweet-toothed** *a.* (later example).

1826 *Blackw. Edin. Mag.* XX. 138/1 A bright-eyed, round-limbed virgin of sweet sixteen. **1898** J. Thornton (*song-title*) When you were sweet sixteen. **1977** *Grimsby Even. Tel.* 5 May 12/3 Unfortunately everybody can't be sweet 16 and there are many shops catering for the older woman. **1887** J. R. Lowell *Credidimus* in *Atlantic Monthly* Feb. 251 Who knows but from our loins may spring (Long hence) some winged sweet-throated thing. **1928** W. B. Yeats tr. *Sophocles' King Oedipus* 5 What message of disaster from that sweet-throated Zeus? **1975** *Times* 31 May 7/2 The puddings, often a weakness in French restaurants from a sweet-toothed British customer's point of view.

Sweet Adeline (swīt æ·dĕlәin). *U.S.* A name in a popular close-harmony song (see quot. 1903), used in *pl.* to denote a group or organization of female barber-shop singers (cf. *barber-shop 2 b). Also *attrib.* in *sing.*

[**1903** Armstrong & Gerard *You're the Flower of my Heart, Sweet Adeline* (song) 5 Sweet Adeline. For you I pine.] **1947** *Harmonizer* Nov. 37/1 The Sweet Adelines, women's quartet organization, held their first convention and contest in Tulsa in October. **1958** *Music Jrnl.* Nov.–Dec. 67/2 A Sweet Adeline Chapter..is a chorus and usually meets once a week. **1969** *Pitch Pipe* Summer 7/1 We as Sweet Adelines have a..responsibility to present ourselves as an 'in' group..capable of presenting.. choruses and performances. **1972** *Music Educators Jrnl.* Dec. 71/1 'The Sweet Adelines', an organization for women, have been largely responsible for keeping alive this style of singing. **1979** *Tucson* (Arizona) *Citizen* 20 Sept. (Old Pueblo Suppl.) 2/4 On the bill for the 5 p.m. show are the Arizona Opera Puppets..the Old Pueblo Sweet Adelines.

So as *v. intr.*, to sing in barber-shop style; hence **Sweet A·delin(e)ing** *vbl. sb.*

1949 *Educational Music Mag.* Nov.–Dec. 38/3 So—get fourboys 'Sweet Adelining' around the place if you want to start a real male section to your choir. **1961** *Pitch Pipe* Aug. 13/2 I've just completed one of the most fabulous weekends in my career of 'Sweet Adelining'. **1966** *Ibid.* Spring 6/1 Sweet Adeline-ing is almost as habit-forming as a drug!

sweet and sour, sweet-sour, *adj. phr.* [f. the adjs.] **1.** = Sour-sweet *a.* Also, alternatively sweet and sour.

1594, 1707 [in Dict. s.v. Sweet *a.* and *adv.* C. 2 b]. *c* **1879** G. M. Hopkins *Poems* (1967) 179 When the air was sweet-and-sour of the flown fineflour of Those goldnails and their gaylinks that hang along a lime. **1909** E. B. Titchener *Text-Bk. Psychol.* I. xxxvii. 141 The mixed sweet-sour stimulus affects only the sour-sensitive bulbs. **1959** *Vogue* Dec. 120 This sweet-and-sour glimpse of the cultured Sahib chez soi continually excites by its freshness. **1967** P. D. James *Unnatural Causes* III. iv. 211 Dalgliesh could smell his breath, the sweet-sour trace of too much drinking. **1975** *Chem. in Brit.* XI. 18/3 If it [*sc.* the chemical industry] is subjected to short term sweet-and-sour treatments of restraints, constraints, stimuli, and instant statute, it could very easily be damaged beyond repair.

2. *Cookery.* Cooked in or flavoured with sugar and vinegar or lemon. Now esp. of Chinese food. Also *absol.* as *sb.*

1723 J. Nott *Cook's & Confectioner's Dict.* sig. L l 2, To make a sweet-sour-tart. Boil..Sugar in..Verjuice, or Lemon Juice. **1932** L. Golding *Magnolia Street* v. 103 Mrs. Emmanuel brought in some fish cooked in sweet-and-sour sauce. **1951** *Good Housek. Home Encycl.* 360/2 The soup should have a pleasant 'sweet-sour' taste. **1959** E. Mannin *Blue-Eyed Boy* I. xi. 129 'A little more sweet-and-sour?' 'A lot,' said Len. He added, 'It's tasty, I must say.' **1961** [see *chow mein]. **1977** *Times* 7 May 9 The usual chow meins and sweet-and-sours can be had. **1978** *Texas Highways* Feb. 16/2 Hong Kong chicken in a sweet-sour sauce. **1982** C. Thomas *Jade Tiger* 5 The German Chancellor's top adviser..struck down by sweet and sour pork.

sweeten, *v.* Add: **7.** Also with *up*.

1875 'Mark Twain' in *Atlantic Monthly* Aug. 195/1 Stephen sweetened him up and put him off a week. **1971** 'E. Lathen' *Ashes to Ashes* x. 99, I know Unger is just trying to sweeten us up... Maybe we should be trying to sweeten him up.

8. c. *Oil Industry.* To free (petroleum products) from sulphur or sulphur compounds.

1924 *Industr. & Engin. Chem.* Nov. 1113 Although naphthas and kerosenes have been sweetened by the sodium plumbite method for many years, the process is entirely empirical. **1975** W. G. Roberts *Quest for Oil* (rev. ed.) ix. 92 The lighter distillates, liquid petroleum gas, gasolenes and kerosenes, can be sweetened by simple chemical treatments which either remove the sulphur compounds or turn them into harmless and non-smelly forms.

sweetened, *ppl. a.* (Further example.)

1924 *Industr. & Engin. Chem.* Nov. 1113 The reactions involved..have furnished explanations of the various complications which appear in sweetening, including.. the sourness developed in rerunning a sweetened oil.

sweetener. 2. c. Delete † and add: (Further examples.) Also, a bribe; a concession or appeasement (esp. in politics, business, etc.). Cf. Douceur 3.

1829 P. Egan *Boxiana* 2nd Ser. II. 415 As a reward, or sweetener for his numerous defeats,..the above unexpected victory has put Sampson once more into good humour with himself. **1847** A. Harris *Settlers & Convicts* vi. 89 The handsome 'sweeteners' (bribes) which old D—'s profits enabled him to give the constables. **1955** *Times* 24 May 16/2, I suggest that what you got from Carroll Levis was a sweetener or a bribe. **1959** *Economist* 28 Mar. 1176/1 The main attraction of the Kennedy Bill is its 'sweeteners' in the form of amendments, made to the order of the labour leaders, to the basic Taft-Hartley Act regulating trade union activities. **1960** *Wall St. Jrnl.* 26 Sept. 11 The State Department responded..by permitting the imports but removing the sweetener—the premium that other sugar suppliers enjoy in their sales to the U.S. **1975** *Times* 10 Apr. 8/2 Mr Nixon used the threat of renewed bombing as a sweetener to get the reluctant President Thieu to sign the agreements. **1979** G. Hammond *Dead Game* x. 138 Everybody gives 'sweeteners' of some kind or another, even if it's only a bottle at Christmas.

3. b. (Earlier example.)

1823 in *Spirit of Public Jrnls.* (1825) 508 Here the music of bidding grows loud and more loud—Here the sweetener is conning his hints for the day.

sweetening, *vbl. sb.* Add: **1. e.** *Oil Industry.* The process of freeing petroleum products of sulphur or sulphur compounds.

1924 *Industr. & Engin. Chem.* Nov. 1113 Sweetening consists in the removal of hydrogen sulfide and of alkyl mercaptans which are the only compounds responsible for sourness. **1959** H. M. Noel *Petroleum Refinery Man.* v. 153/1 Kerosene stocks which are too low in smoke point to be finished by simple sweetening. **1970** C. L. Thomas *Catalytic Processes & Proven Catalysts* xix. 199 Mercaptans in gasoline have an objectionable odor ('sour' gasoline). By converting them to disulfides which have less odor, a 'sweet' gasoline is produced; hence the term 'sweetening'.

sweet-field: see *sweet-veld.

sweet-grass. Add: **b.** *S. Afr.* = *sweet-veld.

1812 A. Plumptre tr. *Lichtenstein's Trav. S. Afr.* I. II. xv. 204 On the high hills, sweet grass grows in tolerable plenty. **1838** W. B. Boyce *Notes S. Afr. Affairs* 186 Men should be sent from..the sweet-grass and karoo farms. **1897** [see *num-num]. **1913** [in Dict.].

c. *N. Amer.* One of several scented grasses, esp. *Hierochloë odorata*, used in basket-making.

1926 *Daily Colonist* (Victoria, B.C.) 24 Jan. 20/1 From making sweet-grass baskets on the shores of the lake of Bays to singing before royalty in the Albert Hall is a far cry. **1968** E. Buckler *Ox Bells & Fireflies* xv. 227 Her contentment grows as the sweet-grass basket fills. **1973** A. H. Whiteford *North Amer. Indian Arts* 43 Sweet grass is widely used in coils.

sweetheart, *sb.* Add: **1.** Also used ironically or contemptuously.

1941 B. Schulberg *What makes Sammy Run?* iv. 51 (*addressing a man*) 'Hiya, sweetheart,' he said. **1977** F. Parrish *Fire in Barley* viii. 82 Try harder, sweetheart, or I'll plug you in the guts.

d. *N. Amer.* Anything especially good of its kind. Cf. *honey *sb.* (*a.*) 5 b.

1942 *Amer. Speech* XVII. 105/1 *Sweetheart*, piece of equipment which performs well. **1970** *Globe & Mail* (Toronto) 28 Sept. 27/7 (Advt.), 68 Renault R10, deluxe, radio, a little sweetheart. **1978** *Detroit Free Press* 2 Apr. 15F/4 (Advt.), Lovely 3 bedrm brick ranch, 1½ baths, re rm, a sweetheart for $45,900.

5. A variety of *Rosa wichuraiana* developed by M. H. Walsh about 1903 which bears clusters of small pink flowers; also = *sweetheart rose*, sense 6 b below.

1905 *Country Life Amer.* VII. 625 Sweetheart..delicate blush. **1920** R. Pyle *How to grow Roses* 106 Some roses have acquired new names... Sweetheart P. Mlle Cecile Brunner. **1955** H. van P. Wilson *Climbing Roses* v. 75 Sweetheart (1901)... Rose-pink buds open to very double, 2½-inch, white flowers that are richly fragrant.

6. a. *attrib.* or as *adj.* Designating a contract, agreement, etc., arranged privately (i.e. without genuine collective bargaining) by trade unions and employers which is beneficial to themselves but prejudicial to the interests of the workers; hence applied to persons, etc., prone to such collaboration. Also *transf.* *colloq.* (orig. and chiefly *U.S.*).

1959 *Washington Post* 5 Feb. A2/2 The Administration's ban..would stop an honest union from picketing a shop that had made a substandard 'sweetheart' deal, recognizing a racket union. **1962** N. S. Falcone *Labor Law* xi. 321 Some employers engaged in collusion with unions and paid union officials to get 'sweetheart' contracts. **1965** *Wall St. Jrnl.* (Eastern ed.) 23 Sept. 1/6 The mine manager is a 'sweetheart' operator... In the classic 'sweetheart' situation, corrupt union leaders accept or extort payoffs from employers in exchange for assuring labour peace or winking at contract violations. **1967** G. Tyler *Labor Revolution* xi. 243 The contract is a 'sweetheart agreement' to give the union heads an income, to give the employer relief from a real union, and to give the workers nothing. **1974** *Australian* 12 Nov. 3 Miss Martin said Mr Jones' description of the..award as a sweetheart agreement was farcical. The award had been decided by arbitration, not by negotiation between Qantas and the unions. **1975** *Publishers Weekly* 14 July 54/2 She takes us to three factors, one unorganized, a second with a sweetheart union, the third with an excellent local. *Ibid.* 24 Nov. 53/1 Caffery, a 35-year-old hockey star... Keeping his medical problem secret Caffery negotiates a sweetheart contract to jump league to Texas. **1977** *Time* 1 Aug. 32/2 William Safire..raised the question of whether the $3.4 million loan that was granted on Jan. 7, after Lance had accepted the sensitive OMB job, was a 'sweetheart oan'. **1979** *Times* 21 Nov. 20/3 What are known as 'sweetheart' transactions (when [supermarket] checkout operators reduce the bill for those they know). **1981** *Times* 30 Nov. 15/1 Mobil has accused US Steel of an illegal 'sweetheart deal' with Marathon board members at the expense of the shareholders.

b. Special Comb.: **sweetheart neck(line),** a heart-shaped neckline on a dress, blouse, etc. (see quot. 1968); **sweetheart plant,** either of two species of *Philodendron*, *P. cordatum* or *P. scandens*, epiphytic herbs of tropical America which have large heart-shaped leaves; **sweetheart rose** *U.S.*, one of several roses having small pink, white, or yellow flowers, particularly attractive as buds, esp. the climbing polyantha Cécile Brunner; see also sense 5 d above.

1965 *Housewife* Jan. 16/1 She has a great feeling for a return to the late forties. 'Wide shoulders, sweetheart necks.' **1968** J. Ironside *Fashion Alphabet* 54 *Sweetheart neckline*, a neckline cut in front in two almost semicircular curves, like a heart. **1974** *Country Life* 17 Jan. 106 Sweater with a sweetheart neckline. **1980** B. Bainbridge *Winter Garden* xvi. 129 Enid..sauntered through the cool reception hall in her pink summer dress with the sweetheart neck and emerged into the evening sunshine. **1981** *Daily Tel.* 21 May 17/2 The bride, of course, was a stunner —all demure in white broderie anglaise with a sweetheart neckline. **1963** *Reader's Digest Compl. Libr. of Garden* II. 658/1 P[*hilodendron*] *scandens* (sweetheart plant) origin: Puerto Rico, Panama. A popular and attractive climbing plant. **1981** *Times* 28 Mar. 11/4 A 6½ ft sweetheart plant..cost £29. **1936** J. H. Nicolas *Year in Rose Garden* xv. 72 Cécile Brunner (Sweetheart Rose): Light pink tea-like flowers. **1976** *Columbus* (Montana) *News* 27 May 6/4 She carried a bouquet of yellow sweetheart roses.

sweetie. Delete 'and chiefly' and add: **1. b.** *sweetie-shop*.

1928 J. Buchan *Runagates Club* ii. 85 Some biscuits which I bought at a sweetie shop. **1980** *Times* 11 Dec. 11/2 Cheery old Mrs Mutterance has a Battersea sweetie shop.

2. *colloq.* (orig. *U.S.*). **a.** A sweetheart, a lover; a lovable person. Also as a term of endearing address.

1778 [see Yankee 1 a]. **1925** F. Scott Fitzgerald *Great Gatsby* ii. 42 Tom's the first sweetie she ever had. **1932** Wodehouse *Hot Water* xv. 248 'I'll drop down off the balcony with the stuff.' 'You won't hurt yourself, sweetie?' **1949** A. Christie *Crooked House* vi. 34 The poor old Sweetie... He..was just on ninety. **1958** *Listener* 3 July 31/1 His *fiancée*, Julia, who is a sweetie. **1964** G. McDonald *Running Scared* i. 14 'Where is Dad?' 'He's in Washington, I think, Sweetie.' **1975** *Times* 19 Sept. 9/3 Karen Black as the steely sweetie on the way up. **1977** N. Marsh *Last Ditch* ii. 40 'Sweetie,' Julia cried extravagantly, 'you *are* such heaven!'

b. *Comb.*, as *sweetie-pie*.

1928 Wodehouse *Money for Nothing* iv. 76 'Hello, sweetie-pie,' said Miss Molloy. **1937** D. B. Wyndham Lewis in L. Russell *Press Gang!* 239 Follies show-girl Gladileen ('Sweetie-Pie') Kisse. **1955** Ld. Winterton *Fifty Tumultuous Years* 28 She is not his daughter; as I tell you, she his girl; how you say, his 'sweetie-pie'. **1957** E. Hyams *Into Dream* I. 77 'I think they're all perfect sweetie-pies,' Barbara said. **1977** 'L. Egan' *Blind Search* v. 83 He's..the kind of man who calls anything female 'honeybunch' and 'sweetie-pie'.

sweetikin. Delete '† *Obs. rare*[−1]' and add later example. Also *sweetiekins*.

1974 I. Murdoch *Sacred & Profane Love Machine* 81 Oh my sweetikin, how can such a love as ours stop? **1978** C. MacLeod *Rest you Merry* (1979) vi. 56 Next time you drop one of your time bombs into the punch bowl, sweetiekins, you clean the bathrooms.

sweet John. Delete '? *Obs.*' and add later example.

1911 C. Mackenzie *Passionate Elopement* xxix. 257 The very heart of high June and hot July dwelt in that fragrant enclosure. Sweet Johns and Sweet Williams with Dragon flowers and crimson Peaseblossom.

sweetmeat, *sb.* Add: Now chiefly *arch.* **3.** *sweetmeat glass, shop* (example).

1897 A. Hartshorne *Old English Glasses* xviii. 299 The bowls of the cut sweetmeat glasses have the edges engrailed, vandycked, or faceted. **1971** *Country Life* 9 Sept. 639/2 Exquisite sweetmeat glasses with elaborately cut bowls and sturdy facet-cut stems were made between 1740 and the 1780s. **1857** Dickens & Collins in *Househ. Words* 10 Oct. 338/1, I see a sweetmeat shop.

sweetness. Add: **1. a.** *sweetness and light*, now usu. in trivial (freq. ironic) use, under influence of senses 6, 7: pleasantness, good will.

1927 Wodehouse *Meet Mr Mulliner* vi. 186 He had been all sweetness and light and had not done a thing to them. **1949** N. Balchin *Sort of Traitors* xi. 191 You know how it is when you've got to poke about round somebody else's work—it's not all sweetness and light as a rule. **1953** P. Wentworth *Anna, where are You?* vii. 45 A desire to spread sweetness and light. **1968** G. Jones *Hist. Vikings* II. iii. 106 Anskar, the monk of Corbey,..whose sweetness and light were probably much lightened and sweetened by his biographer Rimbert. **1974** *Times* 16 Jan. 16/5 When this Act was introduced it was done..to create sweetness and light between management and unions. **1982** *Sunday Tel.* 12 Dec. 14/5 Hell hath no fury like a peace-woman scorned, by comparison with whom even a Cruise missile becomes a soft symbol of sweetness and light.

b. *spec.* molasses. *Canad.*

1912 N. Duncan *Best of Bad Job* xxi. 143 T' beg a barrel o' flour an' a gallon o' sweetness. **1920** W. T. Grenfell *Labrador Doctor* viii. 164 The fact that we were without butter, and that 'sweetness' (molasses) was low, was scarcely even noticed.

sweet pea. Add: **b.** The scent of the sweet pea, esp. as used in cosmetics, etc.

1890–1 *T. Eaton & Co. Catal.* Fall & Winter 42/2 Colgate's perfumes—white rose, sweet pea, Cashmere bouquet. *c* **1938** *Fortnum & Mason Catal.* 54/1 Soaps.. sandal wood..sweet pea..verbena. **1972** [see *lilac 2 c].

sweet singer. Add: **2. a.** A religious poet.

1560, etc. [see Singer[1] 2]. **1892** J. Julian *Dict. Hymnol.* 1284/2 William Williams, of Pantycelyn, was the Sweet Singer of Wales. **1933** *Sign* July 92/2 The sweet singer, Christina Rossetti.

b. A popular, esp. sentimental, writer or singer.

[**1878** J. A. Moore (*title*) Sweet singer of Michigan.] **1936** *New Statesman* 25 Jan. 113/2 Kipling..was a sweet singer to the last. He could bring home the colours and savours of many distant places... But he was not a faultless writer. **1958** *Listener* 4 Dec. 913/1 The latest hit of one of the sweet singers of Hong Kong, Li Li Hua or Yao Lee.

sweet-sour: see *sweet and sour, sweet-sour *adj. phr.*

sweet-talk (swī·t tǭk), *v. colloq.* (orig. and chiefly *U.S.*). [f. (as) next.] **a.** *trans.* To cajole, flatter, persuade. Cf. *smooth-talk *v.*

1936 M. Mitchell *Gone with Wind* xlvii. 836 Don't try to sweet talk me. **1955** T. Williams *Orpheus Descending* II. iv. 80 I'd say a peculiar slew-footer that sweet talks you while he's got his hand in the cashbox. **1965** *Listener* 27 May 791/1 There she worked her life away on the edge of poverty, sweet-talking her customers as she lathered them. **1970** J. H. Gray *Boy from Winnipeg* 199 The 'puller' would come out and sweet-talk them into the store. **1981** *Observer* 17 May 19/6 Many have tried over the years to sweet-talk Walsh into selling, but he remained strongly independent until the last.

b. *intr.* To talk persuasively or flatteringly.

1956 H. Gold *Man who was not with It* (1965) iv. 33, I would just have to sweet-talk a little. **1968** L. Deighton *Only when I Larf* viii. 102 He'll switch on the charm and sweet-talk so hard that I am throwing my arms around him.

Hence **swee·t-talker; swee·t-talking** *ppl. a.* and *vbl. sb.*

1946 Mezzrow & Wolfe *Really Blues* p. vi, To the sweettalkers, the gumbeaters, the highjivers. **1956** R. Ellison in *New World Writing* IX. 230 Now he ain't like that ole clarinet; clarinet so sweet-talking he just *eases* you in the dozens. **1966** J. B. Priestley *Salt is Leaving* vi. 81, I still say, my sweet-talking friend, that..you'd have..forgotten me. **1979** *Arizona Daily Star* 1 Apr. H 1/1 Why did he let himself be wheedled out of a lifetime job? Some sweet-talker, that Lyndon Johnson. **1981** P. Niesewand *Word of Gentleman* xxxii. 221 We tried diplomacy and sweet talking.

swee·t talk, *sb. colloq.* (orig. *U.S.*). [Sweet *a.*] Endearment, blandishment, flattery.

1945 L. Shelly *Jive Talk Dict.* 35/1 *Sweet talk*, endearing terms. **1968** S. Ellin *Valentine Estate* III. viii. 163 'And stop calling me baby!' she said with sudden heat. 'It's not Tinpan Alley sweet talk, the way you say it... It sounds full of contempt.' **1979** J. W. Wainwright *Tension* xlv. 142 A touch of sweet-talk and a winning smile.

sweet-veld. Now the usual form of Sweet-field. Substitute for def.: In South Africa, an area of land providing good nutritious grazing; also, the vegetation of an area of this kind. (Further examples.)

1852 M. B. Hudson *S. Afr. Frontier Life* I. 137 The sheep from the sweet veld fall sick on karroo. **1896** R. Wallace *Farming Industries Cape Colony* 82 Animals brought from sweet veld suffer from what is termed veld sickness. **1937** *Handbk. Farmers S. Afr.* 381 The types of grass found in the sweet veld maintain their feed value after maturity. **1948** H. V. Morton *In Search of S. Afr.* 86 There is sweet veld and sour veld, high veld and low veld. **1972** *Even. Post* (Port Elizabeth) 19 Feb. (Weekend Mag.) 2 Somerset East was, with its wonderful water supply and sweet veld, a choice place for men and beasts to rest.

swell, *sb.* Add: **3. a.** Also *spec.* in *Meteorol.* and *Oceanogr.*, wave movement persisting after the wind causing it has dropped, or due to disturbance at a distance. Constrasted with *sea *sb.* 5 d.

1930 *Meteorol. Gloss.* (Meteorol. Office) (ed. 2) 188 Swell is wave motion in the ocean persisting after the originating cause of the wave motion has ceased or passed away. **1957** *Encycl. Brit.* XXIII. 442A/1 When wind-raised waves travel out of a storm area they advance as 'swell', and after having travelled large distances become a series of long, low and fairly regular undulations. **1977** [see *sea *sb.* 5 d].

4. a. Also, a similar feature on the sea bed; a relatively elevated part of a lithospheric plate.

1963 G. L. Pickard *Descriptive Physical Oceanogr.* ii. 10 The characteristic features [of the deep-sea bottom] are.. either basically long and narrow..or of roughly equal lateral extent (swells and basins). **1971** *Nature* 30 Apr. 555/1 Many areas such as Kenya mark igneous provinces of characteristic per-alkaline magma..which are upswollen portions ('swells') of the African plate some 1,000 km across.

9. (Earlier example.)

1786 *Sessions Papers* 13 Dec. 92/2 Here is a *swell* a coming. What is the meaning of that?—I do not know what meaning they give to it, without it is a gentleman.

swell, *a.* Add: Now chiefly *U.S.* **a.** (Later examples.) More recently, also in weakened use as a general expression of approval.

1926 Maines & Grant *Wise-Crack Dict.* 13/1 *Swell dish*, very beautiful girl. **1951** M. McLuhan *Mech. Bride* (1967) 60/2 He was a swell kid. **1977** I. Shaw *Beggarman, Thief* II. iii. 141 That's great. She's swell, a real lady. What a difference between her and some of the dames we had to put up with on the boat.

b. Also similarly weakened: 'great', 'fine', etc.

1930 E. H. Lavine *Third Degree* xi. 128 The swell time he had with the swell broads in the swell musical comedy company. **1947** A. Miller *All my Sons* II. 62 We're eating at the lake; we could have a swell time. **1952** S. Kauffmann *Tightrope* viii. 142 A play like this, with a swell part for her..all that may not come along again for five years. **1968** *Amer. Speech* XLIII. 223 It was a swell date. **1978** J. Krantz *Scruples* iii. 77 All in all, a swell arrangement, and Spider learned a great deal during the year he was Levy's assistant.

c. *swell mobsman* (earlier example). Now *Obs.* or *Hist.*

1843 *Sessions Papers* 6 Jan. 38, I have heard..that the prisoner is a swell mob's man.

d. *predic.* Most pleasant or kind; very effective; 'splendid'. *U.S.*

1926 *Scribner's Mag.* Aug. 198/2 He also knew that the yeggs were not trained fur-thieves... 'They were swell on safes, but a bum would have showed better judgment on furs.' **1931** H. Crane *Let.* 2 June (1965) 370 Moisés has been swell to me. **1942** Wodehouse *Money in Bank* (1946) ii. 16 You eat vegetables and breathe deep and dance around in circles. It's supposed to be swell for the soul. **1965** A. Lurie *Nowhere City* IV. xxi. 237 Yeah; that'd be really swell, if you would.

e. *int.* As an expression of satisfaction.

1930 D. Hammett *Maltese Falcon* xvii. 201 'She's full of gas and ready to go.' 'Swell.' **1935** Wodehouse *Luck of Bodkins* xxii. 289 'Swell,' said Mabel, placing the document in her vanity-bag. **1976** *Daily Record* (Glasgow) 22 Nov. 10/3 My fellow Scot agreed that you could call it that. 'Swell,' said the reporter.

swell, *v.* Add: **10.** (Earlier example with *it*.)

1841 *Punch* 23 Oct. 178/2 Father Thames..has been 'swelling it'..through some of the streets of the metropolis. As if to inculcate temperance, he walked himself down into public-house cellars, filling all the empty casks with water.

swell-. Add: *swell-head* (earlier U.S. examples).

1845 J. J. Hooper *Some Adventures Simon Suggs* iv. 46 As for the present directory, they're all a pack of d—d swell-heads. **1867** G. W. Harris *Sut Lovingood* 61 Wif an onintemitant attack of swell-head.

swelled, *ppl. a.* Add: **b.** *swelled head*: also, a person affected with 'swelled head'.

1862 *Harper's Mag.* June 33/1 He was set down as a born aristocrat and 'swelled head'. **1900** *Times* 7 July 10/1 The Queen's-hall was filled with swelled heads, and, judging from your correspondent's note, the swelled heads elected one of their own body.

swellishness. (Earlier example.)

1863 W. H. Knight *Diary of Pedestrian in Cashmere & Thibet* v. 186 One..group of Mahomedan exquisites.. had, in addition to their heavy swellishness, an air of Eastern listlessness.

swelp. (Earlier and later examples.)

1894 [see *dicken, dickin *int.*]. **1937** N. Marsh *Vintage Murder* x. 112 It's true... S'welp me. **1981** J. Barnett *Firing Squad* vii. 74 'Think again—harder.' 'Swelp me, Mr Smiff—.'

swept, *ppl. a.* Add: **1.** *swept-up*: *spec.* of hair, brushed up towards the top of the head.

1948 'J. Tey' *Franchise Affair* xviii. 217 With her hair swept up and some make-up on, she would look quite different. **1959** *News Chron.* 18 Aug. 6/7 Swept-up hair styles which straggle down the neck. **1973** M. Woodhouse *Blue Bone* vi. 58 She was about five feet six, with butterfly glasses and swept-up hair.

2. *Electronics.* Of (the frequency of) a signal: increased (or decreased) through a range of values, usu. rapidly and repeatedly.

1965 *Wireless World* Aug. 384/1 A random vibration testing technique which was similar to a swept sinewave frequency test except that the single frequency was replaced by a narrow band of noise. **1980** *IEEE Trans. Microwave Theory & Techniques* XXVIII. 792/1 An automated swept-frequency absorption spectrometer.

3. Special collocations: **swept-back** *a. Aeronaut.*, (of a wing) having its leading edge angled backwards (cf. *sweepback* s.v. *sweep- 3 and *delta wing* s.v. *delta 4); also *transf.*; **swept valley** *Building* [Valley *sb.* 4] (see quot. 1964); **swept volume** *Mech.*, the volume through which a piston or plunger moves as it makes a stroke; **swept wing** *Aeronaut.*, a swept-back wing; freq. *attrib.*; also as *sb.*, a swept-wing aircraft.

1914 *Aeroplane* 26 Feb. 213/2 Swept-back wings with negative tips must always have their centre of side pressure farther back relatively to their centre of lift than normal wings. **1951** *Engineering* 20 Apr. 474/3 The third type of British swept-back 'delta'-wing experimental aircraft. **1959** *Ibid.* 16 Jan. 95/1 At each side of the column just below the engine are 'swept back' service ducts extending to the cell walls. **1976** B. Jackson *Flameout* x. 169 Fast aircraft with swept-back wings are susceptible to dutch rolls. **1926** G. Allen *Smaller House of Today* vi. 96 Swept valleys are very suitable for slated and stone roofs. **1951** N. Wymer *Village Life* iii. 64 A particularly unusual feature of the Cotswold roof is the 'swept valley'. **1964** J. S. Scott *Dict. Building* 326 *Swept valley*, a valley formed of shingles, slates, or tiles cut or made to a taper so as to eliminate the need for a flexible-metal valley. A tile-and-a-half tile is used and cut to shape so that its tail is narrower than its head. **1918** W. E. Dommett *Dict. Aircraft* 45 *Swept Volume.* The volume swept by the piston equals area of piston multiplied by the stroke. **1930** *Flight* 24 Jan. 144/2 The engine is of the five-cylinder radial type of 150 cub. ins. swept volume. **1971** B. Scharf *Engin. & its Lang.* xiii. 193 Volumetric efficiency. This is the ratio of the actual volume discharged [from a pump] (capacity) to the displacement or swept volume of the cylinder(s). **1947** *Jrnl. R. Aeronaut. Soc.* LI. 15/2 Whether or not the delta wing is a better compromise than the swept wing..must await the verdict of appropriate researches. **1955** *Times* 25 June 6/2 The R.A.F.'s latest type of Hawker Hunter swept-wing fighter, the Mark IV, is being used for the first time. **1978** A. Welch *Bk. Airsports* i. 9/2 In between are swallow-tails, swept-wings without tails and even the occasional biplane.

swerve, *sb.* Add: *Cricket.* (Earlier example.) Also *attrib.*, as *swerve-bowler*, *-bowling*.

1900 A. W. Pullin *Talks with Old Eng. Cricketers* 125 One hears occasionally of swerving balls, but the swerve depends very much on the air. **1930** C. V. Grimmett *Getting Wickets* iii. 67 In swerve bowling, like other branches of the art, it must be the bowler's object so to regulate his swerve that the ball will hit the wicket. **1944** E. Blunden *Cricket Country* iii. 37 A large wrathful swerve-bowler using the wind..to the immediate..destruction of all.

swerve, *v.* Add: **7. b.** Also *intr.* Of a delivery, to deviate in the air. Of a bowler: to bowl with a swerve.

1894 *Cricket Field* 437 Lockwood was bowled by a ball that swerved considerably in the air. **1903** C. B. Fry *Let.* Sept. in P. F. Warner *How we recovered Ashes* (1905) ii. 15 Much will depend on how you work your bowlers. I wonder which of your 'swervers' will swerve best in Australia?

swerver. b. (Earlier cricketing example, of a bowler.)

1903 [see *swerve *v.* 7 b].

swerving, *ppl. a.* Add: (Earlier example in *Cricket.*)

1900 [see *swerve *sb.*].

swidden (swi·dәn). *Agric.* [f. *swidden*, var. Swithen *v.* (see *Eng. Dial. Dict.*: also, as a place-name element in Yorks.); in mod. use, a conscious readoption of the dialect word (see sense 2, quot. 1951).] **1. a.** An area of land that has been cleared for cultivation by slashing and burning the vegetation cover. Formerly only *north. dial.* (see quot. 1868).

1868 J. C. Atkinson *Gloss. Cleveland Dial.* 514 *Swidden*, any place on the moor from which the Ling and other herbage has been burnt away, and which still shows signs of burning. **1957** *Proc. 9th Pacific Sci. Congress* (1958) XX. 127/1 They maintain permanent villages.., constructing temporary simple houses in their swidden, where at least part of the family lives during those times of the year when the swidden requires a great deal of care. **1961** *Current Anthropol.* II. 27/2 The specific form that a system of swidden agriculture may exhibit..depends on..the dispersal of swiddens... Swiddens may or may not be fenced. **1972** *Nature* 3 Mar. 41/1 In one case a specific tree is found growing in the new swidden.

b. *ellipt.* for *swidden cultivation.*

1955 *Proc. Prehistoric Soc.* XXI. 45 Even if *swidden* (clearance of woodland by burning) was not widely practised before Neolithic times, [etc.]. **1971** D. J. Robinson in Blakemore & Smith *Latin Amer.: Geogr. Perspectives* v. 191 Swidden appears to have formed the basis of the subsistence agriculture of..a part of the tropical zone. **1977** J. J. Fox *Harvest of Palm* i. 38 The Timorese have been forced..to rely even more heavily on swidden.

2. *attrib.* = *slash-and-burn *attrib. phr.* (The principal use.)

1951 K. G. Izikowitz *Lamet Hill Peasants in French Indo-China* 7 This is a book about the Lamet, swidden cultivators in the northern part of Laos. [*Note*] The primitive system of farming which involves clearing and burning the forest... In English it is sometimes called 'shifting cultivation' or 'slash and burn'. There is no single word in ordinary English which covers the meaning, since the method is no longer used in England... In searching for an English word I have taken..a dialect word, *swidden*. **1957** *Proc. 9th Pacific Sci. Congress* (1958) XX. 127/1 We have swidden cultivators who are sedentary in Southeast Asia and other parts of the humid tropics. **1965** G. A. Collier *Fields in Tzotzil* iii. 60 Virtually all are subsistence corn farmers who utilize the slash-and-burn or 'swidden' system of agriculture. **1971** *Sci. Amer.* Sept. 101/2 There is a structural similarity between a swidden garden and a tropical rain forest. **1978** Kunstadter & Chapman in P. Kunstadter et al. *Farmers in Forest* i. 3/2 Swidden fields are usually located at some distance from markets, generally on land that is considered marginal... Swiddening is often carried out primarily as a subsistence operation..rather than as a source of cash crops.

Hence (as a back-formation) *v. trans.*, to cultivate by the swidden method; **swi·ddener, swi·ddening** *vbl. sb.*

1971 *Sci. Amer.* Sept. 119/1 Between one month and four months after clearing begins..the felled litter on the site is burned. This is a step of considerable importance in the swiddening regime. **1975** J. Nance *Gentle Tasaday* xvi. 282 Swiddeners did not uproot the growth, but burned it over and planted within it. **1978** Kunstadter & Chapman in P. Kunstadter et al. *Farmers in Forest* i. 7/2 The land that is swiddened may or may not be claimed by a village unit as a whole.

Swiderian (swidi°·riăn). *a. Archæol.* [ad. F. *swidérien*, G. *swiderien*, f. *Swidry*, the name of an archæol. site near Warsaw (see quot. 1936): cf. -an.] Of, pertaining to, or characteristic of a (principally) mesolithic culture in Poland and neighbouring countries, or its artefacts. Also *absol.* as *sb.*

[**1922** *Wiadomosci Archeologiczne* VII. 96 Dans la formation III on trouve des documents archéologiques se rapportant aux industries magdalénienne moyenne, swidérienne, azilienne.] **1936** J. G. D. Clark *Mesolithic Settlement N. Europe* ii. 62 Two alternative names have been put forward..to label a culture..centering on the valleys of the Vistula and the Bug–Swiderian, after the site at Swidry..and Chlebowician... Numerous Swiderian sites are known from Poland. **1939** V. G. Childe *Dawn Europ. Civilization* (ed. 3) i. 4 The *Swiderian* culture, represented by assemblages of small flint tools collected from sand-dunes in Russia and Poland, sometimes under fossil turf-lines of Atlantic age, is characterized by small asymmetrically tanged-points..used presumably as arrow-heads. **1948** A. Kroeber *Anthropol.* (ed. 2) xvi. 270 *Swiderian*, Poland, Rumania; smallish, tanged blades; early Mesolithic—in fact apparently late Palaeolithic also. **1951** [see *Lyngby]. **1960** C. Winick *Dict. Anthropol.* 518/2 *Swiderian*, a culture found in Poland, with the tranchet ax a typical tool. Its remains, mostly kitchen middens, resemble the Campignian culture, which is found further south.

swift, *a.* (*adv.*) Add: **C.** Combinations, etc. **3.** *swift-darkening*, *-eddying*, *-falling*, *-flashing*, *-moving*, *-pursuing*, *-sprung*, *-striding*.

1933 W. de la Mare *Fleeting* 33 Even the wise..Have smiled with swift-darkening eyes. **1923** H. Belloc *Sonnets & Verse* 13 Anchor hold against swift-eddying time. **1791** Blake *French Rev.* in *Compl. Writings* (1972) 141 Aumont, whose chaos-born soul Eternally wand'ring a Comet and swift-falling fire, pale enter'd the chamber. **1951** W. de la Mare *Winged Chariot* 38 Swift-falling flower, slowly fretting stone Clock on unheeded those who lie alone. **1855** W. Whitman *Leaves of Grass* 62 The great gay-pennanted..steamboat.., with her..delicate swift-flashing paddles. **1930** Blunden *Summer's Fancy* 44 With swift-flashing hope. **1872** W. Whitman *As Strong Bird on Pinions Free* 4 Thee as another equally needed sun, America-radiant, ablaze, swift-moving, fructifying all. **1955** J. R. R. Tolkien *Return of King* v. i. 19 He wondered if he was..still in the swift-moving dream in which he had been wrapped. **1785** T. Dwight *Conquest of Canaan* VIII. 188 Once hast thou fled the swift-pursuing spear, But fled'st in vain. **1948** R. Graves *Coll. Poems 1914–47* 231 The swift-pursuing reed. **1935** Kipling *King & Sea* in *Times* 17 July 19/4, I opened him all the guile of the seas—Their sullen, swift-sprung treacheries. **1929** —— *Poems 1886–1929* III. 341 One silent, swart, swift-striding camel, oceanward wending.

swift, *v.*[2] Add: (Later example.) Now only as *nonce-usage*.

1935 R. Macaulay *Personal Pleasures* 195 There goes the Atalanta among cars; see how it swifts along, passing all others.

swifter, *sb.* Restrict *Naut.* to senses in Dict. and add: **2.** *N. Amer. Logging.* A cable or spar used to secure a raft of logs.

1870 *Overland Monthly* 5 July 58/1 In a 'square' raft, long, slender spars, called 'swifters', are placed. **1975** H. White *Raincoast Chron.* (1976) 150/1 They were using a hand winch to pull the swifters across that locked the logs in place.

swiftie (swi·fti). Also **swifty.** [f. Swift *a.* + -y[6], -ie.] **1.** A fast-moving person: a rapid runner, a quick thinker. Also *ironically*. *colloq.*

1945 *Sun* (Baltimore) 24 Feb. 9/1 Dan Ferris..says that the Swedish swiftie's provisional entry still is among the 36 hopefuls in the 3-mile run. **1946** J. Irving *Royal Navalese* 170 *Swifty*, a derisive nickname for any particularly lugubrious and slow-moving man. **1969** N. Freeling *Tsing-Boum* xvii. 126 Make no mistake about those feminine nails: a swifty.

2. An act of deception, a trick or sleight; = *roughie 3. Also in phr. *to pull a swiftie* (cf. *to pull a fast one* s.v. *fast *a.* 11, and *pull *v.* 19 e). *Austral. slang.*

1945 Baker *Austral. Lang.* xv. 265 *Swiftie*..will..be heard in male conversation to describe a joke or trick that is either agreeable or disagreeable. **1953** 'Caddie' *Sydney Barmaid* 224 'You didn't work a swiftie on them, did you?' I asked suspiciously. For I was already aware that Bill was collecting three doles for himself. **1962** R. Tullipan *March into Morning* 43 If these mugs hadn't pulled a swifty they wouldn't have been working for me at all. **1969** *Sunday Truth* (Brisbane) 23 Mar. 28/4 Police ..arrested him for his Sydney swiftie. **1976** *Sydney Morning Herald* 9 Apr. 6 The Queensland Premier..is now worried that the Federal Treasury may be trying to pull a swiftie.

swig, *sb.*[3] Add: **2.** (Earlier U.S. example: a punning use of Swig *sb.*[1])

1849 H. Melville *Redburn* I. ix. 94 Every once in a while, the men went into one corner, where the chief mate could not see them, to take a 'swig at the halyards', as they called it;..'to taper off'.

swig, *v.*[3] Add: **3.** Also *intr.*, to pull *on* a rope (see quot. 1961).

1917 A. T. Quiller-Couch *Mortallone & Aunt Trinidad* ix. 77 He had now to hoist sail; which he did very leisurably..swigging on the uphaul till he had it chock-a-block. **1939** A. Ransome *Secret Water* xxi. 250 'It's just the wind we want,' panted Daisy swigging on her halyard. **1961** F. H. Burgess *Dict. Sailing* 203 *Swig*, to swig on a rope is to take half a turn with one hand, whilst heaving and taking up the slack with the other.

swiggle, *v.* **2.** Delete ? *U.S.* and add later example.

1907 J. M. Synge *Playboy* II. 39 To think of you swaying and swiggling at the butt of a rope.

3. Also with vessel (spec. a beer glass) as obj.

1943 *Pub & People* (Mass Observation) vi. 185 Some people have a habit of what may be called 'swiggling' their glasses, which consists in moving them round and round in circles, either on the bar counter or table top, or up in the air.

Hence **swi·ggling** *vbl. sb.*

1948 L. A. G. Strong *Trevannion* xiii. 229 There was a wild splashing; Trevannion, craning forward, saw the gleam of a silver belly, and heard a madly energised swiggling and slithering. **1971** *Weekly Guardian* 2 Jan. 19/3 Such categories of pub behaviour as 'Swiggling'— the habit of moving a beer glass round and round between sips.

swile (swail). *Newfoundland.* Also **swoil(e.** Irregular var. of Seal *sb.*[1] Cf. Soile.

1802 J. Murphy *Old Sealing Days* (1916) 2 [J]ars, Doaters and Gunswoils and many others brew upon the rocks. *c* **1845** in *Dict. Newfoundland Eng.* (1982) 450/2 When we got into the jam the swoiles were very thick. **1878** in C. Hallock *Hallock's Amer. Club List & Sportsman's Gloss.* p. xi/2 Swile. **1907** J. G. Millais *Newfoundland* ii. 39 Swoiles (seals) was much to us in the spring, for it meant 'bout what we lived on. **1924** F. Baird *Parson John of Labrador* iii. 64 It's t' good Lard as does it,.. as made t' harbours for we, an' sends t' fish, an' t' swiles. **1969** H. Horwood *Newfoundland* xii. 83 Seals on the north-east coast are called swiles, and the guns used for hunting them are swilin' guns. **1974** F. Mowat *Boat who wouldn't Float* vi. 58 A number of swile guns—longbarrelled, smooth-bore guns intended for killing seals.

Hence as *v. intr.* = Seal *v.*[3]; **swiling** *vbl. sb.*

c **1894** in *Dict. Newfoundland Eng.* (1982) 455/2 Ma shall have a new silk dress, When Da comes home from swoiling. **1897** B. Willson *Tenth Island* 110, I was no good for 'swilin'' any more. **1905** N. Duncan *Dr Grenfell's Parish* 40, I been swilin'..in these seas every spring for fifty-seven years. **1906** J. Lumsden *Skipper Parson on Bays & Barrens of Newfoundland* 90 If the Canadians come down here to take our country I'll get down my 'swiling gun', and we'll go out and meet 'em. **1969** [see the sb. above].

swiler (swәi·lәɹ). *Newfoundland.* [f. *swile *v.* + -er[1].] **1.** = Sealer *sb.*[2] 2.

1883 Hatton & Harvey *Newfoundland* 88 The roads.. begin to be enlivened by the appearance of the sealers, or, as they are called in the vernacular, 'silers', their enterprise being designated 'swile huntin''. **1927** in *Dict. Newfoundland Eng.* (1982) 455/1 We are swoilers fearless, bold. **1958** M. Harrington *Sea Stories from Newfoundland* 118 She sailed..with Skipper Ned Dower in command, and a crew of able 'swoilers'. **1976** *Globe & Mail* (Toronto) 27 Nov. 35/3 It's the swoilers of Newfoundland and the pea-soupers of Quebec, symbols of cultures that in their turn have been despised for different reasons.

2. = Sealer *sb.*[2] 1.

1897 B. Willson *Tenth Island* 110 When the 'swiler' came to start I give my place to another man. **1900** in Oliver & Burke *People's Songster* 46 The interest of all the people was centred on the 'swoilers'. **1959** in Ryan & Small *Haulin' Rope & Gaff* 70 You'll need no Daylight Bill When a 'swiler' first is sighted From the tower upon the 'Hill'.

swill, *sb.*[2] Add: **2. b.** *six o'clock swill*, the customary bout of hasty drinking in public houses at the end of the working day, occasioned by the former six-o'clock-closing regulations. *Austral.* and *N.Z. colloq.*

[**1951** A. W. Upfield *New Shoe* 93 It wanted ten minutes to the fatal hour of six, and the enforced National Swill was in full flood.] **1955** A. Ross *Australia 55* 81 This evening ritual, known amongst Australians as the 'six o' clock swill'. **1961** F. Hardy *Hard Way* 73 The [prison] yard was filling steadily, mostly with drunks, and a few victims of the six o' clock swill. **1970** D. Horne *Next Australia* 160 The 'six o'clock swill' before the lavatory-tiled bars closed was one of the continuing tests of masculinity.

3. *swill-barrel*, *-bucket*, *-pail* (earlier example); † **swill-milk** *U.S.*, inferior milk produced by cows fed entirely on swill (*obs.*).

1869 Mrs. Stowe *Oldtown Folks* xxxvi. 469 The wasteful excesses she had seen in the minister's swill-barrel. **1932** Kipling *Limits & Renewals* 311 Enoch sat helpless on a swill-bucket. **1975** *Country Life* 13 Mar. 666/1 Those happy-go-lucky swill-bucket days. **1853** *Hunt's Merch. Mag.* XXVIII. 684 The whole business [is] in the hands of the swill milk manufacturers. **1894** P. L. Ford *Hon. Peter Stirling* 72 The press began, too, a crusade against the swill-milk dealers. **1741** *Boston News-Let.* 12 Feb. 2/1 Taken up by John Morey, Esq...a Swill-Pale, otherwise called a Hog-Pale.

swiller[2]. *north. dial.* [f. Swill *sb.*[1] + -er[1].] One who makes swills or baskets.

1859 W. Dickinson *Gloss. Words & Phrases Cumberland* 116 *Swiller*,..a swill-maker. **1901** C. W. Bardsley *Dict. Eng. & Welsh Surnames* 522/2 In Ulverston registers to this day a maker of *swills* (i.e. baskets) is set down as a *swiller*. **1949** K. S. Woods *Rural Crafts of Eng.* III. viii. 142 In Furness the baskets are known as swills, and the craftsmen as swillers. Whether the word is a form of scull or scuttle, or whether it means swaler, is not known. **1972** *Daily Tel.* 5 Aug. 9/4 The Lancashire mountains near Ulverston, home of the 'swillers', or basket-makers. *Ibid.*, With a short and very sharp knife the swiller slices his oak into ribs which he fixes across a hazel rim.

swim, *sb.* **5. a.** (Earlier example.)

1764 J. Wesley *Jrnl.* 16 Jan. (1914) V. 44 My mare lost both her fore feet, but she gave a spring, and recovered the causeway; otherwise we must have taken a swim, for the water on either side was ten or twelve feet deep.

10. a. Simple attrib. 'Worn while swimming', as *swim-cap*, *-pants*, *-shorts*, *-trunks*, *-wear*. Cf. *swimming *vbl. sb.* 6.

1964 *Harper's Bazaar* Nov. 102 Black and white felt, close as a swimcap. **1942** N. Last *Diary* 5 June in *Nella Last's War* (1983) 207 Arthur stripped off to a pair of swim-pants, to get sun-browned. **1977** J. D. MacDonald *Condominium* xxxiv. 328 He wore brief turquoise swim pants and large, very dark sunglasses. **1973** G. Beare *Snake on Grave* iv. 22 All he wore was swim-shorts and leather sandals. **1959** *Spectator* 21 Aug. 223/1 Several were wading about in the water. Two were braving it out in swim-trunks. **1979** G. Mitchell *Mudflats of Dead* iii. 35 He..put on his swim-trunks, and slung a towel around his shoulders. **1935** A. P. Herbert *What a Word!* iv. 115, I have been implored by many to attack 'neck-wear', 'foot-wear', 'sleep-wear', and 'swim-wear'. **1962** *Punch* 23 May p. xiii/1 Harvey Nichols have a new range of Californian swimwear. **1976** J. Archer *Not Penny more, not Penny Less* x. 104 I'll never get into the swimwear I'm..modelling next week.

b. Special combinations. **swim-feeder,** in coarse fishing: a short length of perforated plastic tube about an inch in diameter, used to contain maggots, which escape gradually once it is sunk in the water; **swimgloat,** Logan Pearsall Smith's term for the enjoyment of brief social success without becoming corrupted by it; **swim-hole** = *swimming hole* s.v. *swimming *vbl. sb.* 6; **swim-pool** = *swimming-pool* s.v. *swimming *vbl. sb.* 6; **swimsuit,** a (woman's) bathing costume; hence **swim-suited** *a.*

1958 F. Oates *Coarse Fishing Baits* ix. 68 Another method of ground baiting is by the use of a new gadget called a 'swim-feeder'. **1981** B. Walsh *Live Bait* v. 33, I used a paternoster rig, with a swimfeeder and a coffin leger to hold the bottom. **1943** J. Lees-Milne *Jrnl.* 5 Sept. in *Ancestral Voices* (1975) 236 He [*sc.* Logan Pearsall Smith] calls Stuart's social success a 'swimgloat'. **1974** *Times Lit. Suppl.* 11 Oct. 1112/3 Logan Pearsall Smith coined a word for the buoyant negotiation of the vanities and temptations of society..: 'swimgloat'. It is a term which suggests the eternal resilience of the picaresque hero. **1924** Kipling *Debits & Credits* (1926) 321 There was a wet ditch at the bottom that I had wanted..to dam up to make a swim-hole for Mrs. Bevin's ducks. **1958** J. Kerouac *On Road* I. i. 10 My boyhood in those dye-dumps and swim-holes. **1964** C. Barber *Ling. Change Present-Day Eng.* ii. 21 Recently I have seen..*swim-pool* in a high-class newspaper. **1970** *New Yorker* 10 Oct. 80/1 (Advt.), Two swim pools. **1977** *Lancashire Life* Mar. 115/1 Britain has some of the finest swim pool engineers in the world. **1934** *Times* 18 July 17/6 The one-piece swimsuits with attached skirt are still the most popular. **1948** J. Betjeman *Coll. Poems* (1958) 148 Don't hang swimsuits out on sills (A line has been provided at the back). **1980** B. Castle *Castle Diaries* 151 To the disapproval of the department I insisted on taking an hour off on my way to the office to try to buy a swimsuit for my holiday. **1955** *New Statesman* 16 July 66/2 Brutally honest was the Visual Arts float: the Visual Arts..were..represented by a number of swim-suited young women. **1979** 'J. Ross' *Rattling of Old Bones* iii. 32 She was all fresh and rosy and swimsuited.

swim, *v.* Add: **I. 1. c.** Also, *to swim against the tide*.

1705 Ld. Fermanagh *Let.* 18 Nov. in M. M. Verney *Verney Lett.* (1930) I. xiii. 229, I fancy Mr. Gape may lose it... Its hard Swimming against the Tyde. **1971** *Nature* 22 Oct. 515/3 The Sira Institute seems to be swimming against the economic tide.

II. 13. b. Delete † *Obs.* and add later examples.

1939 A. Ransome *Secret Water* xxvi. 315 You'll just have to lie on your back and keep still, and I'm going to swim you ashore. **1953** *Sun Mag.* (Baltimore) 25 Oct. 29/1 The gun fires and the bay dog is over with a splash. Exultantly he swims the dead game back to his master.

14. c. (Earlier example.)

1794 M. Parry *Jrnl.* 23 May in *Kentucky Hist. Soc. Register* (1936) XXXIV. 380 Forded Buffaloe Creek, at the mouth, which did not quite swim them [*sc.* the horses].

† **15.** To carry (a publication) to success. *Obs.*

1870 'Mark Twain' *Lett. to Publishers* (1967) 45 Launch a book right on our big tidal wave and swim it into a *success*. **1890** G. Meredith *Let.* 19 Nov. (1970) II. 1012 If clogged with the letter-press, I should have my doubts of success, even with his name to swim the book.

swimathon (swi·măþǫn). Also **swim-a-thon.** [f. Swim *v.* + *-athon.] A long-distance swimming race; a marathon (often sponsored) swimming event.

1968 *Telegraph* (Brisbane) 20 Dec. 2/2 Six Gold Coast girls will take part in a swimathon at Southport's Olympic Pool. **1976** *Estevan* (Saskatchewan) *Mercury* 23 June 16/3 A swim-a-thon will be held at Woodlawn Swimming Pool June 27... Proceeds will go toward the aquatic club.

swim-in (swi·min). [f. Swim *v.* + *-in[3].] A form of protest or recreation at which a number of people swim together. Cf. *sit-in *sb.* 1.

1960 *Daily Progress* (Charlottesville, Va.) 2/3 Other white bathers cleared out of the immediate vicinity of the swim-in. **1977** *Navy News* Sept. 25/2 Other events on the social programme have included a barbecue and a 'swim-in' at the local pool.

swimmable, *a.* Add: Also, suitable for swimming in. Also *swimmable-in*.

1963 P. McCutchan *Man from Moscow* ix. 91 The sea's swimmable-in, if you're a Spartan. **1966** *Telegraph* (Brisbane) 3 Feb. 18 (*caption*) Bare midriff camisole tops are the latest on the patio this summer. Worn with snug hip-hugger jams in nylon knit. Both are completely swimable. **1976** *National Observer* (U.S.) 13 Mar. 7/2 Congress poured money into it to help cities do their part in achieving 'swimmable, fishable' waters.

swimmer. Add: **7.** *slang.* A swimming costume. Now (*Austral.*) *pl.* const. *sing.* Cf *bather 3.

1929 *Daily Tel.* 3 June 7/1 Two coloured swimmer with brassiere attached. **1967** *Sunday Truth* (Brisbane) 23 July 1/1 Bikini girls at Parliament House..when a parade of new season's swimmers..will be on show. **1978** *Courier-Mail* (Brisbane) 22 Feb. 1/9, I am not an exhibitionist and if I go swimming on the main beach, I would wear swimmers.

8. Special combination. **swimmer's itch** *Med.*, a painful dermatitis caused by the cercaria of certain species of blood flukes, notably *Schistosoma mansoni*, which penetrate human skin (or mucous membrane) during swimming.

1928 *Minnesota Med.* XI. 573/1 There has been reported from several lake regions in Minnesota a peculiar type of skin eruption locally called 'swimmer's' itch. **1969** *Trans. R. Soc. Trop. Med. & Hygiene* LXIII. 557 Visitors to that camp suffered severe swimmer's itch when bathing in one of the rock pools.., and subsequently developed schistosomiasis.

swimming, *vbl. sb.* Add: **6.** *swimming costume, suit, trunks*; *swimming bath* (examples in *pl.*); **swimming hole** chiefly *U.S.*, *Austral.*, and *N.Z.*, a bathing place in a stream or river; **swimming pool,** an artificial pool designed for swimming in.

1868 A. J. Symonds *Let.* 29 July (1967) I. 828, I went.. to the Victoria Swimming Baths, as I occasionally do, to smoke my cigar & to learn the secrets of Form. **1982** *Financial Times* 9 Dec. 9/1 Proposals are being investigated for private sector school meals and cleaning, the running of swimming baths, [etc.]. **1904** R. Thomas *Swimming* 112 It is very difficult to get photographs of amateur ladies in swimming costume. **1962** F. C. Avis *Swimming Dict.* 95 *Swimsuit*, a superior or elegant swimming costume, with particular reference to the female bather. **1977** N. Slater *Crossfire* iii. 58 A twenty-nine-year-old married woman..who wore a bathing cap and a one-piece swimming costume. **1867** G. W. Harris *Sut Lovingood* 25 He wer aimin fur the swimin hole in the krick. **1912** J. H. Moore *Ethics & Educ.* 128 The boy's love for the water, his affection for the old swimming-hole. **1928** [see *bogy[2], bogey[2]]. **1975** D. Bagley *Snow Tiger* ii. 33 The bluff..projected into the river..and that was where they had their swimming hole. **1899** *Scribner's Mag. Advertiser* Jan. 26/2 You can enjoy..a plunge into the great marble swimming pool, where the water is tempered according to season. **1921** A. Huxley *Crome Yellow* iii. 19 The stone-brimmed swimming-pool. **1972** *Punch* 1 Mar. 266/3 Our goals are increasingly the same —a bigger car, an expense account, and a swimming-pool in every back garden. **1926** E. Hemingway *Sun also Rises* II. xix. 245, I found my swimming suit, wrapped it with a comb in a towel. **1971** 'D. Halliday' *Dolly & Doctor Bird* vi. 83 My swimming-suit, helmet and towel. **1943** *New Yorker* 22 May 26/1 He was big, stalwart, and dressed only in swimming trunks. **1978** I. Murdoch *Sea* 70 Shall I come and bring my swimming trunks?

swimmy, *a.* Add: † **b.** Graceful, elegant. *Obs. nonce-use.*

1827 Coleridge *Let.* 2 June (1971) VI. 687 A fine, tall, slim, swimmy, glidy lass.

c. Of the eyes: watery, tearful. Also, of tears.

1936 J. B. Priestley *They walk in City* vii. 178 She had a round moist face, with swimmy eyes. **1978** J. Irving *World according to Garp* xvii. 358 The woman's.. face, dissolving before him in his own swimmy tears.

Swinburnian (swinbɔ̄·ɹniăn), *a.* [f. the name of the English poet Algernon Charles *Swinburne* (1837–1909) + -ian.] Of, pertaining to, imitative or characteristic of Swinburne or his poetry. Hence **Swinbu·rnianism, Swi·nburnism.**

1867 E. B. Lytton *Let.* 25 Jan. in *Lett. R. Lytton* (1906) I. 207 The 'Gyges and Candaules' have some dangerous supersensual lines which I advise you to reconsider. It will not do for you to be 'Swinburnian'. **1868** A. J. Symonds *Let.* 24 Apr. (1967) I. 803 Courthope..is full of the gall of bitterness against the Apostles of Swinburnism. **1892** W. B. Scott *Autobiogr. Notes* I. xxii. 300 When the Swinburnian passion for French things..had infected nearly all our young writers. **1920** *Glasgow Herald* 30 Dec. 4 The 'Various' verses show now and then a Swinburnian touch. **1931** G. K. Chesterton *All is Grist* xxxviii. 212 Something that is connected not only with Swinburne but with Swinburnianism. **1949** A. Huxley *Let.* 6 Apr. (1969) 595 Any equivalent in English becomes automatically Swinburnian, that is to say rich without the weight ..which Latin imposes. **1960** J. Betjeman *Summoned by Bells* vii. 75, I was released Into Swinburnian stanzas with the wind. **1974** E. Hardwick *Seduction & Betrayal* 109 A Swinburnian mood of spankings and teasing degradation. **1976** *Times Lit. Suppl.* 26 Nov. 1495/2 [Gilbert Murray's] translations of Greek tragedies are still to be found on the shelves of college bookstores today, in spite of all the rude things that have been said about their Swinburnianism.

swindle, *sb.*[3] Add: **1. a.** (Earlier example.)
1883 in A. Bunn *Stage* (1840) I. 134 There was a universal cry of 'off-off'—'swindle-swindle'.

3. Special combination. **swindle sheet** *slang* (chiefly *U.S.*), an expense account; also (*joc.*) in extended use, of other documents which conceal (or reveal) fraudulence and other 'swindles', as a log-book or time sheet.
1923 *N.Y. Times* 9 Sept. vii. 2/3 *Swindle sheet*, the advance agent's expense account. **1934** J. O'Hara *Appointment in Samarra* ii. 42 The Apollo [hotel] got a big play from salesmen who had their swindle sheets to think of. **1936** *Times Lit. Suppl.* 15 Feb. 125/3 The 'swindle-sheet' for the average motor-car shows that 40 per cent of the fuel energy goes into the cooling water. **1960** H. L. Lawrence *Children of Light* v. 77 The fare's ten bob... Put it on the swindle sheet. **1971** M. Tak *Truck Talk* 161 *Swindle sheet*, the daily log book, mandatory for all drivers.

swindler. (Earlier example.)
1774 W. Hawke (*title*) The life, trial, &c. of William Hawke... To which is added a full description of the impositions and deceptions practiced by the swindlers, sharps, gamblers..in and about London.

swindlery. (Later example.)
1869 Dickens in *All Year Round* 2 Jan. 109/2 Swindlery in doubtful boots, on the sharp look-out for any likely young gentleman.

swindling, *vbl. sb.*[2] (Earlier example.)
1788 *Gentl. Mag.* LVIII. 1154/2 As *swindling* is a word that occurs not in our dictionaries, and yet we often meet with it in modern writers.., we should be obliged to any gentleman among your correspondents.., to define it; or..inform us what..distinguishes it from other modes of fraud and imposition.

swine. Add: **2. b.** Of a thing: = *pig *sb.*[1] 1 c. *slang.*
1933 Dylan Thomas *Let.* Oct. (1966) 31 This method of letter writing..is very satisfying, but it's a swine in some ways. **1938** N. Marsh *Artists in Crime* iii. 38 'It's a swine of a pose, Miss Troy.' 'Well, stick it a bit longer.' **1967** K. Giles *Death in Diamonds* ii. 41 The Inspector groaned. 'Could be heroin. That's a swine.' **1976** H. MacInnes *Death Reel* iii. 19 This car's..a swine to drive at slow speeds.

4. *swine-fat*; *swine-headed* adj. (later example).
1922 Joyce *Ulysses* 468 Her odalisk lips..smeared with salve of swinefat. **1922** Swineheaded [see *dog *sb.* 17 c].

5. a. swine-chopped *a.*, of a hound: having the lower jaw projecting forward of the upper one; so **swine-chop,** a malformation of this kind; **swine erysipelas,** an infectious, sometimes fatal, disease of pigs, caused by the bacterium *Erysipelothrix rhusiopathiæ*, and characterized by fever, reddish spots on the skin, and general debility; **swine flu** = *swine influenza* below; **swine-hound** *slang rare*, tr. G. *schweinehund* *schwein(e)hund (quot. in Mil. context); **swine influenza,** an infectious virus disease of pigs, esp. young ones, characterized by fever, coughing, and difficulty in breathing; also, influenza in man caused by the same (or a closely related) virus; **swine vesicular disease,** an infectious virus disease of pigs (similar to foot-and-mouth disease) characterized by mild fever and blisters round the mouth and feet.
1962 *Times* 9 June 11/4, I have seen..puppy show prizes awarded to young hounds with swine-chop. **1930** Kipling *Thy Servant a Dog* 20 Moore-man lifted Ravager's head and opened his mouth... 'Look, m'lord. He's swine-chopped.' **1965** D. Moore *Bk. Foxhound* ii. 29 The forehead and nose merge invisibly, giving always a rather stupid expression, and sometimes accompanying a swine-chopped mouth. **1898** M. M. Hayes tr. *Friedberger & Fröhner's Vet. Pathol.* 72 Swine erysipelas (or swine measles)..is a specific septicæmia produced by a minute bacillus. **1922** A. T. Kinsley *Swine Practice* xii. 338 Swine erysipelas is an infective disease of swine characterized by a high temperature, cerebral disturbances and discoloration of the skin. **1970** W. H. Parker *Health & Dis. in Farm Animals* x. 141 A disease which can easily be confused with swine fever is swine erysipelas. **1921** *Wallace's Farmer* 25 Feb. 371/1 So-called 'swine flu', a name which, while it became quite popular thru its association with the human disease, is nevertheless a misnomer, is primarily a bronchitis. **1976** *National Observer* (U.S.) 21 Aug. 2/2 The swine-flu insurance bill was signed by President Ford, clearing the way for mass inoculations in about six weeks. **1981** *Sci. Amer.* Oct. 46/2 Epidemiologists determined..that recipients of the swine-flu vaccine were developing Guillain-Barré syndrome at a rate several times the usual one. **1916** 'Boyd Cable' *Action Front* 245 'Sulky, eh, my swine-hound!' said the officer. 'But I think we can improve those manners.' **1922** *Jrnl. Amer. Vet. Med. Assoc.* LXI. 178 We must be able to differentiate between hog cholera, necrotic enteritis,..broncho-pneumonia or swine influenza, and many others. **1935** *Lancet* 11 May 1123/2 It seems to me..exceedingly probable that the virus of swine influenza is really the virus of the great [influenza] pandemic of 1918 adapted to the pig and persisting in that species ever since. **1969** C. W. Schwabe *Vet. Med. & Human Health* (ed. 2) vii. 216/2 Swine influenza was not known before the human influenza pandemic of 1918. **1976** *Globe & Mail* (Toronto) 26 Mar. 1/1 Ontario residents probably will be vaccinated against a deadly swine influenza virus, Alan Backley, Ontario's deputy health minister, said yesterday. **1972** *Guardian* 16 Dec. 1/8 The outbreaks of suspected foot-and-mouth disease in the Midlands have turned out to be a rare virus which affects only pigs. Its new name, invented by the Ministry of Agriculture yesterday, is swine vesicular disease. **1981** *Vet. Rec.* 30 May 468/3 The relative decline in the number of cases of swine vesicular disease this year suggests that the campaign against the disease is achieving worthwhile results.

swing, *sb.*[2] Add: **I. 6. d.** A swift tour or journey (*through* a place) involving a number of stops or visits. Now *spec.* a political campaign tour; also, *swing around the circle*, a campaigning tour of a constituency or larger area. Cf. *swing *v.*[1] 11 c. *U.S.*
1860 H. J. Hawley *Jrnl.* 22 Apr. in *Wisconsin Mag. Hist.* (1936) XIX. 330 Had a fine time a nice swing . saw sights and returned. **1905** *Springfield* (Mass.) *Weekly Republican* 6 Oct. 1 Will the appropriated money be available for campaigning swings around the circle? **1929** *Sun* (Baltimore) 23 Oct. 2/7 Mr. Hoover has undergone much strain on this swing and he showed it as he waved a weary adieu to the hospitable Ohioans. **1949** *Manch. Guardian Weekly* 1 Sept. 2 Before his recent swing around the Marshall countries. **1967** *Boston Globe* 5 Apr. 51/1 Dizzy Dean's wife once traveled with him on an Eastern swing by the Cardinals. **1972** *Even. Telegram* (St. John's, Newfoundland) 27 June 5/3 Senior citizens can call a number and a van will come by to pick them up on swings through the city. **1978** L. Heren *Growing up on The Times* iii. 92, I..was sharing a room..with Colin Reid of the *Daily Telegraph*, who was on a swing from Beirut.

e. A worker's rest period between duties; a shift system which incorporates such breaks. Also, time off work as leave, furlough. Cf. *swing-shift* s.v. *swing- 2 a. *U.S. slang.*
1917 D. C. Roper *U.S. Post Office* 353 *Swing*, period of time within the day's tour or 'trick' when an employee is temporarily off duty. **1918** *Outlook* (N.Y.) 17 July 443/2 [On a street railway] The 'swings', or free time between runs, are for the most part so arranged as to be inconvenient for going home. **1945** *Transit News* (Capital Transit Co., Washington, D.C.) 15 June, A 'Swing' works during the morning rush hour, and the operator is then off until time to start the evening rush hour. **1972** J. Mills *Report to Commissioner* 129, I went on my swing after that.

f. In colloq. phr. *to go with a swing*: said of a lively, successful party or other entertainment or undertaking.
1976 *Bridgwater Mercury* 21 Dec. 9/3 Families may soon be enjoying a tipple at the bar to make their parties held in the community centre go with a swing.

7. b. *Boxing.* A punch delivered with a sweep of the arm; a swinging blow. Also in gen. colloq. use, esp. in phr. *to take a swing at* (someone).
1910 [see *hook *sb.*[1] 13 b]. **1962** *Times* 28 Apr. 3/5 Barlow came rushing in attempting to land with right swings to the head. **1983** W. Winward *Last & Greatest Art* 211 If I stand here much longer I'm going to be tempted to take a swing at you.

8. g. *Cricket.* A curving deviation of a ball from a straight line of flight on delivery, occasioned by a combination of the angle of its seam and the relative smoothness of the leather each side of this.
1906 *Cricket* 29 Nov. 450/1 He and Raynor..were two of the best boy bowlers I have seen, and the latter had more break—not swing—in the air than anyone else I ever saw. **1920** Lyttelton & Wilson in P. F. Warner *Cricket* (ed. 2) 270 Baker did not swing too much..and he combined swing with length. **1976** J. Snow *Cricket Rebel* 101 It is not often in Australia that the atmospheric conditions encourage swing.

h. An observable movement in general opinion away from one position towards another. Hence *spec.* in *Pol.*, a change in the relative distribution of popular support for political parties, often measured in terms of percentage gains or losses by each party at a poll. Formerly also † *swing of the pendulum.*
1899, 1912 [in Dict., sense 8 a]. **1933** D. W. Brogan *Amer. Polit. System* x. iv. 368 The strength of American parties is, as a rule, too sectional, too much divorced from any current national controversies, for there to be anything like our 'swing of the pendulum'. **1940** *Economist* 5 Oct. 421/2 The swing in American public opinion has been such that the Nazis may well despair of keeping the United States out of a long war. **1945** *Times* 27 July 4/1 When the votes were counted..it was revealed that the Government formed by Mr Churchill on the break-up of the Coalition had been decisively beaten by a surging swing of opinion to the Left. **1955** *Times* 26 May 10/1 The absence of any pronounced 'swing' towards the Government. **1960** *Where?* III. 17 *'Swing', the*, jargon for the relatively recent tendency among sixth form pupils to specialize in science rather than arts subjects. *a* **1974** R. Crossman *Diaries* (1975) I. 493 It was only a 3·1 swing, and by God a 3·1 swing can become a 4·0 counter-swing very quickly indeed. **1976** H. Wilson *Governance of Britain* ii. 38 We had bad county council results in April, including Lancashire though with a favourable swing in Greater London.

i. *Electr.* An increase or decrease in the magnitude of a current or voltage, the difference between its greatest and smallest values.
1908 *Rep. Brit. Assoc. Adv. Sci. 1907* 622 These [oscillations]..are transferred..into a closed air-condenser circuit, which, when its swings reach a maximum, overflows into the coherer. **1957** *Practical Wireless* XXXIII. 562/2 It is possible to increase the anode voltage swing and the anode peak current. **1978** *Sci. Amer.* Dec. 54/1 Load-following generators are started daily and run most of the time to cope with daily swings in the load; they may be shut down at night.

j. *Psychol.*: see *mood swing* s.v. *mood *sb.*[1] 3 f.

k. *Bridge.* The difference between the total scores of two teams of two pairs playing the same deal at two tables, each team having north-south positions at one table and east-west at the other.
1945 S. J. Simon *Why You lose at Bridge* 24 In Room 1, North-South bid six Spades and made five. In Room 2, North-South stopped in 4 spades and declarer, playing for safety, made three. No swing! **1949** *Contract Bridge Jrnl.* Feb. 5/2 On the very next hand the Scots repaid the compliment; at this stage they were going great guns, and on Board 54 came the biggest swing to-date. **1961** *Listener* 10 Aug. 222/3 The swing on the board was 2,080, or 11 match points.

10. b. *Mus.* A quality of jazz, dance music, etc., that has a flowing but strongly compelling rhythm; since the mid-thirties (esp. for a decade), applied to a variety of big dance-band music played in this style. Cf. *swing- 2 d.
(*a*) **1899** H. H. Mincer (*song-title*) Virginia. Two-step & hot rag swing. **1917** *Sun* (N.Y.) 5 Aug. 3/7 Jazz is based on the savage musician's wonderful gift for progressive retarding and acceleration guided by his sense of swing. **1924** (*music-title*) Lou'siana swing [performed by Piron's New Orleans Orchestra]. **1932** 'Duke' Ellington (*song-title*) It don't mean a thing (If it ain't got that swing). **1939** —— in *Melody Maker* 15 July 8/3 No notes represent swing. You can't write swing because swing is the emotional element in the audience and there is no swing until you hear the note. **1954** *Grove's Dict. Music* (ed. 5) IV. 600/1 'Swing'..can only be said to designate the regular but subtle rhythmic pulsation which animates 4–4 time and must be present in every good jazz performance. Swing is essentially the performer's concern: it cannot be indicated in musical notation except implicitly.
(*b*) **1936** *Delineator* CXXIX. 10/1 This swing, it's nothing more or less than jazz, is it? **1937** L. Armstrong *Swing that Music* xiv. 117 Even now, thirty years after Swing was born, this book is the first history of swing music, and of the men who made it, to be published in the English language. **1943** D. Welch *Maiden Voyage* xiv. 110 'What kind of records have you got?' 'There's plenty of swing.' **1957** R. Hoggart *Uses of Literacy* v. 129 The emotional patterns bodied out by 'swing' are quite close to those of the older, waltz-derived, styles: in fact, 'swing' has been adapted and assimilated; a modern 'swing' song and an old-fashioned waltz tune live together with ease. **1973** J. Wainwright *Pride of Pigs* 61 It wasn't jazz. Not *real* jazz... Swing..that's what they'd called it, when [Artie] Shaw had introduced it in the 1930's.

II. 11. b. Colloq. phr. *to gain on the swings and lose on the roundabouts* and varr., according to which one's losses in one quarter balance one's gains in another. Also *allusively.*
1912 P. R. Chalmers *Green Days & Blue Days* 20 For 'up an' down an' round,' said 'e, goes all appointed things, An' losses on the roundabouts means profits on the swings! **1927** *Times* 24 Mar. 15/5 By screwing more money out of tax-payers he diminishes their savings, and the market for trustee securities loses on the swings what it gains on the roundabouts. **1944** G. B. Shaw *Everybody's Political What's What* xv. 121, I was taxed at a higher rate than my fellow capitalists who had smaller incomes. But then I had to pay at a lower rate than others who had bigger incomes. Whether I lost on the swings what I gained on the roundabouts I do not know. **1964** *English Studies* XLV. (Suppl.). 98 The more a word loses in meaning, the more it gains in functional, as distinct from semantic, importance. What we lose on the swings we win on the roundabouts. **1976** *Listener* 18 Nov. 641/1 There is a certain rough justice in charging for the possibility of using the [broadcasting] service... Swings and roundabouts.

12. d. Also *swing-steer.*
1869 [see *lead *sb.*[2] 11 b].

swing, *v.*[1] Add: **4.** Also, *to swing Kelly* (or *Douglas*), to wield an axe, to do axework. *Austral. slang.*

1945 Baker *Austral. Lang.* i v. 78 *Kelly* and *douglas*, an axe (from the names of makers), with their derivatives *to swing kelly* or *douglas*, to do axework. **1966** 'J. Hackston' *Father clears Out* 98 The scholars..could have passed with honours in such subjects as milking, swinging Douglas, panning off.

7. a. Also, *to swing the lead*: see *lead *sb.*[1] 6 b.

11. c. *to swing around the circle*, to make a political tour of a constituency or larger area. *U.S.*
1866 E. McPherson *Polit. Man.* v. 58 We swing around the circle of the Union with a fixed and unalterable determination to stand by it. **1871** G. W. Peck *Adventures Terence McGrant* iv. 27 Until me Cousin Ulissis gets through swinging around the circle. **1887** *Chicago Tribune* 2 Oct., President Andrew Johnson originated the phrase 'swinging round the circle' on the occasion of his famous tour to Chicago..in September, 1866. **1910** *N.Y. Evening Post* 29 Oct. 2 To stem the rising tide against him, Col. Roosevelt is to swing around the circle in Brooklyn to-night.

d. *Cricket.* Of a bowler: to impart swing to the ball on delivery. Also with the ball as subj. Cf. *swing *sb.*[2] 8 g.
1900 P. F. Warner *Cricket in Many Climes* 84 Morton ..has a beautiful natural action, and swings in the air with his arm. *Ibid.* 179 Rowe..has, too, a very good fast 'yorker' which swings in the air. **1952** *M.C.C. Cricket Coaching Bk.* ii. 37 The farther up the ball is pitched, the more 'room' it has in which to swing. **1977** *World of Cricket Monthly* June 30/1 Bowling medium-pace, he got the ball to swing in the heavy atmosphere.

e. Of a spacecraft: to pass *by* a planet using its gravitational field to change course.
1967 [implied at *swing-by* s.v. *swing- 2 a]. **1970** *Nature* 1 Aug. 434/2 The spacecraft will be launched in the autumn of 1973, swinging by Venus at a distance of 3,000 miles. **1976** *Sci. Amer.* May 116/2 These two spacecraft are scheduled to be launched in 1977 and to swing by Jupiter in 1979.

12. c. (Earlier example.)
1819 J. G. Lockhart *Peter's Letters* (ed. 2) III. lxix. 203 The balls..being swung to and fro in a terrific manner, by means of long queues with elastic shafts.

d. In fig. phr. *to swing it on* or *across* (someone) = *to put it across* s.v. *put *v.*[1] 35* a (*b*).
1923 *Daily Mail* 16 June 11 Too experienced to let even a thundering smart girl swing it on him as easily as that. **1943** N. Marsh *Colour Scheme* iv. 64 You saw Questing swing it across me. **1950** T. E. Lawrence *Mint* 39 'Swinging it on the..rookies, they are, the old sweats' grumbled Tug.

e. *to swing the gate* (see quot. 1933). Cf. *drag *v.* 9 b and *swing-gate* s.v. Swing- 2 a. *Austral.* and *N.Z. slang.*
1933 L. G. D. Acland in *Press* (Christchurch, N.Z.) 16 Dec. 21/8 *Swing the gate*, to be the fastest shearer in the shed. **1941** [see *drag *v.* 9 b]. **1965** J. S. Gunn *Terminol. Shearing Industry* ii. 12 A ringer is..said to 'swing the gate', presumably because he keeps the catching-pen gate swinging.

f. To turn a starting-handle in order to start (a motor vehicle, its engine). Also with *over. colloq.*
1927 R. Lehmann *Dusty Answer* iii. 164 It took ten minutes to get the car started, with Martin and Roddy madly swinging her by turns. *a* **1938** in T. E. Lawrence *Lett.* (1938) 495 S[haw] was asked to swing the car for the old boy. **1957** L. F. R. Williams *State of Israel* iv. 42 Two men break off for a moment from swinging the engine of a tractor. **1977** *Daily Tel.* 12 Jan. 10/2 Attempting to 'swing over' modern high-compression engines would tax the strength of all but the most muscular.

g. *Cricket.* Of a bowler: to bowl (the ball) with swing. Cf. *swing *sb.*[2] 8 g.
1948 [see *seam bowler* s.v. *seam *sb.*[1] 10].

14. b. To bring (something uncertain) about; to contrive or manage; to 'wangle'. Freq. with *it. colloq.*
1934 E. Pound *Let.* 7 Jan. (1971) 250 A guy named Collis... Wants me to edit a mag again. I have replied that..I wd. edit an annual... If he swings it, I shd. want to see a batch of yr. mss. in say about 6 months' time. **1937** Wodehouse *Summer Moonshine* (1938) i. 14 'The idea is to get him to trim the thing a little.' 'How do you expect to swing that?' **1941** B. Schulberg *What makes Sammy Run?* vi. 104 And Julian actually has a real job?.. How the hell did you swing it? **1955** 'J. Christopher' *Year of Comet* ii. 77 I'm not promising anything, but there's a chance I may be able to swing something useful there. **1962** 'K. Orvis' *Damned & Destroyed* x. 71 Phil had gotten himself a white nest-egg. Now how..could a half-broke addict-musician have swung that?' **1975** M. Bradbury *History Man* viii. 138 You can't con me, but you might swing it with someone else.

16. *Mus.* **a.** *intr.* To play jazz music with swing (see *swing *sb.*[2] 10 b). Also, *to swing it.*
[**1918** (*music-title*) Swinging along. **1928** (*music-title*) Swing on the gait.] **1931** (*music-title*) Swing it. **1933** [see *get *v.* 62 l]. **1934** *Esquire* Feb. 96/2 This still leaves a comfortable margin of popular acclaim for the boys who couldn't read it, but who, in the parlance of *hot*, knew how to *swing* it. **1935** *Swing Music* Nov.–Dec. 248/2 In the Duke's band the brass section may swing while the rhythm-section and reed-section provide a harmonic..background. **1937** L. Armstrong *Swing that Music* xiii. 114 A lot of Americans in Paris came to hear me swing. **1955** in Shapiro & Hentoff *Hear Me Talkin' to Ya* xviii. 289 Don't let Benny scare you, you're a *piano player*, Johnny—and you *swing*. **1966** T. Pynchon *Crying of Lot 49* iii. 48 The early crowd tends to dig your Radio Cologne sound. Later on we really swing. **1977** J. Wainwright *Do Nothin' till you hear from Me* viii. 125 He sometimes plays pure 'Palm Court'.., and without that extra lift which can make a band swing.

b. *trans.* To play (a tune) with swing.
1936 (*music-title*) Swingin' them Jingle Bells. **1938** *Times Herald* (Dallas) 1 Apr. iii. 11 The Detroit station pull[ed]..Tommy off the air for 'swinging' Loch Lomond. **1947** *Penguin Music Mag.* II May 28 His instructions in the introduction to the score are that these are to be slightly 'swung', and he admits the influence upon his music of all Negro spirituals. **1954** *Grove's Dict. Music* (ed. 5) 600/2 A score can at most be more or less susceptible to being 'swung'. One band may swing an arrangement while another may play the same arrangement without a touch of swing. **1968** *Blues Unlimited* Nov. 23 The waltz, swung so gently and delicately by the cajuns, is in constant demand.

c. *intr.* To enjoy oneself, have fun, esp. in pursuit of what is considered fashionable or in a manner free of conventional constraints; to be up to date. Also of a place, to provide lively enjoyment.
1957 N. Mailer in *Dissent* Summer 288 Still I am just one cat in a world of cool cats, and everything interesting is crazy, or at least so the Squares who do not know how to swing would say. **1966** *Reporter* 24 Mar. 22/1 Surprising nightlife. Amsterdam *swings*. **1967** *Wall St. Jrnl.* 24 Jan. 30 He has to really swing: Motor-cycle racing, free-fall parachuting, [etc.]. **1975** D. Lodge *Changing Places* ii. 59 Jane Austen and the Theory of Fiction. Professor Morris J. Zapp... 'He makes Austen swing,' was one comment. **1983** *Times* 25 Oct. 10/1 The fashion collections..are supposed to have proved..that 'London swings again'.

d. To engage in (promiscuous) sexual intercourse; *spec.* to advocate or engage in group sex or swapping sexual partners. Also, *to swing both ways*, to enjoy both heterosexual and homosexual relations. *slang.*
1964 W. & J. Breedlove *Swap Clubs* iii. 73 Almost everyone in the group knows one or more couples with which they swing who were not accepted by the recruitment committee. **1970** E. M. Brecher *Sex Researchers* ix. 251 If only one-tenth of one percent of married couples (one couple in a thousand) swing, however, the total still adds up to some 45,000 swinging American couples. **1972** J. G. Vermandel *Last seen in Samarra* xxii. 153 As for the mystery that still surrounded Robin Aseltine's death, the police had picked up and questioned several former girl and boy friends, Robin having been found to swing both ways.

e. Of a party: to go with a swing (see *swing *sb.*[2] 6 f). *colloq.*
[**1963** *Amer. Speech* XXXVIII. 171 [Kansas University slang.] A particularly rough and noisy party..*swinger*.] **1975** D. Lodge *Changing Places* ii. 87 The party's beginning to swing. **1978** J. Anderson *Angel of Death* xii. 128 They were trying hard to make the party swing, but.. there seemed a forced air about the revelry.

swing-. Add: **1. c.** With advbs. forming attrib. phrases in sense 'that swings in the direction specified', as (hyphened) *swing-away, -down, -out.* See also sense 2 a below.
1965 *Wireless World* July 3 (Advt.), Swing-away, lift-off mounting (optional). **1949** *Archit. Rev.* CV. 241 A slightly less conventional example is the swing-down metal wash-basin with which the Viking is equipped. **1977** *Times* 29 Apr. 13/4 There are 156 A class cabins each with two sofa beds, swing-down bunks, lavatory and shower. **1967** K. M. Smith *Insect Virol.* v. 103 In this gradient a discrete band was obtained after 60 minutes centrifugation in a swing-out (Spinco SW25) head at 24,000 rpm.

2. a. swing-back, (*b*) the backward swing of a body, weapon, etc.; back-swing; (*c*) a movement of reaction to(wards) a previous state; (*d*) applied *attrib.* to a style of coat or jacket cut to swing as the wearer moves; **swingball,** a game of table-skittles in which a suspended ball is thrown to hit the skittles on the return pass; also (*U.S.*), a larger-scale version of the game played in a doorway; see also quot. 1980; **swing bowler** *Cricket*, a bowler who makes the ball swing; also **swing bowling; swing-by,** a change of course made by a spacecraft by using a planet's gravitational field (see also quot. 1967); **swing-coat,** a fashionable coat cut to give a swinging motion when the wearer moves (cf. *swing-back* (*d*) above); **swing-gate** (earlier N.Z. example); **swing hand** *Bridge*, a hand which proves to be decisive for a team in the overall result of a rubber or match; **swing label** = *swing ticket* below; **swing man,** (*a*) *U.S.* = Swing *sb.*[2] 12 e; (*b*) *Mus.*, a jazz musician who plays swing music (see also sense 2 d below); (*c*) *U.S. Sports slang*, a versatile player who can play effectively in different positions; (*d*) *slang*, a drug pedlar; **swing mirror** = *swing-glass*; **swing needle,** a sewing-machine needle which can move sideways to the direction of work to accommodate another needle or to form zigzag or patterned stitches; freq. *attrib.*; **swing-over,** a change to a contrastive state or opinion; **swing pass** *U.S. Football*, a short pass to a back running to the outside; **swing room** *U.S.*, a room in which employees may relax while (temporarily) off duty; **swing-round,** a striking change or reversal of direction (in quots., *fig.*); **swing set,** a set of children's play equipment, including one or more swings, supported by a rigid frame; **swing-shift** *U.S.*, a work shift between the standard day and night shifts, esp. from the afternoon to late evening; applied to other irregular shift arrangements; **swing-stoppered** *a.*, applied to a bottle whose stopper is clamped in place by a wire mechanism about the bottle-neck; **swing-tail,** restrict † to sense in Dict. and add: (*b*) *Aeronaut.*, a hinged rear section of a fuselage which can be swung to one side to facilitate the loading of large items of cargo; freq. *attrib.*; **swing-ticket,** a tag or label which carries a guarantee or other information, and hangs loosely from the article to which it is attached; **swing vote(r)** *U.S.*, the independent vote(r) that often decisively influences the result of a poll; also, a casting voter; **swing wing,** an aircraft wing whose sweep can be increased at high speeds to delay the development of shock waves and decreased at low speeds to provide more lift; freq. *attrib.* (with hyphen); cf. **variable sweep*.
1890 H. G. Hutchinson et al. *Golf* iv. 98 It is an effect of stretching after an artificially long swing back. **1924** *Public Opinion* 4 July 16/2 The swing-back to biblicism appears as an accomplished fact. **1945** *N.Y. Times* 12 Aug. IV. 6/2 Legislation will be necessary to tide over those men who are unemployed while the gigantic swing-back to peacetime industry is being accomplished. **1952** W. Cunnington *Eng. Women's Clothes in Present Cent.* viii. 280 Coats were swing-back, flared or tiered [in 1945]. **1972** *Daily Tel.* 1 Mar. 5/1 The firm has charted a remarkable swing-back among its African personnel from rejection to timid acceptance. **1973** *Country Life* 15 Mar. 723/1 Swing-back jacket in showerproof Terylene/cotton twill. **1935** *Popular Mechanics* Dec. 925 (*heading*) 'Swing Ball' table top action game of skill. **1955** D. A. Hindman *Handbk. Indoor Games & Stunts* xii. 188 *Swingball bowling*... The player takes the ball and carries it any desired distance away from the doorway. **1977** *Sci. Amer.* Dec. 39/2 They range from simple board games [to].. indoor versions of miniature golf, swingball bowling (the ball is tethered to the top of a doorframe). **1980** *Trade Marks Jrnl.* 23 July 1316/2 Swingball... Games (other than ordinary playing cards) and playthings incorporating the use of balls. Dunlop Holdings Limited,..London, SW1Y 6PX; a holding company. **1958** *Times* 11 Nov. 15/2 He made an uppish defensive shot against medium-paced swing bowler, Strauss. **1953** *Times* 27 Aug. 3/7 Wind tunnel experiments at this university have shown that spin plays only a secondary part in swing bowling. **1963** A. Ross *Australia 63* iii. 87 This was swing bowling of the kind Statham does not often manage. **1967** *Britannica Bk. of Year 1966* 804/3 *Swing-by*, an interplanetary mission in which a space vehicle utilizes the gravitational field of a planet near which it passes for changing course (a *swing-by* through the gravitational field of Venus on the way to Mars). **1970** *Nature* 1 Aug. 434/2 The next opportunity to make a similar swing-by flight to Mercury will not occur until 1982. **1935** *Times* 4 Nov. 7/1 There is a new flat swing-coat in shower-proof Llamavel curl. **1939** *Country Life* 11 Feb. p. xxxvii/2 (Advt.), The three-quarter 'swing' coat of dyed baby sealskin obtainable in black, brown, or cafe. **1878** E. S. Elwell *Boy Colonists* 214 This was something like a 'race' for drafting sheep, with a swing gate. **1960** T. Reese *Play Bridge with Reese* x. 41 Playing in a team-of-four match against strong opponents, I pick up this swing hand. **1968** J. Ironside *Fashion Alphabet* 100 *Swing label*, the cardboard label hanging from a garment giving name of manufacturer, size, price, etc. **1903** A. Adams *Log of Cowboy* iii. 20 The herd trailed along behind the leaders..guarded by outriders, known as swing men. **1936** *Delineator* CXXIX. 10/3 There have been many other great *swingmen* whose names have become tradition. **1957** D. Hague in S. Traill *Concerning Jazz* 123 Many years ago the best alto player among the swingmen was Johnny Hodges—and to-day he is still tops! **1969** *Daily Progress* (Charlottesville, Va.) 6 Aug. A6/1 As the swingman last year the former Lane High All-Stater was used as a replacement for either of the Browns' starting offensive guards. **1972** T. A. Bulman *Kamloops Cattlemen* xii. 72 Another rider, called the swing man, cut in about the middle of the bunch. **1972** *Sunday Sun* (Brisbane) 2 July 14/2 Now he [drug supplier] is called the connection, the bagman, the swing man, the dealer. **1973** J. Wainwright *High-Class Kill* 157 Tell us about all the dope he pushed... He was taking from *his* swingman. **1930** *Heal & Son Catal.: Matter of Taste in Furnit.* (1972) 11 Toilet Table with two drawers and oval swing mirror. **1978** *Cornish Guardian* 27 Apr. 10/4 (Advt.), Mahogany swing mirror. **1954** M. B. Picken *Singer Sewing Bk.* (ed. 2) 246/1 The twin needles provided for the Swing Needle Machine allow you to do beautiful double stitching, using two different-colour threads. **1959** R. P. Giles *Needlework* i. 6 The more recently introduced swing needle machines..are able to stitch automatically many embroidery stitches. **1961** *Observer* 28 May 33/1 Swing-needle (zigzag) machines ..range from £50 upwards. **1927** *Daily Tel.* 1 Nov. 13/5 Harden was twitted with the violence of his swing-over. **1977** G. Clark *World Prehistory* (ed. 3) ii. 56 The most striking change in respect of animals was a swing-over from heavy emphasis on gazelle to sheep and goat. **1960** *Washington Post* 3 Jan. C4/2 They prattle knowingly of splits and gaps,..of flare passes and swing passes. **1979** *Tucson* (Arizona) *Citizen* 20 Sept. 1D/1 The hardest thing on a linebacker is the swing pass. **1917** D. C. Roper *U.S. Post Office* xxv. 291 The modern 'swing' rooms of many large post offices..have been made ideal club rooms. **1973** 'E. McBain' *Hail to Chief* iv. 56 Patrolman Gomez..was watching television in the swing room on the ground floor. **1940** W. Empson *Gathering Storm* 71 The swing-round of the Trade Unions to rearmament. **1959** *Times* 16 Jan. 14/6 (*heading*) Swing-round in Paris markets. **1951** *Sears, Roebuck & Co. Catal.* Spring and Summer 958/2, 3-Stunt Swing Set. Non-tilt enameled wood swing seat..wood trapeze bar..metal trapeze rings, wood grips. **1978** J. Irving. *World according to Garp* xii. 230, I can travel across lawns, over porches, through swing sets. **1943** *Sun* (Baltimore) 26 Mar. 1/5 (*heading*) Swing-shift workers cross border for 15-cent highballs after California bars close. **1974** *Spartanburg* (S. Carolina) *Herald-Jrnl.* 20 Apr. B5/1 (Advt.), Baby sitter..needed for swing shift in Pacolet area. [**1957** *Encycl. Brit.* I. 225/1 Swing lever stoppered bottles.] **1972** E. Fletcher *Bottle Collecting* iii. 51 In 1894, swing-stoppered bottles were introduced for sterilized milk. **1959** *Wall St. Jrnl.* (Eastern ed.) 20 Feb. 12/2 Feature of the cargo planes is a 'swing tail', which permits the whole aft section of the fuselage to swing aside. *Ibid.*, Although there are other aircraft with rear-loading doors, the flight characteristics of the new swing-tail planes would be considerably better. **1963** *Economist* 21 Sept. 1013/2 The swing-tail version of the Bristol Britannia. **1980** *Jane's Encycl. Aviation* II. 370/1 Commercial Forty-Fours.. were built with swing-tails for straight-in loading as CL-44D4s. **1962** *B.S.I. News* July 10/1 Many of the chromium-plated goods on show carried the now familiar swing ticket indicating that the chrome conformed to British Standard. **1972** *Times* 27 June 11/4 Size and price can both go on swing tickets. **1970** *New Yorker* 12 Dec. 63/3 Sellers told him that Fong was one of the swing votes. **1978** H. Kemelman *Thursday Rabbi walked Out* (1979) iii. 21 Blair and Mitchener will vote for it... So that leaves Cunningham. He's the swing vote. **1966** *Economist* 5 Mar. 898/2 He is expected to join Mr Daane as a 'swing voter', leaving Mr Martin with only one conservative colleague. **1965** *New Scientist* 22 Apr. 217/1 One new project in view is the development..of a swing-wing aeroplane. **1976** *Farnborough Internat. Exhibition* (Official Programme) 8/1 Swing wings..permit Tornado to achieve its best performance in all sections of its flight. **1978** G. Vidal *Kalki* i. 5, I persuaded Boeing to drop the variable-geometry (or swing-wing) aircraft in favor of the fixed delta-shaped wing and tail plane.

d. *Mus.* The jazz sense of 'swing' (*swing *sb.*[2] 10 b) used *attrib.* and in *Comb.*, as *swing band, craze, music*, etc.; *swing-minded* adj.
1933 *Fortune* Aug. 90/3 The best white ensembles usually compromise by playing both *sweet* and *hot* music. This is true of Ben Pollack's excellent *swing* band of Chicago (with Trombonist Teagarden and other crack soloists). **1935** (*title*) Swing music. **1937** L. Armstrong *Swing that Music* xiv. 117 People were beginning now to understand more clearly the difference between a swing orchestra and an ordinary popular orchestra. **1938** *Sat. Even. Post* 7 May 112/2 If any one musician brought about the Swing Age, it is Benny Goodman. **1939** A. Huxley *After Many a Summer* i. xiii. 178 Real romance, like in the pictures, with moonlight, and swing music. **1941** *Melody Maker* 12 July 4/2 Which would you say is the most swing-minded provincial town in the British Isles? **1945** Koestler *Twilight Bar* 11 The swing-band at the Ritz is also on strike, so they play for them. **1947** R. de Toledano *Frontiers of Jazz* v. 68 A combination of events set off the 'swing' craze. **1949** L. Feather *Inside Be-Bop* i. 3 The swing era brought jazz to the attention of the public in the 1930's. **1952** A. Lomax *Mister Jelly Roll* 292 Jelly Roll tried to compete with the swing bands. **1956** M. Stearns *Story of Jazz* (1957) xvi. 189 It was this style, made famous by Benny Goodman and brought to a peak by the Count Basie Band, that characterized the Swing Era. **1968** *Blues Unlimited* Nov. 23 It features a superb vocal with encouragement from someone in the band in the true hot western swing tradition of Bob Wills. **1976** A. Murray *Stomping Blues* vii. 107 (*caption*) The Savoy, the most famous ballroom in Harlem during the so-called Swing Era.

3. In *attrib.* or semi-adjectival use.
a. The electoral sense of 'swing' (*swing *sb.*[2] 8 h) applied to a marginal constituency, state, etc.
1964 *Economist* 4 July 44/2 That interesting phenomenon, a 'swing' state. **1974** *Times* 2 Mar. 4/5 The two major parties have very efficient organizations, as would be expected in a swing constituency. **1980** *Washington Post* 19 Oct. A5/5 An effort to improve his chances of carrying the 26 electoral votes of that swing state.

b. Designating a nation that has the capacity to adjust oil production according to demand; also applied to the oil itself.
1973 *Synagogue Light* Sept. 76/2 U.S. Treasury Deputy Secretary William H. Simon has identified Saudi Arabia as the 'swing nation', capable of a huge increase in its oil production. **1975** *Offshore Engineer* Sept. 24/1 Acting as a 'swing producer', Saudi Arabia has absorbed the biggest drop in oil income. **1980** *Times* 5 Feb. 18/2 Oil is the present 'swing' or 'balancing' fuel. Its flexibility of marketing and supply allows it to be easily taken up or cut back according to demand.

swinger[3]. Add: **I. 2. b.** See also *jim-swinger.

c. *Cricket.* A ball that swings in the air on delivery; an inswinger or outswinger.
1920 Lyttelton & Wilson in P. F. Warner *Cricket* (ed. 2) 266 He bowled a swinger, an off break, and a fast ball, which went with his arm. **1948** *Sporting Mirror* 21 May 2/3 Heath bowls medium fast swingers and opens the bowling. **1966** [see *cutter *sb.*[1] 5 b]. **1977** *Listener* 11 Aug. 182/4 Waving at a late swinger outside the off stump.

d. A gramophone record with an eccentric spindle-hole.
1935 H. C. Bryson *Gramophone Record* vi. 147 The central hole has to be made perfectly true, for were it the least eccentric with the grooves, the records produced from it would be swingers. **1961** E. N. Bradley *Records & Gramophone Equipment* i. 22 The most likely cause of wow is a swinger—a record whose spindle hole is not exactly central and so turns eccentrically as a result. **1981** *Hi-Fi Answers* Apr. 74/2 If you press the grooves off-centre relative to the centre hole it sounds terrible. A swinger that would just be okay at 33 will not do at 45.

2*. a. *Mus.* A musician who plays jazz with swing.
1934 in B. Rust *Jazz Records 1897–1942* (1969) 1516 (recording artists) The Six Swingers. **1958** K. Goodwin in P. Gammond *Decca Bk. of Jazz* xiii. 151 There are some *real* swingers on the coast, among them a young coloured pianist—Hampton Hawes. **1962** *Sunday Times* (Colour Suppl.) 10 June 3 Unexcelled as a technician and swinger, Baker is said by some to lack a musical heart and personality of his own.

b. A lively person who keeps up with what is considered fashionable; one who is 'with it'.
1965 P. Kael *I lost it at Movies* 19, I think in treating indiscriminateness as a *value*, she has become a real swinger. **1966** *Economist* 11 June 1240/3 No attempt has been made to attract the wilder London 'swingers' of *Time*-fame. **1967** H. Kemelman *Nine Mile Walk* (1968) 149 In the parlance of the undergraduate..Professor John Baxter Bowman..was a swinger, with a taste and interest in clothes not usually associated with the professoriat. **1972** J. Gores *Dead Skip* (1973) xiv. 96 The Dukum Inn.. looked..like an aging swinger getting up in the morning with his teeth still in the water glass. **1977** M. French *Women's Room* (1978) i. 14 I'd meet some middle-aged swinger with a deep tan and sideburns.

c. A person who is sexually promiscuous; *spec.* one who advocates or engages in group sex or the swapping of sexual partners. Also, a homosexual. *slang.*
1964 W. & J. Breedlove *Swap Clubs* i. 37 We will on occasion utilize 'swinger' and 'swinging' to describe the advocate of sexual partner exchange and the exercising of that practice. **1966** T. Pynchon *Crying of Lot 49* vi. 147, I had a date last night with an eight-year-old, And she's a swinger just like me. **1972** G. Baxt *Burning Sappho* iii. 47 Flo pondered the invitation... 'You sure you ain't no swinger?' 'I assure you my dear,' said Lady Molly..'I am *not* a womaniser.' **1977** *Time* 4 July 38/2 Some operators have converted nudist colonies into 'swinger camps', the new rural retreats for the randy.

swinging, *vbl. sb.* Add: **5.** Indulgence in sexual promiscuity; *spec.* engaging in group sex or the exchanging of sexual partners. *slang.*
1964 [see *swinger[3] 2* c]. **1967** W. & J. Breedlove *Swinging Set* v. 65 The act of prostitution is separate from 'swinging'. **1970** E. M. Brecher *Sex Researchers* ix. 250 What happened during the 1960's was that group sex in public—swinging—emerged from the brothels and became an established though minor feature of American urban and suburban life. **1973** *New Society* 24 May 437/1 'Swinging' is extra-marital sex by both spouses, at the same time and usually in the same place.

swinging, *ppl. a.* Add: **3. b.** *Mus.* Applied to a musician who plays jazz with swing; also, to the music itself. Cf. *swing *sb.*[2] 10 b.
1955 in A. J. McCarthy *Jazzbook 1955* 31 It has been satisfying to witness the renewed success within the past two years of Count Basie's orchestra, as the swinging spearhead of coloured jazz. **1958** K. Goodwin in P. Gammond *Decca Bk. of Jazz* xiii. 153 Mel Lewis—.. easily the most swinging drummer ever to work with the Kenton band. **1956** B. Holiday *Lady sings Blues* (1973) xxiii. 189 They were the swingingest cats I ever heard.

c. Uninhibited, ignoring conventions; lively and up to date: applied to persons, places (*swinging London*), etc., and *spec.* to the 1960s (*swinging Sixties*). Also, as a general term of approval: fine, splendid, 'great' (temporarily contrasted with *dodgy*). *colloq.*
1958 *Publ. Amer. Dial. Soc.* xxx. 47 *Swingin'*, the highest term of approval. May be applied to anything a jazzman likes, or any person. **1959** *Manch. Guardian* 25 June 8/7 [She] informed him that she wants a large place 'in a swinging part of town'..so he is looking around in Chelsea and Knightsbridge. **1962** J. Baldwin *Another Country* (1963) ii. iii. 299 'You feeling all right?'..'He's going to feel just swinging.' **1964** N. Vaughan in *T.V. World* 24 Sept. 48 When people ask me how I feel about the months ahead, I tell them: 'Sometimes it's a bit dodgy, but most of the time it's swinging!' **1965** *Weekend Telegraph* 16 Apr. 12/2 Diana Vreeland..editor of *Vogue* ..has said simply 'London is the most swinging city in the world at the moment'. **1966** *Time* 15 Apr. 11/3, I know this world, this swinging London... But I wouldn't describe myself as a swinger. **1967** *Listener* 19 Jan. 107/1 He does not fit into the *Zeitgeist* of the swinging 'sixties. **1967** F. Mullally *Prizewinner* iii. 41 The swinging London Percy had read so much about. **1971** H. Wilson *Labour Govt.* xxxvii. 766 The press publicized what they called the new swinging style of the Downing Street receptions. **1976** P. Cave *High Flying Birds* iii. 25 Young people from all over the world—draft-evading Americans, poker-faced Germans, swinging Swedes and the comic-clown Dutch. **1980** M. Sellers *Leonardo & Others* x. 56 Zuleika lived life to the full. She was a product of the swinging sixties. **1982** S. Brett *Murder Unprompted* v. 51 The British film industry..was committed to making zany films about Swinging London.

d. Of or pertaining to one who engages in promiscuous sexual activity (esp. group sex or the swapping of sexual partners). *slang.*
1964 W. & J. Breedlove *Swap Clubs* ii. 43 A 'swinging couple'. **1978** *Bulletin* (Sydney) 11 Apr. 6/2 'Swinging couples' are no longer addicted to square dancing but to the less innocuous pastime of wife-swapping.

4. swinging bridge, (*b*) (earlier Amer. example).
1708 in *Rec. Early Hist. Boston* (Boston Registry Dept.) (1883) VIII. 52 The way leading from Madam Butlers Corner..to the Swinging Bridg.

swingle (swi·ŋg'l), *sb.*[2] *N. Amer. slang.* [Blend of *swinging *ppl. a.* 3 c, d and *single *sb.* 5 c.] A 'swinging' single or unaccompanied person; *spec.* one in search of a sexual partner.
1967 *Glamour* June 82 Hilton Swingles Week. We created a week for people like you: Swinging Singles. **1973** *Newsweek* 16 July 53 The sheer number of singles, meshed with the media's seductive imagery (singles who swing are jauntily dubbed 'swingles'), is gradually revising society's view of its unwed members. **1978** *Chatelaine* (Canada) Dec. 106/3 When she went out with her women friends for an evening, their husbands felt she was luring their wives into swingles bars and white slavery.

swingometer (swiŋǝ·mɪtǝɹ). [f. *swing *sb.*[2] 8 h + -ometer, after *barometer*, etc.] A device consisting of a dial with a movable pointer, used to demonstrate (esp. on television) how a likely or observable 'swing' should influence the outcome of an election. Also *transf.* and *fig.*
1965 *B.B.C. Handbk.* 36 (*caption*) Robert McKenzie demonstrating the Swingometer. **1969** D. Widgery in Cockburn & Blackburn *Student Power* 128 Eventually the 'swingometer' which the UCL NUS Committee had installed in their Gower Street office moved slowly in favour of the ISC and the Executive. **1974** *Daily Tel.* 22 Oct. 18 After five or six results, Robert McKenzie's famous swingometer accurately showed what was to be in the event a majority of 40 or so for Labour over Conservative. **1978** *Sunday Times* 19 Mar. (Weekly Rev.) 37 Guardians of the social swingometer..have been drawing attention to punk rock. **1979** H. Wilson *Final Term* v. 84 This figure headed the election night screens, until the 'swingometer' working on the first declarations rapidly moved into a much more moderate posture.

swingster (swi·ŋstǝɹ). *slang.* [f. *swing *sb.*[2] 10 b + -ster.] = *swinger[3] 2* a.
1937 *Nebraska State Jrnl.* (Lincoln, Nebraska) 22 Aug. CD-9/4 Swingsters got the best touch of feet-itch of the season. **1946** *Jazz Writings* 19/2 Holmes' jazz is 'grown-up' jazz—as opposed to the 'adolescent' jazz of the swingsters. **1952** B. Ulanov *Hist. Jazz in Amer.* xxii. 307 It was an old war in jazz; it had not been declared by the swingsters or the boppers.

swingy (swi·ŋi), *a. colloq.* [f. Swing *sb.*[2] + -y[1].] That swings; characterized by swing. In various senses. **1.** Of music: see *swing *sb.*[2] 10 b.
1933 *Melody Maker* 25 Nov. 3/1 (*song-title*) Swingy little thingy. **1956** E. Delaney in S. Traill *Play that Music* 54 Do you remember the Gerry Mulligan sounds?—easy, swingy and very *quiet*. **1968** *Melody Maker* 30 Nov. 22/2 The arrangements are tight and swingy. **1973** J. Wainwright *Pride of Pigs* 46 The trombonist..improvised a tidy, swingy, four-bar lead-in.

2. Of garments, esp. skirts.
1937 *Evening News* 1 Feb. 1/3 Skirts will be shorter and swingier.., in keeping with swing music. **1960** *She* Dec. 8 It's the swingiest thing! Lister 'Crimplene' in party casuals. **1981** *Times* 28 Apr. 10/6 Chic tweed suits with swingy skirts.

3. Of movement, gait, etc.
1943 J. Steinbeck *Once There was a War* (1959) 22 A band of pipers marches out in kilts, with bagpipes and drums and the swingy march of pipers. **1944** D. Burley *Handbk. Jive* 85 You bend your knees halfway and rock back and forth on your heels and toes with a swingy sway.

Swinhoe (swi·nhǝʊ). The name of Robert *Swinhoe* (fl. 1862–3), British consul in Taiwan, used in the possessive, as **Swinhoe's pheasant** (occas. **kaleege** [Kaleege, kalij]), to designate *Lophura swinhoei*, a brightly coloured pheasant native to Taiwan, where he first collected it in 1862. Also *ellipt.* as **Swinhoe's.**
1863 *Proc. Zool. Soc.* 119 (*heading*) Swinhoe's pheasant. **1921** W. Beebe *Monogr. Pheasants* II. 78 We have no definite information as to the distribution of Swinhoe's kaleege, except that it is not found near the coast of Formosa, but only in the..interior. *Ibid.* 80 The price for the first pair of Swinhoes was between four and five thousand francs. **1951** J. Delacour *Pheasants of World* 162 Swinhoe's Pheasant has developed an interesting mutation in captivity. **1965** P. Wayre *Wind in Reeds* xv. 212 The male Swinhoe's is..magnificent.., his neck, underparts, rump and outer tail feathers being a dark metallic blue; a white crest tops his head and there is a white patch on his back; his scapulars are bright maroon. **1973** *Sci. Amer.* June 40/1 The birds are Swinhoe's pheasant, the mikado pheasant, [etc.].

swipe, *sb.*[2] Add: **II.** [Miscellaneous senses of uncertain affiliation.] **4.** A groom or stable-boy. *U.S. slang.*
1929 S. Anderson in *Mercury Story Bk.* 221, I had taken a job as swipe with one of the two horses Harry was campaigning. **1954** W. Faulkner *Fable* 178 He hasn't

got any money... What little there might have been, that cockney swipe threw away long ago on whores and whisky.

5. An objectionable person; also, such persons considered *collect. slang.*
1929 D. H. Lawrence *Pansies* 138 And do you think it's my business to be handing out money to a lot of inferior swipe? **1944** J. Devanny *By Tropic Sea & Jungle* xviii. 163 Some swipe has lost the fishing lines. **1951** R. Park *Witch's Thorn* xiv. 177 His tormentors leapt off him... 'Bloody little swipes!' said Mr Mate Solivich.

6. The penis. *slang* (*U.S. Blacks*).
1967 'I. Slim' in T. Kochman *Rappin' & Stylin' Out* (1972) 389 Slim, pimping ain't no game of love, so prat 'em and keep your swipe outta 'em.

swipe, *v.* Add: **2. c.** (Earlier example.)
1851 W. Clarke in W. Bolland *Cricket Notes* vii. 148 Some would shut their eyes at a fast one, but might perchance swipe away a slow one for four.

4. For *U.S.* read *slang* (orig. *U.S.*) and add earlier and later examples.
1889 *Seattle Post-Intelligencer* 5 Dec. 8/1 'By adopting this method,' said the merchant, 'we can stand back and laugh at their vain attempts to 'swipe' our goods.' **1936** Wodehouse *Laughing Gas* xxii. 238 You expect me, do you, not only to act as a stooge for you in front of the camera, but to sit smiling in the background while you horn in and swipe my interview. **1946** 'S. Russell' *To Bed with Grand Music* ii. 27 Is there another drink going before you swipe the lot? **1970** T. Roethke *Let.* (1970) 10 June 263 That beautiful Greek anthology you sent me some student swiped. **1982** *Verbatim* Autumn 3/1 The hero gallantly sets out to recover the item, which he does after much derring-do—climbing walls, crawling through windows, swiping addresses out of locked desk drawers.

swiping *vbl. sb.* (earlier example).
1833 in G. W. Ormerod *Ann. Teignbridge Cricket Club* (1889) 14 And when he's in the swiping mood, My stars! how Johnny works 'em!

swiper. 2. (Earlier example.)
1853 F. Gale *Public School Matches* 59 Swiper has the ball; now, if there is one ball which Swiper hits harder than any other, it is an on[-side] long hop rather wide to the leg.

swipey, *a.* (Earlier example.)
1821 P. Egan *Life in London* ii. ii. 181 If the latter are caught in any ways inclined to roosting from being swipy, the young buzzmen will make them pay dearly for the few winks they may enjoy.

swirl, *sb.* Add: **1. b.** A fairground roundabout with freely-pivoted cars drawn by a spider frame. *slang.*
1962 *Sunday Express* 4 Feb. 1/4 She had four rides on the merry-go-round, two trips on the ghost train, and rides on the 'swirl' and the dodgems. **1968** D. Braithwaite *Fairground Architecture* vi. 107 In the 1920's Savages of King's Lynn produced a ride known as the 'Womp'. This was a variant of the 'Whip'... Re-named the 'Swirl' by showmen, this ride was considerably refined by Thurston, Thurston and Lakin's Patents of 1929 and became one of the fastest rides on the fairground.

2. b. *Engin.* A circular motion imparted to the mixture entering the cylinder of an internal-combustion engine. Freq. *attrib.*
1926 *Engineering* 27 Aug. 279/1 It is possible with sleeve valve operation to provide a high degree of swirl in the cylinder, which..serves to bring the air to the fuel. **1940** C. B. Dicksee *High-Speed Compression-Ignition Engine* vii. 170 The swirl ratio, i.e. the ratio between the rate of air swirl and the rotational speed of the engine, varies in different designs. *Ibid.* The effect of the squish is..to increase the swirl already present. **1979** P. J. Bowyer *Boat Engines* ii. 48 The mixing of air and fuel is all important so that the design of the inlet and exhaust ports, as well as the combustion area, is arranged to cause phenomena such as 'squish' or swirl.

4. Special Comb.: **swirl chamber** *Engin.*, a chamber in an internal-combustion engine fashioned so as to impart a whirling motion to the mixture passing through it into the cylinder; **swirl skirt,** a skirt cut circular or with many gores, so as to swirl when the wearer walks.
1934 *Proc. Inst. Mech. Engineers* CXXVIII. 169 In some types of engine a very distinct improvement had been effected by making an additional passage from the cylinder head into the swirl chamber, so that the 'squish' was allowed to interfere..with the swirl. **1976** *Daily Tel.* 3 Nov. 12/3 The swirl chamber in the 2068cc Rekord engine is specially designed for the best mixture and combustion of the fuel. **1962** *Harper's Bazaar* Aug. 32 Dashing young suit with a swirl skirt. **1976** *Morecambe Guardian* 7 Dec., Wearing a Russian-style fur hat, boots and a warm-coloured burgundy coat with swirl skirt, the Princess was escorted..to the private dining room at Cartmel College.

swirly, *a.* Add: **2.** = Swirling *ppl. a.* 2. Also *fig.*
1912 W. R. Titterton *From Theatre to Music Hall* ii. i. 117 Viennese operetta, luscious and swirly. **1939** A. Ransome *Secret Water* xxvii. 317 The water would be a good deal higher.., and already felt swirly and strong. **1979** R. Jaffe *Class Reunion* (1980) i. ii. 38 Her favorite black taffeta dress with the swirly pleated skirt.

swish, *sb.*[1] Add: **1. c.** A rough hiss heard at each revolution of a faulty gramophone record.
1949 G. A. Briggs *Sound Reproduction* xxi. 130 A background noise of even volume..is much more tolerable than a sudden or changing sound such as the click of a damaged surface or the swish of a warped record. **1978** *Gramophone* Jan. 1307/1 Background noise can be at remarkably low levels on disc—though admittedly in only the best examples, and with an ever-present risk of warps, swishes and other annoyances.

5. A male homosexual; an effeminate man. *U.S. slang.*
1941 B. Schulberg *What makes Sammy Run?* iv. 71 If..that fat swish lets the producer know he did all the writing, you're dead. **1967** L. Forrester *Girl called Fathom* xiv. 178 'I think he's a swish.' A—what?' 'Faggot. Queer.' **1975** J. F. Burke *Death Trick* (1976) iv. 62 [He] dresses mod, and he talks like some kind of a swish.

6. *Cricket.* A rapid or careless attacking stroke. *colloq.*
1963 *Times* 25 Feb. 4/1 He resorted at last to the swish, an invitation to the disaster which presently befell him. **1977** *Daily Mirror* 15 Mar. 31/3 The striking sequence that whistled young Hookes from 36 to 56 was as follows: An enormous one-bounce slog over mid-off; a swish to long leg [etc.].

swish (swiʃ), *a. colloq.* [Perh. f. Swish *int.* or *adv.* and *sb.*[1]] Smart, elegant, fashionable.
1879 *N. & Q.* 5th Ser. XI. 116 Provincialisms..in the neighbourhood of Lydford... *Bain't you swish?*= How smart you are. **1922** E. Raymond *Tell England* II. xi. 269 Really, under these conditions, the Peninsula, we felt, would be quite 'swish'. *Ibid.* xii. 273 'If I'm killed you can put those lines over me.'.. 'They are rather swish,' I murmured. **1933** Auden *Witnesses* in *Listener* 12 July (Suppl.) p. ii/1 He was born in a palace, his people were swish. **1960** *Guardian* 14 July 7/7 A party at a swish place with the best people. **1972** *Daily Tel.* (Colour Suppl.) 7 Jan. 7/1 He..is a lover of the sea, food, lilies, the Old Vic, and swish cars: he is contemplating the purchase of a £5,380 Mercedes 350SL Coupé. **1974** P. Dickinson *Poison Oracle* ii. 60 The architects..had made their name running up swish hotels in Beirut.

swish-swish: see Swish *sb.*[1] 1 b.

swishy, *a.* Add: **2.** *slang.* Characteristic of a male homosexual; effeminate. Also as *sb.*
1941 G. W. Henry *Sex Variants* II. 1177 *Swishy*. The reference is to the peculiarly effeminate walk of many male homosexuals. **1954** C. Isherwood *World in Evening* ii. 125 You thought it meant a swishy little boy with peroxided hair, dressed in a picture hat and a feather boa, pretending to be Marlene Dietrich? Yes, in queer circles, they call *that* camping. **1959** J. Osborne *World of Paul Slickey* i. v. 48 He's a cad... He's contrary, he's a swishy. **1968** *Globe & Mail Mag.* (Toronto) 13 Jan. 6/3 Though they ordinarily despise swishy gestures, even masculine homosexuals will sometimes camp (exhibit feminine mannerisms).

Swiss, *sb.* and *a.* **A.** *sb.* **2.** For *rare*[−0] read *rare* and add later example.
1949 J. C. Herold *Swiss without Halos* i. 15 There are several popular misconceptions concerning the language situation in Switzerland. Some believe that the Swiss speak a language called Swiss. **1972** L. P. Johnson in M. Pasley *Germany* i. 19 Alsatian, Swabian and Swiss are forms of Alemannic.

3. Short for *Swiss muslin.* Freq. *dotted Swiss.*
1895 *Montgomery Ward Catal.* Spring & Summer 19/1 Imported Curtain Swiss, white with woven coin spots. **1897** *Sears, Roebuck Catal.* 214/1 *Shirt*.. With wide pleated puff bosom of snow white dotted Swiss. **1909** *Public Ledger* (Philadelphia) 24 June 5/7, 50c. for 75c. to $1. Dressing Sacques: Lawns and Swisses. **1924** C. E. Mulford *Rustlers' Valley* vi. 68 He thought he could make out an oval face drawing back from the dotted Swiss. **1948** E. B. White *Let.* June (1976) 294 Mrs. Dow has just entered this room bearing fresh dotted Swiss curtains. **1978** J. Updike *Coup* (1979) iv. 157 An overheated room with..dotted Swiss curtains.

4. A tournament, usu. of bridge or chess, played in accordance with the Swiss system (see sense 1 b of the adj. below).
1953 *British Chess Federation Year Bk. 1951–52* 150 Do not expect the Swiss to do more than it is capable of doing. **1965** *Listener* 29 Apr. 651/2 It is usual, for publicity purposes, to pair the favourite with one of the local hopes in the first round of a Swiss. **1975** *Games & Puzzles* June 4/2 A 5-round Swiss would make an excellent final.

B. *adj.* **1. a.** Also *Swiss-French, -German* (later examples) adjs. and sbs., (designating) the dialect of French or German spoken in Switzerland, or a speaker of this. Cf. *French-Swiss, German-Swiss* at first element in Suppl.
1941 M. F. K. Fisher in *As they Were* (1983) 58 She.. stood close against the stone, saying, 'Oh, you are adorable, adorable'..in..Swiss-French. **1961** L. F. Brosnahan *Sounds of Lang.* vii. 166 [pf,ts] and [kx] are still almost exclusively characteristic of High German, the former two in the standard language, the last confined to a few of the Swiss-German dialects. **1964** M. A. K. Halliday et al. *Linguistic Sciences* 83 Germanic speakers in Switzerland regard themselves..as speaking a distinct 'Swiss-German'. **1969** R. Petrie *Despatch of Dove* i. i. 19 Her French was fluent, without the savoyard singsong you could so often detect in Swiss-French. **1970** *Guardian* 2 June 15/4 Mr Schwarzenbach is a Swiss-German. **1979** T. Barling *Olympic Sleeper* i. 17 Her accent could have been Swiss-French.

b. Designating a system of organizing tournaments, usu. of bridge or chess, under which each player or team is matched, in each round except the first, against an opponent with a similar score, but no two opponents may meet more than once.
1953 *Brit. Chess Federation Year Bk. 1951–52* 150 The Swiss system in the last few years..has become increasingly popular. **1964** Frey & Truscott *Official Encycl. Bridge* 599/1 *Swiss system* (for multi-session team-of-four events), a method which has been used successfully for many years in major chess tournaments. When insufficient time is available for a complete round robin, a partial round robin is played. **1965** *Listener* 29 Apr. 651/1 The most popular method of deciding a large tournament is now the Swiss system, a hybrid of the conventional all-play-all event and a knock-out. **1973** *Jewish Chron.* 19 Jan. 43/2 This is the first time the English Bridge Union have held a 'Swiss' teams event.

2. *Swiss chalet, clock, franc, lace* (earlier example), *milk* (earlier example), *muslin* (earlier examples), *watch*; **Swiss bank,** a bank in Switzerland, often chosen by international clients, whose anonymity and security are preserved by a system of numbered accounts (see *numbered *ppl. a.* 2); also **Swiss banker, banking; Swiss chard** = *silver beet* s.v. *silver *sb.* and *a.* 21 e; **Swiss cottage,** a chalet; hence, a type of tent; **Swiss cream,** a type of trifle; **Swiss file** (see quot. 1964); **Swiss Itch** *U.S. slang* (see quot. 1967); **Swiss steak** *U.S.*, a steak (usu. round) cooked by dipping in flour, pounding and braising, and served with vegetables; hence, a steak (usu. a less tender cut) suitable for cooking in this way.
1949 M. Cranston *Introd. Switzerland* v. 54 To have one's money in a Swiss bank is to have it somewhere secure against wars and revolutions. *Ibid.* Capital flowed into Switzerland during the two decades between the wars, and Swiss bankers were able to take advantage of this. **1962** Sayers & Linder in R. S. Sayers *Banking in Western Europe* 188 The Swiss banker does not regard bonds as a really attractive use of resources. **1981** P. O'Donnell *Xanadu Talisman* ix. 187 Your Swiss banker telephones confirmation of receipt. **1982** D. Wiltse *Wedding Guest* xvi. 215 The reputation of the entire Swiss banking community, depends on reliability in following procedures the client stipulates. **1879** I. L. Bird *Lady's Life in Rocky Mts.* v. 58 A small house, which bore a delightful resemblance to a Swiss châlet. **1970** 'D. Halliday' *Dolly & Cookie Bird* iii. 33 The story..about Diana doing her Swiss chalet housekeeping last winter. [**1731** P. Miller *Gardeners Dict.* s.v. *Beet*, The Swiss or Chard Beet.] **1832** Swiss chard [see Chard[3]]. **1900** L. H. Bailey *Cycl. Amer. Hort.* I. 289/1 This vegetable is also known as Sea-kale Beet and Swiss Chard. **1950** *N.Z. Jrnl. Agric.* Jan. 10/1 Silver beet also called Swiss chard and sea kale beet is grown for its foliage. **1980** *Times* 23 June 16/5 The Swiss chard, a spinach-like green, was developing brown dry spots on the leaves. **1897** M. H. Kingsley *Trav. W. Africa* i. 16 Manchester cottons and shawls, Swiss clocks, and.. vividly coloured china. **1982** R. Ludlum *Parsifal Mosaic* ix. 306 'Pretty punctual, huh?' 'Like a Swiss clock.' **1820** M. Wilmot *Let.* 1 Aug. (1935) 75 A beautiful Swiss Cottage, built in the most correct manner. **1884** T. H. Lewin *Fly on Wheel* iv. 117 The 'Swiss cottage' tent, on which I decided [in 1862], as it had the advantage of being divisible into two compartments, with, in addition, a small bathing-room, and large outer flaps which served as shelter for my servants. **1968** *Sat. Rev.* (U.S.) 27 July 40/1 Take the omnibus..to the Swiss Cottage—the play-chalet which Prince Albert designed for his children. **1971** A. D. Gorwala *Queen of Beauty* 70 In the snug well-lighted Swiss Cottage tent that was his itinerant home and office for weeks on end while on tour. *Ibid.* 77 There were quite a number of places at which the old Swiss cottage proved useful. **1845** E. Acton *Mod. Cookery* xx. 427 (*heading*) Swiss cream, or trifle. (Very good.) **1861** Mrs. Beeton *Bk. Househ. Managem.* 748 *Swiss cream*.. ½ lb. of macaroons or 6 small sponge-cakes, sherry, 1 pint of cream, 5 oz. of lump sugar, 2 large tablespoonfuls of arrowroot, the rind of 1 lemon, the juice of ½ lemon, 3 tablespoonfuls of milk. **1903** Joyce *Let.* 26 Feb. (1966) II. 31 Today for dejeuner I had some cold ham, bread and butter, Swiss cream with sugar. **1960** E. L. Delmar-Morgan *Cruising Yacht Equipment & Navigation* xiv. 163 The Swiss files are most useful for repair jobs. **1964** S. Crawford *Basic Engin. Processes* i. 8 Needle files or Swiss files, small fine-cut files of various cross-sections, used for instrument work and in the match-making industry. **1934** Webster, Swiss franc. **1938** M. Muggeridge *In Valley* ix. 62, I wrote a cheque for twenty Swiss francs. **1973** 'G. Black' *Bitter Tea* i. 11 She had just made quite a killing..and had invested her take in Swiss francs. **1959** *Life* 7 Dec. 51/1 The technique..was ..old stuff to mature Americans who in Prohibition days had used it, complete with salt, as the safe way to take bathtub gin, then called 'Swiss Itch'. **1967** *Amer. N. & Q.* June 152/1 '*Swiss Hitch*'.., I believe the correct form is Swiss Itch, and I suspect that there is more than one recipe, but I have always heard the term applied to the process by which one places a pinch of salt on the back of the hand, then licks it off, and takes immediately a jigger of tequila, and follows *that* by immediately biting into a segment or a slice of lime. **1865** *Reading Industr. Exhib. Catal.* in *Reading Mercury* (1968) 17 Aug. 11 Swiss lace, tambour muslin, leno and every description of curtain materials. **1889** *Girl's Own Paper* Summer No. 28/1 A small tin, which everyone supposed was Swiss milk, but which proved to be a shilling tin of cream. **1832** T. S. Fay *Dreams & Reveries* I. 155 If I had anything to say about bobbinet or Swiss muslin collars, I should at least wait till he had reached some passage not particularly remarkable for beauty. **1932** E. Craig *Cooking with Elizabeth Craig* 175 Swiss Steak... Take a..2

pound slice of steak. Sprinkle thickly with flour. Pound. ..Brown steak on both sides. 1947 L. P. De Gouy Gold Cookery Bk. vi. 345 Swiss Steak. The original name of this recipe was 'Schmor Braten.' It is three centuries old. 1973 Black Panther 12 May 10/1 Safeway was charged with..mislabeling swiss steaks as round steak for an extra profit of ten cents per pound. 1885 C. M. Yonge Two Sides of Shield I. iii. 28 Her mother's little Swiss watch. 1977 H. Kaplan Damascus Cover (1978) xvi. 174 The Colonel had honed the movements of the two Israeli agents with the precision of a Swiss watch.

Swiss cheese. 1. (In Dict. s.v. Swiss a. 2.)
2. fig., with reference to the pitted or honeycombed structure of some varieties of Swiss cheese. Freq. attrib.
1924 Novak & Martzloff in Amer. Jrnl. Obstetrics & Gynecol. VIII. 387 We are accustomed to speak of the endometrium as presenting a 'swiss cheese' pattern in these cases. Ibid. 409 The glands are of the 'swiss cheese' pattern, large dilated glands being found side by side with glands which are small and narrow. 1929 Hall & Niles One Man's War i. 5 Passing out of the up-rush, I dropped into what was then known as a Swiss cheese section of air. 1949 Bull. Geol. Soc. Amer. LX. 1290/1 Collapsed pumice fragments, usually less resistant than the matrix, weather to give the rock a Swiss-cheese appearance. 1968 Nature 10 Feb. 513/2 Such a universe can be simulated by a 'swiss cheese' model in which spherical perturbations..are distributed through space. 1970 J. R. Lincke Jenny was no Lady vi. 81 The planes ran into 'Swiss-cheese' air, downdrafts..and violent gusts. 1974 A. Lurie War between Tates i. 13 Their friendship now is full of Swiss-cheese holes in which sit things which cannot be discussed.
3. Special combination: **Swiss cheese plant,** an evergreen climbing plant, Monstera deliciosa, of the family Araceæ, native to central America and often cultivated as a house plant for the sake of its large ovate perforated leaves.
1946 M. Free All about House Plants xviii. 267 One of the most interesting [aroids] is the..Swiss Cheese Plant, because of the holes naturally formed in the much-divided leaves. 1955 W. E. Shewell-Cooper Pot Plants i. 15 Monstera deliciosa..is usually called the Swiss Cheese plant. 1970 Sunday Tel. 3 May 19/2 A variation of the 'Swiss cheese plant' has golden marking besides the familiar 'holes' in the leaves. 1981 'J. Ross' Dark Blue & Dangerous xxiii. 130 An immense Swiss Cheese plant climbed a cement pillar.

switch, sb. Add: **I. 3. b.** (Earlier example.)
1865 W. H. Preece Railway Electric Signalling 16 The instrument which is employed to raise and lower the signal is called a 'Switch'.
d. fig. or in fig. contexts, esp. with reference to railway or electrical switches. U.S. colloq.: asleep at the switch, etc., negligent of or oblivious to one's responsibility, off guard.
1898 G. B. Shaw Let. 16 Mar. (1972) II. 16, I am very cross and incommoded..by having to adapt myself [to a new secretary]... For three sentences, I feel resentful.. and quite put out. At the fourth the switch operates and I am on to the new line as if I had never dictated to anybody else. 1906 H. Green At Actors' Boarding House 368 Snow..awoke the startled Williams, asleep at the switch. 1932 W. Faulkner Light in August viii. 161 Mind and body as if on the same switch, believing that he had seen a movement among the shadows. 1958 Observer 19 Oct. 18/4 [The television play], though a bit slow off the switch, scored well over half-marks for sincerity and realism. 1966 C. Achebe Man of People iv. 51 We must not let up. We just must not be caught sleeping on the switch again.
e. Computers. A program instruction that selects one or other of a number of possible paths according to the way it is set.
1951 M. V. Wilkes et al. Preparation of Programs for Electronic Digital Computer 167 Numbers at one end of a permitted range can be detected by adding a constant and testing the sign, and then the result of the discrimination may be used to operate a multiway switch. 1962 R. S. Ledley Programming & utilizing Digital Computers vi. 227 The switch designator is of the form NAME[N] where NAME is the name of the switch corresponding to the switch declaration and N is an integer telling which label of the switch declaration to use. 1970 O. Dopping Computers & Data Processing v. 89 A switch can consist of a branch instruction, the address part of which can be altered by the program.
4. (Earlier U.S. example.)
1870 L. M. Alcott Old-Fashioned Girl xi. 223 So much hair of her own, that she never patronized either rats, mice, waterfalls, switches, or puff combs.
7*. a. A change from one state or course to another; an alteration of position, policy, etc.
1920 Ade Hand-Made Fables 27 A switch had to be made. The Wholesaler..wished him on to the Banker. 1941 News Rev. 14 Aug. 4/2 The Soviet's entry into the war against Nazi Germany meant a switch in the Communist Party's home policy as well as its foreign outlook. 1951 M. McLuhan Mech. Bride (1967) 151/2 In the space of six months it recently shifted a large section of its enterprises from murder to love comics. The combined attacks of Dr. Frederic Wertham, Mr. G. Legman, and others suggested the advisability of a partial switch from Death to Love. 1960 Economist 15 Oct. 260/2 Large-scale cultivation of wheat in the new areas would make possible a switch to industrial crops in the older agricultural parts. 1977 'E. Crispin' Glimpses of Moon viii. 151 If you're thinking I could have done some sort of a switch at some stage, you can put the idea out of your mind straight away. 1981 F. Hoyle Ice x. 158 It is satisfactory that both of the switches, to and from an ice-age condition, can arise from the same kind of cosmogonic event.

b. Bridge. A change of suit either in bidding or play. Cf. sense 7 b of the vb.
1921 A. M. Foster Auction Bridge 70 The take-out or switch. 1923 [see *ASSIST sb. a]. 1939 N. De V. Hart Bridge Players' Bedside Bk. iv. 38 It was a clever switch, and at once turned the hand into a difficult problem. 1952 I. Macleod Bridge is Easy Game xii. 141 Here you dare not concede the opening trick for a Heart switch will surely defeat you. 1980 R. Markus Bridge-Table Tales vi. 19 Declarer was forced to win East's king for fear of a spade switch.
c. An exchange; spec. a substitution which involves criminal deception. colloq. and slang.
1935 Wodehouse Luck of Bodkins xiv. 144, I plunged into..your state-room..and gave the sleeping figure..a hearty wallop..and it was Gertrude... 'What's the idea? Why the switch?' 1938 F. Chester Shot Full xxv. 302 Another of Lewis's rackets was to pose as a buyer of loose diamonds, and then substitute glass for the stones... This form of robbery is known as 'the switch'. 1955 W. Gaddis Recognitions II. ii. 369 Somebody pulled the old twenty-dollar-bill switch on her, Ellery said looking up from his magazine.
III. 8. switch box; **switch base** (see quot. 1940); **switch-blade,** (b) a pocket knife with a blade released by pressing a button or similar device on the handle (cf. flick-knife s.v. *FLICK sb.[1] 4); in full, switch-blade knife; **switch cane,** a large bamboo, Arundinaria gigantea subsp. tecta, native to southern N. America; **switch dealing** Econ., purchase and resale, or sale and repurchase, of a commodity in order to profit by differential values of currency; repurchase or resale through a third party; hence **switch deal, dealer; switch dollar** Econ. (see quots.); **switch gear,** the assembly of switching devices and associated equipment used in the generation and transmission of electric power; **switch gene** Genetics, a gene whose presence or absence determines whether a group of other genes is expressed; **switch-girl** Austral. = switchboard girl s.v. *SWITCHBOARD b; **switch-hitter** U.S. Baseball, an ambidextrous batter; also transf. (colloq.) in sporting and gen. contexts; slang, a bisexual; also **switch-hitting** ppl. a. and vbl. sb.; **switch hook** Teleph., the hook or support in a telephone set which operates the circuit switch when the receiver is placed upon or removed from it; **switch-knife** = switch-blade (b) above; **switch-light** U.S. = switch-lamp, -lantern; also transf. (see quot. 1960); **switch mechanism** Genetics, the mechanism by which a switch gene operates; **switch-reference** Linguistics (see quot. 1972); **switch-room** (earlier example); **switch selling,** a sales technique whereby cheap goods are displayed in order to lead the consumer to buy similar but more expensive items; also transf.; hence **switch-selling** ppl. a. and (as a back-formation) **switch-sell** v. intr.; **switch-tender** (earlier example); **switch-tower** (earlier example); **switch trading** Econ., international trading in commodities conducted through media other than currency (cf. switch deal above); **switchyard** U.S., (a) an area of a railway taken up by points, and in which trains are made up; also transf.; (b) an enclosed area of a power system which contains the switchgear.
1940 Chambers's Techn. Dict. 826/1 Switch-base, the insulating base on which a switch is mounted. 1967 M. Chandler Ceramics in Mod. World iv. 114 For telegraph or telephone insulation, as for domestic switch-bases, fuse-holders, bulb-sockets, and so on, almost any kind of insulating material will work. 1932 L. Hughes Negro Mother 13 'Cause I carries a switch-blade And I swing it a-hummin', And if I don't get you goin', I'll cut you down comin'. 1950 Patterson & Conrad Scottsboro Boy II. ii. 96 He put the shears in his pocket and went to his cell. I had a switch-blade knife. I went looking for him. 1957 New Yorker 5 Oct. 64/1 A fist fight between two champions, but there are emotional complications, and the switch-blade knives are put to work. 1975 P. Theroux Great Railway Bazaar xxx. 342, I would have plotted myself into danger: Sadik would have had a switchblade and gold teeth. 1940 Chambers's Techn. Dict. 826/1 Switch-box, an enclosure housing one or more switches operated by means of an external handle. 1978 W. F. Buckley Stained Glass xxii. 217 Blackford walked to the switch box. 1845 W. T. Porter Big Bear Arkansas 132 They circled about among the switch-cane and priscimmon bushes a long time. 1954 W. Faulkner in Holiday XV. 36/3 The Natchez doctor was clearing the land fast now, plowing under the..switch cane of the creek and river bottoms. 1973 'D. Jordan' Nile Green xxi. 85 KK hovered in his office, too, doing a quick switch deal in forward dollars. 1967 Economist 14 Jan. 143/1 When the Russians don't wish to take up a consignment of Moroccan oranges to which they are committed under a bilateral trade agreement, they go to a specialist known as a switch dealer in one of Europe's financial centres, and he arranges a resale to someone else, at a discount. That, in a nutshell, is the mysterious art of switch trading. 1957 Ibid. 21 Dec. 1082/1 Commodity shunting in general virtually stops when the margin between transferable sterling and official sterling is a little less than three per cent.

But 'switch' dealings in platinum are possible at a narrower margin. 1964 Times Rev. Industry Sept. 17/1 All purchases of foreign exchange for investment in non-Sterling Areas are subject to control... Direct investment projects..may be financed..either by borrowing abroad or by using the non-sterling currency proceeds of the sale of foreign securities..; i.e., so-called 'switch dollars'. 1978 J. Paxton Dict. European Econ. Community (rev. ed.) 236 Switch dollar market. Investment in foreign securities by United Kingdom residents is not normally allowed.., but existing holdings may be realised and the proceeds switched into (i.e. used to buy) other securities, or sold..to other United Kingdom residents who wish to purchase foreign securities. For convenience, such funds, whatever the currency, are expressed in terms of United States dollars called Switch, security or investment dollars. 1901 Switch gear [in Dict.]. 1930 Times 29 Mar. 19/4 Our metal is now being adopted as a substitute for non-magnetic iron in many instances, such as in the large casings for metal-clad switchgear. 1958 Optima Sept. 130/2 The electrical industry uses platinum for switchgear contacts in such equipment as traffic lights, telephone exchanges, radio stations and generating stations. 1978 Jrnl. R. Soc. Arts CXXVI. 609/2 Continued improvements in control circuitry and microprocessors are likely to.. further increase the utilization of transformers, switchgear and circuits. [1941 Mather & de Winton in Ann. Bot. V. 310 The more rigorous the selection of illegitimacy to outbreeding conditions the more efficient it is as an in-breeding mechanism when the switching genes are changed.] 1942 Nature 14 Nov. 564/1 Mather and de Winton have recently spoken of such genes as 'switch genes'. 1968 R. D. Martin tr. Wickler's Mimicry in Plants & Animals vii. 82 Polymorphism is sometimes controlled by single genes, sometimes by groups of genes, and..switch genes (as explained for Papilio dardanus) may also play a part. 1943 K. Tennant Ride on Stranger xi. 120 Some of them would be asking for letters at the switch-girl's desk. 1969 Southerly XXIX. 93 The tea-lady panders to the biological necessities of life, the switchgirl makes communication easier. 1948 L. Durocher Dodgers & Me vii. 49 Cullenbine, a switch hitter, and Steve Rachunok..were two athletes we had picked up..from Detroit. 1956 H. Kurnitz Invasion of Privacy iii. 25 A free-swinging roundhouse slap..landed high on his cheek... 'What do you know!' he said softly. 'A southpaw!' 'Wrong,' said the girl. 'A switch-hitter.' 1960 Wentworth & Flexner Dict. Amer. Slang 534/1 Switch-hitter, a bisexual person. 1972 Pussycat XXXIII. LIX. 8/1 The buddy would shove cock to me. I can still remember the first switch-hitter. 1938 Philadelphia Rec. 5 Feb. 15/6 A signed contract has been received from Emmett Mueller, switch-hitting rookie whom the Phils rescued from the Cardinal chain gang this winter. 1952 Sun (Baltimore) 25 Feb. (B ed.) 14/6 He also picked up a switch-hitting style from baseball. Gordie is the only player..who has mastered the art of switching hands on his stick, so that he can shoot from either his right or left side without warning. 1970 N.Y. Times 16 Aug. II. 1/1 Chock full of scenes of what people apparently want to see today..lesbianism, switch-hitting, group gropes. 1922 Telegr. & Teleph. Jrnl. VIII. 82/2 If a subscriber leaves his receiver off the switchhook..the switching equipment is automatically released after a certain interval. 1975 D. G. Fink Electronics Engineers' Handbk. XXII. 4 In the common-battery [telephone] set..a circuit closure, activated by the switch hook, serves to alert the central office. 1955 Time 6 June 27/3 They manufacture pistols, carry switchknives and use them. 1957 Wodehouse Over Seventy xv. 144 At Eightieth Street he produced a switch-knife... 'This is a stick-up', he announced. 1892 Harper's Mag. Dec. 80 He saw the station agent running down the tracks with the red switch-light. 1929 W. Faulkner Sartoris IV. 350 Along the tracks green switch-lights were steady in the dusk. 1960 Listener 18 Aug. 250/2 When a hungry young boomer came in and demanded..'a couple of switch lights in the fog'..what he really wanted was..two fried eggs with the grease poured over them. 1941 Ann. Bot. V. 308 The switch mechanism at the S,s locus offers the possibility of a very different adjustment to changed breeding conditions. 1953 J. S. Huxley Evolution in Action I. 29 Some genetic differences act as a switch mechanism, turning on a whole battery of further processes. 1967 W. Jacobsen in Hymes & Bittle Stud. in Southwestern Ethnolinguistics 238 This paper discusses a device for pronominal reference, denominated 'switch-reference', which is found, with considerable differences of detail, in three languages of the Hokan-Coahuiltecan group. 1972 D. Hymes in M. E. Smith Stud. in Linguistics in Honor of G. L. Trager 105 The use of separate forms of third person to keep track of discourse is best treated under the general heading of 'switch reference'... It may be preferable to abandon use of the term 'fourth person' altogether, speaking simply of 'inclusive', 'obviative'.., and 'switch reference'. 1978 Language LIV. 220 The presence of switch-reference morphemes..also appears to be an areal feature in parts of California. 1885 List of Subscribers, Classified (United Telephone Co.) (ed. 6) 8 Each subscriber is furnished with a set of instruments..which is connected with a wire communicating with the Exchange or Switch Room nearest his address. [1930 Amer. Speech VI. 128 To switch a customer is to quote to him a low price on an article to inspire him, and then to direct his interest to another article.] 1960 Guardian 30 Nov. 2/7 The practice of switch selling of sewing and other machines..from misleading advertisements. 1965 E. Gundrey Foot in Door ii. 20 The fast-talking, switch-selling, hard-pressing salesman. Ibid. xxxviii. 219 It should..be made illegal to 'switch-sell'. 1971 H. Wilson Labour Govt. xix. 361 Mr Kosygin..was escorted throughout the day by..the Secretary of State, who had been told about the American exercise in switch-selling the night before and had been asked to watch out for any signs of reaction. 1853 Putnam's Mag. July 34/2 We went roaring, rushing, screaming, up the valley of the Susquehanna, occasionally passing a switch-tender with his white lights. 1897 Kipling in Scribner's Mag. Aug. 146/1 They were at the far north end of the yard, now, under a switch-tower, and looking down on the four-track way of the main traffic. 1967 Switch trading [see switch dealer above]. 1974 Harper's Mag. Dec. 54 What Intertel does is ..advise on geopolitical 'switch-trading opportunities'.

1888 Austin (Texas) Statesman 1 Nov. 6/6 In the switch yards of the Chicago & Alton..nearly all the men reported for duty this morning. 1943 J. S. Huxley TVA 85 The transformers and switchyard..are not applied to a predetermined structure, they are part of it. 1956 H. Gold Man who was not with It (1965) x. 70, I was at the switch-yards, still running, and then I was clambering in the coupling of a moving freight. 1969 Daily Colonist (Victoria, B.C.) 16 Oct. 40/7 Work is to begin immediately on placing the..generators at the underground powerhouse.. and the switchyard and central control building on the surface. 1971 Sci. Amer. June 60 Near the end of its trip the electron beam passes through a 'beam switchyard' before reaching the target areas.

switch, v. Add: **1. c.** I'll be switched, a mild indication of exasperation, denial, or surprise. N. Amer. colloq.
1838 U.S. Mag. I. 427, I'll be switched if I do. 1841 J. B. Jones Wild Western Scenes xiv. 178 I'll be switched if many folks lives in higher houses than I does. 1901 Daily Colonist (Victoria, B.C.) 4 Oct. 3/7 'Well, I'll be switched!' ejaculated the chatterer. 1941 L. I. Wilder Little Town on Prairie ix. 99 'Well, I'll be switched!' said Pa... It takes you to think up a chicken pie, a year before there's chickens to make it with.
6. b. (Earlier example.)
1853 'Mark Twain' Let. in Iowa Jrnl. Hist. (1929) XXVII. 413 Our train ran back half a mile and switched off another track, and stopped.
7. b. Substitute for def.: To change or transfer from one thing to another; to alter to another state or activity. Also with advbs. spec. in Bridge, to change to another suit in bidding or in play (see sense 7*b of the sb.) (Later examples.)
1921 A. M. Foster Auction Bridge 32 Your partner.. can support your call or switch into another bid. 1932 Daily Tel. 8 Oct. 15/5 At Contract he has the..duty of raising the opener's bid..and, if he switches, of deciding whether to make a pre-emptive bid or not. 1952 I. Macleod Bridge is Easy Game xiv. 189 Switching to Diamonds declarer made her contract. 1980 R. Markus Bridge-Table Tales iv. 16 West won and switched to a spade.
1923 H. Crane Let. 20 Jan. (1965) 117, I..urged him not to 'waste his time' on any magazine project. But after his visit here last summer I quickly switched about. 1930 H. Zink City Bosses in U.S. x. 207 He bolted the regulars and switched to reform groups. 1954 J. Steinbeck Sweet Thursday v. 34 He knows when high-school boys have switched from gin to marijuana. 1962 Rep. Comm. Broadc. 1960 239 in Parl. Papers 1961–2 (Cmnd. 1753) IX. 259 Viewers who did not switch would find themselves exposed at some time of the evening to informational material. 1978 M. Amis Success ix. 173 Some dead-end toiler asked to switch from one equally meaningless chore to another.
c. trans. To exchange (items), esp. with intent to deceive.
1897 Columbus (Ohio) Dispatch 18 June 5/2 An opportunity presented itself to 'switch' the bottles. 1917 Dialect Notes IV. 330 Switch, v.t., to exchange, esp. surreptitiously. 'I thought I was getting title to this land, but they switched deeds on me in the office.' Neb. 1948 C. L. B. Hubbard Dogs in Britain III. xv. 130 A business in which dogs have been 'switched' (and doped) and the results manipulated is questionable. 1978 F. Weldon Praxis x. 75 Praxis managed to switch envelopes so that an empty one was dispatched instead.
d. To change or alter (from one thing to another); to transfer. Also, with items involved in the change as plural obj.
1919 Wodehouse My Man Jeeves 157 It struck me that I'd no right to butt in on his secret sermons, so I switched the conversation. 1931 W. G. McAdoo Crowded Years x. 157 Sullivan switched the fifty-eight votes of Illinois from Clark to Wilson. 1957 A. C. Clarke Deep Range xxi. 188 The very idea of switching our entire herds to milking instead of slaughtering is just crazy. 1959 Daily Tel. 15 Oct. 12 Among those who have switched offices, Mr. Watkinson's is perhaps the most surprising translation. 1963 Listener 28 Feb. 363/2 The government was forced to switch the full campaign towards the less flexible statutory committees. 1975 D. Lodge Changing Places vi. 229 Philip switches channels until he hits the transmission of the Plotinus March.
8. a. More recently, also of a radio or television set, etc.; to turn out (an electric light). Also, to change the state of (a two-state device).
1935 Radio Times 13 Sept. 4/3 If you were to switch on your set..you would have no difficulty in distinguishing ..who was speaking or singing. 1954 I. Murdoch Under Net iii. 53, I didn't switch out the light, but covered the lamp up again with gauzy stuffs until it gave only a faint glow. 1960 Haley & Scott Analogue & Digital Computers vii. 188 The core is switched from the 1 to the 0 state. 1964 F. L. Westwater Electronic Computers iv. 79 This is..got round by first switching a wound core..and then allowing the read current in this core to be used to write in the appropriate row and column. 1983 J. Fuller Convergence xix. 210 It is no sweat. Easy as switching on the old FM.
(ii) to switch in: to bring into a circuit by the operation of a switch; similarly to switch out.
Cf. quot. 1891, sense 8 in Dict.
1957 Practical Wireless XXXIII. 734/1 A resistor could be switched in initially to limit the maximum possible current flowing to 10 mA. 1970 J. Earl Tuners & Amplifiers iii. 67 The loudness control filter..can be switched out allowing the volume control to work in the ordinary, uncompensated manner. 1978 SLR Camera Aug. 90/1 To switch in the automatic exposure control system all

he need do is turn the shutter speed setting dial to the position marked 'Auto'.
(iii) intr. Of a two-state device: to pass to the other state. Of its state: to change.
1964 F. L. Westwater Electronic Computers iv. 77 The resulting change of flux as the core switches will cause an electromotive force in the read wire. 1981 J. D. Lenk Handbk. Digital Electronics ii. 41 Inputs cause the state of the circuit to switch, reversing the output.
b. intr. or absol. To turn on (or off) a radio or television set, or other device. Cf. *TURN v. 73 h, 74 a.
1932 Even. Standard 21 Jan. 3/3 The best plan is to tell listeners what is going to happen and let them decide whether they switch off or not. 1951 'J. Wyndham' Day of Triffids xi. 206, I could not hear above the noise of the engines. We both switched off. 1958 Listener 20 Nov. 849/3 Many viewers may have missed it by switching off in fatigue. 1975 Ibid. 9 Jan. 38/2 They do it because someone's just switched on. 1977 Rep. Comm. Future of Broadcasting (Cmnd. 6753) iii. 19 Viewers and listeners cannot ..express..disapproval, except by switching off.
c. intr. To change over to another state by means of a switch; spec. to alter the receiving channel of a radio or television set.
1937 Discovery Nov. 348/2 By switching over from white light to black an entire scene can be changed instantaneously. 1940 N. Mitford Pigeon Pie v. 85 It would be difficult to do better, for an account of the Wig Inquest than to switch over, as they say on the wireless, to the columns of the Evening Runner. 1958 Sunday Times 26 Jan. 6/5, 200 pages of mumbo-jumbo which would make anyone switch over to another programme. 1961 S. Price Just for Record i. 13 The phone hasn't rung all day because I've switched over to the answering service.
d. trans. To turn off (a television or radio programme, or its content).
1947 G. B. Shaw How to become Musical Critic (1960) 321 [The B.B.C.'s] worst concessions to popular bad taste..are very horrible. I switch them off so promptly that I am hardly qualified to condemn them. 1962 Listener 18 Oct. 633/3 The archness of the dialogue had to be heard to be switched off.
e. To direct (a telephone link) through to a subsidiary receiver by means of a switch.
1971 'S. Smith' Grave Affair xii. 181 The telephone had not been switched through to my study deliberately. 1976 J. Tate tr. A. Bodelsen's Operation Cobra xvii. 83 They switched the telephone through and went on up.
9. transf. and fig. **a.** To turn on or off, as if by means of a switch.
1929 W. J. Locke Ancestor Jorico viii. 111 Without great discourtesy one couldn't switch off Binkie. 1934 Discovery Sept. 259/2 In this way she succeeded in switching off any unpleasant dream. 1966 Listener 24 Mar. 426/1, I have always found it very easy to 'switch on' emotion. 1967 B. Patten Little Johnny's Confession 54 Those couples who Having been switched off permanently, Are so very still. 1980 Nature 27 Mar. 379/2 The prose style is guaranteed to switch off all but the most ardent student.
b. intr. for refl. With off. Of persons: to cease listening, to lose concentration; to become bored or inattentive.
1921 G. B. Shaw Back to Methuselah III. 94 Dont switch off. Listen. This American has invented a method of breathing under water. 1928 [see *EASY adv. 4 b]. 1955 Times 22 June 11/5 Does he seriously maintain that in a class of 24 boys, where 23 are working keenly and well, it is invariably the master who is to blame because No. 24 always 'switches off'? 1976 J. I. M. Stewart Memorial Service vii. 108 He was heavy alike with his years and his whisky and wine, and he may simply have switched off. 1980 D. Bloodworth Trapdoor xx. 121 For some reason he could not fathom she had switched off. Her love had died.

switchable (swi·tʃăb'l), a. [f. Switch v. + -able.] Capable of being switched between different positions or modes of operation. Freq. in techn. contexts.
1961 [see dual-standard s.v. *DUAL a. (sb.) 3]. 1970 J. Earl Tuners & Amplifiers vi. 142 Signal in the i.f. channel also operates the tuning meter, switchable to a.m. by the f.m./a.m. changeover switch. 1977 Nature 6 Jan. 92/1 The range is −0·05 to 1·999A with a switchable decimal point and double over-range indication. 1982 Sunday Times 31 Oct. 57/1 Switchable pension... Switching is allowed among six funds (corresponding text).

switchback, a. and sb. Add: **A.** adj. (Earlier example.) Also fig. and in extended transf. uses.
1887 R. Fry Let. 21 Nov. (1972) I. 117 Some of us went on a switchback railway (the sensation of which I thought very pleasant). 1908 F. W. Lanchester Aerodonetics 30 A magnificent flight, remarkable 'switch-back' flight path, distance, relative to the wind, probably over 600 yards. 1912 G. Mackenzie Carnival ix. 90 Jenny thought what horrible places they were, these sweeping moorland wastes..with switchback stone walls. 1961 Daily Tel. 5 Sept. 12/2 Where Mr Hill made much more sense was in criticising what he called the Government's 'switchback economies'. 1965 Daily Mail 28 Oct. 5/3 A dangerous 'switchback' course can build up as the plane descends. 1978 S. Wilson Dealer's Move vii. 116, I got hung up behind a couple of lorries... It was switch-back country and there was no way you could see what was coming.
B. sb. (Earlier U.S. and later fig. examples.) transf. (later examples); also applied in N. Amer. to a tight bend on an ascending road or trail.

1863 Harper's Mag. Sept. 465/1 We descend from our high elevation by gravity, changing our direction at various points by means of what is called a Switch-back. 1933 Nat. Geogr. Mag. Feb. 196 (caption) An excellent highway climbs by turns and switchbacks through natural timber to the top of the towering dome. 1934 [see giant racer s.v. *GIANT sb. 6]. 1965 Daily Mail 28 Oct. 5/3 If the Vanguard..was on the down-slope of a switchback when the talk-down was ended,..probably no further cause for the crash need be sought. 1969 Islander (Victoria, B.C.) 9 Nov. 7/1 Steep grades are not the only challenge on this road. Several of the switchbacks are so tight that much manoeuvring is required to get around them. 1976 J. Snow Cricket Rebel 61 Each time I was to find the county side at a lower ebb, with the team on a down slide of the switchback we rode in the 1960s.
switchback v. (later examples); **swi·tchbacked** ppl. a., **swi·tchbacking** vbl. sb. and ppl. a.
1913 Outing Jan. 498/1 Switch-backing or zigzagging up a hill is simply striking off to the right, for instance, at an angle and then turning off to the left. 1930 J. Collier His Monkey Wife vi. 72 Electric light cables..switch-backed along the undulating coast. 1963 Times 16 Feb. 11/1 Another short drop leads to the edge of the lake with its bathing station and a surrealist structure built for high diving and switchbacking into the water. 1972 Daily Tel. (Colour Suppl.) 13 Oct. 86/2 A switchbacking lane over the heath to Studland. 1976 B. Bova Multiple Man (1977) vi. 65 My rented car climbed the switch-backed driveway.

switchboard. Add: **a.** (Earlier examples.)
1873 Harper's Mag. Aug. 349/2 The switch-board..is the central ganglion of the whole [telegraphic] system. 1879 Nature 11 Sept. 461/2 The switch-board at the central office.
b. attrib., esp. as switchboard girl, operator.
1903 P.O. Telephone Service 77 The current from the 'busy-back' and 'don't answer' commutators is not led directly to the switchboard terminals. 1925 F. G. C. Baldwin Hist. Telephone in U.K. xi. 301 Switchboard cable containing 42 wires was employed between the test board and the first switch section. 1952 Traffic Apr. 34 (Advt.), Suppliers to the World for all Telecommunication products..Automatic telephone equipment for main and satellite exchanges..Telephone instruments and accessories. Switchboard lamps (Hivac Ltd.). 1961 M. Kelly Spoilt Kill II. 105 He just wouldn't have anything to do with her. A girl from the works, the switchboard girl! 1967 N. Freeling Strike out where not Applicable 111 'All right,' said the switchboard girl indifferently. 1974 A. Morice Killing with Kindness ii. 18 The call came through a switchboard operator.

switched, a. and ppl. a. Add: **2.** Also, of an egg. rare.
1931 A. J. Cronin Hatter's Castle III. ix. 602 I'm to have a switched egg.
3. a. Of a mechanism: turned on or off by a switch. Also fig.
1962 Listener 28 June 1131/1 His characters were understandably so permanently switched on that their moments of crisis were brought about by the small talk of others. 1968 P. Marlowe Hire me a Hearse vii. 101 He hung up the picture with the switched-off mike. 1968 Listener 25 July 127/2 The screen of a switched-on television. 1974 J. Wainwright Evidence I shall Give i. 9 The switched-on light would emphasise the possible importance of..the one room..with the light still burning. 1977 Times 15 Aug. 2/1 Switched-off heating in every unoccupied room.
b. switched-on: aware of all that is considered fashionable and up to date. Cf. turned on s.v. *TURNED ppl. a. 8. Less frequently, in contrast, switched-off. slang.
1964 House & Garden Nov. 78/2, I..want..to open a department store which caters for switched-on people. 1966 Punch 29 June 946/1 But nowhere have I come across a word of guidance for the 'out' crowd—the vast, non-swinging, switched off, palateless, utterly without-it lot who dominate the community. 1967 N. Fitzgerald Affairs of Death viii. 141 They must be more switched on than I gave them credit for being. 1970 D. Devine Illegal Tender ii. 25 Her mother wasn't switched on, she knew nothing of modern fashion. 1972 C. Weston Poor, Poor Ophelia xxv. 153 The fine beautiful free life. The switched-on scene. 1979 'A. Hailey' Overload IV. iii. 302 She had delivered the tapes to that switched-on black woman who worked for a newspaper. 1982 London Rev. Bks. IV. xxiv. 7/2 What Amis's sprezzatura is saying is that most of his readers are out of touch, old fogies, Prufrock retreads, switched-off.
4. a. and ppl. a. Electr. Having a switch; obtained by switching; subjected to switching.
1961 IBM Jrnl. Res. & Devel. V. 93 A phase reversal data transmission system is described, capable of operating at 2000 bauds over private telephone lines and at 1200 bauds over switched networks. 1971 Gloss. Electrotechnical, Power Terms (B.S.I.) III. vi. 12 Switched beam direction finder. 1971 Physics Bull. Oct. 612/3 The instrument can be used for checking all types of digital logic... Three switched frequency ranges are standard (0–50 kHz, 0–500 kHz, 0–5 MHz). 1974 Harvey & Bohlman Stereo F.M. Radio Handbk. v. 106 (caption) Mullard switched decoder.

switchel. a. For U.S. read N. Amer. and add earlier and further examples.
1790 P. Freneau in Daily Advertiser (N.Y.) 22 Mar. 3/1 Not wretched switchel and vile hogo drams. 1843 Family Herald 29 July 183/1 The drinks ain't good here;..no white noses..switchel-flip..or nothin', but that heavy, stupid, black, fat porter. 1925 Dialect Notes V. 344 [Newfoundland] Switchel, a drink of water and molasses. 1959 W. R. Bird These are Maritimes vii. 190 She kept the jug in the cellar and boy when you came in and had a mug of that switchel it was worth while. 1977 New

Hampshire Times 27 July VII. 20/3 Switchel was a concoction of cold water, sugar, ginger and vinegar, and..it was more or less thirst-quenching.
b. Newfoundland. (A drink of) tea, esp. amongst fishermen and sealers (see quots.).
1897 Jrnl. Amer. Folklore X. 211 Switchel, a mug of weak tea given to the sailors between meals when at the seal fishing. 1924 G. A. England Vikings of Ice 50 Some were devouring beans and salt meat; others, gulping tea that steamed. 'Switchel', this tea was; that is, boiled-over tea whereto now and again fresh leaves are added. 1963 Amer. Speech XXXVIII. 300 Switchel, cold tea. 1974 National Geographic Jan. 114/2 [We] snugged down in the cabin for a 'cup o' switchel', as they call strong tea.

switcher. Add: **d.** One who changes or transfers something to another position; a person who exchanges items, or substitutes one for the other. slang and colloq. (orig. U.S.).
1914 [see *FLOPPER 2]. 1958 Wall St. Jrnl. 3 Nov. 1/1 Almost a third of the voters who plan to vote mostly for Republican candidates tomorrow are recent 'switchers', who until a few days ago either had planned to vote Democratic or were still undecided. 1978 Economist 1 Apr. 26/1 Singapore, South Korea and the Philippines as well as eastern European countries are said to be among the more determined switchers out of dollars.

switcheroo (switʃĕrū·). colloq. (chiefly U.S.). [f. Switch sb. + *-eroo.] = *SWITCH sb. 7*a, c; a change of position or an exchange, esp. one intended to surprise or deceive; a reversal or turn-about; spec. an unexpected change or 'twist' in a story. Also attrib., reversible, reversed.
1933 Forum Dec. 372/2 We'll pull a switcheroo. We'll use olives instead [of cherries]. 1941 B. Schulberg What makes Sammy Run? iv. 81 All you gotta do to that story is to give it the switcheroo. Instead of the minister you got a young dame missionary, see. 1949 Sun (Baltimore) 22 Sept. 7 (Advt.), Girls' 'switcheroo' jacket. One side's red or green corduroy and..the other side's a gay..wool plaid. 1953 C. M. Kornbluth Syndic v. 52 Two strapping girls..began to tear his clothes off, laughing at their switcheroo on the year's big gag. 1961 N.Y. Times Bk. Rev. 21 May 6/3 In Chapter X, then, with a neat whodunit switcheroo, Radin puts the finger on that most obvious suspect who, it appears, was cleared in too much haste. 1970 'A. Gilbert' Death wears Mask vi. 102, I ought to have suggested it was a switcheroo. You know—criminal makes the discovery and informs the police. 1980 Fortune (Chicago) 7 Apr. 44/3 The arbitrator..turned out to be Daniel Collins,..who had upheld Equity in the 1976 row... Collins this time came down on the side of management. It was one of those great switcheroo endings.

switchfoot (swi·tʃfut). Surfing. Pl. -foots. [f. *SWITCH v. 7 + Foot sb.] (See quot. 1970.) Also switch-footer.
1970 Studies in English (Univ. of Cape Town) I. 30 The few surfers who are able to ride with equal skill with either right or left foot forward, are known as switchfoots. 1971 Ibid. II. 26 Finally, surfers differentiate among themselves between naturals,..goofies..and switch-footers, who can switch stance. 1978 G. Wright Illustr. Handbk. Sporting Terms 147/4 Switchfoot, a surfer who can ride with either his left foot or right foot forward.

switching, vbl. sb. Add: **4. b.** Changing or transferring from one position to another; exchanging.
1904 'No. 1500' Life in Sing Sing 253/1 Switching, transferring; passing to another. 1957 Publ. Amer. Dial. Soc. 1956 XXVI. 40 Precision would thus require us to distinguish three stages in diffusion: (1) switching, the alternate use of two languages, [etc.].
c. Stock Exchange. The purchase (or sale) of one stock, and the sale (or purchase) of another stock, at a stipulated price difference.
1932 Literary Digest 30 Jan. 49/1 (title) 'Switching' in a bear market. 1936 Economist 1 Feb. 248/2 The available evidence suggests that 'switching' has not greatly affected the past year's results. Some trusts..have increased their American holdings. 1960 Ibid. 15 Oct. 288/3 Buying in the gilt edged market increased..and demand from both home and continental buyers, including some switching and investment buying, remained high. 1981 Times 18 Aug. 18 The shares managed a 16p rise..with heavy switching from the ordinary into the NV.
5. (b) switching yard = switchyard (a) s.v. *SWITCH sb. 8; (c) Also, pertaining to the switching of electrical apparatus or electronic devices, as switching centre, circuit, speed, station, theory, time.
1894 Daily Ardmoreite (Ardmore, Okla.) 28 Mar. 1/8 There came very near being a disastrous collision..in the upper switching yards of the Santa Fe. 1939 H. J. Reich Theory & Applic. Electron Tubes xii. 459 Two high-vacuum amplifier tubes..are alternately overbiased by the voltage drop through the anode resistors of the switching circuit. 1959 J. M. Pettit Electronic Switching, Timing, & Pulse Circuits iii. 73 The switching speed of a triode in ordinary circuits is impaired by capacitances rather than by transit time of electrons in the tube. 1960 McGraw-Hill Encycl. Sci. & Technol. XIII. 357/1 The bulk of switching theory is concerned with circuits made of binary (two-valued) devices, since these are most common. 1960 R. S. Ledley Digital Computer & Control Engin. xxi. 697 The value of R_s can be estimated by means of considerations concerning the switching time of the core, the time τ required to switch or flip a core. 1962 B.B.C. Handbk. 113 The EBU is responsible for the co-ordination of the programme, legal, and technical aspects

of Eurovision, and operates the switching centre in Brussels. 1968 Switching station [see *FEEDER 10 a]. 1969 Jane's Freight Containers 1968–69 116/3 Progress has been made in reducing the time cars spend in switching yards. 1973 Times 30 Oct. 1/2 Engineers who man power stations and switching centres will refuse to turn out if a breakdown or other difficulty arises while they are off duty. 1977 Sci. Amer. Sept. 212/3 Switching theory, which was developed to help design the relay-operated switching networks of automatic telephone systems, provided guides that enabled a designer to formulate a network with the minimum number of relays for accomplishing some given logical operation. 1978 Ibid. Mar. 61/2 (caption) Supervisory and switching circuits in the central office connect the two sets for the conversation and disconnect them when the call is over.

swi·tch-off. [f. vbl. phr. to switch off: see Switch v. 8.] The turning off of an (electrical) power supply, television set, etc., by means of a switch. Also attrib. and transf.
1947 Times 11 Feb. 2/3 Liverpool electricity undertaking..reduced its load by about 55 per cent, during both the morning and afternoon switch-off periods. 1966 Listener 7 July 9/2 Television is continually at the mercy either of switch-off or switch-over. 1974 Times 15 Jan. 14/3 A mass switch-off of electrical appliances. 1978 Nature 5 Jan. 10/3 This suggests that the switch-off of interferon production is due to cessation of mRNA synthesis as well as to its inactivation. 1980 Daily Tel. 10 Mar. 13/2 The home viewer who cannot cope with BBC 2 or the switch-off button is going to feel swamped by the Olympics.

swi·tch-on. [f. vbl. phr. to switch on: see Switch v. 8.] The switching on of an (electrical) power supply, light, etc. Also transf.
1950 Sun (Baltimore) 4 Mar. 2/7 A new electric 'pick-proof' motor car lock which..permits switchon, and starts with a single key, is being manufactured in Stockholm, Sweden. 1976 Sunday Mail (Glasgow) 28 Nov. 11/6 Councillor Brian Meek..will attend the district council's tree switch-on. 1978 Nature 5 Jan. 10/3 Thus the switch-on of interferon synthesis requires new transcription.

swi·tch-over. [f. vbl. phr. to switch over: see *SWITCH v. 8 c.] A switch or change from one state or course to another; a change-over.
1928 Daily Express 12 Nov. 10/2 The opening left by America's switch-over to the 'talkies' can be brilliantly exploited. 1937 Essays & Stud. XXII. 148 The cross-currents, switch-overs, throw-backs, and quasi-automatic tags of The Waste Land. 1941 'R. West' Black Lamb & Grey Falcon I. 159 He believes that any moment the whole process of life may make a slight switch-over and that every thing will be agreeable for ever. 1952 S. Kauffmann Philanderer (1953) iv. 66 So six months after we start, we've got the name of the book banged into the public's head, and maybe..we've held on to a lot of the confession readers in the switch-over. 1962 Rep. Comm. Broadc. 1960 210 in Parl. Papers 1961–2 (Cmnd. 1753) IX. 259 The method of changing the line standard... It considered the possibility of 'a simultaneous change-over throughout the country'—that is, a 'switchover'. 1979 Financial Rev. (Sydney) 6 July 46/4 Each of the..pulse generators across Australia required manual switchover in the event of failure.

switzerite (swi·tsəraɪt). Min. [f. the name Switzer (see quot. 1967) + -ite[1].] A hydrated phosphate of manganese and iron, $(Mn, Fe)_3(PO_4)_2.4H_2O$, found as pink to brown monoclinic crystals.
1967 Leavens & White in Amer. Mineralogist LII. 1595 The name switzerite is proposed for the mineral in honor of George Switzer, Chairman, Department of Mineral Sciences, Museum of Natural History, Smithsonian Institution. 1978 Rocks & Minerals LIII. 160 A number of new minerals..from the Quarry include switzerite, eakerite and brannockite.

swivel, sb. Add: **4. a.** swivel rocker, -seat.
1916 Blackw. Mag. Oct. 475/1 He walked unsteadily across the room and sat down on a swivel-seat. 1975 Swivel rocker [see saddle brown s.v. *SADDLE sb. 10].
b. swivel eye (earlier example, applied by synecdoche to a dog); **swivel hips** Trampolining, an exercise consisting of a seat drop followed by a half-twist into another seat drop (constr. sing.); also transf.
1765 S. Cibber Let. 3 Oct. in Private Corresp. David Garrick (1831) I. 201, I hope you remember that I have lost poor little swivel-eye, that was blind, and also that you promised me a dog that could see. 1943 L. Griswold Trampoline Tumbling 49 As the legs are swung through the vertical position, the twist is made to right or left and the hips are flexed to assume the sitting position for landing. The movement performed by the hips is called 'swivel hips'. 1948 —— Ibid. (ed. 2) vi. 46 This exercise—popularly called 'swivel hips'—consists of a seat-drop take-off, a half twist, and a seat-drop landing. 1964 Trampolining ('Know the Game' Series) 22/1 The first one [sc. bounce] to practise is the seat bounce with half twist known as the swivel hips. 1966 Rote & Winter Lang. Pro Football III. 141/1 Swivel hips, elusive ball carrier who fakes potential tacklers by shifting hips from side to side. 1980 Sci. Amer. Mar..118/2 An astronaut in space could easily reorient himself in any direction with swivel hips and tuck drops.

swivel, v. Add: **4.** [For an equivalent change of initial shr- to sw-, compare U.S. dial. swimp shrimp.] intr. To shrivel. Also const. up. U.S. dial.

1898 'R. Sanders' Sh. Country Life xxv. 155 Sometimes I think to myself if Christmas didn't come reglar onest a year..this old world would soon swivel and swink up and die out with the dry rots. 1957 W. Faulkner Town (1958) vii. 103 Old hermits setting on rocks out in the hot sun..watching their blood dry up and their legs swivelling.
swivelled ppl. a. (examples in above sense).
1898 'R. Sanders' Sh. Country Life viii. 53, I..filled my pockets full of scalybarks and peanuts and some swivelled up apples of my own raisin. 1938 M. K. Rawlings Yearling xvii. 204 The one we cain't spare was the one was takened... And him a swivveled, no-account thing, too. 1975 E. Wigginton Foxfire 3 258 It'll be a little bitty old swivelled up thing.

swivet (swi·vėt). dial. (chiefly U.S.). Also swivvet, swi(v)vit. [Origin unknown.] A state of agitation; a fluster or panic. Also, a hurry. Freq. in phr. in a swivet.
1892 Dialect Notes I. 232 Swivet (swivit),..'Don't be in such a swivet.' a 1904 in Eng. Dial. Dict. (1904) V. 893/2 What a swivit ee's in. 1913 H. Kephart Our Southern Highlanders xiii. 294 When a man is..in a hurry, he is in a swivvet. 1917 Dialect Notes IV. 418 [N. Carolina] Swivvit, n., hurry. 'He's always in a swivvit.' Also La. 1933 I. S. Cobb Murder Day by Day xvi. 209 And Hilda, so Verity said, was in quite a swivit over the prospect of being interviewed again. 1955 N.Y. Sunday News 27 Mar. 100/1 She does not get in 'swivets' or 'tizzies', either, and she does not often sulk. 1962 M. Carleton Dread Sunset (1963) v. 81 'Don't get into a swivet,' Ellen soothed. 1978 C. Macleod Rest You Merry xxiv. 168 Jemina was always in a swivet about something.

swizz (swiz). slang. Also swiz. [Shortened f. *SWIZZLE sb.[2]] A disappointment or 'swindle.' Freq. in the exclamation 'What a swizz!'
1915 W. Owen Let. 19 Mar. (1967) 328 What a swizz about Harold! 1921 V. Brittain Let. Nov. in Testament of Youth (1933) x. 513 What a swiz for all the people who swore that there was nothing in it between Ramage and Cathleen Nesbitt. 1932 G. Clark Mistress II. v. 186 They want us to go lunch. Just round the corner here... Bit of a swiz, isn't it? I did my best to get out of it. 1937 S. Smith Good Time was had by All 38 The people say that spiritism is a joke and a swizz. 1959 R. Fuller Ruined Boys II. ix. 144 He's given him not out. What a sodding swiz. a 1974 R. Crossman Diaries (1976) II. 208 We were drinking cheerfully when up came that phrase Quintin Hogg is always using: 'Really, it's only a swizz.'
Hence as v. trans., to trick by swindling, to subject to disappointment (in quot., pass.).
1961 H. & M. Williams Irregular Verb to Love in J. C. Trewin Plays of Year XXIII. 84, I..felt I'd been swizzed —not just of sex though that was part of it.

swizzle (swi·z'l), sb.[2] slang (chiefly School-children's). [Prob. altered f. Swindle sb.[2]] = *SWIZZ.
1913 A. H. Dawson Dict. Eng. Slang & Colloquialisms 139 Swizzle. (1) Any sort of drink. (2) A swindle, fraud. Also a verb in both senses. 1931 C. Mackenzie Buttercups & Daisies v. 59 'What a swizzle you can't eat rats,' Roger sighed. Ibid. xviii. 229 What a swizzle it's so late. 1950 A. Buckeridge Jennings goes to School i. 12 It was a rotten swizzle, sir, because we flew through low cloud and we couldn't see a thing. a 1976 A. Christie Autobiogr. (1977) IX. v. 476 This place is awful, Mother... As for those bathrooms,..it's an absolute swizzle! They're never used.

swizzle, v. Add: Hence **swi·zzled** ppl. a., drunk, 'sozzled'; influenced or induced by heavy drinking.
1843 Knickerbocker XXII. 366 We were never 'groggy', ..'swizzled' or 'tight', but once. 1888 Texas Siftings 14 Jan. 8/2 Old Shep, with a swizzled intuition, would darkly imagine that the singers were alluding to his calcium nose. 1918 G. Frankau One of Them ix. 65 Some quaff th'embittered cocktail, or the rum Whose swizzled headaches heavy on to-morrow weigh. 1934 Amer. Spectator July 2/3 The editors of The American Spectator got somewhat swizzled one night last week and didn't feel so good the next day.

swi·zzler. dial. and slang. Now rare. [f. Swizzle v. + -er[1].] **a.** A drunkard. **b.** A swindler.
1876 F. K. Robinson Gloss. Whitby 192/1 Swizzler, a drunkard. 1936 'N. Bell' Crocus ix. 235 Oh, he didn't diddle me... I knew him for a swizzler from the word go. 1938 J. W. Day Dog in Sport xvii. 233 Lights gleamed in a building ashore. The Cockney swizzlers were still at it.

swizzle-stick. (In Dict. s.v. Swizzle sb. b.) Add: **1.** (Earlier and later examples.) Also, a rod used to stir a mixed drink, or to flatten the effervescence of a cocktail, etc.
1879 J. W. Boddam-Whetham Roraima & Brit. Guiana xii. 129 The revolutions of a peculiar instrument called the swizzle-stick. 1899 C. J. C. Hyne Further Adventures Capt. Kettle v. 135 Shout for your boy to bring the cocktail... Where's the swizzle-stick? 1951 N.Y. Herald-Tribune 9 Mar. 18/3 Under Otto Preminger's direction it is all as frothy and inevitable as the action of a swizzle stick in a champagne glass. 1964 Wodehouse Frozen Assets ii. 36 The way the mere sound of her voice got inside one and stirred one up as with a swizzle-stick. 1976 J. I. M. Stewart Young Pattullo iii. 69, I was being officiously counselled not to commit the solecism of using a swizzle-stick too soon.
2. transf.
1962 J. Glenn in Into Orbit 44 A simple little rod... It is ten inches long, has a hook on the end of it for pulling

at levers and a stub for pushing at buttons. You grasp it in your glove if you know you are not going to be able to reach something with your fingers.. We call it, naturally, a 'swizzle stick'. 1977 Lancet 16 Apr. 836/2 The swizzlestick consists of a handle fastened to a small stainless steel circular platform to which is also fastened a stainless steel displacement probe having a diameter slightly less than the bore of a capillary tube and a volume equal to the volume of blood required for the assay.
3. Comb., as **swizzle-stick tree,** a small aromatic evergreen tree, Quararibea turbinata, of the family Bombacaceæ, found in the W. Indies and tropical South America.
1943 Record & Hess Timbers New World 98/1 All specimens of Quararibea without distinct heartwood... Common names: Garrocho, swizzle-stick tree. 1951 E. Mittelholzer Shadows move among Them I. xvi. 153 Wild cacao and swizzle-stick trees and ferns..grew out of reddish sand.

swollen, ppl. a. Add: **2. c.** swollen head: excessive pride, or a person suffering from it; also, a hangover. colloq. Cf. Swelled ppl. a. b.
1899 N. Gould Landed at Last vi. 59 You have got a swollen head this morning... Had too much to drink last night. 1922 F. Hamilton P.J., Secret Service Boy vi. 242 You don't strike me, somehow, as being liable to swollen head. 1928 Daily Express 23 July 9 British film-producers..are..annoyed with me for saying that their swollen-headed outlook was the root-trouble... I feel certain that the swollen heads will bring about many crashes in British film-production.
3. a. swollen-eyed, -headed adjs.
1930 E. Pound XXX Cantos xv. 67 'Ἠέλιον τ' Ἠέλιον' blind with the sunlight, Swollen-eyed, rested. 1977 N. Sahgal Situation in New Delhi xvi. 153 The girl raised her head, swollen-eyed. 1928 Swollen-headed [see sense 2 c above]. 1983 D. Francis Danger i. 13 Chasing personal glory. Stupid, swollen-headed, lethal human failing.
b. Special combination. **swollen shoot,** a virus disease of cocoa trees, spread by mealy bugs and distinguished by swelling of the young shoots, leading to the death of infected trees.
1936 W. F. Steven in Gold Coast Farmer V. 144/1 A new disease of cocoa..has provisionally been named 'Swollen Shoot and Die-back'. 1950 Times 2 Feb. 9/2 We shall test our proposed method for the prevention of swollen shoot disease of cocoa in the Gold Coast, using systemic insecticides. 1972 P. F. Entwistle Pests of Cocoa x. 170 At the time of discovery of swollen shoot disease planting in the Eastern Region was still expanding.

swollenness (swōᵘ·l'n,nės). [f. prec. + -ness.] The state or appearance of being swollen.
1902 E. Selous Thought Transference in Birds (1931) 14 In a very little while..this swollenness subsides and there is the same average appearance of the birds.

swooner (swū·nəɪ). [f. Swoon v. + -er[1].]
1. One who swoons or faints, or pretends to do so.
1911 K. D. Wiggin Mother Carey xxx. 263 Nancy had secretly trained Peter so that he was the best swooner of the family. 1951 M. McLuhan Mech. Bride (1967) 28/2 There's no need to reassure the swooner that Lana is human flesh and blood. 1966 New Statesman 19 Aug. 269/3 Olga Ferri accepted the chance to appear more of a queen, less of a lyrical swooner.
2. U.S. One who sings in a manner which resembles crooning. Also (nonce-wd.) swooner-crooner.
1944 Amer. Speech XIX. 102/1 Swooner-crooner is a characterizing term in current use among journalists and humorists for the singer Frank Sinatra. It made its advent in late 1943. 1952 B. Ulanov Hist. Jazz in Amer. xxi. 268 His voice..never falls into the whispering faint that makes listening to the swooners and crooners so disturbing.

swoony, a. Add: **1.** (Example.)
1978 M. Dickens Open Book (1980) ii. 26 For one of her heroes, in the style she originated as a swoony girl in Dulwich, Fanny wrote: [etc.].
2. Inducing a swoon; hence, distractingly attractive, delightful. colloq.
1934 in Webster. 1960 Wentworth & Flexner Dict. Amer. Slang 534/2 Swoony,..adj., attractive. Teenage use, c1940. More often in movies and stories about teenagers than used by teenagers. 1973 T. Pynchon Gravity's Rainbow I. 57 Those eyes she could never quite see into were so swoony. 1974 'R. Tate' Birds of Bloodied Feather ii. 26 Champers and strawberries and Ronald and swoony lanes on the way back. 1976 P. Flower Crisscross i. 11 Their kiss was long and deep and swoony.
Hence **swoo·niness,** a quality suggestive of a swoon.
1909 R. Bridges in R. W. Dixon Poems p. xxx, The faintness and swooniness is in some sort akin to the remoteness and misty atmosphere of antiquity.

swooping, ppl. a. Add: **2. b.** Of a surface: sloping sharply or steeply.
1956 G. Durrell My Family & Other Animals viii. 103 One of the first to arrive was Zatopec, an Armenian poet, a short, stocky individual with a swooping eagle nose. 1979 Jrnl. R. Soc. Arts Nov. 744/1 The gently upturned, swooping roofs.

swoose (swūs). [Blend of Swan sb. and Goose sb.] A bird that is the offspring of a swan and a goose. Also transf.
1920 Daily Mail 13 July 7/5 A bird prodigy of evil and hybrid character is the despair of a Norfolk farmer. It rejoices in the name of the 'swoose', a portmanteau word indicating its origin, for its father was a swan and its mother a goose. This ill-assorted pair had three children —three 'sweese'. 1927 Daily Express 12 July 8/4 The swoose is a cross between the goose and the swan. 1954 Sun (Baltimore) 27 Apr. 1/7 A Democratic swan..who fell in love with a common farmyard goose today became the father of a 'swoose'. 1964 Sunday Mail (Brisbane) 24 May 29/2 By salvaging parts from the damaged planes, airmen resurrected a few bombers like the Swoose. The name Swoose, for part-swan, part-goose, reflects the plane's patchwork rebirth. 1976 Sydney Sun 20 Aug. 16/2 Like the 'swoose' (a cross between a swan and a goose) the 'churkey' is a mythical bird.

swop (swǫp), informal contraction of 'so help': see S'elp, Swelp.
1890 P. H. Emerson Wild Life 46 Swop my bob. 1912 W. Deeping Sincerity xxviii. 214 Swop me bob, somebody else will be callin' for the police.

sword, sb. Add: **1. e.** pl. One of the four suits in packs of playing-cards used in Italy and in Spanish-speaking countries, and in tarot packs. Cf. Spade sb.[2]
1816 G. W. Singer Researches into Hist. Playing Cards I. 17 The four suits, Spade, (swords,) Coppe, (cups,) Denari, (money,) and Bastone, (clubs,) adopted both by the Italians and Spaniards, were probably the suits of the Eastern game. 1848 W. A. Chatto Facts & Speculations on Origin & Hist. Playing Cards iv. 191 The earliest writers who mention Tarocchi as a kind of cards, always speak of them as consisting of four suits,—Swords, Cups, Batons, and Money. Ibid. 227 The cards most commonly used in Italy in the latter part of the fifteenth century, were those which had..Swords, Cups, Batons, and Money, —as the marks of the suits. 1892 'Papus' Tarot of Bohemians v. 44 When we consider the four colours of the Tarot, new deductions will be called forth... The Sword represents the union of the two by its crucial form. 1911 A. E. Waite Pictorial Key to Tarot I. iv. 36 We must forbear from saying, for example, that the Conditions of Life correspond to the Trumps Major..and the conditions of life to Swords. 1934 J. D. Carr (title) The eight of swords. 1952 V. Wilkins King Reluctant III. iii. 230 He..produced two [tarot] cards... 'The King of Swords and the King of Cups!' he said. 1978 Jrnl. Playing-Card Soc. Feb. 90 It comprises 52 cards, with suits of Swords, Batons, Cups and Pomegranates.
2. b. at the sword's point: under pressure of a threat or an urgent demand; at swords' points: in a state of open hostility. Cf. Dagger sb. 2.
1895 A. Beardsley Let. Nov. (1970) 104 The dreadful thing was a blaze up with Lane-cum-Mathews, and a drawing to be produced at the sword's point. 1909 Webster, At swords' points. 1963 M. McCarthy Group x. 214 Mrs Hartshorn and her dead husband had had a running battle over Wilson and the League, and now Priss and Sloan were at swords' points over Roosevelt and socialized medicine.
3. c. Contrasted with ploughshare (in allusion to Isaiah ii. 4 and Micah iv. 3), as types respectively of war and peace: see Ploughshare 1. Esp. in phr. to beat swords into ploughshares.
1924 L. P. Smith S.P.E. Tract XVII. 38 We must take them [sc. words] as they come to our hands; if they are ploughshares which have been beaten into swords, tools which have been made into battle-axes, they are tools nevertheless for which we have no substitutes. 1976 N. Thornburg Cutter & Bone x. 242 You know the old phrase about beating swords into plowshares—well I think you've beaten your grief into a sword.
5. a. (Further examples.)
1942 R. Davey Measurement of Trees ii. 28 When a tree lies on the ground, there may be some difficulty in passing the tape beneath it. A flat piece of metal with a hook at one end, called a 'timber sword' may be used for this purpose. 1953 H. L. Edlin Forester's Handbk. xiv. 213 Find the mid-point of the log, and pass a girthing tape around it; with large logs, the device called the timber-measurer's sword will be of assistance.
6. a. sword-clash, -flash, -hate, -rust. **c.** Objective, sword-rusting adj.; similative, sword-keen adj.
1946 R. Campbell Talking Bronco 45 Amidst the sword-clash of the reeds. 1969 G. M. Brown Orkney Tapestry 74 It was a long stern battle, hurling of missiles and sword-clash. 1874 R. Buchanan Poet. Wks. III. 228 Feeble as a maid who hides her face In terror at a sword-flash. 1912 E. Pound Ripostes 29 Disease or oldness or sword-hate Beats out the breath from doom-gripped body. 1901 Kipling Kim viii. 209 He caught Mahbub's sword-keen glance. 1923 D. H. Lawrence Birds, Beasts & Flowers 54 Think of it, from the iron fastness Suddenly to dare to come out naked, in perfection of blossom, beyond the sword-rust. 1930 T. S. Eliot tr. St.-J. Perse's Anabasis 47 In the mirror of our dreams, the sword-rusting sea.
d. sword-mat (earlier example); **sword-rattling** a. fig., that threatens military action; aggressive, pugnacious; also as sb. = sabre-rattling vbl. sb. s.v. *SABRE sb. 4 a; **sword-work** = Sword-play 1; also fig.
1851 H. Melville Moby Dick II. v. 31 Queequeg and I were mildly employed weaving what is called a sword-mat, for an additional lashing to our boat. 1914 Bulwark June 84/2 No doubt the Germans will know how to make

their Prince acquainted with the unpopularity of his sword-rattling swagger. 1955 Times 12 May 1 (heading) Afghan 'sword rattling'. Ibid., I would like to ask my Afghan friends whether they really think such a sword-rattling and offensive attitude is going to help them. 1978 Guardian Weekly 29 Jan. 6/3 There are 1,500 British troops in Belize.., as the result of Guatemala's sword-rattling last July. 1913 Nation 28 June 484/2 Not only has he [sc. Sir John Simon] shown his greatest skill in this sword-work [etc.]. 1977 P. Scupham Hinterland 58 Boughs come adrift Over the splayed sword-work of spring flowers.
e. sword-bean (earlier example); **sword-fern:** also, Polystichum munitum, native to western North America; (later examples).
1875 Encycl. Brit. III. 460/2 Beans or pulse, of no small importance as articles of diet, such as the..sword bean of India. 1899 E. Cotes Path of Star i. 4 The bunch of sword-ferns..grew beside the door. 1932 J. Steinbeck Pastures of Heaven vi. 126 Swordferns grew rankly under the alders. 1976 Islander (Victoria, B.C.) 4 Jan. 5/1 Most of the park is forest, the damp coastal rain forest of huge sword-ferns and gigantic cedar trees.

swot, swat, v. Add: intr. Also, to 'bone up'. trans. (examples); more rarely, without up.
1901 Chambers's Jrnl. July 445/2 Dick was 'swotting' blue china for all he was worth, at the British Museum and elsewhere. 1913 Wireless World I. 37/2 There will be a chance for fellows like me, who have been swatting up Fleming's books. 1931 R. Campbell Georgiad i. 18 All who..of despair have baulked the yawning precipice By swotting up his melancholy recipes For 'happiness'. 1955 Times 26 May 13/2 Mr. Forester must have 'swotted up' the subject of wartime Atlantic convoys just as he 'swotted up' the subject of the Navy in Nelson's time. 1967 K. Giles Death in Diamonds vi. 114 Been swatting the maps, I see. 1977 N.Y. Rev. Bks. 23 June 8/2 Our culture hound..swots up in the Encyclopedia before distinguished guests arrive.

swotter (swǫ·təɪ). slang. [f. Swot v. + -er[1].] = Swot sb. 2. Also swotter-up.
1919 in Cassell's New Eng. Dict. 1925 Times Lit. Suppl. 26 Mar. 219/1 If we allow contempt to confuse thought, the 'swotter', for all his dullness, will have us on the hip. 1931 R. Campbell Georgiad iii. 62 Swotters-up of philosophic blisses.

swotting (swǫ·tiŋ), vbl. sb. slang. [f. Swot v. + -ing[1].] = Swot sb. 1; hard work at one's studies.
1873 Punch 11 Jan. 19/2 For downright hard 'swotting' there's no place like School. 1959 I. & P. Opie Lore & Lang. Schoolch. x. 179 'Swotting' or 'mugging up' is only considered good form if a person is on the point of taking an exam. 1974 'J. Herriot' Vet in Harness v. 37 He had been blessed with the kind of brain which made swotting irrelevant.

swozzle, var. *SWAZZLE.

swung, ppl. a. Add: **3.** swung dash, a curved dash ~, used in dictionaries to stand for the headword of an entry or for a specified part of it.
In Oxford dictionaries first used in the first edition of The Little Oxford Dictionary (1930) but there called a tilde.
1951 Conc. Oxf. Dict. p. iii, In this edition..the swung dash has been freely employed. 1975 Amer. N. & Q. XIV. 60/1 ER, like most dictionaries, uses a swung dash to denote the entry word.

swy (swaɪ). Austral. slang. Also swi. [ad. G. zwei two.] **1.** Two; spec. a two-shilling coin or a two-year prison sentence.
1924 Truth (Sydney) 27 Apr. 6 Swy, two. 1941 Baker Dict. Austral. Slang 75 Swy, the game of two-up. (2) A sentence of two years' gaol. (3) A florin. 1983 Age (Melbourne) 15 Dec. 13 (caption) Exhibition of used coin of the realm: bank notes, collector's items, swys, deaners, zacs, treys, brass razoos.
2. The game of two-up (see also quot. 1950). Also swy-up.
1940 Bulletin (Sydney) 17 Jan. 34/3 The crown-and-anchor seminary he avoids; When swi-up's on, a different direction He takes. 1950 K. S. Prichard Winged Seeds 63 What set the whole town agog, though, was their attempt to visit the 'swy': the famous two-up ring on a sand hill near the old Rising Sun Inn. 1953 R. Braddon in I. Bevan Sunburnt Country 127 Swy is a game of chance, requiring the tossing of two or three pennies into the air and the betting of those who watch their rise and fall on whether they come down heads or tails. 1969 Courier-Mail (Brisbane) 24 Sept. 1/1 He said two-up (or swy) was Australia's national game. 1976 Sunday Mail (Brisbane) 24 Oct. 16/4 The police know they will never stop gold-fielders playing swy.
3. Comb., as **swy game,** a game of two-up; **swy school,** a group of persons who have gathered to play two-up.
1950 Austral. Police Jrnl. Apr. 118 Swi, 2s., but a swi-game is a two-up game. 1953 K. Tennant Joyful Condemned xxix. 284 There's all these little crims in the swi-game and the S.P. betting. 1969 Telegraph (Brisbane) 14 May 8/4 Otherwise they blow it at the pub, or at the swy game. 1921 Aussie 15 Mar. 54 Just done me last dollar up at the swi school. 1956 S. Hope Diggers' Paradise 59 Neither shalt thou play two-up for lucre in the street, nor attend such swy schools in any private or public premises.

swyneyed, var. *SWEENIED a.

-sy. Add: In adjectival formations expressing a degree of mocking contempt, as *artsy-and-craftsy, artsy-fartsy, backwoodsy, bitsy, booksy, folksy, itsy-bitsy, teensy*, etc., the suffix may be considered to represent a nursery form (cf. -Y[4]), or the *pl.* (or even a singular ending) in -*s* + -Y[1].

sycon (saiˑkǫn). *Bot.* [a. Gr. σῦκον fig.] † **1.** = SYCONIUM. *Obs.*

1845 *Encycl. Metrop.* VII. 51/1 A sycon is a fleshy, concave receptacle surrounding the fruits.., which are numerous, small, and distinct. 1900 B. D. JACKSON *Gloss. Bot. Terms* 262/1 Sycon..a multiple hollow fruit, as that of the fig.

2. [Adopted as a generic name by A. Risso, *Hist. Nat. Europe Méridionale* (1826) V. 368.] A calcareous sponge of the genus of this name; also, a stage in the development of sponges in which flagellated chambers are developed and lined with choanocytes.

1882 W. J. SOLLAS in P. M. Duncan *Cassell's Nat. Hist.* VI. 326 A transitional series of species can be shown to exist between a simple Ascon and a Sycon in which radiate buds have all united..to form a complex tubulated wall. 1912 *Phil. Trans. R. Soc.* B. CCII. 170 The normal young Sycon has a beautiful double ascular crown of long monaxons. 1932 BORRADAILE & POTTS *Invertebrata* iii. 113 The three grades of sponge structure ..are known as the 'Ascon', 'Sycon', and 'Leucon' grades.

sycophantish, *a.* (Earlier example.)

1821 R. LEE *Diary* 20 Oct. (1897) 26 Mr. L. said that although he admired Sir W. Scott's talents, still there was something about him which he did not like—a sneaking, flattering, sycophantish manner.

Sydama, var. *SIDAMO.

Sydenham (siˑdĕnăm). *Path.* [The name of Thomas *Sydenham* (1624–89), English physician, who described the chorea in *Schedula Monitoria de Novæ Febris Ingressu* (1686).] *Sydenham's chorea*: a self-limited disorder of childhood or pregnancy that is a neurological manifestation of rheumatic fever, affecting the motor activities of the nervous system and characterized by involuntary movements.

1892 *Med. Record* (N.Y.) XLI. 285/2 There are many cases of Sydenham's chorea in which voluntary effort arrests the movements. 1954 *Handbk. for Mental Nurses* (ed. 8) vi. 162 Sydenham's Chorea. This, also known as St. Vitus's Dance, is..much commoner in girls than in boys. 1976 SMYTHIES & CORBETT *Psychiatry* vii. 128 Sydenham's chorea occurs in younger people, there is no family history, no dementia and the course is not progressive.

Sydnæan, var. *SIDNEIAN *a.*

Sydney (siˑdni). [The name of the capital city of New South Wales.] **1.** In *Austral.* colloq. phr. *Sydney or the bush*, all or nothing. Cf. BUSH *sb.*[1] 9.

1924 *Truth* (Sydney) 27 Apr. 6 *Sydney or the bush*, all or nothing. 1930 E. SHANN *Econ. Hist. Austral.* 365 'Sydney or the bush!' cries the Australian when he gambles against odds. 1970 R. BEILBY *No Medals for Aphrodite* 34 'Here we go,' Turk murmured grimly, climbing in behind the wheel. 'It's Sydney or the bush! Keep your fingers crossed.'

2. Special combinations. **a. Sydney-side** [SIDE *sb.*[1] 15 b], Sydney and the surrounding area; also as *adj.*; **Sydneysider,** a resident or native of Sydney or of New South Wales.

[1872 W. M. HUGO *Hist. First Bushmen's Club in Austral. Colonies* 108 Very frequently, however, they are not allowed to proceed so far as the city, but get 'bailed up', as they call it on the Sydney side, before they reach their destination.] 1888 'R. BOLDREWOOD' *Robbery under Arms* I. i. 1 My name's Dick Marston, Sydney-side native. 1928 'BRENT OF BIN BIN' *Up Country* v. 67 She was..supposed to be..a descendant..of the famous Sydney-side sire 'Clifton'. 1941 BAKER *Dict. Austral. Slang* 75 *Sydneyside*, originally the area which is now N.S.W. Later, especially the area of Sydney. 1865 H. KINGSLEY *Hillyars & Burtons* III. xiii. 144 The Sydney-siders' loss is considered by him to have been far greater. 1931 *Times Lit. Suppl.* 1 Oct. 738/1 A fear of its [*sc.* Melbourne's] writers..echoed the nationalist emotion, but in abstract terms that lacked the appeal of the Sydney-siders. 1980 N. MARSH *Photo-Finish* vii. 199 He was a self-made man, a Sydneysider.

b. *Nat. Hist.* In the names of plants or animals associated with the region, as **Sydney blue gum,** a flooded gum, *Eucalyptus saligna*; **Sydney golden wattle,** a shrub or small tree, *Acacia longifolia*; **Sydney silky** (also **silkie**), a small stocky terrier of the breed so called, with long, silky, grey-blue fur and tan markings.

1932 R. H. ANDERSON *Trees N.S.W.* v. 101 Sydney Blue Gum... A tall-growing, shaft-like species. 1933, 1965 Sydney blue gum [see *FLOODED *ppl. a.*]. 1909 A. E. MACK *Bush Calendar* 20 Flowers blooming [in September]. *Acacia longifolia.* Sydney golden wattle. 1976 *Hortus Third* (L. H. Bailey Hortorium) 6/1 Sydney golden wattle..flower heads in loose spikes. 1945 Sydney silky [see *AUSTRALIAN *sb.* 3]. 1965 *Austral. Encycl.* III. 265/2 The Sydney silky, classed as a toy dog,..has a coat of steel-blue, silky hair up to 6 inches long, with a tan face, legs and points. 1977 *N.Z. Herald* 8 Jan. 4-9/7 (Advt.), Sydney silkie dog pups, 6 wks old.

Sydnian, var. *SIDNEIAN *a.*

Syenna, obs. var. *SIENA.

syenodiorite (saiˌēnodaiˑǒrait). *Petrogr.* [f. *syeno-*, comb. form of SYENITE + DIORITE.] A plutonic rock of a kind intermediate between syenite and diorite, containing both alkali feldspar and plagioclase.

1917 A. JOHANNSEN in *Jrnl. Geol.* XXV. 89 Syenodiorite, syenogabbro, and granogabbro are introduced as new terms. 1940 *Bull. Geol. Soc. Amer.* LI. 1592 The laccolith near the abandoned wax factory on Fresno Creek..is composed of a striking augite syenodiorite. 1977 A. HALLAM *Planet Earth* 162/1 Color index rises to about 50 in the gabbros and is between 20 and 50 in the diorites and syenodiorites.

Sykes (saiks). The name of William Henry *Sykes* (1790–1872), English soldier and naturalist, used in the possessive in **Sykes'(s) monkey,** to designate *Cercopithecus albogularis*, a blue-grey guenon native to East Africa.

[1831 *Proc. Zool. Soc.* 105 Major Sykes subsequently called the attention of the Committee to a Monkey presented by him to the Society.] 1864 *Ibid.* 709 Sykes's Monkey. 1905 [see MONKEY *sb.* 1 b]. 1914 R. C. F. MAUGHAM *Wild Game in Zambezia* xi. 252 Sykes' Monkey is a comparative rarity. 1932 S. ZUCKERMAN *Soc. Life Monkeys & Apes* xi. 185 Loveridge found that Sykes' monkeys were plentiful at Morogoro in East Africa. 1963 A. SMITH *Throw out Two Hands* xiii. 132 We had initially been concentrating on a group of Sykes's monkeys.

syllabic, *a.* and *sb.* Add: **A.** *adj.* **1. c.** (Earlier examples.)

1804 J. BARROW *Trav. in China* vi. 270 [The Manchu writing-system] is alphabetic, or, more properly speaking, syllabic. 1838 P. DU PONCEAU *Chinese System of Writing* p. xii, Syllabic alphabets, besides, have considerable advantages over those that we make use of.

d. Of verse or metre: based upon or determined by the number of syllables in a line, etc.

1923 L. ABERCROMBIE in *Times Lit. Suppl.* 12 Apr. 247/1 English metre, according to many theorists, is neither syllabic nor quantitative, but simply accentual. 1965 A. F. SCOTT *Current Lit. Terms* 282 The determining feature of syllabic verse is the number of syllables in the line, not the stress nor the quantity. 1970 G. S. FRASER *Metre, Rhyme & Free Verse* iv. 50 Purely syllabic metrics seems..not suitable to the prosody of English as a natural language.

B. *sb.* **4.** *pl.* Syllabic verse.

1964 *Times Lit. Suppl.* 16 Jan. 53/4 Syllabics are as legitimate a metrical device as any other. *Ibid.*, Syllabics accommodate speech rhythms... MacBeth and..B.S. Johnson, independently discovered this quality of syllabic metre a few years ago. 1977 *Ibid.* 8 Apr. 428/2 The line in Bridges's use of neo-Miltonic syllabics is fundamentally of twelve syllables.

syllabicate, *v.* Delete *rare*[−0] and add examples. Also *intr.*, to form or construct syllables; to divide a word or passage into syllables.

1654 J. BROOKSBANK *Plain, Brief, Rules for Syllabication Eng. Words* 27 To Syllabicate, which is to find out a word by its syllables. 1831 J. BOADEN *Life of Mrs. Jordan* II. xx. 178 He did not syllabicate, his notion of a word was often caught from vulgar speakers. 1902 H. BRADLEY *Let.* 9 Jan. in *Corresp. Bridges & Bradley* (1940) 9 If the Greeks syllabicated like this..a syllable ending in *one* or more consonants is long. 1971 *Language* XLVII. 138 *Perpetual* is syllabicated as *per. pety. u. al.*

syllabication. Add: (Later examples.)

Tending to give way to *syllabification*.–Ed.

1933 L. BLOOMFIELD *Language* vii. 121 The ups and downs of *syllabication* play an important part in the phonetic structure of all languages. 1971 *Language* XLVII. 138 The rule for the devoicing of liquids follows syllabication.

syllabicity (silăbiˑsĭti). [f. SYLLABIC *a.* + -ITY.] = SYLLABICNESS.

1933 L. BLOOMFIELD *Language* viii. 130 Syllabicity determined also by manner of articulation. 1944 L. M. HARTMAN in *Language* XX. 33 One of these [morphophonemic changes] is the loss of syllabicity either by this or by the preceding syllable. 1952 A. COHEN *Phonemes of Eng.* iii. 62 There does not seem to be any need for assuming a special phoneme of syllabicity. 1968 F. G. LOUNSBURY in J. A. Fishman *Readings Sociol. of Lang.* 53 What we have accomplished is to suppress from our transcription the representation of features of the acoustic stimulus (voicing, syllabicity, laryngeal order, and position of the accent) which do not serve as cues for differential responses on the part of the native subjects. 1977 *Archivum Linguisticum* VIII. 87 There is no experimental evidence for, and some experimental evidence against, the necessary presence of such pulses as physiological correlates of syllabicity.

syllabification. (Later examples.)

1972 *Webster's New World Dict.* (Delux Color ed.) p. x, The syllabifications used in this dictionary are in the main those in general use by printers since the 18th century. 1977 *Archivum Linguisticum* VIII. 87 Such questions, he states, are now 'reduced to practical matters of articulatory adjustment in particular languages'..—which would seem to imply that syllabification rules are part of particular phonologies. 1979 *Collins Eng. Dict.* p. x, Syllabification breaks are shown for all headwords. 1980 *Verbatim* Spring 19/1 Lexicography is not simply adding one good point to another to make an ideal dictionary, but balancing the saving of space against fullness of information, the amount of information against cost, a more exact pronunciation guide against added difficulty for some users, the addition of extra information (pronunciation or syllabification) in the headword at the expense of its clean appearance.

syllaˑbify, *v.* Delete *rare*[−0] and add examples. Also *intr.*

1926 [see *SYLLABIZATION]. 1954 F. G. CASSIDY *Robertson's Devel. Mod. Eng.* (ed. 2) xii. 381 Though the American may syllabify more fully where the Englishman elides, he also slurs more. 1972 *Language* XLVIII. 357 If one assumes that ø is the alternative to syllabifying -*s*, then one can establish a graded gamut of markedness among the three alternants.

syllabizaˑtion. [f. SYLLABIZE *v.* + -ATION.] = SYLLABIFICATION.

1926 H. W. FOWLER *Mod. Eng. Usage* 590/2 A verb & a noun are clearly sometimes needed for the notion of dividing words into syllables. The possible pairs seem to be..syllabify..syllabification [etc.]... The best thing would be to accept the most recognized verb *syllabize*, give it the now non-existent noun *syllabization*, [etc.]. 1929 *S.P.E. Tract* XXXIII. 436 Under *syllabize etc.* he [*sc.* Fowler] exposes a want in our vocabulary, which perhaps indicates a general lack of interest in *syllabization*. 1976 *Archivum Linguisticum* VII. 181 In Adrados's explanation, these forms without colouring of the vowel occur in a different syllabization.

syllabize, *v.* Add: **1.** Also **syˑllabized** *ppl. a.*

1957 A. ORAS in N. Frye *Sound & Poetry* 112 Milton's growing dislike of syllabized -*ed* endings. 1969 *Computers & Humanities* III. 257 The latter is based upon successive scanning of the syllabized text in groups of four, three, two and one characters.

syllable, *sb.* Add: **1. c.** Colloq. phr. *in words of one syllable*, in simple language.

1922 F. H. BURNETT *Head of House of Coombe* xvii. 206 The French Revolution..—the cataclysms of agony—need not have been, but they *were*. To put it in words of one syllable. 1941 V. WOOLF *Between Acts* 218 Let's talk in words of one syllable, without larding, stuffing or cant. 1966 'E. LATHEN' *Murder makes Wheels go Round* xxi. 166 'John,' he said breathlessly, 'would you please explain in words of one syllable.' 1970 *Guardian* 9 Mar. 24/1 Why don't they tell us precisely, in words of one syllable, how they would behave if they were in our place?

4. *syllable-count, stress; syllable-counting, -final, -initial* adjs.; **syllable-timed** *a.*, of or having a rhythm in which syllables occur at roughly equivalent time intervals; opp. *stress-timed* adj. s.v. *STRESS *sb.* 10; hence **syllable-timing.**

1969 *Language* XLV. 250 The text itself is composed in syllable-count verse forms. 1983 *Listener* 6 Jan. 21/3 Pop lyrics writers throw in an 'oh yeah' or a 'baby' wherever the syllable-count needs padding out. 1959 *PMLA* LXXIV. 588/2 This has been done on strictly accentual (plus syllable-counting) principles. 1978 *Early Music* Oct. 587/3 He describes in detail three kinds of relationship..between words and music in the period—the metrical..the accentual..the syllable-counting (the characteristic mode of Christian and much courtly poetry). 1964 B. MALMBERG in D. Abercrombie et al. *Daniel Jones* 116 Many languages have an opposition between explosive (syllable-initial) and implosive (syllable-final) consonant. 1978 *Language* LIV. 23 Durand.. points out that the [t] in *petit orage* 'little storm' is syllable-initial, while the [t] in *petite orange* 'little orange' appears to be syllable-final for most speakers. 1924 H. E. PALMER *Gram. Spoken Eng.* I. 6 *Word-stress* (in the opinion of the author the term *syllable-stress* would be more appropriate). This term is used with reference to a syllable. 1964 W. S. ALLEN in D. Abercombie et al. *Daniel Jones* 14 These remarks on English verse are intended only to apply to the 'syllable-stress' metres. 1947 K. L. PIKE *Phonemics* I. ii. 13/1 In English one tends to hear stress-timed rhythm in contra-distinction to a syllable-timed rhythm. In the syllable-timed type the syllables themselves tend to be more or less equally spaced... As a result of the syllable timing the vowels are likely to be clear cut. 1980 *English World-Wide* I. 1. 108 This, as well as the syllable-timed rhythm, gives rise to the staccato impression often noticed by outsiders. 1964 M. A. K. HALLIDAY et al. *Linguistic Sci.* 72 The English type of rhythm is known as 'stress-timing', by contrast with the 'syllable timing' of French.

syllabus. Add: **1. b.** *spec.* a statement of the subjects covered by a course of instruction or by an examination, in a school, college, etc.; a programme of study.

1889 *Rep. Higher Educ. in London* p. ix, in *Parl. Papers* (C. 5709) XXXIX. 323 The colleges having no *locus standi* to make representations to the authorities of the university either as to the settlement or alteration of the 'syllabus' by which the course of the examinations is regulated. 1955 E. BLISHEN *Roaring Boys* III. 117 The history syllabus for the school had been drawn up by Mr Benson. 1972 *Daily Tel.* (Colour Suppl.) 1 Dec. 15 Schools should allow pupils to determine syllabi.

c. *fig.*

1938 AUDEN *Commentary* in *Journey to War* (1939) 290 And the young emerging from the closed parental circle, to whose uncertainty the certain years present their syllabus of limitless anxiety and labour.

syllid (siˑlid), *sb.* and *a.* [ad. mod.L. family name *Syllidæ*, f. generic name *Syllis* (J. B. P. A. de M. de Lamarck *Hist. Nat. Animaux sans Vertèbres* (1818) V. 317) + -ID[3].] **A.** *sb.* A small errant polychæte worm of the family Syllidæ, distinguished by three tentacles on its head and found on rocky shores. **B.** *adj.* Of or pertaining to an animal of this kind. Also † **sylliˑdian** *sb.*

1888 ROLLESTON & JACKSON *Forms Animal Life* (ed. 2) 607 The parent-form in these Syllidians remains non-sexual. 1910 *Encycl. Brit.* V. 793/1 There are even dimorphic forms among the Syllids. 1928 RUSSELL & YONGE *Seas* ii. 52 The little syllid worms break up..into fragments of a few segments, each of which..develops into a full-sized worm. 1930 *Q. Jrnl. Microsc. Sci.* LXXIII. 651 (*heading*) On a new Hermaphrodite Syllid. 1963 R. P. DALES *Annelids* i. 30 The nephridiostome remains as a recognisable notch or pocket in the larger coelomostome, as it does..in some syllids. 1971 *Oxf. Bk. Invertebrates* 96 Syllids usually cling to sea-vegetation or nestle in empty shells and crevices. *Ibid.* 96/2 *Syllis* shows well the typical syllid processes arising from the sides of the body.

Sylow (sīˑlǫf). *Math.* The name of P. L. *Sylow* (1832–1918), Norwegian mathematician, used *attrib.* and in the possessive to designate concepts in group theory propounded by him (*Math. Annalen* (1872) V. 584), as **Sylow (*p*-)subgroup,** a subgroup whose order is the largest power of the prime *p* which divides the order of the group; **Sylow's theorem** (see quots. 1897, 1975).

1893 *Proc. Lond. Math. Soc.* XXV. 14 It is then shown that Sylow's theorem leads to relations between the numbers of operations of different orders which it is impossible to satisfy. 1897 W. BURNSIDE *Theory of Groups of Finite Order* vi. 91 We shall divide the proof of Sylow's theorem into two parts. First we show that, if p^α is the highest power of a prime *p* which divides the order of a group, the group must have a sub-group of order p^α; and secondly that the sub-groups of order p^α form a single conjugate set and that their number is congruent to unity, mod. *p*. 1905 *Messenger Math.* XXXV. 48 A group..all of whose Sylow subgroups are cyclical. 1975 I. STEWART *Concepts Mod. Math.* vii. 104 The best that can be said in general is Sylow's theorem; if *h* is a power of a prime and divides the order of a group *G*, then *G* has a subgroup of order *h*. 1976 *Nature* 20 May p. vii (Advt.), The classification of nonsoluble groups with abelian sylow 2-subgroups.

sylph. Add: **3.** *sylph-like* (earlier example).

1801 C. WILMOT *Let.* 13 Dec. in T. U. Sadleir *Irish Peer* (1920) 15 Madame, their Mother, was too much en bon point to have such a sylphlike appearance as her daughters.

Sylphon (siˑlfǫn). Also **sylphon.** [Invented word.] A proprietary name (see quots. 1906, 1916, 1933) used esp. to designate concertina-like metal bellows and devices employing them.

1906 *Official Gaz.* (U.S. Patent Office) 3 Apr. 1643/1 Heat-regulators for use on boilers, furnaces, and stoves... *Sylphon.* 1916 *Ibid.* 25 July 1432/1 *Sylphon*... A hollow expansible and contractible corrugated tubular metal device. 1933 *Trade Marks Jrnl.* 2 Aug. 925/2 *Sylphon*... Valves, hot and cold water mixers and dampers all being parts of steam boilers... The Fulton Sylphon Company.., Knoxville, Tennessee. 1937 *Jrnl. Psychol.* IV. 281 The essential unit is a small capacity sylphon or thin-gauge metal bellows enclosed in an airtight metal housing. 1938 *Jrnl. R. Aeronaut. Soc.* XLII. 1072 The valves of these tanks may be operated by a sylphon bellows. 1945 H. D. SMYTH *Gen. Acct. Devel. Atomic Energy Mil. Purposes* x. 110 The pumps used were sylphon-sealed reciprocating pumps.

Sylvaner (silvāˑnəɹ). Also **sylvaner.** [a. G. *silvaner, sylvaner*: cf. SYLVAN, SILVAN *sb.* and *a.*]

1. A variety of vine first developed in German-speaking districts, the dominant form bearing white grapes; a vine or grape of this variety. Also *attrib.*

1928 P. M. SHAND *Bk. French Wines* vi. 201 The Gutedel..Ortlieber, Burger, Sylvaner, and Klevner are grown besides [in Alsace]. 1963 *Times* 17 Jan. 4/6 It has the typical flavour of a wine made from Sylvaner grapes. 1965, 1976 [see *PINOT]. 1981 T. MCLEAN *Medieval Eng. Gardens* ix. 256 The Müller-Thurgau vine..is a cross between a Riesling and a Sylvaner.

2. The white wine made from the Sylvaner grape.

1958 A. L. SIMON *Dict. Wines, Spirits & Liqueurs* 152/2 *Sylvaner*, a free-bearing white-wine grape grown extensively in Germany and in Alsace. Much Alsatian wine (white) made from Sylvaner grapes is marketed under the name of *Sylvaner*. 1961 *Spectator* 24 Nov. 756 [The wines] are labelled according to the grapes used, as in Alsace... The commonest names are Riesling..Sauvignon, Sylvaner, [etc.].

sylvatic, silvatic, *a.* Restrict *rare* to **silvatic** and sense in Dict. and add: **2.** *Med.* Also (*rare*) **selvatic.** Applied to certain diseases (as rabies, yellow fever, plague, and Chagas's disease) when contracted by wild rather than domesticated animals, and to the pathogens causing them. [ad. F. *selvatique* (R. Jorge *Les Rongeurs & leurs Puces dans la Propagation de la Peste* (1928) ii. 36); cf. L. *silvāticus* wild.]

1931 C. O. STALLYBRASS *Princ. Epidemiol.* ix. 310 In this way arise two types of epizoötic [plague]... One.. among wild rodents, spreading slowly from colony to colony, independent of man's lines of communication..; to this type of epizoötic Jorge (1928) has given the title selvatic plague. 1935 *Jrnl. Amer. Med. Assoc.* 17 Aug. 535/2 The invasion of sylvatic plague among the ground squirrels of the foothills of the Sierras and Cascade Mountains creates a widening menace in the United States. 1936 WU LIEN-TEH in Wu Lien-Teh et al. *Plague* vi. 195 Jorge..distinguished between the pandemic plague introduced..by..'domestic' rodents, and selvatic plague, dangerous to man only when he invades the remote endemic areas populated by wild rodents. 1970 *Sci. Jrnl.* Apr. 35/1 There has been a steady and alarming increase in rabies in wild animals—so-called sylvatic rabies. 1978 *Nature* 27 Apr. 820/1 We have identified distinct sylvatic and domestic strain-groups of *T*[*rypanosoma*] *cruzi*, apparently circulating independently and transmitted by different vector species.

sylvestrene (silveˑstrīn). *Chem.* [ad. G. *sylvestren* (A. Atterberg 1877, in *Ber. Deut. Chem. Ges.* X. 1203), f. L. *sylvestr-is* found in woods (f. *silva*: see SYLVA, SILVA), specific epithet of the Scots pine, *Pinus sylvestris*: see -ENE.] A liquid monocyclic terpene, $C_{10}H_{16}$, known in two optically active forms and formerly believed to be a natural constituent of pine oil, but now recognized as a product of the extraction process.

1877 *Chem. News* 6 July 7/1 A. Atterberg has examined the crude '*Wood-Spirit from Norwegian Pines*', and found in the higher boiling portions..a new turpentine, to which he assigns the name *sylvestrine.* 1931 [see *ISOPRENE b]. 1952 TURNER & HARRIS *Org. Chem.* xix. 317 Simonsen..showed that the precursor of sylvestrene is the naturally occurring (+)-Δ^3-carene.

Sylvian, *a.*[2] (Earlier example.) *fissure of Sylvius* (examples).

1828 J. QUAIN *Elem. Anat.* ix. 613 The angular part of the anterior lobe..is included between the internal termination of the fissure of Sylvius, the longitudinal fissure, and the commissure of the optic nerves. 1839–47 [see *ROLANDO a]. 1849 S. G. MORTON *Illustr. Syst. Human Anat.* 547 The insula [of Reil] consists of five or six small convolutions grouped and concealed within the Sylvian fissure. 1888, 1939 [see *REIL]. 1980 A. SILVERSTEIN *Human Anat. & Physiol.* xiii. 278/2 Viewed from the side, the cerebrum looks something like a large mitten, with the wrist at the back and the fingers at the front of the head. The 'thumb' of the mitten is separated from the remainder by another prominent groove, the lateral fissure or fissure of Sylvius.

sylvics, silvics. Add: (Further examples.) orig. and chiefly *U.S.* (The form *sylvics* is *rare*[−0].) Cf. *SILVICAL *a.*

1946 *Jrnl. Forestry* XLIV. 965/1 The forester's knowledge of the silvics of the species was pretty thin. 1948 H. J. OOSTING *Study of Plant Communities* (ed. 2) xii. 317 An important part of a forester's training is forest ecology, or silvics, in which he learns the scientific background upon which silvicultural practices are based. 1975 *Agriculture Handbk.* (U.S. Dept. Agric.) No. 486 (*title*) Quaking aspen: silvics and management in the Lake States.

sylvinite (siˑlvinait). *Min.* [ad. G. *sylvinit*, f. G. *sylvin* SYLVINE: see -ITE.[1]] A commercial name for a mixture of sylvite and halite (the form in which sylvite commonly occurs).

1896 A. H. CHESTER *Dict. Names Minerals* 263 *Sylvinite*, the commercial name for sylvite. 1962 *Economist* 31 Mar. 1274/3 The D'Arcy Exploration Company..found ..potash-bearing brine and sylvinite (a mixture of potassium chloride and salt) in a boring near Whitby. 1980 H. BLATT et al. *Origin Sedimentary Rocks* (ed. 2) xv. 558 Sylvinite is composed of sylvite (KCl) and halite (NaCl).

sym-. Add: **symphalaˑngism** *Anat.* [L. *phalang-*: see PHALANX], a condition in which the middle phalanx of a finger or toe is properly developed in length but its proximal (or distal) joint is imperfect or absent; **syˑmphile** *Ent.*, an insect that lives with ants or other social insects as a guest in a relationship of symphilism; hence **symphiˑlic** *a.*, pertaining to or being a symphile; also *fig.*; **symphilism,** add: [ad. G. *symphilie* (M. E. Wasmann 1896, in *3me Congr. Internat. Zool.* 412)]; **syˑmphylan, syˑmphylid** *adjs.* and *sbs. Ent.* [mod. L. *Symphyla*, name of a class of arthropods (J. A. Ryder 1880, in *Amer. Naturalist* XIV. 376), f. Gr. φυλή tribe: so called from their combining characteristics of several other classes] (of or pertaining to) an arthropod of the class Symphyla, the members of which resemble centipedes, having soft bodies and many legs; **syˑmport** *Biochem.* [after TRANSPORT *sb.*], flow of two substances through a membrane in the same direction in which the rate is increased by a cooperative effect.

1916 H. CUSHING in *Genetics* I. 91 This paper will present a much more complete family record of an inherited trait... The malformation will be designated symphalangism. 1943 *Jrnl. Heredity* XXXIV. 344/1 Similar reports of abnormalities of hands and feet including symphalangism, syndactylism, and polydactyly..seem to agree that many anatomical anomalies may be inherited as single dominant traits. 1965 *Arch. Internal Med.* CXV. 580/1 Symphalangism (congenital fusion of the phalanges) is occasionally associated with brachydactyly. 1910 Symphile [see *synœkete* s.v. *SYN-]. 1960 H. OLDROYD tr. *Jeannel's Introd. Entomol.* viii. 212 The greater number of symphiles are beetles, cherished by the ants, and carried with them wherever the nest is moved, but nevertheless terrible enemies of the colony because of the great damage they do to it. 1971 E. O. WILSON *Insect Societies* xx. 403/1 Many of the better-integrated symphiles dispense attractive substances to their hosts from epidermal glands. 1919 W. OSLER *Old Humanities & New Sci.* ii. 12 This attention is what our symphilic community—to use a biological term—bestows on you. 1927 H. ST. J. K. DONISTHORPE *Guests of Brit. Ants* p. xvi, They mostly possess characteristic or 'symphilic' colours and texture—a yellow-red, with an oily looking surface. 1971 E. O. WILSON *Insect Societies* xx. 403/2 A large percentage of the symphilic beetles..possess peculiar tufts of red or golden hairs. 1898 A. S. PACKARD *Text-bk. Entomol.* 21 He..believed that the Symphyla are the forerunners of the myriapods, and not of the insects, his genealogical tree representing the symphylan and thysanuran phyla as originating from the same point. 1964 U. LANHAM *Insects* i. 19 Two of these classes [of many-legged arthropods]—the pauropods and the symphylans—are small, obscure creatures... The other two—centipedes and millipedes—are larger, more conspicuous. 1979 W. D. RUSSELL-HUNTER *Life of Invertebrates* xvi. 301 In some structural features, symphylans resemble the centipedes and in others the apterygote insects. 1936 *Trans. Soc. Brit. Entomol.* III. 14 The contention that the opisthogoneate condition in insects has been derived from Symphylid stock. *Ibid.* 16 The heart, haemocoel, fat-body and anal glands have all been inherited from the Symphylids. 1973 *Nature* 16 Nov. 128/1 It is of interest that certain of the symphilids carry styli on the base of the second and third pairs of legs. 1974 *Encycl. Brit. Macropædia* XII. 771/1 Symphilid species are small, fragile, and lacking pigmentation. 1963 P. MITCHELL in *Biochem. Soc. Symp.* XXII. 148 Over a certain range of concentration, the asymmetry of distribution of the molecules of one substrate across the membrane gives rise to an increased flow of the second substrate in the same direction. We will call this type of coupled movement symport. 1978 *Nature* 2 Mar. 97/1 There are discussions of algal ion transport and of Na^+/organic solute cotransport (symport).

Symbionese (siːmbiˌǒnīˑz), *a.* [f. SYMBIO(SIS + -*n*- + -ESE, after group and people names in -*nese* (*Chinese, Lebanese*, etc.): see quot. 1974.] *Symbionese Liberation Army*, the name adopted by a socialist revolutionary group active in the United States in the mid-1970s.

1973 *N.Y. Times* 10 Nov. 1/4 Two days ago, a group calling itself the Symbionese Liberation Army took the responsibility for the killing. 1974 *Black Panther* 23 Feb. 11/2 The Symbionese Liberation Army is made up of the aged, youth and women and men of all races and people. The name Symbionese is taken from the word symbiosis and we define its meaning as a body of dissimilar bodies and organisms living in deep and loving harmony and partnership in the best interest of all within the body. 1975 *Times* 20 Nov. 1/3 Miss Patricia Hearst, the runaway heiress..was arrested..[in] a routine check on the movements of people associated with the Symbionese Liberation Army (the SLA). 1976 M. J. LASKY *Utopia & Revolution* (1977) 603 Six leading American members..of the so-called Symbionese Liberation Army were killed in Los Angeles in a gun fight [in 1974] with the local police.

symbiose (siˑmbəiˌǒuz), *v.* [Back-formation from next.] *intr.* To live as a symbiont.

1960 *McGraw-Hill Encycl. Sci. & Technol.* XI. 546/2 Strains of each species show marked host specificities in their abilities to symbiose with the plants within each group. 1971 M. ALEXANDER *Microbial Ecol.* xi. 266 A single fungus can apparently symbiose with dissimilar species of algae.

symbiosis. Add: Pl. **symbioses. 1.** Delete † *Obs. rare*[−1] and add later examples.

1910 *Spectator* 30 July 173/2 The savage with his.. sense of 'participation', of 'symbiosis'. 1920 *Q. Rev.* July 164 So long as the people concerned can talk freely together, they form one spiritual symbiosis, and their culture will be the same.

2. a. (Further examples.) Also more widely, any intimate association of two or more different organisms, whether mutually beneficial or not.

1909 tr. *Warming's Oecol. Plants* xxv. 84 Parasitism is a form of symbiosis. 1941 H. KIRBY in Calkins & Summers *Protozoa in Biol. Res.* xix. 891 De Bary..used symbiosis as a collective term, the subdivisions of which include parasitism and mutualism; he recognized two main categories, antagonistic and mutualistic symbiosis. 1953 [see *SYMBIOTE]. 1953, etc. [see *MUTUALISM 2]. 1973 R. G. KRUEGER et al. *Introd. Microbiol.* xxxi. 748/1 Three or more different kinds of organisms are involved in some symbioses. 1977 R. L. SMITH *Elem. Ecol. & Field Biol.* x. 268/1 Mutualism is often termed symbiosis. Actually symbiosis..includes mutualism, commensalism, and parasitism.

b. *transf.* and *fig.*

1921 G. B. SHAW *Back to Methuselah* II. 79 Let the Creator say, if you like, 'I will establish an antipathetic symbiosis between thee and the female.' 1955 *Bull. Atomic Sci.* Apr. 143/2 Two world wars predetermined the henceforth inevitable symbiosis of scientific activity and political decision. 1963 *Listener* 28 Feb. 386/1 The agreement between Castro and the Communist Party early in 1958..began the process of symbiosis which worried many of the more thoughtful *fidelistas*. 1967 M. J. RUGGLES in D. H. Perman *Bibliogr. & Historian* (1968) II. 22 A symbiosis between scholar and librarian is necessary. 1976 *New Yorker* 17 May 127/1 In the symbiosis that will link the candidates and the press throughout this election year, many representatives of each are out in Iowa. 1982 *Listener* 23 & 30 Dec. 29/2 The politician and the journalist exist in a state of uneasy symbiosis.

symbiote, (*b*) = SYMBIONT; also *fig.*; **symbiotic** *a.* (further examples); also *transf.* and *fig.*

1923 *Anat. Rec.* XXV. 2 Portier believes that the 'symbiotes' are especial microorganisms found in great abundance in nature. They are constantly entering and leaving the host organism. 1925 *Jrnl. Infectious Dis.* XXXVI. 94 The intracellular bacteria have been designated as 'symbiotes'. 1953 R. P. HALL *Protozoology* x. 528 Endoparasites which participate in symbiosis, an association involving mutual benefits to host and parasite, are known as symbiotes. 1953 [see *MUTUALISM 2]. 1970 *Times Lit. Suppl.* 14 Aug. 899/5 His suggestion of a future man as a bio-mechanical symbiote. 1951 R. FIRTH *Elem. Social Organization* i. 10 It is most evident in the case of an African tribe having its members living intermingled with those of other tribes and in symbiotic relationship with them. 1956 *Psychiatric Research Rep.* No. 3. 8 A therapeutic move of considerable importance in such a situation is for the physician to function as the other half of the patient's 'symbiotic' system. 1962 *Lancet* 19 May 1033/2 The human infant in its first year is more precariously placed than has hitherto been appreciated since mother and child form a symbiotic union. 1970 *Nature* 6 June 905/1 Throughout its auspicious history the Botanical Society of Edinburgh has had a symbiotic relationship with the Royal Botanic Garden. 1979 W. STYRON *Sophie's Choice* vi. 150 Höss eventually developed what might be called a fruitful—or at least symbiotic—relationship with the man who was to remain his immediate superior.

symbiotrophic (siːmbəiˌotrǒuˑfik, -trǫˑfik), *a.* *Ecol.* [f. SYMBIO(SIS + *-TROPHIC.] Obtaining nourishment through symbiosis.

1905 B. D. JACKSON *Gloss. Bot. Terms* (ed. 2) 358/1. 1974 D. H. LEWIS in Carlile & Skehel *Evolution in Microbial World* 386 Chemoheterotrophs, including animals, may derive nutrients in the free-living state (saprotrophic) or following intimate contact with other organisms (symbiotrophic). 1978 *Proc. Indian Acad. Sci.* B. LXXXVII. x. 243 Despite high salinity and acidity an acid sulfate soil harboured N_2-fixing symbiotrophic organisms with appreciable efficiency.

symbol, *sb.*[1] Add: **4.** *symbol-maker, -making, -object, -system, -user*; *symbol-making, -minded, -using* adjs.

1981 F. INGLIS *Promise of Happiness* iii. 85 The nineteenth-century novelists were the symbol-makers for a new order. 1953 R. LEHMANN *Echoing Grove* 28 Its one round turret, its weather-cock and flag-pole all supernaturally designed in the last sun's last symbol-making glow. 1962 W. NOWOTTNY *Lang. Poets Use* viii. 180 A kind of linguistic ambiguity..seems frequently to occur in poems bent on symbol-making. 1936 O. NASH *Primrose Path* 55 Still, I think, a pig's a pig—Ah, there, symbol-minded Sig! 1977 *N.Y. Times* 20 Jan. 4/3 Ever since he walked home from his inauguration, Mr. Carter has presented the country with a symbol-minded Presidency. 1913 L. BLOOMFIELD in C. F. Hockett *Leonard Bloomfield Anthol.* (1970) 43 This symbol-object is..the word: without it no concept of action, quality, or relation can exist. 1964 E. BECKER in I. L. Horowitz *New Sociol.* 119 Man..possesses both thing-objects, like all other animals; and, uniquely, symbol-objects. 1946 F. P. CHISHOLM in W. S. Knickerbocker *Twentieth Cent. Eng.* II. 183 The communication process involves both speakers and listeners, writers and readers, using a socially-constructed symbol-system, in whose structure 'reality' must be represented. 1964 R. H. ROBINS *Gen. Linguistics* 13 Among symbol systems language occupies a special place. 1946 F. P. CHISHOLM in W. S. Knickerbocker *Twentieth Cent. Eng.* II. 172 Our distinguishing human characteristic is that we are symbol-users. 1951 J. HOLLOWAY *Lang. & Intelligence* vi. 95 Intelligence displayed in a symbol-using planning sequence sometimes enables us to reduce the sequence of actions to a sequence of routines. 1977 R. HOLLAND *Self & Social Context* i. 18 Ethnomethodology..embraces a phenomenological concern for the experiencing, symbol-using self.

symbolic, *a.* (*sb.*) Add: **2. c.** *symbolic logic*, logic that employs a special technical notation of symbols; formal or mathematical logic (see *MATHEMATICAL *a.* 1 e). Hence *symbolic logician.*

1856 A. DE MORGAN in *Trans. Cambr. Philos. Soc.* IX. 83, I think it reasonably probable that the advance of symbolic logic will lead to a calculus of opposite relations, for mere inference, as general as that of + and − in algebra. 1881 [in Dict., sense 2 a]. 1903 B. RUSSELL *Princ. Math.* ii. 10 Symbolic or Formal Logic—I shall use these terms as synonyms—is the study of the various general types of deduction. The word *symbolic* designates the subject by an accidental characteristic, for the employment of mathematical symbols, here as elsewhere, is merely a theoretically irrelevant convenience. *Ibid.* vi. 74 By symbolic logicians..this will be felt as a reactionary view. 1933 C. A. MACE *Princ. Logic* iv. 64 The fact that symbolic logicians have not generally recognized this form compels us to introduce a symbol that is not in common use. 1941 [see *mathematical logic* s.v. *MATHEMATICAL *a.* 1 e]. 1958 *Times Lit. Suppl.* 19 Dec. 729/3 Professor Sparshott quotes the dying symbolic logician—'complete rigour at last!' 1968 *Brit. Med. Bull.* XXIV. 239/2 The final study to be reviewed concerns diagnosis by the computer using a combination of symbolic logic.. and similarity coefficients. 1973 *Sci. Amer.* Apr. 101/3 First Frege, then Peano and finally Russell turned to symbolic logic as a potential source of the fundamental notions necessary for a theory of natural number.

d. *symbolic address* (Computers), an address consisting of a symbol chosen by the programmer for its convenience; so *symbolic addressing*.

1953 *Trans. IRE Professional Group on Electronic Computers* Mar. 10/1 Programs for automatic calculators can be written with symbolic addresses instead of actual addresses. 1970 O. DOPPING *Computers & Data Processing* xix. 308 In automatic coding..each data item receives a name, or symbolic address. 1977 *Gloss. Terms Data Processing* (*B.S.I.*) VII. 13/1 Symbolic addressing. 1981 M. E. WALSH *Understanding Computers* iii. 48 This process of using mnemonic instructions..and symbolic addressing and having them translated into machine language is called assembling a program.

3. c. *symbolic interaction* (Social Psychol. and Sociol.), the sharing and use of common symbols in human communication; freq. *attrib.*; also *symbolic interactionist*, an adherent of the theory that the child is formed into a social being through learning the common meaning attached to symbols by its group; also *attrib.* or as *adj.*; hence *symbolic interactionism*.

1937 H. BLUMER in E. P. Schmidt *Man & Society* 153 The group of social psychologists who may be conveniently labelled 'symbolic interactionists'. *Ibid.* 174 It is clearly an instance of the symbolic interaction. *Ibid.* 191 The stimulus-response approach is interested in *reaction*; the symbolic interaction view in *action*. 1961 D. MARTINDALE *Nature & Types Sociol. Theory* xiv. 339 The symbolic interaction school took shape in America, primarily under the influence of pragmatism. 1967 *Sociol. Q.* VIII. 149 (*title*) On the edge of rapprochement: was Durkheim moving toward the perspective of symbolic interaction? 1969 H. BLUMER (*title*) Symbolic interactionism. *Ibid.* i. 1 George Herbert Mead who, above all others, laid the foundations of the symbolic interactionist approach. 1972 S. MENNELL in Cox & Dyson *20th-Cent. Mind* III. v. 160 Another kind of social action theory has also been influential, especially in the last decade. It is usually known as 'symbolic interactionism', and has deep roots in American sociology. 1977 J. A. KOTARBA in Douglas & Johnson *Existential Sociol.* ix. 272 The concept of illness as deviant behavior..is built upon the labeling theory of the symbolic interactionist perspective. 1979 *Human Relations* Sept. 803 Symbolic interaction stresses the personal definition of the situation, while frame analysis seeks to uncover the background assumptions within which interaction takes place. 1982 *Jrnl. Learning Disabilities* XV. 347 Using a symbolic interaction perspective, the study focused on the extent of agreement..in referring children..to a university clinic for psycho-educational assessment.

symboling: see SYMBOLLING *vbl. sb.* in Dict. and Suppl.

symbolist. 2. c. (Earlier example.)

1888 G. MOORE *Confessions of Young Man* vi. 147 Like a white flag fluttering faintly, Symbolists and Decadents appeared.

‖ **symboliste** (sænbǫlīst). Also with capital initial. [Fr.: cf. SYMBOLIST.] = SYMBOLIST 2 c (*b*). Chiefly *attrib.*

1925 [see *CROCEAN *a.*[2]]. 1957 J. HOLLANDER in N. Frye *Sound & Poetry* 67 Professor Knight has elevated his rather *symboliste* construction of the word 'music' to the heights proclaimed in Verlaine's manifesto. 1966 *Listener* 17 Mar. 378/1 T. S. Eliot..found the clues he needed..in the French *Symbolistes* like Laforgue. 1980 A. ALPERS *Life K. Mansfield* vii. 135 A little *symboliste* prose-poem.

symbolled, *a.* (Example of spelling *symboled*.)

1935 DYLAN THOMAS in *New Verse* Aug.–Sept. 3 The invalid rivals, Voyaging clockwise off the symboled harbour.

symbolling, *vbl. sb.* Add: In U.S. usu. with spelling **symboling.** Also, the use of symbols in communication.

1974 H. G. BURGER in *Gen. Systems* XIX. 64/1 What appears to be non-language..is probably a precisely coded symboling of a non-standard dialect. 1977 —— in B. Bernardi *Concept & Dynamics of Culture* 419 Between the gross human ability of symboling, so well known, and the peculiarly human institutions, also well known, lie special symboling processes. 1977 *Dædalus* Summer 62 Process theory is no longer linked, as in its earlier heyday, with Gumplowicz's notion that 'man's material need is the prime motive of his conduct'; it now recognizes the critical importance of meaning and symboling.

symbolling, var. *SEMBLING.

symbolo-, combining form of Gr. σύμβολον SYMBOL *sb.*[1], as in **symbolo-fideism** (siːmbǒlo-fəiˑdiˌiz'm) [F. *symbolo-fidéisme*], the theory that symbols are of the essence of religious dogma, and that the attitude of faith has priority over intellectual belief (see quot. 1921); hence **syːmbolo-fiˑdeist,** one who holds this theory; **syːmbolomaˑnia** *nonce-wd.*, excessive use of symbols.

[1897 A. SABATIER *Esquisse d'une Philosophie de la Religion* III. vii. 406 En combinant les vues de M. Ménégoz et les miennes qui se complètent en effet réciproquement, on a pu baptiser la conception nouvelle de *symbolo-fidéisme*.] 1903 *Hibbert Jrnl.* I. 555 In these two principles, —the symbolic, pictorial character of all the concepts and terms of religion, and the distinction just mentioned between faith and belief,—we have the germs of Symbolo-fideism. The name appears to have been given to the school by an anonymous writer in 1894. 1921 *Encycl. Relig. & Ethics* XII. 151 Symbolofideism is the name given to the theology taught in the second half of the 19th cent. at the Protestant Faculty of Paris by Professors Auguste Sabatier and Eugène Ménégoz. 1949 E. L. MASCALL *Existence & Analogy* v. 93 Modern Roman Catholic theologians..under the stress of their controversy with symbolo-fideists and modernists of various kinds, have discussed the doctrine of analogy at great length. 1970 *Nature* 4 Apr. 47/2 Such a hornet's nest of symbols, wiggly lines,..and the like, that the reader, despairing of ever reaching the goodies obscured by the fog of symbolomania, may come to think that this new approach to mathematics is not worth while.

symbology. Add: **symbological** *a.*, **symbologist**: for *rare*[−0] read *rare*; (examples).

1924 *Glasgow Herald* 4 Apr. 13 Professor Stern returns to the attack upon this theory-mongering, always recalling the symbologists and complex-jargonists to a consideration of practical realities as a test of their deductions. 1976 *N.Y. Rev. Bks.* 15 Apr. 29/2 He considered the Papal Bull of 1950 declaring the Assumption of the Virgin an article of faith to be the most important symbological event since the Reformation.

symmachy. For † *Obs. rare*[−0] read *rare* and add example.

1911 C. PHILLIPSON *Internat. Law & Custom Ancient Greece & Rome* I. ix. 222 The relationship existing between those cities which constituted the military symmachy in Italy.

symmetallism (simˌmeˑtăliz'm). Also **Symmetallism.** *Econ.* [f. SYM- + BI)METALLISM.] A proposed monetary system based on the use of an amalgam of gold and silver as a standard (see quot. 1979). Hence **symmetaˑllic** *a.*; **symmeˑtallist** *sb.* an advocate of symmetallism (also *attrib.* or as *adj.*).

1895 F. Y. EDGEWORTH in *Econ. Jrnl.* V. 443 The arrangement that there should be a *joint demand* for gold and silver money might, perhaps, be called *symmetallism*, to distinguish it from the arrangement that there should be a *composite supply* which is called bimetallism. *Ibid.* 444 Suppose England with India adopts one symmetallic ratio. 1897 *Daily News* 30 Nov. 4/6 [A man] may be a Symmetallist, and believe that standard coins should be made from a mixture of silver with gold. *Ibid.*, Whether a symmetallist coinage be possible or not, it is certain that we have a symmetallist Administration. 1923 A. MARSHALL *Money, Credit & Commerce* 64 Although coinage of gold and silver at a fixed ratio causes movements of prices to be governed chiefly by the production of gold and silver alternately, a plan can be devised which would make the two metals work together: it may be called Symmetallism. 1934 *Sun* (Baltimore) 24 May 10/2 It [*sc.* the President's proposal] does not involve either bimetallism or symmetalism [*sic*]. 1979 *Econ. Jrnl.* LXXXIX. 29 Consider now Marshall's..proposal for a 'stable bimetallism', which is usually called symmetallism. Under this system the central bank does not attempt to stabilise the price of either gold or silver separately, but rather pegs the price of a reserve unit that corresponds to a specified combination of the two metals. 1980 *Internat. Econ. Rev.* XXI. 675 Under a symmetallic standard..the monetary authority does not set prices for individual commodities.

symmetric, *a.* Add: **1. b.** Math. and Logic. *symmetric difference* (see quot. 1936); *symmetric group*, the group of all the permutations of a set of unlike entities.

1936 *Trans. Amer. Math. Soc.* XL. 38 The Union (modulo 2), or symmetric difference, of two classes is the class of objects belonging to one or the other, but not to both, of those classes. 1971 J. H. CONWAY in Powell & Higman *Finite Simple Groups* vii. 225 If some non-empty ℭ-set has fewer than five elements, every set of the same cardinal would be a ℭ-set,..and by taking symmetric differences we should obtain every two-element set..as a ℭ-set, which cannot be. 1897 W. BURNSIDE *Theory of Groups of Finite Order* viii. 139 The group of order *n*! which consists of all the substitutions that can be performed on *n* symbols is called the symmetric group of degree *n*. 1955 L. MIRSKY *Introd. Linear Algebra* ix. 257 In addition to the symmetric group..there are other groups of permutations, all of them naturally subgroups of the symmetric group.

c. *Physics.* = *SYMMETRICAL *a.* 2 c.

1935 CONDON & SHORTLEY *Theory of Atomic Spectra* iii. 165 If the atom is at a certain moment in a symmetric state it will always remain in a symmetric state. 1965 H. MUIRHEAD *Physics Elem. Particles* ix. 369 It is apparent from this equation that parallel spin states..are symmetric. 1979 *Nature* 29 Mar. 404/2 Certain baryons.. contain three identical quarks with parallel spins in a state which is symmetric (the wave function does not change sign) to the interchange of a pair of quarks.

2. *Logic.* Of a binary relation: such that when two terms for which it is true are interchanged, it remains true.

1933 *Mind* XLII. 34 Not even God himself can make men into women by shifting words, or make what we call 'implies' symmetric by changing names. 1968 *New Scientist* 16 May 339/1 Equality is symmetric as well as being reflexive. 1979 K. J. DEVLIN *Fund. Contemp. Set Theory* i. 14 A binary relation on a set is an equivalence relation just in case it is reflective, symmetric, and transitive.

symmetrical, *a.* Add: **2. b.** Also in *Logic*, = *SYMMETRIC *a.* 2.

1903, 1937 [see *REFLEXIVE *a.* 7]. 1954 I. M. COPI *Symbolic Logic* v. 141 Various symmetrical relations are designated by the phrases: 'is next to', 'is married to', and 'has the same weight as'. A symmetrical relation is one such that if one individual has that relation to a second individual, then the second individual must have that relation to the first. 1979 GEORGACARAKOS & SMITH *Elementary Formal Logic* ix. 329 When a relational expression has this property, we say that it is symmetrical.

c. In *Physics* also applied to a state represented by such a wave function.

1930 P. A. M. DIRAC *Princ. Quantum Mech.* xi. 201 It is quite possible for only symmetrical or antisymmetrical states to occur in nature... One assumes the symmetrical states for photons. 1963 R. P. FEYNMAN et al. *Feynman Lect. Physics* I. xi. 2 The laws of physics are symmetrical for translational displacements,..in the sense that the laws do not change when we make a translation of our coordinates. 1973 B. H. BRANSDEN et al. *Fundamental Particles* iv. 79 The triplet spin state is symmetrical.

d. Math. and Logic. *symmetrical difference* = *symmetric difference* s.v. *SYMMETRIC *a.* 1 b.

1978 C. H. GREENSTEIN *Dict. Logical Terms & Symbols* 172 Symmetrical difference. 1979 KANDEL & LEE *Fuzzy Switching & Automata* ii. 53 The symmetrical difference (or Boolean sum) of two fuzzy sets.

symmetricalness. (*b*) (Earlier example.)

1858 W. BAGEHOT in *National Rev.* Oct. 460 The mode in which those opinions are expressed, and..the mode in which they are framed, affect us..with a sensation of symmetricalness.

symmetrize, *v.* **1.** For *rare*[−1] read *rare* and add earlier example.

1749 J. CLELAND *Mem. Woman of Pleasure* II. 233 An air of becoming manliness..that symetriz'd [*sic*] nobly with that air of distinction..with which nature has stamped it [*sc.* his face].

2. Also *absol.*

1973 *Sci. Amer.* Jan. 111/2 This leaves one column and one row, with the poison piece at the vertex... From now on the first player 'symmetrizes'. Whatever his opponent takes from either line, he takes equally from the other.

Hence **syˑmmetrized** *ppl. a.*

1966 *Math. Rev.* XXXI. 36/1 (*heading*) Matrix applications of a quadratic identity for decomposable symmetrized tensors. 1979 *Nature* 29 Feb. 597/2 It is the interference between the two parts of the symmetrised wave-functions..that leads to the intensity interference.

symmetrodont (simeˑtrǒdǫnt), *sb.* and *a.* [f. mod.L. order name *Symmetrodonta* (G. G. Simpson 1925, in *Amer. Jrnl. Sci.* CCX. 560), f. SYMMETR(Y + Gr. ὀδούς, ὀδοντ- tooth, in allusion to the form of the teeth (see quot. 1979[2]).] **A.** *sb.* A fossil mammal of the order Symmetrodonta, known from remains found in North America and Europe. **B.** *adj.* Of or pertaining to an animal of this kind or the order including it.

1933 A. S. ROMER *Vertebr. Paleont.* xii. 260 The symmetrodonts seem to have been somewhat off the main evolutionary line. 1950 *Nature* 21 Oct. 696/2 The specimen can easily be described as a lower symmetrodont cheek tooth. 1977 A. HALLAM *Planet Earth* 223 Triconodont and symmetrodont mammals died out during the Cretaceous. 1979 R. C. FOX in Fairbridge & Jablonski *Encycl. Paleont.* 429/2 Symmetrodonts were small shrew-sized mammals, probably having insectivorous food habits. *Ibid.*, Symmetrodont molars are highly characteristic: both upper and lower crowns formed simple occlusal triangles... The lower molar triangles are reversed in respect to the uppers and occlusion was alternate in the sense that each molar occluded within the embrasure between two successive molars on the opposite jaw.

symmetry. Add: **3. b.** (*a*) More widely, a property by virtue of which something is effectively unchanged by a particular operation; an operation or set of operations that leaves something effectively unchanged; in *Physics*, a property that is conserved (cf. *symmetry operation*, sense 4 below). (Further examples.)

1908 H. HILTON *Theory of Groups of Finite Order* iv. 42 If a movement (other than identity) brings every point of a figure *F* into the position previously occupied either by itself or by some other point of *F*, *F* is said to possess symmetry. 1941 BIRKHOFF & MACLANE *Survey Mod. Algebra* vi. 122 The algebra of symmetries has its genesis in the fact that we can multiply two motions by performing them in succession. 1965 R. P. FEYNMAN et al. *Feynman Lect. Physics* III. xvii. 8 Symmetry with respect to displacements in time implies the conservation of energy; symmetry with respect to position in *x*, *y*, or *z* implies the conservation of momentum. 1967 *Physical Rev. Lett.* XIX. 1264/2 As far as we know, two of these symmetries are entirely unbroken: the charge *Q*..and the electron number *N*. 1968 M. S. LIVINGSTON *Particle Physics* xii. 201 One consequence of the translational symmetry of space is the invariance of physical laws under translation from one location to another. 1974 FRAUENFELDER & HENLEY *Subatomic Physics* vi. 154 Some of the symmetries are perfect even under closest scrutiny, and no breakdown in the corresponding conservation law has ever been found. Rotational symmetry and conservation of angular momentum are one example.

(*b*) Also in *Logic* (cf. *SYMMETRICAL *a.* 2 b).

1950 [see *REFLEXIVITY]. 1967 S. C. KLEENE *Math. Logic* iii. 158 Sometimes 'equality' is used in a different sense, so that it possesses only the first three properties (reflexivity, symmetry and transitivity).

4. *attrib.* and *Comb.*, as *symmetry principle, property*; **symmetry-breaking** *ppl. a.* and *vbl. sb. Physics*, (causing) the absence of manifest symmetry in a situation despite its presence in the laws of nature underlying it; **symmetry group,** a group (GROUP sb. 5 a in Dict. and Suppl.) whose elements are all the symmetry operations of a particular entity; **symmetry operation** *Physics*, an operation or transformation that leaves something effectively unchanged.

1961 M. GELL-MANN in Gell-Mann & Ne'eman *Eightfold Way* (1964) We attempt..to treat the eight known baryons as a supermultiplet, degenerate in the limit of a certain symmetry but split into isotopic spin multiplets by a symmetry-breaking term. 1977 *Dædalus* Summer 29 As a result of this symmetry-breaking, the quanta of the weak interactions are predicted to acquire a mass approximately forty or more times heavier than that of a proton. 1981 *Nature* 10 Dec. 522/1 The usual analogy used for spontaneous symmetry breaking is ferromagnetism. Maxwell's equations are rotationally invariant; however, below the Curie temperature the rotational invariance of a ferromagnet is spontaneously broken when the magnetization chooses a specific direction. 1956 *Ibid.* 10 Mar. 458/1 To-day the instinctive reaction of every theoretical physicist, confronted with an unexplained regularity in the behaviour of elementary particles, is to postulate an underlying symmetry-group. 1975 I. STEWART *Concepts Mod. Math.* vii. 97 Every shape has a symmetry group. 1981 *Sci. Amer.* Apr. 50/2 The $SU(2) \times U(1)$ theory is only a partial unification because it still includes two distinct forces, each with its own symmetry group and its own coupling constant. 1952 H. WEYL *Symmetry* 27 For forms fixed to the bottom of the ocean the direction of gravity is an important factor, narrowing the set of symmetry operations from all rotations around the center P to all rotations about an axis. 1973 B. H. BRANSDEN et al. *Fundamental Particles* iv. 56 The symmetry operations with which we are concerned are transformations of the dynamical variable that leave the Hamiltonian operator unaltered. 1968 M. S. LIVINGSTON *Particle Physics* xii. 201 It is possible that the number of such symmetry principles is limited and that they are interrelated. *Ibid.*, One of the most basic symmetry principles is that of the homogeneity of space and the associated symmetry of time. 1977 *Dædalus* Fall 31 Some theorists turned to the study of symmetry principles and conservation laws, which can be applied to physical phenomena without detailed dynamical calculations. 1935 PAULING & WILSON *Introd. Quantum Mech.* xiv. 388 The symmetry properties of molecular wave functions. 1968 M. S. LIVINGSTON *Particle Physics* iii. 58 The type of quantum statistics which applies to a system of particles (all of one kind) is related to the symmetry properties of the wave function describing this system of particles.

symmography (simǫˑgrăfi). [f. SYMM(ETRY + -OGRAPHY.] = *string art* s.v. *STRING *sb.* 32. Also **syˑmmograph,** a pattern or picture made by symmography; **symmograˑphic** *a.*

1971 L. KREISCHER *Symmography* 4 Symmography is an art form using yarn, wood, and nails as the media. *Ibid.* 5 The materials you need to begin a symmograph are basically quite simple. *Ibid.*, The nails I use for my symmographic creations are..bright steel wire. 1975 *String Art Encycl.* 9 Whereas originally string-craft creations were often symmographs—art works in which string was wound attractively and symmetrically around nails in a board—this book deals with string in other artistic forms as well.

Symondite (siˑmǫndait). Now *Hist.* [f. the name of Rear-Admiral Sir William *Symonds* (1782–1856): see -ITE[1].] A small warship designed by Sir William Symonds in his capacity as surveyor to the Royal Navy.

1927 B. M. CHAMBERS *Salt Junk* iv. 27 The *Eurydice* was what was known as a Symonite [*sic*] or Jackass frigate, i.e., something between a sloop and a frigate. 1932 A. H. LONG *Round the Bill* 9 She was a good little boat, about seven feet beam, drew three feet six inches, and had a regular Symondite bottom, like the *America*. 1935 H. I. CHAPELLE *Hist. Amer. Sailing Ships* 156 As a class, the Symondites were very unsteady gun-platforms. 1957 *Mariner's Mirror* XLIII. 337 For rolling, pitching and lee-lurches the Symondites beat the lot.

sympathectomy. Add: (Later examples.)

1903 *Med. Record* LXIII. 875/2 So far as the question of choice of operation between hemisection and sympathectomy went, he believed that the Jennesco operation gave better results. 1936 *Q. Jrnl. Med.* XXIX. 438 Of all the 'sympathectomies' which have been proposed and tried, 'ganglionectomy' is the only one really worth doing. 1955 *Sci. News Let.* 22 Oct. 262/1 The nerve-cutting operation, called sympathectomy, is to dilate arteries that have been stopped. 1968 G. MAXWELL *Raven seek thy Brother* ii. 29 There was no alternative, he said, to lumbar sympathectomy. 1979 *Molecular Pharmacol.* XV. 35 Microsomal preparations derived from several peripheral organs of cats or rabbits following chemical sympathectomy.

Hence **sympatheˑctomized** *a.*, that has undergone sympathectomy.

1928 *Amer. Jrnl. Physiol.* LXXXV. 493 Table 3 shows the changes produced in the relative mononuclear count in sympathectomized animals. 1970 H. SHANDS *Semiotic Approaches to Psychiatry* xxiii. 396 He [*sc.* the schizophrenic] thrives (relatively speaking) when, like Cannon's sympathectomized cats, he is never exposed to normally expectable variation.

sympathetic, *a.* (*sb.*) Add: **A.** *adj.* **1. e.** Also *spec.* in Mus., *sympathetic strings*: (see quot. 1960).

1884 F. NIECKS *Dict. Mus. Terms* s.v. *Viola d'amore*, a bow stringed instrument a little longer than the viola, with seven (sometimes fewer) catgut strings about the fingerboard, and seven sympathetic wire strings below it. 1888 HIPKINS & GIBB *Mus. Instruments* 53 In the beautifully carved and inlaid instrument here drawn, a perfect viola d'amore in form.., the sympathetic strings are absent. 1908 L. J. DE BEKKER *Stokes' Encycl. Mus. & Musicians* 706/2 The sympathetic strings give a beautiful effect. 1928 E. BLOM *Romance of Piano* x. 178 In the treble, the sympathetic strings of the Blüthnor piano are tuned in unison with the ordinary strings. 1940 C. SACHS *Hist. Musical Instruments* xvi. 365 Sympathetic strings had come to England from the Near East, apparently in the sixteenth century. Praetorius related that the English used sympathetic viol strings. 1960 H. HAYWARD *Antique Coll.* 297/2 *Viola d'amore*, a musical instrument.. notable for its system of 'sympathetic' strings... Although out of reach of the bow and fingers these strings vibrate freely in sympathy with the notes played and produce a peculiarly ethereal effect. 1966 *Melody Maker* 7 May 10 The sympathetic strings [in a sitar] vibrate when the main strings are played, giving an answering drone. 1976 *Early Music* July 303 This viol still bore twelve wrestpins in the end block which would have originally carried sympathetic strings added in the 18th century. *Ibid.* 305 A viola bastarde..with six sympathetic strings beneath the six bowed strings.

2. b. Tending to elicit sympathy (senses 3 b, d) or to induce a feeling of rapport; also *loosely*, pleasant, likeable. Cf. *SYMPATHIQUE *a.*

1900 M. BEERBOHM in *Sat. Rev.* 10 Mar. 295/2 The true Don Juan..is..not a 'sympathetic' part. 1926 FOWLER *Mod. Eng. Usage* 590/2 Macbeth..is not made sympathetic, however adequately his crime may be explained & palliated, by being the victim of a hallucination. 1965 *Listener* 23 Dec. 1045/1 Being a lover of the south, I personally found it [*sc.* a novel] more sympathetic. 1976 A. EDEN *Another World* iv. 54 It was not a sympathetic house and the furnishing and pictures were ugly.

3. b. *sympathetic strike*, a strike by workers in support of the action of strikers in another union, industry, etc.

1901 [in Dict.]. 1913 in J. O'Connor *Hist. Ireland 1798–1924* (1925) II. xvii. 192 They followed by a somewhat lame conclusion that the 'sympathetic strike was being met with the sympathetic lock-out.' 1958 *Times Rev. Industry* Aug. 7/2 The merest murmur of the words 'sympathetic strike' will command the dockers' attention.

sympatheticotonia, -ic: see *SYMPATHICOTONIA.

sympathico-: see *SYMPATHO-.

sympaːthicotǫˑnia. *Physiol.* Also anglicized as **sympaːthicotoˑny.** [f. *SYMPATHICO- + *-TONIA.] The state or condition in which there is increased influence of the sympathetic nervous system and heightened sensitivity to adrenalin. Also **sympatheːticotoˑnia, -toˑnus,** in the same sense.

1916 J. P. STEWART *Diagnosis of Nervous Dis.* (ed. 4) xx. 356 Vago-tonus and sympathetico-tonus.—Individuals may be classified into two great vegetative types, according as their autonomic sensitiveness prevails over their sympathetic, or *vice versa*. 1923 *Handbk. for Mental Nurses* (Medico-Psychol. Assoc.) (ed. 7) ix. 375 The condition is then known as sympathetico-tonus or vagotonus, as the case may be. 1930 J. E. NICOLE *Psychopathology* ix. 77 The characteristics of vagotonia might be due to the thymus and pituitary glands, while the adrenals and thyroid would account for sympatheticotonia. 1948 A. BRODAL *Neurol. Anat.* xi. 371 Frequently persons are also met with in whom only one organ reveals a clear-cut parasympathetic or sympathetic dominance (local vagotonia or sympathicotonia). 1977 *Lancet* 12 Nov. 1027/2 During sympathicotony the organism is already making use of Nature's reserve supply of lachrymal fluid.

Hence **sympatheːticotoˑnic, sympaːthicotoˑnic** *adjs.*, displaying or promoting sympathicotonia; also as *sbs.*, a sympathicotonic person.

1916 J. P. STEWART *Diagnosis of Nervous Dis.* (ed. 4) xx. 357 Sympathetico-tonic individuals are specially sensitive to adrenalin which exaggerates all their characteristics. 1930 J. E. NICOLE *Psychopathology* viii. 70 The sympatheticotonics..have dry skins, prominent eyes, dilated pupils, and are possessed of great energy, both mental and physical. 1944 L. J. BENDIT *Paranormal Cognition* iii. 47 She was of the sympathicotonic type, given to attacks of vomiting when emotionally upset. 1954 S. DUKE-ELDER *Parsons' Dis. Eye* (ed. 12) xxi. 235 It is seen particularly in those who are highly strung, anxious in disposition and sympatheticotonic in type. 1975 *Year Bk. Ear, Nose & Throat* 273 Sympathicotonic influences might cause hypertonicity of the cricopharyngeus muscle.

sympaːthicotropic (-trǒuˑpik, -trǫˑpik), *a.* *Pharm.* [f. *SYMPATHICO- + *-TROPIC.] Possessing an affinity for the sympathetic nervous system. Also **sy:mpathotro·pic** *a.*, in the same sense.

1914 *Jrnl. Amer. Med. Assoc.* 22 Aug. 619/2 Epinephrin ..affects especially the sympathetic... It is therefore spoken of as a sympathicotropic drug. **1964** *Internat. Jrnl. Neuropharmacol.* III. 217 When evaluating the effect of indirectly acting sympathotropic substances one must first ascertain whether the tissue stores endogenous catecholamines. **1975** *Acta Biol. Med. Germanica* XXXIV. 661 (*heading*) The action of sympathicotropic substances upon liver microsomes in vivo and in vitro.

sympathin (si·mpăþin). *Physiol.* [f. *SYMPATH(O- + -IN[1].] A hormone which acts as a mediator of nerve impulses at sympathetic nerve synapses; now effectively a disused synonym of *NORADRENALINE.

1931 CANNON & BACQ in *Amer. Jrnl. Physiol.* XCVI. 411 Because the substance is derived from structures under sympathetic control, when they are influenced by sympathetic impulses, we suggest that it be called sympathin. **1938** *Nature* 12 Feb. 266/2 The authors reject Bacq's rather factious criticisms of their theory of sympathins *E* and *I*. **1946** *Ibid.* 20 July 88/1 In Cannon's remaining active years he was largely concerned with evidence as to the nature of the sympathetic transmitter 'sympathin', which he believed to be not identical with adrenaline. **1971** *Ibid.* 2 Apr. 340/2 In an attempt to demonstrate this increase the transmitter 'sympathin' liberated from sympathetic nerve to the spleen was examined.

‖**sympathique** (sæ̃patik), *a.* [Fr.: see SYMPATHIC *a.*] Of a thing, place, etc.: agreeable, to one's taste, suitable. Of a person: likeable, en rapport with one, congenial. Cf. *SYMPATHETIC *a.* 2 b.

1859 QUEEN VICTORIA *Let.* 27 Apr. in R. Fulford *Dearest Child* (1964) 187 The sight of a professor or learned man alarms me, and is not sympathique to me. **1865** —— *Let.* 30 Dec. in *Ibid.* 52 Oh if only Antoinette was in Ali's place! She is so much more *sympathique* and *grande dame*. **1869** W. JAMES *Let.* 1 Nov. in R. B. Perry *Thought & Char. W. James* (1935) I. 308 England is evidently *sympathique* to you. **1897** A. BEARDSLEY *Let.* Apr. (1970) 305 The Baronne Dufour came to see us today... How sweet and sympathique she is. **1930** E. WAUGH *Vile Bodies* vii. 116, I do think, when you get to my age, dear, there is something *sympathique* about a wig, don't you? **1960** *Harper's Bazaar* July 25 A warm and *sympathique* personality. **1975** D. GRAY *Ride on Tiger* ii. 14, I find you *sympathique*.

‖**sympathisch** (zümpā·tiʃ), *a.* Also *erron.* sympatisch. [Ger.: see SYMPATHIC *a.*] = *SYMPATHIQUE *a.*

1911 R. BROOKE *Let.* 13 Dec. (1968) 325, I find that Creative Artists are so particularly *sympathisch*. **1922** D. H. LAWRENCE *Let.* 25 Oct. (1932) 559 But it [*sc.* New Mexico] isn't *sympatisch* like Australia. **1937** AUDEN in Auden & MacNeice *Lett. from Iceland* viii. 100 You I find sympatisch, a good townee. **1976** P. HENISSART *Winter Quarry* I. vii. 72 Another lie? It's what makes you so *sympatisch*, isn't it? **1982** *N.Y. Times Mag.* 17 Oct. 100/5 The personality of the singer himself—warm, intelligent, *sympathisch*, recognizable.

sympatho- (si·mpăþo), combining form of SYMPATHETIC *a.* (*sb.*), used to form terms relating to the sympathetic nervous system; also **sympa·thico-**; **sy:mpathogonia** (-gōu·nia) *sb. pl. Med.* [ad. G. *sympathogonien* (H. Poll 1906, in O. Hertwig *Handb. d. vergleichenden und exper. Entwickelungslehre d. Wirbeltiere* V. III. i. 460), f. Gr. γόνος offspring, begetting], undifferentiated embryonic cells of the sympathetic nervous system which give rise to sympathoblasts; also used as *sing.*; **sy:mpathogonio·ma, sympa:thicogonio·ma** [*-OMA], a malignant tumour composed chiefly of sympathogonia.

1934 *Jrnl. Path. & Bacteriol.* XXXIX. 28 The sympathogonia from which the medulla of the suprarenal takes origin, as first described by Wiesel (1902), began to invade the anlage of the adrenal cortex. **1966** *Pharmacol. Rev.* XVIII. 659 The common progenitor, called sympathogonia, is a small lymphocyte-like cell with a dense, chromatin-rich, spherical or pyriform nucleus and a scanty rim of clear, poorly-staining cytoplasm. **1934** *Jrnl. Path. & Bacteriol.* XXXIX. 28 Those formed of sympathogonia have been classified as sympathogoniomas. **1966** *Pharmacol. Rev.* XVIII. 659 Each of these three types of sympathetic cell may give rise to a tumor: the sympathogonia to a sympathogonioma; the sympathoblast to a sympathoblastoma; and the ganglion cell to a ganglioneuroma. **1974** *Oncology* XXIX. 521 Tumour biopsies of a..sympathicogonioma..were obtained when the tumours were removed surgically.

sy:mpatho-adre·nal, *a. Physiol.* [f. *SYMPATHO- + *ADRENAL *a.*] Pertaining to or involving the sympathetic nervous system and the medulla of the adrenal gland, and their activity. Also **sympa:thico-adre·nal** *a.*, in the same sense.

1928 *Amer. Jrnl. Physiol.* LXXXIV. 560 Previous investigations..have emphasized the emergency functions of the sympathico-adrenal mechanism. **1949** KOESTLER *Insight & Outlook* v. 59 The whole complex of sympathico-adrenal excitation which characterizes aughter is not only absent in crying, but is replaced..by parasympathetic excitation, or..by types of reaction, for example, fatigue, which are the direct opposites of sympathetic excitation. **1965** *Jrnl. Physiol.* CLXXIX. 290 Three drugs which are known to have various actions on the sympatho-adrenal system were tested. **1974** *Jrnl. Appl. Physiol.* XXXVI. 183/1 Sympathicoadrenal medullary secretion of catecholamines is increased during acute cold exposures. **1979** *Med. Hypotheses* V. 317 Some disorders in which excessive sweating..is a symptom are also characterized by increased sympatho-adrenal activity.

sympathoblast (si·mpăþoblæst). *Med.* [f. *SYMPATHO- + -BLAST.] A small, relatively undifferentiated cell formed in the early development of nerve tissue which develops into a sympathetic neurone. Also **sympa·thicoblast**, in the same sense.

1927 *Amer. Jrnl. Path.* III. 212 These lesions represent a tumor of a more primitive type of cell (sympathicoblast) than the sympathetic neuroblast. **1934** *Jrnl. Path. & Bacteriol.* XXXIX. 28 Different stages in development may be present in the same tumour *e.g.* sympathogonia, sympathoblasts and ganglion cells. **1966** *Experientia* XXII. 297 In the developing CNS [*sc.* central nervous system] and the spinal ganglia, only the sympathoblasts of the primary and secondary trunk of chick embryos contain a varying amount of catechol amine-containing granules.

Hence **sympa:thicoblasto·ma,** (less commonly **sy:mpathoblasto·ma**) [*-OMA], a malignant tumour composed chiefly of sympathoblasts.

1927 *Amer. Jrnl. Path.* III. 213 The sympathicoblastomas have often been described as consisting of two types of tissue. **1934** *Jrnl. Path. & Bacteriol.* XXXIX. 29 Sympathicoblastomata are the commonest sympathetic tumours found. **1960** *Hirosaki Med. Jrnl.* XII. 92 (*heading*) A case of sympathoblastoma, suspected to be Ewing's tumor. **1974** *Oncology* XXIX. 521 Tumour biopsies of a sympathicoblastoma..were obtained when the tumours were removed surgically.

sy:mpatholytic (-li·tik), *a. Med.* [f. *SYMPATHO- + *-LYTIC.] Annulling or opposing the transmission of nerve impulses in the sympathetic system. Also **sympa:thicoly·tic** *a.*, in the same sense.

1947, 1948 Sympathicolytic [see *ADRENO-]. **1951** Sympatholytic [see *DIOXAN]. **1952** *Acta Endocrinol.* IX. 116 The alarm reaction caused by adrenaline..can be counteracted by using a sympathicolytic agent. **1954** *Brit. Jrnl. Pharmacol.* IX. 236 The assay of sympatholytic (anti-adrenaline) drugs. **1961** *Lancet* 26 Aug. 475/1 Failures of wholly different origin are those due to administration of sympatholytic or ganglion-blocking drugs. **1977** *Ibid.* 19 Mar. 650/2 The medical treatment of essential hypertension is currently based almost exclusively on sympatholytic drugs of one kind or another on the assumption that the disease is caused by over-activity of the sympathetic nervous system.

sy:mpathomime·tic, *a.* (and *sb.*) *Pharm.* [f. *SYMPATHO- + MIMETIC *a.* (and *sb.*).] Producing physiological effects characteristic of the sympathetic nervous system (as raised blood pressure and rate and depth of breathing, decreased secretion and tone of smooth muscle) by promoting stimulation of sympathetic nerves. Also as *sb.*, a substance which does this. Also **sympa:thicomime·tic** *a.* (and *sb.*), in the same sense.

1910 BARGER & DALE in *Jrnl. Physiol.* XLI. 21 A term at once wider and more descriptive than 'adrenine-like' seems needed to indicate the type of action common to these bases. We propose to call it 'sympathomimetic', a term which indicates the relation of the action to innervation by the sympathetic system. **1949** KOESTLER *Insight & Outlook* xx. 281 In contrast to the sympathomimetic hormones, the vagus substance is rapidly destroyed. **1956** *Nature* 7 Jan. 44/2 The presence of sympathomimetic activity in adrenergic nerve tissue has been demonstrated. **1964** W. G. SMITH *Allergy & Tissue Metabolism* ix. 91 Sympathomimetic amines and theophylline are believed to work by a bronchodilator action. **1966** *Acta Physiol. Scand.* LXVII. 482 Peripherally they are generally classified as indirectly acting sympathomimetics, i.e. they depend on an intact sympathetic nervous system for their activity. **1958** *Dis. of Chest* XXXIII. 18 (*heading*) Depressed response to intravenous sympathicomimetic agents in humans during acidosis. **1970** PASSMORE & ROBSON *Compan. Med. Stud.* II. ix. 6/1 Many of the substances most useful in asthma, e.g. sympathomimetic drugs.., act not as specific agents but by producing an opposing, and often overriding, effect on the bronchi. **1973** *Brit. Jrnl. Hosp. Med.* IX. 21/1 Inhalation challenge using agents producing a type I skin response often induce immediate airways obstruction with asthma, reversible by sympathomimetics. **1983** *Amer. Rev. Respiratory Dis.* CXXVII. 413 Airway resistance..and lung volume were assessed before and after inhalation of a β_2-sympathicomimetic.

sympathy. *sb.* Add: **4.** *Comb.* **sympathy card**, a printed card expressing condolence on a bereavement; **sympathy strike** = *sympathetic strike* s.v. *SYMPATHETIC *a.* 3 b; hence **sympathy striker.**

1967 'T. WELLS' *Dead by Light of Moon* (1968) x. 102 'Sympathy cards? Oh yes.' I remembered now. She wrote greeting card verses. **1976** *Billings* (Montana) *Gaz.* 27 June 3-D/4, I left it in a phone booth while I was writing a sympathy card to be mailed. **1937** *Sun* (Baltimore) 19 Mar. 2/3, 200 women in the South Unit sewing department..struck because of a wage dispute. Some 280 other women seamstresses in the North Unit staged a one-hour 'sympathy strike'. **1973** *Morning Star* 28 Aug. 3 (*heading*) Chrysler hit by sympathy strikers. **1981** *Sunday Tel.* 22 Mar. 6/6 The first sympathy strike by students of an American university has been organised in support of demands made by students on strike at a brother-campus in Britain.

sympatisch, erron. var. *SYMPATHISCH *a.*

sympatric, *a.* (In Dict. s.v. SYM-.) Substitute for def.: Occurring in the same geographical region, or in overlapping regions. Opp. *ALLOPATRIC *a.* (Further examples.)

1904 E. B. POULTON in *Trans. Entomol. Soc.* V. p. xc, Forms found together in certain geographical areas and not in other areas. Such groups may be called *Sympatric*. **1942** E. MAYR *Systematics & Origin of Species* vii. 149 The gaps between sympatric species are absolute, otherwise they would not be good species; the gaps between allopatric species are often gradual and relative. **1953, 1958** [see *ALLOPATRIC *a.*]. **1974** *Nature* 16 Aug. 540/1 The two species are sympatric throughout much of their range. **1978** *Ibid.* 21 Sept. 256/1 White makes a good argument for sympatric speciation on small oceanic islands with many species and also for allochronic speciation.

Hence **sympa·trically** *adv.*; **sympatry**, substitute for def.: the occurrence of sympatric species or forms; (further examples).

1904 E. B. POULTON in *Trans. Entomol. Soc.* V. p. xc, The occurrence of forms together may be termed Sympatry, and the discontinuous distribution of similar forms Asympatry. **1968** *Amer. Mus. Novitates* No. 2349. 6 The grasslands of this region are generally similar to those of the area of sympatry west of Bahía Blanca. **1970** *S. Afr. Jrnl. Sci.* LXVI. 392/1 The two species have been found to occur sympatrically over a depth range of 14 to 33 metres. **1973** *Nature* 9 Feb. 406/2 The planting of dense agricultural stands of larval foodplants..[is] believed to have affected the species' geographical ranges and abundances, causing extensive sympatry over much of the eastern United States. **1975** *Jrnl. Zool.* CLXXVII. 330 True polymorphism is thus restricted to multiple forms of a species which regularly occur sympatrically (and synchronically) within a population.

symphonette (simfǒne·t). *rare.* [f. SYMPHON(Y + -ETTE.] A popular musical composition in classical symphonic form (cf. *symphonic jazz* (*a*) s.v. *SYMPHONIC *a.* (*sb.*) 3); a short symphony.

1947 A. EINSTEIN *Mus. Romantic Era* xi. 131 *Overture, Scherzo, and Finale*.., a work that Schumann in all seriousness wanted to bring out as his Second Symphony, or at least as a 'Symphonette'. **1955** L. FEATHER *Encycl. Jazz* 201/1 Completed mambo symphonette in three movements, March 1955.

symphonic, *a.* (*sb.*) Add: **3.** (Further examples.) *symphonic ballet*, a ballet choreographed to the music of a symphony, with an emphasis on pattern rather than plot; *symphonic jazz*, (*a*) jazz influenced by the form and instrumentation of classical music; (*b*) classical music scored and performed in jazz style.

1886 A. L. ALGER tr. *Reissman's Life & Works R. Schumann* iii. 57 Schumann took an important step forward in the path of his progress... Thus arose: 'The Carnival'.., the 'Symphonic Studies' (Op. 13, 1834) [etc.]. **1913** *Times* 3 Oct. 8/5 The two new works—Sir Edward Elgar's symphonic study for orchestra, *Falstaff*, and Mr. Hamilton Harty's setting..of..'The Mystic Trumpeter'. **1926** WHITEMAN & MCBRIDE *Jazz* iii. 58 Symphonic jazz had proved so successful that the Alexandria's cover receipts had risen from $300 to $1200 a day. **1929** *Metronome* Jan. 32/1 Whiteman put jazz in its Sunday dress and made it respectable. He applied the jazz treatment to the classics and established symphonic jazz which could be scored on paper. **1934** S. R. NELSON *All about Jazz* v. 101 Grofé has added 'Grand Canyon Suite'..to his personal contribution to the field of symphonic Jazz. **1936** *Times* 24 June 14/3 'Les Presages', M. Massine's first symphonic ballet..was revived at Covent Garden last night. **1947** W. MELLERS *Stud. Contemp. Mus.* xi. 176 The string quartet Variations, and even Rawsthorne's biggest and most important work, the Symphonic Studies, are more freely based on the same notion of the variation of form. **1958** G. LASCELLES in P. Gammond *Decca Bk. Jazz* viii. 104 Paul Whiteman absorbed the nucleus of the Goldkette Orchestra into his own symphonic jazz group. **1964** RAFFÉ & PURDON *Dict. Dance* 487/1 Most choreographers who have attempted symphonic ballet have..used the music as a basis for vague generalisations by way of theme, while avoiding the technical aridity of 'abstract ballet'. **1976** *New Yorker* 26 Jan. 96/3 The big symphonic ballet in the Allegretto is a space-filling geometrical composition.

symphonically (simfǫ·nikăli), *adv.* [f. SYMPHONIC *a.*: see -ICALLY.] In a symphonic manner; as or like a symphony. Also *transf.*

1854 H. F. CHORLEY *Mod. German Mus.* v. 274 We may arrive at some canons of dramatic orchestral effect, not easy to reconcile with the practice of those writers who have treated Opera symphonically. **1923** G. SAINTSBURY in *Times Lit. Suppl.* 4 Jan. 2/1 There undoubtedly is room for ametric and unrhymed but symphonically rhythmed verse. **1927** R. VAUGHAN WILLIAMS in *Radio Times* 3 June 440/3 The words as well as the music are treated symphonically. **1929** *Sunday Dispatch* 13 Jan.

16 Paul Whiteman records have a wonderful following—chiefly because he can treat jazz symphonically. **1972** *Human World* Feb. 3 Hence the aptness of the symphonically deployed arch-motif in the imagery [of Racine's *Phèdre*]—the repeated reference to monsters. **1977** *Gramophone* Mar. 1457/1, I like both symphonies very much indeed even though I am not sure that they work symphonically.

symphonism (si·mfǒniz'm). [f. SYMPHON(Y + -ISM.] Music of a symphonic kind; symphonies collectively.

1965 *Listener* 27 May 805/2 Operatic music, which has to encompass drama and accommodate it as a further musical element, needs a more flexible technique than the pure music of symphonism. **1973** *Radio Times* 15 Nov. 60 A series of 13 concerts. 7: 'Cyclic' Symphonism.

symphonist. 2. Restrict † *Obs.* to sense in Dict. and add: Also, a player in a symphony orchestra. *rare.*

1964 M. MCLUHAN *Understanding Media* (1967) II. xxiii. 378 The satisfactions are just as few for the.. symphonists, since a player in a big orchestra can hear nothing of the music that reaches the audience.

symphonize, *v.* Restrict 'Now *rare* or *Obs.*' to sense 1–3 and add: **3. b.** *trans.* To accompany musically.

1801 C. WILMOT *Let.* 29 Nov. in T. U. Sadleir *Irish Peer* (1920) 4 During the dinner..we were symphoniz'd by republican tunes, play'd outside the window. **1802** —— *Let.* 19 Oct. in *Ibid.* 103 A Gothic Castle..symphonis'd by the music of the waters.

4. To give the character or style of a symphony to (a piece of music), to render symphonic.

1932 *Amer. Speech* Apr. 241 Jazz is meant for the mass, it isn't meant to be symphonized, and all attempts at symphonization have been no more than negligible.

Hence **symphoniza·tion**; **sy·mphonized** *ppl. a.*, composed in the manner of a symphony.

1932 Symphonization [see sense 4 of the vb. above]. **1946** R. BLESH *Shining Trumpets* i. 14 A spate of symphonized jazz and pseudo-jazz master-works.

symphony. Add: **5. d.** *ellipt.* for 'symphony orchestra'.

1926 WHITEMAN & MCBRIDE *Jazz* xiv. 287 The unknown composer has to pay to get his compositions played by a good symphony. **1934** S. R. NELSON *All about Jazz* v. 87 Symphony work, although of the highest *ton*, is not very lucrative, and most players have additional sources of income. **1968** *Globe & Mail* (Toronto) 17 Feb. 23/4 The former manager of the Vancouver Symphony. **1977** *Times* 23 Apr. 11/3 The seven arias skimpily supported by the Barcelona Symphony.

6. *attrib.* and *Comb.*, as (sense 5 b) *symphony concert, form, orchestra.*

1863 *Dwight's Jrnl. Mus.* XXIII. 110/3 Our concern now is with the concerts... To begin with the most important, those of the Orchestra, the so-called 'Philharmonic', or Symphony concerts. **1919** *Daily Mail Year Bk.* 200/1 Conductor of the Promenade Concerts since 1895, the Queen's Hall Symphony Concerts. **1956** A. H. COMPTON *Atomic Quest* ii. 68 On one occasion, Mrs. Edward Ryerson saw me as I was seeking a little relaxation at a symphony concert. **1911** *Contemp. Rev.* May 615 The idea that the symphonic poem is a further development of the symphony form. **1881** in Grove *Dict. Mus.* (1884) IV. 43/1 Orchestra to be permanent, and to be called The Boston Symphony Orchestra. **1932** *Daily Tel.* 8 Oct. 1/6 London Symphony Orchestra. **1978** *Ann. Reg. 1977* 404 Their success went a good way towards discounting the much-publicized theory that the conventional symphony orchestra now exists only for the purpose of playing music from the past.

symplasm (si·mplæz'm). *Biol.* [f. SYM- + PLASM.] **a.** *Bacteriol.* A group of bacterial cells that have coalesced into one amorphous mass. ? *Obs.*

1916 LÖHNIS & SMITH in *Jrnl. Agric. Res.* VI. 680 Type D is in most cases the dissolution product either of the large forms [of *Bacillus azotobacter*]..or of the small cells. .. As it is made up by a thorough mixing or melting of a frequently large number of cells, spores, or gonidia, the term *symplasm* or *symplastic stage* seems to be a correct and convenient name. **1923** *Anat. Rec.* XXVI. 69 The bacteria coalesce and resolve into a sort of plasmodium. This plasmodium is the symplasm. Later, in the completion of the life-cycle, bacteria are again formed by the breaking up of the symplasm. **1934** A. T. HENRICI *Biol. Bacteria* ix. 153 Such symplasms are found in old cultures and they probably represent masses of gum secreted by the bacteria, or more likely, masses of débris formed from dead and dissolved bacterial cells.

b. *Bot.* The cytoplasm of a symplast (sense *b); an interconnected mass of cytoplasm.

1948 *Recueil d. Travaux botaniques Néerlandais* XLI. 5 Up until now, only a few publications have dealt with estimations as to what extent the symplasm is permeable for solutes. **1954** *Nature* 31 July 223/2 The transport in the cytoplasm from cell to cell..without loss to the outer solution indicates that plasma connexions between the cells must exist. The cells behave like a 'symplasm'. **1973** *Planta* CXII. 293 Stelar tissues only accumulate ions when these are supplied through the root symplasm.

Hence **sympla·smic** *a.*, of or pertaining to (a) symplasm.

1923 *Anat. Rec.* XXVI. 70 The symplasmic stage in the life-cycles of bacteria appear[s] to be universal. **1971** *Protoplasma* LXXII. 315 The concept of symplasmic transport between plant cells must take account of the possible role of plasmodesmata.

symplasma (simplæ·zmă). *Med.* Pl. **-plasmata.** [mod.L., coined in Ger. (R. Bonnet 1903, in *Monatsschr. f. Geburtshülfe u. Gynaekol.* XVIII. 8): see SYM- and PLASMA.] A mass of cell nuclei and cytoplasm regarded as formed by the breaking down of the cell walls of the outer layer of the placenta.

1908 *Q. Jrnl. Microsc. Sci.* LIII. 134 The maternal tissue..is universally recognised to undergo catalytic changes, and to pass into a symplasma, towards the composition of which superficial epithelium, proliferated epithelium of crypts and glands, subepithelial connective tissue, leucocytes, and blood have all largely contributed. **1910** F. H. A. MARSHALL *Physiol. of Reproduction* x. 414 After the destruction of the epithelium, the villi penetrate into the deeper tissues of the mucosa by gradually absorbing the symplasmata, and branch to form secondary and tertiary villi. **1923** *Q. Jrnl. Microsc. Sci.* LXVII. 146 A degenerating syncytium is called a symplasma.., a term which can be correctly applied only to maternal structures of a degenerate nature contained in the plasmodium. **1973** BOVING & LARSEN in Hafez & Evans *Human Reproduction* vii. 149/1 By this stage, the rabbit uterine epithelium has become converted into a 'symplasma' or multinucleated syncytium through the disappearance of the cell membranes between the epithelial cells.

symplasmatic (simplæzmæ·tik), *a.* [f. prec. after *plasma, plasmatic.*] **a.** *Med.* Of or pertaining to a symplasma. **b.** *Bot.* = *SYMPLASMIC *a.*

1923 *Q. Jrnl. Microsc. Sci.* LXVII. 156 In the symplasmatic zone of the diploplasma are to be found remains of maternal nuclei, maternal blood corpuscles and various granules. **1974** *Planta* CXIX. 47 Transients are observed in mutated cells when the illuminated green leaf sample also comprises normally green cells and there is a symplasmatic connection between the 2 types of cells.

symplast (si·mplɑst). *Bot.* [f. SYM- + -PLAST.] † **a.** [ad. G. *symplast* (J. von Hanstein 1880, in *Bot. Abh.* IV. II. 9).] A multinucleate cell created either by the fusion of cells into one cytoplasmic mass, or by the division of the nucleus of a single energid. *Obs.*

1894 *Jrnl. R. Microsc. Soc.* 376 Klemm objects to the term 'unicellular' as applied to *Caulerpa prolifera* and similar organisms. They should be regarded rather as 'symplasts', composed of a number of energids. **1900** *Ibid.* 475 When the polyplasts are so completely fused together that their cytoplasms form a single mass in which a number of nuclei are imbedded, Hanstein's term 'symplasts' may be applied. **1912** L. A. BORRADAILE *Man. Elem. Zool.* vi. 116 Groups of similar, unseparated energids are known as syncytia. They may be plasmodia, formed by the union of free energids, or symplasts, formed by the division of the nucleus of a single energid.

b. [ad. G. *symplast* (E. Münch *Die Stoffbewegungen in der Pflanze* (1930) 73).] A continuous network of interconnected plant cell protoplasts.

1938 *Amer. Jrnl. Bot.* XXV. 529/2 Studies on the occurrence of plasmodesmata in living tissues show that where pits occur the protoplasts are commonly connected by these strands. Consequently the 'symplast'..of the root must constitute an interconnected protoplasmic unit. **1976** B. E. S. GUNNING in Gunning & Robards *Intercellular Communication in Plants: Studies on Plasmodesmata* I. 2 Following the evolution of plasmodesmata, the plant body is..composed of two major compartments, for which the terms apoplast and symplast are convenient (Münch, 1930)... The term symplast refers to the interconnected protoplasts, all bounded by a continuous plasmalemma.

Hence **sympla·stic** *a.*, of or pertaining to a symplast or symplasm; *symplastic growth*, the expansion of a common wall between adjacent plant cells during cell enlargement.

1916 [see *SYMPLASM a]. **1930** J. H. PRIESTLEY in *New Phytologist* XXIX. 132 It is proposed to call this alternative method of growth now described symplastic growth. **1981** J. R. BARNETT *Xylem Cell Devel.* ii. 63 This symplastic growth hypothesis..could not explain satisfactorily the type of growth in which an enlarging cell, such as a fibre, increases the number of cells with which it is in contact as it grows.

symplectic, *a.* and *sb.* For **a, b** read **A, B** and add: **A.** *adj.* **2.** *Petrol.* Of a rock or its texture: exhibiting an intimate intergrowth of two different minerals, esp. one where one mineral has a vermicular habit within the other as a result of secondary action. [ad. G. *symplektisch* (C. F. Naumann *Lehrb. der Geognosie* (1850) I. 667).]

1916 J. J. SEDERHOLM in *Bull. de la Comm. Géol. de Finlande* No. 48. 46, I..take the liberty of proposing that the term symplektic, or symplektitic should be used preferably as a designation of secondary intergrowths of two different minerals. **1949** F. H. HATCH et al. *Petrol. Igneous Rocks* (ed. 10) iv. 281 A characteristic feature of certain noritic rocks is the development of symplectic intergrowths along intercrystal boundaries. **1971** *Nature* 3 Dec. 251/3 The decomposition of a fayalitic olivine (Fe_2SiO_4) to symplectic metallic iron and cristobalite provides confirmation of a very low oxygen fugacity.

Hence **symple·ctite**, an intergrowth of this kind; **symplecti·tic** *a.*

1916 J. J. SEDERHOLM in *Bull. de la Comm. Géol. de Finlande* No. 48. 46, I..propose to use for these intergrowths of two minerals plaited together, and generally of second origin, the common designation symplektites (or symplectites). **1949** F. H. HATCH et al. *Petrol. Igneous Rocks* (ed. 10) iv. 282 (*caption*) Myrmekite-like symplectites of orthopyroxene and plagioclase are lobed into the labradorite. **1976** *Nature* 22 Apr. 673/2 Rare symplectites of spinel and pyroxene may result from the dehydration of amphibole. **1979** *Ibid.* 5 Apr. 512/2 Symplectitic diopside is not different from coarser, recrystallised diopside.

symplesite (si·mplĕzəit). *Min.* [ad. G. *symplesit* (A. Breithaupt 1837, in *Jrnl. f. prakt. Chem.* X. 501), f. Gr. πλησιάζειν to bring together (in allusion to its relations to other minerals): see SYM- and -ITE[1].] A hydrated ferrous arsenate, $Fe_3(AsO_4)_2.8H_2O$, found as green triclinic crystals (altering to blue), usu. in aggregates having a coarsely fibrous radial structure.

1844 J. D. DANA *Syst. Min.* (ed. 2) VI. 532 Symplesite... Occurs at Lobenstein in Voigtland, with cobaltic pyrites and dolomite. **1968** I. KOSTOV *Mineralogy* 453 Symplesite is found in spherical aggregates, light green to indigo-blue when oxidized.

symploce. Add: Also **symploke** (-ðkī). (Later example.)

1952 J. D. DENNISTON *Gr. Prose Style* v. 90 Occasionally repetition occurs both at beginning and at end of clause, anaphora being combined with antistrophe. This is the figure known as symploke.

symposiast. 2. (Later examples.)

1930 *Time & Tide* 16 May 638 This new sally is directed against the Religious Symposiasts of the popular press, against the well-known writers who take part in those series called 'Is Prayer Answered?' [etc.]. **1978** *Social Sci. & Med.* XII. 185 The symposiasts insist that their analyses are stimulated by the cultural science, yet cannot specify just how.

symposium. Add: **2. b.** A book consisting of essays on various aspects of a subject contributed by a number of different authors.

1946 *Nature* 19 Oct. 534/1 Advances in biological sciences in the U.S.S.R. within the recent 25 years, 1917–1942. Symposium. Editor-in-chief: L. A. Orbeli. (In Russian.) Pp. 356. **1969** *Listener* 15 May 696/1 A symposium on 20th-century music, published in 1960, contained a fulsome and over-extended reference to a then almost unknown French composer. **1972** *Daily Tel.* 30 Mar. 6 This generously illustrated symposium, by contributors of different denominations, covers a worldwide range of Christian art and architecture. **1979** *Nature* 1 Mar. 102/1 Symposia are at present, perhaps, an over-popular form of publication: in many of them the thread of supposed common interest which binds the essays together is far too tenuous, and indeed, in the case of some complimentary volumes such as this, completely non-existent.

sympotic (simpǫ·tik), *a.* [f. as next.] = SYMPOTICAL *a.*

1972 P. M. FRASER *Ptolemaic Alexandria* I. x. 565 In other fields Asclepiades shows himself an innovator in his adaptation of existing poetical genres to the epigram. This is clearly shown in his sympotic epigrams, which form a major category of his work. **1981** *Times Lit. Suppl.* 6 Nov. 1307/5 The archaic age was the great age of sympotic pottery: potters and painters became rich and famous, producing shapes and painting designs which echoed the sympotic preoccupations of their aristocratic patrons.

sympotical, *a.* For *rare*[−1] read *rare* and add later example.

1981 *Times* 5 Aug. 12/6 The sympotical form is still quite distinctive of British culture from pubs to clubs.

symptom, *sb.* Add: **1. a.** (Further examples.) Esp., in mod. use, a subjective indication, perceptible to the patient, as opposed to an objective one or sign (*SIGN *sb.* 7 e).

1842, etc. [see *SIGN *sb.* 7 e]. **1869** S. FENWICK *Med. Diagnosis* i. 2 Diseases are distinguished from each other either by such alterations in the organs themselves, or their secretions, as can be ascertained by the senses of the observer (physical signs); or by changes in the functions of the parts affected (symptoms). **1922** *Amer. Jrnl. Med. Sci.* CLXIV. 684 The first sign noticed was cyanosis and the first symptom shortness of breath on exertion.

b. Also *Comb.*, as *symptom-free* adj.

1962 *Lancet* 27 Jan. 212/2 Most remain symptom-free, apart from aching calves, thighs and backs. **1980** *Recent Advances in Surgery* X. 396 Only about 45 per cent of patients achieve a perfect, symptom-free, Visick grade I result.

syn-[1]. Add: **1. synanthro·pic** *a.* [ANTHROPIC *a.*], living in habitats made or altered by man; **syna·pomorphy** *Taxonomy* [f. *apomorphy*, f. APO- + Gr. μορφή form], the possession by two organisms of some character (not necessarily the same in each) that is derived from one character in an organism from which they both evolved; also = next; so **syna·pomorph,**

any such derived character; **syncyano·sis** *Bot.* (pl. **-o·ses**) [ad. G. *syncyanose* (A. Pascher 1914, in *Ber. d. Deutsch. Bot. Ges.* XXXII. 340)], the relationship between a unicellular blue-green alga and a host within which it lives symbiotically; also *concr.*, the organisms themselves; **syne·chthran** *Ent.*, an insect that lives with ants or other social insects as an unwelcome guest in a relationship of synechthry; **synechthry**, add: [ad. G. *synechthrie* (M. E. Wasmann 1896, in *3ème Congr. Internat. Zool.* 412)]; **sy·nform** *Geol.*, a fold that is concave upwards, irrespective of the chronological sequence of the strata; cf. SYNCLINE in Dict. and Suppl.; **synkinema·tic** *a. Geol.*, formed or occurring when moving or as an accompaniment to motion; **synneu·sis** *Petrol.* [Gr. νεῦσις swimming], the clustering together of crystals of a mineral in a rock; freq. *attrib.* in *synneusis texture*; **synœkete** (sinī·kīt) *Ent.* [ad. Gr. συνοικέτης house-fellow, f. συνοικεῖν to live together (f. οἶκος house): cf. G. *synœkie* (M. E. Wasmann 1896, in *3ème Congr. Internat. Zool.* 412)], an insect that lives with ants or other social insects without either benefiting or harming them; **synoroge·nic** *a. Geol.* [cf. G. *synorogenese* sb. (H. Stille *Grundfragen d. vergleichenden Tektonik* (1924) 16)], formed or occurring during a period of orogenesis; **sy:nsedime·ntary** *a. Geol.*, formed or occurring at the time of deposition of (the) sediment; **sy:ntecto·nic** *a. Geol.*, formed or occurring during a period of tectonic activity; hence **syntecto·nically** *adv.*; **sy·nteny** *Genetics* [Gr. ταινία band, ribbon], the condition (of genes) of being on the same chromosome; hence **synte·nic** *a.*

1936 *Discovery* Mar. 89/2 There can be no question of post-glacial colonisation of Iceland by other than synanthropic insects. **1971** *Countryman* Summer 187/1 This is probably a yellow slug, *Limax flavus*, a synanthropic species, which lives in and around houses, cellars and old garden walls. **1969** E. MAYR *Princ. Systematic Zool.* x. 202 Derived characters (synapomorphs of Hennig) shared with a more recent ancestor. **1966** DAVIS & ZANGERL tr. *Hennig's Phylogenetic Systematics* ii. 90 It makes no difference whether the synapomorphy consists in the fact that an apomorphous character (a') is present identically in all species..or whether it is present in different derived conditions (a' and a"). **1979** *Nature* 18 Jan. 176/1 This inference is drawn from the fact that lungfish and cows share derived characters (synapomorphies such as internal nostrils, an epiglottis, a two-chambered auricle and so on..) not found in salmon. **1945** F. E. FRITSCH *Structure & Reproduction of Algae* II. 878 A different relation is seen in the association of certain Myxophyceae of small dimensions with Monads or Bacteria (syncyanoses of Pascher). **1967** *Jrnl. Phycol.* III. 37/2 *Cyanophora* is one of the few forms among the syncyanoses thus far found which has been thoroughly studied. **1978** *Bio Systems* X. 74/2 Lee suggested that members of the group originated from the union of a non-photosynthetic cryptomonad stock with cyanobacteria, resulting in an early 'syncyanosis' similar to that seen in *Cyanophora paradoxa* today. **1910** W. WHEELER *Ants* xxi. 382 In the United States *Megastilicus formicarius*.., which is not uncommon in the large mound nests of *Formica exsectoides*, is..a typical synechthran. **1967** J. H. SUDD *Introd. Behaviour Ants* vi. 127 Synechthrans are always treated with hostility by the ants and usually they, in turn, prey on the ants. **1937** BAILEY & MCCALLIEN in *Trans. R. Soc. Edin.* LIX. 81 In the following pages: Antiform means a fold that closes upwards. Synform means a fold that closes downwards. **1978** *Nature* 12 Oct. 539/1 Preserved in a large secondary synform, there occurs a sequence, several hundred metres thick, consisting of ribbon cherts, bedded jasperites, [etc.]. **1932** *Mineral. und Petrogr. Mitt.* XLII. 475 The older Archaean granites of Fennoscandia.. belong to characteristically synkinematic intrusive complexes from an early stage of an orogenic cycle. **1952** T. F. W. BARTH *Theoret. Petrol.* III. 243 Synkinematic granitization is probably responsible for the majority of the large granodiorite and granite batholiths. **1973** J. T. RENOUF tr. *Didier's Granites & their Enclaves* 7 The orogenic granites are classically divided into synkinematic (=synorogenic or syntectonic) and post-tectonic types. **1921** J. H. L. VOGT in *Jrnl. Geol.* XXIX. 321 The individuals of a mineral, segregated from a magma at an early stage, frequently swam together to assemblings or aggregates, the result of which is a structure, for which I propose the term together-swimming structure or synneusis structure. **1959** W. W. MOORHOUSE *Study of Rocks in Thin Section* xi. 241 Probably related to the banded character of the basic complexes is a texture, sometimes called 'synneusis' texture, in which the dark minerals.. tend to occur as lenticular clumps or aggregates. **1967** *Amer. Mineralogist* LII. 529 The preferential character of synneusis for several common individual minerals and mineral pairs. **1973** J. T. RENOUF tr. *Didier's Granites & their Enclaves* xiv. 368 When two rocks contain the same volumetric percentage of phenocrysts, synneusis is greatest in that with the smaller crystals and thus with the greatest number. **1910** W. WHEELER *Ants* xxi. 381 The symphiles represent the élite,..and number hardly more than 300 to 400 species, whereas the synœketes are much more numerous. **1971** E. O. WILSON *Insect Societies* (1972) xx. 390/2 Most of the time..the *Cremastocheilus* have the status of synoeketes, that is, they are simply ignored and allowed to wander through the nest without interference. **1936** *Bull. Amer. Assoc. Petroleum Geologists* XX. 853 Synorogenic movements dating from this time are to be recognized everywhere in this continent. **1971** I. G. GASS et al. *Understanding Earth* xx. 292/1 The synorogenic sediments that accompany mountain building. **1974** *Nature* 4 Oct. 382/2 In Africa the Kibaran belt experienced major tectonism about 1,300 Myr BP..with the subparallel Irumide belt undergoing synorogenic events about 1,100 Myr ago. **1960** *Gloss. Geol.* (Amer. Geol. Inst.) (ed. 2) Suppl. 65/1 Synsedimentary. **1976** *Jrnl. Geol. Soc.* CXXXII. 124 In sheet III the lower contact is a sharp, curved slide plane with occasional synsedimentary striations. **1979** *Nature* 9 Aug. 483/2 This sealing apparently results from a synsedimentary permineralisation caused by colloidal silica. **1942** M. P. BILLINGS *Structural Geol.* xv. 297 Syntectonic intrusives are always forcefully injected bodies, because the magma was moving under the influence of orogenic pressures. **1974** *Nature* 22 Mar. 325/2 In coastal Liberia the geological evidence of actual faulting that could definitely be said to be syntectonic with rifting is lacking. **1956** L. V. DE SITTER *Structural Geol.* xxvi. 391 The syntectonically metamorphosed mica-schists and migmatites. **1979** *Nature* 25 Jan. 290/1 A phase of upright asymmetric folding..with the steep limbs overturned to the northwest took place syntectonically with major brittle thrusting..of all units. **1971** J. H. RENWICK in *Ann. Human Genetics* XXXV. 80 If the inversion and a marker locus studied in the pedigree are syntenic (lying on the same chromosome pair), the marker may be on either side of either breakpoint and the linkage..to one of them may be close and may have a good chance of being detected. **1978** *Nature* 13 July 161/1 Five genes in the mouse..are syntenic and their human homologues have been assigned to human chromosome 1. **1971** J. H. RENWICK in *Ann. Human Genetics* XXXV. 83 The prior probability of the hypothesis of synteny—i.e. that the autosomal marker locus is somewhere on the chromosome pair that bears the inversion—is A/T. **1974** *Sci. Amer.* July 39/1 Assaying a number of clones for various human enzymes therefore provides information on the synteny of genes.

2. *Chem.* Designating geometrical isomers of organic compounds containing C=N or N=N in which the principal atoms or groups attached to the doubly bonded atoms are on the same side of the plane of the double bond; usu. italicized. Also without hyphen as an independent word. [Introduced in Ger. by A. Hantzsch 1894, in *Ber. Deut. Chem. Ges.* XXVII. 1702.]

1894 *Jrnl. Chem. Soc.* LXVI. I. 454 Such diazo-compounds as exist in the form of rings, due to the formation of inner anhydrides..must be syn-compounds. **1913** T. H. POPE tr. *Molinari's Treat. Gen. & Industr. Org. Chem.* 568 It forms a mixture of phenyldiazonium hydroxide..and syn-diazobenzene hydroxide. **1938** R. L. SHRINER et al. in H. Gilman *Org. Chem.* I. iii. 385 The amine oxide structure does not aid in accounting for the *syn* and *anti* forms of these oximes, but is necessary to account for the tautomerism of these isomers. **1978** *Nature* 9 Feb. 494/2 The intense sweetness of the *α-syn*-oxime of perillartine was first reported in 1920.

syn-[2] (sin), comb. form of SYNTHETIC *a.*, used to form words denoting synthetic products, as **sy·ncrude**, a synthetic product made from coal in imitation of crude oil; also as *adj.*; **sy·nfuel**, any fuel made from coal, oil shale, or the like as a substitute for a petroleum product; **sy·ngas**, a mixture of carbon monoxide and hydrogen, esp. when produced from coal; **sy·njet**, jet fuel derived from synthetic crude oil (syncrude); **sy·noil**, synthetic oil; **sy·nroc** [Roc(K *sb.*[1]], any of various synthetic crystalline materials composed chiefly of oxides of metals and semimetals and devised as sufficiently stable to contain radioactive waste in solid solution deep underground.

1971 *Kirk-Othmer Encycl. Chem. Technol.* (ed. 2) Suppl. 189 Both oils are subsequently hydrotreated to produce a syncrude oil. **1976** *Times* 9 Dec. 27 Looking at the alternative power sources for private transport, the survey reckons that the most likely ones are a synthetic liquid fuel (such as methanol or syncrude) derived from coal, or electricity stored in batteries. **1980** *McGraw-Hill Yearbk. Sci. & Technol.* 303/2 Salable by-products of ammonia, sulfur, and phenols are produced by several of the SNG and syncrude processes. **1976** *Dallas Morning News* 22 Sept. 2-D/3 Whatever has happened to all the synthetic fuel we were supposed to get to ease the oil and gas shortage? Now, three years later, we've still done very little towards the development of 'synfuels'. **1980** *Science* 16 May 740 Certain processes for developing some U.S. oil shales may generate more CO_2 per unit of usable energy produced than any other synfuel development. **1982** *Sunday Times* 9 May 54/6 Multi-billion investment—the basic fee to gain entry to the synfuels game—cannot be justified. **1975** *N.Y. Times* 24 Mar. 20/2 Much of the Western coal has been planned for conversion at the mine to synthetic pipeline gas... The 'syn-gas' is to replace natural gas from wells, the fuel that is expected to be in the most critical depletion by 1985. **1980** *Prospects for Petrochemicals in W. Europe* (Shell Internat. Petroleum Co.) 8 By the middle of next century it is possible that the petrochemical industry could even be sustained very largely, if not entirely, on syngas and methanol derived from coal and methane. **1983** *New Scientist* 28 Apr. 207/2 Syn-gas is also made from natural gas..by the related reaction $CH_4+H_2O=CO+3H_2$. **1979** *Ibid.* 7 June 818 In the long term, the choice must be between jet fuel derived from synthetic crude (synjet) or a wholly new type of fuel. **1980** *Times* 21 Feb. 20/4 The quick and easy solution, which is 'synjet'—kerosene made from coal, shale or tarsands. **1976** *Time* 1 Mar. 47 So far several plants have been..designed to turn 2,700 tons of high-sulfur Illinois coal into 22 million cu. ft. of 'syngas' and 3,000 bbl. of 'synoil' each day. **1978** *Nature* 3 Aug. 413/1 Whereas glassified waste may devitrify when exposed to ground water at high temperature and pressure, thus exposing a large surface area for the dissolution of the radionuclides in the glass, the new mineral—'synroc'—should be as stable as a natural rock. **1980** *New Scientist* 3 July 9/2 In the Synroc process the radioactive wastes are trapped in the crystal lattices of the minerals of the synthetic rocks and so are completely immobilised. **1982** *Nature* 9 Dec. 470/3 The plant..will make Synroc-C, which consists of 60 per cent titanium dioxide, with an admixture of barium oxide, calcium oxide, zirconia and alumina.

synæresis. Add: **2.** *Physical Chem.* The contraction of a gel accompanied by the separating out of liquid.

1864 T. GRAHAM in *Proc. R. Soc.* XIII. 336 In the jelly itself, the specific contraction in question, or synæresis, still proceeds. **1937** *Jrnl. R. Aeronaut. Soc.* XLI. 535 The material in tension might be explained by assuming (in accordance with the phenomenon of syneresis) that the solid portion of the isogel is in a state of contraction relative to the less condensed portions. **1974** *Encycl. Brit. Macropædia* IV. 857/2 A flocculated paste, or suspension of very fine particles, often behaves as a gel... The systems are often thixotropic... They show syneresis.

synæsthesia. Add: **1. c.** (Further examples.)

1895 *Amer. Jrnl. Psychol.* VII. 90 The study of the varying forms of persisting abnormal association, usually known as 'colored-hearing' and 'forms', but grouped together by Theodore Flournoy, under the convenient name *Synæsthesia*, has hardly..completed the stage of scientific observation. **1935** *Brit. Jrnl. Psychol.* XXV. 31 The most interesting phase of M's synaesthesia is the tendency to see the features of people in different colours. Her acquaintances were not only assigned particular colours, but they were remembered in terms of this colour. **1958** *New Scientist* 6 Feb. 29/3 Synaesthesia is not a commonly reported psychiatric symptom. **1971** *Daily Tel.* 21 Aug. 7/3 Synaesthesia (in his case 'colour-hearing') was observed among his blind patients by an English oculist. **1979** C. PRIEST *Infinite Summer* 40 In the morning my synaesthesia seemed to have receded again.

2. *Lit.* The use of metaphors in which terms relating to one kind of sense-impression are used to describe sense-impressions of other kinds; the production of synæsthetic effect in writing or an instance of this.

[**1901** H. OERTEL *Lect. Study of Lang.* v. 327 The second class of metaphors which ought to receive an exhaustive treatment is the transfer of terms from one sense sphere to the other. These..are illustrated by phrases like 'a sharp tone', 'loud colors'... The phenomenon of synaesthesia has received rather full treatment at the hand of the psychologists, but its reflection on language has not yet received adequate treatment by lexicographers.] **1932** G. STERN in *Göteborgs Högskolas Årsskrift* XXXVIII. I. 323 Synaesthesia is especially common among adjectives ..but there are numerous instances of nouns..: *The sound and light of sweeter songs* (Swinburne). **1936** W. B. STANFORD *Gr. Metaphor* 59 Synaesthesia..amongst certain schools of poetry became almost a major element in the technique of sense-expression. **1960** E. H. GOMBRICH *Art & Illusion* x. 366 What is called 'synesthesia', the splashing over of impressions from one sense modality to another is a fact to which all languages testify. **1977** *N.Y. Rev. Bks.* 24 Nov. 11/1 No child who has attempted a list like Whitman's or a synesthesia like Rimbaud's or a colloquy with the sun like Frank O'Hara's is likely to forget the parent-poem. **1978** *Times Lit. Suppl.* 1 Dec. 1406/4 Synaesthesia is a common technique, even a theme, in his work. **1982** *N. & Q.* June 194/2 The 'inevitable' complement to the serene synaesthesia of passages like the Hawkshead dedication.

3. *Linguistics.* **a.** The expression of more than one kind of sense-impression in the same word. **b.** The transfer of the meaning of a word from one kind of sensory experience to another. **c.** The relationship between speech sounds and the sensory experiences that they represent.

1946 A. G. ENGSTROM in *Philological Q.* XXV. 10 Traces of synaesthesia are as clear in language as in laboratory records... Hornbostel cites a Negro tribe that has a separate word for seeing, but employs a common term for hearing, tasting, smelling, and touching. **1946** S. D. ULLMANN in *Word* II. 114 What Wundt and his disciples term 'complicative change of meaning' is known to the vast majority of other students as 'synesthesia'. **1956** J. WHATMOUGH *Language* x. 191 There is some evidence to indicate that synesthesia such as associates the meanings of colour and sound under a single word may extend to smaller linguistic units. **1957** S. POTTER *Mod. Linguistics* vii. 154 By *synaesthesia* or *intersensory transfer* a word may be given a new sense. **1972** HARTMANN & STORK *Dict. Lang. & Linguistics* 229/1 *Synaesthesia*, the association of a particular sound or group of sounds with a particular meaning, e.g. *fl-* in *flare, flicker, flame*, [etc.]. **1977** *Word* 1972 XXVIII. 309 Phonetic symbolism, described as the appropriateness of some phonemes to nonauditory experience, falls under the general heading of synaesthesia or, in psychological terminology, crossmodal association. *Ibid.*, As a result of the clustering, forced-choice testing yields congruent information not only in synesthesia studies but in phonetic symbolism and semantic differential tests as well.

synæsthesis. Add: **b.** (See quots.)

1922 C. K. OGDEN et al. *Foundations of Aesthetics* 76 Synaesthesis..covers both equilibrium and harmony.

1923 OGDEN & RICHARDS *Meaning of Meaning* vii. 267 We cannot enter here into the details of what, from the standpoint of more or less conventional psychology, may be supposed to happen in these states of synaesthesis. **1943** J. T. SHIPLEY *Dict. World Lit. Terms* 327/2 *Synæsthesis*, the harmonious and balanced concord stimulated by art, as posited in the definition of beauty advanced by Ogden, Richards, and Wood in The Foundations of Aesthetics. **1949** WIMSATT & BEARDSLEY in *Sewanee Rev.* LVII. 40 Among these [types of aesthetic theory] the theory of synaesthesis (Beauty is what produces an equilibrium of appetencies) was the one they themselves [*sc.* Ogden, Richards, & Wood] espoused.

synæsthetic (sinísþe·tik) *a.* (*sb.*) Also synesthetic. **1.** [f. SYNÆSTHESIA, after *anæsthetic.*] Of, pertaining to, or exhibiting synæsthesia. Also *absol.* or as *sb.*, a synæsthetic person. So **synæsthe·tically** *adv.*

1910 *Mind* XIX. 296 Sense-experiences synaesthetically aroused. **1920** R. H. WHEELER *Synaesthesia of Blind Subject* 54 Synaesthetic phenomena in the field of imagery..reveal the same characteristics as do the same phenomena in the field of perception. **1925** *Amer. Jrnl. Psychol.* XXXVI. 530 The process of perceiving a synaesthetically colored month as an emotion,—by which we mean that the emotional response is represented in the various qualities of the colored imagery. **1935** *Brit. Jrnl. Psychol.* XXV. 37 Every case of synaesthesia..consists essentially of a parallel arrangement of two gradient series. They may be series of pitches, intensities..or anything else in keeping with the interests..of the synaesthetic. **1936** W. B. STANFORD *Gr. Metaphor* 47 We shall call.. transferences from the sphere of one sense to that of another *synaesthetic* or intersensal metaphor. *Ibid.* 61 Writers like Poe..and..Ayala affect the same kinds of synaesthetic phrases. Edith Sitwell has 'creaking light' and 'dawn...whining'. **1942** *Jrnl. Gen. Psychol.* XXVI. 213 Such results emphasize the continuity between synesthetic phenomena and more general phenomena of language and "thinking. **1949** KOESTLER *Insight & Outlook* xxiii. 320 It is obvious that such 'synaesthetic' metaphors greatly facilitate the sharing by the reader of the teller's vision. **1951** S. D. ULLMANN *Princ. Semantics* iv. 219 Gombocz developed these distinctions..redefining the essence of synaesthetic transfer which, contrary to Wundt and Roudet, he included among cases of affective sense-similarity. **1977** *Word* 1972 XXVIII. 306 On investigation, a group of phonetic-symbolism, synesthetic, and semantic-differential studies was shown to have produced two groups of semantic qualities which were internally coherent and mutually exclusive. **1979** C. PRIEST *Infinite Summer* 39, I was still affected by the enemy's synaesthetic gas I had inhaled. My perception was disturbed.

2. [f. SYNÆSTHESIS, after *æsthetic.*] Of or pertaining to synæsthesis.

1922 C. K. OGDEN et al. *Foundations of Aesthetics* 91 What we have called the synaesthetic character of the experience.

synagogue. Add: Also (*U.S.*) **synagog.**

1929 *Lit. Digest* 2 Nov. 24/1 New York now has..the largest synagog in existence—the new Temple Emanu-El on a site overlooking Central Park. **1963** R. I. MCDAVID *Mencken's Amer. Lang.* 491 The 1962 Style Book, p. 63 specifies the following:..synagog. **1976** *National Observer* (U.S.) 22 May 16-A/5 Usually he was paid by the synagog and served in various capacities as rabbi, cantor, or schoolteacher.

Synanon (si·nănǫn). [See quot. 1965[2] for the supposed origin.] The name of a U.S. foundation concerned chiefly with the rehabilitation of drug addicts through group therapy; also (with small initial) the method of psychotherapy practised in its centres (see also quot. 1963). Freq. *attrib.*

1961 *Time* 7 Apr. 33/1 Synanon offers more than a few cures. *Ibid.*, The Synanon system cannot work until the addict really decides..to kick the habit. **1963** *Amer. Jrnl. Sociol.* LXIX. 135/1 The free, unrestricted interaction in small groups called 'synanons'. **1965** L. YABLONSKY *Tunnel Back* p. viii, Synanon is a community of former addicts and criminals. *Ibid.*, The word 'synanon' originated with a newly arrived addict... In his attempt to say two 'foreign' words, 'symposium' and 'seminar' in the same breath, he blurted out 'synanon'. *Ibid.* vi. 137 The small-'s' synanon is the group psychotherapy of the total Synanon social structure. **1969** *Guardian* 16 Aug. 7/3 Synanon's communities are not unlike Socialist communes... Everything is free, everyone 'mucks in'. *Ibid.* 21 Aug. 8/4 Spinrad's latest novel..examines the process by which psychotherapy has become a religious experience, spawning synthetic cults like scientology..and the Synanon game. **1976** J. ROWAN *Ordinary Ecstasy* iv. 44 The essence of the Synanon approach is direct aggressive confrontation of the one group member by one or more other members.

synaphe. (Earlier example.)

1740 J. GRASSINEAU *Mus. Dict.* 250 *Synaphe*, a Greek term which signifies, according to Boetius..conjunction; a chord is said to be conjoint, when so placed between two fourths.

synapomorphy: see *SYN-[1] I.

synapse (sinæ·ps), *v. Anat.* [f. the sb.] *intr.* Of a nerve-cell or axon: to form a synapse.

1910 *Practitioner* July 98 The rubro-spinal portion (Monakow's bundle) connects the red nucleus with the opposite side of the spinal cord, probably terminating by synapsing round the anterior horn cells. **1963** R. P. DALES *Annelids* vi. 119 The axon is T-shaped, the cell body lying ventrally or ventro-laterally at the bottom of the T and the tips of the arms synapsing with those of the next neuron. **1979** *Sci. Amer.* Sept. 84/1 Many such cells do not themselves make contact with a motor neuron; they synapse instead on yet other neurons of the great intermediate net.

Also **syna·psed** *ppl. a. Genetics* [cf. SYNAPSIS 2 in Dict. and Suppl.], (of chromosomes) in a state of synapsis.

1931 *Amer. Jrnl. Bot.* XVIII. 370 The synapsed spireme strands traverse the nucleus freely. **1946** *Nature* 21 Dec. 912/1 Perhaps such chemical agents in the egg help in separating the synapsed X-chromosomes. **1974** *Ibid.* 12 Apr. 566/2 The X element contains three synapsed chromosomes.

synapsid (sinæ·psid), *a.* and *sb.* [a. mod.L. *Synapsida* (H. F. Osborn 1903, in *Mem. Amer. Mus. Nat. Hist.* I. 455), f. SYN- + Gr. ἁψίς, ἁψιδ- arch: see -ID[3].] **A.** *adj.* Of or pertaining to the subclass Synapsida, which includes fossil reptiles having a single temporal opening on each side of the skull. So **syna·psidan** *a.* **B.** *sb.* A fossil reptile of the subclass Synapsida, showing a skull structure with some mammalian characteristics.

1903 *Mem. Amer. Mus. Nat. Hist.* I. 460 In all Synapsidan types above the Cotylosauria the squamosals and pro-squamosals early coalesce. **1910** *Bull. Amer. Mus. Nat. Hist.* XXVII. 114 The Synapsid reptiles..may conveniently be approached by a cursory review of the reptilian orders. **1933** A. S. ROMER *Vertebr. Paleontol.* vi. 128 Forms with one opening (in which it was presumed that the two openings had fused into one), [were termed] 'synapsid' (fused-arched) reptiles. **1956** —— *Osteol. Reptiles* II. 473 The mesosaurs..are associated with the synapsid orders. *Ibid.* 474 The synapsids..seem to be a very natural assemblage. **1974** D. & M. WEBSTER *Compar. Vertebr. Morphol.* v. 102 The presence of this bone in both monotremes, where it is apparently functionless, as well as in metatherians is particularly confusing since there is no indication of a marsupial bone in synapsid reptiles. **1980** *Nature* 24 Jan. 378/2 The mammals arose from advanced synapsids in the Upper Triassic. *Ibid.*, An alternative view of the origin of the synapsid skull is that the ancestral condition of the temporal region consisted of a large supratemporal bone which was in contact anteriorly with the postorbital.

synapsis. Add: **2.** Substitute for def.: orig. (see quots. 1895, 1905); in mod. use, chromosomal pairing during the zygotene stage of meiosis.

Delete 1892 reference in Dict. (which is spurious). Quot. 1900 there represents a misunderstanding.

1895 J. E. S. MOORE in *Q. Jrnl. Microsc. Sci.* XXXVIII. 296 The transformation of the cells of the first spermatogenetic period into those of the second, which I have termed the synapsis,..is marked by a peculiar evolution in the chromatin with the formation of peculiar nucleoli.. and by the formation of an archoplasmic constituent round the centrosomes. **1905** —— in *Ibid.* XLVIII. 490 Synapsis represents that series of events which are concerned in causing the temporary union in pairs of premaiotic chromosomes, previously to their transverse separation and distribution, in their entirety, between two daughter nuclei. **1912** *Jrnl. Exper. Zool.* XIII. 348 A number of writers have suggested that the term synapsis..should be abandoned in favour of some less ambiguous word (such as Haecker's term 'syndesis') because it has so frequently been applied to the contraction-figure ('synizesis' of McClung). I am, however, in favor of the retention of the word, for the ambiguity has arisen simply through a misunderstanding of Moore's meaning. He applied the term 'synaptic phase', or 'synapsis', to the series of changes following the last diploid division..in the course of which the apparent number of chromosomes is reduced to one-half. **1960** L. PICKEN *Organization of Cells* iv. 137 Given the mitotic apparatus, the special features of meiosis might follow from the one act of synapsis—the pairing of homologues. **1978** M. W. FARNSWORTH *Genetics* vi. 123 During the zygotene stage homologous chromosomes begin to pair lengthwise with one another, a process called synapsis.

† **synaptenic** (sinæptī·nik), *a. Cytology. Obs.* Also **-tænic.** [ad. F. *synaptène* (H. von Winiwarter 1900, in *Arch. de Biol.* XVII. 54), f. Gr. συν- SYN-: see *-TENE, -IC.] Epithet of the stage of meiosis now known as zygotene.

1900 *Jrnl. R. Microsc. Soc.* 654 The reticulum gives rise to a chromatic thread (deutobroch stage), which at first fills the nuclear cavity (leptotænic stage), and later forms a central dense mass (synaptænic stage). **1922** F. H. A. MARSHALL *Physiol. of Reproduction* (ed. 2) iv. 155 The nucleus enters upon the synaptenic condition, which extends over a somewhat longer time.

synaptic, *a.* (Examples in *Anat.* and *Cytology.*)

1895 J. E. S. MOORE in *Q. Jrnl. Microsc. Sci.* XXXVIII. 287, I therefore propose the term Synaptic phase to denote the period at which this most important change appears in the morphological character of reproductive cells. **1974** D. & M. WEBSTER *Compar. Vertebr. Morphol.* ix. 182 Each terminal bouton contains mitochondria and submicroscopic, membrane-bound spheres called synaptic vesicles. **1976** SMYTHIES & CORBETT *Psychiatry* iv. 42 When a nerve-impulse passes down an axon and reaches the terminal a chemical is released which crosses the synaptic cleft and causes depolarization of the next neurone. **1981** *Sci. Amer.* Oct. 122/2 The active form of the transmitter molecule is stored in the sacs called synaptic vesicles until the nerve cell is called on to release it.

synaptinemal, var. *SYNAPTONEMAL *a.*

synapto- (sinæ·pto), ad. Gr. συναπτ-ικός, connective, used as comb. form of SYNAPSE, in various terms in *Physiol.*, as **synaptoge·nesis**, the formation of synapses between nerve cells; **synapto·logy**, the study of the structure and operation of synapses; **syna·ptosome** [*-SOME[4]], a presynaptic nerve ending which, when isolated, seals up to form an intact sac; hence **synaptoso·mal** *a.*

1967 D. P. PURPURA in A. Minkowski *Regional Devel. of Brain in Early Life* 131 We started this morning with looking at myelinogenesis, moved to cytoarchitectonics, and started talking about the probable growth of dendrites. You have now moved us into a fourth area of maturational considerations—that of synaptogenesis. **1979** *Experientia* XXXV. 207/1 Intracerebellar connections are gradually established as the synaptogenesis proceeds. **1962** *Anat. Rec.* CXLII. 332/2 (*heading*) An electron microscope study of the stratum radiatum of the rat hippocampus..with emphasis on synaptology. **1965** *Sci. Amer.* Jan. 56/3 Sir Charles Sherrington..laid the foundations of what is sometimes called synaptology. **1975** *Nature* 8 May 176/2 There have been great advances in knowledge of synaptology from electron microscopic studies of the retina. **1970** *Neurosciences Res.* III. 6 There is more than a sixfold increase in the ATPase activity of the rat brain nerve-ending fraction from prenatal to the 10-day-old animal, the enzyme apparently residing in the synaptosomal limiting membrane. **1978** *Nature* 17 Aug. 706/2 The crude synaptosomal pellet was resuspended in 0·32 M glucose..and equilibrated..in a rotary waterbath. **1964** V. P. WHITTAKER et al. in *Biochem. Jrnl.* XC. 293/1 The club-like presynaptic nerve endings resist disruption and are snapped or torn off from their attachments to form discrete particles (nerve-ending particles) in which all the main structural features of the nerve ending are preserved. For these particles we propose the name 'synaptosomes' in order to emphasize their relative homogeneity and their resemblance in physical properties to other subcellular organelles. **1973** *Nature* 9 Mar. 122/1 Isolation of intact synaptic nerve endings (synaptosomes) has made it possible to investigate transport across synaptic membranes.

synaptonemal (sinæptonī·măl), *a. Cytology.* Also **synapti-**. [f. Gr. συναπτ-ικός connective or *SYNAPTO- + Gr. νῆμα thread.] *synaptonemal complex*: a set of several parallel threads seen adjacent to and coaxial with pairing chromosomes in meiosis.

1958 M. J. MOSES in *Jrnl. Biophysical & Biochem. Cytol.* IV. 637/1 The term 'chromosomal core' was applied to the axial complex when it was first described... A more precise term would..indicate that the structure is associated specifically with chromosome pairing, and that it is thread-like. Unwieldy though it is, *synaptinemal complex* is more accurately descriptive. **1969** —— in *Genetics* LXI. Suppl. 50 It is proposed that 'synaptonemal complex', because it is similar to the original term and has been employed in the literature to refer to the structure in meiotic bivalents, henceforth be used in place of 'synaptinemal complex'. *Ibid.*, The synaptonemal complex (SC) is a regularly occurring, coplanar set of parallel strands (usually three), coaxial to meiotic bivalent chromosomes. Presence of this linear complex is prerequisite to, but not alone sufficient for chiasma formation (and hence, crossing-over). **1971** *Nature* 3 Sept. 48/1 Chromosome synapsis mediated by the synaptonemal complex seems to be non-specific, for it can also pull together non-homologous chromosomes. **1978** *Bio Systems* X. 111/1 A clear prediction of my phylogeny is that meiotic synaptinemal complex proteins should be homologous in all organisms.

sync, synch (siŋk). orig. *U.S.* Also **sink.** Colloq. abbrev. of SYNCHRONISM, SYNCHRONIZATION, SYNCHRONIZE *v.*, etc. in Dict. and Suppl. Cf. *lip-sync(h)* s.v. *LIP *sb.* 7. **a.** In technical senses, esp. in *Cinematogr.* and *Television.* Cf. *POST-SYNC(H.

1929 *Photoplay* Apr. 31/2 *In sink*, in synchronism; picture and sound perfectly timed together. **1939** *Reader's Digest* Mar. 41/1 When your [television] set is out of synchronization the image sort of bobs and weaves; it is then 'out of sync'. **1943** *Gloss. Terms Telecomm.* (*B.S.I.*) 77 *Synchronising signal* (*Sync pulse*), a signal sent out periodically by the transmitter in order to keep the receiving system in synchronism. **1945** F. HAMANN *Air Words* 52 *Synch, to*, to synchronize. **1954** *Proc. IRE* XLII. 106/2 With a strong (clean) sync signal, the color-carrier reference signal may be maintained as closely accurate as desired, independent of other factors. *Ibid.* 116/1 The composite system functions as a form of automatic frequency control system when out of sync and as an automatic phase control system when in sync. **1960** *How TV Works* 16/1 A 'line sync' is..the jargon for the timing signal which is given at the end of each separate line of a television picture. **1962** A. NISBETT *Technique Sound Studio* xii. 219 Mixing the output of two gramophone records or tapes playing almost in sync. **1963** MALEY & EARLE *Logic Design of Transistor Digital Computers* x. 275 If they [*sc.* pulsed-circuit flip-flops] are synced with a clock and thus with each other. **1965** *Wireless World* Aug. 389/1 The conditions for direct sync can now be examined. **1966** *Listener* 4 Aug. 160/1 The introduction of new lightweight sync-sound equipment. **1972** M. MUGGERIDGE *Green Stick* ii. 69 The sync frequently went awry, with the words of the song and the movements of the singers lips not tallying. **1973** C. BONINGTON *Everest* xv. 238 Graham and I played with the little synch-sound super 8 mm. cine camera, trying to make a documentary of what life was like at Camp 4. **1977** *Rolling Stone* 5 May 31/1 It was Mercury who would blow a fuse if the lights were out of sync or the PA system malfunctioned. *Ibid.* 16 June 12/3 They wanted it synched to within one frame. **1978** *SLR Camera* Aug. 88/2 For electronic flash the camera is in sync at all speeds from 1/25th sec downwards and with expendable flash bulbs, at all speeds from 1/30th down. **1979** *Mod. Photogr.* Dec. 192/1 Connect an electronic flash to the camera with the proper sync terminal.

b. *gen.* Esp. in phrs. *in sync, out of sync.* Also *fig.*

1961 J. Steinbeck *Winter of our Discontent* II. xiv. 278 Something's going on... I just feel it... Everybody's a little out of synch. **1964** 'R. Macdonald' *Far Side of Dollar* (1965) xxvi. 225 We could step up our schedule and synch our watches, eh? **1966** E. West *Night is Time for Listening* vi. 200 No cops, no State Department. Are we in sync? **1968** T. Wolfe *Electric Kool-Aid Acid Test* xi. 147 Somehow this ties in, *synchs*, directly with what Kesey has just said. **1974** *Times Lit. Suppl.* 8 Nov. 1247/4 Worldly success depends on being, as it were, in sync with the contemporary scene, and it was at this point that Fleming began to get out of sync, never to get properly in again. **1977** *Time* 17 Oct. 42/3 The next thing will be to bring the players' uniforms into sync with the floor design. **1978** J. Irving *World according to Garp* xvii. 352 His watch..was several hours out of sync with the United States; he had last set it in Vienna. **1978** *English Jrnl.* Dec. 50/1 Or is the teaching 'out of synch' with the cognitive development..and the intentions of the learner? **1982** M. Millar *Mermaid* x. 110 She..sensed his uneasiness, his awareness that he was out of sync, out of tune.

syncategorematic, *a.* Add: (Later examples.) Also in extended uses in linguistic analysis.

1931 [see *autosemantic *a.* (*sb.*)]. **1957** G. Ryle in M. Black *Importance of Lang.* (1962) 159 This is what Mill had said of the syncategorematic words. **1966** J. J. Katz *Philos. Lang.* v. 312 Since the meaning of 'good' cannot stand alone as a complete concept, we shall say that the meaning of 'good' is *syncategorematic.* **1972** *Language* XLVIII. 351 Syncategorematic features such as abrupt/non-abrupt and strident/mellow... By this term I mean features which necessarily occur only in conjunction with certain other features. Besides the abrupt/continuant vs. strident/mellow example, voiced/voiceless vs. tense/lax appear to be syncategorematic, as do compact/non-compact vs. diffuse/non-diffuse in vowels. **1975** *Ibid.* LI. 32 Russell's contextual or syncategorematic definition of definite descriptions is equivalent to the conjunction of three propositions, one of which embodies a uniqueness claim.

syncategorematical, *a.* Delete † *Obs.* and add later example.

1935 H. Straumann *Newspaper Headlines* 72 The distinction resembles that of E. Husserl's (categorematical and syncategorematical words).

syncategorematically *adv.* (later example).

1975 *New Left Rev.* Nov.–Dec. 55 Philosophy has no object, in that it is its task to analyse concepts which can only be used syncategorematically, i.e. under some particular description, in science.

synch: see *sync, synch.

synchisite. Add: Also **synchysite** and with pronunc. (-zəit). (Earlier and later examples.) [ad. G. *synchisit* (G. Flink 1901, in *Bull. Geol. Inst. Univ. Upsala* V. 82).]

1901 *Jrnl. Chem. Soc.* LXXX. II. 663 Synchysite.—This new name is applied to a mineral from Narsarsuk, in South Greenland. **1965** *Bull. Geol. Survey Dept. Malawi* No. 15. 124 Concentrations of bastnaesite and synchysite occur in the central core of sideritic carbonatite at Chilwa Island. **1975** [see *syntaxy].

synchro (si·ŋkro). [f. Synchro(nous *a.*] = *selsyn. Freq. *attrib.*

1943 *Appl. Electronics* (Mass. Inst. Technol. Dept. Electr. Engin.) vi. 316 When designed so that the rotor may turn or be turned freely, the device is given various trade names, such as *Selsyn, Synchro,* or *Autosyn.* **1958** W. G. Holzbock *Automatic Control* vii. 122 There are different synchro components, such as the synchro transmitter, synchro receiver, synchro control transformer, etc., which are combined in control circuits in various ways. **1980** J. D. Lenk *Handbk. Controls & Instrumentation* x. 289 A receiver synchro is limited to light loads such as moving a pointer across a scale to indicate the angular displacement of some device operating a transmitting synchro.

synchro- (si·ŋkro), comb. form repr. Synchronous *a.* and related words, as in **sy·nchroflash** *Photogr.*, a flash whose operation is synchronized with the opening of the shutter; **sy·nchro-su·nlight** *Photogr.*, used *attrib.* to designate the use of flash to supplement sunlight; **sy·nchro-swim(ming)** = *synchronized swimming* s.v. *synchronized *ppl. a.*; also *ellipt.* as **synchro.**

1940 A. L. M. Sowerby *Dict. Photogr.* 626 *Synchro-flash photography,* the taking of photographs with a flashbulb synchronised to the shutter of the camera. **1952** *Sci. News Let.* 24 Dec. 416/1 Synchroflash testing device enables both the professional and amateur photographer to check his equipment. **1974** *Encycl. Brit. Macropædia* XIV. 324/1 In the early days of *Life* and *Look,* photographers made great use of so-called synchroflash. **1940** F. J. Mortimer *Wall's Dict. Photogr.* (ed. 15) 316 Synchro-sunlight technique is chiefly of use in connection with figure subjects, where it gives a well-lit figure against a much less well-lit background. **1981** G. L. Wakefield *Beginner's Guide Photogr.* vii. 137 Synchro-sunlight photography has to be done carefully because if the amount of extra fill-in is excessive the flash takes over from the sunshine and the effect is completely false. **1968** G. Rackham *Synchronized Swimming* i. 27 Being so diverse in character Synchro provides a wide field of related activities. **1974** *Observer* (Colour Suppl.) 17 Mar. 75/1 A member of the Great Britain Synchronised Swimming Team..has been doing synchro for 11 years. **1976** *Star* (Sheffield) 3 Dec. 28/7 Eight Nalgo SC swimmers passed their respective synchro-swim grade examinations at Heeley Baths last night. **1976** *Milton Keynes Express* 11 June 11/1 It is hoped to bring both synchro-swimming and water polo to the city in the near future.

synchrocyclotron (si:ŋkro,səi·klǒtrɒn). *Physics.* [f. *synchro- + *cyclotron.] A particle accelerator similar to a cyclotron in which the frequency of the accelerating electric field is decreased as the particles gain energy so as to allow for the concomitant increase in mass and enable greater energies to be achieved.

1947 *Times* 7 July 3/3 One machine being considered for Brookhaven National Laboratory on Long Island, New York, is a huge synchro-cyclotron. **1956** *Nature* 3 Mar. 397/2 These exchanges of views led to the decision to fix the energy of the proposed synchro-cyclotron for the international laboratory at 600 MeV. and that of the proton-sychrotron at greater than 25 GeV. **1971** *New Scientist* 2 Sept. 510/1 The University of Chicago synchrocyclotron..has been shut down. **1973** L. J. Tassie *Physics Elem. Particles* 221 In circular accelerators, such as synchrocyclotrons and synchrotrons, the particles are confined to circular or spiralling paths by magnetic guide fields so that they pass one or several radiofrequency sources a large number of times.

synchromesh (si·ŋkromeʃ). *Mech.* Also **syncro-**. [f. *synchronized mesh.*] **a.** A mechanism that facilitates gear-changing in a motor vehicle by automatically causing gear-wheels to rotate in synchronism before they engage. Freq. *attrib.*

1929 *Amer. Motorist* Oct. 35/2 (Advt.), Syncro-mesh silent shift transmission. **1931** *Automotive Industries* 24 Oct. 644/1 Constant-mesh gears made synchro-mesh transmissions and free wheeling feasible. **1932** *Oxford Times* 23 Sept. 22/3 Free-wheeling and syncro-mesh gears have for some time been almost universal on cars built in the United States. **1933** *Motor* 10 Oct. 525/3 The Citroën Co. was also early in the field with synchromesh. **1950** *Engineering* 22 Sept. 255/3 Synchromesh engagement is provided for second, third and top gears. **1962** *Which? Car Suppl.* Oct. 129/2 These two cars had no synchromesh on first gear. **1976** P. R. White *Planning for Public Transport* iii. 59 A synchromesh or semi-automatic transmission is quite adequate, as steady running replaces the frequent stopping and starting of urban operation.

b. *fig.*

1966 *Listener* 11 Aug. 204/2 They are vision, sound, decor, lighting, and a sense of style. When all these are in synchromesh, as in a well designed gearbox, the show may be a success. **1977** *Guardian* 5 May 14/2 Somewhere, somewhere, waits the perfect partner, the soul-mate, the sexual syncromesh. **1982** *Church Times* 23 Apr. 7/3 *The Flowers and Fruits of the Bible*..is a lovely book, if slightly out of syncromesh.

Synchromism (si·ŋkrǒmiz'm). [f. Syn- + Gr. χρῶμα colour + -ism.] A movement in art resembling Orphism, founded by the U.S. painters Stanton Macdonald-Wright (b. 1890) and Morgan Russell (1886–1953), with emphasis on the abstract use of colour. Also, *loosely* = *Orphism 2. Cf. *synchronism 4.

1912 M. Russell in G. Levin *Synchromism & Amer. Color Abstraction 1910–1925* (1978) ii. 20 This is cubisme, Futurisme, Synchromisme and any isms possible for many years, perhaps centuries. **1913** *Forum* Dec. 768 This brings us to the latest phase of this chaotic and polyglot age of painting—Synchromism, sired by two Americans, S. Macdonald-Wright and Morgan Russell, which seems destined to have the most far-reaching effects of any art force since Cézanne. **1923** J. Gordon *Mod. French Painters* xiv. 149 Synchromism, Simultaneism..and so on, are merely various more or less pretentious methods adopted by artists..to say that they are going to do just as they like. **1937** T. H. Benton *Artist in America* ii. 38 My old friend Wright came back to America before the gathering of the war clouds in Europe. He came back, the founder of a new school, synchromism, which he had flung in the face of Paris. **1958** M. L. Wolf *Dict. Painting* 288 Synchromism usually displayed its purposes in pictures of huge size, the colors forming prismatic patterns. **1978** G. Levin *Synchromism & Amer. Color Abstraction 1910–1925* ii. 20 The fact that the Delaunays had used closely related terminology, possibly at a slightly earlier date, does not, of course, make Synchromism a direct outgrowth of their art.

Hence **Sy·nchromist** *sb.* and *a.*, **Synchromi·stic** *a.*

1913 *Forum* Dec. 769 The Synchromists claim to have discovered the secrets of color. **1916** *Ibid.* Apr. 461 His later paintings have undergone somewhat the Synchromistic vision. **1923** [see *simultaneist]. **1936** *Cubism & Abstract Art* (N.Y. Mus. Mod. Art) 74 The first large Synchromist exhibition was held in Munich in June 1913. **1958** M. L. Wolf *Dict. Painting* 202 Known also as the Synchromist School, it [*sc.* Orphism] was essentially an abstract style. **1970** *Oxf. Compan. Art* 1118/2 Arthur Burdett Frost..helped to spread the ideas of the Synchromists in America. **1974** *Encycl. Brit. Micropædia* VI. 439/3 Although he [*sc.* Macdonald-Wright] denied any connection, his Synchromist theories were also influenced by the contemporary Parisian movement of Orphism.

Synchromy (si·ŋkrǒmi). Also **synchromy.** [f. as prec. + -y³, after *symphony.*] An abstract painting of a type characteristic of Synchromism.

1916 *Forum* Feb. 213 Why not hang a Pre-Raphaelite-Moreau work of Claude Buck beside an ultra-modern Synchromy. **1936** *Cubism & Abstract Art* (N.Y. Mus. Mod. Art) 74 The first purely abstract 'Synchromy' was not shown until the exhibition in Paris in the autumn of that year [*sc.* 1913]. **1974** *Encycl. Brit. Micropædia* IX. 737/3 The two artists were living in Paris, painting abstract works they called 'synchromies'.

synchroneity (siŋkrǒnī·iti, -ē·iti). Chiefly *Geol.* [f. Synchron(ous *a.* + -*eity*, after *simultaneity, spontaneity,* etc.] = Synchronism 1 a.

1909 *Cent. Dict.* Suppl., *Synchroneity,* synchronism; the character or fact of being synchronous; specifically, in *geol.*, supposed synchronism in time of deposition of strata. **1945** *Bull. Amer. Assoc. Petroleum Geologists* XXIX. 427 Facts bearing on synchroneity. **1968** R. G. West *Pleistocene Geol. & Biol.* xii. 286 There remain very many problems of chronology, in particular the synchroneity of pollen zones. **1979** *Nature* 11 Oct. 431/1 Their data indicate an approximate synchroneity of dinosaur and foram extinctions just below anomaly 29, the maximum error being about 100,000 years.

synchronic, *a.* Restrict *rare* to senses in Dict. and add: **3.** *Linguistics.* [tr. F. *synchronique* (F. de Saussure *a* 1913, in *Cours de linguistique générale* (1916) iii. 117).] Pertaining to or designating a method of linguistic study concerned with the state of a language at one time, past or present; descriptive, as opposed to historical or diachronic. Also *transf.* in Anthropology, etc.

1922 L. Bloomfield in *Classical Weekly* 13 Mar. 142/1 One is glad to see, therefore, that Dr. Sapir deals with synchronic matters (to use De Saussure's terminology) before he deals with diachronic. **1927**, etc. [see *diachronic *a.* 2]. **1937** [see *Saussurean *a.*]. **1946** [see *onomatopy]. **1954** [see *process *sb.* 5 b]. **1968** *Jrnl. Assoc. Teachers of Russian* XVII. 8 A synchronic study of a language studies the language of a particular period without reference to what went before or came after, and in practice the period in question is generally our own. **1975** *Listener* 20 Mar. 367/3 Though the 'synchronic' approach of the semiologists is for the moment more fashionable, it is impossible not to be interested in the history of social myths.

synchronical, *a.* Add: **3.** *Linguistics.* = *synchronic *a.* 3.

1949 *Oxf. Classical Dict.* 971 If we accept J. B. Hofmann's distinction of the 'diachronical' and the '(idio)-synchronical' types of grammar..then Kühner-Blass belongs to the synchronical. **1956** *Archivum Linguisticum* VIII. 174 Particularly out of place in a synchronical book are some assumptions concerning Old Polish.

synchronically *adv.* (later examples relating to sense *3 of the adj.).

1935 *Year's Work in Eng. Stud.* 1933 XIV. 48 The seven explanatory aspects include semasiology and morphology treated synchronically or diachronically, and also diachronic phonology. **1947** *Essays & Studies* XXXII. 79 There is, however, one grammar..that feels the need of changing the method, of recording the facts first synchronically, then, in the second part, diachronically. **1968** J. Lyons *Introd. Theoretical Linguistics* i. 46 It does not matter by what route (the number, nature or order of the moves) the players have arrived at the particular state of the game: this state is describable *synchronically* without reference to the previous moves. **1979** *Dictionaries* I. 6 One may wonder whether words like *fro* or synchronically unproductive suffixes like *-ure*..are really indispensable elements of the defining vocabulary.

synchronicity (si:ŋkrǒni·siti). [f. Synchronic *a.* + -ity.] The name given by the Swiss psychologist, C. G. Jung (1875–1961), to the phenomenon of events which coincide in time and appear meaningfully related but have no discoverable causal connection.

1953 *Jrnl. Soc. for Psychical Res.* XXXVII. 28 Synchronicity, he [*sc.* Jung] explains, is not just synchronousness. In a 'synchronicity phenomenon', as he uses the phrase, two contemporaneous events are linked together in a meaningful manner. **1955** R. F. C. Hull tr. *Jung & Pauli's Interpretation of Nature & Psyche* i. 27, I have picked on the term 'synchronicity' to designate a hypothetical factor equal in rank to causality as a principle of explanation. **1963** *Punch* 25 Dec. 910/2 Hauntings.. magical coincidence ('synchronicity') the lot. **1974** *Sci. Amer.* Jan. 113/2 The Wilhelm-Baynes volume includes the famous foreword by Jung in which he explains the oracular power of the *I Ching* by his theory of 'synchronicity'. **1980** C. Fitzgibbon *Rat Report* vi. 112 A thought-transference has also no mass, but very considerable energy and therefore 'travels'..through the time element called synchronicity.

synchronism. Add: **3. b.** *spec.* in *Cinematogr.* and *Television.* Cf. *synchronize *v.* 2 c.

1904 *Billboard* 27 Aug. 13/4 The motor of the cinematograph is absolutely dependent on the movement of the phonograph axle, and perfect synchronism must be had..in order to render the illusion as perfectly lifelike

as possible. **1928** *Television* Mar. 37 Thus ensuring synchronism between the transmitter and the receiver. **1957** Manvell & Huntley *Film Music* iii. 75 Nothing..can be more vulgar than music synchronism in films. **1967** *Electronics* 6 Mar. 78/2 (Advt.), The display cathode ray tube on which this output is viewed is scanned in synchronism.

4. = *Synchromism. Cf. *Simultaneism 1.

[**1914** M. Russell *Let.* 12 Mar. in G. Levin *Synchromism & Amer. Color Abstraction 1910–1925* (1978) ii. 20 Please don't say Synchromism which does not express the idea which we called it—'chrome', 'color'.] **1961** M. Levy *Studio Dict. Art Terms* 109 Synchronism, used as alternative expression for Orphism. **1972** C. W. E. Bigsby *Dada & Surrealism* ii. 20 In some ways it was part of that artistic re-examination which spawned such schools as impressionism, cubism, futurism, and more exotically, suprematism, rayonism, plasticism, vorticism and synchronism.

5. *Linguistics.* = *synchrony 2.

1962 [see *diachronism 2].

synchronistic, *a.* Add: **2.** *Linguistics.* = *synchronic *a.* 3.

1937 J. Orr tr. *Iordan's Introd. Romance Linguistics* 284 Internal linguistics is static or synchronistic. **1949** *Archivum Linguisticum* I. 127 On the synchronistic plane, homonymy seems..to preclude the existence of any intrinsic link between form and meaning. **1951** [see *diachronistic *a.*]. **1962** L. J. Cohen *Diversity of Meaning* i. 5 The synchronistic and diachronistic enquiries—studies of a single period and studies through several periods, respectively—can and should complement each other.

3. Pertaining to or having the quality of synchronicity.

1955 R. F. C. Hull tr. *Jung & Pauli's Interpretation of Nature & Psyche* i. 10 Synchronistic events rest on the simultaneous occurrence of two different psychic states. **1972** Koestler *Roots of Coincidence* iii. 95 The precognitive experiences are 'evidently synchronistic', since they are experienced as psychic images in the present as though the objective event already existed'. **1979** G. Adler *Dynamics of Self* 10 Synchronistic phenomena, and in particular those of ESP, convinced Jung of the existence of a transcendental 'absolute knowledge'.

synchronistically *adv.*: also, in accordance with synchrony (sense *2) or synchronicity.

1949 *Archivum Linguisticum* I. 128 Is there any intrinsic and synchronistically valid reason for it [*sc.* a name] to have that form and no other? **1980** C. Fitzgibbon *Rat Report* vi. 112 My communication reaches you synchronistically at the same time as all the other rat reports which have been sent out every five hundred years.

synchronization. Add: *spec.* in *Electr. Engin.* and in other technical uses. Cf. *synchronize *v.* 2 c.

1913 *N.Y. Times* 28 Feb. 3/2 Mr. Edison was looking for perfect synchronization of record and film. **1932** *Radio News* (U.S.) Nov. 867/1 Mr. de Forest has solved the secret of the 'talkie movie' with perfect synchronization. **1938** *Manch. Guardian Weekly* 17 Aug. 134/4 The sound is transformed into light and recorded on the margin of the film in automatic synchronization with the movement of the lips. **1939** *Discovery* July 215 Lodge had shown the importance of tuning or 'synchronization'. **1940** *Amateur Radio Handbk.* (ed. 2) xix. 274/1 When the vision signals are..subject to serious interference which tends to upset synchronization. **1958** *Newnes Compl. Amat. Photogr.* 37 Shutter synchronisation. *Ibid.*, The flash synchronisation may have only an X setting. **1962** S. A. Choudhury in G. A. T. Burdett *Automatic Control Handbk.* iv. 28 If, before the supply is switched on, the rotors are 180 degrees displaced, no synchronisation will take place when the supply is switched on.

synchronize, *v.* Add: **2. c.** In technical senses: to cause to coincide in time; to operate simultaneously or in synchronization. Also *intr.*

1920 *Chambers's Jrnl.* Mar. 206/1 The movements of the mouths of the characters on the scene synchronise with the utterance of the phonograph. **1928** *Radio News* (U.S.) Nov. 867/1 (*heading*), Dr. De Forest demonstrates his invention for synchronizing speech with movies... You.. heard the tone which, to a musically trained ear, synchronized perfectly with every movement of the bow. **1934** C. Lambert *Music Ho!* iii. 196 René Clair would not dare to synchronize one of his crowd scenes to a record of a real bal-musette band. **1940** F. J. Mortimer *Wall's Dict. Photogr.* (ed. 15) 315 Focal-plane shutters can be well synchronized with the flash on small cameras. **1956** *Focal Encycl. Photogr.* 483/1 The peak time of the flash discharge is always shorter than the fastest shutter speed with which it can be synchronized. **1957** *Encycl. Brit.* XXI. 912 B/1 It [*sc.* a video signal]..must have added to it the timing pulses needed to synchronize the receiver. **1962** S. A. Choudhury in G. A. T. Burdett *Automatic Control Handbk.* iv. 39 A recent development which enables the receiver to accelerate from rest and automatically synchronize by simply putting the excitation on the selsyn through a three-pole switch. **1977** J. Hedgecoe *Photographer's Handbk.* 34/2 Cheaper cameras which only take bulbs or cubes are synchronized at low speeds, usually around 1/25 sec.

3. *gen.* To combine or co-ordinate.

1973 *N.Y. Law Jrnl.* 3 Apr. 4/5 The law is probably the only profession that must be synchronized with another profession—writing. **1976** *Time* 27 Dec., facing p. 36 (Advt.), Both media synchronize national interests with multinational scope. **1977** *N.Y. Rev. Bks.* 14 July 33/2 Silberman quotes an unreported speech given by the foreign minister, Milos Minic, which alleges not only that Western intelligence is involved with fascist exiles but also that Western press coverage of Yugoslavia is synchronized to discredit the country.

synchronized, *ppl. a.* Add: (Examples.) **synchronized swimming,** a form of swimming which involves a display of ballet-like routines performed to music (often as a competitive sport); hence **synchronized swimmer.**

1927 *N.Y. Times* 28 Aug. VII. 4/1 During the other portions of the reel there will be a synchronized orchestral score. **1932** *New Yorker* 9 Apr. 51/1 The new Ford has synchronized gear-shifting and a silent second-speed. **1942** *R.A.F. Jrnl.* 16 May 15 The normal armament is two 7·7 mm. synchronised guns. **1950** B. Spears (*title*) Beginning synchronized swimming. **1960** C. H. Gibbs-Smith *Aeroplane* iii. 69 Fokker's monoplane..with its fixed synchronised gun. **1968** *Courier-Mail* (Brisbane) 1 Feb. 11/8 Both women returned to Brisbane last week from the New South Wales synchronised swimming championships... Synchronised swimming is the initiate's term for water ballet. **1979** *Oxf. Compan. Sports & Games* 1014/1 As well as having the endurance of a trained racing swimmer, the synchronized swimmer must have the skill and artistry of a ballet dancer, and the grace, rhythm, and acrobatic ability of the gymnast. **1979** A. Fraser *King Charles II* (1980) xi. 193 There were no fewer than seven clocks in his bedroom (their ill-synchronized chiming drove his successor to distraction). **1981** J. Barnett *Firing Squad* xvi. 221 A synchronized rattle of rifle bolts. **1984** *New Yorker* 19 Mar. 114/2 It is normally very difficult to get a new sport accepted for the Olympics. Synchronized swimming will be on the docket this summer.

synchronizer. Add: Also in other technical senses, esp. in *Cinematogr.* and *Photogr.* Cf. *synchronize *v.* 2 c.

1924 S. R. Roget *Dict. Electr. Terms* 253/2 *Synchronizer.* Apparatus for indicating whether two alternating current machines or circuits are in synchronism. **1931** B. Brown *Talking Pictures* i. 5 The synchronizer itself consisted of a box having a transparent disc, bearing an indicating spot, and coupled by gearing to the movement of the cinema projector. **1940** F. J. Mortimer *Wall's Dict. Photogr.* (ed. 15) 314 A sharp distinction should be drawn between the flashgun and the simpler so-called 'synchronisers' which open the shutter, set at 'Bulb' or slow snapshot speed, before the flash begins, and allow it to remain open till the flash is over. **1949** *Proc. Inst. Mech. Engin., Automobile Div.*, 1947–8 III. 98/2 Constant load synchronizers are generally used, except for Vauxhall ..who use the baulked synchronizer. **1957** Manvell & Huntley *Film Music* i. 17 Throughout the silent period various gramophone synchronizers were developed for use with films. **1970** K. Ball *Fiat 600, 600D Autobook* vi. 53/1 This shaft carries the fourth-speed driven gear and synchronizer ring. **1973** *Times Educ. Suppl.* 21 July 27 The sound..is first recorded on the tape. The tape is then passed through the recording head of the synchronizer... Subsequently, when the tape is played back, the control pulses are passed..to the projector..and each automatically initiates a change of slide at the predetermined point.

synchronizing, *vbl. sb.* and *ppl. a.* (Later examples in technical senses.)

1906 *Encycl. Brit.* III. 136/1 In the latter case the synchronizing signals must be transmitted over a channel separate from the picture channel [in phototelegraphy]. **1928** C. F. S. Gamble *Story N. Sea Air Station* iv. 68 The 'synchronizing gear', enabling a machine-gun to fire through the tractor air-screw. **1931** B. Brown *Talking Pictures* i. 3 The earliest practical attempts at synchronizing, i.e. keeping speech and gesture perfectly in phase. **1940** *Gloss. Terms Telecomm.* (B.S.I.) 55 *Synchronizing,* the adjustment of the frequency of the time base to bear an integral relationship to the frequency of the phenomenon under investigation. **1962** G. Millerson *Technique Television Production* ii. 23 The component parts of the video signal..synchronizing level.

synchronous, *a.* Add: **2. a.** (Further examples.) Cf. sense 2 d below.

1972 *Sci. Amer.* Apr. 45/1 In all cases where this effect is significant these same tides will have 'despun' the satellite to synchronous rotation, that is, the satellite's period of rotation around its own axis equals its period of revolution around the planet and it always presents one face to the planet.

b. (Earlier and later examples.)

1908 A. Hay *Princ. Alternate-Current Working* vi. 88 By a synchronous motor is meant one whose speed bears a definite ratio to the periodicity of the alternating current. **1930** *Whitaker's Electr. Engineer's Pocket-bk.* (ed. 4) 279 If..the converter is in parallel with other synchronous machinery. **1950** *Engineering* 25 Apr. 534/3 (*heading*) Hydrogen-cooled synchronous condenser. **1962** J. Bell in G. A. T. Burdett *Automatic Control Handbk.* iv. 77 Torque synchros or synchronous links (Magslips).

c. *Computers* and *Telecommunications.* Of apparatus or methods of working: making use of equally spaced pulses that govern the timing of operations.

[**1947** A. W. Burks et al. in J. Von Neumann *Coll. Wks.* (1963) V. 68 Since the timing of the entire computer is governed by a single pulse source, the computer circuits will be said to operate as a synchronized system.] **1954** *Trans. IRE Prof. Group Electronic Computers* June 14/2 Because the system being designed was centrally synchronous, over-all timing considerations now came to the fore. **1962** Y. Chu *Digital Computer Design Fundamentals* v. 161 The binary state of the signals in logic circuits can be represented by either of two voltage levels or by pulses... A synchronous computer also requires clock pulses. **1971** I. H. Gould *IFIP Guide Concepts & Terms Data Processing* 76 Synchronous working and asynchronous working often coexist in different parts of a computer system. For example, in many computers the central processor is synchronous, but the operation of peripheral equipment is only initiated by signals from the central processor and thereafter proceeds asynchronously. **1982** Heap & Martin *Introd. Digital Electronics* iii. 73 In the worst case the problem of interfacing two independent synchronous systems which are operating at different clock rates may occur.

d. Of a satellite: rotating round the parent planet at the same rate as the planet rotates. Of an orbit: such that a satellite in it is synchronous.

1961 *N.Y. Times* 30 July iv. 9/8 Synchronous satellites will require bigger boosters to reach their higher altitudes. **1964** *Daily Tel.* 4 Mar. 14/6 Three satellites in this synchronous orbit would give a complete global system of communications except for small regions round the North and South Pole. **1967** *Technology Week* 20 Feb. 4/1 There is substantial agreement that a synchronous satellite is desirable for air traffic control. **1970** *Nature* 2 May 503/1 Only one orbit exists which is at the same time equatorial, circular and synchronous. **1978** *Daily Tel.* 11 July 2/5 'Charon', which brings to 35 the number of known moons in the solar system, appears to have a synchronous orbit around Pluto of 12,000 miles, which means that it always stays over a fixed spot over Pluto.

3. *Linguistics.* = *synchronic *a.* 3.

1936 [see *diachronous *a.* 2].

synchronously, *adv.* Add: **3.** *Linguistics.* According to the methods or conclusions of synchronic linguistics.

1923 L. Bloomfield in *Mod. Lang. Jrnl.* VIII. 318 At any given time ('synchronously'), the language of a community is to be viewed as a system of signals.

synchrony. Add: **2.** *Linguistics.* A synchronic method of linguistic study; synchronic treatment.

1931 L. H. Gray in *Amer. Jrnl. Philol.* LII. 77 Synchrony must determine the nature of these categories. **1955** *Word* XI. 630 The Saussurean antinomy between diachrony and synchrony. **1965**, **1965** [see *diachronic *a.* 2]. **1972** *Language* XLVIII. 438 It has recently been claimed that Georg von der Gabelentz anticipated a number of Saussurean concepts, in particular his dichotomies of langue-parole and synchrony-diachrony.

synchroscope (si·ŋkroskōᵘp). [f. *synchro- + -scope.] **1.** *Electr. Engin.* An instrument for indicating any difference in frequency or phase between two alternating voltages.

1908 V. Karapetoff *Exper. Electr. Engin.* xxi. 494 It only remains..to bring the machine into phase with the voltage at the bus-bars. This is done either by means of properly connected incandescent lamps..or special instruments, so-called synchroscopes. **1953** H. P. Banks *Electricity* I. xvi. 268/2 A..usually adopted method of paralleling two alternators is by means of the rotary synchroscope. **1962** A. T. Wells *Electr. Power Technol.* xix. 364/2 Observe the phase angle between E_a and E by means of a synchroscope.

2. *Electronics.* (See quot. 1945.)

1945 *Electronic Industries* Sept. 126 *Synchroscope,* an oscilloscope on which recurrent pulses or wave-forms may be observed, which incorporates a sweep-generator that produces one sweep for each pulse, regardless of frequency, thus allowing no more than one cycle to be viewed on the screen. **1947** R. Lee *Electronic Transformers & Circuits* ix. 257 This curve is obtained by connecting the vertical plates of a synchroscope..across the transformer output winding. **1952** *Electronic Engin.* XXIV. 229/1 Another specialised synchroscope..is provided to display all details of the television waveform.

synchrotron (si·ŋkrotrɒn). *Physics.* [f. *synchro- + *-tron.] An accelerator in which electrons or protons gain energy from an alternating electric field that is kept in synchronism with them as they travel round a closed orbit in a magnetic field, the strength of this field and (in the case of protons, the frequency of the electric field) being increased to keep the radius of the path constant as the particles gain mass relativistically. Also *transf.*

1945 E. M. McMillan in *Physical Rev.* LXVIII. 143/2 (*heading*) The synchrotron—a proposed high energy particle accelerator. **1947**, etc. [see *proton synchrotron* s.v. *proton 3]. **1950** *Engineering* 24 Mar. 323/1 A new electron synchrotron..at work on problems of photodisintegration and pair production. **1970** *Sci. Amer.* July 79/1 The Crab Nebula is a cosmic synchrotron, permeated by electrons with energies of 1,000 billion electron volts or even higher. **1977** J. D. Lawson *Physics of Charged-Particle Beams* ii. 79 Large synchrotrons consist of a sequence of magnets arranged in a ring separated by 'straight sections', which may not all be of the same length.

2. *attrib.* and *Comb.*, as **synchrotron emission, mechanism, process; synchrotron radiation,** polarized radiation emitted by a charged particle as it spirals at high speed in a magnetic field, as in a synchrotron; the emission of this.

1962 C. Susskind *Encycl. Electronics* 275/2 The visible continuum of the Crab nebula has been accepted as synchrotron emission. **1978** Pasachoff & Kutner *University Astron.* xxiv. 594 Continuum radio radiation can be generated by any of several processes. One of the most important is synchrotron emission, the process that produces the radiation from Taurus A. **1956** *Astrophysical Jrnl.* CXXIII. 550 The synchrotron mechanism..beautifully explains the radiation and polarization of the continuum of the Crab Nebula. **1962** C. Susskind *Encycl. Electronics* 275/2 As few as 10^{-4} relativistic electrons per

cubic centimeter, emitting by the synchrotron process in a field of 10^{-8} gauss, can explain the observations. **1975** *Sci. Amer.* Dec. 38/1 When the electrons spiral along the lines of force of the star's magnetic field, they radiate by means of the synchrotron process, emitting radio waves, visible light and X rays. **1956** *Astrophysical Jrnl.* CXXIV. 416 (*heading*) On synchrotron radiation from Messier 87. **1981** J. B. Adams in J. H. Mulvey *Nature of Matter* vii. 165 The large size of LEP [*sc.* a synchrotron] is due not to its particle energy but to the need to reduce synchrotron radiation losses and to economize on electrical power consumption.

syncline. Add: (See also quot. 1972.) Cf. *synform* s.v. *syn-¹ 1.

1937 *Trans. R. Soc. Edin.* LIX. 81 In common tectonic practice, an anticline has come to be understood as a fold with a core of previously underlying rocks, and a syncline as a fold with a core of previously overlying rocks. **1972** *Gloss. Geol.* (Amer. Geol. Inst.) 718/2 *Syncline,* a fold, the core of which contains the stratigraphically younger rocks; it is concave upward.

syncopate, *v.* Add: **3.** Also *transf.*

1928 *Sunday Express* 27 May 15 Her eager feet, that used to patter back and forth in happy household duties, now syncopate to the beat of drums and the clashing of cymbals. **1966** *Listener* 28 July 142/3 At the back of Albéniz's mind there is generally..a dancer whose castanets are always syncopating against each other. **1983** P. Inchbald *Short Break in Venice* xx. 190 They passed a lighthouse syncopating white above with green below.

syncopated, *ppl. a.* Add: **2. b.** Applied to modern popular music played or composed in the manner typical of ragtime and jazz.

1908 *Catal. Copyright Entries* (U.S. Libr. Congress) 1069/2 Floreine waltz; syncopated, by Ernest J. Schuster. **1929** W. Thurman *Blacker the Berry* 120 They muddled their words and seemed to impregnate the syncopated melody with physical content. **1969** E. Roth *Business of Music* x. 247 Apart from syncopated rhythms, jazz proved unfruitful ground for serious music.

c. Designating an orchestra, composer, etc., associated with popular syncopated music.

1927 [see *cross-rhythm* s.v. *cross- B]. **1928** *Grove's Dict. Mus.* (ed. 3) V. 243/1 Dance bands are frequently spoken of as 'Syncopated Orchestras', less because their music employs syncopation than because their constitution with saxophones, percussive instruments, etc., is designed to emphasize the effects essential to dance music of the American type. **1934** C. Lambert *Music Ho!* III. 222 The composer of highbrow jazz must obviously extend his harmonic vocabulary beyond the somewhat narrow range of the syncopated kings.

d. *fig.*

1924 Wodehouse *Bill the Conqueror* iii. 62 The breeze was stronger now, and it ruffled the surface of the water, so that the goldfish had for the moment a sort of syncopated appearance. **1950** 'D. Divine' *King of Fassarai* xvi. 128 A regular syncopated pattern of shifting light. **1964** E. J. Hobsbawm *Labouring Men* 133 The oddly syncopated rhythm of the European trade-union 'leaps' between 1889 and 1914. **1974** M. Cecil *Heroines in Love* vi. 155 Eventually Jizabel awoke from her syncopated dreamland. **1979** *Jrnl. R. Soc. Arts* Nov. 751/2 This last element [*sc.* a colonnade] modulates back and forth in a rather jerky and syncopated manner.

syncopation. Add: **3. b.** Music characterized by a syncopated rhythm, *spec.* dance music influenced by ragtime.

1921 *Chambers's Jrnl.* Jan. 23/1 The pulsating sound [of the drum] heightens excitement to the verge of frenzy, and indicates the direct origin of the orgiastic African syncopation to which the wives and daughters of the conquering Anglo-Saxon race dance with their men-partners, retained by arrangement, in the aristocratic dance-clubs of London and New York. **1928** *Grove's Dict. Mus.* (ed. 3) V. 243/1 Syncopation has become a general term for all that class of 20th-century dance music which has sprung from the American adoption of rag-time. **1962** Charters & Kunstadt *Jazz* vi. 73 Even the Clef Club Orchestra was advertised as a jazz band, with '50 Joy Whooping Sultans of High-Speed Syncopation'. **1968** P. Tamony *Americanisms* (typescript) No. 23. 4 *Syncopation* described 'Alexander's Ragtime Band' and other printed music in the first decade or so of this century, *ragtime* becoming old shoe and pejorative.

c. *fig.*

1979 *Jrnl. R. Soc. Arts* Nov. 777/1 The stained glass and colour syncopations in blues and greens.

syncopator (si·ŋkǒpēᵗtǝɪ). Also **syncopater.** [f. Syncopate *v.* + -or.] One who performs syncopated jazz music, usu. in a dance band. Freq. *pl.*, in the name of a band.

1926 *Daily Colonist* (Victoria, B.C.) 13 Jan. 6/3 With Professor Hunt's syncopators providing the dance music programme. **1927** *Daily Tel.* 22 Feb. 12/1 You may see (and hear) the first 'Lady Syncopators',..cutting rhythm into jazz patterns with the best of mere male 'syncopators'. **1930** *Dancing Times* Oct. 117/2 The White Star Syncopators, the Cunard Dance Band. **1952** B. Ulanov *Hist. Jazz in Amer.* (1958) ix. 94 The swinging Vendome syncopaters took over the stage. **1970** P. Oliver (*title*) Savannah syncopators.

syncope, *sb.* **2.** Delete *rare* and add later examples. Also *attrib.* and in *Comb.*

1953 K. Jackson *Lang. & Hist. in Early Britain* II. 614 A Pr[imitive] W[elsh] syncope-form *Car'dig. **1972** *Language* XLVIII. 350 The same syncope rule which is optional in Russian /stl/ and /stk/ clusters is obligatory in /stn/ and /zdn/ clusters. *Ibid.*, Maximal distinctiveness, hence retention of the consonant, is manifested in the explicit subcode of contemporary standard Russian; whereas partial absence of distinctiveness, hence syncope of the consonant, is manifested in the elliptic subcode.

syncretic, *a.* Add: **2.** *Psychol.* Relating to or characterized by the fusion of concepts or sensations. Cf. *syncretism 3.

1932 M. Gabain tr. *Piaget's Moral Judgment of Child* ii. 192 Since every word obtains its meaning as a function of these syncretic schemas, words end by acquiring a substance of their own independently of reality. **1952** Werner & Kaplan *Acquisition of Word Meanings* ii. 48 The conclusion can be drawn..that syncretic concepts are more characteristic of the younger children. **1962** I. Sarnoff *Personality Dynamics & Devel.* vi. 126 One variety of syncretic perception..involves a synthesis of sensations that pertain to several different sense modalities. **1969** T. Freeman *Psychopathol. of Psychoses* viii. 126 This thinking defect consists in the re-emergence of condensing or syncretic trends, fusing concepts that in normal circumstances are discrete and autonomous.

syncretically (sinkre·tikăli), *adv.* [f. Syncretical *a.* + -ly².] In a syncretic manner.

1900 W. James *Let.* 10 June in R. B. Perry *Tht. & Char. W. James* (1935) I. 647 Assuming no duality of material and mental substance, but starting with bits of 'pure experience', syncretically taken. **1957** *Times Lit. Suppl.* 27 Dec. 781/3 But he manages to square his religious views..with a staunch advocacy of anthropology and sociology as ancillary techniques in historical method. To say that this position is syncretically achieved would be something of an understatement.

syncretism. Add: **2.** *Philol.* The merging of two or more inflectional categories.

1909 in Webster. **1933** L. Bloomfield *Language* xxi. 388 Homonymy and *syncretism,* the merging of inflectional categories, are normal results of sound-change. **1949** C. E. Bazell in E. P. Hamp et al. *Readings in Linguistics II* (1966) 225 It may not always be possible to draw a fast line between syncretism proper and the neutralisation of a morphemic opposition. **1957**, **1963** [see *defectivation]. **1968** W. J. Samarin in J. A. Fishman *Readings Sociol. of Lang.* 664 Planned languages reveal many of the features of pidgin languages, namely, lexical syncretism and reduction of redundancy. **1979** [see *syntagma 4].

3. *Psychol.* The process of fusing diverse ideas or sensations into a general (inexact) impression; an instance of this.

1926 M. Warden tr. *Piaget's Lang. & Thought of Child* iv. 130 We can discern in this activity of understanding and invention on the part of the child several of those schemas of analogy, of those leaps to conclusions which are the outstanding characteristics of verbal syncretism. **1963** T. R. & E. Miles tr. *Michotte's Perception of Causality* xvii. 276 It is probable that an extreme 'syncretism' (i.e. an undifferentiated blending) holds sway at this time. **1967** A. L. Baldwin *Theories Child Devel.* xvii. 501 An example of syncretism in normal adult functioning can be seen in the close relationship between taste and smell.

syncretistic, *a.* Add: Also = *syncretic *a.* 2.

1926 M. Warden tr. *Piaget's Lang. & Thought of Child* iv. 132 To this childish form of perception M. Claparède has given the name of *syncretistic perceptions,* using the name chosen by Renan to denote that first 'wide and comprehensive but obscure and inaccurate' activity of the spirit where 'no distinction is made and things are heaped one upon the other'. **1976** S. Arieti *Creativity* ix. 195 The artist or viewer has a syncretistic grasp of the total object. He abandons precise visualization and experiences an unclear vision of the whole.

syncromesh, var. *synchromesh. **syncrude:** see *syn-². **syncyanosis:** see *syn-¹ 1.

syncytium. Add: Hence **syncy:tiotro·phoblast,** the outer, syncytial layer of the trophoblast; also, one of the cells that make up this layer; **syncy:tiotrophobla·stic** *a.*

1926 *Jrnl. Anat.* LXI. (*Proc. Anat. Soc.*) 77 The trophoblast consists of a thick folded cellular layer (cytotrophoblast), on the outer surface of which an irregular and as yet thin deeply staining layer of syncytio-trophoblast is in process of differentiation. **1961** *Nature* 29 July 510/1 In human beings, the syncytiotrophoblasts are the fœtal cells in direct contact with the maternal bloodstream. *Ibid.* 511/1 The evidence favours the interpretation of the binding of tagged globulin from post-partum sera by the syncytiotrophoblastic cytoplasm as an immune phenomenon. **1980** *Sci. Amer.* Aug. 82/2 (*caption*) As the invasion proceeds the trophoblast differentiates into two layers, the outer syncytiotrophoblast, which leads the advance into the endometrium, and the cytotrophoblast, which forms a complex system of projections that eventually push through the syncytiotrophoblast into the pools of maternal blood.

† **syndesis** (sindī·sis). *Cytology. Obs.* [a. G. *syndesis* (V. Häcker 1904, in *Zool. Jahrb.* VII. 200), f. Gr. συν- Syn- + δέσις binding together (f. δεσμός bond, connection).] The pairing of chromosomes in mitosis or meiosis. Cf. *synapsis 2.

1909 *Ann. Bot.* XXIII. 49 Haecker has proposed the word *Syndesis* to apply to the conjugation or association of the homologous parental chromosomes. **1912** [see *synapsis 2]. **1925** E. B. Wilson *Cell* (ed. 3) vi. 503 It is now widely held that reduction is initiated by a preliminary process or synapsis or syndesis in the course of which the chromosomes conjugate.

syndesmo-. Add: **synde·smophyte** [-phyte], a bony outgrowth from an injured joint or vertebra.

1957 in *Dorland's Med. Dict.* (ed. 23). **1966** E. W. Boland in J. L. Hollander *Arthritis* (ed. 7) v. xxxix. 648/1 In contrast to the marginal, heavy osteophytes of degenerative disease of the spine, the syndesmophytes of ankylosing spondylitis begin as linear, poorly defined calcifications adjacent to the margins of the vertebral bodies. **1980** Bluestone & Katich in R. Bluestone *Rheumatology* xxiii. 284 (*caption*) Note mature syndesmophytes outlining annulus of several disks and bridging vertebral bodies.

syndetic, *a.* Add: **b.** *Librarianship.* Pertaining to or designating a catalogue, index, etc., which uses cross-references to indicate links between entries. Also used in automatic data-processing.

1876 C. A. Cutter *Rules for Printed Dict. Catal.* 15 *Syndetic,* connective, applied to that kind of dictionary catalogue which binds its entries together by means of cross-references so as to form a whole. **1958** T. Landau *Encycl. Librarianship* 299/2 *Syndetic,* applied to an alphabetical subject catalogue or dictionary catalogue which includes cross-references as connecting links between subjects. *Ibid.*, Systematic catalogues have no need of such a syndetic apparatus. **1968** T. C. Hines *Vocab. Control in Indexing Lit. of Librarianship & Information Sci.* (ERIC doc. No. ED050742) 16 Perhaps because of the concurrent use of shelf classification, library heading lists (although they include a syndetic apparatus which serves some of the same purposes) do not include the kind of classification of the headings themselves found in some thesauri expressed as 'broader terms'. **1974** *Encycl. Brit. Macropædia* X. 869/1 Provision is made for cross-references from unused terms and from one term to a related one. A catalog containing these entries is known as a syndetic catalog. **1977** A. P. Jensen et al. (*title*) An instructional and research laboratory for syndetic analog-digital computation in science and engineering education. **1981** *Resources in Educ.* Oct. 138/2 This module describes the main subject heading, LC classification numbers which accompany the headings, 'see' references, 'see also' references, subheadings, and other syndetic features of the LC headings.

syndeton (si·ndītǫn). *Gram.* [Back-formation from Asyndeton and Polysyndeton: cf. Syndetic *a.*] (See quots. 1954, 1972.)

1954 Pei & Gaynor *Dict. Linguistics* 210 *Syndeton,* a phrase or construction in which the elements are linked together by connecting particles. **1971** *Computers & Humanities* V. 262 The frequency distribution enabled us to see also the amount of initial syndeton..in each sample. **1972** Hartmann & Stork *Dict. Lang. & Linguistics* 230/1 *Syndeton,* a construction, parts of which are linked together by means of conjunctions or joining words, e.g. in *He came and went again.*

syndical, *a.* Add: **b.** In other collocations: of or relating to syndicalism; organized in unions.

1907 I. Zangwill *Ghetto Comedies* 411 Your only remedy is a general strike. You must join the Syndical Anarchists. **1943** G. Brenan *Spanish Labyrinth* xii. 271 The real strength of the C.N.T. lay..in their powers of syndical resistance. **1955** *Times* 5 Aug. 8/3 General Perón said that the syndical organization of the people fought for ideals and interests.

syndicalist. For 'Also *attrib.*' read 'Also *attrib.* passing into *adj.*' and add later examples.

1962 V. Nabokov *Pale Fire* 77 We find him next.. printing peevish pamphlets, acting as messenger for obscure syndicalist groups. **1974** J. White tr. *Poulantzas's Fascism & Dictatorship* III. iii. 132 At the same time there was the first rupture with the 'left' syndicalist wing of the movement. **1976** *New Yorker* 3 May 89/1 Marcos has said that he wants to encourage trade unionism, but the fact is that, while his government countenances unions, it appears in some ways to be moving toward the creation of a syndicalist state not unlike Mussolini's Fascist corporate state. **1979** *Jrnl. R. Soc. Arts* Nov. 775/2 Here we have then a typical 'vest-pocket utopia' a form of syndicalist and local organization being collaged into the existing fabric, both formally and politically.

Hence **syndicali·stic** *a.*

1912 *Daily News* 20 Mar. 1 There was nothing particularly syndicalistic about a request for a minimum wage. **1919** M. Beer *Hist. Brit. Socialism* I. II. x. 286 The organized working class turned syndicalistic.

‖ **Syndicat d'Initiative** (sæ̃dika dinisyatīv). Also with small initials. [Fr.] In France, an association for promoting tourism; a tourist information office.

1911 W. J. Locke *Glory of Clementina Wing* ix. 128 The quarter of the town on which the Syndicat d'Initiative prides itself. **1926** E. Hemingway *Sun also Rises* II. x. 92 We went..to the local Syndicat d'Initiative office. **1965** *Harper's Bazaar* Jan. 73/2 The local tourist offices—the *syndicats d'initiative* in France. **1968** F. White *Ways of Aquitaine* 170 Almost all towns and many villages have *syndicats d'initiative.* These are information offices, which will give the tourist local lists of hotels and places of interest. **1972** D. Lees *Zodiac* 85 It's not the sort of thing the Syndicat d'Initiative likes to have get around but it does rain in Antibes every now and again. **1981** C. Watson *Bishop in Back Seat* xxxvi. 208, I would go to the Gendarmerie, the Syndicat d'Initiative.

syndicate, *sb.* Add: **3. a.** Also, an association of people joined in a gambling or betting enterprise. In Gameshooting, a group of sportsmen who share rented shooting rights; also in Angling.

1934 D. Teilhet *Talking Sparrow Murders* ix. 138 La Roc? He's with von Lindbrulle in a betting syndicate. **1961** C. Willock *Death in Covert* i. 25 The game book for the past three seasons showed an average of 1,200 pheasants, 75 woodcock,..160 hares,..and 30 partridges per season... To hell with any qualms he felt about the members of the syndicate individually. **1964** *New Statesman* 3 Apr. 525/1 The fashion for office syndicates and 'sweeps' for charity. **1978** *Country Life* 27 July 272/1 The syndicates that form the basis of many shoots. **1979** *Angling* July 54/1 Catching good fish from strictly private or syndicate waters would prove nothing.

b. *spec.* (freq. with def. article and capital initial). In the U.S., a network of criminals controlling racketeering and other organized crime; also = *Cosa Nostra. Cf. *The Mob* s.v. *mob *sb.*¹ 5 b.

1929 Hostetter & Beesley *It's a Racket!* i. 4 Beer and alcohol running, bombing, bank robbery, murder for pay, window smashing, and a score of other crimes that can be carried on successfully only by organized groups or 'syndicates', are all rackets to the police. **1948** E. L. Irey *Tax Dodgers* xiv. 271 The Syndicate was the remnant of the Al Capone mob. **1952** [see *organized *ppl. a.* 4]. **1962** J. D. MacDonald *Key to Suite* (1968) i. 7 The smut-shadow of beard gave him somewhat the look of imported syndicate muscle. **1963**, **1964** [see *Cosa Nostra]. **1969** *Guardian* 24 Jan. 7/6 The Syndicate is increasingly entering legitimate business. **1980** S. Allan *Dead Giveaway* iv. 38 The Syndicate had not been slow in learning of his involvement..and using it. **1982** *Amer. Speech* LVII. 244 Some successful criminals escape getting a monicker, for they, especially top-notch con men and syndicate members, think it adds 'class' to be without one.

c. *Syndicate of Initiative* = *Syndicat d'Initiative.

1930 Kipling *Limits & Renewals* (1932) 325 A syndicate of Initiative has, indeed, approached me to write on the attractions of the district, as well as on the life of Saint Jubanus.

syndicate, *v.* Add: **2.** Also *spec.* in *Horse-racing,* to sell (a horse) to a syndicate.

1973 *Country Life* 6 Dec. 1897/3 American racing seems to have had a prosperous season with..the prices of bloodstock up. Secretariat was syndicated at $190,000 a share. **1979** D. Francis *Whip Hand* xiii. 161 He buys quite good horses...Then he syndicates them.

syndicated, *ppl. a.* Add: (Examples in the context of journalism.) Similarly in Broadcasting. **syndicated crime,** criminal activities organized by a syndicate (sense *3 b).

1928 [see *gramophone *v.*]. **1959** *Times Lit. Suppl.* 11 Dec. 719/2 Mr. Marquis Childs, the well-known syndicated columnist, formerly of the even better-known *St. Louis Post-Dispatch.* **1968** *Globe & Mail* (Toronto) 13 Feb. 8/3 The Roach report..drew a fine distinction between organized crime and syndicated crime. **1972** *Amer. Speech 1968* XLIII. 211 Van Johnson is quoted in Hedda Hopper's nationally syndicated column. **1974** *Howard Jrnl.* XIV. 108 (Advt.), An exposition of the many problems of organized, syndicated or corporate crime. **1976** *Liverpool Echo* 7 Dec. 17/3 The week gave me new experiences of writing. A syndicated article for the country's local newspapers, a particularly difficult article for a sports journal. **1980** *TWA Ambassador* Oct. 85/1 William R. Allen, professor of economics at UCLA, is known nationally for his syndicated radio commentaries.

syndication. Add: **3.** Publication or ownership by a syndicate. Freq. *attrib.*

1925 A. Huxley *Let.* 21 Apr. (1969) 247, I am trying to arrange for syndication of articles in America. **1955** *Times* 2 Aug. 2/5 Syndication is a nuisance to breeders. They have to guess at a horse's ability as a stallion when they take a share in him at the end of his racing days. **1959** R. Condon *Manchurian Candidate* (1960) ix. 137 The paper..offered Raymond fifty per cent of the syndication money. **1973** K. Giles *File on Death* iv. 97 Once the first instalment hits the street and the syndication rights have been arranged..the Establishment will be chary of proceeding. **1980** *Daily Tel.* 20 Mar. 24/5 (Advt.), Syndication Manager of the Daily Telegraph is looking for a secretary.

syndiotactic (sindəiotæ·ktik), *a. Chem.* Also (more correctly) **syndyo-.** [f. Gr. σύνδυο two together + τακτ-ός arranged, ordered + -ic.] Having or designating a polymer structure in which the substituent groups alternate on either side of the backbone of the molecule.

1956 Natta & Corradini in *Jrnl. Polymer Sci.* XX. 262 We propose to call all vinyl polymers with alternating D- and L-configurations of their substituents (like 1,2-polybutadiene) 'syndyotactic' polymers. **1966** *McGraw-Hill Encycl. Sci. & Techn.* X. 478/2 Isotactic and syndyotactic (stereoregular) polymers are formed in the presence of complex catalysts. **1972** *Physics Bull.* Nov. 668/3 The cellulose molecule..is syndiotactic and hence able to crystallize. **1978** *Nature* 9 Feb. 508/2 Commercial atactic polystyrene is 70% syndiotactic.

Hence **sy:ndiota·ctically** *adv.*; **sy:ndiotacti·city,** the property or state of being syndiotactic.

1959 *Jrnl. Polymer Sci.* XXXIV. 9 Syndiotacticity is the corresponding arrangement. We can think of it as composed of positional and structural arrangements identical to those of isotacticity (repetition arrangements) and of a tacticity opposite to that of isotacticity (inversion tacticity). **1964** *Ibid.* B. II. 319 Predominantly isotactic addition may occur on heterogeneous surfaces, accompanied by conversion of the active complexes to form different..catalyst sites, which propagate predominantly syndiotactically. **1974** *Nature* 26 Apr. 758/1 If there is a high degree of syndiotacticity, a structure incorporating four monomer units per fibre repeat may be present.

syndrome. Delete ‖ and add: Now only with pronunc. (si·ndrōᵘm).

2. Restrict † *Obs.* to sense in Dict. **b.** In recent use, a characteristic combination of opinions, behaviour, etc.; freq. preceded by a qualifying word.

1955 A. Huxley *Genius & Goddess* 26 She took a professional interest in caterpillars... It was part of the Gloom-Tomb syndrome. Caterpillars were the nearest approach, in real life, to Edgar Allen Poe. **1958** C. P. Snow in *Times Lit. Suppl.* 15 Aug. p. iii/2 There is a syndrome of attitudes in literature, nearly all quite modern, apparently unconnected, which spring from the same root. **1965** *Harper's Mag.* Feb. 74 A student..explained Albuquerque's all-enveloping friendliness in terms of the Luke Short syndrome. Typically in a Luke Short novel, a cowboy, footsore and weary, comes into town carrying a saddle over his shoulders. Nobody asks any questions. Friendliness is simply his for the asking. **1971** C. M. Kernan *Lang. Behavior in Black Urban Community* i. 16 The demographic statistics of this community, although depicting accurately a cluster of traits which might be labeled those comprising a lower-class poverty syndrome, do somewhat of an injustice to the social structure of the community. **1976** *Globe & Mail* (Toronto) 21 Dec. 7/1 They were working under the old syndrome that we couldn't do anything—the Government would always block us. **1976** J. I. M. Stewart *Memorial Service* xi. 177 His reclusive side—the withdrawn scholar syndrome, it might be called—remained on top. **1980** *West Lancs. Even. Gaz.* 23 Oct. 13 The falling roll syndrome [in schools] was a problem of the greatest magnitude and one never experienced before.

syndrum (si·ndrɒm). [f. Syn(thesizer + Drum *sb.*¹] A drum designed with electronic means of amplification or alteration of pitch, etc.

1979 *Oxford Times* 28 Sept. 22 The drummer used syndrums more inventively than most disco arrangers. **1980** *Musicians Only* 26 Apr. 13/6 There's a Sonor drumkit, syndrums, and a whole range of Latin percussion. **1981** *Guardian* 13 July 9/1 They dressed up the reggae beat with subtle use of electronic 'syn drums' borrowed from the disco world.

synechism (si·nekiz'm). *Philos.* [f. Gr. συνεχής continuous + -ism: cf. Synechia.] The doctrine that continuity is one of the most important principles in scientific explanation. Hence **sy·nechist,** an adherent of this doctrine.

1892 C. S. Peirce in *Monist* II. 534 The tendency to regard continuity, in the sense in which I shall define it, as an idea of prime importance in philosophy may conveniently be termed *synechism.* **1902** J. M. Baldwin *Dict. Philos. & Psychol.* II. 657/1 The synechist maintains that the only..justification for..entertaining a hypothesis, is that it affords an explanation of the phenomena. **1909** W. James *Pluralistic Universe* 398 Peirce meets this objection by combining his tychism with an express doctrine of 'synechism' or continuity. **1937** *Mind* XLVI. 394 Book i sets forth the doctrines of *Tychism, Synechism,* and *Agapism:* that is to say, it attempts to explain the universe by the use of Pure Chance, Continuity, and psychological categories. **1976** *Internat. Philos. Q.* XVI. 228 This difficulty is also found in Peirce's notions of tychism and synechism.

synechthran, -echthry: see *syn-¹ 1.

synecology (siniko·lǒdʒi). Also † **synoekology.** [ad. G. *synökologie* (Schröter & Kirchner *Die Vegetation des Bodensees* (1902) II. II. 63), f. Syn- + *ecology.] The study of plant or animal communities.

1910 *Proc. 3rd Internat. Bot. Congr.* I. 266 M. Shull... Synoekology of particular regions. **1911** A. Tansley *Types Brit. Veg.* 3 The study of synecology is considerably in advance of autecology. **1936** *Nature* 4 Apr. 565/1 Synecology cannot..be properly studied without a good taxonomic knowledge of the local flora. **1957** [see *biocœnology]. **1977** A. Hallam *Planet Earth* 245 As in ecology, the concern is first with the individual or individual species (autecology), and then investigation proceeds to the assemblage as a whole (synecology).

Hence **synecolo·gical** *a.*; **syneco·logist,** a student of synecology.

1922 *Jrnl. Ecol.* X. 14 Up to the present time most ecological work has been of an extensive (synecological) nature. **1938** *Nature* 17 Dec. 1056/1 The synecologist has to name and ecologically to describe and classify the species components of the vegetation. **1940** E. J. Salisbury in J. S. Huxley *New Systematics* 336 Each [species] has its value for synecological diagnosis. **1974** *Nature* 7 June 599/2 Autecological studies pour out increasing quantities of details for synecologists to work on. **1976** *Ibid.* 22 July p.x (Advt.), Duties: To carry out..synecological studies of aquatic plants in irrigation systems.

synectics (sine·ktiks). orig. *U.S.* Also **Synectics.** [f. Synectic *a.*, perh. after *dialectics.*] A method of problem-solving, esp. by groups, which seeks to illuminate and utilize the factors involved in creative thinking.

A proprietary term in the U.S. (see quot. 1966).

1961 W. J. Gordon *Synectics* ii. 34 Synectics is an attempt to describe those conscious, preconscious and subconscious psychological states which are present in any creative act. **1965** *Times* 11 Aug. 11 A new philosophy, 'synectics', which is said to liberate the creative instinct and so stimulate inventiveness, is gaining a following among big corporations. **1966** *Official Gaz.* (U.S. Patent Office) 25 Oct. 191 *Synectics,* for teaching services—namely, the teaching to individuals and groups, techniques for arriving at creative new concepts, products and solutions; and advising businesses and individuals [etc.]. **1973** *Times* 22 Jan. 20/7 Synectics, a widely used technique for problem-solving in small groups. **1975** R. H. Rimmer *Premar Experiments* i. 128 The basic thrust of Synectics is joining people together into problem-stating and problem-solving groups. **1976** S. Arieti *Creativity* xvi. 376 The synectics method started as a group method. But..the occurrence of analogy—that is, the recognition of similarities—is one of the main processes of individual creativity.

synergetic, *a.* Delete *rare* and add: In mod. use, of or pertaining to synergy (sense *c); = *synergistic *a.* 3. Cf. *synergetical *a.*

1960 R. W. Marks *Dymaxion World of B. Fuller* 8/1 An illustration of the synergetic effect is the behavior of metallic alloys. **1969** R. Buckminster Fuller *Operating Man. Spaceship Earth* v. 73 Universe is synergetic. Life is synergetic. **1975** J. De Bres tr. *Mandel's Late Capitalism* viii. 251 The so-called synergetic model of company planning—in which the overall result of various programmes exceeds the sum of the partial results foreseen for each individual programme—is..derived from military programmes.

synergetical, *a.* For † *Obs.* read *rare* and add: Also in *gen.* use. Hence **synerge·tically** *adv.* Cf. *synergistic *a.* 3, *synergistically *adv.*

1960 R. W. Marks *Dymaxion World of B. Fuller* 166 Thus the system joins together 'synergetically' to distribute and inhibit the loads. [**1969** R. Buckminster Fuller *Operating Man. Spaceship Earth* vi. 77 The patron's supine concessions to the nonsynergetical thinking.] *Ibid.* 109 It produced billions of dollars of new wealth through the increased know-how and intelligence thus released, which synergetically augmented the spontaneous initiative of that younger generation.

synergic, *a.* Add: Also in *Chem.*, with reference to the mutual strengthening of sigma and pi bonds.

1960 L. E. Orgel *Introd. Transition-Metal Chem.* ix. 137 We have discussed σ and π bonding independently. While this gives a satisfactory qualitative picture, the synergic interaction between them is most important. **1974** *Encycl. Brit. Macropædia* XVIII. 606/2 Carbon monoxide is able to form carbonyls with transition metals because the bonding of those metals to the carbon monoxide molecule is of a dual or synergic nature.

synergism. Add: **2. a.** The combined activity of two drugs or other substances, when greater than the sum of the effects of each one present alone.

1910 A. R. Cushny *Textbk. Pharmacol. & Therapeutics* (ed. 5) 29 Other examples of synergism are offered by the anæsthetics, for..a mixture of two of these may induce anæsthesia when administered in a dilution far below that necessary if either is employed alone. **1938** [see *antagonism 1 d]. **1961** *Lancet* 12 Aug. 375/2 Combined chemotherapy of acute leukæmia with 6-mercaptopurine plus corticosteroids gave a higher remission-rate than that observed for either of these administered alone... Synergism was not observed with other drug-combinations given for malignant blood diseases. **1972** *Materials & Technol.* V. x. 313 Products of this type are widely used for their ability to work together with other surfactants (synergism), as clarifying and emulsifying agents. **1979** *Buffalo Evening News* 18 May II. 23/2 By combining 2,4-D and silvex, the weed killer became more effective... This is known as synergism.

b. *transf.* and *fig.*

1925 J. Laird *Our Minds & Their Bodies* ii. 26 These various arguments..are often so closely allied as scarcely to be distinguished. There is synergism in all their ramifications... Nevertheless, we must try to discriminate between the different steps and stages in these converging arguments. **1941** Beigel & Kurth tr. *Reik's Masochism in Mod. Man* i. 30 Freud dropped his attempt to reduce masochism to the assumption of a sexual synergism of pain and discomfort in the infantile organism. **1970** *Nature* 4 July 71/1 We wish to describe synergism between two distinct populations of cells in the production of cellular immunity. **1971** K. Chin Wu in W. O. Dingwall *Survey Linguistic Sci.* (1978) vii. 159 Articulating speech was..a function involving a rather delicate synchronization, synergism to use a more technical term, of many muscular activities that had to be learned. **1974** *Jrnl. Amer. Med. Assoc.* 15 Apr. 290 Thus, it seems that the synergism obtainable by a working together of both professions would, most of all, aid the physicians who read the journals. **1977** *New Scientist* 30 June 767/2 We've done some experiments on cells in culture and you don't get any synergism, you get addition. **1978** *N.Y. Times* 30 Mar. D3/2 (Advt.), The result: a synergism where the results are greater than the sum of the parts.

synergist. 2. (Further examples.)

1938 *Brain* LXI. 322 This is not necessarily true for contraction of synergists. **1959** *New Scientist* 13 Aug. 174/3 Chemical research has already provided 'synergists', cheaper materials which, while not insecticidal themselves, are able to make the pyrethrins more effective in greater dilution. **1969** N. W. Pirie *Food Resources* ii. 67 Substances (called synergists) are added which, though not themselves effective insecticides, prevent those metabolic changes taking place which confer resistance.

synergistic, *a.* Add: **2.** (Further examples.) Also more widely.
1962 *New Scientist* 10 May 263/2 Fucidin is 'synergistic' with penicillin and erythromycin—that is, the activity of a mixture is greater than the sum of the individual antibiotics. **1976** *Sci. Amer.* Feb. p. iv/2 (Advt.), The synergistic effect of mixing finely divided titanium dioxide with opacifying dyes permitted us to use a lesser quantity of dyes than if we had used the dyes alone.
3. *gen.* Co-operative, interacting, mutually stimulating.
1965 H. I. Ansoff *Corporate Strategy* v. 76 This step certainly provides for some of the synergistic interactions. *Ibid.* 83 The synergistic effect can be measured in either of two ways. **1970** *Nature* 26 Dec. 1261/2 The synergistic creativity of Wordsworth and Coleridge which produced *The Lyrical Ballads* in 1798 and began the romantic movement in poetry was over by 1805. **1972** M. Crichton *Terminal Man* i. iii. 30 Designing electronic components to be synergistic with the human brain. **1975** J. A. Argüelles *Transformative Vision* i. 7 The ancient *t'ai-chi* of the Chinese..symbolizes not only the synergistic totality of the two modes of consciousness, but also the interaction of day and night, life and death [etc.]. **1980** *Jrnl. R. Soc. Arts* July 497/2 It was intended that..it should consider the synergistic interactions between the different factors.
synergistically *adv.* (later examples in sense *3 of the adj.).
1968 *N.Y. Times* 8 Jan. 141 By this he meant that science and technology had come to the point where the parts fed upon each other continuously and synergistically to enlarge the whole. **1979** *Logophile* II. v. 8/2 Do the 'principles of acceptance' identified in this study operate synergistically?

synergize, *v.* (In Dict. s.v. Synergist 2.) Add: (Examples.) Hence **sy·nergizing** *ppl. a.*
1954 *Times Lit. Suppl.* 12 Nov. 721/1 The illuminating, synergizing word here, without which the rest is nothing but maundering, is..the word sighs. **1962** *Endocrinology* LXXI. 219/1 Prolactin is well known to synergize with growthhormone in the tibial growth test. **1973** *Nature* 13 Apr. 477/2 Progesterone..synergizes with oestrogen to enhance sexual receptivity.

synergy. Add: **c.** Increased effectiveness, achievement, etc., produced as a result of combined action or co-operation.
1957 R. B. Cattell *Personality & Motivation* xvii. 791 Immediate synergy through group membership..expresses the energy going into the group life as a result of satisfaction with fellow members. **1960** R. W. Marks *Dymaxion World of Buckminster Fuller* 8/1 Fuller refers to the integrated behavior patterns as synergy. **1965** H. I. Ansoff *Corporate Strategy* v. 75 We begin to explore *synergy*.... It is frequently described as the '2+2=5' effect to denote the fact that the firm seeks a product-market posture with a combined performance that is greater than the sum of its parts. **1974** M. B. Brown *Economics of Imperialism* ix. 228 The world-wide 'synergy' of the trans-national company is..the logical conclusion of a long historical process of capital accumulation and territorial assimilation. **1981** *Economist* 28 Nov. 19/2 Others, through mergers (eg, research houses into retail brokerage houses), have demonstrated that there is something to be said for synergy.

synesthesia, var. Synæsthesia in Dict. and Suppl. **synesthetic,** var. *synæsthetic *a.* (*sb.*). **synezesis,** erron. var. *synizesis 3. **synform:** see *syn-[1] 1. **synfuel:** see *syn-[2].

syngameon (singæ·miǫn). *Genetics.* [f. Syngamy + *-on[1].] A cluster of species and subspecies between the members of which natural hybridization occurs.
1922 J. P. Lotsy in *Rep. Brit. Assoc. Adv. Sci. 1921* 453 Nature consists of individuals; similar individuals form syngameons, and these have been mistaken for species. **1930** *Svensk Bot. Tidskr.* XXIV. 386, I have..used Lotsy's term 'syngameon' in a rather wide sense, i.e. as a..handy term for any intercrossing population not divided by distinct lines or zones of discontinuity. **1970** *Brittonia* XXII. 335 We realized that syngameon complexes..were currently contributing to the establishment of additional tetraploid populations in neighbouring areas.

syngas: see *syn-[2].

syngen (si·ndʒěn). *Microbiology.* [f. Syn- + Gen(erate *v.*] A group of organisms, esp. protozoans, capable of breeding together.
1957 T. M. Sonneborn in E. Mayr *Species Problem* 201, I propose the term 'syngen' for the potentially common gene pool, for organisms capable of 'generating together'. **1977** *Jrnl. Protozool.* XXIV. 18/1 We have learned so far that there are 'isozymes' of the epiplasmic proteins in *Tetrahymena*; the molecular weights of proteins B and C vary according to species (including syngens).

syngeneic (sindʒenī·ik, -ē[i]·ik), *a.* *Immunol.* [f. Syn- + Gr. γενε-ά race, stock + -ic.] Immunologically compatible; (of a group of organisms) so closely related that their tissues do not act as antigens when transplanted to one another; = *isogeneic *a.*

1961 P. A. Gorer et al. in *Nature* 25 Mar. 1025/1 We suggest the introduction of 'syngeneic' as a synonym for 'isogenic'. If one does not wish to refer to 'intra-strain immunity' one could then use 'syngeneic immunity'. **1977** *Lancet* 8 Oct. 743/2 Some syngeneic grafts have failed, sometimes because they were undertaken when the patient was already seriously ill. **1978** *Nature* 17 Aug. 697/2 When tumour cells are inoculated into syngeneic mice their establishment and growth is subject to regulation by the host animal.
Hence **syngene·ically** *adv.*
1971 *Nature* 18 June 449/2 The tumour..has been maintained syngeneically as a solid as well as an ascites tumour.

syngenetic, *a.* Add: **2.** (Examples.) More widely, characterized by or pertaining to a formation contemporaneous with the enclosing or surrounding rock.
1905 J. Geikie *Struct. & Field Geol.* xvi. 225 Ore-formations may be grouped under these two main divisions:—1. Syngenetic or Contemporaneous, and 2. Epigenetic or Subsequent.... Syngenetic ore-formations..are formations of the same age..as the rocks in which they occur. **1914** [see *epigenetic *a.* 2]. **1962** *Geografiska Annaler* XLIV. 382/2 The ice-wedges..are clearly 'syngenetic'..i.e. the frost cracks have extended upwards successively as new sediments were deposited on top. **1971** *Nature* 12 Mar. 108/2 It remains to be seen whether the small amounts of racemic amino-acids were syngenetic with the meteorite parent body or were synthesized later. **1978** S. S. Peng *Coal Mine Ground Control* v. 117 Syngenetic anisotropy originates during the formation processes of rock materials. Bedding planes and preferred alignment of pores or mineral grains are examples.
Hence **syngene·tically** *adv.*
1951 *Bull. Amer. Assoc. Petroleum Geologists* XXXV. 2226 Apparently the oölites of hematite formed both syngenetically and epigenetically. **1978** *Nature* 19 Oct. 641/1 The Brioverian unicells and colonies are *bona fide* Precambrian fossils:..they were deposited syngenetically with Precambrian sedimentation.

synizesis. Add: **3.** *Cytology.* Also (*erron.*) synezesis. A stage of meiosis in some species in which all the chromosomal material is seen tightly contracted into a clump. Hence **synize·tic** *a.*
1905 C. E. McClung in *Biol. Bull.* IX. 329, I would suggest that..a new descriptive word be applied to the condition of the nucleus in which the chromatin is found massed at one side of the vesicle, without regard to whether it is a normal phenomenon or not. To carry out this idea I shall call this stage the 'synizesis' of the chromatin. **1921** *Ann. Bot.* XXXV. 367 In this paper the term synizesis is adopted for the tightly contracted phase of the nucleus, following the usage which has become customary in the literature of animal cytology. **1931** *Jrnl. Exper. Zoöl.* LVIII. 299 Synezesis stages were present, but mixed with secondary spermatocytes. **1931** *Amer. Jrnl. Bot.* XVIII. 370 The next stage involves a very rapid shortening of the spireme, the opening out of the spirals, and the transition to an interwoven thread system, which ..persists until toward the climax of synizetic sensitiveness. **1933** *Cytologia* IV. 270 By the use of the methods employed in the studies reported here, the chromatin is drawn into a tight knot (synizesis) at the stages at which synapsis takes place. *Ibid.* 271 Synizetic stages were studied in asynaptic plants which showed little associations of homologous chromosomes at metaphase. **1979** *Hereditas* XCI. 87/1 In the zygotene of the arctic brambles there is a polarised stage called a synizesis. The synizesis is a zygotene bouquet in which mass contraction has occurred.

synjet: see *syn-[2]. **synkinematic, synneusis:** *syn-[1] 1.

synodsman. Add: **2.** Also in other Anglican churches, esp. a member of the General Synod of the Church of England.
1970 *Church Times* 6 Nov. 20/3 Sir John Guillum Scott ..read the gospel—the familiar Beatitudes from St. Matthew containing good advice to all synodsmen. **1972** *Times Lit. Suppl.* 20 Oct. 1261/3 Quite beyond the reach of the average synodsman or lay communicant. **1980** *Times* 23 June 18/2 The synodsmen were concerned to see that the new Anglican Liturgy accurately reflected the contemporary face of Anglicanism.

synœkete: see *syn-[1] 1. **synoil:** *syn-[2].

synonym, *sb.* Add: **5.** *Comb.*, as (sense 1) *synonym-pair*; **synonym-compound** (see quot. 1923).
1923 B. Karlgren *Sound & Symbol in Chinese* iii. 32 The additions were of various kinds, the commonest and by far the most important of which was the formation of what may be called synonym-compounds. This consists in coupling together two simple words with the same or at least analogous meanings, words that formerly had been used alone. **1964** *Language* XL. 104 The Chinese..invented tones to keep the monosyllables apart, and then synonym-compounds, further to clear up the difficulties they were now experiencing with their own language. **1980** *Logophile* IV. 1. 28, I have been working for some time on an article about the curious existence in English and French of synonym-pairs.

synonymic, *sb.* (Examples of form *synonymics.*)
1939 W. E. Collinson in *Trans. Philol. Soc.* 54 (*title*) Comparative synonymics: some principles and illustrations. *Ibid.* 58 The first principle I venture to set up in synonymics..[is that] 'one must never be content with studying synonyms as isolated items'. **1962** H. M. Hoenigswald in Householder & Saporta *Probl. Lexicogr.* 103 Some works, like Dornseiff's monstrous *Wortschatz* or like other thesauri of synonymics are arranged according to semantic areas.

synonymize, *v.* Restrict *rare* to senses in Dict. and add: **5. a.** To be synonymous with (a concept, phrase, etc.).
1938 S. Leslie *Film of Memory* v. 144 The old Baroness was very popular with the crowd and synonymised Victorian charity. **1947** Partridge *Usage & Abusage* 44/2 *As to* in such senses..is defensible when it synonymizes *in respect of* or *in the matter of*.
b. To regard (terms, concepts, etc.) as synonymous.
1970 *Nature* 5 Sept. 1065/1 Hill's classification is not wholly in line with recent trends in primate systematics, which is to synonymize the species of *Papio*. **1976** *Ibid.* 5 Feb. 360/2 If we were now to start referring to cyanophytes as 'blue-green bacteria', we would implicitly synonymise the words 'prokaryota' and 'bacteria'.

synopsis. 2. (Earlier example.)
1844 J. Cowell *Thirty Yrs. passed among Players* 1. xxi. 51/2 Snuffed two tallow-candles, and took a synopsis of the floating apartment.

synopsize, *v.* For (*U.S.*) read (orig. *U.S.*) and add later examples.
1959 J. Gill *Council of Florence* p. xv, Very many long speeches occur in the main sources, which I have synopsised. **1974** *Times Lit. Suppl.* 20 Sept. 1018/5 The 'avowal' here is..that of the editors of the volume, in a short preface in which they attempt to synopsize its contents. **1982** *Daily Tel.* 6 Feb. 11/4 Plot is the very least of Hardy. If you were to synopsise the events of this early novel..you would be a laughing-stock.

synoptic, *a.* (*sb.*) Add: **1. a.** *spec.* Depicting or dealing with weather conditions over a large area at the same point in time.
[**1861** F. Galton (*title*) Synchronous weather chart of England.] **1868** *Symons's Monthly Meteorol. Mag.* III. 144 It is now fourteen years since, impressed with the importance of synoptic weather charts,..I attempted to bring out a series of such charts [of the Indian Ocean]. **1887** [in Dict.]. **1909** A. C. Scott *Notes Meteorol. & Weather Forecasting* 1 Within the last 40 years the Synoptic method of weather-charting has been introduced, which has changed the whole aspect of Meteorology. **1939** *Geogr. Jrnl.* XCIV. 135 Synoptic meteorology (i.e. the science of synchronous weather charts). **1963** G. L. Pickard *Descriptive Physical Oceanogr.* vi. 74 It is.. impracticable to obtain a truly simultaneous picture of the ocean, and the synoptic oceanographer has to make the assumption that when he analyses them the data from his cruise or cruises may be considered as simultaneous. **1974** *Nature* 1 Mar. 87/3 Synoptic climatology is essentially a practical subject.

synoptist. 1. (Earlier example.)
1846 Geo. Eliot tr. *Strauss's Life of Jesus* II. ii. vi. 135 The mode in which the synoptists arrange the sayings of Jesus.

synoptophore (sinǫ·ptofoᵊɹ). *Ophthalm.* Also **-phor.** [Syn- + Opto- + -phore.] An instrument for measuring the deviations of the visual axes of eyes not properly coordinated for binocular vision.
1934 M. L. Hine *May & Worth's Dis. of Eye* (ed. 7) xxvii. 416 The synoptophore..is an elaborate development of Worth's original amblyoscope. **1955** P. D. Trevor-Roper *Ophthalmol.* xviii. 281 Various major amblyoscopes are marketed. The pattern that is perhaps the most generally serviceable is the 'Moorfields Synoptophore'. **1975** *Nature* 17 Apr. 613/2 Binocular interaction was tested on the synoptophor.

synorogenic: see *syn-[1] 1.

synovectomy (sinǒve·ktǒmi). *Surg.* [f. Synov(ia + *-ectomy.] Total or partial excision of the synovial membrane of a joint, esp. the knee, or of a tendon sheath, esp. to relieve pain.
1903 *Buck's Handbk. Med. Sci.* (rev. ed.) VI. 519/1 The procedures ordinarily employed are *évidement* of the focus, and in the later cases..ablation of the patella, with synovectomy or arthrectomy when necessary. **1923** *Jrnl. Amer. Med. Assoc.* 10 Nov. 1579/1 Synovectomy..was originally recommended as a surgical treatment for synovial tuberculosis. **1940** B. I. Comroe *Arthritis & Allied Conditions* xvii. 222 Synovectomy of the knee is the removal of the synovial tissue of the knee. **1976** *Proc. R. Soc. Med.* LXIX. 930/2 The indications for synovectomy or patellectomy in either rheumatoid or osteoarthritis of the knee are well known and differ little in the elderly.

synroc: see *syn-[2]. **synsedimentary:** *syn-[1] 1.

synsemantic (sinsimæ·ntik), *a.* *Philol.* [ad. G. *synsemantisch* (A. Marty *Untersuchungen zur Grundlegung d. allgemeinen Grammatik und Sprachphilosophie* (1908) ii. i. 206): see Syn-, Semantic *a.*] Of a word or phrase: having no meaning outside a context; meaningless in isolation; syncategorematic; opp. *autosemantic*. See also note s.v. *autosemantic *a.* (*sb.*).

1929, etc. [see *autosemantic *a.* (*sb.*)]. **1954** *Archivum Linguisticum* VI. 18 These 'synsemantic' words 'adsignify' or contribute only to the sense of the whole group to which they belong. **1960** *Analysis* XXI. 1. 3 According to Brentano 'Paris' is not a genuine constituent of 'I am thinking-of-Paris'. It is in this context, as he sometimes says, a synsemantic expression. As such, it does not refer to anything. **1965** B. Collinder in Bessinger & Creed *Medieval & Linguistic Stud.* 28 The definite article is a synsemantic demonstrative pronoun.

syntactic, *a.* Add: **1.** (Earlier examples.) Also *Comb.*, as *syntactic-semantic* adj.
1807 R. Kirwan *Logick* iv. i. 531 We learn to ascertain the relation of these different parts to each other, according to the syntactic rules peculiar to each language. **1816** P. S. Duponceau *Let.* 31 July in *Trans. Hist. & Lit. Comm. Amer. Philos. Soc.* (1819) I. 402 As I have given to the Chinese and its kindred dialects, the name of *asyntactic*, the opposite name, *syntactic*, appears to me that which is best suited to the languages of the American Indians. **1972** G. Lakoff in *Language* XLVIII. 291 Anaphora..is a syntactic-semantic phenomenon which can, and must, be specified independently of lexical idiosyncrasies. **1978** *Archivum Linguisticum* IX. 79 We will assume that if such commensurability could be established, we would have strong evidence for the existence of a syntactic-semantic component in our overall grammar, rather than separate syntactic and semantic components.
2. Exhibiting or characterized by syntaxy; *syntactic foam*, a plastic foam made by introducing small hollow spheres into a liquid matrix which then solidifies.
1955 *Sci. News Let.* 2 Apr. 213/3 Called 'syntactic foam', by its developer, the Bakelite Company of New York, the new lightweight material is produced by bonding microscopic hollow spheres made of phenolic resin together with phenolic, epoxy or polyester resins. **1970** *Adv. Chem. Ser.* XCII. 150 Syntactic foams..consist of a dispersion of small hollow glass spheres in a continuous phase or matrix. **1974** *Petroleum Rev.* XXVIII. 675/1 Syntactic foam blocks attached to the top of the frame produce a slight positive buoyancy. **1975** C. A. Harper *Handbk. Plastics & Elastomers* vii. 44 Syntactic foams, like syntactic crystalline polymers, are characterized by their high degree of order or structure.
B. as *sb. pl.* (const. *sing.*). *Linguistics.* C. W. Morris's term for that branch of linguistics which is concerned with the formal relations of signs to each other.
1937 [see *pragmatic *sb.* 4]. **1938** C. W. Morris in *Internat. Encycl. Unified Sci.* I. ii. 14 Syntactics is, then, the consideration of signs and sign combinations in so far as they are subject to syntactical rules. **1941** A. Tate in *Southern Rev.* VI. 636 The role of syntactics in the semiotic science remains somewhat obscure; it seems to consist in a number of 'transformation rules'—that is, in formulas by which given expressions in words, numbers, or symbols can be changed into equivalent but formally different expressions. **1945** [see *intra-linguistic* s.v. *intra- 1]. **1964** E. A. Nida *Toward Sci. Transl.* iii. 35 While semantics deals with the relationship of symbols to referents, syntactics is concerned with the relationship of symbol to symbol; for the meaning of expressions is not to be found merely in adding up symbols, but also in determining their arrangements, including order and hierarchical structuring. For example, the constituents *black* and *bird*, when occuring in juxtaposition, may have two quite different meanings. **1969** [see *pragmatic *sb.* 4].

syntactical, *a.* Add: Also *transf.* in reference to logic (see *syntax 2 d).
1937 A. Smeaton tr. *Carnap's Logical Syntax of Lang.* 2 The difference between syntactical rules in the narrower sense and the logical rules of deduction is the only difference between formation rules and transformation rules, both of which are completely formulable in syntactical terms. **1954** I. M. Copi *Symbolic Logic* vi. 184 To characterize the criterion as 'purely formal' is to say that it is syntactical rather than semantical. **1978** *Jrnl. Symbolic Logic* XLIII. 511 We need first two syntactical transformations on formulae.

syntactically, *adv.* Add: Also *transf.* in reference to logical syntax.
1940 W. V. Quine *Math. Logic* vii. 286 The fact that 'Vbl'.., and 'LFmla' are definable in this syntactical notation is perhaps best expressed..by speaking of them as *syntactically definable*. **1967** *Encycl. Philos.* V. 23/1 Deductive systems for sentential logic..serve to characterize logic syntactically. **1971** G. Hunter *Metalogic* iii. 116 PS is syntactically complete..iff no unprovable schema can be added to it..without inconsistency.

syntactician. Add: **1.** (Later examples.)
1904 *Amer. Jrnl. Philol.* XXV. 355 President Wheeler has not been harder on syntacticians than Piron was on grammarians. **1926** [see *rhythm-deaf* s.v. *rhythm *sb.* 9 a]. **1935** *Punch* 9 Oct. 406/2, I remind myself that the budgerigar is that spiteful little creature known..as the 'lovebird'. I recommend it to future Latin syntacticians as an example of the *lucus a non lucendo*. **1970** *Eng. Stud.* LI. 52 On the whole, present-day syntacticians tend to pay too little regard to the patterning of paradigmatic variables, focusing their attention on syntactic structure on the basis of the theory of grammaticality. **1982** *N. & Q.* Oct. 386/2, I hope the compilers will find a title to express the importance for syntacticians and others of what it will contain.
† **2.** = Syntaxian. *Obs.*
1774 H. T. Blount *Diary* in *Publ. Cath. Rec. Soc.* (1972) LXIII. 358 On the 9th August Jas. Hart, Syntactician, went to England.

syntacticist (sintæ·ktisist). [f. Syntactic *a.* + -ist.] = Syntactician 1.
1889 W. G. Hale in *Classical Rev.* III. 168/2 The very phrases comparative grammar and comparative philology are commonly used in a way to leave the syntacticist outcast and alien. **1944** *Mind* LIII. 243 What the syntacticist is interested in is, that..these isomorphs are themselves analytic. **1977** *Daily Tel.* 10 Feb. 12/6 Syntacticists, linguists and psycholinguists are turning greedily to the poetics of fiction in 'Winnie the Pooh' and the psychological processes in the reading of 'Alice'.

syntacto- (sintæ·kto), used as combining form of Syntactic *a.*, as **syntacto-semantic** *a.* = *syntactic-semantic* adj. s.v. *syntactic *a.* 1; **syntactostyli·stics** *sb. pl.* (const. *sing.*), the study of the stylistic implications of syntactic variation.
1972 *Archivum Linguisticum* III. 7 In particular, I shall show how an adequate grammar must indicate not only a variety of syntacto-semantic features of the noun (such as humanness and inherent duality), but also features of any numeral preceding the noun phrase and certain morphological features of both nouns and adjectives. **1977** *Canad. Jrnl. Linguistics 1976* XXI. 84 The second and third axes represent an arbitrary division of the syntacto-semantic domain of the signal, the motivation for which is discussed below. **1969** Syntactostylistics [see *phonostylistics *sb. pl.*].

syntagm. Restrict † *Obs.* to sense in Dict. and add: **2.** *Linguistics* = *syntagma 4. Also *transf.* and *fig.*
1947 R. S. Wells in *Word* III. 8 A compound sign, i.e. an interrupted sequence of morphemes (no two of which occur simultaneously) is called a syntagm. **1959** W. Baskin tr. *F. de Saussure's Course in General Linguistics* (1960) ii. v. 123 In discourse,..words..are arranged in sequence on the chain of speaking. Combinations supported by linearity are *syntagms*. The syntagm is always composed of two or more consecutive units. **1970** E. Leach *Lévi-Strauss* iii. 48 The term *syntagm*, as applied to an assemblage of non-verbal signs, corresponds to *sentence* in a verbal language. **1973** D. Matias tr. C. Metz in *Screen* Spring/Summer 77 The film-maker at each point in the film..has a choice between a limited number of *basic* combinations..the alternating syntagm, the episodic syntagm, the descriptive syntagm, the 'single shot', [etc.]. **1976** *Times Lit. Suppl.* 19 Nov. 1458/3 The syntagm *il ne va pas* is no longer almost identical with *il ne va un pas*. **1978** 'A. Burgess' *1985* 99 Separate the sexual act from love, and the language of love is devalued. An aspect of our freedom is our right to debase the language totally, so that its syntagms become mere noise

syntagma. Restrict ‖ to senses in Dict. and add: **4.** *Linguistics.* [ad. F. *syntagme* (F. de Saussure *a* 1913, *Cours de Linguistique Générale* (1916) ii. v. 176).] A syntactic unit comprising two or more linguistic signs or elements. Also *transf.*
1937 J. Orr tr. *Iordan's Introd. Romance Linguistics* iv. 286 A syntagma is composed of at least two units in sequence. **1946** *Word* II. 117 To the best of our knowledge, there are three such ultimate and irreducible signs: the phoneme, the word, and the construction or syntagma. *Ibid.* 118 The *syntagma* is defined as the sign of the relations into which the referents of words, enter. **1967** *Ibid.* XXIII. 380 As all composites are syntagmas, i.e., grammatical entities, they must be explainable from an underlying sentence whose syntactic relations they mirror. **1974** M. Taylor *Metz's Film Lang.* p. x, A *syntagma* is, consequently, a unit of actual relationship, while a *paradigm* is a unit of potential relationship. **1979** *Trans. Philol. Soc.* 82 The Latin noun declensions provide more than sufficient illustration of syncretism within word paradigms..with the burden of disambiguating relevant properties then being shifted to the syntagma.

syntagmatic (sintægmæ·tik), *a.* *Linguistics.* [ad. F. *syntagmatique* (F. de Saussure *a* 1913, *Cours de Linguistique Générale* (1916) ii. v. 177).] Of or pertaining to the syntactic relationship between linguistic units. Also *transf.*
1937 J. Orr tr. *Iordan's Introd. Romance Linguistics* iv. 333 To the study of the combinations of linguistic signs ..he [*sc.* Sechehaye] gives the name of syntagmatic grammar. **1948** J. R. Firth in *Trans. Philol. Soc.* 129 We generalize syllabic structure in a new order of abstraction eliminating the specific paradigmatic consonant and vowel systems as such, and enabling the syntagmatic word structure of syllables..to be stated systematically. **1959** W. Baskin tr. *F. de Saussure's Course in General Linguistics* (1960) ii. v. 123 The syntagmatic relation is *in praesentia*. It is based on two or more terms that occur in an effective series. **1966** T. Bendor-Samuel in C. E. Bazell *In Memory of J. R. Firth* 37 In the grammatical description, as in the phonological, it is not sufficient to speak of units as comprising a structure of elements arranged in place since there are also syntagmatic features whose domain of relevance extends beyond any of the elements of the structure. **1973** Matias & Willemen tr. M. Cegarra in *Screen* Spring/Summer 144 Because to Metz, the organisation of the cinema is 'manifestly syntagmatic'.. his criticism of the montage films is in fact no more than reproaching them for being syntagmatically bad. **1981** *Word 1980* XXXI. 243 A hierarchy of phonology, morphology, syntax, and semantics related to each other by paradigmatic and syntagmatic rules.
Hence **syntagma·tically** *adv.*
1937 J. Orr tr. *Iordan's Introd. Romance Linguistics* iv. 286 The constituent elements of human language, considered at a particular moment,..are related to each other syntagmatically and associatively. **1951** *Essays & Studies* IV. 126 Throughout his poetry Swinburne lays general constructions alongside each other, syntagmatically parallel collocations are a feature of verse-form and stanza-form. **1961** Y. Olsson *Syntax Eng. Verb* ii. 27 Both collocation and colligation operate syntagmatically, that is, along the line one-after-another. **1973** [see *syntagmatic *a.*]. **1977** *Word 1972* XXVIII. 261 Examples of the alveolarization of the /t/ in word-initial position could be explained syntagmatically (i.e., it might be attributable to the presence of an alveolar consonant in the preceding word).

syntagmeme (sintæ·gmīm). *Linguistics.* [f. Syntagm(a + *-eme.] In tagmemics, a group of tagmemes of one structural level which represents a tagmeme of a higher level (see also quot. 1964).
1958 K. L. Pike in *Jrnl. Amer. Linguistics* XXIV. 273/2 We have..abandoned the term grammeme and replaced it with Bloomfield's term 'tagmeme'. It also appears probable that we should replace the term 'uttereme' (for 'utterance-eme') with 'syntagmeme'. **1962** E. F. Haden et al. *Resonance-Theory for Linguistics* iii. 41 This complex of slot and filler constitutes a syntagmeme. **1964** R. E. Longacre *Grammar Discovery Procedures* 17 Syntagmemes of one structural level manifest tagmemes of the next higher level; e.g., words manifest phrase level tagmemes. But a syntagmeme may manifest a tagmeme of another syntagmeme on the same level; e.g., one phrase may occur imbedded within another phrase.... On occasion, a syntagmeme of a higher level may manifest a tagmeme of a lower level; e.g. a subordinate clause may occur within a phrase. **1971** *Language* XLVII. 739 Syntagmemes are charted for five levels, according to four features: internal constituents, prosody, nucleus and juncture.

syntax. 1. a. Delete † *Obs.* and add later examples.
1959 J. D. Evans *Malta* ii. 67 The decoration [of certain pottery]..derives its general syntax fairly exactly and its patterns in a more general way from the repertoire of the preceding phases. **1965** *Listener* 9 Dec. 965/2 We have to work to reconcile the shiny shoe with the flat red floor or with the absurd loopy shapes of the legs, or the crushed, pulpy mask of the head. For not only is the syntax of the paint disconnected and inconsistent, but the degree of distortion is too. **1967** G. Steiner *Lang. & Silence* 380 A young East German might come to be more at home, in the syntax of his politics and feelings, in Peking or Albania than in Cologne.
2. a, b. Also with reference to programming languages.
1958 *Communications Assoc. Computing Machinery* Dec. 11 In the sequel explicit rules—and associated interpretations—will be given describing the syntax of the language. **1980** P. Cress et al. *Structured Fortran with WATFIV-S* i. 8 WATFIV-S not only compiles the FORTRAN program, but detects errors in syntax while doing so. **1981** R. D. Tennent *Princ. Programming Languages* ii. 25 An abstract syntax tells us what syntactic structures are available in a language, but does not specify which strings of characters are well-formed program texts, nor their phrase structures.
d. *transf.* in *Logic.* The order and arrangement of the words or symbols forming a logical sentence; the rules operating in formal systems. (See quots.)
1922 tr. *Wittgenstein's Tractatus* 59 The rules of logical syntax must follow of themselves, if we only know how every single sign signifies. **1937** A. Smeaton tr. *Carnap's Logical Syntax of Lang.* 1 By the logical syntax of a language, we mean the formal theory of the linguistic forms of that language—the systematic statement of the formal rules which govern it together with the development of the consequences which follow from these rules. *Ibid.* 2 Thus we are justified in designating as 'logical syntax' the system which comprises the rules of formation and transformation. **1937,** etc. [see *metalogic]. **1940** W. V. Quine *Math. Logic.* vii. 286 Discourse which is 'formal' in this sense, and hence translatable into the notation just now described, is called metamathematics, formal syntax, or briefly syntax. **1955** A. N. Prior *Formal Logic* iii. 70 But as it happens—this can be shown from outside the system—no set of axioms and rules for a system containing its own syntax ever *is* 'complete'. **1979** J. A. Robinson *Logic: Form & Function* ii. 8 The predicate calculus has a simple, systematic basic *syntax*, whose principal feature is the characterization of the class of expressions that are its *formulas*.
3. *attrib.* and *Comb.*, as *syntax diagram*, *table*; *syntax-directed* adj.; **syntax language,** the language used to refer to the syntactical forms of an object language; a metalanguage.
1980 L. V. Atkinson *Pascal Programming* i. 10 The syntax of a programming language can be conveniently illustrated by 'syntax diagrams'. **1961** *Communications Assoc. Computing Machinery* IV. 51 (*heading*) A syntax directed compiler for Algol 60. **1972** J. J. Donovan *Systems Programming* vii. 228 A syntax-directed compiler uses a data base containing the syntactical rules of a source language to parse..the source-language input. **1935** Syntax language [see *object language 1]. **1956** A. Church *Introd. Math. Logic* 58 The meta-language used in order to study the logistic system..is called the syntax language. **1961** *Communications Assoc. Computing Machinery* IV. 55/1 The descriptions are added to the syntax tables used for the second phase, which invokes diagram to output the assembly language program.

syntaxic (sintæ·ksik), *a.* **1.** *Cryst.* [f. *syntaxis 3 or *syntaxy + -ic.] = *syntaxial *a.*
1944 [see *syntaxis 3]. **1972** *Acta Crystallogr.* A. XXVIII. 509/1 The syntaxic intergrowths of the rare-earth carbonates.
2. *Psychol.* [f. Syn- + Tax(is + -ic.] A term orig. used by the American psychiatrist H. S. Sullivan (1892–1949), to designate a mode of experiencing or communicating in which objectivity and the use of consensually validated symbols have replaced subjectivity. Cf. *parataxic, *prototaxic *adjs.*

1945 P. Mullahy in *Psychiatry* VIII. 185/2 Consensually validated symbol activity has more recently been called 'syntaxic' thinking by Sullivan. It involves an appeal to principles which are accepted as true by the hearer. *a* **1948** H. S. Sullivan *Interpersonal Theory of Psychiatry* (1955) ii. 28, I shall offer the thesis that these modes are primarily matters of 'inner' elaboration of events. The mode which is easiest to discuss is relatively uncommon—experience in the syntaxic mode. *Ibid.* xi. 183 Syntaxic symbols are best illustrated by words that have been consensually validated. **1969** A. Neel *Theories of Psychol.* xx. 248 The appearance of the syntaxic or reality-oriented period was greatly aided by acquisition of language skills. **1972** L. Saltzman in Freedman & Kaplan *Interpreting Personality* vi. 176 Obviously, the capacity for syntaxic thinking requires comfortable amounts of self-esteem. **1975** *Psychol. Abstr.* LIV. 141/1 Sullivan's theory of syntaxic mode is compared with Peirce's concept of symbolic interaction.

syntaxis. Restrict † *Obs.* to sense in Dict. and add: Pl. **syntaxes** (-tæ·ksīz). **2.** *Geol.* An arrangement of fold axes or mountain ranges showing convergence towards a common point. [tr. G. *schaarung* (E. Suess).]
1909 H. B. C. & W. J. Sollas tr. *Suess's Face of Earth* IV. 289 In the direction of the syntaxis, i.e. towards the west, their strike bends back in an arc. **1933** W. H. Bucher *Deformation of Earth's Crust* iv. 84 The abrupt deflections and the repeated syntaxes of the Alpine system of southern Europe and western Asia. **1952** *Q. Jrnl. Geol. Soc.* CVIII. 23 The Upper Assam valley..terminates in one of the most impressive examples of syntaxis which our fascinating world can offer.
3. *Cryst.* = *syntaxy.
1944 *Amer. Mineralogist* XXIX. 267 Ungemach..has introduced the term *syntaxie* to describe the coalescence of polyptic substances.... The best English equivalent is probably 'syntaxis', the adjective being 'syntaxic'. **1951** *Phil. Mag.* XLII. 1020 The commonly observed coalescence or 'syntaxis'..of various types of carborundum.

syntaxy (si·ntæksi). *Cryst.* [ad. F. *syntaxie* (H. Ungemach 1935, in *Bull. de la Soc. Française de Minéral.* LVIII. 187): see Syntaxis and -y[3].] Crystal growth or intergrowth in which the new material has the same orientation as the parent, although it may differ chemically.
1952 M. I. Goldman in *Mem. Geol. Soc. Amer.* L. 7 Although Royer uses 'epitaxie' to designate continuous crystallographic orientation between added material and its nucleus, etymologically it means merely 'orientation upon'. It is desirable to distinguish between the relation of added crystalline material with the same orientation as the nucleus, for which I propose *syntaxy* and *syntaxial*. **1953** G. & J. D. H. Donnay in *Amer. Mineralogist* XXXVIII. 939 Ungemach's definition of syntaxic intergrowth seems to be unduly restrictive, as this kind of intergrowth is found to occur also with constituent substances that are chemically different. We therefore propose to abandon the condition of identity of chemical compositions. Henceforth we shall use the term *syntaxy* in this extended meaning. **1973** *Jrnl. Solid State Chem.* VI. 396 Ordered syntaxy and polytypism..give a regular repetition of structural or chemical elements over very long crystalline distances. **1975** *Amer. Mineralogist* LX. 351 Intimate syntaxy between parisite, synchisite, roentgenite, and bastnaesite was quite commonly observed even on a very fine scale.
Hence **synta·xial** *a.*; **synta·xially** *adv.*
1952 Syntaxial [see *syntaxy]. **1958** *Liverpool & Manch. Geol. Jrnl.* II. 15 A mosaic of grains can grow by the deposition of material in lattice-continuity with, or syntaxially with..pre-existing free crystal faces. *Ibid.* 27 The syntaxial rim resembles superficially a cement rim. **1972** H. Blatt et al. *Origin Sedimentary Rocks* xiii. 463 Syntaxial overgrowths are large crystals of calcite that have grown in optical continuity with original single crystal grains.

syntectonic to **synteny**: see *syn-[1] 1.

syntexis (sinte·ksis). *Petrol.* [a. Gr. σύντηξις, f. συντήκειν to fuse together, f. τήκειν to melt: see Syn-.] The alteration of magma by the melting or assimilation of another rock.
1911 F. Loewinson-Lessing in *Geol. Mag.* VIII. 295 When the re-melted portion of the crust is composed of different rocks, eruptive, or sedimentary,..the process is rather a 'syntexis', as I have called it, an assimilation which is followed by liquation and differentiation. **1932** F. F. Grout *Petrogr. & Petrol.* iii. 230 Syntexis has been appealed to in explaining how the more siliceous and the more alkalic rock clans can be derived from primary basaltic magma. **1966** *McGraw-Hill Encycl. Sci. & Technol.* X. 84/1 In some instances such endomorphic effects are sufficiently intensive to result in modification of the composition of the magma (syntexis).

synth (sinþ), colloq. abbrev. of *synthesizer 2 (in sense of musical instrument).
1976 *Liverpool Echo* 24 Nov. 5/4 (Advt.), Crumar String Synth, only one year old, perfect working order, cost nearly £500, accept £270 o.n.o. **1977** *Sounds* 9 July 31/6 They range from a scat slant on 'I Got The Music In Me' through Inter City Disco 'Touch Me Up' to the title cut—a seven minute mutha with heavy southern overtones; brassy, rather stringy, not a synth in sight. **1983** *Yellow Advertiser* (Basildon) 4 Mar. 14/3 Singer-songwriters or synth bands.

synthalin (si·nþălin). *Pharm.* [a. G. *synthalin* (E. Frank et al. 1926, in *Klin. Wochenschr.* 5 Nov. 2101/1), f. *synth-etisch* Synthetic *a.* + *-a-* + *insu-lin* *insulin.] A synthetic but toxic aliphatic diguanidine which has the hypoglycæmic effect of insulin when taken orally; decamethylene-diguanidine, $H_2N\cdot C(NH)\cdot NH\cdot(CH_2)_{10}\cdot NH\cdot C(NH)\cdot NH_2$; also called *synthalin A*; *synthalin B*, an analogous compound in which $-(CH_2)_{10}-$ is replaced by $-(CH_2)_{12}-$.
1927 *Chem. Abstr.* XXI. 772 To produce a molecule of min. toxicity and max. insulin activity, further changes were made in the mol...which resulted finally in a deriv. that is called Synthalin. **1936** Hill & Howitt *Insulin* ix. 190 The decided differences between the action of both synthalin and synthalin-B and insulin render even these substances doubtful as effective substitutes for the hormone. **1952, 1961** [see *diguanidine]. **1980** J. Crossland *Lewis's Pharmacol.* (ed. 5) lvi. 878/1 In the early 1920's..it was found that a number of aliphatic diguanidines such as the Synthalins A and B..produced hypoglycaemia in man.

synthase (si·nþē[i]z). *Biochem.* [f. Synth(esis + *-ase.] Any enzyme that catalyses the addition of a group to carbon atoms joined by a double bond, or the converse reaction; also, a synthetase.
1954 Cohen & Hirsch in *Jrnl. Bacteriol.* LXVII. 182/2 This paper describes an enzyme system synthesizing threonine from homoserine; we have called it threonine synthase. **1961** *Rep. Commission on Enzymes* v. 37 Enzymes removing groups from substrates non-hydrolytically, leaving double bonds (or adding groups to double bonds) will be called 'lyases'.... 'Synthetase' will not be used for any enzymes in this class; where it has been customary, 'synthase' will be used instead. **1976** *Ann. Rev. Microbiol.* XXX. 212 An alternate view is that cyst-specific RNA synthesis during encystment is not necessary for the formation of cellulose synthase. **1979** *Science* 7 Dec. 1149/3 The ATP synthase is a chemiosmotic membrane-located reversible ATPase.

synthesis. Add: **1. b.** In philosophical systems influenced by Hegelian ideas, the final stage of a triadic progression in which an idea is proposed, then negated, and finally transcended by a new idea that resolves the conflict between the first and its negation.
The process is often represented as that of thesis, antithesis, and synthesis, although the terms are not Hegel's. The term is freq. used in relation to the political philosophy of Marx, where this process is seen as exemplified in the history of man's social development (see *dialectical materialism* s.v. *dialectical *a.* 1 b).
1896 J. McTaggart *Stud. in Hegelian Dialectic* i. 2 This idea of the synthesis of opposites is perhaps the most characteristic in the whole of Hegel's system. It is certainly one of the most difficult to explain. **1904** N. I. Stone tr. *Marx's Introd. Critique Pol. Econ.* 288 The two systems by mutually modifying each other may result in something new, a synthesis (which partly resulted from the Germanic conquests). In all of these conquests the method of production..determines the nature of the new distribution which comes into play. **1936** S. Hook *From Hegel to Marx* i. 68 A dialectical synthesis is all this and more. Thesis and antithesis are resolved in such a way that..aspects of each are retained or *conserved* in every new whole or situation; and are reinterpreted or *elevated*. *Ibid.*, For Marx..the manner of synthesis depends..upon the shifts and realignments of human interests in time. **1958** P. Heath tr. *Wetter's Dialectical Materialism* i. i. 4 This third phase then figures in turn as the first step in a new dialectical process, leading to a new synthesis. **1963** F. J. Copleston *Hist. Philos.* VII. ix. 177 We have used the word 'synthesis' for the moment of identity-in-difference in the dialectical advance. But..the terms 'thesis', 'antithesis' and 'synthesis' are more characteristic of Fichte than of Hegel. **1978** P. S. Falla tr. *Kolakowski's Main Currents Marxism* I. vii. 152 As private property develops it necessarily creates its own antagonist; this negative force is itself dehumanized, and as its dehumanization progresses it becomes the precondition of a synthesis that will abolish the existing opposition together with both its terms.
8. Special Comb.: **synthesis gas,** a gas used as a feedstock in the industrial synthesis of a chemical, *esp.* a mixture of hydrogen and carbon monoxide.
1941 *Thorpe's Dict. Appl. Chem.* (ed. 4) V. 503/1 Synthesis Gas.—There is required for the synthesis of methanol or the Fischer synthesis of hydrocarbons, a gas in which the ratio of carbon monoxide to hydrogen is 1:2. The ratio of CO/H_2 in normal blue water-gas is 1:1·25. **1965** M. Sittig *Nitrogen in Industry* ii. 31 Synthesis gas, as the term is used here, is the gaseous mixture of one part nitrogen and three parts hydrogen used as a feed material for ammonia manufacture. **1972** *Sci. Amer.* Oct. 28/1 Lurgi has built more than 50 units to provide town gas (for domestic use) or synthesis gas (for making gasoline). **1980** *Prospects for Petrochemicals in W. Europe* (Shell Internat. Petroleum Co.) 8 A more speculative, but nonetheless plausible, prospect..would be the development of a petrochemicals industry based on synthesis gas.

synthesizer. Add: **1.** (Example in sense b of the vb.)
1980 *Sci. Amer.* Apr. 27/1 The fibrous texture of jade may daunt the synthesizers.
2. *spec.* in *Electronics*, one of various types of instrument for generating and combining signals of different frequencies; esp. a computerized instrument used to create music electronically.
1909 [in Dict.]. **1939** H. Dudley *U.S. Patent 2,151,091* 21 Mar. 13/2 Control currents are then passed to the speech synthesizer. *Ibid.*, In the synthesizer described in detail above, the element equivalent to the vocal system is entirely electrical. **1943** H. J. Finden in *Jrnl. Inst. Electr. Engineers* XC. iii. 165 (*heading*) The frequency synthesizer. *Ibid.* 167/2 There is a demand for a precise frequency generator which will give any desired frequency with a pure output. The frequency synthesizer is an attempt to realise this. **1947** *Jrnl. Appl. Physics* July 601 An electronic synthesizer is described for determination of atomic positions in crystals. **1957** *Sat. Rev.* (U.S.) 26 Jan. 56/2 The American school has not yet, so far as I know, made use of the RCA Electronic Music Synthesizer. *Ibid.* 56/3 The perforated tape operates the music synthesizer in much the same way that a music roll 'plays' a player piano. **1958** E. Fischer-Jørgensen in Saporta & Bastian *Psycholinguistics* (1961) 117/2 Of particular interest to the linguist are the various speech *synthesizers* which have been built recently. **1965** *Wireless World* July 62 (Advt.), The new range of MST transistorized receivers uses synthesizers to provide accurate selection of 250,000 frequencies. **1969,** etc. [see *Moog]. **1973** *Melody Maker* 25 Aug. 27 Baker (electronics, bass) came to London from Australia. He's been working with electronics for ten years, concentrating on solo synthesiser performances. **1975** *New Yorker* 5 May 45/1 The synthesizer can produce a ceaseless kaleidoscope of shapes and colors on the screen. **1981** *Oxford Times* 20 Feb. 13/1 He plays acoustic piano as well as imitating steel drums on his synthesiser.

synthetase (si·nþětē[i]z). *Biochem.* [f. Synthet(ic *a.* + *-ase.] = *ligase; also, a synthase.
1947 Cohen & McGilvery in *Jrnl. Biol. Chem.* CLXXI. 132 We wish to suggest the term 'synthetase' for those enzymes creating a new molecule by the elimination of water between two substrate molecules, excluding the formation of phosphote esters and anhydrides. **1961** [see *ligase]. **1961** [see *synthase]. **1972** *Nature* 15 Dec. 377/1 Aspirin and aspirin-like drugs inhibit..the synthetase which synthesizes prostaglandins from the unsaturated fatty acid, arachidonic acid. **1979** *Sci. Amer.* May 78/2 Interferon also induces the manufacture of a second enzyme, a synthetase that catalyzes the polymerization of adenine nucleotides into a long chain of adenine units called 2,5-oligoadenylic acid.

synthetic, *a.* Add: **A.** *adj.* **2. b.** Of a substance: made by chemical synthesis in imitation of a natural product (cf. *syn-[2]). Also, esp. of a man-made fibre or fabric: made from synthetic materials rather than natural ones (cf. *man-made *a.*).
1874 *Chem. News* 12 June 265/1 (*heading*) Synthetic cymol obtained from normal bromide of propyl and crystalline bromtoluol. **1907** *Chem. Abstr.* I. 1179 (*heading*) Synthetic resins. **1907** *Nature* 25 Apr. 614/2 Since 'synthetic' indigo was put upon the market in 1897, some uncertainty has existed regarding its tinctorial value as compared with the natural dyestuff. **1909,** etc. [see *resin *sb.* 2*]. **1932** B. Hedworth *Foolish Pelican* ii. iv. 136 She had discovered..that synthetic stockings wore better than pure silk. **1941** [see *rubber *sb.*[1] 11 a]. **1946** A. J. Hall *Stand. Handbk. Textiles* i. 66 The du Pont company..commenced the manufacture of a synthetic fibre which has since become known..as nylon. **1955** J. G. Davis *Dict. Dairying* (ed. 2) 1005 Synthetic or imitation cream. **1955, 1966** [see *man-made *a.*]. **1973** *Materials & Technol.* VI. 485 The cleaning of man-made fibres is usually a relatively simple operation which involves a treatment with a mild soap or a synthetic detergent solution. **1983** *Sci. Amer.* Apr. 73/3 In the 19th century, before the boom in organic chemistry that followed the discovery of synthetic dyes, many prominent chemists had undertaken analyses of inorganic natural substances.
c. *fig.* Artificial, imitation, invented.
1930 *Daily Express* 16 Oct. 10/5 With the synthetic idiot, Harpo, you must have a vein of the ridiculous in your laughter gland if boredom is to be kept at bay. **1932** *Sun* (Baltimore) 29 Aug. 8/2 A printing press upon which were struck off bogus service certificates for 'synthetic veterans'. **1934** *Amer. Speech* IX. 101/2 Even when launched in a preliminary fashion, with say fifty or a hundred users, the synthetic language will not grow of itself. **1948** 'N. Shute' *No Highway* iv. 92 The synthetic, phoney film business. **1948** *Newsweek* 10 May 34/2 He has been in London long enough to achieve a synthetic British appearance. **1949** *Hansard Commons* 12 Dec. 2417, I have seldom heard such an outburst of indignation.... It seemed to me a little synthetic. **1976** E. Fromm *To have or to Be?* (1979) ii. iv. 92 The learned, synthetic smile of the marketplace.
d. *Aeronaut.* Of training, exercises, etc.: simulating on the ground what is performed in the air; also *ellipt.* Similarly of equipment used in such training.
1942 *Tee Emm* (Air Ministry) II. 93 All sorts of gadgets and synthetic devices are used..from the cine-film assessor..to the Fisher trainer. **1944** *Horizon* Jan. 49 We are now in the middle of 'synthetic'—i.e. doing things on the ground as they will be done from the air. **1948** *Hansard Commons* 15 Mar. 1808 If people can go for an hour or two in the evenings for synthetic training. **1949** *Aircraft Engin.* Apr. 122/2 There is ample mathematical and electric knowledge in existence to-day to construct 'synthetic aircraft' to simulate the flight of any proposed aircraft from the design data. **1956** *U.S. Air Force Dict.* 504/2 *Synthetic*,..artificial or simulated, as in *synthetic combat mission*, *synthetic training device*, etc. **1976** R.

Hurst *Pilot Error* 260 Complementary process of behavioural engineering and the selection and training of pilots..Performance prediction. Synthetic flight training. Performance assessment.
6. (Earlier example.)
1816 P. Duponceau in *Trans. Hist. & Lit. Comm. Amer. Philos. Soc.* (1819) I. 401 The third class [of languages] would..be that in which the principal parts of speech are formed by a synthetical operation of the mind, and in which several ideas are frequently expressed by one word. Such are what are called the Oriental languages, with the Latin, Greek, Slavonic, and others of the same description. These I would call *synthetic*.
9. Special collocations: *synthetic aperture*, a simulated aperture obtained by moving an aerial or detector transversely during reception so as to increase its effective length; usu. *attrib.*, esp. designating radar employing this; *Synthetic Cubism*, that type of Cubism which involves the combination or reorganization of forms, rather than their analysis (see *Cubism); hence *Synthetic Cubist* adj.
1962 *IRE Trans. Military Electronics* VI. 111 (*heading*) Some early developments in synthetic aperture radar systems. *Ibid.* 113/2 Differences between physical and synthetic apertures. **1977** *Sci. Amer.* Oct. 89/1 Since resolution is proportional to the length of the antenna but inversely proportional to the range, for synthetic-aperture radar the two effects compensate for each other... Synthetic-aperture radar thus makes it possible to obtain high-resolution images of terrain many miles away. **1979** *McGraw-Hill Yearbk. Sci. & Technol.* 224/2 Holography has also been applied, in the form of synthetic-aperture techniques, to the B-scan acoustic reflection systems to provide greater detail in the body areas located near the acoustic transducer. **1947** D. Cooper tr. *Kahnweiler's Juan Gris* ii. vi. 89 Synthetic Cubism was built on a lasting foundation. Gris..finally gave up presenting the beholder with a great variety of information..about the objects which he displayed. He now offered a *synthesis*: that is to say, he packed his knowledge into one significant form, an emblem. **1981** *Times Lit. Suppl.* 9 Jan. 24/3 When constructed sculpture came, along with Synthetic Cubism in 1912, it did so with suddenness, *éclat*, and in quantity. *Ibid.* 24/4 It is often forgotten that Synthetic Cubist space without collage was potentially the most flexible and exciting pictorial space since the Baroque.
B. *sb.* A product obtained by artificial synthesis rather than from natural sources; esp. a synthetic fibre or fabric. Chiefly *pl.*
1934 in Webster. **1940** *New Statesman* 16 Mar. 361/1 The scientists could see in such synthetics [*sc.* plastics].. the threat of maladjustments in industry. **1943** *Sun* (Baltimore) 10 Feb. 4/2 The company built the new plant at its own expense in an effort to increase supplies of the badly needed synthetic. **1951** P. Z. Bedoukian (*title*) Perfumery synthetics and isolates. **1957** *Times* 12 Nov. (Canada Suppl.) p. v/3 Trapping becomes less and less profitable as synthetics displace furs. **1972** D. Bloodworth *Any Number can Play* ii. 10 Lightweight suits cut from one of those shiny Japanese synthetics. **1982** *Sunday Times* 9 May 54/5 There was a sudden scramble to get out of synthetics—those expensive 'fuels of the future'.

synthetical, *a.* **3.** (Earlier examples.)
1796 F. A. Nitsch *Gen. View Kant's Princ. concerning Man* 76 This act may be called a synthetical act of the reproductive imagination. *Ibid.* 89 A synthetical judgment *à priori*.

synthol (si·nþǫl). [a. G. *synthol* (Fischer & Tropsch 1923, in *Brennstoff-Chem.* IV. 281/1), f. *synth-etisch* Synthetic *a.*: see -ol.] (See quot. 1938.)
1924 *Chem. Abstr.* XVIII. 459 The preparation of synthetic oil mixtures (synthol) from carbon monoxide and hydrogen. **1926** J. Joyce *Let.* 5 Mar. (1966) III. 138 He is now using some kind of new chemical stimulant (not chemical but acting by purely physical means, recently discovered here [in France], I understand, synthol for massaging the temples and brow). **1938** *Thorpe's Dict. Appl. Chem.* (ed. 4) II. 350/2 Using mixtures of carbon monoxide with excess hydrogen..Fischer..obtained a mixture which he termed 'Synthol', consisting of alcohols .., ketones, aldehydes, acids.., and various esters. *Ibid.* 425/2 If, instead of zinc-chromium oxides, an alkalised iron catalyst is employed, the liquid product, 'synthol', obtained is a mixture of alcohols, ketones and hydrocarbons containing from 2 to about 8 carbon atoms per molecule.

synthon (si·nþǫn). *Chem.* [f. Synth(etic *a.* + I)on; cf. *-on[1].] A constituent part of a molecule to be synthesized which readily lends itself to an operation of synthesis.
1967 E. J. Corey in *Pure & Appl. Chem.* XIV. 22 The term 'synthon' is suggested [for such units]. These are defined as structural units within a molecule which are related to possible synthetic operations. **1977** *Jrnl. Chem. Soc.: Chem. Communications* 497 (*heading*) A synthon for epoxyolefin cyclisation. **1980** *Jrnl. Amer. Chem. Soc.* CII. 5979/1 (*heading*) Allyl sulfones as synthons for 1,1- and 1,3-dipoles via organopalladium chemistry.

syntone[2] (si·ntōᵘn). *Psychiatry.* [Back-formation from *syntonic *a.*[2] 2.] A person having a syntonic temperament.
1940 J. Bowlby *Personality Types & Mental Illness* ii. 23 Syntones therefore are far from having the 'frank open personalities' commonly attributed to them.

syntonic, *a.*[1] Add: *syntonic comma* (examples).
1944 W. Apel *Harvard Dict. Mus.* (1946) 166/2 The *Didymic* (Didymos, Greek theorist, b. 63 B.C.) or *syntonic comma* which indicates the difference between E as the fourth tone of the circle of fifths..and the E of just intonation. **1954** *Grove's Dict. Mus.* (ed. 5) IV. 523/1 The comma of Didymus (commonly called a *comma* without qualification, and sometimes a syntonic comma). **1979** *Early Music* Apr. 239/2 The major third produced by tuning four successive perfect fifths..is wider than the pure interval..by a syntonic comma.

syntonic, *a.*[2] Restrict *Electr.* to sense in Dict. and add: **2.** *Psychiatry.* [f. *syntony 2 + -ic.] Denoting the responsive, lively type of temperament which is liable to manic-depressive psychosis.
1925 A. A. Brill in *Amer. Jrnl. Psychiatry* LXXI. 592 Bleuler proposed for this reaction the name syntonic. *Syntonic* not only signifies 'equally toned' but also means to be 'attuned' and in 'harmony'.... If a person is neither exclusively schizoid, nor entirely syntonic, one can only say that he is preponderatingly schizoid or preponderatingly syntonic.... Thus, if a person shows a manic attack, it means that the syntonic components predominate qualitatively and quantitatively to a morbid degree. **1927** [see *cycloid *sb.* 3]. **1933** *Brit. Jrnl. Psychol.* July 30 Our results would indicate that the connection of cyclothyme or syntonic type with low perseveration..has no experimental support. **1948** Noyes & Kolb *Mod. Clinical Psychiatry* (ed. 3) vi. 98 Bleuler preferred the term 'syntonic' to Kretschmer's 'cycloid' to describe a personality tendency opposed in characteristics to the schizoid. **1969** H. J. & S. B. G. Eysenck *Personality Structure & Measurement* iv. 23 The large number of persons in the centre of the distribution he would call *syntonic* if they were on the cyclothymic side.
syntonically *adv.* (example in sense *2 of the adj.).
1925 A. A. Brill in *Amer. Jrnl. Psychiatry* LXXXI. 592 The affectivity of the person reacting for the most part syntonically harmonizes with the people of his environment.

syntony. Restrict *Electr.* to sense in Dict. and add: **2.** *Psychiatry.* [ad. G. *syntonie* (E. Bleuler 1922, in *Zeitschr. f. d. gesamte Neurol. u. Psychiatrie* LXXVIII. 373).] A syntonic state or condition (see *syntonic *a.*[2] 2).
1925 A. A. Brill in *Amer. Jrnl. Psychiatry* 598 Translating..syntony into Freudian terms we can say that every transference neurotic has also a fragment of narcistic [*sic*] libido.
3. *transf.* and *fig.*
1958 F. Berry *Poets' Gram.* ii. 20 [The Towneley pageant] is not a work wherein 'anachronisms' occur but a poetic drama where syntony, or multiplicity of tenses running together, is basic to its conception. **1973** D Matias tr. C. Metz in *Screen* Spring/Summer 55 Pierre Schaeffer's specific propositions towards a classification of the possible interactions between music and image into four categories ('masks', 'opposition', 'synchronism', 'syntony'.) **1978** J. Wainwright *Jury People* i. 177 There was a link. A basic syntony which each felt for the other. They each recognised in the other a man proud of his own particular skill.

syntrophy (si·ntrǒfi). *Biol.* [ad. G. *syntrophie* (E. Wasmann 1897, in *Zool. Anz.* XX. 173), f. Gr. συν- Syn- + τροφή nourishment.] The continuing relationship between the individuals of two different species or strains of organisms in which one, or more usually both, benefit nutritionally from the presence of the other; *spec.* that between two bacterial strains which are dependent on each other for their proliferation. Also **sy·ntrophism,** in the same sense. Hence **syntro·phic** *a.*
1897 *Jrnl. R. Microsc. Soc.* 283 Wasmann also speaks of the 'syntrophy' of *Lælaps oophilus* Moniez, [a mite] which occurs freely on the surface of the eggs of ants.., but without doing them any damage, apparently depending on the salivary secretion of the ants, which are always licking their eggs. **1946** *Jrnl. Bacteriol.* LII. 503/2 Syntrophism. This is defined as the growth of two distinct biochemical mutants in mixed culture as a result of the ability of each strain to synthesize the growth factor required by the other.... Mutants blocked at different steps in the synthesis of the same factor show syntrophism. **1950** *Experientia* VI. 42/2 Other possible explanations of the requirement of intermediate cultivation in the penicillin method include segregation of mutant and nonmutant nuclei from a multinucleate cell, and a syntrophic effect of the non-viable irradiated bacteria, which would promote sterilization of mutants by penicillin. **1971** M. Alexander *Microbiol. Ecol.* x. 242 Mutual feeding by dissimilar auxotrophs is termed syntrophism, a relationship in which two or possibly more populations are able to develop in nutrient-deficient circumstances not suitable for the proliferation, or allowing for the poor development at best, of either. *Ibid.* 243 The extensive distribution of bacteria both exporting and importing growth factors suggests a widespread occurrence of syntrophy in nature.

synusia (sin[i]ū·siă). *Ecol.* Pl **-iae.** [mod.L., ad. G. *synusie* (H. Gams 1918, in *Vierteljahrsschr. der Naturforsch. Ges. in Zürich* LXIII. 428), f. Gr. συνουσία society, company.] A group of organisms (usu. plants) of one or more species which have similar life-forms, occupy the same ecological niche, and play a similar role in the community which they form. Also **synu·sium.**
1924 *Jrnl. Ecol.* XII. 15 An aggregation of plants which belong to the same 'life-form' and make similar demands upon a similar habitat constitutes Gams' conception of a synusium. **1926** Tansley & Chipp *Study of Vegetation* ii. 25 The individuals composing a synusia may belong not only to different species but to different families or even different higher groups. **1930** *Svensk Bot. Tidskrift* XXIV. 496 The method of dividing each sociation into its elementary one-layered units, or synusiae, and grouping the synusiae of each layer independently of those of the other layers to synusiae of higher rank. **1932** Fuller & Conard tr. *Braun-Blanquet's Plant Sociol.* xii. 302 A cover of crustose lichens, a pure carpet of moss or of dwarf shrubs, the tree layer of a fir stand are ecological synusiae. **1960** [see *niche *sb.* 3 c]. **1965** B. E. Freeman tr. *Vandel's Biospeleology* xvii. 285 Each biotope contains an animal population which is called the synusium. **1975** T. C. Whitmore *Trop. Rain Forests Far East* ii. 12 Species of very diverse taxonomic affinity make up the synusiae.

syph (sif). *slang.* Also **siph.** Abbrev. of Syphilis. Also with def. article. Cf. *siff.
1914 *Dialect Notes* IV. 113 *Syph, abbrev.*, syphilis. **1925** *Amer. Speech* I. 24/2 For 'syphilis', 'pox' was used wiedly many years ago, but has given place more recently to the simple abbreviation 'syph'. **1930** J. Dos Passos *42nd Parallel* i. 108 He got the siph off 'n her. **1947** *Horizon* Sept. 202 We're going to get the syph. **1960** D. Lytton *Goddam White Man* i. 15 Everybody dies of the cough.... Or you get syph as well.... They say you scream like a hound when you have the syph and the cough together. **1969** P. Roth *Portnoy's Complaint* 129 I'll come down with the syph from just touching the ticket. **1980** 'D. Kavanagh' *Duffy* viii. 149 He goes down to the clinic..and finds he's got the worst case of syph they've seen in years.

syphilide. (Earlier example.)
1829 *Glasgow Med. Jrnl.* II. 327 By syphilide is understood every eruption produced on the skin, by the action of the syphilitic virus.

syphilo-. Add: **syphilo·logy,** the study of syphilis; hence **syphilolo·gic, -lo·gical** *adjs.*; **syphilo·logist,** a specialist in syphilology.
1944 J. H. Stokes et al. *Mod. Clin. Syphilol.* (ed. 3) p. iii, The enormous increase in syphilologic knowledge.., and the earthquake of penicillin have made the revising of a book at this moment a hazardous undertaking. **1908** E. L. Keyes *Syphilis* p. v, The facts upon which the volume rests are the classified cases from the private office books covering forty years of continuous work by myself along syphilological lines. **1890** *Lancet* 13 Dec. 1307/1 The Russian Government has appointed a committee, consisting of Professor Tarnovski and other syphilologists. **1910** *Practitioner* Feb. 231 Few..syphilologists..would now venture to give an opinion on a ..lesion without previously examining a scraping. **1890** Webster, Syphilology. **1893** P. A. Morrow et al. (*title*) A system of genito-urinary diseases, syphilology and dermatology. **1969** J. L. Smith *Spirochetes in Late Seronegative Syphilis* ii. 9/2 Many practitioners are not aware of even the basic doctrines of classic syphilology.

syreen, var. *sireen.

Syrette (sire·t). Also **syrette.** [f. Syr(inge *sb.* + -ette.] The proprietary name of a disposable injection unit, comprising a collapsible tube with an attached hypodermic needle and a single dose of a drug (esp. morphine).
1941 *Official Gaz.* (U.S. Patent Office) 9 Sept. 280/2 E. R. Squibb & Sons, New York, N.Y. Filed July 29, 1939. Syrette for injection units containing narcotic, hypnotic, sedative, analgesic, and vasoconstrictive preparations. **1947** *Sun* (Baltimore) 7 June 5/5 Morphine-containing Syrettes, used by the armed forces during the war to relieve wounded men, are finding their way into illegal narcotic channels. **1953** *Trade Marks Jrnl.* 12 Aug. 718/2 *Syrette*.... Pharmaceutical preparations in collapsible tubes fitted with a hypodermic needle. **1953** [see *shoot *v.* 23 h]. **1962** L. Deighton *Ipcress File* vi. 40 Dalby put the used morphia syrette tube into a matchbox. **1976** *Interdisciplinary Sci. Rev.* I. 179/1 It would be simple to mass-produce disposable syrettes containing one unit of the anti-soma.

Syrian, var. *Zyrian *sb.* and *a.* **Syrianian,** var. *Siryenian *sb.* and *a.*

syringa. Add: Also = Lilac 1 a, b. (Later examples.)
1946 T. C. Mansfield *Shrubs* 244 Syringa is the botanical name for Lilac. **1974** R. L. Fox *Variations on Garden* 75 The old still try and call it [*sc.* philadelphus] *Syringa* which, of course, is the proper name for lilac.

syringe, *sb.* Add: **3. syringe passage** a technique for maintaining a strain of micro-organisms or parasitic protozoans by transferring them through generations of laboratory animals by inoculation with a syringe; also (with hyphen) as *v. trans.*, to subject to this technique; **syringe-passaged** *ppl. a.*
1946 *Ann. Trop. Med. & Parasitol.* XL. 270 All the strains [of *Trypanosoma*] having been maintained by syringe passage through small laboratory animals. **1947** *Ibid.* XLI. 29 It is shown from the literature that a strain which is syringe-passaged through mice gradually increases in its sensitivity to arsenicals. **1970** P. J. Walker in H. W. Mulligan *African Trypanosomiases* v. 89 Syringe passage has certain inherent defects. **1980** *Jrnl. Infection* II. 106 They [*sc.* trypanosomes] had been syringe-passaged from rodent to rodent in the laboratory. **1947**

Ann. Trop. Med. & Parasitol. XLI. 27 All the trypanosomes present in the syringe-passaged strain were the long heterozygous form of *T[rypanosoma] rhodesiense*. **1971** P. C. C. Garnham *Progr. Parasitol.* iii. 28 Such trypanosomes lose their polymorphic morphology, just as they do in syringe-passaged strains in the laboratory.

syringo-. Add: **syringobu·lbia,** [L. *bulbus* onion, bulb], the formation of abnormal cavities in the medulla oblongata of the brain (usu. extensions of those of syringomyelia), resulting in symptoms such as paralysis of the palate, pharynx, and larynx.
1908 *Jrnl. Med. Res.* XVIII. 127 The pathological findings have an important bearing upon the explanation of the bulbar symptoms in cases of syringomyelia and syringobulbia. **1964** S. Duke-Elder *Parsons' Dis. Eye* (ed. 14) 343 In syringomyelia cavities form around which secondary gliosis develops in the cervical and upper dorsal cord; in syringobulbia the process extends up to the medulla. **1975** *Neurology* XXV. 875/1 Syringobulbia is an uncommon lesion of the central nervous system, and is particularly rare in children.

Syrjenian, var. *Siryenian *sb.* and *a.*

Syro-. Add: *Syro-Arabian* (earlier examples), *-Chaldaic* (earlier examples), *-Egyptian*, *-Hittite*, *-Palestinian*.
1841 J. C. Prichard *Res. Physical Hist. Man* (ed. 3) III. 6 The name of Syro-Arabians, formed on the same principle as the now generally admitted term of Indo-Europeans, would be a much more suitable expression. *Ibid.*, The Syro-Arabian tribes lost, at an early period, their ascendancy among the civilized nations of the world. **1835** *Q. Rev.* Sept. 307 A remarkable Syro-Chaldaic lectionarium in the Vatican library. **1836** N. Wiseman *Lect. Doctr. Cath. Ch.* II. xiv. 152 In Syro-Chaldaic there is no expression for to accuse or calumniate. **1964** F. F. Anson *Bishop at Large* vii. 217 Rites and ceremonies were performed like those of the Syro-Egyptian church. **1913** *Times Lit. Suppl.* 3 Mar. 176/2 Syro-Hittite seals. **1964** D. Harden *Phoenicians* xiii. 180 Those cylinder-seals and stamp-seals often termed Syro-Hittite, whose motifs and style are so obviously derived from those of Assyria and Babylonia. **1939** L. H. Gray *Foundations of Lang.* 364 [Arabic] was divided into several dialects, of which only that of Mekkah has survived, this being the parent of a large number of modern vernaculars, notably Arabian, Iraqian, Syro-Palestinian, Egyptian [etc.]. **1978** *Times* 31 Jan. 11/2 Israeli intervention... The mere threat of it headed off Syro-Palestinian intervention in the Jordanian civil war of 1970.

syrup, *sb.* Add: **2.** *syrup of violets* (later examples); *syrup of figs*, an aperient prepared from dried figs, usu. with senna and carminatives.
1849 J. Ruskin *Diary* Apr. in M. Lutyens *Ruskins & Grays* (1972) xxi. 188 The landlady, who noticed my illness, made me some syrup of violets. **1893** *Sears, Roebuck Catal.* 30/2 Sears' fig laxative is pleasant syrup of figs for constipation. **1939** A. Huxley *After Many a Summer* ii. 206 The Baby was acting strange... Acting for all the world like one of those advertisements for Sal Hepatica or California Syrup of Figs. **1981** T. Barling *Bikini Red North* i. 29 A special diet of laxative chocolate... And syrup of figs.

Syryane, Syryen, varr. *Zyrian *sb.* and *a.*
Syryenian, var. *Siryenian *sb.* and *a.*

sysertskite (si·sərtskait). *Min.* Also **sis(s)erskite.** [ad. G. *sisserskit* (W. von Haidinger *Handb. der bestimmenden Min.* (1845) iv. 558), f. *Sysert*, name of a city near Sverdlovsk in Russia: see -ite^1.] A native alloy of osmium and iridium; iridosmine; (see also quots.).
1850 J. D. Dana *Syst. Min.* (ed. 3) 547 At a high temperature the Sisserskite gives out osmium, but undergoes no further change. **1938** *Mineral. Abstr.* VII. 169 The natural alloys are divided into three groups: (1) iridium group with 0–35% Os, cubic; (2) nevyanskite with 35–50% Os; (3) syssertskite with 50–70% Os. **1963** [see *nevyanskite]. **1965** G. J. Williams *Econ. Geol. N.Z.* x. 155 In the other [fraction] consisting of flattened grains of light steel-grey colour, Os was found in excess of Ir, and the Ru proportion is higher—siserskite or ruthenian siserskite. **1973** S. E. Livingstone in J. C. Bailar et al. *Comprehensive Inorg. Chem.* III. xliii. 1165 Alloys of osmium and iridium occur in placer deposits. These are known as osmiridium or syserskite—with less than 60% (usually *ca.* 50%) iridium and *ca.* 35% osmium—and iridosmine or nevyanskite—with over 60% (usually *ca.* 70%) iridium and *ca.* 20% osmium.

systatic, *a.* (*sb.*) **3.** For *rare*$^{-1}$ read *rare* and add examples.
1947 G. Every *Byzantine Patriarchate* xiii. 177 The synod decided to ask the Pope for a 'systatic letter'. **1955** S. Runciman *Eastern Schism* ii. 32 His successor, Sergius IV, sent a Systatic Letter to Constantinople.

system. Also with pronunc. (si·stəm).
I. 1. c. With *the*: (*a*) The prevailing political, economic, or social order, esp. regarded as oppressive; the Establishment; any impersonal, restrictive organization. Freq. with capital initial.
1806 C. Wilmot *Let* 23 Mar. in Londonderry & Hyde *Russ. Jrnls.* (1934) II. 223 Dozens of Slaves are waiting.. to greet the Princess... Her Lenity makes their Lot better perhaps than that of others, but that's saying very little for the System. **1855** *Mechanics' Mag.* LXIII. 542 (*heading*) It is the System. *Ibid.*, I have not heard anything of it from that day to this, and must therefore infer that his Lordship was *instigated* by the 'system'. **1906** U. Sinclair *Jungle* xxx. 384 These Western fellows were just 'meat' for Tommy Hinds—he would get a dozen of them around him and paint little pictures of 'the System'. **1911** H. Walpole *Mr. Perrin & Mr. Traill* ix. 178 She suddenly..had a revelation, that it wasn't really any one's fault at all—that it was the system, the place, the tightness and closeness and helplessness that did for everybody. **1965** G. Jackson *Let.* June in *Soledad Brother* (1971) 78 It's frayed nerves, caused by the harsh terms that defeat brought when they went against the system, the same system that runs this place. **1973** *Ottawa Jrnl.* 18 May 16/1 It is the deeply moving, contemporary story of a young man who wouldn't surrender to the System..and the girl who always stood beside him. **1977** *Gay News* 24 Mar. 20/1 No, I accepted the system wholeheartedly—the suit, white stiff collar and tie, night school, the lot. **1981** 'A. Cross' *Death in Faculty* vi. 65 If I hadn't made it quite to Harvard, I might still have thought there was a chance for me in the system. But Harvard—the oxygen was too pure.
(*b*) *spec.* (See quot. 1945.) *Austral. Hist.*
1874 M. Clarke *His Natural Life* (1875) III. iv. vii. 194 You have a future to live for, man. 'I hope not,' said the victim of the System'. **1934** B. Penton *Landtakers* (1935) i. v. 42 Joe's..not the same as other lags... The System soon breaks them up, but Joe is just sets on fire and leaves him as hard as brick. **1945** Baker *Austral. Lang.* ii. 43 The prison at Fremantle was the *establishment*, a term which is fit to rank with *the System*—as transportation in general and the maltreatment of prisoners in particular became known—as notable examples of understatement.
3. c. In fig. phr. *to get (something) out of one's system* and varr.: to rid oneself of some preoccupation or obsession, esp. by indulging in it to a point of satiety. Cf. quot. 1908, sense 3 in Dict.
1900 H. A. Jones *Mrs. Dane's Defence* iv. 80 I'm rather glad he has taken it [*sc.* a disappointment in love] so violently... It means that in six months it will be out of his system. **1911** G. Stratton-Porter *Harvester* xviii. 430 Let me finish... Let me get this out of my system. **1962** P. Green tr. *S. de Beauvoir's Prime of Life* iii. 139 She still saw him occasionally, trying, at one and the same time and with equal lack of success, to win him back and get him out of her system. **1970** *New Yorker* 17 Oct. 40/1 He..said he'd put a couple of drinks under my belt, I worked the whole thing out of my system. **1974** J. Gardner *Return of Moriarty* 28 We had stayed silent, it was better to let the young fool get it out of his system.
4. c. *Geol.* A major stratigraphic division, composed of a number of series and corresponding to a period (Period *sb.* 4 b) in time; the rocks deposited during any specified period.
1840 A. Sedgwick in *Trans. Geol. Soc.* III. 127 The previous statements seem to show, that the system of the new red sandstone could not have been produced by any sudden and transitory agency. **1835** — in *Ibid.* IV. 50 The lowest beds of the carboniferous system in this region. **1835** R. I. Murchison in *Phil. Mag.* VII. 48, I venture to suggest..the term 'Silurian system' should be adopted as expressive of the deposits which lie between the old red sandstone and the older slaty rocks of Wales. **1839** — *Silurian System* xiv. 169, I venture.. to apply to it [*sc.* the Old Red Sandstone] the term system, in order to convey an idea of the importance of its place in the natural succession of rocks. **1882** A. Geikie *Text-bk. Geol.* 636 The Geological Record is classified into five main divisions. These divisions are further ranged into systems, each system into series..or formations, each formation into groups or stages. **1898, 1927** [see *group *sb.* 4 b (iii)]. **1944** A. Holmes *Princ. Physical Geol.* vii. 103 Pebbles of Shap granite..occur in the conglomerates at the base of the Carboniferous system in Westmorland. **1961** *Bull. Amer. Assoc. Petroleum Geologists* XLV. 658/2 The system is the fundamental unit of world-wide time-stratigraphic classification of Phanerozoic rocks... In the Precambrian, systems still have only local significance. **1971** *Nature* 12 Feb. 460/2 In historical geology, the subdivision of periods into epochs and ages (or systems into series and stages) is usually defined by unconformities.
d. The set of the various phases that two or more given metals are capable of forming at different temperatures and pressures. Usu. with qualifying term, as *alloy system*.
1897 *Jrnl. Inst. Metals* V. 127 In the year 1897 the late Sir William Roberts-Austen..published the complete freezing-point curve of the copper-zinc alloys... This diagram was the first attempt to construct what would in present-day terminology be the Equilibrium Diagram of the Copper-Zinc System. **1922** *Encycl. Brit.* XXXI. 227/2 In non-ferrous alloys, considerable attention has been given to the alloys of zinc, a portion of the ternary system copper-aluminium-zinc. **1967** A. H. Cottrell *Introd. Metallurgy* xv. 233 Many alloy systems are complicated by the appearance of several intermediate phases. **1977** *Sci. Amer.* July 108/3 Both cements are based on the ternary system of oxides of calcium, silicon and aluminum ($CaO-SiO_2-Al_2O_3$).
e. *Linguistics.* A group of terms, units, or categories, in a paradigmatic relationship to one another.
1953 R. H. Robins in *Trans. Philol. Soc.* 109 Professor J. R. Firth has recently suggested that the terms 'Structure' and 'System' be kept distinct in the technical vocabulary of linguistic description. 'Structure' might be used to refer to undimensional, linear abstractions at various levels from utterances or parts of utterances... When..categories have been devised by means of which the utterances of the language can be successfully described and analysed, closed systems are formed of these categories. **1956** J. R. Firth in *Trans. Philol. Soc.* 1955 91 Neither the Americans nor the Scandinavians have controlled and distinguished the use of *system* and *structure* as we have in the linguistics group at the School of Oriental and African Studies. **1961** Y. Olsson *Syntax Eng. Verb* ii. 27 Values for the elements are given by *terms* which commute, that is, operate along the line one-instead-of-another; terms constitute *systems*. **1964** R. H. Robins *Gen. Linguistics* ii. 49 It is useful to employ *structure*..specifically with reference to groupings of syntagmatically related elements, and *system* with reference to classes of paradigmatically related elements. **1977** *Canad. Jrnl. Linguistics 1976* XXI. ii. 196 Throughout the late 'fifties and early 'sixties he [*sc.* M. A. K. Halliday] extended J. R. Firth's concepts of 'system' and 'structure' and 'modes of meaning' into what came to be known as scale and category linguistics.
f. *Computers.* A group of related programs; *spec.* = *operating system* s.v. *operating *vbl. sb.* b.
1963 L. Schultz *Digital Processing* xiii. 271 In applications such as were described in Chapter 6, a system of programs rather than a single program is necessary. **1972** *Computers & Humanities* VII. 82 If a package of programs is so tightly integrated that output from one program is automatically input to another program, then it is frequently called a system. **1978** Lynch & Rice *Computers* ix. 407 A system..handles the manipulation of source programs, language translators, input-output and so on.
g. With reference to business and social organizations and the operations or interactions they involve (see also quot. 1967^2).
1963 *Brit. Jrnl. Sociol.* XIV. 38 The idea of 'system' has been used to imply that its parts (organizations or institutions) are interdependent with each other: that the performances of the parts have consequences or functions, consequences for the 'performing' part, consequences for other 'parts', consequences for the whole system. **1965** H. I. Ansoff *Corporate Strategy* (1968) ix. 166 The term 'systems' is becoming popular for describing large-scale non-military industrial projects. **1967** R. Whitehead in Wills & Yearsley *Handbk. Managem. Technol.* iv. 70 The health of the nation is made possible by a number of systems: doctors, nurses, hospitals, pharmaceutical companies, chemists, and, of course, patients. These are not isolated systems but interacting parts of a large and exceedingly complex whole. *Ibid.* iv. 54 The typewriter may be a relatively simple machine but in this context it is a system with a person and a machine coupled together, both interrelated and interacting. **1969** D. C. Hague *Managerial Econ.* i. 17 We have been considering models for analysing business problems. These seek to state the set of relationships—what we shall call the system—within which and about which business decisions have to be taken.
h. Colloq. phr. *all systems go*: everything functioning correctly, ready to proceed; everything fully operational. Chiefly *fig.* (orig. *U.S.*).
1962 [see *go *a.* 1]. **1967** A. Lurie *Imaginary Friends* i. 8 The Seekers were looking for new members, and we should have no trouble making contact. As McMann put it, all systems were Go. **1969** *Times* 22 July (Moon Rep. Suppl.) p. i/1 Neil Armstrong on the porch of the Eagle at 109 hours 19 minutes and 30 seconds to L.O.S., all systems go, over. **1977** *Listener* 7 Apr. 450/1 It was *sportsfest* time again for the BBC last week—all systems go.
i. A prefabricated construction unit used in system building (see sense 11 d below).
1963 [see *industrialized *ppl.a.*]. **1969** H. A. Frey tr. *Schmid & Testa's Syst. Building* 26/2 Building with systems is naturally more compatible with team thinking than with the approach of the isolated independent architect. **1974** *Encycl. Brit. Macropædia* III. 455/2 Basically a modular volumetric unit composed of some combination of walls, roof, and/or floor, the box system is usually prefabricated in a plant.
II. 9. d. Any method devised by a gambler for determining the placing of his bets.
1850 Thackeray *Pendennis* II. xxvi. 262, I won a good bit of money there, and intend to win a good bit more... I've got a system. I'll make his fortune. **1896** [in Dict., sense 9 a]. **1908** Chesterton *All Things Considered* 47 His vanity..remains a mere mistake of fact, like that of a man who..thinks he has an infallible system for Monte Carlo. **1965** J. Symons *Belting Inheritance* iii. 54 He had all sorts of bright ideas that were going to make a fortune. One was..a racing system, something to do with backing second favourites.
e. *System D* [tr. F. *Système D* (also used)], (see quots. 1918, 1970). *slang.*
1918 in C. A. Smith *New Words Self-Defined* (1919) 185 'System D' is coming into play in the United States Army. 'System D' is a bit of French slang. It means to unmix, to disentangle, to go straight through... It comes from the initial letter of the word 'débrouiller'. **1947** M. Laski in *Vogue* Oct. 63/1 That method called by the French *System D*, the phony medical certificate, the faked-up business journey. **1970** N. Freeling *Kitchen Bk.* v. 45 He was a master of the short cut, the easy way out, the system D. D. stands for dé as in débrouiller or démerder—to extricate, and I suppose that in English it is 'I'm all right, Jack'. **1973** 'Trevanian' *Loo Sanction* (1974) 78 MI-6..muddled their way through the Second World War, relying largely on the French organizational concept, 'système D'.
III. 11. a. (sense *4 g), as *system library*, *technology*, etc.
1952 T. Parsons *Social System* 7 The moment even the most elementary system-level is brought under consideration a component of 'system integration' must enter in. **1962** J. Riordan *Stochastic Service Systems* iv. 70 As noted previously, this is a system with limited waiting capacity. If the waiting capacity is K−1, the system capacity is K. **1962** E. Godfrey *Retail Selling & Organisation* xi. 120 Many firms now recognize that system

training needs to be interspersed with periods of practical selling. **1970** *Gloss. Aeronaut. & Astronaut. Terms* (*B.S.I.*) x. 4 *System capacity*, the total power available from the power sources under the prescribed operating and environmental conditions in the aircraft. **1973** C. W. Gear *Introd. Computer Sci.* iv. 156 These built-in subroutines.. form part of what is called the system library. **1976** *Time* 20 Dec., facing p. 2 (Advt.), This new aid for a communication-saturated world is one more example of Toshiba's sophisticated system technology, which brings together technology from many different fields to solve complex problems of today. **1978** J. McNeil *Consultant* ix. 108 The details of his past career..appeared to have involved Webb in the study of system efficiency.
b. (i) *system-builder* (earlier example), *-building* (examples), *-mongering*; also in sense *4 g, as *system contradiction*, *integration* (so *system-integrative* adj.); (ii) in appositive use, as *system-structure*.
1761 Sterne *Tristram Shandy* IV. xvii. 125 But what it is, I leave to system builders and fish pond diggers betwixt 'em to find out. **1911** J. Drummond *Paul* vi. 79 There is no attempt at system-building. **1969** A. Maude *Common Problem* v. 94 The difference between this process [*sc.* the construction of a system by a political philosopher] and the determinist system-building of social scientists today is concerned with the making of ethical choices about ends. **1977** A. Giddens *Stud. in Social & Polit. Theory* ii. 127 By 'system contradiction' I mean a disjunction between two or more 'principles of organization' or 'structural principles' which govern the connections between social systems within a larger collectivity. **1952** System integration [see sense 11 a above]. **1977** A. Giddens *Stud. in Social & Polit. Theory* ii. 123 While the notion of function is redundant to the theory of structuration, that of 'social integration' can still be regarded as a basic one—together with the further one of 'system integration'. **1953** System integrative [see *polar *a.* 7 b]. **1940** *Mind* XLIX. 120 Hegel was wrong in his formal system-mongering which reflects the influence upon his thought of Christian theology. **1978** *N.Y. Rev. Bks.* 23 Feb. 6/1 [Matthew] Arnold frowned on dogmatic religion, puritanism, and system-mongering. **1964** P. Strevens in D. Abercrombie *Daniel Jones* 125 Such disparate bodies of grammatical theory as those which lie behind phoneme-morpheme-syntax grammar..and system-structure grammar. **1975** M. A. K. Halliday in S. Rogers *Children & Lang.* iv. 225 Prague theory, glossematics, system-structure theory, tagmemics, stratification theory and the later versions of transformation theory are all variants on this theme.
c. In pl. *systems*, used esp. in sense *4 g, as *systems approach, manager, theory*, etc. Cf. also sense 11 d below.
1952 *N.Y. Certified Public Accountant* Oct. 604/2 Principles for acquiring specialized knowledge and experience in the systems field. *Ibid.* 605/2 You can rely on a systems consultant whose business it is to devote more time..than you..can afford to give. **1959** *Economist* 11 Apr. 139/1 The American department is relying increasingly on prime contractors (called 'systems managers') to combine the works of many sub-contractors. **1967** *Ibid.* 28 Jan. 328/3 Airlines in general are shifting toward a 'systems concept' which takes charge of the traveller from door to door, not simply between departure and arrival lounges. **1968** *Sat. Rev.* (U.S.) 23 Nov. 32/3 General Motors and Ford can use a 'systems' approach to their global investments. **1969** *Times* 30 Apr. 23/4 (Advt.), In advanced technology. Systems evaluation engineers. Systems trials engineer. Systems test engineers... We require a number of engineers experienced in the assessment, evaluation and/or trials of complex defence weapon systems. **1970** T. Lupton *Managem. & Social Sci.* (ed. 2) iii. 80 An example of a practical application of a systems theory of organization. **1975** *Modeling & Simulation* VI. 795 (*heading*) Are systems scientists not scientists? **1976** J. Lund *Ultimate* i. 11 Fernandos was a systems consultant to a group of supermarket owners. **1977** *R.A.F. News* 22 June–5 July 9 (Advt.), Systems test engineers £3,500–£4,000. **1977** A. Giddens *Stud. in Soc. & Polit. Theory* ii. 115 Von Bertalanffy counterposes the 'mechanistic' views characteristic of nineteenth-century physical science with the twentieth-century perspective of systems theory. **1978** *Times* 2 Oct. 6/8 A new industry, or sub-industry, has emerged, formed on 'systems houses' which buy in the micro components and other hardware, write the software, and design and market the complete systems. **1978** J. McNeil *Consultant* ix. 109 You might have a bit of trouble with my Systems Manager.
d. Special Comb., as **systems** (or † **system**) **analysis,** the rigorous, often mathematical, analysis of complex situations and processes as an aid to decision-making or preparatory to the introduction of a computer; so **systems analyst; system building** *vbl. sb.*, a method of construction using standardized prefabricated components (see sense 4 i above); hence **system builder; system-built** *ppl. a.*; **system(s) design,** the process or task of matching a computer to the situation into which it is to be introduced and determining the procedures that are to be used; **systems engineering,** the investigation of complex, man-made systems in relation to the apparatus that is or might be involved in them; so **systems engineer; system(s) program** *Computers*, a program forming part of an operating system; so **system(s) programmer, programming; system(s) software** *Computers*, system programs collectively.
1950 in J. H. Batchelor *Operations Research* (1959) 769 Notes on $(m \times 2)$ evaluation matrices for special system analysis applications. **1953** *Jrnl. Operations Res. Soc. Amer.* I. 191 Sometimes this broad type of operations research is called 'systems analysis', 'systems planning', or 'market research'. **1966** A. Battersby *Math. in Management* i. 26 This field of application of mathematics has been defined as 'systems analysis', which considers the thing-being-managed as a system subject to control and operating within an environment with which it interacts. **1977** *Time* 4 Apr. 50/1 Systems analysis, which is really good common sense on a grand scale, combines the knowledge of mathematical probabilities with the aim of dealing with problems in their entirety rather than just piecemeal. **1955** *Operations Research* III. 470 How does the systems analyst choose the preferred strategy? **1967** D. Wilson in Wills & Yearsley *Handbk. Managem. Technol.* 47 The macro block-diagrams show the main logic for a particular program and may be prepared by the systems analyst. **1982** M. Duke *Flashpoint* xxvii. 205 From computer programmer to systems analyst. Quite an achievement. **1965** *Times* 4 Dec. 5/7 If you want to give the system-builder a fair chance of developing his system you have got to have continuous production for a number of years. **1973** *Architect* Jan. 4/1 If you require the services of a good system builder... We can manufacture to your own particular design or in a manner which allows the best use of our standard components. **1964** R. M. E. Diamant *Industrialised Building* I. p. viii, System building is particularly well suited to the rapid erection of tall, low-cost blocks of flats. *a* **1974** R. Crossman *Diaries* (1975) I. 80 A brilliant group of young men and women actively at work developing two methods of system-building, 5M and 12M. **1968** *Guardian* 13 Nov. 1/4 The Minister of Housing..made strenuous efforts to halt the collapse of confidence in system-built blocks. **1973** *Archit. Assoc. Q.* V. iv. 8/2 Later models [of bungalow] were supplied with what would appear to be system-built furniture. Described by the architect as 'chair-furniture', it consisted of various components which could be assembled into chairs, stools, tables, etc. **1954** *Trans. IRE Prof. Group Electronic Computers* June 8/2 The necessity for effecting compromises and avoiding conflicts of this kind between the rival claims of operational effectiveness and engineering reliability and economy strongly influenced the system design of the SEAC and DYSEAC. **1960** Gregory & Van Horn *Automatic Data-Processing Systems* xi. 380 Some analysts with an accounting and systems-design background suggest the straightforward approach—simply asking management people what they must have to control operations. *Ibid.* 396 System design is discussed here in terms of fact finding, developing specifications, meeting specifications, and matching equipment with the system. **1980** *Jrnl. R. Soc. Arts* Feb. 147/1 With the advent of micro-electronics and the growth in the field described nowadays as systems design, there is some danger that.. the manager will fail to appreciate the real importance of the design element. **1980** J. McNeil *Spy Game* i. 22 You're the best systems designer in his Division. *Ibid.* 28 Stick to systems design... You make a lousy financial expert. **1955** *Business Week* 15 Jan. 164/3 Nowadays, the systems engineer starts a project by wrestling with the abstruse questions of what elements in the system need accurate measurement, which ones are important to control. **1974** *Encycl. Brit. Macropædia* XVII. 972/1 The first task of the systems engineer is to develop as clear a formulation of objectives as possible. **1952** W. H. Martin in *Proc. 5th Ann. Conf. Administration of Research 1951* 8/1 In our organization [*sc.* Bell Telephone Laboratories] extensive use is made of an analytical procedure which we call Systems Engineering. **1962** A. Battersby *Guide to Stock Control* i. 9 Two types of specialists concern themselves with the study of these communications networks: we may say broadly that the Organization and Methods experts are responsible for the general layout of the network, whereas the accountants are concerned with the messages which flow along them. The two functions are combined in the new specialism called Systems Engineering. **1973** Gottfried & Weisman *Introd. Optimization Theory* i. 5 The techniques of systems engineering (of which optimization techniques constitute an important subclass) are applicable to a very wide variety of physical problems. **1958** *Communications Assoc. Computing Machinery* Aug. 12 A minimum of 'system programming' should be required to produce the system initially. *Ibid.* 16 System programmers writing in UNCOL can use an existing translator to produce their ML system programs. **1960** *Ibid.* III. 537 (*heading*) A list of computer systems programs for the IBM 650, DATATRON 205, and UNIVAC-SS80. **1970** O. Dopping *Computers & Data Processing* v. 92 Nowadays, only some very specialized 'system programmers' write programs in machine code. **1973** Abrams & Stein *Computer Hardware & Software* iii. 14 Software may be divided into..applications programs, which are written to solve users' problems, and systems programs, which are concerned with operating the computer service. **1973** C. W. Gear *Introd. Computer Sci.* ii. 53 These programs, called system programs, will read in our program after it has been punched on cards in a suitable form and arrange for the instruction counter to be set to the address of the first instruction in our program. **1979** R. Bornat *Understanding & Writing Compilers* xiv. 240 Most system-programming languages allow stack pointers to be used with even more freedom than in ALGOL 68. **1971** B. de Ferranti *Living with Computer* 89 *System software*, those programs, usually prepared by the hardware manufacturer, that provide the link between the programs of the user and the hardware. **1980** Palmer & Morris *Computing Sci.* viii. 283 Systems software is written to schedule the various stages in running a program..at the same time making efficient use of the hardware.

systematic, *a.* and *sb.* Add: **A.** *adj.* **3. a.** (Earlier example.)
1789 *Loiterer* 13 June 8 Armour was rarely used in battles where artillery alone could decide..the Day. ..There was seldom any opportunity of signalizing personal courage amidst the regularity of systematic murder.
c. *systematic ambiguity* (Philos.) (see quot. 1933). Cf. *systematically ambiguous* adj. phr. s.v. *systematically *adv.* 1 c.
1910 B. Russell in Whitehead & Russell *Principia Math.* I. Introd. iii. 45 This is due to a systematic ambiguity in the meanings of 'not' and 'or', by which they adapt themselves to propositions of any order. **1933** L. S. Stebbing *Mod. Introd. Logic* (ed. 2) ix. 161 When the same words are used in sentences which express different kinds of propositions, yet in each case the usage is significant, then these words are said to have 'systematic ambiguity'... This ambiguity is *systematic* because it can be formulated according to a rule. **1952** W. V. Quine *Methods of Logic* p. xi, Systematic ambiguities..are essential to the nature of language. **1979** *Proc. Amer. Cath. Philos. Soc.* LIII. 78 In this paper, I want to argue that there is a systematic ambiguity in the concept of person which explains why it has often been used to favor sexist discrimination.
d. *systematic error*, an error with a non-zero mean, so that its effect is not reduced when observations are averaged.
1925 R. A. Fisher *Statistical Methods for Research Workers* vi. 169 It is worth while to consider the effects of two classes of systematic errors, which, although of little or no importance when single values only are available, become of increasing importance as larger numbers of samples are averaged. **1981** *Astrophysical Jrnl.* CCXLVIII. 34/2 Although there is a statistically significant deviation from a Planck spectrum, there are serious limitations to the statistical analysis where systematic errors are likely.
7. *Chem.* Of the name of a chemical species: constructed in accordance with an agreed set of rules so as to represent the detailed chemical structure of the named species (e.g. *N-methylpent-2-ylamine*); so *systematic nomenclature*. Cf. *trivial *a.*
1858 *Rep. Brit. Assoc. Adv. Sci. 1857* ii. 45 The classification on which the author proposes to base a systematic nomenclature for organic compounds, is a modification of that employed by Gerhardt. **1879** Watts *Dict. Chem.* Suppl. I. 705 A systematic nomenclature for the hydrocarbons, which are the fundamental compounds of organic chemistry, is a great desideratum. **1892** *Nature* 19 May 57/2 It is clearly an absolute necessity of the times that every compound should bear a systematic name of such a character that it can be at once translated into the corresponding formula. **1959** R. S. Cahn *Introd. Chem. Nomenclature* iii. 39 There is a fundamental distinction between the use of trivial and systematic names: trivial names refer to compounds, systematic names to structures. **1978** *Nature* 31 Aug. 929/2 Natural products with particular molecular features cannot normally be located by the keyword approach, as such compounds usually have trivial, rather than systematic, names. **1982** J. E. Fernandez *Org. Chem.* iv. 59 An international, systematic nomenclature system now exists and is used by organic chemists throughout the world.
B. *sb.* **2.** Substitute for def.: = Taxonomy. (Earlier and later examples.)
[**1840** W. Whewell *Philos. Induct. Sci.* I. viii. ii. 468 A department of the philosophy of natural history which has been termed by some writers (as Decandolle,) Taxonomy..by some Germans..has been denominated *Systematik*; if we could now form a new substantive after the analogy of the words Logic, Rhetoric, and the like, we might call it Systematick.] **1940** J. S. Huxley *New Systematics* 1 To-day..systematics has become one of the focal points of biology. **1951** G. H. M. Lawrence *Taxon. Vascular Plants* i. 3 In this text the taxonomy of vascular plants includes the systematics of the taxa known as pteridophytes, gymnosperms, and angiosperms. **1969** E. Mayr *Princ. Systematic Zool.* p. vii, Systematics has had a remarkable renaissance during the last generation.

systematically, *adv.* Add: **1. c.** *systematically ambiguous* adj. phr. (Philos.), having an ambiguity that is systematic (see *systematic *a.* 3 c).
1929 C. I. Lewis *Mind & World-Order* i. 11 The adjective 'real' is systematically ambiguous and can have a single meaning only in a special sense. **1943** I. A. Richards *How to read a Page* iii. 52 There are few important words which are not in varying patterns systematically ambiguous; *say* is typical. These *regular* shifts of sense as a rule give us little trouble in reading. **1967** *Philos.* XLII. 208 'Reality' may be systematically ambiguous.

systematicity (sistĕmăti·sĭti). [f. Systematic *a.* + -ity.] The quality or condition of being systematic; systematicness.
1970 Campbell & Wales in J. Lyons *New Horizons in Linguistics* 257 The child first of all does something 'correctly' and then, with every appearance of systematicity, later proceeds to do it 'wrongly'. **1974** R. A. Hall *External Hist. Romance Languages* 239 Meyer-Lübke's work was characterised by sobriety and balance, with exceptional solidity in detail but with an over-all sense of systematicity. **1977** *Language* LIII. 883 He has little time or inclination to follow through his concept of music theory with any thoroughness or systematicity.

systematy (si·stĕmăti). [f. Gr. σύστημα, -ατ- System + -y^3.] Systematic classification; = Taxonomy.
1912 W. L. Balls *Cotton Plant in Egypt* 5 In this brief summary of the few available historical facts, it has seemed advisable to evade systematy. **1921** *Oxf. Bot. Mem.* XI (*title*) Elementary notes on the systematy of Angiosperms. **1929** E. M. Nicholson *Study of Birds* 20 Mechanical and unfruitful as systematy inevitably is, ornithologists ought to be grateful to those who have plodded through it.

T

‖ **Système International** (sistę̄m æ̃ntęrnasyonal). Also (*erron.*) **-nationale.** [Fr.] In full *Système International d'Unités*. The International System of Units (see *international *a.* 1 c).
[**1957**: see *international *a.* 1 c.] **1969** *Symbols, Signs, & Abbreviations* (R. Soc.) 21 (*heading*) The International System of Units (Système International d'Unités —SI). **1971** [see SI s.v. *S 4 a]. **1972** *Physics Bull.* Aug. 461/3 If the precision of the measurement system can lead us to parts in 10^8, it might be possible to change the way in which the electrical units are defined in the Système Internationale.

systemic, *a.* Add: **1. c.** Of a herbicide, insecticide, or fungicide: entering the system of a plant or animal and freely transported within its tissues. Also as *sb.*, a systemic agent.
1949 *Ann. Appl. Biol.* XXXVI. 160 The term 'systemic insecticides' refers only to chemical substances which are absorbed by the plant and translocated to all parts of it, rendering it insecticidal. **1961** *New Scientist* 5 Jan. 50/2 The animal systemics ronnel..and CoRal have been in commercial use for some time. **1964** *Which?* Apr. 114/1 Dimethoate and menazon are unlike the other insecticides in being systemics, i.e. they are absorbed into the plant instead of just being deposited on it. **1971** *Exper. Agric.* VII. 2 Four different systemics were used. **1979** *Radio Times* 5–11 May 13/4 The best control is to spray young foliage thoroughly with Benlate systemic fungicide.
2. Delete *rare*$^{-0}$ and add examples.
1946 C. Morris *Signs, Lang. & Behavior* 104 In the systemic use of signs the aim is simply to organize sign-produced behavior. **1952** A. Cohen *Phonemes of English* 54 It is not as loans, but as residual structural irregularities, which might rather be called 'systemic fragments' than 'coexistent systems'. **1966** S. Beer *Decision & Control* xvii. 439 They will demand that systemic qualities be measured which no-one as yet knows how to measure. **1975** *Times Lit. Suppl.* 17 Oct. 1233/5 A tradition of American political and social thought..that demands social justice without labelling injustice a systemic product. **1977** *Dædalus* Summer 55 Many scholars, especially those whose level of analysis is systemic, implicitly write as if they were addressing themselves to a world government.
3. *systemic grammar*, a method of linguistic analysis developed by M. A. K. Halliday in 1961 in *Word* XVII, based on the ideas of J. R. Firth and others. Similarly *systemic linguistics*. Cf. *system 4 e.
1968 *Computers & Humanities* II. 147 The linguistic description I adopted for my study was systemic grammar. **1971** D. Crystal *Linguistics* iv. 215 More recently, Halliday has developed out of this a concept of *systemic* grammar. **1975** M. Berry (*title*) Introduction to systemic linguistics. **1978** *Language* LIV. 351 The grammar that assigns to sentences structures like the one in Fig. 1 is generative fusion of elements of American-style immediate-constituent analysis (cf. Nida 1960), European-style dependency theory (cf. Tesnière 1959), and British-style systemic grammar (cf. Halliday 1961).

systoflex (si·stofleks). *Electr. Engin.* [f. *systo-*, of unknown origin + *flex *sb.*^{2}] Flexible sleeving for insulating electric wires.
1922 *Wireless World* X. 556/1 One may lose much time trying to push No. 18 wire into systoflex intended for No. 20. **1968** M. Woodhouse *Rock Baby* ix. 93 His study ..smelled..of soldering fluid, swarf, oil, charred systoflex sleeving..the indefinable smell of home-made electronic gear.

systole. Add: **1. c.** Also *Comb.*, as *systole-diastole*.
1924 C. Gray *Survey Contemp. Music* 260 The immediate future..will witness a return to tradition... So it always has been... It is the *systole-diastole* of the world of art. **1946** M. Lowry *Let.* 2 Jan. (1967) 74 Here we come to the heart of the book which..returns..to the uneasy, but healthy, systole-diastole of Hugh.

systrophe (si·strofi). *Biol.* [ad. G. *systrophe* (A. F. W. Schimper 1885, in *Jahrb. f. wiss. Bot.* XVI. 221), f. Syn- + Gr. στροφή, turning.] The clumping together of chloroplasts in a cell when exposed to bright light.
1886 *Jrnl. R. Microsc. Soc.* VI. 642 Very strong irritation of light causes the chlorophyll-grains to collect into one or two lumps, a phenomenon for which Schimper proposes the term systrophe. **1936** *Geogr. Jrnl.* LXXXVIII. 48 Diatoms..sink deeper and display the phenomenon of systrophe. **1966** E. J. Stadelmann in D. M. Prescott *Methods Cell Physiol.* II. vii. 206 Systrophe is a typical and reversible response reaction of the living protoplasm to a variety of stimuli.

Szechuan (setʃwā·n). Also **Szechwan.** [ad. Chin. *Si-chuān*.] The name of a province in south-western China, used *attrib.* (with reversed stress) to designate the distinctively spicy cuisine originating there. Also *Comb.*, as *Szechuan-style* adj. Also *transf.*
1956 Buwei Yang Chao *How to cook & eat in Chinese* I. i. 30 Szechwan cooking has a fine balance of flavours except that hot pepper is added freely. **1974** *Times* 23 Aug. (Hongkong Suppl.) p. x/7 The Chinese food in Hongkong is superb..Peking duck and Szechwan smoked duck. **1977** *Harpers & Queen* Nov. 30/2 A new restaurant serving Szechaun food. **1979** *United States 1980/81* (Penguin Travel Guides) 179 Honolulu also has several Mandarin or Szechwan-style Chinese restaurants. **1980** E. Behr *Getting Even* vii. 89 There was the smell of real Szechuanese cooking, chillies and hot sesame oil... Waiters began serving an elaborate Szechuan meal.

Szechuanese (setʃwanī·z), *a.* and *sb.* Also **Szechwanese.** [f. as prec. + -ese.] **A.** *adj.* Of, pertaining to, or characteristic of Szechuan or its people, or of the Chinese spoken there. **B.** *sb.* **a.** An inhabitant of Szechuan. **b.** The dialect of Szechuan.
1918 *North-China Herald* 19 Jan. 115/2 Szechuanese invasion of Yunnan... It is reported that the Szechuanese have invaded Yunnan. **1937** E. Snow *Red Star over China* v. 199 The Szechuanese are sentimental about their few bridges. **1947** J. Bertram *Shadow of War* x. 336 We gathered for supper at a Szechwanese restaurant and.. tasted..the roast duck of Chengtu. **1947** N. C. Scott in *Bull. School Orient. & African Studies* XII. 197 (*heading*) The monosyllable in Szechuanese. **1966** R. & D. Morris *Men & Pandas* iv. 65 He not only spoke good English, but also knew the Szechuanese dialect. **1972** 'M. Yorick' *Killer for Chairman* ii. i. 127, I could hear voices talking in a Szechwanese dialect... The Szechwanese had a reputation for bloody-mindedness. **1978** A. Grey *Chinese Assassin* xvii. 174 Your Szechuanese accent is a dead give-away. **1980** [see prec.].

Szekel (se·k'l), *sb.* (*a.*) Also in Ger. form **Szekler.** [ad. Hungarian *Székely* (also used).] (A member of) a Magyar people living in eastern Transylvania. Also *attrib.* or as *adj.*
1843 *Penny Cycl.* XXV. 164/2 The nation [of Transylvania] in the political sense of the word is composed of three bodies or 'nations', the Hungarians, the Szeklers, and the Saxons, who have the collective name of the 'Uniti'. *Ibid.*, When a Hungarian or Szekler nobleman of Transylvania settles in Hungary, he is entitled to all the privileges of noblemen in Hungary. **1869** A. J. Patterson *Magyars* II. xxxi. 354, I started for the little hamlet..accompanied by a Székel. **1888** E. Gerard *Land beyond Forest* II. xxxviii. 151 The Szekel villages, of a formal simplicity, are as far removed from the Roumanian poverty as from Saxon opulence. **1910** *Encycl. Brit.* X. 392/2 The isolated groups of Hungarians now found in Transylvania and called Szeklers are considered the purest descendants of the invading Magyars. **1920** *Glasgow Herald* 23 Apr. 8 At the very most 1,900,000 Magyars are lost, of whom over 900,000 (including the Szekels) are in Transylvania. **1934** R. W. Seton-Watson *Hist. Roumanians* ii. 20 Already Koloman and Stephen II in the first three decades of the Twelfth century began to settle Magyar colonists—the so-called Székelys or Siculi. *Ibid.* 21 Transylvania falls into three distinct political groupings—finally crystallised by the events of 1437 into three privileged nations, the Magyars, the Székels and the Saxons. **1956** F. S. Pisky in S. Fischer-Galati *Romania* iii. 54 Although there are no ethnic or linguistic differences, Hungarians make a distinction between the *Szekelys* and *Magyars*. The *Szekelys*, descendants of the free frontiersmen in Transylvania, populate the Odorhei, Cius and Trei Scaune districts. **1974** *Encycl. Brit. Micropædia* VI. 496/3 The Szeklers, meaning Frontier Guards, received their name, it seems, because they were Magyars sent to Transylvania to protect the eastern flank. *Ibid. Macropædia* IX. 31/1 Colonies of Szekels, a people akin to the Magyars who had preceded the latter into the central plains, were settled behind its eastern passes.

Sze Yap (sī yæp). Also **Sze-Yap, Szeyap.** [Chinese.] The name of an area made up of four counties in the south of Guangdong Province in China (see quot. 1973) used *attrib.* and *absol.* to designate the Chinese dialect spoken there.
[**1948** R. A. D. Forrest *Chinese Lang.* x. 200 Cantonese, with reference also to the dialect of Sze-Yap, to the west of the Canton River delta, generally regarded as a minor variety of Cantonese.] **1964** *Asia Mag.* 12 July 22/3 The Chinese [in Hong Kong]..speak no less than seven tongues—Cantonese, Hoklo, Sze Yap, [etc.]. **1971** K. Hopkins *Hong Kong* 235 Cantonese is very much the predominant language but there are minorities who speak..Sze Yap. **1973** R. A. D. Forrest *Chinese Lang.* (ed. 3) xi. 235 Usually reckoned a sub-dialect of Cantonese, though, in the opinion of the present writer, showing enough distinctive features to warrant its separation, is the dialect of Sze-Yap, the 'Four Towns', spoken on the west of the Canton River delta... Like most varieties of Cantonese, Sze Yap has lost all distinction of *s-* and *ʃ-*. *Ibid.* 328 The Sze Yap dialect has regularly *h-* for *t'-*. **1982** *English World-Wide* III. i. 48 Other varieties of Chinese spoken include Szeyap, Chiuchow, Shanghainese, Hokkien and Hakka.

‖ **szlachta** (ʃla·χta). *Hist.* [Polish.] The aristocratic or land-owning class in Poland before 1945.
1883 *Encycl. Brit.* XIX. 285/2 We soon find the following divisions of society among the Poles:—(1) the nobility, *szlachta*, who throughout Polish history constitute the nation properly so-called. **1905** *Cambr. Mod. Hist.* (1907) III. 61. 76 Poland was at this time on the threshold of a period of political transition of an almost revolutionary character, the most remarkable feature of which was the growing preponderance of the Polish *szlachta*, or gentry. **1969** P. Anderson in Cockburn & Blackburn *Student Power* 264 Bronislaw Malinowski, a Polish aristocrat from the Galician *szlachta*. **1978** W. B. Lincoln *Nicholas I* iv. 136 The more substantial portions of the Polish *szlachta* (nobility) had done relatively well under fifteen years of Russian rule.

T

T. Add: **I. 1. b.** (Earlier example.)
1849 [see Cross *v.* 7 a].
c. (Later examples of *to a tee*.)
1873 K. H. Digby *Last Year's Leaves* viii. 302 Then should you scorn such feasts, like me, We've what will suit you to a tee. **1922** Joyce *Ulysses* 58 Simon Dedalus takes him off to a tee with his eyes screwed up. **1966** *Listener* 29 Sept. 480/1 John Hollis had Walter off to a tee. **1973** *Brit. Printer* May 62/3 Edwin Snell of Yeovil has the direct mail touch to a tee.
2. b. *Electr.* A network of three impedances that can be represented diagrammatically as a T in which the stem and each arm is an impedance. Freq. *attrib.*; so *T-connected* adj.
1909 Bedell & Pierce *Direct & A.C. Testing* vii. 248 Transforming from a 2-phase primary circuit to two sets of T-connected secondaries. *Ibid.* 249 The line voltage, thus obtained by the *T*-connection, is accordingly the same as would be obtained by three 3-phase generator coils..in delta. **1934**, etc. [see *lattice *sb.* 2 d]. **1947** R. Lee *Electronic Transformers & Circuits* vi. 150 In the 'low pass' filter T-section of Fig. 115, the inductance arms shown as *L*/2 and the capacitance *C* are made with losses as low as possible. **1973** J. R. Neuenswander *Mod. Power Systems* iii. 32 For a line of medium length, a better approximation is arrived at through either the π or the *T* connection. **1975** D. G. Fink *Electronics Engineers' Handbk.* xiii. 30 While many null network configurations are useful (including the bridged-T and twin-T), the Wien bridge design predominates.
c. *Naut.* In phr. *to cross the T*: of a fleet or ship, to cross ahead of another (enemy) fleet's or ship's line of advance approximately at right angles, thus securing tactical advantages.
1916 'Taffrail' *Pincher Martin* xvii. 323 Sir David Beatty..altered course to the east and crossed the enemy's T,..inflicting terrible damage with his heavy fire. **1953** *Hist. Today* Feb. 114/1 The Japanese main force was able to steam across the head of the Russian line... This manoeuvre, known as 'crossing the T', has been the dream of all admirals once steam tactics were introduced. **1968** D. Thomas *Battle of Java Sea* x. 143 The destroyer *Oshio*..had crossed the Dutch cruisers' T and exchanged a rapid but ineffectual fire with *Java*. **1976** *Oxf. Compan. Ships & Sea* 213/2 The fleet 'crossing the T' has a considerable gunnery advantage.
3. a. *T-bandage* (later example), *-head* (examples), *-joint* (earlier example).
1885 W. H. Coleman *Hist. Sketch Bk. New Orleans* xviii. 187 The Chevalier appeared in the streets wearing what the surgeons call a T bandage about his face and jaw. **1913** W. E. Dommett *Motor Car Mech.* 42 The arrangement shown at E is the most extensively used, the cylinder being said to have a T-head. **1969** *Jane's Freight Containers 1968–69* 179/3 A new Oil Jetty running 217 m.. out into Harwich Harbour..., and having a minimum depth of 33 ft alongside the T-head. **1889** *Cent. Dict.*, T-joint.
b. T account *U.S. Book-keeping*, a standard form of ledger account (see quot. 1976), or a simplified version of this; **T-bar,** a metal bar with a T-shaped cross-section; a T-shaped fastening on a shoe (cf. *T-strap* below); *spec.* a type of ski-lift consisting of a series of T-shaped bars whereby skiers are towed uphill; **T-bone steak** orig. *U.S.*, a beef-steak cut from the sirloin and containing a T-shaped bone; also *ellipt.* as *T-bone*; **T-formation** *U.S. Football*, a T-shaped offensive formation of players (see quot. 1978); **T-junction,** a T-shaped intersection (of pipes, etc.); *spec.* a T-shaped road junction; † **T-light** *Theatr.*, a type of gas lighting-device utilizing a pipe in the shape of a letter T (*obs.*); **T-strap,** a T-shaped instep strap on a shoe; freq. *attrib.*; also *absol.*, a shoe with such a strap; cf. *T-bar* above. See also *T-shirt.
1936 Owens & Kennedy *Accounting* v. 49 Sometimes 'skeleton' or 'T' accounts are used instead of the regular accounts. **1941** L. O. Foster *Introd. Accounting* x. 181 T accounts are accounts without rulings and are used in classroom discussions because of the simplicity of their structure. **1976** D. W. Moffat *Econ. Dict.* 270/1 *T account.* In double-entry book-keeping, each account has the name of the account on a horizontal across the top, and then a vertical line separates the debit entries from the credit entries. The lines form a T. **1889** *Cent. Dict.*, T-bar. **1940** [see *chair-lift* s.v. *lift *sb.*^{1} 10 b]. **1956** [see *section *sb.* 2 p]. **1964** *Woman* 18 Jan. 13 Today's chair-lifts and T-bars mean you'll be sitting as much as ski-ing. **1966** A. W. Lewis *Gloss. Woodworking Terms* 78 (*caption*) T-bar sach cramp. **1972** *Daily Tel.* 24 Jan. 11/7 The little girl's T-bar beach-shoe. **1979** *United States 1980/81* (Penguin Travel Guides) 648 Excellent skiing facilities including a chair lift,..T-bar, beginners' lift, and various snack bars. **1916** *Dialect Notes* IV. 270 *T-steak* or *T-bone-steak*,..so called from the shape of the bone. **1923** N. Anderson *Hobo* i. 38 These bills of fare..displayed ..T-Bone Steak. **1934** E. Newhouse *You can't Sleep Here* xii. 144 When it's a toss-up between buzzards' gizzards and a t-bone.., me for the t-bone. **1959** *Times* 27 Apr. 7/4 Fillet and T-bone steaks were the order of the hour. **1979** R. Rendell *Make Death love Me* xi. 98 He got Marty to fetch in three great hunks of T-bone because Joyce said she liked steak. **1930** R. C. Zuppke *Coaching Football* vii. 208 The 'T' formation..is at its core a strong formation. **1942** L. O. Waldorf *How to play Football* ix. 112 In 1940, Stanford University used the T formation with great success. **1978** G. Wright *Illustr. Handbk. Sporting Terms* 85/3 *T-formation*, one of the basic offensive formations with the quarterback behind the center and the other three backs behind in a row parallel to the line of scrimmage. **1954** *Gloss. Highway Engin. Terms* (*B.S.I.*) 56 *T junction*, a junction shaped like a T. **1956** *Nature* 24 Mar. 561/1 The study of the stresses in the T-junction of a branched pipe. **1958** *Listener* 20 Nov. 835/3 The first T-junction that comes along without a signpost of any kind. **1969** E. H. Pinto *Treen* 380/2 The method of making a right-angle turn, or a T junction, is shown in Plate 408, where the tapered short section is driven into a cone socket in the side of the longer length of elm pipe. **1982** S. Spender *China Diary* 104 He jumped out of the car..and walked..till he came to a T-junction where the [traffic] blockage seemed to originate. **1898** A. W. Pinero *Trelawny of 'Wells'* iv. 171 Just below the footlights is a T-light, burning gas. **1911** [see *du théâtre]. **1933** J. Martin-Harvey *Autobiogr.* iv. 43 The gloomy underground stage unlit by anything but the 'T' light on which a single jet of gas literally made darkness visible. **1963** *Times* 1 Feb. 14/5 Some shoes had..slender T-straps. **1963** *Harper's Bazaar* Apr. 75 The T-strap sandal shown here in patent leather. **1969** *Sears Catal.* Spring/Summer 7 Dashing T-strap with sparkling patent vinyl upper. **1974** *Country Life* 21 Mar. 687/2 A brogued court with T-strap. **1978** *Detroit Free Press* 5 Mar. A19 (Advt.), Popular T-strap slings with open or closed backs and espadrilles.
II. 4. T-model (Ford) = *Model T* s.v. *model *sb.* 7 e.
1932 [see *model *sb.* 7 e]. **1942** Z. N. Hurston in A. Dundes *Mother Wit* (1973) 28/1 Way after a while a T-model Ford came along full of Negroes. **1962** *John o' London's* 11 Jan. 43/3 Driving a T-model Ford over the roof-tops.
6. a. T (*Physics*) = *tera-; T (*Physics*) = *tesla; *t* (Physics), top or truth, a quark flavour; T (*Physics* and *Chem.*) = *tritium; TA (*U.S.*), teaching assistant; T.A., Territorial Army (see also note s.v. TAVR below); T.A. (*Psychol.*), transactional analysis; T.A.B. (*Austral.* and *N.Z.*), Totalizator Agency Board; T.A.B. (*Med.*), a vaccine against typhoid, paratyphoid A, and paratyphoid B; usu. *attrib.*; T. & A., T and A (*U.S.*), tonsils and adenoids; tonsillectomy and adenoidectomy; t. and g., t. & g., (*Woodworking*), tongued and grooved; TAT (*Psychol.*), thematic apperception test; TAVR, Territorial and Army Volunteer Reserve (the name given to the restructured Territorial Army in 1967, but replaced by the name 'Territorial Army' in 1979); T.B., torpedo-boat; T.B., Treasury Bill (cf. *T-Bill*, sense 6* below); T.B., t.b., tuberculosis; *T.B.-tested* adj., (of an animal) tested to establish the absence of tuberculosis; also (*U.S. slang*), a confidence trickster (see quot. 1930); T.B.D., t.b.d., torpedo boat destroyer; tbs., tbsp., tablespoon(ful); TBS, *t*alk *b*etween *s*hips, a short-wave radio apparatus used for verbal communication between ships at sea; TCA trichloroacetic acid (a herbicide); TCB (*U.S. Black slang*), (to) take care of business; T.C.D., Trinity College, Dublin; TCDD = *tetrachlorodibenzodioxin* s.v. *tetra- 2 a; TCNQ (*Chem.*) [f. T(etra- + CN, chemical formula of the cyano group + Q(uinone], 7,7,8,8-tetracyano-*p*-quinodimethane, an organic compound forming salts of unusually low resistivity; T.C.P., the proprietary name of a disinfectant; TCP (*Physics*), time (reversal), charge (conjugation), and parity (conservation); T.D. [Ir. *Teachtai Dála*], a member of Dáil Éireann, the lower house of the Irish parliament; T.D., Territorial Decoration (in the Territorial Army); TD (*U.S.* and *Canad. Football*), a touchdown; T.D.C., Temporary Detective Constable; TDC (*Mech.*), top dead centre; TDE [f. T(wo *numeral a.* + *d*ichlorethane], an organochlorine insecticide (see quot. 1946) formerly used on fruit and vegetables; T.D.R., Treasury Deposit Receipt; t.d.s. (*Med.*) [L. *ter die sumendus*], to be taken three times a day; TEE, Trans Europ (also Europe, European) Express (train); TEFL, Tefl (tefl), Teaching of English as a Foreign Language; TESL (tesl), Teaching of English as a Second Language; TESOL (te·sǫl), Teachers of English to Speakers of Other Languages; TeV, tera-electron volt; T.G., temporary gentleman (see *temporary *a.* 1 b); T.G., thank God (cf. D.G. s.v. D III. 3); TG (*Linguistics*), transformational-generative (grammar) (see *transformational *a.*); TGV [F. *train à grande vitesse*], a type of high-speed French passenger train; T.G.W.U., Transport and General Workers' Union; THC, tetrahydrocannabinol; t.i.d. (*Med.*) [L. *ter in die*], three times a day; T.I.G., Tig (*Engin.*), tungsten inert gas (with reference to welding with a tungsten electrode in an atmosphere of an inert gas); TIR [F. *transport international routier*), international road transport (with reference to an international customs agreement: see quot. 1969); TKO, t.k.o., (chiefly *N. Amer.*), in Boxing, a technical knock-out; also *fig.* and as *v. trans.*; TL, thermoluminescent (dating technique); also *TL-dating*; TLC (*colloq.*), tender loving care; TLC, t.l.c. (*Chem.*), thin-layer chromatography; TLR (*Photogr.*), twin-lens reflex (camera); T.L.S., *The Times Literary Supplement*; TM, trade mark; TM, transcendental meditation; T.M., trench mortar (cf. *toc emma); TMV, tobacco mosaic virus; T.N.T. = *trinitrotoluene; T.O., turn over (cf. P.T.O. = 'please turn over' s.v. P II in Dict. and Suppl.); TOEFL, Test(ing) of English as a Foreign Language; TOFC (orig. *U.S.*), trailer on flatcar (with reference to a type of freight container); TOW, *t*ube-launched, *o*ptically tracked, *w*ire-guided (missile); T.P.R. (*Med.*), temperature, pulse, and respiration; TR (*Electronics*), transmit–receive; usu. *attrib.*; TRF, TRH (*Biochem.*) = *thyrotropin-releasing factor* or *hormone* s.v. *thyro- 2; tRNA (*Biochem.*), transfer RNA; also † T-RNA; T.S. (*U.S. Forces' Slang*), tough shit (also situation, stuff); also used *attrib.* to designate a (real or imaginary) card, etc., allowing the recipient an interview to discuss his grievances with the chaplain; TS (*pl.* TSS), typescript; TSA, Training Services Agency; TSH (*Biochem.*), thyroid-stimulating hormone (cf. *thyroid *sb.* 2 b); tsp., teaspoon(ful); TSS, twin-screw steamer; TSS (see *TS* = typescript above); T.T., t.t., teetotal, a teetotaller; T.T. (*Comm.*), telegraphic transfer; T.T., Tourist Trophy (freq. used *ellipt.* for *Tourist Trophy Race*); T.T. = *tuberculin-tested* ppl. adj. s.v. *tuberculin b; also *transf.*; T.T.F.N. (*colloq.*), 'ta-ta for now' (a catch-phrase popularized by the 1940s BBC radio programme *Itma*); T.T.L. (*Photogr.*), through-the-lens (metering); TTL (*Electronics*) = **transistor-transistor logic* s.v. *transistor *sb.* 3 b; TV (*colloq.*, orig. and chiefly *N. Amer.*), a transvestite; T.V.A. [Fr., *taxe à la valeur ajoutée*] = *V.A.T.; T.V.A. (*U.S.*), Tennessee Valley Authority; TVP, textured vegetable protein (proprietary name); see *textured *a.*; TWA (*U.S.*), Trans World Airlines (formerly, until 1950, Transcontinental and Western Air). See also *TAM, *TANU, *Tass^1, *TEWT, *TIM, *TOPS, *T.U.C., *TV.
1951 *Symbols, Signs & Abbreviations* (R. Soc.) 15 Tera $(\times 10^{12})$ T. **1978** *Guardian Weekly* 27 Aug. 13/3 Sweden's energy requirements..—125 Twh in 1985. **1964** *Internat. System (SI) Units* (*B.S.I.*) 8 The tesla (symbol T) is the name given to this unit in Continental literature. **1973** *Physics Bull.* Sept. 555/3 He used pulsed magnetic fields as high as 30 T. **1978** *Nature* 2 Feb. 407/1 This new quark pair is labelled t and b for 'top' and 'bottom'. *Ibid.* 407/2 The prudish may care to note that t and b are said to stand for truth and beauty, rather than top and bottom, by some physicists. **1984** *Daily Tel.* 5 July 36/4 Discovery of a sub-atomic particle labelled the 'T-top' quark has been announced by scientists at Cern. **1948** Glasstone *Textbk. Physical Chem.* (ed. 2) ii. 172 The ^{3}H isotope, called tritium, symbol T, has also been obtained by the bombardment of nitrogen by neutrons.

1973 *Nature* 3 Aug. 257/1 The square-root law of mass-dependence does not apply to isotopic variants HDO and HTO. 1969 C. Davidson in Cockburn & Blackburn *Student Power* 357 These considerations make the organization of a radical trade union of TAs a crucial part of any strategy for change. 1980 *Berkeley Graduate* Oct. 3/4 Matthew Soyster, a Comparative Literature graduate student, is currently a TA in the Rhetoric Department. 1924 *Regulations for Territorial Army* I. iii. 33 An officer appointed to command a brigade..will..be granted such rank in the T.A. temporarily. *Ibid.* iv. 53 The senior T.A. officer in the locality. 1939 W. S. Churchill *Let.* 30 Aug. in M. Gilbert *Winston S. Churchill* (1976) V. liii. 1106 Would it not be helpful to call up the reserves and mobilize the TA? 1980 *Whitaker's Almanack 1981* 473/2 The Territorial Army (TA) is designed to provide a reserve of highly trained and well equipped units and individuals. 1972 *N.Y. Times Mag.* 19 Nov. 42 Initial capitals are common in the vocabulary of Transactional Analysis, or T.A. 1976 *S. Wales Echo* 25 Nov. 16/7 In the group therapy of TA members can see a wider range of ego states than they could on their own, and the collective framework is thought to aid analysis and change. 1957 *Press* (Christchurch, N.Z.) 19 Nov. 16/1 If people want things like the T.A.B., alcohol and cigarettes. 1969 *Sydney Morning Herald* 24 May 27/1 (*heading*) The Moorebank Handicap, second leg of the TAB double. 1977 *Herald* (Melbourne) 17 Jan. 6/8 A spokesman for the TAB head office. 1929 *Lancet* 9 Feb. 288/1 These T.A.B. vaccine injections..caused remission in the course of general paralysis. 1970 *Guardian* 24 Jan. 17/3 The Department of Health advises *all* travellers abroad to take the precaution of a TAB vaccination. 1981 *Brit. Med. Jrnl.* 18 Apr. 1313/1 We all lined up, hand on hip, to receive the dreaded TAB. 1942 Berrey & Van den Bark *Amer. Thes. Slang* §532/1 *T.&A.*, tonsils and adenoids. 1960 in *Arch. Otolaryngology* Aug. 183/1 Tonsilloadenoidectomy (T and A) is often classified as minor surgery. *Ibid.* 186/1 Immediate hemorrhage in the first 24 hours post T and A continued to be approximately 3%–4%. 1976 *Amer. Speech 1973* XLVIII. 204 Relatively simple operations like a *T and A*..are considered routine procedures on the hospital's *OR* (operating room) *schedule*. 1948 *Archit. Rev.* CIII. 133 Exterior walls are of two by four studs, four by four posts, faced externally with wood sheathing and t. and g. vertical boarding. 1949 *Gloss. Terms Timber (B.S.I.)* 41 *t. & g.*, tongued and grooved. 1946 *Jrnl. Personality* XV. 70 The Thematic Apperception Test (TAT)..is a projective device which purports to reveal the basic personality characteristics of individuals. 1952 [see *global a. 2]. 1964 M. Argyle *Psychol. & Social Probl.* iv. 49 There are various ways of scoring TAT stories for aggressiveness. 1972 *Jrnl. Social Psychol.* LXXXVIII. 191 The standard Murray TATs.. were not considered appropriate. 1967 *Army Q.* XCIV. 36 T.A.V.R. I and II will be adequately equipped with modern weapons and equipment. 1977 *R.A.F. News* 22 June–5 July 18 (*caption*) The Wapinschaw..attracts entries from many regular and TAVR units in the North of Scotland. 1897, etc. T.B. = torpedo boat [implied at T.B.D. below]. *a* 1912 W. T. Rogers *Dict. Abbrev.* (1913) 187/1 *T.B.*,..torpedo boat. 1925 R. H. Bacon *Naval Scrap-bk.* x. 144 In the 1895 manœuvres, when I was a Lieutenant of just under twelve years' seniority, I was appointed in command of No. 94 T.B. 1938, etc. [see M.T.B. s.v. *M 5 a]. 1977 Preston & Brown tr. *Jentschura's Warships Imperial Jap. Navy, 1869–1945* ix. 124 TBs. *1–4* were modelled by Sir Edward Reed on the RN 100ft type. They were assembled in Japan. 1936 *Financial Times* 20 Nov. 1/1, 3 months T.B.—£0 10 6·24 pc. 1971 *Financial Mail* (Johannesburg) 26 Feb. 661/1 National has about R25m of TBs on its book. 1912 D. Lowrie *My Life in Prison* iv. 33 He's doin' 35 years an' has the T.B. 1921 A. Mason *Flying Bo'sun* ii. 19 Their mother died two years ago... The doctor said it was T.B. 1930 J. Lait *Big House* i. 7 A confidence (or 'con') man is a 'T.B.', ('con' is short and slang for consumption, and 'T.B.' is ditto for tuberculosis). 1932 U. Sinclair *Candid Reminiscences* x. 88 The old captain was ill of TB. 1942 C. Himes *Black on Black* (1973) 176 Men..of all stages of deterioration—drifters and hopheads and tb's and beggars and bums and bindle-stiffs and big sisters. 1951 J. Cannan *And all I Learned* x. 180 We've our own cows, T.B. tested and so beautifully kept. 1957 S. Smith *Coll. Poems* (1975) 336, I lay with my young bride in my arms, A girl with t.b. 1974 M. Butterworth *Man in Sopwith Camel* ii. 26 They wouldn't take me 'cos I'd had a touch of TB. 1897 Kipling *Let.* Aug. in C. Carrington *Rudyard Kipling* (1955) xi. 268 Ref: t.b.d. trials. My attention is at present taken up by one small craft recently launched from my own works. 1902 —— *Traffics & Discoveries* (1904) 182 The chief engineer o' the *Djinn*, 31-knot T.B.D. 1919 C. P. Thompson *Cocktails* 249, I passed an enjoyable day giving a T.B.D. lieutenant a headache. 1978 H. Wouk *War & Remembrance* xx. 200 The TBD is a lot slower. 1950 *Good Housek. Picture Cookery* 170/2 Coffee Glacé Icing. 8 oz. icing sugar 2 tbsps. water. Coffee essence to taste. 1974 J. Paxton *Everyman's Dict. Abbrev.* 338/1 *tbs., tbsp.*, tablespoon; tablespoonful. 1977 *Times Lit. Suppl.* 1 Apr. 391/2, I only wish my dog *liked* his food sprinkled with the recommended 2 tbs of the product every day. 1944 TBS [see *gravelly a. 5]. 1946 *Sat. Even. Post* 26 Oct. 66/3 The astounded admiral grabbed the TBS radio and shouted. 1978 H. Wouk *War & Remembrance* xv. 153 Rear Admiral Spruance, issuing order after order on the TBS, finally regained a semblance of control. 1960 *Farmer & Stockbreeder* 22 Mar. 136/3 After T.C.A. treatment 8 weeks should elapse before planting potatoes. 1971 *Arable Farmer* Feb. 15/2 A pre-planting application of TCA to peas may reduce the waxiness of the crop foliage, leading to unexpected damage from dinoseb applied subsequently. 1969 S. E. Henderson in Cook & Henderson *Militant Black Writer* 78 These poems were not intended for white readers and white audiences..their purpose was direct address to the black community, to get us together to TCB. 1973 *New Times* 2 Nov. 41 Where he is always to be found TCB'ing (takin' care of business, an old ghetto phrase which originally meant to copulate). 1831 M. C. Taylor *Let.* 22 June in J. J. Auchmuty *Sir T. Wyse* (1939) ix. 134 The Scholars of T.C.D. do not afford a Protestant, a learned or an independent constituency. 1916 H. Plunkett *Jrnl.* 29 Apr. in M. Digby *Plunkett* (1949) ix. 212 The firing from the other side of the T.C.D. guards was so fierce that we had to turn back. 1939 Joyce *Finnegans Wake* 424 Go o'er the sea, haythen, from me and leave your libber to TCD. 1979 J. Sheehan in J. J. Lee *Ireland 1945–70* 67 A UCD/TCD merger or co-ordination of some sort. 1971 *New Yorker* 14 Aug. 57 The only real question concerning the safety of 2,4,5-T has to do with the TCDD dioxin with which it is contaminated. 1981 *McGraw-Hill Yearbk. Sci. & Technol.* 199/2 It is impossible to substantiate the charge that TCDD has led to an increase in the number of malformed children in Vietnam. 1960 *Jrnl. Amer. Chem. Soc.* LXXXII. 6408/1, 7,7,8,8-Tetracyanoquinodimethane (TCNQ) has been synthesized and found to yield a series of stable anion-radical derivatives. 1979 *Sci. Amer.* Oct. 48/2 Many salts in which TCNQ is combined with other atoms or molecules form linear-chain solids. 1934 *Trade Marks Jrnl.* 22 Aug. 1084/1 *T.C.P.*... Antiseptic and germicide solutions (being disinfectants). British Alkaloids Limited, 104, Winchester House, Old Broad Street, London, E.C.2; manufacturing chemists. 1947 J. Lees-Milne *Diary* 22 Jan. (1983) 128 Cut my mole shaving this morning and thought it would never stop bleeding. T.C.P. finally staunched it. 1959 I. & P. Opie *Lore & Lang. Schoolch.* ix. 171 Earnestly applying T.C.P...in the privacy of their bedrooms. 1981 G. Kaye *Day after Yesterday* ii. 22 You cried when you only scraped your knee..a little scrape and a bit of TCP. 1957 *Physical Rev.* CVI. 385/1 According to a general theorem, invariance with respect to the product *TCP* follows for a wide class of field theories from invariance with respect to the proper Lorentz group alone. 1974 Frauenfelder & Henley *Subatomic Physics* ix. 223 When violation of parity became a possibility the *TCP* theorem suddenly acquired more meaning. 1979 J. C. Polkinghorne *Particle Play* iii. 47 We do not know how to write down any theories which are not invariant under TCP. 1947 S. Malone *Notes on Procedure in Houses of Oireachtas* p. vii, Members of Dail Eirann (Teachtai Dála)—referred to as T.D.'s or Deputies. 1959 B. Chubb in D. E. Butler et al. *Elections Abroad* iii. 187 TD—Teachta Dála; the Irish equivalent of Member of Parliament. TDs are addressed as 'Deputy'. 1979 M. Manning in J. J. Lee *Ireland 1945–70* 51 Its TDs tended to act more as independents than as members of a political party. 1924 *Regulations for Territorial Army* I. x. 145 The letters 'T.D.' will be inserted in the Army List after the name of the officer on whom the decoration is conferred. 1981 *Whitaker's Almanack 1982* 248 *Alport*, Cuthbert James McCall Alport, P.C., T.D. 1953 Berrey & Van den Bark *Amer. Thes. Slang* (1954) §692/2 Touchdown,..*TD, touch*. 1969 *Eugene* (Oregon) *Register-Guard* 3 Dec. 2D/1 Nyseth scored three TDs en route to 137 yards. 1970 *Globe & Mail* (Toronto) 25 Sept. 33/3 Another fumble set up Riverdale's second TD when Bob Nichols recovered the ball in Lions' end zone. 1977 *Detroit Free Press* 11 Dec. 10-D/4 TD passes, home runs, goals or point averages. 1970 G. F. Newman *Sir, You Bastard* ii. 45 TDC Sneed and DC Lambert to watch on Sloane Square to investigate shoplifting complaint. 1978 B. Norman *To nick Good Body* ii. 9 A temporary detective constable..had just brought in the Guv'nor's tea... The Guv'nor waved away the T.D.C. 1938 *Jrnl. R. Aeronaut. Soc.* XLII. 888 T.D.C. or any other reference marks are marked by the discharge flash of a thyratron circuit, controlled by suitable contacts on the engine crankshaft. 1976 *New Motorcycling Monthly* Oct. 34/2 With piston at TDC, pull barrel up mounting studs. 1946 *Nature* 6 July 22/1 'D.D.D.' or 'T.D.E.' The compound 2,2-*bis*(*p*-chlorphenyl)1,1-dichlorethane..has been shown to be about as toxic as D.D.T. to mosquito larvæ in laboratory tests. 1970 *New Scientist* 1 Jan. 16/1 Much of this case also applies..to the other 'hard' organochlorine insecticides: aldrin,..BHC and TDE (Rhothane). 1948 G. Crowther *Outl. Money* (ed. 2) ii. 37 There are four main types of bank assets, which are..bills.., Treasury Deposit Receipts (usually known as TDRs), investments and loans. 1965 J. L. Hanson *Dict. Econ. & Commerce* 383/1 By 1945 T.D.R.s had reached a total of over £1,800 million. 1899 P. G. Lewis *Nursing* ii. 18 Medicines are ordered to be taken either statim (immediately), or t.d.s. = ter die sumenda (to be taken three times a day). 1919 *Jrnl. R. Naval Med. Service* V. 93 He was given quinine 15 gr. t.d.s. 1961 *Lancet* 29 July 238/1 The response of our patients to a course of 20 electroshock treatments or to chlorpromazine 100–400 mg. t.d.s. 1963 *Times* 23 May 13/7 TEE trains now link 90 European cities. 1967 R. Sawkins *Snow in Paradise* iii. 29 The TEE is just about as quick as the plane. 1977 J. Paxton *Dict. Europ. Econ. Community* 246 *Trans-Europ-Express* (*T.E.E.*). Trans-Europ-Expresses connect major cities in nine European countries by a network of very fast and comfortable trains for which frontier formalities have been reduced to a minimum. 1963 *Language Learning* (Univ. of Michigan) XIII. 225 (*heading*) Reflections on Preparation for TEFL. 1968 *Language* XLIV. 206 Any gathering of TEFL leaders today will be sure to include a large number who received their initial experience at Michigan. 1977 P. Strevens *New Orientations Teaching of English* v. 56 The American terms TEFL, TESL, TESOL, TESOLD have no precise counterparts in British usage. 1981 *Guardian* 14 Apr. 21/4 (Advt.), Applications are invited from candidates who have experience in..Tefl. 1967 *Sat. Rev.* (U.S.) 16 Sept. 83 Though a major curriculum emphasis is developing fluency in the English language using the linguistic approach of TESL (Teaching English as a Second Language), the knowledge of Navaho is still essential to many jobs on the reservation. 1980 *Verbatim* Spring 20/1 A dictionary or international English might have remembered TESL and TEFL 'teaching of English as a Second (or Foreign) Language', however. 1969 *Language* XLV. 171 The two halves of the collection, articles on the description of English and articles on TESOL, reflect the sad dichotomy between descriptive linguists and language-teaching methodologists the world over. 1956 *Proc. CERN Symposium* I. 64/2 These machines would have equivalent energies of 1340 Gev, or 1·3 Tev. 1980 *Sci. Amer.* Jan. 32/3 Completion of the full lower ring is now expected toward the end of 1981, and protons at 1 TeV should be delivered to the experimental areas in 1982. 1916 N. Mitchison *Jrnl.* in *All Change Here* (1975) xv. 147 Last night about half a dozen[officers] came into the salon and started a conversation... They were awful TGs mostly... Here am I, sitting..in the middle of the stuffy salon of a third-rate French hotel, being as charming as I can to an audience of TGs, all to give them the memory of a pleasant evening to take back to the trenches at Givenchy. 1934 J. Rhys *Voyage in Dark* I. i. 17 'Only three more weeks of this damned tour, T.G.,' Maudie said. 'It's no life.' 1978 D. Murphy *Place Apart* xi. 226 Isn't it a tough world to be tryin' to raise twelve boys in? But T.G. so far they're good lads. 1968 B. M. H. Strang *Mod. Eng. Structure* (ed. 2) 200 The most eminent TG thinkers are still evolving and modifying parts of their theory. 1971 *Archivum Linguisticum* III. 64 Of recent years it is transformational-generative grammar (TG) that has undoubtedly called the tune. 1977 *Trans. Philol. Soc. 1975* 8 General linguists unsympathetic to TG. 1978 *English Jrnl.* Dec. 52/1 On the other hand, we should not jump indiscriminately into transformational-generative grammar. A full-blown TG grammar is complex. 1980 *N.Y. Times* 19 Nov. 17/1 Among the trains being studied as possible models for Ohio is the French TGV. [1924 G. D. H. Cole *Organised Labour* II. i. 20 The Transport Workers' Federation, which..has lost much of its importance since the formation of the T. and G.W.U.] 1955 *Times* 2 May 12/6 He became general secretary of the T.G.W.U. in 1945. 1957 *Economist* 26 Oct. 292/1 Last July the TGWU, together with the other unions involved, called an official strike of the provincial busmen. 1977 M. Walker *National Front* vi. 157 It was an American executive who flew to secret talks with the TGWU union negotiators at Llandudno. 1968 *Time* 19 Apr. 79 At a Chicago conference on psychedelic drugs, Dr. Donald R. Jasinski..reported that he had produced LSD-like symptons with tetrahydrocannabinol (THC), one of the purified active ingredients of cannabis. 1974 *Times Lit. Suppl.* 8 Mar. 240/5 Marijuana smokers in the United States are 'playing at cannabis use'—in that their daily ingestion of THC is only one-fifth of that of users in India, Egypt or Morocco. 1885 C. S. Weeks *Textbk. Nursing* vii. 105/2 *T.i.d., ter in dies*, three times a day. 1941 *Jrnl. R. Naval Med. Service* XXVII. 301 The course of atebrin tablets, one t.i.d. 1976 *Amer. Speech 1973* XLVIII. 198 Referring to the exact times a patient must have a certain medication are *b.i.d.* for *bis in die* 'twice daily', *t.i.d.* for *ter in die* 'three times a day', [etc.]. 1960 *McGraw-Hill Encycl. Sci. & Technol.* XIV. 467/2 Inert gas shielding is essential with tungsten electrodes, hence the term Tungsten Inert Gas (TIG) welding. 1969 D. K. Allen *Metallurgy Theory & Practice* xix. 612 (*caption*) Photomicrograph of a Tig weld with low voltage electron beam weld in center of Tig nugget to show relative width of heat affected zone in each process. 1975 Bram & Downs *Manuf. Technol.* ii. 55 The T.I.G. process differs from the manual metal-arc technique in as much as the electrode is virtually non-consumable. 1968 E. Pugh *Dict. Acronyms & Abbrev.* 169 *TIR*, Transports Internationale Routiers [*sic*]. 1969 *Jane's Freight Containers 1968–69* 141/1 For road transport, the so-called 'TIR' Convention, concluded under the auspices of the ECE, allows the transport of goods under Customs seal in lorries from the Customs office of departure to the Customs office of arrival. 1980 K. Hagenbach *Fox Potential* xii. 119 We passed a couple of big TIR trucks, each pulling a trailer. 1942 Berrey & Van den Bark *Amer. Thes. Slang* §704/1 Technical knockout,..*T.K.O.* 1956 'T. Betts' *Across the Board* xxi. 296 Endocrinology TKO's Freud in the second round. 1968 M. Richler in R. Weaver *Canad. Short Stories* 2nd Ser. 164 'You lost by a TKO,' my father said. 'Thanks,' my mother said. 1971 *Weekend World* (Johannesburg) 9 May 1/3 Tulwana's fly weight title which Dlamini took on a third round t.k.o. 1975 J. Gores *Hammett* (1976) iii. 28 Revani TKO'd his Filipino opponent after..using his gut as a workout bag in the fourth [round]. 1972 *Oxf. Univ. Gaz.* CII. Suppl. II. 12 Development and application of TL at the National Museum, Edinburgh. 1978 *Times* 11 Nov. 3 TL-dating was developed in the 1960s for dating pottery and other fired materials from archaeological sites. 1960 I. A. Stanton *Dict. for Medical Secretaries* 149/1 *T.L.C.*, abbreviation for tender, loving care. 1973 *Publishers Weekly* 19 Nov. 55/1 The contagious potential of his TLC when he launches a yarn. 1977 *Listener* 12 May 605/3 It is in a nurse's nature and in her tradition to give to the sick what is well called 'TLC', 'tender loving care'. 1961 *Jrnl. Amer. Oil Chemists Soc.* XXXVIII. 316/1 TLC has a number of features which make it an ideal technique for the analysis of these compounds. 1975 Williams & Wilson *Biologist's Guide to Princ. & Techniques Pract. Biochem.* iii. 58 Adsorbents used in t.l.c. differ from column adsorbents in that they may contain a binding agent such as calcium sulphate. 1980 *Nature* 8 May 105/1 To determine the sequence of the two amino acids, the active sample was dansylated, hydrolysed and the dansyl derivative examined by TLC. 1965 M. J. Langford *Basic Photogr.* 376/1 (Index), TLR. 1978 *SLR Camera* Sept. 61/1 Rollei..originally only manufactured top quality TLR cameras. 1979 *Amat. Photographer* 10 Jan. 110/2 I'd recommend the TLR every time if monochrome prints are the objective, more so when the prints are to over 10 × 8in. 1953 R. Crawshay-Williams *Let.* 1 Aug. in B. Russell *Autobiogr.* (1969) III. ii. 91 There are a nice lot of sly digs..the T.L.S. pastiche. 1967 E. Coxhead *Thankless Muse* i. 18 A little advance something in the TLS never comes amiss, does it? 1977 *Bookseller* 4 June 2704/2 The *Times Literary Supplement* (never called the *T.L.S.* then [in 1952]). 1961 Webster, TM, trademark. 1964 *Trademarks in Marketplace* (U.S. Trademark Assoc.) 64 We use the term 'trademark' right under the selected word, or we put a little 'TM' in the place where you normally would put the ®. 1980 *Sci. Amer.* Oct. 138/1 (Advt.), The Clan of the Cave Bear, Earth's Children™, a novel by Jean M. Auel. 1967 *Listener* 7 Sept. 299/1, I hear you're hostile to drugs now and have taken to TM. 1977 E. V. Cunningham *Case of One-Penny Orange* (1978) ix. 110 Topanga Canyon.. had..sensitivity centers and nudist camps and TM temples. 1980 *Times* 27 May 1/8 The Home office does not approve the use of TM in borstals or prisons. 1925 Fraser & Gibbons *Soldier & Sailor Words* 275 T.M. batteries were created for trench warfare and T.M. schools of instruction were established. 1930 Brophy & Partridge *Songs & Slang 1914–1918* 173 When a T.M.

battery had fired a few shots it departed with speed. 1960 *Proc. Nat. Acad. Sci.* XLVI. 636 (*heading*) The amino acid composition and C-terminal sequence of a chemically evoked mutant of TMV. 1974 W. K. Joklik in Carlile & Skehel *Evolution in Microbial World* 298 The fascinating work of Klug and his colleagues with TMV. 1915 D. O. Barnett *Let.* 4 July 203 The yellow muck doesn't choke you, though, like the black greasy smoke (T.N.T.) which they generally have in the 6 and 8-inch shells. 1962 E. Snow *Other Side of River* (1963) xxix. 217 A responsible Western physicist's estimate that the world then possessed a nuclear weapons stockpile roughly the equivalent of forty tons of TNT for each person alive. 1979 O. Sela *Petrograd Consignment* 53 The casing contains a pound of TNT and it can be attached to a timing device. 1853 Mrs. Gaskell *Cranford* v. 68 However, at the foot of the page was a small 'T.O.', and or turning it over, sure enough, there was a letter to 'my dear, dearest Molly'. 1889 E. C. Dowson *Let.* c 23 Jan. (1967) 29 When we will proceed to Pinoli or where you will. T.O. Act à votre discretion in the matter of booking seats. 1981 *Oxf. Dict. for Writers & Editors* 412/2 *TO*, turn over. 1964 *Overseas* Jan. 22 On February 17, 1964, a new English-proficiency test for foreign students will be administered overseas. Called TOEFL for short, this Test of English as a Foreign Language is designed to help..assess the language competence of foreign students. 1972 J. L. Dillard *Black English* 272 It is of primary importance that all such materials, like all TOEFL materials, take the student's native language (here, dialect) into full account. 1954 TOFC [see *pick-a-back adv. phr. (a., sb.) b]. 1964 *Economist* 26 Sept. 1243/1 The [US] railways have introduced TOFC service—trailer-on-flatcar, better known as Piggyback. 1969 TOFC [see *kangaroo sb. 3 h]. 1972 *Guardian* 22 Aug. 3/1 The missiles [are] called TOWs.. TOW stands for tube-launched, optically tracked, and wire-guided. 1976 *N.Y. Times* 28 Mar. 1 The TOW missile can be used offensively from jeeps or armed cars. 1917 V. Brittain *Let.* 5 Dec. in *Testament of Youth* (1933) viii. 395 Morning work—i.e. beds, T.P.R.s (temperatures, pulses, respirations), washings, medicines, etc. 1976 *Amer. Speech 1973* XLVIII. 197 His *vitals* 'vital signs' such as temperature, pulse, and respiration (*TPR*). 1945 *Electronic Industries* Sept. 226 *T R switch*, transmit-receive switch. A switch which prevents the transmitted energy from getting to the receiver, but allows the received energy, which is much weaker, to reach the receiver without appreciable loss. This is necessary when the same antenna is used for both transmission and reception. 1975 D. G. Fink *Electronics Engineers' Handbk.* IX. 23 Such limiters are replacing TR gas discharge tubes in radar. 1959 K. Shibusawa et al. in *Endocrinol. Jap.* VI. 31 We found a thyroid stimulating neurohumor in the hypothalamus... It was provisionally designated by us as TRF (Thyrotropin Releasing Factor). 1972 *Clin. Endocrinol.* (1973) (B.M.A.) 47 Thyrotrophin-releasing factor (TRF), has recently been synthesized. 1968 A. V. Schally et al. in *Rec. Progr. Hormone Res.* XXIV. 449 (*table*) Present name... Thyrotropin-releasing factor. TRF... Proposed name... Thyrotropin-releasing hormone. TRH. 1977 *Proc. R. Soc. Med.* LXX. 698/1 The chief value of the thyrotrophin releasing hormone (TRH) test has been the diagnosis of dysthyroid eye disease. 1962 *Cold Spring Harbor Symp. Quantitative Biol.* XXVIII. 559 The system consisted of *E. coli* ribosomes and high speed super natant which contained transfer RNA (T-RNA), the amino acyl-T-RNA synthetases, and the enzymes involved in the final steps of the synthesis of the polypeptide chain. 1966 *Ibid.* XXXI. 587/1 The elution profiles of noninfected-cell and infected cell arginyl tRNA exhibit unambiguous major differences. 1977 *Time* 4 Apr. 39/2 Aaron Klug..first determined the crystalline structure of transfer RNA (tRNA), the molecule that brings amino acids to the ribosome for assembly into protein. 1944 A. M. Taylor *Lang. World War II* 69 TS Cards: Beachhead chaplains are carrying a special 'tough stuff' ticket these days which they issue to guys with complaints about which nothing can be done. 1944 *Yank* 18 Aug. 16/2, I..will be ever grateful for any possible solution, for I have tried everything I know, even prayer. Still TS. 1946 *Amer. Speech* XXI. 249 A *T.S. ticket* is an imaginary form entitling the bearer to sympathy and nothing else. 'All I can do is punch ya T.S. ticket (or slip) for ya.' 1946 *Amer. Jrnl. Sociol.* Mar. 422 One such expression is 'TS' or 'tough s—'... 'TS', resigned acceptance, said with a bitter smile. 1966 *Sunday Times* (Colour Suppl.) 4 Dec. 73/4 [GI Jargon] *TS*, tough situation. 1942 Partridge *Dict. Abbrev.* 97/1 *t.s.* or *ts.* or *ts*; also *T.S.*, etc., typescript. 1975 *Record* (Oxf. Univ. Press) xx. 24/1, 11 unsolicited poetry TSS in one week. 1975 *Petroleum Rev.* XXIX. 399/2 The TSA has been very active since its inception. 1976 *Even. Post* (Nottingham) 15 Dec. 12/7 The TSA's direct training services which include the provision of specially tailored courses to meet individual requirements. 1941 *Trans. Amer. Assoc. Study Goiter* 161 Media, which originally contained TSH in a concentration equivalent to one unit per cubic centimeter were found to have lost about seven-eighths of their activity. 1983 *Oxf. Textbk. Med.* I. x. 12/2 Hypothyroidism due to TSH deficiency is often mild and easily overlooked. 1950 *Good Housek. Picture Cookery* 161/1 Unboiled fondant.. 1 lb icing sugar..tartar ..1 tsp. lemon juice..1 egg white. 1955 R. J. Schwartz *Compl. Dict. Abbrev.* 178/1 *tsp*, teaspoon. 1973 Rubinstein & Bush *Penguin Freezer Cookbk.* 211, ½ lb. tomatoes, 1 tsp. allspice, 1 tsp. pine kernels. 1935 Duckworth & Langmuir *West Highland Steamers* ii. 107 T.S.S. 'Flowerdale'..came into Mr. MacBrayne's hands as his first twin screw sea-going steamer. 1981 'J. Ashford' *Loss of 'Culion'* viii. 52, I understand you've some information on the sinking of the TSS *Culion*..in the Indian Ocean. [1841 *Niles' Reg.* 21 Feb. 400/3 TTT. They have temperance wagons in the west, marked with three Ts, to denote that the owner is a tee-to-taller.] 1922 Joyce *Ulysses* 159 Selfish those t.t's are. Dog in the manger. 1936 *Punch* 22 July 97/2 Every birthday he gets a magnificent..Gift Of wine... It is sad To add I've brought him up to be T.T. 1975 J. Symons *Three Pipe Problem* v. 35 Can't offer you a beer, strict TT. 1893 R. Bithell *Counting-House Dict.* (rev. ed.) 292 *T.T.*, telegraphic transfer. 1927 *Financial Times* 3 May 1/4 Kobe, T.T. Yen 24·58d. 1940 *Economist* 11 May 862/2 This compares with..¼ per cent. for T.T. redemptions from Palestine. 1966 A. Gilpin *Dict. Econ. Terms* 201 (*heading*) Telegraphic Transfer ('T.T.'). 1913 W. T. Rogers *Dict. Abbrev.* 192/1 *T.T.* (motor), Tourist Trophy machine. *Ibid.*, *T.T. Race* (motor), Auto-Cycle Tourist Trophy Race. 1914 *Autocar* 16 May 948/2 (*heading*) The T.T. race. 1929 *Motor* 2 July 1060/1 A driver who has, after all, driven at Le Mans and in the T.T. 1976 *Southern Even. Echo* (Southampton) 17 Nov. 21/5 The Isle of Man programme will include the usual classic TT. 1927 *Field* 15 Sept. 413/1 There are now a larger number of farmers producing Grade A (T.T.) milk. 1958 *Times* 28 July 11/5 A T.T. dairy farm. 1970 A. Jenkins *Drinka Pinta* x. 105 Scots were particularly keen that it should be Grade 'A' (T.T.) milk. 1948 F. Worsley *ITMA* 21 The beloved Cockney Charlady, Mrs. Mopp (played by Dorothy Summers)..did not make her first appearance..until 10th October, 1941... Another of her famous sayings were the letters 'T.T.F.N.'—a contraction of 'Ta-ta for now' with which she made her exit. 1966 A. Hall *Frost* 19 'See you soon then.' 'T.T.F.N.' 1976 *Observer* 11 Apr. 2/6 JY [*sc.* Jimmy Young] said TTFN to Mr Healey. 1963 *Electronics* 22 Mar. 54/1 Transistor-transistor logic (TTL) offers a saturated-transistor logic that is simple, compact, and has a high degree of design flexibility. 1967 *Ibid.* 6 Mar. 123/1 Litton Industries Inc. had developed the 'Phoenix gate' TTL for the Phoenix missile. 1977 *Sci. Amer.* Sept. 79/3 Compared with the previous family of RTL gates, TTL circuits provide greater output power (so that more gates in the next stage of an array can be driven), less stringent tolerances in manufacturing and greater immunity to spurious voltages. 1968 *Amateur Photographer* 24 Apr. 4/2 (Advt.), The most sought after T.T.L. camera! 1978 *Ibid.* 2 Aug. 101/1 Landscapes with a lot of sky detail in the shot can often be wrongly exposed, particularly if the camera has TTL metering. 1965 *Realist* Mar. 24 TVs are not as feminine as they themselves think they are. 1979 J. Hansen *Skinflick* (1980) x. 81 Spence doesn't want to be fooled. He knows I'm a TV. 1983 *The Magazine* Apr. 24/3 We get a lot of TVs in and a few of the leather boys of course. 1963 *Times* 2 Feb. 9/5 When we enter the Common Market,.. a T.V.A. (tax value added) tax will have to be substituted for purchase tax. 1963 *Economist* 17 Aug. 567/2 The probable impact of a TVA tax on different types of industries. 1965 *Listener* 22 Apr. 585/2 The added-value tax is commonly known as the TVA. 1935 *Harvard Law Rev.* XLVIII. 806 It would seem that if the TVA is in fact unconstitutional, the contracts are subject to rescission. 1936 *N.Y. Herald Tribune* 4 June 36/8 The T.V.A. development. 1943 J. S. Huxley *TVA* 7 TVA stands for Tennessee Valley Authority, and the Tennessee Valley Authority is the outstanding example of democratic planning. 1965 Mrs. L. B. Johnson *White House Diary* 7 Apr. (1970) 256 Lyndon talked about the vast Mekong River project which can provide food and water and power on a scale to dwarf even our own TVA. 1968 *Guardian* 4 July 7/6 What TVP has been created from is mercifully not revealed. 1969 *Official Gaz.* (U.S. Patent Office) 18 Mar. 120/2 *TVP*... For unflavored and meat and poultry flavored vegetable protein food... First use on or before May 2, 1966. 1974 *Observer* (Colour Suppl.) 24 Feb. 14/1 High-protein mock-meat has been in use in a fairly small way for some years... Known as tvp—textured vegetable protein—it comes in 'extruded' chunks, or minced. 1975 *Trade Marks Jrnl.* 21 May 1049/1 *TVP*... Foods prepared from soya bean derivatives and included in Class 9. Archer Daniels Midland Company.., Decatur, Illinois, United States of America; manufacturers and merchants. 1976 *Times Lit. Suppl.* 13 Feb. 166/1 The chunks of fictionalized, texturized social history (which are to drama as TVP to steak). 1933 *Meccano Mag.* Apr. 270/2 T.W.A. state that most of the transcontinental air mail is carried by their machines. 1941 B. Schulberg *What makes Sammy Run?* vi. 93 They were flying back again via TWA. 1960 [see *red carpet* s.v. *red a. 19 a]. 1976 *National Observer* (U.S.) 17 Apr. 20/5 Pressed by banks that had lent TWA millions, he sold his TWA stock for $546,549,171.

b. *Biol.* T designates lymphocytes that are derived from or have been processed by the thymus, which are responsible for cellular immune reactions.

1970 *New Scientist* 7 May 271/1 Some of the lymphocytes ..'stray' into the thymus gland, where they are converted ..into a new sort of lymphocyte, called a T cell. 1973 *Sci. Amer.* July 58/1 *T* cells and *B* cells cannot be distinguished by their form. 1974 *Nature* 8 Feb. 387/2 T lymphocytes in human peripheral blood may be identified by their ability to form rosettes *in vitro* with untreated sheep erythrocytes. 1976 *Path. Ann.* XI. 437 During famine the severity and frequency of diseases kept in check by T-cell function, such as tuberculosis, will increase. 1982 Arms & Camp *Biology* (ed. 2) xxxiv. 541 Lymphocytes may be divided into two major groups, T lymphocytes and B lymphocytes.

6*. In combinations containing the abbrev. T (or extension, as T.D.) followed by a word, as **T-Bill** [Treasury sb.] = *treasury bill* s.v. Treasury *sb.* 6 (cf. *T.B.*, sense 6 a above); **T.D. pipe** *U.S.* [see quot. 1889], a kind of clay pipe; **T-group** *Psychol.* [Training *vbl. sb.*], a sensitivity-training group (see *sensitivity 4); **T-man** *U.S. colloq.* [Treasury *sb.*: cf. *G-man b], a law enforcement officer of the Treasury Department; **T-rule** *Linguistics* [Transformational *a.*] = transformational rule (see *transformational a.); **T scale, score** *Psychol.* [Thorndike-Terman (see quot. 1922)], a method of scaling or scoring a psychological test; **T stop** (*Photogr.*) [Transmission], a measured point on a scale of aperture values based on the actual light transmitted through the lens of a camera; similarly *T number*; **T-unit** *Linguistics* [Terminable *a.*], a minimal terminable unit (see quot. 1965).

1982 *Daily Tel.* 27 Sept. 18 Examples..occur in financial futures, with the difference between the futures price and the price of its underlying cash instrument (cf. gold futures and the bullion price, or T-Bill futures and cash T-Bills). 1880 *Harvard Lampoon* 19 Mar. 26/2 So, after he had taken his breath of fresh air, he filled his T.D. pipe. 1889 *Amer. N. & Q.* II. 114 'T.D. Pipes.'..It is said that they took their name from Timothy Dexter, an eccentric capitalist, who in his will left a large sum of money to be expended in the erection of a factory where cheap clay pipes, such as those that now bear the name of 'T.D.'s', were to be manufactured. [1947 E. H. Paul *Linden on Saugus Branch* 27 Deacon Parker, known to the boys as T.D., because he smoked the one-cent clay pipes of that name.] 1950 *Jrnl. Social Issues* VI. II. 3 Most of the core staff goes through a 'Practicuum in Group Membership' seminar worked out along the lines of the 'T' group training program at the National Training Laboratory in Group Development. 1967 M. Argyle *Psychol. Interpersonal Behaviour* x. 193 T (training)-groups were first developed in the National Training Laboratories at Bethel, Maine, in 1947... The members of a T-group spend their time studying the group and the processes of social interaction that take place in it. 1977 *N.Y. Times* 15 July c 22/2 We already spend far too much time practicing artificial modes of sociability, such as group encounters, sensitivity training, 'T' groups, Rolfing and the like. 1938 *Sun* (Baltimore) 12 Mar. 10/7 Comparatively little has been heard in late times about the Treasury Department's 'T-men'. 1952 *Daily Progress* (Charlottesville, Va.) 6 Feb. 1/5 T-man William Frank says backdating of tax returns is the major irregularity he found. 1951 G. H. Sewell *Amateur Film-Making* (ed. 2) iii. 32 The T number indicates the actual light transmission obtained by measuring instruments. 1961 G. Millerson *Telev. Production* iii. 38 Modern lenses are sometimes marked in 'T' numbers rather than f-numbers. 1964 E. Bach *Introd. Transformational Gram.* iv. 60 The difference between PS rules and T rules can be made clearer.. by the following remarks. 1976 *Language* LII. 108 In 1962, Wolfgang Motsch proposed a T-rule for a class of German adjectives. 1922 W. A. McCall *How to measure in Educ.* x. 299 It is a tribute to their genius to call the proposed unit..a Thorndike–Terman, or, for brevity, a T... Every product scales [*sic*] may be transmuted into T scales, thereby making all scales performance scales. 1954 A. Anastasi *Psychol. Testing* iv. 83 If the normalized standard score is multiplied by 10 and added to or subtracted from 50, it is converted into a *T-score*, a type of score first proposed by McCall. 1970 F. G. Brown *Princ. Educ. & Psychol. Testing* vii. 173 *T* scores were originally defined..with reference to a particular norm group. .. However, as used today, the *T*-score designation generally applies to any normalized standard score system with $\bar{X} = 50$ and $s = 10$. 1956 J. J. Rose *Amer. Cinematographer* (ed. 9) 133 The new method of calibration will be known as the 'T' Stop system, the T denoting transmission of light and the 'T' stops representing absolutely accurate light measurement. 1977 J. Hedgecoe *Photographer's Handbk.* 344 *'T' stops*, more accurate measurement of light entering a lens than 'f' numbers. 1965 K. W. Hunt *Gram. Structure written at Three Grade Levels* iii. 21 These units might be christened 'minimal terminable units', since they would be minimal as to length, and each would be grammatically capable of being terminated with a capital letter and a period. For short, the 'minimal terminable unit' might be nicknamed a 'T-unit'... T-unit will be the name used for it in this investigation. 1975 *Language for Life* (Dept. Educ. & Sci.) iii. 39 Writing of high quality can employ a simple style that would not necessarily yield a high score as measured by the T-unit. 1977 *Publ. Amer. Dial. Soc. 1974* LXI/LXII. 30 The transcripts of these interviews (exclusive of garbles and false starts) were marked off into T-units and the mean length of T-units was computed for each informant.

6.** Used as a symbol. **a.** *Physics.* [Adopted as being the next letter after S alphabetically (cf. *S 7).] *T* is the symbol of the quantum number of iso-spin; = *I 7 b.

1937 E. Wigner in *Physical Rev.* LI. 110/1 The quantum numbers S_1, T_ζ, ζ_Y can be called magnetic quantum numbers. They determine..the μ uniquely. *Ibid.* 117/1 A total isotopic spin T will be a term with the same binding energy for all nuclei with isotopic numbers from −T to T. 1974 *Encycl. Brit. Macropædia* XIII. 339/2 Isospin-equal-one states can exist in the three isobars $^6_2$He, $^6_3$Li, and in $^6_4$Be, whereas the *T* = 0 state can exist only in $^6_3$Li.

b. *Bacteriology.* [Initial letter of *type*.] T followed by a numeral is the designation of certain strains of phages of the bacterium *Escherichia coli* that have been much used experimentally. So *T-even*, designating the strains for which the numeral is even.

1944 *Proc. Nat. Acad. Sci.* XXX. 398 The viruses used were the strains α and γ described by Delbrück and Luria and a new strain..which has been determined to be identical with strain T7 of Demerec and Fano [*reference given to paper* 'in press', *quoted next*:] 1945 Demerec & Fano in *Genetics* XXX. 119 The materials used in our experiments consisted of the same bacterial strain—*E. coli* B—previously used by Luria and Delbrück.., of seven phage strains active on B, and of various strains of bacteria.. resistant to one or more of the phages. The phage strains were indicated as type 1 to type 7 (T1 to T7). 1960 *New Biol.* XXXI. 78 One group of the larger phages..has become the best understood of any kind of virus. These are the closely related T2, T4 and T6, known collectively as the T-even phages. 1968 H. Harris *Nucleus & Cytoplasm* iii. 46 It was known at the time that this work was undertaken that when *E. coli* cells were infected with the T-even bacteriophages net synthesis of bacterial RNA was rapidly inhibited. 1973 R. G. Krueger et al. *Introd. Microbiol.* xiv. 410/2 Studies of T4 mutants have gone a long way towards elucidating the steps by which biological structures, such as the head and tail of the virus, are assembled.

c. *Astronautics.* [Initial letter of *time*.] T represents the time at which a spacecraft is due to be launched.

1959 *Manch. Guardian* 3 Jan. 5 'T-time' is the moment the firing switch is closed to set off a missile. 1970 N. Armstrong et al. *First on Moon* ii. 32 Only in the latter stages of the final countdown does the nomenclature change to T minus hours and minutes.

IV. 9. 2,4,5-*T*: a selective herbicide used esp. for controlling brushwood; 2,4,5-trichlorophenoxyacetic acid, $C_6H_2Cl_3{\cdot}O{\cdot}CH_2COOH$.

1947 *Bot. Gaz.* CIX. 194/2 The use of 2,4-dichlorophenoxyacetic acid (2,4-D) has been emphasized in nearly all investigations.., although the possible use of 2,4,5-trichlorophenoxyacetic acid (2,4,5-T) for similar purposes was recognized by Hamner and Tukey. 1956 *Nature* 3 Mar. 418/1 An investigation into the effect of the arboricide 2,4,5-T on freshly cut stumps of thicket suggests that this substance may be of practical value in preventing regeneration. 1977 *New Yorker* 25 July 37/2 The military had withdrawn huge stocks of Agent Orange—a fifty-fifty mixture of 2,4,5-T and 2,4-D, which was principally used in its herbicidal operations—from Vietnam.

't², dial. shortened form of That *relative pron.* or *conj.* Cf. At, 'at *rel. pron., adv., conj.*

1867 *Our Young Folks* Mar. 130 Jest show me that! Ur prove 't the bat Hez got more brains than's in my hat. 1871 W. Alexander *Johnny Gibb* xv. 108, I ance was neepours wi' a chap 't could 'a deen that. 1887 R. T. Cooke *Happy Dodd* xxvii. 286, I didn't feel real cherk this week, so't I didn't go to sewin' s'ciety.

ta, *int.* Add: Now also commonly in colloq. adult use.

1931 A. Powell *Afternoon Men* xxx. 252 'Will you give him this, Sophy?'..'What did he say?' Sophy said: 'He just said, "Tar".' 1946 K. Tennant *Lost Haven* (1947) iii. 47 Grandfather Starbrace shovelled great handfuls of pink prawns... 'Ta, Nathe,' Mr. Thorne said. 1970 'R. Gordon' *Doctor on Boil* xxiii. 164 'Ta,' he said, slipping the card into the back pocket of his jeans. 1981 D. Clark *Longest Pleasure* vi. 136 'You know your way, don't you?' 'Ta, love.'

taaffeite (tā·fəit). *Min.* [f. the name of Edward Charles Richard *Taaffe* (1898–1967), Bohemian-born Irish gemmologist + -ite¹.] A mauve gemstone similar to spinel, having the composition $BeMgAl_4O_8$ and a hexagonal crystal structure.

1951 B. W. Anderson in *Gemmologist* XX. 76/2 One of the 'spinels' gave a rather high refractive index reading and seemed to show double refraction... The second 'Taaffeite' was found! 1951 —— & Claringbull in *Mineral. Mag.* XXIX. 765 The new mineral..has been named taaffeite in honour of its discoverer. 1967 *Rocks & Minerals* XLII. 803/1 The world's fourth known cut taaffeite was discovered in August of this year. 1974 *Jrnl. Gemmology* XIV. 104 The discovery of the new gem mineral Taaffeite reads like a gemmological fairy tale.

taaibos (tāi·bɒs). *S. Afr.* Also 9 **ta(a)ybosch; taaibosch.** [Afrikaans, f. Du. *taai* tough + *bos(ch)* bush.] Any of several shrubs or trees with strong, pliable branches, esp. any of several species of *Rhus*.

1821 C. L. Latrobe *Jrnl. Visit S. Afr.* 559 Taaibosch—a species of Rhus, of which genus several bear the name of Taaibosch. 1833 *S. Afr. Almanac & Directory* p. xlviii, One of the Cape sumachs (Taaybosch) has been recommended for culture. 1834 *Cape Good Hope Lit. Gaz.* Mar. 41 The extract may be procured..if the tree is treated as recommended for the colonial Taybosch. 1905 F. G. Stow *Native Races S. Afr.* v. 93 They [*sc.* fish baskets] were composed of reeds and twigs of the taaibosch. 1948 *Cape Times* 5 Aug. 8/7 The safer and more effective plants are reeds and many-stemmed shrubs, such as taaibos. 1974 *Stand. Encycl. S. Afr.* X. 396/1 Taaibos. Name generally applied to plants with tough branches and bark, such as *Passerina vulgaris*..but particularly to shrubby *Rhus* species.

taakhaar, var. *takhaar.

tab, *sb.*¹ Add: **I. 1. f.** = *pull-tab* s.v. *pull- 2: used to open a can of beer, etc.

1963 *Wall St. Jrnl.* 1 Oct. 16/1 The beer drinker opens the can by pulling off the tab. 1978 O. White *Silent Reach* xi. 108 The fat man..pulled the tab from a can.

2. d. A coloured tab, esp. a red tab or gorget patch, worn by a senior or staff Army officer; hence formerly, in *Army slang*, such an officer.

1916 J. Buchan *Greenmantle* i. 2 'Try my tailor,' said Sandy. 'He's got a very nice taste in red tabs.' 1917 *B.E.F. Times* 20 Jan. 4/2 Realising Men must laugh, Some wise Man devised the Staff: Dressed them up in little dabs Of rich variegated tabs. 1925 Fraser & Gibbons *Soldier & Sailor Words* 275 *Tab, a*, a Staff Officer. 1948 Partridge *Dict. Forces' Slang* 154 *Red tabs*, red gorget patches, worn by Colonels and above. 'Red-tab' was sometimes used for an officer who wore them. 1977 D. James *Spy at Evening* vii. 51 He had red tabs on his collar. He had authority even over my father.

4. c. *Aeronaut.* A usu. hinged part of a control surface that serves to modify the action or response of the surface.

1934 *Flight* 25 Jan. 75/1 The word 'tab' has been approved by the [U.S.] Department of Commerce as the name for auxiliary control flaps. 1942 'B. J. Ellan' *Spitfire!* p. x, Winding the bias control one way or the other moves the tab and gives port or starboard bias. 1965 C. N. Van Deventer *Introd. Gen. Aeronautics* iv. 95/2 Controllable or fixed tabs may be attached to any of the control surfaces—the elevators, the ailerons, or the rudder. 1983 D. Stinton *Design of Aeroplane* xi. 397 Often trimming is achieved on the ground by bent plate tabs.

5*. An ear. *dial.* and *slang*.

1866 J. E. Brogden *Provinc. Words Lincs.* 202 *Tab*, a piece of leather in the front of a boot, a latchet, the ear. 1959 *New Statesman* 26 Dec. 904/2 Dad was sitting by the fire, behind his paper with one tab lifted. 1977 Scollins & Titford *Ey up, mi Duck!* III. 15 Ah gorra bile be'int me tab.

II. 6. For *U.S. colloq.* read *colloq.* (orig. *U.S.*). (Further examples.) Also, a bill or charge (chiefly *N. Amer.*): see also *to pick up the tab* s.v. *pick v.¹ 20 r. *to keep (a) tab*: now chiefly current in fig. phr. *to keep tabs (on)*. Also *fig.*

1890 B. Hall *Turnover Club* 19 They knocked off and filed out into the deserted streets, while the Purveyor figured up the 'tabs'. 1924 [see *tab v. 2]. 1929 'E. Queen' *Roman Hat Mystery* iii. 37 We got to keep pretty close tabs on the time, and I know it was ten minutes because..it was just the part on the stage when [etc.]. 1932 D. L. Sayers *Have his Carcase* xxvi. 348 The one person..likely to have kept tabs on Mr Perkins..was old Gaffer Gander. 1946 J. O'Hara in *New Yorker* 23 Mar. 25/1 You signed a small tab, sir. 1952 S. Kauffmann *Philanderer* (1953) iii. 52 So all those old ideas are finished and God is dying. There's nobody in Heaven keeping tabs. And there's just going to be less for people to hang on to? 1954 E. B. White *Let.* 9 July (1976) 395, I did a little haying yesterday..and..I am spending today indoors paying the tab. 1963 T. Parker *Unknown Citizen* v. 138 He's antagonistic to anything or anybody who represents authority... He thinks that our main purpose is to keep tabs on him while he's out. 1976 M. Machlin *Pipeline* xii. 144 He started to reach into the pocket of his Arctic down pants for his wallet, but Nick had already paid the tab. 1978 M. Puzo *Fools Die* ii. 18 Jordan knew that Merlyn the Kid kept tabs on everything he did.

7. A cigarette. *north. dial.* and *slang*.

1934 P. Allingham *Cheapjack* iii. 24 ''Ave you got a tab on yer?' The only tabs I knew were connected with the theatre, but I discovered later that 'tab' is a common name in the north for a cigarette. 1948 A. Baron *From City, from Plough* i. 10 'Gie us a tab, Lanky.' He passed his cigarettes round the carriage. 1968 B. Hines *Kestrel for Knave* 71 From various pockets Gryce collected two ten-packets, which rattled when he shook them, a handful of tabs, three lighters and a box of matches. 1980 C. Ross *Case for Compensation* xiv. 68 'Tab?' Duncan looked blank. 'Cigarette?' he said. Duncan accepted. 1983 *New Society* 2 June 333/1 £13.65 a week to pay for..clothes, 'snake bites' (cider and lager), 'tabs' (cigarettes).

8. A tablet or pill, *spec.* one containing LSD or another illicit drug. *slang*.

1961 in Webster. 1968–70 *Current Slang* (Univ. S. Dakota) III–IV. 123 *Tab*, a tablet of sugar or saccharine impregnated with LSD. (Drug users' jargon.) 1971 *Daily Tel.* 18 Sept. 12 Whenever anybody had any money it nearly always went on drugs, with LSD at £1 a 'tab' (tablet). 1973 'J. Marks' *Mick Jagger* (1974) 137 He presses his palm to his mouth and swallows the tab. 1978 M. Walker *Infiltrator* xii. 136 An order for two tabs of acid.

III. 9. Special combination. **tab collar,** (*a*) a shirt collar whose points are held down by buttons or other fastenings (cf. *button-down* adj. s.v. *button sb. 12); (*b*) (see quot. 1957).

1928 *Men's Wear* 21 Nov. II. 5/1 The tab collar is being worn by quite a few of the best dressed eastern university students. It should prove to be an important feature. 1942 B. G. Chambers *Color & Design in Apparel* xv. 498 *Tab collar*. The fronts have loops on the under side with button-holes which fasten on buttons or small rigid stays, that help keep the tie in place at the top of the collar. 1957 M. B. Picken *Fashion Dict.* 75/2 *Tab c[ollar]*, collar cut in tabs, often with two at front. 1979 *Time Out* 4 May 65 The Mod revival hits London in force: each day offers a gig at which parkas, tab collars and fox-tailed Lambrettas would be acceptable.

tab (tæb), *sb.*² *slang.* [Abbrev. of Tabby *sb.* 3.] **a.** An elderly woman. **b.** *Austral.* A young woman or girl.

1909 J. R. Ware *Passing Eng.* 239/1 Tabs (*Theatrical*), ageing women. 1932 H. Simpson *Boomerang* x. 276 We don't need to go mackin' round with Chinks and wimmen's earnings. We pay our tabs..when we want 'em, and tell 'em to get to hell out of it when we don't. 1971 [see *ma-in-law].

Tab (tæb), *sb.*³ *University slang.* [Short for Cantab.] A member of the University of Cambridge.

1914 C. Mackenzie *Sinister Street* II. iii. iii. 555 He will get his blue next term and show the Tabs that he's a jolly good fellow. 1930 A. Alington *Slowbags & Arethusa* i. 4 Then the morning's play is discussed, the loathly Tabs reviled—for the Slows are Oxford to a man.

tab (tæb), *sb.*⁴ *Theatr. slang.* Abbrev. of *tableau curtain* s.v. *tableau 6.

1929 J. B. Priestley *Good Companions* II. vii. 447 The girls here follow him round with their tongues hanging out, as usual—but away from the tabs he's the same as ever. 1936 N. Royde-Smith *All Star Cast* 44 The tab curtains fell together as the girl and the man stood at arm's length from one another. 1946 'Brahms' & 'Simon' *Trottie True* vii. 164 Down came the crimson tabs. Up went the shouting and the cooing. Out tottered Marie [Lloyd] to the public that idolized her. 1957 P. Frankau *Bridge* 59 The dark stage-hand..came through the tabs. 1983 *Listener* 22/29 Dec. 28/1 When she did the last song, she used to do it in front of the tabs.

tab (tæb), *sb.*⁵ *Typewriting* and *Computing.* [Abbrev. of Tabulator b, Tabular *a.*, etc.] A tabulator (key); a tabular stop, used to preset the movement of the carriage, cursor, etc., under the direction of the tabulator.

1916 H. Etheridge *Bar-Lock Typewriter Manual* 45 The Tab. key acts exactly in the same manner as the carriage release lever. *Ibid.*, On releasing the Tab. key the carriage remains at the number on the scale where the first stop has been fixed. 1969 *Sears, Roebuck Catal.* Spring/Summer 1195/2 Automatic key-set tabs, clear key. 1982 Harris & Chauhan *So You want to Buy a Word Processor?* v. 65/1 Not only do tabs have to be set up at appropriate positions, but the facility needs to be reactivated when any of the text involved is subsequently edited.

tab, *v.* Add: **2.** To identify, name, or 'dub'; to label or record. Also, to watch, 'keep tabs on' (formerly also with *up*). *colloq.* (chiefly *U.S.*).

1924 G. C. Henderson *Keys to Crookdom* 420 *Tab*, to name. *To keep tabs on*, to keep in touch with. 1926 J. Black *You can't Win* xix. 283 You are a burglar; you have put in a week 'tabbing up' a residence. 1946 *Sun* (Baltimore) 18 Feb. 11/5 The Navy has tabbed entertainment with the high-sounding name liaison unit. 1954 'J. Christopher' *Twenty-Second Cent.* 86 The doctors have it all tabbed. It's what they call cumulative stress. 1969 *Eugene* (Oregon) *Register-Guard* 3 Dec. 1D/2 Ken Wiedemann of Cal, tabbed as the best defensive back, was sidelined for a major part of the season with a bad knee. 1978 M. Puzo *Fools Die* ii. 33 He had Jordan tabbed as a degenerate gambler.

tabac, *a.* Add: (Earlier example.) Also as *sb.*

1881 *Queen* 18 June 8/1 *Best felt hats.* For Ladies and Gentlemen. The new 'Vicuna' 'Tabac' Brown, and all Colours. 1886 *Graphic* 30 Jan. 123/2 Colours are Black, Brown, Gold, Geranium, and Tabac. 1922 *Daily Mail* 18 Dec. 1 (Advt.), In shades of Coral, Champagne,..Tabac.

‖ **tabac** (tabak), *sb.* [Fr.] In French-speaking countries: a tobacconist's shop.

1918 'K. Mansfield' *Let.* 11 Jan. (1928) 96 The *tabac* woman did not know me and had no tobacco. 1934 H. Miller *Tropic of Cancer* 52 We sat in the back of the little *tabac* called *L'Eléphant*. 1965 P. O'Donnell *Modesty Blaise* vii. 75 He left the boules with the lady who ran the *tabac*. 1980 'M. Harris' *Treasure of Sainte Foy* ii. 18 There is a boulangerie-pâtisserie, a tabac, and the milk bar.

tabanid, *a.* and *sb.* After 'Linnæus' in etym. add: (1736, in *Acta Soc. R. Scient. Upsaliensis* 31). (Further examples.)

1892 *Insect Life* V. 59 An examination showed it to be a true Tabanid. 1931 K. M. Smith *Textbk. Agric. Entom.* xi. 163 Tabanid larvae are whitish and occur in the water or soil. 1967 V. Nabokov *Speak, Memory* vi. 137 Because of our ferocious Russian tabanids, one could not leave a horse haltered in a wood for any length of time.

tabard. Add: **3. b.** A fashionable slimly cut ladies' jerkin or similar garment with short (or no) sleeves; *spec.* one used as a beach-robe.

1923 in C. W. Cunnington *Eng. Women's Clothing in Present Cent.* (1952) v. 175 Evening dress with tabard top. 1959 *Housewife* June 49 A beach tabard in..cotton, over a bikini and bra. 1977 P. D. James *Death of Expert Witness* II. xii. 106 She wore a dress in fine fawn wool, topped with an elaborately patterned, short-sleeved tabard. 1983 *Times* 11 Mar. 8/4 An odd, misshapen tabard, worn with a long slim suede skirt.

tabardillo (tæbɑːdi·lo, ‖ tabardi·lʸo). *Path.* [Sp. *tabardillo* (see quot. 1980).] A fever common in Mexico and S. America; now *spec.* a murine typhus (cf. *murine a. b) found in Mexico which, unlike most forms of murine typhus, can occur as an epidemic.

[1598 W. Phillip tr. *Linschoten's Voy. E. & W. Indies* I. i. 2/1 Hee fell sicke of a disease called *Tauardilha*.] 1624 W. Aston *Let.* 10 Dec. in *Cabala: sive Scrinia Sacra* (1654) I. 166 He hath been held divers dayes with a terrible Calenture, which proved at last a *Tabardillo*. 1853 W. L. Herndon *Exploration Valley of Amazon* I. v. 113 The most common diseases are pleurisies, rheumatisms, and a putrid fever called *tabardillo*. 1944 R. A. Moore *Textbk. Path.* xlii. 459 It was shown before 1930 that Brill's disease, tabardillo of Mexico, and endemic typhus of the southeastern United States are all transmitted by the rat flea..and by the rat louse. 1980 A. L. Smith *Microbiol. & Path.* (ed. 12) I. xxv. 300/2 The endemic typhus fever of Mexico is known as *tabardillo* (from the Spanish word *tabardo*, meaning a coloured cloak, to designate the mantlelike spotted rash of the disease).

tabaret. (Earlier example.)

1790 P. A. Robb *Let.* in M. Dunsford *Hist. Mem. Tiverton* IV. 235 This year [*sc.* 1752] was introduced to Tiverton the manufacture of..camblets, tarborates, damasks.

Tabasco. Add: A proprietary name. (Earlier and further examples.) *fig.*: delete 'a story "highly-spiced"'.

1876 J. Miller *First Fam'lies Sierras* 126 The following popular drinks..were all made from the same decoction of bad rum, worse tabasco, and first-class cayenne

pepper. 1878 *Let. to E. McIlhenny* (E. S. Hyatt & Co.) 16 Dec. (MS.), Please send us by return mail your lowest prices and terms on your Tabasco pepper sauce. 1879 E. C. Hazard *Let. to E. McIlhenny* 30 July (MS.), Would you agree to sell us your entire packing of tabasco sauce? 1902 *Trade Marks Jrnl.* 3 Sept. 1010 Tabasco pepper sauce. Pepper Sauce made from Tabasco Pepper. Edward Avery McIlhenny,..Sauce Manufacturer.—25th July 1902. Mark has been used in respect of the said Goods by the applicant and his predecessors in business since five years before the 13th August 1875. 1923 Wodehouse *Inimitable Jeeves* iv. 41 Little as he might look like one of the lads of the village, he certainly appeared to be the real tabasco and I wished he had shown me this side of his character before. 1924 *Trade Marks Jrnl.* 18 June 1385 Tabasco. 439,246. Pepper Sauce made from Tabasco Pepper. McIlhenny Company... 21st July 1923. 1949 *Amer. Speech* XXIV. 34 Tabasco sauce is acid used in breaking a limestone formation. 1979 *Guardian* 19 July 12/3 Fred Jackson in among the saxophones to add a tabasco spice to the disco bids.

Tabassaran (tæbăsărā·n), *sb.* (and *a.*). Also **Tabasaran.** [Native name for a district in S. Daghestan.] A North Caucasian, Lesghian language of Daghestan, known in both written and spoken forms. Also *attrib.* and as *adj.*

1951 W. K. Matthews *Languages U.S.S.R.* v. 89 Agul has twenty-five, Awar thirty, and Tabassaran thirty-five cases. 1968 [see *Lesghian sb. and a.]. 1971 [see *Lak]. 1977 C. F. & F. M. Voegelin *Classification & Index World's Lang.* 96 Eleven members of the Caucasian groups have official status as literary languages:.. Tabasaran (35,000).

‖ **tabbouleh** (tabū·le). Also **tabbouli, tabbuuli.** [ad. Arab. *tabbūla*.] An Arabic vegetable salad made with crushed wheat.

1955 J. Gulick *Social Structure & Culture Change in Lebanese Village* 42 The other is *tabbuuli*. This has nearly the same ingredients as the salad, but they are chopped up very fine and mixed with cracked wheat which has been soaked in water. 1958 F. Copeland *Land between: Middle East* ix. 99 No picnic is complete without a special salad called *tabbouleh*. 1965 *Times* 31 May 13/6 Tabbouleh is a salad made from crushed wheat known as bourghul, with chopped parsley and mint. 1968 C. Roden *Bk. Middle Eastern Food* 57 *Tabbouleh* is traditionally served in individual plates lined with boiled vine leaves. 1977 C. McFadden *Serial* (1978) lii. 110/2 Marlene must have cooked her head off. The tabbouleh, all that whole-wheat lasagne,..the brown rice and veggies.

tabby, *sb.* and *a.* Add: **A.** *sb.* **3. b.** An (attractive) young woman or girl; = *tab *sb.*² b. *slang*.

1916 C. J. Dennis *Moods of Ginger Mick* 20 Then the tabbies took to screamin'. 1925 Fraser & Gibbons *Soldier & Sailor Words* 275 *Tabby, a*, a girl. 1935 A. J. Pollock *Underworld Speaks* 118/1 *Tabby*, an attractive girl. 1958 J. Wain *Contenders* iv. 88 'I said, is it true what Joe says that you've got yourself fitted out with a tabby?' 'My humble roof,' said Robert..'is shared by a distinguished actress.'

c. tabby weave *Textiles* = *plain weave* s.v. *plain a.¹ and *adv.* C. c.

1906 H. Nisbet *Gram. Textile Design* ii. 6 The 'plain'.. or 'tabby' weave..is the most simple and elementary combination of two series of threads employed in the construction of textile fabrics. 1957 Simpson & Weir *Weaver's Craft* vii. 77 We may weave a binder thread (a row of plain or tabby weave) of very fine material in between the rows of pattern.

Tabele, var. *Tebele.

tabi. Add: Also 9 **tapie**; (anglicized pl.) **tabis.** For def. read: Thick-soled Japanese ankle-socks with a separate stall for the large toe, worn by both sexes. Also *attrib.* (Earlier and later examples.)

1616 R. Cocks *Diary* 23 Jan. (1883) I. 102, 2 peare *tabis* for Jeffrey. 1822 F. Shoberl tr. *Titsingh's Illustrations of Japan* I. 130 The men leave off the *tapie* on the 1st of the third month, but the women wear them all the year round. 1880 I. L. Bird *Japan* I. xiii. 131 On her little feet she wears white *tabi*, socks of cotton cloth, with a separate place for the great toe. 1892 H. Norman *Real Japan* 193 The costume is completed by a pair of *tabi*. 1938 *N. & Q.* 21 May 361/1 The Japanese private..used to put on his *tabis* and get to work on his arms. 1963 R. Godden *Little Plum* 17 'Why, you have made them tanzen—proper Japanese coats—and tabi,' he said, touching the socks. 1975 J. Clavell *Shōgun* II. xi. 229 He wore a belted kimono of the Browns and tabi socks and military thongs.

tabid, *a.* Add: **1.** (Later example.) Also *fig.*

1914 C. Mackenzie *Sinister St.* II. iv. v. 964 He was disappointed to see no cab..merely a tabid woman clothed in a cobweb of crape, asleep over her tray of matches. 1947 M. Lowry *Under Volcano* ii. 58 Outside.. in the backwash of tabid music from the still-continuing ball.

‖ **tabl** (tāb'l). Also **teboul.** [Arab.: see Atabal.] In the Arab world: a drum (usu. played with the hand).

[1777 J. Richardson *Arab. & Pers. Dict.* p. xlv, *Tebl* is a drum, *Teblek* a small drum. 1836 E. W. Lane *Acct. Manners & Customs Mod. Egyptians* II. v. 75 Several kinds of drums, of which the most common kinds are the *tub'l bel'edee* (or country drum, that is, Egyptian drum), and the *tub'l Sha'mee* (or Syrian drum).] 1876 Stainer & Barrett *Dict. Mus. Terms* 415/1 *Tabl*, an Egyptian drum formed from a hollow block of wood, or made of earthenware, with a skin stretched over one end. 1923 *Chambers's Jrnl.* Apr. 307/1 White-collared gentlemen..play strange instruments—a big guitar, a big drum called a *teboul*. 1976 *Guardian* 16 Dec. 16/3 (Advt.), The *tabl*, the larger wooden drum, probably came [to Qatar] from Africa, though there are obvious etymological links with the Persian *duhul* and the Indian *tabla*.

‖ **tabla** (ta·bla). Also 9 **tubla.** [Hind., ad. Arab. *ṭabl* (see prec.).] A pair of drums used in Indian music (esp. that of the northern region), of which the left-hand (bass) head is larger than the other (tenor) head; either of these drums separately (see quot. 1969). Also *absol.*, a tabla-player. Cf. *mridangam, *pakhawaj.

1865 *Proc. R. Irish Acad.* IX. I. 117 (*Tubla*). These drums, tenor and bass, rank with the pukhraj... The *tubla* drums are made of copper. 1888, etc. [see *mridangam]. 1914 A. H. F. Strangways *Mus. Hindostan* ix. 227 The tubla, left and right, are of the shape of a giant tea-cup and coffee-cup respectively. 1927 *Observer* 12 June 14/5 The Hindu drummer's saying that 'the left *tabla* (hand drum) is like the sleeve of my coat, and the right like such embroidery as I may put on it'. 1955 R. P. Jhabvala *To whom she Will* xiii. 92 The musicians.. played with placid expressions ..only the tabla-player smiled. 1969 R. Shankar *My Music* i. 40/2 Of the multitude of drums that are found throughout India today, the most popular variety in the North is the *tabla*, which is actually two drums, each with one skin stretched across the top. The smaller of the two drums is the right-hand *tabla*, and the bass, left-hand drum is called the *bonya*, though the two are called collectively *tabla*. 1973 'D. Jordan' *Nile Green* xxiv. 99 'No evening Raga tonight?' I asked. 'The tabla has got flu so it's postponed.' 1975 I. Murdoch *Word Child* 3 Christopher was learning to play the 'tabla', a dreary little oriental drum.

tablature. 1. For *Obs.* exc. *Hist.* read 'Now chiefly *Hist.*' and add later examples drawn from the revived study and performance of early music.

1969 *Daily Tel.* 12 Nov. 14/4 The procuring of music for the lute presented the greatest difficulty: it is not written in staff notation but in tablature, so Dolmetsch had to decipher this from old MSS in the British Museum. 1977 *Listener* 15 Dec. 796/3 Tablatures..semi-diagrammatic signs that belong to a specific instrument and make no sense in the abstract. 1980 *Early Music* Apr. 250 Our edition includes voice and tablature as well as voice and transcription.

table, *sb.* Add: **I. 5. b.** (Examples of *on the table*.) *to lay on the table* (examples); also, to present for immediate discussion; *to lie on the table* (earlier examples). Cf. sense 4 a of the vb. in Dict. and Suppl.

1730 E. Knatchbull in *Camden Soc.* (1963) 3rd Ser. XCIV. 106 So a division for it [*sc.* a Petition] lying on the table, carried by 163 against 144. 1733 in *15th Rep. R. Comm. Hist. MSS.* App. VI. 107 in *Parl. Papers 1897* (C. 8551) LI. 1 The majority, for laying the Petition on the Table..and not hearing it by counsel, was only seventeen. 1744 *Archives New Jersey* (1882) 1st Ser. VI. 191 The House of Representatives..would not commit it [*sc.* a bill] but ordered it to lie on the table. 1855 Macaulay *Hist. Eng.* IV. xix. 343 Shrewsbury laid on the table of the Lords a bill for limiting the duration of Parliaments. 1915 J. London *Let.* 25 Aug. (1966) 458 It is..on the table whether or not we shall say 'it is I' or 'it is me'. 1923 H. M. Robert *Parl. Law* (U.S.) 63 It is in order for a mere majority to lay on the table the questions that have not been disposed of. 1958 [see *prayer¹ 5]. 1977 *Times* 14 Apr. 1/3 While stating that those proposals should 'remain on the table', he [*sc.* Ian Smith] is now prepared to listen to new ideas.

c. *spec.* The table which stands before the Speaker's chair in the House of Commons, at which the Clerk of the House and his assistants receive motions and questions to ministers, etc., and at which new members are sworn in (cf. also prec. sense).

1675 *Grey's Debates* (1769) III. 129 Mr Stockdale, and some others, setting their feet upon the mace, which lay below the table, in the usual place at Grand Committees. 1771 *London Even. Post* 28 Feb.–2 Mar. 3/1 Upon which Mr. Byng and Mr. Buller, seized him by the collar, and brought him up to the table. 1885 *Encycl. Brit.* XVIII. 312/1 Having first taken the oath himself, he [*sc.* the Speaker] is followed by other members, who come to the table to be sworn. 1958 *Spectator* 11 July 47/1 Mr. Gaitskell's head wagged up and down as if he wanted to punch a hole in the Table with his nose.

d. A surgeon's operating-table; also, a table or slab on which a body is laid for post-mortem examination.

1917 T. S. Eliot *Prufrock* 9 Like a patient etherized upon a table. 1936 G. B. Shaw *Millionairess* II. 166, I should have cut my patients entirely away if the nurse had not stopped me before they died on the table. 1941 A. Huxley *Let.* 17 Nov. (1969) 470 The patient will die on the table if operated—off the table, if not operated. 1977 P. D. James *Death of Expert Witness* II. 226 As for the cause of death..well, you'll have to wait till I get her on the table.

e. Attrib. phr. *under-the-table*: kept secret, hidden, esp. of clandestine deals or payments. Also (unhyphened) used predicatively and as advb. phr. Cf. *under the counter* s.v. *counter *sb.*³ 4 b.

1949 *Sun* (Baltimore) 25 Feb. 19/3 Two home purchasers told a Federal Court jury..that they were required to make under-the-table payments to purchase housing accommodations. 1973 W. H. Hallahan *Ross Forgery* vi. 115 Under-the-table freight rebates reached absurd proportions. 1976 *Listener* 5 Feb. 144/1 Some of the sports do check people's bank accounts to see that they have not got too much money under the table. 1976 G. Seymour *Glory Boys* vii. 85 This bomb that the Israelis keep so much under the table..what state is that in? 1980 *Times Lit. Suppl.* 25 July 839/3 The Rheinmetal company for long refused to pay anything, but eventually arranged for an under-the-table payment of DM2,500,000 (which provided $425 for each of its former slaves).

f. A table around which parties (esp. in an industrial dispute) sit to discuss points at issue; a negotiating table. *round-the-table* adj. phr., designating such discussions; (*up*)*on the table*: see sense 5 b.

1963 [see *recreationist]. 1976 *West Lancs. Evening Gaz.* 15 Dec. 1. 9/4 Transport and General Workers' Union officials want round-the-table talks with the management. 1980 *Times* 6 Feb. 1/1 We hope to get our negotiators around the table as soon as possible.

6. c. Also *the pleasures of the table* [tr. F. *les plaisirs de la table*], good food and drink, considered as a source of enjoyment.

1769 F. Brooke *Hist. Emily Montague* IV. 146, I love the pleasures of the table. 1825 Scott *Talisman* in *Tales of Crusaders* III. xi. 212 Richard..despised the inclination of the German for the pleasures of the table. 1845 [see *Sard sb.² 1]. 1942 G. M. Trevelyan *Eng. Social Hist.* xiii. 408 Eighteenth Century Englishmen were much addicted to the pleasures of the table. 1981 T. Fitzgibbon (*title*) The pleasures of the table.

d. Slang phr. (*to put*, etc.) *under the table*, (to make) drunk to the point of insensibility.

1921 W. S. Maughan *Trembling of Leaf* 28 Walker had always been a heavy drinker, he was proud of his capacity to see men half his age under the table. 1924 D. Marquis *Old Soak's Hist. World* vi. 32 By three therty everybody was under the tabil. 1936 V. W. Brooks *Flowering of New England* v. 95 He was far from sober, or would have been if two tumblers of brandy had been enough to put him under the table.

9. b. *to lay, put* (or *play with*) (*all*) *one's cards on the table*: see *card *sb.*² 2 d.

c. *Bridge.* The hand belonging to dummy.

1959 *Listener* 7 May 808/2 The lead of the Queen from the table allows East's K 9 x to be smothered. 1960 T. Reese *Play Bridge with Reese* 127, I play low from table. 1974 *Country Life* 28 Feb. 453/2 South won with the Ace, crossed to the Spade Ace on the table, and led a Club.

10. f. = *league table* s.v. *league *sb.*² 4.

1951 *Sport* 6–12 Apr. 10/4 Mr. Drake has been the guiding light behind a remarkable revival that has taken the club soaring up the table. 1972 G. Green *Great Moments in Sport: Soccer* v. 62 Around Christmas, they had begun to catch a tide of success as they crept slowly up the table. 1976 *Western Mail* (Cardiff) 27 Nov. 20/2 Newcastle, third in the table thanks to their midweek win over Everton.

III. 21. a. In sense 5, 'of a table': *table-edge*; in sense 6, 'of the dinner-table': *table-waiter*; 'of implements, etc. used at table': *table cutlery, decoration, fork* (earlier example), *mat, runner*; 'for consumption at table': *table fish* (earlier examples), *-grape*; in sense 10: *table look-up* (*look-up 2). **b.** *table-thumping* adj. and sb. **c.** *table-piano(forte)*.

1861 *Chicago Tribune* 10 July 1/9 Crockery Ware, Table Cutlery, Plated Ware, &c. 1946 A. Christie *Come, tell me how you Live* vii. 116 Civilisation's invention of table cutlery presents a perpetual headache to a worried house-boy. 1937 C. Spry *Flowers in House & Garden* 169 Your choice of table decorations is bound to be influenced by..your guests. 1979 I. Webb *Compl. Guide Flower & Foliage Arrangement* vii. 97/2 'Frensham' roses and ivy berries combine to make an arresting table decoration. 1935 H. H. Bashford *Lodgings for Twelve* 87 George Gedge's Wiltshire guile and a miraculous succession of table-edge strokes. 1977 F. Ormsby *Store of Candles* 30 Resumes his beat from table-edge to door, From door to table. 1770 *Boston Gaz.* 15 Jan. 2/3 *Table fish* warranted the very best, To be Sold at the Store the Corner of Kilby-Street. 1872 F. F. Victor *All over Oregon* 63 Besides the salmon of commerce, the Columbia furnishes a great many other species of edible fish..all of which are excellent table-fish in their proper seasons. 1785 *Daily Universal Reg.* 1 Jan. 3/2 Ivory table knives and forks. 1926 *Zionist Rev.* Apr. 144/2 Splendid prospects exist for good table-grapes in those parts of Palestine where the Jewish urban population is growing. 1979 *Tucson Mag.* Apr. 20/2 An attractive feature..is the possibility that vineyards will be a 'dryland crop', using considerably less water than table grapes. 1957 D. D. McCracken *Digital Computer Programming* xvii. 200 The code number is placed in one of the arithmetic registers and a table look-up instruction given. 1967 Cox & Grose *Organization & Handling Bibl. Rec. by Computer* vi. 142 These will be linked with 'table-look-ups' within the output programs to translate each symbol into a full form. 1779 in *Dict. Amer. Eng.* (1938) s.v. *Table* n., Table mat. 1834 Dickens *Bloomsbury Christening* in *Monthly Mag.* Apr. 380 A front drawing-room, very prettily furnished with a plentiful sprinkling of little baskets, paper table-mats, [etc.]..on the different tables. 1965 A. Nicol *Truly Married Woman* 5 She remembered the wine glasses and the beer-advertising table-mats in time and put those under the sofa. 1911 *Daily Colonist* (Victoria, B.C.) 27 Apr. 11/7 The hotel furniture consists of ..blankets, sheets, spreads, pillows, toilet sets in 60 rooms, 1 table piano, card tables, [etc.]. 1952 J. Gloag *Short Dict. Furniture* 468 Table pianos were designed to conceal the fact that they were musical instruments:

when closed they looked like clumsy and ill-proportioned tables. **1976** Early Music Oct. 483/1, I find the choice of cover picture oddly revealing: a small table-piano ordered by an aristocrat for his children. **1851** Official Catal., Gt. Exhib. III. 1225/1 Patent square and console pianofortes; square and hexagonal table pianofortes. **1889** Table runner [see *runner 14 c]. **1939** W. Fortescue There's Rosemary xliv. 259, I cut lengths of brocatello, designed cushions and table-runners, &c. **1967** E. Short Embroidery & Fabric Collage iii. 63 Small mats, table runners, Radio Times covers are quite unnecessary. **1963** Time 2 Aug. 17/2 The changeover from Stalin, the 'oriental despot', to Khruschev, the table-thumping but jolly politician. **1964** A. Battersby Network Analysis ix. 137 The Esso team..do claim with confidence that resources are utilized more effectively than before,..and that, in general, there is less table-thumping. **1928** D. H. Lawrence Woman who rode Away 15 The sister was all that could be desired as..an upper parlour-maid, and a table-waiter. **1975** Budget (Sugarcreek, Ohio) 20 Mar. 1/4 Tablewaiters were David F. Yoder, Susie Bontreger, [etc.].

d. Designed to stand on a table, as table lamp, lighter, model, stand.

c **1849** J. S. Coyne How to settle Accts. with your Laundress 3 Table at back, L., on which is a table lamp. **1854** C. M. Yonge Heartsease I. ii. i. 102 A pretty little rosewood work-table, on which was..a table-stand of books. **1907** Yesterday's Shopping (1969) 1150/3 Folding Music Stands... Table stand... Brass 7/6. **1922** A. Bennett Lilian ii. vii. 119 It was the silver table-lamps..that impressed her. **1929** Radio Times 8 Nov. 437/1 The table model Columbia is..the most advanced radio of the day. **1951** Catal. Exhibits, South Bank Exhib., Festival of Britain 147/2 Shagreen table lighter. **1954** 'N. Blake' Whisper in Gloom i. vii. 99 Applying his cigar to a massive table-lighter. **1962** A. Nisbett Technique Sound Studio i. 30 There are four types of microphone mounting. These are: (i) The table stand, [etc.]. **1967** P. Chambers Bad die Young i. 11 A grateful client had given me a heavy bronze table lighter. **1976** 'W. Trevor' Children of Dynmouth iii. 60 Only a table-lamp burned, its weak bulb not up to the task of fully illuminating the room. **1977** D. E. Westlake Nobody's Perfect 10 He'd cased that TV repair shop—he'd even brought in a perfectly good Sony table model and let them charge him for six new tubes.

e. Designating various games played on a table, which simulate more or less closely the action of some sport, as table football, hockey, etc. See also table-tennis, sense 22 in Dict. Cf. table game, sense 22 below.

1907 Yesterday's Shopping (1969) 1032/2 Wibley Wob or Table Football. A game for 2 or 4 players, to be placed upon an ordinary dining table. **1948** Sporting Mirror 21 May 10/3 (Advt.), Send 3d. stamp for full details of Subbuteo the game of Table Soccer... Played with 22 miniature men, ball and goals. **1949** S. F. Collis (title) Proper channels for the distribution of 'table hockey'. **1956** H. & L. Eizenberg Omnibus of Fun xvii. 343 Table Hockey. This ping-pong blow game can have four teams on rectangular table. **1976** Deakin & Willis Johnny go Home i. 27 The biggest amusement arcade he had ever seen..the metropolitan mecca of pinball and table football.

22. table bed (earlier example); **table bell** (earlier examples); **table carpet**, delete † and add: also, a decorative table-cloth of other material (now Hist.); **table centrepiece**, a decorative piece placed at the centre of a table, esp. one arrayed with flowers, etc.; **table-chair** = chair table s.v. *Chair sb.¹ 15; **table cover** (earlier example); **table-decker**, delete † and add later example of use in the Royal Household (now rare); **table desk**, (a) a desk with a broad, flat top; (b) a kind of folding writing-box that opens to provide a sloping desk-top, for use on a table; **table game**, a game played on a table or similar surface, usu. with balls, counters, or other pieces (and sometimes distinguished from card- or board-games); **table hand**, (a) N.Z. Sheep-shearing: in a wool-shed, one who helps the fleece-picker to skirt and roll the fleeces; (b) Printing, a bindery assistant; **table-hop** v. intr. U.S. colloq., to go from table to table in a restaurant, meeting the diners (cf. island-hop s.v. *Island sb. 4); also **table-hopping** vbl. sb.; **table jelly**, a flavoured jelly served at table as a sweet; a commercial preparation for making this; **table-maid** (earlier example); **table manners** sb. pl. orig. U.S., behaviour or deportment at table, judged according to accepted standards of propriety; **table-money**, (a) (earlier example); (b) (earlier example with reference to dining charges on board ship); **tablemount** Oceanogr. = *Guyot; **table napkin** (later examples); **Table Office**: in the House of Commons, the office in which the civil servants work whose duties include the preparation of the Notice Paper and the Order Book; by extension, the Office personified by its clerks; **table officer** Canad., any of the principal officers in an organization (cf. Board sb. 8); **table-plan**, a seating plan for those attending a formal meal; **table rock** (earlier example); **table-screen**, (a) a trestle table in a wool-shed; (b) Chinese Ceramics (see quot. 1974); **table-service**, (c) (earlier examples); **table-setting**, (a) the activity of setting a table: see sense 21 b in Dict.; (b) the cutlery, napery, etc., required to set a place at table; **table stake** Poker (see quot. 1885); **table tape** Computers, a magnetic tape containing tabulated numerical information for use in computations; **table-tennis** (earlier and later examples); now an international sport rather than just a parlour game; **table wine**, wine suitable for drinking with a meal, esp. plain wine which is not fortified or sparkling; a wine of this class; cf. *Tafelwein, vin de table s.v. *Vin 3.

1714 E. Postlethwayt Let. 5 Mar. in E. Pyle Mem. Royal Chaplain (1905) 33 Pray take care of putting up the Table Bed, put nothing in but what belongs to it. **1779** in Dict. Amer. Eng. (1938), Table bell. **1832** Chambers's Edin. Jrnl. I. 236/2 This minikin table-bell, which I must have unconsciously pocketed. **1967** E. Short Embroidery & Fabric Collage iii. 74 Great families worked their own table carpets in tent stitch on canvas sometimes incorporating their coats of arms into the design. **1917** Harrods Gen. Catal. 882 Table centre pieces and vases. Finest English hand-made cut crystal. **1979** E. Taylor in I. Webb Compl. Guide Flower & Foliage Arrangement viii. 104/3 The table centre-piece holds Norway spruce, variegated holly and berries, pine cones and red ribbons. **1671** in Farm & Cottage Inventories of Mid-Essex 1635–1749 (1950) (Essex Record Office Publ. No. 8) 120 In The Hall—..one Table-chaire. **1836** S. S. Arnold in Proc. Vermont Hist. Soc. (1940) VIII. 125 Father gave me his old table-chair. **1962** 'K. Orvis' Damned & Destroyed v. 35 Shabby men and women sat in white table-chairs. **1848** C. H. Hartshorne Eng. Med. Embroidery 126 The manner commonly used in braiding table covers. **1983** Daily Express 18 Oct. 22/2 Specially trained 'table-deckers' set the places at State banquets. **1904** M. Corelli God's Good Man 503 Placed below this, and slightly towards the centre of the room, was the Bishop's table-desk and chair. **1933** 'A. Armstrong' Ten-Minute Alibi i. 9 Right centre is a flat table-desk with two drawers. **1933** Burlington Mag. June p. xviii/2 The acquisition from the funds of the Murray Bequest of the table-desk associated with Henry VIII. **1965** J. A. Michener Source 799 Gottesmann was surprised, therefore, when this frail child slammed shut the folding table-desk used by the Palmach as its headquarters. **1864** Amer. Boy's Bk. Sports & Games 455 (heading) Table and toy games. **1905** W. Fiske Chess in Iceland 357 We have, as stated, confined ourselves wholly to table-games, that is those which are played on a board or other surface, on which some peculiar design is drawn. **1976** E. Ward Hanged Man xxviii. 180 Burnett..felt helplessness, a toy rabbit running on the magnetized tracks of a table game made for children. **1950** N.Z. Jrnl. Agric. Oct. 311/2 Pressing the fleece wool with all the skirtings, bellies, stains, [etc.]..still adhering..costs the farmer far more.. than if he had hired one or two table hands at shearing to skirt his fleece wool for him. **1955** G. Bowen Wool Away! vii. 92 A common fault is for a wool-table to be too high, which makes harder work for the table hands and the 'fleeco'. **1972** Classification of Occupations (Dept. Employment) III. 172/2 Bindery assistant. Performs, by hand or machine, folding, gathering, collating and/or sewing tasks in binding books, periodicals or stationery and assists bookbinders... Other titles include..Table hand. **1979** West Lancs. Even. Gaz. 12 Oct. 24 (Advt.), Fully experienced tablehand (SOGAT) required in our Bindery. **1958** Time 6 Oct. 16/1 He table-hopped to shake hands. **1977** Time 28 Mar. 28/2 In Charleston, he table-hopped through the cafeteria at the West Virginia State Capitol. **1967** N.Y. Times Mag. 20 Aug. 33 The writers' club..is a place for gossip, banter, flirtation, shoptalk, confidences and compulsive table-hopping. **1895** Army & Navy Co-op. Soc. Price List 16 Table jelly powder. **1917** Harrods Gen. Catal. 1224/2 Table jellies (Spring's). **1975** in T. Steel Life & Death of St Kilda (1977) xi. 176 She had a few table jellies left. **1862** J. Binny in H. Mayhew London Labour Extra vol. 355/2 Table-maids in aristocratic families or at first-class hotels. **1867** Harper's Mag. Sept. 470/1 That upright position which belongs no less to table-manners than to hygiene. **1904** [in Dict., sense 21 a]. **1949** M. Mead Male & Female ix. 187 In cultures where table-manners are the insignia of humanity people may be unable to eat their food at the table with some one who eats differently. **1835** J. E. Alexander Sketches in Portugal vi. 148 A contract was entered into with them.., that they should receive British pay and table-money during the continuance of the war. **1842** G. Parbury Hand Bk. for India & Egypt (ed. 2) 383 Table money, say 25 days, at 3 rupees per diem. **1952** Procès-Verbaux Assoc. d'Oceanogr. Physique v. 71 The term guyot seems unnecessary in view of the more satisfactory term table mount. **1959** Tablemount [see *Guyot]. **1882** Caulfield & Saward Dict. Needlework 468/1 Tablecloths, table napkins, tray ditto [etc.]. **1917** Harrods Gen. Catal. 1448/3, 1 doz. Table Napkins £1 7s. 6d. **1938** John o' London's Weekly 18 Mar. 991/3 To plant palm trees and pampas grass on the Devon hills is like calling a table napkin in an Englishman's dining-room a serviette. **1970–1** Kay's Catal. Autumn–Winter 585 White cotton tablecloth... Matching table napkins available. **1946** 2nd Rep. Sel. Comm. on Procedure p. iv, in Parl. Papers 1945–46 IX. 161 Questions received at the Table Office before the hour of sitting of the House shall be deemed to have been received the day before. **1950** Erskine May's Law of Parl. (ed. 15) xii. 243 The Table Office assists the Clerks at the Table particularly in the preparation of the Notice Paper and the Order Book. **1973** Times 15 May 7/2 The table office at the House refused, after taking advice, to accept the questions. **1968** Daily Colonist (Victoria, B.C.) 9 Nov. 1/6 John Laxton..confirmed that a..meeting of the caucus of MLA's and the provincial table officers..had agreed on the convention date. **1973** Globe & Mail (Toronto) 8 Sept. 8/5 He's been involved in some of the most complex bargaining in that field..and was one of the table officers when construction workers two years ago rejected a back-to-work order. **1911** W. J. Locke Glory of Clementina Wing xxiii. 345 Quixtus at the end of the table... Clementina had thus arranged the table-plan. **1948** G. V. Galwey Lift & Drop v. 93 Dance was..fretting over..the way his table plan had been upset. **1982** K. Follett Man from St. Petersburg xiv. 252 She sent for Pritchard and made the table plan with him. **1817** in Minnesota Hist. Soc. Coll. (1860) II. 36 The mode I adopted to ascertain the height of the cataract, was to suspend a line and plummet from the table rock on the south side of the river. **1881** Table screen [in Dict., sense 21 c]. **1971** Country Life 10 June 1425/3 Several table screens are on view. A rare example ..is made of turquoise matrix carved with an eastern scene. **1972** Trans. Oriental Ceramic Soc. XXXVIII. 112 Table screen painted in blue... Chêng-tê period, 1506–21. **1974** Savage & Newman Illustr. Dict. Ceramics 282 Table-screen, a small rectangular porcelain plaque or tile, usually decorated on both sides, mounted vertically on a stand, and intended to be placed on the scholar's table, probably to protect his work from unwanted sunlight. **1765** J. Wedgwood Let. 25 July (1965) 36 Your Brother Josiah's Pottworks were the subject of conversation for some time, the Cream colour Table services in particular. **1885** List of Subscribers, Classified (United Telephone Co.) (ed. 6) 229 (Advt.), Crystal and Demi-Crystal Table Services and Ornaments. **1955** House & Garden June 74/2 Table mats are a most practical form of table setting. **1967** E. Short Embroidery & Fabric Collage iii. 66 A tablecloth designed with the table setting in mind will enhance the general effect rather than confuse it. **1885** Encycl. Brit. XIX. 283/1 The modern usage is to play table stakes; i.e., each player puts up such an amount as he pleases at the commencement of each deal, and he cannot be raised more than he has on the table; but he has the option of making good from his pocket a previous raise which exceeds his table stake. **1973** T. Pynchon Gravity's Rainbow i. 7 Routine: plug in American blending machine won from Yank last summer, some poker game, table stakes, B.O.Q. somewhere in the north. **1948** Table tape [see problem tape s.v. *Problem 7]. **1956** G. A. Montgomerie Digital Calculating Machines x. 213 Numbers may also be taken from the table tapes as required. **1887** in 75 Years of Fun (Parker Bros., Inc.) (1958) 19 Table Tennis... This game is laid out like a lawn tennis court, played and counted just the same, all the rules being observed. **1977** World Book Encycl. XIX. 4/2 A British firm manufactured table tennis equipment and registered the name Ping-Pong in England in 1900 and in America in 1901. Soon afterward it sold the American rights to Parker Brothers of Salem, Massachusetts. The monopoly of the game by these two companies and their dictation of rules and equipment led to a revolt by internationally organized players in 1921. As a result, the unpatented name Table Tennis was adopted. **1673** J. Ray Observations Journey Low-Countries 340 The red Florence wine is most commended for a table wine of any in Italy. **1827** Disraeli Vivian Grey III. v. iv. 73 Very fair table-wine, I think. **1978** J. Symons Blackheath Poisonings i. 40 Roger poured a red table wine that had been decanted.

table, v. **4. a.** For U.S. Congress read U.S. Pol., and add later examples.

1916 J. B. Thoburn Stand. Hist. Oklahoma II. 715 [The bill] was sent to the council where it was considered, amended, and finally tabled. **1931** H. F. Pringle Theodore Roosevelt i. vi. 71 The resolution had no sooner been offered than..members were..demanding that it be tabled. **1950** W. S. Churchill 2nd World War III. ii. xxxvi. 609 The British Staff prepared a paper which they wished to raise as a matter of urgency, and informed their American colleagues that they wished to 'table it'. To the American Staff 'tabling' a paper meant putting it away in a drawer and forgetting it. **1974** Sumter (S. Carolina) Daily Item 22 Apr. 5A/7 Various plans for fundraising were discussed but it was decided to table any such plans until the fall.

tableau. Add: **2. c.** Also (Theatr.), a representation of the action at some stage in a play, created by the actors suddenly holding their positions or 'freezing', esp. at a moment critical to the plot, or at the end of a scene or act. Also, as a stage direction.

c **1863** T. Taylor Ticket-of-Leave Man i. 22 Brierly is overpowered and handcuffed—Guests rush in and form Tableau. **1866** Black ey'd Susan ii. 9 fo. 30 The door opens. William enters C. Susan. Ah. William! Alive! (Tableau). **1881** P. Fitzgerald World behind Scenes i. 46 The tableaus at the end of each act..were brought about with admirable simplicity. **1982** C. Castle Folies Bergère vi. 221 There are some 45 sets and tableaux.

d. Cards. The arrangement formed by the cards laid out on the table in the game of patience.

1875 A. Cadogan Illustr. Game Patience (ed. 2) 1 Having placed the tableau, take any aces that may appear on the surface of the packets and play them in their allotted spaces. **1913** 'L. Hoffmann' Sel. Patience Games 5 The first step, in the case of most Patience games, is to arrange a certain number of cards face upwards on the table. The cards thus arranged are known as the 'lay-out', or tableau. **1975** Way to Play 147/4 Spaces in the tableau (caused by the removal of an entire fan) remain unfilled.

5. = simplex tableau s.v. *Simplex sb. 4.

1953 A. Charnes in W. W. Cooper et al. Introd. Linear Programming ii. vi. 67 The coefficients of the ε-polynomial multiplying P_j..are given in due order by the entries in the P_j row of the tableau. **1971** D. C. Hague Managerial Economics ix. 186 The rule in the Simplex method is that any variable—that is, any x or any s—which appears on the left-hand edge of the tableau..has a positive value. **1980** A. J. Jones Game Theory iii. 165 The artificial variables have performed their function,

and we can now drop the columns of the tableau associated with a_4 and a_5.

6. Special combination. **tableau curtain** Theatr. (see quot. 1967); cf. *Tab sb.⁴

1881 W. H. Rideing Dramatic Notes 45 It may here be worth mentioning that the handsome 'tableau curtain' made for this occasion..cost £740. **1967** Oxf. Compan. Theatre (ed. 3) 932/1 Tabs (short for Tableau Curtain), used originally of an act-drop which parted and rose sideways towards the outer top corners, and by extension to any front curtain or, mistakenly, to curtain settings on the stage.

table-book. 1. For † Obs. read Obs. or rare and add later examples.

1852 Thackeray Henry Esmond II. v. 76 We were off Finisterre on the 31st of July, so Esmond's table-book informs him. **1937** Blunden Elegy 11 Might Machiavel Now from his table-book communicate Precept or paradox that could do well In the nerve centres of a modern state?

tabled, a Add: **1. b.** Seated at table. rare⁻¹.

1922 Joyce Ulysses 167 He gazed round the stooled and tabled eaters.

tableful. Add: **b.** (Earlier example.) Also, as many (things) as a table is holding, all that is on the table.

1868 H. A. Vaughan Let. 26 Dec. in Lett. to Lady Herbert (1942) 134 Mrs. Vaughan gave us tablefuls of excellent food. **1872** Geo. Eliot Middlemarch II. iv. xxxvi. 228 Lydgate's tableful of apparatus and specimens. **1977** Zigzag June 15/2 He was so pleased to see me that he threw a tableful of drinks over..like in the movies!

tabler². Add: **4.** (Example in sense 4 a of the vb.)

1976 H. Wilson Governance of Britain vii. 141 If the tablers of each of these questions are called by Mr Speaker, no other question is called.

5. With initial capital. A member of the Round Table organization; a Round-Tabler.

1955 [see *Soroptimist a. and sb.]. **1973** Scotsman 21 Feb. 8/4 During lunch with some executives in Rotary and the Round Table..the prominent young Tabler from Lower Yarrow..suggested [etc.]. **1977** Abingdon Herald 10 Mar. 2/5 Between them, the 1,215 Tablers are buying an off-shore lifeboat.

table-rapping. Add: (Earlier and later examples.) Also as adj., and **ta·ble-rapper**, one who practises table-rapping.

1856 Spiritual Herald Apr. 73 The matter has been explained to us thus... Table-turning and table-rapping are designed to call attention to the existence and presence of super-human powers. **1893** Harper's Mag. Feb. 377/1, I could not ring a bell when there was none to ring, as spirits do in table-rappers' closets. **1936** M. Franklin All that Swagger x. 89 Familiar association with bogus lords and parsons, soothsayers, table-rappers, medical quacks, [etc.]. **1973** T. Pynchon Gravity's Rainbow i. 55 The young statistician is devoted to number and to method, not table-rapping or wishful thinking.

table-spoon. Add: Also loosely, = Tablespoonful.

1960 E. David Fr. Prov. Cooking 506 For 2 lb. of fruit add 6 tablespoons of vanilla sugar. **1981** Sunday Tel. 8 Mar. 11/2 Cucumber soup. 1 small onion; 1 clove garlic; 2 tablespoons sunflower oil; [etc.].

tablespoonful. Add: Also fig.

1880 [see *continued ppl. a. 3 b].

tablet, sb. Add: **1. d.** Also, a plaque of pottery; spec. one forming the central part of a chimney-piece. Cf. *Block sb. 12 c.

1768 J. Wedgwood Let. 6 Nov. (1965) 68, I have lately had a Vision by night of some new Vases, Tablets &c with which Articles we shall certainly serve the whole World! **1775, 1875** [see *block sb. 12 c]. **1970** G. Savage Dict. Antiques 462/2 The year 1773 saw the first catalogue of ornamental wares, which included..tablets for chimney-pieces and furniture-mounts.

e. U.S. = Pad sb.³ 4. Cf. Writing tablet in Dict. and Suppl.

1880 Geyer's Stationer 12 Aug. (Advt.), The M. & H. Blotter Tablet... Beware of tablets sold by J. C. Blair, as he is manufacturing without a legal right. **1897** Sears, Roebuck Catal. 349/2 Everything from a 400 page tablet of fair paper for 4 cents, to one of fine cream laid paper. **1934** Chain Store Age (Gen. Merchandise ed.) Jan. 57/2 The customer does not notice that there are three sheets less in a tablet. **1977** Chicago Tribune 2 Oct. xii. 39/1 Cadillac boasts in its 1978 Seville that it has made 'provision for a phone installation, writing pad, and pen'. But it's only a provision—you have to supply the phone, pen, and tablet.

f. A small metal disc similar in function to a 'staff' (*Staff sb.¹ 9 f), and used for working single-track railways.

1897 W. E. Langdon Applic. Electr. to Railway Working vi. 129 When the instruments are in their normal condition, all tablets being in, the very first ring from the station of whom the permission to withdraw a tablet is made..would..be equally serviceable for the purpose. **1950** O. S. Nock Brit. Locomotives from Footplate 183 This train was booked non-stop between Arrochar and Crianlarich, but the slack for tablet exchange made it necessary to pass very slowly through Ardline. **1969** Railway Mag. Feb. 88 (caption) The single-line tablet for the section to Kingussie is being given by the signalman to the driver.

g. A rigid card used in tablet-weaving (see sense 8 below).

1921 M. & H. Peach tr. Pralle's Tablet Weaving 6 The earliest examples found of the little tablets for the weaving were of thin polished wood. **1964** H. Hodges Artifacts x. 137 The tablets were generally oval or rectangular with a hole, or a pair of holes, at each end. **1970** J. P. Wild Textile Manuf. in N. Roman Provinces vii. 73 Each tablet governs the four (or three) warp-threads which are threaded through its holes..; and the pack of tablets is held in the hand like a pack of cards.

3. b. For Sc. read orig. and chiefly Sc. and add earlier and later examples. Also, a piece of this. Orig. a type of candy, but now a type of fudge.

1736 Mrs. McLintock Receipts for Cookery 35 (heading) To make Orange Tablets with the Grate. **1850** Mrs. Dalgairns Practice of Cookery 347 Ginger tablet may be made in this way. **1897** Private Life of Queen xvii. 140 Among the favourites of the Queen..are..tablets, petits fours,..pralines, almond sweets. **1922** 'R. West' Judge i. ii. 56 Here's some tablet for you, lassie. **1948** Good Housek. Cookery Bk. 643 Ginger Tablet, use the same ingredients as for Hazel Tablet, but omit the nuts and vanilla essence and add ½ oz. of ground ginger. **1973** Times 13 Dec. 12/2 Tablet, for those who don't know, is a delicious, crumbly fudge that melts in your mouth—it's a Scottish speciality.

c. A piece of compacted powder of standard size, shape, and composition, ready for subsequent moulding.

1935 [see *preform sb. 1]. **1936** H. W. Rowell Technol. of Plastics xx. 148 A 'tablet' is of the correct weight and density required and is made of suitable diameter and thickness to fit the mould. It is made in a stock size of die and is not preformed to the approximate shape of the moulding. **1947** R. L. Wakeman Chem. Commercial Plastics v. 76 Tablets and preforms fit freely inside the mold ultimately employed. **1974** Gloss. Packaging Terms (B.S.I.) iii. 9 Tablet, a compressed mass of moulding material of prescribed form and mass.

6. (Earlier example.)

1788 [see *Gobelin 1].

8. tablet paper U.S., notepaper taken from a writing-pad; **tablet-weaving**, an early method of weaving, in which warp-threads are passed through holes in a number of parallel tablets, which are then rotated to form sheds.

1964 Mrs. L. B. Johnson White House Diary 25 Feb. (1970) 73 The file..marked 'particularly appealing'. Those were the letters that were taken to Mrs. Kennedy to read. They came written in poetry, they came in barely legible pencil on tablet paper. **1921** M. & H. Peach tr. Pralle's Tablet Weaving 6 Tablet weaving.. is considered the origin of all weaving. Ibid., In the Museum at Copenhagen..is a belt which must have been woven by this tablet-weaving method. **1950** Proc. Prehist. Soc. XVI. 130 The archaeological material..is then reviewed, with a special note on the curious technique of tablet-weaving. **1979** B. Cunliffe Celtic World 60/2 Finer weaving to make braid and a form of tablet weaving are also attested.

tablet, v. Add: **2.** (Examples of trans. use.)

1936 H. W. Rowell Technol. of Plastics xx. 148 Tableting or preforming or pelleting a powder is generally done on automatic machines. **1963** Times 4 May 11/5 In an article in a medical journal some time ago describing the clinical trial of a drug, reference was made to the manufacturer who 'tableted and distributed' the drug. **1973** R. Parkes Guardians ii. 59 This heroin is comparable in quality to that being sniffed by U.S. troops in Vietnam and far superior to that being tableted for U.K. distribution.

Hence **ta·bleted** ppl. a., **ta·bleting** vbl. sb. and ppl. a. (examples in sense 2 of the vb.).

1889 [see sense 2 in Dict.]. **1936** H. W. Rowell Technol. Plastics xx. 148 Tableting machines measure the charge in this way. **1937** Mod. Packaging Oct. 110/1 Small powdered, tableted and similar products. **1947** R. L. Wakeman Chem. Commercial Plastics v. 76 In compression operations, recourse is often had to tableting and preforming in order to speed up molding. **1972** Materials & Technol. V. xxi. 763 The tabletting process consists of feeding free-flowing granules into a..die, and compressing the material. **1983** Glaxo Group News Sept. 4/3 The accuracy and efficiency of single punch tabletting machines have been monitored by strain gauges.

ta·ble-top, a. Also **table top, tabletop.** [f. table-top sb. (see Table sb. 22).] **1.** Of, pertaining to, or designating photography of subjects which can be contained within the area of a table-top; spec. applied to photography of small-scale models which gives the illusion of a larger subject.

1914 S. C. Johnson Saturday with my Camera xliv. 368 We can all enter the lists of table-top photography,..and spend our winter evenings counterfeiting, at leisure, many of the most attractive sights of the world. **1923** Kodak Mag. Apr. 58 (heading) Home-made landscapes—a few words on table-top photography. **1935** News Chron. Amat. Photogr. xiv. 176 Flashlight has special application to table-top photography, now becoming so popular, as the illumination is under complete control, and all the work can be done in the evening. **1956** Focal Encycl. Photogr. 1151/1 There are three different branches of photography open to the photographer who chooses to work within the limits of a table top studio: still life studies, photography of small scale models and creative composition. The last is the true 'table top photography'. Ibid. 1152/2 Most table-top pictures fail either because they include too many items or because they try to represent the subject accurately and in detail instead of simply suggesting it in a broad effect.

2. That is or can be placed, or that takes place, on a table.

1945 F. Brown in Astounding Sci. Fiction Jan. 133/1 There was a boom market in portable and table-top receivers. **1962** Guardian 9 July 5/3 He made table top models of the machines. **1971** Physics Bull. Sept. 513/2 The total number of installed computers, including table top computers, is expected to be 60 000 in 1975 and 96 500 in 1980. **1979** Guardian 31 Oct. 1/8 Fagging.. includes..retrieving the little red ball when it goes under chairs during games of table-top cricket.

tableware. (Earlier example.)

1772 J. Wedgwood Let. 10 Sept. (1965) 134, I think he might by that means sell now and then a sett of it in Tableware.

Tabloid. Add: **1. a.** (Examples of the proprietary term.) Also loosely, (with small initial), a small (medicinal) tablet.

1884 Trade Marks Jrnl. 23 Apr. 334 Tabloid... Burroughs, Wellcome & Company, Snow Hill Buildings, Holborn Viaduct, London, E.C... Chemical substances not included in Class I, used in Medicine and Pharmacy. **1894** Murray's Handbk. India (ed. 2) p. xx, For medicine, plenty of quinine in 2 or 4 grain 'tabloids' or pills. **1895** Army & Navy Co-op. Soc. Price List 695/1 Tabloids—Ichthyol per bott. 0/7½... Tea per tin 0/5. **1904** Official Gaz. (U.S. Patent Office) 18 Oct. 1743/2 Drugs and chemicals for human and veterinary use... Tabloid. **1916** 'Taffrail' Pincher Martin ix. 161 Morphia tabloids were served out to all the officers of quarters for administration to badly injured men. **1938** E. J. G. Forse Ceremonial Curiosities xxix. 149 It is wise to carry a few simple tabloids with you. **1978** Daily Mirror 19 Apr. 24/1, I found a metal box which used to contain 'Tabloid' tea.

b. fig., etc. (Later examples.)

1909 Westm. Gaz. 22 Oct. 5/2 While in literature the trend of taste is all in the direction of tabloids, composers seem ashamed of anything approaching terseness. **1920** R. Macaulay Potterism vi. iii. 232 People might like their science in cheap and absurd tabloid form... The Potter press exulted in scientific discoveries made easy. **1928** Melody Maker Feb. 145/2 Mr. Harold Craxton's playing on the piano of the 'Three Blind Mice'..as a tabloid Hungarian Rhapsody by Liszt. **1935** Brit. Jrnl. Psychol. July 27 Statements of a vague character, which are condensations of complex propositional wholes... To such propositions I have elsewhere given the name 'tabloids'.

2. a. R.A.F. slang. A small Sopwith biplane. (Disused.)

1913 Aeroplane 11 Dec. 635/2 The small speedy Sopwith biplane has been nicknamed the 'Tabloid' because it contains so many good qualities in such small compass. **1915** War Illustr. 20 Feb. 22/2 The 'Tabloid's' supreme value lies in its speed and climbing power. **1925** Fraser & Gibbons Soldier & Sailor Words 275 Tabloid, a, an Air Force nickname for a type of small Sopwith biplane of high speed and rapid climbing powers, a special favourite from its numerous good points, its, as it were, concentrated excellencies. **1928** C. F. S. Gamble North Sea Air Station x. 149 In addition to its maximum speed of 92 miles an hour the 'Tabloid' was remarkable in those days for its great speed range.

b. In full tabloid cruiser. A small cruising yacht.

1930 Yachting Monthly XLIX. 428/1 T's ship, Honora, is, except for her draught, a 'tabloid' cruiser: 19 ft. LOA, with 5 ft. 9 in. beam. **1937** Ibid. LXIV. 17/1 Reflections on an unusual little tabloid. **1938** Ibid. 452/2 A tabloid cruiser that goes foreign ought to be registered.

3. a. A popular newspaper which presents its news and features in a concentrated, easily assimilable, and often sensational form, esp. one with smaller pages than those of a regular newspaper.

1918 W. E. Carson Northcliffe x. 304 Since 1908 Alfred Harmsworth, like his famous 'tabloid', has disappeared from view. **1926** Encycl. Brit. II. 1055/2 The introduction of tabloids may be explained..by the passing remark of Lord Northcliffe, 'If some American does not start one I shall have to come over to do it.' **1928** Observer 5 Feb. 18/1 The..chain now includes 26 papers, in most cases 'tabloids' or papers with a popular appeal. **1934** A. Woollcott While Rome Burns 100, I remembered how confidently, but how inaccurately, the tabloids had prophesied the..divorce. **1949** [see *angel v. 1]. **1957** Listener 31 Oct. 683/2 Newspapers have been allowed to transform themselves into tabloids with gossip columns, adulation of film stars, beauty contests and other requisites of the popular press of the West. **1970** G. F. Newman Sir, You Bastard vi. 174, I presume you've read the tabloids? **1978** Time 3 July 12/1 The National Enquirer, the Florida-based tabloid, dispatched ten reporters and photographers to scour the Riviera in quest of informants on the courtship.

b. attrib., esp. as tabloid newspaper.

[**1901, 1902**: in Dict.] **1918** W. E. Carson Northcliffe x. 299 The New York World made its appearance. Harmsworth had issued the paper in what he called 'tabloid form'. **1926** Amer. Mercury Dec. 462/1 A tabloid weekly theatrical newspaper, published in New York, and filled with ugly type, heavy black advertisements and the most atrocious English ever put into print, was named as co-respondent by his wife. **1938** [see *jazz sb. 6]. **1949** Koestler Promise & Fulfilment ii. ii. 232 To the distant reader of the tabloid Press..it looked as if history had at last met Metro-Goldwyn-Mayer's most ambitious dream. **1962** V. Nabokov Pale Fire 22 He was back in the car, reading a tabloid newspaper which I had thought no poet would deign to touch. **1977** New Yorker 19 Sept. 31/1 Next day, the tabloid Daily Mail gave the hearing

its entire front page, but the Guardian didn't mention it at all.

Hence as v. trans., to express briefly or concisely; to condense. rare.

1933 Partridge Slang To-day & Yesterday i. iv. 36 Much of the best wit, the most delectable humour is couched in slang; for, slang offers no compulsion to think how the happy thought is phrased or, perhaps, tabloided into an expressive adjective, or a second-sighted noun, an unravelling or illuminating verb. **1934** Punch 21 Mar. 329/2 Also there is a certain sketchiness in the tale as tabloided for the two hours' traffic of our stage, and some of the connecting-links seem to have got lost in the process.

tabnab (tæ·bnæb). Naut. slang. [Origin obscure.] A cake, bun, or pastry; a savoury snack.

1933 M. Lowry Ultramarine v. 212 Perhaps he would be able to speak to Andy when he gave him the tabnabs. **1947** —— Under Volcano vi. 172 What the bosun called, with unction, 'afternoon tea'. With tabnabs. The tabnabs were delicate and delicious little cakes made by the second cook. **1962** Punch 10 Jan. 98/1 Tea and tabnabs (seafaring for cake). **1978** K. Bonfiglioli All Tea in China viii. 111 These 'tabnabs' were little gullet-tickling confections... My favourite 'tabnab' was..a little fried potato-cake with a morsel of kari'd mutton inside.

tabo- (tēⁱ·bo), comb. form of Tabes, as in **tabopara·lysis** Med. (see quot. 1972); **tabopare·sis** = prec.

1910 F. W. Mott in Power & Murphy System of Syphilis IV. x. 328 An important point to remember is the frequency with which optic atrophy is followed by taboparalysis. **1972** R. A. & A. T. Willis Princ. Path. & Bacteriol. (ed. 3) xvi. 201 The quaternary syphilitic diseases are (1) general paralysis of the insane or dementia paralytica, and (2) locomotor ataxia or tabes dorsalis. A combination of the two, taboparalysis, also occurs. **1910** Med. Rec. (N.Y.) LXXVII. 219/1 (heading) The pathological prodromes of taboparesis. **1932** W. Boyd Textbk. Path. xxx. 829 There is sometimes a combination of tabes and paresis (taboparesis), with degeneration of the posterior columns [of the spinal cord]. **1980** A. King et al. Venereal Dis. (ed. 4) v. 89 This suggests that there is an element of tabes present (taboparesis).

taboo, tabu, a. and sb. Add: **B.** sb. **1. c.** Linguistics, a total or partial prohibition of the use of certain words, expressions, topics, etc., esp. in social intercourse.

1933 [see sense 3 b below]. **1962** S. Ullmann Semantics viii. 205 Taboo is an important cause of semantic change. Language taboos fall into three more or less distinct groups according to the psychological motivation behind them. **1980** R. A. Hudson Sociolinguistics ii. 53 The distinction between conventional and necessary social restrictions is also interesting in view of the strength of feeling which the former arouse. This is particularly clear in the case of linguistic taboo, such as the so-called 'four-letter words' of English.

3. b. Linguistics. With reference to an expression or topic considered offensive and hence avoided or prohibited by social custom.

1933 L. Bloomfield Language xxii. 396 In America, knocked up is a tabu-form for 'rendered pregnant'; for this reason, the phrase is not used in the British sense 'tired, exhausted'... In such cases there is little real ambiguity, but some hearers react nevertheless to the powerful stimulus of the tabu-word. **1978** Amer. Speech LIII. 16 It may be that taboo terms form a group which is logically akin to, yet separate from, true slang, since many taboo terms are the only ones available to non-academic speakers. **1980** Scottsdale (Arizona) Progress 9 Feb. 12 We now have a set of taboo expressions relating to ethnic groups and individuals.

taboo·ness, the state or condition of being taboo.

1974 Verbatim I. i. 4/1 The tabooness of fuck. **1978** Maledicta 1977 I. 236 Tabooness focuses on the speaker and his/her decision about what can or cannot be said in a given context.

‖ **taboot¹** (tăbū·t). Also **tabut.** [Hindi, a. Arab. tābūt coffin, box, Ark of the Covenant.] A sacred box or coffin; spec. a box, representing the tomb of Husain, which is carried in procession through the streets during the Muslim festival of Muharram.

1622 in W. Foster Eng. Factories India 1622–3 (1908) 94 This daye is heere aryved Sultan (Khus) roues taboots [sic] from Brampore, (which to-) morrowe is to bee dispeeded to H(elobass?) there to bee intered by his mother. **1862** Mrs. J. B. Speid Our Last Yrs. in India 230 Taboots, or tazzias, the representation of Hosain's mausoleum at Kurbulla. **1879** L. Pelly Miracle Play of Hasan & Husain p. xvii, Against the side of the Imambarrah, directed towards Mecca, is set the tabut. **1891** Daily News 9 Sept. 5/4 Immense sums of money are spent upon the taboots..that, carried in these processions, are broken to pieces and buried at the end of the ceremony. **1958** G. E. von Grunebaum Muhammedan Festivals v. 89 The ta'ziya, or Passion play..became the real climax of the Shī'ite Tenth of Muharram celebrations. The stage requires few properties besides a large tabut (coffin),..and Husain's arms and banner. **1975** Indian Express 15 Jan. 5/1 Taboot processions with music will be allowed only on the last day of the Moharrum.

‖ **taboot²** (tā·būt). Also **tabut.** [Arab., abbrev. of tābūt raf' al-miyāh Archimedes screw, or of tābūt al-sāquiya scoop wheel: see prec. entry.] A form of water-wheel used in Egypt.

1836 E. W. Lane Acct. Manners & Customs Mod. Egyptians II. 25 There is a third machine, called taboot, used for the irrigation of lands in the northern part of Egypt. **1841** J. Kitto Phys. Hist. Palestine vii. p. ccxcvii, Another machine used for the irrigation of lands, when it is only necessary to raise the water a few feet..is called the Taboot. **1877** Encycl. Brit. VII. 708/1 The tâboot..differs from the sâkiyeh principally in having a hollow wheel instead of the wheel with pots. **1924** Countries of World III. 1757/1 The primitive but still effective apparatus known as the 'sakieh', the 'shaduf', and the 'tabut'.

‖ **tabot** (tabǫ·t). [Ge'ez: cf. *Taboot¹.] A box, representing the Ark of the Covenant, which stands on the altar in an Ethiopian church.

1682 J. P. Gent tr. Ludolphus's New Hist. Ethiopia III. vi. 294 In the Sanctuary stands the Holy Table, which they call..Manbar... Upon this they place the sacred vessels. First the Tabot, or Chest..an Oblong Quadrangular Table, upon which the Dish and Cup are set. **1710** F. Tellez Trav. Jesuits in Ethiopia III. x. 242 We restore you the Faith of your Fore-Fathers. The former Clergy-Men may return to their Churches, put in their Tabotes, and say Masses. **1834** S. Gobat Jrnl. Three Years' Residence in Abyssinia ii. 243 A church, when there is no 'tabot' in it, is no more to them than a common house. **1923** Blackw. Mag. Aug. 256/2 It [sc. the Abyssinian Church] venerates an object called the tabot, which is the replica of the Ark of the Covenant. **1968** E. Ullendorff Ethiopia & Bible ii. 83 Criticisms levelled against the Ethiopians on account of their tabot-centred worship.

tabouret. Add: **2. b.** U.S. A small table, esp. one used as a stand for house-plants; a bedside table.

1916 Sears, Roebuck & Co. Catal. Fall 1244/2 Tabourets or jardinier stands. **1968** J. Updike Couples iii. 228 One of his flippers kicked over a tabouret holding a crammed ashtray and a small vase of asters. **1984** M. Babson Trail of Ashes iii. 33 Look in the bedside tabouret for a little nightcap.

Tabriz (tâbrī·z). The name of a city in north-western Iran, used attrib. and absol. to designate carpets and rugs made there, the older styles of which often show a rich decorative medallion pattern.

1900 J. K. Mumford Oriental Rugs xi. 168 The model on which the Tabriz rugs were really designed is the ornamental and richly colored fabric of Kirman in southern Persia. **1911** G. G. Lewis Pract. Bk. Oriental Rugs ii. 184 With the Kirman the Tabriz shares the reputation of having the most graceful floral designs. **1931** A. U. Dilley Oriental Rugs & Carpets iv. 104 Modern Tabriz rugs constitute a..revival of weaving that began at least as early as the Caliphate. **1946** Lancet 2 Mar. 322/1 The bedside carpet is important and must be gratifying to the bare feet... A subdued Shiraz will fulfil most people's requirements, though leptoforms may require something with a more stimulating pattern, say a Tabriz. **1962** N. Freeling Love in Amsterdam ii. 79 He bought books and pictures, and had a treasured Tabriz carpet. **1978** S. Wilson Dealer's Move iii. 42 A superb Tabriz hung on the wall.

tab show. U.S. slang. [f. Tab(loid + Show sb.¹] A short version of a musical, esp. one performed by a travelling company.

1951 Green & Laurie Show Biz 571/2 Tab show, tabloid version of a musical. **1953** Sun (Baltimore) 30 Sept. 38/6 He traveled in 'tab shows'—vaudeville and musical comedy—through the South. **1983** N.Y. Times 12 June xxi. 10/4 Most Broadway musicals are simply streamlined for travel, ranging from the 'tab shows' (short for tabloid) that are the trademark of some resorts, to abridged versions in stock or dinner theaters.

tabular, a. Add: **1. c.** Geol. tabular (ice)berg, a flat-topped iceberg which has broken away from an ice shelf.

1840 C. Wilkes Jrnl. 20 Jan. in Narr. U.S.A. Exploration Exped. (1844) II. ix. 315 These tabular bergs are like masses of beautiful alabaster. **1848** C. Tomlinson Summer in Antarctic Regions iv. 114 Westward of this point the Vincennes met a remarkable collection of tabular icebergs. **1905** R. F. Scott Voyage of 'Discovery' I. iv. 118 Cook preserves the name of Ice Island in describing the long tabular berg so typical of the Southern Regions. **1958** [see ice-shelf s.v. *Ice sb. 8]. **1979** C. Kilian Icequake viii. 133 Drifting snow and falling ice masked much of the cliff face, but it did not look like the side of a tabular berg.

tabulating, vbl. sb. (Later examples.)

1921 J. A. V. Turck Origin Mod. Calculating Machines 124 The carriage in the Ludlum machine..offered no solution to the feature of tabulating. **1979** Washington Post 9 June F3 Citizens in West Germany, France, Italy, Belgium and Luxembourg vote Sunday and tabulating begins that night.

tabulator. b. For 'also' read: a part of the mechanism of a typewriter (formerly, a separate attachment) for controlling the movement of the carriage in tabular work, indentation, etc. (Later examples.) spec. in Computing, a machine that produces lists, tables, or totals from the information in a data storage medium, esp. punched cards or tape.

1917 L. R. Dicksee Office Machinery viii. 96 A complete installation, consisting..of three Punches, one Sorter, and one Tabulator, would involve the employment of four operators, none of whom need be skilled accountants. **1922** F. W. Pixley Accountant's Dict. II. 723/1 Accountants should..consider both systems.. The choice will generally depend on the form in which the data is finally required; in other words, the tabulator will usually govern the system adopted. **1949** [see *interpreter 5 a]. **1970** O. Dopping Computers & Data Processing iv. 73 A tabulator usually prints around 150 lines per minute, while a normal speed for a computer line printer is 1,000 lines per minute. **1978** J. Kellock Elements of Accounting xii. 214 The next step is to feed the cards into a tabulator.

Tabun (tā·bun). Also **tabun.** [Ger., of unkn. origin.] The name of an organophosphorus nerve gas, $(C_2H_5O)(CN)((CH_3)_2N)PO$.

1951 Acta Physiol. Scandinavica Suppl. No. 90. 11 The writer's aim has been to synthesize tabun and some of its homologues. **1953** [see *Soman]. **1967** New Scientist 26 Jan. 196/1 The nerve gas Sarin, known as GB, is said to be four times as toxic as the German Tabun of World War II. **1968** [see *Sarin]. **1978** A. Melville-Ross Blindfold iv. 26 A single 'Sarin' or 'Tabun' shell of British manufacture found its way to Egypt. **1980** Sci. Amer. Apr. 35/1 The first of these compounds, called tabun, was discovered in Germany in 1936 in the course of research on insecticides.

tacan (tæ·kăn). Also **Tacan, TACAN.** [f. the initial letters of tactical air navigation.] A navigational aid system for aircraft which measures bearing and distance from a ground beacon. Freq. attrib.

1955 Times 17 Aug. 6/4 The Defence Department has removed security restrictions from the technical details of Tacan (tactical air navigation system) and the Air Navigation Development Board will reveal them in an announcement on August 19. **1956** Electrical Communication Mar. 3/1 Tacan is a system that provides both bearing and distance information on direct-reading instruments in an airplane within 200 nautical miles of a selected ground station. Ibid. 26/1 A tacan ground-installed beacon consists of a receiver-transmitter..and either a shipboard or a shore antenna. **1966** [see *Shoran]. **1977** P. Way Super-Celeste iii. 129 A small, slim fin broke the smooth underbelly of the plane... 'That's either a UHF antenna or a TACAN aerial,' said Bridge.

tacenda, sb. pl. (Earlier examples.)

1843 Carlyle Past & Present ii. x. 125 Willelmus Sacrista, and his bibations and tacenda are..softly yet irrevocably put an end to. **1870** S. H. Hodgson Theory of Practice I. ii. 217 A greater number of things are classed among tacenda.. The French term pudeur seems exactly to express the feeling which is called out painfully or wounded by an lifting of the veil of the tacenda.

tach (tæk), U.S. colloq. abbrev. of Tachometer. Cf. *Tacho, *Tack sb.⁸

1966 Publ. Amer. Dial. Soc. 1964 XLII. 9 Tach.., abbreviation of tachometer. **1974** R. M. Pirsig Zen & Art of Motorcycle Maintenance i. iii. 36 The speedometer needle swings back and forth but the tach reads a steady nine thousand. **1980** Family Handyman Sept. 98/2 If you've used a tach/dwell meter for the point adjustment, leave it in place, and attach a timing light.

Tachai (da·dʒai). Also **Dazhai.** The name of a village in the Shansi Province of the People's Republic of China, used attrib. to designate its model commune or the methods of work, etc., associated with it. Also Comb., as Tachai-type adj.

1969 Observer 16 Mar. 2/7 Under the 'Tachai' system, not only are peasants being rewarded with work points instead of hard cash, but the schoolteachers and the 'barefoot doctors' in the communes are also to be paid principally in work points. **1973** J. S. Aird in Yuan-Li Wu China Handbk. I. xviii. 463 Even in the Tachai production brigade in Shansi, the family planning propaganda program at first elicited 'little positive response'. **1975** A. Watson Living in China i. 22 The Communist Party almost always introduces its new policies and goals by putting forward examples for others to copy. One of the most famous of these is Dazhai Commune. **1977** China Now June 1 Myriads of such Tachai-type people, Tachai-type cadres, their outlook constantly broadened and deepened by Marxist science. **1979** Ibid. Mar./Apr. 3/1 The movement to spread 'Dazhai-type counties' throughout the country is lagging.

tacharanite (tæ·kārănəit). Min. [See quot. 1961 and -ite¹.] A monoclinic hydrated silicate of calcium, magnesium, and aluminium found as white aggregates or masses.

1961 J. M. Sweet in Mineral. Mag. XXXII. 750 It is thought that this mineral is sufficiently distinctive to deserve a name of its own and that tacharanite..from the Gaelic tacharan (a changeling) would be suitable, both from the nature of its behaviour and the wealth of folklore associated with the island [sc. Skye] in which it occurs. **1975** Ibid. XL. 113 Tacharanite has been re-examined... The composition approximates to $Ca_{12}Al_2Si_{18}O_{69}H_{36}$. **1978** Ibid. XLII. 383/1 Tacharanite in fractures in metagabbroic rocks filled by zeolitic assemblages has been found in the Gruppo di Voltri, Ligurian Alps.

tache, sb.¹ **1. c.** spec. in Art, a spot or dash of colour. Also fig. Cf. *Tachism.

1957 Observer 3 Nov. 14/6 The 'tache' is the mark the painter makes on the canvas with his paint-loaded brush, and an emphasis on the freedom and spontaneity of the creative act itself and on extreme sensitivity towards the actual materials of painting is characteristic of the tachists. **1967** J. N. Barron Lang. Painting 188 Tachisme..., a term used to describe a style of painting in which the color is applied in splotches or blots (taches) of color. **1978** G. Greene Human Factor ii. ii. 67 The simple precise words, with the single tache of colour reminded Castle of the local background so often to be found in primitive paintings.

tache, sb.³ Add: Forms: 8–9 **teach.**

1. (Further examples.)

1788 P. Marsden Acct. Island Jamaica 26 The smallest and last copper is called the teach. **1835** in J. H. Ingraham South-West I. 240 In the last kettle—the teach as it is termed..the sugar is concentrated to the granulating point. **1862** Illustr. Catal. Internat. Exhib., Industr. Dept., Brit. Div. II. No. 6139 Stoves, ranges, sugar pans, teaches, or boilers to any pattern or make. **1887** Encycl. Brit. XXII. 626/1 The [cane sugar] juice..is passed from the one [pan] into the other till it reaches the last of the series, the striking teach. **1949** Caribbean Q. I. i. 9 The juice, now reduced to a syrup, was ladled into a final copper, the teache, for a last boiling.

'tache, var. *Tash.

tachinid (tæ·kinid), sb. and a. Ent. [a. mod.L. family name Tachinidæ, f. generic name Tachina (J. M. Meigen 1803, in Mag. für Insektenkunde II. 280), f. Gr. ταχινή, fem. of ταχινός swift: see -id³.] **A.** sb. A small hairy fly of the family Tachinidæ, the larvæ of which are parasitic on other arthropods. **B.** adj. Designating an insect of this family.

1888 Insect Life I. 44 We also reared an undetermined Tachinid. **1901** Knowledge Oct. 234/2 The ants..protect the caterpillars from the attacks of Ichneumon and Tachinid flies. **1925** R. W. G. Hingston in E. F. Norton Fight for Everest 285 Tachinids were common at the edge of a torrent at 17,000 feet. **1954** Borror & DeLong Introd. Study of Insects xxvii. 636 Tachinid Flies... All the tachinids are parasitic on other insects. **1972** Nature 21 Jan. 135/3 Such long-established successes of biological control as that of the..coconut moth in Fiji by the tachinid parasite. **1979** New Scientist 3 May 380/2 Rettenmayer watched the behaviour of the tachinid and conopid flies.

tachist (tæ·ʃist), sb. (and a.). Also ‖**tachiste** (taʃist). Art. [ad. F. tachiste, f. tache stain, spot + -iste -ist.]

†**1.** One who paints by juxtaposing small patches of unmixed colour. Obs.

1891 [see plein-airiste s.v. *plein-air]. **1909** C. E. Hallé Notes from Painter's Life xi. 234 We have even schools which take their names from the manner of using the brush. We have 'Tachists', 'Vibrists', and Heaven knows how many more.

2. a. One who practices tachism (see below).

1954 New Yorker 4 Dec. 99/1 Negatively, it can be said that the unknowns are certainly not Cubists and not tachistes, and not Mondrianesque or Braqueish either. **1957** [see *tache sb.¹ 1 c]. **1960** Guardian 22 Apr. 9/4 The young English tachistes for whom freedom is an engrossing obsession.

b. attrib. or as adj.

1955 New Yorker 31 Dec. 40/3 The car, maybe the vegetables, and certainly the hope of sharing as an artist in the dubious kudos have all been attributed to a tachiste French painter. **1956** Archit. Rev. CXX. 186/1 In his delectable paintings of trout hovering in light-stained water he uses tachist techniques with a consummate professionalism. **1966** 'H. MacDiarmid' Company I've Kept iii. 103 People should not look at his [sc. William Johnstone's] paintings with any preconceived ideas and seek for elements in them which can be labelled..'tachist', and the like. **1972** R. Quilty Tenth Session i. 123 An aggressive twenty square feet of tachist canvas. **1982** S. Spender China Diary 118 The Western artist looks at the model... The first object of his attention is usually the image, even if this is abstract (except for tachiste painting).

Hence **ta·chism** [cf. F. tachisme (also used)] a style of modern painting in which spots or dabs of colour are arranged in apparently random manner to evoke an emotion, scene, etc.; cf. action painting s.v. *Action sb. 16.

1956 Archit. Rev. CXX. 333/1 The same Cézanne picture, considered simply as a painted surface, is one of the finest examples of 'tachism' in the history of art. **1957** Times 28 Nov. 3/4 The Canadian artist, Mr. Austin Cooper can claim to have been among the first in this country to practise what is now generally known as tachisme. **1960** J. Cohen Chance, Skill & Luck ii. 42 Neakes may deserve to be described by the historian of art as the founder of Tachism. **1978** Jrnl. R. Soc. Arts CXXVI. 696/1 Abstract expressionism and tachisme, dead on time, and an explosion of hard-edged colour, produced, he told me, under the razzamataz influence of New York. **1979** E. H. Gombrich Sense of Order ii. 62 Any number of Ph.D. theses await being written about the influence of Cubism, of Tachism, of Op or Pop art on fabrics and wall paper.

tachistoscope. Add: (Earlier and later examples.) Hence **tachistosco·pic** a.; **tachistosco·pically** adv. (also fig.).

1890 Billings Med. Dict. II. 641/1 Tachistoscope. **1903** Psychol. Rev. X. 393 (heading) Tachistoscopic experiments. Ibid. 394 The number of separate objects that can be apperceived at once with the tachistoscope is given as varying from four to five. **1917** Arch. Psychol. XL. 3 The material was presented tachistoscopically with a fixed tempo of presentation. **1931** Brit. Jrnl. Psychol. XXII. 67 The instrument..embodies an attempt to meet the many theoretical requirements of the perfect tachistoscope by constructing one without moving parts. **1949** Jrnl. Personality XVIII. 24 Present tachistoscopically a picture of a love scene, a handsome young Arab leaning yearningly over his beloved on a couch in a Moorish setting. **1969** J. Brunner Plague on Both your Causes xix. 141, I saw a tachistoscopically rapid glimpse of one of the half-tracks [from a helicopter]. **1979** R. Hawkey Side-Effect vii. 54 I'd like to..have you take the tachistoscopic perception test... Look at a series of images we'd put up on a tachistoscope.

tacho (tæ·ko), colloq. abbrev. of Tachometer. Cf. *Tach, *Tack sb.⁸

1964 Motor 13 June 9/1 (heading) Japanese tachos. **1975** G. J. King Audio Handbk. viii. 195 Now, should the motor speed tend to decrease, owing to an increasing load for example, the tacho output also decreases. **1979** Truck & Bus Transportation May 26/1 On the open roads, the rear axle ratio of 4·33:1 made 100 km/h a comfortable cruising speed with a tacho reading of around the 2800 rpm mark.

tacho-ge·nerator (tæ:ko-). [f. tacho(meter)-generator s.v. *Tachometer 2.] An instrument that generates a voltage accurately proportional to the rate of rotation of a shaft or the like.

1952 Electronic Engin. XXIV. 382/1 Factors affecting the linearity of response with speed of D.C. tacho-generators. **1958** New Scientist 4 Sept. 751/1 One such piece of apparatus for keeping watch on the rpm of engines in flight is actuated by a small tacho-generator to measure the rate of rotation and communicate its warning if overspeeding occurs. **1976** Gramophone Sept. 510/1 Speed accuracy is controlled by a new system using a tacho-generator.

tachograph (tæ·kograf). [f. Gr. τάχο-ς speed + -graph.] A device in a motor vehicle for recording its speed, travel time, and other information automatically.

[**1903** Nature 26 Nov. 95/2 On the use of the Schrader tacheograph in hydrographic work.] **1909** Cent. Dict. Suppl., Tachograph, a recording tachometer applied to shafting or wheels to register rotation-speed; a speed indicator. [With reference to prec. source.] **1941** F. D. Jones Engin. Encycl. II. 1258 Some of these recording tachometers or tachographs have a dial in addition to the recording charts. **1968** Guardian 1 Oct. 5/2 Road tanker drivers..are protesting against the proposal..to install a tachograph in lorries—a device which records speeds, length of time taken on journeys, and periods when the vehicle is stationary. **1976** Citizen (Ottawa) 8 Jan. 1/6 The 10 ambulances in Ottawa are equipped with a tachograph that records the speed of the vehicle. **1980** Times 24 Jan. 2/3 The Road Haulage Association acknowledged last night that the tachograph, which records speed, mileage travelled, stopping time and the use of brake and accelerator, could open the way to productivity deals if drivers could show that they were operating more efficiently.

tachometer. Add: **1. a.** spec. One that indicates the speed of an engine in r.p.m. (Further examples.)

1918 Bull. U.S. Naval Consulting Board No. 3. 10 Many new instruments have been devised for aircraft. These include..tachometers, which indicate the engine speed. **1953** C. A. Lindbergh Spirit of St. Louis ii. vi. 187 The tachometer needle shows 1825 r.p.m. **1975** Drive New Year 88/1 This is an important job performed by the rev-counter, or tachometer.

2. Special Comb.: **tachometer generator** = *Tacho-generator.

1946 Shell Aviation News No. 103. 24/1 The gearbox is of Rolls-Royce design... It is mounted on the bulkhead and provides drives, on the forward side, for the air pump and the generator, and on the rear side for the tachometer generator. **1958** W. D. Cockrell Industr. Electronics Handbk. ii. 254 Tachometer generators are used in systems to generate feedback signals in servoloops or used directly with an indicating instrument.

Hence **tachome·tric** a.

1931 S. R. Roget Dict. Electr. Terms (ed. 2) 341/1 Tachometric electrometer. **1967** O. I. Egerd Control Syst. Theory vii. 238 Tachometric feedback.

‖ **ta chuan** (da dʒwan). Also **ta tchuen.** [Chinese dàzhuàn, f. dà big + zhuàn seal character.] In Chinese calligraphy, an early form of script used during the Chou dynasty (c 1028–221 B.C.); 'large seal script'.

1894 T. de Lacouperie Beginnings of Writing in Central & Eastern Asia 194 The Chinese writing exhibits in its history eight successive styles, viz:..(2) Ta tchuen of 820 B.C. **1910** Encycl. Brit. VI. 220/1 Authentic specimens of the..ta chuan, older or Greater Seal writing, are exceedingly rare. **1958** W. Willetts Chinese Art II. vii. 566 Those [characters] then substituted..were what Han scholars called ta chüan or 'Great Curly', and sometimes chou wên after the name of the supposed Annalist. **1966** C. Ch'en Chinese Calligraphers & their Art iii. 24 Scholars of a later day have chosen to group all the different scripts before Li Ssŭ's time as the ta chuan. **1973** T. C. Lai Chinese Calligraphy 12 (caption) A ceremonial basin ta chuan script.

tachy-. Add: **tachymeta·bolism** Zool. (see quot. 1973); hence **ta:chymetabo·lic** a.; **ta:chyphyla·xis** Pharm. [mod.L., ad. F. tachyphylaxie (Champy & Gley 1911, in Compt. Rend. Soc. de Biol. LXXI. 161), f. Gr. φύλαξις protection], a rapidly diminishing response to successive doses of a drug; **tachypnœ·ic** a. Med., exhibiting tachypnœa; **tachyzo·ite** Zool. [*-zoite], one form of the protozoon toxoplasma (see quot. 1973).

1974 Nature 13 Sept. 143/2 Already at this early age the dog is tachymetabolic. **1978** Ibid. 5 Oct. 441/1 The central nervous system (CNS) is very sensitive to elevated temperatures, and consequently, both bradymetabolic and tachymetabolic terrestrial vertebrates have evolved physiological mechanisms which effect localised cooling of the brain. **1973** Bligh & Johnson in Jrnl. Appl. Physiol. XXXV. 954/2 Tachymetabolism: The high level of basal metabolism of birds and mammals relative to those of reptiles and other nonavian and nonmammalian animals of the same body weight and at the same tissue temperature... Synonym: Warm-Blooded. Antonym: Bradymetabolism, Cold-Blooded. **1911** Index Medicus IX. Index of Subjects 214/2 Tachyphylaxis. **1947** F. K. Oldham et al. Essent. Pharmacol. xi. 132 Its [sc. ephedrine's] disadvantages include..the lessened effect of repeated doses (tachyphylaxis). **1979** Nature 29 Nov. 515/2 The response to DAEA showed neither desensitisation during a 3-min exposure period nor tachyphylaxis with repeated applications. **1961** Webster, Tachypneic. **1976** Lancet 13 Nov. 1083/1 He was not cyanotic or tachypnœic. **1973** J. K. Frenkel in Hammond & Long Coccidia 344/1, I am introducing two other terms: 'tachyzoites' for the rapidly multiplying forms of the acute infection, previously called trophozoites, aggregations, and proliferative forms; and 'bradyzoites' for the slowly multiplying encysted forms characteristic of chronic infection, which have been variously called merozoites or just zoites. **1979** Biol. Abstr. LXVIII. 7579/1 Probably most toxoplasmosis infections involve the ingestion of cat feces bearing cysts and oocytes rather than contact with tachyzoites.

tachygraph. Add: **2.** Also, a tachygraphic sign.

1965 E. V. K. Dobbie in Language XLI. 153 The inventory of the allographs (including nonalphabetic allographs, such as tachygraphs).

tachyon (tæ·ki,ǫn). Physics. [f. Tachy- + *-on¹.] A hypothetical particle that travels faster than light and has imaginary mass. Hence **tachyo·nic** a.

1967 G. Feinberg in Physical Rev. CLIX. 1090/1 One description is presented..for noninteracting faster than light particles, which we call tachyons. **1970** Sci. Amer. Feb. 70/2 Hence a tachyon that was losing energy by interacting with matter or by radiating light would speed up, whereas a tachyon that was gaining energy from some outside source would slow down, and its speed would approach c from above rather than below. **1970** Physical Rev. D II. 265/2 (caption) A and B use tachyonic antitelephones to communicate backwards in time. **1974** Globe & Mail (Toronto) 13 Apr. 5/1 The pursuit of the elusive tachyon lures scientists into the realm of complex mathematical abstraction and high-flown theoretical physics. **1978** Pasachoff & Kutner University Astron. xxvii. 695 So far, there is no experimental evidence that tachyons exist.

tachysterol (tækisti^ə·rǫl, -ste·rǫl). Biochem. [ad. G. tachysterin (A. Windaus et al. 1932, in Ann. d. Chem. CCCCXCIX. 188): see Tachy- and *Sterol.] An oily isomer of ergosterol and lumisterol, $C_{28}H_{43}OH$, which will form calciferol when irradiated with ultraviolet light.

1933 Chem. Abstr. XXVII. 729 The addn. compd. from irradiated ergosterol and citraconic anhydride (20 g.) and 75 cc. Ac_2O, warmed 2 hrs., give [sic] 7 g. of tachysteryl acetate citraconic anhydride... Tachysterol..has not been crystd. **1954** A. White et al. Princ. Biochem. l. 1053 Of the series of compounds obtained from the irradiation of ergosterol only calciferol (vitamin D_2) possesses antirachitic activity. However, one of the series, tachysterol, may be catalytically reduced to dihydrotachysterol.., which is antirachitic. **1976** H. Campion et al. in B. E. C. Nordin Calcium, Phosphate & Mineral Metabolism xii. 445 Lythgoe has reported the synthesis of a closely related isomer, tachysterol, by a nonphotochemical pathway.

tack, sb.¹ Add: **I. 2. a.** Also in colloq. phr. to come (or get) down to brass tacks: see *Brass sb. 5 b. See also *tin-tack b.

3. c. Basket-Making. A size of willow rod, usu. 3 ft. long.

1912 T. Okey Art of Basket-Making ii. 6 White and buff rods are sorted into tacks from 2 ft. 6 in. or 3 ft. to 3 ft. 6 in. **1953** [see *long-small s.v. *Long a.¹ A. 18]. **1961** L. G. Allbon Basic Basketry ii. 11 Willow is sold by the bolt... The rods are sorted..on the farm into lengths... Local usage often gives special names to the sizes..such as Tack or Short Smalls (3ft.), Smalls (4 ft.),..and so on. **1973** B. Maynard Mod. Basketry from Start 171 Tacks, term used for 3 ft willow rods.

IV. 10. b. (Later examples.) Used esp. in Printing.

1939 Printing Feb. 27/1 Where excess tack is attributed to these rollers, it is frequently found that the complainant is..referring to natural rubber rollers rather than synthetics. **1967** E. Chambers Photolitho-Offset xvi. 243 Ink of low tack fills in shadow areas more readily, whilst high tack may pull the surface of a coated paper, if the separation is quick. **1971** Engineering Apr. 17/1 A suitable adhesive..to give a reasonable tack. **1972** Physics Bull. Nov. 665/3 Tack, with prepreg materials, the degree of stickiness of the resin. **1979** G. A. Glaister Gloss. Bk.

(ed. 2) 469/1 If an ink has insufficient tack it will not print sharply.

V. 12. a. tack-hammer (earlier example); also as *v. trans.*

1865 *Atlantic Monthly* June 736/2 If she absolutely cannot get a tack-hammer with a claw on one end, she can take up carpet-nails with an iron spoon. **1908** Kipling *Bk. of Words* (1928) 36 The meanest collection of packing-cases that was ever tack-hammered together.

c. in sense 4: **tack weld** *v. trans.*, to join (materials) at intervals with provisional welds in order to hold them in position for subsequent work; hence as *sb.*

1919 *Chambers's Jrnl.* Jan. 60/2 Up the sides the seams were only welded at intervals, or 'tack welded', as it is called. **1950** B. R. Hilton *Welding Design* ii. 42 If tack welds are not to be removed as the welding proceeds, their section should be equivalent to that of the first weld run. **1964** S. Crawford *Basic Engin. Processes* iii. 82 A tack weld is made by applying the flame to the metal until it melts and then adding a little welding rod. **1979** *Financial Times* 22 Jan. 9/7 The top is tack welded, then the bottom, followed by the sides. Clamps and devices are removed and the weld completed.

d. in sense 10: **tack coat** (see quot. 1954); **tack rag** *U.S.*, an impregnated cloth used for cleaning a surface prior to painting or varnishing.

1949 *Sun* (Baltimore) 17 Oct. 26/3 Workmen spread a 'tack coat' of asphalt on the old pavement. **1954** *Gloss. Highway Engin. Terms* (*B.S.I.*) 28 *Tack coat*, a coat of liquid (such as bitumen, road tar, or an emulsion thereof) applied as a thin film to a surface to improve the adhesion of a course laid thereon. **1979** *Civil Engin.* Nov. 27/1 The painting on of a grid of levels on the tack coat by engineers. **1958** *Washington Post* 16 Aug. B 3/6 So-called 'tack' rags are used in factories where dust particles are likely to mar freshly painted surfaces. **1979** P. Wallage *Restoration Post-W.W.II Cars* ii. 25/2 Go over it with a tack rag.

tack, *sb.*[4] (Later example.)

1826 J. O'Keefe *Recollections* I. vii. 268 The young nobleman..when he was the striker, took the nicest pains to place his tack in such a manner, that to hold his adversary's ball seemed a matter of course.

tack, *sb.*[7] Abbrev. of Tackle *sb.* † **a.** In sense 1. *dial. Obs.*

1777 in *Eng. Dial. Dict.* (1905) VI. 3/2. **1879** G. F. Jackson *Shropshire Word-bk.* 428 My tacks bin at Newport, or I'd soon ketch them rots. **1893** J. Salisbury *Gloss. Words & Phr. S.E. Worcs.* 41 *Tack*,..a collection of tools; a razor-grinder's machine is his tack; a smith's box of tools for shoeing horses is his 'shoeing tack'.

b. In sense 6. Also *Comb.*, as *tack room.*

1924 I. Maddison *Riding Astride for Girls* xiv. 226, I will now give a few hints on tack..in the show ring. The tack on a saddle-horse should be as light as possible. **1933** A. Blewitt *Ponies & Children* iii. 34 Any prize rosettes they win are stuck up on their tack-room wall. **1940** *Evening Sun* (Baltimore) 15 Apr. 21/4 Tack is the name for a rider's equipment—saddle, whip, boots, etc., apparently derived from tackle. **1950** J. Cannan *Murder Included* iv. 65 Patricia..was cleaning tack in the stable. **1964** D. Francis *Nerve* xi. 147 It was a tack-room. Every stable has one..the place where the saddles and bridles are kept. **1975** F. Kennedy *Alberta was my Beat* ix. 111 He stood there like a broke saddle horse as we put the tack on him. **1979** J. Johnston *Old Jest* 100 There was a boy who kept the tack, polished the lovely shiny boots. There was a smell of saddle soap and horse dung. The saddles are flaking now, out in the damp tack room.

tack, *sb.*[8] U.S. colloq. abbrev of *tachograph, Tachometer. Cf. *tach, *tacho.

1963 *Amer. Speech* XXXVIII. 46 *Tack*,..the device in the cab which automatically records miles driven, number of stops, speed, and so on, during a trip; short for *tachometer*. **1971** M. Tak *Truck Talk* 162 *Tack*, short for tachometer or tachograph.

tack, *v.*[1] Add: **5.** Also const. *on.*

1908 L. M. Montgomery *Anne of Green Gables* viii. 83 Marilla was as fond of morals as the Duchess in Wonderland, and was firmly convinced that one should be tacked on to every remark made to a child. **1960** C. Day Lewis *Buried Day* i. 17 My father's family name was originally Day, the Lewis being tacked on by a man who adopted his grandfather or great-grandfather.

tack, *v.*[5] *trans.* Abbrev. of Tackle *v.* 3. Usu. with *up*. Also *absol.* Cf. *tack *sb.*[7] b.

1946 M. C. Self *Horseman's Encycl.* 395 To tack up a horse means to put the saddle and bridle on him. **1962** W. Faulkner *Reivers* viii. 178 So we..tacked up and..led the way. **1972** *Islander* (Victoria, B.C.) 26 Mar. 13/1 In addition to being taught how to groom a horse, the new student must learn how to tack-up (that's putting a saddle and bridle on). **1977** *Sunday Tel.* (Colour Suppl.) 1 May 22/3 It is not a bad idea either to acquire a creature that will come when it is called or will at least stand still long enough to get it tacked up for a bit of a ride.

tacketed, *ppl. a.* (Earlier example.)

1864 J. Brown *Let.* Dec. (1912) 234 To-morrow I meant in a pair of tacketed shoon to have explored some Grampian.

tackie (tæ·ki). *S. Afr.* Also **takkie.** [Origin uncertain: perh. rel. to Tacky *a.*[2] App. not Afrikaans.] A rubber-soled canvas shoe; a plimsoll or sand-shoe. Also, a track shoe with a rubber sole. Usu. *pl.*

c **1902** I. Vaughan *Diary* (1958) 60 We all have to wear..white tackies on the feet. **1913** C. Pettman *Africanderisms* 491 *Tackies*, in the border towns of the Eastern Province this is the name given to rubber-soled sand-shoes. **1924** *Ann. Mountain Club S. Afr.* No. 27. 46 Ye who scale with ropes and 'tackies' Cliffs of awe-inspiring grandeur. **1946** *Amer. Speech* XXI. 59 What are known as 'sand shoes' or 'tackies' in English are the same articles which I still backslide into calling 'sneakers'. **1953** M. Murray *Fire-Raisers* xi. 108 He padded over the rocks on his tackies. **1955** D. Jacobson *Trap* 1. 20 He wore canvas *takkies* on his feet, his toes poking through the ends. **1961** *Personality* 16 May 27, I have yet to discover why tennis shoes, which are known in England as plimsoles, are called 'tackies' in South Africa. **1981** A. Paton *Towards Mountain* xvii. 134 Hofmeyr was a camper of the first water. He wore an ancient canvas hat, a khaki shirt and shorts, and discoloured sandshoes, known as tackies.

tackifier (tæ·kifəi,əɹ). [f. Tacky *a.*[1] + -fy + -er[1].] A substance that makes something sticky; an adhesive agent or ingredient.

1942 *Science Illustr.* Apr. 4/2 As processing aids, naval stores products are numbered among the many plasticizers, softeners, and tackifiers. **1958** *New Scientist* 23 Oct. 1110/2 This pressure-sensitive thermoplastic..is so sticky that a tackifier such as resin need not be added. **1963** H. R. Clauser *Encycl. Engin. Materials* 449/2 The liquid nitrile polymer finds use as a tackifier..in molded rubber parts, cements, friction and calendered stocks. **1970** *New Scientist* 5 Nov. 275 (Advt.), Sometimes it's [*sc.* Lorival liquid rubber] a tackifier in ebonite grinding wheels.

tackily (tæ·kili), *adv.*[1] [f. Tacky *a.*[1] + -ly[2].] In a slightly adhesive or sticky manner. (In quot. 1903, *fig.*)

1903 Kipling *Traffics & Discoveries* (1904) 124 The sea..drummed tackily to gather my attention, coughed, spat, cleared its throat. **1971** *Sunday Times* 20 June 42 For every loving Mum who has ever wished there was an easier way of preventing nappy-rash than smearing on..petroleum jelly, Johnson & Johnson is introducing..a melt-on-contact Baby Gel which does the job less tackily.

tackily (tæ·kili), *adv.*[2] *colloq.* [f. Tacky *a.*[1] + -ly[2].] In a tasteless or vulgar style; shabbily, dowdily.

1936 M. Mitchell *Gone with Wind* iv. xxxii. 544 She was ugly and dressed tackily. **1979** P. Driscoll *Pangolin* i. xiv. 115 A square little hovel tackily partitioned into two rooms.

tackiness[2] *colloq.* (f. Tacky *a.*[1] + -ness.] The quality of being cheap or in poor taste.

1977 *Washington Post* 26 Mar. B5 Their visual craftsmanship and polish are compromised by the manifest tackiness of the story material. **1982** J. Fox *White Mischief* i. 19 A provincial tackiness..pervades the residential suburbs of Nairobi.

tacking, *vbl. sb.* Add: **d.** *Comb.*, as **tacking iron** *Photogr.*, a tool used for attaching tissue to a print or mount by the application of heat at chosen points.

1973 *Bodl. Libr. Rec.* IX. 2 The work bench is equipped with tacking irons and a pH meter. **1977** J. Hedgecoe *Photographer's Handbk.* 309 (*caption*) Using an electric tacking iron gently touch the center of the tissue, sticking it to the print.

tackle, *sb.* Add: **2.** In this sense freq. with pronunc. (tē·k'l).

4. b. *to stand to one's tackle* (later example).

1841 C. Brontë *Let.* 1 July in Wise & Symington *Brontës* (1932) I. 234 Mrs. White offered me a week..but I demanded three weeks, and stood to my tackle with a tenacity worthy of yourself.

9. Restrict *Football* to senses in Dict. **a.** (Earlier example.) Also in *U.S. Football.*

1876 in P. H. Davis *Football* (1911) 462 A tackle is when the holder of the ball is held by one or more players of the opposite side. **1898** A. Spurling in W. A. Morgan *'House' on Sport* 170 If you are running after an opponent who has the ball, and find he is gaining on you, don't give up, as he may be checked, and you have the pleasure of making a good tackle.

b. (Earlier example.)

1891 W. Camp *Amer. Football* 41 The tackle is an assistant to both end and guard.

c. The act of tackling in other sports. Cf. *tackle *v.* 5 (*c*).

1930 M. Pollard *Hockey for Women* viii. 106 A defence player can run towards the tackle, but she should never run into it. **1967** J. Potter *Foul Play* x. 120 Basil was out of the goal in a lightning flash, cutting off the pass with a sliding tackle.

10. *tackle-box* (earlier example), *-maker* (earlier example); **tackle-room,** a room in which horse tackle is stored; cf. *tack room* s.v. *tack *sb.*[7] b.

1902 *Chambers's Jrnl.* Oct. 702/2 This will render carrying a tin tackle-box unnecessary. **1832** *Chambers's Edin. Jrnl.* 7 Apr. 87/1 The lines of the angler may be bought from the tackle makers. **1951** *Chambers's Jrnl.* Oct. 587/1 You enter the tackle-room, where surgical harness is stored. **1962** A. Fry *Ranch on Cariboo* v. 53 Like all cabins, [it] was kitchen, dining and living all rolled into one, sometimes even tackle room.

tackle, *v.* Add: **4. c.** Also *transf.*

1920 *Blackw. Mag.* Jan. 105/2 The sort of road that even a Ford would hesitate to tackle.

d. (Earlier example.)

1884 'Mark Twain' *Huck. Finn* xxx. 313 So the king sneaked into the wigwam, and took to his bottle for comfort; and before long the duke tackled *his* bottle.

5. Restrict *Football* to senses in Dict. (*a*) Also in *U.S. Football.* (*c*) In other sports, to obstruct or accost (an opponent) in order to deprive him of the ball or other object of play.

1895 H. F. P. Battersby *Hockey* 98 In defence, they [*sc.* the halves] must tackle everything, and stick to it. **1935** *Encycl. Sports* 701 The referee may also penalise a side if any of the players holds the ball under the water when tackled. **1959** M. Boyd *Lacrosse Playing & Coaching* vi. 70 As soon as their opponents get the ball, attack players must tackle back onto them. **1975** *Oxf. Compan. Sports & Games* 320/2 The supporting backs and line-backers are prepared to tackle the carrier.

tackler. Add: **c.** Also in other sports.

1955 Doyle & Smith *Lifetime in Hurling* xx. 144 A quick hitter..and a fearless tackler.

tackling, *vbl. sb.* Add: **6. tackling bag** *U.S.* and *Rugby Football*, a stuffed bag suspended and used for practice in tackling; **tackling dummy** *U.S. Football* = *tackling bag* above.

1892 *Outing* (U.S.) Jan. 279/2 Their one special piece of apparatus is..the tackling bag, and this is..necessary to the indoor practice of a football team. **1978** *Rugby World* Apr. 25 (Advt.), Order now for 1978–79 pre-season training the new Allander tackling bag. **1904** *Outing* (U.S.) Dec. 367/2 The tackling dummy was used by many squads. **1959** N. Mailer *Advts. for Myself* (1961) 51 The Japs looked like bushes, or like tackling dummies in the evening when practice was over.

tacky, *sb.* and *a.*[1] Delete *local* and add: Also **tackie. A.** *sb.* **a.** (Earlier examples).

1800 W. Tatham *Agric. & Commerce* 81 A horse, a cow, or a little *tackie*, &c. (which last term signifies a poney or little horse of low price). **1839** C. F. Hoffman *Wild Scenes* 117 The land pirates had disappeared, without molesting my tackey.

B. *adj.* For *U.S. colloq.* read *colloq.* (orig. and chiefly *U.S.*). Also, in poor taste, cheap, vulgar. (Earlier and later examples.) Also *Comb.*, as *tacky-looking* adj.

1862 K. Stone *Jrnl.* 16 Feb. in *Brokenburn* (1955) 89 What a weary, bedraggled tacky-looking set they were. **1883** I. M. Rittenhouse *Maud* 262 Two little cards (with his name printed on them in gilt. Tackey? Ugh). **1937** Hart & Kaufman *You can't take it with You* iii. 180 An extremely tacky-looking evening wrap. **1957** M. Kennedy *Heroes of Clone* iii. i. 158 He went again to the window to watch for the arrival of the tacky little car. **1967** N. Mailer *Cannibals & Christians* i. 16 All the tacky doings of each small town. **1971** 'O. Bleeck' *Thief who painted Sunlight* (1972) iii. 30 A tacky-looking bulletin board. **1983** *Listener* 10 Feb. 29/3 They were really very good, putting together a fast and lively show, full of cheerfully dreadful jokes and inventively tacky songs.

taco (tā·ko, tæ·ko). Chiefly *N. Amer.* [Mex. Sp.] **a.** A Mexican snack comprising a fried, unleavened cornmeal pancake or tortilla filled with seasoned mincemeat, chicken, cheese, beans, etc.

1949 *Amer. Speech* XXIV. 235/2 The *touristas* almost always eat in a Mexican restaurant and bravely attempt to order their meals in Spanish. Such meals are (1) [tækoz], a mispronounciation of the Spanish word *tacos* [takos]. **1957** J. Kerouac *On Road* (1958) xiii. 93 We went into a Mexican restaurant and had tacos and mashed pinto beans. **1965** *Austral. Women's Weekly* 20 Jan. 25/1 She would serve up a traditional Mexican dish of unsurpassable excellence, the white meat..rolled up in a delicate crisp pancake, or taco. **1966** *Listener* 4 Aug. 164/2 *Tacos*..are *tortillas* rolled round shredded meat or bird and fried in oil. **1971** *Islander* (Victoria, B.C.) 4 July 2/4 Friends in San Diego had introduced us to tortillas and tacos. **1978** S. Wilson *Dealer's Move* vi. 107 Washing the food down with two strong cups of tea and mopping up with South London's answer to tacos, sliced white loaf.

b. *attrib.*, as *taco joint, sauce, stand.*

1967 *Trans-Action* Apr. 8/1 Time is alive when and where there is action... During the regular school year it may pick up for an hour in the afternoon when the 'broads' leave school and meet with the set at a corner taco joint. **1977** *Time* 28 Nov. 58/1 Some have about as much feeling for a community's sense of itself and its needs as does the imported manager of a franchised taco joint on the highway outside town. **1976** *Punch* 11 Aug. 227/1, I took to enchilada mix, tortillas, taco sauce, and all those Mexican delicacies. **1977** *Daily News* (Perth, Austral.) 19 Jan. 13 (Advt.), Taco—a crisp tortilla filled with beef, onions, lettuce, taco sauce and cheese. **1969** D. MacKenzie *Night Boat to Puerto Vedra* (1970) 172 A few seamen were at the taco stands... He bought himself a cone of maize flour filled with peppered ground meat. **1979** R. L. Simon *Peking Duck* ii. 18 Jogging behind a gas station and a taco stand to a nondescript stucco gate.

Taconic, *a.* Add: [f. the name of the *Taconic* Mountains in New England and New York State.] **a.** (Earlier example.)

1842 E. Emmons in *Geol. N.Y.* II. vii. 135 It has been deemed advisable to annex to the general account of the group of rocks of the northern district, a brief sketch of the services which constitute the Taconic System.

b. Epithet of an orogeny that occurred in Ordovician times in eastern North America.

1908 *Bull. Geol. Soc. Amer.* XX. 503 The other three [emergences]..were of long duration and of great significance. These are: (1) The Taconic revolution..; (2) the Appalachian revolution.., and (3) the Laramide revolution. **1980** *Sci. Amer.* Oct. 136/1 The southern Appalachians have evolved in a series of collisions of fragments of continental or island-arc material at the eastern edge of North America in the Taconic, the Acadian and the Alleghenian orogenies.

taconite (tæ·kŏnəit). *Geol.* [f. Tacon(ic *a.* in Dict. and Suppl. + -ite[1].] **a.** A type of chert used as an iron ore in parts of N. America.

1905 *Econ. Geol.* I. 48 In the Mesabi district the local name 'taconite' is applied to the ferruginous chert. **1951** *Engineering* 22 June 761/2 To be able to take a hard ore, such as taconite, separate the magnetite and silica by grinding the material down to pass a 300-mesh sieve, and agglomerate the concentrates by pelletising, at a price to compete with imported ores, was no mean achievement. **1981** D. R. Coates *Environmental Geol.* v. 104/1 Taconite is a low-grade ore of iron which is the source of most current U.S. iron production.

b. *attrib.*, as *taconite mine, ore, pellet, tailing.*

1974 *Sumter* (S. Carolina) *Daily Item* 23 Apr. 7B/7 Reserve, which..produces 15 per cent of the iron ore used in the nation's steelmaking blast furnaces, also closed its taconite mine at Babbitt. **1975** *Sci. Amer.* Nov. 32/3 The place of the high-grade ores was then taken by taconite ores containing iron in the form of finely disseminated magnetite. **1958** J. Szarkowski *Face of Minnesota* 270 By 1948 the first mass-produced taconite pellets reached the blast furnaces. **1975** *Telegraph* (Brisbane) 18 Mar. 12/4 The refinery's discharge of taconite tailings.

tacouba (tākū·bā). Also **tacooba, tacuba.** [Origin unknown, perh. an Arawakan word.] In Guyana, a tree which has fallen across a river forming a bridge or obstruction. Also *fig.*

1934 E. Waugh *92 Days* ii. 55 In the wet season..you had to crawl across a tacuba leading a swimming horse. **1951** E. Mittelhölzer *Shadows move among Them* ii. iv. 196 'What's a *tacooba*?' 'Indian word. Means a fallen tree or any sort of obstruction in a river or creek that constitutes a menace to navigation.' **1959** P. Capon *Amongst those Missing* 124 He had expected rapids every few miles, numerous tacoubas and a cataract or two. **1965** 'Lauchmonen' *Old Thom's Harvest* v. 58 He was a squat negro, a stumpy little piece of a man, a knotty old tacuba tree-stump. **1974** H. MacInnes *Climb to Lost World* vi. 73 The walking itself wasn't difficult, but there were plenty of streams to cross, some of them bridged by slippery tacoubas, or tree jams.

tact. Add: **II. 5.** *Psychol.* [Final element of *con)tact.*] B. F. Skinner's term for an utterance which is evoked by an object, event, etc., and reinforces the learning of a response. Hence as *v. trans.*, to respond to (a stimulus) with an utterance; as *v. intr.*, to utter words or sounds in this way; so **ta·cted** *ppl. a.*; **ta·cting** *vbl. sb.* Cf. *mand.

1954 *Brit. Jrnl. Psychol.* Aug. 181 Skinner..describes how a child learns to emit 'tacts' (i.e. verbal responses controlled by properties of objects or situations) under the influence of 'generalized reinforcers', particularly approval. **1957** B. F. Skinner *Verbal Behavior* v. 81 There is no suitable term for this type of operant... The invented word 'tact' will be used here. The term carries a mnemonic suggestion of behavior which 'makes contact with' the physical world. A tact may be defined as a verbal operant in which a response of given form is evoked (or at least strengthened) by a particular object. **1959**, etc. [see *mand]. **1964** A. W. Staats *Human Learning* iii. 73/2 The child learns in this manner to tact environmental events as well as objects. **1969** B. F. Skinner *Contingencies of Reinforcement* viii. 254 The close relation between the topography of behavior and the tacted stimulus. **1977** *Listener* 5 May 597/2 What is nowadays called 'tacting' (i.e., verbal behaviour controlled primarily by its antecedents—in this case the sight of the milk)... Victor was evidently conditioned only to tact.

tactic, *a.*[1] Add: **3.** *Linguistics.* Of or pertaining to taxemes, their arrangement or order. Cf. *tactics 3.

1933 L. Bloomfield *Language* x. 166 Combinations of taxemes, or, quite frequently, single taxemes, occur as conventional grammatical arrangements, *tactic forms.* **1962** E. F. Hayden et al. *Resonance Theory for Linguistics* iii. 24 Like beads on a string, each entity in phonotactics has a distinct Form, since no two beads can occur in the same place on the string. This is the tactic form, i.e. the structural form in the sequence. **1966** S. M. Lamb *Outl. Stratificational Gram.* 5 This process of isolating recurrent partial similarities is the basis of tactic analysis. *Ibid.* 54 Thus the analysis (*un true*) (*ly*) fits the simplest tactic description. *Ibid.* 58 The description of a stratal system is probably most conveniently presented in two parts: the tactic description and the realizational description. **1968** P. M. Postal *Aspects Phonol. Theory* viii. 198 There are four distinct strata, each of which is an independent system with its own generative rules (tactic rules)... The four current properly linguistic strata are..the sememic, the lexemic, the morphemic, and the phonemic. It is apparently the function of the tactic rules on a particular stratum X to generate both the class of X-emes and the possible combinations of X-emes. **1969** *Language* XLV. 303 This tactic fact is that..the low vowels /e a o/ can occur only if accompanied by stress; therefore the only unstressed vowels are /i ə u/.

tactical, *a.* Add: **1. b.** Applied to aircraft, bombing, etc., employed in direct support of ground forces. Cf. *strategic bomber, bombing* s.v. *strategic *a.* 4.

1916 F. W. Lanchester *Aircraft in Warfare* vii. 69 The tactical scout or machine for local reconnaissance will require to be furnished..with both offensive and defensive armament. **1922** *Flight* 24 Aug. 488/1 Tactical bombing and 'trench-strafing', etc., in battles in accordance with the plans and under the command of the naval or military officer in charge of the operations. **1941** A. O. Pollard *Bombers over Reich* xv. 208 Tactical bombing replaces..the long-range attacks on objectives far behind the lines. **1942** *R.A.F. Jrnl.* 3 Oct. (*recto rear cover*), The transition to low wing monoplane trainers and tactical ships of advanced type. **1955** *Bull. Atomic Sci.* May 192/2 One of the pitfalls of the atomic age is the use of words that becloud important meaning, rather than clarify it. Take the words 'tactical' and 'strategic', in defining two kinds of bombing. **1958** *Listener* 11 Sept. 376/2, I believe that the initiation by the West of the use of small tactical bombs on a battlefield in Europe would prove disastrous to NATO forces. **1977** *R.A.F. News* 11–24 May 6 (Advt.), A two-seater all-weather tactical interdictor and attack bomber.

c. Designating nuclear weapons intended for short-range use against an enemy's forces. Opp. *strategic *a.* 2.

1957 [see *strategic *a.* 2]. **1968** *Observer* 31 Mar. 25/1 Consider the weapons that have become operational for the first time in the past 20 years. They include the H-bomb and the so-called 'tactical' A-bombs. **1970** *Toronto Daily Star* 24 Sept. 22/5 It is estimated that about 20 lbs. or so would be sufficient for one atomic bomb in the one kiloton range, a so-called tactical bomb with mainly localized effects. **1976** Ld. Home *Way the Wind Blows* xii. 167 The balance of argument through the years moved towards a substantial conventional force, but it was gradually rendered somewhat academic by the introduction of the tactical nuclear weapon. **1979** N. Calder *Nuclear Nightmares* ii. 35 This definition is..not as sharp as the cynic's version: 'A tactical nuclear weapon is one that explodes in Germany.' *Ibid.*, 'Tactical' nuclear weapons can be let off without necessarily signalling a 'strategic', all-out exchange between the Soviet Union and the United States.

tacticity (tækti·sĭti). *Chem.* [f. Tactic *a.*[1] + -ity.] The stereochemical arrangement of the units in the main chain of a polymer.

1959 Natta & Danusso in *Jrnl. Polymer Sci.* XXXIV. 4 The rule, or *taxis*, which characterizes an arrangement partially or completely ordered, or tacticity, may be simple or composed by few simple rules. **1967** Margerison & East *Introd. Polymer Chem.* ii. 63 The tacticity of the chain. **1975** *Nature* 24 Apr. 696/1 Fibrocytes react to some microarchitectural or 'tacticity' difference between homologous copolymeric substrata.

tactics. Add: **3.** *Linguistics.* C. F. Hockett's term for the study of the relation and arrangement of linguistic units, esp. the study of the arrangement of morphemes.

1947 C. F. Hockett in *Language* XXIII. 274 We should therefore expect to find the following topics treated in his book..; (4) tactics—stating the arrangements of morphemes... This term seems simpler than 'taxemics' or 'tagmemics' which one might derive more directly from Bloomfield's labels. **1953** F. G. Lounsbury in *Yale Univ. Publ. in Anthropol.* xlviii. 18 Statements describing the occurrences of morphemes constitute the portion of a grammar called *tactics*... Tactics is not concerned with the phonemic forms of morphemes, whether they are constant or variable. **1966** S. M. Lamb *Outl. Stratificational Gram.* 1 Each of these systems has its own syntax or tactics, so that a linguistic structure as a whole has a series of tactic components rather than just one.

tactile, *a.* Add: **2. a.** Also, characterized or influenced by the sense of touch. Hence *absol.* as *sb.*, one for whom the sense of touch predominates over the other senses.

1892 [see *motile *a.*]. **1917** [see *audile *sb.*]. **1956** [see *audile *a.*]. **1971** A. Montagu *Touching* v. 169 Children who are highly tactile but have no accompanying sexual interest in others.

c. Art. *tactile value*: B. Berenson's term for the illusion of tangibility which a painter can create with regard to the figures and objects he represents; the attribute or impression of a tangible quality. Also *transf.*

1896 B. Berenson *Florentine Painters of the Renaissance* ii. 4 Every time our eyes recognise reality we are..giving tactile values to retinal impressions. **1907** —— *North Italian Painters of Renaissance* 146 In figure painting, the type of all painting, I have endeavoured to set forth that the principal..sources of life-enhancement are *tactile values, movement*, and *space composition*, by which I mean ideated sensation. **1908** E. M. Forster *Room with View* ii. 22 The traveller who has gone to Italy to study the tactile values of Giotto, or the corruption of the Papacy. **1919** A. N. Whitehead *Princ. Nat. Knowl.* 88 This property of 'conveying' an object..is already well-known in the theory of art-criticism, as is evidenced in such phrases as 'tactile-values'. **1938** R. G. Collingwood *Princ. Art* vii. 146 Mr. Berenson..taught his pupils..to look in paintings for what he called 'tactile values'. **1962** *Listener* 15 Nov. 832/1 It [*sc.* a play] is remarkable because of what one might call, after Berenson, its tactile values. **1970** *Oxf. Compan. Art* 1120/1 Berenson was notoriously incapable of appreciating those schools of modern—and ancient—art which subordinate tactile values to other qualities of pictorial design.

d. *Comb.*, as *tactile-visual* adj.

1969 *New Scientist* 27 Mar. 678/1 A tactile-visual system..should provide valuable information concerning such psychological questions as the nature of sensory processing. **1978** *Verbatim* May 16/1 My point is that the oral-aural mode is intricately combined with the tactile-visual mode.

Hence **ta·ctilely** *adv.*

1953 A. C. Kinsey et al. *Sexual Behavior Human Female* xiv. 578 Some areas which are tactilely sensitive..are of no especial importance as sources of erotic response. **1977** *Verbatim* Feb. 8/1 It takes some talent and not much money to design and manufacture a book artistically, one that provides as much aesthetic pleasure visually and tactilely as it does in its reading.

tactily (tæ·ktĭli), *adv.* [Irreg. f. Tact + -ly[2].] = Tactfully *adv.*

1895 G. Meredith *Amazing Marriage* I. iv. 37 She had to warn her brother to preserve his balance. He tactily did so, aware of the necessity. **1929** M. Lowry *Let.* 13 Mar. (1967) 5 The bewildered parent..would be willing to pay you 5 or 6 guineas a week (I should say six personally, but tactily) if you would tolerate me for any period..as a member of your household.

† **tactism** (tæ·ktiz'm). *Biol. Obs.* [f. L. *tact-*, stem of *tangĕre* to touch + -ism.] The motile response of a living organism to an external stimulus.

1902 *Fortn. Rev.* June 1013 By his revelations of the *rôle* of the 'trophisms' and 'tactisms' Dr. Loeb drove boldly into the domain of mental phenomena. **1912** A. Tridon tr. *Delage & Goldsmith's Theories Evol.* 164 Others..attribute differentiation to the influence of the various tropisms and tactisms.

tactoid (tæ·ktoid). *Physical Chem.* Also † **taktoid.** [a. G. *taktoid* (Zocher & Jacobsohn 1929, in *Kolloidchem. Beihefte* XXVIII. 167), f. Gr. τακτ-ός ordered (see Tactic *a.*[1]): see -oid.] A small anistropic, birefringent region in a dilute, isotropic sol, consisting of an aggregate of rod-like particles or macromolecules aligned parallel to one another.

1929 *Chem. Abstr.* XXIII. 2868 In these taktosols..the individual microscopic particles are called taktoids. **1939** *Nature* 14 Jan. 82/1 The formation of tactoids from thixotropic sols,..and the crystallization of proteins are regarded as being typical of unipolar coacervation. **1952** J. T. G. Overbeek in H. R. Kruyt *Colloid Sci.* I. viii. 327 The concentrated phase in the tactoids still contains a great deal of dispersion medium which implies that the particles are comparatively far apart. **1953** S. E. Luria *Gen. Virol.* v. 94 In the liquid phase, the rod-shaped [virus] particles orient themselves sidewise into 'tactoids' which then settle into the liquid crystalline phase. **1978** *Nature* 14 Dec. 666/3 Minton has applied fundamental models for the entropically driven formation of tactoids from long rod-like molecules or particles to the haemoglobin S system.

So **ta·ctosol** [ad. G. *taktosol* (Zocher & Jacobsohn, *loc. cit.*)], a sol containing tactoids.

1929 [see *tactoid]. **1959** *Lancet* 3 Oct. 513/1 Tactosols are colloidal solutions containing non-spherical particles ('tactoids') which are capable of orienting themselves.

‖ **tactus** (tæ·ktŏs). *Mus.* [L.: see Tact.] = Tact 4.

1740 J. Grassineau *Mus. Dict.* 130 *Metron*, Tactus, Mensura, Battuta,—the beating or measuring the time by a motion of the hand or foot. **1786** Busby *Compl. Dict. Mus.* s.v. *Tactus*,..when the time consisted of a breve in a bar, the time-stroke was called *Tactus-Major*; and when of a semibreve in a bar, *Tactus-Minor*. **1959** *Collins Mus. Encycl.* 644/1 A term used for 'beat' by the theorists of the 15th and 16th cent... The semibreve was the normal *tactus* in the 15th cent.; in the course of the 16th cent. the minim became the normal... With the introduction of bar-lines the semibreve became the unit of a bar and the measuring *tactus* was replaced by the metrical beat. **1980** *Early Music* July 310/2 To sing the passage to a tactus, however, is to miss its *raison d'être* and obliterate the most vital element of the madrigal's expressive scheme.

tacuba, var. *tacouba. **Tacully,** var. *Takulli.

tad (tæd). *colloq.* (orig. and chiefly *N. Amer.*) [Orig. uncertain; perh. f. Tadpole[1].] † **1.** (See quots.) *Obs.*

1845 in C. Cist *Cincinnati Misc.* I. 240 Among a certain class in the eastern cities,..the word *Tad*, is applied to one who don't nor won't pay. **1851** B. H. Hall *College Words* 297 At Centre College, Ky., there is a society..composed of the very best fellows of the College, calling themselves Tads. **1890** E. B. Custer *Following Guidon* 213 These youths [*sc.* graduates from West Point] were called 'tads' and 'plebes'.

2. A young or small child, esp. a boy. Occas. used *joc.* of old men.

1877 Bartlett *Dict. Amer.* (ed. 4) 688 Tads, *little tads*, small boys. *Old tads*, graybeards, old men. **1896** Ade *Artie* xi. 98 Oh, he's a great old tad. **1901** F. Norris *Octopus* i. v. 197 There's a little tad that was just born to be a lady. **1904** W. D. Nesbit *Trail to Boyland* 49 That handle has been broken since he was just a tad. **1928** S. Lewis *Man who knew Coolidge* i. 55 One of the bell-boys at the hotel, cute little tad, knew the town like a book. **1935** H. Davis *Honey in Horn* xxii. 370 I've handled more horses than this tad ever heard of. **1949** O. Nash *Versus* 131 The sea lion loves a loveable lad, An urchin, a gamin,

a tyke, a tad. **1974** W. Garner *Big Enough Wreath* vii. 93 Nowadays young tads think they know it all. **1983** *Sunday Times* 3 Apr. 33/2 The nuns picked me out when I was still a tad, groomed me for a scholarship.

3. A small amount; freq. used *advb.* in the expression *a tad*, a little, slightly.

1940 *Amer. Speech* XV. 448/1 *Tad*, a very small amount. 'I want to borrow a tad of salt.' **1969** L. Michaels *Going Places* 159, I tried to smile. 'You come back later, baby. I'm a tad indisposed.' **1976** *Time* 27 Sept. 39/2 'Pull 'er up a tad, please, mister,' said the nonchalant teen-ager pumping gas. **1977** *Time* 14 Mar. 28/3 White House watchers also think they can glimpse a tad of arrogance showing through the good ole boy pose. **1977** *Globe & Mail* (Toronto) 15 Dec. 8/2 Things are a tad hectic. **1979** D. Anthony *Long Hard Cure* xv. 116 Why don't we sit here on the veranda? There's a tad of breeze. **1980** *N.Y. Times* 12 Aug. A18/1 The Mayor's pitch is a tad exaggerated both on the law's certainty and on the roominess of New York's prisons.

taddy (tæ·di). *Sc.* Also **Taddy.** The name of *Taddy* and Co., of London, used *attrib.* and *absol.* to designate snuff manufactured by them. Also *Comb.*, as **taddy-box,** a snuff-box.

1869 A. Macdonald *Love, Law & Theology* viii. 118 Tapping his box, and inhaling a large pinch of his favourite Taddy. **1870** J. Nicholson *Idylls* 46 Some tea to the auld folk, tobacco or taddy. **1872** 'R. F. Bardinarus' *Arn at Flail* 9 But John took up the Taddy pouder. **1881** R. Ford *Humorous Sc. Readings* 67 Three or four heapit ladlefu's o' London taddy. **1897** J. Wright *Scenes Sc. Life* 5 Irish blackguard and taddy snuff mixed. **1907** N. Munro *Daft Days* xxxiv. 278 The Provost, who had just stepped into P. & A.'s for his Sunday sweeties, smiled tolerantly and passed his taddy-box. **1939** J. M. Dallas *Toakburn* 11 He got his best 'sneeshin' from Johnnie Bickles, who kept the genuine 'taddy'.

Tadjik, Tadzhik, varr. *Tajik.

‖ **tædium vitæ** (tī·diŏm vəi·ti, vī·təi). Also **tedium vitæ.** [L.: cf. Tedium.] Weariness of life; extreme ennui or inertia, sometimes regarded as a pathological state.

[**1618** J. Chamberlain *Let.* 14 Oct. (1939) II. 170 The Lord Clifton..tooke his paterne from your Secretarie of Utrecht to stab and mangle himself with a penknife..without any other shew of reason or cause, but even *vitæ tædio* (as he saide himself).] **1759** E. Young *Conjectures on Original Composition* 8 Both These are happy in *this*, that by fixing their attention on objects most important, they escape numberless little anxieties, and that Taedium Vitae which often hangs so heavy on its evening hours. **1796** T. Jefferson *Let.* 24 Apr. (1926) 86 My health has suddenly broken down, with symptoms which give me to believe I shall not have much to encounter of the *taedium vitae*. **1803** S. Owenson *St. Clair* xxi. 89 The dreadful oppression of the *tedium vitae*. **1826** *Reg. Deb. Congress U.S.* 30 Mar. 402 *Tedium vitæ* appears in Sunday Schools. **1855** *Newsp. & Gen. Reader's Compan.* 156 That *taedium vitae*, which springs from a consciousness of talents abused and opportunities lost. **1883** T. S. Clouston *Clin. Lect. Mental Dis.* xvii. 560 A cloud of vague depression rests on the man, who shuns society, falls off in fat, becomes restless and hypochondriacal, and feels strongly the *tedium vitæ*. **1891** O. Wilde *Pict. Dorian Gray* xi. 216 That ennui, that terrible *tædium vitæ*, that comes on those to whom life denies nothing. **1920** J. Huneker *Painted Veils* vi. 251 Her languour had not been dissipated; 'tædium vitæ', the doctor named it. **1940** 'G. Orwell' *Inside Whale* 159 Everyone with a safe £500 a year..began training himself in *taedium vitae*. **1958** L. Durrell *Mountolive* ix. 187 Even these simple motions of joining with the ordinary world of social habit and pleasure, of relieving the *taedium vitae* of his isolation, were all infected by the new knowledge. **1977** V. S. Pritchett *Gentle Barbarian* xii. 201 He..is suddenly attacked by the *taedium vitae*, the disgust with life, as a man who talks too well may easily be.

‖ **tae kwon do** (tē[1], tai kwɒn dō). Also **Tae Kwon Do, taekwondo,** etc. [Korean: see quot. 1967.] A Korean system of unarmed combat resembling karate.

1967 *Karate & Oriental Arts* Sept./Oct. 2 Taekwondo, which is just starting in this country..will be open to the same abuse as karate. *Ibid.* 27/1 Breaking the word Taekwondo down into its three parts we get: Tae—kick, Kwon—fist, Do—art, way, method. **1969** *Melbourne Truth* 12 July 9/4 Rozinsky gained his Black Belt in Tae Kwon Do at the Jidokwan, Seoul, South Korea. **1972** *Sunday Times* (Kuala Lumpur) 18 June 16/7 To unwind and also to keep fit, he sweats it out at tae-kwon-do classes. **1972** C. Weston *Poor, Poor Ophelia* (1973) xvi. 94 The newcomers to whom Tae Kwon Do seemed an impossible skill. **1976** *Eastern Even. News* (Norwich) 9 Dec. 2/6 Taekwondo (Korean karate) training, Duke Street Centre. **1979** *Sci. Amer.* Apr. 110/1 Karate is just one of a wide variety of martial arts that have evolved in the Orient, including tae kwon do, kempo and kung fu.

tael. Add: Also **tahil. 1.** (Further examples.)

1902 *Encycl. Brit.* XXXIII. 813/1 Tahil..Straits Settlements 1⅓ oz. av. = 10 chee = 100 hoon. **1947** R. O. Winstedt *Malays* vi. 112 Soon after the founding of Malacca Chinese annals under 1416 record..that, 'tin..is cast into small blocks weighing 1 *kati* 8 *tahil* or 1 *kati* 4 *tahil* official weight... They use these pieces of tin instead of money.' **1972** *Straits Times* 25 Nov. 15/1 The gold bars, weighing 15 katis seven tahils.

2. b. A Chinese gold coin based on the value of a tael of silver.

1926 E. Kann *Currency China* i. i. 13 Taiping tael gold coin... During the rule of the T'aipings in Nankin a gold coin was issued there..supposed to represent 25 taels of silver. **1962** R. A. G. Carson *Coins* 543 A rare tael in gold was also struck in this issue. **1979** *Courier-Mail* (Brisbane) 9 July 4/3 Some Chinese had sufficient savings tucked away in gold taels, the traditional, and sensible, way of saving adopted by many East Asian societies, to bribe officials or simply to pay for the right to escape.

tænii-. Add: Also (*U.S.*) **tenii-; tæni·asis** (pl. **-iases**) *Path.* and *Zool.* [*-iasis], infestation with tapeworms, esp. adult worms in (or formerly in) the genus *Taenia.*

1896 F. W. Gamble in *Cambr. Nat. Hist.* II. iii. 82 The Jewish observance with regard to swine is the surest preventive measure against taeniasis. **1900** Dorland *Med. Dict.* 675/1 Teniasis. **1969** Edington & Gilles *Path. in Tropics* iii. 173 Taeniasis may occur in all countries where beef or pork are eaten. The beef tapeworm—*Taenia saginata*—has a cosmopolitan distribution... The pork tapeworm—*T. solium*—is also widely distributed and its larval stage..produces cysticercosis in man. **1971** R. A. Marcial-Rojas *Path. Protozoan & Helminthic Dis.* xxxi. 618/1 The pathology and symptomatology of the taeniases in man vary according to the evolutionary stage of the parasite affecting him.

tæniodont (tī·ni,odɒnt). [f. mod.L. order name *Tæniodonta* (E. D. Cope 1876, in *Proc. Acad. Nat. Sci. Philadelphia* XXVIII. 39), f. Tænio- + Gr. ὀδούς, ὀδοντ- tooth.] A fossil mammal of the order Tæniodonta, related to the edentates and known from remains found in North America.

1933 A. S. Romer *Vertebr. Paleontol.* xiv. 278 Taeniodonts..were seemingly never common. **1949** B. Patterson in G. Jepsen et al. *Genetics, Paleontol. & Evolution* xiii. 243 (*title*) Rates of evolution in taeniodonts. **1979** M. J. Novacek in Fairbridge & Jablonski *Encycl. Paleontol.* 441/1 The taeniodonts..evolved long-clawed feet; ever-growing cheek teeth; short, broad skulls; and deep jaws.

tæniolite (tīni,ō[u]·ləit). *Min.* Also **tainiolite.** [f. Gr. ταινία band, ribbon + -o + -lite.] A rare colourless mica containing lithium and magnesium but without essential aluminium.

1899 G. Flink in *Meddelelser om Grønland* XXIV. 116 The name of tainiolite that I have given the mineral is derived from the Greek word ταινία, a band or strip, because the crystals always have the form of bands or strips. **1900** *Amer. Jrnl. Sci.* CLX. 324 Tainiolite (tæniolite) is a kind of mica occurring in elongated colorless crystals. **1938** *Amer. Mineralogist* XXIII. 110 Because of the absence of essential aluminum, taeniolite is unique among the micas. **1962** W. A. Deer et al. *Rock-Forming Minerals* III. 89 A very rare mica, taeniolite (ideal formula $K_2Mg_4Li_2Si_8O_{20}F_4$), has no aluminium... It may be regarded as a magnesian lepidolite although it could be classed also as a lithian phlogopite. **1968** I. Kostov *Mineralogy* 361 Tainiolite.

‖ **Tafelmusik** (tā·fəlmuzī:k). Also **tafelmusik, tafel musik.** [Ger., lit. 'table music'.] **1.** Music so printed that parts can be read from the same page by two or more persons seated on opposite sides of a table.

1876 Stainer & Barrett *Dict. Mus. Terms* 420/1 *Tafelmusik*,..table music. **1907** [see sense 2 below].

2. Music intended to be performed at a banquet or a convivial meal, esp. popular in the eighteenth century.

1880 Grove *Dict. Mus.* II. 400/1 The Tafelmusik, Nachtmusik, etc., [of Mozart] for wind instruments..often present the most extraordinary combinations. **1907** T. S. Wotton *Dict. Foreign Mus. Terms* 193 *Tafelmusik*, ..(1) Music intended to be sung or played at meal times. (2) Music so arranged that two persons seated at opposite sides of a table can sing from the same page. **1961** *Times* 13 Mar. 3/1 Like the more aimlessly gossipy *Tafelmusik* of an eighteenth-century composer. **1969** *Times* 29 May 8/5 The analogy was with contemporary *tafel musik* rather than the wilder shores of radicalism. **1971** G. Steiner *In Bluebeard's Castle* iv. 92 Much of this [eighteenth-century] music was, in fact, conceived as *Tafelmusik* and aural tapestry around the busy room. **1980** *Early Music* July 300/1 If you seek in your Italian madrigal an escape to remote and perhaps picturesque sonorities, then the last thing you want is to understand it well enough to know why it is different from, say, Gregorian chant or rococo *Tafelmusik*. **1980** *Times* 19 Aug. 7/3 One of Telemann's many pieces of *tafelmusik*.

‖ **Tafelwein** (tā·fəlvəin). Also **tafelwein.** Pl. **-e.** [Ger., lit. 'table wine'.] Wine of less than middle quality, suitable for drinking with an ordinary meal; = *table wine* s.v. *table *sb.* 22. Cf. *vin de table* s.v. *vin.

1972 *Times* 27 Nov. (Wines & Spirits Suppl.) p. iii/5 There will be three categories of wine: *tafelweine* for all the *vins ordinaires*; *qualitätswein* for the middle quality wines..; and *qualitätswein mit prädikat.* **1978** W. F. Buckley *Stained Glass* xix. 186 He flashed his light down and saw a half-empty case of white Tafelwein. **1980** A. Scholefield *Berlin Blind* iii. 183 A bottle of *tafelwein* half empty.

Taff (tæf). Abbrev. of Taffy[2]. Occas. applied also to women.

1929 F. Bowen *Sea Slang* 137 *Taff or Taffy*, any Welsh seaman, or one with a Welsh surname. **1943** [see *Aspic]. **1973** M. Amis *Rachel Papers* 81 She, Nanny, wasn't too bad: a red-faced, fat but strong-looking woman of about sixty-five or seventy. A Taff all right. **1977** *Listener* 25 Aug. 235/1 Taffs and Geordies and Scouses who were barely intelligible.

taffeta, taffety. Add: **B.** *attrib.* and as *adj.* **2.** *taffety cream*, substitute for def.: A dish of cream and eggs. (Earlier example.)

1723 J. Nott *Cook's & Confectioner's Dict.* sig. L1, *Taffaty Cream.* Beat the Whites of eight Eggs..with Rose-water,..put it into a Quart of cream.

taffy[1]. Add: **1.** (Later examples.) Freq. used in comparisons as a type of something which yields to pressure or can be stretched out into lengths.

1960 R. W. Marks *Dymaxion World of B. Fuller* 127/2 The wood die rises, actuated by the console controls, while the universal-jointed giant fists stretch the metal gutter piece like *taffy* around the wooden die's elliptical groove perimeter. **1974** K. Millett *Flying* (1975) v. 474 Each wonderful swatch of hair like a chunk of taffy stretched. **1979** *Sci. Amer.* Oct. 117/2 Below the interface the lava is a fluid that yields like taffy when a drill probe is pushed into it.

2. (Earlier example.)

1878 E. L. Wheeler *Buckhorn Bill* 2/1 Don't try to stuff that kind of taffy down me. *I* know better.

3. *taffy-coloured* adj.; **taffy apple,** a toffee-apple; **taffy pull, pulling,** an occasion on which young people assemble to make taffy.

1967 Mrs. L. B. Johnson *White House Diary* 12 Sept. (1970) 567 Stands dispensing hot dogs, Coca Cola, taffy apples, popcorn, and cotton candy. **1978** A. Maling *Lucky Devil* xxiii. 122, I threw the taffy apple stick away. **1939** L. M. Montgomery *Anne of Ingleside* xxxii. 232 Young Mrs David Ransome, with her taffy-coloured hair. **1970** J. Hansen *Fadeout* (1972) xi. 89 A taffy-colored cocker spaniel. **1883** I. M. Rittenhouse *Jrnl.* in *Maud* (1939) 159 We're going to have a taffy pull at our Y.P.T.A. Friday night. **1926** *One Big Union Bulletin* (Winnipeg) 19 Aug. 5/6 We had planned a taffy pull but the ball game was so prolonged that we only succeeded in making fudge. **1982** S. B. Flexner *Listening to America* 138 The taffy pulls..were a suitable face-to-face pastime for courting couples. **1863** M. B. Chesnut *Diary* 18 Dec. in C. V. Woodward *May Chestnut's Civil War* (1981) xx. 507 General Hood..wanted me to go to a taffy pulling at the Prestons'. **1912** *Out West* Mar. 166/2 He wrote with beautiful flourishes, little notes of regret..declining all socials, taffy pullings and croquet parties. **1959** R. Campbell *I Would do It Again* ii. 7 There was taffy pulling and all the other fun that goes with a picnic.

tafia. (Earlier example.)

1763 tr. *Le Page du Pratz's Hist. Louisiana* II. iv. iv. 266 At night you shall have a cup of *Tafia* (or rum) to give you strength and spirits.

tafoni (tæfō[u]·ni), *sb. pl. Geol.* [a. G. *tafoni* pl. (F. C. A. Penck *Morph. d. Erdoberfläche* (1894) I. 214), a. Corsican dial. *tafóni* pl., holes, hollows.] Shallow rounded hollows in rock produced by weathering.

1942 C. A. Cotton *Climatic Accidents in Landscape Making* i. 9 Rarely or never are positive signs of sand-blasting found in association with typical tafoni. **1968** R. W. Fairbridge *Encycl. Geomorphol.* 1103/1 Smaller rounded depressions may diversify the larger tafoni. **1970** R. J. Small *Study of Landforms* ix. 294 Cavity weathering, leading to the formation of rounded hollows ('tafoni') in granite, may also be due to chemical decay, since it occurs commonly in shaded areas where moisture is likely to linger.

tafrogenesis, var. *taphrogenesis.

taft, *sb.* Add: Also Comb.: **taft joint,** a joint between two pipes, made by tafting one pipe, shaping the other to fit into it, and soldering them.

1891 S. S. Hellyer *Plumbing* xvii. 91 There is no form of wiped soldered joints so easy to make as a *taft joint*. **1906** Goodchild & Tweney *Technol. & Sci. Dict.* 742/1 *Taft joint*, a blown joint. **1945** W. J. Woolgar *Pract. Plumber* iv. 102/2 The taft joints are seldom used in plumbing, because of the low tensile strength. **1972** J. Hastings *Plumber's Compan.* 146 A taft joint..requires little skill in the making. **1976** *Pract. Householder* Nov. 46/1 Plumber's solder can, of course, be used especially with the taft joint and finger wipe joint.

tag, *sb.*[1] Add: **8. a.** Also *fig.* = Tab *sb.*[1] 6 in Dict. and Suppl.

1961 *Times* 5 Jan. 4/3 After the interval Surrey drafted in extra men to help Prosser keep a tag on Farooq.

c. *Electronics.* A small metal projection to which a wire may be soldered or attached.

1919 R. Mordin *Strowger Automatic Telephone Exchange* ii. 34 The tags are arranged in ten sets of three rows, and pass completely through holes in the tag board, so that it is possible to wire the tags on either or both sides of the board. **1958** *Practical Wireless* XXXIV. 63/2 All earth leads on the pre-amplifier are taken to one point, actually to a soldering tag on the input coaxial socket. **1971** *Hi-Fi Sound* Feb. 71/1 Never, under any circumstances, solder connections to the tags with them already on the cartridge.

d. (See quot. 1935.) *N. Amer. slang.*

1935 A. J. Pollock *Underworld Speaks* 118/1 *Tag*, an automobile license plate. **1971** *Maclean's Mag.* Sept. 34/1 The license plates ('tags'), laws unto themselves, somehow manage to contradict and complement each other at the same time. **1976** *Billings* (Montana) *Gaz.* 18 June 7C/3 [They] observed a Thunderbird with Louisiana tags circling the block.

e. *Computers.* A character or set of characters appended to an item of data in order to identify it.

1948 *Theory & Techniques for Design of Electronic Digital Computers* (Moore School of Electr. Engin., Univ. of Pennsylvania) IV. xxxix. 1 To introduce..a new element called a stop order tag which may be attached to the words stored in the memory. **1961** Leeds & Weinberg *Computer Programming Fund.* v. 151 Bits 0, 1, and 2 (often called the prefix of the word) and bits 18 to 20 (called the tag) specify the operation. **1963** *IBM Jrnl. Res. & Devel.* VII. 337/2 If it is desired to translate the text with the aid of a microglossary, the text is preceded by a tag specifying the pertinent field. **1978** J. P. Hayes *Computer Archit. & Organization* iii. 149 The processor merely has to inspect the operand tags to determine the specific type of operation to be performed, e.g., a fixed-point double-precision addition.

f. An epithet; a label or popular designation. *colloq.*

1961 in Webster. **1972** *Times* 7 Aug. (Jamaica Suppl.) p. iii/4 The lost goodwill..and the loss of the tag of still being the safest Caribbean country for investment. **1976** *Daily Tel.* 20 July 3/2 The Black Panther tag, probably coined by the press, was the worst of it. **1982** *Oxford Star* 4–5 Feb. 3/2 Cassells doesn't let the tag of Third Division top scorer bother him too much.

g. A price (cf. *price-tag* s.v. Price *sb.* 14); hence, an account or bill. Cf. Tab *sb.*[1] 6 in Dict. and Suppl.

1968 *Globe & Mail* (Toronto) 3 Feb. B2/3 (*heading*) Petrofind raises fuel oil price, bulk gas tag. **1977** *Modern Railways* Dec. 474/2 BR stresses, too, that if there's a gulf between the price of the basic, low-cost vehicles customers have been using in old-style wagonload working and the tag on a late-1970s air-braked, 75 mile/h vehicle, there's a comparable contrast in the service obtainable. **1979** D. Meiring *Foreign Body* xviii. 197 Even if they went broke, the bank would pick up Sagr's crude-oil tag and pay it.

9. c. (Earlier examples.)

1717 J. Gay et al. *Three Hours after Marriage* i. 25 The tag of the Acts of a new Comedy. **1755** C. Charke *Life* 205 Concluding the Play with Jane Shore's Tag, at the End of the first Act of that Tragedy.

d. A musical phrase added to the end of a piece in composition or performance (see also quot. 1978). Esp. in *Jazz.*

1929 *N.Y. Times* 20 Oct. ix. 8/6 *Tag*, ending added to a musical composition. **1932** *Melody Maker* June 507/3 The tag..implies that this is a band record. **1943** *Riverboat Jazz* (Brunswick Records) 7 He comes in to play a tag—just a few notes. **1958** P. Tanner in P. Gammond *Decca Bk. Jazz* xi. 130 A tradition has grown up..of concluding with a short drum break and a tag ensemble coda. **1960** H. O. Brunn *Story Original Dixieland Jazz Band* v. 59 The Dixieland Band's stock ending, the 'dixieland tag', faithfully concluded every number. **1978** *Amer. Speech* 1975 L. 301 *Tag*, added ending of a song, often repeating the final words and designed to make a complete and satisfying arrangement.

e. *Linguistics.* An interrogative formula used to convert statements into questions. Cf. *tag question*, sense 13 below.

1957 *Publ. Amer. Dial. Soc.* xxviii. 17 An understanding of tags implies an understanding of sentence order and the role of accent. **1963** F. T. Visser *Hist. Syntax* I. ii. 175 *The type* 'oh, Biffin told you, *did he*? (or *He did*?)'. This type differs from that illustrated in the preceding section in the fact that statement and tag with *to do* are either both positive or both negative. **1977** *Language* LIII. 742 An auxiliary verb typically can appear in the tag of tag questions. **1973** *Archivum Linguisticum* IV. 69 Tag constructions can convey much to the discriminating listener.

12*. *slang.* A person who follows another as a detective or spy. Cf. *tag *v.*[1] 4 b, *tail *sb.*[1] 6 b.

1966 'A. Hall' *9th Directive* vii. 62 Who were the tags? The thin one, and the one with the splay-footed walk? **1972** J. D. Buchanan *Professional* v. 62 Guerin realized he had a tag... Guerin would walk and stop, the tag would do the same. **1979** 'A. Hall' *Scorpion Signal* xii. 139 Ignator went through the lights at yellow... I don't think he was going through on the yellow because he'd discovered the tag.

13. tag axle *N. Amer.*, a non-powered set of wheels on a truck, etc., attached so as to support extra weight; **tagboard,** (*a*) *U.S.*, a type of strong cardboard, used esp. for making luggage labels; (*b*) *Electronics*, a board of insulating material containing two or more parallel lines of tags (sense 8 d above), so that a component can be mounted between each pair; **tag day** *N. Amer.* = *flag-day* (*b*) s.v. *flag *sb.*[4] 7; **tag-end** (earlier example); **tag line** *U.S.* = *punch line; **tag-phrase,** an automatically repeated or over-used phrase; **tag question** *Linguistics*, a question formed by the appendage of an interrogative formula to a statement; a formula used in this manner (cf. sense 9 e above); **tag strip** *Electronics*, a strip of insulating material on which are mounted a line of tags (sense 8 c above); **tag-tail** (example of sense 'a parasite, a hanger-on').

1971 M. Tak *Truck Talk* 163 *Tag axle*, the hindmost axle of a tandem-axle tractor if that axle serves only to support additional gross weight. **1977** *Telegraph-Jrnl.* (St. John, New Brunswick) 1 June 3/5 He said in an interview that the Motor Vehicles Branch no longer allows extra weight when a third 'tag axle' is added to tandem drive trucks. **1912** *Walden's Directory of Papers* (Eastern ed.) p. liii, Paper and card board..translucents, tag boards, etc. **1952** E. J. Labarre *Dict. Paper* (ed. 2) 301/1 *Tag paper* or board is a very strong and tough product made on the Fourdrinier (Bristol), used for making the well-known luggage and shipping tags. **1956** *Wireless World* Mar. 125/1 A plain tagboard, carrying resistors and capacitors. **1973** G. Davey *Fun with Hi-Fi* iii. 25 (*caption*) Layout and tagboards of Mullard 510 amplifier. **1976** *National Observer* (U.S.) 23 Oct. 20/1 (Advt.), Each issue is 42 or more pages long, bound in sturdy tagboard. **1908** Tag day [see sense 8 a in Dict.]. **1916** *Daily Colonist* (Victoria, B.C.) 4 July 6/3 Friday, August 4, is to be tag day for the Italian Red Cross Society. **1949** *Courier-Journal* (Louisville, Kentucky) 3 Sept. 10/1 The conference agreed [upon]..a tag day on which Boy Scouts and Girl Scouts will solicit funds during the Kentucky State Fair. **1807** C. Wilmot *Let.* 15 May in *Russ. Jrnls.* (1934) ii. 245, I believe..we have been solemnising..the Tag end of those very May Day ceremonies which scandaliz'd *ould Cato* near two thousand years ago. **1926** G. Ade *Let.* 14 Sept. (1973) 113 The prosecutor asks: 'Do you know him?' She studies him carefully and then pulls the tag line: 'No, I don't recognize him at all.' **1941** B. Schulberg *What makes Sammy Run?* iii. 44 One of those long dirty stories for which the only justification would be the tag line at the climax. **1982** *Fortune* 6 Sept. 53/1 One recent ad. shows a stunning model wearing nothing but a solitaire diamond necklace. 'She can't flaunt a fur on the Côte d'Azur,' reads the tagline. **1933** R. Tuve *Seasons & Months* iv. 110 All these uses of the seasons-introduction appear and reappear, sometimes elaborately, sometimes in a mere conventional tag-phrase. **1963** Tag-phrase [see *goon-like* adj. s.v. *goon 5]. **1933** O. Jespersen *Essentials Eng. Gram.* xxviii. 304 Note especially tag-questions..like: He was angry, wasn't he? **1957** R. W. Zandvoort *Handbk. Eng. Gram.* v. ii. 224 A certain type of compound sentence, consisting of a statement followed by an *appended question* (or 'tag question') modelled on the main clause... You are not ill, are you? **1982** *Amer. Speech* LVII. 95 Lakoff..considers tag questions (*He can work, can't he?* and *He is honest, isn't he?*) as declaratives—assertions. **1942** *Electronic Engin.* XV. 238/2 Such tag strips are found in medium wave receivers, as well as in short wave apparatus. **1960** *Practical Wireless* XXXVI. 405/1 A tag-strip provides a convenient anchoring point for leads. **1834** C. A. Davis *Lett. J. Downing* 311 You are surrounded by such a raft of snuffle-nose, scabby set of tag-tails, that I can't have nothing more to do with you.

tag, *sb.*[2] Add: **2.** *Baseball.* The act of putting out a runner by touching him with the ball (or with the gloved hand holding the ball) while he is off base. Also *tag-out.* Cf. *tag *v.*[2] 2 a.

1941 *Baseball Mag.* Sept. 439/1 A big league infielder..confessed..'I've made the tag with the empty glove outstretched.' *Ibid.* 439/3 Some stars..use a two-handed tag. **1952** *N.Y. Herald-Tribune* 16 Aug. 11/1 Only Lockman's cut-off of Hartung's throw and the subsequent tag-out of Mathews at third averted further damage to the home forces. **1971** L. Koppett *N.Y. Times Guide Spectator Sports* i. 21 The rules forbid a runner to leave the 'base-path'—an imaginary direct line between bases—to avoid a tag.

B. as *adj.* Of, pertaining to, or designating a form of professional wrestling between single alternating representatives of two teams (usu. of two men each).

One team-member cannot enter the ring until the other tags or touches hands with him on leaving it.

1955 *Sun* (Baltimore) 16 May 16/7 (*heading*) 6-man tag bout tops mat card. *Ibid.* For the first time in the history of wrestling, a six-man tag team bout will be staged. **1963** *Economist* 7 Sept. 819/1 The confused spectacle of tag wrestling (four in a ring). **1966** *Times* 28 Feb. (Canada Suppl.) p. xiv/7 The average card in Canada has a tag match (two-man teams with the members taking turns to maul each other). **1972** J. Mosedale *Football* viii. 115 He teamed with his old idol Nagurski in tag team matches. **1974** *Greenville* (S. Carolina) *News* 23 Apr. 8/2 In other bouts, Sandy Scott and Johnny Weaver downed Gene Lewis and Bill White in a special tag team event.

‖ **Tag** (tāk), *sb.*[3] [Ger.] = *day *sb.* 9 c. Usu. *der Tag.*

1914, etc. [see *day *sb.* 9 c]. **1914** J. M. Barrie (*title*) Der Tag. **1916** O. Seaman *Made in England* 35 For now the psychologic *Tag* has come To put the final lid on Christendom. **1918** *Times* 9 Dec. 9/3 The wonderful day, the great Der Tag, Which Prussians had vow'd with unmannerly brag Should see Old England lower her flag. **1924** J. Buchan *Three Hostages* ix. 125 We'll fix the 10th of June for *Der Tag*... The round-up of all must be simultaneous. **1939** C. Day Lewis *Child of Misfortune* iii. iii. 290 You're saving it up for Der Tag... A time will come when those persons will be very sorry. **1966** P. Flower *Fiends of Family* xvi. 187 '*Der tag*,' Maggie said. 'At long last, the day of the great adventure.' **1975** tr. *Melchior's Sleeper Agent* ii. 133 When *der Tag* comes, when his usefulness is Kaput, we'll slap him in detention.

tag, *v.*[1] Add: **1. b.** Also *spec.*, to mark and record (animals) so that their migrations can be traced. (Later examples.)

1953 Scott & Fisher *Thousand Geese* vi. 58 The expedition was confined to camp, except for short dashes..to tag a few whooper cygnets. **1974** *Nature* 19 Apr. 642/2 Anglers tagged 954 bass..on the coast of Devon.

d. *Biol.* and *Chem.* = *label *v.* 2.

1939 *Amer. Jrnl. Physiol.* CXXVII. 557 The radioactivity 'tags' the atoms. **1947** *Ann. Rev. Microbiol.* I. 271 The foregoing method is..not limited to 'tagging' the antigen by means of glucosamine analyses. **1969** *Times* 9 Apr. 7/2 DNA sub-units tagged with radioactive marker atoms were fed to bacteria. **1977** *Sci. Amer.* July 46/3 The antigens were first visualized by tagging their antibodies with a fluorescent dye that could be seen under ultraviolet radiation.

e. *Computers.* To label (an item of data) in order to identify it for subsequent processing or retrieval.

1959 M. H. Wrubel *Primer of Programming for Digital Computers* iii. 56 We must..tag the instructions to be modified..so that those instructions and no others will be modified by adding the contents of the loop box. **1971** *Computers & Humanities* VI. 43 It is a simple matter to enter and tag automatically categories of information indicated by font and/or format... Such tagging is a part of the Dissly service. **1983** *Trans. Philol. Soc.* 33 This is a program which identifies and tags idioms which it finds in an Idiom list.

2. Also with *on.*

1916 T. MacDonagh *Literature in Ireland* 150 The first two verses of the better version..are essential poetry; the three that are tagged on in the song-books are no such thing.

4. a. (Further examples.) Freq. const. *after, along,* (*a*)*round, on.* Also *fig.*

1822 G. F. Cooper *Spy* II. xii. 307 Pooh! Pooh!..if you tag after a troop of horse, a small bit of a joke must be borne. **1897** Kipling *Captains Courageous* ix. 214 Don't go taggin' araound after them whose eyes bung out with fatness. **1900** Ade *More Fables in Slang* (1902) 113 The men..wanted to Tag along, but Clara drove them back. **1930** J. Dos Passos *42nd Parallel* ii. xi. 164 She followed his talk breathless the way she used to tag along after Joe and Alec down to the carbarns when she was little. **1933** D. L. Sayers *Murder must Advertise* iii. 41 He used to tag round with that de Momerie crowd. **1946** 'P. Quentin' *Puzzle for Fiends* (1947) ix. 70 So you're ready to tag along with me, eh, Gordy? **1948** C. Day Lewis *Otterbury Incident* 43 Toppy's kid sister..tagged on, which was rather a bore. **1957** *Economist* 23 Nov. 661/2 There is a Yemeni home public. Its upper crust has been most critical of the recent tendency to tag along with Egypt and do deals, including an arms deal, with Russia. **1960** S. Barstow *Kind of Loving* i. vi. 126 Two or three more people sitting outside the room where they actually take the blood. Me and the Old Man tag on to the line. **1960** L. Cooper *Accomplices* i. iv. 37 He was sick of the sight of those damned Batemans... Couldn't we ever go anywhere without them tagging on? **1960** C. Day Lewis *Buried Day* ii. 44, I would tag around with him, hardly understanding a word he said because of his thick East-Anglian dialect. **1973** J. Pattinson *Search Warrant* v. 80, I guess I'll tag along. Just for the ride.

b. Also *spec.*, to follow as a detective or spy. Cf. *tag *sb.*[1] 12*, *tail *v.*[1] 5 b.

1966 'A. Hall' *9th Directive* vii. 63 Why did you tag me here? **1975** —— *Mandarin Cypher* viii. 123 If I thought I was tagged here because Chiang had blown me I was wrong.

6. (Further examples.)

1853 T. D. Price *Diary* 17 Mar. (MS.), Tagged the ewes in the forenoon. **1863** H. S. Randall *Pract. Shepherd* iii. 141 Tagging sheep before they are let out to grass.

tag, *v.*[2] Add: **1.** (Examples.)

1878 F. H. Hart *Sazerac Lying Club* 166 One of them, who had been 'tagged' seven times in succession, got tired, and proposed to change to playing house. **1891** *Jrnl. Amer. Folk-Lore* IV. 222 One player, who is 'it', attempts to tag, or touch, one of the other players. **1969** I. & P. Opie *Children's Games* ii. 64 In Monmouthshire, Gloucestershire, and Oxfordshire, they speak of 'tagging' each other.

2. *Baseball* and *Softball.* **a.** To put out (a runner) by touching him with the ball (or with the gloved hand holding the ball) while he is off base. Also with *out.* Cf. *tag *sb.*[2] 2.

1907 'B. L. Standish' *Dick Merriwell's Magnetism* xxxviii. 243 He tagged Spratt, and this made the second man out. **1944** E. S. Gardner *Case of Black-Eyed Blonde* 64 Keep cutting corners, Mason, and I'm going to catch you off first base one of these days, and then I'll tag you out. **1971** L. Koppett *N.Y. Times Guide Spectator Sports* i. 21 No one is attempting to tag him. **1976** *Billings* (Montana) *Gaz.* 28 June 1-C/2 (*caption*) Dave Konzen, of Buck's Bar slow pitch softball team, is tagged out as he slides against Heidelberg of Tacoma, Wash. **1982** S. B. Flexner *Listening to America* 34 Someone had the bright idea of forcing the runner out by throwing the ball to the base ahead of him or by tagging him with the ball rather than throwing the ball at him.

b. intr. *to tag up*: of a runner to (return to and) touch one's base after a fly ball is caught.

1942 *Baseball Digest* Dec. 52 Fletcher tagged up at third after the catch and then started for the plate. **1971** L. Koppett *N.Y. Times Guide Spectator Sports* i. 20 The runner 'tags up', waits at his base until the ball is caught, and still beats the throw to the next base. **1978** G. Wright *Illustr. Handbk. Sporting Terms* 89/2 If the ball is caught..the base runners, unless tagging up.., may not advance.

c. *trans.* To make a hit or run off (a pitcher).

1961 in Webster. **1974** *Greenville* (S. Carolina) *News* 23 Apr. 8/5 Seaver was taken out of the game after being tagged for hits by the first two batters in the Pittsburgh sixth.

Tagalog (tāgā·lɒg), *sb.* and *a.* Also † **Tagal, -la, -lian, -lic, -loc.** [Tagalog, f. *tagá* native to + *ilog* river; cf. Sp. *tagalo.*] **A.** *sb.* **a.** (A member of) a people living in the neighbourhood of Manila in the Philippine Islands. **b.**

The Austronesian language spoken by this people, an official language of the Republic of the Philippines.

1704 tr. *Careri's Voy. round World* in A. & J. Churchill *Coll. Voy. & Trav.* IV. 430/2 From these are descended the Tagalians, which are the Natives of Manila and the Country about it. *Ibid.* 446/1 The Languages are so numerous, that there are six in the only Island of Manila, which are Tagalian, Pampangan, [etc.]. **1808** *Asiatick Researches* X. 207 The Tagala or rather Tā-Gála or the Gala language is among the Philippines, what the Malayu is in the Malay islands. *Ibid.* 213 With respect to the original literature of the Tāgalás, the accounts of the Spanish missionaries are rather discordant. **1814** J. Maver *Martinez de Zuñiga's Hist. View Philippine Islands* I. p. xi, In respect to the aboriginal language, or Tagalic, very slight attempts have been made to trace it beyond the quarter in which it was found to prevail. Our author..draws the conclusion that the Tagalic language and original population of all the islands westward of the coast of South America derive from that continent. **1834** W. Marsden *Misc. Wks.* 39 Of these dialects six are considered as meriting distinction..; they are the *Tagala* or *Tagalog* [etc.]. **1840** *Penny Cycl.* XVIII. 88/2 The Malays are divided into a great number of tribes, of which that called Tagala occupies the neighbourhood of Manila. **1885** *Encycl. Brit.* XVIII. 753/1 First among these rank the Tagals... Their language (Tagalog) especially has made extensive encroachments on the other Philippine tongues since the conquest. **1919** F. R. Blake in C. F. Hockett *Leonard Bloomfield Anthol.* (1970) 82 This work contains an extended treatment of Tagalog, the most important native language of the Philippine Islands. **1933** L. Bloomfield *Language* vi. 105 Even simpler is the *three-vowel* system which appears in some languages, such as Tagalog. **1974** *Encycl. Brit. Micropædia* IX. 764/3 Most Tagalogs are farmers. **1976** 'G. Black' *Moon for Killers* vi. 83 A verbal exchange..starting in English, shifting to Spanish, and then apparently getting down to the real obscenities in Tagálog.

B. *adj.* Of or pertaining to this people or their language.

1808 *Asiatick Researches* X. 208 The Tāgála alphabet consists of seventeen letters. **1814** [see the sb. above]. **1859** J. Bowring *Philippine Islands* xiii. 219 What is the Tagáloc language? **1906** *Jrnl. Amer. Folk-Lore* XIX. 191 (*heading*) Philippine (Tagalog) superstitions. **1959** N. Mailer *Advts. for Myself* (1961) 131 Miguel said something to the other Filipinos in the Tagalog language. **1978** M. B. Hooker *Conc. Legal Hist. South-East Asia* viii. 215 The *Maragtas* text..is found in a recension dated A.D. 1650 written in the Tagalog script.

ta·g-along, *a.* and *sb. N. Amer. colloq.* Also **tagalong.** [f. vbl. phr. *to tag along*: see Tag *v.*[1] 4 a in Dict. and Suppl.] **A.** *adj.* **a.** Designating that which is towed or trailed behind something else. **b.** Applied to an uninvited follower.

1944 *Sun* (Baltimore) 21 Jan. 2/3 Evidence of trailed, or 'tag-along', bombs still is scanty. **1960** *Newsweek* 20 June 91/1 The tag-along highway trailers are delivered to..the bus terminals. **1973** *Islander* (Victoria, B.C.) 10 June 16/1 The small trailer snug beside it like a tagalong pup. **1974** *Spartanburg* (S. Carolina) *Herald-Jrnl.* 21 Apr. c9 (Advt.), Also all types of trucks and truck tractors, all types of trailers including low-boy and tag-A-long.

B. *sb.* An unwelcome, uninvited, or neglected companion.

1961 in Webster. **1967** *Boston Sunday Herald* 2–8 Apr. 45/1 The whole family suffers when Ernie becomes a dreadful tagalong. **1974** *Publishers Weekly* 28 Oct. 46/1 Seems to have spent most of her life as a 'tagalong' to a man who, as test pilot and astronaut, was seldom home. **1977** *Islander* (Victoria, B.C.) 16 Jan. 8/1 (*heading*) Are Victoria women just tag-alongs in motor sports?

tagetes (tădʒī·tīz). [med.L. (L. Fuchs *De Historia Stirpium* (1542) 48), f. *Tages*, name of an Etruscan deity.] An annual or perennial herb of the genus of this name, belonging to the family Compositæ and native to South and Central America; esp. = Marigold 1 b.

1792 *Curtis's Bot. Mag.* V. 150 (*heading*) Spreading Tagetes, or French Marigold. **1895** W. Robinson *Eng. Flower Garden* (ed. 4) 780/2 There are also perennial Tagetes, but they are not hardy enough to make satisfactory plants out-of-doors. **1926** *Contemp. Rev.* Feb. 233 A thick mass of petunia and tagetes..in blossom. **1962** R. Page *Educ. Gardener* vii. 220 French gardeners are used to cultivating huge batches of..several varieties of tagetes. **1975** *Country Life* 13 Feb. 388/2 Tagetes marigolds are not at all bad in shade. **1980** L. Mantell *Murder or Three* ii. 23 Small shrubs..and an edging of velvety gold and brown Tagetes.

tagged, *ppl.* a. Add: **1.** **d.** Of an animal: marked to help study of its habits or migrations.

1927 *Zoologica* IX. 204 Every tagged frog was given a new page. **1979** *Fisherman's Weekly* 21 June 4/3 More than 400 of the tagged brown and rainbow trout released into Draycote Water by fisheries officers of the Severn-Trent Water Authority have already been notified.

e. Fastened *on*, appended. Cf. *tag *v.*[1] 2.

1982 *N. & Q.* Feb. 80/1 A tagged-on chapter on 'Critical History' runs only to six pages.

f. *Computers.* Marked or labelled with a 'tag' (*tag *sb.*[1] 8 e).

1983 *Trans. Philol. Soc.* 29 A tagged corpus..provides a head start for anyone undertaking more advanced linguistic analyses of the corpus.

6. *Biol.* and *Chem.* = *labelled *ppl. a.* d.

1945 *Jrnl. Sci. Instruments* XXII. 23/1 Tagged atoms are used to enable the investigator to see where the rest of his material is going. **1955** *Sci. News Let.* 2 July 15/1 When a plant is supplied with isotopically labeled nitrate ..the 'tagged' element rapidly spreads throughout the tissues and is incorporated into all the major nitrogen fractions.

taggeen (tăgī·n). *Anglo-Ir.* [ad. Ir. *taidhgín.*] A small cup or glass (of spirits); a 'dram'.

1899 Somerville & 'Ross' *Some Experiences Irish R.M.* i. 10 'There's no bath in the house, sir.. but..would ye like a taggeen?' This alternative proposal proved to be a glass of raw whisky. **1936** M. Franklin *All that Swagger* iii. 35 Doing the dirty work of some cowardly crawler, who's bought you for a plug of tobacco and a taggeen of rum.

tagger[1]. Add: **4.** Now also *sing.*, chiefly in *attrib.* use.

1938 *Shelf Appeal* July 26/1 The tagger-top in its present form, with a cutter in the lid. **1959** *Gloss. Packaging Terms* (*B.S.I.*) 55 *Lever ring and tagger*, a lever type closure comprising a diaphragm of tagger tinplate or aluminium secured (together with the ring) to the body... *Cutter lid*, an outer lid containing a steel cutter with which the thin tagger tinplate top of the body is pierced and cut away.

tagger[2]. (Example.)

1969 I. & P. Opie *Children's Games* i. 23 One person is the tagger and has to count to thirty.

tagging, *vbl. sb.* Add: **a.** Also *spec.* (*a*) the marking of animals as an aid to study of their migrations; (*b*) *Computers*, the action of *tag *v.* 1 e.

1927 *Zoologica* IX. 201 (*title*) Frog tagging: a method of studying anuran life habits. **1953** Scott & Fisher *Thousand Geese* vi. 65 We worked away in camp at journals, nest records, tagging data. **1960** *Guardian* 25 Oct. 8/4 To study the movement and growth rates there is an elaborate system of fish tagging carried out at sea. **1972** *Even. Telegram* (St. John's, Newfoundland) 24 June 11/1 Tagging studies have shown most salmon intercepted are headed for the northern river spawning beds.

1948 *Theory & Techniques for Design of Electronic Digital Computers* (Moore School of Electr. Engin., Univ. of Pennsylvania) IV. xxxix. 20 This tagging is done by the little cycle..until the sentinel..trips on a coincidence and causes the control to go on to the next stage of computation. **1972** *Computers & Humanities* VII. 5 The study of *Automatic Grammatical Tagging of English*.. describes the theory and method of parts-of-speech tags, procedures used in tagging, and the context frame test employed. **1983** *Trans. Philol. Soc.* 33 The tagging of the LOB Corpus is due for completion by September 1983.

Taghairm (tŏ·yĕrm, tŏ·'ĕrm). *Sc.* [Gael.] A method of divination formerly practised in the Scottish Highlands (see quots.).

1774 T. Pennant *Tour in Scotl. & Voy. Hebrides 1772* 311 A vast cataract, whose waters falling from a high rock, jet so far as to form a dry hollow beneath... One of these imposters was sowed up in the hide of an ox, and.. was placed in this concavity: the trembling enquirer was brought to the place, where the shade, and the roaring of the waters, encreased the dread of the occasion. The question is put, and the person in the hide delivers his answer, and so ends this species of divination styled *Taghairm*. **1810** Scott *Lady of Lake* iv. iv. 146 Brian an augury hath tried, Of that dread kind which must not be Unless in dread extremity, The Taghairm call'd; by which, afar, Our sires foresaw the events of war. **1906** *Athenæum* 2 June 668/3 Another saying, 'Keep the cat turning', refers to the horrid practice of the Taghairm, or divination by the cat. **1953** *Scots Mag.* Dec. 223 Taghairm was, indeed, a magical means of compelling spiritual presences to grant desirable and valuable boons to the sorcerer who invoked them.

tagliarini (talyarī·ni). [ad. It. *taglierini* sb. pl. (also used); cf. *tagliatelle.] Egg noodles cut into very narrow strips.

1846 [see *ravioli]. **1899** J. Ross *Leaves from our Tuscan Kitchen* 69 Repeat the alternate layers of tagliarini, cheese and butter, until the dish is full. **1943** A. Simon *Conc. Encycl. Gastron.* IV. 64/1 Italian pastes..known.. by different names such as Macaroni, Vermicelli,.. Tagliarini, Tagliatelle, [etc.]. **1964** E. H. & M. O. Knopf *Food of Italy* ii. iii. 184 To make tagliarini. Proceed as for lasagne. **1982** G. Bugialli *Classic Techniques Italian Cooking* vi. 138 The finely cut pasta, *taglierini* almost as fine as angel's hair, is appropriate to very delicate sauces and to broths and soups.

tagliatelle (talyate·le). Also **tagliatelli** (-ī). [It., sb. pl., f. *tagliare* to cut.] Egg noodles cut into ribbons. Also *fig.*

1899 J. Ross *Leaves from our Tuscan Kitchen* 69 (*heading*) Tagliatelle with ham. **1926** R. Hall *Adam's Breed* i. iii. 21 There were paste from Naples..Tagliatelle, Gnocchi, [etc.]. **1934** [see *pasta]. **1957** G. Smith *Friends* 9 In Rome, where the *tagliatelli* had disagreed with him. **1967** [see *rigatoni *sb. pl.*]. **1977** *Times Lit. Suppl.* 4 Feb. 120/2 All those tapes, those monstrous forkfuls of magnetized tagliatelle..would lead to definite strangulation. **1980** T. Holme *Neapolitan Streak* 100 She ladled *tagliatelle*..on to the plates of her family. **1983** *Listener* 13 Jan. 5/2 That sentence would have wound up on the cutting-room floor, another piece of inedible Grundig tagliatelle.

tagma. Add: **2.** *Zool.* Each of the morphologically distinct regions, comprising several adjoining segments, into which the bodies of arthropods and some other metamerically organized animals are divided. Chiefly *pl.* Hence **tagmo·sis,** the formation of tagmata.

1902 E. R. Lankester in *Encycl. Brit.* XXV. 691/2 It is convenient to have a special word for..regions of like meres, and we call each a tagma (τάγμα, a regiment). The word 'tagmosis' is applicable to the formation of such regions. **1935** R. E. Snodgrass *Princ. Insect Morphol.* iv. 80 Tagmosis is more variable in the Crustacea (than in insects); in the Chilopoda and Diplopoda it results only in the formation of a head, including the gnathal segments, and a body. **1980** C. Gillott *Entomol.* iii. 54 The basic segmental structure is frequently obscured as a result of tagmosis. In insects three tagmata are found: the head, the thorax, and the abdomen.

3. *Linguistics.* **a.** A feature of grammatical arrangement or syntax.

1949 *Archivum Linguisticum* I. 1 Such distinctions as that of morpheme and 'tagma' as the constituents of the syntagm provide the necessary correction. But there is a different sense in which morpheme and tagma may share in the expression of the meaning of a syntagm: a feature of meaning may be distributed over both.

b. In tagmemics, the smallest meaningful unit of grammatical substance (contrasted with tagmeme).

1964 R. E. Longacre *Grammar Discovery Procedures* i. 46 Copy the data from the filing slips onto charts: (a) There should be a column for each tagma, i.e. for each tentatively identified function-set correlation. **1969** W. A. Cook *Introd. Tagmemic Analysis* vii. 187 Tagmemics is a grouping process, which involves human judgment, an attempt to group tagmas into units essential to the language, as the language appears to the native speaker.

tagmeme (tæ·gmīm). *Linguistics.* [f. Gr. τάγμα arrangement (see Tagma) + *-eme.]

1. The smallest meaningful unit of grammatical form.

1933 L. Bloomfield *Language* x. 166 In the case of lexical forms, we have defined the smallest meaningful units as morphemes, and their meanings as sememes; in the same way, the smallest meaningful units of grammatical form may be spoken of as tagmemes, and their meanings as episememes. **1950** S. Potter *Our Language* 86 Beginning with the phoneme, philologists pass on to speak about morphemes, taxemes or tagmemes.

2. The correlate of a grammatical function and the class of items which can perform it.

1943 K. L. Pike in *Language* XIX. 69 Somewhat diffidently I suggest the following classifications and relabelings as perhaps being a bit easier to handle than Bloomfield's... *Tagmeme*, a composite view of the basic composite taxemes of a linguistic form, at any one specific layer of structure. E.g. the total arrangement features of the form *duchess* considered as a single entity. **1957** —— in *General Linguistics* III. 29 In future work, therefore, we are adopting the term *tagmeme*. It should be noted, however, that our definition of this term is sharply different from Bloomfield. **1968** *Language* XLIV. 190 Another basic concept in tagmemic analysis is the consistent distinction observed between obligatory and optional tagmemes. **1969** S. Potter *Changing English* viii. 163 Every sentence is a frame into which syntactic units, or tagmemes, are fitted. When a word is forced into an unusual tagmemic slot, it is said to undergo grammatical conversion or functional shift. **1973** *Amer. Speech 1970* XLV. 135 The highest level tagmeme operates at the level of the T-unit. **1981** *Word 1980* XXXI. 232 The infrastructure of English contains seven slots or tagmemes and that of German six.

tagmemic (tægmī·mik), *a. Linguistics.* [f. *tagmeme + -ic.] Of or pertaining to tagmemes or tagmemics.

1958 K. L. Pike in *Internat. Jrnl. Amer. Linguistics* XXIV. 275 In order to demonstrate another crucial difference between our tagmeme and that of Bloomfield it is necessary to indicate the manner in which slot and distribution class are relevant to our tagmemic view. **1964** R. H. Robins *Gen. Linguistics* p. xx, P. Postal..sets out a vigorous criticism of both immediate constituent and tagmemic analysis. **1968** *Language* XLIV. 190 Tagmemic theory provides a tight hierarchical scheme for grammatical description. **1969** [see *tagmeme 2]. **1978** *English Jrnl.* Dec. 66 A clause analysis technique reflecting a sketch of the core grammatical system of English based on a tagmemic model. **1981** *Word 1980* XXXI. 231 He.. adopts the generative tagmemic approach as the basis of his work.

tagmemicist (tægmī·misist). *Linguistics.* [f. *tagmemic(s + -ist.] A student or exponent of tagmemics.

1965 *Language* XLI. 640 It would be easy to go farther and say that transformationists represent the extreme of preoccupation with linguistic theory, while tagmemicists represent the extreme of practical concern. **1972** *Times Lit. Suppl.* 22 Sept. 1116/2 Like the units of the tagmemicists, Šaumjan's categories are a combination of form and function. **1977** *Language* LIII. 247 The linguistic study of discourse or intrasentential relations has been carried out by such diverse scholars as..tagmemicists (e.g. Pike 1967).

tagmemics (tægmī·miks), *sb. pl.* (const. as *sing.*). *Linguistics.* [f. *tagmeme: see -ic 2.] The study and description of language in terms of tagmemes; *spec.* a school of linguistic analysis, based on the work of Kenneth L. Pike (b. 1912), which stresses the functional and structural relations of grammatical units.

1947 C. F. Hockett in *Language* XXIII. 274 This term [sc. *tactics*] seems simpler than 'taxemics' or 'tagmemics', which one might derive more directly from Bloomfield's labels. **1958** K. L. Pike in *Internat. Jrnl. Amer. Linguistics* XXIV. 273 Tagmemics, as I see it, should work with neither of these schematic views by itself. **1964** *Language* XL. 314 The similarity of Firthian linguistics to American slot-and-filler grammatical description, notably tagmemics, has already been noted. **1967** R. H. Robins *Short Hist. Linguistics* viii. 212 In thus employing semantics diagnostically, and in severely modifying immediate constituent structures in syntax, tagmemics marks its major divergencies from 'Bloomfieldian' grammatical analysis. **1975** M. A. K. Halliday in S. Rogers *Children & Lang.* iv. 225 With the now general recognition of the basically tri-stratal nature of the linguistic system (and Prague theory, glossematics, system-structure theory, tagmemics, stratification theory and the later versions of transformation theory are all variants on this theme), the semantic perspective has been restored. **1981** *Word 1980* XXXI. 230 He makes use of the slot-and-filler infrastructure, characteristic of tagmemics.

tag-rag, *sb.* **a.** (Later example.)

1870 J. Patrick *Let.* 10 Nov. in D. O. Hunter *Life Marquess of Bute* (1921) v. 96 At the funeral the Rothesay tag-rag outside cheered me as I left the churchyard.

‖ **taharah** (tahărā·). Also 9 **tohoro; tahara.** [a. Heb. *ṭohŭrāh* purification, cleansing.] A Jewish ceremony of washing a corpse before burial.

1819 L. Alexander *Hebrew Ritual* 188 Those who are drawn by lot..to attend, in order to wash the corpse, put on the shrouds... This ceremony is called..*Tohoro*, that is, the cleansing. **1902** *Laws & Bye-laws Burial Soc. United Synagogue* 49 (Index), Tahara men, dismissal of... Tahara women. **1932** C. Roth *Hist. Marranos* vii. 190 The traditional *taharah*, or ritual laving of the body. **1964** H. Rabinowicz *Guide to Life* iii. 38 The utmost respect must be shown to the body during *Taharah*... The body is laid on the *Taharah* board... Warm water..must then be poured down the body. **1974** *Jewish Chron.* 1 Nov. 10/1 (*heading*) Tahara helpers required. *Ibid.*, The ministers.. had expressed their willingness to be trained in tahara and to help out.

tahil, var. Tael in Dict. and Suppl.

‖ **tahina** (tahī·nă). Also **tahine, tahini,** etc. [Arab., f. *ṭaḥana* to grind or crush.] A paste or sauce made of sesame seeds, much eaten in the Middle East.

1950 E. David *Bk. Mediterranean Food* 146 *Tahina* is a thick white oil made from pounded sesame seeds. It is served in a bowl and eaten by dipping bread... You buy the white *tahina* from the grocer, and it is then thinned with water. **1968** C. Roden *Bk. Middle Eastern Food* 35 *Tahina* itself is a paste made from sesame meal, and can be found in all Greek stores. *Ibid.*, Serve the *tahina* cream in a bowl. **1971** W. Tucker *This Witch* (1972) ii. 14 The Arab cook had added a small dish of taheena to whet my appetite. **1972** *Vogue* Feb. 33/1 Moroccan tahine dishes with cone lids for couscous, £1.10. **1976** *Ibid.* Jan. 88/2, I like them [sc. chick peas] best made into *hummus*, mixed to a smooth paste with *tahini*, lemon juice and garlic. **1976** *Islander* (Victoria, B.C.) 1 Aug. 5/1 They saw a tahina factory where sesame oil is made, the peanut butter of the Middle East. **1979** *Guardian* 8 June 9/4 In Morocco it is either couscous or tajine. **1980** C. Smith *Cut-Out* xvi. 120 Both Palestinians ate sparingly..mining modestly into the *hommas* and *tehineh* dips with their *pitta* bread.

Tahiti (tahī·ti). The name of an island in Polynesia used *attrib.* in **Tahiti arrow-root,** a starchy powder made from the tubers of *Tacca pinnatifida*; **Tahiti chestnut** = *ivi.

1861 Tahiti arrow-root [see *Otaheitean *a.* and *sb.*]. **1884** *Encycl. Brit.* XVII. 664/2 Tahiti chestnut. *Inocarpus edulis*. S. Sea Islands. **1974** G. Usher *Dict. Plants used by Man* 319/1 Polynesian Chestnut, Tahiti Chestnut... The seeds are..eaten raw or cooked. *Ibid.* 564/1 The tuber yields a rather indigestible arrowroot (Fiji Arrowroot, Tahiti Arrowroot,..).

Tahitian (tahī·ʃăn, tahī·ti,ăn), *a.* and *sb.* Also **Tahitan, † Taitian, † Taitienne.** [f. *Tahiti + -an; cf. *Otaheitean *a.* and *sb.*] **A.** *adj.* Of or pertaining to Tahiti, its inhabitants, or their language.

1822 tr. *Malte-Brun's Universal Geogr.* i. xxiii. 572 Tagalic, Taitienne languages, &c... The Taitian is distributed through all the small islands of the Great Ocean. **1825** W. Ellis *Jrnl. Tour Hawaii* 244 Both in the Hawaiian and Tahitian languages, every syllable, and every word, ends with a vowel. **1847** *Dublin Rev.* Dec. 357 Numerous other matters of European manufacture.. were strewn about among..the ordinary furniture of a Tahitian dwelling. **1852** J. Crawford *Gram. & Dict. Malay Lang.* I. p. cclix, The names of the three plants are exactly according to the Tahitian pronunciation. **1921** tr. *Rathenau's New Society* iv. 26 When a European artist writes or paints in Tahiti, what he produces is not a work of Tahitian culture. **1958** [see *cargador]. **1980** *London Mag.* July 23 The ochre Tahitian soil of Gauguin's paintings.

B. *sb.* **a.** A native or inhabitant of Tahiti. **b.** The Polynesian language spoken in Tahiti.

[**1822**: see the adj. above.] **1825** W. Ellis *Jrnl. Tour Hawaii* 49 He [sc. Ellis] could not help stating to them the striking identity between theirs [sc. their traditions] and those of the Tahitians. **1854** J. Crawfurd in C. Bunsen *Outl. Philos. Universal Hist.* I. 427 A sentence in the Maori and Tahitian can be written in words common to both. **1859** N. Wiseman *Twelve Lect. Sci. & Rev. Relig.* (ed. 6) I. 186 Charlevoix observed it among the Esquimaux,.. Wallis among the Tahitans. **1914** R. Brooke *Let.* Apr. (1968) 572 My Greek is something rusty. Had it been Tahitian now, or Fijian. **1918** L. Huxley *Life Sir J. D. Hooker* II. 483 He gratified Banks's philanthropic zeal by leaving in his care two Tahitians and two Maoris. **1957** P. Worsley *Trumpet shall Sound* i. 30 Sects including more or less of these elements have also appeared amongst the Tahitians. **1969** J. H. Vance *Deadly Isles* (1970) iii. 22 He would forget the songs and his few words of Tahitian. **1980** *London Mag.* July 24 Stevenson..and two Tahitians lolling among a pile of coconuts.

tahr (tāɹ). Now the usual spelling of Tehr. Substitute for def.: A wild goat of the genus *Hemitragus*, found in mountainous regions of southern Asia or Arabia, esp. *H. jemlahicus*, which has long brown fur and is native to the Himalayas. (Later examples.)

1902 *Little Folks* Apr. 282/1 The tahrs are true goats, though..they have no beard. Their home is..high up in the Himalaya Mountains. **1939** *Proc. Prehistoric Soc.* V. 52 A muscular development and agility in the feet commonly met with in the chamois,..the tahr and others of their kind. **1959** W. Thesiger *Arabian Sands* xiii. 256 The Arabian tahr had never previously been seen by a European. **1972** T. McHugh *Time of Buffalo* xvii. 200 And goats, sheep, and Himalayan tahrs stamp their front hoofs under similar circumstances [sc. in rut].

‖ **tahsil** (tāsī·l). Also **tehsil, † tuhseel.** [Urdu; cf. Tahsildar.] In India and Pakistan, an administrative division comprising several villages; formerly *spec.* a division made for purposes of revenue administration.

1846 *Directions for Collectors of Land Revenue N.W. Provinces* (India) App. VI. p. xxx, Comparing the tuhseel monthly accounts..with the cancelled dustuks. **1881** *Encycl. Brit.* XII. 769/2 Broadly speaking, the subdivision is characteristic of Bengal..and the *tahsil* of Madras. **1921** *Glasgow Herald* 19 Dec. 12 All district Congress Committees..must send out parties of about 20 Volunteers as a patrol daily in every town, tahsil, and village. **1944** Visct. Wavell *Let.* 7 Mar. in *Transfer of Power in India* (Foreign & Commonw. Office) (1973) IV. 784 Whilst I was at Nagpur I visited a village, a rural tahsil office, and a small irrigation work. **1954** O. H. K. Spate *India & Pakistan* p. xxiii, The States of India are divided (if large enough) into Divisions and these into Districts... Districts are subdivided into *taluks* (*taluqs*) or *tahsils* (*tehsils*) normally from 3 to 8 to a district. **1962** *Courier-Mail* (Brisbane) 14 Aug. 2/7 The four-tier scheme of basic democracies—at the village, tehsil, district and division levels. **1968** N. Mitchell *Sir G. Cunningham* v. 100 From 14th to 18th November he was touring on horseback in the Nowshera and Swabi Tahsils. **1971** *Illustr. Weekly India* 18 Apr. 7/2 A very senior and respected elder of our *tehsil* tried to stop me to ask me something. **1975** *Bangladesh Observer* 25 July 4/1 (Advt.), For Tahsil copy in the left hand side of the volume there will be a pocket.. for keeping maps.

tahsildar. Add: Now freq. with spelling **tehsildar.** (Later examples.) Also, the chief official of a tahsil.

1940 *Geogr. Jrnl.* XCV. 426 Khan Sahib Afraz Ghul Khan..is now tehsildar at Gilgit. **1954** J. Masters *Bhowani Junction* xxii. 191 That was a message from the tehsildar in Pathoda. **1972** *Times of India* 28 Nov. 1/2 The Mulki rules will apply only to non-gazetted posts and posts of tahsildar and civil assistant surgeons in the Telengana region. **1978** 'M. M. Kaye' *Far Pavilions* vi. 101 Ash had been given lodging for the night in the house of the *tehsildar*.

Tahunian (tahū·niăn), *a.* Also **Tahounian.** [ad. F. *Tahounien* (R. Neuville 1934, in *Revue Biblique* XLIII. 255), f. the place-name *Tahouneh*: see -ian.] Of, pertaining to, or designating a neolithic culture of Palestine represented by remains found at Tahouneh. Also *absol.*

1936 J. Garstang in *Annals Archaeol. & Anthropol.* XXII. 168 There seems to be no doubt that we are in the presence of a distinctive culture of the neolithic period. Whether it will be classed finally as Tahunian II is a matter for experts; meanwhile, as the Tahunian specimens are surface finds..we propose the more descriptive title 'Neo-Tahunian'. **1949** [see *Maglemosian *a.* and *sb.*]. **1952** V. G. Childe *New Light on Most Ancient East* xi. 225 By 1934 assemblages of flints from caves..had enabled Neuville to define a 'Tahunian' industry which could be classified as Neolithic. **1960** K. M. Kenyon *Archæol. in Holy Land* ii. 46 The flint industry which has for long been accepted as the classic Neolithic industry of Palestine is called the Tahunian. **1961** G. Clark *World Prehist.* iv. 82 The lithic industry in the upper Jericho level, with its pressure-flaked tanged arrowheads.. compares closely with that of the Tahounian. **1977** *Ibid.* (ed. 3) ii. 51 The Tahunian lithic component shows marked continuity, but one notable innovation in the form of flake arrows with side-notches.

tai (tai), *sb.*[1] [Jap.] Also **† tay.** A Pacific sea bream, *Pagrus major*, of the family Sparidæ, eaten as a speciality in Japan.

1620 R. Cocks *Let.* 10 Mar. in *Diary* (1883) II. 311 Dried fish lyke a breame, called heare *tay*, in abundance. **1727** J. G. Scheuchzer tr. *Kæmpfer's Hist. Japan* I. i. 135 *Tai*, is what the Dutch in the Indies call *Steenbrassem*. This is very much esteem'd by the Japanese as the King of Fish. **1795** tr. *Thunberg's Trav.* IV. 39 Among their valuable fishes is what they call the *tay*. **1884** tr. *Rein's Japan* I. vii. 192 The Tai proper is a beautiful deep-red to brown-red gold-bream. **1920** [see *sashimi]. **1965** W. Swaan *Jap. Lantern* iii. 41 The deep red and rather bloody-looking *tai* (a type of sea bream).

Tai (tai), *sb.*[2] and *a.* Also **T'ai.** [Native name.] **A.** *sb.* **a.** (A member of) a group of peoples of southeast Asia which includes the Lao, Shan, and Thai; also *spec.* = *Thai *sb.* b. **b.** A group of languages including Thai (Siamese), Lao, Shan, and other languages of southeast Asia, regarded by some as belonging to the Sino-Tibetan family; also *spec.* = *Thai *sb.* a. Also *Comb.*, as *Tai Dam*, *Tai-Shan*, *Tai-Chinese* adj. and sb., *Tai Yai*.

1693 A. P. tr. *S. de la Loubère's New Hist. Relation Kingdom of Siam* I. i. ii. 6 The Siameses give to themselves the name of *Tai*, or free, as the word now signifies in their language. **1798** *Asiatick Researches* V. 227 The first dialect is that of the kingdom of Siam, the most polished people of eastern India. They called themselves to me simply Tai. **1811** *Ibid.* X. 241 He divides them into two races, the Tai and the Tai Yai. **1837** *Jrnl. Asiatic Soc. Bengal* VI. 18 The Ahom is a branch of the Tai language, which is spoken, with some variations, by the Khamtis, the Shyāns, the Lāos, and the Siamese, all of whom designate themselves by the general appellation of Tai. Among the Ahoms, or the portion of the Tai race inhabiting *Assām*, the language is nearly extinct. *Ibid.*, The sound of the French *u*..is..common in the Tai. **1844** *Chinese Repository* XIII. 169 The inhabitants of this country are not called Siamese but T'ai. **1887** Tai-Shan [see *Mon-Khmer]. **1902** *Census of India 1901* XII. viii. 119 We are..practically where we were ten years ago in respect of our acquaintance with the early beginnings of the Tai. **1934** Webster, Tai-Chinese, *adj.* **1939** L. H. Gray *Foundations of Lang.* 389 The group [sc. Sino-Tibetan] falls into three great divisions: Yenisei-Ostyak, Tibeto-Burman, and Tai-Chinese. *Ibid.* 390 The other members of the Tai-Chinese family are Si-lo-mo.., Karen .., and Tai, the latter sub-divided into south-eastern, eastern, and northern. To the south-eastern division belong Siamese, Lao, Lü, and Khün. **1948** R. A. D. Forrest *Chinese Lang.* v. 100 There is evidence that T'ai in an older phase used a system of prefixes and infixes in word formation. **1956** J. Whatmough *Language* ii. 32 Note also Tai (Siamese). **1977** Tai Dam [see the adj. below]. **1978** *Amer. Poetry Rev.* Nov./Dec. 15/1 Judith Gautier's informant and lover was a Tai and often himself did not understand the Chinese.

B. *adj.* Of or pertaining to the Tai peoples or languages.

1837 [see the sb. above.] **1883** A. P. Phayre *Hist. Burma* i. 12 People of the Tai race were..in the country of the..river..Sālwin; and there is evidence of an irruption of that people into the country of the Irāwadi. **1892** *Census of India 1891* IX. viii. 167 The Tai language, of which there are numerous dialects, is essentially a Polytonic language. *Ibid.* 202 A great wave of Tai migration descended. **1902** *Census of India 1901* XII. viii. 119 The classification of the Tai races is a task of far greater magnitude than appeared when the last census was taken. **1933** L. Bloomfield *Language* iv. 69 The second branch of Indo-Chinese is the *Tai* family, which includes *Siamese*. **1948** R. A. D. Forrest *Chinese Lang.* v. 98 The T'ai languages are remarkably uniform over their wide area. **1977** *New Yorker* 5 Sept. 40/3 Of the three groups..the best off were the Tai Dam..members of a Tai racial group that had settled in China many years ago.

‖ **taiaha** (ta,i·aha). *N.Z.* [Maori.] A long-handled Maori club with a sharp tip. Also *fig.*

1845 E. J. Wakefield *Adventure N. Zealand* I. 140 The *taiaha* is rather a long-handled club than a spear. It ..is about six feet long. **1863** A. S. Atkinson *Jrnl.* 8 Sept. in *Richmond-Atkinson Papers* (1960) II. 61 The first of them was an old man with a handsome taiaha. **1894** *Westm. Gaz.* 29 Aug. 2/1 He looked his best in a picturesque native robe, with lurid feathers in his hair, and a 'taiaha', or spear, in his hand. **1938** R. D. Finlayson *Brown Man's Burden* 10 'Ae, we are one people!' he cries, brandishing a taiaha. **1949** P. Buck *Coming of Maori* (1950) ii. xi. 280 The *taiaha* head with its projecting tongue functioned as a stabbing point. **1963** S. Ashton-Warner *Teacher* 110 How can I protect my beautiful Matawhero from the taiaha of prejudice? **1974** *N.Z. Listener* 20 July 13/4 The Maori bus driver was 'threatened' with a friendly poke of the taiaha carried by another Maori.

‖ **T'ai Chi** (tai tʃī). Also **Tai Chi, t'ai chi,** etc. [ad. Chinese *tàijí*, f. *tài* extreme + *jí* limit.]

1. In Taoism and Neo-Confucianism, the Supreme Ultimate (see quot. 1955). Also, the symbol which represents this.

1736 R. Brookes tr. *Du Halde's Gen. Hist. China* III. 54 They give the first Principle of all things the Name of *Tai ki*. **1845** *Encycl. Metrop.* XVI. 568/2 *T'hai-ki* (the Great Summit) the soul of the universe, when in motion.. produced *Yang*, the living principle; when at rest, *Yn*, the dead principle, the one perfect and male, the other imperfect and female; from the union of which the elements sprang. **1914** D. T. Suzuki *Brief Hist. Early Chinese Philos.* 161 The term, *T'ai Chi*, first appears in one of the Confucian Appendices to the 'Yi Ching'. 'In the system of the Yi there is the Great Ultimate (or source or limit, *t'ai chi*).' **1931** A. U. Dilley *Oriental Rugs & Carpets* (caption to Pl. 63), The centre circle or Tae-keih (Great All) contains Yin (female) and Yang (male). **1955** E. Herbert *Taoist Notebk.* 3 *T'ai Chi* was presented as the starting-point, which was also the finishing-point, of a cosmic process: a cyclic process of constant change, in the course of which were produced in turn the linked principles of *Yang* and *Yin*..the *Wu Hsing* or Five Elements ..and all forms and existences in the material world. **1960** C. Winick *Dict. Anthropol.* 523/1 *T'ai chi*, in Chinese art, the symbol of the Great Absolute. It consists of a wavy or double curved line bisecting a circle, one half of which is red..and the other black. **1962** E. Snow *Other Side of River* (1963) l. 338 Tao is the Absolute that contains the total life force, or T'ai Chi.

2. In full *T'ai Chi Ch'uan* [Chinese *quán* fist], a Chinese martial art, believed to have been devised by a Taoist priest in the Sung dynasty (960–1279), promoting meditative as well as physical proficiency.

1962 E. Snow *Other Side of River* (1963) ii. 24 Servants don't spend their idle time playing mah-jongg now but sit by the bell boards studying English or Russian, or other textbooks, getting ready for after-hours classes; or they do t'ai chi ch'uan calisthenics. **1968** *Times* 22 Nov. 9 Embrace Tiger and Return to Mountain it is called, this being the name of one exercise in T'ai-chi, a Chinese system of callisthenics claimed to produce pliability, health and peace of mind. **1972** Da Liu *T'ai Chi Ch'uan & I Ching* p. v, The movements of T'ai Chi Ch'uan and the hexagrams upon which they are based are both methods of describing the circulation of psychic energy in the body of the meditator. **1979** P. Driscoll *Pangolin* I. xx. 147 Kids were doing Tai Chi exercises and playing soccer.

Taig: see *Teague 2.

taiga (tai·gă). [Russ.] The swampy coniferous forest area of Siberia; also, the zone of temperate coniferous forest stretching across Europe and North America.

1888 *Encycl. Brit.* XXIII. 70/2 They [sc. the Altai] are chiefly hunters, passionately loving their *taiga*, or wild forest. **1920** J. Ritchie *Animal Life Scotl.* vi. 329 The typical pine forest region, or taiga. **1946** F. E. Zeuner *Dating the Past* iii. v. 122 Stunted forest of the taiga type may have played a larger part in preglacial Europe than is commonly assumed. **1957** *Times* 12 Nov. (Canada Suppl.) p. xvi/1 Northward..the timber attenuates into sub-Arctic forest (taiga) and finally gives way to the true Arctic tundra. **1964** *Listener* 12 Nov. 747/1 A huge artificial lake has been created, inundating thousands of square acres of the Siberian taiga, the endless forest of birches and firs and pines that covers southern Siberia. **1969** *Beaver* Summer 5/1 Stunted taiga forest, lakes, yellow-green marshes. **1974** T. P. Whitney tr. *Solzhenitsyn's Gulag Archipelago* I. i. ii. 24 *Before* it came the wave of 1929 and 1930, the size of a good River Ob, which drove a mere fifteen million peasants, maybe even more, out into the taiga and the tundra. **1980** *Jrnl. R. Soc. Arts* Feb. 140/1 These are generally described in terms of bioclimatic zones—arctic, tundra, taiga, boreal forest, temperate deciduous forests, prairies, desert savanna, and rain forest.

‖ **taihoa** (ta,iho·a), *int. N.Z.* Also **taiho.** [Maori.] Wait a bit; by and by; presently. Occas. *attrib.*

1842 W. R. Wade *Journey in N.Z.* iii. 66 'Taihoa.' This word has been translated, By and by; but in truth it has all the latitude of directly,—presently,—by and by,— a long time hence,—and nobody knows when. **1851** J. C. Richmond *Let.* Apr. in *Richmond-Atkinson Papers* (1960) I. ii. 90 Glad we were to pay off our Maori lad & be done with their provoking 'taiho!' 'waiho' (*presently wait*). **1881** J. L. Campbell *Poenamo* v. 87 That irritatingly provoking word, 'taihoa'. **1905** J. M. Thomson *Bush Boys N.Z.* xii. 170 Taisho, Mac. I'll be there in a minute. [Note] The bush-boy corruption of the Maori 'Taihoa', 'Wait a bit'. **1910** A. A. Grace *Hone Tiki Dialogues* 4 There is too much taihoa about you Maoris. **1921** H. Foston *At Front* 188 Taking twelve years instead of five... It was described as a Taihoa policy. **1965** S. T. Ollivier *Petticoat Farm* i. 3 'Taihoa,' Harry said (it was the only Maori word he knew). 'Taihoa, I've not the money yet: wait until I have the money.'

tail, *sb.*[1] Add: **2. m.** The rear part of an aeroplane or air-balloon.

Except in the case of quot. 1804, the 19th-century examples refer to projected not actual aircraft.

1804 G. Cayley in C. H. Gibbs-Smith *Sir George Cayley's Aeronautics* (1962) vi. 18 This rod..supported a tail, made of two planes crossing each other at right angles... The tail could be set to any angle. **1835** *Nautical Mag.* IV. 612 An internal balloon is fitted for the purpose of ascending and descending at will, and the whole is intended to be propelled by fins, paddles, or wings we may call them... Finally the creature enjoys the important appendage of a tail abaft. **1848** *Chambers's Edin. Jrnl.* 6 May 302/2 There was also a tail, which, turning on a joint, was to direct the Ariel's flight. **1909** [see *feathering *ppl.a.* c]. **1913** A. H. Verrill *Harper's Aircraft Bk.* xi. 120 The parts of an aeroplane are mainly the frame, or 'chassis'; the body, or 'fuselage'; ..the rudder and tail;..and the control system. **1915** D. O. Barnett *Let.* 13 June 176 Up went his tail, and he began going down in spirals. **1959** *Chambers's Encycl.* I. 99/1 Streamlining eliminates this feature of bluff sections, a narrow wake forming only as the tail is approached. **1978** J. Gardner *Dancing Dodo* iv. 24 One [body] had been found towards where the tail and elevators should have been... The other had been taken from..the wreckage of the tail cone.

n. *Math.* An extremity of a curve, esp. that of a frequency distribution, as it approaches the horizontal axis of a graph; the part of a distribution that this represents.

1895 K. Pearson in *Phil. Trans. R. Soc.* A. CLXXXVI. 397 We require to have the 'tail' as carefully recorded as the body of statistics. Unfortunately the practical collectors of statistics often..proceed by a method of 'lumping together' at the extremes of their statistical series. **1930** E. Rutherford *Coll. Papers* (1965) III. 235 It is seen that the curve is very nearly symmetrical, but that there is a small 'tail' on the low-velocity side. **1980** K. Randsborg *Viking Age in Denmark* vii. 157 The Russian and Scandinavian finds of the ninth century have long tails of older coins.

3. Also (in *sing.* or *pl.*), the back part of a man's shirt that reaches below the waist. Also, in *pl.*, a tail-coat; a dress suit with tail-coat.

1845, etc. [see *shirt-tail* s.v. *shirt *sb.* 3 c]. **1857, 18..** [in Dict.]. **1915** Mrs. H. Ward *Eltham House* ii. 23 You made up your mind from the time you got into tails at Eton. **1932** S. Gibbons *Cold Comfort Farm* i. 10 Charles looked well in tails. **1958** B. Nichols *Sweet & Twenties* 110 Young men wore tails and white ties as a matter of course. **1960** *Guardian* 16 Dec. 8/3 At balls, even in the London season, tails are not uncompromisingly de rigueur. **1965** R. P. Jhabvala *Backward Place* iii. 166 He ran after her into the street, the tails of his crumpled shirt flying as he ran.

4. c. Also *spec.*, the non-combatant personnel of an armed service or of a military unit.

1946 *Hansard Commons* 30 Oct. 690 Our job must be to secure an efficient fighting force in which the tail is kept as short as possible, and the teeth as long and as keen as possible. **1950** *Ibid.* 26 July 555 If one is to provide an operational division,..the tail cannot be avoided, otherwise the division is not operational at all. **1961** B. Ferguson *Watery Maze* vii. 159 As 'Teeth' troops (to use a phrase which was then [sc. in 1942] both new and picturesque, but has long since become a *cliché*) there was little to equal them; but they lacked a 'Tail'—those ancillaries which in modern war virtually wag the dog **1972** D. Bloodworth *Any Number can Play* xiii. 116 When a soldier moves, all his basic needs are looked after by a vast administrative tail that..clothes him, feeds him, transports him. **1977** *R.A.F. News* 30 Mar.–12 Apr. 7/2 It is possible to continue trimming the so-called 'tail' by successive cuts in defence expenditure.

† m. A small evening party, subsequent to a dinner or a ball. *Obs.*

1837 C. Ridley *Let.* in *Cecilia* (1958) 26 We went to Lady Domville's—the nicest ball I have been at this year. .. We afterwards went to a tail where we saw a collection of unwashed uncombed philosophers. **1912** G. W. E. Russell *One Look Back* viii. 164 'Tails', as the name implies, were little parties tacked on to the end of big dinners, where a few people looked in, rather cross at not having been invited to dine, or else in a desperate hurry to get on to a larger party or a ball.

n. *Phonetics.* (See quot. 1922.)

1922 H. E. Palmer *Eng. Intonation* iv. 10 Any syllable or syllables following the nucleus in the same Tone-Group is termed the 'Tail' of the group. The Tail-syllable or group of syllables following the Falling Nucleus..is pitched on the low level. **1965** *Amer. Speech* XL. 72 Word order affects intonation in the tail, head, and nucleus.

o. The rear part of a motor vehicle.

1928 E. Wallace *Double* xiii. 187 Outside he saw five police cars parked bonnet to tail. **1975** *Drive* New Year 106/3 The car's tail tends to drift out of corners at lower speeds than earlier models.

5. a. Also in U.S. dial. and colloq. use.

1935 J. T. Farrell *Judgment Day* in *Studs Lonigan* (1938) iv. 86 This idea of sweating your tail off with work ..is the undiluted crap. **1942** W. Faulkner *Go down, Moses* 229 This is the first time you've had your tail out of that kitchen since we got here except to chop a little wood. **1969** *New Yorker* 14 June 72/3 Go out there and work your tail off. Don't wake up tomorrow morning regretting that you didn't give a hundred per cent. **1976** *Billings* (Montana) *Gaz.* 1 July 4-E/1, I worked my tail off to help win a pennant for the Dodgers.

c. (Later examples.)

1972 F. Warner *Lying Figures* III. 17 Give her her head..and she'll give you her tail. **1977** *Transatlantic Rev.* LX. 78 He had been after her tail for months, but Judy, being an old-fashioned girl, declined his advances.

d. *slang.* † (i) a prostitute (*obs.*); (ii) women regarded collectively (by men) as a means of sexual gratification; sexual intercourse; a sexual partner. Freq. in phr. *a piece* (or *bit*) *of tail*. Cf. *piece *sb.* 3 d.

1846 *Swell's Night Guide* 58, I takes my pitch last night on Fleet pave, then..a swell was sweet on me for a tail. **1869** F. Henderson *Six Yrs. in Prisons of Eng.* vii. 76 He meant a 'flash-tail', or prostitute who goes about the streets at nights trying to pick up 'toffs'. **1933** M. Lowry *Ultramarine* ii. 67 It's not as though you were a bloody man who'd been having a bit of tail. **1942**, etc. [see *piece *sb.* 3 d]. **1951** J. D. Salinger *Catcher in Rye* xiii. 109 Innarested in a little tail t'night? **1953** H. Miller *Plexus* (1963) xi. 391 He's at loose ends. Hates his work, loathes his wife, and the kids bore him to death. All he thinks of now is tail. And boy, does he chase it! **1967** J. Potter *Foul Play* xiii. 157 Where's all the tail today? No Hermione, no Bunty, no Christabel. **1976** 'R. Gordon' *Doctor on Job* vi. 59 Even if it was deciding whether to go out on the booze at night or have a bit of tail off of the wife. **1977** *Transatlantic Rev.* LX. 39 He would yell, 'How y'all doin, chief? Gettin much tail?'

6. b. A person (as a detective or spy, etc.) who secretly follows and observes another. Also *collect.*, people in the act of following. Cf. *tag *sb.*[1] 12*, *tail *v.*[1] 5 b. *colloq.* (orig. *U.S.*).

1914 [see *tail *v.*[1] 5 b]. **1933** A. Merritt *Burn Witch Burn!* (1934) xii. 181 One of the tails—one of the lads who's been looking—meets up with me. **1940** R. Stout *Over my Dead Body* xiv. 215 'You were having Miss Lovchen followed?' 'Yes, a double tail... Their instructions are to report in every two hours.' **1955** J. Cannan *Long Shadows* iii. 63 I'd like to put a tail on the lady. **1962** 'K. Orvis' *Damned & Destroyed* v. 42, I realized almost at once I'd picked up a tail. The two shadowing me..were..obvious. **1978** M. H. Clark *Stranger is Watching* xxvi. 112 We'll have a loose tail on you—an agent following you from a distance.

8. a. *spec.* in *Cricket*, the lower end of the batting order, comprising the weaker batsmen in a team; (earlier and later examples). Also *fig.*

1851 J. Pycroft *Cricket Field* xi. 221 Never put in all your best men at first, and leave 'a tail' to follow. **1879** *James Lillywhite's Cricketers' Ann.* 17 The tail was again weak, the last five wickets only adding 16 runs. **1913** J. B. Hobbs *How to make Century* xii. 82 The fast bowler.. was bowling far too accurately for 'tail' batsmen to do much with him. **1926** C. E. Montague *Rough Justice* iii. ix. 125 They seemed to be talking about the conflict then arising between the House of Lords and the..House of Commons. 'If it comes to a Test Match,' said Wynnant, 'we'll lose. Too long a tail to our team.' **1955** *Times* 4 July 3/2 Due..to the obstinate wriggling of the tail, the last four Cambridge wickets more than doubled the score. **1977** J. Laker *One-Day Cricket* 67 Marsh, with no support at all from the tail, was left high and dry with 52 not out.

11. b. (Earlier *fig.* example.)

1842 F. A. Kemble *Let.* 6 May in *Rec. Later Life* (1882) II. 218 She has scornfully..departed with her tail over her shoulder, leaving the behind scenes of Her Majesty's Theatre with their tails between their legs.

c. *to get one's tail down* and varr., to become dispirited; *to have one's tail up* and varr., to be in good spirits.

1853 'P. Paxton' *Stray Yankee in Texas* 97 To use an expressive Westernism, 'Dave's tail was up', and every possible preparation was made to preclude a failure. **1874** Hotten *Slang Dict.* 318 *Tail-down*, 'to get the tail down', generally means to lose courage. When a professional at any game loses heart in a match he is said to get his *tail down*. 'His *tail* was quite *down*, and it was all over.' **1917** G. S. Gordon *Let.* 26 Apr. (1943) 75 We were getting jaded till this touch of spring came, and now we have our tails up again, and are prepared to attack anything. **1921** *Punch* 12 Jan. 23, I must try and keep my tail up. **1923** Galsworthy *Captures* 190 He was a Northumbrian.. and his 'tail still up', as he expressed it. **1928** *Sunday Dispatch* 15 July 14, I sincerely hope that..standard producers..will not get their tails down over this 'cheap record boom'. **1933** Wodehouse *Mulliner Nights* iii. 93 'Tails up, Uncle Theodore, tails up!' 'Tails up!' repeated the Bishop dutifully, but he spoke the words without any real ring of conviction in his voice. **1941** C. Morgan *Empty Room* ii. 88 May be a snag somewhere. Usually is when one gets one's tail up about an idea. **1960** [see *balance *sb.* 15 c]. **1978** R. Mark *Office of Constable* xv. 187 Nevertheless, in dealing with the worst forms of crime our tails were well up.

f. *two shakes of a lamb's tail* (and varr.): see *shake *sb.*[1] 2 h.

a **1855** J. F. Kelly *Humors of Falconbridge* (1856) 137 In the wag of a dead lamb's tail. **1901** *Dialect Notes* II. 142 'I'll do it in three jerks of a lamb's tail,' i.e., very quickly. *Ibid.* 429 She got all cleared up in the whisk of a lamb's tail. **1917** *Ibid.* IV. 402 *Two jerks of a lamb's tail*, *n. phr.*, an instant, a jiffy.

g. *the tail wags the dog*, the less important or subsidiary factor dominates the situation; the proper roles are reversed.

[**1907** M. A. von Arnim *Fräulein Schmidt* xxvi. 84 Isn't it rather weak to let yourself be led round by the nose..? It is as though instead of a dog wagging its tail the tail should wag the dog.] **1935** F. Scott Fitzgerald *Let.* 11 Mar. (1964) 260 This letter is a case of the tail (the parenthesis) wagging the dog. **1945** *Jrnl. R. Aeronaut. Soc.* XLIX. 463/1 The aeroplane developing an undamped short period oscillation in which rapid movement of the rudder from side to side plays an essential part—the tail wagging the dog. **1956** W. H. Whyte *Organization Man* ii. 19 The tail wagged the dog in this case and it still often does. **1968** *Listener* 4 Jan. 23/3 Most producers are going to continue resisting..indulgence in an academic exercise. There's a danger of the tail wagging the dog. **1980** *Truck & Bus Transportation* (Surry Hills, New South Wales) Feb. 26/2 Tractor response during a lane-change manoeuvre shows how the externally-applied force through the fifth wheel induces tractor lateral motion. This is better known as 'tail wagging the dog'.

h. *to be on someone's tail* and varr., to follow or pursue someone closely (see also quot. 1925). Also *fig.*

[**1865** 'L. Carroll' *Alice's Adventures in Wonderland* x. 151 There's a porpoise close behind us, and he's treading on my tail.] **1925** Fraser & Gibbons *Soldier & Sailor Words* 275 *Tail, to get on the*, an Air Force expression for an attack on the rear of an opponent. **1937** Partridge *Dict. Slang* 860/2 *Tail, be*—gen. *shall* or *will be*—*on a person's*, to look for, to pursue, a person with a view to punishing or severely scolding him: C. 20. **1962** 'J. le Carré' *Murder of Quality* iv. 54, I rather gathered ..that his Chief Constable was treading on his tail, urging him to scour the country for tramps. **1971** B. Malamud *Tenants* 71, I wouldn't want anybody else on my tail or in my hair, with or without cause. **1971** M. Tak *Truck Talk* 154 *Stay on his tail*, to follow another truck closely. **1981** *Sunday Times* 1 Feb. 63/5 Sir Hugh thought the Lonrho boss had put a private eye on his tail.

i. *to chase one's tail*, to indulge in a futile pursuit; to go round in circles.

1963 *Times* 14 May 8/4 'We have been chasing our tails overlong,' he said. 'Given a Labour Government committed to the principles of equity and justice, a coordinated wages policy may be possible.' **1973** *Archivum Linguisticum* IV. 35 Is anything indeed to be gained from hunting for some notion embodying the cumulate surface exponency of..transitive and perfective..? It is all too easy at times to chase our conceptual tail.

12. a. *tail half* (later example), *-wind* (later *Aeronaut.* examples).

1970 T. Hughes *Crow* 15 He stuffed the head half head-first into woman And it crept in deeper and up to peer out through her eyes Calling its tail-half to join up quickly. **1927** C. A. Lindbergh *We* iii. 39, I left Texarkana with a strong tail wind. **1976** *Evening Times* (Glasgow) 1 Dec. 5/3 Tail winds across the Atlantic knocked up to an hour off the flying times of some transatlantic flights.

13. a. *tail-tuft*; **b.** *tail-buffeting, -chasing* (examples), *-dangler, -wagger*; **c.** *tail-down* adj. and adv.; *tail-first*: also as adj.

1931 *Flight* 30 Jan. 90 To the new phenomenon the sub-committee gives the name 'tail buffeting'. **1947** *Times* 8 Feb. 2/5 There was tail-buffeting within a certain speed range in very bumpy conditions. **1921** J. D. M. Rorke *Musical Pilgrim's Progress* iii. 49 The excitement and tail-chasing demonstrations of a dog at the home-coming. **1957** R. H. Smythe *Conformation of Dog* 123 Tail-chasing, spinning and walking in circles. **1922** Joyce *Ulysses* 646 It [sc. a horse] was a..taildangler, a headhanger. **1916** H. Barber *Aeroplane Speaks* 87 An inclinometer..which will indicate a nose-down position by increase in air speed, and a tail-down position by decrease in air speed. *Ibid.* 113 If the angle of incidence..is too great, it will produce an excess of lift, and that way..result in a tendency to fly 'tail-down'. **1935** P. W. F. Mills *Elem. Pract. Flying* vii. 103 When brought too quickly into tail-down attitude their wings retain an uncomfortable degree of buoyancy for some little time. **1914** H. M. Buist *Aircraft in German War* v. 101 The latter quality lead to the original example of this tail-first machine being purchased by the Rumanian Army. **1945** *Sun* (Baltimore) 7 Feb. 7-0/4 (*heading*) New 'tail-first' fighter plane appears to fly backward. **1910** W. de la Mare *Three Mulla-Mulgars* xvii. 224 They sat, with tail-tufts over their shoulders. *a* **1930** D. H. Lawrence *Last Poems* (1932) 260 The two lions who devoured one another, and left the tail-tufts wagging. **1948** B. Vesey-Fitzgerald *Bk. of Dog* i. 114 Organisations, such as the Tail Waggers Club, undertake to provide discs that can be attached to the collar. **1952** *Chambers's Jrnl.* Apr. 239/1, I reckon that about 3,000,000 folk would have to look elsewhere for their bread and butter if there were no trawlermen—or fish. We mustn't forget the tail-waggers. **1982** L. Cody *Bad Company* iii. 26 'What's this then? The Tail-Waggers Club?' he asked as he..fended off the retriever's enthusiastic welcome.

14. tail-area *Statistics*, an area under the curve of a frequency distribution lying between one end of the curve and any ordinate on the same side of the mode; **tail assembly** [*assembly 1 c] *Aeronaut.* = *empennage; cf. *tail unit* below; **tailback:** in U.S. football, the player stationed farthest from the forwards; **tail boom** *Aeronaut.*, one of the main spars of the longitudinal framework carrying the tail of an aeroplane when not supported by the fuselage; **tail comb,** a comb with a tapering tail or handle used in styling to lift, divide, or curl the hair; **tail cone** *Aeronaut.*, the conical rear end of the fuselage of an aircraft; **tail-dragger** *Aeronaut.*, an aeroplane that lands and taxis on a tail wheel or tail skid, its nose off the ground; **tail fin,** (*a*) the caudal fin of a fish; (*b*) *Aeronaut.* (see quot. 1940); (*c*) an upswept ornamental projection forming a continuation of the fender line at the rear of a motor vehicle; **tail-flap,** (*a*) the tail of a crustacean; (*b*) *Aeronaut.*, an adjustable control surface on the tail of an aircraft; **tail gas** (see quot. 1967); **tail gunner** = *rear gunner* s.v. *rear *sb.*[3] (and *a.*[1]) 9; **tail-heavy** *a.*, of a motor vehicle, boat, etc.: having a tendency for the rear end to bear down more than the front; hence **tail-heaviness** (used esp. with reference to aircraft); **tail-light** (examples with reference to motor vehicles and aircraft); **tail parachute** *Aeronaut.*, a deceleration parachute attached to the tail of an aircraft; **tail-pin,** (*d*) *Mus.* (i) (see quot. 1961); (ii) a metal spike attached to the cello and other instruments to support them at the correct height from the ground; **tailplane** *Aeronaut.*, the horizontal stabilizing surface of the tail of an aircraft; **tail-pole,** a wooden lever or turning beam by means of which a post- or windmill is turned to the wind; **tail-rime, -rhyme** (examples); **tail rotor** *Aeronaut.*, an auxiliary rotor at the tail of a helicopter designed to counterbalance the torque of the main rotor; **tail-shaft** (earlier example); **† tail-shot** = *tail-ill* (*obs.*); so **† tail-shotten** *a.*; **tail skid** [*skid *sb.* 2 f] *Aeronaut.*, that part of an aircraft's landing gear which supports its tail; **tail-slide** *Aeronaut.* (see quot. 1969); **tail-twisting,** (*c*) in gen. *fig.* use, harassment or malicious annoyance; **tail unit** *Aeronaut.* = *empennage; **tail-walking,** the movement of fish over the surface of water by means of propulsion with the tail; hence (as a back-formation) **tail-walk** *v. intr.*; **tail wheel** *Aeronaut.* = *tail skid* above.

1957 Kendall & Buckland *Dict. Statistical Terms* 290 Tail area (of a Distribution). **1971** D. C. Hague *Managerial Economics* vii. 153 If we want to take the probability of there being less than 2 in of rain, we take the area of the first two bars [of the histogram], and so on. If we do this, we are said to be considering tail areas. **1968** *Globe & Mail* (Toronto) 3 Feb. 1/2 The wreckage was a compacted heap of rubble... Only the tail assembly was intact. **1977** J. Cleary *High Road to China* iv. 128 The plane quivered.. then the nose came up, the quivering slid out through the tail-assembly. **1930** R. C. Zuppke *Coaching Football* vii. 208 The tail-back is four and one-half yards back of the scrummage line and directly back of the fullback. **1980** *Washington Post* 10 Oct. c6/5 Of the six Rattler touchdowns Keith pointed out FAMU 'earned' only one: the 69-yard first-quarter run by tailback Archie Jones. **1913** *Flight* 23 Aug. 927/1 One of our sketches shows the method of joining the struts to the tail booms. **1969** K. Munson *Pioneer Aircraft 1903–14* 142/1 The three tubular steel tailbooms formed a triangular section, and the tail control wires were led through the uppermost boom, which also acted as a propeller bearing. **1782** J. Woodforde *Diary* 24 Apr. (1926) II. 19 To a Tail Comb and another Comb for Nancy of Baker pd. 0. 0. 10. **1855** F. Duberly *Let.* 22 July in E. E. P. Tisdall *Mrs. Duberly's Campaigns* (1963) v. 153 Oh, please will you send me a tail comb in the box. **1907** *Yesterday's Shopping* (1969) 110/1 Tail or curling combs—buffalo horn. **1930** V. Sackville-West *Edwardians* i. 38 Don't drag my hair back... Give me the tail comb... It wants more fullness at the sides. **1976** J. Grenfell *Joyce Grenfell requests Pleasure* xvii. 246 Her dark hair was kept neat in a fine net... A tail-comb raised the waves. **1944** H. F. Gregory *Anything a Horse can Do* xxi. 216 The tail rotor and approximately the last four feet of the tail cone were broken completely. **1978** Tail cone [see sense 2 m above]. **1971** *Flying* Apr. 39/2 If you trace the 172 back to the rag-wing 170 taildragger of 1948. **1981** *R.A.F. News* 14 Jan. 12/3 The Chipmunk is well suited to the unit's role because, as a taildragger, it introduces characteristics that 'sort out the men from the boys'. **1681, 1835–6** Tail fin in Dict., sense 13 a]. **1940** *Chambers's Techn. Dict.* 333/1 *Fin*, in an aeroplane, a fixed vertical surface giving lateral stability of motion; usually placed at the tail, then sometimes called a tail fin. **1945** W. Langewiesche *Stick & Rudder* vii. 115 The purpose of the horizontal tail fin is not to hold the tail up, but to hold it down; it is a sort of wing, but a wing set at a negative Angle of Attack. **1954** *Wall St. Jrnl.* 22 Oct. 16/6 Its [sc. the car's] high fender-line sweeps backward in a straight line but is slightly lower at the tail fins than at the headlights. **1974** P. Dickinson *Poison Oracle* i. 22 The plane lay still... The symbol of the rising sun stared from the tall tail fin. **1982** *Quarto* Mar. 7/4 The American family car was a 425-horsepower, twenty-two-foot-long Buick Electra with tail fins in back. **1847–8** Tail-flap [in Dict., sense 13 a]. **1913** A. E. Berriman *Aviation* p. xxiv, The glide..as the pilot switches on at the last moment and cocks up the tail flap to flatten out ere touching the ground. **1980** J. Ditton *Copley's Hunch* ii. i. 115 The tail-flaps were working all right, because he zoomed up and over to gain height. **1948** *Economist* 31 July 193/2 Tail gases..carried ..by pipe-line..will replace some of the coke at present used..for the production of ammonia, methanol and petrol. **1967** *Gloss. Terms Gas Industry* (*B.S.I.*) 12 *Tail gas*, refinery gas which is not required for further processing in the refinery. **1939** *War Illustr.* 29 Dec. 539/2 The tail gunner reported 'Fighters on our tail'. **1971** P. O'Donnell *Impossible Virgin* xii. 246 A bloke called Worsfold, tail-gunner in a Lancaster during the war..fell over seven thousand feet... Only broke a leg and a few ribs. **1919, 1930** Tail-heaviness [see *nose-heaviness* s.v. *nose *sb.* 18]. **1977** D. Beaty *Excellency* vi. 83 The tail-heaviness had been deliberate..this ingenious way of getting rid of him. **1916** H. Barber *Aeroplane Speaks* 110 The aeroplane will, in flight, be nose-heavy or tail-heavy. **1923** G. Sturt *Wheelwright's Shop* 223 *Tail-heavy*, the opposite fault to fore-heavy. In a tail-heavy cart the tendency was to lift the horse off the ground. **1957** [see *scorch *v.*[1] 3]. **1978** R. V. Jones *Most Secret War* xvi. 131 The weight of two cameras, about 120 lbs., would pull back the centre of gravity of the aircraft making it 'tail heavy' and dangerous to fly. **1937** *Esquire* Jan. 64/3 He turned and watched the red tail-light sink into the distant darkness. **1946** R. A. McFarland *Human Factors in Air Transport Design* xii. 610 The pilot..had.. mistaken the taillight of the stationary D.C.-3 for one of a row of ..boundary lights. **1978** S. Brill *Teamsters* vii. 286 Only one man was working the night shift, replacing some tail lights on a trailer. **1937** *Jrnl. R. Aeronaut. Soc.* XLI. 731 The Russian plane A.N.T.6 which was the first to land at the pole was provided with a tail parachute, which was released as soon as the skis touched the ice. **1978** A. Welch *Bk. of Airsports* ii. 29/2 Tail parachutes are 'one-shot' drag producers and are more useful as an emergency aid. **1884** E. Heron-Allen *Violin-Making* xi. 105 The *Tail-pin*..is the peg of ebony or box-wood, which is firmly fixed into the bottom block..to which is fastened the loop..of the tail-piece. **1923** E. van der Straeten *Technics Violoncello Playing* (ed. 4) iii. 18 The use of the tail pin is now generally adopted, and offers the double advantage of steadying the instrument and strengthening its tone. **1946** R. Alton *Violin & 'Cello Building* vii. 60 The tail-rest..over which the tail-gut passes on its way to the tail-pin, must now be inserted. *Ibid.* xv. 147 With a tapered reamer fit the tail-pin into its place, gradually enlarging the hole until the tail-pin fits. **1961** A. Baines *Mus. Instr. through Ages* 358 *Tailpin*, the button let into the bottom block of a violin, etc., to which the tailpiece is attached by a gut loop. **1978** *Early Music* Oct. 530/2 My own contribution to this debate..is concerned with thicknesses and struttings, lengths and positions of necks, bridge heights and string angles and tailpin hitches. **1909** A. Berget *Conquest of Air* ii. iv. 189 Tail planes. **1911** [see *empennage]. **1948** 'N. Shute' *No Highway* i. 8 It had only been necessary to break one of these expensive tailplanes for the strength tests for the airworthiness of the machine. **1979** D. Kyle *Green River High* xvii. 219, I tested the tailplane's firmness to be sure it would take my weight. **1945** *Archit. Rev.* XCVIII. 71 This 'winding' of the mill was first accomplished by pushing the whole body of a post mill round by means of the 'tail pole', which projected downwards through the ladder. **1968** J. Arnold *Shell Bk. Country Crafts* 170 The problem of keeping the sweeps or sails into the wind was originally met by manual labour at the 'tail-pole', or turning beam. **1838** E. Guest *Hist. Eng. Rhythms* II. iv. i. 289 This, like the interwoven and tail-rhime, seems to have been first used by the Latinist. **1916** J. E. Wells *Man. Writings Middle Eng.* I. 86 Lines 3411 to the end are in tail-rime stanzas. **1945** Tail-rhyme [see *rime *sb.*[1] 2 c]. **1982** *N. & Q.* June 242/2 With certain common patterns, of couplets, quatrains, and versions of the tail-rhyme stanza, predominating. **1944** H. F. Gregory *Anything a Horse can Do* x. 107 The control stick..would decrease the pitch of the blades on the right horizontal tail rotor. **1979** *Jrnl. R. Soc. Arts* CXXVII. 571/1 The helicopter for replacement of Sea King is rather a noisy beast, in that it has a tailrotor. **1888** Kipling *Day's Work* (1898) 277 When d'ye ship a new tail-shaft? **1790** J. Woodforde *Diary* 5 Feb. (1927) III. 169 My poor Cow rather better this morning, but not able to get up as yet, she having a Disorder which I never heard of before or any of our Somersett Friends. It is called Tail-shot, that is, a separation of some of the Joints of the Tail about a foot from the tip of the Tail, or rather a slipping of one Joint from another. **1798** *Ibid.* 1 Aug. (1931) V. 130 She is tail-shotten, & hath something of the Gargut. **1913** A. E. Berriman *Aviation* iii. 25 The tail-skid is comparatively an insignificant member of the design: provided it serves its purpose as a protection. **1973** J. D. R. Rawlings *Pictorial Hist. Fleet Air Arm* ii. 18 The fourth broke his tailskid and had to abort the sortie. **1916** H. Barber *Aeroplane Speaks* ii. 73 Should the surface tend to assume too large an angle..the pressure D decreases, with the result that C.P. moves forward and pushes up the front of the surface, thus increasing the angle still further, the final result being a 'tail-slide'. **1969** *Gloss. Aeronaut. & Astronaut. Terms* (*B.S.I.*) ii. 2 *Tail slide*, rearward motion of an aircraft along its longitudinal axis from a vertical or near vertical, stalled attitude. **1887** Kipling *Plain Tales* (1888) 77 The Colonel's Wife..went away to devise means for 'chastening the stubborn heart of her husband'. Which, translated, means, in our slang, 'tail-twisting'. **1937** E. Linklater *Juan in China* ii. 58 He had no reason to feel friendly..and the idea of a little tail-twisting was pleasant. **1982** W. J. Burley *Wycliffe's Wild-Goose Chase* vi. 110 If there is any attempt at tail twisting you can rely on me to see 'em off. **1926** *Chambers's Jrnl.* Aug. 580/1 In every aeroplane the tail unit..comprises the rudder [etc.]. **1977** D. Beaty *Excellency* i. 8 A lot of junk..six DC6 wheels, a Viscount tail unit. **1971** W. Hillen *Blackwater River* viii. 72 The trout leaped, tail-walked, shook himself, leaped again, and ran past the raft for deep water. **1979** *Angling* July 53/2 A fish hits the bait. It runs, leaps, tail-walks. **1946** *Richmond* (Va.) *Times-Dispatch* 16 June 12-B/2 Oh yes, there are certain salt-water fish which do a certain kind of tail-walking, but the way the bass performs these antics is peculiar to himself. **1970** *Islander* (Victoria, B.C.) 25 Oct. 3/3 Out in the salt-chuck where he [sc. a salmon] has a whole ocean to play in you can expect to see some fancy tail-walking. **1910** R. Ferris *How it Flies* xx. 472 *Tail wheel*, a wheel mounted under the rear end of an aeroplane as a part of the alighting gear. **1933** *Jrnl. R. Aeronaut. Soc.* XXXVII. 29 But with the advent of tail wheels, that difficulty should not arise. **1981** *Pilot* Jan. 12/2 A 110 hp tailwheel model.

tail, *sb.*[2] Add: **3. b.** *tail male, female*: also *transf.*, the line of descent of dogs or horses, considering either the male or female ancestors.

1926 Earl Bathurst *Breeding of Foxhounds* vii. 96 The top line perhaps may be considered important, for it represents the descent in tail-male. *Ibid.* 99 The Bruce-Lowe system..is..the importance of the female line, or 'tail-female'. **1931** *Times Lit. Suppl.* 23 Apr. 325/2 His blood is to be found in most of our 'classic' winners, and in tail female it never waned. **1957** C. Leicester *Bloodstock Breeding* ix. 144 This..leaves untouched the tail female line, i.e. the dam, grandam, etc. of the animal under investigation. **1972** *Country Life* 10 Feb. 332/1 One of Whipcord's descendants was the famous Four Burrow Pleader '38, whose ancestry can be traced..on his tail female to Mr. Darley's Damsel..and on tail male (through Whipcord) to the Brocklesby Bumper, 1748.

tail, *v.*[1] Add: **I. 5. a.** Also in N.Z. and with reference to horses.

1852 J. R. Clough *Jrnl.* 29 Feb. in J. Deans *Pioneers of Canterbury* (1939) 291, I have had to tail the cattle on foot this five weeks as I have had no saddle. **1871** C. L. Money *Knocking about in N.Z.* ix. 133 The horses, after being 'tailed', or shepherded, all day by one of us.., were tied in rows..for the night.

b. To follow someone closely; *spec.* to follow secretly as a detective or spy, etc. Cf. *tag *v.*[1] 4 b, *tail *sb.*[1] 6 b. *colloq.* (orig. *U.S.*).

1907 *Everybody's Mag.* Mar. 341/2 Detectives were assigned to 'tail' him. **1914** Jackson & Hellyer *Vocab. Criminal Slang* 83 *Tail, verb.* General circulation. To trail; to follow. Used as a noun in the same sense. **1925** E. Wallace *Strange Countess* ix. 81 'What's your idea in tailing me?'..' "Tailing"? Oh, you mean following you, I suppose?' **1950** D. Hyde *I Believed* viii. 88 For some months I was tailed by a curious assortment of police agents. **1956** S. Plath in *Granta* 20 Oct. 22/2 Ben tailed us out to the kitchen, where the black old gas stove was, and the sink, full of dirty dishes. **1966** T. Pynchon *Crying of Lot 49* v. 130 Oedipa gave him half a block's start, then began to tail him. **1978** S. Brill *Teamsters* iv. 127 I'm not gonna let you tail me like some kinda cop. **1978** G. Greene *Human Factor* v. iii. 278 Castle led the way down the stairs to the cellar. Buller followed him and Mr Halliday tailed Buller.

9*. *slang.* To copulate with (a woman).

1778 in Weis & Pottle *Boswell in Extremes* (1971) 248

When we talk of *pleasure*, we mean sensual pleasure. When a man says he had pleasure with a woman, he does not mean conversation, but that he *tailed* her. **1846** *Swell's Night Guide* 133/2 *Tail*, to cohabit with women. **1973** J. Wainwright *Devil you Don't* 51 So, I tailed his wife.. So what?

III. 19. c. (Earlier example.)

1830 A. Sedgwick *Let.* 21 Nov. in J. W. Clark *Life A. Sedgwick* (1890) I. 366 Many men will tail off, if they have an excuse.

21. The vb.-stem in Comb. **tail-back,** a queue of stationary or slowly moving motor vehicles; **tail-off** *colloq.*, a decline or tapering off of demand, etc.; a period of this.

1975 D. Lodge *Changing Places* v. 188 They hit a tailback of rush-hour traffic in the Midland Road. **1978** *Times* 26 July 8/3 One of the worst traffic jams in living memory with tailbacks of several miles. **1975** D. Francis *High Stakes* vii. 109 There would be at first a patch of sporadic success..and then a long tail-off with no success at all. **1984** *Times* 15 Feb. 20/7 Laurie Millbank does not envisage any tail off in demand.

tail, *v.*[3] (Earlier Canad. example.)

1770 G. Cartwright *Jrnl.* 27 Aug. (1792) I. 30, I tailed a couple of traps for otters, but did not find many rubbing places.

tail-end. Add: **1. a.** (Earlier and later examples.)

1747 H. Glasse *Art of Cookery* ix. 92 Take a large Eel, ..cut it into four Pieces, take the Tail-end, [etc.]. **1917** 'Contact' *Airman's Outings* viii. 214 V., my pilot and flight-commander, was given to a quick dive at the enemy,..and another dash to close grips from an unexpected direction, while I guarded the tail-end.

d. *Cricket.* = *TAIL *sb.*[1] 8 a. Freq. *attrib.*

1888 A. G. Steel in Steel & Lyttelton *Cricket* iii. 176 The tail end of a team are usually victims to a good straight fast bowler. **1904** [in Dict., sense 1 a]. **1930** *Morning Post* 16 July 11/5 He had batted on three different days, and had shown ability and courage. He can never in future be regarded as a tail-end batsman. **1955** *Times* 13 July 3/2 With the first ball of his next Smales bowled Smith, who had..looked the one man likely to deprive Nottinghamshire of a win with tail-end wickets falling fast. **1976** *0–10 Cricket Scene* (Austral.) 15/1 He.. then routed Victoria's tail-end to win another close encounter.

4. tail-end Charlie, a tail-gunner; the last aircraft in a flying formation (*Services' colloq.*); also *transf.* and *fig.*, one who comes last or behind, a tail-ender.

1941 *Illustr. London News* CXCIX. 579 (*caption*) The 'tail-end Charlie' of a 'Halifax' gives the 'thumbs up' sign just before his machine takes off. **1942** *Sun* (Baltimore) 8 Aug. 3/1 Lessig crossed the channel with the RAF, flying a Spitfire in the 'tail end Charlie' position—the last plane in a flight of four. **1956** P. Scott *Male Child* I. ii. 40 My brother..was killed in the war... A tail-end Charlie. **1961** *Times* 7 June 5/7 The Spaniard, Goyeneche was *lanterne rouge*, the cyclists' equivalent of tail-end-Charlie. **1976** *Daily Mail* 4 Oct. 3/3 The average lifespan of a 'Tail-end Charlie' was reckoned as ten 'ops.'

transf. and *fig.* **1962** A. Sampson *Anat. of Britain* xxxiv. 550 The essential services may, as in America, become regarded as the tail-end Charlies, the forgotten drudges under the pavements and pit-heads. **1969** *Daily National* (Nairobi) 31 Oct. 35 (Advt.), Congratulations to Joginder Singh and Ken Ranyard on their magnificent drive in car No. 46, starting as 'tail end Charlie'. But finishing with the major honours. **1973** *Listener* 15 Nov. 661/3 On tours, when I used to go with my parents..a sort of tail-end Charlie. **1978** A. Price *'44 Vintage* x. 131 The jeep behind them was closing up... The Sergeant was taking his tail-end Charlie role..seriously. **1980** *Outdoor Life* (U.S.) (North-east ed.) Oct. 53/3, I found myself on a hillside where the birds were flushing below, but then there was one tail-end Charlie who went up the hill.

tail-ender. Add: (Further examples.) Now used esp. in Sport.

1885 *Sydney Mail* 28 Feb. 451/4 Garrett and Evans, the 'tail-enders', established themselves..firmly at the wickets. **1915** *Lit. Digest* 21 Aug. 360/3 The St. Louis Cardinals, whom the writer designates as 'almost chronic tail-enders', are, in regard to the amounts spent for new players, just about the same. **1955** *Times* 23 June 17/1 It has certainly been focused so far on the leaders rather than the tail-enders in the various markets. One possible brake on the rise..is an increase in the corpus of securities **1961** *Sunday Express* 7 May 1/6 Both men were lapping the tailenders now. **1977** *World of Cricket Monthly* June 24/2 Eric was also the better batsman, Alec being more of the hard-hitting tailender type. **1980** A. Crawley *Dial 200-200* ix. 98 'You might have been killed yourself.' 'Not much chance; the raid had already gone past us. It would have had to be a tail-ender, like the one that got the maid.'

tailer. Restrict *Angling* to sense in Dict. and add: † **2. a.** A follower or hanger-on. **b.** *spec.* on the U.S. Stock Exchange (see quot. 1900). *Obs.*

1838 Disraeli *Let.* in Monypenny & Buckle *Life Disraeli* (1912) II. i. 20 Two of the greatest ruffians in the House... They are 'Tailers'. **1899** G. B. Shaw *Let.* 30 Dec. (1972) II. 127 Though the old favorites would get in on both sides, there would be a real contest between the outsiders and tailers. **1900** S. A. Nelson *ABC Wall St.* 161 Tailer. Big operators have a following of little traders who tail-on a bull or a bear movement on the theory that to make money it is a good thing to follow in the wake of the successful men. **1903** F. Norris *Pit* viii. 269 The 'tailers'—the little Bulls—were radiant.

3. *Austral.* One who follows, drives, or tends sheep or cattle; also, a straggling animal.

1893 K. Mackay *Out Back* (ed. 2) III. iii. 233 Fitzspats was absent, having gone out with the 'tailers' that morning. **1927** M. M. Bennett *Christison of Lammermoor* xxvi. 237 The cattle that had come in were watered and handed over to the tailers' mob. **1959** J. Wright *Generations of Men* (1960) ix. 107 Silent dogs at the heels of the tailers.

4. A device with a metal loop used for landing large fish by the tail.

1962 *Times* 31 Mar. 11/5 Some fishermen prefer the tailer to the gaff at all times. **1974** *Country Life* 14 Mar. 599/3, I was not optimistic enough to bring with me a tailer or gaffe or net, but..my fly was taken by an eight-pounder.

5. tailer-out, one who guides timber as it comes off a saw. *Austral.* and *N.Z.*

1907 G. B. Lancaster *Tracks we Tread* vi. 87 The bench sawyers felt it, and the trolley-men; and each tailer-out and engine driver. **1930** W. Smyth *Wooden Rails* ii. 32 She came upon the sawyer and his mate, the tailer-out. **1950** *Landfall* IV. 125 The planer..spits out faced boards for the tailer-out to stack by the goose-saw. **1971** *N.Z. Listener* 8 Nov. 15/4 He worked all day as a tailer-out in the mill.

taileron (tēi·lĕrǫn). *Aeronaut.* [Blend of Tail *sb.*[1] and *aileron.] A horizontal control surface on an aircraft which can function as both elevator and aileron, moving either in unison with its partner or in opposition to it.

1966 D. Stinton *Anat. Aeroplane* viii. 163 The tailerons of the BAC-TSR 2..were slab surfaces that moved either together, as pitch controls, or independently for additional control in roll. **1975** *Flight* 16 Oct. 569 Roll and pitch stabilisation operates by moving the tailplane surfaces as tailerons, leaving the lateral spoilers and ailerons purely under the pilot's control.

tail-gate. 1. (In Dict. s.v. Tail *sb.*[1] 14, sense (*a*).) (Later example.)

1983 G. Swift *Waterland* v. 29 The lighters are approaching. Dick is opening the tail-gates.

2. (In Dict. as above, sense (*b*).) Also, a tail-board or back on a lorry, etc., hinged or removable to facilitate the loading of goods; a hatchback door on a car. orig. *U.S.*

1868 *Oregon State Jrnl.* 28 Nov. 2/3 The whole charge.. [passed] through the tailgate of the wagon. **1886** [in Dict. s.v. Tail *sb.*[1] 14]. **1909** Webster, *Tail gate*,..a heavy wooden panel pivoted to the end of a railroad car to form an incline from the car bottom to the rails. **1940** W. Faulkner *Hamlet* IV. i. 246 The wagon moved gradually backward until the head of the first horse was snubbed up to the tail-gate. **1956** *New Yorker* 1 Dec. 196/2 This year's crop of friction-motor automobiles includes..a ten-inch Country Squire station wagon, with a tail gate that can be opened and closed. **1963** *Guardian* 13 Mar. 5/4 The one-piece tailgate, which is counter-balanced, rises to 5 ft. 10 in. from the ground, providing protection for both load and loader against the rain. **1967** *Financial Times* 21 Apr. 9/8 Hi-pope vertical tailgate equipment for fitting to lorries. **1974** *Daily Tel.* 22 Oct. 10/7 The styling is angular, but pleasant, and features a large rear tailgate for access to the luggage compartment behind the rear seats. **1978** J. Irving *World according to Garp* xv. 309 She felt her way along the truck toward the tailgate. **1980** *Times* 28 May 3/1 BL's long-awaited new small car, the Mini Metro,..is a front-wheel-drive model with two side doors and a tailgate.

B. *attrib.* or as *adj.* **1.** Used to designate a style of jazz trombone playing characterized by improvisation in the manner of the early New Orleans musicians. [From the traditional position of the trombonist at the rear of the wagon in parades, etc.]

1946 R. Blesh *Shining Trumpets* ii. 32 Long glissandi.. heard in the 'tailgate' or circus-style trombone of jazz. **1959** 'R. Gant' *World in Jug* 26 Vic was our trombonist... He had a real tailgate style—that comes from the days when the trombonist sat at the back of the wagon so that he did not push out the eyes of the other bandsmen. **1973** *Times* 25 Jan. 18/6 It needed the utmost in timing and execution, as many would-be tailgate trombonists have since proved by default.

2. Applied to refreshment stops, etc., made during the course of a journey or outing and arranged at the open tail-gate of a parked car.

1970 [see *potlatch b]. **1980** L. Birnbach et al. *Official Preppy Handbk.* 102/2 Tailgate picnics, whiskey sours in the stadium, and the general complexity of the sport guarantee that nobody knows what is going on.

ta·il-gate, *v.* *colloq.* (orig. *U.S.*) Also unhyphened. [f. the sb.] **1.** *intr.* To drive too close behind another vehicle.

1951, 1955 [implied in *vbl. sb.* below]. **1962** 'F. & R. Lockridge' *Murder has its Points* xiv. 160 The police car they followed knew its way, and Weigand tail-gated. **1964** *Punch* 23 Sept. 442/3 'Don't tailgate!'..meaning don't drive on the other man's tail. **1976** *Good Motoring* May 32/1 In the dangerous sphere of motorway driving, for example, they would not tailgate at speeds where if the man in front stopped suddenly they could not..help but stop in exactly the same place on the road.

2. *trans.* To follow (a motor vehicle) excessively closely in another vehicle.

1967 *Lebende Sprachen* XII. 73/2 The use of the verb (which is a recent accession) no longer requires that the car ahead does in fact have a tailgate. One can tailgate a VW. **1968** *National Observer* (U.S.) 8 Apr. 5/4 Negro cabbie John W. Smith, whose arrest for 'tailgating' a police car..helped spark five days of rioting.., was found guilty of assaulting a policeman. **1970** *Daily Tel.* (Colour Suppl.) 9 Oct. 25 The cruise cars are programmed on an intricate shuttle, one tailgating the other, so that no more than 20 seconds can..pass between a radio alarm and the arrival of a car. **1982** H. Kissinger *Years of Upheaval* vii. 228 We took off in a motorcade traveling at a speed of close to 100 miles per hour with cars tailgating each other.

3. *fig.*

1977 *Times Lit. Suppl.* 20 May 618/3 Pictures tailgate each other, wall to wall, and floor to ceiling, in the authentic eighteenth-century manner. **1978** *Saturday Night* (Toronto) Apr. 5/2 One takeover scenario has tailgated another: in 1969 it was Time Inc. muscling in.

Hence **tai·lgater; tai·lgating** *vbl. sb.*

1951 *Amer. Speech* XXVI. 309/1 *Tail-gating, part. phr.*, a bad practice of following too close to the tail gate of the truck ahead. **1955** *Ibid.* XXX. 93 Twenty-two..[lorry] drivers agreed that tailgating means riding too closely behind the vehicle ahead. **1957** *How to Drive* (Amer. Auto. Assoc.) viii. 71 Expressway 'tailgating' is suicidal. **1968** H. McCloy *Mr. Splitfoot* (1969) xvii. 195 Another car passed him and slipped in between his car and Folly's. One of those eager tailgaters who cannot bear to see a few inches between two cars ahead of them. **1970** V. Johnston *Phantom Cottage* xxi. 160 'So if you will just let me keep following your car—'..'All right. But no tailgating.' **1976** *National Observer* (U.S.) 13 Mar. 8/6 In informal testing by The Observer, a Cyberlite appeared to reduce 'tail-gating' behind the test vehicle. **1978** *Telegraph* (Brisbane) 18 Jan. 2/1 A spider on the boot is a lot less dangerous than a tailgater on the bumper bar. **1980** *West Lancs. Evening Gaz.* 21 May 1/1 In a statement today the AA said poor driving, including the 'often fatal practice of "tailgating" ', was responsible for a big increase in serious accidents.

tailing, *vbl. sb.*[1] Add: **1.** (Further example.)

1848 H. W. Haygarth *Recoll. Bush Life Australia* vi. 56 When cattle are first brought to a new country they are subjected to a process called 'tailing', which consists in watching them with horsemen by day, and driving them into their enclosures every night.

5. *tailing ground, pile, yard.*

1878 E. S. Elwell *Boy Colonists* 67 He had caught sight of a native hanging about the 'tailing-ground'. **1897** 'Mark Twain' *Following Equator* lxviii. 687 The gold fields of the world now deliver up to fifty millions dollars' worth of gold per year which would have gone into the tailing-pile under the former conditions. **1934** I. W. Hutchison *North to Rime-Ringed Sun* vi. 54 Across the entrance of the valley..stretched the heaped 'tailing-piles', tippings of the huge gold-dredges. **1930** A. W. Groom *Merry Christmas* xx. 158 The cattle could be seen moving quietly to the tailing yards. **1963** W. E. Harney *To Ayers Rock & Beyond* v. 45 During my early cattle days all mustering was done into drafting yards... They did not alter the method until the drafting-yards was superseded by the 'tailing-yard' with bronco-panels and twisted greenhide ropes with a leather 'hoonda' for the ring.

tailism (tēi·liz'm). *Pol.* [f. Tail *sb.*[1] + -ism.] In Communist jargon, the fault of accommodating policy to the wishes of the masses, thereby following in their wake rather than taking an active revolutionary role.

[**1933** tr. *Lenin's What is to be Done?* ii. 52 It would be more correct to describe its tendency not as opportunism, but *khvostism* (from the word *khvost*)... [Note] *Khvost* is the Russian word for tail.] **1948** J. Towster *Political Power in U.S.S.R.* ix. 180 A double injunction against either 'commanding' or 'tailism' (following, instead of showing initiative). **1957** *Economist* 26 Oct. 320/1 After three months, all the crimes in the jargon book of communist heresy—including such esoteric offences as.. 'tailism' ('refusal to lead the masses')—have been hurled at the hundreds of eminent non-party rightists uncovered or named during..the recent disharmony. **1966** tr. *Quotations from Chairman Mao Tsetung* xi. 124 The reason why such evils as dogmatism, empiricism, commandism, tailism, sectarianism, bureaucracy and an arrogant attitude in work are definitely harmful..is that they alienate us from the masses. **1971** R. MacFarquhar in S. E. Fraser *Educ. & Communism in China* vi. 352 They read aloud the slogans... 'Do you listen to Chairman Mao, or to doctrinairism? Shameful tailism!'

taille. Add: **3.** *Mus.* (See quot. 1944.)

1842 J. A. Hamilton *Dict. Two Thousand Musical Terms* 85 *Taille* (French), the tenor voice or part. **1876** Stainer & Barrett *Dict. Mus. Terms* 420/1 *Taille* (Fr.), (1) the tenor voice or tenor part, (2) the tenor violin, the viola. **1889** Grove *Dict. Mus.* IV. 52/2 The tenor violoncello clef was originally appropriated to the Taille. **1932** C. S. Terry *Bach's Orchestra* v. 98 His players were certainly never provided with an instrument capable of sounding g, except the taille. **1944** *Harvard Dict. Mus.* 731/2 *Taille* [F.], old name for a middle voice, particularly the tenor. The term was also used for instruments performing such parts, e.g., *taille de basson*, tenor oboe; *taille de violon* or simply *taille*, viola.

taille-douce. Delete *Obs.* and add later examples.

1897 O. Firth *Postage Stamps* ii. 7 The original example of line-engraved stamps, or stamps 'engraved in *taille-douce*'. **1924** F. J. Melville *Compl. Philatelist* vi. 83 Most of the stamps produced after this portrait were surface-printed, but the Falkland Islands and the Turks and Caicos Islands presented it in *taille-douce* engraving. **1955** Boggs & Strange *Foundations of Philately* xi. 128 Line engraving is a classic process which was used for the first adhesive stamps issued in Great Britain in 1840... This process is also known as intaglio, recess printing, or

taille douce. **1969** F. L. Wilder *How to identify Old Prints* v. 77 Line-engraving (taille-douce) had become the principal form of engraving in France and it was said that the art was almost born and died with him [*sc.* Jacques-Philippe Le Bas], shortly before the Revolution. **1975** W. Finlay *Illustr. Hist. Stamp Design* ii. 21/2 The paper usually has to be dampened..and then is laid on top of the plate and forced down under great pressure, so that the plate bites into the paper. The paper squeezed into the grooves picks up the ink; this is what gives stamps and banknotes printed in this fashion their characteristic ridged surface. This process is known as *intaglio, taille douce*, recess printing or direct plate printing. Philatelists often use the term 'line engraving'.

‖ **tailleur** (tayör) [Fr.] A woman's tailor-made suit.

1923 *Weekly Dispatch* 29 Apr. 15 New, indeed, is the sunshade composed of chrome leather, designed specially for use with the morning *tailleur*. **1945** N. Mitford *Pursuit of Love* xviii. 158 The spring *tailleurs*, the summer *imprimés*, the autumn *ensembles*, and the winter furs. **1982** T. Fitzgibbon *With Love* I. ii. 18, I pressed the black *tailleur*, bought a gay scarf..and went off to look for a job.

tailor, *sb.* Add: **5.** *tailor-man* (earlier example), *-shop*.

1882 'Mark Twain' *Prince & Pauper* xiii. 154 Noble large stitches..that do cause these small stingy ones of the tailor-man to look mightily paltry. **1916** G. Frankau in *Wipers Times* 3 July 7/2 Oh! where is Caw-Caw the Captain bold, The pride of the tailor-shop? **1979** *Maledicta* III. 20, I played a lot with Mezzrow. And with Sidney Bechet in his tailorshop in Brooklyn.

6. a. tailor-fashion *adv.* = *tailor-wise* adv. below; **tailor tack(ing)** = *tailor's tack*, sense 6 b below; **tailor-wise** *adv.*, in a cross-legged position.

1877 Ruskin *St. Mark's Rest* II. iv. 45 A curly-haired personage..sitting in an absurd manner, more or less tailor-fashion. **1902** R. P. Browne *Pract. Work of Dressmaking & Tailoring* iii. 80 '*Tailor Tacking*'—This stitch is used to trace the seams, &c., through to the second side of the cloth—following the lines which have been marked with tailor's chalk. **1979** M. McCririck *Better Dressmaking* iv. 35 *Tailor tacker*, for working quantities of tailor tacks on a thick pad of foam rubber... *Marking set*, for transferring single pattern marks to both sides of fabric at the same time as an alternative to tailor tacking. **1885** Tailor-wise [see Tuck *v.*[1] 6]. **1913** W. de la Mare *Peacock Pie* 20 To see them squatting tailor-wise Around a keg of rum. *c* **1973** J. Cholerton *Acrobatic Enchainements* (Assoc. Amer. Dancing) (ed. 7) 3 Lower (side view) to tailor-wise sit.

b. (Occas. *tailors'*.) **tailor's chalk,** hard chalk or soapstone used in tailoring, etc. to make eradicable marks on fabric as a guide to fitting; **tailor's dummy** (lit. example); **tailor's tack** (see quot. 1975); usu. in *pl.*; similarly **tailor's tacking** (cf. *tailor tack(ing)*, sense 6 a above); **tailor's twist** (example).

1881 C. C. Harrison *Woman's Handiwork* III. 167 Grass and iris were sketched on the blue surface with tailor's chalk. **1932** D. C. Minter *Mod. Needlecraft* 107/2 Almost indispensable to successful dressmaking are..a yard stick, tailors' chalk. **1966** *Olney Amsden & Sons Ltd. Price List* 36 Tailor's chalk.. Loose boxes of 100 pieces square or triangle. **1977** A. Scholefield *Venom* v. 204 A maze of sewing machines and tailors' dummies. **1927** *New Butterick Dressmaker* x. 98 Tailors' *tacks*,—after cutting out a garment..mark with tailors' tacks the perforations at 'Outlet' or 'Let-Out' seams. **1964** *McCall's Sewing* ii. 32/2 *Tailor's tack*, method of marking pattern symbols. **1975** C. Calasibetta *Fairchild's Dict. Fashion* 488/2 *Tailor's tacks*, large stitches taken through two thicknesses of fabric with a loop left between the layers which are later cut apart, leaving tufts in each piece; used for guide marks in tailoring. **1952** E. King *Successful Home Dressmaking* iv. 22 *Tailor's tacking*, suitable for all fabrics, but specially for woollens, crêpes, lace, velvets and loosely-woven or flimsy goods. **1873** *Young Englishwoman* Mar. 150/2 Work the button-holes with tailors' twist, which is sold..at one penny per dozen lengths of one yard.

tailor, *v.* Add: **5.** *fig.* To design or alter (something) to suit specific needs; to adjust or make suitable. orig. *U.S.*

1942 *Sun* (Baltimore) 23 Oct. 6/2 Maryland farmers will tailor next year's crops to a size which can be worked by their individual families. **1950** *Engineering* 9 June 655/3 To secure maximum performance the apparatus should be 'tailored' for each application. **1951** M. McLuhan *Mech. Bride* (1967) 98/2 To the mind of the modern girl, legs, like busts, are power points wich she has been taught to tailor. **1959** *Listener* 2 July 35/1 A writer who soberly tailors a piece to fit into sixty minutes. **1961** *New Scientist* 23 Feb. 484/1 Derivatives containing carbon can be 'tailored' to have suitable handling properties. **1964** *Observer* 28 June 23/3 Of course, the story of 'Hiroshima Pilot as Mental Patient' was at once tailored to fit the headlines. **1982** G. F. Newman *Men with Guns* p. vi, The shotguns..the pair of Luigi Franchi double twelve-gauge..he had had tailored in London.

tailored, *ppl. a.* (In Dict. s.v. Tailor *v.* 2.) Add: **b.** *fig.* Made to suit particular needs; adjusted. orig. *U.S.*

1942 *Sun* (Baltimore) 12 Aug. 5/4 Gasoline rationing officials..said that motorists coming into a rationed area from a non-rationed area on a pleasure trip can be supplied with a 'tailored' ration book to fit their needs. **1954** *Economist* 10 Apr. 141 A specially tailored compound containing the radio-active element. **1956** *B.B.C. Handbk.* 1957 38 The *North American Service* produces specially 'tailored' programmes to be rebroadcast by American and Canadian stations and networks. **1963** *Daily Tel.* 14 Mar. 15 (*heading*) 'Tailored' driving for Sunbeam Alpines.

tailoring, *vbl. sb.* Add: **d.** *fig.* The act of adjusting or producing to suit specific needs. orig. *U.S.*

1943 *Sun* (Baltimore) 5 Feb. 4/5 Thirty-seven ration boards in the State now are completing the tailoring of ration books. **1951** *Times* 21 Sept. 1/5 (Advt.), High-polymer chemist for applied research on the 'tailoring' of linear macromolecules required by an important, very modern works in S.W. England. **1979** *United States 1980/81* (Penguin Travel Guides) p. xii, Very precise editing and tailoring keeps our text fiercely subjective.

tailorless, *a.* (Earlier example.)

1876 J. A. H. Murray *Let.* in K. H. E. Murray *Caught in Web of Words* (1977) x. 192 We are not quite *tailorless* and so not obliged to go *trouserless*.

tailor-made, *a.* Add: Also **tailormade. 1.** (Later examples of the sb.)

1932 Auden in *Rev. Eng. Stud.* (1978) Aug. 284 She's been having her tailormades altered. **1981** A. Lurie *Lang. Clothes* viii. 222/1 The ordinary woman..might wear..a.. wool or linen suit (the 'tailor-made') with a shirtwaist.

3. Made to answer a specific demand or requirement; perfectly suited *for* a particular purpose.

1896 G. B. Shaw in *Sat. Rev.* 7 Nov. 494/1 The public were tired of tailormade plays, and were ripe for a revival of colour and costume. **1897** —— in *Ibid.* 1 May 470/2 A theatre which is panelled, and mirrored, and mantelpieced like the first-class saloon of a Peninsular and Oriental liner..is no place..for anything except tailor-made drama and farcical comedy. **1938** *Cine-Kodak News* Aug. 10/1 (*heading*) Tailor-made showings. **1946** *Richmond* (Va.) *Times-Dispatch* 24 Nov. 12B The work is another venture in the science of redesigning plants and animals through genetics to bring forth 'tailor-made' species. **1953** *Economist* 30 May (Suppl.) 9/1 The tailormade molecules which form the long fibres used in weaving Terylene. **1958** *Spectator* 14 Feb. 194/3 A seat tailor-made..for the Liberals to dance further fandangoes on the carefully laid plans of their rivals. **1963** [see *peg *sb.*[1] 1 e]. **1970** G. F. Newman *Sir, You Bastard* i. 14 It's a job almost tailor-made for you. **1972** [see *inside *a.* d]. **1980** *Jrnl. R. Soc. Arts* Mar. 185/2 The tailor-made building had arrived.

4. Designating a ready-made (as opp. hand-rolled) cigarette. Freq. *ellipt.* as *sb. slang* (orig. U.S.).

1924 'Digit' *Confessions 20th Cent. Hobo* 12 *Tailor-mades*, ordinary ready-made cigarettes. **1930** J. Devanny *Bushman Burke* 88 He smoked 'tailor-mades' now. **1945** *N.Z. Geographer* I. 23 The background of all this is tobacco. There are but few pipes, and 'tailor-made' cigarettes are only a luxury. 'Roll your own' is on most lips. **1952** E. Wilson *Equations of Love* 29 Just a match. .. I don't smoke tailor-mades. **1955** *People* (Austral.) 1 June 8/1 Leopold..was..told he had been reported for possessing contraband, a 'tailor-made' cigarette. Leopold didn't even know what a 'tailor-made' cigarette was. **1962** N. Freeling *Love in Amsterdam* III. 145 Martin stayed quiet after distributing his last tailormades. **1974** D. Sears *Lark in Clear Air* iii. 40, I went and bought a package of Turret tailor-made cigarettes and I lit one.

tai·lor-make, *v.* orig. *U.S.* [Back-formation f. *tailor-made *a.* 3.] *trans.* To design (something) according to specific requirements.

1946 *Sun* (Baltimore) 1 July 8/2 Under ideal weather conditions that seemed to have been tailor made for the occasion, the atom bomb was successfully exploded. **1953** *Ibid.* 14 Apr. 3/4 It may be possible to tailor-make drugs which would prevent each type of dangerous virus from getting the electrical charges it needs in order to attack the target cell. **1959** *Listener* 18 June 1052/2 What will happen when Franco does disappear from the political scene? Some think he is trying to tailor-make a monarchy. **1967** M. Chandler *Ceramics in Mod. World* vi. 168 Alumina ceramics can be tailor-made to meet a wide range of industrial requirements. **1973** *Daily Tel.* 9 Aug. 3 (Advt.), If you haven't yet got central heating, an Esso Chartered Installer will tailor-make a complete system to suit your needs. **1981** *Arts Alert* Oct. 3/1 We were asked to tailor-make the hall for the LSO's needs.

tail-piece. 1. (Further examples.)

1723 J. Nott *Cook's & Confectioner's Dict.* sig. KK3v, Draw your Sturgeon;..cut your first and second Rand very fair, cutting the Tail-piece least. **1786** Busby *Dict. Mus.* s.v. *Tail-piece*, the thin, broad piece of ebony horizontally suspended over the lower end of a violin, and to which one end of the strings is attached.

tail-pipe. Add: Also **tailpipe. 2. a.** *Aeronaut.* (See quot. 1933.) Chiefly *U.S.*

1922 L. S. Marks *Airplane Engine* xvii. 423 For durability the muffler should be attached to the end of a tail pipe 6 or 8 ft. long which will cool the gases sufficiently **1933** *Brit. Standards Inst. Specif.* CLXXXV (Gloss. Aeronaut. Terms) vi. 53 *Tail pipe*, a pipe which leads exhaust gases away from a manifold. **1956** C. W. Smith *Aircraft Gas Turbines* viii. 167 The exhaust gas pipe from the turbine is conducted to the propulsion nozzle through a duct (often called *tailpipe*). **1956** [see sense 2 b below]. **1973** *Physics Bull.* Dec. 728/1 This difference is often referred to as 'excess' or 'tailpipe' noise. Work to date has identified this noise source as being associated with the aerodynamic environment in the core exhaust system.

b. *U.S.* The exhaust-pipe of a motor vehicle.

1956 W. A. Heflin *U.S. Air Force Dict.* 511/1 *Tail-pipe*.., an exhaust pipe for escape of gases generated in an internal combustion engine; specif. in a jet engine, the pipelike structure aft of the exhaust nozzle. **1961** E. A. Venk *Automotive Fundamentals* (ed. 2) xi. 179/2 The tail pipe is a long winding pipe which is connected to the muffler... Two tail pipes are used in a dual exhaust system. **1964** S. Bellow *Herzog* 121 Unless you remembered to bear right the tailpipe would scrape on the rocks. **1973** *Sunday Bulletin* (Philadelphia) 14 Oct. (Parade Suppl.) 16/3 Unfortunately the tailpipe didn't clear the underside of the car. **1979** D. Anthony *Long Hard Cure* iv. 33 Lorraine's car was parked..with the tailpipe backed into a hedge.

tail-race. Add: **b.** (Earlier examples.)

1863 *App. Jrnls. House Reps. N.Z.* D. vi. 14 Where the water is heavy, and there are no means of cutting a tail-race, water-wheels have been erected, with Californian pumps attached. **1874** [see *paddock *sb.*[2] 3].

c. The watercourse leading from the turbine of a power-station or dam, etc.

1953 *Times* 4 Aug. 3/4 An electric screen has been devised which is successful in preventing salmon and sea trout from swimming into the tailrace (the outflow from a water turbine) of a power station. **1974** *Progress* (Easley, S. Carolina) 24 Apr. II. 10/2 Rainbows are the most common species stocked into tailrace waters. **1978** *Texas Parks & Wildlife* July 13/3 We've dealt mainly with tailraces below large flood-control and power-generating dams.

tail-rope. 2. a. Delete † *Obs.* and add later N. Amer. examples.

1916 F. W. Wallace *Shack Locker* (1922) 81 Make th' tail rope fast. *a* **1932** L. S. Tawes *Coasting Captain* (1967) 27 We had a long tail rope to slack the staysail over with. **1942** *Amer. Neptune* II. 234 'Tail rope' is a short line made fast to the after end of the fore-staysail boom for the express purpose of backing the jumbo.

tailspin (tēi·lspin), *sb.* Also **tail spin, tail-spin.** [Cf. Spin *sb.*[1]] **a.** *Aeronaut.* A downward movement of an aircraft in which the tail describes a spiral.

1917 V. W. Page *Gloss. Aviation Terms* 23 Tail spin. **1919** in C. A. Smith *New Words Self-Defined* 201, I remember when I thought it was time to try a *vrille* or tail-spin. **1926** *Daily Colonist* (Victoria, B.C.) 17 Jan. 2/5 Collison's airplane went into a tail spin when his motor failed to work. **1953** C. A. Lindbergh *Spirit of St. Louis* II. vi. 263 He'd tried to land out of a tailspin—that was asking for a crash.

b. *fig.* A state of chaos, panic, or loss of control.

1928 R. E. Byrd *Let.* 24 July in *K. W. Rendell Autograph Catal. No. 34* (Kingston Galleries) (1968) 2, I am pretty much in a tail spin which I don't expect to get out of before the main part of the expedition leaves. **1929** *Times* 2 Nov. 12/6 The cyclic tailspin which has occurred in the 11th year of each of the four great previous periods of commercial prosperity. **1935** Wodehouse *Luck of Bodkins* xv. 183 A dashed shame, he considered, that things had gone into a tail spin for him like that. **1954** A. Huxley *Let.* 5 Dec. (1969) 717 A child stoked with fats will find it hard, because of adrenalin, to digest and will probably go into a bad psychological tailspin in consequence. **1967** E. S. Gardner *Case of Queenly Contestant* vii. 84 [He] went into a tailspin. He was afraid of the responsibility. He was afraid his father would find out. *a* **1974** R. Crossman *Diaries* (1977) III. 867 After my denial of the story that appeared in his Manchester lecture, the poor man came along in a terrible tail-spin to see me. **1982** *Daily Tel.* 16 June 19 The Argentine economy was already in a tail-spin before the Falklands invasion.

tai·lspin, *v.* Also **tail-spin.** [f. the sb.] *intr.* To perform or go into a tailspin.

1927 [see *barrel roll* s.v. *barrel *sb.* 11]. **1936** F. Clune *Roaming round Darling* xvii. 167 We slithered in the mud, barrel-rolled, tail-spinned, sank to our hocks, and became part of the everlasting scenery. **1972** M. J. Bosse *Incident at Naha* iv. 204, I had never seen Virgil so moved or so bitter. It was unlike him to tailspin into such a downer. **1973** *Daily Colonist* (Victoria, B.C.) 21 June 1/5 The red and white air craft was about 400 feet in the air a mile from the airport when it suddenly went into a nose dive, then tailspun to the pavement.

tainiolite, var. *tæniolite.

taint, *sb.* Add: **C. 1. c.** An unpleasant scent or smell. Cf. *taint *v.* C. 4 c.

1927 H. Williamson *Tarka the Otter* i. 5 Mingled with the flower odours..was the taint that had given her a sudden shock..; the taint most dreaded by the otters..—the scent of Deadlock, the great pied hound. **1951** 'J. Wyndham' *Day of Triffids* xi. 205 On the higher ground there was still little taint in the fresh air.

taint, *v.* Add: **C. 4. c.** *trans.* To drive *out* (rabbits) from their burrows by the introduction of an offensive smell.

1909 O. Jones *Ten Years Game-Keeping* ii. 22 Gipsies are a help to the keeper..when he has a difficulty in tainting out a colony of rabbits. **1972** *Young's Sporting Appliances* (S. Young & Sons Ltd.) ii. 13 Proved to be best after exhaustive tests over many years for tainting out rabbits to lie out for shooting.

'taint (tēint), *v.* Also **taint, t'ain't,** etc. Dial. and vulg. contraction of *it ain't*: see Ain't *v.*; 'T.

1839 [see *snum *v.*]. **1859** A. J. Symonds *Let.* Sept. (1967) I. 206 You will think I am fallen desperately in love. Yet it taint so. **1919** W. Deeping *Second Youth*

xxiii. 194 'Taint like 'im. 'E used to be sensible. **1942** Z. N. Hurston in A. Dundes *Mother Wit* (1973) 225/1 'T'ain't nothing to you, nohow. **1974** H. R. F. Keating *Underside* viii. 77 'I'm sure whatever they say's undeserved.' 'No, t'ain't. You ninny.'

taintless, *a.* Add: Hence **tai·ntlessness,** the quality of being taintless.

1811 Shelley *Let.* 26 Nov. (1964) I. 144 The first words you spoke to me..are eternal earnests of your taintlessness and sincerity. **1963** *English Studies* XLIV. 22 Taintlessness and incorruptibility seem to depend not so much on the predominance of blood stains as on outlook.

‖ **tai-otoshi** (tai,otǫ·ʃi). *Judo.* Also **Tai-otoshi, tai-o-toshi,** etc. [Jap., f. *tai* body + *otoshi* the act of dropping.] The body drop throw.

1950 E. J. Harrison *Judo* iii. 39 The most suitable moment for attempting the Taiotoshi is when your opponent with unbent legs, his body somewhat stiff, has leaned forward a little with his weight resting on his right leg. **1957** Takagaki & Sharp *Techniques Judo* iii. 24 If *tai-otoshi* fails and the opponent pulls back. **1964** Leggett & Watanabe (*title*) Championship judo: tai-otoshi and o-uchi-gari attacks. **1978** D. Starbrook *Judo* iv. 40/2 The tai-o-toshi is a hand throw, and its great advantage over so many other throws is that it can be performed against opponents who are either stationary or moving.

taipan[1] (tai·pæn). Also 9 **taepan, typan.** [Dial. var. of Chinese *dàbān*.] **a.** A foreign merchant or businessman in China. **b.** The (foreign) manager or head of a firm in China, esp. Hong Kong. Also *fig.* Hence **tai-panism.**

1834 *Canton Reg.* 28 Oct. 170/2 The election of a temporary Chief for the Superintendence of British affairs, until the appointment of one from England, who must be a *taepan* or Merchant, as before and not a Government Officer. **1892** in K. Lentzner *Dict. Slang-English Austral.* 91 My *typan* must make fun of me, When all his crowd can see—Ah! well, perhaps they do not care For a little clerk like me. **1921** *North-China Herald* 24 Dec. 815/1 What is 'Taipanism as seen in China', which Mr. Ku discusses in a recent issue of the 'Evening Standard'?.. Mr. Ku finds that 'Taipanism' is the spirit of respect for the sacred rights of property and vested interest. **1922** W. S. Maugham *On Chinese Screen* xv. 63 With the elderly, but single, taipan of an important firm, what she simply loved was a game of golf. **1957** R. Mason *World of Suzie Wong* I. i. 9 Chinese taipans, who made the richest Europeans seem like paupers. **1972** *Times* 21 Oct. (Hongkong Suppl.) p. i/4 Opium can no longer be indulged in as it was with restraint by the *taipans*, or merchants, in old Shanghai. **1977** W. Tute *Cairo Sleeper* i. 14 Ambassadors..mingled with other taipans of the higher civil service... Most members of the club worked in Whitehall. **1983** *Sunday Tel.* 10 Apr. 20/6 The Keswicks of Dumfries married into the Jardines in the middle of the last century. Four of them, including Henry, became *taipan* or head of the house.

taipan[2] (tai·pæn). Also **Taipan.** [Aboriginal name.] A large dark brown venomous snake, *Oxyuranus scutellatus*, of the family Elapidæ, native to northern parts of Australia.

1933 D. F. Thomson in *Proc. Zool. Soc.* 858 The name 'taipan', by which *O. scutellatus* is known to the aborigines of Cape York Peninsula, is an excellent vernacular name for the species. *Ibid.*, The taipan frequents the open country of the coast, as well as the inland plains and savannah forests. **1953** P. Brickhill in I. Bevan *Sunburnt Country* 96 The long-fanged taipan..grows eleven feet long, and nearly as thick as a man's arm... Only two men are known to have survived a taipan bite. **1966** G. Durrell *Two in Bush* v. 159 To have kept and bred something as rare and shy as a Taipan is a very great triumph. **1982** *Daily Tel.* 18 Feb. 3/1 A western taipan—one of the world's deadliest snakes—was bitten by a mouse and is fighting for its life in the Darwin Museum, Australia.

tai-ping. (Earlier example.)

1853 *North-China Herald* 7 May 158/1 'Conquering the rivers and mountains', the expression by which the 'holy warriors' of Tae-ping designate their enterprise.

‖ **taipo** (tai·po). *N.Z.* Also **taepo, Taipo, typo.** [Origin uncertain: see quots. 1891, 1946.]

1. An evil spirit.

1848 R. Taylor *Leaf from Nat. Hist. N.Z.* 43 (Morris), *Taipo*, female dreamer; a prophetess; an evil spirit. **1883** W. Colenso in E. E. Morris *Austral Eng.* (1898) 454/2 Taepo means to visit or come by night,—a night visitant, —a spectral thing seen in dreams,—a fancied and feared thing, or hobgoblin. **1886** *N.Z. Country Jrnl.* X. 262 His wife became seriously affected, declaring that *Taipo* had entered into her. **1891** E. Tregear *Maori-Polynesian Comparative Dict.* 440/1 *Taepo*, a goblin, a spectre. Cf. *tae*, to arrive; *po*, night. **1921** H. Guthrie-Smith *Tutira* xi. 91 This crossing has always been known in my time as the 'Taipo'—goblin—crossing, a name probably given because of a totara block which used to lie there hewn roughly to the similitude of a man's head. **1946** *Jrnl. Polynesian Soc.* June 150 *Taipo*, supernatural being; goblin: used by the Maori believing it to be Pakeha, and by the Pakeha believing it to be Maori; often spelt taepo, which also is not a Maori word: so taipo is a word coined by no one knows whom. **1968** *N.Z. Listener* 15 Mar 6/5 He hurriedly looked both ways and took to the scrub as if a taipo were after him. **1971** *Ibid.* 1 Mar. 13/2 As for dreaded taniwhas and taipos, why, I could take you to the home of some.

2. = *weta.

1928 J. Devanny *Dawn Beloved* I. vii. 47 The very apogee of excitement would be reached when a 'typo' was discovered. Especially if it happened to be a big fat male. **1946** F. Sargeson *That Summer* 176 But the wetas come out at night... The Maoris call them taipos. **1966** *Encycl. N.Z.* III. 636/2 The tree or ground wetas and the 'taipos' of the West Coast of the South Island, the name of which to the Maori means 'the devil who comes by night'.

tairoa, var. *toheroa.

tait (tēit), *sb.* *Austral.* [Aboriginal name.] = *honey possum* s.v. *honey *sb.* (*a.*) 7.

1894 R. Lydekker *Hand-bk. Marsupialia* 121 Known to the natives by the name of Tait, and Nulbenger, the Long-snouted Phalanger..is generally found..from Swan River to King George's Sound. **1941** E. Troughton *Furred Animals Austral.* 81 The quaint little animal had quite a variety of names in the native vocabularies.., the one favoured as a popular name being 'Noolbenger', and others including 'Ait' [*sic*] and 'Deed'.

Taittinger (ta·tæ̃ʒe). Also *erron.* **Tattinger.** The proprietary name of a champagne manufactured and shipped by the firm of Taittinger in Rheims.

1949 *Trade Marks Jrnl.* 6 Apr. 304 Taittinger... Champagne. Etablissements Taittinger Mailly & Cie (a Societe Anonyme organised under the laws of France). **1964** A. Launay *Caviare & After* xv. 106 Unlike other wines, Champagne is known under the name of the shipper... The best known are Boulinger.., Pommery, Taittinger and Veuve Clicquot. **1967** A. Arent *Grave-digger's Funeral* xii. 189, I will buy you a jereboam of Tattinger. **1971** R. Temple *Schulsinger Affair* i. 12 A bottle of Taittinger Blanc de Blancs, well iced. **1978** R. B. Parker *Judas Goat* xii. 77 Hawk had filled the sink with ice and put..another bottle of Taittinger champagne in to chill. **1978** D. A. Stanwood *Memory of Eva Ryker* xxv. 238 A magnum of Tattinger '05.

Taiwanese (taiwăni·z), *sb.* and *a.* [f. *Taiwan*, the name of a large island off the south-east coast of China + -ese.] **A.** *sb.* A native or inhabitant of Taiwan. **B.** *adj.* Of or pertaining to the island of Taiwan or its inhabitants.

1942 A. J. Grajdanzev *Formosa Today* iv. 53 A rise in the price of Taiwanese rice. *Ibid.* vi. 86 The Taiwanese are a fish-eating people. **1962** E. Snow *Other Side of River* (1963) lxxi. 549 People talk about giving the Taiwanese a plebiscite on whether they want to join China or not. **1969** *Times* 9 Dec. (Taiwan Suppl.) p. v/8 Sixty-three per cent of local government officials are natives of Taiwan. All local magistrates are Taiwanese. **1970** D. Dodge *Hatchetman* i. 22 We've got Army liaison..Taiwanese liaison, British liaison, Old China hands. *Ibid.* ix. 109 Attempts have been made on my life before, but by Taiwanese; my people, not yours. **1978** *Chicago* June 225/1 The simple storefront does a great job with a menu that includes..unusual Taiwanese squid and cuttlefish dishes. **1979** *Pacific Affairs* LII. 455 There was almost no contact between Red Army units and Taiwanese in this period.

taj (tādʒ). Also tuj. [Arab. (Pers.) *tāj* crown.] A crown or head-dress of distinction (see also quot. 1877).

1851 *Illustr. Catal. Gt. Exhib.* IV. i. 918/2 Crown, or tuj, as worn by the King of Oude; without jewels. **1877** *Encycl. Brit.* VII. 113/2 The tâj, or white cap, with the proper number of terks, or sections, belonging to the order [of dervishes]. **1886** J. Atkinson tr. *Firdausi's Shāh Nāmeh* 92 He also gave him a taj, or crown of gold, which kings only were accustomed to wear.

Taj: see *Taj Mahal a.

Tajik (tādʒi·k), *sb.* and *a.* Also **Tadjik, Tadzhik,** etc. [a. Pers. *tājik* one who is neither an Arab nor a Turk, a Persian.] **A.** *sb.* **a.** A people of Iranian descent inhabiting Afghanistan and the Turkistan region of Central Asia; now also *spec.* a native or inhabitant of the Tajik S.S.R. **b.** The Persian dialect spoken by this people. **B.** *adj.* Of, pertaining to, or designating this people.

1815 [see *Hindki]. **1836** *Penny Cycl.* V. 71/2 The Tadjicks consider themselves as the aborigines of the country [*sc.* Bokhara], and as the descendants of the ancient Sogdi and Bactrians. **1871,** etc. [see *Sart *sb.*[2] and *a.*]. **1911** *Encycl. Brit.* XXVI. 365/1 The Tajik population of the richly-cultivated districts north of Kabul. **1920** *Christian World* 26 Aug. 9/2 The fanatical Tadjiks and Sarts of that portion of Turkestan. **1949** F. Maclean *Eastern Approaches* I. x. 152, I had no sooner got to sleep than they dug me in the ribs and introduced themselves as Tajiks. **1953** O. Caroe *Soviet Empire* iii. 33 With sedentary and town populations speaking Tajik it can as safely be accepted that they are of Iranian stock-origin. **1959** E. H. Carr *Socialism in One Country* II. xx. 268 The Tajik population, the only non-Turki-speaking group in the area. **1964** H. H. Paper tr. *Shafeev's Short Gram. Outl. Pashto* 1 Until 1936 the official language of Afghanistan was Kabuli, one of the dialects of Tajik. **1970** *Times* 24 Mar. 7/1 The least emancipated women seem to be found among the Tadjiks, an Iranian people living in the wild mountainous regions at the Soviet Union's extreme south-eastern tip. **1974** *Listener* 17 Oct. 494/2 Travelling through the Hindu Kush..with three Tajik tribesmen. **1977** Yin Ming *United & Equal* 4 There is hardly any place along China's long border without its communities of minority nationalities. Among them are..the Kazakhs, Uighurs, Khalkhas and Tajiks in Sin Kiang. **1978** *Times* 18 Oct. 16/8 [In] Uzbekistan.. many..speak Tadzhik..and other Asian languages. **1979** A. Henning tr. *Myrdal's Silk Road* i. 5 Our hosts..are Tadzhiks... This is the Tashkurghan Tadzhik Autonomous County in the far west of China. *Ibid.* 8 We are sitting on the starkly beautiful, thick, Tadzhik carpet of felted wool. **1980** G. Richards *Red Kill* xxv. 204 Children had come from all corners of the Soviet Union... Azerbaijanis, Tadjiks, Ukrainians.

Taj Mahal (tādʒ, tāʒ māhā·l). [perh. a corruption of *Mumtaz Mahal* (see below) under the influence of *taj; cf. Mahal.] The name of a mausoleum built at Agra by Shah Jahan in memory of his wife known as Mumtaz (Pers. 'chosen one') Mahal (d. 1631), used: **a.** *ellipt.* as **Taj.**

1858 W. H. Russell *Diary* 14 Oct. (1860) II. 279, I had seen that Pearl of architecture, the wonder of the world—The Taj of Agra. **1887** Kipling *Lett. of Marque* (1891) i. 2 He saw from the train the Taj wrapped in the mists of the morning. **1912** E. Lutyens *Let.* May in M. Lutyens *Edwin Lutyens* (1980) vii. 104 The Taj and some other of the tombs have charm. **1978** 'M. M. Kaye' *Far Pavilions* xiv. 225 Shah Jehan's Empress the lady of the Taj.

b. *fig.* denoting that which is excellent or surpassing of its kind. Also occas. in *transf.* use.

1895 *Q. Rev.* Apr. 353 Stevenson has set up and decorated with every precious stone a building so magnificent, that it deserves to be called the Taj Mahal of our prose literature. **1950** Partridge *Name into Word* 429 One not seldom meets with such phrases as 'the Taj Mahal of architectural elegance' or 'the Taj Mahal of romantic architecture'. **1980** D. Pownall *Between Ribble & Lune* i. 20 In 1906..his lordship ordered the structure as a Taj Mahal for his dead wife.

taka (ta·kā). Also **Taka**; pl. **-(s).** [Bengali *ṭākā*.] The basic monetary unit of Bangladesh, equivalent to one hundred paise; also, a banknote of this value.

Quot. 1975 illustrates the idiomatic use of *taka* with *crore* (= 10 million).

1972 *Guardian* 22 Aug. 10/4 The Bangladesh taka, officially at par with the Indian rupee, is being freely offered..at 40 per cent discount. **1975** *Bangladesh Times* 21 July 1/3 He said if the factory could produce..30,000 tons of pulp annually that would meet the home demand as well as fetch about Taka six crore in foreign exchange by exporting pulp abroad. To meet the home demand.. Bangladesh now had to import pulp worth seven to eight crore Taka annually. **1976** M. S. Hoque *Hunger* I. i. 7 Nuribow opens the knot at the corner of her sari, takes a Taka therefrom and pays. **1976** *Sci. Amer.* Oct. 32/1–2 (*caption*) The Bengali writing offers a '250-taka prize' (about $17) to anyone who reports a smallpox case to a health office. **1979** *Church Times* 26 Oct. 13/1 Until two years ago, the family's income consisted entirely of Appavo's wage—a princely thirty takas a week (about a pound).

Taka-diastase (tækă,dəi·ăstēiz). Also **taka-.** [f. the name of J. Jokichi *Taka(mine* (1854–1922), Japanese-born biochemist and industrialist + Diastase.] A preparation containing a variety of enzymes which is obtained after the treatment of rice or bran with the mould *Aspergillus oryzæ*; now a proprietary name.

1896 *Jrnl. Amer. Med. Assoc.* XXVII. 374 Notes on taka-diastase. **1928** [see *pyrophosphatase]. **1928** *Official Gaz.* (U.S. Patent Office) 6 Nov. 10/2 Takamine Ferment Company, New York... *Taka-diastase*... For koji, moyashi, diastase, ferments, and converting agents. Claims use since 1895. **1955** *Trade Marks Jrnl.* 9 Mar. 258/2 *Taka-diastase*... Diastase for pharmaceutical purposes. **1960** [see *koji]. **1969** G. Smith *Introd. Industr. Mycology* xv. 312 Takamine introduced into commerce.. products of high enzymic activity, particularly suitable for the dextrination of starch and the desizing of textiles. These products have been sold under the names 'Taka-diastase'..and 'Oryzyme'. **1976** *Ann. Rev. Microbiol.* XXX. 8 During our studies of adenylate deaminase, we observed that the enzyme from Takadiastase..would deaminate approximately 50% of the adenylate added to the reaction vessel.

takahe (ta·kahi). Also **Takahe.** [Maori.] = *notornis.

1851 G. Mantell *Petrifactions* ii. 128 No one had seen such a bird, but all agreed that it was the traditional Moho or Takahé, which they believed was utterly extinct. **1898** *Daily News* 22 Oct. 2/2 The other day a specimen of the takahe..was found in the South Island. **1915** *Chambers's Jrnl.* May 318/2 There are the flightless kiwi, weka, and kakapo parrot; also the very rare takahe. **1966** G. Durrell *Two in Bush* iii. 103 Then, quite suddenly, from behind a large clump of snow grass, a Takahe appeared... I was imagining something about the size of an English Moorhen..but there stood a bird the size of a large turkey. **1978** *Nature* 9 Feb. 507/2 Take, for example, the takahe (*Notornis mantelli*), a flightless gallinule endemic to New Zealand.

Takali, var. *Takulli.

takapu (ta·kāpu). *N.Z.* [Maori.] The Australian gannet, *Sula serrator*.

1842 W. Colenso in N. M. Taylor *Early Travellers N.Z.* (1959) 54 The natives often take this bird... They call it Takapu. **1882** W. L. Buller *Man. N.Z. Birds* 91 (*heading*) Gannet. Takapu. **1966** R. A. Falla *Field Guide Birds*

N.Z. 60 Australian Gannet..Local name: Takapu... Mainly white; crown and nape pastel yellow.

Takayasu (takaya·su). *Path.* [The name of Michishige *Takayasu* (1872–1938), Japanese ophthalmologist, who described the disease in 1908.] *Takayasu's disease*: a chronic arteritis leading to obstruction of blood-flow, esp. in the vessels arising from the aortic arch; pulseless disease.

1952 *Amer. Heart Jrnl.* XLIV. 629 Pulseless or Takayasu's disease is considered by Japanese ophthalmologists to be a rare but definite clinical entity. **1969** Edington & Gilles *Path. in Tropics* viii. 335 At one time Takayasu's disease was thought to affect only young females and the pathological lesions to be restricted to the aortic arch and its major branches. **1978** *Central African Jrnl. Med.* XXIV. 144/2 Takayasu's disease (pulseless disease) is a disease of unknown aetiology first described in Japan.

take, *v.* Add: **II. 2. a.** Also in *Criminals' slang*, to break into in order to burgle, to rob.

1926 J. Black *You can't Win* xxi. 331 After gathering every scrap of information available, I was sure I could 'take' the spot if I got a fair break on the luck. **1930** D. Runyon in *Liberty* 8 Nov. 24/2 Someone takes a jewellery store in the town.

c. (Earlier examples.)

1846 W. Denison *Cricket* 71 The greatest number of wickets he succeeded in taking in one match was 11. **1870** *Times* 11 July 10/5 Mr. Law was taken easily at the wicket with the score at 22.

6. a. Also, in *Med.*, of animal tissue, etc.: to continue in a healthy state after being transplanted.

1875 *Lancet* 23 Jan. 124/2 The transplanted pieces of skin..were found to have 'taken' remarkably well. **1936** *Anat. Rec.* LXIV. 167 Young donors supply material that is more likely to 'take'. **1977** *Time* 7 Mar. 43/2 Odds that a transplanted cadaveric kidney will 'take' are usually no better than 50%.

c. Of ice: to form (esp. in a lake, river, etc.). Cf. sense 44 d below. *dial.* and *N. Amer.*

1825 *Kingston* (Upper Canada) *Chron.* 4 Feb. 3/2 On Saturday night last, the ice took between Kingston and Long Island. **1877** E. Leigh *Gloss. Words used Dial. Cheshire* 206 'The ice is taking' means it is beginning to freeze. **1881** *Edmonton Bull.* 28 Mar. 1/2 Ice took in the Saskatchewan on the 19th of November. **1931** G. L. Nute *Voyageur* 79 Seines were set in the water just before the ice 'took' on the lake or river.

d. Of a lamb: to be accepted by a foster mother in place of her own dead lamb.

1874 Hardy *Far from Madding Crowd* I. xviii. 204 Mistress and man were engaged in the operation of making a lamb 'take', which is performed whenever an ewe has lost her own offspring, one of the twins of another ewe being given her as substitute.

8. c. *slang.* To swindle, cheat, or deprive of money by extortion. Freq. const. *for*.

1927 [see *clip *v.*[2] 9]. **1930** D. Hammett *Dain Curse* xii. 122 They landed Mrs Rodman... They took her for one of her apartment buildings. **1956** S. Bellow *Seize the Day* i. 9 They make millions. They have smart lawyers... Whereas I got taken. **1968** 'L. Marshall' *Blood on Blotter* xxvii. 183 'How much did you take him for?' 'Slade? Plenty.' **1970** *Washington Post* 30 Sept. B12/4 It looks to me like you're fixin' to git took for the dollar an' thirty cents, Shuffy. **1978** J. B. Hilton *Some run Crooked* ix. 86 It wasn't enough for Julie just to admit she'd been taken. **1982** 'E. Lathen' *Green grow Dollars* xiv. 112 'I told Mary to take them for every penny she could get,' he said stoutly.

d. *Motor-racing.* To overtake (a competitor).

1977 *Custom Car* Nov. 14/2 Jimmy Smith..finally took Falcone, who had developed trouble, and stayed ahead to win the race. **1978** *Guardian Weekly* 12 Mar. 23/5 The South African Grand Prix... Peterson (Lotus) shadowed the leader right to the end, taking him on a bend in the last lap for victory.

10. b. Also without const.

1969 'E. Ferrars' *Skeleton Staff* iii. 61 'Not enormously taken, are you?' 'Not bowled over.' **1978** P. H. Johnson *Good Husband* iii. 24 But about Ann...you were very taken, weren't you?

c. (Later examples.)

1842 J. A. Kasson *Let.* 22 Nov. in *Virginia Mag. Hist. & Biogr.* (1948) LVI. 418 A person, male or female, that relishes society and can *talk*, will take well. **1858** G. Meredith *Let.* 28 Apr. (1970) I. 35 Translate that placard. It would take. **1963** *Listener* 14 Mar. 457/1 Jazz has 'taken' in Africa. **1981** D. Martin in Martin & Mullen *No Alternative* ii. 19 The appeal to primitive practices can obscure the pressures of today which make such practices 'take' with a section of the clergy.

11. a. Delete now *rare*. (Later examples.)

1941 B. Schulberg *What makes Sammy Run?* xi. 203 She was married... The year she came out. But it didn't take. **1978** D. Bloodworth *Crosstalk* xv. 123 [Operation] Crosstalk can do no good whatsoever unless it takes, and ..this move against Sviridov..shows it has taken.

c. Also *fig.*

1906 E. Dyson *Fact'ry 'Ands* iii. 29 Fuzzy's love was the mysterious and unhallowed growth of a moment. Sarah..had beguiled him with her Ethiopian grin and glances of matured coyness... In the words of Benno the wise, 'It took like er vaccination'. **1951** G. Greene *End of Affair* v. iv. 201 'He did it there and then.'..'Did what?' 'Baptized her a Catholic... I always had a wish that it would 'take'. Like vaccination.'

III. 14. c. To possess sexually.

1915 D. H. Lawrence *Rainbow* i. 14 Whether he were going to take her out of inflamed necessity. *Ibid.* viii. 216 Even if he did not take her, he would make her relax, he would fuse away her resistance. **1930** A. Huxley *Brief Candles* 280 She kissed him again. 'Take me.' **1948** G. Vidal *City & Pillar* I. vi. 133 He wanted to throw her on a bed and take her against her will, violently. **1962** I. Murdoch *Unofficial Rose* xiii. 122 'Well, it's up to you too, my queen,' said Randall. 'You want to be—taken, don't you?' **1978** T. Allbeury *Lantern Network* viii. 110 She lay with her eyes open as he took her.

15. d. (Earlier example referring to a newspaper.)

1798 J. Woodforde *Diary* 6 Jan. (1931) V. 92 Crouse's Norwich Paper which we used to take, did not arrive.

17. b. *to take the Fifth Amendment* (U.S.): to appeal to Article V of the ten original amendments (1791) to the Constitution of the United States, which states that 'no person.. shall be compelled in any criminal case to be a witness against himself'; hence, to decline to incriminate oneself. Usu. *ellipt.*, *to take the Fifth.*

1955 *U.S. News & World Report* 22 July 36/2 In the armed services, let a man take the Fifth Amendment and his military career is virtually doomed. **1967** *N.Y. Times* 22 Jan. IV. 10/1 (*heading*) Law: taking the Fifth and making a living. **1972** J. G. Vermandel *Last seen in Samarra* xx. 133 'You can hardly have in mind to cast me as a villain because of that.'..Alex nodded. 'Right... If you want to take the Fifth, maybe Derek will settle it for us?' **1976** *Times Lit. Suppl.* 12 Nov. 1413/2 To do what I did not want to do: take the Fifth Amendment. **1978** S. Brill *Teamsters* Pl. 4 (*caption*) The former gym teacher took the Fifth Amendment when asked about the millions of dollars in insurance he had sold to the Teamsters health and welfare funds.

19. a. Also, to answer (a telephone call).

1970 P. Moyes *Who saw her Die?* iii. 37 The shrilling of the telephone provided a welcome release... Dolly said, 'I'll take it.' **1976** G. Sims *End of Web* i. 13 'Sorry, I'll have to take it. Might be a friend I was trying to contact this morning.'..He picked up the phone. **1979** C. MacLeod *Luck runs Out* iv. 37 The telephone rang. 'I'll take it,' said Shandy.

IV. 25. d. *intr.* *to take and* = *to go and* s.v. Go *v.* 32 c. *dial.* and *U.S. colloq.*

1836 *Southern Lit. Messenger* II. 388/2 If you do so I will take and tell father. **1859** T. Hughes *Scouring of White Horse* vi. 129 This here..maypowl wur the last in all these parts..but..the Uffington chaps cum up, and tuk and carried 'un down ther'. **1876** 'Mark Twain' *Tom Sawyer* i. 8 I'll take and bounce a rock off'n your head. **1901** J. Barlow *From Land of Shamrock* 17 Her cherished Nellie 'took and died on her' of some mysterious malady. **1925** W. Faulkner *As I lay Dying* 44 'She's gone,' Cash says. 'She taken and left us,' pa says. **1977** 'L. Egan' *Blind Search* viii. 133 Poor soul, this awful cancer. She took and died inside of three months.

26. e. *slang.* To confront, attack; to overcome, defeat; to kill.

1939 'E. Queen' in *Blue Bk.* Oct. 17 Seems to me the champ ought to take this boy Koyle. **1956** E. L. Perry in A. Hitchcock *Stories for Late at Night* (1962) 273 Let's take him... That fat guy looks really loaded. **1963** D. Cory *Hammerhead* xi. 161 There were two men now in the doorway, both with pistols... One of them Fedora might have taken; but not, he reluctantly decided, both. **1965** I. Fleming *Man with Golden Gun* vii. 97 It had been damned fine shooting... How in hell was Bond going to take him? **1976** *Publishers Weekly* 1 Mar. 93/3 They broke their tie with the Giants and went on to take the Tigers in seven wild World Series games. **1979** E. Bercovici *Wolftrap* 41 The man who tried to take me was Martinez... Next time I am going to kill him.

28. a. *to take (one's) time* (later examples).

1788 W. Cowper *Let.* 18 Aug. (1904) III. 303, I took my own time to return, and did not reach home till after one. **1796** [see Time *sb.* 8 a]. **1873** Hardy *Pair of Blue Eyes* III. i. 21, I don't press you for an answer now, darling... Take your time. **1912** W. B. Yeats *Land of Heart's Desire* (ed. 7) 11 It's precious wine, so take your time about it. **1925** W. Faulkner *As I lay Dying* 246 'Let him take his time,' I said. 'He ain't as spry as you, remember.' **1946,** etc. [see *sweet *a.* 8 d]. **1966** A. Higgins *Langrishe, go Down* iii. 28 Taking her time, Helen cycled slowly by the wall of the Charter School. **1981** 'E. Ferrars' *Experiment with Death* iv. 68 Emma suggested that Sam had probably gone to the lavatory. 'If so, he's taking his time,' Roger said.

c. *to have (got) what it takes*: to possess the necessary attributes or qualities, esp. those needed for success. *colloq.* (orig. *U.S.*).

1929 *Amer. Speech* IV. 357 To avoid using the word *money*, the well-informed user of slang may use..*the needful, the wherewithal*,..or *what it takes*. **1933** F. Baldwin *Innocent Bystander* ix. 186 Angela, who has planty of what it takes, is the White Hope of the arty crowd which gathers at her penthouse. **1944** M. Laski *Love on Supertax* iv. 49 Only maturity's got what it takes. **1947** D. M. Davin *For Rest of our Lives* 335 The cheap verses had everything it takes to make a soldier's song. **1956** B. Holiday *Lady sings Blues* (1973) iv. 41 Sometimes I wonder how we survived. But we did. If we didn't have what it took at the beginning, we picked it up along the way. **1972** J. Wambaugh *Blue Knight* (1973) xiii. 225 He's got everything it takes but guts. **1977** *Zigzag* Apr. 26/1 They've got the right idea and what it takes.

d. *it takes all sorts to make a world*: see *sort *sb.*[2] 11 d.

e. *to take one all one's time*: see *time *sb.* 8 d.

29. b. *to take it from there*: to take over or continue from the point or situation described.

[**1948** *Radio Times* 19 Mar. 5/3 A new weekly comedy series, *Take It From Here*, will make its appearance..on Tuesday evening.] **1959** *Internat. Celebrity Reg.* 430/1 Miss Shearer informed the studio of her find. They took it from there. **1960** Wodehouse *Jeeves in Offing* xix. 188 His future hangs on this speech, and we've got it and he hasn't. We take it from there. **1973** *Ottawa Jrnl.* 14 July 24/3 They interrupt each other and talk until the breath gives out and then another one cuts in and takes it from there. **1975** N. Luard *Travelling Horseman* vi. 167 I'd tell him what I'd found out and he could take it from there.

V. 33. a. Also in phr. *to take a letter*: to write a letter down in shorthand from another's dictation.

1943 K. Tennant *Ride on Stranger* x. 110 He seated himself at his table... 'Will you take a letter, please?'.. Her pencil travelled quite speedily after his words. **1961** *Times* 7 June 2/5 Director of general publishing house.. needs an assistant-cum-secretary. Will be expected to 'take letters'.

b. Also with reference to cinematographic filming. Occas. *intr.* Cf. *take *sb.* 9 a.

1917 *N.Y. Times* 25 Feb. 4/1 Two thousand persons participated in the coronation, which required two full days to 'take', despite the fact that it remains on the screen only three minutes. **1929** H. B. Abbott *Motion Pictures with Baby Ciné* ii. 4 It has already been stated that the motion picture is made, or 'taken', in a special camera, and that the medium upon which the picture is made is a celluloid film coated with a sensitive emulsion. **1954** N. Bau *How to make 8mm. Films* 99 (*caption*) Hold the camera absolutely steady while taking. *Ibid.* 100 If you are taking a hand-held shot, hold the camera as steady as possible. **1974** *Daily Tel.* 2 May 3/4 Using a friend's projector and screen, he ran a short colour film taken at the wedding.

VI. 34. d. *take that!*: (*a*) said as an accompaniment to the delivery of a blow; (*b*) used, with a suggestion of challenge or defiance, to emphasize a foregoing statement.

a **1425** [in Dict., sense 34 b]. **1805** C. Wilmot *Let.* 7 Dec. in *Russ. Jrnls.* (1934) II. 209, I don't pity you in the least. Take that for asking me to write you 'beautiful Russian storys'. **1846** [in Dict., sense 34 c]. **1932** Kipling *Limits & Renewals* 81 'Then take that!' and he smacked the brute's head. **1942** Berrey & Van den Bark *Amer. Thes. Slang* § 158/8 Take that and see how you like it! **1983** A. Olcott *May Day in Magadan* xiv. 249 His pride was stung. 'They want me..' he said, with an unthinking 'take that!' tilt of his nose.

39. a. Freq. in phr. *take it or leave it* and varr., expressing indifference or a refusal to bargain, compromise, etc. Cf. *take-it-or-leave-it* adj. s.v. Take-.

[**1576** W. Lambarde *Perambulation of Kent* sig. 2D3v, I..doe leaue the Reader to his free choice, to take or leaue the one, or the other.] **1664** T. Killigrew *Thomaso* in *Comedies, & Tragedies* I. IV. ii. 361 That is the price, and less I know, in curtesie you cannot offer me; take it or leave it. **1762** J. Wesley *Let.* 21 May (1931) IV. 182 As to that particular expression, 'Dying at the feet of mercy', I have only farther to add, I do not care as it is not a scriptural phrase, whether anyone takes or leaves it. **1809** B. H. Malkin tr. *Le Sage's Gil Blas* x. x. 39, I will give forty [pistoles] at a word; take them or leave them. **1898** W. S. Churchill in R. S. Churchill *Winston S. Churchill* (1967) I. Compan. II. 917 The tremendous & unchallenged power of the Trust—enabled it to dictate wages to its workmen & prices to its customers. 'Take it or leave it' it said 'This is a free country.' Thereat the oil-mechanic had to accept the offered wage or find another trade and the customer to buy the oil at the offered price or wait in the dark. **1929** D. H. Lawrence in *Forum* Jan. p. l/3 The hen knows she is unanswerable. ..There it is, take it or leave it! **1953** A. Upfield *Murder must Wait* xi. 105, I cock a snook at you... You can take it or leave it. **1962** Wodehouse *Service with Smile* x. 151 Her air was that of somebody who, where Ickenhams were concerned, could take them or leave them alone. **1977** P. G. Winslow *Witch Hill Murder* II. xv. 206, I didn't want to..say I'd gotten married and he could take it or leave it, because I was afraid he'd leave it.

b. In extended use, of a woman. *rare.*

1932 W. Faulkner *Light in August* x. 212 There were white women who would take a man with a black skin. **1941** N. Mailer *Advts. for Myself* (1961) 36 When I take a man, and I may take him for a lot of reasons, in back of it all is the feeling..that that is something I can do better than any other woman.

41. b. Freq. in phr. *take it from me*: believe me, take my word for it, be assured.

1672 Wycherley *Love in Wood* (Dedication) sig. A2v, Madam, take it from me, no man..is more dreadful than a Poet. **1829** G. Griffin *Collegians* I. v. 101 Who should walk in the door to him, only his dead wife..! Take it from me he didn't stay long where he was. **1889** [in Dict.]. **1902** H. James *Wings of Dove* i. 20 You may take it from me once for all that I won't hear of any one of whom *she* won't. **1938** A. Christie *Death on Nile* II. xvii. 178, I think you must take it from me, Mr. Pennington, that we have examined all the possibilities very carefully. **1957** D. Robins *Noble One* xix. 177 You can take it from me that *I* don't believe a word of it.

42. a. *to take it on the chin*: see *chin *sb.*[1] 1 d; *to take it lying down*: see Lie *v.*[1] 21 d; *to take it in one's stride*: see Stride *sb.* 3 d.

b. *to take things as one finds them* (earlier examples); also *to take* (people) *as one finds them*: to judge people without preconceptions; to accept people as they are, esp. by expecting no special preparations for one's entertainment, etc.

[**1548** E. Hall *Union Lancaster & York* fol. ccxliiv, Myne aduise is, let all men trust them, as thei fynde them.] **1580** A. Munday *Zelauto* sig. H2v, In the meane whyle, take as you finde. **1596** J. Harington *Metamorphosis of Ajax* sig. B4v, We must now take him as we finde him, with all his faults. **1638** W. Chillingworth *Relig. Protestants* I. v. 241 But reall externall deeds doe take things in grosse as they find them, not separating things which in reality are joyned together. *c* **1807** J. Austen *Watsons* (1954) 351, I am one of those who always take things as they find them. I hope I can put up with a small apartment for two or three nights. **1825** in H. Wilson *Mem.* I. 147, I could have been a little romantic about you, it is true; but I always take people as I find them. **1868** Dickens in *Our Young Folks* May 260 We have but a simple joint..but if you will take us as you find us it will be *so* kind! **1886** G. B. Shaw *Cashel Byron's Profession* xiv. 148 You can either take me as you find me, or let me alone. **1903** A. Bennett *Leonora* ii. 47 She's gotten sausages for you..though I told her you'd take us as you found us. **1912** A. Lang *Shakespeare, Bacon & Great Unknown* xii. 247, I am only taking Ben as I find him and as I understand him. **1943** K. Tennant *Ride on Stranger* vi. 49 All these go by wearing the peevish expression of a housewife who, not having time to make the beds, grumbles: 'You must take us as you find us.' **1980** T. Barling *Goodbye Piccadilly* vii. 129 'Do we phone ahead in the name of protocol?' 'Hell, no. We take them as we find them.'

c. *to take a joke*: to be able to bear teasing or amusement at one's expense; usu. in negative.

1780 J. Woodforde *Diary* 28 Mar. (1924) I. 276 Poor Sam cant take a Joke. I forgot what I said to disoblige him. **1838** C. Fox *Jrnl.* 4 Apr. in *Memories Old Friends* (1882) iv. 27 Speaking of Dr. [John] Dalton, he said he could not take a joke at all. **1863** M. B. Chesnut *Diary* 14 Dec. in C. V. Woodward *M. Chesnut's Civil War* (1981) xx. 505 When he saw how angry I was, he said, 'Can't you take a joke?' **1921** E. O'Neill *Diff'rent* I. 223 *Mrs. Crosby*... Shet up your foolin', Jack. *Jack*... Nobody in this house kin take a joke. **1972** D. Delman *Sudden Death* (1973) ii. 59 It was a joke. Hell with anybody who can't take a joke.

d. *to (be able to) take it*: to have the capacity to endure punishment, affliction, etc.

1862 in H. Mayhew *London Labour* (ed. 2) III. 387/2 That first flogging made me ripe. I said to myself, 'I can take it like a bullock.' **1914** O. W. Holmes *Let.* 24 Sept. in *Pollock-Holmes Lett.* (1942) I. 222, I value everything that shows the quiet unmelodramatic power to stand and take it in your people. **1941** W. S. Churchill in *Unrelenting Struggle* (1942) 190 If the storm is to renew itself, London will be ready, London will not flinch, London can take it again. **1952** *Chambers's Jrnl.* Apr. 196/2 But as soon as I hadn't got Derek—well, I just couldn't take it. **1976** C. Bermant *Coming Home* I. vi. 87 A slogan, like 'Britain can take it'.

e. *to take things* (or *it*) *as they* (or *it*) *come(s)*: to deal with events as they arise, without anticipating difficulties.

1509 A. Barclay tr. *Brandt's Ship of Fools* fo. 266 *recto*, That man folowes hye wysdome Whych takys all thynges lyke as they come. **1611** J. Davies *Scourge of Folly* 170 Take all things as they come, and bee content. So many whores do, and yet pay their Rent. **1863** 'Ouida' *Held in Bondage* I. ix. 203 The true secret is to take things as they come. **1926** [see *bottom *sb.* 11 c]. **1979** V. Kelleher *Voices from River* iii. 34, I was trying not to think... I kept telling myself, take it as it comes.

43. (Later examples with reference to motor transport.)

1972 M. Kenyon *Shooting of Dan McGrew* xxii. 184 He took the corner like a rally driver. **1976** 'B. Shelby' *Great Pebble Affair* 181, I took the lakeshore S curve designed for thirty mph at fifty-five.

44. d. (N. Amer. examples in sense 'to become frozen'.) Occas. *pass.* Cf. sense 6 c above.

1781 *Quebec Gaz.* 11 Jan. 2/1 It has not been known to take so early as the month of December. **1820** G. Simpson *Jrnl. Occurrences in Athabasca Dept.* (1938) 100 This is an unusual late season as the Lake usually takes from the 15th to the 20th Oct. **1830** J. Macmillan Let. 15 Dec. in G. P. T. Glazebrook *Hargrave Corr. 1821–43* (1938) 58 We had a very mild fall. The river was not taken before 6th of Decr. **1871** *Scribner's Monthly* II. 458 When the rivers are beginning to 'take' or freeze.

VII. 46. a. *to take (someone's) point* (and variants): see *point *sb.*^{1} 28 c.

47. e. *to take* (something) *as done*: to consider an omission not to have occurred; *to take* (something) *as read*: see *read *v.* 11 e.

1893 E. F. Benson *Dodo* I. i. 9 You haven't congratulated me. Never mind, we'll take that as done.

48. a. Freq. in phr. *what* or *who(m) do you take me for?* said as a challenge to a derogatory implication, as of foolishness, dishonesty, etc.

1847 A. & H. Mayhew *Greatest Plague* vii. 87, I wanted to ask her who the dickens she took me for. **1892** Kipling & Balestier *Naulahka* xvii. 202 'You won't get the chance,' said Tarvin unshakenly... 'What do you take me for?' **1912** C. Mackenzie *Carnival* xxx. 293 'What do you take me for?' enquired Irene. 'I take you for what you are—a rotter.' **1921** W. J. Locke *Mountebank* xiii. 164 'You haven't given me away?' 'My good girl,' I protested, 'what do you take me for?' **1927** W. S. Maugham *Constant Wife* III. 186 But, my poor John, whom do you take us for? Am I so unattractive that what I'm telling you is incredible? **1939** G. B. Shaw *Geneva* II. 32 Then you went to school, did you? *Begonia.* Well, of course: what do you take me for? **1983** 'R. B. Dominic' *Flaw in System* xx. 129 What do you take me for? A simp?

51. c. *to take care of*: see also Care *sb.*^{1} 4 b in Dict. and Suppl.

VIII. 52. c. *to take five* (or *ten*): to have a five- (or ten-) minute break. Also *loosely*, to relax. *U.S. colloq.*

1929 *Amer. Speech* V. 147 If the miner craves a rest while on the job, he *takes five*, a long enough period for a smoke. **1943** *Yank* 7 May 3 Six members of a reconnaissance group 'take 10' at a railroad station. **1961** G. T. Simon *Feeling of Jazz* 30 Man, I'm glad they said to take five, because this next arrangement looks rough. **1973** W. Sheed *People will always be Kind* vi. 60 'Could you go a little faster, Fatman?'..It was difficult making jokes... 'O.K. Fatstuff, take five, I was only kidding.'

d. *to take a fall* (U.S.): (*a*) *slang*, to be arrested or convicted of a crime (cf. *fall *v.* 23 f, g); (*b*) *colloq.*, to suffer a fall; similarly *to take a spill*; also *fig.*, to fall for (cf. *fall *v.* 59*).

1942 Berrey & Van den Bark *Amer. Thes. Slang* § 353/7 Fall in love...take a fall. *Ibid.* § 500/6 Be arrested...take *or* have a fall. **1953** W. Burroughs *Junkie* iii. 34 Jack had taken a fall on a safe job and was in the Bronx County jail awaiting trial. **1958** S. J. Perelman *Most of S. J. Perelman* 35, I took a rather nasty fall over a wastebasket. **1962** D. Lessing *Golden Notebk.* iv. 474 Molly rang late—says that Jane Bond has 'taken a fall over' Mr Green. **1968** *Globe & Mail* (Toronto) 15 Jan. 29/6 Even the best skier can take a spill. **1973** *Times* 9 Feb. 12/2 Michael Fish took a couple of falls.

IX. 57. a. Also with *over*, to conduct through or show around (a building, garden, etc.).

c **1810** W. Hickey *Mem.* (1918) II. xix. 251 She..took me over the house, which was as complete a one as ever I saw. **1837** C. Fox *Jrnl.* 15 May in *Mem. Old Friends* (1882) iii. 16 Took them all over the Grove Hill gardens. **1880** Trollope *Duke's Children* III. xix. 215, I want to take her all over the house. **1911** *Rep. Labour & Social Conditions in Germany* (Tariff Reform League) III. 166 [He] was able yesterday to take a small deputation..over the 'Triumph' works.

d. In colloq. phr. *you can't take it with you*, in allusion to the impossibility of benefiting from earthly wealth after death.

1841 Marryat *Masterman Ready* II. ii. 22 He was very fond of money; but that they said was all the better, as he could not take it away with him when he died. **1923** G. Arthur *Let.* 16 Sept. in *Further Lett. Man of no Importance* (1932) 153 Mr. Gladstone, when a dead millionaire was held up for his admiration because he had left large sums for charities, said, 'Thank him for nothing; he was obliged to leave it somewhere as he couldn't take it with him.' **1937** Kaufman & Hart (*title*) You can't take it with you. **1952** A. Christie *Mrs. McGinty's Dead* vii. 48 'They inherited a little money when Mrs. McGinty died.'..'Well, that's natural enough... You can't take it with you.' **1977** J. Porter *Who the Heck is Sylvia?* x. 87 You're not short of the odd penny... And you can't take it with you, can you?

58. b. Also, *to take one's (own) life*: to kill oneself, commit suicide.

1920 D. H. Lawrence *Women in Love* xv. 211 It was not a question of taking one's life—she would *never* kill herself. **1965** *Amer. Speech* XL. 301 This person may indeed take his own life. **1981** *Daily Tel.* 18 June 19/2 A note left by them made it clear that they wanted to take their own lives and also wished to be buried in the same grave.

c. (Further examples.) Also *euphem.* in *pass.*, to die.

1749 Gray *Let.* 7 Nov. (1900) I. 204 He who has preserved her to you so many years..has taken her from us to Himself. **1809** J. Porter *Scottish Chiefs* xxxiv. 257 If all whom I love be lost to me here, take me then to thyself, and let my freed spirit fly to thy embraces in heaven! **1920** E. O'Neill *Beyond Horizon* II. i. 69 It was God's will that he should be taken. **1977** [see *pass *v.* 65 c].

XI. 74. take to—. h. *N.Z. slang.* To attack, usu. with fists.

1911 'Kiwi' *On Swag* iii. 9 Take to him, Bill. **1960** N. Hilliard *Maori Girl* II. xiv. 159 When we got home he really took to me. That was when I lost a lot of my teeth.

XII. 76. take aback. (Earlier *fig.* example.)

1829 F. Marryat *Frank Mildmay* I. ix. 266, I was so taken aback with the sudden appearance and address of this beautiful vision, that I knew not what to say.

76*. take about. *trans.* To conduct on a round of sight-seeing or on excursions, etc.

1823 P. Panam *Mem. Young Greek Lady* 117 If you wish for any thing speak to him; he will take you about everywhere. **1894** E. Fawcett *New Nero* Proem 8 He..took him about for almost an hour, showing him a good many places. **1903** A. W. Patterson *Schumann* 113 He seems to have taken the Laidlaw ladies about a good deal.

77*. take apart. *trans.* **a.** To dismantle or take to pieces; also *fig.*, to search thoroughly; to demolish or wreck.

1936 C. Sandburg *People, Yes* 60 Let's take it apart to see how it ticks. **1958** M. Allingham *Hide my Eyes* xv. 150, I am going to take this shed apart if it costs me my ticket. **1968** 'E. Peters' *Grass Widow's Tale* xi. 140 It has to be somewhere here. Stands to sense. Go take that little front room apart, Skinner. **1969** *Oz* Apr. 25/1 There will be a lobby of Parliament which far from pleading with MPs will probably take Whitehall apart. **1974** D. Seaman *Bomb that could Lip-Read* xxiv. 243 There is going to be one God-awful search for the man... They will take this hamlet apart. **1978** M. Puzo *Fools Die* xv. 161 The new kids were wilder and started taking everything apart.

b. To thrash or beat soundly; also *fig.*, to attack with argument or criticism.

1942 N. Balchin *Darkness falls from Air* v. 94 Supposing I went round and took him apart? **1963** *Listener* 21 Feb. 350/3 The Labour Party's new leader was taken apart with the sort of cheerful and dedicated venom hitherto reserved for Tory Cabinet ministers. **1969** 'J. Ashford' *Prisoner at Bar* xii. 117 And don't get funny with Bladen..or he'll take you apart at the seams. **1971** S. E. Morison *European Discovery Amer.: Northern Voy.* vii. 242 Manuel C. Baptista de Lima..has politely taken me apart and argued for the 1492 date. **1976** *Birmingham Post* 16 Dec. 12/2 League leaders Liverpool were taken apart by the speed, skill and determination of the entire Villa side.

78. take away. a. Also = *put away* s.v. *put *v.*^{1} 38 f (*b*). *U.S.*

1919 E. O'Neill *Where Cross is Made* in *Moon of Caribbees* (1923) 16 They say for his own good he must be taken away.

d. (Earlier example.)

1838 C. Waterton *Ess. Nat. Hist.* p. xxv, After eluding him in cover for nearly half an hour, being hard pressed, I took away down a hedgerow.

80. take down. b. (*e*) (Earlier example.)

1840 M. Edgeworth *Let.* 30 Dec. (1971) 573 Sir John Campbell took me down to dinner and I was seated of course beside him.

c. *to take* (a person) *down a peg*: see Peg *sb.*^{1} 3.

e. Also, with person as obj.: to write down the words of, to take dictation from.

1883 'Mark Twain' *Life on Miss.* xxii. 247, I enlisted a poet for company, and a stenographer to 'take him down'. **1928** D. H. Lawrence *Woman who rode Away* 18 She certainly didn't want to take him down in short hand.

f. *spec.* To record a contentious statement made in a legislative assembly with a view to invoking disciplinary procedure.

1784 *Universal Mag.* Jan. 45/1 Gen Conway said that he was ready to maintain what he had said. Let the right hon. gentleman move to take down his words, and he would make his charge. **1863** *Illustr. Times* 20 June 422/2 Mr Cox had..insinuated that..Lord Ranelagh wished to have power to flog volunteers; and on Monday Mr. Ormsby Gore rose and denounced these words as 'scandalous and unfounded'. Whereupon Sir Robert Jackes Clifton jumped up and moved that the words were taken down. **1934** *Sun* (Baltimore) 3 May 1/4 Representative Pettingill ..threatened to invoke disciplinary procedure against Mr. Britten by means of what is known in the House as 'taking down' his words.

g. To cheat, trick, swindle. *Austral. slang.*

1895 *Argus* (Melbourne) 5 Dec. 5/2 [The defendant] accused him of having 'taken him down', stigmatised him as a thief and a robber.

82. take in. e. Also *spec.*, to lead in (to dinner). (Further examples.) Cf. sense 80 b (*e*), *85 c.

1863 A. J. Munby *Diary* 3 June in D. Hudson *Munby* (1972) 165 The new Lord of the Admiralty..and his wife: whom I took in to supper. **1887** M. Monkswell *Jrnl* fo. 266 *recto*, 25 May in *Victorian Diarist* (1944) 132 We dined with the Dean [of Hereford] that very evening. He took me in.

k. (Earlier and later examples in sense 'to include in a journey or visit'.) Also *loosely*, to go to. No longer restricted to the U.S.

1755 in *Essex Inst. Hist. Coll.* (1916) LII. 80 In our way by the Skuylkill rd. took in ye prop[rieto]rs Gardens. **1880** 'Mark Twain' *Tramp Abroad* iii. 42 An owl that come from Nova Scotia..took this thing in on his way back. **1925** *New Yorker* 7 Mar. 19/1 There's no use me asking you if you took in all the revues. **1940** 'N. Shute' *Landfall* 26 He might pick up Matheson or Hooper and take in a movie. **1968** *Globe & Mail* (Toronto) 17 Feb. 32 (Advt.) Even take in breakfast at Le Drugstore..and head home again on the return flight. **1977** D. Bagley *Enemy* i. 12 We took in more theatres, an opera, a couple of ballets.

p*. *Stock Exchange.* To receive contango on (stocks or shares); to accept (stocks, etc.) as security for a loan. Cf. *give on* b s.v. *give *v.* 61.

1893 R. Bithell *Counting-House Dict.* (ed. 2) 292 The term [taken in stock] is applied solely to stocks taken in for fortnightly or monthly loans on the Stock Exchange. **1911** W. Thomson *Dict. Banking* 503/1 In connection with the Stock Exchange settlements, a 'taker-in' is a broker who lends money against stock (i.e. 'takes in' stock) to a broker who requires to pay for a purchase. **1912** *Q. Rev.* July 102 The dealer says that he will 'take them in', which means that he will lend the money until the settlement following that for which the original bargain was effected. **1928** *Morning Post* 19 Nov. 3/3 If the other man..prefers to take a rate of money rather than to accept the cash which delivery of the shares would produce, he will 'take them in'—the opposite operation to 'giving on'. **1934** F. E. Armstrong *Bk. Stock Exchange* vi. 108 When no 'takers' can be found someone has to provide the cash, and firms known as money brokers frequently agree to 'take in' the securities purely as a money-lending proposition. **1955** *Beginners, Please* (Investors' Chron.) ii. 44 In normal market conditions it is probably easier to 'take-in' shares, i.e., carry over a sale to the next settlement, than to 'give on' shares, i.e., carry over a purchase. This is because generally there are more bulls than bears. Under such conditions the 'giver' pays a rate of interest to the 'taker' for the accommodation provided.

p.** *slang.* To take into custody, arrest. Cf. *pull in* s.v. *pull *v.* 25 e.

1942 Berrey & Van den Bark *Amer. Thes. Slang* § 500/5 Arrest..take, take in or up. **1978** J. B. Hilton *Some run Crooked* xiv. 138 You can tell me now, or I'm taking you in to help. **1979** J. van de Wetering *Maine Massacre* iii. 26 You're not taking me in, sheriff.

s. *N. Amer. dial.* To open, begin, *esp.* of a school term. Cf. sense 90 r below.

1876 'Mark Twain' *Tom Sawyer* 162 She could hardly wait for school to 'take in'. **1906** *Dialect Notes* III. 160 School takes in early and takes out late, seems to me. **1942** *Post* (Morgantown, W. Va.) 14 Sept. 4 An obligation ..upon drivers to be careful of children, esp. in the hours that school takes in and lets out. **1956** W. R. Bird *Off-Trail in Nova Scotia* iii. 99 One girl turned to me and declared she had seen him with it before school took in.

83. take off. a. (*e*) *U.S. Blacks.* To rob or burgle; to 'hold up'. Cf. *to rip off* (ii), (iv) s.v. *rip *v.*^{2} 4**.

1970 C. Major *Dict. Afro-Amer. Slang* 113 *Take off*... to rob or hurt. **1972** J. Hudson in T. Kochman *Rappin' & Stylin' Out* 413, I can't go no place expecting to take off some fat sucker if I look like a greaseball. **1973** *Black World* Jan. 56/1 He and Cecil B were to take off a supermarket in San Jose.

c. (Earlier example.)

1836 Dickens *Pickw.* (1837) ii. 7 Here, No. 924, take your fare, and take yourself off.

e. Also in *Cricket*, to remove (a bowler) after a spell of bowling in order to replace him.

1851 W. Bolland *Cricket Notes* iv. 75 Do not..refuse to bowl any more; neither grumble nor growl if you are taken off. **1921** G. R. C. Harris *Few Short Runs* xi. 280 Don't turn sulky because after bowling five consecutive maidens you are taken off. **1977** *Times* 17 Jan. 7/1 When Greig took him off after 95 minutes his figures for the morning were 10–5–7–1.

n. (*a*) (Later examples in sense 'to go off', etc.)

1959 I. & P. Opie *Lore & Lang. Schoolch.* x. 193 Juvenile language is well stocked..with expressions inviting a person's departure, for instance:..take off, [etc.]. **1968** *Listener* 19 Dec. 809/3 I'm not stopping here,..no matter what they say or do... I'm taking off tonight. **1972** J. Philips *Vanishing Senator* (1973) III. iii. 147 You'd better take off. I've just got to get some sleep. **1978** M. Duffy *Housespy* vii. 178 Danny Oldfield's taken off. I'll let you know when I find her.

(*d*) *Aeronaut.* Of a pilot, plane, etc.: to perform the operations involved in beginning flight; to become air-borne. Also *transf.* of a bird.

?**1849** G. Cayley *Let.* in C. H. Gibbs-Smith *Sir George Cayley's Aeronautics* (1962) xlii. 136 It is absolutely necessary that the tail be securely braced up a little, and that the centre of gravity be made to act steadily on the bulk of the surfaces so that when weighed up to the weight of the person trying the wings—should it take off, they would skim and not either rise up hill or sink down hill. **1918** *Punch* 3 Apr. 222/2 Yes, he crashed a few days ago—in his first solo flip, taking off. **1927** C. A. Lindbergh *We* ii. 19, I taxied to one end of the field, opened the throttle and started to take off. **1936** G. B. Shaw *Simpleton* II. 69 All I want is a parapet to take off from. **1951** A. C. Clarke *Sands of Mars* I. 1, I once took-off standing up, just for a bet. **1973** *Sci. Amer.* Dec. 102/1 If the birds are pursued, they take off, but they do not fly far before they land again.

(*e*) *fig.* Of prices, costs, etc.: to rise steeply or suddenly. Of a scheme, project, etc.: to be launched (successfully), to become popular.

1963 J. N. Harris *Weird World Wes Beattie* (1964) xv. 184 Minerva took off, as we say, on a famous Friday the thirteenth... The stock rose from nineteen cents to over a dollar in the last half-hour of trading. **1970** *Melody Maker* 12 Sept. 33/3, I shall be pretty sick if Andy Williams' record takes off and mine dies. **1971** *Physics Bull.* Oct. 590/2 Prof. E. C. Cherry..devised an arrangement which resulted in reduction in bandwidth requirements... This likewise has not taken off so far although much more interest is now being shown in it. **1976** *Ibid.* Sept. 401/1 Production and salary costs 'took off'. **1978** *Detroit Free Press* 5 Mar. b 12/2 They had best seller hopes for the book, but it hasn't really taken off. **1981** *Church Times* 10 Apr. 9/5 Frank Scuffham has hopes of his committee, but acknowledges that it has not taken off yet. **1983** *Times* 20 Jan. 15/3 Sales of existing properties have taken off during the last few months.

o. *U.S. dial.* To absent oneself from work, school, etc.

1935 W. Faulkner *As I lay Dying* 115 You take off and stay in the house today. **1936** W. Greene *Death in Deep South* (1937) 61 She thought she'd be off in the afternoon and she said she'd take off anyway if she wasn't.

84. take on. d. (*b*) To engage (someone) in a fight, contest, argument, etc.

1885 [in Dict.]. **1915** E. Corri *30 Yrs. Boxing Ref.* 150 Instead of going for what the boxers call the 'easy money', Basham took on Matt Wells. **1928** *Daily Tel.* 24 Apr. 12/6, I saw the Sopwith take him on, and whilst I was changing drums I was attacked again in front by a Roland. **1930** G. B. Shaw *Apple Cart* I. 26 In this conflict we are the challengers. You have the choice of weapons. If you choose scandal, we'll take you on at that. **1976** *Morecambe Guardian* 7 Dec. 8/3 Micky Taylor earned the spotlight with a brilliant, cheeky dribble in which he took on and beat four men.

f. Restrict † *Obs. rare* to sense in Dict. and add: (*b*) To pretend, affect.

1858 Dickens in *Househ. Words* Xmas No. 20/1 This gent took on not to know me.

85. take out. c. Also, to lead (a woman) in (to a formal dinner). Cf. sense 80 b (*e*), 82 e in Dict. and Suppl.

1876 Trollope *Prime Minister* III. x. 166 John Fletcher took her out to dinner and Arthur did not sit near her. **1880** —— *Duke's Children* II. xx. 240 It was of course contrived at dinner that Lord Popplecourt should take out Lady Mary. **1905** J. H. Choate *Let.* 27 Jan. in E. S. Martin *Life J. H. Choate* (1920) II. viii. 272 The King took Mama out to dinner. **1913** in C. Seymour *Intimate Papers Col. House* (1926) I. vii. 188 He considered taking a duchess or royalty out to dinner was hard sledding.

j. *intr.* To go away, make off, start out. *U.S.*

1855 in *Montana Hist. Soc. Contrib.* (1940) X. 137, I took out in order to give them the slip. **1896** 'Mark Twain' in *Harper's Mag.* Aug. 355/1 Out jumps four men and took out up the road as tight as they could go. **1929** W. Faulkner *Sound & Fury* 310 They'll have to hitch up and take out to get home by midnight. **1938** M. K. Rawlings *Yearling* i. 11 How come you to take out such a fur piece?

k. *Bridge.* To remove (one's partner) from his situation in the auction by changing the suit of the probable contract or by bidding in response to his double. Also *into* (the fresh suit), with bid as obj., and *absol.*

1917 E. Bergholt *Royal Auction Bridge* (1918) 88 How am I to know..whether you are taking me out from strength or from weakness? **1921** A. M. Foster *Auction Bridge* 38 If your partner takes you out from weakness into a suit call you are likely to be fined. **1956** Mollo & Gardener *Bridge for Beginners* vii. 75 Responder may have a feeble five or six-card suit and nothing else. Then he takes out the double. **1977** *Homes & Gardens* Feb. 17 If..you held hand II, then it would be correct to take out into Two Hearts. *Ibid.* 14 Most players would take their partners out into Four Hearts on both of these hands.

l. To kill, murder; to destroy or obliterate (a specific target). *slang.*

1939 R. Chandler *Big Sleep* ii. 26 I'll take him out... He'll think a bridge fell on him. **1955** *Times* 28 June 4/4 The purpose of the attack was to 'take out'—as the strategist's jargon has it—the docks. **1962** L. Deighton *Ipcress File* xviii. 109 In terms of destructive area, this is a bomb that would take out a whole city. **1967** J. M. Fox *Dead Pigeon* 170 'He took out two people who could have involved him.'..'Took out? You mean he killed them?' **1977** *Times Lit. Suppl.* 15 Apr. 464/4 A sudden air attack, which would take out London, on a scale comparable with the attacks on Dresden or Hiroshima in 1945. **1978** M. Duffy *Housespy* v. 124 He was taken out yesterday... They ran him down. **1982** *Daily Tel.* 14 June 4/8 For several hours, as a commanding officer and his officers tried to 'take out' the sniper with machine gun, rifle and artillery fire, his bullets ricochetted off rocks above our heads.

m. *Austral.* and *N.Z. colloq.* To accept as a punishment, reward, etc.; to win.

1943 K. Tennant *Ride on Stranger* xvi. 176 George Benson told her briefly he would see her husband had a lawyer. He would probably get a month at the most and he'd better 'take it out'. **1976** *Australian* 15 July 2 Helen Morse..takes out the Australian Film Institute's top actress award tomorrow night. **1977** *N.Z. Herald* 8 Jan. 1. 6/8 The Games we play..can't..end, till Someone takes them out.

86. take out of. g. *to take one out of oneself*: to distract one's attention from one's own concerns; to amuse, divert or occupy (a person).

1848 G. Jewsbury Let. 4 Oct. in *Lett. to Jane W. Carlyle* (1892) 257 There are no bothering algebraical calculations as far as I went, but glimpses, as it were, into the 'everlasting universe of things', till one is taken out of oneself completely. **1908** A. Bennett *Old Wives' Tale* IV. iv. 531 Dr. Stirling wished to practise his curative treatment of taking the sisters 'out of themselves'. **1929** J. B. Priestley *Good Companions* II. iii. 301, I haven't enjoyed anything so much, I don't know when..they're so good they've taken me right out of myself. **1941** A. Christie *Evil under Sun* xii. 218 Poirot had..dwelt on the advantage it would be to Linda to have something to take her out of herself. **1958** P. Marris *Widows & their Families* ii. 21 My sister..took me out for walks. It's wonderful how it takes you out of yourself. **1974** [see *outside interest* s.v. *outside B. 6].

86*. take out on. *trans.* In phr. *to take it out on* (someone or something): to vent one's anger, frustration, etc., on an object other than the cause of it.

1840 H. Cockton *Valentine Vox* xxi. 158 P'r'aps you'd like to take it out on me, 'cos if yer would, yer know, why ony say so. **1903** 'C. E. Merriman' *Lett. from Son* vi. 72 Milligan..came around to take your cussing of him out on me. **1926** G. Hunting *Vicarion* xviii. 311 Make some records of me, and take it out on them. **1947** A. Huxley *Let.* 9 Mar. (1969) 567 He can't associate sex with respectability, but he has to take it all out on tarts or housemaids. **1958** *Daily Sketch* 2 June 12/6 You may be irritable at work, but don't take it out on your colleagues. **1967** *Listener* 11 May 611/2 The country took out its frustrations on Congress. **1978** P. Marsh et al. *Rules of Disorder* ii. 39 My brother..was a troublemaker and now they're taking it out on me.

87. take over. b. Also *absol.* Also *to take over from*: to relieve, take the place of, succeed.

1916 'Boyd Cable' *Action Front* 182 The colonel was severely wounded and had sent for the second in command to take over. *Ibid.* 234 Riley..explained the position to the subaltern who took over from him. **1946** D. C. Peattie *Road of Naturalist* i. 20 A ranker, branching dandelion took over from the desert dandelions. **1978** J. Gardner *Dancing Dodo* xiv. 101 Terry Makepiece was not going to take over on this. He would see it through himself.

90. *take up.* **c.** (*a*) (Further examples.) Also, † to make (a further hole) in order to shorten a strap. To shorten or tighten (a garment, pattern, etc.), esp. by hemming or tucking.

1818 C. Brown *Let.* 7 Aug. in *Lett. J. Keats* (1958) I. 361, I must have another hole taken up in the strap of my Knapsack. **1916** L. I. Baldt *Clothing for Women* ix. 186 To shorten pattern... Lay fold at same point, to shorten length, unless a great deal has to be taken up, in which case some could be taken from the bottom. **1937** P. H. Richards *Dress Creation* XIII. 113 The quantity taken up in the tucks should amount in all to the distance between A and C. **1972** A. Ross *London Assignment* 28 The trousers were a fraction long, and would need to be taken up.

d. (*a*) Also, to apply for or claim. Cf. *take-up *sb.* (*a*.) 6.

1971 *Guardian* 15 Apr. 1/1 A major campaign to persuade people to take up their welfare and social security benefits has been launched by the Government.

(*c*) (Earlier example in sense 'to subscribe for (shares)'.)

1869 *Bradshaw's Railway Man.* XXI. 402 Of 100,000 new 10*l.* shares..84,837 have been taken up.

(*d*) (Earlier examples.) Also *fig.*

1849 E. Davies *Amer. Scenes* 42 While they were singing Brother such-a-one would 'take up the collection'. **1880** 'Mark Twain' *Tramp Abroad* ix. 88 She became a sort of contribution box. This dear young thing in the theatre had been sitting there unconsciously taking up a collection [of fleas].

f. (*a*) Also *absol.* (see quot. 1974).

1960 E. L. Delmar-Morgan *Cruising Yacht Equipment & Navigation* vii. 86 The planks and timbers will dry out... When they are once again waterborne they will leak until the wood 'takes up'. **1974** J. Keats *Of Time & Island* xi. 177 The [fibreglass] boats did not have to be put into the river to soak, or take up, as the people said.

(*b*) *Engin.* To accept, absorb, or assimilate (by gearing, etc.).

1921 *Conquest* Oct. 510/2 It appears to have solved the problem generally of how gradually and smoothly to take up and transmit the power of a prime mover or motor. **1966** *Listener* 24 Nov. 773/1 Although the paint is applied neatly, there are slight irregularities... These slight irregularities help the colours to engage with each other,.. rather as the slightly abrasive surface of a clutch-plate takes up the transmission.

h. (*c*) *to take* (a person) *up on* (something): to accept an offer, invitation, etc. *colloq.*

1914 S. Lewis *Our Mr. Wrenn* v. 63 'We'll go Dutch to a lodging-house.'.. 'All right, sir; all right. I'll take you up on that.' **1948** 'N. Shute' *No Highway* vii. 192 It's just an estimate... I didn't want people to take me up on it like this. **1961** J. Stroud *Touch & Go* iv. 45 'Tell her not to hesitate to ask.' 'Thank you... I might take you up on that.' **1974** 'E. Ferrars' *Hanged Man's House* xv. 149 I'll go over to see Mrs Bayne and take her up on her invitation to lunch. **1979** B. Parvin *Deadly Dyke* xxiv. 134, I must be going. I'll take you up on that coffee later.

n. (*a*) For *Obs.* exc. *dial.* read: Now *U.S.*, of a horse. Also *intr.* of a rider, to rein in.

1942 *Sun* (Baltimore) 20 Oct. 15/1 Fogoso..cut sharply in front of Sunset Boy, causing Jimmy Berger to take up. **1946** *Ibid.* 2 Oct. 15/2 Red Tag ran into tight corners at the head of the stretch and was forced to take up. **1950** *Ibid.* 20 May 11/1 Queen May, ridden by Joe Culmone, was not to get through... Culmone was forced to take up.

r. For † *Obs.* read: *Obs.* exc. *intr.* in *U.S.*, (esp. of a school term) to begin, start up. Cf. sense 82 s above.

1871 E. Eggleston *Hoosier Schoolm.* xii. 104 Meetin's took up. **1903** J. Fox *Little Shepherd* iii. 42 When school 'took up again', Chad was told to say them aloud in concert with the others. **1949** 'J. Nelson' *Backwoods Teacher* 51 Four other children..trooped in, having belatedly heard that school was taking up today. **1961** M. Beadle *These Ruins are Inhabited* (1963) iii. 46 Red's school took up in two days.

x. (Later examples.)

1936 M. Mitchell *Gone with Wind* xii. 234, I knew you were doing it just to take up for me. **1977** *New Yorker* 6 June 85/1 'Wouldn't it embarrass *you*, hearing that *your* daddy spent a night in jail?' And Henry said no, it wouldn't—not if he knew his daddy had been taking up for someone.

z. (*a*) Also, to become friendly with, to form a relationship with.

1957 R. Hoggart *Uses of Literacy* iii. 76 The woman he 'took up with' was likely enough to be married herself and of roughly the same age as his own wife. **1963** *Australasian Post* 14 Mar. 44/1 Miss Dolly has 'taken up' with a poor but respectable cabinet-maker and his wife... She sells her stolen nag to help them out. **1977** *Daily Express* 29 Jan. 7/2 The story is of a poor but pretty girl..who breaks her engagement to a morose butcher..and takes up instead with a feckless punter.

XIII. 91. *take* and 25 d, *t* as it comes 42 e, *t* as done 47 e, *t* as one finds 42 b, *t* a fall 52 d, *t* the Fifth Amendment 17 b, *t* five 52 c, *t* it 42 d, *t* it from me 41 b, *t* it from there 29 b, *t* a joke 42 c, *t* or leave 39 a, *t* a letter 33 a, *t* one's life 58 b, *t* over 57 a, *t* a spill 52 d, *t* ten 52 c, *t* that 34 d, *t* with one 57 d.

take, *sb.* Add: **2. a.** (Further examples.) Also *spec.*, personal income or earnings (*U.S. colloq.*).

1850 N. Wiseman *Let.* 9 Dec. in *Dublin Rev.* (1919) Jan. 9 On Sunday the church was open... You may judge of the crowds when I tell you that the *take* was £94. **1937** *Sun* (Baltimore) 9 Feb. 11/2 There is the case of the check-room girl in a hotel, who receives tips for each garment checked, but turns her 'take' over to the management. **1943** *Ibid.* 25 Feb. 12/1 They will seek to increase their take by selling whisky on the side. **1966** P. O'Donnell *Sabre-Tooth* v. 83 It was a big enough take for her to make a once-only comeback from retirement. **1970** 'B. Mather' *Break in Line* v. 59 The luggage coolies..who kicked back half of their take to the Pathan hall porter. **1978** *Detroit Free Press* 5 Mar. b7/6 Isley will have to make do with his take from the Tractor Pull.

b. *Criminals' slang* (chiefly *U.S.*). Money acquired by theft or fraud.

1888 J. Greenwood *Policeman's Lantern* 69 A tidyish 'take' brought about by what he called the 'sweetstuff lay'. **1927** C. F. Coe *Me—Gangster* xiii. 228 After the stick-up..Carrots..can watch the take till I send the porter over after it. **1934** *Sun* (Baltimore) 14 July 3/5 A self-confessed confidence man..testified that he always handed his take to Graham, who..kept fifteen per cent for providing police protection. **1955** *Publ. Amer. Dial. Soc.* XXIV. 194 The day's *take* is the *knock up*, and the mob usually pools expenses for the day..and this amount is taken off the *top* (the total *take*). **1963** G. J. McCall in A. Dundes *Mother Wit* (1973) 422 The multimilliondollar 'take' of the hoodoo complex.

c. *U.S. colloq.* A percentage of a sum of money which is deducted, as for tax or other levies.

1935 *Sun* (Baltimore) 18 Jan. 14/7 Once the mutuel 'take' is raised over 7½ per cent. Maryland no longer can compete with other Eastern tracks. **1942** *Ibid.* 20 Mar. 14/2 What is the whole take of the Lewis dues collectors? How much of that take is a compulsory tribute through strong-arm imposition of closed-shop contracts? **1975** *Lamp* (Exxon Corporation) Winter 11/2 A recent act of Parliament imposes a special tax on revenues from British fields; combined with royalties and corporate income taxes, it raises total government 'take' to as much as 75 per cent, depending on the size of the field.

7. (Earlier examples.)

1847 J. S. Robb *Streaks of Squatter Life* 54 Here was a '*take*' in the book of human nature, which was most 'fair copy'. **1853** 'Mark Twain' *Let.* 26 Oct. (1917) I. i. 26 When one gets a good agate take, he is sure to make money.

8. a. *Med.* A successful inoculation with a vaccine. **b.** *Agric.* Successful germination and growth of seed. **c.** *Med.* An acceptance by the body of tissue foreign to the site or to the individual.

1909 in *Cent. Dict. Suppl.* **1914** *Q. Jrnl. Med.* VII. 284 Revaccination on an individual who has been vaccinated many years previously usually results in what might be termed a genuine 'take'. **1921** H. Guthrie-Smith *Tutira* xix. 165 Between the isolated plants of the miserable 'take' of seed there was ample space left for the germination of undesirables. **1924** *Surg., Gynecol. & Obstetr.* XXXVIII. 101/2 A temporary take [of grafted skin] occurred but did not thrive well. **1940** R. G. Stapledon *Re-grassing* 21 In cases where the 'take' had been exceptionally good, and where there has been no immediate return of undesirable species, excessive early growth is far from disastrous. **1965** *Nursing Times* 5 Feb. 180/1 Persistent negative nitrogen balance results in..poor graft take. **1977** *Lancet* 13 Aug. 356/2 In studies with live vaccines patients showing serological or virological evidence of virus 'take' are usually compared with the placebo group and/or with vaccine recipients who show no evidence of infection.

9. a. *Cinematogr.* A continuous section of film photographed at one time; an instance of such filming. Also preceding a numeral to distinguish individual sections of film. (In quot. 1922 used *collect.*) orig. *U.S.*

1922 *Opportunities in Motion Picture Industry* (Photoplay Research Soc.) 50 When the daily 'take' is handed in, does Mr. Director look for 'action'; does the technical man look to see how his pet scene photographed? **1928** *Sunday Dispatch* 2 Sept. 5/4 Notwithstanding the fact that the director knows that certain 'takes' are useless and need not be printed. **1937** H. G. Wells *Brynhild* ix. 144 Very few of the players..realized that a movie take was afoot outside the marquee. **1947** *People* 22 June 5/3 Micky and..Dave Crowley did the same fight 25 times before the final take was okayed. **1962** *Movie* Sept. 19/1 This conversation occupies part of a long take in a medium shot which favours neither point of view. **1972** *Listener* 21 Dec. 852/1 Sequence of calls before a shot. Production Assistant: 'Quiet. Going for a take.'.. First Assistant: '245, Take 5.' **1976** H. R. F. Keating *Filmi, Filmi, Inspector Ghote* vi. 53 The scrawled chalk figures on the black board must indicate which scene and 'take' this was.

b. A sound recording; the act of making such a recording. orig. *U.S.*

1926 Whiteman & McBride *Jazz* xii. 248 At 12, a rehearsal or phonograph take. **1946** R. Blesh *Shining Trumpets* ix. 209 The results are..a tribute to the recording engineer who supervised the 'take'. **1965** *Melody Maker* 3 Apr. 10 Of added interest is the fact that these takes..have not been issued in Britain before. **1972** *Daily Tel.* (Colour Suppl.) 17 Nov. 9/3 Barkworth once had a one-word voice-over: 'Maltesers', which required 42 takes. Listening to the playback of the tape, he got the giggles. **1977** *Rolling Stone* 13 Jan. 48/2 Still such gems as 'All the Way from Memphis' and 'All the Young Dudes', along with a different take of 'Roll Away the Stone' and two previously unheard cuts,..give this absorbing group a belated last testament.

10. *on the take*: taking bribes; capable of being suborned. *slang* (orig. *U.S.*).

1930 *Liberty* 29 Nov. 70/1 There are men and women ready to boost the thief's game. The steer guy finds him work, dicks on the take protect him. **1935** J. O'Hara *Appointment in Samarra* i. 27 There was a councilman who was not on the take. Ed for some reason hadn't been about to get to him with a dime, not a dime. **1942** Berrey & Van den Bark *Amer. Thes. Slang* § 374/5 *Bribable*, approachable, fixable, on the take. **1956** B. Holiday *Lady sings Blues* (1973) xxi. 169, I had heard of city cops taking plenty of money, but I never heard of a Treasury agent on the take since long before my time. **1967** *Boston Sunday Globe* 23 Apr. 18/5 In an unguarded public moment [he]..said, 'Half the people in Philadelphia are on the take.' **1975** *Listener* 16 Jan. 67/1 No matter how many Government departments were on the take, it was also evident that..the law enforcers themselves were bent. **1980** R. L. Duncan *Brimstone* v. 90 'I'm not on the take,' he said.

take-. Add: **take-all**, substitute for def.: a disease of wheat and other cereals caused by the fungus *Ophiobolus graminis*, which produces a foot rot, yellowing of the plants, and stunted growth; (later examples); **take-apart** *a.*, capable of being taken to pieces and reassembled; **take-charge** *a. colloq.* (orig. and chiefly *N. Amer.*), pertaining to or characterized by leadership or authority; cf. *take-hold* adj. below; **take-down,** (*a*) (later examples); (*b*) (a rifle with) the capacity to have the barrel and magazine detached from the stock; usu. *attrib.* or as *adj.*; (*c*) *Austral. slang*, a deceiver, cheat, or thief; **take-hold** *a.* (*U.S. colloq.*) = *take-charge* adj. above; **take-home** *a.* (orig. *U.S.*), that may be taken away home; esp. as *take-home pay*, that portion of a person's earnings which is left after deductions of tax, insurance, etc.; **take-it-or-leave-it** *a.*: also as *sb.* in various senses (cf. *take it or leave it* s.v. *take *v.* 39 a); **take-with** *a.* (*U.S. colloq.*), that may be taken along with one; *spec.* (see quot. 1941); also applied to prices; cf. *cash and carry* s.v. *cash *sb.*^{1} 2 g.

1912 *Bull. Misc. Inf. R. Bot. Gardens Kew* 436 In the condition known as 'Take-all', the plants are attacked seriously at an early stage of growth and become yellow, and often die before the stem is formed. **1950** H. J. Massingham *Curious Traveller* viii. 150 The weather.. has certainly nursed the spread of take-all. **1978** *Times* 7 Aug. 3/2 'Take-all'..can make a wheat plant yield a stunted and useless ear. **1966** J. S. Cox *Illustr. Dict. Hairdressing & Wigmaking* 147/1 Take-apart wig. **1979** *Nature* 5 July p. viii/3 Equipment is housed in a practical, take-apart cabinet that provides a support surface and slip-fit vacuum connection for various manifold arrangements. **1954** *Sun* (Baltimore) 1 Apr. 20/4 [The team] lacks a take-charge guy... Neither..players..have those take-charge qualities. **1965** *Economist* 23 Jan. 339/1 Mr Wilson and Mr George Brown, the two take-charge men. **1970** [see *pivot *sb.* 3 c]. **1976** M. Machlin *Pipeline* xlii. 456 Usually that take-charge tone of voice would send the adrenalin racing through Larry's system, setting up deep currents of psychological resistance. **1897** *Sears, Roebuck Catal.* 578/2 All rifles of this model can be furnished with pistol grip, with take down and all other extras. *Ibid.* 579/2 No other lengths made in Take Down style. **1901** *Kynoch Jrnl.* Aug.–Sept. 136/1 This is not a 'take-down'..but the barrel is detachable in the true sense of the word. **1905** A. M. Binstead *Mop Fair* vii. 144 There are more takedowns at this game of racing than are suspected. **1906** E. Dyson *Fact'ry 'Ands* xi. 143 Well, iv she ain't a fair take-down! **1920** G. Burrard *Notes Sporting Rifles* 15 A rifle on this principle cannot be cleaned from the breech unless it is a take-down model. **1926** J. Doone *Timely Tips for New Australians* 19 *Take-down*, a slang word for thief. A cheat. **1934** *Bulletin* (Sydney) 31 Jan. 42/2, I could learn something from a cool-headed young take-down like you. **1971** L. Koppett *N.Y. Times Guide Spectator Sports* xix. 240 A 'takedown' gives 2 points to the wrestler who puts the other down. **1976** *Shooting Times & Country Mag.* 16–22 Dec. 7/2 (Advt.), Beautiful supple quality leather take down style gun cover... Barrels and action in separate compartments inside. **1977** *Rolling Stone* 24 Mar. 67 It survives the knocks and rough handling of countless loadings, unloadings, road bumps, set-ups, and take-downs. **1973** *N.Y. Law Jrnl.* 2 Aug. 16/3 (Advt.), Top salary, benefits for take-hold person with excellent skills, initiative. **1977** I. Shaw *Beggarman, Thief* I. ii. 21 Your brother Rudy is one hell of a man. A take-hold guy. **1943** *Sun* (Baltimore) 4 Dec. 12/1 (*heading*) Steel workers 'take home' pay rises 55%. **1951** *Manch. Guardian Weekly* 8 Feb. 15/4 It used to be that trainmen were classed first or second in take-home pay. **1968** J. D. Watson *Double Helix* xvi. 111 Though the theoretical basis for many of their conclusions was shaky, the take-home lesson was obvious. **1973** *Times* 15 Mar. 23/6 Licences should be granted to 'any retailer whose character and premises satisfy certain minimum standards, at least as far as the take-home trade is concerned. *Ibid.* 23/7 Take-home beer sales. **1976** *National Observer* (U.S.) 28 Aug. 6/3 A total of 823 cadets took the take-home exam.., and many cadets, their attorneys, and others suspect that as many as half the class collaborated. **1977** *R.A.F. News* 11–24 May 1/3 Between £1.68 and £5.29 a week is added to the take-home pay of married men with two children. **1933** Wodehouse *Heavy Weather* xvi. 280 The gallant nonchalance of that take-it-or-leave-it of his..had sent Lord Tilbury scrambling for his cheque-book. **1940** C. Morgan *Voyage* III. v. 267 [He] was a little distinguished among Thérèse's..adorers..by..his attitude of take-it-or-leave-it even towards Thérèse herself. **1942** 'G. Orwell' *War-Time Diary* in *Coll. Essays* (1968) II. 415, I believe, however, that in spite of the 'take it or leave it' with which our government started off, the terms will actually be modified. **1972** *Adoremus* Jan. 19 Devotion to her..is more than a matter of personal taste, a take-it-or-leave-it of the Christian religion. **1930** Godley & Kaylin *Control Retail Store Operation* xviii. 214 For a 'cash-take-with' register transaction, no salescheck is used. **1941** Duncan & Phillips *Retailing Princ. & Methods* xx. 720 Sometimes the terms 'take transaction' or 'take-with' are used to identify those sales in which merchandise is given to the customer at the close of the sale rather than having the goods delivered by the store. **1970** *Redbook* Jan. 97/2 Another feature..of the..Diet is take-with lunches for the working woman... For five days of each week's menu there are lunches that can be eaten at home or packed to take along to work. **1977** *Chicago Tribune* 2 Oct. IX. b (Advt. Suppl.) 2/2 Chain saws are take-with priced.

ta·ke-away, *sb.* and *a.* Also **Take-away, takeaway.** [f. vbl. phr. *to take away*: see Take *v.* 78.] **A.** *sb.* **1.** *U.S.* (See quot.) *rare.*

1931 *Amer. Speech* VII. 52 The train that takes the logs to the mill is the 'takeaway'.

2. *Golf.* The initial movement of the club at the beginning of a backswing.

1961 [see *lightning-quick* s.v. *lightning 3 c]. **1976** *Sunday Mail* (Glasgow) 21 Nov. 39/5 Jack Nicklaus.. gives his advice today on another part of a good golf swing—the take-away.

3. A shop which sells take-away food (see sense 1 of the adj. below).

1970 *Cape Times* 28 Oct. 18/1 (Advt.), Are you interested in a take-away..or supermarket? **1974** *Times* 7 Oct. 8/6 There is just as likely to be a chop suey bar or a chippy or a take-away in..Bognor Regis..as in any big city. **1976** J. Fraser *Who steals my Name?* xi. 134 [He] drove to the Chinese Take-Away on the outskirts. **1981** M. Hardwick *Chinese Detective* xiv. 134 Proprietor of..a small string of burger eateries and takeaways.

B. *adj.* **1.** That may be taken away; *spec.* designating cooked food sold to be eaten away from the premises of sale.

1964 *Punch* 15 Apr. 572/3 Posh Nosh..was serving take-away venisonburgers. **1970** *Final Exam. Hons. Eng. Lang. & Lit.* (Univ. Newcastle upon Tyne) 1 (*heading*) Take-away paper. **1971** *Guardian* 27 Mar. 11/1 We sent out to the Chinese restaurant for a Chinese take-away curry. **1974** *Times* 7 Oct. 8/5 British people buy their take-away meals with convenient regularity. **1975** *Times* 18 Aug. 2/4 Second-class travellers will be able to buy a full meal on a take-away tray which will not slip off the tables in their saloons. **1976** *Nature* 18 Mar. 213/2 The takeaway message of the Dunbars' monograph is that superficially similar social systems may be the product of different behavioural arrangements. **1982** *London Rev. Bks.* IV. xxiv. 3/2 As a takeaway sample of what he had in mind, Alvarez contrasted the horses of Larkin's poem 'At Grass'..with the 'urgent' horses of Ted Hughes's 'A Dream of Horses'.

2. Of, pertaining to, or characterized by the selling of cooked food to be taken away.

1971 *Guardian* 18 June 11/5 Every take-away pieshop and baker sell a Cornish pasty. **1973** *Times* 3 Feb. 13/5 Leslie's also do a take-away service. **1977** *N.Z. Herald* 5 Jan. 2–15/8 (Advt.), Takeaway bar. **1978** *Cornish Guardian* 27 Apr. 14/4 (Advt.), Lucrative beach café.. good take-away business, ice cream servery. **1981** B. Knox *Killing in Antiques* vii. 157 [They] made an expedition..to the nearest Chinese take-away restaurant and brought back enough food.

Takelma (tākeˑlmă), *sb.* and *a.* Also 9 **Takilma.** [ad. Takelma *dāgelmáˀn* those dwelling along the river.] **A.** *sb.* **1.** A Penutian language (now extinct) of south-western Oregon. **2.** The North American Indian speakers of this language. **B.** *adj.* Designating this people or their language.

1882 *Mag. Amer. Hist.* Apr. 258 Phonetically, Takilma is more..vocalic than Kúsa. **1891** J. W. Powell *Indian Linguistic Families* 121 The Takilma formerly dwelt in villages along upper Rogue River, Oregon... They are now included among the 'Rogue River Indians'. **1907** E. Sapir in *Jrnl. Amer. Folklore* Jan.–Mar. 33 The following notes regarding the ideas of the supernatural world held by the Takelma Indians were obtained..from..one of the very few full-blood survivors of the Takelmas. **1909** *Publ. Pennsylvania State Univ.: Anthropol.* II. I. 5 The Takelma language represents one of the distinct linguistic stocks of North America. It is..a source of congratulations that enough of the folk-lore of the Takelmas could be obtained to enable one to assign these Indians a definite place in American mythology. **1912** E. Sapir (*title*) The Takelma language of southwestern Oregon. **1941** C. F. Voegelin in L. Spier et al. *Lang., Culture & Personality* 23 When Sapir studied Takelma in 1906 there were only a few speakers of the language remaining. **1965** *Canad. Jrnl. Linguistics* Spring 124 Some statement should also be made concerning languages for which no speakers were found... Inland, no..Takelma, Molale, or Cayuse were found. **1974** *Encycl. Brit. Micropædia* VII. 859/2 Six of these [Penutian] languages—Costanoan, Cayuse, Molala, the Yakonan languages..and Takelma—are extinct.

taken, *ppl. a.* **b.** *taken-for-granted* (earlier example).

1895 G. B. Shaw in *Liberty* 27 July 2/1 The Impressionist movement..was evidently destined to improve pictures greatly by substituting a natural, observant, real style for a conventional, taken-for-granted, ideal one.

take-off, *sb.* and *a.* Add: Also **takeoff. A.** *sb.* **2.** Freq. in literary or theatrical use, a skit or parody. Const. *of* or (chiefly *U.S.*) *on*. (Earlier and later examples.)

1846 *Knickerbocker* XXVII. 457 Whittier will smile at the following 'take-off' of his spirited 'Songs of Labor'. **1930** C. Wittke *Tambo & Bones* iv. 157 The take-offs on theatrical stars..often displayed rare powers of mimicry. **1951** *Manch. Guardian Weekly* 4 Jan. 15/4 Finely written take-off on New York theatre society. **1967** J. Philip et al. *Best of Granta* II. 103 A favourite ploy was to devote a whole issue to a take-off of a national magazine. **1976** *New Yorker* 15 Nov. 4/2 This musical, with the indicated twist, is a foolish takeoff on foolish musicals of the thirties. **1983** *Listener* 30 June 14/3 As well as being a take-off of Italian opera..*The Beggar's Opera* is a parody of the pastoral mode.

3. a. (Further examples in Sport.)

1887 M. Shearman *Athletics & Football* v. 153 If the 'take-off' is..so slippery as to make the jumper nervous of falling, he may..jump into the bar instead of over it. **1904** R. Thomas *Swimming* (ed. 2) 50 *Takeoff*, the board, side of bath or any standing place whether free from spring or not, from which a leap, header, feet first or other spring into water is made or taken. *Ibid.* 409/2 The one 'ready to dive' should have his feet over the edge of the takeoff. **1920** Naylor & Temple *Mod. Physical Educ.* 189 The 'take-off' may be made from one foot. **1951** *Times* 3 Jan. 4/5 Search the rinks of today for a take-off of beauty such as that of Bernard Adams..and you seek in vain. **1951** *Swimming* (E.S.S.A.) iii. 49 The take-off in the back-crawl is immediate, since there is no over-balancing to contend with as in the front-crawl or the breast-stroke. **1977** *Arab Times* 14 Dec. 10/3 Both high and long jump events were noticeable for their absence of the basic essentials; in the former, jumpers threw themselves at the bar with gay abandon without the slightest modicum of lift at take-off.

b. *Aeronaut.* The act of becoming airborne (see *take *v.* 83 n (*d*)). Also *transf.*

1904 *Aeronaut. Jrnl.* VIII. 56/1 The incline is one in two, the lower end forming a curve... The 'take-off' is on an upward incline of one in ten. **1914** in C. F. S. Gamble *Story North Sea Air Station* (1928) 70 That take-off of his was worth going a long way to see. **1918** in *Ibid.* xx. 353 The boat..taxied slowly along the water until the desired position for the 'take-off' was reached. **1929** *Sat. Even. Post* 14 Dec. 13/2 A group of news camera-men were setting up to catch the take-off of the seaplane. **1942** [see *briefing *vbl. sb.* 2]. **1951** [see *blast-off]. **1966** *Electronics* 17 Oct. 107 Lengthy preflight tests increase the probability that the equipment will fail before takeoff. **1974** *Daily Colonist* (Victoria, B.C.) 28 Dec. 4/3 When the [hydrofoil] boat rises on its struts, it is called 'take-off'. **1977** D. Anthony *Stud Game* xii. 69, I have a witness who says you couldn't have made the aeroplane ride... Grant's take-off that night is a matter of record.

5. *fig.* The act of starting off (on a journey, etc.); a departure.

1928 H. Crane *Let.* Dec. (1965) 332 [I] think I'm going to like London entirely too well for an early take-off to Spain. **1965** J. Pollard *Surfrider* ii. 20 Another thing you have to watch is the 'late take-off', catching a wave at the last minute as it begins to break. **1973** *Black Panther* 13 Oct. 17/1 The little green Fiat conveniently parked on the corner for what was to have been a speedy take-off.

6. *fig.* The beginning of (a new phase of accelerated or increased) growth or development. *spec.* in *Econ.*

1953 W. W. Rostow *Process Econ. Growth* i. 17 The term 'take-off' is here used to describe the transition of a society from a preponderantly agricultural to an industrialized basis, or, more generally, a sustained rate of increase in output *per capita*. **1957** *Listener* 10 Oct. 554/2 Development capital and trading conditions which facilitate their take-off into industrialisation. **1964** M. McLuhan *Understanding Media* (1967) II. xiv. 147 'Backward' countries take a long time to reach economic 'take-off'. **1973** *Daily Tel.* 26 Feb. 17/6 The long-expected take-off in the sale of telephone facsimile machines will not happen before next year at the earliest. **1979** *Dædalus* Spring 1 In a country like France, where the postwar 'take-off' was particularly painful.

7. See *power take-off* s.v. *power *sb.*^{1} 18 f.

B. *attrib.* or *adj.* **1. b.** In sense *3 b of the *sb.*, as *take-off area, run, speed*, etc.

1943 *Yank* 16 July 10 The planes..proceed down the roadway to the take-off strip. **1958** *Chambers's Techn. Dict.* Add. 1019/1 *Take-off rocket*, a rocket, usually jettisonable, used to assist the acceleration of an aeroplane. **1958** [see *right of way 3 a]. **1960** *Guide to Civil Land Aerodrome Lighting* (B.S.I.) 7 *Take-off area*, an area on the ground of specified dimensions which abuts the end of a strip. **1968** Takeoff area [see *sand-bar* s.v. *sand *sb.*^{1} 10 a]. **1969** *New Yorker* 12 Apr. 100/2 The experiments, after they are set up, will be out of range of the blast of the LM's takeoff rocket. **1973** D. Kyle *Raft of Swords* (1974) II. xv. 162 He made take-off speed bumping disconcertingly. **1976** P. Cave *High Flying Birds* iii. 37 Before strapping in, I walked to the edge of the take-off area and cast my eyes around for any possible snags. **1981** C. Potok *Bk. of Lights* (1982) v. 129 They landed in late afternoon... As they walked, a huge aircraft went into its takeoff run.

c. Of or pertaining to a starting-point or point of development, increase, etc. Cf. senses 5, 6 of the sb.

1947 *Radiology* XLIX. 304/2 Prior to death, the heart may be injured, as shown by a lowering of the take-off level of the T-wave. **1962** M. McLuhan *Gutenberg Galaxy* 79 The great medieval invention of typography that was the 'take-off' moment into the new spaces of the modern world. **1962** E. Snow *Other Side of River* (1963) xxiii. 172 By 1962 natural catastrophes and disastrous mistakes in take-off phases of the communes had cruelly combined to expose fantastic overclaims for agricultural output in 1958. **1965** J. A. Michener *Source* (1966) 840 At eight o'clock all units were in take-off position. **1968** B. Magee *Aspects of Wagner* i. 21 This combination of poetry and symphony that provided the take-off point for Wagner. **1974** *Times* 21 Jan. 6/2 Nineteen sixty-six was the take-off year for population.

2. a. Also used of an appliance which removes something.

1945 *Richmond* (Va.) *Times-Dispatch* 8 June 13 For hair-free and satin smooth legs, use take-off hair pads.

b. That may be taken off; designed to be readily put on and taken off.

1950 *N.Y. Times* 29 Nov. 42/6 (*heading*) Take-off attire for spring shown. Bathing suits, town costumes and evening gowns have parts to be removed. **1977** *Lancashire Life* Feb. 20/2 Minty (..showing their latest range of suites with take-off covers).

ta·ke-out, *a.* and *sb.* Also **take out, takeout.** [f. vbl. phr. *to take out*: Take *v.* 85.] **A.** *adj.* **1.** Chiefly *N. Amer.* Designed or made to be taken out. **a.** Applied to a mechanical device that may be pulled or folded out as required.

1908 *Sears, Roebuck Catal.* 94 Our new model long body runabout... Very large, roomy seat, small take out seat in rear. **1982** *Motor* 3 July 55/1 A glass tilt or take-out sunroof.

b. Applied to prepared food sold for consumption elsewhere. Cf. *take-away *a.* 1.

1968 *Globe & Mail* (Toronto) 17 Feb. 47/4 (Advt.), Soft ice cream and take out food. **1970** *Times* 29 Jan. 27/3 One of New York's finest restaurants will provide gourmet 'take-out' lunches for the hard-pressed executive. **1972** B. Garfield *Line of Succession* (1974) I. 73 Lunch in the office..had been dreary with takeout food. **1975** *New Yorker* 21 July 83/2 Ken and Eve do a pretty good take-out-sandwich business at noon with people who work in the neighbourhood.

2. Of, pertaining to, or characterized by the sale of prepared food for consumption elsewhere. Cf. *take-away *a.* 2. orig. *U.S.*

1941 J. M. Cain *Mildred Pierce* ix. 177 Pies she hoped to sell to the 'take-out' trade. **1947** *Sun* (Baltimore) 2 July 30/1 One chain of sandwich shops, which does a large volume of 'take out' business. **1962** *Advance* Mar. 7/1 If you deliver take-out orders for restaurants, the pay is $20 a week and two meals a day. **1970** *New Yorker* 15 Aug. 22/1 Dialogue overheard in the take-out line of a midtown coffee shop. **1972** J. Wambaugh *Blue Knight* (1973) xi. 183 He boxed some chicken..for a take-out customer. **1975** *Times* 8 Feb. 7/1 Simple fish-and-chip shops with a take-out counter and a few tables. **1977** *Belfast Tel.* 17 Jan. 15/7 (Advt.), Capable person to organise and run Safari Take-Out Cafe.

3. *Bridge.* Designating a bid or call that takes the bidder's partner out (see *take *v.* 85 k); *take-out double* = *informatory double* s.v. *informatory *a.* b.

1945 Phillips & Reese *How to play Bridge* 62 When a double is made for that reason it is called an 'informatory' or 'take-out' double. **1959** *Listener* 15 Jan. 146/2 A take-out bid might induce partner to bid too many diamonds. **1962** *Ibid.* 3 May 790/2 The take-out call. **1964** [see *informatory *a.* b]. **1967** R. L. Frey et al. *Bridge Players' Encycl.* 298/2 Negative double, the original name for a take-out double, in general use from 1915 to 1930, about which time the term 'informatory' became current. **1972** *Times* 23 Sept. 9/4 East read more in North's take-out double than in his partner's show of strength.

B. *sb.* **1.** *Bridge.* An act of taking out (see *take *v.* 85 k).

1917 E. Bergholt *Royal Auction Bridge* (1918) 88 The 'weakness take-out' or 'rescue' is obligatory in all suits, but the 'strength take-out' only in hearts or spades. **1927** [see *denial 6]. **1945** Phillips & Reese *How to play Bridge* 62 If partner has not spoken, or has simply passed, a double of One or Two of a suit is for a take-out. **1962** *Times* 14 Nov. 17/2 The Double which may be either for a penalty or a take-out. **1977** *Times* 10 Dec. 13/4 Opponents double for a take-out whenever they can afford the risk.

2. *U.S.* A tax deducted from winnings on a horse-race.

1946 *Sun* (Baltimore) 18 Aug. 8/3 In some states as much as twelve per cent is deducted from every dollar won to cover breakage, track takeout and state and Government taxes. **1954** *Ibid.* 17 Feb. 20/5 With the present 10-cent take-out Maryland mile tracks are producing the highest revenue in proportion to the population in the area of any state in the nation. **1971** L. Koppett *N.Y. Times Guide Spectator Sports* x. 183 There is a 'take-out'—a tax—that is removed from the pool before winnings are returned. **1978** *N.Y. Times* 30 Mar. D17/4 Plans to reduce the parimutuel takeout at New York Racing Association tracks.

3. In *Bowls*, the knocking of an opponent's wood away from the jack; in *Curling*, the striking of an opponent's stone out of play.

1959 *Times* 19 Aug. 4/5 Either by direct scoring, or through judicious take-outs, he swung the outcome by 15 shots on six of the seven ends. **1961** J. S. Salak *Dict. Amer. Sports* 441 *Take out* (curling), striking a stone hard enough to remove it from rings. **1962** *Times* 16 Aug. 3/5 A. R. Allen..successfully essayed take-outs of varying strengths. **1964** [see *Canuck].

4. *U.S.* A special article in a newspaper or journal printed without a break in successive columns or pages so that it can be easily removed.

1961 in Webster. **1980** Collins & Lapierre *Fifth Horseman* I. 37 When..doing a major take-out on violent crime in the city. **1981** B. Granger *Schism* xv. 135 The phenomenon is the story here. UPI already has quite a long takeout on it.

5. A shop selling prepared food or drink for consumption away from the premises. Cf. *take-away *sb.* 3. orig. and chiefly *N. Amer.*

1970 *Globe & Mail* (Toronto) 25 Sept. 36/7 (Advt.), Chicken takeout and snack bar, close to city. **1972** *Even. Telegram* (St. John's, Newfoundland) 29 June 22/6 (Advt.), Commercial site. Formerly used as a Pizza Take-Out. **1972** *Guardian* 5 Sept. 17/6 Pizza parlours in Paddington and chop-suey 'takeouts' in Chorlton-cum-Hardy. **1978** *Detroit Free Press* 16 Apr. f 8/10 (Advt.), Bar & Grill. Liquor, beer & wine takeout.

ta·ke-over. orig. *U.S.* Also **take over, takeover.** [f. vbl. phr. *to take over*: see Take *v.* 87.] **1.** An act of taking over (see Take *v.* 87 b in Dict. and Suppl.). Also, that which is adopted or transferred.

1917 *Acts State New Jersey* xiv. 33 *Take Over*, the action by the department in assuming the control and maintenance of any part or parts of the State Highway System. **1921** W. S. Churchill *Let.* 1 June in M. Gilbert *Winston S. Churchill* (1977) IV. Compan. III. 1489 The whole principle of the 'take over' was to show the actual net cost of Mesopotamia in one vote. **1928** *Amer. Jrnl. Psychiatry* VII. 885 The accessory left ear movement may be a takeover of the same activity during nursing. **1930** J. B. Priestley *Good Compan.* III. i. 479 They want four thousand, lock, stock, and barrel, except the usual takeovers. **1932** M. Joynt tr. *Gougaud's Christianity in Celtic Lands* iii. 63 A direct take over of the monastic doctrines of the East. **1946** *Sun* (Baltimore) 10 Aug. 4/7 Special consideration was being given to the planning of a smooth take-over of UNRA facilities. **1954** *Newsweek* 11 Oct. 66/1 A solid hour without the relief of a teammate's take-over..seemed..a strain at times. **1964** M. McLuhan *Understanding Media* (1967) I. v. 64 Joyce's Bloom is a deliberate takeover from [Charlie] Chaplin. **1968** *Times* 16 Dec. 7/1 An attempt at a Ministry takeover and a threat to a much valued independence. **1980** *News & Observer* (Raleigh, N. Carolina) 28 Oct. WA-5/8, 3 bedrooms, 1½ baths, good loan takeover, nice patio overlooking woods.

2. A (usu. forcible) assumption of power or government; a military coup.

1957 *Economist* 28 Sept. 1023/2 The change in military personnel in Syria has not been followed yet by any complete communist take-over. **1966** *Listener* 10 Mar. 339/1 It is just over a week since the military take-over in Ghana and the dismissal of Dr Nkrumah. **1977** *Arab Times* 14 Dec. 3/1 The black takeover in Zambia. **1980** *Sunday Times* 21 Sept. 18/1 The military takeover in Turkey nine days ago.

3. *Econ.* The assumption of control or ownership of a business concern by another company, esp. by the acquisition of the majority of its shares, either by agreement or after a take-over bid.

1958 Bull & Vice *Bid for Power* 13 A take-over is intelligible only in the light of subsequent developments. **1959** *Punch* 16 Sept. 177/2 A surge of sentiment for Harrods has set in since the Fraser take-over. *a* **1974** R. Crossman *Diaries* (1976) II. 201 This afternoon Tony Wedgwood Benn made a Statement on the Chrysler take-over of Rootes. **1980** D. Williams *Murder for Treasure* xvi. 156 The alarms and excursions that could attend the last stages of a take-over.

4. a. *attrib.*, as *take-over activity, battle, plan,* etc.

1946 *Sun* (Baltimore) 25 Sept. 12/1, I am giving this 'take-over' plan the pitiless publicity it deserves. **1948** *Times* 13 Mar. 7/2 The discount on the shares narrowed slightly to 2½ per cent. to 2 per cent. below take-over values. **1954** R. Sutcliff *Eagle of Ninth* i. 11 After the formal take-over ceremony in the forum, the old garrison marched out. **1957** *Economist* 28 Dec. 1150/2 This is a legitimate dramatic view of a takeover struggle. **1972** *Accountant* 17 Aug. 211/1 Take-over activity serves a dual purpose. **1981** *Times* 13 Oct. 16/4 Takeover fever produced some bright spots in banks.

b. Special Comb.: **take-over bid,** an attempt or offer to gain a controlling interest in a business concern sufficient to take it over (Take *v.* 87 b); also *transf.* and *fig.*; hence **take-over bidder.**

1953 *Times* 10 Nov. 5/9 A certain type of financial operation described in general terms by Lord Hacking—the recent epidemic of 'take-over bids'. **1959** *Listener* 24 Sept. 501/1 The take-over bid [for the television audience] was made by A.B.C. who put back their usual production time by an hour. **1965** E. F. Russell *Somewhere a Voice* 88, I could make a takeover bid for the chief of police and preside over the search for myself. **1979** L. Meynell *Hooky & Villainous Chauffeur* ii. 27 A whole series of trusts, transfers, holding companies and take-over bids had put him close to his million [pounds]. **1982** *Times Lit. Suppl.* 19 Nov. 1281 There is, after all, no limit to the potential fissiparity of national feeling in a world where the United Nations exists to discourage takeover bids. **1958** *Punch* 27 Aug. 287/1 The voice of the take-over bidder has recently been heard loud, clear and sometimes discordant over the market. **1966** *Observer* 13 Nov. 7/6 With take-over bidders out in force last week, the stock market had more excitement than for some time past.

taker. Add: **2. f.** Also in extended use, one who accepts an offer, suggestion, etc.

1897 'Mark Twain' *Following Equator* xxxvii. 333 A youth staked out a claim and tried to sell half for £5; no takers. **1968** *Listener* 25 July 107/3 'If anybody fancies he's better off jumping, he'd better go now.' There were no takers. **1979** J. Thomson *Deadly Relations* vi. 76 Whoever killed her must have..laid her down fairly carefully. Any takers so far?

h. *Stock Exchange.* (See quot. 1979.)

1934, 1955 [see *take *v.* 82 p*]. **1979** G. Cummings *Investor's Guide to Stock Market* 104 *Taker/Taker-In*, a seller of shares previously paid for who is prepared to 'take-in' the shares and receive a rate under a contango instead of delivering in the normal way and receiving payment. Also a speculator who has sold short and is not able to cover his position by the end of the Stock Exchange account by making delivery, and who is ready to take a contango rate from a 'giver'.

4. a. taker-in (examples in Stock Exchange use).

1911 [see *take *v.* 82 p*]. **1928** *Morning Post* 19 Nov., The operation can be repeated as long as the client, the broker and the taker-in mutually consent. **1979** [see sense 2 h above].

take-up, *sb.* (*a*) Add: **3. a.** (Earlier example.)
1850 *Rep. Comm. Patents 1849* (U.S.) 186 Improvement in the delivery and take-up motion of Looms.

b. *Cinematogr.* The apparatus for gathering up film after exposure in a projector or camera.
1915 B. E. Jones *Cinematograph Bk.* 162 The take-up or driving mechanism of the bottom spool. **1931** B. Brown *Talking Pictures* 181 This is threaded through the projector..and down to the take-up. **1940** *Chambers's Techn. Dict.* 832/2 *Take up, take up reel* (Cinema), the drive and the reel which is necessary to accept the cinematograph film after exposure in the gate of a camera or projector.

5. *Engin.* The action or process of taking up (see *Take *v.* 90 f (*b*)).
1912 [see *clutch *sb.*[1] 6 a]. **1927** *Daily Tel.* 21 June 7 The clutch too is light in action and positive,..its 'take-up' is smooth and without jerk.

6. The acceptance of something offered; *spec.* the claiming of benefits provided by the Welfare State. Cf. *Take *v.* 90 d (*a*).
1961 *Economist* 22 Apr. 347/3 Recent studies have suggested a similar social gradient in the use of certain infant welfare services and in the 'take-up' of welfare foods. **1971** *Times* 23 Jan. 18/5 The same low take-up has been shown to apply to free prescriptions. **1975** *Language for Life* (Dept. Educ. & Sci.) xxv. 371 The take-up of longer courses in our sample was very low. **1981** *Daily Tel.* 14 May 36/3 Mr Norman Buchan..put to his 'favourite social services minister'..that not one welfare benefit had 'a take-up' of more than 80 per cent.

7. *spec.* in *Finance*, the action of paying in cash for stock originally bought on margin.
1976 *Glasgow Herald* 26 Nov. 11/9 The market also faced repayment of moderate Bank of England loans.., Treasury Bill take-up, maturing local authority bills in official hands. **1979** *Irish Times* 28 Sept. 15/4 The market had a small net take-up of Treasury Bills to finance.

8. *attrib.*, as *take-up lever, reel, spool*, etc.
1884 Take-up motion [in Dict., sense 3]. **1904** *Dialect Notes* II. 391 *Take-up screw*,..a kind of screw for iron pull-rods or wire rope. **1927** T. Woodhouse *Artificial Silk* 123 Then the cloth goes over the emery take-up roller, and finally on to the cloth beam. **1931** B. Brown *Talking Pictures* 133 Only one spool-carrying spindle is employed, which takes both feed and take-up spools side by side. **1940** Take-up reel [see sense 3 b above]. **1943** *Gloss. Terms Telecomm.* (*B.S.I.*) 85 *Take-up reel*, [on a fire-alarm] a clockwork driven reel, which maintains the tension and winds-up the tape released by the register. **1954** *Trans. IRE Audio* II. 15/1 The tape is fed over another non-rotating tape guide combined with a compliance arm and fed to the take-up reel. **1961** *Which?* Nov. 277 (*caption*) Take up lever. **1972** *Sci. Amer.* Jan. 8/3 The new camera takes the film in a direct line back to the take-up magazine, thus providing an instrument that has only one turn for the film. **1977** J. Hedgecoe *Photographer's Handbk.* 12 The cassette is loaded into a compartment, and film threaded across the back of the camera into a take-up spool. **1978** D. A. Stanwood *Memory of Eva Ryker* I. xx. 191 The last of the seven hours of tape flipped onto the take-up reel of the recorder.

takhaar (tā·khāɹ). *S. Afr.* Also **taakhaar, takhar,** and with capital initial. Pl. **-e** or **-s.** [Afrikaans, f. Du. *tak* branch + *haar* hair.] A rustic or unsophisticated person (with derog. implication of unkempt appearance). Also *attrib.* or as *adj.* and in *transf.* use.
1899 *Graaff Reinet Advertiser* 20 Nov. (Pettman), There are several other poems, one of which urges the Takhaar Boers to 'Fight, fight, fight!' **1906** A. R. Colquhoun *Afrikander Land* 217 With the passing of the old *Taak haare* the little bit of picturesqueness will be gone. **1910** J. Buchan *Prester John* ii. 44 The place..gives the ordinary man the jumps... It may be the natives, or it may be the *taakhaars*, or it may be something else. **1931** T. J. Haarhoff *Vergil in Exper. of S. Africa* 15 The early consuls were described as *capillati*, and the trekkers as *takhare*. **1931** *Times Lit. Suppl.* 3 Dec. 979/2 This motive leads him [*sc.* Haarhoff] to look for special points of resemblance..between the *capillati consules* and the *takhaar* Voortrekkers. **1942** S. Cloete *Hill of Doves* xxvii. 378 A couple of Takhaars. Two backvelders who said I was a spy. **1971** *Rand Daily Mail* 25 May 11 Commentator X (well known for his scathing comments on British 'takhare'). **1972** *Het Suid-Western* 16 Mar. 2 A political meeting of takhare in the deep north.

‖ **takht** (takt). [Pers.] A sofa or bed. So **takhtrawan** (also **taktrevan**) [*rawān* pres. pple. of *raftan* to proceed, travel], a litter or Sedan chair.
1786 S. Henley tr. *Beckford's Vathek* 97 Four of the most amiable, placed the Caliph, on a magnificent takt-revan. **1870** R. Anderson *Hist. Missions Amer. Board* III. xi. 171 Had not divine mercy stayed them just there, takhtrawan, bearers, and occupant would have been dashed down the precipice. **1979** V. L. Pandit *Scope of Happiness* vi. 43 There were also low divans called *takhts* in the living rooms. **1981** S. Rushdie *Midnight's Children* I. 19 She sips fresh lime water, reclining on a takht.

Takilma, var. *Takelma.

taking, *vbl. sb.* Add: **6.** *taking in* (later examples), *out* (later example), *over*.
1902 E. Banks *Newspaper Girl* 193 They prosper exceedingly and their takings-in at the end of the week are apt to be very large. **1983** *Sunday Tel.* 21 Aug. 11/8 Detection first determines a garment's original shape through successive takings-in and lettings-out. **1924** R. Macaulay *Orphan Island* xiii. 160 'If you have nothing to contribute, sir,' he whispered, 'kindly pass the plate, which is for puttings in, not takings out.' **1917** *Acts State New Jersey* xiv. 29 For any road in the State Highway System prior to its taking over as a State Highway [etc.].

7. *taking lens* (see Take *v.* 33 b in Dict. and Suppl.).
1951 [see *finder 3 d]. **1961** G. Millerson *Telev. Production* iii. 28 (*caption*) Small viewfinder kinescope showing TV picture (optically magnified) seen through taking lens. **1962** M. L. Haselgrove *Photogr. Dict.* 187 *Taking lens*, the lower lens of a twin-reflex camera, which forms the image actually falling on the film.

takkie, var. *tackie.

takovite (tæ·kovəit). *Min.* [ad. Serbo-Croat *takovit* (Z. Maksimović 1957, in *Zapisnici Srpskog Geol. Društva za 1955 God.* 219), f. *Tákovo*, name of a place in Serbia: see -ite[1].] A bluish green clay-like mineral that is a rhombohedral hydrated basic aluminate and carbonate of nickel.
1957 *Zapisnici Srpskog Geol. Društva za 1955 God.* 224 (*heading*) Takovite, hydrous nickel aluminate, a new mineral... This occurs in Takovo, Serbia, on the contact of limestone and metamorphosed serpentinite. **1977** *Amer. Mineralogist* LXII. 463/1 The formula of the Australian takovite, for which only kaolinite is a significant impurity, is established with greater certainty.

Takulli (tăkɒ·li). Also 9 **Taculiy, Takali** etc. [a. Carrier *dakełne* (pl. of *dakeł*) Carriers, Indians; lit. 'people who go by boat on the water.'] A name used for the Carrier Indians of British Columbia: at first used only for the eastern Carrier, but later extended to include the Babine Indians of Babine Lake and the Bulkley River.
1820, etc. [see *Carrier 2 b]. **1846** H. Hale *Ethnol. & Philol.* 201 The country of the *Takkali* (or Tacullies) includes the region north of the Oregon Territory, termed by the English New Caledonia. **1932** D. Jenness *Indians of Canada* xxii. 363 The Carrier..had no common name for themselves, only names for the independent subtribes into which they were divided. In the nineteenth century, however, they adopted for themselves the obscure title Takulli, bestowed on them apparently by Europeans. **1974** *Encycl. Brit. Micropædia* II. 590/3 They [*sc.* the Carrier] also assumed the name of Takulli (People Who Go Upon the Water), a name of obscure origin.

takyr (tā·kiᵊɹ). Also **takir.** [a. Turki, Chagatai *taḳir* f. *taḳ* smooth.] In Russian Central Asia, any of the wide expanses of clay which are covered with water in the spring and are dry in summer.
1864 A. Vámbéry *Trav. Central Asia* 91 By degrees the sand disappeared, and about midnight we had so firm a clayey soil under us, that the regulated tread of the distant camels echoed as if some one was beating time in the still night. The Turkomans name such spots Takir. **1902** *Encycl. Brit.* XXIII. 511/2 Large areas amidst the sands are occupied by *takyrs*. **1961** L. D. Stamp *Gloss. Geogr. Terms* 444/2 *Takyr*.., area of barren alkaline soil with heavy unstructural clay soil.

‖ **tala** (tā·lă). *Indian Mus.* Also **tal.** [Skr. *tāla*, Hindi *tāl* hand-clapping, musical time.] Musical time or rhythm; one of a series of traditional metrical patterns.
1891 [see *jati, jāti]. **1921** [see *dhrupad]. **1927** *Grove's Dict. Mus.* (ed. 3) II. 704/2 The principle of the time-units within the bar follows the varieties of prosodic feet. These are of one, two, three, or four syllables, and the times (*tāl*) are decided by the number of units (*mātra*) in each syllable (*akshara*) or beat (*tāl*). **1961** '*Gramophone' Long Playing Classical Record Catal.* Sept. 212/2 *Indian Music..Ragas and Talas.* Indian Instrumentalists. **1967** Singha & Massey *Indian Dances* v. 65 While the musicians play the syllabic beats, the dancer executes patterns within the tal or time-measure (pronounced 'taal'). **1977** Y. Menuhin *Unfinished Journey* xii. 258 The *tala* is the rhythm. Dozens exist.

Talaing (talai·ŋ), *sb.* and *a.* Also 8–9 **Talain;** 9 **Talien.** [Native name.] = *Mon *sb.*[2] and *a.* Cf. *Peguan *sb.* and *a.*
1798 F. Buchanan in *Asiatick Researches* V. 235 This people are named *Talain* by the *Burmas* and *Chinese* of *Yunan*. **1800** M. Symes *Embassy to Ava* v. 183 He has abrogated some severe penal laws imposed by his predecessors upon the Taliens, or native Peguers. **1835** *Penny Cycl.* IV. 438/1 Here and there, on the immediate banks of the river, are a few villages of Talain fishermen. **1844** *Jrnl. Asiatic Soc. Bengal* XIII. 43 The city of Pugan was taken, the Talaings were overawed. **1854** *Jrnl. Amer. Oriental Soc.* IV. 282 In its vocables, the Talaing is the most isolated language in Farther India. **1881** C. J. F. S. Forbes *Compar. Gram. Lang. Further India* iv. 51 The British province of Pegu, representing the old Mon or Talaing kingdom, has become practically as Burman..as Ava itself. **1904** G. A. Grierson *Linguistic Survey India* II. 1 The Mōn or Talaing spoken in Pegu. **1948** D. Diringer *Alphabet* vii. 408 When the ancestors of the modern Burmans came to the Irrawaddy basin, they found the people whom they call Talaings well established in southern Burma. **1957** *Encycl. Brit.* XXI. 759/1 From the Telingas, whose culture they [*sc.* the Mon] took and whose foreign blood they absorbed into their own stock, came the name Talaing. **1973** [see *Peguan *sb.* and *a.*].

‖ **talak** (talā·k). Also **talaq.** [a. Arab *ṭalāq* divorce.] In Muslim law: (a method of) divorce by the husband's mere verbal repudiation of his wife in a set form of words (see quots.). Cf. *get *sb.*[3].
1791 C. Hamilton tr. *Hedàya* I. iv. 200 Talâk, in its primitive sense, means *dismission*:— in law it signifies the dissolution of a marriage, or the annulment of its legality, by certain words. **1861** in E. F. Moore *Rep. Cases Appeal from E. Indies* VIII. 395 A divorce by *Talâk* is the mere arbitrary act of the husband, who may repudiate his wife at his own pleasure, with or without cause. **1917** *Law Rep. King's Bench Div.* I. 649 In my judgment Dr. Mir-Anwaruddin has made such a marriage, as he was legally entitled to do, to which Talak has no application. **1931** S. Vesey-Fitzgerald *Muhammadan Law* ix. 73 Talaq..is a generic name for all divorce, but is specifically applied to repudiation by or on behalf of the husband. **1962** *Times* 2 Feb. 3/6 A husband domiciled in a Mohammedan country could validly divorce by talak a wife whom he had married in England. **1970** *Daily Tel.* 31 Oct. 3/3 A Pakistani doctor's 'Talaq' divorce was held.. to be valid in England. Talaq is the Moslem form which involves saying 'I divorce you' three times... The Talaq was contained in a letter.

talayot (tālā·yɒt). *Archæol.* [a. Cat. *talaiot* small watch-tower, ad. Arab. (Muslim Spain) *ṭāli'āt*, pl. of *ṭāli'a* watch-tower; cf. Arab. *ṭalī'a* with similar meaning.] A Bronze Age stone tower found in the Balearic Islands, usu. circular with a corbelled roof, used for residential or defensive purposes. Hence **talayo·tic** *a.* Cf. Nuragh.
1893 *Athenæum* 2 Sept. 328/1 Like the Sardinian nuragh, the talayot is essentially a vaulted tower of extra-massive proportions. **1927** *Daily Tel.* 12 July 5/4 The Talayots are round mounds, which appear to have served as sepulchres. **1932** M. Murray *Cambr. Excav. in Minorca: Trapuco* I. 6 Taulas are always found in connection with *talayots*, circular buildings of rough masonry. **1939** V. F. Childe *Dawn Europ. Civilization* (ed. 3) xiv. 249 There is no obvious break between the 'Copper Age' culture represented in the rock-cut tombs and that represented in the 'talayots'. **1950** *Antiquity* XXIV. 154 'Taulas'..occur in association with the strong circular towers known as 'talayots'. **1974** *Encycl. Brit. Macropædia* XIX. 279/1 In the Balearic Isles the Bronze Age corresponds to the 2nd millennium BC and is designated 'talayotic' from the name of the talayot, a megalithic monument in the form of a round or quadrangular tower. **1979** Service & Bradbery *Megaliths* vii. 131 In Majorca, the talayotic site called Ses Paisses..includes a central talayot tower, outer walls and four megalithic gateways around the dwellings in the enclosure.

Talbot (tǫ·lbǫt). *Optics.* [The name of W. H. Fox *Talbot* (1800–77), English polymath: cf. Talbotype.] *Talbot's law*, the law that a flickering source of light, varying in either colour or intensity, will be perceived as if it were a constant light source exhibiting the mean value of the varying quantity, provided that the frequency of flickering exceeds the flicker fusion frequency of the eye; also called the *Talbot–Plateau law* [J.A.F. *Plateau* (1801–83), Belgian physicist].
1895 E. C. Sanford *Course Exper. Psychol.* I. vi. 146 (*heading*) The Talbot-Plateau law. **1906** *Bull. Bureau of Standards* (U.S.) II. 2 Talbot's law is thus a statement of physiological rather than of physical phenomena, and depends for its explanation on the action of the eye. **1929** L. T. Troland in C. Murchison *Found. Exper. Psychol.* iv. 187 The Talbot-Plateau law..has been established very accurately. **1943** C. T. Morgan *Physiol. Psychol.* x. 198 As the [flicker] rate is lowered, the subjective brightness of a flickering light may considerably exceed that expected from Talbot's law. **1974** *Sci. Amer.* Apr. 93/1 The law of color fusion, also known as Talbot's law (although it actually goes back to Isaac Newton), enables us to predict what color will be perceived when two colors are mixed.

Talbotype, *sb.* (Earlier example.)
1844 *Times* 9 Sept. 1/3 (Advt.), Claudet's Daguerrotype and Talbotype Portraits... Mr. Claudet continues to take portraits by both the above processes... In Talbotype.. persons may be furnished with any number of copies on paper.

talc, *sb.* Add: **2. d.** [*ellipt.* for *talcum powder* s.v. Talcum.] Talcum powder, esp. as a cosmetic and toilet preparation. *colloq.*
1938 *Vogue Beauty Bk.* 16 Feb. 24/2 Sifter top talc, 3 at 1s. 6d. **1949** *Heiress* Aug. 82 (Advt.), Fragrant talc cool-silky-perfumed by the master Goya. **1966** P. O'Donnell *Sabre-Tooth* vii. 105 Two small bars of soap, a tin of talc, and some body mist. **1977** D. Cory *Bennett* ii. 37 Hunter contrived to take a shower..to apply to himself..a dusting of powdered talc.

talc, *v.* Add: Also, to dust (the skin) with talcum powder. **talced** (tælkt) *ppl. a.* (later example).
1976 M. Nelson *Crusoe Test* iii. 40 She sprayed and talced her body. **1976** L. Deighton *Twinkle, twinkle, Little Spy* xvii. 175 His..face talced like a..cottage-loaf.

talcky, *a.* Add: The usual form is now *talcy*.
a. (Later example.)

1965 G. J. Williams *Econ. Geol. N.Z.* x. 162/2 A large land-slip extends over a good deal of the lens, and material in the slip contains large blocks of talcy serpentine.

b. Of or pertaining to talcum powder, in toilet and cosmetic use. Cf. *talc *sb.* 2 d. *colloq.*
1972 *Daily Tel.* 11 July 13 The faint, talcy smell of babies in the bathroom.

talcum. Delete ‖ and add: *talcum powder*: now *spec.* applied to perfumed or medicated talc for general cosmetic and toilet use. Also used *absol.*
1908 *Sears, Roebuck Catal.* 797 Violet and white rose talcum powder. *Ibid.*, Just the thing for holding your face powders or talcums. **1927–28** *T. Eaton & Co. Catal.* Fall & Winter 367 Djer-Kiss Talcum Powder is low-priced. *Ibid.*, Pompeian Talcum is a fine white powder with the pleasing Pompeian odor. **1949** *Heiress* June 41 (Advt.), For cool, satin-like Comfort Dubarry Talcum is the dainty woman's necessity. **1968** [see *bath oil* s.v. *bath *sb.*[1] VI]. **1981** *Good Housekeeping* Apr. 166/2 *Talcum powder* is closely related to face powder; the components are similar but the aim of body powder is to provide 'slip', a cooling effect and good absorbency.
Hence as *v. trans.*, to treat with talcum powder; **ta·lcumed** *ppl. a.*
1923 H. Jerome *Secret Woman* ix. 106 It is only the American man who smells sweet and soapy, as though he had just been shampooed and talcumed. **1943** G. Greene *Ministry of Fear* I. iii. 47 Her face was talcumed and wrinkled. **1952** C. Armstrong *Black-Eyed Stranger* vi. 47 Charles Salisbury had a clean and 'talcumed' look. **1970** *Daily Tel.* 9 Nov. 18/6 Researchers..concluded that the entire traditional marketing approach to toilet paper was wrong. The result is the soft, talcumed, flower-scented creation now on offer.

talcy: see Talcky *a.* in Dict. and Suppl.

tale, *sb.* Add: **3. c.** Proverbial phr. *dead men tell no tales* (earlier and later examples).
[**1560** T. Becon *Wks.* II. 97 He that hath his body laden with meat & drinke is no more mete to prai vnto god then a dead man is to tel a tale.] **1664** J. Wilson *Andron. Comn.* I. iv. 14 'Twere best To knock 'um i' th head, and give it out The Soldiers did it... The dead can tell no tales. **1681** Dryden *Span. Fryar* IV. i. 48 There is a Proverb..which saies, that Dead-men tell no Tales; but let your Souldiers apply it at their own Perils. **1702** G. Farquhar *Inconstant* v. 76 Ay, ay, Dead Men, tell no Tales. **1850** C. Kingsley *Alton Locke* I. iv. 67 Where are the stories of those who have not risen..who have ended in desperation?..Dead men tell no tales. **1974** 'M. Innes' *Appleby's other Story* xv. 122 'There was only one sure way to do it.' 'To kill him?'..'Yes. Dead men, they say, can tell no tales.'
e. *tale of woe*: see *woe *sb.* 1 a.
III. 10. *tale-writer* (earlier example).
1845 Poe in *Broadway Jrnl.* 7 June 354/2 If we except.. Mr. Hawthorne..and..one or two others..there is not even a respectably skilful tale-writer on this side the Atlantic.

‖ **talea** (tā·lia). *Mus.* Pl. **taleæ.** [L., lit. stick, cutting.] A repeated rhythmic pattern in late-medieval isorhythmic motets.
1944 W. Apel *Harvard Dict. Mus.* 367/1 The repeated scheme of time-values which is used in the sections A, B, C and..in D, is called *talea* in 14th-century treatises. **1960** *New Oxf. Hist. Music* III. v. 145 The rhythmical pattern of the first section or *talea* serves to determine the formal structure of the whole... The melodic development of the individual *taleae* is now entirely subordinate to a rigid framework dictated by rhythm. **1963** *Listener* 17 Jan. 141/1 The structural skeleton of the movement is sixteen rotations of the melody, containing fifteen rotations of the rhythmic pattern or *talea*. **1974** *Early Music* Oct. 220 In some [motets] the isorhythmic voices are constructed not only with a *talea* (the rhythmic pattern) but also a *color*.

Talensi, var. *Tallensi.

talent, *sb.* Add: **III. 6. d.** No longer rare as sing. (later examples.). (*Administration of*) *All the Talents* (earlier examples).
1928 E. Blom *Limitations of Music* 139 Honegger is a Swiss and a great talent to boot. **1958** *Spectator* 4 July 14/1 The studio, with its presiding talent, Lee Strasberg. **1977** *Rolling Stone* 24 Mar. 74/1 The record's not great, but the lady's a real talent.
1807 *All Talents in Ireland!* 10 The general impression upon the public mind, relative to the recent change in administration, seems to be, that the downfall of 'All the Talents' was occasioned by the unbending perverseness of my Lord H-w-k..and the deference which Lord G-n-lle paid to Lord H-w-k. **1837** G. W. Cooke *Hist. Party* III. xviii. 460 The administration, which was ironically designated by its opponents as 'All the Talents'.
e. Frequenters of the underworld. *Austral. slang.* Now *Obs.* or *rare.*
1882 [see *forty *sb.* 7]. **1928** 'Brent of Bin Bin' *Up Country* x. 151 The elder won by telling his son he could use the Waterfall stallion as a saddle-horse in the off season, and have him for his own in place of Black Belle, on condition that he left the talent of Eagle Hawk Gullies strictly alone. **1953** D. Cusack *Southern Steel* 31 He'd learn responsibility quicker married than he would knocking about the ports with the rest of the talent.
f. The women of a particular locality collectively (as *sing.*), judged according to attractiveness and sexual promise, esp. as *local talent* (see *local *a.* 2 d). Also applied occas. to men. *slang.*
1947 [see *local talent* s.v. *local *a.* 2 d]. **1950** J. Cleary *Just let Me Be* 115 [He] looked after her, and Harry grinned at him. 'Not bad, eh?' he said... 'That's a bit of the local talent.' **1963** *Sunday Times* (Colour Suppl.) 1 Sept. 8 You can take a turn on the [sea-]front and see what the talent is like. **1969** J. Fowles *French Lieut.'s Woman* xxxix. 292 Far duller the customers—the numerically equal male sex, who, stick in hand and 'weed' in mouth, eyed the evening's talent. **1972** 'M. Yorke' *Silent Witness* ii. 24 No chance had come her way... 'Your charms are waning,' Liz had said dryly. 'There isn't any talent,' Sue had answered.
8. **talent agency**, an organization which seeks to place talented amateurs in the world of professional entertainment; **talent money** (earlier example); **talent scout** = *scout *sb.*[4] 2 e; so **talent-scouting** *vbl. sb.*; **talent show**, a show or competition consisting of performances by a series of promising entertainers, esp. ones seeking to enter show business professionally; **talent-spotter** = *talent scout* above; also **talent-spot** *v. trans.* and *intr.*, **talent-spotting** *vbl. sb.*
1956 B. Holiday *Lady sings Blues* viii. 95 Nobody was in a position to push a hotel chain, a broadcasting network, and the talent agency around. **1977** *Rolling Stone* 24 Mar. 48/5 Christine and a girlfriend/singing partner snuck away from their strict parents in Birmingham and visited every talent agency they could find in London. **1859** *Bell's Life* 14 Aug. 8/2 When the time arrived for drawing the stumps. Both [players]..were loudly cheered during the presentation of the 'talent' money. **1936** *New Republic* 28 Oct. 351/2 Paramount's 'Big Broadcast of 1937'..(Paramount talent-scouts: there's a joker here somewhere.) **1939** N. Monsarrat *This is Schoolroom* III. xvii. 383 He was appraising the women present, as if he were a talent scout who only recognised one talent. **1952** Wodehouse *Pigs have Wings* x. 202, I understand that he's always being approached with flattering offers by the talent scouts of Colney Hatch and similar institutions. **1976** A. Powell *Infants of Spring* x. 170 Lyall worked intermittently as a film actor, consequence of a talent-scout seeing him making faces in a restaurant. **1934** M. H. Weseen *Dict. Amer. Slang* 154 *Talent scouting*,..seeking new actors. **1955** F. G. Patton *Good Morning, Miss Dove* 70 She had won a talent show and gone to New York. **1977** *Detroit Free Press* 11 Dec. 11-B/3 After that he landed parts in theater productions, ice shows and talent shows. **1937** *Boy's Own Paper* 2 Nov. 80/2 When talent-spotting, the thing he looked out for in a half-back was the ability to deliver an artistic pass. **1968** 'D. Torr' *Treason Line* 69 He had also to prod her into making the best of her mother's party to talent-spot possible agents. **1979** A. Boyle *Climate of Treason* x. 324 George Blake, a Royal Navy lieutenant whom he had 'talent-spotted' as a possible SIS recruit for counter-espionage work in Germany. **1944** *Gen* 15 Jan. 27/2 The B.B.C. talent-spotter is touring the Midlands. **1954** I. Murdoch *Under Net* xiv. 197, I hope that the eye of the talent-spotter has lighted favourably upon you. **1978** L. Meynell *Papersnake* vii. 88 It's punk..no action, what you keep these lousy talent-spotters for I can't imagine. **1957** *Observer* 3 Nov. 9/5 Competitions are an effective method for talent-spotting, an encouragement to architects and a means, sometimes, of acquiring a masterpiece. **1978** J. Pearson *Façades* iv. 69 Thanks to the talent-spotting skill of Richard Jennings..'Drowned Suns' was published in the London *Daily Mirror*.

Talgai (tælgai·). The name of a farm in Queensland, Australia, used *attrib.* in **Talgai boy, man, skull,** etc., to designate the fossil remains of a form of *Homo sapiens* found there in 1884.
The remains were presented to the University of Sydney in 1914.
1918 S. A. Smith in *Phil. Trans. R. Soc.* B. CCVIII. 355 A comparison of these radiographs with those of the Talgai skull show that..an exposure..was only just sufficient for the bone. *Ibid.* 370 The palate of the Talgai boy approaches that of the reconstructed *Eoanthropus*. **1931** A. Keith *New Disc. Antiquity of Man* xix. 303 The exact antiquity of the Talgai skull is still a moot point. *Ibid.*, The Talgai canines projected beyond the level of their neighbours. **1931** *Times Lit. Suppl.* 23 Apr. 317/1 Australia, having scored once with the help of the Talgai man, looks..for a no less successful second innings. **1977** G. Clark *World Prehistory* (ed. 3) xi. 459 The Talgai group,..although clearly sapient, retains a number of archaic features.

talh(a) (tæl, tæ·lā). [a. Arab. *ṭalḥ*, collect. pl. of *ṭalḥa*: see Talca gum.] A small flat-topped spiny tree, *Acacia seyal*, of the family Leguminosæ, native to north-eastern parts of Africa; also, the exudate of this tree, used as a substitute for gum arabic. Also *attrib.*
1857 H. Barth *Trav. & Disc. N. & Central Afr.* II. xxvii. 224 The forest..here consisted principally of.. talha-trees. **1867** [see Talca gum]. **1875** [see *ithel]. **1920** *Nature* 12 Aug. 757/1 The reports..have references to..distillation trials with talh wood (*Acacia Seyal*) from the Sudan. **1977** *Time* 3 Jan. 50/1 We bumped along in darkness looking for spots where talha trees or hills would provide protection.

‖ **taliq** (talī·k). Also **talik, ta'liq,** etc. [Pers., Arab. *ta'līq*, lit. 'hanging'.] A medieval Persian cursive script characterized by sloping rounded forms and exaggerated horizontal stroke, replaced by *nastalik in the fourteenth century.
1771 [see *naskhi *sb. pl.*]. **1795**, etc. [see *nastalik]. **1849** F. Madden tr. *Silvestre's Universal Palæogr.* I. 52 In general, especially in fine copies of poetical works, this writing takes a sloping direction, whence it has obtained the name of *taalik*. **1885** T. P. Hughes *Dict. Islam* 690/2 We now turn to the Oriental style, where we meet again with a bipartition, viz. into the Eastern Naskhī, as it is written in Arabia itself, Egypt, and Syria, and the..Ta'liq, current in Persia, India, and Central Asia. **1962** D. Diringer *Writing* vi. 143 In the course of time the Naskhi script became the parent of innumerable styles and varieties including the *ta'liq* (with its seventy or so secondary forms). **1978** Y. H. Safadi *Islamic Calligr.* 27 Ta'liq..became established as a defined script after the invention of Riyāsī in the ninth century.

Talisker (tæ·liskəɹ). The name of *Talisker* on the island of Skye, used to designate a variety of Scottish malt whisky manufactured at the distillery there, founded in 1831–32. Also, a drink or glass of this.
1883 *Trade Marks Jrnl.* 9 May 254 Talisker. Roderick Kemp & Co., Talisker Distillery, Skye... Whisky. 31,901. 2nd Apr. 1883. **1926** A. A. MacGregor *Over Sea to Skye* xiv. 102 Carbost with its far-famed Talisker distillery is in Bracadale. **1951** R. B. Lockhart *Scotch* I. ii. 27 He feels..unable to decide between Talisker and Clynelish, each of which would be put first by its devotees. **1977** D. Bagley *Enemy* xxxiv. 278, I ordered two large Taliskers.

talk, *sb.* Add: **I. 1. c.** An informal lecture or address; *spec.* = *radio talk* s.v. *radio *sb.* 5 b.
1859 A. J. Munby *Diary* 2 May in D. Hudson *Munby* (1972) 32 Went to the W. M. College to hear Ruskin's 'talk' about Switzerland... His lecture was historical & geographical chiefly—without book, he standing before the fire. **1900** S. Hale *Let.* 25 Mar. (1919) x. 360 She is giving three 'talks' here in Syracuse. **1923** *Radio Times* 28 Sept. 9/3, 8.45.—A Short Talk by the Rev. W. A. Studdert-Kennedy. **1942** E. Waugh *Put out More Flags* i. 58 He had..given the first in what was intended to be a series of talks for the B.B.C. **1962** A. Nisbett *Technique Sound Studio* 273 A 'talk' is a programme or programme segment which consists of one person talking at the microphone, usually from a script. **1977** *Rep. Comm. Future of Broadcasting* (Cmnd. 6753) ii. 12 Radio 3 continued to broadcast some specialised drama, poetry and talks in the evenings.
d. *pl.* Applied *attrib.* to a department of the B.B.C. concerned with the production of radio talks; also to its officials, programmes, etc., and *ellipt.*, the Talks department (with initial capital as a title).
1927 *B.B.C. Handbk. 1928* 124/1 The Talks Department is responsible for the news service, the S.O.S. service,.. Government department talks, and all the sporting, humorous, travel, literary, and general talks. *Ibid.* 125/2 We have evidence..of an increasing demand for the Talks programme and the Talks pamphlet. **1933** J. Reith *Diary* 5 May (1975) ii. 159 He might have done as a talks director.., but not for the big job. **1938** *Times* 5 Feb. 10/3 Two talks studios, music library, listening hall, [etc.]. **1940** R. S. Lambert *Ariel* ii. 49 After the 'hiving off' of Talks, Adult Education and even School Broadcasts, he was left with..Religion and the Children's Hour. **1942** 'G. Orwell' *Let.* 16 Oct. in *Coll. Essays* (1968) II. 246 Yours sincerely, (Geo. Orwell) Talks Producer Indian Section. **1954** W. K. Hancock *Country & Calling* vii. 189 She became a talks producer on the Overseas Service of the B.B.C. **1962** A. Nisbett *Technique Sound Studio* i. 31 Talks studio tables often have perforated, i.e. acoustically transparent, surfaces. **1978** F. Maclean *Take Nine Spies* vii. 228 Guy Burgess..was appointed to the Talks Department of the BBC. **1980** P. Fitzgerald *Human Voices* ii. 35 No-one could have any [brandy]—a disappointment to everybody except Talks, whose allocation..had already run out.
2. a. Also *spec.* in *pl.*, formal discussions, as between representatives of different countries, or between both sides in an industrial dispute; *talks about talks*: preliminary discussions held before entering into formal negotiations.
1952 *Ann. Reg. 1951* 202 The talks broke down on 21 June when it became clear that no agreement was possible. **1971** H. Wilson *Labour Govt.* vi. 78 This time they were more wary, and after some weeks were ready, more for appearances' sake it seemed, to enter into 'talks about talks'. **1977** *Whitaker's Almanack 1978* 590 Greek and Turkish officials began talks in London to try to settle their dispute over rights in the Aegean.
4. Also *talk of the town* = Town-talk.
1624 J. Chamberlain *Let.* 5 June (1939) II. 561 The disgrace that would follow in beeing made *fabula vulgi* and the talke of the towne. **1912** J. N. McIlwraith *Diana of Quebec* xiv. 205 But it would be the talk of the town within twenty-four hours, should even one person chance to see me in pursuit. **1983** *N.Y. Times* 4 Sept. vi. 18/2 The talk of the town this summer, the advertisement was more a lecture than a letter.
II. 5. *big talk* (earlier example).
1860 J. G. Holland *Miss Gilbert's Career* x. 173 He sort o' stands round, and spreads, and lets off all the big talk he hears.
7. † **talk-film** *temporary* = *talkie; **talk-master** *U.S. colloq.*, one who hosts a talk show on radio or television (cf. *quiz-master* (*b*) s.v. *quiz *sb.*[2] 2 b); **talk shop** *colloq.* = *talking shop* s.v. *talking *vbl. sb.* b; **talk show** chiefly *U.S.*, a television programme in which guests are

interviewed by the host or 'talk-master'; a television discussion or 'chat' show; also (more *rarely*), a similar programme on radio.
1929 *Morning Post* 24 May 12/7 He had been booked.. for a talk-film on his voice alone. **1932** *Oxford Times* 23 Sept. 22/5 After a good deal of experimenting I have come to the conclusion that..the best needles to use with a pick-up are those designed for talk-film operation—'talkie' needles. **1975** *Publishers Weekly* 1 Dec. 60/2 Tom Westbrook is the master of radio talkmasters. **1958** *Times* 12 Aug. 7/7 Many Doubting Thomases who regard the United Nations as a mere 'talk shop'. **1973** C. Mullard *Black Britain* xi. 133 They have broken away from the c.r.c., believing that it is a white man's talk-shop about black immigrant problems. **1965** *Times Lit. Suppl.* 25 Nov. 1042/2 There are now literally thousands of talk-shows. **1977** R. Ludlum *Chancellor Manuscript* vi. 78 They don't want you giving those interviews or going on talk shows.

talk, *v.* Add: **I. 1. a.** *talk about* (earlier and later examples of colloq. use); *don't talk to me about* (something), an exclamation against some new topic of conversation of which one has bitter personal experience.
1863 *Fraser's Mag.* Nov. 667 'Talk about women talking!' says a lady of our acquaintance, herself by no means deficient in eloquence. 'Why, look at the debates in the House of Commons, the public dinners, the vestry meetings, and, above all, the gossip, gossip, gossip at those horrid clubs!' **1885** F. Anstey *Tinted Venus* xiv. 167 *Well*, Miss Collum, talk about jealousy! **1940** Wodehouse *Quick Service* xii. 134 Did she mention her views on poor relations?..She believes in treating them rough. Talk about oppressed minorities. **1958** P. Marris *Widows & their Families* viii. 118 'Don't talk to me about shoes,' said the mother of two small children ruefully. 'My little boy just ripped the sole off one pair.' **1973** W. M. Duncan *Big Timer* xxi. 134 Talk about trouble! Goodness knows what Frank will say. **1980** R. Hill *Spy's Wife* vi. 33 'We've had a lot of rain,' said Molly. 'Don't talk to me about rain! You should have been here... I've never seen rain like it.'
c. Of a ship, etc.: to communicate by radio.
1912 *Chambers's Jrnl.* Jan. 61/2 When several ships are 'talking' to the shore-station, some delay may arise in getting a message through. **1927** *Pictorial Weekly* 3 Mar. 117/1 Ships of all nationalities 'talking' with shore stations.
2. *talk of*: also, = *talk about* in colloq. use, sense 1 a in Dict.; *talking of*..: also *joc.* introducing an unconnected subj. (earlier examples).
1831 M. Edgeworth *Let.* 30 Apr. (1971) 531 'Talking of coincidencies' as Mr. Ward would say..that was something of a coincidence. **1840** *Spirit of Times* 23 May 133/2 By the way, 'talking of guns', we shall take it as a great favor if our correspondents will send their orders for English Magazines, papers, etc., direct to Messrs. Wiley and Putnam, instead of ordering them from this office. **1851** E. B. Browning *Let.* 12 Nov. (1897) II. vii. 29 Talk of English comforts! It's a national delusion. **1876** *Atlantic Monthly* Dec. 684 This gives Pulcheria time to murmur, 'Talk of snub-noses!' **1950** F. Stark *Traveller's Prelude* xv. 191 She..asked: 'Were you in a very *small* hospital?' 'I thought so.' Talk of cats!
3. a. *to talk to* (earlier example in sense 'to rebuke, scold'); (*U.S.*) *dial.*, to court or woo (a woman); also *to talk up to*. Of a woman: to accept a man's attentions.
1860 E. M. Cowell *Diary* 17 Feb. in M. W. Disher *Cowells in Amer.* (1934) 18 Arrived there, we were set upon by sleigh conductors, one of whom Sam had also to 'talk to' for using bad language. **1895** *Dialect Notes* I. 374 Judge Jackson's has been talkin' to my daughter nigh on a year. **1905** in *Eng. Dial. Dict.* VI. 22/2 Pat is talking to Kate this six months, they'll soon be married. **1906** *Dialect Notes* III. 160 *Talk up to, v. phr.*, to court, to woo. 'Bud's talkin' up to her.' **1951** L. Craig *Singing Hills* xii. 110 The old man and me are powerful glad Ikey's made up his mind to talk up to a woman. We've been scared he'd be an old bachelor-man. **1951** H. Giles *Harbin's Ridge* xviii. 157 She mentioned that Faleecy John was talking to Jenny Clark a right smart. *Ibid.* 159 I'd not heard of her talking to anybody but Faleecy John.
c. *to talk big*: see also Big *a.* 8 b; *to talk down* (*to*) (examples); also *transf.* (in quots., of writers); *to talk through one's hat*: see *hat *sb.* 5 c; *to talk through* (*the back of*) *one's neck*: see *neck *sb.*[1] 3 e; *to talk turkey*: see Turkey[2] 2 d in Dict. and Suppl.
1856 C. M. Yonge *Daisy Chain* II. xiv. 498 'Say it again—what you said about the sea,' said Mary, more comforted than if Ethel had been talking *down* to her. **1919** H. Walpole *Jeremy* ii. 43 He always talked down to us as though we were beings of another and inferior planet. He called it, 'Getting on with the little ones.' **1954** M. F. Rodell *Mystery Fiction* i. 1 This does not mean that mystery fiction need be hack work; nor that the authors of it must 'talk down' to their audiences. **1970** *Sci. Jrnl.* Apr. 84/2 The authors for future titles all seem to be practising scientists. The problem will be whether they can achieve the necessary clarity of style without 'talking down' to their new audiences.
d. *to talk at* (earlier example); also, *to talk over* (another person): to override or talk simultaneously with (another speaker) on a tape recording, broadcast programme, etc.
1789 *Loiterer* 14 Feb. 11 Sometimes they talk to me, and sometimes at me. **1962** A. Nisbett *Technique Sound Studio* vii. 126 Superimpositions. Two people talking over each other are irritating. **1975** D. Pitts *Target Manhattan* (1976) xxviii. 111 Would you please announce your name and station first and try not to talk over other people.
e. To say something to the purpose, esp. in colloq. phr. *now you're talking*. Also *fig.* of money: see *money *sb.* 6 a.
1841 Dickens *Barnaby Rudge* lii, in *Master Humphrey's Clock* III. 238 That's the kind of game... Now you talk, indeed! **1884** J. Hay *Bread-Winners* x. 149 Now you're talkin'. **1920** 'O. Douglas' *Penny Plain* ii. 21 'I'm going to Scotland.' 'Ah,' said James Lauder, 'now you're talking.' **1936** L. Hellman *Days to Come* I. 26 'I was in Akron.' That's the job I wanted. 'Now we're talking... That's money for you.' **1952** A. Christie *Mrs. McGinty's Dead* viii. 59 'A few days later Mrs. McGinty was murdered.' 'Now you're talking.' **1974** M. Hastings *Dragon Island* xix. 174 'Now you're talking.' Jukes said approvingly... 'That's the best bit of sense you've spoke today.'
f. In various colloq. phrases stating or implying that someone is in no position to criticize another, exemplified by the types: *you can't talk!*; (*look, hark*) *who's talking!*; *who am I to talk?*; *I should talk!*
1847 Thackeray *Van. Fair* (1848) xli. 371 A person can't help their birth... I am sure Aunt Bute need not talk: she wants to marry Kate to young Hooper, the wine-merchant. **1895** Kipling *Day's Work* (1898) 200 'All you other men think of is to give him an absurd nickname.'..'Well, *you* can't talk, William. You christened little Miss Demby the Button-quail.' **1938** N. Marsh *Artists in Crime* ii. 24 You're paying his fare Home, of course. Well, I suppose I can't talk as you've given me the run of your house. **1945** A. Kober *Parm Me* 62 'Look who's talking!' said Pa Gross, glaring at his wife. **1949** D. Smith *I capture Castle* ii. 12 She has had that dressing-gown so long that I don't think she sees it any more... But who am I to talk—who have not had a dressing-gown at all for two years? **1962** J. Cannan *All is Discovered* i. 11 Anyway Daddy can't talk. Whenever she's out he..spends hours with Maria. **1968** M. Ross *Gasteropod* iv. 99 'You should have hated me... Most men would have done.' Max laughed. 'Hark who's talking! What did you do?' **1979** D. Gurr *Troika* viii. 55, I was rough on you... And stupid—I should talk!
g. *to talk back*: to answer back; to indulge in 'back-chat'. *N. Amer. colloq.*
1869 'Mark Twain' *Innoc. Abr.* xii. 112 There was no 'talking back', no dissatisfaction about over-charging. **1939** L. M. Montgomery *Anne of Ingleside* iv. 22 Dad would never let anyone 'talk back' to Aunt Mary Maria. **1955** L. Hughes in *Oliphant Q.* Apr.–June 136 All over the world today folks with not even *Mister* in front of their names are raring up and talking back to the folks called *Mister*. **1977** M. French *Women's Room* (1978) ii. 117, I can't stand it when they start to talk back, be fresh.
h. With alcoholic drink as subj.: used to excuse or explain uncharacteristic sentiments supposedly brought on by the drink consumed. Chiefly in *pres. pple. colloq.*
1922 Joyce *Ulysses* 742 Paying his compliments the Bushmills whisky talking of course. *a* **1953** E. O'Neill *Touch of Poet* (1957) I. 42 But you understand, it was the liquor talking, if I said anything to wound you. **1982** R. Lewis *Gathering of Ghosts* iii. 102 It was all a bit stupid, you know. Beer talking, you know how it is.
i. To disclose information, *spec.* to the police (or another authority), esp. incriminating oneself or others; to confess; to turn informer or 'squeal'. *slang.*
1924 G. C. Henderson *Keys to Crookdom* 420 Talk. See squeal, beef, spiel, chew rag. **1952** M. Allingham *Tiger in Smoke* xv. 216 They've been through it today, but they're not talking. Why should they? **1959** W. Golding *Free Fall* vii. 144 'I wont' talk. I know nothing.' 'Talk. Yes, that is the word. At some point, Mr. Mountjoy, you will talk.' **1976** *Times Lit. Suppl.* 21 May 605/3 He is, as they say, not talking, and refused to be interviewed by the authors of this book.
j. With advb. accusative. *to talk one's way in*: to gain admission by persuasion. Similarly with *out*.
1973 *Ottawa Jrnl.* 3 Feb. 6/1 Their length of hospital stays, during which they tried to talk their way out, ranged from seven to 52 days; the average stay was 19 days. **1978** 'D. Kyle' *Black Camelot* xv. 237 If Rasch could talk his way in..the task would be simple.
5. (Earlier example, of a boat.) Also, of an anchor which is dragged (see Drag *v.* 1 c).
1793 Wordsworth *Evening Walk* 319 The talking boat that moves with pensive sound. **1914** 'Bartimeus' *Naval Occasions* xxiii. 224 'I think the starboard anchor is "talking".'.. A dull metallic sound detached itself from the sibilant rushing of water. **1917** J. S. Margerison *Sure Shield* 86 It was the ring chain that had worked slightly loose and which was allowing the five-ton mass of cast iron to slide three inches each way as the ship rolled or pitched, and the creaking sound of which had given rise to the phrase 'the anchor's talking'. **1962** W. Granville *Dict. Sailors' Slang* 118/1 A ship's anchor is said to come home, or 'talk', when it is dragging.
II. 7. *to talk a good game*: to discuss a topic convincingly (with implication that no action is taken). Also, *to talk a great ball game*. *U.S. colloq.*
1972 'H. Pentecost' *Birthday, Deathday* iv. 39 Hollywood address... Talks a great ball game to the hotel people out there about his big film deals. **1973** *Philadelphia Inquirer* (Today Suppl.) 14 Oct. 17/1 She still talks a good game. To hear her tell it, she's the Henry Kissinger of consumerism. **1977** *New Yorker* 18 July 52/1 While they talk a good game, their considerations are consolidating their power around the world.
8. (Further examples.)
1711 Pope *Ess. Crit.* 36 No Place so Sacred from such Fops is barr'd,.. Nay, run to Altars; there they'll talk you dead. **1738** Johnson *London* 4 And here a female Atheist talks you dead. **1920** R. Macaulay *Potterism* v. iii. 175 He..used to talk one sick about how little scope he had in his parish.
9. a. *to talk down*, (*a*) (earlier example, with impersonal obj.); (*b*) to reduce or diminish by talking; to denigrate or belittle; (*c*) *Econ.*, to depress the value of (a currency) or the price of (a commodity) by making tactical public statements; similarly, *to talk lower*; (*d*) see sense 9 d below; *to talk out of*: also, to extract from (a person) by persuasion; *to talk over* (earlier example); *to talk* (a person) *through* (something), to provide with a commentary on (some event); to take through with helpful explanation; *to talk up*, (*a*) (later example); (*b*) to discuss favourably; to stimulate interest in by talking, esp. exaggeratedly; to praise or advocate (chiefly *U.S.*).
1814 Jane Austen *Mansfield Park* I. xiii. 269 She started no difficulties that were not talked down in five minutes. **1832** J. S. Mill *Let.* 25 Jan. in *Wks.* (1963) XII. 94 Most men in this country have a strong prejudice against any attempt to *talk them over* as the vulgar say. **1863** Trollope *Rachel Ray* I. viii. 151 Mrs. Butler Cornbury..talked her young friend up to the top of the tree. **1872** *Rep. Vermont Board Agric.* I. 679 This little conversation led me to talk the matter up with the marble dealers. *a* **1882** Trollope *Autobiogr.* (1883) I. v. 108, I received £20... The money had been 'talked out of' the worthy publisher by..my brother, who made the bargain for me. **1931** *Daily Express* 5 Sept. 10/5 Amsterdam is taking a bearish view of Royal Dutch and is 'talking' the shares £2 lower. **1932** W. Faulkner *Light in August* iii. 56 The old men and the old women trying to talk down his gleeful excitement. **1933** *Sun* (Baltimore) 12 July 3/2 Thus far the dollar has been 'talked down'. **1938** *Richmond* (Va.) *Times-Dispatch* 10 Nov. 1/7 (*heading*) Victorious Taft talks down 1940 chances. **1956** H. Gold *Man who was not with It* (1965) ii. 17 You talk it up like a longtime grifter. **1962** *Spectator* 28 Dec. 999/3 Most professionals [on the Paris Stock Exchange] are talking their market lower. **1967** *Economist* 4 Mar. 797/2 The non-mandatory system..will allow the Government, in American style, to 'talk down' a price. **1970** *New Eng. Bible Prov.* xix. 26 He who talks his father down vexes his mother. **1976** J. Snow *Cricket Rebel* 27 A year later when I made my Test debut against New Zealand at Lord's. Ted [Dexter] virtually talked me through the game. **1978** *Daily Tel.* 16 Mar. 11/1, I am fed up with people talking down the NHS, failing to put things into perspective by pointing to the successes as well as the difficulties. **1980** *Times* 28 Jan. 19/2 For years the Silver Users Association..has been successfully talking the price of silver down. **1982** *Nature* 13 May 91/1 True, the Western media are delighted to talk up 'Star Wars' fantasies, the US shuttle, Ariane or Soviet space weapons. But they have overlooked the main substance of the Soviet [space] programme.
d. *to talk down*: to provide (an aircraft) with directions by radio communication which enable it to land, esp. in overcast or emergency conditions. Also with *in*: chiefly applied to ships seeking landfall. Hence with the pilot or navigating officer as subj. Occas. with other advbs. and preps.
1943 *Plane Talk* June 28/3 The bombardier talks the pilot 'in', telling him which way to turn. **1945** *Sci. News Let.* 25 Aug. 127/1 A blindfolded pilot..was 'talked-down' to the runway by a control operator. **1946** *Jrnl. Inst. Electr. Engin.* XCIII. III. A. 1–4 124/2 If we make these aircraft carry simple radio receivers, and pin-point them with our accurate radar, it is quite possible to 'talk them down' by giving detailed instructions to the pilot. **1955** *Sun* (Baltimore) 12 June (B ed.) 3/4 The ground controlled approach equipment—through which a plane is 'talked in' to a landing. **1957** *Times* 24 Aug. 4/1 From information telephoned to Ford by police officers it [*sc.* the helicopter] was 'talked' to where two young girls and a young man were clinging to an overturned dinghy. **1960** 'N. Shute' *Trustee from Toolroom* 97 They get us on the radar screen and talk us down on to the runway. **1962** J. Dill in *Into Orbit* p. xviii, Shephard was seated before a console..ready to talk John Glenn back to earth. **1967** *Observer* 2 Apr. 10/6 Ships could be advised of optimum routes and speeds as they approach port, and even 'talked in' through fog. **1970** *Daily Tel.* 21 Dec. 2/6 It is a 'blind landing system' in which a ground controller talks down an airliner in bad weather by watching its approach on a radar screen.
10. Colloq. phrases. *to talk the hind leg off a donkey* (*horse*, etc.), applied to one who: (*a*) talks with unflagging and wearying persistence, or: (*b*) is said to have the power to persuade another by eloquent or charming speech; *to talk* (someone's) *ear off*: to talk incessantly or until one is tired of listening (*U.S.*).
1808 *Cobbett's Weekly Pol. Reg.* XIII. II. 47 The old vulgar hyperbole of 'talking a horse's hind leg off'..will find its verification in the American Congress. **1861** *Temple Bar* I. 414 One has heard of persons who could 'talk a donkey's hind leg off'. **1879** Trollope *John Caldigate* III. ix. 122 She'd talk the hind-legs off a dog, as we used to say out there [*sc.* in Australia]. **1915** W. S. Maugham *Of Human Bondage* lxxxviii. 459 'Doesn't she look like Rubens' second wife?' cried Athelny. 'Wouldn't she look splendid in a seventeenth-century costume? That's the sort of wife to marry, my boy. Look at her.'

'I believe you'd talk the hind leg off a donkey, Athelny,' she answered calmly. **1935** G. & S. Lorimer *Heart Specialist* i. 9 An American will talk your ear off about his sport with a little encouragement. **1942** G. H. D. & M. Cole *Toper's End* IV. ix. 138 You can talk the 'ind leg off any donkey. **1976** *National Observer* (U.S.) 10 Apr. 9/5 Heck! I could talk your ear off. But let me just say that in all my 40 years of organizing and escorting tours, I haven't found a better one than this one.

talkable, *a.* Add: **b.** (Later examples.) *rare.*
1859 Mrs. Gaskell *Lett.* (1966) 546 If Papa gets overbusy; & not talkable as to-day! **1899** H. Van Dyke *Fisherman's Luck* III. i. 54 A person who has the rare merit of being talkable. 'Talkable' is not a new adjective. But it needs a new definition, and the complement of a corresponding noun... *talkability*. **1963** R. Symons *Many Trails* iii. 31 A grown woman yearns for somebody more 'talkable' than a thirteen-year-old.

talkathon (tǫ·kăþǫn). *colloq.* (orig. *U.S.*). [f. Talk *v.* + *-athon, after *walkathon*.] **1.** An interminable session of talk or discussion; *spec.* a prolonged debate in a legislature or similar body, a filibuster.
1934 *Amer. Speech* IX. 76/2 Apropos of the walkathons, ..a contributor suggests that talkathons will be longer lived, especially in legislative halls. **1948** *Times-Dispatch* (Richmond, Va.) 3 Aug. 1/2 Filibustering Dixie Senators won a major round today in their effort to talk the anti-poll tax bill to death. The presiding officer..decided that an effort to curb the debate was in conflict with Senate rules, and so the talkathon continued. **1957** *Economist* 7 Sept. 820/2 The regular Assembly session..will begin with the annual talkathon, oddly styled the general debate. **1969** *Daily Colonist* (Victoria, B.C.) 3 May 2/7 Rene Matte..dashed any hopes of government supporters that the Creditistes were tiring of their talkathon. **1978** D. A. Stanwood *Memory of Eva Ryker* xix. 181 No talkathons tonight, Eva... We'll get a fresh start in the morning.
2. A prolonged broadcast in which a political candidate is interviewed or questioned (*U.S.*). Also applied *transf.* to a lengthy radio or television discussion programme or 'talk show'. Cf. *radiothon.
1952 *Economist* 6 Sept. 555/2 The latest American political novelty, the radio 'talkathon', in which..[Mr. Schmitt] answers questions for 25 hours at a stretch. **1961** *N.Y. Times Mag.* 29 Oct. 24/1 (*heading*) Political walkathons and talkathons. **1965** *Punch* 4 Aug. 176/1 The three-hour and twenty-five minute talkathon on BBC 2, was held by chance on the very day that Mr. Callaghan most heavily stamped on the economic brakes.

talkation. (Earlier example.)
1781 H. Walpole *Let.* 9 Sept. (1955) XXIX. 152 This is the sum total of what I can learn; matter enough to dissert upon if such were my propensity! but besides not loving talkation, it is painful for me to write.

talk-back (tǫ·kbæk). Also **talkback, talk back.** [f. Talk *v.*] **A.** *attrib.* Designating apparatus and facilities for two-way communication by loudspeaker, usu. between one who gives and one who receives instructions; *spec.* that connecting a studio and a control room.
1933 *B.B.C. Yearbk. 1934* 402 The 'talk back' facilities need setting up to suit the particular conditions of studio, listening point, etc., whereby the person listening to the rehearsal may speak over a microphone circuit and by means of a loudspeaker give instructions to those rehearsing. **1959** H. Barnes *Oceanogr. & Marine Biol.* 197 During the descent instructions are always relayed from the lower laboratory to a man at the winch by a microphone and talk-back system, and particular care is taken when nearing the bottom. **1971** M. Lee *Dying for Fun* xxiii. 107 The sound radio producer..put down his talk-back key and said: 'We won't start till you're ready.' **1982** J. Sherwood *Shot in Arm* xiii. 126 He pressed the talk-back key. 'Anthea..could you make it a bit lighter and more informal?'
B. *sb.* A talk-back system.
1957 *Listener* 11 July 53 Large numbers mean more widespread buildings and (in spite of a public address system, talkbacks, and so on) it is not always easy to get the people or things one wants as quickly as one wants them. **1962** A. Nisbett *Technique Sound Studio* vi. 107 The studio manager announces over the studio talk-back (which also goes to the recording room), 'We'll be going ahead in ten seconds from... now!' **1979** *Daily Tel.* 23 Apr. 14/4 Next door again is the studio, with Robin Day, that morning's speaker and, between them, a girl wearing headphones who passes on the caller's name and location over the talk back from the cubicle.

talk-down. [f. vbl. phr. *to talk down*: see *talk *v.* 9 d.] The action or process of talking down an aircraft or a pilot. Chiefly *attrib.*
1946 *Engineering Jrnl.* Apr. 233/1 It is felt that the 'talk down' feature will not be well received in airline work. **1948** *Shell Aviation News* No. 123. 11/2 G.C.A. is a 'talk down' system and where such installations are available.. no equipment in the aircraft is needed other than efficient radio telephony. **1955** *Times* 20 June 3/5 The more complicated problems of refuelling in mid-air, baling out on to snowy tundra slopes, and landing at night by 'talk-down'. **1959** K. Henney *Radio Engin. Handbk.* (ed. 5) xxvi. 91 (*caption*) Information displayed to operators of precision indicators of GCA 'talk-down' radar landing system. **1963** L. Deighton *Horse under Water* xx. 89 A Viscount came down the GCA talkdown, its white, red and green lights peep-boing the traffic pattern.

talked, *ppl. a.* Add: **2.** *talked-about* (chiefly with qualifying advb.): discussed.
1919 C. S. Parker *Amer. Idyll* xvi. 178 He was the most talked-about man at the Convention. **1928** *Manch. Guardian Weekly* 7 Sept. 184/3 A striking proof of the reality and significance of the much-talked-about new entente. **1951** *Rochester* (N.Y.) *Democrat & Chron.* 12 Sept. 14/1 The fight, one of the most talked-about in years, will not be on television or radio. **1968** J. D. Watson *Double Helix* ix. 65 The wine turned the conversation to the currently talked-about Cambridge popsies. **1982** J. Fox *White Mischief* v. 59 She was already a talked-about social success.

talkee-talkee. Add: Also *spec.* (usu. spelt *Taki-Taki*; also with small initials) an English-based creole language of Surinam; = *Ningre Tongo, *Sranan.
1932 A. G. Barnett in *Amer. Speech* VII. 394 In Paramaribo, this speech of former slaves has degenerated into 'Talkee-talkee', which is loaded with a heavy percentage of Dutch. **1933, 1939** [see *Ningre Tongo]. **1955** *Caribbean Q.* IV. II. 167 Translation in the Rural Creole of Surinam (Taki Taki) by Albert Helman. **1961** F. G. Cassidy *Jamaica Talk* i. 8 It [*sc.* Macca] got into the common speech and has remained in Taki-taki and in Jamaican. **1970** *Language* XLVI. 409 Saramaccan..is only partly intelligible to speakers of Sranan or Taki-taki.

ta·lkfest. *slang* (chiefly *U.S.*). [f. Talk *sb.* or *v.* + *fest.] A session of lengthy discussion or conversation, a 'talkathon'. Cf. *gabfest.
1906 S. Ford *Shorty McCabe* ii. 36 Perhaps it'll be the grand annual ball of the Truck Drivers' Association, or just one of them Anarchist talkfests in the back room of some beer parlor. **1928** W. A. White *Masks in Pageant* 247 He stepped naturally into supremacy at that talkfest [*sc.* a political convention] because he had been training for his famous speech. **1948** *Sun* (Baltimore) 18 June 19/4 (*heading*) County boy, girl win in 4-H 'talkfest'. **1961** B. Pike *Robert Musil* viii. 167 Action is somehow made to seem irrelevant and trivial in this enormous talkfest. **1972** J. Potts *Trouble-Maker* vii. 50 She wanted no part of the talkfest that was sure to occupy her fellow guests.

talkie (tǫ·ki). *colloq.* (orig. *U.S.*). Now *Hist.* [f. Talk *v.* + -ie, after *movie.] **a.** A talking film, as opp. to a silent film. Freq. in *pl.* (esp. *the talkies*). Cf. *speakie.
1913 *Writer's Bull.* Mar. 9 The silent 'Movies', so popular to-day, will become tame in comparison with the 'Talkies'. **1921** *Daily Colonist* (Victoria, B.C.) 2 Apr. 12/2 All have seen the movies, now people are to have the opportunity of seeing and hearing the 'Talkies'... The author..of the remarkable speaking photoplay, 'Shell Shocked' is in the city. **1928** *Daily Express* 6 June 3/1 Marvel of the 'talkies'. **1930** E. Waugh *Vile Bodies* vi. 94 She said, 'You're much later than you said. It's so boring to be late for a talkie.' **1939** M. S. Rice *Working-Class Wives* v. 110 Another Leeds woman says 'never get out except to shop; have never been to the talkies'. **1955** *Times* 25 May 5/5 It was not until the talkies came that the cinema divorced itself from reading. **1962** *Movie* Dec. 31/3 Part talkie with music and sound effects. **1978** E. Malpass *Wind brings up Rain* xix. 184 It was about this time that the first Talkies came to Ingerby: a fact that gave Benbow the courage..to ask Ulrike to go to the pictures with him.
b. *attrib.*
1913 *Technical World Mag.* Aug. 815 One hundred other 'talkie-parties' are beginning in different parts of the city. **1922** *Radio News* (U.S.) Nov. 867/1 Mr. de Forest has solved the secret of the 'talkie movie' with perfect synchronization. **1932** L. Golding *Magnolia St.* III. iii. 512 An engagement at a talkie-theatre was about all Wilfred could hope for. **1936** Auden *Look, Stranger!* 34 By cops directed to the fug Of talkie-houses for a drug. **1960** *Times* 19 Feb. 5/1 In 1936 he had again the chance to make a feature film—a rather conventional 'talkie' adaptation of a stage success.

ta·lk-in. [See *-in[3].] **a.** A gathering or meeting for discussion; a conference.
1966 *N.Y. Times* 5 Oct. 46 LeRoi Jones, poet, playwright and polemicist, sustained each aspect of his reputation..at a reading at the Village Theater... It was the second in a series of talk-ins presented at that house. **1970** *Times* 7 Sept. 18 At the end of this month the insurance industry starts its talk-in with the Monopolies Commission over fire insurance. **1976** *Cumberland News* 3 Dec. 18/5 Members..attended a very interesting and comprehensive 'talk-in' on the best way to prepare and present exhibits for showing. **1980** J. Drummond *Such a Nice Family* ii. 16 It's for the patients themselves to choose. Why don't you ask them tonight, at the talk-in?
b. *spec.* as a form of political protest (esp. by students) in which the matter at issue is discussed.
1967 *Time* 30 June 28 Last week 180 Free University students staged a 45-hour hunger strike and talk-in..to demand the release of a jailed anti-Shah demonstrator. **1977** D. James *Spy at Evening* xx. 159, I was in a student meeting... It was a talk-in on Vietnam.
c. A radio or television discussion programme, esp. one broadcast live, and in which the audience may participate.
1971 *Guardian* 14 Dec. 2 Part Two of the Tuesday doomwatch..charts the putative dangers of spreading nuclear power: then a talk-in asks if we should all keep quiet about it. **1977** D. Parker *Radio* 154 The first few months of Capitol Radio and LBC (a station mainly concerned with news, but actually including every kind of 'chat' from author-interviews to interminable talk-ins and phone-ins) were dreadful beyond recall.

talking, *vbl. sb.* Add: **a.** *talking to* (earlier example). Also in colloq. phr. *to do the talking*.
1873 Hardy *Pair of Blue Eyes* III. iv. 91 Knight did most of the talking along the journey. *c* **1875** 'Brenda' *Froggy's Little Brother* (new ed.) iii. 35 I'd give him such a talking-to as never he had in his life before! **1903** *Red Book* Aug. 367/2 The girl did the talking, but for nearly an hour the head of the detective bureau sat silent, impassive as the sphinx. **1948** G. Vidal *City & Pillar* x. 303 Bob did most of the talking and Jim listened.
b. **talking blues**, a blues song in which the lyrics (usu. narrative) are articulated in a rhythm approaching that of speech; **talking cure**, colloq. name for psychoanalysis or psychotherapy which relies on verbal interaction; **talking point**, a topic suitable for or inviting discussion or argument; **talking-shop**, a centre for idle and unconstructive talk; (derog.) applied to a parliament, *spec.* the House of Commons.
1969 *New Yorker* 1 Nov. 25/1 Alice's Restaurant—Arthur Penn's extension of Arlo Guthrie's talking-blues record. **1979** M. Boyce *I was There!* 10/2 My first attempt at song writing took the form of humorous verse (in 'talking Blues' fashion). **1910** tr. *Freud's Outl. & Devel. Psychoanal.* in *Amer. Jrnl. Psychol.* XXI. 184 The patient herself, who at this time of her illness..spoke only English, gave this new kind of treatment the name of 'talking cure'. **1977** R. Holland *Self & Social Context* viii. 240 At first sight nothing would seem more obvious than the dominant role of language in therapy—the talking cure. **1922** S. Lewis *Babbitt* iv. 47 A broker..who understood Talking Points, Strategic Values, Key Situations, Underappraisals. **1979** B. Parvin *Deadly Dyke* vi. 28 Quite a change of climate..but a talking point. **1912** C. Mackenzie *Carnival* xliv. 385 Village! Talking-shop, I should say. **1922** A. M. Hyamson *Dict. Eng. Phrases* 337/1 *Talking mill (shop), the*, (contemptuously) the House of Commons. **1942** R. G. Collingwood *New Leviathan* 213 Contemptuous language about the 'talking-shop at Westminster'. **1944** F. A. Hayek *Road to Serfdom* v. 46 Parliaments came to be regarded as ineffective 'talking shops'. **1963** V. Gielgud *Goggle-Box Affair* v. 49, I was an M.P. then... God knows why, damned talking-shop! **1979** R. Blythe *View in Winter* i. 72 The forge is the classic talking-shop of the village.

talking, *ppl. a.* Add: **2.** Comb. *talking book* (orig. *U.S.*), a sound recording of a book, for use by the blind; *talking clock* = *speaking clock* s.v. *speaking *ppl. a.* 1; cf. *tim; *talking doll*, a doll capable of emitting elementary sounds or words when activated; also *transf.* (of a person); *talking drum*, one of a set of drums, each pitched differently, which are beaten to transmit the words of a message in a tonal language, chiefly in W. Afr. (usu. in *pl.*); *talking head* (slang), a television presenter or interviewer who is shown on the screen talking directly in close-up to the camera; freq. in *pl.*; also *transf.*; *talking machine* (chiefly *U.S.*), (*a*) a machine designed to imitate human speech; *spec.* (the vocal mechanism of) a talking doll (*rare*); also *fig.*; (*b*) a phonograph; a gramophone which reproduces human speech (now only *Hist.*); *talking phonograph*: see Phonograph *sb.* 3 a.
1932 *Ann. Rep. Amer. Foundation for the Blind* 2 The development of books on phonograph records, or 'Talking Books', will be one of the greatest boons ever conferred on the blind. **1960** *Guardian* 22 July 3/4 The blind in Britain are to have 'talking books' incorporating the most advanced tape recording and play-back techniques... The present talking-books are long-playing records. **1978** D. Bloodworth *Crosstalk* xxi. 168 Extra-large cassettes ..from the library of the National Institute for the Blind. .. He examined the talking books. **1936** *Discovery* Oct. 315/1 The 'talking-clock' from which, merely by dialling T-I-M, every telephone user can now learn the exact time. **1969** N. Freeling *Tsing-Boum* xiii. 94 One could phone the talking clock. **1925** C. Morley *Thunder on Left* xix. 241 There was the bleat of one of the talking dolls. '*Maaa-Maa!*' it cried. **1974** H. R. F. Keating *Underside* vii. 66 He had been able to regard her completely in the light of some talking doll, irritating but unconnected..with the reality of life. **1977** *Detroit Free Press* 11 Dec. 15-A/3 (Advt.), The many phrases this big, beautiful talking doll 'Baby Sez' will make your daughter the happiest kid on the block. **1897** M. Kingsley *Trav. W. Afr.* vi. 114 In the street one sees the characteristic standing drum..and one or two talking-drums besides. **1923** R. S. Rattray *Ashanti* iii. 104 The big talking drums were immediately behind him and beat out: The King has sat down. **1966** C. Achebe *Man of People* xiii. 166 These were the same people that only the other day.. praise-singers followed with song and talking-drum. **1977** *Rolling Stone* 24 Mar. 48/2 Even drummer Mick Fleetwood finally ventured out from behind his drum kit to play the African talking drum on 'World Turning'. **1983** *Guardian* 19 July 9/4 A talking drums player from Ghana who seemed to have wandered on stage uninvited. **1968** *Punch* 24 Apr. 617/3 Indifference to people who weren't talking-heads?..If they were so monumentally unmemorable shouldn't they have been skipped altogether? **1976** *Listener* 15 Jan. 52/1, I begin to get very tired of the 'important' interview—indeed, of all talking heads—on television. **1980** *Ibid.* 4 Sept. 299/1, I looked at the talking heads, telling their stories directly to the camera. **1844** *Quincy* (Illinois) *Herald* 9 Feb. 3/2 A German, named Faber,..in New York, has invented and

brought to perfection a talking machine. It is played on by keys like a piano, and can be made to say any thing, in any language, that its inventor desires. **1844** *Picayune* (New Orleans) 18 Mar. 38/2 Why *don't* you go see the Talking Machine, and take little Matilda along with you? **1846** N. P. WILLIS *Compl. Works* III. 674/2, I had a half-hour's interview with the *talking machine* this morning, and found him a more entertaining *android* than most of my wooden acquaintances. **1850** CARLYLE *Latter-Day Pamph.* No. I. 52 A redtape Talking-machine and unhappy Bag of Parliamentary Eloquence. **1891** *Appleton's Ann. Cycl. 1890* 709/1 In 1886, J. S. Taintor, working along the lines followed by Mr. Edison, produced a talking machine, which was called the graphophone, or phonograph-graphophone. **1897** R. STUART *In Simpkinsville* 110 The little talkin' machine inside it has got out o' fix ..an' it don't say 'papa' an' 'mama' any more. **1930** W. FAULKNER *As I lay Dying* 181 Cash aimed to buy that talking machine from Suratt with that money.

b. *talking film, movie, picture* (now *Hist.*) = *TALKIE a; also formerly *talking photograph.*

1904 *Science Siftings* 26 Mar. 353/1 The inventor believes that in a short time these talking photographs will supersede the phonograph. **1908** *Variety* 16 May 11/3 (*heading*) Natural voice talking pictures... The idea is to have a capable company behind the screen and take up the cue as each character comes in view. **1921** [see *CINEPHONE]. **1927** *N.Y. Times* 28 Aug. VII. 4 If a 'talking movie' throughout were being made, it would be necessary to film a thousand feet at a time. **1978** *Lancashire Life* Oct. 97/3 Talking-pictures were introduced under the management of the late Ignatius Cullen, whose daughters continue to run the Civic Hall Cinema..showing 'wholesome family films' only.

ta·lk-out. *colloq.* [f. vbl. phr. *to talk out*: see TALK *v.* 9.] **a.** A 'talking out' of a bill in Parliament, a filibuster. (In quot., with pl. *talks out.*) *rare.*

1884 E. W. HAMILTON *Diary* 25 Mar. (1972) II. 583 One cannot help thinking that the Speaker..has missed an opportunity of applying the closure to prevent those purely obstructive 'talks out', which morning sittings are so well adapted to promote.

b. An exhaustive discussion, in which a matter is 'talked out'.

1965 *Listener* 16 Sept. 431/1 The series contained a fair number of..'talk-outs', as I must shudderingly report I have heard them called, on music, drama, poetry. **1967** *Telegraph* (Brisbane) 10 Apr. 7/6 A 'marathon talk-out' or similar scheme for mass support would pressure the Government to solve the present education problem in Queensland. **1978** *New York* 3 Apr. 58/2 At the conclusion of every Wednesday talkout, I was shocked, saddened, and dismayed.

talky, *a.* Add: *colloq.* **a.** (Earlier example.)

1815 BYRON *Let.* 31 Oct. (1975) IV. 326 Like other [dinner] parties..it was first silent, then talky, then argumentative,..then drunk.

b. Of a play, book, etc.: wordy, long-winded; containing verbose or tedious dialogue.

1937 *Partisan Rev.* Dec. 54 The producers, apparently still under the spell of the old superstition that a play cannot be 'talky', proceeded to excise almost all of the hero's connected conversation. **1967** *Times* 23 Nov. 8/7 The action scenes..do a lot to make up for a slow and talky opening. **1976** *Publishers Weekly* 23 Aug. 60/1 Once past the rather talky and confusing opening, readers will find Price's spy novel a corker.

talky-talky *sb.*, trivial conversation; loquacity; similarly, **talky-talk, talki-talk,** trivial conversation, talk for talking's sake.

1870 *Punch* 22 Oct. 173/1 Perhaps this Lighthearted Gallant doesn't have no end of talky-talkies with all the.. six-foot nothings in jack-boots all over the shop. **1884** 'MARK TWAIN' *Huck. Finn* xxvi. 258 All that kind of humbug talky-talk, just the way people always does at a supper, you know. **1907** G. B. SHAW *John Bull's Other Island* IV. 110 Why cant you say a simple thing simply, Larry, without all that Irish exaggeration and talky-talky? **1928** *Daily Express* 16 July 10/2 A..narrative in which motives and character are revealed in deeds and not in talky-talk. **1937** G. M. YOUNG *Daylight & Champaign* 134, I heartily agree that a large part of the talki-talk about influences and relationships could be with great advantage thrown into a single book. **1938** J. CARY *Castle Corner* 546 The poor child had been upset by all the talky-talky. **1953** W. REICH *Murder of Christ* xii. 111 The admirers feel uncomfortable.., not being free..to engage in small chit-chat and talki-talk. **1965** 'W. HAGGARD' *Hard Sell* xvi. 177 Away with this talky-talk, this stylized sparring between..officials.

tall, *a.* Add: **II. 6. a.** Also in proverbial phr. *tall, dark, and handsome*, denoting a type of attractive man (see also quot. 1965).

1906 R. E. KNOWLES *Undertow* xi. 135 He was tall—and dark—and handsome. **1940** *Chatelaine* Dec. 55/3 One Squadron Leader tells of filling an 'order' for 'three tall, dark and handsomes to go dancing'. **1958** M. STEWART *Nine Coaches Waiting* vii. 93 Tall, dark and handsome—the romantic cliché repeated itself in my head. **1965** T. WOLFE *Kandy-Kolored Tangerine-Flake Streamline Baby* (1966) ix. 178 It was Cary Grant that Mae West was talking about when she launched the phrase 'tall, dark and handsome' in 'She Done Him Wrong' (1933). **1978** 'H. CARMICHAEL' *Life Cycle* v. 64 If she felt like leaning on his shoulder it was certainly not because he was tall, dark and handsome.

7. a. (Later examples of ships, *spec.* square-riggers.)

1715 POPE tr. *Homer's Iliad* I. III. 192 When thy tall Ships triumphant stem'd the tide. **1726** —— tr. *Homer's Odyssey* I. IV. 201 From his tall ship the King of men descends. *a* **1865** SMYTH *Sailor's Word-bk.* (1867) 674 *Tall ship*, a phrase among the early voyagers for square-rigged vessels having topmasts. **1902** J. MASEFIELD *Salt-water Ballads* 59 All I ask is a tall ship and a star to steer her by. **1916** JOYCE *Portrait of Artist* (1969) v. 252 The spell of arms and voices: the white arms of roads..and the black arms of tall ships that stand against the moon. **1975** *Times* 4 July 4/8 The Admiralty Court yesterday granted ..an order that the tall ship, Regina Maris, be appraised and sold by the Admiralty marshal to pay a debt... The 137ft-long ship..competed in last year's Tall Ships Race.

b. *tall copy* (earlier example).

1807 SOUTHEY *Lett. from Eng.* I. xxi. 237 The size of the margin is of great importance. I could not conceive what was meant by *a tall copy*, till this was explained to me. If the leaves of an old book have never been cut smooth, its value is greatly enhanced.

c. (Earlier and later examples.) Cf. *tall-grass* adj., sense C below.

1789 *Ann. Agric.* XII. 441, I was surprized to see no tall oat grass there, the best and most useful of the grasses which meadows can be laid down with. **1979** *United States 1980/81* (*Penguin Travel Guides*) 763 Now scarcely 1% of the original 400,000 square miles of tallgrass remain.

e. *tall timber* (N. Amer.), uninhabited forest. Usu. in phr. *to break* (*strike*, etc.) *for* (*the*) *tall timber*; also *transf.*, to run away, escape. Hence *tall-timbered* adj.

[**1831** *Boston Transcript* 24 June 2/4 Why didn't Van just go and *tell* the old man he wanted to *break for high timber*?] **1845** *St. Louis Reveille* 22 Jan. 1/6 Knowing the direction of the trees that stood in the grove, I 'broke for the tall timber'. **1877** J. M. BEARD *K.K.K. Sketches* 166 The panic-stricken darkies broke across the landscape with a yearning desire for tall timber that was eloquently depicted on every motion of the supple limbs. **1904** [see *SCRATCH *v.* 5 c]. **1914** D. W. ROBERTS *Rangers & Sovereignty* 128 The 'bad men'..began to strike for 'tall timber'. **1921** *Daily Colonist* (Victoria) 22 Mar. 13/1 The northern correspondent of The Colonist has just emerged from the tall timber, where he has been living under canvas for the past two years. **1949** *Skyline Trail* Oct. 18/1, I fell off *three times*; finally the disgusted critter took to the tall timber, leaving me to hike onward and to get across the frigid stream as best I could. **1966** *Times* 28 Feb. (Canada Suppl.) p. ii, Canada is a tall-timbered..rod-and-gun of a country.

f. Of game birds: high-flying. Occas. applied *transf.* to a shot at such a bird.

1913 R. PAYNE-GALLWEY *High Pheasants* iv. 37 The tallest pheasants I know of are at Harpton. *Ibid.* v. 45 These high birds..afford most sporting and tall shots. **1922** H. S. GLADSTONE *Record Bags & Shooting Rec.* 197 Correct judgment of distance is essential for accurate shooting..hence the tall stories of tall birds. **1952** J. W. DAY *New Yeomen of England* xi. 125 It was all done in the sacred service of King Pheasant, by men who laid out their woods in order to show tall birds. **1962** *Times* 28 Apr. 11/4 But equally tall pheasants can and do come over at shoots on ground as flat as a billiard table. **1976** *Shooting Times & Country Mag.* 18–24 Nov. 28/2 Half a dozen superb tall birds came over the gate.

8. d. (Earlier and later examples in *Cricket.*) *tall order* (later examples). Cf. *big* (*large, strong*) *order* s.v. ORDER *sb.* 24 c in Dict. and Suppl.

1864 F. LILLYWHITE *Guide to Cricketers* 95 The..match ..between Surrey and Thirteen of Cambridge University, ..owing to the 'very tall' scoring, was also unfinished. **1891** W. G. GRACE *Cricket* iv. 19 The season, so far, had been dry, and favourable for tall scoring. **1920** C. A. W. MONCKTON *Some Experiences New Guinea Resident Magistrate* xviii. 201, I..told the police we would make the attempt; clearly they thought we were taking on a devil of a tall order. **1946** *Civil & Milit. Gaz.* 26 May 15 (*heading*) Tall scoring by Indians at Lords. **1950** H. READ *Educ. for Peace* iv. 51 It is, to use our slang expression, 'a tall order', but it has been attempted before. **1976** *Norwich Mercury* 19 Nov. 2/1 'You do not by any chance know of anybody with an old lion's skin?' she asked. A tall order indeed.

B. quasi-adv. *to walk tall*: also, to have dignity or self-respect; *to sit tall*, to sit erect, with a straight back (in quot. *fig.*).

1970 *Guardian* 3 June 20/6 Officials gave the doctors folders entitled, 'Walk Tall in Australia'. *Ibid.* 6 Aug. 9/1 Walk tall, sisters... One woman's distinction adds a tiny bit to the stature of every other woman. **1976** *Billings* (Montana) *Gaz.* 6 July 3D/6 'We need to sit tall in the saddle and ride like hell in the right direction,' the governor added. **1980** *Times* 15 Feb. 16/8 'Walk tall,' say the television commercials [in Hongkong], 'report corruption.'

C. *Comb.*: *tall-hatted, -masted* (example), *-stemmed, -wheeled* (example); **tall-grass** *a.*, (esp. of a prairie) characterized by tall grasses (TALL *a.* 7 c); **tall poppy**: see *POPPY *sb.* 1 b.

1825 J. G. WHITTIER *Poet. Wks.* (1898) 522/2 With tall-masted ships on their broad bosoms riding. **1885** KIPLING *Departmental Ditties* (1886) 33, I drive no tall-wheeled traps. **1920** *Carnegie Inst. Washington Publ.* No. 290. 256 The tall-grass prairies are typical of regions in which humid farming prevails. **1922** JOYCE *Ulysses* 250 H. halted and four tallhatted white flagons halted behind him. **1925** J. FERGUSON in *Oxf. Poetry* 18 The tall-stemmed candles brighten. **1951** *Dict. Gardening* (R. Hort. Soc.) II. 921/1 In the tall-stemmed section, such as *G*[*rammatophyllum*] *speciosum*, there are probably six species. **1972** T. MCHUGH *Time of Buffalo* ii. 16 On the lush eastern half of the tall-grass prairie, one buffalo could have lived for a year on about ten acres. **1972** D. LEES *Zodiac* 27 A tall-hatted chef [was] serving a cold lunch. **1980** *Outdoor Life* (U.S.) (Northeast ed.) Oct. 97/1 Oak thickets, tall-grass ridges,..and aspen groves.

Tallensi (tāle·nsi), *sb.* (and *a.*) Also **Talensi, Talenssi,** etc. [Native name.] **a.** An African people of Northern Ghana. **b.** The language spoken by this people, belonging to the Voltaic or Gur group of the Niger-Congo languages. Also *attrib.* or as *adj.*

1920 A. W. CARDINALL *Natives N. Territories Gold Coast* I. 1 The area..is peopled by mixed races, of which the principal tribes are Nankanni, Nabdam, Talansi..and Builsa. **1936** *Discovery* June 169/1 A Tallensi tribesman of the Gold Coast. **1949** M. FORTES *Web of Kinship among Tallensi* i. 1 The Tallensi are typical of the great congeries of Mole-Dagbane-speaking peoples that occupy the basin of the Volta rivers in the French Ivory Coast. **1951** R. FIRTH *Elem. Social Organization* vi. 207 In Tallensi ideology filial piety is an important moral principle. **1952** WESTERMANN & BRYAN *Lang. W. Afr.* v. 65 *Talensi*, own name talene. Language. Spoken by: Talensi (Tallensi, Tallense, Talansi, Talense), call themselves talis, talensi (Sing. talenja, talenga). Where spoken: Gold Coast, Zuarungu District. **1958** D. EMMET *Function, Purpose & Powers* ii. 33 In the West African Tallensi society..the important relations in terms of which behaviour is directed are kinship relations. **1963** *Internat. Jrnl. Amer. Linguistics* XXIX. I. 8 An enumeration of the membership of the Niger-Congo family by tentative genetic subfamilies follows..*Gur*:..Talensi. **1972** J. GOODY in P. Laslett *Household & Family* 106 The Tallensi and the Ashanti of Ghana.

tallied, *ppl. a.* Add: **1. b.** Counted, numbered.

1830 *Dublin Even. Post* 17 July 3/2 Not by 'tallied votes' but by acclamation.

Tallman, var. *TOLMAN.

tallow, *sb.* Add: **4.** (Earlier and later examples.)

1819 M. WILMOT *Let.* 21 Dec. (1935) 42 Wax candles are both bad, and dear. We use them of course, and *tallows* in the nursery and Kitchen. **1980** *Times Lit. Suppl.* 22 Aug. 929/3 He would blow out his tallow behind Coloured glass.

5. c. tallow-bush *U.S.* = *tallow shrub*; **tallow-nut** (examples); **tallow pot** *U.S.* and *Austral. slang*, the fireman on a locomotive engine; **tallow shrub** (earlier example); **tallow-wood** (earlier example).

1835 W. G. SIMMS *Partisan* 387 The prisoners..had been made to file into the groves of tallow bushes. **1791** W. BARTRAM *Trav. N. & S. Carolina* 94 These shelly ridges have a vegetable surface of loose black mould, very fertile, which naturally produces..Tallow-nut, or Wild Lime, and many others. **1884** C. S. SARGENT *Rep. Forests N. Amer.* 34 Wild Lime. Tallow Nut... Common and reaching its greatest development in Florida. **1914** *Dialect Notes* IV. 164 *Tallow pot*,..the fireman of a locomotive. **1929** *Bookman* July 524/1 I'm surprised to find a student tallow-pot up in the cab takin' orders from the bakehead. **1960** *Listener* 18 Aug. 250/2 Firemen are 'tallowpots' or 'bakebrains'. **1968** *Amer. Speech* XLIII. 289 *Tallow pot*, ..originally, before the days of lube oil, a fireman was so-called because he had to get out onto the steam chest of the engine with a can of tallow and hold it so the lubricant would be drawn into the cylinder. **1770** J. R. FORSTER tr. *Kalm's Trav. N. Amer.* I. 192 Tallow shrub, or Candleberry Tree. **1884** A. NILSON *Timber Trees N.S.W.* 67 Tallowwood; Mahogany.—A tall tree, with a persistent furrowed fibrous bark.

tallow-chandlery. a, b. (Earlier examples.)

1815 *Niles' Reg.* VIII. 141 There are..6 tallow chandleries [in Pittsburgh]. **1846** H. MELVILLE *Typee* xxv. 203 Mehevi..looking as if he had..undergone the process of dipping in a tallow-chandlery.

tallower. For *rare*⁻⁰ read *rare* and add example in sense 'a tallow-chandler'.

1946 E. ST. J. BROOKS *Sir C. Hatton* ii. 26 She married.. John Farrington, a tallower of the same town.

tally, *sb.*^{1} **2. a.** Restrict † *Obs.* to sense in Dict. and add: Also, the record of a number.

1951 'J. WYNDHAM' *Day of Triffids* xi. 201 He had taken a tally of the livestock and the number of blind among it. **1976** H. WILSON *Governance of Britain* iii. 55 The prime minister usually keeps a tally of those for and against.

d. *spec.* in sporting use, a total score; also in *Baseball*, a single run.

1856 *Spirit of Times* 27 Dec. 276/3 One of these swiftly-delivered balls, when stopped by a skillful batsman, is sure to give the..striker time to go his rounds in safety, and score one tally as he reaches home. **1868** H. CHADWICK *Game of Baseball* 46 *Tally*, this term applies to the total score of the single innings played, or of the even innings, or of the totals at the close of the match. **1875** *Chicago Tribune* 29 July 5/4 [They] were only two tallies behind at the beginning of the ninth inning. **1949** *Marshfield* (Wisconsin) *News-Herald* 19 July 9/1 Phil Satkowiak homered with none on in the fourth and his teammates added two more tallies. **1976** *Liverpool Echo* 23 Nov. 17/1 Ainsdale marksman Alex Blakeman took his tally to nine goals from his last four games with a brilliant hat-trick. **1977** *Guernsey Weekly Press* 21 July 8/5 They shocked their opponents by scoring four runs on four hits in the top of the first and..holding the red-and-blacks to only one tally in the bottom half.

e. *Austral.* and *N.Z.* (See quot. 1965.)

1881 A. BATHGATE *Waitaruna* xii. 173 There was a rivalry among them [*sc.* shearers] as to who would have the biggest tally. **1908** D. FERGUSON *Bush Life Austral. & N.Z.* (ed. 4) v. 39 Not only did his tallies of 170, 180, and even 190 place him beyond the reach of the keenest competitor, but the quality of his work was far above that

of shearers in ordinary. **1930** L. G. D. ACLAND *Early Canterbury Runs* 1st Ser. vii. 170 The combined shearing tally was..115,000 sheep. **1965** J. S. GUNN *Terminol. Shearing Industry* II. 31 *Tally*, a specialised alternative term for a number of sheep. Each shearer has his personal tally for..the day, progressively to date, and his final tally for the shed... At one time a notch was cut on a tally stick on the call of 'tally' or 'hundred', which indicated a hundred sheep counted.

3. a. (Later example.)

1950 J. AGEE *Morning Watch* (1951) II. 61 Hell of a saint I'd make, he said to himself; and added with cold and level weary self-disgust to the tally of the sins he must soon confess, I swore in Lady Chapel in the presence of the Blessed Sacrament.

4. b. (Earlier examples.)

1864 HOTTEN *Slang Dict.* 253 *Tally*, 'to live *tally*,' to live in a state of unmarried impropriety. **1867** B. BRIERLEY *Traddlepin Fold* 174 Aw'd advise thi t'live tally..if theaw con mak' it reet wi' some owd damsel.

7. a. (Earlier example.)

1851 H. MELVILLE *Moby Dick* I. xli. 303 Tying a lettered leathern tally round its neck.

e. *Naut. slang.* (See quots 1929, 1946.) Also *cap-tally* = *tally ribbon*, sense 9 c below.

1929 F. C. BOWEN *Sea Slang* 138 *Tally*, a name or name-plate of any description. **1945** 'TACKLINE' *Holiday Sailor* i. 9 We queued-up before him to have our cap-tallies—*not* cap-ribbons, we now discovered—secured about our caps with the authentic sailor's-knot. **1946** J. IRVING *Royal Navalese* 172 *Tally*, a sailor's name is his 'tally'—e.g. 'Answer your tally!'

9. a. *tally-book* (examples, also from *2 e), *-keeper* (earlier example). **b.** *tally-system* (earlier example). **c. tally band** *Naut.* = *tally ribbon* below; **tally card** *U.S.*, a score-card; **tally ribbon** *Naut.*, a sailor's cap-ribbon bearing the name of his ship (cf. sense 7 e above); **tally-stick** (earlier U.S. examples).

1977 *Times* 7 Jan. (Royal Navy Suppl.) p. ii/6 That dear little pancake hat with its silk tally band;..I believe that it was because of that little hat that I had joined the WRNS in the first place. **1965** J. S. GUNN *Terminol. Shearing Industry* II. 31 *Tally book*, the official record of what every man has shorn in each run of the day. **1972** T. A. BULMAN *Kamloops Cattlemen* xxxi. 178 Our tally books showed that we were short a bull and six head of cattle. **1909** *Cent. Dict. Suppl.*, Tally card. **1928** *Publishers' Weekly* 14 July 172 Allied with these are tally cards, playing cards, novelties and party favors. **1880** 'MARK TWAIN' *Tramp Abroad* v. 55 The tally-keeper..tallied one for the opposition in his book. **1919** W. LANG *Sea Lawyer's Log* 14 Those three caps, too, look very smart, with the neat, white canvas cover which we wear during the summer months, and the tally ribbon with its tricky little bow on the left side. **1830** *Virginia Lit. Museum* 27 Jan. 526/2 A 'negro boy, with a talley stick was a statesman complete in his school'. *a* **1861** T. WINTHROP *Canoe & Saddle* (1863) vii. 153 She could not tell how many [years], having dropped her tally-stick in the fire..that very day. **1850** J. R. PLANCHÉ *Island of Jewels* II. i. 21 Is it all a trick, you make this mighty splash on, Or, is the tally system here in fashion?

tally, *v.*^{1} Add: **I. 1. d.** *Sport* (chiefly *N. Amer.*). To score (a run, goal, etc.). (*a*) *intr.*

1867 *Ball Players' Chron.* 14 Nov. 2/4 Taylor took his first on a muff by Banker and tallied on passed balls. **1903** *N.Y. Times* 18 Aug. 5/2 The home players tallied only five times during the entire contest. **1931** *Randolph Enterprise* (Elkins, W. Va.) 9 July 5/3 The nine from Randolph had tallied five times in the same frame. **1947** *Sun* (Baltimore) 8 Nov. 12/2 The visitors tallied for the first time in the opening minutes of the game when Ann Worthington sent a hard shot into the corner of the cage. **1968** [see *PERIOD *sb.* 4 f]. **1974** *Cleveland* (Ohio) *Plain Dealer* 13 Oct. C.2/3 Terell tallied on a two-yard run and took a pass from Jeff Starrett 70 yards for a second score.

(*b*) *trans.*

1875 *Chicago Tribune* 24 Aug. 5/6 A furious overthrow by Beals at second..tallied three unearned runs. **1966** *Telegraph* (Brisbane) 22 Jan. 5/2 He started racing in November, and in five starts has tallied a win, second, and a third. **1976** *Billings* (Montana) *Gaz.* 6 July 1-c/6 The Angels tallied their fifth run in the third inning as Bob Brenly came on a Young single to short center.

tally-ho, *int.* and *sb.* Add: Also † **tally-oh.**

1. b. Also *fig.*

1955 *Times* 9 Aug. 9/2 Even the eminent scholar and social worker Liang Shu-ming has been cast out... Throughout China the tally-ho of the party hacks is echoing.

c. (Earlier and later examples.)

1825 H. WILSON *Mem.* (ed. 2) III. 96 A drunken man, in a dashing light green coat, a red waistcoat, and large tally oh! pin in his shirt. **1922** JOYCE *Ulysses* 571 A pack of bloodhounds led by Hornblower of Trinity brandishing a dogwhip in tallyho cap, and an old pair of grey trousers, follows from far.

tallyman. Add: **2.** (Earlier U.S. examples.)

1857 *Spirit of Times* 23 May 190/3 The tallymen were: Olympic, E. W. Cody; Bay State, W. W. Bragg, jr. **1867** 'T. LACKLAND' *Homespun* ii. 155 It may be the vote is very close; in that case the outside counters and tally-men are as much in the dark as the rest.

3. (Earlier example.)

1876 J. READ *They all do It* (song) 3 Mrs Brown says it's a sin, that Mrs Smith drinks gin And harbours tally-men from day to day.

talma. (Earlier example.)

1852 P. S. G. TEN BROECK *Let.* 1 Apr. in T. Donaldson *Moqui Pueblo Indians* (1893) 26 The most beautiful part of their dress is a talma.., which is thrown over the shoulders, fastened in front, and, hanging down behind, reaches halfway below the knee.

talmessite (tælme·səit). *Min.* [a. F. *talmessite* (Bariand & Herpin 1960, in *Bull. de la Soc. franç. de Min. et Crist.* LXXXIII. 120/1), f. *Talmessi*, name of a mine near Anarak, Iran: see -ITE^1.] A hydrated arsenate of calcium, magnesium, and barium, $Ca_2(Mg, Ba)(AsO_4)_2.2H_2O$, found as triclinic crystals.

1961 *Chem. Abstr.* LV. 8186 Analysis of an arsenate of Ca and Mg discovered in 1955..permits the definition of this new mineral as an isomorph of β-roselite... The name talmessite is proposed for the mineral. **1977** *Mineral. Abstr.* XXVIII. 229/2 White fibrous radiating aggregates of talmessite occur linked with the baryte.

‖**Talmid Chacham** (ta·lmid χǫ·χām). Also **talmid chocham, hakham,** etc., and with small initials. [Heb., lit. 'disciple of a wise man'.] One well versed in the Jewish Law; a wise man (see also quot. 1962^1). Cf. *HAHAM.

1863 *Chambers's Encycl.* V. 720/2 Rabbi (Master), *Talmid Chacham* (Disciple of Wisdom), were titles of honour given to those expert in a knowledge of the law. **1905** *Jewish Encycl.* XI. 678/2 *Talmid Ḥakam*..Honorific title given to one well versed in the Law... The talmide ḥakamim formed in Jewish society a kind of aristocracy having many privileges. **1959** D. D. RUNES *Conc. Dict. Judaism* 217/1 *Talmid Chakham*, student of true knowledge as interpreted by the Talmudic sages. **1962** *New Jewish Encycl.* 475/1 *Talmid Ḥakham*, originally this Hebrew term referred to a disciple of a learned man (that is, a teacher of the Oral Law), as its literal meaning implies. Later it applied to any Talmudic scholar of high reputation, and in its present usage it applies to any individual of high Jewish scholarship. **1962** B. ABRAHAMS tr. *Life of Glückel of Hameln* ii. 22 God...does not desert a Talmid Cocham. [*Note*] A wise and clever man, one well versed in the Torah and Talmud. **1973** *Jewish Chron.* 19 Jan. 42/3 A native of Czechoslovakia, he studied at Miskolc, Galanta, and other yeshivot, and was a man of scholarship—a *talmid chacham* in the truest sense of the word.

talmudism. Add: **2.** *fig.* in *Pol.* use [tr. Russ. *talmudizm*], (see quot. 1957). Cf. next.

1957 R. N. C. HUNT *Guide to Communist Jargon* xviii. 65 Dogmatism—or Talmudism, as Stalin at times called it—is defined..as 'the uncritical acceptance of dogma without considering the conditions of its application'. **1965** P. O'DONOVAN et al. *United States* iii. 56/1 Is this the victory of pragmatism over Constitutional talmudism?

talmudist. Add: **2.** *fig.* in *Pol.* use (see quot. 1957). Cf. prec.

1957 R. N. C. HUNT *Guide to Communist Jargon* xviii. 65 The second edition of the *Large Soviet Encyclopædia*.. defines *Talmudist* (figuratively) as 'a pedant, dogmatist, formalist, doctrinaire'. **1965** *New Statesman* 14 May 772/2 Soviet Russia is described as a 'filthy shed'... Its inhabitants are 'slaves' and its rulers 'blockheaded talmudists'.

talmudize, *v.* Add: (Later example.) Hence **talmudiza·tion.**

1927 V. BURCH *Jesus Christ & his Revel.* 13 If the Talmud depraves the values of Jesus Christ with cynicism, and the Kûran talmudises Him so that he becomes an inferior Jewish prophet, then we are guilty of a double talmudisation of the One we are said to follow. **1931** *Times Lit. Suppl.* 19 Mar. 228/2 Extraneous influences, historical development, Talmudization, do not trouble him.

‖**Talmud Torah** (ta·lmŭd tō^u·rŏ, tō^a·rā). [Heb., 'study of the (Jewish) Law'; cf. TALMUD, TORAH.] The study of the Jewish Law. Also freq. used *attrib.* and *absol.* to designate a school run by the community for the instruction of children in the Jewish religion (see also quot. 1962). Cf. *CHEDAR.

1844 *Jewish Chron.* 18 Oct. 7/1 Mr. L. A. Green.. received the rudiments of his education at the 'Talmud Torah School', Bell Lane, Spitalfields. **1881** *Encycl. Brit.* XIII. 686/2 Besides the schools of the Alliance, there are 2287 pupils in the wretched Talmud Torah schools. **1907** I. ZANGWILL *Ghetto Comedies* 387 'We can meet and practise in your Talmud-Torah Hall!' 'The holy hall of study!' gasped the Rabbi. **1932** C. ROTH *Hist. Marranos* 391 A secret society for the purpose of studying the Law—*Talmud Torah.* **1960** L. P. GARTNER *Jewish Immigrant in Eng. 1870–1914* viii. 221 The Jews' Free School in Spitalfields... In its early days..was a Talmud Torah.. for poor children. **1962** *New Jewish Encycl.* 477/1 *Talmud Torah*, the literal meaning of this Hebrew term is 'the study of the Torah' or the 'teaching of the Torah', but it is commonly used as a designation for an elementary Hebrew school. Whereas the traditional *Ḥeder* was a private religious school, the *Talmud Torah* was a public institution maintained and administered by the community in order to provide education for orphans or children of parents who could not afford private teaching. In America, Talmud Torah generally refers to a school which meets in the afternoon after the secular school session is over. Its curriculum includes the study of the Bible, Hebrew, Jewish history, customs and ceremonies, and other phases of elementary Jewish knowledge. **1964** W. MARKFIELD *To Early Grave* (1965) ix. 151 When he wanted he had a marvellous head for Talmud Torah, except that he very seldom wanted and gave the *rebbe* trouble. **1968** M. RICHLER in R. Weaver *Canad. Short Stories* 2nd Ser. 155 My mother came home from her Talmud Torah meeting. **1976** B. WILLIAMS *Making of Manchester Jewry* xii. 322 The ten Corfiote families..contribute[d] collectively in 1872 to the languishing Talmud Torah of their home town.

talnakhite (tælnæ·kəit). *Min.* [ad. Russ. *talnakhit* (Bud'ko & Kulagov 1968, in *Zap. Vsesoyuznogo Min. Obshchestva* XCVII. 63), f. *Talnakh*, name of a locality near Dudinka in northern Siberia: see -ITE^1.] A sulphide of copper and iron, $Cu_9Fe_8S_{16}$, found as yellow, usu. iridescent, crystals of the cubic system.

1969 *Mineral. Abstr.* XX. 148/1 (*heading*) The new mineral talnakhite—the cubic variety of chalcopyrite. **1978** *Canad. Mineralogist* XVI. 29 The talnakhite structure has a broad range of possible compositions including both $Cu_9Fe_8S_{16}$ and $Cu_9Fe_9S_{16}$... The relatively restricted compositions found for natural talnakhite are the result of a change in kinetics of the ordering transformations with composition.

talo-. Add: **talocalcaneal** *a.* (example); also **talocalca·nean** *a.*; **ta:localca:neonavi·cular** *a.*, applied to the joint comprising the rounded head of the talus and the corresponding concavity formed by the navicular bone and calcaneus; **talofibular** *a.* (example); **talonavi·cular** *a.*, applied to the ligament joining the talus and navicular bone, and also to the joint between these bones (part of the talocalcaneonavicular joint).

1979 *Brit. Med. Jrnl.* 15 Dec. 1594/2 Simple rigidity of the talocalcaneal joint increases oxygen consumption during walking by up to 20%. **1900** DORLAND *Med. Dict.* 668/1 Talocalcanean. **1962** *Gray's Anat.* (ed. 33) 559 The lateral talocalcanean ligament..passes downwards and backwards from the lateral process of the talus to be attached to the lateral surface of the calcaneus. **1913** *Ibid.* (ed. 18) 402 (*heading*) Talocalcaneonavicular articulation. **1977** *Bone & Joint Diseases* (Brit. Med. Assoc.) 113 Talipes equinovarus or clubfoot is a limb anomaly that is best regarded as a congenital dislocation at the talocalcaneonavicular joint. **1979** R. H. FREIBERGER et al. *Arthrography* xi. 238/1 The most commonly injured of the lateral supporting ligaments of the ankle is the anterior talofibular ligament. **1907** MORRIS & MCMURRICH *Treat. Human Anat.* (ed. 4) III. 304 The articulations of the anterior and posterior portions of the tarsus..consist of two separate joints, viz., (i) an inner, the talo-navicular, which communicates with the anterior talo-calcaneal articulation; and (ii) an outer, the calcaneo-cuboid. **1977** *Bone & Joint Diseases* (Brit. Med. Assoc.) 3 The talonavicular joint has been replaced experimentally.

talon, *sb.* Add: **II. 4. a.** (Examples.)

1862 C. C. MEEHAN *Law & Practice of Game of Euchre* v. 86 *Talon*, the eleven cards remaining in the pack after the dealer has distributed five to each player and turned up the twenty-first card for the trump. **1921** M. C. WORK *Auction for Two or Three* III. 78 *Talon*..is, in fact, a sort of secondary *Stock* in Russian Bank and a distinguishing term must be used to describe it. **1977** *Jrnl. Playing-Card Soc.* May 25 When the discard is complete, everyone should have 11 cards and the four face-down cards remaining are called the talon.

b. (Later examples.)

1932 *Daily Tel.* 8 Oct. 2/3 Provision was made on May 3 for the conversion of the Austrian share of liability.. into new 4 per cent. bonds. Bonds, Talons, or coupons must now be presented at the Staatszentralkasse, Singerstrasse 17, Vienna, before Dec. 31. **1964** *Lebende Sprachen* IX. 99/2 A *coupon sheet*, consisting of *dividend coupons* and a *talon*, is attached to each share warrant. The dividend coupons are used by the holder to collect dividends. When the last coupon has been cashed, the talon is exchanged for a new coupon sheet. **1979** *Daily Tel.* 26 Nov. 24/6 The Bank of England give notice that new *coupon* sheets for the above-mentioned Loan will be available on and after 17th January 1980 in exchange for *talons.*

5. *talon-like* (earlier example).

1883 'MARK TWAIN' *Life on Mississippi* xxxi. 339 His hand..was talon-like, it was so bony and long-fingered.

talonid (tæ·lǫ̆nid). *Zool.* [f. TALON *sb.* 3 f + *-ID^5.] A flattened cusp on a mammalian lower molar tooth, corresponding to the talon on an upper molar.

1897 H. F. OSBORN in *Amer. Naturalist* XXXI. 1002 It [*sc.* the talon] invariably appeared first in the lower molars (where we may distinguish it as the 'talonid'). **1919** [see *HYPOCONID]. **1922** W. K. GREGORY *Orig. & Evol. Human Dentition* I. 38 The premolars..already possessed incipient talonid spurs. *Ibid.* 39 The talonids had not yet acquired basins or fossæ. **1968** [see *HYPOCONID]. **1979** *Nature* 20 Sept. 213/2 The isolated lower second molar of the right side..has the basic tribosphenic pattern with an elevated trigonid and a low talonid.

talpa^2 (tæ·lpă). *U.S.* Var. CATALPA.

1926 J. MASEFIELD *Odtaa* xvii. 283 Hi could see the fronds of the talpas moving above him. **1933** *Amer. Speech* VIII. 53/1 Talpa, n. Catalpa. This tree..has been planted in some Ozark towns.

tal qual (tæl kwæl), *adv. phr. Newfoundland.* Also **talqual.** [Shortened from L. *talis qualis* such as, of which sort or quality.] 'Just as they come': used with reference to fish sold without sorting.

1732 in *Calendar State Papers, Amer. & W. Indies* (1939) 282 And by carrying a mixt cargoe which is all sold at markett for marchantable fish, when it's only

(what in the stile of the fishermen is called Tal Qual) to the shoarmen. *c* **1894** in *Dict. Newfoundland Eng.* (1982) 557/1 Tal qual, sometimes called *all qualls*, fish bought without culling is clearly the Latin *talis qualis*, 'such as it is'. **1928** in *Ibid.* 557/1 [He] said that tal qual fish was $8.20. **1930** W. F. COAKER *Hist. Fishermen's Protective Union of Newfoundland* 30 Fish would not have advanced beyond $5.30 taqual [*sic*] this season if the F.P.U. did not exist. **1934** *Rep. Newfoundland R. Comm.* 105 in *Parl. Papers 1933–4* (Cd. 4480) XIV. 357 During the war years, quantity rather than quality became the ruling consideration; the 'cull' was therefore dispensed with and fish were bought on what is known as the 'talqual' system, viz., an average price was fixed for the whole of a fisherman's catch without any exact regard to the varying qualities of the fish comprising the catch.

taluk, taluq. Add: Also **taluka.** (Earlier and further examples.)

1793 *Bengal Permanent Settlement Reg.* in *Bengal Code* (1913) I. 9 Dependent *taluk.* **1891** *Rep. Administration Local Boards in Bombay Presidency 1889–90* 1 There are ..39 Taluka Local Boards, presided over by the Assistant and District Deputy Collectors in charge of the tálukas. **1931** *Times Educ. Suppl.* 4 Apr. 123/2 The local municipal *taluka* or district board. **1977** *Lancet* 2 July 39/2 We have in fact already prepared a plan for the treatment of rabies in the taluka (somewhat bigger than a village).

talukdār, taluqdār. (Earlier example.)

1793 *Bengal Permanent Settlement Reg.* in *Bengal Code* (1913) I. 9 Dependent *talukdar.*

talweg, var. THALWEG in Dict. and Suppl.

TAM (tæm). Also **Tam,** etc. [f. the initial letters of *t*elevision *a*udience *m*easurement (see below).] Used in *Comb.*, usu. as *TAM rating*, to denote a measure of the number of people watching a particular television programme as estimated by the company Television Audience Measurement Ltd. Also *absol.*, the company itself.

1958 *Observer* 14 Dec. 15/1 In the telly-dominated world, where everybody from a philosopher to a flat-catcher is judged by his tam-rating, it was encouraging to see originality break through the crust of diffidence. **1959** *Listener* 24 Sept. 501/1 Head-hunting rivalry is understandable, but why does the type of head matter? Egg-shaped or with forehead villainous low, all are equal in the sight of TAM. **1960** *Spectator* 14 Oct. 565 Processed and packaged..down to strip-cartoon versions compatible with the maximum Tam ratings. **1963** *Daily Tel.* 17 Oct. 16/2 According to TAM figures published last Friday, [etc.]. **1966** *Punch* 24 Aug. 302/3 The first episode topped the Tamratings. **1975** LD. HAILSHAM *Door wherein I Went* xxv. 158 They have their own fish to fry, their TAM ratings, their circulations, their Charters, their editorial policies, even their advertisers.

Tamachek, var. *TAMASHEK; **tamain,** var. *TAMEIN.

tamanoir. (Earlier example.)

1785 T. JEFFERSON *Notes on Virginia* vi. 85 Aboriginals of..America..Tamanoir.

tamarack. Add: Also 9 **tamerack, temerack.** Substitute for def.: **a.** Any of several North American larches, esp. the red larch, *Larix laricina*; = HACKMATACK. **b.** The shore pine, *Pinus contorta*, or the lodgepole pine, *P.c.* var. *latifolia*, of western North America; also, the timber of any of these trees. (Earlier and later examples.)

1805 W. CLARK *Jrnl.* 14 Sept. in *Orig. Jrnls. Lewis & Clark Exped.* (1905) III. 66 The Mountains..thickly Strowed with falling timber & Pine Spruce fur Hackmatak & Tamerack. **1810** F. A. MICHAUX *Histoire des Arbres Forestieres* I. 29 American larch..Hacmatack.. Tamarack par les Hollandois du New-Jersey. **1817** W. KEYES *Jrnl.* 25 Aug. in *Wisconsin Mag. Hist.* (1920) III. 351 Crossed a large meadow, a temerack swamp &c. **1894** *Outing* XXIV. 94/1 By vigorous working of three paddles we got up a 'tamarack breeze' that carried us rapidly along. **1947** R. PEATTIE *Sierra Nevada* 160 Lodgepole pine..is plain tamarack to many Californians. **1979** J. VAN DE WETERING *Maine Massacre* xix. 225 The only decorations, hanging from hooks on the low, handhewn tamarack beams, were tools and weapons.

tamari (tamā·ri). [Jap.] A Japanese variety of rich soy sauce. Freq. *attrib.*, as *tamari* (*soy*) *sauce.*

1977 *Spare Rib* Jan. 36/3 For protein use soya bean paste (miso) or tamari soy sauce. **1978** G. DUFF *Vegetarian Cookbk.* 14 Tamari is the genuine traditional Japanese soy sauce made only by natural methods from a mixture of wheat and whole soya beans. **1981** *Times* 22 Jan. 10/8 Root vegetables can be stir-fried..and flavoured with..soy or tamari sauce.

tamarillo (tæmări·lo). *N.Z.* [Artificial name (see quot. 1966); cf. Sp. *tomatillo*, dim. of *tomate* TOMATO.] = *tree tomato* s.v. TOMATO 2 b.

1966 *N.Z. Herald* 6 Sept. 5 The familiar tree tomato will go under the name of tamarillo after January 31, next year. *Ibid.*, It is hoped that export markets for both raw and processed fruit may be more successful with the use of this new, sub-tropical-sounding name. **1969** *N.Z. News* 23 July 4/5 Oranges, lemons,..tamarillos and chinese gooseberries are now in their shortest supply for several years. **1979** *Times* 17 Nov. 26/4 Specialities from New Zealand, including..tamarellos [*sic*].

tamarugo (tæmărū·go). Also *erron.* **tamarugal.** [Chilean Sp.] A small evergreen tree, *Prosopis tamarugo*, of the family Leguminosæ, native to the salt deserts of northern Chile and used to provide fodder in arid regions.

1972 *World Crops* XXIV. 297/1 The tamarugal is a thin-branching tree of 20–25 ft average height, yielding a leaf and a seed which are both nutritious and palatable. **1975** *Underexploited Trop. Plants* (Nat. Acad. Sci., U.S.) 128 In salt-devastated regions of suitable climate, tamarugo, an evergreen plant, could become a year-round fodder supply. *Ibid.* 129 Tamarugo pods need extensive leaching before they can be eaten by humans. **1977** *New Scientist* 17 Mar. 638/1 The leguminous tamarugo tree from Chile..can supply forage in dry, salty regions. **1980** *Times* 5 June 26/9 Where few economic plants could grow, the Tamarugo flourishes and provides good fodder all the year round.

tamasha. Add: **2.** *transf.* A fuss, a commotion.

1882 F. M. CRAWFORD *Mr. Isaacs* x. 213 Mr. Ghyrkins ..wanted to know 'what the deuce all this *tamāsha* was about'. **1923** KIPLING *Land & Sea Tales* 226 Why is there this *tamasha* (fuss)? **1964** A. SWINSON *Six Minutes to Sunset* ii. 24 Stewart..asked 'What's all the *tamasha* about?' **1981** S. RUSHDIE *Midnight's Children* I. 30 Enough of this tamasha! No more of this..tomfoolery!

Tamashek (ta·mășek). Also **Tamachek.** [Berber: see quot. 1896.] The Berber language spoken by the Tuaregs.

1885 *Encycl. Brit.* XVIII. 778/2 The principal dialects [of Berber] are the Kabyle, the Shilha, and the Tuarek or Tamashek, corresponding nearly to the ancient Numidian, Mauretanian and Gætulian respectively. **1896** A. H. KEANE *Ethnol.* xiv. 384 This word [*sc.* *Tamahu*] still exists under various dialectic forms (*Tamahéeg, Tamashek, Tamazigt*) applied collectively to the Hemitic languages of the Sahara and Mauritania. The form T-amazig-t, when stripped of its fem. prefix and postfix particle *t*, is seen to be identical with the *Maxyes* of Herodotus (later *Masices, Mazices*), i.e. Amzigh, pl. Imazighen, 'freemen', the most general name of the Mauritanian Berbers. **1908** T. G. TUCKER *Introd. Nat. Hist. Lang.* 173 Thus old Egyptian *annuk*..='I', for which the Berber Tamashek has *nek*... 'He' is in old Egyptian *entuf*, in Tamashek it is *enta*. *Ibid.* 174 Berber languages are still spoken in the Western Sahara (where is to be found the *Tamashek*, which is usually treated as the most representative dialect). **1974** *Encycl. Brit. Macropædia* VIII. 596/2 Tamashek has several verbal tenses. **1978** D. BAGLEY *Flyaway* xiii. 94 Assekrem is a Tamachek word—it means, 'The End of the World'.

tambala (tæmbā·lā). Also **tambola;** pl. **tambala, -s.** [Nyanja, lit. 'cockerel'.] A currency unit in Malawi, equal to 1/100 of a kwacha. Also, a coin of this value.

1970 *Compton Yearbk.* 188 Malawi's decimal currency was to be introduced in March 1971; the new unit, the kwacha, is divided into 100 tambolas. **1970** *Whitaker's Almanack 1971* 982 Malawi. *Malawi Kwacha* of 100 *Tambala* (from Feb. 15, 1971)... [Coins] *Tambala* 20, 10, 5, 2, 1. **1983** *Times* 19 Dec. 11/2 The Africans think we're mad..but really it wouldn't seem right without.. the five-tambala pieces in the pudding.

† **Tambaroora** (tæmbărū^a·rā). *Austral. Obs.* The name of a town in New South Wales, used to designate a bar game in which the winner buys drinks for the players. Also in *Comb.*, as *Tambaroora muster.*

1882 A. J. BOYD *Old Colonials* 63 It may be that the exciting game of Tambaroora is not familiar to all my readers... Each man of a party throws a shilling, or whatever sum may be mutually agreed upon, into a hat. Dice are then produced, and each man takes three throws. The Nut who throws highest keeps the whole of the subscribed capital, and out of it pays for the drinks of the rest. The advantage of the proceeding lies in this: Where drinks are charged at sixpence, the subscription is double that amount for each... Thus if ten Nuts go in for a Tambaroora, with nobblers at sixpence, the winner pockets five shillings by the transaction. **1895** C. CROWE *Austral. Slang Dict.* 84 *Tambaroora*, a game of a shilling each in the hat and the winner shouts. **1897** *Bulletin* (Sydney) 18 Dec. (Red Page), The essence of a present-day tambaroora is a sweep for the purchase of drinks—frequently on the principle that more liquor can be purchased wholesale for 1s. 6d. than six thirsty people can buy for 3d. each. Hence 'tambaroora muster', when the droughty party musters all the coin it's possessed of, and one individual goes and bargains for the beer. **1901** *Bulletin Reciter* (Sydney) 202 (*poem-title*) Tambaroora. **1945** BAKER *Austral. Lang.* ix. 172 Just as the *shout* is an institution in this part of the world so are the..*Tambaroora muster* and a few other variations on the theme, all of which concern the creation of a jack-pot, usually with the object of buying drinks. The *Tambaroora*—taken from the name of an eastern township—dates from the early 1880s. The idea behind these expressions..is that everyone pays for himself.

tamber (tæ·mbəɹ), var. TIMBRE *sb.*^{3}

[**1920** *S.P.E. Tract* III. 11 Our English form of the French sound of the word would be approximately *tamber*; and this would be not only a good English-sounding word like *amber* and *clamber*, but would be like our *tambour*, which is *tympanum*, which again is *timbre.*] **1923** *Ibid.* XII. 60 Those enchanting qualities, the rhythm, the phrasing, the tamber, and accent of the living voice. **1937** J. R. FIRTH *Tongues of Men* iii. 36 In chiming reduplications you get..likeness of repeated articulation with difference of vowel quality or *tamber difference*—e.g., *see-saw.* **1950** D. JONES *Phoneme* iii. 12 An alphabetic system of phonetic transcription consists of letters representing sound-qualities (tambers) or phonemes. **1973** J. C. WELLS *Jamaican Pronunc. in London* iv. 56 The tamber of nonprevocalic /l/ has been shown only when 'dark'. *Ibid.* 130 Chomsky and Halle call this feature 'tense' rather than 'long', thereby emphasizing the tamber differences.

tambo^1. Add: (Earlier example.) Also, the tambourine played by such a musician.

1848 *New Negro Forget-me-not Songster* 32 We plaid dis song, 'on de banjo', Wid de fiddle and de bones, and ole tambo. **1870** T. A. BROWN *Hist. Amer. Stage* 70/1 George Christy took the bone end, with Lansing Durand as tambo. **1952** [see *QUILL *sb.*^{1} 1 c]. **1958** P. OLIVER in *Decca Bk. Jazz* i. 22 The 'nigger minstrel' troupes comprising banjo players and 'end men' playing 'tambo and bones' being popular in the North throughout the nineteenth century.

‖ **tambo^2** (tæ·mbo). [Sp., ad. Quechua *tampu* wayside hostelry.] In the Andean countries, esp. Peru, a lodging house or inn.

1830 E. TEMPLE *Trav. Var. Parts Peru* ii. 65, I went from *tambo* to *tambo* in search of a lodging. **1854** W. L. HERNDON *Amazon* I. 60 We stopped, at four, at the tambo of Acchahuarcu. **1902** *Encycl. Brit.* XXV. 377/1 One of the most interesting topics of study is the trails along which the seasonal and annual migrations of tribes occurred, becoming in Peru the paved road, with suspension bridges and wayside inns or tambos. **1959** G. WOODCOCK *Incas & Other Men* I. iii. 45 Primitive shelters, still called tambos, which were maintained..where no hotels existed. **1978** D. P. WERLICH *Peru* ii. 36 The Incas constructed granaries to supply the army..and maintained *tambos* to serve travellers.

Tamboekie, var *TAMBOOKIE *sb.* and *a.*, TAMBOUKI *a.* in Dict. and Suppl. **tamboetie,** var. *TAMBOTI(E. **tambola,** var. *TAMBALA.

tamboo^1 (tæmbū·). *Mil. slang.* Also **tambu.** [ad. Pers., Hindi *tambū* tent.] In the war of 1914–18, a temporary rough shelter in a trench.

1916 *Sphere* 19 Feb. 188 *b*/1 The 'Tamboo', as the sleeping apartment [in a dug-out] is called. **1918** W. OWEN *Let.* 28 Sept. (1967) 579 Am still sitting on straw under our Tamboo, for it is raining again. *Ibid.* 10 Oct. 582 The corrugated iron wall of my Tamboo. *Ibid.* 31 Oct. 591 My servant & I ate the chocolate in the cold middle of last night, crouched under a draughty Tamboo, roofed with planks. **1925** [see *BIVVY, BIVY].

tamboo^2 (tæ·mbū). Also **tambou.** [Blend of TAMBO or TAMBOUR *sb.* + BAMBOO *sb.*] In full *tamboo-bamboo.* In the West Indies, a small drum made of bamboo. Freq. *attrib.* Hence **tamboo-bambooist,** one who plays the tamboo.

1942 H. C. GORDON *West Indian Scenes* II. iii. 57 The chief of these [instruments] for beating time was the *tambou*, a small drum. **1955** *New Commonwealth* 28 Nov. Suppl. p. xviii/2 They were not..allowed to make tambour (tamboo) bamboo bands. **1956** *Caribbean Q.* IV. III. & IV. 195 Five to twenty stickmen formed a band.. accompanied by drummers and/or tamboo bamboo bands. **1959** W. A. SIMMONDS '*Pan*'—*Story of Steelband* 8 'Hell-yard', traditional headquarters of the Tamboo-bambooists of downtown Port-of-Spain. **1960** *Times* 17 Sept. 7/6 The three major influences on the development of the modern steel orchestra of Trinidad are the 'tamboo-bamboo' drum, the 'bottle and spoon', and the Indian drum and music of the Hosein festival. *Ibid.*, The tamboo-bamboo requires careful fashioning.

† **Tambookie** (tæmbū·ki), *sb.* and *a.* *Obs.* Also **Tamboekie, Tambouki, Tambuki,** etc. [f. as TAMBOUKI *a.*] **A.** *sb.* The Tembu people. Cf. *TEMBU *sb.*

1786 G. FORSTER tr. *Sparrman's Voy. Cape of Good Hope* II. 147 On the other side of Zomo dwells another nation, who, by the Snese-Hottentots, are called Tambukis. **1792** E. RIOU tr. *J. van Reenan's Jrnl. Journey from Cape of Good Hope* 24 The country of the Tamboekies. *Ibid.* 42 The country of Captain Joobie the Tambookie. **1801** J. BARROW *Trav. Interior S. Afr.* I. iii. 201 With the Tambookies they live on friendly terms. **1824** [see *AMAPONDO]. **1874** *Friend* (Bloemfontein) 2 Apr., We were not sure but that the Tambookies would join the confederacy against us. **1902** *Encycl. Brit.* XXX. 3/1 The Ama-Tembu nation, popularly called Tambookies.

B. *adj.* Designating or pertaining to the Tembu people. Cf. *TEMBU *a.*

1827 G. THOMPSON *Trav. S. Afr.* (ed. 2) II. 336 The *Tambookie* or *Amatymba* tribe. **1835** N. ADAMS *Let.* 5 June in D. J. Kotzé *Lett. Amer. Missionaries* (1950) 75 There is a good waggon road to Natal through the Tambookie country. **1846** J. C. BROWN tr. *Arbousset & Daumas's Explor. Tour N.E. Colony Cape of Good Hope* xii. 93 Lekoro..undertook a military expedition to the Tambuqui country. **1860** W. SHAW *My Mission in S.-E. Afr.* 486 The Rev. John Ayliff, the resident Missionary, was constrained to escape with his family, accompanied by the native inhabitants of the Mission village, and take refuge in the Tambookie country. **1875** *Handbk. S. Afr.* (S. W. Silver & Co.) 460 *Tambookieland*, the country formerly inhabited by the Tambookie tribes of Kaffirs.

tamboora, var. TAMBOURA in Dict. and Suppl. **tamboritsa,** var. *TAMBURITZA.

tamboti(e (tambu·ti). *S. Afr.* Also **tamboetie, tambootie,** etc. [ad. Xhosa *umThombothi.*]

1. A deciduous tree, *Spirostachys africana*, of the family Euphorbiaceæ, native to southern Africa and with dark, rough bark and short spikes of tiny flowers. Freq. *attrib.*

1852 J. S. CHRISTOPHER *Natal* 32 The yellow, assegai, iron,..and Tamboote wood, grow in abundance. **1859** R. J. MANN *Colony of Natal* 157 A dark brown, very hard wood, distinguished by the Kafirs as 'tamboti-wood'..is employed in the construction of axles. **1871** J. MACKENZIE *Ten Years N. Orange River* xxiv. 460 The tall and resinous tambootie tree, which I selected for beams and rafters, was easily split. **1951** *Cape Argus* 27 Oct. (Mag. Sect.) 2/4 Our site was pitched in the shade of a glorious tambotie tree. **1963** H. C. BOSMAN *Unto Dust* 43 Inside were tamboetie wood trestles for the coffins. **1973** PALMER & PITMAN *Trees S. Afr.* II. 1157 The tamboti is widely known for three reasons—its fine wood, its toxic properties, and its 'jumping beans'.

¶ **2.** *tambotie grass* = *tambouki grass* s.v. TAMBOUKI *a.* in Dict. and Suppl.

1866 T. GEAST *S. Afr. Diaries* 42, I observed that almost every stem of the long tambootie grass had a silken filament flying from it. **1899** [see TAMBOUKI *a.*] **1939** tr. *E. N. Marais's My Friends the Baboons* iii. 31 The footpath here passed through a stretch of high tambotie grass.

tambou, var. *TAMBOO^2.

Tambouki, *a.* Add: Also **Tamboekie, Tambuki,** etc., and with small initial. Substitute for def.: *tambouki grass*, one of several tall coarse grasses of southern Africa, esp. one of the genus *Cymbopogon* or *Hyparrhenia*; *tambouki wood*, tamboti wood (see *TAMBOTI(E 1). (Earlier and later examples.)

1837 J. KIRKMAN in F. Owen *Diary* (1926) 158 The mother and child had hidden under the long Tambookie grass. **1910** J. BUCHAN *Prester John* xiv. 230, I was..into a piece of parkland with long, waving tambuki grass. **1963** H. C. BOSMAN *Unto Dust* 119 Her hair was bleached the yellow of tamboekie grass in winter.

tambour, *sb.* Add: **4. c.** (Earlier example.)

1848 [see *PASSING *vbl. sb.* 3].

6*. A sliding, flexible shutter or door on a piece of furniture, made by sticking narrow strips of wood to a backing of canvas.

1940 *Chambers's Techn. Dict.* 833/1 *Tambour* (*Furn.*), a panel of slat-work or pleated textile material. **1952** J. GLOAG *Short Dict. Furnit.* 469 The tambour is used for desk tops and occasionally for doors. **1970** D. ASH *Dict. Eng. Antique Furnit.* 147/1 Tambours were introduced from France where they had been in use since about 1750, and were chiefly popular in England in the last quarter of the 18th century.

7. (in sense *6*) *tambour construction, cupboard, desk, door, front, shutter, slide, top, writing-table*; *tambour-fronted* adj.; **tambour-frame** (earlier examples); **tambour hook** = *tambour needle*; **tambour-stitch** (examples).

1934 *Burlington Mag.* Sept. 213/2 The origin of the tambour construction of doors, roll-tops, etc., which appeared in Europe about the middle of the eighteenth century; the bamboo sticks were split and the halves fastened on a piece of strong canvas, the whole being fitted into grooves. **1918** *Heal & Son Catal.: Cottage Furnit.* 26 Mahogany Inlaid Sideboard, bow front with tambour cupboard in centre. **1797** *Prices Cabinet Work* 57 A Tambour Desk, Three feet long, four long drawers in front. **1803** T. SHERATON *Cabinet Dict.* 316 Tambour doors are often introduced, in small pieces of work, where no great strength or security is requisite, as in night tables, and pot cupboards. **1973** 'K. ROYCE' *Spider Underground* ix. 139 The room was a library with an incongruous television set in one corner... Tambour doors were pulled across the ugly eye of the screen. **1781** in H. M. BROOKS *Olden Time Series* (1886) IV. 52 Isaac Greenwood..makes Flutes,..Tea-Boards, Bottle-Stands, Tamboy [*sic*] Frames. **1782** J. BYNG *Jrnl.* 24 Aug. in *Torrington Diaries* (1934) I. 74 The long gallery is furnish'd with modern frippery, as tambour frames, &c. **1952** J. GLOAG *Short. Dict. Furnit.* 469 A tambour front is shown on the Harlequin Table illustrated on page 283. **1979** *Country Life* 30 Aug. 589/1 The compartment below, tambour-fronted, contains two drawers. **1880** L. HIGGIN *Handbk. Embroidery* v. 52 Irish or Limerick lace..is made on net in the old tambour frames, and with a tambour or crochet hook. **1962** *V. & A. Mus. Internat. Art Treasures Exhib.* 18/2 A George III satinwood secretaire bookcase banded with tulip and kingwood..in the centre a writing desk enclosed by a tambour shutter. **1973** *Country Life* 11 Jan. 91/3 Early-19th-century satinwood bedside cupboard..[with] tambour shutters. **1975** *Ibid.* 9 Oct. (Suppl.) 42/1 Sheraton Period Mahogany Work Table... The top section..is fitted with..two Tambour slides. *c* **1840** LADY WILTON *Art of Needlework* xx. 317 There are tambour-stitch, satin—chain—and queen-stitches. **1953** M. POWYS *Lace & Lace-Making* xi. 179 The outline may be made with tambour or chain stitch. **1797** *Prices Cabinet Work* 57 A Library Writing Table with Tambour Top. **1944** G. HEYER *Friday's Child* vi. 68 They laid the foundations of their future home by purchasing..a tambour-top writing-table, a crystal lustre, and a shaving-stand. **1794** *Cabinet Maker & Upholsterer's Guide* (ed. 3) pl. 69 (*caption*) Tambour Writing Table and Bookcase.

tamboura. Add: Also **tamboora, tambur(a), tanpoora,** etc. Substitute for def.: **a.** A long-necked lute of the Near East and Balkans, with a pear-shaped body and a fretted neck. (Later examples.) Cf. PANDORA^2, *TAMBURITZA.

1909 M. E. DURHAM *High Albania* vi. 141 A man.. came in with a tamboora and played.., his fingers plucking strange trills..from the slim, tinkling instrument. **1937** P. THORNTON *Dead Puppets Dance* I. 19, I contented myself with drumming on the floor..and clapping the rhythms of the tambora players. **1975** L. PICKEN *Folk Musical Instruments of Turkey* III. 220 This orientation..corresponds to one of the orientations adopted in preparing blocks of wood for Bulgarian *tambura.* **1983** *Listener* 20 Oct. 29/1 Saturday night's programme of Turkish music..was a continuous performance by the burbling, throaty flute called *ney* and the harsh, plucked instrument called the *tanbur.*

b. A long-necked fretless instrument of the lute family with a round body and usu. four wire strings plucked by the fingers, and used to provide a drone accompaniment in Indian music.

1864 [in Dict.]. **1875** ENGEL *Musical Instruments* v. 47 The Hindus..have also the divinity Ganesa, the god of Wisdom, who is represented as a man with the head of an elephant, holding a *tamboura* in his hands. **1891** C. R. DAY *Music & Mus. Instruments S. India* vii. 103 The vina, the tamburi or tamburu-vina, and the kinneri still remain just as they are described in the ancient books. **1921** H. A. POPLEY *Mus. India* vii. 110 The *Tambūr* is perhaps the most common stringed instrument in India. **1966** *Punch* 23 Nov. 770/2 The uncanny buzzing drone of a tambura, a four-stringed Indian instrument that looks like a bloated outsize wooden barometer. **1970** *Daily Tel.* 21 Sept. 9/1 She was singing..meanwhile plucking steadily at the drone-strings of her tanpoora. **1971** *Illustr. Weekly India* 25 Apr. 35/1 With Lata Mangeshkar sitting holding the *tanpura*, these inspirations would be shaped and sung and recorded. **1977** *N.Y. Times Mag.* 4 Dec. 41/3 Ram Dass had disembarked barefoot, wearing a long white robe and carrying a tamboura for chanting. **1980** A. DESAI *Clear Light of Day* iv. 178 The *tanpura* player.. strummed the *tanpura* strings as if in a mesmerised state. **1981** LD. HAREWOOD *Tongs & Bones* xvii. 264 He was surrounded by two other vocalists (one also playing the tambura).

tambourer. (Earlier example.)

c **1810** MALTHUS in *Trav. Diaries* (1966) 225 Much deb.[auchery] prevails among the weavers, tambourers and master manufacturers.

tambourinate (tæmbŭrī·nē^{i}t), *v.* [f. TAMBOURINE *sb.* + -ATE^3.] *trans.* To beat (a rhythm) as on a tambourine. (Only in the works of C. Mackenzie.)

1913 C. MACKENZIE *Sinister Street* I. I. iv. 47 He tambourinated upon the window-pane a gay little tune. **1928** —— *Extremes Meet* 63 Waterlow tambourinated with his knuckles on the shop-door the rhythm of the soldier's chorus from *Faust.*

tambu, var. *TAMBOO^1. **Tambuki,** var. *TAMBOOKIE *sb.* and *a.*; TAMBOUKI *a.* in Dict. and Suppl. **tambur(a,** var. TAMBOURA in Dict. and Suppl.

‖ **tamburitza** (tæmbu·ritsă). Also **tamboritsa, tamburica,** etc. Pl. **-n, -s.** [Serbo-Croat.] A stringed musical instrument of the Balkans resembling a guitar or mandoline. Cf. TAMBOURA a.

1941 N. BESSARABOFF *Anc. Europ. Mus. Instruments* IV. 219 The *Tambourica*, as used by the southern Slavs, belongs to a family of instruments similar to the Arabian-Persian *tanbur.* **1961** A. BAINES *Mus. Instruments* ix. 212 The pandoura, or 'long-necked' lute, remains a popular instrument from Persia..to the Balkans (*tamboritsa*). **1969** *Daily Tel.* 5 Nov. 13/6 Dancers from the Northern region of Barania..danced with joyful abandon to the music of tamboritzan—plucked instruments resembling guitars and mandolines. **1970** *Daily Colonist* (Victoria, B.C.) 14 Nov. 22/4 She first played on tamburitzas, Balkan instruments made by her father. **1972** DEAN & SMITH *Wisconsin* 32/2 Then the band starts banging on tamburitzas, and the patrons explode into an intricate folk dance. **1972** *Times* 21 Oct. 7/4 A choir sang Slavonian songs accompanied on the national instruments, *prim* and *tamburica*, like small mandolines. **1979** *United States 1980/81* (Penguin Travel Guides) 263 This is a Serbian restaurant where you can dine to the tune of tinkling tamburitzas.

tame, *a.* **2.** For '*Obs.* in ordinary use since *c* 1650' read 'Now restricted to U.S. use' and add further examples; *tame hay* (examples).

1838 H. W. ELLSWORTH *Valley Upper Wabash* iv. 39 It is very desirable..to get the tame grasses..set as soon as possible. **1857** *Trans. Illinois Agric. Soc.* II. 382 Where tame pasture is resorted to something more needs to be done. **1881** *Rep. Indian Affairs* (U.S.) 107 Some few have raised tame grapes. **1936** *Sun* (Baltimore) 15 Sept. 26/8 Tame hay production..is estimated at 355,000 tons. **1962** A. FRY *Ranch on Cariboo* x. 106 In conjunction with his store, he..put up a big field of good tame hay. **1976** *Billings* (Montana) *Gaz.* 20 June 8-D/8 (Advt.), 320 acres, 148 tillable now in tame grass.

‖ **tamein** (tamə·i·n). *Burma.* Also 9 **tamehn, te-mine; tamain.** [Burmese.] A draped garment resembling a sari, worn usu. by women.

1839 H. MALCOLM *Trav. S.-E. Asia* I. II. iii. 214 Women universally wear a te-mine, or petticoat. **1858** C. T. WINTER *Six Months Brit. Burmah* vii. 56 The *te-mine* is a very scant garment. *Ibid.* viii. 73 The woman's tamehn is a simple piece of cotton or silk. **1863** *Leisure Hour* Oct. 667/1 With their shaven heads..and their scant tameins (petticoats). **1888** *Bow-Bells Weekly* 11 May 293/2 Burmese women..wear of evenings or when visiting religious places, gay-coloured silk 'tameins'. **1908** LADY R. CHURCHILL *Reminisc.* (1973) xiii. 272 As we drove by I saw half a dozen priests in their yellow 'tamains', or robes. **1926** *Chambers's Jrnl.* Feb. 66/1 An old *tamein* of blue cotton check..covered her from armpit to calf. **1950** J. H. WILLIAMS *Elephant Bill* x. 164 She wore her blue *tamain* girdled above her breasts, leaving her beautiful pale shoulders bare. **1984** J. COLENBRANDER *Portrait of Fryn* ix. 133 Mademoiselle Denigré, the blind French silk-weaver of the Royal *tameins.*

tameletjie (tamələ·kʸi, -tʃi). *S. Afr.* Also **tammeletje, tommelaitje,** etc. [perh. f. Afrikaans *tabletje* small cake of chocolate, sugar, etc.] A hard toffee, sometimes containing pine nuts.

1838 T. SHONE *Diary* 6 Aug. in *Voorloper* (1976) 788 In the evening I made the Children some tommelatche as a treat on my birthday. **1862** —— 18 June in *Ibid.*, Made some Tommy Larche for the children. **1862** —— 17 July in *Ibid.*, Mrs. K. gave me a bason of sugar, to make Tomelah. **1904** H. DUCKITT *Hilda's Where is It? of Recipes* 237 *Tamelijtjes* (a favourite Cape sweet). **1926** P. SMITH *Beadle* (1929) 181 Jantje brought with him, secreted about his person, a horrible sticky mess of almond tommelaitjes. **1947** L. G. GREEN *Tavern of Seas* 65 For the children there would always be tameletjies, the sweets made of sugar, water, eggs, naartjie peel and dennebol pits—sweets that were typical of an earlier Cape Town, but which are not made so often now. **1953** *Cape Argus Mag.* 28 Feb. 3/7 Under a large oak in the main avenue sat a friendly, fat Coloured woman selling sweets —the 'Tammeletje Woman' we called her. **1974** *Cape Times Weekend Mag.* 12 Jan. 7 We searched among the pine needles for pips and collected cones. The highlight of our excursions was the tameletjies, a delicious sweet made with butter, brown sugar, syrup, vinegar and water, and the addition of the pips, which my elder sister made for us.

tamerack, obs. var. TAMARACK in Dict. and Suppl.

Tamil, Tamul. Add: The form *Tamil* is now standard. **a, b.** (Later examples of *attrib.* and adjectival use.)

1869 *Chambers's Encycl.* IX. 285/1 The earliest history of the Tamil' country is still involved in obscurity. **1911** *Encycl. Brit.* XVII. 478/2 There has been a great development in agricultural enterprise,..the estates being mainly in the hands of Europeans, and the labour mostly Tamil. **1971** *Ceylon Daily Mirror* 4 Oct. 2/2 There should be healthy rivalry and peaceful co-existence of political parties in Tamil areas.

Tamilian *a.* and *sb.* (later examples of the adj. and earlier and later examples of the sb.); **Tamulic** *a.* (earlier example).

1788 *Asiatick Researches* I. 146 The *Tamulians* (or *Malabars*) having no *h* in their alphabet. *Ibid.*, In their language, which is the *Tamulic*..the place is called *Mâvalipuram.* **1854** [see *MUNDA *sb.* and *a.*]. **1856** [see *MALAYALI]. *a* **1881** [see *BANDYMAN]. **1959** V. CRONIN *Pearl to India* vi. 89 Among the Tamilians *Vêdam* means both the three collections of hymns known as the Vedas, and also religion in general. **1968** P. LAL *Indian Recipes* 57 Pongal..is the Tamilian New Year. **1971** *Illustr. Weekly India* 25 Apr. 4/2 The Tamilian is well known for his remarkable adaptability to his surroundings.

‖ **tamizdat** (tæ·mizdæt). [Russ., f. *tam* there + *izdat*, abbrev. of *izdatel'stvo* publishing house, after *SAMIZDAT.] Russian writings which are published abroad and smuggled back into the U.S.S.R.; also this system of publication.

1974 MOORE & PARRY *Twentieth-Cent. Russ. Lit.* viii. 157 *Tamizdat*, a Russian word of later coinage and less spread than *samizdat*..refers to printed (not typed) material smuggled into the USSR from outside. *Ibid.* viii. 161 Plain mail is frequently used by people sending *samizdat* items out of the Soviet Union and *tamizdat* literature being sent into that country. **1975** *Economist* 11 Oct. 60/1 The volume of *samizdat* (unofficial and uncensored literature) has diminished, but this has been compensated for by the brisk circulation of publications known as *tamizdat* (from *tam*, 'there') printed abroad. **1978** *Observer* 1 Jan. 4/7 The era of *samizdat* is ending, he [*sc.* Georgy Vladimov] says. In its place has arrived the era of what he calls '*tamizdat*'—publication abroad of Russian writings that are then smuggled back into Russia. **1982** *Times Lit. Suppl.* 3 Sept. 950/1 It is thus a combination of *samizdat* and *tamizdat* (i.e., both unofficial Soviet and émigré publications).

tamka, var. TANGA^1 in Dict. and Suppl.

Tamla Motown (tæ·mlă mō^u·taun). The name of two U.S. record labels, *Tamla* and *Motown*, launched in 1960 by Berry Gordy Jr., used *attrib.* and *absol.* to designate a style of music characterized by a heavy beat and influenced by gospel music, which was made popular by the Black artists he employed. Also *ellipt.* as **Tamla.** Cf. *MOTOWN.

[**1964** *Melody Maker* 6 June 13/6 So far, the Tamla-Motown operation has subsisted almost entirely on single record hits.] **1968** P. Oliver *Screening Blues* ii. 46 Rhythm and blues, rock 'n roll, the Tamla-Motown sound and the techniques of the gospel singer. **1970** *Melody Maker* 3 Oct. 25/1 The new record is directly in between reggae and Tamla Motown. *Ibid.* 25/2 I've always wanted to try to get a Jimmy Cliff sound, so I am bound to aim in a Tamla direction. **1977** *Time Out* 28 Jan.–3 Feb. 62/1 (Advt.), All types of sounds wanted—progressive/rock/oldies/middle of the road/jazz/blues/tamla-soul/classical/easy listening/films/shows/budget.

Tammany. Add: **a.** Also applied *transf.* to any similarly corrupt political organization or situation.

1901 'Mark Twain' *Speeches* (1910) 114 Great Britain had a Tammany and a Croker a good while ago. **1910** *Encycl. Brit.* II. 142/2 The spectacle of a Clerico-anti-Semitic tammany in Vienna had strengthened the resistance of the better elements in the country. **1980** J. Barnett *Palmprint* xiii. 137 You're been out in the Caribbean before... Politics here are pure Tammany Hall circa 1900.

b. Tammany tiger, the symbol of the New York Tammany Society.

1871 [in Dict.]. **1953** *Manch. Guardian Weekly* 12 Nov. 2 Wagner, son of the late New Dealer swallowed up his opponents..leaving a broad smile on the face of the Tammany tiger.

tammar (tæ·māɹ). Also **tamma.** [Aboriginal name.] A greyish-brown scrub wallaby, *Thylogale eugenii*, found in south-western parts of Australia. Cf. Paddymelon.

[**1892** A. Zietz in *Trans., Proc., & Rep. R. Soc. S. Austral. 1891–92* XV. 18 *Macropus eugenii.* Dama or Kangaroo Island Wallaby. **1924** F. W. Jones *Mammals S. Austral.* II. 240 The Dama Wallaby group was widely spread over the southern portion of Australia.] **1926** K. S. Prichard *Working Bullocks* iii. 27 The great days they had spent together, as youngsters, hunting.. tammas, in the Paper Bark swamps. **1941** E. Troughton *Furred Animals Austral.* 193 Dama Pademelon; Tammar. **1970** W. D. L. Ride *Guide Native Mammals Austral.* v. 48 Tammars are able to survive on food containing little water. **1979** *Nature* 5 Apr. 549/2, I therefore examined the role of the neural pathway from the mammary gland in tammar wallabies carrying diapausing embryos.

tammeletje, var. *TAMELETJIE.

Tammuz: see Thammuz, Tammuz. **tamongoong,** var. *TEMENGGONG. **tamoure,** var. *TAMURE.

tamoxifen (tămǫ·ksifen). *Pharm.* [f. *t* (perh. f. *T(RANS) + Am(INE + Oxy- + Phen(ol, with alteration of *y* and *ph*.] An œstrogen antagonist, $(CH_3)_2N(CH_2)_2{\cdot}O{\cdot}C_6H_4{\cdot}(C_6H_5)C{:}C(C_6H_5){\cdot}CH_2CH_3$, used to treat breast cancer and infertility in women.

1972 *Approved Names 1970* (Brit. Pharmacopœia Comm.) Suppl. IV. 6 Tamoxifen. **1972** *Clin. Endocrinol.* I. 275 A derivative of triphenylethylene was examined by Harper & Walpole (1967a,b) for its anti-oestrogenic activity in rats. This compound..was subsequently named tamoxifen. **1980** *Brit. Med. Jrnl.* 29 Nov. 1459/1 Lung metastases then enlarged and tamoxifen was substituted but discontinued after only two months because of progressive disease.

tamp, *v.* Add: **3. b.** To pack or consolidate tobacco in (a pipe or cigarette) by a series of light taps. Also with tobacco as obj. and const. *down*. orig. *U.S.*

1920 in Webster. **1939** R. P. Warren *Night Rider* ii. 42 The Captain took out his pipe, tamped it, and with an excess of care lighted it. **1940** *Sun* (Baltimore) 14 Aug. 8/6 The pipe stoppers used to tamp tobacco in the pipe bowl will be on view. **1941** 'A. MacDonald' in *Astounding Sci. Fiction* Oct. 18/2 The man..took out another cigarette, tamped it on one end, turned it and tamped the other. **1959** J. Cary *Captive & Free* xxx. 134 Syson settled himself more comfortably and tamped his pipe with the end of a pencil. **1979** *PN Rev.* No. 9. 35/1 A pipe-smoker Tamps tobacco Down to the base of the pipe bowl. **1981** *Guardian* 12 Oct. 10/4 A local soul, resting from his labours, tamping the dottle in his pipe.

3*. *transf.* and *fig.* To oppress or constrict as by ramming; to subdue or contain by force. Also const. *down*. *U.S.*

1959 N. Mailer *Advts. for Myself* 19 We've all been flattened by the dead air of this time, dinched and tamped into a flat-footed class. *a* **1963** S. Plath *Ariel* (1965) 74 Perfection... Cold as snow breath, it tamps the womb. **1976** *Time* 27 Sept. 27/2 While inflation has been tamped to just over 6%, unemployment is still high. **1977** *Time* 18 Apr. 53/2 Carter may be gambling that..he can tamp down the debate over the safety of nuclear power.

tampan. Substitute for def.: A blood-sucking tick of the genus *Ornithodorus*, esp. *O. moubata*, the vector of African relapsing fever. (Earlier and later examples.)

1861 D. Livingstone *Pop. Acct. Trav. S. Afr.* viii. 120, I dreaded the '*tampans*', so common in all old huts. **1937** *Handbk. for Farmers S. Afr.* 522 The fowl tick, often called 'tampan', is an oval-shaped, slate-coloured tick with light yellow legs. **1971** D. J. Potgieter et al. *Animal Life S. Afr.* 222/1 To control the tampan, the floors of native huts should be well dusted with BHC powder.

Tampax (tæ·mpæks). Also **tampax.** The proprietary name of a sanitary tampon for women; also applied *loosely* to any variety of tampon, and in fig. contexts. Cf. *TAMPON *sb.* 1.

1932 *Official Gaz.* (U.S. Patent Office) 29 Mar. 1063/2 Tampax. For sanitary absorbent tampons. **1935** *Trade Marks Jrnl.* 13 Feb. 187/1 Tampax... Sanitary absorbent tampons. **1955** W. Gaddis *Recognitions* III. v. 884 When we launched the customs almost arrested me, they thought my Tampax was incendiary bombs. **1975** J. McClure *Snake* iii. 35 '*Ach*, come on, Klip—what's got your Tampax in a twist?'..'It's nothing,' he muttered... 'I'm just pissed off.' **1977** M. Drabble *Ice Age* II. 159 She went to the lavatory: there, for the first time for weeks, she found a Tampax machine. **1979** [see *TAMPON 1].

tamper, *sb.* Add: **1.** Also, an instrument or machine used for tamping. (Examples.)

1954 *Highway Engin. Terms* (*B.S.I.*) 49 *Tamper*, a wooden or metal template, beam or frame used for compacting road materials and for shaping the surface. It is manipulated by hand and may or may not carry mechanical means for tamping or vibrating. **1956** *Railway Mag.* May 344/2 It is a four-wheel caravan, painted yellow, providing living accommodation for the two men who operate the Matisa ballast tamper. **1967** *Boston Sunday Globe* 23 Apr. B 63/1 You will also need..a tamper that you can make by nailing handles to a short section of a log. **1976** *Jrnl.* (Newcastle) 26 Nov. (Advt.), Welding equipment, bench drill, bench grinder, wacker tampers, traffic light set.

2. A casing around an atomic bomb which increases its efficiency and decreases the critical mass required for an explosion.

1945 H. D. Smyth *Gen. Acct. Devel. Atomic Energy Mil. Purposes* xii. 126 While the effect of a tamper is to increase the efficiency both by reflecting neutrons and by delaying the expansion of the bomb, the effect on the efficiency is not as great as on the critical mass. **1961** *New Scientist* 26 Oct. 231/1 A tamper, a heavy casing whose inertia resists dispersion in the early stages. **1977** N. Freeling *Gadget* II. 75 If the criticality isn't right then your whole exercise in cores and tampers..stays the way it is.

ta·mperproof, *a.* Also **tamper-proof.** [f. Tamper *v.*[1] + Proof *a.*] Proof against being tampered with; not susceptible to misuse. Esp. of mechanism.

1886 *Time* July (Advt., rear cover), An indicator which records the hours your day or night watchman remains on duty, and is absolutely tamperproof. **1954** *Federal Suppl.* (U.S.) CXVIII. 182/2 Armstrong in part claims a tamper-proof feature on Cel-O-Seal bands since they must be destroyed before being taken off the bottle. **1960** *Times* 3 Oct. (Advt. Suppl.) 1/2 A tamper-proof seal. **1967** D. C. Cooke *c/o American Embassy* (1968) xiii. 129 Timber locks are virtually tamperproof. **1970** *New Yorker* 3 Oct. 41/1 Not altogether tamperproof waiting lists. **1979** C. McCarry *Better Angels* IV. xv. 310 The computer had been designed to be absolutely tamper-proof.

tampon, *sb.* Add: **1.** (Earlier and later examples.) Esp. one inserted into the vagina; now *spec.* one made commercially and bought to provide sanitary protection during menstruation.

1848 C. D. Meigs *Females & their Diseases* xxiii. 432 Having confidence in the power of the tampon to suppress *such* a flooding [*sc.* menorrhagia], I would let her go very far towards a dangerous state rather than subject her to the mortification of the surgical intervention. **1921** B. M. Anspach *Gynecology* xxxix. 709 Tampons are made by placing over a strip of absorbent cotton a smaller strip of lamb's wool, and binding them together. **1932, 1935** [see *TAMPAX]. **1957** T. N. A. Jeffcoate *Princ. Gynaecol.* xliii. 635 Proprietary tampons of various kinds are available but this method of applying antiseptics [to the vagina and cervix] is now rarely used. **1964** *Which?* Mar. 84/1 Medical opinion has always been divided over the use of tampons instead of external towels. **1970** G. Greer *Female Eunuch* 50 The success of the tampon is partly due to the fact that it is hidden. **1979** *Guardian* 27 Mar. 9/4 The tampon market is worth about £14 millions a year. Tampax has 62 per cent.

tamponade. Add: **2.** *Path.* Interference with the action of the heart by an excessive accumulation of blood or other fluid in the pericardial sac.

[**1930** Beck & Cox in *Arch. Surg.* XXI. 1039 The tamponade effect produced by atmospheric pressure may be excited in various types of operation... In the selection of cases for operation, the heart should possess a certain reserve power capable of withstanding this tamponade effect.] **1932** *Southern Med. Jrnl.* (U.S.) XXV. 785/1 Blood is trapped in the pericardial sac, producing heart tamponade. **1962** *Lancet* 8 Dec. 1195/2 In view of the poor response to external massage and the probability of cardiac tamponade, left anterior thoracotomy was undertaken. **1974** Thornton & Levy *Techniques Anaesthesia* vi. 170 Tamponade may develop following open heart surgery.

tam-tam (tæ·mtæm). *Mus.* [Echoic, app. of Creole origin: cf. Fr., Ger. *tam-tam*.] A metal gong of oriental origin, *spec.* a Chinese gong, now used in western orchestras.

1839, etc. [in Dict., s.v. Tom-tom *sb.* 1 b]. **1900,** etc. [see *GONG[1]]. **1933** M. D. Calvocoressi tr. *H. Scherchen's Handbk. Conducting* iii. 132 The tone of the gong does not differ from that of the tamtam but is definite in pitch. Sometimes composers prescribe a gong when they obviously mean a tamtan. **1947** *Penguin Music Mag.* May 85 There is a passage for tubular bells, cymbal, tam-tam, triangle. **1951** E. Paul *Springtime in Paris* vii. 131 The gyrations of a near-eastern dance, to the beat of inaudible tam-tams. **1961** *Radio Times* 20 July 18/3 Six players, their instruments including bongos and maracas..a whip, temple block, four tam-tams, and crotales. **1978** P. Griffiths *Conc. Hist. Mod. Music* xi. 164 A large tam tam (a kind of gong) is activated by two performers with various objects. **1983** *Listener* 15 Sept. 31/4 His orchestra (enlarged by two harps, piano, organ, celesta and a big percussion battery including tam-tam and bells) is handled with extreme refinement and virtuosity.

‖ **tamure** (tamū·re). Also **tamoure.** [Tahitian.] A Tahitian dance, the *ori Tahiti*.

1964 *Wanganui Photo News* 4 July 5 (*caption*) Nurse Anna Paotonu danced the tamoure for the floor. **1970** J. H. Vance *Deadly Isles* iii. 25 Ah! the *tamure*! which was to the hula as whisky to milk. **1976** *Sat. Rev.* (U.S.) 30 Oct. 30/2 Palm trees and girls dancing the sexy *tamure*.

Tamworth (tæ·mwɔ̄ɹþ). The name of a town in Staffordshire, used *absol.* or *attrib.* to designate a pig of the breed of this name, usually red or brown in colour, lean and large in build, and used to produce bacon; also, the breed itself, first developed in the area.

1860 S. Sidney *Youatt's Pig* (ed. 2) iii. 34 The cross of the Berkshire with the Tamworth produces the most profitable bacon pigs in the kingdom. *Ibid.*, The Tamworth Breed is a red, or red-and-black pig,—hardy, prolific,..but slow in maturing. **1886** J. Long *Bk. Pig* ix. 161 A cross between the Tamworth and the Berkshire was considered most valuable. **1897** S. Spencer *Pigs* i. 17 The mahogany or grizzled pig..has acquired the name of the Tamworth. **1950** Carroll & Krider *Swine Production* vi. 83 The Tamworth is a strictly bacon breed and the oldest of domesticated breeds of hogs. *Ibid.* 84 Tamworths are large, rugged hogs. **1967** M. Kenyon *Whole Hog* viii. 93 Most of your Tamworths you call Duroc. They're the red ones—reddish-brown. **1977** *Jrnl. R. Soc. Arts* CXXV. 702/2 Tamworth pigs.., although still relatively scarce, have increased in numbers dramatically. *Ibid.* 704/2 The Tamworth..can adapt itself to a wide range of climatic conditions.

tan, *sb.*[1] (*a.*) Add: **II. 3. b.** (Earlier example.)

1749 J. Cleland *Mem. Woman Pleasure* II. 233 The tan of his travels, and a beard..had..given it [*sc.* his face] an air of becoming manliness.

c. *pl.* Short for *Black and Tans* (see *BLACK *a.* 14 c). Also *sing.*, a member of this organization. Usu. with capital initial.

1932 S. O'Faoláin *Midsummer Night Madness* 19 The Tans in their roaring Lancia patrol cars. *Ibid.* 227 'Shut up, you,' said the Tan angrily. **1951** C. Lynch-Robinson *Last of Irish R.M.s* x. 161, I am sure that even amongst this class of the Tans..there were a number of quite decent fellows. **1962** E. O'Brien *Lost Girl* xii. 132 God Almighty, it reminded me of the tans..the night they burst the door in. **1978** F. Burton *Politics of Legitimacy* iii. 69 Mrs Johnson sees the British troops of today..as latter-day Tans.

B. *adj.* Also, bronzed, sunburnt. Also *euphem.* (*rare*), black, negro.

1950 *Sun* (Baltimore) 13 Sept. 14/2 He has written an article in the 'tan' national magazine *Ebony*. **1963** *Pix* 28 Sept. 63 Her hair should be blonde and bunched and her figure trim and tan. **1974** S. Sheldon *Other Side of Midnight* xiv. 275 He looked tanner and leaner than when she had last seen him.

C. *attrib.* and *Comb.* **1. tan-pot** (see quot. 1978); **Tan war** *Ir. Hist.*, the conflict between the Black and Tans and the Irish Republican Army in 1921; **tan-yard** (later examples).

1946 K. Tennant *Lost Haven* (1947) vii. 96 They could mend nets, boil tan-pots, play football. **1978** *Regional Lang. Stud.-Newfoundland* VIII. 18 A tan-pot or barking kettle is a large cauldron used for 'barking' a fisherman's nets with tan to prevent them from rotting in the salt water. **1968** 'N. Blake' *Private Wound* iii. 41 Flurry and I'd had enough of it after the Tan war. **1981** J. Wright *Devil's Parole* ii. 27 His tales of the Tan War ..his often cruel sketches of the Irish. **1878** Tan-yard [see *hound-pup* s.v. *HOUND *sb.*[1] 7 a]. **1911** J. Masefield *Everlasting Mercy* 18 The tan-yards stank of bitter bark.

‖ **tan** (tan), *sb.*[3] [Jap.] A Japanese unit of arable land or forest, equal to 300 *bu*; in modern use equivalent to approximately 0·245 acres (9·92 ares).

1871 A. B. Mitford *Tales of Old Japan* II. 2 Rice land is divided into three classes; and,..it is computed that one *tan* (1,800 square feet)..should yield to the owner.. five bags of rice per annum. **1914** F. Brinkley *Hist. Jap. People* xxxvi. 527 In Hideyoshi's system,..the rule of 360 *tsubo* to the *tan* (a quarter of an acre) was changed to 300 *tsubo*. **1931** G. B. Sansom *Japan* I. v. 98 The area was two *tan* (1 *tan* is 1,000 square yards). **1964** *Japan* (Unesco) (rev. ed.) i. 19/2 The Imperial Proclamation of the Taika Reformation was announced at the end of the year 646... Taxes shall comprise two large bundles and two small bundles of rice crop on one *tan* (about 0.245 acre). **1970** J. W. Hall *Japan* vi. 54 Strips of one *tan* each (at that time approximately .3 acres).

‖ **tan** (tan), *sb.*[4] [Jap.] A Japanese unit used for measuring cloth, equivalent to about ten yards in length and just over one foot in width; also, a piece of cloth of this size.

1876 W. E. Griffis *Mikado's Empire* (1877) II. 609 A *tan*, or piece of cloth, varies in length from 25 to 30 or more feet. **1909** *Westm. Gaz.* 23 Oct. 13/1, I sentence each one to bring within three days one tan (about twenty-five yards) of cotton cloth. **1931** G. B. Sansom *Japan* III. x. 187 Princes of the blood and ministers of the first rank were restricted to 500 *tan* of cloth. *Ibid.*, One *tan* is about ten yards.

‖ **tan** (dan), *sb.*[5] Also **dan.** [Chinese.] A female character in a Chinese drama or opera; an actor of such a role.

1886 *Jrnl. R. Asiatic Soc.* (N. China Branch) XX. 208 *Tan* is a female character, and varies also according to age and..circumstances. **1917** S. Couling *Encycl. Sinica* 149/1 There are five classes of characters in a play..*shêng* ..*tan*..*ching*..*mo*..and..*ch'ou*. **1937** Arlington & Acton *Famous Chinese Plays* p. xxiv, *Tan*, subdivided into *laotan*, elderly dames with orange bandeau but no make-up; *ch'ing-i*, virtuous maidens and dutiful matrons; and *hua-tan*, vivacious and temperamental—often a courtesan or a piquante little maidservant. **1973** R. F. S. Yang in Yuan-li Wu *China* 739 The 'four great *tan* actors', namely, Mei Lan-fang, Ch'eng Yen-ch'ui, Shang Hsiao-yün, and Hsün Hui-sheng (all were female impersonators). **1975** C. P. Mackerras *Chinese Theatre in Mod. Times* xii. 199 The list of actors who took part in the festival of 1959 is an impressive one. Among exponents of the Peking Opera were three of the 'four great *tan*'. **1978** *Nagel's Encycl.-Guide: China* 201 There exist series of *dan*, or feminine roles, *sheng*, or chief masculine roles, *jing*, or heroic characters with painted faces, and *zhou* or fools.

‖ **tan** (dan), *sb.*[6] [Chinese.] A Chinese unit of weight equivalent to approximately 110 lb. or 50 kg. (formerly approximately 133 lb., 60 kg.)

1911 *Encycl. Brit.* XXVIII. 494/1 *Tan*, China = 25 gallons. Also 133⅓ lb. weight. **1965** T. R. Tregear *Geogr. of China* iii. 108 In terms of food this meant a loss of 250 million *tan* (1 *tan* equals 133 lb.) since one *mow* produces an average of 2½ *tan*. **1973** *Genius of China* 104/2 It is calculated that the price recorded..would in the reign of Wu Ti be approximately the equivalent of over 20 *tan* of grain.

tan, *v.* Add: **2. a.** *fig.* (Later example.)

1979 *Internat. Jrnl. Sociol. of Law* VII. 230 Ehrlich was not interested in the social life but in the 'legal life' of the Bukowina and..his picture of 'legal life' is heavily tanned by the traditional pattern of the Bukowinian society.

3. (Earlier example with person as obj.) Occas. *intr.*

1862 Mrs. H. Wood *Channings* II. ix. 137 'I'll tan you too, Mr. Bywater.'..'Tan away,' coolly responded Bywater. 'I can tan again.'

4. *Photogr.* To harden (gelatin) chemically in proportion to the amount of exposure.

1899 C. F. Townsend *Chem. for Photographers* (ed. 2) iv. 75 Formaldehyde is not the only aldehyde capable of tanning gelatine. **1957** R. W. G. Hunt *Reproduction of Colour* v. 43 Gelatin, in its usual state, is soluble in hot water. But by suitable chemical treatment it can be hardened, or tanned, so that it becomes insoluble. **1979** G. Haist *Mod. Photogr. Processing* I. v. 225 Images produced in the presence of such oxidation products are also tanned.

tana[1]. Add: **b.** Now usu. in form *thana* (see quot. 1961).

1936 W. H. Saumarez Smith *Let.* 5 Dec. in *Young Man's Country* (1977) ii. 45, I am making a tour of all the thanas this fortnight. **1961** L. D. Stamp *Gloss. Geogr. Terms* 450/2 *Thāna*, a political division of a district which is under the jurisdiction of a single police-station so that a thana is really a police-station area. **1975** *Bangladesh Times* 19 July 3/2 Besides forty three members of Jessore district and thana units of the defunct organisation have applied for the membership of BKSAL. **1977** *Ibid.* 19 Jan. 12/3 He made a plea for the abolition of..divisions, districts, sub-divisions and thanas because they are.. ineffective for today's needs.

Tanagra (tæ·năgrā). The name of a city of Boeotia in ancient Greece, used *attrib.* and *absol.* to designate terra cotta statuettes of the 5th to 3rd centuries B.C. found in the neighbourhood. (See also quot. 1899.)

1878 C. Schreiber *Jrnl.* 29 Aug. (1911) II. 195 M. Grean was resettling some of his Greek (Tanagra) specimens. **1890** O. Wilde *Pict. Dorian Gray* (1891) 112 She had all the delicate grace of that Tanagra figurine that you have in your studio. **1899** C. A. Hutton *Greek Terracotta Statuettes* ii. 17 That aspect of individuality which is the great charm of the Bœotian statuettes from the Tanagra district, and which is so characteristic of them that any specially pretty figure, whatever its provenance, is popularly known as a 'Tanagra'. **1915** V. Horsley in S. Paget *Sir V. Horsley* (1919) 309 Some wonderful Greek sculptures small scale, and large Tanagras. **1927** *Glasgow Herald* 22 June 10 The style of the Tanagra heads of ancient Greece. **1960** R. G. Haggar *Conc. Encycl. Continental Pott. & Porc.* 469/1 Tanagra figures have been much admired..and much faked,—many of the fakes appearing..in the late 1870s and early 1880s not many years after the first discoveries of genuine Tanagra statuettes. **1978** 'M. M. Kaye' *Far Pavilions* II. xv. 240 Shushila-Bai was as small and exquisite as a Tanagra figurine.

tanaiste (tǭ·niʃtə). [a. Ir. *tánaiste*: see Tanist.] The deputy prime minister of the Republic of Ireland.

1938 *Éire: New Irish Constitution: Citizen's Manual* 22 The Taoiseach nominates a member of the Government as Tánaiste to act in his place for all purposes, should the Taoiseach die or become permanently incapacitated, or be temporarily absent. **1975** *Irish Times* 10 May 5/4 A tribute to the work of the former members..was paid by the Tanaiste and Minister for Health, Mr. Corish. **1982** M. Wallace *Brit. Govt. in N. Ireland* v. 94 The Labour *Tanaiste* or Deputy Prime Minister.

tanalized (tæ·nălɔiz'd), *a.* [f. *Tanal(ith*, proprietary name of a water-borne preservative for wood + -ize + -ed[1].] Of timber, treated with Tanalith preservative or a similar compound.

1964 *Weekly News* (Auckland, N.Z.) 29 July 42/4, I built in tanalised or creosoted exotic timber which is everlasting and grown for the purpose by our Forest Service. **1967** *Daily Tel.* 15 Apr. 19/7 (Advt.), Garden frames. Made from tanalised timber, never need painting or maintenance. **1980** *Amateur Gardening* 18 Oct. 10/1 To this end, one of the most useful stock items you can keep is a bundle or two of tanalised batten.

Tancook (tæ·nkuk). Now *Hist.* The name of an island at the mouth of Mahone Bay, Nova Scotia, used *attrib.* in *Tancook whaler*, a double-ended schooner, about 45 feet long of a type formerly built there. Also *Tancook schooner*.

1933 *Yachting* Feb. 102/2 It will be the last model of a real Tancook whaler made in Nova Scotia. **1947** *Sun* (Baltimore) 18 Sept. 23/7 The Scotia schooner built to slide over rocks is Nicholas G. Schlegel's Windstark. The type is called a 'Tancook schooner', from the locality in Nova Scotia where she was built. **1951** H. I. Chapelle *Amer. Small Sailing Craft* iii. 166 The Tancook whaler was one of the most handsome of the double-enders used on the Atlantic Coast. **1967** H. F. Pullen *Atlantic Schooners* 33 The Tancook whaler was somewhat similar to the pinky, but with much sharper lines.

tandava (tā·ndāvă). Also **tandav.** [a. Skr. *tāṇḍava*.] The dance of Shiva; a traditional style of dancing in India, of a vigorous and masculine character. Cf. *LASYA.

1924 Ld. Ronaldshay *India: Bird's-Eye View* xxi. 272 A dance of Shiva called Tandava, performed in cemeteries and burning-grounds. **1937,** etc. [see Lasya]. **1967** Singha & Massey *Indian Dances* 22 Tandav was first danced by the god Shiva, Lord of the Dance, who then conveyed this art to mortals through his disciple Tandu. Shiva is the symbol of procreation and it is because of this that tandav is often regarded as a male dance. **1969** *Weekly Mail* (Madras) 26 July 10/4 The same movements performed with the vigour of males is tandava while it becomes lasya when performed by women possessing feminine grace. **1978** G. Vidal *Kalki* vii. 162 Indeed, there is a legend that the end will come when Siva begins the Tandava Dance, or dance of eternity.

tandem, *sb.*[1] and *adv.* Add: **B.** *adv.* Also *in tandem*, arranged one behind the other; also *fig.*, together, in partnership.

1930 [see *CORLISS]. **1949** [see *SCALING *vbl. sb.*[2] 2]. **1957** I. Asimov *Naked Sun* (1958) ii. 30 There were two seats in tandem, each of which could hold three. **1964** S. Lieberson in J. A. Fishman *Readings Sociol. of Lang.* (1968) 553 These measures, used in tandem, provide an instrument for a quantitative approach to a basic sociolinguistic problem. **1974** *News & Press* (Darlington, S. Carolina) 24 Apr. 8/7 Management and labor are learning to work in tandem, as has been evidenced by the sharp drop in strikes.

C. 1. a. *attrib.* and *Comb.* as **tandem axle** (see quot. 1969); usu. *attrib.*; **tandem bicycle** (earlier example); **tandem canoe** (example); **tandem duplication** *Genetics*, an atypical form of duplication (*DUPLICATION 1 e) in which the repeated segments lie immediately adjacent to one another on the same chromosome; **tandem formation** *Tennis*, the position of partners in a doubles match in which server and netman occupy the same half of the court at service; **tandem garage,** a garage with space for two vehicles to be parked one behind the other rather than side by side; **tandem-paced** *a.*, in cycle-racing: involving the use of a tandem for pacing.

1956 in E. Molloy *Automobile Engineer's Ref. Bk.* xxx. § 7 The Eaton type tandem axle unit has been developed as a method of increasing the proportion of payload to gross vehicle weight. **1969** Baker & Stebbins *Dict. Highway Traffic* 230 *Tandem* (*axles*),..an assembly of two axles for the support of the rear of a truck trailer. **1977** 'D. Rutherford' *Return Load* ii. 29 I'd suggest you have one of these new tandem axle Taskvans... They have air-bellow suspension. **1896** G. B. Shaw *Let.* 5 July (1965) I. 634, I will get a tandem bicycle; and we shall ride along over the celestial plains. **1867** J. MacGregor *Thousand Miles in Rob Roy Canoe* (ed. 5) ii. 37 In the Canoe Club we have three 'tandem' canoes, each for two paddles. **1959** C. M. M. Begg *Introd. Genetics* x. 137 A tandem duplication can result from unequal crossing over. **1981** L. L. Mays *Genetics* xii. 529 Tandem duplication of a short region of one chromosome is thought to result from inaccurate crossing over processes. **1967** P. Metzler *Advanced Tennis* ix. 116 When first confronted with the tandem formation..the left-hander is likely to continue hitting his return in its usual direction. **1976** *Tennis Today* Oct. 12/1 The American juniors played the whole match using what they call the Tandem Formation, more normally referred to as the Australian Formation in this country. **1968** *Globe & Mail* (Toronto) 3 Feb. 41/2 (Advt.), Executive home. Seven room solid brick bungalow with tandem double garage. **1974** *Country Life* 7 Mar. (Suppl.) 32 *l*, 2 reception rooms, kitchen/breakfast room. Tandem garage. **1929** *Star* 21 Aug. 17/1 F. W. Southall is expected to make an attack on the one mile tandem-paced record. **1955** *Times* 11 July 5/7 D. Marsh (Shaftesbury C.C.) won the 12-hour tandem-paced cycle race at Herne Hill on Saturday.

b. With reference to systems of work, machines, etc., which involve the use of two or more elements operating together or in series.

1921 *Telegr. & Telephone Jrnl.* VII. 79/2 With a machine switching tandem exchange large groups..can be established between each local exchange and the tandem exchange. **1923** Harbord & Hall *Metallurgy of Steel* II. xiv. 290 Splitting up the mill into a breaking-down mill, which is placed in front of, and feeds, the smaller finishing mill (the arrangement being usually known as a tandem mill) is preferable where there is a sufficient output to justify employing two sets of men. **1924** W. Aitken *Automatic Telephone Systems* III. 79 Tandem working through a tandem office will be introduced into London as quickly as possible. **1933** K. B. Miller *Telephone Theory & Pract.* III. iv. 243 In large cities..tandem operation is especially useful. *Ibid.*, The cost and inconvenience of using the tandem routing must be balanced against the trunk saving which it produces. **1955** *Archit. Rev.* CXVII. 141/2 A motor assembly known as a Two-speed Tandem Motor combines a squirrel cage and slip-ring motor within a single frame. **1959** *Wall St. Jrnl.* 3 July (Eastern ed.) 5/5 The New York State Thruway Authority has approved the use of tandem-trailers over its 559-mile highway system. The tandem vehicles—which consist of a heavy duty truck tractor hitched to two regular size trailer units—have been in experimental use. **1967** *New Scientist* 16 Feb. 397/1 Although the majority of larger telephone exchanges in Britain have direct connections with all other exchanges, many have to route their outgoing calls through central switching exchanges known as tandem exchanges. **1970** *Gloss. Aeronaut. & Astronaut. Terms* (*B.S.I.*) VI. 2 *Tandem boost*, a boost rocket motor assembly which is located at the rear of the missile and co-axially with it. **1971** W. K. V. Gale *Iron & Steel Industry: Dict. Terms* 54 *Continuous mill* (*tandem mill*), any rolling mill in which the stands are arranged in tandem. **1976** *NBR Marketplace* (Wellington, N.Z.) III. 3/4 Tandem blades clog up with shaving foam and don't give as even a shave as the single blade. **1978** *Detroit Free Press* 14 Apr. 16D/2 Carrying 22 extra bolts and a new coupling device, a tandem tanker rumbled down a test track.

2. Passing into *adj.* **a.** In various uses with reference to the occurrence of one thing behind or after another.

1926 *Daily Tel.* 7 Aug. 7/4 Someone has given the appropriate name 'tandem' system to a form of curriculum in which students were limited to one main subject, and one subject only, at a time. **1952** *Mind* LXI. 356 Is minding what one is doing, then, after all, a tandem occurrence? **1968** E. A. Powdrill *Vocab. Land Planning* iii. 54 Two-tier, or tandem, development is merely the utilisation of backland to place one or more houses behind another. **1980** *Archaeology* Nov.–Dec. 37 We took a series of tandem camera exposures from the 300 meter level.

b. Co-operative, joint, dual; involving two persons, organizations, etc.

1962 *Economist* 24 Mar. 1094/1 The future is seen..in terms of a 'tandem' economic partnership between two continental common markets. **1963** *Times Lit. Suppl.* 18 Jan. 44/4 The tandem authors of this study merely add to the confusion. **1976** *National Observer* (U.S.) 21 Feb. 1/5 These tandem operations, Totera and Schuler explain, involve every aspect of IRS forms. **1977** *Time* 26 Sept. 37/3 Kepesh recalls a tandem tantrum he had with his wife.

tandemly (tæ·ndĕmli), *adv.* Chiefly *Genetics*. [f. Tandem *sb.*[1] and *adv.* + -ly[2].] End to end.

1968 *Jrnl. Molecular Biol.* XXXII. 328 Cairns' (1966) hypothesis that the long DNA fibers are composed of many tandemly joined replication sections is proved. **1974** [see *REPLICON]. **1978** *Nature* 5 Oct. 364/3 The histone genes of *Drosophila* represent a tandemly repeated gene family. **1980** *European Jrnl. Cell Biol.* XXII. 430 As Hensen's node regresses, paraxial somitomeres are added tandemly.

tandoori (tændūə·ri), *a.* (*sb.*) Also **tanduri.** [f. as next.] Of, pertaining to, or designating food cooked in a tandour, or this style of cooking. Also *absol.* as *sb.*

1958 R. Howe *Cooking from Commonwealth* 484 Since partition the most popular form of restaurant cooking in India's capital, Delhi, has been tanduri cooking. *Ibid.* 540 Tanduri chicken is always eaten with the fingers. **1961** Mrs. B. Singh *Indian Cookery* I. 18 A tandoor is used for the preparation of tandoori chicken, tandoori fish, seekh kabab, tandoori roti, nan, sheermal etc. **1966** *Daily Tel.* 24 Oct. 11/3 The fabulous *tandoori* cooking from the north-west of India: charcoal-heated clay-oven cooking which ensures food of distinction. **1969** *Guardian* 2 Oct. 17 The restaurant specialises in exclusive Indian cuisine: Tandoories—meat and chicken grilled in clay ovens over a charcoal fire. **1969** *Enact* (Delhi) Nov. 18/2 Bring out the *tanduri* fish. **1970** D. Singh *Indian Cookery* 30 *Tandoori* is food cooked on the spit in a clay oven. *Ibid.* 32 The various kebabs and kaftas used in tandoori, grills and roasts. **1975** *Indian Express* 1 Aug. 4/4 Multi-purpose gadget for tandoori cooking. **1977** D. James *Spy at Evening* viii. 53 Dinner..at..my favourite Tandoori house on Sidney Street. **1979** *Daily Tel.* 11 June 8/8 More rice is being eaten in Britain. Not in the old-fashioned rice pudding, but with curry, chop-suey and tandoori dishes.

tandour. Add: Also **tandoor, tandur. 2.** The form **tandoor** is usual in this sense. [Urdu or Panj. *tandūr* oven.] A clay oven used in northern India and Pakistan; a shop that sells food cooked in this. Also *Comb.*, as *tandoor-cooked* adj.

1840 [in Dict.]. **1925** M. L. Darling *Punjab Peasant* viii. 166 A significant change is that, when the cultivator goes to town, he is not as content as he was with the low-class *tandur*, but has begun to put up at the hotel. **1947** *Civil & Milit. Gaz.* 8 Apr. 11/1 Some 'tandur' shops were partially affected by the fire. **1957** C. Rand *Twain shall Meet* 77 They complain too that there is no room for *tandoors*, special jarlike ovens for baking *chapatis*—wheat pancakes—a Punjab staple. **1967** *Guardian* 8 Dec. 6/4 North Indian or Pakistani [cooking] depends on the clay oven called a tandoor, into which kebabs, chickens marinated in yoghourt and spices, and the special bread called nan are lowered and rapidly cooked. **1974** [see *NAN[2]]. **1977** *Sunday Times* (Colour Suppl.) 27 Nov. 35/3 The tandoor is the traditional oven used in Indian villages. It is fashioned out of clay and left to dry for two days. Then it is seasoned with yoghurt, spinach and oil to smooth it and prevent it from cracking. It is this, combined with the tremendous even heat of the burning coals, that gives tandoor-cooked meat its mild aromatic flavour.

tanewa, obs. var. *TANIWHA.

tang, *sb.*[1] Add: **I. 2. a.** Also in certain firearms.

1805 C. James *Milit. Dict.* (ed. 2), *Tang*, the upper part of the plug, or breech pin. **1869** V. D. Majendie *Milit. Breech-Loading Rifles* 62 The locking arrangements consists of the following parts:— Steel bolt.. Recess in breech tang for bolt. **1909** *Text Bk. Small Arms* I. iv. 35 The rear end of the body is in the form of a tang with sides. **1918** E. S. Farrow *Dict. Milit. Terms* 605 *Tang*, the projecting portion of the breech of a musket, by which the barrel is secured to the stock. **1929** *War Office Textbk. Small Arms* i. 12 On the underside of the cocking-piece is a projection..which travels in a groove, cut for it in the tang of the body. **1965** H. L. Blackmore *Guns & Rifles of World* 100 Butt tang engraved with Royal arms of France. **1976** *Shooting Mag.* Dec. 61/1 (Advt.), Model 801 Luxus O/U shotgun..10 mm wide ventilated barrel rib, sling swivels, top tang safety, double trigger.

3. Substitute for def.: = *surgeon-fish* s.v. Surgeon *sb.* 3 b in Dict. and Suppl. (Later examples.)

1925 D. S. Jordan *Fishes* (rev. ed.) xxxviii. 618 In the next family, *Acanthuridæ*, the surgeon-fishes or tangs, the scales remain small. **1965** Mrs. L. B. Johnson *White House Diary* 3 June (1970) 281 These were lots of little bright yellow fish, young blue tang, they called them—as they grow up they change color. **1980** R. E. Thresher *Reef Fish* xx. 147 The surgeonfish, or tangs, are high-bodied, laterally compressed fish.

4. Restrict *Stereotyping* to senses in Dict. **c.** *Typefounding.* The projection at the bottom of a piece of type which is formed by superfluous metal cooling in the opening of the mould.

1908 *Proc. Institution Mechanical Engineers* Dec. 1034 The gate through which the metal passes into the mould becomes also filled with type-metal and forms a projecting *tang* which must be broken from the type. **1921** W. H. Slater *What Compositor should Know* I. 23 Finishing means breaking off the 'tang' or 'jet' left at the bottom of each letter when this is not done on the machine. This tang occurs on all types cast by hand and all large sizes cast by machine. **1951** S. Jennett *Making of Bks.* ii. 32 When type is cast a fragment of metal, the tang, is left adhering at the base from the orifice in the mould through which the molten metal is injected. This tang is broken off and the resulting roughness of the fracture ground down.

Tang (tæŋ), *sb.*[5] Also **T'ang, Tanga.** [Chinese *táng*.] **a.** The name of a dynasty which ruled in China from A.D. 618 to *c* 906; a ruler belonging to this dynasty.

1669 J. Ogilby *Nieuhoff's Embassy from E.-India Co. to Emperor of China* I. xviii. 282 At last having miserably worried and weakened each other, they were all subdued by the seventh Race called *Tanga*, which seized upon the whole Empire, and reigned with his Posterity till the Year of Christ 618. **1738** J. B. Du Halde *Descr. Empire China & Chinese Tartary* I. 194 (*heading*) The Thirteenth Dynasty, call'd Tang, which had Twenty Emperors, in the Space of Two hundred eighty nine Years. **1788** tr. *Grosier's Gen. Descr. China* II. vi. iii. 209 Under the *Tang*, this superstition still continued. **1837** *Penny Cycl.* VII. 81/1 Ly-yuen..in A.D. 622 founded the dynasty of Tâng. **1925** B. Rackham in R. Fry et al. *Chinese Art* 15 When the classic period of T'ang is reached, the potters are working with the easy mastery of artists in their craft. **1940** E. Pound *Cantos* xiv. 41 Tang rising. And the first Tang was Kao Tseu, the starter. **1979** Mills & Mansfield *Genuine Article* iii. 62 Ming, Sung and T'ang have become names synonymous with the finest ceramics. T'ang was the dynastic name of the pottery of China in the eighth century AD.

b. *attrib.* or as *adj.* Freq. used to designate artefacts, etc., of this period.

1831 *Canton Miscellany* iv. 246 The Tang Dynasty was founded at the commencement of the Seventh century. **1854** *North-China Herald* 17 June 184/1 The first emperor of the T'ang dynasty. **1910** *Encycl. Brit.* VI. 213/2 It is in fact from the early religious schools of Japan that we can best conjecture the grandeur of the T'ang style. **1924** M. Borden *Three Pilgrims & Tinker* ii. 21 The van was full of the favourite belongings of each member of the family: their mother's Tang horses. **1935** *Burlington Mag.* Jan. p. xiii/2 The T'ang and Sung ceramics..should arouse much interest. **1943** D. Welch *Maiden Voyage* xviii. 153 Chou bronzes, T'ang grave figures and Sung porcelain. **1955** *Times* 6 Aug. 7/6 Eight Bodhisattvas, typical of early T'ang painting, and clearly owing their inspiration to Indian models. **1962** E. Snow *Other Side of River* (1963) xviii. 130, I..remembered the description of the scene by the T'ang poet Chen Chang. **1976** 'M. Delving' *China Expert* xiii. 179 How many T'ang horses ..really came from T'ang graves? **1980** *Times* 1 Mar. 13/6 Two hours' drive north-west of Xian is the tomb complex of the Tang emperor.

tanga[1]. Add: Also 9 **tank; tamka, tangka, tenga. b.** (Further examples.)

1876 C. Markham *Narr. Mission George Bogle to Tibet* xiii. 129 The following memorandum of weights used in Tibet is among Mr. Bogle's papers..5 *tanks* make one *nega*. **1889** G. N. Curzon *Russia in Central Asia* vi. 189 At the time of my visit the silver *tenga* was worth about fivepence. **1892** W. W. Rockhill *Jrnl.* 23 July (1894) iv. 253 The chief inquired if I had any Chinese silver or rupees to exchange for Lh'asa tankas. **1904** A. T. de Mattos tr. *Grenaud's Tibet* viii. 301 The commonest coin within the limits of the kingdom of Lhasa is the tangka. **1924** *Glasgow Herald* 30 June 12 Every time I rode through the city [of Bokhara] one of the Cossacks carried a purse with silver 'tengas' (a metal coin worth about sixpence), and distributed them to the..poor. **1970** R. D. Taring *Daughter of Tibet* iv. 44 The *tamka* was then worth about ninepence. **1972** G. Muller tr. *Schön's World Coin Catal. Twentieth Cent.* 826 Tibet..15 skarung = 1 tangka..3 tangka = 1 Indian rupee. **1974** D. Norbu *Red Star over Tibet* i. 34, 670 silver coins called *tamka*.

tanga[2] (tæ·ŋgā). [a. Pg., ad. Quimbundo *ntanga* loincloth.] **a.** (See quot. 1960): the garment is also worn by men. **b.** A bikini made of triangles of material joined by thin ties; *spec.* the lower half of this. Cf. *STRING *sb.* 6 c.

1912 T. A. Joyce *S. Amer. Archeol.* xii. 265 The so-called *tangas*.., triangular in shape, and convex in section,..are found in the burial-urns of women... It has been suggested that they are the 'translations' into pottery of the small triangular leaf coverings worn by many of the women of primitive Brazilian tribes in historical times. **1921** *Museum Jrnl.* (Univ. of Pennsylvania) Sept. 146 Nothing whatever was found on the inside of the burial urns except the so-called 'tangas' or fig leaves supposed to have been worn by the women... The tangas were always well made, hard burned, highly polished, and either in bright red monochrome or painted designs. **1948** B. Meggars in J. H. Steward *Handbk. S. Amer. Indians* III. 157 Tangas, which are found in abundance, are thought to have been worn by the women as a pubic covering. **1948** A. Métraux in *Ibid.* 670 Women..wore a short apronlike (tanga) cotton fringe..or a cotton skirt. **1960** C. Winick *Dict. Anthropol.* 525/1 *Tanga*, a pubic covering worn by Indian women, especially in tropical South America and the West Indies. The most common form of tanga today is a beaded apron. Others consist of a small triangle of inner bark. **1975** *Times* 5 June 12/1 Nylon jersey tanga (or string). **1976** R. Condon *Whisper of Axe* I. x. 60 She had the sort of body that should not..wear anything but a *tanga*, that wonderful Brazilian string bikini.

tangana (tæŋgā·nā). Also **Tangana.** [Origin unknown.] A type of rhythm used in jazz music (see quot. 1952).

1926 A. Niles in W. C. Handy *Blues* 24 The Habañera or tango rhythm... Both the justification for its use in Negro music, and the explanation of its subsequent popularity among the Negroes themselves, are supplied on acceptance of the plausible theory that this is an African rhythm (the native word is *tangana*) and Spanish only by adoption through the Moors. **1952** B. Ulanov *Hist. Jazz in Amer.* (1958) iv. 31 In 1914 Handy published his 'St. Louis Blues' with its provocative Tangana rhythm, which is a kind of habanera or tango beat consisting of a dotted quarter, an eighth-note, and two quarter-notes. **1959** 'F. Newton' *Jazz Scene* iii. 41 An admixture of certain rhythms such as the tangana, or the habanera which..roused a particularly vivid response among continental negroes.

Tanganyikan (tæŋgănyī·kăn), *a.* (*sb.*) [f. *Tanganyik(a* (see below) + -AN.] Of or pertaining to Tanganyika, now the continental part of the E. African republic of Tanzania. Also as *sb.*, a native or inhabitant of Tanganyika. Cf. *TANZANIAN *sb.* and *a.*

1905 *Rep. Brit. Assoc. Adv. Sci.* 430 The total number of Tanganyikan species of fishes amounts to eighty five. **1957** *African Affairs* LVI. 304 He went on to ask..whether the tsetse fly experiments were providing a hope that the scourge of the fly would be overcome... It would make for a great development in Tanganyikan agriculture. **1962** *Sunday Express* 21 Jan. 5/8 We who live in Tanganyika must think of ourselves as Tanganyikans and work for our country. **1971** *Standard* (Dar es Salaam) 7 Apr. 4/5 They could not cope with the then Tanganyikan education curriculum.

‖ **tangata** (ta·ŋata). [Maori: see Kanaka.] **a.** In Maori parlance, a person, a human being. **b.** *tangata whenua* (fenu·a), lit. 'people of the land', local people (e.g. as opp. to *nga manuhiri* the visitors).

1840 W. Deans *Let.* 30 Oct. in J. Deans *Pioneers of Canterbury* (1939) 29 He says they will take no other white man with them and they all want me to go and live there, calling me the *tangata* Widerup or the proprietor of it. **1949** P. Buck *Coming of Maori* (1950) I. v. 65 The Maori people who were in occupation of New Zealand at the time of European contact were the descendants of the intermixture of three successive groups of immigrants: the moa-hunters and the early *tangata whenua* [etc.]. **1974** *N.Z. Listener* 20 July 13/1 The body was lying in state on the stage at the end of the hall and the tangata whenua were seated. **1974** [see *MOA-HUNTER].

tangeite (tæ·ŋge,əit). *Min.* [ad. Russ. *tangeit* (A. Fersman 1925, in *Priroda* No. 7–9. 239), f. the name of the *Tange* Gorge, Tyuya-Muyun, Fergana, central Asia: see -ITE[1].] An orthorhombic basic vanadate of copper and calcium, $CuCa(VO_4)(OH)$, that is a secondary mineral found as green or greenish yellow crystals; calciovolborthite.

1927 *Mineral. Abstr.* III. 234 For the crystalline varieties the name tangeite, from the Tange gorge, is proposed, and for the colloidal variety the term 'Turkestan volborthite' is retained. **1951** C. Palache et al. *Dana's Syst. Min.* (ed. 7) II. 816 Tangeite appears to be identical with calciovolborthite. **1971** *Mineral. Mag.* XXXVIII. 488 The writers..now place on record an occurrence in Leicestershire of tangeite, $CuCaVO_4OH$, and volborthite, $Cu_3(VO_4)_2.3H_2O$, not hitherto described from Britain.

tangelo (tæ·ndʒĕlo). [f. Tang(erine *sb.* 2 a + Pom)elo.] A hybrid citrus fruit resembling a thick-skinned orange, produced by crossing the tangerine, *Citrus reticulata*, and the pomelo, *C. grandis*; also, the tree bearing this fruit.

1904 *Cosmopolitan Mag.* xxxvii. 262 Under the auspices of the United States Department of Agriculture has been evolved the 'tangelo'. **1905** Webber & Swingle in *Yearbk. U.S. Dept. Agric. 1904* 235 The term 'tangelo' is suggested by the writers as a name for this group of loose-skinned fruits, which lie midway between the pomelo and tangerine. **1932** Swingle & Robertson in *Proc. 6th Internat. Congr. Genetics* II. 385 The exhibit shows..tangelo fruits preserved in fluid. **1939** *Times* 23 Feb. 17/4 The basket [of Jamaican fruit] contained uglis, tangelos, grapefruit, oranges, [etc.]. **1969** *Oxf. Bk. Food Plants* 88/2 Tangelos are hybrids between the tangerine and grapefruit. **1970** *Harrod's Summer Food News* 8/1 Canadian tangelo juice. **1977** *N.Z. Herald* 8 Jan. 1-5/6 (Advt.), Fruit trees..tangelos.

tangent, *a.* and *sb.* Add: **A.** *adj.* **1. a.** Also as quasi-*adv.*

1936 A. W. Clapham *Romanesque Archit.* viii. 167 The subsidiary towers..at Mainz and Laach..are set axially and tangent at the ends of the transept. **1977** *Sci. Amer.* Aug. 37/1 The radiation is emitted tangent to these trajectories.

B. *sb.* **1. b.** *spec.* in *Surveying*, a tangent to a curve at a point (*tangent point*) where the curve starts or finishes; freq. *attrib.*, as *tangent distance, length*, the length of such a tangent from the tangent point to its intersection with the other tangent.

1850 T. Baker *Rudimentary Treat. Land & Engin. Surveying* II. ii. 171 Find the radii, as BO, CO′; the tangent point, as C′; and the junction point, as C, with the position of the common tangent at the junction point. **1862** W. J. M. Rankine *Man. Civil Engin.* I. v. 111 The places where permanent marks of the course of the line are chiefly required are on the tangents of curves. **1902** R. E. Middleton et al. *Treat. Surveying* II. v. 61 Find length of tangent for a 3° curve with central angle of 35° 42′. **1978** J. G. McEntyre *Land Survey Systems* x. 329 An easement curve is a transition curve utilized to increase the degree of curvature gradually from the tangent point to the circular portion of a curve. **1983** J. C. McCormack *Surveying Fundamentals* xxi. 384 The tangent distances are taped from the *P.I.* [*sc.* point of intersection] down both tangents to locate the *P.C.* and *P.T.*

ta·ngent, *v. rare.* [f. the sb.] *intr.* To go off or away at a tangent (see Tangent *sb.* 1 c); to fly off at an angle.

1920 F. Niven *Tale that is Told* vii. 44 They are always tangenting away, not from what I have said, but from what they imagine I have said. **1940** 'Gun Buster' *Return via Dunkirk* II. i. 85 The empty sardine tin that the Babe pitched, tangented off the side of his tin hat into the long grass. **1974** K. Millett *Flying* (1975) II. 201 She chuckles and then tangents off to some article..she read.

tangential, *a.* Add: **1. d.** (*c*) Of the pick-up of a record-player: so mounted that it is kept at a tangent to the groove by a rectilinear motion of the arm.

1937 *Electronics* X. 9/2 The whys and wherefores of the 'tangential' type pick-up mount and its effects on distortion and record wear. **1977** *Time* 10 Oct. 43/1 (Advt.), The Beogram 4002 has an electronically controlled tangential arm which plays records in the same way that they were cut: tracing a straight line from the edge of the record to its centre instead of tracing an arc.

2. c. *tangential energy*: in the writings of P. Teilhard de Chardin, the form of energy that is manifest in the workings of the physical world and is described by the laws of thermodynamics. Cf. *radial energy* s.v. *RADIAL *a.* 6. [tr. F. *énergie tangentielle*, introduced *c* 1938 by P. Teilhard de Chardin (*Le Phénomène Humain* (1956) I. ii. 62).]

1959, 1965 [see *RADIAL *a.* 6]. **1969** A. Richardson *Dict. Christian Theol.* 332/2 Tangential energy links units at the same level of organization.

tanger (tæ·ŋəɹ). Also **tenger.** [f. Tang *v.*[1] + -ER[1].] **1.** *dial.* A person who has a noticeable effect on another (see also quot. 1886).

1886 W. Cudworth *Rambles round Horton* 237/1 *Tenger*, a deceiving person. **1914** D. H. Lawrence *Prussian Officer* 221 She looked a tanger sitting there, all like statues, her and the geese. **1922** —— *England, my England* 269 She's a tanger—'s driven the gel to what she is.

2. [Tang *sb.*[1] 2.] One who furnishes implements with a tang.

1921 *Dict. Occup. Terms* (1927) § 279 *Tanger, saw tanger*, places saw blade against gauge in semi-automatic machine, which marks and purchases tang hole in blade and rivets on tang. **1960** *Classification of Occupations* (Gen. Register Office) Index 107/2 Tanger—*cutlery mfr*... *razor mfr*...*saw mfr.*

tangerine, *sb.* Add: **2. b.** Also *Comb.*, as *tangerine-coloured* adj.

1977 M. Kenyon *Rapist* iii. 31 Shovelling sugar into his tangerine-coloured tea.

tangi[1]. Add: (Earlier and later examples.) Also *transf.* and *fig.*

1836 J. A. Wilson *Missionary Life & Work in N.Z.* (1889) III. 34 Here we found many wailing over a dead body... As we passed some left the *tangi* and joined us. **1844** S. Selwyn 10 Apr. in A. Drummond *Married & Gone to N.Z.* (1960) 113 Nothing to disturb us but the incessant tangi of the children at night. **1905** W. Baucke *Where White Man Treads* 278 Our college graduate arrives; the home-coming tangi and nose-greeting is over; the guest meal set out on the floor. **1941** Baker *N.Z. Slang* vi. 56 When we refer to *holding a tangi* about a setback or problem we are putting another Maori term into colloquial use. **1959** Tindale & Lindsay *Rangatira* xviii. 172 The tangi mourning ceremonies. **1963** B. Pearson *Coal Flat* ii. 42 She said to me, 'Come on Joe, we'll have a *tangi* together.'

Hence as *v. intr.* to lament, to mourn; also *transf.* and *fig.*

1844 M. Williams 17 Oct. in A. Drummond *Married & Gone to N.Z.* (1960) 39, I could not think how I had consented, and tangied over his preparations. **1864** A. S. Atkinson *Jrnl.* 19 Apr. in *Richmond-Atkinson Papers* (1960) II. 107 She sat down, began tangi-ing,..& so they remained for some minutes. **1873** J. H. H. St. John *Pakeha Rambles through Maori Lands* x. 168 The old man,..enumerating the different degrees of relationship he stood in to the deceased, and his appreciation of his virtues, '*tangiéd*' again to such an extent, that another relation, affected at his extreme grief, presented him with a horse. **1881** J. L. Campbell *Poenamo* vii. 201 Those who had *tangied* over Ngatai had to come and *tangi* over Te Pirete. **1943** N. Marsh *Colour Scheme* xiii. 224 'She's going to *tangi*.' 'To wail..to lament the dead.'

‖ **tangi**[2] (ta·ŋgi). [Pashtu *tangai*, pl. *tangī*, f. (Pers.) *tang* narrow, *tanga* mountain pass.] A gorge or defile in north-western Pakistan.

[**1854** *Q. Jrnl. Geol. Soc.* X. 467 Mr. Loftus..explains the great clefts or 'Tangs' (as they are termed in Persian) which pass through the elongated limestone saddles. These tangs are very numerous.] **1901** *Mem. Geol. Survey India* XXXI. 188 Narrow gorges or rifts locally called 'tangi'. **1923** *Blackw. Mag.* Feb. 221/2 The river emerged through a narrow Tangi in the hills. **1954** O. H. K. Spate *India & Pakistan* xvi. 425 Tangis or transverse clefts, often only a few yards wide, by which the streams penetrate the longitudinal ridges. **1983** J. Masters *Man of War* x. 125 The Wazirs will try to ambush us... I've been marking my map with every likely gully and *tangi*.

tangible, *a.* Add: **1. b.** *tangible assets*, physical and material assets which can be precisely valued or measured.

1930 [see *INTANGIBLE *sb.*]. **1977** *Time* 8 Aug. 39/3 Recognizing that shares of many companies are selling at far less than the replacement value of their tangible assets, a number of chief executives have been using corporate cash to buy the assets of other companies at a discount. **1982** *Daily Tel.* 2 Sept. 19/2 Alexander Howden's net tangible assets were overstated..according to the audit of the British insurance company.

B. as *sb.* A thing that may be touched; something material or objective. Also *fig.*

1890 W. James *Princ. Psychol.* II. xix. 77 Those things are *tangibles*; their real properties, such as shape, size, mass, consistency, position, reveal themselves only to touch. **1962** Y. Malkiel in Householder & Saporta *Probl. Lexicogr.* 5 Range, i.e. the volume and spread of the material assembled, represents..the most obvious criterion [for classifying dictionaries]; it is also the most objective, involving by definition tangibles alone. **1965** *Economist* 21 Aug. 697/2 He also had some tangibles to offer, in particular a request to Congress to eliminate the import fee on sugar. **1980** I. St. James *Money Stones* II. i. 39 Financial operators like me deal in paper and rarely have first-hand contact with tangibles..reducing..a new jumbo jet or a sugar crop to one common denominator.. Money.

tangiwai (tæ·ŋiwəi). *Min.* [Maori, = tear-water.] A translucent kind of bowenite serpentine found in New Zealand that has droplet-shaped markings when polished.

1863 F. E. Maning *Old New Zealand* 321 This particular *mere* must have been made of the lovely translucent *tangiwai*. **1880** [see *KAWA-KAWA[1] 2]. **1911** [see *BOWENITE]. **1935** *Trans. & Proc. R. Soc. N.Z.* LXV. 201 All the talc-epidote-bearing serpentines examined by the writer are typical *tangiwais*.

tangka, var. Tanga[1] in Dict. and Suppl.

tangle, *sb.*[1] Add: **3. tangle-weed** (earlier and later examples).

1825 G. F. Lyon *Brief Narr. Attempt to reach Repulse Bay* 65 The sea was much agitated, a great quantity of tangle weed floating about. **1870** J. Lauder *Warblings of Caged Bird* 37 Whaur the stanes are green wi' moss And the tangle weeds are plenty.

tangle, *sb.*[2] Add: **1. b.** (Earlier example.)

1882 D. C. Beard *Amer. Boy's Handy Bk.* xi. 88 The Tangle, a name given to tassels of hemp that are often attached to the bottom of the dredge itself or used separately.

3. *tangle-headed* adj. (earlier *fig.* example).

1884 'Mark Twain' *Huck. Finn* xv. 131, I think you're a tangle-headed old fool, Jim.

tangle, *v.*[1] Add: **5. c.** *transf.* To fight, to engage in conflict or argument (*with* or *up with*); also *fig.* and *loosely*, to associate or become involved with. *colloq.* (orig. *U.S.*).

1928 *Amer. Speech* III. 29 Fistic action in large and copious quantities is expected..tonight when Dave Shade tangles with Maxie Rosenbloom. **1929** D. Hammett *Red Harvest* xxiv. 242 While we're tangling, them bums will eat us up. **1929** D. Runyon in *Hearst's Internat.* July 58/2, I remember reading in the paper about a lot of different guys who are considered very sensible until they get tangled up with a doll. **1942** *Sun* (Baltimore) 13 May 15/4 (*heading*) Preakness victor will tangle with old rivals if he runs at Belmont. **1945** L. R. Gribble *Battle Stories of R.A.F.* xxv. 64 There's no better fun in the world than tangling with the Hun. **1953** E. S. Gardner *Case of Green-Eyed Sister* (1959) ii. 17 You tangle up with Brogan..and you'll learn something about the noble art of shake-down. **1957** R. Lawler *Summer of Seventeenth Doll* I. i. 17, I dunno why I always have to get tangled up with little men, just the same. Even Wallie, he was shorter than me. **1958** B. Behan *Borstal Boy* III. 185, I don't like tangling with anyone, but Ickey Summers was the sort of little bastard that would pick a fight with you until he lost and the best thing to do with him was to make sure that he lost the first time. **1960** M. Stewart *My Brother Michael* ix. 118, I didn't particularly want to tangle with Danielle. **1966** P. O'Donnell *Sabre-Tooth* xviii. 244, I fancy we'll tangle in the long run, Willie. But not for a while. **1978** J. B. Hilton *Some run Crooked* xi. 116 There were men here who had nothing particular to hide, but who had learned ..that it was better not to tangle with Kenworthy. **1982** *Times* 10 Dec. 11/5 The mood of the House was sombre, and he had no desire to tangle with the Secretary of State.

tango (tæ·ŋgo), *sb.*[1] Pl. **tangoes, tangos** (*preferred*). [Sp., orig. a Negro or gypsy dance festival.] **1. a.** A Spanish flamenco dance.

1896 H. C. C. Taylor *Land of Castanet* 103 The girls.. dance again, not the vulgar flamenco or tango, but the charming dance of the province. **1931** [see *FARRUCA]. **1967** 'La Meri' *Spanish Dancing* (ed. 2) vi. 83 The Tango (Tango Gitano, Tango Flamenco) is of Arabic origin... It might be well to observe here that the Tango Gitano has nothing whatever in common with the Argentine tango save the name.

b. A syncopated ballroom dance in 2/4 or 4/4 time introduced into Europe and N. America from Argentina, related to the Cuban Habanera but probably of African origin, characterized by a slow gliding movement broken up by pointing positions; a piece of music for this dance.

1913 G. Grossmith in *Daily Graphic* 12 May 9/1 'A Peeress' talks about the Tango. This is a most graceful and beautiful dance. **1915** T. Burke *London Nights* 254 Music, gaiety, sparkle, fine dresses, costume songs, tangos, smart conversation and faces, and all the rest of it. **1921** *Spectator* 19 Feb. 236/3 Instead of a stately waltz at an even pace, Myers had no sooner taken the magnificent lady round the waist, than they were covering the floor with a glorified tango. **1925** C. Connolly *Let.* 8 Apr. in *Romantic Friendship* (1975) 67 Someone is playing tangoes on a guitar. **1947** [see *GUAJIRA]. **1962** *Melody Maker* 7 July 7/5 Kid Thomas..leads his septet through standards, pops and tangoes. **1976** Botham & Donnelly *Valentino* iv. 34 This arrogant and deliciously handsome Continental who danced the tango as though it had been invented for him. **1981** E. North *Dames* viii. 138 The music..played..a tango... He regarded better dancers tangoing.

2. *attrib.* and *Comb.*, as (sense 1 b) *tango band, -dancer, -dancing, foxtrot, music, rhythm, step*; **tango tea,** a thé dansant, usu. at a hotel or other gathering place, arranged for the purpose of dancing the tango (*disused*).

1927 C. Connolly *Let.* 13 Feb. in *Romantic Friendship* (1975) 255 This is a very grand hotel, all lights and carpets and a melodious tango band. **1932** Tango band [see Rumba *sb.* 1 a]. **1974** M. Butterworth *Man in Sopwith Camel* I. 11 Norman the demon tango-dancer of the Palais. **1918** G. Frankau *One of Them* xxxi. 243 She.. knew how wise she'd been to countermand Breastplate and shield which, though they looked entrancing, Would both have been a bore for tango-dancing. *c* **1925** 'H. Nicholls' (*music title*) Sunny Havana: tango foxtrot. **1913** *Sheffield Daily Tel.* 5 Sept. 6/4 Tango music is also expected to rival the popularity of ragtime. **1966** A. Cavanagh *Children are Gone* II. v. 45 The Muzak was dispensing tango rhythms, loud and sinuous. **1977** 'F. Clifford' *Ten Minutes on June Morning* 90 The music came again, tango rhythm. **1981** E. Ward *Baltic Emerald* ix. 64 A new tango step. **1913** *Punch* 6 Aug. 125 No tango-teas shall be given in *this* drawing-room. **1918** G. Frankau *One of Them* I. 14 What time we sang of guns and gore and trenches, Instead of oysters, tango-teas and wenches.

tango (tæ·ŋgo), *sb.*[2] (*a.*) [f. Tang(erine *sb.* 2 + *-o[2], perh. influenced by prec. sb. or by Tan *sb.*[1] 3.] A colour shade of deep orange.

1913 *Vanity Fair* Sept. 59/3 The one shown is of sand color and tango. **1933** *Archit. Rev.* LXXIII. 70 The colour scheme of the shopfront is tango-red throughout. **1958** B. Nichols *Sweet & Twenties* x. 126 Jade green, cerise, tango. **1976** *Burnham-on-Sea Gaz.* 20 Apr. 18/3 (Advt.), 1974 Hillman Hunter GLS 4-door saloon, tango.

ta·ngo, *v.* [f. *TANGO *sb.*[1]] *intr.* **a.** To dance the tango.

1913 *Punch* 10 Dec. 486/1 'Do you tango?' she asked me as soon as we were comfortably seated. **1925** C. Dodd *Farthing Spinster* III. iii. 299 Young Jellis tangoed up to the two ladies. **1941** B. Schulberg *What makes Sammy Run?* xi. 206 Sammy didn't know how to tango. **1952** M. Laski *Village* ii. 42 She..watched her father tangoing ..with Daisy Bruce. **1975** *Times* 4 Sept. 14/6 You have to count anxiously like one learning to tango. **1981** [see *TANGO *sb.*[1] 1 b].

b. *fig.* Freq. in proverbial phr. *it takes two to tango.*

1952 Hoffman & Manning *Takes Two to Tango* (*song*) 2 There are lots of things you can do alone! But, takes two to tango. **1965** *Listener* 24 June 923/2 The President would like to know whom to negotiate with... On this score, the President has a firm, and melancholy, conviction: it takes two to tango. **1970** B. Conacher *Hockey in Canada* (1972) x. 116 Despite all the problems I had had with Imlach, and believe it or not I realize it takes two to tango, I wouldn't have missed playing in the best league in the world. **1973** *Houston* (Texas) *Chron.* (Texas Mag.) 14 Oct. 2/1 An upcoming film of such explicit sexuality it'll have to tango with the new pornography rulings. **1977** *Time* 31 Oct. 48/1 Ellis Rabb can tango with words and he is a sly devil at milking an audience dry of laughter. **1979** *Guardian* 4 Apr. 12/3 It takes two to tango... Mrs Thatcher has turned Mr Callaghan down.

Hence **ta·ngoing** *vbl. sb.*; **ta·ngoist,** an exponent of the tango.

1913 G. B. Crozier *Tango & How to dance It* i. 8 The Parisian version of the Tango..has so much to recommend it that one may..predict for it a permanent place in our affections long after the present craze for 'Tangoing' is over. *Ibid.* ii. 28 Embryonic 'Tangoists' cannot do better than bear that graceful animal [*sc.* the tiger] in mind while attempting to follow their advice. **1928** *Daily Express* 6 July 3/3 This tango is so slow, so smooth, so syrupy. Caterpillars skating over egg-shells could not move more gracefully, more softly, than the contemporary tangoists. **1976** U. Holden *String Horses* vii. 81 They'd show the Camp what real tangoing was.

tangoreceptor (tæ·ŋgorise:ptǫɹ). *Zool.* [f. L. *tang-ĕre* to touch + -o + Receptor.] A sensory receptor which responds to touch or pressure.

1906 C. S. Sherrington *Integrative Action Nervous Syst.* ix. 335 The sessile creature retains..only some gustatory (?) receptors round the mouth, and some tango-receptors..in the tegument. **1937** L. V. Heilbrunn *Outl. Gen. Physiol.* xl. 506 In man and mammals, in addition to ordinary free nerve endings, special types of tangoreceptors are found in the skin and in the viscera. **1980** L. H. Chappell *Physiol. Parasites* ix. 165 The sensory papillae of cercariae are thought to be tango-receptors.

‖ **tangpu** (daŋbu). Also **Tang Pu.** [Chinese *dǎngbù*, f. *dang* party + *bù* office.] The headquarters of the Kuomintang at the central, and various local, levels.

1941 E. Snow *Scorched Earth* VI. ii. 206 The *tangpu* system in China is a product of 'natural' political evolution. **1943** J. T. Pratt *War & Politics in China* xii. 200 Borodin impressed upon Dr. Sun the importance of securing the assent of the people at large to the revolutionary programme... This was to be effected by organizing local branches of the Kuomintang—Tang Pu—and by systematic propaganda. **1972** A. Destenay tr. *Guillermaz's Hist. Chinese Communist Party 1921–1949* vi. 79 Party offices (Tangpu) existed at the provincial, district and local levels.

Tangut (tæ·ŋgut), *sb.* (and *a.*). Also 8–9 **Tangout.** [App. a. Mongol, f. Chinese *Tang-hsiang* (tribal name): (see also quot. 1979).] A Tibetan people who inhabited north-western China and western Inner Mongolia, and formed the independent kingdom of Hsi Hsia from the eleventh to the thirteenth centuries A.D.; the country or language of this people. Also *attrib.* or as *adj.* Also **Tangu·tan** *a.* and *sb.*

1598 Hakluyt tr. W. de Rubruquis in *Voy.* I. 116 Between the foresaid mountaines Eastward inhabiteth the nation of Tangut, who are a most valiant people. **1613** Purchas *Pilgrimage* IV. ix. 337 There were of them divers nations, called by one common name Mogli, which were divided into seven principal tribes, whose names were Tatar, Tangut, Cunat, Talair, Sonich, Monghi, Tebeth. **1795** W. Winterbotham *Hist. & Geogr. View Chinese Empire* 182 Thibet is known under different names, the Chinese call it Tsang; the Tartars, Barantola, Bouttan, and Tangout. **1827** H. E. Lloyd tr. *Timkowski's Trav. through Mongolia to China* I. xii. 442 Tangout is a Mongol word, designating the country which at present forms the whole of the western frontier of China, and is inhabited by the eastern Tibetans. **1876** H. H. Howorth *Hist. Mongols* I. i. 5 This town [*sc.* Ninghia]..was called in the Tangutan language Eyirkai. **1876** E. D. Morgan tr. *Prejevalsky's Mongolia, Tangut*

Country & Northern Tibet II. iv. 109 The Tangutans, or the Si-fan as the Chinese call them, are of the same race as the Tibetans. Ibid. 119 In the Tangut country..the price of brick tea has considerably risen. **1888** Encycl. Brit. XXIII. 343/2 The Tang-chang and Peh-lang tribes boasted also of being descended from a monkey; they were the two great divisions of the Tang-hiang or Tangut, offsets of the same Sien-pi stock as that of the conquerors of Tibet. **1908** J. Curtin Mongols iv. 75 The subjection of the Kirghis and this new victory over Tangut secured the position of Jinghis in Northeastern Asia. **1934** K. S. Latourette Chinese I. iv. 159 The Later Chao was succeeded in the Northwest by a state established by a Mongol people, formerly supposed..to be Tanguts. **1954** Pei & Gaynor Dict. Linguistics 214/1 Tangut, an Asiatic language, a member of the Eastern group of the Mongol branch of the Altaic sub-family of the Ural-Altaic family of languages. **1979** L. Kwanten Imperial Nomads v. 72 This fails to explain how and when the inhabitants of Hsi Hsia became known as Tangut. Ibid., Most scholars remain convinced that the Tangut language is a member of the Sino-Tibetan linguistic family, although..recent linguistic research indicates that there is a distinct possibility that Tangut is either a Turkic dialect or a language heavily influenced by a Turkic dialect.

tangy (tæ·ŋi), *a.* Also † tangey. [f. Tang *sb.*¹ + -y¹.] **a.** Having a sharp, distinct, or spicy taste. Also, characterized by a disagreeable tang or flavour (*rare*).
1875 Ure's Dict. Arts (ed. 7) III. 189 There is a perceptible deficiency in that fine, clean flavour, which is the perfection of a glass of good beer, its place being usurped by a flavour coarse and tangey. **1931** R. Starke Touch & Go xii. 193 The meal was strong and tangy and tough and stringy. **1946** C. S. Forester Lord Hornblower iii. 37 A bite of red cheese, tangy and seasoned. **1958** People 4 May 7/4 (Advt.), You will find Tango the tastiest, tangiest whole orange drink ever! **1966** Woman's Own 22 Jan. 19/1 What all my friends like is this sharp, tangy flavour. **1976** National Observer (U.S.) 17 July 1/3 After a tangy dinner of sausage creole, everybody is lounging about the living room. **1979** J. Wainwright Home at Hunter xxi. 71 The iced cider was a pleasant surprise; tangy, cool.
b. *transf.* and *fig.* Cf. Spicy *a.* 7.
1948 Sun (Baltimore) 3 Dec. 14/1 (heading) The tangy story of the frozen-orange juice industry. **1953** J. Masters Lotus & Wind x. 137 This was heavier, tangier stuff, and it reminded him more of animals than flowers. **1967** Punch 4 Oct. 520/2 Despite a twinkly fondness for waggish puns..his style isn't exactly tangy. **1978** H. Wouk War & Remembrance viii. 78 She knew a tangy tale or two about Madge Knudsen!

tanh (tæn,ʃi·tʃ). *Math.* Abbrev. of *hyperbolic tangent*.
1879 Encycl. Brit. IX. 819/2 Similarly we have the hyperbolic tangent tanh *x*, &c. **1968** E. T. Copson Metric Spaces vii. 107 The restriction..can be removed by considering for example the function tanh *f*(*x*).

tania, tanier, tannier. Add: Also **tannia**. (Later examples.)
1953 Caribbean Q. III. ii. 103 In 1811, Charles Edmonson..reported..: 'The quantity of rice the Bush Negroes raise out of the ground is very considerable independent of yams, tannias, plantains, tobaccos, &c.' **1955** Ibid. IV. ii. 112 Grated tannia fritters..fried in deep fat on a coalpot. **1968** [see *dasheen]. **1979** West Africa 8 Jan. 51/2 In Cameroun..the other, tannia sort [of cocoyam] is called macabo.

tanister. For † Obs. *rare*⁻¹ read *rare* and add later example.
1937 Burke's Landed Gentry (ed. 15) 1496/1 Hugh Vernon Macleod, 11th Chieftain of Talisker, and apparent Tanister of Macleod of Macleod.

‖ **taniwha** (tæ·niwā, ‖ tanifa). *N.Z.* Also † **tanewa, taniwoa**, and with capital initial. [Maori.] A mythical monster supposed to reside in deep water.
1840 J. S. Polack Manners & Customs New Zealanders II. xiii. 126 The additional name of Taniwoa is added, (a water god). **1842** W. R. Wade Journey in N.Z. i. 34 One of our boatmen quickly repeated that the place was tapued for the taniwa (a water demon). **1863** F. E. Maning Old N.Z. ii. 26 Down, villain! down to..the Taniwha cave! **1882** W. D. Hay Brighter Britain II. 214 They (sc. the Maori) have a tale of these taniwha..somewhat parallel to our nursery stories of dragons. **1905** [see *ngarara]. **1921** H. Guthrie-Smith Tutira x. 70 They felt the net being dragged away from them by the taniwha known to haunt the bay. **1938** R. Finlayson Brown Man's Burden 66 The carved taniwha monsters of the meeting-house. **1948** J. K. Baxter Blow, Wind of Fruitfulness 37 Riding the logs upstream, and waiting for the Taniwha. **1966** Encycl. N.Z. I. 483 Places along the sea shore were feared because they were the haunts of the *taniwha*, awesome water monsters with man-killing tendencies.

tank, *sb.*¹ Add: **1. a.** Also in Australia, an artificial reservoir designed to hold water for livestock; *U.S. dial.*, an artificial pond or lake.
1898 D. Carnegie Spinifex & Sand 81, I append a table showing cost and contents of Government tanks excavated at the base of granite rocks between Southern Cross and Coolgardie. **1903** 'T. Collins' Such is Life 265 On a well-managed station..a tank is, whenever possible, excavated on the margin of a swamp. **1911** C. E. W. Bean 'Dreadnought' of Darling i. 7 There is only one boundary rider's hut in it and one 'tank' of water. The tank may have dried up. **1915** Dial. Notes IV. 229 Tank, an artificial lake. 'Most west Texas towns get their water from tanks.' **1936** F. Clune Roaming round Darling xiv. 121, I strongly object to the back country habit of calling holes scooped out of the ground tanks. **1955** W. Foster-Harris Look of Old West ix. 273 Tank is cow country [language] for a small pond, made by damming a ravine or fixing a hollow to catch and hold rain water. **1965** Austral. Encycl. I. 133/2 In Australia, every farmer is interested in constructing and maintaining tanks and dams.
2. a. Also *spec.* a water receptacle (with transparent sides) in which to keep fish; an aquarium.
1854 P. H. Gosse Aquarium i. 3 The tanks in the new Fish House just erected in the [Zoological] Society's Gardens in the Regent's Park. **1890** C. Bateman Fresh-Water Aquaria i. 6 The ordinary oblong tank, containing four glass sides, is both ornamental and useful. **1936** M. C. Elwin First Steps in Aquarium Keeping iv. 27 The tank will look unfinished without a couple of the beautiful Angel fish. **1971** R. F. O'Connell Freshwater Aquarium 127 The breeding tank should be cleaned thoroughly and filled with seasoned water to a depth of 8 inches. **1982** I. Petrovicky Trop. Aquarium Fishes 13 If an aquarium is to be purely ornamental, it is better to select one larger tank.
b. The fuel container of a motor vehicle.
1902 A. C. Harmsworth Motors & Motor-Driving vii. 117 With the gravity-fed carburetter the tank is fitted in the body of the car. **1944** L. D. Kitchin Road Transport Law 19/1 Not more than 60 gallons of petroleum spirit, including that contained in any vehicle fuel tank, may be kept in any one storage place. **1978** H. Ames John's Fling xxv. 124 'Are we low on petrol, Ivor?' 'No, I had a full tank when I picked you up.'
3*. *U.S. slang.* A cell in a police station, *spec.* one in which several prisoners (esp. drunks) are held.
1912 D. Lowrie My Life in Prison iii. 30, I glanced at the number on the cell door. It was..34 Tank. **1933** 'J. Spenser' Limey xvi. 256 In our tank..there were three Chicago gangsters waiting to be returned to that city. **1947** A. R. Bosworth San Francisco Murders 264 The day a police reporter had to pick him out of the collection in the drunk tank. **1951** Life 8 Jan. 24 (caption) Still relatively blissful but due for an unhappy awakening, some of the 1,200 Angelenos charged with drunkenness sleep it off in the tank. **1964** Wodehouse Frozen Assets iii. 50 If it gets boring after a while being shut up in tank, always when one fears nervous feeling that this time the old man won't come through with the necessary bail. **1981** L. Deighton XPD xxv. 210 And then tossed into the drunk tank like a common criminal.
4. (sense 2 a) **tankhouse, -ship, -truck** (later U.S. example); **tank bag**, a receptacle for carrying luggage which fits on to the petrol tank of a motorcycle; **tank circuit** *Electronics*, a resonant circuit placed in the anode circuit of a valve oscillator in order to supply energy to an aerial for transmission; **tank farm** orig. *U.S.*, a collection of tanks for the large-scale storage of oil, etc.; **tank furnace**: substitute for def. (see quot. 1970); (examples); **tankstand** *Austral.* and *N.Z.*, a stand or support for a tank in which water is stored; **tank suit** *U.S.*, a (ladies') one-piece bathing-suit with scooped neck (cf. *maillot 2); **tank top**, a sleeveless upper garment with round neck and deep armholes, freq. of knitted material and similar to the top of a one-piece bathing-suit, worn by men or women; cf. *tank suit* above; **tank town** *U.S.*, a small, unimportant town, orig. one at which trains stopped to take on water.
1974 Cycle World June 24 (Advt.), Rain-proof cycle luggage... Tank bag—straps to gas tank. **1980** Guardian 28 Oct. 9/6 A set of good bike luggage—panniers and top box—is the best solution... A cheaper alternative is a pair of carriers to sling over the seat and a tank-bag. **1931** J. H. Reyner Wireless Princ. & Pract. vi. 183 A slightly different method is that of using a 'tank' circuit, which consists of a low impedance oscillatory circuit connected from the earth end of the aerial inductance to earth. **1959** K. Henney Radio Engin. Handbk. (ed. 5) xviii. 14 Resonant tank circuits are used in power amplifiers to remove the effects of tube and circuit stray capacitances. **1977** Gloss. Electrotechnical, Power Terms (B.S.I.) iii. vii. 24 Tank circuit, a resonant circuit of the final stage of a transmitter which supplies the radio-frequency energy to the aerial or aerial feeder. **1932** Amer. Speech VII. 271 Tank-farm, a group of storage tanks. **1959** Times Lit. Suppl. 5 Nov. 1/3 Chemical flames prevented anyone from getting into the explosion area, which Sears described as a 'tank farm', the storage area of the vinylite or plastics producing section. **1974** Daily Tel. 30 May 8/6 Huge 'tank farms' may be needed in parts of Scotland to store the oil. **1879** Encycl. Brit. X. 659/1 Mr Archibald Stevenson of Glasgow has patented a tank furnace fired by common coal from one end, with working holes on the other three sides. **1908** W. Rosenhain Glass Manufacture iv. 72 The tank furnace utilises the heat of the flame more efficiently. **1970** Gloss. Industr. Furnace Terms (B.S.I.) 10 Tank furnace, a furnace in which glass is melted in a refractory bath. **1941** Sun (Baltimore) 15 Oct. 19/2 It is an unwatched light showing only one dull yellow lamp..a white skeleton tower and tankhouse on concrete piers. **1976** Jrnl. R. Soc. Arts CXXVI. 92/1 An important appreciation is that of titanium blanks for the production of the starting sheets used in copper refinery tankhouses. **1949** Sun (Baltimore) 25 Aug. 9/2 More than 100 tankships, large and small, have been laid..have been cleaned and made gas-free. **1978** M. Dewis Law Health & Safety at Work i. 5 The crew of a British tankship. **1941** Coast to Coast 146 Then she crept off the veranda and went down under the tankstand. The tool under the tank was a rich chocolate brown, and there were drips of water coming from the tap. **1965** S. T. Olliver Petticoat Farm v. 66 Emma carefully retraced her steps down the mud-mill until she reached the..corner of the tank-stand. **1977** C. McCullough Thorn Birds vi. 116 A drover whose cross said only Tankstand Charlie he was a good bloke. **1950** P. Roth Goodbye Columbus ii. 20 She wore a black tank suit and sneaker. **1979** Dancemagazine Feb. 108/3 In *Moth Dance*, the lines of Hermans' tensed, slender body, in tanksuit and reflecting sunglasses, become clearer as the semiclearness grows lighter. **1968** New Yorker 27 Jan. 25 Miss Farrell—a tall, pretty ballerina dressed in a purple tank top and baggy rubber warm-up pants. **1972** Observer 1 Aug. 22/3 A favourite Paris idea is to put little woolly vests or tank tops over shirts and under suit jackets. **1977** Miller & Swift Words & Women 157 Even the latter have given up whalebone corsets and starched winged collars without assuming they have to switch to miniskirts or tank tops. **1910** J. F. Kelly Man with Grip 11 Tank towns are big ones, compared to our route. **1940** R. Chandler Farewell, my Lovely v. 38 You would find them in tanktown vaudeville acts. **1978** Times 25 Mar. 14/4 When vaudeville was in its last death throes, young Donald O'Connor was..going—as the show his legend decreed he should—'two tank towns to another.' **1976** New Yorker 9 Feb. 66/8 It was solved by designing a tank truck that intermittently travelled around the army camps and waited..till the sprinkler was made.

tank, *sb.*⁷ [Special use of Tank *sb.*¹ adopted in Dec. 1915 for purposes of secrecy during manufacture.] **1. a.** An armoured military vehicle moving on a tracked carriage and mounted with a gun, designed for use in rough terrain.
First put into commission on 15 Sept. 1916.
1916 Times 18 Sept. 9/6 'Tanks' is what these new machines are generally called, and the name has the evident official advantage of being quite undescriptive. **1917** A. Macken Terror i. 19 Last summer there were very few people outside high official circles who knew anything about the 'Tanks', of which we have all been talking lately. **1918** Review of Reviews (N.Y.) Oct. 383 The British tanks, as first produced, were of two types, male and female. The male tank was armed with two six-pounder, rapid-fire Hotchkiss guns, and four Lewis machine guns... The female type carried a lighter armament. **1926** Daily Colonist (Victoria, B.C.) 10 Jan. 5/2 Voltaire is said to be the real inventor of the armored tank. In 1756 he designed what was called the 'Assyrian Chariot', which was advanced on the same lines. **1940** Richmond (Va.) Times-Dispatch 10 Aug. 1 (heading) Army irked as news leaks of plans for monster tank. **1957** Encycl. Brit. XXI. 791/1 In 1940..the French alone possessed about 3,600 tanks..superior in armour and fire power to those of the Germans. **1970** Sunday Times (Colour Suppl.) 10 May 56/2 For security purposes the cumbersome metal machines needed a code-name: 'water carriers' was rejected in favour of 'tanks'. **1977** C. McCullough Thorn Birds vi. 34 Went through them big buggers of tanks like a dose of salts.
b. In *pl.*, *ellipt.* for Tank Corps.
1943 J. B. Priestley Daylight on Saturday iv. 21 Her husband was abroad, in the Tanks. **1967** L. Deighton Only when I Larf (1968) vii. 85, I could see that the war had to come, so I..got a commission in the tanks. **1981** A. Price Soldier no More vi. 92 He was in the tanks during the war.
2. *attrib.* and *Comb.*, as *tank battle, Corps, crew, driver, officer, raid, warfare*; *tank-like* adj.; also used of naval vessels designed to transport and put ashore tanks, as *tank-landing craft*, etc. (cf. *landing craft, ship* s.v. *landing *vbl. sb.* 8); **tank buster** *slang*, an aircraft or other device designed to combat tanks (cf. *buster); also *fig.*; **tank-destroyer** *U.S.*, a highly-mobile armoured fighting vehicle equipped with a field gun, designed to combat tanks; **tank-man**, a member of a military tank crew; **tank transporter**, a wheeled vehicle for the transportation of tanks; **tank trap**, an obstacle placed or constructed so as to impede or prevent the progress of a tank; **tank turret**, the rotating structure on a tank on which the gun is mounted; **tank watch**, a gold watch designed by Cartier of Paris, decorated with gemstones, and usu. regarded as a status symbol (designed in 1917; the gold side-panels were held to resemble the wide tracks of the new armoured tanks (see sense 1 a)).
1944 C. Milburn Diary 7 June (1979) 217 A tank battle was raging in one spot and an air battle not far away. **1978** A. Melville-Ross Blindfold xxii. 190 The great tank battles of the Second World War. **1942** Hawker Lambert Hawk 61 Oct. 92/1 The Hawker 'Hurricane' is..proving its superiority in the battle of the Libyan Desert..as a dive-bomber and 'tank-buster'. **1942** J. Sweeney in Murdock & Drake-Brockman Austral. Short Stories (1951) 384 No sooner does the gong go for the third than Irish walks into..a rip-snorting tank-buster that Big Joe had been saving up for a secret weapon. **1967** Electronics 6 Mar. 311/2 (Advt.), Tiny tankbuster. **1917** Army Order 239 28 July, We deem it expedient to authorize the formation of, and to provide rates of pay for, a corps to be entitled 'Tank Corps'. **1976** Listener 20 May 633/3 He had joined the Tank Corps... He was lying in bed in barracks one night, listening to the flow of unremitting obscenity from his fellow tankmen. **a1944** K. Douglas Alamein to Zem Zem (1946) 14 As soon as this was finished I began to make the acquaintance of my tank crews. **1973** A. Price October Men v. 71 Tales of stranded tank crews parboiled. **1941** Sun (Baltimore) 28 Aug. 24/1 The army today demonstrated..the type of unit it hopes is the answer to panzer attack, a 'tank-destroyer' battalion of fast-moving, self-propelled field guns protected to a certain extent by armor. **1961** W. Vaughan-Thomas Anzio v. 76 The American 894th Tank-Destroyer Battalion..attacked again. The tank-destroyers advanced line ahead like battleships of old. **1928** Tank driver [see *caterpillar v.]. **1980** J. Ditton Copley's Hunch I. ii. 42 'You [sc. the RAF] go into action sitting down.' 'So do tank drivers.' **1917** W. S. Churchill in M. Gilbert Winston S. Churchill (1977) IV. Compan. I. 87 In addition a number (say) 50 tank-landing lighters would be provided, each carrying a tank or tanks. **1945** Tank landing ship [see *landing vbl. sb. 8]. **1969** Listener 4 Sept. 304/2 In January 1956 I thought I was going to a reserve fleet, but rather to my delight I was appointed to another command. This was to a tank landing-craft called HMS Redoubt in Malta. **1977** Navy News June 23 Audemer, a Royal Corps of Transport tank landing craft, is in the Review lines. **1916** E. Montagu Let. 31 Oct. in M. Gilbert Winston S. Churchill (1972) III. Compan. II. 1580 Cannot the idea of the Tank be so extended as to use a Tank-like machine to protect our Infantry. **1977** C. Fremlin Spider-Orchid xvii. 111 A sort of monstrous arrogance..driving tank-like over all concerns other than its own. **1934** N. & Q. CLXVI. 73/2 In the tank-man we shall have the steel-clad mediaeval knight back again. **1976** Tankman [see Tank Corps above]. **1949** R. Chandler Let. 24 Jan. (1981) 145 At one [table] sat..a demobbed tank officer with his mother. **1978** A. Price '44 Vintage xi. 136 He certainly didn't intend to let any bloody tank officer..out-crawl him. **1917** 'Contact' Airman's Outings 135 Farther along the road was the scene of the first tank raids. **1943** Times (Weekly ed.) 24 Nov. 6 'Tank Transporter.' To save wear and tear of tracks and to save petrol, tanks are transported over hard roads by huge tank-carrying lorries. **1972** D. Bloodworth Any Number can Play xvi. 149 A respray job... What ran into it, a tank-transporter? **1925** Scribner's Mag. Sept. 234/1 Tank traps, trenches so wide that the little fellows went nose-down into them and stuck, and direct fire from Boche artillery stopped the most of them. **1977** Time 10 Jan. 22/2 The Chileans, bracing for a possible invasion, are mining the desert, implanting tank traps and building fortifications. **1946** E. Linklater Private Angelo xi. 131 Romans..tossed flowers into jeeps and tank-turrets. **1979** D. Graham in K. Douglas Alamein to Zem Zem 10 Sufficient of its character remained, however, for it to move into action..with tank turrets open and umbrellas up. **a1944** K. Douglas Ibid. (1946) 14 It is low-built, which in..tank warfare, is a first consideration. **1977** B. Freemantle Charlie Muffin ii. 29 His absorption in the history of tank warfare. **1976** Vogue Dec. 216 Cufflinks and watches all from Cartier... Tank watch edged with diamonds, £1,800. **1978** T. Gifford Glendower Legacy (1979) 119 She looked at her Cartier tank watch with the sapphire on the stem.

Hence **ta·nker²** *colloq.* = *tank-man*, sense 2 above.
1919 W.R.A.F. on Rhine July 27 Little boys with..tanks..asking the girls to come and play... What jolly boys those Tankers were! **1940** Sun (Baltimore) 23 Sept. 11/4 There are tankers who long ago served in the same regiment when it was fully horsed. **1961** W. Vaughan-Thomas Anzio v. 79 Italian geography is unfriendly to the tank, and there were moments when tankers must have felt that the whole country was one enormous, endless anti-tank ditch. **1974** C. Ryan Bridge Too Far III. ix. 187 Taylor had hoped for the support of the tankers' guns along the fifteen-mile stretch of corridor the Screaming Eagles must control.

tank, *sb.*⁸ *slang.* [Prob. abbrev. of Tankard.] The amount held by a drinking-vessel; hence *loosely*, a drink (usu. of beer). Cf. *jar *sb.*² 2 c, *tank v.¹ 5 a.
1936 O. Nash Primrose Path 46 What can a man.. Ask..More than a pipe..And a modest tank of beer? **1945** Baker Dict. Austral. Slang 75 Tank, a pint of beer. **1958** Spectator 7 Feb. 171/1 Their carousals over a few friendly tanks at the neighbouring Whitehall milk bar.

tank, *v.*¹ Add: **2.** Also, to put into a tank.
1960 Koestler Lotus & Robot I. i. 42 The driver-owners are so poor that they only tank one or two gallons at a time.
3. (Example.)
1906 U. Sinclair Jungle iii. 43 To another room came all the scraps to be 'tanked', which meant boiling and pumping off the grease to make soap and lard.
5. Chiefly *to tank up*. **a.** *intr.* for *refl.* To fill oneself with drink, to drink heavily. Also *refl.* Cf. *tanked *ppl. a.* 1. *slang.*
1902 A. H. Lewis Wolfville Nights xv. 236 Bowlaigs would repair back ag'in to the Major [with the bottle], when they'd both tank up ecstatic. **1920** C. L. Stagg High Speed viii. 142 Both of 'em are tankin' up next door, and layin' for you and the whole bunch. **1925** F. Scott Fitzgerald Great Gatsby (1926) ii. 28, I think he'd tanked up a good deal at luncheon, and his determination to have my company bordered on violence. **1939** A. Huxley After Many a Summer II. iii. 208 She..made him feel good, like you felt when you'd tanked up a bit on Scotch. **1951** W. C. Williams Autobiogr. xxv. 148 Perhaps he was insubordinate or tanked himself up or did something otherwise improper. **1959** A. Christie Cat among Pigeons 18 On Sports Day..Lady Veronica arrived completely sober... But there were times when Lady Veronica tanked herself up. **1974** D. Ramsay No Cause to Kill II. 132 Jessie's a lush... Hardly ever leaves the house..except to tank up at the neighbourhood hangouts. **1980** I. Hunter Malcolm Muggeridge xii. 216 Behan arrived for the interview 'somewhat full' and proceeded to tank up further in the BBC hospitality room.
b. *trans.* To fill the tank of (a vehicle) with fuel; to refuel. Also *absol.*, and *intr.* for *pass. colloq.*
1933 [implied in *tanking vbl. sb.¹]. **1944** 'N. Shute' Pastoral ii. 37 The Bowser was waiting to tank up the Wellington. **1948** —— No Highway ix. 244, I guess we'll make Ivanhoe by sundown... Tank up there, 'n have plenty up at the lake. **1959** Halas & Manvell Technique Film Animation v. 65 A cartoonist may want to give an automobile the characteristics of a dog in its attitude to the fuel that its owner offers it... It shakes its shaggy head in refusal to tank up with the wrong brand of spirit. **1963** D. Irving Destruction of Dresden III. iii. 139 The whole force [of aircraft] had been tanked up with maximum fuel loads, 2,154 gallons of petrol each. **1977** N. Freeling Gadget I. 5 The four cars..stopped once to tank up. **1978** —— Night Lords xxx. 140 At the edge of the service area he stopped..while the car was tanked.
c. *transf.* and *fig.*
1942 Tee Emm (Air Ministry) II. 145 It's no good tanking up on them [sc. vitamins] and hoping you'll be able to detect a black cat at midnight in a Bremen cellar from 10,000 feet. **1959** Word Study Oct. 2/2 We are grateful for them, 'tank up' on their detailed and highly useful messages, and perhaps put them away for future reference. **1975** R. Butler Where All Girls are Sweeter ii. 15 She was twiddling the empty glass... I tanked her up and waited.
6. *intr.* In Lawn Tennis, to lose or fail to finish a match deliberately; to default. *slang.*
1976 Times 30 Sept. 11/5 Too many..singles players do not enter for the doubles. Either that, or they scratch or 'tank' (in boxing parlance, 'take a dive'). **1979** Guardian 13 Jan. 11 But it is ironic that Connors, a player generally considered too honest to 'tank' to anyone, should be the one to suffer.
Hence **ta·nker³**, a heavy drinker; **ta·nking** *vbl. sb.*¹ (in the senses of the vb.).
1891 Tanking [in Dict. s.v. Tank v. 3]. **1918** H. Bindloss Agatha's Fortune iv. 40 When you get the tanking habit such things happen. **1930** Aberdeen Press & Jrnl. 31 Jan. 7/5 Tanking consists of salting ungutted herrings into big tanks for future use. **1932** H. Crane Let. 16 Feb. (1965) 400 Especially with Luz around, who Lisa says is a great little tanker. **1933** Flight 16 Feb. 157/1 Petrol-filling installations, i.e. hand pumps, are now available on all important aerodromes, and the average time spent in tanking is only 45 min. **1935** J. O'Hara Appointment in Samarra ii. 49 But the rest of them! God, what a gang of tankers they were.

tank, *v.*² [f. *tank *sb.*⁷] **1.** *intr.* To proceed or make one's way in a tank. Also *fig.*
1939 H. G. Wells Holy Terror III. ii. 271 The city crowds cheered, the armies went tanking forward. **1945** A. Thirkell Miss Bunting ix. 192 He tanked right over her without so much as noticing her. **1972** R. Poole Towards Deep Subjectivity i. 4 The Russians..shot their way in, they tanked their way in.
2. *trans.* To defeat convincingly, to beat, thrash, or overwhelm. Hence **ta·nking** *vbl. sb.*² Cf. *tonk v. Sc. dial.*
1973 'J. Patrick' Glasgow Gang Observed vi. 60 We were to play football.. ('Uzz Young Team always tank them.'). **1976** Sunday Mail (Glasgow) 26 Dec. 2/1 They..had just come from a party for Rangers F.C., who tanked the local Clachnacuddin side 8–0. **1982** P. Turnbull Dead Knock iii. 56 Glasgow..[is] a good city... The reputation for violence comes from the gangs who give each other tankings.

‖ **tanka²** (tæ·ŋkā). Also **Tanka**. [Jap., f. *tan* short + *ka* song.] A form of Japanese verse which consists of thirty-one syllables, the first and third lines containing five and the other three lines seven syllables.
1877 W. G. Aston Gram. Jap. Written Lang. (ed. 2) x. 197 Tanka..or *mijika-uta*, i.e. 'short poetry', so-called to distinguish it from *naga-uta* or 'long poetry', is by far the commonest Japanese metre. **1899** —— Jap Lit. I. ii. 29 The Tanka is the most universal and characteristic of the various forms of poetry in Japan. **1923** Jun Fujita (*title*) Tanka; poems in exile. **1940** W. de la Mare Pleasures & Speculations 201 A Japanese tanka..on the prolification on the exquisite little cups of the lichen. **1968** Encycl. Brit. XII. 886/1 From the raw material of Chinese poetry came the exquisite haiku and tanka forms. **1982** PN Rev. No. 26. 60/1, I do not think that haiku and tanka are translatable... Fortunately, there is a great deal more to Japanese poetry than tanka and haiku.

‖ **tanka³** (tæ·ŋkā). Also **thang-ka, thanka**, etc. [a. Tibetan *t'án-ka, t'án-ga* image, painting.] A Tibetan religious (scroll-)painting on woven material, hung as a banner in temples and carried in processions.
1925 G. Roerich Tibetan Paintings 17 The most characteristic production of Tibetan pictorial art is the so-called *than-ka*, a word which is commonly interpreted as 'banner'. **1928** 'Ganpat' Magic Ladakh vii. 129 From the beams of the flat ceiling depend painted banners of silk—the gift of various donors. These *tankas*, as they are called, are often very beautiful. **1939** M. Pallis Peaks & Lamas I. vi. 68 We were also shown a scroll-painting of the type found universally in Tibet, and called a *t'hanka*. **1952** A. K. Gordon Tibetan Relig. Art 15 Thang-kas are paintings or, occasionally, embroidered pictures, usually called 'banners'. They..are hung in the temples and at family altars in homes... They portray a deity..or scenes from the life of Buddha. **1969** R. Farre Beckoning Land xx. 242 Thankas are similar to Chinese scroll paintings but the Tibetan ones are always on some religious theme. **1979** Financial Times 7 July 2/2 The imposing Potala Palace is uninhabited with its thousands of priceless gold Buddha statues, rare 'tankas', innumerable Buddhist scrolls and scrips.

tankage. Add: **4.** The fuel-carrying capacity of an aircraft.
1942 W. S. Churchill in Second World War (1951) IV. I. vii. 114 We intend to increase petrol tankage of some Liberator aircraft to give an operational range of 2,300 sea-miles. **1950** Times 17 Feb. 8/5 The maximum still air range with full tankage of 300 gallons and 1,750 lb. of payload will be 1,250 miles and the practical stage length about 850 miles. **1966** M. Woodhouse Tree Frog v. 38 'What really shook me was the tankage.'..'How much fuel does that thing hold?'

tankard. **4.** tankard turnip (earlier example).
1744 W. Ellis Mod. Husbandman Jan. ii. 34, I saw a Farmer transplanting his biggest green tankard Turneps.

tankdrome, var. *tankodrome.

tanked (tæŋkt), *ppl. a.* [f. *tank v.¹ 5 + -ed¹.] **1.** *slang.* Filled with (alcoholic) drink; intoxicated; occas. *transf.*, drugged. Freq. with *up*. Also in phr. *tanked to the wide* (cf. *wide *sb.*) and in developed uses: completely intoxicated.
1893 [see *pub sb. 1]. **1899** A. M. Binstead Gal's Gossip 97 When my male parent, who was a free and frequent librator [sic], came home tanked up. **1917** [see *blotto a.]. **1932** H. Simpson Boomerang ix. 183 Dawlish wrote poetry, and caused acute discomfort by reciting it aloud on starry nights when he was tanked up. **1964** Wodehouse Frozen Assets iv. 77 If a fellow raised from rags to riches at the breakfast table isn't tanked to the uvula by nightfall, it simply means he hasn't been trying. **1968** [see *eyebrow 1 d]. **1977** J. McClure Sunday Hangman xv. 178 He'd arrived half-tanked already.
2. *colloq.* Filled up, fuelled. Also *fig.*
1954 A. Huxley Let. 5 Dec. (1969) 717 A child tanked up with sugar or glucose is likely to get through a party without untoward incidents. **1968** R. Clapperton No News on Monday v. 49, I got the Riley tanked up and started the long haul from Sydney. **1973** J. Drummond Bang! Bang! You're Dead xxxi. 107 We may need the trucks at any time, keep them tanked up.

tanker¹. Delete *colloq.* and substitute for def.: **1. a.** A sea-going vessel fitted with tanks for carrying oil or other liquids in bulk. Cf. *tank-boat, -steamer, -ship* s.v. Tank *sb.*¹ 4 in Dict. and Suppl. (Earlier and later examples.)
1900 Boston Herald 17 Jan. 1/3 The wreck was a tanker. **1920**, etc. [see *oil sb.¹ 6 e]. **1923** R. D. Paine Comrades of Rolling Ocean iv. 73 There was only four of us floated ashore on a capsized boat after the blessed tanker turned turtle. **1950** Sun (Baltimore) 19 Oct. (B ed.) 4/3 It was in 1878 that he [sc. Gustav Conrad Hansen] first put his idea into practice, converting two sailing ships into tankers. **1962** R. H. Brown Dict. Marine Insurance Terms 281 A loaded tanker is usually low in the water. **1977** R.A.F. News 11–24 May 4/4 Two..dinghies attended by a Swedish tanker.
b. A road or rail vehicle with a container designed for transporting fluids in bulk. (Freq. with qualifying word.) Cf. *tank-truck, -wagon* s.v. Tank *sb.*¹ 4.
1927, etc. [see *oil sb.¹ 6 e]. **1947** Times 8 Mar. 4/2 Milk tankers carrying supplies for 400,000 Londoners from creameries in Shropshire. **1951** 'J. Wyndham' Day of Triffids xvii. 300 With a hose from the tanker which held our main petrol supply I filled the half-track's tanks to overflowing. **1960** Farmer & Stockbreeder 2 Feb. 74/1 [He] received fatal injuries..when his car was involved in a collision with a petrol tanker. **1978** J. Sherwood Limericks of Lachasse iii. 35 There were two big road tankers..in the car park. It was used as a safe overnight compound for tankers.
c. An aircraft used for carrying fuel in bulk, esp. for the aerial refuelling of other craft.
1931 Jrnl. R. Aeronaut. Soc. XXXV. 1145 It is..possible that aircraft which had become obsolete as passenger carriers might be utilised as tankers..although..recent experiments indicate that refuelling in the air will best be accomplished by the use of tankers specially designed for their duty. **1950** C. H. Latimer-Needham Refuelling in Flight i. 1 Two aircraft were equipped as tankers..and..were fitted with a 50-ft. length of hose. **1979** J. Barnett Backfire is Hostile xiii. 145 The Russian carrier *Kiev* has flown off a number of strike aircraft at present refuelling from aerial tankers at low level.
2. a. *attrib.* and *Comb.*, as *tanker aircraft, load, train*, etc.
1938 Jrnl. R. Aeronaut. Soc. XLII. 389 The aeroplane would then take off with a small quantity of fuel in the tanks..and the tanks would be filled in the air from the tanker aircraft. **1945** G. Millar Maquis iv. 57 What they don't booze they send off to their factories in tanker wagons to make explosives with. **1953** Times 31 Oct. 11/1 The tanker drivers' strike is out of the way. **1958** Ann. Reg. 1957 v. i. 360 The rise in tanker freights, and the greater use of the Cape route. **1965** D. Francis Odds Against xi. 150 We might not find Smith, the tanker driver. **1973** Country Life 29 Mar. 873/3 A 3,000-hen battery laying unit produces weekly a 600-gallon tanker load of cage manure. **1978** Times 16 Aug. 2/6 The tanker train ran out of control. The tanker guard is blamed. **1981** 'D. Rutherford' Porcupine Basin vii. 126 Two tanker-loading jetties pushed their long arms..out into the water. **1982** Daily Tel. 15 Dec. 24/4 Six TriStar aircraft are to be ordered from British Airways for use as strategic tanker aircraft for the RAF.

b. Special Comb.: **tanker man**, a seaman who is a member of the crew of a tanker (sense 1 a above).
1932 Times 9 Mar. 15/6 What can be done to help these tanker-men to bear or to enrich their isolation? **1974** J. Dyson Prime Minister's Boat xxx. 179 Come sun-up, the tanker men could pick them off at their leisure.
Hence as *v. trans.*, to transport in tankers; **ta·nkering** *vbl. sb.*, the putting (of oil, etc.) into tankers.
1928 Daily Express 20 Feb. 13 Port Said, where the tankering costs the..Company £1,000,000 annually. **1980** Times 8 Jan. 15/4 Airlines..use a complicated logistics operation to 'tanker' fuel around the world to avoid particular airports where it is scarce and expensive.

tankette (tæŋke·t). *Mil. disused.* [f. *tank *sb.*⁷ + -ette.] A small armoured vehicle designed to facilitate the movement of infantry across rough country.
1927 [see *dragon¹ 10]. **1927** Observer 13 Nov. 19/5 The War Office sent a few tanks, tankettes and six-wheelers to perform. **1931** G. le Q. Martel In Wake of Tank xi. 120 He produced a two-man machine with armour protection... These machines were called Tankettes at this time, and they were required to act as scouts. **1942** Times (Weekly ed.) 2 Dec. 12 German barbarians pillaged Mikhailskoye collective farm, while they tied its chairman, an ardent Soviet patriot, to tankettes and tore him to pieces.

tankful. Add: Now usu. with reference to the fuel tank of a car.
1968 'E. Trevor' Place for Wicked ii. 23 He'd..taken her away with the seat-belt on and a warm engine and half a tankful on the dial. **1971** A. Price Alamut Ambush viii. 96 Give him a car and a tankful of petrol. **1978** Detroit Free Press 5 Mar. (Parade Suppl.) 26/1 The result is your average miles per gallon for the previous tankful. The cumulative average, after a few tankfuls, will be an accurate measure of your car's mileage efficiency.

† **tankodrome** (tæ·ŋkodrōᵘm). *Obs.* Also **tankdrome, tanko-drome**. [f. *tank *sb.*⁷ + -o + *-drome, after *aerodrome 2 b.] An area where military tanks are kept.
1918 Illustr. London News 27 July 98 A 'tankdrome' on the Western Front. **1919** C. P. Thompson Cocktails 250, I left the tankodrome and went down with him to the machine. **1920** J. C. Fuller Tanks in Great War 58 A tankodrome (Tank Park) was established at Acheux.

tanky (tæ·ŋki). *Naut. slang.* Also **tankie**. [f. Tank *sb.*¹ + -y⁶, -ie.] The navigator's assistant; the captain of the hold (see quot. 1945).
1909 J. R. Ware Passing Engl. 241/1 Tanky (Navy), foreman or captain of the hold—which looks like a tank. **1921** Blackw. Mag. July 50/1 I'm hanged if I do Tankie any more. **1944** J. Mallalieu Very Ordinary Seaman vi. 91 Draw the meat, spuds, bread, butter, and vegetables from Tanky. **1945** Times Lit. Suppl. 9 June 271/3 The nickname 'Tanky' belonged to his [sc. a navigator's] Yeoman and derived from this man's incidental duty of looking after the freshwater tanks... When refrigerated storage for meat was introduced..it was the practice..to turn the care of these spaces over to 'Tanky'... Thus the lower deck came to connect 'tanky' with fresh meat... There are now, in consequence, at least two 'Tankies' in most ships: the officers' 'Tanky' who provides their bath-water and..the matlows' 'Tanky' who issues the meat and spuds. **1956** H. Tunstall-Behrens Pamir ix. 114 The sharp-witted Amigo had the job of Mate's Tanky.

tanna (ta·nā). Also with capital initial. Usu. in pl. **tannaim** (tanǫ·im); also † **tanaim, tannain**. [Aramaic, = teacher, f. *tĕnā* to repeat, learn, cogn. w. Heb. *šānāh*: see Mishnah, mishna.] One of the Jewish doctors of the law of the first two centuries A.D. whose opinions are recorded in the Mishnah and Baraita. Occas. as **tannaite** (tā·neˌəit) (also as adj., = next); **tannaitic** (tā·neˌi:tik, tā:neˌi·tik) *a.*, of or pertaining to the tannaim.
1718 [see Mishnical a.]. **1845** Encycl. Metrop. IX. 639/2 He [sc. Antigonus of Socho] was the founder of the school of the tannain or mishnical doctors, by which name all the doctors of the Jewish law are distinguished, who lived between the death of Simon [the Just] and the middle of the second century after Christ. **1888** Encycl. Brit. XXIII. 36/1 The Mishnic doctors..were and are called *Tannaim*. **1905** Jewish Encycl. X. 633/1 In rabbinical literature careful discrimination must be made between the tannaitic period and that of the Amoraim. **1911** Encycl. Brit. XXVI. 399/1 The term *tanna* is used in the Talmud of those teachers who flourished in the first two centuries of the Christian era. **1919** H. A. A. Kennedy Theol. Epistles i. 17 These were due to the wisdom of many teachers, of whom the most famous were the so-called Tannaites. **1941** G. G. Scholem Major Trends in Jewish Mysticism ii. 51 The tradition of Tannaitic mysticism and theosophy was really alive among them [sc. the later Merkabah mystics]. **1950** L. S. Thornton Revelation & Mod. World ix. 283 The rabbinical teacher was known as a *tanna*, that is a 'repeater' of the tradition. Ibid., There was a 'tannaite' succession of teachers which traced its genealogy back through successive pairs of rabbis to the men of the Great Synagogue. **1957** Encycl. Brit. XV. 458/1 The Midrash of the schools, often called Halakhic or Tannaite (i.e. Mishnaic) Midrash. **1969** D. Daube Roman Law iii. 158 Tannaitic law, that is to say, the early Talmudic law of, say, 100 BC to AD 200. **1977** New Yorker 17 Oct. 48/3 Rabbi Gabriel used to remind her that, even according to the strictest letter of the law, kissing and embracing are permitted and that tannaim and amoraim frolicked with their spouses in bed.

tannase (tæ·nēⁱz). *Biochem.* [a. F. *tannase* (A. Fernbach 1900, in *Compt. Rend.* CXXXI. 1214): see Tannin and *-ase.] An enzyme that hydrolyses ester linkages in tannins.
1901 J. R. Green Soluble Ferments (ed. 2) x. 169 Tannase attacks not only tannin but the compound of tannin and gelatin, as well as other tannates. **1929** R. P. Walton tr. Waldschmidt-Leitz's Enzyme Actions & Properties 122 Tannase..must be regarded as a specific esterase having a special affinity for the esters of phenol carboxylic acids. **1979** Jrnl. Chromatogr. CLXX. 446 The use of tannase (tannin acylhydrolase..) in localization procedures for enzymatic activity has never been studied.

tanned, *ppl. a.* Add: **5.** *Immunol.* **tanned-(red-)cell**, used *attrib.* to designate a test in which antibodies can be detected by observing the agglutination of red blood cells that have been coated with tannic acid which has then bound with the appropriate soluble antigen.
1956 Jrnl. Immunol. LXXVI. 409/1 The tanned cell hemagglutination test..was also applied to the problem. **1962** Lancet 5 May 951/2 In a series of 78 patients with pernicious anæmia, the tanned-red-cell agglutination test was positive in 24% of males. **1980** Canad. Jrnl. Zool. LVIII. 245 One group [of cattle] infected with only *H[ypoderma] lineatum* was followed using the tanned-cell hemagglutination technique.

tanner¹. Add: **2.** [f. Tan *v.* 2 a.] A lotion, cream, etc., designed to promote a sun-tan when applied to the skin on exposure to the sun; *artificial, man-made tanner*, one which colours the skin brown without the aid of the sun.
1969 Daily Tel. 2 July 15/8 People with sensitive..skins should be wary of all artificial tanners. It's advisable to try any man-made tanner on a small patch of skin. **1972** Vogue June Special 40/2 The new tanners: something here for every kind of skin... Rub your tanner right up into the hairline. **1979** Country Life 31 May 1776/1 There are many artificial tanners on the market; the Charles of the Ritz Self Tanning Foam, £3.95, claims to..tan the skin in a similar way to the sun.

tannia: now the usual form of Tania in Dict. and Suppl.

tannie (ta·ni). *S. Afr. colloq.* Also **Tannie**. [Dim. of Afrikaans *tante*: see *tante 2.] **a.** An informal mode of address used to an aunt or any older woman. **b.** A prim elderly woman. Also *transf.*
1958 L. van der Post Lost World Kalahari i. 16 'Old tannie sea-cow' was our endearing way of naming the hippopotamus, so called because it was there in the surf of the sea to welcome my people when they first landed in Africa. **1958** R. Collins Four-Coloured Flag 16 'Good afternoon, Tannie,' he mumbled, from a distance of twenty feet. **1958** Cape Times 10 Dec. 11/3 The radiologist told him of the shrieks from alarmed tannies in from the country when they found themselves being elevated on high. **1969** I. Vaughan Last of Sunlit Years ix. 77, I am finding that most Afrikaans children call one 'Aunty', or 'Tannie', and are most charmingly co-operative. **1980** Rand Daily Mail 15 Apr. 8/3 So tasteful one feels sure one's most uptight Victorian tannie from Tweetackie-slippersfontein would be pleased to receive her.

tanning, *vbl. sb.* (Later examples.)
1887 J. E. Taylor Tourist's Guide Suffolk 61 Combs..is distant 1 m., well known for the extensive tanning-works of Messrs. Webb. **1899** C. F. Townsend Chem. for Photographers (ed. 2) iv. 75 The 'tanning' of the gelatin or rendering it insoluble. **1935** Discovery July 190/1 Wash-out gelatine relief by development tanning, used for making matrices for obtaining 'imbibition'. **1944** J. S. Huxley On living in Revolution 70 Lamarckism..the inheritance of characters acquired by an individual as a result of changes in the environment, like tanning due to sun. **1963** Listener 17 Jan. 138/1 From winter resort patronized by the pre-1914 crowned heads of Europe to tanning-factory for bikini-clad campers. **1980** B. H. Carroll et al. Introd. Photogr. Theory xi. 227 Tanning development involves the formation of an image in the form of insolubilized gelatin.

tanning, *ppl. a.* Add: **b.** *spec.* in *Photogr.* Cf. *tan v. 4.
1930 G. E. Brown Clerc's Photography xxviii. 224/2 The localized tanning of the gelatine gives rise to unequal contractions of the film during drying... Tanning developers are not used in astronomical work. **1959** Chambers's Encycl. X. 686/2 Most fixing solutions also contain a tanning or hardening agent which unites with the gelatin of the emulsion layer, increases its melting-point and reduces its swelling in water. **1973** D. A. Spencer Focal Dict. Photogr. Technol. 617 A typical tanning developer contains little if any sulphite.

tannish (tæ·niʃ), *a.* [f. Tan *a.* + -ish¹.] Somewhat tan-coloured.
1935 J. T. Farrell Judgment Day xiii. 287 He glanced at a squat fellow in a crimson jersey and tannish knickers. **1947** D. M. Davin Gorse blooms Pale 53 It [sc. the calf] had Rosy's colouring, only at the ends the hair deepened into the Jersey's tannish black. **1961** M. Beadle These Ruins are Inhabited (1963) iii. 40 We ambled into..a tiny Saxon church. Of tannish stone..it was thick-set..and sheltering. **1965** Amer. Philos. Q. II. 320/2 A deep olive green..would be more beautiful..against a tannish pink.

Tannoy (tæ·noi). Also **tannoy**. A proprietary name for electrical apparatus concerned with sound reproduction and amplification. Now used generally, esp. to denote a form of public address system.
1928 Trade Marks Jrnl. 18 Apr. 605 Tannoy 488,958. Electrical Instruments and Apparatus for use in connection with Wireless Telegraphy and Telephony and Electrical Conductors... Guy Rupert Fountain, trading as the Tulsemere Manufacturing Co.... 28th February 1928. **1942** Ibid. 21 Oct. 433/1 Tannoy. Electric signalling and alarm apparatus, telephone systems..television..radio apparatus,..sound reproducing and sound amplifying instruments..; public address apparatus; electric amplifiers. **1944** 'N. Shute' Pastoral ii. 36 He was lighting his pipe..when the Tannoy sounded metallically above his head. **1954** G. Smith Flaw in Crystal xx. 215 The tannoy blared, telling the passengers to go aboard. **1958** M. K. Joseph I'll soldier No More ix. 158 A tannoy-loudspeaker on the deck brought them the hourly news bulletins. **1969** A. Glyn Dragon Variation i. 11 Above them the hoarse voice of the muezzin crackled through the Tannoy, calling the faithful to prayer. **1977** G. Scott Hot Pursuit iii. 34 A call in the Tannoy system drowned my words. **1980** Daily Tel. 12 Nov. 19/3 They were furious after missing a train because of a wrong announcement on the station tannoy.
Hence as *v. trans.*, to call (someone) by public address system; *intr.*, to use a public address system; **ta·nnoyed** *ppl. a.*, transmitted by public address system; **ta·nnoying** *ppl. a.*
1966 Punch 20 Apr. 564/1 The President spoke, and his tannoyed voice boomed back off the blighted trees and the peeling architecture. **1970** B. Turner Another Little Death ii. 14 He held the line while Chief Inspector Rillie was Tannoyed out of the basement. **1976** Times 24 Apr. 4 The Portuguese Socialist leader arrives behind a convoy of tannoying cars and almost bounces into the crowd. **1978** Times 2 Oct. 4/3 A great crowd..being ordered around by Tannoyed voices.

Tanoan (tānōᵘ·ăn), *sb.* and *a.* [f. Sp. *Tano*, ad. Southern Tewa self-designation *tʰá·nu*, + -an.] **A.** *sb.* A family of languages spoken in parts of New Mexico and Arizona by Pueblo Indians; also, the group of people which speaks these languages. **B.** *adj.* Of, pertaining to, or designating this linguistic group.
1891 J. W. Powell Indian Linguistic Families of Amer. 121 Tañoan. **1909** Amer. Anthropologist XI. 564 The Keres (Queres) are introduced among tribes speaking languages of the Tanoan family. **1925** [see *Keres]. **1928** J. P. Harrington Vocab. Kiowa Lang. 1 A brief text [has]..been included, as well as Tanoan etymologies taken from the Tewa dialect spoken at San Juan Pueblo near Santa Fe, New Mexico. Ibid. 11 In the Tanoan languages, several of the consonants have a hard and a soft form. **1941** C. F. Voegelin Language, Culture & Personality 28 Only fragmentary notes and word lists for Tanoan were available. Taos was taken as a type for Tanoan phonetics and morphology. **1950** F. Eggan Social Organization of Western Pueblos 314 The Tanoan groups in the Rio Grande. **1959** G. L. & E. L. Trager in Amer. Anthropologist LXI. 1078 (*title*) Kiowa and Tanoan. **1974** Encycl. Brit. Micropædia IX. 810/2 *Tanoan languages*, the family of Aztec-Tanoan languages spoken in the valley of the Rio Grande in..New Mexico, U.S., including Northern Tiwa, Southern Tiwa, Piro, Tewa, Tano (the type language), and Towa; spoken by about 5,000 people (24,500 in 1680).

tanpoora, tanpura, varr. Tamboura in Dict. and Suppl.

‖ **tansu** (ta·nsu). [Jap.] A Japanese chest of drawers.
1886 E. S. Morse Japanese Homes & their Surroundings iv. 196 The *tansu*—a chest of drawers not unlike our bureau—is often placed within the closet. **1936** K. Nohara True Face of Japan x. 141 The *tansu*, or chest, consists of two, three, or at the most four drawers above each other, which fit exactly into the wall recesses. **1958** M. Joya Things Japanese III. 82 There are unpainted *tansu* or chests, tables, trays and boxes. **1970** P. Zelver Honey Bunch (1971) iv. 16 There was a good modern couch, a Japanese tansu, the art nouveau desk. **1977** South China Morning Post (Hong Kong) 15 Apr. 31/7 Happy Joss has just received a new shipment of tansu, Imari, hibachi and fabrics from Japan.

tansy. Add: **5.** *tansy tea* (examples).
1893–4 R. O. Heslop Northumberland Words II. 718 Tansy-tea, an infusion of the herb. **1902** Spectator 12 Apr. 546/1 Patent pills and soothing syrups have taken the place of calamint and tansy tea. **1965** M. Thomas Grannies' Remedies 26 Hysterics... Strong tansy tea, taken cold and in small quantities, is good.

tant, tanta, varr. *tante 2.

tantadlin, tantoblin. Add: Also **tantaf(f)lin.** **1.** (Later examples.)
1905 in Eng. Dial. Dict. VI. 29/1 (S. Not.) She made cakes an' tantaflin sorts o' things. A bit o' that beef for me; a don't care for non o' yer tarts an' tantaflins. **1911** D. H. Lawrence White Peacock iii. 38 I'm sure you like tantafflins, don't you, Georgie?

tantalian (tæntēⁱ·liăn), *a. Min.* [f. Tantal(um + *-ian 2.] Of a mineral: having a (small) proportion of a constituent element replaced by tantalum.
1930 W. T. Schaller in Amer. Mineralogist XV. 572 Tantalum—tantalian. **1959** [see *ferrian a.].

tantalization. (Later examples.)
1849 C. Brontë Shirley I. viii. 225 Rose had no idea of tantalization, or she would have held him awhile in doubt. **1931** H. S. Walpole Judith Paris iv. 678 From that misery of tantalization he had died. **1983** Listener 28 July 32/3 The human frame is just not conditioned to support such stupendous tantalisation.

tantalus. **2.** (Earlier example.)
1888 G. Grossmith Society Clown 178, I sent him a small souvenir in the shape of a 'Tantalus'.

‖ **tant bien que mal** (taṅ bjeṅ kə mal), *adv. phr.* [Fr., lit. 'as well as badly'.] With indifferent success; moderately well, after a fashion.
1765 Ld. Chesterfield Let. (1774) II. 490 They amuse me, *tant bien que mal*, for an hour or two every morning. **1843** Thackeray Irish Sk.-Bk. II. v. 97 Sketching, *tant bien que mal*, the bridge and the trees..the writer became an object of no small attention. **1890** W. James Princ. Psychol. II. xxvi. 494 We explain the mystery *tant bien que mal* by our evolutionary theories. **1939** Scrutiny VIII. 42 Thus his fleshly desires are satisfied, *tant bien que mal*.

‖ **tante** (taṅt, ta·ntə). Also **Tante**. **1.** [Fr., Ger.: cf. Aunt.] An aunt; also, an older woman who stands in a close relationship. Freq. prefixed to a proper name or as a form of address.
1815 F. Burney Let. 13–18 May (1980) VIII. 129 My tante' made me a long agitated visit. **1929** E. M. Brent-Dyer Rivals of Chalet School i. 18 It was a smiling small face..that was raised to 'Tante Marguérite's' when Mrs Russell drew her close for a kiss. **1932** G. Heyer Devil's Cub i. 15 Aunt Fanny has already warned Tante against your nobody. **1935** Z. N. Hurston Mules & Men II. vi. 287 Ah don't have five nickels, Tante Celestine, but Als'll send a boy to get them. **1941** M. Treadgold We couldn't leave Dinah viii. 130 I lived in Nuremberg... Grandmother and Tante Anna and Tante Frieda were very good to me. **1943** H. T. Kane Bayous Louisiana viii. 166 A nonc and tante..seem far closer than the usual American uncle and aunt. **1964** S. Bellow Herzog 146 Come here, little Moses, and sit on your old *tante's* knee. **1981** B. de Breffny My First Naked Lady III. 147 Anna's mother, tante Rachel, was Turkish.
2. *S. Afr.* With pronunc. (ta·ntə). Also **Tant** (usu. preceding a proper name which begins with a vowel), **tanta**. [Afrikaans, from Du.] = sense 1 above. Also more widely, a term of respect for any elderly woman. Cf. *tannie.
1845 S. Dennison Let. 12 Apr. in D. R. Edgecombe Lett. Hannah Dennison (1968) 205 Pray give my love to Miss Gush and Mrs G and remember me kindly to your good Tant. **1871** H. H. Dugmore Reminisc. Albany Settler (1958) iii. 33 Powers of persuasion had to be employed with oude Tante Nieuwkerk. **1872** Cape Monthly Mag. Sept. 230 And then, does it not sometimes guard the slumbers and share the dreams of our beloved 'tantas'? **1883** [see *oom]. **1900** H. Blore Imperial Light Horseman 162 If a Boer were to be presented at Court he'd offer to shake hands with Queen Victoria, and address her as 'Tante'. All women to whom one wishes to pay respect are called 'Tante'. **1923** O. Schreiner Thoughts on S. Afr. vii. 260 Oom and Tante, I will whisper to you a secret! **1937** C. R. Prance Tante Rebella's Saga 63 Each homestead on a circuit has its own special call..so that every Tante on the line knows..if..someone has something to say to Tant' Emmerentia. **1950** Cape Times Mag. 5 Aug. 3/3 Tanta Theodora thought that her kettle..was not as dazzling as it might be. **1969** Personality 5 June, Always there is the contrast between past and present..the austere black dress of the old tante and the bright mini-wear of a visiting granddaughter. **1974** Panorama Apr. 21 Imagine the surprise of old Tant Alida Prinsloo at finding the contents of her voorhuis (living room) described.

tanteuxenite (tæntyū·ksĭnəit). *Min.* [f. Tant(alum + Euxenite.] A titanotantalate of yttrium, Y(Ta, Ti)$_2$O$_6$, found as brownish or black, tabular or prismatic, orthorhombic crystals.
1928 E. S. Simpson in Jrnl. R. Soc. W. Austral. XIV. 45 Whilst normal euxenite..is a titanoniobate of yttrium, the mineral here described differs from it in being essentially a titanotantalate of yttrium... A new name being required for the species, Tanteuxenite is suggested. **1970** Neues Jahrb. für Mineral.: Abhandl. CXII. 131 Tanteuxenite has been found in rough flat prisms in the alluvials of Liha, together with columbite, euxenite and uranium microlite. **1974** [see *struverite 2].

‖ **tantième** (taṅtiʸɛ̄m). *rare.* [Fr., f. *tant* so much + *-ième*, ending of Fr. ordinal numerals.] A percentage or share, esp. of profits, royalties, etc.
1897 G. B. Shaw Let. 8 Sept. (1965) I. 804 You would simply be robbing the deserving poor if you cut off my tantième. **1911** —— Lett. to Granville Barker (1956) 175 Where should I be if I had to live on my tantiemes.

‖ **tant mieux** (taṅ mjō). [Fr.] So much the better. Cf. next.
1754 Ld. Chesterfield Let. 8 Mar. (1774) II. 354, I really believe [he] will be your friend upon my account; if you can afterwards make him yours, upon your own, *tant mieux*. **1791** H. Walpole Let. 26 May (1944) XI. 272, I am rich in letters from you... You tell me mine entertain you; *tant mieux*; it is my wish, but my wonder. **1830** M. Edgeworth Let. 8 Dec. (1971) 444 He..promised to do all he can for him... If that should be while I am here—tant mieux. **1876** C. M. Yonge Three Brides I. xii. 188 'She seems absolutely repellent.' '*Tant mieux*,' muttered Raymond. **1929** [see next]. **1972** M. Kaye Lively Game of Death (1974) xiii. 74 If your boss can pin his death on somebody, *tant mieux*.

‖ **tant pis** (taṅ pi). Also **tant-pis**. [Fr.] So much the worse. Cf. prec.
[**1768** Sterne Sentimental Journey I. 92 *Tant pis* and *tant mieux* being two of the great hinges in French conversation, a stranger would do well to set himself right in the use of them.] **1782** H. Walpole Let. 25 Mar. (1955) XXIX. 207 The new cabinet are to be Lord Thurlow, Chancellor, (*tant pis*), [etc.]. **1855** A. Thackeray Jrnl. 10 June in H. Ritchie Lett. A. T. Ritchie (1924) v. 68 But tant-pis, when I'm older I hope I shall do it better. **1929** A. Huxley Let. 24 Nov. (1969) 319 If you happen to find people constituted like yourself, tant mieux. If you find them differently constituted, tant pis. **1979** N. Slater Falcon ix. 160 If Hudson wouldn't play ball when the crunch came, then—*tant pis*. Too bad.

Tantric, *a.* Add: Also **tantric, tantrik**. (Later examples.) Also, characteristic of Tantrism. In Western use, also *loosely* denoting the association of spiritual and erotic practices. Occas. as *sb.* (in form *Tantrik*), (*a*) a practitioner of Tantrism; (*b*) a Tantra.
1920 Webster, Tantrik. **1927** A. Huxley Proper Studies 181 The Tantric reformation introduced worship of goddesses, together with a rich collection of magical and erotic rites. **1937** M. Covarrubias Island of Bali (1972) vii. 174 It was Erlangga who instituted Javanese as the official language of Bali. Tantric black magic seems to have played an important part in Erlangga's time. **1954** W. Noyce South Col iv. 61 Erotic paintings of tantric gods and goddesses adorn the walls. **1956** A. Huxley Adonis & Alphabet 59 The Tantriks of northern India and Tibet. **1959** Times Lit. Suppl. 20 Feb. 96/2 It contains elements..which belong to the Tantric Buddhism of India. **1961** A. Huxley Let. 8 Jan. (1969) 902 Pure perceptual receptivity is the basis, incidentally, of many Tantrik exercises. **1966** L. Cohen Beautiful Losers III. 241 An entire cult of Tantric lore perfectionists turned exocentric in their second chance at compassion. **1969** Cultural News from India Nov. 16 But his oils showed an innate understanding of Tantric motifs as well as the..use of material within an arbitrarily fixed boundary. **1977** Time Out 28 Jan.–3 Feb. 65/2 (Advt.), Student of energy release, seeks tantric adept, female. **1977** It May 17/1 The Vak..develops in the course of its manifestation into the seed-sounds of the Tantriks. **1980** Dædalus Spring 103 Repressed Hindus rejoice in myths of extreme forms of..erotic Tantric orgiasticism.

Tantricism (tæ·ntrikiz'm). [f. prec. + -ism.] = Tantrism; also *spec.* Tantric yoga.
1959 Times Lit. Suppl. 20 Feb. 96/2 The secret of Tantricism lies in the control of nervous centres. **1969** Newslet. Tibet Soc. U.K. June 2 Gyudto and Gyudmed, the Upper and Lower Colleges of Tantricism, in its re-organised form.

Tantrism. (Earlier example.)
1877 [see *Sakta].

tantrum. For 'Mostly' read 'Freq.' and add: (Earlier and later examples.) Now often *spec.* a fit of bad temper in a young child.
1714 E. Verney Let. 30 Oct. in M. M. Verney Eighteenth Cent. Memoirs (1930) II. xxi. 18 Our lady has had some of her tanterums as Vapors comeing out etc. **1927** A. Conan Doyle Case-Bk. Sherlock Holmes xi. 283 Sir Robert was in one of his tantrums. **1966** M. Frayn Russian Interpreter xxii. 103 'Let's put all these books away in the case again,' she said coaxingly, as if Proctor-Gould had thrown his toys about in a tantrum. **1979** N. Gordimer Burger's Daughter 329 If we'd still been children, I might have been throwing stones at him in a tantrum.

Tantum ergo (tæ·ntŏm ɜ̄·ɹgo). [First two words of the penultimate stanza, which begins 'Tantum ergo sacramentum Veneremur cernui' (Therefore we, before him bending, This great sacrament revere), of the hymn of St. Thomas Aquinas 'Pange lingua gloriosi Corporis mysterium'.] The last two stanzas of this hymn sung at Benediction; also, a setting of these.
1709 A. Busby in B. Jarrett Eng. Dominicans (1921) ix. 188 Tantum ergo was sung by the Cannons accompanied with Musick wich fild our hearts with joy. **1897** Addis & Arnold Cath. Dict. (ed. 5) 84/1 Next the Te Deum..or some other canticle..is sung, followed by the 'Tantum Ergo'. **1905** J. H. Harting Hist. Sardinian Chapel 38 A *Tantum Ergo* was composed by Alessandro De Angioli for the Sardinian Chapel. **1922** Joyce Ulysses 355 Then they sang the second verse of the *Tantum ergo*. **1976** N. Roberts Face of France vii. 82 The procession to the Altar of Repose, when the *Tantum Ergo* comes rolling out in Gregorian and sonorous Latin.

TANU (tā·nu). Also **Tanu**. [Acronym f. the initial letters of *T*anganyika *A*frican *N*ational *U*nion.] The name of a former political party in Tanganyika (now Tanzania).
1957 Times 15 Feb. 7/2 Speeches made recently by the president of the T.A.N.U. **1957** J. Nyerere Let. in Times 18 Sept. 9/4 For the last three years T.A.N.U. has been trying to get from the administering authority a statement that it intends to develop Tanganyika to become a democratic African state. **1959** New Statesman 12 Sept. 300/1 TANU now has 800,000 members, each of whom pays two shillings entrance fee and six shillings a year. **1967** Economist 18 Feb. 613/1 The constitution of the Tanganyika African National Union (Tanu) lays down [etc.]. **1971** Standard (Dar es Salaam) 7 Apr. 1/4 He asked all members of the corporation to study well the new Tanu Guidelines during their discussions. **1977** Jrnl. Commonwealth & Compar. Politics XV. 247 TANU and the Afro-Shirazi Party of Zanzibar officially merged in February 1977 to form Chama cha Mapinduzi (the Revolutionary Party of Tanzania), abbreviated to CCM. **1978** S. Naipaul North of South II. iii. 163 The peasants here displayed little or no understanding of TANU policies.

Tanzanian (tænzănī·ăn), *sb.* and *a.* [f. *Tanzania* (see below) + -an, -ian.] **A.** *sb.* A native or inhabitant of Tanzania, an E. African state formed in 1964 by the union of the republics of Tanganyika and Zanzibar. **B.** *adj.* Of or pertaining to Tanzania.
1965 Economist 23 Jan. 332/3 Mr Nyerere..apparently decided he must at any cost assert Tanzanian independence from this supposed American intrigue. Ibid. 6 Feb. 519/1 Whatever in the world, Tanzanians are asking, have they done. **1970** Drum (E. African ed.) Feb. 22/3 A gradual increase in standards of health, education and housing..will mean in the end healthier, wealthier, better-fed and happier Tanzanians. **1973** Listener 6 Sept. 301/1 The Chinese-built Freedom Railways reached the Zambian border, its Tanzanian section having been completed. **1978** S. Naipaul North of South II. ii. 158, I watched the Swedish girl weighing the babies... 'We cannot get well-off and educated Tanzanians to come and help us out.' Ibid. vi. 248 The Immigration Officer (Tanzanian) poked a grinning head through the doorway.

tanzanite (tæ·nzănəit). *Min.* [f. as prec. + -ite¹.] A highly pleochroic violet-blue gemstone that is a variety of zoisite in which some of the aluminium is replaced by vanadium.
1968 Wall St. Jrnl. 14 Oct. 4/1 Tiffany & Co. disclosed discovery of a rare gemstone that the firm's vice president, Henry B. Platt, has named tanzanite. The gem, discovered last year in Africa's Tanzania, has the blue coloring and relative transparency of a sapphire. **1975** Nat. Geographic Apr. 490/1, I take a dusty detour to see the mining of glorious gems of transparent purplish blue. Tiffany's has named them tanzanites.

‖ **tao** (tau, d-). Also **Dao, Tao, taou, tau.** [Chinese *dào* road, way.] **1. a.** In Taoism, an absolute entity which is the source of the universe; the way in which this absolute entity functions.
1736 R. Brookes tr. Du Halde's Gen. Hist. China III. 30 Among the Sentences [of Lao Kiun] there is one that is often repeated..: Tao, says he, or Reason, hath produced one, one hath produced two, two have produced three, and three have produced all things. **1797** Encycl. Brit. IV. 676/1 *Tao* is one by nature: the first begot the second; two produced the third; the three created all things. **1868** J. Chalmers Lau-Tsze's Specul. Metaph. p. xii, Existence is..said to be produced from non-existence, and *Tau* is the union of the two. **1904** W. G. Old tr. Laotze's Simple Way iv. 30 Tao is without limitation; its depth is the source of whatever is. **1934** A. D. Waley Way & its Power 50 *Tao* is the way that those must walk who would 'achieve without doing'. But *tao* is not only a means, a doctrine, a principle. It is the ultimate reality in which all attributes are united. **1950** A. Huxley Themes & Variations 172 In China and Japan mountains were taken more seriously. The aspiring artist was advised..to contemplate them lovingly until he could understand the mode of their being and feel within them the workings of the immanent and transcendent Tao. **1957** J. Kerouac On Road (1958) 251 He was reaching his Tao decisions in the simplest direct way. **1963** D. C. Lau Lao Tzu 23 In the *Lao tzu*, the *tao* is no longer 'the way of something', but a completely independent entity, and replaces heaven in all its functions. But the *tao* is also the way followed by the inanimate universe as well as by man. **1971** F. Mann Acupuncture (ed. 2) iv. 47 The root of the way of life (Dao or Tao), of birth and change is Qi. **1980** M. H. Kingston China Men (1981) 96 Bak Goong thought he understood the Tao, which is everywhere and in everything, even in our excrement.
b. = Taoism, Taoist *a.*
1745 tr. J. F. Gemelli Careri in A. & J. Churchill Coll. Voy. & Trav. (ed. 3) IV. II. iv. 313/2 In some of these *Pagods*, religious men and women live in community to serve them; of which there are two sorts, the one of the sect of *Foe*, and the other of that of *Tao*. **1747** New Gen. Coll. Voy. IV. I. v. 214/1 The Sectaries say..that the great Doctrine of *Fo* and *Tau* swallows up all in nothing. **1831** Canton Miscellany I. 28 Hwuy-tsung, an Emperor of the Sung Dynasty, was fond of being a Priest of the Taou Sect. **1959** Listener 26 Feb. 388/2 Tao and Zen. **1980** Jrnl. R. Soc. Arts Feb. 137/2 The garden was seen as the most powerful metaphysical symbol for Tao, Shinto and Zen.
2. In Confucianism and in extended uses, the way to be followed, the right conduct; doctrine or method.
1934 A. D. Waley Way & its Power 30 Each school of philosophy had its *tao*, its doctrine of the way in which

life should be ordered. 1943 C. S. Lewis *Abolition of Man* iii. 30 In the older systems both the kind of man the teachers wished to produce and their motives for producing him were prescribed by the *Tao*—a norm to which the teachers themselves were subject and from which they claimed no liberty to depart. 1966 F. Schurmann *Ideology & Organization in Communist China* i. 50 The Chinese Communists speak of the forces of world history which are universal and cosmic. Though this belief clashes with traditional Confucian beliefs of *tao* (something akin to 'natural law'), it bears certain similarities to the Taoist belief in Heaven as a real force. 1970 H. G. Creel *What is Taoism?* i. 2 Tao at first meant 'road' or 'path'. From this it developed the sense of a method, and of a course of conduct. As a philosophical term it appears first in the Confucian *Analects*. For the Confucians *tao* is the way, the method, of right conduct for the individual and for the state. 1972 F. Fitzgerald *Fire in Lake* i. 11 In analyzing these disasters the emperor blamed them on his deviation from Tao, the traditional way, which was at once the most moral and the most scientific course. 1980 *Dædalus* Spring 34 A Tao of Physics in which the details of modern macrophysics and microphysics are matched to those of the mystical tradition.

Taoiseach (tī·ʃəχ,-aχ). [a. Ir., lit. 'chief, leader'.] The Prime Minister of the Republic of Ireland.

1938 *New Irish Constitution: Citizen's Man.* 17 Dail Eireann is dissolved by the President on the advice of the Taoiseach. 1941 G. B. Shaw *Matter with Ireland* (1962) 285 The Irish Taoiseach (Premier), Mr de Valera, made no move. *a* 1966 'M. Na Gopaleen' *Best of Myles* (1968) 128 You pick up the receiver and say 'Who? The Taoiseach? Oh very well. Put him on.' 1973 *Irish Times* 2 Mar. 9/1 Whoever is going to be Taoiseach is going to have to sweat and work every minute of every day. 1981 *Listener* 1 Jan. 4/2 Mrs Thatcher..permits herself to follow very much the kind of approach the Taoiseach, Mr Charles Haughey, was hoping for.

Taoism. Add: Also **Daoism** and with pronunc. (dau·iz'm). (Earlier and later examples.)

1838 Gützlaff & Reed *China Opened* II. xv. 209 (*heading*) Taouism. 1948 *Mind* LVII. 535 Dr. Fung.. shows how Buddhist philosophy..influenced both Confucianism and Daoism. 1981 *Times* 22 June 6/8 It is not the Vatican which bothers China's leaders most in religious matters—but Daoism (formerly known as Taoism), the only religion truly native to China. *Ibid.* 11 Nov. 6/7 The ancient Chinese religion of Daoism.

Taoist, *sb.* (*a.*) Add: Also **Daoist** and with pronunc. (dau·ist). **a, b.** (Earlier and later examples.)

1838 Gützlaff & Reed *China Opened* II. xv. 209 Such are the better description of Taouists in China. 1971 *Ink* 12 June 8/3 There were many non strict Daoist farmers there who could have been VC. 1981 *Times* 22 June 6/8 Unlike the Buddhists, the Daoists have been granted no licence to continue or revive their practices.

Tao Kuang (dau gwaŋ). Also **Daoguang.** The title of the reign of Xuan Zong (Min-Ning), emperor of China 1821–50, used *attrib.* and *absol.* to designate the period of his reign or pottery and porcelain made at this time.

1927 W. B. Honey *Later Chinese Porcelain* 59 A considerable part of Tao Kuang porcelain was made in revived Yung Chêng patterns. 1951 R. S. Jenyns *Later Chinese Porcelain* ii. 20 A series of 1662 to 1675 wares, often with K'ang Hsi marks, and some marked Shên-tê t'ang, a hallmark which does not occur only on Tao Kuang pieces. 1960 H. Hayward *Antique Coll.* 277/1 *Tao Kuang period*, (1821–50). Characteristic Chinese porcelain wares of this reign are those minutely painted in *famille rose*..style, employing low-toned enamels, and *graviata*, coloured grounds. 1973 *Country Life* 7 June 1680/2 A pair of reticulated ruby ground hexagonal vases..famille rose 19⅞ in, Tao Kuang. 1976 Scott & Koski *Walk-In* (1977) ii. 16 A shabby reproduction, made..in Hong Kong, of a Tao Kuang teapot. 1980 *Catal. Fine Chinese Ceramics* (Sotheby, Hong Kong) 90 Seal mark and period of Daoguang (Tao Kuang).

Taos (taus, tā·os). The name of a town in New Mexico, used *attrib.* (occas. *absol.*) to designate members of a Pueblo Indian tribe living there, or the language of this tribe, a variety of Tiwa.

1844 J. Gregg *Commerce of Prairies* I. 86 A Taos Indian who formed one of the Mexican escort, seeing a gun levelled at his commander, sprang forward and received the ball in his own body, from the effects of which he instantly expired! 1887 *Scribner's Mag.* II. 510 Then the saddle-blanket is laid over his withers, with sometimes a *tilpah*, or parti-colored rug, woven and dyed by the Navajo or Taos Indians. 1939 *Language* XV. 51 The Taos language forms with that of Picuris..the northern branch of Tiwa. *Ibid.*, The old people..speak Taos and Spanish. 1944 B. Johnson *As much as I Dare* 287 Adobe walls around the garden and various nooks and vistas were being built by Taos Indian labor. 1964 *Language* XL. 202 He has published an article describing the application of his system to the Taos language and culture. 1973 A. H. Whiteford *N. Amer. Indian Arts* 28 Taos and Picuris make only unpainted gold-tan pottery. 1978 *Language* LIV. 233 'About the nearest he ever came to having fun' was making charts of the Taos pronoun. 1978 G. A. Sheehan *Running & Being* viii. 111 A Taos Indian chief had once told him that white men were covered with wrinkles because they were crazy.

‖**taotai** (tau·tai, d-). *Hist.* Also **Taotai, tautai,** etc. [Chinese *dàotái*.] The title given to the Chinese provincial officer responsible for the civil and military affairs of a district, abolished shortly after the establishment of the Republic in 1911.

1747 *New Gen. Coll. Voy.* IV. i. vi. 253 To every District there also belongs a Mandarin, called Tau-ti. 1835 *Chinese Repository* Oct. 279 The class of officers next in rank to these are called taou or taoutae: they are not under the orders of the 'two sze', but of the governor and lieut.-governor, and it is their duty to take part in the 'protection' and 'circuit-supervision' of portions of the province. 1848 S. W. Williams *Middle Kingdom* I. vii. 345 The gabel and commissariat are mostly under the direction of officers called *tau*, or *tautai*, sometimes termed intendants of circuit, who have other functions in addition. *Ibid.* 346 The *tautai*..are a kind of deputy of the governor-general and lieutenant-governor, residing in the *tau*, or circuits, into which each province is divided. 1895 *Daily News* 19 Jan. 6/6 A number of Chinese guerilla troops recently tried to enter Neuchwang. The taotai of the city closed the gates, and offered an armed resistance to their entry. 1926 *Blackw. Mag.* Nov. 629/1 A mandarin named Liang was sent to the island as Taotai. 1943 J. T. Pratt *War & Politics in China* xii. 193 When the Revolution came to Shanghai the Taoti—the chief Chinese official—requested the consular body to take temporary charge of the court. 1959 P. Fleming *Siege at Peking* iii. 42 Henceforth Bishops would rank with Governors-General and Governors, Provicaires with Treasurers, Judges and Taotais, and so on down the respective hierarchies.

‖**t'ao t'ieh** (tau tyə). Also **taotie, tao-tieh.** [Chinese *tāotiè*.] The name of a mythical monster, or a mask-design showing its face, found esp. on metalware of the Chou period (1122–221 B.C.). Freq. *attrib.*

1915 R. L. Hobson *Chinese Pottery & Porcelain* II. xvii. 290 This is the face of the t'ao t'ieh (the gluttonous ogre) supposed originally to have represented the demon of the storm. 1933 *Illustr. London News* 9 Dec. (Suppl.) p. i/1 This bell has a *t'ao-t'ieh* design on the upper part. 1958 W. Willetts *Chinese Art* I. iii. 161 *T'ao-t'ieh*..is a device in which two confronting zoomorphs in profile form the left and right sides of an animal mask seen in full face. *Ibid.* 162 Karlgren analyses the *t'ao-t'ieh* motive into six different types. 1965 *New Statesman* 20 Aug. 257/1 Eloquent prose passages like René Grousset's dramatic evocation of the *t'ao-t'ieh* on the sides of Chou vessels. 1973 *Genius of China* 47/2 It is notable that the convention of the face painted on this bowl shows no relation to the *t'ao-t'ieh*, an evil-averting monster mask which pervades the later bronze-age art of central China. 1978 *New Archaeol. Finds in China* II. 29 Some broken pieces of the outer coffin remain; they are carved with a *tao-tieh* (ogre-mask) design in the form of an ox head. 1980 *Catal. Fine Chinese Ceramics* (Sotheby, Hong Kong) 80 A further frieze of upright acanthus leaves around the neck.., the shoulders set with moulded *taotie* (*t'ao t'ieh*) mask and ring handles.

‖**taovala** (tauvala). [Tongan.] In Tonga, a piece of fine matting worn round the waist over a *vala* or Tongan kilt (and without which one is not considered properly dressed).

Traditionally worn by the male (with the exception of the Queen as monarch). It should be torn in several places, to show that the wearer does not set himself above his fellows.

1947 *Pacific Islands Monthly* Sept. 60/3 (*caption*) He wears the 'Taovala' (mat tied with coconut fibre) which is a 'must' with all Tongans who would show respect to their chiefs. 1953 *News Chron.* 2 June 7/1 With him rides a Queen—Queen Salote Tupon of the Tonga Islands. Her ceremonial dress includes a loose blouse and ankle-length skirt, round which is draped a tao-vala—a mat made from brown pandana leaves. 1977 *Daily Tel.* 15 Feb. 17/4 Pretty Tongan girls in white with the Taovala (traditional belts made from tree bark) round their waists kept the Royal party as cool as they could with fans made from the prickly-leaved pandanus tree.

tap, *sb.*¹ Add: **1. c.** *on tap* (later *fig.* examples). Also *spec.* in *Stock Exchange* use, applied to securities which are the subject of a large issue. Cf. quot. 1907, sense 1 b in Dict.

1890 R. L. Stevenson *Vailima Lett.* (1895) 35 The moon is on tap again. 1908 [see *concertina *v.*]. 1923 *Westm. Gaz.* 8 Jan. 4/2 It is some time since 'additional' Treasury Bills have been on 'tap' at so low a rate as.. 1¾ per cent. 1926 L. R. Robinson *Investm. Trust Organization & Managem.* 71 Whether the investment trust should raise its funds by keeping 'on tap' its offerings to the public and 'feeding' them out in response to demand..depends upon a number of factors. 1931 J. Greenhill in *Westm. Bank Guild Lectures 1930–1* III.105 We have not seen Bills 'on tap' for some considerable time past. 1935 A. Huxley *Let.* 5 June (1969) 396 His own left organizations in France will of course be on tap. 1958 *Times* 2 Oct. 3/3 Anything offered by television is on tap. 1965 J. L. Hanson *Dict. Econ. & Commerce* 371/2 Securities are said to be on 'tap' when they are issued in unlimited quantities (though the amounts permitted to each individual may be restricted) and are available for purchase direct from the issuing authority at any time. 1975 J. F. Burke *Death Trick* (1976) v. 82 We'll look into those alibis. Meanwhile, I want you both on tap. Understand?

d. *Electr. Engin.* = *tapping *vbl. sb.*¹ 2 b.

1900 M. A. Oudin *Standard Polyphase Apparatus & Systems* ix. 173 The secondary of each interchangeable transformer has two taps, giving 50 per cent and 86·7 per cent of the full voltage, so that either transformer can serve as the teaser, or supplementary one, by using the proper terminals. 1947 R. Lee *Electronic Transformers & Circuits* viii. 214 To improve the closeness of voltage control, a variable autotransformer has been developed in which the moving tap is a carbon brush which slides over exposed turns of a winding. 1974 C. C. Woodward *Cable Television* vi. 121 A complete new installation.. from the tap to the subscriber's television set.

e. *Stock Exch.* A security which is available 'on tap' (see sense 1 c above).

1948 *Economist* 8 July 772/2 £24½m...was perhaps acquired by original conversion of Local Loan or through the tap; but the additional £55m. was presumably bought on the market. 1960 *Ibid.* 8 Oct. 167/2 The issue price is nominal, since no one expected more than a small fraction of Wednesday's issue of £500 million would be taken by the public. The rest goes into the official tap, and the tap price can of course be adjusted as events dictate. 1967 *Ibid.* 4 Feb. 444/1 Supplies of the long tap (Treasury 6¾% 1995/98), issued only last October 28th, had already run out. 1976 *Southern Even. Echo* (Southampton) 12 Nov. 24/6 Gilts managed to push forward by ½ to ¾ taking the new Treasury 14 per cent. 1982 'tap' up 1/16 to 98 5/16. But the new long 'tap' Treasury 15¼ 1996 stays at 97¼. 1980 *Times* 15 Jan. 18 It would not surprise them to see the authorities issue another tap at the end of this week.

2. a. (Later examples.) Also *spec.* at *Eton College*: († *the*) *Tap*, a place where beer is sold to pupils at Eton.

1865 *Etoniana* 23 The 'Tap' and the Christopher had their earlier prototypes. 1917 A. Huxley *Let.* 30 Sept. (1969) 134 They were regrettably caught at the time just entering Tap. 1980 *Sunday Times* (Colour Suppl.) 14 Dec. 94/3 On the other days I'm free and I go for a drink to Tap.

5*. a. A device by means of which a telephone conversation may be listened to secretly by a third party. Cf. Tap *v.*¹ 2 c.

1923 E. Wallace *Missing Million* xxiii. 181 How did you know where the 'tap' was? 1959 *Washington Post* 26 Oct. A2/1 Law enforcement agencies use the taps even where prohibited by law. 1967 *Times Rev. Industry* Aug. 76/1 Telephones can be tapped so that it is virtually impossible for physical search to locate the tap, and if a searcher came near to it, the tap would automatically destroy itself without trace.

b. The act of listening secretly to a telephone conversation by means of a connection to the wire. Cf. *phone-tap* s.v. *phone *sb.*² 3; *telephone tap* s.v. *telephone *sb.* 3.

1950 *Sun* (Baltimore) 24 Jan. 12/3 The Holmes view has been accepted by the Supreme Court in a series of decisions steadily narrowing the use to which wire-tap material may be put in court. But there is still a shadowland within which the Justice Department feels safe in authorizing use of the wire tap. 1968 W. Garner *Deep, Deep Freeze* iii. 35 He'd made a phone tap, a successful tap, and overheard a reference to an agent..who was being sent to England. 1973 B. Murphy *Business of Spying* viii. 134 As well as being 'bugged', a telephone can be 'tapped'. This permits the recording and/or monitoring of both sides of the conversation. The most basic and easy way to monitor a telephone conversation is to carry out a direct line tap. 1979 *Guardian* 1 Mar. 1/3 There had to be good grounds for suspecting that a tap would be productive.

c. A recording made secretly from a telephone conversation.

1969 L. Sanders *Anderson Tapes* (1970) xxix. 73 Tape SEC. 25 JUN 68... This is a telephone tap. 1978 S. Brill *Teamsters* ii. 62 Telephone taps were played on tape recorders in court.

5.** *on the tap*, begging, making requests for loans. Cf. Tap *v.*¹ 3 in Dict. and Suppl. *slang.*

1932 A. Gardner *Tinker's Kitchen* iii. i. 217 Bob the journalist was, like everyone else at the Cross,..out to get what he could.., in plain words 'on the tap'. 1977 P. Carter *Under Goliath* xii. 61 She was a real moaner and always on the tap, borrowing sugar and milk.

6. (sense *1 e) *tap bill, bond, issue, price, rate, sale, stock*; **tap-changing** *Electr. Engin.*, the process of changing the connection to a transformer from one tap to another so as to vary the turns ratio and hence control the output voltage under a varying load; so **tap-changer**, an apparatus for accomplishing this; † **tap-waiter**, a waiter in a tap-room or tap-house (*obs. rare*); **tap wrench** (later examples).

1957 A. C. L. Day *Outl. Monetary Econ.* xxxv. 443 The British Exchange Equalization Account started operations with large quantities of sterling assets, which it holds in the form of 'tap' Treasury bills. 1942 *Sun* (Baltimore) 9 May 16/1 (*heading*) About $4,500,000 'tap' bonds sold here. 1931 S. R. Roget *Dict. Electr. Terms* (ed. 2) 342/2 Tap changer. 1962 *Newnes Conc. Encycl. Electr. Engin.* 748/2 Where possible, the tap-changer has minimum voltage to earth, and on most high-voltage line transformers it is at the neutral point. 1979 *Railway Gaz. Internat.* Jan. 49/1 As compared with the equivalent ..tap-changer loco, maintenance was halved. 1929 W. T. Taylor *Electr. Supply Transformer Systems* ii. 21 For station and distribution types of transformers, voltage control is now effected by tap-changing on load; several satisfactory designs have been produced which enable tap-changing to be carried out directly on tappings from the main transformers. 1962 *Newnes Conc. Encycl. Electr. Engin.* 748/1 Tap-changing may be done when the transformer is out of circuit. 1926 L. R. Robinson *Investm. Trust Organization & Managem.* 71 'Tap issues' are better fitted for a market in which the investor is learning for the first time the advantages of participation in investment trusts. 1973 *Daily Tel.* 3 Feb. 21/1 During the past fortnight, despite the exhaustion of two sizeable tap issues and the successful launching of a new 9¼ p.c. long-dated stock, the [gilt-edged] market has lacked impetus. 1958 *Times* 21 June 11/3 The strong demand for Funding Five-and-a-Half per Cent., 1982–84,..enabled the 'tap' price (the price at which Government departments are prepared to sell the stock they took up when the original issue was made) to be raised by 1-16 twice during the day. 1922 *Daily Tel.* 12 June 2/1 New second-hand Treasuries were dealt in at 2½ per cent., the 'tap' rate now being 2¾ per cent. 1926 L. R. Robinson *Investm. Trust Organization & Managem.* 71 'Tap' Sales, and occasional flotations. 1958 *Times Rev. Industry* Dec. (London & Cambridge Econ. Bull.) p. x/2 Tap sales have reduced bank liquidity. 1966 *Punch* 9 Nov. 710/3 The Bank will not allow anything like a boom in gilt-edged to develop—and it has tapstocks of its own to sell. 1980 *Times* 15 Jan. 15 Without tap stocks to deter them, gilts climbed briskly. 1835 Dickens *Sk. Boz* (1836) 1st Ser. II. 179 You leave your bag and repair to 'The Tap'... The tap-waiter finds himself much comforted by your brandy-and-water. 1956 H. Townsend in D. L. Linton *Sheffield* xvi. 299 Sheffield plays a large part in the production of drills and tipped cutters,..bit gauges, tap wrenches, pin vices,..and so on. 1964 S. Crawford *Basic Engin. Processes* i. 26 As its name implies the tap-wrench is required to provide a convenient method of revolving the tap in the drilled hole.

tap, *sb.*² Add: **1. c.** = *tap-dancing.

1944 N. Streatfeild *Curtain Up* viii. 97 The same sandals do for everything except tap. 1950 Blesh & Janis *They all played Ragtime* (1958) iii. 57 Chauvin had a fine tenor voice and sang and danced superbly, buck and wing, regular and eccentric tap. 1952 A. Lomax *Mister Jelly Roll* v. 202 Well, I did my tap numbers in a lot of shows after that. 1972 *Guardian* 13 Dec. 9/1 It was quite bad enough doing tap—all the kids at school used to tease me.

d. *Phonetics.* A single momentary contact between vocal organs in the production of a speech sound; the sound produced by such contact.

1952 [see *one-tap* s.v. *one *a.* 32 a]. 1954 Pei & Gaynor *Dict. Linguistics* 214 The Spanish *pero* is pronounced with a *tap r*, but *perro* with a *trill r*. 1964 W. Jassem in D. Abercrombie et al. *Daniel Jones* 339 The assumption that two 'taps' are sufficient for a sound to be labelled 'rolled'. 1977 *Language* LIII. 861 The individual closures of a trill are much more rapid than the single closure of a tap.

e. In fig. phr. *a tap on the wrist*, a mild reprimand. Cf. *slap *sb.*¹ 2 a.

1973 *Black Panther* 20 Oct. 2/1 Forty pages of charges gathered by the Justice Department, and he gets off with a tap on the wrist for income tax evasion. 1974 *Anderson* (S. Carolina) *Independent* 23 Apr. 4A/1 Disrespect for the law and the courts stems from instances..in which the accused have been found not guilty or have received a mere tap-on-the-wrist sentence when it was obvious that all evidence pointed to guilt.

2. (Earlier example.) Also *fig.*, the end.

1824 *H. R. Doc. 18th U.S. Congress 1 Sess.* No. 111. 35 It is his [*sc.* the orderly's] duty..to visit his rooms, at the taps; see that the lights are extinguished; the fires properly secured; the occupants present, and in bed. 1917 A. G. Empey *Over Top* 258 Then it was taps for me.

3. (Further examples.)

1743 J. Hempstead *Diary* 12 Dec. (1901) 418 Nailed on a pr of Tapps on a pr of New Shoes for adam. 1954 J. Steinbeck *Sweet Thursday* 33 Brown calf shoes (heel taps a little run over). 1965 E. Tunis *Colonial Craftsmen* iv. 107 The thick leather for the tap (sole) soaked all day in water. 1976 *National Observer* (U.S.) 6 Mar. 14/6 Amazing polyurethane taps keep heels perfect for months. Attach in seconds to any heel, and no one can tell you're wearing taps.

3*. In negative context: the slightest amount *of work*. Cf. Stroke *sb.*¹ 11 a. *colloq.*

1887 *Lantern* (New Orleans) 22 Jan. 2/2, I understand that Eddie never done a tap of work in his life. 1946 F. Sargeson *That Summer* 185 For several weeks Jack hadn't done a tap of work in the garden. 1952 E. O'Neill *Moon for Misbegotten* i. 10 He's nothing but a drunken bum who never done a tap of work in his life.

4. tap-in *Basketball*, a goal scored by tapping the ball into the basket, usu. when following up an unsuccessful shot; **tap-kick** *Rugby Football*, a light kick given to the ball whereby play is re-started from a penalty and possession retained; also as *v. trans.*; **tap pants** *U.S.*, a type of fashionable ladies' knickers; **tap penalty** *Rugby Football*, a penalty taken with a tap-kick; **tap shoe**, a shoe worn for tap-dancing, having a specially hardened sole or attached metal plates at toe and heel to make a tapping sound.

1948 *Daily Progress* (Charlottesville, Va.) 4 Feb. 9/3 Counting a tap-in one point would help equalize the height advantage, he said, but as in the case of the delayed whistle, what would constitute a tap-in? 1976 *Cumberland & Westmorland Herald* 4 Dec. 12/5 Coward netted a brace with a powerful long-range shot and a tap-in from Martin's cross. 1960 T. McLean *Kings of Rugby* xi. 120 When Hewitt after a tap-kick penalty in the Lions' 25 set off with a tremendous burst of speed..the audience rose to him in wonderment and delight. 1978 *Rugby World* Apr. 6/1 After he had given an indirect free-kick against the French, the Scotland captain, Doug Morgan, dropped a goal direct from hand, without bringing it into play with a tap-kick first. *Ibid.*, Mr Thomas's explanation was that he had been telling the French why the kick had been awarded and that his back was half-turned when Morgan began the run-up to his kick. 'I assumed Morgan had tap-kicked the ball first,' he said. 1977 Tap pants [see *teddy 3]. 1982 *Penthouse* July 26, I..have bought pretty tap pants and knickers from various lingerie establishments. 1976 *Sunday Post* (Glasgow) 26 Dec. 36/3 It only took Gala five minutes to score. From a tap penalty Dickson tore a gash in the defence, and Telfer accepted his scoring pass in the corner. 1978 *Rugby World* Apr. 33/1 He..carries particularly fond memories of the six tries he recorded last season, almost all from tap penalties at close range. 1932 *Boot & Shoe Recorder* 20 Feb. 62/3 A four style range of toe, ballet, acrobatic and tap shoes covers the usual store's requirements. 1936 'Isolde' *Tap Dancing Made Easy* 9 You can practise in an ordinary pair of shoes, but much better results can be obtained when wearing proper Tap shoes. 1980 *Daily Tel.* 10 Mar. 12/6 The musical.. brings out the best in the most bashful of bathroom singers and puts imaginary tapshoes on confirmed wallflowers everywhere.

tap, *v.*¹ Add: **2. c.** (Earlier and later examples.) So *to tap a call, line, message, telephone*, etc.

1869 *Cornh. Mag.* XIX. 759 A favourite plan of the raiders was to 'tap' the wire. 1871 *Q. Jrnl. Sci.* I. 117 For days the unconscious French were sending [telegraphic] messages, which were 'tapped' by the Prussians. 1874 J. H. Bunnel in J. E. Smith *Man. Telegraphy* (ed. 10) p. xv, The means employed to 'tap' a Telegraph line..are very simple. 1878 A. Pinkerton *Strikers, Communists, Tramps & Detectives* xvi. 199 The strikers certainly had some experienced telegraphers..capable of tapping the lines. 1909 G. B. Shaw *Press Cuttings* 3 Why didnt you telephone? *Balsquith.* They tap the telephone. 1911 *World's Work* XVIII. 588/2 Hundreds of amateur installations erected in the vicinity of either station, whereby messages might be tapped or confused. 1957 *Times* 7 June 10/5 (*heading*) Calls tapped on barrister's telephone. Home Secretary questioned. 1972 *Times* 19 Dec. 2/7 He could not prove beyond all reasonable doubt that the call he made to Mr Hope on December 1 had been tapped. 1978 G. A. Sheehan *Running & Being* ii. 32 There is no need to tap my phone or open my mail.

3. (Further examples.) Also *absol.*

1879 A. Pinkerton *Criminal Reminiscences* xiii. 212 In the act of 'tapping' the till of a North Side [of Chicago] German grocery. *c*1926 [see *marble *sb.* 4 c]. 1929 W. R. Burnett *Little Caesar* i. 10 They only bank once or twice a week. They're careless, get that; because they've never been tapped. 1931 T. Horsley *Odyssey of Out-of-Work* xxiii. 247 We'll tap these mansions. 1931 'G. Orwell' *Coll. Ess.* (1968) I. 57 Ginger tapped the local butcher, who gave us the best part of two pounds of sausages. 1935 —— *Clergyman's Daughter* ii. 105 They were begging..'tapping' at every..likely-looking cottage. 1939 J. Worby *Spiv's Progress* iii. 17 Every night he would put on his plimsolls and go tapping. 1979 *Tucson* (Ariz.) *Mag.* Mar. 46/1 Many of the big plush resorts that tap you for $80 to $100 a day.

tap, *v.*² Add: **1. a.** Also *to tap out*, to mark or signify by a tap or series of taps; to cause to be produced thus; *spec.* to type out (a letter, etc.).

1903 R. Langbridge *Flame & Flood* i. 4 He was tapping out a cautious progress towards the women with a stick, letting himself down with a surprised bump upon each step. 1912 *Red Mag.* Mar. 427/2 The clock of the Royal Exchange began to tap out the hour of nine. 1944 in B. A. Botkin *Treas. S. Folklore* (1949) iii. i. 447 He.. tapped out '73', which is the telegrapher's traditional symbol for goodby. 1952 M. Laski *Village* xviii. 248 Gerald tapped out a formal letter on the old typewriter. 1976 J. McClure *Rogue Eagle* ii. 27 Buchanan put down his cup where the writer carriage wouldn't bump it.. and..tapped out the name of his freelance agency.

c. To arrest (someone). Also in phr. *to tap on the shoulder. slang.*

[1785 Grose *Dict. Vulgar T.* sig. Z3, *A tap on the shoulder*, an arrest.] 1859 G. W. Matsell *Vocabulum* 89 *Tap*, to arrest. 1894 J. G. Littlechild *Reminiscences of Chief-Inspector Littlechild* xix. 193 We instructed him.. to hint darkly that he was going to be 'tapped'—i.e. taken into custody on charges connected with the forged cheques. 1968 [see *do *v.* 11 m].

2. a. Also in reduplicated form **tap-tap** and varr., to tap repeatedly (usu. as pres. pple.).

1922 Joyce *Ulysses* 284 A stripling, blind, with a tapping cane, came taptaptapping by Daly's window. 1977 *New Yorker* 6 June 38/2 Two reels of thread escaped from it, rolled along the landing, and went tap-tapping down the stairs. 1982 R. Timperley *Face in Leaves* iv. 34, I heard her typewriter tap-tap-tapping.

d. To sound, esp. as a signal.

1887 A. J. Wilson *At Mercy of Tiberius* xxxiii. 576 Somewhere in the apartment, a bell tapped. *Ibid.* 577 The time has come; the drum taps, I must march away. 1915 C. Johnson *Battleground Adventures* liv. 419 A bell would tap for a waiter to come and take the team.

3. (Earlier and further examples.) Also *absol.*

1746 J. Hempstead *Diary* 1 Jan. (1901) 453, I tapt & nailed Jont. Pierpoints Shoes. 1781 in *Narragansett Historical Reg.* (1882) I. 284 Tapped a pair of shoes. 1852 *Knickerbocker* XL. 149 There is also the shoemaker..who 'taps' for half the city price.

4. To designate or select (a person) for a task, honour, or membership of an organization. *U.S. colloq.*

1952 E. O'Neill *Moon for Misbegotten* i. 55 He was tapped for an exclusive Senior Society at the Ivy university to which his father had given millions. 1972 J. Mosedale *Football* ii. 13 *Sports Illustrated* magazine tapped him..as its 'Sportsman of the Year'. 1977 *Time* 23 May 13/3 Britain's youthful Foreign Secretary David Owen announced last week that he had tapped Jay, at 40, to serve as Ambassador to Washington.

‖**tapa**² (ta·pă). [Sp., lit. 'cover, lid'.] Usu. *pl.* In Spanish bars or cafés, a savoury snack or hors d'œuvre of sausage, cured ham, seafood, potato salad, etc., typically served with glasses of wine or sherry.

The slices of sausage, etc., were originally put on top of the glasses as 'lids'.

1953 C. Salter *Introducing Spain* iv. 36, I should like to draw attention to..the admirable habit of the 'tapa'. In Spain, when you order a drink in a bar.., you will always be given..something to eat. 1959 W. James *Word-bk. of Wine* 186 *Tapas*, small dishes served gratis in boat-shaped saucers with every glass of wine ordered.. in a Spanish bodega or café. 1964 C. Rougvie *Medal from Pamplona* vi. 79 Must be a pub there with tapas..these bits of food they give you free with the booze. 1978 J. Hyams *Pool* vi. 74 She had *tapas* and white wine at Café Monaco with a friend. 1982 D. Serafín *Madrid Underground* 63 It was the hour to take *tapas* or pre-dinner snacks.

tapadero. Add: Also 9 **-daro.** (Earlier and later examples.) Also used elsewhere in North America.

1844 J. Gregg *Commerce of Prairies* I. 213 The stirrups ..over which are fastened the tapaderas or coverings of leather to protect the toes. 1872 'Mark Twain' *Roughing It* xxiv. 178 It was a Spanish saddle, with ponderous *tapidaros*. 1879 *Cimarron* (N. Mexico) *News & Press* 20 Nov. 3/4 New Saddler Shop... Stirrups, Tapaderos, Saddle Bags, etc. 1933 J. Steinbeck *To God Unknown* xxv. 321 Joseph lifted the heavy saddle, and as the tapadero struck the horse's side, it reared. 1975 F. Kennedy *Alberta was my Beat* p. vi, It [*sc.* a saddle] was complete with Tapadero covered stirrups.

‖**tapas** (ta·pǎs). [Skr., lit. 'heat'.] In Hinduism and Jainism, (the practising of) religious austerity or bodily mortification. Also **tapasya.** [Skr., religious austerity].

1810 J. Malcolm in *Asiatic Res.* XI. 267 Na'nac was celebrated for the manner in which he performed *Tapasa*, or austere devotion. 1815 M. Elphinstone *Acct. Kingdom of Caubul* ii. xii. 318 The Tapasya, or ascetic devotion of Gurug. 1924 W. B. Selbie *Psychol. Relig.* 50 The ascetic methods of tapas and yoga, physical practices which belong to a debased mysticism not far removed from magic. 1962 A. Huxley *Island* ix. 137 The real thing only comes to people after years and years of meditation and *tapas* and..well, you know—not going with women. 1969 *Indo-Asian Culture* Oct. 53 In this art form..there is less of the austerity of *tapasya* in the artist's way of working. 1974 *Encycl. Brit. Macropædia* II. 137/1 In India, in the late Vedic period.., the ascetic use of *tapas* ('heat', or austerity) became associated with meditation and *yoga*, inspired by the idea that *tapas* kills sin.

ta·p-dancing. [f. Tap *sb.*²] A form of exhibition dancing characterized by rhythmical tapping of the toes and heels.

1928 *Daily Express* 27 June 9 The inventor of tap dancing. 1934 *Evening News* 1 Mar. 11/2 If the working girl doesn't do her bending and stretching, then she joins a tap-dancing class. 1953 R. Lehmann *Echoing Grove* 33 She wished to study tap-dancing, and to broaden the mind. 1972 *Guardian* 29 Jan. 9/4, I started as a dancer: tap dancing, acrobatic dancing, funny dancing. 1977 D. MacKenzie *Raven & Ratcatcher* ii. 19 Tap-dancing schools.

Also **ta·p-dance** *sb.*; also *fig.* and as *v. intr.* (occas. *trans.*); **ta·p-dancer.**

1927 *New Republic* 12 Oct. 210/1 That fair singer, good tap-dancer, born-to-the-purple, bred-in-the-bone, works-while-she-sleeps comedian, the plump May Barnes. 1929 D. Runyon in *Hearst's Internat.* July 56/2 Miss Billy Perry is worth a few peeks, especially when she is out on the floor of Miss Missouri Martin's Sixteen Hundred Club doing her tap dance. 1931 G. Cadwell (*title*) How to tap dance. 1941 *Penguin New Writing* X. 17 The sergeant.. had been a tap-dancer in civilian life. 1946 R. Campbell *Talking Bronco* 25 The tap-dance of the morning stars. 1950 J. D. MacDonald *Brass Cupcake* (1955) i. 11 He stood up and tap-danced me out to the gate. 1963 A. Lubbock *Austral. Roundabout* 100 The moths tap-danced on the fly-screens. 1972 *Guardian* 28 Jan. 9/1 The lacquered, ringletted monsters who tap-danced their way into the weepies. 1974 *Listener* 17 Jan. 92/2 Old-fashioned, out-dated routines: middle-aged black tap-dancers, a middle-aged blonde. 1977 N. Adam *Triplekip Crackaman* xvii. 171 A larger one [*sc.* table] which would have made a good one-shot tap-dance floor. 1978 W. F. Buckley *Stained Glass* xxi. 209 He could be tap dancing on it and still he'd be a goner.

tape, *sb.*¹ Add: **1. c.** Also used *lit.* or *fig.* in phrases: *to breast the tape*, to reach the finishing-line in a race; *on the tape*, at the very end of a race. Also in *Horse-racing*, a tape or set of tapes suspended across the course at the starting-point of a race; also *fig.*

1903 *Punch* 11 Feb. 103/1 Though a toughish task remains Before I breast the tape, J. Chamberlain, of Birmingham, Will round (or square) the Cape. 1916 J. B. Cooper *Coo-oo-ee* xvii. 270 'They've got me on the tape!' he cried; 'but I'm satisfied.' 1922 Joyce *Ulysses* 608 Judge of his astonishment when he finally did breast the tape and the awful truth dawned upon him anent his better half, wrecked in his affections. 1937 'P. Wentworth' *Case is Closed* vii. 72 He was running in his school sports, winning the hundred yards again, breasting the tape, hearing the applause break out. 1955 *Times* 13 Aug. 4/2 Wheeler..regained the ground he had lost and just robbed the Hungarian of victory on the tape. 1957 D. Francis *Sport of Queens* vi. 122 The six or seven stranded starting tapes familiar in flat racing are not used for National Hunt racing. A single strand across the course is pulled down to a catch at shoulder level, and when this is released..the tape flies up at an oblique angle. 1963 *Times* 21 Feb. 4/5 They were described by an official observer after the last N.E.D.C. meeting as 'just coming up to the tapes'.

d. *Army* and *R.A.F. slang.* A chevron indicating rank worn by a non-commissioned officer on the upper part of the coat-sleeve; a stripe (Stripe *sb.*³ 2).

1943 Hunt & Pringle *Service Slang* 64 *Tapes*, the stripes worn by Corporals, Sergeants, and Flight Sergeants in the R.A.F. and by Lance-Corporals or Lance-Bombardiers, Corporals or Bombardiers, and Sergeants in the Army. 1944 *Gen* 15 Jan. 9/2 That binder's working for his tapes. 1944 *R.A.F. Jrnl.* Aug. 258, I wouldn't leave this unit for three tapes.

2. b. Also used in computing and data processing; = *paper tape* s.v. *paper *sb.* 12.

1945 J. von Neumann in B. Randell *Origins of Digital Computers* (1973) 355 These instructions must be given in some form which the device can sense. Punched into a system of punchcards or on teletype tape, magnetically impressed on steel tape or wire, [etc.]. 1948 *Math. Tables & Other Aids Computation* III. 8 Orders to the machine.. are represented on tape by all combinations of three holes out of six. 1960 M. G. Say et al. *Analogue & Digital Computers* ix. 266 The only problem in fast photoelectric reading arises when the tape has to be set in motion and stopped so rapidly that [etc.]. 1978 D. D. Spencer *Data Processing* v. 105 Data are often hand-sorted before being punched into the tape.

c. = *magnetic tape* s.v. *magnetic *a.* 5. Cf. *steel tape* (*b*) s.v. *steel *sb.*¹ 18.

1932 *Radio Times* 29 July 239/3 The Blattnerphone is an invention for recording magnetically upon steel tape. 1942 *Jrnl. R. Aeronaut. Soc.* XLVI. Abstr. Sci. & Technical Press 68 It consists of recording a sound pattern magnetically on steel tape. The signal is picked up from the tape at frequent split-second intervals. 1953 *Newsweek* 11 May 28/1 It was recorded on tape and was broadcast later that day. 1964 M. McLuhan *Understanding Media* xviii. 295 Tape and the l.p. record suddenly made the phonograph a means of access to all the music and speech of the world. 1982 *Times* 26 Oct. 15/7 Access to specific pieces of information is far faster on a compact, rapidly spinning disk than on a long ribbon of tape.

d. A length or reel of (magnetic or paper) tape; a recording on tape.

*c*1946 [see *subroutine]. 1952 W. Stevens *Let.* 13 May (1967) 750, I read at Cambridge a week or two ago and apparently someone in the audience took a tape. 1956 G. A. Montgomerie *Digital Calculating Machines* x. 213 The instructions are punched in the tapes in a very simple notation. 1966 *Listener* 25 Aug. 287/2 This production came via a tape from the Holland Festival. 1977 *New Yorker* 22 Aug. 56/3 Most thefts of computer tapes are probably not reported to the police. 1978 D. D. Spencer *Data Processing* v. 105 Both the tapes and the tape-producing equipment require less space than punched cards and card-producing equipment. 1983 D. Dunnett *Dolly & Bird of Paradise* v. 54 Kim-Jim loved telly films... I had brought a lot of tapes with me.

e. Used in names designating (paper, transparent film, etc.) tape coated with adhesive and used for fastening packages, etc.; usu. as the final element of a Comb., as *adhesive tape, Scotch tape, Sellotape, sticky tape*: see under first element in Suppl.

1966 A. W. Lewis *Gloss. Woodworking Terms* 99 *Tape*, gummed paper strip used to hold the edges of veneer together while the glue dries.

4. (In sense 2 b) *tape boy, price* (earlier example); (in senses *2 c, d) *tape editing, editor, eraser, head* [*head *sb.* 11 g], *speed, splicing*; *tape-controlled, -playing* adjs.; (in sense *2 e) *tape dispenser*; **tape cartridge** = next (see also quot. 1983); **tape cassette** = *cassette d; **tape-check** *Mus.*: in an upright pianoforte, a type of check (Check *sb.*¹ 10 d) developed by Robert Wornum (1780–1852) and incorporating a tape; also *attrib.* in *tape-check action*; **tape deck** (see *deck *sb.*¹ 3 f); **tape-delay**, the use of a tape recorder to introduce an interval between recording and playing back or transmitting (cf. *delay *sb.* 1 c); **tape drive**, a tape transport or tape deck for use in computing; **tape guipure** *Lace-making* (see quots. 1865, 1881); **tape hiss**, extraneous high-frequency background noise during the playing of a tape recording; **tape loop** = *loop *sb.*¹ 4 k; **tape-measure** (earlier example); also as *v. trans.*, to measure with a tape-measure; hence **tape measurement**; **tape player**, a machine for playing (cassette) tape recordings; cf. *record player* s.v. *record *sb.* 14; **tape punch** *Computers*, a device which punches holes in paper tape in patterns that represent coded information; also **tape-punching**; **tape reader** *Computers*, a device for sensing information recorded by sequences of holes or magnetized areas on computer tape (see *reader 7); also **tape reading**; **tape reproducer**, a machine that plays or reads tapes but does not record or punch them; **tapescript** [after *transcript, typescript*, etc.], a tape recording of the spoken word, esp. in the form of a lesson, interview, etc.; a transcript or text of this; **tape transport**, a mechanism which controls the movement of recording tape past a stationary head; also, a tape deck. See also *tape recorder.

1969 *Listener* 2 Jan. 12/1, I was in the Newsroom:.. where..tape boys..bore in huge foaming trays of paper strip to the duty editors. 1961 *High Fidelity Trade News* Sept. 55/3 Foley Electronics offers..an automatic tape cartridge playback unit employing the endless loop principle. 1972 *Observer* (Colour Suppl.) 22 Oct. 53/2 Tape cartridges are plentiful, even if cassette material is strangely lagging. 1983 D. H. Sandars *Computers Today* vi. 159 (*caption*) Magnetic tape cartridges are used in minicomputers and data entry stations. This tape is ¼ inch wide and varies from 140 to 450 feet in length... Magnetic tape cassettes are used in microcomputers and data entry stations. This tape is either 150 or 300 feet long. 1972 G. V. Higgins *Friends of Eddie Coyle* xv. 88 He opened the glove compartment and removed a tape cassette. 1983 Tape cassette [see *tape cartridge* above]. 1929 *Evening News* 18 Nov. 2/6 Pianette... Iron frame. Tape check action. 1954 *Grove's Dict. Mus.* (ed. 5) VI. 739/1 We must now turn our attention to an action known as the 'tape-check'. 1966 W. L. Sumner *Pianoforte* iii. 66 A later model anticipated the tape-check action, which prevented the hammer from giving unwanted repetition. 1962 *Gloss. Terms Automatic Data Processing* (*B.S.I.*) 98 Tape-controlled carriage. 1949, etc. Tape deck [see *deck *sb.*¹ 3 f]. 1967 *Oxf. Computer Explained* 7 A new configuration..consisting of twelve 30 kc tape decks, a high speed printer, a paper-tape reader, and a monitor printer. 1977 'E. Crispin' *Glimpses of Moon* xii. 242 His stereo tape-deck only a few seconds away from the enormous climax of *Also Sprach Zarathustra*. 1968 *Listener* 12 Dec. 802/3 The music dissolves again, surmounted by quiet seagull sounds produced by high squeaks of feed back multiplied by means of tape-delay. 1983 *Ibid.* 18 Aug. 27/1 The catalyst which sets *Clocks and Clouds* in motion is the gradual de-synchronisation of identical melodic patterns—the classic 'tape-delay' device of electronic music. 1975 *New Yorker* 10 Mar. 31/1 An automatic tape dispenser. 1952 *Proc. Electronic Computer Symp., Los Angeles* (IRE Prof. Group Electronic Computers) 4 (*heading*) Survey of tape drive systems. 1978 J. McNeil *Consultant* x. 114 The computers..showed little signs of life beyond the occasional twitch from the tape drives. 1983 D. H. Sandars *Computers Today* vi. 163 Before the data on a magnetic tape can be processed by a computer, the tape must be placed in a machine called a tape drive or tape transport. 1961 *Times* 17 Apr. 3/1 Knowledge of radio-production and tape-editing. 1973 G. Talbot *Ten Seconds from Now* xix. 239 A wonderland of recording rooms and tape-editing suites. 1959 'F. Newton' *Jazz Scene* x. 169 Recording supervisors, sound engineers and tape editors. 1958 *Oxf. Mail* 27 Aug. (Suppl.) 4/6 (Advt.), The..tape eraser will erase the contents of a spool of tape of any size up to ten inches at one operation. 1865 F. B. Palliser *Hist. Lace* iii. 35 In that class called by the lace-makers 'tape guipure', the outline of the flowers is formed by a pillow or hand-made braid about the eighth of an inch in width, the middle filled in with the needle. 1881 C. C. Harrison *Woman's Handiwork* i. 94 Tape-guipure, made of linen tape twisted and folded into a pattern, held together with bars and then filled in and enriched with needlework. 1882 Caulfeild & Saward *Dict. Needlework* 246/2 An embroidery worked in imitation of the Tape Guipure Laces. 1960 *Practical Wireless* XXXVI. 401/2 (Advt.), Luxury model with press-button inputs to suit any pick-up or tuner and most tape-heads. 1980 *Sunday Times* 24 Aug. 14/7 Magnetic patterns on the tape are translated by a tape-head into electrical impulses. 1962 A. Nisbett *Technique Sound Studio* ii. 38 There may be tape hiss due to setting the gain too low. 1977 *Gramophone* Oct. 638/2 The tape hiss seems now to be higher too. 1962 Tape loop [see *loop *sb.*¹ 4 k]. 1976 W. H. Canaway *Willow-Pattern War* xx. 199 The signal would be going out from a tape-loop. 1873 *Young Englishwoman* Mar. 151/2 Pin your tape-measure down on one of the fronts. 1962 L. Deighton *Ipcress File* xxv. 158 Birth marks..were photographed and tape measured. 1971 Laver & Collins *Educ. Tennis Player* i. 19 Dave Anderson..tape-measured me..and reported that my left forearm is twelve inches around. 1922 Joyce *Ulysses* 523 Tape measurements will be taken next your skin. 1961 *High Fidelity Trade News* Sept. 55/3 (*heading*) Foley presents 'endless loop' tape player. 1977 *New Yorker* 24 Oct. 112/3 Such misdeeds as..having a tape player on too loud. 1962 *Times* 5 July 15/6 Tape-playing equipment to the value of some £160 altogether was being used. 1889 *Hints to Speculators* (G. Gregory & Co.) (ed. 5) 20 Deal at tape prices. 1947 [see *tape reader* below]. 1967 A. Battersby *Network Analysis* (ed. 2) viii. 134 The sheets are then passed to a tape-punch operator who converts the information on them into a punched tape. 1951 M. V. Wilkes et al. *Preparation of Programs for Electronic Digital Computer* 42 (*heading*) Tape punching and editing facilities. 1947 *Math. Tables & Other Aids to Computation* II. 355 In addition to these parts there are.. a drum commutator for operating the relays, a tape reader and a tape punch. 1964 C. Dent *Quantity Surveying by Computer* iii. 26 The program is all ready to be read by the input tape-reader at 300 characters per second. 1972 M. Woodhouse *Mama Doll* x. 143 It's frequency-coded for a tape-reader. 1970 O. Dopping *Computers & Data Processing* xv. 237 Tape reading and tape writing can go on partly simultaneously, if the design of the computer allows it. 1961 Tape reproducer [see *reproducer 2]. 1962 *Gloss. Terms Autom. Data Processing* (*B.S.I.*) 97 *Tape reproducer*, a machine used to copy and edit paper tape. 1962 A. Nisbett *Technique Sound Studio* 246 *Control cubicle* (BBC), the soundproof room equipped with control desk, gramophone and tape reproducers and high quality loudspeaker. 1961 J. S. Holton et al. *Sound Lang. Teaching* 248 *Tapescript*, term used to describe the written program (exercises and instructions) that the teacher has prepared for recording a language laboratory lesson. 1969 *John Edwards Mem. Foundation Q.* V. i. 10 These tapescripts..are resumés of interviews of artists. 1983 *Financial Times* 16 Apr. 14 It is telly docu-drama at its most truth-honouring; based on real tapescripts, speeches and official records. 1954 *Gramophone Record Rev.* Jan. 139/1 With a tape speed of 7½ ins. per second the capstan thus makes about 80 revolutions per second. 1956 *Language* XXXII. 281 These experiments served as a test for two tape-splicing techniques. 1954 *Trans. IRE Audio* II. 23/1 The overall design of a tape transport for a professional tape recording system is very complex. 1981 *Hi-Fi Answers* Nov. 117/1 The tape transport is press button controlled, and operation appears to be achieved by a combination of mechanical and electrical means. 1983 Tape transport [see *tape drive* above].

tape, *v.* Add: **1. a.** Also, to affix or fasten (*up*) with adhesive tape.

1956 A. H. Compton *Atomic Quest* iv. 259 Alvarez taped three copies of this note to instrument boxes. 1972 *Daily Colonist* (Victoria, B.C.) 18 Jan. 17/1 Every girl.. should clip it out and tape it to her bedroom mirror. 1979 R. Jaffe *Class Reunion* (1980) i. i. 36 Daphne had taped up small museum reproductions of famous paintings and prints.

b. *spec.* to bind or gag (a victim, etc.) with adhesive tape. Also with *up*.

1932 'Spindrift' *Yankee Slang* 60 *Tape a guy*, gag a victim with adhesive plaster. 1950 'P. Quentin' *Follower* xix. 135 Once they'd taped his wrists, all hope would be gone. 1962 'H. Howard' *Double Finesse* vi. 69 It shouldn't take Alan longer than that to tape the night-watchman's mouth. 1977 D. Anthony *Stud Game* xiv. 84 He knocks you out..tapes you up neatly, and calls us to come and get you. 1981 P. Mallory *Killing Matter* xv. 157 Tie her and tape her.

2. b. *Gunnery.* To get the range of (a position), by means of a tape-line used in conjunction with a range-finder; hence, to hit and silence (a target).

See also sense 5 below.

1917 A. G. Empey *From Fire Step* xi. 65 Our artillery had taped or silenced them [*sc.* the trench mortars]. *Ibid.* xxi. 146 The German artillery..had us taped (ranged) for fair; it was worth your life to expose yourself for an instant. 1919 J. B. Morton *Barber of Putney* iii. 45 There's a sniper got that corner taped. 1919 G. K. Rose *2/4th Oxf. & Bucks L.I.* 36 The Pozières ridge, whose crest was well 'taped' by the German guns. 1927 E. Thompson *These Men thy Friends* iii. 70 'He tried them at sixteen hundred yards, and got nowhere near them—lengthened the range a thousand, and was still short. But Johnny [Turk] had *us* taped' he added. 'No bothering about mirage for him. He knew the land and the distance of every blotch and pimple on it.'

5. Colloq. phr. *to get* or *have* (someone or something) *taped*: to size up, ascertain, or understand fully (someone or something).

The development of this phrase is unclear. It may have arisen as a figurative use of sense 1 with the idea of 'tying up, having under control or in order' (cf. quot. 1854 in Dict.) or of sense 2 with the idea of 'measuring, assessing'.

1914 Joyce *Dubliners* 210, I never saw such an eye in a man's head. It was at much as to say: *I have you properly taped, my lad.* He had an eye like a hawk. 1919 *War Slang* in *Athenæum* 18 July 632/2 'I got you taped,' an N.C.O. may say to a man, meaning 'I know what you are up to.' 1929 J. B. Priestley *Good Companions* i. iv. 114 We've made a 'ell of a bad break if we tell 'er oo we are and then there's nothing doin'. Got us taped then. 1944 A. E. Coppard in *Wine & Food* xliii. 153, I want to get off the land. Can you find a boat? Not a motorboat, that's noisy and they've got the harbour taped for sure. 1959 *Times Lit. Suppl.* 13 Mar. 142/4 The main part of the book, with its cold effort to get Mencken 'taped'. 1977 *Evening Post* (Nottingham) 27 Jan. 6/5 And so say all of us. Let's hope Portland have illiteracy 'taped'.

6. To record on (magnetic) tape; to make a tape recording of. Also *absol.*

1950 *Senior Scholastic* 1 Mar. 25T (*heading*) We tape it. Recorders produce a transformation in the classroom. 1958 S. Ellin *Eighth Circle* (1959) ii. ii. 41 He's being taped Sunday, so have one of the girls make a transcript of the tape. 1960 *Guardian* 9 Nov. 11/1 One [teenager] with a tape recorder can tape a pile of 'pop' records. 1966 E. McGirr *Funeral was in Spain* 40 Okay, men, let's hear his verbal run through. I understand you didn't tape. 1975 R. H. Rimmer *Premar Experiments* (1976) i. 130 I've tried taping sober, high on alcohol, and stoned on grass. 1978 R. Nixon *Mem.* 501, I was not comfortable with the idea of taping people without their knowledge.

taped *ppl. a.* (examples in sense *6); also with *out*, measured or sized up; fully ascertained (cf. senses *2 b, *5); **taping** *vbl. sb.* (examples).

1929 *Papers Mich. Acad. Sci., Arts & Lett.* X. 329/1 *Taped out*, an expression applied to a strip of land upon which the German gunners had accurately registered distances. 1933 Wodehouse in *Sat. Even. Post* 30 Dec. 58/2 Didn't I tell you that I had everything taped out? 1953 Pohl & Kornbluth *Space Merchants* (1955) v. 55 What..have you got to back that statement up with? Letters? Memos? Taped calls? 1955 I. A. Richards *Speculative Instruments* x. 122 The study of language, even in the most elementary stages, has to be a dependant of that highest generic taping which may be called ethics. 1960 *Daily Mirror* 23 Apr. 18/3 Bobby Darin..left behind a taped Saturday Spectacular... This will be shown on ITV tonight. 1968 H. Waugh *Con Game* xii. 109 A few of the people connected with the show had got together after the taping. 1972 R. Prawer Jhabvala *New Dominion* i. 58 He kept turning on the taped music and the concealed lighting.

‖**tapénade** (tapenad). Also **tapenade.** [Fr., f. Prov. *tapéno* capers.] A Provencal dish, usu. served as an hors d'œuvre, made principally from black olives, capers, and anchovies.

1952 G. Maurois *Cooking with Fr. Touch* iii. 56 Here is a southern (Nice) recipe for *tapenade*, which uses eggs, olives, and anchovies. *Ibid.*, *La tapenade* used always to figure on the list of hors d'œuvres at the old Hotel Victoria in Cannes. 1960 E. David *Fr. Provincial Cooking* 142 To make the *tapénade*, called after the capers (*tapéno* in Provençal) which go into it. *Ibid.*, The *tapénade* is served pressed down into little deep yellow earthenware pots, like a pâté. 1966 P. V. Price *France: Food & Wine Guide* 299 Tapenade or tapanda. Pounded black olives, served on toast or as an *hors d'œuvres*. 1978 *Times* 20 May 10/2 Regular dishes such as tapénade (a Provençal purée of capers, black olives, anchovies, and sometimes tunny fish).

taper, *sb.*¹ Add: **2.** *taper-candlestick, -light* (later example), *-stick* (later examples, as an antique).

1847 Gough & Parker *Gloss. Terms Heraldry* 70 The *taper-candlestick*, borne in the arms of the Founders' Company..has a spike, or..a pricket, upon which the taper is placed. 1913 W. de la Mare *Peacock Pie* 119 Lantern-light, taper-light, Torchlight. 1956 G. Taylor *Silver* v. 114 *Taper Sticks.* Examples do not occur in silver until the later part of the period, and are generally miniature candlesticks. 1982 *Nat. Art-Coll. Fund Ann. Rep. 1981* 39/2 Taper-sticks were made to hold tapers for lighting candles or pipes, and generally have flat circular bases and narrow sockets.

taper, *sb.*² Add: **II. 2.** *spec.* in *Forestry* (see quot. 1957).

1893 P. J. Carter *Treatise Mensuration Timber* i. 5 Long logs should be measured in two or more sections.. the sections increasing..with the taper of each log. 1945 G. B. Grundy *Fifty-Five Years at Oxford* vii. 132 A scale of taper which means the number of inches a tree decreases in girth between its base..and a point in its circumference ten or twenty feet above that. 1957 *Brit. Commonwealth Forest Terminol.* ii. 194 *Taper*, the decrease in diameter of a tree bole or log from the base upwards.

4. Taper-Lock, taper-lock *Mech.*, a proprietary name in the U.S. for a type of tapered bush (Bush *sb.*² 1 b) inserted into a pulley, sprocket, etc., to enable it to be mounted rigidly on a shaft; **taper tap** *Mech.*, a tap (Tap *sb.*¹ 4) tapered lengthways for about two thirds of its length, used to begin the process of cutting a screw thread in a hole.

1954 *Official Gaz.* (U.S. Patent Office) 4 May 31/1 Dodge Manufacturing Corporation, Mishawaka... *Taper-Lock*.. For machine elements to be mounted on shafts and bushings therefor. 1971 *Engineering* Apr. 4/2 Pulleys..complete with Taper-Lock bushes for fast, easy fixing. 1971 *Power Farming* Mar. 75/2 The new pulley illustrated incorporates the quick-fit taper-lock centre. *a* 1877 Knight *Dict. Mech.* III. 2495/1 The process of screw-cutting was greatly improved by Maudslay, who introduced the practice of having three cutting edges, and using three taps, the entering taper tap, the middle tap, and the plug tap. 1964 S. Crawford *Basic Engin. Processes* i. 24 The taper tap has a chamfer or tapered lead for a length of 8–10 threads.

taper, *sb.*³ Add: **a.** (Earlier example.)

1881 *Instructions to Census Clerks* (1885) 68 *Looming and Taping Room* [in Cotton Manufacture]:..Taper. Beam Flanger. Beamer.

b. One who tapes or deals with tape in other occupations.

1921 *Dict. Occup. Terms* (1927) § 304 Coil taper,.. binds coils with tape. 1927 *Daily Express* 2 Dec. 2/4 Taper,..[an] operator in charge of the insulation of armature.

taper, *a.* Add: **b.** † Also, of a person: reduced in funds, short (of money). (Earlier example.)

1789 J. Byng *Torrington Diaries* (1935) II. 88 So now, being taper of the said necessary commodity [*sc.* cash], I was obliged to recruit from M. Oliver.

c. *taper roller bearing*, a roller bearing in which the rollers are tapered slightly and lie at an angle to the axis of the bearing, so as to provide resistance to thrust along the axis as well as at right angles to it.

1930 *Engineering* 7 Feb. 169/3 The driving wheels..are mounted on taper-roller bearings. 1971 *Power Farming* Mar. 50/1 The Benedict Soilmaster takes care of seed bed cultivations—and your tractor... No gears, no cranks and adjustable taper-roller bearings result in minimal maintenance.

taper, *v.* Add: **2. b.** (Earlier *colloq.* example.) Also const. *down*.

1848 J. F. Cooper *Oak-Openings* I. iv. 66 It's hard to give up old habits, all at once. If I could only taper off on a pint a day, [etc.]. 1960 *Wall St. Jrnl.* 18 Nov. 13/1 Carloadings 'taper down' starting in mid-November, when the bulk of Christmas shipping has been completed. 1971 *Daily Tel.* 4 Aug. 2/7 The deal is worth nearly £4-a-week more to the lower grades tapering down to £2 at the top end.

3. b. (Later example.) Also const. *down*.

1971 *Daily Tel.* 2 Aug. 7/8 There is speculation that the Government investment..in tracked hovercraft is to be tapered off. 1977 *Lancet* 23 Apr. 909/2 Oral prednisolone, 1 mg/kg/day, was resumed and rapidly tapered down to 0·5 mg/kg/day.

tape re·cord. *rare.* [f. Tape *sb.*¹ + Record *sb.*] A record(ing) on magnetic or other tape.

1905 *Talking Machine News* III. 57/1 A tape record could be made to be reproduced by either the cylinder of [*sic*] disc type of machine. 1914 *Sci. Abstr.* B. XVII. 372 The author submits some tape records which are reproduced. 1961 Y. Olsson *Syntax Eng. Verb* ii. 17 Speech in its natural environment or in tape-record tapping. 1968 *Listener* 6 June 733/2 Fuzzy telephones or muzzy tape records.

ta·pe-record, *v.* [Back-formation from next.] *trans.* To record (sounds, etc.) on magnetic tape by means of a tape recorder.

1950 *Aviation Week* 6 Mar. 35 (*heading*) Plane-tower talk tape-recorded. 1955 E. Warner *Trial by Sasswood* ix. 177 As though your thoughts..had been tape-recorded and played back to you. 1967 A. Henri in *Penguin Mod. Poets* X. 25, I sit here..trying to taperecord the sound of windflowers and celandines. 1978 N. J. Crisp *London Deal* vii. 110 Could we tape record this?

Hence **ta·pe-recorded** *ppl. a.*

1951 *Time* 25 June 23/2 Testimony continued—part of it played scratchily from long, tape-recorded interviews with addicts. 1973 S. Trueman *Fascinating New Brunswick* xvi. 125 Her tape-recorded folksongs have left a priceless heritage to New Brunswick.

ta·pe reco:rder. Also with hyphen. [f. Tape *sb.*¹ + Recorder¹.] † **1.** A device which records data on 'ticker' tape. *Obs.*

1892 W. P. Lombard in *Jrnl. Physiol.* XIII. 4 The labour of measuring the curves and of computing the total work was so great..that it was found necessary to devise an apparatus which would record automatically the total height to which the weight was lifted. This apparatus, which may be called a tape recorder, consisted of an endless tape. 1922 *Science & Invention* Feb. 935 The accompanying diagram shows a very interesting special arrangement with relay for operating a tape recorder, klaxon or telegraph sounder.

2. An apparatus for recording sounds, etc. on magnetic tape and afterwards reproducing them.

1932 *B.B.C. Techn. Tables & Gloss.* 65/2 Steel Tape Recorder. 1949 *Consumer Reports* Feb. 68/2 The three tape models all proved..substantially more convenient than earlier tape recorders and better than..the wire recorders tested. 1949 *Electronic Engin.* XXI. 369 There will be a selection from the very wide range of G.E.C. sound equipment..and the new G.E.C. Tape Recorder. 1953 M. McCarthy *Groves of Academe* xiii. 252 The psychology student with the tape-recorder. 1964 M. McLuhan *Understanding Media* (1967) v. 63 Radio and gramophone and tape recorder gave us back the poet's voice. 1978 J. A. Michener *Chesapeake* 854 From the moment Amos discovered what those newfangled tape recorders could do, he was satisfied that his..problems were solved.

Also **tape recordist**, one who makes tape recordings.

1960 *Guardian* 9 Nov. 11/1 The tape recordists were clearly making money out of their activities. 1970 J. Earl *Tuners & Amplifiers* iii. 70 If you are a keen tape recordist then the amplifier should certainly be equipped with at least source sockets for tape replay.

ta·pe reco:rding, (*vbl.*) *sb.* [f. Tape *sb.*¹ + Recording *vbl. sb.*, after prec.] A record (of sounds, etc.) on magnetic tape; the process of making such a recording.

1940 *Electronics* May 16 (*heading*) Photo-electric tape recording. *Ibid.* 17/1 The editor of such tape recordings has considerable freedom in arranging the material. 1946 *Electronic Engin.* XVIII. 54 (*heading*) German tape-recording equipment. 1954 A. Huxley *Let.* 12 Dec. (1969) 718, I listened to the tape recording and the foreign language certainly doesn't sound like the gibberish of ordinary glossolalia. 1962 A. Lurie *Love & Friendship* xiv. 286 His voice sounded funny to him, like a tape-recording. 1977 P. Strevens *New Orientations Teaching Eng.* xiii. 163 A set of specialized tape recording/replay machines.

tapering, *vbl. sb.* Add: Also *tapering-off*.

1890 W. James *Princ. Psychol.* I. iv. 124 The question of 'tapering-off', in abandoning such habits as drink and opium-indulgence, comes in here. 1955 Koestler *Trail of Dinosaur* 136 The Jews alone among the varied European immigrant population have resisted this 'tapering off' process.

tapestry, *sb.* Add: **1. c.** Now freq. applied to (pieces of) canvas embroidery executed typically with wool in tent stitch.

1882 Caulfeild & Saward *Dict. Needlework* 473/2 Tapestry worked by the needle..differs but slightly from Embroidery. The stitches are made to lie close together, so that no portion of the foundation is visible. 1955 *Stitchcraft* Mar. 9 The design [for a picture] is worked in tent-stitch by the counted thread..; the chart includes instructions for tent-stitch and hints on stretching tapestry. 1971 *Harrods Magical Christmas* 9 Tapestry Cushion Pack of tramme canvas, wools and needle. *Ibid.*, If desired, we will stretch and mount the finished tapestry on velvet. 1976 P. Clabburn *Needleworker's Dict.* 263/3 Nowadays in Britain, any piece of canvas work, large or small, is called tapestry work, which is a misnomer, while America, although not falling into that particular trap, calls canvas work needlepoint, which is also confusing as that word should apply to lace made with a needle.

2. Also short for *tapestry needle*, sense 3 below.

1895 *Montgomery Ward Catal.* Spring & Summer 88/1 Needle Case... Contains—..Crewel..Tapestry..Bodkin. 1968 J. Ironside *Fashion Alphabet* 94 *Tapestry*, a needle which has a blunt point and large eye, used for embroidery with wool.

3. *tapestry room* (examples), *wool*; **tapestry needle**, a blunt needle with a large eye used in tapestry-making and canvas embroidery.

1888–9 *T. Eaton & Co. Catal.* Fall & Winter 64/2 The Household Needle Case contains darners, glovers, square-carpet, yarn, chenille, tapestry,..and crewel needles. 1967 E. Lemarchand *Death of Old Girl* xvii. 196 Tim Pollard watched her..as she plied a tapestry needle. 1817 M. Edgeworth *Harrington* xviii. 496 Mr. Montenero.. asked, in particular, about a tapestry room,—a picture of Sir Josseline. 1977 R. Player *Month of Mangled Models* vi. 105 The casements of the Tapestry Room were wide span, and the Camelot curtains had been pulled back. 1880 L. Higgin *Handbk. Embroidery* i. 4 *Tapestry Wool* is more than twice the thickness of crewel... Tapestry wool is not yet made in all shades. 1960 G. Lewis *Handbk. Crafts* 36 The most usual wool for this work is that with a slight twist to it called 'tapestry' wool, but other kinds may be used according to the mesh of the canvas.

tapestry-worker. (Later example.)

1908 H. Pentin *Judith* iv. 77 'Judith and Holofernes' was also a favourite subject for tapestry-workers.

‖**tapette** (tapet), *sb.* (and *a.*). [F. *tapette* pederast, homosexual (*slang*), f. *taper* to tap, hit + *-ette*, fem. suffix.] A passive male homosexual; an effeminate man or 'pansy'. Also as *adj.*

1930 E. Waugh *Vile Bodies* ii. 22 My dear, he looks terribly *tapette*. 1936 'R. West' *Thinking Reed* xiii. 455 'It will make my room look as if I were a *tapette*!' exclaimed Marc. 1949 A. Wilson *Wrong Set* 174 She replied '..you do look madly tapette when you're drunk.' 1960 J. Baldwin *Another Country* (1963) ii. i. 183 Yves had lived by his wits in the streets of Paris, as a semi-*tapette*, and as a *rat d'hôtel*. 1978 J. Sherwood *Limericks of Lachasse* xv. 181 My mother..wondered if you were perhaps *tapette*, but my brothers assured her that..you were perfectly masculine.

taphonomy (tæfǫ·nǒmi). *Palæont.* [f. Gr. τάφος grave + -nomy.] The study of the processes by which animal and plant remains become preserved as fossils. Hence **tapho-no·mic, -ical** *adjs.*; **tapho·nomist**, a specialist in taphonomy.

1940 J. A. Efremov in *Pan-Amer. Geol.* LXXIV. 93, I propose for this part of palaeontology the name of 'Taphonomy', the science of the laws of embedding. *Ibid.*, Taphonomical research allows us to glance into the depth of ages from another point of view. 1971 *Nature* 8 Oct. 391/2 There seem to be neither palaeoecological nor taphonomical features of the formation that would preclude *Hipparion* being represented in the assemblage. 1974 *Times* 2 Mar. 14/2 Russian scientists have brought together a team of geologists..and a group of taphonomists: the last belong to a speciality created in Russia for studying the way animals and plants are preserved in their burial sites. 1974 *Nature* 8 Mar. 100/3 Sessions were devoted to..patterns of diversity and implications of taphonomic evidence for behaviour patterns. 1977 Leakey & Lewin *Origins* i. 12/2 Any scientific meeting on our origins nowadays might be attended by archeologists..[and] taphonomists. 1981 *Nature* 10 Dec. 598/3 Palaeontologists are..bringing their subject out of the museum through studies of the processes by which the fossil record forms (taphonomy). *Ibid.*, Much of the book concentrates on the principles, methods of study and results of taphonomic studies of (mainly) African vertebrates.

taphrogenesis (tæfrodʒe·nėsis). *Geol.* Also **tafro-.** [ad. G. *tafrogenese* (E. Krenkel *Die Bruchzonen Ostafrikas* (1922) v. 181), *taphrogenese* (—— *Geologie Afrikas* (1928) II. 636), f. Gr. τάφρος pit: see -genesis.] The formation of large-scale geological structures by high-angle or block faulting, esp. as the result of tensional forces in the crust. Hence **taphroge·nic** *a.*

1923 *Bull. Geol. Soc. Amer.* XXXIV. 200 Faulting through tension [in East Africa] has gone on in compensation for the orogeny elsewhere. Therefore tafrogenesis (from the Greek for rifts or graben) is the counterpart of orogenesis, and East Africa is the type area for tafrogenic structures. 1963 E. S. Hills *Elem. Structural Geol.* xi. 315 Taphrogenic movements—The necessity to recognise a third type of movement is indicated by the tectonic importance of major belts of block faulting, notably the rift valley and graben zones... The term was coined for the East African rifts, and is largely descriptive. In general, however, it implies tensional forces as opposed to horizontal compression for orogeny and differential vertical movements for epeirogeny. 1978 *Nature* 9 Mar. 158/2 Sedimentation has been controlled by NW–SE trending faults in close relationship to the taphrogenesis of the SW–NE trending Benue trough. 1979 *Ibid.* 7 June 478/3 There may be a long time interval between initial taphrogenic activity ('rifting') and creation of ocean floor by spreading ('drifting').

tapidaro, tapidero, varr. Tapadero in Dict. and Suppl.

tapioca. Add: Also 9 tabiaca. (Earlier examples of *tapioca pudding*.)
1837 W. Tayler *Diary* 14 May in J. Burnett *Useful Toil* (1974) II. 181 They had two soles fried with saws.. a tabiaca pudding, cheese and butter. **1859** D. Bunce *Travels with Dr. Leichhardt* 107 Christmas day..tapioca pudding, each man having as much as he could eat.

tapis, *sb.* Add: ‖ **c.** *tapis vert*, a long strip of grass-covered ground; a grass walk. Cf. Carpet *sb.* 3.
1960 O. Manning *Great Fortune* III. 215 They were walking down the main path beside the *tapis vert*. **1965** Mrs. L. B. Johnson *White House Diary* 9 Mar. (1970) 248 He wants to..preserve the tapis vert, the long green ribbon that stretches..from the Capitol to the Lincoln Memorial. **1976** D. Wood *Pract. Garden Design* ii. 42 Two steps down..to the enclosed gardens on either side of the *tapis vert*—'green carpet'.

Tapleyism (tæ·pli,iz'm). [f. the name of Mark *Tapley*, a character in Dickens's *Martin Chuzzlewit* (1843–4) + -ism.] Optimism in the most hopeless circumstances, as expressed by Tapley's determination always to remain 'jolly'. Also **Ta·pleyan** *a.*
1857 B. Smith *Let.* in W. James *Mem. & Stud.* (1911) ix. 246, I have a good share of Tapleyism in me and come out strong under difficulties. **1900** F. W. Maitland *Let.* 22 Jan. (1965) 209 Your letter..told me more than I had learned from any newspapers about the gloom of England, though I had read something between lines which seemed to me Tapleyan. **1972** *Scots Observer* 12 Mar. 2/2 What impressed me..was the grim way they held on to optimism... If that is Tapleyism..then Tapleyism is a fine thing.

tapped, *ppl. a.*[1] (In Dict. s.v. Tap *v.*[1]) (Examples in sense 2 c of the vb.)
1925 P. J. Risdon *Crystal Receivers & Circuits* 9 In the case of a tapped inductance coil, the wire is tapped at every turn for so many turns, for fine adjustment, and then once every few turns for coarse adjustment. **1964** L. Deighton *Funeral in Berlin* v. 37 Transcripts of tapped phone calls. **1978** D. Murphy *Place Apart* ii. 22 He called in a disguised message over the tapped phone that he would be over that evening.

tapped, *ppl. a.*[2] [f. Tap *v.*[2] + -ed[1].] **a.** *Phonetics.* Pronounced with a tap (see *tap *sb.*[2] 1 d). **b.** *tapped penalty* (Rugby Football), a penalty taken with a tap-kick (see *tap *sb.*[2] 4).
1964 R. H. Robins *Gen. Linguistics* 101 A flapped or tapped /r/..when the tongue tip lightly and momentarily touches the alveolar ridge is common between two vowels (as in *merry*). **1966** R. E. Asher in C. E. Bazell *In Memory of J. R. Firth* 17, r is a tapped alveolar consonant. **1977** *Observer* 22 May 23/3 Bevan took a tapped penalty 15 yards from his own line.

tapper[1]. Add: **2. b.** One who 'touches' another for money; a beggar. Cf. Tap *v.*[1] 3 in Dict. and Suppl. *slang.*
1930 G. Smithson *Raffles in Real Life* xiv. 189 He was a hanger-on, a common cadger, a 'tapper'. **1939** J. Worby *Spiv's Progress* iv. 32, I didn't have time to light a cigarette before I was accosted by a tapper. **1962** *John o' London's* 25 Jan. 82/2 One who lives by cadging or begging is a *bummer*, *knocker* or *tapper*.
c. One who taps (Tap *v.*[1] 2 c) telegraph or telephone wires; a phone-tapper, a wire-tapper.
1973 P. Tamony *Americanisms* (typescript) No. 33. 7 Thirty three states legislated total wiretap bans.., while six created partial bans which allowed police to tap.., but forbidding private tappers under any conditions. **1976** *Time* 27 Dec. 42/2 Halperin..was furious at learning that the FBI had tapped his telephone... Last week Morton Halperin won a resounding victory that could cost his tappers, starting with President Nixon, nearly $1 million in damages. **1980** E. Behr *Getting Even* xv. 170 He delivered an oblique message in Chinese... The tappers might not even tell the difference.

Tappertitian (tæpərti·ʃən), *a. rare.* [f. the name of Simon *Tappertit*, a conceited apprentice in Dickens's *Barnaby Rudge* (1841) + -ian.] Characteristic of or resembling Tappertit, esp. in his amorous approaches to Dolly Varden.
1895 G. B. Shaw in *Sat. Rev.* 19 Jan. 94/1 One's gorge rose at the Tappertitian vulgarity and infamy of the thing. **1903** —— *Man & Superman* p. xxviii, I have been proof against the garish splendors and alcoholic excitements of the ordinary stage combinations of Tappertitian romance with the police intelligence. **1949** St. J. Ervine *Craigavon* ii. lxi. 287 Had they been, there would not have been any procession of Tappertitian playboys.

tapping, *vbl. sb.*[1] Add: **2.** (Examples in senses 2 c, 3 of the vb.)
1931 T. Horsley *Odyssey of Out-of-Work* xxiii. 243 You do the tapping; I'll wait for you at the gate. **1955** 'E. Crispin' *Fen Country* (1979) 54 A line from your special switchboard..could be made safe from tapping. **1966** *Times* 14 Nov. 10/6 Mr. Russell Kerr..is to ask the Home Secretary..on how many occasions warrants have been issued for the tapping of M.P.s' private telephones.
b. *spec.* in *Electr. Engin.*, an intermediate connection made in a winding.
1903 K. Edgcombe *Whittaker's Electrical Engineer's Pocket Bk.* 244 The secondary of one of the transformers is wound to give a voltage equal to the required three-phase line-to-line voltage, and is divided into two equal parts by a middle tapping. **1934** *Discovery* Oct. 301/2 Twelve tappings are fitted to give an impedance range of 1·6 to 25,000 ohms. **1950** *Engineering* 22 Sept. 245/3 Tappings from the secondary and filament windings are connected to the electrodes. **1975** G. J. King *Audio Handbk.* iii. 71 A tapping on the resistive element facilitates coupling of the loudness filter.
c. (sense *2 b) *tapping point*; **tapping coil** *Electr. Engin.*, a coil which acts as a tapping.
1933 P. Kemp *Alternating Current Electr. Engin.* (ed. 4) xiv. 188 In order to permit of a slight variation in the secondary terminal voltage, it is usual to provide a number of tapping coils in the L.T. winding. **1958** J. Shepherd et al. *Higher Electr. Engin.* xiv. 344 The tapping coils are placed physically in the centre of the transformer limb to avoid unbalanced axial forces acting on the coils. *Ibid.* 343 If..contact with position 2 is made before contact with position 1 is broken, the coils connected between these two tapping points are short-circuited, and will carry damagingly heavy currents.

tapping, *vbl. sb.*[2] Add: **a.** Also *spec.* tap-dancing.
1944 J. Johns *A.B.C. Tap Dancing* 2 'Tapping' has few stable rules but limitless variations.
c. tapping key *Electr.* = Key *sb.*[1] 12 a.
1916 'Boyd Cable' *Action Front* 173 These [orderlies]..brought them long screeds to be translated to the tapping keys. **1938** *Brit. Jrnl. Psychol.* XXIX. 41 Two tapping keys..were placed below the handles.

tapsell (tæ·ps'l). [Origin uncertain.] *tapsell gate*, a type of churchyard gate peculiar to Sussex, which turns about a central post.
1922 H. Allcroft *Downland Pathways* 114 You approach the building through a 'tapsel-gate'. **1947** E. Meynell *Sussex* vi. 134 One of those curious gates, swinging on a central post, known as tapsell gates. **1957** H. Hall *Some Sussex Sayings & Crafts* 181 Tapsell gate, this ingenious gate is peculiar to Sussex, but nothing is known of its origin or its designer. Its special use is to prevent cattle entering churchyards and to make room for coffin bearers to pass through easily... Today there are only six in the country. **1969** *Daily Tel.* 27 Feb. 21/8 Tapsell gates sometimes have a hook or spring-catch or even are chained and padlocked. **1979** *Country Life* 15 Mar. 726 Tapsell gates, which turn on a central pivot, are peculiar to Sussex. It has been suggested that they are so-called after one Tapsell, a Sussex iron-master.

tap-tap, tap-tapping: see also *tap *v.*[2] 2.

tapu (tā·pū), var. Taboo, tabu *a.* and *sb.* (Largely a regional variation, esp. in *N.Z.*: see note at the dominant form.) **A.** *adj.* **a.** = Taboo *a.* Also (*rarely*) *fig.*
1832 H. Williams *Jrnl.* 18 Jan. in H. Carleton *Life H. Williams* (1874) 114 The canoe was tapu, having conveyed the body of Heagi..to his former place. **1849** W. T. Power *Sketches in N.Z.* p. xliv, A barbarous murder was committed by some of the natives under the protection of Rangihaeta, who refused to give them up; moreover, making the road 'tapu' which communicated between the coast and Wellington. **1873** [see Taboo, tabu *a.* a]. **1902** G. B. Shaw *Mrs. Warren's Profession* p. xiii, Mrs Warren's profession must be either tapu altogether, or else exhibited with the warning side as freely displayed as the tempting side. **1936** *Discovery* Jan. 14/1 The Maoris had found out that the Morioris were a very tapu people. **1967** A. & D. Reid *Paddle Wheels on Wanganui* iv. 33 The bodies were recovered and laid in the old house which was then declared tapu. **1978** P. Grace *Mutuwhenua* ix. 56 Those hills, there are tapu places in them.
B. *sb.* = Taboo, tabu *sb.* 1 a. Also (*rarely*) *fig.*
1833 H. Williams *Let.* July in H. Carleton *Life H. Williams* (1874) 134 Some proposed Paihia, but this we declined, fearing it might bring the tapu upon a considerable portion of the land. **1851** [see Taboo, tabu *sb.* 1 a]. **1872** A. Domett *Ranolf & Amohia* 89 His sole 'tapu' a far securer guard Than lock and key of craftiest notch and ward. **1902** G. B. Shaw *Mrs. Warren's Profession* p. xiii, Each nation has its particular set of tapus in addition to the common human stock. **1938** R. Finlayson *Brown Man's Burden* 41 Uncle Tuna..disapproved of joking about matters of tapu. **1971** *N.Z. Listener* 29 Mar. 11/1 In old New Zealand there were two main causes of sickness and disease. One was the violation of tapu or a tapu place.
Also as *v. trans.* = Taboo, tabu *v.* 1. Now *rare*.
1837 in R. McNab *Old Whaling Days* (1931) xxi. 335, [I] tapued a piece of land of the proprietors, two respectable chiefs, for some blankets and fish-hooks. **1851** V. Lush *Jrnl.* 16 Dec. (1971) 94 A native burial place which the Tryces had been obliged to fence in before the natives would allow them to dwell in peace—so great is the natives' dread lest a spot they have tapued should be desecrated by man or beast. **1863** F. E. Maning *Old New Zealand* i. 14 A good gun... I must have this; I must tapu it before I leave the ship. **1890** *Jrnl. Anthrop. Inst.* XIX. 100 Tapu is an awful weapon. I have seen a strong young man die the same day he was tapued.

Tapuia (tāpū·yā), *sb.* (and *a.*). Also 7 Tapui; 9 Tapuio; Tapuya. [a. Pg. *Tapuia*, Sp. *Tapuya*, ad. Tupi-Guaraní *tapua* savage, slave.] (A member of) a Brazilian Indian people not of Tupi stock. Also *attrib.* or as *adj.*
1613 Purchas *Pilgrimage* ix. vi. 712 Towards the East dwell the Itatini people, which call themselves Garay, that is, Warriours; and others, Tapuis or Slaues. **1860** Mayne Reid *Odd People* 44 Farther down the river, the 'Indio manso' is a 'tapuio', a hireling of the Portuguese, or, to speak more correctly, a *slave*. *Ibid.* 46 By such name is his house (or village rather) known among the *tapuios* and traders of the Amazon. **1910** *Encycl. Brit.* I. 783/2 The name Amazonas arises from the battle he had with a tribe of Tapuya savages. **1944** S. Putnam tr. *E. da Cunha's Rebellion in Backlands* i. 44 The hiding-places of the Tapuia. *Ibid.* ii. 83 The predominance of Tapuia terms in the geographic names of these places—terms that have resisted absorption by the Portuguese and Tupi languages.

tar, *sb.* Add: **1. a.** Also formed in the combustion of tobacco, etc.
1921 *U.S. Patent* 1,398,734 2/1 The catch basins..are adapted to concentrate the heavier particles of tar from smoke. **1932** *Amer. Jrnl. Cancer* XVI. 1513 The tar of cigarette smoke contains nicotine, phenolic bodies, pyridine bases, and ammonia. **1974** M. C. Gerald *Pharmacol.* viii. 155 Some of these are polycyclic hydrocarbons, commonly referred to as 'tars', and are undoubtedly responsible for the disproportionately greater incidence of lung cancer among cigarette smokers.
d. *to beat (knock, etc.) the tar out of*: to beat unmercifully, to reduce to a state of helplessness. Cf. *shit *sb.* 1 g. *U.S. slang.*
1884 *National Police Gaz.* 6 Sept. 11/1 (*heading*) The celebrated New York dubs get the tar knocked out of them. **1916** 'Texas' *Trav. Tourist 'Typo'* 46 The newsboy..whose chief occupation is..to wallop the tar out of smaller newsboys. **1939** D. Trumbo *Johnny got his Gun* ii. 35 Naturally you..wanted Germany to get the tar kicked out of her. **1973** Wodehouse *Bachelors Anonymous* vii. 80 She is a fine upstanding woman, fully capable of beating the tar out of you.
e. Colloq. abbrev. of Tarmac or Tar macadam; a road surfaced with this.
1934 Dylan Thomas *18 Poems* 26 Nor city tar and subway bored to foster Man through macadam. **1971** *E. Afr. Standard* (Nairobi) 13 Apr. 6/5 He thought the tarmac was dangerous anywhere. Next year he wanted to see less tar and higher speeds elsewhere. **1980** G. Lord *Fortress* iii. 26 Both roads were dirt... The tar ended miles back.
4. a. That contains tar: *tar-pill* (earlier example); used for holding, or in making, tar: *tar-bucket* (examples). **b.** Objective, instrumental, etc.: *tar-brand* vb. (earlier example), *-mark* vb., *-painted* (example), *-smelling*, *-streaked* adjs. **c. tar acid**, any of numerous phenolic constituents of coal-tar distillates that react with dilute caustic soda to give water-soluble salts; **tar and feathers** *U.S.* (with reference to the practice of tarring and feathering: see Tar *v.*[1] b); **tar-baby**, (*a*) the doll smeared with tar, set to catch Brer Rabbit (see quot. in Dict.); hence *transf.*, *spec.* an object of censure; a sticky problem, or one which is only aggravated by attempts to solve it (*colloq.*); (*b*) a derog. term for a Black (*U.S.*) or a Maori (*N.Z.*); **tar ball**, (*a*) (in Dict., sense 4 a); (*b*) a ball of crude oil found in or on the sea; **tar base**, any of numerous cyclic, nitrogen-containing bases present in coal-tar distillates; **tar-boiler**, (*a*) (in Dict., sense 4 a); † (*b*) *U.S. slang* = Tarheel (*obs.*); **tar-boy** *Austral.* and *N.Z.*, an assistant hand in a shearing shed who treats injured sheep with tar or other disinfectants; **tar-bush**, one of several aromatic shrubs of western N. America, esp. one of the genus *Eriodictyon*, of the family Hydrophyllaceæ, which includes several sticky or tomentose evergreens; **tar kiln** (later example); **tar-paper** chiefly *N. Amer.* (in Dict., sense 4 a): often used as a building material; freq. *attrib.*; (earlier and later examples); **tar-pavement**, **-paving**, a form of surfacing for roads, pathways, etc., composed mainly of tar; **tar-pot**, (*c*) an opprobrious name for a Black (*U.S.*), or a Maori (*N.Z.*) (cf. *tar-baby* (*b*) above); **tar-sand**, a deposit of sand impregnated with bitumen.
1909 *Chem. Abstr.* III. 1079 (*heading*) Hydrocarbons obtained from the tar acids of petroleum. **1951** M. McLuhan *Mech. Bride* (1967) 91 (Advt.), Coal-Tar Chemicals—Benzol, toluol, naphthalene, tar acids, tar bases, solvents, [etc.]. **1974** Tar acid [see *tar base* below]. **1775** P. V. Fithian *Jrnl.* 8 June (1934) II. 25 He hears many of his Townsmen talking of Tar and Feathers—These mortifying Weapons. **1834** *Southern Lit. Messenger* I. 87 If he remained longer, he was in danger of tar and feathers. **1954** J. Steinbeck *Sweet Thursday* xxxviii. 261 He left town, and just as well. There was talk of tar and feathers. He must have heard. **1982** W. Mankowitz *Mazeppa* vi. 97 The Vigilance Committees..had asserted law summarily with fast necktie parties, rail-rides and tar and feathers. **a 1910** 'Mark Twain' *Autobiogr.* (1924) II. 18 For two years the *Courant* had been making a 'tar baby' of Mr. Blaine, and adding tar every day—and now it was called upon to praise him. **1924** Kipling *Debits & Credits* (1926) 97 Number Five Study..were toiling inspiredly at a Tar Baby made up of Beetle's sweater, and half-a-dozen lavatory towels;..and most of Richard's weekly blacking allowance for Prout's House's boots. **1948** S. Lewis *Kingsblood Royal* 334 'I didn't know she was a tar-baby.'..'Don't be so dumb. Can't you see it by her jaw?' **1959** M. Shadbolt *New Zealanders* 140 'What a hide, though—' 'Those tar-babies and that fellow in the sweater.' **1976** *National Observer* (U.S.) 29 May 15A/3 The troubled U.S. Postal Service is fast becoming the political tar baby of the year. **1978** J. Updike *Coup* (1979) iv. 135 She was one of those white women who cannot leave black men alone... Some questing chromosome within holds her sexually fast to the tarbaby. **1972** *Science* 16 June 1258/2 Crude oil lumps ('tar balls') are now universal constituents of the surfaces of the world oceans. **1921** *Jrnl. Amer. Chem. Soc.* XLIII. 1936 Crude coal tar bases vary greatly..in the nature and in the proportion of the bases which they contain. **1951** [see *tar acid* above]. **1974** *Encycl. Brit. Micropædia* II. 1017/2 Tar bases are the basic constituents of the distillate oils, present after tar acids have been removed. **1845** *Cincinnati Misc.* I. 240 The inhabitants of..N. Carolina [are called] Tar-boilers. **1885** W. Whitman in *N. Amer. Rev.* Nov. 433 Among the rank and file..[in the Civil War] it was very general to speak of the different States they came from by their slang names. Those from..North Carolina [were called] Tar Boilers. **1888** 'R. Boldrewood' *Robbery Under Arms* I. x. 123 There wasn't a man of the lot in the shed, down to the tarboy, that wouldn't have done the same. **1936** A. Russell *Gone Nomad* iii. 19 Then I found myself a tar-boy in the shearing-shed. **1956** G. Bowen *Wool Away!* (ed. 2) 157 Tar-boy, the hand who walks the board where sheep are subject to the fly and who puts a smear of tar on the cuts made on sheep. **1977** C. McCullough *Thorn Birds* x. 231 Luke took himself off on the shearing circuit as a tar boy, slapping molten tar on jagged wounds if a shearer slipped and cut flesh as well as wool. **1878** E. S. Elwell *Boy Colonists* 205 It took a good month to muster and tar-brand all the sheep. **1723** *Amer. Weekly Mercury* 23–30 May 2/1 The forced Men..carryed the Brigantine into Curacao, with the Captains Head in a Tarr Bucket. **a 1909** *Joseph W. Caldwell: Mem. Vol.* (1909) 66 There was a brindled cur dog under the wagon, keeping company with the tar bucket that swung from the coupling pole. **1931** *Sun* (Baltimore) 28 Oct. 13/4 Tall 'tar-bucket' helmets with the black plumes. **1884** W. Miller *Dict. Eng. Names Plants* 134/2 Tar-bush, Californian. *Eriodictyon californicum.* **1902** *Out West* Oct. 452 There were the innumerable cacti with their brilliant flowers, and the tar bush. **1949** *Chicago Tribune* 20 Feb. 30/3 Cedar and mesquite alone are costing Texas ranchers 115 million dollars a year. Add the..blue oak, creosote, tarbush..and prickly pear and the toll is terrific. **1943** H. Parkes in C. Goerch *Down Home* xx. 99 Any one who has ever seen a tar kiln in operation or been around a turpentine distillery will realize what sticky and dirty work it is. **1825** Jamieson *Suppl.* II. 86/1 An old Angus laird,..on observing that one of the young ladies had both earrings and patches, cried out..in obvious allusion to the means employed by store-farmers for preserving their sheep; 'Wow, wow! Mrs. Janet, your father's been michtilie fleyd for tyning you, that he's baith *lug-markit* ye and *tar-markit* ye.' **1918** *Chrons. N.Z.E.F.* 30 Aug. 61 Tar-marking and branding cattle. **1787** P. F. Freneau *Journey from Philadelphia* ii. 14 You tar-painted [*Poems* (1795) vii. 343: Tar-smelling] monster!.. If Snip should be drowned, and lost in the sea, You never once think what a loss it would be! **1891** H. Campbell *Darkness & Daylight* xxxi. 611 If he ventures to charge more, except for a dress..or for tar-paper..he is liable to lose the license. **1919** S. Lewis *Free Air* 122 Then a lonely, tight-haired woman in the doorway of a tar-paper shack waved to her. **1966** D. F. Galouye *Lost Perception* ix. 89 The plane..taxied up to a frame building with a tarpaper roof. **1978** R. Ludlum *Holcroft Covenant* xxxii. 370 The tar paper..bulged slightly next to the wall. **1883** *Proc. Assoc. Municipal Engin.* X. 53 My first experience in tar pavement was in 1850. *Ibid.*, Tar paving had been used to some extent for footways previous to..March 1881. **1967** *Gloss. Highway Engin. Terms* (*B.S.I.*) 47 *Tar paving*, a surfacing of tarmacadam laid in one or two courses for footpaths, playgrounds, and similar areas for pedestrian or very light vehicular traffic. **1728** E. Smith *Compl. Housewife* (ed. 2) 304 The Tar Pills for a Cough. Take Tar and drop it on Powder of Liquorish, and make it up into Pills. **1944** H. L. Mencken in *Amer. Speech* XIX. 174 *Pickaninny* was..used..affectionately. So..was *tar-pot*..signifying a Negro child. **1949** F. Sargeson *I saw in my Dream* xiii. 120 He'd never let the tarpots inside the shed with their lousy sheep. **1899** *Nature* 15 June 159/1 Great trouble has been experienced in the effort to penetrate the 'tar-sands' at the base of the Cretaceous strata. **1978** *Ibid.* 29 June 703/3 The heavier oils will also have to be used as chemical feedstocks in the future: for example, Canada's 'tar sands', now the subject of a multi-million dollar project in Alberta. **1973** R. Adams *Watership Down* (ed. 2) xxiv. 173 He sat..on the bank above the tar-smelling road. **1782** J. Trumbull *M'Fingal* iv. 70 Adown his tar-streak'd visage, clear Fell..th'indignant tear. **1939** S. Spender *Still Centre* iv. 94 You stood once In the tar-streaked drizzling street.

‖ **tar** (tɑːɹ), *sb.*[3] *Anglo-Ind.* [Hindi *tār*.] A telegram.
1893 Kipling in *Harper's Weekly* 30 Dec. 1246/3 My father is at the *tar*-house sending tars. **1901** —— *Kim* xi. 286 Therefore I did not send a *tar* (telegram) to any one saying where the letter lay. **1978** 'M. M. Kaye' *Far Pavilions* iv. xxviii. 411 It would have been a simple matter for the Rana..to arrange for a *tar* (telegram) to be dispatched.

tar, *v.*[1] Add: **b.** (Earlier and later examples.) Also *fig.*
1769 *Boston* (Mass.) *Chron.* 30 Oct. 3/2 A person..was stripped naked, put into a cart, where he was first tarred, then feathered. **1925** A. Huxley *Those Barren Leaves* II. iii. 113 Miss Carruthers, who has a short way with dissenters, would like to see them tarred and feathered—all except pacifists, who, like strikers, could do with a little shooting. **1960** N. Annan in *Victorian Stud.* June 331 The individualist, the eccentric, the man who offends against the trivial rules of the club, are tarred and feathered with gleeful brutality. **1977** *Daily News* (Perth, Austral.) 19 Jan. 6/5 The victims were stripped naked, tarred and shorn of their hair. **1981** A. Price *Soldier no More* 161 The Russians..wouldn't have cared less if we'd tarred and feathered Nasser and run him out of Suez on a rail.

tara, *int.* (Later examples.)
1922 Joyce *Ulysses* 166 Tara tara. Great chorus that. Tara. Must be washed in rainwater. Meyerbeer. Tara: bom bom bom. *Ibid.* 503 Exercise your mnemotechnic. *La causa è santa.* Tara. Tara.

tara (tārā·), *int.*[2] Also **tarra(h**, etc. Colloq. (mainly North.) alteration of Ta-ta, in familiar use.
1958 A. Sillitoe *Saturday Night & Sunday Morning* ii. 27 'See yer't dinnertime, Arthur.' 'Tarr-ar, Dad.' **1967** E. Williams *Beyond Belief* i. i. 6 They are off arm in arm to Belle Vue. Ta-ra-for-now, they call out, the warm casual Lancashire way of saying goodbye, 'ta ra!' **1973** B. Bainbridge *Dressmaker* iii. 39 'Tarrah, Valerie!' called Rita up the stairs. 'Thank you very much for having me.' **1981** B. Hines *Looks & Smiles* 121 'I'm off now, Mum, ta-ra.' 'Ta-ra, love. See you tonight.'

tarabagan, var. *tarbagan.

‖ **tarada** (tarā·dā). [ad. Arab. *ṭarrāda* cruiser, swift war canoe.] A canoe used by the Marsh Arabs of Iraq.
1960 G. Maxwell *Ring of Bright Water* i. vi. 75 We spent the better part of those two months [in Southern Iraq] squatting cross-legged in the bottom of a *tarada* or war canoe. **1964** W. Thesiger *Marsh Arabs* iii. 23 The top part of the ribs was..studded with five rows of flat, round nail-heads two inches across. These decorative nails were the distinguishing mark of a *tarada*..which only a sheikh may own. Years later, in Oslo, I saw the Viking ships preserved there and was at once reminded of the *taradas* in the Marshes. **1974** *Blackw. Mag.* Oct. 341/1 There is a bigger canoe called a *tarada*, graceful and swift, with a fine upswept prow.

Tarahumara (ta:rāhumā·rā), *sb.* and *a.* Also **Tarahumar(e**. [a. Sp., of uncertain origin.] **A.** *sb.* **a.** (A member of) an Uto-Aztecan people of north-western Mexico. **b.** The language of this people. **B.** *adj.* Of, pertaining to, or designating this people.
1874 H. H. Bancroft *Native Races Pacific States* I. v. 609 The *Tarahumares* inhabit the district of Tarahumara in the state of Chihuahua. *Ibid.* III. v. 666 The Tarahumara, which is a more polished language than its neighbors, contains words similar to the Aztec. **1911** J. G. Frazer *Golden Bough: Magic Art* (ed. 3) I. iii. 150 The Tarahumares of Mexico are great runners. **1912** C. Lumholtz *Unknown Mexico* I. viii. 168 Tarahumare pottery is exceedingly crude. **1934** A. L. Kroeber *Uto-Aztecan Lang. Mexico* 13 Varohío..is in all most similar to Tarahumar. But it can hardly be..merely a provincial dialect of Tarahumar. **1963** C. W. Pennington *Tarahumar of Mexico* i. 5 Zapata's résumé of the Tarahumar missions in 1678..indicate that Coyáchic..was an important Tarahumar center. **1964** E. A. Nida *Toward Sci. Transl.* iii. 35 The Tarahumara in northern Mexico have five basic color words, including one term *siyonomi*, which covers both green and blue. **1975** *Language* LI. 798 In Tarahumara, *-tu/-ru* retains the sense 'become' in its use as a derivational suffix. **1979** *Tucson* (Ariz.) *Mag.* June 48/2 Several Tarahumara dwellings and farm buildings have been reconstructed for the exhibit.

tarakihi (taraki·hi). *N.Z.* Also **terakihi** (te·rakī). [Maori.] A marine fish, *Cheilodactylus macropterus*, silver in colour with a black band behind the head, belonging to the morwong group and found off the coasts of New Zealand.
1873 J. H. H. St. John *Pakeha Rambles through Maori Lands* ii. x. 173 Our best fish are the tarakihi, patiki or sole, and whitebait. **1937** *Nature* 7 Aug. 223/1 Tarakihi..flesh was found to contain appreciable amounts of vitamin A. **1959** A. H. McLintock *Descr. Atlas N.Z.* 48 Tarakihi..second only in importance [as a commercial catch] to snapper, is trawled off the east coast..[and] also caught in quantity in deep water along the west coast. **1960** N. Hilliard *Maori Girl* ii. v. 96 She..was preparing to fry a piece of *terakihi* for their tea.

taramellite (tærāme·ləit). *Min.* [a. It. *taramellite* (E. Tacconi 1908, in *Atti dell' Accad. Naz. dei Lincei: Rendiconti, Classe di Sci. Fisiche*, etc. XVII. i. 814), f. the name of T. *Taramelli* (1845–1922), Italian geologist: see -ite[1].] An orthorhombic borosilicate of barium and other metals, chiefly iron and titanium, and usu. containing some chlorine.
1908 *Jrnl. Chem. Soc.* XCIV. ii. 863 The author describes a new mineral, taramellite, occurring in radiating, fibrous, acicular aggregates or slender veins in the calcareous zone of Candoglia-Ornavasso (Val Toce). **1973** *Mineral. Abstr.* XXIV. 433/2 More than 50 minerals have been identified in rocks of this quarry of the Pacific Limestone Products Co. [in California]. They include..15 silicates three of which, celsian, pabstite, and taramellite, are Ba-bearing.

taramosalata (ta:rāmŏsāla·tā). Also **taramasalata**. [a. mod.Gr., f. ταραμᾶς preserved roe (ad. Turk. *tarama* soft roe, red caviare) + σαλάτα Salad.] A Greek fish pâté made (traditionally) from the roe of the grey mullet or from smoked cod's roe, mixed with garlic, lemon juice, olive oil, etc. Also *ellipt.* as **tarama**.
1910 Z. D. Ferriman *Home Life in Hellas* iv. 181 Red caviar..is pounded with garlic and lemon juice into what is called *tarama salata*. **1958** R. Liddell *Morea* ii. iii. 70 A vinegary *taramosalata*, a preparation of salted fish eggs and oil, which is always Lenten food in Greece. **1964** *Spectator* 8 May 645/1 A Greek fish pâté, *taramasalata*. **1972** *Harper's & Queen* Apr. 92/1, I quite often add tarama to go with the avocado. **1978** *Chicago* June 233/1 Dinnertime favorites include saganaki,..taramosalata, red caviar, etc.

Taranaki (tærānæ·ki). The name of a province in New Zealand, used *attrib.* in **Taranaki gate**, a gate made of wire strands attached to upright battens (see quot. 1966).
1937 M. E. C. Scott *Barbara Prospers* 27 She..had the 'Taranaki' gate open in a twinkling. **1948** R. Finlayson *Tidal Creek* 205 Mind you get the taranaki gate properly up. **1953** J. W. Brimblecombe *Shear Nonsense* 110 He can now open a Taranaki gate without rolling himself up in it. **1966** G. W. Turner *Eng. Lang. Austral. & N.Z.* ii. 31 The province gives its name to the *Taranaki gate*, a moveable piece of a wire fence held by a loop of wire at one end to form a makeshift gate. **1968** *Landfall* XXII. 390 Sure enough, we found someone had taken down the Taranaki gate on the side that led to the Oteranika Road.

taranakite (tærānā·kəit). *Min.* [f. prec.: see -ite[1].] A hydrated basic phosphate of potassium (partly replaced by ammonium) and aluminium, found as a soft, whitish or grey clay-like substance composed of minute rhombohedral crystals.
1866 Hector & Skey in *Rep. & Awards of Jurors N.Z. Exhib., 1865* 423 Taranakite, a new Phosphatic mineral..presented by H. Richmond... This singular mineral was mistaken for Wavellite. **1882** *Trans. N.Z. Inst.* XV. 385 Taranakite..a double hydrous phosphate of alumina and potash, part of the alumina being replaced by ferric oxide, was first discovered by H. Richmond, Esq., at the Sugar Loaves, Taranaki. **1976** *Mineral. Abstr.* XXVII. 256/2 Aluminian strengite.., vivianite.., and taranakite..(small yellowish white aggregates) from the Castellane caves, Apulia, southern Italy, occur embedded in small 'terra rossa' deposits, near the contact with superimposed layers of bat guano.

tarantula. Add: **5. tarantula-hawk**: substitute for def.: any of several species of spider wasp of the genus *Pepsis* that occur in the south-western U.S. and kill tarantulas; (examples); **tarantula-juice** *U.S. slang*, inferior whisky; **tarantula-killer** (earlier example).
1861 *Harper's Mag.* Jan. 147/2 Little to drink, except old-fashioned tarantula-juice, 'warranted to kill at forty paces'. **1867** *Amer. Naturalist* I. 137 The large, red-winged 'Tarantula Killer'..is, as far as I know, the largest of the dauber group. **1878** B. F. Taylor *Between Gates* 198 The tarantula hawk..pounces upon his victim and makes a needle-cushion of him. **1932** *Sun* (Baltimore) 8 June 9/1 A large wasp, known..as a 'tarantula hawk', was victor over a tarantula in a battle. **1980** F. H. Wagner *Wildlife of Deserts* 148 Formidable as the tarantula is, it has its own enemies. The large tarantula hawk wasp (*Pepsis*) feeds the spiders to its young. **1939** C. W. Towne *Her Majesty Montana* 38 In the saloons, poisonous liquors are vended to all comers under the name of 'tangleleg', 'forty-rod', 'lightning', 'Tarantula-juice', etc.

† **Tarantulle** (tærăntu·l). *Obs.* [Cf. Tulle.] The proprietary name of a kind of cotton fabric.
1890 *Trade Marks Jrnl.* 3 Sept. 878 Tarantulle 89,034. Cotton piece goods of all kinds. Tootal Broadhurst Lee Company..Manchester..Manufacturers.—10th April 1889. **1915** *Official Gaz.* (U.S. Patent Office) 16 Mar. 1033/1 Tarantulle... Cotton piece goods—viz. Nainsooks, Cambrics, and Madapolams. **1923** *Daily Mail* 9 Jan. 1 (Advt.), Ladies' Night-dress, in standard Tarantulle, trimmed strong Embroidery. **1932** D. C. Minter *Mod. Needlecraft* 253/2 Tarantulle. Dorcas cambric or fine longcloth.

tarara (tārā·rā). Also *redupl.* [Echoic.] = Taratantara i.
1891 Kipling *Light that Failed* ii. 18 Ridin', ridin', ridin', two an' two, Ta-ra-ra-ra-ra-ra-ra, All the way to Kandahar. **1892** —— *Barrack-Room Ballads* 57 You can't refuse when you get the card And the widow gives the party. (*Bugle*: Ta—rara—ra-ra-rara!) **1980** D. Bloodworth *Trapdoor* v. 25 The situation was..saved in the nick of time—*tarara, tarara*—by a little old lady.

Tarascan (tāræ·skăn), *sb.* and *a.* [f. Sp. *Tarasco*, name of a Meso-American Indian language of S.W. Michoacán, Mexico + -an.] **A.** *sb.* **a.** A member of an Indian people of the mountain area about lake Pátzcuara in Michoacán. **b.** Their language. **B.** *adj.* Of or pertaining to this people or their language.
[**1874** H. H. Bancroft *Native Races Pacific States* I. vi. 643 The Michoacagues or Tarascos are warlike and brave. *Ibid.* III. x. 744 The Tarasco, the principal language of Michoacan, can be placed almost upon an equality with the Aztec, as being copious and well-finished.] **1911** C. Thomas *Indian Lang. Mexico & Central Amer.* 51 The Tarascan language is now well known as constituting a separate family. **1914** R. J. MacHugh *Mod. Mexico* i. 2 One hundred and thirty-three separate Indian tribes recognized in Mexico..are arranged under sixteen language groups—the Athapascan..Piman, Tañoan, Tarascan, [etc.]. **1931** S. Chase *Mexico* ii. 31 There was only one basic culture in Mexico and Central America, in which the Mayas, the Toltecs, the Aztecs, the Tarascans, the Zapotecs and various other nations shared. **1948** R. C. West (*title*) Cultural geography of the modern Tarascan area. **1964** E. A. Nida *Toward Sci. Transl.* v. 94 Tarascan, a language of Mexico which has a number of metaphors, does not readily admit new ones. **1974** *Encycl. Brit. Macropædia* XII. 164/2 The relative isolation created by the mountains permitted the Tarascans to work out their own cultural variant.

tarbagan (tā·ɹbăgăn). Also **tarabagan**. [a. Russ. *tarbagán*.] A large long-haired marmot, *Marmota bobak* or *M. sibirica*, found in the steppes of eastern and central Asia; also, the pelt of this animal.
1928 in *Funk's Stand. Dict.* **1930** M. Bachrach *Fur* xii. 136 The other variety [of marmot pelts] is called Tarbagans, or Tarbaganas. **1947** *New Biol.* II. 11 It [sc. the germ of plague] is present..in..tarabagans in China. **1951** Whitby & Hynes *Med. Bacteriol.* (ed. 5) xviii. 303 The more important reservoirs of sylvatic plague include..tarbagans in China. **1962** P. Manson-Bahr *Patrick Manson* xix. 168 The giant marmot..was being hunted for its fur, known as tarabagan skins. **1971** P. C. C. Garnham *Progress in Parasitol.* iii. 32 The infection primarily occurs in a variety of wild rodents..such as the tarabagan in Mongolia.

tar-brush. Add: **b.** (Earlier and later examples.)
1796 Grose *Dict. Vulgar Tongue* (ed. 3), *Blue-skin*,..any one having a cross of the black breed, or, as it is termed, a lick of the tar brush. **1899** C. J. Cutcliffe Hyne *Further Adv. Capt. Kettle* viii. 189 Those snuff-and-butter coloured ladies..ignore their own lick of the tar-brush. **1928** J. Buchan *Runagates Club* i. 18 The Du Preez family had lived..close up to the Kaffir borders, and somewhere had got a dash of the tar-brush. **1970** D. M. Davin *Not Here, Not Now* vii. iii. 340 If he hadn't been an Irishman you'd think he had a touch of the tar-brush himself. **1975** 'S. Marlowe' *Cawthorn Jrnls.* (1976) xix. 163 She was beautiful... High yellow. Some places they would have said she had just a touch of the tar-brush.

Tarbuck knot (tā·ɹbʊk nɒt). *Mountaineering.* [f. the name of the British mountaineer Kenneth *Tarbuck* (b. 1914), who invented it.] An adjustable loop knot (see quots.).
1947 K. Tarbuck in *Wayfarers' Jrnl.* No. 8. 52 The practical advantages of the Tarbuck knot lie in its adjustability. It can easily be slid up or down the standing rope by hand in order to vary the size of the loop. **1950** tr. *Mountaineering Handbk.* (Assoc. Brit. Members Swiss Alpine Club) App. 167 The end of the rope..is tied..with a Tarbuck knot... Although in its general action this knot is similar to the Prusik knot, it has in addition a remarkable shock-absorbing run when subjected to a severe shock load. **1968** P. Crew *Encycl. Dict. Mountaineering* 114/2 Most knots bend the rope into a sharp angle, which reduces the strength of the rope by quite a high percentage—this is avoided in the Tarbuck knot.

tarbuttite (tā·ɹbŏtəit). *Min.* [f. the name of P. C. *Tarbutt* (1874–1943), English mining engineer + -ite[1].] A basic zinc phosphate, Zn_2PO_4OH, found as faintly coloured or colourless triclinic crystals.
1907 *Nature* 27 June 215/1 L. J. Spencer exhibited a suite of beautifully crystallised minerals, presented to the British Museum by Mr. Percy C. Tarbutt, from the Rhodesia Broken Hill mines in north-western Rhodesia... The crystals of this new species, for which the name *tarbuttite* is proposed, are anorthic. **1955, 1974** [see *parahopeite* s.v. *para-[1] 2 c].

Tardenoisian (tāɹdĕnoi·ziăn), *a. Archæol.* [ad. F. *Tardenoisien*, f. *Tardenois* (see below): see -ian.] Of, pertaining to, or resembling the mesolithic culture remains of which were first discovered in Tardenois, dept. of Aisne, France. Also *absol.*, this culture.
[**1912** R. Munro *Palæolithic Man* xi. 277 (*heading*) Tardenoisien flint industry.] **1921** *Glasgow Herald* 16 Feb. 13 Tribes..characterized in their industry by little geometric flints called Tardenoisian. **1939** C. S. Coon *Races of Europe* iii. 56 The cultures of the Mesolithic period may be divided into two elements... One was the intrusive Tardenoisian with its advanced microlithic technique, which came in from the south across the straits of Gibraltar. **1948** A. L. Kroeber *Anthropology* (ed. 2) xvi. 633 As early as 1887, Piette discovered an Azilian period after the Magdalenian, and in 1896 de Mortillet added the Tardenoisian to this. **1951** *Field Archaeol.* (Ordnance Survey) (ed. 3) 13 'Pigmy' flints of the type known as Tardenoisian from the site at Fère-en-Tardenois in Northern France. **1975** J. G. Evans *Environment Early Man Brit. Isles* v. 103 Features of a third group, considered to reflect the continental Tardenoisian, were also incorporated, in particular the use of the chisel-ended..arrowhead.

tardiness. Add: **c.** Lateness in arriving, esp. at a meeting or assembly, a class or school, etc. *U.S.*
1828 Webster, *Tardiness*,..lateness; as, the tardiness of witnesses or jurors in attendance; the tardiness of students in attending prayers or recitation. **1902** J. Corbin *American at Oxford* 17 All this brings recollections of the paternal roof, where tardiness at breakfast meant the loss of dessert. **1930** *Randolph Enterprise* (Elkins, W. Va.) 2 Oct. 5/4 No business enterprise would tolerate the percentage of absence and tardiness experienced in the schools.

tardive, *a.* Add: **b.** *Path.* **tardive dyskinesia**, a neurological disorder, usu. a late-developing side-effect of long-term treatment with anti-psychotic drugs, which is characterized by involuntary movements of the face and jaws.
1964 A. Faurbye et al. in *Acta Psychiatrica Scandinavica* XL. 12 Tardive dyskinesia is first and foremost characterized by the occurrence of dyskinetic movements. **1979** *Nature* 1 Mar. 59/1 Of all the side effects of drugs used to treat psychotic illness such as schizophrenia, chronic tardive dyskinesias are the most disturbing.

tardon, var. *tardyon.

tardy, *a.* (*adv.*). Add: **1. c.** Late for a meeting, assembly, class, school, or appointment. *U.S.*
1638 in *Archives of Maryland* (1883) I. 6 Such as did appeare thoughe tardie should be pardoned. **1843** *Yale Lit. Mag.* VIII. 240 We were 'tardy' at our matins. **1904** *Minneapolis Times* 29 May 6 Don't shoot your husband when he is two hours tardy for supper. **1948** *Daily Ardmoreite* (Ardmore, Okla.) 4 July 21/4 During this time he had been neither absent nor tardy.

tardy, *v.* (*arch.* example.)
1972 Auden *Epistle to Godson* 10 We've had it, are in for a disaster that no four-letter words will tardy.

tardyon (tā·ɹdi,ɒn). *Physics.* Also **tardon** (tā·ɹdɒn). [f. Tardy *a.* (*adv.*) + *-on[1].] A sub-atomic particle that travels at less than the speed of light.
1969 Bilaniuk & Sudarshan in *Physics Today* May 47/2 Let us refer to all subluminal particles as tardyons. **1970** *New Scientist* 10 Sept. 521/1 The number of tachyons in a system may vary from observer to observer—yet another deviation from the conventional world of tardons. **1972** *Nature* 7 Jan. 10/3 This assumes that 'ordinary' particles (called tardyons in this sort of discussion) have a mass *m* such that m^2 is greater than zero. **1975** J. Taylor *Superminds* (1976) vi. 114 Tardons (slower-than-light particles) and tachyons can never interchange roles.

tare, *sb.*[1] Add: **2. c.** *Angling.* (See quot. 1971.)
1971 *Angling Times* 10 June 12/1 Tares: a cereal bait used for roach fishing. **1976** *Reading Chron.* 19 Nov. 26/7 Kennet-style hemp groundbait and caster on the hook failed to get him a bite for the first 90 minutes. Then he switched to floated tares and the roach came thick and fast.

tare, *sb.*[2] Add: **a.** *tare weight* (later examples); also with reference to aircraft.
1944 C. A. Zweng *Aviation Dict.* 329/1 In weighing an aircraft..the weight of any incidental equipment needed, and whose weight is included in the final weight, must be subtracted to obtain the correct weight. This is called the tare weight. **1950** *Gloss. Aeronaut. Terms* (*B.S.I.*) i. 43 *Tare weight*, for design purposes: the standard weight of a type of aircraft complete in flying order but without crew, fuel, oil, removable equipment or payload. **1967** *Times Rev. Industry* May 78/2 Reductions of more than 50 per cent in tare weight..can often be made by using a particular new material. **1977** *Mod. Railways* Dec. 480/2 All timing loads in the working timetables are now calculated for tare weights in tonnes.

tare (in phr. *tare and ages, wounds*): see Tear *sb.*[2] 3 d.

tarentaal (tarĕntā·l). *S. Afr.* Also **tarantal(l)**. [a. Afrikaans.] Either of two guinea-fowl of the family Numididæ, the crowned guinea-fowl, *Numida meleagris*, or the crested guinea-fowl, *Guttera edouardi*, both found in southern Africa.
[**1822** W. J. Burchell *Trav. Interior S. Afr.* I. xv. 364 The missionaries have a few domestic fowls..and Guinea hens or Pintadoes, which are called by the quaint name of *Jan Tadentaal*.] **1827** T. Philipps *Scenes & Occurrences in Albany* 99 Ten guinea-fowl..called here, by the Hottentots, *tarentalls*. **1906** W. L. Sclater *Birds S. Afr.* IV. 228 Crowned Guinea-fowl... 'Tarantal' of Dutch. **1948** H. V. Morton *In Search of S. Afr.* ix. 282 We would walk over the veld watching the guinea-fowl, the *tarentaal*, pour away into the mealies. **1953** U. Krige *Dream & Desert* vii. 187 Great Oupa, whose ear was so acute he could hear from the front stoep the call of a tarentaal against the ridge, slowly opened his eyes. **1958** McLachlan & Liversidge *Roberts' Birds of S. Afr.* (rev. ed.) 100 Crowned Guinea-Fowl. Tarentaal. *Numida meleagris*.. The only Guinea-fowl in our area with a casque on the head.

Tarentine, *a.* (*sb.*) Add: **c.** *sb.* A native or inhabitant of the ancient city of Tarentum (now Taranto), in SE Italy.
1579 T. North tr. *Plutarch's Lives* 443 Pyrrhus..arrived at the length in the city of Tarentum, with twenty thowsand footemen..ioyning thereto to the choycest pyked men of the Tarentines. **1720** H. Eelbeck tr. *Cicero's First Oration Archias* 11, I am of Opinion that the Rheginians, or..the Tarentines, would not have refused the Privilege to this Poet. **1812** C. Kelsall tr. *Cicero's Last Pleadings against Verres* 97 What remuneration..could compensate..the Tarentines, if they were to lose their Europa on a bull..and other works [of art]? **1978** M. Grant *Hist. Rome* iii. v. 77 The Tarentines were governed by a democracy, which..displayed..relative stability.

target, *sb.*[1] Add: **3. a.** Also *transf.*, *spec.* (*a*) a place or object selected for military attack, esp. by aerial bombing or missile assault; (*b*) a part of the body at which a boxer directs his attack.
1902 *Encycl. Brit.* XXXIII. 380/2 Taking range and size of target together, the most advantageous position is to be on the bow of the enemy while he bears abaft your beam. **1914** Hamel & Turner *Flying* xvi. 285 A pilot will have to make three or four attempts before..a bomb can be released in any hope of getting near the target. **1921** J. Driscoll *Text-bk. Boxing* 70 The 'jaw' target is..preferable, if it should happen to be exposed. **1958** F. C. Avis *Boxing Ref. Dict.* 112 *Target*, that part of the boxer's body which may properly be punched, namely, the entire front and side parts of the body above the belt and the head. **1959** *Chambers's Encycl.* XIII. 430/1 The bomber force was sent out and so timed as to converge upon the target and complete the entire attack within a few minutes. **1971** H. Macmillan *Riding Storm* viii. 272 The R.A.F. carried out a number of rocket attacks on military targets.
d. *Physics.* The object or material at which a beam of atomic or sub-atomic particles is directed, as in a cathode-ray tube or particle accelerator.
1915 *Chambers's Jrnl.* Sept. 593/1 This anti-kathode (or target), enabling us to focus the rays, was introduced by Herbert Jackson. **1932** *Proc. R. Soc.* A. CXXXVII. 230 A target, A, of the metal to be investigated is placed at an angle of 45 degrees to the direction of the proton stream. **1953** Amos & Birkinshaw *Television Engin.* I. x. 217 The action of television camera tubes is dependent on an electron beam which is focused on the target and deflected so as to cover it in a series of scanning lines. **1961** G. R. Choppin *Exper. Nuclear Chem.* viii. 111 In a cyclotron, only one target at a time may be irradiated. **1975** D. G. Fink *Electronics Engineers' Handbk.* xi. 61 The target becomes positively charged in proportion to the light intensity.
e. *Biol.* and *Med.* A region in a cell which is especially sensitive to radiation.
1936 D. E. Lea et al. in *Proc. R. Soc.* B. CXX. 56 The hypothesis that the bacterium is uniformly sensitive to radiation throughout its volume raises..difficulties, and attention will therefore be turned to the alternative hypothesis that a target exists which is specially sensitive. **1968** *Brit. Med. Bull.* XXIV. 244/2 The curve can be represented by a model in which each of several targets in the organism must interact with radiation. **1979** I. M. Leahy et al. *Nurse & Radiotherapy* iii. 30 Targets are necessarily very small and are usually assumed to be within the nucleus or the DNA material itself.
f. *colloq.* An amount set as a (minimum) objective, esp. in fund-raising; a result (i.e. a figure, sum of money, etc.) aimed at. Phr. *on target*, on the right track, as forecast. Hence *loosely*, any goal which one strives to achieve.
1942 *N. & Q.* CLXXXIII. 256/1 *Target.* Who invented the ingenious use of this word for the amount aimed at in a public subscription? I think the use has been extended to things like coal consumption. **1943** *Ann. Reg. 1942* 307 The London Warship Week resulted in 146,065,225*l.* being raised as against the original target of 125,000,000*l.* **1951** E. Gowers *ABC of Plain Words* 133 If *target* was to have all the stimulating force it was capable of, it would not do to treat it as a live metaphor, and exhort people to do nothing more exciting..than merely to hit it. So we were offered a great variety of things that we might meritoriously do to our targets. We might reach them, achieve them, attain them or obtain them; we were to feel greatly encouraged if we came in sight of the target to which we were trying to do whatever we were trying to do, and correspondingly depressed if we found ourselves either a long way behind it or (what apparently amounts to the same thing) a long way short of it. **1952** *Sat. Rev.* 20 Sept. 9/2 There are legislative targets, crop targets, charity targets, gross national product targets. **1964** F. Chichester *Lonely Sea & Sky* xxxii. 333, I had failed to beat my 30 day target by 3 days, 15 hours, 7 minutes. **1967** *Time* 18 Aug. 88 Diddy is sure he did it; yet a blind girl near by who hears all and who proves to be on target about everything else, says he never left his seat. **1977** *Times* 10 Aug. 5/5 There has been unrealistic targeteering; the 1960s building target of 500,000 was never required. **1981** *Times* 23 Oct. 22/1 First-half results from Jeavons Engineering are on target at £306,000 pre-tax.
g. *Linguistics.* = *output *sb.* 1 e.
1970 J. Hill in *Linguistic Inquiry* I. 539 The formal statement of the HAB formation rule of Cupeño..is not going to be like the usual..rule involving description of input..; it can instead be visualized as involving first a statement of the target or output, and then a statement for reaching the target. **1977** *Language* LIII. 209 This constraint is a target; i.e., a number of rules of various types conspire to keep the verb in sentential second position.
5. (sense 3) *target-practise* vb., *-seeking*, *-shooting* (example); (appositively) designating an object of attack, as *target area*, *boat*, *vehicle*, etc.; (see also *target ship* in Dict.); *transf.*, esp. of a particular group over which influence is sought, as *target audience*, *company*, *group*, *population*; (sense *3 d) *target nucleus*, *volume*; (sense *3 f, passing into adj.) by which the desired goal is specified, as *target date*, *figure*, *output*, *prize*, *size*, etc.; **target cell** *Biol.* and *Med.*, an abnormal form of red blood cell which appears as a dark ring surrounding a dark central spot in stained blood films; hence *target cell anæmia*, descriptive of any anæmia in which target cells are abundant; **target dialect**, the variety of a language learned as a second dialect; **target indicator**, an object, as a flare, dropped in order to illuminate or delimit a target for aerial bombing; **target man**, (*c*) *Assoc. Football* (see quot. 1978); **target organ** *Biol.*, any organ which responds to a particular hormone or hormones in the body (cf. *target tissue* below); **target program** *Computers* = *object program* s.v. *object *sb.* 10; **target theory** *Biol.* and *Med.* (see quots. and cf. sense 3 e above); **target tissue** *Biol.*, any tissue which responds to a particular hormone or hormones within an organism (cf. *target organ* above).

1936 *Proc. R. Soc.* B. CXX. 57 To prove that the target is a biological reality..the obvious experiment..is to use several different intensities of alpha rays and beta rays and to calculate the target area in each experiment. **1939** W. S. Churchill in *New Statesman* 7 Jan. 6/2, I think a great mistake has been made in spreading our A.R.P. efforts over the whole country, instead of concentrating on what I should call the target areas. **1958** F. C. Avis *Boxing Dict.* 77 *Off the target*, not connecting the opponent in the target area. **1980** J. McClure *Blood of Englishman* xxv. 232 'Target area coming up,' he said, picking up the line of a wriggling dirt road... They were down to about 600 feet above the ground. **1956** *U.S. Air Force Dict.* 513/2 *Target audience*, in psychological warfare, the people at whom propaganda is directed. **1982** *Underground Grammarian* Sept. 2/2 In order to broaden the 'target audience' of your newsletter..I might suggest that such material be written at a lower level of readability. **1934** T. E. Lawrence *Let.* 8 June (1938) 805 At the moment we are all up to the teeth in 5 more target boats. [**1938** A. M. Barrett in *Jrnl. Path. & Bacteriol.* XLVI. 603 They will here be called 'target types of red blood corpuscle', or more briefly, 'target corpuscles'. I have deliberately chosen a name which refers only to their appearance in stained films and not to their three-dimensional form.] *Ibid.* 605 Often the frequency of target cells appeared to be affected by the thickness or thinness of the film. **1940** W. Dameshek in *Amer. Jrnl. Med. Sci.* CC. 445 Since an outstanding abnormality was the presence of large numbers of peculiar erythrocytes designated as 'target cells' by Barrett, the name 'target cell anemia' was adopted for this previously undescribed condition. **1969** Edington & Gilles *Path. in Tropics* x. 353 Excluding films obtained from persons homozygous or heterozygous for haemoglobin C, a high percentage of target cells in normal blood films has been observed in Ghana, Nigeria, and from East Africa. **1977** *Time* 17 Oct. 58/3 The firm that eventually acquires the target company. **1945** W. S. Churchill *Victory* (1946) 108 Full hutting..is nearing completion, the target date for which is May. **1977** *Whitaker's Almanack 1978* 595 The Rhodesia conference in Geneva became deadlocked when leaders of White and Black delegations failed to agree on a target date for legal independence. **1972** J. L. Dillard *Black Eng.* vii. 293 The Network Standard dialect, for which both white and Black speakers have shown marked preferences, is obviously the preferable target dialect. **1978** J. Irving *World according to Garp* viii. 163 Roberta was a target figure; she had made some people very angry. **1972** *Times* 13 Dec. 4/7 It made no discriminations among target groups. **1979** *Bull. Amer. Acad. Arts & Sci.* Mar. 33 The programs to be undertaken in reaching these target groups were to involve workers in both the public and private sectors of health, agriculture, and education. **1944** *Times* 11 Apr. 4/4 The attack began with the dropping of target-indicators through cloud. [**1974** *Times* 23 Feb. 14/8 Even eight, nine and ten-year-olds these days are taught by games masters in terms of 'striker', 'target player'..and the rest.] **1975** *Times* 14 Oct. 10/2 (*caption*) Stuart Pearson, a target man with Manchester United. **1978** *Sunday Times* (Colour Suppl.) 28 May 34/4 *Target man*, forward, usually a large one, used in central positions where colleagues can find him with long passes, usually to his head. **1955** Friedman & Weisskopf in W. Pauli *Niels Bohr* 134 According to this model the effect of the target nucleus upon an incident particle can be described, at least as a first approximation, by an attractive potential. **1947** H. Selye *Textbk. Endocrinol.* 17/1 The so-called 'target organs' or 'end organs' do not necessarily react to hormones under all conditions. **1972** *Sci. Amer.* Nov. 24/1 The pituitary secretes several complex hormones that travel through the bloodstream to target organs, notably the thyroid gland, the gonads and the cortex of the adrenal glands. **1944** *Hutchinson's Pict. Hist. War.* 27 Oct. 1943–11 Apr. 1944. 441 Once more the merchantship target output was achieved. **1971** *Computers & Humanities* V. 292 SPIRES is based on a behavioral science analysis of the information needs of a target population. **1949** H. Preece in B. A. Botkin *Treas. S. Folklore* ii. iv. 341 The rumbling underground is Britt Bailey target-practising for a million years of shooting in hell. **1982** *Sunday Sun-Times* (Chicago) 8 Aug. 9 A witness..allegedly saw Hartmann's widow, Debra, target practicing at a suburban gun shop. **1962** Target price [see *off-farm* s.v. *off- 4 b]. **1969** P. B. Jordain *Condensed Computer Encycl.* 516 The process begins with a source-language program..and ends with a target program. **1979** *Personal Computer World* Nov. 84/1 Any areas of data must be excluded from both and left intact as they are used by both the target program and the trace routine. **1947** *Britannica Bk. of Year* 841/2 *Target-seeking missile*, a missile, equipped with a target-seeking mechanism, which is attracted toward its target when it approaches its vicinity. **1977** *R. Air Force Yearbk.* 4/2 (*caption*) A Harrier GR Mk 3..with laser-ranging and target-seeking equipment in the nose. **1855** Geo. Eliot in *Fraser's Mag.* LI. 706/2, I will tell you of Weimar fairs and target-shooting. **1966** *Observer* 17 Apr. 10/3 Is there any magic in the figure of 30—the target size for classes? **1936** D. E. Lea et al. in *Proc. R. Soc.* B. CXX. 62 That regions of special sensitivity to radiation do exist..has been demonstrated and the additional postulate of the target theory, namely that there is only one such region, in an individual organism, is not improbable for bacteria. **1979** I. M. Leahy et al. *Nurse & Radiotherapy* iii. 30 One theory that has proved to be applicable to radiation biology experimentation is known as target theory. Briefly stated, this theory proposes that if alterations are produced within certain critical molecules in the cell, the loss of vital function that would result would lead to the death of the cell. **1960** Jensen & Jacobsen in Pincus & Vollmer *Biol. Activities Steroids* iii. 162 Information concerning the chemical fate—in the specific target tissues—of physiological amounts of steroid sex hormones should prove of value. **1974** M. C. Gerald *Pharmacol.* xxiii. 409 Whereas growth hormone and thyroid hormone are capable of influencing virtually all the cells of the body, most hormones act rather selectively on specific tissues referred to as target tissues. **1975** *Sci. Amer.* July 94/1 Where trees are concerned one of the target tissues for auxin is the cambium. **1965** *New Scientist* 18 Mar. 701/2 The spacecraft will be manoeuvred by the pilots to approach the target vehicle closely, and finally to dock with it in a firm, mechanical manner. **1946** D. E. Lea *Actions of Radiations* iii. 91 That dose..produces an average of one cluster in a volume equal to the target volume.

target, *v.* Add: **4.** To plan or schedule (something) to attain an objective. Chiefly in *Econ.*
1948 *Observer* 14 Mar. 3/6 Even herrings have targets now: 175,000 tons of fish are being 'targeted' to yield 17,000 tons of oil a year. **1959** *Time* (Atlantic ed.) 17 Aug. 53 Exports of heavy goods..are targeted to rise this year some 40%. **1972** *Newsweek* 7 Aug. 43/3 With test flights now targeted for 1976, the Shuttle is expected to be ready for operational missions in 1978. **1973** *Daily Tel.* 8 Dec. 23/2 Investment income..is targeted to reach £1 million in two years.
5. To aim (a nuclear weapon) at a target. Freq. const. *on*.
1964 *Financial Times* 23 Mar. (Defence Suppl.) 17/4 The Soviet presence..comprises..a force of about 100 MRBMs targeted on Japan. **1972** *Sci. Amer.* Nov. 21/1 Each missile launches a 'bus', which has on board a large number of reentry missiles, each of which can be accurately and independently targeted. **1978** *Observer* 25 June 21/3 There were enough warheads to target some on China as well.
transf. **1973** *Times* 2 Nov. 4/2 The scheme is targeted at those wanting to buy an older property. **1974** *Nature* 1 Mar. 1/1 Research money should be targeted on problems whose solution would have the greatest benefit for society. **1983** *New Scientist* 21 July 208/1 Practical conservation can rarely preserve an entire fauna: rather it is targeted at particular species.
6. To mark out or identify (a place, person, etc.) as a target. Chiefly *U.S.*
1966 *Guardian* 30 Dec. 14/8 US policy is to target North Vietnamese military targets only. **1976** *National Observer* (U.S.) 27 Nov. 5/1 He has no worlds left to conquer, for NCEC has captured all the 'marginal' conservative seats it had targeted. **1978** S. Brill *Teamsters* vii. 297 The airline industry was being targeted for a recruiting drive. **1983** *Listener* 25 Aug. 4/3 They've targeted 22 airlines for special treatment.
7. To direct or aim on a course. Freq. const. *to*.
1974 *Nature* 1 Mar. 16/3 Temperature profiles of the moons of those planets will be helpful in targeting the spacecraft to take a look at the most interesting features. **1976** *Sci. Amer.* June 74/1 The second spacecraft will be targeted to fly past Saturn and on toward Uranus. **1976** *National Observer* (U.S.) 21 Aug. 3/3 Then allocations are made with about one-third going to state governments and two-thirds to local governments, targeted to those jurisdictions with the highest unemployment. **1980** *Sci. Amer.* Aug. 88/2 Highly specialized transport systems that are in effect independent of the tissue through which they convey substances might be exploited as a means of 'targeting' therapeutic drugs for particular organs or tissues. **1981** *New Scientist* 6 Aug. 343/2 Later perhaps it will be possible to target liposomes or red cells ..to whatever part of the body they are needed [*sic*].
Hence **ta·rgeting** *vbl. sb.*
1961 *Guardian* 24 Oct. 8/4 Being forced to rely on so much inspection..that targeting information would be given away to the other side. **1963** *Newsweek* 11 Feb. 23 Planners have recently put forward the notion of city-avoidance, a tacit agreement between potential enemies to arrange their targeting so that missiles are aimed at military objectives rather than civilian populations. **1968** *Economist* 8 June 65/2 A general complaint is that consultants sometimes stick too much to their business precepts, such as 'targeting' and do not bend enough to the particular needs of the company. **1976** *National Observer* (U.S.) 27 Nov. 5/1 NCEC laid out $350,000 for candidates in 1976. That paid for 64 polls in 32 separate congressional districts and for computerized precinct targeting and analysis in more than 40 districts. **1977** *Time* 21 Nov. 24/2 None of these possess as sophisticated a targeting system as the new Soviet model's [*sc.* a T-72 tank]. **1982** *Financial Times* 13 Mar. 14/1 In terms of targeting ability.

targetable (tā·rgĕtăb'l), *a.* [f. Target *v.* + -able.] **a.** Of nuclear missiles or warheads: capable of being aimed at a target. **b.** Of military installations or equipment, etc.: that may be picked out as a target.
1968 *N.Y. Times* 8 Apr. 46 The United States will in the next few years add to its arsenal missiles capable of putting into space a number of individually targetable warheads. **1968** *Economist* 6 July 10/2 Both in submarine-borne and land-based missiles the Americans have established a lead over the Russians in the development of MIRVs (multiple independently targetable re-entry vehicles). **1972** *Sci. Amer.* June 15/3 Land-based intercontinental ballistic missiles..can readily be located with the aid of surveillance satellites, so that they must be regarded as 'targetable' in the event of an enemy first strike. **1981** *Ibid.* Feb. 20/3 Silos are targetable. **1982** M. Duke *Flashpoint* xxi. 151 Minuteman-III, with its multiple independently targetable warheads.

targeted, *ppl. a.* [f. Target *v.* + -ed[1].] **1.** Designated or chosen as a target.
1965 *Economist* 20 Feb. 733 We must..have a short take off and landing (STOL) capability; otherwise the aircraft is tied to targeted concrete and will be destroyed on the ground by the enemy. **1971** *Nature* 22 Oct. 517/3 He labelled sickle cell anaemia a targeted disease for concentrated research. **1974** *Spartanburg* (S. Carolina) *Herald-Jrnl.* 21 Apr. A8/3 Light industry was just meeting a targeted 4 per cent increase. **1979** *Sci. Amer.* Aug. 139/2 In the late 1960's the U.S. Government's 'Operation Shamrock' intercepted international Telex communications to and from 'targeted' invididuals, including antiwar activists. **1983** D. Williams *Treasure Preserved* i. 8 Anyone..who detected Louella engaged in private ombudsman activity had a duty immediately to alert the targeted department.
2. Aimed, directed; given a target.
1969 *Guardian* 23 June 10/2 MIRV (Multiple Independently Targeted Re-Entry Vehicle). **1974** L. Thomas *Lives of Cell* 116 We need more targeted research, more mission-oriented science. **1978** *Dædalus* Spring p. xiv, The distinction between basic and applied or targeted knowledge becomes crucial.

target language. [f. Target *sb.*[1] + Language *sb.*] **a.** The language into which a translation is made.
1953 *Philos. Sci.* XX. 217 Imagine an utterly moronic student without the slightest knowledge of either the source-language or the target-language, i.e., the language into which the given text is to be translated. **1959** [see *language *sb.* 1 d]. **1969** P. B. Jordain *Condensed Computer Encycl.* 515 In assembly and compiler operations, a programmer-oriented language is converted to a target language for execution on the computer. **1976** *Canad. Jrnl. Linguistics* Spring 96 A translator needs to have deciphered the ambiguity in a given sentence..in order to be able to translate it, provided of course that this sentence does not have a syntactic homonym in the target language.
b. A foreign language which it is aimed to learn or acquire.
1965 P. Strevens *Papers in Lang. & Lang. Teaching* viii. 103 The difficulties..vary according to the learner's mother-tongue as well as his target-language. **1973** K. A. Sey *Ghanaian English* ii. 22 Lacking the native speaker's linguistic intuitions, the L_2 speaker has to depend on his limited acquaintance with the target language. **1976** *Word 1971* XXVII. 351 Integratively oriented students are more strongly motivated and more successful in learning the target language than instrumentally oriented students.

tarheel. Add: Also **Tar Heel, Tar-heel, tar-heel.** (Earlier and later examples.) Hence **ta·r-heeled** *a.*
1864 R. E. Park *Diary* 9 Dec. in *Southern Hist. Soc. Papers* (1876) II. 232 A poor, starving Tar Heel at Elmira. **1869** *Overland Monthly* III. 128 A brigade of North Carolinians..failed to hold a certain hill, and were laughed at by the Mississippians for having forgotten to tar their heels that morning. Hence originated their cant name, 'Tar-heels'. **1878** *Scribner's Monthly* Apr. 833/1 A little fellow from North Carolina..announced to the convention he was from 'the tar-heeled state'. **1942** S. Kennedy *Palmetto Country* 260 North Carolina became known as the Tar-heel State. **1959** [see *redneck 1 a].

tariff, *sb.* Add: **4. a.** attrib., *tariff war*; **c.** objective, *tariff adjustment*, *-cutting*. **d.** Special comb., **tariff wall**, a national trade barrier in the form of a tariff; hence **tariff-walled** *a.*
1889 *Puck* (U.S.) XXV. 248 (*heading*) High tariff-wall. **1904** J. Denney *Let.* 4 Aug. (1920) 50 We..have nothing to offer..like a Free Trade Government dealing with tariff-walled nations. **1909** H. W. V. Temperley in *Cambr. Mod. Hist.* VI. ii. 49 The tariff-war was often the precursor of the trade-war. **1932** *Sun* (Baltimore) 17 Sept. 8/3 The proposed policy is variously known as a bargaining or trade-building policy of tariff adjustment. **1934** A. Huxley *Beyond Mexique Bay* 85 Symptoms, such as tariff-wars and armaments. **1935** E. Wingfield-Stratford *Harvest of Victory* i. ii. 19 The combined handicaps of tariff-walled markets, ruined customers, slackening demand for..coal, [etc.]. **1962** *Daily Tel.* 16 Jan. 20/4 The agreement nearing completion in Brussels on a tariff-cutting agreement with the Common Market is satisfactory on the whole. **1964** *Ann. Reg. 1963* 70 Feelings were ruffled by..the Secretary of Commerce's denunciation of a Canadian tariff-adjustment scheme. **1973** *Times* 3 Jan. (Forward into Europe Suppl.) p. xi/2 The tariff walls begin to crumble. **1977** *Whitaker's Almanack 1978* 978 The CET is based on the arithmetical average of those national tariffs it replaced, and after two international tariff-cutting rounds now stands at an average of 6 per cent.

tariffite. (In Dict. s.v. Tariff *sb.*) (Earlier U.S. example.)
1830 *Western Monthly Rev.* III. 376 She is a true tariffite, a hearty and staunch advocate for the genuine American system.

tariff-reform. (Earlier example.)
1859 R. Cobden *Let.* 8 Nov. in F. A. Wellesley *Paris Embassy during Second Empire* (1928) ix. 193 There is no Imperial road to tariff reform, and if he [*sc.* Napoleon III] goes to work à la Villafranca, he will find himself in a supplement of vexations and troubles.

‖ **tarkashi** (tāɹka·ʃi). Also **tar-kashi.** [Hindi *tār-kašī*, lit. 'wire-drawing'.] The Indian craft of inlaying wood with brass wire; the artefacts so produced.
1878 G. C. M. Birdwood *Handbk. Brit. Indian Section* (Paris Universal Exhibition) 79 In Mynpuri work,..we find..wood inlaid with brass wire in various geometrical ..patterns... At Mynpuri,..it goes by the name of *tarkashi*, or 'wire work'; a word which suggests the possible etymology of the word *tarsia*. 1910 E. R. Neave *Mainpuri: Gazetteer* 73 Mainpuri has long been noted for its beautiful wood work inlaid with brass wire, known as *tarkashi* (lit. wire-drawing). The best dark *shisham* is the only wood employed... There are about twenty artisans in the town engaged in the trade. 1979 *Inside-Outside* (Bombay) June–July 51 That was 1963, which you could say was the year that *tarkashi* arrived—in its new incarnation. *Ibid.* 54 The raw material of *tarkashi* used to be brass sheet.

tarlatan. Add: Freq. *attrib.* Also *absol.*, to designate a dress made of this fabric. (Further examples.)
1844 *Lexington* (Kentucky) *Observer* 25 Sept. 1/6 Tarlatan Muslin..will be sold. 1849 *Trelawny* (Jamaica) 24 Apr. 1/2 Rich colored gingham, and tarleton plaid. 1852 Mrs. Stowe *Uncle Tom's Cabin* I. xviii. 309, I was just dying to know whether you would appear in your pink tarletane. 1873 *Young Englishwoman* Jan. 51/3 Does she never go to a ball or dance, and require the extra dress in the shape of a white tarlatan or something of that sort? 1936 M. Mitchell *Gone with Wind* 175 Maybelle Merriwether went toward the next booth..in an apple-green tarlatan so wide that it reduced her waist to nothingness. 1936 N. Streatfeild *Ballet Shoes* iv. 50 When you start on Monday you're having rompers, two each, black-patent ankle-strap shoes, and white tarlatan dresses, two each, with white sandal shoes. 1975 *New Yorker* 29 Dec. 23/3 Sleptsov also found..a tarlatan bag on a collapsible hoop (and the muslin still smelled of summer and sun-hot grass).

Tarmac. (In Dict. s.v. Tar macadam.) Add: Now freq. with small initial. Also designating a surface made of tar macadam; *the tarmac* (colloq.), the airfield or runway.
1905 *Chambers's Jrnl.* 14 Jan. 110/2 The road surveyor ..appears to have almost solved the problem of finding a dustless, a rainproof, and a cheap material by the employment of an iron-slag mixed with tar. This material he calls tarmac. 1919 C. Roberts *Training Airmen* v. 37 An open, wind-swept place... A broad strip of tarmac on which various aeroplanes are receiving the solicitous attention... That is the sight which quickens the cadet's pulse. 1921 *Flight* 11 Aug. 544/2 Aerodrome improvements..are now being carried out on the tarmac. Work has been commenced on the laying of a tarmac road from the sheds to the Customs enclosure. 1931 *Observer* 10 May 5 The lanes that he once used to choose have now been straightened out into fine, noble tarmac highways. 1948 'N. Shute' *No Highway* iv. 109 Samuelson met them on the tarmac. 1970 *Drum* (E. Afr. ed.) Feb. 31/3 One travels on tarmac the whole way to the Kenya border on some of the finest road surfaces on the continent. 1976 *Sunday Telegraph* (Colour Suppl.) 28 Nov. 57/2 A speed established with the car on dry Tarmac. 1979 J. Raban *Arabia through Looking Glass* iii. 67 People in gold-trimmed robes stepped off aeroplanes and were embraced by similarly robed officials who stood in waiting on the tarmac.

ta·rmac, *v.* [f. the sb.] To cover with tar macadam. Chiefly *pass.* or as *ppl. a.*, with spelling *tarmac(c)ed*, *tarmacked*. Hence **ta·rmacing** *vbl. sb.* Cf. *tarmacadam *v.*
1966 C. Wilson *Glass Cage* ii. 90 It was a row of small, semi-detached modern houses with front gardens, and the road had not yet been fully tarmacced. 1972 'R. Gordon' *Doctor on Brain* xiv. 97 All that lies before me is a well-tarmacked dead straight motorway leading to the grave. 1974 *New Society* 14 Mar. 627/3 Ponds which are filled in and reclaimed by farmers, or tarmacked for car parking by the local pub. 1975 *Ibid.* 18 Dec. 663/3 The aesthetic and environmental objections to the tarmacing of 15 odd acres of land. 1977 *Belfast Tel.* 28 Feb. 13/1 (Advt.), Now's the time to have your driveways Bitmaced or Tarmaced. 1981 E. North *Dames* vii. 129 The tarmacked runway.

tar macadam. Add: Now usu. **tarmacadam.**
1959 *Chambers's Encycl.* XI. 724/2 A modification of the tarred macadam road is that known as 'tarmacadam', in which all the pieces of road metal are coated with tar before being spread on the road and rolled. 1965 P. Wayre *Wind in Reeds* xvi. 229 Concrete or tarmacadam paths..were out as far as we were concerned. 1980 *West Lancs. Even. Gaz.* 6 Mar. 17 (Advt.), Tarmacadam—concrete and flagging.
Hence **tarmaca·dam** *v.* (in quots. as *pa. pple.* and *ppl. a.*). Cf. *tarmac *v.*
1910 *Times* 23 July 8/6 The tar-macadamed Madeira-road..proved them to have been pioneers in this matter. 1976 *Glasgow Herald* 26 Nov. 2/7 (Advt.), Driveways excavated, slabbed, tarmacadamed, trees pruned and lopped. 1978 *Morecambe Guardian* 14 Mar. 22/1 (Advt.), Partly tarmacadamed playground.

tarmac(c)ed, tarmacked: see *tarmac *v.*

tarnally, *adv.* (Later examples.)
a 1828 J. Bernard *Retrospections Amer.* (1887) x. 241 May I be 'tarnally starved down for mutton broth, if [etc.]. 1922 Joyce *Ulysses* 419 Tarnally dog gone my shins if this beent the bestest puttiest longbreakyet.

tarnation, *sb.* Delete *rare* and add earlier and further examples.
1790 R. Tyler *Contrast* v. i. 68 Tarnation! That's no laughing matter though. 1830 W. Carleton *Traits Irish Peasantry* I. 49 Tare-nation to the rap itself's in my company. 1922 Joyce *Ulysses* 183 Wall, tarnation strike me! 1938 M. K. Rawlings *Yearling* v. 49 Git away, you blasted bacon-thieves!.. Git to tarnation! 1983 C. MacLeod *Bilbao Looking-Glass* xix. 175 Tarnation! Here comes another o' them mobile camera units.

tarn-cap. (Later example.)
1863 C. M. Yonge *Hist. Christian Names* II. 312 Siegfried, by means of his tarn cap, invisibly vanquished the Valkyr.

ta·rnhelm. Also **Tarn-helm, tarn-helm.** [Ger.; cf. Tarn-cap and Dern *a.*] In Wagner's opera *Der Ring des Nibelungen*, a magic helmet which either secures the invisibility of the wearer or enables him to change his appearance at will; = Tarn-cap. Also *fig.* Hence **ta·rn-helmed** *a.*
1877 A. Forman tr. *Wagner's Nibelung's Ring: Rhinegold* 45 (*stage direction*) He puts the..'Tarn-helm' on his head... His figure disappears; in his place a pillar of cloud is seen. *Ibid.* 57 (*stage direction*) He puts the tarnhelm on again... He disappears: the gods perceive..a toad creeping towards them. 1896 G. B. Shaw in *Star* 22 July 1/7 The magical strangeness of the wishing cap or 'tarnhelm'. a 1930 D. H. Lawrence *Sex, Literature & Censorship* (1955) 84 It is something in her *will*. It is her tarnhelm. 1971 *Daily Tel.* 4 Oct. 13/3 The fateful ring is grabbed by the tarnhelmed Siegfried.

taro. a. (Earlier example.)
1769 S. Parkinson *Jrnl.* 1 Oct. in *Jrnl. Voy. South Seas* (1773) ii. 97 Adjoining to their houses are plantations of Koomarra and Taro.

tarogato (ta·rǫgato). Also **tárogató.** [a. Hungarian *tárogató*.] A Hungarian woodwind instrument with a conical bore, orig. a double-reeded instrument resembling a shawm, but in the 1880s reconstructed with a single reed and fitted with keys. (Now obsolescent in Hungary, and treated as a historical national instrument.)
1907 T. S. Wotton *Dict. Mus. Terms* 195 *Tárogató*, an instrument which has been used in Paris and Brussels etc. to take the *cor anglais* part at the end of Scene 1 Act III *Tristan und Isolde*. 1935 *Swing Music* Mar. 18/2, I had schemes for original instruments—among them the harpsichord..and a Hungarian reed-instrument called a tárogató. 1965 *Listener* 24 June 940/3 The *tárogató*, resembling the clarinet, but essentially an oboe family instrument. 1974 *Encycl. Brit. Micropædia* IX. 828/3 *Tárogató*, single-reed wind instrument, widely played in the folk music of Romania and, especially, Hungary.

tarot. Delete ‖ and add: **a.** (Later *attrib.* examples.)
1928 D. Byrne *Destiny Bay* vii. 319 An old woman crazed by gambling and tarot cards. 1957 L. Durrell *Justine* iii. 180 Justine..would sit cross-legged on the bed and begin to lay out the little pack of Tarot cards. 1972 *Time* 19 June 26/2 The Center also presents tarot-card readings. 1977 *Jrnl. Playing-Card Soc.* May 3 Some Milan card makers reached a high degree of technical and artistic quality, including specialisation in a particular type of Tarot pack, usually with a narrow format.

tarp (tāɹp). Orig. *U.S.* abbrev. of Tarpaulin *sb.*
1906 *Out West* Apr. 319 The men had unrolled their 'tarps' and spread their beds for the night on the ground in front of the little shack. 1919 W. H. Downing *Digger Dial.* 49 *Tarp*, a tarpaulin. 1941 *Times* (Weekly ed.) 15 Oct. 7/3 The gunner had taken the tarp off the seven-pounder forward and was adjusting the sights and oiling the gun. 1964 F. O'Rourke *Mule for Marquesa* (1967) ii. 33 Saddles, blankets, pack cushions, sweat cloths, tarps, ropes. 1971 C. Bonington *Annapurna South Face* 249 Coated nylon tarps..Plastic tarps. 1980 *Christian Sci. Monitor* (Midwestern ed.) 4 Dec. B 32/1 Caked with ice from the violent waves, the tarps were almost unmanageable.

tarpaulin, *sb.* Add: **2. a.** (Later examples.)
1922 Joyce *Ulysses* 610 Chews coca all day long, the communicative tarpaulin added. 1963 *Australasian Post* 14 Mar. 44/1 All the 'tarpaulins' had abandoned their lives of near slavery at sea and with fine wisdom had scattered inland.
4. tarpaulin muster [Muster *sb.*[1]], a collection or pooling of money among seamen; also *transf.* and *fig.*
1889 in *Cent. Dict.* 1904 E. S. Emerson *Shanty Entertainment* 26 Each one in the room to sing, recite, or shout all round, and..a tarpaulin muster every half-hour for drinks, or smokes. 1907 J. Masefield (*title*) A tarpaulin muster. 1920 P. L. Waldron *Afloat & Ashore* vii. 83 The crew had a tarpaulin muster to have a last evening ashore. 1945 E. George *Two at Daly Waters* 102 As she had not brought a town outfit, Daly Waters had what we call in the bush a tarpaulin muster (the loan of everybody's best clothes). 1954 H. W. Edwards *Under Four Flags* xxiv. 125 With the generosity proverbial among sailors, they had a 'tarpaulin muster'.

Tarpeian, *a.* Add: Also *Comb.*, as *Tarpeian-fast* adj. *poet.*
1876 G. M. Hopkins *Wreck of Deutschland* xxix, in *Poems* (1967) 61 The Simon Peter of a soul! to the blast Tarpeian-fast, but a blown beacon of light.

Tarquinian (tāɹkwi·niăn), *a.* [f. L. *Tarquini-us* + -an: ult. Etruscan.] Of or pertaining to either of two kings of ancient Rome traditionally named Tarquinius, or to the dynasty to which these kings belonged.
1600 Index to *P. Holland's Romane Hist.* sig. 6Dv°, Tarquinien gentlemen beheaded in Rome. 1740 J. Dyer *Ruins of Rome* 4 Such the Sewers huge, Whither the great Tarquinian Genius dooms Each wave impure. 1849 D. Spillan tr. *Livy's Hist. Rome* I. ii. iii. 82 Only with the Tarquinian race will kingly power depart hence. 1977 G. Clark *World Prehistory* (ed. 3) iv. 198 The Roman republic dates from the expulsion of the Tarquinian (Etruscan) dynasty in 510 B.C.

tarragon. Add: **3. tarragon vinegar** (earlier example).
1845 E. Acton *Mod. Cookery* v. 163 Tarragon vinegar... Gather the tarragon late in July,..put it into small stone jars..pour in..vinegar to cover.

Tarragona (tæːrăgōᵘ·nă). The name of a town and a province in north-eastern Spain, used *attrib.* and *absol.* to designate any of various wines produced there (see quot. 1958).
1888 *Encycl. Brit.* XXIV. 607/2 In Catalonia there is a much more important wine industry, the district producing what is known in England as Tarragona or Spanish red. 1926 F. H. Butler *Wines & Wine Lands of World* v. 50 From the Catalan country, on the same sea-board but much to the north, come the abundant Tarragona wines. 1958 A. L. Simon *Dict. Wines* 153/2 Tarragona.. is also the name given to the best fortified wines of Catalonia, wines very dark in colour, naturally very sweet, and the fermentation of which is checked by added Brandy—as with Port. Tarragona was for many years the poor man's Port. Its chief merits were its deep colour, its great sweetness, its high alcoholic strength and its low price. 1967 A. Lichine *Encycl. Wines* 504/2 The wine actually named Tarragona is sweet... The appellation is restricted to the dessert and fortified red and white wines produced within a delimited area and matured or prepared in the cellars of Tarragona, or of Reus close by.

tarra(h), var. *tara *int.*[2]

tarrer. (Earlier example.)
1784 Mrs. R. B. Sheridan *Let.* in T. Moore *Mem. Life R. B. Sheridan* (1825) xv. 485 You mistake, if you suppose I am a friend to your tarrers and featherers:—it is such wretches that always ruin a good cause.

tarring, *vbl. sb.* (Earlier examples of phr. *tarring and feathering*.)
1774 T. Hutchinson *Diary* 1 July (1883) I. 164 The committee for tarring and feathering blamed the people for doing it. 1784 [see Tar *v.*[1] b]. 1844 Dickens *Martin Chuzzlewit* xxxiii. 390 He..invariably recommended..the 'tarring and feathering' of any unpopular person who differed from himself.

tarryhooting (tærihū·tiŋ), *vbl. sb.* Chiefly *U.S. dial.* [Prob. var. of *callyhooting* (*Dict. Americanisms*) in same sense.] Going about with much noise and motion; gallivanting. Freq. const. *around*. Also as *v. intr.* (chiefly *pres. pple.*).
1940 *N.Y. Jrnl. & American* (*Amer. Weekly*) 16 June 4/3 Her husband was 'tarryhootin'' around payin' court to gals on both cricks. *Ibid.* 4/4 He took to disappearing & 'tarryhootin'' during the ninth year of the marriage. 1950 R. Moore *Candlemas Bay* iv. 219 'Hi,' Grampie said. 'You're quite a feller for tarryhooting around the woods.' 'Apparently,' Mr. Raymond said. 'Was you chasing Evelyn?' 'Yes... Yes, by God, I was.' 1959 *Spectator* 27 Nov. 779/2 The pundits in Vigo Street.. turned this new venture [*sc.* *The Wind in the Willows*] down... After a certain amount of tarry-hooting around, Mr. Mole..was deposited in the amiable bosom of Sir Algernon Methuen.

ta·r-sealed, *a.* *N.Z.* (and *Austral.*). [f. Tar *sb.* + Sealed *ppl. a.*] Of a road, etc.: surfaced with asphalt. Also as *v. trans.* (chiefly *pa. pple*).
1928 R. G. Stapledon *Tour in Austral. & N.Z.* i. 12 Practically every mile of the road so traversed is 'tar sealed'. 1936 'R. Hyde' *Passport to Hell* iii. 68 The oakum comes in little short rope-lengths, ship-ropes tarsealed, greasy, and hard. 1959 A. H. McLintock *Descr. Atlas N.Z.* 62, 10,384 miles of roads and highways are tar-sealed or concreted. 1960 I. Cross *Backward Sex* i. 14 Across a tarsealed yard was the New Wing. 1963 A. Lubbock *Austral. Roundabout* 10 The bitumen, or tar-sealed, roads are only made over the most frequented highways, and through towns. 1966 G. W. Turner *Eng. Lang. in Austral. & N.Z.* viii. 172 Roads are still 'tar-sealed'. 1977 *N.Z. Herald* 8 Jan. 4-1/3 (Advt.), Situated at Ola Point on the Whangaroa Harbour and gently sloping from tarsealed road frontage to the reserve at Harbour edge.
Hence **ta·r-seal** *sb.*, a road surface made with asphalt; a road so surfaced; also **ta·r-sealing.**

1957 *Numbers* Mar. 14 The tar-seal led purposefully to a wall of..stiff pale grasses. 1959 M. Shadbolt *New Zealanders* 88 They descended..into a lonely part of country. Tar-seal gave way to a road of clay and pumice. 1963 N. Hilliard *Piece of Land* 91 They'd brought in a lot of land around here since the new tarseal went through. 1964 *Evening Post* (Wellington, N.Z.) 10 Mar. 9/4 'Tar-sealing originated in Taranaki,' said Mr. Daniell [of Akura, Masterton]. 'Metal for the roads in the New Plymouth area had to be carted all the way from the Patea River and was, of course, expensive. Traffic threw much of the metal off the roads and one day a New Plymouth councillor suggested that they "seal the metal on the roads" with tar, and so "tarsealing" was born.' 1972 M. Gee *In my Father's Den* 26 Her sandals made a clacking noise on the tar-seal. 1977 *N.Z. Herald* 8 Jan. 5/1 (Advt.), Situated on corner of 600 acres. This is very private yet adjacent to tarseal.

Tarsian (tā·ɹsiăn), *a.* and *sb.* [f. *Tars(us* (see below) + -ian.] **A.** *adj.* Of or pertaining to Tarsus, a Cilician city in south-eastern Asia Minor, and the birthplace of St. Paul. **B.** *sb.* A native or inhabitant of Tarsus.
1907 W. M. Ramsay *Cities St. Paul* iii. 31 Paul was careful to keep within demonstrable law..when he claimed to be a Tarsian citizen. 1914 W. R. Inge in [?] CCXX. 50 The Emperor showed great favour to the Tarsians. 1920 J. A. Robertson *Hidden Romance N.T.* iv. 69 The Tarsian, a diminutive youth, nervous and awkward in manner. 1958 J. P. Arendzen *Men & Manners in Times of Christ* viii. 128 Did St. Paul, by claiming Tarsian citizenship, mean to imply that..he was a man of means?

tarsioid (tā·ɹsi,oid), *sb.* and *a.* *Palæont.* [f. Tarsi(er + -oid.] **A.** *sb.* A fossil primate belonging to the suborder Tarsioidea, of which tarsiers are the only living members. **B.** *adj.* Of, pertaining to, or resembling a fossil tarsioid or a tarsier.
1913 G. E. Smith in *Rep. Brit. Assoc. Adv. Sci. 1912* 585 If it may safely be assumed that the affinities of the Tarsioids ranged from North America to South-eastern Europe. *Ibid.* 590 The factors that..have transformed a Tarsioid Prosimian into an Ape. 1925 *Bull. Geol. Soc. China* IV. 142 Primitive lemuroid and tarsioid fossil remains are widely known from England eastwards to the Carpathians. 1930 F. W. Jones *Man's Place among Mammals* 215 We have further grounds in analogy with the larval types of the Tarsioids. 1968 W. Le Gros Clark *Chant of Pleasant Exploration* iii. 76 A rich assortment of extinct 'tarsioids'..extended their range over considerable areas. 1973 B. J. Williams *Evolution & Human Origins* ix. 224/1 Most forms thought to be ancestral to the higher primates did go through a tarsioid stage of evolution.

tarsonemid (tāɹsŏnī·mid), *a.* (and *sb.*). [f. mod.L. family name *Tarsonemidæ*, f. generic name *Tarsonemus* (Canestrini and Fanzago 1876, in *Atti Soc. Veneto-trentina Sci. Nat.* V. 14), f. Tarso- + Gr. νῆμα thread: see -id[2].] Of or pertaining to a mite of the family Tarsonemidæ. Also as *sb.*
1921 *Nature* 26 Mar. 596/1 A Tarsonemid mite..feeds on the blood of the bee. 1951 *Dict. Gardening* (R. Hort. Soc.) IV. 2082/1 Tarsonemid mites..are of great economic importance owing to the injury caused by them to cultivated plants. 1959 T. E. Hughes *Mites* v. 73 Many other tarsonemids are plant parasites.

tart, *sb.* **2.** Substitute for def.: *fig.* **a.** Applied, gen. (orig. often endearingly) to a girl or woman; freq. in Australia and N.Z. Also in Liverpool dial. (with def. article or possessive pron.): a wife or girl-friend. *slang.*
1864 Hotten *Slang Dict.* 254 *Tart*, a term of approval applied by the London lower orders to a young woman for whom some affection is felt. The expression is not generally employed by the young men, unless the female is 'in' for 'best'. 1898 [see *Sheila]. 1918 *N.Z.E.F. Chron.* 5 July 252/2, I blushes like a 14-year old tart. 1935 [?] (1936) 71 This word [*sc.* tart] seems absolutely interchangeable with 'girl', with no implication of 'prostitute'. People will speak of 'their daughter' as 'a tart'. 1943 L. Esson in *Penguin Bk. Austral. Ballads* (1964) 223 All the tarts is waitin'. In their flashest clobber. 1959 I. & P. Opie *Lore & Lang. Schoolch.* xv. 327 In the south of England a girl is often spoken of as a 'tart' (referred to as such by boys aged 11), and no disrespect is implied by the word. A 'posh tart' is indeed a general term of admiration for a well-dressed, nice-looking girl. 1962 *Guardian* 24 Dec. 4/2 It's the little things at home that start nagging, and the tart's not well. 1966 [see *Judy]. 1980 V. S. Pritchett *Tale Bearers* 84 His mother, a decent, now elderly tart found living with her black servant.
b. A female of immoral character; a prostitute. Also *loosely* as a term of abuse. *slang.*
1887 [?] 1956 Greene *Gun for Sale* iii. 57 A woman policeman kept an eye on the tarts at the corner. 1951 S. Longstreet *Pedlocks* ii. 93 Real fancy night-gowns, pink drawers, black lace. Nothing cheap [?] tarts on Mercury Street. 1966 E. J. Howard *After Julius* ix. 135 People don't..call other people tarts because they go to bed with people without marrying them. 1979 J. Cooper *Class* 17, I evolved a new way of growing [?] high-heeled shoes, tight straight skirts, very very tight cheap sweaters, and masses of make-up... I looked just like a tart.
c. The young favourite of an older man; a catamite. Also *loosely*, a male prostitute. *slang.*
1935 I. Miller *School Tie* ii. ix. 120 Being a tart. The sort of thing you were getting up to with Black last Easter term. 1943 D. Welch *Jrnl.* 23 Feb. (1952) 43 A week afterwards I had a letter from this old boy—quite elaborately romantic... As Geoffrey approached I held the letter down, against me. 'What are you engrossed in?' he jeered. '..A tart-note I bet. You've had a tart-note.' 1952 A. Wilson *Hemlock & After* i. v. 95, I can usually manage a tart's holiday at Cannes or Ischia. 1976 *Times Lit. Suppl.* 30 Jan. 100/3 He neatly loses the boy to a male tart in the city. 1977 *Ibid.* 1 Apr. 401/4 The boys that Isherwood and his friends picked up were not professional tarts only out for what they could get.

tart, *v.*[2] *slang.* [f. *Tart *sb.* 2.] **1.** *trans.* To treat in the manner of a catamite or tart; to favour. *nonce-use.*
1930 Auden *Poems* 31 For where are Basley who won the Ten, Dickon so tarted by the House, Thomas who kept a sparrow-hawk?
2. To dress *up* or adorn (a person), usu. in a showy or gaudy manner; to titivate; also *refl.* and *intr.* for *refl.* Freq. *trans.* and *fig.*
1938 [implied in Tarted *ppl. a.*]. 1952 *Archit. Rev.* CXII. 371/2 Unfortunately these devices to prevent the neighbourhood's slip from showing, have been 'tarted-up' with a variety of recessed panels, pipe ends, scroll banding heads and so forth, which seem to have no function. 1959 *Times Lit. Suppl.* 29 May p. xiv. There was nowadays a disposition to tart up Shakespeare as if he cannot be taken straight. 1961 [see *Pretty *v.*]. 1967 *Spectator* 1 Dec. 690/3 Peacetime seems to have been faced on seducing the daughters of the local townsfolk..or tarting up one's uniform with more feathers or buttons. 1972 J. Wilson *Hide & Seek* ix. 35 You won't be able to tart yourself up like a teenager much longer, Rose. 1976 J. Cooper *Harriet* ii. xiv. 115 They were tarting up in the Ladies. 1978 *Observer* 16 Apr. 18/1 American stores for instance, are often tarted up at suburban auctions at three times the English price.
3. *intr.* **a.** To meet or pursue women.
b. Of a girl or woman: to behave like an immoral woman or a 'tart'; freq. const. *(a)round.*
1948 D. Ballantyne *Cunninghams* 30, I bet he's been tarting. 1949 J. B. Priestley *Home is Tomorrow* ii. i. 47, I know his ways. 1960 K. Waterhouse *Billy Liar* ii. 33, I would fall to wondering whether she was tarting round the town with some American airman. 1960 *Spectator* 18 Nov. 784 The boy would now turn soft and the girl start tarting. 1981 P. Vansittart *Death of Robin Hood* iv. v. 206 All bad tales of adventure... Some claimed to have been tarting. 1980 J. Wainwright *These Evil Ways* ii. 66 Her mother was tarting around with this older bloke.

tartan, *sb.*[1] Add: **1. a.** Also preceded by a clan-name, etc. denoting a particular traditional or authorized design.
1841 D. Stewart *Sk. Highlanders Scotl.* i. ii. i. 129 The pipers wore a red tartan of very bright colours, (of the pattern known by the name of the Stewart tartan). 1897 *Private Life of Queen* xxv. 209 The writing-room is hung entirely with the Balmoral tartan. 1949 'J. Tey' *Brat Farrar* xiii. 114 A frayed Stewart tartan ribbon off a box of Edinburgh rock. 1981 *Times* 3 Feb. 17/8 Streaming from her helmet were two lengths of Colquhoun tartan from the clan of which her father was chief.
c. Used to denote young people who are members of Protestant gangs in Northern Ireland, from their traditional support of Glasgow Rangers Football Club.
1973 *Guardian* 17 Mar. 1/3 The local Protestant street gangs, mostly known as 'Tartans' because of their traditional association with the Rangers Football Club. 1974 *Listener* 14 Mar. 324/2 Until recently these streets were terrorised by Tartan Gangs. Now their place has been taken by these youngsters, acting in the name of the Loyalist cause. Their behaviour is modelled on the Tartans. 1977 P. Carter *Under Goliath* iii. 35 Ever heard of the Tartan gangs?
2. (Earlier examples.)
1827 J. Kirkbride *Northern Angler* 73 What is called the tartan fly kills well in the Highlands at the clearing of the water. *Ibid.* 87 Tartan... 1847 T. T. Stoddart *Angler's Compan.* xiii. 249 Salmon flies... The Tartan..Mottled black and white tail feather from turkey.
3. ¶ (Properly with capital initial.) The proprietary name of a synthetic resin material used for surfacing running tracks, ramps, etc. Usu. *attrib.*, as *Tartan track.*
1964 *Official Gaz.* (U.S. Patent Office) 14 Jan. TM 60/1 *Tartan*. For synthetic resin material for application to various surfaces. To provide a resilient surface thereon. First use Aug. 28, 1962. 1968 *Listener* 10 Oct. 485/2 The 100-metre final is also on Day Three. A fast time with thin air, the 'tartan' track and, maybe, the new brush spike, is inevitable. 1969 *Trade Marks Jrnl.* 22 Oct. 1732/1 *Tartan*... Synthetic resins for use as floor and road surfacing materials. 1972 *Radio Times* 1 June 13/3 Britain's sprint hope... says: 'I've got a good chance, there's a Tartan track two minutes up the road.'
4. c. Used *loosely* in various *transf.* and *fig.* collocations to designate something pertaining to Scotland or which evokes Scottish nationalist fervour.
1954 P. Barter (*title*) Ritchie; or, behind the Tartan Curtain. 1975 *Globe & Mail* (Toronto) 27 Sept. 6/6 The British press has taken extreme care to avoid the suggestion that the activities of the 'Tartan Army' are linked to the legitimate national movement embodied in the Scottish National Party. 1976 *Listener* 28 Oct. 555/2 Radio 3's *Scottish Evening*... Overall there was a blessed absence of..Tartan Romanticism. 1982 *Times* 9 Jan. 6/1 Almost all Scottish MPs..are Scots by birth... It is not simply raw xenophobic tartan nationalism.

Tartar, *sb.*[3] (*a.*) Add: **A.** *sb.*[1] **5.** (Earlier examples.)
1668 [see Scythian *sb.* 2]. 1863 *Jrnl. Amer. Oriental Soc.* 1861 VII. 272 They have by some been spoken of as 'Tartar', by others the 'Finnish', 'Ural-Altaic', 'Mongolian', and 'Turanian'.
C. tartar (also ‖ **-e**) **sauce** [tr. Fr. *sauce tartare*], a sauce made of mayonnaise and chopped gherkins, capers, etc., usu. served with fish.
1855 E. Acton *Mod. Cookery* (rev. ed.) vi. 143 Tartar sauce. (Sauce à la Tartare.)... Tartar sauce..is to be preferred to English for this sauce. 1889 C. Owen *Choice Cookery* 48 Tartare sauce is mayonnaise with the addition of mustard, chives, pickles, and tarragon, chopped. 1927 *Good Housek. Home Encycl.* 677/1 Tartare Sauce..is served with fish, salads, and such vegetables as globe artichokes. 1959 *Good Food Guide* 191 Seafood platter with tartare sauce. 1973 'D. Jordan' *Nile Green* xxiii. 93 The waiter..nearly tipped the tartare sauce down Mara's neck.

Tartary. Add: **3.** *attrib.* **Tartary oat,** a wild oat, *Avena fatua*, which has a loose inflorescence.
1790 S. Deane *New-England Farmer* 193/2 I have lately met with the Tartary oats, which..differ in their appearance from the common sort. 1891 R. Wallace *Rural Econ. Austral. & N.Z.* xvii. 260 The straw is not so long or of such good quality as the straw of the Tartary Oat.

tarted (tā·ɹtĕd), *ppl. a.* [f. *Tart *v.*[2] 2 + -ed[1].] **a.** Of a person: dressed *up* in a showy manner, gaudily adorned. Also without *up*.
1938 E. Bowen *Death of Heart* i. vii. 81 After dark, she [*sc.* London] is like a governess gone to the bad, in a Woolworth tiara, tarted up all wrong. 1952 D. Adams *Murder, Maestro, Please* xvi. 113 [?] tarted-up hussy. 1972 'R. Crawford' *Whip Hand* I. viii. 49 [?] you tarted-up sub. 1979 *Econ. Standard* 15 Sept. 19/6 Prostitutes..are of the conventional kind, in high heeled shoes and characteristically 'tarted' both cosmetically and sartorially.
b. *transf.* and *fig.*
1958 K. Amis *I Like it Here* ii. 21 A collection of tarted-up reviews. 1967 *Spectator* 20 Oct. 455/3 The tarted-up village inn, remodelled with the single aim of attracting motor trade from a distance. 1972 *Where?* Mar. 96/3 ROSLA enthusiasts for tarted-up curricula need to heed the warning. 1981 J. Scott *Distant View of Death* xiv. 182 The tarted panda reversed..and drove in the wake of the quarry. 1981 *Listener* 11 July 31/3 At the other end of the spectrum of the Higher Rubbish—defined for the moment as tarted-up junk—is Elizabeth Taylor in the vastly enjoyable, utterly brainless *The VIPs*.

tarten, *v.* Add: **2.** To affect with sharpness or acidity. *rare*.⁻¹
1925 W. de la Mare *Broomsticks* 58 Hardly had its juice tartened my tongue.

tartine. a. (Earlier example.)
1804 F. Burney *Jrnl.* 1 Oct. (1975) VI. 477, I have given no more medicine—plenty of tisanes &c, & tartines of Honey & salad are all he has.

tartish, *a.*[2] *colloq.* [f. *Tart *sb.* 2 b + -ish.] = *Tarty *a.* (and *sb.*)
1930 C. Connolly *Let.* Nov. in *Romantic Friendship* (1975) 357 We both felt that you thought she was tartish. 1944 E. Bowen in *Penguin New Writing* XX. 62 Collie was wearing that tartish house-coat. 1966 E. Grierson *Second Man* xiv. 149 A brocade house-coat and mules of a pink, tartish shade. 1972 *Daily Tel.* 4 Oct. 13/2 His mother is sluggish, tartish and neglectful.

tartlet. Add: **1.** (Later examples.)
1889 Stevenson & Osbourne *Wrong Box* v. 79 He returned with a large bag of the choicest and most tempting of cakes and tartlets. 1960 [see *Vol-au-vent].
2. A young woman of immoral character, a young 'tart'.
a 1890 in Barrère & Leland *Dict. Slang* (1890) II. 337/2 E'en tartlets are stale, be they ever so tasty—The magic has fled from their tarts, their looks. 1961 *Spectator* 3 Mar. 304/3 Love for a tartlet.

Tartuffe, Tartufe. Add: Hence also **Tartu·ffily** *adv.* (*nonce-wd.*).
1915 [see *Pecksniff].

tarty (tā·ɹti), *a.* (and *sb.*) *colloq.* [f. *Tart *sb.* 2 b + -y[1].] Resembling or suggestive of a 'tart', or women of that character; cheap, gaudy. Occas. *absol.* as *sb.*
1918 G. Frankau *One of Them* xxi. 163 Of that barred citadel whose mincing misses Persuade the chaste to emulate the tarty. 1930 D. H. Lawrence *Pansies* 123, I suppose most girls are a bit tarty to-day. 1969 M. Laski *Love on Supertax* ix. 65 Some very tarty South American perfume. 1980 S. Dickinson *Oxford Folly* xiv. 213 A fat middle-aged woman with henna'd hair and clothes that managed to be at once tarty and tatty. 1978 M. Dickens *Open Bk.* ix. 78 Frank, who was respectably married, was half afraid of Jean, but half delighted. Her tarty teasing made him feel shockingly male.

Tarvia (tā·ɹviă). Chiefly *N. Amer.* [f. Tar *sb.* + L. *via* road.] The proprietary name of a road-surfacing and binding material made from tar. Also (irregularly) **ta·rviate** *v. trans.*; hence **ta·rviated** *ppl. a.*
1912 *Official Gaz.* (U.S. Patent Office) 23 July 1125/1 Tarvia..Pitch prepared from natural or manufactured bituminous oils and tars for road and pavement construction, roofing, waterproofing, and insulating. Claims use since June 1, 1903. 1926 *Daily Colonist* (Victoria, B.C.) 23 July 18/1 There has been a saving, over contract price, of $12,000, in tarviating the twenty-six miles of Island Highway. 1928 *Trade Marks Jrnl.* 18 Jan. 82 *Tarvia*... Raw or partly prepared mineral substances, for use in the manufacture of road-making materials. 1940 *Chambers's Techn. Dict.* 835/1 *Tarviated*.., a term applied to macadam road surfacings in which the stone is bound together with tar. 1947 *Archit. Rev.* CI. 163 A tarvia floor was chosen because of its cheapness and its acoustical properties. 1952 *Jrnl. Acoustical Soc. Amer.* XXIV. 662/1 It is..necessary to record on the identical stretch of tarvia road. 1966 R. H. Rimmer *Harrad Experiment* (1967) 25 A one lane tarvia road between two stone pillars. 1972 *Islander* (Victoria, B.C.) 23 Jan. 16/2 My feet got so sensitive I could sense the difference between tarvia, gravel, or concrete immediately.

tarwinie, var. of *tauhinu.

Tarzan (tā·ɹzăn). The name of a character in a series of novels by the American author Edgar Rice Burroughs (1875–1950), and in subsequent films and television series, who is orphaned in West Africa in his infancy and reared in the jungle by a mother-ape, used *transf.* to designate a person distinguished by physical strength or agility.
[1914 E. R. Burroughs (*title*) Tarzan of the apes.] 1921 *Glasgow Herald* 25 Oct. 5/5 At fruit picking time there is a regular colony of Tarzans disporting themselves in the branches. 1938 M. Allingham *Fashion in Shrouds* vi. 78 Ramillies was ruddy pleased... Saw 'imself a Tarzan. 1946 Koestler *Thieves in Night* 150 Their bodies [are] those of a horde of Hebrew Tarzans roaming in the hills of Galilee. 1960 *John o' London's* 14 Apr. 436 The tough 'Tarzan's' relationship with his landlady..is tenderly portrayed. 1974 H. MacInnes *Climb to Lost World* vi. 85 It wasn't a normal four hour walk—more like an obstacle course for budding Tarzans. 1981 R. Barnard *Mother's Boys* i. 15 Gordon began his morning liturgy of exercises. ..'Bloody Tarzan,' said Brian.
b. Allusively in *attrib.* use.
1932 R. Knox *Broadcast Minds* vii. 161 Though the Tarzan-stuff may make snappy reading. 1941 A. Cotterell *What! No Morning Tea?* 103 Not hothouse gymnasium overdevelopment, but sheer Tarzan physical well-being. 1961 M. Jones *Potbank* xxvi. 114 A remarkably handsome young man with a Tarzan physique. 1974 V. Canning *Painted Tent* ix. 189 Nearly killed myself on the tower ladder today. Saved by a Tarzan act.
Hence **Ta:rzane·sque** [-esque], **Ta·rzan-like** *adjs.*
1933 *Punch* 27 Dec. 712/1 Taken in conjunction with my Tarzanesque agility, They constitute a clue to my athletic versatility. 1943 *Copper Camp* (Writers' Program, Montana) 214 Butt Block gazed pridefull at his partner, smiled and then with brawny fists pounded, Tarzan-like, upon his hairy chest. 1973 C. Bonington *Next Horizon* xi. 158 He loved being the centre of attraction, dropped easily into Tarzanesque poses, and enjoyed showing off the odd feat of strength. 1980 T. Holme *Neapolitan Streak* 160 He had to perform a Tarzan-like operation, lowering himself..and then swinging down.

Tasaday (tasā·dai), *sb.* and *a.* [a. *Tasaday*, prob. f. *tau* person + *sa* (place marker) + *dáya* inland.] **A.** *sb.* **a.** (A member of) a people living on the Philippine island of Mindanao (see note). **b.** The Manobo language of this people.
The Tasaday isolated themselves (most probably in flight from a plague epidemic) some eight hundred years ago, forsaking their skills in rice-agriculture, metallurgy, etc., and taking up a less advanced form of existence (see quots.).
1971 *Guardian* 19 July 3/1 Dark skinned, fruit-eating men, known as Tasadays..near Lake Sebu, in Cotabato Province..south of Manila. The tribesmen..number about sixty... Their isolation was total until 1966. 1971 *Philippine Jrnl. Linguistics* II. 3 A comparison of the lexical items of Tasaday..reveals that the language has most cognates with..B'ht Manobo. 1972 *National Geographic* Aug. 232/2 Igna..translated from T'boli..to Tasaday. *Ibid.* 239 (*caption*) The staple of the Tasaday diet..is a wild yam. 1973 E. Hyams *Final Agenda* ix. 118 The pacific and gentle manners of the Tasaday people, a Stone Age vestige still living on Mindanao. 1975 *New Society* 4 Dec. 559/2 In the early 1960s, Dafal, a wandering hunter, came across a small and timid band of food-gatherers, calling themselves Tasaday, living deep in the forest reaches of southern Mindanao in the Philippines.

Taser (tē[1]·zəɹ). orig. and chiefly *U.S.* Also **taser.** [f. the initial letters of *T*om *S*wift's *e*lectric *r*ifle (a fictitious weapon), after *laser[2].] A weapon which fires barbs attached by wires to batteries, and causes temporary paralysis. Hence **Ta·sered** *a.*, paralysed by means of a Taser.
Developed by Taser Systems Inc., Los Angeles.
1972 *Science* 12 May 615/2 A taser is an instrument that fires a cluster of electrified barbs which become snagged in the victim's clothing and paralyze him until the current is switched off. 1973 *Guardian* 16 Apr. 11 A pan-lethal weapon called the Taser, developed by a California manufacturer... Two electrical wires lash out... The suspect stiffens from shock. His muscles are paralysed. 1975 *Globe & Mail* (Toronto) 4 Oct. 10/1 The Taser Public Defender, as it's called, can penetrate nearly two inches of clothing and give up to a 50,000-volt charge. Taser Systems Inc. of Los Angeles, the manufacturer, says it is not lethal but is designed to stop attackers in their tracks. 1976 *N.Y. Times Mag.* 4 Jan. 13 A powerful transformer within the Taser generates 50,000 volts when a trigger is pressed. This jolt, sent through the wires into the darts, which have been shot into the skin or clothing of the victim, cause him to become 'Tasered'. 1977 *Observer* 21 Aug. 2/7 There was the taser that fired barbs attached to wires into demonstrators to paralyse them with electric shocks.

tash (tæʃ). Also **'tache.** Colloq. abbrev. of Moustache, mustache *sb.* 1 a.
1893–4 R. O. Heslop *Northumberland Words* II. 719 *Tash*, a moustache. 'Him wi' the *tash*.' 1943 Hunt & Pringle *Service Slang* 64 *Tash*, moustache. 1965 R. Simons *Dead Reckoning* iv. 56 'E 'ad a little tash, just under 'is nose. 1968 A. Diment *Great Spy Race* viii. 123 He was..spluttering through his straggly 'tache. 1973 A. MacVicar *Painted Doll Affair* vi. 70 A wee runt wi' a Mexican 'tache. 1980 *Home & Country* Nov. 602/1 (Advt.), 12" male dolls..Painted hair and tash.

Tashi Lama (ta·ʃi lā·mă). Also 8–9 **Teshoo Lama, Teshu Lama;** 9 **Tishu Lama,** etc. [f. *Tashi* Lhunpo, the name of the monastery ruled by the Tashi Lama, + Lama[1].] A title of the Panchen Lama (see *panchen).
1774 G. Bogle *Mission to Tibet* (1876) p. xlvii, There have been two great incarnations of equal rank: the Dalai Lama at Potala..; and the Teshu Lama at Teshu Lumbo, the incarnation of the Buddhisatwa Amitabha. 1784 S. Turner *Let.* 2 Mar. in *Acct. of Embassy to Court of Teshoo Lama in Tibet* (1800) iii. 366 He will ratify the promises made to the former Teshoo Lama, the moment the present Lama is capable of renewing the application. 1811 W. Kirkpatrick *Acct. of Kingdom of Nepaul* 341 Some persons of rank on the part of the Teeshoo Lama, and Sankia Lama, came into the Goorkha camp. 1819 F. Hamilton *Acct. of Kingdom of Nepal* i. i. 57 Still more celebrated is the Tishu Lama, who resides at Degarchi, and is the spiritual guide of the Chinese emperors. 1876 [see Lama[1]]. 1923 *Daily Mail* 18 Apr. 5 The Tashi Lama is of equal rank with the better-known Lama Guru of Lhassa, but he rules over a smaller area, and has not nearly so much temporal power as the latter, though he is regarded as a holier person. 1950 A. De Riencourt *Lost World* viii. 96 His new name was Panchen Rimpoche or Precious Great Sage and his successors became known all over the world as Panchen or Tashi Lamas.

Tasian (tā·siăn, tē[1]·ʃ[1]ăn), *a.* and *sb.* *Archæol.* [f. Deir *Tasa*, the name of a village in Upper Egypt, + -ian.] **A.** *adj.* Of, pertaining to, or designating the pre-Dynastic Neolithic culture represented by remains found at Deir Tasa. **B.** *sb.* A person of the Tasian culture; the culture itself.
1929 G. Brunton in *Antiquity* III. 459 This new cultural phase we have named Tasian from the village of Deir Tasa where the graves were first located. *Ibid.* 466 It may be premature to say definitely that the Tasians preceded the Badarians, but all the evidence points in that direction. 1931 [see *Badarian *a.*]. 1934 V. G. Childe *New Light on Most Anc. East* iii. 52 Remains found at Deir Tasa and other sites on the east bank of the Nile in Middle Egypt..belong to a people who have been termed Tasians. 1939 —— *Dawn Europ. Civilization* (ed. 3) xii. 218 Beaker-like vases decorated with zones of incision which might be clay translations of such basketry vessels occur in Egypt in the early 'Tasian' phase of culture. 1961 G. Clark *World Prehist.* v. 103 Although no radiocarbon or absolute dates are available for the Tasian it is generally held on not very impressive evidence to have preceded the Badarian.

task, *sb.* Add: **I. 2. b.** (Further examples.) *spec.* (*Winchester College slang*) an essay or composition to be written.
1900 J. S. Farmer *Public School Word-Bk.* 201 *Task*, (*subs.*) (Winchester), all kinds of composition other than an Essay or Vulgus. 1980 'T. Hinde' *Sir Henry & Sons* xv. 151 The weekly essay..called a task, is written by every boy in the school.
c. *Psychol.* A piece of work or an exercise given to a subject in a psychological test or experiment. Cf. *aufgabe.
1913 H. Münsterberg *Psychol. & Industrial Efficiency* xviii. 237 We know how the consciousness of the task to be performed has an organizing influence on the system of those psychophysical acts which lead to the goal. 1951 G. Humphrey *Thinking* 99 The energy [for mental operations] may..conceivably originate in..the task or motive. 1972 *Jrnl. Social Psychol.* LXXXVII. 96 Sixty males received..electric shocks of varying magnitude from a confederate during a 10-trial probability estimation task.
III. 6. (senses 2 and 3) *task assignment, performance, role*; *task-directed, -orientated, -oriented, -related* adjs.; **task force** orig. *U.S.*, an armed force organized for a special operation under a unified command; hence *transf.*, any group of persons organized for a special task, esp. an investigative committee; **task group,** a naval task force, or a subdivision of such a force.
1964 Task assignment [see *assignment 13]. 1971 J. S. Bruner *Beyond Information Given* (1974) xvii. 302 The picture of development drawn thus far is much too task-directed, too playless to be characteristic of the first year of life. 1941 *Time* 23 June 41/3 A division of Marines and one of infantry..as a potential A.E.F. 'task force'—for action overseas. 1942 *Jrnl. R. Aeronaut. Soc.* XLVI. 340 Low speed battleships are useless as a constituent of so-called 'task' forces. 1949 *Richmond* (Va.) *Times-Dispatch* 1 Mar. 1/1 The work stoppage resulted from an attempt by the city to try out a 'task force' system of collections. Under this plan, workers are assigned a certain route to be covered each day. When they complete their route, they can go home, regardless of how long it has taken. 1954 *Economist* 9 Jan. 97 The task force appointed..to examine the civil service. 1966 [see *project *sb.* 5 d]. 1971 *Nature* 24 Dec. 435/3 A task force set up under the auspices of the National Heart and Lung Institute. 1980 *Birds* Autumn 13/3 The problem [of bird smuggling] has become so serious in the USA that the Justice Department has been ordered to establish a special task force. Over a period of 12 months, nearly 1,000 birds have been seized by Customs officers. 1982 *Times* 3 June 8/6 The work of getting the components of the task force to sea has often been swift. 1943 *Daily Tel.* 23 Oct. 1/4 Capt. Mackintosh, as the senior commanding officer, commanded a Task Group, which included one of the latest battleships and American destroyers. 1952 [see *kamikaze *sb.* 2]. 1979 *Navy News* Feb. 2/1 Ships in a Royal Navy task group broke off from their work surveying the coast of Iran last month to ferry British and American dependants away from the troubled country. 1953 *Jrnl. Abnormal & Soc. Psychol.* XLVIII. 401 (*heading*) Coding noise in a task-oriented group. 1971 J. Z. Young *Introd. Study Man* xx. 273 Both social and task-orientated behaviours are relatively consistent for both boys and girls from about 4 to 12 years. 1974 tr. *Wertheim's Evolution & Revolution* i. 38 Equally, modernity in political structure is positively related to a task-oriented bureaucracy and a recruitment on the basis of skills. 1970 *Jrnl. Gen. Psychol.* Jan. 91 The findings of this study that the task performance of internals was better. 1956 J. Klein *Study of Groups* viii. 112 If a member proposes that the group shall rehearse a play and another says that he hates play-acting, that is a task-related contribution. 1972 *Accountant* 21 Sept. 357/2 A more task-related analysis might be used. 1967 M. Argyle *Psychol. Interpersonal Behaviour* iv. 71 The task roles of providing 'fuel', putting the 'threads' of the discussion together, and clarification, were generally performed by the same person; the social roles of making tactful comments to heal hurt feelings, and joking, were performed by others.

task, *v.* Add: **2. b.** (Later examples.) Also const. *with*.
1975 *Sentinel* (Ottawa) III. 11. 3/2 Capt. Ditter was tasked to help prepare this issue. 1980 *Oxf. Star* 20 Nov. (Advt.), A small engineering team tasked with the design, building and commissioning of high volume production lines.
5. Delete † *Obs.* (Later examples.) Now const. *with*.
1965 K. Graham *Eng. Criticism of Novel* iv. 117 Trollope is another offender who is frequently tasked with endangering the wholeness of his novels. 1976 *Times Lit. Suppl.* 20 Feb. 197/1 He tasks Taylor with suggesting that Hegel reappeared in Anglo-Saxon thought at the turn of the century.

Taslan (tæ·slăn). [Invented word.] The proprietary name of a process for bulking or texturing synthetic yarns; also, a yarn which has been subjected to this process.
1954 *Trade Marks Jrnl.* 31 Mar. 328/2 Taslan 726,376. All goods included in Class 23. 1954 *Official Gaz.* (U.S. Patent Office) 13 July 263/2 E.I. du Pont de Nemours and Company, Wilmington, Del... *Taslan* for thread and yarn. Use since Jan. 4, 1954. 1957 *Times* 14 Jan. 11 *Taslan*, a process for 'texturing', or 'bulking' synthetic yarns such as acetate, nylon or Terylene to give softer handle and improve draping qualities. 1959 A. J. Hall *Stand. Handbk. Textiles* (ed. 5) iii. 131 Bulk yarns can be produced in various ways..but much success has attended the method used for the production of Taslan yarns. 1960 *Skinner's Silk & Rayon Rec.* Oct. 994/2 Car upholstery is another field in which *Taslan* has found a good reception in the U.S. 1963 A. J. Hall *Textile Sci.* iii. 130 There are various types of textured yarns which have now become available for weaving and knitting into fabrics under branded names such as Agilon, Banlon, Taslan, etc.

Tasmanian, *a.* Add: (Examples.)
1851 *Illustr. Catal. Gt. Exhib.* iv. i. 998/3 Many Tasmanian plants bloom throughout the winter months. 1874 M. Clarke *His Natural Life* ii. v. 104 'And what books do you read?'..' "Blair's Sermons," and "The Tasmanian Almanack".' 1920 *Glasgow Herald* 23 July 6 He..attended the royal meeting of the Tasmanian Racing Club in the afternoon. 1964 W. L. Goodman *Hist. Woodworking Tools* 157 The Sanderson Brothers & Newbould catalogue has a variation of this called the 'Tasmanian' tooth. 1975 *Listener* 7 Aug. 172 The remnants of the four Tasmanian tribes, the last of the 5,000 Tasmanians who were there when the Europeans landed.
B. as *sb.* **a.** A member of the aboriginal people of Tasmania, now extinct.
1842 *Penny Cycl.* XXIV. 90/2 That courage was rewarded by the appointment of Mr. Robinson to the office of 'civilizing' the Tasmanians at Flinders' Island. 1899 J. Milne *Romance of Pro-Consul* viii. 79 The Tasmanians have now been extinct for years. 1918 L. Huxley *Life J. D. Hooker* I. 106 A meagre record of the thousands of native Tasmanians. 1935 Huxley & Haddon *We Europeans* iv. 120 Of all existing men the Arctic Eskimo is the most leptorrhine and the equatorial negro one of the most platyrrhine, but the Tasmanians, who lived in a temperate climate, were also platyrrhine. 1978 *Nature* 18 May 185/2 Within a few decades they were nearly exterminated by European settlers, until the scattered survivors were removed in 1834 to Flinders Island, where the last full-blooded Tasmanian died in 1876.
b. A native or inhabitant of Tasmania.
1934 T. Wood *Cobbers* xiii. 163 Tasmanians, I found, grumble. 1974 *Country Life* 7 Nov. 1388/2 Although.. whaling had started in its adjacent waters, the Tasmanians themselves were prohibited from engaging in the industry. 1978 A. Waugh *Best Wine Last* x. 118 One of my two uncles..married a Tasmanian.

Tasmanoid (tæ·zmănoid), *a.* [f. *Tasman(ian *sb.* + -oid.] Resembling or allied to the ethnological type of the aborigines of Tasmania.
1938 *Skr. Norske Videnskaps-Akademi* (*Mat.-Nat. Kl.*) *1937* 153 It is not inconceivable that an original Tasmanoid population in Australia was driven south. 1943 *Mem. Nat. Mus. Melbourne* XIII. 44 The evidence.. strongly suggests that the earliest migrants belonged to a Tasmanoid (Negrito) race..and that this race..found its way to Tasmania. 1958 F. E. Zeuner *Dating the Past* 281 Wunderly comes to the conclusion that the skull combines Australoid and Tasmanoid characteristics in about equal proportions

Tass[3] (tæs). Also **TASS.** [a. Russ., acronym f. the initial letters of *T*elegrafnoe *a*gentstvo *S*ovetskogo *S*oyuza.] The official Soviet news agency.
1925 *Times* 27 Aug. 11/7 A decree ratified by the Soviet authorities changes the name of Rosta (Russian Telegraph Agency) to Tass (Telegrafnoye Agenstvo Sovietskovo Soyuza = Telegraph Agency of the Soviet Union), as from August 1 next. 1942 *Nature* 25 Apr. 475/1 It is announced by the Tass Agency that [etc.]. 1950 A. Huxley *Themes & Variations* 52 Not..that he had the faintest premonition of Harmsworth or Hearst,..of Goebbels or Tass. 1958 *Spectator* 27 June 824/1 The new Tass statement on the Lebanon. 1974 T. P. Whitney tr. *Solzhenitsyn's Gulag Archipelago* I. i. i. 9 Several days later TASS will issue an angry statement to all the papers. 1981 *Guardian* 27 Apr. 5/2 Tass, reporting from Warsaw, said..that 'revisionist elements in the party' were trying to paralyse it.

tassel, *sb.*[1] Add: **5. tassel-flower,** (*a*) (earlier and later examples); **tassel hyacinth** (examples).
1836 A. Lincoln *Familiar Lect. Bot.* (ed. 5) 83 Tassel-flower; from the East Indies. 1863 'G. Hamilton' *Gala-Days* 10 The scarlet tassel-flower utterly refuses to unfold his brave plumes. 1957 C. O. Booth *Encycl. Ann. & Biennial Plants* 261/2 *E*[*milia*] *sagittata*..is the popular Tassel Flower, or Flora's Paintbrush, a charming half-hardy annual. 1790 *Curtis's Bot. Mag.* IV. 133 (*heading*) Two Coloured, or Tassel Hyacinth. 1865 M. Eyre *Lady's Walks in S. France* xxiii. 251 The starch, and the tassel-hyacinth,..and many others..are all common flowers about Bagnères. 1961 R. Genders *Miniature Bulbs* ii. 165 The 'Tassel Hyacinth' grows a foot high.

tassel, *v.* Add: **2.** (Earlier and further examples.) Also *tassel out*.
1757 in C. R. Woodward *Ploughs & Politicks* (1941) 278 Just before it Tossles it should be plowed & hoed again. 1774 P. V. Fithian *Jrnl.* (1900) 212 The Corn is beginning pretty generally to tassel. 1843 *Amer. Pioneer* II. 83 Corn, if planted, grows a foot high, tassels out and dies. 1887 M. E. Wilkins *Humble Romance* 29 His corn tasselled out..as soon as anybody's. 1966 R. G. Toepfer *Witness* v. 34 Mr. Davis Miller's corn was starting to tassel out and you could pretty near see it grow.

Tassie[2] (tæ·si). Also ¶ **Tassi, tassie.** The name of James *Tassie* (1735–99), Scottish gem engraver, used *attrib.* and *absol.* of replicas of ancient engraved gems or original portrait reliefs made in glass paste by him or by his nephew William Tassie (1777–1860) who succeeded him.
1819 Keats *Let.* 13 Mar. (1958) II. 45 On looking at your seal I cannot tell whether or not it is done with a Tassi—it seems to me to be paste. 1894 J. M. Gray *James & William Tassie* viii. 65 The Shadford Walker Sale included..over a hundred large Tassie medallions of contemporary personages. 1942 E. Blunden *Romantic Poetry & Fine Arts* 10 His [*sc.* Keats's] own particular Tassie was 'a lyre with the strings broken'. 1972 *Times* 1 Aug. 11/5 (Advt.), A collection of paintings..Tassi medallions. 1979 Mills & Mansfield *Genuine Article* vii. 109 Many people fail, when encountering what might be a 'Tassie', to..see if the alleged diamond is backed by metal. 1981 'J. Gash' *Vatican Rip* iii. 36 That glimpse of Mrs Culpepper's 'tassie', as we call such incised semi-precious carvings. *Ibid.* v. 46 The silly bitch laughingly refused to sell me her tassie ring.

Tassie[3] (æ·zi). *Austral. slang.* Also **Tassey, Tassy.** [Hypocoristic, f. *Tasmania* or *Tasmanian *sb.* b: see -ie.] **a.** Tasmania. **b.** A Tasmanian.
1894 *Argus* (Melbourne) 26 Jan. 3 Today Tassy—as most Victorian cricketers and footballers familiarly term our neighbour over the straits—will send a team into the field. 1905 in A. B. Paterson *Old Bush Songs* 51 Once more the Maorilander and the Tassey will be seen Cooking johnny cakes and jimmies on the plains of Riverine. 1915 H. Lawson *Coll. Verse* (1969) III. 154 Fighting hard for little Tassy, where the apple orchards grow. 1936 F. Clune *Roaming round Darling* xix. 188 The Poet says that's the area of Tasmania. I'll take his word for it, as I haven't time to go and measure Tassie. 1938 N. Marsh *Artists in Crime* vi. 76 'Aussie', 'Tassie', 'a goodee', 'a badee'. Pray spare me these bloody abbreviations. 1956 S. Hope *Digger's Paradise* 77 The Hobart–Launceston express has the advantage of allowing the customer to see Tassie's beautiful countryside. 1977 *Herald* (Melbourne) 17 Jan. 14/2 (Advt.), Come to 'Tassie' the Casino State.

tasso (tæ·so). [perh. f. Tasajo: cf. Louisiana French *tasseau* jerked beef.] = Tasajo.
1841 *Southern Lit. Messenger* VII. 77/2 The evening banquet of gumbo, tasso, and beef, in every variety of form, was shortly served up by their attendants. 1934 E. Waugh *Handful of Dust* vi. 336 Mr. Todd..gave him farine and *tasso* and sent him on his journey. 1958 J. Carew *Wild Coast* viii. 106 Enough food to last Hector for a week—unleavened bread and bits of jerked pork in it, strips of tasso, cassava bread. 1959 P. Capon *Amongst those Missing* 168 He bought two sacks of farine, two of tasso, one of Brazil nuts and one of cashew nuts.

taste, *sb.*[1] Add: **II. 3. b.** (Further examples.) Also *spec.*, an alcoholic drink; alcohol. *U.S. slang.*
1919 E. O'Neill *Rope* in *Moon of Caribbees* 202 Will ye have a taste? It's real stuff. 1966 *New Yorker* 25 June 33 Why don't you stop up Wednesday, and we'll have a taste. 1973 T. Kochman *Rappin' & Stylin' Out* 162, I view such terms as 'pluck' for wine and 'taste' for liquor as embodying an action element retained from its more conventional use as a verb. 1976 *New Yorker* 1 Mar. 84/2 He said, 'Take me for a taste.' We went into a bar, and I thought he'd settle down for a few, but he only had two shots. 1978 *Maledicta* 1977 I. 224 Had a complete and unabetting weakness for *taste* (liquor).
5. c. In fig. phr. *a bad* (or *nasty*) *taste in the mouth* and varr., a lingering feeling of repugnance or disgust left behind by a distasteful or unpleasant experience.
1857 Mrs. Gaskell *Life C. Brontë* II. viii. 186 They [*sc.* Balzac's novels] leave such a bad taste in my mouth. 1899 R. Whiteing *No. 5 John St.* ii. xxv. 255 Never before have I heard such a speech... 'Sort o' gives yer a nasty taste in the mouth,' says Low Covey. 1904 [in Dict., sense 5 a]. 1943 *Sun* (Baltimore) 22 Apr. 18/1 A decidedly sour taste was left by the opening number. 1969 R. Harper *World of Thriller* ii. 71 When all the characters are corrupt or shoddy, the reader goes away with a bad taste in his mouth. 1979 R. Perry *Bishop's Pawn* iv. 70 It had taken me nearly an hour to go through the dossier and when I'd finished reading I had a nasty taste in my mouth.
IV. 9. *taste-leader, -maker, -organ*; **taste-blindness** *Biol.* (see quot. 1934); so **taste-blind** *a.*; **taste-bud** (examples); also *fig.*; **taste-test** *v. trans.*, to test (something) by tasting it, to test the taste of (something); also *absol.*; so **taste-tested** *a.*; **taste-tester.**
1934 *Jrnl. Heredity* XXV. 189/2 There is less likelihood of finding a group entirely taste blind. *Ibid.* 190/1 Taste blindness is an inherited inability to taste certain thiocarbamides as crystals or in cold diluted solutions as bitter. 1965 M. A. Amerine et al. *Princ. Sensory Evaluation Food* ii. 112 Based on studies of families and twins, 'taste blindness' was first reported to be a simple recessive character. 1975 *Nature* 6 Feb. 442/1 The designation 'tasters' for the more sensitive individual and 'non-tasters' or 'taste blind' for the less sensitive. 1879 J. Fulton *Text Bk. Physiol.* (ed. 2) xiv. 365 Peculiar structures, known as taste buds, or taste goblets, have been discovered in the circumvallate papillæ. 1951 V. Nabokov *Speak, Memory* ii. 30 It is..to the lowly and ugly agarics, that nations with timorous taste-buds limit their knowledge and appetite. 1963 *Listener* 3 Jan. 40/1 Vested interests and pressure-groups work upon everything from our political opinions to our taste-buds. 1970 T. S. & C. R. Leeson *Histology* (ed. 2) xiv. 274/2 A few taste buds are found also in the palate and epiglottis. 1952 D. Riesman *Individualism Reconsidered* (1954) 207 The problem..of becoming a possible taste-leader. 1961 *New Left Rev.* Jan.–Feb. 34/2 These areas of work are excluded, by the tastemakers, from the concept of 'serious' art. 1978 *Jrnl. R. Soc. Arts* CXXVI. 725/2 Federigo's artistic patronage did prove a tastemaker, though of a limited kind. 1927 Haldane & Huxley *Animal Biol.* i. 25 A number of very small taste-organs are scattered over certain parts of the tongue [of the frog]. 1970 G. Ordish tr. *Chauvin's World of Ants* vii. 175 The taste organ is situated in the antennae. 1979 *Wichita* (Kansas) *Eagle* 23 May 1-B/4 If you've never cooked with fresh ginger.. taste-test before adding more. 1980 D. Williams *Murder for Treasure* xx. 198 He found himself staring into the eyes of a gargantuan dog whose giant tongue was taste-testing his chin. 1960 *Time* (Atlantic ed.) 11 Apr. 58 Critically taste-tested piles of free cigarettes. 1969 *Listener* 24 July 127/3 Game would be virtually tasteless if taste-testers succeeded in eliminating all its off-flavours.

taster[1]. Add: **4.** Also *spec.*, a portion of ice cream served in a shallow glass. *colloq.*
1891 [in Dict.]. 1901 *Daily Tel.* 21 May 10 The irate signor..produced—not a half-penny taster for the policeman but a tattered copy of a work called 'Law without Lawyers'. 1927 W. E. Collinson *Contemp. Eng.* 16 The Italian often known as an ice-cream Jack with his ice-cream barrow still follows his calling and no doubt the youngsters still ask for wafers and tasters.

‖ **tastevin** (tastəvæ̃n, tāt-). [An earlier form of mod.Fr. *tâte-vin*, lit. 'wine-taster', revived as in the title of the *Confrérie des Chevaliers du Tastevin* (founded 1933).] **1. a.** = Taster[1] 3 a. **b.** (With capital initial.) In France, a member of an order or guild of wine-tasters.
1952 A. Lichine *Wines of France* viii. 79 Most tasters use the *tastevin* in Burgundy. 1966 P. V. Price *France: Food & Wine Guide* 147 The three top sketches show a Burgundy tastevin, with irregularly patterned indentations and a thumb-rest. 1969 Dorozynski & Bell *Wine Bk.* 280 Wine-tasting orders, associations, fraternities and clubs are born almost every year... The Tastevins de Bourgogne..are among the best known in France. 1971 *Esquire* July 22/4 The enthronement of the Tastevins takes place in nearby Clos de Vougeot. 1979 *Homes & Gardens* June 129/1 We moved on to the splendid Beaune Boucherottes, its dusky red lights twinkling in the tiny silver *tastevin* that Andre always uses to assess colour.
2. *attrib.* (in sense 1 b), signifying approval by a special committee of wine-tasters.
1964 H. Johnson *Encycl. Wine* 290 System of labelling certain Burgundy wines... Upon payment of a certain sum per bottle, these wines may carry a special, rather elaborate Tastevin label. 1968 *Times* 3 Sept. 20/5 An invitation to attend the Tastevin banquet at the Château Clos de Vougeot. 1977 W. M. Spackman *Armful of Warm Girl* 109 This was a Tastevin bottling which he'd never found before outside of France.

tasting, *vbl. sb.* Add: **2. b.** A gathering for the purpose of tasting and comparing various kinds of drink, usu. wine. See also *wine-tasting* s.v. *wine *sb.*[1] 8.
1959 I. Ross *Image Merchants* viii. 128 'Cognac and coffee' tastings which Kaduson persuaded leading hotels ..to hold. 1963 *Harper's Bazaar* Feb. 22/3 The Lebègue tastings usher in the autumn wine season. 1977 T. Heald *Just Desserts* i. 23 It's their annual tasting..for one or two of the better known wine and food writers. 1978 *Times* 4 Mar. 10/7 The Malmaison Wine Club..holds sit-down tastings, which are often attended by wine trade trainees.
3. *tasting party, room.*
1978 *Times* 4 Mar. 10/3 The 'tasting parties' offered by many [wine] firms..are social rather than studious occasions. a 1966 'M. Na Gopaleen' *Best of Myles* (1977) 122 Gallantry and distinguished conduct in the [ice-cream] tasting room. 1970 *Country Life* 1 Oct. 837/3 It is distinctly cool..in the subterranean tasting room where red and white..wines are sampled.

tasto (ta·sto). *Mus.* [It., = touch, key.] **a.** The key of a piano or the finger-board of a stringed instrument. **b.** Phr. *sul tasto* [lit. 'over the finger-board']: a direction in a musical score that the stringed instrument is to be played with the bow over the finger-board; *tasto solo*: a direction that the bass notes are to be played alone without any harmony.
1740 J. Grassineau *Mus. Dict.* 268 *Tasto*, the touch or part of an instrument whereon, or by means of which its notes are made to sound, be it on the neck, as lutes..; or the front of organs. 1772 W. Tans'ur *Elements of Mus.* v. 221 *Tasto*, the Touches or Keys of Organs &c. *Tasto solo*, to strike such Sounds till other Words, or Parts come on. 1876 Stainer & Barrett *Dict. Mus. Terms* 421/2 *Tasto*,..(1) A key of a pianoforte. (2) The touch of a pianoforte or organ. 1889 Grove *Dict. Mus.* IV. 63/2 'Tasto solo', the key alone, is in old music written over those portions of the bass or continuo part in which the mere notes were to be played by the accompanyist, without the chords or harmonies founded on them. 1946 E. Blom *Everyman's Dict. Mus.* 600/1 *Sul tasto* (It.), a direction indicating that a passage of string mus. is to be played with the bow over the finger-board. 1974 *Sci. Amer.* Jan. 93/2 Farthest from the bridge..the timbre has the gentle character that composers seek by designating *sul tasto*: 'bow over the fingerboard'. 1980 *Early Music* Apr. 150/2 In Rameau's day the single harpsichordist either did not play during the purely orchestral music and choruses or at most played the bass line only, either *tasto solo* (at pitch) or *all'unisono* (doubling an octave above or below).

tasty, *a.* Add: Also † **tastey. 1. b.** (Later examples.)
1899 R. Whiteing *No. 5 John St.* vii. 61 'Nice and tastey,' observes my friend..as he points to a leg that seems to fear nothing on earth..not even Lord Campbell's Act. 1976 *Sounds* 11 Dec. 41/5, I couldn't help wishing Can'd..pump out a tasty three minute song like their last single,..'I Want More'.
c. (See quots.). *slang.*
1975 *Observer* 11 May 2/7 'I got the impression that he with a bit tasty' (i.e., had a criminal record). 1980 *Daily Mail* 21 Mar. 7/2 A 'tasty villain' (a known criminal).
2. b. Fastidious. *rare.*
1905 A. Adams *Outlet* 102 The two strangers were rather tasty, but Siringo ate ravenously.

tat, *sb.*[1] Add: **2.** *pl.* **tats, tatts.** Teeth; now usu. with ref. to a set of false teeth. *slang* (chiefly *Austral.*).
1919 W. H. Downing *Digger Dialects* 49 *Tats*, teeth. 1935 A. J. Pollock *Underworld Speaks* 119/2 *Tats*, teeth. 1962 *Coast to Coast 1961–62* 131 He'd lost his 'tatts' in a brawl in Townsville the night before we left Australia. 1976 *Express* (Austral.) 3 Nov. 2/3 Talking of 'tats' a dental standards official said false teeth for animals are nothing new.

tat, *sb.*[2] Add: **b.** In tea-drying: a tray or shelf, freq. of hessian, on which green tea leaves are spread to wither.
1922 H. J. Moppett *Tea Manuf.* 14 Withering tats must present a smooth even surface free from corrugations or pockets. *Ibid.* 15 Insufficient tat space is a bad

fault. **1935** W. H. Ukers *All about Tea* II. 515/1 *Tat*, a shelf made of wire or Hessian..on which green tea leaves are spread for withering. **1957** *Encycl. Brit.* XXI. 863/2 After plucking, the leaf is withered by being spread on bamboo trays in the sun, or on withering tats within doors. **1958** T. Eden *Tea* xiii. 150 Some factories have 'mobile tats'. They are slung from pulleys travelling on overhead rails such that each individual bank of tats can be removed from its internal location and brought to a loading and weighing platform at any time.

tat, *sb.*[5] Add: Also **tatt. 1. a.** (Earlier and later examples.) Now usu. only in *sing.* Also, poorly made or tasteless clothes. Hence, a shabby person, a slut.

1839 [see *posh *sb.*[2] 1]. **1882** *Sydney Slang Dict.* 9/2 The paper makers get the tats. **1936** N. Coward *To-Night at 8.30* I. 93 You should have seen the company: a couple of old tats got up as Elizabethan pages. **1947** N. Marsh *Final Curtain* iv. 53 Do they think it's any catch living in a mausoleum with a couple of old tats? **1972** D. Goddard *Blimey!* (1974) iv. 43 King's Road beckons the well-heeled traveller into a cloud-cuckoo land of high-priced tat and gear. **1977** M. Drabble *Ice Age* II. 212 She was dressed..in a horrible collection of tat—a long shiny maroon skirt, a baggy flowered blouse, a grey cardigan, and a green cardigan on top of that.

b. Rubbish, junk, worthless goods. Also *transf.* and *fig.*

1951 W. Sansom *Face of Innocence* iv. 55 He was talking of his business in Georgian and early Victorian objets d'oeil. He called it tat. **1958** A. Wilson *Middle Age of Mrs Eliot* II. 151 It was filled..with a jumble of pleasing, valuable antique furniture and hideous, worthless bric-a-brac... 'I like tatt,' he had said. **1967** N. Marsh *Death at Dolphin* ii. 40 A small shop in Walton Street where they sold what he described as: 'Very superior tatt. Jacobean purses, stomachers and the odd codpiece.' **1970** 'D. Halliday' *Dolly & Cookie Bird* iv. 52 Are they selling tat medals as well? **1971** D. Lees *Rainbow Conspiracy* iii. 38 Oh no! Not that load of old tat. We threw it out at afternoon [news] conference. **1976** *New Musical Express* 12 Feb. 26/3 That long deleted album..sounds like a heap of prissy irrelevant whimsical lysergic tat with Disney lyrics. **1981** *Times Lit. Suppl.* 18 Sept. 1060/1 New ways of getting the johns to spend their money on previously unsellable old tat.

tat, *sb.*[6] Delete *Sc.* and add: Also **tatt.** (Later examples.)

1922 Joyce *Ulysses* 423 A slut combs out the tatts from the hair of a scrofulous child. **1968** *Saturday Night* Mar. 34/3 The hair was full of tats so it was easy to find places to stick the flowers.

Tat (tāt), *sb.*[8] Also **Tât.** [a. Russ., from Turkish.] (A member of) an agricultural people perh. related to the Tajiks and living in Azerbaijan and Dagestan; also, the Iranian language spoken by this people.

1834 A. Burnes *Trav. Bokhara* II. 265 We find as great a variety among the citizens of Toorkistan as in the sub-divisions of the Tartars. The aborigines of the country are the Tajiks or Tats. **1888** *Encycl. Brit.* XXIII. 25/1 The Tajaks are known as Tâts on the west side of the Caspian. **1939** H. Field *Contrib. Anthropol. Iran* iii. 157 While..these elements cannot be considered pure Iranian there remain the Talych (91,000), Tat (74,000) and Persian (50,000), all of which are clearly Iranian dialects. **1951** W. K. Matthews *Lang. U.S.S.R.* vii. 104 The Tat of Azerbaijan, north of Baku, is like Talysh, a divergent variety of Persian. **1981** *Jewish Chron.* 24 Apr. 6/5 A Tat, a mountain Jew from Daghestan in the Caucasus, Mr Irmiya Rabayev, 31, has been a refusenik for seven years.

ta-ta, *int.* Add: Also **tata, ta ta**, etc., and with pronunc. (tæ·ta). **1.** Now in gen. colloq. use. Cf. *tatty-bye *int.* and T.T.F.N. s.v. *T 6. (Earlier and later examples.)

1823 S. Hutchinson *Let.* Sept.–Oct. (1954) 261 Baby I believe has not learnt any new words since Mrs M. wrote last, but she has the old ones very perfect—'Gone'—'Ta ta'—'By bye'. **1901** 'M. Franklin' *My Brilliant Career* xxxii. 272 (*heading*) Ta-Ta to Barney's Gap. **1922** T. S. Eliot *Waste Land* II. 26 Goonight Bill. Goonight Lou. Goonight May. Goonight. Ta ta. Goonight. **1934** —— *Rock* ii. 69 Well, tar, tar, boys. **1949** G. B. Shaw *Buoyant Billions* iv. 52, I must go now to see about Father's lunch. Tata. **1951** R. Braddon *Naked Island* II. vi. 153 'All right, Mr. McLeod, fall out.'..'Ta-ta, Rod—see you in Australia.' **1960** L. Reid Banks *L-Shaped Room* xxiii. 297 Charlie'll come up in a few minutes and see how you're getting on. Tata for now. **1983** P. Inchbald *Short Break in Venice* vii. 68 Sod off!.. We'll talk later... Tata.

2. *ta-ta theory* (Philol.), the theory that language originated in an attempt to imitate the body's gestures with the vocal organs.

1930 J. R. Firth *Speech* i. 7 According to the *ta-ta* theory of Sir Richard Paget, the tongue makes the same gesture while saying *ta-ta* or *hither* as would be made by the hand with similar intention. **1939** L. H. Gray *Foundations of Lang.* 40 Language has been traced..by others to sounds produced by the vocal organs when half-consciously imitating the movements of the body in performing some activity (the ta-ta theory). **1972** Hartmann & Stork *Dict. Lang. & Linguistics* 160/1 R. Paget (1869–1955) claimed that language comes from the combination of certain gestures and tongue movements (*ta-ta theory*).

B. as *sb.* A nursery or playful term. **a.** A walk or outing. Also *fig.* **b.** A hat, bonnet, etc.

(*a*) **1886** J. Sully *Teacher's Handbk. Psychol.* x. 185 A child of eighteen months will mentally rehearse a series of experiences, as those of a walk: 'Go tata, see geegee.' **1912** D. H. Lawrence *Let.* 28 Mar. (1962) I. 104 In the evening Diddler took me a tat-tar, and of course got lost. **1930** —— *Nettles* 9 Want to go a little tattah? So it shall... If it's good!.. It shall go a tattah with its Auntie In a motor. **1958** 'N. Blake' *Penknife in my Heart* iii. 45 Sharp at 10.45 every night..he takes his dog out for a ta-ta. **1969** J. Wainwright *Take-Over Men* vii. 121 They're a con man's dream. They're taken for a ta-ta every market day. (*b*) *c* **1910** F. W. Leigh in *Francis & Day's Album of Famous Old Songs* (1956) VIII, Put on your tat-ta, little girlie. **1912** C. Mackenzie *Carnival* iii. 21 The tying on of her 'ta-ta'—at first a frilled bonnet, later on a rakish Tam o' Shanter. **1920** 'K. Mansfield' *Lett. to J. M. Murry* (1951) 506, I put on my ta-ta. **1949** M. Steen *Twilight on Floods* IV. viii. 644 How about having the ta-tas ready to show her?

‖ **tatami** (tata·mi). Also 7 **tatamee, tattami.** [Jap.] A rush-covered straw mat which is the usual floor-covering in Japan and the size of which (approx. six feet by three feet) functions as a standard unit in room measurement.

1614 R. Wickham in *Trans. Asiatic Soc. Japan* (1898) XXVI. 209, I..made Tatamee of Meaco 15§. *Ibid.*, I sold it per 14 Tatamees at 120 Mas per tatame. **1616** R. Cocks *Diary* 23 Jan. (1883) I. 103, 20 *tattamis* for Matingas howse. **1625** Purchas *Pilgrimes* II. v. 326 Hee caused at Ozaca a Hall to bee erected, with a thousand Tatami (very elegant mats). **1880** I. L. Bird *Unbeaten Tracks in Japan* I. ix. 89 Japanese house-mats, *tatami*, are as neat, refined, and soft a covering for the floor as the finest Axminster carpet. **1886** A. C. Maclay *Budget of Lett. from Japan* 42 *Tatamis* are heavy padded mats about seven feet long, three feet wide, and about two inches thick. They are the only covering that the Japanese ever use for their floors... They are manufactured of soft rushes, and are bordered with silken edges. **1909** *Cent. Dict.* Suppl., *Tatami*..2. A Japanese measure of surface, that of a mat 6 shaku in length by 3 shaku in width, or nearly 6 feet by 3 feet. **1924** *Public Opinion* 28 Nov. 527/2 It is a strict rule that tatamis must be kept clean. **1933** R. V. C. Bodley *Japanese Omelette* xii. 116 The dining room floor, instead of being matted with *tatami* as in Japan, was made of some kind of oilcloth. **1957** *New Yorker* 23 Nov. 120/2 *Tatami* cover the floors of nearly all Japanese houses. **1960** B. Leach *Potter in Japan* iii. 68 In twenty years' time, won't the Japanese room with 'tatami' (the thick compressed straw matting) become a luxury as the foreign style is today? **1974** *Encycl. Brit. Micropædia* IX. 837/3 The standardized size of the mat has created an important modular unit in the development of Japanese architecture; for example the *shōji*..are approximately as high as the tatami are long. **1976** P. Quennell *Marble Foot* v. 182 No less beautiful..were the *tatami* that lined our floors, long greenish slabs, that turn with age a dull gold, of finely woven rush-matting. **1981** G. MacBeth *Kind of Treason* ix. 92 He relaxed on the *tatami* and spoke with polite approval of the cousin's *tsuba*.

2. Freq. *attrib.*, as *tatami mat, matting, room.* Also *Comb.*, as *tatami-floored, -matted* adjs.

1947 J. Bertram *Shadow of War* VI. 200 Each man had some two and a half feet by six feet of *tatami* mat. **1962** *Times* 25 Jan. 13/4 The man who comes to lay the *tatami* matting in his house. **1979** S. Coe in I. Webb *Compl. Guide to Flower & Foliage Arrangement* xvii. 231/2 The room..is quite small, about 3×3 m (10×10 ft) or four and a half *tatami* mats. **1979** *Jrnl. R. Soc. Arts* Nov. 749/1 The interior spaces provide everything that the harsh exterior rejects: complex flowing geometries, traditional tatami room, lush furnishing and peaceful, controlled nature. **1980** J. Melville *Chrysanthemum Chain* 127 The tatami-matted floor of Yamamoto's room. **1981** C. Potok *Bk. of Lights* (1982) v. 162 He had a fish dinner in a lovely tatami-floored Japanese restaurant. **1982** *Nature* 20 May 181/1 Thus the popular, but erroneous, Japanese view that they have grown taller by adopting the habit of sitting on Western chairs rather than sitting with legs folded underneath the body on a *tatami* mat.

tater (tēi·təɹ). Also **tator, tatur. 1.** Further dial. variants of Potato *sb.* 2 dγ. Cf. next.

1759 L. Wood *Jrnl.* 27 May in *Essex Inst. Hist. Coll.* (1882) XIX. 65 We travelled..9 miles to Capt Curtises and there we Dined upon codfish and taters. **1911** F. H. Burnett *Secret Garden* xxiv. 252 Anything'll grow for him. His 'taters and cabbages is twice th' size of anyone else's. **1939** F. Thompson *Lark Rise* i. 11 Mother spent hours boiling up the 'little taturs'. **1943** W. Stevens in O. Williams *New Poems 1943* 236 (*title of poem*) No possum, no sop, no taters. **1977** *New Yorker* 27 June 76/3, I et 3lbs. of taters.

2. *attrib.* and *Comb.* **tater-trap** *slang* = *tattie-trap* s.v. *Tattie 2.

1845 W. T. Porter *Big Bear Arkansas* 22 Them ar 'Indian mounds' ar tater Rills. **1846** *Swell's Night Guide* 133/2 *Tater trap*, the mummer, mouth. **1847** J. R. Lowell *Biglow Papers* 1st Ser. iii. 34 He draws his furrer ez straight ez he can, An' into nobody's tater-patch pokes. *c* **1869** Taylor & Dubourg in M. R. Booth *Eng. Plays of 19th Cent.* (1973) III. 303 The coal and 'tatur shed where he worked all the week. **1876** I. Banks *Manchester Man* III. i. 12 Shut up his tater-trap fur him! **1890** P. H. Emerson *Wild Life* 38, I adwised them fellers at the pub ter keep their tater-traps shut. **1902** W. N. Harben *Abner Daniel* 198, I got up on the head of a tater-barrel behind the counter. **1917** 'H. H. Richardson' *Fortunes R. Mahony* I. ii. 20 From the back of the hall came the curt request to him to shut his 'tater-trap'. **1930** *Dialect Notes* VI. 89 *Tater jack*, variety of fermented liquor, made in [lumber] camp from potatoes.

tatie, 'tato. 1. These variants are more widespread than is implied in Dict. Other variants recorded, mostly in dialectical works, include **taty, tautie,** and **tauty.** (The examples that follow supplement those given in Dict. s.v. Potato *sb.* 2 dγ.)

1793 R. Brown *Comic Poems* (1817) 118 Sawt herrings, tawties, water kale. **1812** P. Forbes *Poems* 31 A charger's just a muckle pig, For ha'din' kail or 'tatoes. **1870** D. J. Kirwan *Palace & Hovel* xiv. 214 Guv us a taty, Jenny. **1894** J. Menzies *Our Town* 240 We div look at our tauties on Saubbath. **1920** W. de la Mare *Poems 1901–1919* II. 170 There's goose, baked taties and cabbage. **1979** *Bull. Yorks. Dial. Soc.* Summer 15 The wor acres o gowden corn, taties an sugar beet, peearce an quiet.

2. *attrib.* and *Comb.*, as *tatie pot*; **tatie-bogle** = *tattie-bogle* s.v. *Tattie 2.

1838 J. M. Wilson *Hist. Tales Borders* IV. 306 Ye look mair like a tauty bogle than a Christian man. **1853** S. R. Whitehead *Nelly Armstrong* I. i. 24 It was fitter for a tatie-bogle's back than a leddy's. **1871** J. Richardson *Cummerland Talk* 1st Ser. 7 A dish consisting of beef or mutton, cut into pieces, and put into a large dish along with potatoes, onions, pepper, salt, etc., and then baked in the oven,..is called in Cumberland *taty-pot*. **1893** *West Cumberland Times* (Holiday No.) 5/4 'Begok, it's tatie pot!' says Ben. **1974** *Times Lit. Suppl.* 1 Mar. 215/1 Mr Wyatt soon made friends among the woodmen and farm workers, enjoyed a tatie-pot supper at the nearest inn. **1976** *Cumberland News* 3 Dec. 24/2 Lazonby Methodist Church raised over £70 with a 'tatie-pot' supper. **1978** R. Hill *Pinch of Snuff* xxiii. 234 Going off home for a tatie-pot supper and an early night. **1982** *Sunday Tel.* 21 Feb. 16/7 There await you, in all their rich goodness, leek pie and Mendip snails, Cumbrian tatie pot and tripe,..Dorset sausage and Somerset apple cake.

‖ **tâtonnement** (tatǫnmaṅ). Also **tatonnement.** [Fr., f. *tâtonner* to feel one's way, proceed cautiously.] Experimentation, tentative procedure; *tâtonnement process*, a process of trial and error.

1847 A. De Morgan *Formal Logic* II. 324 M. Gergonne's complex propositions..requiring a separate *tâtonnement* for many things the analogues of which appear as connected results of my system. **1964** W. S. Vickrey *Metastatics & Macroeconomics* i. 21 The postulate that an equilibrium would be reached by a series of adjustments through recontracting, or 'tâtonnements'. **1975** *Jrnl. Econ. Theory* X. 122 W. Neuefeind, A tatonnement process for *N*-person games, *CORE Discussion Paper* 7136. **1978** S. Weintraub *Capitalism's Inflation & Unemployment Crisis* ix. 176 Unbounded pathological Wicksellian-Hayekian cases can be discounted, though some minor (*tatonnement*) imbalances are inevitable.

‖ **tatpurusha** (tatpu·rūʃā). *Philol.* Also **tatpuruṣ(h)a** and with capital initial. [Skr., lit. 'his servant'.] A compound in which the first element qualifies or determines the second, while the second retains its grammatical independence as noun, adjective, or participle.

1846 M. Williams *Elem. Gram. Sanscrit* ix. 157 Native grammarians class compound nouns under five heads: the 1st they call *Tatpurusha* or those composed of two nouns. **1872** [see *bahuvrihi]. **1901** A. A. Macdonnell *Sanskrit Gram.* vi. 159 The past part...gata, 'gone to', is often used at the end of Tatpuruṣas in the sense of 'relating to', 'existing in'. **1946** *Trans. Philol. Soc. 1945* 86 From this combination we get the tatpuruṣa compounds *śilōncha-* and *śilōñchana-*. **1957** S. Potter *Mod. Ling.* iv. 91 *Bookcase* consists of substantival attribute + substantive. It belongs to that class of compounds known as *tatpurusha* to Indian grammarians because the first component determines or qualifies the second. **1969** —— *Changing English* ii. 58 It [*sc.* 'year-book'] is a tatpurusha type of compound that has been in use in English for hundreds of years, written solid by Anglo-Saxon scribes.

tattami, obs. var. *Tatami.

tattarrattat (tæ:tăræ·t). *nonce-wd.* [Echoic.] = Rat-a-tat.

1922 Joyce *Ulysses* 732, I knew his tattarrattat at the door.

tatter, *sb.*[1] Add: **3.** *tatter-eared, -skinned, -tangled* adjs.

1953 R. Graves *Poems* 18 Tatter-eared and slinking alley-toms. **1924** R. Campbell *Flaming Terrapin* ii. 26 And like a leper, faint and tatter-skinned, The wan moon makes a ghost of every tree. **1876** G. M. Hopkins *Poems* (1967) 177 Tatter-tangled and dingle-a-dangled Dandy-hung dainty head.

tatter, *sb.*[3] *slang.* [f. Tat *v.*[3] + -er[1].] A refuse-gatherer, a rag-collector. Cf. Totter (in Dict. s.v. Tot *sb.*[5]). Also **tatterer.**

1890 Barrère & Leland *Dict. Slang*, *Tatter* (tramps), a rag-gatherer. **1910** *Church Times* 15 July, Their occupations being largely that of 'Tatterers'—*i.e.* rag and bone and bottle-gatherers, and casual labourers. **1921** *Dict. Occup. Terms* (1927) § 970 Tatter,..collects [waste] with a hand-pushed barrow or cart. **1969** *Telegraph & Argus* (Bradford) 16 Oct. 9 He was wearing a dark jacket, and light drill trousers. He is believed to be a rag tatter.

tatter, *v.*[1] **b.** (Example.)

1934 Dylan Thomas in *New Verse* Apr. 12 Our strips of stuff that tatter as we move.

Tattersall (tæ·təɹsâl), *sb.* (and *a.*) [The name of Richard *Tattersall* (1724–95), horse-auctioneer.] **1.** Used chiefly in the possessive (occas. abbrev. **Tatt's, Tatts**) to denote: **a.**

The horse-auction market established by him in 1766 at Hyde Park Corner. Also *transf.* and *fig.*

1795 *Sporting Mag.* VI. 5/1 The gentlemen of the turf assembled every sale day..at Tattersalls. **1825** *Monthly Mag.* 1 Mar. 129/2 The sale days, at Tattersall's, formerly on Monday and Thursday. **1834** Carlyle *Sartor Resartus* I. iv. 12/1 He burst forth like the neighing of all Tattersall's. **1846** 'Sylvanus' *Pedestrian & Other Remin.* xxv. 241 They've three or four working the oracle at Tatt's. **1880** *Illustr. Sporting & Dramatic News* 4 Dec. 282/1 There has lately been opened at Christchurch, New Zealand a new 'Tattersall's', the want of which has long been felt. **1882** C. M. Yonge *Unknown to History* II. v. 64 'Paul's Walk' was the Bond Street, the Row, the Tattersall's, the Club of London. **1893** *Cassell's Family Mag.* Aug. 646/1 Ponies and donkeys are here too, for the costermongers. Indeed we might call it the costermonger's horse fair, the Tattersall's of the poorer classes. **1973** *Country Life* 15 Nov. 1547/3 Tattersall's..moved from Hyde Park Corner in 1865, to Knightsbridge Green... Tattersall's continued selling bloodstock and hunters.. for some years after the ring of carriage horses' hooves had ceased. **1982** *Daily Tel.* 4 Mar. 18/3 The [Hunters' Improvement and National Horse Breeding] society.. based..for the past 16 years at Tattersalls.

b. The principal betting enclosure on a racecourse. Also *Tattersall's Ring.*

1836 T. Hood *Let.* 12 Jan. (1973) 211 Tattersall style of betting. **1843** *Illustr. London News* 17 June 418/1 The winners have been haunting Tattersall's..in search of a settlement. **1863** *Observer* 18 Jan. 2/4 The lobby of the hall is like the betting-room at Tattershall's in a low way. **1865** *Once a Week* 28 Oct. 523/1 At one time..rarely did the flood of speculation overflow the dykes of 'The Corner'. Now we have an *al fresco* Tattersall's at nearly every open space in London. **1901** *Cassell's Mag.* Sept. 368/2 Tattersall's Ring at race-meetings and the committee of Tattersall's which rules the betting world, have now no connection with the firm..at Albert Gate. **1922** *N. & Q.* 9 Sept. 206/2 *Outside*, all other enclosures on a racecourse save Tattersall's, which is 'inside'. **1951** E. Rickman *Come racing with Me* xvi. 151 The customary reference to the chief betting 'ring' on any racecourse as 'Tattersalls' or 'Tatts' is a relic of the rough-and-ready days when it was an enclosure used principally by bookmakers and backers who were members of Tattersalls' Subscription Room. **1962** [see *rail *sb.*[2] 2 f]. **1973** [see *silver ring* s.v. *silver *sb.* and *a.* 21 a]. **1983** 'F. Parrish' *Bait on Hook* v. 68 He had no real idea what class of man they were looking for—a denizen of the Members' Enclosure] at £6, Tattersalls at £3.50, the Silver Ring for £1, the course for nothing.

c. The name of a lottery which originated in Sydney in 1881, moved to Tasmania in 1896, and since 1954 has operated from Victoria. Freq. abbrev. in *colloq.* use.

1895 N. Gould *On & off Turf in Austral.* vi. 52 (*heading*) 'Tattersalls' and 'Oxenhams'. *Ibid.* 61 Mr. George Adams..runs his consultations, or sweeps, under the name of 'Tattersall', and they are very popular all over the Colonies. **1945** Baker *Austral. Lang.* xv. 264 *To take a ticket in Tatt's* is to buy a ticket in Tattersall's sweepstakes, Tasmania. *Safe as Tatt's* is synonymous with perfect safety. **1951** J. Frame *Lagoon* 57 His fingers search an envelope for the pink sheet that means Tatts results, ten thousand pounds first prize. **1957** —— *Owls do Cry* xvii. 73 The Art Union? There was a theory that if you bought a ticket up north where the population was thickest you were sure to win a prize. The raffle? Tatts? **1965** *Austral. Encycl.* V. 371/2 The oldest continuing public lottery in Australia is 'Tattersall's',..established at Sydney in 1881... It took its name from Richard Tattersall's horse-auction mart in London. **1969** *Australian* 24 May 40/4 My man asked if he would abolish Tatts, seeing gambling was such a reprehensible thing.

2. *attrib.* or as *adj.* (freq. with small initial). Designating (a fabric with) a small and even check pattern or garments made from such a material. Hence *tattersall-checked* adj. Also *absol.*

From the traditional design of horse blankets.

1891 *Cassell's Family Mag.* Dec. 58/1 All those curious checked cloths which rejoice in the name of 'Tattersall' because, I suppose, they resemble horse cloth. **1951** J. D. Salinger *Catcher in Rye* xii. 103 This..guy, in a grey flannel suit and one of those flitty-looking tattersall vests. **1958** *Vogue* Jan. 35 Tattersall checks of black and caramel. **1963** *Guardian* 2 Oct. 8/5 Simpsons have Tattersall check shirts of woollen fabric. **1967** [see *gun club* s.v. *gun *sb.* 15]. **1972** *New Yorker* 7 Oct. 12/1 (Advt.), Snuggle yourself..inside our tattersall-checked robe. **1976** *National Observer* (U.S.) 2 Oct. 3/3 (Advt.), Also available in neat tattersall checks of rust/green on camel ground. **1978** *N.Y. Times* 30 Mar. c 11/1 The collection's subdued neutral colors, sedate tattersalls and similarly classic patterns. **1980** U. Curtiss *Poisoned Orchard* xii. 126 Fawn corduroys and a yellow tattersall shirt. **1981** *Daily Tel.* 14 Sept. 13/2 Viyella Tattersall check shirt.

tattery, *a.* Add: Also *Comb.*, as *tattery-clothed* adj.

1941 S. O'Casey *Let.* Jan. (1975) I. 874 The half-fed, tattery-clothed, lice-lorn children scattered now over England by the falling bombs.

tattie. 1. This variant of Potato *sb.* 2 dγ is now more widespread than is implied in Dict. See dialect dicts. for fuller documentation. Cf. *Tater; *Tatie, 'tato. Also *fig.*, a stupid person.

c **1800** in F. P. Hett *Mem. S. Sibbald* (1926) 203 Then there's champit tatties, after they are boiled, the water is poured off them then they ar' champet wi' the champer in the pot. **1879** *Forfar Poets* 139 Gae hame, ye simple tattie. **1901** G. Douglas *House with Green Shutters* 234 I'll feenish the tatties at ony rate. **1921** A. S. Neill *Carroty Broon* xv. 202 I'm at the tatties wi' achteenpence a day. **1973** *Stornoway Gazette* 27 Jan. 10/4 Tatties and herrings for supper. **1976** *Sunday Mail* (Glasgow) 26 Dec. 18/1 A tractor ran over her leg as she was picking tatties from a field on the outskirts of Monifeith.

2. *attrib.* and *Comb.*, as *tattie-field*; **tattie-bogle** = *potato-bogle* s.v. Potato *sb.* 7; also *fig.*, a simpleton; **tattie-trap** *slang* = *potato-trap* s.v. Potato *sb.* 7.

1865 *Scotsman* 28 June, First then come the trades' unions and strikes,..as a tattie-bogle..to scare the black and croaking denizens of the wood. **1922** J. Buchan *Huntingtower* xii. 240 There's..me..no more use than a tattie-bogle. **1969** M. Pugh *Last Place Left* xii. 73 The gasworks doctor, you tattie bogle. There's only one doctor. **1979** L. Derwent *Border Bairn* xi. 128 Others I dredged up from somewhere in my head, about tattie-bogles, bubblyjocks and real adventures. **1891** Barrie *Little Minister* III. xxxv. 64 The Retery's in flood..; T'now-dunnie's tattie field's out o' sicht. **1983** *Listener* 18 Aug. 23/2 *Another Time, Another Place* includes a vivid background of wet and windy tattie fields. **1894** J. B. Salmond *My Man Sandy* 175 Juist you keep your tattie-trap steekit. **1899** A. L. Salmon *West-Country Ballads* 74 Cureit's tattie-trap an' muzzle, Like a bwoy's be smooth an' bare.

tatting (tæ·tiŋ), *vbl. sb.*[2] [f. Tat *v.*[3] + -ing[1].] Rag- or scrap-collecting (see also quot. 1926). Cf. Totting (in Dict. s.v. Tot *sb.*[5]).

1926 *Glasgow Herald* 14 Dec. 10/7 The word 'tatting'.. appears..to mean the annexation by dustmen..of stray articles of small value found in dustbins. **1969** *Listener* 6 Feb. 169/2 Now,..scrap-collecting and dealing are the biggest stand-by, with 'tatting' (rag-collecting), [etc.]. **1977** Scollins & Titford *Ey up, mi Duck!* III. 52 *Tattin'*, going round collecting scrap, as a scrap-merchant does.

tattle-tale (tæ·t'l,tēil), *sb.* (and *a.*) *colloq.* (orig. and chiefly *U.S.*). [f. Tattle *sb.* (or *v.*), after *tell-tale.*] **1.** = Tell-tale *sb.* (*a.*) 1 a. Occas. *attrib.* or as *adj.* (cf. Tell-tale *sb.* (*a.*) 3 b).

1889 'C. E. Craddock' *Despot of Broomsedge Cove* 429 I'd strangle that tattle-tale with a mighty good will. **1918** J. G. Thompson *Lest we Forget* 230 An American boy was expelled from a German gymnasium in Berlin, because he refused to 'tattle-tale' on the pupils in his class. **1929** W. Faulkner *Sound & Fury* 91 'I already told on her,' Jason said... 'And see what you got by it,' Caddy said. 'Tattletale.' **1946** C. McCullers *Member of Wedding* I. 45 'Frankie ain't no tattle-tale,' said Berenice. **1962** 'K. Orvis' *Damned & Destroyed* xiii. 89 Helen's tattle-tale scars. **1964** D. MacArthur *Reminiscences* I. 26 Come what may, I would be no tattletale. **1977** D. Bagley *Enemy* v. 37 Nellie is a tattle-tale, isn't she? Too bloody gossipy.

2. A tachograph; also in oil-well drilling (see quot. 1942).

1942 Berrey & Van den Bark *Amer. Thes. Slang* § 516/5 *Tattletale*, an instrument recording the pressure of the bit on the bottom and the work done during a shift. **1949** *Amer. Speech* XXIV. 35 Another characteristic of oil-field language is the predominance of compound words... Compound nouns coined from a verb and its object are *bore hole..swamp pole*, and *tattle tale*. **1962** *Ibid.* XXXVII. 272 *Tattletale*,..a complex device used on trucks, buses, and locomotives which records exact time, distance, and speed on a cylinder, and is used to check on how much speed and how many stops a vehicle made. **1971** M. Tak *Truck Talk* 165 *Tattletale*, a sealed tachograph in a tractor that simultaneously records a truck's speed and the time of day; thus producing a record of driving hours and stops. Installed by large companies to check on driver efficiency and running time, tattletales are, needless to say, unpopular with drivers.

3. *Comb.* **tattle-tale grey**, an off-white colour resulting from inadequate laundering. Also *fig.*

1943 D. W. Brogan *Eng. People* 11 If the victory of the United Nations is a good thing for all..a member of this alliance may be well advised to put up with the fact that the other members' records are not snowy white, but at best tattle-tale grey. **1973** *Houston* (Texas) *Chron. Mag. People, Places, Pleasures* 14 Oct. 11/4 In case all this still leaves the tattle tale grey, one must begin it all over again.

Hence (*rarely*) as *v. intr.*, to tell tales or 'sneak' on (somebody).

1918 J. G. Thompson *Lest we Forget* 230 An American boy was expelled from a German gymnasium in Berlin, because he refused to 'tattle-tale' on the pupils in his class.

tattoo, *sb.*[2] Add: **b.** *tattoo mark* (earlier example).

1892 'Mark Twain' *Amer. Claimant* xvi. 164 His horny hands and wrists were covered with tattoo-marks.

tattooer. (Earlier example.)

1789 *Loiterer* 18 July 7 The most famous Tataower in the Country.

Tatt's: see *Tattersall *sb.* (and *a.*).

tatty (tæ·ti), *a.*[2] *colloq.* [f. Tat *sb.*[5] + -y[1].] **1.** Of a person, an animal: untidy, disreputable, 'scruffy'. Cf. Tatty *a.*[1]

1933 N. Coward *Design for Living* II. iii. 67 Going round in a troupe, with all those tatty old girls. **1951** J. Cannan *And All I Learned* x. 165 You mustn't call Brownie a tatty old trout. **1967** N. Freeling *Strike Out* 38 I've seen the painter..rather a tatty chappy by their standards. **1978** *Lancashire Life* Apr. 36/2 A widower living with his one son and a tatty collie dog, he had been a soldier for many years.

2. Of clothes, decoration, etc.: shabby, tawdry, cheap.

1940 N. Mitford *Pigeon Pie* vii. 117 The 'King's' tatty striped wall-papers. **1951** 'A. Garve' *Murder in Moscow* vii. 84 Ivan pushed up his tatty fur hat. **1959** H. R. F. Keating *Death & Visiting Firemen* xv. 195 You're a man, I can see that, in spite of your tatty old clothes. **1963** *Times* 4 June 14/2 Nineteenth-century-style songs, played by a jaunty orchestra before tatty red-plush curtains and even tattier scenery, accompany the high jinks. **1976** *Sunday Post* (Glasgow) 26 Dec. 29/4 It [*sc.* the car] was a tatty green, so a pal and I painted it navy blue.

3. Of a place or a building: badly cared for, neglected, run down.

1956 L. McIntosh *Oxford Folly* iv. 53 This is Oxford's latest coffee-bar... The others are getting so tatty. **1966** *Listener* 12 May 686/1 Some distance from the edge of the Falls a sizeable crack has opened up... Neither the Americans nor the Canadians can afford to have Niagara looking so tatty. **1978** L. Heren *Growing up on The Times* iii. 63 The car drove through the rather tatty outskirts of Tel Aviv.

4. *transf.* In other miscellaneous uses.

1957 *Listener* 19 Dec. 1026/1 Look what we did to that tatty second act. **1959** *Economist* 28 Mar. 1153/2 The Prime Minister's reply looks like a foretaste of the tattier tactics that will be used by the less inhibited Tories in the election. **1965** *New Statesman* 9 Apr. 585/2 The entire vision's too enormous for accommodation within the tatty ingenuities of the stage. **1975** in R. Crossman *Diaries* I. 376 This was a somewhat tatty account of Labour's first year in Government, prepared in Transport House as a diatribe against the Tories.

Hence **ta·ttily** *adv.*; **ta·ttiness.**

1952 A. Wilson *Hemlock & After* I. v. 93 He rejected the 'tattiness' of dead mullion and withered sycamore berries. **1957** *Observer* 29 Sept. 12/1 The keynote of these tattily exotic revues is imitation. **1959** S. Gibbons *Pink Front Door* ix. 118 I've got you the rooms. Four of them, furnished rather tattily. **1973** J. Wainwright *Pride of Pigs* 8 The impression of tarted up tattiness. **1980** *Times Lit. Suppl.* 3 Oct. 1118/5 The novel is firmly set in the very recent past..and rock music, fashion, the death of Elvis, the tattiness of London are described in detail.

tatty-bye (tæti bəi) (stress variable), *int.* [Fanciful formation: cf. Ta-ta *int.* and Good-bye.] A colloquial form of farewell.

1971 A. Morice *Murder in Married Life* vi. 56 'Ta-ta for now, then.' 'Tatty-bye, Sandy darling.' **1974** M. Babson *Stalking Lamb* xv. 113 I'll say tatty-bye for now then... And we'll see you soon. You know the way, don't you?

‖ **ta tzu-pao** (dā dzə bau). Also **dazebao, dazibao, tatzepao, ta-tzu-pao.** [Chinese *dàzìbào*, f. *dà* big + *zì* character + *bào* newspaper, poster.] In the People's Republic of China, a wall poster written in large characters that expresses a (political) opinion or other message.

The posting of *ta tzu-pao* is no longer encouraged in China.

1960 *Peking Rev.* 5 Apr. 8/2 Criticism and self-criticism through large-scale airing of views and opinions, big debates and putting up *dazibao* (posters in large characters) are carried out in government and people's organizations. **1962** E. Snow *Other Side of River* (1963) xlviii. 368 Around the basketball court and a stage which prisoners had built were bulletin boards posted with *ta tzu-pao* such as you see before any Chinese factory: essays, rhymes, praise and mutual criticism, lists of model workers and their awards. **1967** S. Knight *Window on Shanghai* lviii. 250 Four foreigners in Peking put up a 'dazebao' criticizing the treatment given us by the Chinese. **1968** *Globe & Mail* (Toronto) 5 Feb. 10/2 A tatzepao (big-character poster) reproduced by the Shanghai newspaper claimed that the handful of party leaders now accused of being capitalist roaders exaggerated the role played by technical knowledge. **1973** T. R. Tregear *Chinese* iii. 58 Walls everywhere were covered with *ta-tzu-pao*, big-character newspapers. **1979** *Globe & Mail* (Toronto) 4 Apr. 1/2 Some students at Peking University told Western journalists that the repression was wrong as far as the poster or dazibao writers at Xidan democracy wall were concerned.

tau. Add: **2. b.** (Earlier examples.)

1785 [see *ansated *ppl. a.*]. **1841** J. G. Wilkinson *Manners & Customs Ancient Egyptians* 2nd Ser. II. xv. 283 The sacred *tau*, or sign of Life, was presented to him.

4*. *Particle Physics.* Freq. written τ. **a.** A meson that decays into three pions, now identified with the kaon. Also *tau meson.*

1949 *Nature* 15 Jan. 86/2 We have considered the possible relations of the present results to the particles.. referred to as τ-mesons, evidence for which has been recently reported by Bradt and Peters. **1955** *Proc. Glasgow Conf. Nucl. & Meson Physics 1954* 347 The striking similarity of the masses of the θ° and τ±-mesons. **1968** M. S. Livingston *Particle Physics* vii. 138 The θ decayed into two pions while the τ decayed into three pions. **1973** L. J. Tassie *Physics Elem. Particles* vii. 61 The solution to the θ–τ puzzle was that the θ and τ particles were the same, now called the *K*-meson, and the parity was not conserved in the decay of *K*-mesons. **1974** Frauenfelder & Henley *Subatomic Physics* ix. 205 The decays of the tau and the theta were so slow that they were known to be weak.

b. An unstable heavy charged lepton which has a spin of ½ and a mass of approximately 1780 MeV (3490 times that of the electron)

and which decays into an electron or muon or into hadrons, in every case with one or more neutrinos. Also *tau lepton, particle.*

1977 M. L. Perl in *Proc. Internat. Symposium Lepton & Photon Interactions at High Energies* 146 All..data ..agree on the following points... c. The behavior of these [leptonic] events is consistent with the hypothesis that a new charged lepton, τ, exists with a mass of 1.9±0.1 GeV/c². **1978** Perl & Kirk in *Sci. Amer.* Mar. 50/3 We shall relate here the story of the discovery of the new heavy lepton and its antiparticle, which we have named the tau and the antitau. **1979** *McGraw-Hill Yearbk. Sci. & Technol.* 240/1 It was first discovered through reaction (1), in which a positron (*e*⁺) and electron (*e*⁻) annihilate and produce a pair of τ-leptons of opposite electrical charge. **1980** *Sci. Amer.* July 60/1 More recently a third neutrino flavor has been added to accompany the newly discovered tau particle, which is a massive sibling of the electron and the muon. **1981** D. H. Perkins in J. H. Mulvey *Nature of Matter* iv. 79 The neutrino is a muon-type neutrino and in subsequent interactions will always produce a charged μ, not an e or τ.

tau, obs. var. *Tao.

‖ **taua** (tau·ă). [Maori.] A Maori army or war party.

1858 J. Whiteley *Let.* 19 Apr. in *Richmond-Atkinson Papers* (1960) I. 390 The taua seemed to be mustering their forces last night from the south. **1882** W. D. Hay *Brighter Britain!* I. x. 254 Instead of leading a ferocious taua, he finds himself the venerated pastor of a little community. **1921** H. Guthrie-Smith *Tutira* x. 69 Two *tauas* or war-parties. **1959** Tindale & Lindsay *Rangatira* viii. 79 A taua of fighting-men, about fifty strong, was advancing towards them. **1978** [see *rangatira].

‖ **taubada** (taubā·dā). [Local word.] On the island of New Guinea, used to refer to anyone in a position of authority, esp. as a respectful form of address.

1891 W. D. Pitcairn *Two Years among Savages of New Guinea* iii. 58 If he be a 'Taubada' that is to say a person of importance. **1924** 'R. Daly' *Outpost* xii. 116 Did the Taubada want to make them the laughing-stock of their enemies? **1945** *Coast to Coast 1944* 101 A native voice said, 'One more come, taubada.' **1963** *Times* 3 Dec. 8/1 Some *taubadas* and *mastas* have forsaken the hotels for clubs to which natives are not yet admitted.

Tauberian (taubī°·riăn), *a. Math.* [f. the name of Alfred *Tauber* (1866–?1942), Slovak mathematician + -ian.] Applied to theorems in which the behaviour in the limit of a series or function is deduced from a weaker limiting property together with some additional condition, esp. theorems in which convergence is deduced from summability.

1913 Hardy & Littlewood in *Proc. London Math. Soc.* XI. 411 The general character of the theorems which it [*sc.* this paper] contains is 'Tauberian': they are theorems of the type whose first example was the beautiful converse of Abel's theorem originally proved by Tauber. **1962** D. R. Cox *Renewal Theory* i. 14 A result of this type, enabling the limiting behaviour of $k(x)$ to be deduced from that of $k^*(s)$, is called a Tauberian theorem. **1979** *Nature* 24 May 358/1 Rau is well known and remembered for his valuable contributions to the theory of Tauberian theorems, function-theory and the theory of Dirichlet series.

Tauchnitz (tau·knits, tau·χnits). The name of Christian Bernhard, Baron von *Tauchnitz* (1816–95), publisher of Leipzig, used *attrib.* and *absol.* with reference to volumes in the Collection of British and American Authors, a series begun by him in 1841 for sale on the continent.

1856 Mrs. Gaskell *Let.* 26 Dec. (1966) 430 For 'North & South' I received 600£..having the Tauchnitz profit... I suppose however you would allow me to retain the profits arising from the American & Tauchnitz edit:. **1863** J. Morrel *Jrnl.* 10 July (1963) iii. 94 In a twinkling all our 'Tauchnitzes' became pocket editions. **1895** *Daily News* 15 Aug. 5/1 The excitement lest the Customs' officer were to find the Tauchnitz in the trunk. **1901** *Ibid.* 5 Jan. 7/4 Desolate dwellings, strewn with a few sixpenny magazines and smuggled Tauchnitzes. **1902** H. James *Wings of Dove* III. v. 102 The uncut but antiquated Tauchnitz volume of which, before going out, she had mechanically possessed herself. **1920** Joyce *Let.* 12 May (1966) II. 464 The head of the firm would like to have a copy with a view to including it in the Tauchnitz edition. **1936** E. Ambler *Dark Frontier* vi. 88 A Tauchnitz edition of Butler's *Erewhon* purchased hastily from a station bookstall. **1960** *Times* 7 July 14/7 A library of 'Tauchnitz' books. **1975** T. Allbeury *Special Collection* iv. 125 A handful of Tauchnitz pocket editions.

tauhinu (tauhī·nu). *N.Z.* Also **tarwinie.** [a. Maori.] A health-life evergreen shrub, *Pomaderris phylicifolia*, of the family Rhamnaceæ, native to New Zealand and south-eastern Australia and bearing downy leaves and clusters of small yellow flowers. Also *attrib.*

1848 R. Taylor *Leaf from Nat. Hist. N.Z.* 25/2 *Tauhinu*, a shrub. **1903** B. E. Baughan in Chapman & Bennett *N.Z. Verse* (1956) 54 Briar, tauhinu, an' ruin. **1927** J. Devanny *Old Savage* 47 Behind her, tough shrubs, tarwinie and gorse, mantled the terrace leading back to the low hills. **1950** *N.Z. Jrnl. Agric.* Oct. 297 (*caption*) Note the prevalence of tauhinu scrub. **1964** *Weekly News* (Auckland) 15 Apr. 37/3 Tauhinu was the worst scrub to plough, for it grew outward along the ground.

‖ **taula** (tau·lă). *Archæol.* [Cat., f. L. *tabula* table.] A Bronze Age stone structure found on Minorca in the Balearic Islands, consisting of two slabs forming a T-shaped column, freq. enclosed by a horseshoe-shaped wall and believed to be a place of worship.

1881 R. L. Playfair *Handbk. Mediterranean* II. 459/2 A bi-lithon, or altar, composed of two immense monoliths,..carefully dressed, called *Altar* or *Taula*, altar or table. **1911** M. S. Boyd *Fortunate Isles* xvi. 192 Just behind the talayot..stands another relic of prehistoric times in the shape of a *taula*. **1932** *Discovery* July 238/2 Minorca possesses a form of megalithic structure which does not occur elsewhere. This is the *taula*, a table consisting of a slab of stone set upright in a groove in the flat rock-floor. **1950** *Antiquity* XXIV. 154 These excavations might have solved the problems of the date and purpose of the constructions known as 'taulas'. **1979** Service & Bradbery *Megaliths* vii. 134 The largest taula (the name means 'table' in Catalan..)..is beside the talayot of Trepuco. *Ibid.*, As with all taulas, it is very broad in one dimension, very narrow in the other.

Taung (tauŋ). Also **Taungs.** The name of a town in the northern Cape Province, South Africa, used *attrib.* in *Taung child, skull*, etc., to designate the remains of a fossil hominid, *Australopithecus africanus*, found in a limestone cliff there in 1924. Cf. *Australopithecus.

1931 A. Keith *New Discoveries Antiquity Man* iii. 61 How does the brain development of the Taungs skull fit into the human scheme of growth? **1962** G. H. R. Von Koenigswald *Evol. Man* iii. 61 (*caption*) Front view of original Taungs skull. **1973** B. J. Williams *Evol. & Human Origins* ix. 136/1 The Taung child had an endocranial capacity of approximately 405 cc. *Ibid.* 136/2 Judging the Taung specimen from tooth eruption to be around five years of age, the adult individual would have had a cranial capacity of 440 cc. **1978** P. V. Tobias in C. J. Jolly *Early Hominids in Africa* 45 The *prima facie* case for the likely taxonomic affinities of the Taung skull rests on the concept that two main hominid lineages.. existed side by side in Africa.

‖ **taungya** (tau·ŋyă). Also **taunggya,** † **toungya.** [Burmese, f. *taung* hill + *ya* plot, field.] A temporary hillside clearing. Usu. *attrib.*, designating a method of shifting cultivation practised in Burma, and a system of forest management based on this (see quot. 1938) and employed in tropical countries.

1876 *Encycl. Brit.* IV. 560/2 The system of cultivation known in Bengal as the *jūm*, that is clearing virgin soil by burning, cultivating it for one or two years, and then leaving it again to the jungle, is here [*sc.* in Burma] extensively practised under the name of *toungya* cultivation. **1904** [see *kumri]. **1921** *Times Lit. Suppl.* 8 Sept. 574/3 The best system of raising teak is in taungya plantations. **1926** Tansley & Chipp *Study of Vegetation* xi. 257 Plant succession in deserted taungyas or cultivated land. **1938** H. G. Champion in Champion & Trevor *Man. Indian Silviculture* I. vi. 182 *Taungya* plantation procedure.. depends on getting the cultivator to plant or sow a new forest crop with his food crop, so that when he moves on, useful trees and not weeds will restock the area. **1946** [see *jhoom, jhum]. **1952** [see *shifting cultivation* s.v. *shifting *ppl. a.* 1 b]. **1975** *Daily Tel.* 6 Oct. 9/3 Dr. Earl commends the 'taungya' method of managing eucalyptus plantations in Uganda.

taunt, *sb.*[1] Add: **4.** *Comb.*, as **taunt-song**, used to refer to certain passages in the Old Testament, *spec.* as a rendering of Heb. *māšāl.*

1906 S. R. Driver *Habakkuk* (Cent. Bible) 78 'Parable' in these passages [*sc.* Habakkuk 2: 6] suggests a wholly incorrect idea; and the best rendering is probably *taunt-song.* **1928** C. Gore et al. *New Commentary on Holy Scripture* 444/1 In that day of relief from sorrow and fear a taunt-song will be sung against the King of Babylon. **1959** G. W. Anderson *Crit. Introd. O.T.* v. 109 The second contains a striking taunt-song over the descent of a tyrant ..to the abode of the dead.

tau·ntless, *a. nonce-wd.* [f. Taunt *sb.*[1] + -less.] Lacking or without a taunt (sense 3).

c **1879** G. M. Hopkins *Poems* (1967) 82 Tongue true, vaunt- and tauntless.

Taunton turkey (tǭ·ntən tǭ·ɹki). *U.S.* The name of *Taunton*, Massachusetts, used *attrib.* to designate the ale-wife, *Pomodorus pseudoharengus*, a fish resembling a herring found in marine or fresh water in eastern North America; = Ale-wife[2].

1851 A. Allin *Home Ballads* 18 'Taunton turkeys' are so thick, We sell them by the rod! **1950** *Chicago Tribune* 17 Jan. 14/3 In Massachusetts..the spring herring is known as 'Taunton turkey'.

taupata (tau·pātă). *N.Z.* [a. Maori.] An evergreen shrub or small tree, *Coprosma repens*, of the family Rubiaceæ, native to New Zealand, and bearing shiny leaves and clusters of small white flowers followed by orange-red berries.

1864 J. D. Hooker *Handbk. N.Z. Flora* 268/1 Taupata *Coprosma retusa.* **1906** Laing & Blackwell *Plants N.Z.* 392 C[*oprosma*] *Baueri* is much used for hedges in Wellington and Melbourne. In the former place it is generally known as *taupata.* **1946** *Jrnl. Polynesian Soc.* LV. 159 *Taupata*, a tree..with dark green leaves shining as if varnished. **1960** N. Hilliard *Maori Girl* III. vi. 215 And there's the lights on the leaves of the taupata. **1966** G. Durrell *Two in Bush* ii. 68 The list of plants used in this nest building reads like something out of Lewis Carroll: taupata twigs, scurvy grass and mesembryanthemum. **1982** F. Bream *Island of Fear* 6 The graves..under the taupata trees.

taupe (tōp, tǭp). [a. Fr., f. L. *talpa* mole.] A brownish shade of grey resembling the colour of moleskin. Also *Comb.*, as *taupe-coloured* adj.

1911 *Daily Colonist* (Victoria, B.C.) 5 Apr. 24/1 (Advt.), Important silk purchase..in colors of rose, Persian blue.. taupe, purple, mauve. **1921** *Glasgow Herald* 18 June 4 Pearl, smoke, taupe, mouse and other soft becoming shades of grey. **1955** W. Gaddis *Recognitions* II. ix. 714 The hole in the roof had, of course, been repaired; and the interior done over in taupe and white. **1967** *Boston Sunday Herald Mag.* 26 Mar. 21/1 Use taupe powder to create depth. **1976** L. St. Clair *Fortune in Death* xxi. 215 The subdued taupe of her suit. **1982** M. McMullen *Better off Dead* (1983) I. iii. 19 The velvet was taupe-coloured.

Taurean, *a.* Restrict *rare* to sense in Dict. and add: **b.** Of or pertaining to the constellation or zodiacal sign of Taurus. Cf. *Taurian *a.* b.

1924 C. E. O. Carter *Conc. Encycl. Psychol. Astrol.* 105 Obedience is generally regarded as a Taurean virtue. **1928** W. H. Sampson *Zodiac* iii. 25 The Pleiades are part of the constellation Taurus..more Taurean in nature. **1974** *Woman* 4 May 56/1 A Taurean mother will help her Piscean children to develop their own talents.

B. as *sb.* = *Taurus 1 c.

1911 I. M. Pagan *From Pioneer to Poet* ii. 31 The burlesque Taurean is fat, thick-necked, gross and overfed looking, and often has a great love of low comedy. **1916** K. T. Craig *Stars of Destiny* 37 Mental exactness and persistence are attributed to Taureans. **1969** 'V. Packer' *Don't rely on Gemini* (1970) xiv. 118 Brahms was a Taurus. .. Fred Astaire, Bing Crosby, and Perry Como are all Taureans, too. **1976** *Sunday Mail* (Glasgow) 26 Dec. 27/6, I always believed Taureans were home-lovers.

taureau (tǫ·ro). *Canad. Hist.* Also † **toreau.** Pl. **taureaux.** [a. Canad. Fr., a. Fr. *taureau* bull.] A bag of buffalo-hide for carrying pemmican; also *transf.*, the pemmican itself.

1794 J. MacDonnell *Jrnl.* 14 Jan. in L. E. R. Masson *Les Bourgeois de la Compagnie du Nord-Ouest* (1889) I. 287, I cut 20 sacks or *taureaux* to put pemican in and gave them to Minie to sew. **1795** —— *Jrnl.* 28 Apr. in *Ibid.* 294 Started them for the Forks with 138+137 taureaux of pemican. **1807** W. F. Wentzel *Let.* 27 Mar. in *Ibid.* 90 This is our staple article of provisions when travelling, it is called *taureau* or *Pimecan.* **1821** G. Simpson *Jrnl. Occurrences Athabasca Dept.* (1938) 278 It may be well to have the Toreau in Store. **1890** *Trans. R. Soc. Canada* VIII. II. 104 A sack or 'toreau' of pemmican, as it was called, consisted of nearly equal quantities of tallow and dried meat. **1911** K. Hughes *Father Lacombe* iii. 32 They pounded dried meat to powder in wooden bowls, mixing hot grease and dried berries with it, packing the whole into large sacks of buffalo-hide, called by the Metis—*taureaux*... This was *pimik-kan*, the manna of the Canadian prairies. **1927** A. P. Woollacott *Mackenzie & his Voyageurs* 51 It [*sc.* pemmican] was a staple food among the fur-traders on long journeys, when..game could not be had. Also known as 'taureaux'. **1931** G. L. Nute *Voyageur* 213 The train wound its slow way back to Pembina laden with..228 *taureaux*, or leather bags of pemmican. **1951** W. O'Meara *Grand Portage* xxiv. 139 The pemmican [had been] mixed and stored in shaggy taureaux.

Taurian, *a.* Delete *rare*[−1] and add: **b.** Of or pertaining to the constellation Taurus; characteristic of a person born under the zodiacal sign of Taurus. Cf. *Taurean *a.* b. Occas. as *sb.*, = Taurus 1 c.

1909 Webster, *Taurian*..Of or pert. to Mount Taurus, or the constellation Taurus. **1928** W. Wilson *Astrology* iii. 53 The Taurian life has been spoken of as happier at the end than at the beginning. **1938** D. Anrias *Man & Zodiac* vii. 69 Afflicted Taurians rarely trust men or circumstances. **1972** D. Lees *Zodiac* 15 Zodiac..predicted that Julius would make a tremendous coup on the stock market, provided he got off his lazy Taurian arse and acted immediately. **1972** *Vogue* 1 Mar. 144/4 He's a Gemini... I'd be much better off with a Taurian.

taurine, *a.* Add: Also *spec.*, pertaining to bull-fighting.

1932 R. Campbell *Taurine Provence* 28 Every village westward of Martigues has three or four fêtes a year all accompanied by taurine ceremonies. Nearly every village has its small arena. **1977** *Monitor* (McAllen, Texas) 3 July 6B/3 Longinos Mendoza is also slated to appear on the card having shown exceptional taurine skill during his last performance [*sc.* a bullfight].

tauro-. Add: **tau·robole** [cf. Tauroboly], a bull-slayer; also **taurobo·lic** *a.*, of the nature of tauroboly. (Both *rare.*)

1934 R. Campbell *Broken Record* viii. 183 These two great horsemen are superior equestrian tauroboles to any of the Portuguese, Spanish or Mexican. *Ibid.* iii. 67 A truly taurobolic and Mithraic sensation.

tauroboly. (Earlier example of form *taurobolium.*)

1845 *Encycl. Metrop.* XVI. 114/1 They offered a sacrifice of a bull or ram, (whence the terms *Taurobolium* and *Ariobolium*,) in the blood of which the hierophant was also sprinkled.

taurodont (tǭ·rodǫnt), *a.* [f. Tauro- + Gr. ὀδούς, ὀδοντ- tooth.] Of mammalian molar teeth: having large broad crowns and short roots.

1915 A. Keith *Antiquity of Man* viii. 148 Molar teeth [in Neanderthal man] are large in crown and body and extremely short in root... To [*sic*] this peculiar form of molar tooth..I have proposed the name of 'taurodont'. **1927** Peake & Fleure *Hunters & Artists* 18 This condition of the teeth, known as taurodont, has been found among some representatives of Neanderthal man. **1948** *New Biol.* V. 84 The teeth [of Neanderthal man]..are often of that specialised kind called 'taurodont'—they are stumpy with short roots. **1971** *Nature* 5 Feb. 409/2 The frequent lack of incisors and well worn flattened taurodont molars superficially suggest the crateriform decay of Moon's mulberry molars. **1973** B. J. Williams *Evolution & Human Origins* x. 159/2 The molars and premolars are 'taurodont', meaning that they have an enlarged pulp cavity.

taurodontism (tǭrodǫ·ntiz'm). [f. as prec. + -ism 3.] In certain mammals, the condition of having taurodont teeth.

1913 A. Keith in *Proc. R. Soc. Med.* (Odontol. Sect.) VI. 103 For this condition or tendency Professor Keith proposed the name of 'taurodontism'. **1939** *Nature* 23 Dec. 1055/2 In some peculiarities of the teeth he [*sc.* Sinanthropus] approaches the gorilla..and..the female orang in 'taurodontism'. **1959** J. D. Evans *Malta* i. 36 They are molars, but the roots, instead of being separate, as in normal human teeth, are fused together. This is a phenomenon known technically as *taurodontism*, and it is particularly characteristic of the primitive type of human known as Neanderthal man. **1982** *Times* 27 Nov. 4/1 The molar exhibits marked taurodontism, an enlargement of the pulp cavity extending down into the roots, which..is characteristic of early Neanderthals.

tauromachy. Add: Also sometimes in foreign forms (Gr. and Fr. in quots.). Hence **tauroma·chics** [-ic 2], the business of bullfighting.

[**1910** *Encycl. Brit.* IV. 789/1 *Bull-fighting*, the national Spanish sport. The Spanish name is *tauromaquia.*] **1923** W. J. Locke *Moordius & Co.* xi. 147 After public renunciation of *tauromachie*, gracefully made among a circle of Spanish friends. **1934** F. M. Ford *Let.* 14 Sept. (1965) 234, I don't know why I should deluge you with..tauromachics. **1967** McCormick & Mascareñas *Compl. Aficionado* ii. 35 One has only..to read certain of the early manuals of tauromachia..to realize that..a good deal has been gained through modern changes. **1969** C. Irving *Fake!* (1970) viii. 99 He was to do over two hundred of them [*sc.* drawings], the best of which are probably in the Picasso *Tauromachia* series.

Taurus. Add: **1. c.** A person born under the zodiacal sign of Taurus. Also *attrib.* or as *adj.*

1901 C. A. Walker *Under a Lucky Star* 84 The headworkers of humanity could accomplish but little without assistance from the practical, executive Taurus. **1927** G. Sully *First Princ. Astrol.* iii. 24 Taurus and Gemini make good mates when they set out to help one another unselfishly. **1943** D. Powell *Time to be Born* xi. 272 Her astrologer..failed her by promising a new man, a Taurus with a heart condition. **1964** L. MacNeice *Astrol.* v. 147 A 19th-century astrologer's idea of a Taurus woman. **1971** V. Canning *Firecrest* iii. 35 Henry Martin Dilling, born 1927, the same age as himself; though Dilling was a Leo and he Taurus. **1979** S. Rifkin *McQuaid in August* (1980) ix. 85 I'm Sagittarius. If you're Taurus..we can get a big thing going.

Tau Sug (tau sug), *sb.* (and *a.*) Also **Tao Sug, Tausug, Taw Sug.** [Tau Sug, f. *taw* person + *sug, sulúg* current.] One of the Islamic groups inhabiting the Sulu Archipelago in the Philippine Islands, whose ancestors can be traced back to the Butuan area of north-east Mindanao; the Austronesian language spoken by this people. Also *attrib.* or as *adj.* Cf. *Sulu[1].

1923 [see *Sulu[1]]. **1964** P. G. Gowing *Mosque & Moro* I. 1 Ninety-two percent of all Muslim Filipinos belong to the Tau Sug, Maranao,..and Samal groups. *Ibid.* 2 A vigorous and proud people, the Tau Sug are the backbone of the historic Sultanate of Sulu. **1964** E. A. Nida *Toward Sci. Transl.* ix. 207 In the expression 'sat and begged' as rendered into Tau Sug, a language of the Philippines, one must specify one of these actions as primary and the other as secondary. **1973** J. A. Bruno (*title*) The social world of the Tausug. **1977** [see *Sulu[1]].

taut, taught, *a.* Add: The only current spelling is *taut.* **2. b.** (Earlier and later examples.) Esp. in phr. *taut ship*, a disciplined or strictly run ship. Also *attrib.* and *fig.* Cf. sense 2 c in Dict.

1829 D. Jerrold *Black-Ey'd Susan* III. ii. 43 The trimmest sailor as ever handled rope..give me taut Bill before any able seaman in his Majesty's fleet. **1941** *Time* 29 Dec. 8/1 The promoted admirals were 'taut ship' commanders (meaning rigid disciplinarians, as opposed to 'happy ship' officers). **1970** H. Waugh *Finish me Off* (1971) 106 Yesterday she had been haughty and taut-ship, but today..Mrs. Hardell's position had suddenly become tenuous. **1974** *Progress* (Easley, S. Carolina) 24 Apr. 2/2 His language was salty and he ran a taut ship. He demanded discipline, accuracy, integrity and honesty, as well as good writing. **1977** *Navy News* July 20 All the taut-ship zeal for a tip-top navy, which gave Whale Island its fame as one of the best-known of service establishments, has been redirected. **1980** *Globe & Laurel* July/Aug. 250/1, I found the first half of the book, which describes Trevelyan's efforts to create a taut ship, as he takes Icarus through a NATO exercise, a ship's fire, and a funeral at sea, sketchy and unsatisfactory.

d. *fig.* Of music, literature, etc.: concise, controlled; of the human voice: strained.

1966 *Listener* 10 Feb. 219/2 The music..did not seem to have quite the structural control or clarity of texture one discerned in the piano concerto, where..the thought seems more taut and the tension is therefore more easily maintained. **1972** *Observer* 16 Apr. 33/6 A short, taut, yet circumstantially detailed account. **1976** M. Machlin *Pipeline* lv. 557 Coutts voice was taut with worry. **1978** *Internat. Herald Tribune* 24 July 14/4 Among reviewers, the favored adjectives of the past—trenchant and ironic for books, taut, pert and luminous for theatrical productions—have been overtaken by sentimental.

3. Used adverbially and parasynthetically in *Comb.*, as *taut-necked, -rigged, -stretched* adjs.

1829 D. Jerrold *Black-Ey'd Susan* I. i. 15 There's not so fine, so noble, so taut-rigged a fellow in His Majesty's navy. **1943** D. Gascoyne *Poems 1937–42* 42 The sky's a faded blue and taut-stretched flag Tenting the quadrangle. **1948** L. MacNeice *Holes in Sky* 31 The taut-necked donkey's..lamenting.

tautened, *ppl. a.* Add: Also *fig.*

1929 E. Carfrae *Guarded Heights* xxiii. 202 Carey's voice had a queer little tautened note when he spoke again.

tauto-. Add: **tautophone** = *summator 2; **tau:tosylla·bic** *a.*, belonging to the same syllable.

1940 *Character & Personality* VIII. 216 (*title*) The use of the tautophone ('verbal summator') as an auditory apperceptive test for the study of personality. **1888** J. Wright tr. *Brugmann's Elem. Compar. Gram. Indo-Gmc. Lang.* I. 92 Before the Christian era tautosyllabic *ai* became *ē* in Latin. **1953** *Archivum Linguisticum* V. 22 The lengthening of short vowels before tautosyllabic *s* or [z]. **1978** *Language* LIV. 193 A tautosyllabic consonant.

tautological, *a.* Add: **1. b.** *Mod. Logic.* Characterized by or involving tautology (in sense *f). Hence **tautologica·lity**, the quality of being tautological.

1922 tr. *Wittgenstein's Tractatus* 97 In the one case the proposition is true for all the truth-possibilities of the elementary propositions. We say that the truth-conditions are *tautological.* **1926** F. P. Ramsey in *Proc. London Math. Soc.* XXV. 341 The idea to be defined is one of the essential sides of mathematical propositions, their content, and their form. Their content must be completely generalized, and their form tautological. **1933** *Mind* XLII. 41 Each postulate functions in limiting the ranges of the variables in such a manner that any change in one postulate..involves a reciprocal change in its other parts, which change causes it to remain analytic or tautological. **1936** J. R. Weinberg *Exam. Logical Positivism* ii. 80 The formal property of certain combinations of symbols, which is called tautologicality, is solely responsible for the unconditional truth of the truths of logic. **1950** R. Carnap *Logical Found. Probability* iv. 289 With respect to the tautological evidence '*t*'. **1971** G. Hunter *Metalogic* III. 171 Suppose that A is an instance of a tautological schema of Q.

tautologically, *adv.* (Later example.)

1979 J. A. Robinson *Logic: Form & Function* iii. 43 We can extend this idea to the case when infinitely many sentences together tautologically imply a given sentence.

tautologous, *a.* (Examples in sense of *tautological *a.* 1 b.)

1935 *Mind* XLIV. 195 The tautologous '*p* or not *p*'. **1940** W. V. O. Quine *Mathematical Logic* i. 50 Statements which are true by virtue solely of the truth-functional modes of composition will be called *tautologous.*

tautology. Add: **f.** *Mod. Logic.* A compound proposition which is unconditionally true for all the truth-possibilities of its elementary propositions and by virtue of its logical form.

1919 B. Russell *Introd. Math. Philos.* xviii. 203 The characteristic of logical propositions that we are in search of is the one which was felt..by those who said that it consisted in deducibility from the law of contradiction. This characteristic we may call *tautology.* *Ibid.* 205 The importance of 'tautology' for a definition of mathematics was pointed out to me by..Ludwig Wittgenstein, who was working on the problem. **1922** tr. *Wittgenstein's Tractatus* 97 The tautology..is unconditionally true. **1933** *Mind* XLII. 37 So taken, a postulate is a tautology and cannot be denied. **1959** *Listener* 19 Mar. 510/1 The simplest rigorous proof is tautology. This consists, essentially, of showing that some statement covers all possibilities. **1964** M. Black *Compan. Wittgenstein's Tractatus* xliii. 231 Johnson's..'formal truth' and 'formal falsity'.. seem to correspond exactly to W.'s 'tautology' and 'contradiction'. **1979** J. A. Robinson *Logic: Form & Function* iii. 42 A..general decision procedure for determining whether or not a sentence is a tautology.

tautomerism. Add: *esp.* such a property due to the reversible migration of an atom (esp. of hydrogen) or group within a molecule (see also quots.).

1927 T. M. Lowry in *Chem. Rev.* IV. 233 The necessity for a new definition of tautomerism arises from the fact that Laar embodied in his original definition a theory which is now universally recognized as being incorrect,.. namely, that the various formulae which can be assigned to a tautomerism compound represent 'not isomeric but identical substances'. The new definition [*sc.* that in Dict.] has the advantage that there is no theory behind it, since it is limited to a mere statement of the fact of dual reactivity. **1927, 1936** [see *keto-enol* s.v. *keto- b]. **1937** H. B. Watson *Mod. Theories Org. Chem.* ix. 117 Until a relatively recent date..'tautomerism' was used exclusively to denote the migration of hydrogen. The similar migration of anionic atoms or groups is now recognized, however; this..is included under..'tautomerism'. **1964** [see *prototropy]. **1969** C. K. Ingold *Structure & Mechanism Org. Chem.* (ed. 2) xi. 795 Laar's interpretation was that two such structures did not represent distinct and potentially separable species, but only the end-phases of an intramolecular oscillatory situation in a single chemical species. *Ibid.* 799 Since 1911 there has been no question but that the concept of tautomerism, in terms of which Conrad Laar had incorporated so many scattered observations into a phenomenon, has to be redefined..as meaning reversible isomeric change. Problems of isolation and proof of identity of tautomers are dependent simply on temperature and the available techniques.

Hence **tauto·merize** *v. intr.*, to change into another tautomeric form; **tauto:meri·zable** *a.*, capable of being changed into a tautomeric form; **tauto:meriza·tion.**

1934 Webster, Tautomerize.... Tautomerizable... Tautomerization. **1938** H. Adkins in H. Gilman *Org. Chem.* I. ix. 820 Unsaturated acids tautomerize in the absence of added reagents at temperatures near their boiling points. *Ibid.*, There is little or no correlation between the rate of tautomerization (mobility) and the extent of the reaction. **1962** Tautomerizable [see *Raman]. **1972** R. A. Jackson *Mechanism* iv. 60 This could plausibly lose carbon dioxide to give the carbanion 82 which would rapidly tautomerize to pyridine 83 by a proton shift. **1978** *Jrnl. Amer. Chem. Soc.* C. 4627/2 Activation of the catalyst by tautomerization to a rhodium (I) complex.

tava(h), var. *tawa[2].

tavarish, var. *tovarish, tovarich.

Tavastian (tavæ·stian), *sb.* (and *a.*) [f. *Tavast(ehus*, Sw. name of the Finnish town of Hämeenlinna in the province of Häme + -ian.] A member of one of the major ethnic groups of the Finnish people. Also *attrib.* or as *adj.* Also **Ta·vastlander.**

[**1891** A. Featherman *Social Hist. Races Mankind* IV. 417 The Finns call themselves in their own language Suomalainen (swamp men)... From the dialects which they speak they are divided into two tribes: the Tawaster and the Karelian... The Tawasters..inhabit the central part of Finland.] **1898** J. Abercromby *Pre- & Proto-Historic Finns* I. i. 3 The Finns of Finland (*Suomi*) call themselves *Suomalaiset*, and are broadly divided into two branches, the Tavastlanders (Hämäläiset) and the Karelians (*Karjalaiset*). **1911** Webster, *Tavast*,..*Tavastian*.. *n.*, a Finn of a type characterized by broad, thickset figure and blond complexion. **1934** *Ibid.*, Tavastian, *adj.* **1935** Huxley & Haddon *We Europeans* vi. 181 These characters are seen among Finns, White Russians and.. the Tavastians of Finland. **1957** *Encycl. Brit.* IX. 257/2 The *Finns* proper... There are two principal subdivisions, the Tavastlanders or Hämäläiset in the south and west and the Karelians or Karjalaiset in the east and north. The Tavastlander has a round head, a broad face, concave nose, fair complexion. **1965** E. Jutikkala in Glass & Eversley *Pop. in Hist.* xxiii. 569 The growth was by no means proportional to that of the 11 Tavastian parishes mentioned. **1966** L. Deighton *Billion-Dollar Brain* ix. 79, I am a typical Tavastian... We are Tavastian people from the south and centre of Finland. **1974** *Encycl. Brit. Micropædia* IV. 145/3 The Finns proper, the Tavastlanders (or Tavastians), and the Karelians..had their own chiefs.

Tavel (ta·vel). The name of a commune on the Rhône (department of Gard, France), used *attrib.* and *absol.* to designate a rosé wine produced there.

1875 H. Vizetelly *Wines of World* I. iii. 28 Of the rose-colour wines of the Côtes du Rhône, such as the dry and insidious Tavel,..and the robust Roquemaure,..the majority made default. **1926** P. M. Shand *Bk. Wine* v. 178 On the opposite bank of the Rhône..is the little village of Tavel, which produces the delicious Tavel Rosé (Ancien Vignoble). **1934** J. I. Davis *Beginner's Guide to Wines* 49 Tavel..is a most delicious table wine. **1951** R. Postgate *Plain Man's Guide to Wine* iv. 90 Tavel is both dry and fruity... All rosés are a pretty colour; Tavel is perhaps the prettiest. **1963** N. Freeling *Gun before Butter* III. 153 They drank a Tavel wine from near Avignon. **1974** *Guardian* 24 Jan. 13/5 The best rosé in the world is probably Tavel. **1980** 'M. Harris' *Treasure of Sainte Foy* xviii. 229 He goes on through the small town of Tavel—this is where Tavel rosé comes from.

tavern, *sb.* Add: **1.** (Earlier example referring to *New Inn Hall.*)

1854 'C. Bede' *Verdant Green* II. xi. 103 He seemed to feel that the Dons of his college would look shy upon him, and he expressed his opinion that it would be better for him to migrate to the Tavern... A name given to New Inn Hall, not only from its title..but also because the buttery is open all day, and members of the Hall can call for what they please at any hour.

4. a. *tavern-music* (later example), *-restaurant, -song.*

1979 *Listener* 30 Aug. 275/2 The best tavern music in South London is..jazz. 1880 W. D. Hay *Doom of Gt. City* 46 One place I knew slightly, a tavern-restaurant, where I had occasionally dined. 1973 *Washington Post* 13 Jan. A3/6 Mr. and Mrs. Harry Grenwalt..were celebrating their 42nd wedding anniversary at the tavern-restaurant. 1823 Byron *Don Juan* VIII. lxiii. 142 Without which Glory's but a tavern song. 1917 J. Masefield *Lollingdon Downs* 90 Within the tavern-song, hid in the wine. 1959 I. & P. Opie *Lore & Lang. Schoolch.* xvi. 346 This jingle may be compared with the tavern song which was printed in *Vinculum Societatis*.

taverna (tăvə̆·rnă). [a. mod.Gr. ταβέρνα tavern.] A Greek eating-house.

1914 L. M. J. Garnett *Greece of Hellenes* xii. 151 The typical 'public-house' of Greece is a small tavern... Adjoining many of these humble wayside *tavérnas* are gardens roofed with spreading vines. 1952 [see *bouzouki]. 1963 *Harper's Bazaar* Jan. 59/2 Food in *tavernas* is..not always very good. 1972 *Daily Tel.* (Colour Suppl.) 24 Nov. 20/4 Multitudes of rather second-rate tavernas and pizza-houses [in Australia]. 1977 B. Pym *Quartet in Autumn* xviii. 160 Not Greece of course..—one could hardly imagine Marcia in a taverna, eating octopus. 1978 *Chicago* June 34/2 Penny Evans..is the best taverna singer we've ever heard. 1980 R. Tine *State of Grace* (1982) xix. 169 They won't respect you. In every corner *taverna* [in Sicily] they'll be laughing at you.

Tavgi (tæ·vgi), *sb.* (and *a.*) Also **Tavghi, Tavghy, Tavgy.** [a. Russ.] **a.** (A member of) a Finno-Ugric people (now called Nganasan) living between the Yenisey and Khatanga rivers in north-west Siberia. **b.** The language of this people. Also *attrib.* or as *adj.*, esp. in *Tavgi-Samoyed*.

1886 *Encycl. Brit.* XXI. 251/2 The Tavghi Samoyedes may number about 1000. 1888 *Ibid.* XXIV. 1/1 *Ural-Altaic languages..Tavghi*, between lower Yenisei and Khatanga rivers. 1934 Webster, *Tavgi*, var. of *Tavghi*. 1951 W. K. Matthews *Lang. U.S.S.R.* iii. 17 Tavgi (Nganasan) in Taimyr. 1954, 1967 [see *Nenets]. 1975 G. F. Cushing tr. *Hajdu's Finno-Ugrian Lang. & Peoples* iii. 216 The Nganasans are called Tavgi-Samoyeds. 1977 C. F. & F. M. Voegelin *Classification & Index World's Lang.* 343 Yenisei Samoyed..appears to be transitional between Yurak and Tavgy Samoyed.

taw, *sb.*[2] Add: **c.** (Further examples in *fig.* phrases.)

1840 *Spirit of Times* 7 Mar. 6 We have understood that Boston..will be *en route* for the stable..at Columbia, South Carolina—that is, if Wagner 'comes to taw'. 1868 in *Amer. Speech* (1965) XL. 132 He smiles at all the girls he meets, And you smile at him on the crowded streets, Why don't you make him 'come to taw', I know he wants a mother-in-law. 1904 W. N. Harben *Georgians* xxxii. 292 His wife's a bully woman; she fetched 'im to taw. 1934 D. Runyon in *Collier's* 3 Mar. 41/1 Georges takes a wonderful liking to Princess O'Hara right from taw. 1935 H. Davis *Honey in Horn* ix. 113 The only way Mrs. Yarbro could tell anything was to start from taw. 1956 *Coast to Coast* 183 Starting off from taws with a big load to carry. 1969 *Sunday Truth* (Brisbane) 5 Oct. 14/4 Without a share of overseas star shows, Seven has been battling from taws, but..is..getting stuck into the other networks with a 'super-specials' policy change.

tawa[1]. (Earlier example.)

1839 [see *billy[2]].

tawa[2] (tawā·). Also **tava(h).** [a. Hindi, Punjabi *tavā* frying-pan, griddle.] A circular griddle used in the Indian subcontinent for cooking chupattis and other food.

c 1843 H. Lawrence *Jrnls.* (1980) viii. 134 The *tavah* a convex iron plate on which are [*sic*] baked the thin unleavened bread of the people. 1963 *Guardian* 1 May 6/4 Iron 'tawa' (a baking utensil like the Scots girdle for oat-cakes). 1969 *Eve's Weekly* (Bombay) 20 Dec. 65/3 Make a gash in the flat side of each chestnut and dry roast in the oil in a large tava for a few minutes. 1971 *Femina* (Bombay) 16 Apr. 55/1 Roll out into *chappatis* and bake on a *tawa* on a very low fire till it cooks through.

Tawarek, var. *Tuareg *sb.* and *a.*

tawn, *sb.* For † read *rare* and add later examples. Occas. as *adj.*, tan or tawny-coloured.

1851 H. Melville *Moby Dick* I. v. 48 This young fellow's..cheek is like a sun-toasted pear in hue... That man..looks a few shades lighter... In the complexion of a third still lingers a tropic tawn. 1920 E. Pound *H. S. Mauberley* 27 Tawn fore-shores. 1939 Joyce *Finnegans Wake* 540 Redu Negru may be black in tawn.

tawny, *a.* and *sb.* Add: **B.** as *sb.* **6.** = *tawny port*, sense C. c below.

1929 J. B. Priestley *Good Companions* II. i. 278 'Waiter, I want a bottle of port.'..'Well, we've the Tawny at three-and-nine the bottle.' 1959 W. James *Word-Bk. of Wine* 148 The commonest style of port is tawny, a blend of several vintages between four and ten years old. *Ibid.*, White port is simply port made from white grapes, and is sometimes blended with young red port to make the cheaper sorts of tawny. 1976 *Times* 6 Nov. 13/2 Prices are around £3 for a fine old tawny, about £4 to £5 for a vintage.

C. a. *tawny-eyed, -necked, -stained, -throated.*

1853 M. Arnold *Philomela* in *Poems* (new ed.) 64 Hark! ah, the nightingale! The tawny-throated! 1930 E. Blunden *Poems* 42 And tawny-stained with ruin [the brook] trolls across The tiny village battered into dross. 1940 C. Day Lewis tr. *Virgil's Georgics* IV. 90 A laminated dragon or lioness tawny-necked. 1952 R. Campbell tr. *Baudelaire's Poems* 87 Like angels fierce and tawny-eyed, Back to your chamber I will glide.

c. tawny eagle, *Aquila rapax*, found in Africa and western Asia; **tawny frogmouth** = *podargus; **tawny owl**: see *owl *sb.* 2 b; **tawny port,** a port wine made from a blend of several vintages matured in wood (see quot. 1951).

1859 *Ibis* I. 88 The claim of the Tawny Eagle..to be considered European rests at present solely upon a trophy of the Russian war. 1912 J. Stevenson-Hamilton *Animal Life Afr.* xvii. 286 The tawny eagle is of mottled brown colour above, and tawny chestnut beneath, with yellow legs. 1979 G. & D. Lloyd *Birds of Prey* 82 The Tawny Eagle..of Africa and Asia is 26 to 31 inches in size and is the world's commonest eagle. 1901 A. J. Campbell *Nests & Eggs Austral. Birds* II. 539 (*heading*) Tawny frogmouth. 1933 [see *podargus]. 1968 Breeden & Slater *Birds Austral.* 69 (*caption*) The yawn of this Tawny Frogmouth reveals its enormous gape. 1848 Thackeray *Van. Fair* xliii. 382 The particular tawny port was produced when he dined with Mr. Osborne. 1951 R. Postgate *Plain Man's Guide to Wine* viii. 116 Tawny port is port of various years, blended and matured in cask... Tawny port..soon loses the rich purple colour of vintage port, and is ready to drink much sooner. 1979 *Country Life* 4 Jan. 40/1 (Advt.), Old Tawny Port wines by *Quinta do Noval*..Portugal.

tax, *sb.*[1] Add: **7. a.** *tax bill, bracket, consultant, -defaulter, dodge* (also as *v. intr.*), *fiddle* (colloq.), *return* (earlier example), *year*; *tax-collector* (earlier example), *-fiddler* (colloq.), *inspector*; *tax-free* adj. (later examples). **b. tax allowance,** a sum that is to be deducted from gross income in the calculation of taxable income; **tax avoidance,** the arrangement of financial affairs so as to reduce tax liability within the law; so **tax-avoider, -avoiding** *ppl. a.*; **tax bite** *U.S. colloq.*, a deduction in the form of tax; **tax break** *colloq.* (orig. *U.S.*), a tax advantage or concession allowed by government; **tax code,** a code number representing the tax-free part of an employee's income, assigned by tax authorities for use by employers in calculating the amount of tax to deduct under the PAYE system; **tax credit,** a sum that can be offset against a tax liability; *spec.* one that results in a payment to any person whose liability is less than this sum; **tax-deductible** *a.*, allowable as a tax deduction; so **tax-deductibility; tax deduction** chiefly *U.S.*, an expense that can be deducted from gross income in calculating taxable income; **tax disc,** a circular label displayed in the window of a motor vehicle showing the date up to which motor vehicle excise duty has been paid; **tax dollar** *U.S.*, a dollar paid as tax; **tax-eater, -eating** *a.* (later U.S. examples); **tax evasion** orig. *U.S.*, the reduction of tax payments by misstatement of income or other illegal means; so **tax-evader, -evading** *vbl. sb.*; **tax-exempt** *a.*, free from a liability to be taxed; *sb.*, a tax-exempt security; so **tax exemption; tax exile,** one who lives in a country chosen for its lower taxes on personal income; the state of doing this; **tax haven,** a country that attracts companies or individuals by its low taxes; **tax holiday** *colloq.*, a period of tax exemption or tax reduction, esp. one of fixed duration; **tax-loss,** a loss that can be offset against taxable profit earned elsewhere or in a different period; also *transf.* and *attrib.*; **taxman**: also, an inspector of taxes or similar official; (with *the*) the Board of Inland Revenue, personified; (earlier and later examples); **tax point,** the date upon which value added tax becomes chargeable in any particular transaction; **tax relief** = *relief[2] 7; **tax shelter,** an opportunity for incurring expenses so that they can be used to reduce tax liability; so **tax-sheltered** *a.*, providing such an opportunity; **tax threshold,** the level of income at which tax begins to be payable.

[1935 *Times* 16 Apr. 9/3 The cost of these various amendments in income-tax allowances will amount to £10,000,000.] 1950 *Economist* 22 Apr. 903/2 Statisticians have allowed an increase..to reflect the increase in initial tax allowances on plant purchased from April. *a* 1974 R. Crossman *Diaries* (1976) II. 174 The Chancellor's only votes were gained from those who shared his male views and in particular objected to taking away money from middle-class families by tampering with their children's tax allowances. 1927 *Hansard Commons* 4 July 961, I think that all these devices for tax avoidance ought to be stopped. 1951 L. H. Seltzer *Nature & Tax Treatment Capital Gains & Losses* ii. 43 Wide openings for tax avoidance through so-called reorganization provisions were soon discovered. 1972 *Accountant* 28 Sept. 401/2 This amendment was designed to counteract certain tax avoidance schemes. 1960 *Guardian* 9 July 10/2 Every word of this was fascinating to all tax-payers and tax-avoiders. 1980 *Listener* 1 May 578/3 The tax-avoiding English who have arrived [on the Isle of Man] since the war. 1720 in *Mass. House of Representatives Jrnl.* (1921) II. 284 A Petition..Complaning of the Proceedings of the Court..in their Nulling Three Tax-Bills by them made..[was] Sent up. 1850 R. W. Emerson *Let.* 6 Mar. in R. B. Perry *Thought & Char. W. James* (1935) I. 68 If a good bookseller thinks that such readings in New York will pay my taxbills and bad gardening in Concord, I shall try the experiment. 1978 W. White *W. Whitman's Daybks. & Notebks.* I. p. xii, Tax bills, water bills, subscriptions to daily papers. 1954 *Sun* (Baltimore) 26 Jan. (B ed.) 1/4 The Iowa senator called for..legislation by Congress to put a tax bite on foreign coffee traders operating in this country. 1976 *National Observer* (U.S.) 22 May 2/4 The upshot of the committee's action is a proposal to enlarge the tax bite for some wealthy individuals and contract it for others. 1975 R. Stout *Family Affair* (1976) xviii. 189, I am already in an uncomfortably high tax bracket for the year and would take no jobs anyway. 1968 *Nation* 4 Nov. 463/1 What better way to entice private enterprise than with a tax credit or some other sort of tax break? 1969 *N.Y. Times* 4 Sept. 6/1 Companies will not get the tax breaks they formerly got on amortization of new equipment. 1982 *Economist* 18 Dec. 17/2 Governments should cease to shower capital with tax breaks that artificially lower capital's price. 1961 M. Kelly *Spoilt Kill* III. 163 Writing paper, annual notice of tax code, medical card. 1976 *Star* (Sheffield) 3 Dec. 6/7 We have reached a ludicrous state of taxation when a man on state aid receives enough to exceed his tax allowance but this does not prove that the social security payments are too high, but rather, that wage rates, and especially tax codes are far too low. 1833 J. S. Mill in *Monthly Repos.* VII. 581 These taxes..throw electioneering influence into the hands of the tax-collectors. 1976 J. R. L. Anderson *Redundancy Pay* i. 10 He had..developed a shrewd ability as a tax consultant, particularly in the property market. 1946 H. M. Groves *Postwar Taxation & Econ. Progress* vii. 227 This could be done readily by permitting the taxpayer [with a fluctuating income] to sum his taxes over a period of years, calculate what his tax bill would have been if his income had been distributed evenly among these years, determine the difference between the two, and claim the difference as a refund or tax credit. 1973 *Guardian* 24 Jan. 14 The Green Paper proposes first that most tax allowances..should be replaced by tax credits... Anyone whose tax liability was less than their tax credits would be paid the difference. 1974 *Nature* 10 May 103/3 A company can obtain 'foreign tax credits' (which can be offset against United States tax) in respect of taxes paid to foreign governments. 1980 *Daily Tel.* 23 Feb. 19/2 A final of 8p a share payable on April 3, makes 14.25p net against 10.15p net or 20.36p including the related tax credit compared with 15.15p. 1972 *Accountant* 5 Oct. 422/1 Many captives are established to take advantage of this tax deductibility of insurance reserves. 1954, 1965 Tax-deductible [see *deductible *a.*]. 1977 D. Anthony *Stud Game* vii. 45 Most of Grant's calls were on business, tax-deductible items. 1942 F. W. Marshall *Legitimate Deductions* vi. 40 In enacting provision for income tax deductions, Congress is only interested in determining what part of a company's [or person's] gross income it believes should be treated as net income for the purpose of income taxation. 1971 'O. Bleeck' *Thief who painted Sunlight* (1972) xiv. 122 He can contribute fifty percent of his income each year and claim it as a tax deduction. 1979 *Guardian* 5 July 3/4 The tax deduction for having a company car is a tiny fraction of its real value. 1951 Auden *Nones* (1952) 28 Agents of the Fisc pursue Absconding tax-defaulters through The sewers of provincial towns. 1972 *Times* 3 Oct. 2/8 (*heading*) 'Scrap tax discs' call. 1962 J. Braine *Life at Top* xxiii. 254 The usual tax dodge... It makes me sick to the stomach. 1972 *Listener* 21 Dec. 865/1 He hasn't killed himself yet... He's waiting till 5 April... Some sort of tax dodge. 1976 *Morecambe Guardian* 7 Dec. 17/2 With that film is '*The Swiss Conspiracy*' which is all about people who tax dodge, and blackmail, are blackmailed, and murdered, not necessarily in that order. 1976 *Billings* (Montana) *Gaz.* 17 June 2-E/1 The fate of a plan to use tax dollars to improve off-street parking in downtown Bozeman will be decided July 7. 1965 Mrs. L. B. Johnson *White House Diary* 12 Aug. (1970) 310 We hope for fewer dropouts thirteen years from now, for children able to grow up with a prospect of being responsible citizens, taxpayers, not tax-eaters. 1936 *Sun* (Baltimore) 30 Jan. 1/1 The time has come for a direct attack on the attempt at Washington to substitute a tax-eating bureaucracy for a liberal democratic system. 1927 *Hansard Commons* 4 July 955 What is to be done with the tax evader meanwhile? 1960 Tax-evader [see *snobocracy]. 1971 'G. Black' *Time for Pirates* iv. 78 If this deal went all right..the Hydes could be on their tax-evading boat in a couple of years. 1922 *Hansard Commons* 27 June 1920 That type of company must be perfectly well identifiable when it is seen. You notice the stigmata of tax evasion about it when you see it, not in the mere registration, but in the conduct and carrying on of its business. 1977 Warren & Ponse in Douglas & Johnson *Existential Sociol.* x. 277 It is stigmatized..in the courts of law..and (unlike tax evasion) it is stigmatized morally in the courts of public opinion. 1925 *Contemp. Rev.* June 703 He has a deep resentment against their taking refuge in tax-exempt securities. 1933 *Business Week* 22 Feb. 4/1 Elimination of tax exempts is the object of an amendment to the Constitution offered by Senator Hull of Tennessee. 1966 *Economist* 30 Mar. 78/3 These industrial bonds..have so dogged the market as to increase all borrowing costs for tax-exempts. 1977 *New Yorker* 19 Sept. 27/2 One way to do this is to float a municipal-bond issue, which traditionally pays tax-exempt interest. 1978 G. Vidal *Kalki* v. 117 Any bona fide religion is tax-exempt in the United States. 1927 Bowley & Stamp *Nat. Income 1924* v. 42 Incomes above the tax-exemption limit. 1975 *N.Y. Times* 28 Nov. 37/4 One example of the use of incentives to attract investment is the tax-exemption on municipal bond income. 1969 *Manch. Guardian Weekly* 22 Nov. 11 Has Anthony Grey..joined the ranks of the tax exiles? 1978 J. R. L. Anderson *Death in Greenhouse* ii. 27 Neither of us was

attracted by the prospect of living in tax-exile. 1959 'M. Innes' *Hare sitting Up* I. i. 27 Tax fiddle of some kind? I don't like that sort. 1961 *Times* 31 May 10/4 Inclined to see in every taxpayer a more or less skilful..tax-fiddler. 1917 W. S. Churchill 9 Sept. in M. Gilbert *Winston S. Churchill* (1977) IV. Compan. I. 156, I do not however exclude the possibilities of a special bonus, presumably tax free. 1964 A. Wykes *Gambling* x. 241 Tax-free prizes. 1960 *Business Week* 24 Dec. 32/2 Interest in tax havens is largely due to the fact that U.S. tax law permits a company to accumulate profits abroad tax-free. 1973 *Times* 18 May 29/6 The Briton wanting to minimize his taxes through getting paid in a tax haven. 1950 *Times* 24 Apr. 5/7 The stimulation of enterprise is essentially best organized on a regional footing. At present it is undertaken with varying determination by different colonies. Some grant free import of capital equipment and 'tax holidays' for pioneer industries. 1977 *Time* 10 Oct. 60/1 Haughey's notion of a permanent tax holiday for artists has at least stopped the drain of home-grown talent. 1978 *Jrnl. R. Soc. Arts* CXXVI. 224/2 There was firstly a three-year tax holiday followed by a period allowing accelerated depreciation and gradually increasing rates of tax. 1959 J. Wood *Simple Guide for Taxpayer* iii. 28 Once the form has been completed it must be sent back to the Tax Inspector who sent it out. 1959 *Times* 24 Dec. 7/4 (*heading*) Tax-loss farming. *Ibid.*, Sir,—The so-called tax-loss farmer is generally speaking a man who carries on two businesses, one profitable and the other unprofitable. As a matter of taxation machinery he pays tax in respect of his profitable business and afterwards claims a refund of tax in respect of his unprofitable farming business. 1965 I. Fleming *Man with Golden Gun* vii. 100 Others would want to buy in..cheaply, and use it [*sc.* a hotel] as a tax-loss to set against more profitable enterprises elsewhere. 1970 *Money Which?* Mar. 64/3 Don't wait until the very end of the tax-year before selling shares for tax-loss purposes. 1975 *Times* 14 Jan. 12/6 (Advt.), *Wanted.* Large run-down school... High tax losses an inducement to purchase. 1803 G. Colman *John Bull* I. i. 3 She had disgraced her family by marrying herself to a tax-man. 1968 *Guardian* 22 June 5/5, I don't know what the taxman would say if you tried to get that as an expense allowance. 1970 *Money Which?* Mar. 43/2 You are allowed free of tax..family allowances, most pensions and some other social security benefits. The taxman views all these as earned income. 1972 *Accountant* 21 Sept. 369/2 It might be of interest..to mention the special rule for the tax point of barristers' services. [1916 Tax relief: see *income-tax.] 1931, etc. [see *relief[2] 7]. 1980 *Times* 9 Aug. 16/4, I have been offered remortgage, but my accountant says it will not qualify for tax relief. 1870 'Mark Twain' *Sketches New & Old* (1875) 319 A wicked tax-return..calculated to make a man report about four times his actual income to keep from swearing to a falsehood. 1961 *Guardian* 20 Feb. 16/7 If all the tax shelters were eliminated..the income tax yield would be increased by a third. 1982 *Financial Times* 13 Mar. 14/3 Investors will be seeking to use up the effective tax shelter offered by an appreciation of their assets each year in line with inflation. 1959 *Wall Street Jrnl.* 17 Sept. (Eastern ed.) 21 (Advt.), Tax-sheltered investment. 1976 *National Observer* (U.S.) 17 Jan. 9/6 And how to defer income tax on the interest you get. Plans that offer marvelous tax-sheltered advantages. 1976 F. Zweig *New Acquisitive Society* II. iv. 108 The tax thresholds in real terms have been substantially lowered over the years. 1971 *Money Which?* Mar. 4/1 These taxes are charged for a particular year of assessment, which always starts on 6 April and ends on 5 April in the following year. This is commonly called a tax year.

tax-cart. (Earlier example.)

1796 J. Woodforde *Diary* 31 Oct. (1929) IV. 316 They came in Mrs. Bodhams little Tax-Cart.

taxed, *ppl. a.* Add: **1. b.** Of a motor vehicle: having had excise duty paid for the current period.

1933 *Motor* 2 May (Suppl.) 99/3 (Advt.), Riley... 4-door coachbuilt sunshine saloon.., maroon, taxed. 1976 *Jrnl.* (Newcastle) 26 Nov., (Advt.) Volvo 144 Saloon 1974 N regn, orange, red striped upholstery, taxed Oct. '77.

‖ **taxe de séjour** (taks də seʒur). [Fr., lit. 'tax of visit'.] A tax imposed on visitors to spas or tourist resorts in France and other countries.

1922 *Michelin Guide Gt. Brit.* 764 France... 'Taxe de Séjour.'—The following special taxes are payable. 1979 G. Pottinger *Secretaries of State for Scotland 1926–76* xiv. 151 Another proposal that emerged from the Fraser exercise was to levy a *kurtax* [sic], or *taxe de séjour*, on the continental model, to find funds for tourist amenities.

taxeme (tæ·ksīm). *Linguistics.* [f. Gr. τάξ-ις arrangement + *-eme.] A unit of syntactic relationship, esp. one that cannot be further analysed or lacks meaning by itself, such as word order or stress. Hence **taxe·mic** *a.*; **taxe·mics** *sb. pl.* (const. as *sing.*), the study and description of language in terms of taxemes.

1933 L. Bloomfield *Lang.* x. 166 A simple feature of grammatical arrangement is a grammatical feature or taxeme. A taxeme is in grammar what a phoneme is in the lexicon—namely, the smallest unit of form. 1943 [see *tagmeme 2]. 1947 Taxemics [see *tagmemics *sb. pl.*]. 1950 Webster Add., Taxemic. 1967 M. Schlauch *Language* vi. 127 A taxeme, taken by itself, may have no meaning; when one or more taken together do have meaning the combination is called a tagmeme (for instance, *duck+ess*). *Ibid.* 128 It would seem that the terminology of taxemics requires further study. 1970 G. C. Lepschy *Survey Structural Linguistics* v. 89 Taxemes occur in conventional grammatical arrangements.

taxi, *sb.* Add: Pl. **taxis, †taxies. I. 1.a.** (Examples of *pl.* forms.)

1908 Taxies [in Dict.]. 1911 G. B. Shaw *Getting Married* 285 Me and the beadle have been all over the place in a couple of taxies, maam. 1914 —— *Pygmalion* (1916) III. 128 In future you shall have as many taxis as you want. 1923 A. Bennett *Riceyman Steps* IV. v. 198 Even in the daytime taxies were few in King's Cross Road. 1925 F. Scott Fitzgerald *Great Gatsby* iii. 70 Forms leaned together in the taxis as they waited, and voices sang. 1979 *United States 1980/81* (Penguin Travel Guides) 150 Taxis cannot be hailed in the streets.

b. *colloq.* A (small) passenger aeroplane; also short for *taxiplane* (see sense 4 below).

1911 *Flight* 11 Nov. 986/1 Baldwin, Sabelli and Lieut. Esnie Chinnery were making straight flights on the taxi. 1918 *Atlantic Monthly* Aug. 260 Our hero is..helping his mechanic give the 'taxi' a final looking over. 1919 *N.Y. Times Mag.* 30 Mar. 4 An airplane was..usually [called] a boat, ship, bus, or taxi. 1922 *Daily Mail Year Bk. 1923* 74/2 The 100 miles-an-hour 'aerial taxi'. *Ibid.*, He kept his winged 'taxi' waiting while he transacted urgent business, and then flew on. 1923 *Daily Mail* 7 Aug. 8/2 Companies..which send 'taxis', or aerial tramps, to anywhere between Plymouth and Stockholm. 1943 C. H. Ward-Jackson *It's Piece of Cake* 60 *Taxi*, an aircraft for the conveyance of a small number of passengers.

2. [f. the vb.] An act or spell of taxiing.

1931 *Daily Mirror* 27 Aug. 2/2 The machine..was lost in clouds of spray as it ploughed its way in a graceful 'taxi'. 1965 'J. le Carré' *Looking-Glass War* i. 9 The plane..began the long taxi to the reception point.

II. 3. *U.S. slang.* A prison sentence of between five and fifteen years.

1930 J. Lait *Big House* i. 1 Five to fifteen years, the judge had decreed—a 'stretch' that the convicts call a 'taxi', because most New York cabs bear conspicuously their rates, and are known as 'fifteen-and-fives'. 1935 A. J. Pollock *Underworld Speaks* 119/2 *Taxi*, 5, 10 or 15 years prison sentence. 1962 'D. Shannon' *Extra Kill* viii. 127 Whalen had done a five-to-fifteen year stretch—that's a taxi.

III. 4. *attrib.* and *Comb.*, as (sense 1) *taxi fare, fleet, -horn, man, queue, rank* [*rank *sb.*[1] 1 c], *stand, station*; *taxi-driving* vbl. sb. and ppl. adj.; (sense *2) *taxi apron, clearance, work*; **taxi-boat,** a boat that may be hired like a taxi; **taxi-dance** orig. and chiefly *U.S.*, a dance at which taxi-dancers are available; so *taxi-dance hall*; **taxi-dancer** orig. *U.S.*, a woman (or man) whose services as a dance-partner may be hired; a professional dance-partner; so **taxi-dance** *v. intr.*, **-dancing** *vbl. sb.*; **taxi-driver,** the driver of a taxi-cab; also *transf.*, esp. (*colloq.*) an aeroplane pilot; **taxi-girl,** a young female taxi-dancer; **taxiplane,** a piloted light aeroplane available for public hire; **taxi-ride,** a journey in a taxi; a short distance by car; also *transf.*; **taxi service,** a service providing transport in taxi-cabs (in quot. 1952, a shuttle service of aircraft); **taxi squad** *N. Amer. Football*, a group of players who take part in practices and can serve as reserves for the team; hence **taxi squadder,** a member of such a group; **taxi strip, track** = *taxiway* below; **taxi-truck** *Austral.*, a van, with a driver, for public hire; **taxiway** [after *runway 2 b], a route along which aircraft can taxi on the way to or from a runway.

1978 T. L. Smith *Money War* I. 136 Captain Imman eased the 727 forward on the taxi apron. 1949 *Newsweek* 26 Sept. 44/2 A taxiboat pilot told how, 'as soon as I pulled alongside, the passengers began plummeting down'. 1953 R. Godden *Kingfishers catch Fire* x. 117 Subhan's *shikara* had not cushions and curtains like a real taxi-boat. 1974 *Islander* (Victoria, B.C.) 8 Sept. 6/4 There was a 24-hour taxiboat service supplied. 1966 D. Francis *Flying Finish* ii. 25, I filed my flight plan, checked with the control tower for taxy clearance. 1932 P. G. Cressy (*title*) The taxi-dance hall. 1938 W. Smitter *F.O.B. Detroit* 9 On it there was the name of a taxi-dance place over a store on Woodward Avenue. 1955 N. Marsh *Scales of Justice* ix. 205 [He] cohabited with a so-called Miss Kitty de Vere whom he..met at a taxi-dance. 1972 J. Wambaugh *Blue Knight* (1973) i. 26 She was thirty-eight years old now..and taxi dancing part-time down the street at the ballroom. 1976 *New Yorker* 1 Mar. 80/2 He also played in a taxi-dance-hall band. 1930 *Variety* 12 Feb. 49/4 The Filipino is a much better spender than his white brother and a favorite with the gal taxi-dancers. 1979 P. Driscoll *Pangolin* iii. 36 He would simply pay Mama-san Julie..to take one of her taxi dancers home for the night. 1982 *Times* 8 Feb. 22/7 The ancient (well, 61-year-old) custom of taxi-dancing has returned in style to New York. 1907 Taxi-driver [in Dict.]. 1924 J. Buchan *Three Hostages* vi. 84 It is an outlandish place to get to, but most taxi-drivers know it. 1937 Partridge *Dict. Slang* 868/1 *Taxi-driver*, an aeroplane pilot. 1955 W. Faulkner *Fable* 110 Somebody owes something for that poor bloodstained taxi-driver [*sc.* an aeroplane pilot]. 1971 *Daily Tel.* 30 Jan. 3/3 For the first time separate experiments are being performed by the 'taxi driver', the lonely astronaut in the orbiting Command Module. 1975 J. Meldrum *Semonov Impulse* x. 173 She leaned forward and gave the taxi driver Dörflinger's address. 1909 Taxi-driving [in Dict.]. 1951 Koestler *Age of Longing* II. iii. 234 But there were no grand dukes, taxi-driving generals, or princesses at the Kronstadt. 1932 H. Crane *Let.* Feb. (1965) 401 I'm sure you wouldn't mind advancing the bus and taxi fare, would you? 1965 F. Sargeson *Memoirs of Peon* viii. 264 The taxi fleet was being doubled. 1963 J. Kirkup *Tropic Temper* xv. 163 The professional dance-girls, or taxi-girls as they are sometimes called in Singapore and Hong Kong, sit in a long row in front of the band. 1974 E. Brawley *Rap* (1975) I. xii. 166 Freddy paid a few piastres and danced with the taxi girls on the bandstand. 1943 G. Greene *Ministry of Fear* I. iv. 54 Far away a taxi-horn cried through an empty world. 1909 Taxi man [in Dict.]. 1946 E. O'Neill *Iceman Cometh* II. 111 Sneaking? Why, me and the taxi man made enough noise..to wake the dead. 1982 P. Fitzgerald *At Freddie's* vii. 54 They could all see him..cutting short whatever the taxi-man was saying. 1920 *Daily Tel.* 13 Apr. 1/7 Taxiplanes for any journey. Per mile, 2*s.* 6*d.* 1926 *Bulletin* 6 Aug. 3/1 He made a dash to Constantinople with a taxiplane. 1982 G. Hammond *Game* v. 60, I whistled up the taxi-plane. 1969 G. Lyall *Venus with Pistol* xv. 98 Everybody else in the coach..charged out to get into the taxi queue. 1929 T. S. Moore in *Yeats & Moore: Corr.* (1953) 155 There is a taxi-rank a few steps above the station. 1943 G. Greene *Ministry of Fear* IV. i. 222 A taxi-rank with one cab left. 1917 Kipling *Diversity of Creatures* 333 It demanded Work in the shape of many taxi-rides daily. 1941 B. Schulberg *What makes Sammy Run?* iii. 44 It was a five-dollar taxi ride from the Villa España. 1976 E. Ward *Hanged Man* x. 59 He was no good for the tricky stuff. Just taxi-rides. Light planes over flat country. 1952 *Times* 23 Aug. 4/6 British European Airways..are to make eight flights daily, mainly to Hanover, in addition to a new public 'taxi' service..to west Germany. 1962 L. Davidson *Rose of Tibet* 316 The p.c. questioned the local taxi service. 1981 L. Deighton *XPD* xxvi. 214 London... The parking problem was horrendous, the taxi service inadequate. 1966 Rote & Winter *Lang. Pro Football* III. 141 *Taxi squad*, group of players under contract who practice with team but are not included on official team roster and do not take part in league games. 1967 *N.Y. Times* 8 Dec. 64 The Atlanta Falcons of the National Football League..activated a cornerback..from the taxi squad. 1976 *Globe & Mail* (Toronto) 19 July S9/2 This is the first season the CFL has allowed teams an official taxi squad. 1975 B. Meggs *Matter of Paradise* v. iv. 154 These boys are first string, and a taxi squader like yourself..can get hurt. 1922 M. A. von Arnim *Enchanted April* ix. 135 A taxi stand was at the end of the road. 1982 T. Allbeury *Shadow of Shadows* xxii. 190 He had to walk almost to the bridge before he found a taxi stand. 1930 J. Dos Passos *42nd Parallel* v. 366 Instead she went out to the taxistation. 1943 *Yank* 12 Nov. 6/3 We rode up the taxi strips to our head stand where the crew stood around the ship. 1976 B. Jackson *Flameout* (1977) xii. 209 The small jet began to lose height, and soon airport buildings, a runway, and taxi strips were in sight. 1945 *Tee Emm* (Air Ministry) V. 42 The unit did not use lighted tar barrels to mark taxi tracks. 1966 D. Francis *Flying Finish* x. 127 Patrick moved down the taxy track and turned on to the apron. 1962 *Southerly* XXIII. 98 Taxi-truck. 1974 P. Cave *Dirtiest Picture Postcard* ii. 13 He packed all his books and records into two large cardboard boxes and phoned a taxi-truck. 1933 C. K. Stewart *Speech Amer. Airman* (thesis, Univ. of Akron) 96 *Taxi-way*, a route along the field designated for planes to taxi upon. 1939 *Sun* (Baltimore) 17 Apr. 8/1 The remaining $70,000 would be used for roadways, aprons, taxiways and fences. 1982 I. Johnston *Special Drug Squad* ii. 25 The taxiway..allows aircraft to be towed across the perimeter road to British Airways' maintenance hangars. 1945 *Ann. Reg. 1944* 23 A small land plane of 8,000 lb. to seat eight passengers and suitable for taxi work.

taxi (tæ·ksi), *v.* Also **taxy** (now only in pres. pple.). [f. the sb.] **1. a.** *intr.* Of an aeroplane, etc., or its pilot: to travel slowly along the ground or water under the machine's own power. Also *transf. to taxi in*, to taxi from a runway to a terminal or hangar; similarly *to taxi out*.

1911 [see *remous]. 1914 *Aeroplane* 5 Feb. 140/1 He taxied out to leeward,..turned,..opened out his engine. *Ibid.* 1 July 21/2 The de Bolotoff triplane was 'taxying', but showed no signs of lifting. 1915 [see *come *v.* 24 d]. 1918 H. Barber *Aerobatics* I. 37 In a high wind don't taxy unless necessary. 1927 C. A. Lindbergh *We* ii. 19, I..taxied back to the hangar. 1932 S. Gibbons *Cold Comfort Farm* xxiii. 304 The aeroplane..was taxi-ing comfortably to a standstill. 1955 *Times* 24 Aug. 6/4 Crowds waited at London Airport, and the Canberra was greeted with cheers as it taxied in. 1959 D. A. Bannerman *Birds Brit. Isles* VIII. 280 When well out from the land both birds became silent and 'taxied' heavily over the water until air-borne. 1961 J. Heller *Catch 22* (1962) v. 47 The planes lumbered around and nosed forward lamely..until they taxied into the line at the foot of the landing strip and took off swiftly. 1975 *Daily Tel.* (Colour Suppl.) 4 Apr. 16/2 The ton of fuel..would be burnt while taxiing out to the runway.

b. *trans.* To cause (an aeroplane, etc.) to taxi.

1915 H. Rosher *In R.N.A.S.* (1916) 67, I was taxying my machine to the far end of the aerodrome, to start off into the wind. 1933 *Discovery* Mar. 79 A Moth fuselage is used to 'taxy' a 25 ft. model of a flying boat hull by means of a force-recording undercarriage. 1946 *Proc. IRE* XXXIV. 380/2 (*caption*) General Motors Bug being taxied by radio control from B-23 airplane at Muroc Lake, California. 1977 *R.A.F. Yearbk.* 29 Taxying the aircraft requires a little practice.

2. a. *intr.* To travel in a taxi.

1918 A. Quiller-Couch *Foe-Farrell* xxiv. 397 From Victoria that evening I taxi'd straight to Jermyn Street. 1942 E. Paul *Narrow St.* iv. 33 Anne's father..used to taxi to the rue de la Huchette from the avenue de la Bourdonnais. 1971 L. Blackwell *Blackwell Remembers* xxii. 200 As we taxied down the Mall it was gaily set out with Union Jacks and the Finnish flag.

b. *trans.* To convey in a taxi. Also *transf.*

1973 J. Thomson *Death Cap* ix. 134 I'll get where I want to on my own feet... I don't want no bloody police taxiing me around. 1977 G. McDonald *Confess, Fletch* xxxviii. 178 He taxies Ms. Fryer to her motel. Allows her time to change.

Hence **ta·xiing** *vbl. sb.* and *ppl. a.*, in sense 1 a above; also *fig.*

1916 N. J. Gill *Flyer's Guide* iii. 32 If, however, the machine is subject to a sudden loss of forward way (i.e., taxying over rough ground) the planes then tend to go on. 1946 R. A. McFarland *Human Factors in Air Transport Design* ix. 390 To permit good ground vision during taxiing. 1958 Castle & 'Hailey' *Flight into Danger* vi. 76 One slowly taxi-ing aircraft came to a stop and cut its engines. 1968 J. Ziman *Public Knowledge* i. 10 Greek Science never finally took off from its brilliant taxying runs. 1972 *Guardian* 22 Dec. 4/1 An airliner ran into a taxi-ing plane while taking off. 1982 L. Cook *Under Etna* I. i. 11 The first jarring bounce of touchdown and..the gentle taxi-ing in.

taxi-cab. Add: (Later examples.)

1916 G. B. Shaw *Pygmalion* I. 116 She gets in and pulls the door to with a slam as the taxicab starts. 1978 M. Gilbert *Empty House* xii. 102 An aged taxi-cab parked in the [station] forecourt. 1981 P. Theroux *Mosquito Coast* xxxi. 392 The world..was glorious even here, in this old taxi-cab, with the radio playing.

Hence **ta·xi-cabby,** a driver of a taxi-cab.

1910 'W. Lawton' *Boy Aviators in Nicaragua* 19 The taxi-cabby, like most of his kind, was not averse to making a tip. 1918 G. Frankau *One of Them* xx. 155 Whether five-bob tip to taxi-cabby Presaged the sorting-bells of Hell Fire Abbey?

taximeter. Add: Now only with stress on first syllable. **a.** Also *ellipt.* for *taximeter cab. rare.*

β. 1908 A. Bennett *Buried Alive* iv. 102 He then hailed a taximeter from the stand opposite the Army and Navy Stores.

b. *taximeter-cabriolet.*

1959 P. Bull *I know the Face, But...* xi. 200 Lambert kindly escorted me in a taximeter-cabriolet.

taxis. Add: Pl. **taxes** (-īz). **6.** (Earlier and later examples.) [Introduced in this sense in Ger. by F. Czapek 1898, in *Jahrb. für wissensch. Bot.* XXXII. 308.]

1899 *Jrnl. R. Microsc. Soc.* 180 The phenomena of this [irritability] reaction may be classed under the following heads:—(1) Taxis or movement ('geo-' or 'photo-taxis') [etc.]. 1940, 1955 [see *kinesis 2]. 1973 *Nature* 17 Aug. 468/1 Behaviourism as a general theory of animal behaviour was woefully inadequate—a fact of which anyone must be convinced who tries today to read Jacques Loeb on tropism, taxes and the like.

-taxis (tæ·ksis), the word Taxis (sense 6) used as a suffix in *Biol.*, as in *geotaxis* s.v. *geo-, Phototaxis, etc.

taxogen (tæ·ksŏdʒĕn). *Chem.* [f. Taxo- + -gen.] The monomer in the chain of a telomer.

1948 [see *telomer]. 1974 C. M. Starks *Free Radical Telomerization* i. 2 The term *monomer* is normally employed instead of taxogen, except in the patent literature. 1980 *U.S. Patent 4,183,901* 2 The maleic acid telomers are lower molecular weight polymers formed by reacting a chain transfer agent, or telogen with an olefinic monomer of taxogen.

taxon (tæ·ksǫn). Pl. **taxa.** [a. G. *taxon* (A. Meyer *Logik der Morphologie* (1926) 127), f. *taxonomie* Taxonomy.] A taxonomic group, as a genus or species. Also *fig.*

1929 *Scientific Monthly* Feb. 107/2 Such conceptions as the 'species', or 'taxon' in taxonomy. 1936 *Acta Biotheoretica* II. 180 Meyer's taxa (and phyla) are more valuable concepts in theory than in practice. 1948 *Minutes Utrecht Conf.* 14 June in *Chronica Botanica* (1950) XII. 12 Dr. Lam explained that in order to simplify the wording of the Rules, it is proposed to indicate a taxonomic group of any rank with the term taxon (plural taxa). 1951 [see *endemic *a.* a]. 1953 *Proc. 7th Internat. Bot. Congr. 1950* 465/1 Taxon..was, however, a very convenient word, and after two years [*sc.* by 1950] 60% to 70% of botanists were using it. 1961 *Watsonia* V. 68 Many infraspecific taxa [of *Trifolium repens*] have been described for wild populations. 1971 J. Z. Young *Introd. Study Man* xxviii. 400 There is no fixed typological criterion of an extinct species, genus, family, or other taxon. 1973 *Jrnl. Indo-European Studies* I. 405 A few words should be said about the main taxon in question; namely: Indo-Europeanist. 1976 *Sci. Amer.* Aug. 32/3 His work has revealed a surprisingly wide range of vertebrates: at least a dozen taxa of mammals, 22 of birds, three of fishes and one taxon of reptiles. 1980 *Dædalus* Spring 85 A Renaissance similitude (biblical giants) eventually converted into an Enlightenment taxon (species-giants).

taxonomic, *a.* (In Dict. s.v. Taxonomy.) Add: *spec.* in *Linguistics*, involving or concerned with the identification and classification of the terms into which languages are analysed; esp. as *taxonomic linguistics* (the dominant methodology of the 1940s and 1950s), *phonetics*, etc.

1962 N. Chomsky in *Internat. Preprints Papers 9th Internat. Congr. of Linguists* 556 In the case of perception of language,..the step-by-step analytic models of taxonomic linguistics are not in the least convincing. 1964 —— in *Proc. 9th Internat. Congr. Linguists* 951 Structural linguistics marks a departure from a more traditional point of view... Let us coin the term 'taxonomic phonemics' to refer to this body of doctrine, thus emphasizing its striking reliance..on procedures of segmentation and classification. 1968 P. M. Postal *Aspects Phonol. Theory* p. x, I shall refer..to the dominant conception of phonological structure as 'autonomous phonology' or 'autonomous phonemics', considering this terminology preferable to that of 'taxonomic phonemics' which has been used in the recent past. 1973 *Archivum Linguisticum* IV. 117 He goes on to claim a tacit dependence of transformational grammar on the findings of taxonomic analysis. 1976 *Canad. Jrnl. Linguistics* Spring 128 FBG present excellent discussions of taxonomic structural linguistics, how it is done, what it discovered about language (observing that many of the facts about language structure that taxonomic grammarians set out to capture are real), and the psycholinguistics of taxonomic grammar.

taxonomy. Add: **1.** The systematic classification of living organisms.

2. (With *a* and *pl.*) A classification *of* anything.

1960 *Times Lit. Suppl.* 29 Apr. 277/4 Professor Goldschmidt..has constructed what he calls a 'taxonomy' of human societies. 1971 *Nature* 10 Dec. 319/2 He complained of the taxonomy put forward by Sir Frederick Dainton's committee, with its concepts of basic research, strategic research and technical research. 1972 *Sci. Amer.* Jan. 116/3 His taxonomy of bridge structures before the age of steel and concrete. 1979 *Dictionaries* I. 64 The prefatory notes to the volumes of the *DAE* offer taxonomies of American usages and Americanisms. 1983 *Sci. Amer.* Mar. 102/2 We can outline a taxonomy of chlorite oscillators, and we are beginning to see how they can be related to oscillators of the bromate and iodate families.

taxpayer, tax-payer. Add: **2.** *U.S. colloq.* A building just large enough to provide an income sufficient to meet the expenses it incurs; hence, any small building.

1921 B. Matthews *Ess. on Eng.* vi. 134 A resplendent electric sign on top of a two-story tax-payer. 1950 *N.Y. Times* 28 Dec. (Late city ed.) 39/8 The three-story tax-payer at 288 Jackson Avenue in Jersey City, N.J...has been sold.

taxpaying *sb.* and *a.* (earlier examples).

1832 *Reg. Deb. Congress U.S.* 4 Apr. 2390 As a representative of the tax paying people of the South, I must ask..what becomes of the excessive heavy amount of revenue? 1849 J. S. Mill in *Westm. Rev.* LI. 44 A tax-paying or other property qualification.

Tayacian (tāyē[1]·ʃăn), *a.* *Archæol.* [ad. F. *Tayacien* (H. Breuil 1932, in *Préhistoire* I. 131), f. *Tayac* (see def.): see -ian.] Of, pertaining to, or designating a palæolithic flake industry of which remains were first found at Tayac (Dordogne), SW France. Also *absol.*

1934 *Nature* 7 July 30/1 The Carmel cave series covers from Natufian (Mesolithic) to Tayacian, the recently recognised rough flake industry. 1946 F. E. Zeuner *Dating the Past* ix. 288 The combination of Levalloisian, Acheulian and Tayacian or Clactonian elements which resulted in the Mousterian industry of Europe occurred during the Last Interglacial. 1974 *Encycl. Brit. Macropædia* VIII. 1050/1 The earlier flake tools at this site [*sc.* Fontéchevade] are termed Clactonian tools..., while the later are known as Tayacian tools. 1979 M. Leakey *Olduvai Gorge* ix. 87 The chopper-small-tool complex..is represented in Europe by the Clactonian..and the Tayacian.

tayberry (tē[1]·bĕri). Also **Tay-.** [f. *Tay*, the name of a river in Scotland + Berry *sb.*[1]] A dark purple soft fruit produced by crossing the blackberry and the raspberry, introduced in Scotland in 1977; also, the plant bearing this fruit.

1977 *Ann. Rep. Scottish Hort. Res. Inst. 1976* 42 The new *Rubus* hybrid has been named 'Tayberry' and is being propagated for release. 1980 *Economist* 9 Aug. 63/2 To the strawberry, gooseberry, raspberry, blackcurrant and blackberry cycle is now added a new fruit—the tayberry. 1980 *Amateur Gardening* 8 Nov. 35 The tayberry is deep purple in colour, roughly 1½ in. long and has a refreshing, not-too-sweet flavour. 1982 *Observer* 7 Feb. 43/4 (Advt.), Delicious soft fruits for autumn planting...The new virus-free Tayberry and all soft fruits.

Taylor (tē[1]·ləɪ). **1.** *Math.* [The name of Brook *Taylor* (1685–1731), English mathematician, who published the theorem in his *Methodus Incrementorum Directa et Inversa* (1715).] *Taylor('s) series*, an infinite series of the form $f(a) + hf'(a) + h^2f''(a)/2! + \ldots + h^{n-1}f^{(n-1)}(a)/(n-1)! + \ldots$, where $f^{(i)}(a)$ is the value of the *i*th derivative of a function $f(x)$ at $x = a$; an analogous series for a function of more than one variable; *Taylor's theorem*, the theorem that a function $f(x)$ can be approximated over any interval throughout which its first *n* derivatives exist by the first *n* terms of Taylor's series (with $h = x - a$) plus a remainder dependent on $f^{(n)}(a + \theta h)$ $(0 < \theta < 1)$.

1816 [see Theorem *sb.* a]. 1842 *Penny Cycl.* XXIV. 126/2 Lagrange's paper in the Berlin Memoirs for 1772, in which he proposed to make Taylor's theorem the foundation of the Differential Calculus. *Ibid.*, D'Alembert..gave for the first time..a method of finding the remnant of Taylor's series after a certain number of terms have been taken. 1908 G. H. Hardy *Pure Math.* vii. 255 This expansion of f(a+h) is known as Taylor's series. *Ibid.* 287 (*heading*) Proof of Taylor's theorem by integration of parts. 1968 C. G. Kuper *Introd. Theory Superconductivity* ii. 24 Near the transition temperature, *g* may be expanded in a Taylor series: $g = g_0 + \alpha\zeta + \frac{1}{2}\beta\zeta^2 + \ldots$ 1972 A. W. F. Edwards *Likelihood* v. 72 An alternative method, which readily generalizes to the case of many parameters, is to obtain the Taylor's series approximation to the support curve in the region of the maximum. 1972 M. Kline *Math. Thought* xx. 442 Taylor's theorem for $a = 0$ is now called Maclaurin's theorem. 1979 Page & Wilson *Introd. Computational Combinatorics* ii. 10 We write $E = 1 + \Delta$ and expand the polynomial f by Taylor's theorem.

2. The name of F. W. *Taylor* (1856–1915), U.S. engineer, used *attrib.* to designate the system of scientific management and work efficiency that he expounded.

1911 *Assoc. Machinists Circular* in C. B. Thompson *Scientific Managem.* (1914) 783 The latest danger..is the so-called Taylor system of shop management. 1926 Whiteman & McBride *Jazz* vii. 154 At their work, men and women are the victims of efficiency, the Taylor system, so that humanity itself is being made into machines. 1972 [see *scientific management* s.v. *scientific *a.* 6].

3. Port wine shipped by the firm of *Taylor*, Fladgate, and Yeatman.

1940 M. Healy *Stay me with Flagons* 212 We had some Taylor of the same year, and the connoisseurs usually accorded it a higher place. 1952 H. W. Allen *Sherry & Port* II. i. 126, I tasted some of these wines in the Oporto Lodges, notably Sandeman 1942 and 1945 and Taylor of the same years. 1968 'J. Welcome' *Hell is where you find It* x. 137 Benson brought in the decanter and put it beside me. 'It's the Taylor 47, sir,' he said.

Taylorian (tē[1]lǭ·riăn), *a.* and *sb.* [f. the name *Taylor* (see def.) + -ian.] The familiar name (used as *adj.* and *sb.*) of the Taylor Institution of the University of Oxford, established for the teaching of modern languages from money left for the purpose by Sir Robert Taylor (1714–88), English architect.

1898 *Dict. Nat. Biogr.* LV. s.v. *Taylor, Sir Robert*, The lecture-rooms and library which compose the Taylorian buildings were built in 1841–5. 1913 H. E. Salter *Oxford Deeds of Balliol College* 212 Ball's house was..on the site of the Taylorian. 1932 L. Magnus *Herbert Warren* v. 147 He took much interest as Vice-Chancellor in the Taylorian Institute. 1937 H. Nicolson *Diary* 26 Nov. (1966) 313 Then to the Taylorian where I address a large and interested audience on the German colonies. 1965 Doughty & Wahl *Lett. D. G. Rossetti* I. 47 There is a copy of this letter at Oxford, in the Taylorian.

Taylorism. Add: **2.** [f. the name of F. W. *Taylor*: see *Taylor 2.] The principles or practice of the Taylor system of management.

1928 *Times Lit. Suppl.* 11 Oct. 724/2 The second [essential] was the substitution of exact scientific investigation and knowledge for the old individual judgment or opinion. Mr. Meakin, who speaks somewhat slightingly of 'Taylorism', seems to be unacquainted with this passage. 1952 E. H. Carr *Bolshevik Rev.* II. xvi. 111 The Menshevik journal declared that the Bolsheviks..'are attempting to abolish the eight-hour day and introduce piece-rates and Taylorism'. 1983 *Futurist* June 25/1 Taylorism reduces work to machine-tending that requires little training and effort and that maximizes productivity.

taylorite (tē[1]·ləraɪt). *Min.* [f. the name of its discoverer, W. J. *Taylor* (1833–64), U.S. mineral chemist + -ite[1].] A sulphate of potassium and ammonia found in Peruvian guano beds as yellowish white bitter-tasting orthorhombic crystals.

1868 J. D. Dana *Syst. Min.* (ed. 5) 614 Taylorite... In small compact lumps or concretions. 1968 I. Kostov *Mineralogy* 503 Arcanite and mascagnite are isostructural..and form mixed crystals $(K,NH_4)_2SO_4$ termed taylorite. 1975 *Mineral. Abstr.* XXVI. 353/1 The annual mineral lists for newly recorded Western Australian minerals include..taylorite.

Taylorize (tē[1]·ləraɪz), *v.* Also **taylorize.** [f. *Taylor + -ize.] *trans.* To introduce the Taylor system into (see *Taylor 2); to manage in accordance with this system. Chiefly as **Ta·ylorized, Ta·ylorizing** *ppl. adjs.* Also **Tayloriza·tion,** the action of Taylorizing; Taylorism.

1929 A. Huxley *Holy Face* 64 Machinery, Taylorization..had not yet begun to produce their dehumanizing effects. 1930 *N. & Q.* 26 Apr. 301/2 'Taylorize', then, means 'manage scientifically'. 1939 J. A. Schumpeter *Business Cycles* II. xiv. 783n. A major movement, which however resolves itself into an almost infinite number of small ones, is what may be called Taylorization. Its spread during our period is a typical consequence of the struggle for survival amidst the readjustments of downgrades. 1957 R. Burlingame *Henry Ford* v. 76 In the 'taylorized' industries the pay rise had been more gradual. *Ibid.*, A good many industrialists had repudiated 'taylorization' and called Taylor a crackpot. 1979 *Internat. Jrnl. Sociol. of Law* Feb. 112 Their Taylorizing bureaucracies are only matched by British amateurism..and American razmatazz.

tayn, var. *t'ien.

Tay-Sachs (tē[1],sæ·ks). *Path.* The names of Warren *Tay* (1843–1927), British ophthalmologist, and Bernard *Sachs* (1858–1944), American physician and neurologist, used *attrib.* and *absol.* with reference to a fatal inherited metabolic disorder in which an enzyme deficiency causes accumulation of a ganglioside in the brain and elsewhere, resulting in idiocy and death in childhood (described by them in 1881 and 1887 respectively). [Named in Ger. by H. Higier 1901, in *Neurologisches Centralblatt* XX. 851.]

1907 *Index Medicus* V. 841/1 Hereditary infantile cerebellar ataxy and the Tay-Sachs disease. 1937 [see *amaurotic *a.*]. 1974 *Sci. Amer.* Mar. 63/2 (Advt.), A Tay-Sachs child develops normally for his first six months. Then, as excessive fatty deposits accumulate in his brain cells, he regresses... Usually before his fifth birthday, he dies. 1975 *Nature* 8 May 101/3 Israel's best known ethnic malady is Tay-Sachs Disease, a fatal genetic disorder limited almost entirely to infants whose forebears came here from certain parts of East Europe.

taz (tæz). *colloq.* = *tash.

1951 Partridge *Dict. Slang* (ed. 4) 1198/1 *Taz.* A beard: Cockneys': C. 20... 2. An immature moustache; youthful down, wherever growing: mostly Cockneys': since ca. 1920. 1969 M. Duffy *Wounds* i. 19 He was proud of his little toothbrush taz and elegant white raincoat.

tazetta (tæze·tă). Also **Tazetta.** [mod.L., specific epithet (Linnæus *Species Plantarum* (1753) I. 290), ad. It. *tazzetta* little cup, f. *tazza* (see Tass[2]): see -et.] A fragrant white or yellow polyanthus narcissus, *Narcissus tazetta*, native to the Mediterranean, or any of the numerous varieties developed from it.

1847 *Jrnl. Hort. Soc.* II. 26 Sweet's Hermione Cypri are the produce of poeticus and a white-limbed Hermione, and N[arcissus] bifrons and compressus of Tazetta and jonquil. 1924 L. H. Bailey *Man. Cultivated Plants* 187 They [*sc.* the Poetaz narcissi] are like large-flowered Tazetta. 1956 C. Mackenzie *Thin Ice* iii. 36 The rising sun lighted a green plain covered with tazetta narcissus. 1977 *Chicago Tribune* 2 Oct. XI. 13/2 Midseason... Short-cupped daffodils; poeticus narcissus; jonquils; tazetta daffodils.

‖ **tazia** (tāzī·ā). Also **tazieh, taziyah, tazzia,** etc. [ad. Arab. *ta'ziya* consolation, mourning.]

1. A representation of the tombs of Hasan and Husain (grandsons of Muhammad) carried in the Moharram procession.

1809 T. D. Broughton *Let.* 26 Feb. (1813) 72 There were more than a hundred *Taziyas*, each followed by a long train of *Fuqeers*..beating their breasts. 1862 [see *taboot[1]]. 1885 T. P. Hughes *Dict. Islam* 410/2 Against the side of the Imambarrah, directed towards Mecca, is set the *tabut*—also called *tazia* (*ta'ziyah*), or model of the tombs at Kerbela. 1889 Kipling *In Black & White* 94 Gilt and painted paper presentations of their tombs are borne with shouting and wailing..which fakements are called *tazias*. 1924 *Glasgow Herald* 26 Sept. 10 The procession was a long one, including a number of tazias, or many-storeyed turrets, and followed by 7000 Mussulmans. 1946 *Times of India* 6 Dec. 10/4 Richly decorated in customary Muslim fashion *tazias* looked resplendent.

2. A play commemorating the 'martyrdom' of Hasan and Husain, performed esp. on the anniversary of the event each year.

1893 E. G. Browne *Year amongst Persians* iv. 70 Many people were assembled to witness a *ta'ziya*, or representation of the sufferings of the Imáms Hasan and Huseyn. 1911 D. S. Margoliouth *Mohammedanism* iv. 127 The Indo-Germanic affinities of the Persians have led to the production of miracle-plays, called *ta'ziyah* (consolation), whereby the atrocities are more vividly brought home. 1951 G. E. von Grunebaum *Muhammadan Festivals* (1976) v. 89 At a comparatively recent date..the *ta'ziya*, or Passion play..became the real climax of the Shi'ite Tenth of Muharram celebrations. 1972 *Times* 28 Sept. 9/3 In place of the traditional *ta'zieh* and *ruhozi* performances, a pair of new Iranian plays figured on the main bill. 1974 F. Rosenthal in Schacht & Bosworth *Legacy of Islam* (ed. 2) vii. 335 Numerous librettos for such *ta'ziya* plays have been preserved.

tazza. (Earlier example.)

1824 Lady Blessington *Jrnl.* May in E. Clay *Lady Blessington at Naples* (1979) 86 Antique vases and *tazze*, on which are sculptured bacchanalian orgies.

Tazzie, Tazzy, varr. *Tassie[3]. **T-bone steak**: see *T 3 b.

tch, *int.* Also **tchk, tcht.** A representation of the dental click (freq. reduplicated) used to express vexation (cf. Tchick *sb.*, Tck *int.*, Tut *int.* (*sb.*[3])). Hence **tch** *v. intr.*, to utter this exclamation; also as *sb.*, an utterance of this exclamation.

1898 G. B. Shaw *Mrs. Warren's Profession* I. 176 (Correcting him quickly in a loud whisper) Tch! Nonsense. 1906 N. Munro *Daft Days* ix. 74 'You'll find a curious fearless independence in her.' The twins held up their hands in amazement, 'tcht-tcht-tchting' simultaneously. '*What* a pity!' said Miss Jean, as if it were a physical affliction. 1910 —— in *Blackw. Mag.* Aug. 236/2 Aunt Amelia..tchk-tchked at such preposterous views. 1930 G. B. Shaw *Apple Cart* I. 21 Tch-tch-tch! Gently, Amanda, gently. 1971 *N.Z. Listener* 16 Aug. 50/3 The mind boggles. The dreadful deeds the little monkeys might perpetrate. Tch tch. 1977 *Daily Mirror* 31 Mar. 24 Tch! Of all the times to go down wi' flu! We've got a very important darts match tonight!

Tchaikovskian (tʃaikǭ·fskiăn), *a.* and *sb.* Also **Tchaikovskyan, Tschaikowskian.** [f. the name of Peter Ilyich *Tchaikovsky* (1840–93), Russian composer + -ian.] **A.** *adj.* Of, pertaining to, or characteristic of Tchaikovsky or his style. **B.** *sb.* One who favours or imitates the style of Tchaikovsky.

1937 *Observer* 15 Aug. 3/3 When the time comes for an appraisal..it will not be the Tschaikowskian.., or the Egdon Heath Sibelius..who seems nearest to being a composer of the very greatest rank. 1945 G. Abraham *Tchaikovsky* VII. 138 The G minor andante portion of the ..penultimate scene stands out as truly Tchaikovskian. 1967 *Listener* 12 Jan. 73/3 It reminds one of what Tchaikovsky thought about Mozart, and incidentally of what Stravinsky—a Tchaikovskian to the hilt—thought about musical expression in general. 1973 *Gramophone* 29 June 15/3 The violin concerto is the least Finnish of Sibelius's major works, and its combination of Tchaikovskyan elements with a cosmopolitan concerto-style was emphasised by Pinchas Zukerman's suave, uncommitted manner. 1977 *Ibid.* Dec. 1097/2 The reprise of the Symphony's big tune at the end..lacks the expansive richness that can give the Tchaikovskian a real *frisson* of pleasure. 1979 *Guardian* 23 Mar. 12/7 Tchaikovskian delicacy misses fire in this theatre.

tchaush, tchawoosh, varr. Chiaus.

1819 T. Hope *Anastasius* (1820) II. 377 A Tchawoosh..walked in, and summoned me before the Soo-bashee. 1902 *Encycl. Brit.* XXVII. 213/2 The Sultan's guard consists of ..the 'Tufenkdjis'..the 'Tchaush', of whom there are between 50 and 60, and who are messengers as well as guards [etc.]. 1930 *Observer* 26 Jan. 11 In the last war a peasant's wife, Fatma Hanem, served in the Army as a tchaush, i.e. a sergeant.

tche, var. *se. **Tchehovian,** var. *Chekhovian *a.* and *sb.* **Tcheka,** var. *Cheka. **Tcheremiss,** var. *Cheremis(s. **tchernozem,** var. *chernozem. **tchervonetz,** var. *chervonetz.

tchetvert. (Earlier example.)

1814 *The Commercial Secretary—Il Segretario di Commercio* (Leghorn) 290 Wheat. R 10 1/2 Stock expected to increase to nearly 15/m tschetwer, which embarasses the holders greatly.

tchin. (Earlier example.)

1861 R. Cecil in *Sat. Rev.* 2 Mar. 228/2 The Emperor is practically an absentee landlord, knowing nothing of his estate except what the tchin is pleased to tell him.

tchinovnik, var. *chinovnik. **tchornozem,** var. *chernozem. **tchotchke**: see *tsatske. **Tchuktchi,** var. *Chukchee, Chukchi.

te[1], **ti** (tī). Also **tee.** *Mus.* Now the more usual name, in English-speaking countries, of Si. Cf. *Tonic Sol-fa* s.v. Tonic *a.* 3 b.

1839 S. A. Glover *Scheme for rendering Psalmody Congregational* (ed. 2) 41 *Te* is the *subsemitone* or leading half tone *below* the tonic. 1889 Grove *Dict. Mus.* IV. 144/1 Tonic Sol-fa... The ancient sound-names do, re, mi, etc...are put before a class..in the form of a printed picture of the scale, called a 'Modulator'. For simplicity's sake they are spelt English-wise, and *si* is called *te* to avoid having two names with the same initial letter. 1944 W. Apel *Harvard Dict. Mus.* 690/1 The syllables mostly used today are: *do.., re, mi, fa, sol, la, si* (*ti*). 1969 *Listener* 31 July 162/2 Everything becomes dubious if you suddenly decide to make tee a new doh. 1980 C. Headington *Illustr. Dict. Mus. Terms* 134/1 Solmisation, a system of designating the notes of a diatonic scale by syllables..*do, re, mi, fa, sol.., la, ti.*

‖ **te**[2] (də). Also **Te, teh, tih.** [Chinese *dé* virtue.] **a.** In Taoism, the essence of Tao inherent in all beings. **b.** In Confucianism and in extended use, moral virtue.

1895 G. G. Alexander *Lâo-Tsze: Great Thinker* 123 It is very puzzling to know when '*tih*' is to be treated as a Divine attribute, and when it is to be taken as a moral virtue. 1904 W. G. Old tr. *Laotze's Bk. of Simple Way* li. 114 Tao brings forth, and Teh nourishes. 1912 [see *li[2]]. 1934 A. D. Waley *Way & its Power* 32 Hence *tê* means a latent power, a 'virtue' inherent in something. 1955 E. Herbert *Taoist Notebk.* 18 If allowed free play..these gifts of *Tê* were ample to insure the orderly progression of 'heaven and earth', the disciplined march of 'the ten thousand creatures', all in their ranks and all in step with *Tao*. 1963 D. C. Lau *Lao Tzu* 42 *Te* means 'virtue', and seems to be related to its homophone meaning 'to get'. In its Taoist usage, *te* refers to the virtue of a thing (which is what it 'gets' from the *tao*). In other words, *te* is the nature of a thing, because it is in virtue of its *te* that a thing is what it is. 1975 C.-Y. Chang *Tao* xxxviii. 107 The real meaning of *Tê* is thus the attainment of the self-cultivation of non-discrimination, non-differentiation, and above all, non-willing.

tea, *sb.* Add: **1. c.** Phrases. *given away with a pound of tea*: see *give *v.* 54 a; *not for all the tea in China* (colloq., orig. Austral.): not at any price.

1937 Partridge *Dict. Slang* 148/1 *China!, not for all the tea in*, certainly not!; on no account: Australian coll. from the 1890's. 1943 K. Tennant *Ride on Stranger* ii. 19 I'm not going to stand in my girl's light for all the tea in China. 1958 J. Cannan *And be Villain* vi. 137 She wouldn't get into a sidecar or on a pillion for all the tea in China. 1978 *Radio Times* 11–24 Mar. 25/5, I wouldn't change Newcastle for all the tea in China... It's a lovely place to live in.

2. b. *cup of tea* (colloq. phr.): see *cup *sb.* 12 b.

c. A cupful of tea.

1922 Joyce *Ulysses* 729 We both ordered 2 teas and plain bread and butter. 1976 B. Gibson *Birmingham Bombs* xii. 104 Three teas, two coffees, and a large steak and kidney pie.

d. *one's tea*: what interests or suits one. *rare.* Cf. *cup *sb.* 12 b (ii).

1934 E. Waugh *Handful of Dust* iii. 135 Are you *certain* Jenny will be Tony's tea?

e. *tea and sympathy*: consolation offered to a distressed person. Also *attrib.*

1956 (*film title*) Tea and sympathy. 1958 *Listener* 2 Oct. 537/1 We leave Mrs. Newby enjoying tea and sympathy. 1970 Y. Carter *Mr. Campion's Falcon* i. 7 He was a tea-and-sympathy man, full of tactful advice. 1978 J. Higgins *Day of Judgment* xii. 168 'There may be something I could do.' 'Tea and sympathy... No more than that.'

4. a. (Later examples.) Now usu. a light meal in the late afternoon, but locally in the U.K. (esp. northern), and in Australia and N.Z., a cooked evening meal; in Jamaica, the first meal of the day.

1914 G. B. Shaw *Misalliance* 80 He calls his lunch his dinner, and has his tea at half-past six. Havnt you, dear? *a* 1925 [see *market *sb.* 1 b]. 1938 N. Marsh *Artists in Crime* vi. 81 'We finished tea at half-past eight, about.' 'The gentleman is talking of the evening meal. They dine at noon in the Antipodes, I understand.' 1952 in Cassidy & Le Page *Dict. Jamaican English* (1967) 439/1 Tea—same as chaklata... Tea at 6:30 A.M. 1957 *N.Z. Listener* 22 Nov. 4/3 More than one New Zealander has been invited to 'tea' in England and arrived hours too late, the meal finished and the guests gone. 1968 *Southerly* XXVIII. 5 'What have you got for tea?' he asked... 'It's a coupla nice little bits of fillet Mr Ballard let me have.'

c. *to go (out) for one's tea* (see quots.). *N. Ir. slang.*

1978 F. Burton *Politics of Legitimacy* iii. 78 A Provo would scoff at the Officials' merely elocutionary skills while they were 'going out for their tea' (that is, going on military operations which might result in their death). 1979 *Courier-Mail* (Brisbane) 1 Mar. 5/1 'Going for your tea' in Belfast can be a painful experience—being dragged out by a terrorist punishment squad to get a bullet in the legs.

7. c. Marijuana; *spec.* marijuana brewed in hot water to make a drink. orig. *U.S.*

1935 A. J. Pollock *Underworld Speaks* 119/2 *Tea*, mariahuana; hashish. 1940 [see *ju-ju[2]]. 1950 *San Francisco Chron.* 22 Feb. 20/1 A couple of years ago she started blowing tea. 1957 [see *connexion 6 c]. 1967 *Boston Sunday Herald* 26 Mar. IV. 1/1 Marijuana..when brewed with hot water,..is called 'tea'. 1979 *High Times* Mar. 18/2 Consider the number of words that served for a time and then passed into embarrassed silence. 'Muggles' and 'tea'—words that sound right only in Raymond Chandler novels now.

8. (Earlier example.)

1869 S. R. Hole *Bk. Roses* vi. 77 The autumn leaves.., decayed to mould, are very advantageous to the Teas, Noisettes, and Bourbons.

9. a. *tea-jar* (examples), *-merchant* (earlier Amer. example), *-mug, -slop.* **b.** (Objective and obj. gen.) *tea-strainer, tea-swilling* adj.; (instrumental and parasynthetic) *tea-bathed, -coated, -drowned, -dunked* adjs.; also similative, as *tea-brown* adj.

1922 Joyce *Ulysses* 258 He smiled at bronze's tea-bathed lips, at listening lips and eyes. *Ibid.* 532 A nymph with hair unbound, lightly clad in tea-brown art colours, descends from her grotto. 1953 Dylan Thomas *Under Milk Wood* (1954) 48 Willy Nilly the Postman's dark and sizzling damp tea-coated misty pygmy kitchen. 1882 W. D. Hay *Brighter Britain!* I. 161 What will be the future of these young tea-drowned nations? 1973 M. Amis *Rachel Papers* 68 A small middle-aged man (with unusually big brown ears, like tea-dunked ginger-biscuits). 1870 C. Schreiber *Jrnl.* (1911) I. 74 An old Staffordshire Ware tea-jar. *Ibid.*, Our purchase of the George III tea-jar. 1983 J. Sligo *Concert Masters* iv. 105 The Chinese tea jar on the mantelpiece. 1781 S. A. Peters *Hist. Connecticut* 407 [To] exert themselves..in favour of the Bostonian tea-merchants. 1955 T. H. Pear *English Social Differences* viii. 186 There is said to be a class which considers the tea-mug very *chic*. 1906 Joyce *Let.* 8 Dec. (1966) II. 201 Your friend..ought to get a running kick in the arse for writing his tea-slop about it. 1967 E. A. Gollschewsky in *Coast to Coast 1965–6* 94 Ettie surveyed the tea-table. It was still fairly orderly... No tea-slops in saucers. 1906 *Daily Colonist* (Victoria, B.C.) 26 Jan. 4/6 Kitchen utilities..Tea Strainers. 1970 *Kay's Catal.* 1970/71 Autumn/Winter 895 A stainless steel tea strainer and a decorated ceramic tile are set into a Teak wood base in this contemporary Danish design. 1961 *Times* 2 Oct. 13/4 Arms akimbo, bridling, bristling, and scolding, the tea-swilling Dame would at last be caught in the mangle.

c. tea-bag, (*a*) *Canad.*, a bag for carrying provisions; (*b*) a small permeable bag of paper or cloth containing tea for infusion; **tea ball,** a ball of wire or perforated metal in which tea is placed for infusion; **tea bar,** a bar (Bar *sb.*[1] 28) at which tea is sold as a beverage; **tea basket** (earlier example); **tea-bell** (earlier example); **tea-billy**, also used in New Zealand; **tea-boiler** (example); **tea-bottle**, a bottle containing tea (sense 2 a); also *slang*, an old maid; **tea-box**, (*a*) (earlier example); (*b*) *Canad.*, a box for carrying food and cooking utensils on an expedition; **tea-boy**, (*a*) (later example); also used outside Ireland; (*b*) a youth (occas. a man) employed to serve tea to workers; **tea-break**, an interval, usu. between periods of work, when tea is drunk; **tea-brick**, a brick of compressed tea leaves (cf. *brick-tea* s.v. Brick *sb.*[1] 10); **tea-caddy** (earlier U.S. example); **tea-cake** (earlier examples); **tea-can**, a metal can used for brewing or carrying tea; **tea-canister** (earlier example); **tea cart** *U.S.*, a tea-trolley; **tea ceremony**, in Japan, the preparation and consumption of green tea, according to strict rules of ceremony, as an expression of Zen Buddhist philosophy; **tea-china** (earlier example); **tea-chop** [Chop *sb.*[5] 5], in China, a chop-boat or lighter for the transportation of tea; **tea-cloth** (earlier Amer. examples); **tea-cosy**, (*b*) in full *tea-cosy hat*, a round knitted woollen hat resembling a tea-cosy; **tea dance** = *thé dansant* s.v. *dansant *a.*; also *Canad.*, 'a social gathering held by Indians, so called because in the early days the Hudson's Bay Company contributed tea, bannock, etc.' (*Dict. Canadianisms*, 1967); hence **tea-dance** *v. intr.*; **tea-dancer**; **tea-dancing** *vbl. sb.*; **tea-dust**, tea of inferior quality, often made from leaves broken in the course of production; *attrib.* [tr. Chinese *chá yè mò* tea-leaf dust], used to designate a dark green or brownish (often speckled) glaze on Chinese pottery (see quot. 1899), esp. used on decorative ware; **tea-girl**, a girl who serves tea; **tea-glass**, a glass from which tea (esp. without milk) is drunk; **tea-green**, a shade of greyish green resembling the colour of tea; **tea-head** *slang* (orig. *U.S.*), a habitual user of marijuana (cf. sense 7 c above and *head *sb.* 7 e); **tea hostess**, a woman in charge of serving tea at a tea-party or other social occasion; **tea-hound** [*hound *sb.*[1] 4 e] *U.S. slang* (now *rare*), a man given to frequenting tea-parties; also in extended use, a lady's man (see quot. 1921); **tea-house** (further examples, in oriental contexts); **tea infuser** = *tea-maker* (*c*); **tea interval**, a break for afternoon tea or light refreshment (esp. during a cricket match); **tea-jacket** (earlier example); † **tea-kitchen**, a tea-urn (cf. Kitchen *sb.* 2 b); † **tea-ladle**, a ladle for serving tea; **tea-lady**, a woman who serves tea (esp. in an office); **tea machine**, a machine which makes or dispenses tea; **tea-maker**, (*b*) (earlier example); (*d*) an apparatus incorporating a timer and designed to be kept at the bedside which can be pre-set to make tea automatically at any time (typically on awaking); **tea master**, an expert in the proper conduct of the tea ceremony; **tea money**, money paid by employees for drinks of tea at work (in quot. 1906, money paid by an employer to employees to buy their own tea); also *transf.* (see quot. 1979); **tea olive** [from the Chinese use of the flowers to add scent to tea] = *sweet olive* s.v. *sweet *a.* and *adv.* C. 1 b; **tea pad** *U.S. slang*, a place where one can purchase and smoke marijuana; † **tea-paper**, the ornamental paper used as a wrapper for tea (*obs.*); **tea place** = *tea-shop b; **tea plate**, a small shallow plate for use at tea-time; **tea room**, (*a*) (earlier examples); (*b*) *U.S. slang*, a public lavatory used as a meeting-place by homosexuals; **tea-seed oil** = *tea oil* (*a*); **tea-service** (earlier examples); **tea-set** (earlier example); **tea-ship**, (*a*) (earlier example); **tea-stall**, (*b*) a stall at which tea is sold; **tea-tent**, a tent in which tea is served at an outdoor event; **tea-time**, (*a*) (earlier examples); also *transf.*; (*b*) (*rare*), the time occupied by or allowed for taking tea; **tea-towel** = *tea-cloth* (*a*); **tea-treat**, (chiefly in Cornwall) a publicly provided out-door tea-party for children, esp. of a Sunday-school; also *attrib.*; **tea-trolley**, a trolley (sense *3 c) for conveying tea-things; **tea-urn** (later example); **tea-wagon**, † (*a*) an East Indiaman used to carry cargoes of tea (*obs.*); (*b*) = *tea-trolley* above; **tea-ware** (earlier example); **tea-wrap**, a wrap worn by women and girls at tea (*rare*); **tea yellows**, a deficiency disease of the tea-plant, esp. in Africa, caused by a lack of sulphur and indicated by small, chlorotic leaves, and the eventual death of the bush.

1898 F. Russell *Explorations in Far North* 161 If a crooked knife, a tea bag, or anything that is in the heap is needed, everything is tumbled about until it is found. 1936 K. Conibear *Northland Footprints* p. xii, Give him a large piece of bannock from your tea-bag. 1940 R. Chandler *Farewell, my Lovely* xiii. 102 They put Dad in charge of the Bureau of Records and Identification, which in Bay City is about the size of a tea-bag. 1958 *Sunday Times* 30 Mar. 12/3 The sale of tea-bags is creeping up. 1977 *Lancashire Life* Feb. 19/1 Those who think that tea is grown in teabags will be pleasantly surprised to find the enormous number of teas blended and packaged in Britain. 1895 *Montgomery Ward Catal.* Spring & Summer 187/1 Pure Aluminium, Tea Ball, total length, 7 in. Ball 1¼ in. dia. 1929 *Nation* (N.Y.) 4 Dec. 666 The tea ball enables one to pull the tea out before it has given off its tannin. 1976 *National Observer* (U.S.) 16 Oct. 10/3 Peel and crush six cloves of garlic and tie them in a piece of cheesecloth or put them in a tea ball and add this to the warm liquid. 1952 *Times* 12 Nov. 3/2 Tea bars are increasing. 1976 *Lancs. Evening Post* 7 Dec. 2/2 Mrs. Alice Durdle serves tea to the over 60s at the Lilian Wood Memorial Centre tea bar in Market Street, Preston. 1891 *Queen* 31 Oct. p. xxxvii (Advt.), Drews' Patent En Route 5 o'clock Tea Basket. 1836 *Knickerbocker* VIII. 418 It was nearly time for the tea-bell to ring. 1939 J. Mulgan *Man Alone* viii. 81 Around him were spread his belongings..clothes, boots, two black tea-billies..and a grey blanket. 1839 A. Langton *Jrnl.* in *Gentlewoman Upper Canada* (1950) 101 The pie plates, too, are very nice, and also the little tea-boiler. 1909 J. R. Ware *Passing Eng.* 241/2 *Tea-bottle* (*Mid.-class*), an old maid—from the ordinary drink of spinsters. 1975 B. Meyrick *Behind Light* xv. 198 He unwedged the sought-after tea bottle from its place behind the pipes. 1758 A. Pitt *Let.* 10 Nov. in *Lett. Lady Suffolk* (1824) II. 252 So I design to send it [*sc.* a letter] with a tea box my sister left and does not want. 1972 S. Burnford *One Woman's Arctic* vii. 154 In no time at all had the team hitched up, and his rifle and teabox aboard. 1852 Ld. Granville *Let.* 19 Jan. in E. Fitzmaurice *Life Ld. Granville* (1905) I. iii. 68 The teaboys of our own and our neighbour's establishments. 1954 *Atlantic Monthly* Aug. 35/1 The auction porters ate their dinners off thick white plates brought over by a cross-eyed teaboy from a café down the road. 1963 *Times* 31 May 12/6 Brutus..said that because of the banning order he was no longer able to work as a teacher and had had to take a job as a 'tea boy' with a research worker at the University of the Witwatersrand, for which he got £10 a month. 1977 *Time Out* 28 Jan.–3 Feb. 3/2 He certainly writes with all the flowing panache of a trainee teaboy. 1948 *Brit. Jrnl. Psychol.* Mar. 113 Many reasons were given for the almost universal appreciation of the tea break. 1958 A. Sillitoe *Saturday Night & Sunday Morning* ii. 35 The light flashed: tea-break over. 1981 *Economist* 18 Nov. 17/2 Strikes during the contract period (like the present tea-break strike at BL) would bring heavy damages on the unions. 1962 L. Davidson *Rose of Tibet* v. 87 He had bought tea bricks..and a large cake of yak butter. 1981 *Times* 7 July 12/7 A food shop in Covent Garden..has introduced..tea-bricks, such as Chinese mandarins once used to pay their taxes. 1790 *Pennsylvania Packet* 7 Dec. 3/3 Joseph Anthony, Junior,..Has Imported..Tea cadies, cannisters and salts. 1832 L. M. Child *Amer. Frugal Housewife* 71 There is a kind of tea cake still cheaper. 1844 Dickens *Martin Chuzzlewit* xvii. 216 Tea and coffee arrived (with sweet preserves, and cunning tea-cakes in its train). 1890 H. K. Daniels *Me & Jim* 67 The new plumber he gave him no answer except to drink from his tea-can and go on reading where he'd left off. 1951 J. Fleming *Man who looked Back* x. 120 He picked up his tea-can. 1978 *Lancashire Life* Nov. 75/2 Erect, at Uncle Dan's immediate righthand, was a large, shining tea-can, its lid back in place. 1726 in *N. & Q.* (1942) 24 Jan. 46/1 Bowl & tea canister. 1934 Webster, Tea cart. 1958 P. de Vries *Mackerel Plaza* iv. 56 A teacart hove into view, laden with goodies. 1978 M. Delving *No Sign of Life* i. 15 His wife came into the room followed by the maid pushing a tea-cart. 1886 E. S. Morse *Jap. Homes* iii. 149 The party comes about by the host inviting a company of four to attend the tea-ceremony, and in their presence making the tea in a bowl after certain prescribed forms, and offering it to the guests. 1935 *Burlington Mag.* Mar. 147/2 The tea ceremony, a rite so essentially Japanese that it might be said to epitomize Japanese culture. 1980 J. Melville *Chrysanthemum Chain* 16 A classic four-and-a-half mat tea room with a blond foreigner in full formal Japanese dress performing the tea ceremony. 1790 J. Woodforde *Diary* 15 Dec. (1927) III. 235 My Maid Betty Dade breaking likewise the only Tea China-Slop-Basin..made me more fretful. 1876 F. W. H. Symondson *Two Yrs. abaft Mast* vii. 136 A large 'tea chop' (a tea barge) came alongside. 1886 R. Brown *Spunyarn & Spindrift* xxvii. 328 The river was so swollen by the rains that the tea-chops could not get through Foo-chow bridge. 1770 C. Carroll *Let.* 11 Oct. in *Maryland Hist. Mag.* (1918) XIII. 62 A Hierling..stole a napkin two Towels & a Tea Cloath wh we Recovered. 1881 C. C. Harrison *Woman's Handiwork* I. 49 The beautiful tea-cloth linen, with its firm round thread, the warp and woof of equal thickness, so common in England. 1966 Tea-cosy [see *ensemble *v.*]. 1975 M. Russell *Murder by Mile* x. 101 A scarlet tea-cosy hat perched on top of her hair. 1885 T. Gowanlock *Two Months Camp of Big Bear* 119 When the Indians held their tea-dances or pow-wows in times of peace, the squaws and their children joined in and it was a very amusing sight to watch them. 1916 W. Stevens *Let.* 23 Apr. (1967) 193 People are pretty much dependent on the same things as in New-York: band concerts, tea-dances and..coffee-parties. 1962 A. Buchwald *How much is that in Dollars?* 128 Now you can see why the Patterson-Johansson fight didn't mean much to me. Those kids in the U.S. were just tea-dancing. 1964 *Camsell Arrow* (Edmonton, Alberta) Summer 60/4 High point of their four months in the north was the invitation to join the Indians 'tea dancing' Anne said. 1965 *News of North* (Yellowknife, N.W. Territories) 29 July 5/4 The ceremony was marked by a tea dance, in which everyone joined. 1978 *Lancashire Life* Nov. 129/2 For in a brave gesture of defiance in the punk era, the management has resurrected the Sunday Afternoon Tea Dance. It's all very Palm Court, even if the palms are plastic. 1980 *Radio Times* 29 Nov.–5 Dec. 87/4 So keen are the Tea Dancers that they have picked up all these [dances]. 1946 *New Yorker* 2 Feb. 4 A Melba trio plays in the Café Pierre, where there is tea dancing daily. 1977 *Ibid.* 3 Oct. 95/1 Tea dancing at the Kempinski. This goes on every day. 1899 S. W. Bushell *Oriental Ceramic Art* xviii. 518 One of the best-known glazes..is the *Ch'a-yeh-mo*, or 'Tea-dust' glaze, produced by the insufflation of green enamel upon a yellowish-brown ground, which owes its color to iron. The combination produces a peculiarly soft tint of greenish tone, which was highly prized in the reign of *Ch'ien-lung*. 1909 *Cent. Dict.* Suppl., Tea-dust. 1922 Joyce *Ulysses* 57 Through the open doorway the bar squirted out whiffs of ginger, teadust, biscuitmush. 1945 W. B. Honey *Ceramic Art of China* 145 The 'iron-rust' and greenish 'tea-dust' glazes..are usually of Ch'ien Lung date. 1979 P. Nihalani et al. *Indian & Brit. English* I. 175 Good quality tea packaged in the form of leaf and known as 'leaf tea', and an inferior variety, comparatively inexpensive, called tea-dust. 1980 *Catal. Fine Chinese Ceramics* (Sotheby, Hong Kong) 84 A massive tea-dust bowl..with a finely speckled deep olive-green glaze,..the base covered in an ochre-yellow glaze. 1889 Kipling *From Sea to Sea* (1900) I. 444 The tea-girls giggled. 1976 *S. Wales Echo* 23 Nov. 6/9 Every employee.., from senior executives to tea girls, would be interviewed. 1898 A. Cahan *Imported Bridegroom* xi. 121 Jealousy..of the empty tea-glasses.., of the whole excited crowd. 1979 D. Gurr *Troika* xxxiv. 260 Alexey grabbing at the rail, tea glass dropping from his fingers. 1956 W. Edwards in D. L. Linton *Sheffield* 16 East of the River Trent it [*sc.* the Rhaetic] overlies the 'Tea-Green Marls' at the top of the Keuper. 1967 *Vogue* 1 Mar. (International Collection) 161/1 She loves the colours. White, pink, blue, butterscotch, tea green, [etc.]. 1953 W. Burroughs *Junkie* (1972) ii. 29 Perhaps weed does affect the brain with constant use, or maybe teaheads are naturally silly. 1967 *Guardian* 8 July 6/2 Doctors, commissions, and plain tea-heads have been ready to go on record about the innocence of the weed cannabis sativa. 1970 Tea-head [see *head *sb.* 7 e]. 1976 *Norwich Mercury* 19 Nov. 2/5 Mrs J. Bowhill acted as model for the evening dress... Tea hostesses were Mrs Kedge and Mrs Williams. 1921 *Dialect Notes* V. 111 *Tea-hound*, a lady's man. 1925 *Scribner's Mag.* Oct. 353/2 He was a regular tea-hound, he was seen at so many teas. 1763 J. Bell *Travels from St. Petersburg* II. x. 54 From the temple we went to a publick tea-house, where we saw many people drinking tea [in Peking]. 1959 L.-H. Liang tr. *Ting Yi's Short Hist. Mod. Chinese Lit.* x. 221 In the rear areas, there were other dramatic forms akin to the 'street play', such as the 'tea-house play', 'demonstration play' and the 'lantern play'. 1972 K. Lo *Chinese Food* 1. 50 There are no pubs or bars and most of the informal leisurely drinking takes place in tea-houses. 1889 A. James *Diary* 5 Aug. (1965) 52 A note of farewell from Mr. Godkin with a tea-infuser. 1907 *Yesterday's Shopping* (1969) 188/3 Travellers' Companion... For making tea when touring, boating, &c. ..Comprises kettle,..stand, spirit stove,..and muslin tea infuser. 1960 *Guardian* 4 Jan. 3/1 Collapsible tea infusers. 1923 E. P. Oppenheim *Inevitable Millionaires* xxix. 288 'I haven't done wrong in making the tea, have I?' she asked timidly... 'Of course not,' George Henry assured her. 'The tea interval is an established custom.' 1976 Dexter & Makins *Testkill* 143 In the tea interval..I slipped into the pavilion. 1887 *Girl's Own Paper* 25 June 618/3 New tea-jacket, or *après midi*, for indoor wear. 1770 J. Wedgwood *Let.* 24–26 Dec. (1965) 100 Mr Boulton..shewed me some bodys and necks made of Porcelaine coloured green to be mounted in Ormoleau for Tea Kitchens. 1808 Jane Austen *Let.* 27 Dec. (1952) 243 A silver tea-ladle is also added [to the list]. 1964 *Listener* 13 Feb. 287/1 'Filthy,' said a friend's tea-lady the next morning. 1980 *Times* 13 Nov. 4/8 The tea trolley is being wheeled back... Two years ago, it seemed the ubiquitous tea lady was vanishing beneath a tide of vending machines. This year..automated services are in decline. 1963 *Punch* 8 May 675/1, I..fetched the tea machine into the house. 1972 J. Thomson *Not One of Us* xvii. 227, I kept..some paper cups. I used to nick them from the tea machine at the warehouse. 1814 Jane Austen *Mansfield Park* III. vii. 160 There was..found a chair, and with some hasty washing of the young tea-maker's, a cup and saucer. 1961 'T. Hinde' *For Good of Company* xix. 214 Mary had switched on the bedside tea-maker. 1970 *Sunday Times* 20 Dec. 26/3 When the clock on the tea-maker began to go backwards its owners got their alarm call and a nice pot of tea at three a.m. 1914 Y. Noguchi *Through Torii* 2 It is the high art of the tea-master to make you really taste the water beside the taste of the tea. 1974 *Times Lit. Suppl.* 25 Oct. 1190/3 In Kamakura for the first time a tea master did the tea ceremony for me. 1906 E. Dyson *Fact'ry 'Ands* xvii. 225 We're..puttin' down er mill that'll..never look fer tea money. 1962 L. Deighton *Ipcress File* i. 11 The office tea money. 1979 *Rydge's* (Sydney) Apr. 68/2, I observed a case in Thailand, where payoffs are euphemistically called tea-money. 1952 M. Steen *Phoenix Rising* vi. 117 An overpowering fragrance of tea-olive rose from under her window. 1975 *Country Life* 2 Jan. 39/3 The grassy glade leading from river to house..heavy with the scent of tea olive..and banana tree. 1938 *New Yorker* 12 Mar. 47/1 All tea pads, or marijuana joints, use the blue lamps and nickel machines to induce and sustain the hashish mood. 1963 *Lancet* 9 Nov. 989/2 For a few years the cult of the 'tea pad'..threatened to be imported from the United States. 1814 F. Burney *Let.* 28 Oct. (1978) VII. 488 If you write to me again upon a scrap that can hardly arrive—I shall answer upon a bit of Tea paper. 1884 *Birmingham Daily Post* 23 Feb. 3/4 Lithographic printers. Wanted, a man..well up in Tea-paper and Commercial

Work. 1929 D. H. Lawrence *Let.* 11 Jan. (1932) 780 We were in Toulon yesterday..and went to the same tea-place. 1978 P. Marsh et al. *Rules of Disorder* iii. 72 At the back there you can see down to the tea place underneath. 1862 M. D. Colt *Went to Kansas* iii. 48 Have arranged on them..our five tin plates, two tin cups, one tin tumbler, the nine tea-plates. 1972 *Country Life* 9 Mar. 547/3 These plates were made by the Britannia China Company..between 1895 and 1906. Such plates were known as..tea plates. c 1702 C. Fiennes *Journeys* (1947) iv. 359 Another little closet with the tea equipage and under that was such a little tea roome within the drawing roome. c 1748 Richardson *Let.* in *Corr.* (1804) III. 317 Miss Chudleigh is gone into the tea-room. 1970 [see *nelly[2] 3]. 1976 *New Society* 29 Jan. 227/2 Sentences for what are known in America as 'tearoom' offences—homosexual sodomy or oral copulation—vary. 1884 *Encycl. Brit.* XVII. 746/2 Tea-seed oil is a commercial product in China, where it is used for food, lighting, and soap-making. 1951 E. David *French Country Cooking* 220 Tea seed oil. Much lighter than olive oil and preferred by many for salads. 1809 A. Burr *Private Jrnl.* (1903) I. 253 A splendid tea service of silver and two cups. 1838 J. Romilly *Diary* 26 Feb. (1967) 140 Treated myself with a new tea Service for the occasion (cost 5¼ Gnas). 1786 J. Wedgwood *Let.* 30 June (1965) 297 A single line of colour put on..while it is in the clay state..upon our beer mugs, flower-pots, tea and coffee sets..constitutes *fayence*. 1859 *Harper's Mag.* Sept. 507/2 You might have seen their sign—ay, and their fine staunch tea-ships too—any day you chose to stroll down South. 1889 Kipling *From Sea to Sea* (1900) I. 360 The lower stories were full of tea-stalls and tea-drinkers. 1962 R. Prawer Jhabvala *Get Ready for Battle* iii. 94 A tea-stall under a tree built on upturned kerosene tins. 1890 *Monthly Packet* Christmas 188 She..was not sorry to depart to the tea-tent. 1934 'E. M. Delafield' *Provincial Lady in Amer.* 6 Go with Robert..to..Agricultural Show... We..repair to tea-tent... I drink strong tea and eat chudleighs. 1977 *Oxf. Diocesan Mag.* Oct. 20/2 A tea-tent..apart from affecting the custom of the catering contractors, would give a false picture of the Church as a tea-making machine. 1741 Richardson *Pamela* II. 223, I hope to join you there by your Tea-time in the Afternoon. 1749 J. Cleland *Mem. Woman Pleasure* I. 47 He sat down..and all tea-time kept ogling me. 1936 *Punch* 19 Feb. 204/1 It's still tea-time, you know. 1963 *Times* 31 Jan. 3/3 In the evening of life—or at any rate the tea-time—it is occasionally pleasant to look back. 1863 S. S. Jones *Northumberland* 116 The guid lady shakes her lap an' rubs an' scrapes at her gown wi' the tea-towel. c 1909 D. H. Lawrence *Collier's Friday Night* (1934) ii. 55 *Beatrice*:..You want to wrap it in a damp cloth now. Have you got a cloth? *Ernest*: What?—a clean tea-towel? 1980 *Habitat Catal.* 1980/81 111/1 Honeycomb weave teatowel. Pure cotton. Excellent for easy drying up. c 1748 Richardson *Let.* in *Corr.* (1804) III. 317, I thought..you were of the party at the tea-treats. 1898 C. P. Penberthy *Warp & Woof of Cornish Life* 153 Whas our lil tay-trait to a towser? *Ibid.* 168 Go long up tay-trait field. 1977 *West Briton* 25 Aug. 22/6 Mr. Ken Roberts..said 150 traditional tea-treat buns would be distributed free to children on the estate. 1937 A. Thirkell *Summer Half* x. 275 Mrs. Keith had then bought an excellent tea-trolley with rubber wheels and ball bearings. 1958 J. Cannan *To be a Villain* i. 24 A tea-trolley now stood laden with sandwiches, cakes and buttered buns. 1980 *Times* 13 Nov. 4/8 The tea trolley is being wheeled back. a 1948 D. Welch *Voice through Cloud* (1950) i. 9 This noble room was spoilt by a counter with sizzling tea-urns. 1840 R. H. Dana *Two Yrs. before Mast* xxxiv. 428 Like a true English 'tea-wagon'. 1878 *Appleton's Jrnl.* Jan. 9/2 The good, heavy-bowed, square-countered 'tea-wagons', as the sailors call them, meant for cumbrous freight, heavy stowage, and long passages. 1921 *Daily Colonist* (Victoria, B.C.) 30 Mar. 18/2 A neat Tea Wagon, in walnut finish, fitted with a moveable glass tray top, and mounted on four rubber tired wheels. 1939 J. B. Priestley *Let People Sing* ii. 23 The magazine boy..called to the tea-wagon girl: 'Come and 'ave a look.' 1978 D. Bloodworth *Crosstalk* xv. 122 Don't tell me the tea wagon's come and gone already? 1766 J. Wedgwood *Let.* 15 Sept. (1965) 42 The Teaware, vases, and all other pretty things I shall let alone until I have the pleasure of seeing you here. 1909 H. G. Wells *Tono-Bungay* ii. ii. 176 My aunt too, looking bright and pretty, in a blue-patterned tea-wrap. 1931 *Ann. Rep. Dept. Agric. Nyasaland 1930* 32 Tea Yellows—Investigations into this disease have been carried out. 1958 T. Eden *Tea* ix. 91 Tea yellows, caused by sulphur deficiency, is less severe under shade trees than in open situations.

tea, *v.* Add: **2.** (Earlier example.)
1810 G. Betts *Diary* in K. F. Doughty *Betts of Wortham* (1912) xxix. 286 Mr. Lee..came and *tea-ed* here.
teaing *vbl. sb.* (earlier example).
1845 *Ainsworth's Mag.* VII. 504 During my 'teaing' I was amused with the conversation of my companions.

tea-berry. (Earlier example.)
1818 W. P. C. Barton *Compendium Floræ Philadelphicæ* I. 194 *Gaultheria*..*procumbens*... Mountain Tea. Tea-berry. Partridge-berry. Wintergreen.

teach, *v.* Add: **II. 6. d.** Also without direct object.
c 1863 T. Taylor *Ticket-of-Leave Man* ii. 33 Sam! is it? Confound him! I'll teach him.
e. *teach yourself* (a subject): vbl. phr. used *attrib.* to designate a textbook or manual intended for use without the assistance of a teacher.
The phr. is derived from the titles of books in the Teach Yourself series, published from 1938.
1938 M. Thomas (*title*) Teach yourself embroidery. 1960 G. Butler *Death lives Next Door* He was..going through the Catalogue issued with the Teach Yourself Everything Series. 1961 *Guardian* 4 Feb. 14/6 As I was taught in a teach-yourself book. 1978 P. O'Donnell *Dragon's Claw* ii. 29, I usually spend a few hours with the tape recorder and a Teach Yourself Russian course.

teach (tiːtʃ), *sb.* Colloq. abbrev. of Teacher 2 a.
1958 F. Norman *Bang to Rights* iii. 90 'Now now give him a chance,' said the teach. 1976 A. Hill *Summer's End* i. 6, 'I always suspected it, Hill,' Teach had called across the classroom. *Ibid.* 9 The Teach with the cardboard box stopped in front of each kid and gave him or her a paper bag.

teache, var. Tache *sb.*[3]

teacher, *sb.* Add: **3.** *teacher-factory, -trainee, -trainer, -training* (later examples); appositive, as *teacher-librarian*; *teacher-proof* adj.; also pertaining to each element, as *teacher–pupil* adj. (cf. *pupil–teacher* s.v. *pupil *sb.*[1] 3 b); **teachers' aide**, an assistant employed to help the teaching staff of a school in a variety of duties (see quot. 1967).
1889 'Mark Twain' *Connecticut Yankee* x. 118, I had started a teacher-factory and a lot of Sunday-schools. 1975 *Language for Life* (Dept. Educ. & Sci.) xxi. 304 Except for the Teacher-Librarians' Certificate there have been few opportunities for teachers to acquire help in how to organise and manage a library. 1979 *Jrnl. R. Soc. Arts* July 487/1 A simple handbook for those teacher-librarians in Commonwealth developing countries who are called on to undertake this duty without any previous experience. 1964 P. Strevens *Papers in Lang.* (1965) ii. 32 It is sometimes necessary to prepare 'teacher-proof' materials, if it is known in advance that the proficiency of the teacher is not going to be up to the optimum required. 1965 M. Morse *Unattached* iv. 131 Despite the teacher-pupil relationship the worker could in no way afford to make the girls feel inferior to herself. 1977 *New Yorker* 19 Sept. 44/1 He is a warm, compassionate man, outside the private teacher–pupil relationship in music. 1956 *Sun* (Baltimore) 17 Nov. 6/3 It is another thing when a teacher is also required to be clerk, accountant and nursemaid, as the Parent-Teacher Association of Howard Park Elementary School has recognized in its hiring of two teachers' aides. 1967 *Children & their Primary Schools* (Dept. Educ. & Sci.) I. vi. xxiv. 330 The type of help that is..given by teachers' aides, who ought to have equal status with nursery assistants.., falls into three kinds: (a) Help that amounts to an extra pair of hands for the teacher... (b) help..from those with special skills. This could be available for needlework, art and craft, handicraft... (c) supervising children after school hours while they are waiting for their parents. 1959 *Listener* 12 Mar. 463/1 A graduate teacher-trainee. 1982 *Underground Grammarian* Nov. 6/2 He might actually decide to become a student of something rather than a teacher-trainee. 1964 *Economist* 22 Aug. 709/3 Sending teachers, and teacher-trainers, to the country in question. 1977 P. Strevens *New Orientations Teaching Eng.* vi. 79 Many teacher-trainers regard statements of the kind 'We concentrate on practical teaching—none of this theoretical nonsense!' as if they were robust common sense. 1949 M. Mead *Male & Female* 456 Directed towards particular problems—adolescence, teacher-training, nutrition, housing. 1967 *Listener* 14 Sept. 351/2 You could go to a college of education—they used to be called teacher training colleges. 1975 *Language for Life* (Dept. Educ. & Sci.) xxiii. 331 Our Report emerges at a critical and uncertain time in the development of teacher training.

teacherage (tiːtʃərɪdʒ). *N. Amer.* [f. Teacher *sb.* + -age, after Parsonage (sense 2), Vicarage (sense 3), etc.] A house or lodgings provided for a teacher by a school.
1916 *Boston Evening Transcript* 12 July xi. 3 Educational officials of Monroe County are showing much interest in the new movement for establishing so-called teacherages—cottages for country school teachers—which is spreading rapidly in several Far-Western States. 1959 R. E. Campbell *I would do it Again* vi. 22 There was a teacherage with all the necessaries. 1968 *Globe & Mail* (Toronto) 15 Jan. 26/6 (Advt.), Required immediately, qualified teacher for one-room school... Three-room teacherage available. 1976 T. Walker *Spatsizi* ix. 81 Hungry dogs rushed out to bark, and this brought Lester Dorsey to the door of the teacherage.

teacherly (tiːtʃəlɪ), *a.* [f. Teacher *sb.* + -ly[1].] Of, pertaining to, or characteristic of a teacher; schoolmasterly, schoolmistressy; pedagogic.
a 1683 J. Hull in *Archeologia Americana* (Amer. Antiquarian Soc.) (1857) III. 173 Mr. John Norton..who continued with us three years and upward..laboring in God's work, and joined in a teacherly office with us. 1934 Webster, *Teacherly*,..teacherlike. 1979 *Washington Post* 21 Jan. G1/5 Hesse the German who became a Swiss, the teacherly Peter Pan who hankered after things Asian. 1980 E. Blishen *Nest of Teachers* ii. xii. 124 He knew the teacherly value of conspiracy. 1982 *N. & Q.* Feb. 80/2 A symptom of the book's teacherly liveliness is its wit.

teach-in. orig. *U.S.* [f. Teach *v.* + *-in[3] (after *sit-in*, etc.).] An informal debate (often of some length) on a matter of public, usu. political, interest, orig. between the staff and students of a university. Hence, a conference attended by members of a profession on topics of common concern. Also *loosely*, a lecture or meeting held for the purpose of discussion or disseminating information.
1965 *N.Y. Times* 25 Mar. 9/1 Bomb scares marked the start tonight of a 12-hour series of rallies, speeches and seminars sponsored by some 200 University of Michigan faculty members to protest United States policy in Vietnam... Policemen evacuated..the site of the protest gathering which the faculty members have named a 'teach-in'. 1965 *Economist* 24 Apr. 416/1 Universities all over the country [*sc.* USA] have conducted informal 'teach-ins' on Vietnam, running from eight in the evening to eight the following morning. 1965 *Times* 17 June 8/5 This free-for-all debate..was called by the ugly new jargon name of 'teach-in'—a concept recently invented at Harvard, which has crossed the Atlantic. 1967 McLuhan & Fiore *Medium is Massage* 101 The dropout represents a rejection of nineteenth-century technology... The teach-in represents a creative effort. 1969 *New Scientist* 30 Jan. 219/1 The great Edinburgh Teach-in..on chemical and biological warfare. 1971 *Ibid.* 24 June 741/1 Engineers have run a series of 'teach-ins' to show designers how they wish to apply the new rules. 1973 R. Ludlum *Matlock Paper* iii. 21 Six days of riots on campus. Half a semester lost on teach-ins. 1975 D. Lodge *Changing Places* iv. 138 A two day teach-in on the constitution and scope of the proposed commission.

teaching, *vbl. sb.* Add: **4.** *teaching aid, load, material, post, process*; **teaching hospital**, a hospital at which medical students are instructed; **teaching machine**, a mechanical device for giving instruction in the form of a teaching programme which allows a pupil to progress according to his response to questions of choice.
1966 *Rep. Comm. Inquiry Univ. Oxf.* II. 470 They might even be encouraged to use occasionally the odd teaching aid. 1980 *Underground Grammarian* Dec. 1/2 Think of the audio-visual devices and the teaching aids. 1963 in A. Heron *Towards Quaker View of Sex* 51 All the large teaching hospitals have psychiatric out-patient departments. 1980 *Brit. Med. Jrnl.* 29 Mar. 924/2 The London teaching hospitals, which for so long had served their local population, and which had now been set aside to serve the needs of education, began instead to bear the brunt of the specialised services. 1958, etc. Teaching load [see *load *sb.* 4 c]. 1958 *Science* 24 Oct. 971 (*caption*) Student at work on a teaching machine. 1969 J. Argenti *Managem. Techniques* 215 This method can be used..with a teaching machine. These machines consist of a box like a television set in which there is a film strip. 1972 H. J. Eysenck *Psychol. is about People* iii. 147 Sidney L. Pressey in the mid-1920s designed the precursors of our modern automated teaching machines. 1960 *Tuscaloosa* (Alabama) *News* 20 Nov. 4/4 The student fits teaching material into the box and then uses them [*sic*] at his own speed. 1962 *Sunday Times* (Colour Suppl.) 10 June 4 There have been no sinecures or teaching posts for famous jazzmen. 1975 *Language for Life* (Dept. Educ. & Sci.) xvii. 254 We regard recording as an essential element in the actual teaching process.

teaching, *ppl. a.* Add: **b.** Special collocations, as **teaching elder**: see Elder *sb.*[3] 4; **teaching fellow** *U.S.*, a student at a graduate school who carries out teaching or laboratory duties in return for a stipend, free tuition, or other benefit.
1642 T. Lechford *Plain Dealing* 15 Some Churches have no ruling Elders, some but one, some but one teaching Elder, some have two ruling, and two teaching Elders. 1735 in C. Hazard *Thos. Hazard* (1893) 226 We the Subscribers, Teaching Elders or Pastors of the first gathered..Church in Boston New England. 1936 S. E. Morison *Three Centuries of Harvard* i. 18 There were no funds to maintain more than two teaching fellows. 1979 C. MacLeod *Luck runs Out* (1981) xvii. 169 He'd come there as a teaching fellow... He taught the subject ably.

tea-cup. Add: **a.** (*b*) With reference to fortune-telling by means of interpreting the arrangement of tea-leaves left in a cup. Cf. *tea-leaf 1.
1883 C. S. Burne *Shropshire Folk-Lore* xxi. 277 The apparitions which..nurses used to discover in their teacups when they had..emptied the last remains of the tea in such a manner as to leave the dregs scattered well over the bottom and sides of the cup. 1921 C. Kent *Fortune-Telling by Tea-Leaves* ii. 24 A confused looking tea-cup, without any symbols..is useless for the purpose of divination. 1954 M. Sharp *Gipsy in Parlour* xii. 127 Cook..had an eye for tall dark strangers, who frequently appeared in her tea-cup. 1976 A. E. Lindop in *Winter's Crimes 8* 216 She can 'see' what's best for us... She'll look into our teacups.
c. (Earlier example.) Similarly *tea-cup storm*, etc.
1854 W. B. Bernard (*title*) A storm in a teacup. 1932 *Times Lit. Suppl.* 15 Sept. 639/1 Those old disputes were no teacup squalls. 1951 *Sport* 16–22 Mar. 14/3 Earlier this season a slight 'teacup storm' occurred in Yorkshire Rugby Union circles. 1981 W. Safire in *N.Y. Times Mag.* 15 Feb. 11/1 In the midst of this teacup contretemps came a clear message from John Radosta.

tea-drinker. Add: (Earlier example.)
1737 *London Mag.* Apr. 186/1 Considering the Number of Tea Drinkers, it [*sc.* tea-drinking] has done a great deal more Hurt than Dram-Drinking.
tea-drinking *vbl. sb.* (earlier examples in all senses).
1737 *London Mag.* Apr. 183/2 (*heading*) Of Diet in general, and the bad Effects of Tea-drinking. 1781 A. Storer *Let.* 28 June in *15th Rep. R. Comm. Hist. MSS.* App. vi. 508 in *Parl. Papers* 1897 (C.8551) LI. 1. Lady Craven gave a tea-drinking last night. 1793 W. B. Stevens *Jrnl.* 8 July (1965) 91 Walked with Mrs Cutts, etc...to Schobley Mill, a Tea-Drinking Place.

teaed (tiːd), *a.* *U.S. slang.* Also **tea-d.** [f. *tea *sb.* 7 c + -ed[2].] In a state of euphoria induced by alcohol or marijuana. Usu. with *up*.
1928 L. E. Lawes *Life & Death in Sing Sing* iv. 53 'Didn't alcohol have something to do with your coming here?' 'Yes, sah, dey was bofe considerable teaed up.' 1944 *War Med.* VI. 383/2 Just those thoughts will drive me mad—thinking about my 'boys' all 'tea-d up', and here I am, sitting and thinking about it, and I can't get it. 1966 C. Himes *Heat's On* xvii. 157 The driver was teaed to the gills and on a livewire edge.

Teague. Restrict *Obs.* or *arch.* to sense in Dict. and add: **2.** Usu. in form **Taig** (teɪg). In Northern Ireland, a Protestant term of contempt for a Roman Catholic.
1971 *Times* 13 May 2/6 Taig is Protestant slang for a Roman Catholic. 1973 *Spectator* 3 Mar. 263/2 The Prods are only having their shops blown up and suffering a few slight cases of murder; the Teagues are losing their souls. 1978 D. Murphy *Place Apart* vii. 133 In times of stress Loyalist paramilitaries can easily rouse large mobs and lead them out of the ghettos on Taig-bashing expeditions. 1982 *Observer* 31 Oct. 8/3 This week a new slogan appeared along the Shankill Road, the backbone of Protestant West Belfast. It read: 'All Taigs are targets.'

teak. Add: **1. b.** A fashion shade resembling the colour of teak-wood, a rich reddish brown.
1934 in Webster. 1971 [see *mole *sb.*[1] 6].
3. *teak log, -oil*; *-built* (earlier lit. example), *-panelled, -veneered* adjs.
1835 J. E. Alexander *Sketches in Portugal* viii. 179 In May, the fleet of her Most Faithful Majesty consisted of the following ships:—..50, Don Pedro,..Very strong, teak-built. 1889 Kipling *From Sea to Sea* (1899) I. iii. 220, I saw the elephants playing with the teak logs. 1968 A. Dimont *Bang Bang Birds* vi. 89 We stepped into a small, teak-panelled lift. 1968 J. Arnold *Shell Bk. Country Crafts* 198 For such articles as salad-bowls..teak-oil is used. 1970 *Interior Design* Dec. 753/3 Small teak-veneered tables. 1979 P. Way *Sunrise* xv. 157 Two wrought-iron gates, massively reinforced by teak logs.

tea-kettle. Add: Phr. *ass* (= arse) *over tea-kettle*, head over heels (cf. *arse over tip* s.v. *arse *sb.* 1 b). *U.S. slang.*
1963 T. Pynchon *V.* i. 23 Fast enough..only to send Profane, garbage can and lettuce leaves flying ass over teakettle in a great green shower. 1977 J. Crosby *Company of Friends* vii. 51 Sascha's horse..stopped short... Sascha went ass over teakettle into the brambles.

teal. Add: **1. c.** A shade of dark greenish blue resembling the patches of this colour on the head and wings of the teal.
1923 *Daily Mail* 14 Feb. 10 (Advt.), Jersey frocks... Colours: Teal, Purple and Champ. 1978 *N.Y. Times* 29 Mar. A 14 (Advt.), Both in a delicious new shade of teal... we call it Prussian blue!
3. teal blue, a shade of dark blue tinged with green (cf. sense 1 c above).
1949 *Dict. Colours Interior Decoration* (Brit. Colour Council) III. 26/2 *Teal blue*, a descriptive colour name from the plumage of the small freshwater duck. 1963 *New Yorker* 1 June 75 Sandwich-board sheaths in teal-blue linen are piped down the sides with double rows of lime. 1980 M. H. Clark *Cradle will Fall* iv. 24 Her teal-blue uniform.

tea-leaf. Add: **1.** Also with reference to fortune-telling. Cf. *tea-cup a (*b*).
1883 C. S. Burne *Shropshire Folk-Lore* xxi. 277 A stalk or long tea-leaf floating in the tea was called a 'chap'... was at once taken out and laid on the back of one hand, which was then struck sharply with the palm of the other, in order to see whether the 'chap' would come to the back door or the front. 1931 E. Sackville-West *Simpson* i. 66 Tea-leaves, thought Amy, tell fortunes. 1941 [see *riddle *v.*[1] 2 c]. 1981 *Times* 21 Jan. 11/4 They..read marriage prospects in the tea leaves.
2. Rhyming slang for 'thief'. So **tea-leafing**, thieving.
1899 C. Rook *Hooligan Nights* ii. 23 He could do more than his share at tea-leafing, which denotes the picking up of unconsidered trifles. 1903 C. Booth *Life & Labour of People in London* XVII. ii. 139 'Tea-leaf' is for some inexplicable reason the name used by the police for pickpockets. 1930 A. Bennett *Imperial Palace* v. 20 The badinage..was more picturesque... 'You dirty old tea-leaf.' 1963 J. Prescot *Case for Hearing* ii. 36 Proper tea-leaves they looked, the pair of 'em. 1977 D. Clark *Gimmel Flask* iv. 63 A tea-leaf wouldn't find the key on your person if he broke in.

teallite (tiːləɪt). *Min.* [f. the name of Sir J. J. H. Teall (1849–1924), English geologist + -ite[1].] An orthorhombic sulphide of lead and tin, $PbSnS_2$, found as soft, thin, dark grey crystals having a metallic lustre.
1904 G. T. Prior in *Mineral. Mag.* XIV. 21 (*heading*) On teallite, a new sulphostannite of lead from Bolivia. 1962 W. A. Deer et al. *Rock-Forming Minerals* V. 8 It [*sc.* cassiterite]..has also been recorded from the weathering of teallite and stannite.

team, *sb.* Add: **II. 4. b.** (Earlier examples in Cricket and Football and later gen. examples.)
1846 W. Denison *Cricket: Sk. Players* 32 Hayward..having become a resident at Cambridge, joined the 'team' of that distinguished Club. 1874 *Bell's Life in London* 14 Mar. 5/2 Although the game was won by Scotland..the English team played splendidly. 1921 G. B. Shaw *Back to Methuselah* ii. 53 You will find yourself at the head of a rabble of Socialists and anti-Socialists, of Jingo Imperialists and Little Englanders,..of Syndicalists and Bureaucrats..and the impossibility of keeping such a team together will force you to sell the pass again to the solid Conservative Opposition. 1923 *N.Y. Times* 15 July vi. 1/6 The method of the comedy team remains more or less unvaried. The team is composed, in the first place, of a comedian and a 'straight' man. 1947 *Ann. Reg. 1946* 314 To prevent further clashes General Marshall organised 'teams' composed of an American, a Nationalist, and a Communist member, to visit both parties and to try to create a better spirit [in China]. 1951 *Times* 26 Nov. 2/5 The report..has been compiled by a team from the [metal-finishing] industry which visited the United States last year. 1965 M. Spark *Mandelbaum Gate* iii. 66 Russeifa's one of the most conscientious men in the medical team. 1972 *N.Y. Law Jrnl.* 24 Oct. 15/9 Team I is assigned to preside on the circuit for the October 1972 Term. 1978 *Nagel's Encycl.-Guide: China* 272 The basic cells of agrarian collectivisation are the brigade and the team for the moment.
c. *spec.* A gang. *slang* (chiefly *Criminals'*).
1950 in Partridge *Dict. Underworld* Add. (1961) 814/1. 1955 D. W. Maurer in *Publ. Amer. Dialect Soc.* xxiv. 83 Sometimes a *team* [of pickpockets] is *two handed*, while a *troupe* is *three handed* or larger. 1959 *Observer* 1 Mar. 10/1 Mainly the older brothers in long-resident..families, they are known and feared by other 'teams' (gangs) in North London as the Punchers. 1970 P. Laurie *Scotland Yard* viii. 184 We had a whisper about a team going to do a certain pay van. 1973 'J. Patrick' *Glasgow Gang Observed* ii. 21 The boys themselves never used the word 'gang', always 'team'.
5. c. *fig.* Usually *a whole team*. *U.S. colloq.*
1832 *Polit. Examiner* (Shelbyville, Kentucky) 17 Nov. 4/2 'Whoop! Ain't I a horse?' 'A whole team, I should think,' said Rainsford. 1832 [see *half *sb.* 7 h]. 1843 'J. Slick' *High Life N.Y.* II. 193, I tell you what, he's a hull team, and a horse to let. 1854 *Knickerbocker* XLIV. 416 (Th.), Jump him up when you will, and you'll find him a 'full team' at anything. 1856 G. D. Brewerton *War in Kansas* 270 Avow yourself ready to declare that..a clear-grit Yankee woman quite equal, upon an emergency, to what, in vulgar parlance, is quaintly styled 'a whole team, and *a dog under the wagon*' to boot. 1922 *Dialect Notes* V. 180 *Whole team an(d) little dog under the waggin'*, n. phr. Used facetiously to indicate one's self-importance, energy, etc. Alabama.
V. 11. (sense 4 b) *team-building, leader, manager, player, sport*; (in a team ministry; see below) *team rector, vicar*; **team handball**, a game played by two teams of seven players each on a rectangular court using a ball directed only with the hands; **team honours**, honours awarded to a sporting team; **team-man**, (*a*) (earlier example of *teamsman*); (*b*) with preceding descriptive adj., a member of a sporting team who co-operates (well or badly) with his colleagues; **team-mate**, (*a*) = Teamster (Webster 1934); (*b*) a fellow member of a team; **team ministry**, a group of clergy of incumbent status who minister jointly to several parishes under the leadership of the team rector (contrasted with *group ministry*, in which all members function as equals); the administration of a scheme for such an operation; **team policing**, community policing; **team race**, a race which is won by the team whose members finish on aggregate in higher positions than their opponents; **team spirit**, the spirit of subordination of personal interests to those of the team; **team-talk**, a talk addressed to a team, or a discussion amongst a team; **team-teaching** *vbl. sb.*, the teaching of students by a team of teachers working together; hence (as back-formation) **team-teach** *v. intr.* and *trans.*; **team-work**, (*d*) work done by persons working as a team, i.e. with concerted effort.
1946 *Nature* 12 Oct. 497/1 These are essential conditions for successful team-building and the inherent loyalty it implies. 1970 *Jrnl. Health, Phys. Educ. & Recreation* Mar. 46/1 To the uninformed spectator, the game of team handball would look like a combination of football, basketball, and lacrosse... It is often confused with a popular squash-related sport also dubbed 'handball'. 1978 *Official Associated Press Sports Almanac* 764 The sport spread to gymnasiums throughout Germany and Eastern Europe and full recognition of team handball was achieved when it was included in the 1936 Berlin Olympics. 1928 *Daily Mail* 7 Aug. 15/5 Eagle Road Club secured team honours. 1962 E. Snow *Other Side of River* (1963) lviii. 440 They chose me as team leader. 1977 *R.A.F. News* 11–24 May 8/4 Deputy team leader Chf Tech Mick Young. 1977 *Times* 9 Sept. 3/5 Team leaders, the preferred name for those formerly known as charge hands. 1954 A. G. Moyes *Austral. Batsmen* x. 151 A fighter who was an excellent team-man and therefore most valuable. 1976 J. Snow *Cricket Rebel* 138 His brother Eric [Bedser]..was overheard to say that I was not a good team man. 1926 E. Hemingway *Sun also Rises* xix. 247, I had coffee out on the terrasse with the team manager of one of the big bicycle manufacturers. He said it had been a very pleasant race. 1976 *Evening Post* (Nottingham) 15 Dec. 23/1 Team manager John Sherriff believes the side is now on the right track. 1915 M. E. McLoughlin *Tennis as I play It* xi. 231 Service and the net position go together, the initial stroke giving the server the opportunity to reach the net where his team-mate is already stationed. 1942 Berrey & Van den Bark *Amer. Thes. Slang* § 440/4 *Confederate or partner*..team-mate. 1954 W. K. Hancock *Country & Calling* viii. 223 An orderly row of boxes into which to put the facts that he and his team-mates (for 'team work' is very much the fashion) are collecting. 1977 *Times* 15 July (Motor Racing Suppl.) p. vi/1 James Hunt..was put out of the race by a crash... His German team-mate..had retired three laps earlier. 1964 L. Paul *Deployment & Payment of Clergy* xv. 142 One much-publicised remedy for manpower shortage is a group or team ministry. Caution is necessary here... Group or team ministries in town may cut down 'the plant', or what a tradesman might call the points of service, but they ought eventually to increase the total urban ministry rather than decrease it. 1968 *Pastoral Measure* ii. 13 in *Parl. Papers 1967–68* XVII. 843 A pastoral scheme may provide for the establishment of a team ministry for the area of any benefice, that is to say, for the sharing of the cure of souls in that area by a team of ministers consisting of—(*a*) the incumbent of the benefice which, if it is not or would not otherwise be a rectory, shall be a rectory; (*b*) one or more other ministers who shall have the title of vicar and a status equal to that of an incumbent of a benefice. 1980 *Oxf. Diocesan Mag.* May 15/2 The faint hearts in the Diocese who see team ministry as a threat. 1886 H. Chadwick *Art of Batting* 7 The practical effect of all this is to destroy a batsman's ambition to excel as a 'team player' in batting. 1980 *Newsweek* 17 Nov. 13/3 Reagan wants 'team players' for his Administration—men and women loyal to him personally and to his philosophy generally, willing to argue over policy, but not fundamental ideology. 1977 J. Wambaugh *Black Marble* (1978) iv. 42 Every few years the brass had to come up with some new catchword to justify the budget. 'Team policing.' 1976 *Southern Even. Echo* (Southampton) 10 Nov. 21/7 Southampton were always in control in the team race and were easy team winners from Portsmouth, through Tony Nixon 5th, Bryan Dawkins 8th, and Malcolm Beavis 10th. 1976 *Milton Keynes Express* 11 June 15/1 He will be assisting the Rev Christopher Drummond, team rector from the Christ Church Centre at Stantonbury. 1895 *Wales* Apr. 168/1 He was able to drive a furrow to the delight of even the most envious of surrounding teamsmen. 1928 *Britain's Industrial Future* (Liberal Industrial Inquiry) iii. xvi. 195 The 'fellowship-bonus' system..evokes the team-spirit. 1938 R. G. Collingwood *Princ. Art* iv. 74 These sports, we are told, inculcate a team-spirit. 1976 F. Muir *Frank Muir Bk.* 96 The schools..were sending forth..superbly fit chaps, light on imagination but strong on team-spirit. 1964 G. McDonald *Running Scared* iii. 37 He had never gone out for any team sports. 1947 A. P. Gaskell *Big Game* 12 And then of course, the team-talk on Friday night. The coach would stand on the platform and start on his old game of building us up to fighting pitch. 1960 V. Jenkins *Lions Down Under* viii. 114 The post-mortem at a team-talk in Timaru was a searching one. 1976 *Science News* 28 Feb. 135 Two answers to this problem..are to teach science ethics to college students by presenting them realistic case studies and to bring industrial scientists into the universities to team-teach. 1979 *Maledicta* III. 144 Mary Salawuh Warren, a Yoruba, has team-taught Yoruba and other West African languages with her husband, D. M. Warren, at Iowa State University and in Peace Corps training programs. 1960 *Washington Post* 20 Dec. B2 Principal Harold Wilson..tallied the benefits of team teaching at his school. 1964 *Observer* 13 Sept. 11/8 Team teaching, in which a corps of teachers work with a very large group, already has a long history in the United States. 1976 *Church Times* 8 Oct. 17/5 *Team vicar* required for church of St. Martin, Southdene, to serve large neighbourhood unit and to work as a member of a large established team. 1981 *Ibid.* 10 July 17/4 (Advt.), Applications invited for team vicar to complete established team of three. Priest appointed will have particular pastoral responsibility for four attractively-situated villages. 1909 *World To-day* (U.S.) Sept. 3 (*heading*) Team work in municipal progress. 1954 [see *team-mate* above]. 1977 *Lancet* 23 Apr. 899/2 We need hospitals, hostels, and homes, but we must be sure that by effective communication and teamwork (and these do not cost money), the service we offer is of the highest.

team, *v.* Add: **1. b.** *intr.* Chiefly with *up*: to join together in or as in a team; to ally oneself or get together *with* someone. Occas. *trans.*
1932 W. Faulkner *Light in August* iv. 86 Like man and wife for three years, until Brown and him teamed up. 1932 J. T. Flynn *God's Gold* vii. x. 314 Whetmore was not Rockefeller's agent, but a lawyer and independent promoter who teamed up with the Merritts and worked with them. 1950 D. Hyde *I Believed* ii. 14 The war-wounded were everywhere... Blinded, they teamed up into bands. 1965 J. Lawlor in J. Gibb *Light on C. S. Lewis* 73, I had thought of myself as God's gift to Lindsay's Balliol, with which Magdalen was teamed for scholarship purposes. 1967 M. Chandler *Ceramics in Mod. World* v. 157 In practice two or three refractories may have to be teamed up to do one exacting job. 1978 J. R. L. Anderson *Sprig of Sea Lavender* vi. 94 He seems to have teamed up with Trudi... He was a little in love with Sandra once.
c. *trans.* To use or wear in conjunction *with*. Also *intr.* for *pass.*
1948 M. Laski in *New Statesman* 13 Nov. 417/1 *Team*, vb.: to wear one thing with another; e.g., team your palest grey dress with the subtle flattery of a brief scarlet bolero. 1954 C. L. B. Hubbard *Compl. Dog Breeders' Man.* xx. 203 Well-pressed linen slacks..can look really nice, especially if teamed up with a contrasting blouse, shirt or jumper. 1958 *House & Garden* Feb. 22 (Advt.), [The furniture] will team happily with the pieces you wish to retain. 1960 *Housewife* May 104/2 This sweater teams happily with pants or shorts. 1977 *Jersey Even. Post* 26 July 10/1 A long, tiered empire-line voile dress, made of a yellow and red floral patterned material with a white background. This was teamed with a white floppy hat.
2. For *U.S.* read *N. Amer.* (Later Canad. examples.)

1968 E. Russenholt *Heart of Continent* iii. ix. 162 A sudden freeze-up ends the navigation season, catching many vessels in the ice. Freight which cannot, now, be moved by steamboat, must be teamed. 1951 K. M. Wells *Owl Pen Reader* (1969) iii. 253 He took the road, teamin' hay er cordwood to town.
4. *Comb.* **team-up**, an instance of teaming up (sense 1 b above). *colloq.*
1945 *Richmond* (Va.) *Times-Dispatch* 21 Nov. 15 (Advt.), Santa's a 'good Joe' in their language when he delivers these team-ups [*sc.* a dressing-gown, pyjamas, and slippers]. 1960 *Farmer & Stockbreeder* 8 Mar. 74/1 (*heading*) Poor show but the team-up of American and French manufacturers will be a 'shot in the arm'.

teaman, tea-man. Add: **3.** *U.S. Criminals' slang.* (See quot. 1950.)
1938 *Amer. Speech* XIII. 192/1 *Tea-man*, a reefer-man or marijuana addict. 1950 H. E. Goldin *Dict. Amer. Underworld Lingo* 220/2 *Tea-man*, a smoker or purveyor of marijuana. 1959 in J. E. Schmidt *Narcotics Lingo & Lore* 182.

teamer. Add: **1.** (Earlier U.S. example.)
1778 E. Parkman *Diary* 26 Dec. (1899) 81 May God extend pity to ye miserable poor,—to Sailors, to Soldiers, to Teamers abroad.
2. A member of a team; esp. a member of the first (or second, etc.) team in sport.
1934 R. Macaulay *Going Abroad* ii. 32 Loyal teamers, they agreed that..their team leader's should be the only Guidance sought. 1950 *Sport* 24–30 Mar. 10/1 Phil was in and out of the Liverpool senior side the following season, but was an established first-teamer by 1938–39. 1976 *Evening Post* (Nottingham) 15 Dec. 24/7 Nottingham first teamers Ken MacDonald and Graeme Fraser came back after injury to help Corsairs defeat Nottingham University 36–0 at Beeston last night.

teaming, *vbl. sb.* Add: Also with *up* in senses 1 b and c of *team *v.*
1960 *Farmer & Stockbreeder* 8 Mar. 74/1 The teaming-up of some well-known French manufacturers with..American companies is bound to act as a 'shot in the arm' to the French engineering industry. 1966 *Guardian* 25 Apr. 6/4 A Jaeger shop..in Southport..is making a strong feature of 'teaming up'—that is, exactly matching colours can be found in coats, suits, hats, skirts, slacks, blouses, and knitwear.

teamster. Add: **1.** (Earlier U.S. example.)
1777 in *New Hampshire Hist. Soc. Coll.* (1863) VII. 88 The Committee delivered the Several Teamsters.
2. *N. Amer.* A lorry-driver, a truck-driver; one who drives a truck as his occupation.
1907 J. London *Iron Heel* (1908) x. 182 The teamsters' strike had been broken. 1957 *Economist* 28 Sept. 1024/1 The delegates who are packing their bags this week-end for the convention of the teamsters—as lorry drivers are still called—will also take with them the knowledge that this is likely to be the most momentous meeting in the history of organised labour. 1958 *Daily Express* 29 Aug. 4/2 This type of entertainment is as suspect as the Teamsters' Union. 1978 *New York* 3 Apr. 100/3 The 8,000 sanitationmen represented by a local of the teamsters, usually the most truculent of city unions, are still participating in the coalition talks.

Tean, var. *Teian *a.*

tea party. Add: **2. c.** A gathering at which marijuana is smoked. *slang.*
1944 *War Med.* VI. 383/2 Have you ever been on a 'tea' party? No? You've missed a sensation of a lifetime. 1956 J. Symons *Paper Chase* vii. 32 Used to give tea parties—marihuana. 1972 'J. Quartermain' *Rock of Diamond* i. 7 Jane hadn't taken tea. She..gave no clue..as to what an inhibited Englishman should do at a midtown Manhattan tea-party.
3. *attrib.* of attitudes, behaviour, etc., held to be typical of a tea-party; bland, insipid, trite, trivial.
1961 M. Beadle *These Ruins are Inhabited* (1963) ii. 28, I think he expected the boys to have..tea-party manners. 1962 [see *penguin 2 c]. 1973 C. Mullard *Black Britain* ix. 105 Liberal do-gooders with a tea-party attitude towards race.

tea-planter. (Earlier example.)
1887 Kipling *Plain Tales* (1888) 112 A Subaltern, or a Tea-Planter's Assistant, or anybody who..has no care for to-morrow.

tea-pot. Add: **1. b.** *tempest in a tea-pot* (later example). Also in similar phrases.
1928 R. Campbell *Wayzgoose* i. 16 Storms in a teapot often have occurred. 1942 T. DuBois *Body goes round & Round* xiii. 172 You have been indulging in your favourite occupation of stirring up a tempest in a teapot. 1973 *Times* 1 Aug. 6/5 Senator Ervin said the issue of whether the subpoenas were continuing was 'a difference in a teapot'.
2. *attrib.* and *Comb.*, as *tea-pot stand*.
1895 *Montgomery Ward Catal.* Spring & Summer 439/1 Tea or Coffee Pot Stands. 1968 *Canad. Antiques Collector* June 12/2 A tea service at this period..normally consisted of twelve saucers, twelve cups, twelve coffee cups, a tea pot and cover (occasionally a tea pot stand), a sucrier, [etc.].

tea-pot, *v.* For *nonce-wd.* read *Obs. rare.* and add earlier and later examples.
1842 J. Paget *Let.* 25 July in *Mem. & Lett.* (1901) 117 My pupils have proposed to make a demonstration in my favour, and have asked me if they may open a subscription to 'tea-pot' or commit some similar dignity upon me. 1881 V. Lush *Jrnl.* 26 Oct. (1975) 247, I reminded him that for some time past I had been mulcted of my stipend and that to be 'tea-potted' under such circumstances seemed to me to be a case parallel to that of allowing a man to starve and when he is dead to give him an expensive funeral.

Teapot Dome (tiːpɒt dəʊm). The name of a naval oil reserve in Wyoming, irregularly leased by the U.S. Government in 1922, used *attrib.* and *absol.* to designate the resulting political scandal and, allusively, any similar later scandal.
1936 F. D. Roosevelt *Nothing to Fear* (1947) 64 In spite of all the demand for speed, the complexity of the problem and all the vast sums of money involved, we have had no Teapot Dome. 1973 *New Yorker* 28 Apr. 31/2 Senator Barry Goldwater made his statement: 'The Watergate. The Watergate. It's beginning to be like Teapot Dome.' 1977 *Time* 23 May 54/2 In an era of Teapot Dome and bathtub gin, he [*sc.* Lindbergh] seemed to Americans a cleaner, sharper version of themselves.

tear, *sb.*[1] Add: **1. b.** Also in colloq. phr. *without tears*, without difficulty or distress (freq. used to describe a method whereby some discipline is easily mastered). Also *without-tears* attrib. phr.
1857 F. L. Mortimer (*title*) Reading without tears. 1877 —— (*title*) Latin without tears; or, One word a day. 1896 G. B. Shaw in *Sat. Rev.* 12 Dec. 623/2 (*heading*) Ibsen without tears. 1914 W. Owen *Let.* 1 June (1967) 257, I have a design in sending you this, viz. to keep you hungry to learn French I hope it won't be long before you read such works 'without tears'; at least without tears due to grammatical difficulties. 1932 A. Huxley *Brave New World* xvii. 280 Anybody can be virtuous now. You can carry at least half your morality about in a bottle. Christianity without tears—that's what *soma* is. 1937 T. M. Rattigan (*title*) French without tears. 1956 *New Statesman* 11 Feb. 143/1 The late 19th-century concept of progress without tears. 1962 *Times* 7 June 17/3 It is a without-tears book. 1974 J. I. M. Stewart *Gaudy* i. 19 Charles and Mary..were well-mannered young people, and docile at least to the extent of being resigned to Scrabble as a species of Philology without tears.
6. a. (attributive) *tear-dripping, -flood* (later example), *-tap, -track*; **c.** (instrumental) *tear-bound, -dabbled, -filled* (example), *-logged, -streaked, -strewn, -stuffed, -tricked, -washed* (later examples) adjs.; **d.** (miscellaneous) *tear-tight, -trembling* adjs.
1938 E. Bowen *Death of Heart* ii. i. 178 Her manner..had threatened the afternoon like a cloud that covers the sky but is almost certain never to break. Her eyelids looked rigid—tear-bound, you would have said. 1915 *Pearson's Mag.* Jan. 46/2 She raised a tear dabbled countenance. 1944 W. de la Mare *Coll. Rhymes & Verses* 217 Tear-dabbled cheeks, wild eyes I see. 1893 F. Thompson *Hound of Heaven* in *Poems* 53 And now my heart is as a broken fount, Wherein tear-drippings stagnate. 1951 in M. McLuhan *Mech. Bride* (1967) 11/2 Miss Grable, with tear-filled eyes, showed..a letter she'd received from a soldier's buddy. 1916 R. Graves *Over the Brazier* 21 Till it seemed through a swift tear-flood That dead men blossomed in the garden-close. 1931 R. Campbell *Georgiad* ii. 41 Holding our course among the tear-logged wrecks. 1923 Galsworthy *Captures* 181 The girl's face, tear-streaked, confusedly pretty, had come up before him. 1942 S. Smith *Mother, what is Man?* 76 My reverent reveries and fruitful plod Of tear-strewn steps. 1939 Dylan Thomas *Map of Love* 12 After the feast of tear-stuffed time and thistles. 1922 Joyce *Ulysses* 753 That was the last time she turned on the teartap. 1938 S. Beckett *Murphy* iv. 51 The human eyelid is not teartight. 1965 S. Smith in *Listener* 2 Sept. 347/3 Those awful tear-tracks on her cheeks, As if she had cried a lot! 1916 D. H. Lawrence *Amores* 74 Tear-trembling stars of autumn. 1880 G. M. Hopkins *Poems* (1967) 88 In his hands he has flung His tear-tricked cheeks of flame. 1916 H. G. Wells *Mr. Britling sees it Through* iii. i. 389 Her tear-washed mind became vaguely friendly. 1922 Joyce *Ulysses* 175 Davy Byrne, sated after his yawn, said with tearwashed eyes:—And is that a fact?
e. tear bomb, a bomb containing tear gas; **tear-drop**, (*a*) (see sense 6 a in Dict.) (*b*) *transf.*, freq. *attrib.*, denoting something resembling a tear-drop in shape as: (i) an air cavity in glassware; (ii) a tear-shaped run in paintwork or ceramic glaze; (iii) a streamlined body or component of a vehicle, boat, etc.; (iv) *Surfing* = *pig board* s.v. *pig *sb.*[1] 13; (v) a bead or jewel of tear-drop shape; **tear-gas**, a lachrymatory gas used in warfare or riot control to disable opponents or make crowds disperse; hence as *v. trans.*, to attack with tear gas, to drive *out of* a place with tear gas; **tear-jug** *rare* = Tear-bottle; **tear-shell**, a shell (Shell *sb.*[1] 21 b) containing tear-gas; **tear-smoke** = *tear gas* above.
1929 M. Lief *Hangover* xv. 238 I'm going to have Katie actually taken for a ride..and Rat-Face Walsh's yeggs following..with machine guns and tear bombs. 1953 Wendt & Kogan *Big Bill of Chicago* xxiii. 271 Police squads cruising the city, machine-guns in their laps and tear bombs in their pockets. 1904 *Burlington Mag.* IV. 141/1 Immediately under the bowl at the top of the stem is an air cavity, known as a 'tear-drop'..a frequent form of decoration. 1922 [see *curtain *sb.*[1] 1 e]. 1933 *Burlington Mag.* June 265/1 The presence of 'tear drops' in the glaze [of Chinese porcelain]. 1936 *Times* 29 Dec. 12/6 The new design will allow the manufacturer [of motor vehicles] to indulge in the 'tear-drop' streamlining which has often been discussed in recent years but never achieved. 1948 *Shell Aviation News* No. 115. 6/1 A 25-foot, streamlined, teardrop antenna will project below the fuselage... This will be the main broadcasting antenna. 1962 *Austral. Women's Weekly* Suppl. 24 Oct. 3/4 *Teardrop*, type of surfboard with wide back and pointed front. 1965 *Harper's Bazaar* June 24 Diamond teardrop, £500. 1980 D. Creed *Scarab* i. 9 A large and most marvellous pendant..suspending a teardrop pearl. 1917 W. Owen *Let.* 19 Jan. (1967) 429 It was only tear-gas from a shell, and I got safely back (to the party) in my helmet. 1927 *New Republic* 12 Oct. 202/2 The troopers on the outskirts..hurled tear-gas bombs and charged. 1927 *Daily Express* 16 Dec. 15, I imagine him, first, tear-gassing a river bank and so reducing all the crocodiles to genuinely hopeless grief. 1934 R. Stout *Fer-de-Lance* ii. 25 A gangster had been tear-gassed out of a Brooklyn flat. 1978 R. Ludlum *Holcroft Covenant* xliii. 499 Tear gas and Mace were not unknown in Yakov's line of work. 1869 Tear-jug [see *Pisan *a.*]. 1916 *War Illustr.* 23 Dec. 451/3 Don't you know the scent of tear-shells when you smell it? 1946 F. Burrows *Let.* 22 Aug. in Mansergh & Moon *Transfer of Power* (1979) VIII. 296 He added that the Police had used tear-smoke on crowds frequently. 1949 Koestler *Promise & Fulfilment* i. xii. 136 The boarding party finally gained control of the vessel by using tear-smoke grenades..against them.

tear, *sb.*[2] Add: **4.** Special Comb.: **tear-fault** *Geol.* = *strike-slip fault* s.v. *strike *sb.* 20.
1900 *Proc. Geologists' Assoc.* XVI. 465 It is this ['lag' fault] which gives one a clue as to the nature of some of the most striking 'tear' faults. 1924 J. G. A. Skerl tr. *Wegener's Orig. Continents & Oceans* 58 A lateral displacement of great dimensions, a so-called tear fault. 1957 [see *slice *sb.*[1] 2 b]. 1977 A. Hallam *Planet Earth* 60/1 Another common type of shear—which caused little trouble to miners and so was unrecognised for many years—is variously known as a strike-slip, wrench, tear or transcurrent fault.

tear, *v.*[1] Add: **3. d.** Also with *apart, up*: to render distraught, upset (a person). In *pass.* with *up* (dial. *out*): to be distressed, upset. *N. Amer. colloq.*
1898 F. P. Dunne *Mr. Dooley in Peace & War* 47 They say th' Spanyards is all tore up about it. 1950 R. Moore *Candlemas Bay* 240 Jeb, poor lamb, he was so tore out about you that he never said nothing. 1956 B. Holiday *Lady sings Blues* (1973) xiii. 117 The few I did see when they came back tore me apart. One night..a kid came in to see me... His hair had turned completely white. 1972 *Even. Telegram* (St. John's, Newfoundland) 24 June 9/1 Robert A. Power, who said he was tore up from work in his younger days and was 'no good now to trade and no good to sell.' 1974 K. Millett *Flying* (1975) iii. 323 Pete is too delicate to pattern. Tears him up just to hear Winnie yell.
e. *to be torn between*: to be distracted by (two equal but conflicting desires, emotions, or loyalties).
1871 L. W. M. Lockhart *Fair to See* I. viii. 148 Torn between her desire to underrate Eila and to preserve her own dignity. 1888 Mrs. H. Ward *Robert Elsmere* III. xxxv. 111 Agnes, torn between her interest in what was going on and her desire to get back to her mother, had at last hurriedly accepted this Mrs. Sherwood's offer. 1922 T. Wolfe *Lett.* (1956) ii. 31 The girl Laura, 'torn between' (as the saying goes) love for her beaten father and the blunt young apple grower. 1948 A. Paton *Cry, Beloved Country* ii. viii. 172 Jarvis was torn between compassion and irritation, and he stood and watched uncomfortably. 1971 G. Charles *Destiny Waltz* iii. 104 As usual in such cases he was torn between both sides, angry with Dorn for his patronage of the couple, impatient with them for being what they were.
f. Colloq. phr. (chiefly in *pa. t.* and *perfect*) *to tear it*: to spoil one's chances; to put an end to one's hopes, plans, etc.
1909 'I. Hay' *Man's Man* xvii. 320 'I've fairly torn it, this time!' he reflected morosely. a 1918 W. Owen in *Poems* (1920) 22 First wave we are, first ruddy wave; that's tore it. 1924 Kipling *Debits & Credits* (1926) 258, I expect I must 'ave kept carryin' on, till Headquarters give me that wire from Ma... That wire tore it. 1938 G. Greene *Nineteen Stories* (1947) 77 'I *am* English,' Mr. Calloway said. Even that didn't tear it. 1954 M. Procter *Hell is City* vi. iii. 180 He looked at his watch. 'That's torn it,' he said. 1960 D. Lessing *In Pursuit of English* v. 185 Oh, my God, that tears it, if he's going to start. 1972 D. Delman *Sudden Death* (1973) iv. 110 'Ouch,' she said, grinning... 'Well, that tears that, doesn't it?'
g. In phrases with *up* and indefinite obj., descriptive of unrestrained excitement; esp. in Jazz. *U.S. slang.*
1932 J. Dos Passos *1919* 270 Bud had been tearing things up at the University and was on the edge of getting fired. 1955 Shapiro & Hentoff *Hear Me talkin' to Ya* 204 He had the first big colored band that hit the road and tore it up. 1963 *Listener* 14 Mar. 478/3 The trumpeter Wild Bill Davison, who 'tore it up' with admirable primitivity and sensuality. 1968 [see *identify *v.* 1 b].
h. *to tear down*: to punish; to criticize severely. *U.S. colloq.*
1938 M. K. Rawlings *Yearling* xxix. 381 I'll tear down all two of you. Now git down and pick up ever' one o' them peas and wash 'em off. 1978 I. B. Singer *Shosha* vii. 128 The insolence of a writer tearing down a piece before it's been performed!
i. *to tear apart*, (*a*) to subject to criticism; (*b*) to search (a place) thoroughly.
1953 *Manch. Guardian Weekly* 1 Jan. 13/2 Carefully

tear apart your editorial. 1977 'C. Aird' *Parting Breath* xv. 176 Somebody was ready to tear the place apart. You should have seen Miss Moleyn's house.

5. c. Phrases. *to tear off a strip, tear a strip off*: see *STRIP *sb.*² 1 i; *to tear off a bit, piece* slang (orig. Austral.): to copulate with a woman.

1941 Baker *Dict. Austral. Slang* 76 *Tear off a piece*, to coit with a woman. 1951 S. Longstreet *Pedlocks* iv. v. 222 Look, you come down and tear off a piece anytime. And the wine—*Asti Spumante*—she is ona me. I stand the wine. The girls, that is up to you. 1970 G. Greer *Female Eunuch* 265 The vocabulary of impersonal sex is peculiarly desolating. Who wants to 'tear off a piece of ass?' 1977 *Custom Car* Nov. 67/2 Italian wives must sit and suffer if the men tear off a bit on the sly.

9. b. To make *one's way* violently or impetuously.

1853 C. Kingsley *Hypatia* II. xiv. 328 Furiously..he burst up as if from the ground..tearing his way toward his idol. 1888 Mrs. H. Ward *Robert Elsmere* I. x. 282 A little gully deep in bracken, up which the blast was tearing its tempestuous way.

c. *to tear into*: (*a*) to make a vigorous start on (an activity, performance, or the like); (*b*) to attack vituperatively, reprimand.

(*a*) 1901 M. Franklin *My Brilliant Career* xxxvi. 301 'Syb, I want to speak to you.'..'Very well; "tear into it",' as Horace would say. 1929 W. Smyth *Girl from Mason Creek* xiii. 131 'Three notes a man if we win out!'..'Aw, make it five an' we'll tear into th' job.' 1949 R. Harvey *Curtain Time* 8 Then the lights went down, the baton rapped sharply, and the orchestra tore into the overture. 1961 J. B. Priestley *Saturn over Water* ii. 12, I tore into the business of getting visas.

(*b*) 1934 in Webster. 1946 F. Sargeson *That Summer* 93 You could still hear them tearing into each other. 1954 J. Masters *Bhowani Junction* ii. xi. 94 The sahib tore into me as if I was a little boy he'd caught making a mess on the carpet. 1984 *Miami Herald* 6 Apr. 10A/1 Jackson..tore into both candidates in past debates.

tear, *v.*² Add: **1. c.** (Later examples.) Now chiefly *N. Amer.*

1971 E. Shorris *Death of Great Spirit* i. 16 When your eyes teared and your head fell, I was afraid you were dying. 1980 J. Ball *Then came Violence* (1981) vi. 47 When her eyes teared again, he pulled out his own clean handkerchief.

tear-. Add: **1. tear-away** *sb.*: now usu. (written *tearaway*), an unruly young person, a hooligan, ruffian, or petty criminal (formerly applied *spec.* to a kind of thief: see quot. 1938); **tear-down**, the complete dismantling of a piece of machinery; **tear-off** *a.* (later examples); *sb.* (example); **tear-out**, the action of pulling out the fitments, décor, etc., of a room; **tear-up** *sb.*: also (*slang*), the action or an instance of tearing up; a spell of wild, destructive behaviour; a mêlée; in *Jazz*, a lively, rousing performance (cf. *TEAR *v.*¹ 3 g).

1938 F. D. Sharpe *Sharpe of Flying Squad* i. 15 A type of crime which has almost died out now was that carried out by thieves known as 'Tearaways' who used to hang about outside the theatres after the show and snatch costly brooches from women's dress fronts. 1950 *Observer* 2 Apr. 7/3 He..boasted of being a proper 'tearaway' (one who sticks at nothing). 1958 *Punch* 19 Feb. 263/2 You get some proper tearaways (quarrelsome fellows) at the dogs. 1962 *Observer* 18 Feb. 21/3 My grandfather was a pickpocket, my six uncles were all villains and tearaways, my brothers and friends were thieves. 1978 L. Davidson *Chelsea Murders* xxiii. 140 He had a bit of form..in younger days a tearaway, the odd charge of violence. 1976 *Lebende Sprachen* XXI. 152/2 After engine teardown exercise special care to replace damaged carbon rubbing type seals. 1981 *Pop. Hot Rodding* Feb. 51/1 Installing new pistons means removing the engine, an almost complete tear-down, and then a re-installation job. 1910 Tear-off [see *BLOCK *sb.* 10 c]. 1945 J. Rhys-Williams *Stern Daughter* xxx. 207 The little calendar..was the tear-off kind, with quotations. 1961 'B. Wells' *Day Earth caught Fire* i. 7 An alert..youngish man..was at his desk busily working on tear-offs with swift, practised strokes of his red pencil. 1980 T. Barling *Goodbye Piccadilly* ii. 50 A big tear-off calendar gave today's date. 1976 *Billings* (Montana) *Gaz.* 17 June 3-F/1 (Advt.), Custom Homes. Apartments. Remodeling...'Specializing in Tear-Out.' Every kind of concrete work. 1886 H. Baumann *Londinismen* 206/2 *Tear-up*.., feine(s) Geschäft. 1890 in Barrère & Leland *Dict. Slang* II. 339/2 'What is it this time?'..'Only a tear up.'..Among the readers of the *St. James's Gazette* there may be some who are unacquainted with the accepted method of obtaining a fresh outfit among the casual poor. *Ibid.* 340/1 At his feet, in a heap on the floor, lay some filthy rags,..the remnants of what had recently been his garments... The heap was the result of the tear up. 1958 S. Race in P. Gammond *Decca Bk. Jazz* x. 125 Max Kaminsky..and of course Bunny Berigan all took turns to roughen up that smooth ensemble.., with the historic Berigan tear-ups especially welcome. 1964 E. Parr *Grafters All* xiii. 151 He enters an orgy of crime, more commonly known as 'having a tear-up'. 1974 J. McVicar *McVicar* i. iii. 62, I decided to have a tear-up. 1982 *New Society* 9 Dec. 422/2 We've had a tear-up with the police. 1983 *Listener* 9 June 35/3 The music is not the tear-up associated with jazz at the Phil.

2. tear-arse *slang*, a very active busy person; hence as *v. intr.* (also *U.S.* **tear-ass**), to drive recklessly, rush *around* wildly and rowdily; **tear-sheet** chiefly *U.S.*, a sheet torn from a publication (or, later, separately printed and unbound) to be sent to an advertiser whose advertisement appears on it as proof of insertion; also one containing an article; more generally, a tear-off sheet from a teleprinter or calendar; **tear-thumb** (later examples).

1923 J. Manchon *Le Slang* 308 Tear-arse,..un qui s'échine (à travailler). 1942 Berrey & Van den Bark *Amer. Thes. Slang* § 53/8 *Tear-ass* &c. *around*,..to hasten or rush around. 1954 *Amer. Speech* XXIX. 103 *Tear ass*, v. phr., to drive fast or recklessly. 1968 J. Wainwright *Edge of Extinction* 34 We're the..killjoys. The miserable bastards who won't let 'em tear-arse round the town at sixty miles an hour. 1972 A. Draper *Death Penalty* xix. 119 Some lads were picked up after the cup match for tear-arsing around. 1976 J. Fraser *Who steals my Name?* ii. 17 You'll need to settle down. You can't be a teararse all your life. 1930 H. A. Groesbeck *Practical Photo-Engraving* ii. 4 The 'tear-sheet' of the newspaper advertisement, also the drawing from which it was made, have just been received. 1950 R. Chandler *Let.* 15 Feb. (1981) 210 Herewith the tear sheets from a *Pocket Atlantic* with ..the article you desired. 1962 *Listener* 19 Apr. 672/1 One of the President's advisers came in with a tear-sheet from the news ticker. 1972 M. J. Bosse *Incident at Naha* i. 16 He was slowly flipping the tear sheets of the desk calendar. 1926 *W. Virginia Legislature Hand Bk.* 488 Do you see among those [wild flowers] present, any exotics?..Yes..This bit of smart-weed or tear thumb. 1978 C. B. Dugdale *Mod. Amer. Herbal* II. 124 Tear-thumb; Arrow-leaved Tearthumb... The small cluster of flowers varies in color from pinkish to whitish.

tearer. Add: **1. c.** *tearer-downer* (U.S.), one who tears down, a carping critic (cf. *TEAR *v.*¹ 3 h).

1942 Berrey & Van den Bark *Amer. Thes. Slang* § 421/1 Critic; opposer..tearer-downer. 1944 [see *BUILDER c]. 1955 J. D. Salinger in *New Yorker* 29 Jan. 28/1 I'm just so sick of pedants and conceited little tearer-downers I could scream.

tearing, *ppl. a.*¹ **5.** (Later examples.)

1867 A. Stock *Evidence against & for Walter Tricker* 4 Mrs Hitchins, at the Inquest, says 'It was not ordinary barking. They [*sc.* the dogs] were barking like tearing mad.' 1886 E. L. Dorsey *Midshipman Bob* ii. xi. 219 'Don't you get mad ever, eh?'..'Yes, I'm sorry to say I do—tearing mad sometimes.' 1906 [see *JIM]. 1942 Berrey & Van den Bark *Amer. Thes. Slang* § 284/8 Angry..tearing, tearing angry or mad.

tear-jerker (tɪ°·ɹ,dʒɜ:ɹkəɹ). *colloq.* (orig. *U.S.*). [f. Tear *sb.*¹ + Jerker¹; for the sense of *jerker*, cf. *soda-jerker* s.v. *SODA *sb.*¹ 7.] Something calculated to evoke sadness or sympathy, usu. a sentimental film, play, song, story, etc. Also applied to a person and, *rarely*, to an event.

1921 *Double Dealer* II. 143/2 But no one today, I believe, mistakes his [*sc.* James Whitcomb Riley's] productions for anything but somewhat shallow, fairly easy tear-jerkers. 1935 *Amer. Mercury* Aug. 400/1 A lawyer was imported from California, a magniloquent tear-jerker named Delphin Delmas. 1936 *New Yorker* 7 Mar. 32/2 'Love on the Dole' turns out to be far more than a conventional tear-jerker. 1940 *Manch. Guardian Weekly* 27 Sept. 212 The German description of the torpedoing of the evacuee ship as a 'tear-jerker' recalls Goebbels's clumsy attempt to deny the torpedoing of the Athenia at the beginning of the war. 1948 *Sunday Pictorial* 18 July 11/3 The cameos are linked with a quiet humour and smooth pathos which make the film an A-plus tear-jerker. 1953 'P. Wentworth' *Watersplash* xix. 109 Three copies of the famous *East Lynne*. A notorious tear-jerker. 1958 B. Nichols *Sweet & Twenties* xiv. 187 This number, as sung by Al Jolson, became one of the most efficient tear-jerkers of all time. 1975 *Islander* (Victoria, B.C.) 9 Feb. 2/3 He concluded [his speech] with a real tear-jerker.

Hence (as back-formations) **tea·r-jerk** *sb.*, a sentimental effusion; also as *v. trans.*; **tea·r-jerking** *vbl. sb.* and *ppl. a.*

1940 S. Lewis *Bethel Merriday* xv. 127 You..made me understand how much that poor gutter pup longed for a chance to parade, and yet you didn't do much tear-jerking. 1941 E. Snow *Battle for Asia* iv. 88, I remember a tear-jerking letter from a correspondent appealing for people to boil their garbage and put it beside their ash cans for the hungry—*dogs*. 1953 *Times Lit. Suppl.* 31 July 490/5 The ex-Governor of Illinois does not disdain the obvious 'tear-jerk'. 1961 D. Holbrook *English for Maturity* i. iv. 55 The vague undefined tear-jerk of popular graveyard and funeral verses. 1962 'K. Orvis' *Damned & Destroyed* xvi. 123 She tear-jerked it from a drugstore without a prescription. 1962 Auden *Dyer's Hand* (1963) 430 If Homer had tried reading the Iliad to the gods on Olympus, they would..possibly, even, reacting like ourselves to a tear-jerking movie, have poured pleasing tears. 1965 *Spectator* 5 Feb. 157/3 The prize for tear-jerking seemed..destined for..the *Daily Mail*. 1979 D. Meiring *Foreign Body* xii. 126 Even he [*sc.* God] had sometimes needed a hand, and Hussein had provided that brilliantly, in turn cajoling, tear-jerking, and threatening Americans of huge stature in the oil business. 1981 *Times* 14 Feb. 8/7 An idealized Shavian heroine..the armour-plated, tear-jerking martyr.

tearlet. (Later example.)

1964 V. Nabokov *Defence* x. 162 A warm tearlet would roll down her face.

tea-rose, tea rose. Add: **b.** (Earlier example.)

1872 *Young Englishwoman* Nov. 599/1 The rose colours are *rose frais*..; tea-rose, with yellow tints; and faded rose.

c. A perfume made from or named after this rose.

1897 *Sears, Roebuck Catal.* 19/2 Perfumes..Sweet Pea.. Tea Rose..Tuberose. 1926–7 *Army & Navy Stores Catal.* 486/2 Atkinson's Perfumes..Sweet Pea, Tea Rose, Verbena. 1977 *New Yorker* 10 Oct. 35/3 She..ended with a commercial for her new fragrance, Tea Rose, samples of which her assistants passed around.

teart (tɪ°·ɹt), *a.* and *sb.* [Dial. var. of Tart *a.*] **A.** *adj.* Sour; used of pastures containing an excess of molybdenum. **B.** *sb.* Teart quality in grass; the diarrhœa suffered by cattle grazing a teart pasture.

1850 [see Tart *a.* 2 a]. 1896 *Jrnl. Bath & West Soc.* VI. 207 The herbage possesses the peculiar purging quality known as 'teart'. 1903 *Lancet* 6 June 1590/1 This disease, known as parasitic enteritis, is found to be persistently associated with certain pastures (called 'teart' lands in the West of England) upon heavy moisture-retaining soils. 1939 *Nature* 23 Sept. 532/2 Teart, to which cattle in certain areas are subject, is found to be associated with an increased molybdenum content in the herbage. 1970 W. H. Parker *Health & Dis. in Farm Animals* xiv. 193 The teart pastures of Somerset are on the blue Lyas clay.

Hence **tea·rtness** *sb.* = *TEART *sb.*

1940 *Nature* 15 June 941/2 The cause of teartness is the presence in the herbage of molybdenum. 1979 *Jrnl. Compar. Pathol.* LXXXIX. 495 'Teartness' i.e. the scouring which occurs in cattle but not in horses on 'teart' pastures of high Mo content.

teary, *a.* Add: **1.** Also *transf.*

1941 B. Schulberg *What makes Sammy Run?* iii. 45 Full of teary nostalgia for the glories of his youth. 1976 *Times Lit. Suppl.* 13 Aug. 1010/2 Paul delivers a long, teary monologue about his homosexuality.

Comb. 1949 N. R. Nash *Young & Fair* i. i. 10 But *Patty* is unashamedly teary-eyed. 1960 R. St. John *Foreign Correspondent* iv. 64 We grew teary-eyed trying to fry a fish or a piece of meat over the brazier.

tease, *sb.* Add: **1. b.** *tease number*, a strip-tease act. *U.S.*

1927 *Variety* 13 July 35/5 The four feminine principals alternated in 'tease' numbers with the help of the chorus. 1930 *Ibid.* 3 Dec. 54 With a fair voice, a nice figure and lots of personality, Miss Almond clicked easily in her tease numbers.

2. b. *spec.* = *cock teaser* s.v. *COCK *sb.*¹ 23 (but less coarse). Also *transf.*

1976 *New Yorker* 16 Feb. 107/2 It's easy to get laughs by..showing women..as rich teases, like Mariangela Melato's role in 'Swept Away'. 1978 D. Devine *Sunk without Trace* xxii. 202 Sorry, Ken, but..it's not fair to encourage you to try. I will not be a *tease*. 1979 *Arizona Daily Star* 5 Aug. i. 1/2 Lulu is..a cruel tease to the lesbian countess Geschwitz.

tease, *v.*¹ Add: **1. d.** *U.S. Hairdressing.* = *back-comb* vb. trans. s.v. *BACK- B.

1957 *Amer. Hairdresser* Sept. 66 Pick up one inch of hair and with comb, tease the strand. This creates the lift so necessary to the style. 1962 E. Frank *Best Hairdos* 7 Tease entire head gently for fullness. 1978 J. Updike *Coup* (1979) iv. 171 Her hair bleached platinum and teased to a bouffant mass.

e. *to tease out* (fig.): to extract, get out, obtain, esp. by painstaking effort. Also *to tease on to.*

1959 N. Mailer *Advts. for Myself* (1961) 17 There was a time when Pirandello could tease a comedy of pain out of six characters in search of an author. 1971 *Language* XLVII. 525 It is only by the most careful discrimination that we are able to tease out the critical referential features from the mass of inferential stuff that surrounds them in normal speech. 1974 J. A. Michener *Centennial* x. 580 He was struck with how easy life was in Pennsylvania and how brutally difficult in Colorado, where you had to dig a ditch twenty miles before you could tease a little water onto your land.

2. c. = *strip-tease* vb. intr. s.v. *STRIP-TEASE *sb.* *U.S.*

1927 *Variety* 13 July 35/5 Where they cooch in New York they 'tease' here. 1953 Berrey & Van den Bark *Amer. Thes. Slang* (1954) § 593/22 'Do a striptease.'.. Strip, striptease, tease.

teased, *ppl. a.* Add: **1. b.** Of hair: fluffed out by back-combing. *U.S.*

1965 A. Lurie *Nowhere City* xiv. 147 A waitress appeared in the courtyard, shivering in a teased hair-do and a pink uniform. 1983 J. Valin *Natural Causes* xxix. 200 A pretty nurse with teased brown hair.

2. b. With *out.* Worn out, exhausted. *colloq. rare.*

1943 Hunt & Pringle *Service Slang* 65 *Teased out*, worn out or tired after a long spell of flying or other duty. 1961 D. Moore *Highway of Fear* i. 9 What about you? Still with that teased-out shipping company?

teasel, *sb.* **5. teasel-head** (earlier example).

1743 W. Ellis *Mod. Husbandman* Aug. xvi. 55 One of these stalks has produced..an hundred Teasel Heads.

teaser¹. Add: **1. b.** (Further example.)

1852 Dickens in *Househ. Words* 24 Apr. 118/2 The clay ..is put into mills or teazers, and is sliced, and dug, and cut at.

2. f. A woman who arouses but evades amorous advances; a 'cock-teaser'. *colloq.*

1895 *Cornh. Mag.* Apr. 395 My Joan allus be a teäzer, zur, and when I's wanted to kiss zhe, zhe zes 'Noa, it ain't proper.' 1939 C. R. Cooper *Teen-Age Vice* (1959) iii. 54 The true B-girl is often nothing more than a professional teaser..selling drinks by fraudulent inferences. 1957 J. Braine *Room at Top* vi. 57 She leads young men on and then she turns prim... She's a born teaser. 1980 J. Gardner *Garden of Weapons* i. xi. 111 Martha..sensual in a very obvious way. Herbert always suspected she was a teaser with men.

g. A strip-tease act; a strip-tease artist.

1929 [see *RUNWAY 2 a]. 1930 *Variety* 1 Oct. 49 Miss Dix copped the show from the other femmes with her naughty numbers and teasers. 1931 C. Beaton *Diary* 13 Feb. in *Wandering Years* (1961) 217 There were lots of 'teaser' numbers... The leading lady..tantalisingly takes off one piece of clothing at a time.

3. c. In *Cricket*, a ball that is difficult to play. ? *Obs.*

1856 G. L. H. in V. Dayrell *Weeds from Isis* 69 Your cricketing boy, full of teasers and twisters. 1905 H. A. Vachell *Hill* xii. 249 Fluff's brother bowled slows of a good length, with an awkward break from the off to the leg. 'Teasers,' said the caterpillar critically.

d. *Naut. slang.* A knotted rope's end.

1910 [see *blood-knot* s.v. *BLOOD *sb.* 19]. 1953 J. Masefield *Conway* (rev. ed.) iv. 217 The rope's end, or teaser, made one learn very quickly. 1962 W. Granville *Dict. Sailor's Slang* 118/2 *Teaser*, short length of rope with a 'hangman's knot' at the end, used for chastising *Conway* cadets in the 'tough old days'.

e. *U.S. Theatr.* (See quots.)

1916 A. E. Krows *Play Production in Amer.* xii. 87 The first border (all the borders are numbered consecutively from front to back) is called the teaser. 1923 C. J. De Goveia *Community Playhouse* vii. 80 Just inside the proscenium arch stand two strips of scenery, one on each side of the stage, and usually with a third piece, a border, stretched across the top. The two strips are called *Tormentors* and the particular border the *Teaser*. These pieces are movable. 1933 P. Godfrey *Back-Stage* iii. 34 The 'teaser' and the 'tormentor' are the respective names by which an overhead and side masking arrangement prevents the audience from looking into the wings and the flies.

f. A fisherman's device (orig. live bait) for attracting fish. orig. *U.S.*

1919 Z. Grey *Tales of Fishes* xi. 203 We had three of these flying-fish out as teasers, all close to the boat. 1924 —— *Tales Southern Rivers* 14 The use of teasers..was first used by Avalon boatmen in Marlin fishing. I tried it.., and pronounced it a failure because mackerel, barracuda, and other fish snapped off the cut-bait teasers as fast as they could be put out. 1937 E. Hemingway *To have & have Not* i. i. 17 Eddy put the two big teasers out and the nigger had baits on three rods. 1939 H. Major *Salt Water Fishing Tackle* ii. 69 The first artificial teasers of which I've heard were used by Zane Grey, and I believe he originated them. Most of them are made of wood or metal, brightly colored. 1960 A. Upfield *Myst. Swordfish Reef* vi. 56 To these lines were attached brightly painted cylinders of wood which, when tossed overboard..darted beneath and skimmed over the surface..Teasers, Wilton explained..'the bait-fish and the two teasers look to a shark or swordie just like a small shoal of fish.' 1967 [see *PLUGGER d].

g. A kind of toy pipe with a coil (of paper, etc.) at the end which shoots out when one blows down the stem.

1935 A. J. Cronin *Stars look Down* iii. viii. 554 They had teasers, too, which blew out and hit you as you passed. 1977 D. Jones *My Friend Dylan Thomas* i. 6 Some of them were wearing paper hats..some..blew feather 'teasers' at each other.

4. *Electr. Engin.* **† a.** The shunt winding of a compound-wound dynamo or motor. *Obs.* **b.** The winding or transformer that is connected to the middle of the other transformer in a T-connection. Freq. *attrib.*

1878 C. Brush *Brit. Pat.* 2003 9 This device, which I have called a 'teaser', is used in connection with field magnets..for the purpose of..increasing the magnetic field. *Ibid.*, The teaser wire may be coarser than the principal magnet wire. 1884 S. P. Thompson *Dynamo-Electric Machinery* vi. 92 The shunt part of the circuit, originally called the 'teazer', was adopted at first in machines for electro-plating. 1886 *Ibid.* (ed. 2) x. 238 Brush made the important invention of exciting the field magnets with a compound winding; coarse wire coils being connected in series, with the addition of a so-called 'teazer' coil of finer wire to maintain the magnetism when the main circuit was opened. 1900 —— *Polyphase Electric Currents* (ed. 2) v. 143 The teazer winding is connected with one end to the middle of the main winding. 1937 J. B. Gibbs *Transformer Princ. & Pract.* xi. 82 One transformer, called the 'main transformer' is connected between two of the three-phase lines,..and the 'teaser transformer' is connected between the third line and the 50 per cent tap of the main transformer, using the 86.6 per cent tap of the teaser. 1981 G. McPherson *Introd. Electr. Machines* iii. 221 In the T connection, one transformer has its primary connected directly across two lines. This is called the 'main transformer'. The second transformer is called the 'teaser'.

5. An introductory advertisement, *esp.* an excerpt or sample designed to stimulate interest or curiosity. orig. and chiefly *U.S.*

1934 Webster, *Teaser*, an advertisement meant to arouse curiosity, sometimes by withholding part of the material information (*Trade Slang, U.S.*). 1940 *Chambers's Techn. Dict.* 835/2 *Teaser*, colloquialism for a *trailer* which is intended to advertise films for future exhibition in a cinema. 1960 M. T. Williams *Art of Jazz* 86 Old Town.. was plastered with 'teaser' posters heralding the coming of the famous..Minstrels. 1962 *Daily Progress* (Charlottesville, Va.) 8 Nov. 38/1 A teaser is..a stimulating bit from the story to follow and opens a show. 1962 S. E. Hyman *Tangled Bank* 378 At the end of a lecture, Freud will sometimes tack on a teaser for the next, such as: 'At the next lecture we shall see whether we can agree with the poets in their conception of the meaning of psychological errors.' 1977 'J. le Carré' *Hon. Schoolboy* xii. 270 Our agent asked Ricardo for a teaser so's the information could be evaluated back home.

tea-shop. [f. Tea *sb.* + Shop *sb.*] **a.** A shop where tea (sense 1) is sold. **b.** A café where tea (sense 2 or 4) is served.

a 1745 [in Dict. s.v. Tea *sb.* 9 a]. 1856 A. M. Lang *Diary* (Meean Meer, Punjab) 17 Sept. (MS.), Went to Tea Shop and billiards..at Artillery Mess. 1860 J. R. Edkins *Chinese Scenes* (1863) 153 Drinking tea with about forty nondescript Chinamen... I shall try to give you a little picture of the tea-shop. 1915 W. S. Maugham *Of Human Bondage* lvi. 288 He thought of going to lunch at the tea-shop. 1933 Dylan Thomas *Let.* Oct. (1966) 43 Gower is a very beautiful peninsula,..and so far the Tea-Shop philistines have not spoilt the more beautiful of its bays. 1962 L. Davidson *Rose of Tibet* ii. 48 There were a number of small teashops in the town [*sc.* Kalimpong]; ramshackle sheds with trestle tables containing tea urns and trays of sweetmeats.

c. In the affected spelling **tea-shoppe** [cf. *SHOPPE] applied (freq. disparagingly) to a tea-shop with sham antique decoration.

Second element sometimes pronounced (ʃɒ·pi).

1925 *Amer. Speech* I. 153/1 These names are not intended to be 'quaint' like 'Betty Anne' of Massachusetts and her eternal 'Tea Shoppe'. 1933, etc. [see *SHOPPE]. 1959 *Times* 12 May 3/7 As artificial as a Tudor Tea Shoppe. 1973 Wodehouse *Bachelors Anonymous* v. 50 She had been planning a roll and butter and a cup of coffee at some wayside tea shoppe.

Hence **tea-shoppy** *a.*, characteristic of or resembling a tea-shop (sense b or c).

1931 *Time & Tide* 22 Aug. 992 There are other debatable points, too, in this rather tea-shoppy story. 1959 *Good Food Guide* 52 Portions are ample and not tea-shoppy. 1975 *Times* 27 Dec. 7/3 A restaurant..in a tea-shoppy basement.

teasing, *vbl. sb.*¹ Add: **1. b.** *U.S. Hairdressing.* Back-combing; also, a similar treatment given with a small brush.

1923 F. Korf *Art & Fundamentals of Hairdressing* II. ii. 31 The public seems to fear the back-combing, or as it is often called, teasing of the hair, perhaps with some justification. 1964 D. Z. Hanle *Hairdo Handbk.* vii. 71 Properly done, teasing can play an important part in finishing a hairstyle... Use a small..teasing brush. 1975 C. Calasibetta *Fairchild's Dict. Fashion* 260/1 *Bouffant*, hair exaggeratedly puffed out by means of teasing.

3. (Earlier examples.)

1807 H. Tufts in E. Pearson *Autobiogr. of Criminal* (1930) ii. iv. 292 *Teasing*, whipping. 1821 P. Egan *Life in London* i. 11 The innumerable teazings thou hast book'd.

Teasmade (tɪ·zmēɪd). [perh. f. phr. *tea's made.*] The proprietary name of a brand of automatic tea-maker (see *tea-maker* (*d*) s.v. *TEA *sb.* 9 c).

1938 *Trade Marks Jrnl.* 26 Jan. 93/2 *Goblin Teasmade*. .. Time controlled electric water heating and tea making apparatus. The British Vacuum Cleaner & Engineering Company Limited,..Fulham, London, S.W.6; manufacturers. 1958 *Spectator* 6 June 754/2 The Goblin Vacuum Cleaner, Goblin Washing Machines and the Goblin Teasmades. 1980 A. N. Wilson *Healing Art* 271 The electric clock on the Teasmade in the bedroom. 1983 *Sunday Tel.* 31 July 8/7 He was wearing a watch—one of those that..cannot be used as a calculator and do not double as a Teasmade. It merely gives the time.

tea-spoon. Add: **b.** = Teaspoonful.

1791 J. Woodforde *Diary* 19 Apr. (1927) III. 266, I took half a very small Tea-Spoon of Ether in Water this Evening. 1935 M. Morphy *Recipes of All Nations* 767 A teaspoon of the mixture is wrapped in..a blanched vine leaf. 1963 R. Carrier *Great Dishes of World* ii. 60/2, 1 level teaspoon dried mustard.

teasy, *a.* For *colloq. rare* read *colloq.* and *dial.* and add: **a.** (Earlier and later examples.)

1901 J. H. Harris *Luck of Wheal Veor* 164 A poor woman wethout a man, an' three gert stramming maids to keep, es like a cow wethout a tail when the flies is taisey. 1938 E. Pound *Let.* 8 May (1971) 315, I forget what he and Domenik have to say, but teckon it's teasy.

b. Bad-tempered, irritable, tetchy.

1866 T. R. Higham *Dial. between Two Cornish Miners* 5 My owld 'umman was..so taisy that I cudden live in the house. 1907 A. Quiller-Couch *Major Vigoureux* xvii. 167 He'll be as teasy as fire when he hears about it. 1931 C. C. Rogers *Gwendra Cove* 193 'E was a teasy oald bachelor, I reckon. 1976 J. C. Trewin in D. V. Baker *Cornish Short Stories* 135 'Go 'long, you teasy toad!' shrilled Mrs Bosworthick.

Hence **tea·sily** *adv.*

1928 A. Bennett in *Daily Express* 1 Sept. 5/5 'And what will Mrs. Meadowes say to all this Red politics?' Alan demanded teasily.

tea-table, *v.* [f. the sb.] *trans.* In literature, to treat a dramatic event in a trivial or casual way. Hence **tea-tabling** *vbl. sb.*

1938 C. Isherwood *Lions & Shadows* iv. 175 The accident was to be in the best Forster tradition, 'tea-tabled', slightly absurd. *Ibid.* vi. 258 The murder was cut—'tea-tabled' down to an indecisive, undignified scuffle; and the ending was an apotheosis of the Tea-Table, a decrescendo of anti-climaxes. 1962 *Times Lit. Suppl.* 22 June 460/4 Certain critics have made far too much of Mr. Forster's 'tea-tabling' and of his casual sudden deaths. 1977 *Ibid.* 28 Jan. 97/2 Christina Rossetti's oblique treatment of detail is an early case of what Isherwood, discussing Forster in *Lions and Shadows*, calls 'tea-tabling', the novel's lyrical domestication of disaster.

tea-tree. Add: **2. b.** *tea-tree oil.*

1933 *Bulletin* (Sydney) 12 July 19/2 The distillation of essential oil from *Melaleuca alternifolia* (tea-tree) has become an established N.S.W. industry... Some 40 tea-tree-oil preparations are on the market. 1954 *Econ. Bot.* VIII. 324/1 Tea-Tree Oil... Collection of leaf material differs from that in operation for eucalyptus leaves.

Tebele (tēbī·lɪ). Also **Tabele.** [Native name.] A Niger-Congo language belonging to the South Eastern Bantu group. Cf. *NDEBELE.

1883 R. N. Cust *Mod. Lang. Africa* II. xii. 300 Travellers write about..the Language of the Ma-tabéle, but there is reason to believe that it is Zulu... Syke, of the London Missionary Society, prepared School Books..: this gives the idea that Tabéle is a separate Language. 1897 W. A. Elliott *Dict. Tebele & Shuna Lang.* p. v, The Tebele language is of course only a variety of the Zulu. *Ibid.* p. vi, Different types have been used for the two languages, clarendon for the Tebele, and small capitals for the Shuna. 1919 [see *MATABELE 1]. 1977 C. F. & F. M. Voegelin *Classification & Index World's Lang.* 70 Ndebele = Tabele = Tebele.

Tebeth (te·beþ, te·bet). Also **Tebet, Tevet** (te·vet). [Heb. *ṭēbēṯ*] The fourth month of the Jewish year (though placed tenth in the traditional list of months), corresponding to parts of December and January.

c 1382 Bible (Wycliffe) *Esther* ii. 16 And so she is lad to the priue chaumbre of king Assuer, the tenthe moneth, that is clepid Thebeth, that is, Januer. 1611 *Ibid.* (A.V.), So Esther was taken vnto king Ahasuerus..in the tenth moneth (which is the moneth Tebeth). 1973 *Jewish Chron.* 19 Jan. 22/5 The Fast of Tevet is mentioned in the Talmud and is therefore observed even on Friday.

Tebilized (tɪ·bɪləɪzd), *a.* Also **-ised** and with small initial. [f. the initials of *T*ootal *B*roadhurst *L*ee Company Ltd., the inventors of the process + -ize + -ed¹.] A proprietary name for cotton and other fabrics which have been rendered crease-resistant by impregnation with a synthetic resin.

1937 *Official Gaz.* (U.S. Patent Office) 7 Dec. 36/1 Tootal Broadhurst Lee Company Limited. Manchester, England. Filed Oct. 16, 1937. Tebilized. For Piece Goods of Cotton, Linen, and Artificial Silk. Claims use since May 10, 1934. 1945 *Trade Marks Jrnl.* 28 Mar. 166/2 Tebilized. 1950 *Sun* (Baltimore) 8 May 10/4 This year with all the wonderful 'ized' things (tebilized and sanforized) that have happened to fabrics. 1955 *Times* 4 June 7 Irish linen for dresses or suits. Tebilized for crease-resistance. 1967 H. Thompson *Fibres & Fabrics of Today* 98 The name 'Tebilised' was adopted by Tootal Broadhurst Lee when they introduced this now universal finish.

tec, *sb.* Add: Also † **teck. 1.** (Earlier and later examples.)

1879 *Sessions Papers* 25 July 238 Look out, there is a b— *tec* from Chalk Farm watching. 1909 Galsworthy *Silver Box* ii. i. 47 Drop it, I say, you blooming teck. *a* 1940 [see *PINK *sb.*⁵]. 1977 *Daily Mirror* 30 Mar. 13/4 (*heading*) Porn tec admits bribe plot.

attrib. 1928 D. L. Sayers *Unpleasantness at Bellona Club* viii. 90 He complained of being spied on..like the blighters in the 'tec stories. 1976 *Evening Times* (Glasgow) 1 Dec. 2/1 Softly, Softly: Task Force. Plodding 'tec series which has dragged on too long.

2. Ellipt. for *tec story*, a detective story.

1934 E. Pound *ABC of Reading* ii. 29 Only a very good 'tec' will stand re-reading. 1949 R. Chandler *Let.* 28 Dec. (1966) 76 The mystery and 'tec are on the wane. 1978 S. Hodges *Gollancz* ix. 194 Anthony Price, Nicholas Freeling and Helen McCloy are some of the writers whose 'tecs have been published by Gollancz.

tech (tek), *sb.*¹ *colloq.* (orig. *U.S.*). Also **tec.** Abbreviation for *Technical College, Technical School* (see Technical *a.* 3 a in Dict. and Suppl.), and *Institute of Technology.*

1906 *Dialect Notes* III. 161 *The Boston Tech., n. phr.* Massachusetts Institute of Technology. 'The best engineering school is Cornell; the next best is Michigan; and the next, Arkansas. *The Boston Tech.* aint in it.' 1911 H. Quick *Yellowstone Nights* iii. 63 The insufferable breed of dubs—..who..called an Institute of Technology a 'Tech'. *Ibid.*, I shall have outlived the disgrace of my Tech. training. 1915 E. Wallace *Man who bought London* viii. 79 'I'm off to the "Tec",' he said. *c* 1921 D. H. Lawrence *Mr. Noon* viii, in *Mod. Lover* (1934) 264 He vowed he'd go up to the Tech. with the book. 1947 *Book* (Christchurch, N.Z.) IX. 32, I told him no, I'd have to finish Tech, I wanted to get matric. 1969 D. Collyer in R. Blythe *Akenfield* iv. 94 The young people..go to Ipswich Tec. or even to the university and they learn to talk about anything. 1974 *Sunday* (Charleston, S. Carolina) 7 Apr. 16-C/1 Texas Tech scientists will visit drought-stricken Africa to try to improve its resources and find a cure for famine and poverty. 1980 R. McCrum *In Secret State* x. 86 Rosie's pride would not let her admit that she.. had been to the local Tech.

tech (tek), *sb.*² Slang abbrev. of *TECHNICIAN c.

1942 *Yank* 21 Oct. 15 They're the same kind of tech chevrons. 1953 *Mag. Fantasy & Sci. Fiction* Nov. 30 Let the techs worry about that. 1976 *Amer. Speech* 1973 XLVIII. 194 They used to be called *orderlies*. Today, however, many of these assistants demand the title of *nursing tech* or *emergency room tech*. 1977 *R.A.F. News* 11–24 May 8/4 Chf Tech Mick Young..took part in a cycling expedition...Jnr Tech Rob Patrick..was a schoolboy..cycling champion. 1980 A. Skinner *Mind's Eye* xx. 258 He's got..a thing the other techs made him.

tech (tek), *sb.*³ [Abbrev. of Technology.] **1. high-tech** = *high-technology* attrib. phr. s.v. *TECHNOLOGY 1 d; *spec.* with reference to a style of architecture and interior design that imitates the functionalism of industrial technology. Also (unhyphened) as *sb. phr.* Similarly **low-tech** *attrib. phr.*

1972 *Last Whole Earth Catalog* (Portola Inst.) 247/1 It's the only high-tech home I've found at all lovable. 1978 Kron & Slesin *High-Tech* 1 Some people call this phenomenon 'the industrial style', but we call it 'high-tech'. High-tech..is a term currently used in architectural circles to describe buildings incorporating prefabricated ..building components. 1979 *Jrnl. R. Soc. Arts* Nov. 743/2 Late-Modernism takes the ideas and forms of the Modern Movement to an extreme, exaggerating the structure and technological image of the building beyond the point which the Modernists would have found acceptable, sometimes meriting the label 'High-Tech' as a result. 1980 *New Age* (U.S.) Oct. 36/2 A pocket calculator, a very high-tech gadget. 1981 *Farmstead Mag.* Winter 24/2 These solar greenhouses are often hi-tech. *Ibid.* 26/1 The low-tech greenhouse. 1983 *Times* 25 Jan. 26/7 Mrs Williams touched on such subjects as..the need for more 'high-tech'.

2. Chiefly *attrib.* = Technology 1 b. *rare.*

1982 *Times* 6 May 17/6 (*heading*) Whitehall backs new tech firms. 1983 *New Scientist* 16 June 769/3 (Advt.), Signaal, Philips space-tech company, has developed a meteorological ground station.

tech (tek), *a.* Colloq. (orig. *U.S.*) abbrev. of Technical *a.* **techspeak** [*-SPEAK], technical jargon.

1956 [see *LATENT *a.* j]. 1974 *Some Technical Terms & Slang* (Granada Television), *Tech run*, technical run. A rehearsal by the actors in the rehearsal room to which the technical staff who will be responsible for the programme in the stuido are invited to sort out technical problems. 1974 *Globe & Mail* (Toronto) 20 Feb. 34/9 Will Genge aim at the championship aside from his duties as tech supervisor? 1982 *80 Microcomputing* Feb. 330/1 This is code, jargon, techspeak, whatever.

technetium (teknī·ʃɪ°ŏm). [mod.L., f. Gr. τεχνητ-ός artificial (f. τεχνᾶσθαι to make by art, f. τέχνη art, craft) + -ium.] A dense, refractory, radioactive metallic element, chemically similar to rhenium, which occurs naturally only in trace amounts but is produced in reactors as a fission product of uranium and by neutron irradiation of molybdenum 98 and is used medically as a tracer in scintigraphy. Symbol Tc; atomic number 43. Formerly called *MASURIUM.

1947 Perrier & Segrè in *Nature* 4 Jan. 24/1 We would like to propose the name of 'technetium', from the Greek τεχνητός, artificial, in recognition of the fact that technetium is the first artificially made element. The corresponding chemical symbol should be 'Tc'. 1960 J. Kleinberg et al. *Inorg. Chem.* xxi. 535 The existence of the heptasulfides of technetium and rhenium and their mode of preparation emphasize the high degree of stability of these elements in the +7 state. 1962 [see *MASURIUM]. 1969 *New Scientist* 13 Mar. 564/1 Astrophysicists are puzzled to explain how the element technetium, which must be produced in the hot cores of stars, reaches the surface. 1974 *Encycl. Brit. Micropædia* IX. 859/2 Technetium metal looks like platinum but is usually obtained as a gray powder. 1977 *Lancet* 7 May 1012/1 Individual limbs were scanned 5–6 hr after administration of 10 mCi technetium-99m diphosphonate.

technetronic (te:knītrɒ·nik), *a.* [ult. f. Gr. τέχνη art, craft + *ELEC)TRONIC *a.*] Conditioned, determined, or shaped by advanced technology and electronic communications.

1967 Z. Brzezinski in *New Republic* 23 Dec. 18/2 Our society..is entering a more self-conscious stage; ceasing to be an industrial society, it is being shaped to an ever-increasing extent by technology and electronics, and thus becoming the first *technetronic society*. 1970 D. Gabor *Innovations* i. 7 The 'modern industrial state' or the '*technetronic* society', as it has been variously called, is indeed above the head of the man in the street. 1977 *Time* 8 Aug. 10/2 The old order, based largely on military power and nationalism, is giving way to 'a technetronic age' in which there will be increasing emphasis on economic development and social justice.

technic, *a.* and *sb.* Add: **B.** *sb.* **4.** *U.S.* = Technique.

1908 *Arch. Internal Med.* II. 107 Cunningham's technic was crude. 1922 [see *encephalography* s.v. *ENCEPHALO-]. 1931 [see *PLASMODESMA]. 1943 H. L. Mencken *Heathen Days* vii. 93 The Fourth..went even worse than the Eroica, though it actually makes much less demand on technic. 1954 [see *immunohæmatology* s.v. *IMMUNO-].

technica. (Earlier example.)

1782 [see *CANCRIZANS *a.*].

technical, *a.* (*sb.*) **A.** *adj.* **1.** Delete *rare* and add later examples. Also *spec.* in the official designations of certain ranks in the armed forces of the U.K. and U.S.

1917 'Contact' *Airman's Outings* i. 6 As regards the mechanics, the quality of their skilled work is tempered by the technical sergeant-major, who knows most things about an aeroplane, and the quality of their behaviour by the disciplinary sergeant-major, usually an ex-regular with a lively talent for blasting. 1920 *Army & Navy Jrnl.* (U.S.) 26 June 1338/2 Enlisted men of the 'second grade' will be designated as Technical Sergeants or First Sergeants. 1926 *Brit. Gaz.* 12 May 3/2 At Basingstoke there is a supply of technical engineers available for work elsewhere. 1937 *Discovery* June 168/2 Behind the barge followed the baggage canoe, with two technical assistants and two polers. 1961 G. Millerson *Technique Television Production* 15 The technical director is in charge of the technical operational staff on the show. 1978 J. Irving *World according to Garp* i. 15 Technical Sergeant Garp.. served with the Eighth Air Force.

3. a. *technical college, technical school* (earlier example), *university*; *technical fix* (U.S.), a solution produced by technological means (sometimes used with an implication of superficiality); *technical hitch*, an interruption or breakdown due to mechanical failure; *loosely*, an unexpected obstacle or snag.

1879 Technical university [see *BUFF *sb.*⁷ 1]. 1886 S. A. Barnett in H. Barnett *Canon Barnett* (1918) II. xlv. 246 Relief must..provide training. It may be in technical schools in town. 1886 *Times* 20 May 5/5 Yesterday afternoon the foundation-stone of the Technical College, an extension of the existing Mechanics' Institute, was laid at Keighley. 1940 P. Fleming *Flying Visit* v. 37 [Hitler] remained an equally great man to-day and (despite a technical hitch) equally capable of fulfilling his mission. 1958 'R. Crompton' *William's Television Show* v. 148 Couldn't we say there's been a technical hitch? 1962 D. Lessing *Golden Notebk.* 528 It [*sc.* a film] was running slowly, because there was a technical hitch of some kind. 1971 *N.Y. Times* 8 Sept. 44/2 In recent centuries the scientific revolution has provided much warrant for the notion of the 'technical fix', the idea that scientists or technologists can find an appropriate solution for every problem. 1978 *Nagel's Encycl.-Guide: China* 318 The Technical Universities are in fact like polytechnic or engineering schools, and contain as many faculties as they do specialities. 1980 *Directory of Technical & Further Educ.* (ed. 18) p. xvi, *Technical colleges*, so-called, are usually fairly old-established, with a range of courses for full-time and part-time, day release or block release students, traditionally in engineering, at the levels of Ordinary and often Higher National Certificate..or Diploma. 1980 *New Age* (U.S.) Oct. 30/1 It greatly underestimates the energy savings possible through technical fixes alone—that is, through well-known and presently economic technical measures that would have no significant effect on our lifestyles or economic output.

d. (Later examples.) *technical foul* (Basketball), a foul which does not involve contact between opponents; also *ellipt.* as *sb.*; *technical knockout* (Boxing), the termination of a fight by the referee on the grounds of one boxer's inability to continue (though not counted out), his opponent being declared the winner; abbrev. TKO, t.k.o.: see *T 6.

1921 *Daily Colonist* (Victoria, B.C.) 6 Apr. 10/3 Denial of a report from Saskatoon that in a boxing match in that city on March 24 Bill Barton, of Vancouver, had secured a technical knockout over Billy Mackenzie. 1934 Webster s.v. Foul, *Technical foul*. 1958 F. C. Avis *Boxing Ref. Dict.* 112 *Technical Knock Out*, the decision of the referee when stopping a contest in which one of the boxers, though not knocked out, is scarcely capable of proceeding. 1962 *Times* 2 Aug. 3/3 The ball went to hand off his bat on seven occasions. Most were only technical chances [of a catch], although both Taylor and Swallow..would have held their respective catches more often than not. 1974 *Spartanburg* (S. Carolina) *Herald-Jrnl.* 21 Apr. B1/3 Oscar Robertson sank a free throw on the technical, and Jon McGlocklin hit a long jumper after the Bucks put the ball in play. 1977 *Times* 21 June 10/1 Other technical chances escaped McCosker and O'Keeffe, both at slip, as the ball.. moved about off the seam. 1981 *Washington Post* 17 Feb. D2/2 Before they sat down they had been assessed a technical foul... After Davis missed the first shot of the technical Skipper made the second.

e. So regarded according to a strict legal interpretation. Usu. in phr. *technical assault.*

1911 *Encycl. Brit.* XX. 769/1 Finding himself nonsuited in a court of law he commits a technical assault upon..some high legal functionary. 1914 A. Harrison *Kaiser's War* 126 He [*sc.* an officer] may not accept an apology in the event of a technical assault. Thus a man who on leaving a café, for example, brushes against an officer, is technically liable to be cut down. 1920 Wodehouse *Damsel in Distress* vi. 84 'You ought to have had the scoundrel arrested,' he said vehemently. 'It was a technical assault.'

4. *Finance.* Of, pertaining to, or designating a market in which prices are determined chiefly by internal factors (see also quot. 1962).

1909 in Webster. 1946 *Sun* (Baltimore) 17 July 14/1 A certain amount of support was attracted to individual favorites on the idea the list may have been oversold and was due for a technical comeback. 1962 S. Strand *Marketing Dict.* 732 *Technical position*, a term applied to the various internal factors affecting the market; opposed to external forces such as earnings, dividends, political considerations and general economic conditions. Some internal factors considered in appraising the market's technical position include the size of the short interest, whether the market has had a sustained advance or decline without interruption, a sharp advance or decline on a small volume and the amount of credit in use in the market. 1981 *Times* 22 July 22/6 Most sections of the market staged a technical rally. 1983 *Times* 2 Apr. 10/5 The market remained technical, with positions being covered, and a marked reluctance shown to open new positions.

technician. Add: **c.** *spec.* A person qualified in the practical application of one of the sciences or mechanical arts; now esp., a person whose job is to carry out practical work in a laboratory or to give assistance with technical equipment.

1939 W. A. Richardson *Technical College* xxv. 476 There are certain high-grade technicians, e.g. chemists, physicists, mathematical engineers, who will be engaged in the more elaborate design problems, in industrial research and investigation. 1952, etc. [see *TECHNOLOGIST]. 1954 *Rep. Patent Cases* LXX. 150 The Defendants have a most elaborate system for bringing to the attention of their technicians all technical developments relevant to their work. 1968 *Ibid.* LXXXV. 104 Extracts from laboratory notebooks or other specific records pertaining to the particular strains employed by the laboratory technicians. 1972 *Daily Tel.* 9 Feb. 12 Note that word 'technician'. From today there are no more 'mechanics' at Ford Sure dealers. 'We have changed the name because we think it is time that we built up the image of the man in the workshop.' 1983 *N.Y. Times* 11 Aug. A10/6 French technicians scurry around at the military air base assembling machine guns and helicopters.

technicism. Add: **2.** Technical quality or character; a condition in which practical results or methods are stressed.

1932 tr. *Ortega y Gasset's Revolt of Masses* vi. 61 Three principals have made possible this new world: liberal democracy, scientific experiment, and industrialism. The two latter may be summed-up in one word: technicism. 1951 *Archit. Rev.* CX. 203/2 There is also research into aesthetic functions, and it is due only to this latter, if, instead of arriving at a dry technicism or mere utilitarianism, the modern movement has led to a true new style in architecture. 1977 *Times Educ. Suppl.* 21 Oct. 32/3 Even in the Bullock report..one yet senses, ultimately, a dreary technicism of mind... The report is written in the language of death.

technicist. Add: **2.** *attrib.* or as *adj.* Of or pertaining to technicism (sense *2).

1932 tr. *Ortega y Gasset's Revolt of Masses* 151 We are told quite seriously that the essence of America is its practical and technicist conception of life. 1974 *Times* 16 Mar. 16 A technicist society indifferent to location and continuity will undermine itself by wrecking the structures of meaning on which any society must rest.

technicize (te·knisəiz), *v. rare.* [f. Technic *a.* + -ize.] *trans.* To make technical; to subject to a high degree of technicality. Hence **technicıza·tion**; **te·chnicized** *ppl. a.*

1927 A. Huxley *Proper Studies* 137 The world will become even more completely technicized, even more elaborately regimented, than it is at present. 1959 —— *Let.* 6 May (1969) 869 It dealt with..advancing technicization of everything. 1972 *Listener* 31 Aug. 270/3 The predominant effort..to hang on to and 'technicise' the cult of the socialised warrior—ally the socialised warrior with technology, that is. 1975 J. De Bres tr. *Mandel's Late Capitalism* vii. 243 A far more technicized division of labour now replaces the old factory hierarchy.

technico-. Add: **2.** Forming adjs., as *technico-architectonic, -diplomatic, -economic*, etc.

1951 M. Lowry *Let.* Feb. (1967) 233 Is this a technical technico-architechtonic [*sic*] term? 1970 *New Scientist* 30 July 219/1 Prime Minister Vorster..scored a considerable technico-diplomatic success. 1970 E. Snow *Red China Today* (new ed.) 36 The solution of technico-economic problems of social transformation. *Ibid.*, Centrally led priority programs for essential technico-scientific specialists. 1973 D. Osmond-Smith tr. *Bettetini's Lang. & Technique of Film* i. 45 However, a morphological study of the film, even if conducted on a syntactic level, cannot leave out of account considerations of a technico-grammatical nature.

Technicolor (te·knikɒləɹ). Also **-our. 1.** A proprietary name for various processes of colour cinematography, esp. ones employing dye transfer and separation negatives. Freq. *attrib.*

1917 *U.S. Patent* 1,231,710 Daniel F. Comstock... Assignor, by Mesne Assignments, to Technicolor Motion Picture Corporation, of Boston, Massachusetts. 1929 *Official Gaz.* (U.S. Patent Office) 3 Dec. 24/1 *Technicolor* for motion-picture films. 1930 *Punch* 2 Apr. 385 *Show of Shows* at the Tivoli, the latest and greatest of technicolour talkie reviews. 1932 *Discovery* Dec. 382/2 The subtractive process..has been familiar to picture-goers in the many Technicolor films shown in this country. 1948 *Time* 22 Mar. 85/3 Herbert T. Kalmus,..the co-inventor, developer, majority stockholder and president of Technicolor,..is a graduate of Massachusetts Institute of Technology (after which Technicolor was named). 1950 *Trade Marks Jrnl.* 5 Apr. 314/2 *Technicolor*... Photographic apparatus, cinematographic apparatus, television apparatus, apparatus capable of transmitting and receiving pictures in colour, and talking machines. Technicolor Motion Picture Corporation... 4th August 1949. 1976 L.-A. Bawden *Oxf. Compan. Film* 681/1 A special projector with two apertures—one with a red filter, the other with a green filter—was needed to show early Technicolor

films. *Ibid.* 682/1 Eastman Color, introduced in 1949, changed the whole nature of colour filming and from 1951 onwards Technicolor prints were made, still by dye transfer, almost exclusively from Eastman Color negative. 1978 *Amer. Poetry Rev.* Sept./Oct. 10/2 'On the Towpath' begins in black-and-white and quickly shifts to technicolor.

2. *transf.* and *fig.* (Also with small initial.) Vivid colour characteristic of colour cinematography. Chiefly *attrib.* and in phr. *in glorious* (*gorgeous*, etc.) *Technicolor.*

a 1940 F. Scott Fitzgerald *Last Tycoon* (1941) v. 95 The theme..would come in some such guise as the auto horns from the Technicolor boulevards below. 1946 J. B. Priestley *Bright Day* v. 130 She looked very beautiful, and in glorious technicolour. 1954 M. Procter *Hell is City* i. vi. 35 A natural blonde of Technicolor brilliance. 1960 M. Stewart *My Brother Michael* ii. 25 The heavy Technicolor prosperity of the plains. 1962 N. Del Mar *Richard Strauss* iii. 85 The glorious technicolour of Strauss's orchestra. 1966 *Listener* 13 Jan. 71/2 Illustrations in black and white and gorgeous Technicolor. 1977 *New Yorker* 12 Sept. 32/3 A Technicolor-red leotard that matched her lips and her spike heels. 1979 *Church Times* 2 Nov. 9/1 Evelyn Waugh has described how some glamorise it [*sc.* death] in glorious technicolour.

Hence **te·chnicolored** *a.*

1947 E. F. Russell in Aldiss & Harrison *Decade 1940s* (1975) 166 'Laura [*sc.* a macaw] loves nuts.' 'I know it, you technicoloured bully.' 1954 *Newsweek* 27 Dec. 60/2 The palpable business of this Technicolored and CinemaScope production lies in its physical trappings, its underwater photography, and its action. 1962 *Movie* Sept. 22/1 Setting the events of the previous summer on the Riviera as technicolored flashbacks from the black-and-white Parisian winter. 1982 *New Scientist* 21 Jan. 179/2 Our face fuzz makes a pretty poor showing when compared with that bird's [*sc.* the peacock's] exuberant sensuality: you'd think evolution would have provided us with something more spectacular and 'Technicolored'.

‖ **technicum** (te·knikŏm). Also **tekhnikum.** Pl. **-s,** ‖ **-y.** [ad. Russ. *tekhnikum*, f. mod.L. *technicum*, neut. sing. of *technicus* technical (see Technic *a.* and *sb.*).] In the U.S.S.R., a technical college.

1932 M. Dobb *Soviet Russia & World* ii. 43 To-day there are 663 middle technical schools, or *technicums*. 1957 H. Bower *Short Guide Soviet Life* xiii. 52 Technical Colleges (tekhnikumy) train specialist technicians for a wide range of industries and take pupils from 14 to 30. 1963 *Higher Educ.: Rep. Comm. under Ld. Robbins 1961–3* v. 43 in *Parl. Papers 1962–3* (Cmnd. 2154) XI. 639 In the Soviet Union, the output of *Technicums* is not included. 1974 T. P. Whitney tr. *Solzhenitsyn's Gulag Archipelago* I. i. ii. 72 In Leningrad..the Latvian Technicum, and the Latvian and Estonian newspapers were all closed down.

technification (te:knifikēɪ·ʃən). [f. *techn-* as in Technical *a.*, etc. + -ification.] The adoption or imposition of technical methods. Also **te·chnified** *ppl. a.*, **te·chnify** *v. intr.* (both *rare*).

1959 A. Huxley *Let.* 4 Jan. (1969) 859 In time, I suspect, all fully technified societies will adopt the Russian solution. 1959 *Ibid.* 5 Jan., It may be that the only satisfactory solution..is to accept the inevitability of the technification of everything. 1962 *Spectator* 16 Feb. 199/2 The technification of conditioning minds by emotional images. 1970 J. Cotler in I. L. Horowitz *Masses in Lat. Amer.* xii. 429 Those who might have found employment in the exploitation of sugar and have been unable to do so due to technification, have had to remain in their miserable plots. 1972 *Listener* 18 May 640 This method is designed to be effective against other industrial and technified countries, whose organisation can be so disrupted that they simply can no longer function. 1973 *Harper's Mag.* Nov. 82 They internalized their intelligence activities with headlong speed. They technified senselessly—charts, graphs, bugs, concealed cameras.

technique. Add: **a.** (*a*) Now freq. used of the manner of execution or performance in any discipline, profession, or sport, or of skill or ability in any of these; (*b*) in the sense 'technical or artistic skill', freq. used without article or qualifying word; (*c*) *loosely*, a skilful or efficient means of achieving a purpose; a characteristic way of proceeding; a knack.

(*a*) 1900 [in Dict.]. 1903 [see *ASTROPHOTOGRAPHY]. 1920 Oswald & Pryce *Terra Sigillata* ii. 3 The technique of the green-glazed ware was also to a great extent similar to that of the red-glazed fabric; *e.g.* in the use of moulds with impressed designs..and both techniques were obviously inspired by prototypes in beaten metal. 1932 A. Huxley *Brave New World* i. 15 If they could discover a technique for shortening the period of maturation what a triumph, what a benefaction to Society! 1940 K. Mannheim *Man & Society* v. i. 244 Any deliberate rebuilding of human groups in terms of more elastic organization represents another chapter in the development of social techniques. 1959 C. Porter *Rowing to Win* ii. 40 The measure by which the appearance and technique of a Russian crew differs from that of an American crew is that of the difference in their styles... By style, then, I mean largely the technique of rowing. 1968 J. D. Watson *Double Helix* (1970) x. 63 The witchcraft-like techniques of the biochemist. 1972 M. Kline *Math. Thought* xiii. 282 The more complicated algebraic techniques.

(*b*) 1905 A. Bennett *Sacred & Profane Love* i. ii. 23 A generation of pianists who had lifted technique to a plane of which neither Liszt nor Rubenstein dreamed. 1935 W. S. Maugham *Don Fernando* x. 185 Technique is only the method by which the artist achieves his aim. 1960 P. Goodman *Growing Up Absurd* iv. 89 'The trainee,' says William H. Whyte, Jr., 'believes managing is an end in itself—technique is more vital than content.'

(*c*) 1936 J. A. Lee *Hunted* ii. 49 When brought back they told the story of their venturings, so that northern runaways knew the technique of stealing a passage. 1941 *Punch* 31 Dec. 583/1 There is quite a technique to washing a blanket. 1944 E. S. Gardner *Case of Black-eyed Blonde* ii. 22 Now then, you little son-of-a-bitch,..that technique of planting the diamond pendant is something you used about three years ago on that maid your mother had. 1955 L. P. Hartley *Perfect Woman* xiii. 126 She'll want to know all about you—that's her technique. 1979 D. Halberstam *Powers that Be* (1980) i. v. 273 In 1953 Barth wrote an editorial attacking the FBI's technique of gathering irrelevant information in its investigations and making the information a part of a person's dossier.

b. *spec.* Manner of performance or skill in sexual relations.

1921 M. Arlen *Romantic Lady* 11 She was very close to me, smiling, intimate. Pure coquetry, of course—but what perfect *technique*! 1930 V. Sackville-West *Edwardians* vi. 296 When he chose, his technique could be faultless... He was very gentle with Teresa. 1964 R. Braddon *Year Angry Rabbit* ix. 80 The American servicemen appeared to retain an atavistic but irresistible technique with girls which outraged the local boys. 1974 V. Gielgud *In Such a Night* viii. 71 Alec looks at every woman..like that... He hasn't altered his technique.

techno-. Add: *techno-comme·rcial, -econo·mic* adjs.; **te·chnocomplex** *Archæol.* (see quot. 1968); **te·chnofear** = *technophobia* below; **te·chnofreak** [*FREAK *sb.*¹ 4 c], an enthusiast for technology or for the technical complexities of a particular piece of equipment; hence **techno-frea·kish** *a.*; **technographic** *a.* (earlier U.S. example); **techno-ma·nager**, a person who is both a technologist and a manager; hence **te:chno-manage·rial** *a.*; **technoma·nia**, a mania for technology; hence **technoma·niac**; **te·chnophile**, one who favours technology; **technopho·bia**, fear of technology; so **te·chnophobe**, a person who fears technology; **techno·polis** [-polis], a society dominated by technology; hence **technopo·litan** *a.*; **te·chnosphere** [*-SPHERE], the technological aspect of human activity; **te·chnostructure**, a group of technologists or technical experts that controls the workings of industry or government; **technotro·nic** *a.* = *TECHNETRONIC *a.*

1937 *Discovery* Aug. 254/1 The history of this material [*sc.* synthetic rubber] from the early 'academic synthesis' period, through the 'techno-commercial' period, up to the present time. 1979 J. E. Rowley *Mechanised In-House Information Syst.* ii. 107 Some units..assign equal importance to scientific and technical information and techno-commercial material. 1968 D. L. Clarke *Analytical Archaeol.* viii. 357 *Technocomplex*, a group of cultures characterized by assemblages sharing a polythetic range but differing specific types of the same general families of artefact-types, shared as a widely diffused and interlinked response to common factors in environment, economy and technology. 1976 *Sci. Amer.* Feb. 94/2 What game animals supported the hunters of the Tanged Point Technocomplex? 1980 *Cambr. Encycl. Archaeol.* 69/1 The long, stable period of the Oldowan technocomplex. 1960 Techno-economic [see *FEED *v.* 8 e (ii)]. 1976 *Nature* 5 Feb. 355/2 A team..will be responsible for investigations into the cost-benefit of research done by the BSC and for techno-economic analysis. 1980 *Times* 14 Mar. 20/3 The consumer is still suffering from what many dealers are beginning to call 'technofear'—fear of commitment to purchasing anything in case the technology changes. 1983 *Times* 28 Sept. 3/3 Techno-fear..is defined as 'difficulties in accepting and using high-technology products in the home'. 1973 *Absolute Sound* I. ii. 42 We have always known it [*sc.* the audio industry] was dominated by techno freaks with an unhealthy irreverence for the live sound. *Ibid.* iii. 173 His prose is..so techno-freakish,..so filled with demonstrations that the Great Expert is at work that it is incomprehensible to virtually every informed audiophile we know. 1983 *Austral. Personal Computer* June/July 62/1 A neat piece of technology that..ought to interest any technofreak simply because it's such a good idea. 1891 *Ann. Rep. Smithsonian Inst.* 1889–90 i. 611 The Kunstgewerbe Museum contains much that is like the ethnographic collection, but the reigning concept is technographic. 1962 A. Sampson *Anat. Britain* xiii. 211 It was only after France and Germany had founded their *polytechniques* and *hochschule* [sic] for techno-managers that Britain gradually felt the need to adapt their universities to technical education. 1979 *Times of India* 17 Aug. 12 (Advt.), It has set up a full-fledged Consultancy & Promotional Cell with Indian and Foreign experts for providing technomanagerial Consultancy Services for improving the operational and managerial efficiency of consumer cooperatives. 1969 *Daily Tel.* 29 May 28/3 The days when almost any scientific team could wrest enormous sums from the taxpayer to finance big, glamorous and spectacular projects were coming to an end... 'The era of technomania is passing—and high time too,' Mr. Benn said. *Ibid.* 30 May 20/3 Shall I, who have hated technology all my life... Turn technomaniac myself? 1968 H. Weaver tr. *Ellul's Critique of New Commonplaces* 236 This commonplace is really very common among technicians, technologists, technolasters, technophagi, technophiles, technocrats, [etc.]. 1983 *Daily Tel.* 28 Nov. 12/1 Those technophiles disappointed by the absence of innovative features in IBM's newly announced P. C. Junior home computer have overlooked one splendid novelty. 1965 *New Statesman* 27 Aug. 286/1 Instead of leading us to the golden age, science is dragging us down into a servile, stable hell. Shades of Orwell! Technophobia has struck another good man down. *Ibid.*, The incipient technophobe will rage against the motor-murder of 20 people a day in Britain without once considering that cars also carry 50 million people and their goods. 1965 H. C. Cox *Secular City* 5 We shall make use of a somewhat contrived word, *technopolis*. It will be used here to signify the fusion of technological and political components into the base on which a new cultural style has appeared... it will call to mind the fact that the contemporary secular metropolis was not possible before modern technology. *Ibid.* iii. 63 To say that technopolitan man is pragmatic means that he is a kind of modern ascetic. He disciplines himself to give up certain things... Life for him is a set of problems, not an unfathomable mystery. 1969 Huxley & Nicholson in *Times* 7 Oct. 8/1 The most striking change which it has brought is to create out of a mass of economic, social and technical developments, an entire semi-autonomous new system, which we may call the technosphere, with its own structure and anatomy, its own programmed inputs and outputs, and its accidental or deliberate releases into the biosphere. 1967 J. K. Galbraith *New Industrial State* vi. 71 Management..includes..only a small proportion of those who..contribute information to group decisions. This..group..extends from the most senior officials of the corporation to..blue collar workers... This ..is the guiding intelligence—the brain—of the enterprise... I propose to call this organization the Technostructure. 1978 *Nature* 9 Nov. 147/2 In discussing the origins of the Soviet technical intelligentsia (throughout inelegantly and inaccurately termed the 'technostructure'), Bailes draws on an impressive range of sources, both Soviet and Western. 1969 *Daily Tel.* 23 Apr. 16/3 The 'technotronic society', as the mass technical world is now sometimes labelled, creates its own problems.

technocracy (teknɒ·krăsi). orig. *U.S.* [f. Techno- + -cracy.] The control of society or industry by technical experts; a ruling body of such experts.

Technocracy has been the name of various groups advocating the technical control of society, esp. Technocracy, Inc., established in New York in 1932–3 by Howard Scott.

1919 W. H. Smyth in *Industr. Management* Mar. 211/2 For this unique experiment in rationalized Industrial Democracy I have coined the term 'technocracy'. 1932 *N.Y. Herald-Tribune* 15 Dec. 11/1 Technocracy..the name for a new system and philosophy of government, in which the nation's industrial resources should be organized and managed by technically competent persons for the good of everyone instead of being left to the management of private interests for their own advantage. 1945 C. S. Lewis *That Hideous Strength* xii. 318 The effect of modern war is to eliminate retrogressive types, while sparing the technocracy and increasing its hold upon public affairs. 1947 *Mind* LVI. 164 Such notions as social and economic planning, technocracy,..the denial of natural rights and individual liberties, etc., are due to them [*sc.* French Utopians, St. Simon, etc.] more than to Godwin or the Utilitarians. 1955 *Times* 23 May 3/4 On that unlikely day when England elects a benevolent technocracy to power a Bill will be passed forbidding more than one performance per year per town of such works as *The Messiah*, the *St. Matthew Passion*, [etc.]. 1975 *Political Studies* XXIII. 82 Nevertheless, if technocracy means rule not just by *individuals* who are members of a particular technocratic *élite*, but rule by a technocratic class as such, one has to show that the latter has either a common interest to defend or a common ideology to pursue.

Hence **te·chnocrat**, (*a*) an advocate of technocracy; (*b*) a member of a technocracy, a technologist exercising administrative power; **technocra·tic** *a.*; **techno·cratism.**

1932 *Sun* (Baltimore) 12 Dec. 6/3 The Technocrats, thanks..largely to a peculiarly fetching 'trade label' which embodies in one word two of the most far-reaching of current concepts, technology and democracy, are succeeding in a remarkable degree in breaking down the apathy. 1932 *N.Y. Herald-Tribune* 15 Dec. 11/2 The haunts of technocratic science were situated at numerous places about town, principally in cubbyhole restaurants in Greenwich village. 1933 *Times Lit. Suppl.* 26 Jan. 46/2 An age that was already substituting the technocrat for the monarch. 1945 C. S. Lewis *That Hideous Strength* xii. 318 It was not the great technocrats of Koenigsberg or Moscow who supplied the casualties in the siege of Stalingrad. 1949 *Mind* LVIII. 416 Lersch denies the widely accepted thesis that man's uniqueness consists in his activities (activism, pragmatism, technocratism) since these are characteristic only of the Male's relation to the world. 1957 *London Mag.* Jan. 48 Sprawling in my revolving chair, behind a man-sized desk, I could imagine myself a brisk and efficient technocrat, a kind of highbrow tycoon. 1958 *Times Lit. Suppl.* 26 Dec. 751/1 Either tending towards reliance on a tradition which has been made obsolete..or else attempting a technocratic rule for which no tradition exists. 1965 W. H. G. Armytage *Rise of Technocrats* v. 66 St. Simonians were the first technocrats: apostles of the religion of industry. 1974 J. White tr. *Poulantzas's Fascism & Dictatorship* v. ii. 254 Imperialist ideology in effect represents a displacement of domination within bourgeois ideology itself, from the juridico-political region which was dominant in liberal-bourgeois ideology to economic technocratism. 1980 *Times* 11 Aug. 11/1 Dr Hoss was chosen after the Syrian-imposed end to the civil war in 1976 to head a 'technocratic', ie non-political, government.

technologic, *a.* (Examples.)

1971 *Amer. N. & Q.* Dec. 61/2 The thread of technologic changes is woven throughout the text. 1979 *Bull. Amer. Acad. Arts & Sci.* Mar. 20 The early insights about the 'technologic misfit' and the need for social and technological 'congruence' have given way to more systematic analysis.

technological, *a.* Add: **3.** Pertaining to or characterized by technology; resulting from

1921 R. A. S. MACALISTER *Text-bk. Europ. Archæol.* I. ix. 491 Tectiform devices are essentially shaped like a more or less isosceles triangle, with a vertical line running from apex to base. *Ibid.* 492 The resemblance between such huts and the tectiform devices is striking, but not wholly convincing...I have another suggestion to offer.. namely, that they are intended to represent traps. **1921** M. C. BURKITT *Prehist.* 382 (*caption*) The painted tectiform from La Mouthe. **1962** S. GIEDION *Eternal Present* I. iii. 254 The tectiforms of Font-de-Gaume also belong to the class of symbols with more or less straight lines.

tecto- (te·kto), comb. form of L. *tectum* roof, as in **te·ctocu:ticle** *Ent.* (see quot. 1951); **te·ctospi:nal** *a. Anat.*, applied to a group of nerve fibres which run from the tectum of the midbrain to the spinal cord.

1951 A. G. RICHARDS *Integument of Insects* xvi. 149 The term tectocuticle..is here proposed for any material poured onto the outer surface of the formed epicuticle and hardening there as a reasonably permanent component. **1974** R. H. HACKMAN in M. Rockstein *Physiol. Insecta* (ed. 2) VI. iii. 216 The cuticle is divided into..an inner relatively thick procuticle and a thin outer epicuticle... A cement layer or 'tectocuticle' may also be present. **1916** *Gray's Anat.* (ed. 19) 744 The tectospinal fasciculus originates in the superior colliculus..of the opposite side. **1974** D. & M. WEBSTER *Compar. Vertebr. Morphol.* xii. 290 Another prominent component of the extrapyramidal system is the tectospinal tract.

tectogenesis (tektǫdʒe·nėsis). *Geol.* [ad. G. *tektogenese* (E. Haarman 1926, in *Zeitschr. f. Deutsch. Geol. Ges.* LXXVIII. B. 106), f. Gr. τέκτων, -ον- carpenter, builder: see -GENESIS.] The formation of the highly distorted rock structures characteristic of mountain ranges, as distinct from the formation of mountainous topography itself. Hence **tectogene·tic, tectoge·nic** *adjs.*, of, pertaining to, or involving tectogenesis. Also **te·ctogene** [ad. G. *tektogen* (E. Haarman 1926, loc. cit., 107): see *-GEN 3], a long, narrow belt of downwarping in the earth's crust, said to be an underlying feature of mountain ranges and oceanic trenches.

1937 *Bull. Amer. Assoc. Petroleum Geologists* XXI. 1596 Orogenesis means 'mountain-making', but the term refers only to the production of mountain structure, not to that of mountain topography. Hence, it seems desirable to replace it by 'tectogenesis', as Haarman suggested. *Ibid.*, Tectogenic movements..are incongruent, making structures that vary in the different stories of the crust; and the deformation they produce is wholly irreversible. **1937** *Leidsche Geol. Med.* VIII. 204 (*caption*) Tectogene with root. **1965** *Phil. Trans. R. Soc.* A. CCLVIII. 65 The later Palaeozoic..tectogene stretched in a belt from the Appalachians, through southern Britain and central and southern Europe, to Suess's massive Altaids in the heart of Asia. *Ibid.* 68 Throughout the Caledonides (Spitzbergen, Greenland, Scandinavia and Britain) there seems little doubt that the main tectogenic phase was centred in Silurian time. *Ibid.* 74 Undisturbed marine successions do not necessarily rule out synchronous tectogenesis of a neighbouring region. **1975** *Nature* 24 Jan. 257/1 Sicilian data indicate a time span of only 3–4 Myr for a single 'tectogenetic cycle'. *Ibid.* 10 July 116/1 The radial pattern of transverse folds with respect to the arc..has been taken into account in tectogenetic models of the Alps.

tectonic, *a.* Add: **2.** Also with reference to other planets. Cf. also *plate-tectonic* adj. s.v. *PLATE *sb.* 20. (Further examples.)

1962 F. I. ORDWAY et al. *Basic Astronautics* iii. 65 They [*sc.* the rills on the moon] are apparently the result of tectonic activity. **1976** *Science* 24 Dec. 1386/3 By 2·5 billion years ago the volcanic-tectonic circus on Mars had folded. **1982** *Nature* 28 Jan. 293/2 Additional mapping and analysis based on Voyager images should help resolve many remaining questions about the tectonic evolution of Ganymede.

Hence **tecto·nical** *a.*, in the same sense; **tecto·nically** *adv.*, as regards tectonism; by tectonic processes.

1925 J. JOLY *Surface-Hist. Earth* 191 *Laccolith*, an intrusive mass of igneous rock..generally associated tectonically with a mountain range. **1939** *Geogr. Jrnl.* XCIV. 499 There are also discussions of the tectonical.. and general geological problems of Tanganyika Territory. **1972** *Rep. 24th Internat. Geol. Congr.* III. 3 Among the most important geological (tectonical) factors..are the concentrations of heat flow and juvenile matter ascending from the mantle..along steep, deep-seated tectonic zones. **1974** *Nature* 25 Jan. 194/1 Many investigators interpret ophiolites as slices of oceanic crust which have been tectonically emplaced in orogenic belts. **1976** J. KLECZEK *Universe* iv. 155 On the whole, the Moon is tectonically a very quiet body.

tectonician (tektǫni·ʃăn). *Geol.* [f. TECTON(ICS in Dict. and Suppl. + -ICIAN.] = *TECTONIST 2.

1951 *Amer. Jrnl. Sci.* CCXLIX. 594 For the tectonician, the most useful definition of fold-axis is that given by Wegmann. **1965** *Phil. Trans. R. Soc.* A. CCLVIII. 56 Such tremendous crustal movements within the Alpine fold belt are of a much larger size than those required by the most nappist of tectonicians. **1975** *Nature* 6 Feb. 396/2 The Soviet tectonician Belousov has gone so far as to invoke extensive 'oceanisation' of continental crust to account for the ocean basins.

tectonics. (In Dict. s.v. TECTONIC *a.*) Add: **2.** *Geol.* The structural arrangement of rocks

in the earth's crust (or on another planet); the branch of geology concerned with the understanding of rock structures, esp. large-scale ones. Cf. *plate tectonics* s.v. *PLATE *sb.* 20.

1899 *Q. Jrnl. Geol. Soc.* LV. 399 (*heading*) The tectonics of the district. **1914** J. PARK *Textbk. Geol.* xxxiv. 489 By a series of pressure experiments in 1888, Cadell obtained instructive imitations of the tectonics of mountain-building. **1935** [see *AUTOCHTHONOUS *a.*]. **1976** *Daily Colonist* (Victoria, B.C.) 15 May 5/5 The science of tectonics—the study of the movement of these plates—shows that Africa and Europe are in collision. **1982** *Nature* 28 Jan. 290 (*heading*) The tectonics of Ganymede.

tectonism (te·ktǫniz'm). *Geol.* [f. TECTON(IC *a.* + ISM.] = DIASTROPHISM.

1949 F. J. PETTIJOHN *Sedimentary Rocks* vi. 193 The feldspar content [in sands] is primarily an index of crystal [*read* crustal] instability or tectonism. **1960** *Bull. Amer. Assoc. Petroleum Geologists* XLIV. 1924/2 The San Joaquin Valley was undergoing tremendous tectonism during the Middle Tertiary. **1972** *Gloss. Geol.* (Amer. Geol. Inst.) 726/2 *Tectonism*, a less preferred syn. of diastrophism. **1975** G. ANDERSON *Coring* i. 10 Two basic types of tectonism affect a rock's acceptability and transmissibility of fluids—shearing and flexure folding. **1982** *Nature* 28 Jan. 292/1 Further evidence for distinguishing the style of tectonism on Ganymede is provided by structural features.

tectonist. (In Dict. s.v. TECTONIC *a.*) Restrict † *obs. nonce-wd.* to sense in Dict. and add: **2.** *Geol.* A specialist in tectonics. *rare.*

1933 *Amer. Jrnl. Sci.* CCXXV. 441 Becker's work has been to a large extent discarded by tectonists as being too highly theoretical to be of practical value in the interpretation of rock structure. **1935** [see *HERCYNIAN *a.* 2 c].

tectonite (te·ktǫnəit). *Petrol.* [ad. G. *tektonit* (B. Sander), f. Gr. τέκτων, -ον-, carpenter, builder: see -ITE¹.] A rock whose fabric shows evidence of differential movement during its formation.

1933 *Amer. Jrnl. Sci.* CCXXV. 433 Rocks that owe their present characters to..the integration of differential movements, he [*sc.* Sander] calls tectonites in distinction to non-tectonites, which are formed under conditions involving no differential movement. **1950** *Geol. Mag.* LXXXVII. 331 (*heading*) Note on two lineated tectonites. **1960** TURNER & VERHOOGEN *Igneous & Metamorphic Petrol.* (ed. 2) xxiii. 638, B-tectonites are tectonites in whose fabrics a linear parallelism of elements with reference to the *b* (= B) axis of the fabric is the outstanding structural feature.

tectonization (te:ktǫnəizēⁱ·ʃən). *Geol.* [f. TECTON(IC *a.* + -IZATION.] Modification (of rocks, etc.) by tectonic processes.

1959 *Jrnl. Geol.* LXVII. 26/2 The last schistosity produced was also folded and now preserves the impress of the last phase of tectonization. **1971** *Nature* 2 July 21/1 Part of the central gap may be intermediate crust which has been incorporated, after tectonization, into the neocratons of the Greater Antilles. **1979** *Ibid.* 27 Sept. 267/1 Extensive mélange exists near the base of the Trondheim Nappe, indicating tectonisation of syndepositionally deformed chaotic deposits.

Hence (as a back-formation) **te·ctonize** *v. trans.*, to alter by tectonic processes; **te·ctonized** *ppl. a.*

1970 *Nature* 25 July 351/1 Sedimentary layers have been heavily tectonized. **1975** *Ibid.* 13 Feb. 521/2 Future work should also show whether the distribution of tectonised mantle inclusions in kimberlites is related to tectonic lineaments. **1977** A. HALLAM *Planet Earth* 204 In British Columbia the stratigraphy is less clear in the tectonized zone of the Rockies.

tectono- (tektǫno), comb. form of TECTONIC *a.*, TECTONICS, used in *Geol.*, as in **tecto:nophy·sics**, a branch of geophysics concerned with the forces that cause movement and deformation in the Earth's crust; so **tecto:nophy·sical** *a.*; **tecto:nophy·sicist**, a specialist in tectonophysics; **tecto·nosphere** (see quot. 1926); **tecto:nostratigra·phic** *a.*, of or pertaining to the correlation of rock formations with one another in terms of their connection with a tectonic event; **te:ctonothe·rmal** *a.*, involving both tectonism and geothermal activity.

1960 *Bull. Geol. Soc. Amer.* LXXI. 1255 (*heading*) Tectonophysical investigations. **1979** *Nature* 8 Feb. 495/1 Recently I have been involved in tectonophysical research in the Witwatersrand collar sequence of the Vredefort dome. **1957** *Bull. Geol. Soc. Amer.* LXVIII. 642/1 The various tectonophysicists..have sought to explain the mechanism of the deep-focus earthquakes associated with island arcs. **1978** *Nature* 26 Oct. 733/1 The mechanism by which stress is released in intermediate and deep focus earthquakes is of particular interest to seismologists and tectonophysicists. **1959** *Geosci. Abstr.* July 5/1 The author..theoretically proves the possibility of using scale models in tectonophysics. **1960** *Bull. Geol. Soc. Amer.* LXXI. 1255/1 To Soviet specialists, tectonophysics is a scientific trend dealing with investigations into the physical mechanism of tectonic deformations. Investigators in other countries use the term in a broader, less definite sense. **1971** *Nature* 26 Nov. 185/2 Geophysics as a discipline covers everything from cosmic rays to seismology

by way of..tectonophysics and geomagnetism. **1926** G. W. TYRRELL *Princ. Petrol.* i. 2 The inaccessible heavy interior [of the earth] is known as the barysphere. This is followed outwardly by the lithosphere, the thin, rocky crust of the earth... Other zones have been distinguished and named for special purposes... The zone in which crustal movements originate has been named the tectonosphere by certain continental geologists. **1949** R. W. VAN BEMMELEN *Geol. of Indonesia* IA. iv. 283/2 The tectonosphere comprises three shells: the sial-, salsima- and sima- layers. **1971** M. H. P. BOTT *Interior of Earth* vii. 220 It is a well-established facet of isostatic theory that the weak asthenosphere is overlain by a relatively strong lithosphere (or tectonosphere) about 50–100 km thick. **1971** *Nature* 24 Sept. 246/2 The geology of the Scotia Arc region can be simplified by emphasizing what seem to us to be the fundamental tectonostratigraphic units. **1976** *Ibid.* 9 Sept. 117/2 Taking..the tectonostratigraphic evidence into consideration..an alternative model [of the origin of the Himalaya], based on plate tectonics, involving microcontinents, is suggested here. **1971** I. G. GASS et al. *Understanding Earth* xxii. 323/2 It seems likely that the relative paucity of major deposits of these ores is due to the modifying and/or dissipating influences of younger tectono-thermal activity. **1976** *Nature* 8 Apr. 516/2 It is significant that there is, as yet, no evidence for pre-Dalradian tectonothermal events affecting the Central Highland Granulites.

tectosilicate (te·ktosilikėt). *Min.* Also **tekto-**. [ad. G. *tektosilikat* (H. Strunz 1938, in *Zeitschr. f. ges. Naturwiss.* IV. 189), f. Gr. τεκτο-νεῖα workshops (or τεκτο-νία carpentry), taken as = framework + G. *silikat* SILICATE.] Any of the group of silicates in which the four oxygen atoms of each SiO_4 tetrahedron are shared with four neighbouring tetrahedra in a three-dimensional framework, with a ratio of silicon to oxygen of 1:2.

1947 [see *PHYLLOSILICATE]. **1959** BERRY & MASON *Mineralogy* xv. 471 All the tektosilicates are colorless, white, or pale gray when free from inclusions. **1971** I. G. GASS et al. *Understanding Earth* i. 16/2 The feldspars and quartz are examples of tectosilicates (three-dimensional framework structures).

tectosphere (te·ktosfīᵊɹ). *Geol.* [f. Gr. τέκτ-ων, -ον- carpenter, builder + -O + SPHERE *sb.*] That part of the earth which moves in coherent sections during plate-tectonic activity (see quot. 1979¹). Hence **tectosphe·ric** *a.*, of or pertaining to the tectosphere.

Tectosphere is sometimes confused in dicts. with *TEKTOSPHERE.

1968 *Jrnl. Geophysical Res.* LXXIII. 1980/2 The required strength [to maintain rigidity] cannot be in the crust alone; the oceanic crust is too thin for this. We instead favor a strong tectosphere, perhaps 100 km thick, sliding over a weak asthenosphere. **1969** W. M. ELSASSER in S. K. Runcorn *Applic. Mod. Physics to Earth & Planetary Interiors* 223 Horizontal sliding of the top layer, here called the 'tectosphere', can be more easily achieved than circulation in the material underneath. **1979** *Sci. Amer.* Jan. 76/3 Under the oceans the tectosphere and the lithosphere are..identical in spatial extent... Under the continents, however,..the cratonic tectosphere extends below the lithosphere, perhaps to depths of 400 kilometers or more. *Ibid.*, Tectospheric thickness also correlates with crustal age.

tectum (te·ktǒm). *Anat.* [L., = roof.] **a.** More fully *tectum mesencephali*. The roof of the midbrain, lying dorsal to the cerebral aqueduct.

1907 J. B. JOHNSTON *Nervous Syst. Vertebrates* xvi. 255 It must be remembered always that the tectum opticum is only a part of the tectum mesencephali. **1921** TILNEY & RILEY *Form & Functions Central Nervous Syst.* xxviii. 487 Like other suprasegmental parts of the nervous system, the tectum is capable of great expansion to meet the demands of adaptation. **1979** *Sci. Amer.* Sept. 82/1 The mesencephalon..in mammals includes two pairs of structures that together form a region of four hills known as the lamina quadrigemina, the tectum mesencephali or simply the tectum.

b. More fully *optic tectum* (or *tectum opticum*). That part of the tectum mesencephali concerned with the functioning of the visual system.

1907 J. B. JOHNSTON *Nervous Syst. Vertebrates* viii. 147 Structure of the tectum opticum.—In the lower fishes the tectum contains a large number of cells of several forms. **1926** *Jrnl. Compar. Neurol.* XL. 217 The optic tectum far surpasses the cortical areas of the reptilian hemispheres. **1982** *Sci. Amer.* Mar. 104/3 The optic tectum, also known in mammals as the superior colliculus, is one of the major visual centers of the brain.

Hence **te·ctal** *a.*, of or pertaining to the tectum mesencephali or the optic tectum.

1926 *Jrnl. Compar. Neurol.* XL. 217 A group of nuclei which serve as way-stations between the tectal areas and other centers. **1974** *Sci. Amer.* Mar. 38/2 Recording from individual tectal neurons..tells one how the individual retinal ganglion cells that excite them are reacting. **1975** *Nature* 30 Oct. 738/1 In the vertebrates below mammals, the tectal and subtectal areas are the main centres of termination of sensory pathways.

Ted (ted), *sb.*¹ *Services' slang.* [Abbrev. of TEDESCO *sb.*] A German soldier. *Disused.*

1947 D. M. DAVIN *Gorse blooms Pale* 193 D'you know what those bloody Teds have been up to? They've been bloody well shelling us.

Ted (ted), *sb.*² Also with small initial. Short for *TEDDY BOY. Cf. *TEDDY 4.

1956 in I. & P. OPIE *Lore & Lang. Schoolch.* (1959) vii. 119 Joined the Teds when he was only three, Coshed a cop when he was only four. **1956** *Time* 24 Sept. 28/1 The Ted's notion of sartorial splendor ranges from a caricature of Edwardian elegance to the zoot padding of a Harlem hepcat. **1959** C. MACINNES *Absolute Beginners* 44 Appearing in a telly programme on the Ted question. **1968** *New Scientist* 11 July 64/3 The gangs [of baboons] appeared to carry out his orders, roaming through the troupe like a bunch of leather-jacketed teds. **1977** *Daily Tel.* 19 July 15 A Metropolitan magistrate complained yesterday that he had inadequate power to deal with gangs of 'punk rockers' and 'Teds' who clashed in the King's Road, Chelsea. **1980** *Daily Mirror* 10 Apr. 12/2 The term Ted is a little less popular nowadays, and Rockabilly is Eighties style.

Teddy. Also **teddy**. **1.** Add to first sentence of def.: Short for *TEDDY BEAR 1. Freq. as a proper name for a teddy bear.

1907 *New England Mag.* July 629/2 The Teddy-bear.. suggests to the imaginative owner whatever special being his fancy would have 'Teddy' personify. **1910** *Postcard caption* [to picture of a little girl scrubbing a teddy bear.] You dirty Teddy. **1924** A. A. MILNE *When we were very Young* 90 Then said, 'Excuse me,' with an air, 'But is it Mr. Edward Bear?' And Teddy, bending very low, Replied politely, 'Even so!' **1934** E. TIETJENS in *Child Life* May 214/2, I always find things I'd forgotten, An old brown Teddy stuffed with cotton. **1940** D. WHEATLEY *Faked Passports* xxii. 262 It seemed a rotten business to shoot that harmless Bruin which was so reminiscent of a large teddy in a children's toyshop. **1960** *Sunday Times* 3 Jan. 30/3 My aunt..brought two brown teddies from Vienna in 1904, and in 1905 my mother bought me a white one in Ipswich. **1963** *Sunday Express* 10 Mar. 4/3 Look at teddy—he's got new fur. **1979** *Guardian* 14 June 12/3, I would rather fulfil my role as a mother than have a teddy act as a substitute.

2. [Perh. f. the name of *Theodore* Roosevelt.] (See quot. 1925.)

1917 E. POUND *Let.* 25 Aug. (1971) 118 The *Morning Chronicle* assures me my compatriots are called 'Teddies', which is one in the eye for Mr. Woodie Wilson. **1925** FRASER & GIBBONS *Soldier & Sailor Words* 279 *Teddies, the*, one of the names for the U.S. troops on first landing in France; disliked by the Americans equally with 'Sammies', and soon dropped.

3. orig. and chiefly *N. Amer.* [perh. *transf.* use of sense 1.] A woman's undergarment combining chemise and panties. Also in pl. *teddies*.

1924 H. C. WITWER in *Cosmopolitan* May 122/2 She added..she'd personally get enough enjoyment out of standing before her mirror garbed in a sheer silk teddy to warrant any sacrifice. **1929** *Amer. Speech* IV. 422 There is an article of feminine wearing apparel, a sort of overall piece of underwear, I believe, which is known as a *teddy*. I would suppose that this was so-called from its real or fancied resemblance in general shape (or shapelessness) to the teddy-bear. **1934** J. T. FARRELL *Young Manhood* (1936) xi. 298 Slug whispered to a big..blonde Polack in pink teddies. **1949** GILBRETH & CAREY *Cheaper by Dozen* xvii. 206 Anne..bought silk stockings, two short dresses and four flimsy pieces of underwear known as teddies. *Ibid.* 208 She doesn't even wear a teddy. **1977** *Hartford* (Conn.) *Courant* 6 June 24/4 Teddys are no longer synonymous with teddy bears alone. They also represent the sexiest lingerie around... The teddy is a camisole and tap pants set combined. The chemise bodice, often fashioned after a camisole, unbuttons either in the front or back... The all-in-one feature of teddys has made them more popular as sleepwear. **1978** *Chatelaine* Dec. 72/2 (*caption*) Left: A body-smoothing teddy with deep insets of lace by Emilio Pucci/Formfit Rogers, $30. **1983** *Daily Tel.* 17 Dec. 10/4 The silver satin 'teddy' we picture is one of this Christmas's best-sellers.

4. Short for *TEDDY BOY.

[**1955** *Britannica Bk. of Year* 489/2 *Teddy-boy, Teddy-gang* and *Edwardian* were terms used half derisively to describe youths who affected an Edwardian style of dress and who sometimes formed themselves into hooligan or criminal gangs.] **1956** *Saturday Bk.* 213 Our modern 'teddies' are named after their Edwardian clothes—dress in the manner of the times of King Edward VII—popularly known as 'King Teddy'. **1958** *People* 4 May 12/3 'He'd treat you real good,' said this Teddy. **1960** N. MITFORD *Don't tell Alfred* i. 8 His clothes had been distinctly on the Teddy side. **1963** J. FOWLES *Collector* ii. 165 We saw a group of teddies standing round two middle-aged Indians... The teddies were shouting, chivvying and bullying them off the pavement on to the road. **1968** [see *chukka boot* s.v. *CHUKKA].

teddy bear. **1.** Add to note:

Theodore Roosevelt's bear-hunting expeditions occasioned a celebrated comic poem, accompanied by cartoons, in the *N.Y. Times* of 7 Jan. 1906, concerning the adventures of two bears named 'Teddy B' and 'Teddy G'; these names were transferred to two bears (also known as the 'Roosevelt bears') presented to Bronx Zoo in the same year; finally the fame of these bears was turned to advantage by toy dealers, whose toy 'Roosevelt bears', imported from Germany, became an instant fashion in the U.S.

(Earlier and later examples.) *teddy bears' picnic*, the title of a song (*c* 1932) by Jimmy Kennedy and J. W. Bratton, used allusively to denote an occasion of innocent enjoyment.

1906 *Amer. Stationer* 22 Sept. 18/2 Probably no novelty of recent years has been so popular as the Teddy Bears. **1907** *New England Mag.* July 629/1 The Teddy-bear has come, and one suspects that he has come to stay. **1922**, etc. [see *KEWPIE]. **1927** [see *BONZO]. **1948** *Parents' Mag.* Mar. 8/2 His gently gruff appearance is in best Teddy

She touched her breasts..and Milo said, 'They must have got a heap of exercise since then.'..'If acting as teethers for a middle-aged man with false teeth anyhow is your idea of exercise, they have.' **1974** *Daily Colonist* (Victoria, B.C.) 28 Sept. 21/5 Two of her bells were originally babies' rattles. They are rattle-shaped with tiny bells surrounding the metal body, a teether of either agate or coral as the handle.

teething, *vbl. sb.* Add: **3.** *teething stage* (fig.); **teething powder** (examples); **teething ring,** a small ring or disc for an infant to bite on while teething; **teething troubles** *fig.*, problems arising in the early stages of an enterprise.
1869 *Bradshaw's Railway Man.* XXI. App. 102 Pritchard's teething and fever powders, for..children cutting their teeth. **1872** 'MARK TWAIN' *Roughing It* xv. 125 Soothing-syrup! Teething-rings! **1937** *Jrnl. R. Aeronaut. Soc.* XLI. 917 The dynamometer had now been in use for eight or nine months. It was not quite free from small 'teething' troubles. **1954** 'N. SHUTE' *Slide Rule* iv. 92 These were the inevitable teething troubles of any very large aircraft. **1959** *Observer* 19 Apr. 5/3 They speculated on whether Britain should jump the 'teething' stage of supersonic airliners and concentrate on producing an adult machine in twenty years' time. **1967** *Guardian* 21 Sept. 3/4 It was highly likely that many 'corner shops' still contained stocks of teething powders containing mercury. **1980** 'J. BELL' *Question of Inheritance* i. 18 His favourite toy, the bone ring..a teething ring, with three little bells on it. **1980** *Bookseller* 21 June 2560/1 TBL Book Service was still involved in teething troubles and costs.

† **teetotaciously** (tītotē·ʃəsli), *adv. U.S. dial. Obs.* Also **tetotaciously.** [Fanciful elaboration of TEETOTALLY *adv.*[1]: see -ACIOUS and cf. *BODACIOUS *a.*] = TEETOTALLY *adv.*[1]
1833 J. K. PAULDING *Lion of West* (1954) II. ii. 54 I'm the best man—if I ain't, I wish I may be tetotaciously exfluncticated! **1837** R. M. BIRD *Nick of Woods* I. xvi. 220 If that don't make me eat a niggur, may I be tetotaciously chawed up myself! **1859** T. DE QUINCEY *Coll. Works* X. 247 The slave in Terence, viz., Davus, though otherwise a clever fellow, when puzzled by a secret, or (as in America they say) *teetotaciously exfluncticated*, excuses himself by saying—'Davus sum, non Oedipus.'

teetotalize, *v.* (In Dict. s.v. TEETOTALIST.) Add: Also *intr.*, to practise teetotalism, to abstain.
1898 G. B. SHAW *Let.* 18 Oct. (1972) II. 66 If I began to eat three beefsteaks a day, the ground would..be shifted to the want of stimulants; but as it is,..I am allowed to teetotallize [*sic*] in comparative peace.

teetotally, *adv.*[2] *rare.* [f. TEETOTAL *a.* 1; see TEETOTALLY *adv.*[1] b.] In a teetotal manner, with total abstinence from alcoholic drinks.
1934 H. G. WELLS *Exper. Autobiogr.* II. viii. 533, I lived through my Bohemian days as sober as Shaw if not nearly so teetotally.

teetotum, *sb.*[1] Add: **1. b.** (*a*) For *Sc.* read *Sc.* and *Ir.* (Later example.)
1922 JOYCE *Ulysses* 759 Comical little teetotum always stuck up in some pub corner and her or her son waiting.

teevee (tī:vī·). Also **Teevee, tee-vee.** [A rendering of the names of the letters *T* and *V*.] = *TV.
1949 *N.Y. Mirror* 18 Apr. 10/2 Warners, unworried about teevee, showed a 1948 profit of 3 million. **1959** G. FREEMAN *Jack would be Gentleman* i. 7 It was wonderful what you learned from the tee-vee. **1963** *Guardian* 10 Jan. 6/6 Often..viewers do not know which channel they have been watching, and..will say 'Oh, it was the Teevee.' **1975** *Listener* 11 Sept. 350/3 In California, he achieves success as a low-budget teevee film-maker.

tefillin: see TEPHILLIM, -IN in Dict. and Suppl.

Teflon (te·flǫn). Also **teflon.** [f. TE(TRA- + FL(UOR- + *-on*, arbitrary ending.] **a.** A proprietary name for *POLYTETRAFLUOROETHYLENE.
1945 *Official Gaz.* (U.S. Patent Office) 23 Oct. 531/1 E. I. du Pont de Nemours and Company... *Teflon* for synthetic resinous fluorine-containing polymers. **1951** *Jrnl. Amer. Chem. Soc.* LXXIII. 5195 (*heading*) Tracer studies of oxidation-reduction polymerization and molecular weight of 'Teflon' tetrafluoroethylene resin. **1954** *Trade Marks Jrnl.* 5 May 438/2 *Teflon*... Mouldable plastics in the form of powder, sheets, rods, tubes, tapes, filaments (non-textile) and shaped pieces. **1965** *New Scientist* 11 Mar. 618/2 One end of the blood-carrying teflon tube is joined to a side opening made in the wall of the pulmonary artery. **1970** *Nature* 25 July 382/2 Exudates were..homogenized using a 'Teflon' grinder to rupture any cells present. **1979** A. L. LYDERSEN *Fluid Flow & Heat Transfer* vi. 150 The mesh is available in materials, such as, acid resistant steel, nickel, copper, aluminium, tantalum, hastelloy and Teflon.
b. *Comb.*
1965 *Family Circle* Oct. 92 Won't scratch, scar or mar Teflon coated cookware. **1972** M. CRICHTON *Terminal Man* II. i. 64 The team was now using Briggs stainless-steel Teflon-coated electrode arrays. **1979** A. L. LYDERSEN *Fluid Flow & Heat Transfer* iv. 80 They..have Teflon-covered piston rings.

Tegean (tedʒī·ăn, te·dʒiăn), *sb.* and *a.* [f. Gr. *Τεγέα*, L. *Tegea* Tegea.] **A.** *sb.* A native or inhabitant of the ancient city of Tegea in Arcadia. **B.** *adj.* Of or pertaining to Tegea or its inhabitants. Also **Tegeate** *sb.* and *a.* [ad. L. *Tegeātēs*].
1584 B. RICH tr. *Herodotus' History* I. fol. 17 The Lacedæmonians, hauing escaped a scowring, were triumphant conquerous ouer ye Tegeates. **1709** I. LITTLEBURY tr. *Herodotus' History* II. IX. 377 Next to themselves, the Lacedemonians plac'd the Tegeans, consisting of fifteen hundred Men. **1767** A. STRAHAN tr. *Virgil's Æneid* I. v. 137 He from Arcarnania came, This from Arcadia of Tegæan blood. **1808** [see *MANTINEAN *sb.* and *a.*]. **1858** G. RAWLINSON tr. *Herodotus' History* I. I. lxvi. 204 These persons..measured the Tegean plain as they executed their labours. **1888** *Encycl. Brit.* XXIII. 110/1 At Plataea (479 B.C.) 3000 Tegeans fought the good fight of freedom. **1919** E. POUND *Quia pauper Amavi* 35 Tegean Pan. **1949** *Oxf. Classical Dict.* 881/1 C. 550 B.C. Sparta came to terms with Tegea after a long war; and for two centuries it followed the Spartan lead, though at times unwillingly, for the Tegeans were tough fighters. **1951** [see *MANTINEAN *sb.* and *a.*]. **1952** C. DAY LEWIS tr. *Virgil's Aeneid* v. 101 After him, side by side, Salius and Patron, the one Acarnanian, the other Arcadian, of a Tegaean family.

tegestology (tedʒestǫ·lŏdʒi). [Irreg. f. L. *teges, -etis* covering, mat, f. *teg-ĕre* to cover + *-i* + -OLOGY.] The collecting of beer mats. So **tegesto·logist.**
1960 *Ironmonger* 23 Jan. 123 Tegestologists (beer-mat secreters). **1960** [see *PHILLUMENIST]. **1966** *Punch* 9 Feb. 202/2 'Here let tegestology exact its tribute from a royal embrace' (meaning, pinch a beer-mat from the *Queen's Arms*) is the sort of thing which any rallyist can work out in between emptying the ashtrays and giving the 'V' sign to a fellow competitor. **1977** *Titbits* 20–26 Jan. 17/5 The British tegestologists (a posh name for beermat collectors) are very upset to discover criminals who.. scoop up mats..and decamp.

tegmen. Add: **d.** *tegmen tympani* (examples). Also *ellipt.* as *tegmen.*
1902 D. J. CUNNINGHAM *Text-bk. Anat.* 704 The roof [of the tympanic cavity]..is formed by a thin plate of bone, the tegmen tympani. **1939** JOYCE *Finnegans Wake* 249 There lies her word, you reder!..It vibroverberates upon the tegmen. **1977** *Proc. R. Soc. Med.* LXX. 821/2 In mandibulofacial dysostosis the attic and antrum are typically absent..being replaced..by descent of the tegmen. **1980** *Gray's Anat.* (ed. 36) 312/1 In a young skull the suture between the petrous and the squamous parts of the temporal bone may be visible at the lateral limit of the tegmen tympani.

tegu (te·gu). Abbrev. of TEGUEXIN.
1954 G. DURRELL *Three Singles to Adventure* i. 28 One of the tegus opened his eyes. **1974** D. & M. WEBSTER *Compar. Vertebr. Morphol.* viii. 164 (*caption*) Surface view of a freshly shed skin of a tegu lizard.

teguexin. For *Teius* substitute *Tupinambis* and add: or a similar member of the family Teiidæ.

tegula. Add: **1.** (Later examples.)
1893 A. E. SHIPLEY *Zool. Invertebrata* xix. 376 On the mesothorax [of Hymenoptera] are two small scales known as the tegulae, covering the base of the wings. **1972** M. S. GARDINER *Biol. Invertebrates* xiv. 573/1 Newly hatched wasps..have been made to sting bees near their tegulae.
2. *Archæol.* and *Archit.* A flat roof-tile (see quot. 1964). Cf. IMBREX I. Also *fig.*
1871 R. BURN *Rome & Campagna* p. lxxv, The Roman tiles were of two kinds, flat tiles and smaller curved tiles. The flat tiles had raised rims at the sides... The small curved tiles were..laid over the joined edges..and formed a complete protection for the joint... Tegulæ and imbrices. **1938** in P. E. Thomas *Mod. Building Practice* III. 229 The Double Roman tile..is a descendant of the Roman *tegula* and *imbrex*.., which the Romans used extensively for their villa roofs. **1956** 'H. MACDIARMID' *Stony Limits & Scots Unbound* 9 The gold edging of a bough at sunset, its pantile way Forming a double curve, tegula and imbrex in one. **1964** J. S. SCOTT *Dict. Building* 178 *Italian tiling*.., single-lap tiles which form a roof covering with two different sorts of tiles, the curved over-tile or imbrex and the flat, tray shaped under-tile or tegula. **1977** *Antiquaries Jrnl.* LVII. 264 Bricks and tegulae can frequently be seen in the debris.

tehee, *int.* and *sb.* Add: Now usually in form **tee-hee.** **A.** *int.* (Later example.)
1944 A. HUXLEY *Let.* 24 Feb. (1969) 500 Tee hee, tee hee, oh sweet delight!
C. *attrib.* or as *adj.* **tee-hee farm** (*nonce*), a mental hospital; cf. *funny farm* s.v. *FUNNY *a.* 4.
1955 W. GADDIS *Recognitions* I. v. 172 Everybody knows about Rose, that they've sent her sister Rose back from the tee-hee farm and Esther has to take her in. **1971** *Publishers' Weekly* 1 Nov. 17/2 This accounts for *Newsweek's* rather snide coverage and the tee-hee reports in the press.
tehee *v.* (later example); also as *tee and hee* (*nonce*); **tehee·ing** *ppl. a.*
1904 J. C. LINCOLN *Cap'n Eri* v. 81 'That's it, laff!' almost sobbed Captain Jerry. 'Set there and tee-hee-like a Bedlamite.' **1928** V. WOOLF *Orlando* iv. 163 He teed and heed intolerably. **1935** 'G. ORWELL' in *New English Weekly* 14 Nov. 96/1 Life is full of misery when you believe that the grave really finishes you... Hence the tee-heeing brightness of *Punch*, hence Barrie and his bluebells, hence H. G. Wells and his Utopiæ infested by nude school-marms.

Tehrani (te(h)rā·ni), *sb.* and *a.* Also **Teherani.** [f. *Tehran*, name of a city in northern Iran + *-I.] **A.** *sb.* A native or inhabitant of Tehran, the capital of Iran. **B.** *adj.* Of, pertaining to, or characteristic of the city of Tehran, or of its inhabitants.
1939 W. V. EMANUEL *Wild Asses* xiii. 145 'Assassin's' appearance belied his name; he was a strikingly handsome young Teherani, with a Greek nose. **1941** L. P. ELWELL-SUTTON *Modern Iran* i. 11 The favourite summer resorts of the Tehranis are all along this coast. **1953** A. SMITH *Blind White Fish in Persia* vii. 130 The Tehranis cast soulful glances towards America. **1973** *Times* 22 May (Teheran Suppl.) p. i/1 Few Teheranis..have ever seen a drop of domestically produced crude oil. **1975** P. SOMERVILLE-LARGE *Couch of Earth* x. 178 The Tehrāni police consider it unlikely that you killed Mr Otway. **1977** H. OSBORNE *White Poppy* xii. 96 He was a Tehrani.

tehsil, var. *TAHSIL.

Tê-hua (tē[i]hwā, ‖ dɘhwā). Also (Pinyin) ‖ **Dehua.** The name of a place in the province of Fujian in south-eastern China, used *attrib.* and *absol.* to designate porcelain made there, also known as *BLANC DE CHINE.
1923 R. L. HOBSON *Wares Ming Dynasty* xiii. 173 The term Chien yao was now transferred to a ware..made more than a hundred miles south at Tê-hua (Tehwa)... There are..no lack of actual examples of the Tehwa porcelain. **1945** W. B. HONEY *Ceramic Art of China* II. 133 The whole body of Tê-hua wares, familiarly called Fukien porcelain,..holds together in a remarkable manner. **1953** R. S. JENYNS *Ming Pott. & Porc.* x. 146 Decorated pieces of Têhua, whether painted in blue or enamelled..are of poorer quality than the undecorated pieces. **1980** *Catal. Fine Chinese Ceramics* (Sotheby, Hong Kong) 102 A well modelled Dehua (Te Hua) Seal in the form of a *qilin* (*ch'i-lin*).., the fabulous beast with bushy tail and curly mane shown crouching with horned head sharply turned to one side.

Teian (tī·ăn), *a.* Also **Tean.** [f. Gr. *Τήϊος* of or from *Τέως Teos* + -IAN.] Of or relating to Teos, an ancient Ionian city on the western coast of Asia Minor north of Ephesus.
1646 J. HALL *Poems* 34 Who light'st Love's dying Torch with purer fire, And breath'st new life into the Teian lyre. **1787** J. NOTT *Select Odes from Hafez* p. ix, Whether Anacreon borrowed the gaiety of his Odes from the Persian *Gazel*, or whether Hafez enriched his native language by an imitation of the Teian bard, I will not venture to determine. **1821** BYRON *Don Juan* III. lxxxvi. 47 The Scian and the Teian muse, The hero's harp, the lover's lute, Have found the fame your shores refuse. **1858** [see *ARGIVE *a.*]. **1902** E. R. BEVAN *House of Seleucus* II. xix. 47 Antiochus..backed the envoys of the Teians in other places. **1941** M. ROSTOVTZEFF *Social & Econ. Hist. Hellenistic World* III. 1348 The Tean decree..which may refer to the synoecism of Teos and Lebedos. **1983** R. E. ALLEN *Attalid Kingdom* iii. 50 Teian envoys had been sent to Antiochos.

teichoic (təikō[u]·ik), *a. Biochem.* [f. Gr. *τεῖχο-ς* wall + -IC.] *teichoic acid*: any of various polymers of ribitol or glycerol phosphate that are found in the walls of Gram-positive bacteria.
1958 J. J. ARMSTRONG et al. in *Jrnl. Chem. Soc.* 4346 The general name 'teichoic acid'..is suggested for these polymers of ribitol phosphate with or without other substituents. **1973** *Nature* 4 May 43/2 The wall teichoic acid is responsible for the ability of the cell walls of Gram-positive bacteria to bind divalent cations and the membrane teichoic acid mediates an interaction between magnesium ions..and the cytoplasmic membrane.

Teilhardian (tē[i]yā·ɹdiăn), *a.* and *sb.* [f. the name of Pierre *Teilhard* de Chardin (1881–1955), French scientist and theologian + -IAN.] **A.** *adj.* Of or pertaining to Teilhard de Chardin or his writings, which are noted for their attempt to synthesize science and the Christian faith. **B.** *sb.* An adherent or follower of Teilhard de Chardin.
1967 *Sat. Rev.* (U.S.) 15 Apr. 18/3 Teilhard de Chardin ..restores all the dynamic energies and heroism of Milton's Satan to the Teilhardian Christ. **1970** T. P. O'MALLEY in *Studia Patristica* X. 194 Even if Tertullian was not a Teilhardian, he nevertheless held for a very palpable world here and hereafter. **1971** *Time* 19 Apr. 34 From the Teilhardians, the confidence that God, whoever he is, has something to do with the future and may yet meet man there. **1977** P. JOHNSON *Enemies of Society* ix. 125 Teilhardian phenomenology is a system which enables the more leisured class to accommodate scientific knowledge in a religious setting but which makes no intolerable demands on either flesh or intellect. **1983** KING & SALMON *Teilhard & Unity of Knowledge* i. 1 Teilhardians had long known that May 1, 1981—the centennial of the birth of Pierre Teilhard de Chardin—had to be celebrated in a special way. Two Teilhardian Jesuits would be at Georgetown University on that date.

teindable (tī·ndăb'l), *a. Sc.* [f. TEIND *v.* + -ABLE.] Tithable.
1705 W. FORBES *Treatise on Churchlands* 289 All other Teindable Subjects in Scotland. **1832** *Fife Herald* 14 June, The planted park is valued at £2 10s. sterling, but

is found not to be Teindable. **1924** *Kelso Chronicle* 25 July 4 The mansion house, cottages, joiner's shop, smithy and yard..were not teindable.

teineite (tē[i]·nəˌəit). *Min.* [See quot. 1939 and -ITE[1].] A hydrated sulphate and tellurite of copper, $Cu(Te)O_3 \cdot 2H_2O$, found as blue, prismatic orthorhombic crystals and as fine crusts.
1939 T. YOSIMURA in *Jrnl. Faculty Sci. Hokkaido Univ.: Ser. 4* IV. 465 Professor..Harada collected some.. crystals of a blue mineral... The author proposes the name 'Teineite' from the Teine mine where this new mineral has first been found. **1977** *Mineral. Abstr.* XXVIII. 487/1 A copper-bearing quartz vein cutting phyllites shows..rare azure-blue teineite as single crystals (≤ 1 mm) or crystalline patches (≤ 2·5 mm).

teistie. Add: The form *tystie* is now usual. (Later examples.)
1952 U. VENABLES *Tempestuous Eden* iii. 50 This was an ideal nesting-ground for black guillemots or tysties. **1964** A. L. THOMSON *New Dict. Birds* 69/1 The Tystie (or Black Guillemot)..and its congeners constitute the tribe Cepphini. **1980** *Birds* Summer 51/2 There is a small brackish loch near the shore where tysties display in spring.

tej (tedʒ). Also † **tedge, tedje; tedj.** [Amharic.] A kind of mead that is the national drink of Ethiopia.
1853 M. PARKYNS *Life in Abyssinia* I. xvii. 210 Spirits are of an inferior kind, distilled..from the refuse of the wine or from honey... The grape is called 'wainy', the wine 'wain tedge'. **1877** E. A. DE COSSONS *Cradle of Blue Nile* I. vii. 100 As soon as the baskets of bread had been removed, bottles of tedge were served. **1901** A. B. WYLDE *Mod. Abyssinia* viii. 182 The tedj bearer always pouring out a little of the liquid into..his hand. *Ibid.* xvii. 377 The honey taken from the wanza flowers being greatly prized, as being of a white colour makes very clear tedj. **1925** H. C. MAYDON *Simen* ii. 24 Abdulla, the cook, was ..too apt to celebrate with the local tej and marissa (mead and beer). **1936** E. WAUGH *Waugh in Abyssinia* i. 26 A weekly visit to the cinema, a preference for whisky over *tedj*..were the western innovations that these young men relished. *Ibid.* v. 169 They got very drunk in the *tedj* houses. **1952** [see *POMBE]. **1974** *Country Life* 18 Apr. 938/3 *Tej*, a mead of honey, hops and water. **1981** E. NORTH *Dames* i. 12 It is known as *tedj*... A sort of honey-mead... Some call it *mies*.

‖ **Tejano** (tĕhā·no). [Amer. Sp., formerly written *Texano*, f. *Texas* TEXAS.] A native or inhabitant of Texas, esp. one of Mexican stock; a Texan.
1925 O. P. WHITE *Them was Days* 75 The fear of God, as represented by the wrath of the *Tejanos* (Texans). **1933** H. ALLEN *Anthony Adverse* III. IX. lxiv. 1175 The convoy, when it did start, consisted of about sixty unfortunate mestizos and Indians bound for the mines at Chihuahua as well as the captured Americans, or 'Tejanos', whose cases were to be disposed of by the Viceroy Calleja at Mexico. **1976** *Monitor* (McAllen, Texas) 26 Sept. 7E/2 *Tejanos* more than any other ethnic group in Texas have provided a bridge to Texas' past.

tek (tek), *v.* Also **teck.** Repr. U.S. Black and regional pronunc. of TAKE *v.*
1905 [see *WE *pron.* 1 g]. **1924** M. W. BECKWITH in *Mem. Amer. Folklore Soc.* XVII. 76 Him couldn't get away from de woman until rain tek him in de yard. **1938** C. HIMES *Pork Chop Paradise* in *Black on Black* (1973) 165 W'en de panic cum an' de Lawd tek yo' food..den laff. **1981** *Westindian World* 2 Oct. 4/1 One of London's better known Radio London presenters tek time off te go compere beauty show.

tekhnikum, var. *TECHNICUM.

tekke[1] (te·ke). Also 7 **teke;** 9 **takia, takiya(h;** 9—**tek(k)i(y)eh** (tekī·ye). [a. Turk. *tekke*, Ottoman *tekye* (whence Arab. *tak(k)īya*), ad. Pers. *takya* pillow, place of repose, abode of a fakir.] A monastery of dervishes, esp. in Ottoman Turkey.
1668 P. RYCAUT *Pres. State Ottoman Empire* II. xx. 150 Their poverty..as I have seen in some of their *Tekes* where I have been. **1842** C. MASSON *Baluchistan* II. 278 Many takías are interspersed amongst all the burial-places. **1855** R. F. BURTON *Personal Narr. Pilgrimage to El-Medinah* I. 124 A Takiya is a place where Dervishes have rooms, and perform their devotions. *Ibid.* II. 29 It is flanked on the left..by the domes and minarets of a pretty Turkish building, a 'takiyah', erected by the late Mohammed Ali for the reception of Dervish travellers. **1856** LADY EASTHOPE tr. *Ubicini's Lett. on Turkey* I. 109 He had finished and issued forth from the precincts of the tekieh. **1868** J. P. BROWN *Derwishes* 103, I may here add that of the two hundred, or more, *Tekkiehs* in Constantinople, some fifty only are possessed of sufficient wealth for their support. **1875** *Encycl. Brit.* III. 233/1 The Tekiyeh, or shrine of the Bektash dervishes, on the western bank of the [Tigris] river. **1900** 'ODYSSEUS' *Turkey in Europe* v. 194 Monograms of his name [Ali] are a conspicuous feature in most dervish Tekkes. **1932** G. KAMPFFMEYER in H. A. R. Gibb *Whither Islam?* 167 The *tekkes* or monasteries of the religious orders and the *turbes* or the tombs of the saints are closed. **1977** H. KAPLAN *Damascus Cover* (1978) x. 90 Damascus is overflowing with exotic sights: the tomb of Saladin,..the Tekkiyeh of Suleiman. **1980** A. ALPERS *Life K. Mansfield* xxi. 376 Gurdjieff['s]..'study-house'..resembled..a Dervish tekke.

Tekke[2] (te·ke). [Turkic.] The name of a Turkic tribe inhabiting the Turkmen Soviet Socialist Republic, used to designate a short-piled basically red carpet or rug made by members of this tribe. Also *Tekke Bokhara.*
1900 J. K. MUMFORD *Oriental Rugs* iv. 40 From the rest of it [*sc.* angora fleece] Kashmir shawls are made, and carpet-weaving is illustrated in the finer Tartar fabrics—Tekkes, Yomuds, and Bokhara prayer rugs. *Ibid.* xii. 227 Tekke rugs..will continue to be sold as Bokharas. **1911** G. GRIFFIN LEWIS *Practical Bk. Oriental Rugs* I. x. 135 *Tekke Field Design.*—A repetition of a..Y-shaped design. Found only in the Tekke rugs, especially those of the prayer variety. *Ibid.* II. xvi. 276 *Tekke Bokhara.* Why so named.—Because they are made by the Tekke Turkoman tribes of Nomads. **1931** A. U. DILLEY *Oriental Rugs & Carpets* Pl. 57 (*caption*) Bokhara garden rug. Compare garden designs in Tekke and Saryk prayer rugs. **1967** *Times* 21 Feb. 21/4 (Advt.), A number of fine Turkomans: *Tekke Bokhara.* **1974** *Encycl. Brit. Micropædia* IX. 867/1 *Tekke carpets*... Primarily the rugs that were commonly on the market as Royal Bokharas a generation or two ago... The standard field pattern for the large rugs is a repeat in rows of a particoloured, multifoiled lozenge, a basic motif that serves as the symbol, or gul, of this tribe.

tektite (te·ktəit). [ad. G. *tektit* (F. E. Suess 1900, in *Jahrb. d. K.-K. Geol. Reichsanstalt* L. 194), f. Gr. *τηκτ-ός* molten (f. *τήκειν* to make molten): see -ITE[1].] One of the small, roundish, glassy bodies of unknown origin that occur scattered over various parts of the earth.
1909 [see *BILLITONITE]. **1935** *Times* 28 Jan. 15/4 'Tektites' from Bohemia and Moravia have for more than 150 years been cut as gem stones under the names 'obsidian', 'water chrysolite' and 'moldavite'. **1936** [see *AUSTRALITE]. **1956** *Antiquity* XXX. 70 These stone implements, together with the fossil remains of the stegodon (an extinct elephant), rhinoceros and other mammals, were often found in association with tektites, a form of glass meteorite. **1963** [see *OBSIDIANITE]. **1969** *Times* 29 Aug. 10/3 Tektites, glassy, button-shaped objects which are probably of extraterrestrial origin, have been found in the Libyan desert and dated by radioactive methods to 35 million years ago. **1971** *Sci. Amer.* Oct. 55/2 Well-preserved Australian tektites show clear evidence of aerodynamic shaping.
b. *attrib.*, as **tektite field** = *STREWN FIELD.
1960 *Listener* 22 Dec. 1141/2 Some of the tektite fields are well away from either active or extinct volcanoes. **1968** R. A. LYTTLETON *Mysteries Solar Syst.* vi. 183 There are eight main areas in the world generally accepted to be genuine tektite-fields.

† **tektosphere** (te·ktosfī[ə]ɹ). *Geol. Obs.* Also (in dicts.) **tecto-.** [f. as prec. + -O + *-SPHERE.] = *ASTHENOSPHERE.
1900 J. MURRAY in *Rep. Brit. Assoc. Adv. Sci. 1899* 796 Within [the earth] is situated the vast nucleus or centrosphere; surrounding this is what may be called the tektosphere, a shell of materials in a state bordering on fusion. **1913** [see *geosphere* s.v. *GEO-].

telangiectasis. Add: Also † **teleang-, -ectasia** (pl. **-iæ**). (Further examples.)
1876 VAN DUYN & SEGUIN tr. *Wagner's Man. Gen. Path.* III. 390 Cavernous, or venous tumor. It exists rarely at birth,..and proceeds probably in many cases from teleangiectasia. **1887** T. M'C. ANDERSON *Treat. Dis. Skin* 436 (*heading*) Capillary nævus; mother's mark; port wine or strawberry mark; teleangiectasis. **1948** J. MINCKLER in W. A. D. Anderson *Pathology* xlv. 1390 Telangectasia of cerebral vessels is a fairly common occurrence..and represents a malformation rather than a neoplasm. **1974** PASSMORE & ROBSON *Compan. Med. Stud.* III. xxxiv. 77/1 The disease usually presents in infancy with cerebellar ataxia. Telangiectasiae appear some years later and are distributed over the ears, nose and cheeks.

Telanthropus (telæ·nþrŏpŏs). [mod.L., f. Gr. *τέλ-ος* end, consummation + *ἄνθρωπος* man.] A type of hominid, *Telanthropus capensis*, represented by the fragmentary fossil remains found at Swartkrans near Johannesburg, S. Africa, in 1949.
More recent investigation has led some scholars to believe that Telanthropus properly belongs to the species *Homo erectus*.
[**1949** BROOM & ROBINSON in *Nature* 20 Aug. 323/2 In the cave at Swartkrans..there was found by Mr. J. T. Robinson, on April 29, 1949, the lower jaw of what is fairly manifestly a new type of man... The new type of man represented by this fossil jaw we propose to call *Telanthropus capensis*. We regard him as somewhat allied to Heidelberg man, and intermediate between one of the ape-men and true man.] **1955** *Ann. Reg. 1954* 466 The artefacts were probably made by *Telanthropus* who was either a very advanced Australopithecine or a very primitive true man. **1959** J. D. CLARK *Prehist. S. Afr.* iii. 63 In addition some rather fragmentary remains, considered to be essentially more human-like though still preserving Australopithecine features, have been described by Robinson under the generic name of *Telanthropus*. **1960** W. HOWELLS *Mankind in Making* xii. 178 *Telanthropus*' teeth are smaller. **1977** A. HALLAM *Planet Earth* 286 In addition to these is a jaw which represents another line, closer to modern Man. It has been named Telanthropus.

Tel Avivian (tel avī·viăn), *sb.* and *a.* [f. *Tel Aviv* (see below) + -IAN.] **A.** *sb.* A native or inhabitant of Tel Aviv, the largest city in the state of Israel. **B.** *adj.* Of or pertaining to Tel Aviv.
1939 *Palestine Post* 14 Aug. 2/2 (*heading*) New park benches for tired Tel Avivians. **1941** G. G. SCHOLEM *Major Trends in Jewish Mysticism* 384 A Tel-Avivian scholar, Reuben Margulies, has begun to publish an annotated edition of the Zohar. **1949** KOESTLER *Promise & Fulfilment* II. v. 263 Most Tel Avivians agree that the black-out is logically unwarranted. **1983** *Economist* 30 July 39/3 No fewer than 10 construction companies are vying to provide land- and air-hungry Tel Avivians with a variety of lodgings.

tele (te·li), *sb.*[2] Colloq. abbrev. of *TELEVISION. Cf. *TELLY. Also *attrib.* and *Comb.* (not sharply distinguishable from *TELE- 2).
1936 *Billboard* 14 Nov. 3 (*heading*) RCA–NBC tele progress. **1944** R. E. LEE *Television* vi. 88 The motion picture is a novel; tele is a short-story, or a newspaper article. **1946** [see *lip-read* vb. s.v. *LIP *sb.* 7]. **1956** A. WILSON *Anglo-Saxon Att.* I. iii. 47, I see him on the Tele. **1960** A. KIMMINS *Lugs O'Leary* iii. 33 'Where were you?' 'In my room watching the tele.' **1961** A. WILSON *Old Men at Zoo* iii. 128 Oh, it was on the tele news in the pub I was in. **1964** *Punch* 3 June 833/1 Large sections of the tele-watching population. **1973** J. WAINWRIGHT *Devil you Don't* 5 Sprawling in front of a tele, watching one-day cricket. **1977** *Gay News* 7–20 Apr. 37/2 Hardly home-loving types, likely to be content with baked beans on toast and the tele.

tele (te·li), *sb.*[3] *Psychics.* [a. Gr. *τῆλε* afar, far off.] Psychic affinity between two (or more) people separated by time or space. Cf. *telepsychic* sb. s.v. *TELE- 1.
1937 J. L. MORENO in *Sociometry* I. 16 Tele is defined as a feeling process projected into space and time in which one, two, or more persons may participate. It is an experience of some real factor in the other person and not a subjective fiction... The tele process is..the chief factor in determining the position of an individual in the group. **1952** W. SPROTT *Social Psychol.* ii. 34 A 'monistic origin of life from a common unit' out of which existing networks of 'tele' have been differentiated. *Ibid.* 35 In Moreno's own convention a red line from A towards B represents 'positive tele'.

tele (te·li), *a.*, Abbrev. of TELEPHOTOGRAPHIC *a.*[2] Cf. TELEPHOTO *a.* and *tele-lens* s.v. *TELE- 1.
1979 *SLR Camera* Jan. 42/3 We chose the FD 135 mm f2.5 SC and the FD 200 mm f2.8 SC as being representative of the popular tele range. **1981** *What Camera Weekly* 5 Dec. 3/2 (*heading*) Tele tactics: long lenses with focal lengths of 400 and 500 mm need careful handling.

tele-. Add: **1. te·leba:nking,** a method of effecting banking transactions at a distance by electronic means; **te·le-ca:mera,** (*a*) a telephotographic camera; (*b*) a television camera; **telecentric** *a.* (examples); also *absol.* as *sb.*, a telecentric lens; **teleco·balt,** radioactive cobalt used as a radiation source in teletherapy; usu. *attrib.*; **te:lecomma·nd,** the remote control of machines or the like by electronic means; freq. *attrib.*; **telecommu·te** *v. intr.*, to work from home (esp. at a traditionally office job), communicating with one's place of employment, colleagues, etc., by telephone line or data link; **te·leconne·ction** *Geol.* [tr. Sw. *fjärrkonnektion* (G. De Geer 1916, in *Geol. Fören. Förhandl.* XXXVIII. 18)], the correlation over long distances of varves or other deposits that can be used for dating purposes; also *transf.*; **te:lecontro·l** = *telecommand* above; freq. *attrib.*; **teleconve·rter** *Photogr.*, a camera lens designed to be fitted in front of a standard lens to increase its effective focal length; **te:lecuriethe·rapy** *Med.* [*CURIE] = *teletherapy* below; **tele-diphone** [*Ediphone*, name of a recording machine], a machine for recording speech from a telephone line or radio for subsequent transcription or broadcasting; hence **tele-diphoned** *a.*; **te·le-flash** *U.S.*, (equipment for transmitting) telegraphic news of racing results, odds, etc.; **te:lege·nesis,** the technique of artificial insemination; **tele·gnomy, telegno·sis** *Psychics*, psychic perception of events happening at a distant place; clairvoyance; hence **telegno·stic** *a.*; **telekinesis:** delete ‖ and add later examples; **telekinetic** *a.*: also *transf.* and *fig.* (later examples); **te:lekine·ticist,** one who practises or has the power of telekinesis; **te·le-lens** *Photogr.*, a telephoto lens; **te·leme:ssage,** a form of telegram introduced in October 1981 to replace the inland telegram, and abolished one year later; **te·leordering** *vbl. sb.*, the computerized ordering of books by book sellers from publishers; **telepho·togram,** a message in the form of a picture transmitted by radio or television; **te:lephoto·meter,** an instrument for measuring the brightness of a

distant light source; **te·leplasm** *Psychics*, a hypothetical substance psychically materialized; ectoplasm; hence **telepla·smic** *a.*; **Te·le-player** [*PLAYER[1]], the proprietary name in the U.S. of a device for recording and playing back videotape; **telepro·cessing** *vbl. sb.*, data processing that involves terminals located at a distance from the processor; **telepsy·chic** *sb.*, a medium whose psychical powers are exerted at a distance; *adj.*, pertaining to or involving the exertion of psychic powers at a distance; **te·lepuppet** *colloq.*, a telechiric device, esp. one used in space; **te:leradio·graphy** *Med.*, radiography in which the X-ray tube is placed some distance from the plate in order to minimize distortion; **telera·dium,** radium used as a radiation source in teletherapy; **tele·rgic** *a.*, pertaining to or involving telergy; **te:leroe:ntgeno·graphy** (also **-röntgen-**) *Med.* (chiefly *U.S.*) = *teleradiography* above; **teleseism** (examples); hence **telesei·smic** *a.*; **telesei·smically** *adv.*; **te·lesho:pping,** a method of ordering goods from shops by electronic means; **teleso·ftware** *Computers*, software transmitted by wire or broadcast for use by any number of independent receiving terminals; **te·le-talkies,** cinematographic films broadcast by television (*disused*); **telethe·rapy** *Med.*, radiotherapy using a source of radiation at a distance from the patient; **telethermometer,** substitute for def.: a thermometer that indicates the temperature measured elsewhere; (example); **te:letransporta·tion** = *TELEPORTATION (*rare*); hence (as a back-formation) **teletranspo·rt** *v. trans.*
1981 *Amer. Banker* 18 Feb. 12/3 Consider the things you will be able to do—telereservations, telegames, telebanking, teleshopping. **1910** O. WHEELER *Mod. Telephotography* 68 Messrs. Zeiss also make a special tele-camera. **1951** I. ASIMOV *Stars like Dust* xvii. 162 The movement of the tele-camera can be so adjusted as to counteract the motion of the ship in its orbit. **1960** *Harper's Bazaar* Oct. 82/2 It is not too late even now to bring in the tele-cameras. **1980** T. HOLME *Neapolitan Streak* 42 There were tele-camera teams from the RAI. **1902** MANN & MILLIKAN tr. *Drude's Theory of Optics* I. iv. 75 Certain positions of the iris can be chosen for which the entrance- or exit-pupils lie at infinity... To attain this it is necessary to place the iris behind S_1 at its principal focus... The system is them called telecentric. **1921** *Glasgow Herald* 15 June 7 It was fitted with..a 12 in. Telecentric, and a variant of my 'Dodo' tele-lens. **1973** D. A. SPENCER *Focal Dict. Photogr. Technol.* 619 (*caption*) Telecentric optical system. **1956** C. W. WILSON *Radium Therapy* (ed. 2) 286/1 (Index), Telecobalt therapy. **1959** [see *teleradium* below]. **1980** *Jrnl. R. Soc. Arts* Jan. 95/1 These telecobalt..machines are now part of the routine equipment of most radiotherapy departments. **1972** *Sunday Tel.* 30 Apr. 34/4 He sees on a television screen the view he would have from the driver's seat of the car he is controlling remotely. These 'telecommand' cars are about to be used for complex..handling tests. **1978** *Times* 3 Nov. 27/4 The Post Office itself has listed the main telecommunications services..envisaged for the years 1985 and 2000... By 1985 there will be..viewdata,..telemetry (the radio transmission of measurements), telecommand (remote control of machines). **1980** *Times* 15 Jan. 16 A low-power microcomputer system has been built..and a telecommand receiver has been completed. **1974** *Economist* 5 Jan. 14/1 As there is no logical reason why the cost of telecommunication should vary with distance, quite a lot of people by the late 1980s will telecommute daily to their London offices while living on a Pacific island if they want to. **1975** *Ibid.* 25 Oct. 39/3 Telecommuting is coming. When production is properly automated even in service industries, probably 60% of American breadwinners will be brainworkers. *Ibid.* 43/3 Telecommuter terminals will stop social interaction at the workplace. **1976** *Ibid.* 25 Dec. 56/1 Small ones, employing various piecework-earning telecommuting housewives in their own homes. **1981** *Ibid.* 5 Sept. 20/1 Telecommuters are workers who do not have to travel to their office... They need only their terminal links of today, enhanced by new gadgetry, to make sure they never have to leave their villages. **1982** *N.Y. Times Mag.* 14 Nov. 133 A situation known as 'telecommuting' or, more cozily, the 'electronic cottage'. **1934** G. DE GEER in *Geografiska Annaler* XVI. 3 The general law that the annual amount of meltwater deposits along the ice-border varied congruently is..definitely fixed. Teleconnections were thus, by a great number of close connections,..acting every year for..the whole of Fennoscandia. **1939** G. CLARK *Archæol. & Society* v. 141 Attempts to extend the sequence across the Baltic have not met with general acceptance any more than have the still more ambitious 'teleconnexions' between the Swedish varve-sequence and those in North and South America. **1970** S. THORARINSSON in R. Berger *Sci. Methods Medieval Archaeol.* 325 A young Swedish scientist..has realized my old dream of establishing tephrochronological teleconnection between Iceland and Scandinavia. **1979** *Harvard Mag.* May–June 14 Meterologists have coined the phrase 'teleconnections' to describe the apparent correlation between El Niño [*sc.* an erratically recurring ocean current in the Pacific] and disruptive weather patterns all around the earth. **1983** *Nature* 18 Aug. 583/3 Teleconnection with the Bristlecone pine absolute scale..has already been achieved for Bronze Age varves in south Russia and for tree-rings in Turkey. **1933** *Sci. Abstr.* B. XXXVI. 225 A general survey of the subject of telemeasuring with a brief account of telecontrol systems. **1959** *Times* 30 July 2/3 The installation and commissioning of telecontrol and telemetering systems [for an oil company]. **1974** *Sci. Amer.* Nov. 41/1 The control tasks described so far, including the gathering and presentation of information about the system.., can be realized in principle by analogue control circuits,..telecontrol devices and the like. **1966** 'A. HALL' *9th Directive* ix. 83 A Pentax X-15 35 mm single reflex with a 135 mm lens that took a × 2 Auto teleconverter. **1979** *SLR Camera* Mar. 36/3 A short cut to getting involved in tele photography, where your budget is tight, is to use a tele-converter. **1939** E. LILJENKRANTZ *Cancer Handbk.* ii. 23 Telecurietherapy with 10 gm of radium (a quarter of a million dollars' worth) means usually treatment distance of 15 cm. **1954** *Arch. Otolaryngol.* LIX. 345 Advanced inoperable carcinoma is best treated by telecurietherapy. **1953** *Brit. Jrnl. Psychol.* XLIV. 117 The telediphone records of the [television] programmes were broken up into what appeared to be the principal points contained in the programme. **1973** *Listener* 7 June 757/1 The BBC started making telediphone transcripts of what people actually did say, unscripted, on the air. **1957** *Oxford Mag.* 31 Oct. 70/2, I have before me the B.B.C.'s telediphoned transcript of the discussion. **1937** *Sun* (Baltimore) 16 June 4/5 A 'teleflash' and racing slips taken by police in the tavern were not sufficient for a conviction. **1951** *Ibid.* 23 Mar. 28/1 Equipment of the 'teleflash' type which..was used for announcements of racing results and odds. **1935** Telegenesis [see *EUTELEGENESIS]. **1958** *News Chron.* 4 Feb. 4/8 (*heading*) Telegenesis. **1911** W. F. BARRETT *Psychical Research* xi. 161 Dr. Heysinger..suggests the term telegnosis, or knowing at a distance, instead of clairvoyance. **1932** J. BUCHAN *Gap in Curtain* i. 44 The instinct which had its seat in this cell specialised in time-perception... I had been reading lately about telegnosis. **1962** C. D. BROAD *Lect. on Psychical Res.* viii. 222 The distinction between explicitly referential and merely unwilling telegnosis. *Ibid.* 223 Experiences which are only unwittingly telegnostic. **1906** *Pall Mall Gaz.* 24 Mar. 4 Mr. Grier possesses the faculty of 'telegnomy', which enables him..to perceive..events which are taking place on the other side of the Atlantic. **1962** *Punch* 5 Dec. 805/3 Dusailly..has made a first step towards telekinesis by using the electrical cavity of the brain to operate a switch. **1983** J. MELVILLE *Hand of Glass* vi. 146, I *had* seen it move... If you didn't believe in telekinesis.. then Merry must have practised some form of hypnotism on me. **a 1966** 'M. NA GOPALEEN' *Best of Myles* (1968) 94 An œuvre which would show his telekinetic treatment of over-tonality. **1972** *Countryman* Winter 83 Almost all contemporary investment in the countryside is a telekinetic expression of the distracted town. **1977** A. WILSON *Strange Ride R. Kipling* vi. 291 Trix accumulated clairvoyant, time-travelling, telekinetic and exorcistic powers. **1949** *Startling Stories* May 22/1 'Just what are his potentialities?' queried Shey. 'Is he a hypnotist? A telekineticist?' **1965** J. KINGSTON in J. Carnell *New Writings in S-F* III. 68 Telekineticists..are people who can move things without touching them, change physical states at a distance. **1921** Tele-lens [see *telecentric* adj. above]. **1979** *Amat. Photographer* Feb. 74/2 The modern telelens (and a tele can be as short as 100 mm)..is the biggest boon since sliced bread. **1981** *Times* 20 Oct. 28/4 A new, cheaper form of telegram called the telemessage is to be introduced by British Telecom as an inland service next Monday. **1977** *Bookseller* 14 May 2432/2 A teleordering terminal for bookshops... The value of teleordering from the bookseller's point of view is partly to receive books a few days sooner from publishers. **1929** *Telegr. & Telephone Jrnl.* XVI. 49/1 The transmission of pictures by telegraphic means is coming to the front in the U.S.A... An enterprising firm in New York recently sent out 300 telephotograms of the latest feminine fashions to all parts of the States. **1937** *Times* 30 Oct. 14/2 The London television station transmitted last night the first 'telephotogram' to a ship at sea—a visual message of greeting to the master of the Britannic. **1930** *Monthly Weather Rev.* Nov. 440/2 In the measurements over the sea..the telephotometer..and the theodolite were set up on the point of the mole. **1949** *Proc. Inst. Electr. Engineers* XCVI. II. 456/2 It was possible to calibrate the telephotometer in daylight by reference to tungsten-filament standard lamps of 1-, 2- and 5-kW sizes over ranges varying from 1 500 to 5 000 ft. **1927** *Daily Express* 28 Sept. 9 Teleplasm.. was shown issuing from the face of the tranced woman. **1978** SMYTH & STEMMAN *Mysteries of Afterlife* 225 (*caption*) The teleplasm..is compared with a control sample of ordinary paper. **1930** *Times Lit. Suppl.* 28 Aug. 683/1 Teleplasmic masses resembling arms and hands were seen. **1968** *Daily Tel.* 12 Dec. 25/3 The tele-player will cost about £200 and each tele-cartridge..£20. **1971** *Official Gaz.* (U.S. Patent Office) 23 Nov. TM206/1 *Teleplayer.* For television apparatus... First use Mar. 24 1970. [**1961** *Ibid.* 22 Aug. TM124 *Tele-processing*... For services in organizing, planning, developing, installing, maintaining and operating data processing systems [etc.].] **1962** *Engineering* 8 June 758/2 The development of 'remote computing' or 'teleprocessing' as it is sometimes called. **1970** *Computers & Humanities* IV. 323 Classrooms equipped with voice recorders; and with teletypewriters..for creating perforated paper tape for batch teleprocessing. **1980** R. L. DUNCAN *Brimstone* x. 263 'How will the reprogramming take place?' 'Teleprocessing. Over the telephone lines.' **1914** A. L. TEIXEIRA DE MATTOS tr. *Maeterlinck's Unknown Guest* ii. 63 There are seers, so-called 'telepsychics', who are not psychometers. **1926** F. CAZZAMALLI in *Jrnl. Amer. Soc. Psychical Res.* XX. 1 (*title*) Telepsychic phenomena and cerebral radiations. **1960** *Sci. News Let.* 2 Jan. 4/2 The 'telepuppet', as he [*sc.* F. L. Whipple] called it, would have a little feedback on handling pressure to give the human operator a feel of the object the machine is working on. **1963** *Flight Internat.* LXXXIII. 244/2 It is foreseen that the adaptive machine or 'telepuppet', primitive versions of which are already used in handling radio-active materials, have a key role in space missions. **1973** C. SAGAN *Cosmic Connection* I. viii. 62 There may be telepuppets, devices landed on another planet but fully controlled by an individual human being in orbit. **1909** *Arch. Roentgen Ray* XIV. 38 (*heading*) An instantaneous shutter for teleradiography. **1928** *Brit. Jrnl. Radiol.* I. 368 Arising out of these large milliampereages are the screening stands and radiographic appliances for teleradiography. **1974** *Biol. Abstr.* LVII. 6326/1 Teleradiography and tomography were used to investigate 57 able-bodied male patients. **1937** *Nature* 25 Dec. 1109/1 Teleradium has been practised by several centres in Great Britain over a period of years. **1959** R. W. RAVEN *Cancer* V. 157 Usually a single teleradium or telecobalt field is applied to the undersurface of the chin beneath the tumour. **1909** O. LODGE *Survival of Man* IV. xi. 163 This is the hypothesis of actual telepathic or telergic influence from some outside intelligence. **1912** *Index-Catal. Libr. Surg.-General's Office, U.S. Army* XVII. 712 Teleröntgenography. **1923** R. KNOX *Radiogr. & Radio-Therapy* I. 303 When it is possible to obtain full exposures of the thorax at a distance of 2 metres, then teleröntgenography of the thorax is of decided advantage. **1972** J. E. CULLINAN *Illustrated Guide to X-Ray Technics* i. 3/1 (*caption*) A 72 inch focus-film distance is used for teleroentgenography to minimize geometric enlargement and distortion. **1905** *Rep. Brit. Assoc. Adv. Sci. 1904* 47 [In Italy] there are fifteen first-class observatories provided with apparatus to record teleseisms and local shocks. **1972** J. G. DENNIS *Struct. Geol.* xvi. 363 (*caption*) Teleseism (distant earthquake). **1905** *Rep. Brit. Assoc. Adv. Sci. 1904* 47 Japan has at least five stations for teleseismic observations. **1969** *New Scientist* 25 Dec. 627/1 The so-called teleseismic data.. have provided a rich new fund of research material for analysing the Earth's interior. **1974** *Nature* 23 Aug. 622/3 Nakamura and his colleagues have attempted to determine both P and S wave velocities throughout the lunar mantle from..high frequency teleseismic events and deep moonquakes. **1971** I. G. GASS et al. *Understanding Earth* xxiv. 336/2 A relatively narrow cone at the source can be seen teleseismically. **1981** Teleshopping [see *telebanking* above]. **1983** *Times* 17 Aug. 3/6 The channel will also have the facility for shopping from the armchair at the touch of a switch, now termed 'teleshopping'. **1976** W. J. G. OVERINGTON in *Computing Europe* 4 Mar. 8/2, I have..been theoretically developing a computing system based on Ceefax/Oracle which I call Telesoftware (*ie* software at a distance). **1977** *Wireless World* Sept. 50/2 Perhaps the most marketable use for Telesoftware might be in video games. **1979** *Guardian* 24 Sept. 21/5 Experiments are under way to use Prestel for exchanging software programs and 'telesoftware' is also available for teletext services. **1982** *Datalink* 18 Jan. 5/1 The programme forms only part of the project... There's telesoftware, which uses the BBC's Ceefax teletext service to broadcast software. **1930** MOSELEY & CHAPPLE *Television* viii. 95 Since tele-talkies are sent out in a manner very similar to the transmission of television, they can be received on the identical machine which receives television images. **1913** DORLAND *Med. Dict.* (ed. 7) 946/1 *Teletherapy*, absent treatment. **1929** *Brit. Med. Jrnl.* 11 May 845/1 In teletherapy proper the radium is employed at distances as great as 16 cm. **1945** C. W. WILSON *Radium Therapy* vii. 159 As the name implies, radium teletherapy..is the therapeutic use of a quantity of radium at a distance from the patient. **1974** *Nature* 11 Oct. 521/2 Hyperthermia in conjunction with readily available radiation sources (such as cobalt teletherapy units..) might provide some of the same advantages as heavy particle therapy. **1972** *Science* 5 May 532/3 A thermistor probe which recorded rectal temperatures was connected to a telethermometer. **1968** *Punch* 2 Oct. 488/1 A Royal Martian Vole..teletransported herself to your planet in 1964. **1966** *New Scientist* 20 Jan. 169/3 Each contributing a special faculty such as telekinesis, teletransportation, and so on.
2. [f. *TELE(VISION.] Used to form sbs. denoting activities, persons, things, etc., connected with television (not sharply distinguishable from an attrib. use of *TELE *sb.*[2] **a.** In a virtually limitless range of largely colloq., humorous, or journalistic formations, as *telechair, -course, -drama, -studio*, etc.
1940 *Chambers's Techn. Dict.* 837/2 *Telestudio*,..the enclosure, sound-proofed and treated acoustically, which is used for originating television or broadcasting programmes. **1942** O. E. DUNLAP *Future of Television* vi. 80 The excitement of watching an actual event in progress compensated for any blur or foggy effects, caused chiefly by the tele-eyes' lack of depth and focus. **1953** *Sat. Rev. Lit.* (U.S.) 3 Jan. 3/3 Mr. Sherwood's first tele-drama will be seen in the spring by an anticipating nation. **1953** *Sun* (Baltimore) 15 Dec. (B ed.) 10/2 Mr. Gould's teleplay was the better of the two, although by no means a masterpiece. **1954** *Ibid.* 5 Feb. (B ed.) 8/1 The tele-version..reflected both the assets and faults of the original. **1955** *House & Garden* Apr. 70/1 Yellow appears again on the back of the black-seated telechair. **1957** *Economist* 19 Oct. 226/1 'Tele-courses' [in the U.S.] have in some cases completely replaced conventional classes. **1957** *Cinema* 4 Sept. 3 (*heading*) Tele-movies start in U.S. **1957** P. WILDEBLOOD *Main Chance* 54 Ginny had..blossomed out into a quite new kind of star: the Telepersonality. **1958** *Spectator* 10 Jan. 37/2 The Duke [of Bedford] is so anxious to please the tele-masses that he has taken voice-production lessons. **1962** *Listener* 30 Aug. 327/1 Three of Mr Bowen's teleplays. **1967** *Daily Tel.* (Colour Suppl.) 22 Mar. 27/1 A few swinging teleclerics try vainly to up-date God's image. **1967** *Which?* Oct. 290/1 Telepundits donned ceremonial expressions of awe. **1970** *Times* 25 July 12 The director of telemedicine at the Massachusetts hospital..says that 60 per cent of the patients have found the automated consultations acceptable. **1972** *Observer* 30 Jan. 9/7 They became something that was to be crucial to the development of television—the first tele-journalists. **1978** *Ibid.* 29 Jan. 29/1, I say 'familiar' because teledrama modes are well established. **1983** *Times* 18 Aug. 7/6 We were hanging on the halting lips of all those returning officers..and marvelling at the sharpness and stamina of the telepundits who could divine at the drop of a percentage that the Tories were sweeping the seaside resorts. **1983** *Listener* 22 Sept. 28/3 This was also the week of *The Godfather*, in Coppola's long tele-version, played on BBC1 at 9.25 pm, every week-night but Wednesday.

b. Special Combs.: **te·lefilm,** a cinematographic film shown on television, esp. one made for that purpose; also, the film medium itself; such films collectively; **te:lepo·litics,** political activity conducted through television; **teleroman** (teleroman) [Canad. Fr. *téléroman* (also used): see *ROMAN *sb.*[4]], a French Canadian television soap opera; **te·lescreen,** a television screen; **televa·rsity** [VARSITY], a university that teaches its students by means of television, an open university (*disused*); **tele-vérité** (televerite) (also in Fr. form **télé-vérité**) [f. as *CINÉMA-VÉRITÉ], television broadcasting that presents real life; documentary television; **televersity** [*uni*]*versity*] = *televarsity* above (disused).
1939 *Los Angeles Extended Area Telephone Directory* 1003/1 Telefilm 16 mm Productions Co... 6039 Hollywood Blvd. **1950** *Electronic Engin.* XXII. 8/1 With the advent of television recording or 'telefilm' as it is called, a new tool has been placed in the hands of the television programme builders. **1958** *Times* 20 Nov. 3/4 Plans were announced for the largest Anglo-American co-production scheme yet envisaged in the field of the tele-film. **1975** *New Yorker* 19 May 88/2 It has been translated into telefilm with a greater concern for the Indian position than has been shown by most filmmakers in the past. **1959** *Observer* 4 Oct. 21/3 Lennox-Boyd looked a bit tense, but was certainly controlled. His exit line, to the effect that he had been doing the most wonderful work in the world, deserves a place for itself in the annals of telepolitics. **1975** *Listener* 9 Oct. 479/1 It is a pity telepolitics are so unlike the real thing. **1973** *Globe & Mail* (Toronto) 13 July 13/3 CBC President Laurent Picard's marked liking for the numerous serials seen on CBC French TV, called teleromans. **1942** O. E. DUNLAP *Future of Television* vi. 80 The clarity of the telescreen could not be compared to the sharpness of a newsreel. **1949** 'G. ORWELL' *Nineteen Eighty-Four* I. 6 The telescreen received and transmitted simultaneously. **1979** *Globe & Mail* (Toronto) 13 July 14/3 This wit pales towards the end, as Smith is systematically reduced through the clever interplay of video playbacks (read telescreen for life in Oceania) with O'Brien's stiff and triumphant martinet's voice. **1961** *Economist* 16 Dec. 1105/2 The daytime hours on this network, when the voluntary 'televarsity' students would be at their ordinary everyday work. **1964** *New Statesman* 14 Feb. 264/3 *Télé-vérité* may have reached its apogee.. when a man was tortured to within a few minutes of death in front of the camera. **1976** *Listener* 11 Mar. 310/1 Selected by their daughter's boyfriend, a television producer, as the subject of his *tele-vérité* film. **1950** *Time* 21 Aug. 44/2 *Televersity.* For years, educators have been talking about television as an ideal teaching medium... [The University of] Michigan will start weekly Sunday afternoon telecasts.
3. [f. TELE(PHONE *sb.*] Prefixed to sbs. with reference to a service obtained by means of the telephone, as **te·le-ad,** an advertisement placed in a newspaper by telephone; **te·lebus,** (a service offering) a bus that can be summoned by telephone; **te:lefacsi·mile** (see quot. 1967); **te·lelecture** (see quot. 1969); **te·lesale,** a sale effected by a salesperson who telephones prospective customers.
1976 *Southern Even. Echo* (Southampton) 3 Nov. 3/8 Tele-Ads from telephone subscribers within the 'Southern Evening Echo' circulation area only are accepted. **1977** *Financial Times* 23 Apr. 13/6 Journalists and tele-ad girls should have direct access to the terminals. **1969** Telebus [see *DIAL *v.* 4 b]. **1972** *Daily Colonist* (Victoria, B.C.) 25 Feb. 8/1 The telebus service..uses half-size buses that pick passengers up at their homes and drive them to the nearest regular bus route terminal. **1967** *Britannica Bk. of Year* 804/3 *Telefacsimile*, a system for the transmission and reproduction of fixed graphic matter (as printing) involving the use of signals transmitted over telephone wires (as between libraries). **1968** *Sat. Rev.* (U.S.) 17 Feb. 60 Colleges today are..piping the specialist's voice and face in by telelecture and television. **1969** *Britannica Bk. of Year* 801/1 *Telelecture*, 1. A loudspeaker connected to a telephone line for amplifying voice communication. 2. A lecture delivered to an audience by telelecture. **1981** *Monitor* (McAllen, Texas) 1 Mar. 24/5 A series of telelectures entitled 'Good Health—the key to Happy Living' is continuing at Knapp Memorial Methodist Hospital. **1963** *Spectator* 12 Apr. 478/3 The advantages of 'telesales' over direct mail. **1981** *Event* 16 Oct. 99/3 (Advt.), Dynamic telesales personnel.

tele-ad: see *TELE- 3.

teleangiectasis, -ia, varr. TELANGIECTASIS in Dict. and Suppl.

telebanking, tele-camera: see *TELE- 1.

telecast (te·lĭkast), *sb.* orig. *U.S.* [f. TELE- + BROAD)CAST *sb.*] The action or an act of broadcasting by television; a television broadcast or programme.
1937 *Atlantic Monthly* CLIX. 531/2 He can be assured that any receiver he buys will give him the telecasts sent out by all the major systems of transmission. **1951** M. EHRLICH *Big Eye* i. 11 The Telecast Building way downtown. **1954** 'J. CHRISTOPHER' *22nd Cent.* 21 Within three hours of the Atomics telecast there were riotous assemblies in Canberra. **1961** G. MILLERSON *Technique Television Production* iii. 32 The forms of perspective-distortion introduced by narrow-angle lenses are a familiar feature in telecasts where the camera has to be positioned some way from the subject. **1978** J. IRVING *World according to Garp* xvi. 349 Except for the fact that there had been no nudity in the telecast, the event was an X-rated soap opera from start to finish. **1980** *Daily Tel.* 12 July 5/1 The Minister also said in a telecast that value-added tax would be applied to more products.

telecast (te·lĭkast), *v.* orig. *U.S.* [f. TELE- + *BROAD)CAST *v.* 3: cf. prec.] *trans.* To broadcast by television.
1940 *Topeka* (Kansas) *Daily Capital* 25 Mar. 1/4 Easter Services..were telecast today. **1949** R. GRAVES *Seven Days in New Crete* iv. 44 The garish, raucous, three-dimensioned cartoon-comedies 'telecast every hour in mid-air over the harbour. **1952** *Economist* 26 July 228/1 Old films..are telecast over 49 stations. **1968** *Globe & Mail* (Toronto) 13 Jan. 26/3 In Saint Joan (telecast last month) I searched through myself for parts I could put into Joan. **1978** G. VIDAL *Kalki* viii. 182 Since the networks refused to telecast the Kalki–Arlene Wagstaff interview, Giles was obliged to buy thirty minutes of prime-time television.
So **te·lecasting** *vbl. sb.* and *attrib.* or as *ppl. a.*; **te·lecaster,** one who broadcasts on television.
1937 *Electronics* Sept. 13/1 (*caption*) 'Telecasting' in Great Britain. **1940** A. H. MORTON in Porterfield & Reynolds *We present Television* i. 47 Television standards in the United States must be uniformly adopted by all telecasters. **1945** F. BROWN *Angels & Spaceships* (1955) 90 Every major broadcasting and telecasting station in the world has gone off the air. *Ibid.* 92 With telecasting suspended there were no pictures on their screens. **1951** M. EHRLICH *Big Eye* i. 26 News of the world!..Here is your telecaster—Arthur Morrow! **1957** *Observer* 25 Aug. 11/1 This was a commendable piece of telecasting, though a military exercise unopposed..is always a bit of an anticlimax. **1974** *Times* 8 Jan. 13/7 The 10.30 shut-down has shortened the telecasting day.

telechiric (telĭkəi[ə]·rik), *a.* and *sb.* [f. TELE- + Gr. *χείρ* hand + -IC.] **A.** *adj.* Applied to a device which carries out manipulative operations under the control of a person who is not in the immediate vicinity, but who receives feedback from sensors in the device; also applied to a process or system involving such devices. **B.** *sb.* **a.** *pl.* The branch of technology concerned with telechiric devices. **b.** A telechiric device or system.
1963 J. W. CLARK in *Battelle Technical Rev.* Oct. 3/2 Since the system..can be considered as an extension of man's manipulative and sensory capabilities, even to the use of hand tools, it is well described by the term 'telechirics'. The word is formed from two Greek words—'tele'.., meaning distant, and 'kheir or chir'..which means hand. *Ibid.* 4/2 A well-designed telechiric system replaces man's eyes, hands, and feet with somewhat equivalent mechanical devices. **1968** *Sci. Jrnl.* Oct. 65/2 The commercial cost advantages of the telechiric will displace all competition for work duties undersea. **1970** *Physics Bull.* Oct. 450/1 The telechiric machine requires complete sensing devices, but it presents the sense information to a human as if he were receiving it directly. **1977** *Daily Mail* 11 July 24 (*heading*) Send the telechiric down the pits. *Ibid.* Professor Meredith Thring, Professor of Mechanical Engineering at Queen Mary College, London, is pressing for the development of telechiric mining in Britain. **1978** *Jrnl. R. Soc. Arts* CXXVI. 493/2, I am working on telechirics in mining. It means we shall be able to mine coal in the future without miners going underground, because miners can do their job remotely from the surface.
Hence **telechir** (te·lĭkī[ə]ɹ) = *TELECHIRIC *sb.* b.
1980 *New Scientist* 3 Jan. 5 A telechir is a mobile machine equipped with TV, sensor devices, mechanical arms and hands, and controlled by a skilled human operator situated at the surface.

telecine (telĭsi·nī). [f. *TELE- 2 + *CINE, or f. next.] The broadcasting of cinematographic film on television, or its conversion into television signals; also, apparatus or an organization involved in doing this. Freq. *attrib.*
1935 *Illustr. London News* 23 Feb. 306/1 In the Teleciné apparatus, ordinary standard sound-films may be used. **1937** *Electronics* Aug. 34/3 *Telecine transmission*, the process of transmitting motion-picture film subjects by television. **1938** *Ibid.* July 25/1 A succession of stationary images is projected upon the photoelectric cathode of the pick-up tube in the telecine camera. **1949** *Electronic Engin.* XXI. 194 Two sets of telecine equipment..have just been installed at Alexandra Palace, where they are now being used for televising film programmes. **1959** *Viewpoint* July 32 The telecine operator's in now. Didn't you want to see that piece of film? **1960** D. WILSON *Flight of Dove* 223 Superimpose main title. End telecine (1). **1961** G. MILLERSON *Technique Television Production* vii. 124 Film televising equipment (Telecine) ranges in complexity from slightly adapted cinema projection apparatus, to electronic scanners. **1972** I. HAMILTON *Thrill Machine* xxxix. 187 Joe gave his cues to the men in telecine. **1978** *Gramophone* Aug. 391/3 They have even announced a telecine attachment to use the camera for transcribing home films to video tape. **1983** *New Scientist* 26 May 546/3 The film image is converted to video in the usual way with a 'telecine' machine, which combines a film projector with a video camera.

† **telecinema** (telĭsi·nĭmă). *Obs.* exc. *Hist.* Also **telekinema** and with hyphen. [f. *TELE- 2 + *CINEMA, *KINEMA.] **1.** = prec.
1930 MOSELEY & CHAPPLE *Television* viii. 93 An extraordinary situation in the fascinating history of television was the development of what is now called the tele-cinema.
2. (Also with capital initial.) The name of a building in the Festival of Britain of 1951 in which television programmes could be shown on a large screen as in a cinema; the system or process involved in producing this display.
1951 I. COX *South Bank Exhibition* 83 The Telecinema is the first cinema in the world to be specially designed and built for the showing of both films and television. **1952** *Times* 6 Feb. 4/7 The Minister had suggested that, except for the Telekinema and possibly, also, the Waterloo Road administration blocks, none of the festival buildings should be retained unless the council wished to take them over. **1953** A. K. C. OTTAWAY *Educ. & Society* v. 81 Many new means of mechanical amusement have been created. We have the telecinema; shall we even yet have the 'Feelies'. **1976** *Oxf. Compan. Film* 495/2 *National Film Theatre*, a club cinema established and run by the British Film Institute, took over and rebuilt the Telekinema. **1977** M. STRICKLAND *A. Thirkell* x. 157 Angela treated the events [of the Festival of Britain, 1951] with the greatest scorn, but she agreed nevertheless to be interviewed on the 'telecinema'.

telecobalt: see *TELE- 1.

telecom (te·lĭkǫm). Colloq. abbrev. of *TELECOMMUNICATION. Also (chiefly *attrib.*) in *pl.*
British Telecom, name of a public corporation, formerly part of the Post Office, set up on 1 October 1981 to provide telecommunications and data processing services.
1963 *Telecoms Topics* Aug. 1 This new publication, *Telecoms Topics*,..will contain the latest information about..*G.E.C. Telecommunications.* **1964** D. MACARTHUR *Reminiscences* IX. 331 By 'telecom' I was directed to use the Navy and the Air Force to assist South Korean defenses by whatever use I could make of these two arms. **1970** T. LILLEY *Projects Section* v. 45 That patrol..was now accompanied by telecom and explosive experts. **1981** *Economist* 24 Jan. 100/1 Every big telecoms company is evaluating bubbles (or already buying them) for use in private branch exchanges.

telecommand: see *TELE- 1.

telecommunication (te:lĭkǫmiūnikē[i]·ʃən). [f. TELE- + COMMUNICATION, after F. *télécommunication.*] Communication over long distances, esp. by electrical means such as by telegraphy, telephony, or broadcasting; (usu. in *pl.*) the branch of technology concerned with this. Also *concr.*, a means or channel of such communication. Freq. *attrib.*, esp. in *pl.*
The term *télécommunication* was adopted by the Convention Internationale des Télécommunications at Madrid in 1932 (the official language at the conference was French). The definition then accepted ('toute communication télégraphique ou téléphonique de signes, de signaux, d'écrits, d'images et de sons de toute nature, par fil, radio ou autres systèmes ou procédés de signalisation électriques ou visuels (sémaphore)'), in so far as it includes non-electrical means of communication, is no longer applicable.
1932 *Times* 18 Nov. 13/4 The new convention which is being drawn up by the International Telegraph and Radiotelegraph Conference..will be called International Telecommunications Convention. **1942** *Electronic Engin.* Aug. 128/2 A telecommunication system where intelligence is communicated by means of a radio-frequency carrier. *Ibid.* Dec. 306/1 Rigid frequency control has become a necessity in radio broadcasting and indeed in all forms of telecommunication. **1944** *Times* 21 July 3/4 Resistance groups in Belgium have..been engaged in the systematic destruction of railways, road bridges, telecommunications, [etc.]. **1953** *Science News* XXX. 70 One has only to consider the field of automatic telecommunication to accept the process [*sc.* electrical manipulation of information] as commonplace. **1957** *Technology* July 187/2 The possible applications of solar batteries in telecommunications. **1961** *Engineering* 6 Jan. 33/1 The plan..to put a telecommunications satellite into orbit round the earth. **1971** F. J. M. LAVER in B. de Ferranti *Living with Computer* v. 44 The capture of data at its point of origin, its rapid transmission over telecommunication links, and its filtering and analysis by computers. **a 1974** R. CROSSMAN *Diaries* (1977) III. 583 It was an impressive telecommunications feat, which is why Kennedy decided to do it. **1979** MILLER & CHYNOWETH (*title*) Optical fiber telecommunications.

telecon (te·lĭkǫn). *U.S. Mil.* [f. *teletype* s.v. TELE- + CON(FERENCE *sb.*: see next.] **1.** A device of the U.S. army which sends teletype messages over long distances by means of radio or underwater cable and which typically displays them on a screen; a conference held by this means.
1950 *N.Y. Times* 2 July 8E/1 This was the 'telecon' room, equipped with machines that enable officers in Washington to confer with headquarters overseas... The telecon, essentially, is just a teletype machine, but it has certain modifications which make it especially useful for military communications. **1951** *Sun* (Baltimore) 31 Oct. (B ed.) 7/2 Gen. Douglas MacArthur participated in four history-making telecons. **1969** D. ACHESON *Present at Creation* (1970) xliv. 412 A telecon is a secure device by which a typewriter operated at one end records both there and through a similar machine at the other end.
2. [orig. *transf.* from sense 1; later re-formed as shortening of next.] A long-distance con-

ference held by means of teletype or telephone.

1951 *Sun* (Baltimore) 31 Oct. (B ed.) 7/2 The generals like to impress each other with 'sorry, can't see you then, have a telecon with Washington at that time'. **1981** *Aviation Week & Space Technol.* 15 June 128/3 Reference our telecon regarding the operation of company flights to Antarctica and return nonstop. **1982** *Legal Times* 10 May 14/2 A client..may yelp about a 'six-minute telecon' on his bill that costs $10.

teleconference (te·lĭkǫ:nfĕrĕns). [f. TELE- + CONFERENCE *sb.*] A conference held by people who, though separated physically, are linked by telecommunication devices (e.g. telephones, television screens, etc.). So **tele-co·nferencing.**

1953 *Language* XXIX. 71 A general at a teleconference writes out a message for transmission. **1973** *Times* 4 Oct. 35/5 The prospect of similar teleconference equipment in every main commercial or industrial building—just like the telephone on the business desk—is no longer a pipe dream. **1975** *Financial Times* 21 July 6/2 The psychological and travel-replacement aspects of teleconferencing were underlined in a paper by Quebec University. **1981** *Times* 9 Feb. 20/1 Teleconferencing brings together a number of people in different locations for a meeting by means of a tele-audio link.

teleconnection to **-converter:** see *TELE-1.

telecopier (te·lĭkǫ:piəɹ). Also (*U.S.*) **Tele-.** [f. *TELE- 3 + COPIER.] A facsimile device which transmits and reproduces graphic material over telephone lines.

A proprietary term in the U.S.

1967 *Official Gaz.* (U.S. Patent Office) 24 Jan. TM 152/2 Telecopier. For transmitting and receiving equipment for producing facsimile copies of documents. First use June 13, 1966. **1972** M. GILBERT *Body of Girl* xviii. 160 A photograph..was sent by tele-copier to the Isle of Wight. **1979** *Fortune* 21 May 123/2 He kept a telecopier in his bedroom to take messages and transmit urgent documents.

telecurietherapy: see *TELE- 1.

telediagnosis (te:lĭ,daiăgnōᵘ·sis). [f. TELE- + DIAGNOSIS.] The long-distance assessment of a patient's condition by a doctor using closed-circuit television.

1961 *New Scientist* 7 Dec. 604/1 The most recent use of television in medicine is in 'telediagnosis' now being used in Paris. **1972** D. V. TANSLEY *Radionics* 6 He then experimented with telediagnosis which utilized the overhead telephone wires to link him with the patient sample.

telediphone: see *TELE- 1.

teledu. Add: Also † **telagu.** For *Mydaus meliceps* substitute *Mydaus javanensis*. (Earlier and later examples.)

1821 T. S. RAFFLES in *Trans. Linnean Soc.* XIII. 251 *Mephitis Javanensis* Desm. Telagu of the Malays. **1906** E. INGERSOLL *Life of Animals: Mammals* 176 Two related animals of the East are the teledu, or stinking badger, a small nocturnal burrower of Java and Sumatra ..and the large, long-snouted, piglike sand badgers. **1965** D. MORRIS *Mammals* 292 The Teledu is well able to defend itself by means of the offensive secretions of its large anal glands.

telefacsimile: see *TELE- 3.

teleferic (telĭfe·rik). Also ‖ **teleferica, telepheric.** [ad. It. *teleferica*, f. Gr. τῆλε TELE- + φέρειν to carry + -ικος -IC: see next.] A cableway.

1916 *Windsor Mag.* Oct. 498 We walk a little way and then go up by the teleferic. **1918** W. HUTCHINSON *Doctor in War* (1919) xviii. 262 The miniature cable-railway, or teleferica. **1931** R. H. BEADON *Royal Army Service Corps* II. xi. 336 It was found necessary..to use what was known as the Teleferica, or wire rope carrying cradles, which was extensively used by the Italians. **1964** *Harper's Bazaar* Nov. 140/3 New teleferic extension of the Grandes Rousses lift at Alpe d'Huez. **1973** *Good Motoring* May 31/2 A telepheric takes visitors up part of the hill.

‖ **téléférique, téléphérique** (teleferīk). [Fr., f. as prec.] = prec.

1956 I. BROMIGE *Enchanted Garden* III. 150 We'll go up on the *téléphérique* after tea. **1958** *Times* 15 Nov. 11/5 At present there are no ski-lifts or téléfériques [in the Loetschental]. **1965** G. MCINNES *Road to Gundagai* vi. 97 We..hoped emptily that..a little téléférique would appear to carry us all up high in the clouds. **1977** *New Yorker* 26 Sept. 107/1 The Brévent, to be sure, could be reached in a few minutes by a system of *téléphériques*.

telefilm: see *TELE- 2 b. **teleflash, telegenesis:** see *TELE- 1.

telegenic (telĭdʒe·nik), *a.* orig. *U.S.* [f. *TELE- 2 + *-GENIC b, after *photogenic*.] Of a person or thing: that shows to advantage on television; providing an interesting or attractive subject for a television broadcast.

1939 *Sun* (Baltimore) 16 Oct. 6/8 Judith Barrett, pretty and blonde actress, is the first Telegenic Girl to go on record. In other words she is the perfect type of beauty for television... She is slated for the first television motion picture. **1948** *Daily Tel.* 23 June 6/4 One word that is playing an important part at this gathering is 'telegenic'. With everything that happens..being reproduced on thousands of television screens, the ability of any speaker to look as attractive as he sounds has become an important political asset. **1950** *New Yorker* 26 Aug. 18/3 Korman has spent many hours in the WOR Television studios, experimenting with telegenic properties of various weaves and colors of cloth and styles of suits. **1962** P. FERRIS *Church of England* ii. 35 The letters 'C.R.' after a man's name keep cropping up..in the *TV Times*, television having discovered that the religious communities have some telegenic personalities. **1971** H. WILSON *Labour Govt.* (1974) xvi. 372 We had not yet reached the position where telegenic situations were planned in advance between a television authority and the demonstrators. **1980** *Times* 7 Nov. 12/2 Conscious that jazz by itself is not very telegenic, producers compensate by building shiny sets.

telegnomy: see *TELE- 1.

telegram. Add: (Later *attrib.* example.) Also *transf.* and *fig.*

1908 E. J. BANFIELD *Confessions of Beachcomber* II. i. 244 Telegrams along the line from the sucker [*sc.* a fish] give precise information. **1922** JOYCE *Ulysses* 117 A telegram boy stepped in nimbly, threw an envelope on the counter and stepped off posthaste. **1963** *Listener* 3 Jan. 14/1 A slim, solid brick tower, a kind of telegram about all the best and strongest things in Germany, stands a few feet away.

telegram, *v.* Delete ? *Obs.*, restrict *rare* to *intr.* use, and add: *trans.* (*a*) (later examples); (*b*) to send (news, information, etc.) by telegraph. Hence **te·legrammed** *ppl. a.*, that has been sent by telegraph; **te·legramming** *vbl. sb.*

1952 M. TRIPP *Faith is Windsock* x. 149 In hospital with a broken leg. His mother has just telegrammed the news. **1969** R. MILLAR *Kut* vii. 154 A miserable Aylmer telegrammed Lake with a copy to Townshend. **1970** *Nature* 18 July 225/1 This information was simultaneously telegrammed to doctors. **1972** D. DOUGLASS *Pit Life in Co. Durham* (History Workshop Pamphlets No. 6) 53 The Executive Committee issued telegrammed instructions to the Lodge. **1973** A. BROINOWSKI *Take One Ambassador* ii. 16 The telephoning and telegramming was to begin once more. **1977** C. MCCULLOUGH *Thorn Birds* ix. 201 Mrs. Cleary telegrammed me, a..courtesy I appreciated very much. **1981** *Washington Post* 26 May A15/1 Israeli Premier Menachem Begin's suggestion in telegramming his congratulations for Mitterand's election.

telegraph, *sb.* Add: **1. c.** (Further examples.) See also *bush telegraph* s.v. *BUSH *sb.*[1] 11.

1866 *Tumut & Adelong* (N.S.W.) *Times* 1 Jan. 2/3 They approached to within one hundred yards of the camp unobserved, and then it was apparent that the 'telegraph' had done its work. **1891** 'OLD TIME' *Convict Hulk 'Success'* 20 The 'telegraph' was very extensively worked on board these hulks... The 'telegraph' was a system of speaking from one cell to another by means of tapping on the walls. **1964** D. MACARTHUR *Reminiscences* vi. 206 News of the first such shipment spread rapidly by the 'bamboo telegraph' through the Philippines. **1969** *New Yorker* 14 June 76/2 He would look down at his plate and find two steaks there. He knew what was happening. A message had come from the kitchen, on the Afro-American telegraph.

3. (Earlier example.)

1821 G. GLEIG *Campaigns Brit. Army at Washington & New Orleans 1814–15* vii. 89 We had not proceeded many miles from the river's mouth, when a telegraph from the admiral gave orders for the troops to be in readiness to land.

4. (Earlier example.)

1849 *Sussex Agricultural Express* 8 Sept. 6/4 At the close of the first innings the telegraph showed to the people a score of 61 runs.

5. b. *Austral. spec.* One who warns bushrangers about the movements of police and pursuing troopers.

1864 *Goulburn* (N.S.W.) *Herald* 17 Aug. 2/3 These young scoundrels have got their 'telegraphs' in town, and there is not a stir the police can make but it is known. **1867** *Ibid.* 12 Oct. 4/5 It would make me look a gamer man to the police and other people as has got a down on me for being a telegraph to you chaps. **1908** C. WHITE *John Vane, Bushranger* xv. 76 One of our 'telegraphs' rode up and told us that a party of three police had just gone along the road towards Carcoar.

8. *telegraph house* (earlier and later examples), *line* (earlier U.S. example), *signal*, *station*, *wire* (earlier examples); **telegraph blank** *U.S.* = *telegraph form*; **telegraph coach** = *telegraph-carriage*; **telegraph code** (see quot. 1971); **telegraph editor** *U.S.*, on the staff of a newspaper, one who edits news received by telegraph; **telegraph pole, post** (earlier examples).

1893 S. MERRILL in M. Philips *Making of Newspaper* 99 He struck out the formal matter in the heading of the telegraph blank. **1904** [see *BLANK *sb.* 6]. **1928** F. N. HART *Bellamy Trial* i. 3 [He] had..a good-sized stack of telegraph blanks clasped to his heart. **1812** A. CONSTABLE *Let.* 22 Nov. in J. Constable *Corr.* (1962) I. 85 To Mr. Farrington by last night's Telegraph Coach, a brace of pheasants were forwarded. **1835** *N.Y. Commercial Advertiser* 23 July 4/2 Two Telegraph Coaches will leave Albany every day at half-past 10, A.M...and arrive at Rochester in 44 hours. **1885** *List of Subscribers, Classified* (United Telephone Co.) (ed. 6) 231 Universal Translations Institute,..Specialities:..patent specifications, telegraph codes. **1971** *Gloss. Electrotechnical, Power Terms (B.S.I.)* III. iii. 12 *Telegraph code*, a system of rules and conventions according to which the telegraph signals forming a message, or the data signal forming a block, should be formed, transmitted, received and processed. **1875** C. F. WINGATE *Views & Interviews* 195 Have been continuously employed on the *Missouri Republican* [as] telegraph editor. **1923** G. C. BASTIAN *Editing Day's News* 9 Inside the News Room..[we find the] Managing Editor..City Editor..Telegraph Editor [etc.]. **1981** *N.Y. Times* 15 June A20/3 The telegraph editors of our [*sc.* the Associated Press's] member papers take our word for it and put it in print. **1808** LD. DUNDONALD *Let.* 28 Sept. in *Autobiogr. Seaman* (1860) I. 288 The newly constructed semaphoric telegraphs..have been blown up and completely demolished, together with their telegraph houses. **1923** KIPLING *Land & Sea Tales* 239 My father is at the telegraph-house sending telegrams. **1847** *Michigan Gen. Statutes* (1882) I. 944 The owner of any land through which said telegraph line may pass..having first given consent. **1851** THOREAU *Jrnl.* 12 Sept. in *Writings* (1906) VIII. 497, I instantly sat down on a stone at the foot of the telegraph-pole, and attended to the communication. **1851** —— *Jrnl.* 30 Sept. in *Ibid.* IX. 37 Methinks these telegraph-posts should bear a great price with musical instrument makers. **1821** G. GLEIG *Campaigns Brit. Army at Washington & New Orleans 1814–15* xv. 206 The minds of all were set at ease, as to the place whither we were going, a telegraph signal being made to steer for Jamaica. **1830** M. EDGEWORTH *Let.* 18 Oct. (1971) 419 They use Telegraph signals—flags white—red—and blue —for *all right—moderate speed—stop.* **1971** *Gloss. Electrotechnical, Power Terms (B.S.I.)* III. iii. 12 *Telegraph signal*,..the set of conventional elements established by the code to enable the transmission of a written character [etc.]. **1839** *Knickerbocker* XIV. 187 A recent excursion.. from New-Brighton to the telegraph station. **1973** P. BERTON *Drifting Home* vii. 101 We had stopped at one or more of these solitary telegraph stations whose operators were always fanatically overjoyed to see us. **1848** *Knickerbocker* XXXI. 455 The wrecks of hundreds of little urchins' high-soaring 'hopes'..[hang] on all the telegraph-wires. **1869** *Bradshwa's Railway Man.* XXI. App. 114 Telegraph Wire, Plain or Galvanised, of any length.

telegraph, *v.* Add: **1. b.** (Earlier example in *Cricket*.)

1862 F. LILLYWHITE *Guide to Cricketers* 37 A model of a newly-built covered stand on rollers, with figures for telegraphing on each side.

2. a. (Earlier example.)

1818 'T. BROWN' *Brighton* I. 230 They nod and telegraph to their favourites.

† **3.** *trans.* To send (*esp.* information about police movements) by bush telegraph. *Austral. colloq. Obs.*

1863 *Mudgee* (N.S.W.) *Liberal* 15 Oct. 2/4 The police might have scouts if they would; scouts which would be a match for any system of telegraphing among the bushrangers. **1878** *Australasian Sketcher* 23 Nov. 135/2 The object of the expedition leaked out, and, no doubt, was rapidly telegraphed across the bush to Edward Kelly. **1880** *Victorian Rev.* I. 428 News of the movements of the troopers were 'telegraphed' to them by their confederates.

4. a. In Boxing and other sports: to initiate (a punch, throw, etc.) in such an obvious way as to reveal one's intention. Also in fig. contexts.

1925 J. J. CORBETT *Roar of Crowd* v. 77 Before I would start my right I would, as they say in boxing, 'telegraph' the blow, purposely. **1937** *Daily Mirror* 16 Mar. 30/4 Ford was also landing with some heavy left hooks to the body and although he had never telegraphed his right hand punches..he was now finding Farr's face with such regularity [etc.]. **1945** E. NICHOLS *Hunky Johnny* 68 He telegraphs every curve he throws. **1959** *Charlottesville* (Va.) *Daily Progress* 18 Apr. 10/1 (*caption*), I thought you said he telegraphed his punches! All I saw he delivered personally. **1969** *Wall St. Jrnl.* 1 Dec. 14/1 For competitive reasons, the company won't disclose the nature of the new products now, Mr. Arneson said. 'We're not about to telegraph our punches.'

b. *gen.* To give a clumsily obvious hint or premature indication of (something to come).

1952 *N.Y. Times* 13 Aug. (Late City ed.) 29/5 Inevitably the pay-off gag was 'telegraphed' to the audience far in advance. **1959** *Wall St. Jrnl.* (Eastern ed.) 3 Mar. 12/6 One subplot involving Claudell's mother figures in the story but this development is telegraphed early and does nothing to broaden the book. **1968** *Punch* 16 Oct. 558/3 The exasperating way music [in a film] sometimes not only over-emphasises but even telegraphs effects. **1977** *Time* 7 Nov. 14/2 Young was accurately telegraphing the Administration's view.

telegrapher. 1. For 'Now *rare*' read 'Now chiefly U.S.' and add earlier and later examples.

1851 C. CIST *Sk. Cincinnati in 1851* 51 Telegraphers, 7. **1910** J. HART *Vigilante Girl* 376 This was the room of the man who filled the manifold offices of station-master, ticket-agent, express-agent,..and telegrapher. **1932** E. WILSON *Devil take Hindmost* xvii. 177 The telegrapher ..telephoned ahead along the line to have the niggers taken off. **1955** H. KURNITZ *Invasion of Privacy* (1956) xii. 79 He shared Zorn's hatred of the teletype but..he sometimes operated it himself when the telegrapher was off duty. **1974** T. P. WHITNEY tr. *Solzhenitsyn's Gulag Archipelago* I. I. vi. 268 His White Guard father was just a rank-and-file, unpropertied telegrapher.

telegraphese. Add: **1.** Also *attrib.* or as *adj.*

1951 R. HOGGART *Auden* i. 18 Auden's 'telegraphese' style..is distinguished by its omission of articles, relatives, connectives, personal, demonstrative and other pronouns, and auxiliary verbs. **1978** *Radio Times* 18–24 Mar.

15/1 The actor's opinion hardened into the following telegraphese note: 'Willy beyond question toughest director I've ever worked for.'

telegraphic, *a.* Add: **1. b.** (Earlier example.) Also *telegraphic address*, a brief style of address registered with the postal authorities and designed to reduce the number of words in a telegram.

1839 *Ann. Electricity, Magnetism & Chemistry* III. 442 The telegraph as constructed by Chappe has met with a favourable reception, and since 1793, when the first telegraphic line was established in France, been very generally adopted. **1885** *List of Subscribers, Classified* (United Telephone Co.) (ed. 6) 4 Fraser & Fraser,.. Manufacturers of..Steam Boilers. *Telegraphic address*, 'Pressure, London.' **1910** *Nation* 9 Apr. 54/1 The necessity of being on the right side is self-evident; therefore, when in doubt, write or wire to Mr. Shaw. (Telegraphic address: 'Infallibility', London.) **1930** 'HAY' & WODEHOUSE *Baa, Baa, Black Sheep* I. i. 12 We'll go to the Grotto!..What-ho for the Grott-ho! Telegraphic address —Tighter London! **1963** *B.S.I. News* May 7/2 BSI's telegraphic addresses have been changed. Overseas cables should now be addressed to 'Standards London W 1' and inland telegrams should be addressed to 'Standards Audley London'.

2. d. *spec.* In *Linguistics*, in the context of language acquisition.

1963 BROWN & FRASER in C. N. Cofer *Verbal Behavior & Learning* 192 Young children speak a rather uniform telegraphic English. **1970** D. MCNEILL *Acquisition of Lang.* iii. 20 Telegraphic speech is the outcome of the process of language acquisition. **1973** R. BROWN *First Lang.* I. 143 Telegraphic speech is speech entirely composed of contentive words..and entirely lacking functors or function words. **1978** *Sci. Amer.* Nov. 92/2 Early telegraphic speech is characterized by short, simple sentences made up primarily of content words: words that are rich in semantic content, usually nouns and verbs.

telegraphist. a. (Earlier example.)

1847 *Brit. Patent* 11,926 10 Signals are..quickly received by the mind of the telegraphist.

Telegu, var. TELUGU, TELOOGOO in Dict. and Suppl.

teleguide (te·lĭgəid), *v.* [f. F. *téléguider* (1947), Quemada (1980): see TELE-, GUIDE *v.*] *trans.* To control (a missile, etc.) at a distance or indirectly. So **te·leguided** *ppl. a.*, **te·leguiding** *vbl. sb.* Also *telegui·dance* [F. *téléguidage*].

[**1954** *John o' London's Weekly* 12 Feb. 154/3 The French have shortened 'long distance radio control' to 'téléguidance'.] **1960** *Export Service Bull., Suppl.* 4 June 41 The Bell Aircraft Corporation announced that it had signed an agreement..for the production..of the two teleguided target missiles..produced by the French company. **1964** N. FREELING *Double-Barrel* IV. v. 128 'And the aeroplane?' 'Oh that. I built it. We've used it for various experiments; teleguiding and so on.' *Ibid.* 129 If these toys were that simple everyone could have teleguided missiles. **1969** *Africa Digest* June 51/2 An attempt at subversion teleguided from abroad. **1972** *Daily Tel.* 6 May 5/1 Among the new weapons is a teleguided anti-tank rocket fired from a helicopter, which has never previously been used in combat. **1977** *Dædalus* Fall 150 Jacquart does not celebrate the triumph of small familial agriculture, but rather its defeat by the offensive of the 'large enterprises' teleguided by the urban nobility and the Parisian *notables* of the classical age. **1977** N. FREELING *Gadget* v. 218 He's been thinking all day about teleguidance systems.

telekineticist, telelecture, tele-lens: see *TELE- 1, 3, 1.

telemark, Telemark (te·lĭmāɹk). *Skiing.* [f. *Telemark*, the name of an administrative district in southern Norway, where this originated.] A swing turn, now little used, with the leading ski considerably advanced and the knee bent, employed to change direction or stop short. Freq. *attrib.* Also as *v. intr.*

1904 E. C. RICHARDSON *Ski-Running* 59 The telemark swing..should be practised constantly. **1905** *Ibid.* (ed. 2) 78 We propose to deal with the 'Telemark' first. **1920** A. LUNN *Cross-Country Ski-ing* 88 In powder snow one can Telemark in various undefined ways. **1934** WEBSTER, Telemark turn. **1979** R. FIENNES *Hell on Ice* i. 13 Bending low, I began a curving telemark, leading with the left knee. **1981** *Nordic Skiing* Jan. 49/3 It specializes in teaching downhill technique and telemark turns on some of the finest powder in the Sierra Nevada.

telematics (telĭmæ·tiks), *sb. pl.* [f. *TELE(COMMUNICATION + *INFOR)MATICS; cf. F. *télématique* adj.] (The science of) the long-distance transmission of computerized information. So **telema·tic** *a.*, of or pertaining to telematics.

1979 *Economist* 13 Oct. 52/3 The EEC commission is to launch a programme to help European telematics, the new vogue word for the high-growth industries of telecommunications, computers, microchips and databanks. *Ibid.*, Studies to set up a telematic network for the community institutions. **1981** *Jrnl. R. Soc. Arts* CXXIX. 401/2 This story illustrates the importance of possessing advanced telematic equipment. **1982** *Times* 1 Sept. 3/3 Advances in telematics will be far more significant than aircraft design over the next two decades.

telemessage: see *TELE- 1.

telemeter, *sb.* Add: Also with pronunc. (te·lĭmī:təɹ). **2.** In mod. use, an instrument for measuring a quantity at a distance from the place where the result is displayed or recorded. (Examples.)

1929 *Jrnl. Amer. Inst. Electr. Engineers* XLVIII. 183/1 In addition to surveying the field of application of 'telemetering'..the paper presents several innovations in the types of 'telemeters' available. **1941** T. J. RHODES *Industr. Instruments for Measurem. & Control* viii. 363 One advantage of this telemeter is that the measuring device requires very little power to move the condenser plate. **1947** *Sun* (Baltimore) 10 Nov. 9/2 Electronic gadgets called 'telemeters'..are installed in high-speed missiles and tuned to send back to the ground by radio whatever information the scientists need. **1975** D. G. FINK *Electronics Engineers' Handbk.* XXVI. 43 The passive implant telemeter may contain a resonant circuit in which the resonant frequency, made to vary with the body signal, can be detected with a grid dip meter.

3. (Usu. with capital initial.) A proprietary term for a system of pay-TV involving the use of a coin-box attached to the television set. *U.S.*

1953 *Wall St. Jrnl.* 27 Nov. 18/4 Another pay-as-you-look television set-up makes its debut here... It [*sc.* the International Telemeter Corp.] has installed 'Telemeters' in 78 local homes. **1955** *Amer. Speech* XXX. 232 For *Telemeter*, the subscriber pays for each program by dropping coins into a box which is attached to the receiver. **1961** *Spectator* 24 Mar. 394/3, 5800 telemeter sets were installed [in Canada]. **1961** *Official Gaz.* (U.S. Patent Office) 17 Oct TM87/1 Telemeter. For pay-television systems including a television transmitter.., equipment at a receiver for processing subscription-television signals, and computer equipment for processing records made at each subscription-television subscriber receiver. First use during June 1951.

telemetric *a.* (further examples); hence also **teleme·trically** *adv.*, by means of telemetry.

1957 *Times* 11 Nov. 10/5 The radio transmitters..and the telemetric apparatus on board have..ceased work. **1961** WEBSTER, Telemetrically. **1970** *Sci. Jrnl.* Aug. 7/1 A cosmonaut in the command module working the Vulcan welding equipment telemetrically by cable. **1971** *Nature* 2 July 65/2 The use of telemetric recording of physiological information from free swimming fish in their natural environment. **1974** W. GARNER *Big Enough Wreath* viii. 98 It's definitely a satellite. Multitone telemetric signals. **1977** *Offshore Engineer* May 75/2 Telemetrically operated valves are controlled by two systems in the experiment.

telemeter (te·lĭmĭtəɹ, tĭle·mĭtəɹ), *v.* [f. the sb.] *trans.* To measure (a quantity) and transmit the result to a distant point; to transmit (a measurement or observation). Freq. with *back*.

1929 *Jrnl. Amer. Inst. Electr. Engineers* XLVIII. 185/2 A carrier wave of constant frequency modulated by an audio frequency which varies with the magnitude of the quantity being telemetered. **1941** T. J. RHODES *Industr. Instruments for Measurem. & Control* viii. 356 The pressure or liquid-level indication is telemetered to the operating floor. **1953** A. C. CLARKE *Prelude to Space* xii. 64 'Alpha's' main instrument readings are telemetered back to Earth. **1965** W. L. DONN *Meteorology* i. 4 The radiosonde.. telemeters the most complete information on the vertical distribution of temperature, pressure and humidity to radio receivers in the laboratory. **1969** *Times* 2 May 16/4 Spacecraft designed to telemeter photographs of the moon's surface back to the earth. **1971** G. G. LUCE *Body Time* v. 155 His heart rate was telemetered by a small radio capsule in his undershirt.

Hence **te·lemetered** *ppl. a.*, **te·lemetering** *vbl. sb.* and *ppl. a.*

1929 [see *TELEMETER *sb.* 2]. **1929** *Jrnl. Amer. Inst. Electr. Engineers* XLVIII. 183/1 In the design of a telemetering equipment it is necessary to study the channel limitations. **1953** *Proc. Inst. Electr. Engineers* C. I. 44/2 The rate of change of a telemetered value can be obtained by means of a feedback amplifier. **1955** *Sci. News Let.* 21 May 324/3 Other methods currently used to track missiles are telemetering, in which the missile radios instrument readings to the ground, and radar. **1960** *Brit. Communications & Electronics* VII. 598/1 Telemetering from the digestive tract. **1976** *Offshore Platforms & Pipelining* 211/1 The operator has a complete telemetering system plus sonar and television. **1978** *Broadcast* 5 June 21/2 All the telemetered information is available on..wall mounted key and lamp panels.

telemetry (tĕ-, tele·mĭtri). [f. TELE- + -METRY.] **1. a.** The process or practice of obtaining measurements in one place and relaying them for recording or display to a point at a distance; the transmission of measurements by the apparatus making them.

1885 *Electrician* 9 May 525/1 Telemetry has been practically applied in America..the temperature registrations being exceedingly accurate. **1957** *Economist* 12 Oct. 149 (Advt.), Mullard products are used in almost every electronic application—from radar to radio, from telemetry to television. **1967** *Technology Week* 20 Feb. 10/3 Orbital telemetry indicated that the capsule battery should have sufficient charge to operate the radio beacon and flashing light that serve as recovery aids. **1979** *Sunday Mail* (Brisbane) 7 Jan. 46/5 Occasionally pilots do become disorientated and so a system called air-to-ground telemetry was introduced... Data about the spinning aircraft is transmitted to the ground where another pilot can monitor information and advise the pilot flying how to recover.

b. Apparatus used for telemetry (sense a above).

1958 *Engineering* 28 Feb. 263/3 Instrumentation and telemetry in the satellite are designed to gather and transmit four types of information. **1962** S. CARPENTER in *Into Orbit* 160 Everyone keeps one eye on the telemetry to see how the capsule responds.

c. Telemetered information.

1962 *Flight Internat.* LXXXII. 239/2 These receiving stations have been set up..for receiving telemetry from future topside-sounder satellites. **1979** F. POHL *Jem* i. 8 He had made himself one of the..top experts in reading the telemetry from a tachyon-transmitter probe. **1982** *Daily Tel.* 26 Oct. 28/5 An American spy..gave the Russians details of an NSA satellite system 'rhyolite', which monitored the telemetry of Soviet missile tests.

2. *attrib.*

1953 [see *DOWN-RANGE, DOWNRANGE *adv.*]. **1958** *Times* 10 Oct. 12/7 Telemetry codes provided from the third Russian satellite are expected to permit an analysis of tape recordings made of the sputnik's transmitters. **1962** S. CARPENTER in *Into Orbit* 53 Each time the capsule performs an important function it automatically sends a telemetry report to this effect down to earth. **1976** *Offshore Platforms & Pipelining* 214/2 The latest cable has 65 cores plus telemetry circuits. **1978** R. V. JONES *Most Secret War* xxxviii. 347 Dr. Steinhoff, Head of the Telemetry Department..escaped in his air-raid shelter.

telencephalon (te:lense·fălǫn). *Anat.* [f. TEL(E- + ENCEPHALON.] The anterior of the two vesicles into which the prosencephalon or fore-brain divides in the embryo, or the two antero-lateral vesicles that it gives rise to; the corresponding part of the adult brain, comprising the cerebral hemispheres and the anterior parts of the hypothalamus and the third ventricle.

1897 C. L. DANA *Text-bk. Nerv. Dis.* (ed. 4) i. 3 The anterior vesicle develops two secondary vesicles: the anterior portion of these, including the corpora striata, olfactory lobes, and the cerebral hemispheres, forms the telencephalon. **1934** [see *PALLIUM 3 d]. **1946** B. M. PATTEN *Human Embryol.* v. 111 By the sixth week of development..the prosencephalon has divided to form the telencephalon and diencephalon. **1948** A. BRODAL *Neurol. Anat.* x. 323 In fishes and amphibians the telencephalon is dominated by afferent fibres carrying olfactory impulses to its pallial part. **1977** *Lancet* 9 July 64/1 There was no injury to the mesencephalon in 3 cases, while the telencephalon escaped injury in 2 cases.

Hence **te:lencepha·lic** *a.*, of or pertaining to the telencephalon.

1911 *Jrnl. Compar. Neurol.* XXI. 2 The exact limits of the thalamic and telencephalic gray are discussed later. **1974** D. & M. WEBSTER *Compar. Vertebr. Morphol.* xi. 268 It is not definitely established whether..some of the more superficial avian striatal structures, which are actually large telencephalic nuclei, are homologically related to mammalian cortical structures.

teleo-[2]. Add: **teleo·cracy**, an organization designed to fulfil a specific purpose; hence **te·leocrat, teleocra·tic** *a.*

1973 L. L. & J. M. CONSTANTINE *Group Marriage* xiii. 141 In a teleocratic system, leadership depends on the task at hand; teleocratic means 'purpose-centered'. **1975** *Times Lit. Suppl.* 12 Sept. 1018/4 Two notions, both derived from the Roman law, 'societas' and 'universitas'... The latter applies to teleocratic organizations which are supposed to produce specified outcomes. *Ibid.* 1018/5 The full character of the modern state is a condition of unresolved tension between 'societas' and 'universitas'. Teleocracy has always been one of its aspects... He has permitted the teleocrats to impose their self-image upon him. **1976** F. A. HAYEK *Law, Legislation & Liberty* II. vii. 15, I understand that Professor Michael Oakeshott, in his oral teaching, has long used the terms *teleocratic* (and *teleocracy*) and *nomocratic*..to bring out the same distinction [*sc.* as between an organization and a spontaneous order].

teleological, *a.* Add: (Earlier and later examples.) *teleological ethics* (see quot. 1967).

1798 A. F. M. WILLICH *Elem. Critical Philos.* II. 112 (*heading*) Analysis of the teleological faculty of judging. **1930** C. D. BROAD *Five Types of Ethical Theory* vi. 206, I would first divide ethical theories into two classes, which I will call respectively *deontological* and *teleological*. **1966** F. COPLESTON *Hist. Philos.* VIII. ii. 34 Any form of teleological ethics which interprets the moral imperative as..an assertoric hypothetical imperative. **1967** *Encycl. Philos.* VIII. 88/1 Teleological ethics..is the subordination of the concept of duty, right conduct, or moral obligation to the concept of the good or the humanly desirable. **1973** S. F. COHEN *Bukharin* vii. 228 Rykov.. a perennial foe of grandiose economic projects and teleological planning.

teleonomy (teli,ǫ·nŏmi). *Biol.* [f. TELEO-[2] + -NOMY.] The property, common to all living systems, of being organized towards the attainment of ends (see quots.). Hence **teleono·mic** *a.*, of or pertaining to teleonomy.

1958 C. S. PITTENDRIGH in Roe & Simpson *Behavior & Evolution* xviii. 391 (*heading*) Adaptation: teleonomy versus teleology. *Ibid.* 394 It seems unfortunate that the term 'teleology' should be resurrected... The biologist's long-standing confusion would be more fully removed if all end-directed systems were described by some other term, like 'teleonomic', in order to emphasize that..end-directedness does not carry a commitment to Aristotelian teleology. **1961** *Cold Spring Harbor Symp. Quantitative Biol.* XXVI. 1 (*heading*) The teleonomic significance of biosynthetic control mechanisms. **1976** *Ann. Rev. Microbiol.* XXX. 538 The hoped-for aim of this review is to promote more studies on the diverse glycerol systems in a teleonomic context. **1977** P. B. & J. S. MEDAWAR *Life Sci.* i. 11 Biologists prefer to use the genteelism teleonomy with merely descriptive connotations to signify the goal-directed or 'as-if-purposive' character of biological performances. **1978** J. Z. YOUNG *Programs of Brain* iii. 16 The conception of 'teleology' has been associated with that of a final aim of life, implying metaphysical or religious beliefs. To avoid this, recent authors have used the word 'teleonomy' to describe the directional character of living activities... But words of this sort confuse many people and one is enough, let us keep to 'teleology'.

teleordering: see *TELE- 1.

teleost, *sb.* (In Dict. s.v. TELEOSTEAN *a.* and *sb.*) (Further examples.)

1895 B. DEAN *Fishes* vii. 139 'Teleost' must be used in a popular and convenient..sense..to denote..the modern 'bony fish'. **1955** AUDEN *Shield of Achilles* i. 15 Had He picked a teleost Or an arthropod to inspire, Would our death also have come? **1979** C. E. BOND *Biol. Fishes* i. 6 Arranging the teleosts into lower, middle, and higher levels of organization may be a gross oversimplification.

telepathetic, *a.* For *nonce-wd.* read *rare* and add later example.

1949 KOESTLER *Insight & Outlook* ix. 119 The functional interactions of hypnotic and telepathetic rapports.

telepathize, *v.* Add: Also (te·păþəiz). (Examples.) Also (*c*) *trans.*, to discern by means of telepathy. *rare.*

1895 in *Funk's Stand. Dict.* **1919** E. H. JONES *Road to En-dor* xi. 115 He had put me on parole..not to telepathize with the good folk of Yozgad. **1941** *Mind* L. 315 But this just is telepathy except that you have made the provision that it shall be done by having the same sensation as the person one telepathises. **1954** 'J. CHRISTOPHER' *22nd Cent.* 145 Did they telepathize our coming and retreat to some more civilized solar system? **1963** *Jrnl. Soc. for Psychical Res.* XLII. 33, I have examined a subject telepathized in this way and questioned him. **1980** C. FITZGIBBON *Rat Report* ii. 38 If the rat had telepathized once, then perhaps it would again.

telepheric, var. *TELEFERIC.

telephonable (telĭfōᵘ·năb'l), *a.* [f. TELEPHONE *v.* + -ABLE.] Of a place or person: able to be reached or contacted by telephone.

1908 G. B. SHAW *Lett. to Granville Barker* (1956) 126 She is ill and not telephonable. **1932** *New Statesman* 23 Jan. 88/1 The advertising campaign which is so largely increasing the number of telephonable people. **1973** S. ALSOP *Stay of Execution* (1974) I. 20 Katmandu was not easily telephonable.

telephone, *sb.* Add: **2. b.** *on the telephone* (*a*) (earlier example); (*b*) making a telephone call, ringing up; using or by means of the telephone.

(*a*) **1885** *List of Subscribers, Classified* (United Telephone Co.) (ed. 6) 18 Other people have their own tradesmen, who are doubtless also on the telephone.

(*b*) **1900** C. H. CHAMBERS *Tyranny of Tears* I. 36 (*The telephone bell rings.*).. There's some one on the telephone —forgive me. (*Goes to telephone.*) **1925** F. SCOTT FITZGERALD *Great Gatsby* ii. 35 Mrs. Wilson called up several people on the telephone. *Ibid.* vii. 138 That's Tom's girl on the telephone *Ibid.* ix. 214 You threw me over on the telephone. **1934** G. B. SHAW *Village Wooing* 135 Oh, speak English: you're not on the telephone now. **1963** K. AMIS *One Fat Englishman* xi. 126 Hearing her voice on the telephone in the next room brought an unwelcome reminder of the small hours. **1965** J. H. ROBERTS *'Q' Document* iv. 101 Those moments when he was forced to abandon conversation with Cooper to talk on the telephone.

d. *ellipt.* for *telephone call*, sense 3 below, esp. in *Indian English*. Cf. *PHONE *sb.*[2] 1.

1935 F. W. CROFT *Crime at Guildford* xxi. 298 A telephone to the manager of the hotel produced the needed information. **1979** P. NIHALANI et al. *Indian & Brit. English* I. 176 'Your telephones have not been very clear.' 'He gave me a telephone.'..In BS..the phrase 'telephone call', or simply 'call', would be used.

3. *telephone bell* (earlier example), *call*, *caller*, *cord*, *dial*, *extension*, *line*, *message* (example), *office*, *operator*, *receiver* (earlier example), *survey*, *table*, *transmitter*, *wire*; *telephone-answering* adj.; **telephone bill** = *phone bill* s.v. *PHONE *sb.*[2] 3; **telephone book** = *telephone directory* below; **telephone booth, box** = *phone booth, box* s.v. *PHONE *sb.*[2] 3; **telephone directory**, a book containing an alphabetical list of the names, addresses, and numbers of telephone subscribers; *spec.* (with def. article) such a list covering a particular locality and printed by a telephone company; **telephone girl** (earlier example); **telephone kiosk** = *phone booth, box* s.v. *PHONE *sb.*[2] 3; **telephone number** = *phone number* s.v. *PHONE *sb.*[2] 3; *colloq.*, a large number (esp. with reference to a sum of money or a prison sentence); **telephone pad**, a writing pad for noting telephone messages, etc.; **telephone set**, the assembly of components including a telephone transmitter, receiver, etc., which make up a telephone (sense 2 b); **telephone tapping** *vbl. sb.*, the act of making a connection to a telephone wire so as to listen in on private telephone conversations; cf. *TAPPING *vbl. sb.*[1] a; so **telephone tap**, an instance of telephone tapping; **telephone token**, a small counter designed to operate a public telephone and on sale in certain countries of Europe; cf. *JETON 2.

1963 *Times* 5 Feb. 11/4 This is the telephone-answering machine of the coke department... The machine will record your order or message... Please speak clearly... Please speak now. **1900** Telephone bell [see sense 2 b (*b*) above]. **1935** C. ISHERWOOD *Mr Norris changes Trains* viii. 117 In the present state of Arthur's finances, it was hardly to be expected that he would have settled his telephone bill. **1915** J. BUCHAN *39 Steps* viii. 201, I picked up the telephone book and looked up the number of his house. **1975** D. LODGE *Changing Places* iii. 132, I got out the telephone book and began ringing round. **1895** *Funk's Stand. Dict.*, Telephone-booth. *a* **1910** [see *CALL *v.* 35 h]. **1982** T. HOLME *Devil of Dolce Vita* xxii. 178 There is [in Venice] a plethora of squares, some scarcely bigger than a telephone booth. **1904** *McClure's Mag.* Feb. 405 Golden could snatch only two opportunities to step into the telephone box that morning. **1980** I. MURDOCH *Nuns & Soldiers* i. 44 I'm in a telephone box near Victoria Station. **1885** *List of Subscribers, Classified* (United Telephone Co.) (ed. 6) 9 Telephone Call Rooms have just been opened at Clapham and Kilburn. **1910** 'O. HENRY' *Strictly Business* 13 She pointed out to him clearly how it [*sc.* a play] could be improved by introducing a messenger instead of a telephone call. **1980** I. MURDOCH *Nuns & Soldiers* iv. 243 The Count was sitting ..in torment, waiting for her telephone call. **1948** 'J. TEY' *Franchise Affair* x. 103 Your telephone callers: were they male or female? **1934** WEBSTER, Telephone cord. **1966** 'A. HALL' *9th Directive* xv. 143, I..fiddled with the telephone cord. **1960** COOKE & MARKUS *Electronics & Nucleonics Dict.* 478/2 *Telephone dial*, a switch operated by a finger wheel, used to make and break a pair of contacts the required number of times for setting up a telephone circuit to the party being called. **1972** 'E. MCBAIN' *Sadie when she Died* viii. 81 There is still all day tomorrow to twirl those little holes in the telephone dial and ring up this or that hot number. **1907** *Yesterday's Shopping* (1969) 397/3 Private Telephone Directory..Spanish roan, lettered in gold, 24 leaves, to stand or hang. **1913** W. P. EATON *Barn Doors & Byways* 81 We fail to find this sort of thing any more thrilling or 'literary' than the telephone directory. **1969** B. WEIL *Dossier IX* iii. 20 The *Service* are always up to date with their telephone directories. **1945** C. MILBURN *Diary* 2 Jan. (1979) 261 The telephone handy men had arrived to instal a telephone extension up into Alan's room. **1977** 'J. LE CARRÉ' *Hon. Schoolboy* II. xiv. 322 A telephone extension hung on the wall. **1893** *Chicago Tribune* 2 July 13/3 The telephone girl sits on her high stool..as she produces alternate order and chaos at her switchboard. **1931** G. B. SHAW *Fabian Ess.* p. viii, A couple of pennies to drop into the slot in a telephone kiosk. **1974** M. BABSON *Stalking Lamb* II. xxiv. 178 George had entered the telephone kiosk and could be clearly seen inside the brightly lit box. **1882** T. D. LOCKWOOD *Pract. Information for Telephonists* 163 Now, to consider the possible disturbing influence that electric light wires may exercise upon telephone lines. **1962** A. NISBETT *Technique Sound Studio* 261 *Music line*, broad-band circuit for carrying programme (including speech), as distinct from a telephone line. **1982** A. BROOKNER *Providence* ix. 109 Supposing there is a telephone message waiting for me at the hotel? **1885** *List of Subscribers, Classified* (United Telephone Co.) (ed. 6) 3 Edwin Fox & Compy..Telephone No. 5,110. **1950** T. S. ELIOT *Cocktail Party* I. i. 17 You have the address, and the telephone number? **1963** L. DEIGHTON *Horse under Water* xliv. 171 It looked like he was going up the river for a telephone number. **1878** G. H. LEWES *Jrnl.* 21 Mar. in *Geo. Eliot. Lett.* (1956) VII. 16 We went to the Telephone office to have the Telephone explained and demonstrated. **1894** *Life* 19 Apr. 256/1 One of the young lady telephone operators might be listening to our talk and we don't want our telephone taken out. **1964** M. MCLUHAN *Understanding Media* xxiv. 243 Boors, who inundate defenseless telephone operators. **1923** *Sci. Amer.* Feb. 115/3 When this telephone pad is not in use it rests out of sight beneath..the telephone. **1967** A. WILSON *No Laughing Matter* III. 386 Jack, seeing the telephone pad, did not want to break the mood by asking Marcus if he had rung Gladys. **1884** *List of Subscribers* (London & Globe Telephone Co.) 3 Any form of telephone transmitter or receiver. **1911** *Encycl. Brit.* XXV. 552/1 Each telephone set was equipped with a special key. **1976** P. LOVESEY, *Swing, swing Together* xxx. 146 If anything develops here, you can use the telephone set to leave a message at the Yard. **1976** *National Observer* (U.S.) 13 Mar. 1/6 The Knight-Ridder newspapers asked a similar question in a telephone survey in January and found 81 per cent agreement. **1929** 'E. QUEEN' *Roman Hat Mystery* III. xvii. 251 They..shook out the pages of the telephone-book in the bedside telephone-table. **1977** M. RUSSELL *Dial Death* II. i. 39 The figure slumped across the telephone table. **1958** 'E. MCBAIN' *Killer's Payoff* (1960) xv. 153 There was no intention of maintaining a *telephone tap* in the strictest sense of the word. **1958** *Listener* 12 June 971/1 Some of the evidence had been obtained by telephone tapping. **1978** *Peace News* 25 Aug. 9/1 It is important to note that firstly, the Special Branch 'ambush' was only made possible by some combination of mail interception and telephone tapping. **1963** 'D. CORY' *Hammerhead* x. 127 He..asked for a glass of cognac and a telephone token. **1884** Telephone transmitter [see *telephone receiver* above]. **1937** *Discovery* Jan. 27/2 The use of this material for..telephone transmitter diaphragms, is suggested. **1881** 'MARK TWAIN' *Let.* 31 Jan. in C. Clemens *Mark Twain* (1932) 36 In one place the telephone wire running along six inches above the comb [of the roof] is covered. **1978** 'A. YORK' *Tallant for Disaster* xii. 172, I want a start made on getting the telephone wires back up again.

telephone, *v.* Add: **1. b.** (Earlier example.) Also *fig.*

1878 W. TEGG *Posts & Telegraphs* III. 305 It is said that the results of these experiments were 'telephoned' to the *Boston Guide*. **1908** KIPLING *Lett. to Family* vi. 47, I hear the hard trail telephone a far-off horse's feet. **1923** D. H. LAWRENCE *Birds, Beasts & Flowers* 44 Almond tree... What are you doing in the December rain?.. Do you telephone the roar of the water over the earth?

c. (Earlier example.)

1877 *Telegraphic Jrnl.* 1 Sept. 201/2, I [*sc.* Prof. Graham Bell] telephoned the leader of the band.

telephoner. (Examples.)

1918 A. BENNETT *Pretty Lady* xviii. 116 The telephone-bell rang... The telephoner was Gilbert. **1932** *New Statesman* 23 Jan. 87/1, I have often watched one of these fanatical telephoners sitting opposite the telephone with his hand on the receiver. **1972** *New Yorker* 21 Oct. 31/3 We picked up a mimeographed sheet that tells the telephoners what to say.

telephonically, *adv.* (Earlier example.)

1878 G. B. PRESCOTT *Speaking Telephone* vi. 226 Stations could exchange business telephonically.

telephonist. Add: The pronunc. (tele·fŏnist) is now standard. **a.** (Earlier example.)

1880 *Harper's Mag.* Oct. 723/1 Nor are the Chicago telephonists driven to such an access of rage.

c. = TELEPHONER. *rare.*

1956 [see *CRADLE *v.* 5 b].

telephonitis (te:lĭfōᵘnəi·tis). *joc.* [f. TELEPHONE *sb.* or *v.* + *-ITIS.] A compulsive desire to make telephone calls.

1935 *Even. Sun* (Baltimore) 24 Apr. 4/6 Garrison was suffering from 'telephonitis'. She [*sc.* his wife] has an injunction forbidding him from calling her on the telephone and..Garrison had violated the order because he 'couldn't help it'. **1962** *Punch* 3 Jan. 71 One of the tragedies of telephonitis is that sufferers are unable to help one another. **1979** *Washington Post* 14 Dec. B3/5 Ted [*sc.* Edward Kennedy] has telephonitis and he's on the phone every night.

telephoto[1]. Add: (Later examples.)

1913 KIPLING *Diversity of Creatures* (1917) 290 J'ever see a bird's eye telephoto-survey of England for military purposes? **1920** H. E. IVES *Airplane Photography* xxxi. 383 The telephoto lenses used for spotting would be of long equivalent focus..but of handy size. **1947** A. RANSOME *Great Northern?* i. 17 He would have..a camera with a telephoto lens to take photographs of birds without having to come near enough to disturb them. **1964** M. MCLUHAN *Understanding Media* (1967) II. xxi. 230 The press is now not only a telephoto mosaic of the human community hour by hour, but its technology is also a mosaic of all the technologies of the community. **1977** J. HEDGECOE *Photographer's Handbk.* 119 Some telephoto converters are designed to fit between lens and camera.

B. *sb.* **1.** A telephoto lens or camera.

1904 L. W. BROWNELL *Photogr. for Sportsman Naturalist* v. 72 It is always well, in work with the telephoto, to presuppose that your negative is under-exposed. **1931** O. G. PIKE *Nature Photogr.* iii. 16 A telephoto lens is..a most important part of the nature photographer's equipment... A modern telephoto is a very great advance on those used thirty years ago. **1975** BYFIELD & TEDESCHI *Solemn High Murder* (1976) v. 86, I used a time exposure..with a 135 mm telephoto. They were long exposures.

2. = TELEPHOTOGRAPH *sb.*[2] *rare.*

1974 J. IRVING *158-Pound Marriage* i. 11, I see the close-ups of the shelling of Reims. The telephoto is still unclear.

telephoto[2]. Also *U.S.* **Tele-.** [Abbrev. of TELEPHOTOGRAPH *sb.*[1] or one of its derivatives.] Name for a system of telephotographic transmission.

A proprietary term in the U.S.

1925 *Official Gaz.* (U.S. Patent Office) 3 Feb. 46/1 Telephoto... Photographs. *Ibid.* 7 Apr. 27/2 Telephoto... Machines and apparatus for distance transmission of photographs electrically. **1931** *Daily Express* 13 Oct. 1/6 Picture by telephoto. **1938** F. D. SHARPE *Sharpe of Flying Squad* xxvi. 264 Photographs of finger-prints are greatly reduced in size and are flashed across the sea by the telephoto process to European countries.

telephotogram: see *TELE- 1.

telephotography[2]. (In Dict. s.v. TELEPHOTOGRAPH *sb.*[2]) (Earlier example.)

1892 *Anthony's Photographic Bull.* XXIII. 168 A great deal of discussion is going on at present on the subject of telephotography.

telephotometer to **Teleplayer:** see *TELE- 1. **telepolitics:** see *TELE- 2 b.

teleportation (te:lĭpɔɹtē¹·ʃən). *Psychics* and *Sci. Fiction.* [f. TELE- + TRANS)PORTATION.] The conveyance of persons (esp. of oneself) or things by psychic power; also in futuristic

description, apparently instantaneous transportation of persons, etc., across space by advanced technological means. Cf. *telekinesis* s.v. TELE-, *PSYCHOKINESIS 1.

1931 C. FORT *Lo!* I. iv. 42 Sometimes, in what I call 'teleportations', there seems to be 'agency' and sometimes not... Some other time I may be able more clearly to think out an expression upon flows of pigeons to their homes, and flows of migratory birds, as teleportative, or quasi-teleportative. **1945** N. COLLINS *London belongs to Me* IV. liv. 421 The weekly copy of *The Spirit World* lay on the occasional table... They were wonderful letters— full of glimpses through the veil, and teleportations and proofs of survival. **1951** 'J. WYNDHAM' in *Science-Fantasy* Winter 8 Suppose the Russians..could project things or people here by teleportation. **1960** *Analog Science Fact & Fiction* Nov. 14/1 It took a latent ability to learn teleportation, and some people had it while others didn't. *Ibid.* 36/1 Going from one place to another is teleportation. **1977** 'L. EGAN' *Blind Search* i. 4 Telepathy, telekinesis, teleportation, apports, whatever the hell psychic forces.

Hence (as a back-formation) **te·leport** *v.* (*a*) *intr.*, to convey oneself by teleportation; (*b*) *trans.*, to convey by teleportation; also *absol.*; **te·leporting** *ppl. a.* and *vbl. sb.*; also **te·leport** *sb.*, one who practises teleportation; **telepo·rtage** *rare* = *TELEPORTATION; **telepo·rtative** *a. rare*, pertaining to teleportation.

1931 Teleportative [see *TELEPORTATION]. **1951** 'J. WYNDHAM' in *Science-Fantasy* Winter 5 If there could be teleportation, or teleportage, or whatever it is. *Ibid.* 6 This teleporting guy. **1953** 'T. STURGEON' *More than Human* III. 204 Bonnie and Beanie can't carry so much as a toothpick with them when they teleport, let alone clothes. **1954–5** *Planet Stories* Winter 26 It might teleport him, too, if he attracted its attention. **1955** *Astounding Sci. Fiction* Feb. 11 The Martian was back in the chair again. 'It's not teleportation. We don't teleport.' **1960** *Analog Science Fact & Fiction* Nov. 41/2 After all, he'd found telepaths in insane asylums, and teleports among the juvenile delinquents of New York. **1965** *New Statesman* 5 Nov. 705/2 So we must adjust to instant teleporting from Raspail Métro to the Donnybrook tram by way of Strangeways Gaol. **1967** E. B. NICKERSON *Kayaks to Arctic* ix. 79 It was a steep bushy bank but he [*sc.* a bear] made it as if teleported and did not rustle even a leaf. **1979** B. SHAW *Dagger of Mind* vii. 116 Albert can teleport people... Miss Connie..does it with objects. Psychokinesis.

teleprinter (te·lĭpri:ntəɹ). [f. TELE- + PRINTER.] A telegraph instrument for transmitting telegraph messages as they are typed on a keyboard and printing incoming ones.

1929 *Telegr. & Teleph. Jrnl.* Dec. 42/1 The first Teleprinters to be tried in this country were produced by the Morkrum Corporation under the proprietary name of 'The Teletype'. **1932** *Times Educ. Suppl.* 6 Aug. p. i/4 A service of teleprinters will be opened in London on August 15. **1933** *Post Office Guide* 138 This service is afforded by..the teleprinter, which is a special kind of telegraph instrument operated by a keyboard closely resembling that of an ordinary commercial typewriter, working over the telephone exchange system. **1939** *Daily Tel.* 18 Dec. 12/4 (Advt.), Applicants who are not trained teleprinter operators should possess a typing speed of at least 30 words per minute. **1942** *R.A.F. Jrnl.* 27 June 22 One tribe of Grand Gremlins lives behind typewriter and teleprinter keys. **1957** *Technology* Apr. 68/4 Pushing a button then causes the positions of film-holder and lens to be punched on teleprinter tape. **1959** *Times Rev. Industry* Sept. 34/1 The teleprinter is not a substitute for a telephone. **1964** M. MCLUHAN *Understanding Media* (1967) II. xxv. 263 The teleprinter and the wireless made it possible for orders from the highest levels to be given direct to the lowest levels. **1978** R. V. JONES *Most Secret War* xviii. 148 Grant's interpretation was circulated by teleprinter. *Ibid.* xviii. 150 The teleprinter room into which the messages came was immediately across the corridor from my own.

Hence (as a back-formation) **te·leprint** *v. trans.*, to send or print (a message, etc.) by teleprinter; **te·leprinted** *ppl. a.*

1971 H. WILSON *Labour Govt.* xxvi. 519 On Saturday, 30 March, I was told on the telephone that a very long telegram was coming in from Washington and would be teleprinted to me at Chequers. **1973** G. TALBOT *Ten Seconds from Now* iv. 44 Sheets of teleprinted news 'tape' from the clattering machines. **1980** *Daily Tel.* 4 Aug. 3/3 Some piece of international news,..important enough to rate a teleprinted 'flash'. **1981** *Sci. Digest* Aug. 50/2 Blood samples and X-rays are taken, and the lab data is teleprinted to the team.

teleprocessing: see *TELE- 1.

teleprompter (te·lĭprǫmᵖtəɹ). orig. *U.S.* [f. TELE- + PROMPTER.] An electronic device, placed out of range of the television or cinematographic camera, that slowly unrolls the speaker's script, in order to prompt or assist him.

Formerly a proprietary name in the U.S. The equivalent British proprietary name is *AUTOCUE.

1951 *Life* 12 Mar. 131 Set at the eye level of performers, the Teleprompter unrolls a script whose inch-high letters, printed by special typewriter, can be read 25 feet away. **1953** *Official Gaz.* (U.S. Patent Office) 10 Feb. 232/1 *Teleprompter.* For electrically-operated apparatus for the cuing of speakers and actors by means of scripts advanced in conformity with the action and/or dialogue... Claims use since Sept. 1, 1950. **1958** *Daily Mail* 13 Dec. 8/8 Last night he [*sc.* Lord Montgomery] seemed to be using a teleprompter, so rapidly did the words rattle out. **1961** S. PRICE *Just for Record* ii. 18 That's what the boys in the backroom had written, and it was staring at him..from the teleprompter. **1978** G. VIDAL *Kalki* i. 20 Although most of Arlene's professional career depended upon her ability to read Teleprompters and cue cards, she refused to wear glasses in public.

Hence (as a back-formation) **te·leprompt** *v. trans.*, to assist by means of a teleprompter. Also **te·leprompted** *ppl. a.*, assisted by or by means of a teleprompter.

1956 *Sun* (Baltimore) 29 Aug. (B ed.) 14/2 Two weeks of nominations,..commentaries, teleprompted oratory and gavel-thumping. **1958** *Spectator* 18 July 87/1 Does my conscience need tele-prompting? **1960** *Time* 16 Nov. (Extra ed.) 15/1 The 1960 campaign had been televised, teleguided, teleprompted and telephoned as no other had been before.

telepsychic to **teleradium:** see *TELE- 1.

telerecording (te·lĭrĭkǭɹdiŋ), *vbl. sb.* [f. *TELE- 2 + *RECORDING *vbl. sb.*] A recording of a television programme made while it is being transmitted. Also *occas.*, the action of making such a recording.

1953 *Radio Times* 22 May 44/2 A telerecording of the Abbey Ceremonies and a special Coronation edition of Television Newsreel. **1957** *Times* 16 Nov. 8/4 After the Queen's Christmas Day broadcast, which is to be televised this year for the first time, telerecordings will be flown to Canada and Australia. **1959** *P.O. Telecomm. Jrnl.* Winter 10/1 The signals are..used to operate a slow-speed film telerecording equipment. **1970** A. GLYN *Blood of a Britishman* xvi. 186 A telerecording of a Football Cup semi-final. **1975** *Gramophone* Jan. 1330/1 Next month the BBC will take its cameras to Covent Garden for a tele-recording.

Hence (as a back-formation) **te·lerecord** *v. trans.*, to record (a television programme) during transmission. Also **te·lerecorded** *ppl. a.*

1955 *Radio Times* 22 Apr. 15/2 Monday's telerecorded programme *The Secret Arts.* **1956** *Ann. Reg. 1955* 390 A ..television production of *Romeo and Juliet* was telerecorded and sent to the U.S.S.R. **1960** *News Chron.* 24 June 3/7 The programme was telerecorded this week. **1978** *Times* 4 Sept. 9/4 BBC 2's cameras will..telerecord the new production.

telergic, teleroentgenography: see *TELE- 1. **teleroman, telesale:** see *TELE- 2 b, 3.

telescope, *sb.* Add: **1. a.** Also, an instrument or apparatus that serves the same purpose at other wavelengths of the electromagnetic spectrum.

1948, etc. [see *radio telescope* s.v. *RADIO *sb.* 7]. **1970** [see *light bucket* s.v. *LIGHT *sb.* 16]. **1974** *Physics Bull.* May 208/4 A balloon-borne gamma ray telescope, sensitive to photons with energies greater than 50 MeV, has observed several regions of the sky. **1978** PASACHOFF & KUTNER *University Astron.* xi. 315 Telescopes in orbit that are sensitive to x-rays have detected a number of strong x-ray sources.

2. *telescope-making*; *telescope-bag* (N. Amer. examples), *-table* (earlier U.S. example); **telescope word** chiefly *U.S.*, a portmanteau word.

1885 J. SHORT *Diary* 10 Apr. in *Kingston* (Ontario) *Whig-Standard* (1973) 11 Apr. 29/3 Telescope bags packed, in case we have to start for Calgary in a hurry. **1949** W. FAULKNER *Knight's Gambit* (1951) 110 A tremendous old-fashioned telescope bag, strapped and bulging, sat on a chair. **1881** T. HARDY *Let.* 13 Dec. (1978) I. 97 The telescope-making in the catalogue is also useful. **1937** *Discovery* Nov. 360/1 Amateur telescope-making is a hobby that has found many more enthusiasts in the United States than in the British Isles. **1869** C. L. EASTLAKE *Hints on Househ. Taste* (ed. 2) iii. 67 What is commonly called a 'telescope' table, or one which can be pulled out to twice its usual length, and, by the addition of extra leaves in its middle, accommodates twice the usual number of diners. **1909** *Cent. Dict.* Suppl., Telescope-word. **1933** H. WENTWORTH *Blends in Eng.* 3 *Telescope word* has also been applied to one formed from the first syllables of words. **1977** *Lebende Sprachen* XXII. 9/1 A rather special form of collocation are the so-called blends or telescope words.

telescope, *v.* Add: **1. a.** *fig.* To combine, compress, or condense (a number of things) *into* a more compact or concise form; to combine or conflate (several things, or one thing *with* another); to shorten by compression.

1894, 1909 [in Dict.]. **1911** BEERBOHM *Zuleika Dobson* xvi. 243, I telescoped my toilet and came rushing round to you. **1953** *Essays in Criticism* III. 57 Shelley's mind.. has telescoped the shattered autumn landscape with a stormy Heaven and Ocean. **1958** *Listener* 2 Jan. 13/2 Our own effort at telescoping education is a biological retrogressive step. **1961** *Amer. Speech* XXXVI. 162 Generalizing over all such cases, the linguist can telescope them into one single, economical rule of agreement as a formal requirement for well-formed English sentences. **1965** *Listener* 20 May 756/1 The complex is worked out in the book with poetry and psychological insight... The adaptation had to telescope something of this. **1978** *Jrnl. R. Soc. Arts* Dec. 29/2 And so, telescoping time, I now leap from 1909..to 6th October 1927.

telescopic, *a.* (*sb.*) Add: **1.** (Further examples.) Also *telescopic(-sighted) rifle*, a rifle with a telescopic sight.

1880 W. JAMES in *Atlantic Monthly* Oct. 447/2 An Ajax gets no fame in the day of telescopic-sighted rifles. **1918** E. A. MACKINTOSH *War, the Liberator* 114 With a telescopic rifle he is looking for a Hun. **1936** *Discovery* Aug. 237 His new giant nine-lens aerial camera. On the right is seen the Telescopic viewfinder. **1947** *E. African Ann. 1946–7* 35/2, I use a 12-inch fixed separation telescopic lens for most bird photographs. **1958** *Observer* 10 Aug. 10/2 The use of the Hasselblad long-distance telescopic camera from a concealed position. **1963** 'E. MCBAIN' *Ten plus One* ii. 30 Forrest's murderer must have used a telescopic sight, the distance..being something over a hundred and fifty yards. **1983** J. SLIGO *Concert Masters* v. 109 As Barbarossa pocketed the passports the telescopic lens of a camera focused from among the sand-dunes. *Ibid.* vi. 155 With a telescopic rifle and night sights, he will be a perfect target.

4. (Later examples.)

1931 [see *shock strut* s.v. *SHOCK *sb.*[3] 7 a]. **1962** *Evening Standard* 2 Oct. 7/3 An umbrella, unless you have one of the portable telescopic kind, is best avoided if you have really a long walk ahead of you. **1973** 'A. HALL' *Tango Briefing* x. 124, I..pulled up the telescopic aerial.

telescoping, *vbl. sb.* (In Dict. s.v. TELESCOPE *v.*) (Earlier and later examples.)

1867 *Commercial & Financial Chron.* V. 6/2 There are two principal dangers which have to be guarded against— the 'telescoping' of cars into each other in case of collision [etc.]. **1937** 'M. INNES' *Hamlet, Revenge!* I. ii. 42 Clay's picture of Pepys as Hamlet was..something extraordinary... He had..been examined both in Shakespeare's *Hamlet* and Pepys's *Diaries*. But this sudden telescoping was beyond him. **1951** M. MCLUHAN *Mech. Bridge* (1967) 85/2 The very name 'Wurlitzer', with its telescoping of 'waltz' and 'whirl', conveys the idea of vertigo. **1958** *Listener* 20 Feb. 341 The way in which this telescoping of development and recapitulation is achieved represents the greatest single master-stroke in the work. **1979** *Internat. Jrnl. Sociol. of Law* Feb. 123 Then there is the problem of what has curiously become known as 'telescoping'—the uncertainty and inaccuracy of respondents in identifying precisely the date on which a particular incident took place—which will inflate or deflate the researcher's estimates.

telescreen: see *TELE- 2 b. **teleseismic** to **teleshopping:** see *TELE- 1.

telesis. (Earlier example.)

1896 L. F. WARD in *Amer. Jrnl. Sociol.* II. 248 The only serious lack..is a similar antithetical term to be set over against *genesis*, to denote the distinctively social process which results from the application of the indirect, intellectual or telic method [of human evolution]. In order to supply such a term I propose to revive the Greek form *telesis*.

telesoftware: see *TELE- 1.

telestic, *a.* For † *Obs.* read *rare* and add later example.

1981 *Times Lit. Suppl.* 23 Jan. 79/5 Plato's four types of mania (telestic, or ritual; mantic or divinatory; poetic; and erotic).

tele-talkies: see *TELE- 1.

Teletex (te·lĭteks). [prob. blend of *TELEX and TEXT *sb.*[1]] A proprietary name for a data processing and communication system using interconnected computer terminals.

1978 *Washington Post* 16 Dec. C11/2 It [*sc.* a low cost computer time-sharing network] is directly competitive with the British Viewdata and Teletex system. **1979** *Trade Marks Jrnl.* 16 May 820/1 *Teletex*... Apparatus for the input, output and recording of printed data; printed data storing apparatus and instruments, keyboards for data processing apparatus; [etc.]... Siemens Aktiengesellschaft.; Munich, Federal Republic of Germany; manufacturers and merchants. **1981** *Sci. Amer.* Oct. E5/1 (Advt.), Siemens, for example, is pioneering teletex which is a synthesis of the typewriter and teleprinter and is the first step towards full-scale electronic mail. **1982** *Nature* 27 May 257/1 The information technology community is calling the new service 'teletex'. It is, however, a far cry from teletext, the broadcast information service. Teletex is either a sophisticated form of telex or a standard that allows computer terminals to communicate with each other.

teletext (te·lĭtekst). [f. *TELE- 2 + TEXT *sb.*[1]] A system in which a user's television set is adapted so as to be able to show alphanumeric information selected from displays transmitted using the spare capacity of existing television channels. Cf. *ORACLE *sb.*[1] 9*; contrast *VIEWDATA.

1974 *Wireless World* Nov. 441/1 (*heading*) Teletext to go ahead. **1975** *Electronics & Power* 15 May 548/1 In March 1974, the British Radio Equipment Manufacturers' Association, BBC and IBA reached agreement on a unified standard system, known as TELETEXT, based largely on the CEEFAX system. **1975, 1976** [see *ORACLE *sb.*[1] 9*]. **1980** S. MONEY in K. G. Jackson *Bk. of Video* 89/2 By using the same page format and data coding, viewdata and teletext have been made compatible so that a large part of the decoder can be common for both systems.

teletherapy, telethermometer: see *TELE- 1.

telethon (te·lĭþǫn). orig. and chiefly *U.S.* [f. *TELE- 2 + *-THON.] An especially prolonged television programme used to raise money for a charity or cause; also in extended use, a lengthy television programme for some other purpose.

1949 *Examiner* (San Francisco) 10 Apr. 22/5 'Telethon' nets $702,000. **1952** *Sun* (Baltimore) 23 June (B ed.) 1/6 Bing Crosby and Bob Hope, in a 14½-hour coast-to-coast telethon today raised more than $1,000,000 in contributions and pledges for the United States Olympic fund. **1960** *Daily Tel.* 8 Nov. 1/2 As the climax to his campaign, Mr. Nixon, Republican, answered telephoned questions for four hours in a nation-wide 'telethon'. **1968** *Courier-Mail* (Brisbane) 16 Sept. 6/5 The telethon, now a widely-used fund raising mechanism. **1973** G. W. Hart *Right from Start* iv. 303 Metzenbaum proposed a series of statewide telethons which would provide an opportunity for the Senator to appear on television hook-ups in the key states in a format in which he excels, answering individual citizens' questions. **1982** *Listener* 8 July 3/2 Perhaps we have all been corrupted by the telethons of Vietnam television reporting.

teletransport(ation: see *TELE- 1.

Teletype (te·lītəip), *sb.* Also **teletype.** [f. TELE- + TYPE(WRITER.] **1. a.** A proprietary name for a make of teleprinter. Hence *loosely*, any teleprinter.

1904 *Sci. Amer.* 17 Sept. 193/3 At Brussels it is the telecryptograph of Engineer Malcotti, at Berlin the teletype and the Heljes apparatus. **1922** *Telegr. & Teleph. Jrnl.* VIII. 71/1 The latent innovation in the way of apparatus in the London C.T.O. [*sc.* Central Telegraph Office] is that of the Teletype. **1925** *Official Gaz.* (U.S. Patent Office) 26 May 815/1 Morkrum-Kleinschmidt Corporation, Chicago... *Teletype*... Printing-telegraph apparatus. **1933** *Jrnl. R. Aeronaut. Soc.* XXXVII. 12 America, where the teletype, or electrical distant-controlled typewriter, is such an important part of the weather reporting system. **1952** *Trade Marks Jrnl.* 17 Dec. 1167/2 *Teletype*... Printing telegraph apparatus. **1958** *Times Rev. Industry* July (Suppl.) p. ii/1 The bedrock of the work of most British information posts is the daily Press service sent to them by radio, Morse, Hellschreiber or teletype, from C.O.I.'s telecommunications room. **1968** J. Sangster *Touchfeather* xiv. 161 As security officer, he had no doubt put the details of my visit on the teletype to all branches of the..Corporation. **1976** *Physics Bull.* July 298/2 Teletypes are rapidly being superseded by cathode ray display terminals.

b. A message received and printed by a teleprinter.

1961 in Webster. **1966** 'D. Shannon' *With a Vengeance* xiii. 184 'I'd sent a teletype off.'..He reached into his breast pocket and brought out two folded pages of teletype. **1972** B. F. Conners *Don't embarrass Bureau* (1973) ii. 104 He had read his Bureau mail, received a few teletypes from the Director. **1978** S. Sheldon *Bloodline* iii. 56 The messenger..handed him an envelope. Inside was a teletype from Rhys Williams.

2. *attrib.*, as *Teletype circuit, key, line, machine, message, network, operator, system, terminal.*

1933 *Sun* (Baltimore) 21 July 22/5 At noon..police of all districts were notified on the department's teletype system to pick up Roft. **1934** W. Saroyan *Daring Young Man* 76, I used to sit at a table all day, working a teletype machine, sending and receiving telegrams. **1937** *Sci. Abstr.* B. XL. 302 (*heading*) Teletype network in civil aviation. **1941** C. G. Halpine *Pilot's Meteorol.* iii. 32 Radiosonde observations are made..and the reports transmitted over the teletype circuits. **1950** *Times* 2 Mar. 7/6 The Austrians cannot instal new telephone or even teletype lines to other countries without Allied Council approval. **1962** *New Scientist* 1 Feb. 260/1 A transistorized microwave system for transmitting verbal data and teletype messages. **1971** *Ibid.* 18 Mar. 617/1 Fingers that itch to perform on the teletype keys. **1973** *Physics Bull.* Oct. 632/1 Since January 1973 a 'conversational' reference retrieval system has been available.., providing access to 10 000 references and abstracts through any teletype terminal or visual display unit. **1978** *N.Y. Times* 30 Mar. B-20/1 (Advt.), Shipping co needs expd teletype operator.

teletype (te·lītəip), *v.* [f. prec. sb.] **a.** *intr.* To operate a teleprinter; also, to put in a request *for* by means of a teleprinter. **b.** *trans.* To send by means of a teleprinter.

1924 *Daily News* 19 Dec. 5/4 We must teletype as well as teletalk. **1934** Webster, Teletype *v.t.* & *i.* **1971** 'D. Shannon' *Ringer* viii. 130 Palliser had teletyped an inquiry up to Lompoc... That wasn't a very big town. **1977** D. Anthony *Stud Game* xxvi. 175 Seems he has a record in Texas. We've teletyped for a full report.

So **te·letyped** *ppl. a.*, **te·letyping** *vbl. sb.*

1904 *Knowledge* Feb. 18/2 The Berlin Teletyping Central Station. *Ibid.* 19/2 Teletyping service. **1967** *New Scientist* 28 Dec. 766/2 Professor L. Goldberg and his colleagues..control the course of their observations by daily teletyped instructions to NASA's Goddard Space Flight Center. **1971** *Nature* 25 June 482/3 A human telegraph operator reads the address on an incoming teletyped message, then retypes it in full to send it to its destination.

teletyper (te·lītəipəɹ). [f. as *TELETYPE *sb.* + -ER[1].] A teleprinter.

1904 *Electr. Mag.* I. 64/2 The operation..may..be acquired by anybody in the shortest possible time, the teletyper being nothing else than a teletypewriter. **1948** A. Baron *From City, from Plough* 81 Somewhere in an office..teletypers were clacking their frantic messages. **1965** *Punch* 20 Jan. 76/2 If you can make room between the telephones, dictating machines, office intercoms and teletypers. **1976** *National Observer* (U.S.) 10 Apr. 9/1 The equipment includes telephone-answering devices, burglar-alarm systems, automatic dialers, teletypers.

Teletypesetter (te:lītəi·psetəɹ). Also with small initial. [f. TELE- + *type-setter* s.v. TYPE *sb.*[1] 10.] The proprietary name of an apparatus for the automatic casting and setting of type in response to telegraphed signals recorded on perforated tape; also used loosely for any such apparatus. Hence **te:lety·pe-setting** *vbl. sb.*, the action of a Teletypesetter.

1928 *N.Y. Times* 7 Dec. 1/2 Frank E. Gannett, head of the Gannett newspapers..threw the switch which set in motion the new device known as the Teletypesetter. **1931** *Official Gaz.* (U.S. Patent Office) 17 Mar. 551/1 Teletypesetter. For electrically-controlled typesetting and typecasting apparatus. **1953** *Trade Marks Jrnl.* 22 Apr. 343/1 *Teletypesetter*... Apparatus using perforated tape for the control of type casting machines, and parts thereof included in Class 9. Teletypesetter Corporation.., Chicago 14, Illinois, United States of America; manufacturers. **1961** *Spectator* 14 Apr. 509 He referred to teletypesetting, the process by which the *Guardian* proposes to print in London as well as Manchester. **1965** *Economist* 22 May p. xii/3 The West Coast edition of the *Wall St. Journal*..is being printed with much use of modern long-distance tele-typesetting. **1967** [see *KEY-BOARD *v.*].

telety·pewriter. [f. TELE- + TYPEWRITER.] = *TELEPRINTER.

1904 *Knowledge* Feb. 18/2 The apparatus..is..a tele-typewriter, any letters, figures, or signs of punctuation being printed by pressing down a key corresponding with the signal in question. **1922** *Glasgow Herald* 4 Sept. 11 Another great stride in the advance of civilization is demonstrated by the tele-typewriter. **1954** *Electronic Engin.* XXVI. 260 The effect of one form of telegraph distortion on the teleprinter (British) and teletypewriter (American) will be considered. **1968** *Economist* 21 Dec. 32/2 With time-sharing, he can communicate with a computer by means of a special teletypewriter by his desk. **1979** J. E. Rowley *Mechanised In-House Information Syst.* i. 59 Other devices, such as the tele-typewriter, the visual display unit and the graphical display unit may function both in input and output.

teleutosorus (tĕliū̆tosō°·rŏs). *Bot.* [f. as TELEUTOSPORE + SORUS[1].] A pustule consisting of a group of teliospores (teleutospores) and their supporting hyphæ.

1905 [see *TELIUM]. **1922** H. Gwynne-Vaughan *Fungi* viii. 205 Sooner or later the mycelium of binucleate cells gives rise to teleutospores; these are characteristically grouped together in teleutosori. **1970** J. Webster *Introd. Fungi* ii. iv. 377 *Gymnosporangium* forms teleutosori on *Juniperus.*

televarsity, tele-vérité, televersity: see *TELE- 2 b.

televiewer (te·līviū:əɹ). [f. *TELE- 2 + VIEWER.] One who watches television.

1935 *Discovery* Oct. 285/2 An excellent answer to the questions of the..would-be 'televiewer'. **1937** *Daily Herald* 2 Feb. 3/4 Televiewers will not have to change over a switch on their sets each week to suit the alternate systems of transmission. **1950** *Sun* (Baltimore) 4 Jan. 18/2 Thirty-nine per cent of the entire television and radio broadcast audience during this period were Baltimore televiewers. **1957** E. Hyams *Into Dream* i. v. 40 There was *Dentix*: a half-tone of a face well known to televiewers leered out of the page. **1971** L. Koppett *N.Y. Times Guide Spectator Sports* 3 Many televiewers..had no prior intention to watch a football or basketball game. **1982** *Daily Tel.* 20 Sept. 15/7 Millions of televiewers around the world saw the moving spectacle of Prince Rainier's grief.

So **te·leview** *v. intr.*, to watch television (*rare*); **te·leviewing** *vbl. sb.* and *ppl. a.*

1935 *Times* 15 May 13/2 The German Post Office yesterday opened the first public televiewing post outside Berlin..at Potsdam. **1945** Cooke & Markus *Electronics Dict.* 382 *Teleview*, to watch a scene by means of a television system. **1956** *Encycl. Brit. Bk. of Year* 492/2 *Televiewing*, formerly used as a noun, now appearing as an adjective, as in the phrase 'televiewing families', meaning families habituated to watching television. **1959** *New Statesman* 4 July 10/1 People still like and are permitted to laugh, and occasionally break off from televiewing to have a go at the printed word. **1960** *Twentieth Cent.* Dec. 541 A considerable impression was made..upon the televiewing public. **1976** *Listener* 23 Sept. 366/2 My own loss of televiewing had no effect on the number of times I hit anybody. **1982** *Economist* 5 June 25 By overestimating the numbers who would come, instead of teleview, they left sellers of hot dogs and papal souvenirs bankrupt. **1982** *Nature* 9 Dec. 468/2 Channel 4..seems to have won the allegiance of a mere four per cent of the British televiewing audience.

televisable (te·lī-, telīvəi·zăb'l), *a.* [f. *TELEVISE *v.* + -ABLE.] Capable of being televised, suitable for presentation on television. Similarly **televisible** (telīvi·zīb'l) [after *visible*] *nonce-wd.*

1974 *Times* 18 May 14/7 Unfortunately..Mrs Fawcett [was] eminently less televisable than Mrs Pankhurst. **1975** A. Powell *Hearing Secret Harmonies* ii. 38, I resign St John Clarke to the makers of all things televisible. **1979** *Economist* 3 Nov. 48/1 'Land people' who..have been a deeper tragedy than the more televisable 'boat people'.

televise (te·līvəiz), *v.* [Back-formation from *TELEVISION on the model of verbs that end in -(*v*)*ise* and are related to nouns ending in -(*v*)*ision*, such as *revise.*] **1. a.** *trans.* To transmit (pictures, programmes, scenes, etc.) by television; formerly also, to transmit television pictures of (a person). Also *fig.*

1927 *Glasgow Herald* 14 Jan. 9/1 The distance over which pictures can be televised. **1928** *Television* Mar. 40/3 The subject who is being 'televised' had to face a powerful battery of blinding lights. **1931** *Daily Progress* (Charlottesville, Va.) 2 May 3/4 The first marriage ceremony to be 'televised', if that's the word—is a matter of history to-day. **1939** [see *INTERVIEWER]. **1950** *Sport* 7–11 Apr. 22/2 The Final will definitely not be televised. **1975** *Observer* (Colour Suppl.) 6 Apr. 64 (Advt.), It [*sc.* a series of books on history] is a 'programme' which combines lively and informative reading with vivid illustrations, helpful maps and guides, in such a way that it televises the past for you and your children to understand and enjoy as never before. **1979** S. Brett *Comedian Dies* ix. 95 This..Awards lunch... Big do, being televised. **1983** *Economist* 23 July 24/1 The BBC's Panorama programme on blacks and the police, televised on July 18th.

b. *intr.* for *pass.* To be (well, etc.) suited for television presentation.

1930 *Times* 1 Apr. 28/3 Some faces appear to 'televise' better than others. **1961** G. Millerson *Techn. Telev. Production* 129 Where an iconoscope camera-tube is used as the pick-up device, film shots of dark scenes may televise better in negative form.

2. *intr.* To make a television broadcast.

1948 L. Birch *Something Done* (Central Office of Information) 15 Many performers who are under contract to the big music-halls are not allowed to televise. **1957** [implied at *TELEVISING *ppl. a.*].

Hence **te·levised** *ppl. a.*; **te·levising** *vbl. sb.* and *ppl. a.*

1932 *Jrnl. Television Soc.* I. 107/1 The televising of 'The Man with the Flower in his Mouth', in July, 1930. **1934** *Jrnl. Inst. Electr. Engineers* LXXV. 86/2 The difference in detail between a good Baird televized picture and the cathode-ray picture. **1935** *Times* 1 Feb. 8/4 No doubt the televising of sporting and other public events will have a wide appeal. **1946** *Astounding Sci. Fiction* July 63/1 A man he had seen many times before in televised addresses. **1951** M. Ehrlich *Big Eye* i. 34 The blonde began to take off her robe in a kind of televised strip tease. **1957** D. J. Enright *Apoth. Shop* 221 True pleasure—our moralizing, politicizing and..televising generation has thrown *that* overboard. **1958** *Times Lit. Suppl.* 21 Nov. p. xxix/2 Miss Edwards opens her story with the televising of Punchbowl Farm and the Thornton family. **1978** S. Brill *Teamsters* iii. 80 Fumbling through televised testimony like this to protect Hoffa was worth it to Fitzsimmons.

television (te·līviʒən, telīvi·ʒən). [f. TELE- + VISION *sb.*] **1. a.** A system for reproducing an actual or recorded scene at a distance on a screen by radio transmission, usu. with appropriate sounds; the vision of distant objects obtained thus.

The term normally refers to a system of general transmission over the air, but it also includes systems of restricted transmission to subscribers by wire, such as *cable television*; see also *closed circuit* s.v. *CLOSED *ppl. a.* 3.

[**1904** *Daily News* 3 June 7 Dr. Low talks very modestly of the 'televista' (the name he has given to his 'seeing by wire' invention).] **1907** *Sci. Amer. Suppl.* 15 June 26292/1 Now that the photo-telegraph invented by Prof. Korn is on the eve of being introduced into general practice, we are informed of some similar inventions in the same field, all of which tend to achieve some step toward the solution of the problem of television. **1909** [in Dict. s.v. TELE-]. **1913** *Wireless World* Sept. 353/2 The tele-vision,..being based upon the same principle as photo-telegraphy, is possible in itself. **1926** *Glasgow Herald* 20 Dec. 11/8 Mr. John L. Baird, a native of Helensburgh,..recently invented an apparatus which makes television possible. **1930** J. Buckingham *Matter & Radiation* 122 We have heard so much about Television lately that we are apt to forget that no portion of the apparatus used is novel to scientists. **1942** T. S. Eliot *Music of Poetry* 18 There are words which are ugly because of foreignness or ill-breeding (e.g. *television*): but I do not believe that any word well-established in its own language is either beautiful or ugly. **1948** N. Wiener *Cybernetics* 10 Television was destined to be more useful to engineering by the introduction of such new techniques than as an independent industry. **1957** *Technology* Mar. 9/2 The solution of the major problems in colour television, the public introduction of which is now more a question of economics than of technical difficulty. **1970** *Toronto Daily Star* 24 Sept. 28/1 (Advt.), You can receive Channel 19 by..cable television. **1972** *Times* 21 Jan. 2/5 Cable television was originally introduced in the area in 1962 to provide better reception because Shooters Hill in the south consistently interfered with Television pictures.

b. Organized television broadcasting; the television broadcasting service as a whole or (with defining word) a particular television service. Phr. *on* (*the*) *television.*

1927 [see *HEEBIE-JEEBIE(S]. **1930** N. Coward *Private Lives* ii. 49 Aeroplanes..and Cosmic Atoms, and Television. **1938** *Observer* 26 June 12/6, I reviewed this film three weeks ago when I saw it on television. **1951** *N.Y. Herald-Tribune* 12 Dec. 27/3 Buster is an old playmate and I'm glad to see he..crashed television successfully. **1958,** etc. [see *INDEPENDENT *a.* 5 e]. **1962** *Friend* 1 June 665/1, I have sometimes thought how different life might have been at Haworth if only they had had the television, and Tide, and a Morris Minor. **1965** M. Drabble *Millstone* 194 You could get a job on the television. **1968** *Globe & Mail* (Toronto) 17 Feb. 36 (Advt.), Private balconies,

cable television, off street parking. **1976** *Jrnl. R. Soc. Arts* June 365/2 Are there any recent figures of the cost to every household in the country of the advertising on independent television? **1980** *Private Eye* 26 Sept. 13/1 That ghastly woman with the teeth who's always on the television. **1982** *Listener* 16 Dec. 17/1, I have mixed feelings about cable television's 'autumn of debate'.

c. Television entertainment; television broadcasting considered as a medium of communication or as an art form. Cf. *GOOD *a.* 1 f.

1931 *Daily Progress* (Charlottesville, Va.) 2 May 3/4 The ceremony is television, but in every other way the wedding is absolutely regular. **1957** *Observer* 27 Oct. 17/4 It proved, as discussions on these emotive imponderabilia always do, highly absorbing television. **1977** *Times* 2 Sept. 7/3 Television, the art..that speaks daily to almost everybody. **1982** *Sunday Tel.* 3 Jan. 16/7 Attenborough's 'Life on Earth' was perfect television.

2. A television set.

1955 *Observer* 28 Aug. 7/4 The South London landlady was seeking 'a nice new television'. **1972** *Daily Express* 8 Jan. 12/4 We have never been able to afford a car but we do have a television. **1973** D. Francis *Slay-Ride* viii. 96 Behind me on a wide shelf stood my portable television. **1982** *Sunday Sun-Times* (Chicago) 3 Oct. 72/1 Industry workers last year averaged $75 a month. They buy televisions and send money home to wives.

3. *attrib.* and *Comb.* **a.** In general uses, as *television aerial, antenna, apparatus, box, channel, coverage, dealer, frequency, lounge, receiver, room, screen, service, set, signal, studio, supper, system, theatre, transmission, transmitter, van.*

1940 *Amateur Radio Handbk.* (ed. 2) 306/1 (Index), Television aerials. **1972** J. Porter *Meddler & her Murderer* xi. 136 Rows of ugly little houses, their roofs buckling under a forest of television aerials. **1947** *Electronics* May 96/2 (*heading*) Television antennas for apartments. **1951** W. Faulkner *Requiem for Nun* iii. 246 Lonely farmhouses glittering and gleaming with automatic stoves and washing machines and television antennae. **1930** *Aberdeen Press & Jrnl.* 14 Apr. 6, I do not suppose that many Scottish listeners have yet adopted television apparatus. **1932** A. Huxley *Brave New World* xiv. 234 At the foot of every bed..was a television box. **1950** K. Henney *Radio Engin. Handbk.* (ed. 4) xix. 1024 Allocation of television channels. Figure 40 shows the allocation of twelve 6-Mc channels for television. **1981** *Ann. Reg. 1980* 427 Plans for the setting up of the fourth television channel..went ahead. **1966** *B.B.C. Handbk.* 51 Gaps of the existing BBC radio and television coverage. **1949** *Radio Times* 15 July 42/1 Your nearest Ultra Television dealer. **1955** *Ibid.* 22 Apr. 3/1 The present television frequencies..are in the V.H.F. band. **1970** K. Giles *Death in Church* v. 150 The ladies have arrived... In the television lounge. **1976** W. J. Burley *Wycliffe & Schoolgirls* ii. 42 The door of the television lounge was open and he could see several patients sitting round the set. **1927** Television receiver [see *RECEIVER[1] 7 c]. **1980** *Whitaker's Almanack 1981* 816/1 About 3·8 million television receivers are in use [in the Argentine Republic]. **1959** 'O. Mills' *Stairway to Murder* viii. 95 The Residents' Lounge and Television Room are both at your disposal. **1971** *Country Life* 23 Dec. (Suppl.) 3/2 (Advt.), Mansion. Ideally suitable for institutional purposes...2 television rooms, chapel. **1927** *Pictorial Weekly* 5 Mar. 100/1 These sets will combine a Television screen and loud-speaker. **1973** D. May *Laughter in Djakarta* i. 13 Little figures mouthing words that did not reach him, like a television screen with no sound. **1935** *Times* 1 Feb. 7/4 These first steps being taken towards the establishment of a public television service. **1936** *Radio Times* 30 Oct. 5/3 Television programmes. The BBC Television Service from Alexandra Palace will be opened by the Postmaster-General on Monday. **1966** *B.B.C. Handbk.* 15 *The Northern Echo* [said]..'some parts of the television service are falling asleep'. **1931** *N.Y. Times* 31 May ix. 9/2 The Radio Corporation of America..is concentrating its efforts upon the primary technical developments to be completed before undertaking the manufacture and sale of television sets on a commercial basis. **1976** W. Trevor *Children of Dynmouth* i. 25 Slowly he walked through Dynmouth again, examining the goods in the shop windows, watching golf being played on various television sets. **1927** *Bell System Technical Jrnl.* VI. 560 (*heading*) The production and utilization of television signals. **1935** *Illustr. London News* 23 Feb. 307 (*caption*) The Baird television studios at the Crystal Palace. **1981** S. Brett *Situation Tragedy* iv. 43 He wished he knew a bit more about television studios and their sound systems. **1973** D. Miller *Chinese Jade Affair* xi. 108 This was where the old lady had her television suppers. **1983** *Times* 1 Oct. 8/6 The art of conversation, of manners, of social interplay..cannot be acquired at the nursery table or when eating a solitary television supper with the baby-minder. **1931** *Proc. IRE* XXI. 1655 The experimental television system placed in operation by RCA Victor in..1931..was based on the use of a cathode ray tube as the image reproducing element. **1966** *B.B.C. Handbk.* 35 BBC Television..along with other Western television systems. *Ibid.* 33 There are in the London area six..production studios.., a television theatre, two news studios, and two remote control studios. **1929** *Radio Times* 8 Nov. 412/1, 11.00–11.30 (*London only.*) Experimental television transmission by the Baird Process. **1928** *N.Y. Times* 22 Aug. 1/2 Puppets being used because of the limitations of the television transmitter. **1939** *Electronics* Mar. 26 (*heading*) Television transmitters. **1956** R. Robinson *Landscape with Dead Dons* xiii. 117 'You refer to the television van?' 'The one you told me the little men come in.'

b. Connected with, participating in, or transmitted as part of organized television broadcasting, as *television announcer, audience, broadcast, broadcasting, commercial, crew, critic, discussion, drama, dramatist, film, interview, journalist, magazine, news, personality, play, producer, programme, public, pundit, reporter, serial, series, show, spot, star, version, viewer.*

1938 *Radio Times* 23 Dec. 36/1 It would be nice to say that the television announcers will hang up their stockings. **1972** J. Mosedale *Football* xi. 148 Gifford, then a television announcer, talked briefly with the coach. **1937** *Discovery* Nov. 331/2 Building up a television audience. **1959** *Twentieth Cent.* Nov. 335 Because of its great size, the television audience now closely resembles the population as a whole. **1928** *N.Y. Times* 21 Aug. 26 Hourly television broadcasts over WRNY to aid amateurs and experimenters will begin tomorrow. **1935** *Times* 1 Feb. 8/3 There will be little, if any, scope for television broadcasts unaccompanied by sound. **1928** *Daily Mail* 3 Aug. 9/5 Mr. J. L. Baird, the inventor of television, stated yesterday that television broadcasting would begin in this country in the autumn. **1977** *Rep. Comm. Future of Broadcasting* (Cmnd. 6753) ii. 10 All sound and television broadcasting which uses radio waves for transmission is in the charge of two public Authorities, the British Broadcasting Corporation and the Independent Broadcasting Authority. **1957** Television commercial [see *COMMERCIAL *sb.* 2]. **1975** R. Rendell *Shake Hands for Ever* x. 95 Those children's toys which he had often seen on television commercials. **1964** J. Mitford in *Making of Muckraker* (1979) 90 Our house..was transformed..by television crews filming interviews about the book. **1978** W. F. Buckley *Stained Glass* xviii. 172 All Monday the television crews were at work. **1966** *B.B.C. Handbk.* 13 Television critic Peter Black..asked for a definition of this phrase. **1981** *Listener* 22 Oct. 465/3 He was the best partner I ever had in television discussions. **1949** *Radio Times* 15 July 41/2 A variegated week for television drama. **1973** *Listener* 5 July 27/1 Watergate makes television drama, which rests on an illusion of reality, look pretty thin. **1964** *Ibid.* 16 Apr. 624/2 'Television dramatists'..have the cheek to use television techniques in stage plays. **1951** R. Chandler *Let.* 5 Jan. (1981) 256, I have seen a number of the television films of your stories. **1967** M. Argyle *Psychol. Interpersonal Behaviour* x. 189 A television film is played back to the trainee after his performance. **1964** D. Francis *Nerve* xiv. 192 That was just what I needed... A big race win and a television interview. **1974** *Listener* 29 Oct. 525/1 A bad, sad month for television journalists. **1955** *Radio Times* 22 Apr. 15/2 The fortnightly television magazine *Ulster Mirror*, broadcast since November. **1947** *Billboard* 1 Nov. 16 Television news borrows from the radio, it is related to the newsreel. **1977** D. L. Altheide in Douglas & Johnson *Existential Sociol.* iv. 147 Research on television news. **1958** *Times Lit. Suppl.* 4 Apr. 179/1 The man of letters is no longer a household figure—unless, by coincidence, he is also a television personality. **1978** *Times* 15 Aug. 13/7 Who does qualify for VIP lounges; presumably some television 'personalities' and some entertainers? **1946** *B.B.C. Year Book* 21 The old hand may in time come to ear-mark his evenings primarily for full-length television plays. **1968** Television play [see *PREVIEW *v.* 2 a]. **1951** *Catal. Exhibits, South Bank Exhib., Festival of Britain* 176 *Television Producer* Malcolm Baker Smith. **1982** *Sunday Tel.* 3 Jan. 16/4 Television and film producers..go for their inspiration to the printed word. **1930** *Billboard* 20 Sept. 30/1 Television programs were being broadcast daily from two studios. **1935** *Radio Times* 27 Dec. 3/1 Television programmes from the new station at the Alexandra Palace start next year. **1981** *Ann. Reg. 1980* 428 By far the greatest impact made by a single television programme in 1980 resulted from the screening..of ATV's *Death of a Princess*. **1937** *Discovery* Nov. 332/2 A television public has not been developed at all. **1981** *Listener* 22 Oct. 465/3 Robert McKenzie..was the greatest of all television pundits on politics and elections. **1959** *Housewife* June 32 As a television reporter I've certainly got used to meeting a lot of unhappy people. **1957** M. Summerton *Sunset Hour* x. 133 He hoped to clinch a contract for a part in a television serial. **1965** B. Glanville *Second Home* xii. 301 He had to go up to Birmingham next day for some television series he was directing. **1982** *Sunday Tel.* 3 Jan. 16/7 The most successful television series..are not from books of the very first rank. **1950** R. Chandler *Let.* 22 Nov. (1981) 241 I'd like to have a television show. **1976** W. Trevor *Children of Dynmouth* i. 25 He'd asked Timothy what he found interesting outside the Comprehensive and Timothy had said television shows. **1960** Television spot [see *SPOT *sb.*[1] 8 d]. **1951** A. C. Clarke *Sands of Mars* xiii. 163 A..caricature of a well-known television star. **1982** *Sunday Tel.* 3 Jan. 16/5 The experience of watching the television version [of *Brideshead Revisited*] differed very little from that of reading Evelyn Waugh's masterpiece. **1952** *Sun* (Baltimore) 17 Mar. 1/6 (*heading*) Television viewers are able to see operation in hospital. **1961** *Times* 11 Dec. 13/6 Here is a population consisting almost wholly of industrial workers..car owners, television-viewers.

c. Special Combs. **television camera:** see *CAMERA 3 c; **television camera tube,** an electron tube of the kind used in television cameras for converting a visual image into an electrical signal; **television engineer,** one who designs and maintains the mechanical and electrical processes involved in the transmission and reception of television signals; a television repairman; **television image** = *television picture* below; **television licence,** a licence to use a television set, renewable annually on payment of a fee; **television mast,** (*a*) a tall mast, usu. set up on high ground, carrying a television transmitting aerial; (*b*) = *television aerial*, sense 3 a above; **television network,** a system of television broadcasting stations; a television broadcasting organization or channel; **television picture,** the visual image received on a television screen; **television region,** a region of the country receiving television broadcasts from a local as well as a national transmitting station; **television satellite,** a satellite put into orbit round the earth to reflect back television signals; **television station,** a television broadcasting station (see *STATION *sb.* 13 f); **television tube,** (*a*) = *picture tube* s.v. *PICTURE *sb.* 6 d; (*b*) = *television camera tube* above.

1940 D. G. Fink *Princ. Television Engin.* i. 17 (*caption*) A typical television camera tube, the type 1849 iconoscope, now widely used in television broadcasting. **1974** *Encycl. Brit. Macropædia* XVIII. 112/1 The image orthicon is the most highly developed of the television camera tubes. **1930** *Billboard* 20 Sept. 15/2 Equipment to be used will come from General Electric, under the supervision of Dr. E. F. W. Alexanderson, television engineer. **1978** F. King *Action* x. 31 The failure of the television engineer to call to repair the set. **1933** *Proc. IRE* XXI. 1631 (*heading*) A study of television image characteristics. **1949** *Times* 17 Feb. 5/3 The first hundred thousand mark is about to be reached in..television licences..compared with the eleven million for sound. **1972** C. Drummond *Death at Bar* ii. 52 Jarvis..abandoned the cinema the instant he had paid his first television licence fee. **1958** S. Hyland *Who goes Hang?* i. vii. 39 They could see the enormous, meccano structure of the television mast on top of Sydenham Hill. **1968** M. Allingham *Cargo of Eagles* iv. 54 They drove..through an area of open planned villas, writhing television masts, mini cars. **1947** *Billboard* 1 Nov. 16/2 Now that the Columbia Broadcasting System (CBS) has joined the routes of companies working toward a television network [etc.]. **1974** *B.B.C. Handbk. 1975* 19/1 One..fairly brief consequence of the energy crisis was the decision to close down all television networks at 10.30 pm. **1937** *Chron. & Echo* (Northampton) 8 May 6/1 (Advt.), Real television... Demonstration in advance of how Television pictures should appear when broadcast. **1977** J. Fraser *Heart's Ease in Death* i. 7 The wind shaking the aerial had distorted the television picture. **1974** Television region [see *REGION 5 d]. **1960** *Aeroplane* XCVIII. 419/2 Nowhere will the successful launching of the U.S. television satellite be noted with more interest than in this country. **1976** I. Levin *Boys from Brazil* vi. 180 Speaking to the whole world at once..by television satellite. **1931** *Billboard* 1 Aug. 4/1 As soon as television stations increase in number so that they can't be fitted into low channel, high band will be used for them exclusively. **1980** *Whitaker's Almanack 1981* 816/1 In addition there are 65 television stations, of which 4 are in Buenos Aires. **1943** *Electronic Engin.* XV. 329/2 The modern television tube has many ancestors. We could start its history in 1897. **1975** D. G. Fink *Electronics Engineers' Handbk.* xi. 57 Antimony trisulfide vidicons, lead oxide vidicons, and image orthicons are the work-horse television tubes.

televisionary (te:līvi·ʒənări), *sb.* and *a.* *Humorous.* [Blend of *TELEVISION + VISIONARY *a.* and *sb.*; in adj. use treated as compound with -ARY[1].] **A.** *sb.* **a.** An enthusiast for television. **b.** A television personality.

1928 *Observer* 12 Feb. 11/2 Many 'televisionaries' have spent fortunes in the quest. **1961** A. Clarke *Later Poems* 94 The Pope forbade the clergy..to indulge in daily amusement. He warned them, too, of the danger of becoming televisionaries. **1962** *Listener* 24 May 924/2 With Dylan Thomas and Gilbert Harding gone, Gwyn Thomas has a great future as a televisionary. **1981** Q. Crisp *How to become a Virgin* 156 Impresarios are frequently asked by televisionaries when some scheme or other first occurred to them.

B. *adj.* Of, possessing, or induced by television.

1934 in Webster s.v. *Television.* **1937** E. Blunden *Elegy* 50 The televisionary world to come. **1958** *Times* 5 July 7/2 If, in a televisionary trance, we are induced to buy some commodity for which our waking self has no appetite [etc.]. **1966** *New Statesman* 16 Dec. 917/3 The longish orchestral interludes, which are enough to drive the average televisionary producer to despair. **1980** *Daily Tel.* 5 Sept. 12/1 Televisionary indoctrination in Luanda.

televisioner (telīvi·ʒənəɹ). *rare.* [f. *TELEVISION + -ER[1].] One who watches television.

1928 *Television* Mar. 12/1 (*caption*) Some new uses for television. For the toilet—enabling the televisioner to see the back of his head when brushing his hair.

televisionless (te·lī-, telīvi·ʒənlès), *a.* [f. *TELEVISION + -LESS.] Without a television set; that does not include watching television.

1962 *Guardian* 8 Jan. 2/3 Parents..should provide a warm and undisturbed televisionless place for homework. **1981** *Times Lit. Suppl.* 8 May 516/3 Our customary, television-less routine.

te·levision-wise, *adv.* [-WISE.] In the manner of television; with regard to television.

1962 *Listener* 19 Apr. 702/1 Television-wise, his performance was more convincing. **1968** [see *SCAN *v.* 6 c].

televisor (te·līvəizǫɹ). Also **Televisor.** [f. as *TELEVISE *v.*: see -OR.] **1.** An apparatus for transmitting or receiving television pictures; orig. the name of that designed and patented by John Logie Baird (1888–1946). Now only *Hist.*

1926 *Glasgow Herald* 11 Jan. 7 The Televisor. *Ibid.* 8 Oct. 9/1 The scene in front of the transmitting televisor is turned into electrical impulses. **1927** *Ibid.* 2 Feb. 9 The

cost of televisor equipment for practical use. **1930** Moseley & Chapple *Television* viii. 95 The audience can see on the screen of their 'televisors', and hear the person who is broadcasting from the studio. **1935** H. G. Wells *Things to Come* xii. 98 There is a large televisor disk and telephone and other apparatus on the desk before Cabal. *Ibid.* xiii. 119 A man stands up and switches on a televisor and everybody listens. **1946** *Astounding Science Fiction* July 61/2 The televisor muttered at his elbow and he reached out to flip the toggle. **1984** *Financial Times* 13 June 17/6 A rare John Logie Baird televisor of around 1930 sold for £1,760.

2. A television broadcaster. *rare.*

1942 Berrey & Van den Bark *Amer. Thes. Slang* § 618/3 *Televisor*, the television transmitting apparatus; also a television artist or technician. **1966** *New Statesman* 30 Sept. 497/4 (Advt.), Heather Jenner, televisor, broadcaster and writer.

televisual (te:līvi·ʒiuăl, -vi·ziuăl), *a.* [f. *TELEVISION, after VISUAL *a.*] Of, pertaining to, characteristic of, or appearing on television; suitable for or effective in the medium of television.

1934 in Webster. **1956** *Observer* 15 Jan. 8/5 All day he lies in bed, while televisual phantoms flit across his ruffled cerebral screen. **1959** *Listener* 12 Feb. 303/1 Here was something that the theatre could not do, something essentially televisual. **1960** K. Amis *New Maps of Hell* iii. 82 Televisual views of actual historical scenes. **1973** *Church Times* 16 Nov. 9 Filmic or televisual violence breeds or releases violence in the viewer. **1980** *Times Lit. Suppl.* 25 July 842/2 It is because he keeps in practice with televisual mannerisms that he is so successful a natural in a television-dominated world.

Hence **televi·sually** *adv.*, from the point of view of or as regards television, on or for television.

1957 *Observer* 29 Sept. 13/2 Televisually..there was a surprising amount to be said in favour of this experiment so rich in every kind of producer's death-trap. **1967** *Listener* 9 Feb. 207/2 We had on this uncomfortable occasion not Shakespeare re-worked televisually..but a kind of compromise with what had already been worked in an alien medium. **1979** *Ibid.* 3 May 602/2 A generation ago, we still lived in an age of innocence, televisually, and politicians were apt to come on the box as themselves. **1981** *Times* 27 June 6/2 Who better than Russell, with his televisually perfect mane of white hair?

telex, Telex (te·leks). [f. *TELE(PRINTER + *EX(CHANGE *sb.* 10 c.] **1.** A system of telegraphy in which printed messages are transmitted and received by teleprinters using the public telecommunication lines; the apparatus used in this process. Freq. *attrib.*, esp. in *telex service.*

Despite the frequent use of a capital initial, *telex* in the ordinary uses defined here is not a proprietary term. The names of some specific products (radio equipment, hearing aids, etc.) of the Telex Corporation are registered as trade marks, however.

1932 *Telegr. & Teleph. Jrnl.* Oct. 2/1 In August 1932, came the opening in London of 'Telex' service, otherwise 'teleprinter exchange service'. **1939** *Electrical Communication* Jan. 222/2 Strong grounds exist for the belief that..a spontaneous and universal growth of Telex will follow. **1954** *Communications & Electronics* Nov. 40/1 Telex is an internationally agreed name for a subscriber-to-subscriber public teleprinter exchange service; it is for the printed word what the telephone exchange system is for the spoken word. **1958** *Times* 2 Sept. 5/2 Telex subscribers are linked to a system by which they can exchange typed messages by teleprinter. **1964** C. Dent *Quantity Surveying by Computer* vii. 101 Equipment can be obtained enabling these tapes to be input over the Telex system directly to the computer. **1968** R. V. Beste *Repeat Instructions* xxii. 233 He..was told to ring back at midnight when they should have the answer on the Telex. **1970** *New Scientist* 16 July (Suppl.) 2/1 It is estimated that by 1980 the total of telex and similar machine to machine (including computer) messages in the United States will be just over one billion. **1974** C. Hampton *Savages* (1976) xiii. 64 Installing a telex in the interests of business efficiency. **1981** *Sci. Amer.* Oct. E5/1 (Advt.), Transmission is at 2,400 bit/s and a page of 1500 characters can be sent in ten seconds—20–30 times faster than telex—and received on telexes, electronic typewriters, word processors or other compatible terminals.

2. A message so transmitted.

1978 *Globe & Mail* (Toronto) 12 Oct. 4/4 Here was the only apparent flaw in the Telexes. Mr. Lalonde wasn't elected to the House of Commons and appointed to Cabinet until the following year. **1980** *Daily Tel.* 9 July 16 The first I heard of it was when I got a telex of congratulation.

Hence **te·lex** *v.* (*a*) *trans.*, (i) to contact by telex; (ii) to send (a message) by telex; (*b*) *intr.*, to send a telex message; **te·lexed** *ppl. a.*

1960 *Guardian* 13 Dec. 8/6 In a telexed message to 18 cities and towns Mr. Bevins declared that direct dialling..was planned to start in six months. **1968** *Aramco World Mag.* May–June 25/1 He..waited while the story was telexed—at 60 words a minute—to a communications center in London. **1968** C. Burke *Elephant across Border* ii. 74, I must telex Allard by eleven. **1972** *Daily Tel.* 8 Mar. 17 (Advt.), Write, telephone or telex for information now. **1979** *Courier-Mail* (Brisbane) 4 Oct. 3/6 The managing director of Hudson Conway Holdings Ltd...said yesterday he had telexed the Queensland Opposition offering to open his company's files to them. **1983** *Listener* 28 Apr. 24/2 One didn't know one's every word was being..telexed direct to the Shah for his comment.

telharmonium (te:lhāɹmō^u·niǒm). [f. TEL(E- + HARMONIUM.] An electrophonic instrument, invented by the American scientist Thaddeus Cahill (1867–1934) and designed to produce tones for transmission over telephone wires by means of rotating electro-magnetic generators.

1906 *Chambers's Jrnl.* June 495/1 In the telharmonium the qualifying ripples are ingeniously added to the waves of the fundamental note by a separate contrivance. **1934** *N.Y. Times* 13 Apr. 19/1 Thaddeus Cahill, who invented an electric typewriter and the device for producing music electrically, known as the telharmonium, died suddenly at 10 o'clock yesterday of a heart attack. **1978** P. Griffiths *Conc. Hist. Mod. Music* viii. 111 The pioneer..was Thaddeus Cahill, a Canadian scientist who demonstrated an extraordinarily bulky 'telharmonium' in 1906.

teliospore (tī·liospō°ɹ). *Bot.* [f. *TELI(UM + -O + SPORE.] A spore of the rust fungi (Uredinales) which produces a basidium on germination, often after overwintering; a teleutospore.

1905 [see *TELIUM]. **1970** J. Webster *Introd. Fungi* ii. iv. 369 The teliospores [of *Puccinia graminis*] represent the overwintering stage and only develop further after a period of maturation corresponding to winter dormancy. **1981** *Trans. Brit. Mycol. Soc.* LXXVII. 439/2 If in autumn a rusted rose leaf is shaken gently above a glass slide, many single teliospores fall on to it.

telium (tī·liǒm). *Bot.* Pl. **telia.** [mod.L., f. Gr. τέλ-ος end.] = *TELEUTOSORUS.

1905 J. C. Arthur in *Bot. Gaz.* XXXIX. 222 For the sorus of the third spore-stage, usually..called teleutosorus, I propose *telium*..; derivatives *telial, teliospore*, etc. **1937** *Nature* 8 May 800/2 (*heading*) Production of uredia and telia of *Puccinia graminis* on *Berberis vulgaris*. **1979** *Trans. Brit. Mycol. Soc.* LXXIII. 231/1 Only telia are present on the leaves.., but a few urediniod spores are found in some telia of this specimen.

Hence **te·lial** *a.*

1905 [see above]. **1925** *Jrnl. Agric. Res.* XXXI. 643 [In *Gallowaya pinicola*] there is very little intermingling or intertwining of hyphae as a preliminary to the formation of the telial sorus. **1967** *Trans. Brit. Mycol. Soc.* L. 190 In taxonomic study of *Endophyllum* and *Kunkelia* there are two possibilities. Their sori can either be compared with aecia of other rusts or with telia of other rusts. ..I propose that they should properly be called 'telial aecia'.

tell, *v.* Add: **I. 3. b.** *tell it not in Gath* (earlier *transf.* examples).

1751 S. Richardson *Let.* 11 July (1964) 185 A wise man to be in love! Tell it not in Gath. **1816** M. R. Mitford *Let.* 20 Oct. (1925) 135 My favourite play ('tell it not in Gath!') is the first part of *King Henry the Fourth.*

5. a. Also in phr. *to tell all*, to reveal the whole truth, esp. in a sensational manner (freq. with ref. to the printed word). Now usu. without indirect obj.

1897 B. Stoker *Dracula* v. 57, I *do* so want to tell you all. **1936** *Mademoiselle* Aug. 16 (*heading*) Mrs Chester tells all. **1971** J. Lofland *Analyzing Social Settings* vi. 132 We delude ourselves if we expect very many field workers actually to 'tell all' in print.

7. a. Also with *apart.*

1925 A. Loos *Gentlemen prefer Blondes* (1982) iv. 86 How are we going to tell you gentlemen apart? **1958** K. Amis *I like it Here* xiv. 178 Barbara had complained to him..that she couldn't tell people apart (he found as little difficulty here as he found in telling female film-stars apart). **1982** B. Chatwin *On Black Hill* i. 10 As boys, only their mother could tell them apart.

b. Now freq. in affirmative sentences. (Later examples.)

1920 F. Scott Fitzgerald *This Side of Paradise* i. ii. 56 He was.., as Amory could tell from his general appearance, without much conception of social competition. **1924** 'K. Mansfield' *Something Childish* 124 They're not respectable women—you can tell at a glance. **1936** 'G. Orwell' *Diary* 11 Feb. in *Coll. Essays* (1968) I. 176 You can always tell a miner by the blue tattooing of coal dust on the bridge of his nose. **1963** J. Fowles *Collector* i. 49, I always thought people could tell I lived on my own. **1966** S. Heaney *Death of Naturalist* 15 You could tell the weather by frogs too.

8. a. Also with direct speech as obj. *I told you so*: see also *I TOLD YOU SO.

1904 [see *BOUQUET 1 b]. **1916** H. S. Walpole *Dark Forest* i. v. 135 'I can't marry you,' she told him, 'because I don't love you.' **1936** M. Plowman *Faith called Pacifism* 81 Mr. Lloyd George, as the wild cat of the House of David, said: 'I told you so.' **1943** G. Greene *Ministry of Fear* ii. i. 135 'Mr Digby,' she told him, 'there's a visitor for you.' **1976** H. MacInnes *Agent in Place* xxii. 231 'The police—call the police.' 'It's done,' Tony told her.

c. Const. *on.* To act as informer to (a person) about (another). Cf. sense 16.

1901 M. Franklin *My Brilliant Career* xiii. 107 Now, not a step do you go, my fine young blood, until you pick up every jolly lemon and put them away tidily, or I'll tell the missus on you. **1943** *Crisis* Mar. 78/3 If he told the Big God on them, no telling what would happen.

16. Const. *on* (further examples).

1943 B. Smith *Tree grows in Brooklyn* iii. xxxv. 206 And I didn't tell on you either, the time you made a cigarette out of coffee grounds and when you smoked it the paper caught fire and fell on her blouse and burned a big hole in it. **1955** J. Masters *Coromandel!* 41 Do you think she'll tell on us? **1968** J. Lock *Lady Policeman* xx. 162 They felt they ought not to 'tell on her' unless it was absolutely necessary. **1974** *Age* (Melbourne) 12 Oct. 12/1 Ooh Aah! I'm going to tell on you: I will inform the authorities.

17. a. *to tell the tale*, to relate a false or exaggerated story, esp. in order to evoke a sympathetic response.

1918 W. J. Locke *Rough Road* xi. 133 The temptation to 'tell the tale', to the new-comer was too strong. **1928** *Daily Express* 15 Dec. 7/4 Moneylender at Bow County Court. What did you tell me when you borrowed the money? Debtor: Oh, we all tell the tale when we want money. **1943** J. B. Priestley *Daylight on Saturday* ii. 9 The absentees tell the tale to the National Service Officer, and he tells the tale to Proscot, and nothing's done. **1968** 'J. le Carré' *Small Town in Germany* xiii. 204 He couldn't half tell the tale... He could tell you *any* bloody tale and you believed it. **1979** R. Blythe *View in Winter* iv. 175 I'm not tellin' the tale. We all went to the war.

d. *to tell* (someone) *goodbye, hello,* etc., to say goodbye, hello, etc., (to someone). Chiefly *U.S.*

1859 Bartlett *Dict. Amer.* (ed. 2) 475 To *tell* one good-bye, is the Southern phrase for to bid one good-bye. **1872** E. Eggleston *End of World* xviii. 128 You aren't going without telling me good-by? **1884** [in Dict., sense 4 a]. **1905** B. Tarkington *In Arena* 253 She told me to tell you good-bye. **1973** V. Canning *Flight of Grey Goose* ii. 28 Tell Albert hello and love to you both. **1979** L. Meynell *Hooky & Villainous Chauffeur* viii. 111 Mr Furlong asked me to tell you goodbye.

20. *don't tell me* (later examples); also (with dependent clause) expressing dismay; *do tell!* (earlier and later examples); *to tell the world*, to announce openly; to assert emphatically; *I tell* (or *I am telling*) *a lie*, (in trivial use) I am mistaken (cf. *LIAR 1 a); *you're telling me*, there is no need to tell me; I know that only too well; *to tell* (someone) *what to do with* (something) or *where to put* (something), expressing emphatic rejection with impolite implications; *to tell it like it is*, to relate the facts of a matter realistically or honestly, holding nothing back *colloq.* (orig. *U.S. Blacks'*). Also *tell that to the marines*: see *MARINE *sb.* 4 c; *to tell* (someone) *where he gets* (or *to get*) *off*: see *GET *v.* 62 j.

1861 Geo. Eliot *Silas Marner* i. ix. 143 Not come to live in this house? Don't tell me. **1944** M. Laski *Love on Supertax* iv. 52 Who's your latest pick-up?..Not Sir Hubert Porkington! Don't tell me you've actually hooked him! **1952** H. Garner *Yellow Sweater* 15 'Don't tell me you're in trouble [*sc.* pregnant]?' he asked. **1973** *Farm & Country* 10 Apr. 11/4 'Don't tell me we've got to go through that again,' said one executive member in an audible groan. **1842** J. S. Buckingham *Eastern & Western States Amer.* I. 177 When a person..has concluded his narrative, the hearer will reply, 'Oh! do tell.' **1883** C. F. Wilder *Sister Ridnour's Sacrifice* 138 'Come fur?' 'About eighty miles.'..'Du tell!' **1979** C. MacLeod *Luck runs Out* (1981) i. 17 Do tell. Did she leave any children? **1781** Cowper *Table Talk* in *Poems* (1782) I. 38 And tell the world..That he, who died below and reigns above, Inspires the song, and that his name is love. *a* **1871** T. Carlyle in *Coll. Lett. T. & J. W. Carlyle* (1981) IX. 318 This I *cd* tell the world, you have not had, for 100 years, any Book that came more direct and flamingly sincere from the heart of a living man. **1923** [see *JEEZ(E *int.*]. **1933** *Punch* 11 Jan. 29/3 Say, can he act orr can he act? Ah'll tell de woirrld. **1956** 'C. Blackstock' *Dewey Death* viii. 169 She persistently told the world about her fiancé, her marriage problems, her piano playing. **1925** S. O'Casey *Shadow of Gunman* ii, in *Two Plays* 172 Adolphus..after takin' his tea at six o'clock—no, I'm tellin' a lie—it was before six, for I remember the Angelus was ringin' out. **1956** 'A. Burgess' *Time for Tiger* i. 7 Those Japanese tattooists... I seen one fellow in Jerusalem, wait, I'm telling a lie, it was in Alex,..one fellow with a complete foxhunt on his back. **1968** L. Deighton *Only when I Larf* vii. 89 Six Centurion Mark Fives on that hillock... No tell a lie, one of them is a Mark Two. **1973** J. Mann *Only Security* xii. 172 Oh, it must have been fifteen years or so—..No, I tell a lie, I'm afraid,..can't have been more than five or six years that she was like that, poor lady. **1932** G. Kahn (*song-title*) You're telling me. **1938** M. Allingham *Fashion in Shrouds* xix. 338 'Things are bad enough as they are.' 'You're telling me.' **1954** *Times* 16 July 9/4 When he declares that 'overnutrition has its dangers'..the layman is inclined to reply 'You're telling me.' **1977** 'C. Aird' *Parting Breath* xvii. 194 'Forensic pathologists don't take chances.' 'You're telling me,' said Crosby with audible scorn. **1946** M. Shulman *Zebra Derby* (1947) xxi. 133 Green, an upholsterer, said that he was through with upholstering and had told his old boss what to do with his old job. **1958** M. Dickens *Man Overboard* xii. 176 He had been going to tell Glenn what he could do with his job. **1968** M. Bragg *Without City Wall* ii. xxi. 207 You could tell the people at the hall what to do with that job of theirs. She'd had enough of being a servant. **1977** *Listener* 14 Apr. 483/2 Protestations that if the government did any such thing, the BBC would probably tell it where to stuff such an instruction. **1964** *Down Beat* 19 Nov. 8/2 (*heading*) Mann tells it like it is. **1965** *New Statesman* 1 Oct. 473/1 Although sometimes tardy, all his speeches make their intended points; as they say in Harlem, he tells it like it is. **1969** L. Lokos *House Divided* i. 58 The crowd responded fervently with 'Amen, amen,' and 'Tell it like it is.' **1973** *Field & Stream* Jan. 8/3 Keep this tell-it-like-it-is kind of article going. **1979** *Guardian* 14 Apr. 8/6 'Tell it like it is,' said Hemingway, but that was..before we all became ethnic-conscious.

II. 23. b. (Earlier example.)

1804 J. Whitehouse in Lewis & Clark *Orig. Jrnls.*

Lewis & Clark Exped. (1905) VII. 1. 70 The Captn. form^d. his men On the S.W. Side of the river Missourie and told them off in Sections, from the right.

d. fig. *to tell* (someone) *off*, to scold or reprimand (someone). Cf. *tick off* s.v. *TICK *v.*[1] 3.

1919 *Cassell's New Eng. Dict.* s.v. *Tell, to tell off*, (colloq.) to scold. **1927** A. Christie *Big Four* xiv. 202 They don't like you to notice things—especially if it should seem you were telling them off about it. **1938** G. Arthur *Not Worth Reading* xiv. 216 'It required a very great man,' said F.E. when he emerged from his interview, 'to resist the temptation to tell me off.' **1941** G. Homes *Forty Whacks* ix. 90 The man had just been told off, and told off plenty. **1974** M. Birmingham *You can help Me* iv. 97 She's..telling off the police good and proper... She blames them for all the dirt.

III. 26. (Earlier example.)

1783 *Public Advertiser* 7 Oct. 2/2 Yet strange to tell it, this Distinction, which as the Players call it, tells most forcibly, Garrick overlooked.

tell-. Add: **tell-all** *a.* and *sb.*, (an account which is) revelatory, tending to disclose private information; cf. *TELL *v.* 5 a.

1959 J. Fingleton *Four Chukkas to Australia* (1960) i. 2 This position—in cricket and other sports—has been aggravated by the flood of 'tell-all' books. **1974** *Nat. Geogr. Mag.* Dec. 851/2 A tell-all historian of the time reported that the prince loved to drink and was notably fond of women. **1976** *Publishers Weekly* 7 June 68/1 This man's sometimes engrossing, frenetic, first-person tell-all. **1978** in K. Hudson *Jargon of Professions* iii. 64 A totally engrossing tell-all. Few autobiographies convey so intimately what is involved in creativity.

Tell el-Amarna: see *AMARNA.

tellen. Add: Also **tellin.** (Later examples.)

1901 E. Step *Shell Life* viii. 118 The members of the Tellin family..are distinctly deep burrowers in sand or mud. **1971** S. P. Dance *Seashells* 77 The right and left valves of many tellins are unequal in shape and size.

teller. Add: **2. c.** Also *attrib.* in *teller vote* (U.S.), a vote taken by tellers as members file past them; *spec.* a category of vote in the House of Representatives, in which the tellers record the votes of members but not (until 1970) their names.

1924 *Congressional Rec.* 11 Apr. 6142/1 Is not the teller vote the highest in the committee? **1935** *Sun* (Baltimore) 1 July 1/8 The House is working under a rule which precludes a direct roll-call vote..but a teller vote..is to be taken. **1974** *Encycl. Brit. Macropædia* XIV. 722/1 Voting procedures range from the formal procedure of the division or teller vote in the British House of Commons to the electric voting methods employed in the California legislature. **1972** W. Weaver *Both Your Houses* vii. 99 If the outcome of a division is unsatisfactory to at least twenty members of the Committee of the Whole..they can demand a teller vote.

Teller mine (te·ləɹ məin). Also with hyphen and as one word. [Anglicization of G. *tellermine*, f. *teller* plate + *mine* *MINE *sb.* 3.] A disc-shaped German anti-tank mine containing TNT, used in the war of 1939–45.

1943 *Hutchinson's Pict. Hist. War* 4 Aug.–26 Oct. 11 A huge collection of teller-mines, or disc mines, discovered by the allied forces near Roccalumba. **1945** *Finito! Po Valley Campaign* (15th Army Group) 41 The 10th Mountain Division pushed forward..across a valley studded with Teller mines. **1967** *Punch* 11 Jan. 40/1 We were..taking the fuses out of those wretched Tellermines.

tellina (teləi·nă). Also **Tellina.** [mod.L. (G. E. Rumphius *D'Amboinsche Rariteitkamer* (1705) ii. 145), f. Gr. τελλίνη.] = TELLEN. Also *attrib.* Hence **te·llinoid** *a.*, resembling a bivalve of the genus *Tellina.*

1877 C. M. Yonge in C. Coleridge *C. M. Yonge* (1903) ii. 70 Waves from the Atlantic, bringing in..tellinas of a delicate pink. **1884** G. W. Tryon *Struct. & Syst. Conchol.* III. 170 The branchial apparatus..is..a departure from the Tellina type. **1900** *Proc. U.S. Nat. Mus.* XXIII. 287 The Tellinas are mostly inhabitants of moderate depths. *Ibid.* 291 Shell tellinoid, thin. **1936** A. H. Verrill *Strange Sea Shells* xiv. 132 Why..is the tellina so beautifully colored within its shells.

telling, *vbl. sb.* Add: **1. c.** *that's telling*(*s* (examples of sing. form); similarly *that would be telling*; *to lose nothing in the telling*, (of a story) to become embellished in the course of frequent narration.

1897 'S. Grand' *Beth Bk.* xiii. 112 'May I ask..by whom you were informed?'..'Ah, that would be telling,' said Beth. **1921** S. Kaye-Smith *Joanna Godden* iii. 136 'What sort of surprise?' 'That's telling.' **1930** A. Christie *Murder at Vicarage* x. 78 'When was she talking of earning her own living?'..'That would be telling, wouldn't it?' **1973** G. Mitchell *Murder of Busy Lizzie* xii. 144 'But what could you inform about?' 'That's telling, isn't it?' **1980** A. Price *Hour of Donkey* i. 23 'Are the Germans in Peronne, Dickie?'..'That would be telling!'

[**1721** J. Kelly *Scottish Proverbs* 55 A Tale never loses in the telling. The Fame or Report of a matter of Fact..commonly receives an Addition as it goes from Hand to Hand.] **1906** [in Dict., sense 1 a]. **1914** E. R. Burroughs *Tarzan of Apes* vi. 68 The story of the thunder-stick having lost nothing in the telling during these ten years. **1954** L. P. Hartley *White Wand* 15 No doubt Antonio was telling the story to his fellow-gondoliers at the traghetto, and it would lose nothing in the telling.

2. c. *telling-off*, a scolding or reprimand. Cf. *TELL *v.* 23 d. *colloq.*

1911 Kipling *Diversity of Creatures* (1917) 121 The boys..had had a wildish night..that ended with a telling-off from an artist. **1920** A. J. L. Scott *Hist. Sixty Squadron R.A.F.* 57 He got..a well-deserved and proper 'telling-off' from the Brigadier and Wing Commander. **1959** *Times* 22 July 7/4 Then there was and to some extent still is 'a telling off', sometimes met with in the degenerate form of 'ticking off'. **1974** W. Foley *Child in Forest* ii. 231 Still smarting from my 'telling-off'..and in militant mood.

telling, *ppl. a.* (Earlier example.)

1851 H. Melville *Moby Dick* I. xxv. 180 A staid, steadfast man, whose life for the most part was a telling pantomine of action.

tell-tale, *sb.* (*a.*) Add: **1. b.** (Earlier and later examples.) Also *spec.*, a small hidden object placed so as to reveal a secret intrusion by its disturbance (see quots.).

1754 Richardson *Grandison* II. 295, I was very earnest to know, since my eyes had been such tell-tales, if their brother had any suspicion of my regard for him. **1953** *Sun* (Baltimore) 30 Mar. (B ed.) 26/7 A broken 'tell-tale', one of the little devices policemen set in doorways and fire escapes to help them check on burglars. **1968** 'H. Howard' *Eye of Hurricane* iii. 33 After I'd searched..I was well satisfied that nobody had planted tell-tales anywhere..no microphones, no built-in radio transmitters. **1975** B. Garfield *Death Sentence* (1976) vii. 44 His hands began to sweat: the familiar telltale. **1979** K. Follett *Triple* viii. 184 There were dozens of ways of planting telltales. A hair lightly stuck across the crack of the door..; a lump of sugar under a thick carpet would be silently crushed by a footstep.

2. h. *Building.* A piece of glass or clear plastic, often graduated, fixed over a crack in a building to reveal whether there is further movement in the fabric.

1938 *Times* 16 Feb. 14/1 Of the many tell-tales planted by Mr. Harvey only a few are known to have broken. **1972** *Besselsleigh & Dry Sandford* (Berks.) *Parish Mag.* Oct., The church council has agreed to the architect's suggestion that a few glass tell-tales should be fixed in some of these cracks. **1976** *Daily Tel.* 25 Sept. 21/4 They will hold back on the job while they put up 'telltales' at strategic points.

i. A light on the dashboard of a motor vehicle which shows when the direction indicator or main-beam lights are in use.

1962 *Autocar* Spring 62/2 When any driving lamps are on, a small green warning light appears, and this is supplemented by a blue tell-tale for the main beams. **1966** *Guardian* 17 Oct. 6/3 The winker tell-tale is..self-cancelling. **1970** K. Ball *Fiat 600, 600D Autobook* xi. 138/1 The high beam indicator is on the dashboard behind a blue screen lens, the parking light indicator behind a green and the direction indicator tell-tale behind another green lens.

tellurion. (Example of *tellurium.*)

1905 *Nature* 14 Sept. 493/2 The ordinary wire model or 'tellurium'.

tellurometer (tel^iurǫ·mĭtəɹ). [f. L. *tellūr-em* the earth + -O + -METER.] An instrument that accurately measures distances on land by transmitting a microwave signal and timing the arrival of a return signal that it triggers at the distant point.

1957 T. L. Wadley in *Empire Survey Rev.* July 100 (*heading*) The tellurometer system of distance measurement. **1961** *Aeroplane* C. 63/2 Five of the parties were to do levelling, and the other two distance measuring with tellurometers. **1975** J. B. Harley *O.S. Maps* i. 7 The 'Tellurometer'..has been used to measure distances of 135 km between Great Britain and Ireland as part of the process of checking the triangulation.

telly (te·li). *colloq.* [Shortening of *TELEVISION. Cf. *TELE.] **1.** = *TELEVISION 1. Phr. *on* (*the*) *telly.*

[**1940** *Chambers's Techn. Dict.* 837/2 *Tellies*, colloquialism for cinematograph films with sound; also for television.] **1942** Berrey & Van den Bark *Amer. Thes. Slang* § 618/3 Lookies, tellies, telly. **1957** *Observer* 3 Nov. 4/5 For all practical purposes, if it hasn't been on telly, it doesn't exist. **1957** *Economist* 7 Dec. 842/1 An evening when Sheffield Wednesday were playing Juventus of Milan at football on the telly. **1958** M. Spark *The Go-Away Bird* 152 He said, 'What do you do in the evenings, Lorna? Do you watch Telly?' I did take this as an insult, because we call it TV, and his remark made me out to be uneducated. **1967** E. Williams *Beyond Belief* iii. xix. 204 Tonight, the eve of Christmas Eve,..they are watching telly, nice thriller. **1968** J. Arnold *Shell Bk. Country Crafts* 70 A growing tendency in domestic life of subordinating activities to the 'telly'. **1970** G. F. Newman *Sir, You Bastard* iv. 117 On the news, on the telly this evening. **1977** *New Yorker* 26 Sept. 37/1 His parents lived an isolated life, but now had the telly.

2. = *TELEVISION 2.

1955 M. Allingham *Beckoning Lady* i. 5 He..walked back to the village and the telly. **1957** F. King *Man on Rock* i. 7, I can't even afford to pay the never-never on a wireless, let alone a telly. **1969** A. Glyn *Dragon Variation* vi. 176 An occasional bluish light behind chintz curtains betrayed the night-owls, those who were still glued to the telly, watching the news headlines, the weather forecast. **1978** K. Amis *Jake's Thing* iii. 30 Let's be absolute devils and have the heating on and huddle round the telly.

3. A television performance; a booking or session of filming for this. *Theatr.* and *Broadcasting.*

1963 E. Humphreys *Gift* 8 Every time I did a telly it was a lovely day, while I sweated my guts out under the artificial light. **1979** S. Brett *Comedian Dies* ii. 23 I've got you a telly... It's an *Alexander Harvey Show.*

4. *attrib.* and *Comb.*, as *telly ad, don, mast, -viewer*, etc.; **telly man,** a man who works professionally for a television service.

1958 *Times Lit. Suppl.* 15 Aug. p. xii/3 Turning their backs upon the 'telly' screen, they will strain their eyes in the semi-darkness of the living-room. **1960** C. MacInnes *Mr Love & Justice* 85 His part-time trade of mending radios and telly sets. **1963** *Spectator* 22 Feb. 230/3 The Third Programme..is becoming more and more a private club or experimental research establishment unwittingly financed by the telly-viewers. **1963** *Punch* 2 Oct. 475/1 Small vociferous pockets trying to attract tellymen. **1966** J. Betjeman *High & Low* 4 Slate cottages with sycamore between, Small fields and telly-masts and wires and poles. **1969** Fabian & Byrne *Groupie* (1970) i. 9, I even believe telly ads and things like that. **1971** *Author* LXXXII. 111 Many a paper fills its review columns with inoffensive, but basically uninteresting, books butchered to make a well-known telly don's holiday in the dead summer months. **1977** *Irish Democrat* Mar. 5/2 Surrounded by a regiment of security men, aides, advisers, journalists, tellymen and Unionists. **1977** *Irish Times* 8 June 11/5 But there might have been long telly-watching sessions behind locked doors.

telo-[1]. Add: Also with pronunc. (tī·lo). **telode·ndrion, -de·ndron** (pl. **-dendria** is used for both forms) *Anat.* [Gr. δενδρίον, dim. of δένδρον tree], one of the terminal branches into which the axon of a nerve cell divides; **te·lomere** *Cytology* [*-MERE], the compound structure found at the end of a chromosome in eukaryotes, having only one spindle pole; **telomi·tic** *a. Cytology* [Gr. μίτ-ος thread] = *TELOCENTRIC *a.*; **telope·ptide** *Biochem.*, a peptide which is at or near the end of a polypeptide molecule; **telosyna·psis** *Cytology*, a supposed end-to-end pairing of chromosomes during the zygotene stage of meiosis; = *telosyndesis* below; hence **telosyna·ptic** *a.*, **-syna·ptically** *adv.*; † **telosynde·sis** *Cytology* = *telosynapsis* above.

1899 L. F. Barker *Nervous System* viii. 82 The ultimate terminals (telodendrions) of the axones have been carefully and exactly studied. **1907** I. Hardesty in Morris & McMurrich *Morris's Treat. Human Anat.* (ed. 4) 751 The axone bearing the impulse on approaching its termination loses its sheath and breaks up into its numerous terminal twigs, the final of which are called telodendria. **1949** *New Gould Med. Dict.* 1030/2 *Telodendron*. See *telodendron*. **1966** C. R. & T. S. Leeson *Histology* x. 178/2 In some cases, the telodendria are so numerous as to surround the neuron on which they terminate in a basket-like arrangement. **1940** H. J. Muller in *Jrnl. Genetics* XL. 2 The reconstructed chromosome cannot continue to be transported..unless it happens to be monocentric and—in *Drosophila* at least—ditelic, one centromere and two telomeres being necessary and permanent organelles. **1960** L. Picken *Organization of Cells* vii. 261 Within the nucleus the chromosomes present their telomeres—'centromeres' connected to the chromosomes by terminal filaments—to the aspect of the nuclear membrane nearest to the spindle. **1983** *Nature* 13 Jan. 112/1 Telomeres are very stable as free ends, whereas ends of [DNA] molecules broken *in vivo* tend to stick together irreversibly. **1917** E. E. Carothers in *Jrnl. Morphology* XXVIII. 449 The unusual conditions of the chromosomes in this group have made advisable the introduction of..new terms... Telomitic—a term used to indicate terminal fiber attachment. **1934** L. W. Sharp *Introd. Cytol.* (ed. 3) ix. 116 Supposedly telomitic chromosomes have been shown in some instances to have their attachment region slightly back from the end. **1964** F. O. Schmitt et al. in *Proc. Nat. Acad. Sci.* LI. 494 The term 'end-chains' with its terminal connotation has been replaced by the term 'telopeptides'. **1971** *Nature* 22 Jan. 242/1 Every third residue in the chain is glycine, except in short 'non-collagenous' telopeptides at the N-terminal ends of the chains. **1975** *Ibid.* 10 July 125/1 Rabbit anti-collagen serum is primarily directed to terminal (telopeptide) antigenic sites, and not to helical or central sites. **1909** Telosynapsis [see *parasynapsis* s.v. *PARA-[1] 1]. **1920** W. E. Agar *Cytology* ii. 43 Parasyndesis and Telosyndesis. [*Note*] Called parasynapsis and telosynapsis by cytologists, who employ the term synapsis in the sense in which syndesis is here used. **1945** M. J. D. White *Animal Cytol. & Evolution* v. 79 The old controversy between adherants of the theory of telosynapsis and those who believed in the alternative viewpoint..is only of historical interest, since 'parasynapsis'..is now known to be universal. **1912** *Jrnl. Exper. Zool.* XIII. 394 Attention may be called to the increasing tendency..to reject, or at least restrict, the theory of parasynapsis.., in favor of a telosynaptic conception. **1929** *Jrnl. Genetics* XXI. 47 Digby's interpretation, so much quoted in support of the telosynaptic view.., is due to a misunderstanding of the essential features of meiosis. **1910** Telosynaptically [see *parasynaptically* adv. s.v. *PARA-[1]]. **1926** *Genetics* XI. 274 The third element is joined telosynaptically to the other two chromosomes. **1920** L. Doncaster *Introd. Study Cytol.* v. 68 The end-to-end union [of chromosomes] (formerly supposed to be frequent or universal) is telosyndesis (or telosynapsis).

telocentric (telo-, tīlose·ntrik), *a.* (and *sb.*) *Cytology.* [f. TELO-[1] + *-CENTRIC.] Of a chromosome: having the centromere at the end. Hence as *sb.* Cf. *acrocentric, *metacentric *adjs.*

1939 C. D. Darlington in *Jrnl. Genetics* XXXVII. 349 This does not mean that terminal centromeres or telocentric chromosomes work satisfactorily or survive permanently. *Ibid.* 352 The misdivided chromosomes..will give the functional telocentrics that I have already described at second anaphase. **1949** Darlington & Mather *Elem. Genetics* v. 103 Misdivision of the centromere and Formation of two telocentric chromosomes. **1964** *Hereditas* LII. 209 Most cytologists agree that the centromere in acrocentric chromosomes is less terminal than in telocentric chromosomes. **1971** [see *metacentric *sb.*]. **1976** *Nature* 12 Aug. 580/1 The X chromosome is telocentric (X[1]) in F344 and ACI strains.

telogen (te·lǒdʒěn). [f. Telo-[1] + -gen.] **1.** *Biol.* The stage in the life of a hair or hair follicle following cessation of growth of the hair.

1926 F. W. Dry in *Jrnl. Genetics* XVI. 297 For the present purpose it is convenient to divide the developmental history into the following three phases: (1) The Anagen phase... (2) The Catagen phase... (3) The Telogen phase, the hair having become a dub-hair and not growing further. **1955** *Proc. Soc. Exper. Biol. & Med.* LXXXVIII. 450/1 Telogen is the resting stage of a follicle and it is reached approximately 19 days after plucking. **1980** *Practitioner* Nov. 1161/1 Hair growth..is..phasic, there being a phase of growth (anagen) lasting approximately three to five years in normal scalp hair and a resting phase (telogen) lasting for around three months.

2. *Chem.* A simple compound that provides chain-terminating radicals in polymerization and limits the degree of polymerization.

1948 [see *telomer]. **1974** C. M. Starks *Free Radical Polymerization* i. 2 Chain transfer agents (telogens) are often added to polymerization recipes as molecular weight regulators. **1980** [see *taxogen].

telomer (te·lǒməɹ). *Chem.* [f. Telo-[1] + *-mer.] A low-molecular-weight polymer consisting of a chain of a limited number of units (taxogens) terminated at each end by a radical from a different compound (the telogen).

1948 Hanford & Joyce *U.S. Patent 2,440,800* 1 It has been found necessary to coin new terms to describe the reaction and the participants therein... 'Telomerization' is defined as the process of reacting..a molecule YZ which is called a 'telogen' with more than one unit of a polymerizable compound..called a 'taxogen' to form products called 'telomers' having the formula Y(A)ₙZ. **1951** *Jrnl. Amer. Chem. Soc.* LXXIII. 5197/1 When it [*sc.* the polymerization] is conducted in the presence of any one of a wide variety of organic compounds, telomers containing 1–25 tetrafluoroethylene units combined with one molecule of the chain transfer compound are obtained. **1966** J. A. Brydson *Plastics Materials* xv. 295 With an ethylene–carbon tetrachloride ratio of about 4:1 about 60% of the telomers have 7, 9 or 11 carbon atoms in the molecule. **1980** [see *taxogen].

So **te:lomeriza·tion**, polymerization that is limited by the action of a telogen; also **te·lomerized** *ppl. a.*, **te·lomerizing** *vbl. sb.*

1948 Telomerization [see *telomer]. **1954** *Jrnl. Res. Nat. Bureau Standards* (U.S.) LIII. 122 $YZ + nCX_2{=}CX_2 \rightarrow Y(CX_2CX_2)_nZ$... Such a reaction is called telomerization. **1967** *Gloss. Terms Plastics Industry* (*B.S.I.*) 1. 7 *Telomerized polymer*, a polymer chain-stopped by a telomer. **1968** G. Trappe in P. D. Ritchie *Vinyl & Allied Polymers* xii. 265 Low molecular weight oils, greases, and waxes of polychlorotrifluoroethylene are made by polymerisation in the presence of telomerising agents such as carbon tetrachloride. **1974** H. I. Bolker *Natural & Synthetic Polymers* iii. 136 About 0·5–1·0 mole % of acetic acid is added as a telomerization agent.., which limits the formation of chains of excessively high molecular weight.

telophase. Delete entry s.v. Telo-[1] and substitute:

telophase (te·lǒfē[1]z). *Cytology.* [a. G. *telophase* (M. Heidenhain 1894, in *Arch. f. mikr. Anat.* XLIII. 524): see Telo-[1] and Phase.] The final phase of mitosis and meiosis, following anaphase and preceding interphase, at which the chromatids or chromosomes are at opposite poles of the cell; also, a cell at this stage.

1895 *Jrnl. R. Microsc. Soc.* 35 The fourth chapter discusses the final movements of mitosis (telophases, telokinesis). **1900** [in Dict. s.v. Telo-[1]]. **1934** *Nature* 26 May 800/1 The anaphase and telophase chromosomes thus contain two spiral chromonemata... In late telophase the two threads are found to approximate very closely. **1971** J. Z. Young *Introd. Study Man* xiv. 179 At the end of mitosis (telophase) these enzymes associate with the chromosomes. **1973** *Nature* 6 Apr. 403/2 The DNA determinations were carried out in metaphases and telophases of the original stained lung cultures.

Hence **telopha·sic** *a.*, of or pertaining to telophase.

1907 *Rep. Brit. Assoc. Adv. Sci. 1906* 757 It [*sc.* the nucleolus] lies in the centre of the telophasic figure. **1929** *Bot. Gaz.* LXXXVIII. 360 The foregoing investigators were correct in interpreting some (although not all) of the telophasic aspects they observed as chromosome doubleness. **1976** *Biol. Abstr.* LXII. 693/1 The telophasic plaque in the root meristem of *Triticum vulgare*.

telotaxis (telotæ·ksis). *Biol.* [mod.L., coined in Ger. (A. Kühn *Die Orientierung der Tiere im Raum* (1919) 60): see Telo-[1] and Taxis 6.] Directional movement made by an animal in order to keep a particular source of stimulation acting on its sense receptor(s).

1934 *Jrnl. Exper. Biol.* XI. 129 Those movements in which the animal is truly orientated Kühn calls topotaxis, and these again fall into two main groups: tropotaxis and telotaxis. **1976** D. J. Horn *Biol. Insects* vi. 229 An example of telotaxis that has fascinated researchers is light-compass orientation.

telotype. (Earlier example.)

1850 F. Galton *Telotype* 3 In the Telotype (as our instrument may be termed), by merely touching a key on which any letter is marked that letter is to be printed, almost instantaneously, at the opposite end of the line.

‖ **tel quel** (tęl kęl), *adj. phr.* Also in Fr. pl., **tels quels** (masc.), **telles quelles** (fem.). [Fr.] Just as it is; without improvement or modification.

1876 Geo. Eliot *Let.* 3 Feb. (1956) VI. 220 Don't wear glasses *telles quelles* but consult about getting the right ones. **1903** W. James *Let.* 5 June in R. B. Perry *Thought & Char. W. James* (1935) II. vi. lxxvi. 427 You spoke of publishing these lectures, but not, I hope, *tel quels* [*sic*]. **1967** *Listener* 16 Mar. 368/3 The culled facts have been presented *tel quel*, with inconsistencies noted but not resolved.

telson. Add: Hence **telso·nic** *a.*

1934 in Webster. **1946** *Nature* 28 Dec. 935/2 The caudal segment was probably composed of the sixth abdominal segment with which the telson had become fused, and thus these appendages must be telsonic. **1981** *Israel Jrnl. Zool.* XXX. 115 The more primitive Hyssuridae have no telsonic statocysts.

Teltag (te·ltæg). [?f. Tell- + Tag *sb.*[1]] A label attached to goods manufactured in the U.K. giving information about the size, weight, performance, etc., of the goods.

1965 *Observer* 23 May 9/4 A new label will appear on British goods next year. Called the 'teltag', it has been designed by the Government-backed Consumer Council to give shoppers clear information on the hidden characteristics of their purchases. **1967** *Spectator* 1 Dec. 682/1 So far isn't necessarily very far, since manufacturers are under no obligation to use, and pay for, Teltags. **1971** *Reader's Digest Family Guide to Law* 335/1 The Teltag scheme, devised by the now defunct Consumer Council, is still used to give details of composition, size, performance and use of appliances.

Telugu, Teloogoo, *sb., a.* Add: Also **Telegu. 1.** (Later examples of the form *Telegu*).

1937 W. H. Saumarez Smith *Let.* 31 Jan. in *Young Man's Country* (1977) ii. 54 Kanarese is the vernacular of Mysore, a language allied to Tamil and Telegu. **1957** *Encycl. Brit.* IV. 978/1 Great movements of people: the Telegu- and Tamil-speaking caste groups to the rubber plantations of Burma and Malaya. **1978** 'M. M. Kaye' *Far Pavilions* i. i. 17 He picked up a number of tongues.. Tamil, Gujerati and Telegu from the southerners.

3. (Earlier and later examples.)

1816 A. D. Campbell (*title*) A grammar of the Teloogoo language. **1821** —— *Dict. Teloogoo Language*, Advt., The following Dictionary was commenced soon after the publication of the Author's Teloogoo Grammar. *Ibid.* Many of the most common Teloogoo words are arranged, absurdly enough, in the order of the *French*, and not of the Teloogoo, alphabet. **1891** C. R. Day *Music of S. India* v. 61 They..were probably composed by some Telegu pandit at the court of Mysore. **1909** E. Thurston *Castes & Tribes S. India* VII. 13 The Telugu soldiers might come to be regarded as the Telugus..*par excellence*. **1977** *Listener* 7 Apr. 447 Publishing Telegu books largely for Indians in the Midlands.

temazepam (těmē[1]·zɪpæm). *Pharm.* [f. *tem-* (of unknown origin) + Az(o- + *-ep*(*ine* (see *oxazepam) + Am(ide.] A tricyclic compound, $C_{16}H_{13}ClN_2O_2$, used as a tranquillizer and short-acting hypnotic.

1970 *Approved Names* (Brit. Pharmacopœia Comm.) 74 *Temazepam*. 7-Chloro-3-hydroxy-1-methyl-5-phenyl-1*H*-1,4-benzodiazepin-2(3*H*)-one. **1971** *Pharmacol. Res. Communications* III. 166 Besides the three known benzodiazepines.., in the present experiment a fourth one—temazepam, recently introduced in therapeutic use—was also studied. **1980** *Brit. Med. Jrnl.* 29 Mar. 910/1 Pharmacological differences between 'long'-acting benzodiazepines..and the 'short'-acting rapidly cleared compounds, such as..temazepam. **1984** *Daily Tel.* 11 Feb. 10/2 [He] chose temazepam, a hypnotic drug marketed..for transient insomnia under the trade name of Normison.

‖ **tembe** (te·mbe). [Swahili.] In East Africa, a rectangular house with mud walls and a flat roof.

1887 W. H. G. Kingston *Trav. Dr. Livingstone* vii. 135 Sheikh Said Ben Salim invited him [*sc.* Stanley] to take up his quarters in his tembe, or house, a comfortable looking place for the Centre of Africa. **1896** B. K. Gregory *Story of David Livingstone* xiii. 121 Then the two brave men retired into the doctor's *tembe*, and conversation began, but Stanley could not recollect what it was all about. **1947** *E. African Ann. 1946–7* 93/2 David Livingstone lived for some time in a *tembe* house on the outskirts of Tabora and the house is preserved as a memorial to his name. **1955** *Times* 25 July 5/1 A replica of the tembe (hut) occupied by Livingstone and Stanley, erected near Tabora some years ago and badly damaged by fire recently, is to be rebuilt this year. **1978** J. Updike *Coup* (1979) iii. 111 The space between the square earth houses, the *tembes*, widens.

temblor (temblǫ·ɹ) South-western *U.S.* Pl. **temblores.** [a. Amer. Sp.] An earthquake.

1876 B. Harte *Gabriel Conroy* v. xvii. 288 The temblor has swallowed him. **1896** *Land of Sunshine* July 72 One freshet of one Ohio river a dozen years ago took more lives than all the *temblores* in California in a century and a half have taken. **1932** F. L. Wright *Autobiogr.* ii. 147 One of the features of construction that insured the life of that building in the terrific temblor of 1922. **1950** *Los Angeles Times* 3 Jan. 1/6 The temblor was reported from both Ogden and Logan.

‖ **tembo** (te·mbo). Also **tembu.** [Swahili.] An alcoholic drink made in East Africa, usu. from the sap of the coconut palm; palm wine.

1860 J. Krapf *Travels, Researches & Missionary Labours* i. ix. 149 One of the elders said that it was really true that God loved men, for He gave the Wanika rain, tembo, and clothes. **1937** K. Blixen *Out of Africa* iii. i. 170, I sometimes asked the D.C.'s permission for my Squatters to make tembu, a deadly drink, fabricated from sugar cane. *Ibid.* v. v. 411 The old women took a keen interest in everything that was going on on the farm..; a joke, on a cup of tembu, would make their wrinkled toothless faces dissolve in laughter. **1964** R. M. Gatheru *Child of Two Worlds* vii. 101 Others would be drinking some stinking municipally brewed drink—'Tembo'—at Pumwani as an outlet for their handicapped life.

Tembu (te·mbu), *sb.* and *a.* Forms: 9 **Tymba, 9–Tembu; Thembu.** [Xhosa.] **A.** *sb.* A member of a Xhosa-speaking people of the south-eastern part of South Africa; also, this people collectively. **B.** *adj.* Of, pertaining to, or designating this people. Cf. *Tambookie *sb.* and *a.*

1827 G. Thompson *Trav. & Adv. Southern Afr.* I. i. xvii. 349 A Tambookie Caffer is termed Tymba or Tembu, while the tribe collectively is called Amatymbae. **1874** *Cape Monthly Mag.* IX. 207 He then dilated upon the enormities of Gangelizwe, the Tambookie Chief, and the ability of the present Resident amongst the Tembus. **1912** Whiteside & Ayliff *Hist. Abambo* i. 4 Madikana and his people, at a certain time, were feasting, when they were attacked by a combined force of the Tembus and Xosas. **1927** C. G. Botha in *Wreck of Grosvenor* (Van Riebeeck Soc.) p. iv, The expedition was obliged to turn back owing to the hostile attitude of the Tembu tribe of natives. **1941** C. W. de Kiewiet *Hist. S. Afr.* 73 It was densest in Kafirland, where dwelt the Ama-Xosa, the Tembu, the Pondo, the Xesibe, and the Ama-Baca. **1961** *Guardian* 27 May 6/3 Mr Mandela is..the son of a Tembu chief. **1967** J. A. Broster *Red Blanket Valley* 4 Remember, the Thembu, one of the chief tribes of the Transkei, were never conquered: they asked the European traders to settle. **1971** *Daily Dispatch* (East London, Cape Province) 11 May 2 We have to honour and respect our national leaders such as Faku of the Pondos and Ngubengcuka of the Tembus. **1974** J. Flint *Cecil Rhodes* (1976) ix. 205 The foundations [of peace] were laid by an African..a Tembu man fluent in English and Ndebele. **1976** *Times* 4 Nov. 17/4 Chief Matanzima..is a Thembu leader.

‖ **temenggong** (tẽměŋgǫ·ŋ). Forms: 8 **tamongoong,** 9 **tumângong; temenggong.** [Malay.] In traditional Malay states, a high-ranking official, usu. commanding the army and the police. Also, the title of the rulers of Johore, 1824–85.

1783 W. Marsden *Hist. Sumatra* 285 The *tamongoong*, or commander in the wars. **1859** L. Oliphant *Narr. Lord Elgin's Mission to China & Japan* ii. 23 We ascended to the comfortable bungalow of the Tumângong. **1907** F. Swettenham *Brit. Malaya* iv. 67 A new treaty was made ..between Raffles on the one hand and the Sultan Husein and the Têměnggong on the other. **1964** M. Turnbull in W. Gungwu *Malaysia* ii. viii. 133 This political vacuum encouraged the bid for independence by the subordinate chiefs, the temenggong of Johore and the bendahara of Pahang. **1972** M. Sheppard *Taman Indera* 108 During the reign of Sultan Mahmud of Malacca (1488–1511), Tun Hassan, the Temenggong, who ranked second only to the chief minister, started a new fashion by wearing the sleeves of his tunic long and loose.

temerack, obs. var. Tamarack in Dict. and Suppl.

Temiar (te·mi,āɪ), *sb.* and *a.* [Native name.] **A.** *sb.* An aboriginal people of the Malay Peninsula, also called Sakai; a member of this people. **B.** *adj.* Of or pertaining to this people or their language.

1933 *Jrnl. R. Asiatic Soc.* (Malayan Branch) XI. 291 The Temiars showed an extraordinary desire to sample European medicines. *Ibid.*, The most interesting feature in the Temiar life..was their obvious happiness. **1958** J. Slimming *Temiar Jungle* iv. 54 A Temiar community lives on the centre tongue of land. **1958** *Listener* 13 Nov. 793/3 A Temiar who coveted his wife. **1965** R. McKie *Company of Animals* v. 91 The Temiar are animists who, living among jungle-covered mountains, are perfectly attuned to their environment. **1972** *Times* 6 Dec. (Europe & Third World Suppl.) p. vii/4 (*caption*) Temiar aborigines at the Gombak Aborigine Hospital, Malaysia, for treatment.

te-mine, var. *tamein.

temmoku (te·mokū). Also **tenmoku.** [Jap., ad. Chinese *tiān-mù* eye of heaven (see quot. 1923).] The Japanese name for a type of Chinese porcelain with lustrous black or brown glaze; also, the glaze so used.

1880 A. W. Franks *Jap. Pottery* 5 A tea bowl of porcelain or earthenware (*cha-wan*, or, when of large size, *temmoku*), simple in form, but remarkable for its antiquity or historical associations. **1915** R. L. Hobson *Chinese Pott. & Porc.* I. 31 *Temmoku*..glaze is sometimes flecked with tea green as well as with golden brown. *Ibid.* 131 The Japanese..have always prized the Chien Yao bowls to which they gave the name *temmoku*. **1923** —— & Hetherington *Art of Chinese Potter* 15 The name *temmoku* (*t'ien mu*, or Eye of Heaven) was first given to a bowl, probably of Fukien origin, brought to Japan during the Sung period by a Zen priest from the Zen temple of the *T'ien mu shan* (Eye of Heaven mountain) in the north-west of Chekiang. In later times the generic name of *temmoku* came to be applied to the whole category of wares of this type. **1924** *Trans. Oriental Ceramics Soc. 1923–4* 26 (*title*) The chemistry of the Temmoku glazes. **1934** [see *hare's fur* s.v. *hare *sb.* 6]. **1940** B. Leach *Potter's Bk.* viii. 231 Into the tenmoku I dipped several large jars and bowls. **1958** W. Willetts *Chinese Art* II. vi. 394 *Temmoku* bowls..made to-day at P'êng-ch'êng.., and at T'ai-yüan.., which in many respects can hardly be told apart from their Sung models. **1971** S. Jenyns *Jap. Pottery* iii. 84 The glazes particularly associated with the Toshiros and their successors for tea ceremony wares are of this *tenmoku* variety: 'thick' and semi-transparent but coloured often to opacity by the presence of iron oxide. **1976** *Canadian Collector* Mar.–Apr. 43/1 In Japan this type of black-glazed ware became known as *temmoku*.

Temne (te·mni), *sb.* and *a.* Forms: 8 **Timmany;** 9 **Timan(n)ee, Timmanah, -ee, -eh, -i;** 9– **Temne; Temnee, Timini, Timne, Timni.** [Native name; cf. quot. 1861.] **A.** *sb.* **a.** A people of Sierra Leone; also, a member of this people. **b.** A West Atlantic language spoken by this people. **B.** *adj.* Of, pertaining to, or designating this people or their language.

1791 A. M. Falconbridge *Let.* 8 June in *Narr. Two Voyages to Sierra Leone* (1802) 81 The Burees..were conquered and drove away to other parts of the country by the Timmany's. **1792** —— *Let.* 10 Feb. in *Ibid.* 27 In every house I was accosted by whoever we found at home, in the Timmany language *Currea Yaa*, which signifies—How do you do, mother? **1803** T. Winterbottom *Acct. of Native Africans of Sierra Leone* I. iv. 65 The largest kind [of pepper] is distinguished..in Timmanee, by kik bengbay pootoo; and in Soosoo, by foorootoo bengbay. *Ibid.* v. 80 Timanee towns..have Ro *in general* prefixed to them, as Ro-mak-baynee, a town where they finish things. **1825** [see *Koranko]. **1841** *Penny Cycl.* XXI. 503/1 The Timannees occupy a country extending 90 miles from east to west, and 55 from north to south. **1850** T. E. Poole *Life, Scenery & Customs in Sierra Leone* II. xvii. 274 He gave himself up to the most diligent study of the Timmanah language. **1861** C. F. Schlenker *Collect. Temne Traditions* p. iii, The signification of the term: Témne seems to be 'an old man himself'. They derive it.. from *o-tem*, 'an old man, an old gentleman', to which is affixed the reflexive suffix *-ne*, 'self'; because they believe that the Temne nation will ever exist. **1887** *Encycl. Brit.* XXII. 44/2 The following..can be distinctly classified:—Mandingos, 1190; Timmanehs, 7443 [etc.]. **1903** J. J. Crooks *Hist. Sierra Leone* ix. 121 Many Timini children were carefully brought up according to the Mohammedan creed. **1911** *Encycl. Brit.* XXV. 55/1 Sierra Leone is inhabited by various negro tribes, the chief being the Timni, the Sulima, the Susu and the Mendi. **1916** H. Osman Newland *Sierra Leone* iii. 28 The name of the village is essentially a reminder of the old Temne domination when Kwaia..country embraced all this territory. **1916** N. W. Thomas *Anthropol. Rep. on Sierra Leone* i. i. 7 The remaining languages are Soudanese, and fall into two main groups, prefix and non-prefix tongues. To the former..belong (a) Timne, (b) Limba, [etc.]. **1926** F. W. H. Migeod *View of Sierra Leone* i. iii. 18 The Temne influence is so great that practically all the Bullom north of Freetown speak Temne. **1955** [see *Mande *sb.* and *a.*]. **1957** M. Banton *W. Afr. City* iii. 54 The Temne, conservative as he is in many matters, has a readiness to copy certain new ways. **1957** Ld. Hailey *African Survey 1956* iii. 104 In the Protectorate Mende is the language generally used in education, but Temne and the Kono dialect of Vai are used in mission schools. **1980** E. G. Wilson *John Clarkson* vi. 81 Alexander Falconbridge.. had been prevailed upon..to return to Sierra Leone where his experience with the Temne people would pave the way for the immigrant fleet. *Ibid.* vi. 82 John Clarkson could soothe Falconbridge and win the confidence of the Temne as well.

temp. (temp), abbrev. of L. *tempore* (also used) = in the time of.

1658 W. Dugdale *Hist. St. Paul's Cathedral* 127 Of this, the first mention that I finde, is in that Grant of *Richard de Beaumeir* Bishop of London (*temp.* H.1.) concerning the School. **1767** A. C. Ducarel *Tour through Normandy* 96 Leases of Crown-Lands in Kent, temp. Eliz. **1859** S. A. Allibone *Dict. Eng. Lit.* I. 27 As we have frequently occasion to refer to the reigns of different English sovereigns without specifying dates,—*temp.* Edward II., *temp.* Henry II., &c. **1892** F. W. Maitland *Let.* 6 Sept. (1965) 105 Are you likely to write anything about the frequency of eyres temp. Hen. II? **1932** *N. & Q.* 16 Jan. 46/1 Old registers *tempore* mid-seventeenth century. **1955** *Times* 16 July 7/4 A flying machine, *tempore* Blériot, would almost certainly have crashed in the shrubbery. **1983** W. Blunt *Married to Single Life* i. 2 My great-grandfather, John Blunt of Lindfield in Sussex (*temp.* James I).

temp (temp), *sb.*[1] Colloq. abbrev. of Temperature 7.

1886 R. Roose *Let.* 15 Mar. in R. S. Churchill *Winston S. Churchill* (1967) I. Compan. i. iv. 117 As long as I can fight the temp and keep it under 105 I shall not feel anxious. **1916** G. Bell *Let.* 23 July (1927) I. xv. 383, I shan't be sorry when the temp. drops 20°. **1924** Galsworthy *Forest* ii. ii. 48 Collie's temp.'s down. **1940** M. Dickens *Mariana* v. 170, I wish I had a thermometer with me, I'd take your temp. **1972** J. McClure *Caterpillar Cop* ii. 27 The body temp is above normal. **1977** *Hot Car* Oct. 89/3 The original CF radiator, cooled by a second-hand Kenlowe fan keeps the temps down, no sweat, even in that long hot summer of '76.

temp (temp), *sb.*[2] *colloq.* [f. next or as abbrev. of Temporary *sb.* 4 in Dict. and Suppl.] A temporary employee; esp., a temporary secretary (see *temporary *sb.* 4).

1932 *Amer. Speech* VII. 278 *A temp*, a part-timer [i.e. a part-time non-civil service employee of the U.S. Post Office]. **1967** *Economist* 22 Apr. 374/1 Overstaffing is not solely the result of the unwillingness to use temps. **1970** R. Hill *Clubbable Woman* viii. 238 One of his women, a temp, only comes in at weekends [at a public house]. **1975** *Sunday Sun* (Brisbane) 23 Nov. 51/1 A year ago she sold cattle. Now she sells Temps.

temp (temp), *a.* Colloq. abbrev. of Temporary *a.* 1 a.

1909 in Webster. **1964** in Hamblett & Deverson *Generation X* 173 Between school and university I had several temp. jobs ranging from accounts clerk to van driver. **1968** D. E. Knuth *Art of Computer Programming* I. 632 *Temp storage*, part of memory used to hold a value for a comparatively short time while other values occupy the registers. **1972** *Homes & Gardens* Dec. 22/2 'You do realise it's only temp, dear?' said the Lady Supervisor when I arrived to begin my duties as part-time untrained telephonist. **1977** 'D. Cory' *Bennett* iv. 121 Detective-Inspector (temp.) Eric Hunter.

temp (temp), *v. colloq.* [f. *temp *sb.*[2]] *intr.* To work as a temp (see *temp *sb.*[2]). So **te·mping** *vbl. sb.*

1973 *Times* 15 Mar. 30/5 (Advt.), Bored with temping? We specialise in short term assignments..in the artistic and creative fields. **1974** *Harpers & Queen* Sept. 180/1 You meet such civilized people when temping for Bernardette. **1978** *Times* 2 Oct. (Business Efficiency Suppl.) p. iii/3 Most of the students had given as their explanation for deciding to temp: 'To gain office experience before taking up a permanent job.' **1979** A. Price *Tomorrow's Ghost* i. 9 I'm a bit cheesed off with this temping—I think it's time to dig in somewhere comfy.

tempeh (te·mpe). Also **tempe.** [ad. Indonesian *tempe*.] An Indonesian foodstuff made by fermenting soya beans with Rhizopus and deep-frying them in fat.

1966 *N.Y. Times* 17 Apr. iv. 11/4 A cake of food material..somewhat comparable to Indonesian fermented food called tempeh. **1977** *Daily Colonist* (Victoria, B.C.) 9 Jan. 28/5 Sixty-nine persons died and 265 others in East Java were in hospital for food poisoning after eating tempe, a local dish made from soy beans. **1980** *San Francisco Bay Guardian* 16–23 Oct. 13 (Advt.), Delicious Natural Foods to Eat: Soy Ice Cream, Sushi, Tofu, Tempeh, Mochi, Fresh Juices, Organic Burritos.

temper, *sb.* Add: **III. 12. d.** (See quot. 1975.)

1925 C. E. Guthe *Pueblo Pottery Making* 20 The mixing ..consists of the addition of temper. **1936** K. M. Chapman *Pottery of Santo Domingo Pueblo* 11 The clay and temper are moistened and kneaded together. **1955** Bushnell & Digby *Anc. Amer. Pottery* iv. 32 The temper is normally rather coarse, but in at least one of the three examples..it is very fine. **1975** R. L. Beals *Peasant Marketing System of Oaxaca, Mexico* ii. 18 Temper used in pottery making is a coarse material such as sand or decomposed rock to make the clay more ductile and prevent cracking of the shaped vessel during drying.

IV. 13. *temper-fit, tantrum.*

1884 W. James in *Mind* IX. 199 In injuries to the brain ..we have tears, laughter, and temper-fits, on the most insignificant provocation. **1930** G. C. Myers *Mod. Parent* x. 168 There are vague symptoms of temper tantrum at the age of several weeks when [an infant's] accustomed satisfactions are withheld. **1951** Auden *Nones* (1952) 11 Unable To conceive a god whose temper-tantrums are moral. **1980** *Jrnl. R. Soc. Med.* LXXIII. 217 The affected children themselves are liable to behavioural problems such as temper tantrums.

14. temper-brittleness *Metallurgy*, notch-brittleness produced in certain types of steel when it is held in or cooled slowly through a certain temperature range; hence **temper-brittle** *a.*

1918 *Proc. Inst. Automobile Engin.* XII. 349 If an absolutely unnotched bar is taken and tested under impact conditions, it is frequently found that even if that bar happens to show the peculiar 'temper brittleness' it will bend over without any sign of brittleness in the unnotched state... There is..no difference..between the micro-structure of the 'temper-brittle' steel and the micro-structure of the same steel giving the good impact value. **1930** *Engineering* 24 Oct. 525/3 The metal had been rendered 'temper brittle' by being cooled too slowly in the tempering process. **1967** A. H. Cottrell *Introd. Metallurgy* xx. 384 This temper-brittleness..is associated with fracture along grain boundaries.

temper, *v.* Add: **III. 14. c.** *trans.* To reduce the brittleness in (hardened steel) by re-heating it to a certain temperature and allowing it to cool. Cf. Anneal *v.* 4.

1925 *Jrnl. Iron & Steel Inst.* CXI. 334 Careful observations made on specimens which had been tempered just below 200°C. **1967** A. H. Cottrell *Introd. Metallurgy* xx. 384 If plain carbon or low-alloy steels are tempered below about 250°C they usually remain somewhat brittle.

19. To increase the pliability of straw for corn-dolly making by dampening it with water.

1963 M. Lambeth *Golden Dolly* 11 When plaiting out of season it is necessary to temper the straw. **1976** S. J. Reid *Art of Weaving Corn Dollies* 9 After a period of storage.., straw dries out... To restore it to a supple condition it is necessary to temper (or dampen) the straw.

temperamental, *a.* Add: **2.** Of a person: liable to peculiar moods, having or giving way to an erratic or neurotic temperament. Hence, of a thing: behaving erratically or unpredictably.

1907 *Amer. Mag.* LXIII. 355/2 The Celtic race is above all things temperamental. **1923** E. Wallace *Clue of New Pin* xxix. 255 Tab decided that she was a little temperamental, and loved her for it. **1939** F. Thompson *Lark Rise* iii. 42 A temperamental person was said to be 'one o' them as is either up on the roof or down the well'. **1962** *Amer. N. & Q.* I. 31/1 The horse was particularly suitable in northern Europe where the temperamental climate often made rapid ploughing and planting important. **1965** *Wireless World* Sept. 436/1 He [*sc.* Dr. W. H. Eccles] also started a study of the coherer, the only detector of the period, which led to a better understanding of the action of that temperamental device. **1977** M. Drabble *Ice Age* i. 79 The central heating worked, and he had boosted it with an electric fire, albeit a temperamental electric fire, which needed the occasional kick.

Hence **te:mperame·ntalist**, a temperamental person. *rare.*

1924 *Blackw. Mag.* June 786/1, I was what you might call a temperamentalist, and very easily hypnotised.

temperance. Add: **I. 2. b.** *temperance address, badge, drink* (example), *lecture* (example), *man* (earlier example), *meeting* (examples), *ship, society* (earlier example); **temperance hall,** a building used for public meetings or entertainments at which no intoxicants are sold or provided; **temperance hotel** (examples); **temperance house** = *temperance hotel.*

1831 (*title*) Report of the American Temperance Society. **1833** E. T. Coke *Subaltern's Furlough* ii, in *Waldie's Sel. Circulating Library* 24 Sept. 380/3 Many hotels have 'temperance house' inscribed in large gilded letters over the door or sign. **1833** C. F. Hoffman *Let.* 26 Dec. in *Winter in West* (1835) I. 211 The most devout temperance man could see no harm in that! **1834** J. J. Strang *Diary* 1 July in M. M. Quaife *Kingdom of St. James* (1930) 219, I shall soon have to make the temperance address. **1837** W. Jenkins *Ohio Gaz.* 69 Barnesville..has..2 taverns, and 1 temperance hotel. **1840** R. H. Dana *Two Years before Mast* (1841) xxxi. 118/2 This was a 'temperance ship'; and like too many such ships, the temperance was all in the forecastle... The captain..can drink as much as he chooses. **1841** *Cultivator* VIII. 163, I witnessed..a very large temperance meeting at which I saw 'female influence' fully exerted in a most glorious cause. **1843** Temperance hall [see *perambulatory *sb.* a]. **1862** G. Borrow *Wild Wales* II. xiv. 154, I drew nigh..and read: 'tea made here, the draught which cheers but not inebriates.' I was before what is generally termed a temperance house. **1869** J. H. Browne *Great Metropolis* 327 A large number remain open, pretending to sell nothing but 'temperance drinks'. **1881** I. M. Rittenhouse *Maud* (1939) 17 Went to the Temperance Meeting last night with Emma and Mabel. **1902** O. Wister *Virginian* 454 Once I had to sleep in a room next a ladies' temperance meetin'. **1922** Joyce *Ulysses* 476 Bloom's bodyguard distribute.. loaves and fishes, temperance badges, expensive Henry Clay cigars. **1933** E. O'Neill *Ah, Wilderness!* iv. iii. 153 I'm not going to read you any temperance lecture. **1978** P. Bailey *Leisure & Class in Victorian Eng.* iv. 82 Bolton... A vast new town hall, opened in 1873, looked down upon the Free Library..a Second Temperance Hall. **1978** *Times* 5 Apr. (Tourism in Britain Suppl.) 4/9 For those seeking an unusual experience Birmingham..has a famous temperance hotel.

c. Used as *predic. adj.* Practising or advocating total abstinence.

1907 J. Masefield *Tarpaulin Muster* xiii. 138 They're temperance down at the Point in these times. **1977** H. Fast *Immigrants* iii. 193 You're not temperance, are you?

temperate, *a.* Add: **3. c.** Of food: produced in, or suitable for production in, a moderate climate.

1960 *Economist* 8 Oct. 125/3 The New Zealanders.. have recognised that reconciling their highly competitive exports of temperate farm products with a common European agricultural policy will be an extremely difficult task. **1962** *Listener* 27 Sept. 462/1 The so-called temperate foodstuffs—that's to say corn and meat and dairy produce. **1980** *Nature* 7 Feb. 514/3 In many parts of the Third World people eat temperate vegetables: cauliflower, onions, carrots, potatoes, cucumber, tomatoes.

8. *Microbiology.* Of a phage: not necessarily causing lysis of the host cell, but able to exist as a prophage for a number of generations; giving rise to lysogenic bacteria. [tr. F. *tempéré* (F. Jacob et al. 1953, in *Ann. de l'Inst. Pasteur* LXXXIV. 223).]

1953 *Cold Spring Harbor Symp. Quantitative Biol.* XVIII. 65/1 It has been proposed..to call temperate (as opposed to virulent) those phages which are able to establish the lysogenic condition in their host cells. **1961** M. Hynes *Med. Bacteriol.* (ed. 7) xxvi. 401 The virulent phages are often termed lytic, since they disrupt the bacteria they infect; the symbiotic phages are usually termed temperate. Cultures infected with a temperate phage are confusingly termed lysogenic. **1979** Arms & Camp *Biology* xvi. 244 A virulent, lytic bacteriophage.. can only replicate and cause cell lysis. Other phages, known as temperate phages, may either replicate and lyse the cell they invade, or may instead enter a dormant phase in which the phage DNA is joined to that of the host cell..and replicated with it over many cell generations.

temperature. Add: **10.** *temperature control, -dependence, -dependency, -independence; temperature-controlled, -dependent, -independent, -regulating, -sensitive* adjs.; **temperature coefficient** *Physics*, a coefficient expressing the relation between a change in a physical property and the change in temperature that causes it; **temperature gradient**, a gradient (sense 2 in Dict. and Suppl.) of temperature; **temperature inversion** *Meteorol.*, the phenomenon of an increase of temperature with height above the ground; **temperature-regulation** *Biol.* = *thermoregulation; **temperature-salinity** *adj. phr.*, relating to the temperature and salinity of water; *spec.* applied to a diagram in which both are plotted as a function of depth.

1902 *Encycl. Brit.* XXVIII. 8/1 The quantity *a* is then called the temperature-coefficient, and its reciprocal is the temperature at which the resistivity would become zero. **1962** *Newnes Conc. Encycl. Nucl. Energy* 791/2 For the practical operation of a reactor the temperature coefficient of reactivity should be small so that a steady power can be maintained by moving control rods at a moderate speed. **1923** Glazebrook *Dict. Appl. Physics* III. 582/1 (*heading*) Temperature control. **1959** E. T. Hall *Silent Language* iii. 79 Clothes and houses are extensions of man's biological temperature-control mechanisms. **1935** *Discovery* Nov. 322/1 The centres are passed by an automatic feeding attachment through a curtain of temperature-controlled chocolate. **1970** *Jrnl. General Psychol.* Oct. 163 They were housed in individual cages in a temperature-controlled laboratory. **1946** *Nature* 7 Sept. 333/1 Experiments on the temperature-dependence of the breakdown strength F. **1974** J. W. Drake in Carlile & Skehol *Evolution in Microbial World* 53 The temperature-dependencies of the *Neurospora* and T4 rates differed markedly. **1962** Simpson & Richards *Physical Princ. Junction Transistors* ix. 200 The most important temperature-dependent property of transistors is the collector cutoff current. **1882, 1962** Temperature gradient [see Gradient *sb.* 2]. **1978** *Jrnl. R. Soc. Arts* CXXVI. 683/1 Structures of supersonic aircraft are subject to thermal stresses due to temperature gradients. **1965** *Math. in Biol. & Med.* (Med. Res. Council) vi. 256 The same difficulty arose with the temperature-independence of temporal rhythms. **1946** *Nature* 7 Sept. 333/1 The latter is temperature-independent but increases with the concentration of foreign atoms. **1945** E. Bollay in F. A. Berry et al. *Handbk. Meteorol.* x. 758 (*caption*) Characteristic properties of nonfrontal temperature inversions. **1977** I. M. Campbell *Energy & Atmosphere* viii. 252 A further circumstance of temperature inversion occurs where there is an enclosed valley in which cold air..tends to collect. **1911** J. A. Thomson *Biol. Seasons* iv. 338 It represents an interesting reminiscence of a more primitive physiological state when the temperature-regulating mechanism was not yet well established in the ancestral mammals. **1957** J. S. Huxley *Relig. without Revelation* (rev. ed.) ix. 216 The temperature-regulating mechanism of higher mammals. **1927** Haldane & Huxley *Animal Biol.* xi. 240 The mammals..possess proper temperature-regulation. **1930** *Rep. 'Michael Sars' N. Atlantic Deep-Sea Exped. 1910* I. i. 19 The 'normal' temperature-salinity curve is reproduced..in such a way that the corresponding values of temperature and salinity can easily be read off. **1942** H. U. Sverdrup et al. *Oceans* iv. 141 Water masses can be classified on the basis of their temperature-salinity characteristics. **1959** H. Barnes *Oceanogr. & Marine Biol.* 157 As one passes across the Gulf Stream there is a fairly sharp temperature-salinity boundary between Gulf Stream water and the so-called Slope water lying over the Continental Shelf. **1962** *Science Survey* XX. 308 The surface of the body contains a number of temperature-sensitive patches which produce patterns of nerve impulses related to the ambient temperature.

tempering, *vbl. sb.* Add: **a.** (Examples in sense *14 c of the vb.)

1881 *Encycl. Brit.* XIII. 352/2 The generic phrase 'tempering' is usually applied to mean a combination of the hardening and annealing processes..hardening [the steel] to a red heat and suddenly cooling, and then heating up again to a somewhat lower temperature and allowing to cool slowly. **1941** Jones & Schubert *Engineering Encycl.* II. 1274 The object of tempering, or 'drawing', is to reduce the brittleness in hardened steel.

b. *tempering temperature.*

1910 *Encycl. Brit.* XIV. 808/1 The higher the tempering-temperature, *i.e.* that to which the hardened steel is..reheated, the more is the molecular rigidity relaxed. **1967** A. H. Cottrell *Introd. Metallurgy* xx. 384 Higher tempering temperatures may be used, thereby improving the toughness of the steel.

temperish (te·mpərɪʃ), *a.* [f. Temper *sb.* II + -ish[1].] Inclined to or exhibiting bad temper.

1925 *Chambers's Jrnl.* Nov. 764/1 From these temperish folk arise strange quarrels. **1929** *Times* 16 Jan. 13/5 His [*sc.* Lord Haig's] reply was more than emphatic; it was almost temperish. **1935** [see *riding-blanket* s.v. *riding *vbl. sb.* 5 a].

tempersome (te·mpəɹsəm), *a.* orig. *dial.* [f. Temper *sb.* + -some[1].] Quick-tempered. Also pseudo-*arch.* [after Temper *sb.* 6], displaying extreme conditions of weather. Hence **te·mpersomeness.**

1875 W. D. Parish *Dict. Sussex Dial.* 119 *Tempersome*, hasty-tempered. **1879–81** G. F. Jackson *Shropshire Word-bk.* 434 *Tempersome*, adj., hot-tempered; passionate. **1906** W. De Morgan *Joseph Vance* xliii. 392 He was very tempersome about it. **1909** —— *It never can happen Again* I. xxv. 370 Marianne, for all her tempersomeness and jealousy, loved and reverenced Challis. **1946** M. Peake *Titus Groan* 234 And now that it is so tempersome and cold you are always going out into the nastiness and getting wet or frozen.

tempery (te·mpěri), *a. dial.* [f. Temper *sb.* + -y[1].] Short-tempered.

1905 *Eng. Dial. Dict.* VI. 60/1 Yorks... 'She was a tempery body.' **1951** H. Giles *Harbin's Ridge* xiii. 116 She had tempery ways..and sometimes I felt like I was in a pot of water with it boiling. **1954** *Landfall* VIII. 266, I cursed on like a tempery child deprived Of what he'd loved.

tempest, *sb.* Add: **1. b.** Also *North-eastern N. Amer.*

1877 R. T. Cooke in *Harper's Mag.* Jan. 297/1 Ominous flashes of tempest began to play about the far horizon. **1892** *Dialect Notes* I. 211 *Tempest*, a thunder-shower. [Plymouth, Massachusetts.] **1951** *Amer. Speech* XXVI. 251 Such localized terms as..the southeastern New England tempest (thunderstorm)... Tempest was recorded from a Schuylerville (Saratoga Co.) informant, definitely conscious of his ultimate Nantucket ancestry. **1965** E. Richardson *Living Island* 171 August is also the month of tempests (for here [in Nova Scotia] electrical storms keep the name used by Shakespeare).

5. b. *tempest-charged, -haunted, -shaken, -shattered, -smitten, -throttled, -torn* (later examples), *-troubled* (later example); **c.** *tempest-cleaving, -walking* (example).

1826 J. G. Whittier *Vale of Merrimac* in *Free Press* (Newburyport, Mass.) 29 June 4/1 And the tempest-charg'd vapor their tall tops embraces. **1818** Shelley *Rosalind & Helen* (1819) 77 That a tempest-cleaving swan Of the songs of Albion..Found a nest in Thee. **1880** Longfellow *Ultima Thule* 7 Are not these The tempest-haunted Hebrides, Where sea-gulls scream? **1817** Shelley *Laon* i. 23 The tempest-shaken wood, The waves, the fountains, and the hush of night. **1845** Longfellow *Seaweed* vii, in *Poems* 99 From the wreck of Hopes far-scattered, Tempest-shattered, Floating waste and desolate. **1844** J. G. Whittier *Bridal of Pennacook* in *United States Mag.* Sept. 239 Sometimes The tempest-smitten tree receives From one small root the sap which climbs Its topmost spray and crowning leaves. **1930** R. Campbell *Adamastor* 51 Doomed vessels..Reared to the stars their tempest-throttled cry. **1918** W. S. Churchill in M. Gilbert *Winston S. Churchill* (1977) IV. Compan. i. 418 Prompt and clear must be those great decisions which assign definite limits to the increasing confusion and miseries of the vanquished and above the tempest-torn waters light again the beacons of mankind. **1939** R. Campbell *Flowering Rifle* ii. 64 The wide-winged and wounded Albatross The tempest-torn that rides (and bears) the strife. **1952** —— tr. *Baudelaire's Poems* 183 She sought, with tempest-troubled gaze, the skies Of her first innocence. **1820** Shelley *Prometh. Unb.* i. 35 These are Jove's tempest-walking hounds.

tempest, *v.* **3.** For *? Obs.* read *dial.* or *arch.*

1875 W. D. Parish *Dict. Sussex Dial.* 119 It tempestes so as we're troubled to pitch the hay upon to the stack anyhows in the wurreld. *a* **1907** F. Thompson *Works* (1913) I. 120 Flew spurned the pebbled stars: those splendours then Had tempested on earth, star upon star.

tempest-tossed, *a.* (Later examples.)

1817 Shelley *Prince Athanase* in *Posthumous Poems* (1824) ii. ii. 251 Thus had his age, dark, cold and tempest-tost, shone truth upon Zonoras. *a* **1887** E. Lazarus *New Colossus* in *Poems* (1889) I. 203 Send these, the homeless, tempest-tost to me, I lift my lamp beside the golden door! **1955** E. Pound *Classic Anthol.* i. 8 My heart is no more tempest-toss'd. **1984** *Times* 10 Mar. 17/1 People wrangle and row as though they were part of a tempest-tossed drama by John Cassavetes.

‖ **tempête** (taṅpęt). [Fr., lit. 'tempest'.] An English country-dance (and tune) of the late-nineteenth century.

1873 L. Troubridge *Life amongst Troubridges* (1966) viii. 72 We had a *Tempête*, the Boulanger and 'Wiss'. **1879** *Manners & Tone of Good Society* viii. 139 Country dances, such as the 'Tempête', 'Sir Roger de Coverley', etc., are usually danced at private balls when given in the country. **1923** G. Arthur *Further Lett. from Man of no Importance* (1932) 155 Yes, the Tempête was the popular dance at children's and semi-grown-up dances of our young days, and even later at country balls; fancy its being danced again in Paris.

‖ **tempietto** (tempie·to). Pl. **tempietti.** [It., lit. 'little temple': cf. Templet[4].] A colonnaded building, freq. of circular form, surmounted by a dome.

1896 W. J. Anderson *Archit. Renaissance in Italy* iv. 77 Bramante's *Tempietto* in the Cloisters of San Pietro in Montorio (1502). **1901** M. Carmichael *In Tuscany* 158 Civitali was..a fine and practical architect. He is thought to have built the Palazzo Pretorio, and to him belongs the honour of having built the dome-like chapel (*Tempietto*) in which is preserved the sacrosanct and venerable Volto Santo. **1922** G. Gromort *Italian Renaissance Archit.* iv. 80 Palladio himself, half a century later, made a drawing of Bramante's Tempietto. **1960** E. Bowen *Time in Rome* ii. 47 A *tempietto* in a moist northern garden. **1962** *Listener* 1 Feb. 228/1 The ancient Romans had filled their gardens with statuary as well as with *tempietti*, colonnades, covered galleries, apses, and pavilions. **1971** *Country Life* 7 Oct. 887/1 The Hafeziye [in Shiraz, Iran] is much more than the pleasant municipal garden one might expect, the formal approach, open colonnade and delightful tempietto beyond achieving a simple grandeur that is very memorable.

temping, *vbl. sb.*: see *temp *v.*

temple, *sb.*[1] Add: **I. 1. e.** A Jewish synagogue; now *spec.* the place of worship of Reform (and some Conservative) Jews. Now chiefly *U.S.*

1598 J. Stow *Survey of London* 277 But that this house hath beene a Temple or Jewish Sinagogue..I allow not. **1830** *Monthly Intelligence* May 75/2 There is at Frankfort ..a considerable body of Jews, belonging to what is called the 'New Temple'... Mr. Moritz mentions having visited their Temple. **1850** G. Aguilar *Vale of Cedars* v. 27 The little temple was erected..and the solemn rites of their peculiar faith adhered to. **1914** I. Cohen *Jewish Life in Modern Times* xi. 287 So occidentalized has the Reform temple become that a visitor at first sight can hardly distinguish whether he is in a synagogue or a chapel. **1942** C. Roth in *Menorah Jrnl.* Winter 4 Their place of worship (no longer a homely *Schul* but, with unhappy retrogression, a Temple). **1978** H. Kemelman *Thursday the Rabbi walked Out* (1979) ii. 14 It's the place of women in the temple service I want to talk about, Rabbi. *Ibid.* vii. 45 The synagogue, or as we call it, the temple. **1981** G. V. Higgins *Rat on Fire* vii. 56 Saturdays everybody dressed up and went to temple.

III. 6. a. (senses 1–3) *temple-court* (examples), *-door* (examples), *-priest* (example); (appositive) *temple-pyramid* (examples). **c. temple block,** a percussion instrument of oriental origin consisting of a hollow block of wood which is struck with a drum-stick; also known as a *wood block; usu. in *pl.*; **temple prostitute,** a woman maintained by a temple, who performs rituals of a sexual nature (cf. *devadasi); also *fig.*; hence **temple prostitution; temple-state**: in antiquity, a city-state centred on a temple or similar sacred edifice.

1929 *Melody Maker* Mar. 295/2 The same remark applies to the Temple blocks, and even the tambourine can easily be played too loudly. **1964** J. Carter in Norton & Spacey *Drums & Drumming Today* 40 How I yearn for the days of temple blocks and saucepan lids. **1820** Shelley *Witch of Atlas* in *Posthumous Poems* (1824) 50 And round each temple-court In dormitories ranged..She saw the priests asleep. **1930** R. Graves *Ten Poems More* 13 In every temple-court, for all to see Flourishes one example of each tree In tricunx. *c* **1386** Chaucer *Knight's Tale* (1875, Harl. MS. 7334) 2422 The rynges on þe tempul dore þat hange. **1729** Swift *Directions for Birth-day Song* in *Poems* (1958) 462 What tho for fifteen years and more, Janus hath lock'd his Temple-door? **1921** G. B. Shaw *Back to Methuselah* iv. 178 The temple door is in the middle of the portico. **1941** J. Masefield *Gautama* 31 Like a temple-priest intoning. **1890** A. B. Ellis *Ewe-Speaking Peoples of Slave Coast of W. Afr.* ix. 141 Girls dedicated to a god do not necessarily serve him during the whole of their lives... In Dahomi there seems to be a marked distinction between those who actually minister to the service of the temple, and those who are merely temple prostitutes. **1951** Auden *Nones* (1952) 28 Private rites of magic send The temple prostitutes to sleep. **1980** S. T. Haymon *Death & Pregnant Virgin* xi. 84 Poor old Charlie! Thought he'd recruited a vestal virgin when what he'd got was a temple prostitute! **1912** J. N. Farquhar *Primer of Hinduism* xvi. 194 We are now in a position to realise how it has been possible for the Hindu to admit such things as..cruel torture, temple prostitution. **1961** L. Mumford *City in History* iv. 106 The custom of temple prostitution has not merely been preserved down to our own day.., but the temples of the goddesses of love..were traditionally the favored places of assignation for lovers. **1948** *Nat. Geogr. Mag.* Jan. 127 (*caption*) High and steep were the temple pyramids of the Maya. **1966** M. D. Coe *Maya* v. 94 Towering above all are the mighty temple-pyramids built from limestone blocks over a rubble core. **1920** H. G. Wells *Outl. Hist.* 150/2 There is no temple-state stage, no stage of priest Kings, in the Greek record. **1931** *Times Lit. Suppl.* 1 Jan. 3/2 The Pope's temporal domain..is not a city-state but a temple-state.

temple, *v.* **1.** (Later *poet.* example.)

1936 A. Clarke *Coll. Poems* 98 We saw again How Brigid, while her women slept Around her, temple'd by the flame, Sat in a carven chair.

templed, *ppl. a.* Add: **2.** (Later examples.)

1935 T. S. Eliot *Murder in Cathedral* i. 27 Power obtained grows to glory,..a permanent possession, A templed tomb. **1951** L. MacNeice tr. *Goethe's Faust* 217 Does not your templed home persist!

3. (Earlier example.)

1822 Shelley *Charles the First* in *Posthumous Poems* (1824) ii. 245 Innocent sleep of templed cities and the smiling fields.

templet[1]. Add: The form *template* is now usual, freq. with the spelling-pronunc. (te·mplě[1]t). **2. a.** (Later examples.)

1929 D. S. Stewart *Pract. Design Simple Steel Struct.* I. iii. 24 Templates may be either..the bars which are to be used in the structure or..made from some light and easily worked material. **1942** *Sun* (Baltimore) 25 Nov. 6/4 A ship starts being a ship in the mold loft, where skilled hands make wooden patterns, called templates, from the designer's blueprints. **1967** E. Short *Embroidery & Fabric Collage* iii. 71 The main lines of the design can be chalked in or tacked on to the top fabric, details being put in freely or drawn round a template. **1977** *Early Music* July 443/3 (Advt.), Set of three, fully-explanatory A2 drawings, two templates, [etc.].

c. *Oil Industry.* A frame anchored to the sea-floor to which an offshore platform may be attached.

1975 *Petroleum Rev.* XXIX. 142/1 The system is mounted on a tubular steel frame called a template, which is 124 ft wide and 42 ft high. **1976** *Offshore Platforms & Pipelining* 19/1 The riser..is anchored to a template on the sea floor.

3. Chiefly *Biol.* A molecule or molecular pattern that determines the sequence in which other molecules are assembled into a macromolecule; *spec.* a molecule of nucleic acid that acts thus in the synthesis of nucleic acids or proteins.

[**1904** *Proc. R. Soc.* LXXIII. 542 The protoplasmic complex may be regarded as built up of a series of associated templates which serve as patterns to determine change in the various directions necessary for the maintenance of vital processes and of growth.] **1949** *Q. Rev. Biol.* XXIV. 98/1 If we accept the view that the normal cellular proteins are formed as negative replicas of a positive cellular template, we are confronted with a serious dilemma. **1953** Watson & Crick in *Nature* 30 May 966/1 Previous discussions of self-duplication have usually involved the concept of a template, or mould... Our model for deoxyribonucleic acid is, in effect, a *pair* of templates. **1961** *Ann. Reg. 1960* 402 This theory differed from the older 'instructive' theory in which any cell was able to produce antibody to any pattern using the antigen itself as a template. **1964** *Proc. Nat. Acad. Sci.* LI. 801 (*heading*) A complex of enzymatically synthesised RNA and template DNA. **1970** *Nature* 5 Sept. 1012/2 RNA tumour viruses can act as templates for the synthesis of DNA. **1977** *N.Y. Rev. Bks.* 27 Oct. 17/1 Segments of DNA, selected because they are templates for valuable products such as hormones, antigens or antibodies, might be produced in bulk by multiplying them in culture of *E. coli*. **1980** N. K. Mathur et al. *Polymers as Aids in Org. Chem.* iii. 161 The polymer was prepared with functional groups juxtaposed in an exact, predetermined steric relationship by polymerizing monomers around an optically active template—either D-glyceric acid or [etc.].

4. *transf.* and *fig.*

1965 *Science & Psychoanalysis* VIII. 64 What is established is rather a probabilistic system of implicit or 'unconscious' schemas..which serve as some kind of abstract templates for comparison. **1973** *Computers & Humanities* VII. 159 Each English text to be translated goes through a fragmentation and reordering that allows it to match a template form... The translation into French is then made from the template and the original text. **1976** Nichols & Armstrong *Workers Divided* ii. 143 Their usual point of reference is the old/Northern/real working class. This forms the template against which they judge the modern/militant..generation. **1983** *Microcomputer Printout* Sept. 57/1 Some companies market ready-written models, sometimes called templates on a disk, for standard functions such as a Profit and Loss statement.

tempo. Restrict ‖ to the spec. It. phrases and add: Pl. also **tempos. 1. a.** (Later examples of the plural.) **tempo giusto** (dʒu·sto), strict time; the proper speed that a style of music demands; **tempo rubato** (earlier and later examples).

1740 Grassineau *Mus. Dict.* 283 *Tempo*, or *Tempo giusto*, is often met with after Recitatives, and intimates that the Time be beat equal, which during that recitative was managed otherwise. **1773** C. Burney *Present State of Mus. in Germany* II. 175 It was from her that Quantz first heard what professors call *tempo rubato*. **1810** D. Corri *Singers Preceptor* I. 6 Tempo rubato is a detraction of part of the time from one note, and restoring it by increasing the length of another. **1886** G. M. Hopkins *Let.* 11 Dec. (1935) I. 246 This sonnet shd. be almost sung: it is most carefully timed in *tempo rubato*. **1931** M. D. Calvocoressi tr. *Bartok's Hungarian Folk Music* 23 A few tunes in *tempo giusto*. *Ibid.* 80 In fairly old tunes ..a liking for variable *tempo giusto* rhythm is evinced. **1934** Webster, *Tempo*, n.; pl. *tempi*.., *tempos*. **1956** R. C. Marsh *Toscanini* ii. 83 The earlier performance being somewhat more relaxed and containing some *tempo rubato* that the version of thirty years later lacks. **1967** A. L. Lloyd *Folk Song in England* iv. 312 The group refrains..were always sung plain and in a strict *tempo giusto*. **1980** *Times* 13 May 15/3 Tempos were excellently chosen, most of all perhaps, for the Minuet.

b. *transf.* and *fig.* The rate of motion or activity (*of* someone or something).

1898 G. B. Shaw *You never can Tell* ii. 249 *Again changing his tempo to say to Valentine*..If youll allow me, sir? **1901** *Cassell's Mag.* Sept. 388/2 His *tempo*, to use the expression of our acrobats, is perfect—that is to say, he yields at the proper time and at the proper rate to the descending ball. **1918** A. Gray tr. *Grelling's Crime* II. 239 He describes their readiness 'to retard the tempo of the construction of our warships'. **1925** C. Fox *Educ. Psychol.* 271 Each person is possessed of a vital tempo. **1930** W. K. Hancock *Australia* vii. 139 The State might not have achieved as much if it had been content..to quicken the *tempo* of the economic harmonies—by taxing, by disseminating knowledge, by mobilising credit. **1940** W. Faulkner *Hamlet* 219 But when he stood in the door again, save for the slightly increased rasp and tempo of his breathing, he might never have left it. **1961** A. Christie *Pale Horse* xviii. 198 It's a changing world, Easterbrook... Now the changes come more rapidly. The tempo has quickened. **1974** I. Murdoch *Sacred & Profane Love Machine* 202 The partner who created the confidence and set the tempo was Luca.

temporale (tempǒrā·l). Also ‖ **temporal.** [ad. Sp. *temporal* storm, spell of rainy weather.] A weather condition of the Pacific coast of Central America consisting of strong south-west winds bringing heavy rain.

1853 tr. *F. Gerstaecker's Narr. Journey round World* I. ix. 145 A temporale, however, might happen at any moment. *Ibid.* ix. 160 All the threatened dangers of the cordilleras, of snow-drifts and temporales, were past. **1923** J. Hergesheimer *Bright Shawl* 141 The April temporale lay in an even heat over the city. **1936** L. J. Halle *Transcaribbean* ii. 32 Twice or more in the season there will be 'temporals', when it will rain interminably for a week. **1982** *N.Y. Times* 4 Apr. x. 18/4 There is scarcely one temporale, that sudden downpour of sheet-rain bringing with it a fine red tufa dust.

temporalis (tempǒrē[1]·lis). *Anat.* [L.: see Temporal *a.*[2] and *sb.*[2]] Also *temporalis muscle.* A fan-shaped muscle which closes the lower jaw and which arises from the temporal fossa, passes through the gap between the zygomatic arch and the side of the skull, and is attached to the coronoid process and the anterior border of the ramus of the mandible.

1676 W. Molins *Μυοτομια or Anat. Admin. of Muscles Humane Body* 17 If you throw this Muscle either from his origination or insertion, *Temporalis* will appear in his insertion. **1713** W. Cheselden *Anat. Human Body* ii. ii. 55 Temporalis, arises from the Os Frontis, Parietale, Sphænoides, and Temporis, and..is inserted externally into the Processus Corone of the Lower Jaw which it pulls upwards. **1873** G. Fleming tr. *Chauveau's Compar. Anat.* i. § 3. ii. 223 To dissect the temporalis, excise the external pterygoideus from its inferior border. **1910** *Bull. Amer. Museum Nat. Hist.* XXVIII. 302 The *temporalis* is the most powerful jaw muscle in *Carnivora*. **1938** H. L. Wieman *Gen. Zool.* (ed. 3) iv. 78 A portion of the *temporalis* muscle can be seen between the orbit and the tympanum. **1978** *Sci. Amer.* Apr. 64/2 In 1944 a German surgeon..tried attaching a flap of muscle (the temporalis) from the side of the head to the surface of the brain, hoping that the muscle's blood vessels would join the cerebral arteries and supply them.

temporally, *adv.* Add: **3.** With regard to time.

1961 in Webster. **1971** *Nature* 21 May 172/1 Temporally varying deviations between 40° and 60° were found. **1974** *Sci. Amer.* June 31/2 The pulse, which contains only about 10^{-3} joule of laser energy, is shaped spatially (and also temporally, if desired) prior to amplification. **1982** *Ibid.* Feb. 129/2 Temporally coherent light would show at the sampling point a continuous variation between crest and trough.

‖ **tempora mutantur** (te·mpǒrā miūtæ·ntǝɹ), *Latin phr.* [L., in full *tempora mutantur nos et* (or *et nos*) *mutamur in illis* (also used), times change and we change with them.] A statement emphasizing the inevitability of change in human affairs and customs.

A similar saying *Omnia mutantur, nos et mutamur in illis* is found in *Delitiæ Poetarum Germanorum* (1612) I. 685, where it is ascribed to the Emperor Lothair I by the author (who may be Matthias Borbonius), but there is no evidence that it has a medieval origin. The phrase *tempora mutantur*, however, occurs earlier in other contexts: see quot. *c* 1440.

[*c* **1440** *Gesta Romanorum* (Harl. MS.) x. 28 And þei founde þis y-wreten, Tempora mutantur; Homines deteriorantur;..this is to sey, 'tymys ben chaungid; and men ben hyndred, or turnyd, or I-made worse.'] **1577** W. Harrison *Descr. Britayne* iii. iii. 99 in Holinshed *Chron.* I, Oft in one age, diuers iudgementes doe passe upon one maner of casse, wherby the saying of the poet *Tempora mutantur, & nos mutamur in illis.* [**1666** G. Torriano *Piazza Universale di Proverbi Italiani* 281 Times change, and we with them... [*Note*, 292] The Latin says the same, *Tempora mutantur, et nos mutamur in illis.*] **1855** W. Chappell *Popular Music* I. 309 However unobjectionable this song may have been in Queen Mary's time, the three remaining stanzas would not be very courteously received in Queen Victoria's. *Tempora mutantur.* **1934** Blunden *Mind's Eye* 154 He could not repress an exclamatory *Tempora mutantur.* **1961** *Times* 7 Apr. 20/7 The Rugby-watching public can in no way afford to be smug. *Tempora mutantur* indeed, and..not so many years ago a boo would have caused apoplexy in older Rugby followers. **1978** J. I. M. Stewart *Full Term* iv. 56 'Giles has my old rooms..on the floor above this.' 'More *tempora mutantur* stuff.' **1980** *Times* 7 May 18/2 Charlie Chan was always successful... This was '*because*' he was Chinese... *Tempora mutantur, nos et mutamur in illis*, particularly the progeny of a nasty modern breed of censors... working up a protest against the revival of Charlie Chan.

temporary, *a.* (*sb.*). Add: **A.** *adj.* **1. b.** *temporary hardness*, hardness of water that can be removed by boiling, because it is due to bicarbonates which are thereby precipitated.

1895 H. Collet *Water Softening & Purification* i. 2 The temporary hardness is that due to the bicarbonates of lime and magnesia. **1969** Temporary hardness [see *permanent hardness* s.v. *permanent *a.* (*sb.*) 1 d].

d. Designating one who is commissioned for the duration of a war (esp. that of 1914–18), as *temporary captain, officer*, etc. Also *temporary gent*(*leman*) (in colloq. or pejorative Services' use); abbrev. T.G. s.v. *T 6 a.

1918 Barrie *Echoes of War* 68 Socially he had fallen..; even..in his uniform the hasty might say something clever about 'temporary gentlemen'. **1925** W. Deeping *Sorrell & Son* ii. 21 'My name is Sorrell, Captain Sorrell.' 'You will have to drop the 'captain'. Temporary, I suppose?' **1938** S. Beckett *Murphy* x. 234 He withheld his hand, the little temporary gent and pure in heart. **1958** S. Raven in H. Thomas *Establishment* 72 Temporary Captain C.C. had been in the very first intake at Sandhurst. **1976** *N.Y. Rev. Bks.* 5 Feb. 20/3 The lower-middle-class officer..who in the First World War would have been called a 'temporary gentleman'. **1983** T. Pocock *1945: Dawn came up like Thunder* v. 151 There were two officers' messes—'A' and 'B'; the former primarily for staff officers..the latter, mostly made up of temporary officers.

B. *sb.* **4.** *spec.* a secretary or clerical worker supplied by an agency to cover absences or vacancies for a short period.

1956 C. Blackstock *Dewey Death* vii. 159 Temporaries came, and temporaries went..and the work produced [was] shocking beyond belief. **1957** R. Hart-Davis *Let.* 7 July in *Lyttelton–Hart-Davis Lett.* (1979) II. 123 This might have been possible if my secretary hadn't been on holiday. Instead I was burdened with a pudding-faced 'temporary'. **1970** *New Yorker* 17 Oct. 148/1 (*caption*) But Mr. Clark! I'm just a temporary.

tempore: see *temp. (abbrev.).

temporo-. Add: *temporomandibular* adj. (examples).

1889 A. Macalister *Text-bk. Human Anat.* 243 The temporo-mandibular joint forms a double condylarthrosis. **1974** *Nature* 8 Mar. 165/2 An asymmetrical functional activity of both temporomandibular joint mechanisms must compensate during chewing and non-chewing activities.

‖ **temps** (taṅ). *Ballet.* [Fr., lit. = time.] A term used in the names of various ballet movements in which there is no transfer of weight from one foot to the other (see quots.).

1890 G. B. Shaw in *Star* 21 Feb. 2/4, I do not know.. which particular *temps* is a *battement* and which a *ronde de jambe*. **1922** Beaumont & Idzikowski *Man. Classical Theatr. Dancing* V. 195 Ballonné à trois temps. *Ibid.* 196 Temps de cuisse..Temps levé. *Ibid.* 200 Temps d'Allegre. **1930** Craske & Beaumont *Theory & Pract. Allegro in Classical Ballet* (1960) 9 In a *pas d'élévation*, such as a *Temps de Poisson*, the arms are raised *en attitude*. **1947** N. Nicolaeva-Legat *Ballet Educ.* iv. 72 *Temps lié* is a set combination of steps and arm movements executed to the count of 4 or 2/4. **1948** *Ballet Ann.* ii. 66, I set to work on recording a number of *temps d'allegro* and enchainements. **1957** G. B. L. Wilson *Dict. Ballet* 265 *Temps*..is variously used to describe a movement in which there is no transfer of weight (e.g. temps levé) or the division of a step into a number of movements (e.g. ballonné à trois temps). *c* **1973** J. Cholerton *Theory of Acrobatics* (Assoc. Amer. Dancing) (ed. 12) 9 *Q. What is a Temps lévé?* A. A hop.

‖ **temps perdu** (taṅ pęrdü). [Fr., 'time lost'; used with allusion to Proust: see *recherche du temps perdu.] The past, contemplated with nostalgia and a sense of irretrievability.

1932 S. O'Faoláin *Midsummer Night Madness* 75 Life is too pitiful in these recapturings of the *temps perdu*, these brief intervals of reality. **1942** Partridge *Usage & Abusage* 95/2 One may question whether peace and prosperity will return with or without this word..the 'Sesame' of *le temps perdu*. **1962** *John o' London's* 20 Sept. 287/3 A nostalgic atmosphere, an old-timers' *temps perdu*. **1969** *Listener* 22 May 734/2 The wizened wordsmith..was on the scent of *temps perdu*: in particular those heady years of the early Thirties. **1975** A. Price *Our Man in Camelot* vi. 111 It's a natural human feeling to yearn for the good old days, *le temps perdu*.

temptingness. (Later example.)

1976 *Conc. Oxf. Dict.* (ed. 6) s.v. *Seduce*, Persuade by temptingness or attractiveness.

‖ **tempura** (te·mpurā). [Jap., prob. ad. Pg. *tempêro* seasoning.] A Japanese dish consisting of prawn, shrimp, or white fish, and often vegetables, coated in batter and deep-fried. Also *attrib.*, esp. in *tempura bar, restaurant*.

1920 *Japan Advertiser* 22 Aug. 5 Tempura means a certain way of cooking,—namely, dipping in thin wheat-flour batter and frying in deep oil... The food which forms the base is some kind of fish. **1936** K. Tezuka *Jap. Food* 71 *Tempura* is a characteristic dish of Japan made by dipping fish or shrimps or shell-fish in batter and frying in deep gingelly oil or torreya oil. **1958** *Japan* (Unesco) (1964) 724/1 *Tempura* was adopted from a recipe in Spain and Portugal. **1967** D. & E. T. Riesman *Conversations in Japan* 223 Donald joined us for lunch at a *tempura* restaurant nearby. **1969** *Sat. Rev.* (U.S.) 13 Sept. 62/3 Some of the most popular eateries are *tempura* bars. **1979** *United States 1980/81* (Penguin Travel Guides) 73 This thoughtfully designed Japanese restaurant has separate dining rooms for teppan-yaki, sukiyaki, and tempura.

temura(h (těmū̆rā·). *Jewish Lit.* Also **Themurah.** [Heb. *těmūrāh* exchange.] In cabalistic phraseology, a systematic replacement of the letters of a word with other letters in order to find the hidden meaning of events, laws, etc., in the Torah.

1902 A. E. Waite *Doctrine & Lit. of Kabalah* i. iv. 27 Very curious results were sometimes obtained by these solemn follies which appear so childish and ridiculous at the present day. They comprise: *a. Gematria*..*b. Notaricon*..*c. Themurah*, that is the transposition of letters in a given word or sentence. **1911** 'Sepharial' *Kabala of Numbers* I. iii. 31 The *Temurah*, in which the letters of a word were replaced by others after a definite method. **1974** C. Poncé *Kabbalah* ii. 172 Turning now to Jeremiah xxv. 26,..we discover that Sheshak is none other than Babel. How did the writer of this Biblical passage arrive at Sheshak as another name for Babel? Through the method of *temura*.

ten, *a.*, *sb.* (*adv.*). Add: **A.** *adj.* **2. a.** (Earlier examples in sense 'ten shillings'.)
1837 Dickens *Let.* 10 Feb. (1965) I. 235, I made it five pounds instead of two pounds ten. **1872** Ruskin *Munera Pulveris* p. ix, Worth as many ten-and-sixpences as the impressions which might be taken from the lithographic stones.

b. Also with ellipsis of 'minutes' in phr. *ten past* or *to* or (*U.S.*) *till*, ten minutes after or before the hour; *to take ten* (U.S.): see *TAKE *v.* 52 c.
1852 [see To *prep.* 6 b]. **1937** 'P. Wentworth' *Case is Closed* ix. 95 Then it couldn't have been later than ten past eight when you heard that shot? **1960** S. Barstow *Kind of Loving* I. ii. 53 Nobody ever arranges to meet somebody at ten to and so she must either be late or not coming. **1963** [see *till *prep.* 5 d]. **1979** 'J. le Carré' *Smiley's People* xxiii. 265 The time was ten to eleven.

c. *ten out of ten*, ten marks or points out of ten; hence *transf.* = *full marks* s.v. *Mark *sb.*^{1} 11 g; *the Ten*, (*c*) the group of countries comprising the European Economic Community after January 1981 when Greece joined the existing group of nine countries (the expectation expressed in quot. 1971, but not fulfilled, was that Norway would become a member in 1973 together with Denmark, the Republic of Ireland, and the United Kingdom) (cf. *six *a.* 2 j); *to count ten*: see Count *v.* 1; *spec.*, to do this in order to check oneself from speaking impetuously; also *to count up to ten*.
[The phrase is much older but printed evidence is lacking.] **1981** P. O'Donnell *Xanadu Talisman* iv. 69, I can't claim ten out of ten... I was a bit indecisive. **1981** *Listener* 22 Oct. 481/2 For beating down Whitehall opposition, Mr Sproat gets ten out of ten.
1971 *Guardian* 20 Dec. 10/2 The objective for the Six (and the prospective Ten) should be to bring down the trade barriers. **1981** *Times* 24 Dec. 1/2 The EEC denounced 'the grave violation of the human and civil rights of the Polish people'..and said these were causing 'growing concern' among the people and governments of the Ten.
1817 T. Jefferson *Let.* 12 July in *Writings* (1899) X. 93 When angry count 10. before you speak. **1939** F. Thompson *Lark Rise* xi. 205 Copper-plate maxims..; 'Count ten before you speak', and so on. **1953** E. Simon *Past Masters* ii. 122 Don't say anything. Hold it. Count ten. **1976** R. Perry *One Good Death deserves Another* vii. 116 He counted up to ten before he answered me.

B. *sb.* **1. c.** (Earlier and later examples of '10 A'.) *Number ten*, also *No. 10*: see *number *sb.* 5 e.
1906 H. Müller *Reminiscences* 43 Giovanni endured the punishment that is the Austrian equivalent for '10 A'. **1927** P. Riley *Memories* 74 The drastic punishment, known as '10.A' was introduced into the Service at the same time [*sc.* Jan. 1875].

5. (*b*) (earlier example); (*c*) a ten-dollar note; (*d*) a ten-horse-power car.
c **1863** T. Taylor *Ticket-of-Leave Man* iii. 54 Here are notes—two hundreds—a ten—and two fives.
1829 *Vancouver Herald* (Fredericksburg) 18 Apr. 3/3 The public are cautioned against receiving spurious 5's 10's and 20 dollar bills, purporting to be on the Bank of Virginia. **1907** 'O. Henry' *Trimmed Lamp* 171 He drew out his 'roll' and slapped five tens upon the bar. **1977** J. Crosby *Company of Friends* xxvi. 161 Roger tipped the waiter a ten.
1931 *Daily Express* 16 Oct. 11/2 Cheap 'Tens'. There was a big demand also for cars just above the 'baby' class, the numerous 10's that are cheap to buy. **1942** *R.A.F. Jrnl.* 3 Oct. 24 He would soon be driving around in a family eight or ten. **1968** *Compl. Encycl. Motorcars* 59/1 In 1910 a 1·6 litre 4-cylinder [Austin] Ten was made for export only. *Ibid.* 399/1 That year [*sc.* 1933] Morris's sv 1·3-litre Ten-Four came out as an answer to Austin's Ten and Hillman's Minx.

D. *Comb.* **1. a.** *ten-cent* (examples), *-day* (earlier example), *dollar* (earlier examples), *-figure*, *-shilling* (later examples). **b.** *ten-tongued* (example).
1846 D. Corcoran *Pickings* 26, I gave him a $2 bill, and he only gave me thirteen of these (ten cent pieces) in change. **1873** E. Eggleston *Myst. Metrop.* xviii. 158 The joyful keys and the cheerful ten-cent coins lay in his pocket. **1901** H. Robertson *Inlander* 118 The sleepers in the grass-grown churchyard..had been removed elsewhere to make room for the thriving innovation known as the 'Ten Cent Store'. **1962** E. Snow *Other Side of River* (1963) xxxix. 283 Ten-cent prints are also sold of Italian Renaissance painters and a few French impressionists. **1883** 'Mark Twain' *Life on Mississippi* lx. 582 A ten-day trip by steamer. **1807** *Deb. Congr. U.S.* 19 Aug. (1852) 429, I got two of the notes changed, and one, a ten dollar note, was returned on my hands. **1825** J. Neal *Bro. Jonathan* I. 221 For a 'ten-dollar bill'..Peters would have set fire to it. **1842** *Penny Cycl.* XXIII. 498/1 Nathaniel Roe, 'Tabulæ Logarithmicæ', London. Seven-figure numbers to 100 thousand, ten-figure sines, &c. to hundredths of degrees. **1922** *Biometrika* XIV. 160 It was necessary to calculate π, to eight places, which was done with the help of Vega's ten-figure logarithms. **1959** A. Christie *Cat among Pigeons* iv. 51 She accepted the ten shilling note her mother handed to her. *a* **1974** R. Crossman *Diaries* (1976) II. 279 Then there came fifteen speakers of whom the ten well-informed were all passionately for the ten-shilling unit. **1867** Emerson *May-Day* 86 Speaking by the tongues of flowers, By the ten-tongued laurel speaking.

2. ten-code, a code of signals (all beginning with the number ten) originally used in radio communication by police in the U.S. and later adopted by Citizens' Band radio operators; **ten-eighty** (more commonly **1080**) [see quot. 1945], a formation of sodium fluoroacetate used as a poison against predatory animals; **ten-foot** *a.*: fig. phr. *ten feet tall* used contextually to convey extreme self-assurance or pride; **ten-four, 10-4** *int.*, in the ten-code (see above), the code phrase for 'message received'; used *loosely* as an expression of affirmation; also as *v. intr.*; **ten-gallon** *a.*, that can contain ten gallons; **ten-gallon hat, sombrero,** a high-crowned, wide-brimmed hat of a kind esp. worn in the south-western U.S. (cf. *Stetson); **ten-gauge** *a.*, having a calibre such that ten balls of matching size weigh one pound; also *ellipt.* for *ten-gauge shotgun*; **ten-inch** *a.*, measuring ten inches; *spec.* designating a 78 r.p.m. coarse-groove gramophone record having this diameter; also *ellipt.* = *seventy-eight* s.v. *seventy *a.* 2 b; **ten-minute rule** (also **ten minutes rule**), a standing order of the House of Commons allowing brief discussion of a motion for leave to introduce a bill, each speech being limited to ten minutes' duration; **ten o'clock** (example of the plant); also, a name for *Portulaca grandiflora*, a subtropical annual herb whose flowers open late in the morning; **ten per center** *U.S. Theatr. slang*, a theatrical agent (so called from the ten per cent commission that an agent takes); **ten signal** *U.S.*, any of the signals that form part of the ten-code (see above); **ten-speed**, a multiple-speed set of gears on a vehicle, esp. a bicycle; freq. *attrib.*; **ten-spot** *sb.* (earlier and later examples of both senses); **ten-strike** (earlier example); also *spec.* in *fig.* use, a success, esp. in phr. *to make a ten-strike*, to score a success (U.S. colloq.); **ten tenth(s)** *attrib.* (orig. *Meteorol.*), complete, one hundred per cent; **ten-to-two**, a position of the hands or feet resembling that of the hands of a clock at ten minutes to two, *esp.* a position of the hands on the steering-wheel of a car; freq. *attrib.*
1969 T. E. Drabnek *Lab. Simulation Police Communications Syst. under Stress* 135 Above code corresponds to the 'Ten-Code System' used by many police agencies. **1976** *CB Mag.* June 67/2 Well, it really grabbed me, all this 'good buddy' stuff, the ten-code. **1977** *Rolling Stone* 13 Jan. 45/3 Get a CB and take on a persona, use the 10 code and all the language, and be anybody you want to be. **1945** *Science* 31 Aug. 232 (*heading*) 'Ten-eighty', a war-produced rodenticide. *Ibid.*, One, commonly referred to under its laboratory serial number, '1080',..has been subjected to sufficiently adequate field-testing to warrant the assertion that a promising new rodenticide has been discovered. **1961** *New Scientist* 13 Apr. 17/1 About one and a half grains of 'ten-eighty' are dissolved in a little water and injected into 100 lb of meat. **1971** W. Hillen *Blackwater River* xii. 117 Then Compound 1080 (ten-eighty)..reached British Columbia and created a 'predator control' bureaucracy more difficult to eliminate than wolves. [**1955** (*television film title*) The man is 10 feet tall.] **1962** M. Hastings *Yes, & After* ii. i. 72 You must always be ten feet tall imagining yourself doing this or doing that. **1964** D. Francis *Nerve* xvii. 237 It made me feel warm inside... I felt ten feet tall. **1970** A. Draper *Swansong for Rare Bird* i. 11, I must say I felt 10 feet tall and there was a soppy grin on my face. **1962** *Amer. Speech* XXXVII. 272 *Ten-four* (verb), to understand a message. From the radio code *10-4*, meaning 'I receive you clearly'. **1976** *National Observer* (U.S.) 30 Oct. 5/1 Judge Floyd Smith, a CB operator himself, went by the 'handle' of 'Marryin' Sam', the bride was 'Little Lulu', and the groom was 'Stanley Steamer'. They didn't say 'I do'; they said '10-4'. And the judge didn't pronounce them man and wife; he said, 'Put the hammer down.' **1978** *N.Y. Times Mag.* 23 July 23/2 The CB'ers have a language that's 10-4 with them. **1841** C. Gray *Lays & Lyrics* 241 This song was written on the presentation of a Ten-Gallon China Punch-Bowl...to the Club. **1922** Joyce *Ulysses* 303 Hard by the block stood the grim figure of the executioner, his visage being concealed in a tengallon pot with two circular perforated apertures. **1928** *Daily Express* 7 Oct. 3/7 She instinctively recognized that he was a cowboy, even though he did not wear a ten-gallon hat and a jacket embroidered with Mexican dollars. **1929** T. Wolfe *Look homeward, Angel* (1930) xxvii. 374 He removed from his head the ten-gallon grey sombrero. **1939** *Amer. Speech* XIV. 201/1 In the nomenclature of the South-western cowboy, *sombrero* is used interchangeably for hat, but the qualifying phrase of 'ten gallon' has been arrived at by a mistaken translation of a Spanish word. The word 'gallon'..served to describe the braid with which a vaquero's hat was trimmed..it should have been 'galloon'. **1977** C. McCullough *Thorn Birds* xiv. 331 Only the handful of half-caste aborigines..aped the cowboys of the American West, in high-heeled fancy boots and ten-gallon Stetsons. **1894** Ten-gauge [in Dict., sense D. 1]. **1936** J. Steinbeck *In Dubious Battle* v. 70 'Shot-guns,' he said... 'Soon's somebody sounds off with a ten-gauge, they go for the brush like rabbits.' **1940** W. Faulkner *Hamlet* iii. 194 Looking..into the face which with his own was wedded and twinned forever now by the explosion of that ten-gauge shell. **1908** *Sears, Roebuck Catal.* 201/2 Disc Record Cases... No. 2 holds 50 10-inch disc records. **1959** *Manch. Guardian* 11 Aug. 5/7 Beecham (whose reading has just reappeared on a Fontana ten-inch, KFR 4003). **1979** *Listener* 4 Oct. 461/3 Nearly all the 23 Gillespie tracks..were originally marketed on ten-inch, 78 rpm discs. **1908** A. E. Steinthal tr. *J. Redlich's Procedure House of Commons* III. ii. x. i. 86 Under a standing order passed in 1888, popularly known as the 'ten minutes rule' (Standing Order 11) an abbreviated mode of procedure is authorised for the introduction of bills. **1971** Hindell & Simms *Abortion Law Reformed* xi. 232 Ten minute rule bills can be brought before the House with a short speech each Tuesday and Wednesday after question time, but if the House agrees to the introduction of such a bill *all* further progress to a second reading and beyond depends, in practice, on the Government..giving it parliamentary time. **1826** W. Darlington *Flora Cestrica* 40 *Ornithogalum..umbellatum*... Ten o'clock. Twelve o'clock. Star of Bethlehem. **1953** *Caribbean Q.* III. i. 10 Ten o'clock is a kind of portulaca which blooms in mid morning. **1926** *Variety* 29 Dec. xi. 5/4 Broadway chatter is full of theatrical cracks such as..'ten per center'. **1962** *Punch* 26 Dec. 920/1 A condition of the licence being granted is that the applicant advertises for two weeks in The Stage, stating his intention of joining the ten-percenters. **1951** *Directory Nat. Police Communications Network* (ed. 6) 19 The '10' signals were developed by A.P.C.O...and the system has been widely adopted. **1970** V. A. Leonard *Police Communications Syst.* ii. 34 APCO's Project Series Foundation has produced four nationally recognized projects:..the publication of the APCO Ten Signal Cards. **1971** M. Tak *Truck Talk* 165 *Ten-speed*, a ten-speed Roadranger transmission. **1977** C. McFadden *Serial* (1978) i. 8/2 They spent it rapidly on..twin Motobecane ten-speeds. **1977** *New Yorker* 9 May 34/1 The owner of a ten-speed model asked her why. **1844** 'J. Slick' *High Life N.Y.* II. 215 'Jest so,' sez I, a flingin' down the ten-spot o' clubs. **1848** 'N. Buntline' *Mysteries & Miseries N.Y.* iv. iii. 27 Be quick, and I'll give you a *ten spot!* **1936** E. Culbertson *Contract Bridge Complete* 21 Any six-card suit, even without a ten-spot, is a biddable suit. **1971** B. Malamud *Tenants* 153 Hi, sugar, I took a ten-spot out of your loose change. **1840** *Spirit of Times* 11 July 228/1 [This] he says is an extra touch—a ten strike and two spare balls. **1887** *Scribner's Mag.* May 624/1 But I have got the family to consider, and I am in a position now where I can make a ten-strike for it. **1900** G. Ade *Fables in Slang* 72 He could tell by the Scared Look of the People in Front that he had made a Ten-Strike. **1949** E. Pound *Pisan Cantos* lxxxiii. 124 It comes over me that Mr. Walls must be a ten-strike With the signorinas. [**1945** *Meteorol. Office Air Obs. Handbk.* 34 In estimating the amount of cloud the observer should aim to give the fraction (in tenths) of sky covered by cloud.] **1948** *Daily Tel.* 23 Apr. 6/6 There was ten-tenth cloud at the time. **1973** 'A. Hall' *Tango Briefing* ix. 114 There was a ten-tenths flap on in London so they'd have alerted the whole network. **1977** C. Forbes *Avalanche Express* xviii. 186 One moment he had ten-tenths vision, the next second he was blind. **1979** D. Brierley *Cold War* vii. 57 There was ten-tenths cloud cover, the clouds coming from the north-east..like billowing poison gas. **1961** C. H. D. Todd *Pop. Whippet* iv. 68 It stands with its feet at 'ten to two'. **1962** *Which? Car Suppl.* Apr. 55/2 The steering wheel was..rather high. This made a 'ten-to-two' hand position uncomfortable after a time. **1974** *Drive* Autumn 26/2 We found the square wheel made it difficult for drivers to hold the rim in the ten-to-two position they are taught to adopt.

tenace. Add: Also *Bridge*.
In Bridge, the opponent's card need not be his highest. **1905** 'Cut-Cavendish' *Compl. Bridge Player* 11 *Tenace*, the best and the third best cards of any suit, ace and queen if the king has not been played. **1936** [see *end-play* s.v. *end *sb.* 25]. **1959** *Listener* 10 Sept. 414/1 The K J 8 will still constitute a tenace over the 10 9 x. **1984** *Guardian* 6 Oct. 17/2 The ten of diamonds now end-played West to return a club into declarer's tenace.

‖ **tena koe** (tenā·kwe, te·na ko·,e), *int. New Zealand*. Also 9 **teneako, tenáqui**; (with dual and pl. forms of the pronoun) **tena korua, tena koutou.** [Maori, lit. = 'there you are'.] A Maori greeting. Also **tenako·eing** *vbl. sb.* (nonce-wd.).
a **1842** H. Weekes in Rutherford & Skinner *Establishment of New Plymouth Settlement in N.Z.* (1940) I. 92 The period of *teneáko*-ing, handshaking, etc. *a* **1847** —— in *Ibid.* 119 Came towards me with extended hand and a friendly *Tenáqui*. **1901** A. A. Grace *Tales of Dying Race* 190 'Aaaaah! *tena koutou*,' wailed she, seizing them by the hands. '*Tena koe! tena koe! Katahi te koa!*' she cried, as she rubbed noses with them one by one. **1905** W. Baucke *Where White Man Treads* 170 When I neared the fence I cried the old-time greeting, 'Tena koutou'. **1947** A. P. Gaskell *Big Game* 90 They smiled up at her. 'Tenakoe'. 'A lovely day Miss Brown,' said Mrs. Terari. **1949** P. Buck *Coming of Maori* (rev. ed.) I. vi. 79 The usual Maori greeting of *Tena koe* applies to one person, *Tena korua* applies to two, and *Tena koutou* to more than two. **1960** N. Hilliard *Maori Girl* ii. 92 She could not catch what he said but..she guessed it was 'Tenakoe!'

tenant, *sb.* Add: **4.** *tenant-farmer* (earlier example); *tenant-farming* (earlier U.S. example); hence (as a back-formation) **tenant farm** *v. trans.* (with hyphen) and *sb.*

1949 *Time* 27 June 84/2 The 1,600 acres he tenant-farms. **1979** P. Theroux *Old Patagonian Express* xvii. 263 These are tenant farms..these people own nothing but the clothes on their backs. **1748** Richardson *Clarissa* V. 208 Attended by Susan Morrison, a tenant-farmer's daughter. **1861** *Trans. Illinois Agric. Soc.* IV. 203 On the greater part of this farm are the usual indices of tenant farming.

tenas (te·næs), *a.* (and *sb.*). [Chinook Jargon, ad. Nootka *t'an'as* child.] Small. Also as *sb.*, and in **tenas man** [cf. *klootchman], a child.
1870 *Mainland Guardian* (New Westminster, B.C.) 16 Apr. 3/3 There trip about a few Tenass-men, some with the remains of an old coat and beaver hat, and some [in] almost naked savagedom. **1904** *Wide World Mag.* Sept. 541/2 Klootchmen, tenasses, old bucks, and young hunters, were crowded on the schooner's deck. **1935** H. Davis *Honey in Horn* xxii. 371 Nobody but this *tenas* buck [*sc.* a small Indian] here. **1969** *Islander* (Victoria, B.C.) 19 Oct. 3/2 They were very insolent..laughing at the *tenass* warship for wasting her powder and shot.

Tenby (te·nbi). The name of a town on the coast of Wales, used *attrib.* in **Tenby daffodil** to designate *Narcissus obvallaris*, a small yellow daffodil sometimes found as a wild flower in the region.
1806 A. H. Haworth in *Phil. Mag.* VIII. 130 Truby [*sic*] 6-lobed Daffodil. **1884** J. D. Hooker *Student's Flora Brit. Isles* (ed. 3) 399 The Tenby Daffodil..scarcely differs. **1894** W. Robinson *Wild Garden* (ed. 4) ii. 19 The little Tenby Daffodil is very sturdy and pretty. **1966** J. Bernstein *Wild Garden* iii. 35 The early-blooming *Narcissus obvallaris*, the small Tenby Daffodil. **1981** W. Condry *Nat. Hist. Wales* vi. 171 The Tenby Daffodil..has long puzzled the world.

tend, *sb.* For † *Obs. rare* in Dict. read *rare* and add later example.
1937 V. Woolf *Let.* 30 Apr. (1980) VI. 122 I'm very glad you saw that the tend of the book, its slope to one quarter of the compass and not another, was different from the tend in the long story.

tend, *v.*^{1} **2. a.** For *Obs.* exc. *dial.* read *Obs.* exc. *dial.* and *U.S.* and add U.S. examples.
1880 'Mark Twain' *Tramp Abroad* xx. I got to 'tend to business. **1917** G. B. McCutcheon *Green Fancy* 68 There..paid their regular, 'tended to their own business, and that's all. **1930** W. Faulkner *As I lay Dying* (1935) 155 You got to wait a little while. Then I'll tend to you. **1936** M. Mitchell *Gone with Wind* x. 197 If anybody dares say one little word about you, I'll tend to it. **1973** *Ebony* Nov. The men's linens and jail clothes were sent elsewhere for laundering; the women were expected to tend to their own.

tend, *v.*^{2} **3. b.** (Now usu. in weakened sense: To have a tendency, to be apt or inclined.)
1934 J. Cary *African Witch* ii. 38 Obstinacy and stupidity are things that tend to annoy quick-minded and intelligent people. **1956** H. L. Mencken *Minority Report* 251 Unfortunately, the machine thus devised..tends to go wild. **1976** H. Macklin *Pipeline* xxxiii. 372 After the initial uproar over the Watergate raid, the papers tended to ignore it. **1982** *Daily Tel.* 26 Aug. 11/2 She went to Oxford to read English. 'I didn't really want to, but I tended to do what my parents advised then.'

Tendai (te·ndai). [a. Jap., ad. Chinese *T'ien-t'ai*, the name of the mountain in S.E. China where the doctrines were formulated.] A Buddhist sect introduced into Japan from China by the monk Saichō (767–822), founded by Zhi Yi (515–97) and characterized by elaborate ritual, moral idealism, and philosophical eclecticism.
1727 J. G. Scheuchzer tr. *Kæmpfer's Hist. Japan* I. i. vii. 106 Not far from this hot Bath is a Monastery of the Sect of Tendai. **1833** *Chinese Repository* (1834) Nov. II. 323 There are now in Japan the following sects which are tolerated by government, 1. Zen... 2. Tendai. **1880** E. J. Reed *Japan* I. iv. 91 The Tendai, founded by the priest Saicho, under Kuwammu. **1894** *Trans. Asiatic Soc. Japan* XXII. 318 This comprehensiveness caused the success of the Tendai Sect. **1938** D. T. Suzuki *Zen Buddhism & its Influence on Japanese Culture* i. 23 The philosophy of Tendai is too abstract and abstruse to be understood by the masses. **1973** *Times Lit. Suppl.* 2 Mar. 237/4 Something of the awe Tendai ritual inspired can still be felt by the visitor to the Komponchudo on Hiei-san.

tendency. Add: **1.** Also *pl.* in pregnant use, tendencies towards homosexuality. *colloq.*
1938 J. Betjeman *Oxf. Univ. Chest* ii. 42 Someone who has 'tendencies' as an undergraduate, will in ten years time be settled down to married life. **1958** L. Durrell *Balthazar* ii. 32 Now the Egyptians, they don't give a damn about a man if he has Tendencies.

d. *Pol.* [Infl. by F. *tendance*.] A political association within a larger party or movement, esp. a left-wing group within a socialist party.
1974 J. White tr. *Poulantzas's Fascism & Dictatorship* iv. ii. 171 The communists of the Ruhr, a left tendency, went into combat in isolation in April. **1977** *Politics of Militant* 1 The tendency grouped around the weekly paper 'Militant' has grown considerably in recent years. It absolutely dominates the Labour Party Young Socialists. **1980** *Times* 14 Jan. 1/4 The 'Militant Tendency', a clandestine Trotskyist organization, with its own full-time staff, whose aims are to penetrate the Labour Party. **1981** *Daily Tel.* 10 Dec. 3/7 The arguments of the tendency are about Marxist, Leninist, Stalinist and Trotskyist groups.

3. Also **tendency wit** [after G. *tendenzwitz*].
1916 A. A. Brill tr. *Freud's Wit & its Relation to Unconscious* iii. 138 By virtue of its purpose, the tendency-wit has at its disposal sources of pleasure to which harmless wit has no access. **1954** D. Riesman *Individualism Reconsidered* (1955) xxii. 349 The id expresses its criticism so that Freud called 'tendency-wit', but then turns to its masters with a smile, saying, 'After all, it's only a joke.' **1964** A determination to self-criticism that Freud called 'tendency-wit'.

tendential, *a.* (Earlier example.)
1847 J. D. Morell *Hist. Philos.* (ed. 2) II. vii. 396 He [*sc.* Buchez] has brought to his aid the notion of progress, the logical development of ideas, and the tendential movements of society.

tendentiously (tende·nʃəsli), *adv.* [f. Tendentious *a.* + -ly^2.] In a tendentious manner; with a purposed tendency or aim. So **tende·ntiousness.**
1920 *Glasgow Herald* 9 Apr. 8 This was not due to any tendentiousness on the part of official reports. **1924** *History* Oct. 215 If we could..speak with our enemies in the gate, we should doubtless teach geography..tendentiously. **1966** *Listener* 17 Nov. 745/1 Confessions on the telephone, and witnesses complaining they were not allowed to say what they wanted to say at the trial. An exercise in tendentiousness it would be hard to beat. **1983** *Washington Post* 2 Oct. 4/4 The result is a book of findings, tendentiously presented. *Ibid.* 11 Oct. The tendentiousness of 'The Final Option' may come as a perversely amusing shock.

‖ **tendenz** (tende·nts). [Ger., ad. Eng. Tendence or F. *tendance*.] = Tendency 1 c.
1896 A. W. Small in *Amer. Jrnl. Sociol.* II. 22 Way in *Social Forces* (1935) Mar. 337/2 The connotations are in my mind associated with some 'Tendenz' which is exploited. **1952** A. L. Rowse *England of Eliz.* iii. 379 One sees the *tendenz* of all this... Coke's view was extremely tendentious, but the *tendenz* was good: it was all in favour of the supremacy of law in the State and of the liberty of the subject. **1967** G. Steiner *Lang. & Silence* 336 He argues that the type of '*Tendenz*'..which Engels would find acceptable is..'identical with that "Party element" which materialism..encloses in itself'.
Also **tende·nzroma:n** [G. *roman* novel] = *tendency novel* s.v. Tendency 3, *roman à thèse* s.v. *Roman *sb.*^{3}; similarly, with partial translation, **tendenz novel.**
1855 Geo. Eliot in *Westm. Rev.* July 294 Constance Herbert is a *Tendenz-roman*: the characters and incidents are selected with a view to the enforcement of a principle. **1896** J. Jacobs *Jewish Ideals & Other Essays* p. ix, George Eliot's novels..were to be *Tendenz-Romane*, and we studied them as much for the *Tendenz* as for the *Roman*. **1975** *Listener* 18 Dec. 823/2 Oliver Twist has been..called a *Tendenzroman*..Humphrey House, the eminent Dickens critic, said that it was the closest thing to a tendenzroman that Dickens ever wrote, and yet..very little of that political quality survives.

tender, *sb.*^{3} **3. b.** *tender offer* (U.S.) (see quot. 1979), usu. for the purpose of obtaining effective control.
1968 J. Low *Investor's Dict.* 198 In general when an outside interest makes a tender offer the market price rises close to the tender price. **1979** *Yale Law Jrnl.* LXXXVIII. 510 A tender offer is conventionally defined as a public solicitation of the shareholders of a corporation to tender their shares to the offeror at a specified price.

tender, *a.* (*adv.*) and *sb.*^{4} Add: **A.** *adj.* **I. 1. b.** (Earlier example.)
1727 D. Keys *Let.* 25 Mar. (1971) 105 He has carted at a very unseasonable time when the ground was tender.
II. 3. b. *tender annual*, an annual plant needing the protection of a greenhouse all through its life; cf. *hardy annual* s.v. Hardy *a.* 4 b; *tender plant* (fig.), something needing careful nurture if it is to survive and develop.
1769 Rutter & Carter *Mod. Eden* ii. iv. 218 (*heading*) Of raising tender annuals on hot-beds. **1822** J. C. Loudon *Encycl. Gardening* 1070 The green-house is now filled with tender annuals. **1867** T. Bridgeman *Amer. Gardener's Assistant* iii. 21 Varieties from warm climates ..may with great propriety be treated as tender annuals, by sowing the seed every spring. **1933** *Discovery* Mar. 76/2 The runner bean..of Mexican parentage or origin is here grown as a tender annual. **1969** *Times* 10 Mar. 10/7 These capital sources are conditioned by..the confidence felt in the future profitability of agriculture. That confidence is, at present, rather a tender plant. **1974** J. Warren *Macself's Amat. Greenhouse* (ed. 5) viii. 238 The tender annuals of all kinds should be sown in spring rather than autumn. **1978** *U.S. News & World Rep.* 12 June 36/1 Academic and cultural freedom is a very tender plant, which this country has never cultivated.

III. 8. a. *tender loving care* (colloq.), especially solicitous care such as is given by nurses; also *transf.*; *tender mercies* (occas. *tender mercy*) a Biblical phrase usu. used *ironically* (perh. with spec. allusion to quot. 1611) of attention, care, or treatment thought unlikely to be in the best interests of its object.
1611 Bible *Prov.* xii. 10 A righteous man regardeth the life of his beast: but the tender mercies of the wicked are cruell. **1893** [see Mercy *sb.* 5 c]. **1906** Conrad *Mirror of Sea* xxxi. A ship anchored..is not abandoned by her own men to the tender mercies of shore people. **1925** Galsworthy *Caravan* 571 His feelings revolted against handing 'that poor little beggar' over to the tender mercy of his country's law. **1960**, etc. [see *TLC* s.v. *T 6 a]. **1965** *Listener* 17 June 892/2 Smaller..traders and manufacturers..left to the tender mercies of the open property market. **1973** *Computers & Humanities* VII. 166 The Bernard Quemada *Concordance to Les Fleurs de Mal*, which was perhaps prepared with more tender loving care, corrected such mechanical deficiencies. **1977** *Listener* 12 May 605/1 It is in a nurse's nature and in her tradition to give the sick what is well called 'TLC', 'tender loving care', some constant little service to the sick.

b. *tender-handed*, *-slanted*, *-spirited*; **tender-eared** *a.* (later example); **tenderpad** [f. after *Tenderfoot 2; see Pad *sb.*^{3} 7], a recruit to the Cub Scout movement who has passed the Tenderpad test.
1911 J. Masefield *Everlasting Mercy* (1912) 88 Two hares..Tender-eared..

tenderable, *a.* (Earlier example.)
1868 *Harper's Mag.* Dec. 89/1 The view of Judge Denio that the word 'money' implies the creation of what is tenderable for debts is a much more reasonable..interpretation of the Constitution.

tenderfoot. Add: **2.** In the Scout and Guide movements, a recruit who has passed the enrolment tests (the *tenderfoot tests*); also *tenderfoot badge*, and *ellipt.* = tenderfoot badge, tests.
1908 R. S. S. Baden-Powell *Scouting for Boys* iii. 36 A Tenderfoot is a boy who is not yet a scout. **1910** *Scout Tests* (Boy Scouts Assoc.) 1 It should be noted that Tenderfoots must not wear the button-hole badge until he has passed the Tenderfoot Tests. **1918** R. S. S. Baden-Powell *Girl Guiding* ii. You are a Recruit until you pass your Tenderfoot tests. Then you can go on and rise to the following ranks:—Recruit, Tenderfoot, Second-Class Guide. **1930** *Girl Guide Badges* (Girl Guide Assoc.) 6 No Guider or Guide is entitled to wear the Tenderfoot Badge unless she is a Tenderfoot and has passed the following tests. **1965** G. McInnes *Road to Gundagai* x. 180 ..between..strong scouts and not so strong tenderfoot. **1977** *Times* 13 Oct. 18/7 Mr Bass, who is 6ft 3ins., resigned after passing his Tenderfoot because he did not like wearing short trousers.

tendering, *vbl. sb.*^{1} (In Dict. s.v. Tender *v.*^{1}) (Later example.)
1955 *Times* 24 Jan. 9/4 The President of the Board of Trade..proposed to send to the Commission a second general reference covering 'common prices and level tendering'. **1974** G. L. Rees *Britain's Commodity Markets* vii. 165 For this purpose granaries ('tendering points') have been listed by the Trust.

tenderish, *a.* (Later example.)
1922 Joyce *Ulysses* 436 With a sour tenderish smile.

tenderize, *v.* Add: *spec.* to make (food, esp. meat) tender. Also *absol.* orig. *U.S.* Hence **te·nderized** *ppl. a.*; **te·nderizing** *vbl. sb.* and *ppl. a.*
1934 Webster, Tenderize, *v.t.* **1935** A. P. Herbert in *Punch* 8 May 548/1 He has seen some prunes commended as being 'tenderized by a special process.'..If enough prune-purveyors go on saying 'tenderize' it will be in the next edition of all the dictionaries! **1936** *Amer. Speech* XI. 374/2 Sunsweet Tenderized Prunes, refreshed and pasteurized. **1939** *Sun* (Baltimore) 22 Mar. 3/1 A process of 'tenderizing' meat through the use of ultra violet rays. **1950** [see *pressure cooker a]. **1958** *House & Garden* Feb. 77/2 Wine has a slightly tenderizing effect, so when it is used the meat will cook a little more quickly. **1960** *Times* 24 Sept. 6/6 The new method..ensures that tenderizing liquid gets into the innermost tissues. **1961** *Harper's Bazaar* June 84/2 Diced cubes [of avocado] which have been 'tenderized' and flavoured by the marinade. **1968** L. Durrell *Tunc* v. 235 She's as sweet as a tenderised steak. **1977** *Time* 19 Sept. 61/2 (Advt.), Touchmatic Control Panel—enables you to slow cook, simmer, tenderize and blend flavours.

tenderizer (te·ndəraɪ:zəɪ). [f. Tenderize *v.* + -er^1.] Something used to make meat tender, either (*a*) the enzyme papain, or (*b*) a steak hammer.
1958 *Catal. County Stores, Taunton* June 12 Papaya Juice (meat tenderiser)—a bot. 2/–. **1959** *Housewife* June 75 Steak tenderisers in sycamore. **1969, 1970** [see *meat tenderizer* s.v. *meat *sb.* 5 b]. **1975** A. Ayckbourn *Round & Round the Garden* in *Norman Conquests* 18, I line up the dishes and smash them—slowly—with the steak tenderizer.

tenderloin. Add: **1.** (Further examples.) Also *tenderloin steak*.
1864 *Daily Tel.* 27 Sept. 5/2 The 'tenderloin', the 'porterhouse' steak of America, are infinitely superior to our much-vaunted rump steak. **1875** *Scribner's Monthly* July 274/2 A tenderloin steak,..potatoes, bread and butter, and a cup of coffee will cost fifty cents. **1954** [see *Alaska b]. **1975** *Times* 19 Mar. 16/3 A tenderloin steak made from textured soy protein..sells for 89 cents a pound—as against two dollars for the real thing.
2. (Earlier example.)
1887 *Harper's Mag.* Mar. 500/2 His precinct is known as the 'Tenderloin', because of its social characteristics.

tender-minded, *a.* (and *sb.*). [Parasynthetic, f. *tender mind* (Tender *a.* 8): see -ed^2.] Having a tender mind; sensitive and idealistic (in W. James opp. *tough-minded *a.*). Also *absol.* as *sb.* Hence **te:nder-mi·ndedness.**
1605 [in Dict. s.v. Tender *a.* (*adv.*) and *sb.*^{4} C]. **1907** W. James *Pragmatism* i. 12 You will..recognize the two types of mental make-up that I mean if I head the columns..*The Tender-Minded*: Rationalistic, Intellectualistic, Idealistic, Optimistic, Religious, Free-Willist, Monistic, Dogmatic. *Ibid.* viii. 295 May not the claims of tender-mindedness go too far? May not the notion of a world already saved *in toto* anyhow, be too saccharine to stand? **1924** T. S. Eliot *Homage to Dryden* iii. 45 It is not cynicism, though it has a kind of toughness which may be confused with cynicism by the tender-minded. **1952** H. J. Eysenck in Mace & Vernon *Current Trends Brit. Psychol.* xvii. 210 Idealistic..tender-minded attitudes, such as those approving of church-going and religion, pacifism. **1965** *Listener* 28 Jan. 153/1 According to James, the tender-minded were those who could not endure the violation of a general law... Mathematics, he said, was the typical study of the tender-minded. **1972** H. J. Eysenck *Psychol. is about People* v. 202, I did in fact discover some evidence in favour of the notion that tender-mindedness and tough-mindedness were correlated with personality as hypothesized.

tenderometer (tendərǫ·mɪtəɪ). [f. Tender *a.* + -o + -meter.] An instrument for testing the tenderness of raw peas for picking, processing, etc.
1938 *Encycl. Brit. Bk. of Year* 137/2 The 'Tenderometer' intended to determine the tenderness of raw peas used for canning. **1947** *N.Y. Herald Tribune* 18 May ii. 10 The tenderometer shows when green peas reach their scientific peak of ripeness by registering the amount of pressure it takes for the gadget to shear through a sample pod. **1960** *Farmer & Stockbreeder* 16 Feb. 97/1 The tenderometer readings of the placement peas were 10 better than the broadcast averages. **1971** *Power Farming* Mar. 12/4 Peas can be valued on a tenderometer reading. **1981** *Southern Horticulture* (N.Z.) Spring 36/1 As soon as the tenderometer reading is at its optimum the whole paddock must be cleared within hours.

‖ **tendido** (tenðī·ðo). [Sp., pa. pple. of *tender* to stretch.] An open tier of seats above the barrera at a bull-fight.
1838 *Q. Rev.* LXII. 407 Those whose poverty..consents, sit in the 'tendido', and brave the sun's perpendicular height. **1967** McCormick & Mascareñas *Compl. Aficionado* ii. 59 The toro lopes off from the horse to the sunny side, where it seems to have spotted a man in a blue shirt in the lower *tendido*.

tendinitis (tendĭnəi·tis). *Path.* Also **tendonitis.** [f. med.L. *tendin-em*, *tendōn-em* Tendon + -itis.] Inflammation of a tendon.
1900 in Dorland *Med. Dict.* **1940** B. I. Comroe *Arthritis* xxx. 368 Classified according to location, fibrositis is known as:..Tendinitis—involvement of the fibrous tissue of tendons. **1948** *Nomencl. Disease* (R. Coll. Physicians) (ed. 7) v. 99/1 Calcareous tendonitis. **1972** *Time* 18 Sept. 35/1 Tendinitis of the knees. **1975** *Daily Colonist* (Victoria, B.C.) 9 Jan. 2/2 Tendonitis is a fairly common occurence, especially in the shoulder or elbow or heel. In tendonitis (tennis elbow is an example of it) the joint has been injured or irritated in some way.

tendo. Add: *tendo calcaneus* (also as one word) [L. *calcāneus, -um* heel] = *tendon of Achilles* s.v. Tendon a.
1900 Dorland *Med. Dict.* 674 Tendo calcaneus. **1909** *Gray's Anat.* (ed. 17) 582 The tendo Achillis (tendo calcaneus)..is the thickest and strongest tendon in the body. **1937** J. C. B. Grant *Method Anat.* iv. 302/2 The bellies end at the middle of the leg in a broad aponeurosis that blends with the aponeurosis of the Soleus to form the tendo calcaneus or tendo Achillis. **1967** G. M. Wyburn et al. *Conc. Anat.* vi. 171/1 The muscle groups of the leg are the extensors in front.., the peroneal muscles on the lateral side..and posteriorly the muscles of the calf forming the tendocalcaneus and more deeply the flexors. **1977** *Bone & Joint Diseases* (B.M.A.) 118 At this stage it is essential to elongate the tendocalcaneus through a transverse incision over the heel.

tendon. Add. **b*.** *Engin.* A steel rod or wire that is stretched while in liquid concrete so as to prestress it as it sets.
1958 F. S. Merritt *Building Construction Handbk.* v. 56 After the concrete has attained sufficient strength, the steel is secured to the anchor plates and the jacks are removed. The tendons will tend to shorten and therefore will put compression in the concrete. **1974** [see *pre-stressing *vbl. sb.*]. **1975** [see *pre-tensioning *vbl. sb.*]. **1981** *Sci. Amer.* June 45/3 The prestressed-concrete reactor vessel..is kept in compression at all times by a network of redundant, tensioned steel tendons that can be monitored and retensioned or even replaced if necessary.
c. tendon organ, spindle = *spindle *sb.* 4 e.
1923 V. H. Mottram *Man. Histol.* vii. 225 Similar apparatus is seen in the Golgi tendon organ. **1974** D. & M. Webster *Compar. Vertebr. Morphol.* x. 202 Both the tendon organ and the muscle spindle fire in response to stretch. **1896** Tendon-spindle [see *spindle *sb.* 4 e]. **1930** Maximow & Bloom *Text-bk. Histol.* xiv. 276 Not infrequently, of two branches of the same sensory fiber one supplies a muscle spindle, the other a tendon spindle. **1977** D. P. Winstanley tr. *Leonhardt's Human Histol.* 249 Tendon spindles are situated in the tendon close to its junction with the muscle.

tendonitis, var. *tendinitis.

tendre. Delete Now *rare* and add later examples.
1887 E. Simcox in K. A. McKenzie *Edith Simcox & George Eliot* (1961) 7 Having towards..Garibaldi..perhaps the same sort of *tendre* as that professed by Charlotte Bronte for the Duke of Wellington. **1921** D. H. Lawrence *Sea & Sardinia* vii. 298 She..was relieved to escape the new attachment, though she had a great *tendre* for him. **1980** G. M. Fraser *Mr American* xxii. 439 The cunning old gentleman's reading of her character, and of her supposed *tendre* for Mr Franklin.

tendril, *sb.* Add: **3. a.** *tendril career, finger, hand, -hold*.
1957 C. Day Lewis *Pegasus* 45 Or too much reason chill the air For your tendril career. **1929** *Oxford Poetry* 5 The tendril fingers groping for the bright Eternal beauty. **1939** Dylan Thomas *Map of Love* 14 Shall she receive a bellyful of weeds And bear those tendril hands I touch across The agonized, two seas. **1967** J. Stallworthy *Almond Tree* 11, I am called to the cot to see your focus shift, take tendril-hold on a shaft of sun.

‖ **tendu** (tãndü), *a. Ballet.* [Fr., pa. pple. of *tendre* to stretch.] Stretched out or held tautly, esp. in *battement tendu* (see *battement).
1922 Beaumont & Idzikowski *Man. Classical Theatr. Dancing* I. 34 *Battements Tendus* serve to stretch and strengthen the muscles. **1950** *Ballet Ann.* IV. 129/2 The return, with the Russian masters of 1925–30, of the traditional French style, strictly *tendu*. **1952** [see *battement].

-tene (tīn), f. Gr. ταινία band, ribbon, used in *Cytology* as a formative element of terms denoting stages of the first meiotic division (in nomenclature due to H. von Winiwarter 1900, in *Arch. de Biol.* XVII).

tenebrescence (tenèbre·sěns). *Physics.* [ad. L. *tenebrescens*, pr. pple. of *tenebrescere* to grow dark, f. *tenebræ* darkness.] The property of reversibly darkening and bleaching in response to radiation of different wavelengths (orig. restricted to the property of darkening only). Hence **tenebre·scent** *a.*; **tenebre·sce** *v. intr.*, to darken or bleach thus.
1946 H. W. Leverenz in *RCA Rev.* VII. 199 (heading) Luminescence and tenebrescence as applied in radar. [*Note*] The terms 'tenebrescence' and 'scotophor' are derived..to correspond to the terms 'luminescence' and 'phosphor'... Tenebrescence is any absorption of light not intrinsic to the materials involved;..a scotophor is a tenebrescent material..which may be made to tenebresce reversibly, i.e., visibly darken and bleach (irrespective of chromaticity), under suitable irradiat'ons. **1953** *Amer. Mineralogist* XXXVIII. 919 (*heading*) Composition, tenebrescence and luminescence of spodumene minerals. *Ibid.*, Only non-chromian spodumene is luminescent and tenebrescent. **1970** Dorion & Weibe *Photochromism* i. 9 Tenebrescent and scotophoric materials are crystals that may be colored by radiation such as electron bombardment, X-rays, or light, and that are bleached by other radiation.

tenebrionid (tène·briǫ-nid), *sb.* (and *a.*) [a. mod.L. family name *Tenebrionidæ*, f. Tenebrio 2, adopted as a generic name by Linnæus (*Fauna suecica* (1746) 189): see -id^3.] A dark-coloured beetle of the family Tenebrionidæ, which is widely distributed, esp. in dry regions. Also as *adj.*, of or designating a beetle of this kind. Cf. *meal-worm* s.v. Meal *sb.*^{1} 3 b.
1921 C. A. Ealand *Insect Life* vi. 204 One of the commonest of our sand-loving Tenebrionids is *Heliopathes gibbus*. **1925** A. D. Imms *Gen. Textbk. Ent.* iii. 498 For a bibliography of Tenebrionid larvae vide Graveley. **1942** C. Barrett *On Wallaby* v. 96 Many kinds of Desert beetles are black, notably the large and active Tenebrionids. **1966** [see *carabid *a.*]. **1979** *Jrnl. Arid Environments* II. 265 Such an aposematic function is admirably demonstrated..in a Müllerian complex of American desert tenebrionids.

tenebrity. (Later example.)
1973 E. P. Maten tr. *Budhasvāmin's Bṛhatkathāślokasaṃgraha* i. 32 Light..soiled..by a dense tenebrity.

tenebrosity. (Later examples.)
1815 J. Gilchrist *Labyrinth Demolished* 19 But sure it must be the very essence of tenebrosity to suppose that the hand changes its nature or the name of it its meaning with change of purpose, and application or use. **1922** Joyce *Ulysses* 387 This tenebrosity of the interior..hath not been illumined by the wit of the septuagint. **1924** *Times* 8 Apr. 14/1 Mr. Baldwin was joined by Mr. Asquith in his condemnation of the tenebrosity of the Government statements.

‖ **tenebroso** (tenebro·so), *sb.* and *a.* Also *pl.* (as *sb.*) **tenebrosi.** [It. *tenebroso* dark: see Tenebrous *a.* (*sb.*).] **A.** *sb.* One of a group of early seventeenth-century Italian painters influenced by Caravaggio, whose work is characterized by dramatic contrasts of light and shade. **B.** *adj.* Designating the style of this group of painters.
1886 W. M. Rossetti in *Encycl. Brit.* XX. 532/1 The naturalist school, called also the school of the Tenebrosi, or shadow painters. *Ibid.*, Ribera..had by this time acquired so much mastery over the tenebroso style that his performances were barely distinguishable from Caravaggio's own. **1938** *Burlington Mag.* Feb. 63/1 *The Last Supper*..introduces us to a *tenebroso* effect. **1982** C. Whitfield in Whitfield & Martineau *Painting in Naples* (Catal. of R. Acad. Exhibition) 165/2 Artemisia Gentileschi trained with her father Orazio in the early Seicento, when his style was at its most *tenebroso*.
Hence **te·nebrist** = *tenebroso *sb.*; **te·nebrism,** the style of the *tenebrosi*.
1923 F. J. Mather *Hist. Italian Painting* ix. 454 Both at Rome and Naples swaggering Caravaggio had enormous success... He boasted himself the greatest painter of all time, and he was often believed. From his swarthy tones his entire school took the name, the Tenebrists. **1958** *Archit. Rev.* CXXIV. 56/3 As a belated tenebrist, he [*sc.* Wright of Derby] handles artificial light intelligently without making any visual discoveries of his own. **1959** *Penguin Dict. Art* 313 Tenebrism..is the name given to painting in a very low key, specifically to the works of those early 17th c. painters who were much influenced by Caravaggio. **1978** *Times Lit. Suppl.* 17 Feb. 213/3 Elsheimer's..tenebrism sprang from the same source.

tenement. Add: **5. tenement house** (earlier U.S. example).
1858 W. A. Butler *Two Millions* 47 The Tenement House, o'er which no friendly movement Has waved the Enchanter's wand of 'Modern Improvement'.

‖ **tenente** (teng·nte). [It., Pg.] A lieutenant.
1929 E. Hemingway *Farewell to Arms* iv. 15 'Do they ever shell that battery?'.. 'No, Signor Tenente.' **1969** M. Gilbert *Etruscan Net* iii. i. 215 He sat down, and motioned the Tenente to be seated also. **1970** F. C. Weffort in I. L. Horowitz *Masses in Lat. Amer.* xi. 388 Their [*sc.* the Brazilian middle class's] most radical acts, generally undertaken by young military men—the *tenentes*.

teneral, *a.* (Examples of *lit.* use.)
1900 W. J. Lucas *Brit. Dragonflies* vi. 66 (*heading*) Immature colour. [*Note*] Also called 'teneral'. **1921** G. H. Carpenter *Insect Transformation* ii. 52 For some time after it has acquired the power of flying the dragon-fly has not yet assumed the deep colours and developed pattern that characterize its species; such a relatively newly-emerged insect is defined as teneral. **1957** *Jrnl. Exper. Biol.* XXXIV. 189 The word 'teneral' has been used to describe alate insects at about this time [between emergence and the first flight], although its usage varies between different orders. **1975** *Nature* 15 May 226/1 This [*sc.* a reduction in probing behaviour] was easier to demonstrate in teneral flies (newly emerged flies before their first meal) than in post-teneral flies.

Teneriffe (tenĕrī·f). [The name of one of the Canary Islands.] **1.** A white wine produced on Teneriffe.
1791 J. Woodforde *Diary* 11 Nov. (1927) III. 312 Claret, Teneriffe, and Port Wines to drink. **1833** C. Redding *Mod. Wines* vi. 194 What is called Vidonia is properly the dry Canary wine, best known as Teneriffe. **1855** E. Acton *Mod. Cookery* (rev. ed.) xxxii. 611 A couple of wineglassesful of Madeira (Sherry or Teneriffe will do).
2. Used *attrib.* to designate a kind of lace made in the Canary Islands.
1907 *Yesterday's Shopping* (1969) 3520/2 Teneriffe D'Oyley. **1920** 'K. Mansfield' *Bliss* 137 Now my best little Teneriffe-work teacloth is simply in ribbons. **1969** R. T. Wilcox *Dict. Costume* 342/2 *Teneriffe lace*, a lace of circles and wheels similar to Paraguay lace, made chiefly in the Canary islands.

tenger, var. *tanger. **tengku,** var. *tunku.

‖ **teniente** (tenʸiˌe·nte). [Sp.] A lieutenant.
1798 in *Lett. from Paraguay* (1805) 248 Every thing..is known to the tenientes. **1906** *Soldier Slang* in C. McGovern *Sarjint Larry an' Frinds*, *Teniente*, Spanish for 'Lieutenant'. **1938** *New Statesman* 1 Oct. 488 Aren't we going to eat to-day, teniente? **1979** A. Melville-Ross *Two Faces of Nemesis* xxii. 157 *Teniente* Descola had done his job well enough to be *Capitan* Descola by evening.

tenii-, var. Taenii- in Dict. and Suppl.

‖ **tenko** (te·ŋko). [Jap.] In Japanese prison camps in the war of 1939–45: a muster parade or roll-call of prisoners.
1947 J. Bertram *Shadow of a War* vi. iii. 190 They drilled us by the hour..and firmly broke us in to the sacred mysteries of '*tenko*'—the morning and evening muster parade..that was routine in all prison camps in Japan... In time even *tenko* lost its terrors. **1961** R. Braddon *Naked Island* in *Plays of Year*, *1960* i. iii. 153, I got him a bashing on tenko tonight.

tenmoku, var. *temmoku.

tenné, tenny, *a.* and *sb.* (Later example of *tenney*).
1922 Joyce *Ulysses* 47 On a field tenney a buck, trippant, proper unattired.

tenner. a. (Earlier examples.)
1845 *Ainsworth's Mag.* VIII. 121 The races..he went to as a matter of course, though..he never betted at them beyond a 'tenner'. **1848** *Sessions Papers* 7 Mar. 847 'I was concerned in that affair of Covill's, for which I had a *tenner* out.'.. 'Oh, did they give you 10*l*?'.. I understood by a *tenner*, a 10*l*.-note.

Tennessean (tenĕsī·ăn), *sb.* and *a.* Also **Tennesseean.** [f. next + -an.] **A.** *sb.* A native or inhabitant of Tennessee. **B.** *adj.* Of, pertaining to, or characteristic of Tennessee.
1815 *Niles' Reg.* VII. 373/1 Glory to..the hardy and gallant Tennesseeans, Kentuckians and Louisianians. **1834** [see *hunt *v.* 3 b]. **1853** J. L. McConnel *Western Characters* 269 Its dye a favorite 'Tennessean' brownish-yellow. **1872** 'Mark Twain' *Sketches* 135 The fervent spirit of Tennessean journalism. **1945** H. F. Woods *Amer. Sayings* 274 A Tennessean-born pioneer, he was in the tradition of Daniel Boone and other border heroes of his time. **1959** C. Ogburn *Marauders* (1960) ii. 49 Lieutenant Caldwell..was an angular young Tennessean. **1978** *Times* 24 Oct. 8/6 Her own campaign slogan..[claims] that she 'has not lost touch' with Tennesseans.

Tennessee (tenĕsī·). The name of one of the United States of America, used *attrib.* in **Tennessee Marble,** a kind of marble found in Tennessee and freq. used in building and sculpture; **Tennessee walker, walking horse,** a lightly built horse belonging to a breed developed in the region and distinguished by an easy natural gait.
1875 T. Yelverton *Teresina in Amer.* II. xiv. 177 It is a lofty, domed structure, the dome supported upon pillars of the red Tennessee marble. **1947** J. C. Rich *Materials & Methods of Sculpture* viii. 226 *Tennessee* is the largest producer of marble in the United States... Many sculptors compare Tennessee marble to granite... Tennessee marble is an excellent sculptural stone. **1968** *N.Y. City* (Michelin Tire Corp.) 74 Pierpont Morgan Library..in pink Tennessee marble. **1960** J. W. Patten *Light Horse Breeds* 147 The breed is variously referred to as Tennessee Walkers, Tennessee Plantation Horses, Tennessee Walking Horses. **1979** *Arizona Daily Star* 1 Apr. (Advt. Section) 10/10 Tennessee walker 7 years old, well trained. **1938** *Reg. Tennessee Walking Horse Breeders' Assoc.* I. 5 The Tennessee Walking Horse Breeders' Association of America held its first meeting at Lewisburg, Tenn., April 27, 1935. **1950** *Congress. Rec.* XCVI. App. A 740/3, I [*sc.* Pat Sutton] am delighted today to advise the House that the United States Department of Agriculture in a letter to me dated February 1, 1950, has recognized the Tennessee walking horse as a distinct and standard breed. **1952** J. Sherman *Real Bk. about Horses* iii. 69 The Tennessee walking horse is noted for his three natural gaits—the flat-footed walk, the running walk and the canter. **1976** *Billings* (Montana) *Gaz.* 2 July 11-c/6 (Advt.), Registered Tennessee Walking horse mares, yearlings, and colts.

tennies (te·niz), *sb. pl. U.S. colloq.* [f. *tenn(is shoe* + *ies* (repr. -y^6 + *pl. suff.*).] Tennis shoes.
1969 *Rolling Stone* 28 June 19/3 Electric guitarist James Burton, replete in white turtleneck and matching 'tennies', puts together a lead. **1976** T. Gifford *Cavanaugh Quest* (1977) i. 17 Margaret, one of the cleaning ladies, got out in her green smock and blue shorts... Her costume was completed with blue tennies. **1980** *Outdoor Life* (U.S.) (Northeast ed.) Oct. 63/3 This is good-boot country, so leave your tennies home.

tennis, *sb.* Add: **2. a.** This is now the usual sense. (Earlier example.)
Tennis has replaced *lawn tennis* as the official international name of the sport. Sense 1 is now usually distinguished as *real tennis* (see *real *a.*^{2} 4 e).
1878 Geo. Eliot *Let.* 8 Aug. (1956) VII. 54 My little man..fights resolutely against these ills..having mild games of tennis.

b. *anyone for tennis?*, *who's for tennis*, etc., a typical entrance or exit line given to a young man in a superficial drawing-room comedy, used *attrib.* of (someone or something reminiscent of) this kind of comedy. Also in extended uses.
1953 J. van Druten *Playwright at Work* viii. 99 There is no average Mr. and Mrs. Blank at all. An attempt to draw one..will lead you into the pit of emptiness, and you will emerge with something as unreal as the juveniles in plays who come in impertinently swinging tennis rackets, and when the time for their exit arrives, make it with the remark: 'Tennis, anyone?' **1965** *Listener* 17 June 911/3 One of the panel spoke of 'Who's-for-tennis' comedy,..now a too-familiar pejorative. **1973** *Times* 16 Jan. 11/1 The most unlikely men around London are now dressing as though they might say 'Anyone for tennis?' at any moment. **1974** N. Freeling *Dressing of Diamond* 34 She had seen him..spring up to answer the telephone with an Anyone-for-tennis voice that filled her with pity. **1978** H. MacInnes *Prelude to Terror* ii. 20 He walked over to the small group of staff members... 'Who's for tennis?' he asked, and raised a smile.
c. See *table-tennis* s.v. Table *sb.* 22.
3. b. *tennis apron, dress, frock, -lawn* (earlier U.S. and later examples), *match* (also *fig.*), *netting, partner, party, shirt, shorts, sock, tournament*; **tennis arm** (example); **tennis club:** see Club *sb.* 14; **tennis flannels** (see Flannel *sb.* 2 b); **tennis net,** a net stretched across the centre of a tennis-court, over which the players strike the ball; **tennis pro(fessional,** a tennis player who is paid to act as an instructor and a player at a tennis club, holiday resort, etc.; **tennis shoe,** a light canvas soft-soled shoe suitable for tennis or general casual wear; **tennis stringer** *U.S.*, a person who strings tennis rackets; **tennis whites** (see White *sb.* 9 b).
1880 L. Higgin *Handbk. Needlework* ii. 11 *Kirriemuir Twill*..is good for tennis aprons, dresses, curtains, &c. **1977** *New Yorker* 10 Oct. 123/3 It now sells not only tennis balls, racquets, and apparel but all sorts of knick-knacks—..telephone books, tennis aprons, [etc.]. **1887** *Epoch* 19 Aug. 26/2 The 'base-ball pitcher's arm' as well as the 'tennis arm' are recognized in the medical profession as special diseases. **1894** *Harper's Mag.* June 156/1 The champion player in our tennis club. **1914** L. S. Woolf *Wise Virgins* ii. 41 May was describing the tennis club dance. **1979** K. Conlon *Move in Game* i. i. 14 Why don't you take her with you to the tennis club? **1885** C. M. Yonge *Nuttie's Father* I. xi. 123 Nuttie was very much pleased with her own pretty tennis dress. **1977** J. Didion *Bk. Common Prayer* v. xvii. 258, I never saw her in a tennis dress. **1899** Kipling *From Sea to Sea* I. xx. 404 Member of the Clapham Athletic Club in tennis flannels. **1934** Tennis flannels [see *gravel court* s.v. *gravel *sb.* 9]. **1981** J. Johnston *Christmas Tree* 33 There was a green stain on his tennis flannels, just below the knee. **1934** A. Thirkell *Wild Strawberries* ix. 191 Ursule, in a short silk tennis frock, looked quite presentable. **1882** *Wheelman* (Boston) I. 55 A tennis-lawn..is seldom far removed from the smoke of the town. **1981** T. Thompson *Edwardian Childhoods* v. 130 My grandparents had a big house with a tennis lawn. **1895** E. F. Benson *Dodo* II. xv. 314 A series of tennis matches which he had taken part in a few years ago. **1961** *Listener* 28 Sept. 483/2 A brilliant 'tennis match' between God and Satan. **1979** Reese & Flint *Trick 13* 134, I had a date to play in a tennis match. **1900** C. M. Yonge *Modern Broods* x. 94 Placing tennis nets, arranging croquet hoops. **1977** *Listener* 7 Apr. 450/1 Table tennis..smashing or retrieving a small celluloid sphere over a miniature tennis net. **1915** Kipling *Let.* 22 Aug. in C. Carrington *Rudyard Kipling* (1955) xvii. 436 Don't forget the beauty of rabbit netting overhead against hand-grenades. Even tennis netting is better than nothing. **1934** P. Bottome *Private Worlds* iii. 26 The girl was going to be married to her tennis partner. **1974** E. Ambler *Dr Frigo* ii. 133 My tennis partner at the army communication centre must have been busy. **1887** Kipling *Plain Tales from Hills* (1888) 256 There are garden-parties, and tennis-parties and picnics. **1981** *Times* 24 Mar. 4/4 Sir Roger Hollis..met an MI5 officer at a tennis party and was finally recommended for recruitment. **1942** A. Christie *Body in Library* iii. 31, I do a couple of exhibition dances every evening with Raymond..he's the tennis and dancing pro. **1977** I. Shaw *Beggarman, Thief* iii. vi. 257 A Belgian businessman..had offered him a contract for a year as a tennis pro. **1938** D. du Maurier *Rebecca* v. 52 The tennis professional had complained, the manager has sent a note. **1979** K. Conlon *Move in Game* i. iii. 32 The bronzed tennis professional, who had all the ladies of the club in a lather of longing. **1889** Tennis shirt [see *cellular *a.* (and *sb.*) 2 b]. **1978** *Country Life* 22 June 1841/1 Cotton tennis dress..navy and white tennis shirt..tennis shoes. **1887** Kipling *Plain Tales* (1888) 222 Miss Hollis ..was..five foot seven in her tennis-shoes. **1928** E. Wallace *Flying Squad* xiii. 122 They walked noiselessly, for Mr. Tiser had obligingly supplied them with..tennis shoes. **1975** *Time* (Canada ed.) 22 Dec. 12/3 [He] once flew out to settle a strike at the Vancouver *Sun* wearing tennis shoes and carrying clothes in a Loblaws shopping bag. **1963** D. B. Hughes *Expendable Man* (1964) iv. 122 She was in tennis shorts and a white blouse. **1932** D. C. Minter *Modern Needlecraft* 253/1 Tennis Socks... 3-ply fingering. **1976** *Washington Post* 19 Apr. c15/4 (Advt.), Tennis stringer. Experience preferred but not necessary. **1892** C. M. Yonge *Cross Roads* xii. 127 Miss Clara caught a chill while driving home after a tennis tournament. **1976** *Wymondham & Attleborough Express* 10 Dec. 21/6 Sue Rich..has made great progress in tennis tournaments in several parts of England this year. **1974** M. Ehrlich *Reincarnation* (1975) xxiii. 203 She was in tennis whites now and volleying with the pro.

tennis, *v.* Restrict † *Obs.* to senses in Dict. and add: **2. b.** To play lawn-tennis. Also with quasi-*obj.* *rare*.
1895 Kipling in *Cent. Mag.* Dec. 276/1 They picnicked and they tennised. **1979** *United States 1980/81* (Penguin Travel Guides) 493 Whether you tennis-it at a camp or a clinic, you're guaranteed a certain number of hours of court time every day. **1983** *Washington Post* 15 Aug. c8/6 They'd rather be golfing, or snorkeling, or tennising.

tennis-court. Add: **1.** *Comb.* Also *esp.* in *tennis-court oath*, the pledge given on June 20, 1789, by members of the States General of France that they would not separate before a constitution was granted (see quot. 1911). Also *fig.*
1893 L. Creighton *First Hist. France* xxvi. 231 The Tennis-Court Oath.—Under his [*sc.* Mirabeau's] guidance the Third estate now declared themselves the National Assembly. **1911** H. A. Guerber *Story Mod. France* x. 55 The Third Estate met tumultuously in the Versailles Tennis Court, where..they bound themselves, by the famous 'Tennis Court Oath'. **1959** *Listener* 31 Dec. 1151/1 The integrity of Anatolia was the tennis-court oath of the Kemalist revolution. **1977** *Socialist Press* 2 Mar. 6/3 Faced with procedural fencing on the part of the King's ministers and the Court, they declare themselves a 'National Assembly' and vow (the 'tennis-court oath', June 20th) not to depart until a constitution is drawn up.

tennis-play. 1. (Later example.)
1918 G. Frankau *One of Them* in *Poet. Wks.* (1923) II. xxiv. 143 Who in all Albion on that fateful day..Left not his office-work, his tennis-play, To read black Montmorency's slander-red lines?

tennis-playing, *ppl. a.* [f. as Tennis-player.] That plays (lawn-)tennis.
1956 H. Gold *Man who was not with It* (1965) vi. 58 The long, tennis-playing, suburban legs. **1965** M. Spark *Mandelbaum Gate* ii. 27 Her energetic tennis-playing grandmother..sat on the arm of a chair. **1979** D. Eden *Storrington Papers* iv. 47 'The new governess..arrived this afternoon.' 'Promising?'.. 'No more tennis-playing amazons.'

Tennysonian, *a.* and *sb.* Add: **A.** *adj.* (Earlier example.)
1846 Lytton *New Timon* ii. 51 Where all the airs of patchwork-pastoral chime To drowsy ears in Tennysonian rhyme!
B. *sb.* (Earlier and later examples.)
1850 J. Brown *Let.* Dec. (1912) 116, I am not a Tennysonian, as many are. **1864** H. Sidgwick *Let.* May in A. & E. M. Sidgwick *Henry Sidgwick* (1906) ii. 108 The compressed *inhaltsvoll* classic style of Tennyson and Tennysonians. **1970** T. Hilton *Pre-Raphaelites* vi. 161 William Morris..like all reading men, was a Tennysonian.
Tennysonia·na [*-iana], matters connected with Tennyson; **Tennyso·nianized** *ppl. a.*, rendered in the manner of Tennyson; **Tennyso·nianly** *adv.*; **Tennyso·nianness** = Tennysonianism, Tennysonism; **Tennyso·nize** *v. trans.*, to render in the style of Tennyson.
1866 R. H. Shepherd (*title*) Tennysoniana. **1910** A. D. Godley *Lectures Eng. Lit.* ii, in *Reliquiae* (1926) II. 288 Tennyson is full of reminiscences of the great classics, thoughts and phrases not slavishly copied but Tennysonised—passed through the medium of an art which added beauty to everything it touched. **1915** E. Pound *Let.* Jan. (1971) 49 [In poetry] there must be..no straddled adjectives (as 'addled mosses dank'), no Tennysonianness of speech. **1916** *Ibid.* ? 20 July (1971) 87 Virgil is a second-rater, a Tennysonianized version of Homer. **1932** L. Magnus *Herbert Warren* viii. 215 His own annotated copy of the one-volume edition of the *Memoir* is a veritable treasure of Tennysoniana. **1964** *English Studies* XLV. 73 'Lycidas' once read 'under the glimmering eyelids of the morne', most Tennysonianly. **1975** *Listener* 20 Nov. 685/1 Superintendent Dalziel, on unwanted holiday in sodden fens, meets a Tennysonianly aqueous funeral cortège.

tenon, *sb.* Add: **2. tenon-saw** (later examples).
1898 *Monthly South Dakotan* I. 57 This operation was successfully performed by Dr. Phillips with no further implements at hand than a large butcher's knife and a small tenon-saw. **1979** A. B. Emary *Woodworking* xxix. 125 Saws (a hand saw and a tenon saw) can be stored in the lid.

tenon, *v.* Add: **1. a.** (Later examples.)
1949 H. M. Cautley *Norfolk Churches* 37 A massive sill, frequently unbroken at the entrance to chancel, into which the muntins are tenoned. **1980** *Early Music* Jan. 62/2 At the other end, the neck is tenoned into the post and pegged.
2. b. (Later examples.) Also *fig.*
1935 'E. Queen' *Spanish Cape Mystery* iv. 103 There are a few facts floating about which don't precisely tenon with the psychopathic theory. **1981** *Rescue News* Mar. 8/1 The oak timbers..tenoned and pegged into sole-plates lying on the bottom of the moat.
tenoner (examples).
1944 J. C. Jones in N. W. Kay *Practical Carpenter & Joiner* x. 227/1 The rails are first fed into the tenoner edgeways up for the machining of the haunchings. **1971** *Cabinet Maker & Retail Furnisher* 24 Sept. 532 Mr Taylor has retained in use with the new line a Schwabedissen double end tenoner with overhead beam.

Tenonian, *a.* Add: *Tenon's capsule* (examples); *Tenon's space*, the episcleral space between Tenon's capsule and the sclera.

1868 Hackley & Roosa tr. *C. Stellwag von Carion's Treat. Dis. Eye* I. xi. 434 This anterior part of the sheath of the eye-ball..is also described as Tenon's capsule. **1892** A. Duane tr. *Fuchs's Text-bk. Ophthalm.* II. xv. 285 Exudation into Tenon's space also occurs after it has been laid open by injuries. **1950** *Sci. News* XV. 25 The eye does not form part of a ball-and-socket joint, like the hip joint, but resembles a ball in a sling, the latter..being composed of a thin sheet of fibrous and smooth muscle tissue, called Tenon's Capsule. **1979** G. W. Cibis tr. *Hollwich's Ophthalm.* xvi. 238 The inflammation involves Tenon's capsule in either a serous or a purulent form. As a rule it remains restricted to Tenon's space.

tenor, *sb.*[1] (*a.*) Add: **I. 1. d.** The underlying idea or subject to which a metaphor refers, as distinct from the literal meaning of the words used. Cf. *vehicle *sb.* 3 d.

1936 I. A. Richards *Philos. Rhet.* v. 96 A first step is to introduce two technical terms to assist us in distinguishing..what Dr. Johnson called the two ideas that any metaphor, at its simplest, gives us. Let me call them the tenor and the vehicle. The tenor, as I am calling it—[is] the underlying idea or principal subject which the vehicle or figure means. *Ibid.* 100 The tenor may become almost a mere excuse for the introduction of the vehicle, and so no longer be 'the principal subject'. **1949** *Poetry* (Chicago) Feb. 304 The tenor is the *new* meaning, the vehicle the *old* meaning on which the new meaning is conveyed. **1962** S. Ullmann *Semantics* viii. 213 An important factor in the effectiveness of a metaphor is the distance between tenor and vehicle. **1973** A. Rodway in R. Fowler *Dict. Mod. Crit. Terms* 112 In the phrase 'Now is the winter of our discontent'..discontentedness is the tenor, and an aspect of winter..the vehicle. **1980** G. B. Caird *Lang. & Imagery of Bible* viii. 152 In a living metaphor, although both speaker and hearer are aware that vehicle and tenor are distinct entities, they are not grasped as two but as one.

II. 4. d. (Earlier examples.)

1785 *Daily Universal Register* 1 Jan. 3/2 (Advt.), Mr. Giardini's capital old Violins, Tenors, and Violoncellos for sale. **1833** [see Alto *sb.*[2] 5].

e. *ellipt.* for *tenor saxophone*, sense B. 1 below.

1876 [see *alto *sb.*[2] 6]. **1927** *Melody Maker* Aug. 738 (Advt.), The manufacturers..have been hailed as the saviours of Tenor Saxophonists through their innovation of the astounding B♭ tenor with the extra automatic octave note. **1952** [see *baritone]. **1975** [see *saxist].

B. *attrib.* or *adj.*, and *Comb.* **1.** *attrib.* or *adj. tenor banjo*: see *banjo 1; *tenor clarinet*, an alto clarinet pitched in F; also, one who plays this instrument; *tenor cor*: see *cor[3]; *tenor drum*: see Drum *sb.*[1] 1 b; *tenor horn* = *althorn; *tenor sax, saxophone*, a member of the saxophone family intermediate between the alto and the baritone, usu. pitched in B flat; also, one who plays this instrument; hence *tenor saxist, sax-man, saxophonist*; *tenor violin* (examples).

1802 Tenor violin [see Alto- 1]. **1859**, etc. Tenor horn [see *althorn]. **1865** C. Mandel *Mandel's Syst. Mus.* xvi. 68 There are various kinds of Saxophones. The smallest, or Soprano Saxophone, is in B flat... The.. Tenor Saxophone is an octave lower than the Soprano Saxophone. **1879** Grove *Dict. Mus.* I. 362/2 In F we have the tenor clarinet. **1926** Whiteman & McBride *Jazz* ix. 193 We have computed..that one tenor saxophone equals eight violas. **1927** Tenor saxophonist [see sense *4 e]. **1934** S. R. Nelson *All about Jazz* vi. 127 A brilliant tenor sax was unmistakable. **1938** D. Baker *Young Man with Horn* I. iv. 28 There were five men in Jeff's band—a tenor clarinet, a trombone, a trumpet, traps, and a piano. **1954** *Grove's Dict. Mus.* (ed. 5) II. 326/2 The higher-pitched of the two [alto clarinets] was long known as the 'tenor clarinet' in England. *Ibid.* VIII. 809/1 The true Tenor Violin was the alto of the *viola da braccio* family... The gradual suppression of this instrument in the 18th century was a disaster: neither the lower register of the viola nor the upper register of the violoncello can give its effect. **1954**, etc. Tenor saxophone [see *saxophone sb. 1]. **1955** Keepnews & Grauer *Pict. Hist. Jazz* x. 110 Key members included tenor sax Andy Brown. **1955** Tenor saxist [see *bassist 2]. **1955** Tenor sax-man [see *sax-man* s.v. *sax *sb.*[2] 3]. **1958** T. Hall in P. Gammond *Decca Bk. Jazz* xix. 229 He was mainly featured on an E-flat tenor-horn, which produced a mellophone-like sound. **1963** *Listener* 7 Feb. 264/1 Two virtuoso tenor sax players. **1972** *Guardian* 4 Feb. 10/5 No one strove harder than the tenor saxist John Coltrane. **1979** *Country Life* 12 July 95/2 A..solo by tenor saxophonist Steve Marcus. **1979** *Listener* 4 Oct. 461/3 Charlie Parker..plays tenor-sax on the Miles Davis set.

2. *Comb.* in sense 4 e, as *tenor-man, player, solo, soloist, style.*

1928 *Melody Maker* Feb. 201/2 Quite a few successful tenor players. **1935** *Vanity Fair* (N.Y.) Nov. 38/2 Tenor-men like Hawkins or Fletcher Henderson, are stars in the hot sky. **1943** P. E. Miller *Yearbk. Popular Music* 8/2 He borrowed a tenor from a fellow musician, sat in on a jam session, and from that point forward became a hot tenorman. **1958** R. Horricks in P. Gammond *Decca Bk. Jazz* ix. 118 The prominent tenor soloists outshone even those of the New York scene. **1959** F. Newton *Jazz Scene* ii. 35 A fine tenor player in the Parker tradition. **1962** *Melody Maker* 21 July 7/3 The perfect tenor style for Dixieland jazz. **1966** *Crescendo* Dec. 9/2 Good clarinet, trumpet and tenor solos. **1977** J. Wainwright *Do Nothin'* viii. 125 Tenor men are not *that* hard to find.

te·noring *ppl. a.*

1905 H. G. Wells *Mod. Utopia* iv. 127 It is not only such gross and palpable cases as our blond and tenoring friend. **1930** —— *Autocr. Mr. Parham* II. iii. 119 'But,' said Mr. Mountain in tenoring remonstrance to Sir Bussy, 'doesn't this evening satisfy you, sir?' **1934** —— *Exper. Autobiogr.* II. viii. 602 Bland was a thick-set, broad-faced aggressive man..with a tenoring voice.

‖ **tenore** (tenǒ·re). [It.; cf. Tenor *sb.*[1] (*a.*).] = Tenor *sb.*[1] 4 a and b. Usu. with qualifying adj. or phrase, as *tenore di grazia*, a light or lyric tenor; *tenore robusto* (see quot. 1876), a dramatic tenor.

1740 Grassineau *Mus. Dict.* 272 *Tenore*, the first mean or middle part: or that which is the ordinary pitch of the voice, when neither raised to a treble, or lowered to a bass. **1876** Stainer & Barrett *Dict. Mus. Terms* 432/1 *Tenore robusto*, a tenor singer with a full, strong, sonorous voice. **1889** Grove *Dict. Mus.* IV. 87/1 Hence we have *tenore robusto* (which used to be of about the compass of a modern high baritone), *tenore di forza*, *tenore di mezzo carattere*, *tenore di grazia*, and *tenore leggiero*, one type of which is sometimes called *tenore contraltino*. **1894** G. du Maurier *Trilby* I. i. 25 A voice so rich and deep and full as almost to suggest an incipient *tenore robusto*. **1925** *New Yorker* 19 Sept. 20/2 He is the first *tenore robusto* to emerge since Caruso. **1938** J. Joyce *Let.* 8 Sept. (1966) III. 427 It needs a deep strong voice, not my *tenore di grazia*. **1960** *Times* 4 Mar. 4/1 Mr. Kenneth Macdonald, who has an enviable affinity for *tenore di grazie* singing. **1979** *Times* 19 Nov. 7/5 The..refinements of phrasing and nuance which are hall-marks of a *tenore di grazia*.

tenorino (tenǒrī·no). Pl. **-ini.** [a. It., dim. of *tenore* tenor.] A high tenor; *spec.* a castrato alto.

1867 *Cornh. Mag.* Jan. 32 At present the signor is the pearl of tenorini, and no other artist can match his delivery of the embroidered melodies of the *Cenerentola*, or the *Italiana in Algieri*. **1898** *Harper's Mag.* XCVI. 512, I was..to be the tenor, or rather the *tenorino*. **1980** *New Grove Dict. Mus.* XVIII. 690/2 Another type of light tenor voice was known in the 19th century as the 'tenorino'; such singers were often amateurs who made a speciality of performing love-songs to salon audiences.

tenorist. Add: (Later examples.) Also, one who plays the tenor saxophone.

1958 K. Goodwin in P. Gammond *Decca Bk. Jazz* xiii. 154 Vinegar was also featured..when the tenorist opened a short season at Zardi's. **1962** *Melody Maker* 7 July 9 It is quite remarkable..that so popular a tenorist..has not had a broadcast or TV appearance for more than three years. **1972** *Blues & Jazz* Sept. 11/2 The backing on 'The Fat Man' was provided by the Bartholomew band, including tenorists Herb Hardesty and Red Tyler. **1977** *Listener* 1 Dec. 729/4 Roy Plomley has marooned tenorist Ronnie Scott..on his desert island.

tenoroon. Add: **a.** (Later example.) The use of this instrument has recently been revived in performances of baroque music.

1980 *Early Music Gaz.* Jan. 9/2 William Waterhouse.. performed music by Selma, Böddecker..on racket, two original 18th-century bassoons (Handel period 4-key and Mozart period 7-key, tenoroon and modern Heckel).

tenpence. Add: (Earlier and later examples of *ten pence in the shilling* and varr.) Since 1971, a coin worth ten (new) pence superseding the earlier two-shilling piece or florin; often as two words with pronunc. (ten pens). Also *attrib.* as *tenpence coin, piece.*

1860 Hotten *Dict. Slang* (ed. 2) 235 *Tenpence to the shilling*, a vulgar phrase denoting a deficiency in intellect. **1922** J. Buchan *Huntingtower* vii. 142 There's a certain old lady, an aunt of Mr. Quentin and his sisters, who has always been about tenpence in the shilling. **1936** W. Holtby *South Riding* 9 'Mental?' 'Tenpence halfpenny in the shilling.' **1971** P. Purser *Holy Father's Navy* xxiv. 114, I gave her a ten pence piece and hurried away. **1974** A. Fowles *Pastime* xii. 98 Awkward, that had been, in a phone box. He'd used up two ten pences. **1976** G. Seymour *Glory Boys* xi. 144 He put down two tenpence coins.

Hence **te·npenceworth,** the amount of anything to be bought for tenpence; used contemptuously.

1896 G. B. Shaw *Let.* 16 Nov. in *Ellen Terry & Bernard Shaw* (1931) 124, I have been to Paris, and seen Peer Gynt done in the sentimentallest French style with te npenceworth of scenery.

tenpenny, *a.* (*sb.*) Add: **A.** *adj.* **1. a.** *tenpenny piece* (later examples); also, a coin worth ten (new) pence.

1968 *Guardian* 24 Apr. 3/3, I handed the woman a new tenpenny piece worth 2s. **1971** I. Murdoch *Accidental Man* 305 The room was unheated except for a weak one-bar electric fire which had to be continually fed with tenpenny pieces. **1973** J. Porter *It's Murder with Dover* xiii. 130 MacGregor watched his tenpenny pieces disappearing down the greedy slot of the one-armed bandit.

b. *tenpenny nail* (later examples); in U.S. *spec.* a three-inch nail.

1890 Webster, *Penny*.., denoting pound weight for one thousand;—used in combination, with respect to nails; as, tenpenny nails, nails of which one thousand weigh ten pounds. **1906** *Dialect Notes* III. 146 *Mad enough to bite a tenpenny nail in two, adj. phr.*, very angry. **1909** Webster, *Penny* is used in combination with prefixed numerals..to form adjectives denoting price or value. As applied to nails these adjectives now denote certain arbitrary sizes, though originally, in the 15th century, they designated the price per hundred; as, a *tenpenny* nail, one then costing tenpence per hundred. **1967** *Countryman* Autumn 29 My American companion said: 'I guess we should get some tenpenny nails.' Then he looked at me: 'I suppose you don't know what they are.' I did not. Next day, when we asked for them by that name at the hardware store, we got what we wanted—three-inch nails.

B. *sb.* **1. a.** (Earlier example.)

1822 D. O'Connell *Let.* 13 Apr. (1972) II. 379, I will hug every tenpenny as a link in the chain that is to draw back my Mary to me.

ten-pins. Add: **b.** *ten-pin alley* (earlier examples), *ball* (earlier example), *bowling.*

1835 P. H. Nicklin *Lett. Descr. Va. Springs* 23 The means of amusement at the Warm Springs, consist of a bagatelle table..a ten-pin alley [etc.]. **1842** [see *bowling saloon* s.v. *bowling *vbl. sb.* 3]. **1852** C. A. Bristed *Upper Ten Thousand* v. 117 Perhaps we shall find him at the ten-pin alley. **1870** O. Logan *Before Footlights* 120 Finely cut bits of paper, for fatal snowstorms; ten-pin balls, for the distant muttering of the storm. **1934** *A.B.C. Bulletin* 25 Oct. 11/1 (*heading*) Tenpin Bowling—'The Sport of Kings'. **1960** *Observer* 17 Jan. 3/3 Ten-pin bowling, as its sponsors call it, went to America with the planters and the pilgrims... The game that is being re-imported is hedged about with expensive equipment and social ballyhoo, but is simple enough in itself. **1975** *Oxf. Compan. Sports & Games* 90/1 Tenpin bowling has reached a stage where it can claim to be the largest participant sport in the world.

ten-pounder. Add: **2. a.** (Later examples.)

1844 *Ainsworth's Mag.* VI. 354, I feared I should very soon be obliged to change..my ten-pounder. **1888** C. M. Yonge *Our New Mistress* xii. 109 He took it from me as if I were paying him his wages, and..said..a crisp ten-pounder was a handier thing to drag about than a puling woman.

b. (Earlier example.)

1833 R. Southey *Let.* 13 Jan. in J. Aitken *Eng. Lett. of XIX Century* (1946) 147 The ten-pounders have sent just such members as might have been expected to *Parldemonium* from the great manufacturing towns.

tense, *sb.* Add: **1. a.** (Later example.)

1922 Joyce *Ulysses* 604 To fast and abstain on the days commanded, it being quarter tense or, if not, ember days or something like that.

3. *attrib.* and *Comb.*, as (in sense 2) *tense-aspect, marker, stem, system*; *tense-marking, -modal* adjs.

1892 H. Sweet *New Eng. Gram.* I. 101 By tense-aspect we understand distinctions of time independent of any reference to past, present, or future. **1980** *English World-Wide* I. i. 113 It seems as though the tense-aspect system of English has been restructured. **1971** E. Jones in J. Spencer *Eng. Lang. W. Afr.* 83 Krio is equipped with a range of tense markers, as may be seen from the following set. **1978** *Language* LIV. 84 The advocates of abstract remote structures posit auxiliaries including negative and tense markers as main verbs. **1962** C. L. Barber in F. Behre *Contrib. Eng. Syntax* 27 Any combination of four tense-markings. **1921** E. Sapir *Language* v. 96 Had the statement been made on another's authority, a totally different 'tense-modal' suffix would have had to be used. **1965** *Language* XLI. 173, 1200 adverbial suffixes, partly tense-modal. **1935** T. Hudson-Williams *Short Introd. Study Compar. Gram.* xiii. 72 The endings were added to each tense-stem. **1971** *Archivum Linguisticum* II. 100 The subjunctive is originally independent from the so-called tense stems, as is evident in Celtic and Tocharian and also in Latin. **1951** W. K. Matthews *Lang. U.S.S.R.* iv. 75 The tense system is complicated by being carried into the non-finite grammatical categories, including the gerund. **1963** J. Lyons *Structural Semantics* vi. 112 The 'tense-system' may be set out in terms of the two dimensions of time and aspect.

tense, *a.* Add: **1. a.** Also *spec.* in *Phonetics*, applied to (the articulation of) a speech-sound pronounced with enhanced tension in the muscles of the speech organs. Cf. *lax *a.* 5 c, *slack *a.* 7 e.

1909, etc. [see *lax *a.* 5 c]. **1909** [see *slack *a.* 7 e]. **1918** D. Jones *Outl. Eng. Phonetics* 21 When pronouncing the..tense vowel..the throat feels considerably tenser and is somewhat pushed forward. **1933** L. Bloomfield *Language* vii. 109 In German the tense vowels are longer than the loose; this difference of length is more striking than that of tenseness. **1968** W. S. Allen *Vox Graeca* v. 103 The usually tenser articulation of voiceless plosives might also tend to emphasize the crescendo. **1978** *Canad. Jrnl. Linguistics* 1977 XXII. 211 *Rêve* and *âge* have inherited, underlying tense vowels.

tense, *v.* **a.** Delete *rare* and add: Also *refl.* and with *up*. (Later examples). Also *spec.* of vowel sounds (cf. *tense *a.* 1 a).

1929 P. Gibbs *Hidden City* 1 Rage causes an increase of adrenal secretion, tensing up the nerve cells. **1942** G. Casey *It's Harder for Girls* 130 When his turn came he tensed himself to go through with it. **1951** C. S. Forester *Randall & River of Time* xviii. 263 The constable was tensing himself, ready to restrain him if he should do anything violent. **1978** *Canad. Jrnl. Linguistics* 1977 XXII. 211 Historically, /v/ and /ž/, although lengthening preceding vowels, did not automatically tense them.

b. *intr.* To become tense. Also const. *up*.

1946 *Sunday Express* 31 Mar. 8/2 The court tensed as Ribbentrop gave inside glimpses of events which shaped the war. **1959** *Encounter* Feb. 31, I was tensing for the death-blow. **1973** *Houston* (Texas) *Chron. Texas Mag.* 14 Oct. 2/3 They..feared the kids would tense up if they knew a reporter was in their midst. **1975** I. McEwan *First Love, Last Rites* 42 There was such a sudden ferocity in her silence that I found myself tensing like a sprinter on the starting line.

Hence **te·nsing** *vbl. sb.* (also with *up*).

1921 L. R. Freeman *In Tracks of Trades* 85 There was a sharp tensing of the powerful frame. **1977** *Washington Post* 23 Nov. B2/3 It is the isometric tensing of muscles opposite ones that have been over-developed. **1983** *N.Y. Times* 9 Oct. VI. 56/2 More like a tensing-up that begged for relief.

tensed, *ppl. a.* Add: (Later examples.) Also *fig.* (cf. Tense *a.* 2). Freq. const. *up* and now rarely used *attrib.*

1911 J. London *Adventure* i. 11 The tensed body relaxed. **1934** E. O'Neill *Days without End* I. 29 His eyes fixed before him.., his body tensed defensively. **1952** G. Thomas *Now lead us Home* 191 All tensed up in wait for the hand that will draw some heavenly melody out of them. **1971** S. Hill *Strange Meeting* i. 41 There would be no more anxieties..about how he could bear to sit in the sour-smelling room with the Major, tensed with dread of the night to come. **1980** 'R. B. Dominic' *Attending Physician* xxii. 198 Ben sounds pretty tensed-up to me.

tensed (tenst), *a.* [f. Tense *sb.* + -ed[2].] Having a grammatical tense or tenses.

1972 *Language* XLVIII. 314 The situation of [examples] 34 and 35 is particularly interesting: we do not, in general, find this type of stress in tensed (i.e. non-infinitival) relative clauses. **1978** *Ibid.* LIV. 289 It is also common to most of these proposals to assume that deverbatives are derived from clauses which contain tensed verb forms or their equivalents. **1982** Baker & Hacker in *Language & Communication* II. iii. 240 Animals have immediate purposes and intentions, but long term goals, projects and intentions are available only to creatures who have forms of expression for such things, viz. a tensed language.

tensegrity (tense·grĭti). [f. *tens*(*ional int*)*egrity*.] A stable three-dimensional structure consisting of members under tension that are contiguous and members under compression that are not; the characteristic property of such a structure; also *fig.* Freq. *attrib.*

1959 *Art News* Oct. 29 Of all the ways out of the blind alley in which so much of modern architecture luxuriously relaxes, R. Buckminster Fuller's 'tensegrity' structures seem the most inventive and promising. **1963** R. B. Fuller *Ideas & Integrities* viii. 170 We have in the Geodesic Tensegrity (my name for the discontinuous-compression, continuous-tension structures) the ability to assemble unprecedentedly large, clear-span structures. **1972** *Last Whole Earth Catalog* (Portola Inst.) 4/3 The Universe is a tensegrity. **1976** A. Pugh *Introd. Tensegrity* ii. 11 One of the most impressive Tensegrity figures has six struts which do not touch one another and twenty-four tendons. **1976** H. Kenner *Geodesic Math* p. viii, No useful structures exploiting pure Tensegrity—tension wholly separated from compression—have been built. **1976** *Sci. Amer.* Dec. 144/1 The Kenner book..derives the very nature of a geodesic dome..as the limiting case of a more complex skin of simple tensegrities.

tenseness. Add: Also *spec.* of vowel sounds (cf. *tense *a.* 1 a).

1918 D. Jones *Outl. Eng. Phonetics* 20 The 'tenseness' or 'laxness' of a vowel may be observed mechanically in the case of some vowels by placing the finger on the throat. **1933** [see *tense *a.* 1 a]. **1958** A. S. C. Ross *Etym.* ii. 125 OE *e o* became opened to *ę ǫ* soon after 1200... This loss of tenseness is of great consequence in the development of new diphthongs. **1978** *Canad. Jrnl. Linguistics* 1977 XXII. 211 There appears to have been a recent tendency to attribute tenseness to all lengthened vowels.

tensify, *v.* (Later example.)

1932 V. Woolf *Common Reader* 2nd Ser. 145 That strain of..passion did..not tensify the quiet of the country morning.

tensile, *a.* Add: **2.** *tensile test* (Engin.), a test for determining the tensile strength of a sample of material (usu. metal); so *tensile testing* (also *attrib.*).

a **1877** Knight *Dict. Mech.* III. 2539/1 In the hydraulic tensile testing-machine..the specimen is held by the two clips. **1883** *Jrnl. Iron & Steel Inst.* 98 (*heading*) Results of tensile tests made at University College, London. **1923** Glazebrook *Dict. Appl. Physics* V. 53/1 Two types of testing machine are in use for the tensile testing of fabric specimens. **1953** D. J. O. Brandt *Manuf. Iron & Steel* 362 Fig. 203 shows a tensile testing machine and the method of setting up the test piece. **1973** J. G. Tweeddale *Materials Technol.* I. iv. 78 (*caption*) A tensile test piece.

tensimeter (tensi·mĭtəɹ). [f. Tensi(on *sb.* + -meter.] An instrument for measuring vapour pressure.

1907 *Jrnl. Amer. Chem. Soc.* XXIX. 1055 The hygrometer used above may conveniently serve the purpose of a tensimeter. **1946** J. R. Partington *Gen. & Inorg. Chem.* iii. 54 The dissociation pressure of a salt hydrate is measured in a tensimeter.

tensiometer (tensi,ǫ·mĭtəɹ). [f. Tensi(on *sb.* + -ometer.] **1. a.** An instrument for measuring the surface tension of a liquid. **b.** One for measuring the tension of soil water.

1922 *Jrnl. Exper. Med.* XXXV. 576 The tensiometer was used with an accurately calibrated platinum-iridium ring..and is reliable to ±0·1 dyne. **1936** Richards & Gardner in *Jrnl. Amer. Soc. Agronomy* XXVIII. 352 Rogers.., lacking a more suitable name, has called the combination a soil moisture meter. In the interest of brevity and unambiguity, the name tensiometer is here used. **1973** *McGraw-Hill Yearbk. Sci. & Technol.* 382/1 Tensiometers placed in the active root zone and near the bottom of the root zone..provide information that permits control of deep percolation [in irrigated soil]. **1975** Yong & Warkentin *Soil Properties & Behaviour* iv. 129 Tensiometers are widely used to indicate when irrigation is required.

2. An instrument for measuring the tension in yarn, a rope, etc.

1947 *Textile Research Jrnl.* Jan. 27/1 The thread then passes to a tensiometer,..which instrument feeds voltage proportional to the thread tension into a..chart recorder. **1952** *Electronic Engin.* XXIV. 531 The most common instrument for measuring yarn tension is the pocket-size dial tensiometer. **1978** A. Welch *Bk. of Airsports* vi. 98/2 The tensiometer is important, so that the canopy will never be overloaded by the car being driven too fast in strong winds.

Hence **tensiome·tric** *a.*, **tensio·metry.**

1965 *New Scientist* 18 Nov. 497/1 The rocks are simulated by microscopic glass balls and polymers... They claim that these models enable reliable forecasts of what will happen to the springs if this or that method of mining is adopted. Tensiometry..and ultrasound measurements play an important role in this work. **1968** *McGraw-Hill Yearbk. Sci. & Technol.* 351/1 A tensiometric method utilizes a porous cup filled with water connected by a tube to a vacuum indicator. This approach measures the capillary potential or suction of soil water. **1979** *Acta Protozoologica* XVIII. 64 Two radial measurements by tensiometry.

tension, *sb.* Add: **2. c.** esp. *Psychol.* A condition of strain produced by anxiety, need, or by a sense of mental, emotional, or physical disequilibrium; also *attrib.* or as *adj.*

1884 W. James in *Mind* IX. 12 The states of tension.. have as positive an influence as the discharges in determining the total condition, and consequently in deciding what the *psychosis* shall be to which the complex *neurosis* corresponds. **1925** H. M. & E. R. Guthrie tr. *Janet's Princ. Psychotherapy* iv. 234 Psychic tension [is] characterized by the degree of activation and the hierarchical degree of acts. **1930** J. Riviere tr. *Freud's Civilization & its Discontents* 127 The sense of guilt..is..the ego's appreciation of the tension between its strivings and the standards of the super-ego; and the anxiety that lies behind. **1935** Adams & Zener tr. *Lewin's Dynamic Theory of Personality* ii. 59 A tendency may readily be observed toward immediate discharge of tension (to a state of equilibrium at the lowest possible state of tension). **1958** H. A. Murray in G. Lindzey *Assessment of Human Motives* vii. 194 The concept of human nature..is a concept of perpetually recurrent drives, or tensions.

d. The conflict created by interplay of the constituent elements of a work of art. Used esp. of poetry. (See also quot. 1941.)

1941 A. Tate *Reason in Madness* 72, I proposed..the term *tension*..using the term not as a general metaphor, but as a special one, derived from lopping the prefixes off the logical terms *extension* and *intension*... The meaning of poetry is its 'tension', the full organized body of all the extension and intension that we can find in it. **1949** *Poetry* Feb. 305 *Tension*,..the resultant effectual unity of the poem derived from the operation of such conflict-structures as wit, paradox and irony, slackness being the result of a failure in tension. **1957** N. Frye *Anat. Crit.* 256 It is more likely to be the harsh, rugged, dissonant poem.. that will show in poetry the tension and the driving accented impetus of music. **1975** *Language* LI. 583 Metrical tension can be construed as the degree of difference between underlying and derived metrical patterns.

3. b. Substitute for def.: = Pressure 2 a; now chiefly *Biol.* and *Med.* (Later examples.)

1906 W. Marriott *Hints to Meteorol. Observers* (ed. 6) 69/1 Tension of vapour. **1907** J. H. Parsons *Dis. Eye* ii. 18 The pressure inside the eye is called the intraocular pressure, or the tension, of the eye. **1940** *Jrnl. Bacteriol.* XXXIX. 307 (*heading*) The effect of oxygen tension on the oxygen uptake of lake bacteria. **1971** *Brit. Med. Bull.* XXVII. 55/2 The oxygen tension in the arterial blood may be somewhat lowered. **1972** A. H. Halasa *Basic Aspects of Glaucomas* xi. 97 Low tension glaucoma refers to a condition characterized by a normal intraocular pressure associated with..glaucomatous visual field defects.

c. (Later example.) Also, the degree of tightness or looseness of the stitches in machine sewing or in knitting.

1932 D. C. Minter *Mod. Needlecraft* 199/2 Learn how to regulate machine stitch and tension. **1933** Tillotson & Minter *Compl. Knitting Bk.* ii. 21 The knitted loops, for a correct tension, should just cling lightly and closely to the reader. **1950** J. Norbury *Knitter's Craft* i. 10 A loose tension will produce a flabby, ill-fitting garment. **1973** *Tucson* (Arizona) *Daily Citizen* 22 Aug. 3 (Advt.), Brother sewing machine Lightweight zig zag..fingertip touch tension. **1980** C. Fremlin *With no Crying* x. 61 Alison was concentrating on those first vital rows of her knitting, making sure that she was getting the tension right.

4. (Earlier example.)

1785 G. Adams *Essay on Electricity* (ed. 2) x. 208 The whole energy of electricity depends on its tension, or the force with which it endeavours to fly off from the electrified body.

5. a. *high tension* (earlier examples).

1833 Faraday in *Phil. Trans. R. Soc.* CXXIII. 516, I was anxious..to obtain some idea of the conducting power of ice and solid salts by electricity of high tension. **1877** *Telegr. Jrnl.* V. 289/2 (*heading*) On the effects produced by electric currents of high tension.

c. *transf.* and *fig.*

1898 G. B. Shaw *Candida* III. 148 Eugene, strung to the highest tension, does not move a muscle. **1906** J. M. Synge *Let.* ?6 Nov. (1971) 47, I am working now at very high tension. **1959** D. Cooke *Lang. Music* iv. 183 The high-tension 'current' of Beethoven's emotion, we may say, had to be converted into a high-tension rhythmic energy.

6. (in sense 2) *tension state, system*; *tension-relieving* adj.; **tension bar,** (*a*) (example); (*b*) a metal bar used to apply pressure or exert tension; **tension spring,** (*b*) a spring used to maintain a required degree of tautness; **tension wood** = *reaction wood* s.v. *reaction 5.

1879 *Car-Builder's Dict.* 163/1 *Tension bar*, a bar which is subjected to a strain of tension. **1963** R. A. Higham *Handbk. Papermaking* viii. 212 Tension bars are usually found on calendars, especially when treating light-weight papers, and in action these serve to keep the sheet flat and taut across the working width. **1977** 'E. McBain' *Long Time no See* xiii. 215 The telephone was as vital a tool to policemen as was a tension bar to a burglar. **1949** Koestler *Insight & Outlook* 421 Neglect of the emotional dynamics of laughter, of its tension-relieving aspect. **1974** M. C. Gerald *Pharmacol.* xi. 201 *Quiet World* contains 'special calming and tension-relieving ingredients'. **1966** J. S. Cox *Illustr. Dict. Hairdressing & Wigmaking* 148/1 *Tension spring*, a spirally wound and flattened wire spring which, when stretched returns to its original length... The tension spring is sometimes replaced by elastic. **1970** *Which?* Aug. 238/2 A faulty tension spring on the bobbin case stopped the tensioning adjustment from working properly. **1946** *Mind* LV. 149 We have, therefore, to discover these responses that are the most successful in resolving the personal tension state of which political argument is the expression. **1977** J. D. Douglas in Douglas & Johnson *Existential Sociol.* i. 43 Anomie appears to be a tension state that is produced in the individual by an inability to achieve success by legitimate means. **1936** *Mind* XLV. 248 The technique which seeks to make an undesired goal palatable or a desired goal unpalatable, by linking them up somehow with the 'natural' tension-systems of the child. **1953** M. Horwitz in Cartwright & Zander *Group Dynamics* xx. 371 Individuals develop tension systems coordinated to reaching their own goals. **1924** W. S. Jones *Timbers* iv. 27 'Tension' or 'white' wood differs from 'red' wood in that the cell walls of the tracheids show a well-developed, strongly-lignified, tertiary layer. **1951** McLean & Ivimey-Cook *Textbk. Theoret. Bot.* I. xxi. 907 In conifers the lower wood is reddish, the upper white..the upper wood being called tension-wood. **1972** *Gloss. Terms Timber* (*B.S.I.*) 15 Tension wood. Abnormal wood.. formed typically on the upper sides of branches and of leaning or crooked trunks of hardwood trees.

Hence **te·nsionally** *adv.*, by means of tension, as a result of tension.

1960 R. W. Marks *Dymaxion World of B. Fuller* 195 Magnesium ball-jointed tripods..were tensionally opened by piston-elevated masts. **1975** *New Yorker* 12 May 41/1 Tensionally cohered universe here today and gone tomorrow.

tension, *v.* (Examples.)

1891 *Engineer* LXXI. 120/2 [List of patents.] Tensioning saddles of velocipedes, F. A. Matthews, London. **1950** *Jrnl. R. Aeronaut. Soc.* LIV. 631/1 The 'floating stud'..is a slotted template stud contained in a metal ring, and tensioned by three or four springs. **1975** Kong & Evans *Reinforced & Prestressed Concrete* ix. 196 When the concrete has hardened sufficiently, the tendons are tensioned by jacking against one or both ends of the member.

tensioner (te·nʃənəɹ). *Mech.* [f. Tension *sb.* + -er[1].] A device for applying tension to cables, pipelines, etc.

1950 *Engineering* 5 May 489/3 Heavy spring tensioners are fitted to the front sprockets. **1972** L. M. Harris *Introd. Deepwater Floating Drilling Operations* xiii. 140 Marine-riser tensioning can be provided either by a dead-weight system or by the use of pneumatic tensioner cylinders. **1977** *Austral. Sailing* Jan. 54/1 The luff tensioner..is the most subtle control.

tensometer (tenso·mĭtəɹ). [f. Tens(ion *sb.* + -ometer.] **1.** An apparatus for measuring the tensile strength of a material.

1937 *Nature* 1 May 765/2 The Griffin Gale testing machine enables tensile..tests to be carried out on small samples of metals... In the same class is the Hounsfield tensometer. **1950** *Chem. Abstr.* XLIV. 7 (*heading*) Mechanical tensometer for measurements of deformations on temperature change. **1971** *Nature* 2 Apr. 323/1 The specimens were strained at a rate of 1mm min^{-1}; the tensometer and electrometer head amplifier were enclosed in an earthed brass gauze to screen out extraneous electrical fields.

2. = *tensiometer 1 a.

1941 *Abstr. Sci. & Technical Press* No. 91. 178 The author has designed an interfacial tensometer... In this instrument the force exerted by the interfacial layer on a platinum ring is measured by means of a torsion balance.

3. = *tensiometer 2.

1953 K. H. Inderfurth *Nylon Technol.* vii. 159 It is essential that the yarn tension be frequently checked with a tensometer.

tensor. Add: **2. a.** (Earlier example.)

1846 W. R. Hamilton in *Phil. Mag.* XXIX. 27 Since the square of a scalar is always positive, while the square of a vector is always negative, the algebraical excess of the former over the latter square is always a positive number; if then we make $(TQ)^2 = (SQ)^2 - (VQ)^2$, and if we suppose TQ to be always a real and positive or absolute number, which we may call the tensor of the quaternion Q, we shall not thereby diminish the generality of that quaternion. This tensor is what was called in former articles the modulus.

c. An abstract entity represented by an array of components that are functions of co-ordinates such that, under a transformation of co-ordinates, the new components are related to the transformation and to the original components in a definite way. [This sense is due to W. Voigt (*Die Fund. Physik. Eigenschaften der Krystalle* (1898) p. vi).]

1916 *Monthly Notices R. Astron. Soc.* LXXVI. 701 In the four-dimensional time-space we consider tensors of different orders. The tensor of order zero is a pure number (scalar), the tensor of the first order is a vector, which has 4 components, the tensor of the second order has 16 components, and so on. *Ibid.* 702 If once we have expressed the laws of nature in the form of linear relations between tensors, they will be invariant for all transformations. Thus with the aid of the calculus of tensors Einstein has succeeded in satisfying the postulate of general relativity. **1934** *Nature* 20 Oct. 612 The theory of tensors, so important in physics and geometry on account of their property of vanishing in every co-ordinate system if they vanish in one, was created by Ricci (1887) and his pupil Levi-Civita, although the name *tensor* was not introduced by them. **1943** *Jrnl. London Math. Soc.* XVIII. 109 The study of the particular class of invariants known as tensors goes back to the work of Riemann and Christoffel on quadratic differential forms. **1953** C.-T. Wang *Applied Elasticity* i. 1 Stress is called a tensor, because in addition to its magnitude, direction, and sense, which define a vector, it depends on another vector, which represents the surface upon which it acts. **1970** G. K. Woodgate *Elem. Atomic Struct.* iii. 50 The operator in eqn. (3.95) is a component of a second-rank tensor, the atomic electric quadrupole moment. **1974** G. Reece tr. *Hund's Hist. Quantum Theory* xv. 211 ϕ and χ were scalars, spinors, vectors or tensors.

d. *attrib.* and *Comb.*, as *tensor algebra, analysis, calculus, product*; **tensor field,** a field for which a tensor is defined at each point; **tensor force,** a force between two bodies that has to be expressed as a tensor rather than a vector, *esp.* a non-central force between subatomic particles.

1922 Tensor algebra [see *tensor analysis* below]. **1936** *Electr. Engin.* LV. 1214/1 The object of this paper is to apply tensor algebra to the solution of the circuits of multi-winding transformers. **1971** C. W. Curtis in Powell & Higman *Finite Simple Groups* iii. 142 Form a vector space *M* with basis *X*, and let $\mathfrak{F}_X$ be the tensor algebra over *M*. **1922** H. L. Brose tr. *Weyl's Space–Time–Matter* i. 58 Tensor analysis tells us how, by differentiating with respect to the space co-ordinates, a new tensor can be derived from the old one in a manner entirely independent of the co-ordinate system. This method, like tensor algebra, is of extreme simplicity. **1939** G. Kron *Tensor Analysis of Networks* p. xvi, Tensor analysis may be considered as an extension and generalization of vector analysis from three- to n-dimensional spaces and from Euclidean to non-Euclidean spaces. **1976** *Sci. Amer.* Aug. 98/2 Einstein's ideas were cast in a language very different from even non-Euclidean geometry, called the absolute differential calculus... Einstein used it and changed its name to tensor analysis. **1977** D. Bagley *Enemy* xxxiii. 266 This joker is using Hamiltonian quaternions!.. No one..has used Hamiltonian quaternions since 1915 when tensor analysis was invented. **1922** H. L. Brose tr. *Weyl's Space–Time–Matter* i. 53 The study of tensor-calculus is, without doubt, attended by conceptual difficulties—over and above the apprehension inspired by indices. **1944** G. B. Shaw *Everybody's Political What's What?* ii. 22 Experts in the tensor calculus. **1981** *Sci. Amer.* July 95/1 Tensor calculus..was essential to Einstein's formulation of his general theory of relativity. **1922** H. L. Brose tr. *Weyl's Space–Time–Matter* i. 61 An important example of a tensor field is offered by the stresses occurring in an elastic body. **1934** R. C. Tolman *Relativity, Thermodynamics, & Cosmol.* 36 Tensor fields may..be constructed, in which a value of the field tensor is associated with each point in the continuum. **1948** *Physical Rev.* LXXII. 987/1 The result of the present calculation and that of the proton–neutron scattering, which includes the tensor forces, show that the difference among the three potentials is quite pronounced at these high energies. **1972** *Physics Bull.* June 349/2 The noncentral force causing the anomalies mentioned above is called the tensor force, and it results from a neutron–protonspin–spin interaction. **1964** A. P. & W. Robertson *Topological Vector Spaces* vii. 141 It is essential to form the completion of the tensor product under the correct topology. **1971** E. C. Dade in Powell & Higman *Finite Simple Groups* viii. 252 The tensor product ..is again a finite-dimensional vector space over *F*.

Hence **tenso·rial** *a.*

1934 [see *anti-[1] 2 d]. **1968** C. G. Kuper *Introd. Theory Superconductivity* iv. 58 Since..Pippard's experimental data..do not support the idea of a tensorial anisotropy, these equations have not proved useful.

tent, *sb.*[1] Add: **2. a.** (Later examples.)

1923 T. S. Eliot *Waste Land* iii. 14 The river's tent is broken: the last fingers of leaf Clutch and sink into the wet bank. **1929** *Oxford Poetry* 13 Here in this harbour where straw glows..and overhead The unbroken tent of heaven covers.

f. A plastic or fabric enclosure that can be placed round a patient in bed so that the air he or she breathes can be modified and controlled. Cf. *oxygen tent* s.v. *oxygen 3 b.

1892 J. Carmichael *Dis. Children* xvii. 235 The room should be well ventilated, and the temperature of the tent kept between 65° and 70°. **1941** M. Davidson *Pract. Man. Dis. Chest* (ed. 2) xxxv. 559 Many varieties of tent have been constructed, all of which..demand considerable supervision. **1971** S. M. Bates *Pract. Pediatric Nursing* xii. 237 Both tents are designed to achieve cool super-saturation of the contained air with minimal wetting. **1979** Whaley & Wong *Nursing Care Infants & Children* xxxii. 1201/1 For continuous aerosol therapy a misting device is attached to or incorporated in the mist tent.

5. a. *tent curtain* (examples), *-flap, -frame* (example), *-hand, -mate* (later example), *-picket, -pole* (later example; also *fig.*), *-talk, wagon* (earlier example); *tent-pitching* (examples); (in sense 4) *tent-meeting, -preacher, -preaching* (earlier example).

1835 N. P. Willis *Pencillings by Way* II. xviii. 199 Most of the officers lay asleep on low ottomans, with their tent curtains undrawn. **1926** T. E. Lawrence *Seven Pillars* (1935) VIII. xcvi. 532 He crawled back through the tent-curtain. **1920** *Blackw. Mag.* Sept. 319/2 He paused with his shaking hand on the tent-flap. **1980** D. Hart-Davis *Heights of Rimring* vii. 67 He unrolled the tent-flaps and let them hang down. *Ibid.* vi. 65 The porters.. began to fit tent-frames together. **1938** N. Streatfeild *Circus is Coming* v. 57 The man finished fixing a seat. 'I'm a tent hand.' **1965** H. Gold *Man who was not with It* xiv. 115 A couple of tenthands are taking their flannel shirts off a line. **1972** J. Minifie *Homesteader* xviii. 158, I used the Army-issue straight blade [razor] myself, to the intense admiration of my tent-mates. **1950** *Reader's Digest* Jan. 85/2 Frakes joined the Methodist Church at an evangelistic tent meeting. **1913** S. O'Casey *Let.* 7 June (1975) I. 28 There will be exhibitions of drill, tent-pitching, [etc.]. **1925** G. Bell *Let.* 28 Jan. (1927) II. xxv. 721 The ordinary Scout exercises and tent pitchings—which they did extremely well. **1864** G. M. Hopkins *Poems* (1967) 14 Your hands have borne the tent-poles. **1974** R. Adams *Shardik* xviii. 134 His ugly, unmarriageable tent-pole of a daughter. **1979** *Guardian* 12 June 2/4 These..facts..were..'the fixed and rigid tentpoles' of the whole edifice of the prosecution case. **1966** *Punch* 9 Mar. 362/2 A tent-preacher and healer tells a diabetic woman she is cured. **1977** *Time* 26 Dec. 41/1 The latter include everything from Episcopalians to nearly a million Roman Catholics, to oddball healers and assorted tent preachers. *c* **1795** *Stat. Acct. Scotland* 1791–9 XV. 537 At the celebration of..the Sacrament of the Supper, there is no field or tent preaching..so derogatory from the solemnity of this institution. **1932** Auden *Orators* III. 108 The tent-talk pauses a little till a veteran answers 'Go to sleep, Sonny!' **1819** *Acct. Colony Cape of Good Hope* iii. 118 A light tent waggon, drawn by six or eight horses, constitutes the carriage of the wine boor.

b. tent caterpillar: for 'bombycid moth, *Clisiocampa*' read 'moth of the genus *Malacosoma* of the family Lasiocampidæ'; (earlier and later examples); **tent city,** a very large collection of tents; **tent club**: in India, a club organized for the sport of pig-sticking; **tent coat,** a coat resembling a tent in shape, being narrow at the shoulders and very wide at the hem; **tent-fly** (earlier example); **tent-man,** (*a*) (later example); (*b*) (examples); **tent-master** (later examples); **tent-pin** (earlier example); **tent ring** *Canad.*, a ring of stones used to hold down a tent, teepee, etc.; **tent-sack** (see quot. 1940); **tent show,** a show (such as a circus) given in a tent; **tent-stake** *U.S.* = Tent-peg; also in fig. phr. to *pull up tent-stakes*: to close down a business etc.; cf. Stake *sb.*[1] 1 e; **tent town,** a temporary settlement (as of gold-miners or the like); **tent-trailer** orig. and chiefly *U.S.*, a kind of trailer consisting of a wheeled frame with a collapsible tent cover attached; **tent village,** a small encampment; *spec.* = Douar, dowar.

1854 E. Emmons *Agric. N.Y.* V. 236 To eradicate completely the tent caterpillar, it will be necessary to give attention to the wild cherry trees. **1977** Richards & Davies *Imms's Gen. Textbk. Entomol.* (ed. 10) II. III. 1135 The larvae of *M. americana*..are commonly known as 'tent-caterpillars', their webs measuring 2 feet or more in length. **1934** M. F. K. Fisher in *As they Were* (1983) 64 A tent city, many umbrella and hot-dog concessions. **1980** J. Domatilla *Last Crime* 5 A tent city of tourists on a similar pilgrimage. **1889** R. S. S. Baden-Powell *Pig-sticking or Hoghunting* xvii. 152 Every station near which pig are to be found has its Tent Club. This is an association of the sportsmen of the place for carrying out the preservation of the pig, and successful hunting. **1895** Kipling in *Cent. Mag.* Dec. 271/1 He became a member of the local Tent Club, and chased the mighty boar on horseback. **1920** *Blackw. Mag.* Jan. 105/1 A Tent Club corresponds to a Hunt; the Hon. Secretary to the Master. **1961** *Guardian* 1 Feb. 7/5 A vast tent coat..over an elegant little sheath dress. **1971** *Vogue* 15 Sept. 51 Short tent coat with cape yoke. **1849** T. T. Johnson *Sights Gold Region* 169 The tent-fly is a second roof usually erected over the tent. **1945** *Sun* (Baltimore) 22 Feb. 7/2 Circus men sentenced... Chief tentman, two to seven years in State prison. **1961** *Times* 28 Mar. 12/6 The sweating 'tentman'..will be clown, drummer, lion-tamer rolled into one. **1978** *Illustr. London News* Nov. 32/3 As a kid..I used to take my waddie bottle..full of tea to the tentmen, they were travelling labourers, really. **1864** Tent master [see *tenter *sb.*[2] 2]. **1938** N. Streatfeild *Circus is Coming* v. 57 You'll have the tent master after me. **1763** J. Bell *Trav. from St. Petersburg* I. 323 Our European tents are of little use, as there is no earth in which the tent-pins can be fastened. **1945** *Beaver* Mar. 39/2 When a tent is struck.., the ring of stones which held it down lies there for years to come, and these tent rings are found today scattered even more widely..than the old igloos. **1958** *Even. Telegram* (St. John's, Newfoundland) 29 Apr. 13/3 My large tent needed more rocks to hold it down firmly than the usual tent ring supplied. **1940** F. Smythe *Adventures of Mountaineer* vii. 89 There was little else we could do save..unpack our tent-sack... It was simply a large sack of jaconet waterproof and windproof material intended to shelter climbers in an emergency. **1972** D. Haston *In High Places* iv. 52 As we were sitting comfortably in our tent sacks there was suddenly a loud explosion and a great hissing all around. **1878** M. Long *Life Mason Long* iv. 60, I made the acquaintance of a man named McCoole, who was travelling with a small tent show. **1935** [see *shill *sb.*]. **1973** *Times Lit. Suppl.* 28 Dec. 1593/4 Her life of the tent show, the Black touring company, race labels and buffet flats. **1862** *Rebellion Record* V. II. 156 The desolated, hard-trodden ground, and a few tent-stakes, remain to tell the story. **1923** H. Crane *Let.* 15 Feb. (1965) 123 *Broom*..has busted; N.Y. office closed last Saturday; March issue, the last, to be distributed from Berlin while the tent-stakes are being pulled up. **1956** H. Gold *Man who was not with It* (1965) i. 7 The Popcorn Man was throwing canvas over his machine and had it almost covered when a rube heaved a tent-stake. **1878** J. H. Beadle *Western Wilds* 103 Along the track west of it had sprung up five tent-towns. **1923** H. Steele *Spirit of Iron* 108 He came in touch with all the vice, wretchedness and stark tragedy abounding in the tent-towns and construction camps. **1965** *Globe & Mail* (Toronto) 26 May 3/2 Near this town..there is a tent-town of marginal people and transients. **1963** *Better Camping* Mar.–Apr. 32/2 It is only in the last six years or so that the tent trailer has grown from modest beginnings—little more than a bed on a trailer and just canvas cover enough to turn around in—to fold-out apartment-size dwellings. **1970** *Daily Tel.* 14 Nov. 9 Midway between tent and caravan..come the tent-trailers. **1899** A. E. W. Mason *Miranda of Balcony* x. 139 The Arab..belonged to a douar, a tent village. **1977** *N.Z. Herald* 8 Jan. 1–3/6 They also called for reinforcements for their tent village.

tent, *v.*[6] Add: **1. a.** *spec.* of travelling circus folk.

1875 T. Frost *Circus Life* xvii. 292 During the summer months they 'tented', and in the winter erected temporary wooden buildings in populous towns. **1931** S. McKechnie *Pop. Entertainments* viii. 209 The circus..was purchased by Frederick and Edward, who tented in the summer and spent the winter in..towns. **1952** N. Streatfeild *Aunt Clara* 111 They had been tenting with their mother.

5. To arrange in a shape suggesting a tent; *esp.*, with the fingers as obj., = *steeple *v.* 4.

1966 D. Bagley *Wyatt's Hurricane* ii. 60 He tented his fingers and regarded Wyatt closely. **1977** 'E. McBain' *Long Time no See* xi. 182 She herself sat on the sofa.. pulling her legs up under her Indian-fashion, the caftan tented over her knees. **1980** *TWA Ambassador* Oct. 57/2 Gatmun tented the fat sausages of his fingers.

tentacle. Add: **d. tentacle-feeder,** an invertebrate animal possessing tentacles to trap its food.

1953 J. S. Huxley *Evolution in Action* iii. 73 Tentacle-feeders may either float free in the water or be attached to the bottom. **1963** R. P. Dales *Annelids* 15 The more familiar tentacle-feeders include the terebellids, which have a mass of extensile tentacles.

tentage. (Later examples.)

1905 'L. Hope' *Indian Love* 20 No more the rugged road of Khorasan, The scanty food and tentage of the past! **1948** *Sporting Mirror* 21 May 13/3 (Advt.), Every type of Marquee, tentage and camping equipment. **1978** *B.S.I. News* July 6/1 It is considered unreasonable to deal with the flammability of tentage in isolation from other important features such as resistance to water penetration.

tented, *a.* Add: **1. b.** Of an encampment: consisting of tents.

1872 *Rep. Indian Affairs 1871* (U.S.) 261 Urah..had collected some of the chiefs and headmen of the tribe to receive me in their tented camp. **1955** *Times* 25 May 13/5 The cost of their fares..and their five months' stay in a tented camp in Cyprus will be about £300 each. **1982** *Listener* 23 and 30 Dec. 4/1 On a gentle rise..there is the tented city for the RAF Phantom and Harrier squadrons.

3. b. Of activities, etc.: held in or taking place in a tent or tents.

1898 N. E. Jones *Squirrel Hunters of Ohio* 131 Protracted, tented, or camp-meetings increased,..becoming very popular with preachers and people. **1971** *Morning Star* 22 June 2/5 The percentage of people taking tented holidays in Britain has gone up by 50 per cent since 1965. **1976** J. Wainwright *Bastard* iii. 48 A circus was visiting a neighbouring town. One of the hand-to-mouth, tented circuses. **1976** *Time* 20 Dec. 31/1 Circus Vargas, billed as the world's largest tented show.

tenter, *sb.*[2] Add: **1.** (Earlier U.S. example.)

1846 *Indiana Hist. Mag.* XXIII. 409 The eating hours were the same as those of the tenters.

2. One whose job is to erect and strike tents.

1864 'P. Paterson' *Glimpses of Real Life* xiii. 123 There must be a tent-master and tenters, besides the agent in advance, the members of the brass band, [etc.]. **1979** *Daily Tel.* 23 Apr. 14/8 The slow building and dismantling of the marquee on stage was..a documentary about tenters and their jobs.

tenterhook. **3.** (Later example.)

1980 P. Van Greenaway *Dissident* ii. 41 Alex had listened with tenterhook attention.

tenth, *a.* and *sb.* Add: **B.** *absol.* and *sb.* **4.** The tenth day of the month.

1580 in H. Foley *Jesuits in Conflict* (1873) 105 The tenth of September, 1580. **1868** E. S. P. Ward in *Atlantic Monthly* Mar. 345 (*heading*) The tenth of January. **1951** W. Faulkner *Requiem for a Nun* I. 36 It was barely the tenth of July. *Ibid.* III. 250 On the morning after June tenth.

C. tenth-value *a.*, designating a thickness of material that reduces the intensity of radiation passing through it by a factor of 10.

1955 *Gloss. Terms Radiology* (*B.S.I.*) 17 Tenth-value thickness. **1957** *Effects Nucl. Weapons* (U.S. Defense Atomic Support Agency) viii. 378 For concrete, the tenth-value thickness is..about 48 cm.

tenthredinid, *a.* and *sb.* (Examples.)

1890 *Insect Life* III. 157 Tenthredinid larva on black birch. **1913** *Oxf. Univ. Gaz.* 4 June 954/1 The Tenthredinid *Allantus arcuatus* together with a ♂ humble bee.., a much larger insect. **1970** G. Ordish tr. *Chauvin's World of Ants* i. 61 The biologist Lange put some Tenthredinid larvae..near an ant track.

tenting. Add: **1.** *vbl. sb.*[1] Also (with reference to a touring circus or the like) camping and performing in a tent.

1861 *All Year Round* 16 Nov. 186/1 The tenting system is now so well organised, that everything connected with it is conducted with effect and punctuality. **1878** L. M. Alcott *Under Lilacs* iv. 40 Father..went off sudden.. just before the tentin' season was over. **1931** S. McKechnie *Pop. Entertainments* viii. 222 Bertram W. Mills' Circus and Menagerie..only in its second tenting season..has already revolutionised the status of the circus. **1952** N. Streatfeild *Aunt Clara* 114 We have a little time when tenting is finished. **1971** *Esquire* July 88/2 The Hartford Circus fire of 1944..caused the big show to forswear tenting in 1956 and resolve to play only arena engagements indoors.

2. *sb.* (Later non-attrib. example.)

1936 *Discovery* Dec. 381/1 The second assistant had been carried down the rapids on the tenting.

tenting, *ppl. a.* Add: **1.** (Later example.)

1932 S. O'Faoláin *Midsummer Night Madness* 26 The tenting chestnuts filled the lanes with darkness.

2. Of a circus: that tents (Tent *v.*[6] 1 a in Dict. and Suppl.).

1875 T. Frost *Circus Life* iii. 67 The tenting circuses of those days were on a more limited scale than those of the present time, and were met with chiefly at fairs. **1931** S. McKechnie *Popular Entertainments* viii. 206 The early tenting circuses were unpretentious concerns. **1981** P. O'Donnell *Xanadu Talisman* i. 18 He..owned half a small tenting circus.

tention[3] (**'tention**). Add: Also **ten-shun** (cf. *'shun).

1908 M. Beerbohm in *Sat. Rev.* 26 Sept. 390/1 He.. stood at 'tention to be tapped on the chest by the hero. **1922** Joyce *Ulysses* 417 Get a spurt on. Tention. **1978** J. Blackburn *Dead Man's Handle* viii. 91 He..bellowed an order. 'Ten-shun, Sarn't-Major.'

tenue. Add: (Earlier and later examples.) Also *transf.*

1828 Lady Granville *Let.* 30 Aug. in B. Askwith *Piety & Wit* (1982) x. 154 The tenue, the neatness, the training up of flowers and fruit trees..are what in no other country is dreamt of. **1865** Crown Princess of Prussia *Let.* 12 Dec. in R. Fulford *Your Dear Letter* (1971) 46 She went through it all [*sc.* the marriage].. with the most perfect *tenue*. **1929** R. Fry *Let.* 4 Sept. (1972) II. 641 The building has a certain Florentine *tenue*, very refreshing after the rather sloppy magnificence..of the Venetians. **1956** S. Bedford *Legacy* I. i. 17 A tall, cool..woman,..who had complete *tenu* [sic] and a great deal of character. **1971** A. Fremantle *Three-Cornered Heart* vi. 97 We had quite as many 'love affairs' as girls have now, though we were more reticent about them and carried on our flirtations with a certain amount of *tenue*.

tenurable (te·niŭrāb'l), *a.* Chiefly *U.S.* [f. next + -able.] Of an academic post: subject to tenure (sense *1 c). Of an applicant for such a post: fit to be granted tenure.

1977 *Science* 4 Feb. 440/3 What is their effect on the bright..young scholar-teacher who has not, unfortunately, been productive enough to be undebatably tenurable? **1977** *Nature* 10 Nov. p. xlvi/2 (Advt.), Applications are invited for a full-time, tenurable appointment in the Division of Pharmacy and Pharmaceutics.

tenure. Add: **1. c.** *spec.* (orig. and chiefly *U.S.*) Guaranteed tenure of office, as a right granted to the holder of a position (usu. in a university or school) after a probationary period and protecting him against dismissal under most circumstances.

1957 V. Nabokov *Pnin* vi. 139 Pnin, who had no life tenure at Waindell, would be forced to leave—unless some other literature-and-language Department agreed to adopt him. *Ibid.* 167 'Naturally, I am expecting that I will get tenure at last,' said Pnin rather slyly. 'I am now Assistant Professor nine years.' **1976** *National Observer* (U.S.) 31 July 8/2 Idaho tried to abolish tenure a year ago, but the teachers' lobby was so strong the bill was defeated. **1981** *Listener* 5 Feb. 166/3 Can universities in a time of declining resources still preserve tenure in all its old form?

4. b. In sense 1 c above, as *tenure decision, member, system*; *tenure-heavy* adj.; **tenure track** *U.S.*, an employment structure whereby the holder of a post is guaranteed consideration for eventual tenure, usu. within a stated number of years.

1978 *Chron. Higher Educ.* 2 Oct. 8/3 Some 1,000 complaints of unfair tenure decisions his organization handles each year. **1979** *Yale Alumni Mag.* Apr. 13/1 Faculties are becoming increasingly tenure-heavy. **1960** J. J. Corson *Governance of Colleges & Universities* v. 101 In some institutions only the tenure members of the faculty will be privileged to participate in the school faculty. **1971** *Nature* 31 Dec. 502/2 The tenure system simply allows dead wood to remain in the university. **1979** *Ibid.* 4 Oct. p. xix/1 (Advt.), Two-year appointment with the possibility of tenure track. **1981** *Washington Post* 2 Jan. B16/4 People get into the feeling that they are on a tenure track and that they are unshakeable.

tenured (te·niŭɹd), *a.* Chiefly *U.S.* [f. prec. + -ed[2].] Of an official position, usu. one in a university or school: carrying a guarantee of permanent employment until retirement. Of a teacher, lecturer, etc.: having guaranteed tenure of office.

1969 *Guardian* 24 May 1/6 Left-wing professors, whose only protection is tenured appointment. **1970** 'A. Cross' *Poetic Justice* I. iv. 62 What does she look like?.. I thought I knew all the tenured English faculty. **1975** *Nature* 25 Dec. 653/2 A growing number of French scientists have found themselves having to work on short-term contracts, as tenured posts have dried up. **1976** *Maclean's Mag.* 27 Dec. 46/1 A board's right to fire any teacher, probationary or tenured, who violates the moral principles [etc.].

Hence (as a back-formation) **te·nure** *v. trans.*, to provide (someone) with a tenured post.

1975 *Times Lit. Suppl.* 13 June 639/1 How you propose to recruit, train, tenure and retire faculty [*sc.* in an American university]. **1983** *N.Y. Times* 23 Oct. I. 35/1 We have 22 women who were tenured by this department as a result of evaluations that said they could do the job.

tenuto, *a.* and *adv.* Add: (Examples.) Also *transf.*

1762 Sterne *Tr. Shandy* VI. xi. 50 What *Yorick* could mean by the words *lentamente*,—*tenutè* [sic],—*grave*,—and sometimes *adagio*,—as applied to theological compositions..I dare not venture to guess. **1801** Busby *Dict. Mus.*, *Tenuto*, or *Ten.* (Ital.), a word signifying that the notes are to be sustained, or held on. **1931** D. F. Tovey in Tovey & Craxton *Beethoven Pianoforte Sonatas* (Associated Board of Royal Schools of Music) III. 69/2 *Bars* 1–4.—The *forte* is a matter of string-tone and cantabile... Express the *tenuto* marks without hard accent, and see that in bar 3 the *piano* does not come too soon. **1975** *Gramophone* Nov. 839/1 In the *Meistersinger* piece I..like lots of notes played more *tenuto*.

B. *sb.* Pl. *tenutos*. A note or chord played *tenuto*.

1966 in *Random House Dict.* **1976** *Gramophone* Feb. 1349/3, I should have preferred him not to linger with so pronounced a tenuto on each bar in the bassoon solo. **1977** *Ibid.* Feb. 1279/2 He allows himself few of the momentary tenutos which have become a natural part of phrasing Franck melodies. **1982** *Guardian Weekly* 25 Apr. 20 Variations (with many tenutos Elgar never dreamed of).

Teochew, Teo-chew (tĭ:o tʃiū·). Also **Teochieu, Teochiu, Tiuchiu.** [A place-name in Swatow Chinese, = Putonghua *Chåozhoū*.] (A member of) a people of the Swatow district of Kwangtung in southern China; the dialect spoken by this people. Also *attrib.*

1893 J. D. Ball *Things Chinese* (ed. 2) 229 By..1891 there were 43,791 Teo Chews in the Straits Settlements: Teo Chews is the term applied generally to them in that part of the country, while Hoklo is the name by which they are generally known by the Cantonese speakers in China; the former name being derived from the Departmental city Ch'ao Chao fú (in the local dialect Tíú Chíú fú, or Teo Chew fu). **1927** in R. J. H. Sidney *In Brit. Malaya To-Day* xii. 144 Trouble had been brewing between Hok-kiens and the Teo-chews for some time. **1962** *B.B.C. Handbk.* 109 Adaptations..in the following languages: Chibemba,..Teochew. **1966** M. Freedman *Chinese Lineage & Society* iii. 95 People will assume for all Hakka or Hokkien or Cantonese or Tiuchiu that [etc.]. **1970** M. Pereira *Pigeon's Blood* xv. 164 He was speaking in the Teo-chieu dialect. **1979** *China Now* Jan.–Feb. 10/2 The Teochiu group from one district in Guangdong (Kwangtung) province.

teonanacatl (tĭ:ǒnanaka·t'l). [a. Nahuatl, f. *teotl* god + *nanacatl* mushroom.] Any of several hallucinogenic fungi, esp. *Psilocybe mexicana*, found in Central America. Also *attrib.*

1875 H. H. Bancroft *Native Races Pacific States* II. 360 Among the ingredients used to make their drinks more intoxicating the most powerful was the *teonanacatl*, 'flesh of God', a kind of mushroom which excited the passions. **1915** *Jrnl. Heredity* VII. 294/2 The writer has sought diligently for a fungus having the properties attributed to the *teonanacatl*. **1940** *Amer. Anthropologist* XLII. 439 The identity of teonanacatl was unknown for three centuries. **1953** J. Ramsbottom *Mushrooms & Toadstools* vi. 49 The Aztecs and Chichimecas were the earliest recorders of this teonanacatl. **1966** *Listener* 14 July 47/1 Another Mexican fantasy-producing drug that is used in religious ceremonial comes from the teonanacatl mushroom. **1975** [see *psilocin].

tepa (tĭ·pă). *Chem.* Also **TEPA.** [f. *t*ri*e*thylene *p*hosphor*a*mide.] An organophosphorus compound, $PO(N(CH_2)_2)_3$, used as an insect sterilant and formerly in the treatment of cancer.

1953 *Cancer* VI. 140/2 These observations justify the continuation of studies of the action of tepa in advanced cancer. **1963** *New Scientist* 13 June 603/2 More than 3½ million flies..had been sexually sterilised by dipping their pupae in a 5 per cent solution of tepa. **1973** J. J. McKelvey *Man against Tsetse* iii. 196 [Experimenters] tried to eradicate tsetse from an area in Rhodesia by sterilizing male flies with..tepa.

‖ **tepache** (tepæ·tʃe). [Mexican Sp., ad. Nahuatl *tepiatl*.] Any of several Mexican drinks of varying degrees of fermentation, typically made with pineapple, water, and brown sugar.

1926 D. H. Lawrence *Mornings in Mexico* (1927) 42 *Tepache* is a fermented drink of pineapple rinds and brown sugar. **1930** R. Macaulay *Staying with Relations* xix. 273, I tried for a drink of Mexicali and only got tepache. **1983** M. Gorman *Cooking with Fruit* III. 188 Tepache..is an old Mexican household fruit drink... It is a simple mixture that uses 1 pineapple.

tepal. Delete *rare*[−0] and add: [ad. F. *tépale* (A. P. de Candolle *Organographie Végétale* (1827) I. III. ii. 503).] A segment of a perianth which is not divided into a corolla and a calyx.

1840 B. Kingdon tr. *A. P. de Candolle's Veget. Organogr.* II. ii. 90 It is well to retain for these doubtful cases of a single envelope a particular name... Following the analogy of the terms sepals and petals, I propose..the name of Tepals. **1939** *Rep. Bot. Soc. Brit. Isles* XII. 120 It is proposed to refer to them [*sc.* the perianth segments of *Rumex*] as Inner and Outer Perianth Segments, or Tepals. **1951** G. H. M. Lawrence *Taxon. Vascular Plants* II. 413 Perianth..generally undifferentiated into corolla and calyx, and when so the segments termed tepals. **1968** A. Cronquist *Evol. & Classification Flowering Plants* iii. 87 The tepals of the more primitive angiosperm families are modified leaves. **1972** *Jrnl. R. Hort. Soc.* XCVII. 336 Because magnolia flowers usually have no distinction between petals and sepals the term tepal has been generally accepted for these floral parts.

tepary (te·pări). Also **tepari.** [Origin unknown.] In full, *tepary bean*. An annual legume, *Phaseolus acutifolius*, native to southwestern North America, or a cultivated plant belonging to a variety developed from it and resistant to drought; also, the seed of a plant of this kind.

1912 G. F. Freeman *Southwestern Beans* 582 The name tepary or tepari (Spanish) originated from the Papago. **1912** K. S. Lumholtz *New Trails in Mexico* 318 He had cooked bones of mountain-sheep with tépari beans for us. **1925** K. Bryan *Papago Country* 354 The beans known as tépari..are said to be so resistant to drought that the plants may wither three successive times and then, if enough rain comes, mature a crop. **1942** Castetter & Bell *Pima & Papago Indian Agric.* 92 The cultivated tepary bean antedates the coming of the white man in the southwest. *Ibid.* 191 The Papago made only one planting of teparies. **1972** Y. Lovelock *Vegetable Bk.* I. 56 The Texas or tepary bean..grows wild and is much cultivated in Mexico and the adjoining states of the U.S.

tepee. Add: Also † **teebee, tepe, ti pee; tipi** (sometimes preferred). **1.** (Earlier and later examples.) In extended uses, applied to a similar structure used by peoples of other parts of the world, as a child's toy, or for camping.

1743 J. Isham in *Publ. Hudson's Bay Record Soc.* (1949) XII. 45 A tent Build ti pee. **1847** G. W. Featherstonhaugh *Canoe Voyage* I. xxx. 338 Here, also, were their spring teebees, which they inhabit at that season. **1849** M. H. Eastman *Dahcotah* p. xxii, The messenger enters the wigwam (or *teepee*, as the houses of the Sioux are called) of the juggler. **1910** *Encycl. Brit.* XIV. 469/1 The skin tents or tipis of many of the Plains peoples. **1959** A. Moorehead *No Room in Ark* ix. 211 I liked the Dinka villages... They looked like Red Indian encampments of a century or two ago: tall teepees made of grass and arranged in a rough circle on the bank. **1960** D. Jenness *Indians of Canada* (ed. 5) vii. 90 The plains' area was the home of the tipi, a tent of buffalo hide stretched around a conical framework formed by fourteen to eighteen long poles, whose points radiated like a funnel above the peak. **1970** *New Yorker* 12 Dec. 138/2 A six-and-a-half-foot tepee costs $25 [in a toy-shop]. **1974** N. Gordimer *Conservationist* 42 The newly-ploughed field, where the mealie stalks are piled into tepees. **1976** *Billings* (Montana) *Gaz.* 2 July 8-A/1 (Advt.), You'll love the weekend activities in store for you at four nearby KOA Kampgrounds. All you need to take is your family or friends, sleeping bags and food. The Tepee or Tent will be waiting for you to move right in and there's a grill for cooking.

2. *attrib.* and *Comb.*, as *tepee cloth, cover, pole, trail*; *tepee-like* adj.; **tepee ring** (see quot.).

1877 *Rep. Indian Affairs* (U.S.) 50 Tepee cloth should be discontinued, and..log or frame houses should be substituted. **1890** E. Custer *Following Guidon* 6 The hides were dressed for robes or tepee covers. **1925** *Blackw. Mag.* May 658/1 There were three large tepee-like tents. **1897** J. W. Tyrrell *Across Sub-Arctics of Canada* 68 Instead of Indians, [we] found only old forsaken 'tepee' poles and blackened fire-places. **1899** Tepee pole [in Dict.]. **1956** D. Leechman *Native Tribes of Canada* 10 We find.. on the prairies, tipi rings, circles of stones that once were used to hold down the edges of the skin tents. **1869** *Amer. Naturalist* II. 648 [We] follow upon the dim road or the tepe trail over the broad prairie.

tephigram (te·figræm). *Meteorol.* [f. TE(E *sb.*[1] (*T* being a symbol for temperature) + *PHI (φ being a symbol for entropy) + -GRAM.] A diagram in which one axis represents temperature and another potential temperature (as a measure of entropy), used to represent the thermodynamic state of the atmosphere at different heights.

[*c* 1925 N. SHAW *Sel. Meteorol. Papers* (1955) 226/2, I have found the representation known to engineers as a θ, φ (temperature-entropy) diagram (or, as I shall call it here, a tφ diagram) has the advantage of being..more expressive than the direct pressure-temperature diagram.] 1929 W. J. HUMPHREYS *Physics of Air* (ed. 2) xv. 259 Tephigram... It is convenient, as developed by Sir Napier Shaw and his colleagues, to plot values on a temperature-entropy diagram. *Ibid.* 261 Figure 77 is a tephigram..of a balloon sounding. 1938 *Nature* 29 Oct. 804/1 Daily tephigrams based on aeroplane soundings constitute the most valuable items in forecasting..clearing or persistence of cloud. 1969 *Ibid.* 11 Oct. 170/1 The uniformity of weather conditions over the region in question justified our taking these tephigrams as representative of the state of the atmosphere over the area.

tephillim, -in, *sb. pl.* Add: Now usu. in form tefillin. (Later examples.)

1865 *Chambers's Encycl.* VII. 519/2 Certain strips of parchment, inscribed with certain passages from the Scripture.., enclosed in small cases, and fastened to the forehead and the left arm (*Tefillin*)..in use with the Jews..are..called in the New Testament phylacteries. 1967 C. POTOK *Chosen* I. iii. 61, I got the tefillin and prayer book out of the drawer of the night table and began to put on the tefillin. *Ibid.*, When I finished praying, I took off the tefillin and put them and the prayer book back in the drawer. 1978 J. SACKS in P. Moore *Man, Woman, & Priesthood* iii. 33 They [*sc.* women] are not obliged, as men are, to put on the phylacteries (*tefillin*) or the fringed garment (*tzitzit*).

tephra (te·frǎ). *Geol.* [ad. Sw. *tefra* (S. Thórarinsson 1944, in *Geografiska Annaler* XXVI. 114), f. Gr. τέφρα ash.] Dust and rock fragments that have been ejected into the air by a volcanic eruption. Freq. *attrib.*

1944 *Geografiska Annaler* XXVI. 210 The author suggests (*volcanic*) *ash* or (better) *tephra* as a collective term for all clastic ejectamenta. 1970 *Nature* 25 July 335/1 The maximum thickness of the tephra layer was 7 cm at 15 km from the volcano. 1972 *Nat. Geographic* CXLI. 718/2 Commercial interests are removing this layer—known as tephra—since it makes a highly cohesive and waterproof mortar, serves as an insulating material, and constitutes an important component of cement. 1973 *Nature* 9 Feb. 372/2 Because of its close vicinity to the eruption the town was threatened by tephra fall. 1979 *Sci. Amer.* Dec. 134/2 The cloud of tephra and gas rises high above the volcano, and particles in it are carried downwind, producing a rain of tephra that forms a deposit called a tephra mantle.

tephrochronology (te:frokrono·lŏdʒi). Also tephra-. [ad. Sw. *tefrokronologi* (S. Thórarinsson 1944, in *Geografiska Annaler* XXVI. 6), f. as prec.: see CHRONOLOGY.] The dating of volcanic eruptions and other events by studying layers of tephra. Hence **te:phrochrono·lo·gical** *a.*

1944 S. THÓRARINSSON in *Geografiska Annaler* XXVI. 204 As an international term to designate a geological chronology based on the measuring, interconnecting, and dating of volcanic ash layers in soil profiles the author suggests the term *Tephrochronology*. *Ibid.* (*heading*) Tephrochronological studies in Iceland. 1976 P. FRANCIS *Volcanoes* v. 178 The use of successive pumice or ash deposits in building up a history of the eruptive activity in an area is known as tephrachronology. 1979 *Nature* 25 Oct. 642/1 The tephrochronology of these ashes is well documented, but volcanological interpretations have seldom been attempted. 1979 *Sci. Amer.* Dec. 132/1 A volcano produces successive showers of tephra that fall throughout the surrounding countryside, forming layers that constitute a tephrochronological record of the volcano's activity.

tepid, *a.* **b.** (Later examples of the word applied to persons.)

1926 *Scribner's Mag.* Sept. 259/2 Her smile said that pastels were thin things for tepid people. 1941 A. CHRISTIE *Evil under Sun* x. 197 Some tepid little man, vain and sensitive—the kind of man who broods.

tepidarium. Add: Also applied to a similar room in a Turkish bath.

1969 J. WAINWRIGHT *Take-Over Men* ii. 27, I followed him..into the warm room (the *Tepidarium*)..into the hot room (the *Calidarium*); and finally..into the *Laconicum*. 1975 *Daily Colonist* (Victoria, B.C.) 5 Oct. 4/3 When you are gleaming horribly [with sweat], you go into the tepidarium.

teporingo (tepori·ŋgo). [a. Amer. Sp.] = *volcano rabbit* s.v. *VOLCANO *sb.* 3.

1969 J. FISHER et al. *Red Bk.* 54/2 The *teporingo*.. exists only on the middle slopes of Popocatepetl and Ixtacihuatl and some of the nearby ridges. 1972 [see *park ranger* s.v. *PARK *sb.* 7]. 1980 *Listener* 17 July 90/3 The teporingo appears to be a kind of Mexican rabbit. There aren't many left.

‖ **teppan-yaki** (te:panyā·ki). [Jap.] A Japanese dish consisting of meat, fish, (or both) fried with vegetables on a hot steel plate which forms the centre of the table at which the diners are seated.

1970 P. & J. MARTIN *Japanese Cooking* 80 (*heading*) Teppan-yaki steak. *Ibid.*, Teppan-yaki means literally 'iron plate grilling'. This type of cooking, too, is usually done in front of guests on a large, rectangular griddle. *Ibid.* 81 Teppan-yaki duck. 1972 *Mainichi Daily News* (Japan) 6 Nov. 11/6 (Advt.), A variety of foods including Teppan-yaki (meats roasted before your eyes on hot steel plates). 1979 *United States 1980/81* (Penguin Travel Guides) 73 This thoughtfully designed Japanese restaurant has separate dining rooms for teppan-yaki, sukiyaki, and tempura.

tequila (tekī·lǎ). Also **tequela, tequilla.** [a. Mexican Sp., f. the name of a town which is one of the centres of its production.] **a.** A gin-like Mexican spirit made by distilling the fermented sap of a maguey, *Agave tequilana*; cf. MESCAL.

1849 J. GREGG *Diary & Lett.* (1944) II. 317 So celebrated has this place become, for the manufacture of superior *mezcal*, that that taken from here is known by the name of *Tequila*. 1894 *Harper's Mag.* Feb. 351/2 Between various cigarettes, the last drink of tequela, and the drying of our clothes, we passed the time. 1926 [see *MESCAL 1]. 1941 B. SCHULBERG *What makes Sammy Run?* vii. 110 Burning my stomach with *enchillada* and my brain with more *tequilla*. 1953 W. BURROUGHS *Junkie* (1972) xiii. 129 Every morning when I woke up, I washed down benzedrine, sanicin, and a piece of hop with black coffee and a shot of tequila. 1958 P. HIGHSMITH *Game for Living* (1959) ii. 22 Theodore heard..liquor being poured into a glass, and he knew it would be Lelia's yellowish tequila. 1969 J. MANDER *Static Society* vii. 196 Fuentes had been initiating me into the art of drinking Mexican *tequila* (with salt and lemon). 1977 *Playgirl* May 124/1 For the woman whose liquor larder extends beyond beer and wine, tequila is now a necessity.

b. *attrib.* and *Comb.*, as *tequila sour*; *tequila-based* adj.; **tequila plant,** the maguey from the sap of which tequila is made; **tequila sunrise,** a name given to cocktails containing tequila and grenadine.

1977 T. HEALD *Just Desserts* vii. 156 He was drinking a tequila-based cocktail. 1979 P. THEROUX *Old Patagonian Express* iii. 52, I saw a field of upright swords. It might have been sisal, but more likely was the tequila plant. 1966 T. PYNCHON *Crying of Lot 49* iii. 59 'Who's your client?' asked Metzger, holding out a tequila sour. 1965 O. A. MENDELSOHN *Dict. Drink & Drinking* 336 *Tequila Sunrise*, mixed drink of tequila, lemon juice, grenadine and cinnamon liqueur. 1976 *Daily Tel.* (Colour Suppl.) 11 June 42/3 A Tequila Sunrise has become the 'in' drink at many ski resorts and single bars. It is tequila and orange juice, with half an ounce of grenadine poured on top to filter dramatically down through the drink.

Tequistlatec (teki·stlǎtek). Also **Tequistlate·ca, Tequistlate·co.** [Native name.] (A member of) an Indian people of south-east Oaxaca, Mexico; also, the language of this people. Hence **Tequistlate·can,** the Tequistlatec language or (later) the linguistic family of which it is the principal member; also *attrib.*

1891 D. G. BRINTON *American Race* 112 Quite to the south, in the mountains of Oaxaco and Guerrero, the Tequistlatecas, usually known by the meaningless term, Chontales, belong to this stem. *Ibid.* 148 The only specimen of their idiom which I have obtained is a vocabulary of 23 words... Provisionally, however, I give it the name of *Tequistlatecan*, from the principal village of the tribe. 1902 *Encycl. Brit.* XXV. 374/1 [Linguistic families of Middle America] Tehuantepecan, Isthmus; Tequistlatecan, Oax. 1915 A. L. KROEBER in *Univ. Calif. Publ. Amer. Archæol. & Ethnol.* XI. 279 (*title*) Serian, Tequistlatecan, and Hokan. 1929 E. SAPIR in *Encycl. Brit.* V. 140/2 *Hokan* proper, which includes Seri (coast of Sonora), Yuman (in Lower California) and Tequistlateco or Chontal (coast of Oaxaca). 1965 *Language* XLI. 305 Seri and Tequistlatec, both separate branches of Hokan. *Ibid.*, The Tequistlatecan form seems at least as similar to the Proto-Palaihnihan as the Proto-Shastan. 1974 *Encycl. Brit. Micropædia* IX. 894/1 Tequistlatec, Hokan-speaking Middle American Indians of the Sierra Madre del Sur of Oaxaca, Mex.

ter (tɘɹ), repr. vulg. and dial. pronunc. of TO *prep.*, *conj.*, *adv.* Cf. *ORTER; *OUGHTA, OUGHTER; *USETA, USETER.

1867 *Harper's New Monthly Mag.* Feb. 274/2 This yere is Colonel N—, who wants ter know yer. 1895 W. J. LOCKE *At Gate of Samaria* v. 49 She's bloomin' well got ter. 1926 J. K. STRECKER in J. F. Dobie *Rainbow in Morning* (1965) 61 Ef he sting yo, yo sho gwine ter die. 1934 [see *quarter-turn* s.v. *QUARTER *sb.* 30]. 1944 E. THOMPSON *Robert Bridges* i. 7 One he used in print..that of the cockney who saw on his medicine chart the words *ter die*, and fled in terror to escape his scheduled destruction. 1952 [see *KIN var. CAN *v.*[1]]. 1976 *Southern Even. Echo* (Southampton) 13 Nov. 2/7 You gence wan' anyfing ter drink?

ter-. Add: **2. b.** **termole·cular** *a.*, involving three molecules; **terpo·lymer,** a polymer whose molecule is composed of three different monomers; hence **terpo:lymeriza·tion,** polymerization in which three different monomers go to form a terpolymer.

1901 *Jrnl. Chem. Soc.* LXXIX. I. 229 Even in darkness the action is termolecular. 1931 MAASS & STEACIE *Introd. Princ. Physical Chem.* xii. 199 The reaction between nitric oxide and chlorine is an example of a termolecular reaction. 1974 *Nature* 19 Apr. 666/2 The dimer must be formed by a termolecular collision. 1961 WEBSTER, Terpolymer. 1967 *New Scientist* 18 May 423/3 Terpolymers (e.g. ABS plastics) are in use already. 1975 *Sci. Amer.* Oct. 54/3 The three products that proved to be most broadly successful are generically classified as a urethane, a dimethyl silicone and a styrene acrylonitrile terpolymer. 1964 *Jrnl. Polymer Sci.* A. II. 2740 Penultimate effects in copolymerization can be determined more precisely from composition studies in terpolymerization experiments than in binary polymerizations. 1976 H. SAWADA *Thermodynamics of Polymerization* ix. 262 (*caption*) Heat of terpolymerization for the acrylonitrile-methyl methacrylate-styrene system.

tera- (te·rǎ). *prefix.* [f. Gr. τέρας monster.] Prefixed to the names of units to form the names of units 10^{12} (one million million) times larger (symbol T), as *terabit* [*BIT *sb.*[4]], *-electron-volt*, *-hertz*, *-pascal*, *-second*, *-volt*, *-watt* (hence *-watt-hour*, *-year*).

[1947 *Compt. Rend. de la 14ème Conf.* (Union Internat. de Chimie) 115 The following prefixes to abbreviations for the names of units should be used to indicate the specified multiples or sub-multiples of these units: T tira- 10^{12} ×.] 1951 *Symbols, Signs & Abbreviations* (R. Soc.) 15 Tera (× 10^{12}) T. 1971 *New Scientist* 8 July 80/2 A bigger machine..which will be available next year, will have a 50-terabit memory with only slightly slower cycle time of 100 nanoseconds. 1974 *Sci. Amer.* Feb. 82/2 An energy of 1,000 GeV (one teraelectron-volt, or TeV) does not seem an impossible goal. 1970 *McGraw-Hill Yearbk. Sci. & Technol.* 233/1 M. A. Duguay and coworkers..tuned a mode-locked He-Ne laser..from −45 to +45 GHz, about the central optical frequency of 473·61 terahertz. 1980 *Nature* 21 Feb. 715/1 In relatively simple experiments laser-driven shock waves can be used to study the propagation of shocks in solids for shock pressures up to terapascals (1 TPa = 10^7 bar = 10^{13} dyne cm^{-2}). 1969 *Proc. Geol. Soc.* Aug. 142 Alternatively the second of time may prove to be a more convenient unit as recommended by S.I., thus 1 million years (m.y.) = 31·557 tera seconds (Ts). 1975 *Sci. Amer.* Feb. 40/2 A development program looking toward the creation of a proton beam of about 1,000 GeV, or one teravolt (TeV). 1970 *Britannica Bk. of Year* 322/2 It had an output of 4 trillion watts (4 terawatts). 1972 *Physics Bull.* Mar. 175/2 The terawatt carbon dioxide laser may not be far away. 1979 *Internat. Atomic Energy Agency Bull.* Jan. 7 Let us consider two scenarios which would lead to a total primary energy consumption rate of 50 terawatts (50 000 000 MW) at the end of the next century. 1979 *Times* 11 Dec. 18 The American Department of Energy, Dr Musgrove says,.. could economically produce 500 tera-watt-hours (500 million mega-watt-hours) of electricity. 1980 *Sci. Amer.* Sept. 111/1 From the base year of the IIASA study (1975) to 2030 the total primary-energy consumption rate is projected to rise from 8.2 terawatt-years per year to 36 terawatt-years per year in the high-growth scenario and to 22 terawatt-years per year in the low-growth one.

terai. Add: Also **tarai.** (Examples in the lit. sense (see etym.) and earlier examples of sense in Dict.)

1852 T. SMITH *Narr. Five Years' Residence Nepaul* I. ii. 56 The Terai, or Turay, or Turyanee, is a long strip or belt of low-level-land. 1860 W. H. RUSSELL *My Diary in India* II. ii. 31 This gentleman was one of the unhappy refugees who was sheltered in the terai..and, although he saved his life, he was struck down by terai fever. 1911 *Encycl. Brit.* XIX. 379/1 The low alluvial land of the tarai is well adapted for cultivation, and is, so to speak, the granary of Nepal. 1918 W. BEEBE *Jungle Peace* (1919) xi. 268 The *terai* jungles of Garhwal, the tree-ferns of Pahang, and the mighty *moras*..will stand in silvery silence. 1954 O. H. K. SPATE *India & Pakistan* xviii. 496 Originally the terai covered a zone perhaps 50–60 miles wide... Much of this has been so altered by settlement that the true terai is now confined to a relatively narrow strip. 1981 V. POWELL *Flora Annie Steel* xii. 104 To soothe her fever—terai fever as it was then called—she was given hashish.

1888 KIPLING *Under Deodars* 43 Mrs. Boulte put on a big terai hat. *Ibid.* 73 She was wearing an unclean Terai with the elastic under her chin. 1894 *County Gentlemen's Catal.* 155/2 Soft drab terai double felt hats.

terakihi, var. *TARAKIHI.

‖ **terakoya** (terakō·ya). *Japanese Hist.* [Jap., = temple school, f. *tera* temple + *ko* child(ren) + *ya* place.] In the Japanese feudal period, a private elementary school of a kind established orig. in the Buddhist temples.

1909 D. KIKUCHI *Japanese Education* ii. 33 The name *Terakoya*, or 'House for the Children of the Temple', given to elementary schools up to the beginning of the present era. 1911 *Encycl. Brit.* XV. 220/1 They [*sc.* Buddhist priests] organized schools at the temples.., and at these *tera-koya*..lessons in ethics, calligraphy, reading and etiquette were given to the sons of samurai and even to youths of the mercantile and manufacturing classes. 1938 D. T. SUZUKI *Zen Buddhism & its Influence on Japanese Culture* I. v. 106 The *Terakoya* system was the only popular educational institution during the feudal ages of Japan. 1965 W. SWAAN *Jap. Lantern* xii. 143 The *terakoya*, or 'temple schools' attached to the monasteries, provide the only institution of popular education. 1974 *Encycl. Brit. Macropædia* VI. 342/2 As time passed, some *terakoya* used parts of the houses of commoners as classrooms.

terap (tĕra·p). Also **tarap.** [a. Malay.] An evergreen tree, *Artocarpus elasticus*, of the

family Moraceæ, native to Malaysia and closely related to the bread-fruit tree, also the large edible fruit of this tree or its fibrous bark, which is used to make string or cloth. Also *attrib.*

1839 T. J. NEWBOLD *Straits of Malacca* II. ix. 119 The cloth that encircles their loins is made from the fibrous bark of the Terap tree. 1900 W. W. SKEAT *Malay Magic* v. 225 A string of *tĕrap* bark to tie up the rice that is cut first. 1913 L. W. W. GUDGEON *Brit. N. Borneo* x. 69 The fruit falls in all directions. If it is 'durian' or 'tarap', the size and weight of which are considerable, the Dusuns stand clear. 1935 I. H. BURKILL *Dict. Econ. Products Malay Penin.* I. 248 Every one knows the name 'tĕrap' which is applied to *A[rtocarpus] elastica* by Malays and Sakai. 1940 E. J. H. CORNER *Wayside Trees of Malaya* I. 654 The *Terap* is, undoubtedly, the commonest and best known of our wild species of Artocarpus. 1964 M. E. D. POORE in Wang Gungwu *Malaysia* I. ii. 48 Such occur in..many species of terap or breadfruit (*Artocarpus*).

terato-, comb. form repr. Gr. τέρας, τερατ- monster: **te:ratocarcino·ma** *Path.*, a malignant teratoma containing carcinomatous elements, occurring chiefly in the testis.

1946 FRIEDMAN & MOORE in *Military Surgeon* XCIX. 573 A new term, 'teratocarcinoma', is proposed for the large group of pleomorphic tumours in which both differentiated teratoid structures and histologically malignant elements were present. 1958 *Jrnl. Amer. Med. Assoc.* 28 June 1066/2 A testicular teratocarcinoma occurring in a 35-year-old man was treated initially by surgery. 1975 [see *TERATOMA]. 1979 *Sci. Amer.* Apr. 87/1 As long as the tumors contain embryonal carcinoma cells they continue to grow; such tumors are malignant and are referred to as teratocarcinomas.

teratogen (teræ·tŏdʒĕn, te·rătodʒĕn). *Med.* [f. *TERATO- + -GEN.] An agent or factor which causes malformation of the developing embryo.

1959 *Jrnl. Chronic Dis.* X. 125 Present knowledge of the mechanisms of teratogenic action is meager... The ultimate action of all teratogens seems to be to produce either cell death or an alteration in the rate of cell growth. 1970 G. LEACH *Biocrats* vi. 141 Animals are rarely good models for men when it comes to testing the effects of drugs and other teratogens on the foetus. 1978 *Jrnl. R. Soc. Med.* LXXI. 668 The patient should be seen earlier in pregnancy to help her avoid potential teratogens.

teratogenesis. Add: Also **te:ratogeni·city,** teratogenic property.

1959 *Jrnl. Chronic Dis.* X. 117 More than 20 closely related azo dyes have been tested for teratogenicity in my laboratory. 1964 *Listener* 20 Feb. 311/1 It is apparent that many of the tests that can, in our present state of knowledge, be applied to new drugs to attempt to produce teratogenicity, are neither meaningful nor justifiable. 1981 *Internat. Jrnl. Environmental Stud.* XVII. 10/2 The weak teratogenicity and growth retardative propensity of such a ubiquitous drug as aspirin.

teratologist. b. (Examples.)

1844 *London & Edin. Monthly Jrnl. Med. Sci.* IV. 484 Teratologists are now agreed in referring a considerable number of malformations by defect to the occurrence of an interruption..of natural fœtal development. 1908 *Jrnl. Morphol.* XIX. 51 Teratologists are inclined to read these facts in favor of the germinal origin of monsters, which may even be hereditary. 1973 *Daily Tel.* 13 Jan. 16 Many distinguished obstetricians, pathologists, paediatricians, teratologists and editors were reluctant to accept my hypothesis that thalidomide did cause abnormalities.

teratoma. Substitute for def.: A tumour, esp. of the gonads, characteristically formed of numerous distinct tissues and believed usually to arise from germ cells or their precursors. (Earlier and later examples.)

1879 *Amer. Jrnl. Med. Sci.* LXXVII. 91 (*heading*) Extirpation of teratoma; or, teratoid tumor. *Ibid.* 93 To such tumors Virchow has applied the term teratoma. 1906 [see *LANGHANS]. 1948 R. A. WILLIS *Path. of Tumours* lxi. 940 A teratoma is a true tumour or neoplasm composed of multiple tissues of kinds foreign to the part in which it arises. 1975 *Nature* 6 Nov. 12/1 Teratomas and teratocarcinomas are rare tumours which arise in the gonads, and contain a whole variety of differentiated tissues of ectodermal, mesodermal and endodermal origin (such as skin, nerve, muscle, cartilage, gut and lung), mixed together in a disorganised mass. 1979 *Sci. Amer.* Apr. 87/1 When all the embryonal cells differentiate into various kinds of normal tissue, the tumors stop growing: they are benign and are usually referred to simply as teratomas.

teratomatous *a.* (examples).

1893 *Index-Catal. Library Surgeon-General's Office, U.S. Army* XIV. 896/1 (*heading*) Tumors (teratomatous). 1948 MARTIN & HYNES *Clinical Endocrinol.* ix. 193 A few arrhenoblastomata have been teratomatous, containing cartilage and other tissues. 1962 *Punch* 7 Nov. 658/2 A teratomatous growth of mixed tissues, probably of only low malignancy.

tercio, tertio. Delete 'Now only *Hist.*' and add: **1.** (Later example.) Also used of Spanish units in mod. times.

1938 C. S. FORESTER *Ship of Line* xvii. 236, I am Colonel Juan Claros, of the third tercio of Catalan migueletes... That is to say a thousand men. 1957 P. KEMP *Mine were of Trouble* ii. 19 A Tercio in the sixteenth century was a Regiment of Spanish infantry. The Spanish Foreign Legion is also called *El Tercio*... But the Requetés in the Civil War also organized their fighting units into *tercios*, each approximately of battalion strength. 1965 C. D. EBY *Siege of Alcázar* i. 14 The crack Tercios of the Foreign Legion.

2. *Bull-fighting.* **a.** One of the three parts of a bullfight. **b.** Each of the three concentric circles into which a bullring is technically divided.

1932 E. HEMINGWAY *Death in Afternoon* 331 The bullfight is divided into three parts, the *tercio de varas*, that of the pic, *tercio de banderillas*, and *tercio del muerte* or third of death. 1932 R. CAMPBELL *Taurine Provence* iii. 68 The arena is divided into three areas—tablas, tercios, and medios. It is in the tercios, which extend from a third of the way to the centre until quite near the central area, that the bull is the best to deal with. 1962 J. STEWART tr. *J. Cousseau's Death of Miss Cunningham* 136 The final *tercio* was about to be sounded. 1967 MCCORMICK & MASCAREÑAS *Compl. Aficionado* i. 20 The luring of the bull,..and the ritual staining of the garments of the bridegroom with the bull's blood..aid us in comprehending both the origin of the tercio of the banderillas, and our response to that tercio.

Tercom (tɘ·ɹkɒm). [Abbrev. f. initial letters of *ter*rain *co*ntour *m*atching.] A computerized system for controlling the flight path of a cruise missile which enables it to stay close to the ground.

1975 *Bull. Atomic Sci.* Apr. 13/2 Tercom—Terrain matching device—a system which enables a missile to hug the ground and follow a programmed path. 1980 R. L. DUNCAN *Brimstone* x. 261 The cruise missile would be guided by TERCOM—terrain contour matching.

Terek. For *Terekia cinerea* substitute *Xenus cinereus* and add: Also *Terek sandpiper*. (Later examples.)

1879 *Ibis* XXI. 152 The Terek Sandpiper arrived at our quarters [in Siberia] on the 8th of June. 1915 *Brit. Birds* IX. 69 Three Terek Sandpipers..were shot at Rye in the month of May, 1915. 1956 *Ibis* XCVIII. 161 The Terek Sandpiper thus becomes *Tringa terek* (Latham), the name *Tringa cinerea* being pre-occupied. 1982 *Times* 23 Mar. 10/5 Leicester Museum spent..£400 on a family of [stuffed] Terek sandpipers.

‖ **terem** (te·rĕm). *Russ. Hist.* [Russ., lit. 'tower'.] Secluded separate quarters for women.

1898 G. B. SHAW in *Sat. Rev.* 8 Jan. 42/2 The seclusion of Russian women in the Terem was one of the sacred institutions of his [*sc.* Peter the Great's] country. 1908 *Cambr. Mod. Hist.* V. xvii. 519 The boy soon felt cramped and stifled in the dim and close semi-religious atmosphere of Natalia's *terem*. 1929 S. RUNCIMAN *Emperor Romanus Lecapenus* i. 28 It has been customary to regard the gynaecum as a prison from which Byzantine women never emerged—an exact equivalent of the Russian terem, which most historians say derived from it, forgetting Russia's two and a half centuries of Mongol rule. 1943 E. M. ALMEDINGEN *Frossia* iv. 169 The maiden lived in her *terem*, its windows strictly latticed.

Terena (terē̆·nǎ). Also **Ter(r)eno.** (A member of) an Arawak group of South American Indians of the southern Mato Grosso in Brazil; the language of this group. Also *attrib.*

1891 D. G. BRINTON *American Race* 244 The Terenos.. are members of the Guaycuru stock of the Chaco. 1928 A. R. HAY *Indians of S. Amer. & Gospel* vi. 101 In the Terena tribe we have a typical group of forest Indians who are fast adopting civilized ways. 1932 P. RADIN *Indians of S. Amer.* xi. 204 No evidence exists for its presence [*sc.* class division] in any other of the continental Arawak tribes except the *Terreno*. 1946 *Internat. Jrnl. Amer. Linguistics* XII. 60/1 The basic unit of structure in Terena phonology is the syllable. 1952 E. FISCHER-JØRGENSEN in E. Hamp et al. *Readings in Linguistics II* (1966) 315 It is not at all rare to find particularly nasals entering into the category of phonemes never adjoining the vowel in clusters..; this is the case e.g. in Terena. 1960 *Word* XVI. 349 A phonemic analysis of Tereno establishes as full phonemes a bilabial nasal and an alveolar nasal. 1974 *Encycl. Brit. Macropædia* XVII. 125/1 The Terena..work on cattle breeding farms.

terephthalic, *a.* (Further examples.)

1964 N. G. CLARK *Mod. Org. Chem.* xxiii. 491 Terephthalic acid, after conversion to the dimethyl ester, is an important intermediate in the manufacture of 'Terylene'. 1971 D. POTTER *Brit. Eliz. Stamps* ii. 27 Lettalite B.3 or tere-phthalic acid became known as violet phosphor. 1973 *Materials & Technol.* VI. iv. 326 The starting materials for the manufacture of polyethylene terephthalate are ethylene glycol and terephthalic acid, both obtained from petroleum sources.

terephthalate (also stressed *terephtha·late*) (further examples).

1946, etc. [see *polyethylene terephthalate* s.v. *POLYETHYLENE a]. 1958 *Times Rev. Industry* Aug. 57/3 Mylar polyester film..is a polyester terephthalate film. 1973 *Materials & Technol.* VI. iv. 280 While other fibre-forming polyesters were prepared, none proved to be as satisfactory as the polyethylene terephthalate developed by Whinfield, and this has continued to dominate the commercial fibre-forming polyester field.

teres (te·rīz), *a.* and *sb.* *Anat.* [L.: see TERETE *a.*] **A.** *adj.* In *pronator teres*: a pronating muscle of the forearm that arises from the humerus and ulna, near the elbow, and is inserted into the radius. **B.** *sb.* [sc. *musculus*.] Either of two muscles arising from the shoulder blade and inserted into the upper part of the humerus: the *teres major* draws the humerus towards the body and rotates it inwards; the *teres minor* rotates it outwards and helps steady its head.

1713 W. CHESELDEN *Anat. Human Body* (ed. 3) II. iii. 59 Teres minor, is a small Muscle arising below the former [*sc.* infraspinatus] from the inferior Costa Scapulæ. *Ibid.*, Teres major, arises from the lower Angle of the Scapula. *Ibid.* 66 Pronator Teres, arises from the inner Apophysis of the Os Humeri, and upper and forepart of the Ulna. 1889 J. LEIDY *Elem. Treat. Human Anat.* (ed. 2) v. 295 The greater teres..contributes with the latissimus to form the posterior border of the axilla. 1979 *Sci. Amer.* Dec. 99/3 This feature seems to reflect the strong development in Neanderthals of the teres minor muscle. 1980 *Gray's Anat.* (ed. 36) v. 574/1 The pronator teres rotates the radius upon the ulna, turning the palm of the hand backwards.

tereu. (Later examples.)

1923 T. S. ELIOT *Waste Land* iii. 15 Twit twit twit Jug jug jug jug jug So rudely forc'd. Tereu. 1936 R. CAMPBELL *Mithraic Emblems* 125 Hear how it whistles 'jug, puff-puff, tereu' Better than any nightingale could do.

Tergal (tɘ·ɹgǎl), *a.*[2] and *sb.* Also **tergal.** [a. F. *Tergal*, f. *tér-éphtalique* TEREPHTHALIC *a.* + *gal-lique* GALLIC *a.*[1]] A proprietary name for polyester fibre and fabrics. Cf. *TERITAL.

1954 *Trade Marks Jrnl.* 22 Dec. 1301/2 *Tergal*... Textile piece goods; bed and table covers, curtains; and household textile articles... Societe Rhodiaceta.., Paris VIIIe, France; manufacturers. 1955 *Official Gaz.* (U.S. Patent Office) 26 Apr. TM 192/1 *Tergal*... For..textile fabrics..of synthetic fibers, table cloths and napkins, bed sheets, blankets, and quilts. 1959 *Guardian* 16 Oct. 7/3 Loden or tergal or plaid lined. 1967 *Jane's Surface Skimmer Systems 1967–68* 7/1 The airflow is ducted to ten individual neoprene-coated tergal skirts. 1968 *Economist* 15 June 72/2 Rhone Poulenc had a French monopoly of nylon and terylene (tergal). 1973 'D. RUTHERFORD' *Kick Start* i. 12 My dark blue Tergal trousers.

-teria (tiº·riǎ), *suffix.* orig. and chiefly *U.S.* Also **-eteria.** [Derived from *CAFETERIA by analysis of its components as *café* + *-teria*.] A suffix used commercially to form the names of self-service retail or catering establishments.

1923 *Mod. Lang. Notes* XXXVIII. 188 Every one knows by this time that a cafeteria is a 'help-yourself' restaurant. Apparently in the popular mind the ending *-teria* or *-eteria* has come to indicate just such a process. 1929 *Amer. Speech* IV. 334 To the vast and growing progeny of 'cafeteria' may be added the name given to 'Maxwell's Vegetarian Healthateria', 35 West Van Buren Street, Chicago. 1941 [see *BURGER]. 1959 *Times* 27 Oct. 13/5 To the collection of *-teria* and *-toria* add 'Valeteria' and 'Washeteria'—in Cambridge, Massachusetts. 1965 *Listener* 2 Sept. 339/1 An Italian café-owner..has.. switched his sign from Pizzeria to Pie-teria.

Terital (te·ritǎl). Also **terital.** [a. It. *Terital*, f. *ter-eftalico* TEREPHTHALIC *a.* + *ital-iano* ITALIAN *a.*] A trade name for natural and synthetic (chiefly polyester) fibre, fabrics, and floor-coverings. Cf. *TERGAL.

A proprietary name in the U.S.

1960 *Guardian* 28 Sept. 8/6 Lilion, helion, terital and viscose are blended with wool in textiles by the high fashion houses. 1963 *Official Gaz.* (U.S. Patent Office) 4 June TM 31/2 Società Rhodiatoce S.p.A., Milan... *Terital*... For fabrics (obtained from natural or synthetic fibers). 1963 *Times* 11 June 19/1 Output of 'Rhodia' and 'Albene' yarns was maintained at the 1961 level, while production of 'Nailon' and 'Terital' increased considerably. 1972 *Guardian* 22 Aug. 9/1 See-through blouse.. made in terital which is non-crushable and drip-dry.

‖ **teriyaki** (teriyā·ki). [Jap., f. *teri* gloss, lustre + *yaki* roast.] A Japanese dish consisting of fish or meat marinated in soy sauce and broiled.

1962 M. DOI *Art of Japanese Cookery* 72 In *Teri-yaki*, rich sauce which gives a sheen to ingredients is used as seasoning. 1963 H. TANAKA *Pleasures of Japanese Cooking* iv. 76 Almost as popular as *yakitori* is *teriyaki*, usually fish marinated in a *shoyu* sauce, arranged on long skewers, and then broiled over charcoal. *Teriyaki* means 'glaze broiled'. 1972 A. FOWLES *Double Feature* i. 20 The predictable teriyaki dinner, more edible..than its infra-red mini-grilled BOAC counterpart. 1978 *Amer. Speech* 1975 L. 304 The recipe, one of a number for Japanese dishes, calls for chicken livers marinated in teriyaki sauce, wrapped in bacon, and broiled.

terlinguaite (tɘɹli·ŋgwǎ,ɘit). *Min.* [f. *Terlingua*, name of the village in Texas where it was found + -ITE[1].] An oxychloride of mercury, Hg_2OCl, found as transparent or translucent yellow or greenish yellow monoclinic crystals (see also quot. 1900).

1900 H. W. TURNER in *Mining & Sci. Press* (San Francisco) 21 July 64/1 In addition to cinnabar, mercury occurs in the native form..and as yellow-green crystals. Prof. S. L. Penfield has identified the white coating as calomel or mercury chloride (Hg_2Cl_2), and the greenish

crystals as an oxychloride of mercury, forming a new mineral species, for which I have suggested the name terlinguaite. 1903 A. J. MOSES in *Amer. Jrnl. Sci.* CLXVI. 259 Of the three possibly different substances to which the name terlinguaite has hitherto been applied we have.. 1st. The mineral here described. 2d. The undetermined rough yellow crystals mentioned in No. 5. 3d. The pulverulent yellow masses. *Ibid.*, Terlinguaite.—This name should be limited to the yellow monoclinic oxychloride of mercury here described. 1932 [see *MONTROYDITE]. 1964 *Mineral. Abstr.* XVI. 619/1 The ore deposit is in Upper Cretaceous liparites and tuffs [in Kamchatka, U.S.S.R.]. Brief descriptions..are given for native mercury, calomel, eglestonite, terlinguaite, and mosesite.

term, *sb.* Add: **II. 4. c.** *long-term*, *short-term* adjs.: see as main entries.

5. c. In *pl.* in phrases (esp. *to keep terms*) indicating that a person has attended the required number of lectures at a university, has been in residence for the period of time laid down in the statutes, and has satisfied the authorities in other statutory respects. *N.Z. colloq.*

1959 G. SLATTER *Gun in my Hand* 37 The Old Prof.. gave me 'terms' out of the kindness of his heart, but it was no use. 1962 M. K. JOSEPH *Pound of Saffron* ii. 38 You know the way he barks at you like a sergeant-major and then sees you don't miss terms.

III. 8. b. 3) *to come to terms*: also *fig.* (const. *with*), to reconcile oneself to, to become reconciled with; 7) *terms of reference*: the points referred to an individual or body of persons for decision or report; that which defines the scope of an inquiry.

3) 1923 J. B. PRIESTLEY *I for One* 235 The few [pictures] that it has do not seem so bright, so ideal, but seem to have come to terms with sad reality, showing us the pudding as it is and not as it ought to be. 1934 R. MACAULAY *Milton* vii. 109 He had here come to terms with life, or bravely pretended to have done so. 1965 *Listener* 30 Dec. 1067/1 Kipling, I think characteristically, came to terms with his tormentor. 1970 L. DEIGHTON *Bomber* i. 12 Each of the airmen guests was already coming to terms with the return to duty.

7) 1892 *Daily Graphic* (Suppl.) 30 Dec. 3/1 On the 14th October the constitution of the Commission and the exact terms of reference were made known. 1913 *Rep. Brit. Assoc. Adv. Sci. 1912* 549 The nature of the Inquiry of the Industrial Council is explained in the following 'terms of reference'. 1927 [see *REFER *v.* 8 a]. 1967 G. F. FIENNES *I tried to run Railway* vii. 88 We wrote ourselves new terms of reference in that sense. *a* 1974 R. CROSSMAN *Diaries* (1976) II. 661, I had to point out this was not excluded by the Committee's terms of reference, which had been drafted after consultation with the Foreign Office.

c. *terms of trade*, the ratio between the prices paid for imports and those received for exports.

1923 A. MARSHALL *Money, Credit, & Commerce* III. vi. 161 Illustration of the demands of each of two countries which trade together, for the goods of the other: and the general dependence of the terms of trade on the relative volumes and intensities of those demands. 1942 J. R. HICKS *Social Framework* xvi. 174 *Terms of Trade*, the amount of other countries' products which the nation gets in exchange for a unit of its own products. 1957 A. C. L. DAY *Outl. Monetary Econ.* xxxi. 399 Home terms of trade. [*Note*] An index of the home price of exports divided by the home price of imports. 1976 *Economist* 16 Oct. 22/3 Until exports expand enough, and/or imports fall enough, to offset the terms of trade deterioration, a devaluation makes the balance of payments worse before better.

IV. 11. b. Also, *in* (..) *terms*: in terms of what is designated by (..); *to think in terms of* (colloq.): to make (a particular consideration) the basis of one's attention, enquiries, plans, etc.

1947 MULGAN & DAVIN *Introd. Eng. Lit.* xiii. 164 The impact of Ibsen..did much to revitalize the degenerate English theatre and force it to think in terms of living ideas and contemporary realities. 1959 D. W. BROGAN in F. M. Joseph *As Others see Us* 4, I was predisposed to see American problems in European terms. 1973 'E. MCBAIN' *Hail to Chief* iii. 39 Carella..had suspected the ditch murders were related to organized crime... As it turned out, the cops had been thinking correctly in terms of gang warfare. 1978 *Listener* 26 Jan. 119/1 The hour's delay—a mere hiccup in cricketing terms—was caused by politics.

d. *Physics.* Each of a set of numbers such that lines in the spectrum of an atom have wave numbers given by the differences between two numbers in the set; an atomic state corresponding to one of these numbers, the number being proportional to the binding energy of a valence electron; a symbol representing such a state. Freq. as *spectral term*.

1909 *Sci. Abstr.* A. XII. 20 In any combination formula each of the terms represents the influence of one pole. 1915 *Astrophysical Jrnl.* XLI. 324 The difference in 'wave-number'..between the limit of the series and each member is called the 'term'... The limit itself is commonly a 'term' of some other series. 1922 A. D. UDDEN tr. *Bohr's Theory of Spectra* II. ii. 30 The arrangement of the states in horizontal rows corresponds to the ordinary arrangement of the 'spectral terms' in the spectroscopic tables. 1925 [see *LEVEL *sb.* 3 e]. 1935 [see *STATE *sb.* 4 b]. 1935 W. M. HICKS *Structure of Spectral Terms* i. 1 Any given term in a neutral spectrum is expressible in the form $R/(m+\phi)^2$, where R is a constant..*m* is an ordinal integer and φ a fraction which depends on *m*. 1938 *Nature* 22 Oct. 735/1 Dr. Dobbie has extended the number of classified lines to some 1,700 and has identified 73 terms involving 218 levels. 1967 [see *LYMAN]. 1970 G. K. WOODGATE *Elem. Atomic Struct.* vii. 110 For calcium.. the ^3P and ^1P terms of the configuration $4s4p$ are separated by about 8,000 cm^{-1}.

VI. 17. term paper *U.S. Educ.*, an essay or dissertation representative of the work done during a single term; **term symbol** *Physics*, a symbol of the type ^3P, denoting the values of *L* and *S* for a spectral term; **termwise** *adv.* and *a. Math.*, (carried out) term by term, treating each term separately.

1931 *High School Jrnl.* Jan. 17 A long term paper that will incorporate the results of a semester's reading. 1962 A. LURIE *Love & Friendship* xiv. 281 Students plagiarizing their term papers. 1975 M. BRADBURY *History Man* x. 164 Students..discuss..term-papers, union politics, theses. 1932 BACHER & GOUDSMIT *Atomic Energy States* i. 9 Each of the doublets occurs twice, and it is necessary to distinguish them in the term symbols. 1977 I. M. CAMPBELL *Energy & Atmosphere* viii. 220 The first electronically excited state of the oxygen atom would have in principle the two unpaired electrons of the ground state with opposite spins, producing a singlet state. In fact more detailed quantum mechanical treatment shows that there are two such states, designated by term symbols ^1D and ^1S, with the former the lower in energy of the two. 1912 J. PIERPONT *Lect. Theory of Functions Real Variable* II. v. 180 In this case we can obviously integrate termwise, although the convergence is not uniform. 1930 T. FORT *Infinite Series* vii. 74 Termwise multiplication of series. 1979 *Proc. London Math. Soc.* XXXVIII. 390 We then integrate termwise.

terminable, *a.* (*sb.*) Add: **2.** *minimum terminable unit* (see quot. 1975). Abbrev. *T-unit* s.v. *T 6*.

1965 [see *T-unit* s.v. *T 6*]. 1975 *Language for Life* (Dept. Educ. & Sci.) iii. 39 The minimum terminable unit, or T-unit, is 'roughly any sentence or part of a sentence that is an independent clause, possibly containing, however, one or more dependent clauses'.

terminal, *a.* and *sb.* Add: **A.** *adj.* **3. b.** (Earlier example.)

1869 *Bradshaw's Railway Manual* XXI. 87 This line.. terminates in the city, at a great terminal station in Liverpool Street.

4. b. (*a*) (further examples); (*b*) applied to a patient suffering from such a disease; (*c*) applied to an institution or ward in which such patients are nursed.

(*a*) 1958 A. HUXLEY *Let.* 2 Feb. (1969) 845 The administration of LSD to terminal cancer cases, in the hope that it would make dying a more spiritual, less strictly physiological process. 1961 *Lancet* 2 Sept. 549/1 It is they who have the closest contact with people who are going through 'terminal illnesses'. 1976 *Church Times* 23 July 11/1, I now have several progressive illnesses; one is terminal. 1980 D. LODGE *How Far can you Go?* vi. 160 What would it be like to be told you had a terminal illness, he wondered.

(*b*) 1961 *Amer. Speech* XXXVI. 145 *Terminal* adj., approaching death, moribund. 'The patient looks terminal to me.' 1965 E. M. K. PILLERS in J. S. Mitchell *Treatment of Cancer* 91 Response to this has been disappointing and the patient is now terminal. 1969 *Guardian* 19 Aug. 2/2 We started off with patients who were going to die anyway—terminal patients.

(*c*) 1961 *Lancet* 2 Sept. 548/2 Excellent care of these patients has been carried out not only at St. Joseph's but also in other terminal homes and hospitals. 1974 F. WARNER *Meeting Ends* I. ii. 13 The old lady was taken into the terminal ward, inarticulate, jabbering away.

c. *colloq.* In various *transf.* and *fig.* uses of sense 4 b above (freq. *joc.* or *trivial*).

1973 *Black Panther* 11 Aug. 8/2 The country was plunged into shock and the President faced a terminal crisis. 1975 D. LODGE *Changing Places* iii. 112, I continue to hope that our marital problems are not terminal. 1981 *Daily Tel.* 21 Dec. 2/1 Another contest for Labour's deputy leadership next year could prove 'terminal' for the party, Mr Neil Kinnock..said. 1983 *Times* 23 Sept. 6/4 One commentator said yesterday that his insensitivity was terminal. *Ibid.* 26 Sept. 9/5 A bad case of terminal tiredness had lowered my resistance to every loitering bug.

7. Special collocations: **terminal ballistics,** that branch of ballistics which deals with the impact of the projectile on the target; **terminal guidance** *Aeronaut.* (see quot. 1955); **terminal juncture** *Linguistics*, a juncture (sense *2 c) that occurs at the end of a syntactic unit; **terminal market** *Comm.*, a market that deals in futures; **terminal nose-dive** *Aeronaut.*, a nose-dive during which an aircraft reaches its terminal velocity; **terminal string** *Transformational Gram.*, a string consisting wholly of terminal symbols; **terminal symbol** *Transformational Gram.*, a symbol that denotes a lexical class and cannot be further rewritten; **terminal velocity,** the constant speed of fall that any particular object, given time, will eventually attain, at which the air resistance is equal to its weight.

1947 L. E. SIMON *German Research World War II* vii. 109 Terminal ballistics is concerned with the motion of the projectile, its fragments, and gases in the neighbourhood of the target. 1974 *Encycl. Brit. Macropædia* II. 659/2 A theoretical structure for terminal ballistics is a relatively current development. 1955 A. S. LOCKE *Guidance* i. 19 Terminal guidance is the guidance applied to the missile between the end of the midcourse guidance and contact with or detonation in close proximity to the target. 1979 *Jrnl. R. Soc. Arts* CXXVII. 555/1 Long-range, sea-skimming missiles with terminal guidance. 1956 *Language* XXXII. 653 This set of phonetic phenomena we assign to the terminal juncture. 1975 *Ibid.* LI. 57 The final element, *hacer*, accompanied by terminal juncture, is associated with zero following elements, signals absolute completion, and receives maximum contrast. 1895 Terminal market [in Dict., sense 4]. 1952 *Economist* 22 Nov. 567/1 There is little hope of restoration of a terminal market until the domestic allocation of sugar is freed from rationing. 1962 H. O. BEECHENO *Introd. Business Stud.* xi. 94 Future or terminal markets where goods can be bought and sold in advance. 1933 *Gloss. Aeronaut. Terms* (*B.S.I.*) 9 *Terminal nose-dive*, a dive at terminal velocity. 1956 N. CHOMSKY *Logical Struct. Linguistic Theory* (1975) vii. 174 The mapping in question may rearrange the order of elements of terminal strings and may specify their morphemic shape in various ways. 1967 D. G. HAYS *Introd. Computational Linguistics* vi. 118 A terminal string is composed of a certain number, say n, of terminal symbols. 1964 E. BACH *Introd. Transformational Gram.* ii. 14 Among the symbols of the grammar..there are some which never appear to the left of the arrow in a rule as symbols to be replaced. These are called *terminal symbols*. 1967 D. G. HAYS *Introd. Computational Linguistics* vi. 119 In terms of dependency theory, let the level of a structure be one greater than the number of links from its origin to the terminal symbol furthest removed. 1832 Terminal velocity [in Dict., sense 4 a]. 1910 [see *STOKES[1] c]. 1914 *Aeronaut. Jrnl.* XVIII. 50 He had dived, and had reached a speed so high that he thought it wise to straighten out without waiting to reach the terminal velocity. 1946 T. C. OHART *Elements Ammunition* IV. viii. 199 This theoretical maximum velocity for a given size and shape of bomb is called the terminal velocity; it is really a function of a given design, depending upon the aerodynamic characteristics of a bomb.

B. *sb.* **2. a.** (Earlier example.)

1838 W. STURGEON in *Ann. Electr., Magn., & Chem.* II. 11 That [part] which is connected with the positive pole of the exciting apparatus..may very conveniently be called the 'salient terminal metal', or occasionally the 'salient terminal' only.

d. A device for feeding data into a computer or receiving its output; *esp.* one that can be used by a person as a means of two-way communication with a computer.

1954 *Trans. IRE Prof. Group Electronic Computers* Mar. 2/1 Since the two machines employ the same digital language, this attachment can easily be made through their regular input-output terminals. 1958 *Oxf. Mag.* 29 May 470/1 The 'terminal' equipment, consisting of punched paper tape and a teleprinter, is relatively slow. 1965 *Jrnl. Assoc. Computing Machinery* XII. 350 (*heading*) On a problem concerning a central storage device served by multiple terminals. 1970 O. DOPPING *Computers & Data Processing* ix. 131 An 'impersonal' terminal with card reader, line printer, etc. can be started automatically at the end of the waiting time, but in case of a 'personal' terminal, the computer may send a message to the terminal indicating that the conversation may begin. 1971 *Daily Tel.* (Colour Suppl.) 19 Feb. 5/3 The national police computer with 700 terminals throughout the country opens this year. 1973 *Nature* 12 Oct. p. xxviii/3 (Advt.), There are good in-house computing facilities and a terminal to an IBM 360/195. 1977 *Hongkong Standard* 12 Apr. (Business Suppl.) 5/1 Terminal operators have been responding actively to the encouraging scene. 1979 *Computers in Shell* (Shell Internat. Petroleum Co.) (*recto rear cover*), Types of terminals include card readers, printers, video screens and teletypes. 1981 *Sci. Amer.* Dec. 112/3 The computer executes the operation, simplifies the resulting expression and prints it or displays it on a video terminal.

5. a. For *U.S.* read orig. and chiefly *U.S.* (Later examples.) Hence, in extended use, applied to the terminal point of an airline (= *air terminal* (a) s.v. *AIR *sb.*[1] III. 1), a bus service (= *bus terminal* s.v. *BUS *sb.*[2] 3), or occas. some other transportation service.

1921 *Flight* 16 June 401 (*caption*) Two London–Paris terminals.—The lower photograph shows Cricklewood aerodrome..the upper picture shows the Paris air port. 1922 JOYCE *Ulysses* 699 When citybound frequent connection by train or tram from their respective intermediate station or terminal. 1924 *London Guide No. 3* 152 At all the principal traffic centres and at the route terminals are uniformed 'General' Inspectors [of buses]. 1937 *New Statesman* 25 Dec. 1094/1 A rail-cum-steamer terminal on the Firth of Clyde. 1958 'N. SHUTE' *Rainbow & Rose* vi. 270 Walking from the hostel to the terminal [of an airline]. 1958 *Times* 1 Mar. 7/4 Each city or town would adopt the type of terminal [for helicopters] best suited to its own locality. 1969 *Jane's Freight Containers 1968–69* 113/3 Scheduled national services: door-to-door and terminal-to-terminal. 1975 N. LUARD *Robespierre Serial* xvii. 153 All he could do was head for the bus terminal... The terminal was only twenty minutes by taxi from the hotel. 1980 R. MCCRUM *In Secret State* xiii. 122 Quitman took the Piccadilly line to Heathrow... Soon he was standing on the travolator, riding up towards Terminal Three. 1981 M. MOORCOCK *Byzantium Endures* ii. 50 We eventually arrived at Glavnaya Station, the main terminal of Odessa situated in the heart of the city.

b. An installation where oil is stored, situated at the end of a pipeline or at a port of call for oil tankers.

[1940 *Petroleum Press Service* 19 Apr. 182 This has included laying down a 100-mile 16 inch pipe-line to the coast and constructing ocean terminal facilities at Puerto

La Cruz.] 1947 L. M. FANNING *Amer. Oil Operations Abroad* xi. (*caption*) Oil-loading dock, Puerto La Cruz Terminal. 1948 *Economist* 14 Aug. 259/2 It is obviously difficult to pump oil from an Arab source in Iraq to a terminal in a Jewish-held town. 1976 M. MACHLIN *Pipeline* vi. 70 The 707 could then descend to a relatively low level and follow the route of the pipeline to its terminal.

7. One suffering from a terminal illness.

1960 J. G. BALLARD in *New Worlds* Oct. 95 The terminals sleeping in the adjacent dormitory block attracted hordes of would-be sightseers. 1976 *Church Times* 23 July 11/2 Mr. Rice recently paid a third visit to the nun— who is bedridden and a terminal—and questioned her again, mainly about prayer and intercession. 1982 P. VAN GREENAWAY *Lazarus Lie* vi. 61 'You have maybe a couple of thousand patients.'..'How many terminals?' 'Terminals?' 'Inoperables, end of the liners.'

8. Special Combs.: **terminal building,** a building housing the main facilities for air passengers; **terminal screw** *Electr.*, a screw for fastening an electric wire to the object with the screw hole.

1933 *Jrnl. R. Aeronaut. Soc.* XXXVII. 10 A terminal building will house traffic control and airport administration. 1977 G. SCOTT *Hot Pursuit* iv. 38 At the airport.. I got out of the terminal building and on to the bus. 1931 S. R. ROGET *Dict. Electr. Terms* (ed. 2) 349/2 Terminal screw. 1978 *N.Y. Times* 30 Mar. C-4/1 Aluminum wire is stiffer than copper wire and does not bend as easily when wrapped around small terminal screws on switches and outlets.

terminalia (tɘ̄ɹminē̆ⁱ·liǎ). [mod.L. (Linnæus *Mantissa Plantarum* (1767) I. 21), f. L. *termināl-is* TERMINAL *a.* + -IA[2].] An evergreen tree of the large genus of this name, belonging to the family Combretaceæ, native to tropical or sub-tropical regions, having leaves clustered at the end of branches, and often producing a valuable timber. Also *attrib.*

1830 *Curtis's Bot. Mag.* LVII. 3004 (*heading*) Broad Downy-leaved Terminalia. 1926 *Chambers's Jrnl.* June 411/2 He halted the elephant under one of the Terminalia trees. 1964 C. WILLOCK *Enormous Zoo* iii. 39 The elephants had taken a liking to *terminalia* bark. 1973 *Times* 20 Sept. 8/3 David Dibnah,..convicted of wilfully destroying a flowering terminalia tree, has been ordered by a magistrate to plant another.

terminalization (tɘ̄:ɹminǎlɘizē̆ⁱ·ʃən). *Cytology.* [f. TERMINAL *a.* and *sb.* + -IZATION.] The movement of a chiasma or chiasmata towards the end of a separating bivalent.

1929 C. D. DARLINGTON in *Jrnl. Genetics* XXI. 266 The post-diplotene stages in this species are characterised by a regular movement of chiasmata towards the attachment constriction. The opposite movement, which I will call 'terminalisation', affords a sufficient and indeed the only explanation of the exceptional metaphase configurations found in *Tradescantia*. 1932 *Amer. Naturalist* LXVI. 32 Related species differ in the degree of terminalization of chiasmata. 1979 *Nature* 22 Mar. 349/2 The issue of terminalisation is relevant both theoretically (assessment of chiasma and crossover frequency) and practically (alleged factor in maternal-age-dependent non-disjunction in mammals).

Hence **te·rminalized** *ppl. a.*, (of a chiasma) having moved to, or situated at, the end of a separating bivalent.

1932 C. D. DARLINGTON *Rec. Adv. Cytol.* iv. 103 However many chiasmata are terminalised, the chromosomes remain associated by terminal chiasmata. *Ibid.* 104 (*caption*) Completely terminalised chiasmata. 1959 *Genetics* XLIV. 711 Incompletely terminalized chiasmata were observed in these configurations. 1979 *Nature* 22 Mar. 349/2 Our data do not indicate whether or not chiasmata are terminalised in the mouse.

terminally, *adv.* Add: **2. b.** *Comb.* with an adj. in sense 4 b of TERMINAL *a.*

1973 *Sci. Amer.* Sept. 56/1 One of the terminally sick patients has been given a change of scene by moving his bed to the garden. 1976 M. MILLAR *Ask for me Tomorrow* (1977) iv. 37 This was Aragon's first time in the presence of a terminally ill person.

terminate, *ppl. a.* Add: **1. c.** *Gram.* = *TERMINATIVE *a.* 4 b. Also as *sb.*

Restricted to the writings of G. O. Curme and a few others.

1931 G. O. CURME *Syntax* xix. 385 A large number of simple and compound verbs indicate an action *as a whole*. Such verbs are called terminates. *Ibid.* 386 The terminate aspect has relations also to the durative aspect. 1935 —— *Gram. Eng. Lang.* II. 206 The expanded form often represents the act as a whole, hence it has terminate force: 'I am sorry you doubt my statement. I *am telling* you the truth.' *Ibid.* 237 In terminates it [*sc.* the present participle] represents the act as a whole, as a fact. 1946 [see *PROGRESSIVE *a.* 3 h]. 1972 M. L. SAMUELS *Linguistic Evol.* 161 If a terminate or point-action meaning was required for a majority of its occurrences in the preterite, the short vowel was preferred.

terminate, *v.* Add: **I. 4. c.** In pregnant use: (*a*) to dismiss from employment; (*b*) to assassinate; *to terminate with extreme prejudice*: see *PREJUDICE *sb.* 1 c. *N. Amer. colloq.*

(*a*) 1973 *N.Y. Law Jrnl.* 25 July 13/1 The complainant had been discharged because of an unauthorized absence..and..there was no probable cause to believe that he had been terminated in retaliation for having filed previous complaints against petitioners. 1976 M. MACHLIN *Pipeline* xviii. 228 If the boss didn't care for you very much in the first place, you could be terminated without having any appeal to the union. 1980 R. L. DUNCAN *Brimstone* ii. 36 Adamson's putting pressure on me to terminate you.

(*b*) 1975 N. LUARD *Robespierre Serial* iv. 27 A free-lance agent who'd been given a contract to terminate an individual the Service had declared hostile. 1981 T. BARLING *Bikini Red North* ii. 51 Haddad was terminated by persons unknown.

11. (Earlier example.)

1789 J. WOODFORDE *Diary* 24 Mar. (1927) III. 91 His case is a violent Stranguary and if some remedy or other does not soon, very soon do good, it will terminate fatally to him.

termination. Add: **I. 3. c.** *Chem.* and *Biochem.* The cessation of the building up of a polymer molecule. Freq. *attrib.*

1951 *Jrnl. Amer. Chem. Soc.* LXXIII. 5197/1 It is assumed in the case of tetrafluoroethylene polymerization initiated by inorganic free radicals that chain termination occurs by combination of a polymer radical with either another polymer radical or an initiator radical. 1967 MARGERISON & EAST *Introd. Polymer Chem.* v. 246 Termination may be brought about by many types of reagent. 1968 A. WHITE et al. *Princ. Biochem.* (ed. 4) xxix. 678 The mechanism by which these three codons accomplish chain termination and polypeptide release is not understood. 1978 HAYES & GEORGE in C. E. Carraher et al. *Organometallic Polymers* 16 In vinylferrocene polymerizations, the termination step is monomolecular. 1981 *Sci. Amer.* Aug. 63/2 Two proteins called termination factors are involved, and it appears that UAG, UAA and UGA all serve as termination codons: triplets on the mRNA that cause the ribosome to release the messenger and the newly synthesized protein.

d. The ending of a person's employment; dismissal. Chiefly *N. Amer.*

1961 *Wall St. Jrnl.* 23 Jan. 2/3 They qualify for termination payments and most are eligible for deferred pensions. 1982 *Chicago Sun-Times* 3 Dec. 89 He and fellow workers were informed of the terminations at 10:30 a.m. Wednesday and told to 'pack up and leave immediately'. 1983 M. EDWARDES *Back from Brink* iv. 56 In most cases we allowed the manager to 'resign' but in truth most of these people were dismissed, and were paid termination payments.

e. The ending of a pregnancy before term by artificial means; an induced abortion.

1969 *Times* 3 July 7/6 Women denied a legal abortion commonly seek termination elsewhere. 1973 *Times* 26 Nov. 6/1 The pregnant women walking about the hospital ward were all in for abortions. Or terminations, as they called them—a much nicer word. 1978 F. WELDON *Praxis* xxiv. 256 You can't possibly go through with the pregnancy... If you don't have a termination, you're finished.

f. Assassination (*spec.* of an intelligence agent).

1975 N. LUARD *Robespierre Serial* v. 28 The escort role..was one Carswell had carried out..before, although this was the first occasion where it involved a termination mission. 1980 [see *PREJUDICE *sb.* 1 c]. 1983 G. MARKSTEIN *Ferret* 164 Terminations are no longer as fashionable as they were. Unless the top says so.

terminative, *a.* Add: **4. a.** (Earlier example.)

1860 *Trans. Philol. Soc. 1857* 34 The declension of the personal pronouns [in the Tushi language] is as follows.. Nominative..Genitive..Comitative..Terminative.

b. Applied to an aspect of a verb which denotes a completed action, or its completion.

1911 KRUISINGA & ERADES *Eng. Gram.* (1953) I. II. vii. 257 Verbs of a terminative character, that is such as express the final stage of an activity. 1924 O. JESPERSEN *Philos. Gram.* xx. 273 Lindroth for the first class uses the term 'successive' (with the subdivisions 'terminative' and 'resultative'.) 1930 B. TRNKA *Syntax Eng. Verb* 32 A differentiation between two series of aspects, the *ingressive*, *continuative* and *terminative* (resultative) on the one hand, and the *imperfective* and *perfective*, on the other... The former, common to both Slavonic and Germanic languages, express the ingressive, continuative and terminative stages of the verbal action. 1963 F. T. VISSER *Hist. Syntax Eng. Lang.* III. ii. 1372 There are three kinds [of verbs of aspect]: (1) of egressive or terminative aspect. 1984 *Eng. World-Wide* IV. 208 The *terminative* markers *gaan* and *kom*, which occur only with motion verbs in this *perfective* function, follow the main verb.

terminator. Add: **3.** *Biochem.* A sequence of polynucleotides that causes transcription to end and results in the release of the newly synthesized nucleic acid from the template molecule. Freq. *attrib.*

1969 *Biochemistry* VIII. 4897/1 Would chains bearing such a chain-growth terminator be susceptible to the hydrolytic and pyrophosphorolytic reactions? 1977 *World Bk. Sci. Ann. 1978* 249 To get the gene to work.. certain controlling base sequences had to be added at each end. One end had to have a 'promoter' sequence so transcription could start; the other end had to have a 'terminator' sequence to stop transcription. 1978 *Nature* 30 Mar. 398/1 Analysis of several ρ-independent terminators has revealed that in every case termination occurs distal to a GC-rich region within a run of uridine residues.

terminism. a. (Earlier example.)

1878 S. H. HODGSON *Philos. Reflection* I. i. 66 Nominalism..in its later shape, in which it is opposed to Conceptualism and is then more properly to be called *Terminism*.

terminological, *a.* (In Dict. s.v. TERMINOLOGY.) Add: *terminological inexactitude* (later examples); freq. as a humorous expression for a falsehood.

1926 FOWLER *Mod. Eng. Usage* 444/1 *Polysyllabic humour*... Of the long as distinguished from the abstruse, *terminological inexactitude* for lie or falsehood is a favourable example, but much less amusing at the hundredth than at the first time of hearing. 1940 C. MILBURN *Diary* 19 July (1979) 49, I can't think..why he [*sc.* Hitler] does not choke himself with his 'terminological inexactitudes'! 1976 A. PRICE *War Game* I. viii. 159 It all adds up to a little terminological inexactitude——he was lying through his goddamn teeth.

terminus. Add: **4. b.** *terminus a quo*: also used *spec.* in dating to indicate the starting-point of a period; similarly *terminus ante quem*, *terminus post quem* (= 'term before which', 'term after which') used to indicate the finishing- or starting-point of a period; also *ellipt.*, as *terminus ante*, *terminus post*.

1930 A. H. KRAPPE *Sci. Folklore* i. 14 Generally speaking..a *terminus ante quem* is furnished by the oldest European historical variant. 1936 *Burlington Mag.* Aug. 75/1 The dress the king is wearing did not become fashionable before 1796, a fact which indicates a *terminus post quem*. 1939 *Ibid.* May 218/1 The *termini post* and *ante* of the glass, thus given by the birth of Charles (1500) and the death of Philip (1506). 1953 R. J. C. ATKINSON *Field Archæol.* vi. 166 It must be clearly understood that these *termini post* and *ante quem* are the closest that the archæologist can get (at least with present methods of research) to an absolute date. 1968 *English Studies* XLIX. 455 The chapter..provides a good survey of the evidence..that seems to point to a date around 1200 as the most probable date of composition. However, the *terminus a quo* (1193) that is suggested..on p. 19, seems quite unwarranted. 1973 *Nature* 9 Nov. 77/1 These dates..merely provide, however, a *terminus post quem* for the valley deepening and widening during which cavern opening occurred. 1974 *Bodleian Library Rec.* Dec. 174 Such evidence clearly establishes for the annotations a *terminus a quo* of 1602. 1978 *Maledicta* II. 243, I must now propose 1888 as a conservative *terminus ante quem* for that species, and the evidence at that date points back considerably farther.

6. (Examples of the place at which a tram-line or bus route ends.)

1877 *Tramways Intelligence* 17 The lines of the company have their London termini at Westminster Bridge-road, Blackfriars-road. 1936 J. B. PRIESTLEY *They walk in City* iii. 39 The tram was full, but they pushed their way in... By the time they had arrived at the terminus in Gladstone Lane..a few drops [of rain] were falling. *Ibid.* iv. 85 Taking a bus as far as the edge of the moors, walking over to some place where he could get tea, then walking back to the terminus again. 1975 R. L. BEALS *Peasant Marketing System of Oaxaca* i. 9. Around the peripheries of the district are the termini of most of the bus lines.

termitarium. (In Dict. s.v. TERMITE.) (Later examples.)

1934 *Discovery* Nov. 308/2 A big termitarium..must contain a population of seven million, or even more. 1971 *World Archaeology* III. 124 The large refuse filled pit near the adult burials is at the site of a termitarium dug out for building material. 1981 *Atlantic Monthly* July 49 The nearest thing to a termitarium that I can think of in human behavior is the making of a language, which we do by keeping *at* each other all our lives,..changing the structure by some sort of instinct.

termitary. (In Dict. s.v. TERMITE.) Add: Also *fig.* (Later examples.)

1901 *Jrnl. R. Microsc. Soc.* 32 The Cicindelids select the termitaries as sunny places well suited for play and for hunting. 1935 *Times* 11 Feb. 14/2 Man as a personality is destined to be a free citizen in a free world, not an ant in some human termitary. 1937 *Discovery* Sept. 292/2 One must consider a termitary as a single animal. 1955 J. B. PRIESTLEY in Priestley & Hawkes *Journey down Rainbow* xii. 177 I'd rather stand on the pavement and eat a sandwich than lunch in that underground termitary. 1961 L. MUMFORD *City in History* i. 6 The social functions of the beehive, the termitary, and the ant-hill..have indeed.. many resemblances to those of the city. 1977 *Time* 10 Oct. 51/2 But they linger on paper as the dream architecture of the 20th century. Because these termitaries were never built, they could not be destroyed.

termite. Add: **1. b.** Also *Comb.*, as *termite-proof* adj.; **termite heap, mound** = *termite hill*.

1910 *Dundee Advertiser* 25 Nov. 6 The athletes had to.. jump from a small termite heap. 1920 *Blackw. Mag.* Feb. 205/1 Several enormous termite heaps. 1977 'J. MCVEAN' *Bloodspoor* x. 96 Haston was lying by a termite mound. 1934 WEBSTER, Termite-proof. 1937 *Discovery* Feb. 63/1 Care must be taken to ensure that the timber work is termite-proof. 1971 *Guardian* 18 Nov. 15/1 The immediate problem is to make sure such a building is termite proof.

2. *transf.* (with reference to the destructiveness of the insect).

1943 in S. J. BAKER *Austral. Lang.* (1945) 245 The secretary of the party said 'We can't say too much; there are termites about.' 1949 KOESTLER *Promise & Fulfilment* II. v. 282 One delegation would immediately settle down to silent termite work. 1951 *Economist* 8 Dec. 1402/2 'Any man who betrays the people's trust in a public office is a public enemy..' declared the new chairman..calling on his fellow members to punish the 'termites' relentlessly. 1961 A. MAUND *Worthy Termites* ii. 22 'So you want me to police Great Port for woodpiles which might have somebody besides nigras under them,' Ed said. 'Yes. Look for the *termites*.'

termitologist (tǝ̄ɹmaitǫ·lǒdʒist). [f. TERMITE + -OLOGIST.] One who studies termites.

1936 *Times* 9 June 10/3 Dr. Noyes, of California—a celebrated termitologist—writes doubtfully of *Zootermopsis*. 1971 E. O. WILSON *Insect Societies* vi. 106/2 Termitologists had long looked to the Mesozoic or beyond for traces of a truly archaic termite fauna.

termitophile. (Examples.)

1922 [see *PHYSOGASTRISM]. 1971 E. O. WILSON *Insect Societies* vi. 111/2 Termitophiles, often species-specific and highly modified.

termly, *a.* Delete *rare* and add later examples. Now freq. in the sense: occurring every academic term.

1969 T. FAWTHROP in Cockburn & Blackburn *Student Power* 101 There should be a variety of means by which assessment is arrived at: from termly work standards to dissertations. 1970 M. JONES *Ducal Brittany* vi. 166 The termly sums demanded from individual parishes were always the same. 1983 *Bull. Univ. Coll. London* May 8/2 A termly whole-day inter-disciplinary seminar is proposed.

termorrer (tǝɹmǫ·rǝɹ), repr. vulgar or dial. pronunc. of TO-MORROW *adv.* and *sb.* Cf. *TER.

1898 J. D. BRAYSHAW *Slum Silhouettes* 118 That's ninepence I owes Newsy: must pay that or there won't be no papers to start wiv ter-morrer. 1932 S. GIBBONS *Cold Comfort Farm* xii. 178 Ter-day's dinner... Ter-morrer's too, for all I know. 1974 P. CAVE *Mama* (new ed.) iv. 28 Adolph slipped the merchandise into his pocket. 'I'll do it termorrer,' he vowed.

terna. (Earlier example.)

1885 W. J. WALSH *Let.* 7 Mar. in P. J. Walsh *William J. Walsh* (1928) vii. 163 Then I would, as a matter of course, vote for your Grace, which would put you on the *terna*.

ternary, *a.* Add: **1. b.** Mus. *ternary form.* Substitute for def: The form of a movement which consists of three main divisions, *spec.* one in which the first subject recurs after a contrasting subject. Also *absol.* as *ternary*. (Earlier and further examples.)

1875 F. A. G. OUSELEY *Treat. Mus. Form & Gen. Composition* vi. 41 (*heading*) Of the ternary form. *Ibid.* vii. 44 If the minuet form is adopted for a complete and isolated composition, it should be lengthened considerably, and then both the minuet and trio may be written in the ternary form. 1896 W. H. HADOW *Sonata Form* iv. 29 In its use for purposes of the Folk-song the most primitive Ternary form consists of a melody in three clauses: one of assertion..one of contrast..one of re-assertion. 1931 D. F. TOVEY *Compan. Beethoven's Pianoforte Sonatas* 2 The vital distinction between 'binary' and 'ternary' is that between an aggregate whose members are inseparable and an aggregate containing one or more things already complete. 1938 *Oxf. Compan. Mus.* 334/2 Properly, any composition in which the ear seizes two clear divisions is binary, and any in which it seizes three is ternary... 'Sonata Form'..is often called 'Compound Binary'... Certain text-books..speak of it as 'Ternary'. *Ibid.* 335/1 *Rondo Form*. This may be looked upon as an extension of simple ternary form. 1944 W. APEL *Harvard Dict. Mus.* 88/1 The principle of ternary structure appeared first in the French chansons of the 16th century.... The idea of a contrasting middle section is quite clearly expressed in the shepherd's solo of Monteverdi's *Orfeo*.... Ternary form became clearly established in the da-capo aria, *c.* 1700. 1980 *New Grove Dict. Mus.* XVIII. 694/1 Tripartite musical form designated symbolically as *ABA*. The two elements *A* and *B* are often thematically independent and each is generally a 'closed' structure tonally, so that the interdependence of the two sections characteristic of binary form is not necessarily evident in ternary.

c. Also (of an alloy), composed of three principal metals; (of a mixture) containing three independent components; of or pertaining to such an alloy or mixture.

1889 *Proc. R. Soc.* XLV. 481 Ternary alloys obtained by adding tin to the immiscible pairs of metals, zinc and bismuth, aluminium and lead. 1897 W. D. BANCROFT *Phase Rule* xi. 156 The change of isotherms with the temperature for a ternary system which permits of no compounds and no second solution phase. 1923 GLAZEBROOK *Dict. Appl. Physics* V. 251/2 (*heading*) Methods of representing ternary equilibria. 1969 BENNISON & WRIGHT *Geol. Hist. Brit. Isles* xi. 261 The theoretical succession of deposition of minerals produced by the evaporation of sea-water is supported to a large extent by experimental work... However, the detailed course of crystallization, which varies considerably with different temperatures, can only be adequately represented by ternary diagrams. 1978 P. W. ATKINS *Physical Chem.* x. 305 Consider a ternary solution of composition a_1. This is unsaturated, and is a single phase. 1979 *Sci. Amer.* Nov. 73/3 The new ternary alloys consist of between 68 and 80 percent copper.

g. Nucl. Physics. *ternary fission*, fission of an atomic nucleus into three parts.

1955 *Physical Rev.* XCVII. 748/2 There is some uncertainty regarding the occurrence in ternary fission of a light fragment having a mass and charge greater than an alpha particle. 1979 *Nature* 12 Apr. 615/2 Uranium and plutonium decay by ternary fission with only very low probability.

ternery. Delete *rare* and add examples.

1905 E. SELOUS *Bird Watcher in Shetlands* xxiii. 180, I have mentioned the case of a dog making regular daily expeditions to a ternery, in order to feast upon the eggs. 1932 *Times Lit. Suppl.* 15 Sept. 637/2 The sanctuary at Dungeness, established presumably for the protection of the interesting ternery. 1943 HAGGARD & WILLIAMSON *Norfolk Life* i. 12 There is the azure tide inflowing past the Ternery on the point. 1979 *Woman & Home* June 154/3 The ternery, no longer out of bounds, was uncannily silent.

terotechnology (tiᵊ:ro-, te:roteknǫ·lǒdʒi). [f. Gr. τηρεῖν to watch over, take care of + -O + TECHNOLOGY.] The branch of technology and engineering concerned with the installation, maintenance, and replacement of industrial plant and equipment and with related subjects and practices.

1970 H. P. JOST *Let.* 2 Feb. (in files of *Suppl. to O.E.D.*), Last Saturday a Mintech Steering Committee, which I chaired, took some decisions on a proposed 'Committee on Terotechnology'. We felt that Terotechnology was preferable to Teromechanics, particularly as a good deal of electronics were involved. 1970 *Hansard Commons: Written Answers* 29 Apr. 338 The scope of the subject..includes..installation, commissioning and replacement of plant, machinery and equipment, feedback to designers, and management techniques... It was considered advisable to utilise a name reflecting the wider concept now envisaged. The word Terotechnology has therefore been adopted. 1975 *Daily Tel.* 9 Dec. 11/6 British Leyland has won the award of 'Conservationist of the Year' for its work in saving materials, plant and manpower. The firm estimated that it saved £321,182 on terotechnology—economy in materials, plant and manpower. 1977 *Engin. Synopses* Jan.–Feb. 20 (Advt.), With the continual increase in size and complexity of industrial undertakings terotechnology has emerged as a discipline in its own right. 1980 *Sunday Times* 9 Nov. 19/4 The science of maintenance, or terotechnology if you want to sound knowledgeable about it, is beginning to attract serious interest in industry and universities.

Hence te:rotechnolo·gical *a.*, te:rotechno·logist.

1970 *Chartered Mech. Engineer* June 243/1 According to Mintech, this new high-status role will be filled by none other than the Terotechnologist. *Ibid.*, A Committee of Terotechnology..will advise..on how to introduce appropriate measures to effect terotechnological progress. 1973 *New Scientist* 12 Apr. 95 Terotechnologists quote the case of breakdowns of waste heat boilers for 300 tonne oxygen steelmaking vessels. 1977 *Engin. Synopses* Jan.–Feb. 20 (Advt.), The terotechnologist must be able to call upon a wide variety of other skills and specialisms.

terp¹. Delete ‖ and add: Also applied to similar mounds outside Friesland itself. Now with anglicized pl. *terps* in *Archæol.* contexts. (Later examples.)

1939 G. CLARK *Archæol. & Society* iv. 105 '*Terps*.' Settlement mounds or tells are a commonplace feature of Greek and Middle Danubian prehistory. 1969 G. C. DICKINSON *Maps & Air Photographs* xiv. 217 (*caption*) The villages are built on or (now) around man-made mounds (*terps*) erected as a defence against flooding by the sea.

terp² (tǝ̄ɹp). *Theatr. slang.* [Abbrev. of TERPSICHORE(AN *a.*).] A stage dancer, esp. a chorus-girl; also, a ballroom dancer. Also *attrib.* and *pl.*, dancing. Hence **terp** *v. intr.*, to dance; **te·rping** *vbl. sb.*

1937 *Amer. Speech* XII. 317/2 *Terp*, a dancer. 1937 *Variety* 10 Nov. 58/3 Philly Orch on Thursday (11) night will preem composition of 23-year-old Omaha college soph. Cleffer, titled 'Mystic Pool'... 'Pool' originally composed for his terp. orch. 1942 BERREY & VAN DEN BARK *Amer. Thes. Slang* § 594/1 *Stage dancing*,..terping, terp stuff, toesmithing. 1945 MENCKEN *Amer. Lang. Suppl.* I. v. 338 It [sc. *Variety*] makes verbs of nouns, *e.g.* ..*to preem* and *to terp*. 1951 GREEN & LAURIE *Show Biz* 571/2 *Terps*, dancing. *Ibid.* 572/1 *Terp team*, ballroom dance team. 1952 GRANVILLE *Dict. Theatr. Terms* 183 *Terp*, a stage dancer. 1974 *Spartanburg* (S. Carolina) *Herald* 18 Apr. A6/1 Donna McKechnie is the best dancer in the musical comedy theater (one dance critic tripped over his typewriter when he suggested Donna can't terp).

terpane (tǝ̄·ɹpēᵻn). *Chem.* [ad. G. *terpan* (A. Baeyer 1894, in *Ber. d. Deut. Chem. Ges.* XXVII. 436), f. *terpen* TERPENE: see -ANE.] Any of a class of saturated hydrocarbons related to the terpenes and possessing their carbon skeleton; *spec.* 4-methylprop-2-ylcyclohexane, $CH_3C_6H_{10}CH(CH_3)_2$, a monocyclic liquid.

1902 F. J. POND tr. *Heusler's Chem. Terpenes* 23 Baeyer has..advanced the proposition to designate hexahydrocymene as *terpane*. 1965 *Proc. Nat. Acad. Sci.* LIV. 1412 Peaks [in the mass spectrum] at m/e 191, 203, and 231 probably arise from small amounts of terpane impurities. 1981 *Jrnl. Chromatogr. Sci.* XIX. 156/1 Terpanes and steranes are well-known biological marker hydrocarbons.

terpene. Add: Hence **te·rpeneless** *a.*, rendered free of terpenes.

1921 *Jrnl. R. Naval Med. Service* VII. 89 Terpeneless and sesquiterpeneless oils. 1972 *Materials & Technol.* V. i. 17 'Terpeneless bergamot oil' is used in high-class perfumes.

terpenoid (tǝ̄·ɹpĕnoid), *sb.* (and *a.*) *Chem.* [a. G. *terpenoid* (Vogel & Stohl 1933, in *Ber. d. Deut. Chem. Ges.* LXVI. B. 1066): see TERPENE and -OID.] A terpene in the broadest sense: used when *terpene* itself is restricted to compounds with the formula $C_{10}H_{16}$. Also *attrib.* and as *adj.*

1933 *Chem. Abstr.* XXVII. 4807 The name *terpenoids* is suggested for the resin alcs., resin acids, sterols and xanthophylls, including carotene. 1956 I. L. FINAR *Organic Chem.* II. viii. 250 There is..a tendency to call the whole group terpenoids instead of terpenes, and to restrict the name terpene to the compounds $C_{10}H_{16}$. 1972 *Science* 5 May 512/2 Some species of caterpillars..possess oxidases in their gut that are capable of metabolizing the repellent terpenoids such as pyrethrin from chrysanthemums. 1975 *Nature* 31 Jan. 365/2 Both substances are terpenoid, derived from the essential oils absinthol and cannabinol.

terr (tǝ̄ɹ). *Rhodesian slang.* [abbrev. of TERRORIST.] In Rhodesia (now Zimbabwe) prior to independence, a guerrilla fighting to overthrow the White minority government. Usu. in *pl.*

1976 *Verbatim* Sept. 14/2 Rhodesians, according to a recent news dispatch, now have one [*sc.* a 'clipping'] they could do without, namely *the Terrs* (general for 'the terrorists'). 1976 *Listener* 23/30 Dec. 835/3 It may help ..to know the kind of slang that they [*sc.* Rhodesian Whites] are going to be using. 'Terrs' is short for 'terrorists'... To 'rev' is to shoot somebody. 1980 *Times* 18 Jan. 19/2 Infiltration over the Zambesi River by 'terrs'—or terrorists/freedom fighters, depending on your politics.

terra. Add: **terra alba,** (*b*) pulverized gypsum used industrially. Also used, with qualifying adjectives, in some general expressions, as **terra cognita** [as opp. to TERRA INCOGNITA], *fig.*, familiar territory; **terra ignota** = TERRA INCOGNITA; **terra irredenta** = *IRREDENTA.

1905 H. RIES *Econ. Geol.* vii. 143 Gypsum is also used.. under the name of 'Terra Alba', as an adulterant of foods and medicinal preparations. 1917 J. SHELTON in G. Martin *Industr. & Manuf. Chem.* I. xxxi. 545 Under trade names, such as 'terra alba'.., a certain amount of the purer forms of gypsum is employed by paint manufacturers for admixture with pigments. 1947 J. C. RICH *Materials & Methods of Sculpture* iv. 65 Terra Alba, a finely pulverized raw gypsum powder may be used [as an accelerator in casting plaster of Paris.] 1963 R. R. A. HIGHAM *Handbk. Papermaking* iv. 93 The form containing no water of crystallization..is prepared by calcining the natural gypsum ($CaSO_4,2H_2O$), often referred to as terra alba. 1962 E. SNOW *Other Side of River* (1963) xxxiii. 253 My last remark had put them back on terra cognita and it would have been an appropriate moment to leave. 1975 *Publishers Weekly* 21 July 66/3 But it's all the same old *terra cognita* to those who have read Gerold Frank, Anne Edwards et al. *a* 1925 LD. CURZON *Leaves from Viceroy's Notebook* (1926) vi. 231 The whole country of Annam seems to be almost a *terra ignota* to our countrymen. 1977 *Times* 12 Oct. 9/5 These records [at Greenwich Museum] are largely terra ignota to the outside world. 1934 WEBSTER, Terra irredenta. 1936 *International Affairs* XV. 39 *A terra irredenta*—that is to say, a territory, at present under foreign sovereignty, which, in the claimants' views, ought to be transferred to the sovereignty of his own State and to be incorporated (or re-incorporated) into his own national domain. 1965 *Listener* 20 May 735/1 French Somaliland,..Ethiopian Somaliland, and..a strip of northern Kenya. These represent the Somalis' '*terra irredenta*'.

2. *Science Fiction.* (With capital initial.) The Earth.

1947 E. F. RUSSELL in Aldiss & Harrison *Decade 1940s* (1975) 157 This world..was ten times the size of Terra. But his weight didn't seem abnormal. 1952 P. J. FARMER in —— *Decade 1950s* (1976) 215 It follows the moon around Terra..a much smaller and unseen satellite.

terra a terra. 2. Transfer entry to *TERRE-À-TERRE, q.v.

terrace, *sb.* Add: **1. d.** *Archæol.* = *cultivation terrace* s.v. *CULTIVATION 1 a.

1796, etc. [see *LYNCHET b]. *a* 1964 G. UNDERWOOD *Pattern of Past* (1968) viii. 82 *Terraces*..are found on steep hillsides, and..mark places where a number of geodetic lines run parallel, with wide spaces between them... It seems reasonable to assume that they formed processional ways.

e. At an Association Football or other sports ground, a row for the accommodation of standing spectators; usually in *pl.* Cf. *TERRACING *vbl. sb.* 1 b. Also *attrib.* in *sing.*

1950 *Sport* 7–11 Apr. 2/1 The terrace regulars are..the backbone of many present day clubs. 1959 *Listener* 19 Feb. 332/2 As I saw them from the terraces, I learnt that on top of everything else..they often had to play against their own supporters. 1971 [see *FRIENDLY *a.* 3 c]. 1977 *Times* 6 May 2/5 [The] Minister of State for Sport..imposed his ban on the sale of terrace tickets to Chelsea supporters at away games. 1980 *Observer* 7 Sept. 11/6 It was more like a football terrace than Lord's.

3. Revived in the sense 'balcony in a theatre' or other auditorium (see quot. 1961).

1961 *Ann. Rep. Lincoln Center for Performing Arts* (N.Y.) 10/1 The auditorium's shallow terraces—only six rows deep at the back and two to four seats wide at the sides—surround the orchestra level and flow towards the orchestra platform [in the Philharmonic Hall at the Lincoln Center, N.Y.]. 1963 *Guardian* 5 Mar. 7/3 The music sounds better in the top terrace..than in the lower terraces and orchestra.

7. *terrace cottage, -garden* (later and *attrib.*

examples), *-parapet*, *-walk* (later examples); *terrace-like* adj.; **terrace house,** one of a row of usu. similar houses joined by party-walls.

1973 A. HUNTER *Gently French* iv. 34 Adjacent to the Barge-House were three sad terrace cottages. 1978 *Spectator* 13 May 12/2 Neat little freshly painted two-storey terrace cottages with gardens nearby—already a century old. 1861 QUEEN VICTORIA *Jrnl.* 20 Sept. (1980) 99 The Castle of Auch Mill, which..has traces of a terrace garden remaining. 1931 A. U. DILLEY *Oriental Rugs & Carpets* iii. 58 Many 'tree' and 'landscape' rugs are terrace-garden rugs. 1817 JANE AUSTEN *Sanditon* (1954) x. 413 They were in one of the Terrace Houses. 1922 JOYCE *Ulysses* 697 A terracehouse or semidetached villa. 1958 *Listener* 5 June 947/1 Look at the new hospital-block in Guildford Street, Bloomsbury, and see how well it goes with the old terrace-houses. 1972 *Guardian* 6 Nov. 15/3 The rank and file knew that the real Ulster crisis was happening inside the terrace houses. 1880 'MARK TWAIN' *Tramp Abr.* xxxv. 397 This pile of stone..comes down out of the clouds in a succession of rounded, colossal, terrace-like projections. 1963 *Times* 18 May 5/2 One of the finest things was his terrace-like build-up of the beginning of the allegretto from the Seventh Symphony. 1974 C. TAYLOR *Fieldwork in Medieval Archaeol.* iii. 28 These terrace-like features [*sc.* strip lynchets] on hillsides are the remains of medieval strip cultivation. 1854 DICKENS *Hard Times* II. vii. 207 Tom sat down on a terrace-parapet, plucking buds. 1707 Terrace walk [see *side-wing* s.v. *SIDE *sb.*¹ 27]. 1775 J. WOODFORDE *Diary* 14 Apr. (1924) I. 151 Round it is a fine Terrass Walk which commands the whole City. 1858 M. TUCKETT *Diary* 16 Sept. (*c* 1975) 5 A broad terrace walk goes along the front of the new part of the house.

b. Used to designate a style of women's and girls' clothing suitable for wearing at an informal party.

1963 *Guardian* 2 Feb. 5/2 A series of terrace (ex-casino) dresses. 1965 *Sunday Mail* (Brisbane) 24 Oct. 24 (*caption*) Terrace skirt combines bright red, white, and blue in a lively outfit for girls who want to look graceful at casual parties. 1971 *Rand Daily Mail* (Johannesburg) 4 Dec. 3/8 (Advt.), A fabulous collection of the popular terrace gowns. 1972 *Times* 19 Dec. 11/3, I do not like the idea of little girls in low-cut evening dresses..but I do think that the pale blue terrace two-piece..is a delight.

terraced, *ppl. a.* Add: (Later examples.) Of a house: cf. *terrace house* s.v. *TERRACE *sb.* 7.

1958 *Daily Express* 3 Apr. 7/7 Their tiny terraced home in the back streets of Horden, Durham. 1976 *Milton Keynes Express* 28 May 1/3 She has no intention of leaving her terraced home. 1979 *Guardian* 19 June 4/2 The old-fashioned terraced house..is on the way back.

terrace·ntric, *a.* *nonce-wd.* [f. TERRA + CENTRIC *a.*] Having, or taking, the earth as centre; = GEOCENTRIC *a.* 2.

1932 G. B. SHAW *Adventures Black Girl* 66 As to Bible science..its astronomy is terracentric.

terracette (teräse·t). *Geomorphol.* [f. TERRACE *sb.* + -ETTE.] A small (natural) terrace; *spec.* one of a parallel series on a steep hillside.

1922 H. ODUM in *Dan. Geol. Undersøg.* IV. 1. 27 The phenomenon..occurs on hill-slopes with a steep inclination, and consists in the surface being covered with a number of small terracettes or low ledges, all running parallel, horizontal and at about an equal distance from each other. 1959 G. H. DURY *Face of Earth* ii. 13 On very many steep slopes a kind of ribbed pattern appears on the surface of the creeping waste, with little steps a foot or two in height running horizontally. These steps are called terracettes. An alternative name is sheep-tracks, but this title is grossly misleading. Terracettes can be found where no sheep have ever been. 1971 I. G. GASS et al. *Understanding Earth* xiii. 182/2 Tiny terracettes can be cut into the sides of the ripples [on a beach]. 1974 C. TAYLOR *Fieldwork in Medieval Archaeol.* iv. 85 Terracettes are another form of natural feature likely to be misinterpreted.

terracing, *vbl. sb.* Add: **1. b.** *spec.* = *TERRACE *sb.* 1 e.

1902 [in Dict.]. 1942 *R.A.F. Jrnl.* 30 May 33 It was like the empty terracings of Wembley or Hampden Park after an international. 1976 J. SNOW *Cricket Rebel* 68 The occasional fires on the terracing which are part and parcel of cricket matches in Pakistan. 1978 P. MARSH et al. *Rules of Disorder* iii. 58 Young supporters at every football league ground..have defined sections of the terracing as their own territory.

Terra da (de, di) Sienna, varr. TERRA SIENNA in Dict. and Suppl.

terraglia (terā·lʸă). *Ceramics.* [a. It., = earthenware, china, f. L. *terra* earth.] An (Italian) cream-coloured earthenware, esp. that manufactured from 1728 at Nove, near Bassano, Italy, by G. B. Antonibon and his descendants.

1850 J. MARRYAT *Coll. Hist. Pott. & Porc.* 290 *Terraglia*, the Italian term for fine pottery of hard paste. 1870 W. CHAFFERS *Marks & Monogr. Pott. & Porc.* (ed. 3) 118 Pasqual [Antonibon]..carried on..the manufacture of *maiolíche fine* or fayence, and *terraglia* or terre de pipe. 1877 C. SCHREIBER *Jrnl.* 20 June (1911) II. 28 A very fine Terraglia dish (red ground with figures in white..)..of old Paduan manufacture. 1960 R. G. HAGGAR *Conc. Encycl. Continental Pott. & Porc.* 471/1 *Terraglia*, cream-coloured earthenware.

terrain, *sb.* (*a.*) Add: **A.** *sb.* **2. b.** *fig.*

1860 PRINCESS ROYAL *Let.* 3 Dec. in R. Fulford *Dearest Child* (1964) 288 She will..not [have] such a difficult 'terrain' to work upon as God knows I have here. 1979 *Amer. Poetry Rev.* Mar./Apr. 19/4 He found authors in that terrain of brotherhood and contact the reader calls for and which they, the writers I've named and so many others, gave and go on giving by the courses and the conduct which touch their responsibility as Latin Americans.

4. *attrib.* and *Comb.*, as **terrain-following radar** *Aeronaut.*, a radar system which enables an aircraft or missile to fly at high speed close to the ground by automatically adjusting its altitude in relation to the surface over which it is flying; so **terrain follower.**

1961 *Daily Tel.* 4 Oct. 17/1 A device known as the 'terrain follower', which keeps the plane automatically at a predetermined height above the ground. 1970 *Times* 4 Sept. (Aviation Suppl.) p. ix/8 Advanced avionics technology for the European multirole combat aircraft will be shown, with strike and terrain-following radar. 1980 *Guardian Weekly* 13 July 1/2 Terrain-following radar for low-level penetration of enemy defences.

‖ **terrain vague** (terǣṅ vāg). [Fr. colloq., lit. 'waste ground'.] Wasteland, no man's land (*transf.* and *fig.*).

1920 R. FRY *Let.* 12 Oct. (1972) II. 493 I'm..painting up on Montmartre in a *terrain vague* with the hideous white tower of the Sacré Coeur dominating a jumble of modern houses. 1934 C. LAMBERT *Music Ho!* iii. 207 The same rapprochement between highbrow and lowbrow—both meeting in an emotional *terrain vague*—can be seen in literature. 1957 L. DURRELL *Bitter Lemons* 48 His desk was in the far corner against the wall, and to reach it one traversed a *terrain vague* which resembled the basement of Maple's, so crowded was it with armchairs, desks, prams..and all the impedimenta of gracious living. 1984 *Sunday Times* 26 Aug. 38/8 Alastair Reid occupies a *terrain vague* between reportage and *belles lettres*.

terral. Add to def.: Off the coast of Spain or South America. (Earlier and later examples.)

1863 H. W. BATES *Naturalist on River Amazons* II. ii. 144 Canoes, in descending, only travel at night, when the terral, or light land-breeze, blows off the eastern shore. 1928 *Bryologist* XXXI. 125 The weather was very cold. We had been feeling the intense icy blasts of the *Terral* wind from the North [of Spain]. 1953 W. G. KENDREW *Climates of Continents* (ed. 4) xxxix. 481 The sea-breeze (virazon) and land-breeze (terral) are regular and prominent; the sea-breeze is often so strong on summer afternoons at Valparaiso..that boat-work is stopped. 1961 L. D. STAMP *Gloss. Geogr. Terms* 448/2 *Terral*, the land breeze along the coasts of western Peru.

terramare. (Later examples of *β*: *terremare* is sometimes used as an invariable form.)

1928 C. DAWSON *Age of Gods* xiv. 330 With the Bronze Age there appears in North Italy the new and highly distinctive type of culture known as the Terremare. It receives its name from the peculiar pile settlements which form its characteristic feature. Unlike the lake villages the Terremare were built on dry land and surrounded by an artificial moat and rampart. 1939 *Antiquity* XIII. 490 A terramara is defined by Säflund as 'a station containing the typical bronze-age culture of central and western Emilia'. 1961 L. MUMFORD *City in Hist.* viii. 206 It is doubtful if there is any direct connexion between the *terremare* settlements and the Roman towns.

Terramycin (terămai·sin). *Pharm.* Also **terramycin.** [f. L. *terra* earth + *-MYCIN.] A proprietary name for *OXYTETRACYCLINE.

1950 A. C. FINLAY et al. in *Science* 27 Jan. 85/1 From both cultures of this organism, a crystalline antibiotic was isolated; the name Terramycin has been assigned to this compound. 1953 *Official Gaz.* (U.S. Patent Office) 5 May 22/1 Chas. Pfizer & Co., Inc....*Terramycin* for antibiotic preparation containing oxytetracycline or a derivative thereof. 1954 *Trade Marks Jrnl.* 11 May 479/1 *Terramycin*... Oxytetracycline and derivatives thereof, and substances containing oxytetracycline or derivatives thereof, all prepared for use as antibiotics. Chas. Pfizer & Co., Inc. 1956 A. HUXLEY *Let.* 25 Dec. (1969) 814, I was put on to acromycin, after terramycin had failed to do much good. 1960 *Spectator* 2 Sept. 353 As compensation for the loss of his farmyard freedom he gets a dose of aureomycin, terramycin or penicillin. 1976 *Daily Times* (Lagos) 5 Aug. 14/1 (Advt.), Terramycin eye ointment.

Terran (te·răn), *a.* and *sb.* *Science Fiction.* [f. L. *terra* earth + -AN.] **A.** *adj.* Of or pertaining to the planet Earth or its inhabitants. **B.** *sb.* An inhabitant of the planet Earth.

[1881 W. D. HAY *Three Hundred Years Hence* xi. 267 I am speaking of the Terrane Exodus and the Cities of the Sea.] 1953 *Cosmos Sci. Fiction & Fantasy Mag.* Nov. 83 A chant rose to assail his ears, and the steady beat of a drum thundered in the Terran night. *Ibid.* 78 They set about the tremendous task of educating Martians and Terrans. 1960 K. AMIS *New Maps of Hell* (1961) ii. 63 Any Martian survey team would be well advised to read a sample of the stuff before reporting on Terran civilisation. 1962 E. F. RUSSELL *Great Explosion* i. 13 We shall face them and defeat them as Terrans always have done. 1969 *New Scientist* 23 Jan. 191/3 Like our planet, we Terrans tend to be fat and slow or thin and quick.

terranean, *a.* Add: (Later example.)

1939 JOYCE *Finnegans Wake* 120 Of an early muddy terranean origin.

B. *sb.* Also **terranian.** An inhabitant of the planet Earth. *Science Fiction. rare.*

1913 'SEPHARIAL' *Kabala of Numbers* II. xi. 175 Red is an irritant to all but Martians, and to them green probably means the same thing as does red to us Terranians. 1965 *Punch* 27 Jan. 141/3 When one of the anemone-harnessers..tells a couple of captive 'terraneans', i.e., grown-ups, 'You must be destroyed!' I could sense a tremble of joy in children everywhere.

terrapin. Add: **c.** Terrapin State, a colloquial name for the State of Maryland.

1937 G. E. SHANKLE *Amer. Nicknames* 338/1 Maryland is nicknamed *The Terrapin State* because of the extensive diamondback terrapin farms. 1949 B. A. BOTKIN *Treas. S. Folklore* I. ii. 35 Maryland has had half a dozen or more nicknames since colonial times, but only *Old Line State* and *Terrapin State* have any remaining vitality today.

Terrapin² (te·răpin). Also **terrapin.** The proprietary name of a make of prefabricated building, usu. having one storey and designed for temporary use.

1949 *Trade Marks Jrnl.* 21 Sept. 841/2 *Terrapin*... Portable buildings. Harry Collett Bolt, 21, Kings Gardens, West End Lane, Hampstead, London, N.W... Manufacturer and Merchant. 1962 *Ibid.* 25 July 958/2 Terrapin Minihouses... All goods included in Class 19. Terrapin Limited, Haddon House, 2–4, Fitzroy Street, London, W.1; Manufacturers. 1972 *Guardian* 5 Dec. 17/6 The main concerns of [school] governors tend to be.. asking the Works Department for an extra radiator in an exposed 'terrapin'. 1976 T. SHARPE *Wilt* vi. 56 He looked round the car park at the terrapin huts and the main building.

terrarium. Add: Pl. now usu. **terrariums.**

1. (Later examples.)

1952 M. K. WILSON tr. *Lorenz's King Solomon's Ring* vii. 67 On the table..stands the nucleus of my golden hamster stud, a simple little terrarium. 1976 N. COLEMAN *Shell Collecting in Australia* ix. 149 The larger the terrarium and the smaller the number of snails, the less work will be involved in keeping it sanitary. 1977 *Sci. Amer.* Dec. 36/1 Spiders and scorpions in neat terrariums.

2. A sealed transparent globe or similar container in which plants are grown, usu. for decoration. Cf. *WARDIAN CASE. orig. *U.S.*

1931 *St. Nicholas* Oct. 846/1 Have you ever tried making a terrarium?.. It means a little garden under glass—not a conservatory or a cold frame, but a real little landscape. 1942 *Amer. Speech* XVII. 284/1 He also saw on his rambles a *Terrarium*—a covered glass globe or fish-tank containing flowers and plants to be grown indoors during the winter. 1974 *Evening Herald* (Rock Hill, S. Carolina) 19 Apr. 6/5 (Advt.), Globe shaped terrarium kit... A unique way to display foliage. 1979 *Sunday Mail* (Brisbane) 6 May 5/1 (Advt.), Terrariums add life to any room. In long lasting plastic; each containing an African violet plus three other selected plants. 1982 *N.Y. Times* 11 Apr. II. 33/4 Partridgeberry..is often used in terrariums for its evergreen foliage and red berries.

terra rosa (te·ră rōᵘ·ză). *Painting.* [a. It., lit. = rose-coloured earth.] A light red colour of paint similar to Venetian red.

1897 *Sears, Roebuck Catal.* 360/3 Colors for Artists.. Gold Ochre,..Terra Rosa, Terre Verte. 1935 *Winsor & Newton's Sales Man.* 56 Colours which may be regarded.. inflexible under all conditions of Oil Painting..Terra Rosa. 1973 F. TAUBES *Painter's Dict. Materials & Methods* 224 *Terra rosa* is a very light Iron Oxide Red of relatively slight Tinting Strength. For all practical purposes, this color is interchangeable with *Terra di Pozzuoli*.

terra rossa (te·ră rǫ·să). [a. It., = red earth.] A reddish soil occurring on limestone in Mediterranean climates (see also quot. 1938).

1882 A. GEIKIE *Text-bk. Geol.* 458 Fuchs believes that the 'terra rossa' is only present in dry climates where the amount of humus is small. 1938 *U.S. Dept. Agric. Yearbk.* 991 Many writers have preferred to limit Terra Rossa to soils developed on limestones, while some would have it include any red soil in a Mediterranean climate... At present its only distinction lies in its color. 1956 *Proc. Prehist. Soc.* XXI. 53 Most brown-earths are characteristic of deciduous forests, chernozems of steppe, and terra rossas of Mediterranean forests on limestone. 1975 *Nature* 14 Aug. 566/1 Further evidence for the role of climate is provided by the fact that the limestone soils from Apulia, the Orbetello region and the southern Apennines, which are mainly of the red Mediterranean type (that is, terra rossa) exhibit higher percentage conversions than those from the Central Apennines which are mainly rendzinas.

‖ **terra roxa** (te·ră rǫ·ʃă). [Pg., = reddish-purple soil.] A deep, humus-rich soil of a dark reddish-purple colour on the Paraná Plateau in southern Brazil.

1870 C. F. HARTT *Thayer Exped: Sci. Results Journey Brazil: Geol. & Physical Geogr.* xvii. 514 The terra roxa of Campinas Paulo is..the continuation of the drift-paste of the higher lands and seaward slope of the serra. 1977 *Econ. Geogr.* LIII. 78/2 Areas with larger concentrations of *terra roxas*, the soil of exceptional fertility for coffee, are usually associated with higher levels of production.

Terra Sienna. (Earlier and later examples of *Terra de Sienna* and var.)

1794 A. THOMAS *Newfoundland Jrnl.* (1968) ii. 29 The shores of the Scilly Islands abound with rare and very curious Shells. Some are highly polish'd, of the colour of Terra de Sina [*sic*].., others are striped. 1869 *Bradshaw's Railway Manual* XXI. 460/3 (Advt.), Browns. Vandyke Brown..Terra de Sienna. 1885 *Encycl. Brit.* XIX. 87/1 The well-known ochre Terra da Sienna which in its raw state is a dull-coloured ochre, becomes when burnt a fine warm mahogany brown hue highly valued for artistic purposes.

terra sigillata. Add: **3.** *Archæol.* [Cf. W. Dorow *Opferstätte und Grabhügel der Germanen und Römer am Rhein* (1821) II. 32, etc.] A type of fine Roman pottery made from the first century B.C. to the third century A.D. in Gaul (also Italy and Germany), usu. red in colour and sometimes decorated with stamped figures or patterns. Not the preferred term in English: see *ARRETINE *a.*, SAMIAN *a.* and *sb.* in Dict. and Suppl.

H. Brunsting, in *Overdruk uit Westerheem* (1972) XXI. 252–68, provides a detailed glossary of references to the ware in English and Continental sources. Quot. 1845 below is often mistaken for the first English use of the term: it relates in fact to the medicinal tablet or pastille (sense 1).

[1845 E. B. PRICE in *Gentl. Mag.* Feb. 142/1 It is termed 'Terra Samia sigillata', of circular form, about ⅞ of an inch diameter... It is of a pale dull red colour, and has apparently been made into a paste and dried in a mould.] 1903 *Amer. Jrnl. Archæol.* VII. 485 J. Déchelette publishes the results of a study of the *terra sigillata* ..in the territory of the Ruteni near Millau. 1920 OSWALD & PRYCE *Introd. Stud. Terra Sigillata* i. 1 Next to datable inscriptions, there is, perhaps, no relic of Roman occupation which yields such valuable chronological evidence as *Terra Sigillata*. 1936 *Oxoniensia* I. 50 In and immediately below this layer much pottery was found including Terra Sigillata of the second century. 1959 *Chambers's Encycl.* XI. 761/2 Italian *terra sigillata*, or Arretine ware, was chiefly manufactured at Arretium. 1978 M. GILBERT *Empty House* xi. 94 The stuff we call Samian, or Terra Sigillata, which was manufactured in Central Gaul.

‖ **terrasse** (teras). [Fr.: see TERRACE *sb.*] In France, etc.: a flat, paved area outside a building, esp. a café, where people sit to take refreshments.

1918 A. BENNETT *Roll-Call* I. ix. 197 They sat down on the *terrasse* of a large café near the Place des Ternes. 1930 —— *Imperial Palace* II. lvii. 433 Evelyn, in his big overcoat, was sitting on the *terrasse* of a large café. 1967 R. PETRIE *Foreign Bodies* i. 8 On the *terrasse* Nassim Yussif was turning between his hands a small white envelope. 1979 P. WAY *Sunrise* II. xii. 127 On the hotel *terrasse*, Olsen tipped his chair back and regarded Marriott.

‖ **terra verde,** etc., varr. TERRE VERTE in Dict. and Suppl.

terrazzo (terā·tso). [a. It., = terrace, balcony.] A flooring material made of chips of marble or granite set in concrete and polished to give a smooth surface.

1902 *Encycl. Brit.* XXIX. 340/1 Wood has been superseded by terrazzo... Cracks may be seen in terrazzo floors introduced into several of the recently erected modern hospitals. 1925 *Glasgow Herald* 27 Feb. 7 Terazzo-faced concrete blocks. *Ibid.*, The outer walls being of the terrazzo block. 1958 *House & Garden* Mar. 69/1 The Esse-Dura stove sits on a terrazzo hearth in the living room 1973 [see *RAILED *ppl. a.* 1]. 1983 *Listener* 19 May 16/1 Italians were brought from Italy to lay terrazzo in a new building in West Africa.

‖ **terre-à-terre** (tęr a tęr), *adj.* (and *adv.*) *phr.* The usual form of TERRA A TERRA 2. Substitute for def.: In *Ballet*, applied to a step or manner of dancing in which the feet remain on or close to the ground. In *transf.* use: down-to-earth, realistic, matter-of-fact; pedestrian, unimaginative. Also as *adv. phr.* (Further examples.)

1830 [see *ELEVATION 1 e]. 1884 W. JAMES *Ess. Radical Empiricism* (1912) xii. 266 No seeker of truth can fail to rejoice at the terre-à-terre sort of discussion of the issues between Empiricism and Transcendentalism..that seems to have begun in *Mind*. 1907 —— *Pragmatism* vii. 268 Shutting out all wider metaphysical views and condemning us to the most *terre-à-terre* naturalism. 1915 M. E. PERUGINI *Art of Ballet* iii. 33 To dance, '*terre-à-terre*', that is, with the feet, or one foot at least, on or close to the ground. 1920 *Athenæum* 5 Nov. 617/2 The author of 'Les Baisers' always elegantly *terre-à-terre*, formulates his more concrete desires. 1930 *Time & Tide* 18 Apr. 500/2 He was too frank not to admit that his friend and chief was, intellectually, very *terre-à-terre*. 1941 *Burlington Mag.* Aug. 37/2 The romanticism of this portrait is not the sophisticated, rarefied one..: it is definitely more terre-à-terre. 1961 *Times* 27 May 6/2 He regrets that the Bolshoi ballet seemed to pay so little attention to *terre à terre* dancing. 1981 *Listener* 26 Feb. 284/3 She..was 'a credible girl who suffered from menstrual cramps'... You can't get more *terre à terre* than that. 1983 M. KEYNES *Lydia Lopokova* 59 During the next year, 1912, Lydia.. danced an extremely difficult terre-à-terre 'toe dance'.

‖ **terre cuite** (tęr kwit). [Fr., lit. 'baked (cooked) earth'.] = TERRA-COTTA 1.

1869 C. SCHREIBER *Jrnl.* 14 Oct. (1911) I. 53 We were charmed with four ancient costume pictures of fêtes.. and five terre cuites portraits by J. R. Nini. 1870 *Ibid.* 26 Feb. 76 One of the dealers..took us to her private house to show us some terre cuites. 1882 'OUIDA' *Bimbi* 55 The *terres cuites* of Blasius date from 1560. 1926 R. FIRBANK *Concerning Eccentricities of Cardinal Pirelli* v. 57 A voluptuous small *terre cuite*, depicting a pair of hermaphrodites amusing themselves. 1951 N. MITFORD *Blessing* I. xi. 113 Once you fall into Louis XV you are immediately in the domain of restored terre cuite and broken china.

terrella. Restrict † *Obs.* to sense 2 in Dict. and add: **1.** (Later example.) *rare.*

1959 *Daily Tel.* 23 Feb. 11/8 Col. Steinkamp used the word 'terrella'—a little world or earth—in the context of space flight.

terrenity. For † *Obs.* read *rare* and add later example.

1973 T. PYNCHON *Gravity's Rainbow* IV. 733 Trees creak in sorrow for the engineered wound through their terrain, their terrenity or earthhood.

‖ **terre pisée** (tęr pize). [Fr., lit. 'beaten earth'.] = PISÉ a in Dict. and Suppl. Cf. COB *sb.*² a.

1936 V. G. CHILDE *Man makes Himself* vi. 124 Soon houses built of mud or *terre pisée* were being erected. 1949 W. F. ALBRIGHT *Archaeol. of Palestine* v. 86 A concomitant of the introduction of chariotry into warfare was the spread of the art of building great fortifications of beaten earth (*terre pisée*), usually rectangular in plan. 1972 Y. YADIN *Hazor* 203 As for the claim that the earthen ramparts and *terre pisée* defences were first built in the MB IIA, only little has to be added here.

terrestrial, *a.* and *sb.* Add: **A.** *adj.* **2. e.** *Astr.* Designating planets which are similar in size or composition to the Earth.

1888 C. A. YOUNG *Textbk. Gen. Astron.* xiv. 322 The terrestrial planets are Mercury, Venus, the earth, and Mars. They are bodies of the same order of magnitude.., not very different in density.., and are probably roughly alike in physical constitution. 1926 E. A. FATH *Elem. Astron.* xi. 135 The eight planets fall into two groups, the first usually being termed the terrestrial planets. 1973 *Nature* 17 Aug. 424/2 The terrestrial planets..are solid bodies. 1980 *Sci. Amer.* Jan. 68/2 Io..and Callisto belong to the family of objects designated terrestrial.

terre-verte. (Later examples of variant forms.)

1897 *Sears, Roebuck Catal.* 361/1 Colors for Artists... King's Yellow, Lamp Black, Terra Verta. 1907 *Yesterday's Shopping* (1969) 471/2 Water Colours... Sunny green..Terra vert..Turner brown. 1919 R. FRY *Let.* 27 Oct. (1972) II. 463, I use..burnt umber, indian red and terra verte. Terra verte pure is too bright for the sky. 1944 *Burlington Mag.* Apr. 92/2 According to the literary tradition this fresco was painted in *terra verde*. 1973 *Times Lit. Suppl.* 8 June 634/5 Florence,..where Uccello's fresco of him, in terra-verde, adorns a wall of the Duomo.

‖ **terribilità** (teribilita·). Also **terribiltà.** [It.] **a.** In an artist or work of art: awesomeness of conception and execution, orig. as a quality attributed to Michelangelo by his contemporaries (e.g. by S. del Piombo in a letter of 9 Nov. 1520: see G. Milanesi *Les Correspondants de Michel-Ange* (1890) 24).

1883 *Encycl. Brit.* XVI. 230/2 In it the qualities afterwards proverbially associated with Michelangelo—his *furia*, his *terribilità*, the tempest and hurricane of the spirit which accompanied his unequalled technical mastery and knowledge—first found expression. 1923 A. HUXLEY *Antic Hay* xviii. 253 All this *folie de grandeur*, all this hankering after *terribiltà*..it's led so many people astray. 1948 *Penguin New Writing* XXXIV. 47 The Edwardian Mannerist arcade..with its stork and swiss-roll terribiltà. 1961 *Daily Tel.* 5 Dec. 13/4 In these [*sc.* Piranesi's engravings of prisons] he conveys a degree of terribilità no one else has conferred upon architecture. 1970 *Oxf. Compan. Art* 720/1 That emotional intensity which Michelangelo's contemporaries recognized as his *terribilità* and which earned him the veneration of his juniors.

b. In general use: terrifying or awesome quality.

1957 J. RAYMOND in *New Statesman* 28 Sept. 386/2 Half the horror of Rolfe's life—and its accompanying virtue, his pathetic and gallant attempt to live up to a self-taught conception of honour, *terribilità*, and esteem.. —sprang from the fact that he was a *déraciné* and a homosexual. 1959 *Times* 1 Apr. 11/4 The terribilità has long been drained from air travel. 1975 *New Yorker* 12 May 42/2 Fathers have voices, and each voice has a *terribilità* of its own.

terribility. (Later example.)

1922 JOYCE *Ulysses* 686 The terribility of her.. propinquity.

terrible, *a.* Add: **2. a.** (Later examples.)

1884 QUEEN VICTORIA *Let.* 27 Feb. in R. Fulford *Beloved Mama* (1981) 161 The amount of writing is as they say at Balmoral 'just terrible'. 1924 'K. MANSFIELD' in *Collier's* 5 Jan. 37/2 She leaned against him and looked into his eyes. 'Hasn't it been terrible, all to-day?' said Edna. 'I knew what was the matter.' 1939 G. B. SHAW *In Good King Charles's Golden Days* I. 7 Just as I have my terrible weakness for figures Mr. Rowley has a very similar weakness for women. 1945 [see *SHOW *v.* 25 d]. 1965 E. J. HOWARD *After Julius* vii. 100 Her mother had made his life so terrible—not worth living.

b. Applied to a person who behaves in a shocking or outrageous manner; *terrible child* or *infant* = *ENFANT TERRIBLE; *terrible twins*, applied *joc.* to a pair of associates whose behaviour is troublesome or outrageous.

1859 C. READE *Love me Little, love me Long* I. i. 6 Poor Reginald was not analytical,..like certain pedanticules, who figure in story as children. He was a terrible infant, not a horrible one. 1926 A. HUXLEY *Two or Three Graces* 25 Peddley was not the man to be put out by even the most terrible of terrible infants. 1958 B. BEHAN *Borstal Boy* I. 77 When I was a child, my father used to get the *News of the World*... I used to wonder..why my mother said, half-laughing in spite of herself, that he was a terrible man, because it was banned at home. 1964 in Hamblett & Deverson *Generation X* 47, I used to be terrible. I couldn't stand my girls looking at other men, but I'm different now. 1965 M. SPARK *Mandelbaum Gate* vii. 207 An English female voice..said, 'Oh, look at that terrible man—,' obviously referring to Freddy. 1970 C. HAMPTON *Philanthropist* v. 69 All the men I fall in love with turn out to be such terrible people. 1976 *Evening Advertiser* (Swindon) 31 Dec. 8/1 The 'Terrible Twins' of yesteryear, Mr Jack Jones, general secretary of the Transport and General Workers, and Mr Hugh Scanlon, president of the Engineering Workers, have mellowed. 1978 CADOGAN & CRAIG *Women & Children First* x. 222 Violet Elizabeth, the terrible child of the William books. 1982 *Financial Times* 10 May 10/3 Since the 'terrible twins', as they were dubbed, were both powerful figures, and did not always see eye-to-eye on matters, this rivalry was reflected down the line.

c. As a hyperbolic term of depreciation: of shockingly poor performance or quality; incompetent; defective.

1925 F. SCOTT FITZGERALD *Great Gatsby* ii. 35, I..read a chapter of 'Simon Called Peter'—either it was terrible stuff or the whiskey distorted things, because it didn't make any sense to me. 1946 'E. CRISPIN' *Moving Toyshop* i. 8 Three books..about me (all terrible, but never mind that). 1948 C. FRY *Thor with Angels* 20 As far as he can remember, Though he has a terrible memory for names, His name is Merlin. 1964 in Hamblett & Deverson *Generation X* 88 He was hopeless—you follow me?—terrible in bed. That's why his missus left him..to himself. 1972 [see *LINE *sb.*² 1.e]. 1979 'J. LE CARRÉ' *Smiley's People* (1980) xxi. 257 Grigorieva got herself a driving license two months ago... She's a terrible driver, George. And I mean terrible. 1983 R. RENDELL *Speaker of Mandarin* xvi. 190, I got this Hollywood offer and I went to Hollywood and made that terrible *Mind over Matter*.

3. quasi-*adv.* Now chiefly *dial.* and *U.S.* (Later examples.)

1901 M. FRANKLIN *My Brilliant Career* (1966) xxxi. 197 The old yeos [ewes] looks terrible skinny. 1926 E. O'NEILL *Great God Brown* Prol. 16 My mother used to believe the full of the moon was the time to sow. She was terrible old-fashioned. 1959 L. HUGHES *Sel. Poems* 144 He mistreated her terrible.

terribly, *adv.* Add: **2. a.** (Earlier example as a general intensive.)

1833 DICKENS *Let.* ?Oct. (1965) I. 31, I am terribly out of spirits this morning.

b. Extraordinarily badly; incompetently, feebly. Cf. sense *2 c of the adj. *colloq.*

1930 *Publishers' Weekly* 29 Mar. 1741 (Advt.), About 1,000,000 [bridge players] now play Contract, almost all of them terribly. 1964 J. MITCHELL *White Father* iii. 54 You can sing terribly and get away with it if only you have the right backing.

terricolous, *a.* Add: **2.** = TERRICOLE *a.* 1.

1921 A. L. SMITH *Handbk. Brit. Lichens* 142/1 Terricolous, living on soil. 1959 U. K. DUNCAN *Guide to Study of Lichens* p. xiv, Sand-dunes and heaths are usually rich in terricolous species.

terrier². Add: **3.** (Later examples.)

1915 [see *DO *sb.*¹ 2 b]. 1935 *Economist* 7 Sept. 464/1 This change-over of responsibilities inside the War Office places the 'Terriers' within the same organisation as the 'Regulars'. 1980 *Times* 12 Mar. 12/4 More Terriers. The strength of the Territorial Army on December 31 last year was just under 62,000.

4. terrier-man *Hunting*, a man employed to head the terriers.

1930 C. FREDERICK et al. *Foxhunting* x. 130 If the bag is suspended..by strong pieces of india-rubber..it is more comfortable for the terrier and the terrier man. 1983 *Times* 19 Sept. 4/6 The terrierman..had slashed a fox's paw with a knife before releasing it for young hounds to chase.

terrier (te·riǝɹ), *v. rare.* [f. prec.] *intr.* To burrow in the manner of a terrier; to make *one's way* like a terrier.

1959 R. COLLIER *City that wouldn't Die* ix. 155 Working with hand-shovels and even bare hands, Marotta and his crew began to terrier away. 1965 'J. CHRISTOPHER' *Wrinkle in Skin* iii. 26 He began to terrier his way into the mound.

terrific, *a.* (*sb.*) Add: **A.** *adj.* **2. b.** As an enthusiastic term of commendation: superlatively good, 'marvellous', 'great'. Also *Comb. colloq.*

1930 D. MACKAIL *Young Livingstones* xi. 271 'Thanks awfully,' said Rex. 'That'll be ripping.' 'Fine!' said Derek Yardley. 'Great! Terrific!' 1940 *Chatelaine* Dec. 10/3 But think what it means that they want to come to you. Your bedside manner must be terrific. 1944 *Sun* (Baltimore) 20 Dec. 1/7 Lee McCardell [a reporter] is terrific—first into Metz, first into St. Avold, first into Saarlautern. 1951 'A. GARVE' *Murder in Moscow* iii. 47 Perdita..looked terrific in midnight-blue velvet. 1951 J. D. SALINGER *Catcher in Rye* xii. 103 This..guy had a terrific-looking girl with him. Boy, she was good-looking. 1971 *Farmer & Stockbreeder* 23 Feb. 39/1 He believes the soil is 'terrific' for potatoes and wheat. 1981 *Daily Mail* 14 Oct. 15/1, 'I feel great, really terrific,' said the former Wings guitarist.

terrifyingly, *adv.* (Later examples.)

1908 H. G. WELLS *War in Air* vi. 194 It had crept in upon his mind, chillingly terrifyingly, that these illuminated black masses were great offices afire. 1944 D. WELCH *In Youth is Pleasure* v. 89 He grinned, and then began to make the flesh round his eyes terrifyingly inflamed.

terrifyingness (te·rifǝi,iŋnės). *rare.* [f. TERRIFYING *ppl. a.* + -NESS.] Frightening quality.

1940 *Scrutiny* IX. 294 It is not the terrifyingness of great poetry because it is too exclusively personal.

terrigenous, *a.* Restrict *rare* to senses 1 and 2. **3.** Add *esp.* after 'applied'. (Further examples.)

1884, etc. [see *PELAGIC *a.* c]. 1957 G. E. HUTCHINSON *Treat. Limnol.* I. viii. 550 Most of the calcium in atmospheric precipitation is of terrigenous origin. 1973 *Nature* 27 July 202/1 Deformed early Cainozoic, terrigenous and carbonate sediments form the highest parts of the island.

terrine. Add: **1.** Now, an earthenware or similar fireproof cooking vessel, esp. one in which a terrine (sense *2) is cooked.

1905 A. KENNEY-HERBERT *Common-Sense Cookery* (ed. 2) xxii. 358 Uncooked meats with forcemeat lining.. arranged as within a paste-lined mould, can be baked and finished in a *terrine*. 1914 F. B. JACKS *Cookery for Every Household* 304/2 Compote of Game... Put the joints [of roasted birds] into a fireproof terrine or casserole with the mushrooms and cherries..and leave these until the sauce is prepared. 1960 E. DAVID *French Provincial Cooking* 69 A pâté *en terrine* indicates that the pâté concerned has been cooked and is served in the terrine rather than in a crust. 1979 *Homes & Gardens* June 135/2 The sheets of fat removed from the roast should be thinly sliced and used for lining terrines when making pâté.

2. Delete ‖. In modern use, a kind of pâté cooked in a terrine (sense *1).

1906 A. FILIPPINI *Internat. Cook Bk.* 631 (*heading*) Terrine of duckling. 1914 F. B. JACKS *Cookery for Every Household* 297/2 A terrine like this will keep good for two or three months in cold weather. 1968 D. BRIGGS *Entertaining Single-Handed* iv. 81 *Terrine Andrew*. A terrine is made from he same kind of bits and pieces as a pâté, but..the texture is rougher and it is cooked in a pie-dish or, indeed, a terrine. 1979 REESE & FLINT *Trick 13* 159, I bought an appetizing selection of rough terrine.

territorial, *a.* (*sb.*) Add: **1. d.** *territorial water(s), territorial sea*: the area of sea adjoining the shores of a state and under its jurisdiction (traditionally reckoned as three miles from low water mark, but recently extended by many states). Cf. WATER *sb.* 6 d. Also *territorial limits*, the limits of such water.

1841 J. DODSON in Ld. McNair *Internat. Law Opinions* (1956) I. x. 334 A free permission to Foreign Fishing Vessels so to use the Ports and Territorial waters of our Coasts, would seem likely to lead to constant evasions and violations of the stipulation which prohibits them from fishing within the Limits. 1870 *Act* 33 & 34 *Vict.* c. 90 §2 This Act shall extend to all the dominions of Her Majesty, including the adjacent territorial waters. 1875 [in Dict., sense 1]. 1939 *Daily Tel.* 18 Dec. 1/1 Just beyond the three-mile limit of Uruguayan territorial waters, an unidentified British warship and an Argentine patrol boat had earlier been seen. 1955 *Times* 2 July 6/4 Passage is innocent as long as the vessel does not use the territorial sea for committing acts prejudicial to the security of the coastal State. 1962 *Britannia Bk. of Year* 207/2 Many states had declared, unilaterally, the right to exclusive fishing beyond the territorial limits claimed; *e.g.* Argentina, 3 mi. territorial limits (1869) and 10 mi. exclusive fisheries rights (1907); Thailand, 6 mi. (1958) and 12 mi. (1958). 1976 in R. Crossman *Diaries* II. 71 Since March 1964 pirate radio stations had been transmitting pop music and advertisements, usually from ships anchored outside territorial waters.

e. *Zool.* Of or pertaining to an area defended by an animal or a group of animals against others of the same species; also designating an animal or species that defends its territory in this way; *territorial imperative*, the need to claim and defend a territory.

1920 E. HOWARD *Territory in Bird Life* vi. 228 Do these battles..contribute towards the attainment of the end for which the whole territorial system has been evolved? 1940 *Misc. Publ. Mus. Zool. Univ. Michigan* XLV (*title*) Territorial behavior and populations of some small mammals in southern Michigan. 1961 *Science* 10 Mar. 698/1 The well-defined pattern of year-around territorial behavior of the Uganda kob was discovered in March 1957. 1966 R. ARDREY (*title*) The territorial imperative. *Ibid.* iii. 101 That man is a territorial species has been the conclusion of many a scientist. 1968 K. LORENZ in *Harper's Mag.* May 74 The 'territorial imperative' does much to explain the causes of war, such as the Arab-Israeli dispute, which I consider almost purely territorial. 1971 *Nature* 4 June 295/2 A territorial bull establishes himself as supremely dominant within the confines of his territory. 1980 C. AIRD *Passing Strange* iv. 47 If any one single instinct came to the fore in Superintendent Leeyes it was the territorial imperative. 1981 *Oxf. Compan. Animal Behaviour* 551/1 One benefit of territorial defence is food acquisition.

3. (Earlier example.) Also, of or belonging to one of the 'territories' of Canada (see *TERRITORY¹ 4).

1802 A. GALLATIN *Let.* 13 Feb. in *Deb. Congr. U.S.* 30 Mar. (1851) 1101 If..it is..the interest of the United States to obtain some further security against an injurious sale, under the Territorial or State laws, of lands sold by them to individuals. 1935 *Chambers's Encycl.* II. 703/1 Yukon has a 'Gold Commissioner' and an elected territorial council. 1953 R. MOON *This is Saskatchewan* 18 That day [*sc.* 18 Dec. 1901] the Territorial Grain Growers' Association was formed.

4. b. (Later examples of *Territorial Army.*) In other collocations: of or pertaining to the Territorial Army. *Territorial* as *sb.*: esp. in *pl.* = *Territorial Army.*

The Territorial and Army Volunteer Reserve was a civilian defence force created in 1967 by merging the Territorial Army and the Army Emergency Reserve and was itself renamed the Territorial Army in 1979.

1910 KIPLING *Divers. Creatures* (1917) 315 That was when we found the Territorial battalion undressin' in slow time. It lay on the left flank o' the Blue Army. 1914 G. B. SHAW *Misalliance* 65 *Tarleton*: Why not join the Territorials? *The man*: Because I shouldnt be let. 1938 W. S. CHURCHILL *Into Battle* (1941) 31 Why..are the Guards drilling with flags instead of machine-guns? Why is it that our small Territorial Army is in such a rudimentary condition? 1940 GRAVES & HODGE *Long Week-end* xxvi. 441 Hore-Belisha..called in the Attorney-General, asking him to warn Sandys, who was a Territorial officer,..that he had rendered himself liable to a court martial..for being in possession of confidential data. 1940 J. F. KENNEDY *Why England Slept* vii. 158 From this time on, it was also established that the Territorial Army, which corresponded somewhat to our National Guard, had 'a claim on the same sources and standards of instruction as the Regular Army'. 1962 M. & M. HARDWICK *Sherlock Holmes Companion* 231 He [*sc.* Conan Doyle] campaigned incessantly for the better training of Territorial reservists. 1970 *Daily Tel.* 14 Jan. 16 Trying to get a snappy recruiting message home to the public is a testing business for the TAVR Council now that they have been saddled with the ponderous legal name 'Territorial Army Volunteer Reserve'. *a* 1974 R. CROSSMAN *Diaries* (1976) II. 664 The proposal to disband the Territorials would now naturally be discussed with the Territorial Association.

territorialism. Add: **5.** *Zool.* = *TERRITORIALITY 2.

1933 M. M. NICE in *Fifty Years' Progress Amer. Ornithol.* (Amer. Ornithologists' Union) 89 (*title*) The theory of territorialism and its development. 1969 A. WHEELER *Fishes Brit. Isles & N.-W. Europe* 410 Strong territorialism is shown [in gobies], the males defending a suitable nesting site.

territoriality. Add: **2.** *Zool.* A pattern of behaviour in which an animal or a group of animals defends an area against others of the same species. Cf. prec.

1941 M. M. NICE in *Amer. Midland Naturalist* XXVI. 441 (*title*) The role of territoriality in bird life. 1943 *Jrnl. Mammalogy* XXIV. 346 The more we study the detailed behavior of animals, the larger is the list of kinds known to display some sort of territoriality. 1955 *Sci. Amer.* Oct. 92/3 No room was left for doubting..the territoriality of the owls. 1979 *Nature* 20–27 Dec. 885/1 Territoriality seems to be rare in bees and wasps, with the exception perhaps of males and their mate search behaviour.

Territorian. Substitute for def.: An inhabitant of the Northern Territory of Australia. (Later examples.)

1941 C. BARRETT *Coast of Adventure* 121 Old Territorians, over..a pannikin of tea by the campfire, will tell yarns as long as you'll listen. 1961 J. DANVERS *Living come First* vii. 121 All the people mixed up in the case are far more Territorians than South Australians. 1971 *Southerly* XXXI. 137 I'm a Territorian, Kenny Buckman's my name, n you gohher be good t'survive out there in the desert, I tell you.

‖ **territorium** (teritōᵊ·riŏm). *Rom. Hist.* Pl. **territoria.** [L.] The area of land surrounding and within the boundaries of a Roman city, *municipium*, etc., and under its jurisdiction. Also *transf.* of States having dealings with Rome.

1918 [see *CENTURIATE *v.* 2]. 1926 ABBOTT & JOHNSON *Municipal Admin. Roman Empire* ix. 134 The chief revenues of cities in other parts of the empire came from their *territoria*. 1949 *Oxf. Classical Dict.* 623/2 In Roman territory before 89 B.C. the chief *oppida* were those of the ex-Latin incorporated States. In them was centred the local administration of their former *territorium*. 1962 D. HARDEN *Phoenicians* v. 74 Carthage's fleet was burnt, her domain was henceforth to be confined to her *territorium* in eastern Tunisia.

territory¹. Add: **1. e.** *Zool.* An area chosen by an animal or a group of animals and defended against others of the same species.

1774 O. GOLDSMITH *Hist. Earth* V. 301 All these small birds mark out a territory to themselves, which they will permit none of their own species to remain in. 1914 J. S. HUXLEY in *Proc. Zool. Soc.* 521 There may be hostility between members of one pair and members of another... The only reason I can discover for it is the trespassing of one or both birds of a strange pair upon the 'territory' of another. 1920 E. HOWARD *Territory in Bird Life* i. 3 Securing a territory is then part of a process which has for its goal the successful rearing of offspring. 1933 *Brit. Birds* XXVII. 20 A certain area of land or territory.. extends around the nesting site. 1949 W. C. ALLEE et al. *Princ. Animal Ecol.* xxiii. 412/2 Territories tend to be larger when population pressure is low. 1953 N. TINBERGEN *Herring Gull's World* ix. 82 A Herring Gull returns to the same colony, and often even to the same territory. 1981 *Oxf. Compan. Animal Behaviour* 550/2 Territories range in size from the few millimetres that separate barnacles..on a rock to the distances of several kilometres that separate neighbouring herds of African buffalo.

f. The geographical area within which a firm or salesman operates. orig. *U.S.*

1900 *Cent. Mag.* Feb. 644/1 We've got to begin small. Our territory is Ohio. 1907 F. H. BURNETT *Shuttle* xxxviii. 379 Nick Baumgarten, who having for some time 'beaten' certain streets as assistant salesman..., had recently been elevated to a 'territory' of his own. 1925 *Daily Tel.* 13 May 20/7 Traveller Wanted... Live men can earn £10 a week. Territory given. 1931 *Economist* 26 Dec. 1235/2 A convenient pocket tabulation of the financial results of oil companies, which shows also the area of their territory, the number of wells. 1977 *Evening Gaz.* (Middlesbrough) 11 Jan. 9/1 (Advt.), A career in sales... Local territories available.

3. a. Also in various vague figurative contexts.

1927 *Daily Express* 30 Nov. 3/1, I think it is a fine plan to refuse, if possible, to be affected by an opponent's play... But I am sure I took in entirely too much territory when I said that his work should be ignored. 1971 N. CHOMSKY *Probl. Knowledge & Freedom* (1972) i. 34 To illustrate further, I would like to turn to some still unexplored territory. 1977 J. I. M. STEWART *Madonna of Astrolabe* v. 93 She was frowning now, aware of having got on territory she hadn't designed to tread.

4. Also applied, outside the U.S., to a region administered by a federal or external government, esp. a part of Canada (now only Northwest Territories and Yukon Territory) or Australia (Northern Territory) not yet organized as a province or state.

1897 C. R. TUTTLE *Golden North* 119 Two new provisional districts or territories have been erected in the far northwest by the Canadian government. The first is that called Mackenzie, lying to the north of Athabasca... The second is called Yukon. 1935 *Chambers's Encycl.* II. 699/1 In 1871, after confederation, the population of Canada (including seven provinces and the territories) was 3,689,257. 1936 I. L. IDRIESS *Cattle King* xii. 106 It dribbles south close to the Territory border all in the sand-hill country, until here it crosses the South Australian border. 1957 *Encycl. Brit.* XVII. 12/1 The Territory of Papua in the south-east [of New Guinea], formerly a British protectorate, is administered by Australia under a governor. *Ibid.*, It [*sc.* Norfolk Island] is a dependency of the Commonwealth of Australia, known officially as the Territory of Norfolk Island. 1969 *Northern Territory News* (Darwin) 11 July 3/2 It also has mining interests in the Territory and Queensland. 1979 G. WOODCOCK *Canadians* II. x. 222 Even in Yellowknife, the capital of the [Northwest] Territories, I encountered an astonishing collection of people.

5. *territory-holder, -holding.*

1929 E. M. NICHOLSON *How Birds Live* iii. 31 The solitary territory-holder can only deal with single intruders. 1953 N. TINBERGEN *Herring Gull's World* vi. 55 A territory-holder stretches its neck as soon as a stranger alights in its neighbourhood. *Ibid.* 58 This [fight] happens.. when a territory-holding bird makes a surprise attack 1962 *Science Survey* XV. 238 A 'territory-holding' male robin will attack a bundle of red feathers.

terror, *sb.* Add: **2. a.** Also, this action or quality in fiction, esp. in *novel* (or *tale*) *of terror.*

a 1832 G. CRABBE *Posthumous Tales* xv, in *Poet. Wks.* (1834) VIII. 205 Yet tales of terror are her dear delight, All in the wintry storm to read at night. 1917 D. SCARBOROUGH *Supernatural in Mod. Eng. Fiction* i. 6 And so the Gothic novel came into being. *Gothic* is here used to designate the eighteenth-century novel of terror dealing with mediaeval materials. 1921 E. BIRKHEAD (*title*) The tale of terror: a study of the Gothic romance. 1977 M. ASHLEY *Who's who in Horror & Fantasy Fiction* 103 His masterpiece of terror was *The Castle of Ehrenstein* (1854), a superb portrayal of a ghost-ridden castle.

b. *Trivially.* A person (occas., a thing) fancied to excite terror; esp. a troublesome child; *holy terror*: see *HOLY *a.* 4 c.

1883, etc. [see *HOLY *a.* 4 c]. 1889 *Harper's Mag.* May 933/1 That bright boy..who was a terror six months ago. 1892 LADY R. CHURCHILL *Let.* 10 Jan. in R. S. Churchill *Winston S. Churchill* (1967) I. Compan. I. v. 305 Papa is very well & in good spirits but his beard is a 'terror'. 1900 [in Dict., sense 2]. 1908 G. SANGER *70 Yrs. a Showman* xvii. 58 Brumley..was a bit of a terror in his way, being a drunken bully. 1925 S. LEWIS *Martin Arrowsmith* vi. 63 She's an old terror. If she found a child like you wandering around here she'd drag you out by the ear. 1953 K. TENNANT *Joyful Condemned* xxxii. 311 It wasn't your fault. René was always a terror. You did what you could. 1979 A. MCCOWEN *Young Gemini* 25 At school I was known as a terror and went looking for fights.

4. *Reign of Terror* (earlier transf. examples); *the terror* (later examples applied to any period of remorseless repression). Hence, without article or pl., the use of organized intimidation, terrorism.

Red Terror, White Terror: see also under the adjs. in Suppl.

1831 *Wexford Herald* 11 June 2/3 The reign of terror—of Terryaltism. 1848 GEO. ELIOT *Let.* 8 Mar. (1954) I. 255 The Glasgow riots are more serious, but one cannot believe in a Scotch Reign of Terror in these days. 1920 *Glasgow Herald* 7 May 9 It was admitted that outrages were committed against the Socialists [in Hungary], but it was denied that a 'terror' existed. 1937 KOESTLER *Spanish Testament* vi. 132 They had neither the inclination nor the need to terrorise the population, to make warning examples, to safeguard the territory behind the lines by the application of methods of Terror. 1951 H. ARENDT *Burden of Our Time* I. i. 6 Terror as we know it today strikes without any preliminary provocation. 1966 G. GREENE *Comedians* iii. 100 The Trianon soufflé au Grand Marnier was famous for a time, until the terror started [in

Haiti] and the American Mission left. **1970** G. Jackson *Let.* 4 Apr. in *Soledad Brother* (1971) 212 All times of the day or night our cells were being invaded by the goon squad: you wake up, take your licks, get skin-searched... This treatment, fear therapy, was not accorded to all however... Mostly it came down on us. Rehabilitational terror. **1977** P. Johnson *Enemies of Society* xviii. 241 Thanks to their use of terror, they [*sc.* the Assassins] often controlled local authorities, and forced governments into compliance or impotence. **1978** *Encounter* July 15/1 Anyone who cannot see and appreciate the true difference between Russia today and Russia at the height of the Stalinist terror has a very poor idea of one or other of these phenomena.

5. a. *terror-novel, -romance*; (in sense 4) *terror act, group, organization, régime, tactics.* **b.** *terror-causing, -inspiring* (earlier example). **c.** *terror-stiffened, -stricken* (earlier example). **d.** Special Combs. **terror-bombing**, intensive and indiscriminate bombing designed to frighten a country into surrender; **terror raid**, a bombing raid of this nature.

1946 Koestler *Thieves in Night* 243 While the usual terror acts continued, the Jewish representative bodies issued their usual protests. **1941** *Reader's Digest* June 58/2 It must be remembered that this government today is Hitler, Göring, Goebbels, Himmler and a few others—men who..ordered the terror bombing of Rotterdam last summer and of London last winter. **1945** *Time* 26 Feb. 32/1 Terror bombing of German cities was deliberate military policy. **1959** R. Collier *City that wouldn't Die* ii. 27 To Sperrle the primary consideration was always that the pilot should see his target;..a Nuremberg tribunal absolved him of terror bombing. **1922** Joyce *Ulysses* 384 The terrorcausing shrieking of shrill women in their labour. **1977** P. Johnson *Enemies of Society* xviii. 242 The diabolism of Stavrogin, who preaches the doctrine that the terror-group can only be united by fear and moral depravity. **1839** Poe *William Wilson* in *Gift* 235 In a remote and terror-inspiring angle was a square enclosure. **1917** D. Scarborough *Supernatural in Mod. Eng. Fiction* i. 6 The terror novel proper is generally conceded to begin with his [*sc.* Horace Walpole's] Romantic curiosity, *The Castle of Otranto.* **1972** P. Haining *Gt. Brit. Tales of Terror* I. 117 William Beckford, author of the great Oriental terror-novel, *Vathek.* **1977** *Belfast Tel.* 22 Feb. 5/7 Growing police and Army success against the terror organisations. **1945** *Ann. Reg. 1944* 3 Lord Cranborne..pointed out that the Royal Air Force had never indulged in purely terror raids like those perpetrated by the Luftwaffe. **1977** J. Wainwright *Pool of Tears* 208 Dresden..That was a terror raid... A town turned into a blow-torch. **1952** Koestler *Arrow in Blue* viii. 68 Admiral Horthy established the first semi-Fascist terror régime in post-war Europe. **1931** R. L. Mégroz *Conrad's Mind & Method* x. 237 The 'Gothic' terror-romance of the eighteenth century. **1972** P. Haining *Gt. Brit. Tales of Terror* I. 477 From the later work comes the following grim story which contains much of that chilling atmosphere which made the Gothic terror-romance so widely popular in its time. **1887** Kipling *Departmental Ditties* (1888) 21 Felt the brute's proboscis fingering my terror-stiffened hair. **1831** Poe *Poems* (ed. 2) 75 There the.. clouds do fly,.Through the terror-stricken sky. **1974** *Encycl. Brit. Micropædia* V. 427/2 A 'provisional' wing [of the IRA]..comprising the younger, militant majority committed to the use of terror tactics.

terrorism. Add: **2.** (Later examples; cf. *terrorist 1 b). Also *transf.*

1936 W. H. S. Smith *Let.* 27 July in *Young Man's Country* (1977) ii. 19 The Constitutional League [of India], whose main purpose is to rouse public opinion against terrorism. **1957** L. Durrell *Bitter Lemons* 243 Though his complicity in EOKA was obvious, nevertheless he [*sc.* Makarios] was the only brake to terrorism and the only person who could curb it. **1958** B. Behan *Borstal Boy* III. 271 He said it was the fault of the British boss class that the Irish were forced always into terrorism to get their demands. **1963** *Ann. Reg. 1962* 236 The first half of the year was dominated by the difficulties of obtaining an Algerian settlement and, in particular, by the challenge to the authority of State presented by O.A.S. terrorism. **1973** *Cape Times* 27 Oct. 12 The Minister cannot expect journalists to do violence to the English language..by describing guerilla warfare as terrorism at all times and in all circumstances. **1977** *New Yorker* 24 Oct. 35/1 Last week's manifestations of political terrorism were crowded off the front pages..by more upbeat occurrences.

terrorist. 1. b. (Later examples in general use.)

In modern use the term usually refers to a member of a clandestine or expatriate organization aiming to coerce an established government by acts of violence against it or its subjects.

1947 *Ann. Reg. 1946* 60 The latest and worst of the outrages committed by the Jewish terrorists in Palestine—the blowing up of the King David Hotel in Jerusalem. **1956** H. Nicolson *Diary* 29 Oct. (1968) 311 When people rise against foreign oppression, they are hailed as patriots and heroes; but the Greeks whom we are shooting and hanging in Cyprus are dismissed as terrorists. What cant! **1969** E. J. Hobsbawm *Bandits* viii. 101 The war between police and terrorists is one of nerves as well as of guns. Whoever is more frightened has lost the initiative. **1977** P. Johnson *Enemies of Society* xviii. 240 The Baader-Meinhof gang of ultra-Left terrorists. **1979** *Spectator* 20 Oct. 20/1 (Advt.), In this enthralling autobiography the author of *Maquis*..retravels the course of his life from his childhood to his war-time exploits as a terrorist in the Resistance.

3. (Later examples.)

1937 Koestler *Spanish Testament* vi. 132 The civilian population...whose sympathies they could but alienate by terrorist acts. **1955** *Britannia Bk. of Year* 263/2, 756 Africans executed..incl. 219 for Mau Mau murders and 508 for other terrorist crimes. **1979** R. Perry *Bishop's Pawn* viii. 130 We weren't dealing with ordinary kidnappers. We were faced by a relatively sophisticated terrorist organization. **1983** *Listener* 19 May 8/1 Terrorist theory..says that the brigades should be subdivided into tight terrorist cells.

terroristic, *a.* Add: (Later examples.) Also **terrori·stically**.

1919 M. Beer *Hist. Brit. Socialism* I. ii. ii. 103 The terroristic acts and wars into which that social earthquake had degenerated. **1945** R. Hargreaves *Enemy at Gate* 308 The terroristic procedure associated in these days with Nazism, Fascism and Bolshevism. **1951** McWhiney & Simkins in A. Dundes *Mother Wit* (1973) 590/2 The klansmen used the methods of violence as extensively as any of the other white terroristic organizations. **1972** *Econ. & Polit. Weekly* 1 Apr. 692/1 Consisting almost exclusively of guerilla squads, they (*sc.* the Naxals) moved secretively and acted terroristically. **1977** *Time* 26 Sept. 9/1 The background of terroristic acts is connected with a deep hatred of bourgeois society.

terry, *sb.*[1] and *a.* Add: **A.** *sb.* **1.** In later use = *terry cloth, terry towelling* (see B below).

1895 *Montgomery Ward Catal.* Spring & Summer 24/1 White Turkish Towelling or Terry. The following Terry or Turkish Towelling is for children's cloaking, robes, towels, etc. **1972** *New Yorker* 22 July 74/1 Our new multicolor stripe combines red, navy, gold and sky-blue on white terry. **1981** *Guardian* 19 Oct. 14/5 Having immersed the terries in a steriliser, I give them a short cold water rinse... This regime means that washing terry nappies is no longer a great chore.

B. *adj.* (Later examples.) Now esp. of or pertaining to *terry towelling*, an absorbent cotton or linen cloth used for making towels, beachwear, babies' napkins, etc.; in the U.S. called *terry cloth* (freq. *attrib.*).

1897 H. Neville *Students' Handbk. Pract. Fabric Structure* xii. 136 Beginning with terry-towelling, as the simplest form of looped pile work. **1906** H. Nisbet *Grammar Textile Design* viii. 163 Terry fabrics produced by means of terry motions are exemplified in so-called Turkish towels... The majority of these goods are produced entirely from cotton, although terry towels are sometimes produced either entirely or in part from linen. **1917** *Harrods Catal.* 1440/1 Terry Dusters or Paint Cloths. **1921** *Daily Colonist* (Victoria, B.C.) 15 Mar. 4/5 (Advt.), 54-inch New Terry Cloth $1.69 a Yard. **1937** *Night & Day* 8 July 22/1 Terry-towelling is responsible for a great many irresistible beach affairs. **1944** R. Chandler *Lady in Lake* iii. 18 A big guy in bathing trunks..and a white terrycloth bathrobe. **1959** *Harrods News* Summer 16 Terry towel for bath or beach. **1961** A. Miller *Misfits* iv. 44 He turns and sees Roslyn in a terry-cloth robe emerging from the bedroom doorway. **1961** *Listener* 12 Oct. 558/1 An artificial mother constructed of wire and terry towelling. **1975** *Guardian* 27 Jan. 15/2 A completely new type of tufting machine which is directed specifically at the terry trade...[It] makes a cloth with one face of terry material. **1980** L. Birnbach et al. *Official Preppy Handbk.* 155/1 Need a pair of frog-print slacks or a terry cloth halter jumpsuit? **1981** [see sense A. 1 above].

Terry Alt, Terryalt (te·ri ǫlt). *Irish Hist.* Also *ellipt.* **Terry.** [According to a MS. diary of 1831 quoted in *Times Lit. Suppl.* (1932) 29 Sept. 691/4, *Terry Alts* was the name of an innocent bystander suspected of an outrage on a man.] A member of a secret agrarian association active in western Ireland in the 1830s. Also *attrib.*

1831 D. O'Connell *Let.* 15 May (1888) I. 263 It is probable that, without the aid of the 'Terry Alt' system, he could not poll one hundred votes by all his other exertions. **1931** *Dublin Even. Post* 31 May 3/3 Michael Connelly, a chief leader of the Terry Alts. **1832** *Courier* 17 Feb. 2/5 The Terries in the County of Galway are levying contributions. **1861** W. J. FitzPatrick *Life of Doyle* II. xxxix. 334 He urged the 'Shanavests', the 'Caravats', the 'Terryalts', and the 'Rockites', to abandon their deeds of blood; he implored of the Ribbonmen to cast their evil combinations to the winds. **1898** *Westm. Gaz.* 14 Nov. 4/2 The man was suspected of being a 'Terry Alt', or a member of a local agrarian conspiracy.

tertiary, *a.* and *sb.* Add: **A.** *adj.* **1. b.** Substitute for entry: (i) Applied to compounds regarded as being derived from ammonia by replacement of three hydrogen atoms by organic radicals, and to derivatives of such compounds; also extended to analogous derivatives of other elements, esp. phosphorus. [The sense is due to Gerhardt & Chiozza, who used F. *tertiaire* (*Compt. Rend.* (1853) XXXVII. 88).]

1854 *Q. Jrnl. Chem. Soc.* VI. 195 With regard to the tertiary amides,..their preparation is generally easier than that of the secondary amides. **1888**, etc. [see *primary *a.* 6 f (i)]. **1964** N. G. Clark *Mod. Org. Chem.* xi. 232 The tertiary amines,..with no available hydrogen, are the nitrogen counterparts of the ethers.

(ii) Applied to organic compounds other than amines, etc. (see prec. sense) in which the characteristic functional group is located on a saturated carbon atom which is itself bonded to three other carbon atoms. [Applied orig. to alcohols by H. Kolbe, who used G. *tertiär* (*Ann. der Chem. und Pharm.* (1864) CXXXII. 104).]

1872 *Jrnl. Chem. Soc.* XXV. 295 The oxidation of tertiary alcohols takes place according to a law similar to that which rules the oxidation of ketones. **1932** I. D. Garard *Introd. Org. Chem.* iii. 54 This formation of a ketone having fewer carbon atoms than the alcohol is characteristic of the oxidation of tertiary alcohols. **1964** N. G. Clark *Mod. Org. Chem.* xi. 222 The use of acyl chlorides in the above manner produces hydrogen chloride, which may have a deleterious effect on the compound undergoing acylation, e.g. tertiary alcohols readily give the alkyl chlorides. **1981** Wingrove & Caret *Org. Chem.* x. 435 Under basic conditions, tertiary alcohols do not undergo oxidation.

(iii) Applied to a saturated carbon atom which is bonded to three other carbon atoms; also, bonded to or involving a tertiary carbon atom. Of an ion or a free radical: having (respectively) the electric charge or the unpaired electron located on a tertiary carbon atom.

1903, etc. [see *secondary *a.* 3 i (iii)]. **1972** [see *primary *a.* 6 f (ii)].

c. *Surveying.* Designating triangulation derived by subdivision from secondary triangulation (which in turn results from subdivision of primary triangulation) or points, bench-marks, etc., established by this.

1851 C. Davies *Elementary Surveying* (rev. ed.) iv. i. 181 When the secondary and tertiary triangles have been considerably multiplied, the compass is taken in hand. **1883** J. B. Oliver *Pract. Astron. for Surveyors* ix. 121 The sides of the secondary triangles are from about 5 to 20 miles, and those of the tertiary triangles five or less. **1920** W. N. Thomas *Surveying* xiii. 382 A further subdivision resulted in the 'Tertiary' triangulation. **1965** Bannister & Raymond *Surveying* (ed. 2) iii. 293 The fourth order points give closer spacing in towns—tertiary and higher order points cover almost the whole country at a density of 0·05 trig point per km², with a density of about 0·1 per km² in towns. **1978** [see *secondary *a.* 3 a].

d. *Physics.* Produced by the impact of secondary particles with matter.

1938 R. W. Lawson tr. *Hevesy & Paneth's Man. Radioactivity* (ed. 2) v. 61 On the average 2 or 3 tertiary electrons result from each secondary electron, when the primary β-radiation has a velocity 33 per cent. that of light. **1961** G. R. Choppin *Exper. Nuclear Chem.* iii. 35 Tertiary electrons may be produced by photoemission resulting from the photons of the secondary ionization process.

e. Designating the part of the economy or work-force concerned with services of all kinds, rather than with the production of foodstuffs or raw materials, or with manufacturing.

1940 *Economist* 21 Sept. 363/1 There is a steady tendency for labour to move out of primary production into secondary production (manufacture) and from secondary to tertiary production (all forms of services). **1961**, etc. [see *quaternary *a.* 3 a]. **1974** B. Pearce tr. *Amin's Accumulation on World Scale* I. 16 The sectors of the tertiary part of the economy—transport, trade, financial services—are grafted upon the foreign economy. **1975** *Guardian* 30 Jan. 16/4 Tertiary industries are also being introduced..plants for the preparation of prefabricated houses and timber for construction.

f. *tertiary structure* (Biochem.): the way the helix of a polynucleotide or polypeptide molecule is folded in three dimensions and bound to other helices.

1952, etc. [see *primary *a.* 6 v]. **1964** G. H. Haggis et al. *Introd. Molecular Biol.* iii. 59 The run of the peptide chain through the molecule..is known as the tertiary structure of a protein. **1978** *Nature* 5 Jan. 15/2 Studies on pancreatic trypsin inhibitor and hen egg white lysozyme suggest that at most there are only a limited number of folding pathways to the tertiary structure.

g. *tertiary road* (orig. *U.S.*), a Class III road.

1960 Baker & Stebbins *Dict. Highway Traffic* 114 *Land-service road*, a road which is used primarily to give access to land. Sometimes called: tertiary road. **1971** J. Drummond *Farmer's Party* xxx. 149 We were out on a tertiary road, and more or less alone. **1975** M. Kenyon *Mr. Big* xxi. 214 The secondary road became a tertiary road of muddy craters.

h. *tertiary education*, that which follows secondary education and precedes, includes, or replaces university or professional training; so *tertiary level*; *tertiary college*, one at which such education is provided.

1961 *Mind* LXX. 105 The spread of secondary and tertiary education has created a large population of people..educated far beyond their capacity to undertake analytical thought. **1969** *Guardian* 26 Aug. 16/4 A 'tertiary college' in Exeter where sixth forms are to be merged in the College of Further Education. **1971** *New Scientist* 27 May 513/1 Whenever Britons wrote or talked about tertiary education, they generally meant university education. **1974** *Bookseller* 18 May 2402/1 (Advt.), Can you sell our tertiary-level academic titles to booksellers in Scotland and North-East England? **1981** *New Society* 29 Jan. 192/1 Tertiary colleges—providing everything from A level Russian to pre-nursing courses and apprenticeship courses in motor engineering—are the colleges of the future..A *sixth form college* is for the more traditional sixth form intake. A *tertiary college* provides for all over-16s whatever their needs.

i. *tertiary recovery*, the recovery of oil by

advanced methods after conventional artificial means have ceased to be productive. Cf. *secondary recovery* s.v. *secondary *a.* 3 k.

1975 *Petroleum Economist* Aug. 292/2 Oil produced by tertiary recovery methods, from above the Arctic circle,..could sell at US $8·50 a barrel. **1976** *National Observer* (U.S.) 10 July 13/3 Given that profits hold, what is next for the oilmen? They answer, almost in unison: 'tertiary recovery'.

3. (Later example.)

1967 E. Short *Embroidery & Fabric Collage* i. 11 A mixture of all three primary colours results in tertiary colours. These are the subtle colours such as khaki, various browns, etc.

B. *sb.* **6.** *Gram.* In Jespersen's terminology, a word or group of words of tertiary rank or importance in a phrase or sentence; = *subjunct. Cf. quote 1871, sense A. 1 a in Dict.

1924, **1940** [see *secondary *sb.* 12]. **1959** M. Schlauch *Eng. Lang. in Mod. Times* viii. 221 In this system [of Otto Jespersen's]..the modifier of a modifier (e.g., an adverb) is a tertiary.

‖ **tertium comparationis** (tə·ɹʃiŏm kǫmpæriʃiō"·nis, kɒmpærātiō"·nis). [L., = the third element in comparison.] The factor which links or is the common ground between two elements in comparison.

1922 J. Riviere tr. *Freud's Introd. Lectures Psycho-Anal.* x. 128 In one set of symbols the underlying comparison may be easily apprehended, but there are others in which we have to look about for the common factor, the *tertium comparationis* of the supposed comparison. **1945** *Mind* LIV. 209 A comparison without a *tertium comparationis*. **1956** [J. H. Greenberg in Saporta & Bastian *Psycholinguistics* (1961) 470/1 With what culturally relevant units must one connect a contrast between aspirated and nonaspirated consonants in a given language? Where is the *tertium comparationis*?

‖ **tertium non datur** (tə·ɹʃiŏm nǫn dēi·tŭə). *Lat. phr.* No third possibility exists. Also as *sb. phr.* Cf. *excluded middle, third* s.v. Excluded *ppl. a.* b.

1887 R. Hodgson *Let.* 8 Apr. in R. B. Perry *Th. & Char. of W. James* (1935) I. 642 You are neither empiricist nor transcendentalist; and *tertium non datur*. **1932** tr. *Ortega y Gasset's Revolt of Masses* xiv. 190 The nation is always either in the making, or in the unmaking. *Tertium non datur.* **1948** H. Reichenbach *Elem. Symb. Logic* vi. 227 An example of such a substitution is given by the resulting forms of the *tertium non datur* for higher functions. **1977** *Language* LIII. 319 After all, a pronoun must be either prefixed or infixed; tertium non datur.

tertius. Add: [Earlier Sc. example, appended to the name of an artist.]

1818 *Blackw. Mag.* II. 424/2 Nicol Jarvie, tertius, M.D.

‖ **tertius gaudens** (tə·ɹʃiŏs gau·dĕnz). [L., f. *tertius* third + *gaudens*, pres. ppl. of *gaudere* to rejoice.] A third party that benefits by the conflict or estrangement of two others.

1892 H. Bismarck in *Ann. Reg.* 189 28/4 I should like to interfere in such cases, like a parish beadle bringing peace, and prove that the *tertius gaudens* is the worst enemy. **1933** G. Arthur *Septuagenarian's Scrap Bk.* 39 Having ascertained from M. Bompard..that France would flatly refuse..to be a *tertius gaudens* with Germany and Russia. **1957** R. K. Merton *Social Theory* (rev. ed.) ix. 336 The *tertius gaudens*... **1974** 'M. Innes' *Mysterious Commission* xii. 170 He saw himself as a kind of third force—or even as what the learned would call a *tertius gaudens*, meaning a chap who nips in and does both parties down. **1980** D. Newsome *On Edge of Paradise* v. 160 It would be better for them both to withdraw to allow the election of a *tertius gaudens*.

tertschite (tə·ɹtʃait). *Min.* [ad. G. *tertschit* (H. Meixner 1953, in *Fortschritte der Mineral.*, etc. XXXI. 41), f. the name of H. *Tertsch* (1880–1962), Austrian mineralogist: see -ite[1].] A hydrated calcium borate found as white, fibrous, probably monoclinic crystals.

1953 *Chem. Abstr.* XLVII. 10413 New borate deposits were discovered in 1951... The following types were distinguished:..tertschite, about $Ca_4B_{10}O_{19}·20H_2O$. **1978** *Mineral. Gen.* VI. 261 Tertschite is found only in one locality in the Bigadiç deposits... It is white, contains very fine fibres, shines like silk... Its rare occurrence makes this mineral unique among the other borate minerals.

teruggite (tĕrū·dʒait). *Min.* [f. the name of M. E. *Teruggi*, 20th-c. Argentinian geologist + -ite[1].] A hydrated arsenate and borate of calcium and magnesium, $Ca_4MgB_{12}O_{20}(AsO_4)_2·18H_2O$, found as colourless or white monoclinic crystals.

1968 Aristarain & Hurlbut in *Amer. Mineralogist* LIII. 1815 Teruggite, a new borate mineral, was collected in June, 1967, during a field study of Argentine borates. **1973** *Amer. Mineralogist* LVIII. 1034/2 No chemical analysis was made for the teruggite sample used in the present study. The crystal structure determination confirms the chemical composition reported by Aristarain and Hurlbut (1968) except for the water content which consists of twenty molecules instead of eighteen. **1978** *Mercian Geol.* VI. 264 Teruggite is rare, occurring sporadically at one horizon in the southern basin of the Emet deposits [in Turkey], as very pure white, powdery potato-shaped nodules containing countless minute white euhedral crystals.

Tervueren (tɜɹvū"·rən). Also **Tervuren.** [a. Flemish *Tervueren*, Fr. *Tervuren*, the name of a small town in Belgium, some ten miles east of Brussels.] A fawn, rough-coated, Belgian sheepdog, with dark pricked ears and a black muzzle. Also *attrib.*

1947 C. L. B. Hubbard *Working Dogs of World* ii. 138 The Tervueren is the third main type among the sheepdogs of Belgium. **1964** [see *Groenendael]. **1978** *Detroit Free Press* 2 Apr. 16F/3 Belgian Tervuren Pups, ex show-obedience-guard.

Terylene (te·rilīn). Also **terylene.** [f. *polyethylene terephthalate* s.v. *polyethylene a, by inversion.] **a.** A proprietary name for polyethylene terephthalate used as a textile fibre.

1946 [see *fibre-forming* adj. s.v. *fibre *sb.* 8]. **1946** [see *polyethylene terephthalate* s.v. *polyethylene a]. **1947** *Trade Marks Jrnl.* 23 Apr. 233/1 *Terylene*... All goods in Class 25. Imperial Chemical Industries Limited. **1949** *Official Gaz.* (U.S. Patent Office) 27 Sept. 951/2 Imperial Chemical Industries, Limited... *Terylene*... For synthetic yarns and thread. **1951** *Economist* 22 Sept. 686/1 Dacron, known in Britain as Terylene and made under licence in the United States, replaces light-weight worsted for... **1958** *Sunday Times* 27 Apr. 7/5 New materials, nylon, Terylene and so on bring a certain spick-and-spanness within the reach of all. **1961** *Times* 30 May (I.C.I. Suppl.) p. vi, A conveyor belt made with 'Terylene' synthetic..., out-last, out-wear any other belt. **1976** P. Cave *High Flying Birds* i. 9 The sails are usually made of Terylene.

b. *attrib.*

1951 *Catal. of Exhibits, South Bank Exhib., Festival of Britain* 109/1 Terylene lace, rope, silk, blanket, etc. **1958** *New Statesman* 28 June 831/1 The men who had nylon shirts and terylene suits before those fabrics got into Marks and Spencer's where the rest of us buy our clothes. **1967** E. Short *Embroidery & Fabric Collage* iii. 70 A synthetic interlining such as terylene wadding makes the quilt easy to launder. **1977** B. Pym *Quartet in Autumn* xi. 97 A crisp-looking terylene surplice was suspended from a hanger.

'tes, var. 'Tis in Dict. and Suppl.

teschemacherite (teʃĕmæ·kĕrait). *Min.* [f. the name of E. F. *Teschemacher* (1791–1863), English chemist, who first described it: see -ite[1].] Ammonium bicarbonate, $(NH_4)HCO_3$, occurring as transparent white to yellowish orthorhombic crystals.

1868 J. D. Dana *Syst. Min.* (ed. 5) 705 (*heading*) Teschemacherite. **1968** [see *kalicine]. **1972** *Amer. Mineralogist* LVII. 1305 Teschemacherite, ammonium bicarbonate, was deposited inside the wellhead of the Broadlands [New Zealand] geothermal drillhole BR9 after the well had been shut for several weeks... Teschemacherite has not been reported from other geothermal fields but occurs in some guano deposits.

Teshoo Lama, Teshu Lama, varr. *Tashi Lama.

Tesla (te·slă). [The name of Nicola *Tesla* (1856–1943), Croatian-born American electrical physicist.] **1.** *Tesla coil*, a type of induction coil invented by Tesla, employing a spark gap in place of an interruptor and capable of producing an intense high-frequency discharge.

1896 *Amer. Jrnl. Sci.* CLI. 245 By changing the size of the spark gap in the primary circuit of the Tesla coil one can obtain the range of electrical energy at command. **1930** *Proc. R. Soc.* A. CXXIX. 479 If it is desired simply to obtain the highest possible potentials with the minimum of trouble then the Tesla coil is obviously the ideal solution. **1978** *Bio Systems* VII. 6/2 Thus far, energy has only been fed in by sparking electrodes kept at roughly controlled voltage level with Tesla coils.

2. *Physics.* (Usu. with small initial.) Pl. tesla, teslas. The SI unit of magnetic flux density, equal to one weber (*weber) per square metre; 10,000 gauss. Symbol T.

1960 in Cooke & Markus *Electronics & Nucleonics Dict.* 482/2. **1961** *Symbols, Units & Nomencl. Physics* (Internat. Union Pure & Appl. Physics) 18 The following units of the MKSA system have special names and symbols, which have been approved by the General Conference on Weights and Measures:... tesla (Wb/m²). **1969** *Sci. Jrnl.* June 36/3 The oscillating magnetic field in the radio pulse itself as it leaves the pulsar is probably greater than 10 teslas (10^4 gauss). **1971** *New Scientist* 24 June 737/2 Superconductors cannot yet sustain fields greater than 15 tesla. **1980** J. F. O'Hanlon *User's Guide Vacuum Technol.* ix. 216 Modern [sputter-ion] pumps are constructed..with external permanent magnets of 0·1 to 0·2 Tesla strength.

Teso (te·so). [Native name.] **a.** (Also *Iteso.*) A Nilo-Hamitic people of central Uganda and western Kenya; a member of this people. **b.** (Also *Ateso.*) The Nilo-Hamitic language of this people. Also *attrib.*

1910 *Bible in World* Nov. 323/2 Teso is the speech of one of the Nilotic tribes who are found in the north of the Uganda Protectorate. *Ibid.* 324/1 The Teso language belongs to a group which also includes the dialects spoken by the famous Masai..and the Karamojo tribes. *Ibid.*, Kitching..gave a most encouraging account of his successful work among the Teso. **1915** A. L. Kitching *Handbk. Ateso Lang.* p. v, The Teso are a cheerful, industrious people, amenable to control..yet the name of this tribe does not appear..in any of the works on the peoples of the Uganda Protectorate... The Ateso dialect is spoken by a tribe of some 300,000 people living between Lake Kioga and Mt. Elgon in..the Ugandan Protectorate. **1935**, etc. [see *Sebei]. **1966** [see *Karamojo]. **1973** *Sunday Tel.* 4 Mar. 8/2 The taller and more gaunt appearance of the Nilotic tribes—the Lango, the Acholi and the Iteso.

tesseract (te·sĕrækt). *Math.* Also **tessaract.** [f. Tessara- + Gr. ἀκτ-ίς ray.] A four-dimensional hypercube. Also *fig.* Hence **tessera·ctic** *a.*

1888 C. H. Hinton *New Era of Thought* ii. iii. 118 We call the figure it [*sc.* a cube] traces a Tessaract. *Ibid.* vii. 161 The whole of the 81 cubes make one single tessaractic set extending three inches in each of the four directions. **1919** R. T. Browne *Mystery of Space* v. 134 The hypercube or tesseract is described by moving the generating cube in the direction in which the fourth dimension extends. **1960** *Electronic Engin.* XXXII. 347/1 Fig. 8.. shows a four-dimensional 'tessaract' (the four-dimensional analogue of a cube). **1968** *Listener* 15 Feb. 201 He likes to see A gulping of tesseracts and Gondals in Our crazed search. **1974** S. Sheldon *Other Side of Midnight* xviii. 332 For Catherine time had lost its circadian rhythm; she had fallen into a tesseract of time, and day and night blended into one.

tessitura. Add: (Examples.) Also *transf.*

1884 Grove *Dict. Mus.* IV. 94/1 A term..used by the Italians to indicate how the music of a piece 'lies';..what is the prevailing or average position of its notes in relation to the compass of the voice or instrument for which it is written... 'Range' does not at all give the idea, as the range may be extended, and the general *tessitura* limited; while the range may be high and the *tessitura* low or medium. **1948** *Penguin Music Mag.* Feb. 76 One can hardly blame him, for the *tessitura* is sometimes cruelly high—so many Italian baritone parts seem to have been written for tenors in reduced circumstances. **1956** Auden & Kallman *Magic Flute* (1957) 116 You won't hear a word in Our high tessitura. **1978** *Early Music* Apr. 197/2 He chose singers for whom the resulting tessituras did not mean any strain. **1982** *English World-Wide* II. 136 Tessitura (or the characteristic range of notes, or compass, within which the pitch fluctuation falls) was felt to be generally wider in Br[itish] E[nglish] than in S[ingapore] E[nglish].

test, *sb.*[1] Add: **2. a.** *spec.* (*Austral.* and *N.Z.*) a test for the proportion of butter fat in milk.

1928 *Bulletin* (Sydney) 14 Mar. 32/1 'You should be proud of her,' said I... 'My oath I am!' he made reply—'She gives an eight-five test!' **1950** *N.Z. Jrnl. Agric.* Mar. 270/2 Several [milk] cans can be filled at the same time so that the tests of all cans are, as far as practicable, identical. **1966** G. W. Turner *Eng. Lang. Austral. & N.Z.* iii. 45 A cow with 'a good test', that is, milk rich in butter-fat, may be more valuable than another cow that gives more milk.

c. (Later examples.) Also in *Rugby Football* (cf. *test match*, sense 7 b below); *S. Afr.*, an international match in any of a wide range of games and sports, including Rugby.

1933 M. Nicholls in I. D. Difford *Hist. S. Afr. Rugby Football* xxiv. 336 In the first Test we won 16 scrums to their 36. **1934** [see *Buckley's]. **1954** R. T. Gabe in Wooller & Owen *Fifty Years of All Blacks* i. 14 We travelled over land and a rough sea..to play a Test in Wellington..to lose by 9 points to 3. **1955** [see *deputize *v.* 2]. **1971** *Rand Daily Mail* 4 Sept. 24/7 A series of diving Tests have been arranged against Rhodesia. **1972** *Daily Tel.* 14 Dec. 35/5 A week off before an international, or Test as we call them, is preferable to a mid-week match [New Zealander *loq.*].

4. b. (Later examples.) Also applied to the process or an instance of testing the academic, mental, physiological, or other qualities and conditions of a human subject; in academic and similar contexts usu. implying a simpler, less formal procedure than an examination; freq. as the second element in a collocation or combination denoting a particular kind of test, or used contextually to imply one of these.

A number of other collocations and combinations will be found under the first element in Suppl., as *aptitude, blood, breath, intelligence, means, mental, performance, pregnancy, screen, skin, spot test*; also *road test.

1910, etc. [see *Binet-Simon]. **1918** [see *proficiency test* s.v. *proficiency 3]. **1927** [see *personality test* s.v. *personality 7]. **1928** *Sunday Dispatch* 22 July 4/2 He had had a film test, at the conclusion of which he was told that he filmed remarkably well. **1933** [see *driving *vbl. sb.* 3 a]. *a* **1935** [see *fitness]. **1941** B. Schulberg *What makes Sammy Run?* xi. 198 I'm getting fed up with these floosies you're always promising..a day's work or a test [i.e. a screen test]. **1955** E. H. Clements *Discord in Air* xi. 149 Mummy always drives. I haven't taken my test yet. **1956** [see *nuclear *a.* 3 c]. **1958** *Economist* 8 Nov. 481/2 Russia is trying to make the West agree to a ban on tests. **1959** [see *pass-fail *a.*]. **1960** [see *breathalyser]. **1968** [see *I.Q.* s.v. I. III]. **1968** [see *M.O.T.* s.v. M 5]. **1976** *Star* (Sheffield) 30 Nov. 12/6 Up to £50 paid for scrap and test failure cars and vans.

7. a. (in sense 2 a) *test-sentence, -tree*; (in sense 2 c) *test batsman, captain, cricket, cricketer, team, trial*; (in sense 4 b) *test anxiety, certificate, performance.*

1972 *Jrnl. Social Psychol.* LXXXVII. 155 Few studies have examined the relationship of birth order to test anxiety. **1976** Dexter & Makins *Testkill* 139 The Test batsman, even after net practice, is still forced to use the first few overs in the middle as a warm-up. **1975** *Cricketer* May 8/1 Ian Michael Chappell, the activist of Test captains, has led Australia in 26 Tests in four countries. **1976** *Alyn & Deeside Observer* 10 Dec. 10/2 He did not have an excise licence, a driving licence or a test certificate. **1931** J. Hobbs (*title*) Playing for England! My test-cricket story. **1959** M. Gilbert *Blood & Judgement* iii. 36 In September a test cricketer was still news. **1942** *Mind* LI. 175 A factor which improves certain test-performances when it is not merely absent, but actually negative. **1901** Kipling *Kim* x. 262 Kim repeated the test-sentence. **1977** *Word 1972* XXVIII. 104 There were 15 test sentences in the battery in which the English strongly suggested the use of a diminutive ending in Gaelic. **1955** *Radio Times* 22 Apr. 31/2 The Test Team arrived in this country at the beginning of the week. **1883** G. M. Hopkins *Let.* 25 Oct. (1956) 323 This was the sin of Adam and Eve, who, both in different ways, eat of the 'Test-tree'. **1977** Test trial [see *pencil *v.* 2 c].

b. test ban, a ban on the testing of nuclear weapons; **test bed**, a piece of equipment for testing machines, esp. aircraft engines, before their acceptance for general use; also *attrib.* and *fig.*; **test-body** *Physics*, the imaginary object on which a thought-experiment is carried out; **test card**, (*a*) *Ophthalm.*, a large card printed with rows of letters of decreasing size for use in testing visual acuity (cf. *Snellen); (*b*) *Television*, a diagrammatic still picture transmitted outside normal programme hours and designed for use in judging the quality and position of the image on any particular screen; **test-case** (earlier example); also *transf.* and *attrib.*; **test chart** *Ophthalm.* = *test card* (a) above; **test-drive** *v. trans.* (orig. *U.S.*), to drive (a motor vehicle) in order to determine its qualities with a view to its regular use; **test-fire** orig. *U.S.*, to fire (a gun or missile) experimentally; **test flight**, a flight during which the performance of an aircraft is tested; **test-fly** *v. trans.*, to test the performance of (an aircraft) in flight; hence **test-flying** *vbl. sb.*; **test-hole**, (*b*) = *test well* below; **test letter**, (*b*) (examples); **test-market** *v. trans.* (and *intr.*) (orig. *U.S.*), to put (a new product) on to the market, usu. in a limited area, in order to determine consumers' response to it; also *transf.*; also as *sb.*, an area in which a product is test-marketed; hence **test-marketing** *vbl. sb.*; **test match** (earlier and further examples); also in *Rugby Football* (orig. *S. Afr.*), one of a series of matches between a touring team and teams representing the country of the tour; an international; **test-paper**, (*c*) (example); **test-piece**, (*b*) (see sense 7 a in Dict.); (*c*) a piece performed by each of the competitors in a musical contest to determine which is the best; **test pilot**, one who test-flies an aircraft; also (with hyphen) as *v. trans*; hence **test-piloting** *vbl. sb.*; **test-pit**, (*a*) (see sense 7 a in Dict.); (*b*) *Archæol.*, a pit dug to gain an idea of the contents of a site; **test range**, a range (*range *sb.*[1] 11 c) where missiles are tested; **test-retest** *a. Psychol.*, of or designating a method by which a test is given to a subject on two occasions separated by a lapse of time; **test rig** *Engin.*, an apparatus used for assessing the performance of a piece of mechanical or electrical equipment; **test signal**, a sequence of electrical impulses used for testing purposes in television broadcasting; **test strip**, (*a*) *Cinemat.* (see quot. 1940); (*b*) *Photogr.* (see quot. 1973); **test-type** (examples); **test well** *Oil Industry*, a well made in testing a site for oil; **test-word**, (*a*) *Psychol.*, a word used in a test; (*b*) *Onomastics*, a word used to determine the presence of a particular linguistic form or influence.

1958 *New Statesman* 27 Dec. 898/1 More progress was registered at Geneva last week, when the test-ban conference approved a British draft of Article Four of the treaty. **1971** H. Trevelyan *Worlds Apart* xvi. 177 As we saw it, there were two elements in Soviet thinking about a test-ban. **1979** G. F. Newman *List* vi. 55 Kennedy sees the test ban treaty as a step toward peace. **1914** *Flight* 21 Mar. 312/1 The 120 h.p. engine entered by the Green Engine Co. for the Military Aeroplane Engine Competition is mounted on a tilting test bed. **1924** S. R. Roget *Dict. Electr. Terms* 260 *Test bed*, a base plate or foundation upon which machines may readily be mounted for testing purposes. **1937** *Times* 13 Apr. (Brit. Motor Number) p. xv/4 The car engines undergo a long and thorough trial on the test-bed. **1961** *Aeroplane* CI. 791/1 Two VTOL test-bed aircraft using the G.E. J85-5 fan-lift engine. **1963** *Listener* 28 Mar. 542/2 The Russian leaders ..have spent the last fifteen years on the test-bed of world strategy, feeling the fearful and complex stresses and strains that that involves. **1978** *Sci. Amer.* July 30/1 On test beds turbine-inlet temperatures of well over 1,650 degrees C. have been achieved for at least a decade. **1920** A. S. Eddington *Space, Time & Gravitation* iv. 64 A massive body, such as the earth, seems to be surrounded by a field of latent force, ready, if another body enters the field, to become active, and transmit motion. One usually thinks of this influence as existing in the space round the earth even when there is no test-body to be affected. **1955** L. Rosenfeld in W. Pauli *Niels Bohr* 71 This meant that in studying the measurability of field components we must use as test-bodies finite distributions of charge and current, and not point charges. **1892** A. Duane tr. *Fuchs's Text-bk. Opthalm.* iii. ii. 609 When the visual acuity has become so reduced that the largest letters of Snellen's test-card can no longer be recognized at 6 metres, the patient must go up nearer it. **1935** *Popular Wireless* 16 Mar. 14/2 The 'test cards' radiated recently by the B.B.C. have..been the cause of a lot of correspondence. **1949** H. C. Weston *Sight, Light & Efficiency* vii. 245 External light sources must be relied upon for illuminating the test-cards. **1962** *Which?* Mar. 70/2 To measure the resolution, we used the BS test-slide which has blocks of parallel lines of various thicknesses and spacings, similar to the test card shown to television viewers so that they can adjust their sets for a clear, sharp picture. **1978** S. Wilson *Dealer's Move* iii. vi. 103 A buzzing in my head to match the buzzing of the test card on the screen. **1894** W. Archer in *World* 31 Jan. 25/2 Mr. Gattie is of opinion that the insanity of one of the parties to a marriage should be..a compulsory ground for divorce... He indicts the law by making his hero break it, and showing..that his crime is a law-made crime... He is ..bent upon getting up a good 'test case'. **1911** M. Corelli *Life Everlasting* ix. 205 Because he had seen in me the possibility of a 'test case', Santoris had tried his power upon me. **1959** B. & R. North tr. *M. Duverger's Pol. Parties* (ed. 2) i. ii. 112 Such counts.. presuppose that the leaders of a number of test-case branches..would make a very careful check of attendances over a period of time. **1910** H. C. Parker *Handbk. Dis. Eye* v. 62 (*caption*) Test chart for illiterates. **1978** J. Parr *Introd. Ophthalm.* ii. 64 If a subject's visual acuity is less than 6/60 the distance from the test chart can be progressively reduced down to 1m. **1954** *Sun* (Baltimore) 1 Nov. (B ed.) 9/1 Shaw and his companions were returning from Detroit, where he had test driven a 1955 Chrysler. **1971** *Guardian* 30 Oct. 20/1 Mrs Joy Johnson.. demolished a 'No Entry' sign while test-driving a double-deck bus. **1947** *Birmingham* (Alabama) *News* 27 Oct. 1/2 He stole the automatic pistol from an automobile and test-fired it twice before calling for the cab. **1952** *N.Y. Times* 27 Apr. iv. E5/2 The atomic gun-fired shell..will probably be test-fired in the course of the next year or so. **1960** *Daily Tel.* 8 Jan. 1/3 Russia is to test-fire new heavy rockets, intended for use on inter-planetary flights. **1980** N. Freeling *Castang's City* xxix. 202 We'll have it test-fired tomorrow and the cartridge marks compared. **1912** *Flight* 3 Feb. 106/2 No flying on Friday beyond a test flight by Pizey on the Bristol. **1927** C. A. Lindbergh *We* iv. 59, I took off for a test flight before taking the lady over Pensacola. **1976** *Derbyshire Times* (Peak ed.) 3 Sept. 28/1 The twin-engined Beechcraft monoplane.. was on a test flight at the time. **1936** *Meccano Mag.* Aug. 433/2, I hope it will fall to my lot to test fly these great super-clippers. **1942** W. Simpson *One of our Pilots is Safe* ii. 40 During the day each aircraft received a special check-up and was test-flown by its pilot and crew. **1978** J. A. Michener *Chesapeake* 739 When the time came to test-fly the contraption,..an aviator from Washington.. studied the seaplane. **1928** N. Macmillan *Art of Flying* 7 Immediately after the War, he took up test-flying with considerable success. **1961** Test-flying [see *flight-testing* *vbl. sb.* s.v. *flight *sb.*[1] 15]. **1909** *Chambers's Jrnl.* Mar. 160/1 He [*sc.* the prospector] digs here and there, making test-holes. **1971** *Sunday Australian* 8 Aug. 17/2 The new test hole is sited about 100 miles south-west of Fitzroy Crossing, W.A. **1869** *Trans. 4th & 5th Ann. Meeting Amer. Ophthalm. Soc.* 68 (*heading*) On a new series of test-letters for determining the acuteness of vision. **1970** A. H. Keeney *Ocular Examination* ii. 18/2 Snellen's real contribution was to standardize the size and form of test letters with relation to the distance from the observer. **1958** *Wall St. Jrnl.* 6 Nov. 23/5 A new line of cookingware which is now being test-marketed in three cities of the United States. **1964** *Listener* 12 Mar. 422/1 Many products are produced and tried out in test-markets (usually medium-sized cities or commercial-television areas) for a year or more. *Ibid.*, Decisions about test-marketing..are the cause of severe anxiety. **1972** 'J. Melville' *Ironwood* ix. 154 She had come to me seeking recipes for a new sort of cooking chocolate she was helping test-market in this area. **1862** W. J. Hammersley *Victorian Cricketer's Guide 1861–2* 159 Of the thirteen matches, five only can be termed 'test matches'; the three played at Melbourne, and the two played at Sydney. **1889** *Wisden's Cricketers' Almanack* 162 There was a considerable amount of anxiety as to the result of the first of the three great test matches. **1905** *Westm. Gaz.* 19 Aug. 2/3 Until the year 1894 no one had ever heard of a 'Test' match, but..since that time we have been accustomed thus to speak of an England v. Australia match. **1924** *Times* 15 Aug. 5/4 The British team for the first Rugby Football Test Match on Saturday will be selected [in S. Africa]. **1933** M. Nicholls in I. D. Difford *Hist. S. Afr. Rugby Football* xxiv. 335 We won this fourth Test match by 13 points to 5, and squared the rubber. **1974** *Encycl. Brit. Micropædia* IX. 458/2 Rugby League football... The three principal Test-match series stand as follows. **1926** Kipling *Debits & Credits* 273 To prepare for the Form a General Knowledge test-paper. **1927** *Melody Maker* Aug. 792/1 Some bandsmen tell you that after playing a test-piece for perhaps a hundred times they feel they are only just beginning to appreciate it. **1960** *Times* 23 May 16/6 It would make a good test-piece for an international Eisteddfod. **1917** W. L. Wade *Flying Bk.* 193/1 Now with Parnell and Sons, of Bristol, as chief test pilot. **1927** C. A. Lindbergh *We* iv. 61 The service parachute... gave the test pilot a safe means of escape in most cases when all else had failed. **1947** *Sat. Even. Post* 6 Dec. 78/2 They reminded him of the fiery trail left by the high-altitude jet plane he had test-piloted in the last week of the war. **1978** J. A. Michener *Chesapeake* 740 The trial run was without incident, the test pilot pronouncing the craft airworthy. **1958** *Times Lit. Suppl.* 16 May 274/3 The beauty and immensity of the skies have always been a spiritual bonus added to the satisfactions of test-piloting work well done. **1905** D. Mackenzie *Let.* Sept. in *Observer* (1962) 11 Feb. 11/4 The examination of the later test-pits was reserved for a future time at your own express desire. **1952** V. G. Childe *New Light Most Anc. East* vii. 123 How far other innovations ..coincide with the change in pottery cannot be decided from the limited material furnished by a narrow test pit. **1973** *Lebende Sprachen* XVIII. 72/2 On 5th May..Ariel 3 was successfully launched by a scout rocket from the western test range..at Vandenburg Air Force base, California. **1945** L. Guttman in *Psychometrika* X. 255 (*heading*) A basis for analyzing test-retest reliability. *Ibid.* 266 That the universe of trials be indefinitely large seems part of the *definition* of the problem of test-retest reliability. **1960** F. Land *Lang. Math.* xiv. 253 The 'test re-test' method..involves giving the test and then, after some lapse of time, giving it again to the same group of people. .. A correlation less than 0·9 between the two performances of the same test would indicate that its reliability was below the acceptable level. **1957** *Technology* Sept. 244/3 The..mechanical engineering research laboratory ..developed a new test rig for..fittings for oil hydraulic circuits. **1978** R. V. Jones *Most Secret War* xlv. 435 If only we had complete photographic cover of the Blizna area we could have found the launching site or test rig. **1945** *Daily Herald* 31 Aug. 4/4 The B.B.C. is already sending out test signals on the sound channel. **1975** D. G. Fink *Electronics Engineers' Handbk.* xxi. 29 The use of test signals must not result in significant degradation of the program transmissions. **1940** *Chambers's Techn. Dict.*, *Test strip* (*Cinema.*), the specially exposed unmodulated sound-track which is made to ascertain the current in the exciter lamp of a recording machine which gives the requisite density on the negative, after normal development. **1958** T. L. J. Bentley in *Newnes Compl. Amat. Photogr.* iv. 78 By giving a series of test strips different periods of development and measuring the resulting gammas and plotting them against development times, a curve is produced. **1973** D. A. Spencer *Focal Dict. Photogr. Technol.* 623 *Test strip*, a piece of the sensitised material on which exposure is to be made which is exposed in sections, each receiving a different exposure to enable the correct exposure to be judged by the appearance of the developed strip. Each successive section typically receives twice the exposure of the previous section. **1864** W. D. Moore tr. *Donders' Anomalies of Accommodation & Refraction* ii. 99 We give him small print—I to IV of Snellen's test-types to read. **1907, 1962** Test-type [see *Jaeger[2]]. **1877** *Sci. Amer.* 22 Dec. 387/3 A large number of 'wildcats', or test wells, have gone down off the eastern edge of the defined line, but with very few exceptions they have proved dusters. **1925** A. B. Thompson *Oil-Field Explor.* I. v. 208 The selection of sites for test wells is one of the most responsible duties that devolves on pioneers. **1975** *Offshore* Sept. 91/1 A total of 12 deep onshore test wells have been drilled, all of which have been dry holes. **1905** A. Meyer in *Psychol. Bull.* 15 July 242 The time was measured with a stopwatch from the chief syllable of the test-word to the reaction. **1924** E. Ekwall in Mawer & Stenton *Introd. Survey Eng. Place-Names* iv. 60 Norwegian test-words are *breck, buth* (ON *búð*), *gill, scale, slack*. **1965** G. Kristensson in *English Studies* Apr. 142 This surname [*sc.* Ladyman] is..too unreliable to be used as a test-word for the appearance of OE (*ge*)*lād.*

test, *v.*[2] Add: **2.** Also, to subject (a person) to a test of a particular kind; *to test out*, to put (a theory, etc.) to a practical test. Phrases: *to test* (something) *to failure* or *destruction*; *to test the water* (fig.: cf. quot. 1888 in Dict.).

1926 *Publishers' Weekly* 29 May 1794/2 To test out the value of radio publicity. **1939** *Brit. Jrnl. Psychol.* July 1 The range of chronological age of persons tested was so wide that a special enquiry had to be undertaken. **1957** C. N. Parkinson *Parkinson's Law* (1958) 23 So much time has been spent in studying the art of being tested that the candidate has rarely had time for anything else. **1962** F. I. Ordway et al. *Basic Astronautics* vii. 325 We first select 100 units and test them to failure. **1972** D. Ramsay *Little Murder Music* 62 'If you're attempting to establish a motive—' 'I'm just testing the water,' Meredith said. **1974** *Howard Jrnl.* XIV. 104 Legal philosophers could back up these efforts by testing out some of their theories with research projects. **1978** *Washington Post* 20 Jan. D1 Hepburn had played bit or supporting roles in several European movies..before William Wyler tested her and cast her as the runaway princess in 'Roman Holiday'. **1978** A. Price *'44 Vintage* xviii. 203 Sergeant Winston tested the statement to destruction. **1980** J. Krantz *Princess Daisy* xxv. 443 'I guess it's just..lucky..that Supracorp's such a big business,' Kiki said, testing the waters.

5. *intr.* **a.** To undergo a test. *U.S.*

1934 in Webster. **1961** in Webster, Actors..best suited to the roles for which they tested. *Ibid.*, The great turboprop..was still testing. **1981** *Times* 29 Apr. 12/3, I tested with Jack Nicholson for his own film *Goin' South*. It came down to a choice between myself and Mary Steenburger and she got the part.

b. With phrasal compl. To achieve a rating of (so much) as the result of a test. *U.S.*

1934 Webster, s.v. A compound that tests ten per cent. *a* **1961** R. Benedict in Webster, The eyesight of different peoples may test the same. **1971** 'L. Egan' *Malicious Mischief* viii. 135 They could guess that he might test dull-normal. He was seventeen, not very big and not very bright. **1976** M. Machlin *Pipeline* ii. 32 It tests over two thousand barrels a day.

6. *absol.* or *intr.* To apply or carry out a test.

1961 Webster, s.v., Use the scratch technique in

testing for allergies. **1978** T. Sharpe *Throwback* ix. 87 Then say 'Testing. Testing. Testing' into that little transmitter.

testability (testăbi·lĭti). [f. Testable *a.*[2] + -ity.] The quality or state of being testable (see next).

1936 R. Carnap in *Philos. of Sci.* III. 421 An attempt will be made to formulate the principle of empiricism in a more exact way, by stating a requirement of confirmability or testability as a criterion of meaning. **1945** [see *confirmability]. **1952** C. G. Hempel *Fund. of Concept Formation in Empirical Sci.* 43 Just another formulation of the empiricist requirements of testability. **1968** K. R. Popper *Conjectures & Refutations* (ed. 3) i. 37 The criterion of the scientific status of a theory is its falsifiability, or refutability, or testability. **1981** *Word 1980* XXXI. 151 It is only in this way that the model can attain a high degree of testability.

testable, *a.*[2] Delete *rare* and add later examples; *spec.* in *Philos. of Science*, of a theory: capable of being empirically tested.

1922 *Glasgow Herald* 14 Apr. 8 Japanese history does not become a record of testable facts until the fifth or sixth century A.D. **1945** [see *disconfirm *v.*]. **1959** K. R. Popper *Logic of Sci. Discovery* vi. 112 Theories may be more, or less, severely testable; that is to say, more, or less, easily falsifiable. The degree of their testability is of significance for the selection of theories. **1968** P. A. P. Moran *Introd. Probability Theory* i. 57 The two laws differ in their empirical nature in that the first is empirically testable whilst the second is not. **1973** B. Magee *Popper* ii. 22 Scientific laws are testable in spite of being unprovable: they can be tested by systematic tests to refute them.

testate (te·stēit), *a.*[2] [f. as Testacean *a.*: see -ate[2].] = Testacean *a.*, Testaceous *a.* 2.

1947 *Ann. Rev. Microbiol.* I. 4 In the testate *Pontigulasia vas*, a hypothetical layer of endosomal granules or plaques disappears during the anaphase and reappears during the telophase. **1978** *Bio Systems* X. 79/2 The Foraminifera appear to have had an origin in organisms similar to the testate amoebae.

te·st-cross, *sb.* and *v. Genetics.* [f. Test *sb.*[1] + Cross *sb.*] **A.** *sb.* A cross between an individual whose genotype for a certain trait is unknown and one that is homozygous recessive for that trait, so that the unknown genotype may be determined from that of the offspring.

1934 C. B. Bridges in *Jrnl. Heredity* XXV. 18 The type of cross designated..as the 'backcross' is here to stay... Because of this characteristic we may employ the term 'testcross' wherever 'backcross' has been used in the special meaning. **1979** Arms & Camp *Biology* xiii. 198 If the red-flowered plant of unknown genotype were actually heterozygous (Rr), half the offspring of the test cross would be expected to be white-flowered and half red-flowered.

B. *v. trans.* To make the subject of such a cross.

1950 E. W. Sinnott et al. *Princ. Genetics* (ed. 4) iii. 68 The trihybrid with round and yellow seeds and colored flowers.., when test-crossed to plants with wrinkled, green seeds and white flowers..will produce 8 combinations. **1978** *Nature* 27 July 317/1 A proportion of the males from a wild population in Texas showed low levels of recombination when heterozygous males were test-crossed to homozygous marked females.

teste[2]. **1.** (Later examples.)

1916 G. Saintsbury *Peace of Augustans* iii. 130 Lamb (*teste* Hazlitt..) was inclined to agree with Scott. **1968** *Listener* 6 June 737/2 He tells us, *teste* Evelyn Waugh, of a Sitwellian habit of leaving Sitwell press cuttings (surely not *all* their press cuttings?) in bowls on the drawing-room table.

testee[2] (testī·). [f. Test *v.*[2] + -ee[1].] One who is subjected to a test of his or her health, intelligence, knowledge, etc.

1932 W. S. Duke-Elder *Text-bk. Ophthalm.* I. xxv. 986 The fact that picking skeins of wool does not appeal to the average workman, while the reading of pseudo-chromatic diagrams requires a considerable amount of intelligence, has popularized the adoption of lantern tests wherein coloured glasses are illuminated and the testee is asked to name the colour and (sometimes) to match it with wools or some other coloured material. **1947** *Sci. News* IV. 19 The main difficulty with such tests was that the tester was usually as intoxicated as the testee, and often forgot to press the spindle of his stopwatch, or to take proper notes. **1952** C. P. Blacker *Eugenics: Galton & After* 190 These gaps..make it all the more surprising that so little account was taken of the testee's subsequent services and achievements. **1964** M. Critchley *Developmental Dyslexia* vii. 50 The testee is required to detect which of the simple figures lies concealed or incorporated within the complex design. **1976** K. S. Bowers *Hypnosis for Seriously Curious* iii. 43 Even on the subset of questions for which the testee subjectively feels he is simply guessing, the likelihood is high that he will get more than 25% correct answers. **1983** *Daily Tel.* 22 Sept. 18 Gascoigne..said of his testees [on a quiz programme]: 'They were far more argumentative in the 60s and 70s.'

tester[1]. Add: **3.** *tester cloth.*

1776 in J. S. Moore *Goods & Chattels of our Forefathers* (1976) 270 Bedstead, Green Curtains, Vallens and Head and Tester Cloths etc. **1853** *Heal & Sons Catal.: Bedsteads* 59 Half-Tester Bedstead..Chintz Furniture..fringed and fluted head and tester cloth.

tester[4]. Add: **b.** *Biol.* A stock or strain of organism used to investigate some genetic characteristic of another strain.

1925 *Genetics* X. 421 Two other stocks were needed as 'testers'. **1969** A. M. Campbell *Episomes* iii. 37 Operationally, a bacterial culture is termed F+ if it will mate with an F− tester strain.

testibiopalladite (testi:biopæ·lădait). *Min.* [f. Te(llurium + Stibi(um (or Sti(bium + Bi(smuth) + -o + Pallad)ium[2] + -ite[1].] A mineral, Pd(Sb,Bi)Te, found as minute whitish or grey cubic crystals having a metallic lustre.

1974 *Ti Ch'iu Hua Hsueh* (*Geochimica*) III. 181 Sixteen new minerals..were found from two nickelsulfide deposits in China. The three well studied are: Testibiopalladite, Pd(Sb,Bi)Te. **1978** *Canad. Mineralogist* XVI. 126/1 The first occurrence of testibiopalladite outside China is reported from Kambalda, Western Australia, where it occurs as cores to zoned inclusions of michenerite-testibiopalladite in altaite.

testicular, *a.* Add: **1. b.** *testicular feminization* (or *feminizing*): a familial condition produced in genetically male persons by the failure of tissue to respond to male sex hormones, resulting in a normal female appearance (including external genitalia) but with testes in place of ovaries; usu. *attrib.*

1953 J. M. Morris in *Amer. Jrnl. Obstetr. & Gynecol.* LXV. 1192 Actually these patients present a fairly typical clinical picture. For this reason they have been singled out from the other forms of intersexuality, and we have called the clinical syndrome 'testicular feminization'. **1959**, **1970** [see *feminization 2]. **1974** Passmore & Robson *Compan. Med. Stud.* III. xxix. 2/2 One other specific condition is the testicular feminizing syndrome in which XY individuals with testes nevertheless develop female external genitalia and female secondary development at puberty. **1978** *Price's Textbk. Practice of Med.* (ed. 12) v. 550/2 In the testicular feminization syndrome a different approach is indicated because..'she' may already be married and having satisfactory sexual intercourse.

testimonial, *a.* and *sb.* Add: **B.** *sb.* **6.** (sense 5) = serving as a testimonial or token of esteem, esp. in *testimonial dinner, game, match.*

1851 'Bat' *Cricketer's Man.* (ed. 5) 94 On the 26th July, 1847, the Committee at Lord's got up a testimonial match between Kent and England [for Mynn's benefit]. **1895** *Funk's Stand. Dict.* s.v., A testomonial certificate, benefit, or banquet. **1931** *Daily Express* 21 Sept. 11/5 A testimonial dinner was given to..the millionaire American capitalist. **1972** G. Green *Great Moments in Sport: Soccer* iii. 46 The Russians went to the unusual lengths of giving him [Yashin] a testimonial match at Lenin Stadium. **1977** in *Fremdsprachen* XXIII. (1979) 209/1 The long-serving defender, who collected £35,000 from a testimonial game on Friday, was due to retire. **1979** *Tucson* (*Arizona*) *Citizen* (Weekender Mag.) 28 Apr. 18/1 Sol Stein..was decorating a dais with his presence at a testimonial dinner for a minor television personality.

testimony, *sb.* **5.** Add: except in Evangelical circles. Phr. *to give one's testimony* = Testify *v.* 3 b.

1877 *Independent* 29 Mar. 12/1 A fine-looking young man gave a clear, decided testimony for Christ. **1885** C. T. Studd *Let.* in N. P. Grubb *C. T. Studd* (1933) vi. 53 Now he is just as active for the Lord Jesus as he was formerly for the devil. He has three times publicly given his testimony. **1935** N. L. McClung *Clearing in West* iii. 21 So when an old man who stuttered, was giving his testimony and holding back the meeting with everyone getting impatient, I kept my one eye on the minister and the other one shut. **1966** H. Roseveare *Give me this Mountain* ii. 31 Our leader opened the meeting for testimonies. I didn't know what she meant by a 'testimony' so I waited... She made another attempt to get us talking by pointing out what a blessing it could be to testify to others... 'Well, hasn't anyone anything to tell of what God has done for her during the week?'

testing, *vbl. sb.*[2] Add: **b. testing-ground**, an area used for demonstration and experiment; also *fig.*

1872 *Gentl. Mag.* Jan. 71 Since the battle field is the only thorough testing ground for weapons and military schemes. **1919** H. S. Walpole *Secret City* ii. vii. 213, I turned and devoted myself to Uncle Ivan, who was always delighted to make me a testing-ground for his English. **1943** in R. V. Jones *Most Secret War* (1978) xxxvii. 322 The new weapon is in the form of a rocket which has been seen fired from the testing ground.

testo. Add: **c.** The narrator in an oratorio or similar piece of music.

1947 A. Einstein *Music in Romantic Era* xiii. 177 Through the gradual elimination of the *testo* or narrator, oratorio had approached opera to such a degree as to be confused with it. **1980** *New Grove Dict. Mus.* XVIII. 706/2 The *testo* part was normally set as recitative with continuo accompaniment and sung either by one or more soloists... In secular music the term was occasionally used for the narrator in dramatic dialogues and similar works.

testosterone (testǫ·stĕrōʊn). *Biol.* [a. G. *testosteron* (K. David et al. 1935, in *Zeitschr. f. physiol. Chem.* CCXXXIII. 281): see Testis[2] and *-sterone.] **a.** A steroid hormone that stimulates the development of male secondary sexual characteristics and which is produced in the testes, and, in very much smaller quantities, in the ovaries and adrenal cortex.

1935 *Chem. Abstr.* XXIX. 5165 (*heading*) Crystalline male hormone from testes (testosterone), more active than androsterone prepared from urine or cholesterol. **1944** A. Huxley *After Many a Summer* ii. vi. 234 What a course of thiamin chloride and some testosterone would have made him as happy as a sand-boy. **1947** *Nature* 4 Jan. 15/1 Many mammary cancers would regress when the influence of the female sex hormone was lessened by removal of the ovaries or by injections of testosterone. **1961** *New Scientist* 9 Nov. 340/1 Naturally occurring steroid sex hormones can inhibit ovulation... Testosterone, progesterone and the oestrogens fall into this group. **1969** *Nature* 6 Dec. 945/1 Celibacy apparently has no effect on the androgens, for most of the monks excreted as much testosterone—the most potent naturally occurring androgen—as normal sexually active males. **1976** *Maclean's Mag.* 3 May 60/3 Among women virilized in the womb from an excess of testosterone (the male sex hormone), 60% registered IQs over 110.

b. testosterone propionate, the propionic acid ester of testosterone, given parenterally as a longer-lasting alternative to testosterone.

1937 *Proc. R. Soc.* B. CXXIV. 363 Six normal cyclic rats..were injected daily for 10 days with 0·2 mg. of testosterone propionate. **1941** [see *premenstrual *a.*]. **1970** Passmore & Robson *Compan. Med. Stud.* II. xii. 11/2 Testosterone propionate, given intramuscularly in an oily solution, is active over a period of 1–3 days.

test-tube. Add: **b. test-tube baby**, (*a*) a baby conceived by artificial insemination; (*b*) a baby that has developed from an ovum fertilized outside the mother's body; also *fig.* and in similar Combs., as *test-tube child, pregnancy.*

1935 E. Novak *Woman asks Doctor* xii. 155 There has been..a good deal of unfortunate newspaper discussion on the subject of artificial insemination and 'test-tube babies'. **1945** *Daily Herald* 20 Apr. 3/8 Warning on test-tube babies. Artificial insemination of women is being performed on a small scale in this country. **1958** *Times* 18 Jan. 7/2 A 'test-tube' child cannot grow up feeling normal about his true origin. **1965** *New Scientist* 11 Nov. 399/1 The idea of 'test-tube babies' is no longer something to be woven into the plot of a science fiction novel. Serious-minded scientists are not only thinking about culturing human embryos on the laboratory bench—they are developing the techniques which will make this a practical possibility. **1978** *Times* 26 July 1/1 The world's first test-tube baby, a girl, was born by caesarian section just before midnight at Oldham and District General Hospital, Greater Manchester... The embryo was implanted in Mrs Brown's womb after being fertilized in Mr Steptoe's laboratories. **1982** *New Scientist* 4 Feb. 290 Since the birth of the first 'test-tube baby' three and a half years ago it has become increasingly obvious that fertilising human eggs in the laboratory is not simply a new technique for relieving infertility.

‖ **Tet** (tet). [Vietnamese.] **a.** The Vietnamese lunar New Year. Also *attrib.*

1885 J. G. Scott *France & Tongking* v. 104 The especial great season for every one, rich and poor, is the new year, the Têt, the Annamese new year, which corresponds with the Chinese and falls about the beginning of February. *Ibid.* 105 At a season such as this the evil spirits are particularly active and spiteful on account of the general rejoicing. **1931** H. Norden *Wanderer in Indo-China* iii. 55 Tet is the most important New Year's festival which begins a month later than the Occidental new year. During Tet all work is suspended. **1968** *Times* 30 Jan. 4/1 The United States and South Vietnam authorities announced today that they would not observe the 36-hour Tet (Lunar New Year) truce. **1973** *Times* 28 Dec. 5/1 More than 4,000 civilian and military prisoners still held by the South Vietnamese and the communists are to be released before the Tet (Buddhist new year) celebrations on January 23. **1974** P. Gore-Booth *With Great Truth & Respect* 359 The truce agreed on for the traditional *Tet* (New Year) holiday in Vietnam would start on Wednesday 8 February and finish on Sunday afternoon, 12 February.

b. Tet offensive, in the war in Vietnam, an offensive launched by North Vietnamese and Viet Cong forces on 30 January 1968.

1968 *Times* 17 Feb. 4/4 The Vietcong Tet new year offensive was evidence of the correctness of the United States analysis. *Ibid.* 19 Feb. 1/5 The Vietcong are expected to follow up their Tet offensive. **1977** J. Crosby *Company of Friends* xiv. 95 The CIA had predicted the Tet offensive.

‖ **tetampan** (tĕta·mpan). [Malay.] In Western Malaysia, an ornate shoulder cloth worn by those serving royalty.

1821 J. Leyden tr. *Malay Annals* 342 Tun Sura di Raja..brought the creese from the raja..and covered it with a tetampan. **1909** R. O. Winstedt *Life & Customs* (Papers on Malay Subjects) ii. 90 *Kain tĕtampan*, a shoulder-cloth of yellow silk, embroidered, and with gold or silver fringe, worn by court attendants when waiting on Rajas. **1972** M. Sheppard *Taman Indera* 26 Shoulder cloths of the first grade are called *Tetampan*. These are made of velvet and are usually embroidered with the royal emblem or cypher in gold thread. *Ibid.* 84 A short shoulder cloth of yellow velvet, embroidered with silver thread, called *Tetampan.*

tetarteron (tĭtā·ɹtĕrǫn). *Numism.* [a. Gr. τεταρτηρόν, lit. 'measure of capacity', f. τέταρτος fourth.] A Byzantine gold coin of the 10th–11th cent., a copper coin replacing the old follis from the late 11th cent. (see quot. 1969).

1908 W. Wroth *Catal. Imperial Byzantine Coins in Brit. Mus.* [I. p.l, This..coin is stated to have been called..τεταρτερόν. This was probably..its popular nickname.] *Ibid.* II. 659/1 'Tetarteron', l. **1959** E. Pound *Thrones* xcvi. 12 Here, surely, is a refinement of language. ἢ καὶ νομίσματα ἕξει Wd/ appear to be tetarteron tokens not affecting the aureus. **1969** M. S. Hendy *Coinage & Money in Byzantine Empire 1081–1261* vi. 28 The tetarteron nomisma..originally a gold coin,..was first struck by Nicephorus II, and continued until early in the reign of Alexius I. At some point after this, the name was appropriated to describe a copper coin of similar small, thick fabric, first struck by Alexius as an element of his reformed coinage. This change had taken place by 1097. **1973** P. D. Whitting *Byzantine Coins* iii. 40 A *new* gold coin lighter than the solidus was introduced... The new and lighter coin was called the nomisma tetarteron. These tetartera cannot be distinguished by eye until the later part of Basil II's reign... The name means 'a fourth part', *i.e.* a piece of standard weight diminished by a quarter of a tremissis equivalent to $\frac{1}{3}$ of the whole.

tetched (tetʃt), *pa. pple.* and *ppl. a.* U.S. dial. and colloq. var. of *touched* (see Touch *v.* 23 b); mentally deranged to a slight degree; somewhat mad, crazy, or 'cracked'.

1930–41 in H. Wentworth *Amer. Dial. Dict.* (1944) 657/1. **1983** C. McCarry *Last Supper* III. vi. 333 These people are tetched in the head. **1984** S. Bellow *Him with his Foot in his Mouth* 39 If she had been a little tetched before, melodramatic, in her fifties she seemed to become crazed.

tetchous (te·tʃəs), *a. U.S. dial.* Also **tetchious, tetchus.** [f. Tetchy *a.*: see -ous.] = Tetchy, techy *a.* 1.

1890 *Dialect Notes* I. 66 *Tetchus*,..tetchy. **1893** H. A. Shands *Some Peculiarities of Speech in Mississippi* 62 *Tetchous*.., common among negroes and illiterate whites for *tetchy*. Used also in Kentucky. **1913** H. Kephart *Our Southern Highlanders* xiii. 294 A choleric or fretful person is tetchious. **1948** A. Lomax in A. Dundes *Mother Wit* (1973) 484/1 That's what makes the Negro so *tetchious* till today. **1959** W. Faulkner *Mansion* iii. 58 A respectability that delicate and tetchous that wouldn't nothing else suit.

tête-à-tête, *adv.*, *sb.*, and *a.* Add: **C.** *adj. tête-à-tête set* (example).

1870 L. M. Alcott *Old-Fashioned Girl* viii. 163 Such a cunning teakettle and saucepan, and a tête-à-tête set.

Hence as *v. intr.*, to engage in private conversation (together or *with* another).

1861 Mrs. Gaskell *Let.* 10 June (1966) 657 The reason why she & I were tête à têteing in this way was that Mr Gaskell was gone to Liverpool. **1943** *Two Masques* Nov. 4/2 Maureen O'Hara, Patricia Morison and Martha O'Driscoll are the ladies with whom Garfield goes 'tete-a-tete'ing. **1979** G. Swarthout *Skeletons* 48 I'll tête-à-tête with him, too.

tête-bêche (tẹ̄t,bẹ̄ʃ), *sb.* (*a*). *Philately.* [a. Fr., lit. '(sleeping) head to foot', f. *tête* head + *bêche*, reduced from *béchevet*, lit. 'double bed-head'.] (A stamp) printed upside down relative to the next stamp in the same row or column (see quot. 1913). Freq. *attrib.* in phr. *tête-bêche pair*. Also as *adv.*

1874 *Stamp-Collector's Mag.* XII. 10 The Marquis de L—— has kindly forwarded for notice a reversed 4 centime laureated French empire stamp; technically termed a *tête-bêche*. **1882** E. B. Evans *Catal. Collectors Postage Stamps* 56 One or more stamps upside down,.. forming the varieties termed *têtes-bêches*. *Ibid.*, Varieties 2 and 3 are the result of stamps placed *tête-bêche*. **1891** *S. Gibbons' Monthly Jrnl.* 30 Jan. 153/2 The sheets are composed of four horizontal rows of five stamps,..each row is placed *tête-bêche* to the one below it. **1913** E. B. Evans *Stamps & Stamp Collecting* (ed. 4) 103 *Tête-bêche*. A term applied in French to stamps printed upside down in reference to one another. One such stamp may appear in a sheet, through one of the dies forming the plate being accidentally set the wrong way; this stamp will be *tête-bêche* as regards those surrounding it. Some of the stamps of Grenada were printed with alternate rows reversed, so that the stamps in one row were *tête-bêche* with reference to those in the next. Such varieties must of course be shown in pairs, as the stamps when separated exhibit no peculiarity. **1921** F. A. Bellamy *Oxf. & Cambr. Coll. Messenger Postage Stamps* 14 Balliol, a number of impressions were made one way, then the paper strip was turned round; so a tête bêche pair can be found on each strip. **1971** *Daily Tel.* 16 July 7/6 The 2 annas is known in a tête bêche pair (one stamp upside down in relation to the other).

‖ **tête de bœuf** (tẹ̄t də böf). *Embroidery.* [Fr., lit. 'ox's head'.] Used *attrib.* to designate an embroidery stitch (see quots.).

1882 Caulfeild & Saward *Dict. Needlework* 195/1 *Tête de Bœuf Stitch*. The name of this stitch is derived from its shape, the two upper stitches having the appearance of horns, and the lower ones of an animal's head... The needle is inserted and brought out for the two slanting stitches that commence the next Tête de Bœuf. **1923** *Daily Mail* 19 Jan. 15 *Tete de bœuf stitch*... Two slanting stitches which meet in the form of a V are made, and from the inside of this, at the bottom, is taken a chain-stitch which is caught down with a short over-stitch. **1934** M. Thomas *Dict. Embroidery Stitches* 198 Tête-de-bœuf filling stitch. Single daisy stitches set between two right-angled straight stitches make up this pretty filling.

‖ **tête de cuvée** (tẹ̄t də küve). [Fr., lit. 'head of the vatful'.] A vineyard producing the best wine in a village area; wine from such a vineyard.

[**1833** C. Redding *Hist. & Descr. Mod. Wines* v. 100 The best Burgundies, called *les têtes de cuves*, are from the choicest vines..grown on the best spots in the vineyard, having the finest aspect.] **1908** E. & A. Vizetelly *Wines of France* 122 The finer Volnay, what is called the *tête-de-cuvée* wine, has a most refreshing flavour. **1952** W. Stevens *Let.* 29 Sept. (1967) 761, I sat at lunch with a little Corton (1929, tete de cuvée). **1965** A. Sichel *Penguin Bk. Wines* III. 147 The above listed vineyards are all *têtes de cuvées*, that is the highest class in their village area... It must not be assumed that the *têtes de cuvées* of different villages are equal in quality. Many names of the next category—the *premier cru* or *cuvée*.. may be better.

‖ **tête de nègre** (tẹ̄t də nẹ̄gr'). [Fr., lit. 'Negro's head'.] A dark brown colour approaching black. Usu. *attrib.* Cf. *nigger brown* s.v. *Nigger *sb.* 2 d.

1916 in G. Howell *In Vogue* (1975) 20/1 (Advt.), Tête de Negre..Hat, gold embroidery. **1923** *Daily Mail* 5 Mar. 15/3 A striking gown..is worn over a slip of tete de negre silk. **1973** *Country Life* 22 Feb. 455/1 Design of baskets of spring flowers..on a *tete de nègre* (that is a not dead black) ground.

tête de pont. (Later examples.)

1853 J. H. Stocqueler *Milit. Encycl.* 283/2 In order to add to the defence of *Têtes de Pont*, reduits have been constructed within them. **1918** E. S. Farrow *Dict. Milit. Terms* 613 *Tête-de-pont*, a work thrown up at the end of a bridge to cover communication across a river; a bridgehead. **1926** Fowler *Mod. Eng. Usage* 329/1 The strong tête-de-pont fortifications were rushed by our troops, & a battalion crossed the bridge.

‖ **tête exaltée** (tẹ̄t egzalte), *adv. phr.* [Fr., lit. 'with head elated'.] In an elated or euphoric manner. Also as *sb.*, someone behaving thus.

1841 C. Fox *Jrnl.* 6 June in *Memories of Old Friends* (1882) vii. 128 Carlyle..said, 'Give my love to your dear interesting nephew and nieces!'.. I walked *tête-exaltée* the rest of the day. **1856** C. M. Yonge *Daisy Chain* I. xxii. 232 Flora thought of the words *'tête exaltée'*, and considered herself alone to have sober sense enough to see things in a true light. **1873** —— *Pillars of House* III. xxviii. 128 A pious utterance that only a *tête exaltée* takes literally.

tetel (tẹ̄ⁱ·təl, te·təl). Also **tetl.** [Local name.] = *Tora.

1867 S. W. Baker *Nile Trib.* 308 We had hardly ridden half a mile, when I perceived a fine bull tétel..standing near a bush. **1894** Sclater & Thomas *Bk. Antelopes* I. 16 The Tora or Tétel was confounded by von Heaglin and Sir Samuel Baker, its first discoverers, with the Bubal. **1920** *Blackw. Mag.* Nov. 672/2 A great herd of tetl—big animals the size of a mule—sprang up.

‖ **tête montée** (tẹ̄t mon̂te), *adj. phr.* Also *erron.* **tête monté.** [Fr., lit. 'excited head'.] Over-excited, agitated, worked up. Also as *sb.*, this state of mind.

1825 H. Wilson *Memoirs* I. 12, I had suffered severely from wounded pride, and, in fact, I was very much tête monté. **1836** E. Grosvenor *Let.* in G. Huxley *Lady Elizabeth & Grosvenors* (1965) vii. 160 The tête montée state of the young *Brutus's* and patriots..in France. **1859** Trollope *Bertrams* I. viii. 155 But in truth George was somewhat afflicted by a *tête montée* in this matter. **1882** E. W. Hamilton *Diary* 29 Aug. (1972) I. 328 She regards Davitt as the incarnation of vanity and Dillon as a *tête montée*. **1936** J. M. Keynes *Gen. Theory Employment* vi. 64 It might be, of course, that individuals were so *tête montée* in their decisions as to how much they themselves would save and invest respectively, that there would be no point of price equilibrium at which transactions could take place. **1960** L. Cooper *Certain Compass* 118 Adrian was *tête monté* .., in that slightly exalted state.

tether, *sb.* Add: **5.** *tether-length* (example); **tether-ball** (examples).

1900 *Queen* 29 Sept. (Advt.), Parlour tether ball... This..game consists of a perpendicular pole, to the top of which an india-rubber ball is attached by a cord... Each player is provided with a bat, with which to strike the ball. **1925** T. Dreiser *Amer. Tragedy* I. II. xxxviii. 425 His own mental tether-length having been strained to the breaking point. **1937** J. Bancroft *Games* 632 Tetherball... This is one of the most delightful and vigorous games that is adapted to small playing space. **1973** E. S. Shneidman *Deaths of Man* ix. 95 A..bachelor was found hanging from a tetherball pole.

Tethys (te·þis). *Geol.* [L. *Tēthys*, Gr. Τηθύς, a sea-goddess: see quot. 1893.] The name of a large sea that formerly lay between Eurasia and Africa. Hence **Te·thyan** *a.*

1893 E. Suess in *Nat. Sci.* II. 183 Modern geology permits us to follow the first outlines of the history of a great ocean which once stretched across part of Eurasia. The folded and crumpled deposits of this ocean stand forth to heaven in Thibet, Himalaya, and the Alps. This ocean we designate by the name 'Tethys', after the sister and consort of Oceanus. *Ibid.* 184 The later Tethyan history.. forms certainly one of the most attractive chapters of historical geography. **1931** [see *Laurasia]. **1947** Auden *Age of Anxiety* vi. 133 The Laurentian Landshield was ruthlessly gerrymandered, And there was a terrible tussle over the Tethys Ocean. **1970** R. M. Black *Elements Palaeont.* vi. 52 The rudists occur mainly in the deposits of Tethys. **1971** *Nature* 29 Jan. 311/1 Old ocean floor was subducted into the Tethyan trench. **1972** *Sci. Amer.* June 61/3 In late Paleozoic times a wide tropical seaway, the Tethys, almost circled the globe. The only barrier to the Tethys Sea was formed by the combined land masses of North America and western Europe, which were then connected.

Teton (tī·tǫn). *U.S.* [ad. Dakota *thí* + *hųwą*, dwellers on the prairie.] **a.** (A member of) a Western division of the Dakota or Sioux Indian tribe. Also *attrib.*

1806 *Message from President of U.S., communicating Discoveries made in exploring the Missouri by Captains Lewis & Clark* 32 This trade, as small as it may appear, has been sufficient to render the Tetones independent of the trade of the Missouri. **1840** *N.Y. Mirror* 4 July 12/3 His household was the whole tribe of the Teton Dahcotas. **1873** *Forest & Stream* 9 Oct. 133/1 For several hours we followed on the trail of the Tetons. **1937** R. H. Lowie *Hist. Ethnol. Theory* ix. 133 Boas has trained Miss Ella Delovia to take down Teton stories among her people. **1975** J. A. Hanson *Metal Weapons, Tools, & Ornaments of Teton Dakota Indians* i. 3 The Tetons, who spoke Lakota, took their name from the term *Titonwan*, 'Dwellers of the Prairie'.

b. The dialect spoken by this people.

1911 F. Boas *Handbk. Amer. Indian Lang.* (U.S. Bureau Amer. Ethnol. Bull. No. 40) 880 We give here the description of the Teton as obtained by Dr. Swanton. **1933** [see *Dakota *sb.* 2]. **1976** W. L. Chafe in T. A. Sebeok *Native Lang. Americas* I. 542 There are usually said to be four major Dakota dialects: Santee (Dakota proper), Teton (Lakota), and Yankton and Assiniboine.

tetotaciously, var. *Teetotaciously *adv.*

tetra-. Add: **1. tetra·bolo** [f. *Di)abolo by deliberately false analogy (see quot. 1961)], a polyabolo composed of four triangles; **tetra-co·ccus** (pl. **-cocci**) *Biol.* [Coccus] (see quot. 1968); **tetragamy:** also, marriage with four women simultaneously; **tetralemma** (earlier example); **tetrapare·ntal** *a. Biol.*, (of an organism) produced by the fusion of two embryos; also as *sb.*, a tetraparental individual; **tetrapare·sis** *Path.* [Paresis], muscular weakness of all four limbs; hence **tetrapare·tic** *a.*; **tetrape·ptide** *Biochem.* [ad. G. *tetrapeptid* (see *Peptide)], an oligopeptide in which there are four amino-acid residues in the molecule; **tetrapho·nic** [Gr. φωνή voice, sound], applied to certain forms of quadraphonic recording and reproduction (see quots.); **tetratonon** (earlier example); **tetrawi·ckmanite** *Min.*, a tetragonal polymorph of wickmanite, $MnSn(OH)_6$, found as yellow crystals.

1961 *New Scientist* 21 Dec. 752/3 Mr. S. J. Collins..has experimented with the various plane shapes that can be formed by edgewise joins of four isosceles right-angled triangles; for these he most ingeniously suggests the name 'tetraboloes'. His excuse is that a 'diabolo' has two such triangles in its cross-section (joined pointwise, not edgewise: but no matter!). **1967** Tetrabolo [see *polyabolo* s.v. *Poly- 1]. **1893** W. R. Dawson tr. *Schenk's Man. Bacteriol.* i. 2 Cocci are..found either singly or united in groups... If the elements are joined in pairs and fours we distinguish respectively, according to the number, diplococci and tetracocci. **1907** *Practitioner* Apr. 488 Two diplococci are frequently seen together, giving a tetracoccus form. **1968** M. Hynes *Med. Bacteriol.* (ed. 9) i. 1 Cocci which divide regularly in two planes at right angles to one another result in collections of four organisms, and are known as tetrads or tetracocci. **1907** W. De Morgan *Alice-for-Short* xviii. 203 Charles had said to his friend, jokingly, that if he had fifty sisters, single ones, Johnson was welcome to make offers to them all round.—'But then, my dear Paracelsus, that was to be *if* I *had* fifty. That would leave me forty-nine—or in case of bigamy, forty-eight; or quadrogamy—tetragamy—whatever it ought to be—forty-six. **1918** R. A. Knox *Spiritual Aeneid* x. 166 We could always split the difference between monogamy and tetragamy by having two wives all round. *a* **1856** W. Hamilton *Lect. Metaphysics & Logic* (1860) III. xviii. 352 If it [*sc.* the syllogism] has three, four, or five members, it is called *trilemma* (*tricornis*), *tetralemma* (*quadricornis*), *pentalemma* (*quinquecornis*). **1970** *Nature* 31 Jan. 462/2 Tetraparental mice are produced by fusing two eight-cell stage embryos. *Ibid.*, Allelic differences between the strains produce mosaic patterns in the adult tetraparental which make possible inferences about development. **1971** *New Scientist* 8 Apr. 72/1 One of the most fascinating tools employed in studying these processes [of tissue differentiation]..is provided by 'tetraparental' mice. **1979** *Nature* 11 Oct. 429/1 B. Mintz (Institute for Cancer Research, Philadelphia)..pioneered the use of allophenic mice (formed by aggregating cells from two 8-cell embryos from two different pregnant mice, hence tetraparental). **1972** *New Gould Med. Dict.* (ed. 3), *Tetraparesis*, weakness of all four extremities. **1980** *Brit. Med. Jrnl.* 29 Mar. 902/2 She was anarthric and bedridden with spastic tetraparesis and twitching tremors. **1978** *Jrnl. R. Soc. Med.* LXXI. 449 A woman.. who was tetraparetic following operation for a cerebral tumour. **1906** *Jrnl. Chem. Soc.* XC. I. 810 Triglycylglycine methyl ester, prepared by esterifying the tetrapeptide with methyl alcohol and hydrogen chloride,

crystallises in microscopic, glistening needles. **1927** P. A. Levene *Chem. Relationships of Sugars* 2 in *Contemp. Devel. in Chem.*, A tetrapeptide composed of naturally occurring amino acids is hydrolyzed by trypsin. **1970** R. W. McGilvery *Biochem.* iv. 51 Peptide subunits join to make the tetrapeptide because of interactions between side chains of residues exposed at the meshing surfaces. **1970** *Hi-Fi Mag.* Dec. 23/2 Microphones are placed so channels 1 and 3, and 2 and 4 will operate as pairs to give three walls of sound... In addition, channels 3 and 4 operate as a stereo pair to sharpen the directionality of the reflected sound... When we discovered this..we called it Tetraphonic Sound. **1974** *Wireless World* July 236/2 Gerzon's assertion that 'the optimum characteristic is not known' in regard to a particular tetraphonic technique could be equally well applied to any quadraphonic technique. **1977** *Daily Tel.* 4 May 18 'Tetraphonic' has already been given the technical meaning of a set of signals giving complete first-order directional information including height. **1740** J. Grassineau *Mus. Dict.* 276 Tetratonon: the superfluous fifth may be thus called, as containing four tones. **1973** White & Nelen in *Mineral. Rec.* IV. 24/1 The mineral..was not wickmanite, but its tetragonal dimorph! To emphasize this dimorphic relationship the mineral has been named tetrawickmanite. **1978** *Ibid.* IX. 41/2 The Langban tetrawickmanite occurs as bright yellow euhedra implanted on magnetite.

2. a. te:traalkyllea·d [*Alkyl], any compound in which the molecule consists of four alkyl radicals bonded to an atom of lead; **tetrabe·nazine** *Pharm.* [f. *benzo*(*a*)quinolizin-2-one, f. Benz(ene + *-izine*, denoting two fused rings with a nitrogen atom common to both], a tricyclic compound, $C_{19}H_{27}NO_3$, used in the treatment of chorea; **tetrachloride:** *carbon tetrachloride*, a dense, colourless liquid, CCl_4, used chiefly in the manufacture of refrigerants and aerosols and in fire extinguishers, and also as a solvent; **te:trachlo:rodibe:nzo-p-dioxin**, a polychlorinated tricyclic compound, $C_{12}H_4O_2Cl_4$, that is carcinogenic and teratogenic in animals and is formed in the manufacture of chlorinated phenols; also called *dioxin*; abbrev. TCDD (see *T 6 a); **te:trachlo:rodibe:nzoparadio·xin** = prec.; **te:trachlor(o)e·thane**, either of two isomeric compounds, $C_2H_2Cl_4$, that are dense colourless liquids; *spec.* the symmetrical isomer, which is used chiefly in the manufacture of solvents and is toxic; **te:tradecape·ptide**, a polypeptide having fourteen amino-acid residues in the molecule; **te:tra,e:thylammo·nium** *Chem.* (also † **tetre:thyl-**), the quaternary ion $(C_2H_5)_4N^+$, which is a ganglion-blocking agent and has been used (in the form of its chloride or bromide salt) in the treatment of hypertension; **tetraethyl lead**, a poisonous oily liquid, $Pb(C_2H_5)_4$, which is added to petrol as an anti-knock agent making possible higher compression ratios; = *lead tetraethyl* s.v. *Lead *sb.*[1] 12 b; **tetrae:thyl-pyropho·sphate**, a colourless, hygroscopic liquid, $(C_2H_5)_4P_2O_7$, that is a cholinesterase inhibitor and is used as a garden insecticide and formerly to relieve the symptoms of myasthenia gravis; **tetrahy·drate**, a hydrate containing four molecules of water; **te:trahy:drocanna·binol**, a hydrogenated derivative of cannabinol that is the active principle in cannabis and hashish; **te:trahydrofu·ran**, a colourless liquid, C_4H_8O, used as a solvent for plastics and other polymers and as an intermediate in the manufacture of organic chemicals; **te:trahy:drofu·rfuryl**, the monovalent radical $C_4H_7O.CH_2$—; **te:trahy:dronaphthalene**, any compound derived from naphthalene by the addition of four hydrogen atoms; *spec.* 1,2,3,4-tetrahydronaphthalene, a colourless liquid used as a solvent for fats, resins, esp. varnishes, lacquers, etc.; = *Tetralin; **te:tra,iodothy·ronine** *Biochem.* [*Thyronine] = *Thyroxine; **tetrame·thylene**, the gaseous alicyclic compound $(CH_2)_4$; also, (a compound containing) the bivalent straight-chain radical $—CH_2(CH_2)_2CH_2—$; **tetra·misole** *Pharm.* [f. I]mid(e + *thia*)zole, with alteration of *z*], an anthelminthic drug used in man and animals, usu. as the hydrochloride; 2,3,5,6-tetrahydro-6-phenylimidazo[2,1-*b*]thiazole, $C_{11}H_{12}N_2S$; **tetrapy·rrole**, any compound containing four pyrrole nuclei, esp. when in the form of a ring (cf. *Porphin); hence **tetrapyrro·lic** *a.*

1923 *Jrnl. Amer. Chem. Soc.* July 1821 The method used by Krause..always yields a tetra-alkyl lead compound when an alkyl halide is used. **1978** *Nature* 26 Oct. 738/1 Particulate lead and tetraalkyllead in the atmosphere are due principally to emissions from vehicles fuelled by leaded petrol. **1958** *Federation Proc.* XVII. 404/1 (*heading*) Inhibition of reserpine tranquilizing effects by tetrabenazine, a synthetic tranquilizing agent. **1974** *Lancet* 26 Jan. 107/1 Tetrabenazine is the drug of first choice for the suppression of chorea in patients with Huntington's chorea. **1866** Carbon tetrachloride [in Dict.]. **1930** *Engineering* 26 Dec. 814/2 The use of chemical cleaners, such as..carbon tetrachloride,..is more effective. **1947** J. C. Rich *Materials & Methods of Sculpture* v. 94 The wax is placed in the carbon tetrachloride and set aside for about two days to dissolve, after which it can be used. **1972** *Materials & Technol.* IV. iv. 201 Carbon tetrachloride was originally made from coke and chlorine, carbon disulphide being used as an intermediate, but is now mainly produced by the chlorination of methane, or the chlorinolysis of higher hydrocarbons. **1959** *Jrnl. Pharm. Soc. Japan* LXXIX. 188 (*caption*) 2,3,7,8-Tetrachlorodibenzo-*p*-dioxin. **1970** *Kirk-Othmer Encycl. Chem. Technol.* (ed. 2) XXII. 180 Dioxins, including 2,3,7,8-tetrachlorodibenzo-*p*-dioxin, have been detected as contaminants in samples of 2,4,5-T. **1976** *Daily Tel.* 2 Aug. 11/8 He is to attempt to obtain an experimental patch of land 100 yards square by the introduction of healthy micro-organisms of yeast and mould, which he hopes will 'seed' upon the poisonous substance, breaking down the molecules of TCDD (Tetrachlorodibenzo-dioxin). **1977** *New Yorker* 25 July 30/1 The samples of the chemical used in the experiments had contained uncharacteristically high levels of a toxic contaminant, 2,3,7,8-tetrachlorodibenzo-*p*-dioxin—commonly referred to as TCDD, or, by chemists familiar with the subject, either as tetra dioxin or simply as dioxin. **1978** *Price's Textbk. Pract. Med.* (ed. 12) ii. 284/2 [illegible] tetrachlorodibenzoparadioxin (dioxin) which has produced chloracne in [illegible] workers. **1980** *National Geographic* Aug. 162 TCDD—shorthand for 2,3,7,8-tetrachlorodibenzoparadioxin, frequently simply called dioxin—is the unavoidable by-product of the manufacture of the herbicides 2,4,5-T and silvex. **1871** *Jrnl. Chem. Soc.* XXIV. 191 The former has the composition of tetrachlorethane, $C_2H_2Cl_4$, and is formed, according to the usual reaction of phosphorus pentachloride on aldehydes, by the substitution of 2 at. Cl for 1 at. O in dichloraldehyde. **1922** *Encycl. Brit.* XXX. 35/1 Tetrachlorethane was tried with success, but it proved dangerous to the operatives applying it in enclosed places. Moreover, sunlight decomposed tetrachlorethane. **1933** *Jrnl. R. Hort. Soc.* LVIII. 282 Tetrachlorethane as a Greenhouse Fumigant. **1934** H. Hilton *Notes on Technique of Painting* iii. 235 Dissolve hard copal resin in tetra-chlor-ethan [*sic*]. **1963** A. J. Hall *Textile Sci.* vii. 307 Dry cleaning involves the extraction of dirt, and grease, oil, fat and wax stains from all kinds of textile materials by treatment with a solvent such as..tetrachlorethane. **1974** *Encycl. Brit. Micropædia* IX. 910/3 The other isomer, 1,1,1,2-tetrachloroethane, has no commercial application. **1973** *Science* 5 Jan. 79/1 The linear tetradecapeptide was synthesized by solid-phase methodology. **1852** A. W. Hofmann in *Q. Jrnl. Chem. Soc.* IV. 306 For this [organic] metal I propose..the name Tetrethylammonium..which implies that it is built up by the union of nitrogen with four equivalents of the hypothetical hydrocarbon called ethyl. **1940** *Thorpe's Dict. Appl. Chem.* (ed. 4) IV. 155/1 Tetraethylammonium hydroxide, NEt_4OH, known only in solution. **1946** J. H. Burn *Drugs, Med. & Man* vi. 65 Two American workers attempted to use..tetraethylammonium to reduce blood pressure in patients in 1946, and then two workers in England introduced hexamethonium bromide which was more powerful and acted for a much longer time. **1923** *Jrnl. Amer. Chem. Soc.* July 1821 Tetra-ethyl lead was prepared by P. Pfeiffer by the action of lead chloride on ethyl-magnesium iodide. **1940** *Economist* 30 Mar. 586/2 The 87-octane spirit is generally obtained by the addition of small quantities of tetra-ethyl lead to good-quality 'straight-run' petrol. **1979** *Nature* 14 Mar. 990/1 [illegible] **1949** [illegible] tetraethyl pyrophosphate [illegible]. **1952** H. Beckman *Pharmacol. in Clin. Pract.* 568 Prostigmin may usually be discontinued at this point and tetraethylpyrophosphate cautiously increased..until there is maximal relief of symptoms without toxic effects. **1974** *Encycl. Brit. Micropædia* IX. 923/1 Tetraethyl pyrophosphate is extremely poisonous to humans, the toxic effects being similar to those of parathion. **1886** *Jrnl. Chem. Soc.* XLIX. 418 The thermal reaction..is made up of two distinct quantities—(1) the decomposition of the tetrahydrate, (2) the formation of the tetrahydrate. **1951** [illegible] **1970** [illegible] tetrahydrocannabinol [illegible]. **1970** *Daily Tel.* 19 Sept. 11/2 If the female is pollinated, it uses some of its resinous tetrahydrocannabinol [*printed* -cannibol]—the stimulant in marijuana—to produce seeds. **1908** *Jrnl. Chem. Soc.* XCIV. I. 280 When furan is hydrogenated at 170° by Sabatier and Senderens' method, the principal product consists of tetrahydrofuran. **1956** *Nature* 21 Jan. 128/2 Both the styles 6:6-cyclic monomers ..and the cyclic dimer from caprolactam..were reduced by lithium aluminium hydride in boiling tetrahydrofuran without difficulty. **1978** *Further Perspectives Organic Chem.* (CIBA) 23 Why should the gas phase be a better model for enzymic reactions than a solvent which is more protein-like than, say, water..or tetrahydrofuran? **1928** *Jrnl. Amer. Chem. Soc.* L. 1821 α-Tetrahydrofurfuryl alcohol will shortly be available on a commercial scale from the catalytic reduction of furfural. **1951** Kirk & Othmer *Encycl. Chem. Technol.* VI. 1004 In the United Kingdom, tetrahydrofurfuryl acetate has been used as a lipstick ingredient, based on the high solubility of eosin in it. *Ibid.*, Tetrahydrofurfuryl alcohol is used in the preparation of esters, especially tetrahydrofurfuryl oleate, which is almost colourless and has excellent light and heat stability. **1887** *Abstr. Proc. Chem. Soc.* III. 88 At the next meeting..the following Papers will be read..'Derivatives of Hydrindonaphthene and Tetrahydronaphthalene'. By W. H. Perkin, jun. **1904** [see *Hydronaphthalene]. **1935** *Industr. & Engin. Chem.* (News Ed.) 20 Aug. 331/2 The merit of tetrahydronaphthalene for town gas mains lies in the fact that, being a close chemical relative of naphthalene it has the power of absorbing larger quantities of naphthalene than any of the other solvents in present use. **1964** N. G. Clark *Mod. Organic Chem.* xix. 391 Naphthalene quite readily undergoes addition reactions. For example at 200° catalytic hydrogenation over nickel yields 'tetrahydronaphthalene' ('tetralin'). **1969** N. A. J. Rogers in S. Coffey *Rodd's Chem. Carbon Compounds* (ed. 2) IIIc. xx. 71, 1,4,5,8-Tetrahydronaphthalene, 'isotetralin',..m.p. 58°, may be prepared by reduction of 1,4-dihydronaphthalene by the 'metal in ammonia' method. **1928** Tetraiodothyronine [see *Thyroxine]. **1974** D. & M. Webster *Compar. Vertebr. Morphol.* xiii. 310 The two hormones, triiodothyronine and tetraiodothyronine (thyroxine), are iodinated amino acids. **1885** W. H. Perkin in *Jrnl. Chem. Soc.* XLVII. 806 It was thought that the simplest method would be to regard the saturated hydrocarbons themselves as multiples of methylene, CH_2, and thus name them di-, tri-, tetra-, penta-, &c., methylene, as is easiest seen from the following table:—..Tetra-methylene. **1898** J. Walker *Introd. Study Org. Chem.* xxxv. 219 The tetramethylene compound..is formed by the reduction of ethylene cyanide with sodium amalgam in alcoholic solution. **1909** [see *pentamethylene* s.v. *Penta-]. **1944** S. J. Smith *Princ. Org. Chem.* x. 197 Both their methods of preparation and their paraffins..show them to be cyclic. **1966** [see *Styrene]. **1973** N. L. Allinger et al. *Org. Chem.* x. 193 Frequently occurring hydrocarbon groupings that have more than one site for attachment of substituents..are given special names.. $ICH_2CH_2CH_2CH_2I$, 1,4-Diiodobutane (Tetramethylene iodide). **1966** Thienpont et al. in *Nature* 22 Mar. 1084/1 This article reports the discovery of tetramisole.., a new, potent broad spectrum anthelmintic. **1978** *Ibid.* 22 June 612/1 Tetramisole and/or its levorotatory isomer levamisole is used in many countries against a broad range of nematode infections in birds, pigs, ruminants and man. **1917** *Chem. Abstr.* XI. 452 [illegible] **1968** [see *Prodigiosin]. **1976** *Ann. Rev. Microbiol.* XXX. 410 Relatively general metabolic criteria, such as tetrapyrrole biosynthesis, have proven useful in defining taxonomic relationships. **1944** *Ann. Reg.* 1943 358 Vitamins operate as coenzymes in plant respiration, others of which are the important tetrapyrrolic compounds as are present in animal tissues. **1975** *Nature* 22 May 357/2 The Hans Fischer school which then dominated tetrapyrrolic chemistry.

tetraalkyllead, tetrabenazine: see *Tetra- 2 a. **tetrabolo:** see *Tetra- 1.

tetracaine (te·trăkēⁱn). *U.S. Pharm.* [f. Tetra- + Co)caine.] A compound related to procaine that is used, as a solution of the hydrochloride, as a local anaesthetic, esp. for surface application; amethocaine, $C_4H_9NHC_6H_4COOCH_2CH_2N(CH_3)_2$.

1943 *Dispensatory U.S.A.* (ed. 23) 1121/1 Tetracaine Hydrochloride occurs as a fine, white, crystalline, odorless powder. **1974** [see *spinal anæsthesia* s.v. *Spinal *a.* 6*]. **1979** Snyder & Levinson *Anesthesia for Obstetrics* ix. 110/2 Although tetracaine remains one of the most effective and popular drugs for subarachnoid block, it is a poor choice in epidural analgesia.

tetrachloro-: see *Tetra- 2 a.

tetrachloroethylene (te:traklō^ro,e·pilin). *Chem.* Also **-chlorethylene.** [f. Tetra- + Chloro-[2] + Ethyl + -ene.] = *Perchloroethylene. Also called **te:trachloroe·thene.**

1911 *Chem. Abstr.* V. 1853 In this way the author has obtained..tetrachlorethylene. **1930** A. F. Holleman *Text-bk. Org. Chem.* (ed. 7) i. 137 In contact with water tetrachloroethene reacts with chlorine under the influence of sunlight to form trichloroacetic acid. **1946** J. H. Burn *Lect. Notes Pharmacol.* 99 Hook worms..live in the duodenum. The infestation is treated by carbon tetrachloride and tetrachloroethylene. **1968** A. A. Baker *Unsaturation in Org. Chem.* iii. 27 Faraday..heated the perchloride of carbon to dull redness, finding that chlorine was liberated, and a new compound, which he called protochloride of carbon was formed..known today as tetrachloroethene (C_2Cl_4). **1969** *Times* 20 Feb. 17/5 Some 100,000 gallons of the cleaning fluid tetrachloroethylene were buried in a gold mine, so as to avoid contamination with cosmic rays, and left in place for nearly a year.

tetrachordon. (In Dict. s.v. Tetrachord.) (Earlier example.)

1740 J. Grassineau *Mus. Dict.* 274 Tetrachordon. See Tetrachord.

tetrachoric (tetrăkǭ·rik), *a. Statistics.* [f. Gr. τετράχωρ-ος divided into four (f. χῶρος place: see Tetra-) + -ic 1.] Applied to a table in which data are divided into two according to each of two criteria, and so having four subdivisions; of or pertaining to such a table; applied *esp.* to an estimate of the product-moment coefficient derived from such

a table, and to concepts used in obtaining such an estimate.

1910 P. F. Everitt in *Biometrika* VII. 438 In the present tables the values of the first six τ functions, henceforth to be termed tetrachoric functions, have been computed for values of $\frac{1}{2}(1-\alpha)$ from ·001 to ·500 by successive increments of ·001. **1918** *Ibid.* XII. 95 In the ordinary scheme for a tetrachoric table, the quadrants are denoted by *a, b, c, d*. **1943** M. G. Kendall *Adv. Theory Statistics* I. xiv. 354 Tetrachoric *r* and biserial η. Both these coefficients are, in effect, estimates of a putative product-moment correlation for data which are not specified with the detail of an ordinary bivariate table. **1956** J. Whatmough *Language* 241 The statistical method of the so-called 'tetrachoric R'..is valuable in revealing the degree of correlation..between languages which may be suspected of being historically akin. **1964** [see *Polychoric *a.*].

tetrachotomy. (In Dict. s.v. Tetrachotomous *a.*) Restrict *Zool.* and *Bot.* to sense in Dict. and add: (*b*) *Logic*, a division having four members.

a **1856** W. Hamilton *Lect. Metaphysics & Logic* (1860) IV. xxv. 23 If a division has only two members, it is called a dichotomy;..if four, a tetrachotomy.

tetracoccus: see *Tetra- 1.

tetracosactrin (te:trăkǫsæ·ktrin). *Pharm.* [f. Tetra- + I)cos- + *A(dreno)c(ortico)tr(oph)in.] A synthetic polypeptide (see quot. 1967) which resembles corticotrophin in its action and uses but lacks its antigenic property, and is given (as the acetate) by injection in the long-term treatment of inflammatory and degenerative disorders.

1967 *Brit. Med. Jrnl.* 18 Nov. 391/1 Tetracosactrin (β^{1-24} corticotrophin, Synacthen) is a synthetic polypeptide containing the first 24 amino-acids found in naturally occurring corticotrophin (A.C.T.H.). **1972** *Ibid.* 11 Mar. 680/1 The pathognomic finding is their failure [*sc.* that of urinary and plasma corticosteroids] to show a rise after the administration of ACTH or tetracosactrin. **1979** *Jrnl. R. Soc. Med.* LXXII. 530 A tetracosactrin (Synacthen) stimulation test.

tetracyclic (tetrăsəi·klik, -si·klik), *a.* [f. Tetra- + Cyclic *a.*] **1.** (In Dict. s.v. Tetra- 1.)

2. *Chem.* Of a compound: containing four fused hydrocarbon rings in the molecule.

1928 *Chem. Abstr.* XXII. 2748 (*heading*) Synthesis of tetracyclic compounds and of pyrene. **1977** J. L. Harper *Population Biol. of Plants* xiii. 414 Plants that are only lightly predated contain three or four isomers of lupanine and closely related tetracyclic compounds.

tetracycline (tetrăsəi·klīn). *Pharm.* [f. *Tetracycl(ic *a.* + -ine[5].] **a.** A tetracyclic compound, $C_{22}H_{24}N_2O_8$, which is a broad-spectrum antibiotic (usu. administered as the hydrochloride). **b.** Any of a number of antibiotics structurally related to this compound.

1952 C. R. Stephens et al. in *Jrnl. Amer. Chem. Soc.* LXXIV. 4977/1 Common to both Terramycin and aureomycin is the structure A for which we propose the name tetracycline. **1956** *Nature* 3 Mar. 433/2 (*heading*) Avidity of the tetracyclines for the cations of metals. *Ibid.*, This investigation is now extended to the parent substance, tetracycline,..[and] also includes some new values for the substituted tetracyclines. **1966** I. Jefferies *House-Surgeon* vi. 115 Start her on one of the tetracyclines. **1974** M. C. Gerald *Pharmacol.* xxvi. 457 Tetracyclines are believed to inhibit protein synthesis by blocking the binding of the amino acid-transfer RNA complex to ribosomes. **1978** *Time* 3 July 43/2 Like almost all U.S. farmers, the cattleman is aggrieved... The costs of everything he buys —gasoline, fertilizer, tetracycline for ailing heifers..— have climbed like corn in August.

tetrad. Add: **2. f.** *Ecol.* (See quot. 1976.)

1963 Hawkes & Readett in P. J. Wanstall *Local Floras* 37 We soon realized that it would be impossible to record from every basic square in the county and we modified the method by considering the squares in blocks of four ('tetrads') and selecting one square at random from each tetrad for surveying. **1968** *Watsonia* VI. 351 This involved the detailed survey of 1 km squares as the unit of recording, one square at random being selected from each block of four or 'tetrad'. **1976** J. G. Dony *Bedfordshire Plant Atlas* 10/1 It has become usual in the survey of areas as small as Bedfordshire to divide the ten-kilometre grid squares into 25 smaller squares each 2 km. × 2 km. known as tetrads. Each tetrad has an area of four square kilometres. **1983** *Natural World* Spring 18 Distributional maps based on 2 × 2 kilometre squares, or tetrads.

tetradecapeptide: see *Tetra- 2 a.

tetradic, *a.* **a.** (Later examples.)

a **1914** C. S. Peirce *Coll. Papers* (1935) VI. 222 A tetradic, pentadic, etc. relationship is of no higher nature than a triadic relationship. **1921** [see *Monadic *a.* 1 b].

tetraethyl(-): see *Tetra- 2 a.

te:traflu:oroe·thylene (-flū:ǫro,e·þ-). *Chem.* Also **-fluorethylene.** [f. Tetra- + *Fluoro- + Ethylene.] A dense, colourless gas, $F_2C{:}CF_2$, which is polymerized to make plastics. Cf. *Polytetrafluoroethylene.

1933 *Jrnl. Amer. Chem. Soc.* LV. 3177 There is good evidence that tetrafluoroethylene is produced in small quantities. **1946** *Sci. News* IV. 63 Polymerisation of tetrafluorethylene..yields heat-resistant, chemically inert solids, which can be moulded into any desired shape. **1958** *Times Rev. Industry* Feb. 81/2 A rather typical example of what can be achieved with fluorocarbons is the polymerization of tetrafluorethylene to yield a commercial plastic. **1973** *Materials & Technol.* VI. 544 The monomer tetrafluoroethylene is obtained by pyrolysis from $CHFCl_2$ (..that is, the refrigerants Freon-22 and Genetron-141).

tetragamy: see Tetra- 1 in Dict. and Suppl.

tetragonal, *a.* (*sb.*) Add: **5.** Also applied to (the structure and symmetry of) substances crystallizing in this system. (Further examples.)

1878 Tetragonal symmetry [in Dict.]. **1886** [see *grey tin* s.v. *Grey, gray *a.* 8 c]. **1912** *Dana's Man. Mineral.* (ed. 13) 36 The cross section of a crystal when viewed in the direction of the axis of tetragonal symmetry consists usually of a square or a truncated square. **1937** A. F. Rogers *Introd. Study Minerals* (ed. 3) I. 67 Interfacial angles in the prism zone are the same for corresponding faces of all tetragonal crystals. **1973** H. D. Megaw *Crystal Structures* xii. 307 Several other modifications of the spinel structure have been reported. Cr_2CuO_4 has a different tetragonal structure, with approximately the same unit cell as Cr_2NiO_4.

tetragonally *adv.* (examples).

1963 *Q. Jrnl. Geol. Soc.* CXIX. 326 Tetragonally symmetrical flattened pumice lapilli. **1966** Phillips & Williams *Inorg. Chem.* II. xxv. 248 CuF_2 has a tetragonally-distorted rutile structure.

tetrahedral, *a.* Add: **3.** Math. *tetrahedral numbers*, the series of integers 1, 4, 10, 20,..., the *n*th member of which is the sum of the first *n* triangular numbers.

1939 W. W. R. Ball *Math. Recreations & Ess.* (ed. 11) ii. 59 The sums of consecutive triangular numbers are the tetrahedral numbers. **1983** *Austral. Personal Computer* IV. v. 103/1 The Tetrahedral Numbers..represent the number of identical spheres that can be stacked in a complete triangular pyramid, or tetrahedron.

tetrahedrane (tetrăhī·drẹ̄ⁱn). *Chem.* [f. Tetrahedr(on + -ane.] A compound whose molecule consists of four CH groups forming the corners of a tetrahedron.

1964 *Tetrahedron Lett.* No. 22. 1418 Tetrahedrane (C_4H_4), which has local C_{3v} symmetry, should have a $J_{C\text{-}H}$ of about 225 c.p.s. **1976** *Sci. Amer.* Feb. 106/1 If this carbene were to undergo internal addition, the product would be the hypothetical tetrahedral molecule tetrahedrane (C_4H_4). Although the reaction has been tried many times, tetrahedrane has so far eluded isolation.

tetrahydrate to **-iodothyronine:** see *Tetra- 2 a.

tetrakis- (te·trăkis), formative element [f. Gr. τετράκις four times] used in *Chem.* in the names of compounds to signify four identical groups all substituted in the same way; formerly = Tetra- 2.

1850 [see Tri- 5 a]. **1912** *Jrnl. Chem. Soc.* CI. 2003 The product obtained was tetrakisazobenzene, a deep red substance. **1951** Kirk & Othmer *Encycl. Chem. Technol.* VII. 580 Tetrakis(*p*-dimethylaminophenyl)hydrazine is dissociated to the extent of 10% in benzene solution and 21% in nitrobenzene. **1963** A. J. Hall *Student's Handbk. Textile Sci.* v. 259 The Proban flameproofing process is essentially based on the use of tetrakis-hydroxymethylphosphonium chloride. **1979** G. C. Barrett in D. N. Jones *Comprehensive Organic Chem.* III. xi. iv. 77 Tetrakis(methylthio)methane, $(MeS)_4C$, undergoes exchange with dithiols.

tetralemma: see Tetra- 1 in Dict. and Suppl.

Tetralin (te·trălin). Also **tetralin.** [f. Tetra- + Naphtha)lin(e.] = *tetrahydronaphthalene* s.v. *Tetra- 2 a. (A proprietary name in the U.S.)

1920, etc. [see *Decalin]. **1924** *Nature* 14 June 866/2 Excluding benzol..alternative fuels of greatest promise include 'tetralin' (tetrahydronaphthalene), which, mixed with benzol and alcohol, was used considerably by the Germans during the war. **1944** *Official Gaz.* (U.S. Patent Office) 18 Jan. 350/1 E. I. du Pont de Nemours and Company... *Tetralin* for tetrahydronaphthalene. **1980** *Sci. Amer.* May 89/1 Certain trialkyl tin compounds,.. and acetyl-ethyl-tetramethyl tetralin, a synthetic fragrance at one time used in certain cosmetics, have also been shown to damage myelin preferentially.

tetralogy. Add: **3.** *Med.* A set of four symptoms jointly characteristic of a disorder; chiefly with reference to Fallot's tetralogy (see *Fallot).

1927 [see *Fallot]. **1966** Wright & Summers *Systemic Path.* I. ii. 75/2 The tetralogy..consists of pulmonary stenosis.., ventricular septal defect, displacement of the aorta to the right,..and right ventricular hypertrophy. **1970** [see *Fallot].

tetramer (te·trăməɹ). *Chem.* [f. Tetra- + *-mer.] A compound whose molecule is composed of four molecules of monomer.

1929 *Chem. Abstr.* XXIII. 3213 The dimer and tetramer have also been obtained in cryst. form. **1939** *Jrnl. Amer. Chem. Soc.* LXI. 2320/1 With the aid of a molecular still we isolated the trimer and tetramer of hexatriene. **1966** *New Scientist* 16 June 724/2 The structures of the trimer and tetramer have been determined, and it is known that the trimer is virtually flat, whereas the tetramer is puckered. **1978** *Nature* 6 Apr. 496/2 The haemoglobin molecule is a tetramer of two pairs of identical polypeptide chains, the α and β chains.

Hence **tetrame·ric** *a.*; **tetra:meriza·tion,** the formation of a tetramer from smaller molecules.

1938 *Jrnl. Chem. Soc.* 290 The study of the trimeric and the tetrameric products of acid-catalysed polymerisation. **1962** [see *Depolymerize *v.*]. **1971** *Chem. Abstr.* LXXIII. Subject Index 3626S/3 Tetramerization catalysts. **1974** Gill & Willis *Pericyclic Reactions* iii. 69 In certain circumstances it might then be possible to arrange for a concerted cyclic trimerization or tetramerization to occur. **1976** *Ann. Rev. Microbiol.* XXX. 96 Each tetrameric isozyme is thought to contain the three gene products in the ratio of 2:1:1. **1979** *Nature* 29 Feb. 625/1 Stacked A_4 groups in between the cloverleaf branches are supposed to aid in lining up the protein-complexed individual cloverleaf elements for tetramerisation.

tetramethylene, tetramisole: see *Tetra- 2 a.

tetranu·cleotide. *Biochem.* Also † **-nucleotid.** [f. Tetra- + *Nucleotide.] An oligonucleotide composed of four (unspecified) nucleotides; orig., one composed of one each of four different bases and formerly thought to constitute the nucleic acid molecule (so *tetranucleotide hypothesis, theory*).

1912 W. Jones in *Jrnl. Biol. Chem.* XII. 34 The writer will be permitted to suggest that the nomenclature be made conformable to that which Levene and his co-workers have adopted for the nucleic acids. The term 'tetranuclease' indicates clearly a ferment which exerts its activity upon a tetranucleotide. **1912** A. E. Taylor *Digestion & Metabolism* vii. 429 Mononucleotids and tetranucleotids are definitely known; the most common of the tissue nucleic acids are tetranucleotids. **1914** *Jrnl. Biol. Chem.* XVII. 73 Yeast nucleic acid is a *tetra*-nucleotide and is composed of four mono-nucleotide groups. **1931** Levene & Bass *Nucleic Acids* II. ix. 289 The tetranucleotide theory is the minimum molecular weight and the nucleic acid may as well be a multiple of it. **1952** *Biochem. Jrnl.* LII. 566/1 The core is composed of the tetra- and penta-nucleotides and possibly some of the trinucleotides. **1960** L. Pickin *Organization of Cells* iv. 108 Simultaneously in 1948, several groups of workers independently put forward evidence throwing doubt on the old tetranucleotide hypothesis, according to which the four nitrogenous bases present: adenine, guanine, cytosine, and thymine, were combined in equimolecular proportions in a tetranucleotide. This itself was supposed to consist of four nucleotides, each bearing a molecule of one of the four bases. **1974** *Nature* 26 Apr. 783/1 The tetranucleotide hypothesis..asserted that DNA and RNA had a molecular weight of about 1,300, and contained one of each of the four bases (adenine, guanine, cytosine and thymine in DNA; adenine, guanine, cytosine and uracil in RNA). Because these bases were found to be present in the form of nucleotides (sugar-phosphate + base) the molecule was called a tetranucleotide.

Tetra Pak (te·tră,pæk). Also **Tetra pack** and as one word. [f. Tetra- + Pack *sb.*[1]] A proprietary term in the U.S. for a tetrahedral carton used for packing milk and other drinks. Hence as *v. trans.* (nonce-wd.), to sell in such a pack.

1953 *Official Gaz.* (U.S. Patent Office) 16 June 616/2 Ser. No. 623,384. Aktiebolaget Tetra Pak, Lund, Sweden. Filed Jan. 10, 1952. *Tetra Pak.* **1958** *Mod. Packaging Encycl. 1959* in *Mod. Packaging* Nov. 232 (*caption*) Triangular milk packs are formed, filled from a single roll of paper by this machine. ('Tetra Pak' machine.) **1963** *Economist* 10 Aug. 535/2 Express Dairies only 'tetra-packs' its milk for vending machines and half-pints. **1973** *Times* 29 Oct. 15/8 Just take a look at the Channel Islands, they don't use *any* bottles, not one, they use Tetrapacks; so why can't we. They are cheap, there is no disposal problems [*sic*] as with plastic.

tetraparental to **tetraphonic:** see *Tetra- 1.

tetraplegia (tetrăplī·dʒăi). *Path.* [f. Tetra- + Para)plegia.] = *Quadriplegia.

1911 F. S. Arnold tr. *Bing's Compendium of Regional Diagnosis* IIA. ii. 105 The pyramids are in such close apposition to each other in the medulla, that minimal lesions may cause a condition of tetraplegia. **1964** J. J. Walsh *Understanding Paraplegia* ii. 6 When the arms are involved it is more correct..to use the term 'tetraplegia' or 'quadriplegia', and throughout this book the former word will be adopted. **1974** A. Henry in R. M. Kirk et al. *Surgery* xvi. 349 The incidence of tetraplegia is high in this injury as the spine may have angled acutely ..and thus severely pinched the spinal cord.

Hence **tetraple·gic** *a.* and *sb.* = *Quadriplegic *a.* and *sb.*

1911 F. S. Arnold tr. *Bing's Compendium of Regional Diagnosis* IA. iii. 36 In a complete transverse lesion of the [spinal] cord in the upper cervical region the simultaneous paralysis and anæsthesia affect the four extremities and trunk (tetraplegic type). **1939** W. Haymaker *Bing's Textbk. Nervous Dis.* xix. 523 The contractures give way to flaccid paralyses. The latter may be monoplegic, hemiplegic, paraplegic, even tetraplegic. **1964** J. J. Walsh *Understanding Paraplegia* xvi. 110 Many tetraplegics..are capable of driving a properly converted car with automatic gear box. **1977** *Lancet* 7 May 1013/2 A strain *Ps*[*eudomonas*] *æruginosa* was isolated from a catheter specimen of urine from a tetraplegic patient. **1979** *Daily Tel.* 27 Jan. 18 As a tetraplegic may I thank your Health Services Correspondent..for continued interest in what he describes as the 'shambles' at Stoke Mandeville.

tetraploid (te·trăploid), *a.* (and *sb.*) *Biol.* [f. Tetra- + *-ploid.] (Made up of somatic cells) containing the diploid chromosome complement twice over. Also as *sb.*, a tetraploid organism.

1914 *Proc. R. Soc.* B. LXXXVII. 484 In the tetraploid giants the chromosomes are $2x(24)$ in the gametic and.. $4x(48)$ in the somatic cells. **1921** *Amer. Naturalist* LV. 261 Few-noded tetraploids, however, are less easily distinguished. **1932** [see *Quadruplex *a.* 3]. **1952** *New Biol.* XIII. 38 It is necessary, before tetraploid rye can be introduced for general cultivation, to see that no normal rye is grown in the locality where the tetraploid is to be grown. **1974** E. Stacey *Peace Country Heritage* ii. 90 Clover..breeders also have concentrated some of the better features, such as leaf retention and plant vigor, into 'tetraploids' by doubling the usual number of chromosomes. **1982** *Sci. Amer.* May 118/3 Potato plants are tetraploid.

Hence **te·traploidy,** the state of being tetraploid.

1918 Babcock & Clausen *Genetics* xiv. 263 The doubling of the number of chromosomes typical of the species is known as tetraploidy. **1941** *Amer. Naturalist* LXXV. 321 In many ornamental plants there is a definite need for new types with larger flowers, later blooming habit, and other characteristics usually associated with tetraploidy. **1970** *Sci. Jrnl.* June 78/3 In lower animal forms both triploidy and tetraploidy are compatible with normal development.

tetrapod, *a.* and *sb.* For **a, b** read **A, B** and add: **B.** *sb.* **2.** (See quot. 1962.)

1962 *Newsletter Brit. Petroleum Co. Ltd.* No. 314. 1 An ingenious type of concrete block will next year be helping to protect the harbour at Das Island off the Arabian coast against storms. Known as Tetrapods, these blocks..may be visualised as a central sphere around which are equally spaced four truncated cone-shaped legs. When a number of Tetrapods are placed in position these legs interlock. **1980** *Citizen* (Ottawa) 3 Dec. 43/1 The tetrapods, which look like children's playing jacks, are designed to break up heavy waves in the event of a hurricane.

tetrapyrrole, -ic: see *Tetra- 2 a.

tetrasome (te·trăsōᵘm). *Cytology.* [f. Tetra- + *-some[4].] A chromosome which is represented four times in a chromosomal complement; also, a tetrasomic individual.

1921 [see *hexasome* s.v. *Hexa-]. **1944** *Genetics* XXIX. 232 Through selfing of plants possessing these aberrations, the corresponding nullisomes and tetrasomes have been obtained. **1958** C. P. Swanson *Cytol. & Cytogenetics* vi. 192 Trisomes $(6n+1)$ and tetrasomes $(6n+2)$ have also been found..in *T*[*riticum*] *vulgare.* **1973** [see *Nullisome].

tetrasomic (tetrăsōᵘ·mik), *a.* (*sb.*) *Cytology.* [f. as prec. + -ic.] Of or pertaining to a tetrasome. Also as *sb.*, a tetrasomic chromosome, cell, or individual.

1922 A. F. Blakeslee in *Amer. Naturalist* LVI. 19, I have suggested..the terms disome [etc.]..with the adjectives disomic, trisomic, tetrasomic. **1923** *Bot. Gaz.* LXXVI. 345 Since in the somatic cells of a tetraploid *Datura* each of the 12 chromosomal sets consists of 4 homologous chromosomes instead of only 2 as in diploids, it is obvious that in dealing with the transmission of Mendelian characters we have to do with tetrasomic rather than with the disomic inheritance more familiar to students of heredity. **1937** C. D. Darlington *Rec. Adv. Cytol.* 325 Tetrasomics (whether of fragments or whole chromosomes) are more markedly abnormal..than the corresponding trisomics. **1946** *Nature* 21 Sept. 418/1 Lucerne.., having also given segregation ratios which can best be interpreted as tetrasomic, may be regarded as an autotetraploid. **1961** *Lancet* 7 Oct. 789/1 His father was in effect tetrasomic for that chromosome. **1974** *Nature* 19 Apr. 714/3 The book is then divided into two main parts, the first dealing with trisomics, with some mention of tetrasomics, and the second dealing mainly with monosomics but also mentioning nullisomics.

So **te·trasomy,** tetrasomic state.

1961 *Lancet* 23 Sept. 724/1 Monosomy, trisomy, or even tetrasomy have very little functional effect. **1977** Zellweger & Simpson *Chromosomes of Man* p. xii, Trisomy = 46 + 1 chromosome. Tetrasomy = 46 + 2 chromosomes of the same type.

tetrathlon (tetræ·þlǫn). [f. Tetra- + Gk. ἆθλον contest, after Pentathlon.] An athletic contest comprising a series of four events, esp. one (for juveniles) comprising riding, shooting, swimming, and running.

1959 in *Chambers's 20th Cent. Dict.* Add. **1961** *Times* 29 May 4/4 The 1961 Schools tetrathlon competition, organized by the Modern Pentathlon Association of Great Britain and the R.M.A. Sandhurst, was won by Whitgift. **1973** *Daily Mail* 4 Sept. 33/5 Recently winning the British Horse Society's Pony Club tetrathlon in Warwickshire. **1979** *Daily Tel.* 2 Oct. 19/4 In the recent Pony Club Tetrathlon championships (pentathlon minus fencing) more than 100 girls..competed.

tetratonon: see *Tetra- 1 in Dict. and Suppl.

tetravalent, *a.* Add: = *Quadrivalent *a.* 1; (further example); **tetravalence** (example); also **tetrava·lency.**

1887 *Trans. R. Soc. Edin.* XXXII. 456 The tetravalence of Carbon unsatiated by the bivalence of Oxygen. **1913** *Phil. Mag.* XXVI. 495 The observed trivalency and tetravalency respectively of these elements. **1976** *Sci. Amer.* Dec. 33/1 In the solvents that were used..tetravalent plutonium ions, Pu^{4+} (plutonium atoms from which four electrons have been removed), are soluble. **1982** *Nature* 25 Nov. 386/1 Organic chemistry, thanks to the tetravalency of carbon and the stability of its incestuous bonds, is responsible for most of the compounds.

tetrawickmannite: see *Tetra- 1.

tetrazolium (tetrăzōᵘ·liǫm). *Chem.* [f. Tetrazol(e + *-ium b.] **a.** The ion or radical $N_4CH_3^{(+)}$ derived from tetrazole. **b.** Any of various derivatives of this, esp. triphenyl tetrazolium chloride, a reagent used as a test for viability in biological material. Usu. *attrib.*

1895 *Jrnl. Chem. Soc.* LXVIII. I. 574 (*heading*) Constitution of tetrazolium bases. *Ibid.*, When tetrazolium derivatives are oxidised, the phenyl radicle is not eliminated. **1947** *Nature* 31 May 748/1 (*heading*) Tetrazolium salt as a seed germination indicator. **1969** J. Levitt *Introd. Plant Physiol.* ii. 13/1 Tetrazolium dyes..are converted from the colorless to the colored (e.g., red) form by freshly cut surfaces of living cells. **1980** *Nature* 8 May 80/1 The resulting recombinants are Mal^- when scored on maltose tetrazolium agar.

tetrazotize (tetræ·zǭtəiz), *v. Chem.* [f. Tetra- + Azote + -ize: cf. Diazotize *v.* in Dict. and Suppl.] *trans.* To convert (a compound) into one that contains two diazo groups. Hence **tetra:zoti·zable** *a.*, **tetra·zotized** *ppl. a.*; **tetra:zotiza·tion,** the process of tetrazotizing.

1908 J. C. Cain *Chem. of Diazo-Compounds* 165 Benzidine, when tetrazotized, becomes [etc.]. *Ibid.* 166 This explains why the tetrazotization does not proceed normally. **1933** *Jrnl. Amer. Chem. Soc.* LV. 4540 The technical importance of a method of tetrazotizing *p*-phenylenediamine..in a quantitative way has been shown. *Ibid.* 4541 It appears that *o*-phenylenediamine is tetrazotizable. **1940** [see *Congo 3]. **1972** *Science* 9 June 1132/2 Tetrazotized benzidine spray..aided in the identification of the four major components.

tetrode. Add: **2.** *Electronics.* A thermionic valve with four electrodes.

1919, 1932 [see *Pentode]. **1941** *Electronic Engin.* XIV. 385 The valves in this section are push-pull beam tetrodes. **1943,** etc. [see *Kinkless *a.*]. **1962** D. F. Shaw *Introd. Electronics* xi. 234 The defect in the tetrode characteristics ..is eliminated by the insertion of a third grid, called the suppressor grid, between the anode and the screen. **1976** *Physics Bull.* Aug. 359/3 (*caption*) This component is a new tetrode from Thomson-CSF.

† **B.** *adj. Telegr.* Applied to a mode of multiplex telegraphy by which four messages can be transmitted along a wire simultaneously. *Obs.*

1886 [see *Hexode *a.*].

tetrodotoxin (te:trǭdotǫ·ksin). [a. G. *tetrodotoxin* (Y. Tahara 1911, in *Biochem. Zeitschr.* XXX. 263), f. Tetrodo(n + Toxin.] A poisonous substance found in the ovaries of certain fish of the family Tetraodontidæ.

1911 *Jrnl. Chem. Soc.* C. II. 133 Tetrodotoxin is neither acid nor base, and yields on hydrolysis a base and a crystalline substance. **1938** *Chem. Abstr.* XXXII. 8582 The toxic action of tetrodotoxin is decreased by vitacamphor, coramine,..and cardiazol. **1965** *New Scientist* 18 Feb. 442/3 As deadly as tetrodotoxin from the Japanese globe or 'puffer' fish. **1977** *Lancet* 24/31 Dec. 1331/1 The electrical and mechanical consequences of stimulation are prevented by tetrodotoxin.

tetromino (tetrǫ·mino). [f. Tetr(a- + D)omino by deliberately false analogy: see quot. 1961.] Any of the five distinct planar shapes that can be formed by joining four identical squares by their edges.

1954 S. W. Golomb in *Amer. Math. Monthly* LXI. 678 The checker board cannot be covered with 15 L-tetrominoes and one square tetromino. **1961** [see *Pentomino]. **1979** *Sci. Amer.* Apr. 19/1 It is when we turn to the 4-cell animals (the tetrominoes) that the project really becomes interesting.

tetrose. Delete 'the class of' and add examples.

1904 [see *Diose]. **1916** [see *Biose]. **1963** [see *Erythrose]. **1970** A. L. Lehninger *Biochemistry* xi. 218 If the carbon chains of the trioses are extended by the addition of carbon atoms, we have, successively, tetroses, pentoses, hexoses, heptoses, and octoses.

tetryl. Add: **2.** Also **Tetryl.** [ad. G. *tetril.*] A yellow crystalline nitro-amino explosive, tetranitromethylaniline, $(NO_2)_3C_6H_2N(CH_3)NO_2$, used esp. as a detonator and priming agent.

1909 O. Guttmann *Manuf. Explosives* iii. 47 The Rheinisch-Westfälische Gesellschaft of Troisdorf make now detonators of Tetranitromethylaniline (called Tetryl). **1977** D. MacKenzie *Raven & Ratcatcher* v. 75 A box of detonating-caps, the type with tetryl booster-charges.

tetter, *v.* Restrict † *Obs.* to sense in Dict. and add: **2.** *intr.* To crack, to disintegrate.

1911 J. Masefield *Everlasting Mercy* 30 My mind began to carp and tetter. **1967** T. Keneally *Bring Larks & Heroes* ii. 16 In dutiful vegetable gardens, the leaves of carrots and turnips had tettered and split, shot full of holes by antipodean summer.

tettered (te·təɹd), *a.* [f. Tetter *sb.*, *v.* + -ed[2,1].] Afflicted with tetter (usu. *fig.*).

1906 W. de la Mare *Poems* 84, I marvelled at..this poor creature..tettered with worms of fear. **1908** G. Bottomley *Chambers of Imagery* 2nd Ser. (1912) 19 Iron misused must turn to blight And dwindle to a tettered crust. **1975** J. I. M. Stewart *Gaudy* xvi. 281 The great façade with its massive columns was crumbling, flaked and tettered.

tetterous, *a.* (Later example.)

1977 J. I. M. Stewart *Madonna of Astrolabe* xii. 181 The lizards, darting from crevice to crevice on a crumbling wall, were in process of shedding tetterous skins to reveal a summer green.

tettigonian (tetigōᵘ·niăn). [f. mod.L. *Tettigonia* (see next) + -an.] = next.

1842 T. W. Harris *Treat. Insects Injurious to Vegetation* 183 The Tettigonians, or leaf-hoppers, have the head and thorax somewhat like those of frog-hoppers.

tettigoniid (tetigōᵘ·ni,id). Also **tettigonid.** [a. mod.L. family name *Tettigoniidæ*, f. generic name *Tettigonia* (Linnæus *Systema Naturæ* (ed. 10, 1758) I. 429), f. Tettix + *-onia*: see -id[3].] = *long-horn*(*ed*) *grasshopper* s.v. *Longhorn 4. Also *attrib.*, of or pertaining to an insect of this kind or the family in which it is included.

1921 H. T. Fernald *Appl. Entomol.* xvi. 86 Some of the Tettigoniids are wingless and come out only at night. **1935** *Discovery* Nov. 317/2 Certain long-horned grasshoppers or Tettigonids..are representatives of extensive groups. **1939** M. Barr *Insect Legion* ii. 12 The big Tettigonids..are capable of biting a piece of flesh out of a finger. **1946** F. E. Zeuner *Dating Past* xii. 365 *Platycleis occidentalis jerseyana*.., a tettigoniid grasshopper, is well distinguished in size. **1957** *New Biol.* XXIII. 31 This wing mechanism is also found in the Tettigoniids. **1972** [see *long-horn*(*ed*) *grasshopper* s.v. *Long-horn 4].

teuchter (tiu·χtəɹ). *Sc.* Also **teuchtar.** [Origin unknown.] A Highlander (see also quots. 1962 and 1977).

1940 R. Garioch *17 Poems for 6d.* 13 Thir a glaikit pair o Teuchters, an as Heilant as a peat. **1962** *Scotsman* 26 Jan. 11 There is ample evidence that she referred to him as a 'teuchter', a word which I understand to mean a country bumpkin. **1977** *Times Lit. Suppl.* 9 Sept. 1084/2 For the inhabitants of Harris are mainly what most Scots call 'teuchtars'—a word which I had never heard till I had it applied to me by a teacher in a Glasgow school. What is a teuchtar? It is a Lowland Scots imitation of a Gaelic noise, a term of now genial contempt for a crofter or, more generally, for anyone from beyond the Highland line. **1979** R. Laidlaw *Lion is Rampant* xviii. 137, I look like the archetypal teuchter, right down to the fur-bearing cheeks.

teucrium (tiū·kriǒm). [mod.L., a. Gr. τεύκριον, a name used by Dioscorides.] A herb or shrub of the genus of this name, belonging to the family Labiatæ; = Germander.

1673 J. Ray *Obs. Journey Low-Countries* 257 On the sides of the Mountains..*Teucrium*. **1917** L. H. Bailey *Stand. Cycl. Hort.* VI. 3324/2 The teucriums are little known in cultivation. **1962** R. Page *Educ. of Gardener* v. 151, I would plant grey-leaved shrubs to grow wild and make a thicket: rosemary, cistus, the grey-leaved teucrium so much used in the South of France. **1974** *Country Life* 25 Apr. 997/2 The terraced paths..take you across the hillside..between hedges of rosemary and lavender and teucrium, always aromatic in the sunshine.

teuf-teuf (töf,töf). [a. Fr.: echoic.] An imitation of the repeated sound of gases escaping from the exhaust of a petrol engine. Hence as *v. intr.*, (of a motor) to make such a sound; (of a person) to ride in a chugging motor vehicle. Now usually anglicized as *Tuff-tuff.

1902 *Daily Chron.* 22 Aug. 3/4 The 'teuf-teuf' of the rapid motor is everywhere on the splendid roads. **1904** H. G. Wells *Food of Gods* I. iv. 125 People..used to see him almost daily teufteufing slowly about Hyde Park. **1905** *Westm. Gaz.* 9 Sept. 11/3 The teuf-teufing of the motor is drowned by the tinkle of marriage bells. **1907** G. B. Shaw *John Bull's Other Island* III. 77 His ear catches an approaching teuf-teuf. **1914** T. A. Baggs *Back from Front* xxiv. 121 Suddenly, piercing the night stillness, came the harsh teuf-teuf of a motor-car.

TEUTHOLOGIST

teuthologist. (Later example.)
1982 Sci. Amer. Apr. 82/1 Teuthologists, the specialists who study cephalopods (the group of marine animals that includes the squid, the cuttlefish and the octopus).

teuto-. Add: **1.** (Earlier examples.)
1866 Anthrop. Rev. IV. 62 The Teuto-Celts, under Charlemagne, vanquished the pure Saxons of the fatherland. Ibid. 66 A Teuto-Celtic race extends from the northern shores of the Shetland Isles to the Gulf of Lyons.
2. Teutophobia (earlier example).
1876 H. James Let. 1 Apr. in Parisian Sketches (1958) x. 102 [M. Tissot's] Teutophobia, as an exhibition of vivacity and energy, is really very fine.

Teutonic, *a.* and *sb.* Add: **A.** *adj.* **1.** (Later examples.) Esp., displaying the characteristics attributed to Germans. Cf. Teutonically *adv.*
1925 F. Scott Fitzgerald Great Gatsby i. 3 A little later I participated in that delayed Teutonic migration known as the Great War. 1955 Times 5 July 14/3 The arresting self-portrait of 1914 has a teutonic assurance of manner. 1976 Broadcast Dec. 18/2 He was Teutonic in appearance, and wearing what appeared to be an Army type of tunic, slate grey in colour. 1983 Financial Times 11 Oct. 34/5 Research has shown that Hertz has a rather Teutonic, super-efficient but cold image.
C. *Comb.*, as *Teutonic-Edwardian* adj.
1976 J. Wheeler-Bennett Friends, Enemies & Sovereigns iii. 73 It was hideous, since neither of its previous owners seemed to have had any decorative taste at all, but comfortable in a sort of Teutonic-Edwardian way.

Teu·tonized, *ppl. a.* [f. Teuton + -ize + -ed[1].] Made Teutonic; Germanized.
1866 Anthrop. Rev. IV. 131 The Teutonized Celts of Britain. 1918 Hist. Amer. Lit. I. 357 The Teutonized rhapsodies of Coleridge. 1924 Blackw. Mag. Aug. 280/2 All Germans kept their eyes firmly fixed on a Teutonised Europe.

Tevet, var. *Tebeth. **tevish,** var. *thivish.

tew, *v.*[1] Add: **I. 4. c.** Also *pass.*, to be involved or mixed *up with*.
1890 Kipling Life's Handicap (1981) 67 Happen there was a lass tewed up wi' it. 1904 S. R. Crockett Strong Mac xxxix. 323 Ye were somedeal tewed up wi' a lass, were ye no?

Tewa (tē̆ⁱ·wă), *sb.* and *a.* Also † **Tegua.** [a. Tewa *téwa.*] **A.** *sb.* **a.** An Indian people of the south-western U.S.; a member of this people. **b.** The Tanoan language of this people. **B.** *adj.* Of or pertaining to the Tewa or their language.
1865 Rep. U.S. Bureau Indian Affairs 1864 191 The only reliable, genuine name ascertained is that of the dialect spoken by San Juan, Santa Clara, and others included in that class, which is the Tegua, pronounced Té-wa. 1896 Amer. Anthropologist IX. 345 The Pueblo tribes..embody four linguistic stocks... The Tanoan stock is..composed of five dialectical divisions—Tano, Tewa, Jemez, and Piro. 1910 F. W. Hodge Handbk. Amer. Indians II. 737/2 Tewa ('moccasins', their Keresan name). A group of Pueblo tribes belonging to the Tanoan linguistic family. Ibid., In 1598 Juan de Oñate named 11 of the Tewa pueblos. 1910 Amer. Anthropologist XII. 503 Tewa is rich in sentence-words. 1912 Ibid. XIV. 472 The Tewa-speaking Indians occupy..five villages northwest of Santa Fé. 1914 W. H. Rivers Kinship & Social Organization 53 The Tewa of Hano, a Pueblo tribe, call the father's sister's son tada. 1937 R. H. Lowie Hist. Ethnological Theory ix. 135 In the same category..belongs the Tewa Indian's diary kept at Dr. Elsie Clews Parsons' suggestion. 1959 E. Tunis Indians 115/1 The Hopi still occupy three high mesas in Arizona where they have six towns, plus a seventh occupied by a band of Tewa who have lived with the Hopi for two hundred years. 1980 Smithsonian Oct. 87 Tesuque, a smallish pueblo of some 200 souls, was considered one of the most restive of the six Tewa pueblos north and northwest of Santa Fe.

TEWT (tiūt). *Army slang.* Also **Tewt,** etc. An acronym formed on the initial letters of *t*actical *e*xercise *w*ithout *t*roops, an exercise used in the training of junior officers.
1942 Partridge Dict. Abbrev. 95/2 T.E.W.T., slangily, a tewt or tute. A tactical exercise without troops. 1948 Partridge Dict. Forces' Slang 191 Tewt,..on which junior officers learnt how to be generals. Invaluable according to some authorities (those who set the Tewts), a complete waste of time according to others (those who carried them out). 1952 E. Waugh Men at Arms II. iii. 194 Leonard improvised 'No more tewts and no more drill, No night ops to cause a chill.' 1956 J. Masters Bugles & Tiger viii. 117 Above all, individual training was the time for TEWTs. 1980 Globe & Laurel July/Aug. 206/1 Two TEWTs were laid on for the officers and NCOs.

Tex (teks), *sb.*[1] *U.S. colloq.* [Abbrev. of Texan *a.* and *sb.*] (A nickname for) a Texan.
1909 Cent. Dict. Suppl., Tex, an abbreviation (a) of Texas; (b) of Texan. 1943 R. Vance They made me Leatherneck vii. 29 Call the aborigines 'Tex' and they seem to think that at least you acknowledge Texas to be in the Union and its name well circulated. 1979 P. Theroux Old Patagonian Express x. 140, I could tell you were interested in poetry, Tex. Ibid., That Tex is a real fun guy.

tex (teks), *sb.*[2] [Abbrev. of Textile *a.* and *sb.*] A unit of weight used to estimate the fineness of fibres and yarns.
1953 Textile Research Jrnl. XXIII. 947/1 The Textile Institute recommends the tex and the British Rayon and Synthetic Fibers Federation prefers the grex. 1956 Rev. Textile Progress VIII. 258 A universal system for yarn count in all fibres has been adopted... The system, based on units of grammes per kilometre, is applicable to all types of yarn and is known as the Tex System. 1963 A. J. Hall Textile Sci. iii. 135 This is known as the Tex system and by this the count of a yarn or any other length of fibres in bundle form..is the number of grams which 1,000 metres of the yarn weigh. 1973 Materials & Technol. VI. 263 Silk is a relatively strong fibre, having a tenacity which lies between 3·5 and 4·5 g/denier (31·5 and 40·5 g/tex).

Texan, *a.* and *sb.* Add: **A.** *adj.* (Earlier example.)
1832 W. B. Dewees Lett. from Early Settler Texas (1852) 142 On arriving at that place the Texan troops put to flight seven hundred Mexicans.
B. *sb.* A person or animal native to or inhabiting Texas.
1837 H. Martineau Society in America II. ii. i. 81 If the government wished all possible success to the Texans, it could hardly do better than be quiet. 1868 Trans. Illinois Agric. Soc. (1870) VII. 138 We also put five cows and a buffalo with some Texans about the 20th of June. 1940 W. Faulkner Hamlet iv. i. 246 The Texan..managed to saw the mules about and so lock the wheels. 1974 'R. B. Dominic' Epitaph for Lobbyist ii. 14 He combined formidable intelligence with a Texan's charm.

Texas. 1. Also **texas.** Delete 'pilot-house and' from def. (Earlier simple and attrib. examples.)
1853 Pen & Pencil I. 789/2 The roof of the cabin which offered a splendid promenade, and the spectacle of a second edifice of state-rooms, surrounded by a broad promenade and curiously denominated 'Texas'. 1857 F. L. Olmsted Journey Texas 27 To this Texas, inveterate card-players retire on Sundays. 1875 'Mark Twain' in Atlantic Monthly Feb. 220/2 A tidy, white-aproned, black 'texas-tender', to bring up tarts and ices and coffee.
2. Texas fever, substitute for def.: a North American form of bovine piroplasmosis (red-water) first identified in Texas, indicated by a high fever, reddish urine, and an enlarged spleen, and caused by a protozoan parasite, *Babesia bigemina*, which is transmitted by the cattle tick; (earlier example); **Texas leaguer** *Baseball* (now *rare*), a fly ball that falls to the ground between the infield and the outfield and results in a base hit; **Texas longhorn,** a bull or cow belonging to a breed once common in Texas, distinguished by long horns and able to thrive in dry regions; also *transf.* (see quot. 1908); **Texas Ranger** [Ranger *sb.*[1] 3 a in Dict. and Suppl.], a member of the state constabulary of Texas (formerly, of certain locally mustered regiments in the federal service during the Mexican War); **Texas Tower** [so called from its resemblance to a Texas oil rig], one of a chain of radar towers built along the eastern coast of the U.S.
1866 2nd Ann. Rep. Missouri State Board of Agric. (1867) 16 Another pest..is the Texas fever',..or 'Texas murrain', as it is variously known. 1905 Sporting Life (Philad.) 7 Oct. 9/4 A bit of bad coaching euchered him out of one bingle the other afternoon, when a Texas Leaguer from his bat had to be chalked down a force out instead of a hit. 1935 J. T. Farrell Judgement Day viii. 185 A dumpy texas-leaguer over third base placed runners on first and second. 1977 Verbatim May 5/2 We are no longer besieged with such terms as 'hot corner', 'keystone', 'Texas Leaguer', 'flyhawk', 'maskman', and 'grasscutter'. 1908 Pacific Monthly July 19/1 Pink got here about the same time but he come of old Texas-longhorn stock. a 1918 G. Stuart Forty Years on Frontier (1925) II. 178 None of our cattle were Texas longhorns. 1946 Nat. Geogr. Mag. Jan. 17/1 Cattle then were the rangy Texas longhorns—more head, horns, and tail than thick, juicy steaks. 1972 K. Bonfiglioli Don't point that Thing at Me xiii. 101 The bleached skeleton of a Texas Longhorn..beside a faint track. 1846 Whig Almanac 1847 19/1 Capt. Samuel Walker, at the head of a small company of Texas Rangers, left Point Isabel. 1911 Everybody's Mag. Sept. 354/1 Two Texas rangers faced Antonio Carrasco and his seventeen thieves sometime in December of 1910. 1943 B. House I give you Texas 31 A city was threatened by mob violence, so a telegram was sent to the governor to rush a force of Texas Rangers to the scene. 1980 E. Behr Getting Even x. 114 The Chairman was wearing a Texas Ranger hat the American President had given him. 1954 Tuscaloosa (Albama) News 13 Aug. 3 (caption) Here is a closeup of a section of one of the 'Texas Towers'..being built offshore along the Atlantic coast. Towers, named for oil rigs in the Gulf of Mexico, will be built along the continental shelf. 1971 S. E. Morison European Discovery Amer.: Northern Voy. xix. 653 The Gulf Stream flows within twelve miles of Cape Hatteras, and the counter-currents, strong winds, and shifting sands are a menace to navigation even today. A Texas Tower was established off Diamond Shoals, the most dangerous, in 1966.
3. Used in various depreciatory collocations.
1905, etc. [see Texas Leaguer, sense 2 above]. 1942 Berrey & Van den Bark Amer. Thes. Slang § 926/1 Texas butter, a gravy made with flour and water in meat grease. 1944 R. F. Adams Western Words 164/2 Texas cakewalk, a hanging. Ibid., Texas gate, a makeshift gate made of barbed wire fastened to a pole. 1962 Amer. Speech XXXVII. 266 Arizona stop; Texas stop, n. Slowing down, but not making a full stop at a stop sign. 1968–70 Current Slang (Univ. S. Dakota) III–IV. 125 Texas strawberries, n. Red beans.—New Mexico State. 1969 Britannica Bk. of Year (U.S.) 801/1 Texas toast, a thick slice of bread warmed and covered with butter. 1975 D. Bagley Snow Tiger xi. 97 A Texas nightingale isn't a bird... It's a donkey. This is a similar New Zealand joke. 1976 Boot & Thomas Jamaica 76/2 It certainly had more flair than old LBJ taking a table of journalists and staffers into the men's room, there to reduce them to awe and wonderment at the size of his whopping great Texas trouser snake. 1979 G. Swarthout Skeletons 172 They call it a 'Texas horserace'. Blaise and his deputies sneaked the Mexicans..to the edge of town and told them to hot-foot it for the line. They'd give them an hour's head start. Then they'd come after them, mounted... If Blaise and his boys caught up with them on this side, it was their bad luck... The Mexs didn't make it.

Texel (te·ksəl). The name of an island in the West Frisian group off the northern coast of the Netherlands, used *absol.* or *attrib.* to designate a hardy, hornless sheep belonging to a breed originally developed there; also, the breed itself.
1949 A. Fraser Sheep Husbandry ii. 118 Milch breeds—East Friesian Milch sheep and their strains, the Texel and the West Friesian. 1957 Encycl. Brit. XX. 475/2 The Texel is a medium-wool, white-face, hornless sheep of the Netherlands, well adapted to range conditions and very prolific. 1978 Times 2 Sept. 2/4 (caption) Record prices were paid..at the first sale in Britain of the British Texel Sheep Society.

Texian (te·ksiăn), *a.* and *sb.* Now *rare.* [f. Tex(as + -ian.] = Texan *a.* and *sb.* (See also quot. 1943.)
1835 Franklin Repository (Chambersburg, Pennsylvania) 8 Dec. 1/6 Volunteers are moving from almost every section of the west to the assistance of the Texians. 1836 D. B. Edward Hist. Texas 45 The Texian farmer of the Gulf coast. Ibid. 74 [It] adds to the variety of a Texian landscape. 1943 Sat. Even. Post 11 Sept. 61 Texians are the old rock; Texans, a term which came into use only after the Civil War, are those out of the old rock; the people who live in Texas are those who are wearing the old rock away. 1955 W. Foster-Harris Look of Old West v. 125 The value of the Texian dollar was then descending rapidly and reached an ultimate low of around 2 cents. 1973 R. Symons Where Wagon Led i. vii. 114 All cow people like the Hesters and other Texas folk (or as they said, 'Texians').

Texican (te·ksikăn). [Blend of Texan *a.* and *sb.* and Mexican *a.* and *sb.*] = *Texan *sb.* (sometimes used more narrowly).
1863 Lawrence (Kansas) Republican 16 Apr. 2/4 (heading) 'Texicans' and 'Injuns' again. 1937 D. Coolidge Texas Cowboys x. 149 That's one thing you'll never find around a Mormon town..you'll never find no Texicans. 1969 in Current Trends in Linguistics (1972) X. 596 Texican, a Texan of Mexican background. (Wis.). 1978 Maledicta II. 172 While Texas remained a part of Mexico, Anglo settlers there called themselves Texicans to distinguish themselves from Spanish-speaking Mexicans.

Tex-Mex (te·ks,meks), *a.* and *sb.* [f. Tex(an *a.* and *sb.* + Mex(ican *a.* and *sb.*: cf. *Tex *sb.*[1] and *Mex *a.* and *sb.*] **A.** *adj.* Designating the Texan variety of something Mexican; also *occas.*, of or pertaining to both Texas and Mexico.
1949 Time 14 Feb. 38/1 Fluent in Texmex Spanish, he had been one of the most promising rodeo riders around Tucson, Ariz... The half English, half Spanish patois of the U.S.–Mexican border region. 1973 News (Mexico City) (Vistas Suppl.) 22 July 7 It is a mistake to come to Mexico and not try the local cuisine. It is not the Tex-Mex cooking that one is used to getting in the United States. 1976 M. Machlin Pipeline xx. 246 The voice of Miss Martinez, one of Wilbur's gestures toward Tex-Mex integration, came softly over the intercom. 1977 Time Out 28 Jan.–3 Feb. 8/2 Cooder's current concern is the music of Southern Texas, the 'Tex-Mex' style.
B. *sb.* The Texan variety of Mexican Spanish.
1955 W. Foster-Harris Look of Old West vii. 211 Northern cowboys had their chance to mess up Spanish even more than had the Texas cowhands, with their Tex-Mex, which, incidentally, is a language in itself. 1969 J. Mander Static Society i. 32 A hybrid, like the 'Tex-Mex' spoken in the south-west of the United States. 1981 Verbatim Spring 24/1 The only foreign language she knows is Tex-Mex.

text, *sb.*[1] Add: **7.** *text-editing* vbl. sb., *-figure, -frequency, -processing* ppl. adj. and vbl. sb., *processor, -source, tape*; **text editor,** a machine that permits the user to alter text using a keyboard; also, a program or component for modifying text held in a computer or processor, in accordance with a user's instructions; **text linguistics** [G. *textlinguistik*] (see quot. 1977); hence **text linguist; text paper,** a newspaper containing serious articles.
1972 H. S. Stone Introd. Computer Organization & Structures ix. 208 Another important application of the linked list is text editing. 1975 Business Week 30 June 80 Vydec Corp...soon will add communications to its display text editor. Xerox Corp. will announce the same capability for its automatic typewriter. 1983 I. Flores Word Processing Handbk. vi. 170 If the last word entered does not fit on this line, then the text editor removes that word from the line and puts it at the left of the next line. 1983 Your Computer Sept. 21/1 The M100 runs a full Microsoft BASIC interpreter, appointment scheduler, address filer, text editor and communications utility. 1938 British Birds XXXI. 359 The book is illustrated..by good, if rather infrequent text figures and a coloured plate. 1963 T. G. E. Powell in Foster & Alcock Culture & Environment vi. 169 My thanks are also due to Miss Frances Lynch for preparing the text-figure drawings. 1942 M. Joos in Language XVIII. 33 The Dewey count gives us a statistical picture of text frequencies; the Twaddell count of list frequencies. 1962 P. S. Ray in F. A. Rice Study of Role of Second Languages in Asia, Africa & Latin Amer. (Center for Applied Linguistics) 92 'Text frequency' compares two lexical forms in their repetitions within a body of discourse. 1973 W. O. Hendricks Essays on Semiolinguistics & Verbal Art ii. 53 See Fries..for a discussion of the theme-rheme distinction in text linguistics. 1977 Language LIII. 247 The rapidly growing school of 'text-linguistics'... The general belief shared by these scholars is that the 'natural domain' of linguistic theory consists of discourses, or texts, rather than sentences. However,..this belief is not what distinguishes text-linguistics from other discourse-oriented..trends in linguistics. Ibid. 248 Text-linguistics differs from these approaches in its interpretation of the claim that texts are the natural domain of linguistics. For generative text-linguists, this means that the grammar must actually generate (all and only) possible well-formed texts of the language. 1961 Guardian 30 Jan. 18/2 All possible steps will be taken to make the future of the 'Daily Herald' as a text paper more secure. 1977 Times 5 Sept. 12/6 Tabloid papers sell better than serious text papers. 1968 Jrnl. Assoc. Computing Machinery XV. 8 (heading) Computer evaluation of indexing and text processing. 1980 Lebende Sprachen XXV. 10/2 Other texts..can probably be dealt with more efficiently by an extended text-processing system, than by machine translation as such. 1983 G. Leech et al. in Trans. Philol. Soc. 28 We may..proceed now to consider the kinds of text-processing that can be performed, using a computer corpus as a database. 1970 Technical Disclosure Bull. XIII. iv. 9 A flow chart for a text collection program which operates to collect lines of text for a text processor is described. 1980 Daily Tel. 23 Apr. 3 (Advt.), If you have bought or are about to buy a small computer or text processor, you need Cave Tab to ensure you make the most of it. 1947 A. Einstein Music in Romantic Era xvi. 265 Shakespeare..was no more novel as a text-source for Italian opera than was Sir Walter Scott..or Victor Hugo. 1978 Early Music Oct. 609/1 What is described as a 'text source', the 1545 King's Primer, is also used. 1970 A. Cameron et al. Computers & O.E. Concordances 18 The first thing of course is the production of text tapes and the printing thereof.

text-book. Add: **5.** *attrib.* passing into *adj.* Derived from, dependent upon, or typical of a text-book (sense 2); orig. and still occas. in a derogatory sense, implying mechanical adherence to a stereotype; now freq. used in an approbatory sense of an exemplary or classic instance of something. Cf. Copy-book 2 b in Dict. and Suppl.
1916 [see Middle Western adj. s.v. *middle a. 6]. 1927 M. Sadleir Trollope iv. 183 The presentation of Greshamsbury House..is perfunctory... Trollope..was content to parrot text-book phrases of appreciation. 1939 G. Household Rogue Male 230 To deny..I was uncomfortable, but to produce hypothetical justification for getting more comfort. It was a text-book illustration good enough to take in the foreigner. 1949 'G. Orwell' Nineteen Eighty-Four iii. 260 It was a perfect conversion, a textbook case. 1954 W. Faulkner Fable 33 An authority..among textbook soldiers on how to keep troops fit. 1957 [see *real a.[2] 2 a]. 1963 Times 4 May 3/6 Smith tertius (Edwin, of Derbyshire) hit a swinging six into the Mound stand and followed this with two text-book fours. 1970 Times 13 Feb. 25/7 The..strike..was a textbook example of response to loss of 'our management'. 1979 N. Hynd False Flags xix. 173 He followed an evasive path, a textbook lesson in how to move without leaving a trace.
Hence **te·xt-bookish** *a.*
1914 H. G. Wells Englishman looks at World 84 An educational system..has to be grown; and in the beginning it is bound to be thin, ragged, forced, crammy, text-bookish, superficial. 1951 Sport 27 Apr.–3 May 3/1 Newcastle can be the more brilliant, the more dazzling, the more text-bookish on their day. 1974 Publishers Weekly 4 Feb. 68/2 A textbookish survey of Arab history, religion and culture since the days of Mohammed.

textile, *a.* and *sb.* Add: **A.** *adj.* **1.** Also, of or pertaining to a man-made fibre or filament, not necessarily woven.
1910 Mitchell & Prideaux Fibres used in Textile & Allied Industries i. 8 Textile papers. (a) Spinning fibres in raw state... (b) Cotton or flax fibre previously spun. 1931 K. P. Hess Textile Fibers & their Use v. 232 The fourth method of dissolving cellulose and forming it into fine filaments was worked out... Textile fibres were not developed to any great extent by this method until the close of the World War. 1961 Wall St. Jrnl. 23 Jan. 2/3 DuPont Co. announced it will close its textile rayon operation..by August. 1972 Daily Tel. 20 Nov. 11/4 One single step is..required to convert the chemical raw material of the synthetic fibres into a finished textile cloth, no weaving or knitting being required. 1981 M. L. Joseph Essentials of Textiles ii. 9 Textile fibres..can be manufactured from natural fibrous materials such as wood pulp (rayon) or synthesized from chemicals with no resemblance to fibrous forms (nylon, polyester).
3. *Naturism.* Non-naturist; *spec.* applied to places, etc., prohibited to nudists. Cf. sense *3 of the sb.
1970 Newsweek 25 May 55/2 Its guests follow a daily routine little different from that of the 'textile tourists'—or non-nudists—in nearby hotels. 1979 P. Vallack Free Sun xi. 125 The peninsula that separates textile camping from Funtona Bay designated for nudism.
B. *sb.* **1.** Also, a synthetic material suitable for weaving; any of various materials, as a bonded fabric, which do not require weaving.
1908 A. E. Garrett Fibres for Fabrics iv. 100 Since the Chardonnet silk [sc. artificial silk] is so much more deteriorated by pure water..it matters little, so far as its extended use in textiles is concerned. 1927 M. H. Avram Rayon Industry i. 1 'Rayon'—the first synthetic textile... We shall briefly trace the steps from its conception; through..its struggle to gain a place as a commercially possible textile fibre. 1961 Wall St. Jrnl. 1 Dec. 14/2 Mead,..in cooperation with M. Lowenstein & Sons, Inc...is working on paper 'textiles'. 1962 Z. Travníček tr. Krčma's Nonwoven Textiles i. 11 Nonwoven textiles and, particularly, adhesively bonded textiles can be manufactured by many processes. 1970 Cabinet Maker & Retail Furnisher 23 Oct. 173/2 The original term 'non-woven textiles' used for adhesively-bonded fabrics has grown more and more obscure as novel manufacturing technologies based on mechanical bonding processes have been introduced. Ibid., As a first simplification we can say that 'non-woven' textiles do not comprise traditional textile structures made by processes other than weaving (ie knitting, braiding, lace manufacture, etc). The present meaning of 'non-woven textiles' refers to pliable and porous products from textile materials that are reinforced by mechanical or chemical means.
b. (Later examples.) Also, of non-woven fabric.
1971 N.Y. Law Jrnl. 23 Nov. 1/8 Suskin was a principal in Derby Fabrics, Inc..., a textile converting and jobbing concern. Ibid., Suskin entered a business relationship with Jerry Kassel, Inc..., also a textile converter. 1974 Times 12 Feb. 11 Louis van Praag has a theory that textiles should not be designed by textile designers. 1976 Jrnl. R. Soc. Arts CXXV. 21/1 Most textile conservation begins with cleaning to remove the harmful effects of atmospheric pollution, dust, dirt and undesirable or damaging stains or soiling. Ibid. 24/2 The Textile Conservation Centre came into being primarily to provide the foundation for new textile conservators to base their studies.
3. *Naturism.* A non-naturist; *spec.* one who wears a swimming costume on the beach.
1979 Listener 4 Jan. 20/1 The world's first naturist community..is up for sale and will probably become a holiday resort for 'textiles'—the word naturists use for people who keep their clothes on when they could take them off. 1979 P. Vallack Free Sun vii. 85 What would the sign have to do? Alert non-naturists (textiles) that they will see nude bathers if they continue in that direction. 1983 Times 6 July 32/2 The topless generally inhabit the more remote ends of the beach well away from the 'textiles'.

textless (te·kstlės), *a.* [-less.] Having no text.
1926 United Free Church Missionary Record May 225/1 What a windy textless sermon we got. 1957 J. Hollander in N. Frye Sound & Poetry i. 65 Plato had disapproved of textless music. 1980 Christian Sci. Monitor 12 May B12/4 Another textless wonder, 'Truck' [sc. a book] is a bold and bouncy salute to the open road.

textuality. Add: **2.** (See quot. 1970.)
1970 Babel XVI. 76/1 By textuality, we mean the result of the transformation of the common language of a given type of civilization into the language of a work of literature belonging to that type of civilization. 1976 G. C. Spivak in J. Derrida Of Grammatology p. lxv, Exploiting a false etymological kinship between semantics and semen, Derrida offers this version of textuality: A sowing that does not produce plants, but is simply infinitely repeated. 1979 N. & Q. June 285/2 Glyph is a 'new serial publication' concerned with 'the problems of representation and textuality'.

textura (tekstiū·ra). *Typogr.* [a. G. *textura* (also *textur*), f. L. *textūra*: see Texture *sb.*] One of a group of typefaces first used in the earliest printed books, distinguished by narrow, angular letters and a strong vertical emphasis; also, the manuscript hand on which these typefaces were based. Also *attrib.*
[1922 D. B. Updike Printing Types II. 323/1 (Index), Textur type.] 1929 A. F. Johnson in Library IX. 359 The term which the Germans usually employ is Textur, or Textura, meaning 'woven', from the resemblance of a page in this letter to a woven pattern. 1955 Archit. Rev. CXVIII. 399/3 It [sc. the Gothic letter] is a magnificent letter, both formal textura and Gothic cursive. 1962 [see *lettre b]. 1969 M. B. Parkes Eng. Cursive Book Hands 1250–1500 p. xvii, In the fourteenth century Textura became increasingly more artificial and more difficult to write. 1970 Times Lit. Suppl. 7 Aug. 884/5 The textura types survived in England into the eighteenth century. 1976 [see *rotunda 2*]. 1977 Studies in Eng. Lit.: Eng. Number (Tokyo) 7 Ad is written in a very neat textura hand, which differs from that which copied the rest of the MS.

textural, *a.* Add: **c.** *Mus.* and *Literary Criticism.* See *texture *sb.* 5.
1962 Listener 1 Nov. 735/3 Outward clarity of form, of rhythmic definition, and of textural contrasts, are the most striking features. 1963 Ibid. 3 Jan. 23/1 Joyce's Portrait of the Artist is characterized by organic form, both textural and structural. 1983 Ibid. 10 Feb. 35/4 There are no more than one and a half piano quartets whose great music does not produce textural insuperabilities for the players.
texturally *adv.* (later examples in sense *c of the adj.).
1962 Listener 22 Nov. 885/1 The structurally and texturally elaborate String Quartet. 1976 Gramophone Mar. 1442/3 Missa Salisburgensis..is texturally complex, with its seven 'choirs' of voices and instruments spread over 54 staves.

texture, *sb.* Add: **5.** Also, in Literary Criticism: the constitution or quality of a piece of writing; esp. such perceptible qualities as the imagery, alliteration, assonance, rhythm, etc. (freq. opp. *structure*). In Music: the quality of sound formed by the combination of the different (orchestral, vocal, etc.) parts.
1812 J. Mackintosh in Mem. Life Sir J. Mackintosh (1835) II. iii. 215 This is increased when a few bolder and higher words are happily wrought into the texture of this familiar eloquence. 1895 W. D. Howells My Literary Passions xxxi. 223 All that Mr. De Forest has written is of a texture and color distinctly his own. 1931 Week-End Rev. 3 Jan. 24/2 The texture of the book is much more satisfactory than its theme. 1934 M. Bodkin Archetypal Patterns in Poetry 320 This duality in unity, and harmonized clash, of cosmic and personal that Blake has woven into the texture of his verses. 1941 J. C. Ransom New Criticism iv. 280 The texture, likewise, seems to be of any real content that may be come upon, provided it is so free, unrestricted, and large that it cannot properly get into the structure. One guesses that it is an order of content, rather than a kind of content, that distinguishes texture from structure, and poetry from prose. 1956 M. Krieger New Apologists for Poetry v. 83 The indeterminacies of meaning, into which the poet is forced by his devotion to the determinate sound, constitute the poem's texture [according to J. C. Ransom]. 1956–7 Modern Fiction Studies Winter 209 The birth of Lena's child means more in the texture of the story than a simple event.
1934 C. Lambert Music Ho! iii. 165 The first symphony [of Borodin]..achieves an admirable symphonic texture. 1959 Listener 10 Dec. 1034/1 For a long time now it has been fashionable to cry after new 'textures' in sound. 1963 Ibid. 21 Feb. 354/1 In his last decade as an opera composer Handel..made less use of wind instruments and tended to favour sparser textures. 1980 Dædalus Spring 194 The thinning of texture, and the descending succession of pitches in measures 100 to 103 of Berlioz's melody all seem to foster and presage closure.
6. (Earlier example.)
1845 [see *pearly a. 4].
7. texture brick, a roughened or rough-hewn brick.
1940 Chambers's Techn. Dict. 843/1 Texture brick, a rustic brick. 1961 [see *septic sb. 2].

textured, *a.* Add: Now freq. without specific adj.: Provided with a texture, esp. as opposed to smooth or plain.
1923 Times Lit. Suppl. 28 Jan. 64/1 The method of colour woodcut, with its bold lines and textured tones, suits very well an artist whose painting is apt to be a little thin. 1938 Burlington Mag. Apr. 200/2 Plain or 'textured' weaves. 1943 J. S. Huxley TVA 100 The textured concrete wall finish. 1959 Times 12 Jan. 11/5 Textured: a term frequently met with indicating a process has been used that fluffs up surface or fabric giving greater density, softness of handle and appearance, extra warmth and some degree of absorbency. 1962 Listener 29 Mar. 566/3 They [sc. the collages and picture reliefs] appear to have both an ambivalent scale, a tiny world of textured pleasure inside the larger images they set out to establish, and a somewhat suave finish. 1969 Amateur Photographer 21 May 56/1 These units..have black cases of plastic with simulated textured panelling. 1977 Time 7 Feb. 54/1 Her far-flung locations are not textured settings but flimsy sets where the author vainly attempts to stage her quiet drama of rootlessness and disaffection.
b. *textured yarn*, a yarn which has been modified so as to give a special texture to the fabric.
1960 Which? Jan. 17/2 In recent years..methods of treating continuous filament synthetic yarns have been introduced that modify their properties remarkably. These modified yarns are described as 'textured'. 1964 Ibid. Sept. 285/1 Textured yarns are mainly of two kinds—bulked yarns and stretch yarns. 1975 C. Calasibetta Fairchild's Dict. Fashion 543/2 Textured y[arn]. 1. Man-made continuous-filament yarns permanently heat-set in crimped manner or otherwise modified to give more elasticity, used to make stretch fabrics. 2. Man-made filament yarns processed to change their appearance; e.g., abraded.
c. Designating protein foods derived from vegetables but given a texture that resembles meat, esp. in *textured vegetable protein* (cf. TVP s.v. *T 6 a).
1968 Manch. Guardian Weekly 11 July 12/3 The second exciting stage was launched this May by a Minneapolis manufacturer... TVP (textured vegetable protein) 'could hardly look or taste better..', the makers claim. 1970 New Scientist 24 Dec. 561/2 There is already a big sale for..textured meat analogies. 1977 Times 23 Feb. 4/8 Mincemeat will sometimes be mixed with textured vegetable protein in 800 schools in Kent. 1983 Listener 21 July 23/3 Let us..settle down to textured soya sandwiches for tea.

texturing. Add: Also in other contexts (esp. corresponding to the senses of *textured *a.*).
1958 Listener 18 Dec. 1055/3 Some texturing material such as sawdust, ordinary sand, or silver sand. 1960 Times 4 Jan. 14/1 The two Moores..both avoid the mannered type of texturing of so many of his drawings. 1960 Wall St. Jrnl. (Eastern ed.) 13 Jan. 1/4 This 'texturing' alters the surface of the long, continuous strands of nylon, giving them new properties such as elasticity and bulk without adding weight. 1961 G. Millerson Technique Television Production viii. 151 (caption) The arrangement, distribution and texturing of scenery. 1978 Gramophone June 85/2 It is clear that Lill's stylish performance, with its crystalline texturing and finely pulsing inner voices, is one very much to be reckoned with.

texturize (te·kstiŭrəiz), *v.* [f. Textur(e *sb.* + -ize.] *trans.* To impart a particular texture to (fabrics or food). Also *fig.* Chiefly in *ppl. a.* So **te·xturizing** *vbl. sb.*
1958 Times 26 June 15/3 We have..entered the texturized yarn field with 'Ban-Lon'. 1959 Wall St. Jrnl. 20 Nov. 17/2 Allied Chemical Corp's 'Caprolan' filament nylon is offered to the carpet industry, too. But to achieve the bulkiness of spun yarns, carpet mills have to have 'Caprolan' filament yarn 'texturized', or bulked. 1969 Daily Tel. 24 July 3/2 This involves the design and manufacture of machinery for yarn texturising and the production of texturised yarns, hosiery and knitwear. 1976 Jrnl. R. Soc. Arts CXXIV. 579/1 A great deal of work has been done on converting soyabeans and other high-energy substrates (even oil feedstock) into proteinaceous material that can be spun, like nylon, and given a texture like that of lean meat. This 'Texturized Vegetable Protein' (TVP) has been successfully promoted and seems likely to have a growing impact on the food market. 1976 Times Lit. Suppl. 13 Feb. 166/1 This selfconsciousness distinguishes the whole show from the chunks of fictionalized, texturized social history (which are to drama as TVP to steak) the BBC now seems so casually expert with.

te:xturo·logy. [ad. F. *texturologie*: see -ology.] A term coined by Jean-Philippe-Arthur Dubuffet (b. 1901) for a kind of painting created by him, composed of minute drops of paint entirely covering a flat surface.
1959 J. A. Thwaites in Arts Yearbk. III. 134/2 In the Texturologies..he [sc. Jean Dubuffet] has pulverized the form and color as never before. 1964 New Statesman 1 May 695/2 Dubuffet's finely granulated texturologies. 1973 Art Internat. Mar. 30/2, I don't want to comment here on the nature of Dubuffet's 'texturologies'.

-th, *suffix*[2]. Add: **2.** Used in works of fiction with preceding dash or hyphen to denote an unspecified ordinal number presented as the name of an unspecified or fictitious regiment.
1847 Thackeray Van. Fair (1848) xxxvi. 324 Colonel O'Dowd, of the —th regiment. 1867 'Ouida' Under Two Flags I. v. 101 The —th came back to Brighton and to barracks. 1931 S. Jameson Richer Dust x. 297 Someone asked him if it were true that the —th had run like hell in front of Festubert. 1949 G. Heyer Arabella ii. 33 Algernon..held a commission in the -th Regiment.

thaccy (ðæ·ki), a dial. form of *that*.
Examples of related variants, thac(k, thact, thackey, etc., from 1814 onward, are listed in Eng. Dial. Dict. (Devon, Cornwall, Glos., Wilts.) s.v. Thac(k. See note at Thilk dem. adj. and pron.
1929 H. Williamson Beautiful Years (rev. ed.) i. 21 He produced it [sc. a knife] from his pocket, and opened an enormous blade. ' Not bad, eh?' 'A gude 'un, thaccy!' 1940 J. Cary Charley is my Darling lxi. 332 Tis only booys badness in you and you'll grow out of thaccy.

Thackerayan, *a.* and *sb.* Add: Also with pronunc. (þækərē̆ⁱ·ăn). **a.** *adj.* (Earlier and later examples.) **b.** *sb.* (Examples.)
1857 J. Blackwood Let. 8 June in Geo. Eliot Lett. (1954) II. 344 The harsher Thackerayan view of human nature. 1909 G. K. Chesterton Thackeray p. xxiii, Any Thackerayan must recognize my meaning. 1917 J. B. Cabell Cream of Jest vi. iv. 264, I am thus digressing, in obsolete Thackerayan fashion, to twaddle about love matches alone. 1958 G. N. Ray Thackeray II. vi. 175 Devoted Thackerayans persist in putting it at the top of their favourite's work. 1978 Encounter Feb. 71/1 They have trouble with the inevitable Thackerayan mother-in-law.
Also **Thackeraya·na** [Ana *suff.*], items associated with Thackeray.
1905 Spectator 18 Feb. 256/2 The voracious collector of Thackerayana cannot have too much of a good thing. 1979 Times 27 Dec. 8/1 The cupboard was packed with Thackerayana, early and special editions of his works, bound volumes of the journals he wrote for and a pile of his pictures.

Thai (tai), *sb.* and *a.* Also 9 **T'hai, Thay, T'hay.** [Native name, meaning 'free': the same word as *Tai *sb.*[2] and *a.*] (Occasionally used where *Tai *sb.*[2] and *a.* might be expected.) **A.** *sb.* **a.** The language of the Thai people, a member of the Tai group of languages; Siamese.
1808 Asiatick Res. X. 173 The more ancient eastern languages, are Jawa, Bûgis, T'hay, and Barma. 1880 A. H. Sayce Introd. Sci. Lang. II. viii. 224 In Siamese or T'hai every word which defines another must follow it. 1963 Time & Tide 2 May 23/2 HRH Prince Chula-Chakrabonge of Thailand..has written more than 30 books in Thai and English. 1972 E. A. Nida Bk. Thousand Tongues (ed. 2) 427/1 Thai and related languages are linguistically grouped in a class known as the Tai languages, a class that comprises tongues spoken by at least 40 million people, from Burma to south-eastern China. 1977 Times 15 June 16/6 McGonagall..had just had his prolific collection of bizarre poems translated into Russian, Chinese, Japanese and..Thai.
b. A native or inhabitant of Thailand (called Siam before 1939 and again briefly between 1945 and 1948); a member of the racial group that constitutes the bulk of the population of Thailand. Also, the Thais collectively.
1841 Penny Cycl. XXI. 452/1 The Siamese call themselves Thay. 1939 Times 30 May 11/3 Muang-Thai is the name by which the dominant element in the country, the Thai, call their land... The newcomers amalgamated with their Lao and Thai kinsmen. 1941 Engineer 15 Aug. 99/1 The Thais have always been agriculturists. 1957 Encycl. Brit. XX. 593B/1 Of the total population [of Siam] the great majority (about 75%–80%) belongs to the Thai group of peoples. These may be divided into the southern Thai or Siamese and the northern Thai or Lao. 1962 E. Snow Other Side of River (1963) lxxxv. 681 Cambodians, like the people of Thailand and Upper Burma, are mixed descendants of the same stock as the Thai and other minority peoples of China. 1978 T. Williamson Technicians of Death xi. 90 He was operating his own ship..with a mixed crew of Thais and Filipinos.
B. *adj.* Of or pertaining to Thailand, its people, or its language; *Thai silk*, wild silk woven in Thailand according to traditional designs, often with bright colours; *Thai stick* [cf. *stick *sb.*[1] 11 h], a marijuana cigarette.
1808 [see Siamese sb. 2]. 1939 Times 5 July 15/7 The Siamese Legation, now officially renamed the Thai Legation, issued the following announcement yesterday:—..The word 'Thailand' for 'Siam' and the word 'Thai' for 'Siamese' will be used from now on by the Ministries and Departments of the Thai Government. 1948 D. Diringer Alphabet ii. vii. 413 The thirteenth century witnessed a general advance of the Thai or Shan race, facilitated by the fall of Pagan dynasty. 1955 Times 9 May 8/4 The many millions of people of Thai race considered linguistically and ethnologically now scattered across south-west China, north Viet Nam, and Burma are split into different groups, and in many cases the split dates back for hundreds of years. 1958 A. Toynbee East to West xxviii. 85 A skiff will carry the Thai housewife to a shop-front that could not be reached on foot. 1976 National Observer (U.S.) 25 Sept. 22/2 Cannabis connoisseurs rank Colombian marijuana alongside such Asian types as so-called Thai-sticks from Thailand. 1977 Times 17 May 8/1 (Advt.), Our beautiful slim-bodied hostesses in their glamorous Thai silk outfits. 1978 Chicago June 245/1 Tom yum gung soup..is almost a meal in itself as is Thai fried rice, a combination of green peppers, chicken, and tiny bits of bacon, garnished with cucumber and tomato slices. 1978 [see *stick sb.[1] 11 h]. 1981 Times 22 Apr. 4/3 He had made..money through smuggling Thai sticks.

Thailander (tai·lændəɹ). [f. *Thailand* + -er[1].] A native or inhabitant of Thailand.
1961 in Webster. 1973 P. O'Donnell Silver Mistress v. 81 His personal bodyguard, the silent Thailander who stood two paces away.

Thakali (tăkā·li). [Native name.] **a.** A member of one of the tribes or castes of Nepal, of Mongol origin. **b.** The language or dialect spoken by this tribe. Also *attrib.* or as *adj.*
1928 Northey & Morris Gurkhas xiii. 202 Prosperous, and great traders..the Thakales are of mixed religion and are closely allied to Tibetans. 1961 L. Bajracharya Nepal 1960–61 10/2 Of the trading tribes those next to the Newars are the Thakalis, residents of Thak in Central Nepal. 1974 M. Peissel Gt. Himalayan Passage xvii. 246 Prosperous businessmen..are attempting to obtain for the Thakali people a high rank in the Hindu caste system. 1974 Encycl. Brit. Macropædia XII. 954/1 The languages of the north and east belong predominantly to the Tibeto-Burman family. These include Magar, Gurung,..and a number of Bhote dialects, including Sherpa and Thakali.

Thakin (þā·kin), *sb.* [Burmese.] **a.** A term of respectful address used by the Burmese, formerly in addressing white people. **b.** A member of a militant nationalist movement that arose in Burma during the 1930s; also *attrib.*
1920 Blackw. Mag. June 835/1, I do not know about the deer, thakin. 1934 'G. Orwell' Burmese Days iv. 74 God go with you, thakin. 1942 J. L. Christian Mod. Burma xiii. 238 A current expression of nationalism in Burma is the 'Thakin' movement. Ibid., The Thakins..have done their country little good. 1957 'F. Clifford' Ten Minutes on June Morning (1977) 39 'Who is it?' I asked. ..'A Sikh, thakin.' 1971 W. Laqueur Dict. Politics 66 During the 1930s popular pressure for independence led to anti-British riots, militant student strikes and the formation of political private armies, e.g. the Thakin Army which was trained in Japan. 1974 Encycl. Brit. Macropædia III. 515/1 The young Thakins won the trust of the villagers and emerged as leaders. Ibid. IX. 923/1 Thant was educated at the University of Rangoon, where he met Thakin Nu (afterward U Nu, who became prime minister of Burma in 1948).

thalamo-. Add: **thalamoco·rtical** *a.*, applied to nerves running from the thalamus to the cerebral cortex; **thalamostri·ate** *a.*, connecting or serving the thalamus and the corpus striatum; **thalamo·tomy** *Surg.* [-tomy], an operation to destroy specific groups of cells in the thalamus, used for the relief of pain or for treatment of Parkinson's disease or mental disorders.
1902 D. J. Cunningham Text-bk. Anat. 504 Flechsig divides the thalamo-cortical fibres of ordinary sensation into three sensory systems. 1954 Gray's Anat. (ed. 31) 994 The wealth of thalamo-cortical and cortico-thalamic connexions indicate a very close functional relationship between the two. 1970 Jrnl. Physiol. CCX. 15P The afferent thalamo-cortical pathways to the visual cortex of the cat and monkey have been studied. 1902 D. J. Cunningham Text-bk. Anat. 540 Numerous fibres from the optic thalamus pass into the anterior limb of the internal capsule and enter both the caudate and the lenticular nuclei. These may be termed the thalamo-striate fibres. 1968 Passmore & Robson Compan. Med. Stud. I. xxiv. 73/1 The thalamostriate vein passes forwards between the caudate nucleus and thalamus draining both. 1948 Time 21 June 76/2 Last week they announced first results of their new operation, called thalamotomy. 1955 Sci. News Let. 11 June 381/1 The studies were made on 30 patients who underwent a special brain operation called thalamotomy. In this operation the cutting is done on part of the thalamus, the structure in the brain that serves as the main relay center for feelings of heat, cold, pain and the like to the thinking part of the brain. 1973 Brit. Med. Jrnl. 15 Dec. 666/1 The movements can be abolished only with thalamotomy. 1977 J. N. Walton Brain's Diseases Nervous System (ed. 8) xi. 595 The operations of pallidectomy and ventrolateral thalamotomy.

thalamus. Add: **1.** Now *Obs.* exc. in *spec.* sense, and used without *optic*. (Further examples.)
1902 D. J. Cunningham Text-bk. Anat. 501 The two optic thalami, in their anterior two-thirds, lie close together on either side of a deep mesial cleft, which receives the name of the third ventricle of the brain. 1947 Sci. News IV. 112 There is an anatomically distinct region, the thalamus, deep in the brain-stem which has something to do with the perception of pain and other sensations and the judgment of their quality. 1948 A. Brodal Neurol. Anat. vi. 157 It appears..that the thalamus is not only an important relay station in the large afferent sensory fibre systems and the optic and acoustic systems, but in addition extensive parts of it..also discharge their impulses to the cerebral cortex. 1979 Sci. Amer. Sept. 85/2 The rest of the forebrain is the diencephalon: the upper two-thirds comprises the thalamus (which has numerous subdivisions) and the lower third the hypothalamus.

thalassæmia (þælăsī·miă). *Path.* Also (chiefly *U.S.*) **-emia.** [f. Thalass(o- + Gr. αἷμα blood + -ia[1].] A hereditary hæmolytic anæmia common in malarious (or formerly malarious) areas and caused by the faulty synthesis of part of the hæmoglobin molecule, with symptoms that depend on the part of the molecule affected and on whether the individual is homozygous or heterozygous for the gene concerned; so *thalassæmia major, intermedia, minor.*
1932 Whipple & Bradford in Amer. Jrnl. Dis. Children XLIV. 364 We do not like the term 'erythroblastic anemia'... The disease is limited almost wholly to Italians, Greeks and Syrians, i. e., to the people originating about the Mediterranean Sea. For this reason the term 'thalassemia'..may have an appeal. 1936, etc. [see *Mediterranean sb. 1 b]. 1944 Valentine & Neel in Arch. Internal Med. LXXIV. 196/2 It is suggested, on the basis of the pathologic and genetic evidence, that the full-blown disease be designated 'thalassemia major' and the milder carrier state 'thalassemia minor'. 1954 K. Singer et al. in Blood IX. 1039 We have found the following simple classification very useful; it is based on the alterations of the red cell and hemoglobin levels: 1. Thalassemia major (Cooley's anemia): very severe microcytic hemolytic anemia. 2. Thalassemia intermedia, characterized by a less severe, but still marked anemia. 3. Thalassemia minor: mild anemia. 1962 [see *haemoglobinopathy]. 1972 D. E. Comings in W. J. Williams et al. Hematology xxxi. 332/2 Although the total number of individuals with thalassemia intermedia is relatively small.., it constitutes a clinically important group, since its relatively benign course (compared to thalassemia major) allows affected individuals to live to adulthood. 1973 B. J. Williams Evolution & Human Origins iv. 61/2 Osteoporosis occurs even with thalassemia minor. 1978 Jrnl. R. Soc. Med. LXXI. 465/1 The hepatic cirrhosis of thalassaemia major. 1979 Brit. Med. Jrnl. 17 Nov. 1298/2 The 26-year-old Chinese with beta-thalassaemia who had been transfused with 404 units of blood in his lifetime. His total body iron was so high that it triggered the alarm at an airport security checkpoint. 1982 New Scientist 21 Jan. 164/1 Red blood cells in people with thalassæmia cannot carry oxygen well enough, and patients survive only with regular blood transfusions.
Hence **thalassæ·mic,** a person with thalassæmia.
1974 Nature 8 Feb. 380/1 The δβ-thalassaemics were Sicilian and have been previously reported. 1979 Ibid. 15 Nov. 317/2 In one study, the incidence of HbF Sardinia in β-thalassaemics in Italy was 90%, compared with only 40% in normal Italians.

thalass(o-. Add: **thalassotherapy** (examples).
1910 Index-Catal. Libr. Surgeon-General's Office, U.S. Army 2nd Ser. XV. 362/1 Nyström (O. E.) Några ord om hafskuren eller thalassoterapien. (A few words on seabaths and thalassotherapy.) 8°. Göteborg, 1907. 1966 Punch 5 Oct. 531 Thalassotherapy is sea-bathing in a warm, enclosed and controlled area... Pure sea water is pumped in from several hundred metres out. 1983 Inglis & West Alternative Health Guide 25 Establishments which provide thalassotherapy have been springing up around the continent of Europe..to provide a holiday in which the usual seaside ingredients..are supplemented..by a regime of salt-water treatments.

thalenite (þā·lĕn-, þălī·nəit). *Min.* [ad. Sw. *thalénit* (C. Benedicks 1898, in *Geol. För. Förh.* XX. 308), f. the name of T. R. *Thalén* (1827–1905), Swedish physicist: see -ite[1].] An yttrium silicate, $Y_3Si_3O_{10}OH$, found as translucent monoclinic crystals.
1899 Jrnl. Chem. Soc. LXXVI. ii. 766 Minerals allied to thalénite are yttrialite, rowlandite, and kainosite. 1972 Prof. Papers U.S. Geol. Survey No. 800-c. 63/1 The rarity of thalenite in pegmatites, in which other rare-earth minerals may be abundant, suggests that it can form only under very unusual conditions.

‖ **thali**[1] (tā·li). Also **tali.** [Tamil *tāli.*] A gold pendant that is hung round the bride's neck as part of a Hindu wedding ceremony.
1875 Indian Antiquary IV. 173/1, I am surprised that the opponents of the Kuḍumi have not yet commenced to put down the use of the tâli. This is the Hindu sign of marriage, answering to the ring of European Christendom. .. It is always tied round the Hindu bride's neck. 1957 L. Dumont Hierarchy & Marriage Alliance in S. Indian Kinship (Occasional Papers R. Anthrop. Inst. No. 12) iii. 29 The tying of a string, with or without the well-known marriage badge or tāli, round the bride's neck has certainly a sacramental value. 1963 Guardian 11 Apr. 11/4 At marriage in the wealthier, and even in the not-so-wealthy, families a good deal of gold is passed over from the bride's family to the bridegroom's. And the tokens of marriage are not the miserable rings of the Christian West but gold pendants called thalis which are then hung on solid gold necklaces. 1981 Times 24 Jan. 11/1 The thali is a phallic symbol worn by brides in the Dravidian South.

‖ **thali**[2] (tā·li). [Hind. *thālī.*] In India, a metal platter or flat dish on which food is served.
1969 Times 13 Oct. (Indian Suppl.) p. xx/4 The Apollo room in Bombay's Taj Hotel where a 'thali' platter of assorted spoonfuls of curries and sauces is a good introduction. 1978 F. Olbrich Desouza pays Price xxii. 137 The little serving-boy..brought..a 'thali', a gleaming round metal tray with an assortment of cooked vegetables and pulses arranged in small helpings.

thalidomide (þăli·dŏməid). Also **Thalidomide.** [f. ph*thali*mido*glutarimide*, f. Phthalimid(e + -o + *glutaric* adj. s.v. Glut- + Imide.] **a.** A non-barbiturate sedative and hypnotic, $C_{13}H_{10}N_2O_4$, which was found to be teratogenic when taken early in pregnancy, sometimes causing malformation or absence of limbs in the fœtus.
1958 Lancet 1 Feb. 271 The British Pharmacopœia Commission has issued the following new supplementary list of approved names:..Thalidomide..α-Phthalimidoglutarimide. 1961 Ibid. 2 Dec. 1262/1 We have just received reports from two overseas sources possibly associating thalidomide ('Distaval') with harmful effects on the fœtus in early pregnancy. 1962 New Scientist 28 June 717/1 The tragic cost of the use of thalidomide. 1969 N. W. Pirie Food Resources vii. 167 Thalidomide is the classic example of a substance that passed orthodox tests that were, as it turned out, not relevant. 1978 Dædalus Spring 136 Passage of the 1962 drug law might not have occurred without the public demand for stricter controls over the testing of new drugs following disclosure of deformities caused by Thalidomide. 1979 Nature 29 Nov. 509/1 Other agents, notably Thalidomide, are believed to be teratogenic by virtue of metabolic products rather than the parent compound.
b. *attrib.* and *Comb.*: **thalidomide baby, child,** etc., one born deformed through the effects of thalidomide.
1962 Guardian 31 July 14/2 There is still no information about the number of 'thalidomide babies' in the country. 1962 Lancet 1 Dec. 1155/2 Attempts have been made to lengthen stunted thalidomide arms by grafts from fibula and scapula. 1971 New Scientist 18 Mar. 613/1 The construction of a body harness for armless thalidomide children. 1973 Daily Tel. 9 Jan. 2/2 A Pharmaceutical Society spokesman said..that the scale of the thalidomide compensation would undoubtedly have a 'profound effect' on the manufacture of new drugs in Britain. 1977 J. D. Douglas in Douglas & Johnson Existential Sociol. i. 26 Her highly publicized abortion in Sweden for a thalidomide-damaged fetus. 1979 T. Benn Arguments for Socialism ii. 49 There was also the initial refusal by Distillers to compensate the thalidomide children properly. 1980 Nature 1 May 54/1 The thalidomide tragedy, for example, could have been averted if this synthetic racemate had been separated into its optical isomers, for only the left-handed (s)-(−)-isomer has teratogenic properties. 1981 Daily Tel. 3 Feb. 2/2 This is the first known case of a thalidomide victim becoming a father, although a couple of incidences of thalidomide mothers are known.

thallium. Add: **b. thallium-activated** *a.*, containing a small amount of added thallium so as to make the substance active as a phosphor.
1956 Nature 3 Mar. 413/1 A single-crystal spectrometer (thallium-activated sodium iodide crystal). 1974 Encycl. Brit. Macropædia XV. 397/2 The phosphor in greatest use in scintillation counters is thallium activated sodium iodide.

thalweg. Add: Also **talweg** [after the reformed Ger. spelling]. (Earlier and later examples.)
1831 W. Whewell Let. 22 Feb. in I. Todhunter William Whewell (1876) II. 113 For thalweg and riggin' I do not think you can do better than take daleway and ridgeway. 1937 Geogr. Jrnl. LXXXIX. 260 The development of a terrace..involves two clearly distinguishable phases—firstly, the formation of a continuous flood plain, and secondly, the incision of the talweg below it. 1946 L. D. Stamp Britain's Struct. & Scenery v. 49 Soundings show that the floor of a ria, the old river talweg, slopes steadily seawards and there is no 'lip'. 1966 J. S. Hardman tr. R. Boulanger's Middle East 608 Whilst conducting excavations towards the thalweg of the Kidron valley the Franco-British expedition discovered the remains of a rampart belonging to the Canaanaean (or Jebusite) Jerusalem. 1968 R. W. Fairbridge Encycl. Geomorphol. 1149/1 The opposite of the talweg itself is a divide, i.e., the lines joining all high points in topography.

Thames. Add: **a.** *Thames barge, valley.*
1883 Boats of World 4 Who can mistake the world-renowned Thames Barge, with her long, flat side, picturesque rig, and bright-coloured sails? 1961 F. H. Burgess Dict. Sailing 207 Thames barge, a ketch or yawl-rigged sailing barge with a large spritsail, common on the Thames estuary. 1979 D. May Revenger's Comedy viii. 105 Out on the estuary, a big, red-sailed Thames barge was moving. 1902 Encycl. Brit. XXVIII. 533/2 In the London district the country in the Thames valley..is as largely occupied by flower farms as it is by fruit farms. 1977 D. James Spy at Evening xiv. 113 He..let himself out into the early-morning Thames valley mist.

Thamudic (þămū·dik), *a.* and *sb.* [f. *Thamūd* (Arab. *ṯamūd*) + -ic.] **A.** *adj.* **a.** Of, pertaining to, or designating a class of inscriptions in northern and central Arabia dating from the 5th to the 1st centuries B.C., or the ancient Semitic language of which they are the only evidence. **b.** Of or pertaining to the Thamūd, a tribe that lived in northern Arabia between the 4th century B.C. and the 7th century A.D.
1909 Webster, Thamudic, a. 1937 P. K. Hitti Hist. Arabs vi. 72 The Lihyânites seem also to have held..al-Hijr..once a Thamûdic town. 1951 [see *Safaitic a.]. 1974 Encycl. Brit. Micropædia IX. 921/2 Recent archaeological work has revealed numerous Thamūdic rock writings and pictures not only on Mt. Athlith but also throughout central Arabia.
B. *sb.* The Thamudic language.
1937 F. V. Winnett Study Lihyanite & Thamudic Inscr. 27 In view of the fact that it has the value zāi in both Lihyanite and Ethiopic, there is little likelihood of its having a different value in Thamudic. 1952 Harding & Littmann Some Thamudic Inscr. 47 In Thamudic and Safaitic the verb..often means 'he acquired, he bought'.
Also **Thamudæ·an, Thamu·dene, Thamude·nic** *adjs.*, **Thamudian** *a.* and *sb.*
1909 Webster, Thamudene adj. 1911 Encycl. Brit. XXIII. 956/2 The Thamudaean inscriptions are locally nearer to Phoenicia, and the letters are more like the Phoenician. 1934 J. A. Montgomery Arabia & Bible v. 91 A peculiar and much discussed special type of Arabic inscriptions, the Thamudene, has been discovered. 1936, etc. [see *Lihyanic sb.]. 1948 D. Diringer Alphabet ii. ii. 227 The North Arabian inscriptions..can be separated into three groups: (1) Thamudene or Thamudic... (2) The Dedanite inscriptions... (3) The Safaitic or Safahitic inscriptions. 1981 Word 1980 XXXI. 222 We have here an important isogloss for the chronological division of the Semitic languages into languages with š.., languages with h (Amorite, Hebrew,.. and Thamudian), languages with '.., and languages with y.

Thamudite (þæ·miūdəit), *sb.* and *a.* [f. *Thamūd* (see *Thamudic *a.* b) + -ite[1].] **A.** *sb.* One of the Thamūd. **B.** *adj.* = *Thamudic *a.* b.
1833 A. Crichton Hist. Arabia I. iii. 92 The circumstance of dwelling in caves..was common to other tribes besides the Thamudites. 1881 Encycl. Brit. XIII. 117/2 [The graffiti] are mostly the productions of Thamudite soldiers in the Roman army.

than, *conj.* Add: **2. b.** *different(ly) than* is not uncommon, esp. in the U.S., but continues to be regarded by many as incorrect. (Further examples.)
1857 Trollope Barchester Towers III. xiv. 248 Things were conducted very differently now than in former times. 1912 J. Webster Daddy-Long-Legs (1913) 146 It's different with me than with other girls. 1962 D. Lessing Golden Notebk. 59 Both come from a different world than the housing estate outside London. 1970 Amer. N. & Q. Nov. 39/1 Geoffrey and Erasmus are concerned with classifying metaphors along quite different lines than is Quintilian. 1980 Outdoor Life (U.S.) (Northeast ed.) Oct. 101/1 Mule deer bucks behave differently than whitetails in a few other ways.

thanato-. Add: **tha:natocœno·sis, -cœ·nose** (also *U.S.* **-cen-**) *Ecol.* [a. G. *thanatocoenose* (E. Wasmund 1926, in *Arch. f. Hydrobiol.* XVII. 6), f. Gr. κοίνωσις sharing, as in *biocœnosis], a group of fossils occurring in the same location but not necessarily representing a former biocœnosis; **thanatophi·lia** [*-philia], an undue fascination with death; **thanatopho·ric** *a. Path.* [ad. F. *thanatophore* (P. Maroteaux et al. 1967, in *Presse Méd.* LXXV. 2519), ad. Gr. θανατηφόρος death-bringing] applied to a form of dwarfism that results in death (see quot. 1977).
1953 Amer. Jrnl. Sci. CCLI. 25 The term 'thanatocoenosis' implies a community of death; as used by Wasmund, however, it has come to mean the aggregated remains of organisms that in many cases never constituted a biocoenosis. 1957 Sci. News XLIII. 71 A fossil 'community' (a thanatocoenose or death assemblage) is seldom if ever identical with the original biocoenose. 1967 Oceanogr. & Marine Biol. V. 452 The following (and last) regression..left a very rich fauna which forms most of the thanatocoenoses lying under the present sea level. 1975 Nature 23 Oct. 667/2 It is well known that factors such as habitat preference of the animals in question,..and the environmental setting influence the likelihood of the preservation of thanatocoenoses. 1977 Biotropica IX. 131 (heading) A small-vertebrate thanatocenosis from northern Peru. 1974 Time 28 Jan. 77/2 Romantic cults seem to spring up rapidly round poets who die young. An element of thanatophilia enters into the worship of such poets. 1979 N.Y. Rev. Bks. 25 Oct. 18/4 Many of Sciascia's tales have, at their heart, thanatophilia. 1971 Lancet 12 June 1234/1 An achondroplastic shows some cartilage formation (in fact quite a lot, even in the thanatophoric form). 1977 Ibid. 16 Apr. 854/1 Thanatophoric dwarfism is a congenital chondrodystrophy characterised by short extremities, narrow thorax, a trunk of normal length, and a relatively large head... Affected infants usually die soon after birth.

thanatologist. Add: **a.** (Further examples.)
1972 New Scientist 2 Mar. 497 Thanatologists ask doctors..to help the terminal patient and his family to meet his own death. 1975 Times Lit. Suppl. 31 Oct. 1305/4 Their real subject, as is customary with Signor Manganelli, is death. He has always been proud of introducing himself as the supreme thanatologist. 1983 Oxf. Bk. Death p. xiii, While to 'deny' death would sound as foolish as the lady who told Carlyle she had decided to accept the universe, I cannot say that I share the thanatologists' missionary urge to bring death out into the open.
b. An undertaker.
1972 Daily Colonist (Victoria, B.C.) 1 Mar. 1/8 Quebec's 450 undertakers want to be called thanatologists. 1980 Times 25 Apr. 6/4 He was one of 300 thanatologists, better known as undertakers, gathered in the principality [of Monaco] to discuss death in all its aspects.

thanatology. Delete *rare* and add: (Later examples.) Also (orig. *U.S.*), the study of the effects of approaching death and of the needs of the terminally ill and their families; **thanatological** *a.* (further example).
1912 Jrnl. Amer. Med. Assoc. 27 Apr. 1246/1 There is something more than mere transcendentalism in the Science of Thanatology. 1968 Jrnl. Indiana Med. Assoc. LXI. 1159/1 (heading) Thanatology resurrected. 1969 Courier-Mail (Brisbane) 13 Sept. 12/7 A Foundation of Thanatology is being formed in New York. 1972 New Scientist 2 Mar. 497/2 The most disturbing issue that has arisen anew with thanatology is the problem of what to tell the terminal patient about his illness. Ibid., Another area of thanatological controversy concerns the administration of drugs to relieve the pain of the terminally ill. 1976 Billings (Montana) Gaz. 11 July 3-F/4 Workers in the new field of thanatology are encouraging parents to take their children, even small ones, to funerals. 1977 New York Rev. Bks. 12 May 10/1 There is now a special branch of learning called 'Thanatology', and historians of death, like Philippe Ariès or Michel Vovelle, have suddenly appeared on the scene. 1979 Brit. Med. Jrnl. 15 Dec. 1530/2 The near-dead are not dead; and the dead, whether surviving in some form or not, can be left to thanatology and eschatology.

thanatorium (þænătō°·riŏm). *nonce-wd.* Pl. **-oria.** [Alteration of *sanatorium*, after Thanato-.] An establishment where people are received in order to be killed.
1970 Times 1 May 11/4 We should need public thanatoria, just as we have public crematoria and abattoirs. 1970 New Scientist 24 Sept. 626/2 The Thanatoria, the most negative of all the departments. 1976 Times Lit. Suppl. 13 Feb. 166/1 The violent jerks from excess to excess of the patients at Dr Sacks's pseudonymous New York hospital—'not a sanatorium but a thanatorium', as one of the inmates remarked.

Thanatos (þæ·nătǫs). [a. Gr. θάνατος death.] = *death-instinct* s.v. *death *sb.* 19.
1935 Brit. Jrnl. Psychol. XXVI. 283 Freud's final duality was the division of the mind into two sets of instincts which he termed life instincts and death instincts respectively—or, if one prefers the Greek names, Eros and Thanatos. Ibid. 284 He was inclined..to regard the voice of Thanatos as mute. 1955 [see *defusion]. 1967 [see *Eros 1 b]. 1970 G. Greer Female Eunuch 148 Our life-style contains more thanatos than eros. 1979 H. Segal Klein i. 20 The fundamental conflict, between Eros—life, including sexuality—and Thanatos—self-destruction and destruction—is the deepest source of ambivalence, anxiety and guilt.

thang (þæŋ). Repr. a Southern U.S. pronunc. of Thing *sb.*[1]
1937 Frontier & Midland Autumn 14/2 He done one thang he ought never done. 1941 W. A. Percy Lanterns on Levee xx. 259 Negroes..insisted on going to their [flooded] cabins..to see about their 'thangs'. 1971 in A. Dundes Mother Wit (1973) 319/2 You ain' so bad yourself, girl... I want to help a sweet thang like you all I can. 1973 Black World Sept. 84 Ourselves illusionize About doin our thang.

Thanga, obs. var. *sangha.

thank, *sb.* Add: **II. 5. b.** With intensifying advbs. and phrases, as *thanks awfully, ever so, a lot, a million* (orig. *U.S.*), *very much*, etc. Also used ironically.

1890 A. TUER *Thenks Awf'lly!* i. 11 He at once burst into conversation: 'Thenks awf'lly! I nurly missed the trine.' 1911 D. H. LAWRENCE *Let.* 7 Nov. (1962) I. 84 Dear Garnett: Just got your letter—I am very glad with the *Nation*—thanks very much. 1914 'SAKI' *Beasts & Super-Beasts* 217 If you lend me three pounds that ought to see me through comfortably. Thanks ever so. 1916 E. F. BENSON *David Blaize* vii. 134, I couldn't possibly. But thanks, most awfully. 1936 *Sat. Even. Post* 12 Sept. 10/1 That was a swell lunch. Thanks a million. 1942 N. BALCHIN *Darkness falls from Air* xiv. 237, I gave him a pound and said, 'Thanks a lot.' 1965 WODEHOUSE *Galahad at Blandings* i. 8 The 'Oh, thanks awfully' which betrayed the other's English origin. 1966 H. NICHOLSON *Duckling in Capri* xv. 194 'Spend it on Pam.' 'Shall I? Thanks a million.' 1967 *Plays & Players* Apr. 41/1 *Trebor*: Couldn't we go on an aeroplane, somewhere? *Webster*: No, we couldn't go on an aeroplane. *Trebor*: Thanks very much. 1972 J. MANN *Mrs Knox's Profession* ii. 15 'Thanks ever so,' he said, his voice an octave higher than usual. 1982 'J. BELL' *Innocent* ii. 16 'You'll want a tray, love.'..'Of course, thanks a lot.'

c. *thanks be*: ellipt. for 'thanks be to God', as an expression of relief or satisfaction. *colloq.*

1924 D. MOORE *Fen's First Term* ix. 97 Me 'arf dye, thanks be. 1942 C. MILBURN *Diary* 7 Oct. (1979) 154 Hats are to be fewer—I seem to have many, thanks be! 1963 *Times* 4 Feb. 13/2 And thanks be, that aging design, the longer fitted jacket has not reappeared.

11. (Earlier example of sense 'to say grace'.)

1803 D. WORDSWORTH *Jrnl.* 27 Aug. (1941) I. 269 When breakfast was ended the mistress desired..her husband to 'return thanks'. He said a short grace.

thank, *v.* Add: **3. e.** (Earlier and later examples.) Now usu. ironic, implying a rebuke or command.

1813 I. POCOCK *Miller & his Men* I. iii. 9 Cockatrice!—I'll thank you for that portmanteau. 1907 G. C. WHITWORTH *Indian English* xii. 248 The offence is much mitigated if..the word is followed by 'if' instead of the usual infinitive, as 'I'll thank you to be quiet.' 1930 J. B. PRIESTLEY *Angel Pavement* i. 12 Just say to 'er: 'Mrs. Cross 'as seen the note left.., and..Mrs. Cross'll thank her to keep 'er notes to 'erself in future till they're asked for.' Just you tell 'er that, boy. 1940 H. G. WELLS *Babes in Darkling Wood* II. ii. 160 No decent people are going to bother about it, Mother. And they will thank you not to be bothered about it. 1975 'D. JORDAN' *Black Account* II. xx. 110 I'm here to sell tractors and I'll thank you to remember it.

f. (Later examples.)

1908 A. BENNETT *Old Wives' Tale* IV. v. 559 'Thank you for nothing!' said Dick. 'I don't want it.' 1940 W. S. CHURCHILL *Second World War* (1949) I. II. xxxiv. 548 Sweden will say 'Thank you for nothing' about any offers on our part to defend the Gällivare ironfield. 1975 'R. PLAYER' *Let's talk of Graves* v. 202 I'm not respectable. The Judge has just told everybody that—thank ye' for nothing, my Lord.

g. See also GOD 9 e. Also *thank God for that* (now freq. in weakened use); *thank God hold* (Mountaineering): an easy hold at the top of a difficult climb.

1918 A. P. MCKISHNIE *Willow, the Wisp* xxi. 303 His world was at rest, once more. Thank God for that! 1949 G. DAVENPORT *Family Fortunes* III. ii. 222 'Thank God for that,' he said. 1955 S. STYLES *Introd. Mountaineering* xi. 127 The term *thank-god hold*, which has become part of British climbing jargon, originated on the third ascent of the slab on Route II, Lliwedd East Buttress, when as each climber got his hand over the good knob at the top he expressed his heartfelt gratitude in the same two words. 1978 P. GILLMAN *Fitness on Foot* v. 67 A sense of relief on reaching the top of a difficult climb to discover enormous holds to finish on. These are known as 'thank God' holds. 1978 I. B. SINGER *Shosha* i. 16 Thank God, I found friends among members of the Writers' Club.

h. In negative conditional sentences as an ironical understatement, as *he would not thank you for doing it*, he would be displeased if you did it. Cf. *THANK YOU B. 3.

[1739–40 RICHARDSON *Pamela* (1740) I. xxiv. 65 Now I did not thank her for this, as I told her afterwards (for it brought a great deal of Trouble upon me).] 1873 TROLLOPE *Phineas Redux* (1874) I. iv. 32 His party would not thank him for ventilating a measure which..might well be postponed. 1896 KIPLING *Seven Seas* 148 The things I knew was proper you wouldn't thank me to give. And the things I knew was rotten you said was the way to live. 1970 'A. GILBERT' *Death wears Mask* i. 19 Miss Alice wouldn't thank you for tying her into a chair. 1983 M. HINXMAN *Corpse now Arriving* ii. 14 He was probably in the middle of some world-shattering story and wouldn't thank her for the interruption.

thanka, var. *TANKA[3].

thankful, *a.* Add: **1. a.** Phr. *thankful for small mercies.*

1818 SCOTT *Heart Midl.* in *Tales my Landlord* 2nd Ser. II. xii. 295 'Ye are thankfu' for sma' mercies, then,' said Mrs Howden, with a toss of her head. 1844 EMERSON *Ess.* 2nd Ser. 41, I am thankful for small mercies. 1874 GEO. ELIOT *Let.* 3 Aug. (1956) VI. 72 One has learned to be thankful for sma' mercies in this world of dreadful possibilities. 1947 A. HUXLEY *Let.* 14 Nov. (1969) 576 It is raining harder and harder and Little Rock feels.. remote. However, the Blue Bird is clean and comfortable; so let us be thankful for small mercies. 1950 C. S. FORESTER *Mr. Midshipman Hornblower* viii. 207 Then be thankful for small mercies. And even more thankful for big ones.

thankfully, *adv.* Add: **II. 4.** Let us be thankful (that); one is thankful to say. orig. *U.S.* This use as a sentence adverb, like *HOPEFULLY *adv.* 2, is deprecated by some writers.

1966 in W. Follett *Mod. Amer. Usage* 170/1 The 'suicide needle' which—thankfully—he didn't see fit to use. 1969 *Chatelaine* July 1/1 Thankfully there are fewer movies to endure in which the men have all the lines. 1976 *Shooting Mag.* Dec. 41/2 An alarming safety situation.. caused many a raised eyebrow but thankfully nothing worse. 1980 *New Society* 3 Jan. 33/2 But thankfully social workers will plod on, hopefully with small regard for new fashions. 1982 *Daily Tel.* 30 Aug. 8/4 Thankfully, however, the old style has not entirely disappeared. 1983 *Times* 11 Nov. 2/4 Aldabra Island in the Indian Ocean, where man 'has thankfully failed to establish himself'.

thank-offering. Add: Also **thanks-offering.**

1921 G. O'DONOVAN *Vocations* xi. 171 The united prayers of the nuns were a thanks-offering to God for her. 1952 C. DAY LEWIS tr. *Virgil's Aeneid* VI. 118 A thanks-offering to Phoebus. 1978 *Washington Post* 7 Mar. A13/2 Oberammergau..has performed the Passion play every 10 years for centuries as a thanks offering for the end of the plague.

thanksgive, *v.* For *Obs. rare*[−1] read *rare* and add later examples.

1908 HARDY *Dynasts* III. 353 You almost charm my long philosophy Out of my strong-built thought, and bear me back To when I thanksgave thus. 1938 O. NASH *I'm a Stranger here Myself* 227 And each Thanksgiving I Thanksgive.

thanksgiving. Add: **1. b.** (Later examples of U.S. use in sense *Thanksgiving Day*.)

1930 J. DOS PASSOS *42nd Parallel* I. 87 By Thanksgiving Mac had beaten his way to Sacramento. 1981 *Nordic Skiing* Jan. 50/1 The resort is situated at 7,000 feet..with a ski season extending from Thanksgiving to mid-May.

3. a. *thanksgiving service.*

1902 I. HAMILTON *Let.* 8 June in R. S. Churchill *Winston S. Churchill* (1969) II. Compan. I. 145 We have just had our Thanksgiving Service. 1923 KIPLING *Irish Guards in Gt. War* I. 338 On the 14th a great thanksgiving-service was held in the Cathedral.

b. Thanksgiving day: in the United States, celebrated since 1941 on the fourth Thursday in November; also in Canada, celebrated on the second Monday in October; **Thanksgiving dinner** *U.S.*, a dinner, usu. consisting of traditional dishes, served on Thanksgiving Day; **Thanksgiving turkey** *U.S.*, a turkey served as a traditional part of a Thanksgiving dinner.

1830 *Workingman's Gaz.* (Woodstock, Vermont) 1 Dec. 78/2 They have added to the comfort and happiness of those, whose scanty pittance would hardly allow them to enjoy the luxuries of Thanksgiving dinner. *a* 1892 W. WHITMAN *Daybks. & Notebks.* (1978) I. 89 Took Thanksgiving dinner there Nov 26 '80. 1981 *Washington Post* 22 Nov. K-1/3 Thanksgiving dinner starts with an enormous glut of oysters. 1829 *Virginia Herald* (Fredericksburg) 25 Apr. 4/1 (*heading*) A Thanksgiving Turkey. 1960 *American Home* Nov. 50 Who should know better how to roast a Thanksgiving turkey or bake a mince pie than the women of early America. 1981 *Washington Post* 26 Nov. B1/1 Such a small Thanksgiving turkey.

thank you. For **b** read **B** and add: **A.** *phr.* **1.** Occas. with intensifying advbs. and phrases: cf. *THANK *sb.* 5 b.

1885 A. EDWARDES *Girton Girl* III. x. 182 Oscar Jones looked radiant. 'Thank you, awfully, Miss Bartrand.' 1967 K. GILES *Death in Diamonds* ix. 155 Thank you a million.

2. a. Used to add emphasis to a preceding expression of wish or opinion (usu. one implying a denial or refusal).

1904 E. NESBIT *Phoenix & Carpet* xi. 212 He didn't mean stay and be roasted... No boys on burning decks for me, thank you. 1928 E. O'NEILL *Strange Interlude* IV. 148 No, I've enough guilt in my memory now, thank you! 1940 *Punch* 5 June 624/1, I still have some remnants of self-respect, thank you. 1959 *Times* 27 Apr. 11/3 It was there..that the emissaries of Noah came to give warning of the impending flood, only to be told that the Macneils had a boat of their own, thank you. 1963 N. MARSH *Dead Water* (1964) vii. 170 'Do you mean that you confronted her?' 'Me! No, thank you!' 1974 M. FORSTER *Seduction of Mrs. Pendlebury* x. 105, I don't want to do her good, I just want to keep her well out of sight and mind, thank you very much. 1983 *Listener* 27 Jan. 25/3 Those of us who felt that nuclear weapons were quite enough to be worrying about, thank you very much, were given a nasty jolt by the documentary *Overcast, with Outbreaks of Yellow Rain.*

b. Used in imitation of direct speech to imply self-satisfaction or complacency on the part of a person just referred to; chiefly in phr. *to do very well, thank you* and varr.

1931 S. JAMESON *Richer Dust* xix. 524 He himself was doing very well, thank you. 1969 *Guardian* 4 July 7/1 One of them was Louise Purnell, and you know she's doing very nicely, thank you. 1971 S. JEPSON *Let. to Dead Girl* viii. 85 Merchant bankers..encouraged people like John Kinnon and..did very well out of it thank you. 1972 *National Observer* (U.S.) 27 May 8/4 Pat dresses stylishly, favoring white boots, and gets around just fine, thank you.

B. *sb.* **1.** (Earlier examples.) Also, an unspoken expression of thanks.

1792 F. BURNEY *Jrnl.* May (1972) I. 174 He looked even extremely gratified..& Bowed expressively a *thank you.* 1824 J. KEBLE *Let.* in G. Battiscombe *John Keble* (1963) I. iv. 80 And so with as hearty a thank-you and farewell as ever you received I am your obliged and very faithful John Keble.

3. In negative contexts, used like *THANK *v.* 3 h.

1935 D. L. SAYERS *Gaudy Night* xvii. 365 That's what the man wants. *He* wouldn't say thank you for a critic on the hearth. 1969 W. J. BURLEY *Death in Willow Pattern* v. 56, I wouldn't say thank you for it! 1970 D. BAGLEY *Running Blind* ix. 199 Nordlinger's Chevrolet was too long... I wouldn't have given a thank you for it.

C. *attrib.*, designating something written or done to convey thanks (in quot. 1922, that merits thanks); esp. *thank-you letter, note.*

1912 J. WEBSTER *Daddy-Long-Legs* 57, I meant this to be just a short little thank-you note. 1915 —— *Dear Enemy* 111, I spend my entire time composing thank-you letters that aren't exact copies of the ones I've sent before. 1922 JOYCE *Ulysses* 131 Saving princes is a thank you job. 1939 F. SCOTT FITZGERALD *Let.* 5 Apr. (1964) 55 Got a nice thank-you letter from Frances Turnbull for the check I sent her. 1948 'P. QUENTIN' *Run to Death* x. 83 Vera and I said 'thank-you' speeches to Mrs. Snood and left. 1979 R. JAFFE *Class Reunion* (1980) II. i. 183 After she saw his play she wrote him a thank-you note. 1981 P. DICKINSON *Seventh Raven* xiii. 189 The thank-you party ..for the children—ice-cream and sausage rolls and lemonade.

thanx (þæŋks), commercial and informal spelling of *thanks* (see THANK *sb.* 5). orig. *U.S.*

1936 H. L. MENCKEN *Amer. Lang.* (ed. 4) viii. 406 Such forms as *burlesk..thanx* and *kreem*..are used freely by the advertising writers. 1977 *Zigzag* Apr. 24/1 Thanx for writing.

thar (var. THERE): revived to repr. U.S. pronunc. See *THERE *adv.* (*a.*, *sb.*) A. γ.

thass (ðæs). Also **thas, thash, thazz.** Repr. *that's* in dial. pronunc. or in speech slurred through intoxication.

1919 G. B. SHAW *Great Catherine* II. 138 Thas true. Drungn ruffian... Thas what he said. 1932 S. GIBBONS *Cold Comfort Farm* xvii. 237 Lessee, thass twenty years ago. 1951 'J. WYNDHAM' *Day of Triffids* i. 25 'S that bloody comet... Thash what done it. 1959 E. POUND *Thrones* xcix. 52 Thazz all there is to it. 1973 C. HIMES *Black on Black* 196 Thass 'cause you's a fool. 1981 M. C. SMITH *Gorky Park* III. iii. 341 Wasn't no mink, it was differ'nt. Thass why I took it to town, to fine [*sic*] out what it was.

that, *dem. pron.*, *adj.*, and *adv.* Add: **B. I.** Demonstrative Pronoun. **1. b.** (Further examples of *that's a* —, *that's my* —, used as expressions of commendation or encouragement.) Cf. THERE *adv.* 3 b.

1849 T. ARNOLD *Let.* 10 Aug. (1966) 128 Do you, my dear K, have them sent to me, that's a darling. 1936 [see *BOY *sb.*[1] 2 c]. 1956 M. DICKENS *Angel in Corner* x. 198 'Good girl.' He lay back on the pillow. 'That's my girl,' he murmured. 1964 J. P. CLARK *Three Plays* 32 *Zifa*: He must not see my tears. *Orukorere*: That's my boy. The strong weep only at dead of night. 1973 W. H. CANAWAY *Harry doing Good* II. ii. 139 'Never mind, then,' he said, and kissed her cheek. 'That's my girl.'

4. a. (Further examples, freq. contrasted with *what*.)

1890 W. JAMES *Princ. Psychol.* I. xii. 466 The conception of some object as a whole..points to and identifies for future thought a certain *that*. 1899 F. W. MAITLAND *Let.* 4 Dec. (1965) 205, I wander in a maze of *whiches* and *thats*. 1909 W. JAMES *Pluralistic Universe* 342 All the *whats* as well as the *thats* of reality, relational as well as terminal, are in the end contents of immediate concrete perception. 1933 *Mind* XLII. 27 A fundamental tenet rather insistently taught us..; namely, that things and events, as real, are *thats*, as well as *whats*. 1975 *New Yorker* 5 May 139/1 We wish not to guess but to know more than thises and thats, to know universal truths.

5. a. *that is*: (*b*) accompanying (usu. following) an explanatory limitation or condition of a preceding statement.

1945 N. MITFORD *Pursuit of Love* xiii. 101 I bet the Scotsboro' boys will be electrocuted in the end, if they don't die of old age first, that is. 1965 W. GOLDING *Pincher Martin* x. 155 'I think finally, I shall go into the Navy.' 'You!'..'If they'd have me, that is.' 1958 *Argosy* Sept. 30 The Buttafava household was happy as could be. All, that is, except Fiorella. 1969 R. HUTCHINGS *Lucky in Jeopardy* iii. 99 You'll be tasting it for yourself up at the House this very evening—if you don't go missing another meal there, that is.

b. *all that* (further examples); *and all that* (further examples); freq. implying a diffident or dismissive attitude on the part of the speaker; *and all that jazz*: see *JAZZ *sb.* 3 b; *and that* (further examples); now chiefly in substandard speech or representations of it.

1925 E. P. OPPENHEIM *Wrath to Come* II. xvi. 271 'Glad to see you and all that, Slattery,' he said. 1929 R. GRAVES (*title*) Good bye to all that. 1930 SELLAR & YEATMAN (*title*) 1066 and all that. 1934 J. HILTON *Good-Bye, Mr. Chips* xi. 80 We don't like the fellow a great deal. Very clever and all that, but a bit *too* clever. 1965 *Listener* 2 Dec. 914/1 Having a fag and talking about sex and that just like she was, you know, ordinary. 1968 *Ibid.* 20 June 801/2 Boy: What do you do then? Girl: Well, you know, typing and filing and that. 1971 D. POTTER *Brit. Eliz. Stamps* iii. 43 The Battle of Hastings, 1066 and all that, was given special treatment. 1974 *Economist* 21 Dec. 26/3 Chairman Mao has formally ordered his revolutionary

genie back into the bottle... It sounds like goodbye to all that. 1977 *Listener* 19 May 644/1 They wait outside the pubs for them, and that.

c. (Earlier example.)

1830 *Massachusetts Spy* 28 July 2/3 The march was now hurried on, yet slow at that, for I..could not walk fast.

d. *that's what*: used to add emphasis to a preceding statement: = 'and that is the truth'; *that's that*: indicating that a discussion is closed, a matter settled, a job finished, etc.; similarly *that was that*; *that's so*: that is as you say; that statement is correct; also interrog., (*is*) *that so? that's right*: see *RIGHT *a.* 7 e; *that's it* = *that's that* above. All *colloq.*

1790 F. GROSE *Provincial Gloss.* Suppl., *That's what*, just so; you are right. North. 1813 E. S. BARRETT *Heroine* I. ix. 95 Not a step shall she stir in our cloathes... So that's that. 1857 *Knickerbocker* Jan. 86 The new and popular phrase of '*That's so*', which is working its way into common parlance. 1872 S. BUTLER *Erewhon* vi. 45 'So that's that,' said I to myself, as I watched them scampering. 1891 M. E. RYAN *Pagan of Alleghanies* vi. 93 'That so?' she said. 1914 *Sat. Even. Post* 4 Apr. 10/2 He's a valyable road-kid, that's what, and he ain't for sale. 1924 P. MARKS *Plastic Age* 24 'Well,' he exclaimed, 'that's that. At last I know where I'm going.' 1930 *Times* 26 Mar. 7/2 Martin-Smith and Bond..raced away with 4's, 5's, and 2's; so that was that. 1937 R. MACAULAY *I would be Private* II. v. 196 I'll not be putting up with it. And, that's that. 1967 *Listener* 14 Sept. 326/1, Well, that's it. I don't want to know. 1973 *Ibid.* 15 Nov. 662/3 When I get to bed I absolutely hit the pillow and that's it, I don't know anything until the next morning. 1974 A. FOWLES *Pastime* ii. 14 When she'd gone after the job..and got it, he'd sort of thought that was that and he wouldn't be seeing any more of her. 1976 J. LEE *Ninth Man* I. 79 Ulysses S. Grant..was a war hero, that's what. 1978 B. PARVIN *Deadly Dyke* (1979) v. 25 Alright..that's it Sergeant. Now, where's Alan Tucker's place?

e. *that is* or *was*: added to give emphasis to a statement beginning with those words or the equivalent. *colloq.*

1911 C. E. W. BEAN *'Dreadnought' of Darling* ix. 78 That's exac'ly how it used to be. It's all right, that is. 1911 A. BENNETT *Card* xi. 278 Well, that was a bit of a lark, that was. 1963 *B.B.C. Handbk.* 1964 25 The political world of..'Panorama', or the eventful world of 'That Was the Week That Was'. 1977 *Film & Television Technician* Apr. 4/2 That was the boom that was—and is. 1977 N. MARSH *Last Ditch* ii. 37 He..suddenly ran off down the street. 'That's Master Ferrant, that was,' said Ricky.

II. Demonstrative Adjective. **1. b.** Also *that one*, used disparagingly of a woman.

1848 THACKERAY *Vanity Fair* liv. 486 You don't know how fond I was of that one...Damme, I followed her like a footman. 1922 F. H. BURNETT *Head of House of Coombe* vii. 75 That one in the drawing-room isn't going to interfere with the Nursery. Not her! 1980 J. DRUMMOND *Such a Nice Family* v. 22, I tell you, it's her!..I wouldn't forget that one, not if I lived to be a thousand.

III. Demonstrative Adverb. **a.** For Now only *dial.* and *Sc.* read Now common in *colloq.* use, esp. with a negative: *not* (*all*) *that*, not very. (Later examples.)

1932 R. LEHMANN *Invit. Waltz* I. iii. 39 This weather's that treacherous, you never know. 1937 D. L. SAYERS *Busman's Honeymoon* iv. 85, I was that ashamed I didden know w'ere to look. 1962 *Harper's Bazaar* Aug. 60/3 The Spanish gypsies..hired to do the sweeping were not all that handy with a broom. 1969 J. LEASOR *They don't make Them like That any More* i. 7, I..looked around the stock. It wasn't all that brilliant, I must admit. 1977 *Spare Rib* May 16/1 It's not that easy in a place like Sheffield. 1980 S. BRETT *Dead Side of Mike* xvii. 173 Charles Paris found it difficult to get that excited. 1981 *Listener* 22 Oct. 462/1 The forgiveness of sin isn't just an easygoing matter, as if to say: 'Well, you sinned, but it doesn't matter all that much—I forgive you.'

that, *relative pron.* Add: **I. 1. c.** *that was*: added when a married woman is referred to by her maiden name; occas. also added following the name of a deceased person.

1785 A. SEWARD *Let.* 31 Dec. (1811) I. 97 Miss Jenny Harry that was, for she afterwards married. 1872 GEO. ELIOT *Middlemarch* IV. VIII. lxxiv. 201, I am not so sorry for Rosamond Vincy that was, as I am for her aunt. 1937 D. L. SAYERS *Busman's Honeymoon* 21 Her new ladyship, Miss Vane that was, went down to Oxford the day before. 1970 S. J. PERELMAN *Baby, it's Cold Inside* 178 You remember her, don't you—Luba Pneumatiç that was? 1977 N. MARSH *Last Ditch* v. 135 A..photograph displayed a truculent young woman... 'That's Dulce [*sic*],' said Sergeant Plank. 'That was,' he added.

that, *conj.* Add: **III. 11.** *'that'-clause*: a clause introduced by the word 'that' (as conjunction or, less commonly, as relative pronoun).

1955 J. L. AUSTIN *How to do Things with Words* (1962) vi. 70 Although we have in this type of utterance a 'that' clause following a verb..we must not allude to this as 'indirect speech'. 1964 E. BACH *Introd. Transformational Gram.* v. 102 The rules for *that* clauses..must be made less general. 1978 *Language* LIV. 67 None of the (a) sentences would be appropriate if the proposition expressed in the *that*-clause were part of the undisputed background of already-held belief common to all the participants in the conversation.

thataboy (ðæ·tăboi), *int. slang* (chiefly U.S.). Also **that a boy, thatta boy,** etc. [Corruption of *that's the boy* (cf. *THAT *dem. pron.* B. 1 b), or alteration of *ATTABOY *int.*] An exclamation of encouragement or admiration; = *ATTABOY *int.*

1936 J. DOS PASSOS *Big Money* 287 'All right, let's go,' he said. 'Thataboy,' roared Farrell. 1975 M. BOSSE *Man who loved Zoos* iv. 96 'What should I tell Hopkins?' 'Tell him..I'm not up a blind alley yet.' 'Thatta boy.' 1978 M. PUZO *Fools Die* xi. 114 Frank patted me on the shoulder. 'That a boy,' he said.

that-a-way (ðæ·tăwē[i]), *adv.* Chiefly *dial.* and *U.S.* Also **thataway, that a way,** etc. [f. THAT *dem. adj.* + AWAY *adv.*] **1.** In that direction.

1839 *Southern Lit. Messenger* V. 378/2, I expect, Tommy, you're *a* sparking that *a* way. 1847 *Paddiana* I. 139 It's very careless I hear they are that aways. 1866 H. JACKSON *Gilbert Rugge* II. xii. 174 Down in the marsh lands, that-a-way. 1901 J. PRIOR *Forest Folk* iv. 41 It's out o' my road or I'd show yer; that-away. 1920 M. WEBB *House in Dormer Forest* vii. 89 'I canna see as it's to be found out,' he nodded sideways towards the murmur, 'that-a-way.' 1973 *Washington Post* 13 Jan. B8/7 'Bonanza', the Western series that went thataway a couple of weeks ago after a 14-year ride on the NBC network. 1978 G. MCDONALD *Fletch's Fortune* (1979) xviii. 127 He went thet-away.

2. In that manner; like that.

1887 *Scribner's Mag.* Sept. 366/1, I hadn' 'a' thought ye'd 'a' evidenced agin me that-a-way. 1889 *Spectator* 26 Oct. 549/2 Whin I sees him that a way the second time, your Reverence. 1938 M. K. RAWLINGS *Yearling* v. 46 'You want to tote lunch?' she called after him. 'I'd not insult my neighbors that-a-way. We'll noon with them.' 1959 *Times Lit. Suppl.* 9 Jan. 15/3, I didn't mean to treat her that-a-way. 1973 K. GILES *File on Death* iii. 72 You bloody well don't do it thataway.

thatch, *sb.* Add: **2. b.** orig. and chiefly *U.S.* A matted layer of plant debris, moss, etc., on a lawn; the material of this layer.

1955 *How to install & care for Your Lawn* 59/1 Opening up a thatch of interwoven stolons and stems can be difficult. 1964 *Book of Lawn Care* (N.Y. Times) iii. 15 Because of its rapid growth, this grass has a tendency to form a heavy mat or thatch. 1977 *Western Living* (Vancouver) Apr. 61/1 Power raking for the removal of moss 'thatch' in spring often does harm to the turf. 1980 *Amat. Gardening* 4 Oct. 16/3 Another cause of moss is 'thatch', a layer of dead, moisture retentive grass and debris that builds up on the lawn's surface.

3. b. *U.S.* Tall, coarse grass.

1622 *Relation Eng. Plantation Plimoth, New England* 25 Some of our people being abroad, to get and gather thatch, they saw great fires. 1695 in *Early Rec. Providence, Rhode Island* (1894) VI. 156 That Parcell of Meadow marsh & thatch..belongeth to me. 1797 B. TRUMBULL *Compl. Hist. Connecticut* I. iii. 24 There grew bent grass, or as some called it, thatch, two, three and four feet high. 1863 D. G. MITCHELL *My Farm of Edgewood* 49, I gave them [*sc.* bees] a warm shelter of thatch. 1951 E. GRAHAM *My Window looks down East* iv. 34 Salt hay and thatch, or evergreens, are piled around the houses to insulate against the cold.

4. *thatch-roofed* adj. (earlier example).

1774 J. TRUMBULL *Poet. Wks.* (1820) II. 210 The thatch-roof'd hamlet and defenceless shed..are their fate.

Thatcherite (þæ·tʃərait), *sb.* and *a.* *Pol.* [f. the name *Thatcher* (see def.) + -ITE[1].] **A.** *sb.* One who supports the views or policies of Mrs. Margaret Thatcher (b. 1925), British (Conservative) politician, who became Leader of the Opposition in 1975 and Prime Minister in 1979. **B.** *adj.* Of, pertaining to, or characteristic of Mrs. Thatcher or Thatcherism.

1976 *Economist* 17 Apr. 13/2 Tory constituency rooms were by 1974 fuller of anti-Butler Thatcherites than Mr Heath dreamed. 1977 *Times* 16 May 11/7 The Thatcherite philosophy can be summed-up in two words, 'non-interference'. 1980 J. BOYD-CARPENTER *Way of Life* xxiv. 265 The Thatcherite view accepts the..thesis that 'equality of opportunity means equal opportunity to be unequal'. 1982 *Daily Tel.* 3 Apr. 18/4 The Thatcherites.. are genuinely trying to restore the private sphere, to bring back a world fit for gentlemen. *Ibid.* 11 Aug. 8/1 With the exception of that large part of the Labour party which is now authentically Bennite, we are all, to a greater or lesser extent, Thatcherites now.

So **Tha·tcherism,** the political and economic policies advocated by Mrs. Thatcher, esp. as contrasted with those of earlier Conservative leaders.

1979 *Times* 24 Nov. 2/2 The party was fighting off the shrill divisiveness of Thatcherism, with its simple monetarist policies. 1981 GLYN & HARRISON *Brit. Econ. Disaster* v. 138 Many workers..see Thatcherism as an outmoded nineteenth century ideology with little relevance to contemporary economic reality. 1982 *Daily Tel.* 11 Aug. 8 At heart, Thatcherism is a liberal economic reaction to the collectivism and corporatism of the past 40 years.

thatching, *vbl. sb.* Add: **3. thatching-beetle,** a thatcher's mallet.

1874 HARDY *Far from Madding Crowd* II. vi. 77 Where's your thatching-beetle and rick-stick and spars?

thatchy, *a.* Delete *rare* and add: Also, like thatch. Also *Comb.* (Further examples.)

1944 E. BLUNDEN *Cricket Country* v. 61 The sweetest of hamlets and thatchiest of little old inns. 1952 L. MACNEICE *Ten Burnt Offerings* 30 Like a sick bird..Its thatchy feathers moulting. 1973 T. PYNCHON *Gravity's Rainbow* I. 28 The flooded quarries and logged-off hillsides they'd left..across all that thatchy-brown, moldering witch-country.

Thathanabaing (þaþă:nābai·ŋ). [a. Burmese, f. *thathana* teaching, instruction (f. Pali *sassana*) + *baing* to possess.] The chief Buddhist dignitary in Burma.

1839 H. MALCOM *Trav. S.-E. Asia* I. II. vi. 315 The highest functionary is the *Tha-thena-byng*, or archbishop. 1858 P. BIGANDET *Life Gaudama* 252 In our days, the power of the Thathana-paing is merely nominal. 1912 *Rangoon Gaz.* 31 Oct. 19/1 A rectangular pandal, the central position of which was assigned to the Thathanabaing (Buddhist Archbishop) and his learned sadaws. 1934 'G. ORWELL' *Burmese Days* xv. 234 A big heathen idol..fell down on top of the thathanabaing, that is Buddhist bishop. 1972 A. T. Q. STEWART *Pagoda War* xiii. 151 The *Thathanabaing*, the hierarch whom the English generally referred to as 'the Buddhist Archbishop', had formerly been recognized throughout Burma as the Head of the Buddhist Church.

thattaboy, thatta boy, varr. *THATABOY *int.*

thaumatin (þǭ·mătin). *Biochem.* [f. *Thaumat(ococcus*, mod.L. generic name (f. THAUMATO- + COCCUS) + -IN[1].] Either or both of two related sweet-tasting proteins isolated from the fruit of the African plant *Thaumatococcus daniellii.*

1972 VAN DER WEL & LOEVE in *European Jrnl. Biochem.* XXXI. 221/1 This paper deals with the isolation and characterization of the sweet principles we call thaumatin I and II. *Ibid.* 225/2 The thaumatins are the first sweet-tasting proteins that have been found in nature 1973 [see *MONELLIN]. 1980 *Nature* 24 Apr. 653/2 Tate and Lyle..are to apply recombinant DNA techniques to improve production of the protein thaumatin—a substance 2,500 times sweeter than a 10% sugar solution.

thaw, *sb.* Add: **2. b.** (Earlier example.)

1840 M. EDGEWORTH *Let.* 30 Dec. (1971) 575 Lord Monteagle seated himself..beside Miss Edgeworth who had..made him rather a drawback stand-off curtsey.. He seemed determined there should be a thaw.

c. *Pol.* A relaxation of control or restriction; a lessening of harshness, hostility, etc.; *spec.* that which occurred in the U.S.S.R. after the death of Stalin in 1953.

1950 *Times* 13 June 5/3 The statement on foreign policy is the latest symptom of a thaw in Labour doctrine. 1956 R. MACAULAY *Towers of Trebizond* ii. 19 She had started.. working away at Russian visas..some time before the Great Thaw. 1957 *Economist* 30 Nov. 787/2 When the Polish thaw made emigration again possible, some of these 'autochthons' joined the queue. 1969 A. G. FRANK *Latin Amer.* xxi. 338 In the countries that took the Marxist road there was an increase in freedom or a noticeable thaw after a relatively short period of time. 1971 *Guardian* 13 Sept. 10/1 Krushchev inaugurated the thaw that mitigated some of the harsh intolerance of Stalinist communism. 1981 *Times* 2 Nov. 8/7 Andrei Voznesensky, arguably Russia's greatest living poet.. mirrored the hopes and naivety of the post-Stalin thaw.

3. *thaw-water*; *thaw-cold* adj.

1917 D. H. LAWRENCE *Look! We have come Through!* 156 They are the flowers of ice-vivid mortification, thaw-cold, ice-corrupt blossoms. 1947 K. M. WELLS *Owl Pen Reader* (1969) I. 38 He bumped and slithered over the ice the thaw had laid bare. He splashed through thaw water. 1976 *Times Lit. Suppl.* 23 July 926/3 Lush new green and blue sky reflected in the thaw-waters.

thaw, *v.* Add: **2. a.** (Earlier U.S. example with *out*.)

1835 J. H. INGRAHAM *South-West* I. 33 When vessels in their winter voyages..become coated with ice,..they seek the genial warmth of this region to 'thaw out'.

thawed, thawing *ppl. adjs.* (later examples).

1942 W. FAULKNER *Go down, Moses* 238 Out of the wet and thawing woods. *Ibid.* 240 They plunged down the bank, slipping and sliding in the thawed earth.

thawing, *vbl. sb.* Add: Also *attrib.*, and with *out* (or *up*).

1905 *Kynoch Jrnl.* Oct.–Dec. 200 Many consumers..put frozen cartridges in thawing pans several hours before they are required. 1946 KOESTLER *Thieves in Night* 150 Ellen was engaged in a serious and measured conversation... Dina took no part in the thawing-up proceedings. 1973 'R. MACLEOD' *Nest of Vultures* v. 97 A large whisky gently completing the thawing-out process.

the, *dem. adj.* ('*def. article*') and *pron.* Add: **B. I. 3. b.** (Further examples.) Also forming part of the present and former names of certain countries, as *the Argentine, the Congo, The Gambia, the Lebanon, the Sudan, the Yemen*; with the names of streets, *locally* with ellipsis of the word *Street.*

1853, etc. [see *HIGH *sb.* 1 c]. 1920 G. BELL *Let.* 14 Mar. (1927) II. xviii. 484 On my way home I went to see Frank Balfour..and heard from him the afternoon's news which was that Faisal had been crowned King of Syria and Abdullah King of the Iraq. 1951 DUKE OF WINDSOR *King's Story* xii. 209 Britain had an investment of £400,000,000 in the Argentine. 1959 *Chambers's Encycl.* VIII. 431/2 In internal affairs the Lebanon had to face considerable economic and financial difficulties after the end of the 1939–45 war. *Ibid.* XIV. 796/1 In March 1958 a federal link was established between the Yemen and the United Arab Republic. 1959 *Even. Standard* 31 Dec. 8/6, I am home from the Argentine and would like to link up with some of my old friends. 1975 J. I. M. STEWART

Gaudy xii. 225 The industrious little whirr of his camera was for a moment the only sound in the Broad. *Ibid.* 228, I had crossed Broad Street and was walking down the Turl. 1981 *Church Times* 6 Nov. 14/5 The Hoopoo had nested in his walls when he was in the Yemen. 1984 *Times* 18 Feb. 1/2 Princess Anne's four-day visit to The Gambia brings an extra air of festivity and importance to a tiny African country.

6. Also with names of paintings and sculptures.

1705 ADDISON *Remarks on Several Parts of Italy* 349, I have seen on coins..the Hercules Farnese, the Venus of Medicis, the Apollo in the Belvidere, and the famous Marcus Aurelius on Horseback. *a* 1706 EVELYN *Diary* an. 1693 (1955) V. 147, I..saw & indeede admired the Venus of Coreggio. 1845 *Encycl. Metropolitana* IX. 408 The Apollo Belvidere, the Venus de Medicis, and the Laocoon, have for ages been regarded as the highest possible models of excellence. 1984 *Times* 13 Sept. 13/4 Difficult to think of an art theft with greater sex appeal than that of the Mona Lisa.

7. (Further examples.)

The degree of ellipsis is not easy to determine.

1823 COBBETT *Gram. Eng. Lang.* XIX. 131 It is the same word, you see, in both instances; but you will see it different in the French. 1922 CHESTERTON *Eugenics & Other Evils* I. i. 11, I am content to answer that 'chivalrous' is not the French for 'horsy'. 1934 WEBSTER p. lxxxii/1, The modern descendants of the Latin are called the Romance languages. They include the Italian, the Spanish, the Portuguese [etc.]. *a* 1965 B. HIGGINS *Northern Fiddler* (1966) 34 'I'm corrupt' he said to me in the French, 'I think I live in corruption's stench.'

8. a. (Further examples.)

Still in common use side by side with forms without the definite article.

1961 I. FLEMING *Thunderball* i. 10 His secretary had gone down with the flu. 1972 *Time* 17 Apr. 41/2 Shortly before he was scheduled to make his first space flight aboard Apollo 13 two years ago, the longtime bachelor.. was accidentally exposed to the German measles.

b. With colloq. or humorous names of afflictions, as *the blues, collywobbles, creeps, D.T.'s, habdabs, heebie-jeebies, jitters,* etc., q.v. in Dict. and Suppl. Hence in analogous nonce-expressions.

1976 *Publishers Weekly* 11 Oct. 90/3 The case of the 'cutes' infecting text and pictures. 1976 *Listener* 11 Nov. 626/2 The whole story, like the chateau, has an unmistakable touch of the Enid Blytons.

10. a. (Later examples of formal use with titles of rank. In informal use *the* is usu. omitted.)

1935 C. HAMILTON *Pillion* 25 He was the third son of Colonel the Hon. Almeric Sounds Sharnal Piers Clement Piers, late of the Rifle Brigade. 1939 E. BAX *Miss Bax of Embassy* xviii. 238 Someone is always dashing in to ask me questions like...is Lady V. *The* Lady or only Lady? 1943 H. SAUNDERS *Combined Operations* vii. 52 Admiral of the Fleet Sir Roger Keyes was succeeded as Director of Combined Operations by Captain the Lord Louis Mountbatten. 1981 *Daily Tel.* 5 Nov. 16/2 Her Majesty's Body Guard of the Honourable Corps of Gentlemen-at-Arms under the command of the Lord Denham.

d. (Earlier example.) Also *slang* and sometimes *derogatory*, with a woman's surname or nickname. Cf. *LA, LA.

1730 O. SWINY *Let.* 29 July in R. B. Peake *Mem. Colman Family* (1841) I. 18 If he does not, then we must provide a soprano man, and a contr'alto woman (though the Merighi stays). 1922 *Dialect Notes* V. 143 [At] Somerville..'The Pen' is the Lady Principal, Miss Penrose, 'The Darb', Miss Derbyshire, etc. 1930 WODEHOUSE *Very Good, Jeeves!* iv. 96 The Bellinger..had sung us a few songs before digging in at the trough. 1973 —— *Bachelors Anonymous* xii. 155 The Fitch was at the hair stylist's having a permanent.

15. b. (Further examples.)

1780 COWPER *Progress of Error* in *Wks.* (1905) 29 The creature is so sure to kick and bite, A muleteer's the man to set him right. 1813 JANE AUSTEN *Pride & Prej.* I. xiii. 142, I shall not be the person to discourage him.

16. c. With a sb. characterizing the trade or profession of the person whose name precedes. *local* (esp. in Wales).

1894 SOMERVILLE & 'ROSS' *Real Charlotte* I. iv. 40 Norry the Boat, daughter of Shaunapickeen, the ferryman (whence her title). 1951 W. MORUM *Gabriel* II. vii. 230 He thought Larry the Groan far worse. The effeminate singer..was positively embarrassing. 1974 *Times* 27 Apr. 15/8 The Welsh tradition of referring to people by the names of their jobs, as Jones the Post or Davis the Bread. 1980 R. H. LEWIS *Cracking of Spines* vii. 113 'The prospective client,'..I assumed a Welsh accent. 'Matt the Book.'

II. 21. Also forming phrases with the preposition *on*, as *on the cheap, quiet, sly,* etc., q.v. in Dict. and Suppl.

D. as *sb.* with pl. *thes.*

1882 'MARK TWAIN' *Stolen White Elephant* 269 You [English] say 'out of window'; we always put in a *the*. 1907 —— *Chr. Sci.* II. viii. 239, I uncover to that imperial word... The rare and..exclusive company of the THE's of deathless glory..*the* Saviour..*the* Bible. 1959 *Amer. Speech* XXXIV. 111 The Syrian student tends to put in *the*'s where they are not needed. 1977 *Guardian Weekly* 4 Dec. 4/1 If you are really serious about something and want to be taken seriously yourself, never, ever, under any circumstances, sully its name by putting a 'the' in front of it.

‖ **thé** (te). [Fr., = tea.] † **1.** A tea-party. *Obs.*

1788 H. MORE *Let.* 22 May (1925) 123 A *Thé* is among the stupid new follies of the winter. You are to invite fifty or a hundred people to come at eight o'clock..tea and coffee are made by the company,..and what constitutes the very essence of a *Thé*, an immense load of hot buttered rolls and muffins. 1802 C. WILMOT *Let.* 3 Jan. in T. U. Sadleir *Irish Peer* (1920) 22 We have had.. Plays, Balls, Soirees, Thés, &c.; the first Thé was at Monsieur Amoulin's. 1827 E. GROSVENOR in G. Huxley *Lady Elizabeth & Grosvenors* (1965) vii. 136 On Friday we are to have a thé at the Viceroy's.

2. Phrases. **thé complet** [Fr., lit. 'complete tea'], a light meal including tea and usu. bread and cake; cf. *café complet* s.v. *CAFÉ 3; **thé dansant**: see *DANSANT, DANSANTE *a.*

1951 E. COXHEAD *One Green Bottle* v. 113 She darted away with another *thé complet*. 1967 N. FREELING *Strike out where not Applicable* 113 The Dutch 'thé complet' accessories..sandwiches. Glacé fours. Dry petits fours. Fan wafers.

theatre, theater, *sb.* Add: **2. a.** *saloon theatre* (earlier example) (*obs.* exc. *Hist.*).

1864 G. A. SALA *Robson* 14 Early in 1844 he accepted an engagement at the Grecian Saloon Theatre, in the City Road.

b. *N. Amer.* and *N.Z.* A picture theatre, cinema.

1923 H. CRANE *Let.* 5 Oct. (1965) 149 [Chaplin] is here in New York..to see that the first film he has produced in it [*sc.* a new studio] gets over profitably... It's running now..at the 'Lyric' theatre. 1956 H. KURNITZ *Invasion of Privacy* ii. 20 Do I want to book that man's pictures in my theatres? 1966 G. W. TURNER *Eng. Lang. in Austral. & N.Z.* viii. 176 'Theatre' nearly always a 'picture theatre' or cinema in New Zealand. 1977 *Chicago Tribune* 2 Oct. (TV Week Suppl.) 2/1, I went to the theater and saw George Segal and Goldie Hawn in 'The Duchess and the Dirtwater Fox'.

3. c. (Later examples.) Also, the drama of a particular time or place; dramatic art as a craft, the theatrical profession. Phrases: *theatre-in-the-round*: see *ROUND *sb.*[1] 5 d; *Theatre of Cruelty* [tr. F. *théâtre de la cruauté* (A. Artaud (1932) *Manifeste du théâtre de la cruauté*)], a collective term for plays in which the dramatist seeks to communicate a sense of pain, suffering, and evil through the portrayal of extreme physical violence; *Theatre of the Absurd*, a collective term for plays (chiefly French) portraying the futility and anguish of man's struggle in a senseless and inexplicable world (cf. *ABSURD *sb.*); also *fig.*; *Theatre of Fact*, documentary drama.

1908 E. TERRY *Story of my Life* xiv. 332 The life of an actress belongs to the theatre. *Ibid.* 333, I have had many friends outside the theatre, but I have had very little time to see them. 1938 R. G. COLLINGWOOD *Princ. of Art* xiv. 323 In the Renaissance theatre collaboration between author and actors on the one hand, and audience on the other, was a lively reality. 1955 G. GORER *Exploring English Character* ii. 14 Of the theatre I know of, only the Burmese drama of the second half of the nineteenth century approaches the Elizabethan in its search for horror. 1976 J. ARCHER *Not a Penny More, Not a Penny Less* xvi. 174 Harvey recognized Dame Flora Robson, the actress, who was being honoured for a distinguished lifetime in the theatre. 1977 S. BRETT *Star Trap* IV. xiii. 143 He is a hard-working performer with a great belief in the live theatre.

1954 E. BENTLEY *In Search of Theater* II. vii. 198 Antonin Artaud's 'theatre of cruelty', that theater of Dionysian energy and visionary power. 1958 M. C. RICHARDS tr. *Artaud's Theater & its Double* vi. 79 '*Theater of cruelty*' means a theatre difficult and cruel..on the level of performance, it is not the cruelty we can exercise upon each other..but the much more terrible and necessary cruelty which things can exercise against us. 1964 *Punch* 21 Oct. 627/3 To watch the 'Theatre of Cruelty' season safely on the audience side. 1973 J. ELSOM *Erotic Theatre* x. 190 The one adjective which cannot be used to describe the Theatre of Cruelty evening is, however, *unexpected*.

1961 M. ESSLIN *Theatre of the Absurd* 17 The Theatre of the Absurd strives to express its sense of the senselessness of the human condition and the inadequacy of the rational approach by the open abandonment of rational devices and discursive thought. 1962 [see *ABSURD *sb.*]. 1963 *Sunday Times* 24 Feb. 24/5 They deserved to win, but two of the goals they scored came straight from the theatre of the absurd. 1977 P. JOHNSON *Enemies of Society* xix. 253 We must not be surprised to find that the United Nations..should have become the World Theatre of the Absurd, a global madhouse where lunatic falsehood reigns.

1966 *Punch* 7 Dec. 864/1 Together they make up the most successful example so far of the Theatre of Fact, a gripping story, the clash of widely different personalities and many sharp remarks on the relationship between Science and Government. 1970 *Times* 9 Feb. 5 *Murderous Angels* is another example of the Theatre of Fact... The two main characters are Dag Hammarskjold and Patrice Lumumba. 1974 *Encycl. Brit. Macropædia* XVIII. 232/2 The Brecht approach to stage presentation has something in common with the Theatre of Fact.

e. Without article or pl. (chiefly predicatively). With a descriptive adjective: theatrical or dramatic entertainment (of a specified quality); esp. in *good theatre* (see *GOOD *a.* 1 f); also used *transf.* of an action or work of art that has the quality of (good, etc.) drama or theatrical technique; hence *fig.*, dramatic effect or sensation, spectacle, outward show without serious inward intent.

1926, etc. [see *GOOD *a.* 1 f]. 1927 *Sunday Times* 27 Feb. 6/4 'The Letter' is superb theatre throughout. 1934 *Sun* (Baltimore) 1 June 12/1 Superb tennis and 'goop theater' have never been so generously mixed in the performance of any other player. 1939 A. THIRKELL *Before Lunch* iv. 96 It would have been rather too much theatre to awaken heroine with soft music, don't you think? 1948 A. J. P. TAYLOR *Habsburg Monarchy* i. 12 Austrian Baroque civilisation..was grandiose, full of superficial life, yet sterile within: it was theatre, not reality. 1951 in M. McLuhan *Mech. Bride* (1967) 89 They bring real 'theater' to a sales presentation. 1955 W. W. DENLINGER *Compl. Boston* II. 9 Some of the competition exercises are almost useless; some I consider pure 'theatre' and others are practical. 1958 *Listener* 2 Oct. 499/2 You have to admit that the Old City is good theatre. 1965 *Ibid.* 21 Oct. 630/2 Standing spotlit at the end of a great black-draped room all by itself, it [*sc.* a piece of sculpture]..was above all dramatic. It was, in its way, even great theatre. 1975 J. O'FAOLAIN *Women in Wall* xii. 211 She encourages zeal and all she gets is theatre... This sort of thing was new to the convent.

5. b. A room in a hospital specially designed for surgical operations (orig. one resembling a theatre, for the performance of such operations before observers); = *operating-theatre* s.v. OPERATING *vbl. sb.* b.

1641, 1766 [in Dict., sense 5]. 1823 *Lancet* 5 Oct. 3/1 At half-past Seven this Theatre was crowded in every part, by upwards of four hundred Students, of the most respectable description; in fact we never before witnessed so genteel a Surgical class. 1935 MARSH & JELLETT *Nursing-Home Murder* iii. 38 In the anteroom of the theatre two nurses and a sister prepared for the operation. 1976 J. ARCHER *Not a Penny More, Not a Penny Less* xii. 129 Although the hospital had only some 200 beds, the theatre was of the highest standard.

6. c. A particular region or one of the separate regions of the world in which a war is being fought. Also *theatre of war.*

1914 W. S. CHURCHILL *Let.* 15 Oct. in M. Gilbert *Winston S. Churchill* (1972) III. Compan. I. 193 The hand of war will I expect be heavy upon us in the Western Theatre during the next four weeks. 1928 BLUNDEN *Undertones of War* xv. 160 (*heading*) Theatre of War. 1940 W. S. CHURCHILL *Into Battle* (1941) 261 Far larger operations no doubt impend in the Middle East theatre. 1958 E. BIRNEY *Turvey* vii. 76 Turvey straightened his helmet and marched down the gangplank into the European Theatre of War. 1961 G. F. KENNAN *Russia & West* viii. 118 Real fighting took place between Allied and Bolshevik forces only in one theater, in the Russian north. 1977 C. MCCULLOUGH *Thorn Birds* xv. 352 The biggest and most decisive battle of the North African theater had been fought.

9. a. *theatre audience, bill* (example), *hat* (example), *house* (later example), *man, people, stall, ticket* (examples); *theatre-loving* adj.

1936 *Vogue* 18 Mar. 101/2 The London theatre audience is still all dressed in black and white. 1977 S. BRETT *Star Trap* xiii. 143 It doesn't bear comparison with the contact you can get with a live theatre audience. That's electrifying. 1895 G. B. SHAW *Our Theatres in Nineties* (1932) I. 1 It is not a work of art at all: it is a mere contrivance for filling a theatre bill. 1930 Theatre hat [see *JULIET]. 1977 *N.Y. Rev. Bks.* 29 Sept. 12/4 On the rickety stages of a thousand provincial theater houses, alternative worlds blazed like magic by limelight. 1846 GEO. ELIOT *Let.* 1 June (1954) I. 219 Please to come in a very mischievous, unconscientious, theatre-loving humour. 1933 P. GODFREY *Back-Stage* viii. 112 Every experienced theatre-man knows that there is ample room for criticism inside the theatre. 1961 *Guardian* 6 Mar. 9/4 Some knowing theatre-men say it would have flopped.. even a few years ago. 1952 E. WILSON *Shores of Light* 382, I did not want to see the theater people again; I could not face another evening. 1907 G. B. SHAW *Let.* Dec. (1972) II. 739 With..society out of town during the parliamentary recess, theatre stalls have been empty. 1902 A. C. HEGAN *Mrs. Wiggs of Cabbage Patch* vi. 73 Couldn't you use a whole load [of kindling], if I was to take it out in..theayter tickets? 1980 P. G. WINSLOW *Counsellor Heart* v. 94 Up for a day in town, to get theatre tickets.. and then go shopping.

b. theatre club, a theatre for which tickets are sold only to members (esp. in order to circumvent the censorship of public performances); **theatre-goer** (earlier example); **theatre-going** *sb.* and *adj.* (earlier examples); **theatre-list** *Med.*, a list of patients about to undergo surgical operations; **theatre nurse** *Med.*, a hospital nurse qualified to assist in the operating theatre; **theatre organ** = *cinema organ* s.v. *CINEMA c; hence **theatre organist;** **theatre party**: for *U.S.* read orig. *U.S.* and add earlier and later examples; **theatre-restaurant,** a restaurant where theatrical entertainment is provided for customers; **theatre seat,** (*b*) a seat that may be booked for a performance at a theatre; **theatre sister** *Med.*, in a hospital, nursing-home, etc., a sister qualified to assist in the operating theatre; **theatre suit** *Fashion* (see quot. 1969); **theatre workshop,** a non-commercial theatre company concerned esp. with experimental and unconventional theatrical productions; orig. and *spec.* a company founded by Joan Littlewood and others in 1945 and based in the East End of London from 1953 to 1973.

1961 R. WILLIAMS *Long Revolution* II. vi. 267 The growth of 'free theatres' and theatre-clubs. 1978 R. HOLLES *Spawn* iv. 31 Marianne had met him..at a theatre

club in Notting Hill Gate. 1870 *Boston Transcript* 1 Nov. 2/4 If the theatre is not crowded.., we shall be much disappointed in our estimation of the taste of Boston theatre-goers. 1846 B. I. LANE *Mysteries of Tobacco* 11 The classical theological, feat-haunting, theatre-going, card-playing Reverend Gentleman. 1852 GEO. ELIOT *Let.* 15 June (1954) II. 36 Between theatre-going and proof-reading, my spiritual eyes are burning as dim and bleared as gas-lights. 1964 G. L. COHEN *What's Wrong with Hospitals?* i. 17 A student will undertake the pre-medication of patients on theatre-list. 1934 P. BOTTOME *Private Worlds* xxxi. 302 Matron..is a first-rate surgical nurse... The theatre nurse is about too, in case we want her. 1959 T. S. ELIOT *Elder Statesman* II. 45, I feel in love with him During an appendicitis operation! I was a theatre nurse. 1930 R. WHITWORTH *Electric Organ* xvi. 156 The building of theatre organs has..helped to bring electric..actions to their present state of efficiency. 1977 *Lancashire Life* Nov. 101/1 The story of Ronald Curtis and theatre organs is in effect the chronicle of a love affair which began in his childhood. 1932 R. WHITWORTH *Cinema & Theatre Organ* ix. 105 The cinema or theatre organist fills a very important role. 1883 *Cent. Mag.* Sept. 787/1 A report.. of Mrs. Dash's theater party. 1884 L. TROUBRIDGE *Life amongst Troubridges* (1966) 170 To a theatre party on the 15th and supper after. 1962 J. F. POWERS *Morte d'Urban* viii. 164 The Cathedral curates..wangled an invitation to the Saturday-morning theatre-parties. 1958 *Hotel & Catering Rev.* Oct. 35/1 The only theatre restaurant of its size and type in the world. *a* 1911 D. G. PHILLIPS *Susan Lenox* (1917) II. xviii. 421 A clever play that'll draw the damn fools who buy theater seats. 1982 C. CASTLE *Folies Bergère* vii. 254 As a student..the only theatre seat he could afford was in the gallery. 1935 MARSH & JELLETT *Nursing-Home Murder* iii. 37 Tell the theatre sister I'll operate as soon as they are ready. 1976 C. STORR *Unnatural Fathers* i. 10 The staff nurse on the surgical side who deputised for the theatre sister. 1964 MRS. L. B. JOHNSON *White House Diary* (1970) 202, I changed into my black theater suit en route. 1969 R. T. WILCOX *Dict. Costume* 107/2 *Dinner* or *theater suit*, the feminine 'covered-up' look for evening of the 1930's and '40's, consisting generally of a long black skirt, a delicate blouse, a cummerbund and short jacket. 1945 *Westmorland Gaz.* 4 Aug. 4/9 Addresses were given by Miss Joan Littlewood ..now in Westmorland with the Theatre Workshop, a new venture aimed at furthering the arts in local towns. 1962 *Guardian* 7 Nov. 7/6 A theatre workshop is about to be started in Dublin. 1973 E. BULLINS *Theme is Blackness* 10 Some of the Black Arts approaches and techniques that Marvin X and I had developed in revolutionary theater and literature workshops on the Coast. 1981 *Sunday Tel.* 20 Dec. 16/5 The Arts Council refuses it [*sc.* the D'Oyly Carte Company] a grant—preferring its own East End revolutionary theatre workshops.

c. *attrib.* Designating nuclear weapons for use within a 'theatre' (at present thought of as Europe) as opp. to intercontinental or strategic weapons (cf. *STRATEGIC *a.* 2), or their targets.

1977 *Observer* 3 Apr. 12/4 In a tactical role, Backfire.. is ideally suited to attacking local or 'theatre' targets in Western Europe. 1978 *Orbis* XXII. 309 The United States has deployed a varied array of theater-nuclear weapons and delivery systems in Europe. 1980 *Daily Tel.* 18 June 1/2 Theatre nuclear missiles..have a longer range than battlefield weapons but cannot be fired as far as intercontinental missiles. 1983 *Chicago Sun-Times* 26 Nov. 5/2 'What worries us is the buildup of theater nuclear forces in Europe,' Defense Undersecretary Fred C. Ikle said.

theatrette (þī·ătəre·t). [f. THEATR(E *sb.* + -ETTE.] A small theatre.

1927 *Melody Maker* Sept. 849/3 Things soon went wrong at his Leicester Square Theatrette. 1972 *Malay Mail* 27 May 1/7 The building..will house a theatrette for 200.

theatric, *a.* (*sb.*) Add: **B.** *sb. pl.* **2.** orig. *U.S.* Doings of a theatrical character; theatrical behaviour, effects, or mannerisms; theatricality.

1929 W. FAULKNER *Sartoris* i. 3 With his race's fine feeling for potential theatrics he drew himself up. 1958 A. MILLER *Coll. Plays* iii. 18 Plays..had been written for a theatrical performance, when they should have been written as a kind of testimony whose relevance far surpassed theatrics. 1964 L. HANSBERRY in J. H. Clarke *Harlem* 136 The..little committees..have dragged on their particular obscene theatrics for all these years. 1972 *Time* 2 Oct. 52/2 A desperate device intended to lend a little spine to the sponge-cake theatrics [in a film]. 1977 *Daily Tel.* 12 July 17/6 The 'theatrics' of the Church are important to many Catholics and, in a way, this is what Lefebvre offers. 1983 *Times* 24 Aug. 5/2 Today's so-called peace movement—for all its modern hype and theatrics—makes the same old mistake.

theatrical, *a.* Add: **A.** *adj.* **4.** Special collocations, as *theatrical agency, agent*, an agency, agent whose business is to act as an intermediary between actors and actresses seeking parts and producers offering them.

1825 P. EGAN *Life of Actor* ii. 62 We are engaged at the Harp to meet Mr. Schemer, the theatrical agent, to-morrow night. 1828 J. EBERS *Seven Yrs. King's Theatre* vii. 196 Been actively engaged in theatrical concerns, and the business of theatrical agency. *a* 1911 D. G. PHILLIPS *Susan Lenox* (1917) II. v. 126 She read an advertisement of a theatrical agency. 1973 D. RAMSAY *Deadly Discretion* 111 Why not go to Actors' Equity and theatrical agencies and dance studios? 1978 *Detroit Free Press* 2 Apr. 13c/5 Marco talks Mrs. Hopkins into letting him stay on as boarder by becoming her theatrical agent.

B. *sb.* **1.** *pl.* Now usu. *amateur theatricals.*

1873, etc. [see *AMATEUR 3 a]. 1892 G. & W. GROSSMITH *Diary of Nobody* viii. 118, I..totally disapproved of amateur theatricals. 1965 *Listener* 23 Sept. 462/3 He proved..fond of..amateur theatricals.

2. (Earlier example.) Also *transf.*, the theatrical column of a newspaper.

1763 D. GARRICK *Let.* 8 Oct. in R. B. Peake *Mem. Colman Family* (1841) I. iii. 84 God bless you! my dear Colman, and have a corner of your eye upon my theatricals. 1819 KEATS *Let.* 22 Sept. (1958) II. 176, I purpose living in town in a cheap lodging, and endeavouring, for a beginning, to get the theatricals of some paper.

theatrified (þi,æ·trifəid), *a. nonce-wd.* [f. THEATR(E *sb.* + -IFY + -ED[1].] Deluded by the conventions of popular drama.

1902 G. B. SHAW *Mrs. Warren's Profession* p. xxiv, People with completely theatrified imaginations tell me that no girl would treat her mother as Vivie Warren does.

theatrist (þī·ătrist). *rare.* [f. THEATRE *sb.* + -IST.] A lover or frequenter of the theatre; an expert in theatrical matters.

1889 E. DOWSON *Let.* 24 Dec. (1967) 120 Last night—lo what a theatrist I am becoming—I went to Benson's 'Midsummer Night's Dream'. 1905 M. BEERBOHM in *Sat. Rev.* 13 May 623/1, I do not mean that 'Salomé' has less dramatic than literary fibre. Mr. Wilde was a born dramatist—a born theatrist, too.

theatro-. Add: **theatrophone** (earlier example).

1889 *Telephone* I. 406/1 A 'theatrophone'..is an adaptation of the telephone, by which any one can be put into communication with a certain theatre.

‖ **theatrum** (te,ā·trŏm). [L.: see THEATRE, THEATER.] Theatre, playhouse.

1786 [see *ORCHESTRA 1 c]. 1890 E. DOWSON *Let.* 23 Feb. (1967) 139 We must try & work a theatrum this week. 1967 *Oxf. Compan. Theatre* (ed. 3) 36/1 The line of general development followed the 'Theatrum' of the Lyon Terence, absorbing on the way the lessons learned in the temporary spectacle-theatres.

‖ **theatrum mundi** (te,ā·trŏm mu·ndi). [L., = theatre of the world.] The theatre thought of as a presentation of all aspects of human life; *spec.* (see quot. 1932).

1566 J. ALDAY tr. P. Boaistuau (*title*) Theatrum Mundi, the Theatre or rule of the world, wherein may be sene the running race and course of everye mans life, as touching miserie and felicity, wherein is contained wonderfull examples, learned devices, to the overthrow of vice, and exalting of vertue. 1932 J. NICOLL tr. *von Boehn's Dolls & Puppets* II. i. 261 Gottfried Hautsch, who died in 1703,..constructed in Nürnberg a mechanical automaton with many figures, which was nicknamed his 'little world'. This is a kind of automaton which, to distinguish it from the others, is technically indicated by the term *theatrum mundi*. The *theatrum mundi* for centuries provided the traditional afterpiece of the wandering marionette theatres; by means of small movable figures running on rails it showed a diversity of scenes. 1953 W. R. TRASK tr. *Curtius's European Lit. & Lat. Middle Ages* vii. 140 A *theatrum mundi*, then, with men as actors, Fortune as the stage director, and Heaven as spectator. 1966 H. B. HAWKINS in *Shakespeare Q.* XVII. 174 The idea of the *theatrum mundi* was widely known in the Renaissance, and a number of themes..came to be associated with this concept. The idea that the world itself was God's theater gave cosmic significance to the contemporary stage. 1967 *Listener* 8 June 744/1 Television offers an almost Elizabethan comprehension of the world: it is the new *theatrum mundi*. 1979 C. E. SCHORSKE *Fin-de-Siècle Vienna* v. 227 Klimt..presents the world to us as if we were viewing it from the pit, a *theatrum mundi* in the Baroque tradition... The Baroque *theatrum mundi* was clearly stratified into Heaven, Earth, and Hell.

Theban, *a.* and *sb.* Add: **A.** *adj.* **2.** (Later example.)

1962 E. COLLEDGE *Mediæval Mystics of England* 15 The hermit settlements of the Theban desert.

B. *sb.* **2.** The variety of Greek spoken in Bœotian Thebes. *poet. rare.*

1820 SHELLEY *Oedipus Tyrannus* (1904) II. ii. 451 In plain Theban, that is to say, My name's John Bull.

thebe (þē[i]·be). Pl. **thebe.** [a. Setswana, lit. 'shield'.] A currency unit in Botswana, equal to $\frac{1}{100}$ of a *PULA (sense 2). Also, a coin of this value.

1976 *Botswana: Ten Years of Progress 1966–76* 17 The Bank of Botswana will ultimately perform all of the functions of a modern central bank. Its first task has been to conduct the issuance of the new notes (Pula) and coins (Thebe) in 1976. 1976 *N.Y. Times* 15 Aug. 6/1 The new unit—the pula, which will be divided into 100 thebe—will have the same value, however, as the rand, which ceases to become legal tender in this country as of Aug. 23.

theca. Add: **3. b.** In full *theca folliculi.* A layer of hormonally active cells enclosing a tertiary (vesicular) or a mature (Graafian) ovarian follicle, consisting of an inner, vascular layer (*theca interna*) and an outer, fibrous layer (*theca externa*). [So named in Ger. by C. E. von Baer (*Über Entwickelungsgeschichte der Thiere* (1837) II. III. xv. 23).]

1857 DUNGLISON *Dict. Med. Sci.* (rev. ed.) 400/1 *Folliculi Graafiani*, small spherical vesicles in the stroma of the ovary, which have at least two coats; the outer termed ovicapsule..and theca folliculi. 1859 R. B. TODD *Cycl. Anat. & Physiol.* V. 551/1 The external or vascular coat [of the Graafian follicle]..constitutes the tunic of the ovisac of Barry, the tunica fibrosa, S. theca folliculi of Baer. 1929 [see *LUTEINIZED *ppl. a.*]. 1930 MAXIMOW & BLOOM *Text-bk. Histol.* xxxii. 640 There is no sharp limit between the two layers of the theca or between the theca externa and the stroma. 1966 *McGraw-Hill Encycl. Sci. & Technol.* XI. 474/1 A growing follicle has several layers of follicular cells, or granulosa cells.., and a surrounding capsule of connective tissue, the theca folliculi. 1978 D. B. & W. J. WILSON *Human Anat.* xvi. 388/2 The inner layer is the theca interna, which secretes the hormone estrogen into the vascular system.

4. Special Comb.: **theca cell tumour** *Path.*, an œstrogen-secreting ovarian tumour that consists of cells resembling those of the theca folliculi and is sometimes malignant; = *THECOMA.

1937 *Amer. Jrnl. Obstetr. & Gynecol.* XXXIV. 988 It was not until 1932 that the last member of this interesting group of tumors was reported by Loeffler and Priesel, to which they gave the name of 'fibroma theca cellulare xanthomatodes ovarii', more commonly known as the theca cell tumor. 1974 PASSMORE & ROBSON *Compan. Med. Stud.* III. xxviii. 50/1 Luteinization of the cells of granulosa cell and theca cell tumours may occur.

thecodont, *a.* and *sb.* For '*Thecodontes*, an extinct family of saurians' substitute 'Thecodontia, an extinct order of primitive reptiles'. (Later examples.)

1933 A. S. ROMER *Vertebr. Paleontol.* viii. 170 An overgrown offshoot from the early thecodont stock was.. a large South African Lower Triassic form. *Ibid.*, Confined exclusively to the Triassic, the history of the thecodonts was a brief one. 1980 N. ORRISS tr. *Babin's Elem. Palaeontol.* xviii. 327 In the classical conception, birds originated directly from the Triassic thecodonts.

thecodontian (þīkŏdǭ·ntiăn), *sb.* and *a.* [f. mod.L. order name *Thecodontia* (R. Owen 1859, in *Rep. Brit. Assoc. Adv. Sci.* 163) + -AN: cf. THECODONT.] = prec.

1974 *Nature* 8 Mar. 168/2 The Saurischia and Ornithischia, are usually interpreted as independent derivatives of primitive thecodontian reptiles of the Triassic, but all known Triassic dinosaurs can be distinguished from typical thecodontians. 1979 *Ibid.* 17 May 234/2 Dinosaurs apparently lacked a fenestra pseudorotunda, as did the early thecodontians, eosuchians and captorhinomorphs.

thecoma (þīkō[u]·mă). *Path.* [f. THEC(A + *-OMA.] = *theca cell tumour* s.v. *THECA 4.

1937 *Amer. Jrnl. Obstetr. & Gynecol.* XXXIV. 988 The luteoma, like the thecoma, is also a rare tumor. 1966 WRIGHT & SYMMERS *Systemic Path.* I. xxvii. 869/1 Although they may occur at any age, thecomas are most frequent in women between 50 and 60. 1981 A. D. T. GOVAN et al. *Path. Illustrated* xiii. 685 It has been suggested that fibromas found in ovaries are thecomas which have undergone fibrous degeneration.

thee, *pers. pron.* Add: **3.** (Later examples of use by Quakers or their relatives.) Now *rare.*

1926 *Amer. Speech* I. 638/1 Even in my boyhood in New England I heard very few Quakers who habitually said *thee*. 1950 B. RUSSELL *Let.* 6 Mar. in B. Strachey *Remarkable Relations* (1980) xxi. 312 What thee says about our marriage is very generous. 1964 *Friend* 10 Apr. 453/1 Perhaps thee has noticed the comment on this point in our *Friends Journal* on February 15. 1980 B. STRACHEY *Remarkable Relations* xxi. 314 Alys [Russell (1867–1951)] had been the last of the older ones; the last to say Thee and Thy.

theelin (þī·lin). *Biochem.* [f. Gr. θῆλυς female + -IN[1].] = *ŒSTRONE.

1930 C. D. VELER et al. in *Jrnl. Biol. Chem.* LXXXVII. 357 The isolation of the crystalline hormone seems to justify the selection of a new name... Accordingly, we suggest the term 'theelin'. 1935 [see *OESTRIN]. 1936 *Jrnl. Amer. Med. Assoc.* 10 Oct. 1222/2 The term theelin has not been widely accepted. *Ibid.* 1223/1 The Council [on Pharmacy and Chemistry of the American Medical Association]..decided (1) to adopt the system of nomenclature based on the root *estr-*; (2) to retain *theelin*, *theelol*..as synonyms for..*estrone, estriol.* 1947 *Sci. News* IV. 137 Among the substances administered to plants..are: the juice of leaves subjected to the 'right' length of day; yeast extract;..theelin.

So **thee·lol** [-OL] = *ŒSTRIOL.

1931 DOISY & THAYER in *Jrnl. Biol. Chem.* XCI. 642 We propose now, in view of the facts that the new substance is a trihydroxy compound and that it shows physiological and possibly chemical similarities to theelin, to name it theelol. 1936 [see above]. 1977 *Martindale's Extra Pharmacopoeia* (ed. 27) 1419/1 Oestriol. Estriol; theelol.

theeward, in phr. *to theeward*: see -WARD 6 c.

thegosis (þegō[u]·sis). *Zool.* [f. Gr θηγός sharp + -OSIS.] Tooth-grinding in animals as a means of sharpening the teeth.

1971 EVERY & KÜHNE in D. M. & K. A. Kermack *Early Mammals* 25 While shear edges are used in mastication, they lose their feature which they develop during a powerful and short action of sharpening. This action we call active wear or thegosis. 1974 *Nature* 30 Aug. 730/1 In mammals thegosis wear is accompanied by a second type of tooth wear, dental abrasion, which occurs at food/tooth interfaces during mastication.

theileria (þəili°·riă). Also **Theileria.** [mod.L. (A. Bettencourt et al. 1907, in *Archives R. Inst. Bacteriol. Camara Pestana* I. 343), f. the name of Sir Arnold *Theiler* (1867–1936), South African zoologist + -IA¹.] A tiny, tick-borne protozoan parasite of the genus of this name, which includes those causing theileriasis.

1910 *Parasitology* III. 127 Observations on *Theileria* are fraught with considerable difficulty owing to the minuteness of the parasite. **1927** HALDANE & HUXLEY *Animal Biol.* xii. 279 Smallest parasitic Protozoa (Theileria in ox blood-corpuscle). **1979** *Nature* 5 July p. xiii (Advt.), The research will include the infection and transformation of bovine lymphocytes and other cell types by theileria parasites.

theileriasis (þəilĕrəi·ăsis). Also **theilerio·sis** [f. *THEILERIA + *-IASIS, -OSIS.] An acute, usually fatal, feverish disease of cattle, sheep, and certain other vertebrates caused by a protozoan of the genus *Theileria* or a closely related genus and transmitted by ticks; cf. *East Coast fever* s.v. *EAST D. 1 b.

1944 *Indian Jrnl. Vet. Sci.* XV. 149 (*title*) Control of cattle theileriasis in calves in the Punjab. **1959** *Adv. Vet. Sci.* V. 241 The name theileriosis has come to designate any member of a group of diseases of vertebrates produced by several species of protozoan parasites belonging to the genera *Theileria*, *Gonderia*, and *Crytauxzoon*. **1962** J. A. SMYTH *Introd. Animal Parasitol.* ix. 109 This organism [sc. *Theileria parva*] is the cause of the deadly 'theileriasis' or East-Coast Fever in cattle. **1979** *Protozool. Abstracts* III. 38/1 A good review is given of theileriasis of cattle in India. **1979** *Nature* 5 July p. xiii (Advt.), There are in the laboratory of the Director of ILRAD two vacancies for immunologists to work on theileriosis.

theirn. Add: Also *U.S. dial.*

1836 T. C. HALIBURTON *Clockmaker* 1st Ser. x. 50 When other folks lost theirn from the boys, hisn always hung there like bait to a hook. **1896** 'MARK TWAIN' in *Harper's Mag.* Sept. 532/1, I hain't ever seen eyes bug out..the way theirn did. **1930** *Amer. Speech* V. 267 Such possessive forms as *ourn, yourn, hisn, hern* and *theirn* are almost universal in the Ozarks.

theistical, *a.* Add: **theistically** *adv.* (example).

1881 MAX MÜLLER tr. *Kant's Critique Pure Reason* II. 635 On one side, theistically, that there is a Supreme Being.

thelemic (þelī·mik), *a.* [f. Gr. θέλημα will + -IC, with reference to the abbey of Thelème in Rabelais; see THELEMITE.] That permits people to do as they wish; *spec.* designating the Satanist activities of Aleister Crowley (1875–1947).

1926 T. E. LAWRENCE *Seven Pillars* (1935) V. lix. 335 The Catholic Christians would counter them by demanding European protection of a thelemic order, conferring privileges without obligation. **1951** J. SYMONDS *Great Beast* III. xvi. 151 The intention of these two founders [sc. Sir Francis Dashwood and Aleister Crowley] of Thelemic Abbeys was different. *Ibid.* 152 Five rooms were planned around a central hall, the Sanctum Sanctorum, or the temple, of the Thelemic mysteries. **1956** —— *Ibid.* (rev. ed.) 155 Those *Orgia* which so shocked the readers of the *Sunday Express* and *John Bull*; although through ignorance of magic..these two papers could only hint at the nature of the Thelemic ceremonies. **1973** K. GRANT *Aleister Crowley & Hidden God* v. 73 Elaborate ceremonial and the establishment of fixed Lodges in specific localities would be superseded by a fluid and far-flung web comprised of Thelemic power-zones.

thelemite. (Later example.)

1973 K. GRANT *Aleister Crowley & Hidden God* v. 77 Thelema represents a necessary stage in the spiritual development of the individual. Paradoxically, no one can create or contribute anything original, or bring more to life than he takes from it, unless he is already a Thelemite. The term 'Thelemite' has a wider connotation than its hitherto exclusive use in Crowleian literature might suggest. The artist, the scientist,..the poet, is such only to the degree that he expresses his true will.

them, *pers. pron.* Add: **I. 1. e.** As objective case of *THEY 3 b. Hence phr. *them and us* used *attrib.*

1924 W. HOLTBY *Crowded Street* iii. 27 The magic circle of 'Them', the great ones. 'They' were the élite, the prefects and the games captains. **1945** H. NICOLSON *Let.* 27 May (1967) 465 People feel, in a vague and muddled way, that all the sacrifices to which they have been exposed.. are all the fault of 'them'—namely the authority or the Government. **1957** R. HOGGART *Uses of Literacy* iii. 62 To the very poor, especially, they compose a shadowy but numerous and powerful group affecting their lives at almost every point: the world is divided into 'Them' and 'Us'. **1962** *Listener* 8 Mar. 439/1 It is this feeling of being in a world that belongs to 'them' and not to 'us' that puts a strain on working-class children. **1966** *Guardian* 11 Oct. 3/1 The 'ordinary people' who looked on, who made..the Them and Us division [between cripples and other people]. **1980** A. CORNELISEN *Flight from Torregreca* x. 230 The vicious estrangements of a two-class, a Them-and-us society.

3. a. Also in phr. *them's my sentiments* (now freq. used humorously).

1847 THACKERAY *Van. Fair* (1848) xxi. 179 The sooner it is done the better, Mr. Osborne; them's my sentiments. **c 1864** BROUGH & 'HALLIDAY' *Area Belle* 8 Cold mutton to begin with... Cut near the knuckle, with a little currant jelly if you've got it. Them's my sentiments. **1900** F. NIGHTINGALE *Let.* in C. Woodham-Smith *Florence Nightingale* (1951) xxiv. 590 'Drat' hockey and long live the horse! Them's my sentiments. **1924** E. M. FORSTER *Passage to India* v. 48 We're out here to do justice and keep the peace. Them's my sentiments. **1972** 'J. & E. BONETT' *No Time to Kill* viii. 100 'Them's my sentiments too,' he said. 'As Thackeray wrote,' she exclaimed in delight.

c. As nominative case of sense 1 e above.

1957 R. HOGGART *Uses of Literacy* iii. 62 'Them' is a composite dramatic figure, the chief character in modern urban forms of the rural peasant-big-house relationships. **1962** *Listener* 14 June 1044/2 With their use of Christian names in accusing one another of wilful misrepresentation they impressed me most with being collectively Them trying to get power from Us. **1970** *Guardian* 19 Nov. 1/4 In..the Talk of the Town restaurant, 'them' and 'us' dined last night to earn money for the world's wildlife.

thematic, *a.* (*sb.*) **A.** *adj.* **1. a.** Delete *rare* and add later examples.

1957 N. FRYE *Anatomy of Criticism* 367 Thematic; Relating to works of literature in which no characters are involved except the author and his audience, as in most lyrics and essays, or to works of literature in which internal characters are subordinated to an argument maintained by the author..opposed to fictional. **1974** R. QUIRK *Linguist & Eng. Lang.* iv. 75 There is formulaic and thematic structure..yielding striking if controversial theories about the composition of early English poetry. **1979** *N. & Q.* Feb. 63/2 The orientation of this anthology is essentially thematic.

c. Psychol. *Thematic Apperception Test*: a projective test designed to reveal a person's actual social drives or needs by means of the theme common to the interpretations which he gives to each of a standard series of pictures.

1935 MORGAN & MURRAY in *Arch. Neurol. & Psychiatry* XXXIV. 289 (*title*) Method for investigating fantasies. The Thematic Apperception Test. **1938** H. A. MURRAY et al. *Explorations in Personality* vi. 531 As the subjects who took this test were asked to interpret each picture—that is, to apperceive the plot or dramatic structure exhibited by each picture—we named it the 'Thematic Apperception Test'. **1957** P. LAFITTE *Person in Psychol.* 120 The Thematic Apperception Test is more abstract because of the deliberate vagueness of its pictures as well as the fantastic nature of some. **1981** L. KRISTAL et al. *ABC Psychol.* 189 Two of the best known projective tests are the Thematic Apperception Test..and the Rorschach Inkblot Test.

d. *Philately.* Applied to the collecting of stamps with designs which relate to the same subject, or to such a collection.

1951 R. J. SUTTON *Stamp Collector's Encycl.* 231 Thematic Collecting: Collecting to a theme or subject. **1965** E. H. SPIRE *Adventures in Stamp Collecting* ix. 111 It was ..only a logical development from selected collecting that brought about the advent of Thematic Philately. *Ibid.*, Collections of stamps depicting animals, flowers, ships, railways..and so on, are described as 'thematic'. **1972** *Police Rev.* 1 Dec. 1558 The American Topical Society has recorded more than one thousand subjects for thematic stamp collecting.

e. *Linguistics.* Of, pertaining to, or designating the theme of a sentence: see *THEME *sb.* 1 d.

1959, etc. [see *RHEMATIC *a.* 2]. **1969** K. H. WAGNER *Generative Grammatical Studies in Old Eng. Lang.* i. 52 In interrogative clauses..the initial constituent must be regarded as rhematic rather than thematic. **1977** J. LYONS *Semantics* II. xii. 506 *John Smith I haven't seen for ages.* Here the grammatical subject is 'I', but the thematic subject is 'John Smith'.

3. (Later examples.) Hence, of verb-forms: having a connecting vowel between the verb-stem and the suffixes or inflections.

1894 [see *ATHEMATIC *a.* 1]. **1933** *Language* IX. 82 The athematic verbs were primarily durative in aspect, while the thematic were momentary. **1933**, **1955** [see *NON-THEMATIC *a.* 1 a]. **1972** *Language* XLVIII. 389 Except for certain 'thematic verbs', which are exceptional, the presence of a post-position is mutually implicative with the presence of an ind. obj. morpheme.

4. Of or pertaining to the division of the Byzantine Empire into 'themes' or provinces.

1911 E. FOORD *Byzantine Empire* xi. 203 The army—The thematic system and its development—Organization, arms, equipment, and tactics. **1933** S. RUNCIMAN *Byzantine Civilisation* iv. 90 The thematic tax-gatherers took orders directly from the central government. **1980** C. MANGO *Byzantium* I. ii. 46 The accepted view is that the 'thematic' reform was accompanied by a general fragmentation of the large estates.

B. *sb.* **2.** *Gram.* A thematic verb-form.

1968 *Language* XLIV. 717 The conventional view of the distribution of athematics and thematics seems to be that both types existed even in quite early Proto-Indo-European.

3. *Philately.* A collection of stamps with designs which relate to the same subject.

1972 *Police Rev.* 1 Dec. 1572/3 It was known as United Kingdom Thematics 1972, open to thematic entries from anywhere. **1979** *West Lancs. Even. Gaz.* 6 Apr. 18 (Advt.), Stamp collectors world-wide approvals and thematics.

4. *pl.* const. as *sing.* A body of subjects or topics of discussion or study.

1975 *Amer. Speech* 1973 XLVIII. 125 Conklin's unique credentials..allow him to be catholic in his approach, both in terms of thematics and in his world-wide coverage. **1977** A. SHERIDAN tr. *J. Lacan's Écrits* v. 149 The thematics of this science is henceforth suspended, in effect at the primordial position of the signifier and the signified. **1980** *Encounter* May 34/2 Even if Dr Henry Kissinger's picture of a world of 'multi-polarity' is more a neo-Bismarckian fiction than a reality, the confrontation of two Super-powers describes neither the thematics nor the structure of world politics today.

thematization (þīːmătəizē¹·ʃən). [f. *THEMATIZE *v.* + -ATION.] **1.** The action of *THEMATIZE *v.* 2.

1955 T. BURROW *Sanskrit Lang.* iv. 153 This tendency to thematisation had already been operating in the prehistoric period. **1972** *Language* XLVIII. 399 The thematization process..is triggered by violations of canons of permissible plus and minus values of pronominal features for non-ergator and indirect object. **1977** *Ibid.* LIII. 50 Thus the form **somos* can be regarded as a partial thematization of the copula.

2. The action of *THEMATIZE *v.* 1.

1959 J. FIRBAS in *Brno Studies in English* I. 52 Whereas only two elements ([*the*] *girl* and *broke*) of the three in *The girl broke a vase* allow of thematization, any of the three elements occurring in *The girl broke the vase* can be thematized. **1969** K. H. WAGNER *Generative Grammatical Studies in Old Eng. Lang.* i. 50 In the abstract structure *Th* and *Rh* are empty places, i.e. they will not be expanded by subsequent rules. These places will be filled with constituents from the nucleus by transformation rules which may be termed rules of *thematization* and *rhematization*, respectively. **1977** J. LYONS *Semantics* II. xii. 507 It is certainly true that the processes that different languages make available for the thematization of one expression rather than another frequently involve putting the expression earlier rather than later in the utterance.

thematize (þī·mătəiz), *v. Linguistics.* [f. Gr. θεματ-, stem of θέμα THEME *sb.* + -IZE.] **1.** *trans.* To convert (part of a sentence) into a theme: see *THEME *sb.* 1 d.

1959 J. FIRBAS in *Brno Studies in English* I. 43 When thematic elements..occur in basically transitional or basically thematic positions, they communicatively weaken them, or so to speak, 'dedynamize', 'thematize' them. **1969** [see *RHEMATIZE *v.*]. **1976** *Archivum Linguisticum* VII. 145 The following functional constituents (derived from the mood system) may be thematized by fronting them.

2. To modify (a verb-form) by the addition of a thematic vowel: see THEMATIC *a.* 3 in Dict. and Suppl.

1966 E. P. HAMP in Birnbaum & Puhvel *Anc. Indo-Europ. Dial.* 115 **ĝhē(s)r-om*, thematized from **ĝhēsr*. **1977** *Language* LIII. 50 In addition, Isg. **esmi* was thematized.

Hence **the·matized** *ppl. a.*

1972 *Language* XLVIII. 402 We would expect **n-mi* at the beginning of the relationship terms; instead we get the thematized form, a transitive construction. **1976** *Ibid.* LII. 68 Kuno proposes..that the deep structure of every relative clause in Japanese contains a thematized sentence.

Thembu *sb.* and *a.*, var. *TEMBU *sb.* and *a.*

theme, *sb.* Add: **1. d.** *Linguistics.* That part of a sentence which indicates what is being talked about. Cf. *RHEME.

1959, etc. [see *RHEME]. **1966** J. VACHEK *Linguistic School of Prague* ii. 18 'Functional' elements, the most important of which appear to be the *theme* and the *rheme* (the first being the basis of the statement, known from the context or situation). *Ibid.* v. 89 The *theme*, is that part of the utterance which refers to a fact or facts already known from the preceding context. **1969** K. H. WAGNER *Generative Grammatical Studies in Old Eng. Lang.* i. 48 There is evidence supporting the hypothesis that O.E. is a *theme-rheme* language. That is to say that unless certain factors intervene the most natural order of the elements of a sentence is that progressing from what is known to what is unknown, or rather from what has already been mentioned to what is newly introduced into discourse. **1977** *Language* LIII. 444 Like the article by Cinque, this one gets into the theme/rheme distinction.

3. (Later examples.) For 'now *rare*' read 'now *U.S.*'.

1924 [see *DRIP *sb.* 3 b]. **1955** E. B. WHITE *Let.* 1 Apr. (1976) 406 If you are engaged in writing a theme about my works, I think your best bet is to read them. **1976** *National Observer* (U.S.) 14 Feb. 17/3 In my spare time I go to college and the real reason is that it is here that this small flutter comes alive... Late at night when an English theme, which an hour ago had seemed impossible, starts to jell, I feel it.

8. *themebook*; **theme music**, music which recurs in a film, television programme, or the like; also = *signature tune* s.v. *SIGNATURE *sb.* 9; cf. *theme song, tune* below; **theme park** chiefly *U.S.*, an amusement park organized round a unifying idea or group of ideas; similarly **theme pub, restaurant; theme song, tune,** a song or tune which recurs in a musical play, film, or the like; also = *signature tune* s.v. *SIGNATURE *sb.* 9; also *fig.*; cf. *theme music* above.

1916 JOYCE *Portrait of Artist* (1969) i. 47 Father Arnall gave out the themebooks and he said that they were scandalous. **1957** MANVELL & HUNTLEY *Technique Film Music* 226 Martin and Gaston (1954). 'Theme Music'... Sound-track recording of the music from the English version of the French film on children's drawings. **1967** *Listener* 17 Aug. 222/3 Electronic music..is certainly not

restricted to the novel presentation of sounds in familiar patterns, like the theme music of *Dr Who*. **1976** A. DAVIS *Television: First Forty Years* 136 The commercial was a favourite with viewers and with advertising men. It won awards and its theme music was issued on record. **1960** *Amer. Peoples Encycl. Year Bk.* 881 While most established parks and kiddielands were profitable, the theme parks, seeking to duplicate Disneyland's success, were often in trouble. **1967** *Encycl. Brit. Bk. of Year* 335/2 American-type theme parks around the world included Edenlandia Fun Park, Naples, Italy; Prater Fun Park, Vienna; and a new park, Centro de Diversion, opened at Puerto Rico's Isla Verde. **1983** *Times* 16 Aug. 15/3 The acquisition of a tourist attraction in London and a theme park outside the capital. *Ibid.* 19 July 17 A growth segment of the pub trade is emerging..theme pubs. Their hall mark is a design concept to create a particularly individual atmosphere (the theme) with varying combinations of restaurant, cocktail bar and normal bar service. Various theme restaurants have emerged in the past five years. *Ibid.* 4 Nov. 17/3 Grand Metropolitan's Host Group..is to spend well over £100m over the next three years on converting its outlets to a wide range of theme pubs. **1983** *9,000 Words* 196/2 Theme restaurants that look like railroad cars or Polynesian villages. **1929** Theme song [see *RELEASE *sb.*¹ 7 b]. **1946** KOESTLER *Thieves in Night* 348 The theme-song of all evolution is the trend towards greater articulateness. **1949** 'G. ORWELL' *Nineteen Eighty-Four* II. 149 The new tune which was to be the theme-song of Hate Week..had already been composed. **1977** J. FLEMING *Every Inch a Lady* III. vi. 141 Nathaniel returned to his theme-song..murder must have a plan, a blue-print. **1950** *Sport* 24–30 Mar. 15/4 'This couldn't happen again!' should be the theme-tune of Doncaster Rovers' fans. **1983** *Listener* 21 Apr. 30/3 The furore over the *Today* theme-tune..perfectly illustrates the BBC attitude.

theme *v.* (later example); also **themed** *a.*, having a theme.

1963 *Observer* 29 Sept. 7/4 A themed sequence on summer holidays. **1977** *Broadcast* 28 Nov. 12/2 There are.. possibilities for ethnic themed radio services. *Ibid.* 12/3 He continued the themed service subject. **1979** S. BRETT *Comedian Dies* iii. 32 Great Expectations..was a concept restaurant, themed wittily around the works of Dickens.

themselves, *pron. pl.* **III.** (Later examples.)

a. c **1926** 'MIXER' *Transport Workers' Song Bk.* 92 Their ambition is theirself. **1969** in Halpert & Story *Christmas Mumming in Newfoundland* 159 They used to work theirself from all shapes. They have a couple of pillows up their back and another one on their stomach. **1979** N. MAILER *Executioner's Song* I. xxvii. 422 All they want to do is leave theirself a case for appeal.

γ. **1901** M. FRANKLIN *My Brilliant Career* xxxiii. 277 A new fowl-house which 'Horace and Stanley built all by theirselves'. **1907** G. B. SHAW *Major Barbara* II. 241 Arf the street prayed; an the tother arf larfed fit to split theirselves. **1955** F. O'CONNOR *Wise Blood* x. 167 The unredeemed are redeeming theirselves and the new jesus is at hand! **1965** C. BROWN *Manchild in Promised Land* xiii. 314 Them damn junkies take care of theirselves twice as good as you can.

then, *adv.* (*conj.*, *adj.*, *sb.*) Add: **B. 3. b.** Phr. *and then some*: see *SOME *indef. pron.* 4 f.

4. (Later examples of emphatic use in an apodosis.)

1925 L. ABERCROMBIE *Idea of Great Poetry* i. 8 We have busied ourselves, if not on our own account, then vicariously in the newspapers, with the appreciation of these poets in their several qualities. **1956** A. J. AYER *Probl. Knowledge* i. 7 Can it reasonably be held that knowledge is always knowledge that something is the case? If knowing that something is the case is taken to involve the making of a conscious judgment, then plainly it cannot. **1972** M. KLINE *Math. Thought* ii. 1194 This is the principle of *reductio ad absurdum*. In words, if the assumption of *p* implies that *p* is false, then *p* is false.

10. a. *then-current, -known*; **b. then-clause,** the apodosis in a conditional sentence.

1750 S. RICHARDSON *Let.* 4 June (1964) 161 From robbery to robbery they proceeded, till they had enlarged their den so as to take in the greatest part of the then-known world. **1905** G. B. SHAW *Let.* 28 Sept. (1972) II. 563 She subscribed to the philosophy of a then-current song, 'I Want What I Want When I Want It'. **1927** G. A. GRIERSON *Ling. Survey India* I. i. 376 If the conditional sentence is such a one as we would require the use of 'would' or 'would have' in English, the word *sik* is appended to the apodosis, or then-clause. **1962** *John o' London's* 22 Feb. 188/3 *Would* is often used to express a wish..as in..the *then*-clause of a conditional sentence, as in 'You would enjoy it if you went'. **1976** *Scotsman* 24 Dec. (Weekend Suppl.) 3/2 The military republic of Julius Caesar that ruled the then-known world. **1978** *Detroit Free Press* 5 Mar. c 24/5 (Advt.), The interest..will be recalculated..at the then-current regular passbook interest rate.

thenabouts. Delete *rare* and add further examples. Also (rarely) **thenabout.**

1843 DICKENS *Martin Chuzzlewit* (1844) xiii. 164 Five year ago, or thenabout. **1922** *Times Lit. Suppl.* 19 Oct. 664/2 If Archdeacon Brandon..must go down, it is then or thenabouts that go down he will. *a* **1967** A. RANSOME *Autobiogr.* (1976) i. 18 When I first came to fish the Beela ..in 1930 or thenabouts.

Thénard's blue (te·nāːz blū). The name of a bright blue pigment of considerable stability invented by the French chemist Louis-Jacques *Thénard* (1777–1857), consisting essentially of cobalt aluminate; cobalt blue.

1837 *Penny Cycl.* VII. 301/1 Phosphate of Cobalt..is used in making a pigment known by the name of Thenard's or Cobalt Blue. **1911** *Encycl. Brit.* XXI. 599/1 Several mixed cobalt compounds..represented by cobalt violet and Thénard's blue. **1958** *Listener* 2 Oct. 514/2 With their vivid Thénard's blue, the gleaming black of some of the boots and of Dick Turpin's horse..they have a place with the painted fair-ground horses and roundabouts. **1974** *Encycl. Brit. Micropædia* IX. 930/3 In 1799 he [sc. Thénard] made a discovery that assured him prosperity—Thenard's blue, a pigment used in the colouring of porcelain.

thence-after. (Later examples.)

1921 B. JARRETT *English Dominicans* ix. 180 The boy finished what remained of his noviciate..and thenceafter was no longer interfered with. **1932** BELLOC *Napoleon* iii. 182 Thenceafter, for a week or more, it was intrigue upon intrigue.

theo-. Add: **theoce·ntrism,** theocentric doctrine or belief, also (occas.) **theoce·ntricism; theomaniac** (earlier example).

1925 E. UNDERHILL *Mystics of Church* x. 205 The best traditions of French spirituality, its lofty theocentricism. **1930** *Monument to St. Augustine* viii. 272 The apparent theocentrism of the Calvinist 'glory of God'. **1941** Theocentrism [see *ANTHROPOCENTRICISM, ANTHROPOCENTRISM]. **1863** C. READE *Hard Cash* III. ii. 53 Dr. Wycherley..put down any man a lunatic, whose intellect was manifestly superior to his own... Nor did the dead escape him entirely. Pascal, according to Wycherley, was a madman with an illusion about a precipice... Joan of Arc a theomaniac.

theocrasia (þīokrē¹·ziă). Also **-krasia.** [a. Gr. θεοκρᾶσία, a mingling with God.] = THEOCRASY 1.

1913 *Encycl. Relig. & Ethics* VI. 422/1 The working of the *theokrasia* in the domain of religion and religious art. **1920** H. G. WELLS *Outline Hist.* 496/2 A sort of theocrasia went on between Christianity and Judaism..and other competing cults. **1971** R. E. WITT *Isis* xi. 146 Bast was the intermediary when the process of theocrasia began.

Theocritean (þīǫkritī·ăn), *a.* [f. L. *Theocritus*, a. Gr. Θεόκριτος Theocritus, a Greek poet of Sicily, of the 3rd c. B.C.: cf. SOPHOCLEAN *a.*] Of, pertaining to, or characteristic of Theocritus or his writings, esp. his pastoral poetry; of the style of Theocritus; hence pastoral, idyllic. Also **Theo·critan** *a.*

1846 T. KEIGHTLEY *Notes Bucolics Virgil* 73 He [sc. Virgil] was thoroughly imbued with the Theocritean poetry. **1879** [see *BUCOLICISM]. **1896** *McClure's Mag.* VI. 467/2 With his [sc. Corot's] Theocritan spirit, he could see the fountain of Jouvence in the woods of Sèvres. **1910** *Daily News* 17 Oct. 3 All this is more real and more grim than Arcadia, but even Arcadia had its reality, and there is something of Theocritean sweetness penetrating the intimate truth of this book. **1935** L. MACNEICE *Eclogue by Five-Barred Gate* in *Poems* 24, I am a shepherd of the Theocritean breed. **1969** V. DE S. PINTO *City that Shone* xii. 293 A Theocritean flavour was given to our courtship by the fact that Irène had two white goats.

theodicaea (þīǫdisī·ă). *rare.* [App. an erron. Latinization of Fr. *théodicée* in the title of a work by Leibniz: see THEODICY.] = THEODICY.

1845 *Encycl. Metrop.* II. 659/1 Leibnitz fancied that.. he could construct a *Theodicæa*, in which the doctrines of theology should be reconciled with philosophy. **1883** J. SIBREE tr. *Hegel's Lect. on Philos. of Hist.* 16 Our mode of treating the subject is, in this aspect, a Theodicæa [rendering G. *Theodicee*],—a justification of the ways of God..in indefinite abstract categories. **1974** *Times Lit. Suppl.* 18 Oct. 1161/2 The mysterious 'Spirit' which presides over, or constitutes, reality in the Hegelian scheme of things is supposed..to be pursuing in history a plan of which historical agents are largely or wholly unaware: that is why Hegel can claim to be offering a theodicaea.

theogonical, *a.* (Earlier example.)

1702 *Lives of Anc. Philosophers* p. xxvi, A Magus..was imploy'd to sing a theogonical hymn, as a powerful enchantment.

theologate. 2. (Earlier example.)

1879 H. FOLEY *Records Eng. Province Soc. of Jesus* V. 944 *The College of St. Beuno*... This extensive pile of building was erected in 1848–9 for a Theologate, or House of Divinity, by the late Father Randal Lythgoe, then Provincial.

theologian. Add: **d.** In sense 1 d of *THEOLOGY.

1968 *Listener* 10 Oct. 469/2 To speak well of the past was a mortal sin and got you into trouble with the party theologians and eventually with the police. **1982** 'I. I. MAGDALEN' *Search for Anderson* II. ii. 147 We had the ideologists and the purists and the theologians.

theological, *a.* (*sb.*) Add: **3.** *transf.* In trivial or disparaging use: of, pertaining to, or characterized by dogma or abstract principles (as opposed to practical considerations); doctrinaire, academic.

1959 *Times Lit. Suppl.* 5 June 329/4 The 'theological' approach to Soviet Marxism..proves in the long run unsatisfactory. **1964** S. BRITTAN *Treasury Under Tories* II. vi. 196 The three Treasury ministers played straight into the Prime Minister's hands, by their theological stress on permitting no increase at all on government expenditure for the coming financial year in money terms (which meant cutting it in real terms). **1964** *Listener* 30 July 148/2 Mr Walt Rostow..referred to 'the whole insoluble theological issue' of control... 'Theological' has two senses... He was clearly using the second or vernacular meaning, which is 'as impractical and as irrelevant to the matter in hand as the study of the nature of God'. **1968** *Observer* 28 Apr. 8/3 The doctrinal arguments which used to involve Labour in theological warfare. **1979** H. KISSINGER *White House Years* xviii. 719 It was a reflection on the theological nature of our China debate that many experts still regarded the 'solution' of the UN issue, liturgically, as the absolute precondition of any improvement in our relations with Peking. **1980** *Spectator* 2 Feb. 5/3 The dispute..nominally involves theological distinctions between 'News' and 'Current Events'.

theologically, *adv.* Add: Also in sense 3 of prec.

1973 *Times* 13 Oct. 14 There is a fundamental fascism of the left which is the real problem in the universities: they are theologically right, as they believe, and you are so wrong that you should even be denied the freedom to speak. **1978** *Guardian Weekly* 5 Feb. 5/5 The Government is no longer theologically wedded to it [sc. free trade].

theologico-. Add: *theologico-metaphysical* (earlier example).

1855 H. MARTINEAU *Autobiogr.* (1877) I. 120 Trying my hand at a sort of theologico-metaphysical novel.

theology. Add: **1. a.** *Biblical theology*: orig. theology as a non-dogmatic description of the religious doctrines contained in the Bible, following J. P. Gabler's distinction, in 1787, between biblical and dogmatic theology; now usu. the exposition of biblical texts (both O.T. and N.T.), based on the presupposition that there is a common biblical way of thinking which informs the Bible as a whole.

[**1787** J. P. GABLER *Kleinere Theologische Schriften* (1831) II. 179 (*title*) De iusto discrimine theologiae biblicae et dogmaticae regundisque recte utriusque finibus.] **1846** C. W. BUCH tr. *Hagenbach's Compendium Hist. of Doctrines* I. 5 In our opinion biblical theology is only to be regarded as the *foundation-stone* of the edifice..and dogmatic theology as the builder. **1888** *Encycl. Brit.* XXIII. 264/2 Biblical theology is the delineation of a section of religious ideas,—that section of which the traces and records remain in the Bible. **1904** A. B. DAVIDSON *Theology of O.T.* i. 1 In Biblical Theology the Bible is the source of the knowledge, and also supplies the form in which the knowledge is presented. **1951** H. H. ROWLEY *Old Testament & Modern Study* xi. 312 Since the publication of Gabler's famous address in 1787, the correctness of the distinction which he drew between biblical and dogmatic theology has been accepted as axiomatic by the majority of scholars. **1958** *Listener* 14 Aug. 241/1 A new subject has appeared called Biblical Theology. This means that the Bible, or rather the New Testament, is now subjected to analysis in order to see what is the total message or picture that it contains. **1969** A. RICHARDSON *Dict. Christian Theol.* 36/2 The question ..remains whether there can be a completely presuppositionless interpretation of any historical documents (as tended to be assumed by those who regarded biblical theology as a purely descriptive science).

b. Also *fig.*

1956 P. A. LARKIN *Less Deceived* 36 Our garden, first: where I did not invent Blinding theologies of flowers and fruit, And wasn't spoken to by an old hat.

d. In trivial or disparaging use: a system of theoretical principles; an (impractical or rigid) ideology. Cf. *THEOLOGICAL *a.* (*sb.*) 3.

1962 *Listener* 29 Mar. 551/2, I would also like to see a couple of first-class philosophers, who would..check the new brand of strategists, who have produced a complete theology, not only about the deterrent but about conventional weapons, and about morale and everything else. I think the spectacle of Whitehall trying to keep up with American theologies is too sad for words. **1973** *Times* 3 Oct. 14/4 There were also a few rival interpretations of Marxist theology. *a* **1974** R. CROSSMAN *Diaries* (1976) II. 391 At S.E.P., this morning, we were discussing Paper 105, a brilliant demolition of the theology of PESC. **1980** *Times Lit. Suppl.* 8 Feb. 134/2 Like the farmers of Iowa in relation to China in a later period, they were impatient with the theology of diplomatic recognition.

theomorphism. (Earlier example.)

1822 tr. *Malte-Brun's Universal Geogr.* I. 576 *Theomorphism*, the religion of the Hindoos, is the best supported of all the ancient systems of worship; it still exists.

Theopaschite. Add: Also *attrib.*

1914 W. E. BEET *Medieval Papacy* 15 Uncompromising was the attitude of Hormisdas with reference to the so-called Theopaschite formula. **1971** R. BROWNING *Justinian & Theodora* iii. 102 His [sc. Justinian's] new 'Theopaschite' doctrine in the end contributed nothing to the religious unity of the empire.

theophagous, *a.* Add: (Later example in sense (*b*) of *theophagy*). **theophagy,** (*a*) (earlier example); (*b*) *Anthrop.*, the eating of meals at which the participants believe that they ingest a deity with the consecrated food. (Restrict nonce-wd. to *theophagite* and 'dyslogistic' to definitions in Dict.)

1875 SWINBURNE *Lett.* (1960) III. 49 Would the exalted privilege of theophagy be conceded to a believer in the identity of those two Beings? **1903** G. B. SHAW *Revolutionist's Handbk.* viii, in *Man & Superman* 209 We have relapsed into disputes about transubstantiation at the very moment when the discovery of the wide prevalence of theophagy as a tribal custom has deprived us

of the last excuse for believing that our official religious rites differ in essentials from those of barbarians. **1912** *Encycl. Relig. & Ethics* V. 136/1 The chief among the reasons given for the correlated rite of theophagy. *Ibid.* 137/1 A more detailed account supplies a valuable type of such theophagous ceremonies. **1937** *Jrnl. Theol. Stud.* XXXVIII. 97 Notable also is the demonstration how slight evidence we have, not only that any one ever associated theophagy with the Dionysiac omophagy, but even that any one in the centuries with which we are concerned ever celebrated an omophagy at all. **1956** C. WINICK *Dict. Anthropol.* (1957) 533/2 *Theophagy*, the practice of ingesting the god. It probably stemmed from the ancient habit of eating the sacred animal to secure blessing, grace, and identity with the deity.

theophany. Add: Also *transf.*

1962 AUDEN *Dyer's Hand* (1963) 256 The practical joker desires to make others obey him without being aware of his existence until the moment of his theophany.

theophanism: also, belief in theophanies.

1938 S. BECKETT *Murphy* v. 81 An adherent (on and off) of the extreme theophanism of William of Champeaux. **1970** R. MANHEIM tr. *Corbin's Creative Imagination Ṣūfism* 52 Not to understand..Ibn 'Arabī's conscious intention..of expressing a divine love, would be..to close one's eyes to the theophanism on which this book insists.

theophilanthropist. Add: **theophilanthrope** (earlier example); **theophilanthropical** *a.* (example).

1801 W. DUPRÉ *Lexicographia Neologica Gallica* 275 *Théophilantrope*,..a theophilanthrope. *Ibid.* 276 *Théophilantropique*.., theophilanthropical.

Theophrastian (þīofræ·stiăn), *a.* Also **-an, -ean.** [f. L. *Theophrastus*, a. Gr. Θεόφραστος, a Greek philosopher of Eresus in Lesbos (4th c. B.C.) + -IAN.] Of, pertaining to, or characteristic of Theophrastus or his writings, esp. his *Characters*, a set of thirty sketches on disagreeable aspects of human behaviour. So **Theophra·stic,** † **Theophra·stical** *adjs.*

1662 J. SPARROW tr. J. Boehme in *Remainder of Bks.: Apol. conc. Perfection* 132 Not Tinctured, according to the Cabalisticall, Theophrasticall, Roso-Crucian kind. **1924** *Public Opinion* 18 Jan. 53/2 Some charming little essay or Theophrastan Study. **1926** *Glasgow Herald* 8 Apr. 4 One of the earliest [Characters] which has the true Theophrastian ring. **1928** *Observer* 12 Feb. 4 Some of these Theophrastic 'characters' are very charming. **1962** W. & M. KNEALE *Devel. Logic* iv. 190 Any account of modal syllogisms, either Aristotelian or Theophrastean.

theophylline. (Examples.)

1957 [see *ORAL *a.* 4 b]. **1976** *Lancet* 20 Nov. 1115/2 Theophylline..and caffeine have been shown to be strong prostaglandin antagonists and weak agonists. **1983** *Daily Tel.* 18 Aug. 8/6 The council found that tea has three bracing ingredients—caffeine which stimulates the nervous system, and theophylline and theobromine which relax muscles and stimulate the heart.

theoplasm (þī·,oplæz'm). *rare.* [f. THEO- + PLASM.] (See quot. 1901.)

1901 E. S. HARTLAND in *Folk-Lore* XII. 27 *Tilo*,..like the Siouan *Wakanda*, is found to be theoplasm, god-stuff, not a god fully formed and finally evolved. **1941** R. R. MARETT *Jerseyman at Oxford* xi. 161 My conception of the process whereby both magic and religion had evolved out of the same 'theoplasm or god-stuff', as Hartland was for calling it.

theopolitics, *sb. pl.* (Later example.)

1945 in J. H. Whyte *Church & State in Mod. Ireland* (1980) iii. 71 The Catholic press. A study in theopolitics.

theorbist. (Later examples.)

1976 *Early Music* Oct. 414/2 Quantz wrote that the theorbist should sit behind the second harpsichord, between two cellists. **1980** *Early Music* Jan. 50/1 A lutenist and theorbist are shown in the orchestra in two contemporary drawings of the performance of *Teofane*.

theorem, *sb.* Add: **2.** A stencil. Also *transf.*, a design executed by means of a stencil. *Obs.* exc. *Hist.*

1824 *Federal Gaz.* 29 Apr. 1/5 Theorem painting on velvet..varnished theorems or theorems cut from any design..may be had. **1832** L. M. CHILD *Girl's Own Book* (ed. 4) 137 After all the parts are in readiness, lay your theorem upon your drawing paper, take a stiff brush of bristles..fill it with the colour you want. **1968** *Canad. Antiques Collector* June 21/1 Theorem Painting, designs painted on white cotton velvet, was an art introduced to America from England. Also known as Formula or, if on silk, Poonah painting. **1973** *New Yorker* 3 Feb. 40/3 Old theorems (stencilled paintings or watercolors done on velvet or paper by genteel housebound girls in the nineteenth century).

theoretic, *a.* Add: **5.** As the second element of parasynthetic adjs. formed from compound sbs. of the type *quantum theory*.

1930 *Acta Math.* LIV. 81 (*heading*) A maximal theorem with function-theoretic applications. **1971** E. C. DADE in Powell & Higman *Finite Simple Groups* viii. 249 To use the minimum of ring-theoretic machinery. **1973** *Times Lit. Suppl.* 9 Mar. 267/5 (Advt.), The systems are approached from two directions—proof-theoretic and model-theoretic.

theoretical, *a.* Add: **3. b.** (Further examples.)

1922 *Glasgow Herald* 30 Oct. 10 He was a brilliant theoretical chemist. **1936** *Proc. IRE* XXIV. 353 In our search for good emitters, very little aid can be obtained from the theoretical physicist. **1951** C. P. SNOW *Masters* I. v. 48 One of the earliest theoretical chemists. **1958** J. CLEUGH tr. *Jungk's Brighter than Thousand Suns* i. 17 Rutherford for his part did not hesitate to declare that it was the theoretical, not the experimental, physicists who were to blame for the confusion. **1968** J. J. C. SMART *Betw. Sci. & Philos.* 13 Theoretical physicists have far outstripped philosophers in their imaginativeness. **1980** *English World-Wide* I. 251 The importance of this book is that it is by a scholar who is generally considered to be a 'theoretical' linguist, but who is sympathetic to socio-linguistics and its implications for theory.

5. Used as *THEORETIC *a.* 5.

1920, etc. [see *QUANTUM-THEORETICAL *a.*]. **1934** [see *FIELD-THEORETICAL *a.*].

theoretician. (Later examples.)

1931 *Times Lit. Suppl.* 26 Mar. 255/2 The most articulate theoretician among the Russian film producers. **1954** [see *ACTIVIST]. **1959** K. R. POPPER *Logic of Sci. Discovery* v. 107 The theoretician..shows the experimenter the way. **1970** *Physics Bull.* Apr. 150/2 These results.. provide an incentive for theoreticians to tackle the much more complex problem posed by real finite nuclei. **1980** 'M. FONTEYN' *Magic of Dance* 288 He was by no means a dry, boring theoretician even though he wrote extraordinarily advanced books on dance.

theoreticism (þīore·tisiz'm). [f. THEORETIC *a.* (*sb.*) + -ISM.] (See quot. 1974.) So **theore·ticist** *a.*

1970 B. BREWSTER tr. L. Althusser in *Althusser & Balibar's Reading Capital* 8 One of the theses I advanced as to the nature of philosophy did express a certain 'theoreticist' tendency. More precisely, the definition of philosophy as a theory of theoretical practice..is unilateral and therefore inaccurate. **1974** *Science & Society* XXXVIII. 404 After 1965..Althusser responded to 'theoreticist' and 'positivist' readings of his texts by reformulating in particular his concept of 'philosophy' and its relationship to 'science'. *Ibid.* 421 This dialectical understanding..preserves method against three forms of reductionism and their corresponding ideologies: historical empiricism (historicism)..; structural idealism/empiricism..; and speculative idealism (theoreticism), which radically separates historical and structural analysis. **1981** *Times Lit. Suppl.* 6 Feb. 136/4 They were ..saturated in what would have been called—had the word not borne the double taint of 'jargon' and 'theory'—theoreticism.

theoretico-. Add: **theore:tico-histo·rical** *a.*, pertaining to both the theoretical and the historical sides of a subject.

1922 D. AINSLIE tr. *Croce's Aesthetic* (ed. 2) II. vi. 241 Cesarotti purposed (1762) bringing out a great theoretico-historical book. **1970** B. BREWSTER tr. *Althusser & Balibar's Reading Capital* II. i. 73, I intended to interrogate Marx himself, to see where and how he had theoretically reflected the relationship between his work and the theoretico-historical conditions of its production.

theorism. (Earlier example.)

1820 T. CAMPBELL *Let.* 14 July in W. Beattie *Life Thomas Campbell* (1849) II. 370 At times, perhaps, there is a little German theorism in it [sc. Arndt's conversation].

theorizing, *vbl. sb.* (Earlier example.) Also **the·orized** *ppl. a.*

1817 COLERIDGE *Biog. Lit.* I. xii. 258 The necessity of theorising. **1975** *Amer. Economic Rev.* LXV. 416/1 It is not obvious whether the net effect of all these shortcomings necessarily exaggerates the regression results in favor of the theorized results. **1979** *Internat. Jrnl. Sociol. of Law* VII. 319 The material that is reproduced here would undoubtedly be of considerable use in some more theorized or general analysis.

theory.¹ Add: **4. b.** *theory of knowledge, of logic, of naming.*

1828 J. S. MILL in *Westm. Rev.* IX. 155 A prodigious step in the theory of naming. *a* **1854** —— *Early Draft Autobiogr.* (1961) 135, I pushed on..to try whether I could do anything further to clear up the theory of Logic generally. **1885** *Encycl. Brit.* XVIII. 793/2 Epistemology (theory of knowledge, Erkenntnisstheorie). **1927** B. RUSSELL *Outl. Philos.* xxiii. 248 Descartes..inaugurated two movements, one in metaphysics, one in theory of knowledge. **1966** R. M. CHISHOLM (*title*) Theory of Knowledge.

7. *theory-building* (later example), *-mad* adjs., *-making* adj. and sb., *-ridden* adj.; **theory-laden** *a.*, applied to a term, statement, etc., the use of which implies the acceptance of some theory; contrasted with *theory-free*, *-neutral* adjs.

1964 *Language* XL. 225 Spelling out..how its results have been incorporated into other experiments and theory-building. **1977** A. GIDDENS *Stud. in Social & Polit. Theory* i. 49 As Feigl says, most positivistically inclined authors today..recognize that observation statements cannot be entirely 'theory-free'. **1958** N. R. HANSON *Patterns of Discovery* i. 19 There is a sense..in which seeing is a 'theory-laden' undertaking. Observation of *x* is shaped by prior knowledge of *x*. **1977** A. GIDDENS *Stud. in Social & Polit. Theory* 12 The theory-laden character of observation-statements in natural sciences entails that the meaning of scientific contexts is tied-in to the meaning of other terms in a theoretical network. **1850** E. A. POE in *Sartain's Union Mag.* Oct. 233/1 He must be theory-mad beyond redemption who, in spite of these differences, shall still persist in attempting to reconcile the obstinate oils and waters of Poetry and Truth. **1931** A. HUXLEY *Music at Night* 77 The theory-making mind. **1964** I. L. HOROWITZ *New Sociology* 31 Problems of this kind can be multiplied..in every sphere of sociology from poll-taking to theory-making. **1968** J. J. C. SMART *Betw. Sci. & Philos.* iii. 80 Observation reports can not be couched in theory-neutral language. **1977** A. GIDDENS *Stud. in Social & Polit. Theory* iii. 150 The 'orthodox view' has an answer which Habermas has apparently (although..not finally) rejected: correspondence to sensorily apprehended reality, grounded in the descriptions of a theory-neutral observation language. **1922** R. FRY *Let.* 6 Mar. (1972) II. 522, I don't take it to heart when you say that my pictures are the utterly dismal performances of a theory-ridden painter.

theosis (þi,ōᵘ·sis). *Theol. rare.* [a. med.L. *theōsis*, ad. Patristic Gr. θέωσις deification.] Deification.

1875 J. W. DRAPER *Hist. Conflict between Relig. & Sci.* v. 126 The return of the soul to the universal Intellect is designated by Erigena as Theosis, or Deification. **1934** *Theology* XXVIII. 24 Both natures, therefore, can be correlated positively through the communion of qualities, *communicatio idiomatum*, in the theosis of the created by the Divine. **1967** *Eastern Churches Rev.* I. 246 This is a kind of *theosis* whereby the symbols become life-giving.

Theotiscan (þioti·skăn). *rare.* [f. med.L. (8th–9th cent.) (*lingua*) *theotisca* (the) German (language), reflecting early forms of OHG. *diutisc*, G. *deutsch* (see DUTCH) + -AN.] (See quot.)

1817 COLERIDGE *Biog. Lit.* I. x. 203, I read through.. the most important remains of the Theotiscan, or the transitional state of the Teutonic language from the Gothic to the old German of the Swabian period.

‖ **theotokion** (þīŏtǫ·kiǫn). *Eastern Church.* [eccl. Gr., f. θεοτόκος: see THEOTOKOS.] A *sticheron* or *troparion* addressed to the Mother of God; usu. the last in a series of stanzas.

1850 J. M. NEALE *Hist. Holy Eastern Church* I. 832 The theotokion is simply a sticheron or troparion addressed to the Mother of God. **1880** *Encycl. Brit.* XII. 580/1 A 'theotokion', or ascription of praise to the mother of our Lord. **1961** D. ATTWATER *Christian Churches of East* I. 225 *Theotokos* (Gk., *tokos*, childbirth), the Mother of God. *Theotokion*, a hymn in her honour.

theralite (þī°·rălait). *Petrogr.* [ad. G. *theralith* (H. Rosenbusch *Mikrosk. Physiogr.* (ed. 2, 1887) II. 248), f. Gr. θηρᾶν to hunt, pursue: see -LITE, -LITH.] Any of a group of mafic, intrusive, igneous rocks that contain nepheline and calcic plagioclase.

1898 *Bull. U.S. Geol. Survey* No. 150. 197 This feldspar was determined as in part ordinary sanidine, in part a soda-lime feldspar, and the rocks were therefore made the types of a plutonic rock characterized by the mineral combination nephelite, soda-lime feldspar, and named theralite by Professor Rosenbusch. **1938** A. JOHANNSEN *Descr. Petrogr. Igneous Rocks* IV. 222 Among rocks difficult to place are the shonkinites, theralites, and teschenites. On the basis of their limited extent and mode of occurrence as sills, small intruded masses, and border facies, the rocks are hypabyssal; on the character of their usual textures, they are plutonic. **1978** S. R. NOCKOLDS in S. R. Nockolds et al. *Petrol. for Students* xvi. 177 Theralites appear to be rather rare but occur, for instance, as dykes cutting nepheline syenite in the Khibina complex, Kola Peninsula, U.S.S.R... Theralites are found also in the Lugar sill, Ayrshire.

therapeutic, *a.* Add: **1.** Also *loosely* in weakened use.

1970 *Daily Tel.* 11 Feb. 15 She doesn't get bad-tempered; she merely picks up the piece of patchwork she is working on. 'It is so peaceful and relaxing, quite therapeutic.' **1982** L. CHAMBERLAIN *Food & Cooking of Russia* 253 Bread-making in the last century was a continuous process rather than a therapeutic exercise on a wet afternoon.

3. Special collocations: *therapeutic community*, a residential unit comprising staff and certain classes of mentally or behaviourally disturbed patients run in a deliberately informal manner to encourage social reintegration and rehabilitation; *therapeutic index*, the ratio of the lethal or toxic dose of a drug to the therapeutically effective dose.

1964 G. L. COHEN *What's Wrong with Hospitals?* viii. 167 In the past decade, reformers have gone a step further, attempting to put inmate and authority on the same level: partners in a 'therapeutic community'. This endearing phrase originated at Belmont. **1977** *Lancet* 24/31 Dec. 1344/2 Common-milieu therapy, used by most therapeutic communities, is probably best regarded as re-educative psychotherapy. **1942** H. R. ROSENBERG *Chem. & Physiol. Vitamins* 150 The therapeutic index [of vitamin B_1]..is extremely high. **1973** J. J. MCKELVEY *Man against Tsetse* iii. 200 It had a narrow therapeutic index, that is, a small difference between the 'curative' dose that would kill trypanosomes in human blood and the 'tolerated' dose beyond which the host would suffer damage.

therapeutical, *a.* (*sb.*) **a.** (Later examples.)

1950 G. B. SHAW *Farfetched Fables* Pref. 90 Such a public department should be manned not by chemists analyzing the advertized wares and determining their therapeutical value, but by mathematicians criticizing their statistical pretentions. **1952** E. HOBSBAWM in *Granta* 15 Nov. 12/1 We did not take to politics for therapeutical or aesthetic reasons.

therapist. Delete *rare* and add later examples.

1917 G. B. SHAW in *Eng. Rev.* Dec. 490 A homœopath, or a bonesetter, or a serum therapist. **1937** *Brit. Jrnl. Psychol.* XXVIII. 109 He describes how the therapist.. is able through unconscious observation to conjecture the nature of the patient's unconscious processes. **1978** *Listener* 5 Oct. 430/3, I would describe psychotherapy.. as a treatment in which the doctor or therapist uses talking: first, to establish a relationship with the patient; secondly, to help him understand what is happening to him.

therapsid (þeræ·psid) *sb.* (and *a.*) [a. the name of the order *Therapsida* (R. Broom 1905, in *Rec. Albany Museum* I. 269), f. THERO- + APSIS: see -ID³.] A mammal-like fossil reptile of the order Therapsida; also as *adj.*, of or pertaining to an animal of this kind.

1912 *Rep. Brit. Assoc. Adv. Sci.* 581 The humble Therapsid-like mammal felt the impetus of its new-found power of adaptation. **1933** A. S. ROMER *Vertebr. Paleontol.* xi. 229 The limbs in advanced therapsids are greatly changed from the primitive sprawling position. **1966** E. PALMER *Plains of Camdeboo* vi. 95 The first of the mammal-like reptiles, or Therapsids as these famous fossil reptiles of the Karoo are known, had been discovered in 1838. **1971** J. Z. YOUNG *Introd. Study Man* xxix. 408 Probably we shall never know whether the therapsid reptiles possessed the features of the soft parts that are so characteristic of mammals. **1973** B. J. WILLIAMS *Evolution & Human Origins* vii. 103/1 The therapsids had developed mammal-like features in locomotion and in their dentition. **1979** C. KILIAN *Icequake* i. 14 His sermons on President Wood..had been as boring as his lectures on therapsid endothermy.

therapy. Add: **1.** See also *group therapy* s.v. *GROUP *sb.* 6 b.

2. As the final element in words denoting treatment by means expressed in the first element, as *ACTINOTHERAPY, *CHEMOTHERAPY, PSYCHO-THERAPY in Dict. and Suppl., *RADIOTHERAPY, *roentgenotherapy* s.v. *ROENTGEN-, ROENTGENO-, etc.

Theravada (þerăvā·dă). [a. Pali, lit. 'doctrine of the elders'.] = *HINAYANA.

[**1875** R. C. CHILDERS *Dict. Pali Lang.* 545/1 The adj. *theravādī* (*theravāda*..) means holding the orthodox doctrine.] **1882** W. HOEY tr. *Oldenberg's Buddha* I. i. 75 The Church of Ceylon remained true to the simple, homely, 'Word of the Ancients' (Theravâda). **1923** LD. RONALDSHAY *Lands of Thunderbolt* vii. 48, I shall refer to it as the Thera Vada—'the way of the Elders'—because this is the title which its adherents themselves prefer, the term Hinayāna being objectionable to them. **1959** *Encounter* Jan. 19 Theravada Buddhism stems directly from the Indian tradition. **1978** C. HUMPHREYS *Both Sides of Circle* xii. 132 My own list, however, was far wider than Olcott's 'Fourteen Fundamental Principles', which were largely confined to the Canon of the Southern or Theravada school.

therblig (þɜ·ɹblig). [Anagrammatic formation by partial reversal of the name of its inventor, F. B. *Gilbreth* (1868–1924), American engineer and pioneer of time-and-motion studies.] In time-and-motion study, a unit of work or absence of work into which an industrial operation may be divided (see quot. 1921); a symbol representing such a unit.

1921 F. B. GILBRETH in *Bull. Taylor Soc.* June 128/2 We believe that there are but sixteen sub-divisions of a cycle of motions. They are called therbligs. They are as follows: 1. Search, 2. Find,..13. Transport, empty, 14. Rest for overcoming fatigue [etc.]. **1930** *Movie Makers* Nov. 687/1 The motions of the operator are broken down into fundamental motions known as 'therbligs' (Gilbreth spelled backwards). **1947** [see *MICROMOTION]. **1948** GHISELLI & BROWN *Personnel & Industrial Psychol.* xi. 279 The therblig type of classification of movements is important principally in such problems as changing the sequence of movements and in the elimination of unnecessary movements. **1963** *Engineering* 27 Dec. 826/3 Maynard intended to allocate time values to Gilbreth's therbligs. **1964** A. BATTERSBY *Network Analysis* ii. 13 Two main sets of symbols are used: Gilbreth's 'therbligs' for motion study and the standard ASME (American Society of Mechanical Engineers) symbols for method study. **1975** *Daily Tel.* 11 Dec. 14/5 There are 18 therbligs altogether. Each has its own symbol and colour (e.g. 'search' is black, 'grasp' is red, 'use' is purple) and these can be used to construct simultaneous motion charts. **1976** W. H. CANAWAY *Willow-Pattern War* vii. 82 She was skilled..in lovemaking, and only now and then did you get the feeling that..in her mind there was a stopwatch and a work-study chart covered with therbligs.

there, *adv.* (*a.*, *sb.*) Add: **A.** γ. **thar** (now U.S. *dial.* and *colloq.*: see also sense B. 2 c (*c*) below).

1859 BARTLETT *Dict. Amer.* (ed. 2) 477 A person wishing to imply that he is perfectly at home in any thing, says he is *thar*; a good hunter or fisher is also *thar*. **1885** *Weekly New Mexican Rev.* 29 Jan. 4/5 The Santa Rifles had their first drill at Alhambra hall last night... Nearly all the boys have 'been thar' before, and as a consequence, catch up the command very readily. **1887** [see TCHICK *sb.*]. **1937** W. BLAIR in B. A. Botkin *Treas. S. Folklore* (1949) IV. iii. 645 Hello, thar, gin us 'Forked Deer', old fiddle-teazer. **1980** 'D. SHANNON' *Felony File* i. 27 Thar's a big store, with a lot of different departments.

B. I. 2. b. Now also appended casually to exclamations of greeting, etc., as *hi* (or *hello*) *there!*, with varying purpose: freq. to attract attention or to express cordiality.

1589 [see HOLLO, HOLLOW *int.*]. **1840** [see HALLO, HALLOA *int.*]. **1885** [see *HI *int.* 2]. **1924** *Dialect Notes* V. 270 *Hi there*, (call or warning). **1945** T. WILLIAMS *Battle of Angels* II. i. 33 *A girl*: Hello! *Val*: (amiably) Hello there. **1962** J. BRAINE *Life at Top* xv. 188 'Hello there,' I said, 'What's new?'

c. (*b*) *that there.* Used adjectivally and absolutely, often in *euphem.* reference to sexual activity, esp. in catch-phrase *you can't do that there 'ere* (see quot. 1933).

1819 BYRON *Let.* 26 Oct. (1976) VI. 232 As to 'Don Juan' confess—confess—you dog—and be candid—that it is the sublime of *that there* sort of writing—it may be bawdy—but is it not good English? **1933** SQUIERS & WARK *You can't do that there 'Ere* in *Feldman's 41st Song & Dance Album* 37 As they took a kiss, The keeper shouted this: You can't do that there 'ere, so there! You can't do that there 'ere. You'd ought to know you 'ad, I'm sure, That that there 'ere's agin the law. **1937** *Even. News* 13 Apr. 8/3 The British Government gives vent to a 'John-Bullism', and says, after the abduction of a Hindu girl from within the border, 'You can't do that there 'ere!' **1962** AUDEN *Dyer's Hand* (1963) 406 How suitable, too, for a that-there poet that the room in which his 'Memoirs' were burned should now be called the Byron Room. **1974** P. WRIGHT *Lang. Brit. Industry* xi. 96 Long before the song 'You can't do that there here', Northerners used *that there* as a euphemism for the sexual act. It is a standard phrase in the north when youngsters of both sexes are 'educating' themselves by discussing sex matters.

(*c*) Phr. *there's gold in them there* (freq. *thar*) *hills*, with reference to a potentially profitable enterprise or activity. Also allusively. orig. *U.S.*

1941 C. B. KELLAND *House of Cards* xiv. 159 She heard him chuckle. 'Thar's gold in them thar hills.' **1961** J. L. AUSTIN *Philos. Papers* vi. 129 There is gold in them thar hills. **1965** E. GUNDREY *Foot in Door* xxxiii. 189 There's money in them thar pills—but very little else. **1976** *New Society* 16 Sept. 607/1 There's gold in them there sand-dunes, about 10 million people enjoyed a naturist holiday last year.

3. c. *there is*, usu. contracted to *there's* (with succeeding adj.): used in statements or exclamations in place of standard English *that is* or How *adv.* 7. *Welsh dial.*

1939 R. LLEWELLYN *How Green was my Valley* ii. 23 'Go on, boy,' Cedric whispered, 'there is soft you are to eat old cake.' **1951** E. COXHEAD *One Green Bottle* v. 113 There's tantalising! Plenty of company and no time for a word. **1968** A. LASKI *Keeper* vi. 68 There's sad, about Japhet; that was a good man. **1971** 'H. CALVIN' *Poison Chasers* ix. 123 There's selfish you are, I had him saved up for myself.

7. (Further examples.) Hence *there-there* vb. trans., to soothe or comfort by saying these words.

1798 JANE AUSTEN *Lett.* (1952) 42 There! I may now finish my letter and go and hang myself. **1840** T. C. HALIBURTON *Clockmaker* 3rd Ser. xx. 284 It's no such thing, says mother, quite snappishly; Sam is only twenty-one last Thanksgiving-day, and he was born just nine months and one day arter we was married, so there now. **1875** L. TROUBRIDGE *Life amongst Troubridges* (1966) 101 There now, if I haven't entirely forgotten to say anything about the boys. **1924** R. MACAULAY *Orphan Island* xxi. 280, I suppose you think I'm in love with you. Well, I'm not, so there. **1938** D. RUNYON *Furthermore* viii. 159 He.. starts whispering, 'There, there, there, my itty oddleums.' **1948** 'J. TEY' *Franchise Affair* iv. 39 Only one thing your Aunt Lin makes better than me..hot cross buns, and that's only once a year. So there! **1968** J. SANGSTER *Touchfeather* xv. 180, I was sobbing my heart out on his chest and he was there there-ing me all over the place. **1969** *Listener* 15 May 698/1 But Gwen was going to marry her lecherous tutor, so there. **1977** 'E. CRISPIN' *Glimpses of Moon* xii. 240 There, There, sir. **1977** C. DEXTER *Silent World N. Quinn* 254 Joyce took the baby..and lovingly there-thered his raucous cries.

8. *there and back* (earlier examples); also as a catch-phrase reply (see quot. 1937).

1772 in S. Rosenfeld *Temples of Thespis* (1978) v. 78 Pd Mr. Richards..at 2 Guineas pr Day & expenses there & Back £62.5.0. **1803** G. COLMAN *John Bull* III. ii. 32 Aye, he might have been there and back, over and over again; but my husband is slow enough in his motions. **1830** M. EDGEWORTH *Let.* 18 Oct. (1971) 419 This 'Trip to the Viaduct'..five shillings apiece there and *back*. **1937** PARTRIDGE *Dict. Slang.* 874/2 *There and back*, a c.p. reply to an impertinent or unwelcome inquiry 'where are you going (to)?': late C. 19–20. **1977** *Transatlantic Rev.* LX. 191 'Where are we going?' 'Oh, there and back,' said the cabbie, giggling.

IV. 12. c. *to have been there* (*before*) (colloq.): to have had previous experience of the activity or thing under review; to be fully conversant with or know something at first hand. orig. *U.S.*

1877 *Sat. Even. Post* in J. R. Ware *Passing Eng.* (1909) 24/1 Some reasons why I left off drinking whiskey, by one who has been there. **1913** A. BENNETT *Great Adv.* I. ii. 46 But I'm not a young girl. If it's a question of the male sex, I may say that I've been there before. **1977** J. WAMBAUGH *Black Marble* (1978) viii. 106 Philo Skinner's been in this racket thirty years. Philo Skinner's *been* there, baby!

d. *in there* (U.S. slang): excellent, superlative (esp. of a jazz-musician's performance); well-informed, *au fait*.

1944 D. BURLEY *Orig. Handbk. Harlem Jive* 104 Now, this skull was in there, Jack. **1945** L. SHELLEY *Jive Talk Dict.* 26 *In there*, superlative performance. **1955** SHAPIRO & HENTOFF *Hear me talkin' to Ya* vii. 101 The Lincoln Gardens, of course, was still in there. **1958** J. KEROUAC *On Road* I. i. 6 It took him just a few months..to become completely *in there* with all the terms and jargon. **1962** *Down Beat* 13 Sept. 37 A guy playing a horn has..gotta get in there.

16. a. (*b*) (earlier example); (*c*) = What did I tell you? (*d*) expressing resignation to an unpleasant fact.

1857 DICKENS *Dorrit* II. xxv. 536 All the people who had tried to make money and had not been able to do it, said, There you were! [**1863** H. E. P. SPOFFORD *Amber Gods* 133 She couldn't hire him a nurse, and there he was.] **1883** 'MARK TWAIN' *Life on Mississippi* xlii. 431 The immortelle requires no attention: you just hang it up, and there you are. **1894** A. CONAN DOYLE *Mem. Sherlock Holmes* 142 'There you are!' said Holmes smiling. **1915** CONRAD *Victory* IV. x. 373 'There you are!' Ricardo shrugged his shoulders philosophically. 'Can't be helped.' **1937** M. SHARP *Nutmeg Tree* xix. 250 'We've no business to talk about him. But there you are,' said Julia harshly, 'I'm the sort of woman any one talks to about anything. **1926** S. JAMESON *Three Kingdoms* i. 49 I'm sure that's a revolting sentiment, and revoltingly sentimental, but there you are. **1953** L. P. HARTLEY *Go-Between* xiv. 173 It's a pity we have to shoot so many of them but there you are.

b. *there it is* = sense 16 a (*d*) above. Also with past tense.

1857 TROLLOPE *Barchester T.* III. i. 10 There it is. If they haven't the spirit to enjoy it, the fault shan't be mine. **1884** 'MARK TWAIN' *Huck. Finn* xxxiii. 345 So there it was!—but I couldn't help it. **1904** H. JAMES *Golden Bowl* I. xviii. 311 'It's not, at any rate,' she went on, 'my fault. There it is.' **1932** 'A. BRIDGE' *Peking Picnic* xxiii. 296 He had been hurt hideously, and it made her cry; she was nearly as much surprised as he, but there it was. **1954** R. MACAULAY *Last Lett. to Friend* (1962) 196, I feel a little mean about the dear Chapel, but there it is. **1973** C. SAGAN *Cosmic Connection* xxii. 150 We would not ordinarily consider the flatulence of cattle as a dominant manifestation of life on Earth, but there it is.

V. 17. b. thereamid (later example).

1901 G. GISSING in *Literature* 21 Dec. 572/1 Thereamid stood a girl, her eyes fixed upon the prospect of city roofs.

theremin (þe·rĕmin). Also **thérémin,** and with capital initial. [f. the name of its inventor, L. *Thérémin* (b. 1896), Russian engineer.] An electronic musical instrument in which the tone is generated by two high-frequency oscillators and the pitch controlled by the movement of the performer's hand towards and away from the circuit (see quot. 1971).

1927 *Times* 12 Dec. 11/1 Professor Theremin and his collaborator play duets for two 'theremins' and piano. **1934** S. ROBERTSON *Devel. Mod. Eng.* x. 418 Miscellaneous examples..of the taking over of a surname..and using it as a common noun are *boycott*,..*theremin*, and *zeppelin*. **1950** BLESH & JANIS *They all played Ragtime* x. 199 And then those long metal cylinders, different lengths. The players wore gloves and would pull on them and make weird sounds like a Theremin. **1971** *Daily Colonist* (Victoria, B.C.) 13 Feb. 25/5 Greenway, of Vancouver, will play a theremin. The instrument was developed and introduced to the public in 1920 by Leo Theremin, a Russian scientist. A box-like apparatus, it produced musical tones from two electric circuits running through vibrating radio tubes. The player stands in front of the theremin and moves his hands through the air... The left hand controls volume or tone, and the right hand raises pitch as it nears the instrument and lowers pitch as it moves away. **1974** *Times* 5 Feb. 11 Most illuminating of all were the live performances of Varèse's *Ecuatorial*,.. its original thérémin part now played by two..*ondes Martenot*. **1982** *New Scientist* 16 Dec. 753 Moog recently recorded her playing the theremin.

thereness. Delete *rare* and add later examples.

1929 D. H. LAWRENCE *Paintings D. H. L.* (Introd.) fo. 7 *verso*, All the host of other defiant..cats that have come back..to form and substance and *thereness*, instead of delicious nowhereness. **1958** *Listener* 20 Nov. 822/2 The immense Thereness of someone else. **1976** I. MURDOCH *Henry & Cato* I. 196 All those would-be deep explanations are so abstract and so simple when confronted with the awful complex thereness of a relationship which has gone wrong. **1983** J. JONES *Dostoevsky* i. 7 Its absurd yet maddening thereness like that of the pea under the mattress of the princess.

the-reward, *adv. rare*⁻¹. [f. THERE *adv.* + -WARD.] = THERETOWARD *adv.*

1922 JOYCE *Ulysses* 378 Thereward carrying desire immense among all one another.

therian (þī°·riăn), *a.* (*sb.*) *Zool.* [a. mod.L. *Theria* (Parker & Haswell *Textbk. Zool.* (1897) II. 448), f. Gr. θήρ, θηρίον wild beast.] Of or pertaining to the subclass Theria, one of the four subclasses into which the class Mammalia is commonly divided. Also as *sb.*, a placental or marsupial mammal belonging to this subclass.

1960 *McGraw-Hill Encycl. Sci. & Technol.* XIII. 549/2 Therian mammals are characterized by the distinctive structural history of the molar teeth. **1971** *Nature* 23 Apr. 506/1 The 'cochlea' in birds and reptiles is only slightly curved, in contrast to the tightly coiled cochlea of therian mammals. **1974** D. & M. WEBSTER *Compar. Vertebr. Morphol.* v. 99 The more generalized therians have a clavicle, extending ventromedially and articulating with the anterior portion of the sternum. **1977** *Sci. Amer.* Aug. 79/3 The marsupials have retained the basic ancestral therian reproductive pattern.

theriomorph (pɪ°·riomǭɹf), *sb.* and *a.* [f. THERIO- + *-MORPH; cf. THERIOMORPHIC *a.*] **A.** *sb.* **a.** A representation of an animal form in art.

1913 [see *ANTHROPOMORPH]. **1928** V. G. CHILDE *Most Anc. East* iv. 84 Some theriomorphs are made of just three variegated stones.

b. = THERIOMORPH. Also *fig.*

1920 H. G. WELLS *Outl. Hist.* I. vi. 24/1 These little Theriomorphs, these ancestral mammals, developed hair. **1934** A. J. TOYNBEE *Study of Hist.* III. 194 But he [sc. Wells] comes to grief in the recent annals of our own Western history when he has to size up that singularly etherialized theriomorph William Ewart Gladstone.

B. *adj.* Having the form or characteristics of a beast.

1969 H. ARENDT *On Violence* (1970) 60 Why should we, after having 'eliminated' all anthropomorphisms from man's psychology (whether we actually succeeded is another matter), now try to discover how 'theriomorph' man is?

theriomorphism (pɪ°·riomǭ·ɹfiz'm). [f. THERIO- + *-MORPHISM; cf. THERIOMORPHIC *a.*] The ascription to God or to a god of the form or characteristics of a beast.

1908 *Encycl. Relig. & Ethics* I. 538/1 The Annamese believe in...beings who can pass from one genus or species to another under certain conditions of space and time. Hence theriomorphism and totemism. **1914** H. M. CHADWICK *Heroic Age* vi. 125 It may be remarked in passing that theriomorphism plays a very prominent part in the religious practices and conceptions of primitive peoples, and...we hear not unfrequently of a struggle between a god or national hero and some theriomorphic being whose sanctuary or attributes he appears to have taken over. **1930** A. S. PRINGLE-PATTISON *Stud. Philos. of Relig.* vi. 75 Theriomorphism seems...to precede anthropomorphism, and it is only gradually that the gods are humanized. **1969** *Times Lit. Suppl.* 11 Dec. 1431 The Charolais becomes the focal-point for a mania, of short-lived, cult of theriomorphism. **1982** *Jrnl. Indo-European Stud.* X. 159 Volos is subsumed by theriomorphism...heavenly bodies...fertility...and the realm of life.

therm, *sb.*[3] Add: = *CALORIE b. Now *Obs.*

2. A quantity of heat equal to 100,000 British thermal units, used in Britain as the statutory unit in expressing the quantity of gas supplied.

1920 *Act 10 & 11 Geo. V* c. 28 § 1 (2) A standard or maximum price for each hundred thousand British thermal units (in this Act referred to as 'a therm'). **1922** *Westm. Gaz.* 18 Oct. 8/5 The new method of charging by 'therms'. **1955** *Times* 20 July 8/3 A 'substantial increase' was forecast by the chairman of the South Eastern Gas Board when he announced recently an increase of 2d. a therm in the price of gas. **1982** *Daily Tel.* 26 Apr. 2/8 The average price being paid by British Gas for existing and new supplies is 10p–12p a therm.

therm, *sb.*[3] Add: **2.** In 18th-c. cabinet-making, a rectangular, tapering leg or foot of a chair, table, or the like. Also *attrib.* or as *adj.*

1788 *Cabinet-Makers' London Bk. Prices* (1803) No. 7, If the plinth of the Therms is work'd below [price, extra, 1½d]. **1949** PINKERTON-BRODHURST & LAYTON *Gloss. Eng. Furnit.* 173 *Therm foot*, a rectangular tapering foot to the legs of chairs and tables, also called a spade or taper foot, often used by the brothers Adam and Hepplewhite, and to a lesser degree by Sheraton. *Therm leg*, the taper or therm leg was a favourite feature of Hepplewhite and later designers. **1952** J. GLOAG *Short Dict. Furnit.* 475 (*caption*) Ten designs for therms for claws.

† therm (pɜːm), *sb.*[4] Colloq. abbrev. of THERMOMETER. *Obs.*

1795 J. WOODFORDE *Diary* 11 Dec. (1927) III. 318 It froze all day long even within doors very quick Therm at 52. **1799** MALTHUS *Diary* 21 July (1966) 175 Yesterday his therm was 18, & mine in a dry shade was 71. **1877** W. WHITMAN *Daybks. & Notebks.* (1978) I. 58 Very hot—therm 90–96.

therm (pɜːm), *v.* [f. THERM *sb.*[3]] *trans.* In 18th-c. cabinet-making, to turn (a leg or foot of a chair, table, or the like) to a rectangular, tapering form; also *absol.* Hence **thermed** *ppl. a.*, **the-rming** *vbl. sb.*

1788 *Cabinet-Makers' London Bk. Prices* (1803) No. 12, The Price of Therming in the Legs... The Price of Therming in the Neck... When the legs are therm'd at the top only, the tapering to be paid for extra. **1907** G. O. WHEELER *Old Eng. Furnit.* 461 *Therming*, a process of cornering a delicate taper, especially applied to the feet of chairs, sideboards, and tables of the Sheraton order. **1925** PINKERTON-BRODHURST & LAYTON *Gloss. Eng. Furnit.* 173 *Therming*, a process in use towards the end of the eighteenth century, before circular and band saws were invented, by which the legs of chairs and tables were thermed or tapered, by means of a lathe provided with a cylinder about six feet in diameter, on which the legs were placed and turned down one side at a time. **1952** J. GLOAG *Short Dict. Furnit.* 470 *Taper leg*, a leg of square section, sometimes called a thermed leg, gradually diminishing towards the foot, introduced in the second half of the 18th century for chairs, tables and sideboards.

thermal, *a.* Add: **2. a.** (Further examples.) Also, caused by heat; *thermal agitation*, the motion of atoms or the like due to their thermal energy; *thermal analysis*, analysis of a substance by examination of the way its temperature falls on cooling or rises on heating; *thermal barrier* (Aeronaut.) = *heat barrier* s.v. *HEAT *sb.* 14 d; *thermal bremsstrahlung*, electromagnetic radiation produced by the thermal motion of charged particles in a plasma; *thermal capacity* (earlier example); *thermal column* (Nucl. Physics), a body of moderator inside or projecting from a reactor such that it serves as a source of thermal neutrons for experimental purposes; *thermal cycle*, a cycle in which the temperature of a substance rises or falls and then returns to its initial value; *thermal cycling*, the periodic heating and cooling of a substance; *thermal death point*, the lowest temperature at which a micro-organism is killed under specified conditions; *thermal diffusion*, diffusion occurring as a result of the thermal motion of atoms or molecules, esp. as a technique for separating gaseous compounds of different isotopes of an element (which diffuse at different rates in a temperature gradient); *thermal diffusivity*, the thermal conductivity of a substance divided by the product of its density and its specific heat capacity; *thermal imaging*, the technique of using the heat given off by objects or substances to produce an image of them; so *thermal imager*; *thermal lance* = *thermic lance* s.v. *THERMIC *a.*; *thermal noise* (Electronics), noise arising from the random thermal motion of electrons; *thermal pollution*, the production of heat, or the discharge of warm water, esp. into a river or lake, on a scale that is potentially harmful ecologically; *thermal printer*, a printer having a matrix of fine pins as the print-head, which are selectively heated to form a character on heat-sensitive paper; *thermal runaway* (Electronics), a dramatic or destructive rise in the temperature of a transistor as a result of an increase in its temperature causing an increase in the current through it, and vice versa; *thermal shock* (cf. *SHOCK *sb.*[1] 2); *thermal storage* (later examples); also used *attrib.* to designate appliances which store heat in other ways; *thermal unit* (earlier and further examples); *British thermal unit* is also abbreviated B.T.U., B.t.u.

1853 *Trans. R. Soc. Edin.* XX. 170 The mechanical equivalent of the ordinary thermal unit. **1880** W. THOMSON in *Encycl. Brit.* XI. 558/2 Regnault's measurements of the thermal capacity of water at different temperatures. *Ibid.* 578/1 The thermal conductivity of the substance is not generally the same at different temperatures. *Ibid.* 580/2 It is *kt*, not merely *k*, that expresses the quality of the substance on which the phenomenon chiefly depends. We therefore propose to give to *kt* the name of thermal diffusivity. **1898** *Public Health Papers & Rep.* (Amer. Public Health Assoc.) XXIII. 86 In determining the thermal death point cultures should always be used. **1916** S. CHAPMAN in *Proc. R. Soc.* A. XCIII. 10 We may call D₁, D₂, D₃, and D₄ respectively the coefficients of diffusion, forced diffusion, pressure diffusion, and thermal diffusion. The definition of D₄ agrees with that usually given for the coefficient of diffusion. The other coefficients seem to be defined here for the first time. **1925** *Jrnl. Iron & Steel Inst.* CXII. 489 Thermal analysis was followed by determination of the hardness and a study of the microstructure of the test-pieces. **1925** Thermal shock [see *PULPAL *a.*]. **1927** *Physical Rev.* XXIX. 367 Ordinary electric conductors are sources of random voltage fluctuations, as a result of thermal agitation of the electric charges of the conductor. **1930** F. B. LLEWELLYN in *Proc. IRE* XVIII. 244 The importance of this noise, which will be termed 'thermal noise', in high-frequency receivers... Thermal storage of this kind, when applied to the direct heating of rooms... **1935** *Discovery* July 214/2 Barometric depressions are discussed...in their modern guise as interacting air masses, the forced ascent necessary to give rain being related to dynamical instability instead of to thermal instability as in the older theories. **1936** W. L. NELSON *Petroleum Refinery Engin.* xvii. 504 The thermal decomposition or cracking of oil was called to our attention by Silliman in 1871. **1943** *Gloss. Terms Electr. Engin.* (B.S.I.) 12 A B.t.u. is equivalent to 1054 joules. **1945** H. D. SMYTH *Gen. Acct. Devel. Atomic Energy Mil. Purposes* vii. 79 We are concerned with the second ... the thermal utilization factor, which is defined as the probability that a given thermal neutron will be absorbed in the uranium. **1950** *Sci. News* XV. Plate 16 (*caption*) Slip-lines in pure zinc after exposure to 50 thermal cycles between 30°C and 150°C (×500). **1950** *Canad. Jrnl. Res.* A. XXVIII. 434 The thermal column of the Chalk River pile was used as a large block of scattering material with a high flux of thermal neutrons. **1951** *Jrnl. R. Aeronaut.*

Soc. LV. 757/2 A new barrier is faced after 'climbing' the sonic barrier, namely the Thermal Barrier. **1954** R. STEPHENSON *Introd. Nucl. Engin.* vi. 249 The X-pile has no thermal shield... After 10 years operation there is no material damage of the concrete, which would indicate that the possibility of failure of concrete due to thermal stress may be much less severe than is generally assumed. *Ibid.*, A thermal shield is an inner wall...which is placed between the reactor and the biological shield. Its function is to remove most of the heat energy of the gammas and fast neutrons and thereby protect the biological shield from damage. **1955** *Sci. Amer.* July 38/2 Coal supplied only 34 per cent of the nation's total B.T.U.'s (British thermal units). **1957** *Times* 12 Nov. (Canada Suppl.) p. iii/4 A 1,200,000 h.p. thermal power station is to be built near Vancouver, using natural gas. *Ibid.* 11 Dec. 8/1 It has been necessary to increase the capacity of the distillation unit and provide a thermal cracker. **1958** C. C. ADAMS *Space Flight* 263 The MA-2 is a special suit that is ventilated for travel through the thermal barrier, and the MA-1 is a new ARDC helmet. **1960** CRAIGHEAD & QUARRELL *Physical Examination of Metals* (ed. 2) iii. 183 The simplest method of carrying out a thermal analysis is to place the specimen in a furnace which is arranged to have a negligible temperature gradient over a zone somewhat greater than the length of the specimen. **1961** *Guardian* 2 June 6/6 A domestic thermal storage heater is now available. **1964** *Research* XV. 80/1 Uranium polycarbide...is much more resistant to thermal shock and thermal cycling. **1962** SIMPSON & RICHARDS *Physical Princ. Junction Transistors* ix. 210 In extreme cases this positive feedback may lead to a catastrophic increase in temperature—a phenomenon commonly called 'thermal runaway'. **1963** *Nature* 1 Dec. (Advt.) ... *Energy* 237/2 The separations achieved in a convection-free system are small, and are only used to determine the magnitude of the thermal diffusion effect and to provide information on intermolecular force fields. **1963** B. FOZARD *Instrumentation Nucl. Reactors* xiii. 166 A count rate from the fission chamber of 10⁴ c/s...is produced by a neutron flux, in the thermal column, of about 5×10⁴ neutrons cm⁻² sec⁻¹. *Ibid.* 170 Thus the counter can be operated under virtually isothermal conditions over a wide range of load, with consequent constancy of steam conditions and absence of thermal cycling. **1964** H. S. HVISTENDAHL *Engin. Units* vii. 110 The legal definition of the Btu is the amount of heat required to raise 1lb of water from 60 to 61°F at standard atmospheric pressure. **1965** R. G. KARMAN *Mod. Hydrol.* iv. 109 The so-called thermal pollution of streams has resulted primarily from the installation of steam-electric generating plants along our rivers. **1966** C. R. TOTTLE *Sci. Engin. Materials* v. 127 The mean free path *l* depends on the thermal agitation of the lattice, and in less regular lattices, such as those of amorphous materials, there is a reduced probability of attenuation of the thermal wave, by virtue of the variable distances between the atoms. **1968** R. R. ERNST in LAWRENCE & BLOCK *Disinfection, Sterilization & Preservation* vii. 355 The thermal death point is the lowest temperature at which a suspension of bacteria is killed in 10 minutes. This standard has been almost abandoned. **1969** Thermal pollution [see *POLLUTION 2 a]. **1970** *Nature* 19 Sept. 1185/1 The technique of thermal imaging—picking up infrared radiation from the human body and displaying the resulting thermal image on an cathode-ray tube... has been used to map the flow of warm arterial blood into, for example, tumours and varicose veins. **1971** *Gloss. Soil Sci. Terms (Soil Sci. Soc. Amer.)* 21/1 *Thermal analysis (differential thermal analysis)*, a method of analyzing a soil sample for constituents, based on a differential rate of heating of the unknown and standard samples when a uniform source of heat is applied. **1972** *Sci. Amer.* July 35/3 This combination of conditions is all that is needed to produce the X rays, since accelerated charges are a source of electromagnetic radiation. A compact name for the process is 'thermal bremsstrahlung'. **1972** *Osf. Univ. Gaz.* CII. Suppl. No. 8. 3 Work has begun on the installation of a second thermal storage boiler. **1973** 'K. ROYCE' *Spider Underground* viii. 119 Someone must be on tap to answer awkward questions if Old Bill [sc. the police] arrives. The thermal lance men will be below. **1973** J. G. TWEEDDALE *Materials Technol.* I. iv. 95 Thermal diffusivity (*a*) is important when it is necessary to consider the effects of temperature differences set up in a material during transfer of heat. *Ibid.*, Both thermal conductivity and diffusivity cease to have much meaning for the liquid state since in that state the principal mechanism of transfer becomes convective mixing. **1974** *Sci. Amer.* Dec. 95/2 They are not die-cast in significant quantity now because the thermal shock to the metal components, including the mold, is so severe that the life of the components is short. **1976** *McGraw-Hill Yearbk. Sci. & Technol.* 44/2 Some kind of thermal shield had to be designed, fabricated, and deployed quickly if Skylab was to be saved. **1975** *S. J. King Audio Handbk.* iv. 166 Since *I*₀ and *I*₀₀ are less temperature sensitive on radio transistors, temperature compensation is not necessary, neither can thermal runaway occur, for *I*₀ tends to fall with increasing temperature. **1975** D. G. FINK *Electronics Engineers' Handbk.* xxix. 60 Thermal noise for the most part originates in the first stages of the radio receiver and sets the minimum signal amplitude acceptable for a given signal-to-noise ratio. **1977** F. WEBB *Go for Oil* vii. 125 A modern peterman needs explosives—else thermal lances. **1977** I. M. CAMPBELL *Energy & Atmosphere* v. 86 There is unlikely to be a thermal pollution problem of any importance on a global basis within the forseeable future. **1978** *Jrnl. R. Soc. Arts* CXXVI. 683/1 Structures of supersonic aircraft are subject to thermal stresses due to temperature gradients. **1980** *Times* 4 Aug. 17/2 Thermal imaging devices give an infra-red picture of the scene by detecting and processing the infra-red energy they emit. **1982** *Daily Tel.* 20 Aug. 5/2 Their passive infra-red viewers, image intensifiers, and thermal imagers were excellent. **1982** *Sci. Amer.* Dec. 93/1 Thermal printers, which cost less than $500, burn an image into a special paper at a rate of some 50 characters per second.

b. Nucl. Physics. *thermal neutron*, a neutron which is in thermal equilibrium with its environment (see quot. 1966); so *thermal speed, velocity*, the speed characteristic of such a

neutron; *thermal reactor*, a nuclear reactor in which the fission process relies upon thermal neutrons. Cf. *slow neutron*, *slow reactor* s.v. *SLOW *a.* 13 c.

1936 *Physical Rev.* XLIX. 520/1 It is therefore not necessary to ascribe all large cross sections to neutrons of thermal velocities. **1938** *Ibid.* LIV. 235/1 (*heading*) Collimated, variable energy beam of pure thermal neutrons. **1949** H. ETHERINGTON in S. C. ROTHMANN *Constructive Uses of Atomic Energy* v. 76 In a thermal reactor fission is produced by neutrons that have been slowed approximately to thermal velocity. **1959** *Listener* 19 Nov. 872/1 At slow or 'thermal' speeds neutron capture by nuclei of Uranium 238 is less important. **1966** C. R. TOTTLE *Sci. Engin. Materials* x. 276 Fast neutrons may be moving at speeds of the order of 10¹⁰ cm sec⁻¹ with an energy of 1 to 10 MeV. When slowed down to a similar order of velocity to that of the thermal vibration of atoms (hence called thermal neutrons) the speed is about 10⁶ cm sec⁻¹ at an energy of 1 eV. **1977** *Nature* 21 July 211/1 The companies say that they have between them a good deal of experience in building thermal reactors of several types.

c. Promoting the retention of heat. Usu. of clothes, esp. underwear. *thermal pane* = *Thermopane* s.v. *THERMO-.

1970 *Toronto Daily Star* 24 Sept. 16/2 (Advt.), Quilted thermal units. **1973** *Times* 4 Aug. 16/8 The thermal underwear for use at high altitudes. **1974** R. MACINNES *Climb to Lost World* viii. 122, I had taken the precaution of carrying my pair of calf-length thermal boots with me... The proved very useful in this swampy ground. **1978** *Detroit Free Press* 16 Apr. B 5/1 In addition to long johns, thermal socks and two caps, Harwell wore a tee shirt. **1978** T. GIFFORD *Glendower Legacy* 20 He...glanced out the wide thermal-pane window. **1982** *Oxford Star* 4–5 Feb. 10/7 Tartan Cottage, Oxfordshire's cranked mail order clothing firm, has been rescued by a thermal underwear company.

thermal (pɜ·rmǎl), *sb.* [f. prec. adj.] A rising current of relatively warm air, used by gliders and birds to gain height.

1933 *Jrnl. R. Aeronaut. Soc.* XXXVII. 678 Herr Hirth had gained a great deal of experience regarding 'thermals', that is to say, ascending currents of warm air which can be used for gaining height. **1949** R. M. SAVAGE in *Rep. Brit. Assoc.* ... the centres of thermals are to be seen ... of warm air currents. **1950** 'N. SHUTE' *Town like Alice* 230 She coiled over on her back and watched a seagull soaring in the thermals from the island. **1962** *Amer. Scientist* L. 180 Thermal soaring is the method most commonly used by soaring birds. **1974** 'G. BLACK' *Golden Cockatrice* iii. 60 He had seen high kites riding thermals well beyond the harbour as a bird sees a thermal.

Hence **the-rmalling** *vbl. sb.*, soaring in thermals.

1936 *Archit. Rev.* LXXIX. 255/3 For greater heights the second and more interesting method is employed; what is known as 'thermalling'. This is the utilization of the columns of rising air that are always in existence under certain weather conditions. **1974** *Reader's Digest Frb.* 89 With a Rogallo you can also do another type of soaring, called thermalling, where you circle in chimney-like updrafts of warm air that rise from sun-heated ground.

Thermalite (pɜ·rmǎlait). Also with small initial. [f. THERMAL *a.* + -ITE[1].] The proprietary name of a type of cellular concrete building block with good insulating qualities.

1949 *Trade Marks Jrnl.* 3 Aug. 690/1 *Thermalite*... Concrete products included in Class 19. **1955** *Archit. Rev.* CXVII. 117/2 The construction is a cavity wire with an inner skin of thermalite blocks, with a roof of 1 inch asphalt on rafters at 5 ft spacing, and 2 inch cork insulation. **1960** *Times* 3 Oct. (Suppl.) p. iii/2 External walls are of cavity construction with brick facing and Thermalite inner leaf. **1977** *Reader's Digest Bk. of Do-It-Yourself Skills & Techniques* 201/1 Aerated blocks, e.g. Celcon or Thermalite, are also easy to work and have better insulation qualities [than lightweight aggregate blocks] but are more expensive and are unsuitable for exposed garden walls.

thermalize (pɜ·rmǎlaiz), *v.* *Physics.* [f. THERMAL *a.* + -IZE.] **a.** *trans.* To bring into thermal equilibrium with the environment.

1955 *Ann. Rev. Nucl. Sci.* VI. 317 The coolant...slows down the neutrons, thermalizes them and thereby helps to thermalize the assembly. **1961** R. G. COOPER *Exper. Nuclear Chem.* viii. 216 It is necessary to surround the source with paraffin or water to thermalize the neutrons. **1979** *Nature* 30 Nov. 456/1 The hypothetical Oort cloud should be thermalized by weak stellar encounters on a time scale ~1 Myr.

b. *intr.* Of sub-atomic particles, etc.: to attain thermal equilibrium with their environment.

1966 *New Scientist* 17 Mar. 707/1 It has been predicted that a positron in a metal should 'thermalize'—that is, reach the kinetic energy characteristic of the temperature of the metal—in about 10⁻¹¹ s. **1973** *Physics Bull.* Nov. 652/3 The density was so high that any radiation generated would rapidly thermalize. **1979** *Nature* 1 Mar. 53/2 If the infalling protons thermalize just at the surface of the compact object, the radiation will be in the form of γ rays.

Hence **the-rmalized**, **the-rmalizing** *ppl. adjs.*; also **rhealmaliza-tion**, the process of thermalizing.

1950 GLASSTONE *Sourcebk. Atomic Energy* xi. 294/1 The process of reducing the energy of a neutron to the thermal region by elastic scattering is sometimes called thermalization or, more commonly, slowing down. **1971** *Nature* 16 Apr. 450/1 A more likely possibility...is that electron thermalization during trapping produces local heating of

the matrix. **1971** *Engineering* Apr. 34/2 Moderating materials are often an integral part of the structure of nuclear reactors so that thermalized beams can be obtained directly. **1979** *Nature* 30 Aug. 749/2 The major issue remaining is whether sputtered atoms escape [or are merely supplied to a thermalizing atmosphere for later escape.

thermic, *a.* Add: (Earlier example.) *thermic lance*, a steel pipe packed with steel wool through which a jet of suitable gas may be passed in order to burn away metal, concrete, or the like using heat generated by the burning of the pipe; cf. *LANCE *sb.*[1] 6* **a**, *thermal lance* s.v. *THERMAL *a.* 2 a.

1842 J. F. W. HERSCHEL *Let. Nov.* in *Phil. Trans. R. Soc.* (1843) CXXXIII. 5 If the restriction to these rays of the term *thermic* as distinct from *calorific* be not...a sufficient distinction, I would propose the term *parathermic rays* to designate them. **1970** F. LAURIE *Scotland Yard* x. 251 The thieves...penetrate the wall of the vault with a thermic lance. **1982** *Daily Tel.* 7 Sept. 3/3 He planned to break into the bank with a thermic lance.

Hence **the-rmics** *sb. pl.*, the study of heat, thermodics. *rare.*

1854 [see *palæoneurology* s.v. *PALÆO-, PALEO-]. **1953** *Archit. Rev.* CXIV. 195/2 (*heading*) Kitchen thermics.

Thermidor. Add: **1.** (Earlier examples.)

1801 W. DUPRÉ *Neological French Dict.* 276 *Thermidor*, hot month... The 9th Thermidor answers to the 27th of July. **1802** C. WILMOT *Let.* 30 July in *Irish Peer* (1920) 73 Paris, 30th July, 1802, 9 Thermidor.

2. *lobster thermidor*: see *LOBSTER[1] 5.

3. A moderate reaction following a revolution. Cf. sense 1 in Dict.

1938 C. BRINTON *Anatomy of Revolution* viii. 244 We shall have to call Thermidor a convalescence from the fever of revolution. **1960** *Commentary* June 508/1 The retreat from 'war communism'...did not lead to the revolutionary regime's overthrow by a new Thermidor. **1974** tr. *Wertheim's Evolution & Revolution* 341 Should all such symptoms be taken as evidence that the Cultural Revolution...got stuck in a kind of Thermidor, and that consequently the fight of Mao and his allies against a Thermidor was a last stand, a losing battle? **1981** *Times* 30 June 15/1 Iran is about to enter its Thermidor.

Thermidorian, *sb.* and *a.* Add: **A.** *sb.* **a.** (Earlier example.)

1801 W. DUPRÉ *Neological French Dict.* 276 Un thermidorien, spéculateur révolutionnaire—A thermidorian, and speculator in revolutions.

b. A moderate opponent of a revolutionary movement; a counter-revolutionary.

1981 *Encounter* Dec. 34/1 The Thermidorians...are themselves not easy to define...and the Thermidorian regime...is so fragmented as to defy any simple analysis.

B. *adj.* **c.** Of, pertaining to, or designating a moderate reaction following a revolution.

1938 C. BRINTON *Anatomy of Revolution* viii. 244 The ensuing slow and uneven return to quieter, less heroic times has been well known to French historians as the Thermidorean reaction. **1974** tr. *Wertheim's Evolution & Revolution* 333 It is against the background of these Thermidorian tendencies in the Soviet Union that we have to view the so-called 'Great Proletarian Cultural Revolution' in China. **1978** *Pacific Affairs* LI. 474 For them the Thermidorian effects of the bureaucratic restoration were much mitigated by the emergence at the helm of the enlightened Chou En-lai. **1981** [see sense b of the sb. above].

thermion (pɜ·rmiǫn). *Physics.* [f. THERM(o- + ION.] An electron or ion emitted from an incandescent surface.

1909 O. W. RICHARDSON in *Phil. Mag.* XVII. 814 The substantive Thermionics furnishes naturally the further substantive Thermion. **1932** J. MILLS *Within Atom* vii. 73 An electron which is emitted in this way is sometimes called a 'thermion'. **1973** H. A. ENGE *Introd. Atomic Physics* iii. 81 In 1899, J. J. Thomson showed that the thermions emitted are electrons.

thermionic (pɜ̄mi̞ǫ·nik), *a.* *Physics.* [f. prec. + -IC.] Of, pertaining to, or employing electrons emitted from an incandescent surface; *thermionic valve*, an electronic device consisting of an evacuated envelope containing two or more electrodes, such that a current can flow only in one direction as a result of thermionic emission from one electrode.

1909 O. W. RICHARDSON in *Phil. Mag.* XVII. 814 Here we have two currents: the current used to heat the wire and the thermionic current from the wire to the electrode. **1915** *Electrician* 21 May 241/1 The thermionic current...increased at first. **1917** *Wireless World* June 152 The invention by the writer [sc. J. A. Fleming] of this article of the thermionic detector. *Ibid.* 158 Known by various titles, such as Fleming valve, vacuum valve,... thermionic valve, and audion valve, it is the result of experiments extending over a large number of years. **1920** *Glasgow Herald* 2 July 10 What with the high-frequency alternator, and...the thermionic valve, wireless practice has been very much revolutionised. **1933** *Jrnl. Exper. Biol.* X. 295 (*heading*) A thermionic potentiometer for measuring light intensity with photo-electric cells. **1956** G. A. MONTGOMERIE *Digital Calculating Machines* xi. 216 By using thermionic valves and similar devices, the speed of automatic computors can be made very much higher. **1957** *Times* 26 Nov. 10/6 A new 'thermionic converter'...takes advantage of the fact that electrons can be 'boiled out' of a hot metal surface and used to produce an electric current directly. **1975** D. G. FINK *Electronics Engineers' Handbk.* xxvii. 18 In a thermionic generator...electrons

are emitted from the heated cathode...and collected by a cooler anode.

So **thermio-nically** *adv.*; **thermio-nics** *sb. pl.*, the branch of science and technology concerned with thermionic emission.

1909 O. W. RICHARDSON in *Phil. Mag.* XVII. 814 The author ventures to suggest that the word 'Thermionics' is very suitable for the purpose. **1932** J. MILLS *Within Atom* vii. 74 When electrons are being thermionically emitted from a heated wire. **1933** E. L. CHAFFEE *Theory of Thermionic Vacuum Tubes* i. 8 Early experimenters in the field of thermionics believed that the emission of electricity from hot bodies...was the result of some sort of chemical reaction. **1940** *Times* 12 Aug. 7/4 New branches of knowledge like...photo-electricity, and thermionics sprang up out of the fertile soil. **1967** *Jrnl. Optical Soc. Amer.* XXXVII. 424/2 Associated with this component of the dark current is a shot noise resulting from random thermionically emitted electrons. **1966** *McGraw-Hill Encycl. Sci. & Technol.* XII. 342/1 In thermionics... power densities of 25 watt/cm² at emitter temperatures of 1700°C have been obtained with hardware which is suitable for [space] flight.

thermistor (pɜ̄mi·stɑɹ). [Contraction of *therm(al re)sistor*.] A small piece of semiconducting material the resistance of which falls with increasing temperature, enabling it to be used for the sensitive measurement and control of the latter.

1940 G. L. PEARSON in *Physical Rev.* LVII. 1065/1 Thermistor is a contraction of the words 'thermal resistor' and designates an electrical resistance whose value is markedly dependent on temperature. **1951** *Sci. Amer.* Oct. 50/3 The heart of the instrument is a flake of metallic oxides called a thermistor, whose conduction of electric current is increased when it is heated. An amplifying and recording system translates this conductivity into a temperature reading. **1961** *New Scientist* 5 Mar. 598/3 The thermistor...is a semiconductor the resistance of which varies with temperature. **1972** *Sci. Amer.* Feb. 41/3 [The patient's] ... temperature that sensed the muscle temperature was in the tip of the harpoon and the one that sensed the water temperature was extended from the transmitter, outside the fish's body on the harpoon shaft. **1979** *Guardian* 8 Sept. 20/2 When the water in the panels becomes warm enough, a thermistor produces a signal which is interpreted by the control unit, so that a standard central heating pump takes the hot water from the panel to the bulk storage tank, and from there to the radiators.

thermite. Add. **1.** (Later examples.)

1918 *Nature* 14 Nov. 217/2 Thermit, now an important munition of war, is in a class by itself, for it is used for charging incendiary bombs and sometimes in a kind of shrapnel. **1971** B. SCHAEFER *Faith & the Lamp* xi. 175 The thermite is ignited and the hot metal allowed to flow into the mould, where it fills the gap between the two parts and forms a collar around them. **1973** R. DENTRY *Encounter at Kharmel* xi. 198 The third explosion...set off thermite positioned under the big transceiver.

2. *attrib.* and *Comb.*, as *thermite method*, *reaction*, *shell*, *weld*; *thermite process*, the reduction of finely divided oxides of iron or other metals by means of an exothermic reaction with finely divided aluminium; also *thermite welding*; *thermite welding*, fusion welding in which the heat and the weld metal are produced by the thermite process.

1919 *Times* 16 Jan. 11/4 The thermit method of welding has proved its worth in less-continued use. **1905** *Chambers's Jrnl.* Dec. 78/2 A perfectly successful joint...has been made by utilising the thermite process. **1910** C. H. DESCH *Metallography* vi. 190 The great reducing power of aluminium at high temperatures has been utilised in what is known as the Thermit process to produce metals and alloys free from carbon. **1930** *Engineering* 14 Mar. 349/2 The Thermit process, the oxy-acetylene and similar blowpipes,...have all attained to a facility and convenience applicable...to single jobs. **1938** [see *ALUMINO-THERMIC *a.*]. **1980** *Daily Tel.* 19 Feb. 6/8 British Rail uses two main welding techniques. One, the thermite process used along the Bushey track, involves welding long stretches of rail together in the site. **1973** *Chambers's Jrnl.* July 538/1 The thermite reaction is used largely in the preparation of metals from their oxides. **1923** KIPLING *Irish Guards in Gt. War* I. 219 Oil-drums, gas and thermit shells were added to the trench-mortar range. **1980** Thermite welding [see *ALUMINOTHERMIC *a.*]. **1930** *Times* 19 Feb. 5/5 There are at least 500 track welds, called thermit welds, using this method, on the line that failed. **1906** *Jrnl. Iron & Steel Inst.* I. 452 Experiments with electric and thermite welding for tramway rails. **1927** [see *METALLIC *a.* 1 b]. **1952** FUCHS & BRADLEY *Welding Practice* II. ii. 39 It must be remembered that Thermit welding can only be carried out successfully by specially trained and experienced personnel.

thermo-. Add: **thermocline** (further examples); also, a layer of water marked by such a gradient, the water above and below being at different temperatures; **the:rmocoagula-tion** *Surg.*, the coagulation of tissue, esp. in the brain, by means of heat; so **thermo-coagulated** *ppl. a.*, **-coagulative** *a.*; **thermo-compression**, the simultaneous application of heat and pressure; usu. *attrib.*; **thermo-du·ric** *a.* *Biol.* [L. *dūr-āre* to hold out, last] (of bacteria, etc.) capable of surviving high temperatures, esp. those of pasteurization; **the·rmoforming** *vbl. sb.*, the process of heating a thermoplastic material and shaping it in a

mould; so **the·rmoform** *v. trans.*; **the·rmo-former,** a person who carries out thermo-forming; **thermoha·line** *a. Oceanogr.* [Gr. ἅλς, ἁλι- salt + -INE[1]], of or pertaining to the temperature and salinity of seawater; **thermo-ha·locline** *Oceanogr.* [after *thermocline*], a narrow layer of water separating layers of differing temperature and salinity; **thermo-ha·rden** *v. trans.*, to harden permanently by subjection to heat; **thermo-ha·rdening** *ppl. a.*, rendered permanently hard by heat; **the·rmo-karst** [a. Russ. *termokárst* (M. M. Ermolaev 1932, in *Trudy̆ Soveta po Izuch. proizv. Sil: Ser. yakutsk.* 211)], topography in which the melting of permafrost has produced hollows, hummocks, and the like reminiscent of karst; **thermoly·sin** *Biochem.* [Gr. λύσις a parting], a heat-stable proteolytic enzyme found in some thermophilic bacteria; **thermo-magnetic** *a.* (later example); **thermo-magnetism** (example); **thermona·sty** *Bot.* [*NASTY *sb.*[2]], a nastic movement caused by a change in temperature; so **thermona·stic** *a.*; **The·rmopane** *N. Amer.*, a proprietary name for an openable double-glazed window unit; **thermophy·sics,** the branch of physics dealing with the physical properties of substances at high temperatures; so **thermophy·sical** *a.*; **the·rmo-power** *Electr.* [f. *thermo(electric power)*], the thermoelectric e.m.f. developed by a substance per degree difference in temperature; **thermore·manent** *a.*, pertaining to or being magnetism acquired, esp. by rock, as a result of cooling or solidifying in a magnetic field; so **thermore·manence**; **thermose·nsitive** *a.*, possessing or relating to sensitivity to heat; so **the:rmosensiti·vity**; **† thermo-tank,** a tank containing pipes through which water, air, or the like circulates for heating or cooling, esp. as a heating or ventilating system (*obs.*); **thermoto·lerant** *a.* (see quot. 1940).

1955 *Sci. News Let.* 2 Apr. 217/1 Investigations off the coast of California showed that skin divers can spot thermoclines, the layers of water which mark the sharp change in water temperature, in three different ways. **1973** *Sci. Amer.* Feb. 42/3 The coastal waters of Nova Scotia are characterized by a marked thermocline (a sharp drop in temperature as the depth increases), so that a free-swimming fish might encounter a wide range of water temperature. **1973** P. A. COLINVAUX *Introd. Ecol.* xxxiii. 470 The animals have to feed in the warm surface waters...but they go down to cold water below the thermocline in daytime. **1938** *Yale Jrnl. Biol. & Med.* X. 575 The thermocoagulated layers are completely 'resorbed' within four months. **1933** J. G. DUSSER DE BARENNE in *Science* 2 June 547/1 This method of laminar thermocoagulation of the cerebral cortex, as it might be called, results...in a sharply localized, selective destruction of the nervous elements. **1974** *Nature* 4 Jan. 58/2 The destruction of the area postrema was perfomed by sight through the occipital foramen by thermocoagulation. **1976** *Ibid.* 22 Apr. 660/2 The thermocoagulative lesions were aimed at various limbic tracts, and each estimated at 6 mm in diameter. **1965** *Wireless World* July 337/2 Typically this is done by thermo-compression bonding of extremely thin gold or aluminium wires to the electrodes and terminal posts. **1972** *Physics Bull.* Mar. 154/1 Circuits can now be made with...active components subsequently soldered or thermocompression bonded to the microstrip. **1979** A. L. LYDERSEN *Fluid Flow & Heat Transfer* xi. 323 The waste heat is often available at a temperature which is too low for direct use in the process. However, it may be utilized in conjunction with thermocompression. **1927** *Techn. Bull. N.Y. State Agric. Exper. Station* No. 130. 6 Thermoduric spore-forming bacteria are common types in pasteurized, sterilized, and boiled milk. **1946** *Nature* 23 Nov. 755/1 Working with suspensions of *Staph. aureus*, thermoduric micrococci, and spores of *B. subtilis*, we found that...solutions [of hypochlorite] of low *p*H were more germicidal than at higher *p*H. **1975** CAMPBELL & MARSHALL *Sci. of providing Milk for Man* xxiii. 501 Bacteria that survive specific heat treatments are usually said to be thermoduric (heat-tolerant). **1958** *Times Rev. Industry* Aug. 57/2 Machine...for thermoforming... industrial parts. **1972** *Sci. Amer.* Aug. 9 (Advt.), By helping the thermoformer with mold modifications and adjustments to equipment and operating conditions which enabled him to produce high quality parts economically. **1978** *Detroit Free Press* 5 Mar. c-14/7 (Advt.), Machinist assembler, parts and stock man, and a thermo-former. **1963** SIMONDS & CHURCH *Conc. Guide to Plastics* (ed. 2) vii. 182 There are seven basic techniques for the thermoforming of plastics sheet. **1972** *Sci. Amer.* Aug. 9 (Advt.), By supplying the sheet extruder with a high molecular weight Marlex thermoforming resin ideally suited for the production of large thick sheet. **1978** *N.Y. Times* 30 Mar. B-19/1 (Advt.), We are seeking manager for our model-making dept. in thermoforming. **1942** H. U. SVERDRUP et al. *Oceans* xiii. 509 When examining the circulation [of the waters of the oceans] that arises because of the external factors influencing the density of the surface waters, one must take changes of both temperature and salinity into account, and must consider not the thermal but the thermohaline circulation. **1963** G. L. PICKARD *Descriptive Physical Oceanogr.* vii. 107 The ocean circulation can be divided into two parts, the thermohaline and the wind-driven components. **1978** *Nature* 13 July 151/1 The observed distribution patterns of late Quaternary sapropels favour the hypothesis of periodically altered basin-wide thermohaline circulation entrained by regionally important climatic and eustatic changes. **1964** *Oceanogr. & Marine Biol.* II. 135 When a basin is permanently stagnant, the redox discontinuity may rise to the level of the thermo-halocline, as is well known in the Black Sea. **1976** *Nature* 2 Sept. 23/1 Mechanism and rate of molecular exchange across a well developed thermo-halocline have been studied thoroughly. **1949** R. J. W. REYNOLDS in J. M. Preston *Fibre Science* xvii. 318 The final products may be thermo-hardened by a suitable cross-linking treatment. **1933** *Archit. Rev.* LXXIII. 266/1 The elaborate laboratory researches into the nature of thermo-plastic and thermo-hardening materials. **1961** J. N. ANDERSON *Appl. Dental Materials* (ed. 2) xxi. 220 The Bakelite type of resin is called thermohardening or thermoset as...heat is applied to cure the resin. **1943** S. W. MULLER *Permafrost or Permanently Frozen Ground* 84 Phenomena of thermokarst. a. Cave-in lakes. b. Settling lakes. c. Cave-in and settling funnels. **1970** *Globe Mag.* 17 Jan. 4/3 Even south of the Alaska Range there is much permafrost within the forested areas which will create further problems of heat loss, permafrost melt and thermokarst development. **1965** H. MATSUBARA et al. in *Biochem. & Biophysical Res. Communications* XXI. 242 A proteolytic enzyme with the commercial name 'Thermoase' was isolated by Endo...from cultures of *Bac. thermoproteolyticus* Rokko... It was recently reported that the enzyme had a strong elastase-like activity... We propose the trivial name *thermolysin* for this enzyme. **1979** *Nature* 29 Feb. 667/1 The determination of the three-dimensional structure of the thermostable protease thermolysin showed that heat-stable proteins do not contain unusual structural features absent from less stable proteins. **1954** *Jrnl. Geomagnetism & Geoelectricity* VI. 6 This simple apparatus could be used for the study of the thermomagnetic analysis of ferromagnetic mineral with a fair accuracy. **1828** F. WATKINS *Electro-Magnetism* 22 Experiments in thermo-magnetism teach us that magnetical phænomena will arise from a disturbance in the equilibrium of temperature of metals. **1936** L. J. F. BRIMBLE *Intermediate Bot.* xx. 294 Examples of thermonastic movements are seen in the flowers of the crocus and tulip. **1936** J. B. HILL et al. *Bot.* ix. 228 The rapid opening of certain flowers when brought into a warm room from a cold place is a thermonasty. **1951** Thermonasty [see *NASTIC *a.*]. **1976** BELL & COOMBE tr. *Strasburger's Textbk. Bot.* (rev. ed.) 365 Many flowers... open or close according to the temperature. Such a phenomenon is referred to as thermonasty. *Ibid.*, Repeated thermonastic curvatures may cause an increase in length. **1941** *Official Gaz.* (U.S. Patent Office) 14/2 *Thermopane*. For multiple glass sheet glazing units... Claims use since May 1, 1931. **1968** *Globe & Mail* (Toronto) 13 Jan. 42/8 (Advt.), Large brick and cut stone bungalow...2 fireplaces, drapes, thermopane windows etc. **1974** *Whig-Standard* (Kingston, Ontario) 9 Feb. 15/1 Modernisation of the whole interior of the present building, including modern heating, thermopane, air-conditioning and elevator service. **1978** *N.Y. Times* 29 Mar. B8/1 (Advt.), Builder's custom built 72' hi ranch, 2 acs, circ driveway, thermopane windows. **1957** *Ibid.* 25 Aug. IV. 9/3 Purdue University has established a thermophysical properties research center. **1976** *Physics Bull.* Dec. 561/3 The symposium is concerned with both theoretical and experimental aspects of thermophysical properties of all matter in solid, liquid, gaseous and plasma states. **1962** A. L. KING (*title*) Thermophysics. **1966** G. B. HELLER *Thermophysics & Temperature Control of Spacecraft* p. xi, The modern field of thermophysics rests on some of the oldest branches of physics, namely, thermodynamics, heat transfer, and electromagnetic radiation. **1963** *Canad. Jrnl. Physics* XLI. 1080 The thermopower of the special copper was measured carefully at the low-temperature end. **1976** *Physics Bull.* June 248/2 In figure 4 are plotted the conductivity and thermopower of the magnesium-bismuth alloy as a function of composition. **1938** J. G. KOENIGSBERGER in *Terrestr. Magnetism & Atmospheric Electr.* XLIII. 120 The full apparent remanence acquired by cooling in a given field from *T*ₑ...may be denoted the thermoremanence. **1967** *Nature* 28 Oct. 359/2 The mean directions [of magnetization]...are...fairly well grouped, and are believed to represent the direction of thermoremanence acquired when the rocks cooled. **1951** *Proc. Jap. Acad.* XXVII. 643 The remanent magnetism thus produced has been called the thermo-remanent magnetism. **1958** *Antiquity* XXXII. 124 Measurement of the thermoremanent magnetism in the clay. **1971** *Physics Bull.* Aug. 476/3 Half the papers...were concerned with the analysis of contact printing processes, both the anhysteretic transfer method with $\gamma\ Fe_2O_3$ slave tapes and the thermoremanent method with chromium dioxide tapes. **1975** *Nature* 27 Feb. 701/2 Stable remanent magnetisation discovered in lava samples collected during the Apollo 11 mission has been interpreted as thermoremanent magnetisation acquired when the lava flows cooled through the Curie point 3·6 Gyr ago. **1918** *Jrnl. Exper. Zool.* XXV. 279 The animal is thermosensitive. **1952** *Archit. Rev.* CXI. 278/3 Suitable safety devices operated by a thermosensitive bi-metal strip are fitted. **1978** *Nature* 2 Feb. 470/1 Our results indicate that temperature control of reproduction in an ectothermic thermosensitive species may also be mediated in part by circadian systems. **1918** *Jrnl. Exper. Zool.* XXV. 281 This method... demonstrated the thermosensitivity of the species, because the animals gave an ejection reflex when brought into a region of higher temperature. **1981** *Pflügers Archiv: European Jrnl. Physiol.* CCCXCI. 66/2 It turns out that in the goose a minor fraction only of total body thermosensitivity can be attributed to the spinal cord. **1909** WEBSTER, Thermo-tank. **1920** *Lancet* 25 Sept. 666/2 Eight thermo-tanks. **1928** *Observer* 15 July 9/4 A new Thermo-Tank heating system. **1940** *Chambers's Techn. Dict.* 846/1 *Thermotolerant*, able to endure high temperatures, but not growing well under such conditions. **1964** COONEY & EMERSON *Thermophilic Fungi* 161 [Fungi] which may grow at or near 50°C but which also grow well at temperatures below 20°C, are considered thermotolerant and are excluded from the true thermophilic fungi. **1973** *Nature* 16 Mar. 203/2 Many species of thermophilic and thermotolerant fungi isolated from natural thermal habitats similarly occur in man-made heated habitats.

thermochromism (pɜɹmokrō̆ᵘ-miz'm). [ad. G. *thermochromie* (H. Stobbe 1904, in *Ber. d. Deut. Chem. Ges.* XXXVII. 2239), f. Gr. χρῶμα colour: see THERMO- and -ISM.] The phenomenon whereby certain substances undergo a reversible change of colour or shade when heated or cooled. Also **the·rmochromy,** in the same sense.

1911 *Chem. Abstr.* V. 2087 Characteristics of 'thermochromy'. **1914** *Ibid.* VIII. 2387 The corresponding salts of the thiourethans...are colorless and do not exhibit thermochromism. **1960** *New Scientist* 2 June 1424/1 In all cases where the substance was both thermochromic and photochromic the colour formed either by heat (thermochromism) or by ultra-violet irradiation at low temperature (photochromism) was spectroscopically identical. **1963** [see *PHOTOCHROMY c]. **1965** *New Scientist* 14 Jan. 102/1 Thermochromy...may well be a quite general property of solids containing trivalent chromium ions. **1974** *Inorg. Chem.* XIII. 2512/2 The thermochromism of these compounds involves a gradual change in color from gold to yellow to light green as the temperatures are lowered from 100° down to liquid nitrogen temperature.

So **thermochro·mic** *a.*, of, pertaining to, or displaying thermochromism.

1904 *Jrnl. Chem. Soc.* LXXXVI. I. 672 (*heading*) Thermochromic properties of dibenzylidensuccinic anhydride. **1953** [see *PHOTOCHROMIC *a.*]. **1965** *New Scientist* 14 Jan. 102/1 Heating has the same effect and the higher the chromium content, the lower the temperature required for the 'thermochromic' transition. **1974** *Inorg. Chem.* XIII. 2106/1 This salt is thermochromic: green at 25° and yellow at 80°.

thermocline to **thermocompression:** see *THERMO-.

the·rmocouple. Formerly also **thermo-couple** (with hyphen). [f. THERMO- + COUPLE *sb.*] A thermoelectric device for measuring temperature, consisting of two different metals joined at a point so that the junction develops a voltage dependent on the amount by which its temperature differs from that of the other end of each metal.

1890, 1901 [in Dict. s.v. THERMO-]. **1934** *Jrnl. R. Aeronaut. Soc.* XXXVIII. 618 Twenty-two thermocouples were installed on the heads and bases of all rear-bank cylinders...for temperature tests. **1953** *Brit. Jrnl. Psychol.* XLIV. 41 A thermocouple was used [for recording respiration] It converted temperature variations in front of the nostrils into electrical variations. **1966** *McGraw-Hill Encycl. Sci. & Technol.* XIII. 562/2 Two dissimilar wires welded together at one end form the basic thermocouple. **1977** J. L. HARPER *Population Biol. Plants* v. 145 Emerged seedlings were recorded and marked every week. Surface temperatures were measured with thermocouples.

thermode (pɜ·ɹmōᵘd). [f. THERMO-, after *electrode*.] An object that is introduced into a medium, esp. living tissue, as a means by which heat may enter or leave it.

1938 *Yale Jrnl. Biol. & Med.* X. 573 A simple coagulator was used, the one inconvenience of which was that only the heating surface, the 'thermode' proper, could be sterilized. **1951** *Jrnl. Neurophysiol.* XIV. 424 A metal thermode...was applied on the tongue and kept there at constant pressure during the total experiment. **1967** *New Scientist* 16 Mar. 553/1 About 10 cu. cm of clean mercury is placed in a Perspex boat with mild steel 'thermodes' at either end, one electrically heated, the other water-cooled. **1975** *Nature* 1 May 72/1 Unilateral water-perfused thermodes with thermistors fixed to the tips were placed stereotaxically into the POA [*sc.* preoptic area]. **1978** *Sci. Amer.* Aug. 91/1 In classic experiments conducted by Henry G. Barbour in 1912 silver thermodes were implanted in the hypothalamus.

thermoduric: see *THERMO-.

thermoelectrically, *adv.* (Examples.)

1895 *Electrician* 13 Sept. 637/1 He also considered the possibility of the back E.M.F. being produced thermo-electrically. **1979** *Nature* 11 Oct. 498/1 (*caption*) Photo-multiplier in...thermoelectrically cooled housing.

Thermo-Fax (pɜ·ɹmofæks). Also **Thermofax.** [f. THERMO- + FACS(IMILE *sb.*] The proprietary name of a process for copying documents by means of infra-red radiation, and of a type of overhead projector employing copies made by this process.

1953 *Official Gaz.* (U.S. Patent Office) 17 Mar. 574/1 *Thermo-Fax*. For electrically operated machine employing infrared light source for producing copies of printed or pictorial matter by means of heat-sensitive paper. Claims use since November 1949. **1956** *Trade Marks Jrnl.* 18 Apr. 236/2 *Thermo-fax*... Reproducing (copying) apparatus for office use. Minnesota Mining and Manufacturing Company..., manufacturers. **1962** A. GÜNTHER *Microphotogr. in Library* (Unesco) 16 Some microfilm readers have special accessories for the occasional production of enlarged prints. Either a dry process such as 'Thermofax' is used or a semi-dry process such as diffusion printing (e.g. 'Copy-rapid') or stabilization techniques. **1964** *Times* 7 Feb. (Advt. Suppl.) p. ii/5 This is the situation the Thermo-Fax overhead projector is designed to overcome... It uses large transparencies which are inexpensive and simple to make—a Thermo-Fax infra-red copying machine takes just 4 seconds to produce one.

thermoform, -er, -ing: see *THERMO-.

Thermogene (pɜ·ɹmodʒīn). Also **thermo-, † -gène.** [ad. F. *thermogène* THERMOGENIC *a.*] A proprietary name for medicated cotton wool.

1902 *Official Gaz.* (U.S. Patent Office) 18 Mar. 2357/2 Absorbent wadding. Vandenbroeck & Cie., Brussels, Belgium... *Thermogène*. **1905** *Trade Marks Jrnl.* 22 Nov. 1449 Thermogène absorbent wadding... Medicated wadding for human use. Thomas Other Windsor, trading as the Thermogène Co., Invermay, Lucastes Avenue, Hayward's Heath, Sussex; manufacturer. **1907** *Yesterday's shopping* (1969) 520/1 Thermogene...pkt. 1/o. **1928** A. HUXLEY *Point Counter Point* xviii. 327 What you need...is a good rubbing with camphorated oil and a wad of Thermogene. **1939** M. SPRING RICE *Working-Class Wives* iv. 75 The doctor told her to keep the parts warm, so she used thermogene wool. **1958** W. SANSOM *Cautious Heart* 157 Pinkish brown clouds flew across the cold iron sky like tufts of thermogene loose in the night. **1962** C. WATSON *Hopjoy was Here* xv. 173 I'd been downstairs for some Thermogene.

thermogram. Add: **2.** A photograph or image produced by infra-red radiation emanating naturally from the subject under study.

1957 *Canad. Services Med. Jrnl.* 523 (*caption*) Thermogram lower left showed an area of increased heat. **1964** *Amer. Jrnl. Roentgenol.* XCI. 919/2 The normal breast in the thermogram can be recognized by its size, shape and the heat pattern of the overlying skin. **1967** *Idle Moments* (Austral.) Oct. 5/2 Thermograms...are taken in total darkness, since they are photographic reproductions of infra-red radiations of longer wavelength emitted by the object itself. **1968** *New Scientist* 1 Feb. 263/3 A pair of 'thermograms' of the crater Tycho.

thermograph. Add: **4. a.** = *THERMOGRAM 2. **b.** An apparatus for obtaining thermograms.

1964 *New Scientist* 16 July 163/1 Personal thermographs can sometimes with practice be recognised individually. **1964** *Amer. Jrnl. Roentgenol.* XCI. 919/2 This thermograph has been in daily use in this institution for over a year and has quite recently been replaced by a new model. **1970** *New Scientist* 5 Feb. 260/3 Thermographs of a person's finger before and after smoking show significant changes in the heat pattern.

thermographic, *a.* Add: (Examples corresponding to sense *2 of THERMOGRAPHY.)

1964 *Amer. Jrnl. Roentgenol.* XCI. 925/2 Their experience with thermographic scanning using electrochemical paper was limited, but suggested a promising future. **1975** J. TAYLOR *Superminds* vii. 116 The general features of the aura seem very comparable to pictures obtained by the thermographic camera.

Hence **thermogra·phically** *adv.*, in a thermographic manner; by means of thermography.

1840 *Phil. Trans. R. Soc.* CXXX. 59 The focal image...had acquired the power of imprinting itself thermographically on the paper. **1964** *New Scientist* 16 July 163/2 It now seems clear that most cancers of the breast raise the skin temperature and can be demonstrated thermographically.

thermography. Add: **2.** The taking or use of infra-red thermograms, esp. to detect tumours.

1957 R. LAWSON in *Canad. Services Med. Jrnl.* XIII. 519 It was apparent that 'thermography' or heat imaging by suitable equipment might have a very important place in the early diagnosis of breast lesions. **1963** *Science* 24 May 873/2 The human body is an ideal subject for thermography. **1969** *New Scientist* 8 May 276/1 Thermography is a completely harmless method in which the patient is 'photographed' by her own body heat. **1971** *Daily Tel.* (Colour Suppl.) 22 Jan. 22/2 Aerial thermography has been used to locate old mineshafts, and coal seams burning underground. **1977** *Time* 20 June 48/1 Thermography, or heat scanning, concentrates on looking for infra-red radiation to find tumors.

thermogravimetry (pɜɹmogrǎvi·mĕtri). *Physical Chem.* [f. THERMO- + GRAVIMETRY.] The technique of chemically analysing substances by measuring changes in weight as a function of increasing temperature.

1951 *Chem. Abstr.* XLV. 2274 (*heading*) Thermogravimetry and automatic gravimetry. **1953** *Nature* 22 Aug. 365/1 The advantage of the differential thermogravimetry over differential thermal analysis is that it is quantitative. **1975** H. L. FRIEDMAN in I. M. Kolthoff et al. *Treat. Analytical Chem.* III. III. D-1. 401 In thermogravimetry (TG), one generally records the weight of a sample continually as it is heated through a preselected rate of temperature rise.

Hence **the:rmogravime·tric** *a.*

1953 C. DUVAL *Inorg. Thermogravimetric Analysis* i. 3 The investigations of Honda..., and the construction by Chevenard of automatic recording instruments, have all combined to give birth to a new science which we may perhaps call thermogravimetric analysis. **1972** *Nature* 15 Dec. 418/1 According to our thermogravimetric analyses compounds (III)–(V) contain one molecule of lattice water per formula unit. **1977** *Proc. R. Soc. Med.* LXX. 518/2 If the stone has been passed or removed it should be analysed. This may be by wet chemistry...or by physical methods such as infra-red spectroscopy...X-ray crystallography...or thermogravimetric analysis.

thermohaline to **thermokarst:** see *THERMO-.

thermology. (Earlier example.) **thermological** *a.* (earlier example).

1838 tr. A. Comte in *Edin. Rev.* July 282 The most important and precise laws of thermological phenomena are developed without the slightest enquiry into the intimate nature of heat. *Ibid.* 284 It remained only [for Comte]...to tack to Hydrodynamics the sciences of Magnetism, Electricity, Galvanism and Thermology.

thermolumine·scence. (In Dict. s.v. THERMO-.) Add: (Further examples.) Also, *spec.* used as a means of dating ancient pottery and other material.

1967 *New Scientist* 26 Oct. 206/3 Proposals for thermoluminescence on the Moon are not new. **1968** *New Scientist* 21 Mar. 644/1 Methods of using the natural thermoluminescence of minerals in fired ceramics to find out when they were made have been investigated... for several years. **1968** *Times* 14 Oct. 8/1 Archaeological fakes, some so convincing that they have deceived experts, are being exposed by a new scientific technique developed for dating ancient pottery. The results of one test, known as thermoluminescence dating, are to be used in a court case in America. **1977** G. CLARK *World Prehistory* (ed. 3) p. xx, The degree of thermoluminescence given out by a sample of pottery or stone under heat is proportional to the amount of radiation accumulated since the sample was last fired.

thermolumine·scent, *a.* (In Dict. s.v. THERMO-.) Add: (Further examples.) Also used *spec.* in sense of prec.; abbrev. TL s.v. *T 6 a.

1962 *Oxf. Univ. Gaz.* 9 Mar. 788/2 *Thermoluminescent Dating*. The principle of this technique is that radiation damage accumulates in all clay due to the natural radioactivity of uranium and thorium impurities [etc.]. **1968** *New Scientist* 21 Mar. 644/2 The initial thermoluminescent measurements are capable of good accuracy and reproducibility. **1970** *Daily Tel.* 25 Apr. 11/6 The thermoluminescent method is based on the fact that many minerals when heated to temperatures around 500°C. emit light, additional to the ordinary red-hot glow. **1971** *Physics Bull.* Oct. 579/1 Nearly all acidic rocks and sedimentary carbonates are thermoluminescent.

thermolysin, thermo-magnetic, -ism: see THERMO- in Dict. and Suppl.

thermomecha·nical, *a.* [f. THERMO- + MECHANICAL *a.* and *sb.*] **a.** *Physics.* Designating or referring to an effect observed in helium II in which the liquid tends to flow from a region of lower to one of higher temperature.

1939 H. LONDON in *Proc. R. Soc.* A. CLXXI. 484 By maintaining a temperature gradient along the capillary, it is thus possible to produce a flow of helium against a pressure gradient... The phenomenon which we shall call the 'thermomechanical effect' reveals a new mechanism by which heat can be transformed into mechanical work. **1959** K. R. ATKINS *Liquid Helium* i. 11 The thermomechanical pressure difference Δp arising from a temperature difference ΔT is given by $\Delta p/\Delta T = Q^*/TV$, where Q^* is the heat transfer associated with the transfer of unit mass...and V is the volume of unit mass. **1964** [see *mechanocaloric* adj. s.v. *MECHANO-]. **1974** D. R. & J. TILLEY *Superfluidity & Superconductivity* i. 9 These manifestations of the thermomechanical effect show clearly that heat transfer and mass transfer in the II are inseparable.

b. *gen.* Both thermal and mechanical; *spec.* in *Metallurgy*, involving simultaneous thermal and mechanical treatment to achieve results not obtained when they are applied separately.

1974 *Sci. Amer.* June 34/3 The less energetic ions (thermal deuterons and tritons)...are capable of producing sharp thermomechanical stresses in a thin skin of the first wall. **1975** *Nature* 7 Aug. 455/3 The uncertainty in geochemical characteristics, especially of the radiogenic isotopes, leaves many degrees of freedom in predicting the present thermo-mechanical state of the interior [of Mercury]. **1976** *Ibid.* 22 July p. iii (Advt.), The journal will welcome papers concerned with the relevant areas of materials technology and metallurgy, e.g. thermomechanical treatments.

thermonastic, -nasty: see *THERMO-.

thermoneu·tral, *a.* [f. THERMO- + NEUTRAL *a.* and *sb.*] **1.** *Biol.* Of an environment or its temperature: such that an organism is in thermal equilibrium without thermoregulation.

1961 in WEBSTER. **1966** *Respiration Physiol.* I. 30 The thermoneutral skin temperature zone for fasting adult sheep has been found to be 33–35°C as determined by immersion in a water bath. **1976** *Nature* 13 May 134/1 Neonates and infants were tested using a tight-fitting face mask with minimal dead space while the subjects were asleep in a thermoneutral condition. **1977** *Lancet* 7 May 988/1 The incubator should if possible be kept at the lower end of the thermoneutral range.

2. *Chem.* Of a reaction: accompanied by neither the absorption nor the emission of heat.

1970 *Nature* 12 Sept. 1097/1 Because the translational energies of the reactant ions are approximately thermal, these two reactions must be thermoneutral or exothermic. **1971** *Sci. Amer.* Dec. 57/3 The first and second reactions are strongly endothermic; the third is exothermic; the fourth is essentially thermoneutral. **1977** I. M. CAMPBELL *Energy & Atmosphere* v. 106 It appears probable that this depends upon the attack of radicals like CH upon N_2 in a near thermoneutral elementary reaction.

So **the:rmoneutra·lity**, the condition of being thermoneutral.

1881 [in Dict. s.v. THERMO-]. **1960** K. SCHMIDT-NIELSEN *Animal Physiol.* iii. 43 Thus, man has a narrow range of thermoneutrality between 27°C and 31°C. **1979** *Nature* 24 May 322/1 The abnormal thermoregulatory thermogenesis quantitatively accounts for most of the metabolic efficiency of the obese animals as pair feeding at thermoneutrality rather than at 23°C reduces the excess fat deposited by 65%.

thermonuclear (pɜɹmoniū·kliǎɹ). Also **thermo-nuclear.** [f. THERMO- + NUCLEAR *a.*]

a. Derived from, utilizing, or being a nuclear reaction that occurs only at very high temperatures (such as those inside stars), viz. fusion of hydrogen or other light nuclei.

[**1937** G. GAMOW *Struct. Atomic Nuclei* 232 The first calculations concerning thermal nuclear reactions were carried out by Atkinson and Houtermans] **1938** *Physical Rev.* LIII. 595/1 The behavior of a star with a thermonuclear energy source...is studied. **1942** B. BLIVEN *Men who make Future* xi. 202 This perhaps represents a struggle between two almost unbelievably powerful forces within each star—what is called the 'thermo-nuclear reaction' and gravitation. **1954** *Ann. Reg.* 1953 377 *Pravda* described the results of a 'thermo-nuclear' explosion. **1958** *New Statesman* 1 Feb. 123/2 Thermonuclear energy depends on building up atoms of hydrogen into atoms of helium. **1962** F. I. ORDWAY et al. *Basic Astronautics* vi. 284 A new star not yet hot enough to initiate thermonuclear reactions obtains its luminosity from gravitational contraction. **1964** M. GOWING *Britain & Atomic Energy 1939–1945* ix. 260 Thermonuclear fusion of light elements would provide infinitely more powerful reactions than fission of heavy elements. **1969** *Times* 20 Feb. 17/5 It implied that their calculations of the sun's thermonuclear fuel budget were considerably in error. **1976** *McGraw-Hill Yearbk. Sci. & Technol.* 342/1 In controlled thermonuclear reactors the nuclear reaction $D + T \rightarrow {}^4He + n + 17{\cdot}6$ MeV occurs in a plasma at temperatures of 100–500 × 10⁶ kelvins.

b. Pertaining to, characterized by, or possessing weapons that utilize thermonuclear reactions.

1953 *Time* 19 Oct. 25/3 Secretary of Defense Wilson, at his press conference, cast doubt on a suggestion that the Russians had a thermonuclear bomb 'in droppable form'. **1955** *Ann. Reg.* 1954 169 Mr. Adlai Stevenson... inquired whether the 'New Look' meant leaving the country with 'the choice of inaction or a thermo-nuclear holocaust'. **1955** *Times* 13 July 9/5 They draw attention to the possibility that a thermo-nuclear war might put an end to the human race. **1958** *Listener* 7 Aug. 207/3 So long as Britain makes her own nuclear weapons, and particularly the thermo-nuclear weapon, there is no possibility of dissuading France or Germany or Sweden from developing theirs. **1959** *Times Lit. Suppl.* 16 Jan. 27/3 In the thermo-nuclear age, if civilization is not to disintegrate..., the premises of Gandhi have an immediate relevance. **1959** *Daily Tel.* 4 Mar. 8 With the thermo-nuclear cloud overshadowing the world no statesman, not even a Soviet Prime Minister, can afford to disregard mankind's longing for peace. **1965** H. KAHN *On Escalation* ii. 42 In a thermonuclear balance of terror, both nations will be reluctant to start a crisis. **1966** J. W. BURTON in de Reuck & Knight *Conflict in Society* xxiii. 380 In the relations of the two thermo-nuclear States, tensions arise from unacceptable enactments of their respective roles. **1972** M. H. HALPERIN *Contemporary Mil. Strategy* i. 7 One has only to recall Mr. Khruschev's statements during the Cuban missile crisis about the world being close to thermonuclear war.

Thermopane: see *THERMO-.

thermophil, -phile, *sb.* Add: (Examples.) **thermophilic** *a.* (earlier and further examples); **thermophilous** *a.* (further examples).

1894 MACFADYEN & BLAXALL in *Jrnl. Path. & Bacteriol.* III. 88 To those organisms that grow best at very high temperatures we have applied the name of thermophilic bacteria. **1909** H. W. CONN *Agric. Bacteriol.* (ed. 2) i. 16 A few species...grow best at unexpectedly high temperatures, some having been found flourishing at 140° or even higher. These peculiar bacteria are called thermophiles. **1964** COONEY & EMERSON *Thermophilic Fungi* i. 6 To the algologist a thermophile may have a maximum between 60° and 80°C...to the bacteriologist...55° to 80°C, and the acarologist 35° to 45°C. **1965** BELL & COOMBE tr. *Strasburger's Textbk. Bot.* III. 774 The wood and fruits of thermophilous trees and shrubs have been repeatedly found some hundreds of metres above their present altitudinal limits. **1975** J. G. EVANS *Environment Early Man Brit. Isles* ii. 49 The Allerød Interstadial...was the first zone to see the appearance of thermophilous land snails. **1977** *Time* 26 Dec. 20/3 Their droppings will be placed into fermenter tanks filled with thermophilic (heat-loving) bacteria. **1981** *New Scientist* 10 Sept. 667/2 The plasmid may be of value...for transferring genes into thermophiles (bacteria that like high temperatures).

thermophysical, -physics: see *THERMO-.

thermoplastic, *a.* Delete entry s.v. THERMO- in Dict. and substitute:

thermoplastic (pɜɹmoplæ·stik), *a.* and *sb.* [f. THERMO- + PLASTIC *a.*] **A.** *adj.* Becoming soft when heated and rigid when allowed to cool, and capable of being repeatedly reheated and reshaped without loss of properties; made of such a substance.

1883 [in Dict. s.v. THERMO-]. **1909** *Chem. Abstr.* III. 2063 Thermo-plastic composition containing keratin. Keratin is mixed with β-naphthol and the compound, subjected to heat and pressure, to form a subst. for rubber, celluloid, etc. **1937** *Jrnl. R. Aeronaut. Soc.* XLI. 525 Some kinds of synthetic resins and to a limited extent one natural resin (shellac) have the valuable characteristic of being 'thermosetting', i.e., when once moulded they set to permanently infusible products. Most resins are 'thermoplastic', i.e., they become soft whenever the temperature exceeds a certain value. **1951** [see *benzyl* s.v. *BENZO-]. **1958** *Listener* 4 Dec. 967/1 Thermoplastic tiles and rubber floors. **1976** J. FLEMING *To make an Underworld* xi. 128 These small thermo-plastic boats..were safer than houses, kids could play with them.

B. *sb.* A thermoplastic substance.

1929 *Brit. Plastics & Moulded Products Trader* June 23/1 Thermoplastics are divided into pheno-plastics..and amino-plastics. **1945** *Electronic Engin.* XVII. 516 It is frequently used in association with polythene and other thermoplastics. **1967** *Times Rev. Industry* May 75/3 Other thermoplastics, including polyvinyl chloride (PVC), polypropylene and polyvinylidene chloride comprise about 50 per cent. of the total weight of plastics used in packaging. **1976** *Shooting Mag.* Dec. 23/2 (Advt.), Lightweight multi-type ear protector... Earcups are manufactured from thermoplastic with soft sponge-filled ear cushions.

Hence **the:rmoplasti·city**, the quality of being thermoplastic.

1935 C. ELLIS *Chem. Synthetic Resins* II. xvi. 1150 The thermoplasticity of the resins is a disadvantage. **1962** J. T. MARSH *Self-Smoothing Fabrics* x. 141 Some of these mixed products can be used to reduce the solubility and swelling of the linear polymer but the thermoplasticity will be reduced. **1980** *Brit. Plastics & Design* LVII. 323 (*heading*) Coupling phenomena in thermoplasticity.

thermopower: see *THERMO-.

Thermopylae (þəːmɒ·pɪliː, -pɪlaɪ). The name of a narrow pass on the north-east coast of Greece between Thessaly and Locris, the scene of a battle in 480 B.C. in which a small Greek force temporarily withheld a Persian invasion; used *transf.* and *fig.* with reference to heroic resistance against strong opposition.

1928 A. HUXLEY *Point Counter Point* xxii. 471 E. talked a lot about Thermopylae and the Spartans. But my resistance was even more heroic. Leonidas had three hundred companions. I defended my spiritual Thermopylae single-handed against E. and his Fremen. **1929** J. BUCHAN *Courts of Morning* III. iii. 335 I'm going to try the Thermopylae stunt... Our Thermopylae is going to be a more cunning affair than the old one. **1955** R. Megarry *Second Miscellany-at-Law* (1973) 157 It courageously held the line of reason at the Thermopylae of logic and did not give way at the Gettysburg of fact. **1967** C. SETON-WATSON *Italy from Liberalism to Fascism* iii. 125 A column of 500 Italian troops had been wiped out by several thousand Ethiopians at Dogali, after fighting almost to the last man and the last round. This 'Italian Thermopylae' caught the public imagination. **1974** V. G. KIERNAN *Lords of Human Kind* ... The consequence has been Vietnam's Thermopylae of twenty years.

thermoreceptor (þɜ̄·məʊrɪseptəɹ). *Physiol.* [f. THERMO- + RECEPTOR.] A nerve ending that is sensitive to stimulation by heat and cold.

1937 L. V. HEILBRUNN *Outl. Gen. Physiol.* xli. 506 Animals in general are sensitive to heat and cold but, except in higher animals, thermoreceptors are rare, or at any rate not well known. **1951** *Jrnl. Neurophysiol.* XIV. 423 Knowledge of the intracutaneous depth of the thermoreceptors is of very great importance. **1960** *Lancet* 3 Sept. 610/1 Professor Hensel will lecture on the electrophysiology of thermoreceptors. **1971** D. J. AIDLEY *Physiol. of Excitable Cells* xv. 307 The sense organs themselves can be classified according to the type of stimulus which normally excites them. Thus mechanoreceptors are excited by mechanical stimuli, photoreceptors are sensitive to light, thermoreceptors are temperature sensitive.

thermoregulation (þɜ̄məʊregjʊlēɪ·ʃən). [f. THERMO- + REGULATION.] Regulation of temperature, esp. body temperature in an animal or human.

1927 *Jrnl. Exper. Zool.* XLVII. 156 We deal in nest-building with a behavior pattern definitely adapted to the thermoregulation of the organism. **1932** *Jrnl. Gen. Physiol.* XVI. 9 The machine described above is kept in a thermoregulated room. The thermoregulation of a room for such purposes has been regarded as a difficult and expensive undertaking. **1962** *Lancet* 8 Dec. 1207/1 Most of the research into human problems of thermoregulation in hot surroundings has been sponsored by the Government and the Services. **1972** *Sci. Amer.* June 74/2 Understanding of the physiology of the sphinx moth's thermoregulation requires..an examination of the levels of body temperature maintained during flight.

So **thermore·gulate** *v. intr.*, to regulate temperature, esp. body temperature; **thermore·gulated**, **-regulating** *ppl. adjs.*; **thermore·gulative** (*rare*), **the:rmoregula·tory** *adjs.*, of, pertaining to, or effecting thermoregulation.

1917 T. SOLLMAN *Man. Pharmacol.* 449 (*heading*) The hypothetical thermoregulating centres. **1927** *Jrnl. Exper. Zool.* XLVII. 152 Nest-building may be considered a thermoregulative activity. **1932** Thermoregulated [see above]. **1949** KOESTLER *Insight & Outlook* xi. 168 Animals..develop techniques such as nest building..and, in the case of the termites, even contriving thermo-regulatory devices. **1972** *Science* 12 May 601/3 Bumblebees, which can also thermoregulate, occur on the neotropical and Asian mountains. **1973** *Brit. Med. Jrnl.* 22 Dec. 727/2 Some degree of thermoregulatory failure is common in old age. **1974** M. C. GERALD *Pharmacology* xiv. 265 Aspirin..does increase heat loss from the body by its action on the thermoregulatory centres in the hypothalamus. **1978** *Nature* 19 Oct. 646/2 Reptiles which thermoregulate behaviourally..are subject to considerable daily fluctuations in body temperature.

thermoremanence, -remanent: see *THERMO-.

Thermos. Add: (Further examples.) Hence (freq. with small initial) applied loosely to any vacuum flask. Also *absol.*, and designating a liquid which has been kept in a Thermos flask.

Still (1984) a proprietary term in Great Britain.

1922 S. LEWIS *Babbitt* xx. 235 Say, could I borrow your thermos—just dropped in to see if I could borrow your thermos bottle. **1923** R. FAY *Let.* 21 June (1972) II. 541 I'd got my thermos filled the day before. **1938** C. G. NORRIS *Bricks without Straw* 350 He poured a glass of water from his thermos jug. **1950** *Times* 1 Apr. 24/3 Simon began to pack blankets and Thermoses for a fishing trip. **1960** E. L. WALLANT *Human Season* (1965) v. 55 Let me pour you some coffee.' He poured a cupful from the big Thermos. **1977** *New Yorker* 27 June 53/1 Stoical munching forms in mackintoshes taking a swig of something hot out of thermoses. **1978** P. GRACE *Mutuwhenua* vii. 40 The others had home-made biscuits to offer round, or fruit. Oranges, apples, and feeding bananas. Chocolates, a thermos of soup. **1979** *Church Times* 14 Sept. 12/4 No registration is necessary, but please bring sandwiches and a thermos. **1979** *Nature* 15 Nov. 227/1 Sipping green tea poured out of a large thermos flask, we discussed differences between Chinese and UK science. **1979** *Beautiful Brit. Columbia* Winter 6 Relaxing with a hot cup of thermos coffee. **1980** P. FITZGERALD *Human Voices* ii. 47 Workers off work, each with their own thermos.

thermosensitive, -sensitivity: see *THERMO-.

thermoset (þɜ̄·məset), *a.* and *sb.* [f. THERMO- + SET *ppl. a.*] **A.** *adj.* Incapable of being softened or melted by heat like a thermoplastic; also = *THERMOSETTING *ppl. a.*

1940 [see *CURE *v.*[1] 10]. **1972** *Physics Bull.* Nov. 663/2 Epoxy resins are polymers with one or more epoxide groups... Ones with two such groups..are the monomers which can be converted to a thermoset stage by reaction with appropriate curing agents. **1973** *Sci. Amer.* July 42/3 Today the matrix in glass-reinforced composites may be either a thermoset plastic, such as polyester, phenolic or epoxy, or any of a number of thermoplastic resins, such as nylon, polyethylene or polystyrene. **1973** J. M. G. COWIE *Polymers* i. 19 The thermoset plastics generally have superior abrasion and dimensional stability characteristics compared with the thermoplastics which have better flexural and impact properties. In contrast to the thermoplastics, thermosetting polymers..are changed irreversibly from fusible, soluble products into highly intractable crosslinked resins which cannot be moulded by flow.

B. *sb.* A thermoset substance.

1955 in M. Reifer *Dict. New Words*. **1958** *Times Rev. Industry* Aug. 57/1 Thermoplastics have gained a..market at the expense of thermosets. **1970** *New Scientist* 19 Mar. Suppl. 9/1 Reinforced thermosets are used in many high-strength applications from boat hulls to aircraft radomes. **1975** *Sci. Amer.* July 61/ Nonmetals such as thermoplastics, thermosets, polytetrafluoroethylene and carbon-graphites are successful bearing materials because of their excellent resistance to scoring and erosion.

thermose·tting, *ppl. a.* [f. THERMO- + SETTING *ppl. a.*] Of a plastic: solidifying and becoming thermoset when heated; also, = *THERMOSET *a.*

[**1929** *Brit. Plastics & Moulded Products Trader* June 24/2 Thermoplastic mouldings..comprise substances which go through a process of thermo-setting on the application of pressure and heat.] **1931** *British Plastics Yearbk.* 74 Thermo-setting plastics, those compositions that though thermo-plastic in the first instance, harden off rapidly under the influence of heat. **1937** [see *THERMOPLASTIC *a.*]. **1951** *Archit. Rev.* CIX. 166/1 Hardness and resistance to scratching and heat..is one of the main advantages of the thermosetting plastics. **1973** [see *THERMOSET *a.*]. **1982** M. DUKE *Flashpoint* xviii. 74 The thermosetting polyurethane plastic floor. **1983** *Daily Tel.* 12 Dec. 19/3 ... thermosetting plastics ... is a known and recognised practice for reducing and dissipating exotherm.

thermo-siphon. Add: (Further examples, esp. with reference to internal-combustion engines.) So **thermo-sipho·nic** *a.*

1904 A. B. F. YOUNG *Compl. Motorist* iii. 55 Circulation is maintained either by a centrifugal pump driven by a chain off the engine shaft, or, automatically, by means of what is known as the 'Thermo-Syphon' system, in which advantage is taken of the fact that hot water rises to the top of a tank and cold water sinks to the bottom. **1920** *Autocar* 7 Feb. 251/1 Water is circulated by thermo-syphonic action. **1963** R. F. WEBB *Motorists' Dict.* 210 *Thermo syphon cooling*, a type of liquid cooling for an automobile engine where there is no mechanical assistance for the flow of the liquid through the system. **1968** G. N. GEORGANO *Compl. Encycl. Motorcars* 473 Cooling was by thermo-syphon, the hallmark of the Renault being the huge dashboard radiator. **1982** *Solar Energy* (Shell Internat. Petroleum Co.) 3/2 (*caption*) Typical solar water heater thermo-syphon system.

thermosphere (þɜ̄·məsfɪəɹ). [f. THERMO- + SPHERE *sb.*] **I.** † **a.** (See quot. 1924.) *Obs. rare*[−1]. **b.** The part of the atmosphere between the mesopause and the height at which it ceases to have the properties of a continuous medium, characterized throughout by an increase of temperature with height.

1924 S. N. SEN in *Q. Jrnl. R. Meteorol. Soc.* L. 29 Up to an approximate height of 8 km. above the ground the air density is chiefly controlled by the temperature. The name 'thermosphere' is proposed to denote this layer of the atmosphere. **1950** S. CHAPMAN in *Jrnl. Geophysical Res.* LV. 396, I propose the name..thermosphere for the layer of upward increasing temperature above that level (i.e. of the mesopause). **1967** R. W. FAIRBRIDGE *Encycl. Atmospheric Sci.* 731/2 At the top of the thermosphere, the temperature approaches a constant value of ~1500°K. **1981** *Sci. Amer.* July 46/3 On the earth the thermosphere is present day and night; the large-scale rotation of the atmosphere with the planet carries the heated day-side upper atmosphere to the night side of the planet. On the night side of Venus, however, the thermosphere disappears.

2. The warmer, upper part of the oceans.

1956 *Nature* 16 June 1106 (*in figure*) Thermosphere. **1957** [see *psychrosphere* s.v. *PSYCHRO-].

Hence **thermosphe·ric** *a.*

1971 *Nature* 29 Jan. 333/2 Calculations suggest that the thermospheric winds may produce some net rotation of the atmosphere at low latitudes. **1979** *Ibid.* 8 Feb. 458/2 Rocket measurements of mesospheric and thermospheric nitric oxide concentrations revealed strong enhancements during auroral particle precipitation events.

thermostabile (þɜ̄məstēɪ·bɪl), *a. Biol.* [f. THERMO- + L. *stabilis* STABLE *a.*] = THERMOSTABLE *a.*

1904 *Practitioner* Feb. 249 In the proglottides of tapeworms..there exists a liquid substance..which..is thermo-stabile and is similar to proteolytic ferment. **1947** *Ann. Rev. Microbiol.* I. 92 Motility type strains are more thermostabile than the ringpost type strains.

thermostat. Add: Hence as *v. trans.*, to regulate the temperature of (a substance or a piece of apparatus) by means of a thermostat. Also **the:rmosta·t(t)ed** *ppl. a.*

1940 *Brit. Jrnl. Psychol.* July 69 The tests were done in a roughly thermostated water-bath. **1950** W. J. MOORE *Physical Chem.* xv. 424 The cell must be well thermostated since the conductivity increases with the temperature. **1964** *Plant & Cell Physiol.* III. 212 The temperature of the algal suspension was controlled by flowing thermostated water on the surface of the vessel. **1963** L. C. PICKARD *Descriptive Physical Oceanogr.* vi. 86 This necessitates thermostatting the samples to 0·01°C during measurement. **1967** MARGERISON & EAST *Introd. Polymer Chem.* ii. 94 The data..were obtained using solutions of polystyrene in benzene illuminated by light of 5461Å and thermostatted at 25°C. **1979** *Nature* 25 Jan. 291/2 The complete reaction chamber was heated in a thermostatted oven.

thermo-tank: see *THERMO-.

thermotic, *a.* Add: **thermotics** *sb. pl.* (earlier example).

1831 W. WHEWELL *Let.* 18 Sept. in I. Todhunter *William Whewell* (1876) II. 132 It is very true that we very much want a name for the part of science which treats of light..that which treats of Heat... [In my MSS.] I have called one *Photistics* and the other *Thermotics*.

thermotolerant: see *THERMO-.

thermotropic, *a.* Add: Also with pronunc. (-trǭ·pɪk). **2.** *Physical Chem.* Brought about or effected by a change in temperature: used esp. with reference to mesophases and their phase transitions.

1909 ... in *Jrnl. Chem. Soc.* XCV. ii. ... **1964** G. W. GRAY *Molecular Structure & Properties of Liquid Crystals* ... **1966** ... **1972** *Physics Bull.* May 279/3 ... **1978** *Nature* 13 Apr. 646/1 ...

therophyte (þe·rǭfəit). *Bot.* [ad. Da. *therofyt* (C. Raunkiaer 1904, in *Bot. Tidsskrift* XXVI. p. xiv) f. Gr. θέρ(ος) summer + -PHYTE.] (See quot. 1960.)

1913 *Jrnl. Ecol.* I. 18 Therophytes, or plants of the favourable season, live through the unfavourable season as seeds; hence they are annual plants. **1932** FULLER & CONARD tr. *Braun-Blanquet's Plant Sociol.* i. 13 Many communities of the subtropics begin their annual development with a therophyte aspect. **1952** P. W. RICHARDS *Trop. Rain Forest* i. 10 Therophytes are entirely absent, except in clearings. **1960** N. POLUNIN *Introd. Plant Geogr.* iii. 93 Therophytes (annuals)..complete their life-cycle, from germination to ripe seed, within a single limited vegetative period, surviving the unfavourable times as seeds, spores, or other special (usually resistant) reproductive bodies. They are especially abundant in deserts.

theropod, *a.* (*sb.*) (Examples.)

1933 A. S. ROMER *Vertebr. Paleontol.* ix. 181 The theropods (using this term in a broad sense) include all the carnivorous terrestrial reptilian carnivores of the late Triassic, Jurassic, and Cretaceous. **1970** *Nature* 11 Apr. 109/1 Theropod dinosaurs normally walked on three toes on each hind foot, leaving bird-like tracks. **1981** *Sci. Digest* Aug. 56 (*caption*) The most complete theropod found in China, *Yangchuanosaurus shangyouensis* (discovered 1977, described in 1978).

theropodous, *a.* Add after *Theropoda* in etym. (O. C. Marsh 1881, in *Amer. Jrnl. Sci.* CXXI. 423). For Cope's substitute Marsh's in def.

thesaurosis (þisǭrǭᵘ·sis). *Path.* [f. Gr. θησαυρ-ός store + -OSIS.] A disorder of the lungs caused by the accumulation in them of inhaled material.

1958 *New England Jrnl. Med.* 6 Mar. 475/1 Evidence concerning the etiology of the pulmonary lesion of Case 2 is..less compelling, but we believe that..this, too, was a case of thesaurosis due to hair-spray constituents. **1975** *New Yorker* 7 Apr. 56/3 For at least fifteen years, it has been suspected that the plastic resins in hair sprays cause a restrictive-lung-storage disease called thesaurosis, as well as abnormal lung cells that may be the precursors of lung cancer.

thesaurus. Delete ‖ and add: **2. b.** A collection of concepts or words arranged according to sense; also (*U.S.*) a dictionary of synonyms and antonyms.

1852 [in Dict., sense 2]. **1868** 'MARK TWAIN' *Autobiogr.* (1924) I. 172 The fact that the writer's balance at the vocabulary bank has run short and that he is too lazy to replenish it from the thesaurus. **1942** BERREY & VAN DEN BARK (*title*) The American thesaurus of slang. **1960** W. NAYLOR *Silver Birch Anthol.* 8, I know how you have to polish and repolish, and delve into the thesaurus, change sentences, consult the dictionary and the thesaurus, ... **1975** (*title*) Family Word finder: a new thesaurus of synonyms and antonyms in dictionary form.

c. A classified list of terms, esp. key-words, in a particular field, for use in indexing and information retrieval.

1957 H. PROWNSON in *Proc. Internat. Study Conference on Classification for Information Retrieval* 100 The best ... **1961** *Aslib Proc.* XIII. 265 We decided to..designate the analytical compilation 'a thesaurus'. **1966** *Revue Internat. de Documentation* XXXIII. 21/1 It has become commonplace to refer to ... In this context a thesaurus usually means an arrangement of a vocabulary of terms in each group being connected in some defined way. **1974** *Encycl. Brit. Macropædia* IX. 572/2 The second stage should be a fully automatic selection and matching-process, preferably using a thesaurus.

these, *dem. adj.* Add: **1. f. these days** advb. phr., nowadays, at present.

1936 K. LEHMANN *Weather in Streets* I. v. 97 An estate like this must be a terrible problem these days. **1948** M. DICKENS *Joy & Josephine* I. iv. 132 'Play golf?' Mr. Gray asked George, who answered: 'Not these days,' as if he had never played golf. **1960** ... *Living* vi. iii. 181 He ... picture. ... Dec. 5/1 ... choice ...

Thesean, *a.* Add: Also † **Theseian**. (Earlier example.)

1815 B. R. HAYDON *Jrnl.* 6 Nov. in T. Taylor *Life Haydon* (1853) I. xv. 294 Lord Elgin's steward..thus entirely ruined the moulds of the Thesian bas-reliefs, which had cost Lord Elgin so much.

thesis. Add: **2.** *spec.* in Old English prosody and in the prosody of other Germanic languages.

1870 F. A. MARCH *Introd. Anglo-Saxon* 227 The regular Germanic epic line has four..arses in each section, each of which may have a thesis or not. **1888** A. H. TOLMAN in *Trans. Mod. Lang. Assoc. Amer.* III. 20 Only one accented syllable, out of the first sixteen of this poem [sc. *Beowulf*], has a syllable expressed as its thesis or senkung. **1910** J. SCHIPPER *Hist. Eng. Versification* 28 Syllables with this secondary stress are necessarily in certain cases thesis between the arsis and the thesis. **1938** A. CAMPBELL *Battle of Brunanburh* 18 A disyllabic second thesis seems not to be found in lines of type A. **1942** J. C. POPE *Rhythm of Beowulf* 40 We fill the down-beat or thesis of the first measure.

6. *thesis-monger* (see MONGER[1] 2); **thesis-novel** = *roman à thèse* s.v. *ROMAN *sb.*[4]

1932 *Essays & Stud.* XVII. 75 The aimless burrowings of a thesis-monger. **1959** *Listener* 13 Aug. 255/1 Subjects like the Henrician Reformation..have been far too much in the hands of thesis-mongers. **1934** WEBSTER, Thesis novel. **1954** K. TILLOTSON *Novels of Eighteen-Forties* I. 117 Novelists who..avoided the thesis-novel. **1979** S. WEINTRAUB *London Yankees* vii. 233 Elizabeth Robins.. continued writing thesis-novels on euthanasia, prostitution, women's rights.

Thesmophoric, *a.* Add: Also **Thesmophoria** (see Dict.).

1788 J. LEMPRIERE *Bibliotheca Classica* s.v. *Thesmophora*, The Thesmophoria were instituted by Triptolemus, or according to some by Orpheus, or the daughters of Danaus. *Ibid.*, Such as were initiated at the festivals of Eleusis assisted at the Thesmophoria. **1890** J. G. FRAZER *Golden Bough* II. iii. 26 The casting of the pigs into the vaults at the Thesmophoria formed part of the dramatic representation of Proserpine's descent into the lower world. **1940** M. P. NILSSON *Gr. Folk Relig.* 24 Best known is the festival of the autumn sowing, the Thesmophoria, celebrated by women only. **1978** *Times Lit. Suppl.* 18 Aug. 922/4 The regular female festival of fertility, the Thesmophoria, was entirely confined to women.

thesp (θesp), colloq. abbrev. of THESPIAN *sb.*

1962 *New Statesman* 23 Feb. 274/3 Like all tales about thesps, [it] seems to involve us just that much less in their fate. **1976** *Times Lit. Suppl.* 5 Mar. 262/2 More commonly than not, Shakespeare productions by Eng Lit dons involving undergraduate actors come with terrific programme notes but scrawny thesps. **1978** *Guardian Weekly* 15 Jan. 21/4 Hero and heroine—actors both..a famous pair of Budapest thesps.

Thespianism (θe·spiăniz'm). [f. THESPIAN *sb.* + -ISM.] The art of acting, dramatic art.

1914 C. MACKENZIE *Sinister Street* II. III. viii. 664 Scarcely ever did the Academic Muse enter the O.U.D.S. ..She must greatly dislike Thespianism with all that it connoted of rhetorical balustrades. **1928** *Daily Express* 29 Oct. 9 He still wraps round him..the rags of a tattered toga of Thespianism.

Thessalian (þesēɪ·liăn), *a.* and *sb.* *Gr. Antiq.* [f. L. *Thessalius*, *Thessalus* (Gr. Θεσσάλιος, Θεσσαλός) adj. f. *Thessalia* (Θεσσαλία) Thessaly: see -AN, -IAN.] **A.** *adj.* Of or pertaining to Thessaly, a region in northern Greece.

1590 SHAKES. *Mids. N.* IV. i. 127 Crooke kneed, and dew-lapt, like Thessalian Buls. **1667** MILTON *P.L.* II. 544 As when Alcides..tore Through pain up by the roots Thessalian Pines. **1757** J. DYER *Fleece* II. 56 When, o'er the deep by flying Phryxus brought, The fam'd Thessalian ram enrich'd her plains. **1842** TENNYSON *Talking Oak* 292 Or that Thessalian growth, In which the swarthy ringdove sat, And mystic sentence spoke. **1888** *Encycl. Brit.* XXIII. 299/1 These Thessalian passes were of the utmost importance to southern Greece. **1973** R. LANE-FOX *Alexander* I. iii. 47 Demeratus the Corinthian..had bought the horse from his Thessalian breeder.

B. *sb.* An inhabitant of Thessaly; the dialect of Greek spoken there.

1608 E. TOPSELL *Serpents* 5 One Aleua a Thessalian, who feeding his Oxen in Thessaly..there fell in loue with him a Serpent. **1704** S. PARKER tr. *Tully's Old Age* 34 When Cineas the Thessalian told him of it. **1888** *Encycl. Brit.* XXIII. 299/2 In race, as in geographical position, the Thessalians held an intermediate place between the non-Hellenic Macedonians and the Greeks of pure blood. **1910** C. D. BUCK *Gr. Dial.* I. 3 Thessalian is of all dialects the most closely related to Lesbian, and at the same time shares in some of the characteristics of the West Greek dialects. **1973** R. LANE-FOX *Alexander* I. iii. 60 Philip's heir was ruler of the Thessalians, a people essential for his army. **1978** *Language* LIV. 179 Thereafter Thessalian underwent some influence from Northwest Greek.

theta. Add: **1. b. theta-function**: in def., after 'integral' read '$\int \exp(-t^2)\,dt$'.

2. *Biol.* Used to designate rhythmic activity of the brain recorded by an electroencephalograph and having a frequency of between four and seven cycles per second.

1944 WALTER & DOVEY in *Jrnl. Neurol., Neurosurg., & Psychiatry* VII. 64/1 In the case of the 4–7 c/s. waves the term we suggest is 'theta'. *Ibid.* 65/1 It is suggested that rhythms at about 6 c/s. should be termed 'theta' rhythms and that such rhythms are characteristic of the resting, immature or isolated parieto-temporal complex. **1953** *Brit. Jrnl. Psychol.* XLIV. 320 The excessive theta rhythm ..found in some aggressive psychopaths. **1961** *Lancet* 26 Aug. 465/1 An E.E.G. recorded when fasting showed some abnormality, with rhythmic 6 c.p.s. theta activity in the resting tracing. **1972** *Sci. Amer.* Feb. 85/3 Subjects with a great deal of experience in meditation showed other changes: the alpha waves slowed..and rhythmical theta waves at six to seven cycles per second appeared. **1975** J. TAYLOR *Superminds* vii. 121 In states of extreme emotion, theta waves of four to seven cycles and up to thirty microvolts occur.

3. In Scientology: creative energy or spirit. Also **the·tan**, the embodiment of this spirit in the individual. (See quots. 1965.)

1951 L. R. HUBBARD *Handbk. for Preclears* 77/1 *Theta.* The mathematical symbol for the static of thought. By *theta* is meant the static itself. **1952** —— *Scientology: 8–80*, 83 The thetan, or theta being, takes over a body only a few days or a week before birth. **1957** J. F. HORNER *Summary of Scientol.* iv. 56 In Scientology, the specialized term, 'theta', is used to refer to thought and spirit. The term, 'Thetan', refers to the single unit of beingness which each person is. **1965** L. R. HUBBARD *Scientol. Abridged Dict.* 33 *Theta*, energy peculiar to life or a thetan which acts upon material in the physical universe. *Ibid.*, *Thetan*, the person himself..: that which is aware of being aware: the identity that is the individual. **1971** [see *SCIENTOLOGY]. **1977** C. MCFADDEN *Serial* (1978) xxvii. 60/2 Marlene said Theta taught you how to overcome Specific Negatives. **1977** *Times Lit. Suppl.* 13 May 582/5 The gnosis centres on the 'thetan', the true self... The core task of the cult ought to be cleanse and return the thetan to its immortal, pristine, form.

4. *Chem.* **theta** (or θ, Θ) **temperature**, the temperature of a polymer solution at which it behaves ideally as regards its osmotic pressure; so *theta condition*, *solvent*, etc.

1953 P. J. FLORY *Princ. Polymer Chem.* xii. 533 Frequently it is preferred to use as a parameter the 'ideal' temperature θ. At the temperature T = θ, the chemical potential due to segment-solvent interactions is zero... Hence the temperature θ is that at which the excess chemical potential is zero and deviations from ideality vanish. *Ibid.* xiv. 612 An ideal solvent, or θ-solvent, ... **1973** J. M. G. COWIE *Polymers* viii. 138 Above the theta temperature expansion of the coil takes place, caused by interactions with the solvent, whereas below θ the polymer segments attract one another, the excluded volume is negative, and eventual phase separation occurs. **1974** *Sci. Amer.* Dec. 60/2 At the theta temperature, which varies for different polymers and solvents, the measured properties of the polymer can be usefully compared with those of other polymers at their theta temperature.

5. *Particle Physics.* Freq. written θ. A meson that decays into two pions, now identified with the kaon. Also *theta meson*.

1954 *Physical Rev.* XCV. 1732/1 A large fraction of these θ^0 decays are consistent with the decay scheme $\theta^0 \rightarrow \pi^+ + \pi^-$ ... **1955**, etc. [see *TAU].

6. *theta pinch* (*Nucl. Physics*): a toroidal pinch (PINCH *sb.* 10**) in which the magnetic field follows the axis of the plasma and the current-carrying coils encircle it. [θ, the symbol of the angle of the radius vector of the circular path of the current.]

1959 *Nucleonics* Oct. 82/2 The conference topic that aroused greatest interest was the theta pinch. **1967** *New Scientist* 9 Nov. 369/1 The simplicity of the theta pinch as a means of achieving that sought after goal of power from thermonuclear reactions had led to considerable studies of its real effectiveness in confining and heating up a plasma. **1974** *Nature* 20 Sept. 193/2 In the theta-pinch design, the compression coils are not superconducting and no special measures are taken to reduce neutron heating.

thetatron (þī·tătrǫn). *Nucl. Physics.* Also **Thetatron.** [f. THETA + *-TRON.] A fusion reactor employing a theta pinch in which the plasma is compressed axially by causing the current in the coils, and hence the axial magnetic field, to increase suddenly.

1959 *Nucl. Instruments & Methods* IV. 323/1 At A.W.R.E. both straight and toroidal magnetic compression devices ('Thetatron') have been examined. **1962** W. B. THOMPSON *Introd. Plasma Physics* iv. 63 Much interest has been shown in the axial compression devices, or thetatron, in which a cyclindrical plasma is compressed by a rapidly rising axial field. **1964** *Times Sci. Rev.* Autumn 14/1 Thetatrons produce hot dense plasmas for.. a few microseconds. **1980** W. M. GIBSON *Physics Nucl. Reactors* x. 200 At the Culham laboratory..magnetic mirror and bottle systems have received thorough investigation in experiments with Thetatron and Phoenix devices.

Thetford (þe·tfǒɹd). The name of a town in Norfolk, England, used *attrib.* to designate Saxo-Norman pottery of a type made there and in other parts of East Anglia. Usu. as *Thetford ware*.

1949 *Archaeological News Let.* Feb. 3/1 Until quite recently the origin and affinities of St. Neots (or Thetford) pottery have been misunderstood. **1956** *Proc. Cambridge Antiquarian Soc.* XLIX. 46 There is also evidence that Thetford ware was made at Norwich and Ipswich. **1966** *Daily Tel.* 31 Oct. 14/5 The kilns themselves and the finds of pottery may shed new light on Thetford ware which is known all over East Anglia... This is the first time that a good specimen of a Thetford ware kiln has been available for examination. **1971** *Canad. Antiques Collector* Apr. 17/2 Thetford Ware which is very frequently found at North Elmham is well made, fired hard and has a very characteristic 'feel' which is sandy, but not as friable as the shell-gritted St. Neots Ware also found in this region.

thew, *sb.*[1] Add: **3. b.** Also in *sing.* Also *fig.*

c **1863** E. DICKINSON *Poems* (1955) II. 512 Thigh of Granite—and thew—of Steel. **1873** G. M. HOPKINS *Jrnls. & Papers* (1959) 233 A floating flag is like wind visible and what weeds are in a current; it gives it thew and fires it and bloods it in. **1876** —— *Wr. Deutschland* xvi, in *Poems* (1967) 56 He was pitched to his death at a blow, For all his dreadnought breast and braids of thew. **1887** [in Dict.]. **1930** R. CAMPBELL *Adamastor* 77 A Hercules of matchless thew Whose body is the breath of flowers. **1977** *N.Y. Rev. Bks.* 15 Sept. 40/3 By 'language' he means not the whole body of speech, the thew and sinew of the language..but a precursor's language.

they, *pers. pron.* Add: **I. 1. b.** Also as *theyselves* (var. *theysel(l)s*) for *themselves*.

1882 *Century Mag.* Apr. 892 They're pretty peart at the game theyselves. **1893** P. H. EMERSON *On English Lagoons* xlii. 256 Those gents expect you to keep as clean as theysels. **1901** C. HARE *Dinah Kellow* 255 Passon didn' like for they to be locked in by theysells. **1974** *Black World* Apr. 8 He was presenting the street nigguhs in all they glory without no overt exhortation to them to git theyselves togetha.

c. For † read 'Now *rare*'. (Later examples.)

1889 A. LANG in *Scribner's Mag.* Sept. 265/2 They are small-minded and small-hearted people who are most shocked by what they call 'vanity' in the great. **1899** L. C. CORNFORD *R. L. Stevenson* vi. 140 Alan Breck Stewart is the central figure, and they are his sayings and deeds of arms that go to make the chief interest.

3. b. *colloq.* Used to refer collectively to people in authority, regarded as impersonal and oppressive.

1886 KIPLING *Delilah* in *Definitive Verse* (1940) 7 One day, *they* [sc. people in power] brewed a secret... It related to Appointments. **1939** AUDEN in *Times Lit. Suppl.* 25 Mar. (Spring Books Suppl.) p. i, The legions of cruel inquisitive 'They' Were so solid and strong, like dogs. **1945** H. NICOLSON *Let.* 27 May (1967) 465 People.. believe that 'they' mean the upper classes, or the Conservatives. **1947** 'G. ORWELL' *Eng. People* 24 English political thinking is much governed by the word 'They'. 'They' are the higher-ups, the mysterious powers. **1967** G. F. FIENNES *I tried to run a Railway* iii. 25 'They' shifted me to York. **1976** *Leicester Chron.* 26 Nov., 'They' are always doing you down. **1981** I. BOLAND tr. *Ginzberg's Within Whirlwind* II. v. 228 How could I go and work in a children's establishment, where 'they' would be able to keep tabs on me?

II. 6. As possessive pronoun: = THEIR *poss. pron.* 1. *U.S. dial.*

1928 J. PETERKIN *Scarlet Sister Mary* 162 When dey is worried in dey mind. **1929** W. FAULKNER *Sartoris* i. 23 They was a-settin' behind a table with they pistols layin' on the table. **1935** in Z. N. Hurston *Mules & Men* I. ii. 45 They all brought they rocks and Christ turned 'em into bread. **1974** *Black World* Nov. 58 Just because all the pussy they can snatch is what they can lick off they goddam fingers.

III. 7. As advb.: = THERE *adv.* 4 d. *U.S. dial.*

1874 'MARK TWAIN' in *Atlantic Monthly* Nov. 592/2 Dey was de biggest dey *is*. **1889** J. W. RILEY *Pipes o' Pan at Zekesbury* 41 They's music in the twitter of the bluebird and the jay. **1920** [see *monkey suit* s.v. *MONKEY *sb.* 17 a]. **1949** H. HORNSBY *Lonesome Valley* 185 They's more ways than one to skin a cat. **1973** *Black World* Aug. 61/2 They was ten packets looked like Horse.

thiabendazole (þəiˌăbe·ndăzǭᵘl). *Vet. Med.* and *Pharm.* [f. *thia(zole* s.v. THIO- 1 in Dict. and Suppl. + *ben(zimi)dazole* f. BENZ(ENE, BENZINE + IMID(E + AZO- + *-OLE.] An anthelmintic used in veterinary and human medicine, esp. against intestinal nematodes.

1961 H. D. BROWN et al. in *Jrnl. Amer. Chem. Soc.* LXXXIII. 1765/1, 2-(4'-Thiazolyl)-benzimidazole (I, generic name: thiabendazole) was outstanding in anthelmintic activity. **1970** W. H. PARKER *Health & Dis. in Farm Animals* xx. 265 New worm medicines are appearing frequently. At the present time the choice is likely to be either thiabendazole or tetramisole. **1978** R. B. SCOTT *Price's Textbk. Pract. Med.* (ed. 12) II. 236/2 Thiabendazole has been shown to destroy adult female worms [sc. *Trichinella spiralis*] and larvae but must be used with extreme caution because it may precipitate serious reactions with anorexia, nausea, headache, giddiness, and drowsiness, possibly as a result of allergy to dead larvae.

thiamine (þəi·ămīn). [f. THI(O- + AMINE.]

† **1.** *Chem.* (See quot.) *Obs. rare.*

1886 C. A. BENNERT *Brit. Pat. 13,466* 3 This invention has for its object the manufacture or production of a new class of organic coloring compounds or materials by reactions between certain amines of the aromatic series and the dioxide of sulphur group, which organic colouring compounds or materials so obtained I designate 'Thiamines'.

† **2.** *Chem.* (In Dict. s.v. THIO- 1.) *Obs. rare.*

3. *Biochem.* Also **thiamin**. **a.** Vitamin B_1; a water-soluble, heat-labile, sulphur-containing compound that is present in many foods (esp. whole cereal grains, pork, and liver) but absent from fats and is necessary for carbohydrate metabolism, its dietary deficiency resulting in disturbances of the nervous system.

1937 *Jrnl. Amer. Med. Assoc.* 18 Sept. 952/1 Dr. Jansen has been very cooperative in discussing the matter of nomenclature for vitamin B1 and suggested..that Dr. R. R. Williams..propose a name based on the chemical structure. Dr. Williams..proposed the term Thiamin Chloride. **1939** A. HUXLEY *After Many a Summer* II. vi. 234 With a course of thiamin chloride and some testosterone I could have made him as happy as a sand-boy. **1951** A. GROLLMAN *Pharmacol. & Therapeutics* xxvii. 597 Thiamine contains thiazole and pyrimidine rings. **1952** *New Biol.* XIII. 111 The role of thiamin in cell metabolism is to provide a portion of certain enzyme systems. **1972** *Materials & Technol.* V. xix. 674 In Great Britain, all bread except wholemeal must be fortified with thiamin.. to compensate for the losses of the natural vitamin consequent on milling. **1980** D. MADDEN *Food & Nutrition* i. 23/2 After vitamin C, thiamine is probably the most unstable vitamin.

b. thiamine pyrophosphate, the active form of thiamine in which it acts as a co-enzyme: = *COCARBOXYLASE.

1949 *New Biol.* VII. 108 These include thiamine pyrophosphate, nicotinamide nucleoside and riboflavine, all of which are involved in systems through which carbohydrates are oxidised. **1960** [see *DIPHOSPHOTHIAMINE]. **1973** YUDKIN & OFFORD *Comprehensive Biochem.* (1980) xvii. 297 Transketolase needs as co-factor the vitamin derivative thiamine pyrophosphate, which..is also involved in the transfer of a two-carbon fragment from pyruvate to lipoic acid.

Hence **thia·minase** [*-ASE], an enzyme that destroys thiamine, splitting the thiazole and the pyrimidine rings.

1938 BONNER & BUCHMAN in *Proc. Nat. Acad. Sci.* XXIV. 437 It has been shown..that the pea root synthesizes vitamin B_1..from a mixture of the pyrimidine and thiazole components of the vitamin molecule... This must..be a synthesis in which a specific enzyme, a 'thiaminase'.., takes part. **1972** L. HANCOCK *There's a Seal in my Sleeping Bag* ii. 23 Herring is a fish that contains an enzyme called thiaminase, which breaks down thiamin in the body.

thiazide (þəi·ăzəid). *Pharm.* [f. THI(O- + *AZ(INE + OX)IDE, elements in the systematic name of the parent compound.] Any of a class of drugs derived from 1,2,4-benzothiadiazine-1,1-dioxide that increase the excretion of sodium and chloride and are used as diuretics and as auxiliary hypotensive agents.

1959 *Jrnl. Amer. Med. Assoc.* 22 Aug. 2052/2 When the thiazide derivatives are first administered to hypertensive patients being treated with other drugs, it is advisable to continue giving Rauwolfia derivatives. **1961** *Lancet* 12 Aug. 334/1 The use of guanethidine in combination with one of the thiazide group of diuretics extends the application of these drugs in the management of patients with high blood-pressure. **1974** M. C. GERALD *Pharmacol.* xxii. 399 When used alone, the thiazides have only weak antihypertensive effects. **1980** *Brit. Med. Jrnl.* 18 Oct. 1053 Vasodilators are used mainly as additional treatment in patients who are not controlled by thiazides and beta-blockers.

thiazolidine (þəiˌăzǫ·lidīn). *Chem.* and *Pharm.* [f. *thiazole* s.v. THIO- 1 in Dict. and Suppl. + *-IDINE.] **a.** A liquid, C_3H_7NS, whose molecular structure is that of thiazole with an additional hydrogen atom attached to the nitrogen and each carbon atom. **b.** Any compound containing this ring structure in its molecule.

1916 *Chem. Abstr.* X. 3647/2 (Index), Thiazolidine (tetrahydrothiazole). **1945** *Science* 21 Dec. 628/2 Penicilloic acids are undoubtedly thiazolidines. **1949** E. CHAIN in H. W. Florey et al. *Antibiotics* II. xxii. 823 Penicillamine and its esters react readily with aldehydes or ketones to form thiazolidines. **1951** A. GROLLMAN *Pharmacol. & Therapeutics* xxii. 443 Penicillin is a monocarboxylic acid with β-lactam and thiazolidine rings attached through a CONH linkage to a prosthetic group. **1979** *Nature* 25 Oct. 716/2 The chemical assay based on the behaviour of the fused thiazolidine-β-lactam ring consistently gave higher results than the bioassay which required the full penicillin molecule for biological activity.

thick, *a.* (*sb.*) Add: **I. 3. a.** *a bit thick* (earlier example).

1902 G. W. E. RUSSELL *Londoner's Log-bk.* iii. 46 These manifold exercises of culture are characterized by our curate as 'a bit thick', and he owns himself 'fairly out of it'.

b. *the thick end of the stick = the dirty end* (*of the stick*) s.v. *DIRTY *a.* 1 e.

1957 *Times* 22 Nov. 8/3 Sir Ralph Richardson has the thick end of the stick... He has to represent an ordinary city insurance clerk. **1960** *Woman's Own* 13 Feb. 17/2 I'm the one to get what Father used to call 'the thick end of the stick'.

II. 5. a. Also in colloq. phr. *thick on the ground*: (chiefly of persons) numerous, abundant; closely concentrated or crowded. Cf. *THIN *a.* 2 e.

1893 J. SALISBURY *Gloss. Words S.E. Worcestershire* 42 *Thick on the ground*=crowded. **1919** J. BUCHAN *Mr Standfast* xii. 218, I see you're some kind of general. They're pretty thick on the ground here. **1964** C. WILLOCK *Enormous Zoo* viii. 133 Where animals are thick on the ground as with the herds, often three hundred strong, of topi [etc.]. **1978** 'E. PETERS' *Rainbow's End* i. 24 Willing workers are not so thick on the ground these days.

7. a. Also *dial.* or *colloq.* in phr. (*to be*) *thick o' fog*.

1935 L. LUARD *Conquering Seas* ii. 19 Thick o' fog—can't see whaleback. **1972** E. STAEBLER *Cape Breton Harbour* xvii. 148 We wanted to go back next day but thought we better wait till it was thick-a-fog and nobody'd see us.

III. 9. b. For 'Now *dial.*' read 'Now chiefly *colloq.* of persons' and add later examples. Also emphatically, *as thick as two planks*, etc. Cf. THICK-HEADED *a.* b.

1865 *Harper's Mag.* Dec. 133/2 [He] is nevertheless slow to see the point—in fact, 'thick' otherwise than crosswise. **1961** S. CHAPLIN *Day of Sardine* ii. 53 Free rides on trains and trolleys were routine stuff; and the thickest character in the school could find a buckshee road into a cinema. **1974** G. HONEYCOMBE *Adam's Tale* I. ii. 27 'He must be as thick as two planks,' said Nick. **1976** J. I. M. STEWART *Memorial Service* iii. 40 You might expect to become P.M. if you hadn't been so thick as to accept your idiotic life peerage. **1980** 'J. GASH' *Spend Game* xiii. 130 Rough-mannered and a bit greedy... Corporal's thick as a plank.

V. 12. a. (A further selection of collocations.) *thick-billed* (earlier and later examples), *-fingered*, *-fleeced* (examples), *-legged* (example), *-lensed*, *-lugged*, *-piled*, *-rimmed*, *-shouldered*, *-stemmed* (example), *-walled* (earlier example), *-wooled*.

1770 G. WHITE *Let.* 21 May in *Nat. Hist. Selborne* (1789) II. vi. 131 The bird you kept..abides all the year, and is a thick-billed bird. **1939** F. C. LINCOLN *Migration Amer. Birds* 103 As an exemplar of vagrant migration from south to north, the Thick-billed Parrot may be cited. **1980** CYRUS & ROBSON *Bird Atlas of Natal* 274 Thick-billed Weaver..inhabits coastal bush. **1874** 'MARK TWAIN' *Let.* 9 Dec. (1917) I. xiv. 238, I am so thick-fingered that I miss the keys. **1864** G. M. HOPKINS *Poems* (1967) 136 Thick-fleeced bushes like a heifer's ear. **1924** E. SITWELL *Sleeping Beauty* xxvi. 95 As lovely as the thick-fleeced waters. **1873** J. BROWN *Let.* 23 June (1912) 280 Uig is a pretty snug little bay, with its tidy Inn and its thick-legged, humorous landlord, John Urquhart. **1946** E. O'NEILL *Iceman Cometh* I. 4 He has black eyes which peer near-sightedly from behind thick-lensed spectacles. **1973** J. GOODFIELD *Courier to Peking* ii. 23 A short, squat person with thick-lensed glasses. **1922** JOYCE *Ulysses* 319 The curse of a goodfornothing God light sideways on the bloody thicklugged sons of whores' gets! **1853** M. ARNOLD *Sohrab & Rustum* in *Poems* 6 Upon the thick-pil'd carpets in the tent. **1976** *Sounds* 11 Dec., His hair, short at the sides and thickpiled high on top, makes him look faintly ridiculous. **1976** 'R. GORDON' *Doctor on Job* iii. 18 A small, round, well-scrubbed looking man in a grey business suit and thick-rimmed glasses. **1965** J. A. MICHENER *Source* (1966) 564 And he knelt in the boat, a thick-shouldered, heavy-necked, sandy-haired German seeking God. **1840** EMERSON *Woodnotes* in *Dial* (Boston) Oct. 244 You ask..what guide Me through trackless thickets led, Through thick-stemmed woodlands. **1820** M. EDGEWORTH *Let.* 26 Dec. (1971) 231 Old thick-walled mansions. **1913** W. DE LA MARE *Peacock Pie* 85 Roasting a thick-wooled mountain sheep Upon an iron spit.

b. thick-back, substitute for def.: in full, *thickback sole*; a flat-fish, *Microchirus variegatus*, found in the Mediterranean and off western European coasts; (examples); **thick ear**, an ear swollen or numbed by a sharp blow; usu. in phrases, as *to give* (someone) *a thick ear*; also *spec.* used *attrib.* to designate literature, etc., marked by rough violence and horseplay, or the writers of such material; **thick end**: now also *colloq.*; **thick-knit** *a.*, designating a garment knitted from wool of greater thickness than double knitting; also *absol.* as *sb.*, a thick-knit sweater; **thick sandwich (course)**, a sandwich course (see *SANDWICH *sb.*[2] 1 b) with an extended theoretical component between two periods of practical instruction (see quot. 1978); **thick space** *Typog.*, a third of an em space used in separating words; cf. *thin space* s.v. *THIN *a.* D b; hence **thick-spaced** *a.*; **thick tea** (earlier U.S. example); **thick woods** *Canad.* = *strong wood(s* s.v. *STRONG *a.* 12 b.

1864 J. COUCH *Hist. Fishes Brit. Isles* III. 203 The Thickback seldom exceeds the length of eight or nine inches. **1896** J. T. CUNNINGHAM *Nat. Hist. Marketable Marine Fishes Brit. Isles* 259 The Thickback... Pectoral fins very small. **1925** J. T. JENKINS *Fishes Brit. Isles* 198 The Thickback Sole..is brownish-red, with six or seven dark bands running across the body. **1969** A. WHEELER *Fishes Brit. Isles & N.-W. Europe* 557/1 The thickback sole is found rather more offshore. *Ibid.* 557/2 The thickback makes a very minor contribution to fishery landings of 'soles', but its flesh is of high quality. **1909** J. R. WARE *Passing Eng.* 243/2 Thick ear. **1916** 'TAFFRAIL' *Pincher Martin* ii. 28, I sed I'd give yer a thick ear if yer went on worryin' me. **1922** A. HADDON *Green Room Gossip* ix. 248 'A thick-ear play' was Sir Gerald du Maurier's description of 'Bull-dog Drummond'. **1943** *Gen* 2 Jan. 28/1 A member of the thick-ear fraternity. **1978** *Lancashire Life* Oct. 83/2 Ah geet a reyt thick ear yon time Ah tarned sheets in a tangle! **1981** N. TUCKER *Child & Book* v. 133 One particular favourite type of comic—referred to in the trade as the 'thick-ear market'—is chiefly concerned with crude, knockabout humour. **1938** 'N. SHUTE' *Ruined City* x. 195 It would be the thick end of that sum before we're cracking as a proper yard again. **1965** P. O'DONNELL *Modesty Blaise* xviii. 196 Willie.. tested the weight. 'It's the thick end of a hundredweight .. But I could manage one on me own all right.' **1971** D. LEES *Rainbow Conspiracy* i. 13 It will take them the thick end of half an hour to get to the Travellers from here. **1961** Thickknit [see *COVER-UP]. **1976** J. FLEMING *To make Underworld* xii. 138 The three Irishmen, ill-disguised as sailors or fishermen in their thick-knits. **1962** *Engineering* 13 July 57/2 The 1-3-1 type of 'thick sandwich' course (one year in industry, three years at university, and one year in industry again). *Ibid.* 26 Oct. 555 A pre-university year in industry (as in 1:3:1 thick sandwiches). **1978** *Jrnl. R. Soc. Arts* CXXVI. 549/1 A sandwich course such as the 'thick' sandwich, where you do one year in industry, three years at university and then one year back in industry. **1683** MOXON *Mech. Exerc.* II. 99 Some [letters] are Space thick; that is, one quarter so thick as the Body is high; though Spaces are seldom Cast so,..and therefore..we shall call these Spaces, Thick Spaces. **1808** C. STOWER *Printer's Gram.* iii. 90 Of Spaces... They are cast to various thicknesses... Three to an m—or three thick spaces. **1967** E. CHAMBERS *Photolitho-Offset* ii. 12 The *thick space* and *middle space* are a third and a quarter respectively of the width of the *em quad*. **1824** J. JOHNSON *Typographia* II. 132 A *d* and an *h*..will admit an addition, but not more than a middle and thin space to a thick spaced line. **1893** HART *Rules for Compositors* 22 When the last line but one of a paragraph is widely spaced and the first line of the following paragraph is also more than thick-spaced. **1886** S. COOLIDGE *What Katy did Next* xi. 305 The month's housekeeping wound up that night with a 'thick tea'. **1754** A. HENDRY *Jrnl.* 2 Dec. in *Trans. R. Soc. Canada* (1907) I. II. 343 Strong gale with Snow & Sleet. Obliged to remove into thick woods. **1865** MILTON & CHEADLE *N.W. Passage by Land* xii. 223 We had thirteen horses to pack and drive through the thick woods. **1957** C. HARRIS *Cariboo Trail* 137 The gold-seekers had arrived at the fort after making their way through the thickwoods.

B. *absol.* passing into *sb.* **I. 2. b.** A beverage of thick or heavy consistency, as cocoa, porter, etc. *slang.*

1887 J. W. HORSLEY *Jottings from Jail* i. 26 A somewhat..despairing view of prison life is indicated by 'Lads, your only friend here is your brown lofe [*sic*] and pint of thick'. **1903** FARMER & HENLEY *Slang* VII. 99/1 *Thick*, (common).—porter: ironically said to be 'a decoction of brewers' aprons'. **1923** J. MANCHON *Le Slang* 309 *Thick*, le café, le *jus*. **1947** W. DE LA MARE *Coll. Stories for Children* 222 The mugs of thick proved to be cocoa.

II. *sb.* with *pl.* **6.** For *School slang* read *colloq.* and *slang* (orig. *Schoolboys'*) and add later examples.

1925 S. O'CASEY *Juno & Paycock* III, in *Two Plays* 97 The thick made out the Will wrong. **1960** B. MOORE *Luck of Ginger Coffey* vii. 123 Ha, Ha! cried all the countrified young thicks he had gone to school with. **1970** G. LORD *Marshmallow Pie* iii. 28 Some of those thicks in Earls Court would do it just for the kicks.

7. A thick fog. Cf. sense 7 a of the adj. in Dict. and Suppl. *slang.*

1936 J. BUCHAN *Island of Sheep* ii. 35 Out of the marshes a fog crept which the gunners call a 'thick'. **1961** PARTRIDGE *Slang* Suppl. 1463/1 *Thick, in the*, in, esp. caught in, a thick fog: R.A.F. (operational 'types'): since ca. 1930.

thick, *adv.* Add: **1. c.** (Earlier and later examples.) Also, *to put* (*spread*, etc.) *it on thick*.

1740 *Champion* 29 Jan. (1741) I. 225 You may lay on Honour and Beauty, and all Manner of Virtues as thick as you please. **1865** 'MARK TWAIN' in *Californian* 6 May 9/3 Don't you think he is spreading it on rather thick? **1929** A. CHRISTIE *Seven Dials Mystery* xviii. 148, I thought Bundle was laying it on a bit thick myself... But Codders is such an ass he'd swallow anything. **1955** W. C. GAULT *Ring around Rosa* xiv. 165 Now she was putting it on as thick as a starlet at a producer's party. **1976** *Times* 24 Mar. 3/2 (Advt.), If we are laying it on a bit thick it's only because we want you to volunteer out of a mature realisation of what the Army can be like.

7. Further examples: **a.** *thick-mined*, *-tangled*; **d.** *thick-plotting*, *-scarred*.

1957 C. DAY LEWIS *Pegasus* 35 The rescuer plunging through some thick-mined region Who cannot rescue and is not to die. **1922** JOYCE *Ulysses* 35 They swarmed loud ..their heads thickplotting under maladroit silk hats. **1969** G. MACBETH *War Quartet* 59 Its enormous back, thick-scarred From under-water struggles. **1956** D. GASCOYNE *Night Thoughts* 15 The shadows drift in tattered velvet bunches, Thick-tangled rags of shadow are set swaying.

thick and thin, thick-and-thin, *phr.* **A. 2. b.** *attrib.* or *adj.* (Earlier example.)

1822 M. EDGEWORTH *Let.* 1 Feb. (1971) 339 Mr. Ellice is a *thick and thin* friend of Lord Byron's and defends him..against his wife and all the world.

thickety, *a.* (Earlier Amer. examples.)

1640 in *Maryland Hist. Mag.* (1910) V. 374 The Neck of land..lyeing between thicketty Creek on the North, hog pen Creek on the South. **1740** J. E. OGLETHORPE *Jrnl.* 14 May in *Coll. S. Carolina Hist. Soc.* (1887) IV. 152 They got into such thickety ground that they could not overtake them.

thick-head. 1. (Earlier example.)

1824 H. WILSON *Mem.* I. ii. 41 'Don't you know,' said thickhead, '..that I am blind as well as deaf?'

thickie (þi·ki). *colloq.* [f. THICK *a.* (*sb.*) + -IE.] = THICK-HEAD 1 a. Cf. THICK *sb.* 6 in Dict. and Suppl.

1968 B. EARNSHAW *At St. David's a Year* 52 When I hear that one of our Sixth Form boys Has been kicked-up by a Haverfordwest thickie For stealing his girl friend, I think 'Good'. **1976** J. I. M. STEWART *Memorial Service* v. 74 Just one more grouse-slaughtering thickie like my father. **1983** *Times* 5 Sept. 3/1 Teachers still think that engineering is a subject for 'thickies'.

thicknesser (þi·knėsəɹ). [-ER[1].] A thicknessing machine.

1920 F. T. HILL *Pract. Aeroplane Constr.* 103 Another machine is the surfacer and thicknesser, in which one set of cutters is used for both operations, the top table for the truing-up process..and the bottom table for thicknessing. **1930** *Times Educ. Suppl.* 20 Sept. 400/1 Panel planer and thicknesser. **1959** *Times* 29 Oct. 13/4 A planing machine, which reduces sawn timber to specified dimensions, or sometimes merely smoothes the surface, is frequently referred to as a thicknesser. **1977** *West Briton* 25 Aug. 8/6 (Advt.), Carpenter's universal planer, thicknesser.

thicknessing *vbl. sb.* (In Dict. s.v. THICKNESS.) Hence **thi·ckness** *v. trans.*, **thi·cknessed** *ppl. a.* (back-formations).

1915 Thicknessed [see *spindle machine* s.v. *SPINDLE *sb.* 17]. **1978** *Early Music* Oct. 506 The marks on the inside of the belly..have contributed to the theory that the central strip was first thicknessed and then bent to the long arch, the outer strip being glued to this solid and then carved in the usual way.

thicko (þi·kǫ). *colloq.* [f. THICK *a.* (*sb.*) + *-O[2].] = THICK-HEAD 1 a.

1976 *Oxford Diocesan Mag.* Nov. 8/2 You have mixed ability teaching throughout, which means..having the thickos in with the brainy and the in-betweens. **1981** P. THEROUX *Mosquito Coast* xiv. 176 Where's the camp store, thicko?

thick-set, *a.* Add: **4.** (Later examples.) (This is now the commonest use.)

1893 H. VIZETELLY *Glances back through Seventy Years* I. viii. 165 Captain Marryat was tall,..but broad shouldered and thick set. **1977** N. ADAM *Triplehip Cracksman* xiv. 143 A thickset..guy in a thick woollen polo-neck sweater.

thick 'un (þi·kˌŏn). *slang.* Also **thick one**. [f. THICK *a.* + UN, 'UN[2] (= ONE *pron.*).] Formerly, a gold sovereign (*half a thick 'un*, ten shillings); *to smash a thick 'un* (see SMASH *v.*[2] 2). Also, a crown or five-shilling piece, and rarely in mod. use applied *loosely* to a pound.

1848 *Sessions Papers Cent. Criminal Court* (Kent cases) 7 Mar. 847 Would not a *thick* one or two be very serviceable this cold weather? **1862** *Cornh. Mag.* Nov. 648 If you like..I will send a few thickuns. **1888** J. PAYN *Eavesdropper* II. ii. 79 'Can you smash a thick un for me?' inquired one, handing his friend a sovereign. **1897** HALL CAINE *Christian* IV. iv. 376 A 'thick 'un'? Oh, that was a sovereign, half a thick'un half a sovereign. **1926** 'SAPPER' *Final Count* iii. 66 Done with you, your Graces; a thick 'un it is. **1968** *Gloss. Brit. Argot* (Paramount Pictures), Quid, nicker, thick 'un, one pound.

thicky, *a.* (Later nonce *poet.* example.)

c **1868** G. M. HOPKINS *Poems* (1967) 211 And light us, Lord, with Thy day-break. Beat from our brains the thicky night And fill the world up with delight.

thief. Add: **5.** *thief-den*; *thief-resistant* adj.; **thief-ant**, a small ant of the genus *Solenopsis* which raids the nests of other ants to steal food; **thief-catcher**, (*b*) (earlier example, applied *loosely* to a lantern); **thieves' kitchen**: see *KITCHEN *sb.* 1 b; **thieves' market**, a street market of a type found in many Eastern cities and elsewhere, at which cheap (sometimes, stolen) goods are offered for sale; cf. *flea market* s.v. *FLEA *sb.* 6.

1904 W. M. WHEELER in *Amer. Naturalist* XXXVI. 952 The walls of the galleries in some of the formicaries were tenanted by teeming colonies of the..thief ant. **1971** E. O. WILSON *Insect Societies* xix. 357/1 Colonies of the 'thief ants'..often nest next to larger ant species. **1851** H. MELVILLE *Moby Dick* III. xxii. 150 What art thou thrusting that thief-catcher into my face for, man? Thrusted light is worse than presented pistols. **1844** POE *Marginalia* in *Wks.* (1902) XVI. 20 A race of dolts.. whose clumsily stolen bulls never fail of leaving behind them ample evidence of having been dragged into the thief-den by the tail. **1963** *B.S.I. News* June 9/1 It was the insurance companies and police who first asked the British Standards Institution to lay down a standard for locks for outside doors which really would be thief-resistant. **1968** *Observer* 22 Dec. 22/2 Locks should be built in and made to British Standard 3621, which ensures that they are thief-resistant, although not thief-proof. **1873** TROLLOPE *Eustace Diamonds* II. xlvii. 278 If such a lot of diamonds had been through the thieves' market in London, they would have left some track behind them. [**1927** B. DIQUI *Visit to Bombay* 62 Null Bazaar is..a big market... An interesting section..is the Chor Bazaar. Chor really means 'thieves.' Chor Bazaar, then, means the bazaar of thieves. Probably in the past thieves disposed of their stolen property here. In this bazaar—the Petticoat Lane of Bombay—you can buy secondhand articles of any description.] **1953** S. BEDFORD *Sudden View* I. xi. 103 The Thieves' Market at Mexico City..where thieves offer goods for sale during a limited time to give the owners a chance. **1979** P. DRISCOLL *Pangolin* xiv. 114 Upper Lascar Row, better known as Cat Street, the thieves' market of Hong Kong.

Hence **thie·fwise** *adv. rare* = THIEF-LIKE *adv.*

1898 W. J. LOCKE *Idols* vi. 75 Creeping thiefwise up the stairs. **1904** HARDY *Dynasts* I. IV. iii. 114 Stealing up to us Thiefwise, by our back door.

Thiersch (tīᵊɹʃ). *Surg.* The name of Karl *Thiersch* (1822–95), German surgeon, used *attrib.* and formerly in the possessive with reference to a split-skin graft including only superficial layers, so that regeneration of the donor area can occur (described by him in *Verhandl. d. Deutsch. Ges. f. Chir.* (1874) III. 69).

1890 W. J. WALSHAM *Surgery* (ed. 3) I. 35 New method of skin-grafting (Thiersch's). **1892** KEEN & WHITE *Amer. Text-bk. Surg.* IV. iii. 1095 (*heading*) Thiersch's method of skin-grafting. **1911** F. S. KOLLE *Plastic & Cosmetic Surg.* xi. 180 The remaining raw surface is either allowed to heal by granulation or is covered immediately with Thiersch grafts. **1977** *Brit. Med. Jrnl.* 29 Jan. 278/1 Simple division with a Thiersch graft to the raw areas suffices.

thigh, *sb.* Add: **4.** *thigh-slapping*; **thigh-length** *a.*, (of a garment, boot, etc.) extending down or up to the thigh; **thigh lift**, a dance movement or gymnastic exercise in which the thigh is raised; the lifting of the thigh in this; **thigh roll**, a roll of padding on a saddle, designed to prevent the girths from slipping backwards and to support the rider's legs in jumping and dressage; **thigh-slapper** *colloq.*, an exceptionally funny joke, description, or the like.

1895 Thigh-length [see *knee-length* s.v. *KNEE *sb.* 13]. **1979** S. SMITH *Survivor* xvii. 177, I wore a thigh-length Indian-print chemise. **1949** SHURR & YOCOM *Mod. Dance* v. 113 Do not allow body to tip forward on thigh lift. *c* **1973** J. CHOLERTON *Acrobatic Section Syllabus* (Assoc. Amer. Dancing) (ed. 6) 2 Thigh Lifts.—sit: legs straight in front—hands on the floor behind, lean slightly back. **1963** E. H. EDWARDS *Saddlery* xiv. 101 The Continental panel is similar, but with..the addition of a thigh roll at the rear... The thigh roll is rarely evident to the rider unless it is very heavily stuffed, and its real use is to prevent the girth straps moving back off the flap. **1976** *Horse & Hound* 3 Dec. 31 (Advt.), Colombo the unique jumping saddle... The exterior thigh roll is an unusual but effective feature. **1965** *Wall St. Jrnl.* 13 Sept. 14/4 The thigh-slapper..the President got off to reporters when Lynda Bird showed up in a billowy muu-muu dress. **1932** V. WOOLF *Common Reader* 2nd Ser. 216 [Meredith] overdoes the pith and the sap; the fist-shaking and the thigh-slapping. **1979** *Dance Mag.* Feb. 32/3 'Lovesick Blues' unexpectedly serves as the up-beat, thigh-slapping finale.

thigmokinesis (þigmokəˈni·sis, -kīn-). *Zool.* [f. Gr. θίγμ-α touch + *o* + *KINESIS.] A kinesis in which the stimulus is absence of touch or body contact.

1940 FRAENKEL & GUNN *Orientation of Animals* ii. 23 It seems probable, therefore, that the contact is the positive stimulus involved and that its action is inhibitory. In that case, *thigmo-kinesis* (θίγμα, touch) is the appropriate name for the reaction. **1954** *New Biol.* XVII. 49 There is..a low thigmokinesis, causing the animals to assume positions in which as much as possible of their surface is in contact with another surface. **1961** *Listener* 21 Sept. 439/3 The earwig..displays marked thigmokinesis. **1976** *Acta Neurobiol. Exper.* XXXVI. 579 Crickets attain their ethological praeferendum of resting site owing to their innate photo- and thigmo-kinesis, which are supplemented by hygrophilia and thermophilia.

thill.[1] **a.** (*a*) For (*obs.*) read 'Now only U.S.' and add later examples.

c **1873** E. DICKINSON *Poems* (1955) III. 869 Elijah's wagon has no thill—Was innocent of wheel. **1901** *Century Mag.* Jan. 452/1 I'm like a bronco in a buggy. I want to bust a thill every time I feel the rein. **1944** *Sun* (Baltimore) 16 May 10/3 An old slave cabin and an old ox thill.

thimble, *sb.* Add: **9. a. thimble-berry**: also, any of several other North American raspberries having thimble-shaped fruit; (earlier and later examples); **thimble-screwer** *Criminals' slang*, one who steals watches or 'thimbles' (sense 7); cf. *SCREWER 2.

1789 R. HASWELL *Jrnl.* 16 Mar. in F. W. Howay *Voy. Columbia to Northwest Coast* (1941) 60 We frequently met with gooseberrys rousberrys currants blackberries strawberries and thimble berries. **1847, 1886** [see *BLACK-CAP 5]. **1946** T. M. STANWELL-FLETCHER *Driftwood Valley* 219 Here the devil's-club gave way..to a dense growth of alders, azaleas, and thimbleberry. **1862** H. MAYHEW *London Labour* Extra vol. 25 'Thimble-screwers', those who wrench watches from their guards. **1932** Thimble-screwer [see *RAMPSMAN].

b. More generally applied *attrib.* or as *adj.* to various objects considered to resemble a thimble in size or shape, esp. as *thimble cup*, *glass*. Also *fig.*

1843 *Ainsworth's Mag.* III 470 Having now taken a lump of sugar..and thrown the same into a thimble glass. **1899** H. B. CUSHMAN *Hist. Indians* 501 The monotonous tinkling and rattling of the thimble bells..could be heard. **1907** G. O. WHEELER *Old Eng. Furnit.* xv. 390 The feet [in Sheraton's chairs] were of the thimble-toe or thermed variety. **1933** N. WALN *House of Exile* i. 24 Warm rice-wine, served in thimble cups. **1955** M. ALLINGHAM *Beckoning Lady* xi. 157 The comfortable landlady brought two thimble-glasses, frighteningly overfilled. **1962** *Flight International* LXXXII. 354/1 The new thimble blower is believed to be the world's smallest pressure blower. **1971** *Daily Colonist* (Victoria, B.C.) 24 Nov. 2/1 The thimble-brains who perpetrate the law-breaking know the police are so manacled. **1983** *Times* 16 Dec. 6/4 The thimble measure should be filled to the top.

thimble-eyed, *a.* (Example.)

1815 *Trans. Lit. & Philos. Soc. N.Y.* I. 422 Thimble eyed, bull eyed, or chub mackerel..comes occasionally in prodigious numbers to the coast of New-York in autumn.

thin, *a.* (*sb.*) and *adv.* Add: **A.** *adj.* **I. 1. e.** Phr. *the thin end of the wedge*: see WEDGE *sb.* 2 b.

II. 2. a. Also *thin on top*: of a man, having little hair on the (top of the) scalp, balding. Also, of the hair itself.

1868 TROLLOPE *He knew He was Right* (1869) I. xxxi. 243 'You are not bald at all.'..'I am beginning to be thin enough at the top.' **1921** G. B. SHAW *Back to Methuselah* v. 217 Getting a little hard set and flat-chested and thin on the top, wasn't she? **1933** W. S. MAUGHAM *Sheppey* I. 2 'Air's very dry, sir..getting a bit thin on top. **1950** J. CANNAN *Murder Included* vi. 124 There 'e goes—thin on top, ain't 'e? 1978 L. MEYNELL *Papersnake* vi. 77 At forty-one his hair was definitely receding and getting thin on top.

e. *thin on the ground*: (chiefly of persons) few in number, widely scattered; scarce, and therefore difficult to find. Also of a group, having few members. Cf. *THICK *a.* 5 a.

1951 W. S. CHURCHILL *2nd World War* IV. I. vi. 86 There was very heavy fighting and many craft were sunk, but the Australians were thin on the ground and enemy parties got ashore at many points. *a* 1957 A. BROOKE in A. Bryant *Turn of Tide 1939–43* (1957) ii. 115, I got up early..and started with the 3rd Division, which I found well established but infernally thin on the ground. 1964 'A. GILBERT' *Knock, knock, who's There?* i. 14 The customers were still pretty thin on the ground. 1976 A. HILL *Summer's End* ii. 22 Work was a bit thin on the ground everywhere, wi' long dole queues. 1980 *Times Lit. Suppl.* 25 July 850/1 Even now, when the Anglo-Irish are precariously thin on the ground, people among them who don't like horses can be miserable in certain counties.

f. *Mountaineering.* Of or pertaining to a rock face on which good climbing holds are hard to find.

1955 S. STYLES *Introd. Mountaineering* 144 *Thin*, generally used of steep rock, meaning 'smooth; having few or very small holds'. 1963 A. GREENBANK *Instructions in Rock Climbing* vi. 73 When a guidebook says '*strenuous*', it usually means steep, fierce-looking rock; '*delicate*' or '*thin*', the footholds and/or handholds are tiny. 1970 R. JAMES *Rock Climbing in Wales* 161 Climb this buttress up the L. side, centre and R. side respectively, each giving a thin lower pitch followed by a short artificial section. 1981 *Fell & Rock Jrnl.* XXIII. II. 199 To its left Wafer Thin gives some very thin climbing up flaky pockets to a final smooth slab.

g. *thin red line*: see *RED LINE *sb. phr.* Similarly *thin blue line*: a line of policemen, esp. one which holds back a surging crowd; also *transf.*, the defensive barrier of the law.

1962 *Sunday Times* 16 Dec. 17/2 (*caption*) The 'thin blue line' at an anti-nuclear demonstration. 1970 G. JACKSON *Let.* 17 Apr. in *Soledad Brother* (1971) 222 You've heard the patronizing shit about the thin blue line that protects property and the owners of property. 1979 'M. UNDERWOOD' *Smooth Justice* ii. 45 The sort of protection we can give..isn't even a thin blue line.

3. b. Also in phr. *to vanish* (*melt*, etc.) *into thin air*: to disappear completely from sight or existence (formerly only of spirits). More rarely *to come* (etc.) *out of thin air*. Now chiefly *colloq.*

1610 [in Dict.]. 1671 MILTON *P.R.* I. 499 Satan bowing low His gray dissimulation, disappear'd Into thin air diffus'd. *c* 1800 BLAKE *Vala* v, in *Compl. Writings* (1966) 305 As plants wither'd by winter..Melt into thin air. 1904 CONRAD *Nostromo* I. i. 4 Vapours that..vanish into thin air. 1918 L. STRACHEY *Eminent Victorians* 223 The Ever Victorious Army..was an ill-disciplined, ill-organised body..constantly on the verge of mutiny..and, at the slightest provocation, melting into thin air. 1932 W. FAULKNER *Light in August* ix. 204 Having apparently materialised out of thin air. 1951 *Sport* 7–13 Jan. 16/2 Speed, confidence, shooting ability, all seemed to have vanished into thin air. 1977 'E. MCBAIN' *Long Time no See* xi. 181 The recurring nightmares hadn't come out of thin air.

4. d. Phr. *a thin time*: a wretched period of experience. *colloq.*

1924 A. J. SMALL *Frozen Gold* iv. 108 Yes, sure, you go ..If you don't, she will give me such a thin time. 1935 *Economist* 17 Aug. 326/1 Dairy farming and lumbering and doing poorly; while the mountain peasantry especially are having a thin time. 1955 *Times* 22 Aug. 3/3 The London sides in the Championship had a thin time. Not one of them won.

e. *Econ.* Of or pertaining to a stock market (or to stocks, shares) in which trading is light.

1931 *Economist* 28 Feb. 441/2 Prices were marked up to 10 cents a pound in the hope of attracting buyers who had refused to take metal at 9½ cents, but the market remains thin. 1946 *Sun* (Baltimore) 17 Jan. 12-0/1 Some of the 'thin' shares tacked on around six points. 1964 *Financial Times* 3 Mar. 19/2 Further speculative buying in a thin market led to a fresh rise. 1981 *Times* 30 June 20/1 Dealers described turnover as thin.

D. I. a. *thin-blooded*, *-flanked* (example).

1934 WEBSTER Thin-blooded. 1959 *Times* 10 June 7/3 The rest of the programme, though it sometimes achieved a sort of thin-blooded distinction, was really rather disappointing. 1894 KIPLING *Seven Seas* (1896) 148 Till you married that thin-flanked woman.

b. thin-film *a.*, applied to processes and devices that employ or involve a very thin solid or liquid film; **thin-layer chromatography** *Chem.* [tr. G. *dünnschicht-chromatographie* (E. Stahl 1956, in *Pharmazie* XI. 633)], chromatography in which compounds are separated on a thin layer of adsorbent material such as charcoal or silica gel; **thin-minded** *a. rare*[−1], narrow-minded, prejudiced; **thin section,** a thin, flat piece of rock or tissue prepared with a thickness of about 0·03 mm. for examination with an optical microscope; also, a piece of tissue of the order of 30 nm. thick prepared for electron microscopy; hence **thin-sectioning** *vbl. sb.*, the making of thin sections; **thin space** *Typogr.*, a piece of metal used for separating words, cast five to an em of its own body; cf. *thick space* s.v. *THICK *a.* 12 b.

1956 *Nature* 24 Mar. 571/2 Thin-film lubrication. 1963 *New Scientist* 21 Mar. 632/3 Thin-film memories and logic devices. 1966 D. G. BRANDON *Mod. Techniques Metallogr.* ii. 90 Variations in absorption with crystalline perfection contribute significantly to the contrast in thin-film transmission microscopy. 1970 *Brit. Printer* July 69/2 The advent of thin-film inks gave the screen printer a choice which had not previously existed. 1957 *Chem. Abstr.* LI. 6948 (*heading*) Thin-layer chromatography (the method, affecting factors, and a few examples of application). 1961 *Jrnl. Amer. Oil Chemists' Soc.* XXXVIII. 313/1 Two procedures for the analysis of mixtures of mono-, di-, and triglycerides. One employs.. thin-layer chromatography. 1967 *Oceanogr. & Marine Biol.* V. 267 Thin layer chromatography can pinpoint some inaccuracies in the interpretation of spectra and give a more adequate image of the variety of pigments. 1978 H. H. BAUER et al. *Instrumental Analysis* xxi. 626 Appreciation of the full advantages of planar chromatography then led to thin-layer chromatography (TLC). 1864 TROLLOPE *Small House at Allington* I. ii. 11 Such thin-minded men can hardly go to the proof of any matter without some pre-judgment in their minds. 1858 *Q. Jrnl. Geol. Soc.* XIV. 469 For some purposes, however, thin sections are quite indispensable. 1872 F. DELAFIELD *Handbk. Post-Mortem Examinations* I. 21 The proportion of alcohol is to be afterward increased until the mucous membrane is hard enough to be cut into thin sections. 1916 JORDAN & FERGUSON *Text-bk. Histol.* xx. 734 Thick sections may be obtained from the firmer tissues by freehand sectioning with a razor, but for the satisfactory preparation of thin sections a microtome is a necessity. 1956 *Nature* 14 Jan. 98/1 Although electron microscope contrast may be increased by the use of objective apertures, accurate focusing in thin-section work is still difficult. 1970 *Ibid.* 17 Oct. 251/2 Petrological analysis by thin section has enabled the locality of origin of axes made from hard rock to be identified. 1964 G. H. HAGGIS et al. *Introd. Molecular Biol.* v. 135 The pellet which contains them [*sc.* mitochondria] can be identified, and its purity assessed, by thin-sectioning of the osmium-fixed embedded pellet. 1978 *Sci. Amer.* May 141/2 There are two principal specimen-preparation methods for rendering cells suitable for examination in the electron microscope: thin-sectioning and freeze-fracturing. 1683 Thin space [see SPACE *sb.*[1] 15 b]. 1808 C. STOWER *Printer's Gram.* iii. 90 Of Spaces... Five to an m—or five thin spaces. 1968 J. R. BIGGS *Basic Typogr.* 76/2 Space between words is achieved by means of tiny bricks of metal... They are..thin space..thick space..hair space.

II. thin-clad *a.*: also (*U.S. colloq.*) *absol.* as *sb.*, an athlete.

1947 *Sun* (Baltimore) 30 Oct. 19/2 (*heading*) Maryland thinclads beat navy. 1974 *Anderson* (S. Carolina) *Independent* 24 Apr. 5B/1 Cliff Satterwhite..has been coaching the few Trojan thinclads.

thing, *sb.*[1] Add: **I. 3. b.** With possessive adj. One's particular interest, speciality, or talent. *spec.* in colloq. phr. *to do one's* (*own*) *thing*: to do what one wants, to follow one's interest or inclination.

Evidence for this sense is patchy into the early-twentieth cent. The phrase had become a cliché (often associated with the 'hippie' culture) by the late nineteen-sixties.

1841 EMERSON *Essays* ii. 54 But do your thing and I shall know you. Do your work, and you shall reinforce yourself. 1861 R. W. DIXON *Christ's Company* 98 Go thy way, all things say, Thou hast thy way to go..Do thy thing. 1909 H. G. WELLS *Ann Veronica* xvi. 328 Every human being..exists to do new things... Well, this is *our* thing. 1914 *Egoist* 1 June 216/1, I cannot picture the spring of the editor's actions as being a..desire to do the decent thing. I think, rather, she insists on doing her own thing—what it pleases or suits her to do. 1951 'M. INNES' *Operation Pax* vi. vi. 285 Roof-climbing used to be one of my things, rather. 1962 I. MURDOCH *Unofficial Rose* xxii. 214 Mummy won't be happy, it's not her thing. 1968 [see *FREAK *v.* 3]. 1968 *Melody Maker* 23 Nov. 23/6 No one is right and no one is wrong as long as they say what they feel—as long as they do their thing. 1970 E. BULLINS *Theme is Blackness* (1973) 165 Anything that anybody wants to do is groovy with me... Go ahead and do your thing, champ. 1971 M. SPARK *Not to Disturb* ii. 49 'What are they doing here, anyway in this world?' Heloise, pink and white of skin, fresh from her little sleep, says, 'Doing their own thing.' 1974 K. MILLETT *Flying* (1975) II. 207 She knows her thing. And I am doing mine. 1981 R. BARNARD *Sheer Torture* x. 109 A ghastly warning against..aiming at total self-fulfilment, doing your own thing regardless.

c. *Loosely*, with qualifying adj. or noun (phrase): matter, business; preoccupation (influenced by next sense).

1906 'H. MCHUGH' *Skiddoo!* vii. 94 When it comes to that poetry thing he thinks he can make Hank Longfellow beat it up a tree. 1909 ST. J. LUCAS *First Round* III. xxxiii. 320, I shall have to stay there I suppose; they spoke of giving me a fellowship at Balliol, and of course there is the All Souls thing later on. 1968 T. WOLFE *Electric Kool-Aid Acid Test* i. 11 Thousands of kids were moving into San Francisco for a life based on LSD and the psychedelic thing. 1969 *Listener* 27 Mar. 434/1 The male fashion thing. 1976 *National Observer* (U.S.) 7 Aug. 13/1 You can write the nostalgia thing, but it's been done a thousand times before. It's so easy. 1977 A. SHERIDAN tr. *J. Lacan's Écrits* iv. 131 The psychoanalytic thing has become an accepted thing.

d. A preoccupation or obsession. *spec.* (*a*) *to make a thing about* or *of* (something), to preoccupy oneself greatly with (a matter); to make an issue out of (something), or to (over)-exaggerate its importance; (*b*) *to have a thing about* (occas. *for*) (a person or thing), to be obsessed by (something); to harbour a prejudice or fear about. *colloq.*

[illegible]

4. (*the*) *last thing* (earlier example); also *last thing at night*.

1848 TROLLOPE *Kellys & O'Kellys* II. xii. 281, I must see her the last thing,—about nine. 1935 *Discovery* Apr. 95/2 It [*sc.* the pump] is run last thing at night. 1966 'C. AIRD' *Religious Body* viii. 74, I do a round of doors and windows last thing at night.

II. 7. c. Also (often with initial capital) applied to some particular supernatural or other dreadful monster (i.e. *the Thing*). Hence *transf.* (sometimes *humorously*) of persons.

1888 KIPLING *Smith Administration* (1891) 64 The burning-*ghât*, where a man was piling logs on some Thing that lay wrapped in white cloth. 1917 CONRAD *Shadow-Line* vi. 197 The hair of my head stirred... I could see It—that Thing! 1954 L. M. BOSTON *Children of Green Knowe* 126 The Thing..gave a silent yell... Then it went fumbling round the room. 1973 'B. MATHER' *Snowline* i. 7, I find The Thing hard to take. He's blind,..he can only make mewing noises, and he has no legs and only one arm.

10. a. Also in phrases: *dear old thing*, an expression of affection applied esp. to an elderly person; *old thing*, a jocular or affectionate form of address (not necessarily to an elder). *colloq.*

1852 *Punch* 31 July 55/2 Aunt Ratchet and I had quietly sat down, I to read and she to listen to a new novel—the greatest pleasure the dear old thing can experience. 1864 C. M. YONGE *Trial* II. xiv. 273 I'll do anything for you..you know that, you old thing! 1865 'L. CARROLL' *Alice's Adventures in Wonderland* ix. 130 'You can't think how glad I am to see you again, you dear old thing!' said the Duchess, as she tucked her arm affectionately into Alice's, and they walked off together. 1905 KIPLING *Actions & Reactions* (1909) 18 No, thanks, old thing! Isn't that quite English? 1921 [see *CLEVER-CLEVER *a.*]. 1975 J. DRUMMOND *Slowly the Poison* 13 Don't worry, old thing. It may not be as bad as it sounds.

11. c. (Later examples.) Usu. preceded by possessive pron.

1955 J. P. DONLEAVY *Ginger Man* (1957) vi. 38 Men wagging their things at you from doorways. Disgusting. 1969 L. HELLMAN *Unfinished Woman* ii. 23 One..had opened his pants and was shaking what my circle called 'his thing'. 1981 P. TURNBULL *Deep & Crisp & Even* vi. 110 'His coat was open and his thing...' 'Thing?' 'You know, between his legs... Penis.'

d. With capital initial. Substituted (esp. after a title, as *Miss Thing*, etc.) for the proper name of a person which the speaker cannot recall. Cf. THINGUMMY, WHAT'S-HIS-NAME. *colloq.*

1920 J. M. BARRIE *Kiss for Cinderella* I. 12 She was called something else when she came—Miss Thing, or some such name. 1954 M. RIDDELL *M for Mother* x. 44 Mrs. Thing had absolutely washed her hands of him and my mother was never going to speak to her again. 1977 M. KENYON *Rapist* vi. 70 Keane could not remember the name of..the colonel. Too many names. Colonel Thing.

III. 14. e. (Earlier examples.)

1798 A. F. M. WILLICH *Elem. Crit. Philos.* 21 The position of the sufficient ground, in general, depends.. upon *things in themselves*. 1817 COLERIDGE *Biog. Lit.* I. x. 195 Of this sheet of paper..as a thing in itself, separate from the phænomenon or image in my perception.

f. Also *to tell* (or *teach*) (someone) *a thing or two*; similarly, *to be up to a thing or two*, to be knowing or shrewd.

1816 *Sporting Mag.* XLVIII. 173 The training-groom was up to a thing or two. 1859 HOTTEN *Dict. Slang* 113. 1930 WODEHOUSE *Very Good, Jeeves!* vi. 143 'Listen, Bertie,' said Aunt Dahlia earnestly, 'I'm an older woman than you are—well, you know what I mean—and I can tell you a thing or two.' 1932 L. GOLDING *Magnolia St.* III. ix. 592 It's taught us both a thing or two. 1973 M. BENCE-JONES *Palaces of Raj* xi. 191 Simla could teach Naini Tal a thing or two as regards dances.

g. *one thing..another* (*thing*): see *ONE *numeral a.* 17 c.

h. (*the*) *things of the mind*, matters of a specifically intellectual character. Cf. *life of the mind* s.v. *LIFE *sb.* 12 e.

[1902 H. JAMES *Wings of Dove* II. iii. 44 All the high, dim things she lumped together as of the mind.] 1903 G. K. CHESTERTON *Robert Browning* iii. 61 She..lived her second and real life in literature and the things of the mind. 1965 *New Society* 15 July 10/3 The superiority of the things of the mind over the externals of bodily appearance and success in competitive enterprises. 1980 T. MORGAN *Somerset Maugham* III. 222 Syrie..had no interest in things of the mind. She was the sort who says 'how extraordinary' when a book is being discussed.

i. *of all things*: of all conceivable possibilities (often parenthetically implying that the eventuality is surprising or unexpected).

1925 T. DREISER *Amer. Tragedy* II. II. xii. 170 Well, well, of all things! Well, I'll be damned! 1958 A. HUXLEY *Let.* 22 June (1969) 851 There have been endless contretemps, including, as a last straw, the collapse of the publicity woman with, of all things, chickenpox. 1977 MCKNIGHT & TOBLER *Bob Marley* ix. 110 Keyboard instrument effects..which sound like a harmonium of all things.

j. (*just*) *one of those things*: see *ONE *pron.* 29**f.

k. *like one thing*, 'like anything'. *Austral. colloq.* and *U.S. dial.*

1946 B. JAMES in Murdoch & Drake-Brockman *Austral. Short Stories* (1951) 249 No good for crops,..but it would 'grow cherries like one thing'. 1948 D. BALLANTYNE *Cunninghams* xiv. 75 He saw Phil..and some other..jokers skiting like one thing. 1972 J. S. HALL *Sayings from Old Smoky* 136 *Like one thing*, said of something very well done or in large quantity. 'He can mimic Windy Bill just like one thing.'

l. With reference to a previous statement: *to do that* (*small*, etc.) *thing*, to act in the manner indicated (esp. when taking up a suggestion). *colloq.*

1958 I. MURDOCH *Bell* xiv. 186 'I'm going to have a bath.' 'Darling, you do that small thing!' 1960 K. AMIS *Take Girl like You* vi. 83 'Grab one with us.'..'I'll do that small thing if I may.' 1963 N. FREELING *Because of Cats* i. 23 'I'll plan that.' 'You do that thing.' 1967 N. MARSH *Death at Dolphin* iii. 57 'Will you bear me in mind, then?' 'I'll do that thing,' said Peregrine. 1977 J. TARRANT *Rommel Plot* ix. 89 'I'll be there in twenty minutes.' 'You do that small thing.'

15. a. (Earlier examples of a person's physical condition.)

1832 J. ROMILLY *Diary* 20 Sept. (1967) 19 Better today: tho not quite the thing: dined at home. 1854 C. M. YONGE *Heartsease* I. II. i. 115 And how are you? You don't look quite the thing.

b. Also *colloq.* in weakened use (*the thing is*..), the truth or the fact of the matter.

1971 C. BONINGTON *Annapurna South Face* xiv. 175, I think the thing is that we want to start pushing out the route as fast as possible because the faster we can push the route out the less oxygen we need to use. 1976 'A. HALL' *Kobra Manifesto* xv. 208 They've struck some kind of problem... The thing is they've seized a TWA Boeing.

f. *any old thing*: any thing whatever. *slang* (orig. *U.S.*).

1900 ADE *More Fables in Slang* 205 An Author was sitting at his Desk trying to..grind out Any Old Thing that could be converted into Breakfast Food. 1911, etc. [see *any old*.. s.v. *ANY *a.* 1 e].

18. thing-word, a substantive referring to some material object; after Jespersen, *spec.* a countable noun.

1877 H. SWEET in *Trans. Philol. Soc. 1875–6* 487 'Snow'..is both a thing-word and a noun, 'white' is a quality-word and an adjective, 'whiteness' a quality-word and a noun. 1914 O. JESPERSEN *Mod. Eng. Gram.* II. v. 115 Another difference in the adjuncts of mass-words and thing-words: the former have *what*, the latter *what a* in exclamatory quasi-questions. 1937 A. SMEATON tr. *R. Carnap's Logical Syntax Lang.* v. lxxvii. 297 'Thing' is a universal word (provided that the designation of things constitutes a genus)... 'Moon' is a thing-word..; 'five' is not a thing-word, but a number-word.

For 'all' read 'chiefly' in 'all *rare* or *nonce-wds.*' **thinghood** (later examples); **thingification** = REIFICATION; hence **thi·ngify** *v. trans.*, **thi·ngifying** *vbl. sb.*; **thinginess**, (*a*) (examples); (*b*) (example); **thi·ngism** *Fr. Lit.* [tr. Fr. *chosisme*] (see quot. 1966); **thingling** (later humorous example); **thingly** *a.* (earlier example); **thingness** (later examples).

1919 A. N. WHITEHEAD *Princ. Natural Knowledge* II. vi. 73 Events appear as indefinite entities without clear demarcations and with mutual relations of baffling complexity. They seem..deficient in thinghood. 1950 A. HUXLEY *Themes & Variations* 55 A completer deification of the State, accompanied by a completer reification, or reduction to thing-hood, of individual persons. 1972 L. HUDSON *Cult of Fact* 76 To know about nature, and especially about people, in a way that reduces them to thing-hood, is to pursue knowledge in a way that is inimical to the proper growth of human self-awareness. 1947 *Partisan Rev.* XIV. 456 Everything in this icy landscape must be adapted to the *things* of cold steel. The organic must fuse with the mechanical, the *thingification* of man be pushed to its extreme, and the world of the workers become functional and naked. 1979 E. P. THOMPSON in *PN Rev.* No. 9 (Suppl.) p. xxvi, He had fallen on the 1844 MSS, was high on alienation and reification (which he insisted upon rendering as 'thingification'), and he had put Marx and Freud together in the bed of a single book. 1931 Thingify [see *REIFY *v.*]. 1972 *Guardian* 7 Feb. 12/8 The thingifying of anything else on the road—whether it's another competitor or..a stray pedestrian. 1914 *Morning Post* 26 Feb. 2/2 Let us hear the second-hand eloquence of one of those second-rate authorities in his vain attempt to get at the thinginess of such things. 1962 W. NOWOTTNY *Lang. Poets Use* v. 107 The 'thinginess' even of what we call 'concrete objects' is so inaccessible to the probe of our common language. 1976 *New Yorker* 15 Mar. 118/2 The very thinginess of the contemporary city life in the film shows—the fast cars, the unobtrusive chic, the small cafes—seems to deprive us of any illumination. 1982 T. GUNN *Occasions of Poetry* I. 22 He was in love with the bare fact of the external world, its thinginess. 1961 *Guardian* 7 Feb. 9/6 M. Robbe-Grillet.. hears his method described as 'thingism' because he concentrates on..things. 1966 H. T. MOORE *Twentieth Cent. French Lit.* II. v. 116 They often produce the antiroman (antinovel) or indulge in chosisme (which might be literally translated as thingism)... In their novels and manifestoes, the antinovelists emphasize their escape from the conventional novel's preoccupation with straight-line plot, psychological analysis, and moral involvements. The group of chosistes concentrate on material objects because, in the words of one of their practitioners and spokesmen, Alain Robbe-Grillet, 'things *are there*'. 1950 O. NASH *Family Reunion* 45 I'd rather shake hands with Mr. Ringling And tell him his circus is a beautiful thingling. 1860 J. W. PALMER tr. *Michelet's Love* II. iv. 101 Things have cast off their *thingly* qualities. 1930 'WYNDHAM LEWIS' *Apes of God* IX. 288 Health as intended by Kalman is '*thingness*' right enough! It is vegetable bulk, it is unconsciousness. 1967 S. BECKETT *No's Knife* 34 Into what nightmare thingness am I fallen? 1975 *New Yorker* 2 June 90/1 Clouseau finds himself forced into unkind collision with the thingness of things.

thingum. Add: **thingumajig** (earlier example).

1824 *Casket* June 76 I'd a lot of cousins, that 'com'd all the way down from Varmount to larn the fashions, and to hear and see all the cute and curious thingumajigs of the Old Colony'.

thingummy. Add: Also in extended form **thingummytight** (**-tite**, etc.).

1937 G. FRANKAU *More of Us* xvii. 177 Quick. The small green phial. It's in my bathroom. In the thingummytight—The corner cupboard. 1939 J. CARY *Mister Johnson* 23 What's the trouble? Why, it's thingummytite, aren't you? 1977 D. CLARK *Gimmel Flask* viii. 147 We've got a thingumitite with us...a sort of visionary. Young cops with fantouche ideas! 1980 D. BOGARDE *Gentle Occupation* i. 21 Nothing in the taps of course because the terrorists had buggered up the hydroelectric thingummytites.

thingy, *sb.* (In Dict. s.v. THING *sb.*[1]). Add: **2.** Also **thingie**, (occas. **-ee**). = THING *sb.*[1] (in various senses); cf. THINGUMMY. *colloq.*

1933 GREEN & STEPT (*song-title*) Swingy little thingy. 1968 M. RICHLER *Cocksure* v. 32 It was going to be the rage. A thingee. Like TW3. 1977 *Spare Rib* June 26/3 Then there are those women who make men wear things on their thingies. 1981 J. BARNETT *Firing Squad* xiii. 184 We don't do crime here... Contracts, copyright, companies floated, that's our thingie.

think, *v.*[2] Add: **I. 2. c.** Also, with adj. as quasi-obj. or used quasi-advb., to think in terms of, prefer, have in view (things that are —), esp. *to think big*, to be ambitious.

1960 J. CARSWELL *South Sea Bubble* vi. 100 Even Blunt, though his maxim was certainly to 'think big', would hardly have suggested anything so obviously unacceptable. 1962 A. LURIE *Love & Friendship* xv. 300 Living in a small town had subtly affected my mind, and I had begun to Think Small. 1970 in M. Pei *Words in Sheep's Clothing* ii. 14 For those who think old! 1972 D. HASTON *In High Places* xii. 139 With people pulling off tricks like the West Ridge, Dhyenfurth was not day-dreaming in thinking tall. 1978 *National Geographic* Nov. 615/2 When people think apple,..they usually think red. 1979 *Now!* 21–27 Sept. 134/3 Simple, uncluttered and tubular they illustrate fashion's new mood of 'think thin'. 1981 *Daily Tel.* 21 Dec. 2/3 (*heading*) 'Think British' call to CBI firms.

3. a. *to think for oneself*: to form independent judgements, not to be overinfluenced by preconceptions or received opinions; *to think out loud* = *to think aloud*; *to think straight*: see *STRAIGHT *adv.* 1 g.

1735 BERKELEY *Defence of Free-Thinking in Math.* xix. 19 The only advantage I pretend to, is that I have always thought and judged for myself. 1853 DICKENS *Let.* 28 Nov. (1938) II. 522 One of the great uses of travelling is to encourage a man to think for himself. 1870 J. P. SMITH *Widow Goldsmith's Daughter* vi. 90 The merry mischief in his eyes..made her feel her absurdity in thinking out loud. 1974 *Times* 1 May 6/6 Those matters were thoroughly probed..the President often taking the role of devil's advocate; sometimes merely thinking out loud. 1974 D. L. EDWARDS *What Anglicans Believe* xii. 100 Anglican laymen have been encouraged to think for themselves.

b. *to give* (one) *furiously to think*: see *GIVE *v.* 38.

II. 5. a. *to think that—!* (int.): introducing a statement of a fact thought of as remarkable or surprising.

1906 BELLOC *Hills & Sea* 30 To think that you can get to a place like that for less than a pound! 1919 G. B. SHAW *Heartbreak House* II. 60 And to think that I actually condescended to fascinate that creature there to save you from him! 1946 M. PEAKE *Titus Groan* 337 To think that an hour earlier she had been helping to plait those locks.

b. (Examples of (*it*) *makes you think*.)

1879 J. BLACKWOOD *Let.* 14 Jan. in *Geo. Eliot Lett.* (1956) VII. 94, I have been reading it all with great interest, and it does make one think. 1933 M. LOWRY *Ultramarine* I. 41, I wonder why it [*sc.* a carrier pigeon] had a message from Swansea... Makes you think, that, doesn't it? 1968 G. MITCHELL *Three Quick & Five Dead* ii. 67 'But young Otto is a psychopath!' said Laura. 'Makes you think a bit, that does,' agreed the Superintendent. 1976 'D. HALLIDAY' *Dolly & Nanny Bird* x. 127 Remember how Comer came bursting in one evening?.. It makes you think, doesn't it?

c. Also, *to think back* (*on* or *to*), to recall, reflect on; (*when one*) *come*(*s*) *to think of it*: see *COME *v.* 23 b.

1960 *Times Lit. Suppl.* 17 June 387/1, I think back to the sixth forms of the twenties. 1965 H. GOLD *Man who was not with It* iii. 29 Goombye, I thought back to him. 1976 *Times Lit. Suppl.* 22 Oct. 1327/2 When I think back on it now, that was the best thing I could have done.

g. *to think twice* (const. *about* or *absol.*): to hesitate, change one's mind (about), decide against (something); also, in a negative context: (not) to take any notice of or worry about.

1898 G. B. SHAW *Philanderer* III. 119 He thinks twice before he commits himself. 1910 [in Dict., sense 5 b]. 1936 W. H. SAUMAREZ SMITH *Let.* 21 Nov. in *Young Man's Country* (1977) ii. 43 When they find that I am neither a Blue..nor a bridge-player, they may think twice about offering the appointment to me. 1955 M. HASTINGS *Cork & Serpent* v. 67 Don't worry. Nobody here thinks twice about me. 1956 A. HUXLEY *Let.* 13 Aug. (1969) 805 He thought I had been wise to think twice about talking on TV about mescalin. 1979 R. RENDELL *Make Death love Me* i. 8 The rule was made to be broken and no one ever thought twice about breaking it. 1981 P. SALWAY *Roman Britain* 705 This must make us think twice before attributing reasons to *any* funerary or religious practice from the ancient world for which we have no written evidence.

h. *to think again*: to realize that one is mistaken, to change one's mind, to have second thoughts. Cf. THOUGHT[1] 2 c.

1911 G. B. SHAW *Getting Married* 291 So youre not coming home with me. *Hotchkiss*..: Yes I am. *Mrs George*: No. *Hotchkiss*: Yes. Think again. 1935 C. S. FORESTER *African Queen* ii. 35 It would blow this ole launch..to Kingdom Come. You think again, miss. 1958 P. SHORE in N. Mackenzie *Conviction* 37 Those who imagine that the problem of the public schools will disappear..will have to think again. 1974 M. GILBERT *Flash Point* xiii. 109 It *was* a put-up job. If they think I'm going to sit down under it, they can think again.

7. c. *to think on one's feet*: to react to events, etc., quickly and effectively.

1935 WODEHOUSE *Luck of Bodkins* xvi. 202 PS. Think on your feet, boy! 1960 *Analog Science Fact/Fiction* Oct. 73/2 Your records show that you can think on your feet. 1976 J. ARCHER *Not a Penny More* x. 116 'James,' said Jean-Pierre, thinking on his feet for not the first time in his life. 'You take a taxi immediately.' 1981 T. WRIGHT in *Believing in Church* v. 112 Reports approved by Synod (the Church thinking on its feet).

III. 9. b. *that's what you think* (cliché, with stress on *you*): an expression of emphatic, sometimes scornful, disagreement.

1934 J. O'HARA *Appointment in Samarra* ii. 31 'I can handle that.' 'That's what you think.' 1973 P. MOYES *Curious Affair of Third Dog* xi. 148 'We're going to have the pleasure of your company for several days at least.' 'That's what you think.'

d. *I thought as much*: see As C. 1 in Dict. and Suppl.

e. *I should think*—, introducing emphatic assent: certainly, assuredly, indeed. Also *ellipt.* in neg., *I should think not.*

1894 A. JESSOPP *Random Roaming* iv. 160 Fish? I should think there was fish! There was fish enough to come to at least £15 of our money. 1903 G. B. SHAW *Man & Superman* IV. 167 Promise me that you wont. *Violet* (very decidedly) I should think not indeed. 1944 L. P. HARTLEY *Shrimp & Anemone* iv. 41 'Do you know.. Nancy Steptoe?' 'I should think I did.'

11. a. For '† *on*, † *at*, dial. *to*' read '† *at*, dial. *to*, *on*', and add later examples with *on*.

1974 *Amer. Speech 1971* XLVI. 117 We should think on each student as unique. 1978 R. HILL *Pinch of Snuff* ix. 85 I've known Charlie for years. I asked what he thought on it.

b. (*c*) *imp.* phr. deprecating proffered thanks or apology.

1948 M. ALLINGHAM *More Work for Undertaker* vii. 87 'We did not disturb you, I hope?' 'Think nothing of it,' murmured the torchbearer magnanimously. 1950 [see *PLEASURE *sb.* 2]. 1980 F. OLBRICH *Desouza in Stardust* iv. 41 'Thank you for giving up so much of your time, Mr Chiknis.' 'Think nothing of it, Chief Inspector.' 1982 W. J. BURLEY *Wycliffe's Wild-Goose Chase* i. 17 'Sorry to bother you on a Sunday morning.' ..'Think nothing of it.'

IV. 16*. *think through* = *think out* (sense 15) (*b*) or (*c*).

1922 HARDY *Late Lyrics* 150 I've been thinking it through, as I play here to-night. 1934 T. N. WILDER *Heaven's my Destination* iii. 42 During the journey he.. 'thought through' the matter of capital punishment. 1961 *Observer* 8 Oct. 10/3 It is doubtful if Mr. Gaitskell himself had thought through the problem of inner-party democracy. 1979 B. HEBBLETHWAITE in M. Goulder *Incarnation & Myth* iv. 97 A remarkable attempt to think

through what it means for our concept of God to say that Christ's cross is God's cross in our world.

17. For ?*U.S. colloq.* read orig. *U.S.* and add: to devise, invent, contrive, or produce by thought or cogitation. (Earlier and later examples.)

1855 MRS. STOWE *Tales & Sk. New Eng. Life* 79 Christmas is coming..and I have got to think up presents for everybody. 1872 S. HALE *Lett.* (1918) iv. 83 He asked our plans at once, took right hold and thought up what we had better do. 1901 MERWIN & WEBSTER *Calumet 'K'* vii. 108, I had him pretty busy there for a while thinking up lies. 1930 *G.K.'s Weekly* 15 Nov. 146/1 If Mr. Lloyd George can think up a good ticket. 1956 *People* 13 May 8/6 In America the magic new process—it was thought up over there—is being developed in all sorts of wonderful ways. 1977 J. WAINWRIGHT *Nest of Rats* I. xii. 103 There's a way round it... There *has* to be. Some brainy type thought it up.

V. *Comb.* **18.** (It is not clear in every case whether the verbal or the nominal sense of *think* is dominant.) **think(s) balloon, bubble,** in a comic-strip cartoon, a circle resembling a balloon or bubble floating above a character's head and containing (the word 'thinks' followed by) the character's thought in direct speech; **think book,** a book containing the writer's thoughts, opinions, observations, etc.; one that makes the reader think; **think box** *colloq.* or *joc.*, the brain; **think factory** *U.S. colloq.*, a research institution; **think-fest** [*FEST], an intellectual treat; **think group,** a group of people that meets to thrash out a subject or problem; **think-man** = *idea*(*s*) *man* s.v. *IDEA *sb.* 12; **think-piece** chiefly *Journalism*, a general article containing discussion, analysis, opinion, etc., as opp. to fact or news; **think tank**: see as main entry.

1959 *Spectator* 31 July 133/3 In a 'thinks' balloon are the words: 'Rock Hanson..looks an awful wolf'. 1977 *Times* 31 May 7/6 Roy Lichtenstein's *Girl at the Piano* has a nice verbal irony in the 'thinks balloon' as she muses. 1962 *Listener* 25 Oct. 694/2 People who want a short, quick holiday from newspapers, problems, and 'think' books. 1917 *Dialect Notes* IV. 330 Brain. Also *think tank, think box*. 1937 *Daily Express* 5 Feb. 10/6 I do not believe that their brains, or think-boxes, are of sufficient calibre to understand what they are preaching. 1964 C. HODDER-WILLIAMS *Main Experiment* I. viii. 95 A drawing of a computer with a think-bubble coming out of it with the caption, 'Computers'. 1981 N. TUCKER *Child & Book* v. 141 The self-proclaiming speech styles of the main characters [in comics], and their periodic 'thinks' bubbles. 1959 *Nation* 24 Jan. 62/2 Other think-factories are Johns Hopkins University Operations Research Office... Johns Hopkins thinks for the Army. Stanford Research Institute..does the bulk of its thinking for a variety of government agencies. 1947 AUDEN *Age of Anxiety* (1948) i. 28 Assembled again For a Think-Fest. 1958 *Sunday Times* 21 Dec. 12/3 The most stimulating think-fest in my week was Sir Kenneth Clark's lucid lecture on the revolting subject, 'Can Art Be Democratic?'. 1967 *Guardian* 26 Sept. 8/1 Think groups, in which scientists frighten one another with visions of a not too distant future. 1967 *Economist* 15 July 187/1 Nor is Mr Brezhnev the think-man who throws up bright new ideas to keep his colleagues on the ball. 1947 *Partisan Rev.* XIV. 478 Rapid withering of talent, as shown in slick formula novels or plays or 'think-pieces' for periodicals, has more often than not been the fate of the intellectual. 1966 E. WEST *Night is Time for Listening* iv. 120 I'm not reporting stories... I'm in the think piece business these days. *a* 1974 R. CROSSMAN *Diaries* (1977) III. 546 They are producing various think-pieces, including one on industrial relations, one on social services and one on poverty, discussion papers out of which will be boiled one policy paper to be presented at Conference.

think, *sb.* Add: **2. b.** *to have another think coming*: to be greatly mistaken.

1937 *Amer. Speech* XII. 317/1 Several different statements used for the same idea—that of *some one's making a mistake*..[e.g.] you have another think coming. 1942 T. BAILEY *Pink Camellia* xxvii. 199 If you think you can get me out of Gaywood, you have another think coming. 1979 *Jrnl. R. Soc. Arts* CXXVII. 221/2 Any design consultant who thinks he is going to get British Leyland right by himself on his own has got another think coming.

thinkable, *a.* Add: Also (*rare*) as *sb.*, a thing that can be thought of, a thinkable thing.

1890 W. JAMES *Princ. Psychol.* I. xiii. 529 As 'thinkables' or 'existents' even the smoke of a cigarette and the worth of a dollar-bill are comparable. 1907 —— *Pragmatism* iv. 140 *Absolute* generic unity would obtain if there were one *summum genus* under which all things without exception could be eventually subsumed. 'Beings', 'thinkables', 'experiences', would be candidates for this position.

Hence **thi·nkably** *adv.*, in thought, according to thought; conceivably.

1935 *Mind* XLIV. 325 For finitists, 'to exist' means 'to be thinkably constructible'. 1966 *Listener* 9 June 840/3 Death is thinkably of two sorts—(i) the physical breakup of animate entities,..and (ii) our own projected death.

thinkful, *a.* (Earlier example.)

1864 'MARK TWAIN' in *Californian* 1 Oct. 9/3 There is a handsome portrait in the Art Gallery of a pensive young girl... Says she, 'I like it—it is so sad and thinkful.'

think-in. [f. THINK *v.*[2] + *-IN[3] (after *SIT-IN *a.* and *sb.*, *TEACH-IN, etc.).] A meeting, conference, etc., for thoughtful discussion.

1966 *Newsweek* 19 Sept. 30/3 The think-ins..produced only a few flickers of..anti-war sentiment. 1973 *Belfast Tel.* 23 Feb. 4 The Social Democratic and Labour Party is to have a major 'think-in' this weekend to prepare the party and its supporters for the White Paper.

thinking, *vbl. sb.* Add: **1. a.** *high thinking*, idealistic opinions on or attitudes to social, moral, or religious questions; *good* (or *nice*) *thinking*: an expression of approval of a neat, ingenious, or well-thought-out plan, explanation, observation, etc.

1802 High thinking [in Dict.]. 1910 J. LONDON *Let.* 5 June (1966) 307 Bourgeois circles where he expected to find refinement, culture, high-living and high-thinking. 1959 *Manch. Guardian* 11 Aug. 5/2 For all the high thinking that goes with a branch of the United Nations Association and with folk dancing..there is not much sign of hard living. 1968 *Listener* 26 Dec. 848/2 Marc's *Trendy Ape* saw the final disappearance of high thinking in our new Bloomsbury before the onslaught of the Colonel's cry: 'Good thinking!' 1974 L. DEIGHTON *Spy Story* xx. 214 'They might be security police holding your friend Remoziva in custody.' 'Nice thinking, Pat,' said Schlegel. 1977 D. BAGLEY *Enemy* ix. 65 'We need to keep in his good books.' 'Very good thinking,' said Ogilvie.

3. *thinking-material*; **thinking-box**, *colloq.* (*a*) = *think box* s.v. *THINK *v.*[2] 18; (*b*) a study; **Thinking Day,** 22 February, the joint birthday of the first Chief Scout and Chief Guide, kept by members of the Girl Guides Association for thinking of other Guides all over the world; **thinking distance,** the distance travelled by a motor vehicle from the time when the driver first decides to stop until the time when he begins to apply the brake; cf. *stopping-distance* s.v. *STOPPING *vbl. sb.* 7; **thinking-machine,** (*a*) a person whose thinking consists (merely) in mechanical response to symbols; (*b*) *colloq.*, an electronic computer; **thinking-out** [f. vbl. phr. *to think out* s.v. THINK *v.*[2] 15], the activity of reaching an understanding or a solution of (some problem) by a process of thought; **thinking part** (earlier example); **thinking-through** [f. vbl. phr. *to think through* s.v. THINK *v.*[2] 16*] = *thinking-out* above.

1911 'SEPHARIAL' *Kabala of Numbers* I. vii. 75 The thinking-box of a scientific man [*sc.* Newton]. 1915 GALSWORTHY *Bit o' Love* I. 10 He'm in his thinkin' box. 1951 N. G. ANNAN *Leslie Stephen* i. 29 Stephen wanted to appear..as an athlete who incidentally owned a competent thinking-box. 1927 *Girl Guide Gaz.* Feb. 33/1 At the World Conference in America it was suggested by one of the French delegates that there should be an international 'Thinking Day', on which the Guides of all our different countries should remember each other. 1977 *Guider* July 327/2 The colours of the Retford Unit of Ranger Guides were dedicated on Thinking Day this year. 1947 *Highway Code* (*recto rear cover*), Think in terms of overall stopping distance... Thinking distance = Distance travelled before driver reacts. 1980 J. W. HILL *Intermediate Physics* iii. 20 The Highway Code shows that for a car travelling at 70 m.p.h...the 'thinking distance' travelled is 70ft.. before the brakes are applied. 1943 H. READ *Politics of Unpolitical* iii. 46 We teach them [*sc.* children]..to master abstract symbols and the processes of conceptual thought, and by the age of eleven or twelve we have produced a thinking-machine of sorts. 1948 BLUNDEN *Shakespeare to Hardy* ii. 47 His [*sc.* Francis Bacon's] danger was to turn himself into a thinking-machine. 1950 *Mind* LIX. 436 The present interest in 'thinking machines' has been aroused by a particular kind of machine, usually called an 'electronic computer' or 'digital computer'. 1842 POE in *Graham's Mag.* Jan. 68/2 With the increase of the thinking-material comes the desire..of abandoning particulars for masses. 1971 B. Z. DE FERRANTI *Living with Computer* ix. 80 The amount of 'thinking material' in the brains of a number of fishes was doubled by transplantation from other fishes. 1934 H. G. WELLS *Exper. Autobiogr.* II. ix. 654, I was using my prestige and possibilities as an imaginative writer, to do the thinking-out of this problem of human will and government, under fantastic forms. 1946 R. G. COLLINGWOOD *Idea of Hist.* 196 This thinking-out of the meaning of a concept is philosophy. 1890 'B. HALL' *Turnover Club* i. 17 Then he uses this man to play thinking parts, like the *Bleeding Officer* and the two armies. 1971 *Listener* 16 Dec. 838/2 An inadequate thinking-through of what those fine phrases will mean in practice.

thi·nk-tank. orig. *U.S.* [f. THINK *v.*[2] + TANK *sb.*[1]] **1.** *colloq.* The brain. *U.S.*

1905 A. L. STILLMAN in A. H. Shearer *Little Bk. Rutgers Tales* 51 There's too much scrapping in the Joint... Your Think-tanks are getting to be Air-tight Compartments. 1910 [see *MAVERICK *v.* b]. 1964 *St. Louis Post-Dispatch* 8 May 2A/3 Truman..said he hoped to live to be 90 but only 'if the old think-tank is working'.

2. A research institute or other organization providing advice and ideas on national or commercial problems; an interdisciplinary group of specialist consultants. Also in extended (usu. facetious) use.

1959 *Times Lit. Suppl.* 6 Nov. p. xix/2 Even the Institute of Advanced Studies at Princeton does not quite meet the bill, nor does the 'think tank', the Center for Behavioral Sciences at Palo Alto. 1963 *Business Week* 13 July 61 There are many others..in the special groups—or 'think tanks'—that do analytical work for the armed forces, such as the Air Force's RAND Corp., the Navy's Operations Evaluation Group, the Army's Research Analysis Group, the Defense Dept.'s Institute for Defense Analysis. 1967 MRS. L. B. JOHNSON *White House Diary* 8 Oct. (1970) 577 Mt. Hope Farm..will be the site for the environmental planning center—a sort of a 'think tank' for city-planning experts. 1968 *Sunday Times* 25 Feb. 10 The private research corporations, or 'think tanks' (in the current American terminology) which are paid, mostly by departments of Government, to think about problems. 1968 *Economist* 13 Apr. 29/3 'Think tank' work is usually secret. 1969 *Sunday Mail Mag.* (Brisbane) 4 May 2/7 He's pretty busy as a 'Business Doctor' there—runs a 'think-tank'. 1970 *Daily Tel.* 22 Jan. 3/5 P & O, the world's largest shipping group, have formed a special company to exploit the inventions of their technical staff. A 'think tank' of 40 graduate engineers and naval architects will work on ideas and suggestions sent in by the sea-going and shore staffs. 1971 *Ann. Reg. 1970* 36 Lord Rothschild..described the task of the new organization (or, as it was nicknamed, the 'think-tank') as being to 'weigh up the pros and cons and consequences' of a proposed policy and to advise the Cabinet on it. 1973 *Listener* 26 Apr. 534/3 It has become the fashion among heads of governments and chiefs of state in the parliamentary democracies to equip themselves with a special staff of personal advisers—a think-tank, an entourage..—whose views supplement and sometimes run counter to the processed findings of civil servants. 1976 H. WILSON *Governance of Britain* iv. 95 In addition to establishing the Central Policy Review Staff (the Think Tank),..he [*sc.* Edward Heath] initiated the practice of the Cabinet Office Units, which has been developed since. 1978 R. HILL *Pinch of Snuff* xiv. 146 He leaves the service, possibly under a cloud... The inference in the rugby club think-tank was that the cloud was sexual. 1981 *Daily Tel.* 15 Dec. 13/2 The Rand Corporation, a 'think tank' which undertakes various studies for the Pentagon, completed an analysis of all the Communist-bloc armies. *Ibid.* 29 Dec. 10/4 Lean times are looming for the private consulting firms, 'think tanks' and universities.

3. A meeting or conference of experts, scholars, specialists, etc.

1976 T. SHARPE *Wilt* iv. 39 She had been accepted by people who flew to California or Tokyo to conferences and Think Tanks as casually as she took the bus to town. 1978 *Washington Post* 8 May A14/2 The meeting here, which one official called the NAACP's first 'think tank', was an effort to develop new positions for the association.

Hence **think-tanker,** a member of a think-tank.

1971 *New Scientist* 2 Sept. 536/2 Throughout this century think-tankers have been confidently predicting the imminent exploitation..of the seas. 1975 'J. LYMINGTON' *Spider in Bath* vii. 124 A brief report from our Think Tankers on hypnotising the village people.

thinner. Add: **1.** Also a machine for thinning plants, seeds, etc.

1943 [see *PELLETED *ppl. a.* 2]. 1962 *Times* 21 May 18/4 More use of mechanical beet thinners.

2. (Also in colloq. pl. form.) A liquid used to dilute paint, printing-ink, etc., to a suitable consistency.

1904 *Jrnl. Franklin Inst.* July 17 The painter then adds thinners until the paint will work under his brush. 1958 B. BEHAN *Borstal Boy* III. 338 He..told me where I'd find an extra can of turps, if I wanted thinners. 1967 *Gloss. Paper/Ink Terms for Letterpress Printing* (*B.S.I.*) 9 *Thinner*, a fluid for addition to a printing ink to reduce its consistency. 1973 J. G. TWEEDDALE *Materials Technol.* II. ii. 33 A viscous liquid constituent may present particular problems, perhaps requiring..thinning by solution with a volatile thinner, to make it fluid enough for mixing and pouring. 1980 *New Scientist* 23 Oct. 244/1 This bottle of correction fluid..would be all right if I..added thinners.

thinness. 1. b. (Later example.)

1932 G. GREENE *Stamboul Train* I. i. 5 Her mackintosh showed the thinness of her body.

thinning, *vbl. sb.* Add: spec. in *Forestry*, removal of some of the trees in an immature stand, or removal of part of the crown of an individual tree. Also *attrib.*

1800 W. PONTEY *Profitable Planter* 33 It is not to be wondered at, if, even before thinning, the soil should be nearly exhausted, and the trees checked in their growth. 1822 J. C. LOUDON *Encycl. Gardening* III. iii. 1108 Autumn, or very early in spring, are the proper seasons for thinning where the trees are to be taken up by the root and replanted elsewhere. 1880, 1922 [see *COPPICING *vbl. sb.* b]. 1970 H. L. EDLIN *Collins Guide to Tree Planting & Cultivation* xi. 170 The thinning cycle, or interval of years between thinnings, can be short..or long. *Ibid.*, In regular thinnings a definite proportion of the growing stock of trees is taken out.

thio-. Add: **1. thi·azine** [*AZINE], any of a class of dyes that contain a ring of one nitrogen, one sulphur, and four carbon atoms in the molecule, such as thionine and methylene blue; **thiazole,** also, any of the substituted derivatives of this compound; (further examples); [ad. G. *thiazol* (Hantzsch & Weber 1887, in *Ber. d. Deut. Chem. Ges.* XX. 3118)]; **thiocho·line,** the sulphur analogue of choline, $HS{\cdot}CH_2CH_2N(CH_3)_3OH$, or a derivative in which the hydroxyl group is replaced by an organic radical; **thioe·ster,** the sulphur analogue of an ester, containing the group

—CO·S—; **thio-e·ther,** any compound in which an atom of sulphur is bonded to two organic radicals; **thioglyco·llic** (also **-glycolic**) **acid** [tr. G. *thioglycolsäure* (P. Claesson 1877, in *Ann. d. Chem.* CLXXXVII. 113): see GLYCOLLIC, GLYCOLIC *a.*], a colourless liquid, $CH_2(SH){\cdot}COOH$, that is a strong reducing agent used as a reagent for detecting ferric iron; so **thioglyco·l(l)ate,** a salt or ester of this acid, esp. the sodium salt, used in culture media to produce anaerobic conditions; **thioke·tone,** a sulphur analogue of a ketone, containing the group >CS; **thi·onate,** a salt of a thionic acid; **thionine** (examples); **thionyl** (earlier example); **thiophene**: also **thiophen;** (further examples); **thi:osemica·rbazide** [*SEMICARBAZIDE], a colourless crystalline compound, $H_2N{\cdot}CS{\cdot}NH{\cdot}NH_2$, used esp. as a rodenticide and as a stabilizer in organic liquids; **thi:osemica·rbazone,** any of a class of compounds analogous to the semicarbazones, the oxygen being replaced by a sulphur atom; **thiote·pa, -TEPA** [*TEPA], the thio analogue, $PS(N(CH_2)_2)_3$, of tepa, used in the treatment of cancer; **thiou·racil,** a mercapto derivative of uracil that has been used to depress the activity of the thyroid gland; 4-hydroxy-2-mercaptopyrimidine, $C_4H_4N_2OS$; **thioxa·nthene, † -en,** a tricyclic crystalline compound, $C_{13}H_{10}S$, that is the sulphur analogue of xanthene; also, any of a class of derivatives of this that includes several antipsychotic tranquillizers similar to the phenothiazines.

1893 *Jrnl. Soc. Chem. Industry* 31 Jan. 4 The original colour quickly reappears on exposure to air: Azine-, Oxazine-, Thiazine-, and Acridine-Colours. 1971 R. L. M. ALLEN *Colour Chem.* viii. 130 Thiazine dyes are used on cellulosic fibres, silk, bast fibres, leather and paper. 1888 *Jrnl. Chem. Soc.* LIV. 574 Thioamides condense with α-halogen-substituted ketones to form thiazoles. 1956 I. L. FINAR *Org. Chem.* II. xii. 451 A general method for preparing thiazoles is the condensation between α-halogenocarbonyl compounds..and thioamides. 1929 *Bull. Chem. Soc. Japan* IV. 176 Thio-choline bromide was prepared by heating bromocholine bromide with 2-thio-uracil or 4-methyl-2-thio-uracil with water. 1980 *Sci. Amer.* Apr. 37/3 Nerve gas in the sampled air inhibits the enzyme, just as it would in the human body, resulting in a drop in the thiocholine level, which triggers the alarm. 1952 *Jrnl. Biol. Chem.* CXCVI. 545 These results thus confirm the conclusions of Lynen and Reichert that the acetyl group of acetyl CoA is attached, in thioester linkage, to the thioethanolamine portion of the CoA molecule. 1979 *Nature* 1 Mar. 86/1 The role of ATP and other energy-rich phosphates is considered in detail and this is followed by a study of thioesters including coenzyme A derivatives. 1889 G. M'GOWAN tr. *Bernthsen's Text-bk. Org. Chem.* iv. 94 The Thio-ethers, also termed alkyl sulphides, *e.g.* ethyl sulphide, $(C_2H_5)_2S$, are..neutral volatile liquids. 1979 *Nature* 20–27 Dec. 808/2 Each haem is linked, as in cytochrome *c*, to the apoprotein by two thioether bonds. 1877 *Jrnl. Chem. Soc.* XXXII. 595 Carius obtained an acid of the formula $H.C_2H_3(HS)O_2$, which he called monosulphoglycollic acid. Some uncertainty was attached, however, to the constitution of the product of this reaction, and the author now shows that both thioglycollic acid, $H.C_2H_3(HS)O_2$, and thiodiglycollic acid..are produced. 1980 A. L. SMITH *Microbiol. & Path.* (ed. 12) v. 64/1 Thioglycollate broth, a special medium containing thioglycollic acid, supports the growth of anaerobes.. without special seal. 1877 *Jrnl. Chem. Soc.* XXXII. 595 Potassium thioglycollate..crystallises in masses of small needles, and is readily soluble in water and alcohol. 1976 *Nature* 24 June 652/1 Much early work was concerned with the evaluation of different methods for breaking disulphide bonds, and procedures were developed using.. sodium thioglycollate. 1889 G. M'GOWAN tr. *Bernthsen's Text-bk. Org. Chem.* 542/1 (Index), Thio-ketones. 1965 *New Scientist* 30 Dec. 921/2 The thioketones (compounds containing the >C=S group) are in general red oils with intense nauseating smells. 1878 *Chem. News* 20 Dec. 294/2 (*heading*) Notes on certain thionates. 1938 *Thorpe's Dict. Appl. Chem.* (ed. 4) II. 574/2 All thionates are decomposed by heat, yielding generally sulphates, sulphur dioxide, and, except with dithionates, sulphur. 1886 *Jrnl. Chem. Soc.* L. 53 The addition of strong hydrochloric acid does not turn the solution blue, as is the case with thionine. 1956 *Thorpe's Dict. Appl. Chem.* (ed. 4) XI. 590/1 Lauth had already indicated that bright blue dyes could be obtained by methylating thionine but such a process was not economic. 1976 *Nature* 1–8 Jan. 60/2 Feulgen reaction carried out on the specimens treated with..thionin and exposed to light gave differential staining. 1866 *Chem. News* 9 Mar. 117/1 M. Wurtz presented a note 'On the Synthesis of Chloride of Thionyle'. 1903 A. J. WALKER tr. *Holleman's Textbk. Org. Chem.* 500 Thiophen can be synthesized by various methods, the most important being the interaction of succinic acid and pentasulphide of phosphorus. 1932 I. D. GARARD *Introd. Org. Chem.* xiii. 183 Coal tar benzene always contains thiophene, C_4H_4S, which boils at 85° and is therefore not readily removed by distillation. 1951 *Engineering* 23 Nov. 667/3 Thiophen cannot be removed from [town] gas by any practical chemical method. 1967 M. J. JANSSEN *Organosulfur Chem.* i. 10 Benzene derivatives are much less readily hydrogenated than thiophene. 1894 *Jrnl. Chem. Soc.* LXVI. I. 76 (*heading*) Derivatives of thiosemicarbazide. 1971 *Chem. Abstr.* LXXIV. 74835 Thiosemicarbazide..given i.p. to mice did not affect the incorporation of intraventricularly administered..putrescine-2HCl..into γ-aminobutyric acid in the brain. 1902 *Jrnl. Chem. Soc.* LXXXII. 572 The thiosemicarbazones of aldehydes and ketones readily yield insoluble copper, silver and mercury derivatives, which can be used for the purpose of isolating these compounds. 1979 *Cancer Res.* XXXIX. 4601/1 The isoquinoline thiosemicarbazone derivatives have been shown to be potent inhibitors of ribonucleotide reductase. 1953 *Arch. Internal Med.* XCII. 629 The purpose of this communication is to present our preliminary experience in the treatment of..human leukemias..with triethylene thiophosphoramide (Thio-TEPA). 1976 *Nature* 13 May 135/1 Criticism of the use of the alkylating agent thiotepa (triethylene thiophosphoramide) to sterilise mosquitoes, as part of an eradication programme in India, focused on its toxicity and the possibility that it or its breakdown products would harm other animal components of the food chain. 1905 WHEELER & BRISTOL in *Amer. Chem. Jrnl.* XXXIII. 458, 2-Thiouracil... This compound was first obtained..when pseudoethylthiourea, containing some thiourea, was condensed with ethyl sodium formylacetate. 1977 *Martindale's Extra Pharmacopœia* (ed. 27) 304/1 Thiouracil was formerly used in the control and treatment of thyrotoxicosis and in the preparation of patients for thyroidectomy. 1911 *Jrnl. Chem. Soc.* XCIX. 145 In order to obtain this sulphoxide [*sc.* diphenylmethane *o*-sulphoxide], thioxanthen was oxidised with hydrogen dioxide in acetic anhydride solution. 1924 '*Chem. Age*' *Chem. Dict.* 148/2 *Thioxanthenes*, derivatives of thioxanthene. 1945 *Jrnl. Chem. Soc.* 659 (*heading*) Action of oxygen in sunlight on thioxanthen.

2. thiace·tazone [ACET(YL + *SEMICARB)-AZONE], a semicarbazone used as a bacteriostatic drug in the treatment of tuberculosis and leprosy; 4-acetamidobenzaldehyde thiosemicarbazone, $C_{10}H_{12}N_4OS$; **thi·obacillus** *Biol.* [mod.L., coined in Ger. (M. W. Beijerinck 1904, in *Centralbl. f. Bakteriol.* II Abt. 597)], a rod-shaped Gram-negative autotrophic bacterium deriving energy from the oxidation of sulphur and certain sulphur compounds, and belonging to the genus *Thiobacillus*; **thi·ochrome** *Biochem.* [ad. G. *thiocrom* (R. Kuhn et al. 1935, in *Zeitschr. f. physiol. Chem.* CCXXXIV. 196), f. Gr. χρῶμα colour], a yellow basic solid, $C_{12}H_{14}N_4OS$, that has a strong blue fluorescence in solution and is formed when thiamine is oxidized in a procedure for the estimation of the latter; **Thiokol,** a proprietary name for various polysulphide rubbers and liquids; **thiome·rsal** 'MER(CURY *sb.* + SAL(ICYLATE *sb.*], a bacteriostatic and fungistatic organomercury compound used as a disinfectant for the skin and internally and as a preservative for biological products; sodium ethylmercurithiosalicylate, $H_5C_2{\cdot}Hg{\cdot}S{\cdot}C_6H_4COONa$; cf. *MERTHIOLATE; **thiona·zin** [*pyr*)*azin*(*yl* in the systemic name, f. PYR(O- + AZ(O- + -IN[1] + -YL], an insecticide and nematocide, $(C_2H_5O)_2{\cdot}PS{\cdot}O{\cdot}C_4N_2H_3$; **thioridazine** (-ridǝ̄·zīn) [f. PIPE)RID(INE + *AZINE], a phenothiazine derivative, $C_{21}H_{26}N_2S_2$, that is a white or yellow powder and is given orally as a tranquillizer, esp. in cases of schizophrenia and mania; **thiothi·xene** [f. *thi*(*o*)*x*(*anth*)*ene*], a derivative, $C_{23}H_{29}N_3O_2S_2$, of thioxanthene given orally as an antipsychotic drug.

1952 *Lancet* 1 Mar. 436/2 Para-acetamidobenzaldehyde Thiosemicarbazone. [*Note*] This substance..is marketed under various names;..thiacetazone, &c. 1976 MACGILLIVRAY & HALL in G. S. Avery *Drug Treatment* xiv. 382/2 Massive breast enlargement has been seen with isoniazid regimens containing thiacetazone. 1951 *Biol. Abstr.* XXV. 811/2 It was possible to isolate thiobacilli capable of changing hyposulfides into sulfates. 1973 *Nature* 11 May 99/2 Thiobacilli, iron bacteria and algae can survive in acidic water of about pH 2. 1935 *Chem. Abstr.* XXIX. 6242 When crude lactoflavin is made alk., the fluorescence changes from yellowish green to blue. This phenomenon is due to the presence of a S-contg. pigment for which the name thiochrome is proposed. 1963 STEYN-PARVÉ & MONFOORT in Florkin & Stotz *Comprehensive Biochem.* XI. i. 16 The thiochrome method is based on the observation..that oxidation with alkaline ferricyanide converts thiamine into a compound with intense blue fluorescence: thiochrome... The thiochrome is extracted..and the fluorescence of the extract measured. 1930 *Official Gaz.* (U.S. Patent Office) 20 May 555/2 *Thiokol* for sulfur-containing plastic material used in the manufacture of gaskets,..protective coatings, and like products. 1936 *Industr. & Engin. Chem.* Mar. 275/1 Various olefin-polysulfide reaction products.., under the trade name of Thiokols, have been presented to the industry. 1943 *Trade Marks Jrnl.* 8 Dec. 525/1 *Thiokol*... Thermo-setting or thermo-plastic condensation products of the nature of rubber, being compounds of or containing sulphur, and articles (not included in other Classes) made therefrom. Thiokol Corporation.., Trenton, New Jersey. 1972 *Materials & Technol.* V. xiv. 491 All thiokols are originally obtained in latex form, and as such they have found some industrial applications as an impregnant for textiles and leather. 1958 *Brit. Pharmacopœia* 675 Thiomersal should be protected from light. 1968 WILSON & SCHILD *Appl. Pharmacol.* (ed. 10) xxxvi. 671 Phenylmercuric nitrate, thiomersal (merthiolate), and other organic mercurials, have a better therapeutic index than mercuric chloride. 1964 *B.S.I. News* Mar. 23 Thionazin. 1974 MARTIN & WORTHING *Pesticide Man.* (Brit. Crop Protection Council) (ed. 4) 488 Thionazin is a soil insecticide and nematicide effective against..nematodes,.. root maggots and..aphids. 1959 *Jrnl. Pharmacol. & Exper. Therapeutics* CXXVI. 312 (*heading*) Some neuropharmacological properties of thioridazine hydrochloride (Mellaril). 1965 SIMPSON & IQBAL in *Current Therapeutic Res.* VII. 697 (*heading*) A preliminary study of thiothixene in chronic schizophrenics. 1976 SMYTHIES & CORBETT *Psychiatry* x. 194 Thiothixene is a potent and effective antipsychotic agent in acute and chronic schizophrenia.

3. Used *attrib.* as an independent word (without hyphen), denoting the presence of a sulphur atom, usu. in place of one of oxygen.

1879 *Chem. News* 24 Oct. 204/2 (*heading*) Organic thio compounds. 1926 *Chem. Abstr.* XX. 364 An investigation of the chemistry of the thio ketones as compared with that of the ordinary O ketones. 1955 KIRK & OTHMER *Encycl. Chem. Technol.* XIV. 61 Thio amides may react in either the thiono form, $RC(:S)NH_2$, or the tautomeric thiol form, $RC(:NH)SH$. 1980 J. W. COOPER *Spectroscopic Techniques Organic Chemists* vi. 186 Amino, cyano, and thio groups.

thioctic (þɔi,ǫ·ktik), *a. Chem.* [f. THIO- + OCT(A- + -IC.] *thioctic acid*: any of the sulphur-containing acids with the formula

$$\underbrace{S{-}S{-}CH_2(CH_2)_x}CH(CH_2)_{5-x}COOH,$$

where x = 1, 2, 3, 4, or 5; *spec.* 6,8-dithio-*n*-octanoic acid (x = 1), = *α-lipoic acid* s.v. *LIPO-.

1952 J. A. BROCKMAN et al. in *Jrnl. Amer. Chem. Soc.* LXXIV. 1868/2 The name 'thioctic acid' is proposed for this structure (x = 2), a sulfur-containing organic acid with 8 carbon atoms. 1953 FRUTON & SIMMONDS *Gen. Biochem.* xxxviii. 900 Recent work has assigned to α-lipoic acid.. the structure of 6,8-dithio-*n*-octanoic acid ('6-thioctic acid'). Isomers of this compound such as the 4,8-dithio acid ('4-thioctic acid') or the 5,8-dithio acid ('5-thioctic acid') have less POF [*sc.* pyruvate oxidation factor] activity. 1975 *Sci. Amer.* Mar. 98/2 In the 1950's thioctic acid had been proposed as a remedy for liver damage caused by heavy-metal poisoning.

thioindigo (þɔi,ǫ,i·ndigo). [f. THIO- + INDIGO *sb.* (*a.*).] A red vat dye in which the two imino groups of indigotin are replaced by sulphur atoms; also, any of various derivatives of this also used as dyes.

1906 *Textile Colorist* XXVIII. 321/1 Messrs. Kalle & Co., Aktiengesellschaft, have placed upon the market, under the name of Thio Indigo Red B, a new coloring matter, which like Indigo is admirably suited for dyeing the various textile fibres and for calico printing. 1923 THORPE & INGOLD *Vat Colours* vi. 131 The following aromatic bases have been converted..into the corresponding aromatic thioglycollic acids, which have been transformed..into thioindigos. 1951 KIRK & OTHMER *Encycl. Chem. Technol.* VII. 823 Thioindigo..forms brownish-red metallic crystals when recrystallized from xylene or other organic solvents.... On reduction with sodium hyposulfite in the presence of alkali, it forms a pale yellow leuco compound, Thioindigo White. 1961 COCKETT & HILTON *Dyeing Cellulosic Fibres* v. 185 Thio-indigo is an important basic structure for a number of useful vat dyes.

So **thioi·ndigoid,** any of a class of vat dyes that are substituted derivatives of thioindigo and are used esp. in textile printing; also *attrib.* or as *adj.*

1943 *Thorpe's Dict. Appl. Chem.* (ed. 4) VI. 454/1 This process has been applied..for preparing the orange thioindigoid dye from *p*-phenetidine. 1951 KIRK & OTHMER *Encycl. Chem. Technol.* VII. 824 Although the fastness of thioindigoids does not in general approach that of the anthraquinone vat dyes, the brightness and clarity of shade are in many cases considerably superior. 1952 [see *INDIGOID *a.* (*sb.*)]. 1970 K. VENKATARAMAN *Chem. Synthetic Dyes* III. i. 36 With the exception of halogenated indigo and a few thioindigoids..the indigoid group is steadily declining in commercial importance. 1972 *Materials & Technol.* V. xi. 358 Thioindigoid reds and maroons are vat dyestuff pigments with good light-fastness in full colours and reduced shades and excellent acid and alkali resistance.

thiol (þɔi·ǫl). *Chem.* [f. next.] **a.** = MERCAPTAN.

1900 *Jrnl. Chem. Soc.* LXXVIII. I. 163 Methods used for the preparation of aromatic thiols. 1971 *Nature* 31 Dec. 507/1 This interpretation is supported by the restoration of the equilibrium..when an extraneous thiol, mercaptoethanol, is added.

b. = *MERCAPTO(-) b, SULPHYDRYL in Dict. and Suppl.

1951 C. R. NOLLER *Chem. Organic Compounds* xiv. 265 The —SH group is known as the thiol or sulfhydryl group, or more commonly as the mercapto group. 1973 *Sci. Amer.* Apr. 60/3 All the simplest organic molecules have been found in interstellar space, whereas many of the even simpler nonorganic species such as nitric oxide (NO), sulfur monoxide (SO) and the thiol radical (SH) have not been detected in spite of sensitive searches.

thiol-. Substitute for etym.: [f. THIO- + -OL.]

thiopental (þɔi,ope·ntăl). *Pharm.* Chiefly *U.S.* [f. next + -AL (cf. PENTAL).] = next. Also called *thiopental sodium*.

1947 *U.S. Pharmacopœia* 572 Thiopental Sodium occurs as a yellowish white, hygroscopic powder. 1955 GOODMAN

& Gilman *Pharmacol. Basis Therapeutics* (ed. 2) v. 60/2 Ether (as well as cyclopropane and thiopental) causes renal vasoconstriction.

thiopentone (þaiˌopeˈntōᵘn). *Pharm.* [f. Thio- + *pento(barbit)one.] A sulphur analogue of pentobarbital sodium that is given intravenously as a short-acting general anæsthetic; sodium 5-ethyl-5-(1-methylbutyl)-2-thiobarbiturate, $C_{11}H_{17}N_2O_2SNa$. Also called *thiopentone sodium*. Cf. *Pentothal.

1945 *Brit. Pharmacopœia 1932* Add. vii. 65 (*heading*) Soluble thiopentone. **1952** [see *hexobarbitone]. **1965** J. Pollitt *Depression & its Treatm.* iv. 50 It [sc. electro-convulsive treatment] is usually administered after the patient has been anæsthetised with thiopentone sodium. **1977** *Proc. R. Soc. Med.* LXX. 782/1 Generalized reactions were not seen for the first twenty years after the introduction of thiopentone.

thiram (þaiˑræm). *Chem.* [f. *thi(u)ram* in the systematic name (see def.), f. *thi(o)-ur(ea* s.v. Thio 1 + Carb)am(ic *a.*] Tetramethylthiuram disulphide, $[(CH_3)_2N{\cdot}CS{\cdot}S\text{—}]_2$, used as a fungicide and a seed protectant.

1950 *Phytopathology* XL. 118 The Subcommittee on Fungicide Nomenclature of The American Phytopathological Society, cooperating with the Interdepartmental Committee on Pest Control, has selected common names for five commercially-available fungicidal chemicals... The coined common names and designations are:.. *Thiram* for the fungicidal chemical tetramethylthiuram disulfide. **1962** *Amateur Gardening* 10 Feb. 10 The chemical known as thiram..remarkably increases the percentage of germination of most garden seeds. **1975** *Daily Tel.* 15 Nov. 8/2 A refinement would be to dust them over with antiseptic thiram powder before putting them in a box of peat..to await another spring for replanting.

third, *a.* (*adv.*), *sb.* Add: **I.** *adj.* **1. e.** In proverbial phr. (*the*) *third time('s) lucky*.

[*c* **1840** Browning *Lett.* (1933) 5 'The luck of the third adventure' is proverbial.] **1862** A. Hislop *Proverbs Scotl.* 194 The third time's lucky. **1882** R. L. Stevenson *New Arabian Nights* II. 59 'The next time we come to blows—' 'Will make the third,' I interrupted... 'Ay, true... Well, the third time's lucky.' **1942** N. Marsh *Death & Dancing Footman* vii. 123 It was a glancing blow..but..it might have been my head... One of them's saying to himself: 'Third time lucky'. **1979** J. Tate tr. *Blom's Limits of Pain* ix. 82 Lars Westerberg discovered that the expression third time lucky had something in it.

4*. With following superlative: having two superior in the specified attribute; third in point of quality, position, etc.

1375, 1859 Third best [in Dict., sense 5]. **1938** [see *last *a.* 1 b]. **1962** E. Snow *Other Side of River* (1963) lxvii. 508 In 1960 it was the world's third-greatest reservoir. *Ibid.* lxxv. 577 The most significant additions to China's third-largest educational center are the T'ung Chi Medical College and hospitals. **1979** *Dædalus* Winter 62 Pursuing policies that would be optimal in a first-class world when one actually lives in a..third-best world can be highly inefficient.

5. *third base* (examples), *baseman, cousin* (examples), *-level, realm, -stage, story* (later example); **third dimension**, the dimension of thickness or depth (see Dimension *sb.* 3 a); hence **third-dimensional** *a.*; **third ear** esp. in *Psychoanal.*, a figurative ear which listens intuitively for what lies behind the words heard by the actual ears; **third eye** *Hinduism* and *Buddhism*, the eye of insight or destruction located in the middle of the forehead of the god Siva; hence *transf.*, the power of inward or intuitive sight occasionally gained by humans; **third eyelid**, the nictitating membrane of many animals; **third flute** *Mus.*, a flute pitched a minor third above the ordinary flute (see quots.); **third force, Third Force** [after Fr. *Troisième Force*], a political party or parties standing between two extreme or opposing parties (formerly, esp. between the French Gaullists and Communists); also *loosely*, any neutral power or third body; **third-generation** *attrib.*: see *generation 4 b; **third-grader** *N. Amer.*, a pupil in the third grade (*grade *sb.* 4 c) at school; **third house** (earlier and later examples); **Third International**: see *international *sb.* b; **third man**, (*a*) *Cricket* (earlier example); (*b*) *Lacrosse*, a defence player placed behind the centre; the position occupied by him; (*c*) *Philos.* [Gr. τρίτος ἄνθρωπος], a term from Aristotle (*Metaphysics* Bk. A 990b 17) for a third element (or man) which, in the paradox stated in Plato's *Parmenides*, seems to be needed in arguments from the particular instance (of a man) to the ideal form (of Man); hence *attrib.*, as *third-man argument*; (*d*) *Boxing slang*, the referee; (*e*) an unidentified third participant in a crime; **third market** *U.S.*, trade in stock undertaken outside the stock exchange; cf. *off-board *a.*; **third point**: hence **third-pointed** *a.*; **Third Position**, a name applied to the political stance of Juan Domingo Perón (1895–1974), President of Argentina (1944–55 and 1973–4), being neither capitalist nor communist, but a combination of Fascism and socialism; cf. *Justicialism, *Peronism; **Third Programme**, (from 1946 to 1967, when its name was changed to 'Radio 3': see *radio *sb.* 2 d) one of the three national radio networks of the BBC, broadcasting programmes of a predominantly cultural nature; often used allusively to qualify what is considered intellectually superior or 'highbrow'; **third rail**, (*a*) (i) (earlier example); cf. *conductor rail* s.v. *conductor 12 d; (ii) an additional rail for the accommodation of trains with a wider gauge; (*b*) *U.S. slang*, used *attrib.* to designate highly intoxicating liquor; **third reading, Third Reading**, the third and final presentation of a parliamentary bill after amendments have been made, sometimes allowing for a final debate before it is voted on; cf. Reading *vbl. sb.* 2 c; **third sex**, homosexuals (in quot. 1821, eunuchs) regarded as a separate sexual group; **third slip**: see *slip *sb.*³ 14 c; **third stream** (also hyphenated and with capital initials), a style of music which combines elements of jazz and classical music (see quots.); **third wave** [in allusion to Plato's metaphor (*Republic* 472 a) τὸ μέγιστον τῆς τρικυμίας 'the greatest of the three waves'], the last and most forceful of three successive arguments or propositions; **third way, Third Way**, used in a variety of contexts to designate a third possible ideology or solution to a problem (see quots.); **Third World War**, a hypothetical third war involving the majority of the world's nations (cf. *First World War* s.v. *first *a.* C. 2; *Second World War* s.v. *second *a.* 7 a); **third year**: see also Year 3 b.

1845 in *Appleton's Ann. Cycl. 1885* (1886) XXV. 77/2 A ball knocked outside the range of the first or third base is foul. **1946** *Chicago Sun* 2 July 25/3 He can start a club that would have a Red Sox star at every position except third base and right field. **1857** *Spirit of Times* 7 Feb. 373/1 Mr. Scott, their third base man is always at his post. **1936** O. Nash *Primrose Path* 38 Long have I wondered why a locomotive engineer should be so much nicer than an ambassador or a novelist or a banker or a third-baseman or a quartermaster or a lancer. **1978** *Detroit Free Press* 16 Apr. (Detroit Suppl.) 23/3 Phil spent all of 1977 with Tigers and figures to be the club's 3rd baseman of the future. **1840** Lytton *Money* I. ii. 7 You are very, very, very distantly connected with the deceased—a third cousin, I think? **1921** G. B. Shaw *Back to Methuselah* II. 65 They are all third cousins of somebody with a title or a park. **1858** Third dimension [see Dimension 3 a]. **1923** H. Crane *Let.* 20 Jan. (1965) 116, I prefer Egyptian sculpture to the Greek, and this book makes me feel that the Greeks had more to express in line and design than they had in the third dimension. **1964** M. McLuhan *Understanding Media* (1967) I. i. 28 He acquires the illusion of the third dimension. **1934** H. C. Warren *Dict. Psychol.* 277/1 Third-dimensional. **1937** *Univ. Calif. Publ. Mod. Philol.* XX. 188 Only with such a spray [sc. lipiodol] can the third-dimensional aspect be brought out, giving vivid pictures of the epiglottis and tongue. **1954** *Ann. Reg. 1953* 365 Third dimensional (3-D) or stereoscopic films viewed through polaroid spectacles were no novelty in London. **1907** H. Zimmern tr. *Nietzsche's Beyond Good & Evil* viii. 202 What a torture are books written in German to a reader who has a *third* ear... These were my thoughts when I noticed how..unintuitively two masters in the art of prose-writing have been confounded. **1948** T. Reik *Listening with Third Ear* II. xv. 144 The psychoanalyst has to learn how one mind speaks to another beyond words and silence. He must learn to listen 'with the third ear'. **1979** F. Kermode *Genesis of Secrecy* i. 5 The best psychoanalysts are admired ..for their powers of divination, for the acuteness of their third ear. **1810** E. Moor *Hindu Pantheon* 36 He [sc. Siva] has a third eye in his forehead, pointing up and down. **1921** [see *satori]. **1936** Dylan Thomas *Twenty-Five Poems* 38 No third eye probe into a rainbow's sex That bridged the human halves. **1978** S. Gooch *Paranormal* v. 202 It is the pineal gland to which the Hindu mystics of 3000 years ago gave the name of 'the third eye'—the 'eye' of clairvoyance and second sight. **1822–34** Third eyelid [see Nictitating *ppl. a.*]. **1892** C. S. Minot *Human Embryol.* (1897) xxviii. 727 The third eyelid is well developed in birds, etc., but is rudimentary in man. **1983** *Sci. Amer.* Apr. 86/2 When a cat falls asleep..its eyes close, and the nictitating membrane (the 'third eyelid') covers part of the eye under the outer eyelids. **1876** Stainer & Barrett *Dict. Mus. Terms* 433/2 Third flute. [Terzflöte]. **1906** Goodchild & Tweney *Technol. & Sci. Dict.* 434/2 There is also a flute in E♭ (often spoken of as the third flute in F, but tuned to E♭), which transposes a minor third higher. **1954** *Grove's Dict. Mus.* (ed. 5) III. 168/1 In the 18th century this [sc. the Flute in F] was known as the 'third' flute or 'tierce', since it stood in pitch a minor third above the ordinary flute, whose lowest note at that time was most usually d'. [**1933** *Esprit* 1 Sept. 718 Le projet qui suit a été établi par le Comité économique du mouvement de la Troisième Force et adopté par son Congrès National, à Tours, les 28 et 29 juillet.] **1936** E. Burns tr. *Thorez's France To-day & People's Front* IV. xxv. 228 The 'new economic régime' proposed in the 'Third Force' plan is dressed up in anti-Capitalist garb to make it capable of attracting and winning over the masses. **1951** N. Mitford *Blessing* II. xi. 256 Mr Clarkley, more interested in French politics than English elegance, began asking a few questions about the Third Force. **1955** G. Greene *Quiet American* II. iii. 160 There was always a Third Force to be found free from Communism and the taint of colonialism—national democracy he called it. **1956** *Foreign Affairs* XXXV. 60 An armed 'third force'. **1963** *Listener* 31 Jan. 194/2 Some Europeans have a vision of a great power arising to take its place alongside the Soviet Union and the United States—a third force, possibly armed with a separate European deterrent free of American control. **1971** *Irish News* 31 Aug. 1 What was needed was an immediate increase in the strength of the UDR—or if necessary the formation of a 'third force'. **1974** *Times* 27 Feb. 6/2 A doubling of the vote for the third-force candidates would still leave the relative positions of the Conservative and Labour parties unaffected on current evidence from the polls. **1981** *Daily Tel.* 24 Nov. 1/4 The 'third force' which Loyalist hardliners have formed as their own anti-IRA vigilante group made its first significant appearance on the streets during a commemoration service for terrorist victims. **1962** A. Lurie *Love & Friendship* i. 19 You make me sound like a third-grader. 'I learned simple division, Mummy, and drew a picture of an Eskimo.' **1849** *Alta California* (San Francisco) 31 Dec. 1 The solicitude manifested by the members of the legislature to ascertain where they are to get their mileage and per diem, is a subject of much jocularity among the *third house*. **1950** *Look* 31 Jan. 24/1 In a state where the Third House, the lobbyists,..spend millions every year.., a legislator going on a payroll for 75 bucks a week is looked upon as just a precedent-setting price-cutter, undermining the foundations of a fine profession. **1959** M. Schlauch *Eng. Lang. in Mod. Times* iv. 121 These deviations from strictly completed structure, occurring in formal discourse, are obviously very different from the rambling repetitions, the loose pleonasms and unfinished statements of third-level speech as exemplified in Juliet's nurse. **1975** *Cork Examiner* 30 May 10/4 About 55,200 students were expected to leave the primary, post-primary and third-level education this year. **1801** T. Taylor tr. *Aristotle's Metaphysics* I. vii. 26 Some make ideas of things relative, of which we do not say there is an essential genus, and some assert that there is a third man. **1851** F. Lillywhite *Guide to Cricketers* 23 If Long-slip is required, take the Third man away. **1897** E. T. Sachs in S. Christopherson et al. *Hockey & Lacrosse* 104 In third man I like a powerful player, and a tall. **1916** A. E. Taylor in *Proc. Aristotelian Soc.* XVI. 255 What I propose to show is that the appeal to the regress..is certainly not what Aristotle usually has in mind when he speaks of a certain type of argument as the 'third man'. **1920** S. Alexander *Space, Time, & Deity* I. II. iii. 218 This objection..is analogous to one of the kinds of objection taken in ancient Greece to the Forms under the name of the argument of the 'third man'. **1924** W. D. Ross *Aristotle's Metaphysics* I. 195 Other forms of the 'third man' argument. **1927** J. Palmer *Recoll. Boxing Referee* i. 2, I have acted as third man in the ring on at least three thousand occasions. **1949** G. Greene in *Amer. Mag.* Mar. 142 (*title*) The 3rd Man. *Ibid.* 149/2 And the third man? Who was he? **1954** *Philos. Rev.* LXIII. 342 Plato could neither convince himself that the Third Man Argument was valid, nor refute it convincingly. **1960** M. Golesworthy *Encycl. Boxing* 171/2 Corri..was the third man in the ring for the middleweight bout. **1964** *Lacrosse* ('Know the Game' Ser.) 34/2 Third Man should mark Third Home closely. **1977** M. Green *Children of Sun* (rev. ed.) ix. 434 Kim Philby was finally identified as the 'Third Man', in 1963, when he too fled to Moscow. **1964** *Wall Street Jrnl.* 15 Jan. 1/6 A 10-man Big Board committee..is..studying the expanding role of off-board trading, or the 'third market' as it has come to be known. (The other two are the exchange markets and the over-the-counter market in unlisted securities.) **1868** G. M. Hopkins *Jrnls. & Papers* (1959) 186 The nave is very long, the roof, Third-Pointed, very low... The Third-Pointed altar-screen..and the choir screen..were beautiful in design and proportion. **1953** Third Position [see *Peronism]. **1971** Third Position [see *Justicialism]. **1946** *Times* 1 July 8/3 The future of broadcasting and television was outlined by Sir William Haley... He said that a third programme was planned and awaited only the completion of the Brookman's Park high mast. **1946** *Whitaker's Almanack 1947* 349/2 The *Third Programme*, introduced on Sept. 29, 1946, is broadcast on 203·5 metres and 514·6 metres. **1946** *Lancet* 21 Dec. 921/1 Oh yes, I've met *him*, of course—awfully decent fellow and all that, but frightfully Third Programme! **1951** J. B. Priestley *Festival at Farbridge* II. i. 145 She had fine eyes but a rather ugly despairing sort of mouth, as if she came out of one of those Greek tragedies on the Third Programme. **1960** *Guardian* 22 July 6/3 The lectures—one of the 'Third Programme' ventures that Radio Eireann manages to squeeze in to its narrow broadcasting hours. **1966** H. Ogdon in 'H. MacDiarmid' *Company I've Kept* ii. 56 In England, of course, it [sc. an Indian *naga*] is esoteric, 'Third Programme'; a thesis could be written on it. **1980** *Daily Tel.* 29 May 14/6 MacNeice's most famous two plays..had an impact on a mass Home Service audience before he and his work disappeared into the Third Programme. **1867** *Commercial & Financial Chron.* 29 June 808 It is throughout a double track road, and a third rail is laid..for the accommodation of the wide cars of that line. **1890** *Jrnl. Franklin Inst.* CXXIX. 268 In 1879, Dr. Werner Siemens constructed and operated an exhibition railway... A third rail centrally placed between the other two was used as the outgoing conductor. **1916** *Gazette-News* (Asheville, N. Carolina) 7 Jan. 1/2 This recipe is for fourteen and one-half gallons of the 'third-rail' liquor. **1929** J. Callahan *Man's Grim Justice* i. 4 A shot of the third-rail booze that the Silver Alley joints peddled. **1972** *Modern Railways* Sept. 331/3 Invalides is the terminus of the Western Region 750V third-rail service to Versailles Rive Gauche. *Ibid.* 332/3 From October 1, the third-rail electric trains from Paris St Lazare to St Germain will be replaced by 1500V dc RER trains. ?**1571** *House of Commons Orig. Jrnls.* 14 Apr. II. 16 The Bill for Seweres the

thirde readinge. **1878** W. Stubbs *Constitutional Hist. England* III. xx. 466 It [sc. a bill] is brought up for a third reading, debated again if necessary, read a third time and passed. *a* **1974** R. Crossman *Diaries* (1976) II. 407 Though we had 116 to start with there were only ninety-nine left when the Third Reading vote came. **1908** W. James *Let.* 19 Jan. in R. B. Perry *Th. & Char. W. James* (1935) II. 485 Surely truth can't inhabit a third realm between realities and statements or beliefs. **1957** G. Ryle in M. Black *Importance of Lang.* (1962) 167 It is..positively misleading to speak as if there existed a Third Realm whose denizens are Meanings. **1821** Byron *Don Juan* v. xxvi. 148 A black old neutral personage Of the third sex stept up. **1896** J. A. Symonds *Probl. Mod. Ethics* vi. 78 Burton..was led to surmise a crasis of the two sexes in persons subject to sexual inversion. Thus he came to speak of 'the third sex'. **1924** J. Rivière et al. tr. *Freud's Coll. Papers* II. xviii. 230 A very considerable measure of latent or unconscious homosexuality can be detected in all normal people. If these findings are taken into account, then, to be sure, the supposition that nature in a freakish mood created a 'third sex' falls to the ground. **1977** C. Isherwood *Christopher & his Kind* ii. 27 Their hero, Hirschfeld, was..notorious all over western Europe as a leading expert on homosexuality. Thousands of members of the Third Sex, as he called it, looked to him for guidance and protection. **1981** *Lancet* 8 Aug. 321/1 We had a total of 236 calls, of which 177 were for third-stage complications. [illegible]

II. 8. (Earlier example.)

1768 J. Wedgwood *Let.* June (1965) 66 All our thirds shall be saved for you.

9. c. (Further examples.) [illegible]

10. a. *pl.* Esp. in phr. *on thirds*. An agreement whereby an owner of sheep has them grazed and cared for by another person who in return receives one third of the profits (see quots.). *Austral.* and *N.Z.*

1824 E. Curr *Account Van Diemen's Land* 78 It is a common practice for persons who have not sufficient land, or who cannot attend personally to their flocks, to give them in charge to another party, who receives one third of the increase for his trouble; and if the party taking them [sc. the flocks] for 'the thirds' be careful and trustworthy, it is beneficial to both parties. **1852** G. C. Mundy *Our Antipodes* I. viii. 282 One may buy stock,..or take stock on the system of 'thirds', in which the working partner gets one third of the wool and of the increase, while the proprietary partner..follows some other profession. **1878** E. Jollie *Reminisc.* 18 [Watts]..agreed to take my sheep on 'thirds' for three years. On 'thirds' meant that he was to have one third of the wool each year and I had to have two thirds. **1930** L. G. D. Acland *Early Canterbury Runs* 1st Ser. viii. 206 For five years part of the run and sheep were let on thirds to a man named Thomas.

b. *third(s)-and-fourth(s)*: in cotton and corn farming, a system whereby the tenant contributes towards the cost of seed and fertilizer and the landowner receives a proportion of the crops (see quots. 1964, 1967, 1976). *U.S.*

1940 W. Faulkner *Hamlet* I. i. 8 'What rent were you aiming to pay?' 'What do you rent for?' 'Third and fourths,' Varner said. **1964** *Amer. Folk Music Occasional* I. 62 He could take advantage of the new system of farming on rented land. 'You call this third-and-fourths, now. I do my own furnishing and then the man that owned the land would get every third bale of cotton, every fourth load of corn.' **1967** G. W. Walton in *Publ. Amer. Dial. Soc.* XLVII. 29 *Thirds and fourths*,..a method of tenant farming whereby the following practices are common: the landowner furnishes the land and a house for the tenant; the tenant furnishes his own plow animals and tools and does all the work; the tenant then pays for one-third of the seed and fertilizer for growing the cotton and receives one-fourth of the cotton grown; the tenant pays for one-third of the seed and fertilizer for growing corn and receives one-third of the corn. **1976** C. S. Brown *Gloss. Faulkner's South* 198 *Third and fourths*,..One who pays at this rate is a 'share tenant'... He supplies his own equipment... Then he pays one-third of the seed and fertilizer for cotton, and pays one fourth of his crop as rent.

third class, third-class, *phr.* (*sb.* and *a.*) **1.** (Earlier example.)

1844 *Punch* VII. 258/1 Third class. Make up your mind for unmitigated hail, rain, sleet, snow and lightning... Do not expect the luxury of a seat.

third degree, third-degree, *phr.* (*sb.* and *a.*) **A.** *sb. phr.* **1.** *gen.* The third step or stage in succession, intensity, or amount.

1578 [see Degree *sb.* 6 c]. **1601** Shaks. *Twel. N.* I. v. 145 For he's in the third degree of drinke: hee's drown'd: go looke after him. **1716** Pope tr. *Homer's Iliad* II. v. 48 He got Orsilochus, Diocles He, And these descended in the third Degree. **1966** Tacheron & Udall *Job of Congressman* [illegible] 250 House Rule XIX provides amendments in the third degree. An amendment to an amendment is permitted, but not an amendment to an amendment to an amendment.

2. *Freemasonry.* The highest grade in freemasonry, that of master-mason. Cf. Degree *sb.* 7 b.

1772 W. Preston *Illustrations of Masonry* 203 (*heading*) A charge, to be delivered at Initiation in the Third Degree. **1865** J. How *Freemason's Man.* (ed. 2) 138 The Third Degree is the cement of the whole. *Ibid.* The Fellow-Craft who is duly qualified by time, on presenting himself as candidate for the Third Degree, has to submit himself to an examination of his qualifications as a Craftsman. **1901** [in Dict. s.v. Third *a.* 5].

3. *U.S. Law.* In defining the extent of criminality, the least serious grade of a particular crime. Cf. Degree *sb.* 6 d.

1865 *Penal Code State of New York* xv. i. 192 Maliciously burning in the day time a building, the burning of which in the night time would be arson in the second degree, is arson in the third degree. *Ibid.* ii. 195 Every person who breaks into any dwelling house in the night time, with intent to commit a crime, but under such circumstances as do not constitute the offense of burglary in the first degree, is guilty of burglary in the third degree. **1949** Branham & Kutash *Encycl. Criminology* 203 Assault in the third degree.

4. In many classifications of burns, the deepest variety, resulting in the death of all layers of skin.

Today a twofold classification tends to be used (see quot. 1961).

[**1832** G. Dupuytren *Leçons Orales* I. xvi. 209 Nous avons divisé les brûlures en six degrés ainsi caractérisés: ..3° destruction d'une partie de l'épaisseur du corps papillaire.] **1866** C. H. Fagge tr. *Hebra's On Diseases of Skin* I. xiii. 57 For all practical purposes, the three grades which I have described are sufficient. **1899** *Allbutt's Syst. Med.* VIII. 694 It is the third degree burns which Dupuytren and others have spoken of as burns of the fourth, fifth and sixth degrees. **1930** J. J. Moston in E. A. Graham *Surgical Diagnosis* III. 236 Injuries of the first and second degree will leave practically no scarring but serious deformities may result from the third degree burns. **1961** *Brit. Med. Dict.* 231/2 Classification of burns. Dupuytren's classification: 1st degree..3rd degree..6th degree... Modern classification: superficial burn or partial thickness skin destruction; deep burn or whole thickness skin destruction.

5. An interrogation of a prisoner by the police involving the infliction of mental or physical suffering in order to bring about a confession or to secure information. orig. *U.S.*

[**1880** *Harvard Lampoon* 6 Feb. 166/1 He met the large and celebrated brother of one of the Seniors. He stopped to greet him, and was surprised at receiving a clip over the head from the brother's cane. This was followed by a personal chastisement in the third degree.] **1900** *Everybody's Mag.* [illegible] claims to have had the Third Degree administered to him. **1930** G. B. Shaw *Apple Cart* I. 15 *Boanerges*. What do you mean? put me through it? Is this a police office? *Pliny*. The third degree is not unknown in this palace, my boy. **1976** T. Sharpe *Wilt* xiii. 136 'You don't think they're giving him third degree or anything of that sort?' 'My dear fellow, third degree? You've been watching too many old movies on the TV. The police don't use strong-arm methods in this country.'

B. *attrib.* or *adj.* Of or belonging to the third step or stage (in the senses above).

1926 *Scribner's Mag.* Aug. 193/1 Everybody at Police Headquarters agreed that it was ominously dull that night. There were scarcely a third-degree assault to disturb the city. **1930** J. J. Moston in E. A. Graham *Surgical Diagnosis* II. 236 Second and third-degree burns... **1976** Miller & Keane *Encycl. & Dict. Med. & Nursing* 155/1 Third-degree burns damage the epidermis, dermis and subcutaneous tissue. **1976** 'E. McBain' *Guns* (1977) iv. 86 They'd charged Colley with..second-degree assault. Even better than that, Colley's lawyer thought, would be for him to plead guilty to the lesser charge of third-degree assault.

Hence **third-degree** *v. trans.*, to subject to an intensive or violent interrogation; **third-degreeing** *vbl. sb.*

1928 W. Gillette *Astounding Crime Torrington Rd.* v. 260 They third-degreed Jimmy Dreek good and plenty. *Ibid.* 269 The fools in Boston had third-degreed an innocent man for six months. **1944** 'G. Orwell' in *Horizon* Oct. 237 The third-degreeing of the gangster. **1979** C. Watson *Blue Murder* xiii. 111 He third-degreed me about Bridie... The man just goes on and on.

thirdness. *Philos.* [f. Third *a.* + -ness.] The quality or state of belonging to a third category or of being a third element, *spec.* in the philosophy of C. S. Peirce (1839–1914), that which connects, mediates between, etc., the ontological categories that he designated as firstness and secondness (cf. *secondness).

c **1875** C. S. Peirce *Coll. Papers* (1931) I. §337. 170 (*heading*) Examples of thirdness. *Ibid.* 171 Continuity represents Thirdness almost to perfection. **1934** W. De Morgan in *Mind* XLIII. 210 Thirdness includes the meaning of signs, the conception of general laws, 'infinity, continuity, diffusion, growth, and intelligence'... It is the category that is concerned with connecting, and it is involved in all reflective thought. **1978** *Sci. Amer.* July 153/1 Peirce's Thirdness concerns two things 'mediated' by a third. For example an apple falling from a tree. The tree and the apple are linked by the relation 'falling from'.

third party. Add: **2.** *spec.* **a.** Used *attrib.* to designate insurance arranged against injury to persons other than the insured.

1901 C. H. Green in A. W. Tarn *Insurance Guide & Handbk.* (ed. 5) xix. Insurance of Third Party Insurance, dealing with the Common Law Liability of the individual with regard to the general public and their property. **1910** E. M. Forster *Howards End* x. 84 As we've insured against third-party risks, it won't so much matter. **1931** *Daily Express* 1 Sept. 9/1 Each 'banger' was £15 with an additional £2 for Dompson for driving without a third party insurance. **1978** *Dumfries Courier* 13 Oct. 11/4 Clark was fined £5 and licence endorsed in two other charges—using the car without a third party risks insurance and using it while the handbrake was defective.

b. *third party adoption*, an adoption of a child arranged by a third party who is a private individual rather than an adoption agency.

1965 Hall & Howes *Church in Social Work* iii. 57 That social worker's bane, the 'third party' adoption.

third-rate, *a.* Add: **b.** *absol.* (Earlier examples.)

1814 *Theatrical Inquisitor* IV. 357 Allusions, which have long contributed the commonplaces of poetry among third-rate makers of verse. **1816** Jane Austen *Emma* II. i. 2 In danger of falling in with the second rate and third rate of Highbury.

Third Reich (þɜːd raiç, raik). [Partial tr. of (med.) G. *drittes Reich*: see *Reich.] The German state under the rule of Hitler and the Nazi party, 1933–45; the regime of Hitler.

Of the sequence First, Second (etc.) Reich the Third Reich forms part of recognized English historical terminology.

[**1923** A. Moeller van den Bruck (*title*) Das dritte Reich.] **1930** *Times* 26 Sept. 12/2 Asked to give some idea of the 'Third Reich', Herr Hitler said the old Germany was a State of great honour and of great greatness, but the conception of 'the people' was not the central pillar of the structure. The second State had placed democracy and pacifism in the centre. They hoped for the Third Reich, which would have as its keystone the conception of the people and the national idea. **1933** L. Strowa *Nazi Germany means War* i. 9 On October 18th, four days after he had led the Third Reich out of the League of Nations ..Adolf Hitler made the following peace declaration before eight hundred of his party leaders in Berlin. **1946** J. Flanner *Janet Flanner's World* (1980) i. 105 It was odd to..hear young Dr. Horn..state that it might have been the International London Naval Conference of 1935 that drove Hitler's Imperialistic Third Reich into rearmament. **1966** *Listener* 3 Nov. 659/2 In the early years of the Third Reich, certain ideological ambiguities in Mann's attitude aroused Lukács's apprehension. **1981** S. Dunmore *Ace* 6 The good old days when the flyers of the Third Reich could do no wrong.

Third World, third world, *sb.* (and *a.*) [tr. Fr. *tiers monde.*] The countries of the world, esp. those of Africa and Asia, which are aligned with neither the Communist nor the

non-Communist bloc; hence, the under-developed or poorer countries of the world, usu. those of Africa, Asia, and Latin America. Cf. *Second World* s.v. *second *a.* 7 a. Also *attrib.* or as *adj.*, and in extended use.

[**1956** G. Balandier *Tiers Monde* 369 La conférence tenue à Bandoeng en avril 1955, par les délégués de vingt-neuf nations asiatiques et africaines..manifeste l'accès, au premier plan de la scène politique internationale, de ces peuples qui constituent un 'Tiers Monde' entre les deux 'blocs', selon l'expression d'A. Sauvy.] **1963** *Economist* 26 Oct. 353/1 Relations between Europe and the third world nowadays. **1964** *Ibid.* 18 Jan. 178/2 The ingredients common to most 'third world' countries (poverty, ignorance, love-hate of the former colonial powers). **1967** A. A. Mazrui in *Jrnl. of Politics* XXIX. 792 The concept of the Third World in the sense of the economically underprivileged sector of mankind must include Latin America, as well as the Asian-African countries. But in this paper we use the word, 'the Third World' in a more restrictive sense, meaning the world of the new states. **1969** *Wall St. Jrnl.* 15 May 14/2 By 'Third World' students Jerry means Orientals, Latins and American Indians. **1970** D. Caute *Fanon* v. 65 The 'Third World'..means 'positive neutralism' and 'non-alignment' between the Western and Soviet camps. **1974** *Globe & Mail* (Toronto) 29 Jan. 13/1 The First World Development was Capitalist... The Second World was Communism, in particular Russian Communism. The Third World takes in all the other countries that are not developed. Everyone counts China in the Third World... It includes the whole of Africa, Asia, Latin America. It has to do with income and low standard of living and so this takes in such countries as Greece, Yugoslavia, and some include Spain and Portugal. **1978** *Poland* May 1/2 You will find statements and articles written by Poles, people from other socialist and from capitalist countries as well as from the countries of the Third World. **1978** *Listener* 14 Sept. 322/1 The long ride into town underlines just how Third World, poor and underdeveloped Vietnam still is. **1980** *Times* 6 May 12/8 Andalusia, often described as Spain's 'third world', with its high crime rate and unemployment.

Hence **Third Wo·rlder**, an inhabitant of the Third World; **Third Wo·rldism**, an ideology or policy of support for the Third World.

1970 *New Scientist* 29 Oct. 227/2 Skills which the average 16-year-old western youth would have little difficulty in mastering apparently pose almost insoluble problems for the average Third Worlder. **1970** *New Yorker* 26 Dec. 46 The revolutionary Third Worldism of large sections of Sweden's politically active youth and intellectuals. **1975** *New Left Rev.* Nov.–Dec. 13 It comes down to little more than a more sophisticated justification of romantic nationalism, now transformed in 'Third Worldism'. **1978** G. Vidal *Kalki* v. 119 We have a superb military machine second to none, not to mention a standard of living that is the envy of every commie and the despair of every Third Worlder. **1980** *Encounter* Nov. 40/1 Third-Worldism will not continue to be a one-sided problem for the West..but will also..be embarrassing..the East.

thirst, *sb.* Add: **3.** *thirst-mad, -making* adjs.; **thirstland** (earlier examples).

1878 P. Gillmore (*title*) The great thirstland: a trek through Natal, Transvaal, Orange Free State and the Kalahari Desert. **1889** Farmer *Americanisms* 532/1 The region of extinct lakes and inland seas of Southern Nevada and South-eastern California is the great thirstland of the continent. **1969** G. MacBeth *War Quartet* 68 They.. lived As beasts, thirst-mad. **1952** J. Cannan *Body in Beck* ii. 41 Will you excuse me if I beetle through for another pint?..That's a thirst-making rock if ever there was one.

thirsty, *a.* Add: **2. b.** Of a motor vehicle, engine, etc.: that has a high fuel-consumption rate.

1977 *Jrnl. R. Soc. Arts* CXXV. 364/1 Larger and quieter aeroplanes and less thirsty engines. **1980** *Daily Tel.* 9 July 12/5 It should..appeal to motorists wishing to move up from the normal run of mass-produced saloons without ..running a bigger and thirstier model.

thirteen. Add: **A. 1, 2.** (Examples referring to the original thirteen states (previously, colonies) of the U.S.A.)

1776 *Declaration of Independence*, The unanimous Declaration of the thirteen united States of America. **1776** in *Huntington* (N.Y.) *Town Rec.* (1889) III. 6 Yesterday the Freedom and Independence of the Thirteen United Colonies was..proclaimed. **1834** H. M. Brackenridge *Recollections* vii. 69 Fort Fayette, surmounted by the stripes and stars of the old thirteen. **1904** *Hartford* (Conn.) *Courant* 30 Aug. 10 We want to see the Old Thirteen draw closer and closer together. **1941** S. V. Benét *Listen to People* (1942) 471 There are the pretty girls with their hair curled Who represent the Thirteen Colonies. **1950** *Chicago Tribune* 23 Feb. 4/4 Our 13 original states found that survival and progress depend on closer association and common effort.

4. a. **thirteen-year cicada, locust**, a periodical cicada that reappears every 13 years rather than every 17.

1846 *Dollar Newspaper* (Philad.) 17 June 2/3 The locusts are said to be thirteen years' locusts, having made their appearance before this time in 1833. **1964** Borror & DeLong *Introd. Study Insects* (ed. 2) xx. 204 There are at least 13 broods of 17-year cicadas and 5 of 13-year cicadas.

thirteener. Add: **2. b.** Also in the game of bridge.

1914 M. C. Work *Auction Developments* 611 *Thirteener*, the last card of any suit. **1964** Frey & Truscott *Official Encycl. Bridge* 614/2 *Thirteener*, the card remaining in a suit when all other cards in that suit have been played on the first three tricks of the suit.

thirty, *a.* and *sb.* Add: **A.** *adj.* **1. c.** *like thirty cents* and varr., cheap, worthless. *U.S. slang.*

1896 [see *Reub, Rube]. **1906** J. London *Let.* 24 Nov. (1966) 225 You made my exposition look like thirty cents. **1944** *Chicago Daily News* 31 July 3/6 (*heading*) Sues to make Uncle Sam feel like a 30-cent refund. **1973** T. Tobin *Lett. G. Ade* 2 Feeling 'like thirty cents' and 'the cold gray dawn of the morning after' became part of the American idiom.

B. *sb.* **2. b.** *attrib. spec.* Of, pertaining to, or characteristic of the 1930s.

1967 *Observer* 10 Sept. 24/3 Heaven knows, you can peg people by their opinions—'thirties communist' or '*New Statesman* type' seems as hard a definition as 'whisky priest' or 'teacher's pet'. **1969** 'J. Munro' *Innocent Bystanders* xiv. 205 The whole thing was as English as a Thirties farce: sandwiches and tinkling spoons. **1971** G. Charles *Destiny Waltz* v. 149 It was..furnished in a heavy, thirties style. **1976** S. Hynes *Auden Generation* iii. 82 *New Signatures*..was the first anthology of 'thirties poets. **1981** C. Leopold *Night Fishers of Antibes* ii. 15 His Thirties forehead with the thin black hair brushed back from a parting precisely dead centre.

3. *U.S.* Also in journalism, broadcasting, and wider slang use. (Examples.)

1929 *Amer. Speech* IV. 290 '30' or 'Thirty' indicates the end of a shift or of the day's work, and has come to mean, also, death. **1938** *Sun* (Baltimore) 20 Jan. 2/8 Newsmen.. mourned today at the bier of Edward J. Neil,..who was killed by shrapnel while covering the civil war..in Spain. Prominent..was a shield of white carnations with a red-flowered figure '30'—the traditional 'good night' in the lore of the fourth estate. **1941** J. Smiley *Hash House Lingo* 58 30, end of anything. **1945** J. O'Hara in *New Yorker* 27 Jan. 22/3 'I say thank you and thirty.' This last, the word 'thirty', is the traditional signing-off signal of the newspaper business. **1973** R. Ludlum *Matlock Paper* xxix. 251 The number 30 at the bottom of any news copy meant the story was finished. **1978** G. Vidal *Kalki* iv. i. 88 'When we know those two things, it's fat thirty time.' Bruce had obviously been impressed by journalism school.

4. *thirty-eight*, a revolver of ·38 calibre; ammunition for such a revolver; *thirty-three* (*and a third*), 33(⅓), 33⅓ revolutions per minute; a gramophone record to be played at this speed; *thirty-two*, also a revolver of ·32 calibre.

1942 L. Hughes *Shakespeare in Harlem* 3 Gonna go get my pistol, I mean thirty-two. **1951** Sackville-West & Shawe-Taylor *Record Guide* 716 While we in England cannot say how bad were the worst of the early Columbia 33s, or how good the best of the Victor 45s, the difference would have to be great to justify..the Victor system. **1953** W. Burroughs *Junkie* i. 20 Jack's voice..went on and on... 'Give me a thirty-eight every time. Just flick back the hammer and let her go. I'll drop anyone at five hundred feet.' **1959** I. Jefferies *Thirteen Days* iv. 46 Mostly I filled up with nine-milli..but I threw in some thirty-eights and three-oh-three. **1968** *Melody Maker* 22 June 2 This is the EP which is recorded at 33⅓ and plays for 18 minutes. **1974** R. B. Parker *Godwulf Manuscript* iii. 17 The girl's voice..was thick and very slow, almost like a 45 record played at 33. **1978** R. Thomas *Chinaman's Chance* III. xxix. 291 We need a couple of pieces... Revolvers. No smaller than a thirty-two, no larger than a thirty-eight.

C. *Comb.* **a.** *thirty-foot* (examples); *thirty-five millimetre, mm., mil(l)*. (photographic film, camera).

1880 J. F. Carll *Geol. of Oil Regions* III. 197 Sand shells and slate, '30' Rock'. **1938** R. M. Fanstone *Colour Photogr.* i. 17 Agfacolor... 35 mm. film for miniature cameras. **1969** 'H. Pentecost' *Girl Watcher's Funeral* (1970) III. i. 123 'What kind of a camera was it, Morrie?' I asked. 'Leica—thirty-five millimeter,' he said. **1971** O. Norton *Corpse-Bird Cries* vi. 116 'He couldn't have—well, turned the film back, or anything?' 'On a thirty-five mil Paxette? No, he couldn't.' **1972** I. Hamilton *Thrill Machine* xxii. 102 It's a low-quality blow-up from thirty-five mill. movie film. **1978** F. MacLean *Take Nine Spies* vi. 196 His visitor brought out two rolls of 35 mm film. **1978** S. Sheldon *Bloodline* xxxviii. 341 The thirty-foot police boat..had been built for service, not comfort.

b. **Thirty-nine Articles**: clauses of a doctrinal statement drawn up by the Church of England in the sixteenth century, to which those taking orders in that Church have to assent; **thirty-year rule**, a rule that public records should normally be open to inspection after a lapse of thirty years from their compilation.

1607 T. Rogers *Faith, Doctrine, & Relig. in Realme of Eng. expressed in 39 Articles* 3 The purpose of our church is best knowne by the doctrine which shee doth professe; the Doctrine by the 39. Articles established by Act of Parliament. **1739** (*title*) Thirty nine articles of constitutions & canons of Church of England. **1903** G. B. Shaw *Man & Superman* viii. 211 Straightforward public lying has reached gigantic developments, there being nothing to choose..between..the clergyman subscribing the thirty-nine articles, and the vivisector who pledges his knightly honor that no animal operated on in the physiological laboratory suffers the slightest pain. **1969** A. Richardson *Dict. Christian Theol.* 336/1 In their revised form the Thirty-nine Articles were passed by Convocation in 1571 and the text finally determined in 1604... Subscription is still required from clergymen on their ordination. **1966** *Times* 11 Aug. 13/2 In two years' time the 30-year rule will be operating. **1979** *N.Y. Rev. Bks.* 25 Oct. 52/2 British Foreign Office papers recently opened under the thirty-year rule verify Butterfield's point.

thirtyish (þɜ̄·ɹti,iʃ), *a. colloq.* [f. Thirty *a.* + *-ish¹ 4.] **1.** Of about thirty years of age.

1925 F. M. Ford *No More Parades* I. i. 20 A very thin man; thirtyish. **1926** A. Bennett *Lord Raingo* II. lxxi. 320 The..little thirtyish nurse. **1979** 'J. Ross' *Rattling of Old Bones* vii. 63 'How old was he?' 'Thirty-ish. Perhaps less, perhaps more.'

2. Characteristic of or reminiscent of the 1930s; = *thirty *sb.* 2 b.

1962 *Times* 8 Mar. 16/7 Stuart Davis's hard, bright, 'Thirtyish' cubism. **1976** S. Hynes *Auden Generation* x. 355 The Berlin that their lives express..belongs to the 'thirties... And what makes it 'thirty-ish is that it can define its hell in political terms. **1978** *Broadcast* 20 Nov. 8/1 The decor is full of thirty-ish touches.

this, *dem. pron.* and *adj.* Add: **B. I.** Demonstrative pronoun. **1. b.** Now also indicating a person speaking or (interrog.) being spoken to on a telephone, etc.

1947 *Sun* (Baltimore) 8 Jan. 17 (*caption*) Very well, we'll expect you at nine this evening. Who did you say this was?

c. (Further examples.)

1954 G. Kersh in D. Knight *100 Yrs. Sci. Fiction* (1969) 223 So you came back to life—more than four hundred years ago! Is this right? **1965** *Times* 16 Mar. 13/4, I cannot refrain from a violent protest against the ever increasing use of 'this' instead of 'that': e.g., 'Will you come to supper tomorrow?' Answer: '*This* would be very nice.' **1970** *Nature* 4 Apr. 47/2 The reader..may come to think that this new approach to mathematics is not worth while. This would be a shame.

e. (Later examples.)

1922 Yeats *Player Queen* i. 20 The basket-makers and the sieve-makers will be out by this. **1971** in *Sc. Nat. Dict.* (1974) IX. 283/2 I'll hae plenty adee atween this and Whitsunday.

3. *this, that, and* (or *or*) *the other*, every sort (of), every possible or imaginable.

1824 Scott *St. Ronan's* II. i. 24, I am sure I aye took your part when folk miscaa'd ye, and said ye were this, that, and the other thing. **1918** *Nation* (N.Y.) 7 Feb. 161/1 They..offered us a contract in this, that, or the other company, whose dividend-paying record had been thus and so. **1938** N. Marsh *Artists in Crime* xvii. 255 It's a bit awkward what with this and that and the other thing.

II. Demonstrative adjective. **1. d.** Also, the present or existing. *colloq.*

1785 Boswell *Jrnl. Tour Hebrides* 86 We were told this Mr. Waller was a plain country gentleman. **1788** H. Walpole *Let.* 26 July (1918) Suppl. II. 24 Do you know the medals of gold belonged to this Lord Pembroke's grandfather? **1933** *O.E.D. Suppl.* s.v., [*Mod.*] I knew the last doctor very well. I don't get on with this one.

k. In unliterary narrative: referring to a person, place, etc., not previously mentioned or implied. orig. *U.S.*

1922 S. Lewis *Babbitt* viii. 116 Did you read about this fellow that went and paid a thousand dollars for ten cases of red-eye that proved to be nothing but water? **1946** K. Tennant *Lost Haven* (1947) ii. 41 They dug this great big trench with bull-dozers. **1969** Fabian & Byrne *Groupie* (1970) xvi. 111 The rest of the letters were all written on small sheets of blue notepaper in this really childish handwriting. **1976** *Drive* Nov.–Dec. 24/1 It was on the Chester road, in Birmingham. I saw this car with the keys in the ignition.

2. (Further examples, in attrib. phr.)

1930 R. Graves *Ten Poems More* 11 Neat this-way-that-way and without mistake. **1937** C. Day Lewis *Starting Point* I. iii. 51 The field was scored..with streaking,..incessant this-way that-way movement.

III. this child: see Child *sb.* 7 b.

this, *adv.* Add: **2. b.** Now also qualifying adjs. and advs. other than those of quantity, and no longer chiefly in *this much*. (Grading into an intensive.)

1932 J. Leatham *Fisherfolk* 13 A'm this aul', an' I never had a sy-ystem! **1967** *Boston Sunday Herald* 30 Apr. (Mag.) 34/2, I have a stack of telegrams this thick. **1971** *Where* Dec. 376/3 Yet the picture is usually not even this good. Most teachers..talk much more than half the time in their classes, and the time that is left is not all used for children talking. **1972** *Real Estate Rev.* Winter 8/2 Keep in mind, however, that no existing property is this typical. **1976** *Woman's Day* (U.S.) Nov. 154/2, I haven't felt this well in years.

this-a-way (ðiˑsăwēⁱ), *adv. dial.* and *U.S.* Also **thisaway, this a-way, this-away, thisserway.** [Repr. pronunc. of This *dem. adj.* + intrusive *-a-* + Way *sb.*¹: cf. *that-a-way *adv.*] **1.** In this manner or respect.

1834 S. Lover *Legends & Stories of Ireland* 2nd Ser. 54 Don't ruinate me this-a-way. **1901** A. C. Hegan *Mrs Wiggs* v. 58 Did I ever tell you 'bout how Jim brought our other hoss to town?..It was this a-way. **1938** M. K. Rawlings *Yearling* xxiv. 310 He's mean, at best. He's been this-a-way ever since Oliver takened his gal away from him. **1939** *Best Short Stories* 53 Why can't it last? We're sitting pretty, thisserway. **1956** H. Gold *Man who was not with It* (1965) xix. 166 Boy, if I wanted to tell you that, I would say it this-away: Git! My dotter's too green for screwing! **1978** J. A. Michener *Chesapeake* xii. 728 'I jes' cain't believe it,' he said... 'Goddamnit, he's my own brother-in-law and he hadn't oughta behave thisaway.'

2. In this direction.

1903 *Dialect Notes* II. 333 Was he coming this-a-way when you seed him? **1955** F. O'Connor *Wise Blood* iii. 57 I'm going thisaway too. **1959** *Observer* 11 Oct. 6/4 Politically, Chataway can run thisaway or thataway.

thistle, *sb.* Add: **1. c.** Hence, *Knight of the Thistle*, a member of this order.

1732 *Gentl. Mag.* June 827/1 The E. of Portmore, made a Kt of the Thistle in the Room of the E. of Loudoun, dec. **1828** N. H. Nicolas *Statutes of Order of Thistle* 21 After its [sc. Order of the Thistle's] Revival by King James, the Knights of the Thistle were Installed in the Chapel of Holyrood House. **1911** J. Warrack *Knights of Most Noble Order of Thistle* 29 The King, after consulting the Chapter of the Knights of the Thistle, ordered a letter to be sent. **1963** *Times* 30 Apr. 10/7 Sir Robert Menzies, the Prime Minister, is expected to visit Britain and the United States in June,..later going to Edinburgh to be installed as a Knight of the Thistle.

3. Californian thistle *N.Z.* = *Canada thistle.*

1891 R. Wallace *Rural Econ. Austral. & N.Z.* xxii. 310 One of the most recent importations..is that of the 'Canadian' or 'Californian' thistle. **1948** D. W. Ballantyne *Cunninghams* I. ii. 10 It's been awful with Californian thistle up there.

4. *thistle-clock* [Clock *sb.*¹ 8]; **thistle cup**, a silver cup with an outward-turning rim, of a type formerly manufactured in Scotland; **thistle glass**, a drinking glass with a round bowl and an outward-turning rim.

1948 C. Day Lewis *Poems 1943–47* 63 Thistle-clocks fly. **1947** W. C. Wallis *Silver, Glass & Pott.* 4 Another type of vessel, unique in Scotland, which made its appearance during the last twenty years of the seventeenth century, is the little mug known, from its supposed resemblance to a thistle, as a 'thistle' cup. **1968** *Canadian Antiques Collector* July 17/1 Late in the seventeenth century appear two forms of secular silver unique to Scotland. One is a drinking cup of a type often called a 'thistle' cup, though it has little resemblance to a thistle. **1935** M. Mitchison *We have been Warned* II. 138 Alex.. got a bottle of hock and poured it out into thistle glasses. **1973** *Times* 20 Oct. 14/3 The 'thistle' glass, with its outward-angled rim.

this world. Add: **this-worldliness** (further examples); **thi·s-worldly** *a.*, concerned with the things of this world or the present state of existence; **thi·s-worldness** = This-worldliness.

1883 'Mark Twain' *Life on Mississippi* xlviii. 480 The guests were always this-worldly, and often profane. **1928** C. H. Dodd *Authority of Bible* IV. xii. 268 Hard experience revealed the insufficiency of the robust 'this-worldliness' of the classical Hebrew religion. **1930** G. Greene *Two Witnesses* 92 His sensitively spiritual soul could make no truce with any thisworldness. **1944** J. S. Huxley *On Living in Revol.* xv. 187 The Churches..feel themselves threatened by the rise of an outlook more concerned with social planning for this-worldly improvement than with individual concern for other-worldly salvation. **1957** *Times Lit. Suppl.* 20 Dec. 776/1 Our civilization acquired its secular, scientific, anti-theological character, its 'this-worldliness'. **1978** J. Skorupski in Hookway & Pettit *Action & Interpretation* 83 It treats the magical and, by and large, the religious practices as 'instrumental': as attempts to control the course of events in such a way as to bring about this-worldly ends which the actors seek.

thivish (þaiˑviʃ). *Anglo-Ir.* Also **tevish.** Pl. **-es**; also **tevishies, thevshi.** [ad. Ir. *taibhse*, pl. *taibhsí.*] A ghost, apparition, or spectre.

1852 W. Wilde *Irish Pop. Superstitions* i. 14 Thivishes or thoushas (shadowy apparitions) are literally ghosts. *Ibid.* iii. 71 (*heading*) Reminiscences of the West.—The Welshes.—The Thivish or Fetch. *Ibid.* 111 'Mother,' said he, gazing steadily upon the pale anxious face that was bent upon him, 'I've seen the *thivish*.' **1888** W. B. Yeats *Fairy & Folk Tales of Irish Peasantry* 128 Ghosts, or as they are called in Irish, *Thevshi* or *Tash* (*taidhbhse, tais*), live in a state intermediary between this life and the rest. They are held there by some earthly longing or affection. **1892, 1963** [see *sowlth].

thixotropy (þiksǭtrǫpi). [ad. G. *tixotropie* (T. Péterfi 1927, in *Arch. f. Entwicklungsmech.* CXII. 689), f. Gr. θίξ-ις touching + -o + Gr. τροπ-ή turning: see -y³.] The property of certain gels of becoming fluid when agitated and of reverting back to a gel when left to stand.

1927 *Chem. Abstr.* XXI. 1391 Thixotropy is the phenomenon that a coagulated sol can be liquefied by merely shaking and always again coagulated to a gel by stopping agitation. **1949** P. C. Carman *Chem. Constitution & Properties Engin. Materials* xiii. 379 This type of reversible gelation is known as thixotropy and is particularly strongly marked in montmorillonite clays. **1971** *New Scientist* 19 Aug. 435/2 How to demonstrate thixotropy with custard.

Hence **thixotro·pic** *a.*, exhibiting or pertaining to thixotropy; **thixotro·pically** *adv.*

1927 *Chem. Abstr.* XXI. 1391 (*heading*) Thixotropic behavior of aluminum hydroxide gels. **1947** *Nature* 11 Jan. 70/2 The range of thixotropic materials extends from the hardest solids, through doughs and pastes to liquids such as blood or milk. **1958** *Woman* 22 Feb. 11/2 Thixotropic paint is the non-spill type. **1963** *Geol. Mag.* C. 209 Nodules..produced by allowing a thin layer of sand to sink into thixotropically mobilized mud. **1971** *Nature* 30 July 328/1 A thin layer of grey silt covered a 20 cm layer of black thixotropic mud.

tholeiite (þōᵘ·liəit). *Petrogr.* Also 9 **tholeite.** [ad. G. *tholeiit* (J. Steininger *Geognostische Beschreibung des Landes zwischen der unteren Saar und dem Rheine: Nachträge* (1841) 26), f. *Tholei* (now *Tholey*), name of a village in N.E. Saar, W. Germany: see -ite¹.] Formerly, a basaltic rock containing plagioclase feldspar, pyroxene, and glass, with little or no olivine, and having an intersertal texture; in recent use, any basaltic rock typically containing augite and a calcium-poor pyroxene (pigeonite or hypersthene), and distinguished from alkali basalts by a higher silica and lower alkali content.

1866 P. H. Lawrence tr. *von Cotta's Rocks Classified & Described* II. i. 138 Steininger has given the name of Tholeite to a rock found at the Schaumberg near Tholei, which he took for a compound of albite and titanite. But according to Bergemann's analysis this rock consists of 70 labradorite, 5 augite... It must therefore from its composition be considered a dolerite or basalt unless indeed it be considered as plutonic and classed with melaphyre. **1893** [see *intersertal *a.*]. **1922** *Q. Jrnl. Geol. Soc.* LXXVIII. 229 The rocks were described as being.. of one type: namely, olivine-free dolerites or tholeiites. **1924** Thomas & Bailey in E. B. Bailey et al. *Tertiary & Post-Tertiary Geol. Mull* xxv. 280 Olivine-free and olivine-poor plagioclase-augite rocks with intersertal structure are classed by Rosenbusch as tholeiites. **1962** *Jrnl. Petrol.* III. 352 The two planes identified in Fig. 1.. divide the basalts into five unique groups..: 1. Tholeiite (oversaturated): normative quartz and hypersthene. 2. Tholeiite (saturated; hypersthene basalt): normative hypersthene. 3. Olivine tholeiite (undersaturated): normative hypersthene and olivine. 4. Olivine basalt: normative olivine. 5. Alkali basalt: normative olivine and nepheline. **1967** *Geol. Mag.* CIV. 337 Significant use of the term tholeiite dates from 1887 when Rosenbusch redefined Steininger's Schaumberg, Saar tholeiite (1840) from Tholey as a melaphyre with intersertal texture... Bailey and Thomas (1924) in setting up their non-porphyritic central magma type of Mull recognized its 'tholeiitic' character using..Rosenbusch's definition of 1887... A more precise petrographic and chemical definition was to follow in Kennedy's important papers of 1931 and 1933 in which the non-porphyritic central magma type was redefined as the Tholeiitic Magma Type... Paramount in this definition was the recognition of the presence of an enstatite-augite or pigeonite series pyroxene in the ground-mass of tholeiitic lavas. **1978** *Nature* 13 July 128/1 The two principal Icelandic basalt types are evolved tholeiites, mostly quartz-normative, with $Mg/Mg+Fe^{2+}$ atomic ratio of 0·40 to 0·50, associated with fissure swarms and central volcanoes; and second, olivine tholeiites with $Mg/Mg+Fe^{2+}$ in the 0·60–0·70 range and predominantly associated with monogenetic shield volcanoes.

Hence **tholeii·tic** *a.*

1922 *Q. Jrnl. Geol. Soc.* LXXVIII. 237 The tholeiitic intrusions have produced little thermal alteration of the rocks into which they have been ultimately injected. **1933** [see *plateau basalt* s.v. *plateau 4]. **1933** *Amer. Jrnl. Sci.* CCXXV. 247 The general thesis is..advanced that the olivine-basalt magma-type is the parent of the alkaline line of descent while the tholeiitic magma-type occupies a similar position with respect to the calc-alkaline rock suite. **1967** [see tholeiite above]. **1970** *Nature* 12 Dec. 1030/2 Geochemists have known for some time that [ocean] ridge basalts tend to be tholeiitic whereas alkali basalts are more likely to be found away from ridges. **1972** F. H. Hatch et al. *Petrol. Igneous Rocks* (ed. 13) vii. 364 Alkali olivine-basalt magmas can differentiate towards alkali-enrichment; but tholeiitic magmas differentiate towards silica-enrichment.

tholoid (þōᵘ·loid). *Geol.* [f. Gr. θόλος (see Tholus) + -oid.] A dome-shaped, steep-sided extrusion of hardened lava plugging the vent of a volcano.

In quot. 1912 repr. a Ger. form.

1912 *Jrnl. Geol.* XX. 85 'Tholoides' have slopes of over 35° and are convex upward. **1939** *Nature* 25 Nov. 913/2 Of the latter [sc. new volcanoes] there were two:.. (*b*) Iôzima-Sintô, south of Kyûsyû.., which was of the tholoid type. **1976** P. Francis *Volcanoes* iv. 151 Even when the tholoid is shrouded in mist and invisible, as it often is in the afternoon, the dry clattering continues, with every now and then a much larger collapse taking place.

tholos, var. Tholus in Dict. and Suppl.

tholus. Add: **b.** *tholos* is the usual form. (Later examples.) Also *tholos-tomb*.

1921 *Discovery* Feb. 33/1 The principle of the tholos-tomb was most in use in Mycenæan times. **1957** [see *beehive tomb* s.v. *beehive 3]. **1975** *Times Lit. Suppl.* 14 Mar. 282/1 The megalithic architecture of Britain.. had no conceivable ancestry in the Mycenaean *tholoi*. **1983** *Times* 10 Feb. 1/3 The underground *tholos* tomb, shaped like a giant beehive, lies in a fifteenth-century BC cemetery..three miles south of Argos.

Thomas. Add: **3.** (Earlier *absol.* example.)

1888 Kipling *From Sea to Sea* (1899) I. 185 Every Thomas is interesting, except when he is too drunk to speak.

5. *Surg.* The name of H. O. *Thomas* (1834–91), English surgeon, used *attrib.* and in the possessive to designate a splint that he invented for immobilizing the hip, consisting of a rigid bar that extends from the back to the calf and is bandaged to the leg, and with rings attached that partly encircle the chest and leg; also (now the usual sense), a splint consisting of a soft ring encircling the thigh from which two rigid rods extend on each side of the leg and meet beyond the foot, allowing traction to be applied to the leg via the cross-piece or the knee to be immobilized.

1884 W. Pye *Surg. Handicraft* xxiii. 291 There are many other ways of treating acute hip disease... By Thomas' splint. **1940** N. Mitford *Pigeon Pie* vi. 102 If.. real casualties were brought in and found all the personnel tied up in Thomas's splints. **1961** *Countryman* LVIII. III. 600 The M.O...fixed my fractured leg in a Thomas splint. **1974** Passmore & Robson *Compan. Med. Stud.* III. iv. 11/2 The Thomas' splint is designed so that when the traction tapes are tightened over the end of the splint, a counter thrust is exerted through the padded ring against the bony prominence of the ischium.

6. The name of S. G. *Thomas* (1850–85), English metallurgist and inventor, used *attrib.* to designate a steel-making process like the Bessemer process but using a converter with a basic instead of an acid lining, so that phosphorus is removed (invented by Thomas in 1878). Also *Thomas–Gilchrist* [P. *Gilchrist* (1851–1935), cousin and collaborator of Thomas].

1881 *Encycl. Brit.* XIII. 346/1 Owing to the success of these operations, the 'basic' process has been more frequently spoken of as the 'Thomas-Gilchrist process'. **1925** *Jrnl. Iron & Steel Inst.* CXII. 523 Notwithstanding prognostications as to the supersession..of the Bessemer acid and basic (Thomas) process by the open-hearth process, it is certain that..the Bessemer process will long continue to hold its own. **1948** H. W. Baker *Mod. Workshop Technol.* I. 22 In the Thomas process the necessary amount of lime..is charged into the converter..before the iron.. is poured in. **1973** [see *Siemens a].

Thomas–Fermi (tǫmăsˌfɜ̄·ɹmi). *Physics.* The names of L. H. *Thomas* (b. 1903), English physicist, and E. *Fermi, used *attrib.* with reference to a model of the electronic charge distribution in an atom in which the electrons are treated as a gas of independent particles obeying Fermi–Dirac statistics and the exclusion principle is taken into account, proposed by them in 1927 and 1928 respectively.

1931 *Phil. Mag.* XII. 111 The Thomas–Fermi method of approximating to the atomic charge distribution. **1955** J. Lindhard in W. Pauli *Niels Bohr* 190 If we ask for the behaviour for heavier atoms,..a Thomas–Fermi treatment should again be preferable. **1970** G. K. Woodgate *Elem. Atomic Struct.* vi. 101 The Thomas–Fermi potential does serve as a trial potential for self-consistent field methods.

Thompson (tǫ·msən). *Mil.* (orig. *U.S.*). The name of John T. *Thompson* (1860–1940), U.S. general, used *attrib.* and *absol.* to designate a type of sub-machine-gun which was conceived by him and financed by his company, and named after him in 1919 at the insistence of its designer, O. V. Payne. Cf. *tommy gun.

1920 *Army & Navy Jrnl.* 2 Oct. 120/1 Colonel Thompson is now connected with the Auto-Ordnance Corporation of New York City, which has put the Thompson sub-machine gun on the market. **1921** M. Thompson *Let.* 26 Apr. in J. Helmer *Gun that made Twenties Roar* (1969) iv. 69 We came back with a tentative agreement on their part to purchase 50,000 Thompsons. **1933** 'J. Spenser' *Limey* ii. 27 There was a Thompson sub-machine gun, universally known in gangland as a 'Tommy gun'. **1946** D. M. Ward *Other Battle* xxi. 140 An order was placed in America for Thompsons, but by the time the first few were delivered it was realized that not enough ·45 ammunition could be made. **1970** E. K. Walker in W. King *Black Short Story Anthol.* (1972) 54 Captain Bull..carried a Thompson submachine gun at high port. **1978** D. Murphy *Place Apart* ii. 22 Kerins had a Thompson under the bed but he never had a chance to use it.

Thompsonian (tǫmsōᵘ·niăn), *sb.* and *a.* [f. the name *Thompson* (see below) + -ian.] **A.** *sb.* An admirer of the work of Francis Thompson (1859–1907), English poet and writer. **B.** *adj.* Of, belonging to, or characteristic of Thompson or his work.

1913 T. Hardy in V. Meynell *Francis Thompson & Wilfrid Meynell* (1952) xiii. 198 You may be sure I am a Thompsonian. ?**1921** J. Thomson *Remarks on Francis Thompson's Hound of Heaven* 9 Nearer than either to the Thompsonian *Hound of Heaven*..comes the *heaven's winged hound* of Shelley. **1927** *Observer* 19 June 8 The next in the series is a choice from Francis Thompson's prose... This Thompsonian addition to the 'handy Harraps' is no end of a shillingsworth. **1948** *Tablet* 30 Oct. 282/2 Answering the letters of the growing army of Thompsonians all over the world. **1962** J. Brodrick in F. Thompson *St Ignatius Loyola* p. xii, He made one precious contribution to Thompsonian psychology.

Thomson (tǫ·msən). Also (*erron.*) **Thompson.** **1.** *Physics.* [The name of Sir William *Thomson*: see *Kelvin, kelvin.] *Thomson effect*: the effect an electric current has, when flowing in the direction of a temperature gradient, of absorbing or giving out heat independently of the Joule heating; so *Thomson coefficient*, a numerical measure of this effect for a material.

1878 *Encycl. Brit.* VIII. 98/1 This anomaly led Tait to the discovery..that the Thomson effect in iron changes its sign..at a temperature near low red heat. **1906** [see *Seebeck]. **1930** *Engineering* 9 May 596/2 The Thomson and Peltier coefficients were given values of the right general order of magnitude. **1966** C. R. Tottle *Sci. Engin. Materials* vi. 130 By choosing appropriate values of the Peltier and Thomson coefficients, two dissimilar materials can be arranged to produce a substantial e.m.f. if one junction is maintained at a high temperature and the other at a lower one. **1975** D. G. Fink *Electronics Engineers' Handbk.* xxvi. 3 At large values of ΔT the Thomson effect may be comparable with the Seebeck effect and must be taken into account in the design of generators and refrigerators.

2. *Zool.* The name of Joseph Thomson (1858–94), Scottish explorer, used in the possessive (rarely *attrib.*) to designate an East African gazelle with a broad black lateral stripe, *Gazella thomsoni*, first collected by him and named in his honour by A. Günther in 1884 (*Ann. & Mag. Nat. Hist.* XIV. 428).

1897 *Proc. Zool. Soc.* 454 Thomson's Gazelle does not..extend beyond a few miles north of Lake Nakuru. **1906** [see Tommy 3 c]. **1915** Roosevelt & Heller *Life-Hist. Afr. Game Animals* II. xviii. 600 The Thomson gazelle is essentially a highland antelope. **1969** *Daily Tel.* 13 Aug. 17/3 A Thomson's gazelle, born at Whipsnade Zoo, brings the number in the herd..to 20. **1980** E. W. Hayman tr. *Haltenorth & Diller's Field Guide Mammals Afr.* 94 Thomson's Gazelle..horns lyre-shaped and weakly S-formed.

3. *Physics.* [The name of Sir J. J. Thomson (1856–1940), British physicist.] *Thomson scattering*: scattering of light by free charged particles, *spec.* electrons, in accordance with classical mechanics.

1935 Compton & Allison *X-Rays in Theory & Exper.* (ed. 2) iv. 198 This classical scattering from a free electron is often called 'J. J. Thomson scattering'. *Ibid.* 822/2 (Index), Thomson scattering. **1964** *Sci. Survey* III. 125 Knowing the brightness of the corona, and from the theory of Thomson scattering, the proportion of sunlight scattered towards the observer by each electron, we can calculate the number of electrons per cubic centimetre. **1978** *Nature* 19 Jan. 220/1 One..expects that up to a certain radial distance from the accretion column the neutron star surface is covered by a dense atmosphere and represents a 'tarnished mirror', reflecting quasi-isotropically by Thomson scattering.

-thon, *suffix.* Var. *-athon used in some words, as *Telethon.

1954 *Amer. Speech* XXIX. 229 The word *moviethon* is merely the most recent in what promises to be a long list of words created arbitrarily by means of the *-thon* suffix. **1963** R. I. McDavid *Mencken's Amer. Lang.* 228 *Rockerthon*, *poolathon* and *pianothon* appeared in Canada in 1955.

thong, *sb.* Add: **1. d.** *spec.* A root or root-cutting of horse-radish or sea-kale.

1927 *Smallholder* 26 Mar. 105 Plant [horseradish] each year..fresh pieces..made from the side roots of thongs. *Ibid.*, It is now time to plant out thongs of seakale. **1951** *Dict. Gardening* (R. Hort. Soc.) IV. 1916/2 Cuttings [of seakale], or thongs as they are frequently called, are clean straight pieces of the side roots. **1961** *Amat. Gardening* 21 Oct. 9/3 The thick roots [of seakale] or 'thongs'..are not needed for forcing.

f. *Austral.* and *U.S.* = *flip-flop 1. Cf. *thong-sandal*, sense 2 below.

1967 *Coast to Coast* 1965–67 Her feet, in scuffed leather thongs, were now too clean. **1976** *New Yorker* 17 May 35/2 Please, no clogs, Earth Shoes, or thongs. **1981** H. Kemelman (1982) xxv. 188 I slipped on a pair of Japanese thongs. I slipped them on and felt the skin between my first two toes protest.

2. thong sandal *Austral.* and *U.S.* = sense 1 f above; **thong weed** = *sea thong* s.v. Sea *sb.* 23 e.

1965 *Times Lit. Suppl.* 25 Nov. 1057/2 Supporting activities, like teaching and editing and selling thong-sandals. **1973** J. Aiken *Butterfly Picnic* iii. 58 She wore a magenta trouser shirt..burnt-orange shorts, and local-made thong sandals. **1958** *Listener* 31 July 179/1 With *I-Spy at the Seaside* I shall look for..some thong weed. **1966** *Ill. Elementary Plants* 41 *Himanthalia elongata* ('Thong Weed') is to be found attached to rock surfaces.

Thonga, var. *Tsonga[1].

thonged, *ppl. a.* (In Dict. s.v. Thong *v.*) Add: esp. *thonged sandal*.

1958 N. Marsh *Singing in Shrouds* (1959) ix. 189 She had high-heeled thonged sandals on her feet. **1972** D. Bloodworth *Any Number can Play* xv. 135 Fashionable accessories, including thonged sandals. **1982** J. Elliott *Country of her Dreams* ii. 15 Rosa Treadwell, in thonged sandals and sweat-stained smock, flushed beyond her years.

thora·cically, *adv.* [f. Thoracic *a.*: see -ically.] In the thorax.

1901 W. James *Let.* 13 Apr. (1920) II. 143, I find myself much more comfortable thoracically already than when I came. **1977** *Archivum Linguisticum* VIII. 87 Syllables with long vowels are 'thoracically arrested'.

thoraco-. Add: **thoracolu·mbar** *a.*, pertaining to the thoracic and lumbar parts of the spine; *spec.* an epithet of the sympathetic nervous system (see quot. 1948); **thoracotomy** (examples).

1918 Stedman *Med. Dict.* (ed. 5) 999/1 Thoracolumbar. **1935** J. C. White *Autonomic Nervous System* iv. 52 The tendency to asphyxia, acidosis, dehydration, and loss of body heat which follow general anaesthesia and prolonged operations are all combated by the thoracolumbar division of the autonomic nervous system. **1948** A. Brodal *Neurol. Anat.* xi. 320 The preganglionic efferent neurons of the sympathetic nervous system in man have their perikarya in the spinal cord, more precisely in all the thoracic and the uppermost two lumbar segments... Synonymous designations for the sympathetic and parasympathetic system therefore are the thoraco-lumbar and cranio-sacral systems. **1957** *Jrnl. Nervous & Mental Disease* CXXV. 462/2 Cannon's distinction between the thoraco-lumbar and the cranio-sacral division of the autonomic nervous system deals with the same sort of temporal division. **1967** G. M. Wyburn et al. *Concise Anat.* i. 2/2 Latissimus dorsi; arises from the spines of the lower six thoracic vertebrae, the thoracolumbar fascia, [etc.]. **1974** Passmore & Robson *Compan. Med. Stud.* III. ix. 3/2 The thoracic spine is relatively immobile... The most mobile part..is the thoracolumbar junction and this is damaged most commonly. **1944** *Lancet* 26 Aug. 265/1 A right anterior thoracotomy exposed the bleeding-point. **1976** *Proc. R. Soc. Med.* LXIX. 851/2 Subcutaneous midline sternotomy is a method whereby an upper abdominal vertical incision may be extended into the chest without performing a formal thoracotomy and without opening the pleura.

Thorazine (þǭ°·rǎzīn). *Pharm.* Also **thorazine.** [f. parts of the systematic name, 2-chloro-N,N-dimethyl-10-H-phenothiazine-10-propanamine, rearranged: see Chloro-[2], Thio-, *Azine.] A proprietary name for *Chlorpromazine.

1954 *Official Gaz.* (U.S. Patent Office) 9 Mar. 299/1 Smith, Kline & French Laboratories... *Thorazine* for central nervous system depressant. **1968** J. Hudson *Case of Need* ii. x. 164 Thorazine is a tranquilizer universally used as an antidote to LSD and employed to end bad trips. **1972** *Trade Marks Jrnl.* 20 Sept. 1873/2 *Thorazine*... Pharmaceutical and veterinary preparations and substances, all consisting of or containing azine compounds. Smith Kline & French Laboratories Limited. **1979** *Time* 2 Apr. 46/2 The stronger antipsychotic drugs like Thorazine are useful for handling schizophrenics, whose behavior is characterized by hallucinations and severely disordered thinking.

thoreaulite (þǭ·rəlait). *Min.* [a. F. *thoreaulite* (H. Buttgenbach 1933, in *Bull. Soc. géol. Belgique* LVI. 328), f. the name of J. Thoreau, 20th-c. Belgian geologist: see -ite[1].] A monoclinic oxide of tin and tantalum, $SnTa_2O_6$, found as rough brown prismatic crystals that are transparent in thin splinters.

1934 *Chem. Abstr.* XXVIII. 6658 Thoreaulite (new) is monoclinic with high n and birefringence. **1959** *Mineral. Abstr.* XIV. 107/1 Two flat veins break off from the north of the dyke of pegmatite at Manono (Katanga)... The heavy minerals are: cassiterite, tantalo-columbite, thoreaulite, lillingite?, pyrite, galena, etc. **1974** *Amer. Mineralogist* LIX. 1036/1 It is necessary to justify the choice of $SnTa_2O_6$ as the ideal composition of thoreaulite, since the ideal formula is usually given as $SnTa_2O_6$.

Thoreauvian (þǭrǭ°·viǎn), *sb.* and *a.* [f. *Thoreauvius*, Latinized form of *Thoreau* (see below) + -ian.] **A.** *sb.* One who admires the writings or shares the philosophy of Henry David Thoreau (1817–62), U.S. naturalist and writer. **B.** *adj.* Resembling or characteristic of Thoreau's writing or philosophy.

1927 *Observer* 14 Aug. 8 There was a fair sprinkling of conscious or sub-conscious Thoreauvians among bygone-seemers. **1954** *Sat. Rev.* (U.S.) 10 Oct. 79/2 Thoreauvians favor solitude. **1977** D. Conover *One Man's Island* 142, I am a Thoreauvian by thought and by deed. **1975** *Yankee* Oct. 113/1 In case any reader wishes to test his own ability to distinguish the true Thoreauvian style from the 'ghost-written'. **1977** *Time* 21 Feb. 59/1 It is a graceful Thoreauvian style of a year spent dismantling and rebuilding a 1920s Dodge pickup.

thorian (þǭ°·riǎn), *a.* *Min.* [f. Thor(ium + *-ian 2.] Of a mineral: having a (small) proportion of a constituent element replaced by thorium.

1930 *Amer. Mineralogist* XV. 572 Thorian. **1974** A. R. Philpotts in H. Sørensen *Alkaline Rocks* iv. vi. 295/1 Hydrothermal activity along fractures, producing biotite and enrichment of the carbonatite in thorian pyrochlore.

thoriated (þǭ°·riē¹tėd), *a.* [f. Thor(ium + *-ate[3] + -ed[2].] Of tungsten, or a valve filament made of tungsten: containing a proportion of thorium, e.g. to enhance electron emission in a valve.

1922 *Encycl. Brit.* XXXII. 1025/1 A thermionic valve of the latter type comprises a highly exhausted glass bulb having in it a filament of tungsten, or thoriated tungsten. **1951** *Engineering* 12 Oct. 459/1 The final..output stage consists of two water-cooled thoriated-filament triodes. **1973** J. G. Tweeddale *Materials Technol.* II. v. 118 Commonly, a non consumable electrode is made of tungsten or a tungsten alloy (notably thoriated tungsten).

thorium. Add: **2.** Special Combs.: **thorium lead,** (*a*) the isotope lead 208, which is the final decay product of the series of radioactive transformations beginning with the common isotope of thorium; (*b*) used *attrib.* (with hyphen) to designate a method of isotopic dating, and results obtained with it, based upon measurement of the relative amounts in rock of thorium 232 and its ultimate decay product, lead 208; **thorium series,** the series of isotopes produced by the radioactive decay of thorium 232 (the major natural isotope), each member resulting from the decay of the previous one.

1914 *Phil. Mag.* XXVIII. 827 It may be concluded with reasonable certainty [*though erroneously*] that thorium lead is unstable. **1946** F. E. Zeuner *Dating Past* x. 319 The thorium-family begins with the element thorium (atomic weight 232). In the course of its disintegration, 6 atoms of helium are given off, and thorium-lead remains. **1955** *Bull. Amer. Geol. Soc.* LXVI. 1141/2 The low thorium-lead age for the zircon could have been the result of addition of thorium to the mineral. *Ibid.*, The analysis..demonstrates that small amounts of uranium lead and major amounts of thorium and thorium lead are very loosely bound, chemically. **1971** I. G. Gass et al. *Understanding Earth* ii. 46/1 (*heading*) Uranium-lead.., lead-lead..and thorium-lead..methods. **1913** *Chem. News* 28 Feb. 97/2 The parent of ionium, the product in the uranium series corresponding with radio-thorium in the thorium series, is still experimentally unknown. **1955** I. Kaplan *Nucl. Physics* x. 207 In the actinium and thorium series, the mass numbers are given by the expressions $4n+3$ and $4n$, respectively. **1973** J. Yarwood *Atomic & Nucl. Physics* viii. 246 There are three main series of radioactive elements: the uranium, thorium, and actinium series, leaving out, for the present, the comparatively recently discovered neptunium series.

thorn, *sb.* Add: **I. 2.** (Earlier example of *thorn in the side* and later examples of *to be on thorns* (also *a thorn*) and *thorn in the flesh.*)

1611 Bible *Numbers* xxxiii. 55 Those which ye let remaine of them, shall be..thornes in your sides. **1853** Mrs. Gaskell *Cranford* vii. 100 Peggy wanted now to make several little confidences to her, which Miss Barker was on thorns to hear. **1913** D. H. Lawrence *Sons & Lovers* xiii. 379 He was on thorns to be gone from so trying a situation. **1923** —— *Stud. in Classic Amer. Lit.* ii. 21 Probably I haven't got over those Poor Richard tags yet. I rankle still with them. They are thorns in young flesh. **1924** E. M. Forster *Passage to India* iv. 34, I can be a thorn in Mr. Turton's flesh, and if he asks me I accept the invitation. **1929** J. Buchan *Courts of Morning* ii. iii. 187 You've given me a thorn to lie on, just when I was feeling comfortable. **1946** W. S. Maugham *Then & Now* xxxi. 187 The family that had been for so long a thorn in the flesh of the Vicars of Christ. **1977** E. Quinn tr. *Kung & Lapide's Brother or Lord* 36 Jesus was undoubtedly a thorn in the flesh for many Saducees.

3. d. *thorn needle* = *fibre needle* s.v. *Fibre *sb.* 8. (*Disused.*)

1950 *Vogue* Aug. 98/2 Intellectuals often have an E.M.G gramophone..and they play with thorns, not steels. **1973** *Amateur Photographer* 3 Jan. 33/2 A 'thorn' needle was composed of some soft woody or fibrous substance, which was ground to a point in a special machine.

IV. 8. a. *thorn fence* (examples), *forest* (examples), *jungle* (examples), *scrub* (examples), *-twig*, *woodland*; *thornproof* adj. (also as sb., *sc.* 'material'). **b. thornberry,** (the fruit of) the hawthorn; **thorn-bill,** (*b*) any of several small warblers of the genus *Acanthiza* or a closely related genus, found in Australia, New Guinea, and New Zealand; **thornveld** *S. Afr.*, veld in which Acacias predominate.

1766 Ld. Fife *Let.* 30 Nov. in A. & H. Tayler *Lord Fife & his Factor* (1925) ii. 36 Tell Thos. Reid that his Information as to there being no Thornberrys this season is wrong. **1886** Britten & Holland *Dict. Eng. Plant-Names* 467 Thornberries. Fruit of *Cratægus Oxyacantha.* **1934** E. Reynard *Narrow Land* v. 248 The Dover cliff was a thornberry scratch compared with what befell Cape Cod. **1911** J. A. Leach *Austral. Bird Bk.* 141 These birds..have been called Thornbills by Mr. A. J. North. **1933** *Bulletin* (Sydney) 5 Apr. 27/1 The yellow-tailed thornbill constructs a double nest, the lower cavity..containing the eggs. **1964** *Courier-Mail* (Brisbane) 17 Oct. 2/1 There is a species or more of Thornbill in every mainland State. **1975** I. Rowley *Bird Life* iii. 40 The real diminutives forage..by rapid and nearly continuous searching of ground or shrub layer as by wrens and thornbills. **1843** *Farmers' Cabinet* 15 Jan. 184/1 Our fences are either the worm, post-and-rail, or thorn. **1946** L. G. Green *So Few are Free* 226 Deep in the mountains they discovered a high thorn fence, obviously a man-made obstruction. **1903** W. R. Fisher tr. *Schimper's Plant-Geogr.* i. iii. 260 The Thorn-forest..is very rich in underwood. **1960** N. Polunin *Introd. Plant Geogr.* xiv. 442 Tropical thorn-forests..are usually still more xerophilous. **1913** 'Saki' *When William Came* vi. 102 We have somewhere to go to..better than the scrub and the veldt and the thorn-jungles. **1936** *Discovery* Nov. 337/1 The City of the Lake, buried deep in thorn jungle, through which we cut a path. **1955** W. Gaddis *Recognitions* iii. iv. 846 Engulfed in the flow of a tartan lap robe and folds of Irish thorn-proof, he stared fixedly at an open book. **1978** *Birds* Spring 3/2 (Advt.), *Gamefair Jacket*... In natural olive Beacon Thornproof. **1903** Kipling *Five Nations* 54 The thickets dwined to thorn-scrub, and the water drained to shallows. **1974** R. Adams *Shardik* lviii. 496 This is a country of thorn-scrub and fine, blowing sand.

1895 G. B. Shaw *Let.* 31 Aug. (1965) I. 556, I lay there looking up peacefully at the moon through..the laced thorntwigs of the briar. **1878** A. Aylward *Transvaal of To-Day* xii. 246 Four young men, all Africanders, nearly lost their lives in the Speckboom thornveld. **1936** L. Herrman in N. Isaacs *Trav. & Adventures Eastern Afr.* I. ii. 19 His 'panthers' are the small dark-skinned leopards of the thornveld. **1972** Palmer & Pitman *Trees S. Afr.* I. iii. 81 In the thornveld of Zululand, *Acacia karoo*, *Acacia nilotica*, *Acacia caffra*,..and *Acacia tortilis* subsp. *heteracantha* are frequent. **1903** W. R. Fisher tr. *Schimper's Plant-Geogr.* iii. iv. 492 Thorn-woodland appears..on very permeable, dry, sandy soil. **1960** N. Polunin *Introd. Plant Geogr.* xiv. 442 Grasses are often lacking in the drier thorn-woodlands.

thornless, *a.* (Later examples.)

1900 L. H. Bailey *Cycl. Amer. Hort.* I. 164/2 The Thornless or Mountain Blackberry..is not in cultivation. **1980** M. Spiller *Growing Fruit* vii. 159 The thornless form of the parsley-leaved blackberry..can be planted a little closer than other varieties.

thorn-tree. Add: (Earlier mod. and later examples.) In southern Africa, usually an acacia.

1785 G. Forster tr. *Sparrman's Voy. Cape of Good Hope* I. ix. 324 Being once upon a plain under the shelter of a few scambling thorn-trees, (*mimosa Nilotica*) he thought he should steal upon an elephant that was near the spot. **1798** Lady A. Barnard *Jrnl.* 13 May in *Lives of Lindsays* (1849) III. 440, I plucked from the great thorn-trees some of their prickles. **1970** *Stand. Encycl. S. Afr.* I. 10/2 In South Africa the indigenous members of the genus [*Acacia*] as a whole are generally referred to as 'thorn-trees' or 'acacias'.

thorny, *a.* Add: **5. b. thorny devil** = Moloch 2.

1899 *Strand Mag.* June 653 The Western Australians..describe it [*sc.* the moloch lizard] familiarly as the 'thorny devil'. **1932** *Discovery* Nov. 364/2 The Thorny Devil,..a sturdy creature about a foot long, covered with horny spikes and knobs, killed a puff adder with its tail. **1975** H. G. Cogger *Reptiles & Amphibians Austral.* 226/1 Thorny devil or Moloch... An unmistakable lizard, unique in scalation and form.

thoron (þǭ°·rǫn). *Chem.* and *Physics.* [a. G. *thoron* (C. Schmidt 1918, in *Zeitschr. f. anorg. Chem.* CIII. 114), f. Thor(ium + *-on[2].

Quot. 1920 represents an independent coinage.]

A radioactive isotope of radon, atomic weight 220, that is a gaseous decay product of thorium, being formed by the decay of radium 224; thorium emanation.

1918 *Jrnl. Chem. Soc.* CXIV. ii. 306 A rational system of nomenclature for the radioactive elements and their degradation products is suggested. Radium emanation is given the name Radon, Ro... The other emanations become Thoron, To, and Acton, Ao. **1920** E. Q. Adams in *Jrnl. Amer. Chem. Soc.* XLII. 2206 The names 'radium emanation', 'actinium emanation' and 'thorium emanation' have been shortened, respectively, to 'radon', 'actinon' and 'thoron', names which suggest that the element in question is an inert gas. **1938** R. W. Lawson tr. *Hevesy & Paneth's Man. Radioactivity* (ed. 2) xi. 132 The emanation of thorium (thoron) decays with a half-value period of 55 seconds. **1961** *New Scientist* 2 Nov. 290/3 The filter is placed in a lead cask and, after allowing four days for the decay of the 'daughter' products of natural radon and thoron, the total beta-activity of the sample is recorded. **1981** *Indian Jrnl. Earth Sci.* VIII. 1 (*heading*) On the behaviour and measurement of thoron (Rn^{220}) in soil. **1983** *Canad. Mining Jrnl.* Mar. 34/1 Measurements were carried out to estimate the radon and thoron daughter levels.

Thorotrast (þǭ°·rotrast). *Med.* Also **thorotrast.** [a. G. *thorotrast* (A. Weiser 1930, in *Wiener med. Wochenschr.* 25 Oct. 1428/2), f. *thoro-* Thoro- + *kon*)*trast* Contrast *sb.*] A colloidal solution of thorium dioxide formerly used as a contrast medium in radiography.

A proprietary name in the U.S.

1932 *Official Gaz.* (U.S. Patent Office) 5 Apr. 14/2 Heyden Chemical Corporation, New York... *Thorotrast* for medicinal preparation finding its application in the photography by X-rays for medicinal and similar purposes. **1933** [see *Angiogram]. **1947** *Radiology* XLIX. 362/2 The fact that 'thorotrast' (a colloidal suspension of thorium dioxide) is frequently used in human diagnostic work. **1976** P. Collard *Devel. of Microbiol.* x. 136 Animals were injected intravenously with particulate materials, indian ink, colloidal iron or thorotrast. **1977** *Lancet* 18 June 1297/1 The main long-term sequelæ of thorotrast are local effects and tumours at the site of injection and/or deposit..and fatal blood dyscrasias, including leukæmia and aplastic anæmias.

thorough, *adj.* **2. b.** (Earlier example with reference to servants.)

1822 M. Edgeworth *Let.* 26 Feb. (1971) 359, I have engaged a ladys maid..and a *thorough* maid for cooking and brooming.

thorough-. Add: **2. thorough-draught** (earlier example); **thorough-souled** *a.*, to one's inmost soul, downright.

1853 C. M. Yonge *Heir of Redclyffe* II. ix. 141 Three rooms..opening into each other..so that it was possible to produce a thorough draught. **1842** Poe *Lett.* (1948) I. 193, I cannot bring myself to like that man... He is too thorough-souled a time-server.

thoroughbrace. Add: **†b.** A vehicle whose body is supported on thoroughbraces. *Obs.* **thoroughbraced** *a.* (earlier example).

1865 *Harper's Mag.* Nov. 700/1 Preference to be given to a thorough-braced ambulance of Concord manufacture. **1886** *Leslie's Pop. Monthly* Dec. 722/1 The mustangs looked worse than the thorough-brace itself. **1930** A. W. Groom *Merry Christmas* xv. 111 The heavily laden thoroughbrace was hitched behind seven lively horses.

thoroughfare, *sb.* (*a.*) **1. a.** (Earlier example of *no thoroughfare.*)

a **1817** Jane Austen *Persuasion* (1818) III. xi. 235 The 'no-thoroughfare of Lyme'.

thoroughgoing, *a.* Add: (Earlier example.)

1800 M. Edgeworth *Parent's Assistant* (ed. 3) VI. 168, I am a *thorough-going* friend at any rate.

thoroughgoingness (earlier example).

1838 C. Gilman *Recoll. Southern Matron* xxviii. 194 The gentleman had even the thoroughgoingness to request that my brother's large, stout new slate might be exchanged for a recently-invented tablet.

thoroughwax. (Later examples.)

1925 E. Mellor tr. *Bonnier's Brit. Flora* 76 *Bupleurum* ..Leaves oval, perfoliate..Throw-wax. **1971** *Country Life* 4 Nov. 1192/1 Our cornfield weed thorow wax..is virtually extinct in Britain.

thoroughwort. Add: Also used as a name for other species of *Eupatorium.* (Earlier and further examples.)

1814 J. Bigelow *Florula Bostoniensis* 190 Thoroughwort..has acquired great medicinal reputation. **1842** [see *number six* s.v. *Number *sb.* 19]. **1906** *Harper's Mag.* Oct. 712 The boggy place where she came..for..the wild marsh-marigold, good for greens, thoroughwort, and the root of the sweet-flag. **1968** Peterson & McKenny *Field Guide to Wildflowers* 46 The thoroughworts..are composites of late summer and fall with numerous small fuzzy heads in rounded or flat-topped clusters.

thortveitite (þǫ·ɹtvaitait, -vē¹tait, tǫ·ɹt-). *Min.* [ad. G. *thortveitit* (J. Schetelig 1911, in *Centralbl. f. Min.* 721), f. the name of O. Thortveit of Norway, its discoverer: see -ite[1].] A silicate of scandium, $(Sc,Y)_2Si_2O_7$, found as colourless or greyish monoclinic crystals.

1912 *Jrnl. Chem. Soc.* CII. ii. 56 (*heading*) Thortveitite, a new mineral. **1963** *Prof. Papers U.S. Geol. Survey* No. 475-B. 11/1 The association of thortveitite with fluorite in the deposit at Crystal Mountain represents a new type of occurrence, which contrasts strongly with the occurrences in granitic pegmatites of Norway and Madagascar.

tho't, thot, repr. a U.S. pronunc. of *thought.*

1879 W. Whitman *Daybks. & Notebks.* (1978) I. 161 *Very bad spells*, unable sometimes to walk a block (sometimes tho't it all nearing the end). *a* **1886** E. Dickinson *Poems* (1955) III. 1148 The right to perish might be tho't An undisputed right. **1888** W. Whitman *Daybks. & Notebks.* (1978) II. 453 The Calming Tho't of all. **1971** *Black World* Mar. 53/1 She wuzn't sure but she thot it had. **1975** *Budget* (Sugarcreek, Ohio) 20 Mar. 12/2 Thot: Troubled waters cleanse the garments best.

thou, *sb.* Add: Also **thou.** (with point), **thou'.** (Earlier and later examples.) Also *spec.* a thousandth of an inch; (*U.S.*) a thousand dollars.

1867 'Ouida' *Under Two Flags* I. vi. 113 Losing "long odds in thou' " over the Oaks. **1924** Galsworthy *White Monkey* iii. ix. 276 If he did take a few thou. under the rose, he took 'em off the Huns. **1934** *Practical Motorist* 19 May 94/2 The width of the gap..should not be more than 3 to 5 'thous'. **1952** M. Tripp *Faith is Windsock* xi. 173 We're below ten thou..; you can take off your oxygen masks. **1965** *New Yorker* 20 Feb. 34/3 The gesture cost me a cool ten thou, but I didn't begrudge it. **1975** *Hi-Fi Answers* Feb. 36/1 The AT21X carries an elliptical stylus of 0·3 × 0·7 thou. dimensions.

though, *adv.* and *conj.* Add: **A.** *γ.* The form *tho* has been used in the U.S. as a reformed spelling, and (like *tho'*) is used informally as an abbreviation of the word.

1796 R. Barrie *Let.* 12 Oct. in N. Tolstoy *Half-Mad Lord* (1978) ii. 35 Tho: he sometimes might act imprudently his conduct never merited the ignominious punishment he receiv'd. **1818** M. Edgeworth *Let.* 8 Sept. (1971) 84 The library *tho* magnificent is a most comfortable..room. **1842** Tennyson *Poems* II. 91 Tho' much is taken, much abides. **1849** G. Gray *Let.* 22 June in M. Lutyens *Ruskins & Grays* (1972) xxiii. 217, I have now taken the opportunity..tho' without alluding to your Letter, of asking her how it was. **1879** *Proc. Amer. Philol. Assoc.* 6 The committee now present the following words as the beginning of such [a] list [of reformed spellings], and recommend them for immediate use:..Tho. Thru. Wisht. **1906** *Simplified Spelling Board Circular* (U.S.) No. 2. 12 Tho... Thru. **1973** *Black World* June 66 Sister Habiba's party was still smokin. Tho all the good food and wine and reefer was gone now. **1982** *N.Y. Times* 22 Sept. c-2/3 Tho' the trip's less than a mile it's still a dreary, cheerless bore.

B. I. *adv.* **b.** *colloq.* Used as an intensive after a question or emphatic statement: indeed, truly.

1905 *Eng. Dial. Dict.* VI. 102/1 'How it do rain!' indicates a heavy shower; but, 'How it do rain though!' marks a much heavier. **1906** [see *Geewhillikins *int.*]. **1912** B. Harraden *Out of Wreck I Rise* viii. 153 'I didn't know that persons who wrote plays made thousands.' 'Don't they, though,' Hailsham answered, laughing. **1929** E. M. Brent-Dyer *Rivals of Chalet School* vi. 83 'We've got more than an hour yet!' 'Have you, though?' said Mrs Maynard's voice just behind her. 'You've nothing of the kind.' **1948** G. Vidal *City & Pillar* i. v. 147 'What a sad story!' said Maria. 'Isn't it, though?' **1974** J. Aiken *Midnight is Place* iv. 128 'I get enough money..for Papa and me.' 'Did you though?' said Lucas..with surprise.

II. *conj.* **1. a.** Proverbial phr. *though I say it that should not* and varr.: see Say *v.*[1] B. 2 b in Dict. and Suppl.

4. b. Also with verb in present indicative.

1963 D. Storey *Radcliffe* xxxvi. 367 It's the sense of imitation that's so forbidding... As though it's all a deception, and the only person it doesn't deceive is me.

thought[1]. Add: **1. c.** *train of thought*: see *Train *sb.*[1] 12 b.

d. Also (without defining adj.), that of a named person [cf. G. *denken*].

1903 P. Shorey (*title*) The unity of Plato's thought. **1935** R. B. Perry (*title*) The thought and character of William James as revealed in unpublished correspondence and notes, together with his published writings. **1960** G. Harland *Thought of Reinhold Niebuhr* i. 13 The centrality of Christology in Niebuhr's thought is clear and unmistakable. **1964** S. J. Wilson (*title*) The thought of Cicero. **1968** in Gray & Cavendish *Chinese Communism in Crisis* 222 A force of revolutionised workers, armed with the thought of Mao Tse-tung, has been trained and tempered. **1971** D. McLellan *Thought of Karl Marx* p. ix, An exposition of certain themes central to Marx's thought. **1974** *Encycl. Brit. Macropædia* IV. 395/2 Socialist education at first had a rather abstract quality, because people had to measure their lives against the 'thought of Mao Tse-tung', a slogan that was to grow in popularity.

2. a. (Further examples.)

1967 tr. Mao Tse-Tung (*title*) The thoughts of Chairman Mao Tse-Tung. **1971** [see *red book, red-book 4]. **1977** 'S. Leys' *Chinese Shadows* (1978) i. 11 'We have friends all over the world.' This Thought of Chairman Mao can be seen on many walls. **1982** *Sunday Tel.* 7 Mar. 10/2 Between 1928 and 1941 there were less than 5,000 prosecutions [in Japan] for 'dangerous thoughts'.

b. *thought for the day* (*week*, etc.): a pregnant or gnomic thought (esp. one published or broadcast) to be pondered in the course of the day.

1932 R. Lehmann *Invit. Waltz* i. ii. 6 'Remember what Mother said yesterday.' 'What?' 'She'd have to start calling you herself.' Olivia gave a hoarse chuckle. 'Thought for the day...' **1972** *B.B.C. Handbk. 1973* 82 *Thought for The Day* is broadcast as part of the morning *Today* sequence at 7.45 a.m. **1973** J. Leasor *Host of Extras* iii. 41 Gratitude is sufficiently rare to cause surprise in those who find it, which is my thought for today. **1976** *Listener* 2 Dec. 716/3 So there, for the programme-makers' suggestion box, is a thought for the week. **1978** R. Thomas *Chinaman's Chance* xv. 152 They pay a lot to live here and then they never get up in time to watch the sun rise... Just my thought for the day.

3. c. Phr. *it is the thought that counts* and varr.: the value (to the recipient of a gift) lies in the goodwill, affection, etc., with which it is given.

1934 D. L. Sayers *Nine Tailors* ii. iv. 148 Not that I minded..where my poor little remembrance was placed, for..it is the thought that counts. **1961** C. McCullers *Clock without Hands* iv. 78 A house-warming present..not too modern or attractive, but it's the thought that counts. **1976** L. Thomas *Dangerous Davies* ix. 105 'He's eaten your Smarties.'..'Thanks for bringing them anyway... It's the thought, really.' **1982** *Preview Shopper* (London ed.) Spring 7 It's the thought that matters. When someone you care for has a special occasion to celebrate you want to choose exactly the right gift.

4. b. *lost in thought*: abstracted; absorbed in reverie or contemplation.

1806 J. Porter *Thaddeus of Warsaw* (ed. 4) III. x. 251 Miss Beaufort..was standing by one of the windows, evidently lost in thought. **1863** W. Collins *No Name* i. x. 44/1 He..sat at the table, drawing lines on the blotting-paper with his pen, lost in thought. **1926** B. A. McKelvie *Huldowget* iii. 35 He seemed lost in thought. **1955** L. P. Hartley *Perfect Woman* xxvii. 240 Jeremy stood lost in thought. 'She hasn't been away very long,' he said.

g. Phr. *perish the thought*: see *Perish *v.* 1 e; *it's a thought* (colloq. phr.): it is an idea worth considering.

1967 'S. Mitchell' *Come, Sweet Death* vii. 63 'Possibly he'd had a key cut.' 'It's a thought.' But I gathered from his tone that he didn't think much of it. **1974** M. Hastings *Dragon Island* xiii. 113 'Did they..kill him?' 'Quite a thought. It hadn't occurred to me, but it's a logical explanation.' **1980** J. Ditton *Copley's Hunch* ii. iii. 154 'It's a thought, sir.'..'If so, it doesn't help us.'

h. In negative contexts: *not to give* (something or someone) *a* (or *another*) *thought*, not to think at all (or any more) about, to dismiss from one's mind.

[**1864** [in Dict., sense 2 a].] **1925** F. Scott Fitzgerald *Great Gatsby* iii. 64, I wanted..to apologize for not having known him in the garden. 'Don't mention it,' he enjoined me eagerly. 'Don't give it another thought, old sport.' **1952** M. Allingham *Tiger in Smoke* ii. 50 If it was Martin that was on the tiles I wouldn't give it another thought. **1953** H. Clevely *Public Enemy* xxvii. 214 'After your wife's death, didn't you miss this bag?' 'I didn't even give it a thought.' **1956** M. Dickens *Angel in Corner* viii. 116 There will be plenty of young men in America... You won't give this Joe creature another thought. **1973** W. H. Canaway *Harry doing Good* i. iii. 35 I'll do that. Don't you give it another thought.

6. b. *U.S.* A very short length of time, a moment; usu. in advb. phr.

1912 L. J. Vance *Destroying Angel* xi. 142 Suddenly she turned her head and intercepted his whole-hearted stare. For a thought wonder glimmered in the violet eyes. **1937** in J. S. Hall *Sayings from Old Smoky* (1972) 122 A panther was attracted by the frying venison. In just a thought or two it came out and screamed. **1949** H. Hornsby *Lonesome Valley* 59 Johnny loved to hear the screech owl, except that when the scream came unexpectedly it was enough to scare anybody, for a thought.

7. a. *thought-action*, *-barrier*, *-centre*, *-construction*, *-content*, *-entity*, *-habit*, *-mode*, *-object*, *-picture*, *-process*, *-product*, *-relation*, *-scheme*, *-structure*, *-stuff*. **b.** *thought-destroying*, *-engendering*, *-saving*; *thought block.* **c.** *thought-woven*; limitative, as *thought-tight* [after *airtight*]. **d. thought control,** the control of a person's thoughts; *esp.* the attempt by a government to restrict ideas and impose opinions by such means as censorship and the control of curricula; **thoughtcrime, thought-crime,** in George Orwell's novel *Nineteen Eighty-Four*, the offence of failing in absolute loyalty to the ruling power; hence in any totalitarian system, unorthodox thinking considered as a criminal offence; **thought-experiment** = *Gedankenexperiment; **thought-forms** *pl.*, chiefly *Theol.*, the combination of presuppositions, imagery, vocabulary, etc., current at a particular time or place and in terms of which thinking on a subject takes place; **thought model,** a system of related ideas or images; **thought-pattern,** a set of assumptions and concepts underlying thought; an habitual way of thinking; in *pl.*, thought-forms; **thought police,** in a totalitarian state, a police force established to suppress freedom of thought; *spec.* in pre-war Japan, the Special Higher Police (*Tokubetsu Kōtō Keisatsu* or *Tokkō*); hence **thought-policing** *vbl. sb.*; **thought-provoking** *a.*, prompting serious thought; **thought reform,** a process of individual political indoctrination used in Communist China; also in extended sense; **thought-saver,** a trite expression used to save one the trouble of thinking, a cliché; **thought-stream,** the continuous succession of a person's thoughts, *spec.* as represented in fiction of a certain kind (cf. *stream of consciousness 2); **thought-wave,** (*b*) (further example); **thoughtway,** a customary way of thinking; an unconscious assumption or idea; **thought-word** (earlier example); **thought-world** [cf. G. *gedankenwelt*], the amalgam of mental attitudes, beliefs, presuppositions, and concepts about the world characteristic of any particular people, time, place, etc.

1909 *Encycl. Relig. & Ethics* II. 85/2 Purely mental exercise consists in those 'thought-actions' (*Denkhandlungen* as Eucken calls them) which determine both our mental attitude and our conduct. **1935** Thought-action [see *brain-wave* s.v. *Brain *sb.* 6]. **1958** *New Statesman* 15 Mar. 338/3 This thought-barrier, the difficulty of re-thinking the problems of defence in nuclear terms, is a very real thing. **1969** *Listener* 24 July 98/2 It seems we are again about to ram what C. H. Rolph calls 'a thought-barrier at least as old as the Great Rebellion'. This is the instant assumption of many Englishmen that whatever they dislike ought to be put a stop to. **1965** J. Pollitt *Depression & its Treatment* i. 5 Definite features of schizophrenic illness, e.g. thought block. **1846** E. A. Poe in *U.S. Mag. & Democratic Rev.* Apr. 268/1 We think in cycles, and may, from the frequency or infrequency of our revolutions about the various thought-centres, form an accurate estimate of the advance of our thought toward maturity. **1890** W. James *Princ. Psychol.* I. iv. 115 But our higher thought-centres knew hardly anything about the matter. Few men can tell off-hand which sock, shoe, or trousers-leg they put on first. **1904** Thought-centre [see *Association 9]. **1920** S. Alexander *Space, Time, & Deity* I. 161 In these thought constructions we are dealing all the time with ideas belonging to the empirical world. **1962** *Listener* 15 Mar. 470/2 In science, no thought-construction about the real world can be taken as more than provisionally true. **1916** L. Bloomfield in C. Hockett *Bloomfield Anthol.* (1970) 73 The type of sentence we have so far examined is..often used as the expression of a logical thought-content. **1972** *Jrnl. Social Psychol.* LXXXVI. 258 A 'thought-content' unit refers to all of a subject's utterance which..seems to express a single moral idea. **1935** U. Close *Behind Face of Japan* xxviii. 332 'Thought control' in Japan is strictly constitutional. **1939** R. Lehmann *No More Music* 87 Have you ever tried this healing by thought control?..It seems that if you think right you'll never have an ache or pain. **1945** *Ann. Reg. 1944* 295 Mr. Chen Li-fu, who as Minister of Education had attempted to institute 'thought control' for Chinese students abroad. **1954** T. S. Eliot *Confidential Clerk* i. 33 No, Claude, he only teaches *thought* control, Mind control is a different matter. **1980** 'J. Melville' *Chrysanthemum Chain* 10 A scientist of high intellectual integrity opposed to any form of thought control. **1949** 'G. Orwell' *Nineteen Eighty-Four* i. 22 He had committed..the essential crime that contained all others in itself. Thoughtcrime, they called it. **1954** *Encounter* May 28/1 [The Revolution] first created the 'People's Democracy' of the Terror and of compulsory unanimity, of thought-crimes, and of denunciation as the supreme duty of the citizen. **1968** *Economist* 22 June 19/1 If it were not the habit of Herr Ulbricht's government to put so many people in prison for thought-crime [etc.]. **1909** G. K. Chesterton *Orthodoxy* iii. 62 This..summary of the thought-destroying forces of our time would not be complete without some reference to pragmatism. **1851** H. Melville *Moby Dick* I. xxxiv. 253 How could I—being left completely to myself at such a thought-engendering altitude,—how could I but lightly hold my obligations to observe all whale-ships' standing orders, 'keep your weather eye open, and sing out every time'. **1892** Thought-entity [see Transcendentalistic *a.*]. **1949** *Mind* LVIII. 340 There is present, in addition to the imagery, an entity of another kind, a thought-entity. **1945** M. Wertheimer *Productive Thinking* vii. 180 (*heading*) On movement, on space, a thought experiment. **1965** P. Caws *Philos. of Sci.* xxix. 218 The situation may be illustrated by means of the following thought experiment. **1982** *New Scientist* 14 Jan. 75/2 Bekenstein considered a 'thought experiment' in which a box full of heat radiation was slowly lowered on a rope towards the surface (the horizon) of a black hole. **1890** W. James *Princ. Psychol.* II. xxviii. 664 Kant..insisted on thought-forms with which experience largely *agrees.* **1892** [in Dict.]. **1958** E. L. Mascall *Recovery of Unity* iv. 91 The deadlock between Catholics and Protestants..has been mainly due to their common inheritance of uncriticised..assumptions and thought-forms from the theologically decadent late Middle Ages. **1976** *Times* 2 Aug. 14/8 Bultmann insisted on the task of re-interpreting the substance of the mythological [biblical] materials in terms of thought-forms intelligible and acceptable in the twentieth century. **1939** P. Christophersen *Articles* i. 18 The rise of new grammatical categories must be supposed to result from thought-habits that have become so common and urgent that they demand linguistic expression. **1954** *Essays & Stud.* VII. 66 The common and ancient thought-habit that sight is the chief and most powerful of the senses. **1939** V. A. Demant *Religious Prospect* vi. 145 Dialectical thought has..a kinship with traditional religious thought-modes. **1936** Wirth & Shils tr. *Mannheim's Ideology & Utopia* v. 247 The next factor which may serve to characterize the perspective of thought is the so-called thought-model; i.e. the model that is implicitly in the mind of a person when he proceeds to reflect about an object. **1942** *Mind* LI. 137 It is the perception of spatio-temporal objects, and not the conception of real entities, that is providing the thought-model. **1958** W. Stark *Sociol. of Knowl.* iv. 193 Pareto devalues, and indeed abolishes, the relative in reality; but that means..that he operates with a thought-model which is unrealistic. **1890** W. James *Princ. Psychol.* I. ix. 283 It will show the relative intensities..of the several nerve-processes to which the various parts of the thought-object correspond. **1957** G. Ryle in M. Black *Importance of Lang.* (1962) 166 It is left to philosophy to be the science of this third domain which consists largely..of thought objects or Meanings. **1937** Thought-pattern [see *Pattern *sb.* 8 c]. **1943** *Mind* LII. 123 Those elements in the nineteenth-century thought-pattern, which are frequently referred to as Darwinism. **1962** *N. & Q.* Jan. 33/1 This strenuous attempt to convey the archaic thought-patterns of the New Testament into 'the natural vocabulary, constructions, and rhythms of contemporary speech'. **1977** T. Allbeury *Man with President's Mind* iii. 23 The rigid education..that surrounded all Soviet citizens..led to a thought pattern that automatically rejected anything but the Soviet official position. **1919** W. Deeping *Second Youth* xxix. 243 The arched vestibule..and the figure of the man standing there..reminded Laverach of the picture of the Roman sentinel..at his post in doomed Pompeii, and the..crashing of successive bombs made the thought-picture more vivid. **1963** *Times Lit. Suppl.* 10 May 344/4 Wesley's slowly evolving thought-picture of the nature of sin. **1945** *Sun* (Baltimore) 6 Oct. 4/1 It is an order imposing freedom of speech, thought, religion and assembly on the Japanese people, and requiring the immediate liberation of those imprisoned for political offenses by the so-called 'thought police'. **1949** 'G. Orwell' *Nineteen Eighty-Four* i. 49 He had denounced his uncle to the Thought Police after overhearing a conversation which appeared to him to have criminal tendencies. **1969** *Guardian* 5 Feb. 3/1 The Kremlin's thought-police are moving in slowly, circumspectly, on the Soviet scientific community. **1982** *Sunday Tel.* 7 Mar. 10/2 It may be that the reviewer has confused the latter with the Special Higher Police, or 'Thought Police' as they are sometimes called. **1968** *Listener* 26 Sept. 412/3 To submit to censorship..is to submit to thought-policing, censorship being the prevention of certain thoughts and images from entering your mind. **1973** *Howard Jrnl.* XIII. 268 The attitude develops into official self-protectiveness—restricting law books in case prisoners become litigious, for example—and downright thought-policing. **1889** J. M. Baldwin *Handbk. Psychol.* I. xiv. 271 We are concerned merely with the nature of the thought process—though a full treatment would include also its logic,—its value and bearing in the mental life. **1907** J. London *Iron Heel* i. 18 Each and every thought-process of the scientific reasoner is metaphysical. **1981** 'M. Innes' *Lord Mullion's Secret* ii. 22 This was a well-trodden little path in Honeybath's thought-processes. **1906** J. N. Keynes *Formal Logic* (ed. 4) 6 We may..say that psychology is concerned with thought-processes, logic with thought-products. **1933** *Mind* XLII. 111 The..view..that there must be radical discontinuity between the antecedents of a valid thought and a valid thought-product. **1916** J. Dewey *Ess. Exper. Logic* ii. 84 It..endeavours to define *what* in the various occasions renders them thought-provoking. **1936** *Discovery* Oct. 332/2 Mr Berenson..contributes a thought-provoking foreword. **1983** I. Murdoch *Philosopher's Pupil* 323 This was the most thought-provoking observation John Robert had ever elicited from her. **1959** *Atlantic Monthly* Dec. 75/1 xlviii. 371 In serious cases where criminality is involved..thought reform and punishment are combined. **1964** M. Argyle *Psychol. & Social Probl.* x. 134 Great interest has been aroused by Chinese thought reform, because it has been used on a very wide scale with considerable success and because the methods used are novel. **1966** F. Schurmann *Ideology & Organization in Communist China* i. 47 One of the most important questions..is whether 'thought reform' (*szuhsiang kaitsao*) can produce 'correct' behavior in the individual. **1981** J. Bancroft in Bloch & Chodoff *Psychiatric Ethics* ix. 174 'Thought reform' techniques and aversion therapy. **1887** A. Seth *Hegelianism* i. 36 It does not..follow that the whole external world is nothing more than a complex of thought-relations. **1931** L. Steffens *Autobiogr.* iii. i. 632 They were thoughtless conservatives..whose thought-saver was: 'My father was a Republican, and what was good enough for him is good enough for me.' **1948** E. Gowers *Plain Words* vii. 55 It [*sc.* the word *involve*] is used as a thought-saver because it is so faded. **1963** *Times Lit. Suppl.* 10 May 342/3 Those old thought-savers 'the imagination of England' and 'the American mind'. **1927** A. Huxley *Proper Stud.* 298 There are plenty of people..who feel as much enthusiasm for thought-saving devices as for automatic dishwashers and sewing-machines. **1948** *Mind* LVII. 259 Treating existential intuitions as the perceived convergencies of complementary thought-schemes—the sort of structures that Wittgenstein used to call 'hypotheses'. **1962** *Listener* 15 Mar. 470/2 By purely logical processes of combination, inference, and construction, [mathematics] builds up the most elaborate thought-schemes. **1930** Wyndham Lewis *Let.* 30 July (1963) 191 The Ulyssean 'thought-stream' method is only appropriate to the depiction of children, morons, and the extremely infirm. **1948** E. Bowen *Who do I Write?* 23 But, of course, your monologue isn't simply a thought-stream. **1960** R. St. John *Foreign Correspondent* xi. 225 Could I make it a vital memory for them and part of their thought stream for ever after? **1980** D. Lodge *How Far can You Go?* i. 5 American psychologists have..established..that the thought stream of the normal healthy male turns to sex every other minute between the ages of sixteen and twenty-six. **1931** O. Jespersen in H. N. Shenton et al. *Internat. Communication* iii. 112 Collinson..has been..driven to the view that 'it is precisely through our individual use of and reaction to our mother tongue that we can approach these general and fundamental problems of thought-structures and realize to the full their complexity and subtlety'. **1965** *Eng. Stud.* XLVI. 371 He envisages an extremely..complicated Coleridgean thought-structure which is realized or clothed in a number of images. **1890** W. James *Princ. Psychol.* II. xviii. 58 In some individuals the habitual 'thought-stuff', if one may so call it, is visual. **1915** *New Statesman* 23 Jan. 386/1 Hampered by so much ready-made reach-me-down thoughtstuff. **1913** L. Jerrold *French & English* viii. 153 One is often amazed by..thought-tight compartments in a walled-up mind. **1937** L. Hart *Europe in Arms* xv. 190 Departmentalism tends to thought-tight compartments. *a* **1930** D. H. Lawrence *Last Poems* (1932) 24 A tremendous body of silence Enveloping even the edges of the thought-waves. **1954** L. J. Cohen *Princ. World Citizenship* 4 The middle-class southern English have many thoughtways, like their conception of liberty, which they do not share with Cato. **1976** Nichols & Armstrong *Workers Divided* 19 They provide..ready-made and well trodden thoughtways (so straightaway it appears 'natural' that 'militants' will be 'mindless' [etc.]). **1980** *Times* 13 May 16/4 Their Civil service advisers—whose thoughtways and corporate interest impel them in certain directions. *a* **1866** J. Grote in *Jrnl. Philol.* (1872) IV. 66 Looking at language as it naturally presents itself, its apparently most simple units are what we call words, and therefore I describe a noem as a thought-word. **1947** N. H. Baynes (*title*) The thought-world of East Rome. **1958** *Spectator* 20 June 812/2 The thought-world of the laity, high and low, was in many ways pagan and magical. **1979** J. Hick in M. Goulder *Incarnation & Myth* iv. 78 No Christian who has ever lived within the evangelical thought-world can read without emotion such lines as Cowper's, There is a fountain filled with Blood [etc.]. **1892** W. B. Yeats *Countess Kathleen* 132 The tall thought-woven sails that flap unfurled Above the tide of hours, rise on the air.

thoughtography (þǫtǫ·grăfi). [f. Thought[1] + Phot)ography.] The production of a visible, usu. photographic, image (supposedly) by purely mental means. Hence **thou·ghtograph,** the image produced; **thoughto·grapher,** one who is said to practise thoughtography; **thoughtogra·phic** *a.*

1931 T. Fukurai (*title*) Clairvoyance and thoughtography. *Ibid.* x. 245 The medium can make the thoughtograph of the object presented by the sitter. **1967** *Psychic News* 20 May 4/2 Thus thoughtographic research spans more than half a century. **1968** J. Eisenbud *World of Ted Serios* xiii. 299 These [nightmares] kept up until he began his 'thoughtography'. **1976** C. Wilson *Geller Phenomenon* 33 (caption) Serios producing a thoughtograph. **1978** *Sunday Sun* (Brisbane) 5 Feb. 66/3 Professor Eisenbud..subjected thoughtographer Serios to scientific scrutiny. *Ibid.* 66/5 Ted randomly imprinted on film a thoughtograph.

thousand, *sb.* and *a.* Add: **2. a.** Also *thousand and one.*

1832 F. Trollope *Dom. Manners Amer.* II. xxxiii. 239 Of all the thousand and one towns I saw in America, I think Buffalo is the queerest looking. **1839** E. W. Lane (*title*) The thousand and one nights, commonly called, in England, The Arabian Nights' Entertainments. A new translation from the Arabic, with copious notes. **1842** [in Dict.]. **1910** W. L. Phelps *Essays Mod. Novelists* iii. 63 All the thousand and one details that make up the daily routine of the average person. **1962** J. Wain *Strike Father Dead* iv. 206 Would I be likely to suggest coming along as your manager if I didn't know a thousand and one ways of making myself useful?

b. Phrases: *a thousand times, no*: certainly not; similarly *a thousand times, yes* (rare); *I believe you, thousands wouldn't* (and similar expressions): ambiguous responses to remarks

received with scepticism; *death of* (or *by*) *a thousand cuts*: a succession of minor hurts that are cumulatively very serious or annoying; *a thousand of bricks*: see *Brick *sb.*[1] 5 a.

1896 'M. Rutherford' *Clara Hopgood* v. 57 'No,' said Madge, 'a thousand times no.' **1897** H. James *Spoils of Poynton* xxii. 279 A thousand times yes—her choice should know no scruple. **1926** R. H. Mottram *Crime at Vanderlynden's* 46 'I did twelve months in the line, as a platoon commander. How long did you do that?' 'Twelve months about!' 'I believe you where thousands wouldn't.' **1932** A. Christie *Peril at End House* xvii. 199 Am I sure, myself, about anything at all? No, no—a thousand times, no. **1966** tr. *Quotations from Chairman Mao Tse-Tung* xxvii. 258 'He who is not afraid of death by a thousand cuts dares to unhorse the emperor'—this is the indomitable spirit needed in our struggle to build socialism and communism. **1968** C. Aird *Henrietta Who?* x. 97 'I don't even know..what I don't know.' Bill Thorpe nodded comprehendingly. 'I follow you—though thousands wouldn't.' **1974** D. Seaman *Bomb that could Lip-Read* ix. 73 The head of the rocket..chips off tiny fragments of steel... The poor buggers who get in the way die the death of a thousand cuts. **1980** G. Greene *Doctor Fischer* vii. 39 It had to be the death of a thousand cuts. He told her he forgave her..but he told her also that he could never forget her betrayal. **1980** P. Moyes *Angel Death* xx. 255 I can believe it. Thousands wouldn't. **1981** P. Turnbull *Deep & Crisp & Even* vii. 116 'Don't you think I'm too old?' 'No, a thousand times, no!' **1982** P. Inchbald *Sweet Short Grass* xx. 172 Oh, Franco! *Yes!* A thousand times yes!

3. a. (Further examples.)

1873 Trollope *Phineas Redux* II. xxi. 172 Mere words, supplied at so much the thousand. **1896** G. B. Shaw *Let.* 15 Feb. (1965) I. 597 Men who rattle off their copy at anything from 20/- to 40/- a thousand. **1919** W. S. Maugham *Moon & Sixpence* iii. 14 We would talk..of editors and the sort of contributions they welcomed, how much they paid a thousand, and whether they paid promptly or otherwise.

b. Also (*U.S.*), a thousand dollars.

1919 E. O'Neill *Moon of Carib.* 163 Smith said he would give two thousand cash if I would sell the place to him. **1942** *Amer. Mercury* July 85 He might condence Sweet Back out of a thousand on a plate.

5. *thousand-mile* (example); *thousand-petalled* adj.; **thousand-head(ed) kale,** a branching variety of cabbage, *Brassica oleracea* var. *fruticosa*, cultivated as fodder for sheep or cattle; **thousand island** [f. *Thousand Islands*, name of a large group of islands in the St. Lawrence River], used *attrib.* and *absol.* to designate Russian salad-dressing containing added pieces of garnishing; also **thousand isle; thousand-jacket** *N.Z.* = *Houhere; **thousand-legs** (later example); **thousand-miler** *slang*, a dark shirt that does not show the dirt; **thousand-year(-old) egg,** a Chinese delicacy consisting of a pickled egg that has been kept in earth, lime, and chopped straw for some weeks; **Thousand-Year Reich** [G. *tausendjähriges Reich*], the German Third Reich (1933–45), as a regime envisaged by the Nazis as established for an indefinite period.

1887 *Times* 22 Oct. 8/1 This practice of making thousand-headed kale stand down..on poor land..is likely to come rapidly into favour. **1925** Thousandhead kale [see *marrow-stem* (*kale*) s.v. *Marrow *sb.*[1] 5]. **1929** Oldershaw & Porter *Brit. Farm Crops* v. 235 Thousand-headed kale is a very useful crop to grow both for sheep- and cattle-feed. **1975** Park & Eddowes *Crop Husbandry* (ed. 2) xiii. 294 Marrow stem kale should be used before the new year followed by the hardier thousand head kale. **1916** *Daily Colonist* (Victoria, B.C.) 19 July 6/1 (Advt.), Mrs. Porter's Thousand Island Salad Dressing, bottle 35¢. **1945** J. L. Marshall *Santa Fe* 106 For years, Bill Gardner, steward on the Kansas City–Chicago run, handed out a special '1001 Dressing', an improvement on the usual Thousand Island mixture. **1962** Thousand-isle [see *Roquefort b]. **1981** *Times* 2 Mar. 12/5 In a year or two she will be specifying that the thousand island dressing (a pinkish salad cream with bits of vegetables in it) should be low-calorie. **1888** *Cassell's Picturesque Austral.* III. 210 Toi-toi, supplejack, thousand-jacket, are names of things known well enough to the inhabitants of Napier and Taranaki. **1946** Thousand-jacket [see *Houhere]. **1962** Gordon & Lavoipierre *Entomol. for Students of Med.* vii. 41 The class Diplopoda contains all the millipedes or 'thousand legs'. **1875** 'Mark Twain' in *Atlantic Monthly* Apr. 450/1 The thousand-mile wall of dense forest. **1929** F. C. Bowen *Sea Slang* 139 *Thousand milers*, black twill shirts. **1959** *Washington Post* 8 Oct. c3/3 A thousand-miler is a navy blue shirt which doesn't show the gravy stains and may be worn for days at a time without washing. Slim must have a wardrobe of thousand-milers. **1978** K. Bonfiglioli *All Tea in China* vii. 86 A 'thousand-miler' turned out to be a sort of durable shirt made of black twill; so-called..because it should be washed..after every thousand miles of the voyage. **1951** L. MacNeice tr. *Goethe's Faust* ii. ii. 204 But here at Pharsalus was fought a master model To prove how might opposes greater might and tears To shreds the lovely thousand-petalled wreath of freedom. **1970** *Times* 10 Mar. 17/2 (Advt.), A hitherto unrecorded Baccarat 'thousand-petalled' rose weight. **1961** E.-M. Wong *Chinese Cookery* v. 36 Everyone has heard of 'thousand-year-old' eggs, but in reality these eggs are only a few months old. **1972** K. Lo *Chinese Food* i. 47 Thousand-Year-Old Egg (which, to be more precise, should be called Pickled Eggs)..can be incorporated into a Chinese breakfast. **1980** E. Behr *Getting Even* xviii. 208 Seaslugs, jellyfish and thousand-year eggs appeared on the table. [**1934** *Times* 6 Sept. 12/4 Herr Hitler's proclamation to the rally was read... Herr Hitler declares that 'there will be no further revolution in Germany for a thousand years.'] **1946** A. Huxley *Let.* 27 Oct. (1969) 553 When people think of far-off communist Utopias or Thousand-Year Reichs, they are so much dazzled by the beauty of what they see..in the unknowable future, that they are ready to commit any atrocity in the present. **1970** A. Price *Labyrinth Makers* vi. 89 The Wagnerian last hours of the Thousand Year Reich. **1979** J. Crosby *Party of Year* xviii. 109 Now that the 1,000-year Reich had crumbled, what else was there?

'thout. (Examples.)

1893 H. A. Shands *Some Peculiarities of Speech in Mississippi* 63 *Thout*, Negro for *without.* **1897** Kipling *Capt. Cour.* iv. 91 Don't let's hev another [sad song] 'thout somethin' between. **1917** —— *Divers. Creatures* 341 I'm the only farmer you've got. Nothin' goes off my place 'thout it walks on its own feet. **1935** Z. N. Hurston *Mules & Men* i. x. 205 Nobody can't mention fat 'thout you makin' out they talkin' bout you. **1979** 'E. McBain' *Calypso* iv. 37 You ain't gettin *nothin* 'thout the forty dollars.

Thracian (þrē¹·ʃiăn), *sb.* and *a.* [f. L. *Thrācius, Thrācus*, a. Gr. θρᾴκιος, f. θρᾴκη Thrace: see -an, -ian.] **A.** *sb.* **a.** A native or inhabitant of Thrace, in antiquity a region to the N.E. of Macedonia, and now comprising European Turkey, southern Bulgaria, and the region of Thrace in N.E. Greece.

1569 T. Stocker tr. *Diodorus Siculus' Hist. Successors Alexander* 105 Aboute two thousand Mercenarie Grekes, and so many Thracians. **1618** E. Bolton tr. *Lucius Julius Florus' Roman Hist.* (1636) 176 The Sordiscans were of all the Thracians the most savage. **1875** *Encycl. Brit.* III. 854/1 In the earliest times of history Bœotia was inhabited by various tribes, such as the Aonians, Temmicians, Thracians, [etc.]. **1949** *Oxf. Classical Dict.* 901/2 The Thracians were not without a native culture. **1976** *Daily Tel.* 31 Aug. 6 A major archaeological discovery..has been made in Bulgaria. Knowledge of the Thracians is advanced substantially. **1982** K. Follett *Man from St. Petersburg* ix. 178 'I wonder how the Thracians would feel about all this.' 'They would rather belong to Russia than Turkey.'

b. The language of the ancient Thracians, an Indo-European language thought to be related to Phrygian or Illyrian.

1879 *Academy* XV. 99/1 It is still doubted by..philologists whether Albanian should be classed as an Aryan language... However..., I am quite willing to allow that it is..a descendant of the ancient Illyrian or Thracian, and I will not quarrel with anyone who wishes to call the latter Pelasgian. **1933** C. D. Buck *Compar. Gram. Gk. & Lat.* 14 Thracian is known from proper names and glosses, and there is one obscure inscription believed to be Thracian. **1962** A. J. Beattie in Wace & Stubbings *Compan. to Homer* x. 312 From the Hellespont to Chalcidice most of the inhabitants spoke Thracian. **1972** W. B. Lockwood *Panorama Indo-Europ. Lang.* 172 At the time of its greatest known extent, in antiquity, Thracian was spoken throughout the eastern half of the Balkan Peninsula and stretched northwards into the Central European Plain.

B. *adj.* Of or pertaining to Thrace.

1588 Shakes. *Tit. A.* i. i. 138 The selfe same Gods that arm'd the Queene of Troy With opportunitie of sharpe reuenge Vpon the Thracian Tyrant in his Tent. **1594** Kyd *Cornelia* iii. ii. 49 Stoute Thracian Mars. **1667** Milton *P.L.* vii. 34 The Race Of that wilde Rout that tore the Thracian Bard In Rhodope. **1697** Dryden *Æneis* vi. 877 The Thracian bard..There stands conspicuous in his flowing vest. **1781** Gibbon *Decl. & F.* II. xxvi. 593 Orders were immediately dispatched to the civil and military governors of the Thracian diocese. *a* **1822** Shelley *Cyclops* in *Posthumous Poems* (1824) 343 And when the Thracian wind pours down the snow, I wrap my body in the skins of beasts. **1848** *Rep. Brit. Assoc. Adv. Sci.* 1847 266 The second family is the Thracian or Illyrian, once spread on the Dnieper, the Hellespont, and in Asia Minor. **1920** *Glasgow Herald* 12 May 9 Several of the Thracian harbours now under Greek sovereignty..are to be free. *Ibid.* 27 July 7 To-day at dawn the Thracian Army launched its offensive. **1949** *Oxf. Classical Dict.* 901/2 Greek recruiting officers (especially in the fourth century) enlisted Thracian 'peltasts' or light-armed fighters. **1977** *Jrnl. R. Soc. Arts* CXXV. 485/1 The nearest parallels which we can refer to are two helmets of the so-called 'Thracian' type.

Thraco- (þrē¹·ko), (*rare*) **Thrako-,** used as comb. form of *Thracian *sb.* and *a.*, as in *Thraco-Illyrian* adj., *Thraco-Phrygian* adj. and sb.

1902 *Encycl. Brit.* XXV. 249/2 Albanian is peculiarly interesting as the only surviving representative of the so-called Thraco-Illyrian group of languages which formed the primitive speech of the peninsula. **1924** G. Murray *Rise Gk. Epic* (ed. 3) ii. 40 A great movement of Thraco-Phrygian tribes with eastern linguistic affinities. **1931** *Times Lit. Suppl.* 12 Feb. 116/4 The Thrako-Illyrian stratum which underlies all the races of the Peninsular. **1946** Priebsch & Collinson *German Lang.* (ed. 2) i. 19 He places the prehistoric connexions of the Tokharians with the progenitors of Balts, Slavs, Armenians and Thraco-Phrygians in the steppes of South-East Russia between the Dniepr and the Urals. **1968** D. L. Clarke *Analytical Archaeol.* ix. 391 An older, outer ring of non-Urnfield Indo-European areas—Teutonic and Baltic on one hand and Thraco-Phrygian, Greek, and Hittite on another. **1972** W. B. Lockwood *Panorama Indo-Europ. Lang.* 172 Thraco-Phrygian is the term used to denote a group of languages whose earliest known homeland was South-East Europe. Three languages are distinguished: Thracian, Phrygian and Armenian.

thrash, thresh, *v.* Add: **III. 7. a.** Also *refl.* and *fig.*

1939 T. S. Eliot *Family Reunion* i. ii. 60 The fish Thrashing itself upstream.

8. (Earlier and further examples.) Esp. with *about* or *around.* Also *fig.*

1846 *Boston Courier* 17 June 2/4 Arter I'd gone to bed & heern Him a thrashin round like a short tailed Bull in fli time. **1884** 'Mark Twain' *Huck. Finn* vi. 45 He didn't go sound asleep, but was uneasy. He groaned, and moaned, and thrashed around this way and that. **1962** K. A. Porter *Ship of Fools* 178 He groans and yells and thrashes about at night. **1962** K. Kesey *One Flew over Cuckoo's Nest* iv. 309 It fought a long time against having it taken away, flailing and thrashing around. **1973** *Times* 12 Nov. 11/8 His is in many ways a sad life to watch, as he thrashes around for the opening that will bring him fame. **1978** R. Barnard *Unruly Son* xvii. 190 This little detail panicked you..and then you started thrashing around... You did silly things.

thrash, thresh, *sb.*[1] Add: **3.** A party, esp. one that is lavish or unrestrained. *slang.*

1957 G. Smith *Friends* 120, I think he stole away to London for an occasional thrash when it got too much for him, but in general he was a model pupil. **1968** K. Amis *I want it Now* ii. 68 No quiet family party at all, it had turned out, but a twenty-cover thrash. **1976** *Times Lit. Suppl.* 6 Feb. 131/3 Staggering..from his sick-bed to play host at an enormous black-tie thrash at a Belgravia mansion borrowed for the night. **1980** C. Matthew *Loosely Engaged* 17 Occasionally someone throws a thrash, but most of the time we just bomb round to Wedgies..and have a bit of a giggle.

thrasher[1]. **4. thrasher-whale:** for *Orca gladiator* substitute *Orcinus orca*; (earlier example).

1782 'J. H. St. J. de Crèvecœur' *Lett. from Amer. Farmer* vi. 169 The following are..the various species of whales known to these people... The killer, or thrasher about thirty feet; they often kill the other whales.

thrawnly, *adv.* (Later examples.)

1899 J. Buchan *Grey Weather* 250 'What bird are ye?' he asked thrawnly. **1980** *Times Lit. Suppl.* 28 Mar. 373/1 The Kilbrandon Commission found the stage army of the Scottish good solidly pro-devolution... Only the Labour Party remained thrawnly hostile to the whole idea.

thread, *sb.* Add: **I. 2. f.** *pl.* Clothes. *slang* (orig. and chiefly *U.S.*).

1926 Maines & Grant *Wise-Crack Dict.* 11/2 *New set of threads*, new suit of clothes. **1959** R. Bloch *Blood Runs Cold* (1963) 163 Mitch got into some decent threads—he had this one blue suit and he wore a white shirt and a tie too. **1972** M. J. Bosse *Incident at Naha* ii. 64 *My friends*, who grooved the way I did... I mean, love beads, wild threads, granny glasses..and a bit of grass. **1978** J. Gardner *Dancing Dodo* xxiii. 175 Load it and get in on under that set of executive threads.

5. Also, a similar ridge round the inside of a cylindrical hole, as in a nut or a screwhole.

1677, 1875 [see Tap *sb.*[1] 4]. **1938** [see *self-tapping *ppl. a.*]. **1972** *How Things Work* III. 168 For the majority of screwed work a tap is used for internal threading (Fig. 3, showing the thread being cut in a nut) and a die head is used for external threading. **1977** *Reader's Digest Bk. Do-It-Yourself Skills & Techniques* vi. 175/2 As soon as the tap starts to cut, stop pressing down, and let the tap screw itself into the hole, cutting a thread as it goes.

II. 8. Esp. in phr. *to pick* (or *take*) *up the thread*(*s*) (*of*), to continue (with) after an interruption or separation; *spec.* to resume an interrupted friendship; *to lose the thread*, to cease to follow the sense of what is being said.

1881 R. L. Stevenson *Virginibus Puerisque* 137 We shall..take up again the thread of our enjoyment in the same spirit as we let it fall. **1907** G. B. Shaw *John Bull's Other Island* iv. 95 Eighteen years is a devilish long time, Nora. Now if it had been eighteen minutes, or even eighteen months, we should be able to pick up the interrupted thread, and chatter like two magpies. **1924** A. Christie *Poirot Investigates* v. 125 Philip Ridgeway narrated the circumstances leading to the disappearance of the bonds... When he had finished, Poirot took up the thread with a question. **1929** H. J. Laski in *Holmes-Laski Lett.* (1953) II. 1169, I don't, I suppose, see him more than once in two years; but I always find that we can take up the threads and plunge *in medias res* without any difficulty. **1944** E. S. Gardner *D.A. calls Turn* (1947) xi. 101 If it were true, he'd make some sort of a financial adjustment, but could hardly be expected to pick up the thread of a life where it had been broken ten years ago. **1956** A. Wilson *Anglo-Saxon Attitudes* ii. i. 215 He stopped and, for a moment, he appeared to have lost the thread of his remarks. **1980** D. Lodge *How Far can you Go?* vi. 226 Dennis and Angela picked up the threads of their lives together,.. a little chastened, but both hugely relieved. **1981** A. Schlee *Rhine Journey* xi. 143 He chose..to appear to have lost the thread of the discussion and looked from one to another with a kind of cautious bewilderment.

14. a. (*a*) *threadball.* (*c*) *thread-spinner*; *thread-forming* adj.; *thread-wise* adv.

1896 G. B. Shaw *Our Theatres in Nineties* (1932) II. 252 Peer's wild run through the night over the charred heath, stumbling over the threadballs and broken straws. **1918** G. Frankau *One of Them* xvi. 123 How the three crones must laugh as they entwine Cat's-cradle-wise our mortal threadball's tangle. **1927** T. Woodhouse *Artif. Silk* 34 The tanks which supply the solution to the thread-forming apparatus. **1892** 'Mark Twain' *Amer. Claimant* x. 102 Today, the work of..the 2,000,000 thread-spinners [women] is done by 1,000 girls. **1918** Mrs. Belloc Lowndes *Out of the War?* xx. 255 The narrow, winding road which ran thread-wise on the cliffs.

b. thread bag *Jamaica*, a small cloth bag, tied or drawn closed with a thread or string; **thread belay** *Mountaineering*, a belay in which the rope or sling is passed through a hole in the rock before being secured again to the climber; **thread clips** (see quot. 1964); also *attrib.* in *sing.*; **thread-fin** (examples); **thread-guide** (examples).

1924 M. W. Beckwith *Jamaica Anansi Stories* 35 An' Goat cut her up an' put her in his tread-bag. **1953** R. Mais *Hills were Joyful Together* ii. xii. 226 Her money gone! Somebody had robbed her while she was asleep. She carried it in a threadbag tied with a string around her neck. **1935** *Jrnl. Fell & Rock Climbing Club* X. 236 (*caption*) Thread belay. **1941** C. F. Kirkus *Let's go Climbing* iv. 54 Here you use a thread belay, passing a loop of your rope through a muddy hole behind a chockstone.. and tying it round the stone or on to your waist line. **1965** A. Blackshaw *Mountaineering* viii. 225 Because a thread belay with the main climbing rope is usually very awkward and complicated..slings are normally used. **1958** *Times* 27 Dec. 4/1 Threadclip scissors..are employed in the weaving trade for snipping loose ends during the weaving process. **1964** *McCall's Sewing* v. 62/2 *Thread clips*, a real time-saving little clipper that can be used effectively for snipping threads and making the small clips needed for marking or for curved seams. It has one ring which fits over the little finger, and is operated by squeezing with the palm of the hand. **1896** Jordan & Evermann *Check-List Fishes* 335 Polynemidæ. The Threadfins. **1933** *Bulletin* (Sydney) 5 Apr. 27/3 Threadfins..rarely extend southward to the coast of N.S. Wales. **1979** *Arizona Daily Star* 5 Aug. C-5/1 He was credited with introducing threadfin shad as a forage fish for bass. **1924** Ld. Ronaldshay *India* xiii. 159 The supply from abroad of such things as bobbins, plane tree-rollers..and porcelain thread-guides was cut off. **1964** *McCall's Sewing* v. 69/2 On most machines, the last thread guide will indicate the direction in which the thread must enter the needle.

thread, *v.* Add: **1. d.** Of a man: to have sexual intercourse with (a woman). *slang.*

[**1903** Farmer & Henley *Slang* VII. 109/1 *To thread the needle*, to possess a woman.] **1958** B. Behan *Borstal Boy* i. 15 Sheila would be sorry she did not let me thread her, the night we walked the canal.

10. a. To place the thread, film, or tape in its proper course in (a sewing machine, projector, etc.). Usu. with *up*. Also *absol.*

1873 *Young Englishwoman* Mar. 150/1 Thread up the machine with the same coloured silk. **1913** F. A. Talbot *Pract. Cinematogr.* vii. 85 In threading up the camera it is only necessary to make sure that the image on the negative comes squarely and truly before the window in the gate. **1923** —— *Moving Pictures* 81 Threading the camera, as it is called, completed, the door of the exposed magazine is closed. **1932** Simpson & Weir *Weaver's Craft* x. 92 Threading the Loom.—It is still an advantage for two people to work together for this. **1962** L. Deighton *Ipcress File* xxiv. 155 He threaded up the 16mm projector. **1964** *McCall's Sewing* v. 69/1 Your machine simply won't work if it isn't threaded exactly according to plan. **1970** A. Fowles *Dupe Negative* i. 8 It's [*sc.* the film's] just back from the lab. Take a couple of minutes to thread up.

b. To pass (film, etc.) *through* a projector, recorder, etc., so that it occupies the correct path; = *lace *v.* 4 f.

1915 J. B. Rathbun *Motion Picture Making & Exhibiting* ii. 33 The loading of a motion picture camera is usually no more difficult than threading the film through a projector. **1932** Simpson & Weir *Weaver's Craft* xi. 115 Thread the new piece through the correct heddle and dent of the reed, then wrap the loose end round a pin in the woven fabric. **1959** N. Mailer *Advts. for Myself* (1961) 168 Sam attempts to talk while he is threading the film. **1961** *N.Y. Times* 10 Sept. x. 15/3 The user has to thread the tape through the machine before starting, and rewind the tape after playing or recording. **1972** W. P. Blatty *Exorcist* (1974) iii. i. 279 The priest quickly set up the tape recorder; looked for an outlet; plugged it in; threaded tape.

threaded, *ppl. a.* Add: **1. b.** *Computers.* Of a list or tree: in which items contain a pointer to a preceding node as well as one to the following node.

1960 Perlis & Thornton in *Commun. Assoc. Computing Machinery* III. 196/1 This paper presents an addition to the list structure languages which is expected to add to the above advantages while simplifying machine processing of lists. This is done by the use of threaded lists. A threaded list is a structure in which the last element of each list specifies the location of the head of the list of which it is the terminal member. **1979** Tremblay & Bunt *Introd. Computer Sci.* xi. 583 Given the threaded representation of a binary tree with respect to inorder traversal, it is a simple matter to formulate algorithms for obtaining the inorder predecessor and successor of a designated node.

threading, *vbl. sb.* (Examples corresponding to *thread *v.* 10.)

1913 F. H. Richardson *Motion Picture Handbk.* (ed. 2) 219 It is aggravating for the operator who has to do rapid work in threading up, to be obliged to work with reels in bad condition. **1932** Simpson & Weir *Weaver's Craft* x. 92 For the threading up one person should sit at the back of the loom in such a position that the 'shed' sticks and warp threads can be easily seen. **1933** *Sight & Sound* Spring 34/2 Accessibility of the gates and sprockets for threading up. **1964** *McCall's Sewing* v. 69/1 Always check your instruction manual for proper threading. **1970** *Which?* Jan. 8/1 Threading can be a tedious and sometimes tricky business, especially if the thread has to go *through* other openings as well as the eye of the needle.

Threa·dnee:dle Street. The name of the street in the City of London where the Bank of England is located, used allusively to mean the Bank or its directors. Cf. *the Old Lady of Threadneedle Street* s.v. *lady *sb.* 4 c.

1924 Ld. Birkenhead *Amer. Revisited* i. 10 They [*sc.* U.S. economists and financiers] lack something of the sophistication and age-long sagacity of Threadneedle Street. **1974** G. Vaizey *Tangled Web* ii. 25 It's only a question of time..before it gets to 'the ferret's' ears in Threadneedle Street. Then, heaven help you.

threat, *sb.* Add: **II. 4.** *Zool.* Animal behaviour that keeps other animals at a distance or strengthens social dominance without physical conflict. Freq. *attrib.*

1933 R. W. G. Hingston *Meaning of Animal Colour* v. 119 Whenever a bird has threat-colours on the crown, it either lowers its head so that they show, or erects the feathers..to make them visible above the level of the beak. *Ibid.* x. 291 Song is an exhibition of threat. **1943** D. Lack *Life of Robin* iii. 26 It is..a threat display, serving to intimidate a trespassing robin. **1949** *Brit. Birds* XLII. 234 One female called a peculiar, low, harsh, single note..similar to a harsh growling threat-note. **1966** N. Tinbergen *Animal Behaviour* viii. 177 The signalling movements of higher animals, particularly those used in threat and courtship. **1978** P. Marsh et al. *Rules of Disorder* v. 127 Certain threat signals are evolved so that intra-specific conflicts become ceremonial in character. **1981** *Oxf. Compan. Animal Behaviour* 569/1 The opening of the mouth that precedes biting has evolved into a ritualized baring of the teeth that is characteristic of threat in many mammals.

threaten, *v.* Add: **7.** In weakened use: to express an intention *to do* something, not necessarily evil.

1925 *Dialect Notes* V. 344 *Threaten*, v.i., promise; as, he threatened to give me money. **1928** A. Huxley *Let.* 1 May (1969) 296 [He] was lunching here today and broached a notion about a preliminary limited edition. ..He threatens to come and talk to you about it.

threatened, *ppl. a.* Add: **1. b.** Of a wild animal or plant: in danger of becoming rare or extinct.

1960 *Oryx* V. 381 (*heading*) Australia's threatened mammals. **1966** *Red Data Bk.* II. 2 The object of these lists and sheets of threatened species is not only to draw universal attention to the dangers facing some unique creatures,..but also to provide the factual information necessary for action. **1972** *Ibid.* (new ed.) I. Preamble 1 The threatened species include those that are in immediate danger of extinction (endangered species), those that are likely to enter this category (vulnerable species), and those that are rare and at risk (rare species). **1976** *New Yorker* 12 Jan. 58 Seven hundred and sixty-one plants were designated as 'endangered', meaning their survival was in serious doubt, twelve hundred and thirty-eight were listed as 'threatened' and an even one hundred were declared extinct. **1979** *Birds* Summer 56/1 The most authoritative and comprehensive reference book published on the world's rare, endangered and threatened birds. **1979** *Red Data Bk.* (ed. 2) II. Preamble 3 *Endangered* (E)..Taxa in danger of extinction and whose survival is unlikely if the causal factors continue operating... *Vulnerable* (V)..Taxa believed likely to move into the endangered category in the near future if the causal factors continue operating... *Rare* (R)..Taxa with small world populations that are not at present endangered or vulnerable, but are at risk... A taxon at subspecific level which might qualify for the V..category has not been included if no other subspecies of the species concerned is in a threatened (E, V, R or I) category.

threatful, *a.* Delete *rare* and add further examples.

1895 G. MacDonald *Lilith* viii. 53 The eagle, perched with outstretched wings on the top, appeared threatful. **1922** E. R. Eddison *Worm Ouroboros* xxxi. 391, I have read signs in heaven: nought clear, but threatful unto both you and me. **1923** M. Sadleir *Desolate Splendour* 199 He felt a desire still further to ingratiate himself with this threatful lord. **1932** W. Faulkner *Light in August* ii. 42 It still lingers about her..something dark and outlandish and threatful.

three, *a.* and *sb.* Add: **B. I. 1. g.** (*the three B's*, etc.: examples preceding initial letters other than those of uses referred to in Dict.) *three acres and a cow*: regarded as the requirement for self-sufficiency; *three ages* (Archæol.), the Stone, Bronze, and Iron Ages (see also *three-age*, sense III. 1 a below); *the Three Bishoprics* (Hist.), Metz, Toul, and Verdun; *three cheers*, three successive cheers in unison, freq. *for* someone or something; *three musketeers*, [tr. Fr. *les trois mousquetaires* (title of a novel (1844) by Alexandre Dumas père] three close associates; *the three wise men* = *the three Kings* s.v. King *sb.* 1 c; *transf.*, three men who act as advisers or arbitrators.

1909 J. R. Ware *Passing Eng.* 244/2 *Three B's, the* (*Clerical*), bright, brief, and brotherly—the modern protest against the sleepy nature of a majority of the 19th century church services. **1934** Webster s.v., *Three B's, Music*, the three great composers, Bach, Beethoven, and Brahms. **1969** Three B's [see *cry *ad.* 4]. **1885** Three C's [see C I. 1]. **1976** *Casper* (Wyoming) *Star-Tribune* 29 June 3/3 Q. In money matters, what's meant by 'The Three Cs'? A. That's a term used by the credit experts. Capital, capacity to pay and collectibility. Of the three, capacity to pay is generally thought to be the most important, capital the least. **1955** D. W. Maurer in *Publ. Amer. Dialect Soc.* xxiv. 187 Excuses for missing meets are sometimes delicately referred to as *the three cases: shit, shave, and shine*. **1974** P. Wright *Lang. Brit. Industry* xii. 105 Alliteration, slang and a desire to hide the meaning all contribute to *the three Hs*, standing for 'high, hot, and a hell of a lot', and used when a soft soap and water enema has to be given. **1960** *Guardian Weekly* 20 Oct. 12/4 Women's groups rebelling against the old commandment of the 'Three Ks'—Kinder, Kirche, Küche (children, church and kitchen). **1938** J. Daniels *Southerner* 136 It was impressive how directly the town's merchants made their appeal to poverty with the neatly necessities of living in the three M's, meat, meal and molasses. **1929** F. C. Bowen *Sea Slang* 140 Three Ss, The.. The old naval rule to promotion, to mind your three Ss. That is to say, to be Sober, Silly and Civil. **1885** J. Chamberlain in *Times* 17 Nov. 10/2 This man reported ..that wherever the labourer had land and kept a cow—'three acres and a cow' (loud laughter)..the poor rates were reduced. **1889** G. N. Curzon *Russia in Central Asia* vii. 239 The majority of residents would seem to have attained the ideal of Arcadian bliss expressed elsewhere in the historical phrase, 'Three acres and a cow'. **1964** *English Studies* XLV. (Suppl.) 224 Chesterton the Distributist and advocate of 'three acres and a cow'. **1866** J. Crawfurd in *Trans. Ethnol. Soc. London* IV. 1 The theory which supposes three different ages of civilization, marked respectively by the use of arms and implements of stone, of bronze, and of iron, seems to have originated in the discoveries recently made by the examination of the refuse heaps of Denmark and the pile buildings of the Swiss lakes.. There can be little doubt that the three ages above indicated did really exist. **1944** V. G. Childe *Progress & Archaeol.* i. 5 Of course 'Thomsen's three Ages' are just periods of this relative kind and should be better designated Stages. **1794** A. Young *Trav. France* II. xix. 420 The provinces of Loraine, Alsace, the three Bishoprics, and the West Indies, not included. **1900** H. N. Williams *Henri II* xxi. 271 The princes..authorised him to take possession of the towns of Toul, Metz and Verdun—the 'Three Bishoprics'. **1964** C. Duffy *Wild Goose & Eagle* viii. 111 Conti..might be tempted to harry the Three Bishoprics and the Austrian-garrisoned Duchy of Luxemburg. **1751** Three cheers [see Cheer *sb.* 8]. **1840** *Brother Jonathan* 10 Oct. 4/6 They gave him three cheers. **1907** G. B. Shaw *John Bull's Other Island* ii. 30 Three cheers for ould Ireland, is it? **1970** G. F. Newman *Sir, You Bastard* vi. 174 Three cheers for any good publicity we can get. **1887** Kipling in *Civil & Mil. Gaz.* 11 Mar. (*title*) The three musketeers. **1903** G. B. Shaw *Man & Superman* p. xxviii, He and I and Mr Sidney Webb were sowing our political wild oats as a sort of Fabian Three Musketeers. **1923** Three musketeers [see *blow *v.*¹ 1 b]. **1979** *Nature* 8 Nov. 136/2 Those were the three musketeers who soon agreed that in publishing their joint work they would always share the credit for all ideas, whoever had thought them up. **1867** *Chambers's Encycl.* IX. 419/2 The visit of the three magi or wise men of the East. **1904** Three wise men [used s.v. Magus 2]. **1961** *Ann. Reg. 1960* 467 The conference decided that three wise men (who later became four) should recommend the best means of co-ordinating Western economic policies. **1976** *Hansard Commons* 9 June 1578 The three wise men ..who make up the Programme Complaints Commission. **1979** G. St. Aubyn *Edward VII* iv. 185 The Prince of Wales decided..to submit the negotiations to the Prime Minister, the Lord Chancellor and Lord Hartington... These three wise men produced a memorandum.

3. (Examples short for *three farthings* (now *Hist.*); further examples short for *three inches*.)

1913 J. Vaizey *College Girl* ii. xxvi. 360, I paid eleven-three for it. **1947** *Vogue* Apr. 73/1 Three-and-eleven-three is much less than four shillings. **1962** M. Duffy *That's how it Was* iii. 33 The girls would buy a few yard of stuff at two-and-eleven, three, a yard. **1965** *Canad. Jrnl. Linguistics* Spring 121 A three-by-five card was made out for each address. **1978** W. Stovall *Presidential Emergency* v. 119 Solving his problems on three-by-five index cards.

II. 1. b. Add to def.: Also, a figure resembling that denoting the number three, *esp.* in *Skating*. (Further examples.)

1903 [see *grape-vine 2 c]. **1938** J. Cary *Castle Corner* 378 He cut a three. **1975** *Oxf. Compan. Sports & Games* 522/1 The three, a two-lobed figure, so named because the turn involved at the extreme end of each circle leaves a tracing on the ice resembling the numeral '3'.

2. (Further examples of *in threes*.)

1889 W. B. Yeats *Wanderings of Oisin* 130 Children sing in twos and threes. **1953** R. Chandler *Long Goodbye* xix. 117 Three shots, three misses. I hate it when they come in threes. **1977** 'M. Yorke' *Cost of Silence* xvi. 130 'First Pedro—then Emma Widnes—now Jamie Renshaw. Who'll be next?'..'Things do go in threes, don't they?'

3. c. *Three in One*, (*b*) *attrib.* as *three-in-one*, combining three items, functions, etc., in one whole; (*c*) *Three in One* (attrib. also *Three-in-One*), the proprietary name of a lubricating oil; *three-o(h)-three* (usually printed ·303), a rifle of ·303 calibre; also, ammunition manufactured for use in such a rifle; *three times three* (earlier example).

1909 *Grocery Catal.* (T. Eaton & Co.) 26/2 Three-in-One Hand Saw..combining in one tool a saw, 2 ft. rule and square. **1928** *Trade Marks Jrnl.* 22 Aug. 1354/1 '*Three in One*'... Lubricating oil. Three in One Oil Co..., 130, William Street, City, County, and State of New York,

United States of America; manufacturers. **1931** *Advertiser* (Adelaide) 7 Oct. 10 (Advt.), A three-in-one garment, comprising vest, bloomers, and underskirt. **1962** E. Snow *Other Side of River* (1963) xxxii. 245 We call it a three-in-one technique. It combines, in all grades, teaching, practical research work, and actual production. **1967** N. Marsh *Death at Dolphin* i. 14 The key..refused to turn... 'You want a touch of the old three-in-one.. Oil, mate. Loobrication.' **1970** W. Klatt in D. J. Dwyer *China Now* (1974) xviii. 341 Lin Piao gave high priority for the task of 'struggle—criticism—transformation' which is apparently being carried out by the chief organ of the new order, i.e. the 'three-in-one combination' of the Revolutionary Committees, embracing representatives of the revolutionary cadres, of the People's Liberation Army and of the revolutionary masses. **1977** O. Schell *China* (1978) i. 104 And those old factories with poor conditions and men from three-in-one groups to solve their problems. **1979** T. Gifford *Hollywood Gothic* (1980) xxiv. 240 The lock had gotten rusty, but..Three-in-One oil did the job. **1903** *Kynoch Jrnl.* Aug.–Sept. 128/1 The ·303 Sporting Rifle. The ·303 is used a great deal as a sporting rifle, and being the Government arm is quoted in comparison with Express rifles. **1928** E. Blunden *Undertones of War* xii. 205 I was at least more skilful with the shots of eight grain than with the three-o-three of the small-arms factory. **1959** [see *thirty-eight* s.v. *thirty B. 1]. **1981** J. Barnett *Firing Squad* viii. 79 Firearm certificate holders on 303 Lee Enfields..were mostly Sharpshooters using rifle clubs. **1780** *Loiterer* 19 Sept. 4 My health has been drunk in a bumper, with three times three, by every Club of Tradesmen in the City.

III. 1. a. *three-act* (earlier example), *-age*, *-alarm*, *-bean*, *-blade*, *-car*, *-cent*, *-class* (examples), *-core*, *-dollar*, *-electrode*, *-island*, *-judge*, *-lane* [*lane *sb.* 2 d], *-level*, *-member*, *-mile* (earlier and later examples), *-month(s)* (examples), *-party*, *-person*, *-phase* (later examples), *-pin*, *-place*, *-ply* (later and earlier examples), *-position*, *-shilling* (example), *-speed* (earlier and later examples), *-stage*, *-syllable*, *-term*, *-tier* (examples), *-wheel* (examples), *-word*. **b.** *three-dimensional*, *-engined*, *-fanged*, *-handed* (later examples), *-heeled*, *-numbered*, *-pronged* (later examples), *-storied* (later examples), *-tailed* (earlier example), *-tiered* (later example), *-wheeled* (later examples). **c.** *three-acter*, *-alarmer*, *-hitter*, *-mover* (earlier example), *-tonner* (motor vehicle), *-volumer* (examples), *-wheeler* (later examples).

1825 H. Wilson *Memoirs* II. 76, I..fixed, upon Molière's comedy of the Malade Imaginaire, which I hastily transformed into an English three-act piece! **1948** C. McCullers in *Mademoiselle* Sept. 257/1 By autumn I was writing a three-acter about revenge and incest. **1957** G. Bibby *Testimony of Spade* 31 He [*sc.* Christian Thomsen] was constrained to write a short account of his arrangement of the Copenhagen museum and of his Three Age system. **1970** Bray & Trump *Dict. Archaeol.* 231/2 *Three Age System*, the scheme for dividing prehistory into a stone age, bronze age and iron age. It was first formulated by C. Thomsen 1816–19. **1932** *Amer. Speech* VII. 337 Three alarm fire, used with negative to indicate mediocrity. *a* **1975** Wodehouse *Sunset at Blandings* (1977) xii. 80 Lord Emsworth entered looking like a refugee from a three-alarm fire. **1950** O. Nash *Family Reunion* 80 The author's attention has been called to a type of conflagration known as a three-alarmer. **1976** *Billings* (Montana) *Gaz.* 17 June 7-B/1 Create a picnic-like atmosphere with such favorites as potato salad and three bean salad. **1931** D. Rose *J. de la Cierva's Wings of Tomorrow* vi. 92 For certain purposes the three-blade rotor may prove the most efficient. **1967** *Jane's Surface Skimmer Systems 1967–68* 9 (*caption*) A. Turboméca Artouste IIC drives..two three-blade variable-pitch propellers for thrust. **1944** R. Chandler *Lady in Lake* iv. 23 Outside the wall to the left was the three-car garage. **1980** J. McNeil *Spy Game* ix. 93 There was..a three-car garage. **1851** *Statutes at Large U.S.A.* IX. 587 No ingots shall be used for the coinage of the three-cent pieces herein authorized, of which the quality differs more than five-thousandths from the legal standard. **1898** P. L. Ford *Hon. Peter Stirling* 281 The three-cent papers..abuse me. **1946** *Publ. Amer. Dial. Soc.* vi. 37 To feel like a three-cent piece with a hole in it. (To feel worthless and do-less.) **1910** *Westm. Gaz.* 14 Feb. 6/4 [Germany] The detested three-class system..and..the system of promotion of certain classes of electors from one class to another on other than property qualifications. **1980** *Jrnl. R. Soc. Arts* June 411/1 Gone are..the limitations on political rights of a 'three-class franchise'. **1922** *B.I. Hand-bk.* (Brit. Insulated & Helsby Cables Ltd.) (ed. 3) 106 Three-core cables. **1958** *Spectator* 8 Aug. 190/2, 3-core electric wiring. **1904** Kipling *Traffics & Discoveries* 212 'Heaven is beautiful, Earth is ugly,' The three-dimensioned preacher saith. **1858** J. H. Hickcox *Hist. Acct. Amer. Coinage* 56 Three Dollar gold coins were coined..under an act passed in 1853. **1918** *Wireless World* VI. 144 De Forest was experimenting with a three-electrode valve. **1932** *Discovery* July 216/1 The starting point of modern wireless is what is known as the three-electrode thermionic valve. **1931** *Nineteenth Cent.* Feb. 159 The three-engined types. **1967** *Economist* 16 Sept. 1022/1 What might happen to passenger traffic if one's competitor should advertise 'three engined safety'. **1915** D. H. Lawrence *Let.* 15 July in *Lett. to B. Russell* (1948) 53 Liberty, Equality & Fraternity is the three-fanged serpent. **1792** J. Woodforde *Diary* 10 Feb. (1927) III. 335 After Coffee and Tea we got to Cards to three-handed Cribbage. **1907** W. M. Cockrum *Pioneer Hist. Indiana* xiv. 344 Dancing was the principal amusement..three- and four-handed reels and jigs. **1937** G. Greene *19 Stories* (1947) 57 They had played their usual rubber of three-handed bridge. **1976** 'Trevanian' *Main* (1977) ii. 23 They were playing three-handed cutthroat. *a* **1889** G. M. Hopkins *Poems* (1967) 180 Yet Arthur is a Bowman: his three-heeled timber'll hit The bald..gold. **1976** *Billings* (Montana) *Gaz.* 30 June 3-E/1 Mrs. John Candelaria fired a three-hitter and tripled in two runs during an eight-run first inning. **1930** *Blackw. Mag.* July 1/1 The *Ulidia* was a typical 'three-island' tramp steamer. **1962** A. G. Course *Dict. Naut. Terms* 198 *Three island ship*, a vessel with a raised forecastle forward, a raised bridge deck amidships, and a raised poop aft. [illegible] **1911** *Daily Colonist* (Victoria, B.C.) 14 Apr. 15/4 The schooners were well within the three-mile limit, poaching on the British Columbia fisheries grounds. **1858** E. B. Browning *Seraphim & Other Poems* 160 In the eyes all undefiled Of a little three months' child. **1876** G. M. Hopkins *Wr. Deutschland* ix, in *Poems* (1967) 54 Be adored among men, God, three-numbered form. **1925** J. A. Spender *Public Life* II. xiv. 27 The difficulties of the three-party system. **1978** A. Gilchrist *Cod Wars* viii. 66 As a result of the 1936 election, the conservative-dominated coalition of the Independence and Progressive parties gave way to a three-party coalition from which the Independence Party was excluded. **1922** Three-phase system [see *bias *v.* 5]. **1926** *Jrnl. Iron & Steel Inst.* CXIV. 77 These solutions of the important problem gave a simple and comparatively cheap installation, without transforming the three-phase current to direct current or ..regulating the speed of the generator. **1961** *Listener* 9 Nov. 767/2 The normal three-phase alternating current system. **1868** J. C. Atkinson *Gloss. Cleveland Dial.* 335 *Merls*, sb... Other names are..Five-pin, Nine-pin, Three-pin, Morris or Merels. **1940** *Chambers's Techn. Dict.* 848/1 *Three-pin plug*, a plug with three contact pins, two for the main circuit and one for the earth connexion. **1974** A. Ross *Bradford Business* 75 A length of insulated cable ..snaked across the floor to a three-pin socket. **1947** H. Reichenbach *Elem. Symbolic Logic* § 17. 83 A three-place function is given by the verb 'gives' in the sentence 'Peter gives Paul a book'. **1964** R. H. Robins *Gen. Linguistics* viii. 330 English plosive and nasal consonants fall into a three-place..system, bilabial, alveolar, velar. **1905** *Timber Trades Jrnl.* 21 Jan. 72/1 Date cases, made entirely of three-ply wood. **1910**, etc. Three ply [see *ply *sb.* 1]. **1921** *Daily Mail Year Bk.* 112/2 The new..railway is..equipped with three-position signals. **1971** *Gloss. Electrotechnical, Power Terms* (*B.S.I.*) i. iii. 16 *Three-position relay*, a relay which has one unenergized and two energized conditions. **1944** T. H. Wisdom *Triumph over Tunisia* vi. 54 Jerry..dropped a load of three-pronged spikes on the runway. **1968** N. Mitchell *Sir George Cunningham* vii. 138 A Punjabi called Khurshid Anwar..was on the Hazara border organising a three-pronged drive into Kashmir. *a* **1817** Jane Austen *Persuasion* (1818) IV. vi. 116 She has a blister on one of her heels, as large as a three shilling piece. **1895** *People* 6 Jan. 4/5 The three-speed gear bicycle invented by Messrs. Lindley and Biggs. **1977** *New Yorker* 9 May 34/1 A.. young woman..had just bought a three-speed Raleigh. *Ibid.* 34/2 Why did I buy a three-speed? **1977** *Lancs. Life* Nov. 138/2 Her bike had broken down—something to do with the three-speed. **1980** J. L. Carr *Month in Country* 69 You can have Dad's bike... It's a three-speed and the chain has an oil bath. **1936** *Discovery* Sept. 299/2 The proposed three-stage rocket-ship. **1965** *Language* XLI. 117 A three-stage process of increasing deprovincialization in Russian linguistics. **1939** *Oxoniensia* IV. 127 No. 2 is part of a 'three-storeyed' pitcher, showing a combination of various decorative ideas. **1963** J. Robinson *Honest to God* i. 13 The traditional language of a three-storeyed universe. **1886** *Amer. Jrnl. Philol.* VII. 246 In early Latin this energetic stress-accent was not bound by the three-syllable limit. **1964** W. S. Allen in D. Abercrombie et al. *Daniel Jones* 4 It [*sc.* stress] falls..on a light antepenultimate only because of the overriding three-syllable rule, which will not permit it to recede further. **1718** M. W. Montagu *Let.* 19 May (1965) I. 413 'Tis common for the Heirs of a great three-tail'd Bassa not to be rich enough to keep in repair the House he built. **1957** E. B. Jones *Instrument Technol.* III. ii. 72 (*caption*) Response of a three-term controller to an artificial disturbance. **1977** *Time* 21 Nov. 28/2 In Cleveland, scrappy Dennis Kucinich, 31, a former three-term city councilman, edged out Edward Feighan, 30, the candidate of the regular Democratic organization. **1883** *Heal & Son Catal.: Dining Rm., Libr., & Drawing Rm. Furnit.* 215 Three-tier Whatnot, in Walnut or Ebonised. **1957** Ld. Hailey *African Survey 1956* viii. 467 In form this constituted a 'three-tier' system of Councils, but it was the District Council which was to form the focal point in it. **1977** *Guardian Weekly* 23 Oct. 8/3 The three-tier agreement is to be made up of a treaty limiting the numbers of certain strategic weapons for a period of eight years, a protocol imposing certain limits on other weapons for three years, and a statement of principles looking toward major arms reductions in the future. **1973** 'M. Innes' *Appleby's Answer* xv. 129 A three-tiered contraption loaded with pastries and éclairs. *a* **1944** K. Douglas *Alamein to Zem Zem* (1946) 38 This necessitated sweating about to find a three-tonner or a tank to tow us out. **1971** B. W. Aldiss *Soldier Erect* 39 We marched off the platform in good order..and transferred our kit to a line of three-tonners standing waiting for us outside the station. **1864** G. Meredith *Let.* Oct. (1912) I. 162 My 'plain story' is first to right me and then the 3 volumer will play trumpets. **1927** *Daily Tel.* 27 Sept. 5/1 When the 'three-volumer' went out it was thought we had ceased to ask for literary quantity. **1936** *Discovery* Nov. 351/1 An 1888 Benz three-wheel motor car. **1973** *Times* 30 Oct. 4/1 A new three-wheel car, the Robin, for which an average fuel consumption of 50 miles to the gallon is claimed. **1958** C. Freeling *Hoist before Dawn* iv. 41 She saw Mrs Henderson's miniature three-wheeler drawn up in front of the house. **1975** *Times* 22 Dec. 3/1 Their three-wheeler disintegrated in collision with another car. **1880** 'Mark Twain' *Tramp Abroad* xxx. 321 They have a word here..three-word phrases, filched from the back of the Dictionary. **1978** R. Ludlum *Holcroft Covenant* xiii. 154 He had sent Sam a three-word cablegram from the airport in Lisbon.

2. three-address *a. Computers*, (employing instructions) having three addresses, two that specify the location of the two operands and one that specifies where the result is to be stored; **three-anti** *China* = *Sanfan; **three-axis** *a.*, having or involving an ability to be rotated about each of three mutually perpendicular axes; **three-ball** *a.*, of a golf match: involving three players, each playing his own ball; **three-ball(s)**, a three-ball golf match; **three-banded** (later examples); **three-bar** *a.*, (*b*) of an electric fire: having three heating elements; **three-body** *a. Math.* and *Physics*, involving or pertaining to three objects or particles; **three-body problem** = *problem of three bodies* (see sense 1. 1 g); **three-card** *a.* (later examples); **three-centre** *a. Chem.*, applied to a bond in which the orbital of the two electrons forming it is spread over three contributing atoms; **three-circle diagram**, a Venn diagram in which there are three circles; **three-crop** *a.*, of a ewe: that has borne lambs in three successive years; **three-cushion** *a.*, designating a type of billiards in which the cushion must be struck at least three times by a ball at each play (see quot. 1957); **three-D, 3-D, 3 D** *a.*, three-dimensional, used *esp.* of a stereoptic process of filming; also *ellipt.* as *sb.*, a three-dimensional realization or state; **three-day** *a.*, extending over three days, that takes three days to complete or come to an end, as *three-day event*, a tripartite equestrian competition, usu. with the first day given over to dressage, the second to cross-country riding, and the third to show-jumping in a ring (hence *three-day eventer*, a horse that participates in such competitions), *three-day week*, a reduced working week of only three days; **three-dimensional** *a.*, having, or appearing to have, the three dimensions of length, breadth, and depth (cf. Dimension *sb.* 3 a); = Tri-dimensional *a.*; also *fig.*; hence **three-dimensionality**; **three-dimensionally** *adv.*; **three-figure** *a.*, consisting of three digits; one hundred or more (pounds, runs, miles per hour, etc.); calculated to three decimal places; **three-halves power**, the square root of the cube of a number; in *Electronics* used *attrib.* to designate a law that the anode current of a valve is proportional to the three-halves power of the anode voltage; **Three Hours** (or **Hours'**) **Service**, a devotional service lasting from 12 to 3 o'clock in the afternoon of Good Friday, designed to cover the hours of the crucifixion of Jesus Christ; also *ellipt.*; **three-letter man**, (*a*) *U.S.*, a person awarded a mark of distinction (cf. *letter *sb.*¹ 1 e) in three different sports; (*b*) *colloq.*, an obnoxious person; **three-light**, (*a*) (later examples); also *ellipt.*; **three-line, -lined** *a.*, also, *three-line* (occas. *-lined*) *whip*, a written notice, underlined three times to indicate great urgency, requesting the attendance of members of Parliament at a particular parliamentary session; the discipline of such a notice; (further examples); **three-martini lunch** *U.S.*, a lavish lunch, esp. one charged to a business expense account; **three-minute** *a.*, that occupies, or completes or is completed within, three minutes (in quot. 1833, that completes a mile in three minutes); that indicates the passage

of three minutes; **three-nines** *a.*, (*a*) (see quot. 1927); (*b*) of a telephone call: made to an emergency service, for which in the U.K. 999 is dialled; **three-out**: see *out *sb.* 1 b; **three-pipe problem**, a problem which requires considerable thought (for the duration of the smoking of three pipes of tobacco); **three-putt** *v. intr.* (Golf), to take three putts to hole the ball on a particular green; *trans.* to play (a green or hole) taking three putts; **three-ring, -ringed, circus**, a circus having three rings; hence *fig.*, a showy or extravagant spectacle; a scene of confusion or disorder; cf. *one-ring circus* s.v. *one *numeral a.* 33; **three-sixty**, in various sports, aerobatics, etc.: a turn through three-hundred-and-sixty degrees; **three-space**, three-dimensional space; **three-spined stickleback**: *Gasterosteus aculeatus*; (examples); **three-star** *a.*, having, displaying, bearing as insignia, or being designated by three stars as a mark of quality, rank, etc., usu. in a four- or five-star grading system (see *star *sb.*¹ 10 c, d); *spec.* used to designate: (*a*) a good quality French brandy; (*b*) a highly-rated hotel or restaurant; (*c*) *U.S.*, a lieutenant general (in rank below a general, above a major general); (*d*) a grade of petrol; (*e*) *transf.*, anything of high quality or in a high degree characteristic; also *ellipt.* as *sb.*, three-star brandy, petrol, etc.; **three-striper**: see *striper 1; **three-horned acacia**: = *honey locust* s.v. Honey *sb.* 7 b in Dict. and Suppl.; (earlier example); **three-time** *a.*, that has occurred or been done three times; of a person, to whom something has happened, or who has achieved something, three times; spec. *three-time loser*, a person who has served three prison sentences; **three-valued** *a.*, having three values; *spec.* in *Philos.*, designating a logical system or technique which incorporates a third value such as indeterminacy, uncertainty, half-truth, etc., in addition to the values of truth and falsehood customary in two-valued systems; **three-wire** *a.*, (*b*) applied to a system of mooring used to keep an airship or balloon at a constant height from the ground; **three-wood**, (*a*) *Archery*, a bow made of three pieces of wood; also *attrib.*; (*b*) *Golf*, a wooden club providing medium loft, formerly called a spoon (Spoon *sb.* 4 c).

1948 *Math. Tables & other Aids to Computation* III. 69 The control of this machine is accomplished, for the most part, by means of three-address orders. In contrast, the 'Mark I' at Harvard uses a two-address system. **1970** O. Dopping *Computers & Data Processing* vi. 103 A three-address machine. **1976** Banks & Doupnik *Introd. Computer Sci.* vii. 242 The principal disadvantage of three address instructions is their great length and consequent excessive use of memory space. **1966**, etc. Three-anti [see *Sanfan]. **1975** A. Watson *Living in China* iv. 90 The 'three anti' campaign which opposed the three evils of corruption, waste and excessive red tape in the Party and government. **1962** V. Grissom in *Into Orbit* 78 We had to learn from scratch..how to manipulate the new three-axis control stick and make the precise adjustments in yaw, pitch and roll. **1977** *Dædalus* Fall 52 Some [satellites] are provided..with three-axis stabilization, so that their instruments can be pointed steadily, for long periods of time, to a chosen target. **1839** *Rules of Hon. Co. Edin. Golfers* in C. B. Clapcott *Rules of Ten Oldest Golf Clubs* (1935) 69 In a Three-ball match, the Ball nearest the hole, and within the prescribed distance, must be lifted, if the third party require it, where the Player does so or not. **1890** H. G. Hutchinson *Golf* ix. 241 The three-ball match;..these matches are of two kinds, that wherein each plays against each, and that wherin two are in combination against a third, though each play his individual ball. **1901** *Rules of Golf* 5 Three players may play against each other, each playing his own ball, when the match is called 'a three-ball match'. **1952** *Chambers's Jrnl.* May 299/1 The Major introduced them without enthusiasm and Basil promptly attached himself to the party, much to the Major's annoyance, for he hated three-balls. **1976** *Webster's Sports Dict.* 450/1 *Three-ball*, a golf match in which 3 players compete against each other with each playing his own ball. **1956** G. Durrell *Drunken Forest* iv. 75 Inside the hat, curled into a tight ball, lay..a three-banded armadillo. **1966** E. Palmer *Plains of Camdeboo* xii. 196 A three-banded plover was paddling in the furrow. **1973** 'H. Carmichael' *Too Late for Tears* vi. 81 In the hearth stood a 3-bar electric fire. **1979** T. Wiseman *Game of Secrets* iv. 49 A three-bar electric heater. **1936** *Physical Rev.* L. 638/2 The procedure for the three-body problem. **1968** M. S. Livingston *Particle Physics* iv. 74 Evidence that this is a three-body decay is that the electrons have a wide distribution in energy. **1972** *Sci. Amer.* Jan. 85/1 The Herzberg bands of O_2 and the atmospheric infrared bands probably both owe their origin to three-body association: $O + O + X \rightarrow O_2^* + X$, where X, the third atom..is unchanged in the process. **1920** C. Sandburg *Smoke & Steel* 175 Pickpockets, yeggs, three card men. **1938** [see *broad *sb.* 6]. **1973** *Times* 19 Jan. 3/8 Three-card tricksters are a nuisance. They have someone posted to watch for the police, then they invite people to lay down money on which of three cards..is the 'lady'. **1979** W. H. Canaway *Solid Gold Buddha* xxii. 145 Sam was as confused as a yokel watching a three-card artist. **1954** W. H. Eberhardt et al. in *Jrnl. Chem. Physics* XXII. 989/1 In our approach, the only new, or rather, unfamiliar concept is that which we call the 'three-center bond'. **1978** *Further Perspectives Organic Chem.* (CIBA Symp.) 61 The short S—O bonds in the thioimine may reflect a three-centre bond. **1883** J. Venn in *Proc. Cambr. Philos. Soc.* IV. 51 Both Drobisch and Schröder have used what I have called..the three-circle diagram. **1952** W. V. Quine *Methods of Logic* i. 79 We set up a three circle diagram as usual. **1946** J. Cary *Moonlight* viii. 53 One heard first a single 'aw-aw' from some old three-crop mother, followed at once by a hearty 'mey' from her stout lamb. **1960** *Farmer & Stockbreeder* 1 Mar. 77/3 Mr. McIlwraith also paid..£180 for a three-crop ewe. **1910** *Encycl. Brit.* III. 939/2 There is also Three-Cushion Carom..and the Bank-Shot game. **1957** *Ibid.* III. 569/1 A count is validly made in three-cushion billiards in any one of four ways: (1) when the cue ball strikes an object ball and then strikes three or more cushions before striking the second object ball; (2) when the cue ball strikes three or more cushions before contacting the two object balls; (3) when the one ball strikes a cushion, then the first object ball, then two or more cushions and then the second object ball; (4) when the cue ball strikes two or more cushions, then the first object ball, then one or more cushions and finally the second object ball. **1974** *Mark Twain Jrnl.* Summer 3/1 Cure and Cutler played a game of three cushion billiards, a novelty at the time [*sc.* 1906]. **1952** *Jrnl. Soc. Motion Picture & Television Engineers* Oct. 249/1 Up to now the production of three-dimensional (3-D) films has been sporadic. **1953** *Sun* (Baltimore) 5 Feb. 14/1 We receive with mixed reaction the news that three-dimensional motion pictures, coyly called '3-D', will shortly come into general distribution. **1953**, etc. 3-D, 3D [see D., dimensional, s.v. *D III. 3]. **1955** W. Gaddis *Recognitions* iii. v. 914 She's terrific, even in 3-D she'd be terrific. **1966** *T.V. Times* (Austral.) 7 Dec. 10/2 Three-D Television is now the subject of experiment in several overseas countries, particularly Russia. **1971** 'D. Halliday' *Dolly & Doctor Bird* x. 130 Monopoly would maybe do. Or three-D noughts and crosses? **1983** *U.S.A. Today* 20 May 5D/4 But in a 3-D comic, futility flattens the actors even as the visual gimmick pops them out. **1890** Three-day [in Dict., sense III. 1 a]. **1937** S. Cloete *Turning Wheels* ix. 143 Three day sickness, which as a rule animals recovered from if left alone, meant abandoning beasts since there was no time to wait for them to recover. **1952** *Rules & Reg. governing One-Day Events* (Brit. Horse Soc.) 5 One Day combined Tests.. lead up to the Olympic Three Day Event. **1963** E. H. Edwards *Saddlery* vii. 69 This is not of importance by the time one's horse is sufficiently advanced to perform Three Day Event tests. **1965** *N.Y. Herald Tribune* 18 Apr. 3 An annual three-day..walk. **1974** *Times* 16 Feb. 1/1 The three-day week will carry unemployment to a very high level. **1976** *Times* 21 May 2/5 The committee's three-day conference on negotiated independence..has been postponed. **1976** *Horse & Hound* 10 Dec. 57/1 (Advt.), An ideal type to sire top quality point-to-pointers and three-day-eventers. **1977** M. Walker *National Front* vi. 147 The state of national emergency and the three-day week. **1982** Barr & York *Official Sloane Ranger Handbk.* 152/1 The easiest house parties are for a sport—racing, three-day event, shooting. **1878** Three-dimensional [see Dimension *sb.* 3 a]. **1882** [see Dimensional *a.* 2]. **1920** W. W. Strong *New Philos. of Mod. Sci.* xvi. 142 In gravitational phenomena a small disturbance leaves a circular trajectory finite in a three dimensional space. **1923** H. Crane *Let.* 2 Mar. (1965) 129 O yes, the 'background of life'—and all that is still there, but that is only three-dimensional. **1925** B. Dobrée in W. Congreve *Comedies* p. xvii, Congreve..made his people three-dimensional. **1953** *N.Y. Times* 19 Feb. 20 This much touted picture..is advertised as the first feature made in the three-dimensional Natural Vision process. **1971** A. Drummond *Auckland Jrnls. Vicesimus Lush* 22 His children..emerge from the pages of his journals as sufficiently three-dimensional figures to be of interest to readers a century later. **1926** H. Read *Eng. Stained Glass* i. 11/2 Three-dimensionality. Perspective and shading give the proper spatial relations of the various details represented. **1956** E. H. Hutten *Lang. Mod. Physics* vi. 208 This..suggests that causal action exemplified by the inverse-square law is connected with the three-dimensionality of space. **1977** *Jrnl. Playing-Card Soc.* Nov. 69 The three-dimensionality of 19th- and 20th-century German cards is one of their distinguishing features. **1958** C. Smith in 'E. Crispin' *Best SF Three* 210 Light..allowed the ships to reform three-dimensionally.. as they moved from star to star. **1979** *Nature* 29 Mar. 439/2 The method for producing these three-dimensionally interconnected fibrous structures is described here. **1855** J. Lang *Forger's Wife* xv. 44 'There is not a really good placard on the walls—tens, and fifteens, and twenties; but not a single three-figure gentleman' (he meant £100) 'among 'em.' **1861** C. Knight *Eng. Cycl.: Arts & Sci.* VII. 1007 A. De Morgan. Three-figure logarithms: three figures of numbers to three of logarithm, complete, on a sheet of 7½ by 6 inches. **1929** *Star* 21 Aug. 12/1 A three-figure stand. **1973** J. Wainwright *Devil you Don't* 5 It was a great car—a Jag. Mark II—well capable of three-figure speeds. **1978** R. V. Jones *Most Secret War* xxi. 174 Assuming..that the three-figure entries were bearings. **1920** *Proc. Inst. Radio Engineers* VIII. 70 At low plate voltages..the measured values of the amplification constant are lower and the three-halves power law does not appear to hold. **1963** B. Fozard *Instrumentation Nucl. Reactors* xi. 138 As the anode voltage is raised the equation to the anode characteristic takes the commonly assumed three-halves power law. [**1864** *Guardian* 30 Mar. 299/2 The English Church is indebted to Mr. Mackonochie for the revival of..the admirable ancient Office in Commemoration of the Three Hours.] **1898** (*title*) The three hours' service for Good Friday. **1923** *Spectator* 5 May 753/2 Two hours afterwards I went to the Three Hours at a church in a residential southern suburb. **1976** *Oxford Mission Q. Paper* July/Sept. 5 The little church was packed from beginning to end of the Three Hours. **1929** R. H. Barbour *Tod Hale on Nine* xxiv. 264 He wanted to be a 'three-letter man', and until a few days ago his chance had looked very bright. **1941** *Amer. Speech* XVI. 190 Three-letter man, F-A-G. **1946** J. Irving *Royal Navalese* 81 A three-letter man is a 'cad'. **1972** *Sci. Amer.* Feb. 114/2 A boxer of almost professional caliber; a three-letter man in college, a Rhodes scholar, he passed the bar examination but after only a year of practice decided 'to chuck the law for astronomy'. **1853** in *Notes on Cheshire Churches* (Chetham Soc.) (1894) 10 Each side of the porch having open three-light windows. **1908–9** H. R. Barker *E. Suffolk Illustr.* 330 The east window is a Transitional three-light, and in the side walls are very good two-lights in square heads. **1937** *Burlington Mag.* Mar. 149/1 The three-light Peter de Dene window of the Minster. **1939** W. I. Jennings *Parliament* iii. 78 A 'three-line whip' indicates that all other engagements should be put aside. **1958** *Spectator* 27 June 826/3 A debate sufficiently important to warrant a three-line whip. **1975** J. P. Morgan *House of Lords & Labour Govt.* iv. 127 Labour Peers took their own vote on the question of the vote in the Lords, choosing a free vote, unlike M.P.s who agreed to submit to a three-line whip. **1972** G. McGovern in W. Safire *Polit. Dict.* (1978) 727/1 The rich businessman can deduct his three-martini lunch, but you can't take off the price of a baloney sandwich. **1977** *Time* 26 Sept. 49/2 Carter has railed so vehemently against the 'three-martini lunch' that his staff has to come up with something. **1833** *Knickerbocker* I. 160 The present Mrs. S. admired his three minute roan. **1857** *Uncle Jack the Fault Killer* ix. 131 My three-minute glass lets the sand run through just in three minutes, which is time to boil an egg. **1958** Three minute [see *long-playing *a.*]. **1927** W. Collinson *Contemp. Eng.* 89 The house-agent's repulsive terminology e.g...a three nines agreement (i.e. 999 years). **1982** P. Turnbull *Dead Knock* i. 11 Tango Delta Foxtrot..responded to a three-nines call for a fire appliance. **1891** A. Conan Doyle in *Strand Mag.* Aug. 197/2 It is quite a three-pipe problem, and I beg that you won't speak to me for fifty minutes. **1976** *Lancet* 20 Nov. 1131/2 Appraising and comparing the effectiveness of what we do has certainly up to now proved to be what Sherlock Holmes would have called 'a 3-pipe problem'. **1946** *Sun* (Baltimore) 2 July 17/6 Joe Kirkwood..scored a 74,.. three-putting the last green. *Ibid.*, Lawson Little overshot the greens and three-putted frequently. **1978** *Detroit Free Press* 16 Apr. E7/3 McLendon..three-putted the 18th about the same time. **1904** *Everybody's Mag.* Aug. 161/2 A Barnum three-ring circus compared to Henry H. Rogers's exhibitions. **1914** Kipling *Divers. Creatures* (1917) 394, I can see lots of things from here. It's like a three-ring circus! **1951, 1955** [see *circus 2 c]. **1981** D. Clark *Roast Eggs* viii. 159 Don't m'lud me... You turned my court into a damned three-ring circus. **1898** B. Matthews *Outlines in Local Color* 145 What good is a three-ringed circus to anybody, except the boss of it? **1904** 'O. Henry' in *McClure's Mag.* Apr. 613/2 They commenced to scramble down, and for awhile we had a three-ringed circus. **1927** C. A. Lindbergh *We* v. 82 One of the first lessons was the 'three sixty'—so named because its completion required a total change in direction of three hundred and sixty degrees. **1977** *Skateboard Special* Sept. 7/1 The first really difficult stunt I learned then was a three-sixty. That..is a stunt where you spin the board through a full circle on its back wheels. **1972** *Sci. Amer.* Dec. 102/2 A Möbius strip, for example, has a handedness in 3-space that cannot be altered by twisting and stretching. **1977** *New York Rev. Bks.* 12 May 29/1 A plane is infinite and unbounded. Bend it through 'three-space' (i.e., three-dimensional space) and it can be the closed surface of a sphere. **1769** T. Pennant *Brit. Zool.* III. 217 The three spined s[tickle] back... These are common in many of our rivers. **1836** W. Yarrell *Hist. Brit. Fishes* I. 77 The Three-spined Stickleback was first described by Belon. **1971** *Nature* 23 Apr. 536/2 The three-spined stickleback..found sanctuary in the Atlantic. [**1871** G. H. Lewes *Let.* 27 Aug. in *Geo. Eliot Lett.* (1955) V. 180 Will you meanwhile order for me from the Stores.. 3 bottles of Martorell's three stars Brandy at 5/-.] **1879** R. J. Atcherley *Trip to Boërland* ii. 32 In the up-country towns of the Transvaal..common brandy is retailed at 1s., and 'Three Star' at 1s. 6d. per glass. **1929** *Amer. Speech* IV. 387 A little *three-star Hennessey* brought overland from Detroit. **1931** S. Cooke *This Motoring* xvii. 172 The..Lion at Guildford..is a typical three-star A.A. hotel. **1939** [see S.A.E., s.a.e. s.v. *S 4 a]. **1944** *Mod. Lang. Notes* Dec. 526 Whereas once a sword, a ship, was given an epithet fit for a hero or a goddess, today a warrior is labelled in the manner of a manufactured product:..*3-star general*. **1960** *Harper's Bazaar* July 19/1 The number of three-star restaurants in France has just dropped from 11 to 10. This reduction has been effected by the *Guide Michelin*. **1968** *Listener* 28 Mar. 405/3 Courvoisier V.S.O.P., he croaked, none of your rotten Three Star. *Ibid.* 1 Aug. 159/3 To ask for..two gallons of three-star. **1973** J. Burrows *Like an Evening Gone* xvii. 220 'I'd respect any decent woman.' 'What about Tamara Tayne?' 'That three star whore? That's a different category.' **1973** H. Gilbert *Hotels with Empty Rooms* xiii. 114 He..poured himself a glass of three-star cognac. **1977** *Air Mail* Spring 45/2 (Advt.), Six-berth luxury caravan for hire on three-star site with all amenities. **1977** A. Sampson *Arms Bazaar* xvii. 288 He is a stocky three-star general from Alabama..on his tie was a three-star tie pin. **1979** *Country Life* 13 Sept. 807/2 The typical three-star menu..of foie gras and truffles with everything.. And still, in too many three-stars, foie gras with everything. **1982** S. Wilson *Dealer's Wheels* ix. 85 We filled up with three-star..and I went to check the oil and tyres. **1818** *Mass Agric. Repository & Jrnl.* V. 56 Gleditsia Triacanthos. It is also called Three-Thorned Acacia in the catalogues of nurserymen. **1908** J. Kelley *Thirteen Yrs. Oregon Penitentiary* vii. 81 Pat came back again; he was a three-time loser. **1914**, etc. Three-time [see *loser 4]. **1943** P. Cheyney *You can always Duck* vi. 96 He's a three-time killer. **1979** *Tucson* (Arizona) *Citizen* 20 Sept. 6D/1 The Cats..will be led into tomorrow's meet by three-time All-America selection Thom Hunt and 1978 All-America choice Dirk Lakeman. **1932** Lewis & Langford *Symbolic Logic* vii. 213 One such alternative is the Three-valued Calculus, developed by Lukasiewicz and Tarski. *Ibid.*, If, in addition, the number ½ is taken, then we have the matrix of the three-valued system. **1934**

Mind XLIII. 104 Professor Lukasiewicz is sole author of these systems, having originated the three-valued system in 1920, and *n*-valued systems in 1922. **1946** *Nature* 14 Sept. 356/2 This decisive step opens the way for the construction of a new non-Aristotelian logic, a 'three-valued logic' as it is called. **1965** N. Chomsky *Aspects of Theory of Syntax* 232 Thus we can regard..gender as a three-valued..dimension. **1967** *Encycl. Philos.* VII. 118/1 With this way of reconstructing quantum mechanics, use must be made of a three-valued logic. **1974** tr. *Wertheim's Evol. & Revol.* i. 100 Three-valued prestige models were used by people who placed themselves in the middle class. **1933** Three wire mooring [in *OED Suppl.*]. **1934** J. A. Sinclair *Airships in Peace & War* ix. 186 Then came the three-wire system, which was first employed on the rigid airship No. 9 in 1917... To steady the bow, three wires were taken from the mooring point and attached to three bollards set in a triangle... A three-wire mooring was prepared at Pulham. **1875** *Encycl. Brit.* II. 376/1 Bows..made of three pieces..are called three-woods... Three-wood bows being made a little reflex, should retain their shape. [**1938** R. A. Whitcombe *Golf's no Mystery* xiii. 80 The spoon—or as it is called the No. 3 wood—is one of the golfer's greatest friends.] **1949** B. Hogan *Power Golf* ii. 15 Three wood.. 235 [yards]. **1960** *Times* 24 June 19/2 Second shots with a three-wood and a one-iron at these two holes brought him just short of the green in each case.

three-colour (stress variable), *a.* [Three *a.*] **1.** Utilizing or involving three distinct colours or wavelengths of light, usu. as a means of reproducing any desired colour by a combination of three primary colours in appropriate proportions. Cf. Trichromatic *a.*

1893 *Jrnl. Soc. Arts* 19 May 669/1 This three-colour print, a reproduction of a chromo-lithograph. **1898, 1902** [in Dict. s.v. Three *a.* and *sb.* III. 1 a]. **1906** [see *additive *a.* c]. **1932** R. C. Bayley *Compl. Photographer* xxiv. 291 The amateur who makes his own three-colour prints. **1972** [see *psycho-physical *a.*]. **1978** Pasachoff & Kutner *University Astron.* iii. 55 Thousands of stars have had their colors measured with this UBV set of filters; we call the process three-color photometry... A four-color system, uvby, has ultraviolet, violet, blue and yellow filters.

2. Designating *san ts'ai* ware; also *ellipt.* as *sb.*

1933 *Burlington Mag.* Nov. 211/1 The early Ming three-colour ware. **1959, 1972** [see *san ts'ai].

three-corner, *a.* Add: **three-corner jack** *Austral.* = *three-cornered jack* s.v. *three-cornered *a.* 3; **three-corner-ways** (earlier example replacing quot. 1796).

1919 G. E. A. Russell *Wild Life in Bushland* 32 Springing from the centre or core of each burr are three long spikes or prongs, often half an inch in length—an uninviting thing to sit upon. Most bushmen will recognise the seed—it is the Centralian-famed 'three-corner jack'. **1748** H. Glasse *Art of Cookery* (ed. 3) ix. 199 Toast some thin Slices of Bread cut three-corner ways.

three-cornered, *a.* Add: **3. three-cornered jack** *Austral.*, the spiny burr of the annual weed, *Emex australis.*

1953 A. Upfield *Murder must Wait* xxv. 223 'You lie there.'..'But not on the three-cornered jacks.'.. Her husband..swept the place clean of the skin-piercing burrs.

three-decker. Add: **1.** (Earlier example.)

1792 A. Young *Trav. France* 181 The bason of Toulon, with ranges of three deckers, and other large men of war.

2. a. (Earlier example.) **c.** (Earlier example.) **d.** A three-storey building. *U.S. local.*

1852 A. Mozley in *Christian Remembrancer* July 92 In the midst of the church stands, elaborately carved, the offensive structure of pulpit, reading-desk, and clerk's desk; in fact, a regular old three-decker in full sail westward. **1894** Kipling in *Sat. Rev.* 14 July 44/1 The old three-decker. And the three-volume novel is doomed. **1942** Berrey & Van den Bark *Amer. Thes. Slang* § 83/1 *Three-* (or more) *decker*, a building of three, or more, stories. **1961** L. Mumford *City in History* xv. 465 Vast wooden firetraps called three-deckers in New England, happily blessed with open-air porches. **1978** J. Carroll *Mortal Friends* ii. iii. 151 The flat, the top floor of a Southie three-decker, was large enough.

3. (Further examples, in sense 2 c.)

1926 G. Ade *Let.* 8 Sept. (1973) 110 While some of us have been building chicken coops.., Mr. Dreiser has been creating sky-scrapers. He makes the old three-decker novel look like a pamphlet. **1981** *N. & Q.* June 271/1 The widespread circulation of Evangelical tracts and sermons helped to create a sympathetic readership for the voluminous three-decker novel.

three-double, *a.* (Later examples.)

1728 E. Smith *Compleat Housewife* (ed. 2) 128 Butter Papers three double, one white, and two brown. **1874** Hardy *Far from Madding Crowd* I. viii. 105 'And he's growed terrible crooked, too, lately,' Jacob continued, surveying his father's figure, which was rather more bowed than his own. 'Really, one may say that father there is three-double.'

three-legged, *a.* (Earlier example of *three-legged race*.)

1876 *N.Y. Times* 21 May 2/5 The three-legged race of 100 yards was won by the Brown–Hammond team.

three-masted, *a.* (In Dict. s.v. Three-mast *a.*) (Earlier and later examples.)

1798 *Connecticut Jrnl.* 23 May 3/1 New York... May 14. Loss of the armed three masted schooner Harmony, captain Price, who sailed on Saturday the 5th inst. from this port for Surinam. **1861** *Mitchell's Maritime Register* 28 Sept. 1241/1 On the 21st inst. was launched from the yard of Messrs. Thomas Harvey & Sons, Wivenhoe, a three-masted brigantine. **1970** E. J. March *Inshore Craft* II. iii. 144 Mention should also be made of the big three-masted lugger *New Moon.*

three-master. (In Dict. s.v. Three-mast *a.*) (Earlier example.)

1827 F. Witts *Diary* 26 Apr. (1978) 70 One of these ships was a three-master.

threepenny, *a.* (*sb.*) Add: **3.** *sb.* A length of rod used in basket-making.

1912 [see *long-small* s.v. *long *a.*¹ A. 18]. **1953** A. G. Knock *Willow Basket-Work* (ed. 5) 9 Three feet, Tacks;.. six feet, Threepenny.

three-·piece, *a.* and *sb.* **A.** *adj.* **1.** Of a suite of furniture: comprising three separate items; freq. of a lounge suite: (usu.) comprising two armchairs and a sofa.

1908 *Sears, Roebuck Catal.* 455 Five-piece parlor suite... Three-piece parlor suite. **1952** *New Statesman* 5 July 10/1 Thankfully home, not to sink onto the centre couch of a three-piece suite. **1976** G. Moffat *Short Time to Live* ix. 82 The bathroom held..the usual three-piece suite. **1978** E. Malpass *Wind brings up Rain* xi. 109 A three-piece suite in blue moquette for the front room.

2. Of a suit of clothes: comprising three separate garments; freq. of a man's suit: comprising trousers, jacket, and waistcoat.

1909 in C. W. Cunnington *Eng. Women's Clothing in Present Cent.* (1952) iii. 91 New Three Piece Suit. **1923** *Queen* 26 July p. viii, The three-piece coat-frock. **1965** F. Sargeson *Memoirs of Peon* vi. 134 Three-piece suits were the rule. **1980** *TWA Ambassador* Oct. 84/2 We try to be a little lively, just to get out of that staid, gray, three-piece-suit mold.

3. *Mus.* Of a band: comprising three instruments or players.

1939 C. R. Cooper *Designs in Scarlet* ii. 13 A three-piece string band or a full orchestra. **1959** Wallis & Blair *Thunder Above* xii. 125 The band struck up again. The three-piece combination..played German dance music. **1978** M. Russell *Daylight Robbery* i. 20 The assortment of dance-floor routines that were accompanying the pulsation of a three-piece Latin-American group.

B. *sb.* **1.** A three-piece suit.

1931 W. Holtby *Poor Caroline* 18 Oh, Mums, I *must* show you my new blue three-piece. **1982** Barr & York *Official Sloane Ranger Handbk.* 42/1 There are a few basic lines that continue practically for ever, like..the basic City three-piece.

2. A three-piece suite.

1966 G. Burnett *Dead Account* iv. 27 The room looked more expensive..a grey/red three-piece that wouldn't show much change out of two hundred guineas. **1977** *Times* 24 Dec. 16/6 What suburban child is going to believe that..Santa would be allowed..to put his great sooty footmarks all over the Dralon three piece and the wall to wall?

Hence **three-piecer** *U.S.*, a three-piece suit.

1964 *N.Y. Post* 9 Nov. 13 Orlon® cardigan three-piecer a-glitter with trim. **1976** *Billings* (Montana) *Gaz.* 20 June 5-E (Advt.), Our all polyester knit shirt is great for leisure suits, a three piecer, even tucked into a pair of casual pants.

three-·point, *a.* **1.** Marked with three points; *spec.* designating a grade of point blanket (see *point *sb.*¹ B. 14).

1855 [see *point blanket* s.v. *point *sb.*¹ B. 14]. **1921** [see *point *sb.*¹ A. 32]. **1948** *Beaver* June 21/2 The simple voyageur Leger lost only a three-point blanket..a portage strap, a pair of French shoes.

2. At three points; with contact or support at three points; *spec.* of an aircraft landing: in which all three wheels, or two wheels and the tail skid, touch the ground simultaneously.

1909 *Westm. Gaz.* 9 Nov. 5/1 What is actually achieved by the Rolls-Royce plan is to make a three-point suspension without complication. **1918** R. Frost *Let.* 24 Oct. (1972) 39 You have to learn to make a 'three-point landing' that is on your two wheels and skid simultaneously. **1953** C. A. Lindbergh *Spirit of St. Louis* ii. vi. 266 Boy, *he* always makes 'em three-point! **1960** *Farmer & Stockbreeder* 15 Mar. 103 It connects to any modern tractor by the three-point hitch..and the category shaft. **1969** *Gloss. Terms Dentistry* (*B.S.I.*) 78 *Three-point contact*, a term used to indicate that when the jaws are in eccentric position there is a minimum of three points of occlusal contact, as widely spaced as possible. **1971** *Power Farming* Mar. 15/4 Hitching up an implement to the three-point-linkage could be both difficult and dangerous. **1977** *New Yorker* 4 July 42/1 He came in on final, flared, and made a three-point full-stall landing.

3. In surveying, navigation, etc.: involving the measurement of three known points to determine one's position.

1900 H. M. Wilson *Topographic Surveying* ix. 185 The three-point problem calls for the finding of distances from an unknown and occupied point to three others whose relative positions and distances are known. **1960** E. L. Delmar-Morgan *Cruising Yacht Equipment & Navigation* 18 The best..fix obtainable in coastal navigation is the Three-Point Fix... Here three identified objects are required, and either the sextant is used to measure the angles or, less accurately, by [*sic*] compass bearings.

4. *three-point turn*, a method of turning a vehicle round in a narrow space, whereby the vehicle moves in three arcs, forwards, backwards, then forwards again.

1957 C. Smith *Case of Torches* xiii. 152, I switched on the motor..and did a three-point turn. **1976** 'Z. Stone' *Modigliani Scandal* iii. i. 114 He did a three-point turn on the narrow road.

Hence **three-pointer**, (*a*) a three-point landing; (*b*) a three-point turn.

1932 D. Garnett *Rabbit in Air* iii. 106 This time the wind took off the extra height just as planned and with an almost dead machine made a perfect three-pointer. **1965** P. M. Hubbard *Hive of Glass* i. 12, I did a copy-book three-pointer, drove back to the main street. **1976** B. Lecomber *Dead Weight* i. 21 The Gemini rumbles on to the ground in a neat three-pointer.

three-quarter, -quarters, *sb.*, *a.*, and *advb. phr.* Add: **B.** as *adj.* **b.** (*b*) For '(Also to a lady's coat of similar length.)' read 'Also of a coat, sleeve, etc.: (having) three-fourths of the normal length.' (Further examples.)

1919 in C. W. Cunnington *Eng. Women's Clothing in Present Cent.* (1952) 156 The three-quarter coat is the latest rage. **1940** Graves & Hodge *Long Week-End* xvi. 280 Swagger coats..were of three-quarter length. **1943** P. Cheyney *You can always Duck* v. 85 She is wearin'.. a loose three-quarter length coat. **1960** *Woman's Own* 19 Mar. 42/3 Three-quarter sleeve overblouse. **1960** *News Chron.* 4 Apr. 6/1 The collar sits well away from the face; the sleeves are definitely three-quarter. **1960** *Harper's Bazaar* Aug. 8/2 A brief tartan jacket..which has..three-quarter length sleeves. *Ibid.* Oct. 119 A three-quarter coat in oatmeal tweed. **1972** A. Price *Col. Butler's Wolf* xi. 119 A reversible three-quarter length overcoat.

D. three-quarter bed, a bed intermediate in width between a single and a double bed; **three-quarter line** *Rugby Football*, the row of three-quarter backs aligned (and usu. angled back) across the field, esp. at a set-piece; **three-quarter veneer** *Dentistry* = *partial veneer* s.v. *partial *a.* 3 i.

1919 L. R. Balderston *Housewifery* vii. 196 *Size of Bedspreads*..80 in. × 100 in.—three-quarter bed. **1978** *Lancashire Life* Oct. 141/2 The old four foot wide 'three-quarter' bed—known in the trade as a 'landlady's double' because seaside landladies used it to get two people into what should have been a single room—is rarely seen nowadays. **1960** E. S. & W. J. Higham *High Speed Rugby* vii. 58 The scrum-half's first and foremost duty is to feed the three-quarter-line. **1976** *Eastern Even. News* (Norwich) 9 Dec. 19/8 Norfolk took an early lead when the ball was quickly fed along the three-quarter line to put Hopkins over for a try. **1924** J. F. Hovestad *Pract. Dental Porcelains* xii. 110 (*heading*) The partial coping or the all porcelain three-quarter veneer crown. **1963** J. Osborne *Dental Mechanics* (ed. 5) xxiii. 415 Three-quarter or partial veneer crowns. Such crowns are usually employed as bridge retainers.

threescore, *a.* (*sb.*) (Further examples, esp. in phr. *threescore* (*years*) *and ten*.)

1850 N. Hawthorne *Scarlet Letter* 22 The brave soldier had already numbered, nearly or quite, his threescore years and ten. **1896** A. E. Housman *Shropshire Lad* 3 Now of my threescore years and ten, Twenty will not come again. **1922** Joyce *Ulysses* 683 Evermoving from immeasurably remote eons to infinitely remote futures in comparison with which the years, threescore and ten, of allotted human life formed a parenthesis of infinitesimal brevity. **1977** *Drive* Mar.–Apr. 44/3 Cyril shed..more than threescore pounds of weight from his celebrated 28-stone frame. **1977** *Chicago Tribune Mag.* 2 Oct. 49/1 Those of us who have attained 'three score years and ten' more often now look backward with nostalgia, remembering the years when we were younger.

threesome, *sb.* and *a.* (*adv.*) Restrict 'Chiefly *Sc.*' to sense B and add: **A.** *sb.* **a.** (Further examples.)

1926 [see *foursome *sb.* 2]. **1951** [see *gay *a.* 2 c]. **1959** *Times Lit. Suppl.* 5 June 333/2 The stresses and strains of this uneasy threesome are subtly conveyed. **1972** *Screw* 12 June 33/1 (Advt.), Especially well endowed & very fond of threesomes (with couples or 2 women). **1977** *Gay News* 24 Mar. 14/4 These may include threesomes in which one or both partners bring home a person who is shared in bed. **1977** *New Yorker* 8 Aug. 57/1 A number of them were looking down from the Turnberry Hotel at the Ailsa course, on which the last threesomes were finishing their rounds. **1980** *TWA Ambassador* Oct. 92/1 Robbins's first and better novel, *Another Roadside Attraction* (praised by an odd threesome: Lawrence Ferlinghetti, Graham Greene and Thomas Pynchon). **1981** M. McMullen *Other Shoe* (1982) ii. 22 Justin, why don't you squire Meg?..Threesomes are awkward.

b. A game of golf in which one person plays against two opponents.

1901 *Rules of Golf* 5 A single player may play against two, when the match is called 'a threesome'. **1931** W. Martyn *Scarlett Murder* iv. 48, I was playing a threesome against Sir Griffith Wadham and Lord Berrington.

three-way, *a.* Add: Also, involving three participants; *three-way mirror*, one with three panels to provide a view from three different angles, and often forming part of a dressing-table suite.

1961 Webster s.v., A three-way profit split, a three-way play-off. **1964** *McCall's Sewing* v. 60/2 If you can arrange a three-way mirror it will be even better. **1967**

Economist 29 Apr. 459/1 The slogan of 'three-way alliances' was coined [in China] three months ago. **1978** *Lancashire Life* Oct. 143/2 (*caption*) With a threeway mirror, concealed lighting and a matching stool it is about £154. **1979** *Tucson* (Arizona) *Citizen* 20 Sept. 9D/1 Billie Harper won the second flight with 38, followed by a three-way tie at 41 between Ann Pearsall, Lori Emery and Mary Stewart. **1983** *Daily Tel.* 27 Oct. 21/2 A three-way merger among property groups could result in a new company with combined assets of around £700 million.

b. Of a loudspeaker: having three separate drive units for different frequency ranges.

1960 C. Brown *Introd. Hi-Fi* iv. 91 Small horn-loaded diaphragm loudspeakers are used in some two- and three-way systems of the direct radiator type (*i.e.* all units facing the listener). **1972** *N.Y. Times* 3 Nov. 10/4 (Advt.), One of the finest 3-way speaker systems available. ..Contains a 10" woofer, 5" direct radiating mid-range element and Sonodome ultra-tweeter.

threitol (þrī·itǫl). *Chem.* [f. *Thre(ose + *-itol.] A crystalline tetrahydroxy alcohol, $HOCH_2(CHOH)_2CH_2OH$, formed by the reduction of threose.

1935 *Jrnl. Amer. Chem. Soc.* LVII. 2262/2 Consistency also strongly urges limiting the term 'erythritol' to the natural inactive alcohol.., and using the terms '*d*-threitol' (rotating +4·33° in water) and '*l*-threitol' (rotating −4·4° in water) for the active tetritols derived, respectively, from the two threoses. **1948** *Biochem. Jrnl.* XLII. 330/1 The new product is apparently identical with the substance synthesized by Maquenne from natural xylose and described under the name of '*l*-erythritol'. The name was later changed to '*d*-erythritol' in accordance with the Rosanoff convention for sugars, but recently it has been thought advisable to change the name to D-threitol to relate it more closely to the tetrose threose. **1960** *Ibid.* LXXVII. 272/1 Preparations are given for L-threitol and *meso*erythritol. **1975** *Nature* 11 Dec. 519/1 In adult tenebrionid beetles..an unusual combination of two polyhydric alcohols, sorbitol and threitol, is associated with the ability to tolerate prolonged freezing to at least −50°C. The occurrence of threitol is of particular interest, since this compound has not previously been found in nature.

threne. (Later example.)

1960 R. Eberhart *Coll. Poems 1930–60* 14 The perfect lament, and threne of sorrow's throat.

threonine (þrī·onīn). *Biochem.* [f. *Threo(se + *-n-* + -ine⁵.] A natural amino-acid, α-amino-β-hydroxy-butyric acid, $C_4H_9NO_3$, considered essential for growth and for maintenance of the nitrogen equilibrium in adults.

1936 Meyer & Rose in *Jrnl. Biol. Chem.* CXV. 727 It is proposed that henceforth natural α-amino-β-hydroxy-*n*-butyric acid be known as d(−)-threonine, inasmuch as it possesses a spatial configuration analogous to that of *d*(−)-threose. **1956** [see *plasma 6 b]. **1974** *Nature* 19 Apr. 643/2 Eight essential amino acids—[including].. valine, threonine, phenylalanine and tryptophan—are required by human adults.

threose (þrī·ōᵘz). *Chem.* [a. G. *threose* (O. Ruff 1901, in *Ber. d. Deut. Chem. Ges.* XXXIV. 1364), f. *erythrose* *erythrose by omission and transposition of letters.] A tetrose sugar, $CHO{\cdot}[CH(OH)]_2{\cdot}CH_2OH$, isolated as a hygroscopic solid and existing in two molecular configurations; it differs from erythrose in having the hydroxyl groups on the second and third carbon atoms on opposite sides of the carbon chain.

1901 *Jrnl. Chem. Soc.* LXXX. I. 449 The calcium salt [of *l*-xylonic acid]..was then oxidised with hydrogen peroxide and ferric acetate to *l*-threose. **1963** [see *erythrose]. **1982** T. W. G. Solomons *Fund. of Org. Chem.* xix. 709 One cyanohydrin ultimately yields D-(−)-erythrose and the other yields D-(−)-threose.

threshold, *sb.* Add: Now also with pronunc. (þre·ʃhōᵘld). **2. a.** Also, in an airfield: the beginning of the landing area on a runway. Also *attrib.*

1937 *Jrnl. R. Aeronaut. Soc.* XLI. 295 Sites..for threshold lighting and other signal apparatus required to assist the pilot. **1960** *Guide Civil Land Aerodrome Lighting* (*B.S.I.*) 15 A pilot needs to be given a clear indication of the runway threshold and the addition of wingbars, composed of green lights, is recommended to make the threshold more conspicuous in poor visibility.

c. (i) Also in *Physiol.* and more widely: the limit below which a stimulus is not perceptible; the magnitude or intensity of a stimulus which has to be exceeded for it to produce a certain response. (ii) The magnitude or intensity that must be exceeded for a certain reaction or phenomenon to occur.

1902 J. M. Baldwin *Dict. Philos. & Psychol.* II. 696/2 The least noticeable difference in sensation is called the threshold of discrimination or difference. **1919** W. D. Halliburton *Handbk. Physiol.* (ed. 14) lii. 767 That strength of stimulus which just suffices to evoke a sensation is called..its absolute threshold. **1922** *Electr. Communication* I. I. 45/1 Articulation tests were made upon the..telephone system..when it was set to deliver various intensities from the threshold of audibility to very large values. **1930** *City Noise* (N.Y. Noise Abatement Commission) 34 This means decibels above the threshold of hearing. **1931** *Brit. Jrnl. Psychol.* Jan. 285 There is a definite 'colourless interval' between the 'general threshold', or the intensity which just suffices to produce a sensation of light, and the 'specific threshold', or the intensity at which colour is just noticeable. **1936** G. K. Zipf *Psycho-Biol. of Lang.* 113 Every phoneme must also have a lower threshold below which it cannot pass without strengthening. **1938** *Ann. Reg. 1937* 346 The view [was] advanced that spontaneous mutations are mono-molecular reactions produced by thermal agitation when this oversteps the energy threshold of the chemical bonds. **1941** in M. Gowing *Britain & Atomic Energy 1939–1945* (1964) 403 From..the fact that [uranium] 238 does not give fission with slow neutrons, it is clear that the jump at 1 MeV represents the threshold of 238. The fission which takes place with neutrons of energy less than 1 MeV must therefore be ascribed to 235. **1948** P. M. Morse *Vibration & Sound* (ed. 2) vi. 227 The upper contour is the threshold of pain, above which the sensation is more of pain than of sound (and the result is more or less damaging to the ear). **1949** Koestler *Insight & Outlook* xv. 207 Heightening the threshold of some sensory receptors and lowering the threshold of others. **1949** S. C. Rothmann *Constructive Uses Atomic Energy* 205 The Geiger threshold of a radiation counter tube is the lowest operating voltage at which the charge transferred per isolated count is substantially independent of the nature of the initial ionizing event. **1950** *Gloss. Aeronaut. Terms* (*B.S.I.*) I. 25 *Cruising threshold*, the equivalent air speed giving the lowest comfortable continuous cruising speed. **1955** J. A. Wheeler in W. Pauli *Niels Bohr* 166 A photofission threshold of 5·15 MeV..goes with a half life against spontaneous fission of the order of $10^{16\cdot8}$ years. **1958** *Oxford Univ. Gaz.* 27 Jan. 524/2 (*heading*) Non-random sequences in visual threshold experiments. **1959** *Sunday Times* 5 July 8/6 The absence of a lower threshold for the production of mutations by radiation. **1962** A. Nisbett *Technique Sound Studio* v. 98 At 1,000 c/s the threshold of pain is 110 dB or more above the threshold of hearing. **1963** B. Fozard *Instrumentation Nucl. Reactors* v. 46 The scaling circuit which is used to count the pulses from the G.M. tube has some more or less well defined 'threshold', *i.e.* it accepts only those pulses which exceed a certain amplitude. **1965** *Proc. R. Soc.* B. CLXI. 338 While a climatic change in one area may have produced conditions very favourable for a new species, in another area the same climatic change may have produced conditions only just above the critical physiological thresholds for the existence of that species. **1965** W. Lamb *Posture & Gesture* iii. 44 There has been a lot of investigation of the threshold of fatigue in athletics and the type of training required to push this threshold back is well understood. **1972** J. Mosedale *Football* ix. 124 Performances like Nevers' demonstrates [*sic*] the high threshold of pain common to many athletes. **1973** *Times* 19 Oct. 7/8 A GP who might only see one case of child abuse a year might not have as low a threshold of suspicion as I have. **1983** *Sci. Amer.* Jan. 98/2 Above a certain threshold, known as the critical density, the expansion [of the universe] will eventually cease and contraction will begin.

(iii) In contexts of wages and taxation, in which wage or tax increases become due or obligatory when some predetermined conditions are fulfilled (esp. above a specified point on a graduated scale). Also in more general use in contexts of work. Freq. *attrib.*

1967 L. B. Archer in Wills & Yearsley *Handbk. Managem. Technol.* 131 Usually there is a threshold between 'good enough' and 'not good enough' in respect of each objective, below which a design proposal would not be acceptable. **1971** *Guardian* 7 Sept. 11/2 Mr [Tom] Jackson..argued in favour of a single threshold claim on behalf of all public employees. **1972** *Observer* 13 Aug. 10/8 Threshold cost of living agreements could make things much happier so long as the threshold is put fairly high and/or there is a big reduction in the effective basic level of wage settlements. **1974** *Ann. Reg. 1973* 14 The main features of the incomes plan [of Mr. Edward Heath] were..threshold payments of a maximum of 40p. a week if the retail price index were to rise by 7 per cent [etc.]. **1976** [see *tax threshold* s.v. *tax *sb.*¹ 7 b]. **1979** H. Wilson *Final Term* ii. 42 Viewed with hindsight the thresholds were a disastrous mistake. That does not in fact mean that Mr Heath had been wrong to introduce them in October 1973. **1980** J. Boyd-Carpenter *Way of Life* xiii. 169 The alternative relief was to make a big increase in the level of the 'Thresholds', that is to say the point on the income scale at which people became liable to tax.

3. b. (Having a value or intensity) equal to that of a threshold (sense 2 c in Dict. and Suppl.).

1906 J. R. Murlin tr. *Tigerstedt's Text-bk. Human Physiol.* xvi. 455 In order that an external stimulus may produce a sensation, it must exceed a certain lower limit of strength, which is called, after Herbart, the threshold value of the stimulus. **1921** J. Mills *Within Atom* 215 *Threshold frequency*, the minimum frequency of radiation which will produce photo-electric effects. **1926** J. S. Huxley *Ess. Pop. Sci.* 199 It is needful, not merely that some thyroid secretion should be circulating in the body, but that it should reach a certain definite concentration, a certain 'threshold value'. **1941** in M. Gowing *Britain & Atomic Energy 1939–45* (1964) 400 Neutrons of less than a certain threshold energy..do not cause fission of ²³⁸U. **1959** *Listener* 26 Nov. 929/1 It is possible that the radiation level has to exceed a critical or threshold value before any genetical effects arise. **1964** W. G. Smith *Allergy & Tissue Metabolism* ii. 23 The tissue response would depend upon the number of susceptible cells.. reached by a threshold concentration of histamine. **1971** J. H. Smith *Digital Logic* iv. 69 The device is actuated when the input signal crosses a certain 'threshold' voltage. **1978** J. Paxton *Dict. European Econ. Community* (rev. ed.) 46 Imports were kept up to minimum, or threshold, prices by means of variable import levies.

c. Electronics. *threshold device, element*, etc.: a circuit element having one output and a number of inputs, each of which accepts a binary signal and multiplies it by some factor; the output is 0 or 1 depending on whether or not the sum of the resulting quantities is less than a certain threshold value; *threshold function*, a Boolean function that can be realized by such an element; *threshold logic, switching* (based on such elements).

1960 *IRE Trans. Electronic Computers* IX. 122/1 Another useful logical two-state device is a threshold element. **1960** *Proc. IRE* XLVIII. 1335/3 The increasing use of threshold devices such as magnetic cores and parametrons. **1961** *IRE Trans. Electronic Computers* X. 6/1 Linearly separable switching functions..have been studied under different names, such as..linear-input logic, threshold logic, majority logic, and voting logic. *Ibid.* 798/2 Elementary threshold functions, *i.e.*, functions that can be implemented by a single threshold circuit, are first characterized for the cases of 2, 3, and 4 variables. **1962** *Proc. Internat. Federation Information Processing Congr.* 757/1 A threshold gate determines its output in two steps: a linear summation followed by a discrimination. **1964** H. C. Torng *Introd. Logical Design of Switching Systems* viii. 133 Threshold switching devices are.. extensively used in pattern recognition systems and perception-like automata. **1970** Z. Kohavi *Switching & Finite Automata Theory* vii. 183 One of the limitations of threshold logic is its sensitivity to variations in circuit parameters. **1975** N. N. Biswas *Introd. Logic & Switching Theory* vii. 183 In many cases where the NAND or NOR realizations may require a number of gates, the threshold logic may realize the function by only one gate. **1978** S. C. Lee *Mod. Switching Theory* iv. 117 As another simple example of a threshold function, consider $f(x_1, x_2, x_3) = x_1x_2 + x_3$.

threshold (þre·ʃhōᵘld), *v.* [f. the sb.] *trans.* To alter (an image) by reproducing it in two tones only, each part being dark or light according as the original is darker or lighter than some chosen threshold shade. Hence **thre·sholding** *vbl. sb.*

1968 *Brit. Med. Bull.* XXIV. 262/2 One..comes across objects which have obviously been thresholded at too low or too high a level, resulting in incorrect segmentation. *Ibid.*, Simple thresholding (setting a limit above which everything is considered to be picture and below which everything is considered to be background) seems to work out quite well in coarse density-resolution scanners. **1976** *Physics Bull.* Sept. 381/3 Figure 2 shows the result of magnifying and electronically thresholding a small portion of a LANDSAT infrared image of the UK. **1983** *What's New in Computing* Jan. 16/2 The software modules comprise such algorithms as image thresholding, edge enhancement, [etc.].

thribble, thrible (þri·b'l), dial. var. of Treble *sb.*, *a.* and *adv.*, *v.* Also *spec.* in oil drilling (see quots. 1932, 1975).

1829 J. Hunter *Hallamshire Gloss.* 90 *Thribble*, treble. **1877** *Wide Awake* IV. 348/2 O, let the corn swell till it's three times as bulky as it was in the beginning; that's that's [*sic*] we call 'thribbling'. **1904** E. Nesbit *Phoenix & Carpet* i. 1 The man at the shop said they were worth thribble the money. **1932** *Amer. Speech* VII. 271 *Thrible* .., a *stand* of three joints of pipe. *Thrible-board*.., a platform in the derrick at the height of a *thrible*. **1975** J. Black *Oil* II. ii. 158 Drill pipe came in thirty-foot lengths. These were screwed into sixty- or ninety-foot segments, known to oil field workers as 'doubles' or 'thribbles'.

thrice-cock (þrəi·skǫk). *dial.* [f. var. Thrush¹ + Cock *sb.*¹ 9 b.] = Missel-thrush; cf. *storm-cock* s.v. Storm *sb.* 6 e.

1819 M. Edgeworth *Let.* 26 Jan. (1971) 160 The Thrice-Cock..is the largest kind of thrush. **1913** H. K. Swann *Dict. Eng. & Folk-Names Brit. Birds* 236 Thrice Cock. A Midland and North of England name for the Mistle-thrush. **1965** *Jrnl. Lancs. Dial. Soc.* XIV. 9 Mistle Thrush... Thricecock: Oldham. Cf. Fieldfare.

thrift, *sb.*¹ Add: **3. b.** *U.S.* A savings and loan association.

1981 *Economist* 24 Jan. 28/1 This new charter for the thrifts, as they are called, has not been welcomed by all of them. **1982** *Sunday Sun-Times* (Chicago) 12 Sept. 65 In an effort to keep the funds, banks and thrifts will fire a fusillade of advertising.

5. thrift industry *U.S.*, savings and loan associations as a whole; **thrift institution** *U.S.*, a savings and loan association; **thrift shop** chiefly *U.S.*, **thrift store** *U.S.*, a shop at which second-hand goods (esp. clothes) are sold, usu. in aid of charity.

1981 *Financial Rev.* (Austral.) 1 May 18 When higher interest rates were paid, the thrift industry—building societies, savings banks, credit unions—tended to lose funds to competing institutions. **1982** *Times* 22 May 13/3 The United States House of Representatives..voted to shore up ailing thrift institutions. **1947** S. J. Perelman *Westward Ha!* (1949) xii. 153 A mound of shawls, brocades, bracelets, necklaces, purses, fans, and bric-a-brac resembling the contents of a thrift shop. **1976** *Eastern Even. News* (Norwich) 9 Dec. 2/5 Mums and toddlers and thrift shop, 76, Cadge Road, Community House, 2–4. **1972** T. Ardies *This Suitcase* ix. 85 Someone had probably gone to thrift stores to put together his wardrobe... Even his socks were the wrong size.

thrift, *sb.*² (Later example.)

1969 G. E. Evans *Farm & Village* xiv. 150 The *mill-bill*—it's a kind of steel pick or bill mounted in a *thrift* or handle, made of wych-elm or some other suitable wood.

thriftful (þri·ftfŭl), *a. rare.* [f. Thrift *sb.*¹ + -ful.] Marked by frugality or careful expenditure, thrifty.

1933 V. McNabb *Nazareth or Social Chaos* 75 If..only a country organization is naturally thriftful, it would seem that a town-organization will..end with a famine of real wealth. **1968** P. Falvey *Poems Old & New* 130 No thriftful scrutiny was drawn When, ere creation's mighty dawn, Thou planned'st man's abode.

thrill, *sb.*³ Add: **1. b.** Also, a thrilling experience or incident.

1936 G. B. Shaw *Simpleton Unexpected Isles* I. 48 *The Clergyman*: Yes: I know I should have explained that. But she let me kiss her. *Mrs Hyering*: That must have been a thrill, Mr Hammingtap. Life came to you that time, didn't it? **1947** *Sporting Mirror* 7 Nov. 8/1, I must add that in actual fact there was not much scientific football. But the dizzy paced thrills made up for that. **1951** R. Campbell *Light on Dark Horse* ii. 37 To be driven round in these new horse-less machines was a thrill of which we never tired in those days. **1964** in Hamblett & Deverson *Generation X* 32 Going to a party and being rowdy, dancing to very loud music,..being driven in a very fast car, are all great thrills.

3. a. *Comb.*, as *thrill-seeker*. **b.** *attrib.* passing into *adj.*, of a crime: committed purely for the sake of the excitement experienced in carrying it out, as *thrill hold-up, killing, murder.*

1928 *Daily Tel.* 30 Oct. 11/5 A long series of 'thrill' hold-ups [at Atlanta, Georgia] is cleared up here with the arrest of two Oglethorpe University students. The thrill-seekers are George Harsh and James Gaiogly, both members of good families. **1978** LaRoss & Tanenbaum *Random Factor* [illegible] thrill killings. **1973** R. C. Dennis *Sweat of Fear* [illegible] a thrill murder. [illegible] **1977** W. B. J. Breedlove *Swinging Set* xii. 146 A variety of sexual thrill-seekers.

thrill, *v.*¹ Add: **II. 5.** Also as *pa. pple.*, extremely pleased or delighted (*colloq.*).

1908 E. F. Benson *Climber* vii. 98 Though she would not have dreamed of doing what Elizabeth had done and looked over the letter, she could not but be thrilled by the fact that there were four pages. **1964** in Hamblett & Deverson *Generation X* 153, I adore Nureyev. When he danced on the Palladium show on telly I was thrilled to bits. **1976** A. Miller *Inside Outside* iv. 40 Naturally I was thrilled to bits over the prospect.

c. Now often const. *to.*

1935 *Motion Picture* Nov. 29/2 If you live within range of a national radio network, you've thrilled to their voices. **1940** J. Buchan *Memory Hold-the-Door* 48 Stevenson..thrilled as we did to those antecedents—the lights and glooms of Scottish history. **1943** T. Pyles *Words & Ways Amer. Eng.* iii. 34 Generations of European children have thrilled to the novels of J. F. Cooper.

thriller. Add: (Examples in general use.) Also *spec.* applied to a film.

1934 C. Lambert *Music Ho!* v. 301 The opera *Wozzeck* is on paper a soberly planned symphony, but in performance a 'thriller' of the most theatrical order. **1950** *Sport* 24–30 Mar. 3/2 That was in 1946 when the 'Bishops' were beaten 3–2 by Barnet in a Stamford Bridge thriller. **1968** M. Richler in R. Weaver *Canad. Short Stories* 2nd Ser. 186 'My mother made me promise that one day I would make a prayer in Israel'. 'Did she specify a sexy thriller?' **1976** *New Yorker* 16 Feb. 14/3 The thriller of the afternoon occurred when Redundancy came up in the last stride to beat Summertime Promise by a nose in the Columbiana Handicap.

b. *Comb.*, as *thriller-writer, -writing.*

1925 J. M. Robertson *Mr. Shaw & 'The Maid'* ix. 85 Villains there are in plenty, though those shaped by the thriller-writers are apt to be improbable. **1983** *Listener* 20 Jan. 23/3 The pseudonymous A. J. Quinnell belongs to the generation of good thriller-writers who specialised in South-East Asia. **1958** *Times Lit. Suppl.* 17 Jan. 33/1 It is a fair guess that thriller-writing led to his first interest in the precious stones which are the lure for so many crimes of real life.

Hence **thri·llerdom** [-dom], the world of thrillers or exciting, sensational novels; **thri·llerish** *a.* [-ish¹], suggestive of such a novel.

1922 *John o' London's* 4 Jan. 18/2 The first three-quarters of the play were so good anyway, simply on the level of first-rate thrillerdom. **1957** *Times Lit. Suppl.* 28 June 395/1 When, in the 1930s, one of his best novels, *La Condition Humaine*, was translated under the thrillerish title of *Storm in Shanghai*, many young people must have opened it in the hope that they were going to read a thriller.

thrillingly, *adv.* (Earlier example.)

a **1822** Shelley *Posthum. Poems* (1824) 220 The liquid voice Of pipes, that fills the clear air thrillingly.

thrilly, *a.* Delete *rare* and add further examples.

1924 R. Frost *New Hampshire* 68 A likeness to surprise the thrilly tourist. **1947** *Sun* (Baltimore) 1 Jan. 13/7 The thrilly spot was at Pasadena where the Rose Bowl job was unfolded for the fans. **1967** *Listener* 16 Feb. 239/2 It [*sc.* a story by Conan Doyle] was a chilly, thrilly piece not at all in the Holmes tradition.

thrin, *sb.* (Earlier example.)

1838 Thoreau *Jrnl.* 14 June (1949) I. 51 Truth, Goodness, Beauty,—those celestial thrins.

thrip, *sb.* (Earlier U.S. example.)

1834 W. G. Simms *Guy Rivers* II. 108 Whom he rewarded with a *thrip* (the smallest silver coin known in the southern currency—the five cent issue excepted).

thrive, *v.* Add: **1, 2.** Freq. const. *on.*

c **1862** E. Dickinson *Poems* (1955) II. 403 The Hemlock's nature thrives—on cold. **1907** [see sense 1 b in Dict.]. **1930** G. B. Shaw *Apple Cart* p. xx, The armament firms thrive on war scares [illegible]. **1940** J. Buchan *Memory Hold-the-Door* viii. 84, I throve on it as a diet of oatmeal, mutton and strong tea. **1961** J. Heller *Catch-22* (1962) iii. 35 He thrived on good wit and stimulating intellectual conversation. **1972** *Sci. Amer.* Aug. 73/1 Patent rapport and cooperation thrived on specific instructions.

throat, *sb.* Add: **I. 2. b.** A sore throat. *colloq.*

1885 A. Edwardes *Girton Girl* I. iii. 68 That reasonless creature..has one of her throats again, and I did so want her to take some of my work. [illegible] **1979** M. Soames *Clementine Churchill* xiii. 201 In the last year she had been subject to 'throats' and coughs.

3. a. *to jump down one's throat*, substitute for def.: † (*a*) to be excessively attentive to one; also, to accept one with alacrity as prospective husband (*obs.*); (*b*) to reprimand or contradict one fiercely; (earlier and later examples).

1871 *Monthly Packet* Sept. 287 The small boat held only three. 'Just as well,' Hugh said. 'We don't want all to jump down one another's throats'. **1879** Trollope *Cousin Henry* I. iii. 52 [illegible]

b. *to be at each other's throats*, to quarrel violently; *to have (got) the game* or *it by the throat* (Austral. slang), to have the situation under control.

1947 J. Morrison *Sailors belong Ships* 15 We're sailors, we are! [illegible] We got the game by the throat. **1949** D. M. Davin *Roads from Home* [illegible] **1978** I. Singer *Shosha* 265 The women are at each other's throats.

5. c. *throat-clearing vbl. sb.* (later examples); *throat-catching adj.*; **throat-mane**, a growth of hair on the front of an animal's neck; **throat microphone**, (*colloq.*) **mike**, a microphone attached to a speaker's throat and actuated directly by his larynx.

1958 *Times Lit. Suppl.* 1 Aug. 438/4 Everything about [illegible]

throat, *v.*¹ Delete † *Obs.* and add later examples.

1908 A. S. M. Hutchinson *Once aboard Lugger* v. iii. 304 'Barley water!' Mr. Marrapit throated. 'Barley water!' **1929** S. Leslie *Anglo-Catholic* xx. 116 Music was being throated from a reed organ.

throat-cutting, *vbl. sb.* Also *fig.*, mutually destructive competition in trade. Cf. *to cut one another's throats* s.v. Throat *sb.* 3 d.

1888 E. Bellamy *Looking Backward* xxii. 323 Your contemporaries, with their mutual throat-cutting, knew very well what they were at. **1931** L. Steffens *Autobiog.* II. III. xxxv. 609 It was not exactly a pool, but there had been a lot of throat-cutting in the trade; the competitive bidding had cut prices down till no man could make any profit.

throater. (Earlier example.)

1846 *Knickerbocker* XXVII. 511 The 'throater', the 'header', the 'splitter' take stations at the splitting-table.

throa·tful. [f. Throat *sb.* + -ful.] As much as the throat can hold at once.

1920 D. H. Lawrence *Lost Girl* vii. 139 Geoffrey gulped beer in large throatfuls.

throb, *v.* **3.** (Further example.)

1939 T. S. Eliot *Family Reunion* I. ii. 59 The cold spring now is the time For the ache in the moving root.. The slow flow throbbing the trunk.

throbber (þrǫ·bəɹ). *rare.* [f. Throb *v.* + -er¹.] A person or thing that throbs.

1890 G. Meredith *Let.* 10 Jan. (1970) II. 988 How glad is your poet that he secured his forgiveness before the Voice was..sharpened to hush the Thousands, and whip or curb their hearts. Else—but this poor throbber would have broken. **1934** Webster, *Throbber*,..one who or that which throbs; esp. (*Colloq.*), one whose emotions are easily moved. **1983** *Maledicta 1982* VI. 23 Erection,..throbber.

Thro:gmo·rton Street. The name of the street in the City of London where the Stock Exchange is located, used allusively for the Stock Exchange or its members.

1903 A. Conan Doyle *Green Flag* 243 What could Worlington Dodds know at Dunsloe which was not known in Throgmorton Street? **1953** *Economist* 15 Sept. 658/2 If prosperity is to return to Throgmorton Street the Stock Exchange Council will have to attract more investors.

thrombo-. Add: **thrombasthe·nia** (also thrombo-) [ad. G. *thrombasthenie* (E. Glanzmann 1918, in *Jahrb. f. Kinderheilkunde* LXXXVIII. 28), f. Gr. ἀσθένεια (see Asthenia)], a condition in which the number of platelets is normal but their clotting power is defective; so **thrombasthe·nic** *a.*; **thrombe·ctomy** *Surg.* [*-ectomy], surgical removal of a thrombus from a blood vessel; **thro:mbo-angiitis obli·terans** (-ændʒi,ai·tis) [L. *obliterans* Obliterating *ppl. a.*: see Angio-] = *Buerger's disease* s.v. *Buerger; **thrombocythæmia** (-(saiþī·mia) [ad. G. *thrombozythämie* (E. Epstein 1929, in *Zeitschr. f. Stomatologie* XXVII. 377): see Hæmo-, -hemo-], thrombocytosis, esp. when it is a persistent or primary condition; **thro:mbocytope·nia** [ad. G. *thrombocytopenie* (H. Eppinger in L. Langstein et al. *Enzykl. der klin. Med.* (1920) v. 295): see *-penia], a reduced number of platelets in the blood; hence **thro:mbocytope·nic** *a.*; **thrombocyto·sis** [-osis], a significantly increased number of platelets in the blood; **thrombo-e·mbolism**, embolism of a blood vessel caused by the dislodgement of a thrombus from its site of origin; hence **thromboembo·lic** *a.*; **thrombo-e·mbolus**, an embolus consisting of a thrombus that has become dislodged from its site; **thro:mboendarter(i)e·ctomy** *Surg.* [f. End(o- + Arter(y *sb.* + *-ectomy], an operation to remove a thrombus and part of the inner lining of an obstructed artery; **thrombope·nia** [*-penia] = *thrombocytopenia* above; hence **thrombope·nic** *a.*; **thrombopla·stic** *a. Med.*, causing or promoting the clotting of blood; **thrombopla·stin** *Med.* [-in¹], a natural thromboplastic substance; now *spec.* an enzyme converting prothrombin to thrombin during the early stages of blood coagulation; **thrombosthe·nin** *Biochem.* [Gr. σθένος strength], a contractile protein or mixture of proteins found in blood platelets; **thrombo·xane** *Biochem.* [f. Ox- + -ane], any of several compounds formed from prostaglandin endoperoxides which, when released from blood platelets, induce platelet aggregation and constriction of arterial muscle.

1935 L. E. H. Whitby *Disorders of Blood* xiv. 276 (*heading*) Hereditary hæmorrhagic thrombasthenia. **1962** *Lancet* 22 Dec. 1316/2 This thrombasthenia is a familial hæmorrhagic disease in which the platelet-count is normal but [illegible]. **1974** Passmore & Robson *Compan. Med. Stud.* III. xxi. 45/1 Hereditary thrombasthenia or Glanzmann's disease is characterized by abnormal platelet aggregation and clot retraction. **1928** *Jrnl. Laboratory & Clin. Med.* XIII. 319 Chronic hereditary thrombasthenic purpura.. **1979** *Nature* 6 Dec. 622/1 T-transferase activity was measured in lysates obtained from platelets isolated from a thrombasthenic and a Bernard-Soulier patient, respectively. **1920** *Lippincott's New Med. Dict.* 998/1 Thrombectomy. **1945** *Urologic & Cutaneous Rev.* XL. 672/2 Thrombectomy is an extremely difficult and daring operation..but..in thrombosis of the renal vein prompt surgical intervention offers the only hope for a cure. **1973** D. K. Black *Renal Dis.* (ed. 3) vi. 165/1 Clinical improvement has been documented..with thrombectomy. **1908** L. Buerger in *Amer. Jrnl. Med. Sci.* CXXXVI. 567 (*heading*) Thrombo-angiitis obliterans: a study of the vascular lesions leading to presenile spontaneous gangrene. *Ibid.* 580 Taking the true nature of the lesion into consideration, I would suggest that the names 'endarteritis obliterans' and 'arteriosclerotic gangrene' be discarded in this connection, and that we adopt the terms 'obliterating thrombo-angiitis' of the lower extremities when we wish to speak of the disease under discussion. **1914** [see *Buerger]. **1955** *Sci. News Let.* 11 May 377/2 The relation of cigarette smoking to thromboangiitis obliterans..is well established. **1974** Passmore & Robson *Compan. Med. Stud.* III. xvii. 18/2 In thromboangiitis obliterans the upper limbs are more frequently affected.

1966, 1972 Thrombocythæmia [see *thrombocytosis* below]. **1977** *Lancet* 9 Apr. 775/1 Thrombocythæmia predisposes patients to thrombosis. **1923** *Arch. Internal Med.* XXXII. 939 This constant diminution of the platelets without any known cause..has given rise to the modern name of essential thrombopenia..or better still, thrombocytopenia. **1977** Thrombocytopenia [see *thrombocytosis* below]. **1925** *Jrnl. Amer. Med. Assoc.* 20 June 1888 (*caption*) Blood platelet variations following splenectomy in the thrombocytopenic group of Banti's disease. **1978** *Detroit Free Press* 14 Apr. 4B/1 Andra had a rare blood disease called thrombocytopenic purpura, in which a deficiency of platelets causes bleeding. **1936** *Jrnl. Amer. Med. Assoc.* 21 Mar. 1005/2 (*heading*) Leukemia with thrombocytosis. **1966** Wright & Symmers *Systemic Path.* I. iv. 187/2 In disease, the number of circulating platelets may be greatly raised (thrombocytosis, or thrombocythaemia). **1972** W. J. Williams et al. *Hematology* lxxxiii. 704/1 Whereas the symptomatic rise in the platelet count termed thrombocytosis may be substantial, it is temporary and self-limited. In thrombocythemia the platelet counts are higher..and persistently elevated: the condition is self-perpetuating and must be regarded as neoplastic. **1977** *Lancet* 9 Apr. 774/1 Thrombocytopenia caused by alcohol is reversible after alcohol withdrawal, and is followed by rebound thrombocytosis. **1940** *Acta Chir. Scand.* Suppl. LXI. 37 The first signs of the thrombo-embolic disease itself..can be venographically determined. **1981** *Brit. Med. Jrnl.* 7 Feb. 466/2 The best approach to the prophylaxis of thromboembolic disease is through low-dose heparin. **1907** *Jrnl. Amer. Med. Assoc.* 27 July 360/1 (*heading*) Postoperative thrombo-embolism. **1941** *Archives Surg.* XLIII. 462 In heparin there is available an almost infallible prophylactic against thromboembolism. **1970** *Daily Tel.* 17 July 2/8 Investigations showed that the increased risk of thrombo-embolism declined rapidly after the patient stopped taking the pill. **1955** *Sci. News Let.* 17 Sept. 183/3 The primary reaction is the formation of plugs in arteries and veins. The plugs, or thrombo-emboli as they are known technically, are made up of blood platelets stuck together. **1977** *Lancet* 29 Jan. 251/2 Fulton and Duckett report a significant correlation between high plasma-fibrogen levels..and thromboemboli. **1948** *Index Medicus* XLIV. 1204/2 Aneurysmal development after dos Santos thromboendarteriectomy. **1974** J. D. Maynard in R. M. Kirk et al. *Surgery* xi. 235/1 Thrombo-endarterectomy is successful in over 90 per cent of patients with disease above the inguinal ligaments. **1915** *Index Medicus* XIII. (Subject Index) 166/1 Thrombopenia. **1922** *Nature* 20 May 666/1 The absence of the fat-soluble vitamin from the diet leads, in the rat,..to a progressive diminution in the number of blood-platelets known as thrombopenia. **1981** *Cancer* XLVIII. 198/2 All staging systems isolate a high-risk group of patients defined by anemia and/or thrombopenia. **1934** *Lancet* 21 Apr. 845/1 This thrombopenic hæmorrhagic diathesis occurs regularly when the bone-marrow with its megacaryocytes has been extensively damaged by proliferating lymphadenoid..or neoplastic tissue. **1981** *Cancer* XLVIII. 202/1 The evidence was strong enough to justify putting anemic and thrombopenic patients aside. **1911** W. H. Howell in *Amer. Jrnl. Physiol.* XXIX. 189 They [*sc.* tissue extracts] furnish a substance, which may be designated as a thromboplastic substance or thromboplastin. **1981** *Obstetrics & Gynecol.* LVII. 490/2 Acceleration of the rate of clotting of whole blood is due to the thromboplastic activity of amniotic fluid. **1911** Thromboplastin [see *thromboplastic* adj. above]. **1979** R. Hawkey *Side-Effect* xi. 86 We used a thromboplastin preparation. .. They're hardly likely to test for..thromboplastins. **1961** Bettex-Galland & Lüscher in *Biochim. & Biophysica Acta* XLIX. 537 We have named this protein 'thrombosthenin', firstly because of its role in thrombocyte function, and secondly because of its properties, which in many respects let it appear distinct from muscle actomyosin. **1974** *Encycl. Brit. Macropædia* II. 1107/2 There are about 250,000 platelets per cubic millimetre of blood... They..contain..a contractile protein (thrombosthenin) that allows platelets to extend and retract long footlike projections called pseudopodia. **1975** *Nature* 3 July 14/1 This new intermediate does not have a classical prostaglandin structure and has been named 'Thromboxane'.., because it is a very potent platelet aggregating agent. Since it is the first member of a new series of compounds and contains two double bonds it was further designated 'Thromboxane A₂': its metabolite..becomes 'Thromboxane B₂.' **1979** *Ibid.* 6 Sept. 14/3 Because of their fish diet, Greenland Eskimos have high plasma levels of eicosapentenoic acid which is the precursor of the three series of endoperoxides—prostaglandins, thromboxane and prostacyclin.

thrombocyte (þrǫ·mbosait). *Biol.* [ad. G. *thrombocyt* (M. C. Dekhuyzen 1892, in *Verh. d. Anat. Ges.* 94): see Thrombo- and -cyte.] **a.** A spindle-cell of the lower vertebrates, responsible for the clotting of blood.

1893 *Jrnl. R. Microsc. Soc.* 25 In analogy with Löwit's nomenclature he uses the ending 'blast' for young forms, and 'cyt' for those which are adult. He distinguishes.. thromboblasts and thrombocytes, as the 'spindles' of Eberth and Schimmelbusch may be called. **1910** *Jrnl. Morphol.* XXI. 273 The spindle cells or thrombocytes of certain amphibian blood have a cytoplasm which stains in the same way as does that of the megakaryocyte. **1979** *Nature* 1 Mar. 13/1 The rapid adhesion of platelets (or of their non-mammalian counterpart, the thrombocytes) to the vascular wall when the endothelial lining is breached has been a subject of interest since Wharton-Jones showed in 1851 that thrombocytes accumulated at points of local damage in blood vessels of the frog's foot.

b. A blood platelet.

1907 E. A. Schäfer *Essent. Histol.* (ed. 7) 35 These often seem to radiate from minute round colourless discoid particles less than one-third the diameter of a red corpuscle... These are the..blood-platelets, or thrombocytes. **1938** W. Magner *Textbk. Hematol.* i. 1 The cellular elements of the blood are of three types: red corpuscles or erythrocytes, white corpuscles or leukocytes, and platelets or thrombocytes. **1977** J. Raynor *Anat. & Physiol.* x. 245 There are about 150,000–300,000 thrombocytes per cubic millimeter of blood.

thrombolytic (þrǫmboli·tik), *a.* and *sb. Med.* [f. Thrombo- + *lytic *a.*] **A.** *adj.* Pertaining to or causing the dissolving and breaking down of a thrombus. **B.** *sb.* A thrombolytic agent.

1962 A. P. Fletcher et al. in *Amer. Jrnl. Med.* XXXIII. 738/1 The adjective 'thrombolytic' will be used to designate biochemical moieties capable of inducing thrombolysis. **1965** *Zeitschr. f. d. Ges. Innere Med.* XX. 720/2 The thrombolytics really have enriched the palette of antithrombotics. **1971** *Times* 6 Aug. 4/1 Deaths among patients admitted to hospital with coronary thrombosis have been cut by a third in a trial of a compound, streptokinase... The thrombolytic treatment was assessed in 700 patients in eight hospital centres. **1974** R. M. Kirk in R. M. Kirk et al. *Surgery* ii. 14/1 The thrombus may be removed surgically; at present this is less expensive than giving thrombolytics.

thrombose (þrǫmbōᵘ·z), *v. Path.* [Back-formation from Thrombosis.] **a.** *trans.* To cause thrombosis in (a blood vessel). Cf. Thrombosed *a.*

1910 *Practitioner* June 779 Acute endometritis... When sufficiently severe..to thrombose the endometrial capillaries.

b. *intr.* To become occupied by a thrombus.

1938 *Arch. Path.* XXV. 486 When the hemorrhage occurs into the deeper intimal layers, the capillaries adjacent to the point of rupture may thrombose. **1977** *Proc. R. Soc. Med.* LXX. 401/1 Small blood vessels thrombose but larger vessels appear to be undamaged.

Hence **thrombo·sing** *ppl. a.*, undergoing or causing thrombosis.

1923 *Surg., Gynecol. & Obstetr.* XXXVI. 313/1 The thrombosing part can sometimes be palpated. **1965** *Revue Roumaine d'Inframicrobiol.* II. 71 Caudal thrombosing vasculopathies can be considered as manifestations of a latent pararickettsial infection.

thrombosis. Delete ‖ and add: Also *fig.* with reference to traffic congestion.

1959 *Times* 27 Nov. 8/4 It was clear that the heart of London had traffic thrombosis, said Mr. Ernest Marples.. at a Press conference yesterday. **1975** *Times* 9 June 12/4 In the big cities expansion of car ownership has brought inevitable thrombosis.

thrombus. **b.** Substitute for def.: A clot which forms on the wall of a blood vessel or a chamber of the heart, often impeding or obstructing the flow. (Further examples.)

1961 R. D. Baker *Essent. Path.* v. 82 There is danger of a portion of the thrombus breaking loose and passing as an embolus to the pulmonary artery and lungs. **1970** Passmore & Robson *Compan. Med. Stud.* II. xxvi. 3/1 In large vessels, the thrombus usually remains plastered as a plaque against the wall of the vessel, whereas in small arteries continuation of the process may lead to an occlusive thrombus which blocks completely the direct blood flow.

throne. Add: **1. d.** *fig.* A lavatory bowl and pedestal or other supporting structure. *colloq.*

1922 Joyce *Ulysses* 39 In a Greek watercloset he breathed his last... With beaded mitre and with crozier, stalled upon his throne. **1941** F. Thompson *Over to Candleford* vi. 95 The commode turned out to be a kind of throne with carpeted steps and a lid which opened. **1960** J. J. Rowlands *Spindrift* 52 Our plumber..revealed that the water level in the 'throne' works just like the old glass water barometer. **1981** S. Rushdie *Midnight's Children* I. 62 A wooden 'thunderbox'—a 'throne'—lay on one side, empty enamel pot rolling on coir matting.

2. b. Phr. *the Great White Throne*, used of the throne of God with allusion to Revelation xx. 11. Also *fig.*

1850 Browning *Christmas-Eve & Easter-Day* 116 Is Judgment past for me alone?—And where had place the Great White Throne? **1873** C. M. Yonge *Pillars of House* III. xxxii. 212 It was his first mountain... He raised his hat with an instinct of reverence..then murmured, 'One seems nearer the Great White Throne!' **1922** E. E. Cummings *Enormous Room* vii. 155 The Mecca of respectability, the Great White Throne of purity.

8. *throne-room* (further example, in sense *1 d); **Throne Speech** *Canad.* = *Speech from the Throne* s.v. *speech *sb.*¹ 8 d.

1941 W. Fortescue *Trampled Lilies* xxv. 247 Could I bear to walk through the kitchen to reach the only bath and throne-room? **1955** *Toronto Daily Star* 2 Feb. 6/2 An experimental program for the treatment of drug addicts was announced in this year's Throne speech at the opening of the British Columbia legislature. **1972** *Farm & Country* 19 Dec. 1/3 Informants..say they will be 'very surprised' if the Throne Speech does not contain new provisions to help farmers transfer their properties to following generations.

throng, *sb.* Add: **II. 3.** (Later examples.) Also *transf.*, of impersonal objects.

1912 'Saki' *Unbearable Bassington* x. 170 The Rutland Galleries were crowded..by a fashionable throng of art-patrons. **1955** S. Wilson *Man in Grey Flannel Suit* xxxiv. 256 He joined a throng of men pushing to get aboard the train. **1971** H. Wouk *Winds of War* iv. 53 Was Nazi Germany as strong as the ever-marching columns in the streets, and the throngs of uniforms in cafés, suggested? **1977** P. L. Fermor *Time of Gifts* ii. 35 A throng of villagers had assembled round an enormous bonfire. **1980** D. Adams *Restaurant at End of Universe* xvi. 83 'The End of the Universe is very popular,' said Zaphod threading his way unsteadily through the throng of tables.

throng, *v.* **3.** (Later examples.)

1947 P. Larkin *Girl in Winter* II. vi. 150 Besides—the impossibilities thronged upon her—she was sixteen, while Jane was twenty-five, middle-aged, and foreign, too. **1969** M. Puzo *Godfather* II. xii. 163 The young beautiful girls thronged through the city like lemmings, lasting one year, some two. **1979** A. Fraser *King Charles II* I. ii. 25 This maddened mob was thronging round the palace of Whitehall. **1981** A. N. Wilson *Who was Oswald Fish?* xi. 121 A hundred half-memories of childhood thronged back: the smell of baking from the house next door—the Trenimans.

thronged, *ppl. a.* **2. b.** (Later non-*dial.* example.)

1943 *R.A.F. Jrnl.* Aug. 15 Members of the R.A.F. who in the midst of their thronged days find time to encourage and assist the Air Training Corps squadrons.

throttle, *sb.* Add: **4. a.** (Further examples.) Also, the throttle-pedal.

1957 A. C. Clarke *Deep Range* I. iv. 44 Franklin pressed down the throttle and felt the surge of power as the torpedo leaped forward. **1966** T. Wisdom *High-Performance Driving* viii. 74 You brake with the ball of your foot and blip the throttle with your heel or the side of your foot. **1983** *Listener* 28 July 13/1 Mine was no longer with a functioning throttle, gear change or front-brake on arrival.

b. Phrs.: *to cut* or *chop the throttle*, to close the throttle in order to slow down or stop; *(at) full, half, part*, etc., *throttle*, (at) maximum, etc., power or speed (also *fig.*).

1936 *Motor Man.* (ed. 29) ii. 26 When the throttle is lying flat in the direction of the gas flow, the engine is running 'full bore', the term generally used for this being 'full throttle'. **1948** *N.Y. World-Telegram* 30 Dec. 11/8 The pilot, coming in, doesn't chop the throttle. The jet pilot 'turns down the wick'. **1958** [see *cut *v.* 21 f]. **1969** J. Argenti *Managem. Techniques* viii. 50 Once one has grasped the principle behind Cost-Benefit..one can use the technique at quarter throttle, so to speak. *Ibid.* 51 The results will be less impressive than when an expert uses it at full throttle. **1973** *Daily Tel.* 9 Jan. 1/5 The gunboat, believed to be the Odinn, avoided the ramming by sailing away at full throttle. **1977** J. F. Fixx *Compl. Bk. Running* iii. 42 Even in a race there's no need to run at full throttle if you don't want to.

5. throttle ice (see quots.); **throttle jockey** *slang* (see quot. 1946), **throttleman**, one who controls the throttle(s) of an engine.

1942 *S.A.E. Jrnl.* Jan. 22/1 Ice which collects in the induction system was divided into three classes: impact ice, throttle ice, and fuel evaporation ice... 'Throttle ice' is that which is formed at or near the throttle when the throttle is in a part-closed position due to the cooling effect of the increase in kinetic energy of the air in the restricted flow region. **1972** *Gloss. Aeronaut. & Astronaut. Terms* (*B.S.I.*) xv. 12 *Throttle ice*, ice formed in or near the engine throttle by the cooling due to isentropic expansion of the inspired air in the temperature range of 0°C to 5°C. **1946** *Amer. Speech* XXI. 310/2 *Throttle jockey*, a pilot. **1947** *Seafarers' Log* 25 Apr. 13/2 How could you crush a seamen's strike without captains and throttle-jockeys? **1904** *Everybody's Mag.* X. 663/1 When the officials came out and stood around the engine, there were throttle-men on waiting locals. **1973** H. Gruppe *Truxton Cipher* (1974) xiv. 140 The throttleman nervously wiped his sweating hands on a hank of oily cotton waste. **1982** *Fortune* 22 Mar. 172/2 The throttleman then has to reduce power because the boat's propellers are out of the water and meeting no resistance.

throttle, *v.* Add: **4. a.** Also const. *down.*

1914 Hamel & Turner *Flying* 134 Nearer and nearer we approach and now our pilot throttles down the engine.

b. *absol.* in phrs. *to throttle back, down*, to close the throttle in order to slow down or stop.

1932 D. Garnett *Rabbit in Air* III. 82 The altimeter was at 3000. I throttled back. *Ibid.* 91, I turned over the cement works, flew her level, and turned again by the river, throttled down and made my approach. **1953** C. A. Lindbergh *Spirit of St. Louis* II. vi. 188 The air speed's still over 100 miles an hour... I throttle down to 1750. **1973** R. Rosenblum *Mushroom Cave* (1974) 101 The pilot throttled back to float the helicopter over a large network of paths. **1979** 'K. M. Peyton' *Marion's Angels* viii. 130 He throttled down sharply for the turning to the church.

throttleable (þrǫ·t'lăb'l) *a.* (of an engine) that can be controlled by means of a throttle.

1960 *Aeroplane* XCVIII. 261/2 The Thiokol XLR-99 'throttleable' rocket engine..has completed preliminary static tests and will shortly be installed in an X-15. **1969** *New Scientist* 1 May 243/2 The rotors could be fitted with small, throttleable rockets on their tips.

through, *a.* Add: **1. a.** *spec.* in *Carpentry*, of a housing: running through the whole thickness of the member, not stopped.

1934 [see *stopped *ppl. a.* 9]. **1979** A. B. Emary *Woodworking* iii. 18 (*caption*) Through housing.

b. *through booking* (earlier example), *carriage, passenger* (example); **through traffic**: also, traffic which passes through a particular town, etc., rather than stopping there.

1848 *Amer. Railroad Jrnl.* 29 July 482/1 A *through* passenger in the 9 and 4½ o'clock lines, pays more than $1 25, for each of those parts of the line. **1869** *Bradshaw's Railway Man.* XXI. 43 Through-booking arrangements with the Scottish North Eastern. **1891** S. J. Weyman *New Rector* I. iii. 28 Oh, dear, they are in a through carriage... I would rather go in another carriage and change. **1944** *Sun* (Baltimore) 16 Feb. 9/1 A mid-town express highway as a through-traffic route. **1961** Through-traffic [see *route *v.*]. **1976** *Alyn & Deeside Observer* 10 Dec. 1/6 The new by-pass will provide an additional crossing of the River Dee that will enable much through traffic to avoid the city. **1978** O. S. Nock *Great Western* 110 In such conditions the through carriage was an inestimable boon.

through, *prep.* and *adv.* Add: **A.** β. **thro** (later examples); **thru**: now used informally as a reformed spelling and abbreviation (chiefly) in *N. Amer.*

1878 W. Whitman *Daybks. & Notebks.* (1978) I. 122 Sent piece 'Three Young Men's Deaths' $12 to Mr John Frazer, Tobacco Plant, Liverpool—thro Josiah Child. **1879**, etc. [see *though *adv.* and *conj.* A. γ]. **1904** R. Garnett *Let.* in A. Mizener *Ford Madox Ford* (1971) ix. 96 If Conrad..paid £3 a week thro Pinker it would be a very considerable help. **1917** E. E. Cummings *Let.* *c* Nov. (1969) 40, I see the thing thru, alone. **1921** *Jrnl. Nat. Dental Assoc.* VIII. 609/1 As we look thru our daily papers and our magazines. **1971** *Black World* Mar. 57/1 When she wuz little and she had stuttered thru a sentence. **1977** *Hot Car* Oct. 11/1 Available for S types right thru to Mk 10s it retails for 26 notes.

B. I. *prep.* **1. b.** *through-the-lens* adj., used with reference to light measurement in which it is the light passing through the lens of the camera that is measured (the same light that would form the image).

1965 *Focal Encycl. Photogr.* (rev. ed.) I. 554/1 Through-the-lens exposure measurement has the advantage..that the meter cell receives light from exactly the same subject field as is taken in by the lens. **1977** J. Hedgecoe *Photographer's Handbk.* 14/3 Solid state through-the-lens metering, zoom lenses, motor-drive, are all part of an ever-widening 'system' built around the single lens reflex body. **1984** *What Video?* Aug. 59/2 Fair picture, basic colour temperature controls, through-the-lens viewfinder.

h. (Later example.)

1932 H. S. Walpole *Fortress* III. 562 The mist immediately surrounding him was..so wetting that he was already soaked through and through his clothes.

5. b. Also used in sequences or lists, without necessarily denoting consecutive development.

1938 M. K. Rawlings *Yearling* xi. 110 His wares included the necessities and scanty luxuries of the whole country-side, from plows, wagons, buggies and implements, through food staples to whiskey and hardware, dry goods and notions and medicines. **1962** *Listener* 26 July 130/2 Rents range from just over £3 a month for a small flat, through about £14 for a two-bedroom house, to £23 for the most elegant apartments. **1975** *Nature* 10 Apr. 501/2 Nine recognised glaze types, ranging in colour from pale blue, through green, to yellow, brown and red.

d. *U.S.* Up to (a date, a number, a specified item, etc.) inclusively, up to the end of, up to and including, to, until; often correlative to *from.*

1798 T. Holcroft *Jrnl.* 4 Aug. in *Mem.* (1816) III. 31 Continued the opera through scene 9, Act 3. **1930** H. Brown (*title*) Rabelais in English literature through Sterne. **1932** *Atlantic Monthly* May 538 Mr. Heffernan was mayor for four years, from 1927 through 1931. **1942** M. Kraitchik *Math. Recreations* vi. 130 Poisson calculated this probability, taking into account the cards dealt in the first hand. His result does not differ through the third decimal place. **1950** H. Craig *Hist. Eng. Lit.* 250 Spenser treats of England from the Reformation through the reign of Queen Elizabeth. **1967** *N.Y. Times* (Internat. ed.) 11 Feb. 1/6 At a background briefing early in November, the American command made available infiltration figures covering the year through Sept. 30 and a rough estimate for October. **1971** *Physics Bull.* Dec. 738/1 In the review copy pages 1469 through 1472 are already loose which does not say too much for the quality of the binding. **1977** *Time* 8 Aug. 19/3 We will continue to govern through the end of our term. **1981** L. Deighton *XPD* xliii. 342 A..notice stating that deliveries were only accepted between eight and eleven Monday through Friday.

II. *adv.* **3. b.** (Earlier example of passing an examination and further examples of gen. sense, esp. of persons.) Also, defeated, having no further prospects, no longer friends or associates, outmoded, 'done for'; *to be through with*: also, to be tired of, to have had enough of; (further examples).

1849 Thackeray *Pendennis* I. xxi. 196 'This man has passed,' he thought, 'and I have failed!'..'Good bye, Spavin,' said he. 'I'm very glad you are through.' **1887** *Scribner's Mag.* May 622/2 He..then..scrawled a dash underneath. 'There! I'm through!' he said. **1897** J. L. Allen *Choir Invisible* ii. 22, I was through with the lessons. **1902** W. N. Harben *Abner Daniel* vii. 55, 'I don't understand you.' 'Well, you will before I'm through with you.' **1930** J. B. Priestley *Angel Pavement* x. 508 'You're through then, eh?' 'All I can do to-night, Mr Smeeth. One or two things I've had to leave till tomorrow morning.' **1931** H. F. Pringle *Theodore Roosevelt* I. iii. 37 He..was through with breakfast by 8.30. **1934** G. B. Shaw *On Rocks* II. 271 We were born into good society; and we are through with it: we have no illusions about it, even if we are fit for nothing better. **1939** I. Baird *Waste Heritage* ii. 23 Now when we get down to the hall we're through, see? I don't want nothin' more to do with you, I don't even know your name. **1942** E. Paul *Narrow St.* xxxi. 281 An outsider not familiar with French politics might have thought that Daladier was through. Not at all. He got himself elected by the Popular Front..in April, 1938. **1956** B. Holiday *Lady sings Blues* (1973) xix. 154 It was only a few weeks before, people had been telling me I was through in the United States, that the public would never accept me. **1969** A. Lurie *Real People* 108, I hope you don't think anyone who doesn't paint soda-pop bottles and stripes is through artistically. **1970** *Wall St. Jrnl.* 30 Mar. 1/1 An executive with two dependent children earning the equivalent of $24,000 a year is left with $14,300..after the Board of the Inland Revenue is through.

c. Of a telephone call or caller: connected.

1929 *Telegr. & Teleph. Jrnl.* XVI. 47/2 'You are through' (which is the English way of saying 'here is your party'). **1932** D. L. Sayers *Have his Carcase* iii. 33 The grocer announced that Harriet's call was through... 'Hullo!' she said. **1954** F. P. Keyes *Royal Box* i. 14 Won't you ever learn that when an English operator asks you if you're 'through', she doesn't mean have you *finished*, she means have you got your connection all right? **1977** *Rolling Stone* 30 June 80/3 Directly getting Honolulu information, I got a number for Wiley Hampson and presently was through to him at his home in Hawaii Kai.

5. c. *through-and-through sawing, sawn* = *plain sawing* vbl. sb., *plain-sawn* ppl. adj. s.v. *plain *a.*¹ and *adv.* C. c. Also *through-and-through method.*

1966 [see *plain-sawn* ppl. adj. s.v. *plain *a.*¹ and *adv.* C. c]. **1963** *Gloss. Terms Timber* (*B.S.I.*) 14 *Through-and-through sawing*, a method of converting..logs by parallel cuts in the general direction of the grain. **1979** A. B. Emary *Woodworking* i. 9 Most of the timber at a merchant's will be from logs cut by the through and through method.

through-, in combination. Add: **1. through-compo·sed** *pa. pple.* and *ppl. a.* = *durchkomponiert *a.*

1884 F. Niecks *Conc. Dict. Mus. Terms* 122 A durchcomponiertes Lied, 'a through-composed song', is a song of which each verse has a setting of its own. **1947** A. Einstein *Music in Romantic Era* x. 117 In this prologue there is no longer any spoken dialogue; the scene is 'through-composed'. **1962** *Listener* 11 Jan. 105/2 The opera opens with a Prologue of the spirits..which is the only 'through-composed' portion. **1981** *Times Lit. Suppl.* 20 Feb. 203/2 The first 'Razumovsky' [Quartet].. abandoned the normal first-movement repeat, had the rarity of a through-composed scherzo (without trio or 'da capo'), and only repeated the exposition of the finale.

2. (Some could equally well be placed s.v. *through *a.*) **throu·gh-ball** *Assoc. Football*, etc., a forward pass which goes through the other team's defensive formation; **through deck**, a flight deck which runs the full length of a ship; **throu·gh-deck crui·ser**, a type of lightly-armed aircraft-carrier (see quots.); **throu·gh-draught**: see *thorough-draught* s.v. Thorough- 2; **throu·gh-feed**, in centreless grinding, movement of the work-piece right through the space between the two wheels (cf. *in-feed); also *attrib.*; **throu·ghflow**, the flowing of a fluid, air, etc., through something; also *attrib.*; **through-lounge** (stress variable): in a private house, a lounge that extends from the front to the back of the house; **throu·gh-pass** = *through-ball* above; **throu·gh-valley** (see quot. 1972).

1969 *Punch* 12 Feb. 248/4, I wish I could recall the lingo—the through-balls, high crosses, work-rates and searching diagonals. **1977** *Times* 28 Feb. 8/5 Another through ball and Souness was racing away to make a cross which Armstrong converted. **1969** *Times* 30 Oct. 2/7 Such an aircraft..would be operated from a cruiser-type ship with a newly designed 'through deck'. **1971** *Guardian* 22 Nov. 12/4 The Royal Navy is now considering the operation of Harriers from a new class of flat topped 19,000-ton ships euphemistically known as 'Through-deck cruisers'. **1980** A. Preston *Warships of World* 29/2 The troubled political background accounts for the ludicrous nomenclature applied to the class. First of all *Invincible* was a 'through-deck' cruiser to disguise the flight deck, then a command cruiser, and only now..an aircraft carrier. *Ibid.* 221/2 *Through-deck cruiser*, cumbersome term concocted by the Royal Navy to obtain political and Treasury approval for a ship designed to operate helicopters and V/Stol aircraft. The 'through deck' was a euphemism for a full-length flight deck. **1905** Through-draught [see *thorough-draught* s.v. Thorough- 2]. **1976** B. Lecomber *Dead Weight* vi. 73, I..opened the pilots' side-windows in the pious hope of creating a through-draught of fresh air. **1937** Colvin & Stanley *Grinding Pract.* v. 73 There are two primary methods of grinding the work. One is the through-feed method in which the work passes axially from one side of the machine to the other. **1963** Jones & Schubert *Engin. Encycl.* 215 There are three general methods of centerless grinding which may be described as through-feed, in-feed, and end-feed methods. **1967** M. E. Hale *Biol. Lichens* vii. 96 Stemflow on trees..has been shown to be enriched, relative to throughflow, with potassium and calcium. **1974** *Country Life* 17 Jan. 103/1, 2 litre family saloon..throughflow heating and ventilation. **1962** Jackson & Marsden *Educ. & Working Class* v. 157 The interview took place in the through-lounge. **1976** *Evening Post* (Nottingham) 15 Dec. 15/5 (Advt.), Modern well situated detached house consisting of through lounge, fully fitted kitchen, [etc.]. **1937** F. N. S. Creek *Association Football* iii. 58 Short passing should always consist of 'through' passes so that the ball is sent behind an opponent and never..across his front. **1967** J. Potter *Foul Play* (1968) ii. 29 If you're playing on the left wing I'll feed you with through-passes to the corner flag. **1976** *Denbighshire Free Press* 8 Dec. 24/7 Ian McCarter..got to a splendid through pass from Wildermuth. **1905** *Bull. Geol. Soc. Amer.* XVI. 233 In discussing this paper at the Geological Society meeting in Philadelphia, Professor Davis applied the very descriptive name of 'through valleys' to this condition of valleys connected across lowered divides. **1969** G. C. Dickinson *Maps & Air Photographs* xiii. 208 A valley from one system may meet one from another system 'back-to-back' so that together they form a through valley with scarcely any feature separating the headwaters of the two systems. **1972** *Gloss. Geol.* (Amer. Geol. Inst.) 739/1 *Through valley*, a flat-floored depression or channel eroded across a divide by glacier ice or meltwater streams.

through other, through-other, *adv. phr.* and *adj.* Add: Also 9 **throughther.**

a **1889** G. M. Hopkins *Poems* (1967) 97 For earth her being has unbound; her dapple is at an end, astray or aswarm, all throughther, in throngs.

throughout, *prep.*, *adv.* Add: Also (chiefly *N. Amer.*) **thruout, thru-out** (see *thru* s.v. *through *prep.* and *adv.* A.β).

1922 *Proc. IRE* X. 260 During the past winter an amateur spark station located at Cleveland, Ohio..was received nightly at Yonkers, New York..with sufficient intensity to enable the signals to be read thruout the room. **1968** *Globe & Mail* (Toronto) 17 Feb. 46/6 (Advt.), 6 room brick bungalow with 3 finished rooms in basement, new broadloom thruout. **1976–7** *Sea Spray* (N.Z.) Dec./Jan. 119/1 (Advt.), All kauri ply and timber used exclusively thru-out.

throughput (þrū·put). [f. Through-: cf. Input *sb.*, Output *sb.*] **1.** *Sc.* 'Energy, activity, capacity for or progress at work' (S.N.D.).

1808 Jamieson *Etym. Dict. Sc. Lang.* I. s.v. *ithand*, He has nae great throw-pit, but he's very eident. **1845** *Chambers's Jrnl.* 28 June 412 They'd hang as long as I like at the plough-tail, but I want through-put; and so commend me to my own men and reasonable hours. **1923** G. Watson *Roxburghshire Word-bk.* 309 *Throw-pit*, capacity for accomplishing work.

2. The amount of oil or other raw materials, etc., processed by an industrial plant; *transf.*, the amount or number of units passing through a system, the amount of data processed by a computer (over a stated time period). Also, processing, production, or handling capacity. Also *attrib.*

1915 J. Wilson *Lowland Scotch Lower Strathearn* 273/1 *Throo·put*, *n.* out-turn. **1922** *Daily Mail* 15 Nov. 3 Throughput of oil will necessarily be interfered with temporarily. **1930** *Daily Express* 22 May 15/2 The larger throughput was sufficient to..yield the handsome margin of £89 per ton of ore. **1945** H. D. Smyth *Gen. Acct. Devel. Atomic Energy Mil. Purposes* xi. 113 By the middle of January 1942, a run had been made with a reasonable beam strength and an aggregate flow or through-put of appreciable amount which showed a much improved separation factor. **1950** *Engineering* 20 Jan. 69/1 Plant and machinery made ready to handle maximum throughputs. **1958** *Spectator* 1 Aug. 178/2 The Magheralin Creamery, N. Ireland, had a considerably higher throughput of milk. **1959** *Times Rev. Industry* Mar. 24/1 Coal throughput rates. **1961** B. Fergusson *Watery Maze* ii. 59 The drains..were unable to cope with the 'through-put', as industrialists say, of more than eighteen lavatories. **1962** *Economist* 20 Oct. 215/1 With Stonehenge and the Tower already showing maximum throughput. **1965** *New Scientist* 24 June 883 Time sharing (or multi-programming) is already well established as a means of increasing the throughput and utilization of a computer. **1967** *Listener* 31 Aug. 263/3 Russian economists now discover that pre-packaging..not only makes for better hygiene, but trebles a shop assistant's throughput. **1968** P. Dickinson *Skin Deep* i. 6 Estate agents must be doing nicely down this way, with a constant through-put of moneyed youngsters moving in with one kid and moving out with three. **1969** J. Argenti *Managem. Techniques* 168 By re-siting the factory entrance gates at a cost of a few hundred pounds it was possible for one company to increase the through-put of customers' lorries by 50 per cent. **1970** D. Kut *Warm Air Heating* xx. 338 The grade of oil fuel best suited for a particular installation depends on the type of oil burner and on the hourly through-put of oil. **1972** *Daily Tel.* 11 May 2/4 Prince Philip said..that the criterion for a university was the quality and standard of its life and work, not the total through-put of students. **1974** *Information Handbk. 1974–5* (Shell Internat. Petroleum Co.) 105 The graphs show that costs increase as throughputs exceed or fall below the 'optimum', that is the throughput at which the total unit cost reaches the lowest point. **1976** P. R. White *Planning for Public Transport* iv. 76 An alternative means of maximizing station throughput is to build a station in which a single track carrying a one-way flow bifurcates to form a loop around an island platform. **1978** *Jrnl. R. Soc. Arts* CXXVI. 425/1 Increasing the throughput of vehicles on an existing highway.

throughway (þrū·wēⁱ). Also (*N. Amer.*) **thruway,** and with hyphen. [f. Through- + Way *sb.*¹] **1.** *N. Amer.* An expressway; a large toll road.

1934 in Webster. **1946** *Sun* (Baltimore) 25 Mar. 24/9 The proposal..for a throughway across a filled-in inner harbor. **1951** *Economist* 22 Sept. 685/3 Plans for a 'Thruway' from New York City..to Buffalo on Lake Erie, are well advanced. **1969** A. Lurie *Real People* 42 Down below the rose garden I can hear cars whirring past on the thruway. **1976** T. Heald *Let Sleeping Dogs Die* ix. 176 The Dog Centre..was no more than three and a half hours' fast driving on the thru-way from Kennedy Airport.

THROW

2. *gen.* A way through; a means of passage through or between.

1935 C. Day Lewis *Time to Dance* 22 Speak up, speak up, you skyward man, Speak up and tell us true; To east or west—which is the best, The through-way of the two? **1977** 'E. Crispin' *Glimpses of Moon* xii. 233 The cradle jolts and jerks under their combined assault, but remains obligingly stuck on its through-way.

throw, *sb.*² Add: **I. 2. a.** Also, the extent through which a switch or lever may be moved.

1975 *Gramophone* Sept. 533/3 These are toggle type switches,..and protrude quite appreciably from the facia. They therefore have quite a long throw. **1979** *SLR Camera* Jan. 41/2 The wind-on lever has a short throw of around 120 degrees.

b. (Examples.)

1890 E. Atkinson tr. *Ganot's Elem. Treat. Physics* (ed. 13) x. ii. 794 When a current of very small duration is passed through a galvanometer, a momentary deflection or swing or throw of the needle will be produced. **1931** L. B. Loeb *Fund. Electr. & Magnetism* xviii. 272 Even the first throw of the galvanometer has not the true value which it would have had in the absence of damping.

9. A decorative piece of fabric used as a casual covering for furniture, as a rug, counterpane, etc. Also, a shawl or stole. *N. Amer.* (chiefly *U.S.*).

1895 [see *drape *sb.* c]. **1913** F. H. Burnett *T. Tembarom* xi. 524 They..transformed the cot into a 'couch' by covering it with what Tracy's knew as a 'throw'. **1936** W. Greene *Death in Deep South* xi. 142 Her last summer's scarf made a throw over the pine table. **1952** E. Ferber *Giant* ii. 14 I'm only going to buy a little white mink cape throw. **1963** G. S. Maxwell *Navajo Rugs* iii. 20 Continuing south we come to the city of Gallup, a center for the inexpensive rug known as a 'Throw'. **1980** M. McMullen *My Cousin Death* (1981) xvii. 195 She..brought a plaid throw, and tenderly tucked it in around him.

10. *colloq.* (orig. *U.S.*) A 'go' at anything; freq. in phr. — *a throw*, preceded by a specified sum of money to denote 'so much a go' or 'so much apiece'.

1898 F. P. Dunne *Mr. Dooley in Peace & War* 101 Smaller this New York, but th' livin' was cheaper, with Mon'gahela rye at five a throw. **1930** 'D. Stiff' *Milk & Honey Route* 127 Beer or wine at a jitney a throw. **1948** [see *pork sb. 3 a]. **1958** B. Malamud *Magic Barrel* 30 A column..inviting contributions in the form of stories at five bucks the thousand-word throw. **1966** N. Freeling *Dresden Green* i. 38 Coffee-table books..at a hundred and forty francs the throw. **1975** *Listener* Winter 353 The cost of research... The BBC Archives charge £2 a throw.

throw, *v.*¹ Add: **II. 10. c.** (Earlier example.)

1768 W. Musgrave *Let.* 12 Feb. in *15th Rep. R. Comm. Hist. Manuscripts* App. vi. 245 in *Parl. Papers* 1897 (C. 8551) LI. 1 But if they will be artful enough to throw their votes so as to choose one of your candidates, it is my opinion we ought to remain contented for the present.

12. (Earlier example.)

1777 J. Woodforde *Diary* 8 May (1924) I. 203 Bill caught only one little Miney but he did not throw above four times.

15. a. Also, to project (the voice); also, *spec.* as in ventriloquism. Cf. sense 44 c below.

1962 A. Nisbett *Technique Sound Studio* 263 In dead acoustics the ratio of direct to indirect sound cannot be varied, as indirect sound must be kept to a minimum. In this case it may help if an actor 'throws' his voice, simulating raising it to talk or shout from a distance. **1972** A. Price *Col. Butler's Wolf* xx. 222 He threw his voice past Ryleiev into the mist. **1976** *Listener* 23/30 Dec. 830/1, I can throw my voice. I could make a fortune as a medium.

17. a. Delete † *Obs. rare.* Now usu. in phr. *to throw a punch*, to deliver a blow with the clenched first; occas. with fist as obj.

1923 H. C. Witwer *Fighting Blood* xi. 348, I set myself, took careful aim and threw my right at his chin. **1950** J. Dempsey *Championship Fighting* xvi. 92 You're throwing perfect punches. **1976** J. Lewis *Shadows of Death* iv. 54 Maybe the kid had a hammer in his glove; surely he couldn't have thrown a punch hard enough to hurt him like that one had. **1983** *Daily Tel.* 3 Feb. 3/3 Mr Oatway.. threw punches at the second bandit.

18. a. *to throw a fit.* For *U.S. slang* read *slang* (orig. *U.S.*). (Earlier and later examples.) Chiefly *fig.*

1896 S. Crane *Maggie* (rev. ed.) iii. 22 Deh ol' woman 'ill be trowin' fits. **1926** *S.P.E. Tract* xxiv. 126 Father threw a fit when I came home drunk. **1930** *Observer* 4 May 15 Caesar throws his fit, off stage. **1954** Koestler *Invis. Writing* 172 One day this was discovered by Zsuzsa, who threw a fit. **1973** 'R. Macleod' *Burial in Portugal* v. 97 'Please, Jonathan. If I am late—.' 'The management will throw a fit,' he completed for her.

b. To give or hold (a party), esp. one of an informal or impromptu nature. *colloq.* (orig. *U.S.*).

1922 S. Lewis *Babbitt* xxxix. 339 Saturday night, when they would..'throw a party'. **1937** *Even. News* 6 Mar. 11/5 Anona Winn threw a party a few nights ago at her flat in Maida Vale. **1960** *Sunday Express* 13 Mar. 12/7 She..threw a champagne and scampi party at a nearby pub. **1978** *Detroit Free Press* 5 Mar. B 2/2 When all else fails, throw a party.

III. 19. c. To lose (a contest, race, etc.) deliberately or by corrupt prearrangement. *colloq.* (orig. *U.S.*).

1868 H. Woodruff *Trotting Horse Amer.* xxxi. 263 It was..very unjust to charge Mr. Nodine with throwing the race. **1876** 'E. Queen' in *Blue Bk.* Oct. 27/1 'Brown threw the fight?' asked..a member of the Boxing Commission. **1951** *Manch. Guardian Weekly* 1 Mar. 15/1 Baseball games had been 'thrown' by bribed players. **1959** *News Chron.* 19 Aug. 6/3 ..This was the equivalent of arranging that one side in the Cup Final should throw the game. **1978** *Times* 9 Jan. 8/3 During the Chancellorship of Mr Roy Jenkins, Lord Allen had to 'throw' his position..and publicly support the policy.

25. (Later examples of *transf.* sense.) *U.S.*

1907 [see *puke-out* s.v. *puke *sb.*¹ b]. **1934** *Amer. Ballads & Folk Songs* 14 They had mooched the stem and threw their feet.

V. 30. b. *to throw oneself at* (earlier example).

1789 H. More *Lett.* (1925) 127 The women all threw themselves at his head.

f. *colloq.* To engage (the clutch or gears) of a motor vehicle. Also *transf.* with the vehicle as obj. Usu. with *in, into*.

1904 A. B. F. Young *Compl. Motorist* vii. 176 The mighty engine is fretting and heating itself with impatience, and the clutch is continually being thrown in and out. **1969** J. T. Story *Dishonourable Member* iii. 6, I was forced to rev my engine and throw the gears into reverse. **1970** N. Slater *Falcon* x. 177 He threw the cruising Alfa into third gear and powered away.

g. *trans.* = *throw down*, sense 40 d in Dict. Also *absol.*

1923 A. L. Simon *Supply, Care & Sale of Wine* xvi. 111 If red wines be shipped and bottled too early, they will throw a heavy sediment in the bottle instead of lees in cask. **1956** S. M. Tritton *Amateur Wine Making* iii. 89 The wine..should never be allowed to throw..a deposit. [illegible]

32. b. To operate (a switch), esp. by moving a lever. *colloq.* (orig. *U.S.*).

1930 E. B. White *Let.* July (1976) 94 One of the men [..] threw the switch. **1978** *Times* 20 Nov. 2/3 More than 14,000 civil servants are engaged on government computer work. Those with the direct power to throw the switches..number 1,728.

32*. *b.* To disconcert or confuse (someone), to disturb, upset. Cf. sense 44 l in Dict. *colloq.* (orig. *U.S.*).

1844 R. Browning *Lett. to M. R. Mitford* (1983) II. 431 He appeared to me far more thrown by this intellectual adversity than he ever was by the death of his Katy. **1941** B. Schulberg *What makes Sammy Run?* vi. 104 Don't let Julian's worries throw you. **1950** L. Kaufman *Jubal's Children* xxiii. 247, I knew my way around as a restaurant and a bill of fare. Sometimes, even those French dishes didn't throw me. **1961** C. Willock *Death in Covert* ix. 172 Miche refused to be thrown. 'I rather like enthusiasms,' she said gallantly. **1964** Mrs. L. B. Johnson *White House Diary* 16 June (1970) 169 Although I was a bit thrown by the mix-up over the two Mrs. Does, I felt this was one of those times that shows whether or not you've got poise and inner calm. **1978** *Browning Inst. Studies* VI. 72 One might almost suspect that Browning was trying to 'throw' his reader. **1981** P. Dickinson *Seventh Raven* iv. 47 It'll throw those kids if they have to make the change at the last moment.

e. To break or render inoperable (something mechanical). *colloq.* (orig. and chiefly *U.S.*).

1954 *Amer. Speech* XXIX. 103 *Throw a clutch,*..to break a clutch, usually in speed-shifting. 'On the fourth run we threw a clutch, so that was the end.' **1976** *Billings* (Montana) *Gaz.* 1 July 4-E/1 A truck for Springfield, Va., threw its transmission near Towson. **1980** *Dirt Bike* Oct. 43/2 Suzuki teammate Barnett threw a chain while running fourth early in the race.

34. a. (Earlier example.)

1847 C. Brontë *Jane Eyre* III. viii. 184 And try to restrain the disproportionate fervour with which you throw yourself into common-place home pleasures.

35. *throw about.* **a.** (Examples with reference to aircraft and motor vehicles.)

1942 *Tee Emm* (Air Ministry) II. 85 The operational fighter pilot..wants an aircraft..easily manœuvreable so that he can throw it about when necessary. **1959** *Motor* 11 Nov. 524/1 On the confined test ground it seemed easier to 'throw about' than the big B.M.W., but neither car could show its paces in the space available.

37. throw away. c. Also *spec.* in *Cricket*, to lose (a wicket) through careless play.

1898 K. S. Ranjitsinhji *With Stoddart's Team* (ed. 4) xii. 237 Many wickets were thrown away by the batsmen at critical periods by careless and hasty strokes. **1904** P. F. Warner *How we recovered Ashes* ix. 185 Braund, Bosanquet, and Rhodes literally threw their wickets away. **1977** *World of Cricket Monthly* June 42/3 Once in Lahore, Pakistan, he threw away his wicket so that Australia could win.

e. *Theatr.* To deliver (lines) in a casual manner; to underemphasize or play down (usu. for increased dramatic effect). Also *absol.* and *transf.* Cf. *throw-away B. 3.

1934 J. Agate in *Sunday Times* 18 Nov. 4/1 The spectator becomes aware of a fixed determination on the actor's part to make as little as possible of anything that can be called the orthodox 'acting' of the part, to throw away—in the actor's sense—everything except the highest of its poetry and the most sensitive of its philosophy. **1957** *Times Lit. Suppl.* 11 Oct. 611/3 Mr Fleming can be exuberant; but he prefers, in the stage term, to 'throw away', something he does just as neatly and wittily as du Maurier used to do it in the theatre. **1959** *Times* 26 May 13/4 In this part he again covered up for his author by charmingly throwing away as many lines as possible. **1959** *Listener* 14 May 861/1 The acting was deliberately played down for microphone purposes... This no doubt necessary business of 'throwing it away' must involve some losses as well as some gains.

40. throw down. i. *Cricket.* To knock down (a wicket) with a throw-in from the field, with the intention of dismissing the batsman.

1866 *Baily's Mag.* Sept. 419 John Lillywhite..from long leg..threw down the wicket, and Mr. Davidson was thus run out. **1912** P. F. Warner *England v. Australia* v. 44 Hobbs throwing down Kortlang's wicket from cover-point. **1962** E. W. Swanton in Altham & Swanton *Hist. Cricket* (new ed.) II. xii. 244 All seemed over when Solomon from 25 yards range and square with the wicket on the leg-side threw down the stumps to run out Davidson.

41. throw in. d. (*d*) (Examples in Cricket.)

1816 W. Lambert *Cricketer's Guide* (ed. 6) iii. 43 Long Stop. This man..should be one who is not afraid of the Ball,..and who can throw in well. **1938** *Times* 16 Apr. 8/1 Those 'girls' of the Oval..threw in from the boundary with an accuracy which would have done credit to a University side. **1976** J. Snow *Cricket Rebel* 118 My back troubled me from the strain of bowling on the harder Australian wickets for four months and my right shoulder had 'gone' when it came to throwing the ball in.

e. Also *intr.* (chiefly *U.S.*), *to throw in (with).*

1923 *Century Mag.* Oct. 829/1 Lead me to them humans, and I'll throw in with 'em. **1940** W. Faulkner *Hamlet* 359 When we threw in together that day..he didn't know how long he had been on the road. **1958** J. Carroll *Mortal Friends* i. ii. 25 The important thing was that Jim Brady's best boy—a strong and more than usually efficient young man..has thrown in for good with his own people. **1981** M. Moorcock *Byzantium Endures* xi. 262 We should have stayed with the peasants and not thrown in with Russians and Jews.

g. *to throw in one's hand:* (*a*) to retire from a card game, esp. poker; (*b*) *fig.*, to give up a contest or struggle. *colloq.*

[1904 R. F. Foster *Pract. Poker* 49 Players should be careful not to throw their hands into the deadwood until they have seen openers.] **1923** *Daily Mail* 3 July 8 Our plucky farmers are not 'throwing in their hands'. **1926** *Auction Bridge Mag.* July 119/1 People get so tired of throwing in hand after hand that they come in, regardless of their position. **1928** S. Maugham Ashenden..It needs great self-control to throw in hand after hand. **1938** G. March-Phillipps *Ace High* ii. ii. 34 She turned it up and saw the six of Hearts... 'Bitched again!' Bobby said, and threw in his hand. **1967** *Economist* 5 Oct. 59/2 An international understanding outside Egypt is needed before the board can throw in its hand. **1962** D. Francis *Dead Cert* ii. 20, I threw in my hand. I pushed the four chips across to him. **1973** J. Ashford *Double Run* ii. 9 Nina Ryan had thrown in her hand without bothering to buy cards. She wasn't really fond of poker. **1978** 'S. Woods' *Exit Murderer* 14 Sykes looked at the cards for a long moment..and then suddenly threw in his hand. 'She knew,' he said positively.

42. throw off. e. (Earlier *intr.* examples.)

1789 *Loiterer* 14 Feb. 8 No sooner had the hounds thrown off than my horse grew..hot. *Ibid.* 11 Apr. 5, I have been assured by very experienced Hunters of Tufts, that they never threw off earlier than twelve.

44. throw out. c. Also, to project (the voice), esp. in singing. Cf. sense 15 a above.

1792 H. Newdigate *Let.* Feb. in A. E. Newdigate-Newdegate *Cheverels* (1898) ix. 123 Mortellari..is giving her an Artful Manner of throwing out her Voice to be heard in public.

g. (Earlier example of sense 'to put forward tentatively'.)

1824 M. Wilmot *Let.* 26 May (1935) 214, I only throw out this idea to shew I am ready to act on it.

h. (Earlier *trans.* example.)

1798 Jane Austen *Let.* 1 Dec. (1952) 34 He wants my mother to look yellow and to throw out a rash, but she will do neither.

l. (Earlier example.) Also *transf.*, of a plan, calculation, etc.

1824 H. Davy *Diary* July (1836) II. v. 211 My *senza cura* servant, threw me out by not putting powder enough in my horn. **1892** A. W. Pinero *Magistrate* I. 24, I took five years from my total... It has thrown everything out. As I am now thirty-one, instead of thirty-six as I ought to be, it stands to reason that I couldn't have been married twenty years ago.

45. throw over. b. (Earlier and later examples.) Also, to cancel (an appointment) with someone, to put (someone) off.

1835 Dickens *Let.* 5 Nov. (1965) I. 88, I will throw Bell's life over, altogether. **1891** O. Wilde *Pict. Dorian Gray* iii. 62 'Are you disengaged Tuesday?' 'For you I would throw over anybody, Duchess.' **1903** G. B. Shaw *Man & Superman* I. 40 Tavy will kiss; and you will only turn the cheek. And you will throw him over if anybody better turns up. **1908** E. F. Benson *Climber* x. 146 He had another engagement, and though I urged him not to throw it over when I heard that, he really insisted on coming.

47. throw together. a. (Later examples.)

1878 I. L. Bird *Lady's Life Rocky Mountains* (1879) iii. 37, I threw a few things together and came here. **1967** E. Short *Embroidery & Fabric Collage* i. 18 A bundle of threads or fabrics accidentally thrown together may suggest an exciting scheme.

c. (Earlier example.)

1818 Scott *Heart Midl.* I. viii. 226 The circumstances of their families threw the young people constantly together.

d. To prepare (a snack, meal, etc.) hastily or in an improvised manner.

1962 'E. Ferrars' *Busy Body* i. 13 If we haven't eaten she'll throw something together. **1980** P. G. Winslow *Counsellor Heart* iv. 64 She hadn't been listening much, throwing a quick supper together.

48. throw up. b. (Further examples.) Also *intr.* (now the usual use). Now chiefly *colloq.* or *slang.*

1793 *Morning Chron.* 20 Feb. 3/1 In what odd ways we taste misfortune's cup—While France throws *down* the gauntlet—Pitt throws *up.* **1887** V. Martin *Let.* in M. Collis *Somerville & Ross* (1968) iii. 44 To show that he was quite unembarrassed he began to play with the favourite pug, finally dancing it round on its hind legs. It immediately threw up and that I think ends the story. **1895** Herron & Bacon in A. Dundes *Mother Wit* (1973) 367/2 He made a tea which acted as an emetic and the patient threw up a variety of reptiles. **1934** T. N. Wilder *Heaven's my Destination* 6 He thought he was going to throw up. **1956** [see *showbusiness 1 b]. **1970** G. F. Newman *Sir, You Bastard* i. 24 What if the scene was a man with the front wheel of a bus on his chest? Wasn't the calm, inspiring copper entitled to throw up? **1977** *New Yorker* 19 Sept. 49/3 I'm not sure anyone else even noticed, but it upset me so much I threw up. **1980** A. E. Fisher *Midnight Men* viii. 102 Ogy got drunk and threw up in the backyard.

h. Also without *it* and with personal object: to hold (someone) up as an example, object of reproach, etc. (Earlier and later examples.)

1815 R. Findley *Let.* 6 Dec. in N. E. Eliason *Tarheel Talk* (1956) 300 Betsey..throwed up to me that I made a better bed for Sally then her little Betsey. **1870** 'Mark Twain' in *Galaxy* July 139/1 He [*sc.* Benjamin Franklin] would work all day and then sit up nights..so that all other boys might have to do that also or else have Benjamin Franklin thrown up to them. **1957** R. Lawler *Summer of Seventeenth Doll* I. i. 22 Every time he's away and we have a row, Emma throws him up at me like a dirty dish-cloth.

i. (Earlier example.)

1832 *Q. Rev.* XLVII. 237 For heaven's sake, take care of my hounds in case they may throw up in the lane.

k. *trans.* To produce or provide. *colloq.*

1963 *Guardian* 23 Sept. 3/5 Their memories of this year's fortnight may last even longer than anything Uncle Ted can throw up from his family joke repertoire on Christmas Day. **1981** M. Moorcock *Byzantium Endures* xii. 287 Russia was throwing up better women than men at that time. All the worthwhile men had been killed.

VII. 49. *throw the* *book *at, a* *monkey-*wrench into the machinery, a* *spanner *in the works, one's* *weight *about.*

throw-. Add: **1. b. throw cushion** *N. Amer.* = *scatter cushion* s.v. *scatter *v.* 7 b; **throw-net,** a fishing net cast out by hand; **throw pillow** = *throw cushion* above; **throw weight** (see quot. 1982¹). **c.** (sense 9 of *throw *sb.*²) *throw rug; throw-style, type* adjs. Cf. *throw cushion, pillow* above. *N. Amer.* (chiefly *U.S.*).

1970 *Toronto Daily Star* 24 Sept. 28/6 (Advt.), Throw cushions, chests, dressers. **1931** *Times Educ. Suppl.* 21 Mar. (Home & Classroom Suppl.) p. ii (*caption*), The photograph shows the different kinds of 'nets' used. On the left is the throw-net. **1979** *Field* 9 May 975/4 For smaller fish the locals used circular throw-nets some eight feet in diameter. **1974** *Progress* (Easley, S. Carolina) 24 Apr. III. 5/2 (Advt.), Decorator throw pillows. **1952** *Better Homes & Gardens* Feb. 173/1 How to vacuum your throw rugs. **1978** R. Ludlum *Holcroft Covenant* xi. 132 She leaned forward on the hassock, picking at an imaginary piece of lint on the throw rug beneath his feet. **1974** *Spartanburg* (S. Carolina) *Herald* 18 Apr. (Kmart Advts. Suppl.) 8 Tailored, throw-style bedspread, elegantly puff-quilted to the floor. **1970** *Toronto Daily Star* 24 Sept. 37/4 (Advt.), Quilted throw type bedspreads. **1969** *Sci. Amer.* Aug. 19/2 Parity..is clearly not numerical equality in the number of warheads or in the number of megatons or in the total 'throw weight'. **1982** *Times* 10 May 22/1 The second stage would seek to achieve equal ceilings on ballistic missile 'throw weight' at less than current American levels. 'Throw weight' is the term used to describe the weight of the warheads which missiles can carry onto a target. **1982** H. Kissinger *Years of Upheaval* vii. 264 The Soviets were ahead in numbers of land-based missiles and throwweight.

2. throw-down, (*b*) *Austral.* and *N.Z.*, a type of small firework, a squib; **throw-in,** (*a*) (earlier examples); (*b*) in Bridge, an end play in which the declarer throws the lead to an opponent who has to play into a tenace combination; (*c*) in Polo, the act of throwing the ball (by the umpire) between opposing ranks of players, each team being on its own side of the line of the throw, in order to (re)start the game; (*d*) in Baseball, a throw made by an outfielder to an infielder; **throw-out,** (*b*) in Cricket, the act of throwing out a batsman (see Throw *v.*¹ 44 n); (*c*) *Bookbinding* (see quot. 1976); (*d*) the mechanism by which the driven and driving plates of a clutch in a motor vehicle are separated; usu. *attrib.*

1896 E. Turner *Little Larrikin* xxvi. 321 Lol was.. projecting jumping Jacks and throwdowns on the floor, and keeping the cook..on the table. **1948** D. W. Ballantyne *Cunninghams* xiv. 75 Kids were exploding throwdowns on the footpaths. **1864** *Baily's Mag.* July 173 What a recreation!—wherein..the hands, dropping for a while the pen.., can..deliver a 'throw-in' decisive of all argument. **1891** [see *goal-kick* s.v. *goal *sb.* 6]. **1898** T. B. Drybrough *Polo* xi. 245 Players should leave room for the ball to pass between them. In case of overlapping, the umpire may delay throwing-in or recall the throw, and he may recall a faulty throw-in. **1935** [see *exit *sb.* 2 b]. **1937** *Times* 12 July 5/2 Captain Robinson got away from the throw-in but broke his stick. **1940** *Sun* (Baltimore) 13 May 13 (*caption*) Dolph Camilli..slides safely into third..after a throw-in from outfielder Chuck Klein. **1952** *Ibid.* 2 Oct. 20/1 The Yankees might have had a big inning on the way since Martin had taken second on the throw-in. **1959** M. Gilbert *Blood & Judgement* x. 108 He settled down with *Reese on Play*, to study the tactics of the Throw-in. **1964** *Official Encycl. Bridge* 619/1 In a throw-in play, an opponent gains the lead, but it costs him a trick (or more) to do so. **1977** *Navy News* Sept. 38/4 (*caption*) All eyes on the ball at a throw-in during the Rundle Cup polo match between the Navy and Army at Tidworth. **1928** *Weekly Dispatch* 24 June 20/5 That 'throw-out' which has to be executed so quickly at cover-point. **1953** *Vocab. Bibliothec.* (Unesco) 190/1 Throw out (for maps, illus., etc.). **1966** *New Statesman* 18 Mar. 393/1 (Advt.), Illustrated and with 4 coloured throwout maps. **1969** *Catal. Austral., N.Z., & Pacific* (O.U.P., Melbourne) 16, 20 half-tone plates, 2 maps, end-paper map, 1 coloured throw out. **1970** K. Ball *Fiat 600, 600D Autobook* v. 46/2 (*caption*) Diagrammatic view of clutch throwout mechanism. **1976** *Gloss. Documentation Terms* (*B.S.I.*) 68 *Throw-out*, a leaf, usually bearing illustrative material, bound in at one edge and designed to fit the book when folded. **1979** P. Wallage *Restoration of Post-War Cars* 6 Clutch release bearing, throwout bearing.

throwable (þrōᵘ·ăb'l), *a.* [f. Throw *v.*¹ + -able.] Capable of being thrown.

1888 Lees & Clutterbuck *Ramble in Brit. Columbia* xxiii. 264 We had thrown every throwable article at him. **1966** J. Derrick *Teaching English to Immigrants* iv. 172 If it is difficult to find throwable objects, balls can be used instead..to represent them. **1977** G. V. Higgins *Dreamland* iii. 27, I reached for a throwable life preserver cushion.

throw-away. (In Dict. s.v. Throw-.) Add: **A.** *sb.* **1. a.** Now usu. with reference to ephemeral material distributed free of charge, as pamphlets, advertising leaflets, certain newspapers, etc. (Later examples.)

1922 Joyce *Ulysses* 149 A sombre Y.M.C.A. young man..placed a throwaway in a hand of Mr Bloom. **1944** *Sun* (Baltimore) 1 June 16-0/3 The small throw-aways contain the latest authentic reports of the progress of the war. **1954** [see *junk mail* s.v. *junk *sb.*² 5]. **1965** *Newsweek* 21 June 70/2 The advertisements in the two editions of the weekly shopper's throwaway. **1973** D. Ramsay *Deadly Discretion* 102 Here's our throwaway. It's also the ad we run in the papers.

b. More generally, anything designed to be thrown away after use; *spec.* a disposable container. Cf. sense 2 a of the adj. below. *colloq.*

1953 *Sun* (Baltimore) 17 Feb. (B ed.) 30/7 The group of county delegates submitted a bill..which would ban the dispensing of alcoholic beverages in throw-aways. **1976** *Monitor* (McAllen, Texas) 29 Oct. 3A/3 Consumers could save millions of dollars a year in lower prices if soft drinks and beer were sold in returnable containers instead of throwaways.

2. An act of throwing away, or that which is thrown away, in various senses. Also *fig.* Cf. Throw *v.*¹ 37 in Dict. and Suppl.

1911 G. B. Shaw *Lett. to Granville Barker* (1956) 178 If the attempt proves a throw away, it is only a throw-away of the chance I promised him, not of the play. **1922** Joyce *Ulysses* 223 A skiff, a crumpled throwaway..rode lightly down the Liffey. **1955** *N.Y. Times* 29 May VI. 15/1 Generally, the program opens with a line of girls in two or three minutes of fast-stepping, high-kicking precision dancing. This is a throwaway, designed to get late-comers settled into their seats before the real show starts. **1960** *Twentieth Cent.* Aug. 137 Each is a finely polished stylist: let no one be deceived by the easy, laconic throw-away of Ada Leverson. **1976** B. Jackson *Flameout* vi. 115 It was a pity that the best question was a throwaway to the other reporters: they didn't deserve it. **1983** *Listener* 13 Oct. 21/1 Even in films they hiss 'What did he say?' at a throwaway of dialogue, thus ensuring that the next few minutes are lost to all around.

B. as *adj.* **1.** Of prices: so low as to represent virtually no return for the goods sold; 'give-away'.

1924 A. J. Small *Frozen Gold* xiii. 288 With a modicum of luck they might even be able to record every claim they had pegged—and then get rid of them at throw-away prices. **1967** *Spectator* 14 July 53/3 At throwaway prices everyone can afford the latest Camp, and there will be something new coming along next month. **1976** [see *saturate *v.* 2 d].

2. a. Designating something designed to be thrown away after use; disposable.

1928 *Weekly Dispatch* 13 May 17 You can..clean your face at intervals with those throwaway hankies you buy from any chemist. **1945** *Forbes* (N.Y.) 15 Oct. 16/1 'Throw-away' towels will arrive soon. **1958** *Engineering* 7 Feb. 192/3 The butane comes from a throwaway cartridge. **1970** *Worship* Jan. 41 Already one hears of loose-leaf prayer books and throw-away hymn books. **1982** J. Hansen *Gravedigger* iv. 33 Two plastic-handled throwaway razors.

b. Pertaining to or characterized by the use of disposable goods or those with a short life-span.

1969 *New Scientist* 25 Sept. 648/1 We will undoubtedly have a formidable litter problem in our 'throw away' world..from..household equipment with built-in obsolescence. **1977** M. Drabble *Ice Age* II. 114 She thanked God that she lived in a consumer throw-away flush-away advertising society. **1980** *Jrnl. R. Soc. Arts* Mar. 188/1 At the same time the 'throw away' attitude developed in society.

3. Underemphatic or casual in style or technique; understated (usu. for increased effect). Cf. *throw *v.*¹ 37 e.

1955 *Time* 4 Apr. 77/2 It takes a certain nerve for a comedian to try a throwaway line. **1958** M. Dickens *Man Overboard* vii. 102 He was more cunning than he seemed with that throw-away sixth-form voice. **1961** *John o' London's* 25 May 591/4, I remember his beautiful throw-away performance in *Mr. Deeds goes to Town*. **1969** *N.Y. Rev. Bks.* 30 Jan. 27/1 He will..carry us with him, a little breathless perhaps, and dizzy with his throw-away allusions and polyglot versatility. **1972** *Daily Tel.* 29 June 7/7 You can carry your enthusiasm..into casual slouchy nonchalance, and the outstanding collection of Stephen Adnitt had plenty of this throwaway chic. **1980** *Times Lit. Suppl.* 11 July 786/3 The style of the narrative is measured but evocative; a little throw-away, a little affected by the insidious influence of Peter Fleming.

throw-back. Add: **3.** (Earlier example.) Also, a reversion to the technique or methods of an earlier period. Also applied to a person using such techniques.

1888 Kipling *Plain Tales from Hills* 209 The queer, savage feeling..must be a 'throw back' to times when men and women were rather worse than they are now. **1930** E. Blunden in *Nation & Athenaeum* 6 Dec. 327/1 The Canterbury Poets were a throw-back to Cooke's little volumes. **1938** *Sun* (Baltimore) 19 July 8/3 His flight was harebrained and foolhardy... It was an unnecessary throwback to the romantic era of long-distance aviation. **1949** *Ibid.* 14 July 23/1 On the whole it was wartime baseball, a throwback to the hilarious 1945 World Series between the Tigers and the Cubs. **1976** *UCT Stud. in Eng.* (Univ. of Cape Town) Oct. 13 If..modern literature is distinctive in its claim on our moral attention, Stevens is distinctly un-modern, a transcendental throwback.

throwed, pa. t. and pa. pple. of Throw *v.*¹ (Later U.S. dial. examples.)

1914 *Sat. Even. Post* 3 Oct. 20/3 This should ought to of gave me a record of 16 wins and 0 defeats because the only games I lost was throwed away behind me. **1930** [see *honky-tonk 1]. **1949** [see *grain *v.*¹ 8]. **1968** E. J. Gaines in A. Chapman *New Black Voices* (1972) 91 He wiped his mouth and throwed his cup on his bunk.

thrower. Add: **II. 3. a.** (Earlier example in *Cricket.*)

1832 P. Egan *Bk. Sports* 344/1 Mr. K–ngsc—e comes next... An excellent thrower—a hundred yards clear.

b. *thrower-out* (later example).

1963 N. Streatfeild *Vicarage Family* vi. 75 Some of the members of his men's society..are coming by train as throwers-out. If there is a rumpus it will be all over almost before it begins.

throwing, *vbl. sb.*¹ Add: **III. 5. b.** *throwing arm* (properly *ppl. a.*). **c. throwing power,** the ability of an electrodepositing solution to produce an even coating on an irregularly shaped object.

1972 J. Mosedale *Football* ii. 19 Thanks to his accurate throwing arm, New Orleans was..one of the darkest cities in America. **1977** *World of Cricket Monthly* June 47/2 He had one of the strongest throwing arms ever seen in Australian cricket. **1922** *Trans. Amer. Electrochem. Soc.* XLI. 363 The object of this study was the development of a zinc plating bath of high 'throwing power'. **1932** *Metal Industry* XL. 501/2 He had to undertake the task of determining the throwing power of several selected plating baths. **1966** D. G. Brandon *Mod. Techniques Metallogr.* 5 Electropolishing..may even give a poor macropolish as a result of..poor throwing power in the electrolyte.

thrown, *ppl. a.* Add: **II. 4.** *thrown-away, -back* (earlier example), *-together.*

1778 'J. H. St. John de Crèvecœur' *Sk. 18th-Cent. Amer.* (1925) 323 Your neighbours won't thank you for this thrownaway humanity. **1890** W. James *Princ. Psychol.* II. xxv. 484 The frowning brow, the thrown-back shoulders, and clenched fists of rage. **1934** Thrown-together [see *red-hot *a.* (and *sb.*) 4]. **1973** J. Leasor *Host of Extras* v. 69 The beach was littered with..thrown-away cigarette packets.

throw-off. a. (Earlier example.)

1843 *Ainsworth's Mag.* III. 144 The throw-off will be unusually great to-morrow.

throw-over. Add: **throw-over switch** *Electr. Engin.* (see quot. 1943).

1902 [in Dict.]. **1919** *Wireless World* VII. 40/1 A simple differential indicator operated by a throw-over switch. **1943** *Gloss. Terms Electr. Engin.* (*B.S.I.*) 52 *Throw-over switch*, a switch for changing over from one set of connections to another set of connections.

thrum, *sb.*² Add: **7. a.** *thrum mat* (later examples). **b. thrum cap,** (*b*) *Canad.* (*obs.* exc. in place-names), a small island with a conical shape suggestive of a thrum cap.

1832 T. Baillie *Acct. New Brunswick* 120 Opposite to this point an islet or thrum-cap..was once considered available for the purpose of drawing fish. **1903** G. S. Wasson *Cap'n Simeon's Store* xi. 248 They had..taken to their boat and pulled for Thrumcap Island Light. **1966** T. H. Raddall *Hangman's Beach* II. xiv. 217 'Why is this called the Thrum Cap?' 'Thrumb's a coarse kind of wool. ..Sailors used to wear knitted caps of red thrum... You notice the red bank of the knoll standing up like a thrum cap.' **1883** *Man. Seamanship Boys' Training Ships* (Admiralty) (1886) 184 A thrum mat is made by cutting a certain number of yarns of equal length and reeving them through holes made in the mat, both ends to come through on one side. **1961** F. H. Burgess *Dict. Sailing* 208 *Thrum mat*, a piece of canvas or other coarse material into which thrums are inserted, either roughly for chafing purposes and collision mats, or in decorative pattern with materials suitable for homes.

thrumming, *vbl. sb.* Add: (Examples in sense 2 of Thrum *v.*³)

1941 D. C. Peattie *Road of Naturalist* (1946) i. 9 Unleashed for the long stretch, the motor took up a loyal thrumming. **1969** *Daily Tel.* 14 Apr. 19/6 The arrival of a hovercraft..brings a thrumming which rattles the window. **1977** *Islander* (Victoria, B.C.) 21 Aug. 10/1 Switch on, and the Yanmar diesel barked into life, warming up slowly to a soft thrumming.

thrump (þrɒmp). [Echoic.] The sound of a blow, heavy fall or beat, etc. Freq. redupl.

1871 *Daily News* 25 Jan., The heavy thrump, thrump of the mitrailleuse. **1886** *Century Mag.* Feb. 520/1 The banjo's thrump and strum. **1903** *Westm. Gaz.* 5 Oct. 2/1, I awoke with the music of marching men's feet in my ears—thrump, thrump, thrump, thrump.

† **thrums** (þrɒms). *Obs. slang.* Also quasi-*sing.* thrum (7 thrumm). [Repr. colloq. or dial. pronunc. of *thrup(p)ence.] = Threepence.

1699 B. E. *New Dict. Canting Crew, Thrumms*, threepence. **1846** *Swell's Night Guide* 78 There is a hanger on here who teaches the art of self defence—thrums (three pence) a lesson. **1865** *Leaves from Diary of Celebrated Burglar & Pickpocket* 108/2 His first putting-up place was at a low padding ken in St. Giles's, where he paid a 'thrum' per night for share of a 'doss'. **1880** W. H. Patterson *Gloss. Words Antrim & Down* 106 *Thrum*,..a threepence. A commission of three pence per stone of flax, paid by a flax-buyer to a person who brings the buyer and seller together in open market. **1933** *Bulletin* (Sydney) 23 Aug. 10/2, I haven't encountered a crook thrum yet.

thrup(p)ence, thrup(p)enny (þrɒ·pĕns, þrɒ·pĕni; þru·-). Repr. colloq. or dial. pronunc. of Threepence, Threepenny *a.* (*sb.*).

1895 H. Nevinson *Neighbours of Ours* v. 142 They paid 'er fivepence for doin' each large flag... So she sublets to Ginger at thruppence a flag. **1962** *Spectator* 27 July 117 A letter with a thrupenny stamp.

thrush¹. Add: **2*.** *fig.* A female singer. *U.S. slang.*

1946 *Amer. Speech* XV. 205/1 *Thrush*, a songstress. **1966** *Crescendo* Oct. 31/2 She has established herself as one of the best female thrushes in this area. **1982** B. Fantoni *Stickman* iv. 38 The band's thrush and Moses sang so long over the tempo to play 'Lover Man' we ditched it.

thrush². Add: **1.** Also, an infection of any other part with the same fungus (now called *Candida albicans*), esp. of a woman's vagina. (Further examples.)

1967 *Current Medicine & Drugs* Dec. 4/1 The occurrence of Thrush Bowel Infection after antibiotics is..argued by the makers of Nystatin. **1970** Passmore & Robson *Compan. Med. Stud.* II. xviii. 19/2 Vaginal thrush is fairly commonly associated with pregnancy. **1977** *Spare Rib* Jan. 36/1 Thrush is very irritating and can make you extremely sore if it's allowed to develop.

thrust, *sb.* Add: **I. 3. a.** (*b*) (Further examples.) (*e*) The propulsive force developed by a jet or rocket engine.

1870 *4th Ann. Rep. Aeronaut. Soc. 1869* 9 The thrust of aerial screw propellers. **1933** *Aircraft Engin.* Jan. 22/1 These series of tests were undertaken to determine how much the reaction thrust of a jet could be increased by the use of thrust augmentors. **1950** *Sci. News* XV. 24 Since news of Whittle's jet engine was released there has been a popular misconception that if there is no atmosphere for the exhaust gases to push against there will be no thrust. **1977** *R.A.F. Yearbk.* 32/1 Recovery..involves unstalling the wing and re-establishing lift rather than blasting the aircraft out with thrust. **1982** *Daily Tel.* 14 Jan. 16/5 The engine was a Russian-built Nene of higher thrust (5,000 lb) than the original models. **1983** D. Stinton *Design of Aeroplane* viii. 297 A propeller consists of a number of wing-like aerofoils designed to convert torque into thrust.

II. 6. c. The principal theme or gist (*of* remarks, an argument, etc.); a point, aim, or purpose. orig. and chiefly *U.S.*

1968 Mrs. L. B. Johnson *White House Diary* 17 Apr. (1970) 667 He spoke well... His 'thrust' was that we 'make open spaces and recreation facilities a part of the daily..environment of people'. **1973** A. Chapman *New Black Voices* 275 The Institute of the Black World in Atlanta... Its central thrust is towards the creation of an international center for Black Studies. **1973** *Globe & Mail* (Toronto) 1 Aug. 6/3 The thrust of your editorial..is premised on the discredited 'compact theory'. **1977** *Guardian Weekly* 28 Aug. 18/4 That was the thrust of the exclusive story in the New York Times on March 10. **Ibid.** 6 Nov. 16/1 The postwar thrust of U.S. policy..has been to enlist Thailand in an anti-Communist alliance. **1982** *Church Times* 15 Jan. 12/4 A major part of the thrust of my article was to dissuade others from proposing such a cutback.

6*. *Geol.* = *thrust-fault* s.v. Thrust *sb.* 7 in Dict. and Suppl.

1888 *Q. Jrnl. Geol. Soc.* XLIV. 420 Outliers of the 'Fucoid-beds' and Serpulite-grit are found,..separated from each other by major thrusts. **1910** *Ibid.* LXVI. 593 Thrust is here employed in the sense of a fold-fault replacing the lower limb of an overturned anticline. Lag..is employed in the sense of a fold-fault replacing the upper limb of an overturned anticline. **1926** *Ibid.* LXXXII. 315 The Creag-an-Lochan Thrust. [*Note*] Equivalent to the 'lag' of E. B. Bailey [*preceding quot.*]; but the term 'thrust' is preferred and used throughout this paper for all structures indicating differential resistance to folding forces at a comparatively early stage. **1934** B. & R. Willis *Geologic Structures* (ed. 3) vii. 153 The term 'thrust' too often connotes the idea of an overthrust, whereas the structure may be an underthrust. **1942** E. M. Anderson *Dynamics of Faulting* i. 1 Overthrusts, or more simply thrusts, are faults which are inclined, in theory, at well under 45° to the horizon, and in field experience it is found that they are sometimes nearly horizontal. **1942** [see *overthrust *ppl. a.*]. **1971** C. R. Twidale *Structural Landforms* iv. 98 In a normal thrust the upper block rides over the lower.. but in a lag thrust the lower block is thrust forward and upwards beneath the upper.

7. thrust augmentor *Aeronaut.*, a procedure or modification used with a jet engine to increase its thrust; so **thrust augmentation; thrust-bearing** (earlier example); **thrust-box** (example); **thrust chamber** *Astronautics* (see quot.); **thrust-fault:** in mod. use, a low-angle reverse fault; also, any low-angle fault; = sense 6* above; (further examples); hence **thrust-faulted** *a.*, **-faulting** *vbl. sb.*; **thrust reverser** *Aeronaut.*, a device for reversing the flow of gas from a jet engine so as to produce a retarding backward thrust; **thrust spoiler** *Aeronaut.*, a device for deflecting the flow of gas from a jet engine so as to reduce the thrust quickly without reducing the engine power; **thrust vector,** a vector representing the direction (and magnitude) of the thrust produced by a jet engine, propeller, etc.; **thrust washer,** a washer (Washer *sb.*²) against which a thrust-bearing rests.

1956 W. A. Heflin *U.S. Air Force Dict.* 524/2 Thrust augmentation for jet engines is accomplished by afterburning, reheating, water injection, etc. **1967** N. E. Borden *Jet-Engine Fundamentals* 126 On some jet engines, it is advantageous to provide a means of thrust augmentation during take-off on warm or hot days. **1933** Thrust augmentor [see sense 3 a (*e*) above]. **1947** *Jrnl. R. Aeronaut. Soc.* LI. 79/1 The pumping could be directly produced by a ducted fan..[illegible] **1918** *Blackw. Mag.* Aug. 291 Pretty drawings in colour of such things as thrust-boxes and oil-pumps. **1962** F. I. Ordway et al. *Basic Astronautics* ii. 413 The rocket thrust chamber is a device into which propellants are injected and burned to form gases. The basic components of the thrust chamber are the injector, the combustion chamber, and the exhaust nozzle. **1959** *Sci. Amer.* Feb. 86 The fan... **1969** [illegible] It was suggested that a thrust-fault through the Mendip axis carrying up the upper portion northward. **1951** C. Schuchert in Pirsson & Schuchert *Text-bk. Geol.* I. xiv. 344 Reverse faults, having a gently inclined fault-surface are known as thrust-faults or simply thrusts. **1982** *Earth-Sci. Rev.* [illegible] **1973** J. G. Dennis *Structural Geol.* xii. 271 Since thrust faults were originally considered a class of reverse faults, they should bring older rocks over younger. This is..[illegible]. **1980** *Sci. Amer.* Oct. 127/2 The thrust sheets and folds indicate that the rocks were much compressed in the horizontal direction. **Ibid.** The Valley and Ridge province is characterized by folded and thrust-faulted strata of mostly unmetamorphosed sedimentary rocks formed between 600 million and 300 million years ago. **1912** *Q. Jrnl. Geol. Soc.* LXVIII. 59 The occurrence of two lines of block-faulting. **1936** *Geogr. Jrnl.* LXXXVII. 214 The Purari Plateau is characterized by..elevation uplifting with the accompanying development of block-faulting, probably more normal faulting than thrust-faulting. **1954** *Flight Handb.* (ed. 5) xi. 164 (*caption*) A turbojet thrust-reverser developed by the American Boeing company. The jet is deflected by a W-shaped pair of clamshell doors. **1958** B. Jackson *Flameout* (1977) x. 182 At his low altitude there was simply no way to..thrust reversers..[illegible] **1967** *Gloss. Aeronaut. & Astronaut. Terms* (*B.S.I.*) 57 Thrust spoiler... **1947** *Jrnl. R. Aeronaut. Soc.* LI. 679/2 The thrust vector..[illegible] **1961** *Flight* Jan. [illegible] **1962** *Flight Internat.* LXXXII. 393/1 In the pioneer SC.1 the lift units are arranged in two pairs mounted on lateral trunnions to pivot some 25° fore and aft, in order to provide longitudinal thrust components to assist transition to and from wing-supported flight. This idea has now given way to a fixed installation with thrust-vector control. **1975** *Offshore Engineer* Dec. 54/2 (Advt.), The whole unit, and thus the thrust vector, can be directed through 360°, which means that it has been possible to optimise the nozzle and propeller for one main flow direction. **1954** Thrust washer [see *space *sb.* 1 a]. **1962** Gas turbine... **1970** K. Ball *Fiat 600, 600D Autobook* ix. 106/2 If the line is out of center, other parts must be checked.

thrust, *v.* Add: **IV. 8. c.** *Comb.*, as **thrust stage** *Theatr.*, an open stage that projects into the auditorium so that the audience is seated around three sides.

1968 *Sat. Rev.* (U.S.) 1 June 22 The Fine Arts Theatre.. is a compact, multipurpose amphitheater seating 600, which can be utilized for conventional theatricals, as a thrust stage, or even—with the built-in pit—for musicals and intimate opera. **1969** *Guardian* 28 Oct. 7/3 Knighted actors argue the merits of a 'thrust' stage for the costly new Sheffield Theatre. **1977** *Times* 25 Aug. 15/4 Kate went to the University of Toronto where they had just built a superb thrust-stage theatre.

thrust, *ppl. a.* [See the vb.] With adverbs, as *thrust-out* adj. = Out-thrust *ppl. a.*

1872 R. W. Buchanan *St. Abe & his Seven Wives* 153 And with thrust-out jaw and set Teeth, the Yankee threatens yet. **1976** [see *snake-hipped* s.v. *snake *sb.* 10].

thruster. Add: **2. b.** *fig.* One who pushes his way; an aggressive or go-ahead person. Also *spec.* with reference to driving.

1925 Fraser & Gibbons *Soldier & Sailor Words & Phr.* 280 *Thruster, a*, an obnoxious, pushing person. A 'bounder'. **1927** *Observer* 6 Nov. 13/1 Those who described their more successful fellows as 'thrusters' and 'climbers'. **1927** *Morning Post* 28 Nov. 10/4 The 'road thruster', or the man with the 'passing' mania, is usually a nuisance. **1960** *Times* 7 June 13/7 The ordinary lorry or van driver.. is often the worst thruster of all. **1964** C. Willock *Enormous Zoo* ii. 21 Bere..was known as a thruster when it came to administration. **1974** R. Harris *Double Snare* xxvi. 197 Robert's car has fallen back, displaced by an Italian thruster's enormous red tourer. **1982** Barr & York *Official Sloane Ranger Handbk.* 106/1 She will not lend her support to any ambitious young thruster or leak the firm's secrets to a rival.

3. a. *Astronaut.* A small rocket engine on a spacecraft for providing the thrust needed to alter or correct its flight path or its attitude.

1962 J. Glenn in *Into Orbit* 146 This would include warming up the hydrogen peroxide thrusters so that they would go to work without delay when I activated the controls. **1965** *Newsweek* 13 Dec. 60 In past missions, excitement, the thump of the thruster rockets and a busy schedule all conspired to deprive the astronauts of sleep. **1969** *Guardian* 22 July 18/3 A 46-second firing of the reaction control thrusters altered Eagle's orbit. **1973** *Times* 28 Nov. 8/2 Small thrusters were fired yesterday which put the 570lb craft on course so that it will fly to within 81,000 miles of Jupiter on December 3. **1977** *Sci. Amer.* Feb. 61/1 The satellite is carefully tracked by ground stations, the propulsion vector needed to compensate for drift is computed and the appropriate thrusters are fired by radio command.

b. *Oil Industry.* Each of several jets or propellers on a drill ship or offshore rig, used for accurate manœuvring and maintenance of position.

1972 L. M. Harris *Introd. Deepwater Floating Drilling Operations* i. 5 Several vessels have been built, or are in the design and construction stage, that depend entirely upon thrusters, or powered units, to position the vessel dynamically. **1975** *Gen. Electric Investor* Winter 18/2 In the several thruster systems called 'dynamic positioning', grabbed above have thrusters propelled keeps the rigs positioned over the drilling hole despite wind, waves and currents. **1983** *New Scientist* 2 June 637/1 When it is semi-submerged, a 'dynamic positioning' system, thrusters controlled by computer, keeps the ship within two metres of a programmed position in the face of strong winds.

thrustful, thrustfulness. (Later examples.)

1963 *Economist* 22 June 1236/1 A Butler government would, one imagines, be less thrustful. **1966** *Times Lit. Suppl.* 1 Sept. 1/4 Few men have a good word for the Cecil cardinals... They are the automatic targets of the thrustfulness of young radicals.

thruway, var. *throughway.

thucholite (þᵘ·tʃolait). *Min.* Also thucolite. [f. Th, U, C, H, O, symbols of the constituent elements + -lite.] A naturally occurring brittle, highly lustrous, black mixture or complex of carbon and hydrocarbons with uraninite.

1928 H. V. Ellsworth in *Amer. Mineralogist* XIII. 66 A remarkable carbon mineral, which will be described in another paper under the name of thucholite, is sometimes very intimately associated with both uraninite and samarskite. **1938** R. W. Lawson tr. *Hevesy & Paneth's Man. Radioactivity* (ed. 2) v. 61 Fig. 12..shows the Z-spectrum of the mineral thucolite. **1950** G. J. Williams *Econ. Geol. N.Z.* xiii. 207/2 The radioactivity in the carbonaceous matter is due to very minute grains of uraninite, 1 to 5 microns in size, distributed at random or as strings of granules. **1980** *Mineral. Abstr.* XXXI. 487/2 Assemblages of thucholite in sandstone, dolomite, and shale of Polish Zechstein rocks are described. The thucholite..forms disseminated spheroidal or irregular bodies.

thuck (þɒk). [Echoic.] The sound of a missile, as an arrow, bullet, etc., hitting a target.

1948 F. Blake *Johnny Christmas* i. 46 Knifings, head-blows muffled by sand, the *thuck* of arrows striking home. **1979** P. Cosgrave *Three Colonels* 202 The whine of the rifle bullet and the thuck of its striking home.

Thucydidean (þiū·sidĭd·iăn), *a.* Also † -æan, -ian. [f. L. *Thūcȳdidēs* (Gr. Θουκυδίδης), name of a Greek historian of the fifth century B.C. +

-ean; cf. *Thūcȳdidēus* adj. (Cicero).] Of, pertaining to, or characteristic of Thucydides or his work.

1752 *Phil. Trans. R. Soc.* XLVII. 385 The European plagues are much more violent than the eastern; those being really the Thucydidian, which sweep all away. **1826** K. H. Digby *Morus* 125 Having no character of solemn reserve or Thucydidean dignity of style to support. **1834** —— *Mores Cath.* v. vi. 183 The Thucydidæan expression. **1888** *Encycl. Brit.* XXIII. 326/2 The best clue to Thucydidean bibliography is in Engelmann's *Scriptores Graeci.* **1911** *19th Cent.* Apr. 697 He even heightens the pathos of the Thucydidean original. **1945** E. K. Chambers *Eng. Lit. at Close of Middle Ages* iii. 130 It was not long before, in Thucydidean phrase, he had won his way to the mythical. **1977** *Trans. Philol. Soc. 1975* 128 The Hellenizing and specifically Thucydidean tradition of Sallust and Tacitus.

thuddingly, *adv.* Add: Also *fig.*

1976 *Daily Tel.* 16 Dec. 10/5 The man who shot her, incidentally, is called Lord Lichfield—just one of the names dropped thuddingly at every opportunity. **1979** *Ibid.* 18 July 16/3 'Human stories'..even more thuddingly boring than the well-boiled cabbage-slabs of opinion.

thug, *sb.* **b.** Delete Now *U.S.* and add further examples.

1881 R. L. Stevenson *Virginibus Puerisque* 164 Sometimes it [*sc.* death] leaps suddenly upon its victims, like a Thug. **1958** P. Gibbs *Curtains of Yesterday* xxvii. 216 'Isn't he a madman?' he asked. 'Isn't he raising an army of young thugs, his brutish young Brownshirts? Haven't they been fighting and brawling in Bavaria?' **1967** S. Faessler in *Atlantic Monthly* Apr. 103/2 The old man ducked for cover.., but not my father. Unarmed he stood up to the thugs, and was cracked over the head for it. **1982** *Daily Tel.* 7 Jan. 16/6 A plea..that it [*sc.* corporal punishment] should be retained as the final deterrent for school 'thugs'.

thug, *v.* Add: **2.** *intr.* To be a thug (sense b). *U.S.*

1937 *Sun* (Baltimore) 6 May 7/2 When I was thugging in Harlan county,..Merle Middleton was the chief of the gang. **1965** W. Soyinka *Road* 22 *Chief*:..Are you..one of the boys? *Samson*: I won't thug for you if that is what you mean.

thuggee. (Earlier *attrib.* example.)

1858 J. S. Mill *Memo. of Improvements in Admin. of India during Last Thirty Years* 46 The work of the Thuggee Suppression Department was nearly completed.

thuggery. (Later *transf.* examples.)

1930 E. D. Sullivan *Chicago Surrenders* xiv. 239 Thuggery..has met no serious rebuff on any front. **1959** *Economist* 29 June 1154/2 The police are having an extremely difficult task combating thuggery. **1973** *Nation Rev.* (Melbourne) 31 Aug. III. 1445/2 Envoys confronted with unfamiliar thuggery can be as naive as Birmingham businessmen in the newer industrial countries.

thuggish (þɒ·gɪʃ), *a.* [f. Thug *sb.* + -ish¹.] Resembling a thug. Also *Comb.*, as *thuggish-looking* adj.

1953 W. Burroughs *Junkie* viii. 75, I..got in conversation with a thuggish-looking young Italian. **1976** *Daily Tel.* 25 Oct. 2/4 Thuggish youths streaming from the rival supporters' clubs. **1981** *Sci. Amer.* Apr. [illegible] The Nationalists..had allowed some of their..thuggish.. [illegible]

‖ **thugyi** (þɒ·dʒi). Also 9 thoogyee. [Burmese.] The headman of a village.

1863 C. Williams *Jrnl.* 3 Feb. in *Through Burmah* (1868) 57 On the Thoogyee sending for some fishermen, he found the characters to be Nagiri. **1887** *Rangoon Gaz. Weekly Budget* 7 Jan. 8/4 On 18th a thugyi successfully attacked dacoits and recovered dacoited cattle. **1934** 'G. Orwell' *Burmese Days* xxv. 369 Old thugyis murmured to one another. **1948** F. S. V. Donnison *Public Admin. Burma* iv. 33 In the villages, the thugyis or headmen, chosen by the villagers and appointed by the government, have limited magisterial powers and collect the revenue. **1980** J. Silverstein *Burmese Politics* ii. 30 The thugyi was directly responsible to the deputy commissioner.

thuja. Add: Hence **thujaplicin** (-plai·sin) [L. *plic-ātus*, pa. pple. of *plicāre* to fold], any of three isomers of isopropyltropolone, $C_3H_7C_7H_5O_2$, that have fungicidal properties and occur in the conifer *Thuja plicata*.

1948 Erdtman & Gripenberg in *Nature* 8 May 719/2 The compound, m.p. 82°, is termed γ-thujaplicin because of the isopropyl group occupies the γ-position in the cycloheptatrienolone. **1964** *New Scientist* 4 June 613/1 The tropolones beta thujaplicin and gamma thujaplicin were highly toxic to four of the test fungi. **1978** *Further Perspectives Org. Chem.* 189 The three isomeric thujaplicins..comprise natural tropolones—the thujaplicins—by ring expansion in some unknown way from a benzene ring and an attached carbon.

thula (þu·lă). [ON *þula*.] A metrical list of names or poetic synonyms assembled in categories (orig. for oral recitation) to preserve traditional knowledge.

1936 K. Malone *Widsith* 1 Three name-lists (and one fragment of such a list) make up what we might call the thula.. taken from the Icelandic. **1937** —— in *Angl. Beibl.* XLVIII. 221 In an old thula-fragment quoted in the Icelandic..Wearne by..[illegible] **1963** Brown & Foote *Early Eng. & Norse Stud.* 113 It is undoubtedly because of these lists of names that the poem is called a *þula*. **1974** *Eng. Stud.* LV. 507 The first and third thulas are..parallel in introducing kings uninterruptedly.

Thule. Add: **1. b.** *ultima Thule*: in *fig.* use, also the lowest limit, the nadir.

1954 M. Lowry *Let.* 22 May (1967) 370 Before you write off that behaviour as being the ultima thule of ingratitude..try to understand the effect your news..had on me. **1976** L. Davidoff et al. in Mitchell & Oakley *Rights & Wrongs of Women* iv. 157 The one who had 'fallen' out of the respectable society..to the *ultima Thule* of prostitution.

2. *Archæol.* (with pronunc. þūl, þiūl). Used chiefly *attrib.* to designate a prehistoric Eskimo culture widely distributed from Alaska to Greenland *c* 500–1400 A.D. [From the name *Thule* (now Dundas), a settlement in N.W. Greenland.]

1927 T. Mathiassen *Archæol. of Central Eskimos* II. i. 2 We have..found remnants of an older culture which, after a locality outside the Central Eskimo territory, in North Greenland, we have called the Thule culture. **1935** *Nature* 3 Aug. 188/1 He [*sc.* Mathiassen] regards the Thule culture as originating in Asia. **1956** G. Freeman tr. *Malaurie's Last Kings of Thule* I. vii. 87 The Thule Culture—neo-Eskimo—would be derived from an anterior continental culture—palaeo-Eskimo. **1962** *Times* 4 Aug. 7/7 Within recent historical times Greenland has known three Eskimo cultures. The most recent is that of the Thule people who moved across Canada from Alaska. **1972** *Country Life* 12 Oct. 880 About AD 1100 a second Eskimo culture, known to archaeologists as the Thule period because it was first identified at Thule in Greenland, spread eastward from Alaska. *Ibid.* 881 The Thule Eskimos lived in stone houses. **1974** *Encycl. Brit. Macropædia* I. 1130/2 The spread of Thule..has been traced eastward from Alaska, arriving in Greenland about 1200. .. Later there was a resurge of Thule back toward the west, reaching all the way to Bering Strait. **1977** G. Clark *World Prehist.* (ed. 3) IX. 411 The Neo-Eskimo bearers of the Thule culture, immediate forebears of the existing population.

Thulean (þiū·liăn), *a. Geol.* [f. Thule + -an.] Of, pertaining to, or designating a region of Tertiary volcanic activity including Iceland and parts of Britain and Greenland.

1925 *Glasgow Herald* 10 Nov. 11/5 Jan Mayen, he said, formed a part of the Thulean or Brito-Arctic petrographical province. **1938** A. K. Wells *Outl. Hist. Geol.* xvii. 206 Evidently Britain lay on the fringe of a vast North Atlantic (or Thulean) volcanic province. **1976** P. Francis *Volcanoes* vii. 219 The Giant's Causeway in Antrim, and Fingal's Cave on the island of Staffa both originated when the Thulean plateau was being formed.

thulia (þiū·liă). *Chem.* [mod.L., f. next after Thoria, Yttria, etc.] The sesquioxide of thulium, Tm_2O_3, a dense white powder.

1886 [see *fractional *a.* b]. **1924** J. W. Mellor *Compreh. Treat. Inorg. & Theoret. Chem.* V. xxxviii. 698 Cleve's preparation of thulia was purified..[illegible] **1964** *Chem. Abstr.* [illegible] Should the demand arise, thulia, Tm_2O_3, could be isolated readily, as one of a number of individual rare earth by-products from the current (1967) commercial production of yttria by ion exchange, at a rate exceeding 1000 pounds per month.

thulium (þiū·liəm). *Chem.* [mod.L., coined in Fr. (P. T. Cleve 1879, in *Compt. Rend.* LXXXIX. 480).] A rare metallic element of the lanthanide series that forms pale green salts in which it is trivalent. Atomic number 69; symbol Tm.

1879 P. T. Cleve in *Chem. News* 12 Sept. 126/2 For the radical of the oxide placed between ytterbia and erbia, I propose the name of Thulium..derived from Thule, the ancient name of Scandinavia. **1923** U. R. Evans *Metals & Metallic Compounds* II. 236 Thulium is only known in very small quantities. **1956** *Nature* 17 Mar. 494/2 Iridium-192 continues to be the only widely used radioisotope in industry for gamma-radiography... Thulium-170 gives very low energy radiations, and it has been successfully used.. **1970** [illegible] Thulium has limited..[illegible] Autonomous sophisticated speculators have already shifted into antimony and silver while as *anned gadds* group is believed to have plunged heavily on thulium.

thumb, *sb.* Add: **5. h.** Also in mod. use (with significance the reverse of that in the ancient amphitheatre): *thumbs down, up*, gestures made with the fingers closed and the thumb pointing vertically downwards (indicating disapproval or rejection) or upwards (as a sign of approval, acceptance, encouragement, etc.); also *attrib.* and *fig.*

1906 Kipling *Puck of Pook's Hill* 180 We're finished men—thumbs down against both of us. **1917** A. G. Empey *Over the Top* 311 *Thumb up*, 'Tommy's expression which means "Everything is fine with me".' **1920** A. C. & C. Edington *Studio Murder Myst.* iii. 26 The irrevocable 'thumbs down' on a lovely female actor, because certain people turned their thumbs down on her in the front. **1946** *War Illustr.* 4 Nov. 3, iii/1 French peasants now return the 'thumbs up' gesture with which the jeep and the British troops are now received in the front. **1946** *Sunday Dispatch* 8 Sept. 3/1 Not a single man gives the thumbs-up sign. **1951** *Sport* 7–13 Jan. 16/3 The London team has been given the thumbs down sign by a meeting of 1st division promoters. **1951** S. Spender *World within World* v. 275 Our chief comedian was Buckfast... Everything about him suggested a 'thumbs up' attitude. **1954** R. Sutcliff *Eagle of Ninth* iii. 27 He laughed, and made the 'thumbs up' to his troops, calling 'Well done, lads!' **1961** *Guardian* 25 Mar. 6/7 The Chancellor of the Exchequer's thumbs-down to a National Theatre. **1967** *Technology Week* XX. 95/2 Giving a final 'thumbs up' on the rocket's readiness. **1971** *Sunday Times* (Johannesburg) 28 Mar. 5/1 She said the thumbs-down vote was not unanimous. **1976** *Scotsman* 25 Nov. 3/7 The market yesterday gave Sir Hugh the thumbs-up. The Fraser shares went up 3p to 58p on the report, which was apparently better than expected. **1979** R. Fiennes *Hell on Ice* i. 14 Both drivers gave a 'thumbs up'. **1982** *Daily Tel.* 5 Mar. 17/1 (*heading*) Baldwin statue gets thumbs down from Foot.

i. *to have a green thumb*: see *green *a.* 1 k; *to stick out like a sore thumb*: see *sore *a.* 9 f.

6. c. *thumb-latch* (earlier example), *-wheel* (examples). **d.** *thumb-sucking* (earlier example) *-twiddling* (cf. Twiddle *v.*¹ 2 c); *thumb -stained, -worn* (earlier example). **e. thumb-pad** (later example); **thumb paper** *U.S.*, a paper or card inserted in a book at the bottom of a page to protect it from thumb-marks; **thumb piano** *Mus.* = *sansa; **thumb pick** *Mus.*, a kind of plectrum; **thumb print:** also *fig.*; **thumb-stick,** a tall walking-stick with a forked thumb-rest at the top; **thumb-sucker,** (*a*) a child who habitually sucks his thumb; (*b*) *Journalists' slang* (see quots. 1974, 1980); **thumb-tack,** add to def.: *N. Amer.* = *drawing-pin* s.v. Drawing *vbl. sb.* 6 b; (earlier example); also as *v. trans.*; hence **thumb-tacked** *ppl. a.*

1761 *Essex Inst. Hist. Coll.* (1912) XLVIII. 96 Hinges, thumb latches, hammers. **1965** Lee & Knowles *Animal Hormones* iii. 53 At sexual maturity in male frogs (for example *Rana temporaria*) there is hypertrophy of the muscles of the forearms and thickening of the thumb-pads. **1843** B. R. Hall *New Purchase* I. xxx. 286 To have used..any other than the thumb-paper just named would have been considerably worse than ridiculous. **1888** E. Eggleston *Graysons* viii. 79 Fervid little love-notes.. were folded like the 'thumb-papers' that served to protect their books. **1942** F. Warnick *Dial. Garrett County, Maryland* 15 *Thumb-paper*,..a small piece of paper used to protect the pages. **1952** R. A. Waterman in *Proc. 29th Internat. Congress Americanists 1949* II. 212 Melodic instruments..are utilized for their percussive value, as in the case of 'thumb pianos', [etc.]. **1974** *Encycl. Brit. Macropædia* I. 250/1 The *mbira* is also known as..thumb piano, and by other regional names. The common term *sansa* is not correct; it is not found in Africa. **1969** *John Edwards Mem. Foundation Q.* V. 15 Riley used a thumb pick to achieve the heavy bass runs. **1973** Thumb pick [see *pick *sb.*² 5 c]. **1967** G. Steiner *Lang. & Silence* 66 Thumb-print..[illegible] the name and nature of the modern poet. **1979** *Times* 30 July 12 Caddell's thumb-prints are now on the energy speech that Carter delivered to the nation Sunday after returning to Washington. **1976** *Guardian* 14 Aug. 8/4 Fettered in The haives..[illegible] **1945** *Sun* (Baltimore) 25 Oct. 4/1 Believing the thumbstick to be important during the..[illegible] on at a forty year-old youth. **1974** R. Adams *Shardik* xi. 79 Bel-ka-Trazet walked with the help of a long thumb-stick which Kelderek remembered to have seen him trimming the evening before. **1982** *Church Times* 2 Apr. (Advt. Feature) p. iv/4 Whether it is..a military swagger cane, a stick you have whittled, a Shepherd's crook or a thumb stick for the country walk. **1891** 'Mark Twain' tr. *Hoffman-Donner's Slovenly Peter* (1935), Lld. Ed.] 25 Story of the thumb-sucker. **1957** M. Argyle *Psychol. & Social Probl.* ix. 111 There is also some evidence that children who have been given the breast for as long as they want it, or use a dummy, are more likely to become thumb-suckers. **1974** *Bell System Technical Jrnl.* [illegible] Walter Lippmann wrote the best straight think-pieces, or thumb-suckers as they are known in the trade. **1980** W. Safire *On Language* [illegible] *N.Y. Times Mag.* 11 May 12/1 Slurs like 'paper pusher' for bureaucrat, or 'thumbsucker' for columnist. **1858** *Vermont Herald* [illegible] thumb-tacks in my picture by myself. **1951** R. Mayer *Artist's Handbk.* (1965) 475 A much better way to preserve unstretched pictures..is to thumb-tack them face down to sheets of wallboard. **1978** *N.Y. Times* 14 Sept. x. 1/2 Thumbtacked to the bulletin board was a color snapshot. **1966** D. Francis *Flying Finish* ii. 24 We heard the..thumb-tacked weather report. **1920** *Times* 26 Mar. 14/1 Continuous thumb-twiddling, idleness and aimless grumblings degenerate into dreary thumb-twiddling. **1964** M. McLuhan *Understanding Media* viii. 78 More aesthetic than thumb-twiddling, less expensive than smoking. **1967** *Electronics* 6 Mar. 129/1 High and low limits can be set separately on the computer by..thumbwheel switches on the front panel during routine testing [etc.]. **1976** *Sci. Amer.* Jan. 120/3 There are now correct..thumbwheels for feeding in ambient air conditions in order to get standard-ized results on the digital display; they affect only the fourth digit and beyond. **1980** *Nature* 1 May p. xii/3 (Advt.), A continuously variable thumbwheel control for a wide range of illumination. **1851** H. Melville *Moby Dick* II. 191 A thumb-worn old thing.

thumb, *v.* Add: **3. b.** = *thumb-read* vb. s.v. Thumb *sb.* 6 e; freq. const. *through*. Also, to turn (pages) with or as with the thumb in glancing through a book, etc.

1930 D. HAMMETT *Maltese Falcon* xvi. 186 He took a battered memorandum-book from a vest-pocket, licked his thumb, thumbed pages, and held the book out open to Spade. **1934** WEBSTER, *thumb*..*v.*,..to run over the pages of, (a book, periodical, newspaper, pamphlet, or the like), as by turning them rapidly with the thumb. **1966** G. GREENE *Comedians* I. v. 140 He sat on the sofa and thumbed through *Paris-Match.* **1966** S. SMITH *Frog Prince* 37, I dare say he had thumbed a book about it. **1976** J. ARCHER *Not Penny More* v. 62 Stephen left his study for the Senior Common Room where he thumbed through the latest copy of *Who's Who* and found the noble lord.

4. c. *to thumb one's nose*: see *NOSE *sb.* 8 f.

5. To seek or get (a ride or lift) in a passing vehicle by signalling with one's thumb the direction in which one hopes to travel (also *fig.*); to signal to (a driver or vehicle) with the thumb. Also *intr.*, to make *one's way* by thumbing lifts, to hitch-hike. orig. *U.S.*

1932 *Sun* (Baltimore) 4 Oct. 15/8 He was 'thumbed' into picking up two lads. **1933** *Ibid.* 26 Aug. 6/7 New England..is filled with young men and young women who are continually thumbing their way from one camp to another. **1934** *Amer. Speech* IX. 111/1 Those not fortunate enough to possess a car of their own stand by the side of the road and attempt to thumb a ride. **1939** N. MONSARRAT *This is Schoolroom* xii. 250, I thumbed my way across England..spending..four-and-sixpence and walking about thirty miles out of the hundred and fifty. **1944** H. NICOLSON *Diary* 1 May (1967) 369 Eventually an American lorry came along. We thumbed them. They stopped, and jumped off and with many jokes mended the tyre for us. **1952** J. CANNAN *Body in Beck* vii. 135 He had been thumbed for a lift by a desperate man. **1958** *Landfall* XII. 32 When a likely lift came by, Pat would..thumb it with a slow impressive sweep of his arm. **1958** *Oxford Mail* 15 Feb. 1/5 Photographed thumbing a lift near Wolverhampton are two..boys..who hitch-hiked to see the Wolves cup-tie with Darlington at Molineux. **1959** *News Chron.* 14 Aug. 7/5 The only Government-sponsored effort has been a plan to 'thumb a lift' in American rockets for British-made instruments. **1960** O. MANNING *Great Fortune* II. 146 He..had been 'thumbing' his way through Galicia when war broke out. **1975** D. NOBBS *Death of R. Perrin* 154 Reggie stood at the entrance to the lay-by and tried to thumb a lift. **1979** *Listener* 1 Mar. 314/2 Like many students..I had thumbed my way through France.

6. *intr.* To gesture with the thumb; *esp.* to signal with the thumb in the hope of getting a lift in a passing vehicle.

1935 G. STEIN *Let.* Dec. in R. L. White *S. Anderson/G. Stein* (1972) 99 Yesterday an American described thumbing on the roads. **1951** E. PAUL *Springtime in Paris* xvi. 309 Gilles thumbed over toward the Abbot. 'His Nibs should have given us the list in advance.' **1955** *Times* 18 Aug. 10/7, I thumbed for four hours without stopping a single vehicle. **1966** R. PRICE *Generous Man* (1967) ii. 142 He turned to Yancey.., thumbing to the house—'Is that all the house old Rooster can afford?' **1976** N. THORNBURG *Cutter & Bone* viii. 191 He was on the freeway entrance ramp, thumbing with his usual touch of calculated restraint.

thumber (þʌ·məɹ). *N. Amer. colloq.* [f. *THUMB *v.* 5, 6 + -ER[1].] One who 'thumbs' a lift, a hitch-hiker.

1935 *Even. Sun* (Baltimore) 8 Feb. 39/8 Chief of Police..has turned 'thumbs down' on the 'thumbers'. **1973** *Daily Colonist* (Victoria, B.C.) 29 June 31/5 For the hitch-hiker, Canada's roads hold little but terror and horrors, say many thumbers from Toronto.

thumbful (þʌ·mful). [f. THUMB *sb.* + -FUL.] As much as a thumb can hold.

1930 E. POUND *XXX Cantos* xxii. 109 He..pulled out his snuff-box, And sniffed up a thumb-full. **1957** [see *RIFFLE *v.* 3 b].

thumb-mark, *v.* (Earlier example.)

1891 J. L. KIPLING *Beast & Man in India* xviii. 400 St. Peter thumb-marked the haddock when he took from its gills the providential tribute money.

thumb-nail. Add: **2.** (Earlier and later examples.)

1852 E. E. HALE in *Sartain's Mag.* Jan. 39 (*heading*) The old and the new, face to face. A thumb-nail sketch. **1911** R. D. SAUNDERS *Col. Todhunter* 125 A full-length 'character-cartoon' of the Colonel surrounded by 'thumb-nail' impressions of his face and bodily pose. **1968** R. GITTINGS *John Keats* xi. 148 He wrote a brilliant thumb-nail sketch of Oxford.

3. *thumb-nail scraper* (Archæol.), a kind of microlith made for scraping.

1937 GARROD & BATE *Stone Age of Mt. Carmel* I. I. iii. 31 *Thumb-nail scraper*..a very well-made minute round scraper. **1977** G. CLARK *World Prehist.* (ed. 3) v. 226 Late Stone Age assemblages including..microliths and thumb-nail scrapers, which in this part of Africa [*sc.* Nigeria] were usually made of quartz.

thumby (þʌ·mi), *a. colloq.* [f. THUMB *sb.* + -Y[1].] **1.** Soiled by thumb-marks.

1900 *Daily News* 11 Jan. 7/2 The report books look as prosaic as any ordinary account books, only very black and 'thumby'.

2. Clumsy, 'all thumbs'. Cf. THUMB *sb.* 5 c.

1909 R. A. WASON *Happy Hawkins* 103 One day we was kiddin' him about bein' so thumby. **1915** *Pearson's Mag.* XXXIX. 28 You have no idea how thumby your fingers are when fixing a bike under shrapnel fire. **1939** X. HERBERT *Capricornia* ix. 122 The box was set down, the stiff buckles of its mildewed straps tackled by a dozen thumby hands. **1974** P. WRIGHT *Lang. Brit. Industry* vi. 59 Their efficiency is affected when..they are known to be ..awkward,..numb-pawed, or thumby.

thump, *sb.* **3.** In Yorkshire (esp. Halifax): a local festival; a feast, wake, etc. **Thump Sunday,** the Sunday of the annual fair or festival week.

1884 *Folk-Lore Jrnl.* II. 25 Last Halifax Thump, a teetotaller..was punished, according to custom, by the company laying him face downwards and beating him on the back of the body with a heated fire-shovel. **1916** J. HARTLEY *Seets i' Yorks. & Lancs.* ii. 19 It'll be five year sin come Halifax thump Sunday. **1930** *Brit. Weekly* 4 Sept. 448/4 A correspondent sends us a description of 'Deanhead Thump Sunday', the..annual musical festival. **1976** H. WILSON *Governance of Britain* ii. 40 A prime minister must, and if he is a northerner usually does, understand the complex of Wakes Weeks and Feast Weeks, to say nothing of Longwood Thump.

thump, *v.* Add: **1.** (Earlier and later examples with a person as obj.)

1746 *Exmoor Scolding* 6 Chell vump tha. **1803** G. COLMAN *John Bull* III. ii. 35 If he don't behave himself, I'll come in and thump him blue. **1848** THACKERAY *Vanity Fair* viii. 67 Don't you remember..how she was always thumping Louisa? **1960** J. RAE *Custard Boys* II. xii. 145 If you interrupt me again, Felix, I'll bloody well thump you. **1978** D. DEVINE *Sunk without Trace* xxi. 194, I saw red. If I didn't get out, I would thump him.

b. Also, with *out*: to produce (a tune, beat, etc.) by thumping.

1929 T. WOLFE *Look Homeward, Angel* xxiii. 315 She..thumped out popular tunes on a battered piano. **1974** C. RYAN *Bridge Too Far* III. i. 134 The bass drummer..thumped out a symbolic beat in Morse code: three dots and a dash—V for victory.

e. To express by thumps.

1928 *Manch. Guardian Weekly* 26 Oct. 335/3 His [*sc.* a dog's] tail..thumped a welcome.

thumper. Add: **1. b.** *Geol.* A device for creating artificial seismic waves in the earth.

1962 *Times* 18 Apr. 4/7 Another device carried is a hydroprod, more familiarly known as a 'thumper'... It transmits a powerful sound impulse which penetrates the sub-surface strata of the sea bed, the return echo recording the depths of the different layers of sea bed material. **1977** *Sci. Amer.* July 65/2 (Advt.), Seismic excitation—a shock wave from an explosive charge or a mechanical 'thumper'—is applied to the earth.

thumpingly, *adv.* Add: **a.** (Later examples); **b.** *colloq.*, very, exceedingly.

1923 *Chambers's Jrnl.* Apr. 211/1 [He] gripped my hand—Shook it thumpingly. **1948** *Manch. Guardian Weekly* 11 Nov. 3 There was a thumpingly false assumption made about the American farmer. **1977** *Times* 17 Feb. 8/5 A book of thumpingly high entertainment value. **1983** *N. & Q.* Feb. 85/2 Even the thumpingly main statement of the last couplet of Spenser's 'tradefull Merchants' sonnet can be regarded as a mere gesture.

thu·mp-up. *slang.* [f. THUMP *v.* + UP *adv.*[1]] = *PUNCH-UP.

1967 H. W. SUTHERLAND *Magnie* vi. 80 There'd be a killing... But need it get as far as that?.. Just a thump up, maybe. **1978** *Maledicta* 1977 I. 130 Teacher: 'What would you have if you had 10 apples, and the boy next to you took 6 apples from you?' Boy: 'A thump-up (fight), Miss.'

thumri (thu·mrī, þ-). [a. Hindi *ṭhumrī*.] A light classical form of North Indian vocal or instrumental music; a piece in this form. Also *attrib.*, designating this style of music.

1834 N. A. WILLARD *Treat. on Mus. Hindoostan* 89 *Thoomree.* This is an impure dialect of the Vrujbhasha. The measure is lively, and so peculiar, that it is not mistaken by one who has heard a few songs of this class. **1914** A. H. F. STRANGWAYS *Music of Hindostan* vi. 165 'Hādi e illah' is a *Thumri* from Benares. **1964** AHMED ALI *Ocean of Night* II. v. 71 The poets were writing songs, sonnets, free verse, and no one cared for the *ghazal, thumri* or *mustezad.* **1972** P. HOLROYDE *Indian Music* iii. 97 Thumri style of singing evolved from the more austere forms of dhrupad and khayal. **1981** LD. HAREWOOD *Tongs & Bones* xvii. 257 The sarangi player..finished with a gentle and rather sentimental thumri.

thunbergia. Add to etym. after mod.L.: (A. J. Retzius 1776, in *Handl. K. Physiografiska Sällsk. Lund* I. 163). Also, a plant of this genus. (Earlier and further examples.)

1797 H. ANDREWS *Botanist's Repository* I. 123 Twining Thunbergia... The Thunbergia, here figured, is a native of the East Indies. **1867** H. KINGSLEY *Silcote of Silcotes* I. xi. 110 Thunbergias, when clumsily gathered, are apt to come up by the root. **1935** H. NICOLSON *Let.* 22 Feb. (1966) 198 On the wall..is a vast *Thunbergia* with blue morning-glory flowers. **1977** M. ALLAN *Darwin & his Flowers* xii. 215 The woody shoots of the wisteria move faster than those of the flexible Morning Glory and thunbergia.

thunder, *sb.* Add: **3. d.** Fig. phr. *to steal* (someone's) *thunder*: to use the ideas, policies, etc., devised by another person, political party, etc., for one's own advantage or to anticipate their use by the originator.

Derived from the utterance of John Dennis (1657–1734), 'Damn them!..they will not let my play run, but they steal my thunder,' on hearing the stage thunder produced by a method designed for his own play of *Appius & Virginia* being used for a performance of *Macbeth.* (Spence quoted in W. S. Walsh *Lit. Curios.* (1893) 1052; cf. Pope's note on *Dunciad* II. 223.)

1900 E. E. PEAKE *Darlingtons* iii. 23 You must all remember that papa had stolen my thunder. **1911** M. BEERBOHM *Zuleika Dobson* ix. 144 'Happy maid!' he murmured. Zuleika replied that he was stealing her thunder: hadn't she envied the girl at his lodgings? **1931** *Time & Tide* 12 Sept. 1049 Sir Oswald Mosley's exploit was to steal a little of the protectionist thunder temporarily abandoned by the Conservatives. **1937** 'G. ORWELL' *Road to Wigan Pier* xii. 222 It is important..to disregard the jealousy of the modern literary gent who hates science because science has stolen literature's thunder. **1973** A. BROINOWSKI *Take One Ambassador* ii. 19 He would have been watching the returns in the Senate elections I guess. This'll steal a bit of their thunder, that's for sure.

4. (Further examples, including earlier and later examples of *in thunder*.)

1826 *Massachusetts Spy* 23 Aug. (Th.), The bull roared like thunder! I split like lightning! **1834** C. A. DAVIS *Lett. J. Downing* xxxiii. 274 He turned..and giv me a look as black as thunder. **1841** H. GREELEY in R. W. Griswold *Passages from Corr.* (1898) 94 Why in thunder did you go off on Saturday without seeing me? **1852** MRS. STOWE *Uncle Tom's Cabin* I. vii. 95 Go to thunder, gal! **1854** M. J. HOLMES *Tempest & Sunshine* xv. 204 Don't none on you tread on my corns for thunder's sake. **1867** H. J. DANIEL *Muse in Motley* 25 He'll screech like thoonder, iss he will. **1876** E. W. HEAP *Diary* 24 Nov. in *Publ. Amer. Dial. Soc.* (1969) LII. 55 Every paper around is giving the road Thunder. **1916** G. B. SHAW *Pygmalion* v. 188 Of course they do. Then what in thunder are we quarrelling about? **1920** E. O'NEILL *Beyond Horizon* I. ii. 48 You kin go to thunder, Jim Mayo! **1927** —— *Marco Millions* II. i. 102 War is a waste of money which eats into the profits of life like thunder! **1940** W. FAULKNER *Hamlet* I. iii. 77 What in thunder are you fellows up to over at Varner's?

5. a. *thunder-burst, -colour, -crackle, -quake, -rain* (later example), *-sound, -throne.* **b.** *thunder-throning* adj. **c.** *thunder-cloven, -scathed* (earlier example) adjs.; *thunder-heavy, -stormy* adjs. **d.** *thunder-browed, -coloured* adjs.; *thunder-purple, -red* adjs.

1913 J. MASEFIELD *Daffodil Fields* 44 Full of wrath and thunder-browed. **1882** *Imperial Dict.*, Thunder-burst. *a* **1910** 'MARK TWAIN' *Autobiogr.* (1924) II. 176, I can remember those awful thunder-bursts and the white glare of the lightning yet. **1939** JOYCE *Finnegans Wake* 362 Thunderburst, ravishment, dissolution and providentiality. **1851** H. MELVILLE *Moby Dick* I. xxvii. 197 The barest..most thunder-cloven old oak. **1873** G. M. HOPKINS *Jrnls. & Papers* (1959) 232 The others [*sc.* pigeons] are dull thundercolour or black-grape-colour. **1907** R. BROOKE *Let.* Sept. (1968) 106 We have been sitting at an evil café sipping thunder-coloured coffee from glasses. **1941** L. MACNEICE *Plant & Phantom* 20 Thunder-crackle and the bounce of hail. **1922** BLUNDEN *Bonadventure* xii. 68 After the storm, the air was thunder-heavy all that day. **1879** G. M. HOPKINS *Poems* (1967) 80 The thunder-purple seabeach plumèd purple-of-thunder. **1940** J. BETJEMAN *Coll. Poems* (1958) 58 Not Satan's thunder-quake Can cause the mighty walls of Heaven to shake. **1926** D. H. LAWRENCE *David* xi. 78 Till they drop in drops of blood, like thunder-rain, and the land is red. **1949** BLUNDEN *After Bombing* 15 And foam, pearl-pink and thunder-red. **1826** J. G. WHITTIER *Writings* (1888) IV. App. 303 Where the thunder-scath'd peaks of Helvetia are frowning. **1886** W. B. YEATS *Mosada* 7 The faint far thunder-sound. **1930** J. DOS PASSOS *42nd Parallel* II. 149 Hot thunderstormy Washington summers. *a* **1974** R. CROSSMAN *Diaries* (1975) I. 197 A fairly restful Easter weekend of mixed, blowy, brilliant, shiny, thunder-stormy weather. **1876** G. M. HOPKINS *Wreck of Deutschland* xxxiv, in *Poems* (1967) 62 Mid-numberèd he [*sc.* Christ] in three of the thunder-throne! *a* **1918** W. OWEN *Poems* (1963) 135 That columnar, thunder-throning cloud.

6. *thunder-ball,* (*b*) (later examples); **thunderboat** *U.S.*, an unlimited hydroplane; **thunder-box** *slang*, a portable commode; by extension, any lavatory; **thunderbug** *dial.*, (*a*) *U.S.*, a horse-fly; (*b*) a midge; **thunder-drum,** (*a*) (later example); **thunder egg** *N. Amer.* and *Austral.*, a geode, esp. of chalcedony; **thunder-flash**: in military use, a harmless, very noisy, form of explosive; a firework imitating this; **thunder-head,** (*b*) *nonce-use*, a large head, as a whale's head; **thunder-mug** *slang* = CHAMBER-POT; **thunder-pumper,** (*a*) (earlier example); **thunder-rod** (earlier example); **thunder run** *Theatr.*, two wooden troughs down which iron balls were rolled to imitate thunder; **thunder-sheet** *Theatr.*, a piece of sheet metal shaken to imitate thunder; **thunder-shower,** now chiefly *U.S.* (later examples); **thunder stick,** a name said to have been given to a rifle or cannon by peoples who did not possess firearms; **thunder-trunk** *Theatr.*, a trunk in which iron balls were rolled to imitate thunder.

1889 W. B. YEATS *Wanderings of Oisin* II. 30 Trembling, on the flags we fall, Fearful of the thunder-ball. **1942** L. BENNETT *Jamaica Dial. Verses* 41 Wen..Him tun roun..Him se de sinting two yeye dem A roll like tunder-ball... It was a rollin' kealf. **1967** *Compton Yearbk.* 153/2 Through the previous 20 years, only three 'thunderboat' drivers had died in races. **1976** *Popular Mechanics* June 61/1 Officially, they are..hydroplane racing boats... To their hundreds of thousands of fans, they are unlimited hydros, thunderboats, gold cuppers, or just unlimiteds—the fastest racing machines afloat. **1939** AUDEN & ISHERWOOD *Journey to War* vii. 182 We should wash the dishes and clean the thunder-boxes. **1952** E. WAUGH *Men at Arms* II. ii. 178 'If you *must* know, it's my thunderbox.'.. He..dragged out the treasure, a brass-bound, oak cube... On the inside of the lid was a plaque bearing the embossed title *Connolly's Chemical Closet.* **1955** N. FITZGERALD *House is Falling* xi. 188 When the plumber called for instructions, Hapleigh chose the ground floor for the new thunder-box. **1980** *Daily Tel.* 18 Oct. 18 Life in India was ..coping with the indignities of the 'thunder box' (a portable earth commode) and searching sponges for stealthy scorpions. **1837** J. L. WILLIAMS *Territory of Florida* 71 Horse Fly.—..Of these there are five kinds.—1st. the large black, called thunder bug, an inch long. **1875** W. D. PARISH *Dict. Sussex Dial.* 66 Those thunder-bugs did kiddle [*sc.* tickle] me so. **1974** P. HAINES *Tea at Gunter's* xx. 214 Outside the air was still heavy; there were thunderbugs everywhere... I felt them settle on my skin, my hair. **1967** *Stage* 2 Mar. 4/2 (Advt.), Thunder drums, bells, chimes, gongs and effects of every description. **1951** W. F. HEALD *Scenic Guide to Oregon* 25 Agate- and opal-filled nodules called 'Thunder Eggs' can be found near Madras. **1962** E. LUCIA *Klondike Kate* ix. 187 She never returned empty-handed, hauling back..petrified woods, agates..thundereggs, [etc.]. **1973** *Sunday Mail Mag.* (Brisbane) 25 Feb. 14/5 The individual bays of the caravan park are marked off by 'thunder eggs' (round stones a foot and more in diameter, many of which contain fossilised fish). **1977** *Trailer Life* July 16/3 'Thunder eggs', the agate-colored nodules familiar to many rockhounds, are a variety of spherulite. **1943** C. C. KNIGHTS *What H.G. needs to know about Explosives* 10 The only 'fireworks' issued to the typical H.G. unit are crackers and thunder-flashes used to give 'an air of verisimilitude to an otherwise bald and unconvincing' exercise. **1959** G. ADAMSON *Let.* 19 Mar. in J. Adamson *Born Free* (1960) 139, I went to visit Elsa... I let off three thunder flashes..and..she suddenly appeared. **1977** 'E. CRISPIN' *Glimpses of Moon* xi. 210 And we did ought to have..used firecrackers and thunderflashes and horns and whistles. **1851** H. MELVILLE *Moby Dick* II. lxxiii. 59 Throw all these thunder-heads overboard, and then you will float light and right. **1890** BARRÈRE & LELAND *Dict. Slang* II. 347/2 *Thunder-mug* (American low), a chamber utensil. **1942** D. GILBERT *Lost Chords* 6 His room furnishings were meager—a rag carpet,..a bowl and pitcher on a washstand whose closet concealed a chamber, or 'thunder mug'. **1966** 'L. LANE' *ABZ of Scouse* II. 108 *Thundermug*, a chamber-pot. **1877** *Scribner's Monthly* July 285/2 The natives call these bitterns by the very appropriate if not euphonious name of 'thunder-pumper'. **1784** G. ADAMS *Ess. Electricity* ix. 154 When lightning strikes a tree..or a thunder-rod, it is not because these objects are high..but because they communicate with..the surface of the ground. **1944** *Archit. Rev.* XCV. 135/2 Archaic devices like the 'thunder run', the 'sloat' system of raising scenery, the 'drum and shaft' method of hanging it, still survive at Bristol. **1976** *Early Music* Oct. 401/1 The thunder simulated at the beginning and end of the Cave scene must be..baroque-artificial—for preference made by cannon-balls in a thunder run. **1913** 'V. D. BROWNE' *Secrets Scene Painting & Stage Effects* 66 Hung from flies. A thunder Sheet. **1939** JOYCE *Finnegans Wake* 503 Raindrum, windmachine, snowbox. But thundersheet? **1967** *Oxf. Compan. Theatre* (ed. 3) 947/2 The noise of thunder is usually produced off-stage by the shaking of a suspended iron sheet known as the Thunder Sheet. **1856** E. B. BROWNING *Aurora Leigh* IV. 174 Softly, as the last repenting drops Of a thunder-shower. **1947** S. BELLOW *Victim* i. 5 A thundershower began when he approached the outside door. **1980** *News & Observer* (Raleigh, N. Carolina) 28 Oct. 2/3 Clouds will prevail across much of North Carolina today, with some showers or thundershowers possible through Wednesday. **1918** E. R. BURROUGHS *Tarzan & Jewels of Opar* (1919) xvii. 157 The ape folk fear the thunder-sticks of the Tarmangani. **1947** I. L. IDRIESS *Isles of Despair* xxxiv. 229 A puff of smoke belched from the brig... They had expected resistance, but had hoped the vessel was too small to carry the 'big thunder sticks'. **1965** *Canad. Geogr. Jrnl.* Apr. 115/1 The white man came to shatter the silence of the wilderness with his thunder stick. **1767** D. GARRICK *Peep Behind Curtain* I. 22 Ladies, you can't possibly have any thunder and lightning this morning; one of the planks of the thunder-trunk started the other night. **1830** G. COLMAN *Random Rec.* I. vii. 229 For then did my Evil Genius enthrone himself upon a thunder-trunk, with a roll of play-bills in his hand.

thunder, *v.* **2.** (Further examples connoting movement.)

1934 J. B. PRIESTLEY *Eng. Journey* i. 4 The children of these fist-shakers now go thundering by in their own huge coaches and loll in velvet as they go. **1946** *R.A.F. Jrnl.* May 169 Lancasters..thundered through the night to pinpoint their objectives. **1951** 'J. WYNDHAM' *Day of Triffids* i. 9 The westbound buses thundered along trying to beat the lights. **1960** C. DAY LEWIS *Buried Day* ii. 38 We thundered down the steep hill into the centre of the town, the squawking hens bouncing up and down on the flat cart, straw and feathers flying.

thunder and lightning. Add: **6.** *Angling.* A variety of artificial fly.

1910 *Encycl. Brit.* II. 26/1 In most fly-books great variety of patterns will be discoverable, while certain old standard favourites such as the Jock Scott, Durham Ranger, Silver Doctor, and Thunder and Lightning will be prominent. **1972** *Country Life* 23 Mar. 697/3 Three large salmon were clearly visible... I changed to a tiny low water Thunder and Lightning.

7. *attrib.* Melodramatic, startling, violent.

1892 'MARK TWAIN' *Amer. Claimant* ix. 71 Take What's-her-name, that plays those sensational thunder-and-lightning parts. **1981** J. WAINWRIGHT *All on Summer's Day* 12 Some thunder-and-lightning speed merchant on the opposing side bowled a yorker and chipped a bone in the batsman's ankle.

thunderation. For *U.S. slang* read *slang* (orig. and chiefly *U.S.*) and add earlier and later examples.

1836 *Crockett's Yaller Flower Almanac* 21, I don't know as I can say he was so all darned thunderation fat. **1939** JOYCE *Finnegans Wake* 245 Bing. Bong. Bangbong. Thunderation! **1949** *Sat. Even. Post* 9 July 86/2 Why in thunderation does he have to play such a hay-in-the-hair?

thunderbolt, *sb.* Add: **2. c.** In *Sport*, a fast, hard-struck shot or stroke.

1959 *Times* 29 May 4/7 [Lawn Tennis] Maloney, with his 'thunderbolts' made no mistake in the next for the match. **1977** *Times* 7 Feb. 7/2 Heighway, at full steam, lashed a thunderbolt past Latchford from the edge of the box.

3. d. (Further examples.)

1884 A. LANG *Custom & Myth* i. 10 Village wisdom determines that the wedge-shaped piece of metal is a 'thunderbolt'. **1949** 'J. NELSON' *Backwoods Teacher* vi. 57 We spoke of lightning and 'thunderbolts'. Of these latter, Fritz Baily said his uncle used to 'gather them up—and we still get lots kickin' 'round the barn.'.. He promised to bring me one. (Next morning he did—a meteorite the size of his fist.)

5. thunderbolt attack, raid, a short-lived but heavy air-raid.

1943 *Hutchinson's Pictorial Hist. War* 25 Nov. 1942–16 Feb. 1943 240/1 At night Lancasters and Halifaxes carry out a large-scale 'thunderbolt' raid on Duesseldorf, dropping several hundred tons of bombs in a 20-minutes attack. **1944** H. HAWTON *Night Bombing* v. 66 There is no necessary connection between concentrated and precision bombing, but it would be quite wrong to think that the 'thunderbolt attack', as it is sometimes called, lacks exactness. The Renault factory was almost completely demolished in an attack of short duration.

thunder-clap. b. (Later example.)

1924 R. CAMPBELL *Flaming Terrapin* iv. 65 And steepled cities stun the hollow sky With thunderclaps of bells as they go by.

thunderer. 2. (Earlier example applied to *The Times.*)

1830 *Morning Herald* 15 Feb. 3/1 Any person or persons..may receive further particulars..by application..at the office of *The Thunderer*, Printing House-square.

thundering, *ppl. a.* (*adv.*) **4. b.** Delete quot. 1839 and add earlier examples.

1809 *Salmagundi* 7 Mar. 95 He..prefers..telling his story among cronies of his own gender..and thundering long stories they are. **1839** *Havana* (N.Y.) *Republican* 25 Dec. (Th.), He is thundering shy of me.

thunk (þʌŋk), *sb.*[1] Joc. var. THINK *sb.*

1922 JOYCE *Ulysses* 503 Have a good old thunk.

thunk (þʌŋk), *sb.*[2] (*int.* or *adv.*) [Onomatopœic.] A sound of an impact, either dull or plangent. Also *int.* or as *adv.*

1952 B. HARWIN *Home is Upriver* xviii. 178 He heard the dull thunk of wood against wood and felt the planking jar over his head. **1958** 'W. HENRY' *Seven Men at Mimbres Springs* vi. 70 Presently the sodden 'thunk!' of an ax blade caving in barrel staves echoed wetly. **1968** W. GARNER *Deep, Deep Freeze* ix. 109 The door said *thunk* in a well-bred whisper. **1970** M. CHISHOLM *McAllister says No* x. 93 The bullet tore through the canvas of the cover and went *thunk* into a barrel. **1971** A. ROSS *Huddersfield Job* 57, I heard the triple *thunk* of the undercarriage locks. **1979** *Herald* (Melbourne) 23 Apr. 2 The familiar 'thunk, zing, ding' of a pinball machine.

thunk (þʌŋk), *v.* [f. *THUNK *sb.*[2]] *intr.* To make a thunk; to fall or land with a thunk.

a **1963** S. PLATH *Johnny Panic & Bible of Dreams* (1977) I. 133 With shovels and picks they crawled through the attic trapdoor and soon great masses of snow were thunking from the roof into the yard. **1972** *Daily Tel.* (Colour Suppl.) 14 Jan. 21/4 A quoit which thunks into the clay just to the left of the hob, leaning towards it, bevel downwards, is called a pot. **1976** *New Yorker* 3 May 44/3 Last night, I slept nine hours, rain thunking on the tent.

thunk (þʌŋk), dial. and joc. pa. t. and pa. pple. of THINK *v.*[2] Cf. *THUNK *sb.*[1]

1876 C. C. ROBINSON *Dial. Mid-Yorks.* p. xlii, Think.. (Thuongk) The last form is less employed participially than in the past, in which tense it is of constant occurrence. **1887** *Lantern* (New Orleans) 15 Oct. 3/2 Who'd a thunk it? **1908** N. DUNCAN *Every Man for Himself* ii. 60 Leastwise, he *thunk* so, admittin' 'twas open t' argument. **1939** JOYCE *Finnegans Wake* 504, I then tuk my takenplace lying down, I thunk I told you. **1967** T. SAVAGE *Power of Dog* xiii. 240 Phil had most excellent use of the hides after all. Who'd a thunk it!

Thurberesque (þɜ·ɹbəre·sk), *a.* [-ESQUE.] Of or pertaining to the American cartoonist and writer James *Thurber* (1894–1961), the characters in his work, or his style of writing or drawing.

1954 *Encounter* June 88/1 The essentially sexual (we might almost say Thurberesque) nature of true comedy, its concern with the war of men and women. **1958** *Times Lit. Suppl.* 31 Oct. 626/4 A very amusing Thurberesque anecdote by Miss Rebecca West about meeting Pirandello, without knowing it, on a wild night out in New York in the 1920s. **1972** *Listener* 6 Apr. 458/1 A Thurberesque doodle. **1980** *Washington Star* 23 Oct. D6 Nabokov is in it as kind of a Thurberesque stage manager character.

thuringer (þiū̇ə·riŋəɹ). Also **thüringer.** [ad. G. *thüringer*, lit. = next.] Summer sausage.

1933 *Sausage Man. & Text Bk.* (Oppenheimer Casing Co.) 70 (*heading*) Thueringer (fresh Summer sausage) 120 lbs. regular beef trimmings, 40 lbs. beef heart, 60 lbs. chuck, 80 lbs. regular pork trimmings, [etc.]. **1938** *Packer's Encycl.* III. vii. 123 Hang thuringer on clean sticks and place in cooler. **1965** [see *summer sausage* s.v. *SUMMER *sb.*[1] 6 a]. **1978** *Chicago* June 241/2 The restaurant serves a variety of sandwiches..thuringer, and delicious homemade soups.

Thuringian (þiuri·ndʒiăn), *a.* and *sb.* [f. the name *Thuringia* (see below) + -AN.] **A.** *adj.* Of or pertaining to (the inhabitants of) Thuringia, a region of central Germany, in mediæval times a principality. **B.** *sb.* A native or inhabitant of Thuringia.

1607 E. TOPSELL *Foure-footed Beasts* 293 The Thuringean horsses are neighbors to Hessis. **1618** SELDEN *Hist. Tithes* vi. 90 As in the examples which wee anon have of the Turingians, and those of Holtz. **1812** C. BUTLER *Hist. Revolutions of Empire of Germany* App. 35 The electorate was successively enjoyed by Frederick the warlike, and Frederick the wise, of the *Thuringian* branch of the Wittikindian stem. **1839** J. F. STANFORD *Jrnl.* 21 Sept. in *Rambles & Researches in Thuringian Saxony* (1842) 2 Old Palace of Friedenstein..commanding the finest view of the Thuringian Forest. **1839** —— *Let.* Aug. in *Ibid.* 158 In physical development the Thuringians..are..fine powerful men. **1881** C. C. HARRISON *Woman's Handiwork* III. 232 Small plates for tea or dessert, in Thuringian ware, imitating old Dresden. **1974** P. GORE-BOOTH *With Great Truth & Respect* 41 The Münchs had living with them..a brother of Frau Münch's, a general who..spoke such total Thuringian dialect that I never learned to understand a word he said. **1975** F. HEER *Charlemagne & his World* ix. 121 Hessians and Franconians were divisions of the Austrasian sub-kingdom... The Thuringians were a distinct group, but fairly well assimilated into France.

Thurstone (þɜ·ɹstən). *Psychol.* The name of the American psychologist, Louis Leon *Thurstone* (1887–1955), used *attrib.* to denote tests or methods devised by him, esp. for the measurement of mental abilities and attitudes, for factor analysis, and the study of personality.

1935 R. R. WILLOUGHBY in C. Murchison *Handbk. Soc. Psychol.* xii. 502 Perhaps the most definite evidence is that secured..with personality inventories of the Thurstone type. **1954** A. ANASTASI *Psychol. Testing* ix. 229 Another abbreviated adaptation..is to be found in the Thurstone Test of Mental Alertness. *Ibid.* xx. 538 The resulting inventory is known as the Thurstone Temperament Schedule. **1958** M. ARGYLE *Relig. Behaviour* viii. 97 A group of eighty-three delinquent girls scored higher on Thurstone scales measuring attitudes towards Sunday observance. **1972** *Jrnl. Soc. Psychol.* Dec. 243 Attitude toward reclaimed water for noningestive, close-contact use was assessed by a Thurstone-type scale.

thus, *adv.* **1. e.** *thus and thus*: delete † *Obs.* and add later examples.

1892 KIPLING & BALESTIER *Naulahka* xviii. 211 Now we are come to our Kingdom, And the State is thus and thus. **1909** H. G. WELLS *Tono-Bungay* II. iv. 200 Nobody, no book, ever came and said to me, thus and thus is the world made and so and so is necessary. *Ibid.* 225 Thus and thus it was the Will in things had its way with me. **1942** R. CHANDLER *Let.* 15 Mar. (1981) 20 The reader expects thus and thus of Chandler because he did it before.

g. *thus and so* = SO-AND-SO *a.*, *adv.* 2. *dial.* and *U.S.*

1824 W. CARR *Craven Dial.* i. 6 Hees lang been vara indifferent, and hees now nobbud thus an seea. **1901** F. E. TAYLOR *Folk-Speech S. Lancashire* s.v. *Thus an'-so*, 'Heawsto bin gerrin' on?' 'Well, nobbo thus an' so.' **1904** *N.Y. Even. Post* 23 Apr., The statement that matters will result thus and so 'if the crops turn out all right'. **1924** R. M. OGDEN tr. *Koffka's Growth of Mind* iii. 100 The present situation appears..not as one that is constituted thus-and-so. **1932** *Atlantic Monthly* Apr. 407/1 We know why we stand thus and so in the sample of conflicting faiths.

thusly, *adv.* (Earlier examples.)

1865 *Harper's Mag.* Dec. 133/2 It happened, as J. Billings would say, 'thusly'. **1876** [see *DINGUS].

thwack, *sb.* Add: Also as *int.*

1908 L. M. MONTGOMERY *Anne of Green Gables* xv. 156 And then—Thwack! Anne had brought her slate down on Gilbert's head. **1976** *National Observer* (U.S.) 14 Aug. 6/3 Thwack! Boston's Jim Rice sends the first pitch sailing over the left-field wall.

thwartly, *adv.* **2.** (Later example.)

1914 HARDY *Satires of Circumstance* 192 Then grinned the Ancient Briton from the tumulus treed with pine: 'So, hearts are thwartly smitted In these days as in mine!'

thwartness. (Later example.)

a **1907** F. THOMPSON *Works* (1913) III. 66 It is full of thwartness and eating and drinking, and selffulness.

thylakoid (þəi·lăkoid) *Bot.* [a. G. *thylakoid* (W. Menke 1961, in *Zeitschr. f. Naturforsch.* B. XVI. 335/1), f. Gr. θυλακοειδ-ής pouch-like, f. θύλακ-ος pouch: see -OID.] Each of the flattened, fluid-filled, membranous sacs inside a chloroplast in which photochemical reactions take place.

1962 *Ann. Rev. Plant Physiol.* XIII. 35 These are designated discs by Gibbs, flattened vesicles by Mühlethaler, and thylakoids by Menke. **1967** [see *GRANUM]. **1974** R. Y. STANIER in Carlile & Skehel *Evolution in Microbial World* 231 The enclosing thylakoid membrane provides an external constraint to prevent disaggregation.

thymallus. Add: = GRAYLING 1 a. (Later example.)

1921 *Chambers's Jrnl.* Jan. 114/2 We have fished that river again, and succeeded in capturing good baskets of thymallus.

thyme, *sb.* **3. thyme-leaved** *a.* (later examples).

1859 D. BUNCE *Trav. Dr. Leichhardt* iv. 27 *Decaspora disticha* and *Thymifolia*, the first two-leaved, and the latter the thyme-leaved *Decaspora.* **1972** Y. LOVELOCK *Veg. Bk.* III. 336 Small or thyme-leaved mint..is only found in Tasmania.

thymectomy, thymectomize. (Later examples.)

1963 *Lancet* 5 Jan. 43/1 In the adult animal, thymectomy is associated with little or no significant depression of the immune response. **1970** *Nature* 26 Sept. 1353/1 This preliminary observation on the long term persistence of T cells has been extended by thymectomizing CBA/Lac mice. **1974** R. M. KIRK et al. *Surgery* ii. 33 Neonatal thymectomy suppresses the immune response in experimental animals. **1976** *Ann. Rev. Microbiol.* XXX. 584 Some adjuvants fail to work in thymectomized mice treated with antilymphocyte (anti-T) serum.

thymidine (þəi·midīn). *Biochem.* [f. THYM(INE + *-IDINE.] A pyrimidine nucleoside, $C_5H_9O_3{\cdot}C_5H_5N_2O_2$, in which the base is thymine and the sugar deoxyribose, and which is obtained by the partial hydrolysis of DNA.

1912 *Jrnl. Biol. Chem.* XII. 414 Barium salts of a hexocytidine diphosphoric acid and a hexo-thymidine diphosphoric acid were obtained. **1931** LEVENE & BASS *Nucleic Acids* 335/1 (Index), Thymidine. *See* Thymine desoxyribonucleoside. **1963** *New Scientist* 21 Mar. 616/2 Before a cell can divide, it has to duplicate its own normal complement of DNA and this it synthesises from many different starting products, one of the most important being thymidine. **1978** *Bull. Amer. Acad. Arts & Sci.* Feb. 17 If, during the culturing process which forms an essential feature of the preparation of cells for examination of their karyotype, tritiated thymidine (a radioactive precursor of DNA) is added, the resultant metaphase chromosomes become radioactively labeled to a degree that is a function of their recent activity in synthesizing DNA.

thymidylic (þəimidi·lik), *a. Biochem.* [f. prec. + -YL + -IC.] *thymidylic acid*: any phosphoric acid ester of thymidine, $C_5H_8O_2{\text -}(PH_2O_4).C_5H_5N_2O_2$, one or other of which is one of the four nucleotides present in most DNA.

1951 *Jrnl. Bacteriol.* LXI. 41 Samples of thymidylic acid, desoxyadenylic acid..and desoxyguanylic acid.. were made available. **1968** A. WHITE et al. *Princ. Biochem.* (ed. 4) ix. 186 Deoxyguanylic, thymidylic, and deoxycytidylic acids esterified at positions 3′ and 5′ are also found in hydrolysates of DNA. **1972** *Sci. Amer.* Dec. 85/1 Until well into the 1930's DNA was generally thought to be merely a tetranucleotide composed of one unit each of adenylic, guanylic, thymidylic and cytidylic acids.

Hence **thymi·dylate,** the anion derived from thymidylic acid by the loss of a hydrogen atom from the phosphate group.

1959 FLAKS & COHEN in *Jrnl. Biol. Chem.* CCXXXIV. 2981/1 We are designating the latter catalytic activity.. as 'thymidylate synthetase'. **1974** J. W. DRAKE in Carlile & Skehel *Evolution in Microbial World* 51 The *td* gene encodes the viral thymidylate synthetase.

thymin. Add: **1.** Now always **thymine.** (Further examples.)

1954 *New Biol.* XVI. 15 In desoxyribose nucleic acid the purine is either adenine or guanine, the pyrimidine either cytosine or thymine. **1976** *Sci. Amer.* Jan. 64/3 The DNA molecule consists of two long chains of nucleotides wound in a double helix... Each nucleotide consists of a deoxyribose sugar, a phosphate group and one of four nitrogenous bases: adenine, guanine, thymine or cytosine.

2. *Biochem.* = *THYMOPOIETIN.

1968 G. GOLDSTEIN in *Lancet* 20 July 122/1 'Thymin'.. seems an appropriate name for this substance, which not only is present in normal thymus but also appears to be secreted normally and to have an effect in physiological concentrations on neuromuscular transmission. **1974** [see *THYMOPOIETIN].

thymocyte (þəi·mosəit). *Histology.* [f. THYM(US + -O + -CYTE.] A lymphocyte-like cell derived from the thymus gland; a T-lymphocyte.

1929 E. S. SCHAFER *Essent. Histol.* (ed. 12) xxii. 265 Besides lymphocyte-like cells (thymocytes) it [*sc.* the thymus] contains peculiar granular cells. **1976** *Nature* 13 May 139/1 Differentiation of T cells is generally depicted as involving transformation of prethymic cells to immunoincompetent thymocytes, which themselves are precursors of functionally mature, immunocompetent T cells.

thymoleptic (þeimole·ptik), *a.* (*sb.*) *Pharm.* [f. Gr. θυμός soul, spirit + λῆψις seizing: see -IC.] (Of or pertaining to) a psychic energizer (see *PSYCHIC *a.* (*sb.*) 1 a).

1959 *New Scientist* 31 Dec. 1351/2 So novel are these [psychotherapeutic] drugs that even the scientific terms used to describe their action—ataractic, thymoleptic—have been coined within the past few years. *Ibid.* 1353/1 The psychotherapeutic drugs fall at present into two classes:..the major tranquillizers..and those that relieve psychotic depression—the psychic energizers or thymoleptics. **1967** W. L. L. REES *Short Textbk. Psychiatry* xxxii. 281 The tricyclic (thymoleptic) group of antidepressants are quite dissimilar in chemical structure to the monoamine oxidase inhibitors. **1971** H. B. MURPHREE in J. R. DiPalma *Drill's Pharmacol. in Med.* (ed. 4) xxiii. 437/2 A term used in Europe, thymoleptic..has not gained currency in North America.

thymoma (þəi·mōᵘmă). *Path.* [ad. G. *thymome* (F. Grandhomme *Ueber Tumoren des Vorderen Mediastinums* (1900) 43): see THYMUS, *-OMA.] A rare, usually benign tumour arising from tissue of the thymus gland and often associated with myasthenia gravis.

1919 J. EWING *Neoplastic Dis.* xlvi. 894 Lymphosarcoma or thymoma is the most frequent form of thymus tumor. **1961** R. D. BAKER *Essent. Path.* xviii. 510 Thymomas may at times be lymphomas, and be associated with myasthenia gravis. **1974** J. D. MAYNARD in R. M. Kirk et al. *Surgery* x. 219 Thymoma is a soft lobulated tumour situated in the thymus or close to it.

thymo-nucleic, *a.* Add: *thymo-nucleic acid* (examples); now identified with DNA.

1904 *Jrnl. Chem. Soc.* LXXXVI. I. 127 (*heading*) Oxidation of thymonucleic acid with calcium permanganate. **1942** *Nature* 17 Jan. 66/1 To these fibres..are attached desoxyribose-, or thymo-, nucleic acid which is responsible for the specific aldehyde reaction given by the chromosomes in Feulgen's test. **1951** *New Biol.* X. 48 Phosphorus-32 uptake is found to be intimately related to the process of synthesis of thymonucleic acid.

thymopoietin (þəimopoi,e·tin). *Biochem.* [f. THYM(US + -O + Gr. ποιητικ-ός (see POIETIC *a.*) + -IN[1].] A polypeptide hormone secreted by the thymus which stimulates the development of thymocytes.

1974 BASCH & GOLDSTEIN in *Proc. Nat. Acad. Sci.* LXXI. 1477/2 It has been brought to our attention that our term thymin for this polypeptide hormone is being confused with the base thymine... We therefore intend to change our terminology in the future and use the term thymopoietin. **1980** H. N. EISEN *Immunology* (ed. 2) xviii. 416/2 Thymopoietin, a small protein (49 amino acid residues), has been sequenced and synthesized. It causes pre-T cells..to become somewhat sensitive to mitogens, but not to become immunologically competent.

thymosin (þəi·mŏsin). *Biochem.* [f. Gr. θύμος THYMUS + -IN[1].] An extract of the thymus gland which has a stimulating effect on the immune system (see quots.).

1966 A. L. GOLDSTEIN et al. in *Proc. Nat. Acad. Sci.* LVI. 1010 We wish to report the preparation..from calf thymic tissue of a product which stimulates incorporation of H³-thymidine into mesenteric lymph node cells. The lymphocytopoietic factor, which we term thymosin, is active..*in vivo*, as well as..*in vitro.* **1980** H. N. EISEN *Immunology* (ed. 2) xviii. 416/2 Thymosin, a mixture of many active low-molecular-weight polypeptides.., probably comes closer than many others to behaving like a natural hormone... It restores immune function to neonatally thymectomized mice.

thymus. Add: **1. c.** *thymus nucleic acid* = *thymonucleic acid* s.v. THYMONUCLEIC *a.* in Dict. and Suppl. Now *Hist.*

1904 *Jrnl. Chem. Soc.* LXXXVI. I. 837 (*heading*) Thymus nucleic acid. **1938** [see *DEOXYRIBONUCLEIC ACID].

3. *Comb.*, as *thymus-dependent, -derived, -independent* adjs.

1963 *Jrnl. Nat. Cancer Inst.* XXXI. 1466 In the thymectomized newborn mouse..re-establishment of immunologic competence..is also known to be thymus-dependent. **1977** *Proc. R. Soc. Med.* LXX. 524/2 This change may be responsible for the diminished response to thymus-dependent antigens that is seen in subjects with chronic liver disease. **1970** *New Scientist* 7 May 271/2 The thymus-derived T cells apparently move through the body more actively than the B cells. **1974** *Ciba Symposium* XX. 114 They suggest that, by one means or another, trypanosome infections break the control link between thymus-dependent lymphocytes (T cells) and thymus-independent lymphocytes (B cells).

thyratron (þəi·rătrŏn). *Electronics.* [f. Gr. θύρα door + *-TRON.] A thermionic valve utilizing an arc discharge in mercury vapour or low-pressure gas and having a heated cathode and at least one grid.

1929 *General Electr. Rev.* Apr. 213/1 The thyratron is an electrically controlled arc rectifier. **1948** *Jrnl. R. Aeronaut. Soc.* LII. 216/1 An automatic speed control has been provided in which the voltage from a tachometer generator driven by the fan shaft is composed with a pre-set standard voltage, and the difference amplified by thyratrons is fed back to an auxiliary field in the generator. **1981** [see *THYRISTOR].

thyristor (þəiri·stəɹ). *Electronics.* [f. *THYR(ATRON + *TRANS)ISTOR.] A three-terminal semiconductor rectifier made up of four layers, *p-n-p-n*, so that when the fourth is positive with respect to the first, a voltage pulse applied to the third layer initiates a flow of current through the device which continues as long as it is greater than some minimum value.

1958 MUELLER & HILIBRAND in *IRE Trans. Electron Devices* V. 2/2 Because of its thyratron-like properties, the device has been named a Thyristor. **1965** *New Scientist* 19 Aug. 446/2 Its transmission system eliminates the commutator by employing thyristors (silicon-controlled rectifiers) as inverters under electronic control. **1981** J. SEYMOUR *Electronic Devices & Components* iv. 185 Industrial process control, the control of electrical machines and high voltage d.c. transmission are among the many areas in which thyristors are applied. They have superseded the thyratron..in all but the highest voltage applications.

thyro-. Add: **2. thy:rocalcitonin** (-kælsitōᵘ·nin [*calcitonin* f. L. *calx, calc(i)-* lime + TON(IC *a.* and *sb.* + -IN[1]], a polypeptide hormone secreted by the thyroid gland which reduces the levels of calcium in the blood; also called *calcitonin*; **thy:roparathyroide·ctomy,** excision of both the thyroid and the parathyroids; so **thy:roparathyroide·ctomize** *v. trans.*, **-e·ctomized** *ppl. a.*; **thy·rotoxico:sis,** a disorder in which there is an excessive amount of thyroid hormones in the blood; **thyrotro·p(h)ic** *a.* (also † **thyreo-**) [*-TROPHIC, *-TROPIC], applied to a hormone secreted by the pituitary which regulates the activity of the thyroid gland; hence **thyrotro·p(h)in,** thyrotropic hormone; *thyrotrop(h)in-releasing factor, hormone*, a hormone secreted by the hypothalamus which stimulates the release of thyrotropin from the pituitary.

1963 P. F. HIRSCH et al. in *Endocrinology* LXXIII. 252/2 Whether or not the thyroid hypocalcemic agent that we have demonstrated is the same as..calcitonin..it is of considerable pharmacological interest. As a tentative name for the substance we propose 'thyrocalcitonin'. **1974** D. & M. WEBSTER *Compar. Vertebr. Morphol.* xiii. 311 The parafollicular cells of the thyroid gland produce quite a different hormone, called calcitonin or thyrocalcitonin. **1956** *Nature* 21 Jan. 138/1 Six animals were adrenalectomized..: the remaining six were thyroparathyroidectomized. **1932** *Amer. Jrnl. Physiol.* C. 262 (*heading*) Heterotopic bone formation in thyroparathyroidectomized dogs. **1976** H. CAMPION et al. in B.E.C. Nordin *Calcium, Phosphate & Mineral Metabolism* xii. 466 Thyroparathyroidectomized (TPTX) rats had their serum phosphate levels manipulated by dietary means. **1920** *Nature* 9 Dec. 488/2 The effect of thyroid-feeding and of thyroparathyroidectomy upon the pituitrin content of the posterior lobe of the pituitary. **1956** *Ibid.* 21 Jan. 138/1 The effect of thyro-parathyroidectomy on the blood changes induced by injected cortisone..has been studied. **1911** STEDMAN *Med. Dict.* 887/1 *Thyrotoxicosis*, exophthalmic goiter. **1912** *Jrnl. Amer. Med. Assoc.* 3 Aug. 328/1 A patient 23 years of age having an adenoma has a definite fixed chance of developing thyrotoxicosis during her thirty-seventh year. **1977** *Lancet* 27 Aug. 438/2 Simpson hypothesised an autoimmune basis for myasthenia gravis on account of its association with disorders thought to have an autoimmune ætiology—such as..thyrotoxicosis. **1930** CREW & WIESNER in *Brit. Med. Jrnl.* 26 Apr. 777/1 (*heading*) On the existence of a fourth hormone, thyreotropic in nature, of the anterior pituitary. **1957** *New Biol.* XXIII. 117 The pituitary gland exerts some control over the thyroid gland, as it secretes a 'thyrotropic' hormone. **1965** LEE & KNOWLES *Animal Hormones* ii. 21 Adrenocorticotrophic hormone..and thyrotrophic hormone..are also secreted by the adenohypophysis. **1944** *Hackh's Chem. Dict.* (ed. 3) 856/2 Thyrotrophin. **1952** *New Biol.* XIII. 65 Activity of the thyroid is maintained by the stimulus of a hormone, thyrotropin. **1959** [see *TRF* s.v. *T 6]. **1968** [see *TRH* s.v. *T 6]. **1976** *Nature* 8 Apr. 480/2 The α subunit of HCG..is nearly identical to the α subunits of thyroid-stimulating hormone (thyrotropin) ..and luteinising hormone. **1980** *Brit. Med. Jrnl.* 29 Mar. 895/1 Two patients..had impaired responses of serum thyroid stimulating hormone to thyrotrophin releasing hormone.

thyroid, *sb.* Add: **2. b. thyroid-stimulating** *a.* = *thyrotrop(h)ic* adj. s.v. *THYRO-.

1941 *Trans. Amer. Assoc. Study Goiter* 159 Attempts have been made to demonstrate the thyroid stimulating hormone in the urine of patients with thyroid disease. **1974** D. & M. WEBSTER *Compar. Vertebr. Morphol.* xiii. 307 The thyroid-stimulating hormone, TSH, is produced by basophilic cells [within the pituitary].

thyroidectomize *v.* (examples); **thyroide·ctomized** *ppl. a.*, deprived of the thyroid; **thyroi·dic** *a.* and *sb.*, (designating) a person with a disordered thyroid gland; **thyroidless** *a.* (examples).

1932 J. S. HUXLEY *Probl. Relative Growth* vi. iv. 183 Groups of these [albino rats] were thyroidectomized. **1974** *Nature* 5 Apr. 525/1 All rats were ovariectomized and thyroidectomized on day 1 of the experiment. **1899** Thyroidectomized [in Dict.]. **1946** *Nature* 19 Oct. 557/2 Another series of experiments was carried out with thyroidectomized mice. **1922** G. B. SHAW in S. & B. Webb *English Local Govt.* VI. p. lxiii, By all means let the endocrinists go on dividing abnormal people, in prison and out, into hyper and sub pituitaries and thyroidics and

adrenals. **1965** M. BRADBURY *Stepping Westward* i. 31 This was James Walker, a stout, slightly thyroidic, very shambling person. **1908** Thyroidless [see *anti-thyroid* s.v. *ANTI-[1] 3 b and c]. **1946** *Nature* 26 Oct. 590/1 During the third month, mortality was as high as 75 per cent in thyroidless animals, 60 per cent in the controls, and only 15 per cent of the hyperthyroid animals.

thyronine (þəiᵊ·rŏnīn). *Chem.* [f. THYRO- + -*n*- + -INE[5].] The amino-acid $HOC_6H_4OC_6H_4CH_2CH(NH_2)COOH$, of which thyroxine can be regarded as a formal derivative (see quots.).

1928 C. R. HARINGTON in *Biochem. Jrnl.* XXII. 1430 In order to lessen the clumsiness of the systematic nomenclature of thyroxine derivatives it is proposed to call the amino-acid, desiodothyroxine, 'thyronine'..so that thyroxine would be '3:5:3':5'-tetraiodothyronine'. **1970** R. W. MCGILVERY *Biochemistry* xxiii. 562 (*caption*) The active thyroid hormone, thyroxine, is in a formal sense the tetraiodo derivative of an amino acid called thyronine.

thyroxine (þəiᵊrǫ·ksīn). *Biochem.* Also **-in.** [f. THYR(O- + OX(Y- + IN(DOLE *sb.*, after the original (erroneous) description of its chemical composition: see quot. 1918.] A hormone secreted by the thyroid gland which increases the metabolic rate and regulates growth and development in animals; tetraiodothyronine, $HO{\cdot}C_6H_2I_2{\cdot}O{\cdot}C_6H_2I_2{\cdot}CH_2CH(NH_2)COOH$.

1918 E. C. KENDALL in *Endocrinology* II. 90 It appeared desirable to emphasize the presence of the oxy-indol nucleus and it appeared equally desirable not to emphasize the presence of iodin. The substance was therefore named 'Thyro-oxy-indol', which has been shortened to 'thyroxin' for every-day reference to the substance. **1926** C. R. HARINGTON in *Biochem. Jrnl.* XX. 294 It is impossible to accept the formula proposed by Kendall, and..the constitution of thyroxine must be regarded as not proven. **1969** *Daily Tel.* 19 Dec. 11/7 Thyroxine, given to thyroid-deficient babies and adults, saves them from cretinism and myxoedema. **1979** ARMS & CAMP *Biology* xxx. 501 If the tail is removed from a tadpole and placed in a bath containing thyroxin, the white blood cells in the tail will digest it.

ti. Add: (Earlier example.)

1832 G. BENNETT in *London Med. Gaz.* 22 Sept. 795/2 *Dracæna indivisa. Ti* of the natives. This species of Dracæna..attains an elevation of ten or twelve feet,.. The leaves form an excellent food as sea stock for cattle, &c.

b. *ti-palm* (earlier example), *-tree* (earlier example). Cf. TEA-TREE 2.

1851 V. LUSH *Jrnl.* 25 Sept. (1971) 86 Planted 14 Ti-palms in various parts of my grounds. **1864** R. ANDERSON *Hawaiian Islands* 134 Then came..an extensive tract covered with the *ti* trees.

ti: see *TE[1], TI.

Tiahuanaco (tī:āwānā·co). *Hist.* The name of a ruined ceremonial site south of Lake Titicaca in Bolivia, used *attrib.* and as *adj.* with reference to a pre-Incan culture, esp. notable for its stonemasonry and distinctive pottery, which flourished in South America in the first millennium A.D. Hence **Ti:ahuana·coid** *a.* [-OID].

1892 C. R. MARKHAM *Hist. Peru* i. 19 The work of the builders of the Tiahuanaco period is met with in other parts of Peru. **1926** *Brief Guide to Peruvian Textiles* (Victoria & Albert Museum, Dept. Textiles) 6 The culture of the highlands has been termed 'Tiahuanaco' after the place..where stands the ruined archway. **1957** *Encycl. Brit.* II. 259 V/2 Nazca ceramics and textiles and Tiahuanaco stone carving. *Ibid.*, concomitant with this change is the Peruvian-wide diffusion of an art style referred to as 'Tiahuanacoid'. **1973** D. MENZEL in D. R. Gross *Peoples & Cultures Native S. Amer.* I. ii. 20 The Inca-period vogue of antiquarianism which revives post-Tiahuanacoid styles first appears at a time just preceding the Inca conquest of the area. **1976** *Times* 16 Nov. 19/2 The Tiahuanaco culture..flourished from about AD 130 to 1170.

Tia Maria (tī·ă mărī·ă). Also **tia maria.** [Sp., lit. 'Aunt Mary'.] The proprietary name of a coffee-flavoured liqueur based on rum, made originally in the West Indies. Also, a drink or glassful of this.

1948 *Trade Marks Jrnl.* 29 Sept. 794/2 *Tia-Maria*... Liqueurs (alcoholic). **1951** E. DAVID *French Country Cooking* 190 A tablespoon of rum liqueur such as Courantin or Tia Maria. **1954** *Official Gaz.* (U.S. Patent Office) 31 Aug. 968 Tia Maria. For liqueurs. **1957** J. FRAME *Owls do Cry* 127 Tim has said something about drinks, a liqueur, benedictine, or tia maria. **1967** C. DRUMMOND *Death at Furlong Post* vi. 80 'Justifies a taxi back..,' said Hart eventually, after a Tia Maria. **1981** 'D. KAVANAGH' *Fiddle City* vii. 133 Soft, package-tour airports where bandits swirl through the green channel in a bustle of tired perms and duty-free Tia Maria.

tiang (ti,æ·ŋ). [Dinka.] A small dark brown antelope belonging to a race of the korrigum, *Damaliscus lunatus*, found in the Sudan and neighbouring parts of Ethiopia.

1894 SCLATER & THOMAS *Bk. Antelopes* I. 63 The Tiang, as the well-known German traveller and naturalist Theodor von Heuglin proposed to call this Antelope, after its native name, is a representative form of the Korrigum in the upper valley of the Nile. **1920** *Blackw. Mag.* Nov. 668/2 Herds of hartebeeste and tiang. **1969** *Times* 30 Jan. (Ethiopia Suppl.) p. iii/3 The south-west corner of Ethiopia is an area inhabited by Sudanese lowland fauna, including..tiang.

‖ **tiare** (ti,ā·re). Also **tiara, tiaré, tiari.** [a. Fr. *tiare* tiara.] In Tahiti, one of several species of *Gardenia* bearing fragrant white flowers. Also *attrib.*

[*a* **1771** S. PARKINSON *Jrnl. Voy. S. Seas* (1773) 37 Plants of Use for Food, Medicine, &c., in Otaheite..E teea-ree Gardenia-florida.] **1888** W. HILLEBRAND *Flora Hawaiian Islands* 171 Cultivated species: G[ardenia] *Tahitensis*—Tiara [etc.]. **1891** D. HORT *Tahiti* ii. 25 The houses are..surrounded by umbrageous trees, and..gardenias—called there *tiari.* **1914** R. BROOKE in *New Numbers* Aug. 110 With the starred *tiare's* white.. Mamua, your lovelier head! **1919** W. S. MAUGHAM *Moon & Sixpence* xlix. 214 Tiaré..the white, scented flower which..will always draw you back to Tahiti in the end. **1931** A. WAUGH *Most Women* iv. 71 On her face was a look of serene contentment, and behind her left ear was the white tiare flower. **1960** H. E. BATES *Aspidistra in Babylon* 222 She was wearing in her hair not the big customary hibiscus flower but a little cluster of *tiare*, not more than six or seven blooms of small wax-white stars. **1980** *Daily Tel.* 26 Jan. 16 The local *tiaré* flower..is similar to a tiny gardenia.

tiarella (tiăre·lă). [mod.L. (Linnæus *Dissertatio Botanica qua Nova Plantarum Genera* (1751) 29), f. L. *tiara* turban + dim. suffix *-ella.*] A small perennial herb of the genus of this name, belonging to the family Saxifragaceæ, native to North America and Asia, and bearing basal, lobed leaves and clusters of small white or reddish flowers. Also *attrib.* Cf. *foam flower* s.v. *FOAM *sb.* 5 b.

1759 P. MILLER *Gardener's Dict.* (ed. 7) s.v. Tiarella, Tiarella with Heart-shaped leaves. This is the Mitella Americana. **1871** *Scribner's Monthly* II. 470 Tiarella leaves just tipped with claret colour. **1887** *Harper's Mag.* July 303/1 The tiarella sent up feathery spikes of white. **1944** T. C. MANSFIELD *Border in Colour* 224 Tiarellas.. are, fortunately, indifferent as to soil. **1976** J. BERRISFORD *Backyards & Tiny Gardens* vi. 47 Solomon's seal, epimediums and tiarella will also do well in shady places.

‖ **tibbin** (ti·bin). Also **tibben, tibn.** [Arab. *tibn.*] Hay or chopped straw.

1900 A. CONAN DOYLE *Green Flag* 271 Each camel provided with his own little heap of tibbin laid in the centre of the tablecloth. **1909** T. E. LAWRENCE *Let.* 29 Aug. (1938) 77 A bed for the night in a threshing floor, on a pile of tibn, chopped straw. **1923** *Blackw. Mag.* Nov. 692/2 Their sister-craft in the Ægean..filling up with stores for the Army—tibben (or Egyptian hay), barley, firewood, eggs,..and pigs. **1958** L. DURRELL *Balthazar* iv. 73 River-craft moved about their task of loading *tibbin* (corn).

Tibet. Add: The spelling *Thibet* is now obs. **Tibet dog, mastiff** = *Tibetan mastiff* s.v. *TIBETAN *a.* 2.

1845 W. YOUATT *Dog* ii. 18 The colour of the Thibet dog is of a deep black slightly clouded on the sides... He has the broad short truncated muzzle of the mastiff. **1884** G. STABLES *Our Friend the Dog* xxv. 245 He was called a Thibet Mastiff.

Tibetan, *a.* Substitute for entry:

Tibetan (tibe·tăn), *sb.* and *a.* Also † **Thibetan. A.** *sb.* A native or inhabitant of Tibet; also, the language of Tibet, a member of the Tibeto-Burmese sub-family of the Sino-Tibetan language group.

1822 tr. *Malte-Brun's Universal Geogr.* I. 571 The stock or family of the languages of Eastern Asia..differs entirely from that of the Indo-Germanic languages. It comprehends the *Thibetan*, the *Chinese*, the *Burman*, [etc.]. **1842** *Penny Cycl.* XXIV. 429/1 The Tibetans belong to the Mongol race. **1891** W. W. ROCKHILL *Land of Lamas* 97 It was with him..that I commenced studying Tibetan. **1962** L. DAVIDSON *Rose of Tibet* ii. 48 Caravan teamsters strolled everywhere; but..he noticed no Tibetans. **1979** A. HENNING tr. *Myrdal's Silk Road* (1980) ix. 71 Large steles..inscribed in Han, Manchu, Oirat, and Tibetan.

B. *adj.* **1.** Of or pertaining to Tibet, its inhabitants, or their language.

1828 *Asiatick Res.* XVI. 410, I have added a few words from the Tibetan vocabularies of the *Asia Polyglotta.* **1888** *Encycl. Brit.* XXIII. 343/1 The centres for *Tibetan* trade. *Ibid.* 843/2 The Tibetan race is not thoroughly homogeneous. **1942** M. CABLE *Gobi Desert* 32 Flowing beards made from the soft white tail of the Tibetan yak. **1960** [see *BHUTANESE *sb.* and *a.*]. **1974** *China Reconstructs* July 35/3 It used to be thought that the Tibetan plateau had no coal... For generations the Tibetan serfs used butter lamps and pine knots for lighting.

2. Special collocations: **Tibetan cherry,** a white-flowered cherry tree, *Prunus serrula*, native to western China; **Tibetan mastiff,** a large black-and-tan dog with a thick coat and drop ears, belonging to the breed of this name; **Tibetan spaniel,** a small white, brown, or black dog with a silky coat of medium length, belonging to the breed of this name; **Tibetan terrier,** a grey, black, cream, or parti-coloured terrier with a thick, shaggy coat, belonging to the breed of this name.

1948 C. INGRAM *Ornamental Cherries* II. 138 In cultivation the Tibetan Cherry tends to lose its squat, compact habit of growth. **1982** *Times* 20 Nov. (Saturday Suppl.) 3/3 A tree which likes to be stroked is the Tibetan cherry. .. The bark has probably the richest shade of all the coloured-bark trees—a striking mahogany is discovered when the outer bark peels away. **1852** T. SMITH *Narr. Five Years' Residence at Nepaul* II. 295 Young Porcupine. Tibetan Mastiff. Common Hare of central region. **1905** P. LANDON *Lhasa* I. xi. 403 The so-called Tibetan mastiff.. is a great shaggy creature, with a very massive head. **1976** T. HEALD *Let Sleeping Dogs Die* ix. 186 A more than generous helping of Tibetan mastiff, so fierce a dog that Aristotle thought it half tiger. **1930** *Observer* 9 Feb. 13/2 The foreign classes..will contain such rarities as Lhasa terriers, Thibetan spaniels, [etc.]. **1970** *Times* 5 Feb. (Pedigree Dog Suppl.) p. ii/2 Recently there was a market tip for Tibetan spaniels, golden-coated, lion-like dogs of pleasing temperament. **1905** P. LANDON *Lhasa* I. xi. 387 The typical Tibetan terrier, a long-coated little fellow with a sharp nose, prick ears, and..black from muzzle to tail. **1976** T. HEALD *Let Sleeping Dogs Die* i. 12 The latest is a Tibetan terrier in Tokyo.

Tibetian (tibe·tiăn, -ʃiăn), *sb.* and *a.* Now *rare.* Also † **Thibetian.** [f. TIBET, THIBET + -IAN.] = *TIBETAN *sb.* and *a.*

1747 *Astley's New Gen. Coll. Voyages* IV. II. iv. 451/2 The Mogul's Empire; called by the Tibetians, Anonkek, or Anonjen. **1790** *Asiatick Res.* II. 32 We know, that rolls of Tibetian writing have been brought even from the borders of the Caspian. **1841** G. BORROW *Zincali* II. III. 108 Many of the principal languages of Asia are..of the great Tartar family, at the head of which there is good reason for placing the Chinese and the Tibetian. **1889** J. J. REIN *Industries Japan* IV. iii. 517 Thibetian cats. **1973** *Times* 12 Apr. 8/6 [The Chogyal's] American wife.. is also said to be a devotee of Tibetian culture.

Tibeto- (tibe·to), combining form of TIBET, = 'pertaining to Tibet and —', as **Tibeto-Burman**: see below as main entry; **Tibeto-Burme·se** *sb.* and *a.* = *TIBETO-BURMAN; **Tibeto-Chine·se** *a.* = *Sino-Tibetan* adj. s.v. *SINO- 2; also as *sb.*; **Tibeto-Himala·yan** *a.* and *sb.*, (*a*) *adj.*, pertaining to Tibet and the Himalayas; (*b*) *sb.*, a branch of the Tibeto-Burmese sub-family of the Sino-Tibetan language group.

1954 PEI & GAYNOR *Dict. Linguistics* 217 Tibeto-Burmese (Tibeto-Burman). **1974** M. PEISSEL *Great Himalayan Passage* xv. 230 Tibetan is the root of the Tibeto-Burmese languages. **1910** *Encycl. Brit.* XIV. 384/1 Of the Tibeto-Chinese family, the Tibeto-Burman sub-family..is spoken from Tibet to Burma. **1960** C. WINICK *Dict. Anthropol.* 537/2 *Tibeto-Chinese*, a language family that is both agglutinative..and isolating... It has two sub-families: Tibeto-Burmese and Siamese-Chinese. **1961** L. F. BROSNAHAN *Sounds of Language* viii. 179 The Tibeto-Chinese languages..have neither palatalisation nor simple accentuation. **1875** *Encycl. Brit.* II. 684/1 Tibeto-Himalayan mountains. **1939** Tibeto-Himalayan [see *TIBETO-BURMAN *a.* and *sb.*].

Tibeto-Bu·rman, *a.* and *sb.* [f. *TIBETO- + *BURMAN *a.* and *sb.*] **A.** *adj.* Pertaining to Tibet and Burma; *spec.* designating or belonging to a group of languages spoken in Asia, belonging to the Sino-Tibetan family, or the peoples speaking any of these languages. **B.** *sb.* The Tibeto-Burman group of languages.

1878 R. N. CUST *Sk. Mod. Lang. E. Indies* 4 The great Tibeto-Burman sea. *Ibid.* 93 The Bhramu speak a purely Tibeto-Burman language. **1880** *Encycl. Brit.* XII. 777/2 The early peoples of India belonged to three great stocks, known as the Tibeto-Burman, the Kolarian, and the Dravidian. **1895** E. W. HOPKINS *Relig. India* (1896) xviii. 525 The native wild tribes of India (excluding the extreme Northern Tibeto-Burman group) fall into two great classes. **1939** L. H. GRAY *Foundations of Lang.* 389 *Tibeto-Burman* is divided into *Tibeto-Himalayan*, consisting of Tibetan..; and of Himalayan, [etc.]. **1976** W. H. CANAWAY *Willow-Pattern War* xv. 156 Thupten's own language was some impenetrable offshoot of Tibeto-Burman. **1982** *Whitaker's Almanack 1983* 803/2 The indigenous inhabitants who entered Burma from the north and east are of similar racial types and speak languages of the Tibeto-Burman, Mon-Khmer and Thai groups.

Tibetology (tibetǫ·lŏdʒi). [f. TIBET + -OLOGY.] The study of Tibetan culture. Also **Tibeto·logist,** one who specializes or is expert in this branch of study.

1964 *Bull. Tibetology* I. 5 Tibetology, that is, study of culture or cultures expressed through the medium of Po Key (Bod Sked = Tibetan language), is not confined to the geographical boundaries of Tibet. *Ibid.* 6 The very first difficulty which a Tibetologist faces is that of non-availability of literary data. **1974** M. PEISSEL *Great Himalayan Passage* iv. 84 Professor Tucci, the noted Italian Tibetologist, wrote of Nepal. **1982** *Bodl. Libr. Record* X. 371 The modern discipline of Tibetology.

tibicinist. For *rare*⁻¹ read *rare* and add earlier example.

1776 J. HAWKINS *Hist. Music* I. p. xlvi, A scene in an ancient comedy, in which a tibicinist is delineated standing on the stage.

tic. 3. For 'see' read '=' and add examples.

1896 [see TICK *sb.*⁴ 2]. **1927** F. M. FORD *Let.* 28 Mar. (1965) 172, I have such a tic against writing letters that I cannot do it. **1960** *Twentieth Cent.* Apr. 361 This is an irritating tic of the British Left, this substitution of moral gestures for practical policies. **1978** C. P. SNOW *Realists* vi. 176 He had the tic, common to many writers, of insisting that the table be kept pernicketily tidy.

ticarcillin (tikāsi·lin). *Pharm.* [f. *ti-* (of unknown origin) + *CAR(BOXY- + *-cillin*, after *PENICILLIN.] A semisynthetic penicillin antibiotic, (6R)-6-[2-carboxy-2-(3-thienyl)-acetamido]penicillanic acid, usu. administered as the disodium salt, $C_{15}H_{14}N_2Na_2O_6S_2$.

1972 in *Approved Names* (Brit. Pharmacopoeia Comm.) Suppl. v. 4. **1974** *Jrnl. Clin. Pharmacol.* XIV. 172/1 Ticarcillin..is a new semisynthetic penicillin, with a wide spectrum of activity against gram-positive and gram-negative bacteria. **1980** *Brit. Med. Jrnl.* 24 May 1220/2 Bacteriological monitoring of seriously ill patients may be rewarding if combined with early and energetic use of suitable antibiotics such as tobramycin, plus ticarcillin.., when septicaemia is clinically suspected.

tice, *sb.* **a.** (Earlier examples in *Cricket.*)

1843 'WYKEHAMIST' *Pract. Hints Cricket* 15 The 'Tice', which may be described as a short pitched full-pitch..is one of the most destructive Balls that can be bowled. **1869** L. M. ALCOTT *Little Women* II. ix. 128 The phrases, 'caught off a tice', 'stumped off his ground',..were as intelligible to her as Sanscrit.

tich (titʃ). *slang.* Also **Tich, titch.** The stage name Little *Tich* of the dwarfish music-hall comedian Harry Relph (1868–1928), who was given the nickname as a child because of a resemblance to the Tichborne claimant (see below), used as a name for any small person. Cf. *TITCHY *a.*

Arthur Orton (1834–98), the Tichborne claimant, claimed in 1866 to be Roger Charles Tichborne (1829–54), the heir to an English baronetcy, who was lost at sea. Orton was finally discredited and imprisoned in 1874.

1934 W. POLLOCK *Cream of Cricket* xi. 66 'Tich' Freeman would toss up his slows with that maddening lure to come and commit cricket suicide which so many misguided batsmen cannot resist. **1959** I. & P. OPIE *Lore & Lang. Schoolch.* ix. 169 A chap who has got duck's disease is most often labelled 'Tich' in a friendly manner, or 'squirt'. **1960** D. ABSE in J. C. TREWIN *Plays of Year 1960–61* XXIII. 147, I vowed to work harder. To make more money. For you and the titch. **1963** *Daily Mail* 7 Mar. 12/5 (*caption*) This will prove we're not 'tiches' in spare communications.

Ticinese (titʃinī·z), *sb.* and *a.* Pl. Ticinese, ‖ -esi. [a. It. *Ticinese*: see -ESE.] **A.** *sb. collect.* The natives or inhabitants of Ticino, an Italian-speaking canton in southern Switzerland. **B.** *adj.* Of or pertaining to Ticino or its inhabitants.

1961 'W. HAGGARD' *Arena* x. 84 Walter didn't much care for Ticinesi, but it was certain they weren't Swiss-Germans. **1964** *Language* XL. 287 The article gives a detailed description of Ticinese customs. **1974** *Sat. Rev. World* (U.S.) 19 Oct. 42/2 The Ticinese can manage most of the other tongues in current employ in Switzerland. **1978** [see *SCHWEIZERDEUTSCH, SCHWYZERTÜTSCH].

tick, *sb.*¹ **1. b.** Delete † and add later *colloq.* examples. Freq. as *little tick*.

1909 WODEHOUSE *Mike* xli. 231 Can't you see that..we've got a chance of getting a jolly good bit of our own back against those Downing's ticks? **1928** J. VAN DRUTEN *Young Woodley* I. 17 Milner: 'Cope, your presence is urgently desired... Scrimshanking, the little tic.' **1952** E. O'NEILL *Moon for Misbegotten* I. 17 Everyone says you're a wicked old tick, as crooked as a corkscrew. **1973** R. FULFORD in D. Pryce-Jones *Evelyn Waugh & his World* ii. 17 How often in those early days did I hear those ominous words 'that awful little tick Waugh'.

c. Phr. *as full (or tight) as a tick*: full to repletion, esp. with alcoholic drink.

1678 J. RAY *Coll. Eng. Proverbs* (ed. 2) 284 As full as a pipers bag; as a tick. **1822** *Yankee Phrases* in *New Jersey Alm. 1823* (Elizabethtown, N.J.) 31 Though of love I'm as full as a tick. **1911** L. STONE *Jonah* 226 'Ard luck, to grudge a man a pint, with 'is own missis inside there gittin' as full as a tick. **1933** M. LOWRY *Ultramarine* IV. 177 He was tight as a tick so couldn't tell the difference. **1952** E. O'NEILL *Moon for Misbegotten* IV. 168 You must have seen how blotto I was.' 'I did. You were as full as a tick.' **1980** A. PRICE *Soldier no More* v. 59 He was drunk as a lord..tight as a tick.

3. *tick-infested adj.*; **tick-bean** (earlier and later examples); **tick-bird** (earlier example); **tick-borne** *a.*, transmitted by ticks; *tick-borne fever*, a mild, transient, rickettsial, febrile disease of sheep, cattle, and goats; **tick paralysis**, paralysis caused by neurotoxin in the saliva of certain biting ticks; **tick pyæmia**, a type of blood-poisoning in sheep, esp. lambs, caused by *Staphylococcus aureus* and leading to lameness or death; **tick-trefoil** (earlier example); **tick typhus** = *Rocky Mountain fever* s.v. *ROCKY *a.*² 1 b.

1744 W. ELLIS *Mod. Husbandman* Feb. ii. 17 Chiltern Farmers can get a full Crop of Horse or Tick-beans. **1969** *Oxf. Bk. Food Plants* 40 Prehistoric specimens are all small-seeded forms—even smaller than the 'Horse bean' or 'Tick bean' varieties grown as food for livestock in modern times. **1850** T. E. POOLE *Life, Scenery, & Customs in Sierra Leone & Gambia* II. xiv. 220 Perched upon these animals [*sc.* cattle], which did not seem in the least to mind them, were a species of birds called 'Tick-birds', from the circumstances of their feeding upon certain insects of that name, which they find in great numbers on these beasts. **1921** *Indian Med. Gaz.* LVI. 368/1 It [*sc.* Brill's disease] has no epidemiological relationship whatever with the Rocky Mountain fever which is tick-borne. **1932** W. S. GORDON et al. in *Jrnl. Compar. Path. & Therapeutics* XLV. 121 A disease characterized by a low mortality, with an incubation period of about four days, followed by a sharp rise in temperature and a period of fever.. We have..shown..that this reaction is a 'tick-borne fever'. *Ibid.* 301 This condition we have..named 'tick-borne fever'. **1970** W. H. PARKER *Health & Dis. in Farm Animals* xvii. 241 In areas of late lambing, abortions in ewes are sometimes attributable to tick-borne fever. **1973** J. M. KELLY *Man Against Ticks* i. 42 Dutton died at Kasongo in February 1905 of tick-borne relapsing fever. **1932** C. FULLER *Louis Trigardt's Trek* 128 Our small stock were so tick-infested that we despaired of rearing them. **1960** *Times* 1 Oct. 7/7 Tick-infested hinterland. **1914** E. NUTTALL in *Parasitology* VII. 307 (*heading*) Tick paralysis. **1962** GORDON & LAVOIPIERRE *Entomol. for Students of Med.* xlii. 260 In Australia, North America, South Africa and South Eastern Europe several species of ticks..produce a type of ascending motor paralysis known as 'tick paralysis'. **1946** *Nature* 27 July 132/2 The sheep tick, *Ixodes ricinus*, is involved in the transmission of..tick-pyæmia. **1970** W. H. PARKER *Health & Dis. in Farm Animals* xviii. 241 Tick pyæmia is caused by the ubiquitous bacterium *Staphylococcus*. **1853** THOREAU *Jrnl.* 31 July in *Writings* (1906) XI. 350 *Desmodium nudiflorum*, naked-flowered tick trefoil, some already with loments round-angled. **1960** *Internal. Med. Gaz.* LVI. 370/2 (*heading*) Tick typhus. **1973** J. D. HARLAN *Wild Ozarks* vi. 55 Tick typhus, or spotted fever..is..in that area.

tick, *sb.*³ **b.** (Further examples.)

1853 *Heal & Son Catal.: Bedsteads* 3 Best Grey Goose.. in Fine Linen Ticks. **1968** L. MORRISON *Heart of Grass* 2/1 [?] She made her bed less successfully, for she had never learned the art of wrestling with a feather tick. **1981** *People* 3 June 6/8 [*Advt.*] Pillow ticks black white striped. **1980** J. C. OATES *Bellefleur* (1981) IV. 319 A plain four-poster with white ruffled skirts, a cornhusk tick and feather bed on top.

tick, *sb.*⁴ Add: **3. c.** A ticked item on a list, esp. a list of birds to be observed. Also *Comb.*, as *tick-hunter, -hunting*.

1975 W. CONDRY *Pathway to Wild* vi. 93 R. S. Thomas..saw it [*sc.* foreign travel] as an opportunity of adding to his life-list of birds. 'Tick-hunting' is what bird-watchers call it. You carry a card with a list of all the birds on it and you..tick them off as you spot them. *Ibid.* 94 We saw a signpost on our right, 'La Route des Lacs', and what tick-hunter short of waterbirds could resist a lake-side road? **1981** *Birds* Autumn 60/3 Their [*sc.* the Country Life team's] ticks..included glossy ibis, spoonbill, Savi's warbler, [etc.].

4. (Further examples.) *on* or *to the tick*, exactly at the appointed time, punctually; cf. *on the dot* s.v. *DOT *sb.*¹ 4 d.

1900 [see *CONNIVER]. **1907** To the tick [in Dict.]. **1913** A. BENNETT *Regent* ix. 262 If you don't clear out on the tick I'll chuck this cop and saucer down into the stalls. **1927** *Daily Express* 6 July 1/3, I am always here on the tick myself, and I do not see why jurors should not do the same. **1963** T. PARKER *Unknown Citizen* i. 38 Won't be a tick, don't go away. **1973** J. WILSON *Hide & Seek* i. 18 Just wait till I get these grotty old school things off. Won't be a tick. **1973** P. WHITE *Eye of Storm* ii. 83 Shan't be a couple of ticks, love. **1983** E. BAVERLY *In Good Faith* vii. 194 Just wait a tick while I tell George where we'll be, and then we can go down together.

tick, *sb.*⁵ Delete 'Rarely tic.'

tick, *v.*¹ Add: **1. c.** *trans.* = TIC *v.* 2.

1913 A. G. CATON *Romance of Wernal* viii. 69 One out of the corner..would ticket over it [?]..Then a chase over the country began between these two. **1969** I. & P. OPIE *Children's Games* ii. 64 In the west midlands they 'tick' him, and he is then said to have been 'took', 'tocked', or sometimes 'tucked'. **1983** T. THOMPSON *Edwardian Childhoods* iii. 83 We used to play..tick... You had..to tick somebody.

2. d. *intr.* with *over*. Of an internal combustion engine: to run or work with the propeller or gears disengaged, or at a low rate of revolutions; to idle. Also *transf.* and *fig.*, to function (merely); to work or operate continuously, esp. at a low capacity. Chiefly in *pres. pple.*

1916 H. BARBER *Aeroplane Speaks* 50 The engine is awake again and slowly ticking over. **1934** *Humorist* 28 July 38/2 How shall I know when the influence is ticking over? **1950** *Sport 7–13* Apr. 22/4 It is the money in the pockets of the man-in-the-street which keeps sport, the cinemas and the B.B.C. ticking over. **1962** A. BEVAN in *Planet of Fear* iv. 70 Old out-of-date steel plants were kept ticking over by means of bank overdrafts. **1963** C. A. LINDBERGH *Spirit of St. Louis* i. 9, I..pull back my throttle until the propeller is just ticking over. **1966** M. CRONIN *Begin with Gun* vii. 81 Just the way you said, chief. All ticking over nicely. **1977** F. WEBB *Go for Out* v. 97 The car engine fired. He let it tick-over for a moment, then switched it off.

e. *intr.* Of a taximeter (cab): to make a ticking sound while recording the fare due for a period of hire, esp. while waiting; also quasi-*trans.* with complement. With *up*, to record an increasing fare.

1926 W. S. MAUGHAM *Constant Wife* III. 208, I don't want to hurry you, but the taxi is just ticking its head off. **1930** E. P. OPPENHEIM *Million Pound Deposit* xi. 104 'Got a car?' she enquired. 'No, a taxi, ticking up like blazes.' **1938** E. BOWEN *Death of Heart* III. vi. 438 Taxi ticked outside. **1940** DYLAN THOMAS *Portrait of Artist as Young Dog* 155 The taxi was ticking away, and that worried Beatrice and Betti, and at last the sisters and the cousin and Mary drove together to the church. **1954** T. RATTIGAN *Sleeping Prince* I. i. 44 Mary.. The taxi isn't ticking up, is it? *Regent.* No. They will tell us when it arrives. **1966** A. L. COBURN *Autobiogr.* vi. 72 He..whisked away in the cab which he had kept ticking at the door. **1979** J. GRIMOND *Memoirs* vi. 93 The General was alarmed to find a taxi waiting, the clock on it ticking up from £15.8.6 to £15.8.9.

f. *intr. fig.* To work, function, operate; *what makes* (someone) *tick*, what motivates (a person). *colloq.*

1931 E. F. BENSON *Mapp & Lucia* i. 26, I want to get round up again and find out what made to tick. **1947** AUDEN *Age of Anxiety* (1948) ii. 43 They watch others with a covert but passionate curiosity. What makes them tick? **1957** *Listener* 3 Oct. 541/1 Television could show the minds ticking; no need here for those stage directions. **1964** H. E. B. JOHNSON *Whole Human Diary* 6 Jan. (1970) 31 Then came the big event of the day—the White House staff reception... We would be meeting..everybody who makes the house tick. **1971** A. PRICE *Alamut Ambush* iii. 151, I still don't quite know what makes Razzak tick. You have going to find out about. **1975** *Nature* 24 Apr. 659/2 The first step to correct this source of insecurity and fear is to learn what makes it 'tick'.

g. *intr.* with *by* or *away*: (of time, events, etc.) to pass, come to an end. Cf. sense 2 b in Dict.

1932 C. ODETS *Golden Boy* 42 You don't know what it means to sit around here and watch the months go ticking by! **1974** *Publishers Weekly* 30 Sept. 35 [*Advt.*], Their father, his own life ticking away after a freak accident, must prepare his children for the pending battle ahead. **1981** G. BOYCOTT *In Fast Lane* xii. 92 A statement was expected by the hour but each hour ticked away without any news.

3. a. (Earlier and later *fig.* examples.)

1864 DICKENS *Hard Times* I. xiv. 106 He was not sure that if he had been required..to tick her off into columns in a parliamentary return, he would have quite known how to divide her. **1932** A. HUXLEY *Let.* 14 Oct. (1969) 355 All that stupid unreal rhetoric of tension... It's manifestations..have to be ticked off and different manifestations.., by Tolstoy. **1966** N. MAILER *Cannibals & Christians* (1967) 31 God you tick off a few of the major issues you think will be in the campaign against the Democrats?

c. To reprimand or scold. Cf. *to tell off* s.v. *TELL *v.* 23 d. *colloq.* (orig. *Mil. slang*).

1915 W. OWEN *Let.* 1 Nov. (1967) 365 He has been 'ticked-off' four or five times for it; but is not yet shot at dawn. **1926** [see *ESSENCE *sb.* 8]. **1937** *Listener* 29 Aug. 227/1 'Ticked off' by one of the boys for leaving his car unlocked and complete with ignition key. **1978** K. AMIS *Jake's Thing* xvii. 182 He'd ticked Ed off without being told to.

d. To annoy, anger; to dispirit. Cf. *TICKED *ppl. a.* c. *U.S. slang.*

1975 *Washington Post* 19 Feb. c 12/7 We got hit something! Hence: 'It really ticked me off'—That's what ticks me off. **1979** R. L. SIMON *Peking Duck* xvi. 117 Shit, it ticks me off? I spent all the money on this tour and look what happens.

tick, *v.*² Add: **1. b.** (Later Austral. and N.Z. examples.) Also const. *up*.

1926 'MIXER' *Transport Workers' Song Bk.* 42 You've never 'ticked' a penny Whilst you worked. **1947** M. MORRIS in 'B. James' *Austral. Short Stories* (1963) 355 Best be off soon. No use ticking things up. **1966** 'J. HACKSTON' *Father clears Out* 114 Going on the slate and other payments.

ticked, *ppl. a.* Add: **c.** *ticked off*: angry, annoyed, 'fed up'. Cf. *TICK *v.*¹ 3 d. *U.S. slang.*

1959 *a.* *mer.* *Speech* XXXIV. 156 When one is angry, he's *ticked* or *tick off*. **1973** 'J. ROSS' *Don't Lie to Me* (1974) v. 54 Now you can see why Grazko is so ticked off. **1977** C. MCFADDEN *Serial* (1978) xxxii. 72/1 Joan was beginning to get ticked off.

ticker³. Add: **1. a.** (Earlier example, in sense 'a watch'.)

1821 P. EGAN *Boxiana* III. 622 To nail the ticker..or to mill the cly.

c. *slang* (orig. *U.S.*). The heart; also *U.S.* and *Austral.*, courage, spirit, 'guts'. Cf. HEART *sb.* 11.

1930 J. TAIT *Big Howe* 7 Because the heart is a 'ticker'. **1935** D. RUNYON *Money from Home* 87, I never see a guy with a better ticker than Sherman. **1950** *Chambers's Jrnl.* Mar. 144/1 Then I heard my mother say it had met my eyes made the old ticker miss more beats than it had done when Martin clamped his gun on the back of my neck. **1979** *Sunday Sun* (Brisbane) 4 Nov. 54/1 The lady has ticker.. She didn't opt for the soft life. **1980** J. CARTWRIGHT *Horse of Darius* viii. 106 Put something at the bottom about your heart. Say, 'The ticker seems to be a little dodgy at the moment.'

2. Someone who ticks off items in a list, etc.; *spec.* = *TWITCHER 4.

1980 *Guardian* 25 June 12/5 'Twitchers' or 'tickers'—

the serious ornithologists' somewhat disparaging term for those bird watchers whose main interest in their hobby is adding new species to their lists. **1982** *Birds* Spring 70/2 Bird tickers contribute little to the well being of the environment and often do little but disturb it.

3. Special Comb.: **ticker tape**, the paper strip on which telegraphic messages are recorded in a tape-machine; this or similar paper material thrown from windows as a form of greeting for a celebrity; also *attrib.*

1902 H. L. WILSON *Spenders* 407 For two days he clung to the ticker tape as to a life line. **1957** *Listener* 10 Oct. 556/1 A traditional ticker-tape reception is to be accorded. **1972** D. E. WESTLAKE *Cops & Robbers* (1973) x. 135 The Wall Street ticker-tape parade is a tradition. **1976** H. WILSON *Governance of Britain* iii. 51 On the afternoon of the second day, the ticker-tape carried a story that President Truman had said that General MacArthur, supreme commnder in Korea, had the authority to use the nuclear weapon there, without reference to the President. **1980** L. ST. CLAIR *Obsessions* ii. 48 Tomorrow's ticker-tape welcome to Commander Richard E. Byrd.

ticket, *sb.*¹ Add: **2. b.** An official documentary notification of an offence, esp. in connection with traffic regulations. Cf. *parking ticket* s.v. *PARKING *vbl. sb.* 3 b. orig. *U.S.*

1930 *Outlook* 12 Feb. 249/1 He wrote the young professor a ticket for speeding. **1935** J. O'HARA *Appointment in Samarra* vii. 201 It was two blocks from the hotel, and he might get a ticket for parking, but if he couldn't get the ticket fixed it was worth the two-dollar fine to have things straightened out with Harry. **1956** S. BELLOW *Seize the Day* (1957) ii. 46 Some fool puts advertising leaflets under your windshield wiper and you have heard failure a block away because you think you've got a ticket. **1964** M. BANTON *Policeman in Community* iii. 62 A driver who made an illegal turn against a red light was 'given a ticket' (i.e. a citation was issued against him). **1981** C. DEXTER *Dead of Jericho* xxxii. 176 Cheque for £6 being the penalty fixed for the traffic offence detailed on the ticket.

c. *big* (or *large*) *ticket item*, something expensive. Cf. *price ticket* s.v. *PRICE *sb.* 14. *N. Amer. colloq.*

1970 *Globe & Mail* (Toronto) 25 Sept. B2/2 Buying plans for big ticket items are up since the previous survey in May and June. **1972** *Mod. Law Rev.* XXXV. 20 Legal aid does not seem to have made much difference, except with regard to large ticket items in middle class communities. **1975** *Washington Post* 28 Jan. A19/3 His proposed tax rebate obviously is designed to stimulate consumption of what is known in this week's argot as 'big ticket items' like cars.

4. b. *spec.* an airman's or seaman's certificate of qualification (further examples).

1907 M. ROBERTS *Flying Cloud* 7 Seventeen years before he got his 'ticket', his second greaser's, second mate's ticket, he served in the foc'sle before the mast. **1910** *Flight* 26 Nov. 970/1 He did rolling practice in the morning, straight flights before luncheon, circuits in the afternoon, and qualified for his 'ticket' before dark. **1947** M. LOWRY *Under Volcano* iv. 111 If all goes well I'll be sailing from Vera Cruz in about a week. As quartermaster, you knew I had an A.B.'s ticket, didn't you? **1977** 'E. CRISPIN' *Glimpses of Moon* xiii. 268 George..had eventually got his Mate's ticket.

† **c.** A certificate given to children at Sunday school recording their progress in religious instruction, esp. their readiness for confirmation. *Obs.*

1838 J. ROMILLY *Diary* 18 July (1967) 153 George went..to St Mary's vestry to be examd by Mr Carus:—he only asked him..'the meaning of "Sacrament"', & gave him his ticket. **1879** C. M. YONGE *Burnt Out* i. 1 Mother! mother! where's my ticket bag? Oh! my tickets, my tickets and my Bible and all my prize books!

5. a. *meal ticket*: see *MEAL *sb.*² 4.

c. *to have tickets on* (a person or thing), to have a strong liking for; esp. *to have tickets on oneself* and varr., to be vain, to be conceited. *Austral. slang.*

1908 W. H. KOEBEL *Anchorage* viii. 140, I don't know whether she's got any tickets on me. **1938** 'R. HYDE' *Nor Years Condemn* ix. 179 You must have tickets on her, Starkie. **1941** K. TENNANT *Battlers* 20 'Arr,' the busker said disgustedly, 'you've got tickets all over yourself.' **1951** CUSACK & JAMES *Come in Spinner* iv. 32 If people have got any tickets on themselves, Blue don't get nowhere with them. **1970** J. HIBBERD *Plays* 227 You're the bastard that's always been smug and had tickets on himself.

d. *to write one's own ticket*: to be able to stipulate one's own conditions, to be in an advantageous position. *colloq.*

1928 WODEHOUSE *Money for Nothing* v. 94 'But Oil's the stuff, and if you want to part with any of that Silver River of yours, Tom,' he said, 'pass it across this desk and write your own ticket.' **1961** C. COCKBURN *View from West* vii. 75 A prelate in the Archbishop's position can..write his own ticket as to what is in the mind of God. **1981** A. PRICE *Soldier no More* xvii. 246 He could make his own terms, and write his own ticket.

e. A (counterfeit) pass or passport. *slang.*

1969 R. AIRTH *Snatch!* ii. 23 A small but select stock of tickets—Ziggy sold only the best, no London-issued Lithuanians for him. **1973** G. M. FRASER *Flashman at Charge* 164 Russia—where everyone has to show his damned ticket every few miles.

f. A piece of paper impregnated with lysergic acid diethylamide (see quot. 1969). *slang* (chiefly *Austral.*).

1969 *Pix* 19 Apr. 11/1 It [*sc.* LSD] is sold usually in absorbent paper in a portion of 120 micrograms known as a ticket. When you take a ticket you are on a trip. 70/2 Bird tickers [?]

6. a. Also, any certificate of discharge from service, prison, etc.; freq. in phr. *to work one's ticket*, to obtain (by scheming) one's discharge.

1869 *Temple Bar* XXV. 217 'Coiners'..as a rule returned to their profession as soon as they got their 'ticket'. Prison is..a great punishment to such men. **1899** H. WYNDHAM *Queen's Service* xxiii. 231 It is a comparatively easy matter for a discontented man to 'work his ticket'. **1958** M. ALLINGHAM *Tiger in Smoke* iv. 77 He ..attempted to work his ticket to one of these new-style open prisons. **1960** W. SMITH *Gold Mine* xxv. 56 My boss boy has worked his ticket.. Can you see that I get a good man to replace him?

b. (Earlier example.)

1840 in *Occasional Papers Univ. Sydney Lang. Res. Centre* (1981) No. 19. 3, I have this day given the prisoner named in the margin a pass to proceed to Bathurst as he wished to have his Ticket issued for that District.

7. b. = *pawn-ticket* s.v. PAWN *sb.*² 4.

1835 DICKENS *Sk. Boz* (1836) 1st Ser. II. 152 You leave your ticket here, will you, with Jenny and your wife for them two places. **1863** E. BARLEE in N. Longmate *Hungry Mills* (1978) viii. 112 [He] owned straight home, made up the fire and burnt every blessed ticket. **1899** KIPLING *Stalky & Co.* 45 Why, last month you and Beetle sold mine [*sc.* a watch]! 'Never got a ticket'.

8. For 'In U.S. politics' read 'In politics (orig. U.S.)' and add further examples. Also, the subject or theme of an election campaign; the principles of a political party as presented for an election.

1863 *Clare Jrnl.* 28 May 2/4 We venture to tell Mr Vereker and Captain Knox that they need not..attempt to go into Parliament on the Conservative 'ticket'. **1899** G. B. SHAW *Let.* 30 Dec. (1972) II. 127 Suppose we do run a ticket, how is it to be done? The [Fabian] Society would not vote a ticket except as between two rival tails to the Exec. **1927** A. HUXLEY *Let.* 23 Mar. (1969) 286 How are you going to make a strong working government from a body of people elected on a great variety of different tickets? **1962** *Listener* 22 Mar. 505/2 Lloyd George had actually fought the election of 1929 on the ticket 'We can conquer unemployment'. **1968** *Globe & Mail* (Toronto) 3 Feb. 3/8 Mr Woodcock..is running on a youth ticket. **1974** *Argus* (Cape Town) 2 Aug. 8/7 In 1966, the then Mayor..had to contest an election... He won by topping the poll on a ticket of 'Vote for me'. **1977** *Grimsby Even. Tel.* 14 May 7/4 I did not ask Mr. Mungeridge for permission to put his name on the ticket. **1979** H. KISSINGER *White House Years* ii. 24 Dwight Eisenhower had been elected on the Republican ticket, but he owed little to the Republican Party.

10. a. *ticket-pocket* (later examples), *stub*; **b.** *ticket agency, agent* (examples), *booth, counter, hall, machine, money* (examples), *office* (examples), *wagon, wicket, window*; **d.** *ticket-collector* (examples), *-dispenser, -examiner* (examples), *-holder, punch, -seller, -writer*; *ticket-collecting* (earlier example), *-issuing* (examples), *writing* (examples).

1953 *Tatler & Bystander* 18 Mar. 424/1 He..walked across the floor and kissed her, taking the ticket. [?] ..in the basis of a night work bay. **1975** R. HOBAN *Turtle Diary* iv. 26 A lady from the ticket agency where Miss Neap had worked. **1861** *Richmond* (Va.) *Examiner* 6 Dec. 3/3 Mr. John M. Parker, for several years the efficient General Freight and Ticket Agent of the Richmond and Petersburg railroad. **1976** SCOTT & KOSKI *Walk-In* (1977) ii. 20 The ticket booth sat to the..right. **1935** *Kauffman* [?] 'Do you ever go to the Ticket-counter of the..' **1926** E. HEMINGWAY *Sun also Rises* xiii. 186 The ticket-booth was at a side of the room... I made a kind of..ticket-collecting. **1850** F. B. HEAD in E. R. *Pike Human Doc. Victorian Golden Age* 97 The ticket collector at Camden station. **1857** W. BARNARD *Blood Brotherhood* ii. 18 The Bishop bestowed his ticket on the ticket-collector. **1862** *Railway Traveller's Handy Bk.* 68 An elderly lady presents herself at the ticket counter. **1962** A. LURIE *Love & Friendship* xv. 293 Will..leaning on the ticket counter below the boarded window. **1977** G. SCOTT *Hot Pursuit* iii. 54 When I got to the airport I..walked over to the ticket counter. **1979** P. CAVE *High Flying Birds* ii. 17 He turned the handle on his little ticket dispenser and delivered my receipt. **1943** K. TENNANT *Ride on Stranger* v. 48 He dived under the seat as they drew in at a station. The ranks drew together and out-stared the ticket-examiner. **1950** *Sunday Times* (Johannesburg) 31 Mar. 1/4 Mr. Manaka reported this to the ticket examiner. **1960** *Listener* 8 Mar. 293/3 The miners sat about the job of undertaking a very large ticket hall, designed to become the busiest in London. **1978** H. R. F. KEATING *Long Walk to Wimbledon* x. 165 She made her way..to the station entrance... There were yet more people inside the old ticket hall. **1930** J. ROBERT-HOUDIN *Memoirs* I. viii. 154 Suppose the ticket-holder declined, he was not admitted, and when matters came to that pass, people always paid. **1979** D. HURD *End to Promises* i. 13 To avert violence these rallies were to be for ticket-holders only. **1951** *Evening News* (Port of Spain, Trinidad) 11 Jan. 2/1 Recently 'Tim', a ticket-issuing machine. **1980** *Times* 20 Jan. [?] machines now. **1964** A. WYKES *Gambling* x. 237 Mechanical ticket-issuing machines. **1977** *Transport* 29 May (Suppl.) p. viii/4 Another relic of postwar contempt for the passenger is the refusal to install at Underground stations that have automatic ticket machines a single machine that will enable the passenger without small change to obtain a ticket so that he can operate the ticket machine. **1979** *Listener* 18 Oct. 520/1 Ten-pence pieces have to be hoarded for..the ticket machine at St James's Park tube station. **1827** L. T. REDE *Road to Stage* 56, I remember Miss S—, at Drury, from neglecting this precaution, having to pay

one hundred and ninety-eight pounds, out of her ticket money alone, to her co-partner in the benefit. **1902** 'MARK TWAIN' in *Harper's Weekly* 6 Dec. 4/2 The man could not get back the ticket-money. **1667** PEPYS *Diary* 4 Jan. (1974) VIII. 4 My lord Brouncker went away after dinner to the Ticket Office. **1835** J. H. INGRAHAM *South-West* I. 217 A noisy crowd was gathering around the ticket-office. **1980** J. O'FAOLAIN *No Country for Young Men* iii. 56 'No point travelling First,' the man at the Euston ticket office had advised. **1934** L. A. G. STRONG *Corporal Tune* ix [?] the ticket was left in the ticket pocket. **1978** F. MACLEAN *Take Nine Spies* vii. 266 Kim Philby..was ..carrying a Soviet Secret Service cipher in the ticket pocket of his trousers. **1866** *Outing* 7 Feb. 588/1 Conductor..H. is unlocking his little corner cupboard and taking therefrom his punch [I mean ticket-punch, of course]. **1978** E. MALPASS *Wind brings up Rain* ix. 59 The girl sat down, fiddling with her ticket punch. **1844** J. COWELL *Thirty Years passed among Players* iv. 65 John Blake I appointed secretary of the treasury and principal ticket-seller. **1929** 'E. QUEEN' *Roman Hat Mystery* i. 14 He'd be looking for ticket-stubs... Anything resembling half a ticket. **1979** 'M. HEBDEN' *Death set to Music* v. 50 'But I was in Paris!' 'We have only your word and two ticket stubs to confirm that.' **1895** McClure's Mag. June 15/1 The band-wagon and the chariots, the calliope, the chimes, the oil-tank, the sprinklers, the ticket-wagon..have arrived. **1946** E. O'NEILL *Iceman Cometh* i. 60 Thinking of the old ticket wagon brings those days back. **1892** KIPLING & BALESTIER *Naulahka* v. 49 Tarvin..stepped out through the ticket wicket into Rajputana. **1964** M. LAURENCE *Stone Angel* v. 124 The ticket wicket's still there. **1865** *Harper's Mag.* May 816/1 [He] asked me to wait my return while he crowded to the ticket-window and procured tickets for both. **1979** P. THEROUX *Old Patagonian Express* xiii. 201 Ticket windows were often open only a few hours before the train was to go. **1922** JOYCE *Ulysses* 471 Export bottlers, fellmongers, ticketwriters, heraldic seal engravers. **1962** E. GODFREY *Retail Selling & Organization* vii. 57 A ticket-writing department, where price tickets, display and sales placards are written or printed. **1979** *Arizona Daily Star* 8 Apr. 22/1 Tucson travel agents and the University of Arizona are at loggerheads over the proposed creation of a taxpayer-financed ticket-writing service that would be used by the university to spend some of the $2 million the school expects to spend on air travel this year.

11. *ticket barrier*, the point at a railway station beyond which one cannot proceed without a ticket; **ticket chopper**, (*a*) also, a similar device used in cinemas; (*b*) (earlier and later examples); **ticket fine**, a fine imposed on a motorist for violation of traffic regulations by the issuing of a ticket (sense *2 b) rather than by prosecution in court; **ticket-scalper** (example); **ticket-splitting** *vbl. sb.* and *ppl. a.*, *U.S.*, the practice of voting for candidates of different political parties in the same election; hence **ticket-splitter**; **ticket tout**, one who obtains tickets for sporting, theatrical, or other events and attempts to resell them at more than the published price.

1939 AUDEN in *New Writing* Spring 2 Crowds round the ticket barrier. **1981** J. B. HILTON *Surrender Value* iii. 29 The man who stood beyond the ticket barrier, scanning the boat-train. **1888** C. B. DAVIS *Borderland of Society* 90 She took up with ticket-chopper on the elevated road. **1919** J. B. RATHBUN *Motion Picture Making* 119 To prevent the tickets from being used a second time a 'ticket chopper' may be used that mutilates the ticket in such a way that it is impossible to present it without detection. **1932** Ticket-chopper [see *movie house]. **1959** *Daily Tel.* 18 Dec. 1 (*heading*) Ticket-fine system for drivers opposed. **1979** T. SEVERN *Changing Image of Magistracy* vii. 89 Further relief came in 1960 with the introduction of 'ticket fines' for illegal parking and some other minor offences. **1879** W. N. BRYANT *Railroad Guide* 13 [?] It would prevent the deception daily practiced upon this class by ticket scalpers. **1972** DE VRIES & TARRANCE *Ticket-Splitter* i. 12 We will examine the ticket-splitting voter..as a product of..Republican politics. **1978** *Times* [?] one of the three states in which ticket splitters. **1957** *Amer. Pol. Sci. Rev.* LI. 308 (*heading*) Other motives and ticket splitting. **1972** *Times* 13 Oct. 9/3 Mr Nixon..is obviously favoured by the prospect of ticket-splitting, under which Democrats are being invited to salve their consciences by voting for their local party candidates while giving their presidential vote to Mr Nixon. **1976** *National Observer* (U.S.) 22 May 5/1 This is the damndest crossover, ticket-splitting state in the nation. **1980** *Spare Rib* 29/3 Mar. 26/4 A band that ticket touts. **1982** *Times* 11 Mar. 11/2 *Cats*..is still giving the ticket touts an excellent living.

ticket, *v.* Add: **4.** *trans.* To attach a parking ticket, etc. to (a vehicle); to serve with a ticket for a traffic or other offence. *U.S.*

1955 V. NABOKOV *Lolita* (1959) I. xxiii. 97, I should explain that the car in the parking lot..had been ticketed by the police for..having ticketed the illegally parked cars in a precise line two blocks down the grade. **1966** *Cavalier Daily* (Charlottesville, Va.) 8 Feb. 4/1 If you don't park next to the curb, you're still liable to be ticketed. **1976** *Billings* (Montana) *Gaz.* 30 June 1-B/3 Two dog owners were ticketed by animal wardens because their animals were allegedly involved in a wasting food melee are planning to fight back. **1979** R. BALLANTINE *Richard's Bicycle Bk.* (rev. ed.) I. vii. 122 Cyclists have been ticketed for causing an obstruction by riding too far to the right.

ticketing, *vbl. sb.* Add: **1. c.** The buying and selling of (airline) tickets. Freq. *attrib.*

1962 *Flight International* LXXXII. 382/2 If after expiry of a 'ticketing time limit' a provisional reservation is cancelled by a passenger, his booking deposit..must be forfeited. **1972** *Accountant* 28 Sept. 386/2 An International airline employing very many highly-qualified people—engineers, cabin-crew, pilots, ticketing and reservation clerks. **1977** 'O. JACKS' *Autumn Heroes* iii. 48 'Get the ticketing under way. Sixty-two.. Can you get me sixty-two air tickets?' **1983** *Jetaway* (Air New Zealand) Sept.–Oct. 26 (*caption*) Artist's impression of the completed 'ticketing street' in the West Terminal at Los Angeles International Airport.

ticketless, *a.* (Later examples.)

1960 F. WYLIE *Let.* 31 Aug. in Mansergh & Moon *Transfer of Power* (1979) VIII. 369 Simultaneously a regular campaign of ticketless travel on the railways will be started. **1979** P. NIHALANI et al. *Indian & Brit. English* i. 179, 2500 cases of ticketless travel and un-booked luggage were reported last month by the railway authorities.

ticket of leave. Add: **a.** (Earlier *Austral.* and further *gen.* or *transf.* examples.)

1801 *Hist. Rec. New S. Wales* (1896) IV. 300 All prisoners whose terms of transportation is [*sic*] not expired and are off the stores, or those with settlers, are to attend at the Secretary's office at Sydney, to receive their tickets of leave. **1876** *Vale Rev.* XXV. 789 Those [slaves] who went visiting, came for a 'ticket of leave',..stating in a line or two the name of the person, and where he was going. **1917** 'CONTACT' *Airman's Outings* v. 111 On the last occasion when I was let loose from the front on ticket-of-leave, I added twenty-four hours to my Blighty period. **1938** [see *CONVOY *sb.*].

b. *ticket-of-leave man* (earlier example).

1807 *Hist. Rec. New S. Wales* (1898) VI. 292 A considerable injury to the colony had crept in: that of ticket-of leave [hyphenation *sic*] men—men that were taken off the stores, and permitted to work for themselves.

tickety-boo (ti·kĕtibū·), *a. colloq.* Also **tickety-boo, tiggity-boo**, etc. [Etym. obscure: perh. f. Hindi *ṭhīk hai* all right; cf. also TICKET *sb.*¹ 9.] In order, correct, satisfactory.

1939 N. STREATFEILD *Luke* 186 Things ought to have shaped right.. Couldn't have looked more tickety-boo. **1947** *Amer. N. & Q.* Sept. 94/1 Lord Mountbatten, new Governor General of India, is credited in the *New York Times Magazine* (June 22, 1947, p. 45) with 'giving currency' to the phrase 'tickety-boo' (or 'tiggerty-boo'). This Royal Navy term for 'okay' is derived from the Hindustani. **1954** 'G. CARR' *Death under Snowdon* xi. 143 'All tiggity-boo.' 'Tiggity—? Never mind, Sergeant. Go on.' 'Everything's jake, sir.' **1957** J. BRAINE *Room at Top* xxi. 179 Everything was tickety-boo again. **1960** D. FRANCIS *Murder-on-Thames* xviii. 168 'I never liked her. It's Ewans either'. 'Then that's all tickety-boo.' **1977** *Listener* 28 Apr. 450/3 Attempting vainly to get everything tickety-boo for the Big Day. **1981** S. RUSHDIE *Midnight's Children* i. 97 Everything's in fine fettle, don't you agree? Tickety-boo, we used to say.

ticking, *vbl. sb.*¹ Add: **4.** *ticking-off*, a scolding or reprimand: see *TICK *v.*¹ 3 c. Cf. *telling-off* s.v. *TELLING *vbl. sb.* 2 c.

1950 J. CANNAN *Murder Included* ii. 16 Iona's a little beast, but she knows how to take a ticking off—she learned that at St. Olaf's. **1960** *News Chron.* 23 July 5/5 The machine shop inspector..expects a 'ticking-off' when he goes back to work. **1977** E. AMBLER *Send no More Roses* ii. 36, I gave him a ticking-off. Not that he cared. Too clever by half.

5. *ticking-over*: the idling of an engine; also *transf.* See *TICK *v.*¹ 2 d.

1972 J. WAINWRIGHT *Requiem for Loser* i. 9 Originally the talk..had been meant as the first of a quartet of 'stop gap' lectures..a ticking-over of the association's activities until the end of the holiday season. **1973** —— *Pride of Pigs* 88 The youth..revved the engine, then quietened it down to the soft ticking-over.

ticking, *ppl. a.*¹ Add: **2.** *ticking-over*: (merely) working or functioning; unproductive. See *TICK *v.*¹ 2 d.

1960 *Guardian* 31 Dec. 6/3 It seemed to be rather a ticking-over year so far as new buildings..were concerned. **1963** *Times* 13 June 8/6 As long as we have a 'ticking over' laity who are still living in the Victorian era and don't want to be shaken out of their complacency, so long will the ministry remain a reflection of the body of laity from which they came. **1974** 'J. ROSS' *Burning of Billy Toober* xvi. 149 Waiting like a ticking-over computer to be programmed.

3. Special collocation: **ticking bomb** = *time bomb* s.v. TIME *sb.* 51 a.

1960 WODEHOUSE *Jeeves in Offing* i. 13 But while equipped with eyes like twin stars.. Bo Wickham had also the disposition and general outlook on life of a ticking bomb. **1980** G. M. FRASER *Mr American* ii. xvii. 222 Mr Asquith..would find himself out of office, and the ticking bomb of Ireland could be hastily passed to his successor.

ticklace, var. *TICKLE-ACE.

tickle, *sb.*² Add: **2.** *Criminals' slang.* A successful deal or crime. Cf. *TICKLE *v.* 6 e.

1938 F. D. SHARPE *Sharpe of Flying Squad* 333 *Tickle*, a successful deal. **1955** D. WEBB *Deadline for Crime* i. 13 If there's a good tickle, say for as much as £10,000, which is as much as anyone got from it, it's soon gone to the birds.. the bookmakers, the hangers-on. **1960** [see *GRAFT *v.*²]. **1979** 'P. O'CONNOR' *Into Strong City* i. xiv. 48 Keeps me going till the big tickle comes along.

tickle, *v.* Add: **II. 3.** (Examples of *to tickle to death*.) Also in colloq. phr. *to tickle pink*, to delight; to overcome with pleasure or amusement. Cf. sense 5 in Dict.

1886 [?] A. DAVIS *Let.* J. *Downing* xxv. 188 It has tickled me my most to death. **1907** *St. Nicholas* May 607/1 I'm tickled to death to find some one who think they call human emotions. **1922** 'G. EMERY' in A. H. Quinn *Contemporary Amer. Plays* 138 He'll be tickled pink. **1939** W. IREMONGER *There's Rosemary* xiv. 266 Knowing the great artist, he had hopes that my rather cheeky suggestion might 'tickle him to death'. **1948** F. A. IREMONGER *William Temple* xxiv. 416 An American delegate who sat opposite Temple at the table—'Archbishop, you tickle me pink!' **1950** WODEHOUSE *Nothing Serious* 29 Your view, then, is that he is tickled pink to be freed from his obligations? **1976** *Scottish Daily Express* 21 Dec. 8/7 We are tickled pink that we were able to come home to do the concert at Liverpool Philharmonic Hall. **1977** E. LEONARD *Unknown Man* i. 24 He said he'd been tickled to death I'm talking to you,' Mr. Perez said..smiling into the telephone.

5. c. Also, to play or operate (the keys of a keyboard instrument or machine); esp. in phr. *to tickle the ivories* (*IVORY 5 d). *colloq.*

1919 H. CRANE *Let.* 13 Dec. (1965) 28 Tickling the typewriter keys is a stiff proposition. **1930** S. SASSOON *Mem. Infantry Officer* VIII. ii. 291 He lost no time in telling us that he had discovered a place where we could 'buy some bubbly and tickle the ivories'. **1940** N. SAYLER *Fanny by Gaslight* ii. 371 Chunks..enchanted to the pianist to tickle the ivories. **1980** *Times* 1 Oct. 12/6 The 24-year-old virtuoso who tickles the very keys once played by Reginald Dixon.

c. (Earlier example.)

1833 C. MATHEWS *Let.* 11 Oct. in A. Mathews *Mem. Charles Mathews* (1839) IV. x. 208 If you do not tickle up my matter for me after I have put it down, I will not contrive my 'Life'.

d. (See quot. 1967.)

1919 C. P. THOMPSON *Cocktails* 257 We had got out to his cycle, and he bent to tickle the carburettor. **1967** *Dr. About the Practical Gerber Chemists* viii. 33 Often the float chamber has a little plunger on top, and by pushing it you can sink the float a little and allow petrol to run through into the carburettor. People call this 'tickling the carburettor'.

e. *Criminals' slang.* To rob or burgle. Esp. in phr. *to tickle the peter*, to rob the till or cash box; also in extended use. Chiefly *Austral.* and *N.Z.* Cf. *TICKLE *sb.*² 2.

1945 [see *PETER *sb.*² 6 b]. **1950** *Austral. Police Jrnl.* Apr. 119 Tickle the peter, to embezzle or steal funds, usually by the servant of an employer. **1973** *Courier-Mail* (Brisbane) 14 Mar. 14/9 Senator Georges..was accused in State Parliament last night of having 'tickled the peter' when he was in. **1978** P. GREENLAND *Misericordia Deepl.* 1. 44 Get a Portuguese villain to hide the place.

f. *Cricket.* Of a batsman: to deflect (a delivery) with a light stroke or glance. (In quots., with bowler as obj.)

1934 *Times* 1 Mar. 4/1 Dowling, who..is probably New Zealand's finest batsman..today tickled Trueman round the corner. **1977** *Sunday Times* 3 July 28/6 At last, however, Brearley tickled Doshi away behind the wicket for three.

tickle-ace (ti·k'l,ēs). *Newfoundland* and *Labrador*. Also **ticklace, tickle-ass, tickle-lace**, etc. [perh. imit. of the bird's cry.] The kittiwake.

1819 in H. J. PADDOCK *Languages in Newfoundland & Labrador* (1977) 11 Tikle-[?]ase. **1880** J. P. HOWLEY in G. M. Story et al. *Dict. Newfoundland Eng.* (1982) 562/1 We shot a few Gulls, Terns, Pigeons and Tickelaces. **1909** P. W. BROWNE *Where Fishers Go* 207 The Kittiwake..is known to fishermen as 'Tickelelse ('Tickelers). **1932** J. BARBOUR *Forty-Eight Days Adrift* 43 The only thing alive, to be seen, was a tickalass, and we tried with poles to kill it, to eat. **1951** PETERS & BURLEIGH *Birds of Newfoundland* 235 Atlantic Black-legged Kittiwake..Local names: Tickle-ace, Tickle-lace, Tickle-ass. **1966** A. R. SCAMMELL *My Newfoundland* 26 [?] Parsons was out shooting tickle-aces yesterday. **1974** F. MOWAT *Boat who wouldn't Float* xviii. 219 She [sc. a boat] lay lightly on the harbour looking as pretty as a tickle-ass. **1975** T. RUSSELL *Chron. Uncle Mose* 23 There'd been a few ticklases around the fishing-rooms..and being as how I liked ticklace soup as well as the next man, I wished I'd had my breech-loader and a few cartridges.

tickler. Add: **1. b.** A pianist. Cf. *TICKLE *v.* 6 a. *slang.*

1980 *N.Y. Age* 9 Oct. 2/7 The Round Mr. Fletcher Butler, the distinguished Chicago piano tickler.] **1962** *Down Beat* 16 Aug. 26/3 He had a complicated attack.. combined with the gaiety and shy humor that one looks for in a true 'tickler'. **1975** J. MCCLURE *Snake* vi. 86 'Bet he's a tickler.' Pianist. Y'know.

2. j. Also *transf.*, anything intended to serve as a reminder.

1905 CALKINS & HOLDEN *Art Mod. Advertising* 351 A tickler is any small piece of printed matter sent out to keep open a prospective sale on the part of the inquirer. **1970** *New Yorker* 3 Oct. 34/3 The new lineup of teams can be easily remembered with the aid of a handy mental tickler.

k. A stern or severe letter. Cf. TICKLE *v.* 6 b. *rare*.

1846 LD. PALMERSTON *Let.* 27 Sept. in H. L. E. Bulwer *Life Palmerston* (1874) III. viii. 299 Do not mention it to anyone; but the Queen has written the King of the French a tickler in answer to a letter he sent her. **1902** G. W. E. RUSSELL *Onlooker's Notebook* xix. 102 An Illustrious Personage wrote to the Dean [of Windsor] suggesting that, as Mr. Gladstone was engaged in violent attacks upon the Government, it might be better if his visits to the Deanery were discontinued. 'Whereupon,' said the stout old Dean.. 'I wrote her a tickler'.

l. An inductance coil in the anode circuit of a valve, giving positive feedback through another coil in the grid circuit.

1922 [see *REACTION 3 c]. **1931** MOYER & WOSTREL *Radio Handbk.* VII. 335 A tickler..usually more than makes up for this decrease in detection coefficient by increasing the strength of the impressed radio signal. **1948** [see *RADIO *sb.* 2 a].

m. A hand-rolled cigarette or the tobacco from which it is made. *Naut. slang.*

1929 F. BOWEN *Sea Slang* 141 *Ticklers*,..is also a name for cigarettes made from the monthly issue of naval tobacco. **1964** J. HALE *Grudge Fight* I. ii. 29 Brooks rolls and lights a tickler. **1977** G. MELLY *Rum, Bum & Concertina* iv. 54 'Tickler' is naval slang for duty-free tobacco.

4. *attrib.* and *Comb.*, as (sense *2 j) *tickler card, file, system, telegram*; **tickler coil** = sense 2 l above.

1931 *New Yorker* 10 Oct. 21/1 He ran the tickler cards in their credit department. **1940** *Chambers's Techn. Dict.* 850/2 *Tickler coil*.., an inductance coil included in the anode circuit of a thermionic valve and magnetically coupled to the grid circuit to obtain reaction. **1975** R. L. SHRADER *Electronic Communication* (ed. 3) xviii. 433/1 To increase the sensitivity and selectivity of the grid-leak detector, a plate-circuit tickler coil can be used. **1939** C. MORLEY *Kitty Foyle* (1940) xix. 174 You might remind me of that sometimes. Put it in the tickler file. **1972** D. BAGLEY *Enemy* xiii. 105 Authority had lost interest in him and he would exist only in a tickler file to remind someone to give an annual check. **1962** 'A. A. FAIR' *Stop at Red Light* iii. 56 Real estate, subdivisions..figuring interest, keeping a tickler system for time payments. **1938** *Sun* (Baltimore) 21 June 6/7 One father..remembered last year to send admonitioning letters to each member of his family and follow them up with 'tickler' telegrams.

tickling, *vbl. sb.* Add: **4. tickling stick** *joc.*, a feather duster or similar device used as a comedian's prop.

1969 *Listener* 6 Mar. 299/2 Beneath your tickling stick and candy floss came Heath and Wilson and Woodcock. **1980** *Times Lit. Suppl.* 24 Oct. 1198/5 Hints of the tickling-stick [in a performance of *Peer Gynt*] vulgarized the first meeting with Solveig. **1983** *Sun* 8 June 4/4 Ken Dodd painted his famous tickling stick blue yesterday for a Liverpool walkabout in support of Mrs Thatcher.

ti·ck-off. *slang.* [f. vbl. phr. *to tick off*: see TICK *v.*¹ 3.] Fortune-telling; a fortune-teller. *to work the tick-off*, to practise fortune-telling.

1934 P. ALLINGHAM *Cheapjack* ii. 18 Several show-people were in the bar. 'You're working the tick-off, aren't you?' said one of them. *Ibid.*, I discovered that 'tick-off' was the fair-ground slang for fortune-teller. **1966** *Punch* 3 Aug. 193/1 No palmists, tick-offs, character readers, mock auctions, pick-a-straw.

ti·ck-over. [f. vbl. phr. *to tick over*: see *TICK *v.*¹ 2 d.] **1.** The running of an internal combustion engine while out of gear or at a low rate of revolutions. Cf. *ticking over* s.v. *TICKING *vbl. sb.*¹ 5.

1931 *Flight* 2 Oct. 990/2 All engines were tested on a hangar with a propeller fitted before going away, for opening and for tick-over. **1960** *Times* 27 Sept. 5/3 The tick-over is inevitably 'lumpy' and uneven. **1981** P. O'DONNELL *Xanadu Talisman* iii. 60 The tick-over became a roar as a motor was revved.

2. *attrib.*, as *tick-over speed*.

1965 [see *IDLE *v.* 4 a]. **1978** *Country Life* 10 Aug. 393/1 A Mercedez-Benz dieselcar..at tickover speeds..almost runs on pure air.

tick-tack, *sb.* Add: **1. c.** Chiefly *N. Amer.* A contrivance, such as a button on a piece of thread, spun to make a clattering sound against a window or door as a practical joke, esp. at Hallowe'en.

1884 I. M. RITTENHOUSE *Jrnl.* in *Maud* (1939) 288, I formed plan after plan to frighten them. Finally a 'tick-tack' was decided on. **1947** *Sun* (Baltimore) 7 Oct. 16/3 The Park has decided to have an old-time Hallowe'en, with the old boys puttin' tick-tacks on windows.

d. *transf.*

1927 D. H. LAWRENCE *Mornings in Mexico* 63 Seeing the white monkeys for ever mechanically bossing, with their incessant tick-tack of work. **1934** S. BECKETT *More Pricks than Kicks* 133 'God' he exclaimed, executing a kind of passionate tick-tack through his pockets.

2*. = TICK-TACKER.

1918 G. FRANKAU *One of Them* xxi. 159 Silent the tic-tac's tell-tale Semaphore: On thousand tracks, unridden, ..Hay waves.

tick-tack, *v.* Also tic tac, tic-tac. [f. the sb.] **1.** *intr.* = *TICK-TOCK *v.*

1842, *a* **1847** [implied at TICK-TACKING *ppl. a.*]. **1859** MRS. STOWE *Minister's Wooing* ii. 17 The solemn old clock that tick-tacked in the corner.

2. *intr.* and *trans.* To produce a whirring, clattering sound by spinning a tick-tack (sense 1 c) against a window, etc., as a practical joke. *dial.* and *N. Amer.*

1901 F. E. TAYLOR *Folk-Speech S. Lancs.* s.v. *Ticktackin*, a boys' practical joke. See Window-tackin'. **1970**

J. H. GRAY *Boy from Winnipeg* 188 We got tired of the project and abandoned it in favour of ringing doorbells and tick-tacking windows.

3. *trans.* and *intr.* To signal (information) by means of tick-tack telegraphy.

1907 *Favourite* 30 Nov. 9/3 Kilbeg was 'tick-tacked' out at 4 to 1 by the private clerk of one particular firm. **1908** *Tatler* 3 June 247 The above system of signalling, which is known as tick-tacking, may be seen on any racecourse. **1927** *Observer* 27 Mar. 18/6 A man in the body of the hall was detected tictacing to Labour supporters and guiding the uproar. **1937** L. MANN *Murder in Sydney* xxv. 273, I also noticed Leon Caspar ticktacking to the girl in response to which the girl challenged two of those called on the panel. **1972** *Guardian* 11 Aug. 8/6 The policeman tic-taced to the judge what the punishment should be.

tick-tack-toe. Add: (*b*) *U.S.* = *noughts and crosses* (see NOUGHT *sb.* 7 c; OUGHT *sb.*³); also the cross-shaped frame in which this game is played; also *fig.*

1960 S. PLATH in *Sewanee Review* LXVIII. 604 The jacket is patterned with brown squares the size of cigarette packs, each square boldly outlined in black. You could play tick-tack-toe on it. **1975** *Nat. Geographic* Apr. 500 (*caption*) Tick-tack-toe of a new apartment complex rises amid mud-and-wattle houses in Zanzibar town. **1976** N. THORNBURG *Cutter & Bone* xi. 266 A tick-tack-toe form filled with zeros. **1978** G. VIDAL *Kalki* i. 8 Just past the tall sick palms at the edge of the pool, the exhaust of a half-dozen jets was making a kind of tick-tack-toe in the dusty brown sky over Los Angeles. **1980** *Dædalus* Spring 46 A computer designed only to issue the company's pay-checks might stalemate me perpetually in tic-tac-toe.

tick-tock (ti·k,tǫ·k), *v.* [f. the sb.] *intr.* Of a clock, etc.: to make a rhythmic alternating ticking sound. Hence **tick-tocking** *vbl. sb.*

1921 H. S. WALPOLE *Young Enchanted* III. iii. 274 The gaudy clock..now tick-tocked along in amiable approval of them both. *a* **1947** F. THOMPSON *Country Calendar* (1979) 201 The loud tick-tocking of the clock in the hall. **1950** G. BARKER *Dead Seagull* ii. 89, I heard the murmur in the distance and then the rumbled tick-tocking and then the appalling cacophony all around as the approaching train swept up and past us. **1962** L. DEIGHTON *Ipcress File* ix. 55 The clock tick-tocked on, adding a second or so to its seventy years of tick.

ticky-tacky (ti·kitæ:ki), *sb.* and *a.* orig. *U.S.* Prob. redupl. f. TACKY *sb.* and *a.*¹]

A. *sb.* Inferior or cheap material, esp. that used in uniform suburban building.

1962 M. REYNOLDS *Little Boxes* (1964) (*song*) 3 And they're all made out of ticky tacky, And they all look just the same. **1973** *Newsweek* 30 July 71 The real point is, will..Watchung Pharmaceutical get those 250 unspoiled acres around Howard's tree farm which have been zoned for a park, there to produce more poppable pills and sprinkle company ticky-tacky over the landscape? **1978** M. BUTTERWORTH *X marks Spot* iii. 26 A large and gloomy Victorian pub, heavily overlaid with up-to-date ticky-tacky.

B. *adj.* Made of ticky-tacky; cheap, in poor taste.

1969 *Sat. Rev.* (U.S.) 10 May 19/1 Men who desecrate the landscape with hundreds of ticky-tacky houses and.. call themselves developers. **1970** O. NORTON *Dead on Prediction* iii. 46 The house was one of those ticky-tacky semi-bungalows, which are given a false air of expensiveness. **1977** *Jrnl. R. Soc. Arts* CXXV. 119/1 Critics of the private sector complain of ticky tacky little boxes spread all over the country.

tidal, *a.* Add: **1. b.** (*b*) (Earlier examples.)

1870 'MARK TWAIN' *Lett. to Publishers* (1967) 45 'We'll have somebody standing ready to launch a book right on our big tidal wave and swim it into a success. **1875** —— *Sketches New & Old* 213 A great tidal wave of grief swept over us all.

2. b. *spec.* Of (esp. rush-hour) road traffic, its flow, or a road carrying it: that uses the same lane(s) for travel in opposite directions, depending on time and conditions.

1954 *Highway Engin. Terms (B.S.I.)* 55 *Tidal traffic*, traffic on a two-way road proceeding predominantly in one direction or the other according to time or recurrent circumstances. **1955** *Times* 25 Oct. 9 The reversible lane on tidal highways. **1960** *News Chron.* 26 Feb. 5/1 Tidal-flow traffic was introduced on Chelsea Bridge last year. **1960** *Guardian* 7 June 1/2 The 'tidal flow scheme'..has already been tried out in London. **1969** *Soviet Weekly* 13 Sept. 2 All the flyovers, underpasses, tidal flows and one-way streets the authorities organized only eased the problem without curing it.

3. a. *tidal train* (earlier example).

1855 DICKENS in *Househ. Words* 29 Sept. 194/1 The South Eastern Company..with their tidal trains and splendid steam-packets.

b. (Earlier example.)

1866 DICKENS in *All Year Round* Christmas No. 18/1 A return pass by South-Eastern Tidal, to go right through ..to Marseilles.

tiddle, *v.* Add: **2. b.** To move potteringly.

1881 W. D. HOWELLS *Modern Instance* (1882) I. x. 173 Mr. Macallister, a slight little straight man..tiddled farcically forward on his toes. **1970** *Times* 6 July 6/8 You can't just tiddle up to the town hall to see the man.

tiddled (ti·dl'd), *a. slang.* [f. *TIDDLY *sb.* + -ED².] Drunk, tipsy; = *TIDDLY *a.*¹

1956 G. DURRELL *My Family & Other Animals* xii. 163 'I've got the most splitting headache.' 'I'm not surprised; you were as tiddled as an owl last night.'

tiddler¹. Add: Also applied to other small fish, as a minnow. Hence, a child; any small person or thing.

1927 R. LEHMANN *Dusty Answer* IV. 294, I remember you hated lightning when you were a tiddler. **1937** G. FRANKAU *More of Us* vi. 69 When father's under par, Mother suggests a tiddler at the bar. **1966** F. SHAW et al. *Lern yerself Scouse* 33 *Tiddler*, silver threepenny piece. **1971** *Daily Mail* 18 Feb. 5/6 They will scrap the ½p coin—the 'tiddler'—when they change to decimals. **1976** *Courier-Mail* (Brisbane) 26 Aug. 9/6 Pastime anglers would not be allowed to keep 'tiddlers'. **1980** E. BLISHEN *Nest of Teachers* I. viii. 43 A couple of days with Class 1A and.. he will know a deuce of a lot..about the little tiddlers.

ti·ddler³. *U.S. colloq.* [f. TIDDL(YWINK + -ER¹.] A tiddlywinks player.

1958 *N.Y. Times* 9 May 28 An Oxford University tiddler..says there's a 100-year-old controversy over proper spelling of the sport. **1962** *Boston Globe* 14 Oct. 81 The Crimson tiddlers winked their way to a 23 to 12 victory over a green Purple team.

tiddly (ti·dli), *sb.* and *a.*¹ *slang.* Also 9 **titley**, 9- **tiddley**. [Origin uncertain: cf. TIDDLYWINK 1 in Dict. and Suppl.] **A.** *sb.* Drink; an alcoholic drink, esp. a 'short'.

1859 HOTTEN *Dict. Slang* 109 *Titley*, drink. **1864** *Ibid.* 258 *Titley*, drink, generally applied to intoxicating beverages. **1895** *Punch* 18 May 230/2 It took two 'ot tiddleys to warm 'er. *Ibid.* 12 Oct. 180/1 A helderly humorous gent, on the tiddley. **1923** E. P. OPPENHEIM *Inevitable Millionaires* xxiv. 259 Just a tiddley to drink success to the club. **1930** E. V. LUCAS *Down Sky* 222 It wasn't oysters that she really wanted, but..tiddley.

B. *adj.* Drunk, tipsy.

1905 *To-day* XLVI. 182/2 If ever you was tiddly in crossing the old 'un [*sc.* a bridge], it was as easy as anything to fall into that blarsted river. **1909** *Chambers's Jrnl.* 17 Apr. 316/1 Mind you don't get tiddley and blow the gaff. **1930** W. S. MAUGHAM *Cakes & Ale* xxiii. 232, I don't say he ever got tiddly, but he used to like to sit in the bar and talk. **1958** B. NICHOLS *Sweet & Twenties* xvi. 208 No more wine, George, thank you. I shall be quite tiddly. **1979** 'J. SCOTT' *Angels in your Beer* xxv. 254 Yvonne giggled. 'I do believe I'm tiddly,' she said.

tiddly (ti·dli), *a.*² *colloq.* Also 9 **tidly**; **tiddley**. [var. *TIDDY *a.*] Very little, tiny. Freq. in phr. *tiddly bit*.

1868 TROLLOPE *He knew he was Right* (1869) I. xxiii. 183 The smallest little 'tiddly' things do so often turn up trumps. **1885** E. C. SHARLAND *Ways & Means in Devonshire Village* 42 I'd only got but a tiddly bit o' that mutton left. **1888** B. LOWSLEY *Gloss. Berks. Words* s.v. *Tidly*, I had un in my arms when a was a tidly little chap. **1937** *John o' London's Weekly* 5 Feb. 762/2 When there was a bad harvest he hung the little tiddley sheaves o' corn on his garden railing to let God Almighty know how badly he'd been treated! **1950** WODEHOUSE *Nothing Serious* 106 A mere tiddly seaside competition. **1956** 'N. SHUTE' *Beyond Black Stump* viii. 237 If I have one will you have just a little bit, in your coke? **1978** J. GOODMAN *Last Sentence* iii. 114 The whole bally case for the prosecution is built on tiddly bits of non-evidence.

tiddly (ti·dli), *a.*³ *slang* (orig. and chiefly *Naut.*). Also **tiddley**. [perh. f. TIDY *a.*] **1.** Smart, shipshape, spruce.

1925 FRASER & GIBBONS *Soldier & Sailor Words* 281 *Tiddly*, smartly dressed... Also applied to a ship of smart appearance. **1942** G. HACKFORTH-JONES *One-One-One* i. 7 Like all sailors he took great pleasure in keeping his surroundings what he described as 'tiddley'. **1960** 'N. SHUTE' *Trustee from Toolroom* ii. 25 We'll have to get everything all tiddley. **1973** R. DOUGALL *In & out of Box* 119 The aim was to achieve as pale a blue on the collar as possible—this was considered 'tiddley', which in Naval parlance means smart or *comme il faut*. It was also tiddley to have smart horizontal creases in the bell-bottom trousers like a concertina.

2. Special collocation: **tiddly suit**, one's best suit of clothes.

1943 'TAFFRAIL' *White Ensigns* 33 'Tiddley' suits with light jumpers and bell-bottomed trousers cut far wider than the regulation permitted. **1951** C. CAUSLEY *Farewell, Aggie Weston* 9 Farewell, Aggie Weston, the barracks at Guz, Hang my tiddley suit on the door.

tiddly-om-pom-pom (tidli,ǫ·mpǫ·mpǫ·m). [Imit.: cf. *POM-POM 2.] Representing the sound or regular beat of brass-band or similar music. Also **tiddy-pom** *a.*, with a simple beat or tune, trite.

1909 J. A. GLOVER-KIND *I do like to be beside the Seaside* (1970) (*song*) 4 The brass bands play tiddely-om-pom-pom. **1937** *Scrutiny* VI. 333 The tiddly-pom guitar accompaniment. **1958** *Times* 30 Dec. 7/4 The brass band certainly plays tiddly-om-pom-pom with any amount of spirit at Association football matches. **1973** T. HEALD *Unbecoming Habits* iii. 76 The piano had a tiddly pom, tin pan alley, sort of complacency.

tiddlypush (stress variable). *colloq.* A meaningless word substituted for the name of a person or thing which the speaker has forgotten, does not know, or is unwilling to mention. Cf. THINGUMMY.

1923 J. MANCHON *Le Slang* 342 And tidderly push, et patati et patata. **1934** S. BECKETT *More Pricks than Kicks* 168 He wore a belt Whenever he felt A pain in his tiddlypush. **1939** WODEHOUSE *Uncle Fred in Springtime* ii. 27 It simply says 'The marriage arranged between George Tiddlypush and Amelia Stick-in-the-mud will not take place.' **1962** *Guardian* 17 July 16/4 Mr Randolph Churchill..said..he had shown that if he wanted to 'get rid of anyone—from Lord Salisbury down to Mr Tiddlypush—he does so'.

tiddlywink, *sb.* Add: **1. b.** *Rhyming slang.* A drink.

1880 [see *pig's ear* s.v. *PIG *sb.*¹ 13 c]. **1960** J. FRANKLYN *Dict. Rhyming Slang* 129/1 *Tiddly wink*..is applied more frequently to 'shorts'..than to beer.

2. a. (Earlier example.)

1857 'DUCANGE ANGLICUS' *Vulgar Tongue* 43 At *knock'emsdown* and tiddlywink, To be a sharp you must not shrink.

b. orig. **Tiddledy-Winks.** (Earlier examples.) Also *attrib.* Also used *fig.* of a useless or frivolous activity; *esp.* in phr. *to play tiddlywinks*, to waste time on trivia.

1889 *Trade Marks Jrnl.* 15 May 476 Tiddledy-Winks... Toys or games. Joseph Assheton Fincher. **1890** *Amer. Stationer* 18 Sept. 691 In 'Tiddledy Wink Tennis' E. I. Horsman..has brought out a very pretty and lively parlor game. **1892** E. LYTTON *Let.* 24 Apr. in E. Lutyens *Blessed Girl* (1953) vii. 97 We all played the most exciting game that ever was invented, called Tiddleywinks. It consists in flipping counters in a bowl. **1895** *Montgomery Ward Catal.* Spring & Summer 236/2 Tiddledy Winks may be played by any number... Each player is provided with four to six counters..and one larger one..to press to the edge of the smaller one and..cause it to jump into the cup..in the centre of the table. **1919** *Collier's* 8 Feb. 7 There's trouble down there and I've been playing tiddledy-winks on Broadway! **1947** *Economist* 18 Oct. 626/1 The storm was long predicted..yet when its first icy gust blew in the windows of the Cabinet room.., it found the Ministers playing tiddleywinks. **1964** *New Yorker* 4 Apr. 147 Others seem to take little interest in the organized activities, describing them as 'make-work' or 'tiddlywinks'. **1975** *Way to Play* 135/1 Tiddlywinks golf sets, with tiddlywinks, greens, obstacles, and holes, are produced by various toy manufacturers. **1980** *Disarmament Times* 6 Oct. 4/4 This is not a game of diplomatic tiddlewinks. It is, rather, the game of human survival.

c. Any of the counters used in the game of tiddlywinks (sense 2 b); hence, a similar counter used in other games.

1891 J. K. BANGS *Tiddledywinks Tales* 35 Jimmieboy thought a great deal of his Tiddledewinks and had been playing with them nearly all that day. **1939** J. STEINBECK *Grapes of Wrath* iii. 22 His front wheel struck the edge of the shell, flipped the turtle like a tiddly-wink,..and rolled it off the highway. **1949** 'G. ORWELL' *Nineteen Eighty-Four* III. 296 Soon he was wildly excited and shouting with laughter as the tiddleywinks climbed hopefully up the ladders and then came slithering down the snakes again. **1977** B. JEWELL *Sports & Games* 109 The object was to flip the tiddlywink into one of the window openings and so ring the bell.

Hence **ti·ddlywink** *v. intr.*, (*a*) to flip like a counter in tiddlywinks; (*b*) to play tiddlywinks; **tiddlywinker**, (*b*) a tiddlywinks player; **tiddlywinking** *vbl. sb.*, (*b*) the activity of playing tiddlywinks.

1958 *Sports Illustrated* 7 Apr. M5 Each tiddlywinker plays with two large and four medium-size winks. **1965** *Northeastern Reporter* 2nd Ser. CCVI. 847/2 We have raised a lot of manhole covers, and you never walk across them to see if they are going to tiddlywink. **1971** *Ottawa Citizen* 6 Feb. (Canad. Mag.) 24 He loves the game so much that in 1967 he played in a 67-hour tiddlywinking marathon. **1975** *Milwaukee Jrnl.* 25 May IV. 2/2 Dean, a high school mathematics teacher, is certainly Britain's top Tiddlywinker. **1977** P. DICKSON *Mature Person's Guide* 162 The general consensus in tiddlywinking circles is that all of this might change if the game was given a new name. **1980** *Milwaukee Sentinel* 11 Feb. 1. 6/4 (*heading*) Fame opens the eyes of those who tiddlywink.

tiddy (ti·di), *a. dial.* or *nursery*. [Origin obscure. (The spelling *tidy* in quot. 1781 is not recorded elsewhere and may be an error.)] Small, very small, tiny. Also (*redupl.*) **tiddy iddy** *a.*

1781 J. HUTTON *Tour to Caves* (ed. 2) 98 *Tidy*, small. **1869** W. S. GILBERT *Bab Ballads* x. 171 A tiddy iddy daughter, and a tiddy iddy son! **1896** KIPLING *Seven Seas* 191, I got me a tiddy live 'eathen... Doll in a teacup she were. **1907** *Daily News* 4 Feb. 2/5 It was only a 'tiddy' pup. **1958** M. KELLY *Christmas Egg* III. 99 Do you know this Richborough?.. There's a tiddy railway, power cables, and the castle. **1960** 'N. SHUTE' *Trustee from Toolroom* vi. 133, I filled out the buttocks a tiddy bit on this one.

tide, *sb.* Add: **I. 6.** (Later examples in the names of saints' days.) *St. Andrew's tide*: delete †.

1957 *Oxf. Dict. Chr. Ch.* 49/2 In the Anglican Communion, St. Andrewstide is widely observed by intercessions for foreign missions. **1975** *Church Times* 15 Aug. 2/3 Last week—St Laurence-tide—all the churches supported a flower festival in the chapel. **1976** *Ibid.* 15 Oct. 5/1 It has become the custom at St. Luke's-tide for the Church to pray for doctors.

II. 9. (Earlier and later examples with reference to the 'turning' of the tide.)

1593 [see TURN *v.* 20]. **1781** C. WESLEY *Protestant Association* in *Poet. Wks.* (1870) VIII. 464 His faithful troops from every side Are brought to turn the rapid tide.

1843 Lytton *Last of Barons* I. i. vii. 117 This speech turned the tide. **1915** Mrs. Belloc Lowndes *Let.* 10 Mar. (1971) 57 There is an invasion scare but I don't believe in that.. I do think the tide has now turned. **1935** E. Waugh *Edmund Campion* i. 24 That generation was inured to change; sooner or later the tide would turn in their favour again. **1941** P. Carr *English are like That* i. 18 He must have patience—patience..in the face of misrepresentation, patience to wait for the turn of the tide. **1982** *Church Times* 5 Feb. 8/3 After that [donation] who will be surprised to hear that the tide shows signs of turning at St. Christopher's? **1982** D. Fraser *Alanbrooke* iv. 79 In the autumn of 1918 the tide finally turned.

V. 16. a. (*a*) *tide-edge, -print, -stream* (earlier example); (*b*) *tide-flat*; (*c*) *tide-master, -turner*; (*d*) *tide-beat* (example), *-borne, -carved, -hoisted, -looped, -tongued, -traced* adjs.

1807 J. Barlow *Columbiad* vii. 272 Two British forts the growing siege outflank, Rake its wide works and awe the tide-beat bank. **1957** R. Campbell *Coll. Poems* II. 99 The swirl, the spray, the nimbus, and the wave Of tide-borne lust and beauty. **1897** Tide-carved [see *pediment¹ 1 b]. **1976** *National Observer* (U.S.) 4 Sept. 7/3 We clambered down to tide-carved caverns. **1931** W. Faulkner *Sanctuary* xxiii. 221 A world left stark and dying above the tide-edge of the fluid in which it lived. **1859** C. Kingsley *Glaucus* (ed. 4) 146 The tide-flats below are still unfinished, dry land in the process of creation. **1929** W. Faulkner *Sound & Fury* 211, I saw the last light supine and tranquil upon tideflats. **1936** Dylan Thomas *Twenty-Five Poems* 41 The winder of the clock-wise scene..threw on that tide-hoisted screen Love's image. **1939** —— *Map of Love* 4 Or like the tide-looped breastknot reefed again. *Ibid.*, The silent tide Lapping the still canals, the dry tide-master Ribbed between desert and water storm. *Ibid.* 9 Moonfall and sailing emperor, pale as their tide-print. **1795** *Essex Inst. Hist. Coll.* (1918) LIV. 101 To compensate for any supposed inconveniences that may attend a tide stream. **1934** Dylan Thomas *New Verse* xii. 11 Among the rabble Of tide-tongued heads and bladders in the deep. **1936** —— *Twenty-Five Poems* 44 Cartoon of slashes on the tide-traced crater. **1922** D. H. Lawrence *Fantasia of Unconscious* xv. 272 The moon is the tide-turner.

b. tide-land(s) *N. Amer.*, land(s) covered by the tide; *tide-land spruce* = *Sitka spruce* s.v. *Sitka; **tide-lock** (earlier example).

1787 W. H. Siebert *Loyalists in E. Florida, 1851–94* (1929) II. 239, 200 acres of rich tide land well dam'd. **1884** C. S. Sargent *Rep. Forests N. Amer.* 206 Tide-Land Spruce... A large tree of great economic value. **1891, 1895** [in Dict., sense 16 a]. **1969** R. C. Hosie *Native Trees Canada* (ed. 7) 68 Sitka Spruce. Tideland Spruce... Produces a long, branch-free, cylindrical trunk. **1975** *N.Y. Times* 25 Feb. 16/3 Gas resources did not become a controversial issue until the mid-nineteenthirties when oil companies began drilling wells in the tidelands. **1808** B. H. Latrobe *Let.* 16 Mar. in *Niles' Reg.* (1818) XV. 54/2 It would be necessary to place the tide lock as far out as possible.

tide-mark. Add: *fig.* (examples); *spec.* a line of dirt left on a surface, esp. at the limit to which water has reached (cf. *high-water mark c).

1865 Geo. Eliot *Ess.* (1884) 203 A particular class of facts..in their relation to certain grand tide-marks of opinion. **1907** N. Munro *Daft Days* xxii. 190 With a smut on your nose and tide-marks on your eyebrows. **1928** *Daily Express* 20 Dec. 5/3 Your fur collar leaves a horrid tide-mark after it has been worn for a time. **1934** A. Huxley *Beyond Mexique Bay* 197 A tide-mark of grit ..running round the bath. **1957** J. M. Mackintosh in 'C. H. Rolph' *Human Sum* ix. 177 The succession of tide-marks round the necks of my young patients. **1973** C. Egleton *Seven Days to Killing* ii. 27 She..casually lit a cigarette; pink lipstick formed a tide-mark around the filter tip. **1976** T. Heald *Let Sleeping Dogs Die* i. 21 From the hooks..she took three cups, chipped..and with tidemarks of tea clearly visible an inch below the rim. **1977** A. Carter *Passion of New Eve* ix. 113 Dead peonies in enormous glass jars streaked with tide marks where the water had evaporated long ago.

tide-mill. 1. (Earlier examples.)

1640 *Essex Inst. Hist. Coll.* (1863) V. 169/2 Captane Traske hath leave to sett up a tyde myll upon the North River. **1755** *Mass. Prov. Acts* (1878) III. 810 Tide-mills that have or shall be set up on across the mouth of rivers.

tide-water. Add: Also **tidewater. 2.** (Earlier examples.)

1772 *Va. Statutes at Large* (1821) VIII. 564 The extension of the navigation of James river..will be greatly promoted by cutting a canal..from the Westham to the tide water. **1832** J. F. Watson *Hist. Tales N.Y.* 38 [Hudson City] is deemed at the head of tide water and ship navigation.

b. *attrib.* Designating or pertaining to regions situated on tide-water or affected by tides, *esp.* (also with capital initial) eastern Virginia. Also *ellipt.*, a tide-water region, esp. that of eastern Virginia.

1832 J. P. Kennedy *Swallow Barn* I. xviii. 179 The tide-water country of Virginia. **1835** *Southern Lit. Messenger* I. 662 The tranquil and affectionate hearths of tide-water Virginia. **1868** [in Dict.]. **1884** *Century Mag.* Apr. 825/2 Mr. Jones..[has] one of those thin, mournful faces common to tide-water Maryland. **1936** W. Faulkner *Absalom, Absalom!* vii. 222 He didn't listen to the vague and cloudy tales of Tidewater splendor that penetrated even his mountains. **1943** *Sun* (Baltimore) 14 June 10/1 A comprehensive program to increase the annual harvest of the oyster industry has been submitted to conservation authorities in the various tidewater States. **1944** L. Mumford *Condition of Man* vi. 216 The stately tide-water mansions of Virginia. **1949** B. A. Botkin *Treas. S. Folklore* I. ii. 32 Within his state the Southerner thinks in terms of Eastern Shore or Tobacco Country, Tidewater or Valley, [etc.]. **1976** *Amer. Speech 1974* XLIX. 45 The Delmarva area..comprises southern Delaware, portions of the eastern shore of Maryland, and the peninsula and islands of tidewater Virginia. **1979** D. Anthony *Long Hard Cure* xxi. 166 A hardnosed captain of industry who wanted a pretty mannequin from tidewater aristocracy. **1980** *Amer. Speech* LV. 285 It is common in the Virginia piedmont and tidewater.

c. *attrib.* Designating bodies of water affected by tides.

1835 J. Martin *Descr. Virginia* 40 A tide water river, or more correctly a bay, the Chowan. **1888** [in Dict.]. **1939** *WPA Guide to Florida* (1984) II. 177 Daytona Beach ..is a city with a triple waterfront, one on the Atlantic Ocean and one on each side of a tidewater lagoon.

tidied (tai·did), *ppl. a.* [f. Tidy *v.* + -ed¹.] That has been made tidy; esp. with *up*.

1922 E. O'Neill *Hairy Ape* v. 49 A general atmosphere of clean, well-tidied, wide street. **1959** *Daily Tel.* 28 July 5/3 (*heading*) Russians get 'tidied-up' version of Nixon clash.

tidier (tai·diəɹ). [f. Tidy *v.* + -er¹.] One who makes (something) tidy. Also **tidier-up,** (*colloq.*) **tidier-upper.**

1923 *Public Opinion* 15 June 565/3 He became a champion tidier from that moment on. **1961** J. Wilson *Reason & Morals* i. 8 The philosopher seems to be not so much a universal aunt as a sort of universal tidier-up. **1963** *Times Lit. Suppl.* 25 Oct. 849/3 Among scientists there are collectors, classifiers, and compulsive tidiers up. **1976** Culross & Robb *Leaving Home* iv. 32 Nancy was the neat sort..and..an habitual tidier-upper.

tidingman, U.S. var. Tithingman¹ c.

1703 *Early Rec. Groton, Mass.* (1880) 123 For tiding men [for the year 1703] Joseph gilson Benjmen farnworth. **1878** Mrs. Stowe *Poganuc People* vi. 63 They're goin' clean agin everything—Sunday laws and tiding man and all.

tidy, *a.* Add: **5.** *tidy-mindedness*; (in sense 3) *tidy-sized* adj.; **tidy bin,** a bin into which things may be discarded or tidied away, a waste-bin.

1972 *House & Garden* Dec. 84/3 Colourful bathroom accessories, including shelf units, tidy bins, mirrors. **1978** *People's Friend* 13 May 3/3 Lifting some toffee papers and preparing to transfer them to the tidy bin, she spotted the torn photograph. **1951** *Essays & Studies* IV. 21 Too much tidy-mindedness and love of classification. **1975** J. P. Morgan *House of Lords & Labour Govt.* vi. 163 Even if administrative tidy-mindedness..was the motive behind the Seats Bill.., its consequences would still transcend any administrative convenience. **1922** Joyce *Ulysses* 628, I want to see everyone..having a comfortable tidysized income. **1945** W. de la Mare *Scarecrow* 128 'Lor bless me,' said Alice. 'The questions he asks!'..'And that's a tidy-sized one too!' said Alice, smiling at him again.

B. *sb.* **2.** An act of making tidy; a period of time devoted to tidying. Freq. with *-up*.

1909 E. Nesbit *Daphne in Fitzroy St.* ix. 122 The dreadful neatness that follows a 'good tidy-up'. **1915** Kipling *Diversity of Creatures* (1917) 428 We'll pull up the blinds and we'll have a general tidy. **1949** N. Streatfeild *Painted Garden* xiii. 140 Rachel was going on to tell Jane to give her hair a tidy. **1970** C. Whitman *Death out of Focus* ix. 141, I bustled around..giving my flat a rough tidy-up. **1971** H. Wilson *Labour Govt.* xxxvi. 756, I went for a quick wash and tidy-up. **1980** 'T. Hinde' *Daymate* i. vi. 53 [She] is coming to give his house its Saturday morning tidy.

tidying, *vbl. sb.* Add: Also with *-up*.

1959 N. Marsh *False Scent* (1960) iv. 102 When they arrived..the tidying-up process had considerably advanced. **1964** M. McLuhan *Understanding Media* ii. xx. 197 This immense tidying-up of our inner lives..has had its obvious parallels in our attempts to rearrange our homes and gardens. **1975** S. Brett *Cast, in Order of Disappearance* vi. 47 He'd..even done a token tidying-up of his room.

tie, *sb.* Add: **4. a.** In mod. use the tie or neck-tie is usu. distinguished from the cravat; also, a bow-tie.

b. A lady's ornamental necklet or scarf.

1860 C. M. Yonge *Hopes & Fears* I. ii. iii. 204 Ladies affected coats and waistcoats..both cousins..wearing.. black ties round the neck. **1895** *Montgomery Ward Catal.* Spring & Summer 79/2 White Hemstitch Lawn Ties, embroidered ends. (Size 4½ × 44 inch.) **1919** *Queen* 4 Oct. 5 A..Mink Tie beautifully worked in three strands. **1930** *Daily Tel.* 8 Apr. 9/5 Wherever fashionable women may meet this Easter most assuredly will you mark the popularity of the Fur Tie. **1973** *Country Life* 22 Nov. (Suppl.) 721 Important auction sale... Mink & Astrakhan fur coats and ties.

6. c. The locking together of dog and bitch during copulation.

1951 E. F. Daglish *Dog Breeder's Man.* xi. 102 Penetration by the dog is usually followed by the 'tie'.. usually considered evidence of a successful union. **1969** M. Roslin-Williams *Dual-Purpose Labrador* iv. 52 When the mating is effective and normal, the 'tie' will be so strong that the dog can be turned carefully round.

8. d. Delete '?*dial.* or *colloq.*' (Examples.)

1868 J. C. Atkinson *Gloss. Cleveland Dial.* 534 T' au'd lady's a gret age. She'll be a desper't *tie* on em. **1928** A. Huxley *Point Counter Point* xix. 343 Free, without ties, unpossessed by any possessions, free to do as one will, to go at a moment's notice wherever the fancy may suggest—it is good. **1960** R. Collier *House called Memory* iii. 45 We'd love to do an evening show sometimes but the children are *such* a tie.

¶ *to ride in tie* (earlier example).

1870 G. T. Curtis *Life D. Webster* I. 37 As Mr Webster once humorously expressed their frequent interchange of study and labour for their joint support, as they had but one horse between them, they 'rode in tie'.

e. *Logic.* Something that unites the elements of a linguistic construct, e.g. the verb 'to be'.

1918 W. E. Johnson in *Mind* XXVII. 14 In order to understand the verbal juxtaposition of substantive and adjective, we must recognise a latent element of form in this construct... This element of form constitutes what I shall call the characterising tie. **1921** —— *Logic* I. i. 10 The general term 'tie' is used to denote what..is involved in understanding the specific form of unity that gives significance to the construct. **1923** C. D. Broad *Sci. Thought* ii. 75 Take first a very simple characterising judgment, like '3 is a prime.'..We might say that the first judgment is about the number 3 and the characteristic of primeness, and asserts that they are connected by the characterising tie. **1959** P. F. Strawson *Individuals* v. 168 To the characterizing tie between Socrates and the universal, *dying*, there corresponds the attributive tie between Socrates and the particular, his death.

tie, *v.* Add: **1. f.** *intr.* for *pass.*

1842 *Amer. Pioneer* I. 274 A pair of buckskin leggins,.. made to fit the leg and tie in at the ankle with the moccasins. **1924** A. D. Sedgwick *Little French Girl* ii. iii. 114 Straightly falling dress,..tying at the breast with tassels and at the waist with a loosely knotted sash.

2. *to tie to the stake*: delete † *obs.* and add later examples; *tie that bull outside* or *to another ashcan* (U.S. slang): I do not believe you; 'tell me another'; *to tie a can to* (or *on*) (slang): to reject or dismiss (a person); to stop (an activity); *to tie one on* [cf. *to tie a bun on* s.v. *bun *sb.*⁵] *slang* (chiefly *U.S.*): to get drunk.

1605 Shaks. *Lear* iii. vii. 54, I am tyed to' th' Stake, And I must stand the Course. **1921** J. Dos Passos *Three Soldiers* iv. i. 212 'Fellers, the war's over!'..'Tie that bull outside,' came from every side of the ward. **1922** H. Crane *Let.* 10 Dec. (1965) 108 Life is meagre with me. I am unsatisfied and left always begging for beauty. I am tied to the stake—a little more wastefully burnt every day of my life. **1926** Wodehouse *Heart of Goof* viii. 265 What caused the definite rift was Jane's refusal to tie a can to Rodney Spelvin. **1928** C. Sandburg *Good Morning, America* 16 They got a fat nerve to try to tie a can on you. **1932** J. T. Farrell *Young Lonigan* ii. 60 Three-Star told Vinc to tie his bull to another ash can. **1933** E. O'Neill *Ah, Wilderness!* i. 27 Aw say, you fresh kid, tie that bull outside! **1942** Wodehouse *Money in Bank* (1946) xix. 163 Tie a can to the funny stuff, see? If I want to laugh, I'll read the comic strip. **1951** *Western Folklore* X. 82 The Act of Drinking:..to swill one down; to tie one on. **1959** *Listener* 4 June 971/1 That was what lost Mr. Acheson votes when he was tied to the Senatorial stake. **1962** J. Onslow *Bowler-Hatted Cowboy* xix. 186 You used to tie one on with the boys. **1972** Wodehouse *Pearls, Girls, & Monty Bodkin* v. 65 I'm warning you to kiss her goodbye and tie a can to her. Never marry anyone who makes conditions. **1982** A. Mather *Impetuous Masquerade* vii. 107 He had..tied one on, if you know what I mean.

3. a. Also with *into*, = *to tie in to* (see sense 10* a below).

1913 in *O.E.D.* s.v. *Tie* vb. 3 e. **1969** D. Acheson *Present at Creation* xliv. 402 The white telephone tied into the White House switchboard was used sparingly by considerate associates. **1974** *Sci. Amer.* Oct. 113/1 We began work at the complex by establishing over the target area a submerged grid of 10-meter squares that was tied into the Greek ordnance survey grid ashore.

f. *intr.* and *trans.* Of a dog or bitch: to remain linked (with) for a period during copulation.

1910 J. S. Turner *Kennel Encycl.* III. 919 Occasionally a dog does not tie in the normal fashion... A dog that regularly ties is preferable. **1934** F. W. Cousens *Dogs & their Managem.* v. 83 When the stud dog is unable to 'tie' a bitch, he is unable to remain sufficiently long in position to impregnate the bitch properly. **1952** C. L. B. Hubbard *Pembrokeshire Corgi Handbk.* v. 60 Not all breeds tie. **1968** J. F. Gordon *Pet Library's Beagle Guide* x. 149 Once the pair have tied, they can be steadied, and..left to complete their task.

5. f. To impose conditions on (foreign aid), esp. by restricting its use *to* purchases from the source country. Cf. *tied *ppl. a.* 2 c.

1965 *McGraw-Hill Dict. Mod. Econ.* 515 A considerable part of U.S. foreign aid has been tied. **1965** *New Statesman* 18 June 945/3 This strain [on the balance of payments] can be reduced by 'tying' aid—insisting that it be spent on British exports. **1976** *New Internationalist* Jan. 7/2 Virtually all aid from the USSR is tied to the purchase of Russian goods or expertise. **1980** *North-South* (Rep. Independent Comm. Internat. Devel. Issues) xii. 198 When they tie aid to their own sources the donor countries greatly limit choices and discourage local initiatives.

7. d. (Further examples.) Now chiefly *N. Amer.*

1966 *N.Y. Times* (Internat. ed.) 22 Apr. 12/5 Real Madrid tied Internazionale of Milan, 1–1, last night. **1968** *Globe & Mail* (Toronto) 17 Feb. 39 If Canada ties Russians, Swedes beat Czechs—Russia wins gold on goal spread, Canada takes silver. **1977** *Arab Times* 13 Dec. 9/1 The American Embassy is currently in second place and needs a victory to tie Dresser and force a play-off for the League Championship.

e. *N. Amer.* To match or equal (an existing record or score); colloq. phrases *can you tie that?, tie that!*, expressions of surprise or amazement. Cf. *beat *v.*¹ 10 g.

1918 *Collier's* 11 May 46/3 The French won't even admit he's dead yet—they call this joint the Invalides, which is only concedin' that he's *sick!* Can you tie that? **1930** *Sat. Even. Post* 28 June 162 'Can you beat that?' he muttered. 'Can you even tie it?' **1932** W. Faulkner *Light in August* viii. 172 Well, say. Can you tie that. **1946** *Sat. Even. Post* 30 Mar. 46/2 Mr. Carter got hold of Blily Rose and offered him $1000 a day for 100 days if he would come to Fort Worth and put on a show that not only couldn't be beat but couldn't be tied. **1948** Wodehouse *Uncle Dynamite* vi. 83 Tie that for a disaster, Uncle Fred. **1968** *Globe & Mail* (Toronto) 15 Jan. 21/3 Willie Turner, a young sprinter who has yet to reach his peak, tied a world indoor standard Saturday. **1974** *State* (Columbia, S. Carolina) 3 Mar. 6-d/1 Cincinnati tied a school record by hitting 16 of 17 free throws. **1978** *Detroit Free Press* 5 Mar. c2/1 Connors, a 25-year-old lefthander from Belleville, Ill., quickly served a love game to tie the score.

9. Also, to get to work vigorously on; to tuck into (food). (Further examples.) *U.S.*

1912 R. A. Wason *Friar Tuck* xiv. 99 They girded up their loins, an' tied into him a little harder. **1948** 'J. Evans' *Halo for Satan* ix. 130 She put her head back and tied into her drink with the easy grace of a practiced drinker. **1965** M. Bradbury *Stepping Westward* v. 238 I'm going to take a peanut-butter sandwich..but I want to see these important men tie into something really good.

II. 10*. tie in. a. *trans.* To connect or join *to* an existing structure or network.

1793 [in Dict., sense 3 a]. **1914** *Dialect Notes* IV. 164 *Tie in*, in surveying, to join or connect up. 'We'll run over to the monument and *tie in* this survey.' **1943** J. S. Huxley *TVA* xi. 95 The framing to the exit..neatly ties in the air exhaust trough at the bottom of the walls. **1975** *North Sea Background Notes* (Brit. Petroleum Co.) 30 It is not impossible that a branch line from another nearby oilfield may be tied in to the Forties line in the future. **1978** *Lancashire Life* July 37/3 Instead of being tied-in to the building next-door this 19th century addition was simply slapped-up alongside it.

b. *intr.* To accord or be consonant (*with*); to be connected or associated (*with*).

1938 S. Chase *Tyranny of Words* viii. 91 This ties in with Korzybski's central idea of knowledge as structural. **1954** 'A. Garve' *Riddle of Samson* x. 97 There's another thing that ties in rather neatly, too. **1959** H. Nielsen *Fifth Caller* xiv. 216 A stranger? That didn't tie in with the words Dr. Whitehall was quoted as having used in greeting. **1967** *Sci. Amer.* Sept. 276 The problem ties in with the discussion of Pascal's triangle. **1972** D. Lodge *20th Cent. Lit. Crit.* 174 Jung's theory of the Collective Unconscious tied in neatly with the anthropological study of primitive myth and ritual, initiated..by Sir James Frazer in *The Golden Bough*.

c. *trans.* To associate or connect (*with*).

1958 'A. Bridge' *Portuguese Escape* ix. 146 How can they have tied the Monsignor in with the Duke's house? **1959** *Listener* 26 Feb. 364/1 Nowadays, more emphasis is placed on teaching foreign languages phonetically and on trying to tie lessons in with exchange visits of pupils abroad. **1972** 'T. Coe' *Don't lie to Me* (1974) xi. 102 The detectives on the case think the two things are tied in. The killing and the acid. **1972** J. L. Dillard *Black English* iv. 140 Pidgin has been tied in historically with a lot of regrettable racial and economic policies. **1982** R. Leigh *Girl with Bright Head* xix. 131 'Just tell me what Mrs Storm wanted with you.' 'Not unless you can tie her in with the murder.'

10. tie off. a.** *trans.* To close (a tubular vessel) by tying something round it. Also *transf.*

1903 J. J. McGrath *Surg. Anat. & Operative Surg.* i. 13 In resecting portions of the alimentary canal the mesentery or omentum that carries the blood-supply to the parts must be tied off. **1973** 'D. Halliday' *Dolly & Starry Bird* ii. 27 You must have Digham tied off... I won't have you become preggy.

b. *trans.* To secure or make fast (a rope or line); also *fig.* Also *absol.*

1928 *Amer. Speech* IV. 69 [Stage-hand language.] The lines pass..down to a fly-floor,..where they are tied-off, or belayed. **1933** P. Godfrey *Back-Stage* vii. 88 Stage-hands are speaking strangely cryptic phrases to people overhead... 'Up on yer long—dead it—tie off at that—mark yer new set.' **1952** R. Bissell *Monongahela* xix. 217 While the deckhand is tying off you jump down out of the brain box and knock the face wires loose. **1973** J. Thomson *Death Cap* x. 143, I like all the ends tied off and *Finis* written on the file. **1974** H. MacInnes *Climb to Lost World* xi. 193, I..asked him to tie-off the bottom end of Joe's rope.

11. tie up. a. Also *intr.* for *pass.*

1865 'L. Carroll' *Alice's Adv. Wonderland* xi. 172 A large canvass bag, which tied up at the mouth with strings.

c. Also, to hold up; to keep busy or occupied. Chiefly *pass.* orig. *U.S.*

1887 C. B. George *40 Years on Rail* vii. 140, I ran into a snow-storm that tied us up until we were six days making the run. **1907** *Springfield* (Mass.) *Weekly Republican* 10 Oct. 16 Traffic west of Springfield was tied up until about midnight. **1935** D. L. Sayers *Gaudy Night* xiv. 295, I meant to come round yesterday evening, but I got tied up with people. **1941** B. Schulberg *What makes Sammy Run?* iii. 53 He was tied up in a story conference. **1959** W. D. Pereira *North Flight* ii. 29 Sir Arthur's terribly tied up at the moment and regrets he cannot speak to you. **1973** *New Yorker* 24 Feb. 36/1 The *World Almanac and Book of Facts* is a small buoy indeed but one that, whenever we stop to read it, ties us up for several hours. **1978** *Nature* 21 Sept. p. xii/2 The computer or scope is tied up only a fraction of a second while the exposure is made. **1980** D. Lodge *How far can you Go?* iv. 125 She sent her apologies, but she's tied up organizing some bazaar.

i. *to tie* (a person) *up in(to) knots* (or *a knot*): see *knot *sb.*¹ 10 a.

j. *intr.* To associate or unite oneself or one's interests *with* (or *to*). Also *trans.*, to associate (one thing) *with* another. orig. *U.S.* Cf. *tie-up *sb.* 7 b.

1888 *Texas Siftings* 3 Mar. 3/1 He's all O.K. There is no subterfuge about him... He is a man who will do to tie up to. **1903** *N.Y. Even. Post* 5 Dec. 1 It becomes his first interest to make business for that yard. He can best do this by tying up with the other navy yard representatives on the committee. **1904** *Indianapolis News* 21 June 6 The assurance that Captain New is to have a good post may be the reason that so many fellows want to tie up to him. **1925** *Round Table* June 593 It is clearly to South Africa's interest to tie up definitely either with sterling.. or with gold. **1928** *Daily Express* 13 June 3/4 Registered readers..have..'tied up' with the newspaper which.. offers the best..insurance benefits. **1943** [see *generation 2 b]. **1958** *Times Lit. Suppl.* 19 Sept. 526/2 [He] does not rest solely on his spade but takes every opportunity of tying up archaeological discoveries with references obtainable from written authorities.

k. To bring to a satisfactory conclusion.

1954 'R. Crompton' *William & Moon Rocket* i. 25 Taking that lorry's number and giving a description of where the shed was on the Minster road. Tied things up a treat, that did. **1959** *Listener* 12 Feb. 305/1 When the play ends..one is left intentionally with the feeling that not everything has been tied up. **1973** A. Broinowski *Take One Ambassador* vii. 90 The trade mark of the few in the know. That ties it all up. **1980** S. Brett *Dead Side of Mike* xiii. 147 It all fits in... It just ties up the whole package.

l. *intr.* = sense 10* b above.

1959 M. Gilbert *Blood & Judgement* xiii. 138 'That would make him..in his late fifties now.' 'Which ties up all right with our man.' **1968** *Listener* 20 June 799/2 This may well tie up with the fact that he was an intensely religious person who believed in people going to hell and being saved. **1974** J. Aiken *Midnight is Place* iv. 130 'He had had two men sent to jail for protesting.' 'Yes, that seems to tie oop with what we had heard.'

tie-. Add: **1.** *tie fabric*; **tie-break, -breaker,** a means of deciding a winner out of two or more contestants who have tied; also *fig.*; so **tie-breaking** *ppl. a.* and *vbl. sb.*; **tie-clasp, clip,** a small ornamental clasp for securing the ends of a tie to one's shirt; **tie press,** an instrument for pressing ties; **tie rack,** a rack on which to hang neckties; **tie silk,** a strong silk fabric used esp. for ties and clothing; cf. Foulard 1; **tie tack,** a two-part ornamental fastener for a necktie, one part of which is worn under the shirt to receive the point of the part worn on the tie.

1970 *Times* 5 Mar. 13 In principle, the tie-break is an undesirable expedient, but there is a case for it in indoor tournaments confined to one court. **1974** *Observer* 1 Sept. 18/6 In the tie break Miss Mappin led 4–1. **1979** *Daily Tel.* 10 Dec. 19/1 Nigel..failed in a tie-break to win the British Chess championship in August. **1961** Webster, Tie-breaker. **1970** *New Yorker* 10 Oct. 179/1 There are several species of tie-breakers, but the one that Bill Talbert, the tournament director, selected..was the nine-point sudden-death variety. **1971** *Computers & Humanities* VI. 68 The identifiers will be indexed and will serve as ultimate tie-breakers in all sorting operations. **1979** G. Hammond *Dead Game* xiv. 188 At the end of the quiz, honours were even..and the chairman asked for a tie-breaker from the audience. **1982** *Daily Tel.* 21 Sept. 16/4 [Rifle-shooting.] Belither..beat Paul Kent..by a single point on a tiebreaker. **1970** *Times* 1 Oct. 10/6 Okker, of the Netherlands, had to battle through two tie-breaking sets to beat El Shafei..7–6, 7–6. **1971** *Jrnl. Gen. Psychol.* LXXXV. 265 The stratification procedure entails a large number of random assignments and tie-breakings. **1978** R. Nixon *Memoirs* 85 His only important functions were to cast occasional tie-breaking votes in the Senate. **1955** W. Gaddis *Recognitions* ii. i. 285 They..fastened monogrammed tie-clasps the more firmly. **1971** 'D. Shannon' *Murder with Love* (1972) ix. 153 His dapper tailoring, gold cuff links and tie-clasp. **1898–9** *T. Eaton & Co. Catal.* Fall & Winter 16/1 Tie clips... For holding the tie in place on shirt or blouse fronts, 2 for 5c. **1913** [see *innovation 6]. **1976** 'R. Boyle' *Cry Rape* xii. 62 It wasn't an elegant tie-clip. Not the kind with a diamond or emerald in it. **1931** *Fairchild's Fabrics Buyers' Guide* I. 143 (*heading*) Tie and muffler fabrics. **1977** *Man-Made Textiles in India* XX. 92/3 The attributes which make a tie fabric satisfactory in use are similar to those in a dress fabric. **1926–7** *Army & Navy Stores Catal.* 757/3 'Watts' tie press..9½ in. by 3½ in. *a* **1974** T. R. Dennis in J. Burnett *Useful Toil* (1974) iii. 354 A very cheap watch from my parents..a tie-press from a friend. **1916** *Daily Colonist* (Victoria, B.C.) 1 July 9/3 (Advt.), Pipe Racks, Tie Racks and Collar Bags, regular to 75c. **1974** L. Deighton *Spy Story* ii. 22 Rummaging through the wardrobe I..noticed the tie rack had been moved. **1920** M. S. Woolman *Clothing* iv. 53 Ribbons, velvets, tie silks..and knitting silks are made of this fiber. **1934** *Vogue* 30 May 96 (*caption*) Jenny makes a tailored suit of checked tie silk. **1961** *Guardian* 30 Mar. 9/4 Charming and practical tie-silk shirtwaisters. **1961** Webster, Tie tack. **1962** 'D. Shannon' *Extra Kill* xi. 171 He'd always wear a tie clasp, or one of those new tie tacks. **1970** *New Yorker* 10 Jan. 56/3 (Advt.), Peace Tie Tac. **1980** *Outdoor Life* (U.S.) (Northeast ed.) Oct. 154/3 (Advt.), Detective profession. Easy home study; free tie tack or lapel pin.

2. tie-back: also, a device for holding a drawn curtain back from the window; **tie-down,** the state (of an aircraft, etc.) of being tied down or otherwise kept on the ground; also, a device to or with which something may be tied down; freq. *attrib.*; **tie-on** *a.* (later examples.) See also *tie-in, *tie-off.

1927 *Ladies' Home Jrnl.* Dec. 35/3 Flat festoons of green used for tie-backs on the curtains. **1961** *Times* 14 Jan. 9/7 Brass tie-backs in various shapes for holding the drawn but voluminous folds of the mid-Victorian drawing room curtains became a period 'must'. **1982** Barr & York *Official Sloane Ranger Handbk.* 148/1 Sloane windows need curtains with a capital C: with pelmets, twiddly bits, bands, tassels, tie-backs, edging. **1952** *Sun* (Baltimore) 4 Aug. 1/5 $1 tiedown fees for [flying] saucers less than 1,000 feet in diameter. **1955** *Ibid.* 11 Feb. 2/4 Exhaustive 'tie-down' ground tests are scheduled for the XV-3..before actual flight tests begin. **1956** W. A. Heflin *U.S. Air Force Dict.* 525/1 *Tie-down*, a ring, hook, stake, or the like to which something is secured; a tie-down fitting. **1969** *Jane's Freight Containers 1968–69* 418/1 Forklift entries and tie-down inserts are provided. **1971** *Flying* Apr. 13/1 Big tie-down areas for smaller airplanes. **1974** *Union* (S. Carolina) *Daily Times* 22 Apr. 7/7 (Advt.), Mobile homes anchored: Storm tie-downs to guard you against wind damage. **1978** F. Mullally *Deadly Payoff* xii. 171 He let the loosened end of the tie-down rope fall to the ground and kicked the chocks away from the front wheels. **1982** *Chicago Sun-Times* 31 Oct. 8 Another 30 residences..have planes with tie-downs instead of hangars. **1949** Tie-on [see *occluder]. **1967** E. Short *Embroidery & Fabric Collage* iii. 80 Tie-on cushions for dining chairs. **1971** 'D. Halliday' *Dolly & Doctor Bird* xii. 161 My skin became brown..between my tie-on tops and my hipsters.

3. tie belt, a belt which is fastened by tying; hence **tie-belted** *a.*; **tie-dye** *sb.* = *tie-and-dye*, sense 4 below; freq. *attrib.*; also as *v. trans.*, to dye by this process; also *absol.*; hence **tie-dyed** *ppl. a.*, **-dying** *vbl. sb.*; **tie game,** a game in which the result is a tie; **tie-neck,** a collar attached at the back of the neck but left loose in front so that the ends can be tied; hence **tie-necked** *a.*; **tie-post** (earlier example); **tie rail** orig. *U.S.*, a rail or railing to which horses may be hitched; **tie rod,** (*a*) (further examples); (*b*) a track rod, or one of the rods of which it is composed, in the steering gear of a motor vehicle; **tie-teeth** *W. Indian colloq.*, any sticky foodstuff which is difficult to chew; **tie-tie** (earlier and later examples).

1964 *McCall's Sewing* xii. 227/1 A tie belt, without stiffening, becomes a string in no time. **1977** *Daily Mirror* 16 Mar. 16/2 (Advt.), The tie-belt style is 12–18. **1976** *Woman's Weekly* 6 Nov. 4/2 Tunic top is hip-length, tie-belted and tie-necked. **1904** G. Watt *Indian Art at Delhi* vii. xxxi. 255 From Chamba State has been received..a most remarkable..cotton fabric woven in alternate bands of cotton and gold thread, the cotton being tie-dyed so as to show large wavy formations. **1926** *Daily Colonist* (Victoria, B.C.) 22 July 8/7 (Advt.), Marvelous tie-dye patterns, glorious color-blendings are amazingly simple to make. **1951** A. N. Gulati *Patolu of Gujarat* 18 The first essential step, therefore, is to tie-dye both warp and weft in conformity with the proposed design in the fabric. **1956** J. Irwin in *Textiles & Ornaments India* 29 The so-called *ikat*-technique is another kind of tie-dye. **1970** *Time* 20 Apr. 72 The stars fussed with their see-through dresses, tie-dyes and black ties and then paraded up a red-carpeted walkway. **1971** *New Yorker* 4 Sept. 61 What the boys who tie-dye and the grandmothers who rug-hook are doing is, in effect, as mechanical as anything done by a machine. **1975** *Advocate-News* (Barbados) 28 June 1/7 Organza flowers, soft toys, tie-dye and other craft work will be displayed by the YWCA craft group. **1977** *Guardian* 10 Jan. 8/1 The pallid youth in the tie-die shirt with a sewn-on picture of Marx. **1904** G. Watt *Indian Art at Delhi* vii. xxxi. 257 The beauty of these warp and weft tie-dyed textiles. **1978** J. Updike *Coup* (1979) vi. 236 Their countrymen wearing cowboy hats, blue jeans, tie-dyed T-shirts. **1904** G. Watt *Indian Art at Delhi* vii. xxxi. 252 The once famous Bandana handkerchiefs may be given as the best known example of tie-dyeing. **1939** G. Clark *Archæol. & Society* iii. 63 The tunics of the notables, which were.. coloured by the tie-dying method. **1970** *Time* 26 Jan. 40 The art is almost as old as India—where it is called *bandhnu*. It is as new as the boutiques that blossom along Sunset Strip and Madison Avenue—where it is called tie-dying. **1742** in H. T. Waghorn *Dawn of Cricket* (1906) 12 That played the tie game the beginning of the season. **1832** P. Egan *Bk. of Sports* xxii. 347/2 In the first innings Woking gained 71 runs, Shiere then went in and got 71. Second innings, Woking 71; ditto, Shiere 71; it was consequently a *tye*-game, under circumstances unprecedented in the annals of cricket-playing. **1928** *Collier's* 29 Dec. 17/4 A tie game in football is certainly more thrilling..than a one-sided game. **1960** *Washington Post* 18 Oct. a18/1 It was Eddie Erdelatz, the Navy football coach, who once described a tie game as an unsatisfactory experience that permitted no enthusiasm. 'A tie game is like kissing your sister,' he said. **1968** J. Ironside *Fashion Alphabet* 54 Tie neck. **1983** *Daily Tel.* 28 Feb. 24/5 The Queen's outfit was a navy-and-white jacket and dress with ..a tie neck. **1973** *Country Life* 22 Feb. 490/3 A tie-necked silk shirt. **1977** *Harpers & Queen* Sept. 44/1 A tie-necked champagne blouse in washable crepe. **1861** *Harper's Mag.* Feb. 424/2 He alighted,..throwing the reins over a tie-post. **1920** C. E. Mulford *Johnny Nelson* 238 He'll never forget my kickin' him off'n th' tie-rail. **1970** *Sunday Mail Mag.* (Brisbane) 9 Aug. 14/2 The stranger dismounted and hitched his horse to the tierails in front of the pub. **1910** J. Gunn *Practical Design Motor Cars* ix. 230 The rod which transmits the motion of the steering gear to the front wheels should be connected at

the front end..to the tie-rod. **1922** Joyce *Ulysses* 698 Water closet..provided with opaque singlepane oblong window, tipup seat, bracket lamp, brass tierod brace. **1966** *McGraw-Hill Encycl. Sci. & Technol.* XIII. 640/1 In pressure piping, large forces are produced between connected parts. The pipes or parts are constrained by tie rods. **1976** *Jrnl.* (Newcastle) 26 Nov. (Advt.), Viva, 1969, white, taxed 11 months,..new balls and tie rods. *c* **1915** in Cassidy & Le Page *Dict. Jamaican Eng.* (1967) 444 Tie-teeth. **1953** *Caribbean Q.* III. i. 9 *Tie-teeth* (candy, sweets, or other very sticky food). **1975** E. L. Ortiz *Caribbean Cookery* (1977) 266 If it is overcooked the mango paste turns into what Jamaicans graphically call tie-teeth. **1774** E. Long *Hist. Jamaica* II. 427 The Negroes seem very fond of reduplications..as..*tie-tie, lilly-lilly, fum-fum.* **1827** *Hamel, Obeah Man* II. 257 Stretching what they called a *tie-tie* of tent-ropes, hempen cordage, mahoe bark, and bush ropes, all spliced together, to form..a guide for those who could be induced to cross the bridge. **1931** *Discovery* May 153/1 The whole of the frame-work [of a Nigerian house] is secured with what is known in pidgin English as *tie-tie* which is fibre from certain plants. **1958** C. Achebe *Things fall Apart* i. vii. 47 There were little holes..in the upper levels of the wall, and through these Okonkwo passed the rope, or *tie-tie*, to the boys and they passed it round the wooden stays and then back to him; and in this way the cover was strengthened on the wall.

4. Phrasal Comb.: **tie-and-dye,** a technique for producing a mottled appearance in dyed cloth by folding it and tying it before it is put in the dye bath; a garment or piece of cloth so dyed; freq. *attrib.* and unhyphened.

1886 *Jrnl. Indian Art* I. 117 The wonderfully constructed patterns of Patolo weaving with 'tie and die [*sic*]' warp and woof..testify..to the skill achieved by Indian dyers and weavers. **1928** *Daily Express* 21 May 5/2 The 'tie and dye' process—an old craft which gives a charming hazed effect. **1937** M. Covarrubias *Island of Bali* i. vii. 196 The Balinese often decorate pieces of silk by the tie-and-dye process. **1976** *Billings* (Montana) *Gaz.* 7 July 9-b/7 Demonstrations in the hotel courtyard will include ..silver casting, tie and dye, silk screening and oil painting.

tied, *ppl. a.* Add: **2. c.** Of an international loan, etc.: given subject to conditions as to its use (see *tie *v.* 5 f).

1958 C. N. Henning *Internat. Finance* iv. xxi. 441/1 The so-called 'tied loan' principle. **1961** *Ann. Reg. 1960* 170 The issue of tied grants and credits was the subject of some dispute at international meetings. **1965** *McGraw-Hill Dict. Mod. Econ.* 515 The advantages of tied loans are that they stimulate employment and income in the creditor nation and do not affect the balance of payments of that country adversely. **1976** *New Internationalist* Jan. 7 For the Third World, 'tied' aid generally means having to pay between 20% and 50% more for goods than the competitive world market level.

d. Of a retail garage: of which the tenant is bound to receive fuel from a particular supplier.

1957 *Economist* 7 Dec. 885/1 About a third of these 'tied garages' have been signed up for periods of five years or less; the other two-thirds, which sell over half the petrol sold through dealers, are tied to their suppliers for longer periods of up to 20 years. **1965** [see *solus *a.* 3 b].

3. b. *tied-back*: held back by tying.

1895 M. Beerbohm in *Yellow Book* IV. 280 The women wore jerseys and tied-back skirts. **1979** A. Buck *Dress in Eighteenth-Century England* 32/1 (*caption*) The tied-back hair in a bag.

tied (təid), *a.* [f. Tie *sb.* + -ed².] Wearing a tie.

1911 G. K. Chesterton *Innocence of Father Brown* iv. 105 The red-tied youth. **1976** Scott & Koski *Walk-In* (1977) ii. 17 One clean-shaven, suited and tied, scrubbed Asian.

tie-in (tai·in). orig. and chiefly *U.S.* [f. vbl. phr. *to tie in*: see *tie *v.* 10*.] **1. a.** A connection or association *with*; a link-up.

1934 in Webster. **1941** W. Kozlenko *100 Non-Royalty Radio Plays* 535/2 That's the tie in. He killed Tom and after that decided to really cripple the Dominion team so that we wouldn't have a chance at winning. **1949** M. Mead *Male & Female* vii. 152 The tie-in with birth can be close. **1965** *Listener* 10 June 875/1 The London studio tie-in with the pictures from America was also very successful, with John Tidmarsh getting lucid technical comments out of Geoffrey Pardoe at each stage of the operation. **1972** 'T. Coe' *Don't lie to Me* (1974) vii. 66, I didn't know the museum had a tie-in with City College. **1973** *Black Panther* 21 July 2/3 Could you say something about the tie-in of David Hilliard's case with Watergate?

b. Used *attrib.* with reference to sales that are made conditional upon the purchase of some additional item or items from the same supplier.

1943 *Amer. N. & Q.* July 54/2 The phrase 'tie-in sales' (referring to those whereby tradesmen are obliged to buy unwanted stock in order to get even a small amount of a scarce item) appears to have established itself..about the first week of June (1943). **1946** *Sun* (Baltimore) 8 Feb. 2/6 This [liquor] industry can put an immediate stop to tie-in selling. **1980** *Times* 13 Sept. 18/7 It was information from an individual 'mole', working temporarily in a discount store, that first alerted the National Consumer Council to..'tie-in sales' (a stipulation that a buyer must purchase part or all of his requirements of a second (tied) product from the supplier of a first (tying) product).

c. An association between two publicity campaigns in the form of a theme common to both, or an advertisement that appears in two different media.

1949 *Newsweek* 28 Nov. 70/1 By next May, Paramount expects to have spent $1,000,000 calling attention to the show by way of..tie-ins with fashion designers and department stores who will make and sell women's clothes influenced by the 'Minoan period' costumes [in a film about Samson and Delilah]. **1959** I. Ross *Image Merchants* ix. 149 Ryland [of NBC] also spends a good deal of time working up promotional tie-ins with manufacturers, department stores..and publications. It is a major coup when a national magazine can be persuaded to incorporate a TV plug in its own vast promotional outlay. **1962** *Economist* 10 Feb. 521/1 'Calories Don't Count' has been offered to the public as a promotional tie-in to encourage the purchase of safflower oil.

d. A book, film, or the like published to take advantage of the appearance of the same work in another medium.

1962 *Publishers' Weekly* 23 Apr. 43/1 Students show an interest in books which have been made into successful movies. Watch for these tie-ins. **1976** *Ibid.* 29 Mar. 42/1 The paper edition of Ibsen's 'Hedda Gabler and Other Plays' even has a movie tie-in cover with a still from the recent Glenda Jackson film. **1981** *Times Lit. Suppl.* 17 Apr. 425/2 The 1970s was the decade of the 'tie-in'—the almost simultaneous film of the novel, novel of the film, TV series of the novel.

2. (The making of) a connection between two pipelines or sections of pipeline. Freq. *attrib.*

1975 *North Sea Background Notes* (Brit. Petroleum Co.) 32 The line was welded, trenched and buried as quickly as possible, with only tie-in ends being left temporarily exposed. **1975** *Offshore Engineer* Nov. 18/2 The towing method..could be developed for installing a pipeline across the Norwegian trench in section, with tie-ins between them. **1976** *Ibid.* July 5/3 The 406mm line from SPAR to Brent B is being completed by Serra-Comex tie-in barge *Sandokan*. **1977** *Ibid.* Apr. 9/2 Three wells have been completed on the field and await tie-in when the production decision is made.

tie-line. [f. Tie- + Line *sb.*²] **1.** (In Dict. s.v. Tie- 3.)

2. *Teleph.* A line connecting two private branch exchanges. Also *transf.*

1923 T. E. Herbert *Telephony* xvii. 437 Should an answering plug be inserted into a jack of the tie line, and the corresponding calling plug be placed into an exchange jack, [etc.]. **1955** E. F. Russell *Somewhere a Voice* (1965) 107 Man is born of Earth and needs a tieline to Earth. **1969** J. Martin *Telecommunications & Computer* v. 91 Many companies have a leased system of telecommunications lines with switching facilities. To telephone a person in a distant company location, an employee must first obtain the appropriate tie line to that person's private branch exchange. **1978** *Broadcast* 29 May 9/2 Dring got a phone call out [of Zaire] via the BBC's New York office on tie line to the TV Centre just three minutes before the nine o'clock bulletin went out.

3. A line in a phase diagram joining two points that each represent the composition of two phases in equilibrium with one another.

1924 A. E. Hill in H. S. Taylor *Treat. Physical Chem.* I. ix. 400 No solution therefore can exist having composition indicated by points within the area *A–B–C–B'–A'*, but such compositions can lead only to two conjugate solutions of composition lying upon the curves at the intersections with the isothermal tie-lines. **1935** *Amer. Jrnl. Sci.* CCXXIX. 174 All liquids *a–b* are in equilibrium with olivine, each with an olivine of different composition as indicated by the tie-lines. **1961** *Geol. Mag.* XCVIII. 336 (*caption*) Tie-lines joining co-existing, chemically analysed pyroxenes from igneous rocks. **1979** *Nature* 3 May 53/2 (*caption*) Miscibility gap between carbonate and silicate melts... Conjugate liquids at 0·7 kbar and 7·6 kbar are joined by tie-lines.

4. A pipeline or transmission line connecting two distribution systems or two parts of a single system.

1949 *Sun* (Baltimore) 5 July 10/4 With the completion of a 48-inch tie-line..the city Bureau of Water Supply believes it could supply 4,000,000 gallons daily. **1962** *Newnes Conc. Encycl. Electr. Engin.* 348/1 If two machines [*sc.* a.c. generators] are to be coupled together by a tie-line perhaps 50 or 60 miles in length, there may be an appreciable impedance presented to the flow of the synchronizing current. **1974** *Sci. Amer.* Nov. 40/3 Its objective is to maintain the frequency within an area served by several generating stations and to maintain the sum of all active tie-line power exchanges between the area and its neighbors.

‖ **t'ien** (tiˀe·n). Also 7 tayn, 8 tyen, 8– tien. [Chinese *tiān.*] In Chinese thought: Heaven; the Deity.

1613 Purchas *Pilgrimage* iv. xvi. 373 All being a rude and vnformed Chaos, *Tayn* (say they) framed and setled the Heaven and Earth. **1710** *Memoires for Rome concerning State Christian Relig. China* iii. 71 He would not retract, nor acknowledge that *Tien*, that is to say, the visible Heavens was the God of the Christians. **1747** *Astley's New Gen. Coll. Voy.* IV. i. v. 202/1 It appears.. that this *Tyen*, or first Being, is the Creator of all Things. **1788** tr. *Grosier's Gen. Descr. China* II. vi. i. 186 Between the *Tien* and man there is a relation. **1878** *Jrnl. North-China Branch R. Asiatic Soc.* XII. 122 Even in the time of Confucius the change from the personality of T'ien-tsze to the abstract idea of the modern T'ien does not seem to have been complete. **1904** *Athenæum* 17 Sept. 373/2 The Chinese Emperor is not regarded as other than the regent of the empire appointed by *t'ien* or heaven. **1940** E. Pound *Cantos* liv. 39 The rites of *Tien*, that is Heaven Were ploughing and the raising of silk worms. **1958** W. Willetts *Chinese Art* I. ii. 92 After the Chou displaced the Shang-Yin, they apparently came to equate the old Shang Ti with their own chief deity T'ien.

tienda (tie·nda). orig. and chiefly *U.S.* [a. Sp., = tent, awning, shop: see Tent *sb.*¹] In the south-western U.S.: a shop or stall, esp. a draper's or general store.

1844 G. W. Kendall *Santa Fé Exped.* II. ii. 38 Standing in front of a small tienda, or store. **1870** J. C. Duval *Adv. Big-Foot Wallace* xxxviii. 235, I searched in vain every shop and 'tienda' in the city for even a pair of No. 11's, though 12's fit me best. **1912** C. F. Saunders *Indians of Terraced Houses* 71 Our proximity to the pueblo was indicated by our meeting Indians..on their way to the trader's *tienda* beneath the shady cottonwoods at Algodones. **1927** *Blackw. Mag.* Nov. 658/2 Its *tienda* with long counter, handsome shelves, and fabulous profits. **1948** F. Blake *Johnny Christmas* i. 10 Across the plaza, too, in the sprawling Martinez tienda, were articles that would trade or sell high in northeastern Texas.

‖ **tiens** (tiæ̃), *int.* [Fr., imp. sing. of *tenir* to hold.] An expression of surprise.

1932 G. Heyer *Devil's Cub* xi. 169 '*Tiens!*' said the Duchess with polite interest. 'My son is then a house-breaker.' **1958** L. Durrell *Mountolive* viii. 179 *Tiens!* I forgot. Here is the thousand I promised you. **1975** A. Christie *Curtain* xiii. 133 Did he take it by accident or intention. *Tiens*, his fingerprints are not on the bottle.

‖ **tienta** (tie·nta). [Sp., lit. 'probe'.] In Spain, an occasion at which young bulls in the field are tested for spirit as prospective stud and fighting bulls.

1909 J. Villiers-Wardell *Spain of Spanish* vi. 110 It is necessary for the *ganadero*—an owner of cattle—to test his young bulls while they are still running wild... This testing of the young bulls is called a *tienta*. **1932** E. Hemingway *Death in Afternoon* xii. 124 The strain of fighting blood..can only be kept pure by conscientious testing in the tientas. **1957** R. Campbell *Portugal* 109 His name..is given him along with his branded number and noted down on the day of the *tienta*. **1967** McCormick & Mascareñas *Compl. Aficionado* ii. 55 Memory of the *tienta* stirs and fires him to a prompt charge, his weariness forgotten.

‖ **tiento** (tie·nto). *Mus.* [Sp., lit. 'touch, feel'.] In sixteenth- and seventeenth-cent. Spanish music: a contrapuntal piece resembling a *ricercar*, orig. for strings and, later, organ.

1905 C. F. A. Williams *Story of Organ Music* xi. 163 The compositions of Cabezon's collection consist of nine practice pieces,..short preludes called Versos or Versillos, ..[and] longer preludes called Tientos. **1947** A. Einstein *Music in Romantic Era* xvii. 294 The 'schools' of the ricercar, whether called fugue as in Germany, *tiento* as in Spain, or fancy as in England, were different; but the spirit and form were the same. **1976** D. Munrow *Instruments Middle Ages & Renaissance* ix. 74/4 The solo *tiento* for harp or organ by Alonso Mudarra, published in his *Tres Libros* (1546). **1980** *Early Music* Apr. 248/1 Some polyphonic works of the period in the tradition of the *tiento* offer most remarkable variety within their stylistic limitations.

Tientsin (tyen(t)si·n, tin(t)si·n). The name of a city and port on the East coast of north China, used *attrib.* and *absol.* to designate carpets made or shipped from there.

1904, 1913 [see *Pekin 1 b]. **1922** Kendrick & Tattersall *Hand-Woven Carpets* I. i. vi. 66 Many carpets find their way to Tientsin for export, and on that account the term 'Tientsin Carpet' has come into use. **1980** D. Creed *Scarab* ii. xiv. 132 Deep-piled Tientsin carpets covered the tiled floor.

tie-off (təi·ǫf). [Tie- 2.] **1.** *Show-jumping.* = *jump-off 3.

1958 *Listener* 2 Oct. 536/3 We could see the tie-off for the Grand Prix between the Italian and German champions.

2. *Mountaineering.* (See quot. 1968.)

1968 P. Crew *Encyclopædic Dict. Mountaineering* 117/2 *Tie-off*, a method of reducing the leverage on a piton which has not been fully inserted, by tying a short loop of rope to the piton blade, close against the rock face. **1971** C. Bonington *Annapurna South Face* 247 We used huge quantities of tape for everything from tie-offs on pitons to belts for trousers. **1974** H. MacInnes *Climb to Lost World* x. 162 It took him most of the day to put a couple of bolts in, with some weird tie-off pegging to do.

Tiepolesque (tie:pole·sk), *a.* [f. *Tiepolo* (see below) + -esque.] Characteristic of or resembling the work of Giovanni Battista Tiepolo (1696–1770), or of his son Domenico (1727–1804), Italian painters famous esp. for (ceiling) frescos. Often used somewhat *loosely*.

1895 G. B. Shaw *Our Theatres in Nineties* (1932) I. 159 Goddesses in a Tiepolesque ceiling. **1934** *Burlington Mag.* Feb. 91/1 The purely Tiepolesque conception of the war-horse. **1958** *Times* 17 Oct. 17/6 The dash of Murillo in 'The Guerilla taking leave of his Confessor', or the much less happy Tiepolesque pose of the figure straining at the wheel of the gun in 'The Defence of Saragossa'. **1974** *Times Lit. Suppl.* 20 Dec. 1449/5 Her inability to understand Visconti's Tiepolesque conception of *Iphigénie en Tauride* is significant: she wanted to look Greek, primeval and savage whereas Visconti wished upon her the bejewelled artificiality of a Venetian fresco.

tier, *sb.*¹ Add: **1. e.** A range or line *of* contiguous lots, townships, counties, or states. *U.S.*

1693 in *Connecticut Hist. Soc. Coll.* (1912) XIV. 212 One lyeing in the Same Teere of lotts abutting on a Highway. **1720, 1722** [in Dict., sense 1 a]. **1824** in S. C. Cox *Recoll. Early Settlem. Wabash Valley* (1860) iii. 18 The land is sold in tiers of townships. **1856** *Spirit of Times* 18 Oct. 113/1 The great varying hare..is no longer to be found in our state,..until we reach the northern tier of counties, on the Canada line. **1949** *Ward County* (North Dakota) *Independent* 21 July 1/3 Each of the big wheat states in the tier from Texas up through North Dakota appears to be coming up with a crop just under that state's all-time record.

f. A mountainous scarp; a mountain. *Tasmania.*

1850 T. Arnold *Let.* 29 Sept. (1966) 55 The next day.. we all walked up a 'tier' (Tasmanian for hill) near the house. **1902** *Encycl. Brit.* XXXIII. 185/2 The marginal crests of this mountain table-land, together with its upper surface,..are known locally as 'Tiers'. **1965** *Austral. Encycl.* VIII. 425/2 Along the north-western coast [of Tasmania] there is a strip of rich, undulating land, climbing steadily to the base of steep escarpments called 'tiers'.

g. Each of a number of successively overlapping ruffles or flounces on a garment.

1934 in Webster. **1938** F. P. Walkup *Dressing Part* xv. 354 Tiers of ruffles, side pleats, and diagonal layers were introduced, for variety. **1978** *Detroit Free Press* 5 Mar. d12/1 Tiers will fall from little squared yokes, freely.

3. tiersman *Tasmanian colloq.* (see quot.); cf. sense 1 f above.

1941 Baker *Dict. Austral. Slang* 76 'Tiersman': one who lives in the mountains. Tasmanian slang.

tier, *sb.*² Add: **3.** *spec.* one for tying a sail.

1860 G. S. Nares *Naval Cadet's Guide* 81 The sail is then secured to the yard with tyers. **1873** 'Vanderdecken' *Yachts & Yachting* 265 Let the gaskets, or as they are sometimes called, the tyers, which confine the mainsail in its furl, be taken off. **1939** A. Ransome *Secret Water* iii. 44 In a minute or two, he had bundled the sail along the boom and put a couple of tiers to hold it there. **1947** —— *Great Northern?* viii. 101 The next few minutes were full of the regular drill of getting under way. Tyers were cast off the sails.

tierce, *sb.* **8.** *tierce de Picardie* (examples).

1849 J. A. Hamilton *Celebr. Dict.* 117 *Tierce de Picardie* (French), a term applied to the concluding chord of a piece of music in a minor key, when its third is made major by an accidental sharp or natural. **1889** Grove *Dict. Mus.* IV. 114/2 The Third, thus made major by an accidental sharp or natural, is called the 'tierce de Picardie'. **1940** *Scrutiny* IX. 128 Beneath the suspended F sharp of the concluding tierce de picardie, trombones, violas, and horns emphatically sing. **1959** D. Cooke *Language of Music* ii. 57 For centuries, pieces in a minor key had to have a 'happy ending'—a final major chord (the 'tierce de Picardie') or a bare fifth. **1978** *New Universities Q.* XXXII. 288 The final cadence achieves, and deserves, the bliss of a tierce de Picardie.

tiercé, *a.* Restrict *Her.* to sense in Dict. and add: **B.** *sb.* Also **tierce. a.** A method of betting in which the first three horses in a race have to be named in the correct order. **b.** A French horse-race at which this method prevails. **c.** *fig.*

1964 A. Wykes *Gambling* viii. 195 The French have developed a special way of betting... It is called 'Tiercé' and involves the selection of three horses to finish first, second, and third in specified races held on Sundays and holidays. **1966** *Economist* 24 Dec. 1336/4 The French happily bet vast sums on the *tierce*. **1974** *Times Lit. Suppl.* 6 Sept. 941/4 She had achieved a far more considerable success, having managed both to break up her sister Eliza's marriage..and, better still, indirectly to cause the death of her infant niece... This was not a *tiercé*, but it was a considerable double triumph. **1981** *Times* 30 June 6/5 Frenchmen had to do without the 'Tiercé'..yesterday. *Ibid.* 6/6 It was the third time since ..May, 1968 that the Tiercé fixtures had not taken place.

tiered, *a.* Add: Now freq. not in Combs.

1930 *Antiquity* IV. 422 He regards the tomb of Cyrus as a reproduction in stone of such a tiered temple. **1939** M. B. Picken *Lang. of Fashion* 155/1 Tiered skirt. **1951** *Good Housek. Home Encycl.* 276/1 The tiered steamers have..three separate pans which fit on top of each other over a base pan. **1952** C. P. Blacker *Eugenics* 294 Certain professions—the Church, the law, the fighting Services—have a hierarchical structure, and success can be gauged by the level attained by an individual in such a tiered system. **1955** V. Cronin *Wise Man from West* 92 Among the limitless paddy-fields and tiered hills. **1971** *Daily Tel.* 28 July 11 The skirt tiered all the way down. **1978** *Morecambe Guardian* 14 Mar. 16/3 She was attended by her sister..who wore a pale blue chiffon tiered dress.

tiering (tiᵊ·riŋ), *vbl. sb.* [f. Tier *sb.*¹ + -ing¹.] Arrangement in tiers; the formation of tiers.

1969 T. E. B. Howarth *Culture, Anarchy & Public Schools* v. 78 Local authorities are required under circular 10/65 to plan a complete system of comprehensive education for their areas either by the erection of all-through comprehensive schools or by various devices of amalgamation or tiering ('middle schools', sixth form colleges and so on). **1971** *Daily Tel.* 28 July 11 Cardin has run amok.. with the tucking and tiering attachments on his sewing machine. **1978** *Detroit Free Press* 5 Mar. d12/1 Tiering [of dresses] is a continuation of the peasant theme that has been with us for what seems like a long, long time. **1980** *Amer. Film* Oct. 25/1 Viewers are willing to pay generously..for more programming. There is a term for this program proliferation: 'tiering'. When a cable system adds another channel of programs for an extra fee, there is a new tier of service to offer subscribers.

‖ **tiers monde** (tyε̄r mǫ̃nd). Also **Tiers Monde.** [Fr.: see *Third World.] = *Third World.

1963 *Listener* 28 Feb. 386/1 The book exemplifies the need for utter empiricism in our approach to the *tiers monde*. **1964** *Financial Times* 11 Feb. 7/6 France is strongly placed to make a success of her championship of the 'Tiers Monde'. **1966** *Listener* 18 Aug. 229/1 What has commonly come to be called the *tiers monde*, the uncommitted nations, Africa and Asia, have a position in the world.

Tietze (tī·tsə). *Path.* [The name of A. *Tietze* (1864–1927), Polish surgeon, who described the condition in 1921 (*Berlin. klin. Wochenschr.* LVIII. 829).] *Tietze's disease, syndrome*: a condition in which there is painful swelling of one or more costal cartilages without evident cause.

1932 *Index Medicus* XII. 1146/2 Tietze's disease; dystrophy of costal cartilage. **1942** *Brit. Med. Jrnl.* 19 Sept. 352/1 Among the many manifestations of subnutrition recently seen by us have been two cases of swelling of the costal cartilages, associated with weakness and lethargy, which conform to the description of Tietze's disease. **1945** *Canad. Med. Assoc. Jrnl.* LIII. 572/2 It is not suggested that these rib changes are necessarily due to Tietze's syndrome. **1977** *Daily Colonist* (Victoria, B.C.) 20 July 2/1 Tietze's disease can be mistaken for angina.

tie-up, *sb.* (*a.*) Add: **I. 2. b.** A (makeshift) garter.

1970 R. Hill *Clubbable Woman* i. 5 One stocking was down. His tie-up hung loose round his ankle. **1976** E. Dunphy *Only a Game?* v. 139 Laying the shirts out before the game, making sure the lads have chewing gum, tie-ups, tea at half-time.

3. b. A building or stall in which cattle are tied up for the night. orig. and chiefly *U.S.*

1851 J. S. Springer *Forest Life* 82 At the further end of the 'tie-up' he thinks he hears a little clattering noise. **1883** *Rep. Maine Board Agric. 1882* 49 Those who have not the convenience for a barn cellar can save the manure very well by a tight floor in the tie-up. **1930** W. Faulkner *As I lay Dying* 117 The cows were still in the tie-up. **1952** E. Caldwell *Lamp for Nightfall* iii. 32 They constructed a chicken house..and installed six tie-ups in the dairy barn. **1960** *Farmer & Stockbreeder* 2 Feb. 17/1 The property..includes a farmhouse, two cottages and farm buildings in good condition with tie-up accommodation for 80 cows.

II. 6. (Later examples.) Also a stoppage of transport, a traffic hold-up. orig. and chiefly *U.S.*

1904 *N.Y. Tribune* 29 Oct. 1 An accident to one of the motor cars caused a tie-up of the southbound trains. **1923** C. R. Cooper *Under Big Top* xi. 227 Add to all this the handicaps of weather, of railroad tie-ups [etc.]. **1943** *Coast to Coast 1942* 181 Old McAlister had made a reference to some trouble on the Brisbane waterfront that threatened a temporary tie-up in the shipping of produce. **1962** E. Snow *Other Side of River* (1963) xxiv. 183, I myself saw new motors and parts piled up unprotected in the weather and damaged or ruined because of transport tie-ups or lack of storage facilities. **1977** *Time* 18 July 14/2 Tie-ups extending for 30 miles are almost normal.

7. b. A connection or association. Cf. *tie *v.* 11 j.

1927 *Daily Express* 7 Mar. 11/5 There is a tie-up, too, over this firm with the gramophone records. Every record of the 'Happiness Boys' is an advertisement for Happiness Chocolates. **1938** F. Scott Fitzgerald *Let.* 18 Apr. (1964) 28, I had made the mental tie-up that work equals something unpleasant. **1945** E. Daly *House without Door* xi. 118 It's quite an interesting tie-up... The Locke case and the Gregson murder case. **1960** *Guardian* 27 Sept. 8/4 The Labour Party..has..an out-of-date tie-up with the trades unions. *a* **1963** L. MacNeice *Astrol.* (1964) vii. 220 The modern astrological tendency..to seek tie-ups between astrology and psychological knowledge. **1974** S. Gulliver *Vulcan Bulletins* 32 'How the hell did he get in with the Libyans in the first place?' I asked. 'Old school tie-ups,' said Selby.

c. A telecommunication link or network. *U.S.*

1927 *Sci. Amer.* July 37/2, 27 stations..were connected to the battery of microphones in front of the Capitol. This record tie-up of transmitters was surpassed..when President Coolidge addressed a joint session of Congress.. through a network of 42 broadcasters. **1939** *Sun* (Baltimore) 24 Aug. 3/5 (*heading*) Western Union speeds deliveries with tieup. **1940** *Nature* 20 July 91/2 It is reported from Pittsburgh, Pa., that the Bell Telephone Co. has applied for a permit to install transmitters and receivers on what is called a 'tie-up' with existing telephone facilities.

Tiffany² (ti·fǎni). The name of Charles L. *Tiffany* (1812–1902), goldsmith and founder of a fashionable New York firm of jewellers *Tiffany and Co.*, and of his son Louis C. *Tiffany* (1848–1933), Art Nouveau decorator noted for his iridescent glassware, used: **a.** *attrib.* (sometimes with small initial) to designate objects made or designed by either of these (but esp. the son) or imitating his style, esp. as *Tiffany glass, lamp.* **b.** Allusively to denote organizations, etc., considered equal to Tiffany and Co. in high-quality craftsmanship or exclusiveness. Also *Comb.*, as *Tiffany-style, -type* adjs.

1895 *Montgomery Ward Catal.* Spring & Summer 153/1, 1K Diamond. Tiffany Setting. **1907** Kipling *Let.* 10 Dec. (1983) 56 A slippery slidy red leather box—like a huge Tiffany jewel case. **1930** E. Waugh *Labels* vii. 175 Gaudi alone was able to use it [*sc.* a mosaic of broken china and pebbles embedded in cement] with precision.. and make of it the craft which, in New York, is reverently known as 'Tiffany bathroom'. **1936** J. Dos Passos *Big Money* 307 Gladys kept..tiffanyglass bowls full of freezias. [**1964** M. J. Simon *New York This Way* iii. 31 Tiffany & Co., celebrated American jewellers whose name often is used colloquially to signify highest quality of product or service in any field.] **1969** A. Lurie *Real People* 89 The Tiffany lanterns [had been] lit beside the front door. **1971** P. A. Whitney *Listen for Whisperer* v. 95, I switched on the Tiffany lamp. **1973** *Synagogue Light* Sept. 4/2 By maintaining a small, specialized firm, we have been able to attain a consistently high calibre level of financial performance... It is the quality we are noted for, a Tiffany type organization. **1977** *Lancashire Life* Mar. 56/3 (*caption*) His policy is to supplement 'the world's finest exhibition' of Tiffany glass with a monthly change of exhibitions. **1977** *Texas Highways* Dec. 33/1 Tiffany-style lampshades. **1978** M. Puzo *Fools Die* xiii. 144, I had become the Tiffany's of bribe takers, with rich, trusting customers. **1979** *United States 1980/81* (Penguin Travel Guides) 425 This one..has Tiffany lamps, antique tables, and a very friendly atmosphere. **1980** 'E. McBain' *Ghosts* iv. 75 The Tiffany-style lamp on the end table.

tiffin, *sb.* Add: **b. tiffin-carrier,** a tiered container for transporting meals.

1960 R. P. Jhabvala *Householder* i. 13 He always brought his breakfast with him in a tin tiffin-carrier.

tig, *sb.*¹ Add: **3.** *colloq.* (orig. *Sc.*). A fit of bad temper.

1773 R. Fergusson in *Weekly Mag.* XXII. 209/1 What tig then takes the fates, that they can thole, Thrawart to fix me i' this weary hole? **1895** R. Ford *Tayside Songs* 71 She left me in a tig. **1934** N. Marsh *A Man lay Dead* xii. 206 'You shall have every opportunity,' soothed Alleyn. 'What a tig you are in, to be sure!' **1943** —— *Colour Scheme* xii. 217, I must say..that I can't see why you're getting into such a tig over it. **1962** *Punch* 24 Oct. 587/1 The spectacle of a man in a tig, even of two men in a tig, is not as a rule wholly entertaining.

tiger, *sb.* Add: **2. a.** (Earlier and later S. Afr. examples.)

1708 tr. F. Leguat in R. Raven-Hart *Cape of Good Hope 1652–1702* (1971) II. 431 The Company gives twenty Crowns to anyone that kills a Lion, and ten to him that kills a Tigre. **1907** P. FitzPatrick *Jock of Bushveld* (1909) 252 Tigers—as they are almost invariably called, but properly, leopards—were plentiful enough.

c. (*a*) (Earlier example.)

1829 H. Widowson *Present State Van Dieman's Land* xviii. 179 The hyena, or as it is sometimes called, the tiger, is about the size of a large terrier; it frequents the wilds of Tasmania.

4. a. Also *spec.*, a native of the Fens. In full, *fen tiger. colloq.*

1893 [in Dict.]. **1963** 'C. Marchant' *Fen Tiger* ii. 40 The term 'fen tiger'... Andrew explained it was the name given to a type of fen man, now almost extinct but not quite, for here and there a descendant of the type of man who had lived deep in the trackless, treacherous fenland, and who fought against the land being drained with cunning, craftiness, and even murder, was still to be found. **1971** *Country Life* 28 Oct. 1128/3 A scattered crowd of rough 'Fen Tigers' in corduroy trousers. **1981** S. Marshall *Everyman's Bk. Eng. Folk Tales* 13, I was.. not accorded the welcome I would have expected to be given to a fen-tiger returning home from choice.

d. *colloq.* (chiefly *Austral.* and *N.Z.*). One who has an insatiable appetite *for* something. Cf. Glutton *sb.*

1896 *Bulletin* (Sydney) 24 Oct. (Red Page), His father thought a lot of Henry; he used to call him a tiger for work. **1927** R. Lehmann *Dusty Answer* ii. 98 Martin.. was still..a tiger for raw vegetables. *Ibid.* iv. 259 You're a tiger for conversation, aren't you? **1935** W. Hatfield *Black Waterlily* 15 'Tiger for work, aren't you?' he smiled. 'A good fault, of course, if you don't carry it to extremes.' **1972** P. Newton *Sheep Thief* xx. 170 Don't tell me you're up to your capers again... You're a tiger for punishment.

e. A sportsman or climber of outstanding skill and confidence. Cf. *rabbit *sb.*¹ 2 a. *colloq.*

1929 E. Bowen *Joining Charles* 166 They may be tigers at ping-pong. **1935** D. Pilley *Climbing Days* ii. 27 Wet ground, where most climbers—bar the latest 'tigers'—find that they slip. **1941** R. R. Marett *Jerseyman at Oxford* ix. 138, I was never really worth more than bare scratch, and clean outside the 'tiger' class [in golf]. **1957** Clark & Pyatt *Mountaineering in Britain* x. 178 'Tiger' is the word used to describe the climber whose abilities are outstandingly in advance of his generation. **1974** *Times* 23 Feb. 13/3 There is a third [golf] course strictly for tigers; rabbits should try the excellent par three to seawards of the big course. **1979** *Country Life* 24 May 1674/2 Odon has less of a reputation as a tiger on difficult climbs than his father.

6. b. *Naut. slang.* A captain's personal steward.

1929 F. C. Bowen *Sea Slang* 141 *Tiger, the*, the steward who acts as personal servant to the captain of a liner. **1936** E. T. Britten *Million Ocean Miles* iii. 30 Croughan is my 'Tiger', as the Captain's steward is called at sea. **1961** 'R. Gordon' *Doctor on Toast* x. 87 In the old days, you could have swapped the Captain's tiger for the butler in any stately home in the kingdom, and no one would have been the wiser. 1982 *Times* 11 May 6/6 Captain Jackson's 'tiger'—the merchant navy equivalent of a batman.. was married after the weekend.

8. (Earlier example.)

1845 *Florence de Lacey* 28/1 Nine cheers for old Tip—one, two, three, four, five, six, seven, eight, nine, and a tiger.

11*. a. In proverbial phrases: *to ride a tiger* and varr. [after the Chinese proverb 'He who rides a tiger is afraid to dismount' (W. Scarborough *Coll. Chinese Proverbs* (1875) xvi. 388)]: to take on a responsibility or embark on a course of action which subsequently cannot safely be abandoned; *to have a tiger by the tail* and varr.: to catch a Tartar (see TARTAR *sb.*[2] 4).

1902 A. R. COLQUHOUN *Mastery of Pacific* xvi. 388 These colonies are..for her [*sc.* France] the tiger which she has mounted (to use the Chinese phrase), and which she can neither manage nor get rid of. 1940 *Daily Progress* (Charlottesville, Va.) 27 Nov. 1/7, I believe that Hitler is riding a tiger in trying to keep all Europe under control by sheer force. 1969 *Guardian* 7 July 9/5 All African politics to-day is concerned with the art of riding this terrible tiger [*sc.* tribalism]. 1972 'E. LATHEN' *Murder without Icing* (1973) iii. 30 Convulsions..could be expected... The Sloan Guaranty Trust..might well have a tiger by the tail. 1979 P. DRISCOLL *Pangolin* xii. 101 You're taking on an organization with..reserves you know nothing about. How do you know you won't be catching a tiger by the tail? 1981 W. H. HALLAHAN *Trade* iii. 79 It was done. They were all riding the tiger now.

b. *to put a tiger in one's tank* [after an Esso Petroleum Co. advertising campaign of 1965]: to invest one with energy or 'go'; also in similar allusive phrases.

1965 *Guardian* 31 May 4/7 Esso's tiger has pounced on to the national consciousness within two months. The phrase 'Put a tiger in your tank' has become part of everyday conversation. 1967 *Listener* 22 June 835/2 Westin and Friedman are young men with ideas of their own... They are the tigers in the Ford [Foundation] tank. 1973 P. GEDDES *Ottawa Allegation* iii. 32 Lorimount..began pouring tea... The movements were brisk and purposeful. No safety belts worn here, they said, there's a tiger in the tank. 1981 *N.Z. Tablet* 10 June 10/4 Young girls must be made to realise that boys of the same age have a 'tiger in their tank' as far as sexual desire goes.

c. *paper tiger*: see *PAPER *sb.* 12.

12. a. *tiger cage, country, -pit* (examples), *-skin* (earlier example), *trap*.

1970 *Guardian* 8 July 1 (*caption*) Political prisoners peering up out of a 'tiger cage' in Con Son prison in South Vietnam. 1982 *Times* 28 Sept. 3/4 (*caption*) An apprentice animal trainer, in the tiger cage with six Bengal tigers. 1931 E. A. ROBERTSON *Four Frightened People* v. 178 This was tiger country, she knew, but she had never yet seen one of those animals. 1978 'M. M. KAYE' *Far Pavilions* xxv. 369 Biju Ram would only have had to wait until they were in tiger country—preferably..where there was known to be a man-eater. 1936 T. S. ELIOT *Coll. Poems 1909–35* 153 The tiger in the tiger-pit Is not more irritable than I. 1970 *Daily Progress* (Charlottesville, Va.) 8 July 5A/3 Harkin said more than 200 men, crammed three to five in 86 5-by-8-foot tiger pits in one building, were unable to stand because they had been there so long. 1763 J. BELL *Trav. from St. Petersburg* xi. 162 There appeared two troops of Tartars, clothed in coats of tiger-skins. 1934 M. MITCHELL *Warning to Wantons* x. 324 They were like two big game-hunters whose elaborate tiger-trap has netted..a domestic cat! 1980 N. FREELING *Castang's City* viii. 47 She was extremely sharp. One kept falling..into tiger traps full of pointed bamboo stakes. One got little out of her.

13. a. tiger barb, any of several brightly coloured freshwater fishes of the genus *Barbus*, esp. *B. tetrazona*; **tiger maple** *N. Amer.*, a kind of maple-wood with strongly contrasting light and dark lines in the grain; **Tiger Milk**, a name given to Yugoslavian dessert wine made from over-ripe grapes; **tiger-nut**: also eaten locally as a sweetmeat by children; (later examples); **tiger prawn** *Austral.*, a large prawn marked with dark bands, *Penæus esculentus*; **tiger salamander**: delete *l* from *Amblystoma*; (examples); **tiger-snake**, (*a*) for *Hoplocephalus curtus* substitute: of the elapid genus *Notechis*, esp. *N. scutatus*; (earlier and later examples); (*b*) a slightly venomous southern African colubrid snake of the genus *Telescopus*, esp. *T. semiannulatus*; **tiger-stripe(d)** = *TIGER CAT d; **tiger suit**, a striped combat uniform worn as camouflage in jungle warfare; **tiger-ware**, substitute for def.: sixteenth- or seventeenth-century German stoneware with a mottled brown glaze, or English stoneware made in imitation of this; (examples).

1951 R. DUTTA *Right Way to keep Pet Fish* xviii. 155/2 Tiger barbs. 1962 *Listener* 22 Nov. 852/2, I brought home a tiger barb, round and flat with bold orange and black stripes. 1976 *Norwich Mercury* 19 Nov. 4/8 (Advt.), This week's Fish Centre offers: Neons..Silvertips..Tiger-barbs. [1952 J. DOWNS *Amer. Furnit.* p. xxxii, [In] Queen Anne maple furniture..the curly figure is produced by fibers which develop spirally, without any known reason, giving a tiger-stripe pattern much prized by collectors.] 1961 WEBSTER, Tiger maple. 1967 *Canad. Antiques Collector* Apr. 4/1 (Advt.), Canadian Tiger maple desk,..circa 1830. 1978 *Times* 13 Mar. 20/4 Another American Chippendale piece was a tiger maple desk and bookcase. 1961 *Guardian* 21 Nov. 16/5 Yugoslavia is now exporting..'Tiger Milk',..an excellent dessert wine. 1977 T. HEALD *Just Desserts* vii. 172 Not just claret..but ..Tigermilk (or Ranina Radgona Spatlese). 1927 W. E. COLLINSON *Contemp. Eng.* 18 Bull's eyes.., acid drops, fondants..are still in demand, though the popularity of monkey-nuts and tiger-nuts has somewhat waned. 1957 J. KIRKUP *Only Child* ix. 122 We knew..the illicit joy of spending our Sunday school collection money on 'tiger-nuts' and coconut ice. 1972 *Country Life* 30 Nov. 1481/3 The sort of boy who would..find such delight in munching tiger nuts. 1893 J. D. OGILBY *Edible Fishes & Crustaceans N.S.W.* 203 This is the 'Tiger Prawn' of the Sydney fishermen. 1952 W. J. DAKIN *Austral. Seashores* xv. 176 The tiger-prawn is a large northern species that.. has dark vertical bands on its body. 1978 O. WHITE *Silent Reach* vi. 72 It could be arranged..for a marine biologist..to complete his thesis on the breeding habits of..the tiger prawn. 1926 J. K. STRECKER in J. F. Dobie *Rainbow in Morning* (1965) 63 In the plains region of Western Texas, the large tiger salamander is a common animal. 1966 R. C. STEBBINS *Field Guide Western Reptiles & Amphibians* 33 Tiger salamander... A large stocky salamander with small eyes. 1869 Tiger-snake [see *brown-banded snake* s.v. *BROWN *a.* 7]. 1910 F. W. FITZSIMONS *Snakes S. Afr.* iii. 54 Tiger Snake... Average length 2 feet to 2 feet 6 inches. 1941 K. TENNANT *Battlers* xviii. 193 The driver of the car..very efficiently despatched a large tiger snake. 1947 J. STEVENSON-HAMILTON *Wild Life S. Afr.* xxxvi. 330 The tiger snake..is a yellowish snake spotted with brown. 1966 *Southerly* XXVI. 109 A fisherman had been bitten by a tigersnake there and had died. 1974 *Stand. Encycl. S. Afr.* X. 504/1 Tiger-snake..is conspicuously marked throughout its length with alternate black and yellow to reddish brown cross-bands. 1977 *Time* 31 Oct. 49/1 His tabby—a tiger-stripe he calls Dr. Carleton P. Forbes—has amassed $3,000 worth of 'cat toys' by filching checks from Steve's mailbox. 1981 P. MALLORY *Killing Matter* ii. 23 The cat.. was a big grey tiger-stripe. 1965 F. MANOLSON *C is for Cat* 187 A striped cat (even if it's the result of the mating of a Tiger striped with a Tabby striped) is either one or the other. 1970 A. MARIN *Rise with Wind* xx. 241 The soldier was dressed in a tiger suit. 1977 M. HERR *Dispatches* (1978) 5 [He] took his pills by the fistful, downs from the left pocket of his tiger suit and ups from the right. 1874 C. SCHREIBER *Jrnl.* (1911) I. 325 A grand old cruche of Tiger Ware, with Royal Arms of England, and date 1604. 1928 *Daily Express* 5 June 4 There are few [objects] which exercise a stronger fascination over collectors than old stone wine jugs known as tiger-ware. High prices—up to £1,500—have been paid for these 'stone pottes garnished with sylver'. 1983 *Country Life* 1 Dec. (Suppl.) 85/1 A rare Elizabeth I Norwich Tigerware jug.

b. Tiger's Milk = *Tiger Milk*, sense 13 a above.

1959 W. JAMES *Word-bk. Wine* 155 *Ranina*, the 'Tiger's Milk' wine of Radgona, in Yugoslavia, a sweet and strong dessert wine made from late-gathered grapes. 1965 O. A. MENDELSOHN *Dict. Drink & Drinking* 277 *Ranina*, Yugoslavian (Radgona) dessert wine. Syn. 'Tiger's Milk'.

tiger, *v.* Restrict *nonce-wd.* to sense in Dict. and add: **2.** *trans.* To mark like a tiger with lines or streaks of contrasting colour.

1930 R. CAMPBELL *Adamastor* 50 Striped with the fiery colours of the sky, Tigered with war-paint.. The green waves charged the sunrise. 1934 —— *Broken Record* iii. 74 She [*sc.* a dog] was tigered with wounds from head to tail. 1960 T. HUGHES *Lupercal* 56 Pike, three inches long, perfect Pike in all parts, green tigering the gold.

Hence **ti·gered** *ppl. a.*, striped or broken into stripes; **ti·gering** *vbl. sb.*, a striated condition (see quot. 1961).

1961 R. D. BAKER *Essent. Path.* xiv. 344 The yellowness of the heart muscle may be diffuse or concentrated in narrow stripes forming a peculiar and distinctive pattern especially along the papillary muscles and the inside of the ventricles... This striated appearance has suggested the descriptive designation of 'tigering' or 'thrush breast'. 1969 *Burpee Catal.* 50/2 *Calceolaria*... Many [flowers] are attractively tigered, blotched, spotted and laced in most unique patterns. 1980 J. O'FAOLAIN *No Country for Young Men* ix. 197 Tigered light which fell slantwise through a Venetian blind.

tiger-cat. Add: **a.** Also *fig.*

1863 M. BRADDON *Aurora Floyd* II. v. 112, I should get nothing—but my revenge upon a tiger-cat. 1959 *Times* 7 Jan. 13/5 Whether as a tiger-cat, a hard-headed business woman..her command of the character was superb. 1979 P. ALEXANDER *Show me a Hero* xx. 208 If she's a trained resistance girl she'll be a tiger-cat.

d. = *tabby cat* s.v. TABBY *a.* 2, esp. a tabby cat with vertical stripes in its markings.

1903 F. SIMPSON *Bk. of Cat* xix. 216/1 The term 'tiger cat' is, I believe, often used in America, and it well describes the true type of a brown tabby. 1915 *Century Mag.* Sept. 673/2 She has..a nice purry tiger cat asleep on a braided rug. 1939 I. M. MELLEN *Pract. Cat Bk.* i. 4 The striped tabby or tiger cat..is the most ancient type of cat known to science. 1969 R. LOWELL *Notebk. 1967–68* 52 A first tiger cat stationed on the record-player spies on a second.

tiggy (ti·gi). [Perh. rel. to dial. *tig* little pig (*Eng. Dial. Dict.*)] = HEDGEHOG 1.

[1905 B. POTTER *Tale Mrs. Tiggy-winkle* 85 Mrs. Tiggy-winkle was nothing but a Hedgehog!] 1938 T. H. WHITE *Sword in Stone* xxi. 303 Us be'nt no common tiggy Mëaster, for to be munched and mumbled.

tight, *a.* (*adv.*) Add: **A.** *adj.* **2. d.** *spec.* Uncommunicative; secret; *Oil Industry*, applied to a well about which little information is released.

[1661: in Dict.] 1949 *Amer. Speech* XXIV. 34 If information about the *venture* is withheld from the public, then it becomes a *tight well*. 1966 *Natural Resources Jrnl.* (Univ. New Mexico) VI. 55 If the draining well is what is commonly known in the industry as a 'tight hole', the information concerning its performance is probably more closely guarded than most national defense secrets. 1976 M. MACHLIN *Pipeline* ii. 32 'Who knows about this [*sc.* an oil-strike]?' 'Nobody but me and a couple of guys here on the platform know for sure... Communication is lousy here, but rumor travels faster than radio waves.' 'Okay. Shut it off..and see if you can keep it as tight a hole as possible.' 1977 B. FREEMANTLE *Charlie Muffin* v. 59 The British..[have] gone completely silent... The British Embassy is tighter than the Kremlin itself.

5. Also in U.S. fig. use, on terms of close friendship, intimate. Cf. THICK *a.* (*sb.*) 10. *slang*.

1956 B. HOLIDAY *Lady sings Blues* (1973) ii. 23 Blue.. had me busted... He and Bub were real tight with the cops. 1971 *Current Slang* (Univ. S. Dakota) VI. 10 *Tight*, very much in love; very friendly. 'John and Mary were really tight for awhile but they seem to have drifted apart.' 1977 *Rolling Stone* 5 May 55/1, I was very tight with him for a long time.

6. c. Of an organization or group: strict, disciplined, well co-ordinated; *spec.* of a pop group or an individual member of it.

1968 L. DEIGHTON *Only when I Larf* vi. 80, I ran a tight unit, and if that meant repeating my lecture every week, then I'd do that. 1971 *Melody Maker* 9 Oct. 21/3 Keef can be proud of his [jazz-rock] band... It's well rehearsed, tight and above all fun. 1977 *Ibid.* 26 Mar. 46/6 (Advt.), Wanted. Good tight drummer for funky group. 1980 MOTSON & ROWLINSON *European Cup 1955–80* ix. 190 This tight triumvirate, Smith in the boardroom, Robinson at the administrative helm, and Paisley on the training ground, headed an Anfield staff which worked as efficiently..as the team.

7. (Earlier examples.)

1830 [implied at *TIGHTISH *a.* 3]. 1840 in *Amer. Speech* (1951) XXVI. 184 After supper I got tight, sick with oysters, and slept.

8. b. Of ground: allowing (vehicles) little room for manœuvre. Of a turn, curve, etc.: having a short radius.

1937 *Sun* (Baltimore) 20 Apr. 4/2 He expressed a hope the airport work would be completed as rapidly as possible, pointing out that Logan Field was 'rather tight' for large transports. 1947 A. C. DOUGLAS *Gliding & Advanced Soaring* i. 24 He based this opinion on the belief..that they [*sc.* contemporary airplanes] could not be turned in tight circles like the birds. 1958 *Times* 19 Feb. 5/4 She [*sc.* an aircraft] started to turn to starboard, and it seemed clear that the turn became tighter and tighter. 1969 *Times* 23 May 1/3 The L.M. was due to spend about an hour in a tight orbit approaching within eight nautical miles of the surface. 1979 *Beautiful Brit. Columbia* Fall 19/1 The highway narrows down to one lane which clings in tight curves around a sheer mountainside.

c. Applied to persons: tough, hard, unyielding; also, aggressive, 'stroppy'. *U.S. dial.* or *slang*.

1928 R. FISHER *Walls of Jericho* 306 *Tight*, tough; redoubtable; hard. 1950 PATTERSON & CONRAD *Scottsboro Boy* I. iii. 30 'You'll get it [*sc.* a bath] Saturday,' he said. Saturday came and he put me off... I got tight with him. 'I got to have a bath!' *Ibid.* 31, I was a tight guy who would not show people tears, but I felt the water behind my lids. *Ibid.* II. vii. 129 There were guys there [*sc.* in a prison], they made reputations for themselves as tight guys and killers just from defending themselves against the insane. 1960 L. BUCKLEY *Hiporama of Classics* 16 He was a hard, tight, tough Cat.

10. b. (Earlier example.)

1805 LEWIS & CLARK *Orig. Jrnls. Lewis & Clark Exped.* (1904) III. 278 They are tite Deelers, value Blu and white beeds very highly, and sell their roots also highly.

c. Also of a person: in financial straits, hard up. *dial.* or *slang*.

1859 HOTTEN *Dict. Slang* 109 *Tight*,..hard up, short of cash. 1864 J. S. LE FANU *Uncle Silas* II. xvi. 247 It is a hard case, Miss, a lad o' spirit should be kept so tight. I havn't a shilling. 1892 'T. COBBLEIGH' *Gentleman Upcott's Daughter* ix. 173 Any man might find himself tight—temporarily.

d. *Journalism.* (See quot. 1970.) Hence also of (a day of) restricted newspaper space.

1927 *Amer. Speech* II. 241/2 If advertising crowds out news, the paper is said to be 'tight'; if advertising is scant, the paper is 'wide open'. 1927 *New Republic* 12 Oct. 202/1 Possibly space was 'tight' that day, and the newspapers didn't have room for this minor angle of the story. 1928 *Amer. Speech* IV. 135 The 'desk' must know whether 'room' is 'tight', 'fair', 'good' or 'wide open'. If news is 'heavy' on a 'tight day' and is permitted to 'run' in length practically as written, 'oversets'..may result. 1970 R. K. KENT *Lang. Journalism* 133 *Tight.* 1. designating a newspaper that has little room for news because there is a great deal of advertising: opposite of *open*. 2. designating a newspaper on a day when there are a great many newsworthy events to record, and hardly enough space to cover them all.

11. a. Also, of a group or formation: having the individual members positioned close together. Freq. in *Sport*; also *transf.* (esp. in *Cricket*), that allows the opposition little chance to score, etc.: *tight bowling, fielding*, etc.

1942 *R.A.F. Jrnl.* 13 June 22 They lived in dread of our fighters, and normally kept a tight formation. 1961 F. C. AVIS *Sportsman's Gloss.* 138/2 *Tight field*, the fieldsmen when drawn closely round the wicket, so preventing the easy scoring of runs. 1961 *Times* 12 May 4/1 Surrey's bid for quick runs..was foiled by tight bowling. 1965 *Daily Express* 13 Aug. 15/5 *Tight position*, an area of the field in which there are a large number of players, both attacking and defending. 1968 I. URE *Ure's Truly* xvii. 116 Let's have expressions such as 'a steady defence' rather than a side being described as 'tight at the back'. 1976 *Milton Keynes Express* 11 June 41/5 Farnham Royal found it difficult to score against the tight Wolverton bowling. 1977 *Arab Times* 13 Dec. 9/6 Tight fielding by the Airlines prevented easy scoring and the FSC batsmen had to rely on quick singles and doubles.

b. Also in general, of literary, artistic, or intellectual work: kept within strict limits; pared to essentials; disciplined, taut, not loose or diffuse.

1958 C. A. LARSON *Who: Sixty Yrs. Amer. Eminence* 75 Quaint little items and details were often inserted in these early biographical sketches which would scarcely survive the tight editing of a modern Marquis editor. 1962 [see *PILOT *sb.* 1 f]. 1979 *Sci. Amer.* Aug. 24/1 The author..seeks the answer in this small book of tight argument.

e. Of a schedule or timetable: packed with engagements, leaving little free time. Also applied to a space of time which is limited or restricted.

1959 J. POPE-HENNESSY *Queen Mary* III. ii. 386 Princess May concentrated on seeing as many of the wonders and beauties of the spacious old Imperial city as she could crowd into four days and a tight social schedule. 1968 P. G. HOLLOWELL *Lorry Driver* ii. 31 The older drivers are constantly aware that schedules are getting tighter. 1971 J. SANGSTER *Your Friendly Neighbourhood Death Pedlar* vii. 187 It was unlikely that anything would happen that night, and the following morning was going to be awfully tight for time. 1972 M. CRICHTON *Terminal Man* I. v. 42 'I can't see her to-day,' Morris said, 'and to-morrow is tight.' 1976 *New Yorker* 1 Mar. 30/3 'How about seven at O'Hoolihan's? I'm going there with Pat and Betsy.' 'Seven's a little tight.' 1981 P. HARCOURT *Turn of Traitors* ix. 82 Time's going to be tight, so take my car.

13. a. Delete *rare* and add later examples.

1939 *Daily Tel.* 18 Dec. 11/1 They were better served by their forwards in the tight. 1979 *Times* 12 Dec. 9/1 They outscrummaged their opponents in the tight.

b. An awkward situation, predicament, 'tight corner' (TIGHT *a.* 9). Usu. in phr. *in a tight*; occas. const. *for*. *U.S. dial.* or *colloq.*

1896 in *Dialect Notes* (1916) IV. 348 *Tight*, n. (From *tight place*.) A difficult or precarious position. 1902 W. N. HARBEN *Abner Daniel* xxi. 182 It would tempt five men out of ten if they were inclined to go wrong, and were in a tight. 1930 W. FAULKNER *As I lay Dying* 29, I tell him again I will help him out if he gets into a tight, with her sick and all. 1938 M. K. RAWLINGS *Yearling* xv. 177 Jody's in a tight for a name for the new Baxter. 1950 PATTERSON & CONRAD *Scottsboro Boy* II. xi. 175 'You scared that man almost to death.' 'I was in a tight jam then, Warden. I was trying to get out of a tight.' 1979 G. SWARTHOUT *Skeletons* 18 Pat Garrett..had said of him in public: 'I would rather have Wood with me in a tight than any man I know.'

B. *adv.* **1.** Delete 'Now *dial.* and *U.S.*' Now chiefly in colloq. phr. (*good night*) *sleep tight*, a conventional (rhyming) formula used when parting for the night or at bedtime. Also in slang phr. *blow me tight*: see *BLOW *v.*[1] 29.

1933 E. O'NEILL *Ah, Wilderness!* III. ii. 101 Good night, Son. Sleep tight. 1957 [see *NIGHTY-NIGHT *int.*]. 1960 D. LESSING *In Pursuit of English* iii. 99 When we left him, she patted his shoulder with triumphant patronage, and said: 'Sleep tight. And keep your dreams clean.' 1976 'R. BOYLE' *Cry Rape* xxi. 94 Goodnight, Anne. Sleep tight.

2. b. (Earlier example of sense 'to maintain one's position'.)

1890 G. B. SHAW in *Star* 27 Nov. 2/7 I, therefore, again urge Mr Parnell to 'sit tight'.

4. *as tight as* ( ): as quickly or rapidly as ( ). Cf. TITE *adv.* *U.S. dial.*

1833 S. SMITH *Life & Writings J. Downing* lix. 200 The President shook hands with all his might an hour or two until he couldn't hardly stand it... I..stood behind him and reached my arm round under his, and shook for him for about a half an hour as tight as I could spring. 1867 W. L. GOSS *Soldier's Story of his Captivity* 185 Captain Sherman..was making for Macon as 'tight as he can come'. 1884 'MARK TWAIN' *Huck. Finn* xix. 180 A couple of men were tearen up the path as tight as they could foot it.

5. Close *up to, after*, or *on*. *dial.* or *colloq.*

1886 F. T. ELWORTHY *W. Somerset Word-bk.* 756 The bitch was tight arter'n. 1901 *Century Mag.* May 123/1 They was tight up t'me all the way. 1919 J. C. SNAITH *Love Lane* xxx. 160 He lived to be tight on ninety.

C. 2. *tight-fitting* (earlier example), *-packed* (examples).

1846 E. A. POE in *Godey's Lady's Bk.* Nov. 216/1 He had on a tight-fitting parti-striped dress. 1918 G. FRANKAU *One of Them* xxix. 223 Tight-packed as, face to tail and tail to face, Bristle in Watson's tins the silvery 'Skippers'. 1950 PARTRIDGE *Slang To-day & Yesterday* (ed. 3) III. iii. 257 The ten-page, tight-packed chapter entitled 'War Words' in Collinson's Contemporary English.

3. tight back *Bookbinding*, a book cover which is stuck directly on to the spine; cf. *fast back, fastback* s.v. *FAST *a.* 11; **tight end** *N. Amer. Football*, an offensive end (*END *sb.* 3 g) who lines up close to the tackle; the position occupied by this player; **tight-fisted** *a.*: hence **tight-fistedness**; **tight head** *Rugby Football*, (the position of) the prop forward supporting the hooker on the opposite side of the scrum from the loose head; *to win a tight head* = *to win the ball against the head* (see *HEAD *sb.* 26 c); **tight junction** *Cytology*, a specialized connection of two adjacent animal cell membranes such that the space usually lying between them is absent.

1913 *Funk's Stand. Dict.* II. 2518/1 *Tight back* (Bookbinding), a back that clings to the signatures or to the paster attached to them: distinguished from *loose back* or *spring back*. 1929 A. J. VAUGHAN *Mod. Bookbinding* I. 2 (*caption*) A limp paper book bound with a tight back. Effect of a tight back binding upon stiff paper. 1957 E. A. CLOUGH *Bookbinding for Librarians* vi. 60 Because the tight back bends with the spine of the book, there is a tendency for the tooling on the spine to crack. 1963 HUFF & SMITH *Defensive Football* vi. 72 When playing the tight end head to head, the linebacker must be aware [etc.]. 1972 J. MOSEDALE *Football* ii. 29 With Fears spread out on one side and Shaw in close, the 'three end offense' was born. Today the positions are called split end, tight end and flanker. 1978 J. IRVING *World according to Garp* viii. 162 A standout tight end for the Philadelphia Eagles. 1975 *Church Times* 25 Apr. 2/3 Let it be seen..that we have asked the very necessary questions about the liberality or tight-fistedness of the [European] Community's policy on trade with the under-developed countries. 1959 *N.Z. Listener* 28 Aug. 7 You can't afford to give tight heads in your own 25 when you've got fast backs like that against you. 1960 V. JENKINS *Lions down Under* xiii. 170 Dawson won six tight-heads to three in the scrums. 1969 *Advanced Coaching* (Rugby Football Union) 73 On the tight head it may still be preferable to hook with the nearside foot, *but* with the body facing the *loose* head side so as to hook the ball with the *inside*.. part of the foot. 1978 *Rugby World* Apr. 17/1 There seems nothing to stop Graham Price reigning for many years to come as the world's outstanding tight-head prop. 1961 *Jrnl. Exper. Med.* CXIV. 706 The normal slits as well as the tight junctions have structural features reminiscent of usual epithelial desmosomes. 1982 *Nature* 1 Apr. 464/1 Our evidence, which is based on direct rapid freezing of newly formed tight junctions between rat prostate epithelial cells, indicates that individual tight junction strands are pairs of inverted cylindrical micelles sandwiched between linear fusions of the external membrane leaflets of adjacent cells.

ti·ght-ass, *sb.* (and *a.*). *slang* (orig. and chiefly *U.S.*). Also **tight-arse**. [Back-formation from next.] An inhibited or strait-laced person; occas. a stingy person, a skinflint. Freq. *attrib.* passing into *adj.* = next; occas. applied (*lit.*) to clothes: fitting tightly around the buttocks.

1969 P. ROTH *Portnoy's Complaint* 79 A really constrained and tight-ass human being. 1970 J. G. VERMANDEL *Dine with Devil* iii. 11 A well-muscled, virile-looking type with a neat little beard, who favors tight-ass pants and rather beautiful, flowing shirts. 1972 'E. MCBAIN' *Sadie when she Died* x. 105 'Don't be such a tight-ass,' the girl said. 'I'm thirsty as hell here.' 1972 *Southerly* XXXII. 103 Jenny was very friendly but very tightarse inhibited headinnamess. 1981 M. GORDON *Company of Women* II. iv. 190 My three o'clock class was an incredible bummer. Bunch of tight asses. Tight-ass kids. 1982 J. SHERWOOD *Shot in Arm* x. 96 As though any policeman in his senses would pocket cash..with a virgin-faced tight-arse like Verney looking on.

ti·ght-assed, *a. slang*. Also **tight-arsed**. [f. TIGHT *a.* + Ass, vulg. and dial. spelling of ARSE *sb.* + -ED[2]: see prec.] **a.** Of a woman: (see quot. 1903). **b.** Unwilling to relax or enjoy oneself, full of inhibitions or constraints; occas. stingy, mean.

1903 FARMER & HENLEY *Slang* VII. 126/1 *Tight-arsed*, chaste; close-legged. 1961 PARTRIDGE *Dict. Slang* 885/2 *Tight-arsed*, stingy. 1967 J. ORTON *Diary* 28 July in J. Lahr *Prick up your Ears* (1978) 16, I hate this tight-assed civilization. 1974 D. SEARS *Lark in Clear Air* iv. 50 Rough quarterings of pine that even the tight-assed mill owners considered too lean to..ship to Toronto. 1976 *Listener* 15 July 54/4 Me and Gambaccini would like tight-assed British radio to relax a bit. 1981 T. HEALD *Murder at Moose Jaw* ix. 112 I'd sooner have an entertaining shit than a tight-arsed saint.

tighten, *v.* Add: **2.** Also *intr.* for *pass.* Freq. const. *up*, esp. in *fig.* use.

1933 P. GODFREY *Back-Stage* vi. 86 As the day of the dress rehearsal draws near the acting tightens up. 1947 *Milwaukee Jrnl.* 25 Apr. 2 His stomach tightened up on him. 1959 *Listener* 2 July 24/3 We should alter the tax laws to tighten up on tax-dodging. 1973 *Daily Tel.* 9 Feb. 2/8 Yellow line no-waiting regulations are to tighten up in London.

tightish, *a.* Add: **3.** Somewhat drunk. Cf. TIGHT *a.* 7. *colloq.*

1830 H. LEE *Mem. of Manager* I. iii. 110 'I think they be getting on pretty tightish!' 'What do you mean, getting drunk!'

ti·ght-li:pped, *a.* [f. TIGHT *a.* + LIPPED *ppl. a.*.] Having the lips firmly closed, esp. as a sign of determined suppression of emotion; also *transf.* and *fig.* Also, determinedly reticent or uncommunicative.

1876 [in Dict. s.v. TIGHT *a.* C.1]. 1918 H. G. WELLS *Joan & Peter* iii. 62 Her pride was white and tight-lipped. 1936 P. FLEMING *News from Tartary* v. ii. 194 This, I know, is my cue for tight-lipped heroics. 1952 [see *INSTALMENT[2] 3]. 1958 *Observer* 15 June 15/2 Joan Mitchell's rather beautiful painting 'Hudson River Day Line' (which has a sensitive, tight-lipped, almost Slade School quality). 1970 J. SANGSTER *Touchfeather, Too* ii. 32 The Russians, notoriously tight-lipped normally, had been approached through tortuous channels. 1979 A. PRICE *Tomorrow's Ghost* iii. 37 She smiled her careful tight-lipped smile. 1981 D. HOPKINSON *Edward Penrose Arnold* iv. 40 Here is that melancholy resignation which Matthew Arnold so often..conveys... But his finest poetry..is stoically tight-lipped. 1983 *Times* 2 Apr. 10/2 All parties are keeping tight-lipped. A spokesman for DTR issued a firm 'no comment'.

tightness. Add: **3.** (Earlier example.)

1861 E. COWELL *Diary* 1 Jan. in *Cowells in Amer.* (1934) 234, I congratulated him on his remaining free from 'tightness' after so many calls.

4. (Earlier example.)

1847 *Punch* XIII. 77/1 There is a tightness at present in the Omnibus Market.

5. As an artistic quality: (*a*) crampedness, lack of freedom, constraint; (*b*) sense of control, rigorousness.

1933 *Burlington Mag.* Jan. 22/1 The effort to achieve a difficult and unfamiliar piece of modelling gives the penwork a certain tightness. 1959 *Listener* 26 Mar. 542/1 Lack of balance between voice and accompaniment, acoustic 'tightness', restricted and uneven frequency-range. 1973 *Art Internat.* Mar. 73/2 Erlebacher's tightness makes her poetic allegories more like kitsch.

tight rope, tight-rope. Add: Now freq. as one word. **1.** Also *fig.*

1934 *Essays & Stud.* XIX. 123 He moves with complete security on the tight-rope of serio-comic wit. 1959 *Daily Tel.* 30 Nov. 1 For 35 minutes the Deputy Leader balanced himself on a verbal tightrope which purported to bridge the awesome gap within the party. 1979 *Sci. Amer.* Nov. 126/3 The nocturnal *S. laevistriatus* beetles are on an energy tightrope.

2. *Comb.*, as *tight-rope dancer, dancing, walk, walker, -walking* vbl. sb. and ppl. adj.

1824 *Advt.* (Theatre-Royal, Worcester) in *Henry Bristow Ltd. Catal.* (1973) No. 205, The celebrated Mr. Wilson, the tight rope dancer. 1890 Tight-rope dancer [in Dict.]. 1800 W. DYOTT *Diary* July (1907) I. 138 Besides rural sports in the gardens, such as gipsies guying, lofty tumbling and tight-rope dancing. 1890 Tight-rope-dancing [see *billiard-playing* s.v. *BILLIARDS 2]. 1952 R. KNOX *Hidden Stream* vi. 55 We, in this tight-rope-walk business of trying to live our lives..want more than a metaphysical conviction that God exists. 1869 *Atlantic Monthly* July 83/2 This tight-rope walker was one of the most exemplary domestic little bodies imaginable. 1910 *Encycl. Brit.* IV. 77/1 *Blondin* (1824–1897), French tight-rope walker and acrobat. 1979 R. JAFFE *Class Reunion* (1980) II. xi. 287 When he was drunk he had a slow and precise quality, like a tightrope walker. 1982 C. CASTLE *Folies Bergère* i. 24 Tightrope walkers, magicians and sleight-of-hand artists..attracted Parisians and tourists. 1958 *Spectator* 30 May 675/1 Admiral Auboyneau, who had been doing some tightrope-walking of his own in the previous weeks, finally came out for the rebels. 1981 *Times Lit. Suppl.* 3 Apr. 368/2 Ours is the age of heartless efficiency and tight-rope-walking virtuosity in music.

tights, *sb. pl.* Add: **a.** (Earlier example.)

1827 M. WILMOT *Let.* 24 May (1935) 260 His [*sc.* a jockey's] shirt-collar was open, so were the knees of his 'tights'.

c. A woman's or girl's one-piece stretchable garment covering the legs and body up to the waist, worn in place of stockings; formerly, an undergarment taking the place of knickers and stockings.

1897 *Sears, Roebuck Catal.* 241/2 Ladies' summer drawers and tights..Ladies' fine jersey ribbed fast black open summer tights,..knee length. 1908 *Ibid.* 962 Wool and fleece lined underwear for children..Union suits, tights and sleeping garments. 1929 R. S. & H. M. LYND *Middletown* xii. 159 Today flannel underwear is almost as obsolete as the long black equestrian tights, high-necked, long-sleeved nightgowns for women, and the heavily-lined trousers of the working men of a generation ago. 1965 P. O'DONNELL *Modesty Blaise* ix. 105 Modesty wore ..a full black skirt, with black stretch tights. 1967, 1970 [see *panty-hose* s.v. *PANTIES 3]. 1977 C. MCCULLOUGH *Thorn Birds* xviii. 469 A five-guinea pair of tights.

tightwad (tai·t,wǫd), *sb.* (and *a.*). *slang* (orig. and chiefly *U.S.*). Also **tight-wad**. [f. TIGHT *a.* + WAD *sb.*[1] 2 b.] A miserly person; one who keeps his wad of paper money tightly rolled. Also *attrib.* and (*rarely*) as *adj.* Also *fig.*

[1900 ADE *More Fables* 30 Henry was undoubtedly the Tightest Wad in the Township.] 1906 S. FORD *Shorty McCabe* ii. 32 Keep these, and found a home for Incurable Tight-Wads. *a* 1911 D. G. PHILLIPS *Susan Lenox* (1917) I. xv. 255 You've forgotten what a lot of tightwads and petty swindlers they are. 1914 *Wells Fargo Messenger* III. 9 (*caption*) Tight-wad Tim. 1934 *Punch* 26 Dec. 715/3 'A spendthrift', countered Chloe, 'makes a much more satisfactory husband than a tight-wad.' 1945 S. LEWIS *Cass Timberlane* xviii. 107 Hey, don't be so tight-wad with that hootch. *Ibid.* xxxii. 228 The man said to his wife, 'Our friend here has made a pretty good joke?' She said, 'Come on now—don't be a tightwad—what's his pretty good joke?' 1959 'A. GILBERT' *Death takes a Wife* xv. 188 He may be a bit of a tightwad. 1971 *Sunday Express* (Johannesburg) 28 Mar. (Comic Suppl.) 1/2 Blondie, I have a great idea to get a raise out of old tightwad Mr. Dithers. 1976 *National Observer* (U.S.) 26 June 10/2 When hard times hit..salesmen put the blame for their dwindling commissions on everybody else: the manufacturers, tightwad consumers [etc.]. 1977 *Sunday Tel.* (Colour Suppl.) 31 July 19/4 Bleeding tightwad! You'd think with all that cash he'd take a taxi.

‖ **tignon** (tī·yǫn). [Louisiana Fr., f. F. *tigne*, dial. var. of standard F. *teigne* moth.] A handkerchief worn as a turban head-dress by Creole women.

1884 *Daily Inter-Ocean* (Chicago) 2 July 1/5 [The women's heads were] adorned with the traditional head handkerchief of the tignon. 1935 M. MORPHY *Recipes of All Nations* 664 The old Creole women..with their brightly coloured bandana *tignons* or head-dress. 1961 M. G. EBERHART *Cup, Blade or Gun* i. 3 A..Negro woman, in a dark dress and snowy white apron and *tignon*. 1979 —— *Bayou Road* xix. 187 Liss's tightly wrapped tignon, her dark face.

tigon (tai·gǫn). Also **tiglon, tigron.** [f. TIG(ER *sb.* + LI)ON *sb.*, etc.: cf. *LIGER.] The offspring of a tiger and a lioness.

1927 G. JENNISON *Nat. Hist. Animals* 57 It should be noted particularly that the markings of the Tigon are not stripes, but rhomboids, almost like the markings on the Clouded Leopard. 1932 *Times Educ. Suppl.* 16 Jan. p. iv/2 The male parent of the 'tigon' was a tiger, the female an Indian lioness. 1938 *Times* 28 May 7/7 The name liger is given to the offspring of a male lion and tigress, the opposite cross being called a tigron. 1947 *Partisan Rev.* XIV. 395 Mr. Gielgud's gravity, his sensitive, melancholy profile, here becomes exquisitely comic—he looks like a tiglon with a heart. 1964 *Sunday Mail Mag.* (Brisbane) 8 Nov. 4/6 For many years the London Zoo possessed a magnificent 'Tigon'—a tiger-lion hybrid. 1972 *Times* 15 Feb. 6/5 A 'tigon' has been born in a Calcutta zoo from cross-breeding a tiger with a lioness. 1976 *Observer* (Colour Suppl.) 5 Sept. 42/1 (*caption*) This animal, one of the world's rarest cats, is a tiglon, a cross between a tiger and a lion. 1981 M. DUFFY *Gor Saga* v. 214 'Can you combine the sexual elements from two different species?' 'Like tigers and lions making tigrons you mean?'

Tigray (tigrai·). Also † **Teegray.** An alternative name for *TIGRINYA.

a 1860 W. C. PLOWDEN *Trav. Abyssinia* (1868) i. 10 They call their language Teegray. 1939 [see *TIGRINYA]. 1954 PEI & GAYNOR *Dict. Linguistics* 217 *Tigray*, an alternative name for *Tigriña*.

Tigre (tigre·), *sb.* and *a.* Also **Tigré.** [Native name.] **A.** *sb.* A Semitic language spoken in northern Ethiopia and adjoining parts of Sudan (distinguished from *TIGRINYA, spoken in Tigre itself). **B.** *adj.* Of or pertaining to the province of Tigre in northern Ethiopia. Cf. *TIGREAN *sb.* (*a.*).

1878 *Encycl. Brit.* VIII. 612/2 There are at least two modern languages which have sprung from the ancient Geez, distinguished in modern philology by the conventional names of Tigriña and Tigrê. 1908 [see *TIGRINYA]. 1933 L. BLOOMFIELD *Language* iv. 66 Ethiopian... The present-day languages of this group are Tigre, Tigriña, and Amharic. 1972 *Bk. Thousand Tongues* (rev. ed.) 429/1 Tigré is spoken by about 175,000 Muslims living in northern Eritrea and spreading into Sudan. 1978 *Observer* 29 Jan. 10/7 Tigre nationalism: The Tigre people, who inhabit Northern Ethiopia want to redress the balance of power lost with the defeat of the ancient kingdom of Gondar towards the end of the last century.

Tigrean (tigrē̆·ăn, -ī·ăn), *sb.* (*a.*). Also † **Tigrian.** [f. *Tigre* (see prec.) + -AN; cf. -IAN.] A native of the Tigre province in northern Ethiopia. Also *attrib.* or as *adj.* Cf. *TIGRE *a.*

1842 ISENBERG & KRAPF *Jrnl.* 22 Apr. (1843) 500, I must confess that I had conceived a more favourable idea of the hospitality of the Tigrians. 1901 [see STEW *v.*[2] 2 c]. 1960 E. ULLENDORFF *Ethiopians* iii. 37 A greater measure of non-Semitic ingredients in the Amharic language may accurately reflect a lesser degree of ethnic Semitization among the Amharas than is the case with the Tigreans. 1977 *Daily Tel.* 4 May 1/6 The secessionist Eritreans and Tigreans in the North. 1980 *Observer* 21 Sept. 7/5 Guerrillas of the Tigrean People's Liberation Front.

Tigrinya (tigrī·nya). Also **Tigriña, Tigrine.** [Native name.] A Semitic language spoken in the Tigre province of Ethiopia. Cf. *TIGRAY.

1878 [see *TIGRE]. 1908 T. G. TUCKER *Introd. Nat. Hist. Lang.* 172 Along with Amharic (of South-West Abyssinia) go the contaminated dialects Tigre (of the north), Tigrine (of the centre), and Harari (of Harai, east of Gallaland). 1939 L. H. GRAY *Foundations of Lang.* 364 Ethiopic..still serves as a learned language, though its true linguistic successor is Tigriña or Tigray, which is written by few except officials of the Italian colony of Eritrea. 1955 F. R. PALMER in E. P. Hamp et al. *Readings in Linguistics II* (1966) 341 The paradigms that may be set up for the nominals in Tigrinya..consist of two members. 1976 D. TOPOLSKI *Muzungu* iii. 35 She smacked my hand sharply and said something loudly in Tigrinya.

tigron, var. *TIGON.

‖ **tika** (tī·kā). Also **tikka.** [Hindi *ṭikā, ṭikkā*: cf. *TILAK.] Among Hindus, a mark on the forehead indicating caste, status, sectarian affiliation, etc., or worn simply as an ornament. Also *attrib.*, as *tika mark*, etc.

1884 in J. STORMONTH *Dict. Eng. Lang.* 1053/2. 1960 Z. EGLAR *Punjabi Village in Pakistan* xi. 114 A *tika* is a gold ornament worn on the forehead. 1963 S. S. IKRAMULLAH *Purdah to Parliament* xvi. 137 The use of *tikka* (a red spot on the forehead, originally a caste-mark among Hindus but now used more or less for its decorative effect). 1965 P. ROBINSON *Pakistani Agent* iii. 46 On her forehead was the red *tika* dot of marriage. 1973 *Australian Women's Weekly* 27 June 75/4 This colorful tika adds to the Indian woman's adornment. 1979 P. MATTHIESSEN *Snow Leopard* i. 22 The vermilion tikka dot on the women's foreheads. 1980 *Old Lady* June 76/3 We were invited to witness the *tika* ceremony... Granny.. had given the first *tika* mark to her eldest son, Naina Singh.

tiki (ti·ki). Also 8 **tigi.** Pl. **tiki, tikis.** [a. Eastern Polynesian *tiki* image; cf. *HEI-TIKI.] A large wooden image of Tiki, the creator and first created being of the Maoris and Polynesians, or of an ancestor; also, a small, usu. greenstone, image of the same, worn as a charm or ornament. Also *attrib.*

1777 D. SAMWELL *Jrnl.* 24 Feb. in Cook *Jrnls.* (1967) III. II. 1001 They [*sc.* the Maoris] had brought many Articles of Trade such as Ahoos, green Images called Tigis, Stone Adzes &c. 1840 J. S. POLACK *Manners & Customs New Zealanders* II. xxi. 178 Around the neck, similar kinds of ornaments are worn, but the principal favourite is the *Tiki*. 1889 HOCKEN *Catal. N.Z. Exhib.* 81 (Morris), Wooden Tikis, some of immense size, usually represented the ancestors. 1921 *Outward Bound* June 46/2 The beautiful and valuable greenstone from which the Maoris fashion all their..tikis. 1955 W. J. PHILLIPPS *Maori Carving Illustrated* 12 The body is U shaped, humanised below after the manner of a Maori tiki. 1963 *Times* 26 Feb. 8/5 Gas-flared tiki torches. 1974 T. HEYERDAHL *Fatu-Hiva* iii. 123 Some large wooden images, *tiki*, barred the entrances. 1975 *Times* 17 Mar. 1/6 (Advt.), By the flickering light of a Tiki lamp, you'll delight in sampling exotic dishes. 1977 *Chicago Tribune* 2 Oct. IV. 13/2 Papeete, the main town of Tahiti, generally isn't considered much of a shopping port, but it can bring surprises. Carved wooden, bone, and shell tikis and other jewelry; [etc.].

tikka[1], var. TICCA.

1888 KIPLING *Soldiers Three* (1889) 10 That *tikka*..has been *owin'* an' *fere-owin'* all over the bloomin' *maidan*. 1911 R. E. VERNEDE *An Ignorant in India* i. 5 He had collected enough porters to get it into a tikka ghari. 1928 *Blackw. Mag.* Apr. 498/1 Tikka-gharris, bullock-carts, hand-trucks, and coolies combined to raise..an excruciating din. 1978 'M. M. KAYE' *Far Pavilions* ix. 151 The drivers of phaetons and *tikka-gharis*, tongas and *ekkas*.

tikka[2] (ti·kā). [a. Hindi *ṭikka*.] In Indian and Pakistani cookery, (a dish of) small pieces of meat or vegetable marinaded in spices and cooked on a skewer. Freq. with qualifying word indicating the type of meat, etc. used. Also *attrib.*

1955 *Times* 28 May 5/1 Enormous dishes of *tikka kabab* —pieces of lamb deliciously roasted on skewers. 1960 *Dawn* 17 Apr. 12/5 President Nasser relished 'kabab' and 'tikka' dishes. 1961 *Guardian* 16 Feb. 12/4 You will hear all the gossip..over chicken tikka at Farooq's. 1980 J. PASSMORE *Indian Cookery* 110 A whole, smallish bird will take a maximum of 30 minutes, while smaller 'Tikkas' (pieces) only a few minutes. 1982 L. CODY *Bad Company* xxx. 200 She set out to find a chicken tikka and some live music.

tikka, var. *TIKA.

Tikopian (tikōᵘ·piăn), *sb.* and *a.* Also 9 **Tucopean.** [f. *Tikopia* (see below) + -AN.] **A.** *sb.* A native or inhabitant of Tikopia, one of the Solomon Islands. **B.** *adj.* Of or pertaining to Tikopia.

1832 A. EARLE *Narr. Res. N.Z.* 207 Savages are not much affected by music; but these two Tucopeans were excited to a most extraordinary degree. 1874 C. M. YONGE *Life J. C. Patteson* II. xii. 456 Three Tikopian giants had made a visit at Mota. 1951 W. J. GOODE *Relig. among Primitives* 234 The Tikopian emphasis is clearly on the culture-hero.

'til (til), var. of TILL *prep.*, *conj.* or short for UNTIL *prep.* and *conj.*

1939 P. G. PERRIN *Index to Eng.* 606 *Till, until,* (*'til*), these three words are not distinguishable in meaning. Since *'til* in speech sounds the same as *till* and looks slightly odd on paper, it may well be abandoned. 1956 H. KURNITZ *Invasion of Privacy* iii. 27 You sit on your tail 'til they come up with another story. 1978 M. DUFFY *Housespy* v. 126, I shall have to tell the P.M. but not 'til I'm sure.

tilak (ti·læk). Also **tilaka, tilka.** [a. Hindi *tilak*, Skr. *tilaka*, f. Skr. *tila* sesamum.] = *TIKA.

1879 E. ARNOLD *Light of Asia* II. 72 Slender hands and feet new-stained With crimson, and the tilka-spots stamped bright. 1895 *Funk's Stand. Dict.*, Tilak. 1961 *Spectator* 3 Mar. 281 The traditional welcome with music and *tilak* (a red mark placed in blessing on the honoured guest's forehead). 1969 'R. FARRE' *Beckoning Land* xiii. 151 On his forehead was a tilak mark showing that he was a follower of Shiva. 1974 *Encycl. Brit. Micropædia* IX. 1009/1 Among Śaivas (followers of Lord Śiva), the *tilaka* usually takes the form of three horizontal parallel lines across the forehead, with or without a red dot.

tilapia (tilē̆ⁱ·piă). [mod.L. (A. Smith: see quot. 1849), perh. f. Gr. τίλων, a fish name used by Aristotle + ἄπος distant.] A freshwater fish of the genus of this name, belonging to the family Cichlidæ, native to Africa but introduced elsewhere as a food fish or an aquarium species.

1849 A. SMITH *Illustr. Zool. S. Afr.: Pisces* 5 The *fossæ* or caverns connected with the gills are very indistinct, yet such traces of them exist as appears to warrant our regarding *Tilapia* as a fish of an aberrant form. 1901 *Trans. Zool. Soc.* XV. 18 In these *Tilapiæ*..the function of protecting the eggs devolves on the male sex. 1932 *Discovery* Feb. 43 (*caption*) The photograph shows a catfish with a *Tilapia* from its stomach. 1954 V. BARTLETT *Rep. from Malaya* iii. 56 A fish imported from Egypt, the tilapia,..hatches its eggs in its mouth and provides a refuge there for its young in moments of danger. 1965 A. J. MCCLANE *Standard Fishing Encycl.* 947/2 Tilapias have been experimentally bred in the southern United States (Alabama and Florida) where they have been stocked and caught in public waters. 1971 *Daily Tel.* 13 May 13/5 Families [in Uganda] are widely encouraged to cultivate their own stocks of tilapia, a fat, flat-bodied fish common throughout Africa.

tilasite (tī·lăsait). *Min.* [ad. Sw. *tilasit* (H. Sjögren 1895, in *Geol. Föreningens i Stockholm Förhandlingar* XVII. 291), f. the name of D. *Tilas* (1712–72), Swedish mining engineer: see -ITE[1].] A fluor-arsenate of calcium and magnesium, $CaMg(AsO_4)F$, that is isostructural with sphene and occurs as translucent monoclinic crystals that are colourless, green, or greyish.

1897 *Jrnl. Chem. Soc.* LXXII. II. 325 Tilasite or fluor-adelite from Långban... Irregular grains of this new mineral are found with hausmannite, berzeliite and calcite in a grey limestone. 1968 [see *ISOKITE]. 1979 *Mineral. Abstr.* XXX. 293/1 Tilasite crystals < 2·5 mm across have been found with friedelite and baryte at the Sterling Hill mine, Ogdensburg, Sussex County, New Jersey.

tilde. Add: **1. a.** Also, the mark placed in Portuguese above the letters *a* and *o* to indicate nasalization. Used simularly in systems of phonetic transcription.

1915 G. NOËL-ARMFIELD *General Phonetics* xvi. 88 The phonetic symbol for nasalised vowels is [˜] (the Spanish *tilde*) placed over the vowel symbol. 1958 J. L. TAYLOR *Portuguese-Eng. Dict.* (1959) p. x/2 If a word bears a tilde..stress the syllable so marked. 1974 *Encycl. Brit. Micropædia* VIII. 147/2 Typical of the Portuguese sound system is the use of nasal vowels, indicated in the orthography by *m* or *n* following the vowel..or by the use of a tilde..over the vowel.

b. *Palæography* and *Early Printing*. The diacritic mark ~ placed above a letter to indicate contraction of a following *n* or *m*.

1959 *N. & Q.* Feb. 77/2 From c. 1560 the tilde was used in printed English dramatic texts only over the vowels *a, e, o*, and *u*. 1975 J. BUTCHER *Copy-Editing* xi. 205 Superscript letters and tildes (nunnation marks) in contractions are normalized to modern usage unless there are good reasons to the contrary.

2. Used as a symbol in *Math.* and *Logic*, chiefly to indicate negation.

1958 *New Scientist* 10 July 364/2 If A is a matrix it is usual to denote the transpose of A by A′ or A*, and the trace of A by tr A. However, the author plays the part of 'the odd man out' by placing a tilde (~) over A to denote the transpose of A. 1971 [see *HOOK *sb*[1] 10 d]. 1979 D. R. HOFSTADTER *Gödel, Escher, Bach* viii. 212 With the tilde in front, the whole statement is denied. 1982 *S. Afr. Jrnl. Philos.* I. 117/1 The standard interpretation of the propositional calculus identifies the tilde with the English word *not*.

tile, *sb.*[1] Add: **1. a.** Now freq. made of concrete.

c. (*b*) *transf.* Regularly-shaped pieces (often squares) of floor- or wall-covering made of some other material, as *carpet* (*cork*, etc.) *tile*.

1960 *Mrs. Beeton's Cookery & Househ. Managem.* 44 (*heading*) Cork tiles. 1975 *N.Y. Times* 6 Apr. II. 40/4 Carpet tiles are installed by starting at the center of the room. 1976 *Evening Post* (Nottingham) 15 Dec. 20/1 (Advt.), Super savings on all branded Axminsters, Wiltons, Foambacks, Cords and Carpet tiles. 1982 *Habitat Catal.* 1982/83 140 Wipe clean natural cork tiles, pre-sealed with polyurethane varnish for protection.

h. *on the tiles*: [after the nocturnal activities of cats] on a spree, on a debauch. *slang*.

1887 H. BAUMANN *Londinismen* 125/2 *On the tiles*, auf dem Nachtbummel. *c* 1906 GALSWORTHY *Silver Box* (1910) 7 Been on the tiles and brought 'ome some of yer cat's fur. 1948 'J. TEY' *Franchise Affair* xi. 119 I'd say she was what is known as 'out on the tiles', sir. A very cool customer she was. 1977 C. MCCULLOUGH *Thorn Birds* xviii. 458 They all went out on the tiles... It was some night.

3. (Earlier example.)

1813 M. EDGEWORTH *Let.* 1 May (1971) 33 A number of *Fellows* and scholars with black tiles on their heads.

4. b. A thin flat piece used in a game, esp. in mah-jong or Scrabble.

1923 [see *MAH JONG *v.*]. 1973 *Times* 17 Nov. 2 A mah-jong set with ivory tiles. 1976 'M. ALBRAND' *Taste of Terror* ix. 56 The original..was..printed in red block letters. They seemed to think that tiles from a scrabble set had been used.

6. *tile-hat, -top*; *tile-hatted, -roofed, -topped* adjs.; **tile-and-a-half tile**, a tile one and a half times the width of the tiles used with it; **tile-drainage**, drainage constructed of tiles; **tile game**, a game played with flat pieces; **tile-hanging**, tiling fixed vertically to an outside wall, for its weather-resisting and decorative properties; hence **tile-hung** *a.*

1940 *Chambers's Techn. Dict.* 850/2 *Tile-and-a-half tile*.., a purpose-made tile of extra width, used to form the bond at a laced valley. 1866 MRS. GASKELL *Wives & Daughters* II. i. 4 He had taken the lead among the neighbouring landowners, when he first began tile-drainage. 1971 *Power Farming* Mar. 36/4 First-time sub-soiling over an existing satisfactory tile drainage system. 1950 E. CULBERTSON *Culbertson's Hoyle* p. xiii, Tile Games: Mah Jongg..Dominoes. 1974 *Encycl. Brit. Macropædia* II. 1149/2 Board and tile games are games played with a number of pieces on a specially constructed or marked board or with marked pieces (tiles) on a tabletop or other flat surface. 1932 *Times Lit. Suppl.* 7 July 494/3 Suffolk, Essex and Norfolk, counties remarkable for their brick-work, tile-hanging and weather-boarding. 1977 M. GIROUARD *Sweetness & Light* viii. 202 Stucco was replaced by red brick and tile-hanging. 1937 PARTRIDGE *Dict. Slang* 886/1 *Tile*,..extant as *tile-hat*, esp. in Glasgow. 1976 C. BERMANT *Coming Home* I. iii. 40 The topper, or a tile-hat as it was known in Scotland..was virtually the badge of office of the Rabbi. 1924 *Glasgow Herald* 24 Dec. 6 There is something as Christmas-like as snow in the sight of a tile-hatted gentleman purchasing a sausage-balloon. 1948 J. BETJEMAN *Coll. Poems* (1958) 140 Gabled lodges, tile-hung churches, catch the lights of our Lagonda. 1977 FEDDEN & JOEKES *National Trust Guide* (ed. 2) v. 375 Filled with brick nogging and tile-hung. 1962 E. SNOW *Other Side of River* (1963) xxxi. 239 On the Sungari River at Harbin I saw a tile-roofed structure really elaborate enough to be called a palace. 1977 H. FAST *Immigrants* II. 86 Seven thousand dollars for the tile-roofed, tile-floored house..was a tremendous bargain. 1907 *Yesterday's Shopping* (1969) 129/1 Bamboo tile top table ..12/4. 1931 'G. TREVOR' *Murder at School* xiii. 253 A sort of lounge, fitted up with tile-topped tables and deep armchairs. 1979 J. LEASOR *Love & Land Beyond* iii. 52 He sat down at a tile-topped table.

tilekum, tilicum, varr. *TILLICUM.

tiling, *vbl. sb.* Add: **2. b.** = *tile-draining* vbl. sb. s.v. TILE *sb.*[1] 6.

1943 J. W. DAY *Farming Adventure* xii. 138 The land was drained by tiling and moling, given three ploughings, and by the end of July sown with wheat. 1960 *Times* 5 July (Agric. Suppl.) p. v/3 More tiling has been undertaken.

tilka, var. *TILAK.

till, *sb.* Add: **4. till-roll**, a roll of paper recording an account of the transactions made at the till to which it is attached.

1972 *Times* 18 Oct. 4/4 The butcher..examined his till roll, and there was no record of anyone having paid that amount.

till, *prep.*, *conj.*, *adv.* Add: **A.** *prep.* **II. 5. d.** = To *prep.* 6 b, in stating the time of day. *U.S.*

1949 [see *QUARTER *sb.* 8 c]. 1962 M. & G. GORDON *Journey with Stranger* (1963) iv. 36 'Ten till,' he said... 'I'll go in first.'

B. *conj.* **1. g.** Indicating purpose: in order that (one) may, (*loosely*) 'and let me'. *Sc.* and *Ir.*

1881 A. MACKIE *Scotticisms* 18 Give me a match till I light the gas. 1904 W. B. YEATS *Pot of Broth* 78 Give me some vessel, till I give this sky-woman a taste of it. 1931 A. J. CRONIN *Hatter's Castle* III. iv. 525 Come till I give ye a grand, big hug. *a* 1966 'M. NA GOPALEEN' *Best of Myles* (1977) 57 Will yez all come in here..till I show yez me new picture.

tillage. Add: **1. d.** *fig.* Sexual intercourse (with a woman). *poet.*

c 1600 SHAKES. *Sonnets* (1609) sig. B1ᵛ, For where is she so faire whose vn-eard wombe Disdaines the tillage of thy husbandry? 1918 E. POUND *Pavannes & Divisions* 41 Some men will live as prudes in their own village And make the tour abroad for their wild tillage.

tiller, *sb.*[2] Add: **4. tillerman** *U.S.*, a fireman who controls the rear portion of a fire-engine; **tiller soup**, the minatory wielding of a tiller by the coxswain to encourage his boat's crew.

1968 L. LOKOS *House Divided* vii. 226 A hook and ladder truck went to answer the alarm without the tillerman aboard to control the rear portion. 1929 F. C. BOWEN *Sea Slang* 141 *Tiller soup*, the man-handling with the tiller by threat of which a coxswain encourages his boat's crew. 1947 *Sea Breezes* III. 151 We boys got a few gentle taps with the boat's tiller, too. We called it 'tiller soup'.

tilleyite (ti·li,ait). *Min.* [f. the name of C. E. *Tilley* (1894–1973), Australian-born English petrologist + -ITE[1].] A silicate and carbonate of calcium, $Ca_5Si_2O_7(CO_3)_2$, found as white monoclinic crystals.

1933 LARSEN & DUNHAM in *Amer. Mineralogist* XVIII. 469 An examination of blocks of the material from the Wet Weather quarry..has revealed the presence of a new lime silicate-carbonate mineral, with distinctive optical and chemical properties. For the new mineral the name tilleyite is proposed in honor of Professor C. E. Tilley of Cambridge University.., in recognition of the contributions he has made to the study of metamorphism. 1977 *Mineral. Abstr.* XXVIII. 71/1 At Kushiro, Tojo-cho, Hiroshima Prefecture, various rare skarn minerals occur in the contact zone between dioritic rock and limestone. These include..tilleyite.

Tilley lamp (ti·li læmp). Also (*erron.*) **Tilly lamp**, and with small initial. [f. the name of the manufacturing company.] The proprietary name of a type of portable oil or paraffin lamp in which air pressure is used to supply the burner with fuel. Also *ellipt.* as *Tilley*.

1932 R. F. FORTUNE *Sorcerers of Dobu* iv. 181 My great shadow cast..by my Tilley lamp. 1948 *Trade Marks Jrnl.* 17 Mar. 201/2 *Tilley*... Lighting and heating lamps employing liquid fuel. The Tilley Lamp Company Limited,.. Brent Works, Brent Street, Hendon, London, N.W.4; manufacturers. 1953 G. M. DURRELL *Overloaded Ark* vi. 115, I left the Tilly lamp in the middle of the swaying hypnotised circle. 1960 A. WESKER *I'm talking about Jerusalem* I. 23 Light the tilly lamp for me. 1966 *Guardian* 22 Dec. 5 Get that 'Tilley' goin', Gary. 1981 J. B. HILTON *Surrender Value* v. 42 Men filed out with Tilley-lamps, wicker baskets and rod-bags.

tillicum (ti·likǫm). *N.W. Amer.* Chiefly in *pl.* Also **tilicum, tilekum.** [a. Chinook Jargon *tilikum* people, ad. Chinook *tilxam*, f. *t-* pl. prefix + *ilxam* village.] **1.** A member of one's own tribe or people; also *pl.*, the people, common people.

1847 J. PALMER *Jrnl. Trav. Rocky Mts.* 105 A long time ago the Great Spirit became angry with them, set the mountain on fire, destroyed their towns, turned the *tiye* (chief) and *tilicums* (people) into stone, and cast them in the ocean outside of Cape Lookout. 1859 *Brit. Colonist* (Victoria, B.C.) 30 Apr. 2/2 Four Haidah Indians..came to the house of Mr. Oxner, Colquitz Farm, and after agreeing to work for him, one left, when the rest said they preferred to go with tilekum. 1922 E. P. JOHNSON *Legends of Vancouver* 35, I saw her graceful, high-bowed canoe heading for the beach that is the favourite landing-place of the 'tillicums' from the Mission.

2. A friend.

1869 *Mainland Guardian* (New Westminster, B.C.) 9 Oct. 3/2 The rescue from the courthouse at Kootenay, by his tillicums, of the Indian taken there by the Prices when caught in the act of robbing their sluices. 1906 *Daily Colonist* (Victoria, B.C.) 5 Jan. 2/4 Mr. French expects to resume his citizenship of Princeton after about four years' absence, his present hurried trip precluding a visit with many old tillicums here about. 1969 *Islander* (Victoria, B.C.) 30 Nov. 3/2 Thiepval was in Hakodate when we were there and during our stay in that port we had all become good tillicums.

tillite (ti·lait). *Geol.* [ad. G. *tillit* (A. Pencke 1906, in *Geogr. Zeitschr.* XII. 608): see TILL *sb.*[2], -ITE[1].] A sedimentary rock composed of glacial till compacted into hard rock.

1907 E. BLACKWELDER in B. Willis et al. *Res. in China* (Carnegie Inst. Publ. No. 54) I. I. xii. 267 (*heading*) Basal quartzite and glacial tillite. [*Note*] Tillite, a term recently proposed for consolidated till. 1944 A. HOLMES *Princ. Physical Geol.* xxi. 500 Tillites of the same age have been found in the north of Angola, in the eastern Congo, in Uganda, and in Madagascar. 1977 A. HALLAM *Planet Earth* 87 (*caption*) Striations on smaller rock bodies may be used to identify those as tillites (glacial till converted to rock) rather than tilloids, which are superficially similar.

tilloid (ti·loid). *Geol.* [f. TILL *sb.*[2] + -OID.] A sedimentary rock which is similar in appearance to a tillite but is not known to be of glacial origin; a sedimentary rock composed of non-glacial deposits.

1931 E. BLACKWELDER in *Bull. Geol. Soc. Amer.* XLII. 903 For these reasons the scarcity of striated stones in the McGee tilloid* need not be given undue weight. [*Note*] *Meaning a till-like deposit of doubtful origin. 1957 F. J. PETTIJOHN *Sedimentary Rocks* (ed. 2) vi. 265 The non-glacial conglomeratic mudstones or tilloids vary from a chaotic unassorted assemblage of coarse materials set in a mudstone matrix to a mudstone with sparsely distributed cobbles. 1966 *Earth-Sci. Rev.* II. 251 When there is any doubt about the origin of a till-like deposit, we encourage the use of tilloid, as a non-genetic term. 1977 [see *TILLITE].

tilly (ti·li), *sb.*[2] [f. Ir. *tuilleadh* an additional quantity, a supplement.] In Ireland and places of Irish settlement, an additional article or amount unpaid for by the purchaser, as a gift from the vendor.

1922 JOYCE *Ulysses* 15 She poured again a measureful and a tilly. 1958 M. & P. COLUM *Our Friend James Joyce* II. ii. 120 As the price of the little volume [*sc.* *Pomes Penyeach*] was a shilling, one expected to find twelve poems in it, one for each penny. Actually there were thirteen, the additional one being named 'Tilly'. In this Joyce was being obscurely local: the extra half-cup of milk that the milkman left in the Dublin householder's jug in the morning was a 'tilly'—something unpaid for. 1975 *Canadian Antiques Collector* Mar.–Apr. 22/2 [In Newfoundland] we have still in common use such Anglo-Irish terms as:..tilly (a small amount over and above what is purchased).

Tilly, tilly: see *TILLEY LAMP.

tilorone (ti·lŏrōᵘn). *Biochem.* [f. *tilor-*, of unkn. origin + -ONE.] An aromatic amine which induces the production of interferon and acts as an anti-viral agent.

1970 *Science* 18 Sept. 1213/1 Tilorone hydrochloride, the orange, water-soluble dihydrochloride salt of 2,7-bis[2-(diethylamino)ethoxy]fluoren-9-one, is a broad spectrum, orally active antiviral agent. 1978 *Nature* 29 June 760/1 Tilorone, statolon and Newcastle disease virus (NDV), all potent inducers of interferon in mice, induced a marked increase in spleen cell cytotoxicity.

Tilsit (ti·lsit). Also **Tilsiter, tilsit(er).** [The name of a town in East Prussia (now Sovetsk, U.S.S.R.); *Tilsiter* f. Ger. (= of Tilsit).] In full *Tilsit cheese*. A semi-hard cheese orig. made at Tilsit.

1936 *N.Y. Herald-Tribune* 8 Aug. 12/7 First on his list is tilsit, a sharp cheese from Denmark. 1950 J. G. DAVIS *Dict. Dairying* 130 *Tilsit cheese*,..an East Prussian hard-pressed cheese made from cow's whole milk. 1961 WEBSTER, Tilsiter. 1965 *Harrods Food News* May, Tilsiter (German—per lb. 6/6). 1971 J. AIKEN *Nightly Deadshade* iv. 46 They go at once to the larder and eat two pounds of Tilsit cheese.

tilt, *sb.*[1] Add: **3.** Also for a motor vehicle.

1976 *Milton Keynes Express* 16 July 31/3 (Advt.), 1975 Mini Pick-up, green, one owner, fitted tilt. 1977 'D. RUTHERFORD' *Return Load* iv. 77 One of his employees was fastening the blue canvas cover to the hooks on the side of one of the big, steel-framed tilts that were in such great demand for Continental journeys.

4. Also, a lean-to shelter. (Earlier examples.)

1612 in G. M. Story et al. *Dict. Newfoundland Eng.* (1982) 567/2 They had made a tilte with a sayle, that they got from some Christian. 1819 L. A. ANSPACH *Hist. Island of Newfoundland* 468 They call *tilts* temporary log houses, which they erect in the woods to pursue there their winter occupations.

tilt, *sb.*[2] Add: **II. 4. d.** *Television* and *Cinematogr.* (See quot. 1959.)

1959 HALAS & MANVELL *Technique Film Animation* 342 *Tilt*, the upward or downward pivoting movement of the camera across the screen. 1963 D. BOTTING in A. Smith *Throw out Two Hands* App. I. 266 A Miller tripod with a fully fluid head..giving smooth, controlled pans and tilts even with extreme long-focus lenses.

e. *fig.* An inclination; a bias.

1975 *N.Y. Times Bk. Rev.* 11 May 14 The contribution to the American language of other cultures has long been acknowledged..but it is unscholarly to insist on a 'tilt' toward minority contribution to satisfy resentment over past neglect. 1978 *Time* 18 Dec. 40 The pro-Soviet tilt of the new rulers in Kabul, the Afghan capital, is already stirring some recriminations in Washington.

10. tilt cab, a cab of a lorry, etc., which can tilt forwards; **tilt-cart** (earlier example); **tiltmeter** *Geol.*, an instrument for measuring changes in the steepness of a slope; **tilt-top** *a.*, having a top that tilts; **tilt-wheel**, (*b*) *U.S.*, a steering wheel that tilts; **tilt-wing** *a.* and *sb.*, (designating) an aircraft with wings that tilt.

1963 *Lebende Sprachen* VIII. 166/1 Tilt cab. 1977 *Horse & Hound* 10 June 42/1 (Advt.), Container 21ft Williams with Luton for tilt cab, metal framed and in sound order. 1834 J. B. BUCKSTONE *Wreck Ashore* II. iii. 40, I ha' just brought him home from the Physickiners in a tilt cart. 1937 *Nature* 10 Apr. 616/1 In both the Tango and Ito districts, tiltmeters..were erected. 1980 *New Scientist* 26 June 388/2 Everything that is happening to Mount St Helens is a 'classroom' experience for geologists and scientists scrambling to gather as much data as they can with seismic recording instruments, tiltmeters, and water level gauges. 1940 I. CRUMP *Our Airliners* vii. 138 For this disassembling operation, the engine is bolted on a portable tilttop table. 1973 *Canadian Antiques Collector* Jan.–Feb. 12/1 A birch tilt-top candlestand with an oval top. 1974 *State* (Columbia, S. Carolina) 15 Feb. 19-B/7 (Advt.), Power steering, power brakes, tilt wheel, vinyl roof, rally wheels. 1953 W. A. SHRADER *Fifty Years of Flight* 124/2 Spratt Aircraft Co...test-flies another in a series of tilt-wing flying boats designed by George Spratt. 1963 *Times* 2 Dec. 9/7 Two prototypes of a twin-engined 'tilt-wing' short and vertical take-off and landing transport aircraft. 1970 *New Scientist* 23 Apr. 173/1 Several tilt-wings have flown satisfactorily.

tilt, *v.*[1] Add: **II. 4. a.** Also *fig.*

1976 *Globe & Mail* (Toronto) 12 Nov. 1/3 The UN [*sc.* Union Nationale] will have taken seats from the Liberals and acted as spoiler in many other ridings, tilting the victory toward the Parti Quebecois. 1979 *N.Y. Post* 8 Aug. 2/5 Yesterday Israeli Foreign Minister Moshe Dayan charged that the U.S. was tilting its policy toward the PLO to appease Saudi Arabia and insure a steady flow of oil at reasonable prices.

b. Also *fig.*

1967 *Globe & Mail* (Toronto) 10 Nov. 2/9 During the India-Pakistan war, President Nixon ordered assistance to one side... Press reports at the time said Mr. Nixon wanted to 'tilt' in favour of Pakistan. 1978 *Guardian Weekly* 22 Jan. 17/3 Officials tend to tilt toward secrecy from a parochial view of their responsibilities.

d. *Television* and *Cinematogr.* To move (a camera) in a vertical plane.

1915 A. Lockett in B. E. Jones *Cinematograph Bk.* iv. 26 Tilting the camera causes convergence of upright lines.

tiltable (ti·ltăb'l), *a.* [f. Tilt *v.*¹ + -able.] Able to be tilted.

1934 in Webster. 1955 *Sci. Amer.* Mar. 42/3 Their tiltable antenna has four such units, each 320 feet long and 40 feet wide. 1979 *Observer* (Colour Suppl.) 22 Apr. 94 (Advt.), The steering column is tiltable.

tilting, *vbl. sb.*¹ Add: **2. c.** *Television* and *Cinematogr.* Movement of a camera in a vertical plane.

1938 G. H. Sewell *Amateur Film-Making* ix. 80 Movement [of the camera] in the up and down direction is known as 'tilting'.

TIM (tim). Also Tim, etc. Repr. pronunc. of *TIM* (the first three letters of Time *sb.*), the dialling code formerly used to obtain the telephone service giving the correct time in words; hence, this telephone service itself.

1936 [see *talking clock* s.v. *talking *ppl. a.* 2]. 1939 *Times* 23 Mar. 15/5 Public appreciation of the service was shown by the 18,000,000 or so calls a year that were made on 'Tim'. 1952 W. de la Mare *Winged Chariot* 7 Though every minute of your life's your own,..you ring up TIM; consult the telephone. 1978 E. St. Johnston *One Policeman's Story* ii. 41 Always receptive to innovations and new ideas, he [sc. Sir Donald Banks] introduced TIM, 999, Greetings Telegrams and the neo-Georgian style of post office buildings.

Tima(n)nee, var. *Temne.

timar. For *Obs.* read *Hist.* and add later examples.

1877 J. Baker *Turkey in Europe* 157 A Timar contained from three to five hundred acres of land. 1974 *Encycl. Brit. Macropædia* XIII. 776/2 The newly conquered lands were assigned to their commanders in the form of timars.

timariot. For *Obs.* read *Hist.* and add later examples.

1913 A. H. Lybyer *Government of Ottoman Empire* 101 The Zaims and Timariotes..were a class of country gentlemen. 1981 A. Toynbee *Greeks & their Heritages* 183 Ottoman Muslim timariots.

timbal, tymbal. Restrict 'Now *Hist.* or *arch.*' to sense in Dict. and add: **2.** = Timbale 1.

1929 J. G. Myers *Insect Singers* vi. 77 The essential elements of the [sound-producing] apparatus are the tymbals and the tymbal-muscles. 1969 R. F. Chapman *Insects* xxviii. 585 The tymbal is [illegible] segment of *Platypleura* (Cicadidae) there is an area of very thin cuticle supported by a thick cuticular rim... This area of cuticle forms the tymbal.

timbale. 2. (Earlier examples.)

1824 Byron *Don Juan* xv. lxvi. 58 Then there was God knows what 'à l'Allemande', 'timballe', and 'Salpicon'. 1866 Mrs. Gaskell *Wives & Daughters* I. xv. 178 Mr. Gibson had to satisfy his healthy English appetite on badly-made omelettes, rissoles, vol-au-vents, croquets, and timbales.

timbales (timbă·lez), *pl.* [a. F. *timbales*, pl. of *timbale* (or a. Sp., Pg., pl. of *timbal*): see Timbal, tymbal.] Two single-headed drums played as a pair with drumsticks (a percussion instrument of Caribbean or African origin). Also *attrib.*

1928 [see *maraca]. 1955 [see *conga 2]. 1966 *Crescendo* Dec. 27/1 L/A bands heavily feature bongoes, timbales and conga drums. 1974 *Nation* (Barbados) 3 Mar. 5/1 The BRC were doing a calypso and cut for drummer 'Boo' and stand-in timbales player L.O.D. to go into a frenzy of madness.

timber, *sb.*¹ Add: **4. a.** *tall timber*: see *tall *a.* 7 e.

c. *int.* The warning call of the feller when a tree is about to fall.

1912 J. Sandilands *Western Canad. Dict.* (ed. 2) 47 'Timber-r-r' the long-drawn melodious warning call of the sawyers in a lumber camp when a tree is about to fall. 1935 'C. Ford' *Burn Forever* 56 There was a stentorian shout: 'Timber!' 1968 *Islander* (Victoria, B.C.) 25 Dec. 2/1 The sharp ring of Father's axe echoed in the icy air, and we cried 'timber' as our tree fell.

5. (Earlier example in sense 'a wicket'.) Also (*rare*), an arrow.

1840 [see *screw *sb.*¹ 11 c]. *c* 1879 G. M. Hopkins *Poems* (1967) 180 Yet Arthur is a Bowman: his three-heeled timber'll hit The bald and bold blinking gold when all's done.

7. Preceded by a qualifying word: suitable quality or character for the specified office, etc. Cf. material *sb.* 6*. Chiefly *U.S.*

1892 *Chicago Tribune* 4 Apr. 4/5 Senator Cullom of Illinois is better Presidential timber than was generally supposed. 1914 *Emporia Gaz.* 13 Jan. 2/1 He is everlastingly..N.G. as gubernatorial timber. 1954 *Sat. Even. Post* 6 Nov. 64/4 CIA recruits many employees from our colleges and universities through a process beginning even before individual students realize that they are being singled out as possible CIA timber. 1967 R. S. Churchill *Winston S. Churchill* II. vi. 193 His parliamentary stature had grown and he had proved that he was of Cabinet timber. 1975 *Times Lit. Suppl.* 13 June 661/2 My concern [illegible] [sc. J. F. Kennedy] was potential Presidential timber.

9. a. *timber-claim* (earlier example), *growth, harvest, jinker* (Austral.) (Jinker²), *-land* (examples), *management, -market* (later examples), *-merchant* (later example), *preservation, production, -raft* (earlier example), *-slide* (earlier example), *-trade* (earlier example), *truck, value.* **b.** *timber feller* (later example), *-harvester; timber-harvesting.* **c.** *timber-lined* (earlier example) adj.

1857 *Lawrence* (Kansas) *Republican* 4 June, Timber claims..may be purchased on better terms than in any place of equal distance from Lawrence. 1922 E. M. Forster *Life to Come* (1972) 74 There was a tolerable [illegible] timber-fellers. 1968 *Ceba* XIV. 29 (*heading*) Forecasting timber growth by the point center extension modification of the Bitterlich system. 1969 *U.S. Forest Service Research Bull.* [illegible] (*title*) 1968 Washington timber harvest. 1965 *Canad. Geogr. Jrnl.* Sept. 86/2 The economic arguments of timber-harvesters. *Ibid.* 86/1 It may also be that park administrators..apply the terminology and techniques of timber harvesting to areas in which the primary use is recreational. 1916 J. B. Cooper *Coo-oo-ee* i. 19 [illegible] Quality trucks at lowest prices..also semi trailers, semi tippers, low-loaders, timber jinkers, tippers. 1654 *Suffolk Deeds* (Boston, Mass.) (1883) II. 55 Howses fence..gardens, Tymber Lands broaken & unbroaken. 1804 P. Gass *Jrnl.* 29 Aug. (1807) ii. 31 We passed some timber land on the south side. 1981 *Bull. Yale Univ. School Forestry* No. 92. 15 Ownership of timberlands by the forest products industry..grew by [illegible]. 1939 W. Whitman *Leaves of Grass* (ed. 2) xii. 225 You timber-lined sides! You distant ships! 1952 *U.S. Forest Service Research Paper* No. 50 40. 1 (*title*) 25 years of selective timber management on the Crossett Experimental Forest. 1981 *Bull. Yale Univ. School Forestry* No. 92. 39 This model will probably be implemented empirically on a [illegible] timber markets were available. 1946 A. R. M. Lower *Colony to Nation* 169 The timber merchants..bought square timber and deals..to ship them to England. 1966 *Encycl. N.Z.* I. 724/1 As building authorities have rightly been unwilling to accept this non-durable sap-timber, a suitable timber preservation industry has grown up. 1968 *Wisconsin Agric. Exper. Sta. Research Bull.* No. 272. 1 (*heading*) Mycorrhizae: their role in tree nutrition and timber production. 1818 F. Hall *Trav. Canada & U.S.* iii. 116 The frequent sail, or heavy timber-raft, 'floating many a rood'. 1836 *Bytown* (Ottawa) *Gaz.* 21 July 2/5 This improvement with many of others (amongst the rest a timber slide at the Chats) the country owes to..George Buchanan. 1732 in *Calendar of State Papers Colonial Series, Amer. & W. Indies 1731* (1939) XXXIX. 243 Abundance of saw-mills are erecting for the timber trade. 1859 D. Bunce *Travels with Dr. Leichhardt* ii. 23 These pipes..afforded excellent substitutes for bridges, wherever it was necessary for a road to be made for the timber-trucks. 1976 M. Birmingham *Heat of Sun* vi. 75 The timber trucks—great articulated monsters each carrying thirty-odd tons of timber, usually in the form of three huge logs. 1927 *Chambers's Jrnl.* Jan. 4/2 What bearing has the presence, or the increase, of timber-wolves upon the problem of timber values? 1981 *Bull. Yale Univ. School Forestry* No. 92. 33 Timber values must be foregone to obtain additional nonmarket values.

10. timber beast *N. Amer.*, a logger; **timber berth** *Canad.*, a tract of forested land the bounds of which have been established by the government, which leases or sells the rights to fell and remove timber; **timber carriage** = *timber-cart*; **timber cruise** *N. Amer.* = *cruise *sb.* 2; **timber-cruiser** *N. Amer.* [*cruiser 3], a timber prospector; hence **timber cruising**; **timber-doodle** (examples of *U.S. local* sense and earlier example of *slang* sense); **timber drive** *N. Amer.*, an organized floating of loose timber down a waterway; a quantity of timber so floated; **timber due** *Canad.*, a tax paid to the government on each tree taken out of a timber berth; **timber-frame,** (*c*) *attrib.* = *timber-framed* adj.; **timber-framing,** the construction of buildings having frames of timber; **timber-getter** *Austral.*, a lumberman or logger; **timber-head,** (*b*) *slang* (*rare*) = Blockhead 2; **timberjack** *N. Amer.*, a lumberman or logger; **timber jam** = *log-jam 1; **timber-jumper** (earlier example); **timber licence** *Canad.*, a licence to cut timber on a timber berth on payment of dues to the government; **timber-limit** *Canad.*, (*a*) (earlier example); (*b*) any tract of forested land suitable for lumbering; (*c*) = *timberline; **timber-line:** delete and see as main entry; **timber rattler, rattlesnake,** a venomous snake, *Crotalus horridus horridus*, found in the northeastern United States and marked with dark bands or blotches; **timber-tug** (later example); **timber-wolf,** for *Western U.S.* read *N. Amer.* and add earlier and later examples.

1919 *Camp Worker* 26 Apr. 5/2 A large number of our city folk imagine that a 'timber beast' has just about as much need for brain as a Canadian soldier in Siberia has for refrigeration machinery. 1975 J. Gores *Hammett* (1976) xiii. 93 They thought he was a timber beast out of Seattle. 1837 *Times* (Halifax, Nova Scotia) 17 Jan. 22/1 The selling of Crown lands by auction—and the disposal of the timber berths. 1957 *Camsell Arrow* (Edmonton, Alberta) Christmas 68/3 The mission bought a sawmill and set it up on a timber berth just north of the Sunchild reserve buildings. 1747 E. Purefoy *Let.* 12 May in *Purefoy Lett.* (1931) i. 41 This day I ordered the Timber Carriage to be set up by Simon Hobcroft the Carpenter. 1901 'L. Malet' *Hist. Sir R. Calmady* ii. iii. 110 A miller's tented waggon,..a timber-carriage, and a couple of spring-carts. 1933 E. Merrick *True North* 319 The people in Mud Lake remember the lumbermen by..their timber cruises. 1949 *Boston Globe* 17 July (Fiction Mag.) 8/1 Hard years in mine and timber cruise had given Flood a certain steadiness and maturity. 1956 T. Raddall *Wings of Night* (1957) iii. 33 Someone offered me a job on a timber cruise up in the north Ontario bush. 1894 *Century Mag.* Mar. 671/2 The timber-cruiser is a hero... The location of a choice tract of timber is a secret to be guarded with his life. 1981 *Publ. Amer. Dial. Soc.* LXVII. 5 In 1890 the area was inhabited only by a few transient timber cruisers and mineral prospectors. 1933 *Meccano Mag.* Mar. 195/1 Another activity confined largely to the Eastern Lines is 'timber cruising', which consists of the surveying and mapping out of various forest areas. 1956 T. Raddall *Wings of Night* (1957) xviii. 143 Winter was a good time for timber cruising. A pair of snow-shoes would carry you anywhere. 1842 Dickens *Amer. Notes* I. iii. 141 Mint Julep, Sherry-cobbler, Timber Doodle, and other rare drinks. 1856 *Spirit of Times* 25 Oct. 129/1 While we have been dosing timberdoodles with infinitesimal blue pills, they shall have..been doctoring bruins. 1979 *Globe & Mail* (Toronto) 24 Oct. 41/1 Any hunter who has ever overheated his shotgun barrel trying to down the elusive timberdoodle knows..of the erratic flight of the woodcock. *a* 1861 T. Winthrop *Life in Open Air* (1863) 23 The head-driver of a timber-drive leads a disorderly army. 1920 *Blackw. Mag.* Nov. 616/1, I caught sight..of a second log, followed by a third and yet others in an apparently endless procession. I had never encountered a timber drive before. 1957 B. Hutchison *Canada* 101 He had heard only vague rumors of the old timber drives in the days of Peter Emberley. 1883 J. Fraser *Shanty Life Backwoods of Canada* 87 How easily this could be balanced in the treasury accounts by the smallest additional fraction upon timber dues. 1936 A. R. M. Lower *Settlement & Forest Frontier in Eastern Canada* 77 To Crown timber dues was added 'timber licence', 'timber-limit', or 'timber-berth' arrangement. 1967 *Times Rev. Industry* Apr. 32/2 Timber-frame houses are composed of lightweight sections and therefore are easy to erect. *Ibid.*, Timber-framing has completely overcome the postwar 'pre-fab' image of industrialized building. 1912 *Contemp. Rev.* Aug. 248 The professional timber-getter is a Southern miscreant. 1970 M. Kelly *Spinifex* vi. 103 The word's Timber Getter, not lumberjack. 1849 H. Melville *Redburn* 37 You timber-head..take this bucket here, and go up the rigging. 1916 A. Bridle *Sons of Canada* 5 He was a timberjack in the hardwood bush of western Ontario. 1953 D. Cushman *Timberjack* 127 You boys hired out to be timberjacks. 1888 Lees & Clutterbuck *B.C. 1887* 186 On one of the huge timber jams which so often occurred we passed close to a wolverene. 1937 Kipling *Something of Myself* iv. 101 The removal of the key-log in a timber-jam starts the whole pile. 1832 *Q. Rev.* XLVII. 237 'Now for the timber-jumper,' cries Osbaldeston, pleased to find himself upon Clasher. 1910 J. London *Lost Face* 133 Crossing a timber jam on the frozen bed of the Teelee, the sled suffered a wrenching capsize. 1921 *Daily Colonist* (Victoria, B.C.) 16 Mar. 5/2 It was a wide open invitation to the speculator to buy up large numbers of timber licences in arrears. 1966 *Canad. Forest Industries* Nov. 55/1 Amendments made to the Forest Act in 1965 now permit: Application of the cost of timber sales to the timber licence as a whole, or to cutting permits issued pursuant to the licence. 1854 T. C. Keeper *Ottawa* 56 No timber limits are without water—for it is by water alone that the timber can reach its market. 1890 *Grip* (Toronto) 8 Feb. 83/2 A Journal..is..agitated lest, by disputing our timber-limits.., the Ontario Government shall bring the province to direct taxation. 1898 W. T. Jennings *Rep. Routes to Yukon* 9 The whole valley and slopes to the timber limit are clothed with cotton-wood, spruce and alder trees. 1914 H. Bindloss *Intriguers* 108 We want to get as far north as the timber limit. 1960 *Ottawa Citizen* 18 June 38/5 A Blackfoot Indian band in 1892 surrendered a timber limit in Alberta. 1936 E. G. Barnard *Rider of Cherokee Strip* 51 Some called it the timber rattler or the black diamond rattler. 1974 A. Dillard *Pilgrim at Tinker Creek* xiii. 223 The only other poisonous snake around here is the timber rattler. 1950 *Chicago Tribune* 16 Mar. ii. 12/2 With him to the zoo went five timber rattlesnakes. 1982 J. S. Borthwick *Case of Hook-Billed Kites* (1983) xlviii. 171 He swept from the timber rattlesnake..to the western diamondback. 1977 N. Freeling *Gadget* iii. 142 Sturdy horses could haul carts, timber-tugs, sleds in winter. 1860 *Nor' Wester* (Red River Settlement) 28 Feb. 4/2 We also saw a large timber wolf (not a wolf made of wood, but a gentleman who inhabits prairies and wooded country). 1936, 1964 [see *grey wolf* s.v. *grey, gray *a.* 8 b]. 1980 *Beautiful Brit. Columbia* Spring 19 In the warehouse, red and silver fox furs hang beside the sleek pelts of lynx, timber wolf and beaver.

ti·mberline. Chiefly *N. Amer.* Also **timber line, timber-line.** [Timber *sb.*¹] **1.** On a mountain, the line or level above which no trees grow. Freq. with omission of *the*. Also *attrib.*

1867 *Harper's Mag.* June 17/2 A high mountain ridge divided into innumerable peaks, all of which tower above the timber-line. 1874 [in Dict. s.v. Timber *sb.*¹ 10]. 1904 *Bull. Geol. Soc. Amer.* XIV. 557 On the mountains of central Idaho, the cold timberline is sharply drawn at an elevation of about 10,000 feet, while the dry timberline, equally well defined, has an elevation of about 7,000 feet. 1936 *Scrutiny* IV. 443 Timber-line settlers. 1961 R. M. Patterson *Buffalo Head* ii. 40 The timberline country I would learn to call it in the years to come. 1966 *Encycl. N.Z.* I. 730/1 Upper timber-line belts are largely composed of broadleaf shrubs. 1980 *Outdoor Life* (U.S.) (Northeast ed.) Oct. 100/2 Hunters working from 8,000-feet to timberline.

2. In the northern hemisphere, the line north of which no trees grow.

1896 C. Whitney *On Snow-Shoes to Barren Grounds* 287 How well I remember that birch-tree! It was the only birch-tree we saw on the whole trip.. I was so delighted, for I knew by that sign the timber-line was very close. 1930 H. Godsell *Arctic Trader* 288 We had a warm camp that night, as we were still within the timber line. 1977 A. Hallam *Planet Earth* 88 In practice we find that the zone of continuous permafrost is limited to those areas where the mean annual temperature is −15°C. and this also roughly coincides with the timberline, the northernmost point where trees exist.

timberman. Add: **1. c.** *Canad.* An owner or manager of a company engaged in lumbering.

1889 W. H. Withrow *Our Own Country—Canada* 157 The trees, where the timbermen have not culled out the finest, are most picturesque. [illegible] 1964 W. Lindsay *B.C. Outlaws* 7 Among them is..a prominent timberman who recently became interested in timber limits at [illegible].

timbo (timbo·). Also **timbó.** [a. Tupi.] **1.** Any of various South American woody vines cultivated as a source of fish poison and the insecticide rotenone, esp. those of the genus *Lonchocarpus* (family Leguminosæ); also, the poison itself.

1725 H. Sloane *Voy. to Islands of Madera, Jamaica* II. 40 Timbo, a sort of Withe, intoxicates Fishes. 1930 *Jrnl. Econ. Entomol.* XXIII. 868 Rotenone also occurs in 'cube' roots (*Lonchocarpus nicou*) and in timbo, haiari, and other members of the genus *Lonchocarpus.* 1949 *Thorpe's Dict. Appl. Chem.* (ed. 4) IX. 306/2 South American sources of rotenone..have been developed recently, chiefly the cubé of Peru, and the timbó of Brazil. 1971 *Nat. Geographic* Sept. 433 Like most Brazilian Indians, they usually fish with the sap of a vine called timbo.

2. A South American timber tree of the genus *Enterolobium* (family Leguminosæ) from which a soft red wood is obtained and used for making furniture.

1924 Record & Mell *Timbers Trop. Amer.* 205 'Timbó' is well known to the trade in Argentina..in the form of squared logs and because of its softness and ease of working is used as a cheap substitute for cedar. 1960 T. H. Everett *Living Trees of World* 156/2 *Enterolobium.* Ten New World species comprise this genus. The best known are the timbo..(*E. contortisiliquum*) and the guanacaste or elephant's ear (*E. cyclocarpum*).

timbre, *sb.*³ Delete ‖ and add: Now also with pronunc. (tæ·mbəɹ).

timbred (tɛ̃·bər'd, tæ·mbəd), *a.* [f. Timbre *sb.*³ + -ed².] Having a timbre of a specified kind.

1944 C. Himes *Black on Black* (1973) 183 His voice, timbred with a quality of raw intensity, encompassed them both. 1969 'E. Lathen' *When in Greece* xvi. 182 Her splendidly timbred voice.

timbrous (tɛ̃·brəs, tæ·mbrəs), *a.* [f. Timbre *sb.*³ + -ous.] Sonorous, resonant.

1929 W. Faulkner *Sartoris* iv. 283 The other one bayed her single timbrous note. 1977 T. Wicker *Facing Lions* 67 The low, timbrous voice was as resilient and its words as rambling and loosely collected as those of the farmers who preceded him.

Timbuctoo (ti:mbʌktū·). The name of a town (now officially spelt *Timbuktu*) on the edge of the Sahara in West Africa, used as the type of the most distant place imaginable.

1863 L. Duff-Gordon *Lett. from Egypt* (1969) 104 It is growing dreadfully Cockney here [sc. Cairo]. I must go to Timbuctoo. 1935 A. Christie *Death in Clouds* xiv. 148 'She's very——' '—but she may be going to Timbuctoo' finished Norman. 1974 V. Canning *Painted Tent* ix. 182 Trevor Green wished Sinter in Timbuctoo, but since that couldn't be arranged, he just had to put up with him. 1982 *Times* 19 Feb. 2/4 Electors promising to 'follow their leader' [sc. Shirley Williams] to Timbuctoo if she should choose to go there.

Hence **Ti:mbuctoo·t** *a.* (fanciful *nonce-wd.*), of or pertaining to Timbuctoo, foreign, outlandish.

1930 D. H. Lawrence *Nettles* 16 And the world it didn't give a hoot If his blood was British or Timbuctoot.

time, *sb.* Add: **I. 1. a.** *in no time*, etc. (earlier example).

1832 [see *no *a.* 3 b].

2. (*for the*) *time being* (later examples).

1883 'Mark Twain' *Life on Mississippi* xliii. 440 It's human nature—human nature in grief. It don't reason, you see. 'Time being, it don't care a dam. 1913 *Granta* 22 Feb. 225/1 When listening to a singer of such extraordinary natural gifts as Madame Melba, one loses the faculty of criticism for the time being. 1948 V. Massey *On being Canadian* xi. 176 It is perhaps natural that there should be a period of pause; that we should not for the time-being be 'on the march'. 1977 J. Crosby *Company of Friends* xx. 127 The pilot's one of ours—for the time being.

3. e. (*there's*) *a good time coming.*

1817 Scott *Rob Roy* III. v. 149 'There's a guid time coming.' 'No time like the time present, Mr Campbell.' 1846 C. Mackay *Voices from Crowd* (1851) 9 There's a good time coming, boys, A good time coming: We may not live to see the day. 1873 Trollope *Lady Anna* (1874) I. iii. 33 She did not believe in the good time coming as did her mother.

5. a. *behind the time(s)* (examples); *to move with the times*: see *move *v.* 19 c.

1846 Dickens *Dombey & Son* (1848) ix. 87 I'm old-fashioned, and behind the time. 1921 E. O'Neill *Diff'rent* ii. in *Emperor Jones* 244 You needn't think we're all so behind the times here. [illegible] 1937 G. Orwell *Road to Wigan Pier* viii. 163 Here I shall be accused of being behind the times, for I was a child before and during the war and it may be claimed that children nowadays are brought up with more enlightened notions.

b. Also used *attrib.* to designate typefaces designed for *The Times.*

1932 S. Morison in *Monotype Recorder* XXXI. 22 (*heading*) The Times New Roman. *Ibid.* 13/1 A set of drawings was completed some two years ago, and [illegible] at the Monotype Works, Redhill, in April, 1931. 1965 *Times* 4 June 8/4 The weekly news-magazine..appears this week with a changed type face—Times Roman instead of the former old style. 1972 *Times* 9 Oct. 14 After 40 years *The Times* today appears in a new typeface, Times Europa. This has been designed for *The Times* to suit changing printing methods and largely replaces Times Roman.

c. *big time*: see *big *a.* B. 2. See also *small time, small-time sb.* and *a.* (*phr.*).

6. *to make a time* (examples); *to have the time of one's life* (examples); *to have one more time* (U.S. colloq.), to have a very good time [illegible]. Also (without specification, depending on context) a good or bad experience lasting some time; *N. Amer.* a social occasion, a party.

1860 H. J. Hawley in *Wisconsin Mag. Hist.* (1936) Mar. 331, I had a time getting the chairs up. 1878 in G. M. Story et al. *Dict. Newfoundland Eng.* (1982) 568/2 But..while on a visit to Bett's Cove [he] got on a time and let the cat out of the bag. 1882 J. M. Kettlewell *Maud* (1939) 149 I've had a 'time' with Mr. [illegible]. 1883 A. Pinkerton *Spy of Rebellion* xx. 328 While there I met some of the boys, and we had a little 'time'. 1887 [see *life *n.* 8 a]. 1888 *Boston Jrnl.* 31 July 2/5 She doesn't weep at the parting or make a time over it. [illegible] 1932 R. Hichens *Spirit of Time* 186 Arab chieftains..having the time of their lives in the redecorated hotels. [illegible] 1933 Greenleaf & Mansfield *Ballads & Sea Songs Newfoundland* p. xxii, To raise money for the schoolhouse and the church, the Sally's Cove people held a 'time' on Orangemen's Day, which took the form of an all-day fair and was held in the school-house. [illegible]

7. b. (*a*) (Later example.) (*d*) (Earlier example.) (*f*) (Further examples.) (*h*) The prescribed duration of opening-hours at a public house; the moment at which this ends; also *ellipt.* as the signal for closing-time.

(*a*) 1966 P. Wrightson *Thirteen-Bitten Portrait* [illegible]

(*d*) 1837 Dickens *O. Twist* (1838) I. xviii. 306 His 'time' was only out an hour before.

(*f*) 1829 P. Egan *Boxiana* 2nd Ser. II. 519 The Gas was defeated, [illegible] 1833 *Sporting Mag.* Aug. 354/2 On time being called, [illegible] went in again. 1892 J. Hibson *Hist. Salford Football Club* 52 A fight began, and the game was brought to an abrupt termination about five minutes before 'Time'. 1926 [see *end *sb.* 22 f]. 1976 *Sunday Mail* (Glasgow) 28 Nov. 44/6 Scorers were Martin, in the first half, and Johnston just before time.

(*h*) 1912 G. Frankau *One of Us* xi. 104 Lingered a remnant; querulous to these, one spake unceasing: 'Gentlemen! Time, Please!!' 1922 T. S. Eliot *Waste Land* ii. 23 Hurry up please its time. 1932 L. Golding *Magnolia St.* i. iii. 47 Collecting empty glasses and shouting, 'Time, gentlemen, please!' 1953 J. Mortimer *Like Men Betrayed* v. 87 It's not very comfortable in our pub... They're always shouting 'time' and turning the lights on and off. 1979 'C. Brand' *Rose in Darkness* ii. 20 Soon he must turn her out..five minutes to Time. 1984 *Daily Tel.* 4 Jan. 15/7 A car crashed into the Mermaid public house..and ended up in the bar just as the landlord..was calling time.

8. b. (Earlier example.)

1842 *Congress. Globe* 2 Mar. App. 188/2 A single horse in a sulky would..be able to make..even better time, with the letter mail alone.

c. *pressed for time*, short of time, in a hurry.

1817 [see Press *v.*¹ 6 c]. 1833 [in Dict., sense 8 a]. 1942 W. Faulkner *Go down, Moses* 274 Not impatient but just pressed for time.

d. *it takes* (a person) *all his time*, it presents great difficulties to, it requires great effort from. *colloq.*

1900 R. Guthrie *Kitty Fagan* 208 We've a ticklish job on hand, an I'm bound to say, it's taken us all our time. 1905 in *Eng. Dialect Dict.* VI. 151/1 It'll tak him all his time to mak their theirs public pay it ways. 1941 E. Carr *Klee Wyck* 85 It took Jimmie all his time in the shallows to keep us in the channel.

e. Imp. phr. *give* (a person) *time*, be patient with (another), in expectation of some future change of attitude, competence, etc.

1902 [see *smile *v.* 8 b]. 1940 W. Faulkner *Hamlet* i. iii. 60 He'll pick it up though..just give him time. 1962 J. F. Powers *Morte d'Urban* vii. 151, I don't say the present population wants it, but give them time.

f. With a quantifier: *to have no* (*a lot of*, etc.) *time for*, to have no, etc., respect or admiration for. *colloq.*

1911 E. M. Clowes *On Wallaby* vi. 166 The merely fictitious value of age here [sc. Australian youth] 'have no time for'. 1922 C. Wilson *Rambles in Bookland* i. I never had much time, to use an effective colloquialism, for the list of 'the best hundred books'. 1930 N. Marsh *Artists in Crime* xi. 256 The only one they seemed to have much time for was the Honourable Basil Pilgrim. 1952 A. Grimble *Pattern of Islands* ix. 177 He never had much time for pen-pushers, as he called them. 1966 J. Cleary *High Commissioner* xi. 247, I don't think he'd much time for him. [illegible] 'Yes, I've got a lot of time for Lester,' the Vicar continued.. 'He'll always lend a hand if he's at home.'

g. Time available for a certain purpose, *spec.* for an advertising broadcast.

1930 *Daily Express* 6 Sept. 4/6 To the big advertiser the broadcasting stations came with an offer to 'sell time' to pay the cost of broadcasting programmes. 1935 S. Lewis *It can't happen Here* iv. 43 Father Charles Coughlin, of Detroit..first thought out the device of freeing himself from any censorship of his political sermons..by 'buying his own time on the air'. 1967 *Boston Sunday Herald* 26 Mar. ii. 11/1 CBS-TV explained..that its policy 'prohibits..the sale of time for the presentation of views on current issues other than in connection with elections'. 1979 *Daily Tel.* (Colour Suppl.) 18 Sept. 30 The world's 1300 or so professional astronomers who obtain 'time' on the big telescopes all have different programmes in different parts of the sky. 1977 *Zigzag* June 10/3 A friend of mine who was an engineer rang up to see if I had any songs I wanted to cut, and if he could get some cheap time.

h. *to take* (*one's*) *time*: see Take *v.* 28 a; *lost time*: see Lost *ppl. a.* 9.

9. (Earlier and further examples of sense 'pay', esp. at the termination of a period of employment.)

1887 *Courier-Journal* (Louisville, Kentucky) 12 Jan. 6/3 All that remained for the brakemen and switchmen to do was to go to the office, and get their 'time'. [illegible] railroad parlance as their 'time'. 1902 S. E. White *Blazed Trail* viii. 56 So Pat and Henry's [illegible] not discharged—were not instructed to 'get their time'. 1926 J. Black *You can't Win* xx. 297 He threw down his shovel, walked over to the boss, and demanded his 'time'. I heard the foreman say: 'All right, you're no good anyway. I was going to fire you to-night.' 1935 A. J. Cronin *Stars look Down* iii. xiv. 668 It broke his heart to give the boy his time, to send them to join the six hundred men from the Neptune already on the dole.

10. b. *time and a half* and varr.: one and a half (or one and a quarter, etc.) times the usual rate of pay.

1888 [in Dict.]. 1921 *Daily Colonist* (Victoria, B.C.) 16 Mar. 3/4 The extra payment was due to the time and a half allowance for overtime. 1932 *Economist* 12 Mar. 552/2 Extra pay for night duty is to be reduced from 'time and a quarter' to 'time and an eighth', and for duty on Sundays, Christmas Day and Good Friday from 'time and a half' to 'time and a third'. 1970 *Daily Tel.* 9 Jan. 9 The claim..is for payment of time and a quarter for the first six hours' overtime, time-and-a-half for the next six, and double time payments thereafter. 1976 *Ibid.* 12 Nov. 21/7 They want holiday pay, at present single time, increased to time and a third. 1980 M. Kenyon *Deep Pocket* ii. 28 Tell the men I'm paying time and a half for every forty yards dug by the weekend.

11. *to mark time*: see Mark *v.* 10 b.

II. 13. d. *Stock Exch.* The account.

1901 W. G. Cordingley *Dict. Stock Exch. Terms* 89 Time Bargains refer to those speculative transactions which are made for settlement on the next Account. They are made 'for time', and are 'Bought for the Account' or 'Sold for the Account'. 1928 *Daily Mail* 13 Aug. 18/2 Dealing for 'new time',..the new Stock Exchange account.

15. b. (Later examples in the sense 'time of childbirth'.)

1841 Thackeray *Gt. Hoggarty Diamond* xii, in *Miscellanies* (1857) IV. 428 My poor wife, then very near her time, insisted upon accompanying me. 1931 H. S. Walpole *Judith Paris* iii. 582 Judith was very near her time, and, in consideration..that this was her first child, it had been wiser of her perhaps not to have come. 1980 R. Butler *Blood-Red Sun at Noon* (1981) i. i. 19 She..became pregnant... What she called 'her time' approached.

16. (Earlier example of *now's your time.*)

c 1810 W. Hickey *Mem.* (1960) xix. 309 Now's your time, Hickey. That beast Mordaunt was called away..so that you will have a couple of days' enjoyment together.

17. a. With a price: (so much) *a time*, on each occasion, (*colloq.*) for each item.

1718 R. Grosvenor in C. T. Gatty *Mary Davies* (1921) II. 205 One that is grown pretty rich by his attendance upon Patients in Garrets at Half-a-Crown a time. 1976 *West Lancs. Evening Gaz.* 13 Dec. 7/5 Buying..cashmere scarves at £15 a time.

c. *every time*: see *every *a.* 1 e.

18. *many a time and oft* (later example); *times* (later examples) also, sometimes, at times; *many's the time.*

1920 E. O'Neill *Beyond Horizon* ii. i. 67 Many's the time I've said to her [etc.]. 1936 Wodehouse *Laughing Gas* xxiii. 247 You have many a time and oft referred to her as a piece of cheese. 1938 M. K. Rawlings *Yearling* xvii. 213 Seems to me, times, hit ain't done nothin' to you but sharpen your tongue. 1956 A. Wilson *Anglo-Saxon Attitudes* i. iii. 48 Many's the time Sir Beerbohm Tree's stood outside the theatre. 1968 C. Aird *Henrietta Who?* xiv. 130 Times, it's a bit quiet at Holly Tree. 1980 P. G. Winslow *Counsellor Heart* iii. 52 There's one that likes a joke. Times I've had her in fits. 1982 S. Johnson *Of Wilful Intent* i. 13 'And you say this has all been reported before?' the sergeant asked him. 'Times,' came the despondent reply.

19. Also *attrib.* with *table* or *ellipt.*, designating the multiplication table of the preceding cardinal number. Cf. *times table*, sense 52 below.

1906 Kipling *Puck of Pook's Hill* 38, I don't know my Nine Times—not to say it dodging. 1973 J. Wainwright *Touch of Malice* 124 A long-suffering father explaining the two-times-table to his dull-witted son. 1976 D. Storey *Saville* xi. 133, I want you to recite the two times, the three times, right through to your twelve times. 1982 *Sunday Tel.* 2 May 11/1 (*heading*) Know your 6-times table.

III. 24. b. *time will tell* (and varr.); *borrowed time*: see *borrowed *ppl. a.* 1; *the* (*very*) *nick of time*: see Nick *sb.*¹ 9.

[1539 R. Taverner tr. *Erasmus's Proverbes* f. 37 *Tempus omnia reuelat.* Tyme discloseth all thynges. 1616 T. Draxe *Bibliotheca Scholastica* 205/2 Time reuealeth all things.] 1771 C. Stuart *Let.* 15 Apr. in *Publ. Miss. Hist. Soc.* (1925) V. 50 Time only will shew how far those Informations have been well founded. 1863 C. Reade *Hard Cash* I. v. 164/2 She shall speak as distinctly to music as you do in conversation. *Sampson*.... Time will show, madam. 1913 E. H. Porter *Pollyanna* xxiii. 234 The doctor had looked very grave..and had said that time alone could tell. 1957 A. Huxley *Lett.* (1969) 839 It may turn out, of course, that the experts are right and that their play is better... Time will show. 1971 D. Eden *Afternoon Walk* vii. 94 Time will tell, Mrs. Simpson. 1980 J. Gardner *Garden of Weapons* iii. xi. 332 'Big Herbie gone over, has he?' Fincher said that time would tell. 1983 *Sunday Tel.* 20 Feb. 16/4 Whether this general mania for physical purification extends also to schoolgirls we are not told. Time alone will tell.

c. *it's* (*only*, etc.) *a question* (or *matter*) *of time*: said of an event that is thought certain to happen sooner or later.

1867 E. A. Freeman *Hist. Norman Conquest of England* I. iv. 251 The definitive alliance of Rouen and Paris fixed the extinction..of the royalty of Laôn. It was a question of time. 1928 E. O'Neill *Strange Interlude* iii. 94 I'm making good, all right..since I got married—and it's only a question of time. 1960 S. Barstow *Kind of Loving* ii. viii. 273 It's just a question of time now, apparently. Making all the arrangements and all that. 1960 D. Storey *This Sporting Life* ii. v. 236 Mrs. Hammond was in a coma. It seemed only a matter of time before she died. 1963 'J. le Carré' *Spy who came in from Cold* viii. 81 It was only a matter of time before it packed up. 1982 *N.Y. Times* 30 Sept. c 18/4 It's only a matter of time before something terrible happens.

d. *to kill time*: see Kill *v.* 5; *time hangs heav(il)y*: see Hang *v.* 15 b; *to redeem the time*: see Redeem *v.* 8.

27. (Further examples.) *standard time*, a standard system of reckoning time adopted throughout a country or region, now based on the time zone in which it is situated; cf. *zone time* s.v. *zone *sb.* 9 a. With preceding place-name or possessive pronoun, the time as reckoned at the place referred to (normally differing from one's own). Cf. *Greenwich time* s.v. Greenwich, Summer time 2, *railway time.

1840 *Minutes Board of G.W.R.* in *Railway Gaz.* (1935) 30 Aug. (G.W.R. Suppl.) 7/2 Outside clock to be provided for each station so as to be seen by passing trains, in order to ensure punctuality. London time to be adopted at all stations. 1841 *G.W.R. Timetable* in *Ibid.*, London time is about 4 min. earlier than Reading time. 1847 H. Booth *Uniformity of Time* 4 The managers [of the Post Office]..are quite aware of the advantages of one uniform system of time..; accordingly all their movements are regulated by 'London Time'. 1847 *Minutes Railway Clearing House Committee* 22 Sept. in *Vistas in Astron.* (1976) XX. 221 That it be recommended to each company to adopt Greenwich Time as soon as the Post Office permits them to do so. 1847, etc. [see *railway time]. 1863 Dickens in *All Year Round* 2 May 232/1 They don't keep 'London time' on a French railway. 1879 S. Fleming *Papers on Time-Reckoning* 13 On a journey from Paris to Vienna..the standard time employed by the railways changes frequently. 1883 *N.Y. Times* 19 Nov. 5/2 Standard time clocks were set to correspond to the new signals. 1917 *Whitaker's Almanack* 90/1 Since the year 1883 the system of Standard Time by Zones has been gradually accepted, and now the majority of the countries of the world use as Standard Time the time of some meridian which differs from that of Greenwich by a multiple of 15°. 1924 J. C. W. Reith in *Radio Times* 4 Jan. 42/3 We broadcast standard time. 1935 *Cook's Continental Time Table* Mar. 102 Moscow time is two hours later than that of Greenwich. 1941 *Ann. Reg. 1940* 126 A state of war existed..as from 9 a.m. (Eastern standard time). 1948 A. N. Keith *Three came Home* xviii. 295 He telephones me from Australia... We speak at twelve midnight, my time. 1958 'N. Shute' *Rainbow & Rose* vi. 229, I would hand over to him at two in the morning, Honolulu time. 1974 *Encycl. Brit. Macropædia* XVIII. 415/1 All clocks in the United States were kept one hour ahead of standard time for the interval February 9, 1942–September 30, 1945... Since then, the time in a large part of Europe has been kept one hour ahead of standard zone time without any change during the summer. 1979 P. Hill *Washermen* xxxiv. 81 [He] arrives at Kai Tak Airport, Hong Kong, 3 p.m. tomorrow afternoon our time.

IV. Phrases.

*With another sb.

28. time of day. a. *not to give* (a person) *the time of day* (colloq.), not to help or cooperate with (a person) at all, to be surly or mean towards. Cf. salutations at sense 28 b.

1951 N. Mailer *Advts. for Myself* (1961) 146 You don't even give me the time of day. You're the coldest man I've ever known. 1979 A. Maling *Koberg Link* (1980) xxiii. 123 You've come to the wrong place. Paul Carmichael won't give me the time of day. 1982 A. Price *Old 'Vengeful'* vii. 112 Lippy wouldn't have given Danny the time of day on a wet Sunday afternoon, not if he'd have come to him on bended knees.

b. *to pass the time of day*: see also Pass *v.* 52 c.

28*. time of life, the age of a person, *esp.* middle age, the menopause.

1764 [in Dict., sense 13]. 1838 Mrs. Gaskell *Let.* 17 Aug. (1966) 25 We agreed..that when people are come to *yr time of life*, there is no use having long engagements. 1971 'E. Ferrars' *Stranger & Afraid* vi. 100 Whatever's wrong with a woman over forty, it seems to me, people say it's her Time of Life. 1981 J. Mann *Funeral Sites* xxii. 132 Aidan has already threatened me with psychiatrists. He says it is 'my time of life'.

30. time and tide. (Further examples.)

1639 J. Clarke *Parœmiologia Anglo-Latina* 233 Time and tide tary on no man. 1796 'A. Barton' *Disappointment* ii. iv. 50 Let's step into the state-room, you know the old saying, 'Time and tide waits for no one'. 1935 J. Masefield *Box of Delights* i. 21 Time and Tide and Buttered eggs wait for no man. 1979 'C. Aird' *Some die Eloquent* x. 112 Time and tide and newspapers wait for no man. 1983 *Out of Town* Dec. 19/2 Only two years ago it [sc. The National Trust] completed a major repair to the tiny and beautiful Mullion Harbour on the Lizard Peninsula. Time and tide wait for no man.

**With a following adv.

33. a. Also, **time and time again.**

1887 J. Hartley *Halifax Orig. Illuminated Clock Almanack* 48 He's browt us in a bit o' dinner time an' time again. 1957 E. Waugh *Ordeal of Gilbert Pinfold* i. 2 He would dearly have liked to revise it, envying painters, who are allowed to return to the same theme time and time again. 1977 *It* May 29/2 Time and time again we have been told of the desperate need to coordinate squatting activities.

35*. time off, a break from one's occupation, absence from work, school, service, etc. (cf. *off *adv.* 4 d); also, remission of part of a prison sentence.

1930 H. Crane *Let.* 30 Sept. (1965) 356 I'm taking 'time off' to answer in the hope that you'll write me more news. 1951 'J. Tey' *Daughter of Time* i. 9 Benny would get time off for good behaviour. 1954 *Spectator* 10 Dec. 736/2 Theorists who indulge the undemocratic vice of taking time off to think. 1977 *Whitaker's Almanack 1978* 580/2 The *Financial Times* was not published because of a dispute between management and N.G.A. compositors over time-off.

35. time out, time-out, timeout** (orig. and chiefly *U.S.*). **a.** (Usu. as one word.) In various games: a deduction of playing time for a stoppage; a (usu. brief) break in play called by a coach, referee, or player.

1896 Camp & Deland *Football* vi. 61 *Time out*, time taken out by the referee when play is not actually in progress. 1930 *Sun* (Baltimore) 26 Dec. 11/7 The game..takes up about two and a quarter hours, when one allows for the intermission between halves and the innumerable 'times out'. 1946 [see *Dolly Varden]. 1972 J. Mosedale *Football* v. 61 We'd just stopped them on our one-yard line and called time-out. 1979 *Arizona Daily Star* 5 Aug. c 1/2 Clark's directive created a difference of opinion during a Tucson timeout in the opening game. 1981 *Times* 11 Nov. 8/8 Experts said Korchnoi might postpone the fifteenth game, which is due on Thursday. Each man has used up two of the three timeouts allowed each player under the championship rules.

b. A break from one's occupation.

1939 I. Baird *Waste Heritage* vi. 76 An' I took time out to tell you why we got to have those rules. 1962 J. D. MacDonald *Girl, Gold Watch, & Everything* ix. 115 Everybody in such a damn hurry, sugar, it's good for them to take a little time out. 1978 *Chicago* June 131/2 Sandwiching Sunday-morning Mass between an appearance on *Meet the Press* and a press conference back at the Ramada Inn was the only time-out Schtafly took in Houston.

***With a governing preposition or adverb.

35*. about time,** approximately the right time; usu. *iron.*, long past the right time; (also with *too*) this should have happened much earlier, this is long overdue.

1920 E. O'Neill *Beyond Horizon* ii. i. 67 It's about time you put a stop to his nonsense. 1931 [see *curve *sb.* 2]. 1940 W. Faulkner *Hamlet* iv. i. 285, I reckon it's about time to get dinner started. 1952 M. R. Rinehart *Swimming Pool* xxvii. 240 'It's about time,' he said, pushing aside the junket Jennie had served him. 1977 A. Clarke *Letter from Dead* ix. 103 'Now you're talking,' said Jill, 'and about time too.'

36. against time. (Earlier and later examples.)

1835 Dickens *Sk. Boz* (1836) 1st Ser. II. 178 The kennels seem to be doing matches against time. 1933 D. L. Sayers *Hangman's Holiday* 37 It must have been put in the wrong way round... You know, sir, we often have to work against time, and I suppose—but it's very careless. 1935 'E. Queen' *Adventures* 86 What would you gentlemen expect a thief, working against time, to do under these circumstances? 1975 *Economist* 1 Feb. 16 Sheikh Mujib's 'second revolution' last weekend was his personal answer to this race against time. 1982 *Washington Post* 11 Nov. d 11 Whether it realizes it or not, the government of Japan is in a race against time.

42. in time. d. *what* (*why*, etc.) *in time*.., 'what (etc.) in the world..' or 'on earth..'. *U.S. colloq.*

1849 J. T. Fields *Let.* 28 Feb. in R. W. Griswold *Passages from Corr.* (1898) 250 Why in Time don't you come our way and see the boys? 1883 *Harper's Mag.* Jan. 212/1 He wondered what In time made 'em keep the cars so hot. 1904 J. C. Lincoln *Cap'n Eri* vii. 125 What in time did you tell the Doctor that she was a relation of mine for?

42*. not before (*dial.* **afore**) **time,** not soon enough, almost too late, long overdue.

1905 in *Eng. Dialect Dict.* VI. 15/1 Ah see they're beginnin' ti mend rooad, an nat afoor time. 1955 'N. Shute' *Requiem for Wren* v. 144 She got her clothes brush from her quarters and gave him a grooming with it, not before time. 1967 *Listener* 26 Oct. 553/3 The Minister..is right: 'the licensed victualler must now recognise that he has to provide a different kind of social life in his pub.' It is not before time. 1972 *Observer* 16 July 13/6 It all points to a wind of change blowing in the direction of the Ordinary shares..: and not before time either. 1974 *New Statesman* 29 Nov. 766/1 It's..goodbye to cheap sugar—and perhaps not before time so far as the developing countries are concerned.

43. on time. a. For 'Chiefly' read 'Formerly chiefly'. (Earlier example.) Also (with hyphen) *attrib.* (see *on-¹ 4 b).

1821 R. Cadell *Let.* 28 Nov. in *Times Lit. Suppl.* (1933) 7 Sept. 592/2 In order to effect this and *on time* we have resolved [etc.]. 1965 *Economist* 13 Feb. 675/1 Exact scheduling..and perfect coordination to assure on-time completion of the project. 1967 R. J. Serling *President's Plane is Missing* (1968) i. 13 As my airline friends would say, I prefer on-time departures.

c. On credit. *N. Amer.*

1840 *Spirit of Times* 15 Aug. 277/1 *On time*, the prices would at once be enhanced. 1873 W. Mathews *Getting on in World* xix. 316 We need not expect that the practice of selling goods on time will ever be abandoned. 1925 *Sat. Even. Post* 10 Oct. 133/1 It's like peddling lots on time, instead of selling and developing acreage. 1972 J. M. Minifie *Homesteader* vi. 44 Everything was bought 'on time', hardly any transactions involved cash. 1979 R. L. Simon *Peking Duck* xx. 144 On the table with Harvey's Sony tape recorder was a Nikon FT... I wondered if he had bought it all on time.

44. out of time. c. *out of one's time*, in an era unsympathetic to one's attitudes, aspirations, etc.; at the wrong season.

1950 'D. Divine' *King of Fassarai* xvi. 125 Kellie was born out of his time. Last piece of history he could have flourished in was the Alaska rushes. 1958 B. Behan *Borstal Boy* iii. 334 It was a little undersized goat born out of its time, and it was so small now that it wouldn't be any bigger than a lamb at Christmas when we put on the play. 1973 R. Lewis *Of Singular Purpose* vi. 130 'Major Cornelius Van Rijk.' He laughed shortly. 'A man out of his time.' 1976 L. Henderson *Major Enquiry* xii. 78 You know, Mildred, you were born out of your time, you really belong to the naughty nineties.

44*. over time, gradually, during a period of (past or future) time.

1966 *Rep. Comm. Inquiry Univ. Oxf.* II. 46 The proportion from independent schools has fallen over time. 1973 *New Society* 1 Nov. 258/3 Like the Foot-Steel proposals, these would be introduced over time.

****With a verb.

50*. *to make time* (*with*): see *make *v.*¹ 66.

IV*. Ellipt.

50.** as *conj.* At or by the time that; as soon as; when. *U.S. dial.* and *colloq.*

1919 E. O'Neill *Moon of Caribbees* 6 It *was* in New Guinea, time I was shipwrecked there. 1938 M. K. Rawlings *Yearling* iv. 30 You'll likely not be so merry, time the day be done. *Ibid.* ix. 78 You git on to the sink-hole, son, and I'll foller time I've skinned out your 'coon hide. 1950 R. Moore *Candlemas Bay* 13 Time Joel Walls had his net, one night he caught seven hogsids.

V. Combinations.

51. a. (*a*) *time-behaviour, -consciousness, -co-ordinate, -cycle, -depth, -dimension, -direction, -displacement, -evolution, -factor, -flow,*

-foot, -gap, -horizon, -integral (earlier example), *-interval, -measure, -order, -pattern, -period, -perspective* (earlier example), *-plane, -process, -ratio, -relation, -rhythm, -scheme, -sense* (earlier example), *-sequence, -shift, -slip, -span, -sphere, -stream, -succession, -unit, -word*; (*b*) *time-budget, -chart*; (*c*) *time-bomb* (*fig.* and *attrib.* examples).

a 1974 R. Crossman *Diaries* (1975) I. 259 A mere day and a half was a crazy time-allocation if only because the eight new clauses and fifty amendments..would take all the time available. 1955 Friedman & Weisskopf in W. Pauli *Niels Bohr* 153 We can then examine the time behaviour of the outgoing parts of the wave packet as they pass a given radius. 1977 *Jrnl. R. Soc. Arts* CXXV. 765/2 Other types of lasers..can be controlled in either frequency or time-behaviour with the limits set only by the Uncertainty Principle. 1939 Dylan Thomas *Map of Love* 21 Strike in the time-bomb town. 1941 B. Schulberg *What makes Sammy Run?* vi. 128 Sammy Glick is a time bomb in my brain and it's going to go off. 1966 N. Nicolson *Diaries & Lett. H. Nicolson* 275 Harold Nicolson had long been aware that a constitutional time-bomb was ticking beneath the throne. 1981 'M. Underwood' *Double Jeopardy* xxiii. 181 His official diary could become a time bomb. 1948 Time-budget [see *participant *sb.* 1]. 1976 P. R. White *Planning for Public Transport* ii. 46 Within a fixed time-budget, as work trips become longer, less time is available for shopping and recreational trips on weekdays. 1934 *Burlington Mag.* Jan. 50/1 The author establishes his time-chart, proving conclusively that the wide early influence..has no foundation in fact. 1958 *Times Lit. Suppl.* 17 Jan. 33/2 Mr Sullivan..gives a much more orderly unfolding of the time-chart of discovery because he sticks to a straightforward chronology. 1890 W. James *Princ. Psychol.* I. xv. 632 Th. Waitz is guilty of similar question-begging when he explains our time-consciousness. 1963 H. Lindenberger *On Wordsworth's 'Prelude'* vi. 199 The dissolution of the traditional literary genres and the increasing eccentricity of structure..have proved coincidental with..the development of time-consciousness among writers. *a* 1942 B. Malinowski *Sci. Theory of Culture* (1944) iii. 20 In order..to make an historical process..significant in terms of explanation or analysis, it is..necessary to prove that we are, along the time coördinate, linking up phenomena that are strictly comparable. 1903 A. W. Patterson *Schumann* 7 That the mind should work in a regular time-cycle, passing from one phase of sentiment to another with almost mechanical exactness. 1968 *Brit. Med. Bull.* XXIV. 197/2 All disease, whether it be physical or emotional, appears to have its time-cycles. 1957 P. Worsley *Trumpet shall Sound* 266 There is time-depth to all social action. 1978 *Archivum Linguisticum* IX. 76 The ultimate sources of the verbal root *es- can never be definitively known because of the huge time-depth involved here. 1937 *Mind* XLVI. 162 It is rather unfortunate that philosophers..should have paid little attention to the problem suggested by Minkowski's imaginary time-dimensions. 1982 M. Duke *Flashpoint* vii. 46 Shmuel let his mind slip into a new time-dimension. The near future looked good. 1890 W. James *Princ. Psychol.* I. ix. 283 With each prolongation of the scheme in the time-direction, the summit of the curve of section would come further towards the end of the sentence. 1937 *Mind* XLVI. 177 We may agree..to regard as the time-direction that in which the number of beats registered by the clock is increasing. 1890 W. James *Princ. Psychol.* I. xi. 411 The cases he [sc. Wundt] describes are really cases of anachronistic perception, of subjective time-displacement, to use his own term. 1901 Time-displacement [see *attentional *a.*]. 1937 R. A. Wilson *Birth of Language* ii. ii. 79 To one..who tries to work out a concrete philosophical view of the world..on the basis of a time-evolution of all its forms from matter to man, the mechanistic hypothesis appears..to obscure the real problem of the beginnings of life. 1911 *Aeronaut. Jrnl.* XV. 66 This switch has a 'time-factor' approximating to that of the motor. 1976 M. & G. Gordon *Ordeal* (1977) xxix. 198 He worked fast, conscious of the time factor. 1936 J. Kantor *Objective Psychol. Gram.* xvii. 240 Grammarians mean by time the abstract points in a field-spread or an equally abstract time-flow. 1956 E. L. Mascall *Christian Theol. & Natural Sci.* iv. 134 This continuous activity of God is not to be thought of as if it were the insertion of the creatures into a time-flow which existed antecedently to them. 1883 Time-foot [see *rhythm-foot* s.v. *rhythm *sb.* 9 a]. 1890 W. James *Princ. Psychol.* I. ix. 237 Interruptions, time-gaps during which the consciousness went out altogether to come into existence again at a later moment. 1978 *Early Music Gaz.* Oct. 15/2 Leonhardt emphasises the large time-gap between the two books of the so-called '48'. 1965 H. I. Ansoff *Corporate Strategy* iii. 40 To make this concept meaningful, we need the idea of the *time horizon* of a firm—the period over which the firm seeks to optimize its resource conversion efficiency. 1983 *Listener* 8 Dec. 23/3 The time-horizon over which policy is formulated would become markedly biased towards the short-term. 1873 J. C. Maxwell *Electr. & Magn.* II. 186 The time-integral of a force is called the Impulse of the force. 1871 *Cornh. Mag.* July 58 The imagination is wholly unable either to conceive the duration of the time-intervals..occupied by these wonderful processes. 1975 *Language for Life* (Dept. Educ. & Sci.) xxv. 503 We had to ask whether the width of the time intervals had forced the pattern of responses and rendered suspect our method of calculating average times. 1911 W. James *Some Probl. Philos.* xi. 179 Mr. Bertrand Russell..treats the Achilles-puzzle as if the difficulty lay only in seeing how the paths traversed by the two runners..should have the same time-measure if they be not themselves of the same length. 1943 *Mind* LII. 61 From no sort of correlation between space-measures and time-measures can the obliteration of the ontological distinction between space and time be validly inferred. 1890 W. James *Princ. Psychol.* I. iii. 88 The whole succession is so rapid that perception seems to be retrospective, and the time-order of events to be read off in memory rather than known at the moment. 1960 *Colston Research Soc. Symposium* XII. 90 These paranormal cognitive powers, it seems, are indifferent in some degree to a physical time order, which of course raises frightful difficulties. 1946 R. Blesh *Shining Trumpets* ii. 42 The..sequential time-patterns of human or divine speech. 1968 D. L. Clarke *Analytical Archaeol.* vi. 254 The time pattern regularities ..in the trajectories and traditions of many quite different cultures. 1894 *Jrnl. Inst. Electrical Engineers* XXIII. iii. 295 It was due to synchronism between the changes of load *on the engines* and the time-period of the governor. 1953 *Scottish Jrnl. Theol.* VI. 162 We believe that patterns and expanding purposes have been established through such time-periods. 1965 *Canad. Jrnl. Linguistics* Spring 125 Seven languages..died out during the twenty-three year period ending in 1964... In this same time-period Coos was reduced to a single informant. 1890 W. James *Princ. Psychol.* I. xv. 639 In hashish-intoxication there is a curious increase in the apparent time-perspective. 1969 Bennison & Wright *Geol. Hist. Brit. Isles* ii. 26 Lines indicating contemporaneity, so-called 'time-planes', are normally presented diagrammatically as horizontal. 1977 G. Clark *World Prehist.* (ed. 3) i. 24 Among the factors that caused peoples living on the same time-plane to retain or discard old forms while adopting new ones were ..variations in the environment to which they had to adapt. 1887 A. Seth *Hegelianism* v. 170 The time-process of the finite world is..the reality with which we are immediately acquainted. 1938 E. Bevan *Symbolism & Belief* iv. 114 The time-process goes on throughout a universe of which our planet is, spatially, only an infinitesimal part. 1964 W. S. Allen in D. Abercombie et al. *Daniel Jones* 3 Whereas in modern verse the rhythms are marked by 'stress', the classical rhythms were expressed solely in terms of time-ratios. 1965 *Wireless World* July 336/2 The same idea of time-ratio control can be used in regulated power supplies. 1924 R. M. Ogden tr. *Koffka's Growth of Mind* iii. 118 When a pianist..articulates a series of muscular innervations..fixed time-relations are determined in the series of sound-waves. 1962 D. Nichols *Echinoderms* v. 71 There is no clear-cut evidence as to the origin of the echinoids. Time-relations do not allow their derivation from the Ophiocistioidea of the Silurian. *c* 1873–4 Time-rhythm [see *beat *sb.*¹ 5]. 1934 J. J. Hogan *Outl. Eng. Philol.* iv. 29 Quantity or time-rhythm, consisting of the alternation of longer and shorter, not of stronger and weaker beats, is the rhythm of music. 1904 *Mind* Oct. 468 The distribution of terms in our inner time-scheme and space-scheme must be an exact copy of the distribution in real time and space of the real terms. 1978 *N. & Q.* Feb. 55/1 Given Sterne's complicated time-scheme..such inconsistencies are surprisingly rare. 1890 W. James *Princ. Psychol.* I. xv. 611 The units of duration ..which the time-sense is able to take in at a single stroke, are groups of a few seconds. *Ibid.* II. xxviii. 671 The principle [of causality] expresses a demand for *some* deeper sort of inward connection between phenomena than their merely habitual time-sequence seems to us to be. 1974 G. Jenkins *Bridge of Magpies* viii. 135 What the time sequence of events was in regard to the two killings I'd never know. 1933 F. M. Ford *Let.* 24 Aug. (1965) 222 To them, on account of the 'time-shift'..they [sc. novels] must be quite incomprehensible. 1958 *Times Lit. Suppl.* 18 July 414/2 This device, moreover, involves Mr. Young in irritating time-shifts and flash-backs and rather strained symbolism. 1978 Cadogan & Craig *Women & Children First* xi. 268 The technical problems involved in the time-shift structure have simply failed to interest the author. 1983 *Listener* 24 Nov. 37/2 The whole business of recording broadcasts and watching them later is known to the trade as 'time-shift'. 1952 P. Wentworth *Brading Collection* xii. 72 It brought a horrid feeling that there had been a kind of time-slip—that they had been caught back again, she and Charles, to where they were three years ago. 1974 *Bookseller* 10 Aug. 999/2 (Advt.), Four children, a disused railway line, a time-slip to an Edwardian scene—this enchanting fantasy [etc.]. 1981 V. Glendinning *Edith Sitwell* xv. 195 Old Beau Nash, in her *Bath*, sees the past float by... This conceit, or technique of the time slip was not unique. 1933 A. N. Whitehead *Adventures of Ideas* vi. 98 The recent shortening of the time-span between notable changes in social customs is very obvious, if we examine history. 1979 A. Storr *Art of Psychotherapy* ii. 11 Attention is difficult to sustain without a break beyond a time-span of forty-five to fifty minutes. 1928 H. Poutsma *Gram. Late Mod. Eng.* (ed. 2) I. i. i. 43 *Should*, as a modal verb, is a preterite conditional, used irrespective of the time-sphere of the predication. 1957 R. W. Zandvoort *Handbk. Eng. Gram.* iv. 61 The perfect tense usually denotes an action that falls within the time-sphere of the present. 1937 R. A. Wilson *Birth of Lang.* ii. i. 69 In his emergence to consciousness man rose above the time-stream of sense. 1978 P. G. Winslow *Coppergold* 125 At some point a man begins to feel out of place in the time stream. 1890 W. James *Princ. Psychol.* II. xx. 147 There enters thus an element of time-succession into our perception of ourselves which transforms the latter from an act of intuition to one of construction. 1922 A. S. Eddington *Theory of Relativity* 18 It [sc. the relativity theory] fully recognizes that the chain of events in such a time-succession is a series of an entirely distinctive character from the succession of points along a line in space. 1925 J. Joly *Surface-Hist. Earth* v. 79 Our time-units have become millions of years. 1968 R. A. Lyttleton *Mysteries Solar Syst.* iv. 133 The time-unit of the abscissa i. 50,000 years. 1933 W. A. Russell *Devel. Art of Lang.* viii. 57 The thought of an action is intimately associated with the thought of time; so much so that some grammarians have called the verb the *Time-word.* 1973 *Archivum Linguisticum* IV. 5 The distinction between /bin/ and /dɔŋ/ is clear when we consider..the time words with which they collocate.

b. *time-waster* (later examples); *time-allocation, -consuming, -measuring, -reckoning, -saving, -wasting* (examples) adjs. and sbs.

1951 Parsons & Shils *Toward General Theory of Action* ii. iii. 143 A compulsive fixation on time-allocation is a familiar phenomenon. 1890 *Proc. Soc. for Psychical Research* Dec. 654, I dropped my inquiries..for a period of about two years,..being over-freighted with time-consuming duties. 1978 R. Mitchison *Life in Scotland* viii. 161 There were a great many strenuous, unpleasant and time-consuming tasks to be done in any house. 1890 W. James *Princ. Psychol.* I. xiv. 557 The time-measuring psychologists of recent days have tried their hand at this problem. 1959 *Publ. Mod. Lang. Assoc. Amer.* LXXIV. 589/2 Music—or at least music with bar-lines—is precisely a time-measuring notation; it divides the time into equal intervals. 1974 tr. *Wertheim's Evolution & Revolution* 363 The relatively rapid and consistent process of evolution—slow as it was in terms of time-measuring as applied by mankind. 1920 A. S. Eddington *Space, Time & Gravit.* ii. 31 Observers with different motions use different space- and time-reckoning. 1964 L. MacNeice *Astrol.* iv. 112 The skies were observed..for the old time-reckoning reasons. 1891 A. James *Diary* 24 June (1965) 216 A restricted nature, not admirable or generous in its impulses, but highly practical and time saving. 1960 *Farmer & Stockbreeder* 15 Mar. 140/2 The light tractor with all the features for time-saving. 1977 *Listener* 10 Nov. 607/2 The amounts that air travellers would be willing to pay for the time-savings that it [sc. Concorde] made possible. 1982 R. Ludlum *Parsifal Mosaic* xiv. 215 It's basically an economic, time-saving decision. 1930 J. Bailey *Let.* 5 Apr. (1935) 311 All these new inventions are time wasters for people like you and me. 1980 M. Drabble *Middle Ground* 118 Sally was a moaner and a timewaster. 1845 Geo. Eliot *Let.* c 16 Apr. (1954) I. 187, I am..full of hope that..I shall be able to ward off these time-wasting visitations. 1853 C. M. Yonge *Heir of Redclyffe* I. xv. 258 Abstaining from the time-wasting that might have tempted him if he had had plenty of money to spend. 1976 S. R. Simpson *Land Law & Registration* viii. 144 Sporadic survey is expensive and time-wasting. 1981 'J. Ross' *Dark Blue & Dangerous* x. 58 The time-wasting had gone on long enough.

c. *time-blurred, -bound* (later examples), *-conditioned, -constrained, -controlled, -dulled, -eaten* (for *a* 1849 in Dict. read 1831), *-hallowed, -limited, -obsessed* (also *absol.*), *-ridden, -sanctioned, -shaken, -stained, -tested, -tormented* adjs.

1916 A. Huxley *Burning Wheel* 24 Some lover of an older day Has carved in time-blurred lettering One word only:- 'Alas'. 1924 R. Graves *Mock Beggar Hall* 79 Neither eternal nor time-bound, Not certain, nor in change. 1978 *Dædalus* Summer 168 Our ideas about childhood..are very much time-bound and culture-bound. 1951 R. A. Knox *Stimuli* i. 3 We force our time-conditioned minds, once a year, into an artificial mood of expectancy. 1967 A. Battersby *Network Analysis* (ed. 2) ix. 146 If we simply say that the project must be completed as quickly as possible, we have what is called a 'time constrained' network. 1979 *Jrnl. R. Soc. Arts* Dec. 9/1 Colleges will be asked..to specify the learning objectives which they wish to assess by time-constrained examination. 1954 *Gloss. Highway Engin. Terms* (*B.S.I.*) 58 *Time controlled traffic signals*, signals in which the aspects are displayed for fixed periods which are determined by manual or time-clock adjustment of the controller. 1960 *Farmer & Stockbreeder* 23 Feb. 100/2 Light for time-controlled poultry-house lighting. 1971 P. C. Sylvester-Bradley in I. G. Gass et al. *Understanding Earth* ix. 124/1 Evolution is a time-controlled process. 1922 Joyce *Ulysses* 238 Stephen Dedalus watched..the lapidary's fingers prove a timedulled chain. 1749 W. Collins in R. Dodsley *Collection of Poems* IV. 65 Where..some time-hallow'd pile, Or up-land fallows grey Reflect it's last cool gleam. 1959 J. L. Austin *Sense & Sensibilia* (1962) vi. 61 His wholesale acceptance of the traditional, time-hallowed, and disastrous manner of expounding them [sc. arguments]. 1974 Dawa Norbu *Red Star over Tibet* i. 30 The nomads were given a time-hallowed concoction. 1947 Partridge *Usage & Abusage* 51/1 Turn those sentences into *to be* equivalents: 'To receive wounds is no fun'; 'To become a casualty is no fun'; 'To be wounded is no fun': crisp, clear-cut, single-action, time-limited connotations. 1977 *Jrnl. R. Soc. Arts* CXXV. 462/2 The price he pays for that time-limited monopoly is to give up any other advantages which the usual rights of exclusive ownership..might otherwise confer on him. 1945 Auden *Coll. Poetry* 12 These only feared another kind of Death To which the time-obsessed are all condemned. 1951 S. Spender *World within World* 137 No wonder that the literature of this period is time-obsessed, time-tormented, as though beaten with rods of restless days. 1936 T. S. Eliot *Burnt Norton* in *Coll. Poems 1909–35* 188 A flicker Over the strained time-ridden faces. 1838 J. S. Mill in *London & Westm. Rev.* Aug. 469 The inconsistencies and absurdities of time-sanctioned opinions and institutions. 1949 Dylan Thomas in *Botteghe Oscure* IV. 399, I who hear..the notes on this time-shaken Stone. 1835 W. C. Bryant in *N.Y. Mirror* 19 Sept. 92/1 How the time-stained walls That earthquakes shook not from their poise, appear To shiver. 1904 W. S. Kennedy *Walt Whitman's Diary in Canada* p. v, The transcribing of these out-door notes from the worn and time-stained fragments of paper. 1930 *Times Lit. Suppl.* 10 July 566/3 The kind of faith which cheerfully believes things about the East which time-tested experience has proved to be untrue. 1977 *Daily Express* 29 Jan. 7/1, I will defend.. the right of the ingenious and time-tested Mr. [Peter] Hall to spend my taxes any way his fancy takes him. 1947 Auden *Age of Anxiety* v. 125 Transpose our plight like a poignant theme Into twenty tongues, time-tormented But His People still. 1951 Time-tormented [see *time-obsessed* adj. above].

d. *time-based, -centred, -conscious, -dead, -dependent, -faced, -independent, -kept, -lost, -old, -pressed, -served* (later examples), *-varying* adjs.

1976 P. R. White *Planning for Public Transport* vi. 123 The classification of time-based, mileage-based and peak-vehicle-based costs..was adopted. 1964 I. L. Horowitz *New Sociol.* i. 25 We can draw..upon the information available from the historian and the journalist to forge a time-centered sociology. 1977 P. Johnson *Enemies of Society* iii. 33 Against this background of a time-centred religion, there were also solid economic reasons why the fulcrum of progress would shift northwards across the Alps. 1934 A. Huxley *Beyond Mexique Bay* 218 What causes a people..to become as acutely time-conscious as the priestly mathematicians of the Maya Old Empire? **1962** J. Glenn in *Into Orbit* 43 All of us wear very exact watches... As you can see, we are extremely time-conscious during a mission. **1923** L. Hughes in *Crisis* (N.Y.) Feb. 174/2 When Susanna Jones wears red, A queen from some time-dead Egyptian night Walks once again. **1955** O. Klein in W. Pauli *Niels Bohr* 113 We need a time-dependent operator. **1974** J. W. Drake in Carlile & Skehel *Evolution in Microbial World* 53 When *E. coli* is maintained in the chemostat, mutant accumulation is generation-dependent when growth is limited by glucose, but becomes time-dependent when growth is limited by amino acids. **1936** Dylan Thomas *Twenty-five Poems* 38 Now Jack my fathers let the time-faced crook..Sneak down the stallion grave. **1953** *Physical Rev.* XCI. 740/1 Initial and final currents that are time-independent with respect to different reference systems. **1970** G. K. Woodgate *Elem. Atomic Struct.* ii. 14 The value of *E'* has to be found from eq. (2.22)..which is a time-independent eigenvalue equation. **1934** T. S. Eliot *Rock* i. 7, I journeyed to London, to the timekept City. **1930** L. Hughes in *Crisis* (N.Y.) July 235/1 Subdued and time lost are the drums. **1861** J. R. Lowell *Washers of Shroud* in *Poems* (1912) 476 The time-old web of the implacable Three. **1922** D. H. Lawrence *England, my England* 8 The wide, black, time-old chimney. **1886** Hardy *Mayor of Casterbridge* I. ix. 110 The bow-windows protruded like bastions, necessitating a pleasing *chassez-déchassez* movement to the time-pressed pedestrian at every few yards. **1960** *Times* 18 Feb. 3/3 Only time-served engineers..will be considered. **1979** *Navy News* Feb. 9/1 To say that every seaman is a good seaman by virtue of being a time-served man would be very much open to question. **1962** W. B. Thompson *Introd. Plasma Physics* vii. 151 A particle going past a point target at a distance *l* produces a time-varying field at the target. **1981** *Word 1980* XXXI. 172 The time-varying spectral pattern of the processed stimuli differed radically from that of the natural speech.

52. time-and-motion, used *attrib.* to designate a study, person, etc., concerned with the measurement of the efficiency of an industrial or other operation; **time-average** *Physics* and *Math.*, an average evaluated over a period of time; hence **time-averaged** *a.*; **time-barred** *a.*, disqualified or invalid by reasons of arriving or being presented after the expiry of a statutory time-limit; **time capsule,** a container used to store for posterity a selection of objects thought to be representative of life at a particular time; also *fig.*; **time-change,** (*a*) change that takes place with the passage of time; (*b*) the difference in standard time between widely separated localities, as experienced by travellers; **time-charter** *v. trans.*, to hire (a vessel) under a time-charter agreement; **time check,** (*a*) *Canad.* a chit from a foreman stating the number of hours for which a man is due to be paid; (*b*) the act of ascertaining or stating the exact time; **time clock,** (*a*) a clock with a mechanism for recording the time on time-cards pressed into it; (*b*) a clock which can be set to switch an appliance on or off at specified times; **time-constant,** substitute for def.: the time taken by an exponentially varying quantity to change by a factor $1-1/e$ (approximately 0.6321), regarded as a parameter of the system in which the variation occurs; more widely, a time taken as representative of the speed of response of a system; (examples); **time-course,** (*b*) the period of time in which something happens, the length of time taken; **time-delay** = *time-lag*; used chiefly *attrib.* of a mechanism, system, etc., into the operation of which a time-lag has been deliberately introduced; **time deposit** (orig. *U.S.*), a sum placed in a bank at interest and not to be drawn before a set maturity date; **time derivative** *Physics* and *Math.*, a derivative of a variable with respect to time; **time difference,** (*a*) the difference between the lengths of time taken by different operations or processes; (*b*) the difference in standard time between widely separated localities; **time differential,** (*a*) = *time difference* (b) above; (*b*) the difference in the length of time taken by a process in different places or at different stages; **time dilatation** or **dilation** *Physics*, the apparent slowing down of the passage of time in a frame of reference moving relative to the observer, a relativistic effect analogous to the increase in mass and the Lorentz contraction of length; **time–distance,** used *attrib.* of the relation (*esp.* as expressed in graphs) between time and distance; **time division** *Telecommunication*, allocation of transmission time to each of a number of signals in quick rotation, so that all can be transmitted over the same channel if the sampling rate is sufficiently high; usu. *attrib.*; **time-element,** (*a*) time conceived as the natural element of temporal beings; (*b*) time as a factor to be taken into consideration; **time-expired** *a.*, (*a*) also of convicts: whose term of sentence has expired; (*b*) of perishable goods: of which the term of safe storage (before sale or use) has expired; **time-frame,** a limited and established period of time during which an event, etc., took place or is planned to take place; a schedule; **time-lag** (examples); **time lapse,** lapse of time; *spec.* (usu. with hyphen) *attrib.*, designating or pertaining to a technique of taking a sequence of photographs at set intervals to record changes that take place slowly over time; **time-line,** (*b*) (earlier example); (*c*) a schedule, a deadline; **time-lock** (earlier example); **time machine,** an imaginary machine capable of transporting a person backwards or forwards in time; **time-of-flight** *a. Physics*, designating techniques and apparatus that depend on the time taken by particles to traverse a set distance, e.g. in the separation of ions according to their mass; **time-payment** (later examples); **time pencil,** a type of delayed-action firing-switch or detonator for setting off explosive devices; **time-release** *a.* = *slow-release* adj. (b) s.v. *slow *a.* 16 d; **time-resolved** *a. Chem.* and *Physics*, produced by or pertaining to a spectroscopic technique in which the spectrum is obtained at known times after excitation; **time reversal** *Physics*, a transformation in which the passage of time (and so all velocities) is imagined to be reversed; *time reversal invariance*, invariance of laws of nature under this transformation, so that all processes allowed by them are also allowable when all motions in them are reversed; so **time-reverse** *v. trans.*, to subject to time reversal; **time-reversed** *a.*; **time-sampling,** the collection of data or observation of events at given times or intervals or within given periods of time; **time-scale,** (the relative length of) the period of time in which a sequence of events takes place, the successive stages of a process, operation, etc.; a representation or exposition of the stages of such a sequence, etc.; **time-series,** (*a*) the sequence of events which constitutes or is measured by time; (*b*) a series of values of some quantity obtained at successive times (often with equal intervals between them); *time-series analysis*, the statistical analysis of such series; **time-signal,** (*b*) a signal transmitted to indicate the exact time of day, *esp.* that broadcast by the BBC at certain hours; **time slice,** (*a*) a short period in the continuum of time; (*b*) *Computers*, each of the short intervals of time during which a computer or its central processor deals uninterruptedly with one user or program, before switching to another; so **time-slicing** *vbl. sb.*, the division of processor running time into a succession of short intervals that are allocated in turn to different users or programs; **time slot,** (*a*) a portion of time allocated to a purpose or person, *esp.* to an individual broadcast programme; (*b*) *Computers*, = *time slice* (b) above; **times table,** a multiplication table; cf. sense 19 above; **time-stamp,** a mechanical device for stamping letters, tickets, etc., with the date and time of receipt; hence as *v. trans.*; **time-stamped** *ppl. a.*; **time step,** an even-timed basic tap-dancing step; **time study,** a time-and-motion study; the close observation of an industrial or other process with a view to time-saving alterations in procedure; also *attrib.*; **time switch,** a switch that acts automatically at a set time; **time term,** a term of an equation in which time is the main variable; **time train** = *going train* s.v. Train *sb.*[1] 15; **time travel** = *time-travelling* vbl. sb. below; also (hyphened) as *v. intr.*, **time-traveller,** one who practises time-travelling; **time-travelling** *vbl. sb.*, the imagined activity of travelling into the past or future, hypothetical movement through time; also *fig.* and as *ppl. a.* (also = 'extending through time'); **time trial,** a test of individual speed over a set distance, a race in which competitors are separately timed; **time wage** (see quot. 1892); **time warp** *Science fiction*, a distortion of space-time that is conceived as causing or enabling a person to remain stationary in time or to travel backwards or forwards in time; also (with hyphen) as *v. trans.*, to transport in a time warp; **timewise** *adv.*, with regard to time; **time-worker** (examples); **time-zone** (earlier and later examples).

1932 C. Reynolds *Production Planning* ix. 87 (*heading*) Time and motion studies. **1959** *Listener* 5 Nov. 762/2 That sinister figure, the man with the stop-watch, the time-and-motion expert, disliked by union men the world over. **1966** *Punch* 20 July 127/1 Time-and-motion study techniques, applied to American office workers, enable employers to reorganise office procedures and streamline routine chores. **1973** M. Woodhouse *Blue Bone* vi. 61 We have decided to divide the job according to the best work-study, time-and-motion principles. **1980** *Daily Tel.* 11 Sept. 29 (Advt.), Knowledge of..time and motion studies is very advantageous. **1875** *Encycl. Brit.* III. 39/1 In a material system in a state of stationary motion the time-average of the kinetic energy is equal to the time-average of the virial. **1914** *Phil. Mag.* XXVIII. 826 The ratios have been calculated using the time-average values. **1965** Phillips & Williams *Inorg. Chem.* I. i. 11 Time-average potential and kinetic energies. **1946** Time-averaged [see *space-averaged* adj. s.v. *space *sb.*[1] 19]. **1957** *Financial Times* 23 Mar. 4/3, I would appreciate advice..whether such a claim is now time-barred or not. **1971** *E. Afr. Standard* (Nairobi) 10 Apr. 1/3 The fourth car in the Datsun 240Z team..was time-barred at Korogwe. **1978** *Observer* 4 June 16/6 Employees must make their claim..within six months of ceasing work. Failure to do this results in employees being 'time-barred'. **1938** *N.Y. Times* 19 Aug. 21 A record of the world of the present era ..will be buried on the site of the World's Fair in the hope that it will give to historians 5,000 years hence a picture of the middle twentieth century... The record will be contained in a 'time capsule', a specially devised container of metallic alloy of high corrosion resistance. **1947** Campbell & Robinson *Skeleton Key to 'Finnegans Wake'* 8 The *Wake*..is a huge time-capsule, a complete and permanent record of our age. **1965** *Christian Century* 27 Oct. 1313/1 With the cooperation of a hospital physician, the hospital administrator, a funeral director and a local commercial firm which happened to be building 'time capsules' or cryogenic (low temperature) storage units.. preparations were made for this pioneering effort. **1973** *Art Internat.* Mar. 56/1 This image of Venice, as a waterlogged time capsule, is very much the creation of outsiders. **1982** *Daily Tel.* 6 July 12/4 George Howard, the BBC chairman, is..asking around for corporation 'artefacts' to be sealed into an age-proof time capsule, so future generations may be able to discover the wonder of our broadcasting services. **1937** Blunden *Elegy* 31 To haunt and cling To this one ground, whatever closed Of strange power, or time-change. **1941** *Mind* L. 182 A person as a 'self' exhibits beside its individuality and its identity in time-change the peculiar character of self-assertion through vicissitudes. **1969** N. Denny tr. *Veraldi's Spies of Good Intent* xi. 177 I'm suffering from the time-change. With me it's three in the morning. **1976** 'M. Delving' *China Expert* i. 9 I'd been in Honolulu two days before, and when I got back to New York I'd forgotten about the time change. I lost eight hours. **1963** *Times* 10 June 21/5 The companion for our new ship will be time-chartered.. until we decide about building another. **1974** *Information Handbk. 1974–5* (Shell Internat. Petroleum Co.) 100 Ships time-chartered by Shell companies total 235 of 20·6 million dwt. **1911** *Daily Colonist* (Victoria, B.C.) 26 Apr. 2/4 He cashed a time check after working a short time on a log drive. **1937** *Printer's Ink Monthly* May 45/1 *Time check*, synchronizing the time pieces of all concerned in a broadcast. **1968** M. Woodhouse *Rock Baby* xv. 149, I.. got a time-check to correct my watch. **1887** C. B. George *40 Yrs. on Rail* 56 [He] pulled a wire leading to a time-clock. **1930** *Engineering* 1 Aug. 130/1 Special terms for night service has encouraged the use of heat-storage ovens... In France, on weekdays, the luncheon interval lasts two hours, and that gives a pause in the factory power consumption which, with the aid of time-clocks, allows of a valuable heat storage period. **1943** Time-clock [see *punch *v.*[1] 3 c]. **1961** *Listener* 19 Oct. 629/3 Running costs also tend to be higher but a judicious use of time clocks can keep these in check. **1976** *Billings* (Montana) *Gaz.* 4 July 6-c/4, I am not the nine-to-five cat—couldn't punch a time clock if my life depended on it. **1869** Ld. Rayleigh in *Phil. Mag.* XXXVIII. 4 There is for every conducting circuit a certain time-constant which determines the rapidity of the rise or fall of currents, and which is proportional to the self-induction and conductivity of the circuit. Thus, to use Maxwell's notation,.. the time-constant is $L/R=\tau$. **1892** O. Lodge *Lightning Conductors* xxvi. 297 A column shows the time taken for the current amplitude to decay to one-millionth of its initial value, *i.e.*, 14 times what is ordinarily called the 'time constant'. **1943** *Electronic Engin.* XV. 346 The amplifier..is a standard three-stage..circuit. The time constant of the stages is 6 seconds. **1962** A. Nisbett *Technique Sound Studio* v. 94 The PPM is a special type of voltmeter... It has a rapid rise characteristic (the BBC version has a time constant of 2·5 milliseconds; this gives 80% of full deflection in 4 milliseconds). **1977** *Nature* 1 Sept. 11/1 The radiative time constant is the time taken for a mass of air to warm up or cool by radiating in the infrared portion of the spectrum. **1971** *Jrnl. Gen. Psychol.* Jan. 38 The time course for recovery from these effects was found to be slower for the subfusional stimulation effect than the suprafusional stimulation effect. **1977** J. L. Harper *Population Biol. Plants* 678 The way in which the time course of the risk of death is related to the time course of producing offspring. **1959** H. Barnes *Oceanography & Marine Biol.* 183 Instead of using a bottom release to start the camera, a time-delay mechanism, set for the depth to which it is intended to work, is used. **1963** *Times* 15 Feb. 7/6 The levels from which different frequencies are reflected are thus obtained from the time-delay between outgoing pulse and received echo. **1978** *Tucson Mag.* Dec. 30/1 A digital time-delay system ..that brings 'concert hall sound' into the living room.

By delaying the impulse coming out of the back speakers ..by something under one one-thousandth of a second, this tiny wonder simulates the spatial quality of sound present in large auditoriums. **1851** C. Cist *Sk. Cincinnati in 1851* 90 Their policy of taking *time* deposits and allowing eight and ten per cent interest..[has] attracted public attention. **1930** J. M. Keynes *Treatise on Money* II. xxiii. 7 In the United States the law requires that the amounts of Time Deposits and Demand Deposits respectively shall be separately published. **1982** *Bank of England Q. Bull.* Dec. 519/1 Sight deposits grew by 28% and time deposits by 30%. **1956** *Nature* 11 Feb. 267/1 The passage of electrolytic current through the first coil induces its time-derivative in the second one. **1909** E. B. Titchener *Text-Bk. Psychol.* I. 134 In ordinary life, these time-differences escape notice, so that we may regard two tastes as occurring together when really they occur in succession. **1953** [see *capitalize *v.* 1 c]. **1981** 'W. Haggard' *Money Men* iii. 38 Seven-thirty from Schiphol and an hour for the flight. Time difference at the moment one hour. **1968** J. Sangster *Touchfeather* xi. 112 With the time differential on my side, I was back in Los Angeles by three-thirty p.m. local time. **1974** tr. *Wertheim's Evolution & Revolution* i. 33 The differential in space-time between more or less parallel processes in the past..might..imply time differentials in future developments. **1980** L. St. Clair *Obsessions* xvii. 291 Due to the time differential between New York and Paris, which he had left at sundown, he arrived at La Guardia shortly after 10 p.m. **1973** Time dilatation [see *relativity 2]. **1934** R. C. Tolman *Relativity, Thermodynamics & Cosmology* ii. 24 This time dilation and the conclusions as to the setting of clocks..are to be regarded except for experimental difficulties as an entirely verifiable mutual property of systems of clocks in relative motion. **1968** *Guardian* 28 Dec. 9 After two years the spacecraft's velocity would be such for 'time dilation' to have an effect. In other words time would be slowing down on board the craft and, to those on board, a single lifetime would be longer than for those on earth. **1981** *Sci. Amer.* Feb. 108/1 The relativistic effect of time dilation prolongs the life of pions and kaons that are particularly energetic. **1936** Time-distance [see *P III 3]. **1965** *Math. in Biol. & Med.* (Med. Res. Council) v. 230 The question whether short distances are correlated with short times is represented by the third degree of freedom, the interaction of the table, and for this reason this component can be referred to as a space-time, or time-distance, interaction. **1905** Time-division [see *Skiddavian *a.*]. **1938** *Proc. IRE* XXVI. 56 Many proposals have been offered for multiplexing. These divide naturally into two major categories; (1) frequency division and (2) time division. *Ibid.* 57 The major disadvantage of time-division multiplex..is that provision must always be made to insure accurate timing of the channel assignments. **1947** [see *quantization b]. **1975** D. G. Fink *Electronics Engineers' Handbk.* xxii. 35 In time-division multiplexing message information from many channels is sampled briefly in time sequence. **1831** Time element [in Dict., sense 51 a]. **1936** J. M. Keynes *Gen. Theory Employment* xxii. 317 The explanation of the *time-element* in the trade cycle. **1979** D. Anthony *Long Hard Cure* ix. 77 The time element fits... You heard the shot about nine forty-five. That gave him half an hour. **1931** *Times Lit. Suppl.* 21 May 402/4 The kindly Egyptian prince who..helped time-expired convicts to find honest employment. **1972** *Times* 17 Oct. 14/7 Return of time-expired stock like sausages and yoghurt. **1973** [see *hot-house, hothouse *sb.* 3 b]. **1974** *Ciba Symposium* XX. 253 Those of us who have used 'time-expired' human blood from blood banks in culture media have frequently found that some batches are highly toxic to trypanosomes. **1964** *Sunday Times* 5 July 14/4 All three considerations argue for abandoning the artificial time-frame which the Johnson Administration has now set. **1969** *Ottawa Commons Debates* 24 July 11573/1 In a timeframe of less than seven decades in length,..man has ceased to remain earthbound. **1976** 'R. B. Dominic' *Murder out of Commission* xx. 179 Somebody had..shot him. The time frame was pretty well established: between twelve o'clock, when he had phoned the desk, and twelve fifteen, when Ben and Tony arrived. **1980** D. Bloodworth *Trapdoor* xii. 71 It's right outside our brief, and our time-frame would not allow it. **1983** *Listener* 16 June 37/2 The time-frame is one intense, meandering weekend. **1892** O. J. Lodge *Lightning Conductors* vii. 48 Lord Rayleigh.. thinks these induced peripheral currents competent to explain magnetic time-lag in every case. **1939** *John o' London's* 7 Apr. 46/2 There is often a necessary time-lag between discovery and application. **1979** *Bull. Amer. Acad. Arts & Sci.* Mar. 40 Even if it were granted that a given intervention would..produce a given result..there would inevitably be time lags..which would diminish the efficacy of the remedy. **1937** *Mind* XLVI. 169 The duration which will be indicated as a 'time-lapse' in the standard clock which S carries. **1937** *Discovery* June 192/1 Experimenters in time-lapse photography and nature photography in general. **1956** *Kenyon Rev.* XVIII. 418 The time-lapse between any two primary stresses tends to be the same irrespective of the number of syllables and the junctures between them. **1957** [see *high-speed *a.*]. **1957** *New Scientist* 19 Sept. 14/1 The study of living cells by time-lapse cinematography... Special ciné-cameras are used to take microphotographs..at the rate of about eight photographs per minute. **1974** *Sci. Amer.* Apr. 123/2 In that industrial valley, the time-lapse photographs show, exposure from A.D. 1702 up to 1908 induced only light damage. **1983** *Listener* 3 Nov. 35/3 Time-lapse, slow-motion, stop frame, microscopic, infra-red and underwater techniques are part of the stock-in-trade of the TV wildlife producer. **1890** W. James *Princ. Psychol.* I. iii. 86 Another electric pen..traces alongside the former line a 'time-line' of which each undulation or link stands for a certain fraction of a second. **1967** *Sci. Year* 51 The *time line* (sequence of the mission) and trajectory of the specific mission are analyzed. **1976** *New Yorker* 30 Aug. 59 They were well behind the timeline, for the ground had allotted only ten minutes to move from one experiment to the next. **1871** *Rep. Comm. Patents 1869* (U.S.) II. 224/2 *Time-Lock*... The combination of the shaft [etc.]. **1895** H. G. Wells (*title*) The time machine. **1944** C. Day Lewis *Poetry for You* ii. 12 How *Poetry Began*. To find this out, we'll have to jump into a Time Machine, put its gear-lever into reverse, and race backwards through many thousands of years into prehistoric time. **1960** *Guardian* 7 Oct. 16/4 Nigel and Wendy arrive by time-machine in the Garden of Eden. **1978** I. Watson in C. Priest *Anticipations* 22 The assumption that a time machine should proceed to its destination *instanter* instead of at a snail's pace. **1945** H. D. Smyth *Gen. Acct. Devel. Atomic Energy Mil. Purposes* xii. 131 One elegant scheme for studying the effects of neutrons of a single, arbitrarily-selected velocity is the 'time of flight' method. **1948** S. A. Goudsmit in *Physical Rev.* LXXIV. 622/2 (*heading*) A time-of-flight mass spectrometer. **1969** Egelstaff & Poole *Exper. Neutron Thermalisation* iv. 74 Measuring the velocity of the scattered neutrons by the time-of-flight technique. **1975** *McGraw-Hill Yearbk. Sci. & Technol.* 189/2 Both types exceed a mass resolution *M/ΔM* of 1000,..and in addition the time-of-flight version is completely free of artifacts. **1927** *Ladies' Home Jrnl.* Dec. 45/2 Chrysler dealers are in a position to extend the convenience of time-payments. **1955** *Times* 25 Aug. 6/6 The Government had been greatly concerned at the rapid expansion of time payment and hire purchase and its effects upon the economy. **1955** J. Thomas *No Banners* xix. 171 Two sticks of plastic, a primer bound with a length of instantaneous fuse, and a ten-minute time pencil. **1977** J. Hutchison *That Drug Danger* xi. 88/1 Having pressed the time pencil which set off the explosive a quarter of an hour or one or two hours later. **1977** H. J. Eysenck *You & Neurosis* iv. 135 They were told that this was a peripheral-acting time-release muscle-relaxant known to be effective. **1980** *Holistic Health News* Sept./Oct. 2/1 This treatment is called 'Magnetroph' and was developed by Dr. Ernest Pescetti, renowned as 'the father of the time-release principle'. **1956** *Nature* 4 Feb. 222 (*caption*) Optical arrangement for spectrochemical analyses with time-resolved spectra. **1977** *Ibid.* 17 Feb. 659/2 (*heading*) Time-resolved resonance Raman spectroscopy of bacteriorhodopsin. [**1922** I. Fisher *Making of Index Numbers* iv. 64 The time reversal test... The index number reckoned forward should be the reciprocal of that reckoned backward.] **1955** *Proc. Glasgow Conf. Nucl. & Meson Physics 1954* viii. 341 The superselection rule on the parity of spinor-particle number is derived from a consideration which does not depend on double time-reversal. **1958** *Physical Rev.* CX. 783/2 We see that at least one prediction of time reversal invariance is very well fulfilled. **1979** *Sci. Amer.* June 116/3 The discovery of an electric dipole moment of the neutron would reveal a violation of the physical principle known as time-reversal symmetry. **1971** *Ibid.* Dec. 97/1 Time-reverse all motions and the three will return at the same instant to the starting point. **1962** M. A. Preston *Physics of Nucleus* xvi. 475 The wave function of a time-reversed state is obtained by taking the complex conjugate of the wave function of the original state and reversing all spin directions. **1981** T. D. Lee *Particle Physics & Introd. Field Theory* xiii. 283 If the T-invariant classical system consists of a large number of particles, although the time-reversed sequence is always possible, it is in general improbable. **1960** J. B. Carroll in *Encycl. Educ. Research* (ed. 3) 746/1 Careful time-sampling and situation-sampling designs seem to be in order if one wants purely normative or 'typical' data. **1973** *Jrnl. Genetic Psychol.* Sept. 99 Time sampling has been demonstrated to be an effective observational technique. **1979** J. Jaffe et al. in Aaronson & Rieber *Psycholinguistic Research* 404 Assume the two states are digitized as 'one' or 'two' by discrete time sampling at intervals that must be shorter than the expected minimum time the system spends in any state. **1890** W. James *Princ. Psychol.* I. ix. 283 We make a solid wooden frame with the sentence written on its front, and the time-scale on one of its sides. **1923** N. Shaw *Air & its Ways* v. 46 The time-scale of the operations which tend to cause deviation is so large that the course of the operations escapes observation. **1934** *Discovery* Aug. 227/1 The constructing of a geological and meteorological time-scale for the Southern Hemisphere. **1958** F. E. Zeuner *Dating Past* 5 All these studies aiming at the establishment of absolute time-scales for the past are comprised by the term *Geochronology*. **1972** *Country Life* 23 Mar. 672/2 The timescale of planning is a long one. **1979** *Jrnl. R. Soc. Arts* CXXVII. 409/1 We must also bear in mind the timescale involved in exploitation of oil and gas discoveries. **1892** J. Royce *Spirit Mod. Philos.* 431 This transcending of a time-series..is in fact what one might call the soul of the natural order. **1919** *Rev. Econ. Statistics* Apr. 123/2, *x* and *y* represent the deviations of the items of two time series from their respective linear secular trends. **1928** *Jrnl. Amer. Statistical Assoc.* XXIII. 407 The inferential possibilities of time series analysis are contingent upon segregating from the specific historical the repeatable, recurrent element. **1978** *Nature* 19 Oct. 630/2 The resulting time series is then smoothed by using a 13-month running mean average to remove seasonal variations. **1979** *Daily Tel.* 6 July 16 He has confirmed his original time series analysis with a more recent cross-sectional study. **1979** J. Hicks *Causality in Economics* v. 64 There are two kinds of time-series, with different kinds of time reference. In one of them each item relates to a point of time, in the other to a period. **1853** *Rep. Brit. Assoc. Adv. Sci. 1852* II. 131 On the 5th of August 1852, the first time-signal passed;..the clock at Greenwich..originates the signals. **1862** *Monthly Notices R. Astron. Soc.* XXII. 119 At any time, day or night, when the wire is not wanted for ordinary work, London can receive time-signals from Liverpool every minute. **1923** [see *news *sb.*(*pl.*) 5 c]. **1972** P. Lively *Driftway* v. 59 Through the atmospherics there came the time-signal, and then a man reading the news. **1965** *Language* XLI. 193 To the comparativists of the 1870's..a 'synchronic' view could be nothing more than taking a relatively thin time-slice and doing the best one can with it. **1973** Murrill & Smith *Introd. Computer Sci.* i. 31 A user receives the processor's undivided attention for one time slice but then receives no attention over the next three time slices. **1981** *Kilobaud Microcomputing* June 35/2 At the end of the current user's time slice, he is put on hold by the scheduler, and the next user gets to run his program. **1967** J. Martin *Design of Real-Time Computer Systems* 628/2 (Index), Time-slicing. **1978** W. S. Davies *Information Processing Systems* xv. 328 The whole purpose of the time-slicing approach..is to prevent any single user from dominating the system at the expense of everyone else. **1962** Time slot [see *programme planning* s.v. *program, programme *sb.* 4]. **1967** J. Martin *Design of Real-Time Computer Systems* ix. 130 When the time-slot..for one user ends, it may be necessary to bring in a completely new set of data and programs for the next user. **1979** *Arizona Daily Star* 5 Aug. 1. 7/5 It..will be interesting to learn what the viewer response is to a program concept of this dimension when aired at an odd afternoon time slot. **1980** Pužman & Pořízek *Communication Control in Computer Networks* iii. 168 Time division multiplexing means that a certain time slot is to be assigned for a station for the time during which it can exchange data with another station. **1982** 'W. R. Duncan' *Queen's Messenger* vi. 49 People were units to be fitted into Sir John's available time slots. **1960** N. Hilliard *Maori Girl* I. ii. 14 Netta was supposed to learn her words and the older ones their times-tables. **1983** *Cotswold Life* Nov. 19/1 Here the infants practised wobbly letters, began to chant the Times Tables, and spilt paint water all over the floor. **1984** *Reader's Digest* Jan. 81/2 Children..slipping from ignorance to knowledge.., until one day finally they know their times tables. **1892** *Work* IV. 75/3 The time-stamp is altered every minute. **1963** L. Meynell *Virgin Luck* viii. 181 Each slip is time-stamped as it comes from the telephonist who takes the bet. **1973** *Times* 6 Dec. 9/8 Waiter's [*sic*] checks are time-stamped on receipt... This way, the clerk knows exactly when each item was ordered. **1962** *Times* 16 Apr. 11/3 The 'limpet' meter for issuing time-stamped tickets. **1929** Time-step [see *stooge *sb.* 1]. **1956** B. Holiday *Lady sings Blues* (1973) iii. 32, I knew exactly two steps, the time step and the crossover. **1975** Time step [see *pull-back 1 d]. **1911** F. W. Taylor *Princ. Scientific Managem.* 75 A careful time study of men working under these conditions will disclose facts which are ludicrous as well as pitiable. **1928** *Britain's Industrial Future* (Liberal Industr. Inquiry) III. xvi. 194 There should be means of revising rates..in consultation, with all the facts and time-study figures on the table. **1944** *Jrnl. R. Aeronaut. Soc.* XLVIII. 257 Mr. Westbrook asked whether time study was used sufficiently to convince shop superintendents of the amount of labour necessary for any job. **1970** T. Lupton *Managem. & Social Sci.* (ed. 2) i. 14 The organization as a whole, or at least those members of it with whom they have close contact, the supervisor, the time study man, and so on. **1979** J. Harvey *Plate Shop* iii. 15 He was the best plater in the shop, and the Time Study would never believe he was working as fast as he could; and it was true, he always did work slow when studied, out of loyalty to his mates. **1902** *Specifications of Patents* (U.S. Patent Office) 14 Oct. 1560/1, I, William B. Coulter,..have invented certain new and useful Improvements in Time-Switches for Electric Lights. **1963** *Listener* 3 Jan. 47/1 Portable time-switches vary in elaborateness and price. *a* **1977** *Harrison Mayer Ltd. Catal.* 64/2 A time switch can be installed on any kiln. **1920** A. S. Eddington *Space, Time & Gravit.* vi. 103 Leaving aside now the time-term as sufficiently discussed, we consider the space-terms alone. **1974** *Nature* 15 Mar. 204/1 If..the true P_n velocity is 8·0 km s^{-1}, then the average time terms are slightly reduced and the resulting crustal thickness estimates are about 5–10% lower. **1965** E. Tunis *Colonial Craftsmen* vi. 146 Both the time train and the striking train of a Terry clock are driven by weights wound up by a key. **1977** *Lancashire Life* Jan. 39/2 One of the chief technical advances in the making of watches..was the improved design of the escapement, the device which secures uniformity in the rate of movement of the time-train. **1953** A. C. Clarke *Prelude to Space* xxvi. 137 That lurid magazine..that goes in for hyperspace time-travel. **1969** *Punch* 1 Jan. 35/2 This intelligent and ingenious story..has a fascinating climax where people from both sides time-travel back to see the Passion and Crucifixion. **1975** G. Ewart *Be my Guest!* 1. 23 Belief in Time Travel and supernatural facilities. **1930** *Wonder Stories* Nov. 489 We have purposely allowed our time travellers to become known to the people of the eras that they visit, for in this way the great drama of the story becomes apparent. **1934** C. Lambert *Music Ho!* ii. 69 The most successful time traveller of our days was undoubtedly Serge Diaghileff. **1972** M. Crichton *Terminal Man* iii. i. 106 Slowly..he seemed to emerge like a time-traveler advancing through the years. **1871** Swinburne *Songs before Sunrise* 171 Mother of man's time-travelling generations,.. The temples and the towers of time thou breakest. **1895** H. G. Wells *Time Machine* iv. 28, I am afraid I cannot convey the peculiar sensations of time travelling. They are excessively unpleasant. **1934** C. Lambert *Music Ho!* ii. ii. 73 By his adoption or even invention of the particular type of present-day pastiche that can conveniently be described as time travelling Diaghileff immediately established a position of mastery again. **1961** *Guardian* 16 June 9/5 This musical time-travelling is presumably deliberate. **1981** Craig & Cadogan *Lady Investigates* viii. 152 In children's fiction in general..magic and time-travelling were acceptable forms of the supernatural; ghosts were not. **1954** *Amer. Speech* XXIX. 103 *Time trial*,..a competitive event..in which cars are separately timed for top speed over set distances. **1976** *Scottish Daily Express* 24 Dec. 13/6 Ten days ago an unknown sprinter ran an impressive time-trial against local man Jim Smith. **1977** *Wandsworth Borough News* 16 Sept. 10/4 The Wandsworth and District Cycling Club Open 10 miles' Time-trial..resulted in a win for Eddie Adkins.. in the fine time of 22 mins, 4 secs. **1887** Moore & Aveling tr. *Marx's Capital* II. vi. xx. 553 The converted form under which the daily, weekly, &c., value of labour-power presents itself, is hence that of time-wages, therefore day-wages, &c. **1892** D. F. Schloss *Methods Industrial Remuneration* i. 11 The two leading forms of Industrial Remuneration under the wage-system are time-wages and piece-wages... The employee engaged on time-wage sells to his employer the labour which he shall perform within a given period, irrespective of the amount of labour performed within that period. **1954** W. M. Miller in *Fantastic* Jan.–Feb. 34/1 They showed me a dozen pictures of moppets with LTR-guns, moppets in time-warp suits, moppets wearing Captain Chronos costumes, [etc.]. **1954** *Sociological Rev.* Dec. 242 Instead of the 'dream' to move us out of lived time, it [*sc.* science and fantasy fiction] uses some machine to 'move across

the "time-warp"', 'to cut across the 4th dimension'. **1971** *Guardian* 17 June 10/5 The time warp effect was..intensified by having David Frost—essentially an early sixties figure. **1974** *Times* 22 Aug. 6 He pauses in his narrative and time-warps it back to South Staffs. **1976** N. Thornburg *Cutter & Bone* i. 13 A sensation that always made him feel as if he had been time-warped back into wet diapers. **1983** *Listener* 3 Nov. 32/3 Molly Kean's images are of psychic rather than physical decay, of families able to live beyond their means because they are trapped in a peculiar time-warp. **1953** *Sun* (Baltimore) 27 Aug. 21/3 Of course, we [*sc.* railroads] can't meet the planes timewise. **1981** J. D. MacDonald *Free Fall in Crimson* xx. 229 'How far are you from a pay phone, timewise?' 'Ten minutes.' **1917** W. S. Churchill in M. Gilbert *Winston S. Churchill* (1975) IV. iv. 63 These advances arose out of an intention to remedy the contrast between the wages of skilled time workers in certain munitions industries which had grown up during the war. **1971** *Daily Tel.* 15 Sept. 7/1 Representatives of nearly 3,000 'timeworkers' at the Austin-Morris car body plant..have said that [their] pay offers..are not acceptable. **1892** E. Noel *Internat. Time* 4 The country [*sc.* United States of America]..is divided into time-zones, each stretching over fifteen degrees of longitude, and differing one hour in time from the zone on either side of it. **1929** [see *diachronous *a.* 1]. **1976** M. Machlin *Pipeline* lv. 553 Due to the time zone differences, they would have plenty of time to make the six o'clock news coast to coast. **1982** *Daily Tel.* 24 Apr. 9/1 They conclude..that rapid travel through time zones precipitates psychiatric illness in people already predisposed to it.

b. Comb. with *time's*: **time's arrow,** the temporal asymmetry whereby many macroscopic phenomena do not occur in reverse even though such an event would not conflict with the laws of nature.

1928 A. S. Eddington *Nature of Physical World* iv. 69, I shall use the phrase 'time's arrow' to express this one-way property of time which has no analogue in space. **1937** *Nature* 27 Feb. 356/2 Eddington's statement..that the second law of thermodynamics holds 'the supreme position among the laws of nature'. His reason is that this law alone reveals 'time's arrow'. **1972** S. Weinberg *Gravitation & Cosmol.* xv. 597 It is the expansion of the universe that, by providing a heat sink, sets the direction of time's arrow in thermodynamic processes.

time base. *Electronics.* [f. Time *sb.* + Base *sb.*[1]] **a.** A line on a cathode-ray tube display representing the time axis, usu. horizontally. **b.** A signal for uniformly and repeatedly deflecting the electron beam so as to produce such a line, usu. consisting of a saw-tooth waveform. **c.** Also *time-base generator*. A circuit for generating such a signal; a sweep generator.

1925 *Proc. Physical Soc.* XXXVII. 167 Circular and elliptical time-bases were not nearly so useful as a linear and unidirectional base obtainable from a special type of triode oscillator circuit. **1933** R. A. W. Watt et al. *Applications Cathode Ray Oscillograph* ii. 44 The general nature of the e.m.f. desirable as a linear time-base for use with a cathode-ray oscillograph. **1942** *Electronic Engin.* XV. 102/2 An outstanding case where stroboscopic viewing is a necessity is in conjunction with a spiral timebase. *Ibid.* 664 The mechanism of the time-base circuit. **1959** Davies & Palmer *Radio Studies of Universe* iii. 43 As the speed of radio waves is 3×10^5 km/s, the horizontal trace of the tube, called the time base, can be calibrated in range directly. **1960** *Practical Wireless* XXXVI. 408/2 A lead from the timebase is brought out to the front panel of the oscilloscope. **1962** Simpson & Richards *Physical Princ. Junction Transistors* xiii. 309 This form of operational amplifier..has been used extensively as a time-base generator. **1977** *Broadcast* 7 Feb. 5 (Advt.), VPR-1 systems, with..timebase correction..for installation in studio.

Time-ese, Timese (təimī·z), *sb.* and *a.* [-ese.] (Characteristic of) the prose style of *Time* magazine.

1958 C. Logue in *Times Lit. Suppl.* 15 Aug. p. xxiii/1 They can model their syntax on the new 'international-English' style—a compound of Hemingway and *Time*-ese. **1967** *Time* 17 Mar. 7 The humor of a *Playboy* cartoon is often more sophisticated than the cleverness of Timese. **1973** *N.Y. Times Mag.* 8 July 4/2 Mr. Kanfer's prose: part showbiz, part Time-ese, and all posturing absurdity. **1977** *Globe & Mail* (Toronto) 15 Mar. 6/3 Elaine Dewar's review..is..bad book reviewing with a Timese style.

timeist (təi·mist). Now *Obs.* or *rare*. [f. Time *sb.* + -ist.] = Time-keeper, timekeeper 3.

1830 W. T. Parke *Musical Mem.* II. 320 This habit [of playing in concert] made him a good timeist. *a* **1837** E. C. Knight *Autobiogr.* (1861) I. i. 7 The dancing-master was ..graceful without affection, a good time-ist, and..a good domestic character.

time-keeper, timekeeper. 2. c. (Later examples.)

1829 P. Egan *Boxiana* 2nd Ser. II. 128 *Maurice*.. repeated the signal for fighting. *Randall* was the time-keeper. **1896** [see *corner *sb.*[1] 13 d]. **1950** *Sport* 7–11 Apr. 7/1 Did the timekeeper and second..want to catch the early train home?

timeless, *a.* Add: **2. a.** Esp. in phr. *timeless moment.*

1942 T. S. Eliot *Little Gidding* i. 9 Here, the intersection of the timeless moment Is England and nowhere. Never and always. *Ibid.* v. 15 History is a pattern Of timeless moments. **1957** L. MacNeice *Visitations* 22 A timeless moment where the nether blue meets the upper blue. **1981** J. Brabazon *Dorothy L. Sayers* xix. 238 The timeless moment after death, when choices made while living are seen in the light of eternal reality.

ti·me-like, *a. Physics.* [f. Time *sb.* + -like.] Being or related to an interval between two points in space-time that lie inside one another's light cones (so that a signal or an observer can pass from one to another).

1914 [see *four-vector* s.v. *four C. 2]. **1920** [see *space-like *a.*]. **1955** O. Klein in W. Pauli *Niels Bohr* 112 We divide the *k*-states into those belonging to time-like *k*-vectors..and those belonging to space-like *k*-vectors.., a division which is relativistically invariant. **1959** [see *null *a.* 4 d]. **1978** Pasachoff & Kutner *University Astron.* xxvii. 694 The world lines of light divide space-time into spacelike and timelike regions.

timenoguy (ti·menǫgi). Delete †*Obs.* and add: Also **timm(e)y-nog(gy). 1.** (Further examples.) In extended sense, a gadget.

1880 W. Bottrell *Traditions W. Cornwall* 3rd Ser. 198/2 *Timmy-noggy*, a notched square piece of wood, used to support the lower end of the Vargord. **1886** [see *gadget]. **1925** Fraser & Gibbons *Soldier & Sailor Words* 282 *Timmynoggy*,..a name given to various strop and toggle gadgets on board ship... A device more or less makeshift, to enable something to be done more expeditiously than would be possible in the ordinary way. **1963** R. M. Nance *Gloss. Cornish Sea-Words* 160 *Timmy-noggy*,..may be a nautical substitute word like 'thingum-a-bob'. **1976** *Oxf. Compan. Ships & Sea* 870/2 *Timenoguy*, ..more recently it was a rope made fast in the mizen rigging with a thimble in the end through which passed the hauling part of the mainbrace.

timer. Add: **3. d.** An instrument for automatically timing a process or activating a device at a set time or set times; a time-switch; *egg-timer*: see *egg *sb.* 7; *oven timer*: see *oven *sb.* 4.

1908 *Sears, Roebuck Catal.* 266/1 Stem wind, jeweled horse timer. This timer is operated from the crown by merely pressing down the..mechanism. **1939** C. Morley *Kitty Foyle* i. 9 They worked right round the clock..making some kind of timers for shells. **1953** R. Chandler *Long Good-Bye* ii. 25, I stirred the coffee and covered it. I set my timer for three minutes... The bell of the timer went just as I got back. **1961** *Which?* Dec. 335/1 Two.. were clockwork dial timers, similar to kitchen timers, which have to be wound up by rotating the dial. **1972** *Jrnl. Social Psychol.* Dec. 225 The timer was preset so that the signals for all shock options were of three seconds duration, with a 15-second intertrial interval. **1977** *New Yorker* 27 June 89/1 Attached to the timer of the bomb was a note. **1982** *Listener* 15 July 34/3 These timers are perfectly adequate for some domestic tasks, like..switching lights on and off in a house to fool..burglars.

4. See also *two-timer.

ti·me-sharing, *vbl. sb.* [f. Time *sb.* + Sharing *vbl. sb.*[2]] **1.** *Computers.* The automatic sharing of (central) processor time so that a computer can serve two or more users or devices concurrently, switching between them rapidly and automatically so that each user has the impression of continuous exclusive use. Also *attrib.*

1953 *Digital & Analog Computers* (Amer. Soc. Mech. Engineers) 41 'Time-sharing' can be relatively simple. Nearly all the machines have convenient features for changing the problems easily. **1958** *Computer Jrnl.* Apr. 4/1 If time-sharing is being employed it can be arranged that the control unit proceeds to another branch of the program which is not held up in the same way. **1960** *Information Processing* 336 Time sharing, in the sense of causing the main computer to interrupt its program to perform the arithmetic and control operations required by external or peripheral equipment, has been used on a limited scale for a long time. This paper explores the possibility of applying time sharing to a large fast computer on a very extensive scale. **1964** *Discovery* Oct. 56/2 The advent of time-sharing machines. **1965** *Math. in Biol. & Med.* (Med. Res. Council) 295 Without time-sharing, the 'on-line' use of a fast modern machine would be unthinkably costly. **1969** *New Scientist* 2 Jan. 16/1 With a time-sharing system, the user can now call the system over the ordinary GPO network using an adapted telephone. **1980** P. Way *Icarus* iii. 20 'Where is this computer?' 'Texas... You dial a telephone number. Ring it up. Bounces off a satellite. It's called time-sharing.'

2. The ownership or right to use of a property (esp. as a holiday home) for a fixed limited time each year. Also *attrib.* orig. *U.S.*

1976 *Time* 30 Aug. 67 In exchange for guaranteed occupancy over an extended period time-sharing resorts offer low prices, [etc.]. **1977** *Detroit Free Press* 11 Dec. 17-C/1 Time-sharing..is proving to be a popular concept for vacations. **1980** *Country Life* 27 Nov. 2038/3 The concept of time-sharing (that is, buying an annual choice of weeks in a holiday home instead of buying the home outright).

Hence **ti·me-share** *sb.*, (*a*) = *time-sharing *vbl. sb.* 2 (freq. *attrib.*); (*b*) a share in a property under a time-sharing scheme; **ti·me-share** *v. trans.*, to use concurrently with others on a time-sharing basis; **ti·me-shared** *ppl. a.*; **ti·me-sharer** *sb.*, a participant in a time-sharing scheme.

1954 *Jrnl. Assoc. Computing Machinery* I. 136/1 It seems reasonable to investigate the possibility of time-sharing one or more integrators in such circuits. **1962** *Spring Joint Computer Conf.* 307/1 By having these programs time share areas of memory, the program may be executed. **1973** *Lebende Sprachen* XVIII. 72/2 Subsequent data processing can then be carried out with a small and cheap dedicated computer, by a time-shared processor or off-line on a large computer. **1977** *Sci. Amer.* Sept. 135 (*caption*) One head can serve for both reading and writing on a time-shared basis. *Ibid.* 177/1 At a terminal an operator is connected to the computer and time-shares the computer with other operators. **1978** *Detroit Free Press* 2 Apr. 6F/1 The Minnesota resort is among more than 150 time share facilities in 26 states. **1980** R. Rejnis *Her Home* 73 A time-share can be very inexpensive.. compared to hotel costs. **1980** *Times* 13 Oct. 24/4 (Advt.), Time-share ownership of luxury 8-berth yacht in Mediterranean. **1981** *Times* 31 Jan. 17/5 An organization.. which offers an exchange service with other timesharers elsewhere. **1981** *Sunday Express Mag.* 14 June 33/2 It is important for the public to be aware that time-share is a property-based purchase.

time-spa·ce, *sb.* and *adj. phr.* [f. Time *sb.* + Space *sb.*[1]] **A.** *sb.* = *space–time *sb.*

1916 [see *metric *a.*[1] 2]. **1942** E. Waugh *Put out More Flags* i. 86 Metaphysical war, war in time-space, war eternal..all war is nonsense. **1978** *Church Times* 15 Dec. 6/5 Using the discoveries of Einstein concerning the nature of time-space, Professor Macquarrie sketches in a theory of both individual and cosmic hope which depends upon the idea that God occupies every point in space-time.

B. *adj. phr.* = *space–time *adj. phr.*

1951 N. M. Gunn *Well at World's End* xxviii. 264 Giving and receiving are aspects of the same thing; as time and space are aspects of the one time-space continuum, according to the latest mode of scientific utterance. **1961** G. Clark *World Prehistory* 4 It is important to avoid European or other regional bias, yet it would be pointless to try and achieve a kind of time-space parity for its own sake. **1968** D. L. Clarke *Analytical Archaeol.* viii. 327 Strongly localized and smaller time-space entities..by contrast suggest culture group transforms in time depth. **1977** *Brit. Med. Bull.* XXVII. 20/1 One of the most interesting features has been the recognition in some areas of time–space clusters.

timetable, *sb.* Add: **1. c.** (Earlier example.)

1820 *Edin. Monthly Rev.* Jan. 15 One of the most striking epochs in music, as we conceive, was that of the invention of the time-table.

2. *timetable motion* = *guillotine motion* s.v. *guillotine 4 a.

1976 S. Lloyd *Mr Speaker, Sir* iii. 76 Peyton protested about the timetable motion, and asked why aid to Norton-Villiers had suddenly become so urgent. **1977** *Whitaker's Almanack 1978* 349/2 A third timetable motion to limit debate on Lords' amendments to the Bills..was carried by 312–296.

ti·meta:ble, time-table, *v.* [f. the sb.] *trans.* To schedule, to plan or arrange according to a time-table, to include in a time-table.

1917 'Contact' *Airman's Outings* v. 111 The leave train at Arrière was time-tabled for midnight. **1939** G. Greene *Confidential Agent* i. i. 29 'My employers..wouldn't understand the delay.' They would have time-tabled his movements. **1970** *Times* 20 Oct. 4 Next year the school will admit 15 musically talented children who will be separately timetabled to give full scope for their abilities. **1974** *Times* 3 June 7/5 Headmistresses..were adamant that non-examinable, creative subjects must be time-tabled in girls' schools.

Hence **ti·meta:bled** *ppl. a.*; **ti·meta:bling** *vbl. sb.*

1960 *Farmer & Stockbreeder* 22 Mar. 47/1 Apart from complex timetabling, his work included liaison with police ..and dovetailing turn-rounds. **1963** *Times* 24 May p. xii/5 Regular timetabled lorry services between the depots and Acton. **1969** *Guardian* 24 Oct. 9/2 Some will protest that the division of school time into 40-minute periods is essential for time-tabling. **1975** *Language for Life* (Dept. Educ. & Sci.) xxv. 403, 93 per cent of classes..were taught English as a separately time-tabled subject. **1977** *Modern Railways* Dec. 481/3 The obsolete layout at Liverpool Street is a major timetabling handicap particularly in that there has to be a comfortable margin at Platform 9 between the 00.30 Norwich departures and the 00.39 arrivals. **1978** P. Bailey *Leisure & Class in Victorian England* vii. 167 Cutting down ad libs and encores..helped ensure the predictable time-tabling of acts. **1982** *Daily Tel.* 12 Feb. 2/5 While many timetabled [rail] services will be axed, staff shortages mean there will be few cuts in services operated.

timing, *vbl. sb.* Add: **2. a.** (Earlier example in *Cricket.*)

1863 *Boys' Jrnl.* I. 264/1 He..showed us all the power and velocity resulting from hitting the balls at the right time. He called this the art of *timing*.

3. In an internal-combustion engine, the times when the valves open and close, and the time of the ignition spark, in relation to the movement of the piston in the cylinder. Freq. *attrib.*, as *timing diagram*; **timing chain,** the chain that drives the camshaft of an engine from the crankshaft, part of the timing gear; **timing gear,** the mechanism by which the valves of an internal-combustion engine are made to open and close at the right moment; **timing mark,** each of the marks on an engine that are used as guides when

assembling the timing gear or altering the valve timing.

1915 V. W. Pagé *Automobile Repairing made Easy* iii. 304 In these diagrams the timing used is: Inlet opens at 8 degrees past the upper center; [etc.] *Ibid.* 306 Two typical valve timing diagrams, one for a four cyclinder engine having a flywheel diameter of 15¾ inches, the other of a six cylinder engine are given. **1929** —— *Ford Model A Car* vii. 270 (*caption*) Diagram showing timing marks on crankshaft and camshaft timing gears of Ford model A engines. **1935** F. J. Camm *Practical Motorist's Encycl.* 319/1 Timing chain. **1970** K. Ball *Fiat 600D Autobook* i. 20/1 Replace and secure timing gear gasket, cover and crankshaft oilseal. *Ibid.*, Install camshaft, crankshaft sprockets with timing chain so that timing marks are adjacent. **1977** *Hot Car* Oct. 81/1 When setting the timing on a Mini engine with a stroboscopic light, it is often difficult to see the timing marks. **1977** 'J. Fraser' *Hearts Ease* ix. 107 That lorry..needed the tappets adjusting. And I don't think the timing was right.

Timini, Timmanah, Timni, obs. varr. *Temne *sb.* and *a.*

timolol (ti·mǫlǫl). *Pharm.* [f. *tim-* (of unknown origin) + *-olol*, after *propranolol.] A β-adrenergic blocking agent used in the treatment of hypertension; (−)-1-*tert*-butylamino-3-(4-morpholino-1,2,5-thiadiazol-3-yloxy)-propan-2-ol, $C_{13}H_{24}N_4O_3S$.

1973 *Arch. Internat. de Pharmacodyn. et de Thérapie* CCV. 92 As a β-adrenergic blocking agent, timolol was found to be more potent than propranolol. **1976** *Nature* 22 July 307/2 Neither phentolamine..nor timolol, a specific blocker of beta-receptors..had any effect.

Timor (tī·mǫɹ). The name of an Indonesian island off the north-west coast of Australia, part of which was before 1976 a Portuguese colony, used *attrib.* in **Timor pony,** a small, stocky horse belonging to a variety first found there. Also *absol.*

[**1841** G. Grey *Jrnls. Two Exped. N.-W. & W. Austral.* I. iv. 68 The vessel could then proceed..to the Island of Timor, to procure the requisite number of ponies for our expedition.] **1895** A. B. Paterson *Man from Snowy River* (1896) 4 He was something like a racehorse undersized, With a touch of Timor pony. **1928** 'Brent of Bin Bin' *Up Country* xv. 253 He rode a yellow bay with a dash of Timor, hardy and sure-footed as a goat. **1933** *Bulletin* (Sydney) 27 Sept. 21/2 Those Timor ponies..date back to 1840. **1965** *Austral. Encycl.* IV. 552/1 The Timor pony possesses remarkable hardiness and stamina.

Timorese (tīmorī·z), *sb.* and *a.* [f. *Timor* (see prec.) + -ese.] **A.** *sb.* (A member of) the indigenous people of Timor, of Indonesian-Malay stock. **B.** *adj.* Of, pertaining to, or characteristic of Timor or its inhabitants.

1869 A. R. Wallace *Malay Archipelago* I. xiii. 290 The native Timorese predominate. **1875** [see *Papuan *a.* 1]. **1964** M. Dickson *World Elsewhere* iii. 122 The standard of the game was mainly due to the exemplary keenness.. of the Timorese and Dusan opponents. **1979** *Times* 28 Dec. 16/2 Left-wing Timorese students..began to return to the island. *Ibid.*, Not a few Timorese are in somewhat uneasy cooperation with the Indonesians.

timothy. 1. b. *timothy hay* (examples).

1772 *Pennsylvania Gaz.* 16 Apr. 4/3 (Advt.), Timothy and blue grass hay to be sold. **1911** *Daily Colonist* (Victoria, B.C.) 25 Apr. 6/1 (Advt.), *Horses*. Try some of our Washington Timothy Hay—big load just in.

timpani (ti·mpăni), *sb. pl.* [a. It., pl. of *timpano* kettledrum (also used), f. L. *tympanum* drum: cf. Tympanum, Tympany 3.] **a.** The kettledrums. (The preferred spelling: see also *tympani.) **b.** Timpani-players, timpanists.

[**1557**: see Tympany 3.] **1740** J. Grassineau *Mus. Dict.* 283 *Timpano*. See *Tympanum*. **1876** Stainer & Barrett *Dict. Mus. Terms* 435/1 Timpani, (*It.*) kettle-drums. **1906** [see Tympanist]. **1947** A. Einstein *Mus. in Romantic Era* ii. 14 One of them, that in G minor, renounces trumpets and timpani. **1962** N. Del Mar *R. Strauss* v. 161 It takes the form of a powerful pedal-point reinforced by regular strokes on the timpani. **1970** J. Blades *Percussion Instruments* xii. 270 The soft pulsation of the C timpano. **1977** *Listener* 27 Jan. 128/3 BBC Northern Orchestra..requires: Timpani and Percussion... Salary £3,876 per annum.

Hence **ti·mpanist,** one who plays the kettle-drums.

1939 J. Harrison *Brahms & his Four Symphonies* vi. 79 The Timpanist reiterates those challenging Cs. **1948** *Penguin Music Mag.* June 40 The timpanist's job is extremely responsible. **1978** *Washington Post* 28 Mar. c2/3 The kid snuck down to the front and got a wink from the timpanist in the orchestra.

timps (timps), *sb. pl.* Colloq. abbrev. of prec.

1934 S. R. Nelson *All about Jazz* ii. 50 There were drummers who could already read and play timps, etc. **1948** *Penguin Music Mag.* June 40 He himself was a 'timps' player as well as conductor. **1978** R. Donington in J. M. Thomson *Future of Early Music in Britain* 15 The baroque orchestra..should have a basic need for.. trumpets and timps as required.

Timurid (ti·miurid), *a.* and *sb.* [f. the personal name *Timur* (see below) + patronymic suff. *-id*.] **A.** *adj.* Descended from Timur Lenk (see Tamerlane, Tamburlaine); of or pertaining to the Turkic dynasty which ruled in Central Asia after the death of Timur in 1405 until 1506. **B.** *sb.* A descendant of Timur Lenk; a member of the Timurid dynasty.

1889 G. N. Curzon *Russia in Central Asia* vii. 216 The *Koktash*, or coronation-stone, of the Timurid sovereigns. **1908** C. Eliot *Turkey in Europe* (ed. 2) iii. 87 Shah Shaitan..and his descendants struggled with the Timurids for the supreme power in Transoxiana. **1934** A. Toynbee *Study of Hist.* I. ii. 369 The Timurid Empire had held together..for half a century longer. **1947** Auden *Age of Anxiety* v. 110 The Timurids and Torguts. **1958** O. Caroe *Pathans* x. 138 His claim to the Delhi throne was based on his Timurid ancestry. **1975** [see *Safavid *a.* and *sb.*].

tin, *sb.* **1. b. block tin,** substitute for def.: tin of second quality cast into blocks; solid tin as distinct from tin plate; a receptacle made from this; (further examples).

1836 Dickens in *Bell's Life in London* 17 Jan. 1/1 The little block-tin temple sacred to 'baked 'taturs'. **1852** —— *Bleak Ho.* (1853) xxvi. 259 He could play 'em a tune on any sort of pot you please, so as it was iron or block tin. **1879** M. E. Braddon *Vixen* I. xiii. 255 The silver kettle..was conducting itself as spitfireishly as any blackened block-tin on a kitchen hob. **1910** G. B. Shaw *Let.* 21 Mar. (1972) II. 915 You inherited from your father a sense of the importance of block-tin piping.

2. a. (Earlier examples in first two senses.)

1795 S. Martin *New Experienced Eng.-Housekeeper* v. 73 Butter the tins, and bake them in a pretty quick oven. **1861** Mrs. Beeton *Bk. Househ. Managem.* 100 Many cooks use the tinned turtle..preserved in hermetically-sealed canisters... The cost of a tin..is about £2.

c. Cricket colloq. *the tins*, rectangular metal pieces each with a single white number painted on a black ground, set on the score-board or 'telegraph' to show the score, etc., during a match. Phr. *on the tins*, on the score-board.

1903 D. L. A. Jephson in H. G. Hutchinson *Cricket* iv. 97 Poor old Surrey in the soup again!.. The mouldy eight runs on the tins were only hoisted there by a mighty effort. **1944** E. Blunden *Cricket Country* i. 19 The call from the pavilion..sent the tins hustling up on the score-board.

d. *Squash Rackets.* A strip of metal or other material fitted along the bottom of the front wall of the court, which resounds when struck by the ball, showing it to have dropped out of play. With *the*.

[**1926** C. Arnold *Game of Squash Rackets* i. 1 On the front wall is fixed the playboard or tell tale. This consists of a piece of boarding backed with tin extending to a height of 19 inches from the ground. A ball striking this surface would not count, hence the name and the tin backing which sends forth a metallic clang when struck.] **1933** *Times* 18 Nov. 5/7 Time after time he got his opponent out of position and then, in too great a hurry to finish off the rally, put the ball on to the tin. **1960** *Times* 29 Nov. 17/4 Gordon..cast away his chances into the tin. **1973** M. Russell *Double Hit* xxv. 187 The boy aimed a stroke which missed. The fourth he returned into the tin.

3. b. *the Tins*: a nickname of the Household Cavalry (from their cuirasses).

1918 G. Frankau *Poet. Works* (1923) 181 Why ride the Tins in full review-array? 'Tis Hazeline Tredither's wedding-day! **1947** *Times* 16 Sept. 5/4 The Household Cavalry are the 'Tins', in allusion to their cuirasses; the shrapnel helmet of our day is a 'tinhat'. **1982** Barr & York *Official Sloane Ranger Handbk.* 90/1 Household Cavalry (the Tins: the Life Guards, and the Blues & Royals).

c. The badge or shield of a policeman. *U.S.*

1949 Partridge *Dict. Underworld* 725/2 *Tin*,..as 'a sheriff's badge', it is American s[lang]. **1956** 'E. McBain' *Cop Hater* (1958) xiii. 109 They reached for the leather cases to which their shields were pinned... They pinned the tin to their collars. **1975** 'S. Marlowe' *Cawthorn Jrnls.* (1976) ii. xix. 170 Mason Reed flashed the tin. 'Police officer. March right out of here.'

4. a. *tin box* (earlier example), *roof*, *trunk*, *ware* (earlier example); *tin-shop* (examples).

1723 J. Nott *Cook's & Confectioner's Dict.* sig. 1 i 8, You may boil it [*sc.* spinach] in a Tin-box, which shuts so close, that no Liquor can get in. **1912** E. Lutyens *Let.* May in M. Lutyens *Edwin Lutyens* (1980) vii. 105 It is.. very English!—to have a capital as Simla is entirely of tin roofs. **1982** M. Duke *Flashpoint* x. 69 Untidy shanties with tin roofs. **1851** H. Mayhew *London Labour* I. 336/1 The street-sellers of that order are supplied at the 'tin-shops'. **1979** *United States 1980/81* (Penguin Travel Guides) 546 The bakery, tin shop and garden house look ..as if they were still open for business. **1922** E. H. Young *Bridge Dividing* ii. i. 79 Henrietta went to her room to unpack the brown tin trunk which contained all her possessions. **1981** R. Grayson *Death of Abbé Didier* xiv. 125 He could see in the bedroom two large tin trunks. **1812** 'H. Bull-Us' *Diverting Hist. John Bull & Brother Jonathan* xiii. 91 These people are also very ingenious in making tin ware, brooms, cider-brandy, wooden bowls, and tallow candles.

b. Freq. in phr. (*little*) *tin god* (later examples). Also *tin Jesus* (only in the work of G. B. Shaw).

1909 *Our German Cousins* xv. 89 In Prussia alone there are 492 Landräte—a sort of district commissioner—all Government officials or directly in touch with the central government, and all little tin gods in their own district. **1917** S. Lewis *Job* xiii. 193 If they'd work like sixty they might get to be little tin gods on wheels like himself? **1928** R. Campbell *Wayzgoose* ii. 55 Of Tin Gods you may oft have heard or read But this one was entirely made of lead. **1930** G. B. Shaw *What I really wrote about War* xii. 368 The victorious Chauvinists..derided him [*sc.* Woodrow Wilson] as 'a tin Jesus'. **1951** E. Coxhead *One Green Bottle* v. 115 We economists are going to be the little tin gods of this generation. **1978** 'M. M. Kaye' *Far Pavilions* vi. xxxvi. 534 With luck the 'Tin Gods' who had banished him to Gujerat..would leave him alone.

c. *tin-pedlar* (earlier example), *smelter*, *-smelting* (example), *-stamping* (example); *tin-roofed* (earlier example).

1812 J. K. Paulding *Beauties Bro. Bull-Us* 53 Feather-merchants, rag-men, tin-pedlars, and horse-jockies. **1882** J. G. Whittier in *Atlantic Monthly* Feb. 145 Its tin-roofed chapel stood Half hid in the dwarf spruce wood. **1885** *List of Subscribers, Classified* (United Telephone Co.) (ed. 6) 211 Tin Smelters... Redruth Tin Smelting Co. **1977** *Whitaker's Almanack 1978* 757 Some of the country's more important industrial installations include..a tin smelter. **1885** *List of Subscribers, Classified* (United Telephone Co.) (ed. 6) 151 Anglo-American Tin Stamping Co., Limited.

d. Applied disparagingly to buildings (esp. Nonconformist churches) made partly of corrugated iron: *tin chapel*, *tabernacle* (cf. Tabernacle *sb.* 6 b), etc. Also, *tin town*.

1884 *Lichfield Diocesan Mag.* Jan. 11/2 It was decided to build 'a little bit of a tin tabernacle'. **1886** Marquess of Bute *Let.* 17 Apr. in D. H. Blair *John Patrick 3rd Marquess of Bute* (1921) ix. 154 The persistent wish of my Lord of Argyll to have what he calls an 'opening' of the tin temple in August. **1897** E. Edwards *Journey through S. Afr.* viii. 48 It would not be out of place to refer to Kimberley as a 'tin town'. **1919** A. T. Bassett *S. Barnabas' Oxford* iv. 36 This was before the 'tin' church at Cowley S. John existed. **1929** J. B. Priestley *Good Companions* i. vi. 242 That's the Station Refreshment Rooms, a tin place, just opposite. **1934** Dylan Thomas *Let.* Oct. (1966) 143 We made a tour of the pubs..drinking to the..destruction of the Tin Bethels. **1937** *New Statesman* 13 Nov. 802/2 The several designs of late-Victorian tin-chapel in the slums of a northern industrial town. **1962** 'J. le Carré' *Murder of Quality* x. 108 That parson man from the tin tabernacle. **1979** 'P. O'Connor' *Into Strong City* ii. xxix. 103 *Being born again* had become no longer a derisive tin chapel slogan but a phrase to describe what was happening to me.

5. tin-arsed *a. Austral.* and *N.Z. slang*, very lucky; **tin-back** *Austral. slang*, a very lucky man; **tin can,** (*a*) (see sense 4 a in Dict.) (earlier example); (*b*) *slang* (chiefly *U.S.*), a warship, esp. a destroyer (often, one of an older design); also applied to a submarine; **tin-canning** *N.Z.*, a greeting or serenading on a special occasion by beating tin cans; hence **tin-can** *v. trans.* (cf. Tin-kettle *v.*); **tin disease** = *tin pest* below; **tin ear,** (*a*) *slang* = *cauliflower ear* s.v. *cauliflower 2; (*b*) *colloq.* (usu. with indefinite art.), tone-deafness, aural insensitivity, esp. in phr. *to have a tin ear*; also *fig.*; hence **tin-eared** *a.*; **tin-enamel,** white tin-glaze decorated in enamel colours; hence **tin-enamelled** *a.*; **tin fish**: see *fish *sb.*[1] 1 h; **tin-glaze** (examples); **tin hare** *slang* (chiefly *Austral.*) = *electric hare* s.v. *electric *a.* 2 b; also fig. (in quot. 1941 a nickname for a train); **tin helmet** = *tin hat 1 a; hence **tin-helmeted** *a.*; **Tin Lizzie**: see *Lizzie 2; **tin-mouth** (earlier example); **tin pest,** the crumbling of pure tin that occurs at low temperatures as the ordinary white allotrope changes to grey tin; cf. *grey modification* s.v. *grey, gray *a.* 8 c; **tin printing** (see quot. 1957); **tin wedding** orig. *U.S.*, the tenth anniversary of a wedding (cf. Wedding *vbl. sb.* 2 b).

1937 Partridge *Dict. Slang* 888/1 *Tinny*, adj...Occ. tin-arsed. **1971** R. F. Brissenden *Winter Matins* 25 This tin-arsed character Hasn't been there six months before he starts To fidget, gets to grizzling in his beer. **1897** W. T. Goodge *Hits! Skits! & Jingles!* (1899) 150 And a 'tin-back.' is a party Who's remarkable for luck. **1770** G. Washington *Diary* 18 Nov. (1925) I. 442, I was to pay 6 Dollars and give them a Quart Tinn Can. **1937** *Sun* (Baltimore) 20 Apr. 15/3 Of the forty-seven destroyers out with the United States fleet..thirty-nine are the 'tin cans' of the World War days. **1957** J. Frame *Owls do Cry* ii. xix. 84 We tin-canned them and threw rice at them. **1959** *N.Z. Listener* 10 July 5/3 With us were the first warships I had seen in the heavy seas, and some of the American tin cans they gave us under Lease Lend. **1974** H. Gruppe *Truxton Cipher* iii. 31 He had noticed the Admiral's wince at Pozo's use of the archaic phrase 'tin cans' to denote destroyers. **1981** G. Markstein *Ultimate Issue* 27 'Boy, you must have been cramped in that sub.'..'Plenty of space. You'd be surprised how roomy those tin cans are.' **1926** A. F. Webb *Miss Peters' Special* vii. 62 A promoter of most of the tin-canning parties when anyone got married. **1953** M. Scott *Breakfast at Six* (1960) ii. 19 The chaps are coming up tonight. Tin-canning. The usual thing. Thought I'd better warn you. **1908** H. C. Cooper tr. *Holleman's Text-bk. Inorg. Chem.* (ed. 3) 275 It [*sc.* white tin] turns very slowly into gray tin, falling to powder, probably because of the increase in volume (this phenomenon is called the 'tin-disease'). **1965** Phillips & Williams *Inorg. Chem.* I. vii. 246 In cold countries the disintegration of organ pipes and other tin objects has..been observed, and the phenomenon was known as tin disease. **1923** *Dialect Notes* V.

239 *Tin ear on, to put a,* v. phr. To strike or beat, especially, about the head. **1935** *Peabody Bull.* (Baltimore) Dec. 42/2 A player has a 'tin ear' when his intonation is poor and his playing is mechanical. **1958** *Times Lit. Suppl.* 24 Jan. 39/2 He gives the, possibly false, impression, that he has a 'tin ear', as his countrymen put it, for many of the popular art-forms he discusses. **1962** Young & Willmott *Family & Kinship in E. London* (rev. ed.) I. vi. 101 A man with skill as a boxer, and a 'tin ear' (cauliflower ear) to prove it, had..prestige. **1978** *Times* 17 Apr. 14/4 Manson had a tin ear but..the Beach Boys recorded at least one of his songs. **1981** L. Deighton *XPD* xviii. 159 'Do you play the piano?'..'My wife insisted I get it for Billy, but that kid's got a tin ear.' **1983** *New Yorker* 28 Apr. 130/1 Who but a tin-eared organ fancier..can bear to listen to elaborate contrapuntal textures sounded in consecutive fifths? **1900** F. Litchfield *Pottery & Porcelain* ii. 14 Stanniferous or tin-enamel. **1964** H. Hodges *Artifacts* ii. 51 Today the terms faience, majolica and tin-enamel glaze are all variously applied to mean wares with a red body covered with a tin-opacified lead glaze which has been coloured with over-glaze designs. **1981** 'J. Gash' *Vatican Rip* x. 86 Forged nineteenth-century tin-enamel porcelain maiolicas. **1933** *Burlington Mag.* July 16/1 The tin-enamelled ware, façon de Pise, especially for the shelves of pharmacies and still-rooms. **1974** *Country Life* 5 Dec. 1728/1 Faience is tin-enamelled earthenware, not porcelain. **1897** C. F. Binns *Ceramic Technol.* ii. 17 These wares were uniformly coated with opaque tin glazes. **1975** *Times* 18 Feb. 13/1 Tin glaze earthenware shelve. **1934** Webster, Tin hare. **1941** K. Tennant *Battlers* 159 The 'Tin Hare's' whistle was heard in the distance. **1969** *Northern Territory News* (Darwin) Focus '69 109/1 Many top notch tin hare chasers tried at open coursing are 'left for dead' by very ordinary live hare chasers. **1934** Webster, Tin helmet. **1942** E. Waugh *Put out More Flags* i. 40 A man in a tin helmet shouted..'Take cover, there.' **1980** 'T. Hinde' *Sir Henry & Sons* i. 10 He mounted a tin helmet on the top of a rifle. **1939** N. Blake' *Smiler with Knife* xi. 156 The 'Riddle of Third Mile' i. 10 The tin-helmeted head spattered with blood. **1878** C. Hallock *Sportsman's Gaz.* 378 Sand Perch, or Bachelor Perch; called also 'Tin-Mouth'. **1900** H. C. Cooper tr. *Holleman's Text-bk. Inorg. Chem.* 301 [Tin (i.e. white tin)] turns very slowly into gray tin, falling to powder (this phenomenon is called the tin-pest). **1933** *Jrnl. R. Aeronaut. Soc.* XXXVII. 540 This change does not proceed with disintegrating effects until considerably lower temperatures, when the 'Tin-Pest', experienced in organ-pipes, during cold winters on the Continent, occurs. **1960** E. S. Hedges *Tin & its Alloys* i. 2 There seems to be a paucity of very ancient objects made entirely of tin—a fact which is sometimes laid at the door of the disintegration of tin through 'tin pest'. **1887** *Amer. Lithographer & Printer* 2 Apr. 192/3 Could you give me something practical on tin printing? **1957** *Encycl. Brit.* XIV. 214/1 Tin Printing, which was introduced about 1875, is the application of the lithographic process to the decoration of metal plate. A substantial percentage of canlabel work formerly done by the paper lithographer has in recent years gone to the tin lithographers.. Sheets of prepared tin..are fed into the press and are then oven dried in a continuous path, the procedure being repeated for each additional colour. **1968** Gloss. *Terms Offset Lithogr. Printing* (B.S.I.) 33 *Tin printing machine*, (designated) a machine for printing on sheet metal. **1863** *Harper's Mag.* Nov. 856/2 Mr Jones's people made him a tin-wedding visit on the tenth anniversary of his marriage. **1981** *N.Y. Times* 19 July II. 25/4 A tin wedding bouquet and a brooch of jewelled flowers set side by side.

‖ **tinaja** (tina·xa). [Sp.: see Tinage.] **1.** In Spain: a large earthenware jar used to hold wine, oil, olives, or salted fish or meat; in parts of Spanish America, such a jar used for storing water.

1676, 1845 [see Tinage]. **1885** *Encycl. Brit.* XIX. 629/1 The earliest kinds now existing of Spanish pottery without either enamel or glaze are chiefly large wine-jars, 'tinajas', about 3 or 4 feet high, of graceful amphora-like shape, stamped with simple patterns in relief. **1914** Gorgas & Hendrick *William Crawford Gorgas* v. 179 An assault on the water barrels, cisterns, tinajas, and dish pans of the cities of Colon and Panama involved greater difficulties. **1949** *Jrnl. N.Y. Bot. Garden* Mar. 59 The women look like animated tea cozies..loaded down..with *tinajas* (jars) or rolls of mats made of reeds. **1971** L. Boger *Dict. World Pott. & Porc.* 343/1 Tinajas were produced in all parts of Modern Spain with..little variety in design.

2. *South-west U.S.* A rock hollow where water is retained; hence, any temporary or intermittent pool.

1835 T. Coulter *Notes on Upper California* 65 The only water to be had is found..in excavations called Tinajas, made by the Indians. **1857** A. Schott *Obs. on Country along Mexican Boundary* 69 Permanent water is found under a cleft of igneous rocks, and does not properly deserve the name of a spring, but is rather a tinaja supplied by water trickling through the rocks from water-holes above. **1896** *Science* 3 Apr. 494/1 Knowledge of the few widely separated tinajas and springs was bought at the price of many lives. **1958** 'W. Henry' *Seven Men at Mimbres Springs* 55 The wells were pothole water tanks, rock *tinajas*.

tincalconite (tinkæ·lkŏnait). *Min.* [f. Tincal + Gr. κονία powder + -ite¹.] A hydrated basic borate of sodium, $Na_2B_4O_5(OH)_4.3H_2O$, occurring as a fine white powder with rhombohedral symmetry and formed by the dehydration of borax.

[**1878** C. U. Shepard in *Bull. Soc. Franç. de Minéral.* I. 144 Tincalconite (Shepard). Borax pulvérulent et efflorescent.] **1892** J. D. Dana *Dana's Syst. Min.* (ed. 6) 887 Tincalconite. **1930** *Prof. Papers U.S. Geol. Survey* No. 158. 164/2 Tincalconite can readily be made by boiling a solution of borax until crystallization ensues. **1977** *Mineral. Abstr.* XXVIII. 280/2 Tincalconite and meyerhofferite are thought to be metastable, and unsuccessful attempts to bring about their dehydration to kernite and colemanite are reported.

tinctorially, *adv.* (Earlier example.)

1895 *Sci. Progress* II. 418 In 'acid' solutions the staining principle is the acid although the dye may be a chemically neutral salt; tinctorially it reacts as a free acid.

tincture, *sb.* Add: **7. c.** An alcoholic drink, a 'snifter'. *colloq.*

1914 Joyce *Dubliners* 115 Weathers made them all have just one little tincture at his expense. **1980** Ingrams & Wells *Dear Bill* 36 Rough diamond, especially after a tincture or two.

tinder, *sb.* Add: **d.** *tinder-lighter.*

1915 V. Asquith *Let.* 16 Nov. in M. Gilbert *Winston S. Churchill* (1972) III. Compan. II. 1272 Is there anything you *haven't* got for the Front? Compass? Luminous wristwatch? Muffler? Tinderlighter? **1977** 'J. Gash' *Judas Pair* ii. 25 Flintlocks..the standard tinder-lighter of history.

tinderish *a.* (earlier Canad. example).

1837 A. Langton *Jrnl.* 8 July in *Gentlewoman in Upper Canada* (1950) 15 From her gingham never having been washed I suppose it was more tinderish than my sister's and mine.

tine, *sb.*¹ **1.** (Later examples.)

1968 J. Arnold *Shell Bk. Country Crafts* 92 The larger, called a drag rake, carrying about thirty tines compared with fifteen for the garden rake. **1978** *Cornish Guardian* 27 Apr. 10/4 (Advt.), 60in rotavator with new tines. **1979** P. Theroux *Old Patagonian Express* (1980) xiv. 289 The man jerked the tines of his fork into a slab of ham.

tinea. Add: **tinean** *sb.* (example), **tineid** *a.* and *sb.* (examples).

1842 T. W. Harris *Insects Injurious to Vegetation* 361 The Tineans..have four short and slender feelers. **1888** *Insect Life* I. 191 These insects..are cloth-feeding Tineids. **1890** *Ibid.* II. 330 The Tineid Leaf-miner..affects the younger leaves only. **1924** J. A. Thomson *Science Old & New* x. 55 There is a very interesting Tineid caterpillar, found in the tree-nest of one of the Termites. **1964** Edwards & Heath *Princ. Agric. Entom.* xiii. 285 Corn moth..is one of the most common Tineid moths which attacks grain.

tined, *a.* **a.** (Later examples.)

1886 R. E. G. Cole *Gloss. S.-W. Lincs.* 154 He was charged with stealing a steel-tined fork. **1971** *Farmers Weekly* 19 Mar. 84 There was plenty to interest traditionalists, particularly among tined implements.

ting, *sb.*¹ Add: **b.** Also *ting a ling ling* and (*rarely*) *tingating.* Cf. *tink-a-tink* s.v. Tink *int.* and *sb.*

1922 Joyce *Ulysses* 734 And he so quiet and mild with his tingating zither. **1932** T. S. Eliot *Sweeney Agonistes* 12 *Telephone*: Ting a ling ling. Ting a ling ling.

‖ **t'ing** (tiŋ), *sb.*² Also **ting.** [Chinese *tíng*.] In China: a small open pavilion, esp. in which one may rest or enjoy the landscape.

1853 *North-China Herald* 7 May 159/2 Another accommodation for travellers, called *ting*, are of more frequent occurrence. **1947** *Archit. Rev.* CII. 12/2 An island in a lake will have its t'ing, a bridge spanning the water is crowned by a t'ing, and a t'ing will invariably mark any particularly charming viewpoint. **1958** W. Willetts *Chinese Art* II. viii. 701 A small Chinese open pavilion (*t'ing*) of traditional form at Fuchow in Fukien, dating from the nineteenth century.

Ting (diŋ, tiŋ), *sb.*³ Also **Ding. a.** The name of a county in Hebei province, China, used *attrib.* to designate a type of white porcelain first made there during the Tang dynasty and perfected during the Song dynasty. **b. Ting-yao,** the name of a kiln in this county, used *attrib.* and *absol.* to denote the porcelain made there.

1904 E. Dillon *Porcelain* v. 67 In the Ting yao of the Sung dynasty..we have the oldest type of an important class of porcelain. **1915** R. L. Hobson *Chinese Pott. & Porc.* I. iv. 51 Many of the white Ting wares are thin enough to be translucent. **1933** *Burlington Mag.* June 265/1 The standard Ting ware was white..porcelain, which was either perfectly plain or decorated with free-hand carved designs. **1953** B. Gray *Early Chinese Pott. & Porc.* v. 31 To return to the Ting wares. The most characteristic Ting shape is a conical bowl on a small foot. **1958** W. Willetts *Chinese Art* II. vi. 446 North of Ting Chou and the district where Ting wares are supposed to have originated. **1971** L. A. Boger *Dict. World Pott. & Porc.* 343/2 As a rule the Ting bowls had a raw edge, as though placed in the furnace in an inverted position. **1972** *Times* 30 May 11/2 (Advt.), A carved ting yao plate. **1980** *Catal. Fine Chinese Ceramics* (Sotheby, Hong Kong) 32 A small Ding (Ting) Ware bowl with curved sides, freely carved with a lotus blossom in the interior.

‖ **ting** (diŋ, tiŋ), *sb.*⁴ [Chinese *dǐng*.] An ancient Chinese vessel, usu. bronze, having two looped handles and three or four legs (see quots.).

1904 S. W. Bushell *Chinese Art* I. iv. 80 The word *ting* is occasionally rendered 'tripod', but this is hardly applicable to a second not uncommon form which has a rectangular body of oblong section supported by four legs. **1958** W. Willetts *Chinese Art* I. iii. 138 The Han dictionary *Êrh ya* defines the *ting* as a *li* with solid legs. **1959** G. Savage *Antique Collector's Handbk.* 40 The *ting* is a bowl of hemispherical shape with three legs and two upstanding handles. **1973** *Genius of China* 12/1 In 219 BC the Ch'in emperor tried to recover from a river the nine *ting* tripods on which the power of the Chou king over his feudal subordinates was said to depend.

ting, *v.* Add: **1. c.** To announce (a person) by 'ringing in' (see Ring *v.*² 7 c). *rare.*

1880 Hardy *Trumpet-Major* II. xxiii. 157 'There, they be tinging in the passon!' exclaimed David,..as the bells changed from chiming all three together to a quick beating of one.

tingle, *sb.*¹ Add: **2. b.** A sheet of metal, usu. copper, used for making temporary repairs on a small wooden boat when it has been holed.

1909 in Webster. **1932** F. B. Cooke *Cruising Chats* xxv. 228 The best material for a tingle is a piece of thin sheet lead. **1961** B. Fergusson *Watery Maze* ix. 222 Plans should never be regarded as immutable. If they leak when first floated, it is no good patching them up until they are all tingles and no hull; scrap and start again. **1969** *Beaver* Spring 30/2 It has been suggested that these pieces [of sheet copper] may be nothing more than a tingle or patch on a ship's boat.

tingle, *sb.*³ [Abbrev. of *whelk-tingle* s.v. Whelk¹ d.] Any of several marine molluscs, esp. the rough tingle, *Ocenebra erinacea*, the smooth tingle, *Nucella lapillus*, or the American tingle, *Urosalpinx cinerea*, all of which bore holes in the shells of oysters and other molluscs.

1930 *Essex Naturalist* XXII. 299 In the autumn of 1928 samples of living tingles dredged in the River Blackwater were forwarded to me. **1959** *Times* 25 Aug. 5/6 The investigation was designed to control the tingles. **1974** P. R. Walne *Culture of Bivalve Molluscs* vi. 125 The introduction of the American slipper limpet..and the American tingle..on to the south-east coast of England..are well-documented examples.

tingler. (Earlier example.)

1829 P. Egan *Boxiana* 2nd Ser. II. 703 Johnson fell from a tingler on the left lug.

ti·ngle-ta·ngle². Also **tingel-tangel.** [ad. G. *tingeltangel* (with orig. reference to Berlin café chantant music); cf. Tingle-tangle¹.] A cheap or disreputable music-hall or night-club, esp. in Germany; cabaret.

1911 *Mariner's Mirror* I. 190/1 Those sing-song houses of ill repute, which in German and Scandinavian ports are called 'tingle-tangles'. **1939** Adeler & West *Remember Fred Karno?* 71 The music halls in Glasgow in those days were pretty rough houses. There was one called The White Bait, where the artistes were all girls, as in the Continental Tingel-Tangels. **1948** [see *peck horn]. **1972** *Sat. Rev.* (U.S.) 25 Mar. 68/2 Cabaret in Germany never managed to be counted as a major or serious artistic venture... People came to refer to cabaret by a term whose unimportance needs no translation: *Tingel-tangel*.

tingly, *a.* Delete *rare* and add later examples.

1945 B. Macdonald *Egg & I* (1946) 185 There was already beginning to be a tingly feel of autumn in the air. **1975** R. H. Rimmer *Premar Experiments* (1976) iii. 219 Arguing with Bren is like standing under a cool needle shower spray. He forces you into an alive, tingly state.

tinguaite (ti·ŋgwăˌəit). *Petrogr.* [ad. G. *tinguait* (H. Rosenbusch *Mikrosk. Physiogr. der massigen Gesteine* (ed. 2, 1887) II. 628), f. Serra de *Tinguá*, name of a spur of the Serra do Mar, W. of Rio de Janeiro: see -ite¹.] A hypabyssal rock similar to phonolite and nepheline-syenite, composed essentially of alkali feldspar, nepheline, and ægirine (acmite), the last occurring usu. as acicular crystals and conferring a greenish colour.

1893 *Mineral. Mag.* X. 173 More markedly porphyritic types are a rock with porphyritic elæolite, an 'elæolite tinguaite' with porphyritic orthoclase. **1965** G. J. Williams *Econ. Geol. N.Z.* xi. 167/2 Among these rocks is a tinguaite consisting of a network of aegirine crystals with phenocrysts of anorthoclase in a groundmass of anorthoclase, cancrinite and nepheline. **1978** S. R. Nockolds in S. R. Nockolds et al. *Petrol. for Students* xv. 173 Some tinguaites have small amounts of a sodalite mineral, of analcite.., or of cancrinite. More rarely, one or other of these becomes an important constituent.

tin hat. *slang.* **1. a.** A metal hat or helmet; *spec.* a steel helmet worn for protection against shrapnel. Chiefly *Mil.*

1903 A. M. Binstead *Pitcher in Paradise* viii. 194 A Tommy in a tin hat as I squared with a couple o'blow. **1917** W. E. Molesworth *Let.* Mar. in A. J. L. Scott *Sixty Squadron R.A.F.* (1920) iii. 38 We managed to collect some tin hats bombs,..and a few other odds and ends. **1923** *Sci. Amer.* Nov. 360/2 The trench hat, 'the old tin hat', is coming into quite extensive use as a means of head protection against small falls of rock in mines. **1932** G. Campbell *Number Thirteen* vii. 107, I happened to be dressed in a rather extraordinary rig, consisting of a tin hat, a naval monkey-jacket, grey flannel trousers,

and puttees. **1940** *War Illustr.* 5 Jan. 563 'Tin Hats' for the Heads of Britain's Defenders. **1961** Joswick & Keating *Combat Cameraman* xiv. 124 An upside-down tin hat lay nearby. **1976** 'A. Hall' *Kobra Manifesto* vi. 86 We haven't got a single tin hat in the place... Not even a blinking first aid kit!

† **b.** A general officer. Cf. *brass-hat* s.v. *brass sb.* 7. *Mil. Obs.*

1919 *Athenæum* 18 July 632/2 May I add one or two more army slang terms? 'Tin hat' or 'brass hat' for general officer.

2. Used predicatively, usu. in *pl.*: drunk.

1909 J. R. Ware *Passing Eng.* 246/2 Tin hat (Anglo-Port Said), drunk—two tin hats very drunk—three, incapable, and to be carried on board. **1926** 'Taffrail' *Pincher Martin* ii. 25 'No, sir, not drunk, only a bit shaky like,' I set, though I knowed orl the time I'd bin properly tin 'ats.' **1919** W. Lang *Sea-Lawyer's Log* 69 If you do come off tin 'ats (i.e. inebriated), go quietly below to the Mess Deck.

3. Phr. *to put the tin hat on* (*it, things,* etc.): to bring something to a (usu. unwelcome) close or climax. Cf. *to put the lid on* s.v. *lid sb.* 1 e.

1919 *Athenæum* 8 Aug. 727/2 The shrapnel helmet was invariably a 'tin hat', and 'to put the tin hat on it' is..'to ko-kush it.' **1927** 'Sapper' *Saving Clause* i. 22 This second exhibition of cowardice had put the tin-hat on. **1933** Wodehouse *Mulliner Nights* vii. 225 It was the limit, he felt, the extreme edge. It put the tin hat on things. **1943** C. Dickson' *She died a Lady* xx. 177 Next came the point that put the tin hat on it. **1977** J. M. Johnson in Douglas & Johnson *Existential Sociol.* viii. 244 He reflected that, at the time, he thought his efforts had 'put the tin hat on it'.

Hence **tin-hatted** *a.*, wearing a metal helmet or helmets.

1926 *British Worker* 10 May 2/3 At the Iron Bridge, at Canning Town, I met a half company of soldiers, tin-hatted, and with rifles and packs, marching into the docks. **1940** Hargreaves & Marsh *War begins at Home* iii. 96 A tin-hatted policeman began to push the crowd back. **1978** E. Malpass *Of Human Frailty* xiv. 232 A tin-hatted Air Raid Warden.

tinhorn (ti·nhǭn), *a.* and *sb. slang* (orig. and chiefly *U.S.*) Also **tin-horn.** [f. Tin *sb.* + Horn *sb.*; cf. quot. 1931, sense A 1 below.]

A. *adj.* **1.** *tinhorn gambler*: a cheap gambler, esp. one who acts showily.

1885 *Weekly New Mexican Rev.* 26 Feb. 4/2 We have been greatly annoyed of late by a lot of tin horn gamblers and prostitutes. **1912** *Maclean's Mag.* Mar. 478/1 He says he aint no piker, and he is a game loser, and nobody can walk around his collar, and he begins to put us all like a tin horn gambler. **1931** G. F. Willison *Here they dug Gold* 216 Chuck-a-luck operators shake their dice in a 'small churn-like affair of metal'—hence the expression, 'tinhorn gambler', for the game is rather looked down upon as one for 'chubbers' and chuck-a-luck gamblers are never admitted within the aristocratic circle of faro dealers. **1958** P. Berton *Klondike Fever* 6 A circus parade of camp-followers crowded in upon them, saloon-keepers,..tinhorn gamblers and three-card monte men. **1963** *Punch* 17 July 102/2 A Western..with..tinhorn gamblers, fisticuffs, guns and so on.

2. Interior, contemptible; pretentious, flashy. Cf. Tin-pot *a.*

1886 *San Juan* (Colorado) *Prospector* 4 Sept. 3/7 The Silverton vigilantes have notified the tin-horn element to evacuate. **1903** A. Adams *Log of Cowboy* xii. 80 A tin horn lawyer. **1935** E. Pound *Let.* 25 Sept. (1971) 276 All American Communists are, as far as I can discover, absolute bonehead, tinhorn repeaters. **1950** R. Stout *Crime & Again* xiv. 104 'You tin-horn Casanova,' she said.. 'Hitching to me that you had her, and I knew all the time you didn't.' **1977** C. Weston *House Demon* xxiii. 111 This godforsaken tinhorn paradise.

3. *tinhorn sport*: a contemptible person.

1903 Mrs. C. Foote *Shocky McCabe* ii. 34 He wasn't no Johnnie, and he wasn't no tinhorn sport. **1925** S. Lewis *Arrowsmith* v. 47 I'm a—I'm a—Martin, I'm a tin-horn sport! **1958** 'W. Henry' *Seven Men at Mimbres Springs* x. 120 The stage roads of this whole plateau are littered with the bones of tinhorn sports who didn't have the brains to fort up before morning. **1975** R. Davies *World of Wonders* (1977) I. vi. 57 Swifty Dealer, the village tin-horn sport.

B. *sb.* A poor or contemptible person, esp. one who is pretentious or flashy; *spec.* one who gambles for low stakes.

1887 F. Francis *Saddle & Moccasin* 225 The tin-horns were there in a body, with a few stacks of chips, playing light. **1908** S. E. White *Riverman* vi. 55 You ain't a tin-horn yourself? **1922** S. Lewis *Babbitt* ii. 22 I'll bet I make a whole lot more money than some of those tin-horns that spend all they got on dress-suits. **1949** *Penguin New Writing* XXXVI. 91 A guy got off at the next stop and came back for the change. A tin-horn. **1962** E. Lucia *Klondike Kate* iii. 65 Conditions produced the greatest opportunity for the tinhorns in the history of nineteenth-century gold strikes. **1977** D. Anthony *Stud Game* xiv. 180 Tony Hunter called me... 'Greetings, Tinhorn,' he said.

tink (tiŋk), *sb.*³ Chiefly *Sc.* Colloq. abbrev. of Tinker *sb.* 1 b; hence, a foul-mouthed, brawling, or disreputable person.

1857 J. Stewart *Sketches Sc. Character* 74 Nae swearing tink', nor beggar body That tak's a glass. **1894** J. B. Salmond *Dr. Bouddes* (1895) iv. 36 To sleep on the Common amon' the tinks. **1924** R. B. Cunninghame Graham *Sc. Stories* 19 Ca' yer man a gentleman? I just ca' him naething better than a tink. **1939** J. M. Caie *Twixt Hills & Sea* 58 There's kindly, honest ,eident fowk There's kyaards an' tinks forbye. **1968** A. MacLeod *Dam* i. 15 How disgusting it was to..back up a drunken tink like Sorley. **1980** D. K. Cameron *Willie Gavin* viii. 73 A fear that had driven her..in the hope that she might spot..another human soul (a tink on the road, some shepherd walking the hill fields).

tink, *v.*¹ Restrict ? *Obs.* to senses 1 and 2 and add: **3.** (Modern example.) Now *arch.*

1958 H. Hinks *Rented for Knave* 168 The glass shone. He licked it with his tongue, tapped it with a knuckle, then repeated it with his knuckles.

tink (tiŋk), *v.*³ Repr. dial. or foreign pronunc. of Think *v.*²

1767 'A. Barton' *Disappointment* I. ii. 53, I put too much confidence in dose I tought my friends, and dey deceib'd me. **1801** T. Tenney *Female Quixotism* II. xi. 117 How cou'd I tink, ma'am, it was John, in massa chamber? **1821** J. F. Cooper *Spy* II. xii. 186 'I don't tink I look a bit like me,' said Caesar. **1936** E. O'Neill *Bound East for Cardiff* in *Provincetown Plays* 1st Ser. 7 Yust tink of it! **1933** M. Lowry *Ultramarine* i. 10, I tink you are very much English all the same. **1944** in H. Wentworth *Amer. Dial. Dict.* 637 'I never tought it would happen.'..'I have to keep thinkin' about de enough I'm gettin', mumbled Bill in his best Brooklynese. **1973** *Nation* (Barbados) 23 Nov. 1/1 Yuh tink it easy?

tinker, *sb.* Add: **1. d.** *not to be worth a tinker's* (*damn* (earlier example); also *not to be worth* (etc.) *a tinker's cuss* and (ellipt.) *a tinker's*.

1839 Thoreau *Jrnl.* 25 Apr. in *Writings* (1906) VII. 78 'Tis true they are not worth a 'tinker's damn'. **1865** [see *till v.*¹]. **1891** Kipling *Light that Failed* vii. 137 The real peaceful doesn't care a tinker's—doesn't care a bit. **1947** [see *hoot *sb.*³]. **1973** *Jewish Chron.* 2 Feb. 19/3 It doesn't matter a tinker's cuss whether you amend the constitution to call the chairman president. **1973** J. Symons *Name of Annabel Lee* ii. viii. 139, I don't give a tinker's if you'll forgive the old fashioned way of putting it, who killed Ira Wolfdale.

e. A rascal, a persistently naughty child. As a term of mild contempt, usu. familiarly or playfully. Cf. Beggar *sb.* 6.

1925 R. Rees *Lake of Enchantment* 50 I'll soon settle the young tinker if he's up to his tricks. **1953** R. Lehmann *Echoing Grove* 133 Oh, you little tinker. **1974** *Tenant Joyful Condemned* viii. 124 She's a little tinker..Even you couldn't do anything with her. **1960** J. Stroud *Shorn Lamb* xxii. 247 I'm not so sure about Clement, he's a bit of a tinker at the moment. **1972** G. Sims *Deadhand* ii. viii. 141 Did the boys scare you? I expect they did. **1982** 'Saki' killed a duck and a tinker.

5. c. (Earlier example.)

1771 G. Cartwright *Jrnl.* 1 June (1792) I. 128 They killed a duck and a tinker.

5. tinker-bird, any of several African birds having a call like repetitive hammering, esp. a barbet of the genus *Pogoniulus*; **Tinkertoy** orig. and chiefly *U.S.*, the proprietary name of a type of child's construction set; a toy made of this; also *fig.*

1884 E. Ayres in R. B. Sharpe *Layard's Birds S. Afr.* (1884) 175 The note of this curious little bird so much resembles the tapping of a hammer on an anvil (having that peculiar metallic ring) that it is called in Natal the tinker bird. **1960** C. Dunsell *Zoo in my Luggage* 12 A tinker-bird was giving its monotonous cry, toink, toink.. tink.. like someone beating forever on a tiny anvil. **1914** *Official Gaz.* (U.S. Patent Office) 3 Feb. 254/2 Charles H. Pajeau, Chicago, Ill. Filed June 27, 1913. Tinkertoy. Particular description of goods.—Games, Toys, and Children's Building-Blocks. Claims use since June 12, 1913. **1935** *Trade Marks Jrnl.* 22 Dec. 1279 Tinkertoy.. Toys. Charles Hamilton Pajeau, McCormick Building, 332, South Michigan Avenue, Chicago, County of Cook, State of Illinois, United States of America; manufacturers. **1978** *Sears, Roebuck Catal.* Fall 829/1 Tinker Toy is the Wonder Builder. **1938** *Harper's Mag.* Dec. 77/2 At the Tinkertoy factory in Evanston, Illinois, I was shown a Japanese imitation of the construction sets faithful even to the trademark drawing and two typographical errors in the instruction sheet. **1972** *Newsweek* 19 June 23/3 McGovern will go into the campaign against Nixon with those Tinkertoy proposals of his.

tinker, *v.* Add: **1. b.** (Later examples const. *with.*)

1903 G. B. Shaw *Man & Superman* 193 Parliaments and synods may tinker as much as they please with their codes and creeds. **1936** *Discovery* Sept. 273/2 When the harbour is reached he will be at liberty to tinker with it [sc. a boat] to his heart's content. **1955** *Times* 2 May 13/7 Nobody is prepared to tinker with a social structure that has withstood every kind of outside pressure. **1977** J. L. Houlden *Patterns of Faith* iii. 39 Matthew often 'tinkers' with Mark's work as he received it.

2. a. (Earlier example.)

1769 J. Wedgwood *Let.* 23 Feb. (1965) 71, I have settled a plan..to Tinker all the black Vases that are crooked.

tin-kettle, *sb.* Add: **tin-kettle** *v.* (earlier example); **tin-kettling** *vbl. sb.* (earlier example).

1881 A. Bathgate *Waitaruna* xvi. 234, I was wakened by the din caused by a lot of the diggers tin-kettling the newly-married pair. **1894** B. Rock *Where Three Creeds Meet* (1897) 103 What cheering and tin-kettling Had they after at the 'settling'.

tinkle, *sb.* Add: **d.** *colloq.* A telephone call. Usu. in phr. *to give* (*someone*) *a tinkle.* Cf. *ring *sb.*³ 3 c.

1938 F. D. Sharpe *Sharpe of Flying Squad* xxiii. 241 They know as soon as we find 'em I'll give you a tinkle on the blower. **1939** N. Monsarrat *This is Schoolroom* ix. 207 Shall I give you a tinkle later? **1949** S. Ginsburg *Matchmaker* 51 And then not another word for three weeks! Not even a tinkle to ask if my cold was better! **1959** 'D. Buckingham' *Wind Tunnel* xix. 153 Shall we give Robin a tinkle and tell him that you're home? **1960** H. E. Bates *Aspidistra in Babylon* 143 Give us a tinkle. **1980** B. Bainbridge *Winter Garden* xii. 89 'Next time you're in London,' advised Ashburner, 'give me a tinkle and I'll take you to my Oxfam shop.'

e. *colloq.* An act of urination. Cf. *tinkle v.* 6.

1965 J. R. Hetherington *Selina's Aunt* 54 Tinkle (have a)... No. 1. **1974** E. Brawley *Rap* (1975) II. xiv. 239 And went over and had a tinkle. **1978** C. MacLeod *Rest you Merry* (1979) vi. 57, I was having a tinkle.

tinkle, *v.* Add: **6.** *intr.* To urinate. Cf. *tinkle *sb.* e.

1960 Wentworth & Flexner *Dict. Amer. Slang* 547/1 *Tinkle*..v.i., to urinate. Common usage by small children; humorously used by adults. **1972** *Sat. Rev.* (U.S.) 17 June 77/2 The handy man..picked the wrong moment to urinate on the roses. 'He's been tinkling on the roses for twenty-five years.' **1976** 'E. McBain' *Guns* (1977) vii. 198 I'm looking for the loo... 'I really have to tinkle.'

tinktinkie (tiŋkti·ŋki). *S. Afr.* [Afrikaans.] = *tinker-bird* s.v. *tinker sb. 5.

1874 [see *bushman, Bushman 3]. **1908** F. C. Slater *Sunburnt South* 186 The little mouse-coloured tinktinkie is a most mischievous little creature. **1956** A. G. McRae *Hill called Grazing* iii. 24 Big birds and small birds, from huge vultures..to the tiniest tinktinkies piping their thin little reeds of song.

tinman. Add: **2.** Comb. **tinman's solder,** a common low-melting solder composed of tin and lead in similar proportions, suitable for joining either of those metals; **tinmen's snips** = *tinsnips.

1937 *Archit. Rev.* LXXXI. 272/1 Solders vary in their proportions, fine solder or 'tinman's' consisting of equal parts of lead and tin and 'plumbing solder' two of lead to one of tin. **1976** *Pract. Householder* Nov. 46/1 With a soft flamed blow lamp..tin the tube end with tinmans solder. **1950** *N.Z. Jrnl. Agric.* June 563/2 Tinmen's snips are used by almost every beekeeper, as they are necessary for cutting sheet metal or wire gauze. **1974** G. Stokes *Jewelry Making* vi. 76 A small pair of tinmen's snips for shaping sheet metal.

tinned, *ppl. a.* Add: **2. a.** (Earlier example.) *tinned dog* (Austral. slang), canned meat.

1861 Mrs. Beeton *Bk. Househ. Managem.* 100 When live turtle is dear, many cooks use the tinned turtle. **1895** *Bulletin* (Sydney) 17 Aug. 27 We gave him some 'tinned dorg' and a drink. **1950** G. Casey *City of Men* 196 We'd be living in a tent and eating tinned dog. It's no place for a woman.

b. *tinned air*, air supplied by an artificial ventilation system (*Naut. slang*); also (chiefly *oc.*) air sealed in a tin for sale.

1913 *Rep. Brit. Assoc. Adv. Sci.* 1912 635 The fresh air driven in by fans through the metal conduits..is spoken of by the officers (in the battleship)..as 'tinned' or 'potted' air. **1929** F. Bowen *Sea Slang* 142 *Tinned air*, artificial ventilation. **1969** *Daily Tel.* 12 Dec. 13/3, I shouldn't think even the American who thought Brooklyn Bridge would fall for tinned air as a serious buy.

c. Of music: = *canned *ppl. a.* 2. Cf. *potted *ppl. a.* 3 b.

1924 J. Reith in *Radio Times* 23 May 1/3 The sound is metallic and unsatisfying, and..we do not sufficiently use 'tinned' music. **1976** A. Christie *Autobiogr.* (1977) iv. v. 196 There was..no 'tinned' music in those days: no broadcasting, no tape-recorders, no stereophonic gramophones.

tinny, tinnie, *sb.* Restrict *Sc.* to sense in Dict. and add: **2.** *Austral. colloq.* A can of beer.

1974 *Telegraph* (Brisbane) 2 Mar. 6/5 In olden days audiences took the equivalent of a cut lunch and a few tinnies to the theatre and expected to be entertained for hour after hour. **1978** *Sydney Morning Herald* 20 Feb. 1 Next time you feel inclined to toss that scrap of paper or empty tinny out of the car window..pause a thought. **1980** *Truck & Bus Transportation* Feb. 34/3 We doubt if the driver would have enough room on board to stow his lunch box or a couple of tinnies.

tinny, *a.* Add: **2. a.** (Earlier example of a sound.) Also applied dismissively to (a device which produces) sound of poor quality from which the lower frequencies are largely missing; cheaply contrived.

1884 *Encycl. Brit.* XVII. 831/1 The tone tends towards a certain quality which may be described as 'tinny' or metallic. **1926** *Encycl. Brit.* III. 281/2 When the low notes are dropped out, the result is 'tinny'—high-pitched, shrill, mechanical, lacking in body. **1953** A. Huxley *Let.* (1969) 377 The particular nature of the device gives to the brevity something rather tinny, something (in an undesirable sense) artificial. **1961** *Lancaster Seward's Folly* vi. 66 A tinny radio was playing pop music.

b. (Earlier example.)

1873 'S. Coolidge' *What Katy did at School* ii. 30 The cans gave the oysters a curious taste,—tinny, or was it mere like solder?

4. *Austral.* and *N.Z. slang.* Lucky. †on

the tinny luck: by a lucky chance. Cf. *tin-arsed* adj., *-back* s.v. *tin sb.* 5.

1918 *Chrons. N.Z.E.F.* 7 June 205/1 Remarks are heard on the 'tinny' luck. **1919** W. H. Downing *Digger Dialects* 50 *Tin*, *Tinny*, lucky. **1947** L. Doolan *Opportunity in Australia* 90 Tinny—lucky. **1951** D. Stivens *Jimmy Brockett* x. Landfall V. 168 And this one's yours, Edith. Hey, you're tinny, aren't you? **1969** G. Slatter *Gun in Hand* xvii. 229 He's tinny because some people are tinny and always win. **1978** O. White *Silent Reach* xvii. 173 You'll have to be pretty tinny to pin down those blokes.

Hence **ti·nnily** *adv.*, with a tinny sound.

1927 J. Masefield *Midnight Folk* 198 He had no sooner wished, than invisible someones came silently, blocked up the approach to Otter's lair, tinnily reported, 'Entrance blocked securely', and disappeared. **1954** M. Sharp *Gipsy in Parlour* iii. 133 A bell above my head rang tinnily. **1982** A. Desai *Clear Light of Day* iv. 171 Teacups clinked on the saucers, tinnily.

tin pan. (In Dict. s.v. Tin *sb.* 5.) Add: **1.** (Earlier examples.)

1806 *Austin Papers* (1924) I. 102, 1 doz. Tin pans. **1843** *Knickerbocker* XXII. 50 With discordant fife and old tin-pans for drums.

2. A cheap, 'tinny' piano. Cf. *tin-panny a.* *U.S. slang.*

1882 C. Parker *Amat. & Prof. Stage Life* viii. 156 It was now Linwood's turn, and with a wail, that sounded like 'Oh, if I only had a decent piano!' he went back to the old 'tin-pan' again.

3. Special Comb. **Tin Pan Alley** *colloq.* (orig. *U.S.*), the world of the composers and publishers of popular music; also applied *loosely* to a district where song publishing houses abound, *spec.* (formerly) in New York in 28th Street and in London around Denmark Street (see *Denmark).

1908 *Hampton's Broadway Mag.* Oct. 456/2 Oh it's a world in itself, is Tin Pan Alley. It has its laughter and its tears. **1909** *Busy Man's Mag.* Jan. 48/1 Down Twenty-eighth Street, which is known as 'Tin Pan Alley', a dozen music publishing houses grind out new songs 'by hand' daily. **1926** Whiteman & McBride *Jazz* viii. 161 Like everybody else, I think of the Alley as a 'Tin Pan Alley'. As a matter of fact, Tin Pan Alley exists now only as a tradition. **1934** *Des Denmark Street* s.v. *Denmark]. **1945** A. Bellow *Dangling Man* 132, I guess she sees herself in Tin Pan Alley, her face streaked with tears. **1962** Radio & B. Janis *They all played Ragtime* (1958) xi. 220 While Tin Pan Alley was squeezing ragtime dry, a few people made a lot of money and a great number of people made a little. **1979** P. O'Connor *Into Strong City* I. xvii. 61, I found Seven Dials... Then Denmark Street. Tin Pan Alley it belongs in the pages of the Melody Maker.

tin-panny *a. U.S.*, of a piano: tinny-sounding. Cf. *tin pan 2.

1904 J. C. Lincoln *Cap'n Eri* ii. 30 On the platform of one [shop] a small crowd was gathered, and from the interior came shouts of laughter and the sound of a tin-panny piano. **1931** G. G. Russell *Speech & Voice* iii. xv. 158 The high partials become 'metallic' like the tin-panny piano.

tin-pot. 4. (Earlier example.)

1838 *Remarks* G. *Taylor's Factory Strike* 5 in *Pattie's Mod. Stage* II, Mr. Taylor, is a patriot in his little tin pot way.

tinsel, *sb.*² and *a.* Add: **5.** Also similative, as *tinsel-pink*, *-violet.*

1920 E. Sitwell *Wooden Pegasus* 49 As I, a puppet tinsel-pink, Leap on my springs. **1966** D. Barnham *One Man's Window* vi. 67 The hills are tinsel-violet with distance, encrusted with the Valetta buildings and almost encircled by the blue waters of Grand Harbour.

6. b. Special Comb.: **Tinseltown,** a nickname for Hollywood; also *transf.*, the supposedly glittering world of Hollywood cinema; the Hollywood 'myth'.

1975 *Bookseller* 16 Aug. 1305/1 The tinseltown stuff when Wodehouse won the applause of the theatre-going fans. **1984** *Times* 3 Mar. 8/7 When a filmmaker starts cherishing the natural roar of traffic on the soundtrack..you know the believes in Tinseltown.

tinsnips, *sb. pl.* Also **tin snips.** [f. Tin *sb.* + Snip *sb.* 8 (*pl.*).] A pair of hand-held clippers used for cutting (sheet) metal. Cf. *tinmen's snips* s.v. *tinman 2.

1944 *Living off Land* vii. 131 Materials required are: four sound kerosene tins, tin snips, soldering-iron and flux. **1947** J. Conroy *Midland Humor* 182 Ain't you got no tinsnips in this town?.. Get a pair of tinsnips, extra large. **1964** C. Willock *Enormous Zoo* v. 85 Our work..consisted of cutting off lengths of heavy gauge wire with tin snips. **1976** *Conservation News* Sept./Oct. 24/1 As a parent I am very wary of allowing a child not conscious of danger at fixing an electric plug but keen about with tin snips.

tint, *sb.*¹ Add: **1. c.** *Hairdressing.* An artificial colouring, less permanent than a dye, applied to enhance the colour of the hair; an application of this.

1921 [see *tinter d]. **1957** *Encycl. Brit.* VII. 496a/2 The tint..is only temporary and is not a dye in the true sense. **1979** 'M. Hebden' *Pel & Faceless Corpse* x. 109 What is it you wanted? Tint? Shampoo? Set? Cut?

tint, *v.* Add: **d.** *trans. Hairdressing.* To colour (the hair) with a tint. See *tint *sb.*¹ 1 c.

1921 [implied at *tinter d]. **1966** J. S. Cox *Illustr. Dict. Hairdressing* 149/2 Tint, to dye. When the many-hued fashion... to one of the many euphemisms employed in the hairdressing craft. **1977** A. Morice *Scared to Death* xvii. 119 I'm going a bit grey... So I have it tinted three or four times a year.

ti·ntable *a.* [-able], capable of being tinted.

1974 *Spartanburg* (S. Carolina) *Herald* 18 Apr. (Kmart Advt. Suppl.) 10 Washable latex acrylic is tintable to hundreds of colors! **1979** *Chatelaine* (Canada) Jan. 95/2 Tintable, so the bifocals with no lines. They're featherweight and tintable.

tin-tack. Add: **a.** (Earlier example.)

1839 Dickens *Nicholas Nickleby* xxxv. 346 A..parcel of tin tacks and a very large hammer.

b. Colloq. phr. *to come* (*or get*) *down to tin tacks* = *to come* (*or get*) *down to brass tacks* s.v. *brass sb. 5 b. (Found only in the work of G. B. Shaw.)

1932 G. B. Shaw *Pen Portraits* (1932) 183 Keats..had he lived, would no doubt have come down from Hyperions and Endymions to tin tacks as a very full-blooded modern revolutionist. **1949** —— *Buoyant Billions* (1950) iii. 45 Let us get down to tin tacks. Is Clemmy going to marry him or is she not?

tinted, *ppl. a.* Add: **a.** (Earlier *fig.* example.) Also, coloured in a manner specified by defining word in Comb.

b. Coloured so as to reduce the strength of light, e.g. *tinted glass*.

1756 *Crit. Rev.* II. 340 The French author is not much oblig'd..to his English translator. We meet with..tinted ideas, propell'd, devastated, bilious, and many others of this kind. To what language these most properly belong the translator best knows, most certainly not to the English. **1804** Southey *Poet's Pilgr.* i. iv. 92 The autumnal-tinted groves. **1905** *Proc. R. Soc. Vic.* LXIV. 528 A similar experiment was tried on eyes with 3-ply red-tinted glasses. **1909** *Westm. Gaz.* 1 July 14/1 Frowning heights the outline of which stood out dark and desolate against the orange-tinted sky. **1931** *Index Catal. Libr. Surg.-Gen.'s Office, U.S. Army* XVI. 272 (*heading*) Spectacles (tinted) etc. **1977** E. Lemarchand *Let or Hindrance* xiii. 160 A small brisk woman with auburn-tinted hair. **1978** *Newcastle Guardian* 14 Mar. 32/6 (Advt.), 1974 Ford Capri..black vinyl roof, tinted glass.

tinter. Add: **d.** *Hairdressing.* One employed to tint hair. Cf. *tint *v.* d.

1921 *Dict. Occup. Terms* (1927) § 920 Tinter..washes and applies tint or colour to human hair..or to one of the heads or models of wig maker. **1966** J. S. Cox *Illustr. Dict. Hairdressing* 10 Hg. Words denoting a person engaged in some aspect of the Hairdressing or Wigmaking crafts... tinter.

tinticite (ti·ntikəit). *Min.* [f. *Tintic* (see quot. 1946) + -ite¹.] A hydrated basic ferric phosphate, $Fe_6(PO_4)_4(OH)_6.7H_2O$, found as whitish masses of submicroscopic crystals.

1946 B. Stringham in *Amer. Mineralogist* XXXI. 395 A creamy white clay-like substance with unusual physical and chemical properties was found as a vault coating in a limestone cave near the Tintic Standard Mine in the Tintic Mining District, Utah. The chemical analysis and x-ray comparative data show the mineral to be new and it has been named tinticite. **1967** *Virginia Jrnl. Sci.* XVIII. 160 Secondary minerals, formed primarily by weathering processes, include aragonite..rozenite..and tinticite.

tintinnid (tinti·nid). [a. mod.L. family name *Tintinnidæ*, f. generic name *Tintinnus* (F. von P. von Schrank *Fauna Boica* (1803) III. 305): see Tintinnabulum and -id³.] A ciliated protozoan of the family Tintinnidæ or the order Tintinnina, often distinguished by a bell-shaped test. Also *attrib.*

1945 M. F. Glaessner *Princ. Micropalæont.* ii. 22 Apart from their stratigraphical value..the Tintinnids..are used by stratigraphy..they are of some interest in view of the stratigraphical analogy between *Calpionella* and the chitinous tintinnid tests suggestive of taxonomic relations. **1953** Shrock & Twenhofel *Princ. Invertebr. Paleontol.* (ed. 2) ii. 67 Fossil tintinnids have been found in Pleistocene peat bogs. **1970** *Nature* 25 July 381/1 Micropaleontological..dating... In..Calpionellid... **1974** *Nature* 8 Feb. 382/2 The pattern of deposition..with respect to tintinnid..were fine, well preserved..zooplankton. **1983** *Jrnl. Protozoology* XXVI. 415/1 The process of lorica building by tintinnids is poorly understood.

tinto, *sb.*¹ Delete † *Obs.* In recent use, short for *vino tinto* s.v. *vino 3 or [Pg.] *vinho tinto* s.v. *vinho (= red wine); a glass or drink of this.

1958 K. Amis *I like it Here* x. 128 What about some wine? Will you have the *branco* or the *tinto*? **1978** M. Walker *Infiltrator* iii. 35 We'll get to one of the Galician bars..and I ordered two tintos.

Tintometer. Add: This is a proprietary term. (Further examples.)

1893 J. W. Lovibond *Measurement of Light & Colour Sensations* 3 This work is a record of some investigations on light and colour carried out whilst..I found it necessary to provide a colour-meter, which the author was perfecting a colorimeter, to which he terms 'The Tintometer'. **1957** G. E. Hutchinson *Treat. Limnol.* I. vi. 416 Kalle (1938), in his studies of the color of the sea, has introduced a tintometer which permits the color to be expressed as a single equivalent wave length. **1976** *Trade Marks Jrnl.* 27 Aug. 1188/1 Tintometer. B856,033. Colorimeters. The Tintometer Limited... 1st November, 1963.

tintype. Add: **a.** (Earlier example.)

1864 E. W. Pearson *Lett. from Port Royal* (1906) 243 You will probably in due course..see the tin-types of Rose and Flora.

b. Colloq. phr. (orig. *U.S.*) *not on your tintype*, certainly not. Cf. *not on your Nelly* s.v. *nelly².

1900 Ade *Fables in Slang* 78 Oh, rats! Not on your Tintype. **1918** *Christ. Herald* 23 Sept. (1969) 52 'Not on your tintype', he said: as Uncle George ecstatically would have said. **1934** C. Stead *Seven Poor Men of Sydney* iii. 100 Does the Pope keep beggars, or the Vatican police hand out alms? Not on your tintype! Not by Joe. **1963** P. H. Johnson *Night & Silence* xvii. 209 'No, no... she couldn't make a bracelet for three gentlemen...' 'I got standards... Two gents, just about. Three, not on your tintype.' **1970** S. J. Perelman *Baby, It's Cold Inside* 118 Let's eschew all pious cant to the effect that it turned to ashes in my mouth. Not on your tintype—it was nectar.

tiny, *a.* (*sb.*) Add: **A.** *adj.* **b.** *tiny garment*, an article of clothing made for an expected baby.

1925 'E. Queen' *Fourth Side of Triangle* i. 12 Her 'needlework'..consisted of 'tiny garments', prepared for a lay sisterhood which aided 'unfortunate' young women. **1978** R. Freeling *Night Lords* i. 9 She..had managed to get herself pregnant..but there was no display of tiny garments.

c. *tiny mind*: esp. when used self-deprecatingly, a term of abuse suggesting an absence of common sense. Chiefly in colloq. phr. *out of one's tiny mind*, an emphatic form of *out of one's mind* s.v. *mind sb.* 19 a.

1959 'W. Haggard' *Mad Tick* xii. 68 Why don't you use your tiny mind? **1970** K. Benton *Sole Agent* vii. 71 We'd had a row. I was out of my tiny mind. **1976** *New Yorker* 23 Feb. 98/1 What doesn't my tiny mind? **1977** D. Beaty *Excellency* xix. 225 Everyone with the possible exception of H.E. was scared out of their tiny British minds.

B. *sb.* **2. a.** (Earlier example.)

1797 F. Burney *Jrnls. & Lett.* July (1973) III. 326 He..hesitated before he could persuade himself to give at once to my bigger Children, if they came accompanied by tinies.

b. A nickname for a very large or tall man. Cf. *tich.

1924 *Literary Digest* 18 Apr. 40 A big fat guy will be called 'Babe' or 'Tiny'. **1976** *Burnham-on-Sea Gaz.* 20 Apr. 12/8 He is 6ft 7in tall and not unnaturally, is known as 'Tiny'.

tinzenite (ti·nzĕnəit). *Min.* [ad. G. *tinzenit* (J. Jakob 1923, in *Schweiz. mineral. und petrogr. Mitteil.* III. 234), f. *Tinzen*, name of a mountain in Graubünden, Switzerland: see -ite¹.] A basic silicate of calcium, manganese, aluminium, boron, and iron that occurs as triclinic yellow crystals and is now regarded as a member of the axinite group.

1924 *Chem. Abstr.* XVIII. 3337 Tinzenite: yellow; radial platy masses in quartz. **1968** *Amer. Mineralogist* LIII. 2099 The name 'tinzenite' must be used for those axinites in which Ca is less than 2Mn and Fe (but usually Mn > Fe). At this stage it is impossible to say whether tinzenite is an independent mineralogical species or only a variety.

Tio Pepe (ti·o pe·pe). [a. Sp., lit. 'Uncle Joe'; cf. *Tia Maria.] The proprietary name of a dry Spanish fino sherry. A glass or drink of this.

1886 *Trade Marks Jrnl.* 15 Sept. 953 Tio Pepe... Gonzalez, Byass, & Co., London, E.C., Wine and Spirit Merchants. **1907** *Yesterday's Shopping* (1969) 951 'Tio Pepe', very dry, delicate—per doz. bots.. 48/6. **1920** G. Saintsbury *Notes on Cellar-Bk.* ii. 18 Some of the finer kinds are really supernacular—the best 'Tio Pepe', for instance. **1943** G. Greene *Ministry of Fear* ii. i. 135 He..took out a glass and a bottle of sherry... 'Tio Pepe,' Digby said. **1955** R. Fitzgerald *House is Falling* xi. 177 They had brought in a bottle of Tio Pepe to keep them company. **1968** *Times* 18 Dec. 18 (Advt.), Tio Pepe Apr. 141/2 Gonzalez Byass & Co. Limited... Tio Pepe... First use 1905. **1978** R. MacInnes *Agent in Place* xi. 187 Get Mackenzie a Tio Pepe.

tip, *sb.*¹ Add: **2. a.** (Earlier example with reference to shoes.)

1840 H. Moseley *Let.* 11 Feb. in D. Mozley *Newman Family Lett.* (1962) III. 86 They danced very prettily, though he had 'tips'.

c. Used in *pl.* to denote the leaf-buds used in tea-making, preceded by an adj. or trade-name to designate a particular brand of tea.

1897 *Sears, Roebuck Catal.* 8/1 Golden tips. **1952** 'W. Cooper' *Struggles of Albert Woods* II. iv. 98 He said: 'I give him tips'. **1978** *Listener* 16 Nov. 642/3 (caption) Come back to my pad, man, I've got some amazing PG tips.

f. Formerly, a band of (gold, etc.) paper round a cigarette at the end held by the lips; now, = *filter *sb.* 3 c. Cf. *gold-tipped* adj. s.v. *gold¹ 10 a.

1897 Kipling *Captains Courageous* v. 107 Cigarettes with gold-leaf tips. **1981** *Times* 25 July 3/8 Filter cigarettes were..assumed to be safer than those with plain

5. c. *arse over tip*: see *arse *sb.* 1 d.

6. e. *Aeronaut.* with reference to the extremity of an aerofoil, as *tip loss, speed, stall, stalling, tank;* **tip-touch** *v. trans.*, to touch with the tips of one's fingers.

1938 *Jrnl. R. Aeronaut. Soc.* XLII. 39 The engine r.p.m. can with advantage be increased until the tip speed of the airscrew approaches the speed of sound, at which speed there are serious tip losses which reduce the thrust. Thin bladed metal airscrews show less tip loss due to high speed than the thicker sections. **1966** *Gloss. Aeronaut. & Astronaut. Terms* (B.S.I.) iv. 15 *Tip loss*, loss of lift. A failure of the tip of an aerofoil associated with the formation of tip vortices. **1921** R. M. Pierce *Dict. Aviation* 231 Tip speed, the oscillatory speed of the tip of a reciprocating wing; the up-and-down velocity of a wing-tip in flapping flight. **1960** *Jrnl. R. Aeronaut. Soc.* XLV. 505, A measure of the high tip speed of the windmill beat our photographer. **1969** *Gloss. Aeronaut. & Astronaut. Terms* (B.S.I.) *Sect.* 1, 18 *Tip speed*, the velocity of the tip of a rotor multiplied by the rotor radius. **1946** *Jrnl. Brit. Interplanetary Soc.* VI. 55 The phenomenon of tip stall is brought about by spanwise drift in the boundary layer over a swept wing. **1957** *Jrnl. R. Aeronaut. Soc.* XLI. 205, Increasing wing-tip loading..one of the most efficient means known at present to prevent tip stalling of highly tapered wings. **1933** *Wall St. Jrnl.* 25 Aug. 5 First..for extra fuel the Sabrejet carried standard as auxiliary fuel containers for the early-day F-80 jet fighters. **1977** *R.A.F. Yearbk.* 11/1 Max range with tip tanks, 900 mls. **1902** Joyce *Ulysses* 330 Must I tiptouch it with my nails? **1976** H. Gold *Man who was not with It* (1965) i. 6 Trickster with her pretty little hand tip-touching the black-and-blue spot.

tip, *sb.*³ Add: **A.** Also, a present of money given to a schoolboy by an older person.

1810 W. Hickey *Mem.* (1960) x. 164 Joseph Pott... whom I had frequently called upon, and tipped at Eton School. **1909** G. B. Shaw *Genevra* ii. 38, I have exchanged twenty words with the boy since I tipped him when he was going from Eton to Oxford.

tip, *sb.*⁴ Add: **d.** (Earlier example.)

1847 *Punch* 16 Oct. 138/1 You attach him for making himself conspicuous at the sale of Shakspeare's house. You seem to think he has missed his tip.

e. *tip-sheet* (orig. *U.S.*), *-slinger* (*Austral. slang*).

1943 *Sun* (Baltimore) 21 Feb. 12 (*caption*) Tip sheet. **1955** *Sci. News Let.* 20 Aug. 126/3 Tip sheets may feature a rectangular description of the long-shot horse's heart before long. **1972** *Daily Tel.* 14 Nov. 18 A tip sheet on ways of fitting in smoothly in America has been handed to the 2,000 guests..the tipping term for race-course tipster. **1944** *Ban Bulletin* (Sydney) 3 Mar. 27/3 He's a tip-slinger as a regular patron..up to the races. Not hurt..some. **1934** *Bulletin* (Sydney) 15 Aug. 49/1 By their conversation most of them were tipslingers or urgers.

tip, *v.*¹ Add: **1. b.** Also *spec.*, to glance or touch with the edge of the bat. *tip-and-run*: (earlier examples); also *transf.* in *attrib.* use, esp. to designate short, sudden attacks in war; also as *v. intr.*

1846 W. Lambert *Cricketer's Guide* (ed. 3) iii. 43 It is..to such [balls] as are just tipped with the edge of the Bat. **1968** P. Thomson *Cricket* xii. 215 His first play was tip and run. **1887** R. H. Lyttelton *Cricket* 74 Mr. J. Pycroft *Cricket Field* x. 185 Put in two skeletons 'tip and run'. **1858** 'G. Forrest' *Playground* ix. 137 If you only tip the [ball] ball, it will go to fair for chucking out you. The trouble of striking it. **1918** *Chambers's Jrnl.* June 391/2 Where was that 'tip-and-run' raid—the enemy's favourite 'tip-and-run' daily raids. **1922** *Rep. Commissioner Police Metropolis* 1926 6 The sudden rise in..'tip-and-run' raids on jewellers' shops. **1942** *R.A.F. Jrnl.* 13 June 26 The Italians, with their half-hearted enthusiasm and their 'tip-and-run' type of bombing. **1946** J. W. Day *Harvest Adventure* xvi. 273 At Mersea Island..we enjoyed the Committee's folly in placing a machinery dump within a hundred yards of the sea and tip-and-run raids.

c. *U.S. Sport.* To hit (a ball, puck, etc.) into the net or goal with a (light) touch.

1958 G. F. Pinholster *Encycl. Basketball* ix. 111 The player with the best position tries to tip in the goal as the other two players block for him. **1965** T. J. A. Lionne *How to play and teach Basketball* vi. 131 The tip-in shot is used when a player jumps in the air and tips the ball into the basket. **1976** *Sunday Mail* (Brisbane) 31 Oct. 57/1 Mulder times his jump so that he can lightly tip the ball with the fingers of his right hand. **1968** [see *rap *v.*¹ 2 a].

c. *to tip one's hand*(*s*) (or *mitt*): to disclose one's intentions inadvertently. *slang* (orig. and chiefly *U.S.*).

1917 G. Ade *Let.* 8 July (1973) 67 For a time in the play it should appear that the plans of the smooth citizen are working out perfectly. He becomes confident and even praises himself, 'tips his hands', so to speak. **1929** *Sat. Even. Post* 16 Feb. 6/3 Carter did not tip his hand. **1930** In an effort to tip the balance, New Zealand began to take British shopgirls and hairdressers. **1935** *Jrnl. Amer.* 25 Apr. 34 I'd tip my mitt. **1948** *New Republic* 26 Oct. 33/1 That would be a certain tip off. That guy's probably got a red under his coat. **1938** *New Republic* 26 Oct. 33/1 That would be tipping our hand. **1966** M. Woodhouse *Tree Frog* xvii. 133 We couldn't very well oppose it without tipping our hand. **1979** *Economist* 17 Nov. 12/1 Mr Hunt will not tip his hand on the price at which he will buy more bullion.

d. *Bookbinding.* *to tip in*, to attach a single leaf, often an illustration, to the neighbouring leaf of a book by a thin line of paste down its inner margin.

1928 S. Unwin *Truth about Publishing* v. 131 Should an extra page..be needed, it may have to be separately printed and specially 'tipped' or 'pasted in' as a frontispiece often is. **1949** Melcher & Larrick *Printing & Promotion Handbk.* 289/2 The leaf to be tipped in is first run with a narrow line of paste along its back edge. **1962** H. Williamson *Methods Bk. Design* (ed. 2) xix. 322 So too engravings..are tipped..separately in the text is concerned, the plates must be separately tipped in. **1976** N. Wright in *Times Lit. Suppl.* 6 Jan. 1/2 Hackett's..magnificence being..were tipped..in. **1974** Tipped in here is a clipping from a magazine, with a notation in the margin in WW's hand.

5*. To dispose of or kill (a person). Also *fig.* Cf. sense 9 and *to bump off* s.v. *bump *v.*¹ 2 c.

1930 W. Camp *Football without Coach* vii. 129 Time after time methods such as these have 'tipped off' keen football players and have spelled the failure of good plays. **1958** *Evening News* 18 Aug. 11/5 Jake's sort o' done me a good turn, getting himself tipped.

tip, *v.*⁴ Add: **1. b.** (Earlier example with *up*, and later *absol.* example.)

1829 P. Egan *Boxiana* 2nd Ser. II. 13, I shall expect, before we part, that you will tip up my half of the prize. **1865** *Sunday Times* (Colour Suppl.) 7 Nov. 41/2 For 'first two year the tipped up, she give me her wage packet and I give her her spending money.

c. (Earlier and later examples of presenting a schoolboy with money.)

1810 W. Hickey *Mem.* (1960) x. 164 Joseph Pott... whom I had frequently called upon, and tipped at Eton School. **1909** G. B. Shaw *Genevra* ii. 38, I have exchanged twenty words with the boy since I tipped him when he was going from Eton to Oxford.

tip, *v.*⁵ Add: **1.** (Earlier example of the simple vb.)

1886 E. Dowson *Let.* 16 Nov. (1967) 117 Ye gods what of the 'Marchester Nov.' I have been tipped (i) Lady Rosebery (ii) Goldseeker (iii) Phil—(by you).

2. Also, to warn, alert, or inform (a person); to make known or give away (someone or something). Freq. const. *off. slang* (orig. *U.S.*).

1893 L. W. Moore *His Own Story* xxi. 292 This was 'tipped off' to me on Thursday, and also that the arrest of the whole party was to be made. **1904** *Ibid.* xxxii. 445 He must have been tipped off. **1909** 'I. Y. Ryan' *Poker* 12 'I have he had 'tipped me off' to bet, I said, 'Look at me, for I am the best'. **1930** C. H. Sewell *Dossin Feb.* 7/1 I'd 'tipped off' the situation made Mrs. Jones particularly downhearted. **1899** S. Crane *Monster* xxix. 76 I told him to keep tipped..shut. You know how he'll go all over town yapping about the thing. I thought I'd better tip you. **1939** 'S. Luck' *Wake Gangs* 2nd ser. *London* xv. 136 He was doing badly and was tipped off with three or four times. **1958** *Harper's Mag.* Feb. 70/2 Marks that have tipped. **1955** R. Gilbert *Sky High* xiii. 184 That one [crime] we were tipped off about that put a dragnet. **1960** M. Spark *Bachelors* i. 163 'Someone has tipped the police,' said Mike Garland. **1964** McLuhan *Understanding Media* (1967) 111. xxxi. 339 There could be no better tip-off..the Wild Swan have been pointed out by others. **1966** S. Gilbert *Sky High* xiii. 184 That one [crime] we were tipped off about but put a dragnet. **1975** T. Allbeury *Special Collection* xiv. 96 Was there any mileage in tipping them off? [sometimes pr. **1977** *Sunday Times* Mar. 7/5 Ring from tipping..to manage the money. **1978** G. McDonald *Fletch's Fortune* xix. 130 Who tipped you off? The one that you about the editorial, and the campaign?

tip-. Add: **tip-horse** (example).

1912 A. Bennett *Matador of Five Towns* 207 The old horse-car..climbing hills with the aid of a tip-horse and a boy perched on the back thereof.

tip-in. [f. vbl. phr. *to tip in*: see *tip *v.*¹ 1 c and *v.*¹ 2 d.] **1.** *Bookbinding.* = *paste-in sb.*

1958 G. F. Pinholster *Encycl. Basketball* ix. 111 The player with the best position tries to tip in the goal as the other two players block for him. **1965** T. J. A. Lionne *How to play and teach Basketball* vi. 131 The tip-in shot..is used when a player jumps in the air and tips the ball into the basket.

2. *U.S. Sport* (esp. Basketball). A score made by tipping a rebound into the basket or net.

1958 G. F. Pinholster *Encycl. Basketball* ix. 110 Tipping a basketball into the basket is a spectacular feat of timing, coordination, and jumping ability... Several tip-ins in a game can mean the difference between victory and defeat. **1963** [see *tip *v.*¹ 1 c]. **1966** [see *lay-up 3]. **1980** *Washington Star* 17 Dec. 83 Theus..led all scorers with 23 points, including a tip-in with 1:51 [left] that put the Bulls in front.

tipiti (tipiti·). [a. Tupi.] A strainer used by Amazonian Indians for expressing the poisonous juice of the cassava.

1960 Mayne Reid *Odd People* 52 A long elastic cylinder-shaped basket or net, of the bark of the 'jacitara' palm (*Desmoncus macracanthus*). This is the tipiti. **1866** Lindley & Moore *Treas. Bot.* I. 396/1 Indians use strips of the [*Desmoncus*] to make their tipitis or strainers used for squeezing out the poisonous juice of the mandioc root. **1952** G. Sarton *Hist. Sci.* I. i. 5 The South American tipiti is an elastic plaited cylinder of jacitara-palm bark which is used to express the juice of the cassava.

tiple (ti·ple). [a. Sp., lit. 'treble'.] Any of various high-pitched stringed instruments played in Spain and the Spanish-speaking parts of the Americas, each resembling a small guitar.

1942 N. MacDonald *Orchid Hunters* xx. 125 He strummed lightly on the tiple, changed to one of the quick-tempoed, popular Colombian dances. **1965** A. Marcuse *Mus. Instruments* 525/2 *Tiple*... 1. syn. of guitarrillo 2. In Cuba a small bandurria with 5 pairs of strings..tuned. **1976** *New Yorker* 12 Apr. 111/1 He played the tiple (a stringed instrument between the guitar and the ukulele).

ti·p-off. Also **tip off, tipoff.** [f. vbl. phr. *to tip off*: see *tip *v.*⁵ 2.] **a.** Information, a 'tip', esp. about criminal activity; a hint or 'give-away'. *slang* (orig. *U.S.*).

1901 J. F. Willard *World of Graft* 164 'So much down now,' he said, 'and so much when the show's over.' Otherwise it's a tip-off and pinch.' **1928** H. C. Witwer *From Baseball to Boches* iv. 145 Nobody knew we was coming—I'd tip to front... The first real tip off was when they served out the identification tags. **1928** E. Wallace *Sooper* xii. 236 Now he had something under his hat; a tip-off straight from headquarters, news of high international importance. **1945** *Richmond* (Va.) *Times-Dispatch* 23 May 8 One of the first tipoffs on this came from the press. La Bacall followed 'The Leer' to New York was more months ago. **1955** W. Tucker *Wild Talent* ix. 151 Karen was driving the tip-off car. **1982** *Observer* 24 Jan. 5/2 There was a tip-off available about when it [sc. a bank raid] was going to be stacked up with cash. **1977** *Birds* Winter 15/2 There was a tip-off about the robbery and RSPB staff lay in wait near the nest.

b. A person supplying information, esp. in connection with criminal activity. *slang.*

1941 M. Allingham *Traitor's Purse* xiv. 157 Or the other little syndicate..must have a tip-off in the police somewhere. **1962** J. Welcome *Beware of Midnight* xiii. 166, I don't know which of the servants here is a tip-off to the Secret Police, but one of them is. **1977** R. Busby *Pattern of Violence* vi. 90 They've got a good tip-off on the inside.

2. [f. Tip *v.*¹ 1 a, after *kick-off*, etc.] *Basketball.* A method of (re)starting play, in which two opposing players contest a jump-ball; an instance of this.

1924 [see *jump-ball, jump ball s.v. *jump-]. **1937** F. C. Allen *Better Basketball* 181 On the tip-off, a player should jump for the ball. **1964** and the other..tip-off. **1977** *Evening Gaz.* (Middlesbrough) 11 Jan. 13/9 Loughborough are the visitors to Newton Aycliffe on Sunday for a 3 p.m. tip-off.

tippable (ti·păb'l), *a.* [f. Tip *v.*² and *v.*⁴ + -able.] [Tip *v.*⁴] Designating one who may be tipped, or who is open to tips or *douceurs*. Occas. *absol.* as *sb.*

1907 *Sat. Rev.* 21 Dec. 752/2 The great aim of the tippable is to squeeze. **1923** S. Graham *Europe—Whither Bound?* 27 He is..plucked staring upon a queue of all the rabble of the East.

2. [Tip *v.*²] Of seats, etc.: that can be tilted or tipped up.

1936 *Times* 15 Oct. 8/4 The Austin Seven seems fresher than ever... In its latest form the front tippable seats have a rearward slide.

tipped, tipt, *ppl. a.*¹ Add: **4.** = *filter-tipped* adj. s.v. *filter *sb.* 5. Also *ellipt.*, filter-tipped cigarettes.

1964 M. Drabble *Garrick Year* ii. 27 'I don't smoke tipped', I said. *Ibid.* 28 He let me have the other Gauloise, and smoked the tipped cigarettes himself. **1972** *Guardian* 24 June 9/6 Have you got any cigarettes?.. Anything tipped. **1977** 'A. Hall' *Beauty Gentleman of Road* ix. 81 The gamut of grades of the cigarette-ends, ranging from..the complete cigarette left in a throw-away packet..to the abhorred tipped variety and the trodden-out stub.

5. *Bookbinding.* Of a leaf, plate, etc.: inserted in a book by attaching to another leaf with a narrow strip of paste at the inner edge. Usu. with *in*. Cf. *tip *v.*¹ 2 d, *tip-in 1.

1912 A. J. Philip *Business of Bookbinding* 217/2 *Tipped-in*, when a leaf, illustration, map, etc., is pasted in without guarding it it is said to be tipped in. **1958** J. Carter *ABC for Book-Collectors* 175 *Tipped in*, slightly attached by a thin strip of paste at the inner margin. **1960** J. Betjeman *Summoned by Bells* v. 47, I wished the book to have a more tipped-in colour plate. **1962** H. Williamson *Methods Bk. Design* (ed. 2) xix. 322 Unless the leaves to be..tipped..separate from the plates..will..tipped. **1977** W. Matheson in *Q. Jrnl. Libr. Congr.* July 233/2 The collector's directions in the endpapers, when tipped-in plates or any of the plates by Augustus Peck are sometimes a tipped-in leaf.

tippee. Add: Also **tipee. 2.** [f. Tip *v.*⁵] One who receives inside information about a company or business enterprise and uses it to trade profitably in stocks and shares. orig. and chiefly *U.S.*

1961 L. Loss *Securities Regulation* (ed. 2) III. ix. 1451 To hold 'tippees' liable under Rule 10b-5 when they had no reason to suspect that their informant was an insider might result in an unreasonable entrapment of innocent persons. **1967** *Federal Suppl.* CCLVIII. 284/2 This is strong circumstantial evidence that Darke must have passed the word to one or more of his 'tippees' that drilling on the Kidd 55 segment was about to be resumed. **1973** *N.Y. Law Jrnl.* 23 July 5/5 New rules for tippors [*sic*] and tippees. **1978** *Times* 12 Oct. 29/1 What about so called 'tipees'—people who come by price sensitive information often because of a breakdown in security by a professional adviser or within the company? **1980** *U.S. Reports* CCCCXLV. 242 It [*sc.* the SEC] did not hesitate to extend *Cady, Roberts* to reach a 'tippee' of a Government insider.

tipper[1]. Add: **2. a.** (Earlier example.)
1861 *MacMillan's Mag.* Dec. 143/2, I got some work in Sussex, as a 'tipper'.

c. (Examples without comb. element.) Freq. *attrib.*
1920 *Glasgow Herald* 18 Apr. 10 The farmer can get on with his work, and the waggon which, in preference to being a 'tipper' would have a moving lattice floor and a removable drum..would spread the manure direct on the fields. **1950** *Engineering* 3 Feb. 140/1 Specialised vehicles, such as tippers. **1955** *Times* 14 June 4/5 There had been a shortage of tipper vehicles for emergency purposes. **1977** *Jersey Even. Post* 26 July 25/7 (Advt.), Tipper lorry (high sided), immediately available for contract work, etc. **1979** *West Lancs. Even. Gaz.* 10 Sept. 12 (Advt.), Tipper driver required. **1983** *Truckin' Life* Aug. 70/3 I've watched some poor little tipper struggling up a 5 per cent grade in second gear with two or three tonnes of topsoil.

tippet (ti·pėt), *v.*[2] [perh. alteration of TIPTOE *v.*] To move on tiptoe. Also *transf.* and *fig.* Hence **ti·ppeting** *vbl. sb.*
1916 W. DE LA MARE *Songs of Childhood* (new ed.) 3 See they're tippeting at the door; Their wee feet in measure falling. **1932** E. BOWEN *To North* vii. 72 Then someone's wife opened a cold piano: she tinkled, she tippetted, she struck false chords and tried them again. **1934** —— *Cat Jumps* 84 Her affronted, muddled and rather tippeting manner. **1944** R. LEHMANN *Ballad & Source* 258 Their wives tippet about on the bank in high heels.

tipping, *vbl. sb.*[1] Add: Also **tippen. 2. b.** = TIPPET *sb.* 4 a.
1881 W. GREGOR *Folk-Lore N.E. Scotl.* 52 Lines, hair for tippens, hooks. **1924** *Chambers's Jrnl.* Oct. 710/1 Many of the hooks have been torn from their tippings.

4. *Bookbinding.* (See quot. 1931.) Usu. with *in.* Cf. *TIP *v.*[2] 2 d; *TIP-IN 1.
1931 A. ESDAILE *Man. Bibliogr.* 183 *Tipping-in*; pasting the edge of a single leaf to the next leaf. **1963** W. CLOWES *Guide to Printing* i. 8 Sometimes..tipping-in might not give sufficient strength. **1966** [see *TIPPED, TIPT *ppl.a.*[1] 5].

tipping, *vbl. sb.*[2] Add: **1.** (Earlier example.)
1853 DICKENS in *Househ. Words* 7 May 218/2 'Tippings'..denotes the spiritual movements of the tables and chairs.

3. d. *tipping machine.*
1877 *Scotsman* 1 Sept. 4/7 Tipping machine.

tipple, *v.*[1] Add: **5.** *intr.* To rain heavily; to gush, to pour. Freq. const. *down.*
a **1930** D. H. LAWRENCE *Last Poems* (1932) 204 Now it is almost night, from the bronzey soft sky Jugfull after jugfull of pure white liquid fire, bright white Tipples over and spills down. **1968** J. PORTER *Dover goes to Pott* xiv. 177, I saw her from the office window. No hat, no coat, nothing... It was tippling down too, absolutely tippling. **1971** *Country Life* 27 May 1283/1 After getting out of bed on the wrong side because the day is overcast or rain is tippling down.

tippy, *a.*[1] (*sb.*) Add: **1. b.** (Earlier example.)
1790 A. M. WOODFORDE *Let.* 3 Sept. in *Parson Woodforde Soc. Jrnl.* (1972) V. III. 55 Your Bonnets are quite the *Tippy.*

† **c.** as *sb.* A dandy. *Obs.*
1798 *Monthly Mag. & Brit. Reg.* VI. 173/1 His dress..will be, elegant; exhibiting no articles of apparel but such as are 'All the rage', he is 'Quite the tippy'. **1844** 'J. SLICK' *High Life N.Y.* II. 92 You wouldn't ketch one of our York tippies at that, let alone a ginuine Lord.

tippy, *a.*[2] Add: (Later example.) *U.S.*
1923 E. F. WYATT *Invisible Gods* III. i. 93 A tippy, wire-legged table.

tippy-toe (ti·pi,tǝᵘ), *sb.* (*adv.*, *a.*). Also **tippi-toe.** [Alteration of TIPTOE, TIP-TOE *sb.* (*adv.*, *a.*): cf. -Y[6].] **A.** *sb. pl.* The tips of the toes. Usu. in phr. *on* (one's) *tippy-toes.* Occas. also as collect. *sing.*
1899 *Century Mag.* Nov. 47/2 The whole court now stood on its tippy-toes. **1965** *New Statesman* 3 Dec. 897/2 Illustrations..show the dear little mite standing on tippy-toe to feed famished and deserted nestlings. **1980** *Dirt Bike* Oct. 68/1, I stood on tippi-toes to watch.

B. *adv.* Short for *on tippy-toes* (see sense A above).
1901 'ZACK' *Dunstable Weir* 216 The rocking stone stud tippy-toe above his girt shadder. **1975** R. HELMS *Tolkien's World* iv. 126 A rather vulgar sugar-iced concoction, with a doll tippy-toe on its pinnacle as the Fairy Queen.

C. *adj.* Standing or walking on tiptoe. Also *fig.*
1951 J. STEINBECK *Log from 'Sea of Cortez'* p. xli, Ed would be smiling and doing his tippy-toe mouse dance. **1968** *Courier-Mail* (Brisbane) 8 Nov. 1/5 He [*sc.* Richard Nixon] is not as cautious and tippy-toe as he appeared to many voters. **1980** S. T. HAYMON *Death & Pregnant Virgin* ix. 71 When Jack Ellers was excited he rose up on the balls of his feet... Now he came into the dreary room..all tippy-toe.

ti·ppy-toe, *v.* Also **tippie-toe.** [f. prec.] *intr.* To go on tiptoe, to move lightly. Also *fig.* Cf. TIPTOE *v.* 2.
1901 'ZACK' *Dunstable Weir* 232, I tippy-toed back to the fire. **1942** C. MORLEY *Thorofare* xxxvi. 169, I tippy-toed down one side to join the Confed'racy while my old man was pretending to look up the other. **1974** *Globe & Mail* (Toronto) 29 Oct. 6/6 Did he tippy-toe across the press? **1980** *Daily Tel.* 2 Sept. 11/2 The roly-poly little girl who tippy-toed through Sandy Wilson's enchanting 'Big Best Shoes' number in 'Valmouth' all those years ago.

Hence **ti·ppy-toed** *ppl. a.* (in quot. as quasi-*adv.*).
1938 M. K. RAWLINGS *Yearling* iv. 31 A deer track'll prove the same. A deer or bear that's fat and heavy'll sink in that-a-way [at the heel]. A lettle ol' light doe or yearlin' 'll walk tippy-toed, and you'll not see more than the front of their hooves.

tipster. Add: **1. a.** Also *attrib.*, as *tipster sheet* (U.S.). Cf. *tip-sheet* s.v. *TIP *sb.*[4] e.
1933 *Sun* (Baltimore) 17 July 1/2 Warning against the use of 'tipster sheets' and 'market service' by stock market and other investment patrons was issued today.

tip-tap *sb.* (*a.*, *v.*). Add: **d.** as *adv.* With a tapping sound.
1911 H. S. WALPOLE *Mr. Perrin & Mr. Traill* vi. 104 He came tip-tap across the floor to him.

tip-top, *sb.* Add: Also **tiptop. 3.** *Angling.* A line guide on a fishing-rod. *N. Amer.*
1961 *Washington Post* 5 Feb. c 6/6 A tiptop for flyrod and casting rod. **1971** *Islander* (Victoria, B.C.) 5 Sept. 5/3 Many cases of broken lines can be traced to cracks or nicks on a fishing rod's guides or tiptop.

Hence **ti:p-to·pmost** *a. colloq.*, (*a*) highest; (*b*) best; **tip-topper** (earlier example of a person).
1937 G. FRANKAU *More of Us* ii. 26 Clashed home the gates. Slow to tip-topmost storey Groaned lift. **1960** *Guardian* 22 Apr. 8/5 All their tip-topmost British merchandise. **1829** P. EGAN *Boxiana* 2nd Ser. II. 239 Some *tip-toppers* on the *Corinthian list* were witnessed getting over the ground as 'gaily as larks'.

tipulid, *sb.* (Examples.)
1951 C. N. COLYER *Flies Brit. Isles* xxv. 317 It will seem a far cry from the large, long-winged, long-legged Tipulids. **1976** *Nature* 22 Jan. 251/2 There is no evidence that hen grouse do eat tipulids during incubation.

tip-up, *sb.* and *a.* Add: **3.** (Examples.)
1850 S. F. COOPER *Rural Hours* 42 The boys call these contrivances 'tip-ups', from the bit of stick to which the line is attached, falling over when the fish bite. **1880** *Harper's Mag.* Mar. 517 With baited lines and tip-ups set, we waited. **1923** H. E. WILLIAMS *Spinning Wheels & Homespun* 247 Fishing through the ice... With the aid of what they call 'tip-upses' several lines can be used simultaneously. **1978** *Globe & Mail* (Toronto) 1 Feb. 34/1 Anglers using jigs were more successful than those waiting by a tipup.

4. A tip-up seat (see sense B in Dict. and Suppl.).
1966 'A. HALL' *9th Directive* xxvi. 234 The other sat on the tip-up behind the front seat.

B. *adj.* Esp. as *tip-up seat* (later examples); also with reference to a method of opening, as *tip-up door.*
1936 N. STREATFEILD *Ballet Shoes* xiii. 203 Mr. French pulled down the tip-up seat next to him. **1959** *News Chron.* 7 July 6/5 A local vicar, whose garage had tip-up doors. **1973** J. DRUMMOND *Bang! Bang! You're Dead* xxxvii. 127 Garages with tip-up doors. **1980** *Oxf. Diocesan Mag.* May 8/1 They are halls with tip-up seats designed for secular entertainment..as well as worship.

tirade, *sb.* **2.** (Earlier example.)
1806 SCOTT *Let.* Sept. (1932) I. 321 Tales they had heard in infancy with here & there a tirade really taken from an old poem.

tire, *sb.*[2] Add: In the twentieth cent. the revived spelling *tyre* has become standard in the British Isles, whilst American English retains the traditional *tire.* **3.** *tire pressure, repair, track*; **tire chain,** a metal chain designed to be attached to the tyre of a motor vehicle to prevent skidding on snow or ice; **tire-iron**: also (*N. Amer.*), a length of steel flattened at one end, used as a lever for removing tyres from wheel-rims.
1917 *T. Eaton & Co. Catal.* Spring & Summer 282 Every car owner should carry a pair of tire chains. **1980** 'E. MCBAIN' *Ghosts* ii. 20 Carella could hear the sounds of tire chains jangling. **1952** R. ELLISON *Invisible Man* xxv. 426 You could hear that gun striking that ole shield like somebody dropping tire irons out a twelve-story window. **1976** *Globe & Mail* (Toronto) 16 Feb. 10/1 Two 19-year old brothers have been arrested after a man, his wife and son were assaulted with tire irons in a Towers Department Store parking lot. **1920** *T. Eaton & Co. Catal.* Spring & Summer 224 Tire pressure gauge. Accurate and reliable for correct air pressure. **1895** *Montgomery Ward Catal.* Spring & Summer 556/3 Pneumatic tire repair outfit. **1975** J. GRADY *Shadow of Condor* viii. 128 The man..stowed the tire-repair items..in the trunk. **1947** E. S. GARDNER in *Amer. Mag.* Aug. 150/3 Mason, studying the tire tracks, said, 'It was an automobile and a horse trailer.' **1973** T. PYNCHON *Gravity's Rainbow* I. 113 The Dutch resistance will then 'raid' this site, making a lot of commotion, faking in tire-tracks and detailing the litter of hasty departure.

tired, *ppl. a.*[1] Add: **1. a.** Also in phr. *to make* (someone) *tired*: to get on the nerves of, irritate. *slang* (orig. *U.S.*).
c **1883** C. H. HOYT *Bunch of Keys* I, in *America's Lost Plays* (1941) IX. 13 That makes me tired! **1904** S. E. WHITE *Blazed Trail Stories* iv. 65 Such talk made Daly tired, and he said so. **1925** W. J. LOCKE *Great Pandolfo* xiii. 172 'Women like you,' said Myrtilla a trifle sourly, 'make me tired.' **1950** P. WOODRUFF *Island of Chamba* viii. 124 They make me tired... Things are bound to get worse.

c. *tired Tim* (or *Timothy*), usu. associated with *weary Willie*: the names of two tramps, characters in the comic magazine *Illustrated Chips*; hence both used as nicknames for tramps or other work-shy people. Also *attrib.*
1906 *Daily Chron.* 15 Feb. 3/5 Heroes of the Tired Timothy stamp. **1927** W. E. COLLINSON *Contemp. Eng.* 27 Comic papers..brought home to us the picturesque language of Weary Willy and Tired Tim (the genial tramps—whence these words are frequently used as appellatives for 'tramps' in general). **1930** H. HERD *Diagnosis of Mental Deficiency* 10 Mental defectives are the 'weary Willies and tired Tims' *par excellence.* **1932** W. S. CHURCHILL *Let.* 6 Feb. in Ld. Boothby *Recoll. Rebel* (1978) vi. 86 These two old tired Tims of the Commons have ceased to command my allegiance. **1972** J. PORTER *Meddler & her Murder* xii. 157 With Miss Jones in..her Tired-Tim-and-Weary-Willie mood, there was no temptation to linger.

d. *the tired business man*: a cliché, often used with satirical allusion to the short working hours and pleasure-loving habits popularly ascribed to business men.
The phrase is said to have been used by Mark Twain in 1896.
1913 *Vanity Fair* Nov. 37/2 'I mean simply this, my dear,' replied the Tired Business Man. **1927** A. HUXLEY *Proper Studies* 186 From the fetish-worshipper to the metaphysician, from the tired business man to the mystic..every type of human being can find in Catholicism the spiritual nourishment which he or she requires. **1940** I. BROWN in *Best One-Act Plays of 1939* 141 Oh, how I've longed to be a Tired Business Man once more—office at ten, out at twelve-thirty, back at three, sign the letters, off home! **1969** *Listener* 20 Mar. 399/1 We often use the cliché of the tired businessman to define the low response..that sustains leg-shows.

2. a. Also, of language, literature, etc.: hackneyed, trite.
1951 *Chambers's Jrnl.* Sept. 521/1 The start of the paper was promising enough, for Greenwood collected a group of writers around him equal to deserving that tired word 'brilliant'. **1956** *Sat. Rev.* (U.S.) 30 June 34/1 It [*sc.* a book] is (to use a tired phrase) history made interesting. **1966** *Listener* 28 Apr. 630/3 *The Pipeline*..turned out to be as tired and cliché-ridden a spy story as any I have heard.

b. Of food, flowers, etc.: limp with long exposure, no longer fresh. Of clothes: crumpled, shapeless, or baggy with long wear.
1897, 1909 [in Dict.]. **1933** *N. & Q.* 26 Aug. 130/1 To-day people speak of stale vegetables or fruits as 'tired'. **1934** E. BOWEN *Cat Jumps* 252 Tired dance dresses. **1947** H. NICOLSON *Diary* 9 May (1968) 97 The spring-garden has lost its early bloom... The primroses are looking a trifle tired. **1958** *Spectator* 15 Aug. 222/1 The last batch of eggs I got from the local grocer turned out to be very tired. **1963** *Times* 11 June 10/6 In the present heat, merchants are reluctant to pay high prices for 'tired' fish. **1974** A. LURIE *War between Tates* (1977) v. 117 He is ill-dressed in a tired grey turtleneck sweater and sagging work pants. **1977** G. MARTON *Alarum* 16 Chris read..occasionally munching on a very tired cheese sandwich.

tiringly (tǝiᵊ·riŋli), *adv.* [f. TIRING *ppl. a.*[1] + -LY[2].] In a tiring manner, to a wearisome degree.
1894 E. FAWCETT *New Nero* xx. 219 'It's a trifle tyrannical, is it not?' 'Yes; amusingly so.' 'Never tiringly so.'

Tir-na-nog (tiᵊ·ɹnanǭᵘg). *Irish Mythol.* Also **Tir-nan-og, Tir-n-an-oge,** etc. [a. Ir. *Tír na nÓg* land of the young.] A fabled land of perpetual youth, an Irish version of Elysium. Also *transf.*
1889 W. B. YEATS *Let.* 29 July (1954) 132 The Irish peasant's notion that Tir-n-an-oge (the Country of the Young) is made up of three phantom islands. **1898** E. C. BREWER *Reader's Handbk. Allusions* (new ed.) 590/2 The ancient inhabitants of Erin had..the vague belief that there somewhere existed a land where people were always youthful..and lived for ever. This country went by various names, as *Tir-na-nóg*, etc. **1906** P. E. MORE *Shelburne Ess.* 4th Ser. 245 It is the Tir-nan-og of the Celts, the country of the young. **1938** L. MACNEICE *I crossed Minch* II. x. 155 Don't talk to me about the Isles of Youth. These are the Isles of Senescence, of Inactivity. ..I do not want to sleep or dream of Tir n'an Og. **1955** *Bull. Atomic Sci.* Mar. 82/3 Without a better balance, the science-created *Tir na nog*, the legendary Irish land of youth on earth, must result in the rigid Malthusian principle which population experts sternly predicted.

1980 *London Mag.* Mar. 23 The old Irish myth of Tir nan-Og, the land of everlasting youthfulness.

tirodite (tirǭᵘ·dǝit). *Min.* [f. *Tirodi*, name of a village in Madhya Pradesh, India + -ITE[1].] A monoclinic mineral, $Mn_2Mg_5Si_8O_{22}(OH)_2$, of the amphibole group which forms a series with dannemorite ($Mn_2Fe_5Si_8O_{22}(OH)_2$); also, any member of this series having more magnesium than iron.
1938 DUNN & ROY in *Rec. Geol. Survey India* LXXIII. 295 (*heading*) Tirodite, a manganese amphibole from Tirodi, Central Provinces. **1973** *Indian Jrnl. Earth Sci.* I. 38/2 The colour of tirodite is distinctive as opposed to the vivid colours shown by other manganese-bearing amphiboles occurring in the area.

tirra-lirra. Add: Hence also as *v. intr.*, to sing tirra-lirra.
1879 G. MEREDITH in *New Q. Mag.* July 83 Duchess Susan was distinguished coming across a broad, uncut meadow, tirra-lirraing beneath a lark.

‖ **Tirthankara** (tiᵊɹþa·ŋkarā). Also **Tirthankar, tirthankara, Tirthanker.** [Skr., lit. 'maker of a ford', f. *tīrthá* ford, passage + *kará* maker.] In the Jain region, one of the twenty-four founding prophets or Jinas, venerated as having successfully crossed the stream of time and having made a path for others to follow.
1835 J. WILSON *Let.* 13 Mar. in G. Smith *Life of John Wilson* (1878) vi. 205 In the inferior parts there are the images of all the twenty-four Tirthankars. **1881** *Encycl. Brit.* XIII. 543/2 The Jains count twenty-four such prophets, whom they call Jinas, or Tirthankaras, that is, conquerors or leaders of schools of thought. **1901** KIPLING *Kim* vi. 159, I'd give a month's pay to hear how he explained it all at the Tirthankers' Temple at Benares. **1961** A. J. TOYNBEE *Between Oxus & Jumna* xii. 37 At Ludra..there is a Jain temple containing a hallowed image of the last Jain tirthankara but one. **1971** *Illustr. Weekly of India* 11 Apr. 8/1 (*caption*) Rishabha and Vardhamana are the most honoured among the twenty-four Tirthankaras.

'tis. Add: Also in dial. and colloq. use. (Later examples.) Also **'tes.**
1896 I. T. THURSTON *Well Won* iv. 39 'Henderson, is that true?' demanded Gordon sternly... For once, Henderson absolutely looked ashamed of himself..as he said sulkily, 'Yes, 'tis.' **1922** JOYCE *Ulysses* 58 'Lovely weather, sir.' ''Tis all that.' **1922** E. O'NEILL *Anna Christie* III. 177 'Tis quare, rough talk, that—for a dacent girl the like of you! **1932** S. GIBBONS *Cold Comfort Farm* iv. 53 'Tes the cowshed! 'Tes our Feckless openin' the door fer me! **1977** P. HILL *Liars* (1978) xii. 154 'Tis your business if'n you want to waste your time.

tisane (tiza·n). [Mod. re-adoption of Fr. *tisane*: see PTISAN, which it has largely supplanted.] A medicinal tea or infusion made from herbs.
1931 W. CATHER *Shadows on Rock* I. iv. 29 He kept them away from doctors,—gave them tisanes and herb-teas and poultices. **1941** W. FORTESCUE *Trampled Lilies* v. 52 A communicating room could be used as a kitchen..where hot chocolate, coffee, and *tisanes* could be prepared for the men. **1959** *News Chron.* 6 July 6/5 Old ladies..drink herb teas in France, where they are called tisanes. **1965** *Punch* 7 July p. xii/2 The health food shop with..lime flower tisanes and heather honey. **1981** M. GEE *Dying, in Other Words* xlix. 114 And the tea, the lime-flower *tisane* which was good for her chest and smelled citrous and fresh, singing to her when she drank it of blue summer skies over yellow-green lime trees.

Tiselius (tisē·liŭs, tiz-). *Biochem.* [The name of A. W. K. *Tiselius* (1902–71), Swedish biochemist.] *Tiselius* (*electrophoresis*) *apparatus*: an apparatus in which electrophoresis is carried out in free solution in a U-tube (see quot. 1964).
1939 *Jrnl. Franklin Inst.* CCXXVIII. 798 (*heading*) U-tube portion of the Tiselius electrophoresis apparatus. **1946** *Nature* 13 July 41/2 The Tiselius electrophoresis apparatus is now established as an essential part of the equipment of protein chemists... Its first appearance.., in 1937, was the result of a careful technical study. **1964** G. H. HAGGIS et al. *Introd. Molecular Biol.* ii. 23 In the Tiselius apparatus a potential is applied across a boundary between the solution containing the proteins and a protein-free buffer solution (moving boundary electrophoresis).

‖ **Tisha b'Av** (ti·ʃā bǫv). Also **Tisha be-Ab, Tisha Bov,** etc. [Heb. *tiš'āh bǝ'āḇ.*] The ninth day of the month Av, on which both the First and the Second Temples are said to have been destroyed: observed by Jews as a day of mourning.
1938 *Vallentine's Jewish Encycl.* 2/1 The 9th of Ab (*Tisha be-Ab*..) is a fast day commemorating the destruction of the 1st and 2nd Temples. **1958** A. L. EISENBERG *Story Jewish Calendar* 19 Later on, the three months in which Hannukah, Purim, and the fast day of Tisha B'Av occur were also included. **1970** *New Yorker* 19 Sept. 32/2 My grandfather always fasted on Tisha b'Av and slept with his head on a stone. **1973** *Synagogue Light* Sept. 12/1 On Tish B'Av [*sic*], the traditional day of mourning, we take off our shoes. **1978** I. B. SINGER *Shosha* I. ii. 10 The day that Zelig and Bashele moved..was like Tisha Bov for me.

tisicky (ti·ziki), *a.* Also **tissicky, tizzicky.** [dial. var. of PHTHISICKY *a.*] Wheezy, asthmatic; also *transf.*, delicate, squeamish.
1905 E. PHILPOTTS *Secret Woman* II. x. 193 Once a labourer have gone in the back an' thighs, an' growed tisicky in the breathing parts—then [etc.]. **1924** *Western Daily Press* (Bristol) 18 Mar. 8/3 A person troubled with a slight but frequent and annoying cough is said to be Tissicky. **1961** 'K. NORWAY' *Waterfront Hospital* iii. 44 Men are more tissicky than women when it comes to telling people the worst. They're kinder by nature. **1969** J. CLARKE *Foxon's Hole* iii. 23, I can see you've a good appetite. None of Frances's tizzicky ways.

'tisn't (ti·zěnt), dial. or colloq. shortening of *it isn't* (= it is not); see IT *pron.* A. γ; NOT *adv.* 2, 3. Cf. 'TIS.
1803 G. COLMAN *John Bull* IV. i. 42, I be but the guide, and 'tisn't for I to go first. **1888** KIPLING *Under Deodars* 93 Remember, Bobby, 't isn't the best drill..it's the man who knows how to handle men. **1924** M. KENNEDY *Constant Nymph* II. viii. 117 ''Tisn't yours,' cried Antonia... 'It ought to be mine.' **1972** P. CLEIFE *Slick & Dead* I. i. 14 Oh, come now—'tisn't always like that.

tissue, *sb.* Add: **6. a.** (Later examples.)
1880 J. DUNBAR *Pract. Papermaker* 32 (*heading*) Lilac tissue, deep shade. **1937** E. J. LABARRE *Dict. Paper & Paper-Making Terms* 244/1 *Tissue* or tissue-papers are fine, thin, soft papers made of strong materials such as rag and hemp fibres... They are usually unsized, nearly transparent, chiefly used for wrapping and protective purposes. **1977** J. HEDGECOE *Photographer's Handbk.* 309 Carefully trim the print, with its attached tissue.

b. *Racing.* A sheet of paper showing the 'form' of the horses competing in a race (see also quot. 1866).
1866 *Daily Tel.* 24 Feb. 3/4 A 'tissue' is a slip of paper written for a telegraph company, showing results of betting transactions and accounts. **1914** JOYCE *Dubliners* 59 No one knew how he achieved the stern task of living, but his name was vaguely associated with racing tissues. **1972** G. F. NEWMAN *You Nice Bastard* ii. 83 Manso quickly got a bet on the fifth and sixth, and studied the tissue for the previous races.

c. A piece of soft absorbent paper used as a handkerchief, for drying or cleaning the skin, etc. Hence as *v. trans.*, to wipe with a tissue.
1929 *Punch* 10 Apr. p. xv (Advt.), Two or three times every day you should massage the hands with Ponds' Cold Cream, removing the cream after a minute or so with a Ponds' Cleansing Tissue. **1938** [see *COMPLEXION *sb.*]. **1958** M. DICKENS *Man Overboard* x. 162 Ben grabbed a make-up stick and scrawled it [*sc.* an address] on the side of a box of tissues. **1960** *Woman* 25 Apr. 2/1 Pond's Cold Cream..goes on moisturising long after you tissue it off. **1976** M. & G. GORDON *Ordeal* (1977) 142 Sniffling, he asked Penny for a tissue. **1981** *Economist* 8 Aug. 79/1 The battle against the common cold may not be over... So do not throw away your tissues yet. **1983** *Harrods Mag.* Spring/Summer 72 Yellow Herbal Astringent is sprayed on..then tissued dry.

d. A cigarette paper. *Austral.* and *N.Z. slang.*
1952 *Here & Now* (N.Z.) Jan. 32/2 Better go and see if the parole-jumper in Number 8 has got any tissues left. **1966** G. W. TURNER *Eng. Lang. Austral. & N.Z.* viii. 164 In Hobart the [expression]..'Got a tissue, mate?' [is commoner than elsewhere]. A *tissue* is a cigarette paper.

9. b. *tissue-specificity*; *tissue-dwelling, -specific* adjs.; **tissue-bank** [*BANK *sb.*[3] 7 f], a place where a supply of human or animal tissue for grafting is stored; **tissue culture,** a culture [*CULTURE *sb.* 3 c] of cells derived from tissue; the practice of culturing such cells; **tissue fluid,** extracellular fluid which bathes the cells of most tissues, arriving via blood capillaries and being removed via the lymphatic vessels; **tissue type** *Med.*, a class of tissues all of which are immunologically compatible with each other; **tissue-type** *v. trans.*, to determine the tissue type of; **tissue typing** *Med.*, the assessment of tissue in order to predict its immunological compatibility with other tissue, esp. prior to transplantation.
1968 *Punch* 14 Feb. 239 Donald Pleasence plays the night attendant at a central tissue-bank in Montreal... A stupid, illiterate man with inexplicable operatic aspirations, he thinks that if he can only get the right larynx he will be able to sing. **1971** *New Scientist* 8 Apr. 101/2 Tissue Banks where human and animal tissues could be readily obtained. **1912** *Anat. Rec.* VI. 91 The character of the growth in tissue cultures varies primarily with the kind of tissue used. **1926** J. S. HUXLEY *Ess. Pop. Sci.* 283 A fundamental experiment from which sprang the whole sub-science of tissue-culture. **1955** *Sci. News* XXXVI. 8 It is interesting to compare the events in regeneration with what happens in tissue culture. **1975** *Daily Tel.* 8 Sept. 8/4 At present, if a dog or other animal is sick or dies it takes several days to grow the virus in tissue culture to be sure rabies is to blame. **1964** M. HYNES *Med. Bacteriol.* (ed. 8) xxviii. 443 The tissue-dwelling parasites which cause relapses [in malaria] are not affected. **1974** *Ciba Symposium* XX. 309 Few drugs have any significant action against its tissue-dwelling amastigotes. **1900** E. H. STARLING *Elem. Human Physiol.* (ed. 4) vii. 292 This absorption depends on the small proportion of proteid contained in the tissue-fluid as compared with the blood-plasma. **1954** S. DUKE-ELDER *Parsons' Dis. Eye* (ed. 12) i. 4 It [*sc.* the cornea] has no blood vessels with the exception of minute arcades, about 1 mm. broad, at the limbus so that it is dependent for its nourishment upon the diffusion of tissue-fluid from the vessels at its periphery and materials from the aqueous humour. **1976** D. JENSEN *Princ. Physiol.* ix. 524/1 The interstitial (or tissue) fluid forms the actual internal environment of the body. **1962** *Sci. Survey* III. 224 This type of change may be associated with the changes in tissue-specific antigens. **1932** J. S. HUXLEY *Probl. Relative Growth* VI. iv. 177 The tissue-specificity is apparently the same..in both sexes. **1967** *Science* 25 Aug. 942/1 The first two explanations should be tested more critically if applied to a single tissue type. **1968** *Times* 7 Nov. 3/2 In a year or two it might be possible to store human hearts for a period of hours; this would enable donors and recipients to be tissue-typed on an international basis. **1969** *Private Eye* 6 June 3/2 Experts from Guys Hospital came to tissue type her to see if she was a 'suitable donor'. **1971** H. FESTENSTEIN et al. in R. Y. Calne *Clin. Organ Transplantation* vi. 158 It may be possible to tissue type potential recipients from several hospitals in one central laboratory. **1973** *Daily Tel.* 27 Feb. 2/7 Simon has a tissue type shared by only one in 50,000 of the population. **1965** *Israel Jrnl. Med. Sci.* I. 498/2 This seems..a hopeful avenue toward the goal of tissue typing. **1967** *Observer* 26 Nov. 1/5 Research on tissue-typing has reached the stage where tissues from different people can be matched (just as blood can be matched) so that grafts will 'take' without resort to drugs to suppress the immune mechanism. **1971** *New Scientist* 8 July 63/2 One or two of these cases, particularly when recipient and donor have been well 'matched' by tissue typing, have been spectacularly successful.

tissuey (ti·siui, ti·ʃ¹ui), *a.* [TISSUE *sb.* + -Y[1].] Having the quality or texture of tissue.
1867 G. MEREDITH *Vittoria* III. xlii. 196 Letting her..crumble the black tissuey fragments to smut in her hands. **1965** P. WYLIE *They both were Naked* I. i. 4, I could see that overcast less than a hundred feet above..its tissuey substance was rolling and boiling. **1974** J. HELLER *Something Happened* 210 The..silken feel of the tissuey things between her legs the first time she let me touch her there.

tiswas (ti·zwǫz). *slang.* Also **tis-was, tizz-wozz,** etc. [Perh. a fanciful enlargement of *TIZZ, TIZ.] A state of nervous agitation or confusion; occas. a state of physical disorder or chaos.
1960 M. CECIL *Something in Common* xvii. 195 Gets you all of a tiswas, when he's up the wall. **1974** *Observer* 27 Oct. 5/5 A young man rang up in quite a 'tis-was. **1980** *Encounter* May 7 She doesn't clean, but circumvents the dirt. Chairs stand on tables—'All of a tizz-wozz.' Has that been spelt before?

tiswin. (Earlier example.)
1877 *Rep. Indian Affairs* (U.S.) 162 Addicted..to the use of intoxicating liquors, 'tiswin', which they manufacture from corn, and whiskey obtained from strangers.

tit, *sb.*[2] Add: **1. tit for tat.** Also used as rhyming slang for 'hat'. Cf. *TITFER.
1925 FRASER & GIBBONS *Soldier & Sailor Words & Phrases* 285 *Tit for tat*, hat. (Rhyming slang). **1930, 1937** [see *TITFER].

3. tit-tat-toe, (*b*) *dial.* or *U.S.* = *noughts and crosses* (see NOUGHT *sb.* 7 c); cf. *TICK-TACK-TOE (b).
The precise nature of the activity referred to in quot. 1865 is uncertain and cannot be determined from the context.
1865 TROLLOPE *Can you forgive Her?* II. xxi. 164 The signing-clerk's clerk..playing tit-tat-to by himself upon official blotting-paper. **1888** B. LOWSLEY *Gloss. Berks. Words & Phrases* 164 *Tit-tat-toe*, the first game taught to children when they can use a slate pencil, the words 'Tit-tat-toe, My first go', being said by the one who first makes three crosses, or noughts in a row. **1898** A. T. SLOSSON *Dumb Foxglove* 11 Checkers, and tit-tat-toe, and fox-and-geese, and set down games like those. **1961** *New Scientist* 9 Nov. 367 Noughts and Crosses (known in America as Tit-Tat-To). **1973** J. SCARNE *Encycl. Games* 583 Tit-tat-toe. This simple game, also called Noughts and Crosses in Great Britain, is played on diagrams consisting of intersecting parallel lines.

tit, *sb.*[3] Add: **2.** Also applied indiscriminately to women of any age. ?*dial.*
1922 E. R. EDDISON *Worm Ouroboros* xxxi. 397 The Demons,..since they had a strong loathing for such ugly tits and stale old trots, would no doubt hang her up or disembowel her. **1932** S. O'FAOLAIN *Midsummer Night Madness* 62 I'm sorry for his two tits of sisters, though. **1969** H. E. BATES *Vanished World* ix. 87 'The old tit doddered forth... I see her as a kind of..diminutive nun, untouched and unprotected.

3. c. tit-bell, a bell-shaped container filled with seeds, fat, etc., and suspended out of doors to supply food to tits and other birds of similar habits.
1934 J. M. CROSTHWAITE in H. M. Batten *Our Garden Birds* 184 Mr. Mortimer Batten has..invented and developed many ingenious and artistic feeding devices... The tit bell is filled with melted fat, which is allowed to set, after which the bell is hung. **1976** *Southern Evening Echo* (Southampton) 3 Nov. 12/3 Another useful device for feeding tits and woodpeckers is to make a 'tit bell'.

tit, *sb.*[6] [Var. of TEAT.] **1. a.** See TEAT 1 a for dial. uses.

b. *pl.* A woman's breasts. Also in *sing. slang* (orig. *U.S.*).
1928 in A. W. Read *Classical Amer. Graffiti* (1935) 80 A girl may sit & finger her tits and play with her cunt all day. **1947** C. WILLINGHAM *End as Man* 93 'Well,' said Munro.

'That girl ought to go to Hollywood.' 'She wouldn't make it out there,' blushed Wilson. 'No tits.' **1962** J. HELLER *Catch-22* xviii. 181 How do you expect anyone to believe you have a liver condition if you keep squeezing the nurses tits every time you get a chance. **1969** *Oz* May 40/2 Mary Anne Shelley, with the best tits off-off-Broadway. **1980** J. BARNES *Metroland* I. xi. 63 Tits? I asked myself in furtive panic. Well, you couldn't really see, not with that dress.

c. *to get on one's tits* or (occas.) *tit*: to irritate intensely, get on the nerves of. *slang.*
1945 BAKER *Austral. Lang.* vi. 121 Someone or something disagreeable is said *to get on one's..tit.* **1966** 'L. LANE' *ABZ of Scouse* 40 *Gets on me tits*, annoys me very much. **1967** N. FREELING *Strike out where not Applicable* 114 Those women, who even wear a corset under riding breeches..they're the ones who get on my tits. **1973** P. WHITE *Eye of Storm* vii. 304 Much as she disliked men, Sister Manhood began to think women got on her tits as badly. **1977** J. WILSON *Making Hate* xiii. 153 This Sherlock Holmes act of yours gets right on my tits.

d. *tits and ass* or *arse*: slang phr. used to denote crude sexuality. Similarly *tits and bums.* Also *transf.*, a magazine containing photographs of nude women; also called *tit mag*(*azine*).
1972 R. A. WILSON *Playboy's Bk. Forbidden Words* 288 The late Lenny Bruce once suggested that 'Tits and Ass' would be the most accurate advertisement for most night-club acts. **1975** *New Society* 3 July 26/3 His lascivious sisters in the tit mags who part their legs and leer. **1975** *Wentworth & Flexner's Dict. Amer. Slang* Suppl. 750/1 *Tits and ass* [taboo] adj., of, being, or pertaining to commercial photographs of nude young women. **1976** N. THORNBURG *Cutter & Bone* i. 24 A tits-and-ass independent, you might call him. **1977** D. FRANCIS *Risk* xii. 150 On Wednesday, paragraphs in all the dailies... 'Fun Jock Twice Removed?' from a tits and bums. **1977** *Zigzag* Apr. 34/1 Not unless you look at some jerk-off magazine, a tit-and-ass magazine disguised as some junior hippy kind of thing. **1978** K. AMIS *Jake's Thing* (1979) v. 49 A keen buyer of tit-magazines. **1978** *Globe & Mail* (Toronto) 4 Nov. 13/2 Victor Matthews, chairman of Express Newspapers..put his people to work on plans for a new tabloid 'with plenty of tits and bums'. **1980** in S. Terkel *Amer. Dreams* 1 There are certain images that come to mind when people talk about beauty queens. It's mostly what's known as t and a, tits and arse. No talent. **1982** *Sunday Times* 2 Sept. 29/1 Ugly George, America's prime TV porn artist (who invites women to undress for his video camera), with his 'tit n' ass' cable channel.

2. = TEAT 2; *spec.* a push-button, esp. one used to fire a gun or release a bomb. orig. *Forces' slang.*
1942 J. GLEED *Arise to Conquer* iv. 30 Pull the tit. [*Note*] This is the emergency control which, by driving the supercharger at its very maximum pace, gives the aeroplane considerable extra speed. **1943** 'T. DUDLEY-GORDON' *Coastal Command at War* xvii. 165 It was time to release the depth-charges... I pressed the tit and that was the last I saw of it [*sc.* the bomb] for a bit. **1972** A. PRICE *Colonel Butler's Wolf* xii. 135 They've built this mock-up in the Museum... You press the tit, and the lights go out. **1976** 'J. ROSS' *I know what it's like to Die* xxi. 136 He pressed the tit of the bell push and she opened the door.

tit, *sb.*[7] *slang.* [Of uncertain origin: perh. f. *TIT *sb.*[6]; cf. TIT *sb.*[3], *TWIT *sb.*[1] 2 b.] A foolish or ineffectual person, a nincompoop.
1947 *Landfall* (N.Z.) Dec. 290 Why didn't Lachlan go, the silly tit? **1965** M. FRAYN *Tin Men* (1966) xv. 69 'Who are all these people?' they shouted at one another. 'All which people?' 'All these tits in tweed sports jackets.' *Ibid.* 70 'Peculiar friends he has.' 'Tits, a lot of them.' **1968** *Listener* 19 Sept. 370/2, I don't think much of this little tit Hitler, do you, ducky? **1978** S. WILSON *Dealer's Move* vii. 122 We always took a gun, and it kept me quite alert, not wishing to make a tit of myself in front of the laird.

titania (tǝitē¹·niā). *Chem.* [f. TITAN(IUM + -IA[1], after YTTRIA, etc.] = *titanium dioxide* s.v. *TITANIUM b.
1922 *Jrnl. Amer. Chem. Soc.* XLIV. 387 Titania was prepared by dissolving the oxide in hot conc. sulfuric acid, diluting with 10 volumes of water and precipitating as with alumina. **1971** *Materials & Technol.* II. v. 326 Titania and its compounds..are important for high-capacity condensers, and have considerably displaced the natural product mica.

titanian, *a.*[2] Restrict † *Obs.* to sense in Dict. and add: **b.** *Min.* [*-IAN 2.] Of a mineral: having a (small) proportion of a constituent element replaced by titanium.
1930 *Amer. Mineralogist* XV. 572 Titanium—titanian. **1944** [see *PLUMBIAN *a.*]. **1967** *Amer. Mineralogist* LII. 780 Zr also has been reported in amounts up to a few weight percent in titanian andradite (melanite).

Titanic, *sb.* [f. TITANIC *a.*[1]] The name of a giant British liner which sank on its maiden voyage in 1912 after collision with an iceberg; used allusively or as a metaphor for a vast and supposedly indestructible organization fated to disaster. Also *Titanic clause* (see quot. 1915).
1915 *N.Y. World* 3 Aug. 5/4 When he executed his will on March 16, 1914, Joseph E. Greenfield inserted what is known as a 'Titanic' clause, which anticipated the possibility of the testator and his wife meeting death together in a catastrophe. **1975** S. LAUDER *Killing Time on Corvo* x. 91 It was some horrifying *Titanic* disaster. **1975** *Times* 3 Sept. 10/6 The implications of the final song, that England is a Titanic with a crew composed only of vagabonds and privileged yachtsmen. **1976** *Times* 13 May 1/3 The hapless President's campaign manager..sounded..fatalistic... 'I'm not going to do anything to re-arrange the furniture on the deck of the Titanic,' he said. **1980** K. HAGENBACH *Fox Potential* vi. 57, I wanted to leave England... I did not intend to be aboard when that particular Titanic finally foundered in a sea of bureaucracy.

Titanism. b. (Earlier example.)
1851 H. MELVILLE *Moby Dick* II. xliv. 297 Where infantileness of ease undulates through a Titanism of power.

titanium. Add: **b. titanium dioxide,** the oxide TiO_2, occurring naturally as the minerals rutile, anatase, and brookite, and used esp. as a white pigment and opacifying agent; **titanium oxide,** any oxide of titanium, esp. the dioxide; **titanium sponge,** titanium in a porous form; **titanium white,** a white pigment consisting chiefly or wholly of titanium dioxide.
1877 *Jrnl. Chem. Soc.* I. 688 Sulphate of titanium dioxide, $TiS_2O_8 + 3H_2O$, is a yellow resinous mass. **1963** R. R. A. HIGHAM *Handbk. Papermaking* iv. 94 Titanium dioxide has the property of extreme chemical inertness, i.e., it is not affected by acids, alkalis, or the common solvents at standard temperature and pressure and is insoluble in water. **1982** *Sci. Amer.* Oct. 58 (Advt.), Titanium dioxide makes the plastic of your coffee cup opaque and the color of your telephone deep and bright. **1885** *Jrnl. Chem. Soc.* XLVIII. I. 640 The author describes the hydrated titanium oxide with phosphoric acid and various earths from the diamond diggings of Diamantina, in Brazil. **1955** *Sci. News Let.* 9 Apr. 233/1 A liquid at ordinary temperatures, titanium tetrachloride changes to smoky fumes of titanium oxide when air touches it. **1977** *Whitaker's Almanack* 1978 152/2 The M stars, like Betelgeuse, show very complex molecular spectra, chiefly of titanium oxide. **1950** *Metal Industry Handbk. & Directory* (ed. 39) I. 39/1 The resultant mixture of molten magnesium chloride, unused magnesium and titanium sponge is allowed to cool..and the product is bored out as chips. **1978** *Jrnl. R. Soc. Arts* CXXVI. 679/1 The extraction method normally used now is to chlorinate rutile (TiO_2), turning it into titanium tetrachloride ($TiCl_4$) liquid, which is then reduced with magnesium or sodium to produce titanium sponge which can subsequently be melted and cast into ingots. **1920** *Chem. Abstr.* XIV. 355 (*heading*) Titanium and titanium white. **1934** H. HILER *Notes on Technique of Painting* ii. 101 Titanium white is the oxide of a metal which until lately was considered as a curiosity in the laboratory.

titano-[2]. Add: **titanau·gite** *Min.* [ad. G. *titan-augit* (A. Knop *Der Kaiserstuhl im Breisgau* (1892) ii. 72)], a variety of augite containing titanium; **titan(o)hæ·matite** (also **-hem-**) *Min.*, a variety of hæmatite containing titanium dioxide in solid solution; **ti:tanomaghe·mite** [*MAGHEMITE] *Min.*, a titanian variety of maghemite; **titanoma·gnetite** *Min.* [ad. G. *titanomagnetit* (P. Groth *Tabellarische Übersicht der Min.* (ed. 4, 1898) 79)], a variety of magnetite containing titanium.
1933 *Zeitschr. für Kristallogr.* LXXXVI. 112 The titanaugite in question forms a small patch or segregation within a sphene-rich, plagioclase-diopside-hornfels xenolith in the Haddo norite. **1963** W. A. DEER et al. *Rock-Forming Minerals* II. 109 Titanaugites are the typical pyroxenes of basic alkaline rocks, *e.g.* teschenite, essexite and nepheline dolerite. **1970** *Nature* 28 Nov. 850/2 The principal minerals in the rock are zoned plagioclase..; nepheline; and titanaugite, grading at the edges of the crystals into aegirine. **1938** A. B. EDWARDS in *Proc. Australasian Inst. Mining & Metallurgy* No. 110. 42 This 'white ilmenite' is quite distinct from ordinary hematite... The name 'titanhematite' is here suggested to indicate its difference from pure hematite. **1945** *N.Z. Jrnl. Sci. & Technol.* B. XXVI. 299 A range from pure titanomagnetite to pure titanhæmatite is..present in a small percentage of the iron-ore grains. **1971** I. G. GASS et al. *Understanding Earth* xvii. 255/1 Ultimately, in the highest state of oxidation.., the original titanomagnetite has been converted mainly to pseudobrookite (Fe_2TiO_5) and titanohaematite (Fe_2O_3) containing a little titanium. **1953** E. Z. BASTA *Mineral. Aspects of System FeO–Fe₂O₃–TiO₂* (Ph.D. Thesis, Univ. of Bristol) vi. 71 For those minerals with composition approaching (Fe, Ti)$_3$O$_4$ (e.g. the Bushveld maghemites) I propose the new name 'titanomaghemite'. **1971** *Nature* 5 Mar. 28/1 Under conditions of low temperature and high oxygen fugacity..titanomagnetite tends to oxidize to an equilibrium mineral assemblage of rutile and haematite, with intermediate formation of titanomaghemite, ilmenite and iron-rich titanomagnetite. **1900** *Mineral. Mag.* XII. 393 Titanomagnetite... Titaniferous magnetite. [(Fe,Ti)O$_3$]$_2$Fe. **1945** [see *titanohæmatite* above]. **1962** W. A. DEER et al. *Rock-Forming Minerals* V. 68 A considerable amount of Ti can enter the magnetite structure, there being a continuous relationship between magnetite and the ulvöspinel molecule, Fe_2TiO_4... The term titanomagnetite is best restricted to those specimens where the presence of an ulvöspinel phase can be demonstrated. **1971** I. G. GASS et al. *Understanding Earth* xvii. 255/1 The mineral which accounts for the magnetic properties of most rocks, and especially basalts, is titanomagnetite.

tit-bit, tid-bit. Add: **tid-bit** is now chiefly *N. Amer.* **a, b.** (Later N. Amer. examples.)
1906 U. SINCLAIR *Jungle* xiv. 162 Things..went into the sausages in comparison with which a poisoned rat was a tidbit. **1941** AUDEN *New Year Let.* I. 26 Add his small tid-bit to the rest. **1968** *Globe & Mail* (Toronto) 17 Feb. 28 An unusually good selection of hot and cold tid bits. **1976** *Time* 27 Dec. 49/3 There were enough tidbits of good news last week to soothe the fears of some Ford Administration economists.

titch, var. *TICH.

titchy (ti·tʃi), *a. colloq.* [f. *titch*, var. of *TICH + -Y[1].] Insignificantly small, diminutive, tiny.
1950 A. BUCKERIDGE *Jennings goes to School* vii. 139 Well, anyway,..there'll be a titchy hunk all round, so no one'll have any reason to grumble. **1958** *Spectator* 13 June 768/2 Towering six foot three inches over a titchy Laertes. **1967** J. PORTER *Chinks in Curtain* ii. 20 Titchy little automatics. **1978** *Lancashire Life* Sept. 96/1 'E 'olds a titchy rod an' line An' angles in a pond.

titfer (ti·tfǝɹ). *slang.* Also **titfa, titfor.** [Shortened from *tit for tat* used as rhyming slang: see *TIT *sb.*[2] 1.] A hat.
1930 BROPHY & PARTRIDGE *Songs & Slang Brit. Soldier* 171 *Tit-for*, tit-for-tat, i.e. hat. **1937** N. GAY *Me & my Girl* I. i, in J. Franklyn *Dict. Rhyming Slang* (1960) 172 *Duchess*: I hope you enjoyed your drive. *Bill*: Not 'arf—but I nearly lost my titfa! *All*: Titfa? *Bill*: Me tit for tat. *All*: Tit for tat? *Bill*: My Hat! **1939** J. B. PRIESTLEY *Let People Sing* x. 257 I'll see Billy Fitt, with me titfer in me 'and. **1943** HUNT & PRINGLE *Service Slang* 67 *Tin titfor*, steel helmet. **1952** M. TRIPP *Faith is Windsock* x. 151 I've got a lucky scarf too, so's Jake. Dig always takes his titfer, and Arthur's got a brassiere. **1960** *Observer* 20 Mar. 10/3 Last week I told you about the time I popped my titfa. **1976** U. HOLDEN *String Horses* viii. 102 The old lady made a show... Lil Pratt forgot to fill her mouth... She'd not seen a titfer like that since the film of mountain people in the Dardanelles, made after World War one.

tithe, *a.* and *sb.*[1] Add: **B.** *sb.* **1. a.** Also, in recent use, in certain religious sects: a tenth part of an individual's income which is pledged to the church. Cf. *TITHE *v.*[2] 1 b, 2.
1965 M. J. C. CALLEY *God's People* ix. 106 The 1960–1 income of a London congregation of the New Testament Church of God which claimed fifty-nine members in 1961 consisted of £900 from tithes and £200 from free-will offerings.

4. b. *tithe-accounts, -audit, -campaign, -dinner, -map* (earlier example), *-payer* (later example), *-war*; *tithe-free* adj. (later example).
1781 J. WOODFORDE *Diary* 6 Nov. (1924) I. 329 Being obliged to go to Lenewade Bridge to settle Dr. Bathurst's Tithe accounts. *Ibid.* 4 Dec. 333, I asked them to dine with us..this day being my Tithe Audit. **1878** F. KILVERT *Jrnl.* 5 Feb. (1977) 302 Today was the Tithe audit and tithe dinner to the farmers, both held at the Vicarage. About 50 tithe payers came. **1832** COBBETT *Weekly Reg.* 21 Apr. 134 Look at the tithe-campaign preparing for Ireland; the tithe-war, indeed. **1878** Tithe dinner [see *tithe-audit* above]. **1960** *Farmer & Stockbreeder* 29 Mar. 31/3 Gentleman's tithe-free residential farm. **1895** *Law Jrnl. Rep., Queen's Bench* LXIV. 159/2 The late Master of the Rolls held that a tithe-map was not evidence of boundaries between two adjoining owners. **1878** Tithe-payer [see *tithe-audit* above]. **1832** Tithe-war [see *tithe-campaign* above]. **1979** V. BOGDANOR *Devolution* v. 123 The 'tithe war', under which tithes were withheld.

tithe, *v.*[2] **1. b.** Delete † *Obs.* and add: (Later examples.) Also *gen.*, to pledge or contribute as a levy.
1967 *Observer* 6 Aug. 4/5 A reply sent to a young member by the sect's letter-answering department was more precise: 'A person working for wages is to tithe one-tenth of the total amount of his wages before income tax, national health, or other deductions are removed.' **1976** *Billings* (Montana) *Gaz.* 20 June 6-c/1 Former Southern officers prospered and tithed up to 50 percent for Civil War II, which never came.

2. Delete † *Obs.* Revived in recent use in connection with voluntary church giving. Cf. *TITHE *sb.*[1] 1 a.
1942 *Esquire* Sept. 174/2 They went to the Six Hickories church—tithed—and behaved themselves. **1968** N. GIOVANNI in W. King *Black Short Story Anthol.* (1972) 23 He quit church after a couple of months, but he continued to tithe every month faithfully and never drank again.

Hence **ti·thing** *ppl. a.*
1965 M. J. C. CALLEY *God's People* ix. 111 Everybody [in the congregation]..gives generously, probably more than the tenth required by tithing sects.

tither (ti·ðǝɹ), *sb.*[2] [Of obscure origin; cf. Hampshire dial. *to be on tither-thorns* 'to be tremulously anxious' (*Eng. Dial. Dict.*) and DITHER *sb.*] A state of feverish excitement.
1960 V. JENKINS *Lions Down Under* vi. 91 His amazing side-stepping and running had the crowd in a tither. *a* **1974** R. CROSSMAN *Diaries* (1977) III. 640 He adored discussing the Health Service, he was all of a tither and quiver of excitement at having the Secretary of State there.

tithing, *vbl. sb.* Add: **a.** (Later U.S. example: cf. *TITHE *v.*[2] 2.)
1929 R. S. & H. M. LYND *Middletown* xxii. 356 Traditionally every Christian 'returns a tenth of his substance to the Lord'. A few families in Middletown continue this practice of tithing, but..the great majority contribute far less than a tenth.

d. *tithing-system.*
1904 F. W. MAITLAND *Let.* 19 May (1965) 305 We still want a little more light on the tithing system. **1978** *Daily Mirror* 12 Jan. 6/6 Much of their wealth comes from the use of the 'tithe-ing' system—members of the religion are required (like the Mormons) to donate one-tenth of their income to the funds.

Tithonian (tǝiþǭᵘ·niăn), *a. Geol.* [ad. G. *tithonisch* (A. Oppel 1865, in *Zeitschr. der deutsch. geol. Ges.* XVII. 535), f. L. *Tīthōnus*, Gr. *Τιθωνός*: see -IAN.] Designating a stage of the European Upper Jurassic, thought to correspond to the Portlandian, or the Portlandian and part of the Kimmeridgian, in Britain; also *ellipt.*
1871 *Q. Jrnl. Geol. Soc.* XXVII. 208 The deposition of the Wealden strata..commenced before the close of the Oolitic period; it continued during the whole of the Tithonian. **1882** A. GEIKIE *Text-bk. Geol.* 800 At the top of the Alpine Jurassic series an important group of deposits occurs to which the name of Tithonian stage was given by Oppel. **1975** *Nature* 13 Mar. 108/1 Two different directions have been obtained from sediments in the lower and upper parts of the Morrison formation (Kimmeridgian–Tithonian respectively).

titi[2]. Add: Now the usual spelling of TEETEE[1]. Substitute for def.: A small long-coated monkey of the genus *Callicebus*, native to the tropical forests of S. America. (Later examples.)
1927 *Ann. Mag. Nat. Hist.* 9th Ser. XIX. 509 (*title*) On the Titi Monkeys. *Ibid.*, The British Museum has received ..some further specimens of..the Yellow-handed Titis. **1963** *Mammalia* XXVII. 3 Other names..have been applied to the titis of eastern Brazil. **1976** *Nature* 23 Sept. 321/1 Titi monkeys..remain paired throughout the year. **1978** *Ibid.* 18 May 193/2 Titis and siamangs carry the infant(s) for much of the day.

Titian. Add: (Later example in sense 'a picture by Titian'.) Also, a person with Titian or bright auburn hair. As *adj.*, also 'painted by or in the style of Titian'; (earlier examples); freq. in *Comb.*, as *Titian-haired* (occas. with small initial).
1841 M. E. LUCY *Diary* 10 Mar. (1983) 66 Lord Byron's favourite, Countess Guiccioli was there; she had..reddish auburn hair..looking very much like a Titian Magdalene. **1892** S. WATERLOO *Man & Woman* xiii. 97 A setter, with Titian hair and big eyes, which slept on the clover beside him. **1903** H. JAMES *Ambassadors* III. vii. 86 Standing with his fellow-visitor before one of the splendid Titians. **1923** *Times* 3 May 14/6 (Advt.), Tecla pearls..are equally becoming whether worn by blondes, brunettes or Titians. **1934** *Times Lit. Suppl.* 25 Oct. 732/2 His Titian-haired wife. **1959** W. BURROUGHS *Naked Lunch* 77 Titian-haired Venetian lads. **1982** 'D. SERAFIN' *Madrid Underground* 103 The tall, titian-haired girl.

Titius (ti·ʃǒs). *Astr.* [The name of J. D. *Titius* (1729–96), German astronomer, who published the law in 1766, six years before Bode.] *Titius–Bode law*: = *Bode's law* s.v. LAW *sb.*[1] 17 c.
1954 H. ALFVÉN *Origin of Solar System* viii. 128 The so-called Titius-Bode's law gives an empirical formula for the distance of the planets from the sun. **1972** *Nature* 21 Apr. 374/2 The chief facts which have to be explained are..the peculiar regularities in the planetary and satellite spacings, summed up by the Titius-Bode law.

titivate, tittivate, *v.* Add: ¶ **2.** Used by confusion for TITILLATE *v.* 1.
1915 [see *EROTICIZE *v.*]. **1933** DYLAN THOMAS *Let.* Sept. (1966) 23 Even now twelve heartfelt pages are titivating the senses of a Dead Letter superintendent. **1976** *Telegraph-Journal* (Saint John, New Brunswick) 27 Aug. 2/4 What would the exhibition be without the midway, with its sounds, sights and smells to titivate the senses?

titivating *ppl. a.*, **titivator**: also in similar erron. use; hence **ti·tivatory** *a.* ¶ = TITILLATORY *a.*
1928 GALSWORTHY *Swan Song* I. iv. 29 The papers were like cocktails—titivators mostly of the appetite and the nerves. **1964** E. HUXLEY *Back Street New Worlds* ix. 95 Displaying a shapely but naked midriff, eyelids kohl-ed, hands henna'd, perfumed with eastern essences as titivating to the senses as they were no doubt unsettling to the aldermen. **1975** *Time Out* 24 Jan. 5/2 All the magazine lacks is a titivatory piece on 'what they do in bed'.

title, *sb.* **1.** Restrict † *Obs.* to sense in Dict. and add: **c.** A piece of written material introduced into a film or television programme to explain action or represent dialogue; a caption; cf. *SUB-TITLE *sb.* 3. Also, a credit title (see *CREDIT *sb.* 13 f).
1905 *Billboard* 21 Oct. 42 All our films come with red titles, and show our trade mark. **1909** *Moving Picture World* 10 July 57 We make film titles, 5 feet for 50 cents in any color desired. **1922** [see *CREDIT *sb.* 13 f]. **1929** I. MONTAGU tr. *Pudovkin's On Film Technique* III. 45 *Scene I.*... A passer-by, coming towards the waggon, pauses... The driver turns to him. Title: '*Is it far to Nakhabin?*' The pedestrian answers, pointing with his hand. **1958** *Punch* 27 Aug. 285/3, I shall remember [this film] as the first exception I have noticed to the rule that amusingly well-designed titles..mean a good film. **1961** G. MILLERSON *Technique Television Production* xix. 358 Roll titles give us a continuous, unbroken stream of information. **1964** T. RATTIGAN *Heart to Heart* in *Coll. Plays* III. 498 Cut, sound. Start titles... Cue grams.

3. c. (*a*) Chiefly in *Publishing*, a book, a magazine, a newspaper; (*b*) a gramophone record.
1895 *Montgomery Ward Catal.* Spring & Summer 62/1 Burt's Library of the World's Best Books... This series comprises titles selected from the standard works of the world's literature. **1908** *Sears, Roebuck Catal.* 200 Columbia P Records. Your own selection of subjects, any of the titles shown on the list. **1935** A. C. BAUGH *Hist. Eng. Lang.* viii. 246 In England over 20,000 titles in English appeared by 1640, ranging all the way from mere pamphlets to massive folios. **1953** J. MORTIMER *Like Men Betrayed* v. 83 I'd never read any titles by Dickens, but we're thinking of bringing out a Victorian Omnibus so I read one. **1958** G. BOATFIELD in P. Gammond *Decca Bk. Jazz* xxiv. 313 A 1938 session with Pete Brown on alto and flamboyant trumpeter Charlie Shavers produced eight titles. **1977** *Times* 10 Sept. 2/5 The *Daily Express*..Mr Matthews thought 'had lost its way'... Mr Matthews is..fairly satisfied with Beaverbrook's other titles, the *Sunday Express* and *Evening Standard.* **1979** P. THEROUX *Old Patagonian Express* xix. 301, I..introduced myself as the author of the three titles I had seen in the bookstores in Tucuman. **1982** *Times* 4 May 15/2 The latest casualty is the IPC romantic weekly for teenage girls, Love Affair... The title is no longer profitable.

5. b. *Sport.* The championship or supremacy in a contest or competition; the game or contest in which this is decided.
1922 *Encycl. Brit.* XXXII. 566/1 J. J. McDermott won the [golf] open tournament both in 1911 and 1912. Travers defeated Anderson for the amateur title in 1913. **1930** *Amer. Speech* VI. 121 *Title*, championship: Al Brown Signed For Title Battle. **1939** *Encycl. Brit. Bk. of Year* 117/1 Joe Louis reigned as world heavyweight champion and defended his title three times. **1955** R. BANNISTER *First Four Minutes* vi. 59 To win the 100 and 200 metre titles in the World Student Games. **1971** *Rand Daily Mail* 4 Sept. 3/6 The visit of South African squash players to Hobart to compete in the Australian squash titles. **1973** P. EVANS *Bodyguard Man* iv. 33 He goes straight into the Fiorentina first team, in his first year helps to win the League title for his new club.

11. *title-leaf* (later example), *-registration, -search, -searching*; **title catalogue** *Librarianship* (see quots.); **title entry** *Librarianship*, an entry for a book in a library catalogue made under the title (as opp. under the author); **title fight** *Boxing*, a match held to decide the championship; **title-holder,** (*a*) one who holds title-deeds; (*b*) *Sport*, the reigning champion in a particular field; **title insurance** *U.S.*, insurance protecting the owner or mortgagee of real estate against lawsuits arising from defective title; **title-music,** music played during the credits at the beginning of a film or television programme; **title-part** (examples); **title-piece,** an essay, piece of music, etc., giving its name to the collection of which it forms part; **title song, -track,** the song or track giving its name to a long-playing record.
1876 C. A. CUTTER in *Public Libraries in U.S.A.* xxvii. 528 *Title-catalogue*, one in which the entries are arranged alphabetically according to some word of the title, especially the first, (a dictionary of titles). **1910** A. E. BOSTWICK *Amer. Public Library* 175 If they [*sc.* entries] are arranged alphabetically by the chief word in the title, it is a title catalogue. **1968** P. QUIGG *Theory of Cataloguing* vi. 63 The *author catalogue* is a catalogue with, in the main, authors' names... The entries will, however, usually include..for certain works..title entries. Added entries for significant titles are usually included..so that..the form of the catalogue should be designated as an *author/title catalogue.* **1875** C. A. CUTTER in *Nation* 4 Mar. 151/1 especially impressed with the usefulness of title-entries. **1935** *Library Q.* V. 459 He..had obtained permission to change entries in the university library catalog for publications of corporate bodies from title entry to entry under their names. **1969** P. S. DUNKIN *Cataloging U.S.A.* iii. 46 So much for author entry, title entry, and arbitrary entry and the heading which introduces each. **1951** *Sport* 7 Jan. 14 The forthcoming feather-weight title fight between champion Ronnie Clayton and veteran Al Phillips. **1973** 'S. HARVESTER' *Corner of Playground* I. viii. 71 She went away, walking on her heels like a boxer after thirteen rounds of a title fight. **1904** Title-holder [in Dict.]. **1938** *Encycl. Brit. Bk. of Year* 113/1 Three of the title-holders, Louis (heavy), Lewis (light-heavy), and Armstrong (bantam), are negroes. **1978** H. COOPER *Great Heavyweights* 86 Willie Pastrano, the then world light-heavy titleholder and a boxer of beautiful science. **1902** C. J. PIDGIN *Stephen Holton* 260 That was a mighty good idea of yours, Mr. Lethbridge—telling me to go to a title insurance company. **1942** *Federal Reporter* (U.S.) CXXXII. 44 The contention of the appellant is that premiums paid for title insurance are earned when received. **1979** *Arizona Daily Star* 5 Aug. D 2/4 The firm has been a division of First American Title for 20 years, offering title insurance and escrow services nationally. **1936** *Discovery* Dec. 384/2 The booksellers also displayed the title-leaves of new works as advertisements. **1977** *Gramophone* Apr. 1555/3 The opening and closing are of great impact (like the title-music for one of the more dramatic of those films). **1898** G. B. SHAW in *Sat. Review* 5 Feb. 171/2 Miss Irene Vanbrugh, in the title part..vanquishes it easily and successfully. **1927** F. HARRIS *My Life* III. xix. 334 'Poil de Carotte', (Carrots!) I think it was, with Madame Nau in the title part. **1927** *New Republic* 12 Oct. 211/1 He has possibly scored some moderate hits: in 'Manhattan Mary', 'Broadway', 'The Five Step'.., a curiously constructed sob-song called 'Memories', and the title-piece. **1936** in A. Huxley *Olive Tree* (dust-jacket), This is one of the best collections of essays that Mr. Huxley has ever made. The title-piece is a completely new departure in technique. **1968** ROBERTS & MOORE in D. H. Lawrence *Phoenix II* p. xii, 'The Gentleman from San Francisco' appeared as the title-piece of a collection of Bunin's stories. **1971** A. AXELROD et al. *Land Transfer & Finance* 693 Patton has been a strong proponent of title registration. **1965** *Amer. Bar Assoc. Jrnl.* LI. 1071/1 In the second step the contract is drawn up by the lawyer and he handles the closing, but the title search is conducted by full-time, salaried employees of the title company. **1980** *Daily Tel.* 16 Jan. 23/3 They will ask you to complete a 'title search' on the new property to make sure that all deeds and papers are in order. **1899** G. B. SHAW *Let.* 1 Aug. (1972) II. 95 After much title-searching, I have resolved to give that play..the ugly but arresting name 'Captain Brassbound's Conversion'. **1971** A. AXELROD et al. *Land Transfer & Finance* 499 This loss of title searching and examination illustrates the vulnerability of lawyers in private practice to competition from specialized high volume businesses and professions. **1961** *New Musical Express* 6 Jan. 4/2 *Am I that easy to forget*..is the title song of a soft-sung album by Debbie Reynolds. **1970** *Melody Maker* 21 Feb. 21/3 It's hard to believe that the same man who could write and play the extraordinary title track could also be responsible for 'Spirits' and 'Search'.

titling, *vbl. sb.* Add: **c.** The action of providing a film, television programme, or photograph with captions or titles (sense *1 c).
1913 *Moving Picture World* 4 Oct. 25 The perfect picture tells its story without any titles, but as there are very few perfect pictures good titling becomes a necessity. **1958** *Punch* 1 Jan. 80/2 This horrible cliché of TV drama presentation, known to the trade as delayed titling. **1966** *Ibid.* 22 June 922/1 We may have seen an unsuccessful copy [of the Yugoslavian film *Covek Nija Tica*], and certainly the titling wasn't very efficient. **1970** *Amateur Photographer* 11 Mar. 61/1 Positive in the camera comes into its own for titling.

titlist (tǝi·tlist). *U.S. Sport.* [f. TITLE *sb.* + -IST.] A title-holder or champion.
1924 [see *STRING *sb.* 12 c]. **1955** M. REIFER *Dict. New Words* 209/2 *Titlist*, one who has won a title, usually in a sport or game contest. **1973** *Internat. Herald Tribune* 15 June 15/2 Defending titlist Manuel Orantes of Spain joined other top seeds on the sidelines of the German tennis championships. **1976** *Billings* (Montana) *Gaz.* 11 July 7-G/1 Carner, the 1971 titlist, a winner three times on the LPGA tour this year.

Titoism (tī·toiz'm). [f. *Tito*, name adopted by Josip Broz (?1892–1980) premier of Yugoslavia from 1945 + -ISM.] The ideas or policies associated with Marshal Tito; *spec.* a form of communism which concentrates on the national interest without reference to the Soviet Union.
1949 *Economist* 2 July 5/1 During the year which has followed the famous resolution of the Cominform expelling Jugoslavia from its membership, Titoism has become an international phenomenon. **1949** [see *TITOIST *sb.* and *a.*]. **1958** F. W. NEAL (*title*) Titoism in action. **1962** [see *MAOISM]. **1965** *Listener* 30 Sept. 480/1 Where communism cannot be avoided the best hope is to steer it towards affluence and Titoism. **1973** R. J. ALEXANDER *Latin Amer. Polit. Parties* xx. 397 Titoism was more attractive to some of the Latin American Socialist parties than to the Communists of the area. **1978** *Time* 3 July 7/2 Exactly 30 years after fiercely independent Yugoslavia was expelled from Joseph Stalin's Cominform for what became known as 'Titoism'.

Titoist (tī·toist), *sb.* and *a.* [f. as prec. + -IST.] **A.** *sb.* A follower or adherent of Titoism. **B.** *adj.* Of, pertaining to, or resembling Titoism.
1949 *Newsweek* 15 Aug. 9 Titoism has now infected the Communist Party in the Middle East, a split is developing between Stalinists and Titoists in Beirut. **1949** *N.Y. Times* 30 Oct. I. 17/1 The delegation denied any intention of starting a Titoist 'movement' in Italy. **1951** *Round Table* CLXIII. 247 (*heading*) The Titoist schism. **1954** KOESTLER *Invisible Writing* ii. 21 These [*sc.* communist cells] are living, pulsating units..susceptible to various diseases—to the Titoist virus, to bourgeois infection or Trotskyist cancer. *Ibid.* xiv. 155 Every Communist who had lived in the Soviet Union for some length of time, returned to his country as a Titoist at heart. **1961** *Listener* 9 Nov. 754/1 This aspect of Titoist reforms looms large in official literature. **1971** 'P. KAVANAGH' *Triumph of Evil* iii. 25 Dorn, a Croat, had spent the war years with Ante Pavelic, killing Serbs and Titoist partisans. **1977** *Time* 27 June 12/3 My son recently wrote to me that he is a liberal conservative. I replied, 'In Yugoslavia, that means a Titoist.'

Titoite (tī·toǝit), *sb.* (and *a.*) [f. as prec. + -ITE[1].] = *TITOIST *sb.* Also *attrib.* or as *adj.* Usu. with derog. implication.
1955 H. HODGKINSON *Doubletalk* 129 Soviet writers have not thought fit to elevate Tito's heresy to the rank of an 'ism'... He and his colleagues were..referred to abusively as Titoites. **1961** *Spectator* 9 June 829 He was..not even a Titoite or fellow-traveller. **1973** R. J. ALEXANDER *Latin Amer. Political Parties* xx. 396 (*heading*) The Titoite heresy. *Ibid.*, In the Communist-controlled countries of Eastern Europe, the Titoite dissidence had considerable support.

titrate, *v.* Add: Also with pronunc. (tǝi-). Hence **titra·table** *a.*, capable of being titrated.

1919 *Jrnl. Physiol.* LIII. 189 Duplicate estimations were made..of titratable acidity and ammonia by the method of Folin. 1929 *Amer. Jrnl. Physiol.* LXXXVII. 538 During the experiments no marked change occurred in the titratable acid concentration of the urine. 1963 [see *END-POINT 1]. 1973 *Nature* 14 Dec. 425/1 Calculation of the total titratable acid in the collected effluent.

titrator (təi-, titrē^{i}·təɹ). *Chem.* [f. TITRATE *v.* + -OR.] An apparatus for automatically performing a titration.
1948 *Analytical Chem.* XX. 288/1 The recorder switch should be set so that when the titrator stops finally the potential of the indicator electrode will coincide with the equivalence point potential of the particular titration. 1964 *Oceanogr. & Marine Biol.* II. 103 Many commercially produced titrators would doubtless prove suitable for routine work of moderate accuracy. 1979 *Nature* 25 Jan. p. xx/3 The range, consisting of Karl Fischer, endpoint, *p*h stat and recording titrators, all incorporate a unique photoelectric counting device.

titre, titer. Add: Also with pronunc. (təi·təɹ). (Further examples.) In *Med.*, the concentration of an antibody, as measured by the extent to which it can be diluted before ceasing to give a positive reaction with antigen.
1868 *Chem. News* 13 Mar. 132/1 Provided the 'tinctorial' power, and consequently 'titre' of the ammonia standard be correctly ascertained in terms of the iodine solution, the former may be dispensed with. 1918 [see *PEPTIDASE]. 1947 *Ann. Rev. Microbiol.* I. 337 If soluble antigen has been released it may be detected in relatively high titer in the supernatant fluid. 1958 *Spectator* 19 Sept. 379/1 The shot I'm going to give you..ought to moderate the symptoms until you've developed a high anti-body titer of your own. 1979 *Jrnl. R. Soc. Arts* CXXVII. 421/1 As the anti-body titres waned they would go through a succession of mid-gestation abortions.
2. (See quots.)
1895 J. LEWKOWITSCH *Chem. Analysis Oils, Fats, Waxes* iv. 100 The temperature will continue to fall, but then it will rise suddenly..and reach a maximum, remaining thereat stationary for some little time before it falls again. This point is called the titer or solidifying point. 1951 KIRK & OTHMER *Encycl. Chem. Technol.* VI. 153 (*heading of table*) Fat or oil... Titer range, °C. 1972 *Materials & Technol.* V. viii. 193 Fats containing fatty acids with a titre above 40°C are known as tallows. (The titre of an oil is the highest temperature reached when the liberated water-insoluble fatty acids are crystallizing under controlled conditions. It is generally taken to be the solidification point of the fatty acids.)

ti-tree. (Earlier and later examples.)
1864 R. ANDERSON *Hawaiian Islands* vii. 134 Then came gigantic ferns, and an extensive tract covered with the *ti* trees. 1881 [in Dict. s.v. TI b]. 1981 M. GEE *Meg* xiii. 116 A ti-tree prop.

tit-tat-toe: see *TIT *sb.*^{2} 3 in Dict. and Suppl.

titter, *sb.*^{3} *slang.* [Of uncertain origin: cf. TIT *sb.*^{3}, *TIT *sb.*^{6}] A young woman or girl.
1812 J. H. VAUX *Vocab. Flash Lang.* in *Mem.* (1964) 274 *Titter*, a young woman or girl. 1845 E. J. WAKEFIELD *Adventure in N.Z.* I. xi. 319 A chief was called [by whalers] a 'nob'; a slave, a 'doctor'; a woman, a 'heifer'; a girl, a 'titter'. 1882 *Sydney Slang Dict.* 6/2 Nark, to watch, to look after; 'Nark the *titter*', watch the girl. *a* 1890 in Barrère & Leland *Dict. Slang* (1890) II. 356/2 Only a glass of bitter! Only a sandwich mild! Only a stupid titter! Only she's not a child! 1953 *Landfall* (N.Z.) Sept. 179 Boys, she's a larky little titter.

tittery (ti·təri), *a.* [f. TITTER *v.*^{1} or *sb.*^{2} + -Y^{1}.] Of laughter, remarks, etc.: having a nervous, tittering quality.
1936 M. MITCHELL *Gone with Wind* xxxv. 591 The tittery cackling laugh which she always found so annoying. 1962 E. O'BRIEN *Lonely Girls* ii. 23 He would have seen me..saying foolish tittery things to amuse the others. 1983 *Financial Times* 15 Sept. 13 After a richly funny beginning the characters and story fracture into a series of fluttery, tittery vignettes.

tittup, *v.* Add: (Later examples with reference to a person's gait.) Also *fig.*
1910 E. M. FORSTER *Howard's End* xiv. 121 No one felt uneasy as he titupped along the pavements. 1968 B. HEALEY *Murder without Crime* vii. 126 Like benevolent Mr. Pickwick he tittuped along beside us. 1972 N. FREELING *Long Silence* I. 51 'We're not very happy about art,' Van der Valk tittupped on.

titty, *sb.*^{3} Add: Also **tittie.** Now *colloq.* or *slang* and also as dim. of *TIT *sb.*^{6} **1.** (Later examples.) Chiefly *pl.* Applied both to a nipple (sometimes of a boy) and to a woman's breast including the nipple; *tough titty*: see *TOUGH *a.* 6 d.
1922 JOYCE *Ulysses* 738 Yes I think he made them a bit firmer sucking them like that so long he made me thirsty titties he calls them. 1940 C. MCCULLERS *Heart is Lonely Hunter* I. iii. 39 His little titties were like blue raisins on his chest. 1957 J. FRAME *Owls do Cry* vii. 33 She had pink bulges where Daphne had mere tittie dots. 1972 *Screw* 12 June 10/2 Man, those firm nice buttocks and titties filled that bikini to overflowing. 1976 M. MACHLIN *Pipeline* iv. 46 Man, that *is* cold. My titties feel like a pair of Pecos strawberries.
2. *Comb.*, as **titty-bag,** a sweetened object given to a baby to suck; **titty-bottle,** a baby's feeding bottle with teat.
1923 J. MANCHON *Le Slang* 314 *Titty-bag*, un suçon. 1976 A. HILL *Summer's End* x. 147 A titty-bag was a piece of rag with sugar poured inside, then the rag was tied up with string and the sugar-lump stuck into the blarting mouth. 1871 B. BRIERLEY *Cotters of Mossburn* iv. 46 He's suckin' th' sofy bowster i' th' bar an' doesno' know but it's a titty-bottle. 1920 D. H. LAWRENCE *Lost Girl* xi. 280 'Eh, tha can ta'e th' titty-bottle wi' thee,' said the labourer.

titty (ti·ti), *a. dial.* and *colloq.* [f. TIT *sb.*^{3} + -Y^{1}.] Diminutive, insignificant.
1884 *Rep. & Trans. Devonshire Assoc.* XVI. 118 A titty piece of cake. 1943 J. W. DAY *Farming Adventure* iii. 41 War Agricultural Committee officials, whom he described as 'titty little bits on motor-bikes—never got their feet wet yet'. 1967 K. GILES *Death & Mr. Prettyman* viii. 156 One of those titty little bikes with a one-horse engine. 1969 E. MCGIRR *Entry of Death* iii. 38 It was a sliver of card... 'This titty little bit of card could be anythink.'
Hence in reduplicated form **titty-totty** *a.* (*sb.*) (*dial.*).
1893 H. COZENS-HARDY *Broad Norfolk* 56 *Titty totty*, extremely tiny. 1943 J. W. DAY *Farming Adventure* v. 62 He hed a little owd titty-totty boy from Tollesbury as a hand—a furriner! 1970 *Morning Star* 28 Mar. 2/8 It is a titty-totty of a tree, a crab apple, a tree nonetheless.

titular, *a.* and *sb.* Add: **A.** *adj.* **1.** *titular abbot*, one holding the title of abbot from a monastery that no longer exists as a religious community; *titular bishop* (earlier example).
1767 A. BUTLER *Short Acct. Life & Virtues of Mary of Holy Cross* p. xviii, He repeated this Charge..to his Coadjutor and Successor the Right Reverend Benjamin Petre, titular Bishop of Prusa. 1934 WEBSTER, Titular abbot. 1977 *Church Times* 1 July 14/2 Coventry. Benedictine anniversary... The Titular Abbot of Westminster will preach.

titulature (ti·tiŭlătiŭɹ). *Anc. Hist.* [f. late L. *titulātum*, pa. pple. of *titulāre* to give title to: see -URE.] The set of titles borne by an official; the form of title by which an official is known.
1893 E. G. HARDY in *Classical Rev.* VII. 49/1 It is well known how carefully the proconsulare imperium was omitted in the imperial titulature. 1971 R. BROWNING *Justinian & Theodora* iii. 97 The memory of it [*sc.* the city Justiniana Prima] remained, and its name appeared in the titulature of certain Serbian archbishops down to 1718. 1973 J. BRISCOE *Commentary on Livy Books XXXI–XXXIII* 5 The titulature of the governors of the Spanish provinces is a more complex problem. L[ivy] describes them variously as praetors, propraetors and proconsuls.

‖ **titulus** (ti·tiŭlŭs). [L.: see TITLE *sb.*] An inscription on or over something; esp. the inscription on the Cross.
1918 *By an Unknown Disciple* xx. 238 He ordered the centurion to have it so inscribed on the Titulus. 1927 A. H. MCNEILE *Introd. N.T.* 10 There was a deep irony in the mockery by the soldiers, and in the *titulus* on the Cross. 1963 *N. & Q.* May 166/2 The best-known type [of 'illustrated' poem] is the inscription-poem used as a *titulus*, carved or painted on tombs or the walls of buildings.

‖ **ti-tzu** (di·dzə). *Mus.* Also **Ti tzu.** [Chinese *dízi*.] A type of Chinese bamboo flute (see quots.).
1874 *Jrnl. N. China Branch R. Asiatic Soc.* VIII. 109 The modern *Ti-tzu*..has seven holes besides the embouchure. 1917 *Encycl. Sinica* 389/1 The *Ti tzŭ* is a very popular flute, about 26 inches long, formerly with 11 finger-holes, one of which was covered with membrane, but now having 6 finger-holes and a 7th covered with membrane. 1954 *Grove's Dict. Mus.* (ed. 5) II. 235/2 *Ti* (to-day usually *ti-tzu*), a transverse flute, the common or popular flute of China... A bamboo tube..pierced with 8 holes (6 fingerholes, 1 blown across and 1 covered with a paper membrane). 1975 C. P. MACKERRAS *Chinese Theatre in Mod. Times* 16 The *K'un-ch'ü*..found most of its admirers among the gentry, officials and scholars. Its music was softer and more melodious than that of the popular drama and it was accompanied principally by the *ti-tzu*, or Chinese transverse flute.

Tiuchiu, var. *TEOCHEW.

Tiv (tiv), *a.* and *sb.* [Native name.] **A.** *adj.* Of, pertaining to, or designating a people of central Nigeria, who live on either side of the Benue River, or the language spoken by them. **B.** *sb.* **a.** A member of this people; also *collect.* **b.** The Bantu language of this people.
1939 R. M. EAST tr. B. Abiga (*title*) Abiga's story: the Tiv tribe as seen by one of its members. 1957 W. M. HAILEY *Afr. Survey* (rev. ed.) iii. 101 The languages of the Tiv group are largely used in mass literacy schemes in the Benue Province. 1960 *Guardian* 15 July 14/4 The important smaller peoples like the Tivs, Kanauris, Binis and Ibibias. 1962 *Listener* 25 Jan. 187/3 Artists and critics among the Tiv of Nigeria. 1973 T. KOCHMAN *Rappin' & Stylin' Out* 83 [Charles Keil] studied Tiv language, culture, music, and aesthetics, and Yoruba urban music. 1976 *Nigeria Herald* 21 May 2/2 News in Kanuri, Tiv and Yoruba. 1977 *Language* LIII. 290 Loanwords are not all lumped into one class (as they tend to be in Tiv). 1982 B. EMECHETA *Destination Biafra* i. 8 In Nigeria..there are smaller groups..surrounding each major tribe... Around the Hausas you have the Tivs, the Fulanis.

tizz, tiz (tiz), shortened form of *TIZZY^{2}.
1954 B. BOLAND *Return* II. iii, in *Plays of Year* IX. 326 *Peter*: Nothing matters. *Angela*: Practically nothing. All the things people get in such a tizz about—. 1958 'N. BLAKE' *Penknife in my Heart* iv. 52 Miriam was as usual in a tiz. 1967 R. RENDELL *New Lease of Death* vi. 54 When Burden looked murderous, he added, 'Don't get in a tiz with me.' 1978 *Illustr. London News* Nov. 97/1 The people of Morecambe were thrown into a tizz by this idea of a barrage [across Morecambe Bay].

tizz-wozz, var. *TISWAS.

tizzy^{1}. Add: Also *Comb.*, as **tizzy-snatcher** *Naut. slang*, an assistant paymaster.
1914 'BARTIMEUS' *Naval Occasions* xiii. 107 'Bloomin' tizzy-snatcher' he muttered slipping the coins into his trousers pocket. He referred to the A.P. (Assistant Paymaster, who had mulched him of sixpence). 1916 'TAFFRAIL' *Pincher Martin* v. 74, I cursed them for a couple of tizzy-snatchers. 1946 J. IRVING *Royal Navalese* 176 *Tizzysnatcher*, a disillusioned Nor' Easter's name for the Paymaster. The derivation is the Cockney 'tizzy' meaning sixpence.

tizzy^{2}. *colloq.* (orig. *U.S.*). [Of uncertain origin.] A state of nervous excitement, agitation or worry, a 'flap'; esp. in phr. *in a tizzy.*
1935 *Amer. Speech* X. 192/1 The *tizzy* in which a huge wedding kept society columnists for weeks. 1938 *Ladies' Home Jrnl.* Oct. 14/2 Maybe it's better for the future of the race to live from daze to daze in a perpetual tizzy like Alix. 1952 A. WILSON *Hemlock & After* II. iii. 170 Politics and the sun together always put me into a mad tizzy. 1958 N. MARSH *Singing in Shrouds* (1959) v. 83 Gets in a tizzy over details. 1967 *Spectator* 10 Nov. 582/3 John Whiting's play..about the tizzies of English gentlefolk involved in the Napoleonic invasion scare. 1974 *Courier-Mail* (Brisbane) 14 Sept. 10/6 A small band of private fliers has had the RAAF base at Amberley in a tizzy. 1983 *Daily Tel.* 8 June 20/3 He hopes this mass production of original art may throw 'into a state of total tizzy' an art world where 'more and more money is being made by less and less people'.

tjaele (‖ʃē·lə, tʃē^{i}·lə). *Geol.* Also **taele, tjäle.** [a. Sw. *tjäle* ice in frozen ground.] Frozen ground; also, permafrost. Freq. *attrib.*
1924 *Geogr. Jrnl.* LXIII. 210 The arrangement of materials can usually be traced right down to the tjaele. *Ibid.* 211 The tjaele ice was still very close to the surface. 1937 *Nature* 4 Sept. 410/1 The Third Glaciation is represented in the north-west by the Irish Sea Drifts,..and by taele-gravels and melt-water flood-gravels. 1939 *Geogr. Jrnl.* XCIV. 451 Near the edge of the snow the *tjäle* approaches the surface and underneath the snow the soil is completely frozen. 1960 B. W. SPARKS *Geomorphol.* xiv. 314 Solifluxion gravels of various sorts have long been recognised and mapped in Great Britain, where they are known by a variety of names, e.g. head, coombe rock, taele gravel..warp and trail. 1970 R. J. SMALL *Study of Landforms* iv. 124 A climatic refrigeration, allied to the onset of periglacial conditions and the formation of permafrost or annual taele.., is capable of rendering permeable rocks impermeable.

tjalk. (Earlier example.)
1861 *Mitchell's Maritime Reg.* 1417/1 Eja, Dutch tjalk, Bronuma, from London for Amsterdam, was totally lost Nov. 1 at West Kapelle.

tjurunga, var. *CHURINGA.

‖ **tlachtli** (tlæ·tʃtli). Also **tlaxtli.** [Nahuatl.] The ceremonial ball-game of the Aztecs; = *POK-TA-POK. Also *attrib.*, as *tlachtli-court, -field.*
1875 H. H. BANCROFT *Native Races Pacific States N. Amer.* II. viii. 297 The national game of the Nahuas was the *tlachtli*, which strongly resembled in many points our game of football. 1914 T. A. JOYCE *Mexican Archaeol.* vi. 165 But the national game, tlaxtli, was closely connected with the worship of the gods, and the tlaxtli-courts..were generally associated with temples. 1959 *Times* 27 Apr. (Rubber Industry Suppl.) p. v/1 The sacramental Tlachtli played by Aztecs in the temple courts of Mexico. 1963 [see *POK-TA-POK]. 1968 H. HELFRITZ *Mexican Cities of Gods* xv. 157 The playing area, or *tlachtli*-field,..is 199 ft. wide.

Tlapanec (tlæ·pānek), *sb.* (and *a.*) Also **Tlapaneco, Tlappanec.** [ad. Sp. *tlapaneca, tlapaneco*, ad. Nahuatl (Aztec) *tlapanecatl.*] **a.** An Indian people of south-west Guerrero, Mexico. **b.** The language of this people, formerly classified as Hokan but now regarded as Otomanguean. Also *attrib.* or as *adj.* Also in *Comb.*, as *Subtiaba-Tlapanec* (see *SUBTIABA).
1875 H. H. BANCROFT *Native Races Pacific States N. Amer.* I. vi. 677 The Tlapanecs, Coviscas, Yopes, Yopis.., Chochos,..or Popolucas are one and the same people, who by different writers are described under one or other of these names. *Ibid.* III. x. 752 Several tongues, of which.. I find nothing mentioned but the names;..further there are mentioned the Chatino, Tlapanec, and Popoluca. 1900 F. STARR *Notes Ethnography S. Mexico* 71 The name Chocho is said by Orozco to be applied to a language in Oaxaca, while to the same language in Puebla is given the name *popoloco*; it is also the *tlapaneco* of Guerrero..and the ancient *Yope.* 1911 THOMAS & SWANTON *Indian Languages Mexico & Central Amer.* 53 Sahagun..says the

Tlapaneco language is precisely the same as those called Tenime, Pinome..in the singular Pinotl.. This brings Tlapaneco into the same relation as that given by Orozco y Berra. *Ibid.* 54 The Tlapanec group is located by Orozco y Berra in Guerrero. 1925, etc. [see *SUBTIABA]. 1940 J. A. MASON in *Maya & Their Neighbors* v. 61 According to Sapir..Subtiaba in Nicaragua differs hardly more than dialectically from Tlapanec in Guerrero. 1972 *Bk. Thousand Tongues* (Amer. Bible Soc.) (rev. ed.) 430/2 The Tlapaneco Indians..live in Guerrero, Mexico. Tlapaneco, spoken with numerous dialectal variations, is a Hokan language, related in Mexico to Seri and Chontal of Oaxaca. 1974 *Encycl. Brit. Macropædia* XI. 960/1 The Tlapanec complex was first correctly identified by Walter Lehmann..in 1920. *Ibid.*, He [*sc.* Sapir] believed Tlapanec to be Oto-Manguean.

Tlingit (kli·ŋkit, kli·ŋgit; also, incorrectly, tl-), *sb.* and *a.* Also **Thlinget, Thlinkeet, Tlinget, Tlinkit,** etc. [a. Tlingit *ɬingít*, *ɬi·ngít* person, Tlingit.] **A.** *sb.* **a.** (A member of) an Indian people of the coasts and islands of south-eastern Alaska. **b.** The language spoken by this people.
1865 M. MACFIE *Vancouver Island & British Columbia* xvi. 452 The 'Clingats', which name is applied to all the northern tribes, relate the following tradition. 1876 *Encycl. Brit.* V. 187/2 This [*sc.* exogamy] is illustrated in the case of the Thlinkeets, or Kolosches, who inhabit the coasts and islands from Mt. St Elias to the River Nass. 1901 G. W. JAMES *Indian Basketry* 50 In Alaska the chief basket-makers are the Thlinkets and Haidas. 1908 *26th Ann. Rep. Bureau Amer. Ethnol.* 472 In Tlingit..*p* and *b* do not occur in words of native origin. *Ibid.* 407 The Tlingit..trace the origin of nearly all their clans..to the neighbourhood of the mouth of the Skeena river. 1921 E. SAPIR *Language* iv. 84 A good example of such a pitch language is Tlingit, spoken by the Indians of the southern coast of Alaska. 1951 R. FIRTH *Elem. Social Organiz.* iv. 145 The *potlach* of the Haida, Tlinkit and other Indians of the American North-West coast. 1965 *Canad. Jrnl. Linguistics* X. 97 To this Athapaskan-Eyak unit, Tlingit seems related definitely but remotely. 1970 *Language* XLVI. 401 Similar problems are presented by the so-called 'third modals' of Tlingit. 1978 *Amer. Poetry Rev.* Sept./Oct. 18/4 The Tlingit of Alaska prescribe the wearing of broad-brimmed hats.
B. *adj.* Of, pertaining to, or designating these Indians.
1881 [see *ALEUT]. 1908 *26th Ann. Rep. Bureau Amer. Ethnol.* 463 Each Tlingit shaman was guarded by a number of helpers and possessed a number of masks. *Ibid.* 472 The Tlingit language tends to shorten its vowels. 1932 D. JENNESS *Indians of Canada* xxi. 331 Feasts and ceremonies occurred constantly in all the Tlinkit villages. 1963 *Times* 29 Apr. 5/5 A Thlinget Indian from Alaska. 1966 [see *CHILKAT]. 1976 *Times* 25 Sept. 12/8 The Tlingit screen..was used as a partition in the house of Chief Shakes of the Tlingit tribe in Wrangell, Alaska.

to, *prep., conj., adv.* Add: **A.** *prep.* **I. 4. a.** (Earlier and later U.S. examples.) Cf. *HOME *sb.*^{1} 13*.
1795, etc. [see *HOME *sb.*^{1} 13*]. 1801 J. QUINCY in *Proc. Mass. Hist. Soc.* (1888) 2nd Ser. IV. 130 Mr. William Hammatt and Mr. Josiah Barker..called and invited us to a party they had made for us to the East end of the Island. 1818 L. D. CLARK *Jrnl.* 10 Sept. in *Firelands Pioneer* (1920) XXI. 2321 Stayed to Canfields all night. 1977 *New Yorker* 15 Aug. 37/2 Suzanne said, 'What about Sunday? We could do something in the afternoon. Were you ever to the Botanic Gardens?'
b. *to work*: at work, working. *U.S. colloq.*
1776 *Proc. Mass. Hist. Soc.* (1886) 2nd. Ser. II. 304 [We] met some people to work on the High: way. 1827 S. S. ARNOLD *Proc. Vermont Hist. Soc.* (1940) VIII. 111 Her husband..had died instantly in the barn, where he was to work. 1834 C. A. DAVIS *Lett. J. Downing* 116, I have been to work on it ever since we was at the Rip-Raps. 1858 *Rome* (N.Y.) *Sentinel* Sept., The boiler.. passed through the main building..without injuring the workmen there, although men were to work on each side of where the boiler passed. 1949 *N.Y. Herald Tribune* 6 Dec. 1 Some 450,000 miners were back to work today. 1978 M. Z. LEWIN *Silent Salesman* xxvii. 146 He's to work... Don't rightly know what time he'll be back.
II. 6. b. (Earlier and later examples of use with ellipsis of the hour.)
1842 TENNYSON *Walking to the Mail* in *Poems* II. 47 *James.* The mail? At one o'clock. *John.* What is it now? *James.* A quarter to. 1968 'R. PETRIE' *MacLurg goes West* II. vii. 60 'I thought we might just catch you before dinner,' said Mrs. Robbins to them quickly. 'It's twelve minutes to.'
III. 8. d. Indicating the crop with which ground is planted. Chiefly *U.S.*
1799, etc. [see *PLANT *v.* 6 a]. 1833 S. SMITH *Life & Writings Major J. Downing* 22 [He]..planted the ground all over to corn, and potatoes. 1848 F. A. DURIVAGE *Stray Subjects* 21 Having laid down a few acres to oats. 1902 [in Dict., sense 8 a]. 1945 B. MACDONALD *Egg & I* (1946) I. iii. 45 The garden..was planted to peas, beets, beans, corn, Swiss chard, lettuce, cabbage, onions, turnips, celery, cucumbers, tomatoes and squash. 1980 *Daily Tel.* 17 Sept. 8/3 The area sown to winter barley was greatly increased.
V. 15. a. (Further examples.) Now only indicating food taken in addition to a dish or meal, and in this use *dial.*
1792 W. COWPER *Let.* 30 Nov. in J. A. Roy *Cowper & his Poetry* (1914) viii. 166 It is impossible any longer to find a pound of butter or cream to our tea in all the country. 1916 'TAFFRAIL' *Pincher Martin* vii. 107 My poor John was fond of a hegg to 'is tea. 1925 V. WOOLF *Let.* 20 Sept. (1977) III. 213, I am growing old, and want more mustard to my meat.

b. Also indicating the tune to which words are set.
1591 [see TUNE *sb.* 2 a]. 1611, 1702 [see GO *v.* 17]. 1825 C. WATERTON *Wanderings S. Amer.* iv. 279 There is an old song, to the tune of La Belle Catharine. 1906 BELLOC *Hills & Sea* 116 The two trumpets of the battery sounding the call which is known among French gunners as 'the eighty hunters', because the words to it are '*Quatre-vingt, quatre-vingt..chasseurs*'.
17. c. In colloq. phrases with *there is* and a quantitative or pronominal expression: belonging as a quality, attribute, or capacity to (someone or something, freq. *it*); *that is all there is to it*: it is that and nothing more; *there's nothing to it*: see *NOTHING *sb.* 10 c. orig. *U.S.*
1880 'MARK TWAIN' *Tramp Abroad* ii. 36 There's more *to* a blue-jay than any other creature. 1883 —— *Life on Mississippi* xlv. 459 The steamboat shoved out up the creek. That was all there was 'to it'. 1895 KIPLING *Day's Work* (1898) 83 'That's all there is to it,' seethed the white water roaring through the scuppers. 1903 A. H. LEWIS *Boss* 14 Tell me what there is to this shindy. 1914 V. CASTLE *Mod. Dancing* 44 Simply *walk* as softly and smoothly as possible... This is the One Step, and this is all there is to it. 1936 L. C. DOUGLAS *White Banners* xvi. 343 He's a wonderful person, you know. There's a lot to him that doesn't show up on the surface. 1974 J. THOMSON *Long Revenge* iii. 40 He had the feeling that there was a great deal more to it [*sc.* a case] than he had so far discovered. 1976 *New Yorker* 26 Apr. 38/1, I thought she had a lot to her, a lot to offer.
VI. 19. a. (Earlier example with reference to votes.)
1836 J. ROMILLY *Diary* 30 Nov. (1967) 109 The grace.. was thrown out in the White hood house by 30 to 21; it past in the black by 23 to 20.
VIII. 34. *Book-keeping.* Placed before debit entries, and followed by particulars of the goods or services for which money has been paid, or by the name of the account containing the corresponding credit entry. Cf. BY *prep.* 37.
1772 in *Country Life* (1973) 7 June (Suppl.) 104 To mending a Waiter & Candlestick & a Sauceboat 5s. 1803 G. COLMAN *John Bull* III. i. 31 These charges are brought in like a bill!—To attending your ladyship at such a time —to dancing down twenty couple at another. 1876 *Encycl. Brit.* IV. 46 To J. Bevan and Co., for Bales, *ex* 'Mary Jane' £2349 0s. 0d. 1901 *Jrnl. R. Microsc. Soc.* 109 The Treasurer's Account for 1900... To Balance from 1899..£195 11s. 3d. 1968 G. M. WHITEHEAD *Book-Keeping made Simple* v. 79 Whenever a debit entry is made on an account we begin with the word 'To' and follow with the name of the account where the other half of the double entry is to be found. 1978 J. KELLOCK *Elements of Accounting* i. 11 In many accounting text books the words 'To' and 'By' are used to preface debit and credit entries respectively in the ledger... These prefixes are now being discontinued in modern accounting systems.
35. Preceding the name of persons or groups who use a specified name or expression: in the language or usage of.
1922 P. S. O'HEGARTY *Terence MacSwiney* ii. 3 Terence James MacSwiney on the baptismal register, but Terry always to his friends and to Cork generally, was born in Cork City on March 28th, 1879. 1941 *Poor Souls' Friend* June 111 Her father, Edmund William Roe (Ted to his friends) was a man of character and great individuality. 1956 J. BRODRICK *St Ignatius Loyola* i. 12 In the Basque countries (to the Basques Euskalerria). 1970 *Outlook* Mar. 34 Owen Glyn Dwr—Glendower to the Anglo-Saxon—was the Welsh prince who made most of the mischief. 1977 *Transatlantic Rev.* LX. 118 Lindy (Miss Hoffmann to the kids) had to glide it back down to them.
B. To before an infinitive.
11. a. (*c*) *to let* used *absol.* as *sb.* Freq. applied *attrib.* to a board, sign, etc., indicating that premises are offered for rent.
1886 F. H. BURNETT *Little Lord Fauntleroy* xi. 174 He stopped opposite the empty house..staring at the 'To Let', and smoking his pipe. 1894 A. MORRISON *Martin Hewitt, Investigator* ii. 80 The three shops..appeared not yet to have been occupied. A dusty 'To Let' bill hung in each window. 1903 A. BENNETT *Truth about an Author* xv. 206 A To-let notice flourished suddenly in my front-garden. 1936 A. CHRISTIE *ABC Murders* vi. 46 A 'To Let' sign appeared in the windows. 1938 G. GREENE *Brighton Rock* III. iii. 122 A vista of To Let boards. 1976 J. BINGHAM *God's Defector* v. 54 One day they have hope, a basement, a letter-head, and the next their place is occupied by a 'To let' sign.

‖ **to** (tǫ), *sb.* Now *rare.* Pl. **to.** [Jap.] A Japanese unit of capacity equal to ten *sho*, equivalent to approximately 3·97 gallons (18·0 litres) or 0·496 bushel.
1871 A. B. MITFORD *Tales of Old Japan* II. 2 Each of these bags holds four tô (a tô is rather less than half an imperial bushel). 1884 *Murray's Handbk. Japan* (ed. 2) 18, 10 *shō* = 1 *to*. 1901 F. BRINKLEY *Japan* II. iii. 118 At the close of the sixteenth century,..the measure of capacity was exactly fixed, and its volume was called *tō*; ten *tō* (i.e. a sheaf of grain) being called a *koku*. 1956 R. J. SMITH in Cornell & Smith *Two Japanese Villages* 90 The most expensive *hōji*..costs a minimum of 4,000 *yen* (one *koku* of rice). The least expensive costs 1,000 *yen* (1 *tō* of rice). 1959 R. K. BEARDSLEY et al. *Village Japan* 488/2 *Tō*, measure of volume; about 4 gallons.

toa. (Earlier example.)
1792 W. BLIGH *Voy. South Sea* xii. 148 A wooden spear ..pointed with the toa wood. *Ibid.* xvi. 213 The soil is little other than sand, yet it produced small toa-trees.

‖ **toa**^{2} (tō^{u}·ă). *N.Z.* [Maori.] A brave warrior.
1860 A. S. ATKINSON *Jrnl.* 25 Dec. in *Richmond-Atkinson Papers* (1960) I. 671 A remark of Manuka's was rather good—Maori chiefs (leaders) were toas & went out at the head of their men, Pakeha chiefs stayed in Town & ate biscuit. 1881 C. W. HURSTHOUSE *Let.* 12 June in *Ibid.* (1960) II. 486 Tuninia was a celebrated toa, and knew it. 1901 A. H. GRACE in D. M. Davin *N.Z. Short Stories* (1953) 19 You can imagine what joy it is to become the wife of such a brave *toa*. 1949 P. H. BUCK *Coming of Maori* (1950) III. i. 338 Those who had distinguished themselves in battle were termed *toa* and were usually of the *rangatira* class, for ruling chiefs had to lead in war as well as in peace.

toad, *sb.* Add: **3. b.** Applied to children. Cf. *TAD 2.
1836 T. C. HALIBURTON *Clockmaker* 1st Ser. xxvii. 178 Two little orphan children, the prettiest little toads I ever beheld. 1897 *Private Life of Queen* xi. 93 Jonathan Mace..had been a day labourer at Frogmore... He always spoke of the Queen's spirited sons as 'rare young toads'. 1954 M. SHARP *Gipsy in Parlour* IV. xxiv. 234 Why shouldn't 'ee wed the poor toad? 1958 'MISS READ' *Storm in Village* iii. 38 If our Billy has the nightmares, I shan't wonder! Poor little toad, and him so high strung! 1981 V. CANNING *Boy on Platform One* vii. 108 I'll show you. Never seen a salmon! You poor little toad.
4. (Later example.)
1959 I. & P. OPIE *Lore & Lang. Schoolch.* x. 191 One who makes up to a teacher is recognized as being in a slightly different category from an outright sneak, although almost as nauseous. The usual epithets are 'toad' or 'toady', 'worms', 'crawler', or, in Camberwell, 'grease boy' or 'grease rat'.
6. a. For '*toad in a hole*: see quots.' read *toad in the* († *a*) *hole*: meat, now usu. sausages, baked in batter. (Later examples.)
1927 [see *SCHOONER *sb.*^{1} 1 b]. 1934 *Cassell's Mod. Practical Cookery* 145/1 Put in the sausages. Season the batter,..and pour it over them. Put the toad-in-the-hole into a good, moderately hot oven. 1934 T. S. ELIOT *Rock* i. 40 Restaurants where you can get..sausage and mashed or toad-in-the-'ole for twopence. 1943 L. CHATTERTON *Mod. Cookery* 167/1 Toad in the hole without eggs... Milk,..flour,..bicarbonate of soda,..vinegar,..salt,.. sausages,..dripping... Small sausages are best.., or.. steak cut in two-inch pieces may be used. 1959 B. NILSON *Penguin Cookery Bk.* 330 Toad-in-the-hole is made like Yorkshire pudding, but 1lb. skinned sausages is heated in the fat for 5 minutes before adding the batter. 1971 *Guardian* 18 June 11/6 Although 'toad-in-the-hole' is an unprepossessing name it must be one of the few dishes which sounds even worse in French. I am sure no one would eat 'Le crapaud au trou'.
b. *toad in the hole*: a name applied to various games, esp. a form of hide-and-seek and a game in which lead discs are thrown at holes in a wooden structure.
1930 J. DOS PASSOS *42nd Parallel* 381 They got tired playing toad in the hole in the deep weeds. 1969 I. & P. OPIE *Children's Games* iv. 154 When played after dark, as is not unusual, it [*sc.* Hide-and-Seek] may have a special name, such as ..'Toad in the Hole' (Forfar). 1969 E. H. PINTO *Treen* 229 Toad-in-the-hole..probably originated in England in Tudor times. Since then, it has been played in many parts of the world, including Argentina, where it is known as *Sapo*. 1970 *Daily Tel.* 18 Mar. 15/6 Mr. P. N. Barnard's letter..about the 'charity game' known as toad in the hole..has provided me with a translation into English of this very old game which used to be played a great deal in France, where it is known as *tonneau*. 1975 *Country Life* 11 Dec. 1677/4, I am..looking for examples of the following regional inn sports:..twister (Suffolk), and toad in the hole (Sussex)... Every one of the games I have mentioned is actually played in English pubs today.
7. a. *toad-bellied* (later example), *-spotted* (later example).
1922 JOYCE *Ulysses* 465 Beside him stands Father Coffey, chaplain, toadbellied, wrynecked. 1915 W. OWEN *Let.* Apr. (1967) 331 It's Measles!... Bloody eyes—toad-spotted, raw-meat-coloured skin.
b. **toad-cheese**: (earlier example of form *toad's cheese*); **toad-frog,** (*b*) *U.S. dial.* = sense 1 a in Dict; † **toadskin** *N. Amer. slang* (*obs.*), (*a*) a five-cent stamp; (*b*) a banknote; **toad-stabber** *slang* (chiefly *U.S.*), a large pocket-knife or jack-knife; **toad-sticker** *U.S. slang*, a large knife; formerly also, a sword; **toad-strangler** *U.S. dial.*, a heavy downpour of rain.
1853 J. LOUSLEY *Let.* 9 Jan. in *N. & Q.* (1962) Mar. 84/1 Toads cheeses are the poisenous Fungusses which grow in our hedgerows and woods. 1861 *Harper's Mag.* Aug. 421/1 Every body is a pitching into this matter like toad-frogs into a willow swamp. 1913 H. KEPHART *Our Southern Highlanders* xiii. 295 In the Smokies a toad is called a frog or a toad-frog. 1964 J. H. CLARKE *Harlem* 276 She just stood..swellin' up like a big toad frog. 1981 *Amer. Speech* LVI. 45 Toad-frog is Southern and Midland. 1867 F. H. LUDLOW *Little Brother* 251 'Why, ma, don't you know what a toadskin is?' said Billy, drawing a dingy five-cent stamp from his pocket. 'Here's one... And don't I wish I had lots of 'em!' 1912 J. SANDILANDS *Western Canad. Dict. & Phrase-Bk.*, *Toadskin*, a dollar bill. Originally, in the States, a toadskin meant a five-cent stamp, and of a mean, grasping person it was said 'His purse is made of toad's skin.' 1926 Toadskin [see *IRONMAN 1 c]. 1885 G. SWEETMAN *Gloss. Wincanton, Toad-stabber*,..a bad knife. 1915 S. LEWIS *Trail of Hawk* x. 102

Carl..pried open a class-room window with his large jack-knife..known as a 'toad-stabber'. 1938 W. SMITTER *F.O.B. Detroit* 48 'There you are,' said Russ, snapping the blade open. 'A regular toad-stabber of a thing.' 1858 *Calif. Spirit of Times* 7 Aug. 1/8 The Judge put his toad sticker atween his teeth, tuk a pistol in won hand, and a slung shot in the other, an sez thru his nose, 'cum on'. 1944 J. S. PENNELL *Hist. Rome Hanks* 293, I must have picked up this old toadsticker. 1938 M. K. RAWLINGS *Yearling* xix. 228 Hit's a toad-strangler of a rain. 1980 *Knoxville* (Tennessee) *News-Sentinel* 6 Apr. c4/5 'We say toad-strangler for a hard rain around here,' Farley said.

toad-fish. **a.** Substitute for second part of def.: also, a fish belonging to any of numerous other species of the family Tetraodontidæ, many of which are poisonous. (Further Austral. examples.)
1871 G. BENNETT in *N.S.W. Med. Gaz.* I. 176 (*title*) On the 'toad fish'..of New South Wales. 1923 *Med. Jrnl. Austral.* 1 Dec. 572/1 The toad fish belongs to the genus Tetraodon of which a number of species are known to be poisonous. 1974 [see *TOADO].

toadless (tō^{u}·d,lĕs), *a.* [f. TOAD *sb.* + -LESS.] Devoid of toads.
1911 *Chambers's Jrnl.* July 435/2 When the garden was dug..it was toadless. 1922 M. TEMPLE *Shallowdale* iii. 39 No dog can be thoroughly happy in a toadless garden.

toado (tō^{u}·do). *Austral.* [f. TOAD(-FISH + *-O^{2}.] A poisonous puffer-fish of the family Tetraodontidæ.
1943 G. P. WHITLEY in *Bull. Council Sci. & Industrial Res.* (Australia) CLIX. 8 Australia alone has more than 30 different species of Toadoes. 1953 *Copeia* I. 32/2 The terms 'tetrodon', fugu, globefish, toad-fish, toado, and puffer poisoning are synonymous. 1965 *Courier-Mail* (Brisbane) 11 Sept. 2/1 Few fish in the sea..pack such deadly poison as the Toadoes. 1974 J. M. THOMSON *Fish of Ocean & Shore* xv. 162 The common toadfish or toado..has a splotchy brown or green coat. *Ibid.*, The silver-cheeked toadfish or giant toado..attains a metre in length.

toa·dstool, *v. rare.* [f. the sb.] *intr.* To grow up like a toadstool; to expand or increase rapidly and objectionably. Cf. *MUSHROOM *v.* 4.
1939 R. CAMPBELL *Flowering Rifle* I. 14 As limply fungoid in the idle rich As when it grimly toadstools from a ditch. 1971 M. MCCARTHY *Birds of America* 60 New little houses had toadstooled; they passed a trailer camp.

toa·dstooled, *a. rare.* [f. as prec. + -ED^{2}.] Overgrown with toadstools.
1910 KIPLING *Rewards & Fairies* 282 They hit an old toadstooled stump.

toadyish (tō^{u}·di,iʃ), *a.* [f. TOADY *sb.* + -ISH^{1}.] Characteristic of or resembling a toady; meanly servile.
1909 in WEBSTER. 1955 M. COOPER in H. Van Thal *Fanfare for E. Newman* 50 A particularly toadyish begging letter. 1977 B. PYM *Quartet in Autumn* vi. 57 Only Marya, toadyish with her murmurs of 'such delicious coffee', accepted the offer.

toa grass, var. of *TWA(A-GRAS(S.

to and fro, *phr.* **D.** *adj.* (Earlier example.)
1749 J. CLELAND *Mem. Woman of Pleasure* I. 212 The sweet urgency of this to-and-fro friction.

toast, *sb.*^{1} Add: **1. a.** For 'Now *rare* or *Obs.* except as in b' read 'Now *rare* or *Obs.* except in India' and add later examples.
1838 DICKENS *Let.* 1 Feb. (1965) I. 366 We have had for breakfast, toasts, cakes, a yorkshire pie [etc.]. 1978 *Vishveshvaranand Indological Jrnl.* XVI. 218 He had stopped taking cereals after the age of sixty but after 85 he had to re-start on medical advice taking two toasts or some cornflakes.
b. (Earlier example of phr. *as warm as toast.*)
1855 A. S. STEPHENS *Old Homestead* i. 16 Every thing nice and warm as toast.
2. a. *French toast*: see *FRENCH *a.* 3.
c. Add to def.: *to have* (one) *on toast* (colloq.), to have (a person) at one's mercy or 'where one wants him'; to subject to anxiety (later examples). Also with other verbs.
1896 B. L. FARJEON *Betrayal of John Fordham* III. 288 'It's my night,' I sed. 'Didn't I tell yer? I've got 'im on toast.' 1916 E. F. BENSON *David Blaize* xiv. 285 To think that half an hour ago that little squirt thought he had us on toast. 1929 D. H. LAWRENCE *Pansies* 127 But Tolstoi was a traitor To the Russia that needed him most... He shifted his job on to the peasants And landed them all on toast. 1942 'R. CROMPTON' *William Carries On* v. 119 Well, let's have 'em on toast for a bit wonderin' what's happened to him. 1964 J. CREASEY *Guilt of Innocence* xvii. 151, I think the time has come to tell the Press we want to interview him... That will get 'em both on toast. 1981 'J. ASHFORD' *Loss of Culion* xix. 151 'You've been positively identified by Mr Barnard.'..'Then he's having you on toast.' 'He has no reason for lying.'
4. *toast-crumb* (example); **toast Melba**: see *MELBA.
1872 G. M. HOPKINS *Let.* 22 Mar. (1956) 55 If you say the Mahābhārata is your toast-crumb ordinary breakfast book I am jaundiced all marigold under the eyes.

toast, *sb.*^{2} Add: **3. toast-master('s) glass,** a drinking-glass having a thick bowl on a tall stem and thus giving the impression of having greater capacity than it really has; **toast-mistress,** a female toast-master.
1916 J. H. YOXALL *Collecting Old Glass* ix. 63 (*heading*) Toastmaster glasses. 1919 M. PERCIVAL *Glass Collector* 162 Toast-masters' [*sic*] glasses are found in many varieties. 1969 *Canad. Antiques Collector* Oct. 25/1 The tiny, clear toastmaster glass was usually solid except for a narrow v-shaped depression at the top capable of holding a bare half-ounce of liquor. 1921 *Daily Colonist* (Victoria, B.C.) 7 Apr. 7/5 The toast mistress, Mrs. Sutton, referred in very complimentary terms to the naval lads and their splendid services during the war. 1979 *Arizona Daily Star* 5 Aug. J 6/1 She..has been picked as 'Toastmistress of the Year'.

toast (tō^{u}st), *sb.*^{3} Chiefly *U.S.* (and *W. Indies*). [Perh. f. TOAST *sb.*^{2}] **1.** A type of long narrative poem recited extempore by American and Caribbean Blacks.
1962 R. D. ABRAHAMS in A. Dundes *Mother Wit* (1973) 300/1 Many of them [*sc.* insults] take the form of rhymes or puns, signaling the beginning of the bloom of verbal dexterity which comes to fruition later in the long narrative poem called the 'toast'. 1972 T. KOCHMAN *Rappin' & Stylin' Out* 261 The best talkers from this group often become the successful streetcorner, barber shop, and pool hall storytellers who deliver the long, rhymed, witty narrative stories called 'toasts'. 1978 *Maledicta* II. 290 An extraordinary collection of black American folk poetry (*toasts*) collected by the author from lower-class black males (inmates of county jails, streetcorner gangs).
2. In reggae, a performance by a disc-jockey who speaks or shouts while playing a record.
1980 N. KIMBERLEY in J. Collis *Rock Primer* 249 The wedding of John Holt's sentimental singing and Roy's effervescent toast..show [*sic*] us the new musical idiom in full flower. 1983 *Listener* 19 May 22/3 Loud and bass-heavy 'dub' music with a patois talkover 'toast' booms into the bus.

toast, *v.*^{1} Add: **toasting-fork** (earlier example).
1807 SOUTHEY *Lett. from Eng.* I. xvii. 185 Pocket-toasting-forks have been invented, as if it were possible to want a toasting-fork in the pocket.

toast (tō^{u}st), *v.*^{3} [See *TOAST *sb.*^{3}] *trans.* and *intr.* In reggae, to accompany (music) by speaking or shouting. Freq. **toa·sting** *vbl. sb.*^{3}
1976 *New Musical Express* 17 Apr. 17/5 Another bass riff that cracks foundations, knocks down walls, and brushes aside nine stone weaklings, but this and all the dubwise trickery in Trenchtown can't hide the absolute ordinariness of Woosh's toasting. 1980 N. KIMBERLEY in J. Collis *Rock Primer* 249 Much of the strength of 'Your Ace From Space', 'Version Galore', etc, lies in the original rhythm which Roy toasts. 1980 *Times* 19 May 9/4 A group of young London blacks whose lives centre on their reggae music—the technology of sound systems, the virtuoso techniques of improvisational 'toasting'.

toaster^{1}. Add: **3.** An electric appliance for toasting bread; = *electric toaster* s.v. *ELECTRIC *a.* 2 b. See also *pop-up toaster* s.v. *POP-UP *a.* b.
1913 *Maclean's Mag.* Feb. 163/1 Electric cooking appliances—the shining nickel-plated or aluminum utensils, including coffee percolators, toasters, chafing dishes, each with its long connecting cord and plug for attachment to the electric light socket. 1948 *Clarke County Democrat* (Grove Hill, Alabama) 2 Dec. 4/4 Frigidaire Ranges, Water Heaters, Sunbeam Mixmasters, Toasters..are here. 1962 A. LURIE *Love & Friendship* x. 187 He has to have it quiet so he can fix our toaster for us. 1975 C. FREMLIN *Long Shadow* xi. 82 The second lot of crumpets were out of the toaster.
4. Special Comb.: **toaster-oven,** a small oven suitable for toasting, broiling, and baking.
1961 *Better Homes & Gardens* Sept. 74/2 Toasters come in family-fitting styles... New dual purpose styles: a toaster-oven and a toaster-broiler. 1976 *Woman's Day* (U.S.) Nov. 124 The island's taller side..houses a microwave oven, toaster-oven and..a waffle grill. 1980 *Redbook* Oct. 54/1 In the kitchen, use a small toaster-oven whenever possible rather than the main oven, particularly if the latter is electric.

toaster^{3} (tō^{u}·stəɹ). [f. *TOAST *v.*^{3} + -ER^{1}.] In reggae, one who accompanies music by speaking or shouting.
1976 in Cassidy & Le Page *Dict. Jamaican Eng.* (1980) 507/2 The toaster has the microphone; he introduces the performers, comments, beats time and talks with the music—keeps things lively. 1977 *Melody Maker* 5 Feb. 16/2 New albums by two of Jamaica's most popular toasters. 1980 *Guardian* 6 Nov. 9/1 Blue..is a garage mechanic by day and a 'toaster' for a reggae group by night.

toast-rack. For '(quot. 1905)' read: a vehicle, esp. a tram, having full-width seats and (usu.) open sides; also *attrib.* (Later examples.)
1941 BAKER *Dict. Austral. Slang* 77 *Toastrack*, one of the old-style footboard trams still used in Sydney. 1957 *Railway Mag.* June 427/2 There are now 24 bogie motor cars and 25 bogie trailers. Toastrack and saloon types exist in both categories; the former, as befits a holiday line, are in the majority. 1966 P. MATHERS *Trap* 190 A tram now, it would be a toast-rack with ten or so compartments with the only physical intercommunication along the outside footboards, and the concertina doors. 1970 *Railway Mag.* Oct. 587/2 Only horse traction was used there; there were two four-wheel cars, a closed one for winter service, and an open 'toast-rack', used in the summer. 1976 *Country Life* 22 Jan. 191/4 Single-deck 'toast-rack' trams, so named because they closely resembled that familiar implement on wheels.

toasty, *a.* Add: (Earlier and later examples.) Now usu. *spec.*, warm and comfortable.
1890 BARRÈRE & LELAND *Dict. Slang* II. 357/1 *Toasty* (artists), said of a picture painted in very warm tints. French painters call this *rôti*. 1961 WEBSTER, *Toasty*,..toasted or comfortably warmed. 1970 *New Yorker* 19 Sept. 14/1 (Advt.), A smart, double-breasted coat... In Melton..with..a toasty wool interlining for cold weather. 1977 *Time* 14 Feb. 32/2 His toasty (75°) office on Nevsky Prospekt. 1978 *Nieman-Marcus Xmas Bk.* 80/2 Natural shearling ear muffs keep ears truly toasty in the bitterest weather.

toatoa (tō^{u}·ātō^{u}·ā). *N.Z.* [a. Maori.] = *celery pine* s.v. *CELERY 2.
1831 G. BENNETT in *London Med. Gaz.* 12 Nov. 184/1 Toatoa, of the natives of New Zealand, is an unpublished species of Phyllocladus. 1845 E. J. WAKEFIELD *Adv. N.Z.* I. 20 The *toa toa*..is much prized by the natives for walking-sticks. 1910 L. COCKAYNE *N.Z. Plants* iii. 46 Confined to the north are..the toatoa (*Phyllocladus glaucus*) and some other trees. 1966 *Encycl. N.Z.* I. 730/1 In the same forests are rimu, tanekaha..and, in places, toatoa.

tobacco. Add: **1. b.** A fashion shade; cf. *TABAC *a.* Cf. sense 3 c below.
1923 *Daily Mail* 10 Jan. 1 Becoming Hat in good quality Petersham Ribbon... Colours: Grey, Cherry, Nigger, Tobacco, Peacock. *Ibid.* 5 June 6 In Pale and Mid Fawn,..Sky, Tobacco, Lemon. 1954 [see *ALIZARIN]. 1972 *Country Life* 7 Dec. (Suppl.) 24/2 Haroun Keshan carpet..rich tobacco field. 1980 G. M. FRASER *Mr American* I. iii. 41 Socks, in the fashionable shades of tobacco, Leander, Wedgwood and crushed strawberry.
3. a. *tobacco bag, barn, field, -jar* (examples), *tin.* **b.** *tobacco-chewing* (earlier examples), *-growing* (example); *tobacco-chewer, -planter, -trader* (example). **c.** *tobacco-brown* sb. and adj. (cf. sense 1 b above); *tobacco-coloured* adj.
d. tobacco baron [*BARON 2 b, c], (*a*) *colloq.*, a powerful tobacco merchant or manufacturer; (*b*) *slang*, a prisoner who dominates his companions because he is able to sell tobacco to them (cf. quot. 1950 s.v. *BARON 2 c); **tobacco beetle** (examples); **tobacco dove,** substitute for def.: a small light brown ground-dove, *Columbina passerina*, native to central America; (example); **tobacco fly,** a hawk moth of the genus *Protoparce*, either *P. quinquemaculata* or *P. sexta*, the larva of which feeds on tobacco leaves; **tobacco-grater** (example); **tobacco housing** (see quot. 1965); **tobacco-leaf,** (*a*) sense 3 a in Dict.; (*b*) used *attrib.* to designate eighteenth-century Chinese porcelain decorated with a floral pattern including tobacco-leaves; **tobacco lord** (now *Sc. Hist.*), a wealthy tobacco merchant of Glasgow; **tobacco mosaic virus,** the virus that causes mosaic disease in tobacco and similar effects in other plants, much used as an experimental subject; **Tobacco Road,** the title of a novel (1932) and play by Erskine Caldwell, used allusively with reference to conditions of extreme poverty, esp. in rural districts of the Southern U.S.; **tobacco-root,** also = VALERIAN 1; (earlier and later examples); **tobacco-shop** (earlier example); **tobacco streak,** a streak disease of tobacco (*STREAK *sb.*^{1} 6*); **tobacco worm,** substitute for def.: the larva of the tobacco fly; (earlier and later examples).
1643 R. WILLIAMS *Key into Lang. of America* vi. 44 Generally all the men throughout the countrey have a Tobacco-bag, with a pipe in it, hanging at their back. 1864 M. CUMMING *Jrnl. Hospital Life* (1866) 120/2, I noticed some of the ladies about having tobacco bags made. 1961 L. VAN DER POST *Heart of Hunter* vii. 115 He poured the capsules into an empty canvas tobacco bag. 1877 G. W. BAGBY *Old Virginia Gentleman* (1910) 3 Where is your plank to come from, and your logs for new barns and tobacco barns? 1971 *Country Life* 22 July 214/2 (*caption*) The tobacco barns are a characteristic feature of the landscape. 1961 *Spectator* 7 July 5/1 The Tories and the 'tobacco barons', who had recently raised their prices. 1964 *Daily Tel.* 15 Jan. 15/1 Powers to limit the activities of prison 'tobacco barons' are provided in the modernised prison and Borstal rules. 1896, 1959 Tobacco beetle [see *cigarette beetle* s.v. *CIGARETTE 2]. 1908 *Sears, Roebuck Catal.* 437/1 We can furnish plain brocaded velour in solid color of myrtle green, deep red or tobacco brown. 1940 R. CHANDLER *Farewell, my Lovely* xiii. 99 She was wearing a tobacco-brown suit. 1977 Tobacco-chewer [see *SMOKING *vbl. sb.* 2 c]. *a* 1832 F. TROLLOPE *Dom. Manners Amer.* (1949) 421 Doom to death the spitter and tobacco chewer. *Ibid.*, When a tobacco-chewing age preceded that of Anacreon, my books do not say. 1972 D. BLOODWORTH *Any Number can Play* xiii. 113 A tobacco-coloured dress of [illegible] 1954 SMILEY & WHITE *Hurricane Road* xi. 97 Blackbirds, tobacco doves, and a roseate tern fluttered

about in bewilderment. 1852 J. B. JONES *Adventures Col. Vanderbomb* 46 They rode by a large tobacco field. 1981 B. HEALEY *Week of Scorpion* vi. 112 They turned aside along a quieter lane between tobacco fields. [1688 *Phil. Trans. R. Soc.* XVII. 947 There be various Accidents and Distempers, whereunto Tobacco is liable, as the Worm, the Flie,..and the like.] 1807 Tobacco fly [in Dict.]. 1904 E. GLASGOW *Deliverance* 126 It was..mid-August—the time of the harvest moon and the dreaded tobacco fly. 1962 METCALF & FLINT *Destructive & Useful Insects* (ed. 4) xiii. 594 The parent 'tobacco flies', or hawk moths,..lay the eggs of the hornworm. *a* 1877 KNIGHT *Dict. Mech.* III. 2583/1 *Tobacco-grater*, a machine for grinding tobacco into small pieces suitable for smoking in pipes. 1824 *Deb. Congress U.S.* 13 Apr. (1856) 2324 The effect of this measure on the cotton, rice, and tobacco-growing States will be pernicious in the extreme. 1960 *Encounter* Feb. 31/1 Those [G.I.s] who live in the semi-luxury of on-base 'tobacco' housing. 1965 *New Society* 22 Apr. 5/3 'Tobacco housing' constructed with sterling funds from sales of American tobacco in England. 1775 J. LOVELL *Let.* 26 June in *Essex Inst. Hist. Coll.* (1875) XIII. 186, 1 Tobacco Jar; 1 Large Lead.º dº. 1857 T. B. GUNN *N.Y. Boarding-Houses* 26 Hair-brush and tobacco-jar jumbled among your shirt-collars. 1967 M. KENYON *Whole Hog* xxv. 252 A tobacco jar bounced..to the floor, where it exploded into fragments. 1969 *Times* 25 Feb. 12/5 A magnificent tobacco leaf dinner service of 96 pieces. 1976 *Times* 27 July 14/5 A famille rose tobacco leaf part service, painted with a lady punting a lotus leaf of flowers. 1832 J. CLELAND *Enumeration Inhabitants of Glasgow* 258 When any of the most respectable master tradesmen of the city had occasion to speak to a tobacco lord, he required to walk on the other side of the street till he was fortunate enough to meet his eye. 1975 T. M. DEVINE (*title*) The tobacco lords. 1914 *Bull. U.S. Dept. Agric.* No. 40. 15, 15 healthy tobacco plants..were inoculated with tobacco mosaic virus. 1947 *Ann. Rev. Microbiol.* I. 87 Ordinary tobacco mosaic virus consists of submicroscopic, rod-shaped particles..composed chiefly of nucleoprotein and possessing a high degree of resistance to heat, desiccation and deleterious chemicals. 1970 PASSMORE & ROBSON *Compan. Med. Stud.* II. xviii. 93/2 Plant viruses, such as tobacco mosaic virus (TMV), are more easily studied than animal viruses. 1775 *Amer. Husbandry* I. 66 Those who have dealings with London.. are the tobacco and rice planters. 1838 *Southern Lit. Messenger* IV. 197 A fine old specimen of the real Virginia tobacco planter, a half domesticated son of France. 1937 *Harper's Mag.* Nov. 566/1 Nobody in his senses wants slums, Tobacco Roads, and undernourished, ragged schoolchildren in a land of potential economic plenty. 1961 Tobacco Road [see *OVER-EXPLOIT *v.*]. 1845 J. C. FRÉMONT *Rep. Exploring Exped.* 135, I ate here, for the first time, the *kooyah*, or tobacco root, (*valeriana edulis*). 1919 E. L. STURTEVANT *Notes Edible Plants* 589 Tobacco Root. Valerian. Ohio to Wisconsin and westward. 1974 J. AIKEN *Midnight is Place* v. 145, I have sold some [cigar] ends..to a man in a tobacco-shop. 1936 Tobacco streak [see *STREAK *sb.*^{1} 6*]. 1968 *Times* 3 Oct. 13/5 Tobacco streak virus, so called because of the symptoms it produces in tobacco plants, infects a wide variety of plants, including French beans, peas and clover. 1930 J. S. HUXLEY *Bird-Watching & Bird Behaviour* v. Plate VII, A Black-headed gull contentedly brooding a tobacco-tin which has been substituted for its eggs. 1975 M. BRADBURY *History Man* ix. 153 Ashtrays have been stolen, and replaced by..tobacco tins. 1840 *Picayune* (New Orleans) 13 Sept. 3/1 The same Mac..[is] well known to the Western country Tobacco Traders. [1688 Tobacco worm: see *tobacco fly* above.] 1737 J. BRICKELL *Nat. Hist. N. Carolina* 168 The Tobacco-worm..has two sharp horns on its Head. 1872 *Rep. Vermont Board Agric.* I. 319 The large night-flying moths..produce the large larvæ, as the potato-worm and the tobacco-worm. 1962 METCALF & FLINT *Destructive & Useful Insects* (ed. 4) xiii. 594 The best known of tobacco insects..are the large green tobacco worms with white bars on the sides and a slender horn at the end of the body.

tobacco-box. **2.** For *Pomotis* read *Lepomis*. (Examples.)
1877 C. HALLOCK *Sportsman's Gazetteer* I. 379 Black Perch, sometimes called 'tobacco-box'; found in ponds. 1903 *Outing* Apr. 134/1 He is content to lure to the surface..the 'sunny', 'tobacco box', or 'pumpkin seed'. 1913 A. DOUGLAS *Fast Nine* 150 The commonest and smallest skate of the Eastern coast of the United States is the 'Tobacco Box'.

tobacconalian, *a.* (Earlier example.)
1835 J. H. INGRAHAM *South-West by Yankee* I. viii. 89 Every other gentleman we met was enveloped in a cloud, not of bacchanalian, but tobacconalian incense.

Tobagonian (tŏbē^{i}gō^{u}·niăn), *sb.* and *a.* [f. *Tobago* (see below and etym. of TOBACCO) + -n- + -IAN.] **A.** *sb.* A native or inhabitant of Tobago, an island in the West Indies, part of the nation of Trinidad and Tobago. **B.** *adj.* Of or pertaining to Tobago.
1955 *Caribbean Q.* IV. II. 158 The Tobagonian says *is go ah goin'* (it's go I'm going). 1957 C. MACINNES *City of Spades* II. xiii. 189 Word will reach the ear of this Tobagonian owner and I lose my good job. 1962 *Times* 31 Aug. (Trinidad Suppl.) p. iv/5 There are more Tobagonians living in Trinidad than in Tobago. 1972 [see *ring-play* s.v. *RING *sb.*^{1} 18 a]. 1974 *Trinidad Guardian* (Port-of-Spain) 16 Oct. 24/9 Trinidadians and Tobagonians have an outstanding record in bodybuilding.

to-be. Add: **B.** (Later examples.) Esp. following sbs. of kinship, as *grandfather-, wife-to-be*; see also *mother-to-be* s.v. *MOTHER *sb.*^{1} 16 a.
1930 A. BENNETT *Imperial Palace* lxix. 581 The excited grandfather-to-be. 1969 L. HELLMAN *Unfinished Woman* xii. 174 The so-called good life for us is the to-be-good life for them. 1973 H. NIELSEN *Severed Key* iii. 33 My wife-to-be is going to be fabulously successful.

tober (tō^{u}·bəɹ). *Showmen's slang.* Also **tobur.** [a. Shelta; see TOBY *sb.*^{2}] The site occupied by a circus, fair, or market.
1890 BARRÈRE & LELAND *Dict. Slang* II. 357/2 *Tobur, toba* (showmen, &c.), the ground or field at fairs, hired to put the waggons on for show or circuses, or other *al fresco* entertainments, which does not amount to much, so that a man or manager is considered very hard up if he has not enough to pay the *tobur*. Gypsy *tober*, the road, hence ground. 1933 E. SEAGO *Circus Company* vi. 85 How can I walk about the tober without me trousers, I'd be askin' ye? 1939 J. B. PRIESTLEY *Let the People Sing* x. 256 'It's not a bad tober—but what's 'is bunce?'..'It's all fair-ground slang... Micky..said this wasn't a bad market here, but what had Knocker made?' 1957 *Times Lit. Suppl.* 6 Dec. 742/1 She lived with it [*sc.* a circus] for some time as a privileged outsider, parking her caravan on the tober, as the site is called. 1968 [see *JOINT *sb.* 14 b].

tobermorite (tō^{u}bəɹmō^{ə}·rəit). *Min.* [f. *Tobermory*, name of a village on the Isle of Mull, Scotland + -ITE^{1}.] A hydrated, basic calcium silicate occurring as masses of pale pinkish white translucent orthorhombic crystals.
1880 M. F. HEDDLE in *Mineral. Mag.* IV. 119 Tobermorite. This is a zeolite which I first found filling small druses in the cliffs of the shore immediately to the north of the pier of Tobermory in the Island of Mull. 1962 *Engineering* 3 Aug. 137 All tobermorites, natural or synthetic, are layer crystals having some similarity to vermiculite. 1978 *Mineral. Mag.* XLII. 229/1 Tobermorite minerals vary in some properties, most notably in whether loss of molecular water is accompanied by unidimensional lattice shrinkage... Tobermorites that show this lattice shrinkage have come to be called 'normal' and ones that do not, 'anomalous'.

Tobias night (tobəi·ăs nəit). [tr. G. *Tobiasnacht*, which alludes to Tobit viii. 1–3.] (See quots.) Cf. *Toby-night* s.v. *TOBY *sb.*^{1} 6.
1960 C. WINICK *Dict. Anthropol.* 539/1 *Tobias nights*, in the Catholic church, postponing the consummation of a marriage for several nights. 1975 *Amer. Speech 1973* XLVIII. 73 One need not go all the way back to Semitic antiquity..nor to the derivative nineteenth-century Swiss and German custom of the 'Tobiasnächte', the 'Tobias nights', in which the next of kin slept between the newlyweds for the first three nights of marriage to protect them at a time when their resistance to evil would be at its lowest ebb.

toboggan, *sb.* Add: **2. c.** *U.S. slang.* A rapid decline, a progression towards disaster. Usu. in phr. *on the toboggan.*
1910 E. A. WALCOTT *Open Door* xii. 153 Do you remember the time I got Conny Mulnix off, when the police had him on the toboggan for the Kinsley affair? 1947 *Christian Cent.* 20 Aug. 999/1 The United States is sliding down the toboggan with 75 per cent of the south a negroid population. 1950 J. DEMPSEY *Championship Fighting* 197 A veteran of thirty or thirty-one who is on the 'toboggan'. 1978 J. A. MICHENER *Chesapeake* 853 My daughter Clara's a little younger than you. For three years she's been on one hell of a toboggan.
2*. *U.S.* A long woollen cap. Cf. *toboggan-cap*, in Dict. and Suppl.
1929 *Amer. Speech* V. 152 *Toboggan*, a woolen cap. 'Take off your toboggan.' 1948 *Pacific Spectator* Winter 83 He had on faded overalls with new blue patches on the knees, and a sweater under the overalls, and a knitted blue toboggan on his head, against the cold. 1975 *Raleigh* (N. Carolina) *News & Observer* 6 Jan. 24/4 He [*sc.* a burglar] was wearing a red toboggan and tight pants, police said.
3. *toboggan-cap* (examples), *club* (earlier example), *-sleigh* (example); **toboggan-chute** (examples); **toboggan-slide** (earlier example).
1884 *Brandon* (Manitoba) *Blade* 21 Feb. 9/2 The several Toboggan Slides were illuminated every evening. 1890 *Silverton* (Colorado) *Miner* 1 Mar. 3/2 During the storm, the big tree on Anvil, which was generally known as the starting point for snow shoers and the toboggan club, was blown down. 1902 *Sears, Roebuck Catal.* 1159/3 Toboggan Caps or Toques. 1902 A. C. LAUT *Story of Trapper* xiv. 196 Wrapping her husband in robes on the long toboggan sleigh, the squaw placed her younger child beside him and with the other began tramping through the forest drawing the sleigh behind. 1913 W. P. EATON *Barn Doors & Byways* 223 One road runs along the ridge, the other plunges over it and crosses the intervale like the smooth, straight drop of a great toboggan chute. 1928 *Chicago Tribune* 11 June 10/5 Women and children in winter wore toboggan caps which wrapped two or three times around the neck and hung about a yard down the back. 1936 J. H. STREET *Look Away!* xiii. 91 Tiller wore a faded green coat, woolen stockings, and a toboggan cap. 1964 *Globe & Mail* (Toronto) 15 Dec. 32/3 Winter funseekers will take to..two new toboggan chutes.

toboggan, *v.* Add: (Earlier example.) **tobogganing** *vbl. sb.* (earlier example).
1846 E. WARBURTON *Hochelaga* I. v. 68 They tarbogginned, slid, and trudged about merrily in the deep dry snow. 1849 J. E. ALEXANDER *L'Acadie* I. 186 An amusement of which Canadian boys, and sometimes ladies too, are passionately fond..is called 'toboggoning'.

tobogganer. (Earlier example.)
1878 *Canad. Gentleman's Jrnl.* 8 Mar. 1/5 In fact, there has been so little snow this winter that the tobogganers have not had a good time.

Tobralco (tobræ·lko). [f. the name of *To*otal *Br*oadhurst *L*ee *Co*mpany, Limited, the manufacturers.] The proprietary name of a type of cotton fabric.
1910 *Westm. Gaz.* 25 Jan. 9/3 (Advt.), 'Tobralco', a new material, made in White, Ecru, and Black. 1917 *Trade Marks Jrnl.* 26 Sept. 942 *Tobralco*... Textile fabrics (not included in other Classes) made from substances covered by Class 50, but not including Incandescent Gas Mantles and not including any goods of a like kind... Tootal Broadhurst Lee Company, Limited, 56, Oxford Street, Manchester, manufacturers and merchants. 1932 D. C. MINTER *Modern Needlecraft* 250 *Kimono Pinafore*... Gingham, zephyr, tobralco. 1961 D. STUART *Driven* xx. 200 The agent was a middle-aged man, sharp and alert, clean-shaven, in gaberdines and Tobralco shirt.

tobramycin (tǫbrāməi·sin). *Pharm.* [f. *to-* (of unkn. origin) + L. *tene*)*brā*(*rius* belonging to darkness (see def.), f. *tenebræ* darkness: see *-MYCIN.] An antibiotic related to streptomycin that is produced by the fungus *Streptomyces tenebrarius* and is active mainly against Gram-negative bacteria, being used esp. to treat *Pseudomonas* infections.
1971 *Appl. Microbiol.* XXII. 1147/2 Factor 6 of the nebramycin complex was originally given the generic name of ebbramycin in 1970. Subsequently, its official generic designation was changed to tobramycin. 1977 *Lancet* 19 Mar. 655/2 She had a *Pseudomonas æruginosa* skin infection which responded rapidly to tobramycin and carbenicillin. 1980 [see *TICARCILLIN].

toby, *sb.*^{1} Add: **4.** In full *toby tub.* (Earlier and later examples.)
1842 *London Jrnl. Arts & Sci.* XIX. 35 The printing [of the fabric] is to be done in an ordinary machine or press, the colours being furnished from what is called the 'toby tub'. 1881 *Instructions to Census Clerks* (1885) 43 Toby and Rainbow Tub Maker.
5. (Earlier example.)
1894 T. B. SEARIGHT *Old Pike* 144 They [*sc.* cheap cigars] became very popular with the drivers, and were at first called Conestoga cigars; since, by usage, corrupted into 'stogies' and 'tobies'.
6. *Toby-night* = *TOBIAS NIGHT.
1910 T. M. PARROTT *Chapman's Plays & Poems* I. 699 The custom..is the well-known 'Toby-night', or 'nights', ordained as a rule of the Church by the Council at Carthage, A.D. 398. The rule was authorized by the example of Tobith (Toby), who spent the first three nights of his marriage in prayer.
7. *Austral. slang.* A stick of ochre used for marking sheep which have not been shorn to the owner's satisfaction.
1912 in Stewart & Keesing *Old Bush Songs* (1957) 273 I've been shearing on the Goulburn side and down at Douglas Park, Where every day 'twas 'Wool away!' and toby did his work. 1964 H. P. TRITTON *Time means Tucker* 41 Raddle was a stick of blue or yellow ochre, also called 'Toby'. 1965 J. S. GUNN *Terminol. Shearing Indust.* II. 11 The raddle stick was also called 'Toby', and its improper use was one of the main reasons for the formation of the first Shearers' Union.
8. (With capital initial.) The name of a stock character of American comedy (see quot. 1961), used *attrib.*, esp. in *Toby show.*
1946 *Theatre Arts* Nov. 652/1 Young actors who have played juveniles or ingenues with a Toby show seldom succumb to first-night nerves in later years. 1961 BOWMAN & BALL *Theatre Language* 393 *Toby*,..a comic character type, a boisterous, blundering yokel as the protagonist. Hence *Toby play* (or *show*), a repertory favorite. 1964 *Tennessee Folklore Soc. Bull.* June 49 Bisbee's Comedians..is one of the two surviving Toby Shows left in the entire country. 1967 *Oxf. Compan. Theatre* (ed. 3) 949/1 Most travelling dramatic tent-shows, playing one-week stands in rural communities, feature a Toby-comedian. *Ibid.*, Frederick R. Wilson, member of a touring tent-show company known as Horace Murphy's Comedians, was the first of a long line of actors to specialize in Toby roles. *Ibid.*, Toby-comedy includes generous use of the topical 'ad-lib'. 1978 *Chicago* June 56/2 We thought this [*sc.* donkey baseball] had gone the way of the Toby shows.
9. *Angling.* (With capital initial.) A type of lure used in spinning.
1969 V. CANNING *Queen's Pawn* i. 2 The river would be high... No use for a fly. He wanted..a few small Tobies for spinning. *Ibid.* 3 He bought some..four-gram golden Tobies, and the rod. 1973 A. ROSS *Dunfermline Affair* 139 Bayne's biggest lure—a six-inch metal Toby with a big triple hook.

toby, *sb.*^{2} Add: *toby concern* (example); *low toby* (earlier example); *toby-gill* (earlier example).
1807 *Sessions' Papers* Feb. 133/1 He..asked me if I had any objection of being in a good thing... I asked him when and..he replied it was *low* toby, meaning a footpad [*sic*] robbery. 1811 *Lexicon Balatronicum* s.v. *galloper*, The toby gill clapped his bleeders to his galloper. 1830 LYTTON *Paul Clifford* I. iv. 76, I heered as ow Long Ned started for Hampshire this werry morning on a toby consarn!

toc (tǫk). Also **tock.** Used for *t* in telecommunication codes and in the oral transliteration of coded messages. Cf. *TOC EMMA, *Toc H.
1898 [see *ACK]. 1913 [see *PIP *sb.* 4]. 1944 K. DOUGLAS *Alamein to Zem Zem* (1946) xvii. 100 This means they are

hopelessly broken down and want the technical adjutant, known officially over the wireless as 'Tock Ack', to arrange their recovery.

toccatina (tǫkătī·nă). *Mus.* [a. It., dim. of Toccata.] A short toccata. Also **toccatella** (-e·lă).
1740 J. Grassineau *Mus. Dict.* 284 Toccatina, a small research when we have not time to perform it in all its parts. **1889** Grove *Dict. Mus.* IV. 130/1 Dupont has published a little pf. piece entitled Toccatella. *Ibid.*, The same composer [*sc.* Rheinberger] has used the diminutive term Toccatina for one of a set of short pieces. **1938** *Oxf. Compan. Mus.* 937/1 Widor in his seventh Organ Symphony has a toccatina—a sort of *Perpetuum Mobile*.

toc emma (tǫk e·mă). *Mil. slang.* Also **tock** (and **toch**) **emma** and with capital initials. [Representing T.M. (see *T 6); see *toc, *emma.] A trench mortar. Also *transf.*
1916 *B.E.F. Times* 1 Dec. f. 3/1 Completely oblivious of the dangers I encountered from our own artillery and Tock Emmas! **1918** J. H. Douglas *Captured* ii. 25 He turned out to be Bombardier 'Chuck' Gibson who was with the sixty-pound 'Tock Emma' (Trench Mortar) Battery located on our frontage. **1928** R. C. Sherriff *Journey's End* (1929) ii. ii. 57 Can't have men out there while the toch-emmas are blowing holes in the Boche wire. **1931** [see *emma].

Toc H (tǫk ē̆ɪtʃ). [Representing T.H., initials of *T*albot *H*ouse (see sense 1 below), which was so called in memory of Gilbert W. L. Talbot (d. 1915); see *toc.] **1.** Colloq. abbrev. of the name of *Talbot House*, a rest-house and club for soldiers opened at Poperinghe, 15 Dec. 1915.
1918 in P. B. Clayton *Tales Talbot House* (1919) 138 Owing to the inconsiderate retirement of our old neighbours, the Boche, Toc. H. is in a pretty fix. **1925** Fraser & Gibbons *Soldier & Sailor Words* 286 Poperinghe..was visited by thousands of officers and men, for practically every one of whom 'Toc H', with its unique atmosphere and surroundings, proved alike a club and a home from home.
2. An association, orig. of ex-servicemen, founded by the Rev. P. T. B. Clayton after the war of 1914–18 to embody Christian fellowship and service.
1920 *Christmas Spirit 'Toc. H.' Ann.* 77 (*heading*) Toc. H. *Late* Talbot House. *Ibid.*, To open the club houses, Toc. H. asks for sympathy and help in many practical forms... H.R.H. The Prince of Wales has consented to open the H.Q. Club in London in 1921. **1930** *Toc H Jrnl.* Jan. 3 Toc H will indeed begin..to be..a power making for righteousness. **1954** P. Toynbee *Friends Apart* i. 17, I intended to work in a Toc H settlement. **1981** J. Brabazon *Dorothy L. Sayers* xix. 241 An Anglican priest, chaplain to the Toc H hostel where I was staying.
3. *Toc H lamp*: an oil lamp, an emblem of Toc H, used *iron.* as a type of dimness.
1977 *New Statesman* 9 Sept. 341/1 'He is as dim as a Toc H Lamp'..is not yet rare as a phrase though members of the Toc H organisation may well be thin on the ground. **1977** J. Porter *Who the Heck is Sylvia?* v. 46 Sometimes you can be dimmer than a Toc H lamp.

Tocharian (tǫkē̆ə·riăn, -ā·riăn), *a.* and *sb.* Also **Tokharian.** [ad. F. *tocharien* (or next), f. Gr. Τοχάροι (Strabo) a Central Asiatic people formerly thought to speak Tocharian; see -ian.] **A.** *adj.* Of, pertaining to, or designating an extinct Indo-European language spoken in the latter half of the first millennium A.D., of which remains have been discovered in Chinese Turkestan. **B.** *sb.* This language; also, a member of the people or peoples speaking the language.
Two dialects of Tocharian are recognized: an eastern, *Tocharian A* (= *Turfanian), and a western, *Tocharian B* (= *Kuchaean, Kuchean).
1927 Peake & Fleure *Peasants & Potters* 134 The Tocharian language of parts of Turkestan. **1934** A. Toynbee *Study of Hist.* I. i. iii. 113 One isolated language in the far north-east (the now extinct 'Tokharian', which has become known to Western scholars through the discovery..of documents in this language.) **1950** *Trans. Philol. Soc. 1949* 9 The system of *r*-endings found in the verbal paradigms of various I[ndo-] E[uropean] languages ..is clearly attested in Hittite, Indo-Iranian, Tocharian, Phrygian and Armenian, Italic, Celtic. **1960** Partridge *Charm of Words* 170 The -*k*- variation attested by Lett *aka*, a water-spring, and Hittite *eku*-, to drink, and dubiously Tokharian *yoko*, (a) thirst, should perhaps be aligned with certain OE and ON -*g*- words. **1966** G. S. Lane in Birnbaum & Puhvel *Anc. Indo-European Dial.* 218 If we could ever find out what non-Indo-European influence brought about the distinction in gender in A, we might know considerably more about the wanderings and contacts of the 'Tocharians'. **1975** *Language* LI. 141 Tocharian -*tsi*..is regularly added to a verbal stem, the present stem in East Tocharian and the subjunctive stem in West Tocharian. **1977** *Word 1972* XXVIII. 1 We have on one side Latin and Keltic, on the other Indo-Aryan, Iranian ..and Tocharian.

Tocharish (tǫkā·riʃ, -ē̆ə·riʃ). Also **Tokharish.** [ad. G. *tocharisch*; see *Tocharian *a.* and *sb.*, -ish¹.] The Tocharian language.
1910 *Encycl. Brit.* II. 712/2 Up to 1909 only a preliminary account had been given of Tocharish, a hitherto unknown Indo-European language. **1926** J. R. R. Tolkien in *Year's Work in Eng. Stud. 1924* 27 The traditional Indo-European philology has suffered shocks in recent years, shocks from Tokharish and Hittite that begin at last to be felt even by the inexpert. **1939** [see *Kuchaean, Kuchean]. **1956** J. Whatmough *Language* ix. 179 Irish and Welsh have a middle or passive voice in -*r*, analogies to which are known in Hittite, Phrygian, Tocharish, Latin, [etc.].

tochilinite (tǫtʃi·linait). *Min.* [ad. Russ. *tochilinit* (N. I. Organova et al. 1971, in *Zap. Vsesoyuznogo Min. Obshch.* C. 477), f. the name of M. S. *Tochilin* (1910–55), Russian geologist: see -ite¹.] A mineral that is a complex of iron sulphide and magnesium and iron hydroxides, found as bronze-black grains and fibrous aggregates.
1973 *Mineral. Abstr.* XXIV. 186/2 A new mineral tochilinite..occurs in two habit modifications. **1976** *Papers Geol. Survey Canada* No. 76–1B. 66/1 Tochilinite is associated with clear and white calcites, some of which are coarse euhedral crystals.

tochus (tōᵘ·χəs, tǫ·χəs). *slang* (chiefly *Jewish* and *N. Amer.*). Also **tochas** (-**ess**, etc.), **tuchus** (tu·χəs), -**as**; *Anglicized* **tokus** (tōᵘ·kəs), tocus, etc. [ad. Yiddish *tokhes*, ad. Heb. *taḥat* beneath.] The backside, buttocks; the anus.
1914 *Dialect Notes* IV. 114 *Tookis, n.*, the anus:—said to be of Jewish origin. Also *tukis*. **1930** M. Gold *Jews without Money* 250 I'll spit in his face..and tell him to kiss my *tochess* for his rent. **1934** J. T. Farrell *Young Manhood of Studs Lonigan* i. 11 He was hurtled forwards by three swift kicks in the tocus. **1938** J. Curtis *They drive by Night* xxiv. 269, I could do three months on me tochas. **1951** B. Schulberg *Disenchanted* xvii. 308, I don't go for all these fancy conferences and I don't kiss anybody's tochis. **1952** W. R. Burnett *Vanity Row* v. 43, I was.. getting my tokus pinched all over the place. **1963** 'R. L. Pike' *Mute Witness* (1965) iv. 59 They call this stuff Sun-Bay Tinge... I'd call it Tuchus Pink myself. **1973** *Kingston* (Ontario) *Whig-Standard* 22 Dec. 7/3 Now get your tokus off my beat. You want to get killed go over on the next beat. **1975** R. H. Rimmer *Premar Experiments* (1976) i. 99 Your tuchas is smiling sideways at me.

tock (tǫk), *v.* [Echoic; cf. Tick-tock.] *intr.* To make a sound similar to Tick *sb.*³ 2, but slightly lower and therefore more resonant. Esp. of a clock, and in phr. *to tick and tock.*
1913 J. Masefield *Daffodil Fields* 72 A stately timepiece ticked and tocked. **1917** S. Graham *Priest of Ideal* xxxi. 306 The comfortable grandfather clock ticked and tocked temperamentally. **1961** H. R. F. Keating *Rush on Ultimate* iv. 69 There were tears in her eyes as Sebastian's second ball tocked against the peg. **1967** T. Kenneally *Bring Larks & Heroes* iv. 24 Their ears, drenched by the south wind, tocked like clocks, thumped like sails. **1970** W. Brown in Ramchand & Gray *West Indian Poetry* (1972) 14 The clock tocked and the stable dried.

toco². Substitute for etym.: [ad. Hindi *ṭhōko*, imp. of *ṭhoknā* beat, thrash.] Replace quot. 1823 and add: Also *fig.* and in phr. *to get toco for yam.*
1823 'J. Bee' *Slang* s.v., Yams are food for negroes in the West-Indies,..and if, instead of receiving his proper ration of these, Blackee gets a whip (*toco*) about his back, why 'he has caught *toco*' instead of yam. **1848** J. R. Planché *Theseus & Ariadne* (1859) i. ii. 14 Toco from my father I instead of yam shall get. **1885** W. S. Gilbert *Mikado* i. 16 To embrace you thus, *con fuoco*, Would distinctly be no gioco, And for yam I should get toco. **1910** Kipling *Let.* in Ld. Birkenhead *R. Kipling* (1978) xvi. 252 The Teuton..prepares to give us toko when he feels good and ready. **1921** [see *Bolshy, Bolshie]. **1941** J. Cary *Herself Surprised* lviii. 143 You'd better tell people how I took your trousers down last time and gave you toko.

tocopherol (tǫkǫ·fērǫl). *Biochem.* [f. Toco- + Gr. φέρ-ειν to bear + -ol.] Vitamin E: any or all of a group of closely related fat-soluble compounds that occur esp. in plant oils and are anti-oxidants essential in the diets of many animals and probably of man.
1936 H. M. Evans in *Jrnl. Biol. Chem.* CXIII. 321 For this alcohol we propose the name 'α-tocopherol'. **1956** *Nature* 14 Jan. 86/2 (*heading*), η-Tocopherol (7-methyltocol): a new tocopherol in rice. **1968** Passmore & Robson *Compan. Med. Stud.* I. x. 9/1 Vitamin E is a mixture of tocopherols, which are yellow oily liquids remarkably stable to heat. **1972** *Daily Colonist* (Victoria, B.C.) 13 Feb. 27/4 Glib armchair vitamin experts discuss tocopherol, the chemical name of Vitamin E, as easily as they talked of ascorbic acid and riboflavin two years ago. **1979** *Nature* 19 Apr. 737/2 Vitamin E (α-tocopherol) and vitamin C (ascorbic acid) react rapidly with organic free radicals, and it is widely accepted that the antioxidant properties of these compounds are responsible in part for their biological activity.

tocusso (toku·so). Also **tocussa.** [a. Amharic.] A name used in Ethiopia for finger millet, *Eleusine coracana*, the ear of which is composed of several spikes resembling the fingers of a hand.
1790 J. Bruce *Trav. Source of Nile* V. 79 In place of Teff..there grows a black grain called Tocusso. **1866** Lindley & Moore *Treas. Bot.* II. 1154/2 Tocusso. An Abyssinian corn-plant or millet. **1875** *Encycl. Brit.* I. 63/1 The low grounds produce also a kind of corn known as *tocussa*, of which a black bread is made.

tod (tǫd), *sb.*⁴ *slang.* [Short for *Tod* Sloan (occas. used in full), name of a U.S. jockey (1874–1933), used as rhyming slang for 'own' in the phr. *on one's own.*] *on one's tod*: alone, on one's own. Cf. *Pat Malone.
1934 P. Allingham *Cheapjack* vi. 56 'Are you on your tod?' I gathered that she was asking me if I was on my own. **1956** L. Godfrey in *Pick of Today's Short Stories* 91, I was in a small ward, and one evening some clot turned on the bloomin' wireless, and then went out, leaving me on my tod. **1959** J. Wain *Travelling Woman* 7 Frequent visits to town on your Tod Sloan—no need to account for your doings. Leave her to keep the home fires burning. **1966** T. E. B. Clarke *Wide Open Door* xi. 156 I'm on me Tod 'cept for the baby. **1972** J. Brown *Chancer* v. 64 That left Sonny and me on our tod in the public. **1981** 'G. Gaunt' *Incomer* xiii. 71 Maybe they don't want your company... Never seen you on your tod before.

Toda (tōᵘ·dă), *sb.* and *a.* **A.** *sb.* **a.** (A member of) a people of southern India. **b.** The language of this people, a Dravidian language closely related to Tamil. **B.** *adj.* Of or pertaining to this people.
1864 F. Metz *Tribes inhabiting Neilgherry Hills* 19 The Todas justify their belief in intermediate spiritual agencies by a reference to analogy. *Ibid.* 20 Great sanctity attaches to the person of the Pa'laul in the eyes of his Toda brethren. **1873** W. E. Marshall *Travels amongst Todas* xxiv. 208 Everything with the Toda is taken *au sérieux*. **1900** *Knowledge* 1 Mar. 67/1 On the clearances amid the dense and luxuriant primeval forest, or on the open grass-lands of the hill-tops, dwell a number of interesting aboriginal wild tribes, among whom the Todas and the Kotas are perhaps those whose names are the least unfamiliar to European ears. **1921** *Blackw. Mag.* July 28/1 The Toda puzzles and interests the Occidental because the Toda's origin is undiscoverable. **1938** [see *moiety 4]. **1939** L. H. Gray *Foundations of Language* 386 [Dravidian] falls into four great divisions, the first of which is *Tamil-Kurukh*, comprising Tamil..; Malayāḷam..; Tuḷu..; Koḍagu..; Kanarese, including Toda, Kōta, and Baḍaga ..; and Kurukh. **1955** T. Burrow *Sanskrit Language* viii. 376 Besides the major languages there are numerous minor non-literary Dravidian languages spoken in various parts of India, namely: (i) Southern: Tulu, Coorg, Toda, Kota. **1976** *Language* LII. 259 This latest monograph..is most easily understood when studied in conjunction with his massive earlier tome on Toda songs. **1980** H. Trevelyan *Public & Private* 7 The bee-hive huts of the Todas, the earliest known inhabitants of the [Nilgiri] hills who still lived there.

to-day, *adv.* and *sb.* Add: **A.** *adv.* Freq. in phr. *here today and gone tomorrow*: see *here *adv.* 1 e.
C. *adj. colloq.* Modern, characteristic of or suitable for the present day.
1969 *Harper's Mag.* Oct. 65/2 I'm a today writer. **1976** A. Cross *Question of Max* iii. xiii. 154 It's old-fashioned and sentimental and altogether not 'today' to talk of restitution. **1980** J. Wainwright *Eye of Beholder* 24 The today song-smiths..wrote boy-girl-and-bed words.

Todd-AO (tǫd,ē̆ɪ,ōᵘ). *Cinemat.* [f. the name of Mike *Todd* (1907–58), U.S. stage and film producer, + the initials of *A*merican *O*ptical Co.] The proprietary name (in the U.S.) of a cinematic process producing a wide-screen image. Freq. *attrib.*
1955 *Times* 6 Aug. 3/5 The Todd-AO process uses 70mm. film instead of the standard 35mm. It has six sound tracks running at 30 frames a second as against the usual 24. The image is projected on to a huge curved screen presenting a picture about 25 ft. high and 65 ft. wide. **1955** *Official Gaz.* (U.S. Patent Office) 29 Nov. TM 239/2 *Todd-AO*. For motion picture equipment. Use since August 1953 on motion picture camera equipment... The Todd-AO Corporation, New York. **1958** *New Statesman* 26 Apr. 530/3 There can be no question about it; Todd-AO wipes the floor with Cinerama. **1958** *Observer* 27 Apr. 15/6 The Todd-AO screen is a huge, incurved affair, rather like the other side of a monstrous broken cup. **1976** *Oxf. Compan. Film* 692/2 His [*sc.* Todd's] main contribution to films..was his promotion of a new 70mm wide screen process, called Todd-AO, in the mid-fifties.

toddle, *sb.* Add: **toddlekins** (earlier example).
1852 C. J. Mathews (*title*) Little Toddlekins.

to·ddlerhood. [f. Toddler + -hood.] The condition of being a toddler.
1966 'L. Lane' *ABZ of Scouse* Foreword, In his very toddlerhood his mother *lerned* him. **1967** *Punch* 15 Mar. 377/2 A normal childhood needs a good decade to run its full noisy course from toddlerhood to puberty. **1976** *Word 1971* XXVII. 37 The physical transition from infancy to early toddlerhood is marked by a qualitative transition in the nature of mother-child communication.

toddy, *sb.* Add: **3. a.** *toddy-wine* (later example); **toddy-maker** = *toddy-man* (see also sense 3 b in Dict.); **toddy-tapper** = *toddy-man*; **toddy-tapping,** the collection of toddy from palms.

1821 J. Leyden tr. *Malay Annals* 151 There was a toddy-maker, who went to amuse himself on the sea. **1937** *Discovery* May 143/1 It [*sc.* coconut shell] is an indispensable part of the toddy-tapper's outfit, for it is in a coconut shell that he carries his cinnamon leaf paste and his lime for the purpose of stimulating the reluctant flowers to give up their sweet nectar. **1971** *National Geographic* Mar. 355/2 No Than Shwe, like many men around Pagan, is a toddy tapper. **1968** *Nature* 5 Oct. 493/2 Toddy-tapping is a popular occupation in a large section of the population. **1958** *Contributions to Indian Sociology* II. 14 Toddy-tapping and the taking of animal life are associated with low status. **1971** *National Geographic* Mar. 358/1 Juice collected at 8 a.m. ferments to toddy wine by 5 that evening.
b. **toddy-lifter,** a device used in the manner of a pipette to transfer hot toddy from a bowl to a glass; **toddy-stick** (earlier example).
1840 *Picayune* (New Orleans) 4 Oct. 2/5 A 'toddy stick' is as spirit-stirring an article as any poet can boast. **1923** *Classical Q.* July–Oct. 173 The 'Toddy-lifter', known in Scotch and Irish households during the eighteenth and early nineteenth centuries, a bulbous glass cylinder, is exactly the instrument [*sc.* the clepsydra] described here. **1954** E. M. Elville *Paperweights* x. 107 A toddy-lifter was something like a miniature decanter in shape, with a body large enough to hold a glassful of liquid. **1970** G. Savage *Dict. Antiques* 420/1 The toddy-lifter was dipped into the bowl and filled through the hole in the bottom. The thumb was then placed over the upper orifice, air-pressure keeping the contents from flowing out.

todorokite (todǫ·rǫkait). *Min.* [f. the name of the *Todoroki* mine, Hokkaidō, Japan + -ite¹.] A hydrated oxide of manganese, calcium, and other elements occurring as soft black aggregates of minute, probably monoclinic laths having a metallic lustre; also, any of a group of minerals structurally related to this.
1934 T. Yoshimura in *Jrnl. Faculty Sci. Hokkaidō Univ.* Ser. IV. II. 297 The new mineral belonged to the pyrolusite species of crystalline manganese... As such a mineral had not yet been reported, this mineral was named 'todorokite' after the name of the mine where it had been first noticed. **1982** *Science* 29 May 1024/1 Todorokites are calcium-bearing manganese oxides found in terrestrial manganese ore deposits, in weathering products of manganese-bearing rocks, and in some manganese nodules.

toe, *sb.* Add: **1. e.** *fig.* Speed, energy. *Austral.* and *N.Z. slang.*
1903 *Truth* (Wellington) 8 Oct., Happy Song has a fair share of toe in spite of her nine years and she was flying in fifth place after losing ground at the start. **1969** *Sun* (Melbourne) 22 July 58/3 The North half-forward line... has a ton of toe and could give Richmond's novice half-back line a torrid afternoon.
4. f. In full *the toe of Italy.* The south-western extremity of that country. Cf. Heel *sb.*¹ 6.
1894 in Dict., sense 4 a]. **1941** C. Milburn *Diary* 15 Feb. (1979) 85 We have dropped parachutists...on Italy's toe...near Brindisi. **1974** *Times* 5 Jan. 5 The boy had been kept in various hideouts in the south-western 'toe' of Italy. **1979** R. Perry *Bishop's Pawn* iv. 68 The advancing Allied armies..forced themselves northwards from the toe of Italy.
g. A flattish portion at the foot of an otherwise steep curve.
1940 *Wall's Dict. Photogr.* (ed. 15) 573 The method of speed-measurement used must..depend on the position, not of the extreme under-exposed 'toe' of the curve, but of its straight-line portion. **1962** *Rep. Progress Physics* XI. 284 A pronounced toe can be obtained on a density-development-time curve by adding bromide ions to a hydroquinone developer. **1982** *Sci. Amer.* Apr. 41/1 The design of tension-leg platforms, like the design of guyed towers, is still at the toe of the learning curve and will undoubtedly go through several generations of improvement.
h. *Hort.* A section of a fleshy root.
1952 A. G. L. Hellyer *Sanders' Encycl. Gardening* (ed. 22) 169 Dracaena... Propagation: by cuttings or 'toes' of fleshy roots in sandy peat in spring. **1976** *Billings* (Montana) *Gaz.* 27 June 4-G/6 Rhizomes branching from the old toe will bear flowers next year. **1984** *Gardening from Which?* (Consumers' Assoc.) Mar. 84/1 Remove the offsets..known as yucca toes... Remove the 'toes' if new plants are needed.
5. i. *to tread on the toes of* (earlier example).
1866 Trollope *Belton Est.* (ed. 3) I. iii. 71 'But you mustn't offend my father.'..'I won't tread on his toes.'
l. *on one's toes*: alert, eager.
1921 J. Dos Passos *Three Soldiers* ii. i. 56 If he just kept on his toes he'd be sure to get it. **1938** B. Nichols *Sweet & Twenties* 94 You have to be on your toes to make the right sort of ripostes on such an occasion. **1972** P. Marks *Collector's Choice* ii. 123 Anavi was convinced that he had the right to deduce from the gesture's consequences: he was doing them a service because it kept them on their toes.
m. *toe-to-toe*: (carried on) in close combat, at close quarters; also, neck and neck. Cf. *foot to foot* s.v. Foot *sb.* 26 b.
1942 Berrey & Van den Bark *Amer. Thes. Slang.* § 701/14 *Toe-to-toe*, evenly matched. **1950** J. Dempsey *Championship Fighting* 199 Has each enough confidence in his own punching ability..to engage the other in toe-to-toe exchanges? **1952** *Newsweek* 23 June 21/1 In the toe-to-toe fight for the Republican Presidential nomination, last week's round went to Sen. Robert A. Taft of Ohio. **1958** *Oxf. Mag.* 15 May 429/2 The sense of toe-to-toe negotiation with financial giants. **1971** *Flying* Apr. 42/1 My wife and I landed..to top up the tanks and have a toe-to-toe talk with the weather guys. **1977** *Sounds* 9 July 43/3, I love real eccentric people, getting toe to toe with them.
n. *to have it on one's toes*: to run away. *slang.*
1958 F. Norman *Bang to Rights* 53 They hold us responsible for anyone having it on their toes. **1966** 'P. B. Yuill' *Hazell & Menacing Jester* vi. 67, I had it across the road on my toes.
o. *toes over* (Surfing) (see quots.).
1964 T. Masters *Surfing made Easy* 65 *Toes over*, walking to the very front of the board during a ride on a steep hollow wave. **1965** J. Pollard *Surfrider* ii. 19 Walking the board when you don't wish to put up your toes or you can still put a few over the edge—do a 'toes over'.
p. *a toe in the door*: a position from which progress can be made.
1977 *Times* 7 Oct. 17/2 Gail Sheehy stopped her sample at 50... She says she now has a toe in the door of the 50's and 60's. **1978** *Dumfries Courier* 20 Oct. 6/5 He was only using the application for boating as a 'toe in the door' to sell something else. **1979** D. Sanders *Queen tends for Mrs Chadwick* 11 He'd be thirty-five at the next election. Just the right age to get a toe in the door.
q. *to dig in one's toes*: see *dig *v.* 11 c.
6. **toe-board** (earlier example); **toe brake** *Aeronaut.*, in an aircraft, a brake that is operated with the foot; so **toe braking** *vbl. sb.*; **toe-clip,** (*a*) (earlier example); **toe-cover** *slang*, an inexpensive and useless present; **toe-dancing,** dancing on points; **toe-end** *v. trans.*, to kick with the point of one's foot; **toe-hold,** (*b*) a place of support for the toe (of a boot) in climbing; hence *fig.*, a position of little significance or influence, esp. one seen as providing a base from which they may be increased; **toe-hole** *rare*, a place of support for the toe (of a boot) in climbing; **toe-jam** *slang*, dirt which accumulates between the toes; **toe jump** *Skating*, a jump initiated with the tip or toe of one's non-skating foot; **toe loop,** (*a*) *Skating*, a loop jump that is also a toe jump (see quot. 1979); more fully *toe loop jump*; (*b*) a loop on a sandal through which a toe is placed; **toe-nail** *sb.*, (*a*) (earlier example); so *fig.*; (*b*) an iron nail employed for the toe in shoeing; **toe-puff,** a stiffener for the toe of the upper of a shoe; **toe rake** *Skating*, a set of teeth at the front of the blade of a skate; **toe-ring** (earlier example); **toe-rubber** *N. Amer.*, a rubber overshoe that covers only the front part of a shoe; **toe shoe** *N. Amer. Ballet*, a shoe with a reinforced toe, worn for toe-dancing; a point shoe; **toe-spin** *Skating*, a spin performed on the toe; **toe-strap,** (*a*) (earlier example); (*b*) a strap on a bicycle pedal to keep the foot from slipping off it; (*c*) a band fixed to a boat and serving to hold the foot of someone leaning out; **toe-tapping** *vbl. sb.*, the tapping of feet in time to music; (in quot. 1929, a derogatory term for 'dancing'); *ppl. a.*, that makes one want to tap one's feet; **toe-thong sandal** = *thong sandal* s.v. *thong *sb.* 2; **toe wall** a low wall built at the foot of an embankment to help keep the earth in place.
1892 *Harper's Mag.* Jan. 271/1 The..bag..to put under his feet on the toe-board. **1944** *Jrnl. R. Aeronaut. Soc.* XLVIII. 297 The toe brakes are awkward to operate, and heavy pressure is needed on them to get the desired brake effect. **1976** R. Lecomber *Dead Weight* ii. 32, I stood on the toe-brakes and opened the throttle. **1977** *R.A.F. Yearbk.* 29 Direction is maintained or altered by braking with the toe-brakes. **1896** *Army & Navy Co-op. Soc. Price List* 1379/2 The Courier Toe Clips... For Rat Trap Pedals (adjustable). per pair 1/6. **1948** B. MacDonald *Plague & I* vi. 85 Our Christmas presents..were what Bob called toe-covers. **1983** *Listener* 3 Feb. 22/2 Gifts are given, not only the completely useless trivia of 'toe-covers' which litter the surgery, but more substantial gifts, such as briefcases. **1924** Sharp & Oppé *Dance* 47 Toe-dancing is perhaps the most striking instance of the virtuosity achieved by the ballet-dancers of the last century. **1976** F. Muir *Frank Muir Bk.* 42 About 1820 the ballerina Taglioni popularized toe-dancing, which called for special ballet shoes. **1968** B. Hines *Kestrel for Knave* 98 He pivoted on his left foot and toe-ended a lump of coke back across the asphalt. **1976** *Sunday Mail* (Glasgow) 21 Nov., Jonquin took a free-kick and the inside-right toe-ended the ball into the net. **1880** 'Mark Twain' *Tramp Abroad* xxxix. 379 One man's toe-hold broke and he fell! **1928** *Observer* 20 Nov. 8/6 The enemy retains a toehold in the Rimeuse Valley. **1945** Mencken *Amer. Lang.* Suppl. I. 324 So many novelties swarm in... A large number come and go without the lexicographers so much as hearing of them... At least four-fifths of those which get any sort of toe-hold in the language originate in the United States. **1963** M. I. Finley *Ancient Greeks* ii. 12 Small groups of men began to migrate eastward across the Aegean to find toeholds on the Asia Minor coast. **1965** *Listener* 10 June 869/3 By Carletti's time Europe..retained only a toe-hold on the China trade. **1980** 'M. Fonteyn' *Magic of Dance* 155 A model rock about twelve inches high was dragged onto the stage by the corps de ballet. It had a special toehold into which I had to place my foot and balance for a moment on pointe. **1876** H. Melville *Clarel* I. ii. xix. 224 A ladder of steep stone With toe-holes cut. **1973** R. Campbell *Broken Record* 163 The stale smell of the toe-jam of the shuffling pedestrian Charlot. **1973** *Black World* June 21 If you miss now Picking time Then you collect Three and one half milligrams of toe jam To give it to barbara's cat. **1928** M. V. Vinson *Primer Figure Skating* ix. 129 Another nice toe jump is the 'ballet hop'. **1975** *Oxf. Compan. Sports & Games* 323/1 The split jump, a toe jump in which the skater takes off from a back inside edge, assisted by the toe-point of the free foot, half-turning in mid-air. **1924** J. Ross *Figure Skating for Beginners* ix. 92 The toe loop and double toe loop jumps are the ordinary loop and double loop jumps with the addition of toe-strikes. **1973** K. Mareheva *Nowhere Man* iii. 18 Sandals on her smooth-skinned feet, with thongs and a toe-loop. **1976** *Times* 29 Jan. 10/5 The modern toe-puff makes feeling the position of toes impossible. **1963** T. D. Richardson *New Bk. of Skating* ii. 20 The strike must be from the edge of the blade—and not from the point of the toe rake of the skate. **1973** *Times* 3 Mar. 18/1 Towards the end of the programme..Miss Buck tripped over the toe rake of her skate. **1980** *Radio Times* 16 Feb. 33 The front of the blade has teeth (the toe-rake) to assist with spins, pivots and jumps. **1896** 'Mark Twain' *Diary* 30 Jan. in *Following Equator* (1897) xlv. 403 All the females among them [*sc.* Hindoos] wore clumsy and shapely and showy nose-rings, toe-rings, leglets, and armlets. **1968** *Sun* (Baltimore) 16 Jan. 7 (Advt.), Handy, dual-purpose umbrella that protects you top to toe! Its smart plastic handle holds a pair of excellent quality toe-rubbers that fit any size foot. **1975** *Toronto Star* 25 Oct. H7/1 Who wouldn't develop a sense of humor in a country where some men have to wear toe rubbers half the year. **1949** Chujoy & Manchester *Dance Encycl.* 480/2 Toe-shoes are usually, but not always, made of satin. **1962** H. Meyer *Skating with Bror Meyer* 117 All the toe-spins are beautiful if well executed. **1958** [see *counter *sb.* 8]. **1960** M. V. Owen *Fun Figure Skating* vii. 130 Back toe spins (with the free leg closing in in front) and back sit spins (with the free leg held high in front) should be done by advanced free skating. **1884** *Queen* 29 Nov. (Advt.), Superior polished wood skates with thick toe-strap. **1910** *Cycling* 26 Jan. 66/1 The first time I ever essayed to climb Westerham I had no toe-straps, and I failed. **1948** L. Proctor *Racing Dinghy Handling* vi. 56 At least one foot should be tucked under the canvas toe strap. **1966** [see *cross-bar *sb.* 1 a]. **1968** *Daily Tel.* 29 Jan. 7/8 Dean hit the buoy, and Hinton fell in when his toestrap broke. **1981** B. Wren tr. *Scholl's Sailing Dict.* 257/2 A crew can only sit out effectively if the boat has toe-straps or some other device to enable weight to be placed well outboard. **1929** 'Sakmark' *Down River* iii. 46 You didn't think I wanted to come to your toe-tapping do did you? **1935** *Motion Picture* Nov. 4 (Advt.), Roaring comedy, warm romance, sensational song hits, gay toe-tapping dances. **1966** C. Keil in T. Kochman *Rappin' & Stylin' Out* (1972) 87 The jazz audience now remains immobile save for some head-bobbing, toe-tapping, and finger-popping. **1976** *Broadcast* 1 Nov. 14/1 A charming presentation of music in a toe-tapping reminiscent mood. **1966** M. Laurence *Jest of God* viii. 130 Her feet..slap with the rubbery sound of her royal-blue thong sandals. **1934** Webster, Toe wall. **1947** *Sun* (Baltimore) 5 May 16/6 Concrete toe walls have been installed on both sides of the river. **1975** Westerman & Fang *Foundation Engin. Handbk.* xi. 398/2 In England wide dry-stone toe walls have been used successfully to stabilize cuts in over-consolidated clay.

toe, *v.* Add: **1.** Also with *off*, to complete (a sock, etc.) by knitting the toe and then casting off. Also *fig.*
1856 M. J. Holmes *Homestead* 126 She..was toeing off the stocking only that morning commenced. **1870** G. M. Hopkins *Jrnls. & Papers* (1959) 196 The next morning a heavy fall of snow..It lasted until toed the firs and yews.
2. *to toe the line*, etc. (earlier examples); also *to toe the trig* (Trig *sb.*⁵).
1813 'H. Bull-Us' *Diverting Hist. John Bull & Bro. Jonathan* (ed. 2) xii. 92 He began to think it was high time to toe the mark. **1817** *Deb. Congress U.S.* 30 Jan. (1854) 792 The necessity appeared..of toeing the trig, and standing there at all hazards. **1826** W. N. Glascock *Naval Sketch-Bk.* (ed. 2) I. 27 The brigades of seamen embodied to act with our troops in America, as well as in the recent contest in Spain, contrived to 'ship a bagnet' on a pinch, and to 'toe' (for 'toe the line' was the phrase) 'a tolerable line'. **1905** *Eng. Dial. Dict.* VI. 253/2 The player may 'toe the trig', but may not overstep it.
6. orig. *U.S.* **a.** *intr.* To turn the toes *in* or *out*. Also *fig.*
1877 Bartlett *Dict. Amer.* (ed. 4) 710 *To toe in*, to turn in the toes. **1894** *Vermont Agric. Rep.* XIV. 120 Avoid a horse which toes in or toes out. **1945** B. MacDonald *Egg & I* (1946) i. 16 She toed out and had trouble with her arches. **1950** J. Dempsey *Championship Fighting* 70 If you toe-in slightly with the left foot, you'll get greater freedom in the whirl.
b. Of a pair of wheels: to have a slight forward convergence (*toe in*) or divergence (*toe out*). Also *trans.* (causatively).

1926 J. A. Moyer *Gasoline Automobiles* (ed. 2) i. 25 To facilitate steering, the front wheels of the conventional rear-wheel drive 'toe in' about ¼ to ⅛ inch. **1929** Newton & Steeds *Motor Vehicle* xxvii. 324 The alignment of the wheels should be checked occasionally since if the wheels should get to 'toe-out' the wear on the tyres will be excessive. **1939** *Automobile Engineer* XXIX. 40/1 In addition to a camber change, the wheel is 'toed-in' as it rises or falls in relation to the car. **1962** *Which? Car Suppl.* Oct. 139/1 Front wheel alignment [was] toeing out ⅛ in. instead of toeing in ⅛ in. **1976** Crouse & Anglin *Pocket Automotive Dict.* 101 On a turn, the inner wheel turns, or toes out, more.
toeing *vbl. sb.* (examples in senses 1 and *6 of the vb.).
1876 A. D. Whitney *Sights & Insights* I. 21 It is the 'toeing off' that is the satisfaction, after all, even whilst you knit the stocking. **1891** S. M. Welch *Home Hist.* 116 That peculiar turn of the foot called 'toeing in' which in the white girl would be called 'pigeon toed'. **1904** M. E. Waller *Wood-Carver of 'Lympus* 36 Ther ain't nothin' more ter learn but 'toein' off'. **1928** *Bureau of Standards Jrnl. Res.* (U.S.) I. 24 The common practice of cambering and toeing in of the front wheels of an automobile doubtless influences the tread wear. **1962** R. H. Smythe *Anat. Dog Breeding* 77 Such a dog might show no sign of toeing-in. **1970** K. Ball *Fiat 600, 600D Autobook* vii. 78/1 The final torque loading of the short arm mounting pin nut is determined after the toeing-in procedure.

toeding, var. *toering.

toe-in (tōᵘ·ˌin). [f. vbl. phr. *to toe in*: see *toe *v.* 6 b.] The inclination of a pair of wheels so that they are closer together in front than behind.
1929 Newton & Steeds *Motor Vehicle* xxvii. 324 The distances between the marks at the front and at the rear should then be measured and the amount of toe-in determined. **1979** *Arizona Daily Star* 5 Aug. C 9/2 (Advt.), We'll set caster, camber and toe-in to manufacturer's original specifications.

toeless, *a.* (Examples referring to footwear.)
1942 D. Powell *Time to be Born* viii. 187 Her feet in toeless, heelless sandals. **1952** C. W. Cunnington *Eng. Women's Clothing* vii. 248 (*caption*) Toeless sandal with low square heel.

‖ **toenadering** (tu·nadəriŋ). *S. Afr.* [Du., f. *toe* To *prep.*, *conj.*, *adv.* + *nadering* approach (f. *na* Near *adv.*²).] Rapprochement, esp. between political parties or factions.
1920 S. Black *Dorp* 187 All Oakley saw in any toenadering (coming together) of the bickering factions, was a trick to deprive King George and his heirs of their legitimate ownership of the country. **1947** *Forum* (Johannesburg) 3 May 3/2 The whole question of toenadering with the English-speaking section has..been..an apple of discord in Nationalist-Afrikaner Party circles. **1957** *Cape Times* 18 June 8/7 He must draw a large Nationalist vote if he is to win those English-speaking people who want White *toenadering*. **1971** *Financial Mail* (Johannesburg) 26 Feb. 669/1 Michael Botha..revealed to shareholders.. details of a deal with the Afrikaanse Pers... It certainly is a fairly ingenious bit of *toenadering*. **1973** *Star* (Johannesburg) 16 June 13, I have a feeling there is a good deal of public support for 'toenadering', particularly on the part of the unthinking and the wishful thinkers.

toe-out (tōᵘ·ˌaut). [f. vbl. phr. *to toe out*: see *toe *v.* 6 b.] The inclination of a pair of wheels so that they are closer together behind than in front.
1930 *Flight* 25 Apr. 460 Toe in or toe out of wheels should be carefully avoided. **1970** K. Ball *Fiat 600, 600D Autobook* ix. 112/2 With toe-in or toe-out correctly set, securely tighten the track rod clamps.

Toepler, var. *Töpler.

toe-rag. [f. Toe *sb.* + Rag *sb.*¹] **1.** A rag wrapped round the foot and worn inside a shoe, in place of a sock.
1864 J. F. Mortlock *Experiences of Convict* ii. ix. 80 Stockings being unknown, some luxurious men wrapped round their feet a piece of old shirting, called, in language more expressive than elegant, a 'toe-rag'. **1932** F. Jennings *Tramping with Tramps* vi. 98 Socks are very seldom worn. Instead you get a winding of cotton rag round the ball and toes of the foot as a safeguard against blisters. Toe-rags, the tramp calls them. **1933** 'G. Orwell' *Down & Out in Paris & London* xxvii. 197 Less than half the tramps actually bathed.., but they all washed their faces and feet, and the horrid greasy little clouts known as toe-rags which they bind round their toes.
2. A tramp or vagrant; a despicable or worthless person. Also *attrib.*
1875 T. Frost *Circus Life & Circus Celebrities* xvi. 278 *Toe rags* is another expression of contempt..used.. chiefly by the lower grades of circus men, and the acrobats who stroll about the country, performing at fairs. **1903** 'T. Collins' *Such is Life* (1937) v. 229 'Come over to the wagon, and have a drink of tea,' says I. 'No, no,' says he, 'none of your toe-rag business.' **1912** D. H. Lawrence *Let.* (1962) I. 154 Remember, whatever toe-rag I may be personally, I am the person she livanted with. So you be careful. **1960** H. Pinter *Caretaker* i. 9 All them toe-rags, mate, got the manners of pigs. **1971** 'H. Calvin' *Poison Chasers* xii. 168 Move, ya useless big toerag! **1978** M. Kenyon *Deep Pocket* xiii. 165 Could she have loved this toe-rag sheikh out of the desert? **1980** J. Wainwright *Tainted Man* 171 The Law doesn't differentiate between you and the most miserable towrag [*sic*] on the face of the earth.
Hence **toe-ragger** *Austral. slang* = sense 2 above.
1896 *Truth* (Sydney) 12 Jan. (Morris), The bushie's favourite term of opprobrium 'a toe-ragger' is also probably from the Maori. Amongst whom the nastiest term of contempt was that of *tau rika rika*, or slave. **1919** V. Marshall *World of Living Dead* (1969) 82 Over the way a 'trial' man had tossed a 'chew' to a 'toeragger'. **1953** E. Partridge in I. Bevan *Sunburnt Country* 217 Some of the gold-diggers were tramps,..and several terms connected with them are worth recording—..*toe-ragger*, a deadbeat wanderer. **1966** G. W. Turner *Eng. Lang. Austral. & N.Z.* vii. 144 The battler seems to have been the poorest itinerant. The toeragger was not much wealthier than the battler.

‖ **toering** (tū·riŋ). *S. Afr.* Also **toeding, toudang, tudong.** [Afrikaans, ad. Malay *tudong* (now *tudung*) cover, lid, sun-hat.] A wide-brimmed conical hat of straw, formerly worn by Cape Malays.
1855 J. S. Mayson *Malays of Capetown* 10 The coloured cap, the *tudong* or hat, and the sandals of wood, formerly formed a part of the national dress; but being adopted by Mahometan converts of every class, are now regarded as badges of a common faith. **1909** *Cape* 25 June 9 There was..the 'toeding' (sometimes spelt 'toering'), a conical, wide-brimmed hat of plaited straw. **1913** D. Fairbridge *That which hath Been* 52 The *toudang* of the old Malay coachmen is still to be seen at the Cape, but it is fast disappearing. **1944** I. D. du Plessis *Cape Malays* iii. 48 The *toering* is still worn by Malay coachmen when driving the wedding group. **1965** A. Gordon-Brown *C. W. Smith, Artist at Cape of Good Hope* iv. 18 Very small figures appear in some of the drawings... They are usually Malays in which the 'toering', or conical straw hat worn by the men, is prominent.

toey (tōᵘ·i), *a. slang* (chiefly *Austral.*). [f. Toe *sb.* + -y¹.] Restive, anxious, touchy.
1930 *Bulletin* (Sydney) 8 Oct. 35/2 Wise Force [*sc.* a horse] was 'toey' before the race, and behaved in alarming fashion on his way to and at the post. **1961** *Coast to Coast 1959–60* 47 And the other umpire a bit toey out there at square leg. **1968** K. Weatherly *Roo Shooter* 91 He knew that the roos were toey, and, as they were drinking on the opposite side, they would be gone as soon as he moved a muscle. **1969** C. Drummond *Odds on Death* viii. 175 The horse seemed to him a bit on the toey side. He looked down to see if saliva was dripping. **1974** *Sydney Morning Herald* 1 Jan. 2 He's that toey he's got us all nervous, too. **1981** *National Times* (Austral.) 25–31 Jan. 24/3 Dallas Jongs..had a hotel bouncer friend who could get as toey as a Roman sandal.

toff, *sb.* Add: **toffish** *a.* (earlier example); hence **to·ffishness,** behaviour characteristic of a 'toff'.
1873 J. Greenwood *In Strange Company* 43 Thick slices, bear in mind: anything under an inch thick would be regarded with contempt by the bony young barrow-man, and perhaps with an uncomfortable suspicion that you have designs to inveigle him into the detestable ways of gentility. He calls it 'toffishness'. *Ibid.*, To affect thin bread and butter is undoubtedly 'toffish'. *c* **1876** J. Albery *Dramatic Wks.* (1939) II. 105 But only because his toffishness wexes me.

toff (tǫf), *v. slang.* [f. Toff *sb.*] *trans.* and *refl.* To dress *up* like a 'toff'.
1914 D. H. Lawrence *Widowing of Mrs. Holroyd* i. i. 5 He'd got a game on somewhere—toffed himself up to the nines, and skedaddled off as brisk as a turkey-cock. **1928** *East End Star* Dec. 2/2 Notice the perfect stillness when the 'lovely lidy all toffed up' sings. **1932** L. Golding *Magnolia Street* ii. ii. 298 The fellows come in [to a hair-dressing saloon] when they're on leave. They want to get toffed up for their girls.

toffee, toffy, *sb.* and *a.* Add: **A.** *sb.* **1. c.** A small, shaped piece of toffee, usu. sold wrapped.
1938 G. Greene *Brighton Rock* i. iii. 52 'Have a toffee.' 'It's bad for the figure.' **1984** W. Garner *Rats' Alley* x. 195 He..bought..a box of her favourite toffees from the shop next door.
2. Phr. *not to be able* (to do a thing) *for toffee*: to be incompetent at it. *colloq.*
1914 *Illustr. London News* 12 Sept. 380/1 Their opponents cannot 'shoot for nuts' (or 'for toffee', as one Tommy more expressly put it). **1932** D. L. Sayers *Have his Carcase* xii. 145 The Morgan wouldn't start, not for toffee. **1951** M. Kennedy *Lucy Carmichael* ii. 76 Those dreary girls you get in every Drama School who can't act for toffee. **1977** C. McCullough *Thorn Birds* xiii. 325 You can't kiss for toffee. You open your mouth too wide.
3. Nonsense, rubbish.
a **1930** D. H. Lawrence *Phoenix* (1936) 588 The eternal flame of the high ideal is all my-eye. It's all toffee, my dear sirs. **1957** P. Wildeblood *Main Chance* 220 Working-class to the backbone, just like us... And if he's been filling you up with a lot of toffee to the contrary, more fool you. **1970** M. Tripp *Man without Friends* vii. 77 'It was all a lot of toffee,' I said, 'as Hardacre very well knows.'
4. A medium shade of brown. Cf. sense B below.
1960 *Woman's Own* 19 Mar. 42/2 In stone, toffee, scarlet, green. **1976** *Honolulu Star-Bull.* 21 Dec. A-12 (Advt.), In toffee, green or blue... Jacket with stitched back-belt.
5. *toffee-coloured*, *-like* adjs.; **toffee apple,** (*a*) an apple coated with toffee and mounted on a stick; (*b*) *slang*, a bomb of similar shape that is fired from a trench mortar; **toffee-brown** = sense A. 4 above; **toffee hammer,** a miniature hammer such as may be used to break pieces of toffee; **toffee-nose** *slang*, a snob or supercilious person; also *attrib.*; **toffee-nosed** *a. slang*, snobbish, supercilious; **toffee paper,** a small piece of paper in which a toffee is wrapped.
1917 *B.E.F. Times* 25 Dec. f. 3/2 The planting of Toffee-apples on the border of your neighbour's allotment will seriously interfere with the ripening of his gooseberries. **1930** Brophy & Partridge *Songs & Slang 1914–18* 171 Toffee Apples.—Trench mortar bombs, so called from the haft, like the skewer in a toffee-apple. **1937** 'R. Crompton' *William—the Showman* vi. 127 A little girl was leaning against the wall, eating a toffee-apple on a stick. **1957** *Times* 5 Sept. 11/4 We must kill the idea that Weymouth is just a candy-floss, toffee-apple resort. **1975** P. Fussell *Gt. War & Mod. Memory* (1977) ix. 313 Everything from shovels,..and rolls of barbed wire, to.. the perverse toffee-apple. **1976** *Milton Keynes Express* 16 July 8/2 Toffee apples and ice-cream, sweets and raffles, pony rides and competitions—these were all part of the scene. **1961** M. Kelly *Spoilt Kill* i. 30 Creased forehead, receding toffee-brown hair. **1978** R. Rendell *Sleeping Life* xvi. 129 Malina..wore jeans, of toffee-brown silk. **1948** M. Allingham *More Work for Undertaker* xiii. 167 The clear toffee-coloured pavements. **1979** D. MacKenzie *Raven settles Score* 5 His long toffee-coloured hair. **1958** B. Behan *Borstal Boy* iii. 230, I sometimes saw a fellow wearing overalls and walking round..carrying brushes and paint and sometimes glazing tools; hacking knife, glazing knife, toffee hammer,..and rule. **1978** D. Bloodworth *Crosstalk* vii. 54 Toby jugs and toffee hammers. **1919** Toffee-like [see *raft *v.*¹ 5]. **1944** K. Douglas *Alamein to Zem Zem* (1946) 78 A tin of treacle, which had been well heated, contained a delicious black toffee-like substance. **1943** Hunt & Pringle *Service Slang* 67 *Toffee-nose*, another of the expressions chiefly heard amongst the W.A.A.F. This refers to a snob or someone who considers herself 'superior'. It is very apt since it implies that the nose is kept high to prevent it coming into contact with the mouth. **1958** *Woman* 12 Apr. 69/4 People thought I was a bit of a toffee-nose for the first few months because I didn't speak to them. **1962** *John o' London's* 29 Nov. 506/3 Christian was a gentleman, hence Mr. Brando's toffee-nose accent. **1974** Toffee-nose [see *Jew boy* s.v. *Jew *sb.* 3 a]. **1925** Fraser & Gibbons *Soldier & Sailor Words* 287 *Toffee-nosed*, stuck up. **1928** T. E. Lawrence *Let.* 20 Jan. (1938) 568 A premature 'life' will do more to disgust the select and superior people (the R.A.F. call them the 'toffee-nosed') than anything. **1960** K. Amis *Take a Girl like You* iv. 60 She did not want any more chat, but could not think how to say so without running the risk of sounding both stagey and toffee-nosed. **1978** *Radio Times* 28 Jan.–3 Feb. 17/2 Let Elkan Allan and the rest of the toffee-nosed critics sneer; I shall be watching *Big Jim McLain* this Sunday and so, I am sure, will a lot of other people. **1958** G. Bellairs *Corpse at Carnival* i. 9 A little Manx cat..chasing a piece of toffee-paper. **1983** R. Sutcliff *Blue Remembered Hills* xii. 91 They..flipped screwed-up toffee papers onto the heads of the orchestra.
B. *adj.* Toffee-coloured; medium brown. Cf. sense A. 4 above.
1962 J. D. MacDonald *Key to Suite* (1968) vii. 116 A very pretty slender girl with toffee hair and dark-blue eyes. **1971** *Homes & Gardens* Aug. 57/1 The dining chairs are covered in a toffee and black houndstooth check. **1975** G. Howell *In Vogue* 259/2 (*caption*) A toffee and gold mesh sweater.

Tofranil (tǫ·frănil). *Pharm.* [Of unknown origin.] A proprietary name for the drug imipramine.
1958 *Trade Marks Jrnl.* 4 June 564/2 *Tofranil*... Pharmaceutical preparations for human and veterinary use... J. R. Geigy..Basle, Switzerland. **1958** *Official Gaz.* (U.S. Patent Office) 16 Sept. TM 91/2 Geigy Chemical Corporation, Ardsley, N.Y... *Tofranil* for antidepressants. **1963**, **1965** [see *imipramine]. **1979** *Daily Tel.* 27 Nov. 12/7 Prescriptions for the brand leaders Tofranil and Tryptizol accounted for an estimated £5 million.

to-fro (tū·ˌfrōᵘ), *adj.*, *adv.*, and *sb. poet.* [f. To and fro *phr.*] **A.** *adj.* = To and fro *adj. phr.*
1879 G. M. Hopkins *Poems* (1967) 81 To-fro tender trambeams truckle at the eye. **1936** R. Campbell *Mithraic Emblems* 83 How shrill the long hosannahs of despair With which those to-fro scolopendras bear, Statesmen to conferences, troops to war. **1952** C. Day Lewis tr. *Virgil's Aeneid* xi. 253 It was like the to-fro rhythm of the sea, when a wave runs forward..then rapidly draws away. **1983** T. Hughes in *Listener* 13 Jan. 21/1 The silent to-fro hurrying of nurses, The bowed stillness of surgeons.
B. *adv.* = To and fro *adv. phr.* 1. *rare.*
1920 Blunden *Waggoner* 44 A sharp snatch, swirling to-fro of the line.
C. *sb.* = To and fro *sb. phr.* 1. *rare.*
1937 C. Day Lewis *Starting Point* 200 The rhythmic tap and to-fro of the white ball. **1960** —— *Buried Day* viii. 157 Almost from the start I seem to have been aware of a fidgetiness, and a constant to-fro made up of many individual, desultory movements.

toft¹. 1. (Later examples.)
1870 Lady Verney *Lettice Lisle* xiii. 146, I might ha' been a comfortable man by this; and now I'm like to have neither toft nor croft. **1955** *Times* 19 Aug. 8/5 Even a layman, with guidance, can recognize the signs pointing to medieval occupation: the hollow said to be the main street; the adjoining humps of the house enclosures, each with its 'toft' (garden) and croft, or small holding. **1965**

Auden *About House* (1966) 17 A toft-and-croft Where I needn't, ever, be at home *to* Those I am not at home *with*.

Toft² (tǫft). The surname of a Staffordshire family used *attrib.* to designate (a style of) lead-glazed slipware made there in the late-seventeenth cent., some of the best examples of which bear the name of Thomas Toft (d. 1689) or another Toft, usu. regarded as the maker of the piece.
1878 L. Jewitt *Ceramic Art of Gt. Brit.* I. iv. 103 Another Toft dish..bears a female figure..and the name Ralphoft, or Ralph Toft, the h and t being apparently conjoined. **1900** F. Litchfield *Pottery & Porcelain* ii. 26 Those buff-coloured dishes which we now recognise as 'Toft ware'. **1957** Mankowitz & Haggar *Conc. Encycl. Eng. Pottery & Porcelain* 222/2 The name [of Ralph Toft] occurs on many typical large Toft-style dishes. **1961** L. G. G. Ramsey *Connoisseur New Guide Antique Eng. Pottery* 20 Signed Toft pieces are known dated 1671 and 1674. **1975** *Country Life* 26 June (Suppl.) 56/1 Christie's. .. Fine English Porcelain and Pottery... Toft dated slipware bragget-pot.

tofu (tōᵘ·fū). [a. Jap. *tōfu*, ad. Chinese *dòufu*, f. *dòu* beans + *fŭ* rotten.] A curd made in Japan and China from mashed soya beans; bean curd.
1880 *Trans. Asiatic Soc. Japan* VIII. 399 *Tôfu* is made by pounding the soy beans after soaking in water. **1905** *Bull. U.S. Dept. Agric.* CLIX. 46 The larger part of the leguminous food in the Japanese diet consists of the preparations of soy beans, such as miso, shoyu and tofu. **1934** Blunden *Mind's Eye* 109 Two hawks have raided the *tofu*. **1936** K. Tezuka *Jap. Food* 28 Tōfu (bean-curd) is made by soaking soy beans in water, mashing them, straining the mass through cloth and solidifying with the addition of magnesium chloride. **1979** *Sunset* Apr. 214/2 Arrange all tofu strips in the casserole and cover with ⅓ of the cheese. **1981** *Guardian* 14 Aug. 7/1 In the United States,..tofu has become an 'in' food.

tog, *sb.*¹ Add: **1.** (Earlier examples.)
1708 *Memoirs Right Villanous John Hall* (ed. 4) 10/2 *Togge*, a Coat. **1718** C. Hitching *Regulator* 20 The names of the flash words now in vogue among thieves... Togge, *alias* Coat. **1755** J. Potts *Jrnl.* in R. Price *Howling Arctic* (1970) i. 16 Having no beaver coats in the factory to make their togs, mittens nor caps.
2. a. (Earlier examples.)
1779 J. Wedgwood *Let.* 9 May (1965) 233 He determined to strip off his waistcoat, and put on the togs at once. *a* **1790** H. T. Potter *New Dict. Cant & Flash* (1795) 59 *Toges* or *toggs*, cloaths for both sexes.
c. *Austral.* and *N.Z. colloq.* A swimming-costume.
1930 V. Palmer *Passage* i. x. 83 'You nip in and get my togs.'.. He was much more at ease in his bathing-trunks than in his..suit and slippery shoes. **1935** J. Guthrie *Little Country* xiii. 216 We..tore down to a quiet beach, stripped off our clothes, and plunged in... We didn't bother about togs. **1944** G. Texidor in D. M. Davin *N.Z. Short Stories* (1953) 313 Mum came over and said.. they could put on their togs. But they mustn't stay in for long, it was getting chilly. **1959** M. Shadbolt *New Zealanders* 96 'I forgot my togs. I left them at the other place.' 'Never mind, you can swim in your shorts, can't you?' **1971** *N.Z. Listener* 15 Feb. 14/5 'I haven't got a costume.'..'Go back and get your togs.'
2*. A unit of thermal resistance used to express the insulating properties of clothes and quilts (see quots. 1945, 1978); so *tog rating*, *value*. [Modelled on the earlier U.S. term *clo*.]
1945 Peirce & Rees in *Shirley Inst. Mem.* XIX. 343 So that practical clothing may be described conveniently by a range of small integers, the unit of thermal resistance, to be called the 'tog', is the resistance that will maintain a temperature difference of 0·1°C. with a flux of 1 watt per square metre, or in more practical terms, 10°C. with a flux of 1 watt per square decimetre. This is the resistance of a light summer suit, and 10 togs represents about the thickest clothing..practicable to wear. **1975** *Daily Tel.* 9 Dec. 13/4 White goose down: 10·5 togs (which means that it is extra-warm and light).—Terylene P.3: 8·5 to 9 togs (normal warmth). The heaviest quilt, I am told, gives the same tog warmth as five blankets at less than half the weight. **1977** *Observer* (Colour Suppl.) 25 Sept. 60/1 (Advt.), Genuine continental quilt luxury at bargain prices: Tog rating (warmth factor) 9·5+. **1978** *Textiles* VII. ii. 50/2 The tog value of a textile is equal to ten times the temperature difference between its two faces when the flow of heat is equal to one watt per square metre. One tog is the thermal resistance of a fabric for a conventional man's suiting or of a blanket of medium quality.

togavirus (tōᵘ·găvaiᵊrǝs). *Min.* [f. L. *toga* Toga + Virus.] Any of a group of RNA animal viruses with enveloped icosahedral capsids, many of which are arthropod-borne and including the viruses of rubella, yellow fever, dengue, and several forms of encephalitis.
1970 C. H. Andrewes et al. in *Virology* XL. 1070/1 Togavirus (from the Latin toga = a cloak) is the name now proposed to cover what is likely to prove the great majority of arboviruses having taxonomic characters like those of the A and B groups. **1974** *Nature* 27 Sept. 343/1 We have examined only one virus outside the picornavirus family, the togavirus, Semliki Forest virus. **1980** R. W. Schlesinger *Togaviruses* i. 10 The rubiviruses, pestiviruses and other 'non-arbo' togaviruses are also of enormous medical or veterinary importance.

together, *adv.* Add: **2. f.** Colloq. phr. (*all*) *girls together*: see *girl *sb.* 2 f.
D. *adj.* **a.** Fashionable, up-to-date; hence used as a general term of commendation. *slang.*
1968 *Daily Mirror* 27 Aug. 7/5 No finer honour can be bestowed on a man down the King's Road than to be called a together cat. **1970** E. Bullins *Theme is Blackness* (1973) 176 Honey, with the right clothes and a together front I'd be a knockout. **1971** *Jamaican Weekly Gleaner* 3 Nov. 5/1, I read in the Miami Herald that conditions in the women's jails [are] not so together.
b. Composed, self-assured; free of emotional difficulties or inhibitions. *colloq.*
1969 Fabian & Byrne *Groupie* ii. 19, I reckoned it was no good putting on a together image if you were all screwed up inside. **1971** *New Yorker* 18 Dec. 31 A young lady of twenty-two who's been through what Twiggy has been through has got to be a very together person to survive. **1974** A. Lurie *War between Tates* (1977) iii. 67, I forgot you, and me, and where I was—I felt very calm, very together. **1977** *O.D.* No. 3. 13/3 All free festivals dream of a together stage manager—try your best to get one, as on the day it's all up to him. **1978** I. M. Gaskin *Spiritual Midwifery* (rev. ed.) i. 41, I knew William was together enough to be there through the whole birthing and I was really excited that he was going to get to see such a heavy thing as a birth. **1979** *Amat. Photographer* 10 Jan. 67 (*caption*) Biddy and Eve—a very together cabaret act. **1983** *Times* 25 Mar. 13/3 An amateur flute player, well groomed and articulate, she looks a very together young woman.

togetherness. Delete *nonce-wd.* and add: **a.** (Later examples.)
1909 F. Barclay *Rosary* xv. 156 Having been apart for a little while seemed to make this curious feeling of 'togetherness' deeper and sweeter than ever. **1912** [see *compresence]. **1920** A. S. Pringle-Pattison *Idea of God* 354 Our primitive and basal experience of time is thus characterized by a togetherness of parts or elements. **1953** E. L. Mascall *Corpus Christi* iii. 57 Assuming that the corporateness of the liturgy is produced by a merely geographical togetherness of the worshippers. **1966** J. Porter *Sour Cream* ix. 123 'I thought I'd take Katia out somewhere.' 'How about making up a foursome?' This blasted Russian passion for togetherness! **1971** *Sci. Amer.* May 105/3 Bullheads often form a dense community, composed of hundreds of individuals, that is based not on a hierarchy or a collection of territories but on close togetherness, with the members swimming freely and peacefully throughout the pond.
b. The fact of getting on well together or being well suited to one another; a sense of belonging together, fellowship.
1930 D. H. Lawrence *A Propos of Lady Chatterley's Lover* 58 Class-hate and class-consciousness are only a sign that the old togetherness, the old blood-warmth has collapsed. **1930** *Times Lit. Suppl.* 13 Nov. 925/3 Characters..must also be real in relation to each other... The personages of Tourneur have this togetherness. **1941** Auden *New Year Let.* 34 O cruel intellect that chills His natural warmth until it kills The roots of all togetherness! **1952** C. Bardsley *Bishop's Move* vi. 74, I wish I saw more of this 'togetherness' in church congregations. **1963** *Economist* 9 Mar. 876/2 The new togetherness [in the Ministry of Defence] is unlikely to mean that..controversies..will disappear. **1972** M. Williams *Inside Number 10* xiv. 352 So there we had social class divisions within the organization itself, so one can imagine how much 'togetherness' that encouraged. **1981** G. Clare *Last Waltz in Vienna* ii. 126 What mattered to me was the ideal of scouting, one for all and all for one, the togetherness in a good and just cause.

Toggenburg (tǫ·gənbɜ̄ɪg). Also **Toggenburgh.** The name of a valley in the canton of St. Gall, N.E. Switzerland, used *attrib.* and *absol.* to designate a hornless light brown goat belonging to a breed first developed in the region.
1886 H. S. H. Pegler *Bk. Goat* (ed. 3) iii. 27 The Toggenburgh goat is generally hornless, and of a rather unusual colour, being a pale drab. **1891** *Goat-Keeper* Sept. 7/2 Champion Zampa, Swiss Toggenburg Goat, 5 years old, short-haired, hornless. **1921** *Blackw. Mag.* June 764/2 The white Nanny and her kid are Alpino goats, and the brown lot are Toggenburgs. **1937** E. B. White *Let.* 9 Sept. (1976) 162 He lives there, with a wife, three children, and a Toggenburg goat. **1979** B. Malamud *Dubin's Lives* ix. 356 One of the Toggenburgs was killed in the goat pasture by a dog.

Togger. (Earlier example.)
1891 P. S. Allen *Let.* Oct. (1939) 10, I hope to combine reading and rowing, for both I hugely desire to get into the Eight next summer and also they are rather hard up for men for the Togger.

toggle, *sb.* Add: **2. a.** Also, a short rod attached to one side of a garment to fasten it by being passed through a loop attached to the other side.
1903 W. F. Petrie *Abydos* II. ii. 26/2, 141–3 appear to be toggles for fastening dress through a loop, like the frogs on a modern military cloak. **1916** *Chambers's Jrnl.* Sept. 617/1 He undid the toggles of his thick lammy coat. **1968** [see *duffle coat* s.v. *duffel, duffle 3]. **1982** B. Aldiss *Helliconia Spring* ix. 231 She was buttoning up her tunic, looking down at the toggles.
f. A kind of wall fastener for use on open-backed plasterboarding, etc., having a part that springs open or turns through 90 degrees after it is inserted, so as to prevent withdrawal and aid gripping.
1934 in Webster. **1964** *Practical Householder* Nov. 1369/1, I had an occasion to use Rawlplug ⅜ in. ⅜ in. gravity toggles... If you decide to remove the toggle at a later date, when the burr on the screw comes up against the swivel nut, the whole device will turn. **1977** *Reader's Digest Bk. Do-It-Yourself Skills & Techniques* v. 154 Gravity toggles have a swivel toggle that drops vertically when pushed through a hole bored in the wall... Spring toggles have two spring-loaded gripping arms which expand after the toggle is pushed through a hole.
g. *Electronics.* = *latch *sb.*¹ 3 b. Also *toggle circuit.*
1953 *Proc. IRE* XLI. 1429/1 The toggles or other storage elements hold the accumulated count. **1955** *Sci. Amer.* June 93/2 In the logical circuits of a modern computer the memory units commonly consist of pairs of vacuum tubes connected in a circuit which is called a 'toggle' because of its analogy with a toggle switch. **1962** [see *bistable *a.*]. **1971** J. H. Smith *Digital Logic* iv. 54 The latch or toggle circuit is used to hold signals fed momentarily into a system.
h. *Computers.* A key or command that is always operated the same way but has the opposite effect on successive occasions.
1982 *Personal Computer World* Dec. 138/1, I find that the 'Install' program is unable to make the best of configuring for my printer as Wordstar expects toggles where the Epson has separate control codes for turning on and off certain modes.
3. *toggle fastening*; **toggle-bolt,** (*b*) = sense 2 f above; **toggle switch,** an electric switch operated by means of a projecting lever that is moved with a snap action, usu. up and down.
1934 Webster, *Toggle bolt*, a bolt having a nut with pivoted flanged wings that close against a spring when passed through a constricted passage and open after emerging. **1968** *Trade Marks Jrnl.* 8 May 736/2 *Rawlplug*. .. Bolt anchoring devices, expansion bolts; toggle bolts, wall plugs and sockets. **1976** *Country Life* 29 Apr. 1143/1 Suit with toggle fastenings. **1976** *Woman's Weekly* 6 Nov. 49/2 Plus a zingy crochet jacket in bold bright stripes with toggle fastenings. **1938** *Rev. Sci. Instruments* IX. 86/1 A toggle switch allows application of the input pulses either..to the scaling circuit or to a thyratron pulse sharpener. **1962** *Times* 8 May 16/5 A steering column lever would be handier than the headlight toggle switch. **1976** *Gramophone* Apr. 1687/3 Neat toggle switches are provided for loudness, tape monitor, low and high filters and tone cancel.

toggle, *v.*¹ Add: **1.** (Earlier example.) Also *fig.*
1836 *Knickerbocker* VIII. 207 What,..has the devil toggled you at last, Jacky?

togidashi (tǫgida·ʃi). Also **togi-dashi.** [a. Jap. f. *togu* to whet, grind + *dasu* to produce, let appear.] A kind of Japanese lacquering in which several coats of lacquer, applied over gold or silver designs, are rubbed and ground down to let the underlying picture appear as if floating below the lacquer surface.
1881 *Trans. Asiatic Soc. Japan* IX. 26 For making *Togi-dashi*, gold dust of a slightly coarser quality is used than for ordinary *Hira-makiye*. **1911** *Encycl. Brit.* XV. 189/1 The togi-dashi design, when finely executed, seems to hang suspended in the velvety lacquer. **1911** [see *nashiji]. **1972** *Times* 15 June 21/3 A two-case togidashi inro by Moei at £1,000.

Togolese (togolī·z), *a.* and *sb.* [f. *Togo* + -ese, after F. *togolais*.] **A.** *adj.* Of or pertaining to the state of Togo (formerly Togoland) in W. Africa. **B.** *sb.* The people of Togo.
1957 *Keesing's Contemp. Archives* 27 Apr.–4 May 15511/1 Continued Togolese representation in the French National Assembly. **1962** A. Lejeune *Duel in Shadows* ii. x. 142 The Togolese..are not too fond of Dr. Nkrumah. **1972** *Times* 9 Oct. (Nigeria Suppl.) p. ii/2 General Gowon visited Lome for the twelfth anniversary of Togolese independence. **1983** *Times* 25 Jan. 6/1 Thousands of Togolese and Beninese have already left Lagos.

togt (toχt). *S. Afr.* [a. Afrikaans, a. Du. *tocht* expedition, journey.] † **1.** A trading expedition or venture. *Obs.*
[**1816** C. I. Latrobe *Jrnl.* 6 May (1818) 265 The master ..was about to set off..on a trip..to dispose of it [*sc.* arrack] in barter... They call this, going *op de tocht*.] **1860** *Queenstown Free Press* 8 Feb., Horses have been discovered amongst those of 'smouses' who were returning..after a somewhat successful *togt*. **1862** Lady Duff-Gordon *Lett.* (1921) 105 He has made a fortune by 'going on Togt'.
2. Casual labour, hired for a specific job.
1901 A. R. R. Turnbull *Tales from Natal* 120 The black devils..so often put us about by deserting—without even the possibility of our being able to obtain togt even. **1948** *Rep. Native Laws Comm. 1946–8* (Dept. Native Affairs, S. Afr.) 37/1 Migrant labour tends to be casual and to produce less and earn less than stable labour. The supply of such labour is often badly adjusted to the demand... In Durban it is..a characteristic of so-called togt or daily labour.
3. *attrib.* and *Comb.*, as *togt labour*, *labourer*, *work*; **togt boy,** a casual labourer; **togt-ganger** [ad. Afrikans *togganger* (also used)], a travelling trader; **togt licence,** a licence authorizing the holder to undertake casual labour.

1898 *Port Elizabeth Tel.* (Weekly ed.) 2 Sept., A Chinaman refused to supply a small quantity of bread and sugar to a togt boy on Saturday. 1972 J. McClure *Caterpillar Cop* ix. 139 He had slunk up to the door..and informed the maid he was a *togt* boy. She..said there were no odd jobs going. 1879 *Cape Monthly Mag.* Feb. 88 For a long time he used to accompany the togtgangers (hawkers or traders). 1896 R. Wallace *Farming Industr. Cape Colony* 91 The plant [sc. prickly pear] was first spread in the Colony by transport riders or 'togt-gangers'. 1957 L. G. Green *Beyond City Lights* 31 In slack times the clever speculators known as *togtgangers* would drive out of Paarl with cavalcades of carts and wagons. 1951 *Cape Argus* 5 Jan. 5/7 Durban harbour had been crippled by a shortage of rail trucks and togt (casual) labour. *Ibid.*, A compound capable of housing up to 1,000 togt labourers should be set aside for this purpose. 1960 J. L. L. Sisson *S. Afr. Judicial Dict.* 121 Casual Labourer, in terms of Native Pass Laws, is synonymous with the term togt labourer. 1948 O. Walker *Kaffirs are Lively* 172 A Native is required to carry on his person..one or more of the following documents... 7. A receipt for *togt* (casual labour) licence. 1968 K. L. McMagh *Dinner of Herbs* xiv. 101 Is there work, togt work nearby?

toheroa (tō^u·ərō^u·ă). Also 9 **tairoa.** [a. Maori.] A large edible bivalve mollusc, *Amphidesma ventricosum*, native to New Zealand.

1873 J. E. Tinne *Wonderland of Antipodes* 66 She sent us a present of a basket of tairoas, a large white shell-fish from the coast, which is considered a delicacy... When roasted, or, better still, made into soup, they are not unlike the clams of New England. 1908 A. Hamilton *Fishing & Sea-Foods Anc. Maori* 13 An enterprising firm in Auckland has recently started a factory for canning toheroa. 1934 *Bulletin* (Sydney) 18 Apr. 20/4 It is only quite recently that toheroa soup has had any standing in culinary circles. 1967 K. Giles *Death & Mr. Prettyman* i. 37 The number one has special black caviar, but the two has smoked Spanish swordfish and New Zealand toheroa patties. 1976 *Daily Colonist* (Victoria, B.C.) 1 Oct. 25/5 The toheroa has a distinctive flavor.

tohoro, var. *Taharah.

tohunga. Delete 'of the second rank' in def. and add earlier and further examples.

1831 G. Bennett in *London Med. Gaz.* 12 Nov. 182/1 This species of Asplenium is a sacred plant among the New Zealanders..; it is used by the Tohunga, or Priest, when praying over a sick person. 1843 E. Dieffenbach *Trav. N.Z.* II. 1. iv. 60 If a chief or his wife falls sick, the most influential tohunga..attends. 1928 [see *atua]. 1938 R. D. Finlayson *Brown Man's Burden* 42 She was a witch all right—like her father the tohunga. 1943 [see *rangatira]. 1955 W. J. Phillipps *Maori Carving Illustrated* 4 Some [carvings] were carried out under the instruction of the old tohunga. 1976–7 *Art N.Z.* Dec./Jan. 34/1 The art of kite making and flying played an important role in the lifestyles of the ancient Maoris. Their manufacture was a sacred and time-consuming affair, for, according to tradition, only a tohunga (priest) of some standing in the tribe could prepare them.

toich (toiχ). *Geogr.* [a. Dinka.] In Southern Sudan, a stretch of flat land near a river that is subject to annual flooding.

1948 J. D. Tothill *Agric. in Sudan* vii. 136 The 'toich' lands..are primarily used as grazing lands. *Ibid.* 954 Toiches are a feature of Equatoria Province. 1955 P. A. Buxton *Nat. Hist. Tsetse Flies* ix. 271 The same sharp edge may generally be observed in the southern Anglo-Egyptian Sudan, at the boundary of the 'toich' or grassy flood-plain on cotton soil, with deciduous woodland on ironstone. 1974 *Nature* 10 May 121/2 The site may have been subject to the type of annual flooding and partial drying which now occurs in 'toich' soils which adjoin rivers or 'khors' in the permanent swamps of the Sudd region in the southern Sudan.

Toidey (toi·di). Also **toid(e)y.** The proprietary name (in the U.S.) for a toilet-training apparatus that can be clipped or strapped on to an ordinary lavatory seat. Also *attrib.*

1924 *Official Gaz.* (U.S. Patent Office) 9 Dec. 300/2 Little Toidey... Water-Closet Seats for Infants, Attachable to Ordinary Water-Closet Seats. 1956 *Ibid.* 6 Mar. 42/2 Toidey. The Toidey Company, Gertrude A. Muller. 1963 M. McCarthy *Group* xiv. 323 She had set him on the new toidey-seat strapped to the regular toilet. *Ibid.*, She tried leaving him on the toidey. 1981 D. Uhnak *False Witness* xxiv. 164 It is the current day-by-day life we have to deal with, not mama and the toidy potty and papa and the primal scene.

toil, *sb.*¹ Add: **4.** *toil-bowed* adj.

1890 Kipling *Poems 1886–1929* (1929) III. 289 They strove to stand to attention, to straighten the toil-bowed back.

toil, *sb.*² Add: **3.** (Further *pl.* examples.)

1897 B. Stoker *Dracula* iv. 41, I am surely in the toils. .. In the present state of things it would be madness to quarrel openly with the Count whilst I am so absolutely in his power. 1931 V. Sackville-West *All Passion Spent* I. 69 Their mother quietly disentangling herself from their toils. 1958 E. Birney *Turvey* v. 46 Soldiers in the toils of civilian law for thefts, burglaries, assaults, rapes and the odd murder. 1973 J. G. Farrell *Siege of Krishnapur* vi. 91 How hopelessly Prejudice, on the point of throwing a net over Truth, had become enmeshed in its own toils.

toile. Add: Also 8 **toille. 1. b.** A painting on canvas.

1919 R. Fry *Let.* 6 Oct. (1972) II. 458 Her old studio.. was stacked with her husband's immense *toiles* of Majorca. 1949 N. Mitford *Love in Cold Climate* II. v. 237 The dozens of *toiles* by him..are worthy of the Douanier himself.

2. Also with Fr. defining addition; esp. *toile de Jouy*, a fabric for upholstery or drapery with a characteristic floral, figure, or landscape design, usu. in one colour on a light background, orig. made at Jouy-en-Josas near Paris.

1794 A. Young *Trav. France* (ed. 2) I. xix. 552 The linens..are *toille de menage*; that exported to Spain is.. called *toille de leon*. 1873 *Young Englishwoman* Oct. 506/1 Paletot of grey-coloured toile de laine. 1923 *Weekly Dispatch* 11 Feb. 14, 5,000 Yards of Toile de Chine. Made of finest Spun Silk. 1934 E. Waugh *Handful of Dust* i. 16 She hasn't paid for the toile-de-jouy chaircovers we made her last April. 1958 I. Murdoch *Bell* ix. 135 She saw the flat in Knightsbridge..glowing with stripy wallpaper and *toile de Jouy*. 1983 *Harrods Mag.* Spring & Summer 143 'Toile de Jouy', a French pastoral scene in Deep Pink on Plain Pink.

3. A pattern for a garment made up in muslin, cotton, or the like, for fitting or for use in making copies.

1959 *Guardian* 18 Nov. 6/4 An excellent little collection of models made from toiles bought in Paris. 1982 *Times* 29 June 13/4, I spent seven months of a two-year couture course *just* making toiles for skirts.

‖ **toilé** (twale). [Fr. *toilé*, f. *toile* Toile.] In lace-making, an area with a closely-worked inwrought pattern.

1865 F. B. Palliser *Hist. Lace* iii. 27 The flower or ornament..is called 'toilé', from the flat, close texture resembling linen, and also from its being often made of that material, or of muslin. 1902 [see *Binche]. 1953 M. Powys *Lace & Lace-Making* v. 41 The linen part or Toilé should show the effect of the raised work.

toilet, *sb.* Add: With spelling **toilette** now usu. pronounced (twăle·t).

5. a. (Further examples, with emphasis on washing and grooming.)

1811 B. Wynne *Diary* 18 Sept. in *Wynne Diaries* (1940) III. x. 340 We began our toilette which refreshed us much after the fatigue of having sat up the whole night. 1890 G. Gissing *Emancipated* I. 1. iii. 83 But when at length he appeared at the dinner-table, once more fresh from his toilet, then did a gleam of animation transform his countenance. 1939 T. S. Eliot *Old Possum's Bk. Pract. Cats* 20 They make their toilette and take their repose.

6. (Earlier example.)

1752 A. Hervey *Let.* 27 Nov. in *11th Rep. Hist. MSS. Comm.* App. iv. 380 in *Parl. Papers 1887* (C. 5060–111) XLVII. 309 'Tis so long (tell Lady Caroline) since I have seen so spruce a Toylet as hers.

7. The restricted sense 'a lavatory' is now in widespread use. (Further examples.) Also (contextually), a lavatory bowl or pedestal; a room or cubicle containing a lavatory.

1917 C. R. Wadhams *Simple Directions for Chambermaid* 50 The toilet should be kept absolutely clean. Hot water with washing soda or cleanser is often needed to clean it thoroughly, using the chamber-cloth or toilet brush for that purpose. 1930 F. J. Eble tr. *Grisar's Martin Luther* v. 108 In the second story of this tower there was a so-called hypocaust, i.e., a furnace-room, and beneath it the toilet (*cloaca*) of the monks. 1955 A. Huxley *Genius & Goddess* 109 She..poured the perfume into the toilet and pulled the plug. 1957 J. Braine *Room at Top* xxiv. 196 You could watch me on the toilet. 1959 S. Gibbons *Pink Front Door* xviii. 222 Such a *gentleman*.. always pretended not to see you if he met you coming out of the toilet. 1968 B. Hines *Kestrel for Knave* 193 He struck a match. A moment while it flared, then two urinals, a toilet in a doorless cubicle, and the sink without a tap. 1979 L. & J. Brown *Our Miracle called Louise* i. 16 A harsh voice beat against the door of a toilet at the home. I had bolted myself inside.

9. a. (mainly sense *7) *toilet article*, *bucket*, *jug*, *kit*, *-service* (earlier example), *-set* (further examples), *-tidy*, *ware*; *toilet block*, *bowl*, *lid*, *seat*, *stall*, *tank*.

1868 *Mich. Agric. Rep.* VII. 351 Perfumery toilet articles. 1981 'J. Ross' *Dark Blue & Dangerous* iv. 23 He checked the essentially male shaver and toilet articles. 1976 *Star* (Sheffield) 26 Nov. 26/5 A two-classroom mobile unit and toilet block. 1947 E. Hodgins *Mr. Blandings builds his Dream House* II. xii. 164 One bathroom seemed all but finished... A toilet bowl was in place. 1983 *Out of Town* Dec. 16/3 We could at least instal the odd alligator in the toilet-bowl. 1957 H. Roosenburg *Walls came tumbling Down* 11 A guard would..open one cell, and allow two prisoners to come out, one with the toilet bucket, the other with the water jug. 1873 J. H. Ewing *Flat Iron for Farthing* iv. 37, I fancied that I heard the familiar sound of Rubens lapping water from the toilette jug in my room at home. 1913 C. Mackenzie *Sinister Street* I. II. iv. 198 After trying to soak a shadowy tomcat down below with water from the toilet-jug Michael and Alan would undress. 1922 S. Lewis *Babbitt* xiii. 163 The leather seat piled with dingy toilet-kits, and the air nauseating with the smell of soap and toothpaste. 1982 D. Bagley *Windfall* xxii. 221 Stafford took his toilet kit and went into the bathroom. 1971 J. D. MacDonald *Seven* (1974) vi. 143 Turned toilet lid down. Sat on it. 1941 N. Last *Diary* 4 Aug. in *Nella Last's War* (1983) 165, I remember washing the floor and toilet seat. 1982 J. Aiken *Whisper in Night* 114 A small round lavatory, with a mahogany Victorian toilet seat. 1855 Geo. Eliot in *Fraser's Mag.* June 706/1 A decanter and a sugar-basin or pie-dish, are an ample toilette service for them. 1890 O. Wilde *Picture of Dorian Gray* (1891) viii. 138 A chased silver Louis-Quinze toilet-set. 1977 G. Marton *Alarum* 6 The New Yorkers on top of Claire's toilet set. 1978 R. Ludlum *Holcroft Covenant* xxiii. 265 If they let the weapon through, he was to reassemble it immediately, in the toilet stall of a men's room. 1974 R. B. Parker *Godwulf Manuscript* xvi. 126 There was nothing in the toilet tank. 1912 Toilet-tidy [used s.v. Tidy *sb.* c]. 1921 D. H. Lawrence *Let.* 10 Nov. (1962) 674, I haven't heard from him, so I can't send him..a set of toilet-tidies until I do. 1928 A. M. Davies *Bk. with Seven Seals* xvii. 380 Mary Anne was not inspired with any admiration for the antimacassars and toilet-tidies that she made. 1864 *Hist. North-Western Soldiers' Fair* (Chicago) 168, 1 set fancy toilet ware. 1977 *Western Morning News* 30 Aug. 2/3 Woolland, Son & Manico have received instructions from private vendors to sell by auction..toilet ware.

b. toilet box, a box containing toilet articles; **toilet brush,** (*a*) a brush used in washing and grooming; (*b*) a lavatory brush; **toilet-case** (earlier example); **toilet club** (see quot. 1966); **toilet-cover** (earlier example); **toilet-glass** (earlier example); **toilet humour** = *lavatory humour* s.v. *Lavatory *sb.* 8; **toilet-paper** (examples of use in lavatories); also *attrib.*; **toilet powder,** a form of dusting powder employed in the toilet, talcum powder; **toilet roll,** a roll of toilet paper; also *attrib.*; **toilet tent,** a tent serving as a lavatory; **toilet tissue,** tissue for use as toilet paper; **toilet-training** *vbl. sb.*, the training of a child to adopt acceptable habits of urination and defecation; hence **toilet-train** *v. trans.*, **toilet-trained** *ppl. a.*

1774 J. Wedgwood *Let.* 2 Mar. (1965) 158 We shall send you some Toilet boxes. 1869 Lady C. Schreiber *Jrnl.* (1950) I. 10 A fan-shaped toilet box made of Chelsea china. 1936 *Burlington Mag.* July 26/1 The tall rectangular toilet-box bears the old standard hall-marks of 1723–24. 1897 *Sears, Roebuck Catal.* 33/2 Rubber Toilet Brush. For the nails and hands. 1917 Toilet brush [see sense 7 above]. 1976 *Sunday Mail* (Glasgow) 26 Dec. 5/7 We went into a bar and found a toilet brush in the gents which still had a price label on it. ?1879 S. F. P. *Stepping Westward* 8 My toilet case for my brush and comb. 1884 W. S. Gilbert *Princess Ida* ii. 69 He grew moustachios, and he took his tub, And he paid a gui-nea to a toi-let club—And he paid a gui-nea to a toi-let club. 1966 J. S. Cox *Illustr. Dict. Hairdressing & Wigmaking* 150/2 *Toilet club*, a barber's shop which, in the 19th cent., offered reduced charges to clients who paid a regular quarterly or yearly subscription. 1838 C. Gilman *Recoll. Southern Matron* xxix. 207 The bride's chamber..neatly set off with white curtains and toilet cover. 1818 M. Edgeworth *Let.* 29 Oct. (1971) 132 Long rolling toilette glass and every piece of furniture belonging to better times. 1956 Auden & Kallman *Magic Flute* (1957) 60 Indulged in toilet-humour with his cousin. 1907 *Yesterday's Shopping* (1969) 340/1 Automatic toilet paper rack... Only one sheet of paper can be drawn at a time. 1956 [see *lavatory paper* s.v. *Lavatory *sb.* 8]. 1976 M. H. Kingston *Woman Warrior* (1977) 181 Her sister went into one of the stalls and got handfuls of toilet paper and wiped her off. 1840 *Picayune* (New Orleans) 28 July 4/1 [Merchandise includes]..perfumed toilet and pearl powders. 1898 W. J. Locke *Idols* viii. 107 A delicate odour of toilette-washes and powder hung on the warmth of the room. 1934 A. P. Herbert *Holy Deadlock* 172 Mr. Rigby..propped the door open with a toilet roll. 1961 *Guardian* 20 Jan. 2/7, 17 toilet roll holders..have been broken or stolen. 1982 *Times* 1 June 18/6 There would seem to be very little difference between one soft toilet roll and another. 1969 *Guardian* 2 Sept. 6/7 A crushed marquee, two burnt-out toilet tents, and a partly demolished fence. 1968 N. Giovanni in W. King *Black Short Story Anthol.* (1972) 30 We'd get slightly used toilet tissue with an article on it or brown paper bags with short sayings or just a note to say they dig us. 1982 *Verbatim* Summer 5/1 Take lavatory paper, or, in the genteel euphemism of Adspeak, toilet tissue. 1951 *Med. Jrnl. Austral.* XXXVIII. II. 111/2 Of our 73 children, 43 were toilet trained with impatience. 1961 *Guardian* 8 May 6/1 When a baby shows signs of being 'toilet trained' a mother can look forward to the end of nappy washing. 1980 A. Cornelisen *Flight from Torregreca* x. 202 The long process of toilet training her children. 1940 *Time* 15 Apr. 48/1 Toilet Training. 1955 B. Spock *Baby & Child Care* 177 Sometimes parents make a great fuss about toilet training. 1964 M. Argyle *Psychol. & Social Probl.* i. 19 An example of a Freudian hypothesis which has been definitely refuted is the hypothesis that early or traumatic toilet training should lead to an anal personality. 1973 E.-J. Bahr *Nice Neighbourhood* ii. 19, I hope Women's Lib never gets to hear about it, but..it was like fun to sit and talk about toilet training.

toilet, *v.* Add: **c.** To assist or supervise in using a toilet (sense 7); *refl.* to use a toilet unaided.

1954 F. G. Blake *Child, his Parents & Nurse* v. 150 Some children stay awake after they have been toileted. 1973 *Lancet* 9 June 1301/1 The same type of elderly patient..may be..wheeled to the toilet in sanitary chairs, or toileted on these within the ward. 1976 *Listener* 19 Feb. 206/1 He is 100 per cent physically and mentally handicapped... He cannot walk, talk, feed or toilet himself.

toileted *ppl. a.* (earlier example); **toi·leting** *vbl. sb.*

1864 'P. Paterson' *Glimpses of Real Life* xxx. 289 There is a gay cavalcade of exquisitely-'toileted' ladies. 1954 F. G. Blake *Child, his Parents & Nurse* v. 157 Until the child has thoroughly mastered his impulses to soil and has incorporated his mother's standards into his conscience, successful achievement in toileting is dependent upon his reward (mother's pride and acceptance). 1977 *Lancet* 17 Dec. 1294/2 She was found to be depressed, occasionally aggressive, and totally dependent on others in all self-care activities, including feeding, toileting, and grooming.

toiletry. (In Dict. s.v. Toilet *sb.*) Restrict *nonce* to senses in Dict. and add: **c.** A preparation for use in washing or grooming. Chiefly in *pl.*

1927 *Glasgow Herald* 5 May 8 One really up-to-date shop coins a new and compact name for these indispensable odds and ends and calls them 'toiletries'. 1927 *Hollis St. Theatre Programme* (U.S.) 19 Sept. (Advt.), Her keen individuality finds in the inimitable Djer-Kiss odeur a refreshing complement; she fastidiously insists upon it in *all* her toiletries! 1957 *Observer* 10 Nov. 11/2 [illegible] toiletries' are now accepted here as they have been in America for years past. 1981 M. Kenyon *Zigzag* iv. 23 The hold-all..held clothing and toiletries.

to--infi·nitive. [f. To *prep.* + Infinitive *sb.*] The infinitive form of the verb immediately preceded by *to*.

1946 O. Jespersen *Mod. Eng. Gram.* V. 154 There is no reason to have a separate name for the to-infinitive. 1964 C. Barber *Linguistic Change in Present-Day Eng.* vi. 135 Another auxiliary which is becoming extremely important to be followed by a *to*-infinitive.

toi-toi. Now the usual spelling of Toe-toe.

1843 [see Toe-toe]. 1907 'K. Mansfield' *Jrnl.* Nov.–Dec. (1954) 23 A clump of toi-toi waving in the wind, and looking for all the world like a family of fine gentlemen [illegible] their heads. 1957 J. Frame *Owls do Cry* 13 The place was like a shelf with gold tickle of toi-toi around its edges. 1981 M. Gee *Meg* xxi. 244 Hedges of toi-toi and fields of fat spring grass.

tokamak (tō^u·kămæk). *Physics.* [a. Russ. *tokamak*, f. *toroidal'naya kamera s magnitnym polem*, toroidal chamber with magnetic field.] One kind of toroidal apparatus for producing controlled fusion reactions in a hot plasma, distinguished by the fact that the controlling magnetic field is the sum of a toroidal field due to external windings and a poloidal field due to an induced longitudinal current in the plasma.

1969 *Nature* 1 Nov. 488/1 Measurements have been made of the electron temperature and density of the plasma in the toroidal device designated Tokamak T3, at the Kurchatov Institute. 1972 *Sci. Amer.* July 73/3 The first large Tokamak machine put into operation in the U.S. resulted from a conversion of the Model C stellarator in the Princeton University Plasma Physics Laboratory. 1980 *Ann. Rep. 1979/80* (U.K. Atomic Energy Authority) 31/1 In tokamaks, plasma is heated and confined by an electric current induced by transformer action, while a strong external field stabilizes the plasma. 1981 [see *stellarator]. 1984 *N.Y. Times* 8 Dec. 1. 49/2 In the race to achieve commercial success, Princeton's tokamak (the original Russian acronym for a toroidal magnetic chamber) is pitted against..laser technology.

Tokarev (tō^u·kărĕf). The name of the Russian designer of firearms F. V. *Tokarev* (1871–1968), used *attrib.* and *absol.* to designate any of a range of automatic and semi-automatic firearms designed or developed by him.

1953 W. G. B. Allen *Pistols, Rifles & Machine Guns* iii. 33 Hotchkiss machine guns and the Russian Tokarev rifles provide examples of the gas operation. *Ibid.* xi. 128 Modern manufacturing techniques permit the weight to be no greater than that of an orthodox rifle; the Russian Tokarev weighs 8½ pounds. 1956 'E. McBain' *Cop Hater* (1958) xix. 123, I keep a few guns... There's a Luger, and a Mauser, and I even got a Tokarev. 1981 S. Dunmore *Ace* ii. i. 190 The Russian was dead..a Tokarev pistol in his right hand.

Tokay. Add: Also in Alsace (more fully *Tokay d'Alsace*) the Pinot Gris grape, or the white wine made from this. Also, an Australian vine, grape, and white wine.

1953 W. James *Word-Bk. Wine* 190 The tokay of Alsace bears no resemblance to the Hungarian wine; light and fruity, with sometimes a pink tinge, it is made from the pinot gris... Australia, too, has a vine called the tokay, which yields a good sweet white dessert wine. 1967 A. Lichine *Encycl. Wines & Spirits* 67/2 Wines from the Pinot Gris (which, in this region, has for centuries been called the Tokay d'Alsace) and Gewürztraminer varieties. *Ibid.* 107/1 The versatility of Australian soil has been amply demonstrated..by its ability to produce..from Tokay either a dry white or a luscious dessert wine. 1976 [see *Pinot]. 1983 *Wine Soc. 1982/1983* (Ann. Rep., Internat. Exhib. Co-operative Wine Soc.) 21 These '82s..are good by any standard, the Gewürztraminer and Tokay particularly.

toke (tō^uk), *sb.*¹ *slang.* [Origin uncertain.] (A piece of) bread; also *fig.* (see quot. 1967).

1843 Dickens *Let.* 7 June (1974) III. 503 [illegible] throwed crumbs and such like up in the air over the side. 1905 [see *ground *ppl. a.* 1 a]. 1963 M. Kendon *Ladies College, Goudhurst* 8 Dripping..spread on 'tokes', was eaten for eleven o'clock lunch by schoolgirls for well nigh forty years. 1967 K. Giles *Death in Diamonds* v. 90 'If you fall foul of Tiny Holdsworth he gives you toke.'.. 'In the local dialect..toke used to be poor quality bread, hence toke and water equals punishment.'

toke (tō^uk), *sb.*² *U.S. slang.* [Origin uncertain: cf. *toke *v.*] An inhalation of smoke from a cigarette or pipe containing marijuana or other narcotic substance.

1968 *Harper's Mag.* Mar. 48 If he still took a toke of marijuana from time to time..still! Mailer was not in approval of any drug. 1973 R. L. Simon *Big Fix* xii. 87, I packed my pipe with..hashish... I took a good heavy toke and held it in as long as I could. 1976 *New Yorker* 7 May 34/3 The host shall light up and take the first toke. He will then offer the joint to the first woman on his left. 1980 *London Mag.* Aug.–Sept. 106/1 He takes huge tokes from a home-made hookah.

toke (tō^uk), *sb.*³ *N. Amer. slang.* [Origin uncertain: perh. an abbrev. of Token *sb.*] A gratuity or tip.

1971 *Daily Colonist* (Victoria, B.C.) 22 June 18/4 The prime advantage [of being a waitress] is instant money—tips or 'tokes' as they are known in the profession. 1981 *Miami Herald* 26 Mar. 30A/2 They have just gone in and hassled people on tips and tokes.

toke (tō^uk), *v.* *U.S. slang.* [Origin uncertain: cf. *toke *sb.*²] *intr.* and *trans.* To smoke (a marijuana cigarette). Also const. *up*. Hence **to·ker.**

1952 *Amer. Speech* XXVII. 30 *Toke v.*, to smoke a cigarette; to take a puff of a cigarette. 1973 *Newsweek* 1 Jan. 5 Bill Buckley says he went 'toking' once... [illegible] 1975 *High Times* Dec. 6/2 Thousands of tokers. *Ibid.* 13/2 [illegible] 1979 N. Mailer *Executioner's Song* (1980) I. xxi. 359 He had been over at a friend of his selling drugs, a little crystal, some speed, toked a couple, got blasted.

token, *sb.* Add: **1. f.** *Semiotics*, etc. A particular and individual sign, as opposed to the type of which it is an instance. Cf. *type *sb.*¹ 8 e.

1906 C. S. Peirce in *Coll. Papers* (1933) VIII. 240, I devoted much study to my ten trichotomies of signs... I called..an Actisign a Token, a Famisign a Type. 1955 N. Chomsky *Logical Struct. Linguistic Theory* (microfilm, Mass. Inst. Technol.) i. 31 [illegible] 1967 J. B. Carroll et al. *Word Freq. Bk.* p. xv, A type is a particular word, counted just once, regardless of how many times it occurs; a *token* is any of the individual occurrences of the type. 1979 *Computers & Humanities* X. 133/1 Without further intervention concordances remain concordances of word tokens and not of headwords.

7. b. *Railways.* (See quot. 1936.)

1936 *Gloss. Terms Railway Signalling (B.S.I.)* 51 *Token*, the authority which must be carried by trainmen to permit a train to travel over a prescribed section of a single line. 1966 O. S. Nock *Railway Enthusiast's Encycl.* 273 The tokens are engraved with the stations at each end of the sections to which they apply. 1971 D. J. Smith *Discovering Railwayana* iv. 20 [illegible]

11. b. A voucher exchangeable for goods or services; *book token*: see *book *sb.* 18; *gift-token*: see *gift *sb.* 9 b; *record token*: see *record *sb.* 14. Also, a small disc or other piece representing or resembling a coin, esp. one used to operate a machine or in exchange for goods or services. Freq. with defining word.

1908 R. Brooke *Let.* Mar. (1968) 123 Dear Mother, I am so sorry about the Boots token. [illegible] 1934 Webster, *Token*, ..the metal fare or ticket issued by a transportation company. 1942 Berrey & Van den Bark *Amer. Thes. Slang* § 160/4 Scrip; tokens; coupons; etc. 1954 *Daily Progress* (Charlottesville, Va.) [illegible] 1965 Ayllon & Azrin in *Jrnl. Exper. Anal. Behav.* VIII. 358/2 Special metal tokens were used as conditioned reinforcers. 1966 G. W. Turner *Eng. Lang. Austral. & N.Z.* viii. 174 The *milk tokens*..are put out for milk last thing at night. 1968 *Listener* 29 Aug. [illegible] 1973 *People's Jrnl.* (Inverness) 4 Aug. [illegible] 1976 *Southern Even. Echo* (Southampton) 18 Nov. [illegible] 1977 *Washington Post* 16 June [illegible] 1978 *Times Lit. Suppl.* 1 Dec. [illegible] car will be waiting. 1981 M. Gee *Dying, in Other Words* 111 The milkman, who was dishonest, and sometimes stole Clothilde's token, leaving no milk. 1982 *Christian Sci. Monitor* 15 Nov. 1 A collector of transportation tokens. 1983 *N.Y. Times* 9 Oct. I. 1. 1/1 The price of bus and subway tokens..must be increased.

15. a. Restrict '*arch.* or *dial.*' to *by this* (or *that*) *token* and sense (b) and add: (*a*) (later examples of *by the same token*).

1875 'Mark Twain' in *Atlantic Monthly* Aug. 193/2 By the same token any person can see that seven hundred and forty-two years from now the Lower Mississippi will be only a mile and three quarters long. 1945 B. MacDonald *Egg & I* (1946) I. i. 11 If you marry a doctor, don't whine because he doesn't keep the hours of a shoe clerk, and by the same token if you marry a shoe clerk, don't complain because he doesn't make as much money as a doctor. 1970 'D. Halliday' *Dolly & Cookie Bird* v. 66 I've dined out on a few stories about her. But not ones that matter. By the same token, she could have made quite a good thing out telling how she saw you..that night. 1978 *Jrnl. R. Soc. Arts* CXXVI. 701/1 By the same token, among the most interesting and valuable sections of this book are those which deal with technique.

16. a. token booth *U.S.*, a booth from which tokens are sold, esp. those for obtaining tickets for a subway; **token economy,** in the treatment of behavioural disorders, the principle or practice of rewarding desirable behaviour with tokens which can be exchanged for goods or privileges and punishing undesirable behaviour by withholding or forfeiting such tokens; **token-reflexive** *a.* (*Logic*), denoting words the referent or temporal or spatial orientation of which is contextually determined, e.g. 'I', 'now', 'here', 'today'; also as *sb.*

1970 *New Yorker* 31 Oct. 123/1 Their reptile-papered basement..is a bit bigger than a token booth. 1968 Ayllon & Azrin *Token Economy* ii. 16 We first conceived of the token economy and its use as a motivational system for therapy during the early part of 1961. 1981 W. Reich in Bloch & Chodoff *Psychiatr. Ethics* iv. 59 The development of aversive techniques of control, 'token economies' and other forms of behaviour modification. 1947 H. Reichenbach *Elem. Symbolic Logic* vii. 284 Words which refer to the corresponding token used in an individual act of speech, or writing..may therefore be called *token-reflexive* words. 1949 *Mind* LVIII. 356 Personal pronouns are to be distinguished from personal proper names such as 'Jones', or 'Fleur',..by their different use, the former being 'token reflexive', the latter 'proper name'. 1962 W. & M. Kneale *Devel. Logic* ii. 53 A sentence containing a token-reflexive taken out of context expresses no proposition at all. 1968 A. J. Ayer *Origins of Pragmatism* 156 With the exception of quantifiers and relative pronouns,..designations are token-reflexive. That is to say, their use is determined by the context.

b. passing into *adj.* Serving as a token; pro forma; (purely) symbolic; constituting a gesture (only); minimal, nominal, perfunctory; cf. *statutory *a.* 3 b; **token estimate,** a provisional statement of a sum of money, placed before Parliament to allow discussion to proceed; **token payment,** (*a*) the payment of a small proportion of a sum due, as an indication that the debt has not been repudiated; (*b*) a nominal payment; **token stoppage, strike,** a brief strike to demonstrate strength of feeling only; **token vote,** a vote of money on the basis of a token estimate.

1915 *Political Q.* May 147 For form's sake 'token' estimates were presented, on the basis of £1,000 for each vote and £100 for each appropriation in aid. 1923 *Times* 27 Feb. 18/3 On the Supplementary Vote of £10 for Diplomatic and Consular Services..the anticipated savings under various subheads were rather larger than £155,198, and would..be sufficient to cover the whole amount now asked for; but inasmuch as all but one of the subheads referred to new services, it had been thought right that a token vote of £10 should be put down in order to provide the opportunity for discussing these new services. 1933 *Sun* (Baltimore) 15 June 1/7 The British Government..tendered a partial or 'token' payment of $10,000,000 to the United States 'as an acknowledgment of the (war) debts pending a final settlement'. 1937 *Ibid.* 19 Oct. 6/1 British and French authorities have expressed belief that there are at least 100,000 Italians serving under Generalissimo Francisco Franco and have urged a 'token' withdrawal on that basis as a guarantee of good faith. 1941 *Ibid.* 28 June 6/3 Less than a week after launching its aggression, Berlin has requested other European states to dispatch 'token forces' to the battlefield. *Ibid.* 29 Aug. 4/3 They [sc. the Persians] have insured themselves against this..by making a token resistance and yielding to demonstrated superior force. 1947 *Daily Mail* 22 May 1 Civil Servants in some sections are considering 'token' strikes if their wages claims continue to drag on without result. 1954 *Times* 20 Jan. 6/7 Twenty-six workers employed by a Manchester contractor have been dismissed for participating in Monday's token stoppage. 1958 *Listener* 12 June 978/1 Some London railway workers vote in favour of an unofficial 'token' strike in support of busmen. 1960 *Time* 12 Dec. 56 The schools took in token Negroes. 1962 *N.Y. Times Mag.* 5 Aug. 11 The current notion that token integration will satisfy his people, says Dr. King, is an illusion. 1968 C. Brooke-Rose *Between* 2 More often the bathroom..has a token window on the hotel corridor or no window at all, merely a ventilation shaft. 1970 J. G. Farrell *Troubles* I. 10 For some reason —the poor quality of the soil or the proximity of the sea— vegetation has only made a token attempt to possess them. 1971 H. MacMillan *Riding Storm* xiv. 442, I.. only agreed to a very small, almost a token, delivery of arms to Tunisia. 1972 D. E. Westlake *Bank Shot* ix. 64 He and his wife Linda were the token whites at this dinner party..the three other couples all being black. 1974 *Times* 21 May 7/8 No tightly run business will have 'token' women on the board. Each director must be able to offer some exceptional contribution. 1976 *New Society* 7 Oct.

28/3 The resistance is little more than token. 1979 J. Cooper *Class* iv. 82 'We've even got two Punk Rockers' (rather like token blacks).

tokenism (tō^u·kĕniz'm). orig. *U.S.* [f. *token *sb.* 16 b + -ism.] The practice or policy of making merely a token effort or granting only minimal concessions, esp. to minority or suppressed groups.

1962 *N.Y. Times Mag.* 5 Aug. 11 (*heading*) The case against tokenism. 1963 *Times* 28 May 10/1 Tactics such as 'tokenism', which have been adopted in some southern states seeking to delay the process of desegregation without opposing the original court order. 1972 [see *joual]. 1976 R. Baxandall in Mitchell & Oakley *Rights & Wrongs of Women* viii. 265 Women are not encouraged to become leaders in unions. Tokenism is a standard practice. 1980 *Jewish Chron.* 12 Sept. 27/3 Philip Rosenthal.. waffled on about 'tokenism' in his factory, where two workers sit on the board, as if real democracy had been achieved.

Hence **tokeni·stic** *a.*, of the nature of tokenism.

1976 in *6,000 Words*. 1977 M. Edelman *Political Lang.* vii. 125 Disorder..invites a response that is only tokenistic or symbolic when the protest is narrow in scope and expressed through conventional tactics. 1983 *Daily Tel.* 14 Sept. 32/5 That was just a tokenistic load of knackers about gay rights.

tokenless, *a.* (Later example.)

1969 *Railway Mag.* Feb. 88 (*caption*) Introduction of tokenless-block working over this route will make these [sc. tablet-catchers], and associated lineside equipment, redundant.

‖ **toki** (tǫ·ki). *N.Z.* [Maori.] A Maori war adze or axe, usu. of stone. Freq. with defining addition.

1860 A. S. Atkinson *Let.* 9 Apr. in *Richmond–Atkinson Papers* (1960) I. 559 It was perfectly innocuous, I believe, in the ordinary way of guns and was probably intended not to hit us but to frighten us into a fit state for nobbling with the toki. 1905 W. B. *Where White Man Treads* 93 The Maori was..provided with no more efficient tool.. than the 'toki panehe' (stone axe). 1949 P. Buck *Coming of Maori* II. vi. 188 The ceremonial adze termed *toki pou tangata* was formed of a nephrite adze lashed to a carved haft.

‖ **tokoloshe** (tǫ:kǭlǫ·ʃi). *S. Afr.* Also **thikoloshe, tikolosh(e), tokolosh,** etc. [Sotho *thokolosi*, *t(h)ikoloshi*, Xhosa *uThikoloshe*, Zulu *utokoloshe*.] In African folklore, a mischievous and lascivious hairy dwarf.

1833 S. Kay *Trav. in Caffraria* xiii. 339 *Tikaloshi* also is much more frequently and familiarly talked about than amongst the more southern tribes. 1894 E. Glanville *Fair Colonist* 82 Tikoloshe is supposed to be an evil spirit which takes the shape of a small man. 1911 *Daily Dispatch* (E. London, S. Afr.) 24 Nov. 7 One might be dragged into the watery den of the *tikolosh*. 1927 W. Plomer *Notes for Poems* 30 The water spirits laughed at her, my friends the tokoloshes. 1949 *Handbk. Race Relations in S. Afr.* 561 The most widely believed in amongst many Nguni tribes is *thikoloshe*. 1959 L. G. Green *These Wonders to Behold* 22 Witchcraft has never lost its grip on the African native. Wherever you go, from Algiers to Cape Town..you will find black millions who are still ruled by the fear of djinns and demons,..ngogwe and tokoloshe. 1972 *Country Life* 10 Feb. 348/2, I was looking at a modern bronze..of a Tokoloshe, a Bantu evil spirit. 1974 *Stand. Encycl. S. Afr.* X. 504/2 The tikolosh (or tokolosh) is believed to be a dwarfish being, pitch-black, with a hairy body and baboon-like face..half-human and half-animal, having only one buttock, while the male has an enormous sexual organ.

‖ **tokonoma** (tō^u:kǒnō^u·mă). Also 8–9 **toko, tokko.** [Jap.] In a Japanese house, a recess or alcove, usu. a few inches above floor-level, in which pictures, ornaments, etc., are displayed. Also *attrib.*

1727 J. G. Scheuchzer tr. *Kæmpfer's Hist. Japan* II. iv. 421 In the solid wall of the room there is allways a *Tokko*..or a sort of a cupboard, raised about a foot.. above the floor, and very near two feet deep. 1822 F. Shoberl tr. *Titsingh's Illustrations of Japan* II. 202 Two cakes..which are placed as an ornament within the *toko*. 1871 A. B. Mitford *Tales of Old Japan* II. 127 The *tokonoma*—that portion of the Japanese room which is raised a few inches above the rest of the floor, and which is regarded as the place of honour. 1929 *Periodical* Feb. 25 The whole set [of the *O.E.D.*] of mine is now sitting stately on the 'tokonoma' of my study in my residence by the side of a Japanese 'Oxford Dictionary'. 1957 C. Brooke-Rose *Languages of Love* 45 The recess on the left of the chimney-breast, which Georgina had turned into a *tokonoma*. 1980 J. Melville *Chrysanthemum Chain* 14 In the *tokonoma* alcove a modest flower arrangement stood in a simple bowl.

Tok Pisin (tǫk pi·zin). Also **talk Pidgin.** [Pidgin, = talk pidgin.] A Melanesian pidgin English spoken in Papua New Guinea.

1943 R. A. Hall *Melanesian Pidgin Phrase-Bk. & Vocab.* 52 *Pidgin*,..talk Pidgin. 1976 *Language* LII. 631 We first began working on the problem of relativization in New Guinea Tok Pisin (Melanesian Pidgin English, Neo-Melanesian) in 1972. 1982 *Trans. Philol. Soc.* 103 Unfortunately, data about the use of Tok Pisin by the indigenous population is very scarce.

tola² (tō^u·lă). [a. *ntola*, name used in Zaïre.] = *agba.

1897 M. Kingsley *Trav. W. Afr.* iii. 58 The Bubi... His idea of decoration goes in the direction of a plaster of 'tola' pomatum over his body. 1959 *Archit. Rev.* CXXVI. 313 The table top and chair seats are covered with black imitation leather; the chair backs are of tola. 1962 *House & Garden* Jan. 45/2 (*caption*) 3-drawer desk in tola wood.

tolazamide (tǫlē^i·zămaid). *Pharm.* [f. *tol(uene* s.v. Tolu- + Az(o- + Amide.] A hypoglycæmic sulphonylurea drug given orally in the treatment of diabetes; 1-perhydroazepin-1-yl-3-*p*-tolylsulphonylurea, $C_{14}H_{21}N_3O_3S$.

1963 *Canad. Med. Assoc. Jrnl.* LXXXIX. 669/1 Tolazamide is a new sulfonylurea compound which has hypoglycemic properties said to be six to eight times as potent as tolbutamide. 1974 M. C. Gerald *Pharmacology* xxv. 441 Several oral hypoglycemic sulfonylurea agents are currently used in the United States, including..tolazamide.

tolazoline (tǫlē^i·zǒlīn). *Pharm.* [f. Tol(yl + *imid)azoline*, f. *imidazol(e + -ine⁵.] An adrenergic blocking agent and vasodilator, used esp. in the treatment of spasm of the peripheral arteries; 2-benzyl-2-imidazoline, $C_{10}H_{12}N_2$.

1952 *Martindale's Extra Pharmacopœia* (ed. 23) I. 605 Tolazoline hydrochloride is a sympatholytic and adrenolytic compound which also exerts a vasodilator effect. 1959 S. Duke-Elder *Parsons' Dis. Eye* (ed. 13) 582/1 (*heading*) Tolazoline ('Priscol') injection. 1972 [see *phentolamine].

tolbooth, toll-booth. 1. Delete † and add: *spec.* a booth at which the toll for the right of passage across a bridge, along a road, etc., is collected. (Later examples.)

The spelling *tolbooth* is now *arch.* or *Sc.*

1973 *Times* 8 May (Hong Kong Suppl.) p. iii/9 Fourteen toll booths can be seen from the control room. 1978 D. Devine *Sunk without Trace* xxvi. 243 The car halted at the toll. He jumped out..to question the man in the toll-booth.

tolbutamide (tǫlbiū·tămaid). *Pharm.* [f. *tol(uene* s.v. Tolu- + But(yl + Amide.] A hypoglycæmic sulphonylurea drug given orally in the treatment of diabetes; 1-butyl-3-tosylurea, $C_{12}H_{18}N_2O_3S$.

1956 *Metabolism* V. 801 (*heading*) Hypoglycemic actions of tolbutamide and carbutamide. 1974 M. C. Gerald *Pharmacol.* xxv. 436 Tolbutamide (Orinase) and related sulfonylurea compounds used for the oral treatment of diabetes act by enhancing the release of insulin from the pancreas.

tole (tō^ul), *sb.*² Also ‖ **tôle.** [a. F. *tôle* sheet-iron, f. dial. *taule* table, f. L. *tabula* a flat board.] **a.** Tin-plated sheet-iron which is first varnished and then ornamented by decorative painting. Also in phr. *tôle peinte*, painted sheet-iron.

1946 *National Button Bull.* (U.S.) Oct. 290 Let's consider these buttons of japanned or lacquered metals. First we have those of Tôle, which are really of tin... Tôle is composed of thinly rolled sheets of iron, tinned. 1958 *Times* 29 Nov. 8/6 A Regency library table decorated in *tôle peinte*..made £540. 1973 *Canadian Antiques Collector* Jan.–Feb. 28/2 The French-Canadian workers in 'tole' or sheet iron.

b. *attrib.* and *Comb.*

1948 E. O. Christensen *Popular Art in U.S.* 19 *Toleware Coffee-pot*... Though called Pennsylvania Dutch, it was of English production... The term 'japanned' tinware, or toleware, is applied to..household utensils.. made of sheet iron, covered with a coating of tin. 1960 *Washington Post* 12 Mar. A-6 (Advt.), All metal tole lamps with brass accents. 1973 *New Yorker* 3 Feb. 40/3 Toleware—exquisite hand-painted tinware produced in the nineteenth century in Pennsylvania, New York State, Ohio, and New England. 1975 *Country Life* 11 Dec. (Suppl.) 22/1 Pair of Austrian tôle peinte figures of red Indians..sold at Christie's..for £4,200. 1976 *National Observer* (U.S.) 17 Jan. 10/2 The city's community-education system offers hundreds of classes that vary from appliance repair to tole painting.

tole (tō^ul), repr. a U.S. dial. and Black English pronunc. of *told* pa.t. and pa. pple.

1797 J. Barton *Let.* in *Amer. Speech 1969* (1973) XLIV. 304 If I had a tole any such a story. *a* 1911 D. G. Phillips *Susan Lenox* (1917) I. ix. 152 So you ain't tole her? Well, Keziah, I've been and gone and got married. 1935 Z. N. Hurston *Mules & Men* (1970) I. i. 26 John tole him, 'Massa, he had two great big eyes lak balls of fire.' 1955 F. O'Connor *Wise Blood* iv. 72, I done tole you them tires won't bust. 1979 'E. McBain' *Calypso* ix. 122 She tole me what she'd been doin.

Toledan (tolē^i·dăn, tolī·dăn), *a.* and *sb.* [f. *Toledo* (see Toledo) + -an.] **A.** *adj.* Of or pertaining to Toledo. Cf. Toletan *a.* **B.** *sb.* A native or inhabitant of Toledo.

1846 R. Ford *Gatherings from Spain* iii. 28 The Toledan chroniclers derive the name from Tagus, fifth king of Iberia. *Ibid.* 29 The performance has been contemplated by many *foreigners*, the Toledans looking lazily on. 1914 J. Masefield *Philip the King* 47 Till their Toledan armour was burnt black. 1965 C. D. Eby *Siege of Alcazar* (1966) viii. 156 Once and for all the Toledan nightmare must be ended.

tolerable, *a.* (*adv.*) Add: **5. a.** As *adv.* (Later U.S. example.)

1884 'Mark Twain' *Huck. Finn* i. 3 Her sister, Miss Watson, a tolerable slim old maid, with goggles on.

b. (Earlier example.)

1812 J. Constable *Let.* 16 Feb. (1962) I. 77 Your Father looks well & is very tolerable as to his cough & breathing.

tolerance, *sb.* Add: **1. b.** Also, diminution in the response to a drug after continued use. Also const. *to*. (Further examples.)

1951 A. Grollman *Pharmacol. & Therapeutics* xviii. 362 A certain degree of tolerance to the nitrites is gained by man from their repeated administration. Especially is this true as regards the headache which they often produce. 1974 M. C. Gerald *Pharmacol.* iii. 62 For the heroin addict, tolerance represents a very real problem, for he is obliged to take larger and larger doses to get the same psychological response. 1982 *Sci. Amer.* Mar. 112/3 The body may accumulate the drug or develop a tolerance to it.

c. For *U.S.* read orig. *U.S.* More widely in *Biol.*, the ability of any organism to withstand some particular environmental condition. Const. *to*. (Further examples.)

1932 Fuller & Conard tr. *Braun-Blanquet's Plant Sociol.* vi. 169 The higher plants have a more or less wide pH tolerance. 1939 *Ecology* XX. 71 (*heading*) A study of the tolerance of trees to breakage by ice accumulation. 1953 E. P. Odum *Fund. Ecol.* iii. 29 Trees give way to grassland as the amount of available water drops below the limits of tolerance for forests. 1960 N. Polunin *Introd. Plant Geogr.* xiv. 428 The arborescent species..fall into groups having a particular height-limit and degree of tolerance to shading. 1961 *Biol. Abstr.* XXXVI. 6632/1 Restraint may affect altitude tolerance in the rat by hastening the body temperature fall. 1979 *Environmental Biol. Fishes* IV. 253/1 Cox..found differences in thermal tolerance of large and small 26°C acclimated bluegill sunfish warmed at 0·1 and 1·0° C min⁻¹.

d. *Biol.* The ability of an organism to survive or to flourish despite infection with a parasite or an otherwise pathogenic organism.

1904 *Q. Rev.* July 137 It is probable that the sleeping-sickness parasite flourished innocently in a state of adjustment due to tolerance on the part of the aboriginal men and animals of West Africa. 1951 R. H. Painter *Insect Resistance in Plant Crops* ii. 59 Corn strains that are tolerant to chinch bug infestation under the moisture conditions of Illinois may not show as much tolerance under drier conditions in Kansas. 1976 Gibbs & Harrison *Plant Virology* xv. 226/1 The use of tomato plants containing a single gene for tolerance to TMV resulted in the selection and rapid spread of virus strains virulent for the plants.

e. *Immunol.* The ability to accept without an immunological reaction an antigen that normally produces one.

1951 *Heredity* V. 396 It may seem surprising that the interchange of red cell precursors should confer tolerance upon homografts of, effectively, skin epithelium. 1968 Passmore & Robson *Compan. Med. Stud.* I. xxvii. 21/2 Experimentally tolerance can be induced by exposure to antigens either in utero or..in the neonatal period. 1979 *Nature* 15 Mar. 257/2 It is pertinent to ask whether the induction and maintenance of specific immunological unresponsiveness (tolerance) to foreign antigens is also under genetic control.

4. a, b. More widely, the allowable amount of variation in any specified quantity.

1916 *Yorkshire Post* 28 Mar. 8/1 Permissible margins of error in workmanship are known as tolerances. 1937 *Times* 13 Apr. (Suppl.) p. xii/4 Visitors may see..how the metal cools and can be withdrawn a minute or two later, finally to be machined to within a tolerance of 0·001 in. on the inside and 0·0005 in. on the outside. 1957 R. W. G. Hunt *Reproduction of Colour* xii. 174 With this system, discrepancies..will result only in errors in chrominance and not in errors of luminance. The tolerances thus become slightly larger. 1965 *Economist* 28 Aug. 812/2 The Ministry will be able to tighten up on tolerances in the road building specifications which it is now rewriting. 1973 A. Parrish *Mech. Engineer's Ref. Bk.* III. 17 A geometrical tolerance is applied to a feature when there is a requirement to control its variation of form or position. 1975 D. G. Fink *Electronics Engineers' Handbk.* I. 48 Stations must operate on an assigned carrier frequency.. which must be maintained within specified limits of frequency tolerances.

5. *attrib.* and *Comb.*: **tolerance dose** *Med.*, a dose, esp. of radiation, believed to be received or taken without harm; **tolerance level,** the level that can be tolerated or is acceptable; *spec.* in *Med.* = *tolerance dose* above; **tolerance limit,** a limit laid down for the permitted variation of a parameter of a product.

1925 *Amer. Jrnl. Röntgenol.* XIII. 66/2 We will have then to decide upon a tolerance dose which can be considered harmless for the operator within a certain assumed period of time. 1958 W. D. Claus *Radiation Biol. & Med.* xvi. 390 The concept of 'tolerance dose' has changed somewhat to the thought that there is no such thing as a literally harmless dose of radiation. 1972 H. C. Rae *Shooting Gallery* III. 202 You know what controlled tolerance doses [of drugs] are?..I had it under control. 1947 *Radiology* XLIX. 364/2 What are the first changes produced by exposures just above the tolerance level? 1964 F. G. W. & M. G. Jones *Pests of Field Crops* xvi. 361 The U.S.A. and Canada have laws determining the tolerance levels

for those pesticides that leave residues on or in the crops. 1977 *New Yorker* 19 Sept. 82/2 It's very important to gauge your audience's tolerance level—decide what it's receptive to, what it can take. 1931 W. A. Shewhart *Econ. Control of Quality of Manufactured Product* xvii. 249 The tolerance range for a given quality X is defined as the range between the maximum and minimum tolerance limits specified for this quality. 1963 Begeman & Amstead *Manuf. Processes* (ed. 5) xv. 356 The tolerance limits for a part are placed outside of the control limits.

tolerance (tǫ·lĕrăns), *v.* *Engin.* [f. the sb.] *trans.* To specify a tolerance for (a machine part, etc.). So **to·leranced** *ppl. a.*, **to·lerancing** *vbl. sb.*

1950 W. Staniar *Plant Engin. Handbk.* ii. 45 (*caption*) Quality-control chart—correct tolerancing of operations. 1953 F. Zozzora *Engin. Drawing* viii. 126/2 As a general rule, nonmating members are toleranced bilaterally, while mating surfaces are toleranced unilaterally. 1959 *B.S.I. News* Aug. 13 British proposals on dimensioning and tolerancing of tapers were generally approved. 1971 J. H. Smith *Digital Logic* ii. 19 The designs are well toleranced and the reader will find that almost any small-signal transistor will function quite satisfactorily. 1973 A. Parrish *Mech. Engineer's Ref. Bk.* III. 18 The concept of geometrical tolerancing is complex. *Ibid.* 19 The feature toleranced is indicated by a leader line.

tolerant, *a.* (*sb.*) Add: **d.** For *U.S.* read orig. *U.S.* More widely in *Biol.*, capable of withstanding any particular environmental condition. (Further examples.)

1929 Weaver & Clements *Plant Ecol.* xiii. 321 Tolerant species..retain their branches. 1943 D. V. Baxter *Path. Forest Practice* viii. 478 Certain woody species tolerant of wet soil. 1979 *Austral. Jrnl. Bot.* XXVII. 531 *Coleochloa setifera* is a desiccation-tolerant sedge which becomes yellow during drying. 1980 Spurr & Barnes *Forest Ecol.* (ed. 3) xiv. 380 A forest tree that can survive and prosper under a forest canopy is said to be tolerant.

e. *Biol.* Of an organism: exhibiting tolerance (sense *1 d) to infection.

1904 E. R. Lankester in *Q. Rev.* July 128 A more precise nomenclature would describe the attacked organism..as 'tolerant', for it tolerates the presence and multiplication of the parasite without suffering by it. 1951 [see *tolerance *sb.* 1 d]. 1976 Gibbs & Harrison *Plant Virology* xv. 225/2 In the western U.S.A., where beet curly top virus is widespread, the sugar-beet industry has been saved by introducing tolerant cultivars.

f. *Immunol.* Exhibiting immunological tolerance (sense *1 e). Const. *of*, *to*.

1951 *Heredity* V. 396 Not all dizygotic twins are completely tolerant to grafts of each other's skin. 1969 R. S. Weiser et al. *Fund. Immunol.* xviii. 227 The F₁ hybrid is an example of an allogeneic recipient which for genetic reasons is immunologically tolerant of parental grafts.

toleration. 2. b. (Earlier examples.)

1796 *Rec. Smithtown, N.Y.* (1898) 129 Any person not an inhabitant..taking Soft shelled clams within the limits of said Town shall pay six pence for every bushel as toleration for taking the same. 1881 E. Ingersoll *Oyster-Industry* III. 249 Toleration.—License to gather oysters or operate beds... The money paid is called a Toleration fee.

tolerize (tǫ·lĕraiz), *v.* *Immunol.* [f. Toler(ant *a.* (*sb.*) + -ize.] *trans.* To render immunologically tolerant. So **to·lerizing** *ppl. a.*; also **toleriza·tion,** the action of tolerizing.

1967 *Immunology* XIII. 156 Immunocompetent cells might not have been exposed to the tolerizing antigen. 1973 *Nature* 16 Mar. 161/3 Free IgT complexes are present and these effectively tolerize B cells. 1974 I. M. Roitt *Essent. Immunol.* (ed. 2) viii. 202 It may be that the tolerance-inducing regimen does truly tolerize T cells.. but not all B-cells. 1978 *Jrnl. R. Soc. Med.* LXXI. 161/1 The interaction of bacteria with the adjuvant and tolerizing agents in plaque may induce immune responses which could enhance or inhibit the development of caries. 1979 *Nature* 15 Mar. 258/2 One day after tolerisation the recipient mice plus appropriate controls were sensitised by two paintings with DNFB.

tolerogen (tǫ·lĕrǒdʒĕn). *Immunol.* [f. Toler(ance + -o + -gen.] A substance inducing immunological tolerance.

1967 *Immunochemistry* IV. 180 The multi-chain polymer was a very efficient tolerogen. 1980 *Nature* 28 Aug. 837/2 Both generations, neither of which had been intentionally exposed to the tolerogen, showed a wide range of response, from normal down to undetectable.

Hence **toleroge·nic** *a.*, **to:lerogeni·city.**

1967 *Immunochemistry* IV. 180 The protein carrier is not necessary to endow a molecule with tolerogenic capacity. 1970 *Nature* 11 July 176/1 There is much evidence for a reciprocal relationship between immunogenicity and tolerogenicity. 1979 *Jrnl. Immunol.* CXXII. 1886/2 Polyethylene glycol has been shown to serve as an effective tolerogenic carrier.

tolidine (tǫ·lidīn). *Chem.* [f. Tol(yl + Benz)idine.] A benzidine derivative, $(NH_2 \cdot (CH_3)C_6H_3—)_2$, which is the parent compound of a group of azo dyes and is used (in the *ortho* form) as a reagent in chemical analysis.

1879 *Jrnl. Chem. Soc.* XXXVI. 235 (*heading*) The three isomeric tolidines (diamido-ditolyls). 1935 *Discovery* July 208/1 A minute sample of the bath water is taken and 'doped' by the mixture of few drops of 'O.T.' (Ortho-Tolidine), a chemical which turns chlorinated water yellow. 1964 *Kirk-Othmer Encycl. Chem. Technol.* (ed. 2) III. 414 *o*-Tolidine is used to a rather large extent in qualitative and quantitative analysis. It is employed for the detection..of such substances as chlorine..gold.. and tungsten. *Ibid.* 415 Azo dyes prepared from *m*-tolidine..have little affinity for cotton, but are interesting dyes for wool.

tolite (tǫ·lait). Also **Tolite.** [f. Tol(u- + -ite¹.] Trinitrotoluene used as an explosive.

1909 O. Guttmann *Manuf. Explosives* i. 14 Trinitrotoluene has been introduced into the French Service under the name of Tolite. 1924 *Chem. Abstr.* XVIII. 2604 During the war..use was made of a cold, dil. soln. of Na_2SO_3 for the purification of crude trinitrotoluene (tolite) to produce TNT. 1953 J. Y. Cousteau *Silent World* 39 With one-pound tablets of German tolite explosive the effect was different. 1968 *New Scientist* 7 Nov. 304/3 Molène favours the use of classic explosives such as TNT, Tolite, Melinite or Trinitrophenylamine.

‖ **tolkach** (tǫ·lkatʃ). Pl. **tolkachi.** [Russ., f. *tolkat'* to push or jostle.] In the U.S.S.R., a person who negotiates difficulties or arranges things, a 'fixer'.

1955 H. Hodgkinson *Doubletalk* 129 *Tolkach* (from *tolkat*, to push or jostle), a 'fixer'; the man who knows a man; the man who can get it for you wholesale; who has *blat*. 1957 J. S. Berliner *Factory & Manager in U.S.S.R.* xii. 215 The key figure in financing all these operations of the tolkachi is the accountant. 1963 *Economist* 29 June 1390/2 Plant directors..buy..supplies illegally from each other, through *tolkachi* or spivs. 1977 *Western Political Q.* XXX. 217 The premier practitioner of *blat* is the *tolkach*. He is the plant's representative who travels the country searching for needed supplies or unsnarling bureaucratic bottlenecks.

Tolkienian (tǫlkī·niăn), *a.* Also **Tolkinian** (tǫlki·niăn). [f. the name *Tolkien* (see below) + -ian.] Of or pertaining to the philologist and author of fantasy literature John Ronald Reuel Tolkien (1892–1973) or his writings.

1954 C. S. Lewis in *Time & Tide* 14 Aug. 1083/1 In the Tolkinian world you can hardly put your foot down.. without stirring the dust of history. 1975 C. N. Manlove *Mod. Fantasy* v. 160 Escape merges into another Tolkinian criterion of the higher fairy-tale—Consolation. 1979 J. C. Nitzsche *Tolkien's Art* 4 A pattern emerges upon an examination of the titles of other Tolkienian works. 1980 *Times Lit. Suppl.* 7 Nov. 1258/3 If nursery puddings, Tolkienian fantasy and public school cuddles are anything to do with politics at all, they are slightly more identifiable with the Right.

So **To:lkiene·sque** *a.* [-esque], characteristic of or resembling Tolkien or his writings.

1970 *Nature* 18 July 215/2 Earlier this month it was announced that Loch Morar, too, would be screened for a monster, already christened with suitably Tolkienesque undertones, as Morag. 1977 *Sounds* 9 July 28/3 Tyrannosaurus Rex, a duo small in sound but big in Tolkienesque fantastical imagery.

toll, *sb.*¹ Add: **2. g.** (Further examples, esp. of death, loss, or injury.)

1870 J. C. Duval *Adv. Big-Foot Wallace* p. xv, Wallace joined Colonel Hays's regiment..and was with it at the storming of Monterey, where he says he took 'full toll' out of the Mexicans for killing his brother and cousin.. in 1836. 1927 W. E. Collinson *Contemp. Eng.* 34 The 'toll of the road' (often used by the Daily Mail in 1925–6) has been so heavy that some of the local authorities have adopted the expedient of the white line. 1929 *Daily Express* 7 Nov. 2/3 Miners' members were artists in presenting the toll of the mines in its most impressive form. 1959 *Listener* 6 Aug. 222/1 A thoroughly well-intentioned programme aimed at reducing the toll on the roads. 1962 E. Roosevelt *Autobiogr.* II. xv. 123 The war had taken from France a heavy toll of her young men from 1914 to 1918. 1972 N. Freeling *Long Silence* II. 189 In fact it had been very hardbought, some of the winnings, taking fearful tolls of nerve, straining every atom of him. 1974 F. Forsyth *Dogs of War* xiv. 264 He felt tired and flat; the strain of the past thirty days was taking its toll. 1974 C. Ryan *Bridge too Far* III. ii. 147 Forty-five patients were dead (the toll would increase to over eighty), and countless more were wounded. 1981 *Times* 9 June 6/3 The death toll in the train disaster..could be more than 1,000. 1982 S. Brett *Murder Unprompted* xii. 113 The obscurity of the play, and the..lack of star names—all the elements which pessimists had predicted would work against the show—were now beginning to take their toll.

i. A charge for a telephone call. Usu. *attrib.* (see also *toll call*, sense 3 below, *toll-free *a.*). *N. Amer.*

1886 *Jrnl. Soc. Telegr. Engineers* XV. 275 Another term also very largely used in America is 'toll line working'; i.e., communication between town and town.. which is paid for by tolls per message, and not by annual subscription such as usually occurs in local exchanges. 1912 Thiess & Joy *Toll Telephone Pract.* i. 3 In Bell practice the terms 'suburban' and 'toll' are often used synonymously, but suburban business is always short haul..while toll business may be long haul but not exceeding 100 miles. The term 'long distance' in Bell practice implies any haul exceeding 100 miles and in many cases the toll and the long distance business are handled at different switchboards... The terms 'toll' and 'long distance' do not have this distinction in independent practice and are commonly synonymous. *Ibid.* iv. 43 When a toll subscriber desires to call central, he will operate his generator. *Ibid.* 44 If the jack of the subscriber wanted terminates in the toll position.., the toll operator will simply insert her calling plug and ring. 1921 *Telegraph & Telephone Jrnl.* VII. 180/2 It was eventually decided to remove all the short trunk lines (i.e. up to 25 miles in length) from the Trunk Exchange and to connect them to one or more Toll Exchanges. 1926 *Daily Colonist* (Victoria, B.C.) 11 July 16/1 The British Columbia Telephone Co...built the first toll line to Nanaimo. 1933 K. B. Miller *Telephone Theory & Pract.* III. xi. 466 Telephone traffic between subscribers whose lines are connected to different local exchanges is ordinarily called 'toll traffic'... Toll traffic is generally classified as 'short haul' or 'suburban' for distances up to the neighborhood of about 50 miles, and 'long haul' or 'long distance' for connections between more widely separated points. 1970 N. Armstrong et al. *First on Moon* xiv. 363, I had a telephone call yesterday. The toll wasn't..as great as the one I made to you fellows..on the moon. 1978 *Sci. Amer.* June 90/2 It may also provide paths between trunks, but this task is usually performed by special exchanges called toll exchanges.

3. toll call, orig., † a telephone call which was not a local call and for which an individual charge was therefore made; in later (not British) use, a long-distance call, or a call between different telephone areas; **toll plaza** *U.S.*, a row of toll booths on a toll road; **toll television, TV** = *pay television*, *-TV* s.v. *pay- 4; **tollway** *U.S.*, a highway for the use of which a charge is made.

1912 Thiess & Joy *Toll Telephone Pract.* i. 3 Common usage among telephone men has led to the general classification of telephone service under four headings, as follows: local, suburban, toll and long distance... Any telephone call which is not local bears a special or toll charge and broadly may be termed a toll call; hence it seems proper to use the term 'toll' to embrace all service of this class, whether it be suburban, toll, or long distance in the narrow sense. 1928 E. Wallace *Again Three Just Men* x. 223 The telephone-bell rang. The voice of the porter informed him that a toll call had come through. 1965 S. T. Ollivier *Petticoat Farm* vi. 88 This ring was a toll call and would be costing a fortune. 1977 D. Anthony *Stud Game* vii. 45 Grant must have called the Bishop girl from here. She lives in Santa Monica, which makes it a toll call. 1948 *Sun* (Baltimore) 20 Nov. 14/2 All tolls for travel across the Chesapeake Bay Bridge will be paid at booths on the 1,000-foot toll plaza on the Western Shore approach to the bridge. 1983 *N.Y. Times* 16 July 26/6 Connecticut state police have been stopping northbound trucks at the Greenwich toll plaza on the Connecticut Turnpike. 1956 *Britannica Bk. of Year* 492/2 *Pay T.V., Toll T.V.*,..these phrases having been coined in America during a discussion of the possibilities of providing additional television programmes to viewers willing to pay for them on a subscription basis. 1960 *News Chron.* 13 Oct. 14/2 The pioneer Toll-TV service working in a suburb of Toronto. 1960 *Spectator* 30 Dec. 1039 The principle of toll television is that the viewer should pay only for films..he wants to see. 1955 *Britannica Bk. of Year* 490/1 Modern automobile transport continued to produce new words. Among these were *Tollway*, a modern version of the old toll road. 1958 *Times* 22 Nov. 7/6 To the other points of the compass the expressways—and even newer, faster, tollways—throw out rippling tentacles. 1969 'E. Lathen' *When in Greece* vi. 61 The new tollway along the coast..was..the only high speed road in Greece. 1982 S. Paretsky *Indemnity Only* xvi. 214, I headed back to the tollway and Chicago.

toll, *v.*² Add: **2. d.** To charge a toll for the use of (a bridge, crossing, etc.). Chiefly as *ppl. a.*

1978 *Financial Times* 24 Oct. 11/3 The tolled Humber bridge. 1978 in H. Wilson *Final Term* (1979) 247 Concessions to 'Orange Badge' holders at most tolled crossings.

toll-bait. (Earlier example.)

1870 *Amer. Naturalist* IV. 516 The 'tole-bait' consists chiefly of Menhaden (*Alausa menhaden*) ground very fine.

Tollens (tǫ·lĕnz). *Chem.* [The name of B. C. G. *Tollens* (1841–1918), German chemist.] *Tollens'(s) reagent*: a solution of ammoniacal silver nitrate and sodium hydroxide, used as a test for aldehydes with which it reacts to give a silver precipitate; so *Tollens' test* (described by Tollens in *Ber. d. deut. Chem. Ges.* (1882) XV. 1828).

1904 S. P. Mulliken *Method for Identification Pure Organic Compounds* I. 22 (*heading*) Compounds reducing silver from Tollen's [*sic*] reagent. 1946 F. Schneider *Qualitative Organic Microanalysis* vii. 198 A blank experiment should be run with the Tollens' reagent and the original substance. 1964 N. G. Clark *Mod. Organic Chem.* x. 187 Tollens's reagent..gives a precipitate of silver, often in the form of a silver mirror. 1981 Wingrove & Caret *Organic Chem.* xxi. 867 A simple chemical test to distinguish between aldehydes and ketones is based on aldehydes being readily oxidized to carboxylic acids whereas ketones are not. Tollens' test..is used most frequently.

tollent, *a.* (Earlier example.)

1770 tr. *C. F. von Wolff's Logic* 87 The Tollent mode.

toller², toler. 2. Delete second quot. and add earlier example.

1831 I. T. Sharpless in *Cabinet of Nat. Hist.* I. 43/2 Most persons on these waters, have a race of small, white or liver coloured dogs, which are familiarly called the *toler* breed.

tolley (tǫ·li). [Var. of *taw-alley* (Eng. Dial. Dict. s.v. Taw *sb.*¹): see Ally, alley, alay *sb.*²] = Taw *sb.*² a.

1970 *Times* 18 Feb. 2 Playing marbles requires a player to bend double to flick the tolley. 1972 *Daily Tel.* (Colour Suppl.) 14 Jan. 25/1 You flick the tolley with your thumb.

toll-free, *a.* (Later N. Amer. examples, of telephone calls, lines, etc.)

1970 *Globe & Mail* (Toronto) 26 Sept. 31/1 (Advt), For reservations, call toll-free 368-7474. **1971** *Sci. Amer.* Oct. 7/2 (Advt.), When you buy a '72, you get the name and toll-free number of a person in Detroit. **1976** *National Observer* (U.S.) 13 Mar. 3/6 (Advt.), Write or call now. Our toll-free lines are open 24 hours daily, 7 days a week. **1979** *Arizona Daily Star* 5 Aug. (Advt. Section) 4/3 Make a toll-free call to Bill Jackson. **1984** *Gainesville* (Florida) *Sun* 28 Mar. 2A/3, I have called the same toll-free number I ordered from several times but these people will not return my call.

toll-house. Add: **3.** *attrib.* and *Comb.*: **toll-house cookie** *U.S.*, a biscuit containing chocolate chips.

1973 *Publishers Weekly* 22 Jan. 71/2 Henry begins suffering from dark spots that pop out all over his body; he soon looks like a toll-house cookie. **1978** R. Nixon *Memoirs* 316 After our meeting we had a delicious lunch of steak and fresh corn on the cob, followed by Lady Bird's homemade toll-house cookies.

tolling, toling, *vbl. sb.*[1] **b.** (Earlier example.)

1838 J. J. Audubon *Ornith. Biogr.* IV. 6 The usual mode of taking these birds has been..by *toling*, as it is strangely termed, an operation by which the ducks are sometimes induced to approach within a few feet of the shore.

tol-lol-ish, *a.* (Earlier example.)

1840 H. Cockton *Valentine Vox* xxvii. 210 'And the ladies, how are they?' 'Why, they're only tollolish. You know what women are.'

tolly (tǫ·li). *School slang.* Now *arch.* or *Hist.* [app. f. Tallow *sb.*] A (tallow) candle.

1890 Barrère & Leland *Dict. Slang.* II. 360/2 *Tolly* (public schools), a candle. **1905** *Daily News* 2 Aug. 4 Who does not recognise a living experience in Hugh working after prohibited hours for a scholarship, caught by the master and 'jawed for having a tolly alight'. **1924** E. Marsh tr. *La Fontaine's Fables* 77 The luckless tolly.. Ended as a pool of grease.

Also as *v. intr.* (*Harrow*), to work by candle-light after the lights have been extinguished; to 'burn the midnight oil'; usu. with *up*.

1890 Barrère & Leland *Dict. Slang.* II. 360/2 *Tolly up, to* (Harrow School), to keep a candle alight after the gas has been turned off. **1894** Wilkins & Vivian *Green Bay Tree* I. 73 The process known as 'tollying up', or working by candle-light after the legal hours.

Tolman (tǫ·lmăn). Also **Tallman.** The surname *Tolman* used *ellipt.*, *attrib.*, or in the possessive in **Tolman('s) sweet(ing)**, to designate a yellow-skinned apple belonging to a variety originally developed in Rhode Island; also, the variety itself, or the tree bearing this fruit.

1822 J. Thacher *Amer. Orchardist* 139 Tolman sweeting..is held in much estimation for family use during the autumn. **1838** *Genesee Farmer* 17 Mar. 81/1 Winter Fruit..Tallman Sweeting. **1845** A. J. Downing *Fruits & Fruit Trees Amer.* viii. 137 The Tolman's Sweeting..is one of the most popular orchard sorts. **1867** J. A. Warder *Amer. Pomology—Apples* 557 Tallman's Sweet..has traveled from Rhode Island wherever her hardy sons have gone westward. **1875** J. Burroughs *Winter Sunshine* 155 Now you have got a Tolman sweet. **1878** [see *McIntosh]. **1893** A. M. Diaz *William Henry Lett.* 7 He..set out Baldwins and Tallmans and Porters. **1909** S. B. Green *Pop. Fruit Growing* ix. 169 Some varieties..are adapted to a wide range as..Tolman Sweet. **1928** W. H. Chandler *N. Amer. Orchards* iv. 105 Tolman, the leading sweet apple to reach the market,..but few trees now being planted. **1949** *Boston Globe* 14 Aug. (Fiction Mag.) 11/4 We could look squarely at the tolman sweet standing all alone with the moonlight making its blossoms burn like candles. **1970** [see *russet *a.* 1 b].

toloache (tǫlwæ·tʃi). Also **toloachi.** [a. Mexican Sp. *toloache*, a. Nahuatl *toloatzin*, f. *toloa* to bow the head + *tzin* reverential.] A preparation of a plant of the genus *Datura* used as an intoxicating and hallucinogenic drug.

1894 [see *marijuana, marihuana 1 a]. **1948** A. L. Kroeber *Anthropol.* (rev. ed.) xiv. 567 Southern and south-central California: Initiation of youths with *toloache* or jimson-weed drug (*Datura* species). The narcosis is accompanied by visions, which were considered sacred. **1964** I. Fleming *You only live Twice* vii. 93 Addiction to toloachi, a drink made from *D. tatula*, causes chronic imbecility.

Toltec (tǫ·ltek), *sb.* and *a.* Also 8 **Tolteca,** 9 **Tolteck, Tultec.** [ad. Sp. *tolteca*, ad. Nahuatl *toltecatl*, pl. *tolteca.*] **A.** *sb.* (A member of) a Nahuatl people who dominated the valley of Mexico *c* 900–1150 A.D., before the arrival of the Aztecs. **B.** *adj.* Of or pertaining to this people.

1787, etc. [see *Olmec 1]. **1814** [see *Aztec *sb.* and *a.*]. **1843** W. H. Prescott *Conquest of Mexico* I. i. i. 12 The Toltecs were well instructed in agriculture. **1875** *Encycl. Brit.* I. 696/1 The Toltec and Aztec races. **1939** G. Greene *Lawless Roads* iii. 104 Quetzalcoatl..was the white Toltec god of culture. **1955** *Sci. Amer.* May 82 To be a Toltec in Mexico was to be an exponent of civilization. **1977** *Time* 21 Feb. 19/1 He wrote his mystical novelette about the god Quetzalcoatl, who figures so largely in the Toltec legends of the Mexican people. **1979** P. Theroux *Old Patagonian Express* iii. 52 Towards Tula, a treeless desert ..rose into peaks like pyramids. This was the capital of the Toltecs.

Hence **To·ltecan** *sb.* and *a.*

1839 *Penny Cycl.* XV. 165/1 The older..monuments of Mexico are..the productions..of the Toltecans. *Ibid.* 165/2 The extraordinary vastness of..these..Toltecan constructions.

tolu-. Add: **toluqui·none** [Quinone], the aromatic compound $CH_3C_6H_3O_2$; also, any of the derivatives of this compound.

[**1870** *Jrnl. Chem. Soc.* XXXIII. 135 (*caption*) Trichlorotoluquinone.] **1874** *Index Jrnl. Chem. Soc. 1848–72* 254/1 Toluquinones. **1975** *Nature* 20 Nov. 194/1 The chemicals present in the glandular secretions of insects are often exceedingly diverse... They comprise..a great number of phenolic substances and quinones (such as phenol.. and toluquinone).

toluidine. (In Dict. s.v. Tolu-.) Add: Also **toluidene, -in. toluidine blue,** a thiazine dye, $C_{15}H_{16}ClN_3S$, now used chiefly as a biological stain.

1898 *Philadelphia Med. Jrnl.* II. 343 Toluidin-blue is a member of the aniline group closely related chemically to methylene-blue. **1908** Toluidene blue [see *metachromatically *adv.*]. **1947** *Ann. Rev. Microbiol.* I. 346 Both methylene blue..and toluidin blue..have shown therapeutic activity in experimental rickettsial infections in animals. **1981** J. A. Kiernan *Histol. & Histochem. Methods* xviii. 258/1 The procedure..for demonstrating metachromasia with toluidine blue gives excellent results when used as a Nissl stain.

tom, *sb.*[1] Add: **1. a.** *Tom, Dick, and Harry* (earlier example). Also with *or*.

1734 *Vocal Miscellany* (ed. 2) I. 332 Farewell, Tom, Dick, and Harry, Farewell, Moll, Nell, and Sue. **1762** J. Otis *Vindication House Representatives Massachusetts-Bay* 21 That I should die very soon after my head should be cut off..whether chopped off to gratify a tyrant by the christian name of Tom, Dick or Harry is evident. **1864** Trollope *Can you forgive Her?* I. xxxii. 254 Didn't he want to squander every shilling of the property,..property which I could give to Tom, Dick, or Harry tomorrow, if I liked? **1906** I. Zangwill *Let.* 29 Oct. in K. Gregory *First Cuckoo* (1978) 64 And have these wise and witty ladies less right than Tom, Dick or 'Arry to a direct influence on the government of their country? **1974** *New Statesman* 22 Nov. 740/3 There is no legislation for giving them a licence, so that any Tom, Dick or Harry can work as a guide and give..wrong information.

d. A girl or woman. *Austral. slang.*

[**1882** *Sydney Slang Dict.* 8 *Tom-tart*, Sydney, phrase for a girl or sweetheart.] **1906** E. Dyson *Fact'ry 'Ands* i. 8, I may be wrong in thinkin' your tom was tryin' t' mash ther man shootin' off ther camera. **1951** D. Stivens *Jimmy Brockett* 102 'You did, darling,' one of the little social toms said. She was a nuggety little sheila.

e. A prostitute. *slang.*

[**1914** Jackson & Hellyer *Vocab. Criminal Slang* 84 *Tommy*... A prostitute.] **1941** V. Davis *Phenomena in Crime* xix. 255 *Tom*, old prostitute. **1955** M. Hastings *Cork & Serpent* i. 12 I'll bet she's holding out on us. We know these toms, sir. **1957** H. Williamson *Golden Virgin* ix. 134 'Is Lily a tom?'..'Not within the meaning of the act. She works in Nett's Laundry, on the lower side of Randiswell Bridge. Of course, I don't say she doesn't have a bit of fun at times, but that's her business.' **1977** *Time Out* 17–23 June 18/1 What doesn't appear in the film but is very revealing about police mentality, is the filing room on prostitutes (or Toms as they are called).

f. Short for *Uncle Tom*: a Black regarded, esp. by other Blacks, as behaving in a servile, ingratiating, or complaisant manner towards white people. *slang.*

1959 *Esquire* Nov. 122/1 *Tom*.., a Negro who does not try to maintain his complete dignity before whites. **1968** N. Giovanni in W. King *Black Short Story Anthol.* (1972) 26 Toms, I told him, only have power if we let them have power. I mean, if a tom says get off the streets and you get off the streets, then that's your fault, not his. **1973** R. Ludlum *Matlock Paper* ii. 14 The African studies may be in trouble. That 'Tom' I recruited from Howard turned out to be..a little to the right of Louis XIV. **1975** *Publishers Weekly* 3 Feb. 72/1 By installing 'American Nigger Toms' as the Third World élite, the U.S. has controlled the angry hunger of the poor populace.

7. b. Tom Collins orig. *U.S.*, a cocktail made of gin, lime or lemon juice, sugar, and soda water (cf. sense 5 in Dict. and *Collins[2]); **Tom Walker** *U.S. dial.*, the Devil; also *the Devil and Tom Walker.*

1888 H. Johnson *New Improved Bartenders Man.* (rev. ed.) 227 Tom Collins. **1906** L. Muckensturm *Louis' Mixed Drinks* 99 (*heading*) Tom Collins. **1959** 'M. Ainsworth' *Murder is Catching* xvi. 178 She made us both long cool Tom Collinses, the tumblers frosted with ice-chips. **1979** S. Rifkin *McQuaid in August* (1980) ii. 7 The bartender would make me a tall tom collins without any cherry. **1833** S. Smith *Life & Writings J. Downing* 139 They always would have their way in spite of every body and Tom Walker besides. **1914** *Dial. Notes* IV. 71 He wukked like the Devil an' Tom Walker. **1949** 'T. Nelson' *Backwoods Teacher* xiii. 136, I don't know nary charm, but they's an' old sayin' that some folks says but all it is is, 'Ol' Tom Walker [the devil] under yore hat, God the Father, God the Son, an' God the Holy Ghost.' **1958** *Virginia Q. Rev.* Spring 261 He whispered: 'Old Tom Walker under your hat. Father, son and holy ghost,' the way blue-eyed Dulcie would have done.

c. Tom Pudding *slang*, one of the box-like iron boats that are connected together and towed by a tug to carry coal on canals. (See also sense 8 b in Dict.)

1906 *Westm. Gaz.* 28 Mar. 8/2 Trains of iron compartment boats, known locally as 'Tom Puddings', are towed all the way to Goole. **1949** *Archit. Rev.* CVI. 8/3 On the Aire and Calder, compartment boats, or Tom Puddings, are used. These are oblong iron boxes towed in trains up to 32 in number by steam tugs. **1970** *New Society* 19 Nov. 898/2 If you haven't seen a chain of tom puddings then you've missed one of the sights of England.

tom (tǫm), *sb.*[2] Colloq. abbrev. of Tomato.

1920 *Chambers's Jrnl.* 15 May 384/1 The acreage of 'outside toms' is increasing annually. **1935** [see *cue *sb.* 4]. **1976** *Coventry Evening Tel.* 27 Oct. 9/3 (*heading*) Summer of the giant toms.

tom (tǫm), *sb.*[3] Slang abbrev. of Tomfoolery (sense *2, = jewellery).

1955 P. Wildeblood *Against the Law* 119 Two grand's worth of tom. **1970** G. F. Newman *Sir, You Bastard* ii. 68 What d'you do with the tom and money you had out of Manor Gardens this afternoon? **1980** *Times* 23 Feb. 3/3 One of thieves..told police: 'When we found the tom (Cockney slang: tomfoolery jewellery) in the car we were amazed.'

tom (tǫm), *sb.*[4] Slang abbrev. of *tom-tom *sb.* 1 d.

1970 J. Wainwright *Freeze thy Blood* 11 Fatso grinned and notched the buckle of a tom case. **1975** J. Pidgeon *Flame* v. 65 Around the drum kit he arranged four mikes, one for the bass drum, one for the floor tom, one for the snare, and one overhead.

tom, *v.* Restrict *nonce-wd.* to sense in Dict. (s.v. Tom *sb.*) and add: **2.** *intr.* [f. *tom *sb.*[1] 1 f.] To behave in an ingratiating and servile way to someone of another (esp. white) race. Also *to tom it* (*up*). *U.S. slang.*

1963 L. Bennett in W. King *Black Short Story Anthol.* (1972) 161 They say you are going to chicken out, Papa... They're betting you'll 'Tom'. **1972** M. J. Bosse *Incident at Naha* ii. 94 Virgil just smiled, Tomming it up. **1976** *Public Opinion Q.* XXXIX. 527 The respondent 'accommodates', or to use the colloquial term, 'toms', in order to get through the racial interaction with minimal tension.

3. *intr.* To practise prostitution, to behave promiscuously; also, to have sexual intercourse in such a context. Also *to tom* (*it*) *around. slang.*

1964 Z. Progl *Woman of Underworld* iii. 35 They were perfectly willing to go 'tomming' on the streets to earn a few quid, but I never could. **1968** 'J. Ross' *Diminished by Death* i. 14 She's just tomming around. **1973** J. Rossiter *Manipulators* ix. 102 This woman... Is she tomming it around with the local villains? **1981** A. Sewart *Close your Eyes & Sleep* xviii. 181 What was she doing? Tomming, to put it bluntly. She was having it off with a bloke.

Hence **to·mming** *vbl. sb.*

1968 J. Lock *Lady Policeman* ii. 12 A prostitute was a 'tom'..and to practise prostitution was 'tomming'. **1973** *Black World* May 44 Afrikan People all over the world Conscious, unconscious, struggling, sleeping, Resisting, tomming, killing the enemy. **1981** J. Ross *Dark Blue & Dangerous* ix. 55 His own tomming around had given him a charitable view of casual sex.

tomahawk, *sb.* Add: **1. d.** (Earlier example.)

c **1825** J. Clare in M. Grainger *Nat. Hist. Prose Writings J. Clare* (1982) 88 The hookd bill usd by hedgers & calld by them a tomahawk.

2. *to bury the tomahawk* (earlier and *lit.* examples).

1705 R. Beverley *Hist. Virginia* iii. 27 They use.. very ceremonious ways in concluding of Peace..such as burying a tomahawk. **1775** in *Virginia Hist. Coll.* (1887) VI. 80, I..resolve never to bury the Tomahawk untill liberty shall be fixed on an immovable basis thro' the whole Continent.

tomalley. (Later examples.)

1950 R. Moore *Candlemas Bay* 289 The lobsters boiled to a fine, even red. Grampie ate five. Then he wiped the tomalley off his jacknife. **1981** *Times* 13 June 12/7 The [lobster's] red coral and the creamy green liver, known as tomalley, are delicious.

tomatin (tǫ·mătin). *Biochem.* Also **-ine.** [f. Tomat(o + -in[1].] A steroidal alkaloid present as a glycoside in the stems and leaves of the tomato plant and some other members of the family Solanaceæ.

The distinction made in quot. 1948 was not generally adopted.

1946 [see *lycopersicin b]. **1948** *Arch. Biochem.* XVI. 399 The crystalline compound has very low antibacterial activity..and is designated tomatine to distinguish it from the crude or partially purified tomatin. **1959** H. Martin *Sci. Princ. Crop Protection* (ed. 4) ii. 19 Although tomatine effectively inhibits the growth of *F*[*usarium*] *lycopersici* in pure culture.., no direct evidence has been obtained that it is responsible for wilt resistance. **1973** L. P. Miller *Phytochemistry* I. vi. 160 Lycobiose (4-*O*-β-D-glucopyranosyl-D-galactose) and the trisaccharide lycotriose were obtained from the tetrasaccharide tomatin in tomato leaf. **1980** *Phytochemistry* XIX. 1322/1 Tomatin in neutral or alkaline pH is highly membranolytic..and forms a complex with cholesterol.

tomato. Add: **1. b.** = *tomato-red* sense 3 in Dict.

1920 *Queen* 22 May (front cover), Colours:..Apricot, Ivory, Mastic, Tomato, Suede, and Saxe. **1923** *Daily Mail* 29 Jan. 1 (Advt.), Striped suitings... On grounds of mole, grey, fawn, black and tomato. **1977** M. Kenyon *Rapist* v. 56 Her rosy cheeks turned tomato with indignation. **1978** P. McCutchan *Blackmail North* viii. 93 She got into the tomato Mini and drove away.

c. An attractive girl. *slang* (orig. *U.S.*).

1929 D. Runyon in *Hearst's Internat.* Nov. 74/1 Different guys have different names for dolls, such as broads,..and tomatoes, which I claim are not respectful. **1962** J. Heller *Catch-22* xvi. 153, I can rush back to that night club before Aarfy leaves with that wonderful tomato he's got without giving me a chance to ask about an aunt or friend she must have who's just like her. **1977** H. Fast *Immigrants* v. 303 This tomato is twenty-three years old and she's a virgin.

2. b. tree tomato (later examples).

1944 *Living off Land* ii. 40 Tree-tomatoes will be found as garden-escapees. **1959** *N.Z. Listener* 8 May 22/3 Tree Tomato Sauce. Eight pounds tree tomatoes, 2 large onions, [etc.]. **1966** G. W. Turner *Eng. Lang. Austral. & N.Z.* viii. 172 Tree tomato and Chinese gooseberry seem to be commoner in New Zealand than in other English-speaking countries. **1976** K. Thackeray *Crownbird* v. 91 A separate table bore..tree tomatoes, portions of yellow jackfruit and chilled mountain paw-paw.

3. *tomato can* (examples), *chutney, ketchup* (earlier example), *purée, salad, sandwich, soup*; *tomato-coloured* adj. (earlier example); **tomato hornworm** = *tomato worm* in Dict.; **tomato juice,** the juice from tomatoes; also, a drink of this; **tomato paste,** thick, concentrated tomato purée; **tomato pinworm,** the larva of a small moth, *Keiferia lycopersicella*, which bores holes in the buds or fruit of the tomato plant; **tomato vine** *U.S.*, a tomato plant.

1868 'O. C. Kerr' *Smoked Glass* xviii. 216 What mean these letters which I find imprinted upon..the tomato can? **1914** *Sat. Even. Post* 4 Apr. 11/1 A gay-cat..will turn against a friend when that friend is down to tomato cans. **1855** E. Acton *Mod. Cookery* (rev. ed.) xxxii. 609 Tomata and other chatnies. **1963** A. L. Simon *Guide Good Food & Wines* 133/1 (*heading*) Green tomato chutney. **1869** L. M. Alcott *Little Women* ii. iii. 31 Brown rain, and purple clouds, with a tomato-coloured splash in the middle. **1928** Metcalf & Flint *Destructive & Useful Insects* xvi. 488 The southern or tomato hornworm ranges from the northern states southward far into South America. **1972** Swan & Papp *Common Insects N. Amer.* xix. 263 Tomato hornworm...sometimes called the five-spotted hawk moth. **1935** M. Morphy *Recipes of All Nations* 595 Tomato juice cocktail. **1936** 'R. West' *Thinking Reed* ii. 56 She ordered some tomato juice, for she would never again need a cocktail. **1981** P. Van Greenaway *'Cassandra' Bell* iv. 47, I ordered another tomato juice to calm his nerves. **1845** E. Acton *Mod. Cookery* v. 136 Tomata catsup. *c* **1938** *Fortnum & Mason Price List* 59/1 Tomato Paste. **1979** *Guardian* 25 Aug. 10/4 Pepsi's had been selling cola in Budapest..but all the company had been able to get out of the country was tomato paste. **1931** *Monthly Bull. Calif. Dept. Agric.* XX. 458 (*heading*) Damage to tomatoes in Southern California by the tomato pin worm. **1972** Swan & Papp *Common Insects N. Amer.* xix. 323 The closely related Tomato Pinworm ..bores pinholes in the developing buds, green and ripening fruits of tomatoes. **1877** E. S. Dallas *Kettner's Bk. of Table* 122 Add to the sauce a tablespoonful of tomato purée. **1977** B. Pym *Quartet in Autumn* vii. 64 Tomato purée, stuffed vine leaves..and tapioca pudding. **1877** E. S. Dallas *Kettner's Bk. of Table* 460 For the tomato salad a dash of mustard is not a bad addition. **1980** I. Murdoch *Nuns & Soldiers* 176 Supper consisted of onion soup, black sausage with tomato salad, and a local cheese with herbs. **1911** W. J. Locke *Glory of Clementina Wing* ii. 17 Tomato sandwiches and plum-cake set out for a visitor's tea. **1978** F. Olbrich *Desouza pays Price* xxv. 161 Delicate tomato sandwiches and fragrant Darjeeling tea. **1846** *Jewish Manual, or Pract. Information Jewish & Mod. Cookery* v. 97 Dry tomato soup. **1974** 'E. Lathen' *Sweet & Low* xvi. 154 The problem we have with tomato soup in my cannery. **1876** 'Mark Twain' *Adventures Tom Sawyer* i. 18 [She] looked out among the tomato vines and jimpson weeds that constituted the garden. **1981** G. V. Higgins *Rat on Fire* xii. 92 You..tromp all over the old people's tomato vines.

Hence **toma·toey** *a.*, having the taste or flavour of tomatoes.

1972 *Homes & Gardens* Aug. 101 (Advt.), The result is a tomato juice that's thicker, smoother and more tomatoey than any you've ever tasted. **1982** D. Wiltse *Wedding Guest* xv. 205 The shared appreciation of the wine, the spicy sausage, the tomatoey beans.

tomb, *sb.* Add: **4*.** *The Tombs*: New York City prison. *U.S. slang.*

1840 *Daily Picayune* (New Orleans) 27 Aug. 2/3 Poor Chapman..is in the 'Tombs', charged with false swearing at an election. **1842** Dickens *Amer. Notes* I. vi. 199 What is this dismal-fronted pile of bastard Egyptian?.. A famous prison, called the Tombs. **1935** A. G. Macdonell *Visit to Amer.* iii. 53 A criminal had been brought from the Tombs..to be examined in the 'Line-Up'. **1981** M. C. Smith *Gorky Park* iii. iii. 330 It's the Tombs... Night Court's open now.

5. a. *tomb chest, figure, figurine, furniture, house* (later example), *monument.* **b.** *tomb-haunter.*

1955 M. D. Anderson *Imagery Brit. Churches* ii. ii. 44 The late medieval tomb chests often have small figures arranged in niches all round them. **1925** B. Rackham in R. Fry et al. *Chinese Art* 13 In his wonderful tomb figures.. we come to the very border-line of sculpture. **1970** Tomb figure [see *haniwa]. **1933** *Burlington Mag.* Nov. 233/2 There are several tomb figurines which show vivacity and able characterization. **1976** 'M. Delving' *China Expert* v. 56 Tashjian..had..unloaded an extremely dubious Han tomb figurine on an unsuspecting German dealer. **1908** *Chambers's Jrnl.* July 527/2 We were in the midst of such a medley of tomb-furniture. **1977** *Times* 23 Apr. 13/3 The increasing vogue for tomb 'furniture' among the lower echelons of T'ang society. **1939** W. B. Yeats *Last Poems* 20 What great tomb-haunter sweeps the distant sky. **1975** G. Ewart *Be My Guest!* i. 15 But stay! who is this totally melodramatic opium-smoking tomb-haunter? **1963** E. M. Jope in Foster & Alcock *Culture & Environment* xiii. 338 It is an area where hog-backed tomb-houses are to be found. **1948** D. Diringer *Alphabet* ii. iv. 261 Tomb-monuments in various countries.

tombac. Add: In *Archæol.* usu. **tumbaga, -bago.**

1931 E. Nordenskiöld *Compar. Ethnogr. Stud.* IX. 104 The awl..has been analyzed by Dr. K. G. Almström, and found to contain 33% Au, 12% Ag and 55% Cu. This is the composition generally known in archaeological literature as tumbaga. **1936** *Nature* 4 Jan. 29/1 In Antioquia, in Colombia,..objects are found made particularly of the gold alloy..which is known as 'tumbago'. **1974** S. E. Morison *European Discovery of America: Southern Voy.* vii. 152 As ornaments they [*sc.* Venezuelan Indians] displayed great polished disks made of an alloy of copper and gold that they called *guanin*, and which modern archaeologists have named *tumbaga.*

‖ **tombarolo** (tǫmbărǭ·lo). Pl. **tombaroli.** [It.] A grave-robber.

1973 *New Yorker* 7 Apr. 98/2 A skilled *tombarolo* knows the surface indications of a tomb and pilfers the treasure inside after using an ingenious probing device known as an *asta* ('lance') or *chiave* ('key'). **1975** *Listener* 11 Dec. 793/2 The ware..had been illegally excavated from an Etruscan tomb..by a gang of grave robbers, the notorious *tombaroli.*

tombola. Add: Now usu. with pronunc. (tǫmbǭᵘ·lă).

tombolo (tǫ·mbǭlo). *Physical Geogr.* [a. It. *tombolo* sand dune, tombolo.] A bar joining an island to the mainland.

1899 F. P. Gulliver in *Proc. Amer. Acad. Arts & Sci.* XXXIV. 189 Upon the coast of Italy where island-tying ..is beautifully shown, such a bar is called a tombolo. For convenience in distinguishing island-tying bars from those of other kinds, the writer proposes to call every bar of this kind a tombolo, giving an English plural tombolos. **1937** *Geogr. Jrnl.* XC. 190 The writer goes on to discuss bars, spits, and tombolos along the more sheltered shores of the Firth of Clyde. **1960** B. W. Sparks *Geomorphol.* viii. 201 In the British Isles a fine example of a tombolo is provided by Chesil Beach, an eighteen-mile-long ridge connecting the Isle of Portland to the mainland. **1977** A. Hallam *Planet Earth* 94/3 In high-latitude, previously glaciated areas, tombolos are often formed of shingle, the commonest beach material in such regions.

tombstone. Add: **2*.** *Comm.* An advertisement displaying the names of the underwriters or firms associated with a new issue or the like. Also *tombstone ad*(*vertisement.*

1968 *Times* 27 Feb. 22/3 'Tombstones'..are getting bigger. I am referring, of course, to the new issue advertisements. **1972** *Times* 24 Oct. 2 The Times is now able to offer financial advertisers an exclusive service for the placing of Tombstone advertising (public announcements and notices of redemption) in Europe. **1977** *National Times* (Austral.) 17 Jan. 40/2 The advertising of trusts is limited to 'tombstone' advertising. **1981** *U.S. Banker* Dec. 56/1 The old tombstone ad, promoting a service and basing the appeal largely on price,..is long gone, according to advertising men. **1983** *Marketing* 24 Mar. 39/1 Financial advertising columnage, i.e. tombstones, company meetings, prospectuses, takeovers, etc.

3. (Earlier example.)

1751 B. Lynde *Diary* 16 Oct. (1880) 176 Yesterday Cox and Stacy ½ day abo. Tombstone monument.

tom-cat, tomcat, *v. U.S. slang.* [f. the sb.] *intr.* To pursue women promiscuously for sexual gratification. Freq. const. *around.* Hence **to·mcatting** *vbl. sb.*

1927 *Dial. Notes* V. 478 *Tom catting*, v., to seek illicit sexual adventure. 'Jeff he's out a-tomcattin' roun' some 'ers.' **1932** W. Faulkner *Light in August* xii. 259 So this is where you tomcat to every night. **1939** J. Steinbeck *Grapes of Wrath* xvi. 239, I was goin' out an' dance, an' I was gonna go tom-cattin'. **1953** W. P. McGivern *Big Heat* ii. 23 A guy tom-catting around while his wife's away. **1962** H. Green *Time to pass Over* iii. 42 Don't drink, don't chew, don't tomcat with the women so far as I know. **1975** *New Review* May 20/1 'He had a very strong tendency to be doing a lot of tomcatting.'..Would go out looking for women. **1980** G. Thompson *Murder Mystery* (1981) xxii. 172 A man who's been tom-catting around with three women all day long.

-tome (tǭᵘm), terminal element: (*a*) f. Gr. -τόμον, neut. of -τόμος that cuts (see -tomy), used in names of instruments for cutting, esp. ones used in surgical operations denoted by the corresponding word in *-tomy*, as in *cystotome* s.v. Cysto-, Hysterotome, Microtome; (*b*) f. Gr. τομή a cutting, used in words denoting a distinct section or segment of a body or part, as in Myotome 1, *gonotome* s.v. *gono-.

tom-fool, *sb.* **d.** (Earlier example.)

1762 Sterne *Tr. Shandy* V. xxx. 107 'Twas a Tom-fool-battle.

tomfoolery. Add: **2.** *Rhyming slang.* Jewellery. Cf. *tom *sb.*[3].

1931 C. Rimington *Bon Voyage Bk.* xv. 88 *Tomfoolery*, jewelry. **1943** M. Harrison *Reported Safe Arrival* 52, I wouldn' be surprised if you both done a stretch fer knockin' orf some ole bloke's tom-foolery. **1975** *Sunday Times* 30 Mar. 49/2 He will have contacts in 'tomfoolery', or jewellery outlets.

Tom Jones. The name of the hero of Fielding's novel *History of Tom Jones* (1749), used *attrib.* to designate dress and hair styles represented in the film version of 1963 and considered suggestive of eighteenth-century styles.

1964 *Glamour* May 186 Wonderful way to wear hair that's fit and silky: Brush it back from the brow, catch it at the nape, tie in a Tom Jones queue. **1967** *Observer* 14 May 28/6 A chiffon scarf in a Tom Jones bow. **1971** *Jamaican Weekly Gleaner* 10 Nov. 15/2 Looking..beautiful in belted tan tailored pants with chocolate Tom Jones blouse.

‖ **tomme** (tǫm). Also (erron.) **tome.** [Fr.] The name given to a variety of cheeses made in Savoy, a region of S.E. France.

1946 A. L. Simon *Conc. Encycl. Gastron.* IX. 24/2 [One of] the best known *Tommes* of Savoy [is]..*Tomme au Fenouil.* **1958** *Catal. County Stores, Taunton* June 9 Cheese... Tome de Savoie—each 7/6. **1966** P. V. Price *France* ii. 310 There are several *tommes* in the region and this is the most famous. **1972** *Sat. Rev.* (U.S.) 24 June 77/3 In Savoy I sampled..*reblochon*..and the famous *tomme aux raisins*, a blander cheese coated on the outside with dried grape pips.

tommelaitje, var. *tameletjie.

Tommy. Add: **1. c.** (Earlier example.)

1884 Kipling in L. L. Cornell *Kipling in India* (1966) iii. 83 (*title*) The story of Tommy.

3. b. Substitute for def.: = *tommy bar*, sense 6 below. (Earlier example.)

1843 J. J. Greer *Brit. Patent 9811* (1856) 2 My invention..consists..first, in working a double screw..from a central axis into two separate boxes or cases, either by a lever commonly called a tommy, or by a spanner.

6. tommy-bag (example); **tommy bar** *Mech.*, a short bar that can be inserted into a hole in a box-spanner or screw to assist in turning it; **Tommy('s) cooker** *Mil. slang*, a small portable spirit stove; also, a piece of rolled-up canvas soaked in grease used in place of this; **Tommy Dodd** (earlier example); **Tommy talker** *colloq.* = Kazoo in Dict. and Suppl.

1983 P. Nash *Coup de Grass* ii. 23 He was wearing overalls and carrying a canvas tommy-bag. **1920** Webster, Tommy bar. **1930** *Engineering* 25 Apr. 538/1 The cylinder is removed bodily from its supporting bracket, by unscrewing a hinged fixing pillar by means of a tommy bar. **1953** E. Hyams *Vineyards in England* 226 Pass a tommy-bar through the hole at the top of the screw and spin the screw to bring the piston down on to the grapes. **1973** D. Lees *Rape of Quiet Town* vii. 121 He fixed each jack separately..then gave the tommy bars a few..twists until they were tight. **1917** W. Owen *Let.* 4 Feb. (1963) 430 We had 5 Tommy's cookers between the Platoon, but they did not suffice to melt the ice in the water-cans. **1919** *N.Y. Times* 23 Feb. iv. 12/2 When 4 o'clock came around every manjack of us would take out his Tommy-cooker and begin making his tea. **1948** A. Baron *From City, from Plough* xviii. 165 On the little tommy-cookers that they sheltered between their feet they brewed..tea in their mess-tins. **1870** A. Steinmetz *Gaming-Table* II. 221 Not long ago a returned tradesman ..allowed himself to be induced to play at Tommy Dodd with two low sharpers. **1938** Tommy Talker [see *kazoo]. **1976** S. Barstow *Right True End* i. iii. 38 Learn the piano... Well, the french horn, clarinet, fiddle, trombone; mouth-organ, jew's harp, tommy-talker.

to·mmy[2]. *Austral.* and *N.Z. colloq.* [Shortened f. Tomahawk *sb.*] A hatchet. Also **tommy-axe.**

1873 M. A. Barker *Station Amusements N.Z.* ix. 148, I had to get the tommy (*anglicé*—tomahawk) and *chop* his boots off. **1898** Morris *Austral Eng.* 474/2 *Tommy-axe*, a popular corruption of the word *Tomahawk.* **1939** X. Herbert *Capricornia* xxii. 328 Cutting a strip of bark from the tree with the tommy-axe.

to·mmy-gun. *colloq.* (orig. *U.S.*). Also **Tommy-** and as two words. [f. Tommy, repr. the name of J. T. *Thompson*: see *Thompson.] A Thompson or other sub-machine-gun.

1929 *Sat. Even. Post* 13 Apr. 54/3 There are three types of machine gun used—Tommy guns, Browny guns and Louie guns. **1934** E. Hemingway in *Cosmopolitan* Apr. 23/2 The nigger shot him in the belly with the Tommy gun. **1941** *Times* 21 Nov. 3/4 Twelve of the enemy were killed with bayonets and tommy-guns. **1955** M. Banks *Commando Climber* iii. 40 He survived the war by a combination of good fortune and adroitness with a tommy-gun. **1973** D. Lees *Rape of Quiet Town* vii. 111 A wild assortment of weapons, including a couple of tommy guns and a cavalry sabre.

Hence as *v. trans.*; also **to·mmy-gunner.**

1942 *Times* 3 Oct. 4/1 A party of our tommy-gunners penetrated enemy barbed wire entanglements. *Ibid.* 24 Nov. 4/1 When one Soviet tank was engaged in pursuing the retreating Germans its crew leapt out and tommy-gunned them. **1973** A. Mann *Tiara* iii. 22 Thirty years ago he was the best tommy-gunner in the Appennines. **1978** T. Gifford *Glendower Legacy* (1979) 144 Maybe they're waiting for me to leave so they can come in and tommy-gun you.

tomo (tǭᵘ·mo). *N.Z.* [a. Maori.] A depression or hole in limestone terrain.

1952 *Arena* (N.Z.) xxxi. 3 See him drag a gully full of tomos and patches of scrub. **1961** R. Park *Hole in Hill* (1962) x. 76, I fell down a big hole... Not just a hole, but a *tomo*, a limestone shaft. **1975** *N.Z. Jrnl. Agric.* Sept. 24/2 It may be located in a tomo or natural depression down the back, or in a corner, of a paddock.

tomography (tǒmǫ·grăfi). *Med.* [f. Gr. τόμος slice, section + -graphy.] Radiography in which an image of a predetermined plane in the body or other object is obtained by rotating the detector and the source of radiation in such a way that points outside the plane give a blurred image. Also in extended use, any analogous technique using other forms of radiation.

1935 *Brit. Jrnl. Radiol.* VIII. 750 The most significant field of application for the method of reproducing body layers (tomography) will, in practice, be lung diagnosis. **1949** *Ibid.* XXII. 627/1 Today, sectional röntgenography, commonly called tomography has indeed, in many places, found recognition as a method of radiological diagnosis. **1968** *Brit. Med. Bull.* XXIV. 242/1 With the advent of transverse tomography, it is now possible to get much more accurate estimations of the various volumes and areas of tissues of differing density, in any cross-section of the body. **1977** 'E. Trevor' *Theta Syndrome* iii. 40 The axial tomography revealed bilateral subdural hematomas. **1982** *New Scientist* 21 Jan. 155/1 Paul Carson and others ..have been comparing the new-developed technique of ultrasonic computed tomography (UCT) with pulse echo imaging. **1983** *Ibid.* 17 Mar. 725/3 Tree physiology and dendrochronology are just two of the possible applications for portable computer tomography. **1984** *McGraw-Hill Yearbk. Sci. & Technol. 1985* 335/1 Positron emission tomography (PET scan) now allows scientists to measure noninvasively the functional activity of the living human brain.

So **to·mogram,** an X-ray picture taken by tomography; **to·mograph,** (*a*) a tomogram; (*b*) an apparatus for carrying out tomography; **tomogra·phic** *a.*, **tomogra·phically** *adv.*

1935 *Brit. Jrnl. Radiol.* VIII. 736 The new tomographic method is also based upon the principle of the method already mentioned. *Ibid.* 750 Three tomographs are generally sufficient for lung diagnosis. *Ibid.*, A new device, the tomograph, is described, which permits the radiography of body layers of any size and variable thickness. **1936** *Lancet* 25 July 185/2 For all practical purposes three tomograms are ample for accurate diagnosis. **1949** *Radiography* XV. 242/1 (*caption*) Aneurysm demonstrated by lesion of dorsal vertebræ tomographically. **1950** *Times* 22 Nov. 6/6 The infection was discovered, thanks to the use of a tomograph, a new X-ray apparatus, apparently during Sir Stafford Cripps's last stay at Zurich. **1955** D. B. Fry in B. I. Evans *Stud. in Communication* 156 The internal action of the larynx was for a long time inaccessible to X-ray photography but the development of the tomogram technique has enabled Ardram and Kemp to obtain remarkable pictures of the larynx in action. **1968** Tomographic [see *laminagraphic *a.*]. **1975** *Radio Times* 1–7 Nov. 7/1 The EMI ('Emmy')-scanner is a machine for taking axial X-ray tomograms. **1975** *Sci. Amer.* Oct. 56/3 In most tomographic instruments the X-ray source moves in one direction and the photographic film simultaneously moves in the opposite direction. The patient lies in between. **1976** *Physics Bull.* Oct. 436/3 X-ray tomography has become a popular method of medical imaging over the past few years... A M Cormack and A M Koehler have come up with an alternative idea for improving the density resolution of tomographs. **1982** *Times* 26 July 2/3 Using the tomograms, the interior of the helmet has now been excavated.

tomorrer (tŭmǫ·rəɪ), *adv.* Also **to-morrer.** Repr. a dial. and slang pronunc. of To-morrow *adv.*

1901 M. Franklin *My Brilliant Career* iii. 16 Only I promised to stick to the missus a while, I'd scoot tomorrer. **1921** H. Williamson *Beautiful Yrs.* 81 Go when you like, miboy. Only I shall be a-mowing to-morrer. **1959** [see *sickie 1]. **1970** T. Hughes *Crow* 63 O do not chop his winkle off His Mammy cried with horrer Think of the joy will come of it Tomorrer and tomorrer.

to-morrow, *adv.* and *sb.* Add: Now usu. written **tomorrow. A.** *adv.* **1. c.** *fig.* In the (near) future.

1871, etc. [see *jam tomorrow* s.v. *jam *sb.*[1] b]. **1957** *Listener* 15 Aug. 223/1 An accelerated movement towards independence: Ghana yesterday; Nigeria, French West Africa, the Cameroons, tomorrow.

B. *sb.* **1*.** *fig.* The (near) future. Freq. in the possessive.

1943 J. B. Priestley *Daylight on Saturday* ii. 5 He belonged to tomorrow's new ruling class. **1959** *Brno Studies in English* I. 73 Progressive poets preferred to look forward into distant future and dreamed..of a better to-morrow. **1979** *Guardian* 30 Oct. 32/8 The Prime Minister..told the Wales TUC that British industry was not going to get tomorrow's jobs 'unless we move into tomorrow's world'.

4. *as if there were no tomorrow* and varr., recklessly, with no regard for the future.

1862 Whyte-Melville *Queen's Maries* II. xxii. 10 Why should you thus risk your life as if there was no to-morrow? **1980** *Guardian Weekly* 3 Feb. 1/3 Oil supplies that Americans at home continue to consume as though there were no tomorrow.

5. Proverb. *tomorrow is another* (or † *a new*) *day.*

c **1527** J. Rastell *Calisto & Melebea* sig. C 1[v], Well mother to morrow is a new day. **1603** Florio tr. *Montaigne's Ess.* II. iv. 57 A letter..beeing delivered him..at supper, he deferred the opening of it, pronouncing this byword, *To morrow is a new day.* **1824** Scott *St. Ronan's* III. vii. 192 We will say no more of it at present... Tomorrow is a new day. **1927** P. Green *Field God* i. 148 Go to it, you Mag and Lonie! To-morrow's another day, and you'll need all you can hold. **1956** M. Dickens *Angel in Corner* vi. 90 'You can run along now... Those few letters will keep until the morning.'..'But there will be a whole heap of new ones by the morning.'..'I know, dear... If the letters didn't come, that would be the time to start worrying. But tomorrow is another day.' **1980** B. Pym *Few Green Leaves* xiii. 107 He would probably have said nothing and so missed his opportunity. Still, tomorrow was another day.

tom-tit, tomtit. Add: **1. a.** Also, the great tit, *Parus major.*

1965 *Jrnl. Lancs. Dial. Soc.* Jan. 19 Great Tit... Tom Tit. Accrington. **1972** *Guardian* 23 Feb. 12/2 An octogenarian farm worker..pointed out that, just as one's largest toe is the 'tom toe', so it followed naturally that the largest tit was the tom-tit.

2. A small sailing boat.

1857 A. W. Habersham *My Last Cruise* xvii. 333 Some of us also took the tomtit, (a boat smaller even than the dingy). **1925** A. B. Armitage *Cadet to Commodore* xx. 278 Out with the 'Tom-tits' in the harbour on a breezy Saturday afternoon was another delightful sport.

3. Rhyming slang for 'shit'. Freq. with *the* in *fig.* use: 'the willies'; also, nonsense.

1943 'R. Llewellyn' *None but Lonely Heart* xx. 116 'You're always doing it, you shower of tom tit, you.' **1944** L. Glassop *We were Rats* 67 'Break it down,' said the corporal. 'You'll give these blokes the tomtits before they get their first lot of C.B.' **1967** J. Gardner *Madrigal* i. iii. 54 You can cut the Tom Tit, sergeant. **1970** C. Wood *'Terrible Hard,' says Alice* vii. 91 Perhaps 'e stopped for a tomtit. **1973** P. A. Smith *Barcoo Salute* 14 What's the matter, got the tom tits? **1982** L. Cody *Bad Company* xii. 80, I was just sitting there, trousers round me ankles. .. If I hadn't been doing it already, he'd 've given me the tom-tits.

tom-tom, *sb.* **1. b.** Delete entry and see *tam-tam.

d. A low-toned drum (without snares), used in Western music.

1934 E. Little *Mod. Rhythmic Drumming* (rev. ed.) 26 No outfit is complete without at least one tomtom. **1977** *Rolling Stone* 30 June 97/2 Ringo slams away on his tom-toms.

ton[1]. Add: **4. b.** (Earlier and later examples, in *sing.*)

1770 P. Freneau in Brackenridge & Freneau *Father Bombo* (1975) i. iii. 13 My head stuck a considerable time in a ton of mud. **1899** H. Sweet *Pract. Study Languages* x. 115, I am told that the great English lexicographers of the present day look down with contempt on anything less than a ton of such materials. **1971** *Scope* (S. Afr.) 19 Mar. 38/1 Fine, thanks a ton, Len. I won't be a sec. **1977** *Belfast Tel.* 28 Feb. 20/8 This has brought the lass on a ton.

c. *pl.* As *adv.* qualifying comparative or (*U.S.*) positive adjs.: much; very. *colloq.*

1908 S. Wilson *Let.* 17 Aug. in R. S. Churchill *Winston S. Churchill* (1969) II. Compan. ii. 804. I feel tons better for being in the wonderful air. **1970** 'D. Halliday' *Dolly & Cookie Bird* viii. 127 He was looking tons better, with his ribs done up in crêpe. **1977** *Amer. Speech 1975* L. 68 *Tons adv*, very, extremely. 'Her outfit is tons neat.'

d. Phr. *to come down* (*on* or *upon*) (a person) *like a ton of bricks*: see *come *v.* 56 g.

4*. *transf.* **a.** *colloq.* A score of one hundred in a game, *spec.* in *Cricket* (= *century 3 b) and *Darts.*

1936 R. Croft-Cooke *Darts* vi. 42 *Ton*, the word means simply 100. While in more gentlemanly games they speak of Centuries, in Darts we curtly say 'One Ton'. **1946** J. Moore *Brensham Village* iii. 95 Darts has its own esoteric terminology... A hundred is a 'ton', of course, all over England. **1958** *Punch* 9 July 40/2, I owe everything to Cambridge. I got a ton in the Freshman's Match of 1941. **1973** *Atlantic Monthly* Aug. 73 Now he's averaging 60 or more, frequently throws a 'ton'—a round of 100 or more points—and can put a dart into a fifty-cent piece area every time. **1978** *Lancashire Life* Apr. 41/3 Scoring a century didn't mean a hoot to me then... Now, as an experienced pro, I know I must make a 'ton' and then keep going to get another.

b. *slang.* A hundred pounds.

1946 *People* 7 Apr. 2/6 A red-faced punter..whose conversational powers were limited to..jargon, which translated fivers as 'flims'..; £100 as a 'ton' [etc.]. **1960** 'A. Burgess' *Doctor is Sick* 164 'And what's the first prize?' asked Edwin. 'A ton,' screamed Harry Stone. ''Undred nicker an' a film test.' **1981** P. Turnbull *Deep & Crisp & Even* vii. 131 The old man would charge three ton for this but me and the boys will do it for half-price.

c. *colloq.* A speed of one hundred miles per hour (esp. with reference to motor cycles). Freq. in phr. *to do the* (or *a*) *ton.* Cf. *ton-up *sb.* and *a.*

1954 G. Smith *Flaw in Crystal* iv. 36 At eighty I felt a wild sense of elation... I watched to see if Several would triumphantly lead Teddy onwards at a majestic full ton. **1959** *News Chron.* 17 Dec. 3/1 The dangerous noddles who boast about doing the ton on the public roads. **1964** *New Statesman* 21 Feb. 288/3 We do the ton sometimes, but not where any one's goin' to get 'urt. **1973** *Hansard Lords* 5 Dec. 684 In that case, you must have been doing a 'ton', if very few cars passed you.

d. In other miscellaneous colloq. uses to denote one hundred.

1962 *Electronics Weekly* 21 Nov. 3/1 Elliott reach a ton. The 100th National Elliott 803 computer has been installed. **1970** *Sunday Tel.* 22 Mar. 13/3 Blissful summer breezes..ease the discomfort of temperatures which occasionally threaten to make the ton. **1980** *Financial Rev.* (Sydney) 29 Aug. 29/1 Australians staying at the best capital city hotels..will have reached 'the ton' in their room rates—accommodation will be costing $100 a night.

6. ton-force (pl. *tons-force*), a unit of force equal to the weight of a mass of one ton, esp. under standard gravity; **ton weight,** the weight of one ton; usu. *fig.*

1961 *B.S.I. News* Oct. 26/2 A similar distinction is made between..ton (no abbreviation) and ton-force (tonf). **1972** *Physics Bull.* May 285/1 The 50 tonf dead-weight standard was originally designed to give forces only in units of tons-force. *a* **1855** C. Brontë *Professor* (1857) I. vi. 87 This liability is a ton weight at least. **1893** H. Frederic *Return of O'Mahony* i. x. 83 Then would come..the fierce buffeting of ton-weight blows as the boat staggered blindly at the bottom of the abyss. **1936** *Discovery* Feb. 37/2 The power developed per ton-weight of the engine. **1960** H. Pinter *Caretaker* i. 18 I'll give you a hand. (*They lift it.*) It's a ton weight, en't? **1981** J. Wainwright *All on Summer's Day* 198 She'd been like a ton weight across his shoulders. Her and her infernal daughters.

ton[3]. Delete 'Now *rare*' and add: **a.** (Later examples.)

1939 D. Cecil *Young Melbourne* viii. 220 Some humble country acquaintances and a few persons of ton. **1978** J. Krantz *Scruples* ii. 39 And these Bostonians..did own a gratifying number of mills and plants and banks and brokerage firms. Also they had *ton.*

b. (Earlier and later examples.)

c **1770** in de Vries & Fryer *Venus Unmasked* (1967) 33 Miss P... D... will only..take engagements from billiard table gentlemen, gentlemen of the ton, and young shopmen. **1969** H. Elsna *Abbot's House* 99 A waste, when all the *ton* will flock here for this event. *Ibid.* 103 The *ton* are here in force.

‖ **tonadilla** (tonadī·lyă). [Sp., dim. of *tonada* tune, song.] A light operatic interlude of the mid-eighteenth to early-nineteenth cent., orig. forming an intermezzo between the acts of a serious play or opera, but later performed independently.

1830 W. C. Stafford *Hist. Music* xviii. 263 The Tonadilla, originally a simple and popular song, sung in the *Zarguela* and *Saynette*, now frequently represents an entire action, consisting of a whole scene, or even of an act. **1876** Stainer & Barrett *Dict. Mus. Terms* 435/2 *Tonadilla* (*Sp.*), a short tune, an interlude, ritornello, symphony to a song. **1920** C. van Vechten *Mus. Spain* v. 82 The *tonadilla*..accompanied by a guitar or violin and interspersed with dances, was very popular for a number of years. **1947** A. Einstein *Mus. Romantic Era* xvii. 329 The works decisive in bringing out this national style were..*zarzuelas* in the realistic manner of the *tonadilla.* **1973** *Oxford Times* 30 Mar. 14/4 The scenic tonadilla in the 18th century can be described as miniature comic opera.

tonal *a.* Add: **1. c.** Of tonality; pertaining to music written in keys. Opp. *atonal *a.*

1884 G. Oakey *Text Bk. Harmony* viii. 51 A sequence.. in which the intervals belong to one scale, is termed..a Tonal Sequence. **1922** [see *atonal *a.*] **1957** *Encycl. Brit.* XI. 205/1 With the development of polyphony, tonality becomes as important as the concord-discord system itself; and, indeed, that system could not have existed without tonal guidance at every point. **1978** P. Griffiths *Conc. Hist. Mod. Music* ii. 23 Sibelius's long silence..may suggest the difficulty of maintaining tonal composition in the twentieth century.

3. Pertaining to or characterized by shades of colour or effects of light and shade. Cf. Tone *sb.* 10 a, b.

1910 S. J. Solomon *Practice of Oil Painting* vi. 62 The same method is applicable in arriving at a similar decision with regard to the relation of shadows, all intervening tones, and the general tonal aspect of the whole figure. **1931** J. H. Brown *Water-Colour Guidance* x. 192 Present-day colour work..has tended to divert the colourist's attention from the *tonal* aspect of painting. **1967** E. Short *Embroidery & Fabric Collage* i. 4 (*caption*) As one fabric is used throughout, there are no contrasts of colour or texture, and the design relies for interest on the tonal pattern made by the shadows. **1980** *Economist* 20 Aug. 62/1 Plain and tonal designs—mingled colours, but no pattern—are the fashion. *Ibid.* 62/2 Tonal carpets are what the customer wants.

tonalite. Add: [ad. G. *tonalit* (G. vom Rath 1864, in *Zeitschr. der deutsch. geol. Ges.* XVI. 249).] Later used for any quartz-diorite; now *spec.* a rock in which quartz represents 20–60 per cent of quartz plus feldspars (a higher proportion than in quartz-diorite) and alkali feldspar is less than 10 per cent of total feldspars. (Further examples.)

1913 *Jrnl. Geol.* XXI. 213 Lindgren has defined tonalite (or quartz diorite), as containing less than 8 per cent of alkali feldspar, granodiorite as containing 8–20 per cent of alkali feldspar. **1932** A. Johannsen *Descr. Petrogr. Igneous Rocks* II. 379 Diorites with less than 5 per cent of quartz may be called quartz-bearing-diorites. With more than 5 per cent, they become quartz-diorites (tonalites). **1962** W. T. Huang *Petrology* iv. 91 From granite, granodiorite, tonalite to diorite, the proportional amounts of quartz and alkalic feldspar decrease, while that of sodic plagioclase increases. **1976** *Earth-Sci. Rev.* XII. 14 For field 5 the term *tonalite* is recommended, whether hornblende is present or not, in agreement with Johannsen..; whereas quartz diorite, frequently used for this field, is restricted to field 10*. *Ibid.* 27, 20–60% of light-colored minerals... Plag 90–100% of total feldspar: (5) tonalite. **1977** *Sci. Amer.* Mar. 94/1 The average chemical composition [of gneisses] resembles that of the common igneous rocks diorite and tonalite.

Hence **tonali·tic** *a.*

1963 *Revista de la Asociación Geológica Argentina* XVIII. 97 In the Pampean Ranges, tungsten ore deposits are related to tonalitic-granitic intrusives of the pre-Cambrian tectomagmatic cycle. **1978** *Nature* 16 Mar. 241/1 Although the gneiss belts show great structural complexity..granodioritic and tonalitic compositions greatly predominate.

tonalitive (tonæ·litiv), *a. Mus. Obs.* or *rare.* [f. Tonalit(y + -ive.] Of or pertaining to tonality.

1907 M. H. Glyn *Rhythmic Conception of Music* iii. 64 Nothing would seem more natural than that tonality should suggest 'tonalitive', but the word has not hitherto appeared. **1918** *Mus. Assoc. Proc. 1917–18* 162, I should expect the new tonalitive schemes of such composers as Debussy and Ravel to bring about great changes in composition. **1924** T. H. Y. Trotter *Music & Mind* 237 The old major and minor tonalitive schemes are giving way.

tonality. Add: **1. b.** The principle or practice of organizing musical composition around a key note or tonic.

1932 J. Yasser *Theory Evolving Tonality* 375 *Tonality*, a principle which organically and tonocentrically unites the function of a certain number of systematically arranged sounds..in their melodic and harmonic aspects. **1957** *Encycl. Brit.* XI. 205/2 Palestrina's tonality is one of the most mature and subtle things in music. **1978** P. Griffiths *Conc. Hist. Mod. Music* iii. 39 Berg's opera differs from Schoenberg's monodrama in its direct references back to tonality.

3. *Linguistics.* The differentiation of words, phrases, or syllables by a change in the pitch of the voice.

1948 R. A. D. Forrest *Chinese Lang.* i. 26 All the languages which we group under the term Sinitic have..a tendency to develop significant tonality. **1956** Jakobson & Halle *Fundamentals of Lang.* iii. 29 All the inherent features are divided into two classes that might be termed sonority features and tonality features,..the latter [akin] to the prosodic pitch features. **1973** *Archivum Linguisticum* IV. 23 The 'paratonality system' determines the relationship between sentences and paratone groups in the same way as the tonality system relates clauses and tone-groups.

‖ **tonari gumi** (tonā·ri gū·mi). Also **tonari-gumi, tonarigumi.** [Jap.] A neighbourhood association in Japan, formed of groups of families who assume responsibility for their own community affairs.

1947 R. Benedict *Chrysanthemum & Sword* iv. 82 Japan had, like China, tiny units of five to ten families, called in recent times the *tonari gumi*, which were the smallest responsible units of the population. **1958** W. J. H. Sprott *Human Groups* 93 In Japanese cities they have associations called *tonari-gumi*, made up of from ten to twenty households. **1980** J. Melville *Chrysanthemum Chain* 119 People who lived nearby..members of Murrow's *tonari-gumi* or neighbourhood association.

tondino. Restrict *Arch.* to sense in Dict. and add: **2.** *Ceramics.* Pl. **tondini.** A plate with a wide flat rim and deep centre made of majolica.

1885 [see *Faenza]. **1900** F. Litchfield *Pott. & Porc.* ii. 16 A set of round plates or *tondini.* **1958** M. Wykes-Joyce *7000 Years Pottery* vi. 76 What we may term useless wares, a particular kind of *piatto con fondo*, the *tondino*, which would be sent by a gallant to his current inamorata, filled with candied flower petals... A flirtatious signorina would have some dozen or score of the *tondini* on display. **1971** *Times* 30 Nov. 24/3 (Advt.), A Faenza tondino by the 'Green Man'.

tondo. (Earlier example.)

1877 Geo. Eliot *Let.* 29 Mar. (1956) VI. 359 The little *tondi* on the covers of the 3/6 edition are charming.

tone, *sb.* Add: **I. 2. a.** Also, such a sound produced electrically; cf. *pure tone* s.v. *pure *a.* 1 e; in *Teleph.*, a pure tone or a more complex sound generated automatically to convey to a calling subscriber information about the line or the number required (see *busy, dial, dialling, engaged*, etc., *tone* under the first elements in Suppl.).

1919 J. Poole *Pract. Telephone Handbk.* (ed. 6) xxi. 364 The tones and interruptions required are as follows:— (1) A 'tone' of 24 interruptions per revolution of the armature or 400 interruptions per second, [etc.]. **1958** G. Higgs in E. Molloy *High Fidelity Sound Reproduction* i. 10 The specification of a definite acoustical or electrical level necessarily involves reference to a steady-value test-tone of the stipulated frequency. **1962** A. Nisbett *Technique Sound Studio* v. 103 To calibrate for this, the most accurate method is to replay a reference tone (or some other steady sound). **1973** T. J. Glattke in F. D. Minifie et al. *Normal Aspects of Speech* viii. 329 A series of three tones at 800 Hz..followed by a series at 800, 1,000, and 800 Hz..was differentiated by cats following cortical ablation. **1976** T. H. Flowers *Introd. Exchange Syst.* iii. 67 Each tone is generated by a tone generator common to the whole exchange.

5. d. Also *spec.* in literary criticism, an author's attitude to his subject matter or audience; the distinctive mood created by this.

1929 I. A. Richards *Pract. Crit.* iii. i. 183 A man writing a scientific treatise, for example, will put the *Sense* of what he has to say first... His *Tone* will be settled for him by academic convention. **1950** F. B. Millett *Reading Fiction* 11 This tone, the general feeling which suffuses and surrounds the work, arises ultimately out of the writer's attitude toward his subject. **1959** H. Gardner *Business of Crit.* 40 The tone of the close of the play. **1973** G. W. Turner *Stylistics* vi. 186, I shall use..*tone* for the range of variation reflecting adjustments to an audience. **1977** *N.Y. Rev. Bks.* 15 Sept. 40/2 His practical criticism is not much concerned with the structure of an individual poem except as an embodiment of crisis; it has little to say of diction, the metres, rhythm, syntax, or tone.

6. a. (Earlier example.)

1679 R. Hooke *Diary* 14 May (1935) 412 At Garways, Chinese Language Tones.

III. 11. (sense 2) *tone-quality*; (sense 6) *tone-curve, -group, -mark, -pattern, -sequence, -unit*; *tone-bearing* adj.; (sense 9) *tone-setter*; *tone-setting* adj.; (sense 10) *tone relation, value*; **tone-arm,** †(*a*) the tubular arm connecting the sound-box of a gramophone to the horn (*obs.*); (*b*) = *pick-up arm* s.v. *pick-up *a.* a; **tone burst,** an audio signal used in testing the transient response of audio components; **tone cluster** *Mus.*, a group of adjacent notes on a piano played simultaneously by placing the forearm or flat of the hand on the keys; cf. *note-cluster* s.v. *note *sb.*[2] 21; **tone control,** the adjustment of the proportion of high and low frequencies in reproduced sound; a device or manual control for achieving this; **tone-deaf** *a.*: also *transf.* and *fig.*, insensitive, lacking in perception; hence **tone-deafness; tone-full** *a.*: now usu. *toneful*; cf. Tuneful *a.*; **tone generator,** an apparatus for electronically producing tones of a desired frequency; **tone language** *Linguistics*, a language which uses variations in pitch, in addition to different consonants and vowels, to distinguish words, e.g. Chinese; **tone-on-tone** *a.*, applied to designs, textiles, etc., composed of toning rather than contrasting shades of colour; **tone-painting** (earlier example); hence **tone-painter; tone poem** *Mus.* = *symphonic poem* s.v. Symphonic *a.* (*sb.*) 3; **tone poet,** (*b*) *spec.* one who composes tone poems; hence **tone poetry; tone-row** *Mus.*, the twelve notes of the chromatic scale arranged in a fixed order to form the basis of a composition; **tone sandhi** *Linguistics* [*sandhi], in tone languages: the differences between the tones of words through the influence of contiguous tonal patterns; **tone separation** *Photogr.* = *posterization.

1907 *T. Eaton & Co. Catal.* Spring & Summer 249/1 Columbia Graphophone..patent aluminium tone-arm. **1923** *Gramophone* Apr. p. vii/2 (Advt.), 18 models of Tonearms with and without Goosenecks. **1946** [see *record groove* s.v. *record *sb.* 13 b]. **1981** *Popular Hi-Fi* Mar. 85/3 This is a direct drive, quartz locked, fully automatic turntable with integrated tonearm. **1971** B. Mafeni in J. Spencer *Eng. Lang. W. Afr.* 107 There is a syllabic nasal /N/ [in Nigerian Pidgin] which is tone-bearing and is always homorganic with the succeeding consonant. **1981** *Word 1980* XXXI. 186 Other syllables..may be higher, lower, or on the same level relative to the onset of the tone-bearing syllable. **1967** *Electronics* 6 Mar. 82/1 (Advt.), See the little boxes. See what they can do... Tone burst..trigger..sweep. **1978** *Gramophone* Jan. 1336/1 The toneburst oscillogram..shows that the output across an 8-ohm dummy load is virtually identical with the input signal. **1921** *Freeman* 13 Apr. 112/2 The significance of the tone cluster, like that of the single tone, is to be found in its possibility of combinations with other tone clusters. **1937** N. Slonimsky *Music since 1900* 122 [12 March 1912.] At the San Francisco Music Club Henry Cowell performs for the first time in public, on the day after his fifteenth birthday, piano tone-clusters on white or black keys, struck with the forearm. **1973** *Daily Tel.* 24 Nov. 11/2 He watched the Sinfonietta's resident pianist..elbowing his way through the tone clusters of an early Roberto Gerhard. **1983** *Listener* 28 July 30/3 The music abounds in such special effects as tone-clusters like smudged chords, microtones, fragmentation of the text, whistling, whispers, shouts. **1930** *Electronics* July 195/1 Tone control was the most evident technical idea at the Trade Show of the Radio Manufacturers Association in June. **1934** *Discovery* Nov. 324/2 The models..have effective tone and volume controls fitted. **1974** *Harrods Christmas Catal.* 70/3 Electric Guitar..with volume and tone controls. **1922** H. E. Palmer *Eng. Intonation* i. 3 That part which is concerned chiefly with the tone-curves irrespective of their meanings has been called *Tonetics.* **1953** C. E. Bazell *Linguistic Form* 99 'Questioning intonation' (a special tone-curve) in English. **1932** R. Knox *Broadcast Minds* iv. 85 When we ask him precisely what it is which 'religion' can give us that is inaccessible to a nature..tone-deaf to religion, he has nothing to point to except those moments themselves. **1972** F. Warner *Lying Figures* iii. 35 We are spiritually tone-deaf. Mum's the word! **1884** T. Barr *Man. Dis. Ear* iv. ii. 459 If this partial tone-deafness is not connected with disease of the conducting apparatus, the anomaly is probably due to cochlear disturbance. **1941** F. Matthiessen *Amer. Renaissance* i. iv. 34 The honesty of Whittier's effort was somewhat vitiated by the tone-deafness that robbed his verse of any full variety of cadences. **1973** *Listener* 14 June 786/3 Mr Nixon..has persistently shown ..a disturbing tone-deafness to the legal restraints which ..are built into the American system. **1977** *Proc. R. Soc. Med.* LXX. 134/1 Tone deafness is a defect of pitch discrimination in which the relationship of one musical tone to others cannot be accurately assessed or imitated. **1925** T. Dreiser *Amer. Trag.* I. i. xi. 77 The none too toneful piano. **1927** *Observer* 10 Apr. 24 The short, quick flutter of the wing and the most toneful croak of satisfaction. **1942** *Brit. Jrnl. Psychol.* XXXII. 292 We now have in the laboratory a tone generator capable of sounding tones of any desired harmonic structure. **1980** *Sci. Amer.* Oct. 74/1 Each phoneme is generated by a particular setting of various tone generators, noise generators and acoustic filters. **1922** H. E. Palmer *Eng. Intonation* i. 6 The more serious difficulty is the teaching of the semantic values of the tone-groups. **1977** *Bull. School Oriental & African Stud.* XL. 654/2 The structure of the basic intonational unit, the *tone-group*, consists of an obligatory *tonic*, i.e. the syllable where the pitch movement identifying the *tonic type* begins, and an optional *pretonic* element. **1930** R. Paget *Human Speech* 188 In the Tone-languages, the melody of phonation is tied to the articulation. **1971** G. Ansre in J. Spencer *Eng. Lang. W. Afr.* 157 Most of the languages of the region [*sc.* West Africa] use pitch in their phonological patterning in a way which has earned them the term 'tone languages'. **1978** *Sci. Amer.* Nov. 96/1 Many African and Asian languages are tone languages. **1924** H. E. Palmer *Gram. Spoken Eng.* 6 When tone-marks are provided, the use of the sign ['] may therefore be entirely dispensed with. **1964** M. Schubiger in D. Abercrombie et al. *Daniel Jones* 265 The tone-marks are mine. **1939** *Country Life* 11 Feb. p. xxxviii/2 This Matita two-piece redingote and dress is in a tone-on-tone effect in light and dark grey. **1965** 'L. Egan' *Detective's Due* i. 10 Beige tone-on-tone carpet. **1979** *Arizona Daily Star* 5 Aug. A-12/3 (Advt.), From the tip of its tone-on-tone toe to its sleek, stacked heel, it's everything you'd expect from Evan Picone. **1903** A. W. Patterson *Schumann* 49 How first the pianoforte, next the orchestra, and lastly the string quartet suggested sound pictures to the tone-painter. **1897** *Daily Tel.* 31 Mar. 10/4 Even great musicians do not appear at their best in tone-painting. **1931** T. H. Pear *Voice & Personality* 74 The tone-pattern of the Welsh sentence. **1961** *Amer. Speech* XXXVI. 221 Tone patterns illustrated by Kingdon's tonetic stress marks. **1889** G. B. Shaw *London Music 1888–89* (1937) 68 A long, scrappy movement which is neither bravura nor tone poem. **1942** E. Paul *Narrow St.* xviii. 142 Jacques Benoit-Mechin, who wrote tone poems about South America. **1977** *Gramophone* Apr. 1561/3 Nor does the performance..really project the work as the blazing tone poem that it self-evidently is. **1983** *Listener* 3 Nov. 36/4 At seven and a half minutes it is perhaps a little short-winded for a full-blown tone poem. **1892** *Review of Reviews* Sept. 289/1 A most original tone-poet. **1903** A. W. Patterson *Schumann* p. viii, The writer.. has endeavoured..to let the great tone poet speak to the readers through his own thoughts. **1890** Tone poetry [see *absolute music* s.v. *absolute *a.* 16]. **1934** Webster, Tone quality. **1936** *Discovery* July 224/1 The tone-quality [can] be very considerably altered. **1961** *Times* 10 Mar. 22/2 No conductor in my experience has shaped a melody with more tenderness and lustre of tone-quality. **1903** R. Fry *Let.* 6 Mar. (1972) I. 204 The tone relations are nearer to Moretto's in breadth. **1955** *Times* 9 May 3/5 He was before everything a colourist, and all the machinery of his art—composition, drawing, tone relation, and touch—was organised in the interests of his ruling passion. **1936** *Musical Q.* XXII. 14 (*title*) Schoenberg's tone-rows and the tonal system of the future. **1958** *Times* 6 June 4/3 Composition in tone-rows of 12 notes. **1967** A. L. Lloyd *Folk Song in England* i. 38 The scale of a folk tune is the series of notes used, the tone-row and no more. **1925** E. Sapir in *Language* I. 45 In Sarcee, an Athabaskan language..there is a true middle tone and a pseudo-middle tone which results from the lowering of a high tone to the middle position because of certain mechanical rules of tone sandhi. **1968** P. Kratochvíl *Chinese Lang. Today* ii. 38 One of the factors which cause modifications of these general tendencies of tones in continuous speech is the influence of the tone environment of the given syllable. This is what is known as tone sandhi. **1943** Tone separation [see *posterize *v.*]. **1977** Tone separation [see *posterization]. **1924** H. E. Palmer *Gram. Spoken Eng.* 21 Any pair or more of tone-groups in one sentence constitutes a tone-sequence. **1973** *Archivum Linguisticum* IV. 17 Halliday..though he describes certain tone sequences ..implies that these are no more than chance associations of tones. **1973** *Publishers Weekly* 9 July 44/2 A tone-setter on the field, he contributed to five Packer championships and two Super Bowl wins. **1979** C. E. Schorske *Fin-de-Siècle Vienna* p. xxiii, The intellectual tone-setters among college students. **1962** Y. Malkiel in Householder & Saporta *Probl. Lexicogr.* 11 Many tone-setting Academy dictionaries. **1978** *Language* LIV. 430 Condillac and other tone-setting figures were concerned solely with generalities. **1964** Crystal & Quirk *Prosodic & Paralinguistic Features in Eng.* iv. 50 We come now to the system which has the tone-unit..as its actual matrix. **1981** *Word 1980*

XXXI. 154 The vertical bar marks the 'onset' of the tone unit. **1927** R. H. Wilenski *Mod. Movement in Art* 35 Taught successfully to draw 'by the shadows' and paint 'by the tone values'. **1967** E. Short *Embroidery & Fabric Collage* i. 9 It is easy to assess the relative tone values of strong contrasts, such as black and white.

tone, *v.* Add: **II. 5. a.** (Earlier example, const. *down*.) Cf. sense 6 b in Dict.

1831 J. Constable *Let.* 13 Oct. (1966) IV. 357, I think the *large sail*..much too light. I shall like it toned down very considerably.

c. Also without const.

1976 W. J. Burley *Wycliffe & Schoolgirls* i. 31 The colour scheme was old gold from the carpet to the wallpaper, cushions and curtains. Everything was 'to tone' as Mrs Clarke would..have said.

III. 6. b. (Earlier example.)

1847 Dickens *Dombey* (1848) xx. 197 The Native..handed him..his hat; which..the Mayor wore with a rakish air on one side of his head, by way of toning down his remarkable visage.

d. The vb.-stem in Comb. **to·ne-up,** an act or means of raising to a higher tone; a strengthening or improvement.

1943 W. S. Churchill *Second World War* (1951) IV. 852 It is time to have another tone-up of security arrangements. **1950** *Times* 2 Feb. 2/7 He was a man of 37, and if I had known he was going on this course I should have advised a period of drill training as a tone-up.

toned, *ppl. a.* and *adj.* Add: **I.** *ppl. a.* **3.** *toned-down*, modified, reduced in intensity.

1974 *Listener* 24 Jan. 122/2 The dances are in effect toned-down Bartok. **1981** V. Glendinning *Edith Sitwell* v. 83 Edith published two very similar accounts of Wyndham Lewis..both toned-down versions of a provocative essay she wrote.

toneless, *a.* **2.** (*a*) (Earlier example.)

1833 *Philol. Museum* II. 386 The Old English..the Middle English, and the New, inflect all these verbs in a plain and toneless -ed.

toneme (tōᵘ·nīm). *Linguistics.* [f. Tone *sb.* + *-eme.] A tone or set of tones functioning as a distinctive unit in a tone language (cf. *phoneme 1 b). Hence **tone·mic** *a.*; **tone·mically** *adv.*

1923 D. M. Beach *Phonetics of Pekingese* (London Univ. thesis) ii. 9 Corresponding to the phoneme is the *toneme*, which is a group of tones no one of which may be used in the same position as any other. **1924** —— in *Bantu Stud.* Dec. 90 The key to all tonetic transcription is the principle of the *toneme*,..a group of tones within a given language. **1926** C. M. Doke in *Ibid.* July 218 Words tonemically different and differing in meaning may under certain morphological circumstances become tonemically alike. **1930** J. R. Firth *Speech* iii. 27 A proper understanding and use of the tonemes of these languages would appear to be a *sine qua non*. **1944** *Internat. Jrnl. Amer. Linguistics* X. 123/2 The tonemic changes are by no means always substitution of a high for a mid toneme. **1957** D. Jones *Hist. & Meaning of Term 'Phoneme'* 12 One day—..about February 1921—he [*sc.* D. M. Beach] gave a lecture in the Department of Phonetics at University College, London,..in the course of which he demonstrated that each of the four so-called 'tones' of that language [*sc.* Pekingese] had 'variants' conditioned by the tones of syllables adjoining them in connected speech, and sometimes by other factors. The word 'toneme' was coined at my suggestion: it was readily accepted by Beach and the..staff of the Department. **1965** [see *intoneme]. **1976** *Word 1971* XXVII. 379 The toneme distinction in Swedish (accent 1 and accent 2), differentiating words like *stegen* [ste:gen] with accent 1 'the ladder' and *stegen* [*ste:gen] with accent 2 'the steps', however, is not mastered at an early stage. **1978** *Language* LIV. 245/2 R. W. Wilkinson..argues that mid tone is tonemic.

toner. Add: **2.** *Photogr.* A chemical bath used to change the tone or colour of a (black-and-white) photographic print.

1920 E. J. Wall *Dict. Photogr.* (ed. 10) 661 There are many alternative toners for different colours which may be used alone or in combination on the same print. **1950** O. R. Croy *Compl. Art Printing & Enlarging* iv. 201 Immerse the print in the toner until the desired tone is reached. **1977** J. Hedgecoe *Photographer's Handbk.* 275 (*caption*) The already stark appearance of the print..was further accentuated by processing in a blue toner.

3. Particles of pigment used in xerographic processes to render an electrostatic image visible.

1954 *RCA Rev.* XV. 471 Developing consists in bringing fine positively charged particles of developer powder, or toner, close to the surface, so that they will be attracted to those areas which are still charged. **1977** *Sci. Amer.* Nov. 69/1 (Advt.), Develop in a second or two by a dip through any of at least three makes of commercial electrographic liquid developers that carry positive-charging toner particles.

4. *Hairdressing.* (See quot. 1966.)

1966 J. S. Cox *Illustr. Dict. Hairdressing* 151/1 *Toner*, a substance which when applied to the hair effects a change of tone or an accentuation of an existing tone. **1969** E. Tasho *Hair Styling for Women* x. 182 Temporary rinses can be used as toners on bleached hair. **1976** *Wymondham & Attleborough Express* 10 Dec. 6/3 Brassiness in blondes can be cured by an application of one of the latest ash blonde toners—or by a visit to your hairdresser.

5. = *toning lotion* s.v. *toning *vbl. sb.* and *ppl. a.*

1970 *Vogue* Jan. 5/2 (Advt.), Toners that polish, purify, pep up circulation. **1983** *Harrods Mag.* Spring–Summer 74 Skin care products..Refresher Toner..£15·75.

tonetic (tone·tik), *a.* *Linguistics.* [f. Tone *sb.*, after *phonetic*.] Of or pertaining to the use of tones in languages. So **tone·tics,** the study of tones; **tone·tically** *adv.*

1921 in *Trans. Philol. Soc. 1921–4* (1932) 11 A paper on Tonetics by Mr. D. H. Beach. **1921** H. E. Palmer *Princ. Lang.-Study* i. 37 Learned specialist in 'tonetics' (or whatever the science of tones will come to be called). **1922** —— *Eng. Intonation* i. 6 The teacher articulates one or more syllables and calls upon the students to write down in tonetic symbols what they think they have heard. **1924** [see *toneme.] **1926** *Bantu Stud.* July 198 Phonetics treats of phones, phone-groups, and phonemes: tonetics treats of tones, tone-groups, and tonemes. **1934** Webster Tonetically. **1938** D. M. Beach *Phonetics Hottentot Lang.* ix. 125 In Hottentot, as in Chinese, two roots which are identical *phonetically* may differ *tonetically* when pronounced in isolation. **1958** R. Kingdon *Groundwork Eng. Intonation* p. xxix, The tonetic stress-mark system used in this book was developed..in an endeavour to find the most practical system of marking intonation. **1964** M. Schubiger in D. Abercrombie et al. *Daniel Jones* 256 Even without extra tonetic prominence, it [*sc.* the self-pronoun] increases the weight of the head. **1975** *Language* LI. 561 A considerable residue of cases still remains which must be analysed in terms of underlying homonyms yielding tonetically distinct forms in many environments. **1981** *Word 1980* XXXI. 151 (*heading*) Intonation: tonetic stress marks *versus* levels *versus* configurations.

tonette (tōᵘne·t). [f. Tone *sb.* + -ette.] A simple end-blown wind-instrument resembling a small flute.

1958 E. Birney *Turvey* i. 10 'Plays tonette.' Some kind of a whistle, wasn't it? **1963** *Guardian* 29 Oct. 7/1 [He] is able to produce notes from a small flute, called a Tonette, by sticking the mouthpiece in his nostrils. **1979** *Arizona Daily Star* 1 Apr. K10/1 Tonettes, recorders, triangles and tambourines are the types of instruments needed.

tone-up, *sb.*: see *tone *v.* 6 d. **toney,** var. Tony *a.* in Dict. and Suppl.

tong (tǫŋ), *sb.*² Also **Tong.** [ad. Cantonese *tohng* hall, meeting place.] **a.** An association or secret society of Chinese in the U.S., orig. formed as a benevolent or protective society but freq. associated with underworld criminal activity.

1883 *Harper's Mag.* May 831/1 This burial-place..is parcelled off by white fences into enclosures for a large number of separate burial guilds, or *tongs*, as the Fook Yam Tong [etc.]. **1913** [see *hatchet-man 2]. **1924** *Glasgow Herald* 29 Oct. 8 Rival Tongs, whose principal object seems to be mutual extermination. **1948** P. Johnston *Lost & Living Cities of California Gold Rush* 15/2 Chinese who were members of two tongs, the Sam-yap and the Yan-wo, were working side by side at Two-mile Bar, on the Stanislaus River. **1968** *New York City* (Michelin Tire Corp.) 83 Chinatown was then an area rife with debauchery and vice, the scene of 'tong wars' fought by rival 'tongs' to win control over opium dens, gambling haunts and houses of ill fame. **1972** K. Bonfiglioli *Don't point that Thing at Me* xv. 128 The Chinese Tongs used to favour a six-inch nail, the Japanese use a sharpened umbrella-rib. **1977** *Time* 12 Dec. 28/2 The famous Tongs were something else, more mysterious—secret societies similar to Mafia families. They ran gambling, prostitution, drugs, and offered merchants 'protection'.

b. *attrib.*, esp. in *tong war.*

1927 *Daily Express* 25 Mar. 2/1 Chinese Tong (secret society) warfare broke out at midnight throughout the United States. **1928** H. Asbury *Gangs of N.Y.* 301 The tong wars appeared to have begun about 1899, and..were all caused by conflicting gambling interests. **1950** *Los Angeles Times Home Mag.* 26 Mar. 5/2 The servants of 70 years ago were mostly Chinese whose favorite outdoor sports were tong wars. **1962** 'K. Orvis' *Damned & Destroyed* ii. 20, I was called in by a tong leader. **1966** 'G. Black' *You want to die, Johnny?* iii. 51 The police are not neutrals in the little wars. Our interests are not entirely focussed on Tong feuds. **1972** J. Ball *Five Pieces of Jade* xvi. 220 He had the idea that he could get rid of the two Chinese by..making it look like a ritual killing or a tong murder. **1976** J. O'Connor *Eleventh Commandment* viii. 101 The screws weren't standing for Tong warfare. **1980** G. V. Higgins *Kennedy for Defence* xvii. 149 We are liable to have a nice little tong war on our hands.

tong, *v.*² Add: **d.** To style (hair, etc.) with curling tongs.

1932 'E. M. Delafield' *Thank Heaven Fasting* I. ii. 35 Monica's hair had been tonged into waves. **1953** P. L. Fermor *Violins of Saint-Jacques* 48 His moustache was crisply tonged. **1976** 'D. Halliday' *Dolly & Nanny Bird* iii. 37 She had her hair waved to her ears, and then tonged out sideways.

tonga¹. Add: **b. tonga wallah,** the driver of a tonga.

1942 M. R. Anand *Sword & Sickle* i. 27 A tonga wallah called rudely. **1955** R. P. Jhabvala *To whom she Will* xiv. 98 The tonga-walla in his stained turban cursed and muttered and whipped his horse. **1978** 'M. M. Kaye' *Far Pavilions* ii. x. 155 Tell the tonga-wallah to wait.

Tonga (tǫ·ŋā), *sb.*⁴ and *a.* [Native name: cf. Zulu *i(li) Thonga* member of the Tonga tribe, perh. f. *-thonga* member of a subject race.] **A.** *sb.* The name of several African peoples living chiefly in southern Mozambique, Malawi, and Zambia; a member of these peoples. Also, the Bantu language spoken by them. Cf. *Tsonga. **B.** *adj.* Of, pertaining to, or designating these peoples or their language.

1866 in A. Mackenzie *Memorials Henrietta Robertson* 247 He might make beginnings both in the Amaswazi and Amatonga countries. **1872** *Cape Monthly Mag.* Feb. 117 The Tonga does not own a single head of cattle. **1875** *Jrnl. R. Geogr. Soc.* XLV. 53 The Portuguese..must have found conquest south of the Zambesi an easy matter when the country was entirely peopled by these industrious natives, called generally by the Zulus by the contemptuous title of Tongas. *Ibid.* 93 We found..the poor Tonga Chief at his wit's end. **1910** *Jrnl. Afr. Soc.* IX. 305 The Tonga of the Zambezi is..a different language from the one just mentioned [*sc.* Thonga], and also distinct from the Tonga of Lake Nyasa. **1910** *Encycl. Brit.* III. 360/1 The *Ronga* (*Tonga*) languages of Portuguese South-East Africa..are almost equally related to the *Nyanja* group ..and to *Zulu*. **1929** [see *Nguni *sb.* and *a.*]. **1951** Colson & Gluckman *Seven Tribes Brit. Central Africa* ii. i. 94 The Plateau Tonga, a matrilineal people, occupy a large portion of the Southern Province of Northern Rhodesia. *Ibid.* 95 Within the Tonga group itself slight changes accumulate. **1968** C. Burke *Elephant across Border* iii. 79 Roger broke the silence, translating his discussion in Tonga dialect to Gomez. **1968** *Guardian* 15 Apr. 9/4 In Zambia..there are four strong tribal vernaculars—Bemba, Tonga, Lozi and Nyanja. **1970** [see *Ila *sb.* and *a.*]. **1977** *Times Lit. Suppl.* 30 Sept. 1109/3, I am a chi-Tonga speaker from the lakeside Tonga of Malawi, and I should say that the so-called Tongas of Zambia, Mozambique, Malawi, etc., are one people.

Tongan (tǫ·ŋan), *a.* and *sb.* [f. *Tonga* (see def.) + -an.] **A.** *adj.* Of or pertaining to the island kingdom of Tonga in the south-west Pacific Ocean. **B.** *sb.* A native of Tonga. Also, the Polynesian language spoken in Tonga.

[**1818** J. Martin in Mariner & Martin *Acct. Natives Tonga Islands* p. iii, Having written down sundry examples in English,..I gave them to Mr. Mariner to translate into Tonga.] **1853** J. E. Erskine *Jrnl. Cruise Western Pacific* iv. 119 The group has from time immemorial formed part of the Tongan dominions. *Ibid.* iv. 157 In imitation of the Feejeeans, the Tongans have occasionally practised cannibalism. **1897** S. W. Baker *Eng. & Tongan Vocab.* ii. iii. 35 The present tense in the Tongan language ..for which we have two forms in English... In Tongan.. there is but one form. **1901,** etc. [see *Niuean *sb.* and *a.*]. **1927** J. S. Huxley *Relig. without Revelation* vi. 183 The Tongans became..the most virulent Sabbatarians. **1939** G. Blamires *Little Island Kingdom of South* 13 We had at least five anthems that evening, some of Tongan composition, others European, in the latter case the words being translated into Tongan. **1951** R. Firth *Elem. Social Organization* iii. 105 Every male Tongan is entitled by law when he becomes a taxpayer to receive a residential plot of land in a village or town. **1966** *Listener* 13 Jan. 62/1, I wrote it out in Tongan. **1972** *Vogue* June (Special no.) 135/1 Rate of exchange: 2·14 Tongan dollars to £1. **1976** S. R. Anderson in *Symposium on Subject & Topic* 3 A language in which ergativity is indicated by case marking alone is Tongan. **1978** *Times* 17 Jan. 17/4 Captain Cook..noted that Tongans boxed in much the same way as the English.

tongkang (tǫŋkæ·ŋ). Also **tongkan, tonkang.** [a. Malay.] A sea-going barge used as a cargo boat in the Malay archipelago.

1834 *Singapore Chron.* 2 Jan. 3/2 The fourth [race] was a sailing match, between several tonkangs or cargo-boats. **1858** P. L. Simmonds *Dict. Trade Products* 383/2 *Tongkang*, a kind of boat or junk used in the seas of the Eastern archipelago. **1892** *Nautical Mag.* Dec. 1155 *Hydra*, s.s. and a tongkang, in collision in Singapore Harbour. **1922** *Chambers's Jrnl.* 8 July 503/1 A string of big, heavy *tongkangs*, towed by a puffing launch. **1950** *People* (Austral.) 11 Oct. 16/2 It was a hazardous adventure but they brought it off in spectacular fashion, first by sampan and then by tongkan. **1972** *Straits Times* (Malaysian ed.) 25 Nov. 87/2 Liew Kim..later became a trishaw puller and a tongkang builder.

Tongkinese, var. *Tonkinese *sb.* and *a.*

tongs, *sb. pl.* Add: **3. h.** *Oil Industry.* A large pipe wrench used for making up or breaking out lengths of pipe or casing.

1922 F. M. Towl in D. T. Day *Handbk. Petroleum Industry* I. 411 When the friction becomes so great that this method cannot be used, the tongs are placed on the line. **1972** L. M. Harris *Introd. Deepwater Floating Drilling Operations* v. 46 The normal rig-floor tools, such as, tongs, slips, and small hand tools.

4. tongsman, (*b*) *Oil Industry*, one who handles the large pipe wrench used for making up or breaking out lengths of pipe.

1974 *China Reconstructs* July 47/1 Before long he became a skilled tongsman.

tongue, *sb.* Add: **II. 4. b.** (Further examples.)

1911 H. H. Harper *Bob Hardwick* 88, I was so angry at her that I..made no answer... Presently she said, 'Has the cat got your tongue?' **1940** 'J. Falstaff' *Jacoby's Corners* vi. 69 The cat has got his tongue. **1981** I. St. James *Balfour Conspiracy* vi. 229 Shaughnessy shook his head. 'Cat got your tongue?'

d. Also in phr. *to stick* (or *thrust*) *one's tongue in one's cheek*, as a gesture of sly or † contemptuous humour. Hence *with* (one's) *tongue in* (one's) *cheek*, with sly irony or

humorous insincerity. Cf. *tongue-in-cheek *a.* and *adv.*

1748 [see Cheek *sb.* 2]. **1828** Scott *Fair Maid of Perth* in *Chron. Canongate* 2nd Ser. I. viii. 153 The fellow who gave this all-hail thrust his tongue in his cheek to some scapegraces like himself. **1849** *Blackw. Edin. Mag.* Oct. 450/2 Hows'ever, I just sticks my tongue in my cheek,.. watches my chance, an' off by a track-boat..to New Orleens. **1887** R. H. Roberts *In Shires* i. 10 [He] sticks his tongue in his cheek, and whispers to his neighbour. **1928** *Observer* 19 Feb. 5/1, I must confess my utter inability to grasp what Mr. B. Nicholson is after, though I am loath to believe that he painted his apparently flippant still life arrangements with his tongue in his cheek. **1951** *Sport* 30 Mar.–5 Apr. 9/3 Walsall fans will tell you, with tongue in cheek, that the Fellows Park club is always on the alert where transfer of players is concerned.

e. *with* (one's) *tongue hanging out* and varr., with great thirst or (*fig.*) eager expectation. *colloq.*

1897 Kipling *Day's Work* (1898) 102 They've been waiting for this youth with their tongues hanging out. **1928** Wodehouse *Money for Nothing* x. 222, I should hurry. His tongue was hanging out when I left him. **1967** E. Lemarchand *Death of Old Girl* xii. 141 My tongue was hanging out, so I thought I'd..see if there was any sherry going. **1974** L. Lamb *Man in Mist* ii. 16, I don't have to run round to them with my tongue hanging out the moment I am promised something.

8. c. (Later examples of sense *glossolalia*.) Also *pl.* in *collect.* sense.

1965 *Sunday Mail* (Brisbane) 5 Dec. 31/5 Some parishioners have complained to the Diocesan authorities.. about Mr. Schofield's interest in speaking with tongues. **1972** S. Tugwell *Did you receive Spirit?* v. 40 Some manifestation, usually tongues, is generally expected; indeed, strict Pentecostals demand it. **1976** *Church Times* 5 Mar. 14/2 Tongues is a personal and devotional gift as opposed to the others, which are intended to help people.

III. 13. d. *Geol.* A part of a formation that projects laterally into the material of an adjacent formation, becoming thinner in the direction of its length.

1917 L. W. Stephenson in *Jrnl. Washington Acad. Sci.* VII. 245 It is..proposed that such features as *x* and *y* in figure 1 be designated 'tongues'... A tongue is not a member nor a lentil, either one of which differs lithologically from the typical material composing the formation of which it forms part. **1953, 1970** [see *lentil 4*].

e. *gen.*

1881 E. A. Freeman *Sk. Subject & Neighbour Lands Venice* 207 Columns with richly carved capitals, and..with tongues of foliage at their bases. **1954** F. T. Prince *Soldiers Bathing* 7 Letting the sea-waves coil Their frothy tongues about his feet. **1965** E. L. Myles *Emperor of Peace River* ii. iv. 226 The frantic bawling of a calf in the edge of a tongue of brush near the river's bank. **1966** D. Bagley *Wyatt's Hurricane* v. 129 They emerged on to an open place, an incursive tongue of the countryside licking into the suburbs.

14. o. *Mus.* = *plaque 1 d.

1953, 1957[see *plaque 1 d]. **1977** Goossens & Roxburgh *Oboe* iii. 34 The scraping tongue (or plaque). A flat oval plate of steel... Some players prefer a narrower plate to prevent the knife from coming into contact at the edges.

IV. 15. a. *tongue-position*. **b.** *tongue-wagger*. **c.** *tongue-lashing* vbl. sb.

1881 'Mark Twain' *Prince & Pauper* xix. 222 She promptly brought the King out of his dreams with a brisk and cordial tongue-lashing. **1918** D. Jones *Outl. Eng. Phonetics* vi. 16 We examine the tongue positions of these five classes [of vowels]. **1977** *Word 1972* XXVIII. 321 The most important feature for the correct perception of this phoneme from the viewpoint of the listener is high tongue position. **1913** D. H. Lawrence *Sons & Lovers* i. 16 He was blab-mouthed, a tongue-wagger.

16. tongue-and-groove, applied (chiefly *attrib.*) to boards in which a tongue along one edge fits into a groove along the edge of its neighbour, and to joints, etc., so made; also *fig.*; **tongue-slip,** a slip of the tongue; **tongue-speaker,** one who speaks with tongues (see sense 8 c in Dict. and Suppl.); **tongue-twisting** *a.*, difficult to articulate.

1882 W. J. Christy *Practical Treat. Joints* iii. 52 Joggle Joint.—This term is applied to a square, semi-circular,..or otherwise shaped tongue and groove joint generally of equal depth the full way through. **1929** W. Faulkner *Sound & Fury* 353 He emerged carrying a sawn section of tongue-and-groove planking. **1939** —— *Wild Palms* 19 The flimsy walls (they were not even tongue-and-groove..but were of ship-lap). **1976** *Southern Even. Echo* (Southampton) 11 Nov. (Advt. Suppl.) 4/2 End terr. house, built 1972,..d. glazing, tongue and groove floors, etc. **1977** *Time* 3 Oct. 53/1 Despite its style and tongue-and-groove plotting, *The Honourable Schoolboy* sometimes displays a Balzacian tendency to turn urges into passions. **1913** Tongue-slip [listed in Dict. in sense 15 a]. **1948** *Sunday Pictorial* 18 July 5/2 Freud took up this pioneer work and showed how the half-forgotten world of dreams and tongue-slips could be explored. **1978** *Canadian Jrnl. Linguistics 1977* XXII. 179 The penultimate chapter of AM whips through pauses, tongue slips, and other topics in the science of word-botching. **1910** *Encycl. Relig. & Ethics* III. 370/2 The 'tongue-speaker' needed as his complement the 'interpreter'. **1978** *Amer. Speech* LIII. 59 They..associate these utterances with the inspiration of the Holy Spirit, although tongue-speakers differ in their beliefs about the significance of the gift of tongues. **1949** Koestler *Insight & Outlook* vii. 109 Its name, too, is funny—foreign and tongue-twisting. **1961** E. S. Turner *Phoney War* viii. 109 Each new campaign brought them a crop of tongue-twisting place names.

tongue, *v.* Add: **2. c.** *Mus.* To move the tongue when playing a woodwind instrument so as to interrupt the air flow briefly. Also *trans.*, to produce (a note) repeatedly interrupted in this way. Cf. Tonguing *vbl. sb.* a.

1936 F. B. Chapman *Flute Technique* iv. 18 The student must..ultimately aim at producing notes by multiple tonguing..: he should..be able to tongue them continuously and quite clearly at the rate of nine or ten to the second. **1953** E. Rothwell *Oboe Technique* iii. 30 To 'tongue' a note pronounce the consonant 'T' with your tongue on the reed. **1977** *Early Music* July 343/1 Do not tongue too much or you may dislodge the reed from its staple.

5. Delete '(of ice)'.

1942 *Bull. U.S. Geol. Survey* No. 936. 374 In places a thick shale lens lies within, or tongues into, an ore-bearing sandstone lens. **1973** *Nature* 2 Mar. 41/2 The patch reef, 13 m long and over 2 m high, tongues out to the west. **1980** D. Creed *Scarab* iii. xix. 183 A low spit of land tongued out into the shallow water.

tongued, *a.* (*ppl. a.*) Add: *tongued and grooved*, furnished with a tongue and groove joint (see *tongue *sb.* 16).

1773 *Bristol* (Va.) *Vestry Bk.* (1898) 238 A Dwelling House [shall] be built..[with] Good flouring Plank, well Tong'd & Groved. **1897** F. C. Moore *How to build Home* 15 The sheathing should be tongued and grooved and planed on one side. **1955** J. S. Chappell *Woodworking* x. 136 A..stronger form of tongued and grooved joint is made by ploughing a groove in both edges to be joined.

tongue-in-cheek, *a.* and *adv.* [See Tongue *sb.* 4 d in Dict. and Suppl.] **A.** *adj.* Ironic, slyly humorous; not meant to be taken seriously. Also **tongue-in-the-cheek.**

1933 *Times Lit. Suppl.* 30 Mar. 223/4 *Shooting the Bull* ..is a tongue-in-the-cheek march through newspaperdom. **1937** M. Covarrubias *Island of Bali* xi. 375 A typical tongue-in-cheek Balinese answer to dodge a complicated explanation for outsiders. **1953** *Spectator* 13 Mar. 320/2 This..novel..seems too facile, too tongue-in-cheek. **1959** *Times* 4 Sept. 5/1 Though the piece was energetic and often exuberant it was certainly not tongue-in-the-cheek or humorous in style. **1976** *National Observer* (U.S.) 27 Mar. 10/1, I enjoyed Wesley Pruden's tongue-in-cheek suggestion..that every man, woman, and child in the United States be given a college degree so they 'become equal'. **1982** *Listener* 16 Dec. 28/1 Angela Carter translated Perrault's fairy tales..with absolute fidelity to the understatement, the tongue-in-cheek charm of the originals.

B. *adv.* = *with tongue in cheek* s.v. *tongue *sb.* 4 d.

1934 in Webster. **1976** *Listener* 18 Mar. 334/3 Someone told Muhammad Ali, tongue-in-cheek, that his book made him come over as a 'deep thinker'. **1979** H. McLeave *Borderline Case* xi. 113 'You mean you're a spy.' 'Only for those people who have something sinister to hide,' he said, tongue-in-cheek.

tonguer. (In Dict. s.v. Tongue *v.*) Add: **† 2.** *N.Z.* (See quots.) *Obs.* exc. *hist.*

1836 in R. McNab *Old Whaling Days* (1913) 436 Some mention of what are called tonguers... When the whale is cut in they are entitled to the carcass and the tongue. **1843** [see *cutter *sb.*¹ 1 b]. **1845** E. J. Wakefield *Adventure N.Z.* I. xi. 323 The proper officers have been selected—such as cooper, carpenter, cooks, painter, and 'tonguer'..[who] takes his name from having an exclusive right to the oil obtained from the tongue..in payment of his duty of 'cutting-in', or dissecting, the whale. **1941** Baker *N.Z. Slang* ii. 13 A tonguer was a native or white living in New Zealand who assisted a whaling crew to cut up whales... These men earned their name..from the fact that they were given the whale's carcass and tongue to dispose of as they wished.

toni (tōᵘ·ni). *India.* Also **tonee, tony.** [a. Tamil: see Dhoney, doney.] **a.** See Tony *sb.*² **b.** A dug-out boat. **c.** A ferry boat.

1881 *Naval Encycl.* 811/2 *Tonee*, a canoe formerly used on the coast of Malabar. **1914** *Yachting Monthly* June 83/2 Inshore the flood tide had already turned the tony and bunder boats, the former a canoe-like craft hollowed straight out from the tree. **1917** *Ibid.* Sept. 268/2 The large Tonis are between 25 and 30 feet long. **1946** *Mariner's Mirror* XXXII. 209 The ordinary Malabar dug-out, called *toni* in Bombay. **1978** *Times of India* 15 Jan. 3/5 Contraband radios and cassettes..were seized from a toni, Laxmi, on Thursday at the Ferry Wharf.

-tonia (tōᵘ·niă), also anglicized as **-tony,** terminal element [f. Gr. τόν-ος Tone *sb.* + -ia¹] with the sense 'tone, condition' in terms in *Med.*, as *hypotonia, *sympathicotonia.

tonic, *a.* and *sb.* Add: **A.** *adj.* **2.** *tonic water*, a non-alcoholic carbonated drink containing quinine or another bitter as a stimulant of appetite and digestion; a drink or glass of this; *tonic wine*, weak, flavoured wine sold as a medicinal tonic.

1899 *Graphic* 11 Mar. 320/1 (Advt.), His Holiness the Pope writes that he has fully appreciated the beneficent effects of this Tonic Wine. **1926** *Daily Colonist* (Victoria, B.C.) 6 July 9/6 (Advt.), Schweppes famous British table waters. Soda water,..ginger beer, tonic water. **1958** S. Hyland *Who goes Hang?* xi. 53 'What will there be to drink?' asked Mrs. Kimmis..over the top of a tonic water. **1970** G. Greer *Female Eunuch* 276 Perhaps she can try a glass or two of tonic wine? More likely her G.P. will.. prescribe a happiness pill. **1982** G. F. Newman *Men with Guns* x. 74 He drank gin swamped with Indian tonic water.

4. b. (Earlier example.)

1849 *Jrnl. Indian Archipelago* III. 668 The influence of this habit of the tonic languages is still largely impressed on their Malay-Polynesian and Turonian descendants and congeners.

B. *sb.* **1. c.** Tonic water.

1935, 1949 [see *gin and tonic* s.v. *gin *sb.*² 2 a]. **1972** M. J. Bosse *Incident at Naha* ii. 108 We all had vodka and tonics.

2. b. The principal key of a musical composition or passage; the home key.

1896 G. Grove *Beethoven & his Nine Symphonies* 8 The *Coda* which closes the first movement, after repeating in the tonic the phrase already quoted as No. 5, combines the wind instrument passage with the first subject. **1923** E. Evans *Beethoven's Nine Symphonies* I. 177 At the third portion we have a new treatment of the first part of the same subject..leading to a triumphant cadence in C as tonic. **1961** A. Hopkins *Talking about Symphonies* i. 20 The key you start in is called the 'Tonic'. **1979** D. R. Hofstadter *Gödel, Escher, Bach* v. 130 With the inversion of the theme for our melody, we begin in D as if that had always been the tonic—but we modulate back to G after all, which means that we pop back into the tonic, and the *B*-section ends properly.

Tonica, var. *Tunica².

toning, *vbl. sb.* and *ppl. a.* (In Dict. s.v. Tone *v.*) Add: *spec.* Having or being a colour that certain place in an intonation pattern.

1963 M. A. K. Halliday in *Archivum Linguisticum* XV. 13 Second [is]..the placing of the tonic syllable..—the location, in each tone group, of the pretonic and tonic sections... I propose to call these three systems 'tonality', 'tonicity' and 'tone'. **1966** G. N. Leech *Eng. in Advertising* ix. 88 Devices of graphic emphasis such as underlining and italics can be used to represent special tonicity in speech. **1973** [see *nucleus *sb.* 12 a].

Tonikan, var. *Tunica(n.

toning, *vbl. sb.* and *ppl. a.* (In Dict. s.v. Tone *v.*) Add: *spec.* Having or being a colour that tones in (with something previously mentioned). *toning lotion*, a lotion, usu. slightly astringent, used for cosmetic purposes to refine the texture of the skin.

1960 *Harper's Bazaar* (U.K. ed.) Aug. 22/2 An easy seven-eighths suit, that wraps around a simple toning dress. **1965** *Ibid.* (U.K. ed.) June 40 A toning lotion beneath base provides a protective film beneath make-up. **1970** *Cabinet Maker & Retail Furnisher* 30 Oct. 204/1 It is available in toning shades of browns, blues, red and greens. **1977** *Sunday Times* 6 June 43/1 The attack on the Cooper pores was maintained with Fresh Toning Lotion. **1982** Barr & York *Official Sloane Ranger Handbk.* 31/1 She has, naturally, a kilt, which she wears with toning tights.

tonite (tənəi·t). Simplified spelling of To-night *adv.* and *sb.*, after *nite *sb.*², used chiefly in advertisements.

1968 R. Clapperton *No News on Monday* iv. 35 Another placard announced..'Tonite, for one nite only, the fabulous Lisa Mundt.' **1971** *Black World* Mar. 55/1 They wud be up all nite tonite if they didn't wake up soon. **1976** *Leicester Mercury* 16 July (Advt.), Riverside disco. Tonite. All welcome.

tonk (tǫŋk), *sb.*¹ *slang* (chiefly *Austral.*). [Etym. unknown.] **a.** A term of abuse: a fool, an idiot.

1941 Baker *Dict. Austral. Slang* 77 *Tonk*, a simpleton or fool. (2) A dude or fop. (3) A general term of contempt. **1963** *New Society* 22 Aug. 5/1 'Bleg', 'thick boot', 'tonk', and 'greb' are all of uncertain origin, but probably have euphemistic backgrounds. **1965** [see *nana¹].

b. A homosexual man.

1943 *Penguin New Writing* XVII. 83 The cook got my goat when he started trying to do the same thing. He was a tonk all right, just a real old auntie. **1965** H. Porter *Stars Austral. Stage & Screen* 280 During the last ten years or more, there have been imported a coterie of *untalented* English homosexuals, English tonks unheard of outside their home country. **1970** *TV Times* (Austral.) 15 July 41/3 There was also a homosexual (who was referred to as a 'tonk'—thereby dating Mr Porter rather badly).

tonk (tǫŋk), *sb.*² Colloq. abbrev. of *honky-tonk.

1937 [see *smoke-shop* s.v. *smoke *sb.* 11]. **1948** *Common Ground* VIII. 38 The man who owned the little country Tonk was named Hamp... It was a one-room shanty store that doubled as a country bar room at night. **1960** C. Hamblett in J. Pudney *Pick of Today's Short Stories* XI. 138 None of the other rundown bars and tonks had anyone remotely like Lia.

tonk (tǫŋk), *v.* *colloq.* (chiefly *Sport*). [Echoic.] *trans.* **a.** To strike. **b.** To beat or defeat. Freq. *pass.*

1910 A. A. Milne *Day's Play* 114 Wanting four to win, I fairly tried to tonk the leather. **1926** Galsworthy *Silver Spoon* iii. i. 224 'He seems to enjoy the prospect of getting tonked,' murmured Michael. **1945** Baker *Austral. Lang.* 207 Here are a few general expressions concerned with school life:..*to get tonked*, to receive corporal punishment. **1963** A. Ross *Australia 63* ii. 55 Our spinners have been tonked about yet again by uncouth country batsmen.

Tonkawa (tǫ·ŋkāwā), *sb.* (*a.*) [ad. Sp. *tancagueis, tancahues*, etc., prob. ad. Wichita (Waco dial.) *tonkawéya*, said to mean 'they all stay together'.] **a.** (A member of) an Indian people of Texas. **b.** The language of this people. Also *attrib.* or as *adj.*

1806 J. Sibley in *Message from President of U.S., communicating Discoveries made in exploring the Missouri by Captains Lewis & Clark* 74 Tankaways..have no land, ..but are always moving. **1870** J. C. Duval *Adventures Big-Foot Wallace* xxv. 148, I got it from 'Puppy's Foot', the Tonkawa chief. *Ibid.* xl. 245 My old friend 'Bah-pish-na-ba-hoo-tee' (which means 'Little blue whistling thunder' in the Tonkawa language). **1933** H. Hoijer *Tonkawa* (thesis, Columbia Univ.) p. ix, The Tonkawa appear to have been an important and warlike tribe living in central Texas during most of the 18th and 19th centuries. *Ibid.* p. x, Tonkawa is now spoken by only six persons—all of them past middle age. **1974** *Encycl. Brit. Micropædia* X. 43/3 By the 1970s the Tonkawa reservation in Oklahoma was reported to have a total population of about 60.

Tonkinese (tǫŋkinī·z), *sb.* and *a.* Also 8–9 **Tonquinese,** 9 **Tong-, Tungkin(g)ese.** [f. *Ton-kin* (*Tongking*) + -ese.] **A.** *sb.* **a.** The people of Tongking, a region of northern Vietnam on the border with China; also, a member of this people. **b.** The language of the Tonkinese. **B.** *adj.* Of or pertaining to the Tonkinese.

1726 Swift *Gulliver* II. iii. i. 4 Several sorts of Goods, wherewith the Tonquinese usually trade to the Neighbouring Islands. **1806** J. Barrow *Voy. Cochinchina* ix. 251 The *Tung-quinese*, being in fact of the same character and disposition as the Chinese, were little able to cope with the hardy and disciplined troops. **1845** *Encycl. Metrop.* XXV. 673/1 In self-sufficiency and jealousy of strangers, the Tonkinese do not yield to their neighbours in China. **1884** W. Mesny *Tungking* xviii. 115 The protracted struggle that has been going on for long months is quite as much due to the action of China as to the Tungkingese themselves. **1885** J. G. Scott *France & Tongking* i. 11 Garnier and his lieutenants..enrolled many thousand Tongkinese auxiliaries. **1890** J. Frazer *Golden Bough* I. i. 36 Many other peoples (Tonquinese, Hindoos, Chuwash, etc.) have adopted the same test of a suitable victim. **1926** H. A. Franck *East of Siam* xii. 220 In the mess that followed the Manchu conquest of China, a Tonkinese fisherman founded a new dynasty. **1934** Webster, *Tonkinese*, the Annamese dialect of Tonkin. **1951** M. B. Emeneau *Stud. Vietnamese (Annamese) Gram.* p. v, Tonkinese and Chochin Chinese are slightly differentiated from one another by differences of pronunciation and of vocabulary. **1983** C. McCarry *Last Supper* 172 Christopher met a fellow poet, a Tonkinese who had studied at the Sorbonne... The Tonkinese poet was a female.

tonlet (tɒ·nlėt). [ad. MF. *tonnel(l)et* short, full skirt, (also) *tonlet*, dim. of *tonneau* cask (see Tonnel, -ell).] A short skirt of armour; also, each of the overlapping horizontal bands of which this was sometimes made.

[*a* **1480** *Traictié de la Forme et Devis d'ung Tournoy* in *Œuvres Complètes du Roi René* (1844) II. 11 Le harnoys de corps est come une cuirasse ou comme ung harnoys à pié qu'on appelle tonnellet.] *a* **1486** in *Archaeologia* LVII (2nd Ser. VII. 1900) 43 *To arme a man*... Firste ye muste sette on Sabatones..& þe the breche of mayle And the tonletis And the brest And þe vambras [etc.]. **1894** *Antiquary* Jan. 26/2 Another suit, or rather part of one, of Henry VIII..is that which has been called the tonlet, or, as in the Tower inventories it is written, the trundlet suit. **1910** *Encycl. Brit.* II. 587/2 The surcoat being gone we see him armed in breast and back plate, his loins covered by a skirt of 'tonlets', as the defence of overlapping horizontal bands comes to be named. **1934** G. C. Stone *Gloss. Construction, Decoration & Use of Arms & Armor* 622/2 The tonlet suit was used mainly for fighting on foot, but was sometimes used in place of other leg armor when jousting at the barrier. It had wide, bell-shaped skirts of plate which were often solid and elaborately fluted with deep vertical folds... Sometimes it was made of horizontal plates. **1975** *Country Life* 3 July 45/3 The superb 'tonlet' or skirted armour made for King Henry VIII for foot combat in about 1512. **1983** *Daily Tel.* 18 Oct. 10 (*caption*) Henry VIII's tonlet armour ready to go on view at the Burlington House Fair.

tonne (obs. form of Ton, Tun). Add: Re-introduced from Fr. in mod. use to denote a metric ton of 1000 kilogrammes (Ton¹ 4).

1877 *Rep. Brit. Assoc. Adv. Sci. 1876* ii. 32 The Tonne is the mass or quantity of matter contained in a cubic metre of water, and is very nearly the same as the British Ton. **1930** *Engineering* 25 July 119/3 Each of the two high-pressure turbines takes some 224 tonnes of steam per hour. **1953** *Economist* 28 Mar. 902/1 The country's refining capacity, in terms of crude oil throughput, was about 23 million tonnes a year. **1972** *Which?* May 130/3 The British Steel Corporation, going metric but realising the possible confusion between a ton and a tonne (1,000 kilograms) has directed its staff to pronounce 'tonne' 'tunnie'. **1975** *B.S.I. News* Apr. 5/1 Our units committee has been asked to advise how, in speech, confusion between 'tonne' and 'ton' can best be avoided. Their advice is simply this: when saying the word 'tonne' never say it alone; always say 'metric tonne'. **1977** A. Hallam *Planet Earth* 24/2 Meteorites vary in weight from a few tens of grammes to several tonnes. **1981** *Southern Horticulture* (N.Z.) Spring 3 It should be possible to achieve yields of 5 tonne/ha or more from mature roots.

tonneau. Add: **1.** Also, the rear part of a car with front and rear compartments or of an open car; a car having a tonneau; **tonneau cover,** a removable, flexible cover for protecting the rear or passenger seats in an open car when they are not in use; also *transf.*

1905 A. M. Binstead *Mop Fair* 118 With the entrancing little green *tonneau* which a railway rustic delivered.. next morning, it was entirely different. So winsome was the diminutive car [etc.]. **1931** Garrard & Geddes *Practical Motoring* 643/1 *Tonneau*. The rear part of an open four or five seater motor car body was at one time commonly referred to as the tonneau but this term is now rarely used by itself. When the front seats only are used, a special cover known as the tonneau cover is sometimes stretched across the whole of the rear part. **1976** *Glasgow Herald* 26 Nov. 21/6 (Advt.), 16ft cabin cruiser... Complete with tonneau cover and canopy, £3000. **1978** *Sat. Rev.* (U.S.) 1 Apr. 45 That leaves London [taxis]. Ah, how civilized... There is room in the tonneau for five, with two on jump seats. **1979** J. Leasor *Love & Land Beyond* vi. 88 Victoria brought the plane down... Love helped her batten down tonneau covers over the two cockpits. **1980** *Times* 1 Nov. 14/3 (*caption*) During the 1975 run, an 1898 Daimler wagonette..overtakes a 1903 de Dietrich tonneau. **1981** *West Lancs. Evening Gaz.* 25 Feb. 15 (Advt.), Sports boat.., complete with trailer, tonneau cover, etc.

2. (A gallicism.) A barrel or cask; a measure of capacity for wine, equal to one tun (198 gallons).

1794 A. Young *Trav. France* (ed. 2) I. xviii. 535 Wine has increased in its export to England..; before the treaty it was 8000 tonneaux a year. **1851** [see *queue *sb.* 6]. **1978** S. Sheldon *Bloodline* iv. 69 We should get three hundred thousand francs a *tonneau* for the first pressings.

‖ **tonnelle** (tonɛ̆l). [Fr., = Tunnel *sb.*] An arbour. Also *fig.*

1861 Thackeray *Roundabout Papers* (1863) 219 Those who sit down under my *tonnelle*, and have a half-hour's drink and gossip. *a* **1922** H. Jones *Old Memories* (1923) 160, I can even yet see him sitting peaceably, sheltering from the heat in our vine *tonnelle*. **1947** *Horizon* Feb. 106 The Queen is looking back along the flowery *tonnelle* of her day.

tonner. Add: (Earlier example.) Also, a lorry of (so many) tons weight.

1851 A. O. Hall *Manhattaner in New Orleans* 177 A seven hundred tonner,..full of Dutch emigrants. **1959** I. Jefferies *Thirteen Days* iv. 43 The Arab six-tonner driver. **1978** R. Mark *Office of Constable* iii. 38 Came the great day when the survivors were packed with their kit into a three-tonner en route for Sandhurst.

tono-. Add: **tonofi·bril** *Histology* [ad. G. *tonofibrille* (M. Heidenhain 1899, in *Arch. f. mikrosk. Anat.* LIV. 212)], a bundle of tonofilaments; **tonofibri·lla,** (*a*) *Histology* = prec.; (*b*) *Ent.*, a non-contractile fibril in an insect that passes from a myofibril through the epidermis into the cuticle; **to·nofilament** *Histology*, one of the minute supportive or non-contractile filaments that occur aggregated into networks in the cytoplasm of many epithelial cells, esp. in the epidermis; **tono·logy,** the study of tones or of intonation in speech; hence **tonolo·gical** *a.*; **tonoto·pic, -topical** *adjs.* *Anat.* [Gr. τόπος place], exhibiting a spatial correspondence with the frequency of heard sound; hence **tonoto·pically** *adv.*

1901 *Jrnl. R. Microsc. Soc.* 512 (*table*) Tonofibrils or resistance fibrils, e.g. in intestinal epithelial cells, epidermis cells. **1964** G. H. Haggis et al. *Introd. Molecular Biol.* v. 120 In the stratified squamous epithelium of the skin, tonofibrils arch through all the cells like scaffolding and they are attached to numerous desmosomes over the entire surface of the cells. **1976** *Path. Ann.* XI. 220 At a fine structural level, the cells of thymoma contain tonofibrils and complex desmosomes, but no neurosecretory granules. **1925** E. B. Wilson *Cell* (ed. 3) i. 41 The greater number of writers have..accepted the conclusion..that they are of the nature of supporting or skeletal structures, hence the term tonofibrillæ (Heidenhain). **1935** R. E. Snodgrass *Princ. Insect Morphol.* iii. 63 It frequently appears not only that the tonofibrillae traverse the epidermal layer, but that they penetrate a varying distance into the cuticula. **1969** R. F. Chapman *Insects* xii. 211 In *Musca* each myofibril is attached to the cuticle by about twelve tonofibrillae. **1964** *Jrnl. Investigative Dermatol.* XLIII. 278/1 In pemphigus vulgaris, a severe necrotizing injury of unknown etiology leads to complete destruction of the tonofilaments with ensuing loss of desmosomes. **1978** *Sci. Amer.* May 145/1 The tonofilaments are not contractile but seem to form a tensile, structural framework for the cell cytoplasm. **1980** *Nature* 17 Jan. 249/1 Electron microscopy shows that bundles of keratin tonofilaments often terminate in membrane-bound desmosomes. **1934** Webster, Tonological. **1975** *Language* LI. 565 The nouns in the two classes with L final vowels show tonological behavior parallel to that of the nouns in the two classes with Ø final vowels. **1983** *Word 1982* XXXIII. 230 With regard to other tonological features in the area, one can mention a number of languages with four-tone systems. **1874** H. Sweet in *Trans. Philol. Soc. 1873–4* 98 What is wanted, then, is a comparative 'tonology' of the Danish dialects. **1924** D. M. Beach in *Bantu Studies* Dec. 77 An entirely new field..is lying open before us—the comparative and historical study of tones. This study.. will be called tonology. **1970** *Stud. Afr. Linguistics* I. 100 (*heading*) Nupe tonology. **1978** *Language* LIV. 245/2 There are ten papers on phonology, seven of which deal specifically with tonology. **1942** *Anat. Rec.* LXXXII. 430 In the monkey and chimpanzee..surface positive potentials evoked by various pitches indicate tonotopic localization within the primary auditory cortex. **1983** *Nature* 10 Feb. 463/1 This 'tonotopic' organization is preserved in all levels of the central auditory pathway. **1948** A. Brodal *Neurol. Anat.* ix. 314 Pfeifer (1936)..was led to conclude that if there exists any tonotopical localization in the primary acoustic cortex, tones of the highest pitch must be represented medially, those of lowest pitch laterally. **1963** *Jrnl. Neurophysiol.* XXVI. 294 (*heading*) Tonotopical organization, relation of spike counts to tone intensity, and firing patterns of single elements. **1971** *Brain Res.* XXVI. 402 There is good evidence that cells in the cochlear nucleus, superior olivary nuclei, nuclei of the lateral lemniscus, and inferior colliculus are organized according to their best frequencies, or tonotopically. **1978** *Nature* 9 Mar. 139/2 Spatial analyses of the evoked potentials indicate that the auditory centre in the midbrain is organised tonotopically.

Tonquinese, obs. var. *Tonkinese *sb.* and *a.*

tonsillectomy (tǫnsil,e·ktŏmi). *Surg.* Also **tonsilectomy.** [f. Tonsil + *-ectomy.] Removal of the tonsils.

1899 *Jrnl. Amer. Med. Assoc.* 23 Sept. 768/2 What then are the general results of tonsillectomy, as compared with those obtained by the usual operation of tonsillotomy? **1932** *Oxford Times* 23 Sept. 22/7 It would be a mistake to suppose that tonsilectomy..is indicated only where there is throat trouble. **1961** J. Heller *Catch-22* (1962) xvii. 164 He could come through other people's tonsilectomies without suffering any postoperative distress. **1977** *Rolling Stone* 16 June 43/2 They told him he was going to the circus but instead took him for a tonsillectomy.

tonsillotomy. Add: Usu. applied to partial removal of the tonsils, in contrast to *tonsillectomy. Now *rare.* (Earlier and further examples.)

1876 *Louisville Med. News* I. 280 (*heading*) Dangerous hemorrhage after tonsillotomy. **1899** [see *tonsillectomy]. **1902** C. Joyes (*title*) Tonsillotomy or tonsillectomy, which? **1924** W. D. Harmer in H. W. Carson *Mod. Operative Surg.* II. 272 The question whether it is better to cut away part of the tonsils (tonsillotomy) or to remove them entirely with their capsules (tonsillectomy) has been hotly discussed.

tonsorial, *a.* Add: Hence **tonso·rialist** *humorous*, a 'tonsorial artist', a barber.

1869 *New North West* (Deer Lodge, Montana) 6 Aug. 3/1 Mr. Plummer, the colored tonsorialist..has the misfortune to be a 'bloody Hinglishman'. **1898** A. M. Binstead *Pink 'Un & Pelican* xi. 253 One of them [*sc.* constables]..gazed..at the abstracted sign of the tonsorialist.

tonstein (tǫ·nstəin). *Geol.* [a. Ger., lit. 'clay stone'.] A rock composed mainly of kaolinite which is commonly found in association with certain coal seams, or a thin band of such a rock (see quots.).

1961 I. A. Williamson in *Mining Mag.* CIV. 9 Tonsteins are essentially argillaceous rocks containing kaolinite in a variety of forms together with occasional detrital and carbonaceous material. **1971** *Nature* 6 Aug. 371/2 The thin, curious, kaolinitic bands called tonsteins discovered more recently in the coalfields of Western Europe have provided welcome additional markers for coalfield correlation.

tontine, *a.* (Earlier example.)

1790 J. Woodforde *Diary* 3 Sept. (1927) III. 211 Mr. Custance brought some Papers for me to sign respecting all his Children being put into the new Tontine Annuities.

Tonton Macoute (toñtoñ makū·t). Also **Ton Ton Macoute** and with small initials. [a. Haitian French, of uncertain origin.] **a.** A militia which was formed by President Duvalier in Haiti and became notorious for its brutal and arbitrary behaviour; also, a member of this. Also *ellipt.* as **Tonton.**

1962 S. E. Finer *Man on Horseback* ix. 133 Duvalier took office in October 1957... Instead of relying upon the *Garde Nationale*, he has built himself up a 5,000 strong counterforce of palace guards, civilian militia, and civilian hoodlums called '*tonton macoute*'. **1965** J. E. Fagg *Cuba, Haiti & Dominican Republic* 136 A gang of ruffians known as *Tonton Macoute* murdered or beat citizens who complained. **1966** G. Greene *Comedians* i. i. 16 The Tontons Macoute... The President's bogey men. They wear dark glasses and they call on their victims after dark. *Ibid.* ii. 47 He was believed by some to have connections with the Tontons. *Ibid.*, He exchanged some words with a Tonton Macoute at the door. **1972** *Times* 23 Nov. 10/8 Mr Luckner Cambronne, former Haitian Minister of Defence and of the Interior, who organized the dreaded Ton Ton Macoutes, has sought asylum in the Colombian Embassy in Port-au-Prince. **1976** *Globe & Mail* (Toronto) 5 Nov. 7/2 Rather than fire his father's tontons, (and risk sending a bunch of disgruntled triggermen underground) the young President found them jobs in public institutions such as hospitals. **1981** Plate & Darvi *Secret Police* ii. 47 There is no agreement among scholars of Haiti about what 'Tonton Macoutes' means. Some claim the name refers to a primitive bad figure in Haitian voodoo culture who takes you away when you misbehave.

b. *fig.*

1970 'D. Craig' *Young Men may Die* xi. 83 Our two people..have on Tonton Macoute sunglasses. **1973** *Publishers' Weekly* 8 Jan. 34/1 'Veronica Ganz', the one-girl *ton ton macoute*.

to·n-up, *sb.* and *a.* *slang.* [f. Ton¹ + Up *adv.*² 12 c.] **A.** *sb.* A speed of 100 m.p.h.; a

motor-cyclist who achieves this. Also in the sense of *ton¹ 4*a. Also *fig.*

1961 *Daily Tel.* 11 Feb. 1/2 The term 'Ton Up' is used by young motor-cyclists to indicate doing 100 m.p.h. **1964** *New Statesman* 21 Feb. 288/3 Many made a point of.. assuring me that the ton-ups weren't as black as they thought I'd painted them. **1964** in Hamblett & Deverson *Generation X* 146 Of course, there were the Tonups (now Rockers) who mustn't be forgotten, but then they have always been an untouchable group on their own, kinkily keen on their bikes. **1972** J. Blackburn *For Fear of Little Men* xiv. 147 Eighty miles an hour, ninety, a ton-up—as the motorcycle maniacs call a hundred. **1976** J. Snow *Cricket Rebel* 44 My return read nought for 117. I got a ton-up in my next Test at Headingley as well. **1978** *Gramophone* Aug. 329/3 The sleeve photograph shows Perényi in action, head, hands and cello blurred as if moving at too great a rate for even the fastest stop on the camera. There is indeed an element of the 'ton-up' about his performance.

B. *attrib.* **1. a.** Applied to young motor-cyclists who enjoy travelling at high speed.

1961 *Harper's Bazaar* May 104/2 Gangs, rebels without a cause and ton-up kids. **1961** *Times* 1 Sept. 11/1 A reasoned defence, of the 'throttle-potties 'or 'ton up boys' was submitted to the psychology section. **1965** G. McInnes *Road to Gundagai* xiii. 225 Dad wore leather hip boots and jacket, goggles and a cap with its peak at the rear... The Ton-Up kids on the M1 had nothing on him. **1982** 'C. Aird' *Last Respects* iii. 31 I'm not stupid enough to want that boy Crosby behind the wheel of one of Traffic Division's vehicles... He'd be after a ton-up kid.

b. *fig.* Applied to a person who incongruously imitates the dress or behaviour of such people.

1964 *Economist* 13 June 1246/2 The ton-up type of vicar who is trying to be 'with it'.

2. Achieving a speed or score of 100 in other contexts.

1967 *Daily Tel.* 17 Feb. 1/4 Plans are being made to cut BEA's Manchester-to-London air service..as a result of big passenger losses to British Railways 'ton-up' trains. **1976** *Southern Even. Echo* (Southampton) 6 Nov. 13/4 After his two records for the number of winners trained in a season on the flat, it is the North Country that has the ton-up trainers of the jumping game. **1977** *News of World* 17 Apr. 19/1 'Ton-up' Taylor—he landed 100 winners last season for the first time.

Tony, *sb.*¹ Restrict † *Obs. slang* to sense in Dict. and add: **2.** [The nickname of Antoinette Perry (1888–1946), U.S. actress, manager, and producer, arbitrarily used.] One of the medallions that have been awarded annually since 1947 by the American Theatre Wing (New York) for excellence in some aspect of the theatre. Freq. in *Tony award.*

1947 *N.Y. Times* 7 Apr. 40/1 The award already has been dubbed a 'Tony', as her associates called Miss Perry. **1948** *Ibid.* 29 Mar. 23/6 John Garfield represented the Experimental Theatre in accepting a 'Tony' for 'experiment in theatre'. **1975** *Times* 10 May 9/3 The Tonys have been awarded, and the 1974/75 New York theatre season is over. **1976** *Time* 27 Dec. 5/3 He later starred in several musicals, including his 1963 Tony Award-winning performance in *She Loves Me.*

tony, *sb.*²: see also *toni.

tony, *a.* For '*U.S.* and *Colon. colloq.*' read '*colloq.* (orig. *U.S.*)' and add: Also **toney.**

1. (Earlier and later examples.)

1877 R. J. Burdette *Rise & Fall of Mustache* 177 He's a toney old cyclopedia on the patter. **1880** *Harper's Mag.* Jan. 209/2 He just put on heaps of style..you know—regular tony. **1920** D. H. Lawrence *Lost Girl* xii. 299 The really toney women of the place came to take tea. **1922** Joyce *Ulysses* 158 Theodore's cousin in Dublin Castle. One tony relative in every family. **1959** D. Barton *Loving Cup* i. iii. 60 Have you got your dinner-jacket with you, old man?.. I'm afraid we're very toney these days. We seem to get tonier. **1966** 'J. Hackston' *Father clears Out* 84 Father, dignified and collected,.. entered the calm, cool tony atmosphere of the Commercial Hotel. **1982** A. H. Garnet *Maze* (1983) iii. 14 He was charming..what Cyrus's mother used to call a 'toney fella'.

2. A fashion colour between red and brown; also as *sb.* *temporary.*

1921 *Punch* 4 May 357/1 Ladies' artificial silk stockings. In black, white, nigger, grey and toney. **1927** W. E. Collinson *Contemp. Eng.* 61 Brogues..sometimes of ox-blood or tony red colour. **1965** *Guardian* 31 Mar. 15/1 Toney was a colour of the twenties which died with the twenties.

-tony [-y³], anglicized f. *-tonia.

Tony Curtis (tōᵘ·ni kɔ̆·ɹtis). The film-name of Bernard Schwarz (b. 1925), U.S. actor, used *attrib.* and *absol.* to designate a style of haircut in which the hair at the sides of the head is combed back and that on the forehead is combed forward.

1956 *People* 13 May 10/2 The blokes with crew cuts or Tony Curtises. **1961** J. I. M. Stewart *Man who won Pools* iv. 48 His girl had..made him quit that Duck's Behind for a straight sleeking back with oil. George Pratley had his Tony Curtis still. **1969** *It* 13–28 June, She had seen him..with a well slicked back Tony Curtis style complete with DA at the back.

too, *adv.* Add: **I. 1. a.** The use at the beginning of a clause has been revived, at first in the U.S.

1930 *Publishers' Weekly* 17 May 2514/2 Too, chain store merchandising tactics are the result..of the keenest.. retailing brains in this country. **1956** Gardner & Smith *Geneal. Res. Eng. & Wales* I. iv. 46 Many births and deaths were not recorded in the parish registers of England and Wales. Too, some of the other denominations kept poor records. **1969** *Daily Tel.* (Colour Suppl.) 17 Oct. 59/1 And, too, is there any future for the Dunebuggy in Britain? **1976** *National Observer* (U.S.) 7 Mar. 13/2 Too, supermarket officials note, the projected 10 to 20 per cent saving ..covers only part of the..bill. **1978** R. Ludlum *Holcroft Covenant* vii. 89 Too, the windows were not that close to one another.

b. Used after a vb. to emphasize a reassertion of a denied statement. orig. and chiefly *U.S.*

1914 B. Tarkington *Penrod* xiv. 122 'No, I didn't.'.. 'He did, too! Didn't he, Sam?' **1936** M. Mitchell *Gone with Wind* xlvii. 843 'Surely you can't be thinking of marrying a man who wasn't in the army..?' 'He was, too, in the army.' **1937** Wodehouse *Summer Moonshine* (1938) v. 59 'Do you know the Princess?' 'My stepmother.' 'She isn't!' 'She is, too. I have documents to prove it.' **1939** *Reader's Digest* Dec. 25 'She hasn't got appendicitis.' The husband became even wilder, insisting that she did too have appendicitis. **1963** L. Deighton *Horse under Water* xxi. 92 'How do you think she guessed?' 'No idea,' I said. 'You have too. Please tell me,' said Jean. **1969** tr. Godard's *Masculine Feminine* 60 *Madeleine*: You don't care, but for me my first record is very important. *Paul*: I do too care. **1978** A. Maling *Lucky Devil* xxxiii. 181 'Well, you can't really believe in both,' she said. 'You can too!' Frances said hotly.

2. c. (Examples of *too true*.) *just too bad*: see *just *adv.* 6 c.

1568 [in Dict.]. **1849** [in Dict., sense 5 c]. **1900** C. M. Yonge *Modern Broods* i. 5 'I am considered quite passée—' 'My dear! With your art, and music, and all!' 'Too true!' **1930** 'E. Queen' *French Powder Mystery* xxxi. 261 'The presumption is that he slept home all night and therefore couldn't have committed the crime. Yet physically it was possible.'..'Too true, too true,' murmured Ellery. **1976** N. Freeling *Lake Isle* x. 67 'Rare, that sort of saint.' 'Too true.'

4. c. (Earlier example.)

1881 *Punch* 26 Mar. 138 (*caption*) 'Have you seen the Old Masters at Burlington House?'..'Are they not really quite *too too*!!'

5. b. (Earlier example of *too much for*.) *too much*: also (orig. *U.S.*), excellent, first-rate.

1794 A. Radcliffe *Myst. Udolpho* I. ix. 251 The sight of this poor old woman would have been too much for Emily. **1937** *Metronome* Mar. 55/1 Man, if you didn't you really missed something. That man's too much! What great bass drum work he shows. **1958** G. Lea *Somewhere there's Music* xviii. 155, I want to make it to the City... Man, like the City is too much—and that's where I want to be. **1966** *Melody Maker* 15 Oct. 19, I just can't wait for his Spring return with Earl Hines, Budd Johnson and the rest. This could be too much. **1967** [see *lean *v.*¹ 6 d]. **1968** *Scottish Daily Mail* 3 Jan. 6 They got 'Absolutely divine'; we get 'Too much'... One day 'Too much' will sound as old fashioned as 'ripping'.

e. Also, rather less than; only moderately; not very. Also in other negative contexts, esp. *not too*— (cf. *not *adv.* 15 d).

1842 E. A. Poe in *Graham's Mag.* Feb. 126/2 The mind of the not-too-acute reader. **1866** Geo. Eliot *Felix Holt* I. iii. 86 They were not too hopeful about Protestants who adhered to a bloated and worldly Prelacy. **1866,** etc. [see *not *adv.* 15 d]. **1892** E. G. White *Steps to Christ* (1908) 108 We do not pray any too much, but we are too sparing of giving thanks. **1909** Galsworthy *Fraternity* xxxvii. 313 There were not too many people in London who.. would have behaved with such seemliness—not too many so civilised as they! **1912** J. Sandilands *Western Canad. Dict. & Phrase-Bk.*, *Not too bad*, a characteristic Canadian reply to an inquiry regarding one's health or circumstances. **1947** *Sun* (Baltimore) 5 Nov. 2/7 There is little incentive for him to do more than seek a mere existence for himself and family, without too keen a regard for the plight of others. **1956** *English* Summer 45 The English Association..having survived half a century and two world wars..has not done too badly. **1967** L. Deighton *Expensive Place to Die* iii. 19 'Can I have a shower?' she asked. 'The water's not too warm I'm afraid,' said Byrd. **1984** A. Brookner *Hotel du Lac* i. 10 My intervention did not seem to be too welcome.

h. *too right*: expressing emphatic agreement or assertion. orig. *Austral.*

1926 'J. Doone' *Timely Tips to New Australians*, Too right!—A slang term expressing agreement or corroboration. **1934** T. Wood *Cobbers* v. 76 What I says is, give 'em an axe and send 'em into the bush. Then they'd work, or starve. Too right they would. **1951** J. Fleming *Man who looked Back* xi. 145 'We should have thought of that before we started out.' 'Too right,' Joe agreed. **1961** *Lancet* 5 Aug. 311/2 The chairman agreed it was thumbs down for Dr. Y., too right it was. **1978** P. McCutchan *Blackmail North* viii. 95 'He'll see you now sir.' 'Too right he will.'

toodle-oo (tūd'l,ū·), *int.* *colloq.* [Origin unknown; perh. f. Toot *sb.*²] Goodbye. Cf. *pip-pip.

1907 *Punch* 26 June 465 'Toodle-oo, old sport.' Mr. Punch turned round at the amazing words and gazed at his companion. **1908** T. E. Lawrence *Let.* 16 Aug. (1938) 62 Tootle 'oo. E.L. **1931** D. L. Sayers *Five Red Herrings* vi. 64 Well, toodle-oo! **1960** [see *blot *sb.*¹ 1 c]. **1981** R. Barnard *Sheer Torture* xi. 121 I'll be downstairs. Toodle-oo.

Also **toodle-, tootle-pip.**

1977 A. C. H. Smith *Jericho Gun* v. 67 Well, tootle-pip for now. **1983** *Standard* 26 Oct. 23 (*heading*) Toodlepip to the poor British Exec.

toofer, var. *twofer.

tool, *sb.* **1. b.** Restrict *arch.* to def. in Dict. and add: In *Criminals' slang*, any weapon.

1938 F. D. Sharpe *Sharpe of Flying Squad* xix. 209 'Here they are, boys; get your tools ready.'..As they ran they pulled weapons from under their coats, hatchets, knuckle-dusters, hammers, and bars of iron. **1971** J. Mandelkau *Buttons* i. 28 We grabbed our tools and by then the Mods were at the bottom of the street.

2. b. (Further examples of the sense 'penis'.)

1922 Joyce *Ulysses* 299 The poor bugger's tool that's being hanged. **1966** L. Cohen *Beautiful Losers* (1970) i. 114 You uncovered his nakedness!—You peeked at his tool! **1971** J. Stewart tr. *Simenon's Rich Man* iii. 64 A little slut of a girl..who had not protested when he had put his tool in her hand.

3. c. A pickpocket; the member of a pair or team of pickpockets who actually picks pockets; = Wire *sb.* 13.

1865 *Leaves from Diary Celebr. Burglar* xviii. 62/1 They were getting uneasy about the absence of their 'tool'. **1886** A. Pinkerton *Thirty Yrs. a Detective* 38 The man who is to do the actual stealing is called the 'tool' or 'hook' and the others are known as 'stalls'. **1936** *Evening News* 9 Dec. 8/5 Modern pickpockets are either 'tools' or 'stalls.'..Really clever tools work alone, disdaining the assistance of a stall. **1955** *Publ. Amer. Dial. Soc.* xxiv. 60 The tool selects the mark to be robbed, and actually takes the score.

5. *tool bag* (also *fig.*), *box* (earlier example), *-kit, -maker* (earlier example), *-roll, -room* (*attrib.* examples); **tool-bar,** a frame fitted to a tractor on which interchangeable implements may be mounted; **tool-crib,** a place from which tools or other stores are issued to workmen; **tool-dresser** *Oil Industry* = *roustabout 3; **tool head,** a part of a machine that carries the tool or tool-holder and can be moved to bring the tool to bear on the work; **tool-man,** (*a*) a worker with tools; a toolroom worker; (*b*) *Criminals' slang*, a lock-picker or (*U.S.*) safe-breaker; **tool-pusher** *Oil Industry*, someone in charge of a drilling rig; **tool slide,** a sliding machine part which carries a tool; **tool subject** *Educ.*, a subject taught or studied as a help to a main subject.

1892–3 *T. Eaton & Co. Catal.* Fall & Winter 95/1 Bicycle Accessories... Tool Bags—Flat pouch with fastener. **1970** *New York* 16 Nov. 42/2 Talk is the most unreliable and over-reacted-to weapon in the black revolutionary toolbag. **1960** *Farmer & Stockbreeder* 8 Mar. 74/3 In a great many cases they are designed as units to be carried on an ordinary toolbar. **1832** *Chambers's Edin. Jrnl.* I. 236/2 Lifting his tool-box, and going through all the operations of horse-shoeing. **1936** J. Dos Passos *Big Money* 19 In six years he rose from machinist's helper to keeper of toolcribs. **1973** T. Pynchon *Gravity's Rainbow* i. 160 It must have been the wind that was carrying him down a dirt road..among the shacks and tool cribs to a wire fence with a gate. **1896** B. Redwood *Petroleum* I. v. 258 The drilling 'crew' consists of two drillers and two tool-dressers. **1976** M. Machlin *Pipeline* iv. 53 His Daddy started him as a tool-dresser—same way I started. **1950** W. Cooper in A. W. Judge *Centre, Capstan & Automatic Lathes* I. iv. 212 Independent feed for the tool heads is provided at each [work] station. **1977** *Sci. Amer.* Sept. 188/1 The appropriate toolhead was selected automatically by a punched-paper-tape program that was read by an electronic computer-controller. **1963** A. Lubbock *Austral. Roundabout* 108, I took..a tool-kit, a box of spare parts, two spare wheels, [etc.]. **1977** C. McFadden *Serial* (1978) iv. 14/2 Did she have enough cash in her Swedish carpenter's tool kit? **1844** Mill *Ess. Pol. Econ.* iv. 98 The producer..must set aside a portion of the produce to replace not only the wages paid both by himself and by the tool-maker, but also the profits of the tool-maker. **1909** Webster, *Toolman*, one who works with or makes tools. **1949** W. R. Burnett *Asphalt Jungle* vii. 47 We need an expert toolman. **1970** R. Busby *Frighteners* xvi. 157 The toolman..got his nickname and reputation by proving there wasn't a lock made that he couldn't tickle. **1977** *Whitaker's Almanack 1978* 577 British Leyland was given 28 days to get the striking toolmen back to work. **1979** K. Bonfiglioli *After you with Pistol* xxiv. 149 Every sound, professional team of thieves has..a 'toolman' who knows how to neutralize burglar-alarm systems and to open locks. **1932** *Amer. Speech* VII. 271 *Tool pusher*, a foreman in charge of drilling operations—distinct from driller. **1976** M. Machlin *Pipeline* xlii. 460 Around daylight a tool-pusher comes out and he tells us we better shut down our rig and put out the fire because the crew on the next well's going to change their control head. **1917** *Harrods Gen. Catal.* 1059/4 Motor car tool roll.. containing 19 best quality tools. **1979** W. H. Canaway *Solid Gold Buddha* xxii. 146 Pete spread a toolroll on the spillway. **1937** *Times* 13 Apr. p. xv/2 This checking is the function of the tool room staff in which are to be found the finest craftsmen in the factory. **1963** *Times* 28 May 5/2 More than 1,400 toolroom workers in 10 Birmingham factories of the Joseph Lucas group took part in a 24-hour token strike today. **1976** *Milton Keynes Express* 16 July 13/4 (Advt.), Toolmaker and toolroom miller required. **1919** G. W. Burley *Lathes* vii. 108 In some special forms of vertical turning and boring mills the tool-slide is of the non-swivelling variety,..and only vertical and horizontal movements are possible. **1936** Colvin & Stanley *Turning & Boring Practice* vii. 101 When the cut is completed the spindle stops, the flow of coolant is shut off, and the tool slides return..to the starting point. 1963 Tool slide [see *SEMI-AUTOMATIC sb. 1]. 1934 WEBSTER, Tool subject. 1966 Rep. Comm. Inquiry Univ. Oxf. II. 436 Teaching in any 'ancillary' or 'tool' subject (e.g. languages for historians or mathematics for economists).

tool, v. Add: **1. d.** *trans.* To equip (a factory) with the machine tools needed for a particular product; to provide the tools needed for (a new product); also *intr.* Usu. with *up*. Also *fig.*

1927 Observer 25 Sept. 4 The work of tooling up the Manchester and Cork factories may result in production within the next two months. 1939 Times 7 Apr. 6/3 The United States National Defence Council is taking steps to see that American plants shall not be tooled to fit European needs at the expense of the United States' own later military needs. 1933 Flight 27 Apr. 391 It is standard practice to 'tool up' for a certain type as soon as the size of the order warrants the expenditure on jigs and dies. 1940 E. J. H. JONES Production Engin. i. 6 The expression 'to tool up' a component means to design and supply all jigs, fixtures, cutting tools, and gauges required for the manufacture and inspection of the pieces. 1957 Observer 3 Nov. 11/4 Makers must be given a chance to sell models already tooled-up. 1959 Times Rev. Industry Apr. 37/3 Much expenditure had to be faced for tooling new models. 1962 Listener 13 Sept. 375/1 The automotive factories have tooled up for their new models. 1972 M. KAYE Lively Game v. 23, I saw all of the specs..and I helped to tool it up.

c. *intr.* to *tool up* (*fig.*): to arm oneself. *slang.*

1959 Times 7 Apr. 6/3 There seemed a general agreement that the fashion of carrying dangerous weapons was more widespread to-day than formerly. One read all too often about groups of young men 'tooling up' before setting off for a showdown with a rival gang. 1971 J. MANDELKAU Buttons xiii. 144 We tooled up with pieces of wood and iron bars and hiked over towards their main camp. 1978 H. WOUK War & Remembrance xxxi. 283 We might have closed the Mediterranean and forced England to her knees even while..we tooled up for our summer Caucasus thrust.

2. *slang.* **b.** (Earlier example.) In mod. use, to travel in any kind of vehicle.

1835 DICKENS Let. 11 Jan. (1965) I. 53, I wish..you could have seen me tooling in and out of the banners, drums,..and go-carts. 1923 WODEHOUSE Inimitable Jeeves xiii. 155, I borrowed a bicycle from one of the grooms and tooled off. 1964 Mankind May 134/1, I was tooling home from the Mexican border in a light blue convertible. 1977 D. ANTHONY Stud Game xii. 114, I tooled down the Coast Highway to Sunset.

c. Of a person: to go (or come) in an easy manner; to go off quickly.

1862 A. J. MUNBY Diary 23 Feb. (1972) 116 Near S. Martin's Lane, I met W. M. Thackeray..'tooling' along quietly, alone, with hands in pockets. 1882 Punch 17 Dec. 285/2 Now we'll just tool off to some quiet part of a place where we can divide this 'ere shining swag without fireworks. 1936 WODEHOUSE Laughing Gas xxviii. 293 Well, I know when I'm licked. I tooled straight round to the Temple of the New Dawn and asked for an entrance form. 1937 D. L. SAYERS Busman's Honeymoon 25 The Dowager saw them and was quite nice to them, so they tooled off, fairly happy. 1940 WODEHOUSE Eggs, Beans & Crumpets 38 Bingo was tooling along the road with the Peke in his arms. 1945 'A. GILBERT' Don't open Door! xix. 176 Citizens come tooling up to the Police Station..to give information. 1955 E. WAUGH Officers & Gentlemen 8 Tool off to Headquarters and get the gen about tonight's do. 1977 E. CRISPIN Glimpses of Moon vi. 90 Then along comes the boy-friend..and tools off without the least idea that anything's seriously wrong.

d. To play around; to behave in an aimless or irresponsible manner.

1932 F. ILES Before the Fact v. 77 'Well, anyhow, what are you doing with yourself.' 'Oh, tooling around, you know. Nothing much.' 1957 New Yorker 21 Sept. 37/3 Let him stay parched or get a head cold tooling around in ferryboats. 1973 'A. HALL' Tango Briefing ii. 22 We were tooling around in Malta on a friendly visit. 1981 J. WAINWRIGHT All on Summer's Day 8 Lord around long enough with the paper work and..half the night has slipped by.

tooled, *a.* (Later example.)

1935 [see middle leg s.v. *MIDDLE a. 6].

tooled, *ppl. a.* **2.** *tooled up*: equipped with an offensive weapon. *slang.*

1959 Observer 1 Mar. 10/1 They sit in all-night cafés, 'tooled up' sometimes with knives and 'choppers' and crank handles. 1973 Time Out 2–8 Mar. 13/2 They had knives, chains, house-bricks, iron bars, and everything. We weren't tooled up, because we were coming out of school, and so couldn't get ready properly. 1982 J. BARNETT Marked for Destruction v. 58 Smith brandished the shotgun..to let the minder know he was tooled up.

toolie (tū·li). *Oil Industry. slang.* [f. TOOL sb. + -IE.] = *tool-dresser* s.v. *TOOL sb. 5.

1932 Amer. Speech VII. 271 Toolie..., a tool-dresser—i.e., the assistant of a cable-tool driller, who dresses bits, fires the boilers, and maintains the rig in order. 1976 L. ST. CLAIR Fortune in Death i. 13 He wouldn't be the first wildcatter to end up a toolie, a roustabout.

toolies (tū·liz), *sb. pl. Canad.* [Respelling of *tules*, pl. of TULE.] Backwoods; remote or thinly populated regions.

1961 R. P. HOBSON Rancher takes Wife i. 22 We're plenty far back in the toolies at Batnuni. 1976 M. MACHLIN Pipeline xii. 141 This here's a program they got for people out in the toolies.

tooling, *vbl. sb.* Add: **2. c.** The process of designing and supplying the machine tools needed for a product or model; also *concr.*, these tools collectively.

1939 Daily Tel. 18 Dec. 12/3 (Advt.), Experience should include the setting up and tooling of automatic and turret lathes. 1940 Sun (Baltimore) 9 Dec. 14/8 War plants emerge from the 'tooling up' phase into production. 1958 Times Rev. Industry June 57/2 A minimum of three-year development time; it includes..design, prototypes, testing and tooling. 1963 Wall St. Jrnl. 1 Oct., Dayton Reliable Tool & Manufacturing Co. has switched 90% of its manufacturing to making tooling for turning out pull-open devices. 1967 L. E. ASCHER in Wills & Yearsley Handb. Management Technol. 125 Should we begin to sell at a high price to recover our tooling costs quickly? 1977 C. HIGGINS Dreamland xii. 150, I had earned commissions in hand, from cleaning out Vidalia's obsolete tooling.

toon² (tūn), repr. a dial. pronunc. of TUNE sb.

1901 M. FRANKLIN My Brilliant Career iii. 26 Some of us was always good for a toon on the concertina. 1977 New Musical Express 12 Feb. 27/1 How the mighty are fallen. Shel Talony..is reduced to dealing with a none-too-choice churning out four mock country toons for a 99p line.

toonie (tū·ni). [f. *toon*, repr. Sc. pronunc. of TOWN sb. (in Shetland, spec. = the arable land on a croft) + -IE.] In full *toonie dog*: = *Shetland sheepdog* s.v. *SHETLAND 1 d.

1910 J. A. LOGGIE in J. S. TURNER Kennel Encycl. III. 1425 The 'Shetland Sheep-dog'..was originally known as the 'Shetland Collie' or 'Toonie Dog'; the first of its being used to drive the sheep off the township, crofts, or what is known in Shetland as the 'Toon'. 1958 O. GWYNNE-JONES Shetland Sheepdog Handb. 5 The 'Toonie' dogs ('Toonies' or Shetland Sheepdogs) are of great service to the farmer. 1971 [see *PEERIE sb.].

tooraloo (tūrălū·), *int. colloq.* [Var. *TOODLE-OO int.] Goodbye.

c 1921 D. H. LAWRENCE Phoenix II (1968) 121 So long! See you soon! Too-ra-loo! 1922 JOYCE Ulysses 229 Tooraloo, Lenehan said, see you later. 1974 J. JOHNSTON How Many Miles to Babylon? 22 I'll have to be off... Tooraloo.

toot, *sb.³* Add: **3.** Cocaine; a 'snort' of cocaine. *U.S. slang.*

1977 Maclean's Mag. 2 May 24 They slink into some of the finer furnished bathrooms of the city for a quick toot. 1978 Detroit Free Press 16 Apr. (Parade Suppl.) 21/1 Cocaine—also called 'toot', 'snow', 'coke'. 1979 Daily News (N.Y.) 23 Sept. 5 Each man dipped a spoon into the white powder and got his toot. 1981 W. SAFIRE in N.Y. Times Mag. 15 Mar. The familiar 'to go on a toot', or to drink heavily and thereby lose a weekend, has been replaced by 'to blow a toot', or to inhale a 'line' of cocaine.

toot, *sb.⁵* **a.** For *ruscifolia* substitute *arborea*. See also TUTU.¹ (Earlier and later examples.)

1851 E. WARD Jrnl. 18 Feb. (1951) 133 Found poor Novice had taken the 'toot' and had been very ill. 1949 F. SARGESON I saw in my Dream 126 There wasn't even any fern, only a few pieces of the tutu which everybody called toot.

b. *to eat (one's) toot*: see *TUTU¹ b. *N.Z. slang* (now *Obs.* exc. *Hist.*).

Hence **too·ted** (*ppl.*) *a.* = *TUTUED *a.*; also as *pa. pple.*

1879 in H. Guthrie-Smith Tutira (1921) xvi. 123 Two bullocks dead at Troutbeck's. One 'tuted', the other goodness knows what. 1930 L. G. D. ACLAND Early Canterbury Runs 1st Ser. vii. 169 A travelling showman had the bad luck to get his elephant tooted near the Waitaki.

toot, *v.³* Add: **4.** *trans.* To inhale (cocaine).

1975 High Times Dec. 110/2 Counterculture advocates of cocaine sniffing now have public confirmation of what they've known for a long time: the chief drawbacks to tooting coke are high costs and the law. 1979 Ibid. Jan. 31 You'll feel better knowing that what you toot is cut with the original Italian Mannite Conoscenti.

tooter². Add: **2.** One who proclaims loudly; *spec.* = TOUTER¹. *U.S.*

1863 Rio Abajo Weekly Press (Albuquerque, New Mexico) 19 May 2 The nameless party's tooter speaks confidently of the success of its nominee. 1868 Harper's Mag. Aug. 417/2 The wharf..was alive with vehicles and tooters for the hotels. 1897 R. E. ROBINSON Uncle Lisha's Outing 297 Noisiest of all were the tooters, vociferously proclaiming the wonders of the side shows.

tooth, *sb.* Add: **I. 2. b.** (*b*) *pl.* Denoting the ability to compel or enforce, esp. by the exaction of penalties, etc.

1925 Country Gentleman 25 July 13/1 How many teeth can you put in a grower's contract of membership with a cooperative marketing association? 1931 Week-End Rev. 14 Mar. 369/1 It is even more clear..that..it will lead to the success of the Disarmament Conference next February than to 'give teeth' to the Paris Peace Pact. 1935 Evening Sun (Baltimore) 27 May 13/3 (heading) Coal control bill with teeth studied. 1949 Economist 16 Apr. 694/2 It is well that President Truman should have made quite clear, not only that the Atlantic Pact is meant to have teeth in it, but also what sort of teeth. 1963 Listener 7 Mar. 432/3 It needed guts to fight a battle against an Establishment with teeth. 1964 Daily Tel. 20 Mar. 24 (heading) 'Teeth' put in scheme for fair coal sales. 1976 Howard Jrnl. XV. 1. 29 The Magistrates' Association..asked for an order stronger than a care order to show that 'in the last resort the law has teeth'.

(*c*) *spec.*, denoting the combatant personnel of an armed service or military unit. Cf. *TAIL sb.¹ 4 c.

1946, 1962 [see *TAIL sb.¹ 4 c]. 1962 Daily Tel. 26 Sept. 12/1 There is, indeed, room for a 'teeth' role for certain units (of the Territorial Army). 1967 R. AYCH KRAN Friends, not Masters iv. 2 These changes gave the infantry more teeth and less tail. 1968 Listener 25 July 99/3 As for recruiting, the 'teeth arms' of the three Services are likely to go on attracting young men of high quality. 1977 J. HAINES Politics of Power ii. 24 He knew..that the Labour Government's great 'defence review' had left a swingeing spending—administrative 'tail' as opposed to front-line 'teeth'—almost untouched and unharmed.

II. 3. a. (Later example.)

1966 B. MALAMUD Fixer vii. iii. 239 He combed his hair and beard with the comb and the teeth of the comb fell out.

f. (Earlier example.)

1806 J. DAVIS Post-Captain iv. 19 'She looks, sir, like a whacking frigate.' 'Can you see her teeth?' 'Yes, sir; she has a very heavy tier of teeth.'

6. c. So as to be utterly committed; *up to the teeth*: heavily involved or absorbed.

1934 T. E. LAWRENCE Let. 8 June (1938) 805 At the moment we are all up to the teeth in 3 more target boats. 1974 Spartanburg (S. Carolina) Herald 25 Apr. A11/6 A young farmer who is starting out and he's mortgaged to the teeth at the bank would not look at it the same way.

d. *fed (up) to the (back) teeth*: see *FED *pa. pple.*

7. a. (Earlier *attrib.* example.)

1872 B. JERROLD London xiv. 116 Honourable instinct making a tooth-and-nail fight against adverse circumstances.

8. h. *to get one's teeth into*, to become engrossed in; to come to grips with, to begin serious work on. **i.** *long in the tooth*: see *LONG A¹ 1 c.

1935 D. L. SAYERS Gaudy Night i. 23 If one could work here steadily..getting one's teeth into something dull and durable. 1962 R. FERGUSSON Watery Maze vi. 140 American eagerness to get their teeth into the enemy. 1983 G. MITCHELL Cold, Lone, & Still xi. 111 He's not the man to get his teeth into anything big.

9. a. tooth-comb, (*a*) in *fig.* use (later examples); also *attrib.* and as *v. trans.*, to investigate minutely: cf. *FINE-TOOTH *a.*; (*b*) *Zool.*, a group of procumbent lower front teeth found in tree shrews and lemurs; **tooth fairy**, a fairy believed by children to take away milk teeth and leave a small sum of money; also *transf.*; **tooth-glass**, (*a*) (see quot. 1858); (*b*) a glass used to hold false teeth; **tooth-mug** = prec. (*b*); **tooth-paste** (earlier and later examples); freq. *attrib.* in *tooth-paste tube*; also in *fig.* phr. *to put the toothpaste back in the tube*, illustrating the futility of trying to restore a stable state of affairs in the light of subsequent events; **tooth-root**: for a substitute 3; example; **tooth wash**, a liquid dentifrice.

1918 Daily Chron. 25 Jan. 3/7 The Army behind the front is being tooth-combed of all men fit for the fighting line. 1934 Glasgow Herald 28 Aug. 4/2 She was a strong woman. 1955 Times Lit. Suppl. 1 Oct. 749/3 Whatever the 'tooth-combs' of Dr. Hutton's fellow-scholars may leave of it, he must be congratulated on his discovery. 1958 N.Y. Times Mag. 6 Apr. 66/1 The three officers start their tooth-comb check of their huge ship. 1962 Amer. Jrnl. Physical Anthropol. XX. 128/1 The closely spaced incisors of lemurs seem to scrape the fur rather than comb it. The use of the term 'tooth comb' may be, therefore, objectionable. 1972 Times Lit. Suppl. 11 Aug. 948/1 A novel which has been picked over with toothcombs, in search of clues to 'The Mystery'. 1977 Lancet 20 Aug. 442/1 Decides whether the bid is contentious enough to be toothcombed by the mergers panel. 1962 J. GARDNER Garden of Weapons i. xi. 10 The four men and one woman trained after a toothcomb selection. 1981 Times 16 Feb. 14/4 Tooth combs are found today in tree shrews..as well as in the lemurs and lorises. 1977 Age (Melbourne) 18 Jan. 13/5 Who do you suppose paid the $50 billion difference? The tooth fairy? Hardly. You do. 1977 Rolling Stone 7 Apr. 45/4 Anyone who thinks they acted alone must also believe in Santa Claus, the Easter Bunny and the Tooth Fairy. 1978 J. WYLES Food for Sea 4, 46 Has ceased to believe in miracles at about the same age he stopped believing in the Tooth Fairy. 1858 P. L. SIMMONDS Dict. Trade Products 384/2 Tooth-glass, a toilet water-glass for washing the mouth. 1935 KIPLING Diversity of Creatures (1917) 411 That plate of the four lower ones in the tooth-glass there. 1976 Greener Human Factor v. iii. 186 The toothglasses were swathed in plastic. 1840 Outing (U.S.) Dec. 244/2 Some drank their champagne out of tooth-mugs. 1935 Discovery Apr. 114/1 To this communal tooth-mug debris and food particles get transferred. 1979 A. MORICE Murder in Outline ii. 77 Why not repair to..our rooms, where tooth mugs abound, and push the boat out? 1852 Amer. Railroad Jrnl. I. 607/1 (Advt.), Seidlitz powders, chloride of soda, chlorine tooth paste. 1966 A. SACHS Jail Diary vi. 62, I get up off the floor, fetch my toothpaste-tube. 1975 Listener 9 Jan. 44/3 Haldeman says to him: 'John, you ought to think about that, because once the toothpaste is out of the tube, it is awfully hard to put it back again.' 1978 F. KING Action xxxi. 105 Pinching at an exhausted toothpaste tube. 1818 Tooth-root [see *DENTARIA]. 1871 'MARK TWAIN' in Galaxy Aug. 284/2 He tendered me a tooth-wash atrocity

of his own invention. 1895 Army & Navy Co-op. Soc. Price List 726/2 Tooth-wash. 1949 E. POUND Pisan Cantos lxxx. 104 Pepitone was wasting toothwash.

b. *teeth-grinding adj.* (later example), *-ridge*.

1969 FABIAN & BYRNE Groupie (1970) xvi. 173 Teeth-grinding teenagers from Muswell Hill picking you up in Cortinas. 1928 J. C. WADE Phonetics of Eng. xiii. 117 T and d before r are articulated on the teeth, not on the teeth ridge. 1969 J. DERRICK Teaching Eng. to Immigrants iii. 127 The tip of the blade (i.e. the very front part) of the tongue is just behind the upper gums ('the teeth ridge').

tooth-brush. Add: **b.** *tooth-brush glass, holder, rack.*

1931 H. E. L. MELLERSH Salt of Earth vii. 159 Put them [sc. flowers] in the toothbrush glass or something till May can see to them. 1979 R. JAFFE Class Reunion (1980) i. 19 Go get your toothbrush glasses and we'll lock the door. 1971 T. Eaton & Co. Catal. Spring & Summer 199/2 Combination tumbler and tooth brush holder..nickel plated. 1979 M. MILLAR Murder of Miranda ii. 85 The chrome toothbrush holder was empty. 1926–7 Army & Navy Stores Catal. 1111/2 Bath bracket outfit consisting of tumbler ring, tooth brush rack. 1969 HURD & OSMOND Smile on Face of Tiger v. 164 [He] stood leaning against the toothbrush rack, cup in hand.

Hence **tooth-brushing** *vbl. sb.*

1920 A. HUXLEY Limbo 168 A foam of tooth-brushing. 1976 J. PHILIPS Backlash (1977) ii. 93 Go into the bathrooms... Time for tooth brushing.

toothily (tū·þili), *adv.* [f. TOOTHY *a.* + -LY².] In a toothy manner; so as to display the teeth.

1930 R. MACAULAY Staying with Relations xiii. 193 She stared at her master and smiled toothily. 1939 A. HUXLEY After Many a Summer i. 4 The chauffeur..slightly over-acting the part of an old-world negro retainer, bowed, smiled toothily. 1977 P. D. JAMES Death of Expert Witness iv. 190 A studio photograph of Miss Willard herself, young, toothily coy.

toothless, *a.* Add: **3. c.** Lacking the means of compulsion or enforcement; ineffectual. Cf. *TOOTH *sb.* 2 b (*b*).

1961 in WEBSTER. 1966 Federal Suppl. (U.S.) CCXLIV. 823/2 Congress might as well have legalized the closed shop as have enacted such a cynical and toothless provision. 1971 Nature 23 Apr. 486/1 Reasons for the failure include unenforceable and toothless laws,..foot dragging by local and state authorities, [etc.]. 1973 Guardian 16 Feb. 13 The EEC's social and economic committee..is a toothless organisation... Its views are not seriously taken into account. 1984 N.Y. Times Mag. 22 Jan. 46/4 If we do not do what we propose to do, we shall be reviled as toothless and irrelevant.

toothsomely, *adv.* (Earlier example.)

1871 'MARK TWAIN' Let. 25 Dec. in C. Clemens My Father Mark Twain (1931) 53 Gossip of any kind, and about anybody is one of the most toothsomely Christian dishes I know of.

toothy-peg. Add: Also **toospeg.** (Later examples.) Also used *joc.* in other contexts.

1921 'K. MANSFIELD' Let. 3 Feb. (1928) II. 91 When the time comes just put your toospeg brush, pyjamas and a collar..into a handkerchief. 1931 A. CHRISTIE Sittaford Mystery xxi. 171 [He] took his elephant's trotters and his hippopotamus's toothy pegs and all the sporting rifles and what nots. 1977 J. WILSON Making Hate vi. 71 It's those toothypegs, isn't it, my lovey, those naughty old toothypegs.

tooting, *ppl. a.* (In Dict. s.v. TOOT *v.*²) **1.** See also *rootin' tootin'* s.v. *ROOTING *ppl. a.*² 2.

2. *U.S. slang.* Used, usu. with preceding adv. or adj. (as *damn* or var.), as a strong affirmative or intensive.

1932 Amer. Speech VII. 338 You're damn tootin', emphatic affirmative. 1933 E. CALDWELL God's Little Acre i. 12 'After the albino, Pa?' Buck asked. 'You're durn tooting, son,' he said. 1952 B. MALAMUD Natural 36 You're plumb tootin' crazy. 1970 E. BERCKMAN She asked for It xi. 134 You're goddam' tootin' I'm on that again. Y'say I've been prying, you admit there's something to pry into. 1981 G. MCDONALD Fletch & Widow Bradley xviii. 72, I got pregnant, when you said I wouldn't... You tol' me a tootin' lie.

tootle, *v.* Add: **2.** *trans.* To play music on (a wind instrument). Also *transf.* and with music as direct obj. *colloq.*

1890 J. SERVICE Thir Notandums xiv. 99 Heralds clad in green tootled glorious musick frae their siller horns. 1895 G. MORTIMER Like Stars that Fall iii. 28 'There's no need for the cornet in this piece,' said Jenny. 'No, only Abrahams is so fond of tootling his bloomin' instrument,' said Larpenti. 1939 [see *SMOKER 4 b]. 1978 J. GALWAY Autobiogr. xiv. 164, I had tootled my flute to some purpose with Herbert von Karajan.

3. *intr.* To walk, to wander casually or aimlessly; usu. const. *along, around*, etc. Also *transf.* with reference to motor transport; *to tootle off*, to go, to depart. *colloq.*

1902 Cornh. Mag. July 102, I tootled down to Cooney's a half-hour before time. 1914 M. & J. FINDLATER Crossriggs xx. 149 Take that beast and stop all his work, feed him fat and let him sleep on the rug and tootle around the garden. 1918 Punch 3 Apr. 222 Well, I must tootle off now. 1951 J. B. PRIESTLEY Festival at Farbridge II. ii. 272 You're going to be tootling round to a lot of big houses. 1956 N. COWARD South Sea Bubble II. i. 52 It's getting late. ..It is time for me to tootle off home. *a* 1974 R. CROSSMAN Diaries (1975) I. 532 We had a real honeymoon holiday..tootling round in a Volkswagen which Helga Greene's villainous friend Johnnie in Heraklion had rented to us. 1978 E. O'BRIEN Mrs. Reinhardt 55 He would work for an hour or so and then tootle off. 1983 Listener 20 Oct. 31/3 Veteran car tootles down country lanes.

toots (tuts). *slang* (orig. and chiefly *U.S.*). Also **Toots.** [Prob. abrev. of *TOOTSY 2.] A woman, a girl; freq. used as a familiar form of address, esp. to a female.

1936 Amer. Speech XI. 275/2 Toots used to be used in families here and there as a nickname, or a term of endearment, the vowel sounded as in 'boots'... Is this term the ancestor of the present mode of address in 'O.K., toots', 'Hello, toots!' etc., the vowel shortened into that of 'full'? 1936 Mademoiselle Jan. 60 Everyone who isn't 'Toots' or 'Cookie' is 'Darling'. 1941 H. A. SMITH Low Man iii. 30, I..raised my hand in a clumsy wave and cried out: 'Hiya, toots!'.. I had called J. P. Morgan 'toots' to his face. 1946 E. LINKLATER Private Angelo xii. 143 'Hiya, toots,' repeated the Count. 'I like that. It is the felicitous expression of a young people who are making their own language.' 1951 J. B. PRIESTLEY Festival at Farbridge III. iii. 365 'Wasting it on you, Toots,' Ricky retorted respectfully. 1975 New Yorker 19 Dec. 33/2 'Hi, toots,' Ducky said in Donald's voice a few minutes later to a tiny girl. 1981 G. MARKSTEIN Revenge Game xv. 161 Maybe it's in his mind to come back for you and dig a quiet grave... How does that grab you, Toots?

Tootsie Roll (tu·tsi rǝᵘl). *U.S.* The proprietary name of a type of sweet or candy bar.

1925 Official Gaz. (U.S. Patent Office) 7 Apr. 19/1 The Sweets Company of America, Incorporated, New York, N.Y. Filed Oct. 14, 1924. Tootsie Rolls... Claims use since September, 1908. 1955 W. GADDIS Recognitions I. v. 177 And so brown. Like a tootsie roll. 1969 L. HELLMAN Unfinished Woman ii. 22, I bought a few Tootsie Rolls and a half loaf of bread. 1980 J. KRANTZ Princess Daisy xxiv. 421 Her Tootsie Roll brown eyes sparkling.

toot sweet (tūt swīt), *adv.* Also **toot and sweat, toots sweet**; *compar.* **the tooter the sweeter.** [Repr. colloq. anglicization of F. *tout de suite*.] Straightaway; promptly, quickly; freq. used as *imp.*

1917 A. G. EMPEY Over Top 311 'Toots Sweet.' Tommy's French for 'hurry up', 'look smart'. 1917 Punch 5 Dec. 389 (caption) Tommy (to inquisitive French children): 'Nah, then, alley toot sweet, an' the tooter the sweeter.' 1929 [see *FUCK v. 3]. 1942 'N. SHUTE' Pied Piper v. 109 Get them kids dressed toot and sweet—I ain't going to wait all night. 1959 R. POSTGATE Every Man is God xxiii. 216 'The tooter the sweeter' was an adjuration to do something, usually bringing a drink, more *tout de suite*, more promptly. 1967 Guardian 16 May 2/8 Your two brace of crocodiles[' eggs] have arrived—Yes. I'll get 'em incubated toot sweet. 1978 D. WILLIAMS Treasure up in Smoke xiii. 118 The Governor wanted him toot sweet this morning but he hasn't shown up.

tootsy, tootsy-wootsy. Add: **2.** Also **tootsie, tootsey-wootsey, tootsie-wootsie**, etc. A woman, a girl; a sweetheart; occas. applied to a male lover. Freq. as a familiar form of address. *slang* (chiefly *U.S.*).

1895 W. STEVENS Let. 23 July (1967) 6, I can be your own dearest tootsey wootsey. 1901 'H. MCHUGH' John Henry 88 One of the kind that's anxious to lead you away from your own tootsie-wootsie, in the hope that you may have a spare bunch of sweet talk you can hand her in the quiet. 1905 E. M. FORSTER Purple Envelope in Life to Come (1972) 48 'Well, she's not my idea of a tootsy!' said Howard, and clenched his criticism by a coarse and vapid jest. 1920 D. H. LAWRENCE Lost Girl vi. 114 Underneath the oak-tree nice and shady Calling me your tootsey-wootsey lady? 1930 Sat. Even. Post 5 Apr. 72/4 'Hello, tootsie,' Rusty Charley says. 1938 'E. QUEEN' Four of Hearts ii. 23 The future Mrs. Butcher wouldn't throw her tootsie, would she? 1952 B. WOLFE Limbo xvi. 239 'What's the matter, tootsie?' she whispered. 1952 Sat. Even. Post 1 Mar. 21/2 What about one of those tootsy-wootsies? 1968 G. DE FRAGA Murder at Cookout xv. 71 Don't bother to kiss me. Save that for the little tootsies who think you're as marvellous as you do yourself. 1979 'P. O'CONNOR' Into Strong City xx. 74 Two chicks. One for me... One of the hot-time tootsies.

top, *sb.*¹ Add: **II. 3. a.** Also *pl.*, mountain tops, high moorland, etc.

1930 L. G. D. ACLAND Early Canterbury Runs 1st Ser. viii. 190 A wedge-shaped block of ninety thousand acres of high tops, mostly bush-bound. 1948 A. PATON Cry, Beloved Country I. x. 65 He would tell him of..the mist that shrouded the tops above Ndotsheni. 1951 E. COXHEAD One Green Bottle ii. 45 Cathy saw the great Welsh tops at last..four great blue mountains grouped at its farther end. 1976 Lancs. Evening Post 7 Dec. 8/3 There's no collective name for these tops but I've always known them as the Troutbeck Fells. 1980 J. WAINWRIGHT Kill of Small Consequence xiv. 109 Up on The Tops the first snows of winter had already etched the dry-stone walls.

c. (Examples in *Billiards*.)

1896 W. BROADFOOT Billiards i. 51 McNeil..certainly played the 'top of the table' game better than any of his contemporaries. 1927 Observer 20 Mar. 29 Prior..is essentially an all-round player with a tendency to make the top of the table game his chief scoring medium.

d. In the war of 1914–18, with reference to the parapet of a trench; esp. in phr. (*to go*) *over the top* (at the start of an attack). Also *fig.*

1916 War Illustr. 9 Sept. 80/1 Some fellows asked our captain when we were going over the top. 1917 'CONTACT' Airman's Outings 184 When, at a scheduled time, the infantry emerge over the top behind a curtain of shells, the contact patrol buses follow their doings. 1923 Publishers' Circular 24 Nov. 703/1 If Canada, metaphorically speaking, 'goes over the top', it will be against the wishes of the rest of the Empire and against the wishes of her own authors and publishers. 1933 J. BUCHAN Prince of Captivity II. i. 154 Life's a perpetual affair of going over the top. 1962 [see *AUNTIE, AUNTY b]. 1972 S. HILL Strange Meeting 121 Armstrong went over the top with the first wave and was hit almost at once. 1976 F. WILLIS Buckingham Palace Connection ix. 179 'This is it, then.' 'Yep... Over the top and the best of luck.'

4. a. Also in slang phrases, as *to blow one's top*: see *BLOW v.¹ 24 i; *to be off one's top* (chiefly *Austral.*) = *to be off one's nut* s.v. NUT sb.¹ 7 b; *to do one's top* = *to do one's nut* s.v. *NUT sb.¹ 7 d. Also *up top*, with reference to brains, intelligence.

1916 C. J. DENNIS Songs Sentimental Bloke vi. 48 'E's fair orf 'is top wiv love. 1945 BAKER Austral. Lang. vi. 130 The state of being stupid is described variously as being off one's..tile, top or saucer. 1961 [see *DO v. 26 i]. 1973 Guardian 17 Oct. 18/5 Get Dad, you've got enough up top for both of us. 1977 W. KANNER Lying Figures iii. 33 Money little creature, bless her, not much up top if y'know what I mean. 1977 Shoot 18 June 22 (caption) Always does his top when he misses chances!

8. f. A circus tent. Cf. *big top* s.v. *BIG a. B. 2. orig. *U.S.*

1931 Amer. Mercury Nov. 354/2 Top, a tent. 1942 D. POWELL Time to be Born iii. 29 A perpetual rain cloud seemed like a circus top. 1959 Manch. Guardian 16 July 5/1 He supervises the erection of the 'top'.

III. 11. g. Also, of the lid of other kinds of container, esp. the metal-foil cover of a milk bottle, the colour of which may indicate the kind or quality of the milk, as *gold top, silver top*, etc. **h.** Also of a motor-car (chiefly *U.S.*); see *hard top* s.v. *HARD *a.* 22 a, *soft top* s.v. *SOFT *a.* 27. **j.** Now usu. a blouse or similar upper garment for wearing with a skirt, trousers, etc.; cf. *sun-top* s.v. *SUN *sb.*11 d, *tank top* s.v. *TANK *sb.*¹ 4, etc.

g. 1958 A. SILLITOE Saturday Night & Sunday Morning ii. 35 Screwing the top back on the flask. 1959, 1972 [see *milk bottle* s.v. *MILK sb. 9 a]. 1979 Dairy Mirror Nov. 8/3 (caption) Lisa Faulkner..displays the 10,000th Gold Top Milk Gymnastics Award Scheme double gold certificate. 1980 Ibid. Feb. 1/1 The retail price of a pint of ordinary silver top milk goes up from 15p to 16½p. 1981 J. BARNETT Firing Squad v. 48 What do you think this is, laddie? The top off a Fry's cocoa-tin? **h.** 1910 Sears, Roebuck Catal. 1143/2 Three-bow skeleton automobile top of heavy moroccoline. 1942 D. POWELL Time to be Born ii. 51 Ted would never put the top down when he drove. 1977 H. FAST Immigrants III. 199 They argued about putting up the top. **j.** 1922 JOYCE Ulysses 341 His little man-o'-war top and unmentionables were full of sand. 1949 N. MITFORD Love in Cold Climate II. i. 186 A jersey top, however Parisian, was obviously unacceptable for evening wear in high Oxford society. 1968 Daily Mirror 20 Aug. 9/2 And I got a couple of bright flowery roll-neck tops..and some super things for the beach, stretch bikini bottoms and loose towelling tops in hectic colours.

IV. 14. a. Also in Journalism, Broadcasting, etc., the leading position in a news bulletin, or the top of a column in a newspaper.

1973 L. HEREN Growing up Poor in London vi. 163 The first flashes were coming through on the attempt to get an abandoned ship in tow somewhere in the Atlantic... The story rated a top. 1979 'A. HAILEY' Overload I. iv. 23 On the radio,..a news bulletin. The item Nim had been waiting for was at the top.

b. Also *spec.* in Journalism. Cf. sense 14 a above.

1960 R. ST. JOHN Foreign Correspondent x. 195, I.. dictated a new 'top' for Sunday papers.

c. *ellipt.* for *top sergeant*, sense 32 below. *U.S. Mil. slang.*

1898 E. H. BLATCHFORD Let. 30 July (1920) 53 The 'top' said he wanted us to sign the pay-roll and be back at ten to-night. 1930 T. FREDENBURGH Soldiers March! ii. 12 The Top says he'll pass the word along. 1970 W. JUST Military Men iii. 95 Don't worry, Top.

15. a. Also, *at the top of one's form* (FORM *sb.* 16 a).

1933 A. POWELL From View to Death iii. 89 It had come at a time when he was not feeling at the top of his form. 1947 L. P. HARTLEY Eustace & Hilda vii. 138, I can't pretend that she was at the top of her form.

c. Now in various constructions, esp. *in(to) top*.

1925 Morris Owner's Man. 10 When changing gear up from first to second, or second to top, the clutch pedal should be pressed down. 1932 S. GIBBONS Cold Comfort Farm xvi. 217 They heard him change into top. 1953 [see *CHANGE v. 6 d]. 1958 [see *GEAR sb. 7 b]. 1970 N. FLEMING Czech Point (1971) viii. 107 Melanie rammed the car into top and kept up the acceleration.

d. *Bridge.* (*a*) Either of the two highest cards of a suit; (*b*) the best score made in the play of a particular hand.

1929 M. C. WORK Compl. Contract Bridge Gloss. 246 Tops, Aces and Kings. 1945 'S. J. SIMON' Why you lose at Bridge ix. 103 As the Clubs didn't break, and he took the Heart finesse to try and save something from the wreck, he went six down. A cold top for us. 1958 Listener 23 Oct. 669/2 To ask whether, at match points, East-West should try for Seven Hearts is like asking whether a golfer should play for a birdie or a bogey: it all depends on the state of the game. If they need a 'top' they take the chance. 1977 Hongkong Standard 12 Apr. 10/3 Romik was able to claim all 13 tricks for an outright 'top' on the hand.

16. Also in fig. phr. *on top of the world*, at the peak of well-being, prosperity, or elation; hence *top of the world* attrib. phr. (also with hyphens).

c 1920 D. HAMMETT in W. F. Nolan Dashiell Hammett (1969) ii. 19 A Samuels diamond puts you on top of the world! 1930 WODEHOUSE Very Good, Jeeves! ix. 226 If ever a bird was sitting on top of the world, that bird was Bingo. 1946 E. S. GARDNER Case of Borrowed Brunette xi. 132 This time Gulling, with this new evidence making him feel he's sitting on top of the world, slapped my proposition right back in my face. 1962 D. FRANCIS Dead Cert vii. 79 His eyes were alight with that fantastic, top-of-the-world elation. 1978 D. DEVINE Sunk without Trace v. 51 Last time I spoke to Liz she was on top of the world. 1979 Guardian 12 Jan. 9/8 As Colt's say in their publicity handout: 'This top of the world feeling can now be reproduced in a factory, office or shop.'

17. b. (*d*) In gen. colloq. use (predicatively): the best. Freq. with *the*. orig. *U.S.*

1935 Motion Picture Nov. 41/1 Top Hat is tops—it has everything! 1937 R. STOUT Red Box xv. 249 Your conversation is an intellectual and esthetic delight. It's the tops. 1942 N. STREATFEILD I ordered Table for Six 243 He didn't go near your mother until he was the tops, so to speak. 1948 C. DAY LEWIS Otterbury Incident 94 Toppy is tops at spur-of-the-moment tactics. 1958 Punch 9 July 44/3 Cooney's Cassocks stand the test, Choosy Churchmen say they're best. Sure-fire sermons, never flops; Cooney's Cassocks are the tops. 1976 'W. TREVOR' Children of Dynmouth i. 36 'You're easily tops, lad,' Hughie Green was enthusing, putting an arm round his shoulder. 1979 L. MEYNELL Hooky & Villainous Chauffeur i. 14, I always looked up to him... I just thought he was the tops.

17*. *Particle Physics.* [An arbitrary choice of name.] The name of (a quark carrying) a possible sixth flavour, with a charge of $+\frac{2}{3}$. Freq. *attrib.*

1977 Sci. Amer. Oct. 74/2 The new quarks will apparently be called 'top' and 'bottom', the names being meant to suggest properties surpassing those of the up and down quarks found in ordinary matter. 1978 Nature 2 Feb. 407/2 Similarly if top quarks exist then 'naked top'—or 'topless' states will eventually be found. The prudish may care to note that t and b are said to stand for truth and beauty, rather than top and bottom, by some physicists. It is predicted that..top decays to bottom. 1980 J. S. TREFIL From Atoms to Quarks xii. 184 As I write (spring 1979), there is no evidence to indicate that particles made from a top quark have been seen. 1982 Sci. Amer. Mar. 64/2 The member of the top family that should be easiest to identify would be made up of a top quark bound to a top antiquark. 1984 Nature 12 July 97/1 Last week, the 80-strong collaboration..announced the discovery of the missing sixth quark, called *top*. Ibid. 97/3 The discovery of the *top* quark. Ibid., The discovery of *top*.

VI. Phrases. **21. a. on (the) top of**, (*a*) also, in addition to (further examples); (*b*) too close to; esp. *on top of one another*, in crowded conditions; (*c*) burdensome to, too much for; *to get on top of*: to overwhelm, harass, depress; (*d*) in control of.

1824 M. WILMOT Let. 5 Feb. (1935) 207, I came home hungry, took some hot tea on the top of a cold ice which I got there, got an indigestion. 1947 A. L. ROWSE Tudor Cornwall xvi. 434 There was little privacy, for they lived on top of one another. 1952 Chambers's Jrnl. May 267/1 Our work consisted, mainly, in safeguarding road convoys from attack by hostile tribesmen. By no stretch of the imagination could it have been termed exhausting, but it was always on top of you. 1952 M. ALLINGHAM Tiger in Smoke x. 167 This time there was..no faltering. He was on top of himself and them. 1955 —— Beckoning Lady iv. 55 None of us saw her until she was right on top of us. 1962 B. COBB Murder: Men Only ix. 109 Oh, Kitty, it's Thursday and I know we agreed, but how? With everything on top of me. 1965 New Society 11 Nov. 7/1 People..do not necessarily want to live 'on top of each other'. 1968 Listener 4 July 5/1 On top of all this there are the continuing constitutional negotiations. 1974 A. MORICE Killing with Kindness ii. 21, I didn't mean to be rude. It's all got so much on top of me that I don't know what I'm saying half the time. 1977 M. ALLEN Spence in Petal Park xxxiii. 158 He still lives in Downsea. Near enough for me to babysit but not so close that we're on top of him. 1977 'A. YORK' Tallant for Trouble vi. 87 He really felt he was getting on top of the situation. 1981 Sunday Express Mag. 2 Aug./33 Lord Mackan has had a busy programme of special ceremonial events on top of his normal Household chores.

b. at the top: in a position of power or authority. Cf. sense 14 in Dict. and *room at the top* s.v. *ROOM sb.¹ 2 a.

1936 G. B. SHAW Millionairess I. 145 That's what keeps him at the top in the city. 1962 J. BRAINE (title) Life at the top. 1979 A. FOX Threat Warning Red iii. 41 The machines..hadn't made life easier at the top.

c. over the top, beyond reasonable limits, too far, into exaggeration.

1968 C. WATSON Charity ends at Home x. 129 For instance, you said at our first interview that your wife got so worked up about some things that she was in danger of going 'over the top', as you put it. 1974 Times 6 Mar. 2/8 We agreed to give every possible support to the Labour Government, including not going over the top with wage claims. 1981 'D. JORDAN' Double Red i. 11, I could summon less and less response to Magnus's more rhetorical flights: so here we are, going over the top again, I was thinking.

22. b. off the top of one's head and varr., impromptu, without consideration, superficially; hence *top-of-the-head* attrib. phr.

1939 H. L. ICKES Secret Diary (1954) II. 718 He was impetuous and inclined to think off the top of his head at times. 1959 'E. MCBAIN' 'Til Death xiii. 169 The jokes.. took on an ad lib quality, each prankster..coming up with top-of-the-head advice on the proper hotel-room behaviour. 1967 Listener 20 Apr. 518/2 His [sc. Bertrand Russell's] political activities..are not something that is coming out of the top of his head, they are coming from his nature. 1972 J. RIPLEY My Word you should have seen Us 159 You're talking out of the top of your head, mac. 1977 W. J. BATE Samuel Johnson (1978) xi. 173 London.. seems breezy, as if written off the top of the head..; it lacks the sublime moral elevation of the Vanity. 1981 C. DEXTER Dead of Jericho xxviii. 160 A bit of bread-and-butter investigation was worth a good deal more than some of that top-of-the-head stuff.

VII. 26. *top buggy* (earlier examples), *-wagon* (earlier examples).

1849 Knickerbocker XXXIV. 266 An ordinary 'top-buggy' wagon. 1866 'MARK TWAIN' Lett. from Hawaii (1967) 45 His 'turnout', as he calls a top buggy that Captain Cook brought here in 1778. 1852 C. A. BRISTED Upper Ten Thousand 208, I have a top-wagon. 1880 W. WHITMAN Daybks. & Notebks. (1978) III. 639 Many queer old one-horse top-wagons.

29. (Later examples.) Also, of high standing. See also *TOP PEOPLE *sb. pl.*

c 1926 [see *RUNNING vbl. sb. 2 e (b)]. 1936 Publishers' Weekly 21 Nov. 1965/2 Publishers involved in recent top-seller crises. 1938 E. AMBLER Cause for Alarm xiii. 213 The prisons would be overflowing..and most of the top men would be with them. 1939 Supervision Feb. 1/1 The whole related circle which reaches from top management down to the worker. 1945 Richmond (Va.) Times-Dispatch 25 Oct. 12/1 The A-bomb has aroused so much interest a complex technical tome on that subject is now a top-seller. 1958 Observer 3 Aug. 5/1 Grouse-shooting, it must be conceded, is the top sport. 1965 Mod. Law Rev. XXVIII. v. 587 Corporations, seemingly, will be liable for the acts and omissions of 'top management'. 1972 L. DEIGHTON Close-Up viii. 166 They are going to spend a hundred thousand dollars just on this one story. They'll get a top outside photographer to do it. 1981 R. SAMUEL East End Underworld xi. 133 Cockney Cohen was the favourite; he was regarded as the top man of the two [boxers]. 1982 Lakeland Echo 18 Mar. 6/7 Peter Frankl was going to play there..on May 1, and other top performers were lined up for the future.

31. a. *top lacing*; *top-feeding, -opening* adjs.

1925 F. SCOTT FITZGERALD Great Gatsby i. 8 He seemed to fill those glistening boots until he strained the top lacing. 1926–7 Army & Navy Stores Catal. 409/2 Top-opening handbag. 1933 Sun (Baltimore) 20 Apr. 5/6 Some 200,000 of the American top-feeding minnow species Gambusia were dumped into some of the ponds. 1963 Which? 6 Feb. 36/1 The chest top-opening freezer.

31*. In sense 'highest or first'. **a.** With nouns forming attrib. phr., as *top class, quality, rank*, etc. See also *top drawer, flight*, sense 32 below.

1948 J. TOWSTER Political Power in U.S.S.R. III. xiii. 318 'Stakhanovites', that is, top-efficiency workers. 1950 N.Y. Times 20 Apr. 1/3 Virtually every top bracket job.. could be filled from the proposed register. 1950 Times 23 May 5/6 Time and again one reads..of top grade British films which will never be shown here at all. 1953 Newsweek 30 Mar. 81/2 Higgins-built mine sweepers..became the top-priority ships on the Navy's program. 1959 Times 29 Oct. 2/2 Position calls for top-calibre executive with experience of marketing. 1960 Farmer & Stockbreeder 9 Feb. 74/3 These..would only be interested in top-quality products tailor-made to suit their demand. 1960 Times 12 July 13/4 It isn't only the field events that are a poor show at top-class athletics meetings. 1961 Lancet 9 Sept. 598/1 We have very few top-rate managers. 1962 L. DEIGHTON Ipcress File i. 15 It makes eight top rank Disappearances in..six and a half weeks. 1972 J. AIKEN Butterfly Picnic x. 190 [He] is doing forty years in a top-security prison for handing over state secrets. 1973 Country Life 29 Nov. 1773/1 No champagne is made exclusively from top-price grapes. 1975 G. ST. GEORGE Proteus Pact i. 36 An urgent matter, a top-priority project. 1977 National Observer (U.S.) 8 Jan. 7/1 They conclude that there are no significant differences in intellectual and social development between young children reared at home and those placed in a top-quality day-care center. 1978 K. HUDSON Jargon of Professions v. 122 X is a top-class product manager because his father and mother were top-class product managers. 1979 Jrnl. R. Soc. Arts July 504/2 Something should be done to encourage a really efficient and top-grade display of arts and crafts. 1982 Lakeland Echo 18 Mar. 6/4 Special attention has been paid to acoustics and lighting so that really top-rank artistes can be persuaded to play there.

b. Adverbially with adjs. or ppl. adjs., as *top-ranking, -rated, -secret*, etc.

1936 Time 19 Oct. 67/1 Adapting a story which is to be played by four top-ranking film personalities. 1944 Top secret [see *CLASSIFIED ppl. a. c]. 1946 KOESTLER Thieves in Night 194 Turning to the urgent blue and so to the top-urgent red tray. 1958 People 4 May 19/7 Top-rated American Davey Moore said yes. 1960 Farmer & Stockbreeder 22 Mar. 83/3 Top-priced bull at Hereford last week was Haven Possible. 1962 Guardian 13 July 8/1 Drambuie..ranks with Benedictine and Cointreau among the world's five top-selling digestifs. 1975 Listener 17 July 69/1 Top-paid people should agree to limit their incomes. 1976 Billings (Montana) Gaz. 16 June 3-c/2 The victory by the third-rated Hurons left top-rated Arizona State one defeat from elimination. 1976 Scotsman 25 Nov. 14/5 A commercial paper nowadays would have to be less 'Left-of-centre'..to be read by the top-earning businessmen and stockbrokers who justify expensive advertising. 1976 H. WILSON Governance of Britain iv. 92 The Churchill and Macmillan appointees inevitably had the same access to secret and top-secret documents as any civil service appointee. 1978 N.Y. Times 30 Mar. D22/1 In a postponed first-round match, top seeded Vitas Gerulaitis.. defeated Ray Moore. 1978 Observer (Colour Suppl.) 9 Apr. 30 These two were 'top rankin'' gunmen in the ghetto for Jamaica's two main political parties. 1980 Washington Star 17 Dec. E2/6 Mississippi State jumped into national prominence with its big win over top-ranked Alabama.

32. top banana *Theatr. slang* (chiefly *U.S.*), the leading comic in a burlesque entertainment; also *fig.*; **top board** *Chess*, the principal player of a team in a tournament; **top box**, on a motor-cycle, a carrier box for baggage, etc., placed on top of the cycle behind the saddle (as opp. to panniers at the sides); **top brass**: see *BRASS *sb.* 2 e; **top-coat**, (*a*) (earlier example); (*b*) any of the finishing layers of paint applied after undercoat; **top copy**, the original typescript of a document, of which the under-sheets are carbon copies; also *ellipt.*; **top-cut**, reduction of the strength of the higher-frequency components of a signal; **top cutter** *U.S. Mil. slang* = *top sergeant* below; **top cymbal** *Mus.* = *ride cymbal* s.v. *RIDE *sb.*¹ 7; **top dead centre** (see quot. 1978); **top deck**: see *DECK *sb.*¹ 3 d; **top dollar** *N. Amer. colloq.*, a high price; **top-down** *a.*, (*a*) *Computers*, working from the top or root of a tree towards the branches (with or without backtracking); (*b*) that proceeds from the top downwards; authoritarian, hierarchical; occas. as *adv.*; (*c*) (of planning or design) starting with the overall structure and going on to successively more detailed parts of it; **top drawer**: in *fig.* use freq. with reference to social standing (later example); also *attrib.* or as *adj.*, first-class, of the highest level; **Top End** *Austral. colloq.*, the Northern Territory of Australia; hence **Top-Ender**; **top fermentation** *Brewing*, a process in which the yeast rises to the surface during fermentation; **top flight**, the highest rank or peak of excellence; also *attrib.* or as *adj.*; hence **top-flighter**; **top gear**, (*b*) also *fig.*, a fast pace, full speed; **top-graft** *v.*, substitute for def.: = *top-work* vb. s.v. *TOP *sb.*¹ 32; hence **top-grafting** *vbl. sb.*; **top hand** *N. Amer. colloq.*, a cowboy who is an experienced or first-rate ranch-worker; also *fig.*; **top kick** *U.S. Mil. slang* = *top sergeant* below; also **top kicker**; **top light**, a pane of glass affording illumination from overhead; a skylight; hence **top-lighted, top-lit** *adjs.*; **top loader**, a machine or device designed to be loaded from the top; opp. *front-loader* s.v. *FRONT *sb.* (and *a.*) 14; hence **top-loading** *vbl. sb.*; **top minnow**, a small, often brightly coloured, fish belonging to the family Cyprinodontidæ or Poeciliidæ; **top notch** (*attrib.* examples); **top of the bill** *Theatr.*, the chief place on a bill of entertainment; also (with hyphens) *attrib.* and *fig.* (cf. *to top the bill* s.v. *TOP *v.*¹ 16 a); **top-of-the-line** *a.* (chiefly *U.S.*), designating a commercially produced commodity that is the best, most expensive or luxurious, etc., of its kind; **top of the milk**, the cream that rises to the top of milk when left undisturbed; **top of the pops**: see *POP *a.* (*sb.*⁸) 1 b; **top-score** *v. intr. Cricket*, to make the greatest number of runs of an innings; hence **top scorer**; **top sergeant** *U.S. Mil. slang*, first sergeant; **top-slicing** *vbl. sb.*, (*a*) *Mining*, a method of working in which successive slices up to 12 feet thick are mined from the top of an ore body, working downwards, the material overlying each slice being made to cave after its completion; so **top-slice** *v. trans.*, to work in this way; (*b*) a method of assessing income- or surtax chargeable on a lump sum by averaging it out over the years for which it has accrued and charging tax accordingly; **top-soil**, the cultivatable) surface layer of the soil, as distinct from the subsoil; in *Archæol.*, the soil covering a site being investigated; **top soldier** *U.S. Mil. slang* = *top sergeant* above; **top-spin** = *OVERSPIN *sb.*; see also sense 19 in Dict.; also *fig.*; **top-spinner** = *overspinner* s.v. *OVERSPIN *sb.*; also **top-spun** *a.*; **top-stitch** *v. trans.*, to make a row of stitches on (the right side of a garment or other piece of sewn work), usu. as a form of decoration; so **top-stitched** *ppl. a.*; **top-stitching** *vbl. sb.*; **top table**, at a formal dinner, the table at which the chief guests are placed; also *fig.*, esp. in *Pol.*; **top-tail** *v.*

(example); **top ten** *Popular Music*, the first ten tunes or gramophone records in the popularity charts (*CHART *sb.* 3 c) at a particular time; also *transf.*; similarly **top twenty**, etc.; **top view** = *plan view* s.v. *PLAN *sb.* 6; **top-water**, (*b*) as *adj.* (of a lure) that floats on top of the water; **top-work** *v. trans. Horticulture*, to replace part or all of the top of (a fruit tree) by grafts of another variety; so **top-worked** *ppl. a.*, **-working** *vbl. sb.*

1953 BERREY & VAN DEN BARK Amer. Thes. Slang (1954) § 583/12 *Top banana*, the burlesque comedian who gets top billing. 1956 Picturegoer 21 July 29/3 'Top banana' is the comic-in-chief of a burlesque show. 1974 Time 21 Jan. 53/3 Dentsu Advertising Ltd..has become the new top banana of world-wide advertising. 1978 N.Y. Times 29 Mar. C 27/1 Miss Burnett is a..very, very funny woman. She is a superb top banana. 1910 British Chess Mag. XXX. 463 A top-board winning seven times successively might find himself temporarily or unjustly displaced in the ninth match. 1976 Milton Keynes Express 28 May 55/7 The competition was won..by county top board Norman Stephenson. 1976 Eastern Daily Press (Norwich) 19 Nov. 5/6 (Advt.), 1975 Yamaha FS1E, excellent condition, low mileage, winkers, topbox. 1804 F. ASBURY Jrnl. 18 Apr. (1821) III. 136, I had heedlessly thrown off my top-coat for a few hours, and caught cold. 1959 Sears, Roebuck Catal. Spring & Summer 1182/3 House paint undercoat... Insures longer wear..and a smoother appearance of top coat. 1977 Custom Car Nov. 26/3 After three undercoats and four topcoats of Dulux Golden Yellow Coach Paint,..Chris describes the finish as 'not bad'. 1919 H. ETHERIDGE Dict. Typewriting 68 If an error is made whilst taking carbon copies, it is a lengthy process to make the correction, as, in erasing the original or top copy, the pressure of the eraser will make a bad smudge on the copies. 1967 L. MEYNELL Mauve Front Door vii. 89 If you could possibly do a top and two carbons of these notes. 1979 G. MITCHELL Mudflats of Dead II. xvi. 162 The bill is for typing a top copy and two carbons of a book. 1957 Practical Wireless XXXIII. 706/1 Simple switched bass-boost and top-cut compensation is provided by S1 and S2 respectively. 1962 A. NISBETT Technique Sound Studio ii. 35 There is no worse microphone defect..for emphasizing any slight sibilance which may be present (and you cannot get rid of it by top cut if the emphasis lies in upper middle peaks). 1917 Editor 13 Jan. 33 *Top cutter*, first sergeant. 1930 T. FREDENBURGH Soldiers March! 279 It's a damn good book. Lots of swell dope for Top Cutters in it. 1948 Record Changer July 12/1 The top cymbal has become the main tool of the bebop drummer. 1956 M. STEARNS Story of Jazz (1957) xviii. 234 Clarke made the single right-hand 'ride' or 'top' or 'front' cymbal the rhythmic center... The top cymbal was the only regular and continuous sound made by the drummer. 1924 E. C. M. SHEPHERD Motor Car ii. 23 When a piston is at the top of its stroke..on the point of changing from an upward motion to a downward motion, it is said to have reached top dead centre. 1978 Vocab. Reciprocating Int. Combust. Engines (B.S.I.) (1979) 7 *Top dead centre*, dead centre when the piston is farthest from the crankshaft. 1970 Toronto Daily Star 24 Sept. 15/9 He said Sault residents 'are paying top dollar for a second-rate flight'. 1978 M. PUZO Fools Die xvi. 170 A lot of those guys..had paid him top dollar to buy their enlistment in the six months' program. 1964 Communications Assoc. Computing Machinery VII. 80 [What are your general views regarding the merits of doing the syntax analysis from the 'top down' as against the 'bottom up'?] Ibid., My analyzer is bottom-up and Warshall's is top-down. 1969 R. BLACKBURN in Cockburn & Blackburn Student Power 178 Its officials are robbed of all real initiative by the requirements of rule obedience, top-down control and hierarchy. 1972 O. J. DAHL et al. Structured Programming p. v, Structured programming principles can be equally applied in 'bottom-up' as in 'top-down' program design. 1975 Nature 16 Oct. 548/1 Many somewhat different algorithms are properly classified as top-down parsing algorithms. 1976 Eastern Daily Press (Norwich) 16 Dec. 8/6 You take the familiar top-down view, pointing out the various problems which always beset constitutional changes. 1977 N.Y. Rev. Bks. 13 Oct. 28/1 The students were intent on showing that..'every decision was made top-down by the power structure'. 1979 Personal Computer World Nov. 74/2 The approach we shall take in programming this problem is known as 'Top-Down Design'. 1980 Times 9 Feb. 17/2 The emphasis in the latest public spending round has shifted from 'bottom up' planning, where spending totals are built up from the individual elements in the programmes, to 'top down' planning. 1920 R. MACAULAY Potterism I. i. 10 The Potter family, however respectable now, wasn't really 'top-drawer'. 1946 Sun (Baltimore) 10 Oct. 12/2 The National Bureau of Economic Research, a top-drawer group of research economists. 1958 'A. BRIDGE' Portuguese Escape iii. 42 The composed decision that somehow had so much distinction. 'She *is* out of the top drawer, isn't she?' 1959 Vogue Dec. 61 Vedonis also make ladies' underwear, sweaters, nightwear and bed jackets. They're top-drawer because everything about Vedonis is so good. 1960 Guardian 25 June 4/4 The word 'Hampstead' with all its associations of top-drawer socialism. 1976 Time 20 Dec. 10/3 Tanaka and four other Diet members linked to Lockheed's scheme to buy top-drawer influence and stimulate sales with more than $2 million in bribes were re-elected by loyal rural constituencies. 1977 C. MCCULLOUGH Thorn Birds vii. 149 Quite respectable, socially admissible, but not top drawer. Never top drawer. 1933 F. E. BAUME Tragedy Track 93 She..left again for the more human..regions of the Top End, where at least one could drink fresh water occasionally. 1969 Northern Territory News (Darwin) Focus '69 81/1 Beef roads..will criss-cross the Top End with 665 miles of good bitumen. 1941 C. BARRETT Coast of Adventure 14 The old Top-ender drank beer, which, to the men up there, is more desirable than iced nectar is to gods. 1961 T. RONAN Only a Short Walk 52 Any 'Top-Ender' who wanted..a tip for the races..went to Billy. 1902 Encycl. Brit. XXVI. 367/1 The system is called *top-fermentation*, because the type of yeast employed develops on the surface of the liquid, forming the 'head'. 1905 [see *bottom fermentation* s.v. *BOTTOM sb. 19]. 1974 Encycl. Brit. Macropædia III. 161/2 Top fermentations are usually carried out using selected strains of *Saccharomyces cerevisiae*, botanically identical with bakers' yeast. 1939 R. CHANDLER Big Sleep xx. 150 Top-flight racketeers have business brains. 1958 Times 19 July 3/4 Lifting himself into the top flight of English batsmen. 1959 J. THURBER Years with Ross viii. 138 Reporters.. joined the staff, all of them top flight. 1967 Punch 20 Dec. 951/3 Good though it is, it isn't top-flight. 1979 E. NEWMAN Sunday Punch xv. 123 Every successful fighter when he reaches a point just below the top flight. 1981 Beautiful Brit. Columbia Fall 37 The University of Victoria, with its new, acoustically exuberant auditorium, is the scene now for many top-flight performances. 1950 'M. INNES' Hare sitting Up II. ii. 52 He lives on his nerves, as so many top-flighters do. 1959 J. DEMPSEY Championship Fighting v. 21 If you boast only nine professional fights, there's little danger of your being tossed in with a top-flighter or a champion. 1932 E. BOWEN To North xxii. 235, I can't live at top gear. 1973 Nature 13 Apr. 440/1 The coal industry must now put its research and development programmes into top gear. 1912 F. A. WAUGH Beginners' Guide Fruit Growing i. 13 Trees for top-grafting may be of almost any age. 1975 W. E. SHEWELL-COOPER Compost Fruit Grower viii. 118 Many of the trees consist of quite unsuitable varieties... It is, therefore, worth realising that top-grafting methods may be adopted which will convert one variety into another. 1912 'B. M. BOWER' Flying U Ranch 201 We can both safely consider ourselves top-hands when it comes to lying. 1955 R. P. HOBSON Nothing too Good for Cowboy i. 12 It will be impossible to line up enough top hands to carry on. 1972 T. A. BULMAN Kamloops Cattlemen iii. 19 They were all top hands with either saddle or work horses. 1918 J. E. RENDINELL Diary 28 Mar. in One Man's War (1928) viii. 63 The old top-kick would make a running dive for the dugout. 1976 L. DEIGHTON Twinkle, twinkle, Little Spy vii. 70, I was a gunner, nineteen—youngest top-kick in the group. 1979 Arizona Daily Star 22 July A8/1 The president's appointment of Hamilton Jordan as the White House topkick. 1919 L. L. LINCOLN Company C, Eleventh Engineers 8 Veeder was our first top-kicker. 1843 J. BALLANTINE Gaberlunzie's Wallet ix. 199 The speck of sky overhead looked not larger than a common top-light or cupola. 1873 Young Englishwoman July 342/2 The top-lights were..removed..and whitewashing and painting were done. 1924 GALSWORTHY White Monkey II. ii. 131 A high room with rafters and a top light, and lots of pictures. 1972 P. DIAMAND in D. Sutton Lett. R. Fry I. 60 On the top floor was Roger's studio... It had a top light. 1911 W. J. LOCKE Glory of Clementina Wing xxiv. 374 The room, spacious and top-lighted, was converted into a studio. 1932 F. L. WRIGHT Autobiogr. II. 152 The top-lighted interior created the effect of a great official family at work in day-lit, clean airy quarters. 1962 Times 16 May 5/5 The great, top-lit room of the Whitechapel Gallery. 1979 Jrnl. R. Soc. Arts CXXVII. 655/1 The new galleries should be on the same level as the old and top-lit by natural light. 1968 Which? May 149/1 This machine is a top loader, but has a horizontal stainless steel drum—you have to lift the top lid before being able to open the doors of the wash drum. 1976 Gramophone Dec. 1084/3 Included in the 1977 Tandy catalogue..are..two new stereo cassette decks with Dolby—a front-loader.. and a top-loader. 1978 Nature 18 May p. xviii/3 The Sartorius 3802MP electronic balance is a toploader of large capacity and high readability. 1976 CB Mag. June 59/1 (Advt.), And top loading eliminates vehicle body obstructions, a common problem for base loaded antennas. 1884 Bull. U.S. Nat. Museum No. 27. 471 *Gambusia patruelis*..Top Minnow..Southern United States, from Virginia to Texas. 1962 K. F. LAGLER et al. Ichthyology vi. 180 The mouths are superior in most of the topminnows. 1900 Billboard 29 Dec. 8/1 The last is a top-notch figure, and it is reached no oftener than can be helped. 1928 Amer. Speech IV. 244 Some successful criminals escape getting a monicker, for they, especially top-notch con men and syndicate members, think it adds 'class' to be without one. 1912 Music Hall & Theatre Rev. 7 Mar. 157/1 The divided 'top' of the bill happens with these two artistes. 1933 P. GODFREY Back-Stage xviii. 222 The London theatre queues provide a great variety of performances. At the top of the bill are a few well-organized teams of strolling players. 1965 Times Lit. Suppl. 25 Nov. 1047/4 The..top-of-the-bill entertainer. 1963 Economist 19 Oct. 301/1 The top-of-the-line sporty version [of a car]. 1981 Sci. Amer. Feb. 4/1 (Advt.), The new, top-of-the-line HP 3000 Series 44 computer has up to double the throughput power and memory size of its predecessor. 1942 C. SPRY Come into Garden, Cook v. 51 Make a mixture of tomato sauce..and a little 'top of the milk' cream. 1958 Listener 21 Aug. 287/2 Serve hot or cold, with cream or top of the milk. 1979 A. PARKER Country Recipe Notebk. viii. 103 The milk..pasteurized..has no 'top of the milk'. 1960 J. FINGLETON Four Chukkas to Australia 29 Huntington..top-scored with 73. 1977 World of Cricket Monthly June 26/3 Mohsin Khan batted well to top-score with 55. 1860 Baily's Mag. Aug. 367 The top scorer for the Midland was Mr. J. H. Marshall. 1976 Milton Keynes Express 30 July 41/1 Arnold Mann was top scorer with a patient knock of 24. 1898 J. BOWE Diary 2 June in With 13th Minnesota in Philippines (1905) 12 The top sergeant went around with a lantern. 1969 I. KEMP Brit. G.I. in Vietnam vii. 150 My immediate superior was First (or Top) Sergeant Rutledge, a dour and somewhat autocratic professional soldier in his early forties. 1905 IHLSENG & WILSON Man. Mining (ed. 4) I. iii. 79 (heading) Top-slicing and caving. Ibid. (caption) A system of top-slicing the ore. 1963 Economist 23 Mar. 1141/2 Other taxpayers..deserve some form of relief by 'top-slicing'. 1973 L. J. THOMAS Introd. Mining vi. 209 Top slicing is more suitable for large horizontal deposits. Ibid., Small pillars that could be top sliced are more likely to be recovered by cut and fill methods or to be abandoned. 1983 Sunday Tel. 5 June 28/7 If the recipient pays tax above the basic, 'top slicing' relief is provided to mitigate the effect of taxing the whole gain in one year. 1836, 1850 Top-soil [see *ENCALLOW sb.]. 1868, 1904 [in Dict. s.v. TOP sb.¹ 28]. 1967 Antiquaries Jrnl. XLVII. 188 In 1965 the stripping of the turf and topsoil from the rampart defences..exposed the top of the wall and an internal tower. 1975 J. G. EVANS Environment Early Man Brit. Isles vi. 128 Chalk waste bringing about the burial of topsoil and the destruction of what may have been valuable pasture or arable land. 1926 ANDERSON & STALLINGS What Price Glory? in Three Amer. Plays I. 10 I'm the new top soldier here. 1935 Our Army Nov. 39 Top Soldier Rawhide was sitting in the NCO club. 1913 Daily Mail 7 July 9/2 A good straight ball, with top spin, that comes off the ground very quickly. 1934 Punch 7 Feb. 141/1 She has thrown her husband out of the house sixty-one times, but he always returned. It looks as if she put too much top-spin on him. 1977 Time 4 July 10/3 Guillermo Vilas,..winner of the French Open last month, never could get his big topspin game going on grass. 1980 Times Lit. Suppl. 12 Sept. 983/5 Such existentialist propositions..sound pedestrian when summarized. The kind of intellectual top-spin required to give them philosophic solemnity is supplied. 1921 P. F. WARNER My Cricketing Life x. 194 A. R. Littlejohn..bowled an occasional top spinner which came very quickly off the ground. 1975 Times 13 Aug. 6/8 Intikhab..beat him with a top-spinner and hit his middle stump. 1969 New Yorker 14 June 44/3 He just can't hit a heavily top-spun backhand. 1977 Sunday Times (Perth, Austral.) 16 Jan. 11/4 The large crowd..reserved their warmest applause for some Wilkinson top-spun forehand lobs. 1960 Lebende Sprachen V. 35/3 *Top stitch*, Steppstich, steppen. 1964 McCall's Sewing vii. 100/2 Faced edges should be top-stitched to keep them flat. 1976 N. C. ANDERS Appliqué Old & New v. 104 Bind edge of each potholder with double fold bias tape. Cut two 2¼" strips for loops. Topstitch edges. Attach to potholder. 1934 A. L. HIRD Needlework & Dressmaking v. 100 List of seams..plain lapped = top stitched. 1975 New Yorker 17 Nov. 138/2 Bottega Veneta has some splendid wrist-length styles..in topstitched pigskin. 1947 C. TALBOT Compl. Bk. Sewing xxi. 145/1 Top-stitching is the frank use of stitching on the outside of a dress, suit or coat to emphasize lines that are important in the design. 1979 Tucson (Arizona) Citizen 20 Sept. 2B/1 Remove any top-stitching to 4 inches above the 'new' hemline. 1964 Guardian 7 Oct. 10/1 (heading) At the top table in Washington. *a* 1974 R. CROSSMAN Diaries (1976) II. 125 We found ourselves at the top table. I was sitting opposite the Bishop of London and next to the wife of a City alderman. 1977 'J. LE CARRÉ' Hon. Schoolboy I. viii. 191, I have a standing instruction..to repair our American liaison... 'To get us back at the top table.' 1983 Daily Tel. 1 Mar. 16/4 A late guest [at the St. David's Day banquet] will be Simon Hughes, Liberal victor of Bermondsey—too late to get a place on the top table. 1839 Knickerbocker XIII. 385 'There she top-tails! there she blows!' added he,..after taking a long look at the sporting shoal. 1958 J. ASMAN in P. Gammond Decca Bk. Jazz xiv. 174 Traditional jazz records vie with the accepted 'pop' Top Ten in selling power. 1960 News Chron. 7 May 3/5 Buxton will have to change..to make the tourist top ten. 1979 E. H. GOMBRICH Ideas & Idols 157 It [sc. a Beethoven Quartet] will never belong to the top ten. But it does belong to the canon. 1981 R. D. EDWARDS Corridors of Death v. 22 One of the country's top ten management whizz-kids. 1959 'F. NEWTON' Jazz Scene (1960) xiii. 236 Jazz has until recently simply not been big business in Britain, in the terms in which those who prepare records for the 'hit parade' of the 'top ten' or 'top twenty' think of it. 1962 Listener 20 Sept. 451/3 TAM puts the repeats of *Steptoe and Son* in the top twenty week after week. 1982 Daily Tel. 15 Apr. 16/6 All we need now is a royal baby named George.., and the name may be back in the top 20 once more! 1895 T. S. LAWLEY Lessons in Woodwork Drawing 10 The top view of a penny..placed on a table will be a circle. 1912 V. C. GETTY How to Read a Drawing i. 8 As we were..looking at the top of the object, this view would be known as the top view, or plan view. 1953 A. C. PARKINSON Pictorial Drawing for Engineers vi. 44/1 We commence by drawing a true-shape top view or plan view of the object. 1945 Richmond (Va.) Times-Dispatch 21 Sept. 18/2 It is well to try your top-water lures first, and if they fail, then try the under-water varieties. 1980 Hunting Ann. 1981 36/2 Rather than look for a long stick or get clothing and boots wet and muddy, the hunter can use a multihooked topwater lure and cast for his bird. 1883 Maine Agric. Rep. XXVI. 342 The Bourassa..does well top-worked on a strong stock, and then produces bountifully of apples. 1910 PADDOCK & WHIPPLE Fruit-Growing in Arid Regions ix. 150 It seldom pays to top-work any crab. 1968 Punch 27 Mar. 466/2 Though apples can be 'top worked' by grafting another kind on the sawn off branch ends, no peach will stand this treatment. 1934 WEBSTER, Top-worked. 1974 Country Life 28 Nov. 1660/1 Topworked trees, that is those which are grafted at the top of a standard stem and trunk. 1897 L. H. BAILEY Princ. Fruit-Growing v. 235 Some persons have proposed to sow seeds in the very spot where the trees are to stand, and thereby to raise stocks for top-working without transplanting them. 1946 Nature 28 Dec. 941/2 There is a particularly good chapter on top-working and frame-working, but that on pruning might have been improved.

top, *sb.*² Add: **4.** *top-spinning* sb. (examples), *-string* (example).

1964 Catal. National Mus. Kuala Lumpur 3/1 Dioramas present aspects of Malay dances, Kelantan top spinning, [etc.]. 1979 Arizona Daily Star 5 Aug. B 5/3 For relaxation, the brothers have taken up juggling, motocross bike-riding and top-spinning. 1855 MRS. GASKELL Lizzie Leigh & Other Tales 247 He had been the..Robin Goodfellow of the neighbourhood..whose top-strings were always hanging in nooses to catch the unwary.

top, *v.*¹ Add: **II. 3. b.** *to top and tail*, to wash the face and bottom of (a baby or small child); also *absol.*; hence *top-and-tail* attrib. phr. *colloq.*

1924 H. DE SÉLINCOURT Cricket Match ii. 22 She topped and tailed each small boy with the same rubber sponge. 1931 P. W. YEOMANS Happy Motherhood vii. 61, 5.50 to 6.20 p.m.—Top-and-tail wash, and feed baby. 1964 Guardian 24 June 6/2 Freda showed me how to top and tail (which is done on the lap because these babies do not get enough cuddling). **1983** *Woman's Weekly* 8 Jan. 53/3 There is no need to bath your new baby more than twice a week, 'topping and tailing' on the other days.

6. (Earlier example.) Also simply, to kill (someone); chiefly *refl.*, to commit suicide.

1718 C. Hitchin *Regulator* in F. J. Lyons *Jonathan Wild* (1936) 238 He, being known to be an old practitioner, will certainly be *cast* and *top'd, alias* hang'd for the same. **1958** F. Norman *Bang to Rights* 30 He also took my tie and belt so that I could not top myself. **1961** [see *slag *sb.*¹ 4* (*b*)]. **1983** *Listener* 3 Feb. 18/3, I have to try and get a key to it all, otherwise I'll just top myself. **1984** M. Litchfield *See how they Run* xvii. 157 That shooter..wasn't used to top Frost.

III. 11. For *Obs. rare* read *Obs.* exc. *U.S.* (Later example.)

1959 W. Faulkner *Mansion* i. 14 My young bull topped her last week.

IV. 16. a. Freq. in phr. *to top the bill*: to be at the top of a bill of entertainment (*bill *sb.*³ 8 c); to be the star of a show. Also *fig.* and with the entertainment as object.

1910 Wodehouse *Psmith in City* 3 He is a man of hobbies... When I left the house this morning he was all for cricket... Cricket seems still to be topping the bill. **1933** P. Godfrey *Back-Stage* xiv. 179 The old favourites, when they still topped the bill, had to revise the material they had formerly worked. **1959** [see *bill *sb.*³ 8 c]. **1977** *Sounds* 9 July 4/4 Led Zeppelin remain favourites to top a one-day festival at Wrotham Park.

b. (Later examples in *U.S. Sport.*)

1951 *Amer. Speech* XXVI. 230/2 Dartmouth tops Harvard. **1974** *State* (Columbia, S. Carolina) 27 Feb. 3-B/1 The Panthers demolished both, topping Duquesne, 82–65, and trouncing Davidson in Charlotte, 90–63. **1979** *Tucson* (Arizona) *Citizen* 20 Sept. 8D/3 Boston topped Toronto, 8–0.

VI. Idiomatically combined with adverbs. (See also sense 9 in Dict. and Suppl.)

19. top off. a. *intr.* Of a ship, aircraft, etc.: to fill up or complete a cargo. Cf. sense 20 b below. *colloq.* (chiefly *U.S.*).

1937 G. S. Doorly *In Wake* 22 A tramp steamer.. called in to the Gulf to top-off with sugar. **1950** *Sun* (Baltimore) 3 July 14/2 Ships go to other ports to 'top off'. **1961** *Aeroplane* C. 761/2 Since the passenger carriers ..'top-off' with cargo, it..seems fair and reasonable to permit the all-cargo carriers to carry cargo and to 'top-off' with passengers. **1978** H. Wouk *War & Remembrance* v. 46 We top off, take on provisions and torpedoes, and go.

b. *trans.* To fill up to the top (a tank already partly full) with fuel. *U.S. colloq.*

1943 F. J. Bell *Condition Red* 16 There'll be a fuel barge alongside some time tonight to top us off. **1953** C. A. Lindbergh *Spirit of St. Louis* ii. vi. 182 The fuel tanks would need topping off again. **1970** N. Armstrong et al. *First on Moon* iii. 65 White streaks of vapor were emitted by the fuel tanks—which were constantly being 'topped off'. **1979** *Farmington* (New Mexico) *Daily Times* 27 May 3c/6 If everyone in New Mexico topped off their tank, that would use about 10 million gallons of gasoline.

c. *intr.* = *top out*, sense 20 c below.

1970 *Toronto Daily Star* 24 Sept. 4/2 If wage rates show signs of topping off, the Cabinet can face Parliament. **1976** *Survey* Spring 60 The progressive character of the scale tops off at 3 per cent of earnings for any income over 300 R/mo.

20. top out. a. *trans.* To put the finishing touch to (the roof of a building, etc.), freq. (in modern times) accompanied by some form of ceremony. *colloq.*

1834 W. Sewall *Diary* 22 Dec. (1930) 160 Topped out house chimney, and went to saw mill. **1962** *Engineering* 16 Nov. 640 The dome was 'topped out' on 2 November. **1969** *Daily Tel.* 18 Apr. 27 (*caption*) Ald. Walter C. Dennis, Mayor of Lambeth, toasting the workmen..when the G.L.C.'s..Lambeth Walk development was 'topped out' yesterday. **1979** *Guardian* 25 July 3/2 Britain's most expensive new homes in Knightsbridge, London, were 'topped out' at a champagne reception yesterday.

b. Of a ship: to fill up or complete (its cargo). Also *absol.* Cf. sense 19 a above. *U.S. colloq.*

1940 *Sun* (Baltimore) 16 Apr. 24/6 Preparations were being made to tow her into the stream to 'top out' a 12,500-ton cargo. **1941** *Ibid.* 24 June 22/4 Every ship.. 'topped out' with scrap, if there was any room left.

c. *intr.* To reach a peak, to cease rising. Cf. sense 19 c above and *bottom *v.* 4 c.

1972 *Sunday Tel.* 26 Mar. 30/4 Gilts now look as though they have topped out, and this is another sign that we are in the late stages of this bull market. **1972** *Guardian* 24 June 10/6 World population, he says, will probably top out at 10,000 millions sometime in the twenty-first century. **1979** *Sci. Amer.* Feb. 28/1 From the 10th century to the Mongol Wars, numbers rose, topping out in A.D. 1200.

21. top up. *trans.* **a.** To bring (something) up to its full capacity; to fill to the top (a partly full container, *spec.* (the cells of) a motor vehicle's battery). Used esp. with reference to a drinker's glass, freq. with the person as object. Occas. *absol.* and *transf.*

1937 *Times* 13 Apr. p. xxii/2 In order to help the owner-driver to look after his battery, a combined acid-level indicator, vent plug and filler cup has been introduced, thus enabling the cells to be 'topped up' accurately and visibly, without removing the vent plugs. **1946** *Happy Landings* July 12/1 Failure to..top-up brake pressure.. and to check the voltage readings of batteries, are common examples. **1958** *Times* 1 Mar. 6/3 Liquid oxygen..to top up its [*sc.* a missile's] fuel tanks. **1960** 'N. Shute' *Trustee from Toolroom* ix. 237 We'll need water, and top up with diesel fuel. **1965** *Listener* 18 Nov. 800/3 Tea is expensive.. so you economize by topping up your mug with hot water. **1969** 'R. Petrie' in *E. Queen's Mystery Mag.* Mar. 33/1 Jim Morris tiptoed over to the sideboard for the bottle of brandy... Top him up, he told himself. **1971** 'E. Ferrars' *Stranger & Afraid* iii. 40 She..picked up the glass of sherry that she had started earlier. He said at once, 'Shall I top that up?' and..filled the glass to the brim. **1976** J. I. M. Stewart *Memorial Service* i. 14, I tried to teach him how to translate Tacitus, but had more success in topping him up with madeira. **1981** G. Boycott *In Fast Lane* xi. 79 There was at least three feet of water in the main channel, constantly topped up by torrential showers.

b. *fig.*

1968 *Listener* 27 June 835/3 They..topped up the Welfare State with plenty of money for its more exquisite and bizarre excrescences. **1973** *Times* 20 Oct. 20/3 (*heading*) Topping up a mortgage with a loan from a life office. **1976** *Scotsman* 27 Dec. 1/2 It proposes a Scottish Assembly of 100 members... An Assembly member elected for each of the 71 parliamentary constituencies, 'topped up' by 29 additional members.

topaz. Add: **1. b.** The dark yellow colour of topaz.

1908 *Sears, Roebuck Catal.* 360/2 These colors are.. sapphire blue, emerald green, topaz, etc. **1942** W. Faulkner *Go down, Moses* 237 A horse stands, blinking his sleepy topaz eyes at nothing. **1974** *Times* 2 Dec. (Wines & Spirits Suppl.) p. iii/5 Pale topaz with a gentle grapey aroma..it [*sc.* a wine] costs less than £12 a dozen.

top-boot. 1. (Earlier U.S. example.)

1768 J. R. Peyton *Let.* 10 Apr. in J. L. Peyton *Adventures of my Grandfather* (1867) ii. 17, I found my heavy top-boots of immense service.

top-booted, *a.* (Earlier example.)

1829 G. Griffin *Collegians* I. viii. 169 A stout top-booted elderly gentleman.

‖**topchee** (tōᵘ·ptʃi). Also 7 **topagee, toptchi,** 9 **topechee, topdji, topgi; topgey.** [Hind. *topcī*, Pers. *topchī*, Turk. *topçu* artilleryman, f. Pers., Turk. *top* gun, cannon.] A term used in the former Ottoman Empire for a gunner or artilleryman.

1623 in W. Foster *Eng. Factories India* (1908) 234 They delivered mee..into the hands of the Topagee. **1668** P. Rycaut *Present State Ottoman Empire* iii. x. 200 The Toptchi. These are Gunners, called so from the word *Tope*, which in Turkish signifies a Cannon. **1828** J. B. Fraser *Kuzzilbash* I. 337 The men..bore down like lightning on the topechees. **1854** R. Curzon *Armenia* 73 He brought four guns with him, and a number of topgis, or gunners, to work these instruments of destruction. **1892** P. L. Simmonds *Dict. Trade Products* (new ed.) Suppl. 502/1 *Topdjis*, militia artillerymen in Turkey. **1918** E. S. Farrow *Dict. Mil. Terms* 620 *Topgeys*, the term for Turkish artillerymen or gunners. Also written *topgis*.

top-dressing, *vbl. sb.* **a.** (Earlier example.)

1744 W. Ellis *Mod. Husbandman* Mar. i. 7 The Top-dressing of a powdered Manure is far more preferable on this Account to the Top-dressing of Dung.

topectomy (tǫpe·ktŏmi). *Surg.* [f. Top(o- + *-ectomy.] An operation in which specific areas are removed from the frontal lobe of the cerebral cortex as a treatment for mental illness.

1948 *Newsweek* 29 Mar. 47/3 In the new operation, called a topectomy, the brain fibers are not cut. **1967** [see *gyrectomy].

topgallant, *sb.* and *a.* Add: **B. 2.** *topgallant breeze* (earlier example).

1798 *Authentic Narrative Battle of Nile* 21 The wind was at this time N.N.W. and blew what Seamen call a Topgallant breeze.

top hat. Add: **1. b.** *transf.* A person of the kind or class that wears a top hat; an important or senior person (see also quot. 1938).

1936 *Amer. Speech* XI. 221 The ermines and top-hats, the carriage trade of the early part of the century. **1938** F. D. Sharpe *Sharpe of Flying Squad* xix. 209 It's no good here, boys, there are too many top-hats (detectives). **1974** *Globe & Mail* (Toronto) 4 Oct. 7/1 Mr. White brought Mr. Benoit along to the Albany Club, and the Tory party tophats liked what they saw.

2. *attrib.* **a.** With reference to the wearing of top hats. **b.** Shaped like a top hat.

1902 [in Dict.]. **1958** *Engineering* 14 Mar. 344/2 Shallow 'top-hat' sections are used for the horizontal rails. **1966** Mrs. L. B. Johnson *White House Diary* 13 Dec. (1970) 461 Lyndon made a splendid toast. He said he wanted American art to be enjoyed at the grass-roots level, just as it is by the top-hat crowd. **1967** M. Chandler *Ceramics in Mod. World* ii. 83 (*caption*) A 'top-hat' kiln loaded for firing.

3. Passing into *adj.* Designating insurance or pension schemes devised as a means of deferred payment for senior executives.

1952 *Economist* 8 Nov. 417/1 'Top-hat' contracts for higher executives. **1964** *Daily Tel.* 16 Apr. 23/1 The annual check on senior executives under the 'top-hat scheme', because of the neurosis it created, was a pernicious plan. **1979** *Guardian* 18 June 10/4 It would be good to know that the company car, the top hat pension,.. will not survive the axe next spring.

top-heavy, *a.* Add: **c.** *fig.* Of a business, organization, etc.: (*a*) overcapitalized; (*b*) having a disproportionately large number of people in senior administrative positions.

1934 in Webster. **1945** A. Huxley *Let.* 8 Aug. (1969) 531 A country with a decentralized..economy could not compete as a military power with countries having a top-heavy capital goods industry. **1962** *Rep. Comm. Broadcasting 1960* in *Parl. Papers 1961–2* (Cmnd. 1753) IX. 259 Some witnesses alleged..that the BBC was 'top-heavy': that is, it spent too much on administration at the expense of programme production. **1963** *Listener* 7 Mar. 417/1 Petty officials multiplied till the administration became top-heavy. **1976** M. Machlin *Pipeline* vi. 71 Even now his firm was top-heavy on distribution and transportation.

Hence **to·p-hea·vily** *adv.*; **top-heaviness** (earlier *fig.* example).

1843 H. James *Let.* 11 May in R. B. Perry *Tht. & Char. W. James* (1935) I. 46 Thought heaped up to top-heaviness and inevitable lopsidedness. **1926** Fowler *Mod Eng. Usage* 504/2 What reads well & what reads..jerkily, lopsidedly, topheavily, or otherwise badly. **1947** *Penguin New Writing* XXXI. 172 In the cottage gardens the crumpled, bunchy flowers of double daffodils waved top-heavily in the wind.

topi² (to·pi). [a. Mende.] A glossy dark brown antelope belonging to a race of the korrigum, *Damaliscus lunatus*, found in the coastal region of East Africa.

1894 Sclater & Thomas *Bk. Antelopes* I. 68 The 'Topi', as we propose to designate another local representative of the Korrigum, from the name given to it by the Swahili, has been known for some years. **1910** R. E. Drake-Brockman *Mammals Somaliland* 59 The Topi Hartebeest ..is purplish brown. *Ibid.*, The Topi I have never seen in large herds. **1959** A. Moorehead *No Room in Ark* iv. 91 The topi is a fairly mad animal.., a large brown gleaming antelope with gun-metal blazes on its legs and an air of continual stage-fright. **1976** K. Thackeray *Crownbird* i. 9 They were now doing fifty..and overhauling the topi fast.

topic, *a.* and *sb.* Add: **B.** *sb.* **I. 3. b.** *Gram.* The part of a sentence which is marked as that on which the rest of the sentence makes a statement (comment), asks a question, etc.

Topic sometimes corresponds to *subject*, but the *topic/comment* contrast is not necessarily the same as that of *subject/predicate*.

1958 C. F. Hockett *Course in Mod. Linguistics* xxiii. 201 In English and the familiar languages of Europe, topics are usually also subjects, and comments are predicates. **1972** Hartmann & Stork *Dict. Lang. & Linguistics* 239/1 Some languages, e.g. Japanese, have special particles to mark the topic of the sentence, and for such languages the topic/comment is a more satisfactory analysis than the subject/predicate division. **1976** *Archivum Linguisticum* VII. 123 'Topicalization'..will here be used to denote a process of both foregrounding of information..and selection of the 'topic' of information, that is a process which singles out certain elements in a sentence and makes them the 'topic' on which some 'comment' is made. **1979** *Canad. Jrnl. Linguistics* XXIV. i. 42 Topics are created by a rule of Topic Formation, and preposed by a rule of Topic Preposing.

III. *Comb.*, as (sense 3 b) *topic-neutral* adj.; **topic-(and-)comment,** (based on) the dichotomy in grammar of topic and comment.

1964 E. A. Nida *Toward Sci. Transl.* iv. 66 It has been found that all languages seem to have something equivalent to subject-predicate constructions. These may in some instances be more aptly termed topic-comment, but essentially they are very similar from one language to another. **1978** *Language* LIV. 231 He [*sc.* R. Scollon] then suggests that topic-comment structures themselves may arise from discourse. **1979** *Amer. Speech 1978* LIII. 279, I think the basic type of openness in human language behavior is that of the topic-and-comment pattern. **1951** *Mind* LX. 541 There are..some forms of inference which can occur only in a restricted field of discourse... There are others, depending on the meaning of what Professor Ryle has called 'topic-neutral words', which can occur in the handling of any kind of subject matter. **1961** D. S. Shwayder *Modes of Referring* iii. 81 A distinguishing use may be more or less topic-neutral.

topical, *a.* (*sb.*) **B.** *sb.* Restrict † *Obs. rare*⁻¹ to sense in Dict. and add: **2.** A film dealing with topical events. (Now *disused.*)

1912 F. A. Talbot *Moving Pictures* 123 This point of view is responsible for the apathetic American attitude toward the 'topical', as it is called in Great Britain. **1915** B. E. Jones *Cinematograph Bk.* 33 Something may here be said about topicals or 'newsy' films. **1917** C. N. Bennett *Guide to Kinematogr.* 123 Fourpence or fivepence a foot will be the most a country showman will pay for a local topical. **1976** *Oxf. Compan. Film* 500/2 The early 'topicals' were very short, often less than a minute long, each dealing with a single event. The regular issue of newsreels in the conventional sense—several short items grouped under no general heading other than topicality—was begun by Pathé in 1908.

to·picalize, *v.* [f. Topical *a.* + -ize.] *trans.* To make into a grammatical topic. Usu. *pass.*

1970 *Language* XLVI. 375 In 47, *le tama* is topicalized out of the objective case. **1976** *Archivum Linguisticum* VII. 132 It must be some other quality inherent in *pay*, the 'abstract predication'.., or more precisely, the act, action, or activity, which is topicalized and appears on the surface as *-ment*. **1977** *Amer. Speech 1975* L. 71 One important aspect of this argument is that those proposing it..question the underlying verbal status of *have* and regard it

instead as a surface form that serves to topicalize the possessor, which in other languages would be a locative or dative.

Hence **topicaliza·tion; to·picalized** *ppl. a.*

1967 *Foundations of Lang.* III. 47 Topicalization.. means that some major constituent of a sentence, such as a noun phrase, which is identical with (or has the same referent as) a constituent in the given sentence, may be generated before or after this sentence. **1970** *Language* XLVI. 375 Reversing the order of the two topicalized noun phrases is not possible; the sentence cannot be interpreted 'the girl loves the boy'. **1978** [see *preposing* vbl. sb. s.v. *prepose *v.* 2 a]. **1980** *English World-Wide* I. 249 In the structure, only the plural marker, the locative copula, and the topicalized predicator were found to be of African origin. **1983** *Studies in Eng. Lit.: Eng. Number* (Tokyo) 158 Topicalization is a stylistic rule.

Topkhana (tōᵘ·pkā:nā). Also 7 **Tophana,** 8 **Tope Khonnah;** 9 **Tope Khâna; top-khana.** [a. Pers. and Hind., ad. Turk. *tophane* (also used), f. *top* (see *topchee) + Pers. *khāna* house.] **a.** In Turkey, a gun-factory or arsenal, *spec.* the gun-factories in Galata, Constantinople, during the Ottoman Empire; hence (the current sense) the district of Istanbul adjoining them. **b.** In India, artillery; ordnance department.

1656 W. Jesson *Let.* 4 Dec. in W. Foster *Eng. Factories in India 1618–69* (1921) X. 73 The King, being resolved to renue the warr against Candahor, dispeeded Cossom Ckawn..with the tobeconah [*tōp-khāna*, artillery] for Lahore. **1668** P. Rycaut *Pres. State Ottoman Empire* 200 Their quarters are at Tophana, or the place of Guns in the Suburbs of Constantinople. **1765** J. Z. Holwell *Hist. Events* (1766) I. 96 By the treachery of the *Tope Khonnah Droger*, the cannon were loaded with powder only. **1842** C. Masson *Baluchistan, Afghanistan & Panjab* II. 256 From the court of the Dafta Khâna the Tope Khâna, or artillery ground, is entered. **1901** Kipling *Kim* iii. 67, I have known Him since he was a lieutenant in the *top-khana* (the Artillery). **1969** D. Walder *Chanak Affair* xix. 350 An enormous Turkish flag floated over the Tophana naval depot. **1972** D. K. Palit *Hist. Regiment of Artillery* i. 8 Although Shivaji had a regular department of *Topkhana*..he never had a foundry of his own. He managed to obtain some guns from the foundries at Surat..but by and large his *Topkhana* was described as a collection of 'old and defective' guns. **1974** J. F. Guilmartin *Gunpowder & Galleys* i. 44 The cannon foundries of the *Tophane* were directly dependent upon English tin which they could get only through Genoese entrepreneurs.

topknot. Add: **1. b.** Also *Austral.* and *N.Z.*, wool shorn from the top of the head of a sheep.

1950 *N.Z. Jrnl. Agric.* Oct. 313 (*caption*) Pick over lambs' wool. Pick out stained wool, face pieces, leggings, and top-knots. **1972** J. S. Gunn in G. W. Turner *Good Austral. Eng.* iii. 61 One thing I did notice about shearing was that two terms for the one idea sometimes shared popularity, for example *rouseabout/shedhand*,.. *topknot/wig*, [etc.].

Töpler (tǝ̄pləɹ). *Physics.* Also **Toepler.** [The name of A. J. I. *Töpler* (1836–1912), German physicist, who invented the pump in 1862.] *Töpler* (or † *Töpler's*) *pump*: a pump in which the reservoir to be evacuated is connected by a glass tube containing a valve to a second reservoir, which in turn is connected by a flexible U-tube to a reservoir of mercury, so that lowering the last draws gas out of the first reservoir into the second and raising it expels gas from the second reservoir into the atmosphere. Also *absol.*

1883 *Encycl. Brit.* XVI. 31/1 In Töpler's pump this is attained by using..for the inlet and the outlet vertical capillary glass tubes. **1922** Glazebrook *Dict. Appl. Physics* I. 8/1 The working of a Töpler by hand is extremely tedious, for several hours may be required to reach the limit of pressure. **1932** [see *Sprengel]. **1976** A. Roth *Vacuum Technol.* v. 201 With the Toepler pump, pressures down to 10^{-6} Torr can be obtained, except the mercury vapour pressure which is about 10^{-3} Torr.

topless, *a.* Add: **3. a.** Designating or pertaining to a garment, esp. a (woman's) bathing-suit or dress, having little or no material above the waist; that does not cover the breasts and upper body.

1937 *Time* 21 June 53/1 With another bathing-suit season at hand, local lawmakers are aiming their ordinances at males on the score of topless suits rather than at underclad females. **1964** *San Francisco Chron.* 16 June 4/1 Saigis introduced San Francisco's first topless bathing suit for women. **1964** *Punch* 1 July 20/2 The topless look. **1964** *New Statesman* 24 July 116/1 A girl who wears a topless dress in the streets of Coventry or Nottingham could be doing as much for her sex as any Mrs Pankhurst. **1966** C. Mackenzie *Paper Lives* xii. 166 And those topless dresses they're going in for now aren't nearly as topless as those dresses they were wearing about 1500 B.C. **1971** S. Jepson *Let. to Dead Girl* xvii. 195 Her jeans were covered with dirt, and her shirt torn into topless decolleté. **1978** P. Glynn *In Fashion* iv. 97 Rudi Gernreich's topless bathing suit had appeared in 1964.

b. Of a person (esp. a woman): naked or almost naked above the waist; bare-breasted.

1966 *Observer* 13 Nov. 2/8 The appearance of topless waitresses. **1968** 'R. Raine' *Night of Hawk* xxvi. 125 Various acts, the main one being African girl dancers who perform topless. **1969** *Observer* 7 Dec. 25/3 Topless boys with shoulder-length hair pause as they cycle past you: 'Wanna buy some acid?' **1970** *Daily Progress* (Charlottesville, Va.) 29 May 19/5 (*heading*) Topless girl is acquitted. **1981** *Birds* Summer 55/2 New reserve records included nude female bathers on Loch Garten, a 'topless' woman picking blackberries, [etc.].

c. Applied to a place or area in which women are permitted to appear naked above the waist; esp. of bars, etc., employing bare-breasted waitresses or dancers, and of beaches at which women sunbathe topless. Hence also of entertainments, sunbathing, etc., so conducted.

1967 F. Warner *Madrigals* 30 Draining down screw-drivers in topless Broadway. **1970** G. R. Taylor *Doomsday Bk.* x. 246 Fairbanks has become a boom-town with topless entertainment in the bars and saloons. **1972** G. Baxt *Burning Sappho* ii. 41 For Chrissakes not one of them topless joints. Who wants flabby tits hanging over my shrimp cocktail. **1976** P. Cave *High Flying Birds* iii. 36 This section of the beach was strictly topless. **1978** *N.Y. Times* 30 Mar. B 2/4 He has campaigned against marijuana decriminalization, 'topless' bars, [etc.]. **1979** *Globe & Mail* (Toronto) 26 Jan. 5 Toronto City Council probably will move on Feb. 5 to restrict new bars offering topless or nude entertainment. **1983** *Times* 6 July 32/1 Topless sunbathing is a well-established practice on a great many British beaches.

Hence **to·plessness,** the condition of being topless (sense 3).

1964 *Punch* 22 July 117/3 His pronouncement on toplessness. **1982** C. Castle *Folies Bergère* iii. 127 Today, the question of toplessness..evokes little excitement in an age of naked beaches.

top level. orig. *U.S.* [f. Top *sb.*¹ + Level *sb.*] The highest degree of importance, prestige, or ability; usu. (with hyphen) *attrib.*, designating that which belongs to or takes place at such a level.

1951 *Sci. Amer.* Sept. 43/2 The story is even less favorable when we examine what proportion of able people obtain the Ph.D. degree—today a requirement for many of our top-level intellectual occupations. **1952** *Times* 29 Jan. 3/2 It is estimated by the assistant director of public relations at American University that 'there must be more than 150 ghost writers at the top level in Washington alone'. **1955** *Times* 16 May 4/1 No one can say for certain that the results of a top-level meeting will fulfil our hopes. **1956** Wallis & Blair *Thunder Above* (1959) xii. 116 They'll probably engage you..as a top-level consultant. **1957** *Ann. Reg. 1956* 23 It was announced that a meeting 'at top level' between union officials and the management of Standard Motors had been arranged for 13 May. **1967** C. Berners-Lee in Wills & Yearsley *Handbk. Management Technol.* 15 In due course [a really good technical salesman] will find himself inundated with as much top-level work as he can cope with. **1969** *Daily Tel.* 17 Dec. 2/6 Top-level talks to discuss the threatened unofficial strike at 63 power stations..are being held..between employers and unions. *a* **1974** R. Crossman *Diaries* (1976) II. 125 This Lord Mayor's dinner was the first top-level invitation which Anne and I have accepted since I was a Minister. **1979** *Yale Alumni Mag.* Apr. 8/2 The desire to watch top-level squash competition has also increased.

top line. [f. Top *sb.*¹ + Line *sb.*²] **1.** (In Dict. s.v. Top *sb.*¹ 32.)

2. The head item on a bill of entertainment; the headline of a newspaper; freq. (with hyphen) *attrib.*; also *fig.*

1906 *N.Y. Times* 6 May iv. 2/1 The rumor of a new $30,000,000 vaudeville act—a regular 'top-line' combination—has been giving Broadway a good deal to talk about. **1922** A. Haddon *Green Room Gossip* vi. 138 His..duets with Miss Muriel George have reached a top-line position on the halls. **1928** *Boston Even. Transcript* 30 Mar. 15/6 This big 'top line' caught my eye: 'The Pocasset filicide.' **1928** *Daily Express* 31 July 13/7 The top-line contest on Saturday next is that between Jack Stanley of Deptford and Gipsey Daniels. **1947** [see *care *v.* 4 a]. **1958** *New Statesman* 23 Aug. 211/3 The *Daily Mail* has recently been the most persistent in putting news first in its public appeal and in tying its 'top line' features much more closely to it than most. **1981** B. Healey *Last Ferry from Lido* i. 17 A real, top-line Venetian socialite.

3. *on the top line*: in the highest state of perfection, readiness, etc. *slang* (orig. Naval). Phr. *to sweat on the top line*: see *sweat *v.* 9* c.

1916 'Taffrail' *Pincher Martin* ix. 155 I've 'eard tell, too, that that there Kayser bloke o' theirs 'as gingered 'em up somethin' crool, an' a navy wot's been gingered up must be on th' top line same as us, mustn't it? **1942** *Tee Emm* (Air Ministry) II. 81 There are a lot of ways of navigating these days; you ought to be on the top line in them all. **1958** *Punch* 9 July 57/3 The famous scene in which he has to serve both his masters with dinner at the same moment becomes a juggling turn on the top line. **1972** *Times* 30 Nov. 17/4 It cannot be but harmful to the patient to be attended by a doctor who through sheer exhaustion is not on the top line.

4. *Mus.* In music divided into four parts for singing, the highest (usu. soprano) line.

1965 *Listener* 1 July 33/3 The top line could be sung by sopranos and altos in unison, while the baritones..could dodge from the bass line to the tenor line. **1968** E. R. Buckler *Ox Bells & Fireflies* iv. 73 We sang—with the teacher always taking what was known as 'top line'.

Hence **top-li·ner,** one who or that which appears in the top line, or in the first or principal place.

1901 *Munsey's Mag.* Nov. 247/1 Grand opera in this and other countries sadly needs a new bright particular star, a 'top liner', as they say in the music halls. **1901** *Daily Colonist* (Victoria, B.C.) 1 Nov. 3/1 One of the four schooners which returned to port yesterday morning was the top-liner of the fleet, the Otto. **1928** *Daily Express* 19 June 16/1 Every Ascot race is a top-liner. **1950** *Sport* 7–11 Apr. 9/1 For months past people have been trying to grab tickets for this match, which is always a top-liner. **1970** *Daily Tel.* 29 Apr. 13/1 On the club circuit there are top-liners..making up to £400 a week. **1982** R. Hill *Who guards a Prince* iv. vi. 227 A journalist..a top-liner as well as an old friend.

topman¹. Add: **4. c.** (See quot. 1964.)

1961 *Evening Standard* 3 Aug. 21/4 (Advt.), Topmen reqd. for demolition trade. **1962** Parker & Allerton *Courage of his Convictions* i. 22 He's a top man, and don't you forget it... When derelict buildings were being knocked down he was one of those on the roof. **1964** J. S. Scott *Dict. Building* 94 *Demolisher* or *mattock man* or *topman* or *housebreaker*, a skilled man who pulls down a wall by standing on top of it and breaking pieces off below him, or by pulling a loose wall with a winch and rope, or by means of a concrete breaker. **1973** *Daily Tel.* 7 Mar. 13/2 Pulling down the YMCA dome, 120ft. or about 15 storeys high, had to begin with a topman..picking away with mattock.

topo, U.S. colloq. abbrev. of Topographic *a.* Also *ellipt.* for 'topographic map'.

1970 N. Armstrong et al. *First on Moon* xi. 256 The best we can do on topo features is to advise you to look to the west of the irregularly shaped crater. **1977** *Chicago Tribune* 2 Oct. iii. 10/3 'Topos' show land contours, elevations, marshes, lakes, streams. **1979** *Amer. Alpine Jrnl.* XXII. i. 35 Japan's foremost mountain magazine.. contains..valuable maps, diagrams and topos with English captions. **1981** *Northeast Woods & Waters* Jan. 16/2 You can buy topos most anywhere in Maine, but they are dated and won't show all the logging roads built in the past 25 years.

topo-. Add: **topoce·ntric** *Astronautics*, (of a parameter of a spacecraft or an orbit) measured relative to a point on the earth's surface (rather than its centre); **to·pocline** [*cline *sb.*], a cline associated with variations in the locality of the species concerned; **topo·genous** *a.*, formed as the result of a combination of geographical features; **to:po,inhibi·tion** *Biol.*, the inhibition of cell multiplication by contact with other cells; **topo,iso·merase** *Biochem.*, any enzyme that alters the supercoiled form of a DNA molecule.

1965 P. R. Escobal *Methods Orbit Determination* vii. 241 The topocentric right ascension-declination of the unknown orbit at the three times..can be obtained as follows. **1976** *Sci. Amer.* June 70/2 Since the tracking stations are on the earth's surface, the direct measurements they provide of the spacecraft's radial parameters (range, velocity and acceleration) are topocentric rather than geocentric. **1939** J. W. Gregor in *New Phytologist* XXXVIII. 317 Prefixes can be used to denote clines of different types, for example topocline. **1953** J. Heslop-Harrison *New Concepts Flowering-Plant Taxon.* v. 68 Independent topoclines exist for different morphological features. **1970** *Watsonia* VIII. 140 The two subspecies may be regarded as the relatively extreme end-points of a topocline. **1939** A. G. Tansley *Brit. Islands & their Vegetation* xxxv. 719 Valley bog–topogenous, formed in valleys and depressions where water..stagnates, and bog plants establish themselves. **1975** J. G. Evans *Environment Early Man Brit. Isles* iv. 76 Essentially there are two types of peat, topogenous and ombrogenous. Topogenous peat forms in places of impeded drainage. **1970** *Nature* 22 Aug. 806/1 Topoinhibition is probably an important mechanism regulating cell multiplication in organisms in normal conditions. **1975** *Ibid.* 29 May 371/3 The loss of topoinhibition at wound edges in culture is apparently not due to loss of junctional communication. **1978** *Devel. Biol.* LXIV. 273/2 The enzyme has been referred to as ω-protein.., swivelase.., untwisting enzyme.., relaxing activity.., relaxing protein.., nicking-closing activity.., and DNA topoisomerase. **1979** Wang & Liu in J. H. Taylor *Molecular Genetics* III. ii. 66 We propose that they be called DNA topoisomerases. **1980** *Sci. Amer.* July 109/2 These nicking-closing enzymes, which are also called topoisomerases, generally require no energy source to function.

topochemical (tǫpoke·mikăl), *a.* [f. Topo- + Chemical *a.*] **1.** *Ent.* Of, pertaining to, or denoting an insect's capacity to perceive spatial relationships through the sense of smell (see quots.).

1908 M. Yearsley tr. *A. Forel's Senses of Insects* x. 237 It is..a question of a chemical sense which gives the exact relations between the different parts of space... The faculty of smell-by-contact could be called chemaphesthesia and relational smell topochemical sense. **1967** J. H. Sudd *Introd. Behaviour Ants* ii. 17 Because the antennae are moveable and have sense organs all over the surface it has been suggested that ants can perceive spatial arrangements of chemical stimuli, rather as we can perceive patterns of colour or texture. This is the so-called topochemical sense.

2. *Chem.* [ad. G. *topochemisch* (Kohlschütter & Tüscher 1920, in *Zeitschr. f. anorg. u. allgemeine Chem.* CXI. 193).] Of or pertaining to topochemistry.

1920 *Chem. Abstr.* XIV. 3202 The term topochemical is proposed as a designation of 'locally confined' reactions. **1945** *Electronic Engin.* XVII. 425 Not only pressure and quality of the gas play an important part in vacuum measurement but also its quantity... It may be altered by 'topophysical' or 'topochemical' effects. **1948** *Research* I. 262/2 The topochemical significance is that corrosion is almost always an electrochemical process, in which the anodic localities of corrosion are small surface elements. **1967** [see *biogenesis 3].

Hence **topoche·mically** *adv.*

1962 *Jrnl. Physical Chem.* LXVI. 2442/1 Photochromy in this series is a topochemically determined phenomenon, *i.e.*,..the packing arrangement in the crystal is of importance. **1967** *Science* 26 May 1123/2 When the construction [of the termites] reaches a certain critical density it attracts other termites topochemically.

topochemistry (tǫpoke·mistri). [f. Topo- + Chemistry.] The chemistry of reactions as affected by local variations in the structure of the medium on or in which they occur.

1948 *Research* I. 260/2 Topochemistry, derived from the Greek τόπος meaning 'location',..and topochemical processes are localized reactions in every sense. **1971** *Nature* 16 July 194/1 We have demonstrated a new method for the study of the topochemistry of membrane surfaces. **1975** *Ibid.* 31 Jan. 310/3 Topochemistry is concerned with the nature and kinetics of chemical reactions, including polymerisation reactions, between adjacent molecules in crystals.

to·p-off, *sb. Austral. slang.* [Of obscure origin: cf. Top *v.*¹ 6 and *tip *v.*⁵ 2.] An informer. Also *top-off merchant* (*merchant *sb.* 3).

1941 Baker *Dict. Austral. Slang* 77 *Top-off*, a police informer. **1944** L. Glassop *We were Rats* 133 He pooled me with the Q.M. Just a top-off merchant, that's all he is. **1966** B. Collins *Copper Crucible* 14 About four o'clock in the morning some top-off rings the cops. **1973** *Sunday Mail* (Brisbane) 4 Mar. 4/2 He believed him to be a prison 'top-off'.

to·p-off, *a.* [f. the vbl. phr. *to top off*: see *top *v.*¹ 19 a.] Of a passenger: carried in a freight aircraft that would not otherwise be full.

1961 *Flight* CXXX. 864/2 Seaboard World Airlines is continuing to press hard its proposal to carry 'top-off' passengers on its..cargo flights across the Atlantic. *Ibid.*, 'Top-off' passengers would..afford the necessary extra revenues without interfering with the basic objective of profitable transatlantic freight services. **1962** *Aeroplane* CIII. 5/3 Application to carry 'top-off' passengers on scheduled cargo flights to be rejected.

topograph. Restrict *rare* to senses in Dict. and add: **2.** *Cryst.* A photograph taken in such a way, usu. with X-rays, as to exhibit the variation over the surface of a crystal of some physical or structural characteristic.

1944 G. N. Ramachandran in *Proc. Indian Acad. Sci.* A. XIX. 292 Eighteen such 'topographs' of cleavage plates of diamond..are reproduced. **1963** G. L. Clark *Encycl. X-Rays & Gamma Rays* 1053/2 Topographs may be taken that show, for example, the distribution of optical absorption or optical fluorescence in a specimen. **1971** *Physics Bull.* Sept. 553/2 X ray topographs..were used by many authors to assess accurately the quality of their crystals.

topographize, *v.* **b.** (Earlier example.)

1792 W. B. Stevens *Jrnl.* 30 May (1965) 25 Shaw was topographising around Shenstone Church.

topoi: see *topos.

topological, *a.* (In Dict. s.v. Topology.) Add: **2.** *Math.* Of or pertaining to topology; such as is dealt with by topology; *topological invariant*, something invariant under a topological mapping; *topological mapping* or *transformation* = *homœomorphism 2; *topological space* [tr. G. *topologisch raum* (F. Hausdorff *Grundzüge der Mengenlehre* (1914) vii. 213); the sense is due to M. Fréchet (*Compt. Rend.* (1925) CLXXX. 421)], an abstract space together with a topology (sense *3 c) on it.

1913 *Amer. Jrnl. Math.* XXXV. 189 (*heading*) On some topological properties of plane curves and a theorem of Möbius. **1926** *Proc. Sect. Sci. K. Akad. van Wetenschappen te Amsterdam* XXIX. 462 Any normal, not absolutely closed topological space $\mathfrak{R}$ can be extended to a normal topological space $R = \mathfrak{R} + \xi$ by adjunction of a non isolated point ξ. **1939** M. H. A. Newman *Elem. Topol. Plane Sets of Points* iii. 51 The correlation is called a homoeomorphism between the spaces, or a topological mapping of the one space on the other. **1946** E. Lehmer tr. *Pontryagin's Topological Groups* iii. 53 For any two elements p and q of the group G there exists a topological transformation $f(x)$ of the space G into itself which transforms p into q. **1956** [see *homœomorphism 2]. **1961** A. E. Farley tr. *Alexandroff's Elem. Concepts Topology* 16 A simple closed curve (i.e., the topological image of a circle). **1968** E. T. Copson *Metric Spaces* vii. 92 Properties of a metric space which depend only on its open sets..are called topological properties. **1975** I. Stewart *Concepts Mod. Math.* x. 144 Straightness is not a topological property.

topologize (tǫpǫ·lŏdʒaiz), *v. Math.* [f. Topolog(ical *a.* + -ize.] *trans.* To make into a topological space. So **topo·logized** *ppl. a.*; **topo:logiza·tion,** the process of topologizing.

1946 E. Lehmer tr. *Pontryagin's Topological Groups* iii. 56 The abstract group G admits one and only one topologization under which the system Σ^* is a complete system of neighborhoods of the identity. *Ibid.*, If the group G can be topologized in such a way that [etc.]. **1963** D. Bushaw *Elem. Gen. Topol.* ii. 29 It is no great step to convert each of these methods of topologization into an alternative definition of the concept of a topological space. **1964** A. P. & W. Robertson *Topological Vector Spaces* i. 8 The real numbers and the complex numbers can both be topologised by taking $d(x, y) = |x - y|$. **1979** *Proc. London Math. Soc.* XXXVIII. 231 For any closed subspace S of ER^n, we topologize the space of continuous maps of S into ER^n with the compact open topology.

topology. For **1. c.** read **3. a.** and add: **3. a.** The branch of mathematics concerned with those properties of figures and surface which are independent of size and shape and are unchanged by any deformation that is continuous, neither creating new points nor fusing existing ones; hence, with those of abstract spaces that are invariant under homœomorphic transformations. (Further examples.) [ad. G. *topologie* (J. B. Listing 1847, in *Göttinger Studien* I. 814).]

1929 *Trans. Amer. Math. Soc.* XXXI. 290 Analysis situs or topology is primarily concerned with invariants under homeomorphic transformations of a space into itself. **1952** F. Bagemihl et al. tr. *Pontryagin's Found. Combinatorial Topol.* i. 1 Combinatorial topology studies geometric forms by decomposing them into the simplest geometric figures, simplexes, which adjoin one another in a regular fashion. **1959** E. M. Patterson *Topology* i. 1 Nowadays mathematicians are fairly general agreement that topology is a study of continuity. **1970** *Observer* (Colour Suppl.) 15 Feb. 19/2 Topology is one of the most recent and rapidly advancing branches of mathematics, and is a kind of universal geometry of surfaces. **1972** M. Kline *Math. Thought* l. 1158 Topology, as it is understood in this century, breaks down into two somewhat separate divisions: point set topology, which is concerned with geometrical figures regarded as collections of points ..; and combinatorial or algebraic topology, which treats geometrical figures as aggregates of smaller building blocks. **1975** I. Stewart *Concepts Mod. Math.* x. 146 The basic objects studied in topology are called topological spaces.

b. (The study of) the topological properties *of* something. Also *transf.*

1913 *Amer. Jrnl. Math.* XXXV. 189 An application.. of the transformation by inversion to the topology of plane curves. **1930** *Proc. Nat. Acad. Sci.* XVI. 240 (*heading*) Combinatory topology of convex regions. **1959** *Ibid.* XLV. 1607 (*heading*) On the topology of the genetic fine structure. **1972** *Sci. Amer.* Jan. 65/1 With careful dissection techniques one can expose deep-lying sections of bulk specimens so that their topology can be studied by scanning electron microscopy. **1980** D. L. Cohn *Measure Theory* p. vii, Chapters 1 through 5..presuppose only the familiarity with the topology of Euclidean spaces that a student should acquire in an advanced calculus course.

c. A family of open subsets of an abstract space such that the union of any of the subsets and the intersection of any two of them are members of the family, together with the space itself and the null set.

1946 E. Lehmer tr. *Pontryagin's Topological Groups* iii. 55 A topology can be introduced into any abstract group G whatsoever in such a way that G becomes a discrete group. **1963** M. J. Mansfield *Introd. Topol.* ii. 21 The topologies $\mathscr{S}$ and $\mathscr{U}$ for R..were defined, in effect by specifying neighborhoods for each point and then declaring a set to be a member of the topology if and only if the set contains a neighborhood of each of its points. **1976** *Physics Bull.* Sept. 388/2 A useful way to think of a topology for a space is as a specification of which functions on it are to be continuous.

d. *gen.* The way in which constituent parts are interrelated or arranged.

1967 *Electronics* 6 Mar. 149/1 If consideration is restricted to bipolar gate topologies..there are just three basic forms of IC logic schemes. **1970** *Nature* 7 Nov. 553/1 These data have been used to construct a topology based on the minimal mutation distance method... This topology places castor..closest to sesame.., then mung bean .., then sunflower. **1971** *Physics Bull.* Dec. 717/3 Having an axisymmetric topology permits an easier study [of tokamaks] than, say, stellarators. **1972** *Computer Jrnl.* XV. 204/1 The resulting list structure has the same topology as the old, so that re-entrancy and sharing of common substructure are preserved.

topologically *adv.*, **topologist** (examples in *Math.*).

1915 *Trans. Amer. Math. Soc.* XVI. 153 A manifold.. topologically equivalent to the boundary of an *n*-dimensional complex. **1938** *Mind* XLVII. 126 Tests are gauged 'topologically' by the extent to which they succeed in grouping together men who are also grouped together in respect of their performances in life. **1954** *Sci. News* XXXIII. 56 If you cross the curve..you must go from one part to another—you cannot stay inside or stay outside. I think that anyone who is not a topologist will accept this as a self-evident fact. **1967** G. Steiner *Lang. & Silence* 33, I have watched topologists, knowing no syllable of each other's language, working effectively together at a blackboard. **1969** R. B. Fuller *Operating Man. Spaceship Earth* v. 67 All the system's paths must be topologically and circularly interrelated. **1971** I. G. Gass et al. *Understanding Earth* iv. 77/1 It is not what the topologists call a simply connected body; it is like a Henry Moore statue: it has a hole in it. **1975** I. Stewart *Concepts Mod. Math.* x. 146 The oft-quoted assertion that to a topologist a doughnut is the same as a coffee-cup provides an example. *Ibid.* 156 Now topologically a dog is a sphere (assuming it keeps its mouth shut and neglecting internal organs) because all we have to do is shrink its legs and fatten it up a bit.

toponium (tǫpōᵘ·niŏm). *Particle Physics.* [f. Top *sb.*¹ + *-onium, after *positronium.] A bound state of a top quark and a top antiquark.

1978 *Nature* 28 Sept. 268/2 This gives the prospect of detailed investigation of quark dynamics by studying transitions among these levels of 'toponium' analogous to transitions among levels of positronium and hydrogen in atomic physics. **1984** *New Scientist* 17 May 15/2 Theory suggests that if there is an electrically neutral Higgs boson whose mass is less than that of toponium, then we should be able to observe toponium decaying into a photon and a Higgs.

toponomastic (tǫpǫnomæ·stik), *sb.* and *a.* [f. Topo- + Onomastic *a.* and *sb.*] **A.** *sb.* (also *pl.*) = Toponymy. **B.** *adj.* Of or pertaining to place-names.

1916 T. Taylor *Celtic Christianity of Cornwall* iv. 54 Professor Loth, as the result of a careful study of Breton toponomastic, has arrived at the conclusion that the Armorican parishes were placed as early as the sixth and seventh century under the invocation of the saints.. whose names they still bear. **1922** Joyce *Ulysses* 673 Their..toponomastic, historical and religious literatures. **1971** W. F. H. Nicolaisen in A. J. Aitken et al. *Edin. Stud. Eng. & Scots* 211 There is no basic difference in the proper handling of the Scottish toponomastic material in this respect. **1977** *Maledicta* I. 41 (*heading*) Macedonian toponomastics.

toponym. (in Dict. s.v. Toponymy.) Add: **2. a.** A place-name; a name given to a person or thing marking its place of origin. **b.** = *toponymic *sb.* 2.

1939 *Antiquity* Sept. 311 Important also are certain long lists of personal names followed by those of the cities from which these persons came... The toponyms include Byblus (*G-b-'l*, the *Gebal* of the Old Testament). **1958** [see *anthroponymy]. **1973** *Times Lit. Suppl.* 5 Jan. 3/3 *The Manchester Guardian* (before it dropped the toponym). **1978** *Regional Lang. Stud.—Newfoundland* viii. 3 The surname of the original author of this list, Jeddore, occurs as a toponym in Nova Scotia, and.. Jeddore's Harbour is about 40 miles east of Halifax. **1980** *Sci. Amer.* Feb. 48/1 The Zapotec people also used toponyms, glyphic 'place signs' for important places or landmarks, mountains in particular.

toponymic, *sb.* (In Dict. s.v. Toponymy.) Add: **2.** A descriptive place-name, usu. derived from some topographical feature of the place.

1933 *Times Lit. Suppl.* 20 Apr. 280/2 With a few.. exceptions..all Sussex place-names..ought to be explained as toponymics, i.e., as containing descriptive words. **1956** A. H. Smith *Eng. Place-Name Elements* i. 6 Village names which were originally toponymics. **1957** H. H. Jenkins *Diction of 'Yank'* (Univ. Florida thesis) v. 45 Some of the GI's toponymics are banal and unimaginative; others are witty and colorful. **1977** *Word 1972* XXVIII. 117 (*heading*) Celtic toponymics in Scotland.

topos (tǫ·pǫs). Pl. **topoi.** [a. Gr. τόπος place: cf. etym. note s.v. Topic *a.* and *sb.*] A traditional motif or theme (in a literary composition); a rhetorical commonplace, a literary convention or formula.

1948 L. Spitzer *Linguistics & Lit. Hist.* v. 201 In a proem there is generally present a second topos. **1957** N. Frye *Anatomy of Criticism* 103 The *topoi* or rhetorical commonplaces..are so dull when stated as propositions, and so rich and variegated when they are used as structural principles in literature. **1957** *Medium Ævum* XXVI. 148 We have identified the *topos* of the sixth age of the world (and its approaching end) in our two OE poems. **1962** D. A. Pearsall *Floure & Leafe* 68 The submission formula is a *topos* of classical rhetoric. **1966** *Eng. Stud.* XLVII. 150 There must be few literary historians who have traced with such thoroughness the development of a single *topos* through two centuries. **1976** *Classical Q.* XXVI. 246 Kinds of anecdotes which Herodotus loved to include: raiding parties, espionage.., tales of bravery and cowardice, and other such *topoi*. **1981** *Times Lit. Suppl.* 16 Jan. 60/4 It is a common *topos* to remark that thanks are due to the editor or author for raising weighty questions.

toposcope (tǫ·poskōᵘp). [f. Topo- + -scope.] **1.** A device (as a horizontal circular dial) showing the direction of designated features of the landscape and usu. erected on a hilltop.

1907 [see *indicator 1 a]. **1968** V. Waite *Malvern Country* vi. 77 The direction indicator—or toposcope, to give it the technical name—was set up on the Beacon summit to commemorate the reign of Queen Victoria. **1974** *Victorian* (Victoria, B.C.) 22 Mar. 45/3 Nearby is a panoramic viewing area with a toposcope, giving details of all the islands dotted around in the sparkling blue China Sea 1,800 feet below.

2. *Med.* An instrument used for toposcopy.

1951 Walter & Shipton in *Electroencephalogr. & Clin. Neurophysiol.* III. 282/2 Because it provided a visual display of topographic detail this device was called a Toposcope. **1965** *Math. in Biol. & Med.* (Med. Res. Council) IV. 159 The 'toposcope'..depicts each EEG electrode as a spot on a cathode ray tube. **1977** *Lancet* 21 May 1114/1 He devised the first on-line frequency analyser, which with subsequent modifications led to the 'toposcope'.

toposcopy (tŏpǫ·skŏpi). *Med.* [f. as prec. + -y³.] Examination of the electrical activity at different points in the brain simultaneously by means of a number of electrodes each connected to a separate oscilloscope or the like. Hence **toposco·pic** *a.*

1950 *Electroencephalogr. & Clin. Neurophysiol.* II. 97/2 The split disc neon lamp method of toposcopy is particularly attractive. **1951** *Ibid.* III. 283/1 The mark I Toposcope was too limited in its resolution, but the results obtained with it were sufficiently encouraging to warrant the initiation of a development programme with the aim of investigating the general problem of toposcopic display. **1974** *Ibid.* XXXVI. 566/2 (*heading*) Current thoughts on toposcopy. *Ibid.*, Many of the toposcopic display devices used in EEG are products of the age of analog computers and devices.

topotaxy (tǫ·potæksi). *Cryst.* [f. Topo- + *-taxy*, after *epitaxy.] (See quot. 1959³.) So **topota·ctic, -ical** *adjs.*; **topota·ctically** *adv.*

1959 F. K. Lotgering in *Jrnl. Inorganic & Nucl. Chem.* IX. 115 For these reactions the name 'topotactical reactions'..is proposed. *Ibid.* Plate facing p. 120 Photomicrographs of topotactically oriented $Co_{0.1}Zn_{0.1}Z$. *Ibid.* 123 We propose the term 'topotaxy' for all chemical solid state reactions that lead to a material with crystal orientations which are correlated with crystal orientations in the initial product [*read* substance]. **1969** *Nature* 9 Aug. 609/1 Studies of the topotaxy of solid state precipitation from spinel crystalline solutions. **1976** *Ibid.* 19 Aug. 721/1 Dislocations in molecular crystals..have been studied on the microscale chiefly with a view to establishing their role in..topotactic transformations. **1982** *Ibid.* 22 Apr. 730/1 (*caption*) Electron micrograph of a fault-free β-phase grain, topotactically replacing faulted ring-woodite.

top people, *sb. pl.* Also with capital initials. [Top *sb.*¹ 29.] The aristocracy; leaders and people of rank and influence in the arts, politics, the professions, etc. Occas. *sing.* as *top person.*

The expression gained wide currency from the advertising slogan used by *The Times* in 1957.

1752 in M. M. Verney *Verney Lett.* (1930) II. xxxiv. 250 When they get in liquor they are very troublesome and noisy. They kept it up all night, several of the top people. **1957** *Economist* 21 Sept. 929/3 Like the *Times*, it [*sc. Punch*] has been read by top people for a long time. **1959** C. MacInnes *Absolute Beginners* 103 He had a very sharp top-person suit on. **1960** *Sunday Times* 21 Feb. 9/5 (Advt.), The man who gets on is he who makes himself bigger than his job... Top People take *The Times.* **1963** *Punch* 4 Sept. 356/2 Southerners, satiated with Top People prissiness. **1977** *News of World* 17 Apr. 2/3 The top people's directory, Who's Who. **1981** R. D. Edwards *Corridors of Death* vi. 29 The need for tact and sensitivity in handling the Top People involved.

topper, *sb.*¹ Add: **3.** An action, remark, etc., that puts a finishing touch to what has gone before, *esp.* an outrageous one or one that cannot be capped. Cf. Top *v.*¹ 12 b.

1939 J. O'Hara *Pal Joey* (1940) 98 It was a famous historical topper when Josephine was informed that the poor people did not have any bread and she said 'Why don't they eat some cake'. **1973** *Black Panther* 6 Oct. 6/3 The topper is that Foster completes this vicious cycle with a cold beg..to get his money under the..Law Enforcement Assistance Act. **1977** *Amer. Film* July–Aug. 18/1 The shot was part of a gag..a 'topper' that Keaton used to finish off a duel between two Civil War locomotives. **1977** *New Yorker* 12 Sept. 92/3 'Will you do me a favor and take me over to the Carnegie Deli..?' The driver said sure, swung off his bus route, and deposited Henny at the door. Henny's topper? He got off the bus and said, 'Pick me up in twenty minutes.'

topper, *sb.*² Add: **1. b.** *U.S. Mil. slang.* A first sergeant. Cf. *top sergeant* s.v. *top *sb.*¹ 32.

1918 *Radiator* 22 Aug. 2 Sergeant Hulbert, the 'topper' of 95. **1937** *Our Army* (U.S.) Jan. 19 'I'm sure there's no Lieutenant McGonigle here,' replies the Topper.

2. b. A kind of loose-fitting jacket or short coat worn by women or children. orig. and chiefly *U.S.*

1937 *Los Angeles Times* 2 Sept. 5/1 (Advt.), Soft, fluffy casual toppers tailored in the best British way. **1938** *Sears, Roebuck Catal.* Spring/Summer 20/2 The absolutely perfect topper coat. **1948** *N.Y. Times Mag.* 8 Feb. 38 (Advt.), A boxy topper with cardigan neckline. **1960** *News Chron.* 22 July 6/4 Attractive little top jackets ('toppers'). **1972** *Country Life* 30 Nov. 1533 Printed dress ..worn with a topper jacket in orange wool. **1978** *Detroit Free Press* 5 Mar. D 10/4 (Advt.), There are girls' topper sets and perky dresses.

3. (Earlier examples.)

1785 *Sessions Papers* 6 Apr. 571/2 One of them said, damn his eyes, *give him a topper at once.* **1803** in *Occasional Papers Univ. Sydney Austral. Lang. Res. Centre* (1980) No. 18. 44 He..had 'knocked him down and given him a topper for luck!'

6. (Earlier example.)

1874 Hotten *Slang Dict.* 327 *Topper*, the tobacco which is left in the bottom of a pipe-bowl..; or the stump of a smoked cigar.

Hence **to·ppering** *vbl. sb.*

1829 P. Egan *Boxiana* 2nd Ser. II. 671 The *topper*-ing system was in full practice, till *poor* Cock Robin went down quite exhausted!

toppie (tǫ·pi). [perh. f. Topknot.] A small brown bulbul of the genus *Pycnonotus*, found in southern Africa, either the Cape bulbul, *P. capensis*, or the red-eyed bulbul, *P. nigricans.*

1899 G. Russell *Hist. Old Durban* viii. 176 Doves cooed and 'Toppies' answered each other obtrusively. **1940** A. Roberts *Birds S. Afr.* 225 Cape Bulbul, or Toppie ..is not such a common bird as the Layard's Bulbul found further east. **1951** R. Campbell *Light on Dark Horse* x. 142 The commonest bird in Africa, the Toppie as we call him.., is a great character. **1964** D. Varaday *Gara-Yaka* xv. 126 His best friends became the Toppies, the conical-helmeted small dark birds that disclose the presence of snakes. **1971** J. Drummond *Farewell Party* vi. 35 There were birds..finches and toppies and a collared sunbird.

topping, *vbl. sb.*¹ Add: **1. e.** Also, filling up or bringing to capacity.

1919 *Gloss. Aeronaut. Terms* (R. Aeronaut. Soc.) 54 *Topping up*, the operation of replenishing the balloon with fresh gas. **1935** *Motor Commerce* Jan. 8 (Advt.), Acid-level indicator..shows when the maximum level is reached on 'topping-up', and so safeguards against overfilling. **1941** *Illustr. London News* CXCIX. 308 (*caption*) Interior contents of the rubber dinghy: bailer, 'topping-up' pump, chocolate, paddles, stoppers or plugs, and seadrogue. **1959** *Engineering* 6 Feb. 192/3 An infra-red heater provides a focal point for 'topping-up' if required. **1963** *Motor* 17 July 3/1 The automatic gearbox of my 3.4 Jaguar was in need of topping up. **1972** Hillier & Pittuck *Fund. Motor Vehicle Technol.* 534 A further check is the need for topping up.

g. *slang.* Execution by hanging.

1699 B. E. *New Dict. Canting Crew, Topping cheat*, the gallows. *Topping cove*, the hangman. **1846** *Swell's Night Guide* 134/2 *Topping*, hanging. **1968** *Daily Tel.* 15 Nov. 24/7, I wish they still had topping..because if I get bird.. for this I will get 30 years and I will top myself anyway. **1971** *Times* 6 Oct. 3/8, I deserve topping for shooting a copper.

h. *topping and tailing*, the action or practice of washing a baby's face and bottom; a sketchy wash. Also *top-and-tailing.* Cf. *top *v.*¹ 3 b.

1931 P. W. Yeomans *Happy Motherhood* vii. 69 A.. recommendation of the evening tub comes when the baby crawls and gets really grubby. Topping-and-tailing is not then sufficient. **1941** U. Orange *Tom Tiddler's Ground* xi. 205 The next hour was a busy one, what with Norman's 'topping and tailing', Norman's bottle and Marguerite's bath. **1960** C. Day Lewis *Buried Day* ii. 31 We did not go in for a desperate amount of washing—top-and-tailing twice a day, and a hip-bath once a week. **1972** *Times* 9 Aug. 7/7 Washing and changing a baby..can be a problem... Topping and tailing is all that is required.

i. *topping out* (*ceremony*), (the ceremony accompanying) the finishing of the roof of a building. Cf. *top *v.*¹ 20 a.

1961 *Times* 9 Sept. 16/5 Mr. H. R. Lake, New Zealand Minister of Finance,..performed the topping-out ceremony. **1962** *Guardian* 3 Nov. 6/6 The mystical ceremony of 'topping out' was performed this morning on the domed roof of the new Smithfield Market. A green bough was nailed to the roof..as a means..of warding off evil spirits and protecting the future inhabitants. **1973** *Times* 19 Mar. 11/4 A topping-out ceremony of a new shopping centre being built for Ravenstone Securities..took place in Glasgow. **1977** P. Van Greenaway *Destiny Man* i. 9 For any building, there must be some sort of topping out. **1984** *Daily Tel.* 26 Jan. 6/4 Some concrete words of advice from Mr Gow..as he performed the 'topping-out' ceremony for the International Conference Centre in Westminster yesterday.

4. d. A top layer or garnish put on food, esp. dessert (see also quots. 1926, 1927). Chiefly *U.S.*

1926 *Amer. Speech* I. 653/2 *Toppings*, pastry or cakes. **1927** *Ibid.* II. 389/2 *Toppings* refers to pastry, because it is used to top off a meal. **1950** *Manch. Guardian Weekly* 31 Aug. 5 Sundae 'toppings'. **1978** *Chicago* June 248/2 Toppings—including sausage, pepperoni, green peppers, anchovies, and fresh mushrooms are generously applied [to pizzas]. **1981** *Living Trends* (U.S.) Summer 6 Ice cream plus one or two toppings, such as crushed peanuts, toasted coconut, maple syrup, cherries.

topping, *ppl. a.* **4.** (Earlier examples.)

1815 D. Humphreys *Yankey in Eng.* 30 She's lofty—topping—has her highs—sometimes. **1852** Mrs. Stowe *Uncle Tom's Cabin* II. xxix. 147 She'd..have me know.. that I wasn't going to be so topping as I had been.

topple, *v.* Add: **3. a.** (Later *fig.* examples.)

1951 *Amer. Speech* XXVI. 230/2 California topples Washington. **1970** A. Toffler *Future Shock* (1971) viii. 177 Research topples older conceptions of man and nature. **1976** *Evening Post* (Nottingham) 15 Dec. 23 They beat Scotland 6-3 (one drawn) in the semi-finals and went on to topple England 'A' 6-1 (three drawn) in the final. **1979** *Daily Tel.* 26 May 14/4 A painting by Burne-Jones..made £48,000, toppling the artist's previous best price of £33,000. **1983** *Times* 15 Feb. 7/1 The revolution that toppled the regime of Emperor Haile Selassie eight years ago.

topply (tǫ·pli), *a.* [f. Topple *v.* + -y¹.] Liable to topple over.

1913 J. Vaizey *College Girl* xvii. 236 The screen's..a topply one. **1950** G. Greene *Third Man* ix. 67 The inadequate too-fancy topply table.

toppy, *a.* Restrict 'Now *low*' to senses in Dict. and add: **d.** *U.S.* Of animals: of superior quality.

1893 [in Dict.]. **1927** J. Lomax *Cowboy Songs* 303 I've cut your toppy mounts, boys. **1938** *Sun* (Baltimore) 24 Oct. 13/1 Firm rates were paid for vealers..and indications at times suggested that more of the 'toppy' offerings would have met with good reception. **1960** A. West *Trend is Up* (1961) v. 145 He turned to his favorite page in the book which showed a nice toppy lot of young lambs.

e. *Mus.* Containing too much treble. *colloq.*

1956 B. Edwards in S. Traill *Play that Music* vi. 63, I should want from a side-drum plenty of response and life —without being too 'toppy' so that the drum has a certain amount of 'guts' when it comes to playing in large places. **1969** W. Rutherford *Gallows Set* vi. 85 'What the hell's "toppy sound"?' 'Sound with too much treble in it. It's caused by the camera running slow when the film was taken so when its projected the sound is speeded up.' **1976** *Gramophone* Aug. 360/2 It could be argued that some modern discs are too toppy.

f. *Stock Exchange.* Of a market currency, etc.: high and unstable. *colloq.*

1961 *Spectator* 26 May 774 When markets became 'toppy' he should be busy switching from vulnerable to stronger positions. **1968** *Economist* 5 Oct. 81/2 What could be developing in London is a 'toppy' situation for a few months as the market consolidates the laurels it has earned. **1979** *Daily Tel.* 2 Aug. 17 Sterling suddenly looked toppy. **1983** *Times* 28 Apr. 16/2 If, however, the market looks 'toppy', it is the highest-rated sectors which are likely to suffer most.

TOPS (tǫps). [Acronym f. the initial letters of *T*raining *Op*portunities *S*cheme.] A system of vocational training programmes established in 1972, and organized by the Training Services Agency within the Manpower Services Commission. Also *attrib.*

1975 *Ann. Rep. Manpower Services Comm. 1974–75* 8/3 The government's vocational training programmes, which were to be developed into the Training Opportunities Scheme (TOPS) to meet the needs of individuals. *Ibid.*, The initial target for the TOPS scheme was to increase the number of people trained each year. **1979** *Jrnl. R. Soc. Arts* Dec. 10/1 The Board arranged for TOPS candidates to be provided with special examinations in a limited range of subjects.

topside, *sb.* (*adv.*) Add: **A.** *sb.* **a.** (Later example.)

1980 *Family Handyman* Sept. 85/1 Did you just smear caulk over the topside crack and hope for the best?

c. (Earlier example.)

1896 *Girl's Own Paper* 8 Feb. 295/1 Braised beef.—A piece of 'top-side' is best for the purpose.

f. *Oil Industry.* (See quot. 1948.) Now usu. with reference to the equipment and installations above water in offshore drilling. Freq. *attrib.*

1948 *Dialect Notes* IX. 60 *Top side*, any place above ground or ground level... Borrowed from nautical usage. **1975** *Offshore* Aug. 136/3 The system consists of a neutrally buoyant helmet mounted camera assembly and topside controls. **1977** *Offshore Engineer* Aug. 38/2 The manufacture of topside equipment. **1981** *Daily Tel.* 2 June 2/7 He [*sc.* a diver] complained of 'poor topside management'. **1982** *Sci. Amer.* Apr. 35/1 In designing platforms for North American conditions one can assume extended periods of calm for fastening the platform to the sea floor and erecting the topsides.

g. (With capital initial.) The upper or ruling classes, the Establishment. Freq. *attrib.* passing into *adj.*

1958 J. B. Priestley *Topside* 5 Topside people. *Ibid.* 8 Topside..takes and uses power, controls all patronage, imposes whatever pattern it prefers on the life of the nation. **1959** *New Statesman* 10 Jan. 47/2, I make this criticism as one who, in general, agrees with Priestley. What he calls Topside, what others call the Establishment, and what others, including myself, still call the capitalist ruling-class, does indeed..behave very much as Jordan asserts. **1962** *Guardian* 6 Oct. 5/2 J. B. Priestley, quite an attraction even in Topside Cheltenham. **1973** *Listener* 7 June 742/3 Jazz in the thirties in Topside circles was synonymous with vulgarity.

h. *Meteorol.* The part of the ionosphere above the height at which the concentration of free electrons is greatest, viz. about 300 km. Freq. *attrib.*

1962 *Canad. Jrnl. Physics* XL. 1692 The sounder is part of the 'Alouette' satellite.., which was launched at 0605 GMT on September 29, 1962... Several unfamiliar phenomena appear on the top-side ionograms... The top-side sounder project is an international one. **1965** Heikkila & Axford in C. O. Hines et al. *Physics of Earth's Upper Atmosphere* v. 114 Most recently, the technique of 'topside sounding' from a satellite has been accomplished. *Ibid.*, Better geographical coverage can be obtained now for the topside than for the bottom. **1976** *Nature* 19 Aug. 675/1 It may be possible to consider this process as a means of modifying the topside ionosphere by using a high power, low frequency transmitter at high latitudes. **1979** J. K. Hargreaves *Upper Atmosphere* iii. 39 To study the 'topside', an ionosonde may be carried on a satellite.

B. *adv.* Freq. with reference to the upper deck of a ship. Also, to the top, and in form *topsides.*

1946 P. Carter in Aldiss & Harrison *Decade 40s* (1975) 115 Chief [Navigator] Schmidt..relayed data topside. **1971** H. T. Walden *Anchorage Northeast* i. 27 While the ferry awaited its capacity load of eight vehicles a band topsides serenaded the passengers with martial airs. **1976** L. Sanders *Hamlet Warning* (1977) ii. xix. 164 On the third level below the main deck, they met four

TOP-SIDER

men..fleeing topside in panic. **1977** *New Yorker* 15 Aug. 54/3, I bring two of the sandwiches topside. **1978** *Guardian Weekly* 2 Apr. 24/5 A carrion crow..who, though damaged, can get top-sides of the noisy pack of gulls who winter near here.

To·p-sider. Also topsider. [f. TOPSIDE *sb.* + -ER[1].] A kind of casual shoe, freq. of canvas with a rubber sole.

A proprietary term in the U.S.
1937 *Official Gaz.* (U.S. Patent Office) 13 Apr. 261/1 Top-sider. For boots and shoes made of a combination of rubber or rubber substitute in combination with either fabric or leather or both. **1958** S. A. GRAU *Hard Blue Sky* (1959) 17 He waited perched on the railing picking the shells from the soles of his topsiders. **1968** [see *SPANDY *a.*]. **1977** *New Yorker* 10 Oct. 121/1 Standing in Topsiders and white ducks.

Topsy (tǫ·psi). The name of a character in Mrs. H. B. Stowe's novel *Uncle Tom's Cabin*; used allusively as the type of something that seems to have grown of itself without anyone's intention or direction (see quot. 1851).
[**1851** MRS. STOWE *Uncle Tom's Cabin* in *Nat. Era* 6 Nov. 1/5 Have you ever heard anything about God, Topsy?.. Do you know who made you?' 'Nobody, as I knows on,' said the child... 'I spect I grow'd. Don't think nobody never made me.'] **1885** KIPLING *Let.* in Ld. Birkenhead *Rudyard Kipling* (1978) vi. 81, I have really embarked..on my novel *Mother Maturin*—Like Topsy 'it growed' while I wrote. **1936** C. ROUSE *Old Towns* i. 17 The planning of towns in medieval England can be said to have followed no given rule—they were like Topsy who 'just growed'. **1955** *Times* 30 Aug. 9/7 It may be that political parties must emulate Topsy and just grow. **1967** *Boston Sunday Herald* 9 Apr. 27/4 This practice [*sc.* bugging] has grown like Topsy. **1973** 'J. RYDER' *Trevayne* (1974) xxv. 201 Are you implying that it [*sc.* a business] just grew—a Topsy? **1982** *Oxford Times* 30 Apr. 31/1 The garden, like Topsy, 'just growed'.

topsy-turvy, *adv.* (*a.*, *sb.*, and *v.*) Add: **topsy-turvydom** (earlier example); **topsy-turvying** *vbl. sb.* (later example).
1870 W. S. GILBERT in *Fun* 19 Mar. 15/1, I dreamt that somehow I had come To dwell in Topsy-Turveydom! **1967** [see *SKOOB].

to·psy ve·rsy, colloq. alteration of TOPSY-TURVY *adv.*, after ARSY-VERSY.
1767 D. GARRICK *Peep behind Curtain* (1772) 9 Damn all these new vagaries, that put us all upon our heads topsy versy. **1911** J. MASEFIELD *Everlasting Mercy* 46 Joe, and Si, and Nick, and Percy I rolled together topsy versy.

to·p-up. [f. the vbl. phr. *to top up* s.v. *TOP *v.*[1] 21.] An addition; that which serves to complete an amount or a number; a filling up to the top of something already partly full (esp. a glass of alcoholic drink). Freq. *attrib.*
1967 N. FREELING *Strike Out* 55 Fifty children that already had their top-up shots [against polio] in February. **1968** J. LOCK *Lady Policeman* xvi. 138 Cypriots..predominated, then a fair sprinkling of Italians, with a top-up of every nationality under the sun. **1971** A. MORICE *Death of Gay Dog* vii. 85 'Time for a top-up.' 'No more, thank you.' **1977** *Proc. R. Soc. Med.* LXX. 160/2 Post-operative analgesia occurs for a few hours after short operations, but is otherwise much as after other methods, unless 'top-up' doses are given through a catheter. **1978** A. BAINES in J. M. Thomson *Future of Early Music in Britain* 23 The next stage should be..to..get the Associated Board to accept this instrument—but excluding the crumhorns etc. save perhaps as optional top-ups. **1983** *Truckin' Life* Aug. 66/1 We..were looking for some improvement in consumption at the top-up point, the Ampol station. **1984** *Times* 13 July 2/5 Moneylenders..offer 'top-up' loans..before the original debt has been settled.

toque. Add: **1. c.** *Canad.* = TUQUE.
1890 S. M. ST. MAUR *Impressions Tenderfoot* 265 [He] was..made picturesque by a red cap 'toque', sash and red duffel overall stockings. **1906** G. LAWRENCE *Let.* 20 Jan. in R. E. Watters *Brit. Columbia* (1958) 7 With his toque pulled down..he looked for all the world like one of the small gnomes we children used to see pictures of. **1945** K. M. HAIG *Brave Harvest* 170 They were matched by toques and mittens, and tied in with gay voyageur sashes streaming like banners against the snow. **1972** *Daily Tel.* (Colour Suppl.) 8 Dec. 10/4 He wore a woollen ski toque on his head. **1977** *Westworld* (Vancouver, B.C.) Jan.–Feb. 44/2 A few other clothing essentials to remember are an extra pair of wool socks, some wool mitts, with a waterproof outer mitt, and a woolen toque.

d. A kind of tall white hat worn by chefs. In full *toque blanche*. orig. *U.S.*
[**1965** C. KLEIN *Professional Cook* ii. 28 The origin of the high white hat (*la toque blanche*) that has been the cook's trademark for centuries is in the monastery.] **1966** *McCall's* June 168/4 The symposium ended with a discussion about the shape of the chef's hat, the *toque blanche*. **1975** *N.Y. Times* 1 Nov. 18/1 The wearer of the toque blanche..is among a handful of women now presiding over restaurant kitchens here. **1977** *Guardian Weekly* 9 Oct. 19/1 France's two super-gourmets..found only 21 restaurants worth a chef's toque, their mark of at least qualified approval. **1978** R. CONDON *Bandicoot* i. 2 His cook..was a man of great height, made taller still by the two-foot-high, fluted *toque blanche* he wore to ventilate his head. **1981** *Listener* 17 & 24 Dec. 772/1 When the celebrated Swiss *chef de cuisine*..retired..he handed his toque to another Swiss.

toquilla (tǫkī·lʸa). [a. Amer. Sp.] = *JIPIJAPA a.
1877 *Encycl. Brit.* VII. 647/2 The leaves of the toquilla ..furnish material for the well-known hats. **1924** *Countries of World* III. 1713/1 The toquilla palm yields the leaf-fibre for the famous 'Panama' hats of Ecuador.

tor, *sb.* **3.** Add: **tor grass,** a perennial grass, *Brachypodium pinnatum*; cf. TORE *sb.*[3]
1954 C. E. HUBBARD *Grasses* 71 Tor grass..A worthless grass of neglected open grassland on chalk and limestone. **1976** *Times* 28 June 14/8 Coarse Tor and Erect Brome grasses have supplanted the grazed pastures.

tora (tō^ə·ra). [a. Amharic.] In full, *tora hartebeest*. A light brown hartebeest, *Alcelaphus buselaphus tora*, found in parts of north-eastern Africa.
1873 J. E. GRAY in *Nature* 4 Sept. 364 The British Museum has just received a series of skins of a new Bubale from Abyssinia called Tora... I propose to call it *Alcephalus tora*. **1873** —— in *Ann. Mag. Nat. Hist.* XII. 341 The male of the Tora..has a large, round, convex tuft..of darker hair in front of each eye. **1894** [see *TETEL]. **1912** J. STEVENSON-HAMILTON *Animal Life Afr.* vii. 102 The Tora Hartebeest..is found in Abyssinia and the Blue Nile Valley. **1970** DORST & DANDELOT *Field Guide Large Mammals Afr.* 220 The Tora Hartebeest..is pale tawny.

Toradja (tora·dʒā). Also **Toraja.** [Native name.] **a.** An Austronesian people living in central Sulawesi (formerly Celebes); a member of this people. **b.** The language of the Toradja.
1911 J. FRAZER *Golden Bough: Magic Art* (ed. 3) I. iii. 109 The Toradjas of Central Celebes believe that things of the same sort attract each other by means of their indwelling spirits. **1937** M. COVARRUBIAS *Island of Bali* ii. 16 The ancient inhabitants of the Malay Archipelago were 'Indonesians', also called Malayo-Polynesians,..and so forth. Of these, pure branches are to be found today in the Dyak of Borneo, the Batak of Sumatra, the Toradja of Celebes. **1957** *Encycl. Brit.* V. 89/2 The Toraja are a collection of tribes, living in central, southeast and east Celebes. **1964** E. A. NIDA *Toward Sci. Transl.* ix. 208 In east Toradja, spoken in Indonesia, 'they mentioned his-he-ness' is a way of rendering 'they spoke about him'. **1979** *Radio Times* 5–11 May 61/1 It's in the mountains of central Sulawesi that you find the land of the Torajas.

Torah. Add: **b.** *attrib.* and *Comb.*, as *Torah scroll*; esp. denoting ornaments or accoutrements of the parchment scrolls of the Torah, as *Torah breastplate, crown, curtain, finial, mantle, pointer, wrapper*.
1901 *Ann. Rep. Board of Regents Smithsonian Inst. 1899* 545 *Torah scroll*, parchment scroll of the Pentateuch in Hebrew mounted on wooden rollers. **1941** F. LANDSBERGER in *Hebrew Union College Ann.* XVI. 374 The ritual implements used in the synagogues, above all the Torah crowns..must have been made by Jews. *Ibid.* 398 The desire of German Jews for costly..ceremonial objects was strong, and there were..a few wealthy benefactors who could donate rich Torah curtains, Torah mantles, Torah finials, Torah breastplates, pointers..and the like. **1950** S. S. KAYSER in *Ibid.* XXIII. II. vii. 498 The final benediction of the Brith Milah ceremony and therefore.. the text on Torah wrappers. **1968** *N.Y. City* (Michelin Tire Corp.) 112 Jewish Museum..admirable examples of Torah Scrolls, the sacred texts of the Pentateuch... You can also admire..Torah wrappers, in finely engraved silver, and Torah mantles with silk and gold embroidery, ..and Torah pointers, used by rabbis to follow the text. **1976** Y. L. BIALER *Jewish Life* 100 (*caption*) Torah crown, hammered, gilded silver with precious stones..Poland, 18th century. *Ibid.* 120 (*caption*) Torah breastplate, pierced and engraved silver, Turkey, 19th century.

Toraja, var. *TORADJA.

‖ **torba** (tǭ·ɪbă). [ad. Arab. *turba* dust, earth, soil.] A primitive kind of cement made with broken pottery, traditionally used in Malta for the floors of buildings. Freq. *attrib.*
1910 *Ann. Rep. Valetta Mus. 1909–10* 4 The original 'torba' floor existed at a depth of about 15 cms. **1923** M. A. MURRAY *Excavations in Malta* I. v. 33 The beaten-earth floor, known as *torba*, is made as follows: the pots are broken..or..pounded; [etc.]. **1953** *Proc. Prehistoric Soc.* XIX. 43 They were found only on the lowest torba floor at Santa Verna and Kordin III. **1968** J. D. EVANS in S. Rossiter *Malta* 13 Remains of huts..were found at Skorba... The floors were of beaten earth or *torba* (a plaster made of crushed limestone).

torch, *sb.* Add: **1. a.** Now also = *electric torch* (*b*) s.v. *ELECTRIC *a.* 2 b.
1901 E. W. HORNUNG *Black Mask* xii. 253, I saw Raffles on my right striking with his torch; a face flew out of the darkness to meet the thick glass bulb with the glowing wire enclosed. **1936** W. FAULKNER *Absalom, Absalom!* ix. 370 He..saw the light of the torch approaching along the upper hall. **1967** P. SHAFFER *Black Comedy* 48 The Colonel takes the torch from Harold and shines it pitilessly in Schuppanzigh's face.

b. Also in phrs.: *to hand* (*pass*, etc.) *on the torch* (and varr.), to pass on a tradition, etc., esp. one of enlightenment (after L. *lampada tradere*, Gk. λαμπάδα παραδιδόναι, a metaphor from the ancient Greek torch-race; cf. LAMP *sb.*[1] 1 c); *to carry* (etc.) *a torch for* (someone), to feel (esp. unrequited) love for, to feel lingering affection for.
1887 *Q. Rev.* Oct. 276 Her [*sc.* Italy's] work has been done among the nations, and in their turn France, England and Germany hand on the torch. **1912** E. GOSSE *Portr. & Sk.* p. viii, They were all..engaged in keeping bright, and in handing on unquenched, the torch of literary tradition. **1927** *Vanity Fair* (N.Y.) Nov. 132/3 When a fellow 'carries the torch' it doesn't imply that he is 'lit up' or drunk, but girl-less. His steady has quit him for another or he is lonesome for her. **1932** L. GOLDING *Magnolia Street* I. xi. 189 He had sometimes hoped that in Max a son was born to him who would take the torch from his dying hand and jump on to the platform he had vacated. **1953** L. Z. HOBSON *Celebrity* vi. 78 Jim's still carrying a torch for Roosevelt. **1959** *Manch. Guardian* 16 June 5/2 She was carrying a torch for someone. **1969** J. GROSS *Rise & Fall Man of Lett.* iv. 104 Dante was the poet of Catholicism, who handed over the torch to Shakespeare, the poet of Feudalism, who passed it on to Milton, the poet of Protestantism. **1977** H. FAST *Immigrants* v. 305 Maybe you got a torch for her, maybe not. But we both got her interest at heart.

c. = *blow-torch* s.v. *BLOW-.
1909 WEBSTER, *Torch*, 3... Any of various devices for emitting a hot flame, as for vaporizing oil to start an oil engine, burning off old paint, melting solder, or the like. **1931** *Writer's Digest* Oct. 28 A keister torch is an acetylene torch which can be carried in a suitcase. **1961** *Sheet Metal Industries* XXXVIII. 613/1 The high rating of the water cooled models..is made possible by the design which permits the circulation of cooling water right to the tip of the torch. **1978** S. BRILL *Teamsters* vi. 225 Our theory is that the car was hidden there and then cut up with torches and carted out.

d. An arsonist. *U.S. slang.*
1938 *Reader's Digest* Mar. 71/1 The torch is now serving a 20-year sentence. **1977** *Time* 31 Oct. 28/3 Blazes are set by quasi-professional 'torches' hired by landlords, real estate brokers, store owners, or welfare tenants who want to be relocated.

3. *torch battery, -flare; torch-lighted* ppl. adj.; **torch-carrying** *vbl. sb.* (*fig.*), the harbouring of (esp. unrequited) love (see sense 1 b above); **Torch Commando** *S. Afr.*, (see quots.); **torch igniter** (see quots.); **torch singer** orig. *U.S.*, a singer of torch songs; **torch singing** *vbl. sb.* (orig. U.S.), the singing of torch songs; **torch song** orig. *U.S.*, a popular song on the subject of unrequited love; a sad sentimental or romantic song.
1926–7 *Army & Navy Stores Catal.* p. xxii/3 (Index), Batteries, Torch. **1957** C. SMITH *Case of Torches* x. 128 Bring in the..report on the torch batteries. **1970** E. R. JOHNSON *God Keepers* (1971) iii. 26 When does the torch-carrying stop..and work itself into hate? **1951** *Sun* (Baltimore) 9 Nov. 13/1 South Africa's Torch Commando, an organization of war veterans pledged to uphold the Dominion Constitution, is building up into a potent opposition to Prime Minister Daniel F. Malan's Nationalist Government. **1971** L. BLACKWELL *Blackwell Remembers* xviii. 158 The Torch Commando, a militant organization which strongly opposed the policy of the Nationalist Government relating to the coloured voters at the Cape. **1910** W. DE LA MARE *Three Mulla-Mulgars* xv. 205 All the Men of the Mountains came out with their little ones in the starlight and torch-flare to see them go. **1959** E. POUND *Thrones* cii. 82 Lit by the torch-flare. **1948** *Jrnl. R. Aeronaut. Soc.* LII. 170/2 The simple expedient of fitting an igniter plug in the flame tube was not sufficient to cater for the more arduous duties of ignition, such as under flight conditions at over 20,000 feet. The torch igniter was designed..to get over these problems. This unit was an ordinary igniter plug, with a subsidiary supply of fuel to it. The fuel was injected through a small hole on to the plug points. **1970** *Gloss. Aeronaut. & Astronaut. Terms* (*B.S.I.*) viii. 17 *Torch igniter*, a combined igniter plug and fuel atomizer for initiating combustion when starting the turbine. **1881** 'MARK TWAIN' *Prince & Pauper* xxxii. 365 We find the torch-lighted galleries already filling up with people. **1975** R. H. RIMMER *Premar Experiments* (1976) ii. 164 In every direction you looked, torchlighted faces were swaying to the music. **1934** J. O'HARA *Appointment in Samarra* vii. 210 Taking that dame out, that torch singer. **1973** *Times* 15 Dec. 10/1 She is sometimes a movie vamp, or a torch singer. **1947** E. JENKINS *Young Enthusiasts* 163 Jazz bands, torch singing and swing. **1983** *Listener* 9 June 35/4 If this is 'torch' singing, then Julie London is not a flimsy key-ring flashlight. **1927** *Vanity Fair* (N.Y.) Nov. 132/3 'Sing a torch song' is commonly used in Broadway late-places as a request for a ballad in commemoration of the lonesome state. Tommy Lyman is said to have created the slang and he announced one night: 'My famous torch song: "Come To Me, My Melancholy Baby".' **1939** G. GREENE *Lawless Roads* x. 256, I was grateful for the darkness and the torch songs. **1977** *Listener* 13 Oct. 481/2 The songs are pleasant parodies of Nashville, of torch songs and even of grand opera.

torch, *v.*[1] Add: **1. b.** To set alight, to set fire to, *esp.* in order to claim insurance money. *slang* (orig. and chiefly *U.S.*).
1931 *Writer's Digest* Oct. 29, I had just lit a match to torch the squib when I heard steps behind me. **1971** *Wall St. Jrnl.* 16 Aug. 1/3 Two bombs were planted in a university dormitory, part of a Belfast soccer stadium was torched, and snipers attacked army patrols. **1977** *Time* 31 Oct. 34/1 Griffith relied on an arsonist turned informant..who worked as a 'broker' for landlords eager to torch their property. **1979** *Arizona Daily Star* 22 July E 4/4 BIA police suspected a night of violence..when a group of people ran a car into Pike Creek... The car was then torched. **1983** *Granta* VII. 37 Halfway through the first pint of coffee, I torched a cigarette. Mmm, tasted good.

to·rcher[3]. *U.S. slang.* [f. TORCH *sb.* + -ER[1].] = *torch singer* s.v. *TORCH *sb.* 3.
1940 R. CHANDLER *Farewell, my Lovely* xli. 317 A handsome..torcher who could sing as if she meant it. **1975** J. GORES *Hammett* xxii. 147 A colored band..was backing a torcher.

torch-light. c. (Earlier example.)
1837 *New Yorker* 30 Sept. 441/3 A Loco-Foco 'Torch-Light Meeting', auxiliary to the larger concern in Tammany, was held in the Park on Thursday Evening.

torchon. Add: **torchon lace** (earlier example).
1865 F. B. PALLISER *Hist. Lace* iv. 53 But Venice point is now no more. The sole relic of this far-famed trade is the coarse torchon lace of the old lozenge pattern offered by the peasant women of Palestrina.

torchy, *a.* Restrict *rare* to sense in Dict. and add: **2.** Of, pertaining to, or characteristic of a torch song or torch singer. *colloq.* (orig. and chiefly *U.S.*).
1941 W. C. HANDY *Father of Blues* xxi. 285 The torchiest of all torch songs, *Melancholy Baby*. **1962** [see *SLINKY *a.*]. **1977** *Time* 25 July 60/2 He is married to Actress Diahnne Abbott, whose torchy rendition of *Honeysuckle Rose* in *New York, New York* upstages Liza Minnelli's belting.

tordion: see TURDION in Dict. and Suppl.

toreador. Delete ‖ and add: **c. toreador pants** chiefly *U.S.*, women's tight-fitting trousers, tapering to mid-calf.
1956 E. BAIN *Mugger* x. 83 She wore a white blouse, and black toreador pants, which tapered down to her naked ankles and feet. **1960** 'A. BURGESS' *Right to Answer* ii. 22 Veronica..went off, slim as a blade in toreador pants. **1974** R. B. PARKER *God save Child* vi. 48 She was dressed for a bull fight. Tight gold toreador pants... A ruffled red shirt,..a bronze wide-brimmed vaquero hat.

torenia (torī·niă). [mod.L. (Linnæus *Nova Plantarum Genera* (1751) 45), f. the name of the Rev. Olof *Torén* (1718–53), chaplain to the Swedish East India Company + -IA[1].] An annual herb of the genus of this name, belonging to the family Scrophulariaceæ, native to sub-tropical or tropical Africa and Asia, and bearing racemes of yellow, blue, or purple flowers.
1840 *Curtis's Bot. Mag.* LXVI. 3715 (*heading*) Heart-leaved Torenia. **1902** L. H. BAILEY *Cycl. Amer. Hort.* IV. 1822/1 Torenias are of easy cultivation. **1978** *Detroit Free Press* 16 Apr. (Gardening Guide) 12/1 Torenia or wishbone flower and browallia perform well in considerable shade.

tore-out (tō^ə·raut). [f. *tore*, dial. pa. pple. of TEAR *v.*[1] + OUT *adv.*] A small inferior type of sailing-boat.
1923 *Yachting Monthly* Jan. 145/1 Wave was of the genus usually referred to by East Coast watermen as 'little old toreouts', being, in fact, a small converted ship's boat. **1956** A. DAVISON *My Ship is so Small* ii. 19 Most of the boats in the yard were old tore-outs like the one I lived in. **1979** *Yachts & Yachting* 9 Nov. 1443/3 Let it be hastily explained that the Sirens in the title of this book are all boats: old toreouts, yachts and ultimately a 117ft ex-trading schooner.

torgoch. (Later examples.)
1924 *Glasgow Herald* 8 Feb. 8 Among the fish got in the Clyde at that time (about 1840) was the..torgoch or char. **1936** J. T. JENKINS *Fishes Brit. Isles* (ed. 2) 237 The Torgoch..inhabits two lakes near Llanberis.

torgsin (tǫ·ɹgsin). Also **Torgsin.** [a. Russ., contraction of *vsesoyuznoe ob"edinenie po torgovle s inostrantsami*, the All-Union Association for Trade with Foreigners.] A Soviet trading organization in the 1920s and 1930s which sold goods only in return for foreign currency. Used *attrib.*
1933 *Sun* (Baltimore) 24 Nov. 18/5 Sale of 200,000 pounds of bacon to the 'torgsin' stores, which cater only to foreigners in Soviet Russia, has been announced by Polish exporters. **1934** H. G. WELLS *Exper. Autobiogr.* II. ix. 819 All over Moscow and Leningrad you can bribe with foreign currency because of the absurd *Torgsin* system. **1968** *Listener* 3 Oct. 434/2 It was fairly easy for us because of those Torgsin shops where you could really get everything.

Torgut (tǭ·ɹgut). Now *Hist.* Also **Torgod, Torgot, Torgud.** [Native name.] A migratory Mongol people now absorbed into China; a member of this people. Also *attrib.*
1883 *Encycl. Brit.* XVI. 745/1 It was with no great difficulty..that his brother Ki Wang detached the greater part of the Kerait tribes from his banner, and founded the Torgod chieftainship. *Ibid.*, The position of the Torgod at this time..was rapidly becoming unbearable. **1947** AUDEN *Age of Anxiety* (1948) v. 103 The Timurids and Torguts. **1957** *Encycl. Brit.* VIII. 351/1 Some of them [*sc.* the western Mongols], notably the Torgots (Torgud) migrated as far west as the Volga, whence they returned, and their descendants are settled in the Ili district. **1962** E. SNOW *Other Side of River* (1963) xix. 142 In the past three centuries there had been no similar armed migration of a nation in Asia, with the exception of the amazing Flight of the Torgut from the Caucasus to Mongolia.

toric, *a.* Add: Also with pronunc. (tǫ·rik). (Further examples.) *spec.* in *Ophthalm.*, applied to a lens with one surface curved like part of a torus, the radius of curvature having a minimum value in one direction and a maximum value in the direction at right angles to it; also as *sb.*, such a lens.
1890 *Trans. Amer. Ophthalm. Soc.* V. 708 Such toric surfaces are concave, and, when sunk in a plate of glass, afford typical examples of concavo-plane toric lenses. **1954** S. DUKE-ELDER *Parsons' Dis. Eye* (ed. 12) iv. 41 A regularly astigmatic surface is said to have a toric curvature. **1962** L. S. SASIENI *Princ. & Pract. Optical Dispensing* x. 264 'Commercial' torics of minus powers are usually transposed to the form which provides a plus cylinder. **1973** *Nature* 21/28 Dec. 479/2 A toric segment of bore diameter *d*, generator circle diameter *D* and length *s* measured along the central axis. **1978** J. PARR *Introd. Ophthalmol.* ii. 44/1 A cylindrical curvature can be combined with a spherical curvature in a lens which is then called a toric surfaced lens. In astigmatism the surface of the cornea is toric instead of spherical.

‖ **torii** (tǫ·ri,i). Also **tori, torij.** [Jap., f. *tori* bird, fowl + *i* to sit, perch.] A ceremonial gateway in front of a Japanese Shintō shrine, consisting of two uprights and two cross-pieces of which the lower is straight and the upper usu. curved and projecting. Also *attrib.*
Occas. erron. interpreted as *pl.*
1727 J. SCHEUCHZER tr. *Kæmpfer's Hist. Japan* III. ii. 208 At the entry of the walk, which leads to the temple, stands..a particular fashioned gate, called *Torij*, and built either of stone or wood. **1874** C. HOEY tr. *Humbert's Japan & Japanese* I. ii. 14 A long avenue of fir-trees, headed by a sacred gate called a Tori. **1874** *Trans. Asiatic Soc. Japan* II. 116 The *torii* gradually assumed the character of a general symbol of *Shintō*. **1904** D. SLADEN *Playing the Game* II. v. 231 Tall torii, those mystic arches of Japan. **1911** *Encycl. Brit.* XV. 182 Originally designed as a perch for fowls which sang to the deities at daybreak, this torii subsequently came to be erroneously regarded as a gateway characteristic of the Shintō shrine. **1960** B. LEACH *Potter in Japan* ii. 55 Torii gateway leading to cryptomeria-shaded steps climbing to little empty Shinto Shrines. **1977** *Amer. Speech 1975* L. 69 Mills..must be almost as exotic to most younger Americans as torii or dagobas.

‖ **toril** (torī·l). Pl. **toriles.** [Sp.] One (*spec.* the last) of a series of pens in which a bull at a bullfight is confined before being released into the ring. Also *attrib.*
1893 CHAPMAN & BUCK *Wild Spain* v. 65 The noble bulls will be lured in their company away from their native plains,..to the entrance of the fatal *toril*. **1932** E. HEMINGWAY *Death in Afternoon* vi. 61 The alguacils ride up to under the president's box to ask for the key to the red door of the toril where the bull is waiting. **1961** *Times* 8 July 10/6 The *toril* gates open and out bounds the animal. **1974** *Encycl. Brit. Macropædia* III. 477/1 The *presidente municipal*..throws down to one of the *alguaciles* the key to the *toriles* or bull pens.

Torinese (tǫrinī·z), *a.* and *sb.* Also **Turinese.** [a. It. *Torinese*, f. *Torino* Turin: see -ESE.] **A.** *adj.* Of, pertaining to, or characteristic of the city of Turin in Piedmont, north-west Italy, its natives and inhabitants, or their dialect. **B.** *sb.* **a.** A native or inhabitant of Turin. **b.** The dialect of Italian spoken in Turin.
1883 H. JAMES *Little Tour in France* (1885) xix. 132 The shops are probably better than the Turinese, but the people are not so good. **1886** WEBSTER, Torinese *a.* and *sb.* **1960** A. COLQUHOUN tr. *di Lampedusa's Leopard* i. 19 The Piedmontese... Wouldn't things be just the same? Just Torinese instead of Neapolitan dialect; that's all. **1975** R. COBB *Paris & its Provinces* 2 The famous café.. that unkind Torinese have sometimes called *il caffè della mennapausa*. *Ibid.* 3 Two louts engage the auburn beauty in the coarsest Torinese. **1977** *Times* 26 Mar. 10/1 Turinese cooking. **1980** *Times* 16 Oct. 7/2 As the five acts of the *opera* drifted their way onwards the Torinese decided they had had enough.

‖ **torista** (torī·sta). [Sp.] An enthusiast for bullfighting who is chiefly interested in the performance of the bull.
1957 A. MACNAB *Bulls of Iberia* p. xii, The more solid Spanish *aficionados*, especially those known as *toristas* because their primary interest is in the Bull rather than the Bullfighter. **1967** MCCORMICK & MASCAREÑAS *Compl. Aficionado* ii. 31 The torista exalts the toro over the torero to the point where he will ignore and despise the man's best achievements.

torma (tǫ·ɹmă). Pl. **torma, tormas.** [a. Tibetan.] A sacrificial offering burned in a Tibetan Buddhist ceremony.
1895 L. A. WADDELL *Buddhism of Tibet* xii. 297 Another food-offering is a high conical cake of dough, butter and sugar, variously coloured, named *tormă*.., that is, holy food. It is placed on a metal tray supported by a tripod. **1929** D. MACDONALD *Land of Lama* xvii. 205 Everyone makes offerings in the temples, while the monks of the monastery prepare *torma*, symbolic emblems made of butter, on thin wooden or leather backing, often eight or ten feet high. **1958** *Illustr. London News* 13 Dec. 1041/1 A set of such figures, known as *torma* and representing divine personages, spirits and demons, as well as offerings of various forms, are required for every rite conducted by lamas. **1970** R. D. TARING *Daughter of Tibet* x. 135 On a certain day they burnt *tormas* (cone-shaped religious sweets). **1979** J. NORBU *Horseman in Snow* 46 The grass.. yields a red dye that is used to colour sacrificial offerings called tormas.

tormented, *ppl. a.* Add: **2.** *U.S. slang.* Used adjectivally and adverbially as a mild equivalent of DAMNED *ppl. a.* 4.
1825 J. NEAL *Brother Jonathan* I. 138 They hadn't come such a tormented long piece. **1867** J. R. LOWELL *Biglow Papers* 2nd Ser. p. lix, *Tormented*, euphemism for damned, as, 'not a tormented cent'. **1903** G. S. WASSON *Cap'n Simeon's Store* 86 It don't look right for nobody.. to take and hang on to them tormented ole witch-bridles so-fashion! **1938** M. K. RAWLINGS *Yearling* xi. 119 'Look at him,' she said. 'Tormented Yankee. His feet drag like a 'gator's tail.'

tormentor. Add: **3. f.** Freq. a device for squirting liquid (further examples).
1894 A. MORRISON *Tales of Mean Streets* 34 The ladies' tormentors are larger, and their contents smell worse than at any other fair. *Ibid.* 36 Billy bought a ladies' tormentor and began to squirt it at Lizerunt. **1912** J. MASEFIELD *Widow in Bye St.* 19 One's so safe with such a son to con her Through all the noises and through all the press, Boys daredn't squirt tormenters on her dress.

torn, *ppl. a.* Add: **c. torn-down** (examples in literal sense).
1933 S. SPENDER *Poems* 40 Through torn-down portions of old fabric. **1953** K. REISZ *Technique Film Editing* i. 35 Other fragments of the torn-down statue of the Czar reassembling.

tornadoed (tǫɹnē^ɪ·dō^ud), *ppl. a. nonce-wd.* [f. TORNADO + -ED[2].] Affected by tornadoes. (In quot. *fig.*)
1851 H. MELVILLE *Moby Dick* III. i. 19 Even so, amid the tornadoed Atlantic of my being, do I myself still for ever centrally disport in mute calm.

toro (tǭ·ro, ‖ to·ro). [Sp.] A bull used in bullfighting. Also, a child's bullfighting game.
1660 in T. Mathews *Collection of Lett.* 172 But, in a word, me-thinks, that not onely in their sports of Cannas and Toros, but even in some more solemn and serious things than those, they are not free from having still somewhat of the Moor. **1846** R. FORD *Gatherings from Spain* xxi. 291 The young urchins in the streets play at '*toro*', as ours do at leap-frog. **1932** R. CAMPBELL *Taurine Provence* 66 Novillos (bulls that are full grown but lack the ultimate footsureness of the toros). **1958** L. VAN DER POST *Lost World Kalahari* i. 29 The rhinoceros, angry like a pricked toro with the rosette of blood that comes to it in adolescence vivid on the flank. **1967** MCCORMICK & MASCAREÑAS *Compl. Aficionado* i. 4 He might be compared to the actor but for the fact that he writes his own script, in collaboration with the *toro*.

toroid (tǫ·roid). [f. TOR(US + -OID.] An object having the shape of a torus (sense *4); a toroidal object.
1886 G. S. CARR *Synopsis Elem. Results Pure & Applied Math.* I. II. 932/1 (Index), Toroid. **1903** *Astrophysical Jrnl.* XVIII. 339 Each reflecting surface..would be part of a parabolic toroid. **1916** G. KAPP *Princ. Electr. Engin.* I. xi. 211 The term 'toroid' in this connection means a circular coil of *n* turns of wire wound so that the cross-section is a circle. **1954** *Electronic Engin.* XXVI. 196 Magnetic toroids could..be used..for random access information storage at the rate of one core per digit. **1973** *Nature* 15 June 386/2 This field is supplemented by an additional field produced by a coil wound on the toroid.

toroidal, *a.* Add: (Earlier and later examples.) Also = TORIC *a.* in Dict. and Suppl.
1881 *Phil. Trans. R. Soc.* CLXXII. 609 By 'toroidal functions' are understood functions which satisfy Laplace's equation and which are suitable for conditions given over the surface of tores. **1946** [see *POLOIDAL *a.*]. **1948** *Electronic Engin.* XX. 28 The problem..was solved by using a very few turns of concentrically wound copper tape and achieving the neccessary self-inductance by winding on a high permeability toroidal core. **1962** CORSON & LORRAIN *Introd. Electromagn. Fields* v. 200 (*caption*) Toroidal coil of square cross section carrying a current *I*. **1973** *Sci. Amer.* Aug. 111/1 Optical surfaces are far from spherical these days; even our spectacle lenses are toroidal. **1975** *New Yorker* 12 May 84/2 Wind instruments were bolted to a toroidal buoy.
Hence **toroi·dally** *adv.*
1961 *Technology* Mar. 72/1 A variable-ratio toroidally-wound transformer. **1970** *Nature* 26 Sept. 1299/1 Toroidally contained fusion plasmas. **1978** *Gramophone* Aug. 396/3 To the right of the PCB is a massive, toroidally wound mains transformer.

Torontonian (torǫntō^u·niăn), *sb.* [f. *Toronto*, capital of the province of Ontario in Canada + -*n*- + -IAN.] A native or inhabitant of Toronto.
1875 *United Service Mag.* Dec. 500 'No great thing after all!' exclaimed a Quebecker or Torontonian. **1967** *Economist* 29 July 396/3 Torontonians are often suspected of believing that money decides everything. **1975** *Globe & Mail* (Toronto) 16 July 7/3 Many Torontonians are probably not aware of the Western Guard's propaganda.

torp (tǭɹp), slang abbrev. of TORPEDO *sb.* 2 and *torpedo juice* s.v. *TORPEDO *sb.* 4. Cf. *TORPS.
1929 *Papers Mich. Acad. Sci., Arts & Lett.* X. 330 *Torp*, a torpedo. **1945** J. BRYAN *Diary* 23 Apr. in *Aircraft Carrier* (1954) 193 Someone brought a pint of torp. **1967** B. KNOX *Blacklight* ii. 42 If anyone does find a stray torp, then they'll make damn' sure it stays lost.

torpedo, *sb.* Add: **2.** (Earlier example.)
1776 J. THACHER *Military Jrnl.* (1823) 75 Mr. Bushnell gave to his machine the name of American Turtle or Torpedo.

b. See *aerial torpedo* s.v. *AERIAL *a.* 5. Also without specifying adj.
1922 W. RALEIGH *War in Air* I. 467 The hope of using the torpedo, launched from the air, against ships which are sheltered and protected from naval attack, was never long absent from the minds of those who directed the activities of the Royal Naval Air Service. **1943** *Jane's Fighting Ships 1942* 5, 18 inch torpedoes are used by the torpedo bombers of the Fleet Air Arm.

3*. *slang.* **a.** *U.S.* A professional gunman. **b.** A tablet or capsule of a narcotic drug.
1929 G. L. HOSTETTER *It's a Racket!* 241 *Torpedo*, a professional gunman or bomb tosser. **1940** R. CHANDLER *Farewell, my Lovely* xxxvi. 239 There's yellow cops and there's yellow torpedoes. **1971** *Go ask Alice* (1972) 28 He introduced me to torpedoes on Friday and Speed on Sunday. **1973** P. EVANS *Bodyguard Man* iii. 24 Ask Al Capone. Mention his torpedoes to most people and they conjure up mental sketches of middle-aged heavyweights. **1978** M. RUSSELL *Daylight Robbery* xv. 154 The phial.. contained more tablets... He tried to estimate how long.. it took a couple of the torpedoes to send him off.

3.** = *torpedo-body*. Also, a car with such a body.
1909 *Daily Chron.* 13 Nov. 9/6 There is a general tendency..to utilise the form of body known as the 'torpedo'. **1930** V. PALMER *Passage* III. i. 213 Another car coming! They..watched with strained eyes as the dusty torpedo shot into view. **1968** *Compl. Encycl. Motorcars* 624 *Torpedo*,..an open touring car with an unbroken line from bonnet to windscreen, and from windscreen right through to the back of the car, the seats being flush with the body sides. Bodies of this design began to appear in about 1910.

4. (sense 2) *torpedo bomber, coxswain, gunner, plane, ship*; *torpedo-carrying, -proof* adjs.; **torpedo-body** (example); **torpedo destroyer** (earlier example); **torpedo juice** *slang*, intoxicating liquor extracted from torpedo fuel; any strong home-made alcoholic liquor; **torpedo-tube** (earlier example); **torpedo vessel** (earlier example).
1924 *Motor* 21 Oct. 630 (*caption*) A handsome torpedo body, on a Voisin chassis, by H. J. Mulliner. **1930** *Flight* 16 May 535/2 The aircraft equipment consists of..two flights of two-seater fleet torpedo bombers. **1970** *Times* 24 Mar. 2/4 (*caption*) Using engine cylinders taken from an exhibit at the Imperial War Museum, the last airworthy Swordfish torpedo bomber in the Royal Navy. **1922** W. RALEIGH *War in Air* I. 466 The torpedo-carrying aeroplane or seaplane would outrival the submarine as a weapon of offence against enemy shipping. **1903** *Windsor Mag.* XIX. 6/2 Speakin' as a torpedo-coxswain,..I presume we fall in. **1918** KIPLING *Land & Sea Tales* (1923) 107 If his torpedo-coxswain had ever allowed anyone to look there. **1896** *World* 12 Feb. 29/1 It would not cost us much—not so much, in the long run, as a single torpedo-destroyer. **1903** *Windsor Mag.* XIX. 9/2 What's a torpedo-gunner more or less to a full lootenant? **1928** C. F. S. GAMBLE *Story N. Sea Air Station* xiv. 234 The crew consisted of the pilot, the observer, the torpedo-gunner, and a machine-gunner for the back seat. **1946** *Seafarers' Log* 31 May 13/5, I have known many Navy men who were chronic drinkers at sea as well as ashore. Some have gone blind from drinking torpedo juice. **1961** *Guardian* 26 Sept. 9/5 Torpedo juice is a combination of these [*sc.* bush beer and toddy] and acquires its name from its lethal effect. The original torpedo juice was the neat alcohol extracted from torpedoes during the war by American servicemen and sometimes mixed with local bush beers to soften the blow. **1917** *Flying* 1 May 317/2 Admiral Fiske declared that torpedo-planes in the battle of Jutland would have given a tremendous advantage to the side employing them. **1981** G. MACBETH *Kind of Treason* xi. 106 The *Prince of Wales* and the *Repulse* had been sunk by Japanese torpedo planes. **1914** H. H. ASQUITH *Let.* 27 Oct. in M. Gilbert *W. S. Churchill* (1972) III. Compan. I. 220 Torpedo-proof harbours and refuges. **1873** *Illustr. London News* 29 Mar. 294/2 It is also proposed to build a small torpedo-ship of 214 tons burden by way of trying experiments. **1911** *Q. Rev.* Oct. 476 This gradual merging of the essential features of the gun-ship and the torpedo-ship is now about to find expression in the submarine. **1893** *Souvenir World's Fair: Naval Exhibit*, The battery mounted comprises..two Gatling guns, and six torpedo tubes or torpedo guns. **1877** *Illustr. London News* 16 June 556 Steel Torpedo-vessel used by the Russians on the Danube.

torpex (tǭ·ɹpeks). [Blend of TORPEDO *sb.* + EXPLOSIVE *sb.*] An explosive consisting largely of T.N.T., cyclonite, and aluminium, used for depth charges.
1948 *Jane's Fighting Ships 1947–48* 6/1 These charges could be set to explode at any depth, the former [*sc.* hedgehog] employing torpex, and the latter minol. **1974** [see *RDX].

‖ **torpilleur** (tǭɹpiyö̂ɹ). *Obs.* exc. *Hist.* [Fr., f. *torpille* torpedo.] A torpedo boat or destroyer in the French navy.
[**1894** W. LE QUEUX *Great War in England in 1897* xiv. 88 The reinforcements consisted of the French battleships ..together with nine cruisers, and thirty-eight *torpilleurs de haute mer*.] **1950** *Jane's Fighting Ships 1949–50* 181/2 (*caption*) 3 ex-German Type (Classed as *Torpilleurs*.)

torps (tǭɹps), slang abbrev. of *torpedo-lieutenant* s.v. TORPEDO *sb.* 4. Cf. *TORP.
1914 'BARTIMEUS' *Naval Occasions* xxiv. 237 The Torpedo Lieutenant (hereinafter known as 'Torps') was awakened by the June sunlight. **1943** C. S. FORESTER *Ship* 20 Torps and Lightfoot, the Officers of the Watch.

torque. Add: **c. torque converter,** a device that varies or multiplies torque; **torque meter, torquemeter** = *torsionmeter* s.v. *TORSION 3; **torque motor** *Electr. Engin.*, an electric motor designed to exert a torque without continuous rotation; **torque wrench,** a device to set and adjust the tension of nuts and bolts.
1934 WEBSTER, Torque converter. **1944** *Machine Design* July 117 The Lysholm-Smith torque converter.. incorporates multiple-stage turbine blading. **1970** *Commercial Motor* 25 Sept. 65/3 Its 190 bhp is passed through a two-stage torque-convertor which in its hydraulic stage multiplies the engine torque by 4·7 and in its mechanical stage has a ratio of 1·046 to 1, a slight overdrive. **1911** H. M. HOBART *Dict. Electr. Terms* II. 564/1 Torque meter. **1955** *Electronic Engin.* XXVII. 430/1 Several types of torquemeter are known but..it has sometimes proved difficult to obtain reliable results in this particular application [*sc.* in aircraft]. **1926** *Gloss. Terms Electr. Engin.* (Brit. Engin. Stand. Assoc.) 54 Torque motor. **1946** *Nature* 13 July 54/2 As a torque motor, the machine was required to operate on 333 c./s. and works continuously under standstill conditions, with a temperature rise at 20 volts/phase of about 30°C. **1979** *Engin. Materials & Design* Oct. 43/1 Torque motors are those which are designed to provide their maximum torque under the conditions of 'stall' or 'locked rotor'. **1948** A. W. JUDGE *Automobile Engine Overhaul* (ed. 3) x. 187 It is important to tighten each nut a little at a time... A torque wrench is here an advantage. **1979** G. HAMMOND *Dead Game* v. 55 Janet was still working on the hydraulics of the tractor... She pulled gently on the torque-wrench.

torque, *v.* [f. TORQUE[2].] *trans.* To apply torque to. So **to·rquing** *vbl. sb.*
1954 *Fasteners* IX. v. 3 The subject of bolt torquing. **1960** D. A. HALPERIN *Building with Steel* xii. 163/2 He then checks the bolts to verify that they have been torqued to at least specified minimum tension. **1978** *Nature* 12 Oct. 517/1 Perhaps these radio galaxies are systems where a pre-existing central black hole is gradually being torqued into alignment with the angular momentum of newly supplied fuel that has reactivated the nucleus. **1981** *Ibid.* 24 Sept. 261/2 Drilling conditions were rugged... Cave-ins and torquing finally caused them to abandon the hole well above their oceanic crustal target.

torquey (tǭ·ɹki), *a.* [f. TORQUE[2] + -Y[1].] Of the engine of a motor vehicle: producing plenty of torque; able to pull well.
1977 *What Car?* Apr. 59/1 Tractable, torquey engine. **1981** *Motor* 27 June 9/2 The torquey, sweet engine.

torr (tǭɹ). *Physics.* Also **Torr.** [f. the name of *Torricelli* (see TORRICELLIAN *a.*).] A unit of pressure used chiefly in measuring partial vacuums (see quot. 1958); 133·32 newton/sq. metre.
1949 S. DUSHMAN *Sci. Foundations Vacuum Technique* i. 4 In German literature, 1 Tor (or 1 Torr) is used to designate 1 mm Hg pressure. **1958** *Gloss. Terms Vacuum Technol.* (*B.S.I.*) 7 *Torr*, a unit of pressure defined by the relationship: 760 torr = 1 standard atmosphere (atm) = 1 013 250 dyn/cm² exactly. 1 torr is equal to the conventional barometric millimetre of mercury..within 1 part in 7×10^6. *Ibid.* 20, 1 torr = 1·333 22 millibar (mb) approximately. **1981** J. B. ADAMS in J. H. Mulvey *Nature of Matter* vii. 151 It is necessary to achieve a very low pressure in the vacuum chamber; pressures of 10^{-11} Torr are essential.

Torrens (Tǫ·rĕnz). The name of Sir Robert *Torrens* (1814–84), first Premier of South Australia, used *attrib.* in *Torrens system*, a system of land title registration devised by him, and adopted in Australia and elsewhere outside the U.K.
1863 R. R. TORRENS (*title*) Transfer of land by 'registration of title', as now in operation in Australia under the 'Torrens system'. **1905** J. E. HOGG (*title*) The Australian Torrens system. *Ibid.* p. v, The system of land transfer and registration known as the Torrens system has now been in operation in Australasia more than forty-six years. **1976** S. R. SIMPSON *Land Law & Registration* v. 68 Outside Great Britain the system of registering title..is widely known as the 'Torrens system', for it was Sir Robert Torrens who in 1858 introduced registration of title into South Australia, the first jurisdiction (at least of jurisdictions using English land law) to establish such a system.

torreyite (tǫ·ri,əit). *Min.* [f. the name of John *Torrey* (1796–1873), U.S. scientist + -ITE[1].] A hydrated basic sulphate of magnesium, manganese, and zinc, $(Mg, Mn)_5Zn_2SO_4(OH)_{12}.4H_2O$, found as massive aggregates of white to colourless monoclinic crystals.
1949 J. PREWITT-HOPKINS in *Amer. Mineralogist* XXXIV. 595 A new name, torreyite, is proposed to replace the name delta-mooreite. **1979** *Ibid.* LXIV. 952/2 Torreyite in fine-grained aggregates resembles hardened granular sugar.

Torricellian, *a.* (Earlier example of *Torricellian vacuum*.)
1753 *Phil. Trans. R. Soc.* XLVII. 371 The Torricellian vacuum then occupied a space of about thirty inches.

Torridon (tǫ·ridǫn). *Geol.* The name of Loch *Torridon* on the NW. coast of Scotland, used *attrib.* to designate Torridonian rocks (see next), which are well exposed there.
1873 *Q. Jrnl. Geol. Soc.* XXIX. 334 The lowest conglomerate bed of the Torridon sandstone is seen in fine section in a cliff north-west of Brochel castle. **1896** [see next]. **1930** PEACH & HORNE *Chapters Geol. Scotl.* iii. 83 The various groups of Torridon Sandstone are not equally developed along the belt from Cape Wrath to Skye. **1963** D. W. & E. E. HUMPHRIES tr. *Termier's Erosion & Sedimentation* i. 7 In this field the studies of British scientists.. extend over rocks of all ages back to the Torridon Sandstone (Precambrian).

Torridonian (tǫridō^u·niăn), *a. Geol.* [f. prec. + -IAN.] Of, pertaining to, or designating the later of the two main series of Pre-Cambrian rocks in NW. Scotland, which occur in a narrow belt running from Cape Wrath to Skye and consist chiefly of sandstones, grits, and shales; also the time of their deposition. Also *absol.*, the Torridonian series.
1896 J. W. JUDD *Student's Lyell* xxviii. 435/1 The Torridon Sandstone or Torridonian. *Ibid.* 435/2 The Cambrian strata being found lying on every portion of the Torridonian series. **1934**, etc. [see *MOINE]. **1938**, etc. [see *MOINIAN *a.*]. **1952** *Geol. Mag.* LXXXIX. 70 The possibility of a Torridonian age for this very isolated outlier should not be excluded. **1969** BENNISON & WRIGHT *Geol. Hist. Brit. Isles* iii. 46 Resting with marked unconformity upon the Lewisian of the foreland is the Torridonian Series, an enormous thickness of dominantly arenaceous rocks. **1971** *Country Life* 18 Nov. 1349/3 The fertile oolite soil gives way to Torridonian sandstone just north of James Gillie's croft. **1976** T. R. OWEN *Geol. Evolution Brit. Isles* ii. 19 Radiometric datings (1000–800 m.y.) now confirm that the Torridonian and the Moine are the lateral equivalents of one another.

torry (tǫ·ri), *v.* [ad. Sp. *torear* to fight (a bull), to be a bullfighter.] *trans.* To provoke and fight (a bull).
1936 R. CAMPBELL *Mithraic Emblems* 52 The white Torero—him who took the toss Sky-high upon the black horns of the Cross, For torrying the hornèd prince of Death. **1957** A. MACNAB *Bulls of Iberia* viii. 81 To give passes to a bull is not the same as to 'torry' it (*torear*). *Torear* means to exercise control over the bull at all times, and to work on it as desired. *Ibid.* 263 *Torear*, to 'torry', neologism..adopted from Roy Campbell..who has used it in print for twenty years.

torsade. (Earlier example.)
1872 *Young Englishwoman* Nov. 593/1 The black velvet bonnet is trimmed with a torsade of violet faille ribbon.

‖ **Torschlusspanik** (tǭ^ə·ɹʃluspa:nik). [Ger. lit. 'shut door (or gate) panic'.] A sense of alarm or anxiety (said to be experienced particularly in middle age) caused by the suspicion that life's opportunities are passing (or have passed) one by; *spec.* that manifested in an ageing woman who longs to (re)discover the (sexual) excitement of youth, and who fears being left 'on the shelf'.
1963 P. BRACKEN *I Hate to Housekeep Bk.* ix. 92 The random housewife is often prone to Torschlusspanik, or fear of being locked in the park at night, after the gates are closed. **1977** *Time* 8 Aug. 21/3 She was haunted by *Torschluss-panik* (mid-life crisis). **1980** *Times Lit. Suppl.* 14 Mar. 287/2 She [*sc.* Mme de Staël] is perhaps history's most outstanding case of *Torschlusspanik*: the panic at the shutting of the door.

torsiograph (tǭ·ɹsiograf). *Mech.* [f. late L. *torsiō* (see TORSION) + -GRAPH.] An instrument for measuring torsional oscillations of the crankshaft of an engine.
1930 *Engineering* 25 Apr. 551/3 The handiest instrument, and therefore the most widely used, is the torsiograph. **1950** *Ibid.* 1 Sept. 204/1 It was decided..to carry out torsiograph tests on the engines under specially standardised conditions with specified instrumentation.

torsion. Add: **1. e.** *Math.* The degree to which a curve departs from being planar at any given point, measured by the rate of change of the angle of the osculating plane or the binormal with respect to distance along the curve; *radius of torsion*, the reciprocal of this.
1862 G. SALMON *Treat. Analytic Geom. Three Dimensions* xi. 269 The angle made with each other by two consecutive osculating planes..we shall call the angle of torsion, and denote by $d\eta$. *Ibid.* 270 Following the analogy of the radius of curvature which is $ds/d\theta$, the later French writers denote the quantity $ds/d\eta$ by the letter *r*, and call it the radius of torsion. **1939** BURINGTON & TORRANCE

Higher Math. vi. 711 Torsion is agreed to be positive when the rotation (with *s* increasing) of the binormal increases in the same sense as that of a right-handed screw traveling in the direction of **t**. **1978** E. C. YOUNG *Vector & Tensor Analysis* ii. 106 The torsion of a plane curve is zero, just as the curvature of a straight line is zero.

f. *Zool.* The twisting of the visceral hump of gastropod molluscs through 180 degrees when the embryo reaches a certain stage of development.
1888 ROLLESTON & JACKSON *Forms Animal Life* (ed. 2) 475 In..the *Streptoneura*, the posterior union of the visceral nerves..is situated dorsally to the intestine, and the loop is therefore twisted with the torsion of the visceral dome. **1930** G. R. DE BEER *Embryol. & Evol.* vii. 53 The limpet develops into a more or less symmetrical Veliger larva which suddenly undergoes a twist through 180°, the process of torsion occupying two or three minutes. **1972** M. S. GARDINER *Biol. of Invertebrates* ii. 59/2 Torsion appears to be a reversible process, for in some genera the anus and the organs on either side of it lie posteriorly and the nerve commissures are untwisted.

g. *Chem.* Restricted rotation of an atom or group about a bond joining it to another atom.
1932 *Physical Rev.* XL. 445 (*heading*) The torsion oscillator-rotator in the quantum mechanics. **1978** *Nature* 14 Dec. 674/1 Although in some cases, rotation of a rigid molecular structure cannot lead to superposition, this may be possible as a result of torsion about certain bonds.

3. torsion bar, a bar that is subject to torque; *spec.* one in the suspension of some motor vehicles, fixed to the frame at one end and the wheel assembly at the other so that up-and-down motion of the latter tends to twist the bar and is thereby absorbed; **torsionmeter, torsion meter,** an instrument which measures the torsion in a rotating shaft, thus providing information about the power output of the engine driving it; **torsion test** *Engin.*, a test in which a material is subjected to torsion (see quot. 1936).
1937 *Daily Herald* 15 Jan. 16/6 The action of the torsion-bar controlled shock-absorbers preventing roll and pitch. **1978** L. PRYOR *Viper* viii. 150 He'd nodded and set to work ordering adjustments to the tires, wings and torsion bars. **1983** *Sci. Amer.* Jan. 120/2 The strength of the gravitational force is measured from the magnetic force that is required to prevent the rotation of a torsion bar when an additional mass is brought close to it. **1905** *Engineering* 7 Apr. 440 (*heading*) Denny and Johnson's torsion meter. **1970** *Jrnl. Physics E* III. 105/1 The only practicable way to measure the power [of a ship's main engine] is by the use of a torsionmeter, ie an instrument that measures the twist put in the propeller shaft by the torque it transmits. **1891** W. G. KIRCALDY *Strength & Properties of Materials* vii. 196 Some examples of Twisting, or torsion, tests have been given to show in a graphic way the behaviour of different metals. **1936** P. F. FOSTER *Mech. Testing of Metals & Alloys* vii. 115 Torsion tests are carried out to determine the modulus of rigidity of a material..or to ascertain its ultimate torisional strength. **1969** DIVAKARAN & GARG *Strength of Materials* i. 27 In a torsion test on the same specimen the angle of twist was found to be 0·43°.

torso. Add: **2.** (Earlier example.)
1825 T. MOORE *Life R. B. Sheridan* xvi. 534 And exhibit little more than the mere *Torso* of his eloquence.
3. *Comb.*: **torso-tosser** *slang*, a hootchy-kootchy dancer.
1927 *Vanity Fair* (N.Y.) XXIX. 134/2 A kootch or hootchie kootchie dancer is a 'torso tosser'. **1954** F. P. KEYES *Royal Box* 361 Barbara Villiers, a torso-tosser who got to be no less than the Duchess of Cleveland.

torte. Restrict † *Obs.* to sense in Dict. and add: **2.** Pl. **torten** or **tortes.** [a. G. *torte*, of same origin.] An elaborate sweet cake or tart.
1748 H. GLASSE *Art of Cookery* (ed. 3) viii. 142 To make a Tort. First make a fine Puff-paste. **1957** [see *PASTRY 1 c]. **1967** V. NABOKOV *Speak, Memory* (rev. ed.) x. 205 Wedges of slightly salty mokka *torte* with whipped cream. **1972** F. B. MAYNARD *Raisins & Almonds* 20 Mama produced meringues..puff paste..tortes layered with nut creams and Turkish delight.

tortellini (tǭɹtĕlī·ni), *sb. pl.* Also *erron.* **tortelloni.** [a. It., pl. of *tortellino*, dim. of *tortello* cake, fritter, dim. of *torta*: see TORTE.] Small squares of pasta rolled round a filling and then formed into a ring shape.
1937 M. MORPHY *Good Food from Italy* 5 One of the characteristic features of Italian soups is their garnish ..—*ravioli, tortellini, gnocchi* and plain dumplings or forcemeat dumplings. **1954** E. DAVID *Italian Food* 102 Although tortellini are always to be had in the restaurants of Bologna, in private houses they are still the great dish for Christmas Eve. **1975** J. CLEARY *Safe House* iv. 172 Charlie Lincoln ate the now-just-warm tortelloni he had bought. **1980** *Times* 18 Oct. 11/8 The Bolognese invented tagliatelle, tortellini and lasagne.

tortie (tǭ·ɹti). [dim. of TORTOISE-SHELL.] = *tortoise-shell cat* s.v. TORTOISE-SHELL 4 b. Also *attrib.*
1948 P. M. SODERBERG *Cat Breeding* 312 The three colours required in the Tortie are black, red and cream. **1958** E. F. DAGLISH *Pet-Keeper's Man.* 1 Tortoise shell cats are almost invariably females... Anyone owning a male 'tortie' may pride himself on the possession of a feline rarity. **1960** *Times* 17 Mar. 1/4 (Advt.), Must find home for beautiful tortiecat. **1976** POND & SAYER *Cats* 25 Bi-Colours, bred from Tortie and White mothers,.. could be produced.

tortillon (tǭɹti·lʸǫn). [a. F. *tortillon*, f. *tortiller* to twist, twirl.] = STUMP *sb.*[2]
1885 F. FOWLER *Drawing in Charcoal and Crayon* ii. 12 The other form of paper stump, known as the tortillon, is made of strips of paper rolled to a point like spills. **1895** *Army & Navy Co-op Soc. Price List* 674/2 Stumps for chalk drawing... Tortillons, White. **1970** *Oxf. Compan. Art* 1111/2 *Stump*, also called *tortillon*. A short tapered stick usually of cork or tightly rolled leather or paper, used to soften the edges of a drawing or spread the chalk, crayon, or pencil in shading. It was used in 18th-c. France.

tortoise. Add: **4.** *tortoise-shaped* adj. (later example); *tortoise-like* adj. and adv. (later examples); *tortoise-fashion* adv.; **tortoise core** *Archæol.*, a core (CORE *sb.*[1] 5) resembling a tortoise in shape; **tortoise race,** a race in which the last person home wins.
1919 R. A. SMITH in *Man* July 101 Tortoise-cores have been found on the bank of the ancient river to the south-east. **1972** K. P. OAKLEY *Man the Tool-maker* 52 Viewed on its outer face an oval flake thus detached from a tortoise-core has the appearance of a flat, finely worked hand-axe. **1894** A. BEARDSLEY *Let.* Oct. (1971) 75 The Tannhaüser gets on tortoise fashion but admirably for all that. **1921** W. DE LA MARE *Crossings* 71 Ann slowly thrusts her head out of the snow-house, tortoise-fashion. **1956** P. H. JOHNSON *Last Resort* xxiii. 143 His aged, stilted stride, his tortoise-like out-thrusting of the head. **1982** 'J. ROSS' *Death's Head* iv. 22 The traffic once more moving, though at a tortoise-like crawl. **1913** Tortoise race [listed in Dict]. **1914** ROWE & WEBB *Guide to Study of Eng.* iii. 126 This is a 'tortoise' race, the last man to receive the prize. **1911** *Archæologia* LXII. II. 523 Tortoise-shaped cores. This is perhaps the most striking group in the enormous series from Northfleet.

b. (With capital initial.) A proprietary name for a type of solid-fuel-burning stove.
1884 *Trade Marks Jrnl.* 5 Nov. 1025 The Tortoise... Slow combustion stoves. Charles Portway & Son, 'Tortoise' Stove Works, and High Street, Halstead, Essex; Stove Manufacturers. **1895** *Army & Navy Co-op Soc. Price List* 336 (*heading*) 'Tortoise' heating stoves. **1948** J. BETJEMAN *Few Late Chrysanthemums* (1954) 10 The Tortoise stove is lit again. **1981** *Country Life* 12 Feb. 411/3 (Advt.), The old world appeal of the Tortoise Ornamental Stove... Accepts woods, coal or smokeless fuel.

tortoise-shell. Add: **4. b. tortoise-shell ware** (examples).
1879 [see *agate-ware* s.v. *AGATE 6]. **1975** *Country Life* 9 Oct. 898/2 (*caption*) 19th-century teapot similar to Whieldon's tortoiseshell ware.
5. *tortoise-shell worker.*
1903 [see *CEMENTER]. **1931** P. A. S. PHILLIPS (*title*) John Obrisset Huguenot: carver, medallist, horn and tortoiseshell worker.

Tortolan (tǫɹtō^u·lăn), *a.* and *sb.* [f. *Tortola* + -AN.] **A.** *adj.* Of or pertaining to Tortola, the largest of the British Virgin Islands. **B.** *sb.* A native or inhabitant of Tortola.
1923 C. F. JENKINS *Tortola* xi. 67 Captain Tittley, the commander of the Tortolan navy. **1953** *Caribbean Q.* III. II. 112 Most of Tortola's exports go to St. Thomas and most of her imports come from the island, and Tortolans go frequently by sloop or launch merely to do their marketing. *Ibid.* 113 In the Tortolan countryside..no one is seen at work between ten and four o'clock. **1980** *Washington Post* 20 July K6 Almost certainly, the government worker is a native St. Thomian, the taxi driver a Tortolan by birth.

Tortoni (tǫɹtō^u·ni). Also **tortoni.** The name of an Italian café-owner in Paris in the 18th cent., used *attrib.* and *absol.* to designate a kind of ice-cream.
1911 LEITER & VAN BERGH *Flower City Cook Bk.* xxiv. 128 *Tortoni pudding*, Scald 1 pint milk..; add 1 tablespoon of flour..; 1 egg and 1 cupful sugar... Add ½ pound chopped almonds, 7 macaroons..and a little vanilla. Cool. Add 1 pint whipped cream. Freeze. **1958** *Sunday Times* 27 Apr. 22/4 A raspberry Tortoni (raspberries plus cream, macaroons and a touch of kirsch). **1979** *Tucson Mag.* Sept. 61/2 Then try tortoni (made on the premises).

Tortonian (tǫɹtō^u·niăn), *a. Geol.* [ad. G. *Tortonien* sb. (K. Mayer 1857, in *Verhandl. der allgemeinen schweiz. Ges. für die gesammten Naturwiss.* 171), f. *Tortona*, name of a town in N. Italy: see -IAN.] Of, pertaining to, or designating a stage of the upper (or middle) Miocene in Europe. Also *absol.*, the Tortonian age.
1885 A. GEIKIE *Text-bk. Geol.* (ed. 2) 873 Italy... The Tortonian stage (3) is made up [of] blue marls, reaching the great thickness of 3900 feet. **1931** GREGORY & BARRETT *Gen. Stratigr.* xii. 192 The volcanoes of the Rhine began in the Tortonian. **1974** *Nature* 22 Mar. 312/2 We suggest that the separation of Calabria from Sardinia took place in the middle Miocene, and that Tortonian deformation in Sicily indicates a collision that terminates this phase of separation.

torture, *v.* Add: **3. b.** Also with *into*.
1789 J. MOORE *Zeluco* I. ix. 80 What he said was excusable; to endeavour to torture it into mutiny would be absurd. **1840** POE *Tales of Mystery* (1905) 365 An unredeemed dreariness of thought which no goading of the imagination could torture into aught of the sublime. **1956** E. H. HUTTEN *Lang. Mod. Physics* vi. 232 It is possible to torture almost any statement into the logical form of an implication.

torula. Add: Hence **torulo·sis** *Path.* [-OSIS] = *CRYPTOCOCCOSIS.
1929 *Jrnl. Amer. Med. Assoc.* 9 Feb. 438/1 Acute miliary torulosis of the lungs follows a blood stream dissemination of the torula organisms from some chronic lesion. **1974** S. L. ROBBINS *Pathologic Basis of Dis.* xxxii. 1494/1 European blastomycosis is known more commonly as cryptococcosis or torulosis... It is caused by a blastomyces known as *Cryptococcus hominis* or *Torula histolytica*.

† **torulin.** *Biochem. Obs.* [f. TORUL(A + -IN[1].] = *THIAMINE 3.
1912 E. S. EDIE et al. in *Biochem. Jrnl.* VI. 242 The substance isolated we propose to call Torulin. **1931** SHERMAN & SMITH *Vitamins* (ed. 2) ii. 81 The symptom of opisthotonus in pigeons appears to be due to excess of lactic acid in the brain and is cured most quickly by injecting concentrated preparations of torulin locally.

torus. Add: **4.** In mod. use, a surface or solid conceived of as generated by the circular motion of a circle about an axis outside itself but lying in its plane; also, any body topologically equivalent to this, having one hole in it but not necessarily circular in form or cross-section. (Further examples.)
1958 *Times* 25 Jan. 4/5 The Zeta apparatus is essentially a ring-shaped metal tube, or torus,..containing deuterium gas at low pressure. **1966** E. H. SPANIER *Algebraic Topology* 148 The surface with one handle is topologically the torus. **1976** *Offshore Platforms & Pipelining* 49/1 The base structure consists of a torus through which the piles are placed. **1977** *Time* 6 June 54/3 Tokamaks are toruses, or doughnut-shaped chambers, surrounded by huge electromagnets.

Tory, *sb.* and *a.* Add: **A.** *sb.* **4. b.** During the American civil war, applied in the Confederate states to a Union sympathizer.
1862 *Southern Confederacy* (Atlanta, Georgia) 3 May 3/1 The other prisoners..are all sharp, intelligent-looking men—no hard looking cases like Yankee prisoners, and East Tennessee tories usually are. **1866** W. REID *After the War* 402 Ef you fetch any d— tories heah, that went agin their State, and so kin take the oath,..'twill soon be too hot to hold 'em. **1953** T. C. BRYAN *Confederate Georgia* ix. 152 In the fall of 1864 bands of Tories were plundering northeast Georgia.
C. 3. a. (Earlier example.)
1867 LD. SALISBURY in *Q. Rev.* CXXIII. 539 It was not till the earlier struggles of the session were over..that the project of Tory democracy, which had been so long and so sedulously concealed, was at last given to the world.
b. Tory Democratic *a.* (earlier example).
1885 E. W. HAMILTON *Diary* 15 June (1972) II. 885 It was R. Churchill's way of protesting publicly against a revival of the old Tory Cabinet..without any infusion of fresh (Tory-democratic) blood.

Tosa (tō^u·sā), *sb.*[1] The name of an aristocratic Japanese family of court painters used *attrib.* to designate (the products of) a school of painting characterized by the use of traditional themes and techniques, which flourished from the mid-fifteenth to the late-nineteenth century.
1879 *Trans. Asiatic Soc. Japan* VII. 355 The reputation of the Tosa school was maintained during the progress of the Kano riu. **1909** L. BINYON *Jap. Art* ii. 10 The typical Tosa picture was a long scroll (makimono) portraying scenes of battle, adventure, scenes of court life, or the lives of saints. **1952** L. WARNER *Enduring Art of Japan* vi. 64 It had been appropriate enough, for narrow Tosa scrolls..to use small patches of opaque colour set in cells of black ink. **1972** *Times* 18 May 21/5 An album of hand paintings of the Tosa school..dating from the turn of the seventeenth and eighteenth centuries.

Tosa (tō^u·sā), *sb.*[2] [a. *Tosa*, the former name of a province on the island of Shikoku, Japan.] A black, tan, or brindle mastiff of the breed of this name, originally developed as a type of fighting dog in Japan. Also *attrib.*
1945 C. L. B. HUBBARD *Observer's Bk. Dogs* 191 The Tosa..has been known for at least six centuries. **1966** 'G. BLACK' *You want to die, Johnny?* ii. 39 Taro, my Japanese Tosa hound..is a big brindle fighting dog. **1971** DANGERFIELD & HOWELL *Internat. Encycl. Dogs* 309/2 Little effort has been made to keep the Tosa purebred until quite recently. *Ibid.*, The modern Tosa dogs are about 28 inches tall..and weigh well over 100 lb.

tosh (tǫʃ), *sb.*[3] *slang.* [Cf. *TOSHER[1].] Items of value retrieved from drains and sewers.
a **1852** [see *TOSHER[1]]. **1974** J. AIKEN *Midnight is Place* v. 164, I am at present engaged in fishing for tosh in the sewers of Blastburn.

tosh (tǫʃ), *sb.*[4] *slang.* Also **tush.** Abbrev. of *TOSHEROON. Also used *loosely* for two shillings, money.

1912 J. W. Horsley *I Remember* xii. 253 'Tush', for money, would be an abbreviation of 'tusheroon', which in old cant, and also in tinker dialect, signified a crown. **1937** *Night & Day* 22 July 14/3 A couple of grafters had the courage..to bat for a straight tush. **1961** J. Maclaren-Ross *Doomsday Bk.* I. v. 63 Here's a tosh to buy yourself some beer. **1964** A. Prior *Z Cars Again* ix. 74 'You can give me three blacks for a tush,' he said. 'Two blacks for half a dollar,' was Mr. Thistlethwaite's reply.

tosh (tɒʃ), *sb.*[5] *slang.* [Origin uncertain; perh. f. Tosh *a.* (*adv.*).] Used as a neutral or joc. form of address.

1954 E. Hyams *Stories & Cream* 175 'Ere, tosh, you bin at Cha'ham? **1978** M. Kenyon *Deep Pocket* vi. 75 'Sortin' you out for a start, tosh!' came a voice.

tosher[1]. Add: **b.** One who searches for valuable refuse in drains and sewers.

a **1852** H. Mayhew *London Labour* (1861) II. 150/2 The sewer-hunters were formerly, and indeed are still, called by the name of 'Toshers', the articles which they pick up in the course of their wanderings along shore being known among themselves by the general term 'tosh', a word more particularly applied by them to anything made of copper. **1870** D. J. Kirwan *Palace & Hovel* xxi. 331 These men..search the sewers..for..whatever is of value... They are called 'Toshers' or 'Shore-men'. **1974** J. Aiken *Midnight is Place* v. 154 Gudgeon's your mate, boy, he's my other tosher.

toshing (later example).

1974 J. Aiken *Midnight is Place* vi. 180 You tend to the toshing, let Mester Hobday tend to the dealing.

tosheroon (tɒʃərū·n). *slang.* Also **tusheroon.** [Etym. unknown.] Half-a-crown; a coin of this value (in quot. 1859 erron. said to be a crown).

1859 Hotten *Dict. Slang* 112 *Tusheroon*, a crown piece, five shillings. **1933** 'G. Orwell' *Down & Out in Paris & London* xxix. 214 A tosheroon (half a crown) for the coat, two 'ogs for the trousers. **1960** 'A. Burgess' *Doctor is Sick* xvi. 125 'I haven't got three nicker,' said Edwin, 'nor one nicker, nor half a bar, nor a tosheroon, nor,' he added, 'a solitary single clod. I can't buy anything.' **1978** *Daily Mirror* 18 Feb. 19/1 All sorts of things, places and creatures we believed were everlasting have vanished, like trams, tosheroons and Constantinople.

Tosk (tɒsk), *sb.* and *a.* Also **Toshke.** [a. Alb. *Toskë.*] **A.** *sb.* (A member of) one of the major ethnic groups of Albania, living mainly in the south of the country. Also, the Albanian dialect spoken by this people. **B.** *adj.* Of or pertaining to the Tosks or their language. Cf. *Gheg.

1835 [see *Gheg]. **1900** 'Odysseus' *Turkey in Europe* ix. 397 The Southern Albanians differ from the Northerners. ..Their generic name is Tosk. *Ibid.* 401 The whole of the Tosk country has been strongly influenced by Greece. **1908** T. G. Tucker *Introd. Nat. Hist. Lang.* ix. 195 A number of dialects, usually grouped under the two heads *Ghegh* (to the north) and *Tosk* (to the south), spoken by the Albanians..are admitted to belong to the Indo-European stock. **1939** L. H. Gray *Foundations of Lang.* 331 The language [*sc.* Albanian] falls into two groups, each with a number of sub-dialects: *Geg* (with colonies in Dalmatia) to the north and *Tosk* (with colonies in Greece, Italy, and Sicily) to the south of the Shumbi River. **1958** [see *Gheg]. **1966** E. P. Hamp in Birnbaum & Puhvel *Anc. Indo-Europ. Dial.* 98 Rosetti, however, mistakenly repeats the myth that some Tosk dialects show Geg characteristics. **1980** *Word* 1979 XXX. 27 Within Albania the proportion of northerners (Gegs) to southerners (Tosks) is almost equal. *Ibid.* 41 The imposition of Tosk upon a population that was not entirely Tosk-speaking. **1983** *Ibid.* XXXIV. 26 Arvanítika is the form of Tosk Albanian spoken in Greece; it is closely related to the southernmost variety of Albanian.

toss, *sb.*[1] Add: **3. b.** *to take a toss*, to suffer a fall from a horse; also *fig.*

1917 [see *half *sb.* 7 h]. **1926** Galsworthy *Silver Spoon* I. xiv. 101 The Government had 'taken their toss' over the Editor. **1949** M. Steen *Twilight on Floods* iv. x. 699 By Jove, old boy, she's taken a toss for you! **1966** [see *league *sb.*[1] 1 c]. **1973** 'M. Innes' *Appleby's Answer* v. 47 The red-haired Lady Curricle, who had 'taken a toss', you will remember, over a hedge.

c. *U.S. slang.* A search (of a building or person) conducted by the police. Cf. *toss *v.* 1 c.

1970 L. Sanders *Anderson Tapes* xciv. 220 The author was allowed to attend as an observer but not active participant in the search. The toss of the above premises.. was..conducted with professional skill. **1972** J. Mills *Report to Commissioner* 86 You wanta give her a toss, give her a toss, but let's not stand here all night.

4. b. A spread of fall (of hair).

1946 D. C. Peattie *Road of Naturalist* i. 13 My wife lay hiding from the light in the toss of her hair. **1978** J. Updike *Coup* (1979) vi. 233 Ezana looked at her and saw beyond the brassy toss of her hair.

6. b. *to argue the toss*, to dispute a decision or opinion.

1925 Fraser & Gibbons *Soldier & Sailor Words* 288 *Toss, to argue the*, to dispute: wrangle: to have too much to say. **1945** *Penguin New Writing* XXIV. 84 Poetry was never much in my line, except Shelley, and Terry didn't think much of him, so..we argued the toss about it. **1958** *Economist* 11 Jan. 92/2 The Prime Minister's..venture.. cannot do more than clear the way... More is involved than just arguing the tosses of the moment. **1978** 'M. Underwood' *Crooked Wood* iv. 61 He was not in a strong position to argue the toss.

c. *fig.* In negative contexts: a jot, a whit, a very small amount. Usu. in phr. *not to care* (or *give*) *a toss. colloq.*

1876 [see sense 6 in Dict.]. **1925** P. Gibbs *Unchanging Quest* xviii. 132 She..didn't care a toss what people thought of her. **1973** *Time Out* 2–5 Mar. 13/1, I don't give a toss whether he's black, white or purple. **1979** *Bull. Yorks. Dial. Soc.* No. 26. 11 Ah deean't odd wi them as mooan As prices mak em cross. Then spend ther brass on eaps o things Wat isn't woth a toss.

7. (Earlier example.)

1882 J. L. Burgess *Homing Fancier's Ann.* 11 The intermediate tosses were Redhill, 184 miles from Brussels, and Worcester, 280 miles.

11. toss-off. An act of masturbation. Cf. *toss *v.* 12 d. *coarse slang.*

1735 *Rake's Progress* III. 19 Or loudly sing some bawdy Song, Then drops into St. D—n's C—h, And take a *Toss-off* in the Porch.

12. *Comb.*: **toss pillow** *U.S.* = *scatter cushion* s.v. *scatter *v.* 7 b.

1956 *Sears, Roebuck Catal.* Fall & Winter 897 (*heading*) Toss pillows..new colors..styles..shapes. **1978** *Washington Post Mag.* 26 Mar. 20/2 (Advt.), Matching toss pillows $10 each.

toss, *v.* Add: **I. 1. c.** *U.S. slang.* To search (a building or person) in the course of a police investigation. Cf. *toss *sb.*[1] 3 c.

1939 *Fortune* July 102/2 [He] sent a couple of detectives across town to 'toss their flat' (i.e. search their apartment). **1969** 'P. Kavanagh' *Such Men are Dangerous* vii. 82 They tossed the room while I had breakfast. **1972** B. Garfield *Line of Succession* III. 186 He had been tossed seven times..but no drugs had been found on him. **1980** 'E. McBain' *Ghosts* iii. 56 We ought to try for an order to toss his apartment.

3. e. In cookery, to stir or turn (food) over, esp. so as to coat it with butter, oil, etc.

1723 J. Nott *Cook's & Confectioner's Dict.* No. 107a Artichokes with cream... Toss them up with butter. **1877** E. S. Dallas *Kettner's Bk. of Table* 120 The fillets..are to be lightly tossed in butter, taking care not to colour them. **1913** C. M. Pearse *Kitchen Garden & Cook* 197 Separate the endive into tufts. Toss these in the salad dressing. **1952** G. W. Brace *Spire* (1953) viii. 69 The salad was tossed amid murmurs of pleasure. **1976** 'Trevanian' *Main* (1977) v. 109 You don't stir a salad. You toss it.

III. 9. c. To wager with (a person) on the toss of a coin. Usu. const. *for* (something).

1851 H. Mayhew *London Labour* I. 196/1 To 'toss the pieman' is a favourite pastime with costermongers' boys. ..If the pieman win the toss, he receives 1*d.* without giving a pie. **1858** G. H. Lewes *Sea-Side Stud.* IV. i. 271 We used to 'toss' the pieman for epicurean slices of pudding—a vulgar, but seductive form of juvenile gambling. **1942** Wodehouse *Money in Bank* (1946) xv. 128 He was in the frame of mind when he would have patted a small boy on the head and given him sixpence, though it is probable that a moment later he would have tossed him for it and won it back again. **1983** 'D. Shannon' *Exploit of Death* (1984) i. 18 'I'll toss you for the job.' Conway produced a quarter and flipped it.

d. To release (a homing pigeon) in a race or trial flight. Cf. Toss *sb.*[1] 7.

1882 J. L. Burgess *Homing Fancier's Ann.* 10 Five hundred and eighty-nine birds were tossed at 4 a.m. **1911** *Encycl. Brit.* XXI. 596/2 Training..consists in taking it [*sc.* the bird] out in a closed wicker basket and liberating or 'tossing' it at gradually increasing distances from its loft.

IV. 11*. toss in. To finish, to give up. *N.Z. slang.*

1956 D. M. Davin *Sullen Bell* II. iv. 128 I'd toss it in for tonight, Miss Sim. It's an ugly night and you should get your train home. **1971** *N.Z. Listener* 19 Apr. 56/5 In the end they saw some hogsbacks up above the col so they tossed it in and glissaded down back to their bivvy.

12. toss off. c. To do or make easily, without effort. **d.** *trans.* and *intr.* To masturbate. *slang.*

1874 L. Troubridge *Life amongst Troubridges* (1966) 80 A new rage..for painting the panels of the shutters of our bedrooms..and they only take a jiffy to toss off. **1879–80** *Pearl* (1970) 258, I don't like to see, though at me you might scoff, An old woman trying to toss herself off. **1915** L. Strachey *Let.* 12 Mar. in P. N. Furbank *E. M. Forster* (1979) II. i. 16 À propos of Maurice tossing himself off..you say—'He knew what the price would be —a creeping apathy towards all things.' **1927** Joyce *Let.* 25 July (1966) III. 162 The verb 'to toss off' an expression for 'to masturbate'. **1937** M. Hillis *Orchids on your Budget* (1938) vi. 109 Any man worth anything could toss off a rarebit or an omelet. **1969** *Jeremy* I. III. 22/1 All they want to do is toss off in the cottage while they look at my prick. **1979** *Church Times* 17 Aug. 7/1 [These books] do not provide bibliographical support for the learned references tossed off in the articles. **1981** 'D. Kavanagh' *Fiddle City* iv. 82 Would you like me to toss you off?.. It's ten if you're worried about the price.

tossed, *ppl. a.* Add: Of a salad: stirred or turned, esp. so as to be coated with dressing. See *toss *v.* 3 e.

1947 M. Given *Mod. Encycl. Cooking* II. 1315 Whole meal tossed salad... Wash and clean all vegetables, chill thoroughly. **1962** H. Hood in R. Weaver *Canad. Short Stories* (1968) 2nd Ser. 207 It was a good hot-weather supper, tossed greens with the correct proportions of vinegar and oil. **1978** H. C. Rae *Sullivan* I. ii. 15 Artichokes, tossed green salad and cold lake trout.

tosser. Add: **1. b.** [Prob. f. sense *12 d of the vb.] A term of contempt or abuse for a person; a 'jerk'. Cf. Bugger *sb.* 2 b. *slang.*

1977 *Zigzag* Apr. 40/3 She came on in a big mac and flashed her legs like an old tosser before throwing it off. **1983** P. Inchbald *Short Break in Venice* xviii. 172 It's a right pig's job... Poor little tosser. As if he wasn't suffering enough already.

3. A penny, a coin of small value. Cf. Toss *v.* 9, *toss *sb.*[1] 6 c.

1935 [see *carve-up]. **1964** J. Aiken *Black Hearts in Battersea* (1965) vi. 70 Shall we play for money?..I haven't a tosser to my kick. **1966** *New Statesman* 4 Nov. 662/1 There's no one mumming today whose opinion is worth a tosser on the cut of a dressing-gown. **1980** J. Gerson *Assassination Run* xi. 158 Your people don't give a tosser for me!

tostada, tostado (tɒstā·dā, -o). [a. Sp., pa. pple. of *tostar* to toast.] A deep-fried cornmeal pancake topped with a seasoned mixture of beans, mincemeat, and vegetables.

1945 E. Fergusson *Mexican Cookbk.* (ed. 2) p. v, Mexican food has, even since the 'American Occupation', been a part of the Southwestern diet... In every Southwestern town *tostados* are served with cocktails. **1958** *McCall's Mag.* Aug. 9/2 The tortilla is the basis of many famous Mexican dishes: Enchiladas, Tacos, Tostadas. **1972** *Times* 6 May 12/7 Wait for the crunch of tostados—for the next culinary invasion..will be Mexican. **1975** 'S. Marlowe' *Cawthorn Journals* xxiv. 235 Maruja fed tostadas into the hot splattering oil.

‖ **toston** (tɒstɒ·n). [(Amer.) Sp.: cf. Teston, testoon.] A silver coin formerly in use in various Latin American countries; in Mexico, equivalent in value to half a peso.

1884 A. R. Conkling *Appleton's Guide to Mexico* xv. 61 Mexico has followed the example of Spain in adopting the decimal system of coinage, of which the *real* is the basis. The current coins are as follows: *Silver.* One *peso*, or dollar, containing 8 *reales*, or 100 cents. Four *reales*, or one *toston*, 50 cents. **1932** H. Crane *Let.* 22 Apr. (1965) 412 The telegraph office paid us off in six hundred and some odd 'Tostons' (about like getting it all in dimes) and neither the Ward Line office nor the official Banco de Mexico would accept them. **1947** M. Lowry *Under Volcano* vii. 240 Laying a tostón on the counter. **1980** in S. Terkel *Amer. Dreams* 6 My father's father came from Mexico... He paid a *toston*, a half-dollar. That automatically made him a U.S. citizen.

tosudite (tōᵘ·siūdəit). *Min.* [ad. Russ. *tosudit* (V. A. Frank-Kamenetsky et al. 1963, in *Zap. Vsesoyuz. Min. Obshch.* XCIII. 563): see quot. 1964 and -ite[1].] A blue mixed-layer clay mineral (see quots.).

1964 *Mineral. Abstr.* XVI. 549/2 This newly characterized mixed-layer mineral consisting of unusual aluminian chlorite and montmorillonite is named tosudite, in honour of Toshio Sudo, who described the Japanese occurrences. **1976** *Clays & Clay Minerals* XXIV. 142/1 The name tosudite is usually used for a regularly interstratified mineral with dioctahedral or di-trioctahedral chlorite component.

tosyl (tɒ·sil). *Chem.* [a. G. *tosyl* (Hess & Pfleger 1933, in *Ann. d. Chemie* DVII. 48), f. *to*luol*s*ulfon*yl*: see *Toluol* s.v. Tolu-, *sulphonyl.] The *para* isomer of the univalent radical toluenesulphonyl, $H_3C{\cdot}C_6H_4{\cdot}SO_2$—.

1938 *Jrnl. Amer. Chem. Soc.* LX. 398/1 The unimolar tosylation of α- and β-methyl-*d*-glucosides in pyridine solution with tosyl chloride, followed by acetylation results in the formation of 6-tosyl-triacetyl-α-methyl-*d*-glucoside. **1975** *Nature* 30 Oct. 763/1 There is no evidence for a hydrophobic binding pocket or tosyl hole as seen in the α-chymotrypsin structure.

Hence **to·sylate** *sb.*, an ester of the tosyl group; *v. trans.*, to introduce a tosyl group into (a compound) or add one to (an atom); **to·sylated, to·sylating** *ppl. adjs.*; **tosyla·tion,** the process of tosylating.

1938 *Jrnl. Amer. Chem. Soc.* LX. 1203/2 The structure of this compound may be considered proved since it is identical with that obtained..upon tosylating 2,3,4,2',3'-pentaacetyl-β-methylcellobioside. **1938** [see *tosyl above]. **1963** I. L. Finar *Org. Chem.* (ed. 4) I. xxv. 612 Tosylates are useful for preparing, *e.g.*, ethers. **1972** *Jrnl. Chem. Soc.: Chem. Communications* 1148/1 We have found that *N*-methyl-*N*-tosylpyrrolidinium perchlorate..can be used as a selective tosylating reagent. *Ibid.* 1149/1 We assumed that only nitrogen would be tosylated if both amino- and hydroxy-groups were present in the substrate. **1974** *Jrnl. Org. Chem.* XXXIX. 635/2 A greater chemical shift difference was observed for the C-2 ring protons adjacent to the tosylated nitrogen. *Ibid.*, Tosylation was carried out with tosyl chloride in the usual basic media of pyridine or aqueous sodium bicarbonate. **1978** J. M. & D. J. Cram *Essence Org. Chem.* vi. 150 The name *p*-toluenesulfonyl group ($Ar—SO_2—$) is shortened to the 'tosyl' group, and the ester is ethyl tosylate.

tot, *sb.*[3] Add: Also *gen.*, the total number or amount. (Earlier examples.)

1755 C. Charke *Narr. Life Mrs. C. Charke* 260 The above-mentioned notable Gentleman, with his wife and a young Fellow, besides our two selves, made up the whole *Totte.* **1857** *Londonderry Standard* 26 Feb. 2/2 Forty monopolists whose numerical 'tot' is so oddly coincident with the history of Ali Baba. **1866** *Times* 28 Apr. 5/6 He added up the gross 'tots' of the several poll books himself.

tot, *sb.*[4] Add: **4.** *Comb.*: **tot lot** *N. Amer.*, a playground for small children; **tot system** *S. Afr.*, a system of paying agricultural workers, esp. in vine-growing districts, part of their wages in 'tots' (usu. mugs) of wine.

1944 *Sun* (Baltimore) 15 July 13/3 (*heading*) Pall Mall tot lot open daily except Saturday. **1968** *Daily Colonist* (Victoria, B.C.) 6 Dec. 41/4 Snow fails to stop Gold River children from enjoying recently completed tot-lot behind the community hall. **1977** *Ottawa Citizen* 19 May 2/2 The plan includes..a 'tot lot' for pre-schoolers. **1926** *Eastern Province Herald* (Port Elizabeth) 12 Feb. 7 (*heading*) Liquor bill under fire—evils of the tot system. **1953** P. Abrahams *Return to Goli* II. 77 The vicious 'tot'-system which obtains in the wine-growing Cape valley..is ruining the health..of a very large number of Coloureds. **1974** *Sunday Times* (Johannesburg) 24 Feb. 14 Asked whether he made use of the tot system, Mr. — said: [etc.].

tot, *v.*[3] Add: **totting** *vbl. sb.* (later simple and *attrib.* examples with *up*).

1963 *Guardian* 2 May 5/1 The 'totting up' procedure is a new principle, and Mr Marples explained that its sole purpose was to make anyone who had two endorsements in two years drive extra carefully. **1976** *Deeside Advertiser* 9 Dec. 24/1 Hamilton had been disqualified by Chester City Magistrates in October for six months under the totting up procedure because of previous endorsements on his licence. **1978** R. Mark *Office of Constable* xxii. 273 'Totting up'..is the arrangement under the 1972 Road Traffic Act authorizing the endorsement of a driving licence after conviction for any one of a number of offences. The endorsement lasts three years and two further endorsements during that period mean mandatory disqualification from driving for a minimum of six months.

tot (tɒt), *v.*[4] [Back-formation f. *totting* s.v. Tot *sb.*[5]] *intr.* To pick anything saleable from a dustbin or tip; † to pick up bones.

1884 J. Greenwood *Little Ragamuffins* xiv. 121 'P'r'aps he's going a-tottin'' (picking up bones,) said Ripston. **1922** Joyce *Ulysses* 422 On a step a gnome totting among a rubbishtip crouches to shoulder a sack of rags and bones. **1969** *Guardian* 6 Feb. 5 The right to tot or sell salvage is the cause of a 10-day-old strike of 267 dustmen. **1976** M. Russell *Double Deal* iv. 32, I could earn as much, totting for the corporation.

total, *a.* Add: **2. b.** *total heat* (Physics): = *enthalpy; *spec.* (see quot. 1853).

1851 *Phil. Mag.* II. 4 We often hear of the total heat of bodies, and of gases and vapours in particular, this term being meant to express the sum of the sensible and latent heat. **1853** *Trans. R. Soc. Edin.* XX. 172 If to the latent heat of evaporation at a given temperature, is added the quantity of heat necessary to raise unity of weight of the liquid from a certain fixed temperature (usually that of melting ice) to the temperature at which the evaporation takes place, the result is called the total heat of evaporation from the fixed temperature chosen. **1927, 1962** [see *enthalpy].

c. *total impulse* (Astronautics): (see quot. 1949).

1949 G. P. Sutton *Rocket Propulsion Elements* i. 18 The impulse (often called total impulse) is the integral of the thrust over the firing duration. For a constant thrust it is the product of thrust and duration. **1979** J. W. Cornelisse et al. *Rocket Propulsion & Space Flight Dynamics* vi. 115 The specific consumption is defined as the ratio of propellant weight consumed and the total impulse delivered.

3. a. *total recall*: see *recall *sb.*[1] 2.

c. Complete in nature; involving all resources; manifesting every characteristic or the whole nature of an activity, person, etc.; all-encompassing, all-inclusive; fully co-ordinated or integrated; *total diplomacy*, diplomacy conducted with the consent or participation of all citizens and institutions; *total institution* (see quot. 1962); *total theatre*, (*a*) a theatre designed for maximum involvement of performers and audience; dramaturgy which achieves this; (*b*) theatre involving a wide range of techniques and conventions; *total war*, a war to which all resources and the whole population are committed; loosely, a war conducted without any scruples or limitations; *total woman*, spec. a woman who conforms to the female 'ideal' or stereotype of complete self-abnegation and devotion to the interests of a man.

1935 W. Gropius in S. Giedion *W. Gropius* (1954) I. 61 The aim of this 'Total Theater' is to draw the spectator into the drama. All technical means have to be subordinated to this aim and must never become ends in themselves. **1937** W. L. Shirer *Berlin Diary* (1941) 86 Total war means the complete and final disappearance of the vanquished from the stage of history! **1940** Graves & Hodge *Long Week-End* i. 13 The philosophy of 'total war', that a war can best be won by complete ruthlessness, was of German origin. **1942** Total war [see *people's war* s.v. *people *sb.* 9]. **1950** *World-Telegram-Sun* (N.Y.) 14 Mar. 12 He [*sc.* D. Acheson] defines 'total diplomacy' as the full use of Congress,..government agencies, as well as business, labor and agriculture. **1951** E. A. Walsh (*title*) Total empire: the roots and progress of world communism. **1957** *Sat. Rev.* (U.S.) 26 Jan. 22/1 M. [Jean-Louis] Barrault..describes it as follows: ' "Total Theatre" is simply the true and traditional theatre, the one which makes use of man "in his totality", his gestures, pantomime, dances, breath, cries, articulation, speech, poetry, and singing.' **1957** J. D. MacDonald *Man of Affairs* iii. 43, I did not see how any platonic relationship between Mike and this total woman would be possible. **1957** E. Goffman in *Symposium on Preventive & Soc. Psychiatry* 44 [These institutions'] encompassing or total character is symbolized by the barrier to social intercourse with the outside that is often built right into the physical plant: locked doors, high walls, barbed wire, cliffs and water, open terrain, and so forth. These I am calling total institutions. **1962** —— *Asylums* p. xiii, A total institution may be defined as a place of residence and work where a large number of like-situated individuals, cut off from the wider society for an appreciable period of time, together lead an enclosed, formally administered round of life. **1963** I. Fletcher in B. Sewell *Two Friends* 63 *Spiritual Poems* is a fine example of 'total art', its paper and typography reflecting the eclecticism and scholarly caprice of content. **1966** *Listener* 29 Dec. 959/2, I still think the best kind of Christmas show for children is pantomime. It is a form of total theatre—story, spectacle, ballet, song, revue sketch, comedians' jokes, audience-participation, even circus acts. **1966** Schwarz & Hadix *Strategic Terminology* 132 *War, total*, conflict in which the issue is a threat to survival and in which all weapons of the combatants are used. Many modern definitions of total war used the term for war involving nuclear weapons or the direct confrontation of the great powers on the assumption that it would be unlimited or become unlimited. **1967** *Boston Sunday Herald* 26 Mar. I. 34/2 Rather he is thought of as a human being who needs dentistry badly and one for whom total care is now available. **1969** *Times* 17 Oct. 17/4 Carlo Palazzi, who is a past-master of the total look..had all his men wearing saffron coloured clothes which mixed and, of course, matched. **1972** A. Bryant *Bless this House* iv. 44 Marabel Morgan..talked about her course entitled 'The Total Woman'. **1973** M. Morgan *Total Woman* (1975) iv. 60 A Total Woman caters to her man's special quirks, whether it be in salads, sex, or sports. **1973** *Times* 30 July 20/3 They became convinced that developers and local authorities ought to concern themselves with engineering a 'total' environment for a community, of which the buildings themselves are only a part. **1974** *Howard Jrnl.* XIV. 86 Most total institutions leave their mark on those who devote their lives to them—the colonel, the sea captain, the public school headmaster, the monk and the nun are popular cultural stereotypes. **1975** *Times Lit. Suppl.* 2 May 477/5 The approach throughout is scholarly and thorough, no one period receiving less attention than any other. As one might expect, Roman and medieval features are fully treated, but so too are post-medieval and recent... This is total archaeology at its best. **1977** *N.Y. Rev. Bks.* 27 Oct. 46/3 (Advt.), Professional man, 60, needs slender, total woman over 30 for September, 1978, fortnight Alaskan cruise. **1978** *New Yorker* 7 Aug. 45/1 In the past few years the Dutch have been the most thrilling [soccer] side to watch, playing a running game—'total football', it is called—with players interchanging positions and functions but always pressing forward on the attack. **1980** R. Moody *Devil you Don't* iv. 45 In the fight against Hitler, we progressed to the concept of Total War, no quarter given, no humanity expected, victory at any price.

5*. *total float* (see quot. 1967).

1964 K. G. Lockyer *Introd. Critical Path Anal.* v. 49 *Total float*, the time by which an activity can expand. **1967** S. Woodgate in Wills & Yearsley *Handbk. Management Technol.* 80 Total float is..the maximum amount of spare time which can be made available to any activity.

B. *sb.* **b.** *in total*, all together, entirely.

1965 *Listener* 7 Jan. 3/1 Does the Government mean incomes in total cannot go up by more than production..? Or does it mean that all incomes should go up by the same percentage? **1969** R. Buckminster Fuller *Operating Man. Spaceship Earth* iv. 52 We have not been seeing our Spaceship Earth as an integrally-designed machine which to be persistently successful must be comprehended and serviced in total.

total, *v.* Add: **2.** Also with *out. U.S.*

1966 *Word Study* Dec. 2/2 How long did I control on fly-by-wire?..That is something we want to total out. **1977** *New Yorker* 13 June 30/1 We weren't going to wait until the Police Department could total out what it would cost them.

3. To damage beyond repair (esp. a motor vehicle, in an accident); to destroy, to demolish, to wreck; to kill or injure severely; also *fig.* Also with *out.* Freq. in *pass.* and as *pa. ppl.* Chiefly *N. Amer.*

1895 W. Rye *Vocab. E. Anglia* 232 Totald [sic], killed or injured. **1954** *Amer. Speech* XXIX. 103 Bob totaled his car last night. **1965** *Ibid.* XL. 159 Her son was hospitalized because of an automobile accident and..his car was 'totalled'. **1966** *Newsweek* 13 June 48c/3 Amazingly, no drivers were 'totaled'. **1966** *Current Slang* (Univ. S. Dakota) Summer 4 *Totalled out*, intoxicated... Tom was *totalled out* by midnight. **1966** *Ibid.* Winter 8 *Totalled*, adj., mentally upset... After one semester, he was *totalled.* **1970** E. Segal *Love Story* iii. 24 Did you at least total the guy that hit you? **1971** *Wall St. Jrnl.* 17 Mar. 1/1 He has had 44 planes, three of which were 'totaled' in accidents. **1971** *New Yorker* 28 Aug. 81 Townshend did total his instrument during his last song. **1971** M. Tak *Truck Talk* 169 *Total it out*, to wreck a truck completely. **1972** C. Weston *Poor, Poor Ophelia* x. 52 You think it's a fantasy my car's totaled? **1973** *New Yorker* 16 July 34/3 A streak of sudden tire skids,..a totalled car at the bottom of a ravine. **1974** *Publishers Weekly* 1 Apr. 50/1 Water from fire engines and hydrants cascaded into the burning ruins. Eighteen businesses were totaled. **1974** J. Goldman *Man from Greek & Roman* xxiii. 211 'Totalled out.'.. Big gash along the side, hood all barged up. **1975** *Times Lit. Suppl.* 12 Dec. 1486/5 'The Execution of Lady Jane Grey'... Lady Jane is about to be totalled by the axe. **1977** *Time* 10 Jan. 42/2 He can still total a liquor store in the course of rescuing hostages. **1979** *Yale Alumni Mag.* Apr. (Suppl.) cn10/3 Little Robert was totaled by a bus that ran a red light but escaped with fractures of collar bone and right hand. **1979** G. Swarthout *Skeletons* 98 I'm too totalled to hate anyone... This has been the worst week of my life. **1981** J. D. MacDonald *Free Fall in Crimson* xx. 230 'He's the one that beat the old man to death.'..'They *think* he totaled the movie lady.' **1982** *Guardian* 26 Oct. 8/7 Daddy's BMW which she can drive any time she wants as long as she doesn't total it.

to·talist, *sb.* (and *a.*). [f. Total *a.* and *sb.* + -ist.] One who inclines to treat or regard things as a whole; one concerned with the whole social environment, esp. as a means of thought-control; one concerned with the whole person. Also *attrib.* or as *adj.*

1956 J. S. Bruner et al. *Study of Thinking* v. 128 The totalists have wanted to stay as close as possible to the whole cortex as an explanation, and it is only with the greatest reluctance that they will subtract any of its attributes as irrelevant. **1961** R. J. Lifton *Thought Reform* xxii. 420 Through this milieu control the totalist environment seeks to establish domain over not only the individual's communication with the outside.., but also ..over what we may speak of as his communication with himself. *Ibid.* 422 Ideological totalists do not pursue this approach *solely* for maintaining a sense of power over others. **1964** R. Wilkinson *Gentlemanly Power* xiii. 184 A 'totalist ideology'..refers to any doctrine which attempts a complete, unified explanation of world and society. **1969** *Political Q.* XL. 472 Only that which is known by the 'whole being' is sound and healthy... The origins of this totalist view of knowledge..are no doubt various.

So **totali·stic** *a.*

1932 H. H. Price *Perception* vi. 151 The perceptual act still has this totalistic character. **1942** *Mind* LI. 316 Some writers are frightened by the word *intuition*, and admittedly it has bad associations... Again, under, I believe, Croce's influence, it has come to mean the apprehension of a whole as a whole, a 'totalistic' apprehension. **1976** *Brit. Jrnl. Sociol.* XXVII. 88 A totalistic rejection of the contemporary order is not encountered. **1979** *Jrnl. R. Soc. Arts* Nov. 772/1 The idea of collage city was dualism itself, an incorporation of opposite qualities which Modern city planning in its utopian, or totalistic phase, had denied.

totalitarian (tōᵘtælītēᵊ·riăn), *a.* and *sb.* [f. Totality + *-arian, after It. *totalitario* complete, absolute; totalitarian.] **A.** *adj.* Of or pertaining to a system of government which tolerates only one political party, to which all other institutions are subordinated, and which usu. demands the complete subservience of the individual to the State. Also *transf.* Cf. *total *a.* 3 c.

1926 B. B. Carter tr. *Sturzo's Italy & Fascismo* ix. 220 Anti-Fascism..has, however, a positive sense if it is taken to represent an element antagonistic to the 'totalitarian' and absolute position of Fascism. **1929** *Times* 2 Nov. 7/5 A reaction against parliamentarism..in favour of a 'totalitarian' or unitary state, whether Fascist or Communist. **1936** E. Underhill *Worship* xii. 251 This cultus is, in origin, an acknowledgement of the corporate and totalitarian character of the Christian response to God. **1937** E. Pound in *Germany & You* 25 Apr. 95 (*heading*) Totalitarian scholarship and the new paideuma. *Ibid.* 96/2 In 1937 we are concerned with the reintegration of the arts in totalitarian synthesis. **1940** *Hutchinson's Pictorial Hist. War* 2 Oct.–26 Nov. 183 We have all heard lately about total or totalitarian war. It has been defined as conflict between nations taking the place of armed forces. Every citizen is in a sense a combatant and also the object of attack. **1951** H. Arendt *Burden of Our Time* III. x. 303 Totalitarian movements aim at and succeed in organizing masses—not classes. **1964** H. Marcuse *One Dimensional Man* i. 3 'Totalitarian' is not only a terroristic political coordination of society, but also a non-terroristic economic-technical coordination which operates through the manipulation of needs by vested interests. **1977** M. Walker *National Front* i. 13 The totalitarian society is a single-minded structure. It mobilizes all its resources under one authority to achieve one goal.

B. *sb.* A leader or member of a totalitarian party; an advocate or supporter of totalitarianism.

1938 *Times* 20 Oct. 15/3 The new methods of the totalitarians. **1944** A. Huxley *Let.* 10 Apr. (1969) 504 The Left-wing Intellectuals and the Labour Party are eager totalitarians. **1958** R. Liddell *Morea* II. vi. 149 Sparta has one of the finest romantic backgrounds in Greece; this is not surprising, for only romantics can successfully be totalitarians. **1978** L. Deighton *SS-GB* xiv. 115 The totalitarians of right and left have constantly to describe the faith they have in common.

Hence **totalita:rianiza·tion,** the action or process of rendering totalitarian; the fact of becoming totalitarian.

1941 'G. Orwell' in *Partisan Rev.* July–Aug. 321, I don't believe that the ordinary man cares a damn about the totalitarianisation of our economy. **1954** *Encounter* Dec. 32/2 Professor Hayek originated the phrase, 'the Road to Serfdom', to describe the progressive totalitarianisation of a whole *society* by a government that only wanted at first to control the *economy*, but finds more and more human obstacles to this aim. **1958** M. Fainsod *Smolensk under Soviet Rule* xxiii. 446 The stately procession includes urbanization, industrialization, collectivization, secularization, bureaucratization and totalitarianization.

totalita·rianism. [f. prec. + -ism.] Totalitarian theory and practice; the advocacy of totalitarian government. Also *loosely*, authoritarianism; *transf.* monolithic character.

1926 B. B. Carter tr. *Sturzo's Italy & Fascismo* ix. 233 This would mark the end of Fascist 'totalitarianism' and the renewal of political dualism. **1937** *Times* 9 Nov. 12/6 Nothing could be worse than to introduce totalitarianism into literature and to try to breed in a single country races of men and women fundamentally incapable of understanding one another. **1944** J. S. Huxley *On Living in Revolution* iii. 31 It [*sc.* Japan] has transformed itself from tribal and feudal totalitarianism to a modern technological totalitarianism. **1952** J. L. Talmon *Origins of Totalitarian Democracy* 6 The starting-point of totalitarianism of the Left has been and ultimately still is man, his reason and salvation, that of the Right totalitarian schools has been the collective entity, the State. **1967** G. Steiner *Lang. & Silence* 408 Soviet totalitarianism is most extreme not in the claims it makes on the utopian future, but in the violence it would do to the past, to the vital integrity of human remembrance. **1974** *Guardian* 2 Dec. 14/1 Most Swedish companies must have at least two elected representatives of workers on their boards of directors. .. Yet there was general agreement that the work place is the last bastion of totalitarianism in an otherwise democratic society.

totalizator. Add: **1.** Add to *spec.* def.: also, on each greyhound, etc., in a race; also, a system of betting based on the totalizator.

1910 *Encycl. Brit.* III. 827/1 On all French racecourses .., a system of betting known as the *Pari-Mutuel* or Totalizator, is carried on. **1935** *Encycl. Sports* 321/1 In 1931 and 1932 most of the licensed tracks in England were equipped with the most up-to-date all-electric totalisators; but in the February of 1933, following a High Court decision of December 17, 1932, that a totalisator on a greyhound track constituted a place within the meaning of the Act, these machines ceased to operate. **1975** *Oxf. Compan. Sports & Games* 445/2 Punters betting with a totalisator endeavour to select the dog or dogs they think will either win or obtain a place.

2. *Comb.*: **Totalizator Agency Board,** in Australia and New Zealand, an official organization, with local offices, for off-course betting on horses.

1950 *N.Z. Statutes 1949* 496 There shall be a board to be known as the Totalizator Agency Board. **1957** D. Glover *Since Then* 16 We are equals all in the sight of the Lord And the Totalizator Agency Board. **1957** *Weekly News* (Auckland, N.Z.) 6 Nov. 48/1 With the virtual extermination of the illegal bookmaker in the Dominion and the establishment of the Totalisator Agency Board as a firm medium for off-course bettors. **1983** *Austral. Encycl.* (ed. 4) X. 75 In Australia, the Totalizator Agency Boards are the government agencies which control legal off-course betting in the respective States and in the Australian Capital Territory.

totalness. (In Dict. s.v. Total *a.* and *sb.*) For *rare*[−0] substitute *rare* and add example.

c **1864** E. Dickinson *Poems* (1955) II. 619 All I may, if small, Do it not display Larger for the Totalness—'Tis Economy To bestow a World And withold a Star.

tote, *sb.*[1] **2.** For 'in *Australian colloq.*' read '*colloq.* (orig. *Austral.*)' and add: (further examples); also *loosely*, a lottery; hence also *tote board, double, ticket.*

1926 *Spectator* 9 Jan. 45/2 The 'tote' goes steadily on and the bookies do a roaring trade secretly. **1927** *Glasgow Herald* 30 Apr. 9 Information will be given on the legal and practical aspects of the 'tote'. **1930** *Cambridge Daily News* 25 Sept. 5/7 In the Tote Double on the 2.30..the winning dividend was £10 13s. 9d. **1933** *Sun* (Baltimore) 3 May 14/4 (*heading*) Ticket a second expected of electric 'tote' at racing meet. **1945** *Daily Herald* 31 Aug. 3/4 A fraud by which a considerable sum..was obtained with forged Tote tickets at Harringay Greyhound Stadium.. is thought to have been carefully planned by a gang. **1950** *Amer. Speech* XXV. 304/2 A tote board is a board where odds, payoffs, time of race and numbers of winners are posted in electric lights. **1966** *Listener* 27 Oct. 605/1 Further along there was a board showing the latest stock prices on Wall Street:..Zurichers watch them in much the same spirit as race-goers watch the tote. **1974** *Times* 26 Nov. 16/4 Young women were selling tote tickets through the window. **1975** *Ox. Compan. Sports & Games* 495/2 The Tote also operates 'doubles', 'trebles', and 'jack-pot' prizes for correct forecasts. **1976** *Star* (Sheffield) 30 Nov., The money was raised through totes and the fund is being wound up with a final pay-out because income was not enough to keep it going. **1977** *N.Z. Herald* 5 Jan. 2-8/4 The tote at Ellerslie, in line with the general trend this year, was up 22 per cent.

tote, *sb.*[4] (In Dict. s.v. Tote *v.*) Add: **b.** *ellipt.* = *tote bag* below.

1959 *Sears, Roebuck Catal.* Spring–Summer 68 Cowhide Bag... 2 side zip pockets in this top-zipper tote. Rayon lined. **1967** *Observer* 24 Dec. 15/7 A Twiggy Fashion Tote, 'for shopping and surfing'. **1979** *Kingston* (Ontario) *Whig-Standard* 5 Apr. 24/6 Remember that an open bag, like a tote may make access to its contents easier for you, but it also means access is easier for a pickpocket.

c. tote bag, a large hand-bag or shoulder-bag; **tote box,** a portable box for small items.

1900 in *N. & Q.* (1904) 27 Aug. 162/1 The Watson Tote Bag..best thing..for carrying coat, camera,..lunch, &c. **1969** *Daily Colonist* (Victoria, B.C.) 24 Sept. 2/6 Tote Bags —Great for knitting supplies, shopping. **1982** M. Millar *Mermaid* x. 108 A girl entered, carrying an oversized canvas tote bag with the name Gretchen printed on it. **1917** *Machinery* (N.Y.) July 957/1 The New Britain Machine Co...makes these tote boxes of steel, and they are designed in such a way that they may be stacked up to economize in floor space. **1951** Urquhart & Boyle *Materials Handling Case Bk.* 7/1 If the parts are small and are handled in a tote box, then each operator must: 1. Position the totebox of pieces to be worked [etc.]. **1966** *Guardian* 18 Apr. 6/1 Tote boxes are also available with cushion tops and back cushions.

tote, *v.* For *U.S. colloq.* read '*colloq.* (orig. *U.S.*)' and add: **a.** Also, to wear or carry regularly as part of one's equipment; to take (a person) with one; (further examples); *to tote fair* (earlier example).

1823 J. A. Quitman in J. F. H. Claiborne *Life Quitman* (1860) I. 85 The belles..'tote' their fans with the air of Spanish señoritas. **1828** J. Hall in *Western Souvenir for 1829* 269 This is a poor shooting-iron..it might do for young men to 'tote' in a settlement, but it is of no use in the woods. **1866** C. H. Smith *Bill Arp, so Called* 147, I don't think you tote fair. **1909** R. Parrish *My Lady of South* viii. 95 Thar warn't many Danielses left able ter tote a gun. **1909** H. G. Wells *Tono-Bungay* II. iii 194 The old merchant used to tote about commodities. **1952** C. Day Lewis tr. *Virgil's Aeneid* IV. 89 One who, men say, totes round his home-gods Everywhere. **1975** *Nation* 20 Dec. 659/1 Others wear official-looking uniforms and tote service revolvers. **1977** C. McCullough *Thorn Birds* vi. 110 He toted the infants with easy familiarity. **1979** *Chatelaine* (Canada) Jan. 24/3, I toted a canvas bag over one shoulder. **1983** E. Reveley *In Good Faith* iii. 59 They still tote the two original evangelists around with them but I think that's mostly so's they can keep getting money from the mother church.

totem. Add: **1. d.** *ellipt.* = *totem-pole*, esp. in *low on the totem. colloq.*

1974 K. Millett *Flying* (1975) II. 167 Counting on faculty privilege. Almost too low on the totem even to deserve it. **1977** D. Bagley *Enemy* xviii. 148 'What's your status here?' 'Low man on the bloody totem... I have a line into the Embassy but that's for emergency use only.'

2. *totem ancestor* (examples); **totem-pole,** (*a*) (earlier examples); also *fig.*, esp. in colloq. phr. *low on the totem pole*, of lowly status (see also sense *1 d); (*b*) *Electronics*, an arrangement of two output transistors or valves in which one takes the place of the load of the other, the output being taken from between the two.

1937 *Jrnl. Anthrop. Inst.* XL. 413 It is thus clear that the gi is nothing other than the totem-ancestor. **1949** J. Campbell *Hero with Thousand Faces* 390 An unconscious identification took place, and this was finally rendered conscious in the half-human, half-animal, figures of the mythological totem-ancestors. **1880** S. Jackson *Alaska & Missions on North Pacific Coast* ix. 263 Daylight found us near Fort Tongas... From the water there seemed to be a whole forest of.. *totem* poles. **1897** B. W. James *Alaska* 75 It has ever been an unanswerable question as to the origin of these totem poles. **1940** L. MacNeice *Last Ditch* 18 And under the totem poles—the ancient terror—between the enormous fluted Ionic columns There seeps..The guttural sorrow of the refugees. **1945** *Sun* (Baltimore) 3 Sept. 1/7 The lowest brass to sign the surrender documents was Colonel L. Moore Cosgrave... 'He's low man on the totem pole,' murmured an Australian correspondent. **1967** *Electronics* 6 Mar. 155/2 High leakage of the multiple-emitter transistor may load a circuit excessively. To offset this..the totem pole output stage is used. **1973** 'B. Mather' *Snowline* iii. 36 Just how far up the Departmental totem pole was Hallaby? **1978** *Jrnl. R. Soc. Arts* CXXVI. 456/2 In looking at the heritage of ideas or values we are looking at the totem poles of the heritage, symbols that are of more importance to us for what they represent than for themselves. **1978** D. Bagley *Flyaway* xxxiii. 311 Kissack ..was pretty low on the totem pole—a hired hand. **1981** P. M. Chirlian *Analysis & Design Integrated Electronic Circuits* v. 114 The load resistance of Fig. 5–10 has been replaced by an enhancement MOSFET. This 'load' is called an active load or active pull-up... The circuit is also called a totem pole because the elements are drawn one above the other in the schematic diagram.

totemic, *a.* (Earlier example.)

1846 H. R. Schoolcraft *Notes on Iroquois* 79 It will be necessary to go back, and examine..the curious and intricate principles of the Totemic Bond.

totemistic, *a.* (Earlier example.)

1873 *Fortn. Rev.* May 631 They have lost whatever meaning their totemistic forefathers may have had.

‖ **Totenkopf** (to·tĕnkɒpf). [Ger., = 'death's head'.] Used *attrib.* and *absol.* to designate (a member of) one of the divisions of the SS in Nazi Germany, having a death's head as its badge; *spec.* in the war of 1939–45, designating a unit (*Verband*) of concentration-camp guards.

1943 W. Necker *German Army of To-day* iv. 164 During the war some formations of the *Totenkopf-Verbände* or *Death's-head* (*Skull*) *Detachments* were incorporated into the *Waffen-SS.* Originally, these formations were guards at concentration camps. *Ibid.* 167 The *Totenkopf-Division*, or 'Death's Head' Division. **1953** G. Reitlinger *Final Solution* ii. 43 Later there developed a strong distinction between the *Totenkopfverbaende* and the SS *Totenkopf* Division, which..became a field division like any other. **1975** tr. *Melchoir's Sleeper Agent* II. 81 You and me both know what those *Totenkopf* bastards are. Concentration camp guards, that's what! **1977** D. James *Spy at Evening* vii. 45 Eicke was..a *Waffen SS* divisional commander... He commanded The Death's Head, Totenkopf division. **1981** 'E. Trevor' *Damocles Sword* xv. 149 The Totenkopf and the Gestapo keep extensive records.

‖ **Totentanz** (to·tentants). Also 8 **Toden Tans;** (*hist.*) **Todtentanz.** [Ger.] = *dance of death* s.v. Dance *sb.* 6 c. Also *fig.*

1789 *Emblems of Mortality* p. xxiv, He [*sc.* Holbein] also engraved several things upon wood, among which are his *Scripture Cuts*, and *Dance of Death*, vulgarly called *Toden Tans.* **1937** *Jrnl. Archæol. Assoc.* I. 249 Switzerland, which was once rich in representations of the Dances of Death, has suffered grievous losses. All that remains are a few..fragments in Museums. The earliest of these—the Klingental Totentanz at Basle—..was destroyed.. about 1850. **1950** A. Wilson *Such Darling Dodos* 144 (*heading*) Totentanz. *Ibid.* 163 Your first big reception, duckie, shall be a Totentanz. **1964** W. G. Raffé *Dict. Dance* 50/1 The popular play and mime dance of the *Todtentanz* was then [*sc.* in 1535] still extant in its ritual or miracle play form. **1966** J. Fowles *Magus* xix. 113 Dupes of the reality of war, of the ultimate *Tōtentanz.* **1982** P. Dickinson *Last House-Party* (1983) ii. 18 Time emerged..no friendly old gaffer with a scythe, but close kin to the skeleton reaper of the *Totentanz.*

tother, *pron.* and *a.* Add: **A. 1. a.** Phr. *to tell tother* (or *t'other*) *from which* (joc.), to tell one from the other or (loosely) another; to distinguish or tell apart.

1874 M. Clarke *His Natural Life* (1975) III. xxii. 24 You're so much alike one can't tell t'other from which. **1904** Kipling *Traffics & Discoveries* 258 We've mixed the whole show up..till you can't tell t'other from which. **1979** D. Francis *Whip Hand* ii. 27 He calls them all Tommy, because he doesn't know tother from which.

B. 1. c. *tother school, 'un* (Public School slang), a preparatory school, a school one attended before one's public school.

1880 Trollope *Dr. Wortle's School* (1881) I. ii. 34 The old prescribed form of education..must be followed,—a t'other school, namely, then Eton... Therefore Bowick was chosen as the t'other school. **1940** M. Marples *Public School Slang* 179 'Where's your t'other 'un?' a question generally addressed to new boys. **1958** *Sunday Times* 25 May 8/3 Mr. Kenward's totherun (if the reviewer may be permitted to adopt, for the moment, his own public-school terminology) is named Ripple.

C. tothersider, delete *nonce-wd.* and add: *spec.* of Australia (earlier and later examples).

1896 H. Lawson *Let.* 3 Sept. (1970) 62 W.A. is a fraud. ..The old Sand-gropers are the best to work for or having dealings with. The Tothersiders are cutting each other's throats. **1903** 'T. Collins' *Such is Life* (1944) 276 The ancient t'other-sider [*sc.* Vandemonian Jack] oscillated his frame-saw. **1929** J. Raeside *Golden Days* 224 The population of Hannans, although mostly composed of t'othersiders, included not a small sprinkling of West Australians. **1949** *Geographical Mag.* Feb. 373 *Tothersider*, a Western Australian. **1950** K. S. Prichard *Winged Seeds* 30 Unemployed from all over the country swarmin' here, t'other siders as well as W.A. blokes. **1963** X. Herbert *Disturbing Element* 2 My parents..were what were called T'othersiders, meaning people who had come to West Australia from the other side of the continent.

‖ **totidem verbis** (tɒ·tidem vɜ̄·ɹbīs), *adv. phr.* [L.] In so many words.

1659 N. Hardy *First Ep. John* vi. 101 We do not read (*totidem verbis*) in the Scripture that the Apostle Baptized Infants, yet it is very probable. **1704** Swift *Tale of Tub* 64 'Tis true, said he, there is nothing here in this Will, *totidem verbis*, making mention of Shoulder-knots. **1844** Mill *Ess. Pol. Econ.* ii. 47 This object, under the varying names of an extensive demand, a brisk circulation, a great expenditure of money, and sometimes *totidem verbis* a large consumption, was conceived to be the great condition of prosperity. **1902** L. A. Burd in *Cambr. Mod. Hist.* (1907) I. vi. 202 These [fundamental beliefs or hypotheses] are rarely stated *totidem verbis* in any passage, though implied in nearly all.

totipotent, *a.* Add: Also, able to differentiate into any other related kind of cell; so **totipo·tency** = *totipotence* s.v. Totipotent *a.*; **totipote·ntial** *a.* = Totipotent *a.* in Dict. and Suppl.

1918 *Jrnl. Exper. Zool.* XXV. 500 Totipotency is restricted to those girdle-forming cells which become implanted along with the limb bud. **1934** *Discovery* Aug. 220/1 There must be some power which controls and regulates the powers of these turbulent totipotential cells and this is exactly what Driesch called the 'entelechy'. **1942** M. M. Wintrobe *Clin. Hematol.* i. 30 According to the monophyletic school..the lymphocyte of lymphatic tissue is identical with the primitive blood cell and is thus totipotential, giving rise under proper stimulation to any other type of blood cell. **1959** W. Andrew *Textbk. Compar. Histol.* xii. 458 These cells are 'totipotent' and, according to need, can give rise to any other type of cell in the body of the sponge. **1967** *Amer. Jrnl. Med.* XLII 932/1 Yoffey and Courtice..believe that a major function of the small lymphocyte is that of a circulating totipotential cell. **1979** *Sci. Amer.* Apr. 93/2 Totipotency equivalent to that of early embryonic cells was thereby established unequivocally for individual embryonal carcinoma cells.

toto[1] (tōᵘ·to). [ad. Swahili *mtoto* offspring, child.] In East Africa: a child; a baby; a young animal; a young servant.

1916 *Chambers's Jrnl.* Nov. 719/2 Poor little 'toto', bereft of his mother. **1927** *Ibid.* Nov. 762/1, I was a 'toto' then. How old I cannot say. In my tribe there is no record of birth or death. **1937** K. Blixen *Out of Africa* IV. 336, I was..a long way in front of the waggons, with Farah, my dog Dusk and the Toto who looked after Dusk. **1964** C. Willock *Enormous Zoo* v. 90 At first he ran out in front as is the custom with white rhino *totos.* **1979** *Observer* (Colour Suppl.) 9 Sept. 43/1 We hear goat-bells, and tiny herd-boys emerge cautiously from the bush. 'Give those *totos* a bowl of maize meal,' Thesiger orders.

toto[2] (tōᵘ·to). *Mil. slang* (of the 1914–18 war). [a. Fr. mil. argot.] A louse.

[**1917** G. Clover *Stop at Suzanne's* (1919) 223 They were all covered with lice—*les totos* they call them.] **1918** *Radiator* 30 May 1 Dr. Kent Hagler..saw no evidence of flea or toto. *a* **1919** in E. C. Garrett *Trench Ballads* (1919) 78 Some people call 'em Totos—Some people call 'em lice. **1929** Hall & Niles *One Man's War* 46, I do not know who developed lice first, but I noticed them on that march. We called them 'totos'.

Totonac (totŏnæ·k). Also † **Totonaca.** [ad. Sp. *Totonaca*, f. Nahuatl *Totonacatl*, pl. *Totonaca.*] An Indian people of east central Mexico; a member of this people. Also, their language. Also *attrib.*

1787 C. Cullen tr. *Clavigero's Hist. Mexico* II. viii. 18 Five men..said in Mexican..that they were of the nation of the Totonacas, and sent by the lord of Chempoalla. **1852** B. Mayer *Mexico* I. iii. 29 The Tlascalans were not so easily won as his allies, the Totonacs. **1900** *Proc. Davenport Acad. Sci.* VIII. 187 Some of the..Totonac women in the Plaza..are really gay with hair ribbons. **1908** F. Starr *In Indian Mexico* xx. 245 In Tlaxco, a small village in this *municipio*, four idioms are spoken—Aztec, Otomi, Totonac and Tepehua. **1940** F. Johnson in C. L. Hay et al. *Maya & their Neighbors* vi. 109 The area on the map in which Totonac was spoken remains practically identical with that originally drawn by Orozco y Berra. **1948** A. L. Kroeber *Anthropol.* (rev. ed.) xviii. 794 Along the Gulf coast in Tabasco and Vera Cruz, a series of peoples known as Olmec, Totonac, and Huastec, in order from south to north. **1977** T. A. Sebeok *Native Lang. Americas* II. 153 The Totonac..were the first to become allies of Cortez when he disembarked on the coast of Veracruz and set out to conquer Mexico. **1977** *Language* LIII. 262/2 The publication of these volumes.. makes Totonac an unusually well-documented language.

Hence **Totona·can** *a.*, of or pertaining to the family of languages that comprises Totonac and Tepehua.

1933 J. E. Thompson *Mexico before Cortez* viii. 259 The Tajin pyramid is in Totonacan linguistic territory. **1940** F. Johnson in *Maya & their Neighbors* vi. 109 (*heading*) Totonacan family [of languages]. **1977** T. A. Sebeok *Native Lang. Americas* II. 153 There are..well over 130,000 speakers of Totonacan dialects in Veracruz and Puebla.

totora (totōᵊ·ra). [a. Quechua, Aymara.] A perennial bulrush, *Scirpus totora*, native to alpine lakes in Peru and Bolivia. Also *attrib.*

1936 *Discovery* Dec. 372/2 The 'totora'..is still used by the Indians of Lake Titicaca to make nets and cordage. **1958** A. Toynbee *East to West* vi. 17 The cattle wade out, breast-deep, to crop the tender shoots of the totóra reed. **1971** P. Crampton tr. *Heyerdahl's Ra Exped.* vi. 110, I knew that the *totora* reed in America was capable of long sea voyages. **1974** T. Morrison *Land above Clouds* 142 For centuries, the Indians around Titicaca have used the Totora for making their reed boats.

‖ **tot siens** (tɒt sins). *S. Afr.* Also **totsiens.** [Afrikaans *tot* (*weer*)*siens* 'until we meet again', f. Du. *tot* until + *zien* to see.] A formula of farewell: au revoir, till I see you again.

1937 S. Cloete *Turning Wheels* ix. 142 'Tot Siens. Tot siens,' they cried, their voices growing fainter. **1948** H. V. Morton *In Search S. Afr.* ii. 52 We said good-bye, with a *tot siens* or two and one 's' long folks'. **1963** A. Delius *Day Natal took Off* 72 This is, then, not good-bye, only totsiens. **1974** *Argus* (Cape Town) 30 July 22/9 Tot siens! May you come back to Britain soon.

totsy (tɒ·tsi). *slang.* [f. Tot *sb.*[4] + -sy.] = *totty *sb.* 2.

1938 G. Greene *Brighton Rock* IV. i. 142 The atmosphere of innumerable roadhouses, of totsies gathered round swimming pools.

Tottenham. Add: **b.** *Tottenham pudding*, feed for pigs or poultry, consisting of sterilized kitchen waste.

1944 Halnan & Garner *Princ. & Pract. Feeding Farm Animals* (ed. 2) xvi. 329 In the process of cooking the waste..sets to a firm..'pudding' which finds a ready market for feeding to pigs and poultry. Since this pudding was first produced in..Tottenham..such prepared food is known as 'Tottenham Pudding'. **1966** K. Nicholson *Hook, Line & Sinker* viii. 91 He's a Large White... In the winter, he has his quota of Tottenham Pudding—that's concentrated swill, sterilised by steam. **1980** *Good Housekeeping* Dec. 226/6 The days of muck and mystery, when poultry were fed on 'Tottenham Pudding'—kitchen waste from hotels and restaurants, boiled up and sterilised.

tottle, *v.*[1] Add: **tottlish** *a.* (earlier example); **to·ttly** *a.* = *tottlish* adj.

1835 *Knickerbocker* VI. 6 Had she not been obliged.. to steady her tottleish bark with the paddle which now loitered behind the stern. **1905** *Eng. Dial. Dict.* VI. 203/1 *Tottly*,..ready to fall, unstable. **1910** Kipling *Rewards & Fairies* 155 My legs was pretty tottly, but I made shift to go on deck.

totty, *sb.* Add: **2.** *slang.* A girl or woman, *esp.* a 'good-time' girl.

1890 Barrère & Leland *Dict. Slang* II. 368/2 *Tottie*.., a girl, a fast girl. **1914** Joyce *Dubliners* 29 He asked us which of us had the most sweethearts. Mahony mentioned lightly that he had three totties. **1957** J. Braine *Room at Top* xxviii. 230 She has a pal, some old tottie that lends her a flat. **1968** 'O. Mills' *Sundry Fell Designs* viii. 86 All Dan's Manchester-type ladies were only totties. This marriage wasn't going to make any difference to them. **1977** C. Watson *One Man's Meat* iv. 34 Showing off. Certainly, why not? There were a couple of totties just behind.

3. *Comb.*: **totty-pot** = *potty *sb.* 1.

1966 'L. Lane' *ABZ of Scouse* II. 83 *Potty* or *totty-pot*, a child's chamber-pot. **1971** *Daily Tel.* (Colour Suppl.) 22 Oct. 17/2 Room in boot for pram, pushchair, totty-pots, picnic gear.

‖ **tou** (dōᵘ, tōᵘ). *Chinese Antiq.* [Chinese *dòu.*] A hemispherical pedestalled bowl with a lid of similar shape, used as a container for food.

1899 S. W. Bushell *Oriental Ceramic Art* xvii. 491 The twelve bowls (*pien*) on the right being made of closely woven slips of bamboo, lacquered yellow, the twelve (*tou*) on the left of carved wood, gilded. **1909** B. Laufer *Chinese Pottery of Han Dynasty* iv. 122 The ears are not essential to the type of the *tou*, as there are also bronze *tou* which lack them. **1973** *Genius of China* 52/2 Red Pottery tazza *tou*, with undulating ornament painted in black.

Touareg, var. *Tuareg.

touch, *sb.* Add: **II. 8. c.** (Earlier example.)

1816 Jane Austen *Emma* II. viii. 147 Having so much to ask and to say as to tone, touch, and pedal.

12. (Earlier example.)

1857 [see *play *sb.* 10 d].

III. 13. d. *out of touch* (examples).

1891 G. Moore *Impressions & Opinions* 88 He is out of touch with them; he cannot make them understand. **1969** H. Perkin *Key Profession* i. 4 By then Newman was out of touch with what universities were becoming. **1980** D. Lodge *How Far can you Go?* iv. 121 Most of them had been out of touch with him for many years, but he spoke to them as if it was only yesterday.

18. b. (*fig.* from senses 8 b, 10 b.) A person's characteristic skill or aptitude in any activity, *spec.* a sport; *to lose one's touch, to be out of touch*, not to show one's customary skill; similarly *to be in touch.*

1927 *Sat. Rev.* 9 July 60/1 But in the extended character-drawing of Ferdinand Banting and Tom Lord, who are not indigenous to the King's Cross scene, he seems to lose his touch. **1933** *Times* 18 Nov. 5/7 Success depended on being in touch for his drop shots, and yesterday his touch failed him. *Ibid.* 8 Dec. 6/2 He lost his touch and made but one more ace before the match was over. **1939** *Punch* 4 Oct. 378/1 That fatal hour when Hitler lost his touch. **1955** *Times* 13 July 8/5 It is one of the signs of greatness to be able to stay in for a long time without finding touch, and yet without looking exactly like getting out. **1959** *Times* 29 May 4/2 Nicholls, who has been out of touch, is a tall, stylish player. **1976** *Liverpool Echo* 6 Dec. 17/2 It was..stalemate until Ipswich found the touch which produced the winner 15 minutes from the end. **1977** *World of Cricket Monthly* June 28/3 The presence of Kallicharran in his best touch for the series..encouraged a little West Indian optimism entering the last day. **1979** A. Morice *Murder in Outline* v. 44 She may not be in the pink of health, but she has not lost her touch.

19. b. Also *a touch of the sun*, a mild attack of sunstroke.

1890 Kipling *Life's Handicap* (1891) 165, I judge no man this weather... He had a touch of the sun, I fancy. **1915** R. Brooke *Let.* Apr. (1968) 680 When I had a touch of the sun, in Egypt. **1965** M. Spark *Mandelbaum Gate* v. 118 'A touch of the sun,' Freddy said. Amnesia, was the doctor's conclusion... Nonsense, I'm suffering from sunstroke.

20. b. Add after 'person': esp. by persuasion or glib talk; *to make a touch*, to obtain money thus. (Earlier and further examples.)

1846 *Nat. Police Gaz.* (U.S.) 18 July 390/1 Ingenious Touch... Phillsburg..felt for his money, and..found in its place another pocket-book filled with newspaper instead of money. *Ibid.* 25 July 389/1 The Read Street Touch Case. Ann Henry, the keeper of a den of infamy and..one of her syrens, have been fully committed on the charge of robbing Townsend W. Hetherington. **1865** *Leaves from Diary of Celebrated Burglar & Pickpocket* xv. 48/2 The most splendid 'touch' of the campaign was already in our grasp! *Ibid.* xvii. 58/2 They took a furnished room in..Seven Dials, until a lucky 'touch' came off, when they took larger apartments. **1896** Ade *Artie* v. 43 Next day they had to make a hot touch for a short coin so as to get the price of a couple o' sinkers and a good old 'draw one'. **1914** *Automobile Topics* 4 July 638/3 His story of not being able to find employment..has enabled him to make many a successful 'touch'. **1939** R. Chandler *Big Sleep* xvi. 114, I figure it's a good time to..make a quick touch on the Sternwoods for travel money. **1953** *Essays in Crit.* III. 111 The poet might regard the patron as an intimate,..or simply as a public Maecenas, good for a 'touch' of three guineas a dedication. **1964** C. Chaplin *Autobiogr.* xvii. 299 It seemed obvious from the tone of the letter that it was all leading up to a 'touch'. So I thought I would take along $500.

c. *soft* or *easy touch*: a person easily manipulated; *spec.* one easily induced to part with money; also, a task or opponent easily handled. *colloq.*

1940 J. O'Hara *Pal Joey* 44 You get the reputation of being a soft touch. **1945** *Sun* (Baltimore) 4 Oct. 1/1 Newhouser.., who figured to be the outstanding pitcher of the season, proved..to be the softest sort of a touch for Manager Charlie Grimm's National Leaguers. The second Cub to bat belted him cleanly. **1955** H. Kurnitz *Invasion of Privacy* (1956) xii. 80 Dorsey's appetite for easy money..was honed to a razor edge... He sensed a vast soft touch. **1959** H. P. Tritton *Time means Tucker* vi. 45/1 He was an easy touch for any hard-luck story. **1972** *Police Rev.* 1 Dec. 1562/2, I would also warn any university student regarding the Police as a 'soft touch' for graduates. **1976** *Eastern Even. News* (Norwich) 29 Nov., Caravan dwellers are on the increase and they will keep on increasing while Norwich remains an easy touch; the complacency regarding this problem is alarming.

d. *to cut up* (*old*) *touches*: see *cut *v.* 59 r.

V. 28. *in* or *out of touch*: see also sense *18 b.

touch, *v.* Add: **I. 2. a.** Also = *to touch up*, sense 34 c (*b*) below; *refl.*, to masturbate.

1903 Farmer & Henley *Slang* VII. 177/2 *Touch*... verb... (or *to touch up*), to grope a woman. **1927** F. Harris *My Life & Loves* (1934) IV. ix. 182 You want to know if I have touched myself. Sure, all girls have. **1973** *Family Circle* July 114/1 Little girls are told not to touch or play with themselves, and later their sexual parts are associated with urination and menstruation, which are considered 'dirty'.

5. Also *fig.* in pa. pple., = *touché b. *rare.*

1853 E. Sewell *Experience of Life* xviii. 183 'I dislike this kind of bantering very much, Horatia,' I said... Horatia laughed merrily. 'Touched, I declare!'

7. Also, *pass.*, to be slightly affected by drink. *colloq.*

1834 C. Brontë *My Angria & Angrians* in W. Gérin *C. Brontë* (1967) vi. 84 Two bottles of..ale, and a double quart of Porter..and I'm not a bit touched—only light and smart and active. **1888** 'R. Boldrewood' *Robbery under Arms* II. xi. 180, I wasn't no ways drunk; but I must have been touched more or less, because I felt myself to be so sober.

11. a. Also in legal formula *to touch and stay.*

? **1796** in *Eng. Reports* (1927) CLXX. 471 Liberty to sail to, touch and stay at any port or ports whatsoever on her passage out..without prejudice to the insurance. **1895** W. Gow *Marine Insurance* iii. 59 The liberty to touch and stay is limited by its close application to the main object of the voyage. **1969** E. R. H. Ivamy *Marine Insurance* xiii. 142 It was formerly held that 'liberty to touch and stay' did not permit of trading at the port of call.

16. b. (Examples of sense 'to obtain a loan or gift of money from' without const., and of sense 'to rob'.)

1807 H. Tufts in E. Pearson *Autobiogr. of Criminal* (1930) II. iv. 293 *Touching a cly*, robbing a pocket. **1888** in Farmer & Henley *Slang* (1903) VII. 177/1 A dip [*sc.* pickpocket] touched the Canadian sheriff for his watch and massive chain while he was reading the Riot Act. **1928** [see* knock *v.* 13 d]. **1950** *Austral. Police Jrnl.* Apr. 110 To touch a person is to steal from him, but to touch him for a loan is to ask him for one. **1951** G. Greene *End of Affair* v. iv. 197 'If you would lend me a pound.'..Had she 'touched' Henry once too often? **1963** T. Parker *Unknown Citizen* i. 32 He wants some money... Don't you send it to him, let him touch somebody else for a change.

23. b. Also in phr. *touched in the head* or *the upper story.*

1867 Trollope *Last Chron. Barset* I. xx. 172 We tried to get him through as being a little touched in the upper story. **1902** E. Nesbit *Five Children & It* ii. 61 Touched in the head, eh?.. All the more shame to you boys dragging the poor afflicted child into your sinful burglaries.

III. 30. touch down. b. *Aeronaut. intr.* To alight on the ground from the air; to land; also *transf.* Also (*rare*) *trans.*, to land (an aircraft).

1935 C. Day Lewis *Time to Dance* 41 M'Intosh touched her down. **1938** *Jrnl. R. Aeronaut. Soc.* XLII. 498 A successful flight down the beam..gives the feeling that if the ceiling had been only 50 feet one could have held on.. longer before finally touching down. **1942** P. Brennan et al. *Spitfires over Malta* (1943) ii. 55, I touched down and swung my aircraft away from the pitted landing path, braking violently. **1955** *Times* 22 Aug. 5/4 The first aircraft to touch down brought an official party from Kallang. **1962** *Listener* 8 Feb. 260/2, I send this dove from the ark Where she must never touch down. **1970** N. Armstrong et al. *First on Moon* xiv. 369 At 12.45 a.m. Houston time, Apollo 12's lunar module Intrepid touched down on the moon. **1979** *Arizona Daily Star* 1 Apr. E 9/1 14 persons were injured when a tornado touched down near Glasgow, Ky.

32. touch off. b. (Earlier example.) Also *fig.*, to provoke (a reaction), to spark off.

1884 'Mark Twain' *Huck. Finn* xxviii. 282 It does seem most like setting down on a kag of powder and touching it off. *a* **1934** in Webster s.v., These terms..have become push buttons which touch off emotional reflexes. **1943** *Sun* (Baltimore) 12 Feb. 6/2 Senator Truman touched off the debate with a speech in which he assailed the supplanting of Lou Holland as chairman of the SWPC. **1950** *N.Y. Times* 20 Apr. 1/6 The surprise proposal..touched off several outbursts of denunciation of the Soviet action. **1958** *Listener* 29 Nov. 813/1 The Bundestag declaration has touched off a chain-reaction of inquiry, proposal, examination, plan. **1966** *Ibid.* 10 Feb. 221/2 All these insights the piano touched off with its single hollow note struck over and over again. **1979** *Tucson* (Arizona) *Citizen* 20 Sept. 5c/1 A $1·8 million error may touch off a legal challenge.

34. touch up. b. Also, to exert influence upon; to rouse the emotions of.

1811 Jane Austen *Sense & Sens.* III. iv. 88 We must touch up the Colonel to do something to the Parsonage. **1817** M. Edgeworth *Harrington* I. iii. 55 You will see.. how cleverly I will get myself out of the scrape with her. I know how to touch her up. **1846** Dickens *Let.* 28 Mar. (1977) IV. 528, I hope you mean to go to the General Theatrical Fund Dinner on Monday Week... Let me know, that I may touch up the Committee to place you near me. *c* **1863** T. Taylor in M. R. Booth *Engl. Plays of 19th Cent.* (1969) II. 140 The roughs adore music..and as for sentiment and sensation, if you could hear Miss St. Evremond touch them up with the 'Maniac's Tear', the new sensation ballad [etc.]. **1884** E. W. HAMILTON *Diary* 10 Mar. (1972) II. 573 Slavery is a matter which specially touches up the British public.

c. † (*a*) (See quot. 1785.) *Obs.* (*b*) To finger or caress so as to excite sexually. *slang.*

1785 GROSE *Dict. Vulgar T.*, *To touch up a woman*, to have carnal knowledge of her. **1903** [see sense 2 a above]. **1923** J. MANCHON *Le Slang* 318 *To touch up a woman*, caramboler une femme. **1961** H. S. TURNER *Something Extraordinary* vii. 135 She..went in for a crass practice..known as 'touching up'. It is..a quick flick, in passing, from the crotch upwards... When a girl 'touches up' a boy it seems to be a very casual signal,..nowhere near a definite proposition. **1966** P. WILLMOTT *Adolescent Boys E. London* iii. 49 They would often try to move on from kissing to sexual play: as they put it, they ..went up her skirt or 'touched her up'. **1973** C. EGLETON *Seven Days to Killing* iv. 48 Good-looking tart... I wouldn't have minded her touching me up.

touch-. Add: **1. a.** *touch-sensation* (earlier example), *-stimulus*; *touch-sensitive* adj. **b.** **touch-dancing** orig. *U.S.*, dancing in which the partner is held close; hence (as a back-formation) **touch-dance** *v. intr.*; **touch-finder** *Rugby Football*, one who or a kick which succeeds in driving the ball into touch (TOUCH *sb.* 12); so **touch-finding** *vbl. sb.* and *ppl. a.*; **touch football** *U.S.*, a form of American football in which a player carrying the ball may be stopped simply by touching him, instead of tackling; **touch-kicking** *Rugby Football*, the action of kicking the ball into touch (TOUCH *sb.* 12); hence (as a back-formation) **touch-kick** *v. intr.*; so **touch-kick** *sb.*; **touch-mark,** an official stamp on pewterware, esp. one identifying the maker; cf. TOUCH *sb.* 5 b; **touch pad** *Computers*, a computer input device in the form of a small touch panel; **touch panel,** a panel containing different areas that need only to be touched to operate an electrical device; **touch preparation** *Microscopy*, a preparation made by lightly touching cultured or freshly cut tissue with a slide so that a thin layer of cells adheres to it; **touch rugby** or **rugger,** a version of rugby football in which touching takes the place of tackling; **touch screen** *Computers*, a VDU screen that is also an input device operated by touching it; **touch shot** *Lawn Tennis*, a shot without any force; **touch spot** *Physiol.*, one of the spots on the skin specially sensitive to touch or pressure; **Touch-Tone** *U.S.*, a proprietary name for telephone apparatus in which push-buttons take the place of a dial; **touch-typing,** the art of typing without looking at the keys; hence (as a back-formation) **touch-type** *v. intr.*; so **touch-typist**; **touch-writer** = *touch-typist* above.

1972 *Harper's Bazaar* Oct. 72/1 Dance experts agree that, as the East goes, so goes the nation, and what you've heard by now is true—'touch', 'partner', 'ballroom' dancing is back. *Ibid.* 72/3 Freddie doesn't touch dance at all. **1974** *Courier-Mail* (Brisbane) 23 Feb. 18/4 The latest craze among young people in the U.S. is 'touch dancing', which their mums and dads used to call 'dancing cheek-to-cheek'. **1939** *Daily Tel.* 18 Dec. 11/1 Jenkins..alternated long touch-finders with sliced shots. *Ibid.*, Ellis was allowed a lot of latitude in..putting in touch-finding kicks. **1960** *Times* 30 Nov. 3/6 The small, durable halves were dedicated touch-finders. **1976** *Leicester Mercury* 14 Oct. 46/1 It was a very solid display of good catching, good touch-finding and some probing entries into the line. **1933** *Jrnl. Health & Physical Educ.* Oct. 41/1 Touch football is now a scientific and standardized game. **1951** J. STEINBECK *Burning Bright* i. 37, I was..just playing around with some of the fellows—touch football. **1977** *Transatlantic Rev.* LX. 119 You often see Winterville kids playing touch football along the parkway. **1954** J. B. G. THOMAS *On Tour* vi. 71 They..saved and counter-rushed and touch-kicked with unerring accuracy. **1960** *Times* 31 Oct. 4/4 Long touch-kicks. **1978** *Morecambe Guardian* 14 Mar. 11/3 Glover took a while to find his usual accuracy with his touch kicks. **1936** *Times* 9 Jan. 4/3 Some excellent touch-kicking by Morris forced the Navy back into their own '25'. **1904** H. J. L. J. MASSÉ *Pewter Plate* xiv. 190 The touch-marks usually were the initials of the maker of the pewter, and various other devices such as the Company's quality mark. **1959** L. GROSS *Housewives' Guide to Antiques* viii. 103 Some, but not all, pewter will be found with a touch-mark. **1974** L. KOENIG *Little Girl* iii. 34 A pewter tankard..seemed to demand examination. She ..turned it over to study the touchmark. **1980** *Displays* I. 206/1 These experimental studies, conducted during the development of the Touchpad, demonstrate the viabiity of an off-display touch input device. **1983** *Your Computer* Aug. 32/2 Once out of its package the Wizzard takes on the appearance of a quite simple, compact unit, complete with two joysticks, touch pads and firing buttons. **1974** *Physics Bull.* June 225/2 One of these is a touch panel made of a glass plate on which capacitors are thinly etched in copper. A TV tube behind the panel 'names' the capacitors in an array of 4×4 and these 'buttons' can be 'pushed' by merely touching the panel. **1981** J. B. ADAMS in J. H. Mulvey *Nature of Matter* vii. 156 The operator..can send instructions to any component of the machine by means of a touch panel, which identifies the component, and one knob, which determines the required action. **1956** *Nature* 7 Jan. 47/1 Touch-preparations of spleen and lung on slides were fixed in 95 per cent ethanol at 37°C. for 30 min. **1975** *Ibid.* 17 July 225/2 Spleens were sectioned sagittally and touch preparations were made and stained with the Wright–Giemsa stain or benzidine. **1977** *Arab Times* 14 Dec. 9/7 Both [games] entail constant running, both in defence and attack, especially touch rugby. **1942** C. MILBURN *Diary* 16 Dec. (1979) 161 He talks of hockey, soccer and touch-rugger, describing the latter game. **1974** *Management Informatics* III. 70/1 As a first step, a prototype touch screen was designed and constructed in our Laboratory by Mr. Stephen Salter. **1983** *Austral. Personal Computer* Aug. 60/2 The touch-screen and light-pen both have the limitation that the user must first identify the location that has to be touched, and then a physical movement has to be made... Also, touch screens do get finger-marked. **1865** S. HODGSON *Time & Space* ii. 78 A combination of a whole series of touch-sensation into a solid whole is apparently possible..in grasping a small object, where the fingers meet each other. **1969** *Bull. Radio & Electr. Engin. Div. Nat. Res. Council Canada* July–Sept. 15 (*heading*) An *X-Y* touch sensitive position encoder for computer input. **1979** *Washington Post Mag.* 25 Mar. 5/3 The clavichord changed. Its keyboard widened... 'It got louder and more touch-sensitive,' says Tom. **1983** *Listener* 12 May 3/2 You controlled your route by pressing buttons on a touch-sensitive screen. **1959** Touch shot [see *DINK *sb.*[2]]. **1969** *New Yorker* 14 June 56/2 A loose, liberal, infuriating touch shot. **1906** C. S. SHERRINGTON *Integrative Action of Nervous System* ix. 324 The retina is thus a group of glorified 'warm-spots,' and the cochlea a group of glorified 'touch-spots'. **1927** HALDANE & HUXLEY *Animal Biol.* v. 122 The fineness of discrimination for touch depends mainly on the closeness of touch-spots. **1968** D. F. HORROBIN *Med. Physiol. & Biochem.* xxv. 150/2 There are touch spots, cold spots and warm spots, each particularly sensitive to one modality. **1927** HALDANE & HUXLEY *Animal Biol.* xii. 268 Most..of the group possess nerves, and at least scattered sense-organs for perceiving touch-stimuli. **1962** *Official Gaz.* (U.S. Patent Office) 19 June TM 122 American Telegraph and Telephone Company, New York... *Touch-Tone.* For providing telephone communication service. **1970** O. DOPPING *Computers & Data Processing* xi. 163 If the telephone is of the touch-tone type, the same buttons that are used for dialling can be used also for putting questions, expressed in numerical form, to the computer. **1972** *Sci. Amer.* Sept. 112/3 The first telephone switching systems were actuated by human operators; today the job is done automatically by means of a dial or 'Touch-Tone' terminal on the user's telephone. **1976** *National Observer* (U.S.) 10 Apr. 9/4 Touch-tone converters. This device changes your dial telephone to a touch-tone telephone. **1962** *Punch* 8 Aug. 191/1 If you will learn to touch-type, I will give you a new, feather-light portable. [**1897** in *Story of Typewriter* (1923) 113 Omaha has become the storm centre of the commotion over the touch method of typewriting.] **1947** K. JAEDIKER *Tall, Dark & Dead* viii. 117 All I know about touch-typing is that there are home keys and if you don't put your fingers on them, you go haywire. **1976** 'J. FRASER' *Who steals my Name?* iv. 46 Many coppers had done touchtyping courses. **1929** *Telegr. & Telephone Jrnl.* XVI. 13/1 Attention was concentrated upon touch-typing with the object of turning out highly-skilled touch-typists. **1972** 'J. & E. BONETT' *No Time to Kill* vii. 90 I'm not a touch typist. Very few writers are. **1915** *Literary Digest* (N.Y.) 21 Aug. Advt. p. i, Great numbers were so-called touch-writers—yet there has hardly been a single one who hasn't doubled or trebled his or her speed and accuracy.

2. a. touch-back: (examples); also, a similar action in some other ball games; **touch-last,** a children's game, = TOUCH *sb.* 1 g.

1891 W. CAMP *Amer. Football* 172 A touch-back is made when a player touches the ball to the ground behind his own goal, the impetus which sent the ball across the line having been received from an opponent. **1941** *Daily Progress* (Charlottesville, Va.) 14 Jan. 11 This used to be an automatic touchback and the ball was placed in play on the 20-yard line. **1976** *Webster's Sports Dict.* 455/2 *Touchback.* Speedball. A situation in which the ball is driven over the end line by an offensive player without scoring. **1825** JAMIESON *Suppl.* II. 568/2 *Tig*, a game among children, in which one strikes another and runs off... This game in S[cotland] is the same with Touchlast in E[ngland]. **1902** [see *HE *pers. pron.* 6 b]. **1927** *Sunday Express* 17 July 8/2 The younger and sprightlier guests..played 'touch last' on the lawn. **1951** E. GRAHAM *My Window looks down East* iv. 29 He walkes sideways away from her, like a child playing 'touch-last'.

touchable, *a.* Add: Hence also **touchabi·lity,** suitability to be touched.

1937 L. MACNEICE in *Essays & Stud.* XXII. 157 Spender..believes in 'touchability'... It means..the belief that people in themselves are worth knowing and touching, just as for Auden facts are worth remembering. **1944** [see *MARRIAGEABILITY].

tou·chdown. [TOUCH- 2 a.] **1.** (In Dict. s.v. TOUCH- 2 a.) Also in *American Football.*

1876 in P. H. Davis *Football, Amer. Intercollegiate Game* (1911) 462 A match shall be decided by a majority of touchdowns. **1949** *Desplaines Valley News* (Summit, Illinois) 28 Oct. 7/3 Harvard could not push across a touchdown in the first half. **1977** *New Yorker* 9 May 122/2 A figure holding hands overhead like a referee indicating a touchdown.

2. *Aeronaut.* The action of coming into contact with the ground during landing.

1935 P. W. F. MILLS *Elem. Pract. Flying* vii. 102 [The purpose] of causing the actual touchdown, when it takes place, to take place with the aeroplane in its natural position on the ground. **1948** *Sun* (Baltimore) 3 Nov. 11/3 You are 50 feet above glide path and one quarter of a mile from touchdown. **1961** H. H. KOLBE *Handbk. Astronaut. Engin.* XXVII. 7 The term *landing*, when used in a discussion of space flight, actually can be considered as four phases: i.e., the exit from orbit, the reentry, the letdown, and the touchdown. **1975** *Daily Tel.* 11 Aug. 11/4 One vehicle will make a soft touchdown on Mars while the large spacecraft which carried it on its journey will remain in orbit.

‖ **touché** (tuʃe), *int.* Occas. *fem.* **touchée.** [Fr., pa. pple. of *toucher* to hit.] **a.** *Fencing.* An exclamation used to acknowledge a hit.

1904 *Red Book* Feb. 382/1 '*Touché!*' Jarsac growled sharply. **1958** A. WEST *Princ. & Persuasions* 202 These cosy thrusts will never slip between the ribs into the lungs; the weapons are not rapiers but buttoned foils that will bend double against a jacket and at most produce a murmur of '*Touché*'.

b. A pleasant admission of a valid point or justified accusation made by another person.

1907 *Everybody's Mag.* XVI. 221/1 They did not cry *touché*, but the House cheered to the echo. **1912** E. C. BENTLEY *Trent's Last Case* xv. 322 '*Touché*', Trent said, with a dry smile. **1928** *Sat. Rev.* 17 Nov. 649/1 Touché—I apologize to Messrs. Brown and Phillips for my lack of technical discernment. **1952** H. INNES *Campbell's Kingdom* I. iv. 81 'I'd my own reasons, the same as you have.' ..'Touchée,' she said softly. **1981** A. PRICE *Soldier no More* 50 '*Touché*...' he nodded, accepting the rebuke.

touched, *ppl. a.* Add: **touched proof** (earlier example).

1831 J. CONSTABLE *Let.* 13 Apr. (1966) IV. 348, I send you the *twelve pounds*, and a touched proof of the Heath.

toucheous, var. *TOUCHOUS *a.*

toucher. Add: **1. d.** One who † robs or seeks to obtain gifts or loans of money for himself. *slang.*

1849 G. G. FOSTER *New York in Slices* 25 The other places in the cotillion are occupied by a notorious kracksman [*sic*] with his 'pal'—a celebrated 'toucher'. **1904** *Chicago Tribune* 30 Oct. (Worker's Mag.) 4/2 The salaried clerk who keeps his wife..at a fashionable hotel is, usually, a toucher of the kind that makes a good front. **1919** WODEHOUSE *My Man Jeeves* 91 Many's the time in London, I've hurried along Piccadilly and felt the hot breath of the toucher on the back of my neck. **1961** 'F. O'BRIEN' *Hard Life* xii. 101 The streets aren't crawling with touchers like Dublin.

4. b. *as near as a toucher* (earlier example). Also *within a toucher*, within an inch *of* doing something (only in Wodehouse).

1827 W. CLARKE *Every Night Bk.* 73 The cock which takes your fancy..is..to all appearance, right-thorough bred, or 'as near it as a toucher'. **1932** WODEHOUSE *Doctor Sally* viii. 78, I came within a toucher of saying, 'pause before it is too late!' **1954** —— *Jeeves & Feudal Spirit* xviii. 173 The hand of doom within a toucher of descending.

touching, *vbl. sb.* Add: **1. c.** (Examples with *up*: see TOUCH *v.* 34 in Dict. and Suppl.)

1902 [in Dict., sense 1 d]. **1936** *Burlington Mag.* May 208/1 The artist's later touching-up. **1957** *Practical Wireless* XXXIII. 558/1 'Look Back to Lyttleton' was a novel by Caryl Brahms, which had been laid aside for touching up and taken out as suitable material for a radio play. **1973** C. MULLARD *Black Britain* vii. 87 Both employees had for some months been practising a mild form of homosexuality—'touching up'. **1980** J. SCOTT *Gospel Lamb* iii. 51 Touchings-up were frequent—the girls seemed as eager as the boys.

4. *touching-distance*: hence *touching-distant* adj. (*poet.*).

1881 W. WHITMAN *Leaves of Grass* (new ed.) 352 Thy touching-distant beams.

touch-line. Add: Also **touchline. 3.** Also in some ball games other than rugby football, and *fig.*

1932 AUDEN *Orators* II. 46 The two-faced, the obscure and amazed, the touch-line admirers. **1964** *Sunday Times* 25 Oct. 22/5 A charming touchline companion called the [hockey] match 'grotty'. **1973** PARK & FAHEY *Team Handball* 50 The Boundary-Lines on the long sides shall be termed the Touch-lines. **1973** *Nature* 9 Nov. 108/2 From the touchlines the editor does, however, bias the issue by setting H. G. Haas's article on 'Active Ion Transport' immediately before that on the sinoatrial node.

touch-me-not, *sb.* **3. b.** (Earlier example.)

1817 M. EDGEWORTH *Harington* I. v. 112 Lady de Brantefield, the *touch-me-not* mistress of the mansion.

touchous (tʌ·tʃəs), *a. dial.* Also **toucheous.** [f. TOUCH *sb.* or *v.* + -OUS.] Easily offended, sensitive, touchy.

1867 P. KENNEDY *Banks of Boro* xxv. 190 By the time I got home, however, I was very cross and *touchous.* **1933** C. MILLER *Lamb in his Bosom* iv. 28 The ill-temper worked in her body like a slow fever... Lonzo called her toucheous. **1960** H. LEE *To kill Mockingbird* viii. 72 He said Atticus was still touchous about us and the Radleys and it wouldn't do to push him away. **1973** *N.Y. Times* 3 June L-19/1 [In the Caribbean] an overly sensitive person is 'touchous', not touchy.

toudang, var. *TOERING.

tough, *a.* Add: **1. b.** Phr. *tough as* (*old*) *boots* or *leather.* Freq. *fig.*, implying sense 4.

1843 As tough as leather [in Dict.] **1870** As tough as old boots [see BOOT *sb.*[3] 1 b]. **1946** J. B. PRIESTLEY *Bright Day* iv. 111 Joe Ackworth's more the type. He's as tough as old leather. **1967** *Listener* 7 Dec. 765/1 This is no sweet old dolly... She is tough as old boots, working for a living. **1981** M. HATFIELD *Spy Fever* I. iii. 31 Colonel Theakston was..as the saying goes, as tough as old boots.

5. b. Resolute in dealing with opposition; vigorously uncompromising; severe; esp. in phr. *to get tough* (cf. *GET *v.* 73 d). *colloq.* (orig. *U.S.*).

1906 U. SINCLAIR *Jungle* i. 11 He affects a 'tough' aspect, wearing his hat on one side and keeping a cigarette in his mouth all the evening. **1930** E. H. LAVINE *Third Degree* ii. 17 A conscientious, or 'tough', [police] sergeant was assigned to a west-side precinct. **1935** WODEHOUSE *Blandings Castle* vi. 151 In all villages, of course, there must..be an occasional tough egg. **1938** E. AMBLER *Cause for Alarm* vii. 116 Vagas got tough. They had a showdown. **1964** in Hamblett & Deverson *Generation X* 10 The funniest thing was seeing the cops getting tough. If they want a fight we'll give it to them. **1972** J. SYMONS *Bloody Murder* xii. 159 The behaviour of the private detective may be tough, but is based on ethical standards. **1978** J. IRVING *World according to Garp* i. 14 They initiated a get-tough policy with Jenny Fields. It was a staff decision—'for her own good', of course. **1984** *N.Y. Times* 12 Feb. (Late City Final) I. I. 35/1 My policy is to be tough but fair with the gaming industry.... Federal law-enforcement officials have greater access to data on Nevada.

c. Of laws or rules: strict, inflexible. Of an institution: marked by strict enforcement of discipline.

1961 in WEBSTER s.v., When the law gets too tough the courts don't convict. **1971** J. OSBORNE *West of Suez* I. 42 Father decided I needed 'toughening up' at a really tough school. **1977** *National Observer* (U.S.) 22 Jan. 1/1 Reformers want a tougher code of ethics for Presidential appointees. *Ibid.*, The environmentalists want a tougher line on automobiles that pollute.

6. c. Of circumstances, etc.: imposing hardship, distress, or injustice. *colloq.* (orig. and chiefly *U.S.*).

1890 *Stock Grower & Farmer* 8 Mar. 4/2 The recent blizzard..was pretty tough on range cattle. **1901** S. E. WHITE *Claim Jumpers* 256 I've been a little tough on you occasionally. **1929** WODEHOUSE *Mr. Mulliner Speaking* i. 34, 'I suppose it's because I'm rather an out-size and modelled on the lines of Cleopatra.' 'Tough!' 'You bet it's tough. A girl can't help her appearance.' **1933** P. GODFREY *Back-Stage* xvii. 216 The 'tough breaks' in their gipsy life soon weed out the weaklings. **1942** E. PAUL *Narrow St.* xxix. 265 You know you're likely to be bumped off? ..Things are tough down there, and they won't get any better. **1959** H. P. TRITTON *Time means Tucker* (1965) i. 11 Work was scarce and wages low, and conditions all round were tough. **1962** J. H. CUTLER *Honey Fitz* xx. 291 Joe [Kennedy] made his children stay on their toes... 'He would bear down on them and tell them, "When the going gets tough, the tough get going."' **1982** *Church Times* 30 Apr. 11/1 The life of a nun is extremely tough and involves a lot of physical hard work.

d. *tough luck* (colloq., orig. U.S.), hard luck, misfortune; esp. as an expression of (sometimes ironic) commiseration; also (chiefly *U.S. slang*) *tough shit, stuff*, or *tiddy* (*titty*).

1912 Tough luck [see *old top* s.v. *OLD *a.* 8 a]. **1932** *Kansas City* (Missouri) *Times* 14 Jan. 18 It may be Mr. Hoover's tough luck to be both renominated and reelected. **1934** J. T. FARRELL *Calico Shoes* 143 You have to take your chances, and if you can't swim, you sink. It's just your tough tiddy. **1944** in A. M. Taylor *Lang. World War II* 198 Beachhead chaplains are carrying a special 'tough stuff' ticket these days which they issue to guys with complaints about which nothing can be done. **1946** *Amer. Speech* XXI. 249 [Army vocabulary.] *Tough shit*, something which is unfortunate, but about which nothing can be done. **1958** S. A. GRAU *Hard Blue Sky* ii. 89 'And the whole building near to going down with the next strong wind.' 'Tough titty, man.' **1971** 'A. BURGESS' *MF* ii. 32 [I got] robbed and rumpled.—Tough titty she said with little sympathy. **1974** *Black World* Jan. 10/2 Is Mr. Gayle exasperated by the fact that I do not give clear-cut answers to these questions? Tough luck: I do not have them. **1976** *New Yorker* 1 Mar. 74/2 I'm awfully sorry to hear about your tough luck. **1978** J. CARROLL *Mortal Friends* II. v. 200 Tough shit, Lady! Morning wears to evening and hearts break.

9. Restrict † to sense in Dict. **b.** In an uncompromising, aggressive, or unyielding manner.

1943 R. CHANDLER *Lady in Lake* iv. 25 You fellows [*sc.* cops] ever flash a buzzer—or is acting tough all the identification you need? **1968** *Globe & Mail* (Toronto) 3 Feb. 7/5 Saskatchewan's Premier Ross Thatcher, while he talks tough in private, is apparently willing to make at least a gesture.

9*. As an epithet of commendation: very good, 'great'. *U.S. slang* (orig. *Blacks'*).

1937 [see *CATCH *v.* 35 b]. **1960** R. G. REISNER *Jazz Titans* 167 *Tough*, great. **1965** MRS L. B. JOHNSON *White House Diary* 3 June (1970) 282 'Pat Nugent..he's just tons, Mother—he's a tough guy!' ('Tough' means great, wonderful, nice, attractive, it seems.) **1972** J. HUDSON in T. Kochman *Rappin' & Stylin' Out* 422 Now my singing ain't none too tough, but I can sell some dope.

10. a. tough baby, boy *slang* (orig. *U.S.*), a person given to hard-headed, violent, or lawless behaviour; **tough guy** *colloq.* (orig. *U.S.*), a person not easily injured or thwarted; freq. *attrib.*; **tough movement** *Transformational Grammar*, a transformation applied to a sentence moving words of a certain class (of which *tough* is one), from one part of the sentence to another (e.g. *to convince John is hard*: *John is hard to convince*); **tough nut** *colloq.* (orig. *U.S.*), a person difficult or dangerous to deal with; **tough pitch,** substitute for def.: commercially pure copper in which the amount of cuprous oxide was reduced by poling to the value at which it would produce minimum brittleness; usu. *attrib.* or as *adj.*; (examples).

1932 E. WALLACE *When Gangs came to London* xxiii. 234, I've had real tough babies on their knees to me in a police station, begging me to be put in a cell. **1946** WODEHOUSE *Joy in Morning* ii. 12 Scanning the roster of the females I've nearly got married to in my time, we find the names of some tough babies. **1958** F. NEWTON in P. Gammond *Decca Bk. Jazz* v. 68 It is no use being censorious about the atmosphere of..tough boys and sleezy vaudevilles in which the great blues singers were nurtured. **1974** T. P. WHITNEY tr. *Solzhenitsyn's Gulag Archipelago* I. I. vii. 294 The interrogators and their tough-boy helpers dashed in from the interrogation prison. **1932** Tough guy [see *CLEAN *v.* 6 b]. **1938** L. MACNEICE *Mod. Poetry* viii. 149 E. E. Cummings, the 'tough-guy' American poet. **1946** R. CHANDLER *Let.* 30 May (1981) 75 Bogart, of course, is..much better than any other tough-guy actor... Ladd is..a small boy's idea of a tough guy. **1946** H. CROOME *Faithless Mirror* vii. 75 Tough guys with a heart of gold. **1981** J. DUNNING *Deadline* (1982) xix. 187 At the bottom of that tough-guy facade, you're just like all the rest... Scared to death. **1971** P. M. POSTAL *Cross-Over Phenomena* iii. 27 There is a class of adjectives in English, *hard, tough, easy, difficult, impossible, simple*, which have played a prominent role in discussions of the need for a transformational grammar of English... The contrast between sentences like ..*a* Throneberry is easy to please. *b* Throneberry is eager to please..is by now well known... There is a special rule defined for this class ..which involves the movement of an NP out of the predicate of the complement sentence. Let us refer to this rule as tough-movement. **1977** *Canad. Jrnl. Linguistics* 1976 XXI. 157 Consider (24), resulting from Passive, and (25), resulting from Tough-Movement, as answers to the question 'Why was John arrested?' (24) That he robbed a store was reported in the newspaper. (25) That he robbed a store is hard for us to believe. **1862** in E. W. Pearson *Lett. from Port Royal* (1906) 81 There are a great many men of twenty-five to forty, 'tough-nuts' many of them. **1892** 'MARK TWAIN' *Amer. Claim.* xxv. 263 His father was rather a tough nut. **1922** E. O'NEILL *Hairy Ape* viii. 83 Say, yuh're some hard-lookin' guy, ain't yuh? I seen lots of tough nuts dat de gang called gorillas, but yuh're de foist real one I ever seen. **1950** *Times* 12 May 7/7 For the 'tough nut' the youth club as at present constituted offered no fold. **1977** C. MCCULLOUGH *Thorn Birds* x. 236 Meggie was going to be a tough nut to crack and he couldn't afford to frighten or disgust her... He'd woo her the way she obviously wanted. **1903** *Engineering* 4 Dec. 753/3 When the right amount of oxygen is present, the copper is said to be 'tough-pitch'. **1949** P. C. CARMAN *Chem. Constitution & Properties Engin. Materials* vii. 220 The product is a 'tough-pitch' copper of over 99·9% purity. **1964** H. HODGES *Artifacts* iv. 70 Correctly poled copper, tough pitch copper, still contains a little cuprous oxide.

b. *tough-hided* (in quots. *fig.*), *-necked*, *-skinned* (*fig.* example).

1925 D. H. LAWRENCE *St. Mawr* 158 She felt a peculiar tough-necked arrogance in him. **1930** R. LEHMANN *Note in Music* VI. 249 It would take a good deal..to harm a tough-hided old hippopotamus like Uncle Tom. **1933** C. S. LEWIS *Pilgrim's Regress* VII. v. 146, I always think it is possible for a place to be *too* bracing. They call it the land of the tough-minded—tough-skinned would be a better name. **1964** *Listener* 30 Apr. 731/3 A tough-hided, soft-centred, north-country, working-class dramatist.

B. *sb.* **1.** For *U.S.* read 'orig. *U.S.*' and substitute for def.: A person given to rough or violent behaviour. (Later examples.)

1929 'G. DAVIOT' *Man in Queue* iii. 25 The missing man ..was, in the opinion of the Durham inspector, a tough. **1946** R. LEHMANN *Gipsy's Baby* 145 Can't think how your parents put up with it—all that gang of young toughs in and out all day. **1972** E. GRIERSON *Confessions of Country Magistrate* ix. 86 Certainly the treatment of the teenage tough..is a problem to which no one has ever hazarded an optimistic answer. **1982** I. HAMILTON *Robert Lowell* (1983) ii. 16 He graduated to the status of school tough via a series of spectacular playground victories.

2. A person of uncompromising or aggressive views.

1928 C. CONNOLLY *Let.* July in *Romantic Friendship* (1975) 321, I am becoming a tough, an anglophobe, and reverting to intolerance and intellectual pride. **1931** H. NICOLSON *Diary* 21 Aug. (1966) 89 The latter asked whether Tom would join him and the Tory toughs in opposition. **1980** *Times* 23 June 31/1 The so-called 'toughs' who support Mrs Thatcher's policy—like Sir Keith Joseph.. against the 'wets' led by Jim Prior.

tough (tʌf), *v. slang* (orig. and chiefly *U.S.*). [f. TOUGH *a.*] **a.** *intr. to tough it* (*out*): to withstand (to the end) difficult conditions or adverse circumstances without flinching. Cf. *to rough* (*it*) *out* s.v. ROUGH *v.*[1] 4 b.

1830 *Mass. Spy* 27 Jan. (Th.), Judy with whom he had toughed it three years. **1852** *Knickerbocker* XXXIX. 26 You don't need no medicine; you'll tough it out, I dare say. **1873** C. THAXTER *Isles of Shoals* 64 (Th.), Our brave little schooner 'toughed it out' on the distant ledge. **1939** L. M. MONTGOMERY *Anne of Ingleside* xviii. 121 She darkly opined that it would be a miracle if he toughed it out till spring. **1956** T. RADDALL *Wings of Night* (1957) xxxii. 241 She was a great ol' lady... Just kep' her chin up and..toughed it out to the end. **1982** H. LIEBERMAN *Night Call* xvi. 94 We'll tough it out, but sacrifices will have to be made.

b. *trans.* With obj. in place of *it*: to withstand to the end.

1974 *Newsweek* 20 May 23/2 Everybody..was pressed into service denying that Mr. Nixon planned to quit; his daughter Julie vowed that he would tough out the impeachment process to its end in the Senate. **1979** *Courier-Mail* (Brisbane) 27 Sept. 1/6 Mr. Sinclair signalled he would try to tough out the crisis. **1981** *Observer* 26 Apr. 15/4 Fraser, it is assumed, will tough out this latest crisis.

toughie (tʌ·fi), *sb.* (and *a.*) *colloq.* (orig. *U.S.*). Also **toughy.** [f. TOUGH *a.* + -IE, -Y[6].] **1.** A tough person. **a.** = *tough guy* s.v. *TOUGH *a.* 10 a. **b.** A person of aggressive or uncompromising views.

1929 *Princeton Alumni Weekly* 24 May 981/2 The *toughie* is the man of the hour when the policeman's whistle blows. It is always nice to be close to him when the riot calls are turned in. **1938** *New Republic* 21 Sept. 188/1 Getting the toughies off the street. **1940** R. CHANDLER *Farewell, my Lovely* xxvii. 168 A toughie..came in and showed me a blackjack. **1959** *She* May 65/2 Luxury-lovers had better stick to planes or boats. A trip in the 'Bombay Bus' is definitely one for the toughies. **1960** *Sunday Express* 24 July 16/5 Mr. Butlin is a toughie too... A man who has learned to cater for the mood of the people and take full advantage of their longer purses... I think he will continue to prosper despite credit squeezes, Chancellors, the rain, and other vexations. **1971** J. MANDELKAU *Buttons* x. 121 A group of bikers riding out for kicks every Friday and Saturday night, getting drunk and swinging back and forth over the white line behind the toughie they would call 'our leader'. **1980** I. MURDOCH *Nuns & Soldiers* I. 84 Daisy had women friends.. 'Women's Libbers', and left wing toughies. **1984** *Observer* 8 Apr. 12/4 Mondale may think that he makes a good political toughie.

2. A difficult problem, enterprise, or contest.

1945 *Good Housekeeping* June 230/2 How about the $80 question?..Think now. This is a toughie. **1947** *Sun* (Baltimore) 18 Jan. 1/7 West has another toughie for Congress... Would it be O.K. if the District used tax money to buy uniforms for the policemen's band? **1972** J. L. DILLARD *Black English* vii. 281 Children who speak Standard English may react to words like *island*,..with 'That's a toughie!' **1972** D. LEES *Zodiac* 6 It sounded, as Harry put it, 'a doddle', but even if it had looked a toughy the Riviera would have sold it.

3. *attrib.* or as *adj.*

1962 *John o' London's* 9 Feb. 138/4 The nice, 'toughie' Irishman. **1974** *Times Lit. Suppl.* 14 June 644/3 Admirers of Stephen Becher's urbane *When the War is Over* will be surprised at the toughie metaphysics of the opening of his new novel. **1977** *Film & Television Technician* Jan. 9/3 Bob was one of those 'toughie' production managers, greatly concerned not only with the budget, but with the film and the people working on it.

tough-mi·nded, *a.* In the philosophy of William James: marked by a purely empirical, sceptical, non-metaphysical approach to questions; opp. *TENDER-MINDED *a.* Hence more widely: free from excessive sensitivity, realistic, unsqueamish, etc. Also *absol.*

1907 W. JAMES *Pragmatism* i. 12 You will..recognize the two types of mental make-up that I mean if I head the columns..*The Tender-Minded.* Rationalistic (going by 'principles'), Intellectualistic, Idealistic, Optimistic, Religious, Free-willist, Monistic, Dogmatical. *The Tough-Minded.* Empiricist (going by 'facts'), Sensationalistic, Materialistic, Pessimistic, Irreligious, Fatalistic, Pluralistic, Sceptical. **1927** J. S. HUXLEY *Relig. without Revelation* iv. 116 Youth wakes to the fact of social inequality, and, if not one of the tough-minded, to remorseful distress of its own privileged position. **1945** AUDEN *Coll. Poetry* 123 Tough-minded men get mushy in their sleep. **1952** C. P. BLACKER *Eugenics* 240 The scientist, or at least the scientist's camp-followers, may become tough-minded and contemptuous: the word 'mysticism' expresses what they most passionately abjure. **1960** *Guardian* 17 Dec. 6/2 The tough-minded and unromantic pragmatism of the new President. **1974** *Sci. Amer.* Jan. 113/2 Tough-minded skeptics. **1980** F. K. PROCHASKA *Women & Philanthropy in Nineteenth Cent. Eng.* vi. 191 Reclaiming prostitutes was a daunting prospect for charitable women however tough-minded.

Hence **tough-mi·ndedness.**

1907 W. JAMES *Pragmatism* vii. 267 One misunderstanding of pragmatism is to identify it with positivistic tough-mindedness. **1936** *Mind* XLV. 218 The sort of tough-mindedness which, on principle, excludes ethical, aesthetic and religious considerations from metaphysical thinking, is wilfully blind. **1961** *Guardian* 3 Apr. 12/7 The first fruits of tough-mindedness [in U.S. policy in Laos] are..encouraging. **1975** *Nature* 10 Apr. 470/2 In addition to being careful in this way, we have also to be sensitive and observant, and not to react with a preconceived pattern of tough-mindedness.

‖ **toughra** (tu·grā). Also **toghra, tughra, tuğra.** [a. Turkish *tura, tuğra.*] An ornamental monogram incorporating the name and title of the Sultan.

1888 S. LANE-POOLE *Turkey* 36 [1365] It is said that Murād signed the treaty, for lack of a pen, with his open hand, over which he had smeared some ink, in the manner of Eastern seals. This veritable sign-manual is believed to be the origin of the *tughra* or Sultan's cipher, which has ever since appeared on the coinage and the official documents of the Turks. **1903** *Amer. Jrnl. Numismatics* Jan. 73 The principal device on the gold coins of the present Sultan..is the imperial *toghra.* **1954** *Stamp Lover* XLVI. 136/1 Turkey was introduced by an official document of 1840 bearing the manuscript Toughra or signature of the

Sultan. **1962** R. A. G. CARSON *Coins* 488 The tughra, the monogram of the sultan's names and titles which is a feature of later Turkish coinage.., appeared for the first time under Suleyman I. **1974** *Encycl. Brit. Macropædia* III. 664/2 A distinctive *tuğra* was created for each sultan and affixed to imperial decrees by a skilled calligrapher.

‖ **toujours** (tuʒūr), *adv.* (and *sb.*) [Fr., = always.]

1. *toujours gai* (gɛ), 'always cheerful'; cheerful under all circumstances; also as *sb.*, an unfailingly cheerful disposition. Occas. partially anglicized as *toujours gay.*

1711 ADDISON in D. Piper *Eng. Face* (1957) vii. 163 A certain smirking Air..bestowed indifferently on every Age and Degree... The *Toujours Gai* appeared even in Judges, Bishops and Privy-Counsellors. **1899** KIPLING *From Sea to Sea* I. viii. 263 They [*sc.* prostitutes] spoke of themselves as 'gay'... A night's reflection has convinced me that there is no hell for these women in another world... It was my duty to watch through the night a patient—gay, *toujours* gay, remember—quivering on the verge of the 'jumps'. **1927** D. MARQUIS *archy & mehitabel* xiv. 56 Well archy the world is full of ups and downs but toujours gai is my motto. **1972** M. KENYON *Shooting of Dan McGrew* xxi. 174 He was 'toujours gai' (I wonder is he on drugs?).

2. *toujours perdrix* (pɛrdrī), lit. 'always partridge', an allusive phr. used to imply that one can have too much of a good thing. [For an explanation see A. M. Hyamson *Dict. Eng. Phrs.* (1922) 346/1.]

1818 *Blackw. Mag.* Feb. 569/2 A partridge is a good thing; and yet even '*Toujours Perdrix*' is not to be borne. **1877** L. W. M. LOCKHART *Mine is Thine* (1879) xvii. 163 He wanted a rest, a change from this *toujours perdrix* of ladies' society, polite small-talk, boredom. **1927** D. H. LAWRENCE *Let.* 12 Dec. (1962) II. 1026 I'm sick of Jesus. .. We might have somebody else born for a change. *Toujours perdrix!*

3. Used simply: always.

1902 G. MEREDITH *Let.* 19 Jan. (1970) III. 147 If it is *toujours* Goethe, that is because I share the *culte.*

touladi, var. *TULADI.

Toulousain (tūlūzæ̃n), *sb.* and *a.* Fem. **-aine** (-ɛn). [Fr., f. *Toulouse* + *-ain* -AN.] **A.** *sb.* A native or inhabitant of Toulouse, in SW France. **B.** *adj.* Of, pertaining to, or characteristic of Toulouse.

1883 H. JAMES in *Atlantic Monthly* Oct. 461/1 A big, brown, expansive woman... This terrible Toulousaine of today. *Ibid.* 462/1 Saint-Sernin..dedicated to Saint Saturninus—the Toulousains have abbreviated—is, I think, alone worth the journey to Toulouse. **1970** *Sat. Rev.* (U.S.) 3 Oct. 45/1 A young Toulousaine from the local tourist office. **1972** R. COBB *Reactions to Fr. Revolution* iii. 119 To do the opposite to what the Bordelais did was an ancient rule of Toulousain conduct. **1980** 'M. HARRIS' *Treasure of Ste. Foy* v. 54 Serge Gaspar is a Toulousain, dark, with a Spanish look to him. *Ibid.* 55 She is the same Toulousain type as Gaspar, with dark hair..and a clear olive complexion.

toungya, var. *TAUNGYA.

toup (tūp), slang abbrev. of TOUPEE in Dict. and Suppl.

1959 P. BULL *I know the Face* viii. 138 'Say, Padre, is that a toup?' he naively enquires. **1973** R. HAYES *Hungarian Game* xxxiv. 205 He picked a blond wig... He slipped the toup over his gray hair and adjusted it to cover the high forehead.

toupee. Add: No longer *rare* in the sense 'a patch of false hair or small wig to cover a bald place'. (Later examples.)

1973 M. AMIS *Rachel Papers* 81 My hair hung on my head as if it were a cut-price toupée. **1980** V. S. PRITCHETT *Tale Bearers* 20 He is having his toupee fixed and his hair dyed.

tour, *sb.* Add: **I. 1.** Freq. in *tour of duty.* Also, with pronunc. (tauᵊɹ), in *Oil Industry.* (Further examples.)

1800 WELLINGTON *Suppl. Desp.* (1858) I. 464 This tour of duty to commence at morning parade on halting days. **1903** *Dialect Notes* II. 345 The morning *tour* lasts from midnight until noon. **1939** D. HAGER *Fund. Petroleum Industry* ix. 212 These men work in shifts or 'tours' (pronounced towers) of 6 or 8 hrs. **1946** *R.A.F. Jrnl.* May 153 The existing Editor having performed his tour of duty and taken up other duties in the Service. **1975** L. CROOK *Oil Terms* 60 The Driller is responsible for his crew and the running of the rig during his eight or twelve hour 'tour'. **1981** 'J. ROSS' *Dark Blue & Dangerous* xxvii. 158 He's too soft, and..I don't know how he did the rest of his tour of duty without showing it on his face.

2. b. *Dancing.* Also with pronunc. (‖tūr). In a cotillion, a circular movement by the dancers. In *Ballet*, a turn by a solo dancer; *tour en l'air*, such a turn while leaping in the air.

1841 MRS. GASKELL *Lett.* (1966) 822 The cotillion was so pretty—such amusing & graceful *tours.* **1930** CRASKE & BEAUMONT *Theory & Pract. Allegro Class. Ballet* 94 (*heading*) Series of tours en dedans en diagonale. **1948** A. H. FRANKS *Approach to Ballet* iii. 45 Used sparingly, *tours* can become most effective highlights in a male solo. **1958** [see *PLIÉ]. **1960** M. WOOD *Advanced Historical Dances* 93 The refrain had been replaced by a fixed series of movements called Tours, forming a framework into which the figures were fitted. The Tours in order of performance were these: (1) Grand Rond. All take hands in a ring and go round both ways. [Etc.] **1963** *Times* 29 May 13/4 Mr. Flindt..can produce effortless, waist-high *cabrioles*, yet is often constricted, even rough, in *tours.* **1977** *Times* 5 May 11/8 The skill with which she sustained the series of *tours en l'air* in her solo.

3. a. Also, † an account of such a journey.

1812 [in Dict.]. **1817** JANE AUSTEN *Sanditon* (1925) viii. 110 He read all the Essays, Letters, Tours & Criticisms of the day. **1821** BYRON *Don Juan* v. lii. 161 Nature..Resigns herself with exemplary patience To guide-books, rhymes, tours, sketches, illustrations.

III. 12. *tour bus, director, guide, operator, party.*

1952 *Galaxy* June 56/2 I'm the tour director. Can I help you? **1965** J. A. MICHENER *Source* (1966) 22 That afternoon the first excursionists stopped at the tell, asking to see the Candlestick of Death, and the next morning a tour bus arrived. **1971** M. MCCARTHY *Birds of America* 107 Their tour director, who had met them at Le Havre. **1973** P. THEROUX *Saint Jack* xviii. 219 The tour-guide had started his spiel. **1976** J. SNOW *Cricket Rebel* 47, I could only wait anxiously for the announcement of the names of the tour party to visit the West Indies that winter of 1967–68. **1981** *Sunday Express Mag.* 11 Oct. 9/1 (Advt.), We've taken more British holiday-makers here than any other tour operator. **1981** M. KENYON *Zigzag* xi. 67 He was agonizingly shy... Guiding tour parties petrified him.

tour, *v.* Add: **4.** Also with a performer as obj.

1920 *Glasgow Herald* 10 June 7 Mr. Quinlan..recently toured Madame Tetrazzini and Signor Caruso. **1922** JOYCE *Ulysses* 92 The idea is to tour the chief towns... Mary Anderson is up there now... Louis Werner is touring her.

touring *ppl. a.* (earlier and *Theatr.* examples).

1832 F. TROLLOPE *Dom. Manners Amer.* II. xxxiii. 236 To this frail shelter..nearly all the touring gentlemen.. find their way. **1867** *Harper's Mag.* Dec. 96/1 As railways have multiplied, the formation of what are called Touring-parties, for the purpose of giving concerts and operas in the provinces, has become the business of many. **1895** ROBERTS & MORTON *Adventures Arthur Roberts* xiii. 159 Whether from preference or economy,..this touring company generally slept at night on the beach. **1969** G. GREENE *Travels with my Aunt* I. xi. 105 The touring company..came after my Paris days. It was in Paris that I was spotted by Mr Visconti... 'He was a great amateur of..the stage.' **1983** *Times* 7 Sept. 3/4 Scunthorpe is a regular stop-off point for touring companies.

touraco. Delete ‖ and add: Now usu. spelt **turaco.** (Later examples.)

1932 *Discovery* Jan. 25/1 Colies, turacos, various doves and barbets..throng to the Amani plantations. **1965** G. B. SCHALLER *Year of Gorilla* ii. 69 A black-billed turaco flew up, flashing its brilliant crimson wings. Then this crow-sized bird hopped along a branch, chattering like a squirrel. **1977** *Daily Colonist* (Victoria, B.C.) 9 Oct. 23/4 You will see..exotic birds—bee-eaters, love-birds, sunbirds and turacos.

‖ **Tourangeau** (turaṅʒō), *sb.* and *a.* Pl. **-x.** [Fr.] (A native or inhabitant) of Touraine, a former province of France corresponding more or less to the modern department of Indre-et-Loire, or of Tours, its chief town.

1883 H. JAMES in *Atlantic Monthly* July 25/1 The real Tourangeau will not make an effort..to go in search of a pleasure. **1969** B. ARTHAUD tr. *Martin-Demézil's Loire Valley* 9 The Tourangeaux gave it the wonderful name of '*le Jardin de France*'. **1973** *Listener* 22 Mar. 372/3 Motifs from this Tourangeau château or that, from Florentine palazzi or Classical temple architecture.

‖ **Tourangeois** (turaṅʒwa), *a.* and *sb.* [Fr.] = prec.

1857 C. KINGSLEY *Two Years Ago* III. vi. 155 Balzac's old Tourangeois judge. **1958** C. COCKBURN *Crossing the Line* ii. 37, I took a terrible chance by recommending to him—a Tourangeois—a certain Catalan wine I had discovered.

tourbillion, tourbillon. 2. Delete '*Obs.* exc. as French' and add: Also, an eddy, a whirlpool; also *fig.* (Further examples.)

1891 'MARK TWAIN' *Lett.* (1917) II. xxxi. 557 We were allowed to go through the wrong arch, which brought us into a tourbillion below which tried to make this old scow stand on its head. **1931** R. GRAVES *To whom Else?* 19 Such portents are not to be wondered at Being tourbillions in Time made By the strong pulling of her bladed mind Through that ever-reluctant element. **1972** J. WAIN in Cox & Dyson *20th-Cent. Mind* I. xi. 374 In the history of any art there are unexpected eddies and tourbillions.

3. (Earlier example.)

1749 *Descr. Machine for Fireworks* 15 Tourbillons..88.

tour de force. (Earlier examples.)

1802 LD. ELGIN *Let.* 18 Feb. in *Paget Papers* (1896) II. 41 To exult over what is styled a *tour de force* of the British Influence here. **1802** M. BERRY *Jrnl.* 16 Mar. (1865) II. 137 Women now dance in the style of men, that is to say, with all the difficult steps and *tours de force* possible.

‖ **Tour de France** (tūr də frɑ̃s). [Fr., lit. 'tour of France'.] The name of an annual cycling stage race on the public roads of France (with some crossing into adjoining countries), now typically over about 4,000 kilometres including mountainous terrain.

1922 *Times* 24 July 7/5 The 'Tour de France' cyclist competition..ended to-day. **1926** E. HEMINGWAY *Sun also Rises* III. xix. 247 Bicycle road-racing was the only sport in the world, he said. Had I ever followed the Tour de France? **1967** *Guardian* 14 July 1/2 Tommy Simpson, the British cyclist, died early this evening after collapsing during a mountain stage of the Tour de France. **1978** *Listener* 6 July 15/1 These days find Paris in a sort of limbo... There is, it is true, the Tour de France to distract us in the weeks ahead.

‖ **tour d'horizon** (tūr dorizɔ̃). [Fr., lit. 'tour of the horizon'.] An extensive tour. Usu. *fig.*, a broad, general survey.

1952 *Ann. Reg. 1951* 181 General Eisenhower, returning from his European *tour d'horizon*, was able to give Congress..a report. **1964** *Economist* 5 Sept. 929/1 A fascinating *tour d'horizon* of the main theories. **1979** T. SKYRME *Changing Image Magistracy* xii. 162 Lord Widgery, in the course of a *tour d'horizon* in 1973 said: [etc.]. **1984** *Oxf. Univ. Gaz.* 22 Mar. 584/1, I hope it is not invidious, in an avowedly personal *tour d'horizon*, to single out here the museums for mention.

tourdion: see TURDION in Dict. and Suppl.

tourer (tūᵊ·rəɹ). [f. TOUR *v.* + -ER[1].] **1. a.** A touring-car.

1927 *Sunday Pictorial* 28 Aug. 8/4 Two and five-seater tourers will be £495. **1948** G. H. JOHNSTON *Death takes Small Bites* v. 102 Through a great rent in the tourer's flapping hood she could see Cavendish hunched over the wheel. **1978** *Hot Car* June 91/3 The *Munster Koach*, a wild '27 Model T tourer with a gung-ho Cobra engine.

b. A kind of caravan for touring.

1970 *Guardian* 25 Apr. 15/3 There are three basic types of caravans... There are tourers tacked on to the back of family cars, [etc.].

2. One who tours or goes on tour.

1931 *Times Lit. Suppl.* 1 Oct. 746/3 Mürren..tends to breed racers, the other [*sc.* Malója] 'tourers'. **1981** *Beautiful British Columbia* Summer 5 The first thing that bicycle tourers learn about British Columbia is that it is all up and down.

touring, *vbl. sb.* Add: **b.** *touring bag, centre*; **touring-car** (earlier example).

1903 *Encycl. Amer.* II. s.v. *Automobile*, Gasoline touring cars. **1930** *Cycling* 4 July p. iii, Touring bag. **1978** *Exchange & Mart* (South ed.) 20 Apr. 133/3 (Advt.), Motor cycle tank touring bag. **1981** *Nordic Skiing* Jan. 4/2 A group of Appalachian Mountain Club members went for a hike on trails established by a touring center for cross country skiing.

tourism. Add: Also ‖ **tourisme** (turi·z'm). (Further examples.) Also, the business of attracting tourists and providing for their accommodation and entertainment; the business of operating tours.

1910 *Blackw. Mag.* Feb. 207/2 He thus inaugurated veritable aerial tourism. **1930** *Time & Tide* 2 May 555 The office of the commissioner of *tourisme* in France..was organized to clip as many petty annoyances as possible from the routine activities of visitors. **1954** H. F. M. PRESCOTT *Jerusalem Journey* v. 127 A mosque, to the Christian pilgrim, was forbidden ground, unless..he were ready to risk martyrdom in the cause of 'tourisme'. **1955** *Times* 24 May 9/3 In addition to the importance of the Kariba project for power and irrigation, emphasis is being laid on its potentialities with regard to fishing and tourism. **1976** J. ARCHER *Not a Penny More, Not a Penny Less* xii. 128 Tourism is the chief source of income for the Principality, and the Monégasques take the welfare of their visitors very seriously. **1981** I. MCEWAN *Comfort of Strangers* i. 12 They dutifully fulfilled the many tasks of tourism the ancient city imposed, visiting its major and minor churches, its museums and palaces, all treasure-packed.

tourist. Add: **a.** (Earlier example.) Also *spec.* a member of a touring sports team (usu. *pl.*).

1780 *Ode to Genius of Lakes in North of England* 3 (Advt.), He throws the piece only into the way of *actual tourists.* **1975** *Cricketer* May 11/2 On the fourth day Julien joined the feast, hitting a scintillating 101 and helping the tourists to an 87-run first-innings lead.

a*. *ellipt.* = *tourist class* (see sense b below).

1936 [see *cabin class* s.v. *CABIN *sb.* 8]. **1939** T. S. ELIOT *Family Reunion* 42, I was down in the Tourist..and you could see the corner of the upper deck. **1939** G. GREENE *Lawless Roads* 297 A middle-aged American woman, who should have been travelling tourist. **1976** V. CANNING *Doomsday Carrier* vii. 117 A few men could fly tourist to any country and plague would fly in with them. **1981** J. RATHBONE *Base Case* i. 8 The service in tourist had been appalling.

b. (*i*) *tourist agent, attraction, board, bureau, bus, camp, circuit, hotel, industry, office, resort, route, season, shop, tax, ticket* (example), *trade, traffic, visa*; *tourist-crowded* adj.; **tourist cabin,** (*a*) the tourist-class accommodation on a ship; (*b*) a cabin for tourists; **tourist-car** (earlier example); **tourist card,** an identity card substituting for a passport or visa for a short visit to certain countries; **tourist centre,** a place much frequented by tourists; **tourist class,** a low-charge class of passenger accommodation in a ship, aircraft, etc.; also *attrib.*; also adverbially, in the class of accommodation so designated; **tourist court** *U.S.*, a group

of self-contained living units with service buildings, for tourists; **tourist flight,** a flight in the tourist class of a passenger aircraft, or in an aircraft which has only tourist-class seating; **tourist guide,** (*a*) = *tour guide* s.v. *TOUR *sb.* 12; (b) a guide-book which introduces tourists to a region or locality; **tourist park** *U.S.*, a park or camp-site with facilities for overnight campers, etc.; **tourist track,** (*a*) a route from place to place frequented by tourists; (*b*) *N.Z.*, a track through the bush for walkers; **tourist trap,** (*a*) a type of object sold to tourists at an excessively high price; (*b*) a place where tourists are exploited; **Tourist Trophy,** the name of a trophy awarded to the winner of motor-cycle races held annually on the Isle of Man since 1907; freq. *attrib.*; also *ellipt.* one of these races; usu. abbrev. as TT (see *T II. 6).

1884 *Queen* 16 Feb. (Advt.), Mr H. Laurence..Oculist Optician... Testimonials from..Thomas Cook, Esq., the well-known tourist agent, [etc.]. **1918** E. POUND *Pavannes & Divisions* 41, I knew a tourist agent, one whose art is To run such tours. **1959** A. H. MCLINTOCK *Descr. Atlas N.Z.* p. xix, The..geysers and boiling pools have long been a tourist attraction. **1978** *Times* 5 Aug. 5/6 Belsen had become almost a tourist attraction for visitors to Germany. [**1948** *Wales* (Tourist & Holidays Board Wales) 3 The Tourist and Holidays Board for Wales and Monmouthshire has pleasure in presenting its second Annual National Holiday Guide.] **1957** JACK & BLAIR *Chambers's Guide to Scotland* 13 An excellent publication (2s. 6d.) of the Scottish Tourist Board. **1972** 'R. CRAWFORD' *Whip Hand* I. viii. 44 Ballycroom might be as pretty in summer as the Tourist Board said all Kerry was. **1935** G. GREENE *Basement Room* 108, I opened a tourist bureau. Trips to the London underworld. Limehouse and all that. **1978** J. IRVING *World according to Garp* vi. 122 'I am from the Tourist Bureau,' Father announced. **1964** L. DEIGHTON *Funeral in Berlin* xvii. 105 In Horse Guards Avenue.. tourist buses were parked. **1928** H. CRANE *Let.* Dec. (1965) 332, I have been the only native American in the whole tourist cabin. The rest being Britishers, Canadians, Australians, [etc.]. **1937** A. HUXLEY *Let.* 7 May (1969) 421 Are now at Del Rio, on the Mexican frontier, in a tourist cabin. **1943** J. S. HUXLEY *TVA* ix. 61 Cove Lake Inn, with its group of tourist cabins,..has become..an important overnight stopping place for tourists. **1923** *Outlook* Aug. 591/3 The University of Iowa has published a bulletin on the tourist camps of that State. **1968** C. BURKE *Elephant across Border* vi. 217 Tourist camps with hot and cold running. **1895** J. C. WAIT *Car-Builder's Dict.* 134 *Tourist car*, a car roughly built and furnished for the transportation of men alone, such as bodies of troops, parties of excursionists, emigrants, etc. **1971** *Daily Nation* (Nairobi) 10 Apr. 5/3 Visitors now travelling to several Latin-American countries like Mexico require only a Tourist Card, which is issued by an airline authorised by the respective government. **1922** W. J. LOCKE *Tale of Triona* xxvi. 292 The hiring garages, in anything like tourist centres, found their resources strained. **1978** N. FREELING *Night Lords* xxvi. 123 The Loire country.. a centre for tourism... Normandy..another natural tourist centre. **1962** E. SNOW *Other Side of River* (1963) lxxv. 577 The city is after all on the main 'tourist circuit' —Peking–Shanghai–Hankow–Canton. **1936** *New Yorker* 22 Feb. 63/2 In Tourist Class, too, you find typical American standards. **1939** F. SCOTT FITZGERALD *Let.* Aug. (1964) 107 Only the rich now can do the things you and I once did in Europe—it is a tourist-class world. **1951** *Word Study* Feb. 5/1 People sailed to Europe *cabin class* or *tourist class.* **1952** *Shell Aviation News* No. 164. 4/2 Tourist class fares are being introduced and are going to lead to a doubling of passenger air travel in two years. **1964** MRS. L. B. JOHNSON *White House Diary* 20 Jan. (1970) 55, I left early this morning..tourist class on a commercial airline for New York. **1978** J. A. MICHENER *Chesapeake* xii. 712 At the Sunday meal many of the first-class passengers came down to the tourist class..to urge that Paxmore conduct the services. **1937** *Amer. City* Oct. 115 (*heading*) House trailer and tourist court regulations. **1979** R. THOMAS *Eighth Dwarf* iii. 27 A cluster of fishing shacks..and the odd tourist court. **1872** R. BROUGHTON in *Temple Bar* XXXVI. 340 The great glaring Schweizerhof, with its colonnaded, tourist-crowded porch. **1959** A. HUXLEY *Let.* 9 Jan. (1969) 861 The tourist flights are relatively cheap and I will treat you to the ticket. **1969** G. LYALL *Venus with Pistol* xxxii. 207 Maybe old Georgy-boy used to pop over to Germany on the cheap tourist flights. **1924** R. CUMMINS *Sky-High Corral* 15 You cut me down because you leased eighty acres of Hay Fork Meadow to that measly tourist guide to fence for horse feed. **1925** W. DEEPING *Sorrell & Son* iii. 26 Medlum.. kept the book-shop and sold..pretty-pretty art tourist guides. **1977** J. VAN DE WETERING *Japanese Corpse* (1978) xi. 109 He speaks English fairly well. He used to be a tourist guide. **1927** G. ADE et al. *Let.* 4 Mar. (1973) 118 We went to a most attractive tourist hotel above the town [*sc.* Gibraltar]..surrounded by palms and tropical plants. **1980** R. CONNOLLY *Sunday Kind of Woman* xxix. 194 She was staying in an ordinary tourist hotel just off Gloucester Road. **1938** *United Empire* Sept. 398 With peace achieved, the Government is fostering the tourist industry. **1977** *Times* 5 Apr. 16/6 The stately homes of England have been..making an important contribution to the tourist industry. **1875** *Cooks Continental Timetable* (1973) Mar. p. iv/1 (Advt.), Cook's Waterloo Coach Tickets..may be obtained at any of their tourist offices. **1977** *Times* 14 May 12/5 The tourist office put us in touch with the guest house association. **1927** F. F. VAN DE WATER *Family Flivvers to Frisco* iii. 52 There is money in well-run tourist parks. **1977** *Chicago Tribune* 2 Oct. XII. 27/8 (Advt.), Facility has been updated with boat launch, tourist park and public beach nearby. **1906** 'MARK TWAIN' *Autobiogr.* (1924) II. 215 Mr. Richmond had become possessed of Tom Sawyer's cave in the hills three miles from town, and had made a tourist-resort of it. **1959** A. H. MCLINTOCK *Descr. Atlas N.Z.* 74 New Zealand ..has roads of high quality leading to the principal tourist resorts. **1874** *Cooks Continental Timetable* (1973) Mar. p. ii/2 (Advt.), List of the Principal Hotels on the chief Tourist Routes in Europe. **1976** BOTHAM & DONNELLY *Valentino* xx. 147 He and Natacha followed the tourist routes to Windsor Castle, the Tower of London, Hampton Court, the theatre. **1884** MRS. F. F. MILLER *Harriet Martineau* viii. 135 Hunters of celebrities were wont, in the tourist season,..to walk round her garden. **1980** I. MURDOCH *Nuns & Soldiers* ii. 139 Now the tourist season's starting it's better to have someone there, like a caretaker. **1969** 'E. LATHEN' *When in Greece* x. 112, I call Athens... I have business with the tourist shops there. **1946** P. BOTTOME *Lifeline* vi. 61 It was two years since Mark had seen them, two years of grinding poverty under the [Nazi] Tourist tax. **1963** *Times* 21 Jan. 5/4 The tourist tax proposal appears to have had its brief, controversial highland fling. **1977** *Times* 22 Aug. 13/2 Proponents of a tourist tax really mean for it to be imposed only on foreign tourists. **1887** M. CORELLI *Thelma* I. xiii. 288 She wad send a' her relations there wi' tourist tickets, not available for the return journey. **1912** 'SAKI' *Chron. Clovis* 137 Continental travel..away from the great tourist tracks, was a favoured hobby. **1959** M. SHADBOLT *New Zealanders* 110 Ted..moved about the country from season to season..sheelite-mining in the Alps, cutting tourist tracks at Milford sound. **1969** *Guardian* 7 July 8/3 The Piazza Navona, right on the Roman tourist track. **1936** *Discovery* Sept. 264/1 That trade which cannot survive without peace—the tourist trade. **1946** J. S. HUXLEY *Unesco* ii. 53 An unregulated tourist trade in 'curios'. **1979** V. S. NAIPAUL *Bend in River* xv. 254 London is destroying itself for its tourist trade. **1892** KIPLING & BALESTIER *Naulahka* xiii. 156 Some towns might think we had a little tourist traffic now. **1979** *Tucson Mag.* June 54/1 Some extremists might seek out.. Hell on Earth. But, judging by the light tourist traffic into El Paso, these extremists are a small minority. **1939** G. GREENE *Lawless Roads* iv. 117 Different stations had their different tourist traps—at Apizaco hideous hand-painted clubs..at Rinconada little grey stone mortars. **1942** E. PAUL *Narrow St.* xxvii. 241 The Oubliette Rouge was a small tourist-trap..where..the drinks were watered to such an extent that temperance was automatically accomplished. **1967** O. WYND *Walk Softly, Men Praying* x. 157 The village..[was] now a tourist trap almost entirely given over to eating houses and souvenir shops. **1981** C. STORR *Vicky* xiv. 99 Vicky's eye was caught by something in one of the tourist-trap shop windows. **1907** *Motor Cycle* 1 May 358/1 Cash prizes will be awarded to the drivers of machines taking part in the International Auto Cycle Tourist Trophy Race. **1913** [see TT s.v. *T II. 6]. **1968** S. E. ELLACOTT *Everyday Things in England 1914–68* V. xii. 174 The Isle of Man..Tourist Trophy races.. began in May 1907. **1973** S. JACKMAN *Guns covered with Flowers* iv. 40 The Immigration Officer...issued him with a Tourist visa.

(ii) Of a foreign language: of that degree of proficiency required by a tourist or to communicate with a tourist.

1938 E. AMBLER *Cause for Alarm* ii. 30, I have seen six gentlemen before you. Three of them could speak tourist French and insisted that most Italians would understand it. **1953** S. BEDFORD *Sudden View* I. iii. 36, I had a small deposit of past tourist Spanish..equal to ordering the *comida corrida*, the table d'hôte luncheon. **1975** A. WORBOYS *Lion of Delos* ii. 23 Most of the [Greek] people you will come across talk Tourist English.

Hence **tou·rist** *v. intr.*, to travel for pleasure, as a tourist; **touristy** *a.*, also, designed for or likely to appeal to tourists, consisting of tourists, frequented by tourists; freq. with derog. connotation, superficial, trashy.

1953 E. M. FORSTER *Hill of Devi* 27, I continued my tour—I was touristing—and I did not expect to see my delightful host again. **1971** K. WHEELER *Epitaph for Mr. Wynn* xxviii. 364 He wasn't just touristing around. **1937** A. CHRISTIE *Death on Nile* vi. 86 This trip..feels somehow, so much less touristy—as though we were really going into the heart of Egypt. **1958** *Listener* 25 Sept. 454/1 Its rushing rivers leaping with salmon; its canyons; its glaciers;..steam from thermal springs..all these are perhaps the 'touristy' things. **1961** I. FLEMING *Thunderball* xii. 125 A tepid, touristy breakfast on his balcony. **1967** [see *STRIP *sb.*[2] 1 h.] **1969** E. MCGIRR *Entry of Death* iii. 59 There's a little 'ostelry wot only the helite know of. None of yer touristy trash. **1973** *Daily Tel.* (Colour Suppl.) 16 Feb. 31/3 They've turned out so many phoney things—touristy things about the Civil War, say.

touristic, *a.* Add: **touri·stically** *adv.*, from the point of view of a tourist; as regards tourists or tourism.

1928 *Sunday Dispatch* 16 Dec. 4/2 Asked to show a young French boy of fifteen the 'sights' of London... But he had one ambition, touristically, and one only. **1959** *Encounter* May 30/2 The girls..go anywhere, do anything sexually or touristically interesting. County Dublin, Positano, Ibiza, Tangier. **1971** *Daily Tel.* 10 Sept. 14/4 Replicas of the castle could also be erected not only in London but at touristically under-privileged places like Leeds and Merthyr Tydfil. **1983** *Which?* Nov. 515/1 The qualities I look for in a country house hotel—a tranquil location in a touristically rewarding area, [etc.].

tourmaline. Add: **b*.** (See quot. 1957.)

1957 J. H. F. STEVENSON *Mink in Britain* (ed. 2) 101 *Tourmaline*, EMBA [*sc.* Mutation Mink Breeders of America] brand name for high quality natural pale beige mutation mink skins. **1959** *Vogue* 1 Oct. 188/3 You'd see the all-important minks..from the very dark ranch mink through the lighter, lovely Emba mutations to the pales, which include a pearly blondness, 'Tourmaline'. **1959** *Official Gaz.* (U.S. Patent Office) 27 Oct. TM 127 Mutation Mink Breeders Association... Tourmaline. For Mink Fur Pelts. First use Dec. 11, 1956. **1965** P. O'DONNELL *Modesty Blaise* iii. 32 A Dior-designed mink coat in embl tourmaline.

c. *tourmaline mink.*

1959 *Vogue* Dec. 25 (Advt.), Tourmaline mink coat. **1973** 'R. MACDONALD' *Sleeping Beauty* x. 51 Elizabeth Somerville came to the front door in a tourmaline mink which almost matched her blonde head.

tourmalinize, *v.* (In Dict. s.v. TOURMALINE.) Add: (Examples.) Hence **tou·rmalinized** *ppl. a.*, **tou·rmalinizing** *vbl. sb.*; **tou:rmaliniza·tion,** the process or state of being tourmalinized.

1899 *Bull. Geol. Soc. Amer.* X. 23 As the vein widens the tourmalinizing of the schist becomes less and less marked. *Ibid.*, Fragments of this tourmalinized schist are thickly strewn along both sides of the road. *Ibid.* 24 A more intense tourmalinization is to be noticed in the immediate vicinity of the veins. *Ibid.* 26 The hornfels..has been tourmalinized on both sides of the vein. **1946** *Geol. Survey Nigeria Bull.* No. 17. 56 The altered country-rock on the hanging-wall has been recrystallized and tourmalinized to form a rock composed of brown biotite..and tourmaline. **1982** D. S. SUTHERLAND *Igneous Rocks of Brit. Isles* xxiii. 318/2 The second stage of tourmalinization followed the second stage of greisening, using the same channels. *Ibid.* 524/1 The tourmalinized varieties of the granitic rocks are of such striking appearance that they have given rise to several special names.

‖ **tourmente** (turmɑ̃nt). [Fr.: see TORMENT *sb.*] A whirling storm or eddy (of snow): see TORMENT *sb.* 5.

[**1843** J. D. FORBES *Trav. through Alps* 187 The strong west wind..raises the snow into fearful eddies, called *tourmentes* in the French.] **1847, 1909** [see TORMENT *sb.* 5]. **1924** J. BUCHAN *Three Hostages* xii. 177, I could see that it was blowing hard, for my glass showed me little *tourmentes* of snow. **1928** *Blackw. Mag.* May 601/1 Above the howl of the *tourmente* came the crash of falling rocks.

Tournai. Now the usual form of TOURNAY. Used *attrib.* and *absol.* to designate products of the town, *esp.* the porcelain manufactured there from 1751.

1873 C. SCHREIBER *Jrnl.* (1911) I. 207 He..prides himself upon his Tournai. **1874** *Ibid.* 292 One Tournai plate, gold marked. **1907** C. H. WYLDE *How to collect Continental China* 119 Birds..had always been a favourite subject as a *motif* of decoration on Tournay porcelain. **1959** *Listener* 22 Jan. 179/1 The more familiar tombstone of a bishop in Tournai 'marble'. **1980** *Times* 5 Feb. 16/5 A Tournai shaped oval dish of about 1765..made £11,000.

Tournaisian (tūᵊɹnē¹·ziăn), *a. Geol.* [ad. Fr. *tournaisien* of Tournai: see prec., -IAN.] Of, pertaining to, or designating the lower of the two divisions of the Lower Carboniferous (Dinantian) in Europe. Also *absol.*

1910 *Rep. Brit. Assoc. Adv. Sci.* 1909 187 A minute exposition of the coral sequence in the Tournaisian. *Ibid.*, The Tournaisian Beds of this fine coast section. **1923** L. D. STAMP *Introd. Stratigr.* ix. 129 The lower three zones constitute the Tournaisian or Lower Carboniferous Limestone, the upper zones the Visean or Upper Carboniferous Limestone. **1969** BENNISON & WRIGHT *Geol. Hist. Brit. Isles* ix. 187 The Tournaisian is..exposed in a continuous section in the gorge of the River Avon. *Ibid.* 205 The Upper Old Red Sandstone may be in part of Tournaisian age.

tournament, *sb.* Add: **3*.** *Math.* A set of points each of which is joined to every other point by a line having a direction. Also *tournament graph.*

1959 F. HARARY in *Management Sci.* V. 398 Consider a tournament in which there are n players, every pair of players play each other once, and none of the games ends in a draw. For brevity let us call the resulting digraph D itself a tournament. **1972** R. J. WILSON *Introd. Graph Theory* vii. 108 Let T be a tournament on $n+1$ vertices, and let T' be the tournament on n vertices obtained by removing from T a vertex v and every arc incident to v. **1980** *Sci. Amer.* Mar. 18/3 Tournament graphs provide a convenient means of modeling a person's pairwise preferences for any set of choices, such as brands of coffee or candidates in an election.

Tournay: see also *TOURNAI.

‖ **tournedos** (turnədo). *Gastronomy.* [Fr., f. *tourner* to turn + *dos* back: acc. to Littré and Robert, so called because the dish is traditionally not placed on the table, but is passed behind the backs of the guests (see quots. for this and another account).] A fillet steak of beef with a surrounding strip of fat; *tournedos Rossini*, such a fillet served with a *croûton* and pâté, and a Madeira sauce.

1877 E. S. DALLAS *Kettner's Bk. of Table* 460 A tournedos is a thin collop, which..is done on one side before the cook has had time to turn round. **1937** G. FRANKAU *More of Us* v. 53 Italy! Twice ten thousand special pleaders Present your case—and tournedos Rossini, Whose rich appeal 'spite facial controlling Still sets these eyes in a fine frenzy rolling. **1958** *Times* 29 Nov. 7/6 Tucking into scampi and tournedos. **1966** P. V. PRICE *France: Food & Wine Guide* 124 About [1869]..the word tournedos came into use... Rossini was dining at the Café Anglais and..suggested an alternative method of cutting and preparing the steak... The horrified *maître d'hotel* announced that he could not..present a dish that was..unpresentable. 'Very well,' said Rossini, 'then don't let us see you do it—I'll turn my back' (*tourne le dos*). **1979** C. Curzon *Leaven of Malice* xi. 126 *Tournedos Rossini* with salad and a reasonable Valpolicella.

tournee (tūᵊ·ɹni), ‖ **tournée** (turne). [a. F. *tournée* round.] A round, circuit, tour.
1794 B. Wynne *Diary* 8 Oct. (1952) xiii. 163 We did today what is called the *Tournee* and which is visits to all the ministers and *Grand* Families. **1834** W. F. Tolmie *Jrnls.* (1963) 298 Made an unsuccessful tournee in the S. plain in quest of deer. **1961** *Times* 13 Oct. 10/5 When Louis Armstrong and Dave Brubeck come to Germany their tournées take them to sold-out concert halls.

tournette (tūᵊɹne·t). *Archæol.* [a. Fr., f. *tourner* to turn.] A rotating disc resembling a potter's wheel. Also *attrib.*
1927 Peake & Fleure *Peasants & Potters* iv. 47 The pots were made on a tournette, a slow wheel turned by hand. **1952** V. G. Childe *New Light on Most Anc. East* xi. 235 Centrally perforated stone discs some 20 cm. in diameter have been called 'tournettes'. **1964** H. Hodges *Artifacts* i. 27 Ring or coil-built pots are generally flat-bottomed since they are frequently formed on a turn-table, or tournette. **1977** *Antiquaries Jrnl.* LVII. 317 The upper fills of such pits can often be seen as discarded rubbish relating to the use of clay: pot sherds,..loom-weight, and tournette fragments.

tourney, *sb.*[1] Add: **1. c.** = Tournament *sb.* 3.
1890 J. Rayner *Chess Problems* 15 If..one should creep into a problem deemed by him..to be fit for a tourney, it will be useful..to know that the German school of problematists is less puritanical than the English. *Ibid.* 28 In solution and problem tourneys..it is necessary to throw aside all conventionalities. **1950** *Sun* (Baltimore) 20 June 21/6 It was really rather astonishing to watch this youth club his way through the tourney to a sturdy victory..in a 36-hole grind. **1951** *Sport* 30 Mar.–5 Apr. 10/2 J. Parsons..outscored Billy McHale, newly-crowned Northern Counties A.B.A. champion, in the miners' divisional tourney. **1971** *Rand Daily Mail* 4 Sept. 2/9 The Government's new sports policy..has guaranteed a welcome for all teams for next year's Federation Cup tennis tourney. **1976** *Star* (Sheffield) 3 Dec. 28/8 Last week with the results boosted by the netball tourney..there were 140 results in the Hotline columns.

tournure. Add: **3.** (Earlier example.)
1827 Disraeli *Vivian Grey* III. v. xv. 299 Touched in with freedom—a grand *tournure*—great *goût* in the swell of the neck.
4. Also, a kind of corset. (Earlier examples.)
1831 H. Granville *Let.* Jan. (1894) II. 75 Very fat, but squeezed into a *tournure*. **1872** *Young Englishwoman* Dec. 646/2 The tournure is high indeed behind... It has superseded the crinoline.

Tourte[2] (tūᵊɹt, ‖ turt). The name of the French violin-bow maker François *Tourte* (1747–1835), who perfected the modern bow, used *attrib.* and *absol.* to designate bows made according to his model.
1889 Grove *Dict. Mus.* IV. 155 The Tourte bow greatly facilitated the new development of violin music. *Ibid.* 156 A very fine Tourte has been recently sold for £30. **1896** H. Saint-George *Bow* vi. 51 What a marvellous thing a fine Tourte is! What a revelation the first time a player handles one! **1908** *Sears, Roebuck Catal.* 233/2 This [professional violin] outfit includes:..One Tourte model bow, full German silver trimmings and best quality Brazil wood stick. **1950** *Musical Q.* Jan. 16 French and Italian bows shown in early 18-century sources seem lighter in construction than the Tourte bows. **1980** *Early Music* Apr. 200/1 Illus. 1c shows a genuine Tourte bow and 1d an English bow stamped forster..one of the types of so-called 'transitional' bows in vogue about 1775, some ten years before the invention of the Tourte.

tourtière (tūᵊɹti,ēᵊ·ɹ, ‖ turtyɛ̄r'). [a. Fr., lit. 'tart-tin', f. *tourte* Tourte tart.] **a.** *Fr. Canad.* A kind of meat pie traditionally eaten at Christmas. **b.** A tart-tin or round baking-sheet.
1953 Wattie & Donaldson *Nellie L. Pattinson's Canad. Cookbk.* (rev. ed.) xxv. 467 Tourtière (Pork Pies). These pies are traditional Christmas Eve fare. **1959** J. Donon *Classic French Cuisine* ix. 303 Pour into a well-buttered *tourtière*, or pie plate. **1960** E. David *French Provincial Cooking* 69 *Tourtière*, a shallow tart tin... In former times a *tourtière* was a heavy iron or earthenware dish, much deeper than a tart tin, in which many things besides pastry could be cooked. **1975** *Globe & Mail* (Toronto) 3 Dec. s8/5 A meat pie of Christmas Eve fame in Quebec, the tourtiere is traditional fare at the reveillon after midnight mass. **1978** *N.Y. Times* 29 Mar. c 8/6, I like to bake crust in flan rings set on round black baking sheets called tourtières.

tousle, touzle, *sb.* Add: The spelling *touzle* seems now to be obsolete. **3.** *touslehead.*
1900 M. Hewlett *Life & Death Richard Yea-and-Nay* ii. xi. 364 The townsmen of Gratz, hoarse-voiced touzle-heads mostly, divined her to be an anchoress. **1981** *Sunday Express Mag.* 7 June 12/1 The breathlessness of a touslehead at a school concert.

tously, *a.* Add: Also formerly **touzly.** (Earlier example.)
1832 *Chambers's Edin. Jrnl.* I. 193/2 Ye may be as touzly as ye like i' the outside o' your claes.

‖ **Toussaint** (tusæ̃). [Fr., f. *tous*, pl. of *tout* all + *saint* saint.] The feast of All Saints (1 November).
1930 K. Boyle *Plagued by Nightingale* (1931) xxv. 246 They would linger in the country..perhaps even until the *Toussaint*. **1955** *Caribbean Q.* IV. ii. (*verso front cover*) In many West Indian Islands, especially but not exclusively those under Catholic influence, it is customary to keep the Festival of All Saints, or Toussaint. **1979** N. Freeling *Widow* xxxiv. 210 The flowers for mum's birthday or the Toussaint.

tousy, towsy, *a.* Add: Also **touzie, -y. 1.** Also *transf.* and *fig.*
1873 A. G. Murdoch *Lilts* 57 Tell him, when in the touzie key, A nicht wi' him I wadna gie. **1897** H. Ochiltree *Out of Shroud* xxiv. 331, I was oot gey late ae nicht—a touzie nicht it was. **1925** G. B. Cumming *A'anside Lilts* 71 The times then were touzie to live in. **1955** *Times* 9 May 6/3 A campaign that is already showing signs of developing into a tousy fight. **1972** *Listener* 27 July 104/3 (*heading*) Glory goals rock tousy Rangers.
2. Abundant, prolific; *esp.* in *Comb.*, **tousy tea,** a knife-and-fork tea, high tea.
1835 *Glasgow Jrnl.* 31 Oct. 44 Mrs Stewart had laid what she styled a 'touzie tea'. **1895** H. Ochiltree *Redburn* ix. 90 It's no very great place for yits or barley, but a gye tousie place for gress. **1934** T. Smellie *Tea-Pairty* 12 Next to a touzie tea there's naething like maesic tae soothe a savage beast.

tout, *sb.*[1] Add: **5.** A spy; an informer. Cf. Tout *v.*[1] 2. *N. Ireland* and *Sc.*
1959 I. & P. Opie *Lore & Lang. Schoolch.* x. 189 The tell tale is..a touт, traitor, quisling, or widemouth. **1973** *Times* 6 June 1/5 The body of a young man..was found..shot through the head 800 yards from the southern Irish border... A label with the word 'Tout' written on it was attached to his neck. **1977** W. McIlvanney *Laidlaw* xl. 186 'What's his business?' 'Same as any tout's. Other people's.'

‖ **tout** (tu), *adv.*, *sb.*[4], and *a.* [Fr.] **A.** *adv.* Quite, entirely: *tout au contraire* (tut o koṅtrɛ̄'r), quite the contrary; *tout court* (tu kūr), in short, in little, simply, without qualification or addition; *tout de suite* (tu də swīt) [*de suite* in sequence], at once, immediately; cf. *toot sweet; *tout seul* (tu sōl), quite alone, on its (or his, etc.) own; *tout simple, simplement* (tu sæ̃pl', sæ̃pl'mañ) quite simply, just that.
1841 M. Edgeworth *Let.* 23 Mar. (1971) 590 Scandal but not by any means ill natured tout au contraire. **1982** E. Dewhurst *Whoever I Am* i. 18 'You find it obvious that I've been on the amateur stage?' '*Tout au contraire*... But I know.'
1747 H. Walpole *Let.* 26 June (1955) XIX. 420 My eagle is arrived—my eagle *tout court*, for I hear nothing of the pedestal. **1888** Kipling *Wee Willie Winkie* 38 Judy was officially 'Miss Judy'; but Black Sheep was never anything but Black Sheep *tout court*. **1928** C. Dawson *Age of Gods* xii. 262 There are grave objections to the identification *tout court* of the Nordic race with the Indo-European stock. **1958** *Oxf. Mag.* 15 May 435/1 Hove, instead of asking for Psychology *tout court*, has a course by a Harley Street psychiatrist. **1981** J. Sutherland *Bestsellers* xxiv. 240 Len Deighton's..history *tout court* of the Second World War (*Bomber* and *Fighter*).
1895 E. Dowson *Let.* c 13 Nov. (1967) 319 If you see Moore tell him that I am writing tout de suite. **1971** *Ink* 12 June 14/3 Some of the underwriters quietly told their clients to resell their shares *tout-de-suite*.
1926 H. Crane *Let.* 19 Aug. (1965) 273, I have encountered him in the road, talking again tout seul and examining pebbles. **1954** *Essays in Criticism* IV. 272 The danger in self-exploration *tout seul* is that it can lead to loss of urgency.
1930 *Harvard Law Rev.* XLIII. 881 Strict or liberal construction or interpretation is therefore the ordinary process of interpretation, *tout simple*. **1977** *Times* 14 Apr. 14/6 The event was listed as a variety show, *tout simple*. **1939** *Burlington Mag.* Mar. 142/2 The most probable explanation..is, *tout simplement*, faulty recollection. **1973** E. Berckman *Victorian Album* 114 There it was. There, *tout simplement*, as they say, was my murder.
B. *sb.*[4] and *a.* All: *tout compris* (tu koṅprī), all included, inclusive; *tout ensemble*: see Ensemble *sb.* 1; *tout le monde* (tu lə moṅd), all the world, everyone; (*le*) *tout Paris* (lə tu parī), all Paris, i.e. Parisian society; also *transf.*, of other cities, social circles, etc.
1901 Ld. Milner *Let.* in J. A. Smith *John Buchan* (1979) 34/2 You will have to pay your own way out—about £60 *tout compris*. **1960** *Harper's Bazaar* Aug. 63/1 A day in one of these hotels..can cost under 15 shillings, *tout compris*.
1825 H. Wilson *Memoirs* III. 110 Tout le monde seemed so very much to admire my person. **1944** Auden *Sea & Mirror* in *For Time Being* iii. 29 She invites..just *tout le monde* to drop in at any time.
1894 G. du Maurier *Trilby* III. vii. 15 'Tout Paris' passed them; but they were none the wiser, and agreed that the show was not a patch on that in Hyde Park during the London season. **1921** G. Bell *Let.* 5 May (1927) xx. 480 'Le tout' Bagdad was there—the Arab world. **1965** N. Freeling *Criminal Conversation* ii. xix. 183, I married her..for the introduction she could give me into what I thought of as 'the club'. Le tout-Paris. **1975** P. Moyes *Black Widower* ii. 21 *Tout Washington* tends to arrive late at diplomatic cocktail parties. **1980** T. Morgan *Somerset Maugham* iii. 221 He wanted a hostess, who knew the tout-Londres. **1982** *Times* 14 Jan. 15/3 It is the talk of le tout Paris in the French business world. Who will be getting the plum jobs?

tout, *v.*[1] Add: **3. b.** *trans.* (*a*) To importune (a person) in a touting manner; (*b*) to solicit custom for (a thing), to try to sell; also (*U.S.*) in extended sense, to recommend.
1920 S. Lewis *Main Street* xvi. 199 Why, you're always touting these Greek dancers. **1928** *Daily Tel.* 5 May 9/6 It strikes one as..unfair for bankers to tout their clients for..investment business. **1930** R. H. Mottram *Europa's Beast* vii. 164 He was involved in the ghastly job of touting motor cars. **1948** M. Laski *Tory Heaven* i. 14 Touting vacuum-cleaners at back doors. **1974** *Nature* 11 Jan. 81/1 Such deposits of geothermal energy have long been touted as potential sources of power. **1978** *Detroit Free Press* 2 Apr. 6E/1 Any team that touts Jerry Augustine as the ace of its staff is in serious trouble.
Hence **tout·ed** *ppl. a.* (*U.S.*), (usu. with qualifying advb.) vaunted, extolled.
1953 *Manch. Guardian Weekly* 5 Feb. 3 The much touted Nationalist 'offensive' on the Chinese mainland. **1978** *Sci. Amer.* Aug. 32/2 The highly touted system of separating isotopes by laser excitation.

toutou (tū·tū). [Fr. nursery term.] A pet name for a dog, esp. a lap-dog. Cf. *loulou.
1894, 1916 [see *loulou].

‖ **tou ts'ai** (tū tsai). *Ceramics.* Also **doucai.** [Chinese (Wade-Giles), *duŏcǎi* (Pinyin), lit. 'multi-coloured', f. *tou* many + *ts'ai* colours.] Used *attrib.* and *absol.* of a kind of enamel painting on Chinese porcelain, developed in the reign of Ch'êng Hua (1465–87), and of (pieces of) porcelain so decorated.
1953 S. Jenyns *Ming Pottery & Porcelain* vi. 90 Another problem piece is the famous Kitchener bowl... Brankston is uncomfortably non-committal. 'There are some points', he says, 'on which it differs from other known Ch'êng Hua *tou ts'ai* pieces.' **1960** H. Hayward *Antique Coll.* 285/1 *Tou ts'ai* ('*contrasting colour*') *enamels*, delicate, sparing designs on Chinese porcelain in underglaze blue, set off by transparent enamel colours. **1972** *Trans. Oriental Ceramics Soc.* 109 A globular jar..painted in *tou-ts'ai* style in underglaze blue. **1980** *Catal. Fine Chinese Ceramics* (Sotheby, Hong Kong) 104 A rare doucai (tou ts'ai) Vase, brightly enamelled with a formal pattern of lotus scrolls.

tovarish, tovarich (tovā·riʃ). Also **tav-; -isch, -ishch, -istch, -itch.** Pl. **-i.** [ad. Russ. *továrishch* comrade.] In the U.S.S.R., comrade (freq. as a form of address).
1918 C. E. Russell *Unchained Russia* ii. 95 After the Revolution everybody in Russia was 'tavarisch'. **1930** E. Pound *XXX Cantos* xxvii. 127 And these are the labours of tovarisch, That tovarisch lay in the earth, And rose, and wrecked the house of the tyrants. **1935** N. Mitchison *We have been Warned* iii. 236 I'm rather looking forward myself to the first time someone calls me tovarish... It seems much more romantic in Russian. **1938** E. Hemingway *Fifth Column* iii. ii. 86 Hurry up, Tovaritch, and tape good the mouths. **1968** L. Smith *Fear & Dead Man* x. 80 The Russian grinned slyly. 'Now you know I cannot tell you why I wanted you brought here, *tovarich*.' **1976** M. Barak *Secret List of Heinrich Roehm* xi. 120 The KGB Chairman shrugged. 'Very well... We shall take care of Tovarishch Joe Gonen.' **1977** *Time* 28 Feb. 12/3 To compensate for her lost lover, she found at least one more torrid *tovarish*.

tove (tōᵘv). A factitious word introduced by 'Lewis Carroll' (see quot. 1855[2]).
Quot. 1855[1] also occurs in the first verse of 'Jabberwocky' in *Through the Looking-Glass* (1871) i. 21.
1855 [see *slithy *a.*]. **1855** 'L. Carroll' *Rectory Umbrella & Mischmasch* (1932) 142 Tove, a species of Badger. They had smooth white hair, long hind legs, and short horns like a stag: lived chiefly on cheese. **1928** [see *slithy *a.*]. **1937** G. Frankau *More of Us* 2 While the free-versifier gyres and gimbles The slithy tove—with his own 'private symbols'.

tow, *sb.*[1] Add: **3*.** A bundle of untwisted natural or manmade fibres.
1950 B. E. Hartsuch *Introd. Textile Chem.* viii. 237 The filaments from several coagulating baths or cabinets (acetate) are combined to form a thick strand known as tow. **1969** [see *prepreg *sb.* (*a*)]. **1971** *New Scientist* 8 July 68/2 The material [*sc.* carbon fibres] was in the form of 'tows'—14 inch long bundles containing 10 000 filaments each of 1½ denier. **1973** *Materials & Technol.* VI. iv. 302 In the case of viscose rayon the thick tows are sometimes supplied to mills which desire to do their own cutting into staple lengths.
4. a. *tow-card* (earlier example), *-linen, -sack, string* (examples). **c. tow-head** (examples); *spec.*: for *Mergus* read *Lophodytes*; also (*U.S.*), a sand-bar or other obstruction causing ripples in a river or stream; **tow-headed** *a.* (earlier example).
1655 *Essex County, Mass. Probate Rec.* (1916) I. 201 A pair of tow cards, 1s. **1829** S. Cummings *Western Pilot* 7 There are..a great number of tow-heads and sand-bars. **1830** A. Royall *Southern Tour* I. 92 One insolent little tow-head. **1883** 'Mark Twain' *Life on Miss.* xxiii. 262 A large town which lay shut in behind a tow-head (*i.e.* new island). **1888** G. Trumbull *Names & Portraits of Birds* 75 The name Tow-Head..was heard in one of our Southern States. **1901** A. H. Rice *Mrs. Wiggs of Cabbage Patch* vii. 95 Little Europena, with baby wisdom, put her tow head under the cloth. **1960** *Guardian* 5 Nov. 1/6 The abominable tow-head from Massachusetts. **1977** *Verbatim* Dec. 6/2 Even the most casual student of American literature should have no trouble with..*tow-head* ('sand bar with cottonwoods'). **1850** S. Judd *R. Edney* xlix. 450 Bronze-faced and tow-headed Wild Olive boys. **1779** *New Jersey Archives* (1906) 2nd Ser. III. 154 [A] blue long elk

saddle cloth lined with tow linen. **1884** 'Mark Twain' *Huck. Finn* xx. 196 Some of the children didn't have on any clothes but just a tow-linen shirt. **1921** J. Buchan *Path of King* xiii. 259 He wore an old skin shirt and a pair of tow-linen pants. **1930** W. Faulkner *As I lay Dying* 48 The soaked towsack tied about his shoulders. **1976** J. Lee *Ninth Man* 243 He anchored the tow-sack bundle on his left arm. **1806** *Balance* V. 5/3 Our worthy old friend..sometimes wears a tow string round his hat. **1892** *Harper's Mag.* Mar. 649/2 What was known as the 'tow-string survey' offered him an excellent opportunity for the display of his peculiar talents.

tow, *sb.*[4] Add: **2. b.** (Earlier example.) Also *in tow* (*with*) in extended sense: in company (with), accompanying, following.
1722 *New-England Courant* 17–24 Sept. 1/1 Their eager and amorous Emotions of the Body, occasion'd by taking their Mistresses *in Tow*, they call'd *wild Steerage*. **1907** C. S. Ross *Early Otago* 169 He had got in tow with a young lady. **1937** C. Day Lewis *Starting Point* i. 14 Oh, he's got one of his Swedish blondes in tow. **1965** M. Shadbolt *Among Cinders* xiv. 117 I'm in tow with my parents. **1979** S. Brett *Comedian Dies* i. 18 'Come along, Paul.' And Walter Proud, with his writer in tow, hurried along to join them.
3. Also, a string of barges that is pushed rather than pulled.
1976 *Sci. Amer.* July 124/3 The U.S. has a network of inland waterways that carries roughly a sixth of all the nation's freight in multiple-barge 'tows' that are usually pushed, rather than pulled, by powerful tugs. **1977** *Washington Post* 4 Sept. A12/1 They [*sc.* towboats] push the barges ahead of them. But the nest of barges that is pushed is called the 'tow'.
4. a. *tow hook, -horse* (example), *-truck*; **tow-bar,** a bar used in towing; *spec.* the bar by means of which a trailer, caravan, etc., is attached to the vehicle that tows it; **tow-boating** *U.S.*, the piloting or operating of a tow-boat; **towfish,** a housing with measuring or detecting instruments in it or attached to it and designed to be towed underwater behind a ship; **tow-plane,** an aircraft that tows gliders; **tow-start** *v. trans.*, to tow (a motor vehicle) in order to start the engine.
1956 *Archit. Rev.* CXIX. 259 Small trucks are coupled to it by a quick-action towbar. **1959** '*Motor*' *Manual* (ed. 36) xiii. 270 Don't be misled by the Unladen Weight which is stamped on the towbar. **1960** *Guardian* 19 Sept. 2/4 For sailing families, boat-cradle, mast support, and long tow-bar can be bought as extras. **1980** *West Lancs. Even. Gaz.* 5 June 16/4 (Advt.), Tow-bars supplied and fitted with electrics to suit every caravan and trailer requirement. **1887** *Courier-Jrnl.* (Louisville, Kentucky) 7 Feb. 3/3 Theodore Brooks..will try his hand at tow-boating this season. **1977** *Amer. N. & Q.* XV. 153/1 For more than 300 years its [*sc.* New London's] people were involved in just about every activity related to the sea including fisheries,..towboating, [etc.]. [**1973** *Jrnl. Marine Res.* XXXI. 73 An alternative solution..is the replacement of the weighted fish with a lighter tow body having a controllable fin.] **1975** *McGraw-Hill Yearbk. Sci. & Technol.* 292/1 The..towfish..continuously monitors the depth of a preselected isotherm by towing at about 5 knots..two depth-controlled instrument packages which bracket an isotherm. **1876** Knight *Amer. Mech. Dict.* 2604/2 Tow-hook, an artillery-man's hook, used in unpacking ammunition-chests. **1971** M. Tak *Truck Talk* 170 *Tow hooks*, hooks, generally found on a tractor's bumper, by which it can be towed or pulled. **1978** J. McNeil *Consultant* xx. 180 He unclipped the tow hook from the back of the car. **1865** *Harper's Mag.* Apr. 571/1 It requires as much judgment to drive tow-horses up the Alleghany as to pilot a steamboat down the Mississippi. **1940** *Aeronautics* Nov. 42/1 If the glider is more heavily loaded than the towplane, the latter will unstick first. **1973** 'A. Hall' *Tango Briefing* viii. 107 The change in the engine-note of the tow-plane. **1976** A. Schroeder *Shaking it Rough* i. 4 We passed a trio of youths trying to tow-start an old Chevrolet. **1957** Tow-truck [see *shopping plaza* s.v. *shopping *vbl. sb.* 2]. **1972** *Sat. Rev.* (U.S.) 17 June 6/2 A tow truck..came for the crippled car.
b. With advbs.: **towaway** *U.S.*, the towing away of an illegally parked vehicle; freq. *attrib.* as *towaway zone*, an area from which such vehicles may be towed away; **tow-out,** the action of towing a drilling platform out to an oil-field at sea.
1956 *Sun* (Baltimore) 31 Jan. 32/6 Such cars parked in the 'tow-a-way' zones would be hauled off to the police impounding lot. **1967** *N.Y. Times* 5 Mar. 75 Despite the published warnings about illegal parking—towaways averaged close to 200 cars daily during February. **1975** *Petroleum Rev.* XXIX. 303/1, 56 steel cylinders..provide buoyancy during tow-out. **1977** *Offshore Engineer* July 55/2 Initial plans were for flat tops to the storage tanks, but these were redesigned to increase the deck loading capacity, at tow-out, by 10,000t, to 24,000t.

towable (tōᵘ·ăb'l), *a.* [f. Tow *v.*[1] + -able.] That may be towed.
1927 *Glasgow Herald* Jan. 7 The dock..is towable to any part of Australia. **1967** [see *dracone, Dracone]. **1982** *Social Trends 1983* viii. 114/2 Towable caravans and house-boats.

towel, *sb.* Add: **1. b.** Phr. *to throw* (*chuck*, or *toss*) *in the towel*: to admit defeat. orig. *Boxing.* Cf. Sponge *sb.*[1] 1 c.
1915 E. Corri *30 Yrs. Boxing Ref.* 223 In the nineteenth round Storbeck's seconds 'threw the towel' in literally. **1916** C. J. Dennis *Moods of Ginger Mick* 132 I've done me limit, an' tossed in the tow'l. **1923** Wodehouse *Inimit. Jeeves* xv. 192 He had found the going too hard and had chucked in the towel. **1952** [see *dingo *v.*]. **1979** M. Russell *Touchdown* ii. 90 'Don't give up.'..'Have no fear... I shan't throw in the towel, I promise you.'
2. c. = *sanitary towel* s.v. *sanitary *a.* 3. Also *ellipt.*
1896 *Eng. Illustr. Mag.* Aug. (Advts. Section) 8/2 A sample of the improved 'towel' will be sent free to any lady applying to the Lady Manager. **1907** *Yesterday's Shopping* (1969) 1264 Full-sized towels reduced by pressure, packed in tiny boxes. **1979** *Guardian* 27 Mar. 9/5 A campaign for free sanitary protection through the NHS started in 1973 when the Government imposed VAT on towels and tampons.
4. *towel-covered* adj.; **towel-horse** (earlier example); **towel rail, ring,** a rail or ring on which to hang towels.
1916 H. G. Wells *Mr. Britling sees it Through* i. i. 19 A towel-covered can of hot water. **1947** *Nation* 22 Feb. 214/1 The crooked towel-covered table. **1833** J. C. Loudon *Encycl. Cottage, Farm, & Villa Archit.* 349 A Towel Horse has generally one rail at top... It..should..be painted, for the reasons given when speaking of fixed towel rails. **1961** *Times* 24 July 13/5 The civilized English custom of having heated towel-rails has not reached the United States. **1895** *Montgomery Ward Catal.* Spring & Summer 126/3 Towel Ring. Consists of polished hard-wood ring with brass chain and hook. **1977** *Times* 30 July 10/6 Rough towels..towel rings..and all manner of taps.

towel, *v.* Add: **1. c.** *absol.* for *refl.* Also with *down, off.*
1972 M. Crichton *Terminal Man* iv. ii. 141 One of the girls got out of the pool lithely and began toweling off. **1977** P. Moyes *To kill Coconut* viii. 118 Emmy emerged from the shower, towelling vigorously. **1977** G. Fisher *Villain of Piece* iii. 29, I towelled down, dressed.
2. Also (*Austral.*) with *up*; also *fig.*
1941 Baker *Dict. Austral. Slang* 78 *Towel up, to*, to beat, thrash. **1951** Cusack & James *Come in Spinner* 372, I think you deserve the V.C. for the way you towelled Old Mole up. **1973** A. Buzo *Rooted* 42 Gary got his big serve working, I chipped in at the net and we were laughing. Towelled them up in no time.
Hence **tow·elled** *ppl. a.*, wrapped in a towel.
1920 T. S. Eliot *Ara Vus Prec* 23 Doris towelled from the bath Enters padding on broad feet. **1940** G. Arthur *Concerning W. S. Churchill* 8 Standing at the edge of a deep swimming pool a junior boy mistook a towelled, stocky figure for a contemporary and playfully pushed him into the water. **1978** C. Tomlinson *Shaft* 3 The towelled head next.

towelette (taučle·t). [f. Towel *sb.* + -ette.] A small towel.
1902 in W. Forrester *Great-Grandmama's Weekly* (1980) iv. 84/1 Artmann's Hygienic towelettes. Superior to any other Sanitary Towels. **1926** *Blackw. Mag.* Apr. 528/1 Drying ourselves on pretty hand-woven towelettes. **1981** *Times* 16 Sept. 9/4 On board, someone opened my milk carton for me and someone else explained the uses of my towelette.

tower, *sb.*[1] Add: **I. 1. a.** For earlier examples of *Tower of Silence* see Silence *sb.* 2 c.
3. Freq. in *tower of strength.* See also *ivory tower.
1549 *Bk. Common Prayer* fol. xvᵛ, O lorde..Bee vnto them a tower of strength. **1594** Shaks. *Richard III* v. iii. 12 Besides, the King's name is a Tower of strength. **1852** Tennyson *Ode on Death of Duke of Wellington* 7 O fall'n at length that tower of strength. **1866** Mrs Gaskell *Wives & Daughters* II. xxii. 224 But, my dear Cynthia,—how soon Roger will be back,—a tower of strength. **1956** A. Wilson *Anglo-Saxon Attitudes* ii. iii. 394 She's been such a tower of strength all this time. **1970** *New English Bible Prov.* xviii. 10 The name of the Lord is a tower of strength, where the righteous may run for refuge. **1981** P. H. Johnson *Bonfire* ii. i. 84 He put his arms round Agnes... She thought of him as a 'tower of strength'.
5. d. A railway signal-box. *U.S.*
1900 *Everybody's Mag.* II. 442/2 The tower from which the traffic entering and leaving the Grand Central Station in New York city is directed, is located just outside the station itself. **1910** H. A. Franck *Vagabond Journey* 328 A man in the neighbouring tower opened the block, and the diminutive freight screamed by us. **1946** [see *tower house*, sense 10 a below].
e. = *pylon 4.
1930 *Engineering* 9 May 603/2 There are four standard types of tower for the single-circuit lines. **1946** D. C. Peattie *Road of Naturalist* iv. 42 The car lamps picked up out of vacancy the marching towers of the power lines. **1963** A. Lubbock *Austral. Roundabout* 72 The electric pylons, or towers, as they are called here, stalk up and down great rides cut through the trees, carrying the cable in their upflung arms.
f. See *control tower* s.v. *control *sb.* 5. Also *ellipt.*; *transf.*, the flight-control staff.
1958 'N. Shute' *Rainbow & Rose* i. 7 I'll come up to the Tower when we land. **1971** A. Diment *Think Inc.* xii. 201 Captain Roberts..asked tower, politely, for permission to taxi. **1977** *Time* 11 Apr. 23/2 The tower ordered KLM to taxi the full length of the runway.
g. *ellipt.* = *tower block*, sense 10 e below.
1970 *Times* 6 July 6/5 The towers, cheerless in their four tones of dun-colour. **1975** M. Bradbury *History Man* i. 11 Higher on the hill grow the new concrete towers.
IV. 10. a. *tower house* (later example), *mill, silo.* **e. tower apartment,** an apartment in a tower block, a high-rise flat; **tower block,** a tall block of flats, a high-rise building, a sky-scraper; **tower bolt** = *barrel bolt* s.v. *barrel *sb.* 11; **tower crane** (see quot. 1940); **tower karst** *Geomorphol.* [tr. G. *turmkarst* (H. von Wissmann 1954, in *Erdkunde* VIII. 122/1)], a type of karst characterized by isolated steep-sided hills; **towerman,** one who works in a tower; *spec.* (*U.S.*) (*a*) a railway signalman; (*b*) a look-out for forest fires; **Tower musket** *Hist.*, a tower-proof musket; **tower-shell** = *turret-shell* s.v. Turret *sb.*[1] 5; **tower skull** = *oxycephaly* s.v. Oxy- 1.
1961 Tower apartment [see *maisonnette 2]. **1966** *Atlantic Monthly* Oct. 127 Tower blocks can be accused of leading to eardrum degeneration, owing to constant use of high-speed elevators. a **1974** R. Crossman *Diaries* (1975) I. 82 The jack-block building in Coventry, a fifteen- or sixteen-storey tower block built by a new technique of jacking each storey up after it has been erected. **1982** *Listener* 23/30 Dec. 58/4 Most American film crews refuse to take rooms higher than the second floor of towerblock hotels since this picture. **1911** Tower bolt [see *barrel bolt* s.v. *barrel *sb.* 11]. **1906** *Electr. World* XLVII. 743 An illustrated description of an electrically-operated rotating tower crane for the Dublin docks. **1940** *Chambers's Techn. Dict.* 856/1 *Tower crane*, a rotatable cantilever pivoted to the top of a steelwork tower, either fixed or carried on rails. **1967** *Listener* 27 July 111/1 A tower crane on our university building site. **1946** E. B. Thompson *Amer. Daughter* 124 We climbed the little ladder to the railroad tower house. **1954** *Erdkunde* VIII. 122/1 Of the various formations of kegelkarst, two widely differ in appearance from one another... **2.** A river plain, dotted with groups or swarms of limestone towers or castles... This is the tower karst. **1977** A. Hallam *Planet Earth* 83/2 Hills with slopes of 70° and more occur, the relief being called tower karst. **1895** Tower man [see *roundhouse *sb.* 4]. **1947** *Sun* (Baltimore) 18 Oct. 7/1 Towermen..serve as the eyes of the fire fighters. **1951** Towerman [see *goldfish b]. **1888** Tower mill [see Smockmill]. **1933** *Times Lit. Suppl.* 14 Dec. 891/1 Even in brick or stone tower-mills the sweeps may be caught in the rear..by a suddenly veering storm. **1979** *Jrnl. R. Soc. Arts* Dec. 3/1 The viewer is taken inside one of the last remaining working tower mills. **1832** A. Earle *Narr. Residence in N.Z.* (1966) 170 He had with him a beautiful double-barrelled gun, and a very good Tower musket. **1947** Tower musket [see *dane gun]. **1888** *Cassell's Encycl. Dict.*, Tower-shell. **1927** Haldane & Huxley *Animal Biol.* xii. 300 One fossil tower-shell stands nearly five feet high. **1959** A. C. Hardy *Open Sea* II. v. 118 The tall slender *Turritella*, or tower-shell, is another common gastropod burrowing just below the surface. **1939** J. R. McCalmont *Silo Types & Construction* 2 Silos may be divided roughly into above-ground—tower or upright—and the below-ground—pit or trench—silo, either of which may be built for temporary or continued use. **1982** *Daily Tel.* 19 Apr. 9/3 Tower silos, standing as they do up to some 60 feet high and painted in various..colours are not particularly attractive features of our rural areas. **1905** *Trans. Ophthalm. Soc.* XXV. 364 (*heading*) Oxycephaly or 'tower skull'. **1918** J. H. Parsons *Dis. Eye* (ed. 3) xxxiii. 620 Bilateral proptosis occurs in exophthalmic goitre..as a result of diminished orbital volume in oxycephaly or 'tower-skull' and leontiasis ossea. **1969** Edington & Gilles *Path. in Tropics* x. 379 Patients with sickle-cell anaemia tend to have a certain type of habitus, with tower-skull, parietal bossing, and long slender, limbs.

tower, *v.* **I. 3. c.** (Earlier example.)
1799 Coleridge *Notebks.* (1957) I. entry 564 Partridges towering after being shot is a certain Proof that they are mortaly wounded.

towing, *vbl. sb.*[1] Add: **b.** *towing bracket.*
1959 [see *little man 2]. **1977** *West Briton* 25 Aug. 17/1 (Advt.), All types of caravan repairs undertaken,..towing brackets supplied and fitted.

‖ **towkay** (tau·kēⁱ). [ad. Malay *tauke.*] A Chinese businessman or employer, esp. in Malaysia.
1854 *Jrnl. Indian Archipelago* VIII. 16 Country born Chinese have a club called Sip Gee Seeah; they elect 12 Towkays or trustees. **1900** W. W. Skeat *Malay Magic* v. 253 The Malay *pawang* may squeeze a hundred or perhaps two hundred dollars out of the Chinese *towkay* who comes to mine for tin in Malaya. **1948** *Straits Times* 7 July 4/5 Our lives were probably saved by a Chinese towkay from Karak who used to keep us supplied with fresh fruit and tinned milk. **1966** D. Forbes *Heart of Malaya* iii. 41 Nancy, the fourth daughter of Lee Kwan Bock, the sawmill *towkay*, was a schoolteacher.

towl (taul), *v. dial.* [Imit.] *intr.* To yowl.
1906 Kipling *Puck of Pook's Hill* 283 Dan and Una found a couple of them [*sc.* beagles] towling round the kitchen-garden after the laundry cat. **1930** *Punch* 30 Apr. 478/2 They make reverent overtures to our Siamese Pugsie, who hates them and scurries towling with nerves from their outstretched hands.

town, *sb.* Add: **8. b.** *fellow about town* (earlier example); *man about town* (earlier example); also *woman about town.*
1734 in *15th Rep. R. Comm. Hist. Manuscripts* App. vi. 146 in *Parl. Papers* 1897 (C. 8551) LI. 1 Though being what is called an idle man about Town, I generally read all that is writ on both sides. **1752** M. W. Montagu *Let.* 16 Feb. (1967) III. 6 One of the most disagreable Fellows about Town, as odious in his outside as stupid in his conversation. **1927** *Manch. Guardian Weekly* Jan. 75/1 Another surrender to the woman-about-town who wants a different kind of entertainment. **1979** 'S. Kemp' *Goodbye, Pussy* xii. 160 Zoë had been an 'actress'. Actress, model, woman-about-town.

c. *woman* (also *lady*) *of the town* (later examples).
1873 G. H. Lewes *Diary* 1 Jan. in *Geo. Eliot Lett.* (1956) V. 357 Trollope came to lunch. Told me of his trouble with Harry wanting to marry a woman of the town. **1886** *Lantern* (New Orleans) 20 Oct. 2/2 Orders were issued to the police to remove all women-of-the-town. **1982** C. Castle *Folies Bergère* i. 37 At the back of the stalls..the notorious 'ladies of the town'..plied their trade.
f. *to go to town*: to do something energetically, enthusiastically, or without restraint; *spec.* to make a great fuss. Freq. const. *on. colloq.* (orig. *Jazz slang*). See also sense 8 a in Dict.
1933 [see *get *v.* 62 l]. **1940** E. S. Gardner *Case of Silent Partner* xii. 222 Chocolate creams are one of the fondest things I am of [*sic*]. I was feeling low, and I went to town. **1946** J. B. Priestley *Bright Day* viii. 252 He surveyed me with mock admiration. 'The only writer who ever made..Gruman pay him a royalty on the gross. ..And did we go to town with it, I'll say we did.' **1947** J. Bertram *Shadow of War* 238 'Skeleton's' in a bad mood; he's going to town on 'em. **1958** A. Hocking *Epitaph for Nurse* ix. 159 The local papers naturally went to town over the murder of Sister Biggs. **1960** N. Hilliard *Maori Girl* ii. ix. 128 'It's funny as hell to see girls fight.'..'They're really tough sorts, and boy! do they go to town. And swear! Punching and spitting and pulling hair.' **1972** P. M. Hubbard *Whisper in Glen* vii. 67 Whoever had painted the thing, he had gone to town on his picture. **1980** *Times Lit. Suppl.* 14 Mar. 290/2 Professor MacAndrew goes to town on this novel, deciphering the code which she believes Henry James to have set up.
9. c. *town-builder* (example); *-loving* adj. (examples); *town-dark, -tied* (earlier example) adjs.
1859 S. R. Stumbo *Let.* 11 Jan. in L. R. Hafen *Colorado Gold Rush* (1941) 214 The reports you see in the papers..are put in circulation by town builders for speculative purposes. **1918** D. H. Lawrence *New Poems* 26 Gay birds of the town-dark sea. **1960** R. Williams *Border Country* 10 It was dark..town dark. **1900** F. W. Maitland *Let.* 18 Feb. (1965) 211 The Spaniard of the middle class is a town-loving animal. **1941** *Mind* L. 396 A statement which no purely town-born, town-bred, town-loving person can..verify. **1845** E. A. Poe in *Broadway Jrnl.* 13 Sept. 155 We poor town-tied denizens..can revel in scenes which we may never be able to visit, and snuff up in imagination the incense of the flowers, which only bloom for us through the painter's art.
10. town and country planning, the preparation and construction of plans in accordance with which the development of towns and countryside is to be regulated; cf. Town-planning *sb.*; **town ball** *U.S.*, a game resembling baseball; **town belt** *N.Z.*, a belt of public land reserved chiefly for recreational purposes in or round a town; **town car** *U.S.*, a four-door motor car having a passenger compartment which is permanently enclosed and a driver's compartment which is not; **town centre,** a place or a collection of buildings forming a central point in a town (see *centre *sb.* 6 a); **town clown** *U.S. slang*, a policeman working in a village or small town; **town gas,** gas manufactured and supplied for domestic or commercial use, based on coal gas; **town-head** (example); **townhithe** *rare*[−1], a haven or landing-place in a town; **townhome** *U.S.* = *town-house, town house 2 b; **town-miss,** a young woman who lives in a town: *spec.* a prostitute; **town-mouse** (earlier example); **town-plat** (earlier examples); **town-site** (earlier example); **town trail,** a route through a town for tourists or walkers linking features of interest, which are described and interpreted by explanatory notices, printed leaflets, or a guide; **town-traveller,** a commercial traveller whose operations are confined to the town which is his employer's place of business; **town twinning,** the establishment of regular contacts between two towns in different countries; cf. *twin *v.*[2] 2.
1933 P. Abercrombie (*title*) Town & country planning. **1941** J. S. Huxley in *Times Educ. Suppl.* 6 Dec. 581/2 It is here that adult education, enlightened town and country planning, and deliberate encouragement by the State and local authorities of living art, music, drama and all other branches of cultural life, must be called on to do most of the bridging of the gap. **1972** *Whitaker's Almanack 1973* 1177/2 The Town and Country Planning Act 1971 (consolidating earlier Acts) contains very far-reaching provisions affecting the liberty of an owner of land to develop and use it as he will. **1852** *California Dispatch* 18 Jan. 2/4 A game of 'town ball' which was had on the Plaza during the week, reminded us of other days and other scenes. **1909** *Collier's* 8 May 12/1 In America the corresponding game generally went under the name of 'rounders', and because it was played at the time of town meetings, 'townball'. **1975** E. Wigginton *Foxfire 3* 466 We'd go out and play town ball. **1851** E. Ward *Jrnl.* 3 Jan. (1951) 98 We afterwards went on to the Town Belt and Riccarton. **1889** W. Davidson *Stories N.Z. Life* 61 The native bush which covers a large portion of the 'town belt'. **1907** *Horseless Age* 16 Oct. 589/3 There will be [from Ford Motor Co.] an enclosed town car, which is to be an exact copy..of the Renault town car. **1929** *Vanity Fair* (N.Y.) Mar. 89 (*caption*) The Blackhawk, a smaller and lower edition of the Stutz, is represented here by a town car of dignified proportions. **1968** G. N. Georgano *Compl. Encycl. Motorcars* 621 *Coupé de ville*,..Some 'de ville' bodies had folding rear quarters as in the landaulette. In America they were more often known as town cars. **1932** T. Sharp *Town & Countryside* x. 203 But even if the naturalistic style could be quite perfectly carried out in the perfect replica of a romantic natural scene, what..is the purpose of such a scene in a sensibly-sized town, when..genuine countryside [is] accessible within a few minutes' walk of the town centre. **1966** *Guardian* 10 Sept. 14/1 A recently started town-centre housing scheme. **1980** P. Lively *Judgement Day* iii. 26 The street plan of the town centre, an elongated triangle enclosing an open space. **1927** *Amer. Speech* II. 387/1 The town clown's badge is called a *tomato can*. **1931** 'D. Stiff' *Milk & Honey Route* i. 20 There should always be some retreat, preferably a thicket, into which the hobos can flee, should they receive an unwelcome visit from the 'town clown', or the law enforcer of the community. **1908** F. E. Junge *Gas Power* ii. 31 The price for town gas has been gradually reduced during this period. **1958** *B.S.I. News* Aug. 16 Flexible tubing and connector ends for appliances burning town gas. **1973** C. Callow *Power from Sea* iv. 85 The number of people who have been inconvenienced is small compared with the total number converted from town gas to natural gas. **1805** G. McIndoe *Poems & Songs* 62 Some b—h frae the town head has stown't. **1922** Joyce *Ulysses* 379 Once her in townhithe meeting he her bow had not doffed. **1976** *Washington Post* 19 Apr. c18/2 (Advt.), Rockshire Townhome w/3 bedrms., 2½ baths, English pub rec.rm., den, show well. **1979** *Arizona Daily Star* 22 July H 4/1 Accordingly, when what used to be called 'row houses'—attached houses—became economically desirable, they were at first called 'town houses' and are now in the process of being renamed 'town homes'. **1749** J. Cleland *Mem. Woman Pleasure* II. 98, I was not at all out of figure to pass for a modest girl. I had neither the feathers, nor *fumet* of a tawdry town-miss. **1921** D. H. Lawrence *Sea & Sardinia* vi. 245 Two town-misses in fur coats. **1750** *Student* 31 May 190 Town-mice, he knew, luxurious were. **1656** *Public Rec. Colony of Connecticut* (1850) I. 282 Thos persons that cohabitt in the towne platte. **1723** *Proprietors' Rec. Waterbury, Connecticut* (1911) 121 To settle the old Town platt Lotts. **1821** *Canad. Courant* 17 Jan. 1/2 There are about fourteen acres cleared for a Town site but not a single house in a finished state. **1973** *Nature* 11 May 105/2 The local College of Education has sponsored the idea of 'town trails' in Leicester. **1980** *Jrnl. R. Soc. Arts* CXXVIII. 303/1 Its 140 pages of practical advice on ..town trails, heritage centres and other 'media' are not aimed at the general reader. **1850** Dickens *D. Copperfield* xi. 114 He was a sort of town traveller for a number of miscellaneous houses. **1930** A. Bennett *Imperial Palace* x. 59 A town-traveller in tinned comestibles. **1960** *Sunday Express* 16 Oct. 9/6 Town twinning between cities of highly developed and under-developed countries. **1981** *Times* 23 Mar. 4/4 The question of Dundee's association with Nablus would be raised with the Scottish Town Twinning Association.
11. Townswomen's Guild, an urban organization of women, engaging in educational and social activities.
1929 *Times* 26 Nov. 19/4 Lady Cynthia Colville, the president of the Townswomen's Guild Appeal..spoke of the great need there was in small towns and residential suburbs for the new Townswomen's Guilds, which are to fulfil a role similar to that played by the women's institutes in the rural areas. **1933** *Ludlow Advertiser* 25 Feb. 6/4 The Townswomen's Guild held a whist drive on Monday night in the Guild Room in Broad Street. **1960** J. Stroud *Shorn Lamb* vii. 79 Miss Dashforth stumped the whole area addressing Mothers' Unions, Townswomen's Guilds, Parent-Teacher Associations and so on. **1977** *Belfast Telegraph* 19 Jan. 3/5 Bloomfield Collegiate School—Knock Townswomen's Guild, talk on community relations, 7.45 pm.

townee, *sb.* Add: (Later examples.) Now usu. as distinguished from a country-dweller. Freq. pejorative. Cf. Towny *sb.* 1.
1929 S. Kaye-Smith in H. C. Minchin *Legion Bk.* 195 The æsthetic week-ender is like other townees in that he generally fails to realize that the real country-dweller..is a very mass of conventions. **1939** Auden in *I Believe* (1940) 18 We frequently admire the 'goodness' of illiterate peasants as compared with the 'badness' of many townees. **1976** J. I. M. Stewart *Memorial Service* xv. 245 Janet as a child had been a townee like myself.
B. *adj.* Of, pertaining to, or characteristic of town-dwellers or the town.
1935 H. H. Bashford *Lodgings for Twelve* 110 For the townee, as he called him, and townee pursuits, he had a quite unconcealed if tolerant contempt. **1936** Auden *Look, Stranger!* 55 The identical and townee smartness. **1960** W. Miller *Russians as People* 60 It is all fascinating to the foreigner trying to sniff out 'Russian life', but to the townee Russian it is the shabby side of the familiar. **1972** *Daily Tel.* (Colour Suppl.) 24 Nov. 18/1 The Australians are far and away the most urbanised and townee of all nations.

towney: see Towny *a.* and *sb.* in Dict. and Suppl.

town hall. Add: **2.** *Comb.*: **town-hall clock(s)** = Moschatel.
1900 Dickinson & Prevost *Gloss. Dial. Cumberland* (rev. ed.) p. xcv, *Adoxa moschatellina*. Town-hall clock (Carlisle). **1968** F. Warner *Garland* 13 The red herb-Robert twined a bridge With celandine and town-hall-clocks. **1980** *Country Life* 28 Feb. 589/3 The countryman's name for the four-faced pale green woodland flower moschatel is..'Town Hall Clocks'.

town-house, town house. Add: **1. b.** (*a*) (Earlier and further examples.)
c **1870** in *Dict. Amer. Eng.* (1944) IV. 2341/2 *Town-house*, an almshouse.—Conn. **1889** R. Cooke *Steadfast* 28 Just as soon as the road settled she should 'cart her off to the town-house'.
2. a. (Earlier example.)
1771 Smollett *Humph. Cl.* I. 185 He has his town-house, and his country-house, his coach, and his post-chaise.
b. = *terrace house* s.v. *terrace *sb.* 7. orig. *U.S.*
Often in multiple units designed in a stylish or adventurous manner.
1965 *Daily Progress* (Charlottesville, Va.) 13 July 13 The City Planning Commission..is to receive a proposed ordinance permitting the development of privately owned town houses. **1968** *Globe & Mail* (Toronto) 17 Feb. 5/3 It would include 1,800 dwellings comprising apartments, maisonettes and townhouses for 5,600 people. **1971** *Rand Daily Mail* (Home Owner) 27 Mar. 7/2 City dwellers are gravitating towards high density living (flat complexes, town houses). **1971** *Ideal Home* Apr. 69/2 Something become apparent is that..the modern terrace, even under the pretty name of 'town house', is not popular. **1977** *Telegraph* (Brisbane) 28 Oct. 49/3 The townhouse is a two-storey 'unit' which features a separate courtyard and more privacy than a home unit. **1977** *Detroit Free Press* 11 Dec. 1-B/2 One east side developer,..would like very much to build a community of townhouses along one of the canals that leads to the Detroit river. **1982** *Habitat Catal.* 1982/83 28/3 Room 2 has a 'metropolitan' style associated with townhouse living.

townie: see Towny *a.* and *sb.* in Dict. and Suppl.

town-major. Add: **d.** An officer responsible for liaison between troops stationed in a town and the townspeople. (No longer current.)
1917 A. G. Empey *Over Top* 312 *Town major*, an officer stationed in a French town or village who is supposed to look after billets, upkeep of roads, and act as interpreter. **1919** A. P. Herbert *Bomber Gipsy* 19 Town-major jobs that break men's hearts, and billets at the Base.

townscape. Delete *nonce-wd.* and add: **1.** (Later examples.)
1959 *Sunday Times* 18 Jan. 16/8 These townscapes display, in short, the internal contradictions which also mark this painter's portraits of the aged rich. **1962** *Listener* 4 Oct. 515/2 The tranquillity of his land- and town-scapes.
2. The arrangement and overall appearance of the buildings, spaces, and other physical features of a town.
1937 *Evening News* 23 Apr. 10/4, I prefer a townscape with human figures to a landscape with trees. **1939** *Archit. Rev.* LXXXVI. 235/2 That universal Croydon towards which the townscapes of England are tending. **1953** *Ibid.* CXIV. 33 If I were asked to define Townscape I would say that one building is architecture but two buildings is Townscape. For as soon as two buildings are juxtaposed the art of Townscape is released. Such problems as the relationship between the building and the space between the buildings immediately assume importance. **1972** *Oxford Times* 27 Oct. 2/6 Mr John Ashdown, the city's conservation officer, said the monument would add to the townscape and be particularly attractive when seen by people walking up Turl Street. **1983** *Listener* 20 Jan. 27/3 The tall, usurping factor of Aurungzeb's mosque, dominating the townscape that some remember.
Hence **tow·nscaper,** one who plans townscapes; **tow·nscaping,** the planning of townscapes.
1949 *Archit. Rev.* CV. 249 Though the townscaper may welcome the contribution of street publicity to the urban scene there is one pitfall he must avoid—impropriety. **1953** *Ibid.* CXIV. 251 The townscaper's box of tricks—enclosure, escape, claustrophobia, surprise, delight, relief. **1959** *New Statesman* 16 May 686/2 All over the unbuilt ground surface of the Churchill Gardens estate there is evidence of an attitude indistinguishable from Townscaping. **1961** *Guardian* 18 Jan. 10/6 This cool new piece of townscaping.

Townsend (tau·nzend). *Physics.* The name of Sir John *Townsend* (1868–1957), Irish physicist, used *attrib.* with reference to certain phenomena and concepts related to his work on the conduction of electricity through gases, as **Townsend discharge,** a dark, low-current electric discharge in a gas that depends on an external source of ionization for its continuance.
1932 K. K. Darrow *Electr. Phenomena in Gases* ix. 293 The relative importance of the various processes of ionization and electron-expulsion..which figure in the 'Townsend discharge'. **1956** *Nature* 25 Feb. 391/1 The possibility of using the Townsend electron avalanche process in a gas in a stage-by-stage system to give a highly stable electron-multiplication factor. **1968** Romanowitz & Puckett *Introd. Electronics* xiv. 563 As the voltage is increased from zero, the dark current (called Townsend current..) increases slowly until the voltage approaches that at which the tube operates with a glow discharge. **1978** J. H. Ingold in Hirsh & Oskam *Gaseous Electronics* I. ii. 23 At extremely low currents, on the order of microamperes and less, the discharge is a Townsend discharge, with little or no visible light emanating from the discharge tube. **1982** *Nature* 28 Oct. 774/1 The main agents of discharges are eelctrons that participate in Townsend avalanches.

township. Add: **6. a.** Also in New Zealand.
1857 R. B. Paul *Lett. from Canterbury, N.Z.* iv. 72 Malvern Hills, where Mr Cass thinks there is a site suited

for a township. **1911** W. H. Koebel *In Maoriland Bush* xviii. 241 Half an hour later the street of the township opens out before the rider. **1977** *N.Z. Herald* 8 Jan. 4-1/5 (Advt.), From Henderson Township take Swanson Rd for 1 mile.
b. In South Africa, an area set aside for non-White occupation.
1934 *Lovedale Sol-fa Leaflet* No. 17. 4 When the Bantu Township of Nancefield or Klipspruit (eleven miles West of Johannesburg) was first settled as a Suburb of the Rand Municipality, the late Enoch Sontonga..was a teacher in one of the Methodist Mission Schools. **1946** P. Abrahams *Mine Boy* viii. 98 This side of the township had mostly Coloured people. The other side was where the native people were. **1964** L. Nkosi *Rhythm of Violence* 15 Which black township would you go to? **1971** *Sunday Express* (Johannesburg) 28 Mar. 6/1 The non-Whites..are not going to be satisfied much longer with leading third-rate lives in third-rate townships. **1984** *Observer* 9 Dec. 12/2 The flood [of people] has overflowed the inadequate African townships built by apartheid planners.
9. township trustee (earlier example).
1836 *New-Yorker* 30 Apr. 92/1 The vote (by general ticket) for Township Trustees is stated as follows.

Townsville (tau·nzvil). The name of a town on the coast of Queensland, Australia, used *attrib.* in **Townsville lucerne, stylo** [abbrev. *Stylosanthes* (see below and Stylo-)] to designate an annual or perennial leguminous plant with trifoliate leaves, *Stylosanthes humilis*, now used as a pasture plant in northern Australia and other tropical regions.
1937 *Jrnl. Council Sci. & Industr. Res.* (Australia) X. 201 The so-called wild or Townsville lucerne..was introduced accidentally into north Queensland. **1968** *Times* 23 Jan. (Austral. Suppl.) p. xiv/4 Several decades ago a plant, now called Townsville lucerne, drifted ashore from a South American ship at the port of Townsville and took root. **1977** A. V. Bogdan *Trop. Pasture & Fodder Plants* 402 *Stylosanthes humilis* was known as Townsville lucerne until about 1968–9 when the Queensland Herbage Plant Liaison Committee recommended that the name should be changed to Townsville stylo in order to avoid confusion with species of *Medicago*.

towny, *sb.* Add: **1.** In first quot. for **1828** (ed. 3) 223 read **1827** (ed. 2) 227 and add later examples.
1934 [see *bushy *sb.*]. **1942** C. Barrett *On Wallaby* i. 13 A sundowner's life was better than that of the 'townie' who had to work hard for a living and wear clean collars and shirts. **1959** I. & P. Opie *Lore & Lang. Schoolch.* iv. 62 The 'towney' touches the leaf gingerly. **1972** P. Lively *Driftway* vi. 85 He was a real townie, didn't care for walking at all. **1984** *Times* 13 Feb. 2/8 Farmers know and care far more about conservation than meddlesome townies.
2. a. (Earlier example.)
1852 *Deseret News* (Salt Lake City) 7 Aug. 1/1 'O, nothing,' replied the 'towney'.
b. *N. Amer. Circus slang.* A town-dweller, as opp. to a person travelling with a circus or carnival. Cf. *townee *sb.*
1937 [see *Reub, Rube]. **1951** *N.Y. Times Bk. Rev.* 8 Apr. 7/5 A fight [of carnival workers] with the townies. **1971** *Islander* (Victoria, B.C.) 19 Dec. 6/4 Everything had been set up for the show and tickets were being sold when several 'townies' attempted to crash the gate.
3. (Earlier example.)
1834 *Knickerbocker* IV. 279 Five or six fellows, whom I knew were friends and 'townies' of his.

tow-path. (Earlier example.)
1788 G. Washington *Diary* 2 June (1925) III. 361 A tow path on the Maryland side.

toxaphene (tǫ·ksăfīn). *Chem.* [f. Tox-[1] + *-a-* + *cam*)*phene* s.v. Camph-.] Chlorinated camphene used chiefly as an insecticide for pests of crops and livestock.
1947 *Jrnl. Econ. Entomol.* XL. 79/1 The chlorinated bicyclic terpene, now designated *Toxaphene* is an insecticide developed cooperatively by Hercules Powder Company and University of Delaware entomologists. **1975** *Nature* 9 Oct. 475/2 Over the past ten years toxaphene has been used in the USA in larger quantities than any other insecticide.

toxi-. Add: **toxige·nic** *a.* [*-genic] = *toxicogenic* adj. s.v. Toxico-; so **toxigeni·city,** toxigenic property.
1930 J. A. Arkwright in *Syst. Bacteriol. in Relation to Med.* (Med. Res. Council) I. xi. 344 The new form present may be a variant which makes less growth than a more serviceable strain, or may be less toxigenic though equally prolific. **1979** *Nature* 8 Feb. 453/1 Toxigenic strains of *E. coli* elaborate two types of toxins. **1929** Topley & Wilson *Princ. Bacteriol. & Immunity* II. xlii. 626 The same association of phenomena, which occurs in the case of virulence, may hold in the case of toxigenicity. **1977** *Lancet* 19 Mar. 649/2 A half antitoxin Naglar plate was prepared..for use in toxigenicity testing with specific antiserum.

toxic, *a.* Add: **2. a.** *toxic shock syndrome*, an acute bacterial illness observed esp. in women using tampons, characterized by fever, vomiting, diarrhoea, muscle pain, and some peeling of the skin, and in severe cases followed by shock.
1978 Todd & Fishaut in *Lancet* 25 Nov. 1117/2 The acute illness we have described and called the toxic-shock syndrome seems to affect older children. **1982** *Brit. Med. Jrnl.* 29 May 1586/1 There is no justification at present for any suggestion that women should avoid using tampons, since the risk of developing toxic shock syndrome is extremely small.

toxicant, *a.* and *sb.* Restrict *rare* to the adj. and add further examples of the sb.
1951 *Ann. Rev. Plant Physiol.* I. 311 The injury [to vegetation] has economic significance only in 'heavy' smog when..the concentration of the toxicants is raised materially above the usual 'smog' level. **1982** *Nature* 14 Jan. p. xvii/3 The system measures the effect of toxicants on the light output of a special strain of luminescent bacteria.

toxin. Add: **b.** *toxin-antitoxin* (used *attrib.*).
1904 [see Toxin b]. **1910** Hiss & Zinsser *Text-bk. Bacteriol.* xiii. 204 This work..showed that the element of time entered into the toxin-antitoxin reaction, just as it enters into reactions of known chemical nature. **1923** *Daily Mail* 16 Feb. 5/2 Since May in that borough 250 children have been tested and protection has been conferred on 70 by injection of the toxin-anti-toxin mixture. **1951** Whitby & Hynes *Med. Bacteriol.* (ed. 5) vi. 70 Toxin-antitoxin mixtures become highly toxic if frozen. **1975** E. Nnochiri *Med. Microbiol. in Tropics* iii. 43/1 Large visible floccules may be produced following a toxin-antitoxin reaction.

toxo-[2]. Add: **toxophore** *a.* (earlier and later examples); also as *sb.*, a toxophoric group; [ad. G. *toxophor* adj. (P. Ehrlich 1898, in *Deutsch. med. Wochenschr.* 22 Sept. 599/2)].
1899 Toxophore [see *haptophore]. **1951** Kirk & Othmer *Encycl. Chem. Technol.* VII. 121 In World War I, derivatives of trivalent arsenic received considerable attention, the structure —As= being considered a toxophore.

toxocara (tǫksokæ·ră). *Vet. Sci.* [mod. L. (C. W. Stiles 1905, in *Bull. Bur. Animal Industry* (U.S. Dept. Agric.) No. 79. 150), f. Toxo-[1] + Gr. κάρα head.] **a.** A nematode worm of the genus of this name, which includes species parasitic in cats and dogs; also in *pl.* sense. **b.** = *toxocariasis.
1940 A. C. Chandler *Introd. Parasitol.* (ed. 5) xxii. 388 Toxocara males have a small finger-like process at the tip of the tail. **1962** *Lancet* 6 Jan. 35/1 Duguid has now recorded 28 cases of children with retinal granuloma endophthalmitis due to toxocara. **1968** *New Scientist* 4 Apr. 41/1 The nematode is toxocara, of which there are two important species. **1976** *Milton Keynes Express* 16 July 1/2 Toxocara, a 'rare and horrific' disease caught from dog-dirt, has infected a Bletchley boy.
So **toxoca·ral** *a.*; **toxocari·asis** [*-iasis], infection with *Toxocara* nematodes.
1930 E. C. Faust *Human Helminthol.* 613/2 (Index), Toxocariasis. **1966** *Arch. Dis. in Childhood* XLI. 222/1 The diagnosis of toxocariasis is essentially a clinical one. **1968** *Brit. Med. Jrnl.* 16 Mar. 677/2 Persistent eosinophilia, hepatomegaly, choroiditis, and pulmonary infiltration are mentioned together or separately as indicating possible toxocaral infection. **1976** *Ibid.* 19 June 1486/2 Of human toxocariasis virtually nothing was known until about 12 years ago. **1981** *Ibid.* 18 July 192/1 A case of arthritis and arthralgia associated with toxocaral infestation.

toxophily. (In Dict. s.v. Toxophilite.) For *nonce-wd.* read *rare* and add later examples.
1970 M. Gilmore *World Away* 75 Mervyn had become fascinated by toxophily, and he returned with very beautiful equipment for us all. **1983** *N.Z. Listener* 19 Nov. 67 *Toxophily*,..a very exclusive and fancy word for the shooting of bows and arrows.

toxoplasma (tǫksoplæ·zmă). *Zool.* Pl. **-plasmata.** [mod.L. (coined in Fr. by Nicolle & Manceaux 1909, in *Compt. Rend.* CXLVIII. 371), f. Toxo-[1] + Plasma.] A micro-organism of the genus of this name, which comprises crescentic uninucleate sporozoans that are parasites of vertebrates. Also in pl. sense.
1926 C. M. Wenyon *Protozoology* II. ii. 1042 Mayer..discovered a parasite which appeared to be a toxoplasma in the spleen and liver of a bird. *Ibid.*, Though some of the toxoplasmata may be merozoites of hæmogregarines or coccidia, this cannot apply to such an organism as *Toxoplasma gondii*. **1937** *Science* 2 Apr. 336/1 Toxoplasma have been described as the causative agents of various pathologic conditions in birds and mammals, including man. **1962** *Lancet* 6 Jan. 23/2 Only when a woman is initially infected with toxoplasma during pregnancy can she pass the infection to her fœtus. **1973** *Times* 31 Oct. 14/3 It is thought that toxoplasma infection may occasionally lead to abortion, or to some cases of mental abnormality in the child.
Hence **toxopla·smic** *a.*; **toxopla·smin** [-in[1]], an antigenic preparation of toxoplasma; **to:xoplasmo·sis** [-osis], infection with or a disease caused by toxoplasma, which may vary from symptomless to fatal.
1934 *Biol. Abstr.* VIII. 972/2 This is the first observation of incidence of toxoplasmosis in canary birds in Argentina. **1937** *Science* 2 Apr. 337/2 It is not known how toxoplasmic infection is transmitted in nature. **1948** J. K. Frenkel in *Proc. Soc. Exper. Biol. & Med.* LXVIII. 639/2 The preparation of toxoplasmin, a skin testing antigen made of toxoplasma has been described. **1962** *Lancet* 6 Jan. 23/2 At the end of the third month of pregnancy, all women should be tested by the toxoplasmin skin test, which is cheap and simple. **1971** *Physics Bull.* July 409/2 Recurrent toxoplasmic chorioretinitis, leading to satellite lesions about a scar caused by cysts, presents a serious ocular threat. **1977** *Rolling Stone* 24 Mar. 24/5 Roy Harper, popular British singer and guitarist, almost had to delay a tour of Britain when he caught a rare (in humans), flulike virus called toxoplasmosis while giving mouth to mouth resuscitation to a pregnant sheep.

toy, *sb.* Add: **II. 7. c.** (Earlier example.)
1826 *Sessions Papers* 21 Sept. 546/2 James Boyce..said 'The b—g—r has got no *toy*'; I had no watch.
d. *U.S. slang.* A small tin or jar containing opium; the quantity of opium held in such a container.
1934 *Detective Fiction Weekly* 21 Apr. 114/1 Toy, small receptacle for opium. **1951** *Suggestions for Teaching Nature & Effects of Narcotics* (U.S. Board of Education) 9 It [*sc.* opium] is usually sold in round tin salve containers, about the size of a five-cent piece, and is known as a 'toy'. **1955** *U.S. Senate Hearings* (1956) VIII. 4161 The containers thereof are known as 'toys' (small jars or like containers). **1961** *Dissent* VIII. 349 Opium itself is often available. However, it is expensive ($15–20 for a *toy*, a ball about the size of a large pea).
9. a. (Earlier example.)
1876 *All Year Round* 15 Jan. 377/1 'Toys' repose on velvet cushions.
b. (Earlier example.)
1855 [see *hyacinth 3 b].
10*. *pl.* At Winchester College, a bureau or desk; hence, a cubicle used as a study.
1816 *Hist. Colleges Winchester, Eton & Westminster* 43 Besides his scob, every boy has, in the chamber to which he belongs, another receptacle for his books, with convenience for writing, &c. denominated, in the language of the place, Toys. **1901** *Public School Mag.* VII. 158/1 A series of small compartments, semi-secluded, but answering in their way to private studies. Each of these little dens is known as 'Toys'. **1974** K. Clark *Another Part of Wood* ii. 74 We all sat in the same large enclosure, round the walls of which were small partitions (known as toyes) like uncomfortable polling booths, with just enough room for two shelves, one to serve as a seat and the other as a desk.
III. 11. a. *toy pistol.*
1883, 1889 [see *amorce]. **1978** N. Freeling *Night Lords* ii. 11 The bandits..were pathetic imbeciles armed with toy pistols.
c. (Earlier example.)
1806 M. Lewis *Jrnl.* 1 July in *Orig. Jrnls. Lewis & Clark Expedition* (1905) V. 178 [Barking squirrels] will generally set and bark at you.., their note being much that of the little toy dogs.
d. toy book *U.S.*, a children's book; **toy-getter** (earlier example); *toy-getting* (example); **toy soldier,** a small model of a soldier; also *fig.*; **toy theatre,** a miniature theatre in which the characters are represented by printed pictures mounted on card or wood; also *fig.*; **toy time,** at Winchester College, time allocated for work in toys (see sense 10* above). See also *toytown.
1801 M. L. Weems *Let.* 10 Mar. in E. Skeel *M. L. Weems* (1929) II. 177, I sell the Primers & toy books wholesale at great discount. **1865** Toy book [in Dict.]. **1879** *Macmillan's Mag.* Oct. 502/1 The following people used to go in there—toy-getters (watch-stealers), mags-men [etc.]. **1896** A. Morrison *Child of Jago* xxiv. 239 The gains of the toy-getting trade were poor, except to the fence. **1850** Dickens in *Househ. Words* Extra Christmas No. 291/2 The lazy-tongs that used to bear the toy soldiers. **1922** M. Arlen *Piracy* iii. xi. 232 Poor Hugo..has gone clucking back for to be a toy soldier at Aldershot. **1980** *Listener* 19 June 796/1 A shopful of toy soldiers cast from the same lead mould. **1850** Dickens in *Househ. Words* Extra Christmas No. 292/1 Out of this delight springs the toy theatre,..with its familiar proscenium, and..boxes. **1931** A. C. Ward *Found. Eng. Prose* iii. 98 Stevenson loved to play with toy-theatres, and all his novels, with one exception, are reflected through the toy-theatre temperament: life is not in them. **1978** A. & P. Miall *Victorian Christmas Bk.* 30 The toy theatre..was similar to the kind..still being made by Pollocks of London. The printed figures and scenery were cut out and applied to wooden backings. **1881** W. H. David in C. E. Pascoe *Everyday Life in our Public Schools* 84 The clock marking 7, each junior retires to his 'toys' or bureau, for an hour and a half—during what is known as 'toy-time', when the work of the next morning and the week's composition have to be prepared. **1901** *Public School Mag.* VII. 158/1 Thus we find that from seven o'clock to half-past eight is 'toy time'.

toyable (toi·ăb'l), *a. nonce-wd.* [f. Toy *v.* + -able.] Fit for toying with.
1922 Joyce *Ulysses* 213 Phedo's toyable fair hair.

Toynbeean (toi·nbīăn), *a.* [f. the name of Arnold Joseph *Toynbee* (1889–1975), British historian + -an.] Of or pertaining to Toynbee, his style, or his theories of the rise and decline of civilizations.
1954 W. K. Hancock *Country & Calling* viii. 221 Even when he has attained these two virtues of attachment and justice, the historian still needs a third, which I call *span*. This resembles the Toynbean quality of remoteness. **1962** *Listener* 8 Feb. 240/1, I would translate your words into Toynbeean English, namely an answer to a challenge. **1966** *New Statesman* 18 Feb. 230/3 A remark..that 'the mature societies of Europe..in numerous ways..responded to the challenge' has the authentic Toynbeean stamp. **1975** *Asian Affairs* LXII. 239, I speak of dominance in the Toynbeean sense of a dominant civilization.

toy-shop. Add: **2.** (Earlier example.)
1796 *Boston Directory* 232 Butler, Mary, crockery and toy shop.
3. (Earlier example.)
1813 *Theatrical Inquisitor* II. 124 Her arms..drop inanimate like the..limbs of a toy-shop harlequin.

toytown (toi·taun), *sb.* (and *a.*) Also **toy-town.** [f. Toy *sb.* + Town *sb.*] A model of a town used as a plaything; *fig.* a small or insignificant town; also (with capital initial) the name of a town featured in a series of books and radio plays for children by S. G. Hulme Beaman (1887–1932). Also *attrib.* or as *adj.*
1836 [see Toy *sb.* 11 a]. 1864 R. Browning *Let.* 22 Aug. in G. R. Hudson *Browning to his American Friends* (1965) 145 A toy-town with boulevards traced through the sand-hills. 1897 'S. Grand' *Beth Book* (1898) xxiii. 207 The place..[had] a look of having been..set in order like a toy-town. 1928 S. G. H. Beaman *Tales of Toytown* 53 'Did you tell him I am busy?' the Mayor asked, laying down the copy of the *Toytown News* he had been reading. 1941 *Sun* (Baltimore) 27 Nov. 10/7 The demonstration, in the [chemical warfare] school's 'toytown' buildings which simulate actual city conditions, is a regular part of the two-week course at the school. 1964 W. Markfield *To Early Grave* (1965) xi. 191 Platters of marzipan cookies shone with a toytown brilliance. 1971 R. Falkirk *Chill Factor* iv. 38 Austurvollur Square was still toytown with the little white Lutheran Cathedral. 1972 *Daily Tel.* (Colour Suppl.) 27 Oct. 19/4 The slums of Kingston are horrendous: hovels of cardboard and plywood, tiny packing-case houses like a stricken toytown. 1972 'S. Woods' *They love not Poison* vii. 97 She..was..listening to a Toytown play on the Children's Hour. 1973 *Times* 15 Nov. 25/6 The Treasury are also fairly unimpressed by it; they refer to it internally as 'toytown money'. 1979 *Theatre Australia* Apr. 30/1 A *faux naif* toytown set of kitchen cupboard colours clashed dismally with furniture. 1984 *Times* 13 Mar. 17/1 This toytown situation became the occasion for a number of serious-looking people (all men, as it happened) in serious-looking suits to respond in a serious way to the questions of a sombre moderator.

trabacolo. (Earlier example.)
1800 E. C. Knight *Let.* 9 Aug. in *Autobiogr.* (1960) 221 Had we sailed, as was first intended, in the imperial [Russian] frigate, we should have been taken by eight trabaccoli, which the French armed on purpose at Pisaro.

trabant. Restrict 'Now chiefly *Hist.*' to sense in Dict. and add: **2.** *Cytology.* = *satellite *sb.* 9.
1926 C. D. Darlington in *Jrnl. Genetics* XVI. 248 A portion thus narrower than the main body of the chromosome seems to require the name of satellite or trabant; such an element, having an attraction for the parent body proportionally less than a larger element, is naturally more subject to external forces, hence the common appearance of flying out. 1967 C. P. Swanson et al. *Cytogenetics* ii. 26 The region of the chromosome distal to the nucleolar gap is called a trabant or satellite. 1980 *Caryologia* XXXIII. 207 In three individuals we observed different thickness of the intercalary trabant.

trac (træk). *Basketry.* [Etym. unknown.] In full *trac border.* A basketwork border made by taking the remaining length of an upright and weaving it in and out of the following uprights before repeating the process with the next.
1924 C. Crampton *Cane Work* 13 *Back trac*, an additional border worked with the remaining ends of a three-rod plain border. 1959 D. Wright *Baskets & Basketry* ii. 57 A Foot-Border... This is a trac border used for securing stakes to a wooden base with holes in it... Other tracs may also be used. 1964 H. Hodges *Artifacts* x. 146 In the trac border one stake at a time was bent and woven in completely.

trace, *sb.*¹ Add: **6. a.** Also *to sink without trace*: see *sink *v.* 1 a.
c. Also in *Meteorol.* (see quot. 1930).
1908 *Observer's Handbk.* (Meteorol. Office) I. 35 Falls [of rain] of less than ·005-inch should be noted in the register by entering the word 'trace'. 1930 *Meteorol. Gloss.* (ed. 2) (Meteorol. Office) 177 The word 'trace' is entered in the daily record sheet when some rain (or other form of precipitation) is known to have fallen and the amount in the gauge is not large enough to be measured. 1974 *Nature* 25 Oct. 694/2 The measurements were made in very light snowfall (which never exceeded a 'trace' in equivalent precipitation rate).
d. *Psychol.* A change in the brain as a result of some mental experience; the physical after-effect *of* such.
1690 Locke *Essay Hum. Und.* II. i. 41 The memory of Thoughts, is retained by the impressions that are made on the Brain, and the traces there left after such thinking. *Ibid.* x. 67 There is no reason why the sound of a Pipe should leave traces in their [*sc.* birds'] Brains. 1892 G. F. Stout *Man. Psychol.* I. I. ii. 76 Mental development would be impossible unless previous experience left behind it persistent after-effects to determine the nature and course of subsequent experience. These after-effects are called..traces or dispositions. 1927 G. V. Anrep tr. *Pavlov's Conditioned Reflexes* iii. 39 The stimulus this time is not the actual disappearance of an external agent, but the trace left by the action of the agent on the central nervous system after the agent itself has been removed. 1930 W. Köhler *Gestalt Psychol.* ix. 232 Learning and those processes the traces of which make reproduction and recognition possible. 1940 *Brit. Jrnl. Psychol.* XXX. 193 This process..works in conjunction with and through the trace-column, the masses of traces, ultimately of a chemical nature, left by past experience. 1978 Tarpy & Mayer *Found. Learning & Memory* ii. 22 In trace conditioning, the CS does not impinge directly upon the sense receptors.
e. *Linguistics.* In transformational grammar, a phonetically null element considered to have been left in the position from which another element has been moved by a transformation, and to retain some influence on the resultant sentence.
1975 N. Chomsky *Logical Struct. Linguistic Theory* 22 Transformations that move expressions leave a 'trace' in the position from which the item was moved. 1977 *Stud. in Eng. Lit.* Eng. Number (Tokyo) 95 Traces make it possible to define permissible transformations correctly. 1978 *Language* LIV. 412 S-initial sentential complements are base-generated in topic position,..and are linked to an empty subject position (actually a trace in subject position)..by a general rule of interpretation.
8. b. The luminous line or pattern on the screen of a cathode-ray tube.
1937 G. Parr *Low Voltage Cathode Ray Tube* ii. 28 The effect on the trace on the screen is..to break up the line into a series of light and dark patches. 1966 D. Bagley *Wyatt's Hurricane* vi. 156 He blinked them open again and stared at the radar screen, following the sweep of the trace as it swept hypnotically round and round. 1975 D. G. Fink *Electronics Engineers' Handbk.* vii. 39 This trace is displayed continuously until erased, so long as the flood beam is maintained in operation.
11. *Math.* The sum of the elements in the principal diagonal of a matrix.
1938 A. A. Albert *Mod. Higher Algebra* iv. 80 We call *T*(*A*) the trace of *A*. 1958 *New Scientist* 10 July 364/2 If A is a matrix it is usual to denote the transpose of A by A′ or A^T and the trace of A by tr A. 1972 *Jrnl. Physics* B. V. 990 Evaluating the constant of proportionality by taking the trace of each side of the result for the particular case of complete recapture.
12. a. *Computers.* The detailed examination of the execution of a program or part of one (usu. to investigate a fault) with the aid of another program that can cause individual instructions, operands, and results to be printed as they are reached by the first program; the analysis so obtained; also, a trace program. Freq. *attrib.*, as *trace program, routine.*
1957 M. V. Wilkes et al. *Preparation of Programs for Electronic Digital Computer* (ed. 2) 96 A useful error-diagnosis subroutine..prints the function letters of orders as they are executed... The printed sequence of function letters is sometimes known as a trace. 1960 Gregory & Van Horn *Automatic Data-Processing Systems* iii. 82 (*caption*) Trace of operations in read-write loop using an index register. *Ibid.* viii. 271 A trace routine is used to observe how the object program..operates while it is being executed. 1966 Trace program [see *snapshot *sb.* 2 b]. 1980 N. Rushby in Meek & Heath *Guide to Good Programming Practice* iii. 64 Some debugging compilers provide a trace, which can be used simply to follow the program flow from statement to statement, or can include details of each assignment. 1982 Ghezzi & Jazayeri *Programming Language Concepts* vii. 221 After the program has terminated, a trace and an indication of the cause of failure can be produced.
b. A request for information to be sought concerning a particular person or thing; an investigation which traces this information (freq. to discover the source of a telephone call).
1974 M. Fenovski *Breach of Security* i. 4, I might put in a trace to London to see if any Cities has got anything on him. 1977 S. Seymour *Glory Boys* iv. 49 Very professional. Say possibility of a trace by the time they'd left. 1978 R. Ludlum *Holcroft Covenant* xxviii. 311 There are men following me. I think it's called a 'trace'. Put out by you? 1981 D. Boggis *Time to Betray* xii. 114 He..got the index number... 'Get me a trace through Yard liaison.'
II. 13. *attrib.* or as *adj.* Present or required only in traces.
1956 *N.Z. Jrnl. Agric.* Sept. 195/1 The control..of peat scours by copper in trace amounts. 1956 *Sun* (Baltimore) 1 May 12/7 The oysters..are loaded with trace minerals. 1962 *Listener* 16 Aug. 243/2 It is the selected drug that is actually the effective agent and not trace amounts of some as yet unrecognized contaminant. 1965 G. J. Williams *Econ. Geol. N.Z.* vi. 63/2 To place these figures in genetic perspective we must realize that gold is a trace-metal—even where concentrated in a payable reef. 1978 *Sci. Amer.* Dec. 124/1 It is the trace ions (those at least 1,000 times less abundant than hydrogen and helium) that serve to control the nebula thermostatically, maintaining the nebular temperature generally between 5,000 and 15,000 degrees. 1979 *Brit. Jrnl. Psychol.* 15 Dec. 1529/1 Most trace minerals and other micronutrients required for survival are known.
b. Special Comb.: **trace element,** an element that is present (esp. in the soil) or required only in minute amounts; also *fig.*; **trace fossil** *Palaeont.* [tr. G. *spurenfossil* (R. Krejci-Graf 1932, in *Senckenbergiana* XIV. 21)], a fossil that represents the burrowing or similar activity of an animal rather than the animal itself.
1932 *Yale Jrnl. Biol. & Med.* IV. 501 Investigation as to the occurrence and function of 'trace' elements in both plant and animal life is now very active. 1954 R. L. Parker tr. *Niggli's Rocks & Mineral Deposits* i. 9 Fundamental rock chemistry need consider only comparatively few elements... This does not mean that relatively rare elements or even the so-called trace elements lack importance. 1970 *Nature* 17 Oct. 251 Trace element analyses of flint show statistically valid differences between products of major British and European Neolithic flint mines. 1976 *Church Times* 16 July 7/2, I found little that has not been said already by many radical Christian writers. There are trace-elements of Marxism, but hardly more than that. 1980 *Q. Jrnl. Geol. Soc.* CXII. 475 A consideration of the..morphology of the fossil, permits its interpretation as a trace-fossil resulting from the driving of a system of branching tunnels in the sea-bed sediment. 1974 *Nature* 22 Mar. 328/2 It is generally agreed that the earliest metazoan animals were soft-bodied forms which are rarely preserved but have left tracks, trails and burrows, collectively known as trace fossils.

trace, *v.*¹ Add: **6. b.** (Earlier example.)
1876 *Rep. Vermont Board Agric.* III. 107 The farmer loses sight of the fact that the..calf..may 'trace back', as it is termed, to a remote ancestor.
7. c. To make a tracing of (a listed item); to derive (a tracing) *from* an index or catalogue; see *tracing *vbl. sb.*¹ 1 b.
1905 *N.Y. State Library Bull.* No. 95. 578 See that every secondary card is traced on one or both main cards. 1914 [see *tracing *vbl. sb.*¹ 1 b]. 1926 *Amer. Speech* II. 93 The catalog cards are 'main entry' cards and 'secondary entry' cards, the latter being 'traced' from the former.
8. c. *Computers.* To subject (a program) to a trace (*trace *sb.*¹ 12 a).
1959 M. H. Wrubel *Primer of Programming for Digital Computers* v. 107 When a program is traced, the machine produces a record of each instruction as it is performed. 1967 Klerer & Korn *Digital Computer User's Handbk.* I. 3. 33 The location limits of the program segment to be traced enter as initial parameters to the trace program. 1982 L. A. Hill *Structured Programming in FORTRAN* iii. 73 The program is traced in Table 3-6 with Rule 4 relaxed.

trace, *v.*² Add: **1.** Also with *up*.
1884 *Vermont Agric. Rep.* VIII. 285 The ears thus selected would be 'traced up' and hung away to dry. 1943 *Old Farmer's Almanac* 79 In the early fall the farmers would speak of 'tracing up' the yellow ears of corn to hang from the beams of the woodshed.

traceless, *a.* Add: **2.** *Math.* Having a trace equal to zero.
1966 *Rev. Mod. Physics* XXXVIII. 220/1 To each representation belongs a traceless tensor. 1973 *Nature* 14 Sept. 78/1 A view of the Earth from Polaris, with W representing the pole of the traceless part of the rotation tensor. 1977 J. C. Polkinghorne *Particle Play* iv. 63 There is a triplet representation of SU(3) which is called the fundamental representation because all other representations..can be constructed by mathematical manipulations on these three fundamental objects. The mathematical operations involved are direct products, symmetrizing, and making traceless.

tracer¹. Add: **2.** (Further examples.) **c.** A substance (as a radioactive isotope or a dye) with distinctive properties that is introduced into a system so that its subsequent distribution may be readily followed. Freq. *attrib.*
1938 *Encycl. Brit. Bk. of Year* 220/1 The use of deuterium as a tracer in biochemical studies has been important. 1946 *Nature* 12 Oct. 537/1 The attempt..to correlate by radioactive-tracer techniques the localization of heavy metals in the body and their chemotherapeutic activity. 1952 *New Biol.* XIII. 63 One method..involves injecting into such mammals as sheep and rats a very minute dose [called a 'tracer dose'] of the isotope. 1960 P. Daudel tr. *Eisner's Radioactive Tracers in Chem. & Industry* v. 164 The Russian workers..have investigated the metabolism of nucleotides and carbohydrates by ion exchange using labelled inorganic phosphate as a tracer. 1963 L. C. Pickard *Descr. Physical Oceanogr.* vi. 81 Radioactive materials seem attractive as tracers of water movement... A very convenient artificial tracer is the dye fluorescein. 1976 *Sci. Amer.* Apr. 130/1 (Advt.), Using beryllium-7 as a tracer of stratospheric ozone, our scientists found that such ozone is distributed throughout high pressure weather systems. 1977 *Nature* 26 July 299/1 Sparticles have been used as tracers to determine the rotation rate of the sun.
4. a. Bullets or shells whose course is made visible by the trail that they emit during flight; occas. in *sing.* sense. Orig., the trail produced by these.
1910 *Blackw. Mag.* July 62/2 The projectiles of airship guns may possibly give out a jet of flame and a smoke tracer during flight. 1922 *Encycl. Brit.* XXX. 120/2 For night use, the tracer shows a luminous spark, for day use the tracer gives a smoky trail. 1939 *Times* 16 Apr. 8/6 This was a most spectacular demonstration, the machine-guns using tracer and the new smoke observation projectiles. 1957 P. Kemp *Mine were of Trouble* iv. 173 A minute later bursts of tracer flew over us from high ground on our right. 1967 *Boston Sunday Globe* 23 Apr. 16/4 Helicopter gunships tried to protect the other busy helicopters by circling in pairs, one with a light on to draw a stream of enemy tracers. 1970 L. Deighton *Bomber* xxiii. 335 He was in the nose watching ropes of red and yellow tracer curve towards them and fall away. 1983 'W. Haggard' *Heirloom* xv. 169 He'd seen..appeals for death..that airman with tracer burning his lungs out.

b. *attrib.* and *Comb.*
1916 'Taffrail' *Pincher Martin* xv. 278 A thin trail of dim light climbed skywards in a curve as a tracer shell hurtled its way through the air. 1918 'Boyd Cable' *Air Men o' War* 22 Tracer bullets emit smoke and flame to allow the shooter to follow their flight. 1928 C. F. S. Gamble *Story of North Sea Air Station* xii. 179 When about 1,500 feet below the airship, he fired two trays of explosive and tracer ammunition from his Lewis gun into her. 1943 *Sun* (Baltimore) 3 Aug. 4/6 During this exchange of fire, a Japanese plane had managed to get on the tail of Captain Walter's Warhawk. Tracer bullets were flying past him, but none hit. 1944 *Return to Attack* (Army Board, N.Z.) 18/1 The flash and crack of the high-velocity tank guns, the low parabola of the tracer shell. 1969 G. Macbeth *War Quartet* 40 Tracer-filled In open air-space. 1973 M. Woodhouse *Blue Bone* xii. 135 A machine-gun stammered and tracer bullets began to draw graceful curves in space. 1976 A. White *Long Silence* vii. 58, I saw a lone fighter come in from the west... Sudden streams of tracer fire came from him.

tracery. Add: **2.** *stump tracery*: see Stump *sb.*¹ 18.

tracheide. Add after **-id**: (the usual spelling).
1907 D. P. Penhallow *Man. N. Amer. Gymnosperms* vi. 88 Such tracheids are invariable features of the ray in all the higher Coniferæ. 1910 J. M. Coulter et al. *Textbk. Bot.* I. iv. 241 Tracheids are single cells thus formed. 1948 [see *collapse *sb.* 4]. 1974 *Sci. Amer.* Apr. 59/1 Tracheids predominate in softwoods, which have no vessel cells or libriform fiber cells.

tracheostomy (trē¹ki,ǫ·stōmi). *Surg.* [f. Tracheo- + *-stomy.] **a.** The operation of making an opening in the trachea near its upper end, so that the patient can breathe through it; also, the opening so made.
1945 W. V. Mullin in F. Christopher *Textbk. Surg.* (ed. 4) xix. 774/2 Tracheostomy may be necessary to facilitate the operation. 1961 *Lancet* 7 Oct. 819/2 He spoke of the need for asepsis in managing a tracheostomy, especially during cleaning and replacement of the tube. 1976 *National Observer* (U.S.) 18 Dec. 16/4 Siegel underwent a tracheostomy 2½ months ago. 1977 *Proc. R. Soc. Med.* LXX. 160/1 He was then able to breathe spontaneously and the tracheostomy was allowed to close.
b. **tracheostomy tube,** a curved tube which can be inserted into the trachea via a tracheostomy.
1961 *Lancet* 7 Oct. 819/2 There was less chance of stenosis of the trachea after removal of the tracheostomy tube. 1977 *Ibid.* 19 Mar. 636/2 Indwelling urinary catheters, endotracheal tubes, and tracheostomy tubes put the patient at special risk.

tracing, *vbl. sb.*¹ Add: **1. b.** The procedure of making a list of all the headings under which a given item occurs in an index or catalogue; an entry in such a list.
1905 *N.Y. State Libr. Bull.* No. 95. 582 Make slip under personal name as in 14a, following same method of tracing. 1914 *N.Y. State Libr. School Cataloging Rules* 32 Trace added entries on the back of the main card. Write the tracing for other cards toward what will then be the lower right corner. 1953 R. L. Collison *Indexes & Indexing* I. 73 Since revision [of an index]..may sometimes be necessary, it is a good policy to enter 'tracings' of any references made on the main slip. 1978 Fothergill & Butchart *Non-Book Materials for Libraries* iv. 191 Then the primary name heading and tracings for added entries and references can be given.
c. The following of the course of the cutting stylus by a reproducing stylus; usu. in *tracing distortion*, distortion that occurs when the stylus does not describe exactly the same path as the groove owing to its size in relation to the groove.
1942 *Jrnl. Acoustical Soc. Amer.* XIII. 276/1 (*heading*) Tracing distortion in the reproduction of constant amplitude recordings. 1959 *Listener* 26 Mar. 542/1 Turning now to the gramophone record, we find a sound source which suffers from certain well-known inherent defects, among the most important being..tracing distortion. 1961 G. A. Briggs *A to Z in Audio* 208 These delicately balanced arms which permit good tracking and tracing at extremely light weights give the best available quality from the finest pickups. 1975 G. J. King *Audio Handbk.* viii. 192 Incorrect adjustment of the lateral and vertical tracking of the pickup can also aggravate distortion; but that resulting from 'normal' errors is generally less than tracing distortion.
d. *Computers.* The process of performing a trace (*trace *sb.*¹ 12 a). Also *attrib.*, as *tracing routine.*
1959 M. H. Wrubel *Primer of Programming for Digital Computers* v. 107 Tracing is an important technique in testing programs, but it must be used in moderation. 1967 Klerer & Korn *Digital Computer User's Handbk.* I. i. 23 It is possible to build in a tracing structure into any given program. 1969 P. B. Jordain *Condensed Computer Encycl.* 539 It is important that the tracing routine leave intact the natural operation of the subject program.
5. **tracing table,** a table with a translucent top illuminated from underneath; **tracing-wheel** (examples).
1953 A. H. Robinson *Elements Cartogr.* v. 82/2 (*caption*) A tracing table with fluorescent illumination. 1978 *N.Y. Times* 30 Mar. B-21/8 (Advt.), Nuarc lighted tracing table, 5 drawer blueprint files. 1894 J. E. Davis *Elem. Mod. Dressmaking* i. 7 Tracing the fitting-lines of the pattern through the doubled lining only..is now almost entirely done by tracing-wheel. 1969 *Guardian* 30 Sept. 11/2 Singers..stock everything for dressmaking..hem markers, tracing wheels, fastenings.

track, *sb.* Add: **I. 1. a.** *spec.* in *Particle Physics*, a line marking the path taken by an atomic or sub-atomic particle.
1912 *Proc. R. Soc.* A. LXXXVII. 277 It has now been found possible to photograph the tracks of even the fastest β-particles. 1942 J. D. Stranathan *Particles Mod. Physics* i. 43 The ions formed directly by the alpha particle must all have been of very low speed; otherwise the track would not be as narrow and sharply defined. 1955 *Sci. News Let.* 12 Feb. 103/1 When exposed to the special photographic plates, particles from the radioactive samples leave a distinct pattern, known as tracks, on the emulsion. 1973 L. J. Tassie *Physics Elementary Particles* vi. 50 The forked track, *ab*, in Fig. 23.1 was due to the decay of a heavy neutral particle..into two charged particles.
d. A line on the skin made by the repeated injection of an addictive drug. Usu. *pl. slang.*
1964 H. Rodriguez in Larner & Tefferteller *Addict in Street* 34 Tracks are marks,..like a long black streak coming down your arm directly over your vein; that comes from hitting in the same place so much. 1965 *Life* 26 Feb. 86/4 In summer, they [*sc.* addicts] alone wear long sleeves (to cover their 'tracks'—needle marks). 1972 J. Mills *Report to Commissioner* 104 Whaddya mean, lemme see your tracks? I'm a pros, man, I shoot up in my thighs. 1977 *Rolling Stone* 13 Jan. 14/3 The coroner found four fresh needle marks but no tracks, indicating that Bolan was not a junkie. 1979 R. B. Parker *Wilderness* i. 10 'Junkie,' he said. The white trooper said, 'Tracks?'..The black trooper nodded, 'All up and down her right arm.'
4. d. *Aeronaut.* The projection on the earth's surface of the (actual or intended) course of an aircraft; the representation of this on a chart.
1919 S. F. Card *Air Navigation* i. 6 The straight line on the map or chart joining the two places will be called the desired track. 1943 Redpath & Coburn *Air Transport Navigation* viii. 176 Measurement of the line must give us the groundspeed, since track and groundspeed go hand in hand. 1970 Taylor & Parmar *Ground Stud. for Pilots* II. i. 13 Plot in the places carefully on the chart..and join them up, putting the two arrows on the line..to indicate the Track you wish to follow over the Earth's surface.
e. The plane in which the blades of a propeller are intended to rotate.
1920 W. E. Park *Treat. Airscrews* xii. 206 The relative position of corresponding points in opposite blades..in the side elevation is considered as the 'track' of the blade. 1948 C. E. Chapel *Aircraft Power Plants* xv. 323/2 This and the several other methods of field checking the track of the propeller are rough methods only. 1956 W. A. Heflin *U.S.A.F. Dict.* 81/2 A blade is said to be *in*, or *out of, track.*
f. = *line *sb.*² 19 c.
1931 *Flight* 23 Jan. 73/1 Each fuselage moves along the track to the next [stage]..until at the end of the track the machine is complete. 1979 *Daily Tel.* 3 Aug. 2 The jobs of about 1,000 workers will be affected at British Leyland's Rover saloon car plant..by plans to cut production tracks from three to two. 1981 B. Walsh *Live Bait* ix. 85 Me and Brian work on the track... The assembly line.
g. *U.S. Educ.* = *stream *sb.* 6 d. Usu. *attrib.*
1959 *Washington Post* 17 May E 4/2 The extension of the track system to District junior high and elementary schools ought to benefit most pupils, but the School Board should make certain that Superintendent Carl F. Hansen's 3-track plan for children below the senior high school level is as flexible as possible. 1964 B. Fine *Stretching their Minds* 19 'Acceleration' became fashionable—meaning either old-fashioned 'skipping' or the modern 'multiple-track' plan (in racing terms, a slow track for the average student, a fast track for the superior). 1968 *Economist* 7 Dec. 47/2 It ordered the end of the track system (which divided children according to academic ability, with most Negroes landing inevitably in the lowest track). 1983 *N.Y. Times* 13 Nov. XII. 71/2 There could be different rooms for learners and spurners. If a traditional track system is preferred, there could be a class for those who are on the track and another for those who have derailed.
6. a. Delete '(now *U.S.*)'. *Esp.* a single pair of rails, in contrast to a line (which may denote the route and comprise one or more tracks: cf. Line *sb.*² 26 b). Also (*U.S.*) with following number, denoting the line served by a particular platform or gate.
1869 *Bradshaw's Railway Man.* XXI. 390 The length of this line is 94 miles... Of the whole only 33 are 'double track'. 1911 *Encycl. Brit.* XXII. 820/2 The Stockton & Darlington railway... This line..was in the first instance laid with a single track. 1955 J. L. Austin *How to do Things with Words* (1962) v. 57 Passengers are warned to cross the track by the bridge only. 1967 [see *double-tracked* s.v. *double *a.* C. 1]. 1978 R. Ludlum *Holcroft Covenant* i. 12 He had learned before the announcement that the train for Zurich would leave from track twelve. 1984 *Financial Times* 27 Jan. 2/8 Yesterday, Breton farmers suspended their disruption of rail traffic and removed the barriers they had placed across the tracks in Brittany.
b. (Earlier examples.)
1836 *Spirit of Times* 20 Feb. 5/3 And he will run a match against either, or a sweepstakes with both, one, two, three, or four mile heats, over any good track in East Tennessee. 1851 *Fraser's Mag.* June 657/1 A barouche and four does not differ more from a trottingwaggon..than an English race-course from an American 'track'.
c. The distance between a wheel on one side of a vehicle and the corresponding wheel on the other side.
1850 *Western Jrnl.* IV. 96 This distance will, therefore, vary in different sections of the country according to the usual 'track' of wagons. 1910 J. Gunn *Practical Design Motor Cars* viii. 219 The wheel base and wheel track of a motor car require consideration. 1928 [see *roadability]. 1948 J. D. Rittenhouse *Amer. Horse-Drawn Vehicles* 1 The term 'track' refers to the extreme width of the vehicle as measured from outside rim of one wheel to the outside of the rim of the opposite wheel, measured at the bottom of the wheel. 1969 *Gloss. Aeronaut. & Astronaut. Terms* (*B.S.I.*) v. 12 *Track*, the distance between the outer points of contact of the port and starboard main undercarriages.
d. Each of the endless bands on certain heavy vehicles, esp. tanks, passing round and driven by wheels and facilitating travel over rough or soft ground. Cf. *caterpillar 1 b.
1884 *Patent 269,998* in *Specifications & Drawings* (U.S. Patent Office) 15 Apr. 1384/1 This invention relates to certain improvements in that class of road-engines in which the driving and pilot or guiding wheels are connected by a chain or series of links, which together form an endless track which the wheels traverse in the movement of the engine. 1926 *Encycl. Brit.* Suppl. III. 723/2 The track..was carried all round the tank; this track was driven from the engine through a two-speed gear box. 1931 G. Le Q. Martel *In Wake of Tank* 83 The whole of the engine power could be transmitted to the track on the one or the other side of the tank as desired. 1971 *Power Farming* Mar. 13/2 Before the development of the large rubber tyre it was possible to transmit high power to the soil only through tracks. 1974 'W. Haggard' *Kinsmen* viii. 82 A simple crane on a pair of tracks was well within his modest competence.
e. (Without article.) The branch of athletics in which a running track is used; track athletics, track events; *track and field* (also *attrib.*), athletics in general. orig. *U.S.*
1905 *Outing* XLVI. 490/1 Track and field sport has been working out its own spontaneous solution. 1934 T. V. Wilder *Heaven's My Destination* 66, I was captain of track and basket-ball. 1936 *Nat. Geogr. Mag.* LXIX. 799/2, I progressed the next year to my class squads in football and track. 1964 A. Wykes *Gambling* iv. 102 As for athletics, or 'track and field' sports, there are practically no places where public betting..flourishes to any extent. 1972 *N.Y. Times* 4 June 4/3 The Oregon Track Club is very active in promoting track in the area. 1978 G. A. Sheehan *Running & Being* viii. 107 Despite the detailed and accurate statistics of track and field, the scientists consistently underestimate the human body and its potential. 1979 R. Jaffe *Class Reunion* (1980) II. viii. 265 'Do you have a favorite sport?' 'Track.'
f. A ballroom or dance-hall. *U.S. slang.*
1945 L. Shelly *Hepcats Jive Talk Dict.* 19/2 *Track*, hall for dancing. 1960 Wentworth & Flexner *Dict. Amer. Slang* 553/2 The Savoy Ballroom in N.Y.C.'s Harlem was widely known as 'The Track' to hepsters. 1965 'Malcolm X' *Autobiogr.* xvi. 315, I dig your holding this all-originals scene at the track. 1972 T. Kochman *Rappin' & Stylin' Out* 163 The place where the movement can occur is appropriately termed the 'track', whether the place is a dance hall.., the street.., or, as used figuratively, the life span.
g. A metal or plastic strip designed to carry the sliding fittings from which a curtain is hung, or on which an electric (spot)light may be positioned. Cf. *track lighting* in sense 13 below.
1971 *Guardian* 18 Aug. 9/6 Curtain tracks and pelmets. 1976 *N.Y. Times Mag.* 15 Aug. 47 Installation of the track on the ceiling is tricky, and, in most instances, involves hiding the wires from the ceiling to the light switch. 1979 D. Brierley *Cold War* iv. 39, I checked plugs, sockets..the track for the curtain across the window.
h. *Cricket.* = Wicket 3 c.
1976 J. Snow *Cricket Rebel* 102 Deliveries..that pitched half way down the track and went through above head high. 1977 *Grimsby Even. Tel.* 31 May 12/6 Fast bowler Robert Herkes again gave a good account of himself and his figures of two for 33 off 16 overs was no mean feat on a track that did not give him a shred of help. 1983 *Daily Tel.* 3 Sept. 12 The commentators—particularly the professional cricketers (active or retired)—use a vocabulary peculiarly their own: the pitch is a 'track', good or bad.
8*. a. = *groove *sb.* 2 c (now *rare* or *Obs.*); hence, a single recorded item (esp. of popular music), which on a long-playing record is a band bounded on both sides by an area of widely-spaced grooves.
1904 S. R. Bottone *Talking Machines & Records* 60 We must have some means of controlling or varying the pressure of the stylus of the reproducer on the record, so as to enable it to follow correctly every indentation in the 'track'. 1949 *Playback* Oct.–Nov. 4/2 This took the form of a 10″ record with two 'tracks' or 'grooves' impressed on each side. 1956 *Gramophone* Dec. 265/1 None of the tracks lives up to the promise of the star-studded personnel. 1957 [see *band *sb.*² 9 b]. 1958 *Observer* 28 Dec. 6/7 All the tracks had been released as singles in the era of seventy-eight r.p.m. 1967 A. Diment *Dolly Dolly Spy* xii. 160, I wandered over to the juke box and selected a Dylan track for relaxation. 1970 *Honey* June 53/4 Their first LP is so polished. There are some great original tracks. 1980 *Oxford Times* 1 Feb. 23/3 On tracks like 'Rock Music' they seem to want to be a hard rock 'n' roll band... On several other tracks their aim is vague and visionary.
b. *Cinemat.* = *sound track* s.v. *sound *sb.*³ 7 b.
1931 B. Brown *Talking Pictures* x. 226 To a certain extent surface noise is due to irregularities in the sensitive film used for recording the track. 1976 *Oxf. Compan. Film* 203/1 'Dolbyized' tracks sound 'cleaner' and clearer than ordinary tracks.

c. A lengthwise strip on magnetic tape consisting of a single sequence of signals; more widely, a linear path in any information storage device or medium that accommodates one sequence of signals or corresponds to one head.
Orig. identical with prec. sense.
1947 *Jrnl. Soc. Motion Picture Engineers* XLVIII. 9 A magnetic recording track..on 16-mm film. 1951 *Audio Engin.* Sept. 40/2 The recording medium is an endless polyvinyl-chloride [magnetic] tape with 56 parallel sound tracks spaced at the ordinary rate of four tracks per millimeter. 1951 *Proc. Inst. Electr. Engineers* XCVIII. II. 29/1 As the drum rotates the surface is carried past a fixed magnetic recording and reading head... Many separate tracks can be recorded side by side. 1957 *Practical Wireless* XXXIII. 697/1 Rotation of VR1 should cause it to click in and out as a certain point on the track is passed. 1962 *Times* 5 July 15/6 Some tapes have two tracks, others (in stereo) have four, and a four-track recorder will not produce the best results on a two-track tape. 1969 P. B. Jordain *Condensed Computer Encycl.* 305 Data are addressed on a drum by specifying the track number and word number within the drum. 1970 O. Dopping *Computers & Data Processing* iii. 57 Readers and punches for paper tape can easily be adjusted to different numbers of tracks. 1977 *Time* 4 July 4 (Advt.), There's a film to watch—a recent release—8 tracks of stereo to listen to, free naturally, and plenty of room to stretch out or stroll about. 1983 *Austral. Personal Computer* Aug. 62/1 Files stored on a disk are located by means of a directory set up on a particular grouping of tracks.
9. *in one's tracks* (earlier and later examples); delete (*U.S.*); *on the right track*, having the right idea; heading in the right direction; also *on the* (or *a*) *wrong track*; *the wrong side of the tracks*, the socially inferior part of town; so *to cross the tracks* and similar phrases; *on the track* (Austral.), tramping from place to place in search of work; *on track* (U.S.), on course; achieving or doing what is required; *to comb the tracks*: see *comb *v.*¹ 4 c; *to jump the track*: see Jump *v.* 6 c.
1824 T. D. Arnold in M. James *A. Jackson* (1937) 156 He failed to shoot 'Jackson dead in his tracks'. 1873 J. C. F. Johnson *Christmas on Carringa* 19 'Tis Christmas Eve again to day, and I am on 'the track'. 1886 C. M. Yonge *Chantry House* I. xiii. 116 This had done more to convince my father that he was on the right track than the having found him on his knees. 1889 J. K. Jerome *Three Men in Boat* iii. 37 You know we are on a wrong track altogether. We must not think of the things we could do with, but only of the things that we can't do without. 1896 H. Lawson *While the Billy Boils* 207 I've been knocking round for five years, and the last two years constant on the track, and no show of getting off it unless I go for good. 1915 A. Huxley *Let.* Oct. (1969) 84 These maximal horrors of war are really too unthinkably appalling; but things I trust are on the right track now for health. [1929 T. Smith *Stray Lamb* iv. 29 In most commuting towns..there are always two sides of which the tracks serve as a line of demarcation. There is the right side and the wrong side. Translated into terms of modern American idealism, this means, the rich side and the side that hopes to be rich.] 1945 S. Lewis *C. Timberlane* (1947) xxxiv. 230, I thought at first that she was from the wrong side of the railroad tracks, but she seems to have settled down to being a nice little lady and a good war worker. 1953 'Caddie' *Sydney Barmaid* xliv. 255 It would have been impossible for him to maintain the home on a dole ration... He was going on the track. 1954 I. Murdoch *Under Net* xi. 141 What I saw as I opened the door made me stop dead in my tracks. 1956 W. H. Whyte *Organization Man* xxi. 269 The boy from Shantytown was going to have less chance than ever of crossing over the tracks. 1965 E. Lambert *Long White Night* 12 His clothes clearly proclaimed him as a man who had been on the track, one of that tattered, aimless, wandering band which the Depression threw up. 1973 *Times* 19 May 6/6 'The Government may fall,' Mr Caulfield reportedly said, complaining: 'Everybody else is on track but you.' 1977 *Listener* 13 Oct. 478/2 Eva Duarte Peron..came from the wrong side of the tracks. 1978 *Detroit Free Press* 16 Apr. F 3 (Advt.), We're looking for a professional who can keep us on track by making contributions that improve efficiency. 1978 *Time* 24 Apr. 20/2 If we can reach a salt agreement..that will begin to change the whole character of the relationship, put it on the right track again. 1979 B. L. C. Johnson *Pakistan* xiii. 199/1 The whole area has something of a 'beyond the tracks' character about it. 1984 *Gainesville* (Florida) *Sun* 3 Apr. 10 B/5 Three weeks ago, Mondale won the Illinois primary and said his comeback was on track.
III. 13. (sense *6 e) *track coach, event* (examples), *meet* (*U.S.*), *shirt, shorts, team*; also *track-mounted* adj.; **track-bed** = Bed *sb.* 12 e; **track circuit,** an electric circuit formed by the two rails of a railway line, so that the short-circuit produced by the presence of a train can be used to control the signals protecting it; so **track-circuit** *v. trans.*, to equip with or make into a track circuit; **track circuiting** *vbl. sb.*; **track-in,** the movement of a film or television camera towards the subject; **track-layer,** (*a*) (earlier example); (*b*) one who lays the trail in training dogs to track criminals; (*c*) a tractor or other vehicle which travels on endless tracks (sense 6 d above); **track-laying** *sb.*, (*a*) (earlier example); (*b*) in film editing, the putting together of the sound track that is to accompany a picture; *adj.*, (*b*) (of a vehicle) having endless tracks (sense 6 e above); **track lighting,** lighting in which the lights are fitted on to tracks, allowing variable positioning (see sense 6 g above); **track-man,** (*b*) a track athlete; **track record** (examples); also *fig.*, known facts about past achievements or behaviour taken as a guide to future performance; **track rod,** a rod that connects the two front wheels of a motor vehicle and transmits the steering action from the steering column to the stub axle of each wheel; **track-shoe,** (*b*) = *running shoe* s.v. *running *vbl. sb.* 17 a; **track suit,** a loose two-piece garment (elasticized at the wrists and ankles) worn by athletes while training and before and after contests; hence **track-suited** *a.*; **track system** *U.S. Educ.* (see sense 4 g above).
1962 *Mod. Railways* Apr. 278/2 The jack, which had been left projecting from the track bed, did considerable damage to the gear underneath the cars. 1978 W. Hjortsberg *Falling Angel* xliv. 217, I followed the trackbed of the downtown express, measuring my pace to the spacing of the ties. 1911 *Encycl. Brit.* XXV. 76/2 At points the track circuit is run through a circuit breaker, so that the 'opening' of the points sets the signal for the section. 1931 E. T. MacDermot *Hist. G.W.R.* II. 498 Track circuits, whereby the signals protecting an occupied section of line are electrically locked at Danger, were first introduced in August 1907. 1935 *Economist* 22 June 1419/1 If every mile of line in use were 'track-circuited'..a blunder by a signalman would be impossible; no train could be signalled forward unless the line really was clear. 1983 *Internat. Railway Jrnl.* May 8/1 The Landskut box controls 125 signals, 83 points and 110 track circuits. 1931 *Times Lit. Suppl.* 10 Dec. 999/1 Exactly the same argument might be heard to-day against the introduction of track-circuiting or automatic train control. 1956 *Railway Mag.* Nov. 793/1 Track circuiting at both home and starting signals can be used to give additional safeguards. 1962 A. Lurie *Love & Friendship* x. 199 Hal Humphrey, the track coach. 1977 J. F. Fixx *Compl. Bk. Running* vii. 85 It has been attributed to..Lauri Pihkala, a pre-World War I Finnish runner, and George W. Orton, at one time Penn State's track coach. 1912 *Times* 29 June 13/1 In the track events, all of which must be held in the Stadium..there are 95 entries. 1928 *Daily Sketch* 10 Aug. 2/4 There are bound to be fine finishes in the international track events. 1973 C. Bonington *Next Horizon* ix. 138 He had always been a brilliant natural athlete, excelling at almost every game and track event in which he took part. 1954 *Encounter* Aug. 53/1 Her abject jealous misery has been..conveyed by the camera's slow track-in to close-up of her anguished face. 1961 *Listener* 2 Nov. 716/1 A track-in suggests an increase in intensity. a 1861 T. Winthrop *Life in Open Air* (1863) 234 'Wanted, experienced track-layer!' was the word along the files. 1928 *Daily Express* 19 Sept. 2 Coastguards..acted as track-layers for the open police dog trials. 1934 Webster s.v., The tracklayer is used especially where tractive conditions are poor. 1952 J. W. Day *New Yeomen of England* viii. 96 In all there are 45 tractors (including 6 track-layers), 4 moto-carts for hauling and odd jobs, and 8 combine harvesters. 1971 *Power Farming* Mar. 13/2 Use of the tracklayer was now mainly restricted to heavy clay soils and industrial duties where its higher costs could still be justified. 1857 R. G. Payne *Rep. Condition Railroads Tennessee* 7 The track-laying is progressing from the southern end of the road. 1920 *Sci. Amer.* 2 Oct. 335 (*caption*) Typical tractors of the wheeled and track-laying species now employed for agricultural and other purposes. 1957 *Times Lit. Suppl.* 27 Dec. 781/3 They resemble (as it says on driving licences) a 'track-laying vehicle steered by its tracks'. 1957 Manvell & Huntley *Technique Film Music* iv. 178 The four technical branches of film production, i.e., design..photography..sound..and editing (including assembly cutting, track-laying, laboratory liason, and post-production processes). 1962 A. Nisbett *Technique Sound Studio* xii. 206 Track-laying systems, where a whole series of tracks can be recorded individually on a single broad tape and then scanned together. 1972 *Times* 30 Nov. 18/1 Ceiling mounted spotlights and..track lighting systems. 1980 D. Francis *Reflex* iii. 41 In the sitting room, white walls..track lighting. 1972 *N.Y. Times* 4 June 4/2 An illegal water-jump area has impaired the credibility of the three fastest steeplechase performances by American trackmen. 1977 *Evening Gaz.* (Middlesbrough) 11 Jan. 14/2 Teesside Clarion's top trackmen..were among those honoured at their club's annual presentation in the Normanby Hotel last Friday. 1904 *Cap & Gown* (Chicago) IX. 215 Track Meets and Scores, 1903... Second Annual Interscholastic Meet, at Marshall Field. 1976 *Columbus* (Montana) *News* 27 May 1/4 Absarokee and Columbus scored first and second respectively in the Southern C Divisional track meet on May 20. 1977 *Time* 21 Feb. 34/2 He would be inclined to forgo continued development of a mobile U.S. nuclear missile launcher (the MX) if the Soviet Union will abandon deployment of its track-mounted launcher (the SS-20). 1951 *Publ. Amer. Dial. Soc.* XVI. 66 *Track record*, the best time made by a horse over a certain distance on a certain track. 1965 *Life* 15 Jan. 56 A/1 Wilder has had a series of extremely successful pictures... We were betting on his track record that this one would be too. 1972 *Observer* 30 Apr. 12/5 The airlines have over the years had enough confidence in our track record to be perfectly happy about this procedure. 1974 *Spartanburg* (S. Carolina) *Herald* 18 Apr. C 2/5 Charles Mathis..set a new track record at Northwestern with a 149′6″ in the discus. 1976 *Milton Keynes Express* 25 June 51/3 Houghton Rip..came fourth behind the Irish dog, whose track record was smashed by last year's Derby consolation winner Shamrock Point. 1976 *Time* 20 Dec. 17/1 A lot of the women candidates..have no management track records to be judged on. So they keep being passed over. 1983 *Daily Tel.* 23 Mar. 21 The Trustee Savings Banks, which plan to go public towards the end of next year, badly need to establish a good track record on profits. 1926 *Amer. Speech* I. 686/2 The following list of automobile terms in American and English nomenclature appeared in a 'special' from the Boston News Bureau early in the present year... [American] Tie rod [English] Track rod. 1930 *Engineering* 12 Sept. 326/3 Each pair of steering pivots is connected by a track rod at right angles to the chassis centre line. 1976 *Flintshire Leader* 10 Dec. 25/10 (Advt.), New springs for Land Rovers and most cars, 1935–70, towing brackets,..new kingpins, trackrods. 1977 J. F. Fixx *Compl. Bk. Running* x. 121, I bought them all inexpensive track shoes and University of Southern California track shirts just like Daddy's. 1978 *Detroit Free Press* 5 Mar. A 19/1 (Advt.), Juvenile to teen male track shirts, hooded sweatshirts. 1961 Webster, Track shoe. 1970 G. Jackson *Let.* 28 May in *Soledad Brother* (1971) 261 We're wearing track shoes. 1983 'J. le Carré' *Little Drummer Girl* ii. 45 Kids in summer rig and track shoes. 1946 C. McCullers *Member of Wedding* I. 4 She wore a pair of blue track shorts. 1974 *Index-Jrnl.* (Greenwood, S. Carolina) 23 Apr. 7/6 Smith describes his actions as mild, temporary schizophrenia, or Clark Kent in track shorts. 'When I pole vault..I'm like a complete different person.' 1955 R. Bannister *First Four Minutes* 46 Not having had the importance of warming up explained to me I did not wear a track suit. 1980 *Times Lit. Suppl.* 7 Nov. 1258/4 His..wife..memorably fetching in her pink towelling track suit. 1965 R. T. Bickers *Scent of Mayhem* iv. 42 His sweaty, track-suited figure. 1907 *St. Nicholas* (N.Y.) XXXIV. 693/2 Hammond has a track team, but we have n't. 1976 *Billings* (Montana) *Gaz.* 30 June 7-E/1 Aams was also an outstanding performer on the basketball and track teams at East Bay.

track, *v.*¹ Add: **I. 1. d.** Also (*U.S.*) of a horse: to walk with the fore and hind feet placed in the same straight line. Of the feet: to be placed thus.
1857 R. Glisan *Jrnl. Army Life* (1874) xxvii. 382, I observed..that he does not 'track' (step his hind foot straight after the fore one). 1897 E. Hough *Story of Cowboy* 34 His feet, in the vernacular of the range, do not 'track', but cross each other weakly.
e. *intr. Electronics.* Of a tunable circuit or component: to vary in frequency in the same way as another circuit or component, so that the frequency difference between them remains constant.
1932 [implied at *tracking *vbl. sb.*¹ 3]. 1939 [see *padder *sb.*² 3]. 1948 Slurzberg & Osterheld *Essent. Radio* vi. 271 In order to obtain the maximum fidelity, selectivity, and sensitivity..it is necessary that all the tuning circuits track together over the entire range of the receiver. 1975 D. G. Fink *Electronics Engineers' Handbk.* XIII. 40 The tuned circuits must track across the frequency band, and in the case of the superheterodyne, tracking of the local oscillator is necessary so that a constant frequency difference..is maintained.
f. *trans.* To follow the course of (a distant object) by means of a telescope, radar, or the like.
1950 in Webster Add. 1959 *Listener* 18 June 1057/2 The Jodrell Bank telescope and the smaller one at Bedford, Massachusetts, were tracking the moon. 1966 M. Woodhouse *Tree Frog* viii. 64 In order to track it [*sc.* a pilotless plane]..during flight trials..we've had to fit travelling wave reflection amplifiers under the wings. 1971 *Daily Tel.* 20 July 8/8 Every commercial and military aircraft flying over Europe can be tracked by radar. 1976 *Nature* 16 Sept. 216/1 An ITT FW 130 (S20) photomultiplier..was mounted about 40 feet above the ground on a radar dish programmed to track the star.
g. *intr.* To enjoy a rapport or 'get on' *with* another person; to take things in. *U.S.*
1972 *Newsweek* 17 July 22/3 He tracks better with reporters than did his phlegmatic predecessor. 1977 C. McFadden *Serial* (1978) xvi. 38/1 She's practically out of her mind. Like, she isn't even tracking. 1978 J. L. Hensley *Killing in Gold* ix. 116 Mom didn't track very well after the second stroke... It didn't mean anything to her any more.
2. c. (Earlier and later examples.) Also, *to track up* (a floor, etc.); to bring *in* (dirt, etc.) on one's feet (also const. preps.). Also *fig.*
1838 C. Gilman *Recoll. Southern Matron* xviii. 127 Miss Neely, one buckra woman want for track up all de clean floor. 1866 *Harper's Mag.* Jan. 271/2 The snow had been tracked in till it lay pretty thick on the floor. 1901 Merwin & Webster *Calumet 'K'* vi. 117 There's going to be a law passed about tracking mud inside the railing. 1915 *Century Mag.* Aug. 496/2 A good live boy..is a drug in the market. There seems to be a general feeling that they track in dirt. 1919 J. Reed *Ten Days that shook World* i. 11 The mud underfoot was deep, slippery and clinging, tracked everywhere by heavy boots. 1944 S. Bellow *Dangling Man* 100 Tracked your mat up. I'm sorry. 1950 M. Mead *Male & Female* xvi. 338 Floors do not need to be polished so often when there are no children's feet to track them up. 1980 R. Hill *Killing Kindness* ix. 87, I was trying not to track my work into the house too much. 1981 *Farmstead Mag.* Winter 50/2 Birds really use the trees to nest in, and small rabbits revel in tracking up fresh snow.
3. a. (U.S. examples.)
1843 R. Carlton *New Purchase* xxvii. 254 I'll track round a little—I wants any how to go over to the post-office. 1868 *Putnam's Mag.* June 670/1 We tracked through the dirty streets till we got to the house.
c. *Austral.* To keep company *with* (a person of the opposite sex, esp. a woman); *to track square* (see quot. 1919).
1916 C. J. Dennis *Songs of Sentimental Bloke* 51, I swear I'll never track wiv 'er no more. 1919 W. H. Downing

Digger Dialects 50 *Track square*, to pursue an amorous enterprise with honorable intentions. 1926 K. S. Prichard *Working Bullocks* 47 Combo's what they call a man tracks round with a gin in the nor'-west. 1933 N. Lindsay *Saturdee* 239 Who are you trackin' with now? 1949 A. Marshall *How Beautiful are thy Feet* 64 He wants me to track square with him. To look at him you'd never think he could talk seriously. 1954 T. A. G. Hungerford *Sowers of Wind* 270, I bet it's that cross-eyed harlot he's been tracking with. 1964 G. Johnston *My Brother Jack* 161 He's been at me for years about how irresponsible I am, and the first time I come back with a girl I'm tracking square with, I get hoisted!
d. Of a stylus or pick-up: to follow the waveform of a record groove. Also *trans.*, with the record, the groove, or the sound represented as obj.
1929 Wilson & Webb *Mod. Gramophones* vi. 129 As the needle tracks in the groove it is gradually worn to a chisel point. 1937 *Electronics* Nov. 21/2 The test was stopped when the records had been played 185 times each and the quality had become very bad indeed. The needles still would track the grooves, however, showing that complete breakdown of the walls had not yet occurred. 1850 *Audio Engin.* Aug. 15/2 In ordinary recorded music, the inability of the stylus to track at high groove curvatures leads to objectionable high-frequency distortion. 1957 *Records & Recording* Nov. 20 It is these grooves which must be tracked with absolute accuracy by the pickup needle. 1977 *Gramophone* June 10/2 (Advt.), This cartridge successfully tracks all types of records at forces even lighter than one gram. 1978 *Ibid.* June 128/3 There is a solo flute passage which could only just about be tracked at 1 gram. 1981 *Popular Hi-Fi* Mar. 7/4 The DT1 tracks exceptionally well and retrieves more informations from the grooves.
e. Of a film or television camera, or its operator: to move (esp. *back* or *in*) in relation to the subject being filmed.
1959 *Listener* 30 Apr. 772/1 After we had seen Mac in close-up, the camera suddenly tracked right away. 1960 N. Kneale *Quatermass II* II. 61 Track in on him fast. Fade in end music. 1961 G. Millerson *Technique Television Production* iii. 26 If..he is tracking backwards through an archway at too high an elevation, he might severely injure himself, as more than one cameraman has found. 1962 *Movie* June 5/2 Track into close-up of irrelevant detail; cut to close-up irrelevant detail of new setting; track out and begin sequence. 1975 *Radio Times* 22 May 66/3 Tufano starts a close shot on the broken walls of a bombed house. He then tracks back and pans across the blitzed street.
III. 5. *Comb.* **track-ball** *Computers*, a VDU input device in the form of a small ball that is rotated in a holder to move a cursor on the screen; = *tracker ball* s.v. *tracker¹ 2.
1969 M. H. Mehr in *Internat. Symposium Man-Machine Systems* V, Positioning to 0·1% of the screen diameter could be accomplished in 3–4 seconds which compares favorably with the published track ball data. 1972 *Acta Crystallogr.* A. XXVIII. S 253/2 The operator can interact with the display by means of a track-ball cursor. 1983 *Austral. Personal Computer* Aug. 60/2 Lisa's engineers are sometimes criticised for selecting the mouse rather than other quick data input devices—notably the trackball, touch-screen and light-pen.

trackability (trækăbi·lĭti). [f. Track *v.*¹ + Ability; cf. *-bility.] The ability of a stylus or cartridge to track adequately (*track *v.*¹ 3 d).
1972 Anderson & Jenrick in *Jrnl. Audio Engin. Soc.* XX. 162 (*heading*) A practical high-frequency trackability test for phono pickups. 1978 *Gramophone* May 1970/1 Not only does it check trackability, but also the tone-arm resonance over the frequency range of 4–12Hz.

tracked (trækt), *a.* [f. Track *sb.* + -ed².] **1.** Of a vehicle: having endless tracks (*track *sb.* 6 d).
1926 *Westm. Gaz.* 26 Jan. 6/2 'Tracked' vehicles, or, as most people would say,..'caterpillar' or roadless tractors. 1950 *Times* 17 Feb. 8/5 With a few bulldozers and grabs and tracked trucks, even the intimidating Snowdonian slate-tips could be put on the road to recovery. 1979 *Daily Tel.* 3 Dec. 3/1 A tracked armoured personnel carrier.
2. Of a hovercraft: confined to a fixed track.
1967 *Jane's Surface Skimmer Systems 1967–68* 52 (*heading*) Tracked air cushion vehicles. 1971 *New Scientist* 24 June 756/1 This time it is the £3·5 million tracked hovercraft project which is coming under attack.

tracker¹. Add: **2.** Special Combs.: **tracker ball** *Computers* = *track-ball* s.v. *track *v.*¹ 5; **tracker dog,** a dog trained to pick up and follow a scent, esp. a police dog trained to track people; cf. *sniffer dog* s.v. *sniffer 3 b.
1969 *Advance in Electronics: Proc. 16th Electronics Congr.* 484 The input devices..vary from scheme to scheme but common ones are keyboards, light pens and tracker balls. 1982 *Internat. Conf. Radar-82* (Inst. Electr. Engineers) 306/1 A Maintrace Section..distributes the radar signals to the displays and an Intertrace Section..generates the video maps and provides interactive keyboard/tracker-ball facilities. 1962 'J. le Carré' *Murder of Quality* iv. 51 We've got to rely on laboratories, tracker dogs and nation-wide searches. 1979 *Sunday Express* 16 Dec. 11 Police with tracker dogs will resume the search to-day for farmer's wife. 1984 *Times* 27 Feb. 8/1 Searching for..explosions with the eagerness of a tracker dog from the bomb squad.

tracking, *vbl. sb.* (In Dict. s.v. Track *v.*¹) Add: **I. 1.** (Further examples, corresp. to senses of *track *v.*¹)
1932 *Jrnl. Sci. Instruments* IX. 288 The best tracking conditions are reached by choosing the position of the axis about which the tone arm rotates by the method shown. 1937 *Discovery* Nov. 330/2 The operators of the electron cameras receive their instructions through telephones, so that they can advance or withdraw their cameras (tracking) or swing them sideways (panning) as planned by the producer. 1958 *Listener* 4 Dec. 908/2 The manufacture of the satellite and its instruments, with the associated tracking and computing systems. 1959 *Cambr. Rev.* 24 Oct. 73/1 The technique in this section cuts right across Bazin's distinction between classic Eisenstein montage, and more realistic modern methods of tracking, panning, and so on, on a wide screen. 1969 *Times* 16 July 4/1 The radio tracking instruments spread around the globe for keeping in continuous contact with the spacecraft. 1975 G. J. King *Audio Handbk.* viii. 184 Tracking ability thus takes account of the compliance, effective tip mass and mechanical resistance.
II. Specific senses. **2.** The formation or occurrence of conducting paths for electricity over the surface of an insulating material.
1931 H. Warren *Electr. Insulating Materials* III. xix. 239 The surface carbonization, or what is commonly called 'tracking'.., of these materials is their most serious electrical handicap. 1945 *Electronic Engin.* XVII. 600 Moisture..very soon causes..tracking between the connecting tags. 1967 M. Chandler *Ceramics in Mod. World* iv. 115 The electrical properties that matter most are high volume and surface resistivity, high puncture strength, and good tracking resistance. 1970 K. Ball *Fiat 600, 600D Autobook* iii. 34/1 Examine the distributor cap for cracks or signs of carbonisation (tracking).
3. *Electronics.* The maintenance of a constant difference in frequency between two or more connected circuits or components.
1932 *Electronics* Aug. 250/1 (*caption*) Deviation from exact tracking at various intermediate frequencies. 1971 [see *padder *sb.*² 3]. 1975 [see *track *v.*¹ 1 e].
4. *U.S. Educ.* = *streaming *vbl. sb.* f.
1967 *N.Y. Times* 23 June 36 Tracking can be a useful educational device if tests are frequently administered and if movement from one track to another is made easy. 1974 *Florida FL Reporter* XIII. 29/3 School would simplify their task if they could separate those students who want to learn to speak the new dialect from those who do not... This would not be a matter of 'tracking'.
III. 5. Special Combs.: **tracking error,** the error that occurs in gramophone reproduction when the tone-arm is pivoted, so that in general the axis of the cartridge is not in line with the groove; **tracking shot** *Cinemat.* and *Television*, a shot during which the camera tracks (*track *v.*¹ 3 e); **tracking snow** *N. Amer.*, snow sufficiently deep to enable hunters to track animals; **tracking station,** an establishment set up to track objects in the sky; **tracking weight,** the weight with which a stylus rests on a gramophone record.
1924 *Gramophone* Sept. 129/2 The 'tracking error'. 1930 *Wireless World* 26 Mar. 340/1 The mean between the highest and lowest deviations..gives a slightly smaller maximum tracking error. 1975 *Gramophone* Jan. 1424/1 Tracking error at all points across the record was within 2° overall. 1940 *Chambers's Technical Dict.* 857/1 Tracking shot. 1957 Manvell & Huntley *Technique Film Music* ii. 33 There follows a continuous tracking shot lasting in all for 2 minutes 22 seconds. 1973 D. Osmond-Smith tr. *Bettetini's Lang. & Technique of Film* ii. 95 Another narratively effective element is the 'tracking shot'; this consists of moving the whole cine-camera on a mechanical device known as a 'dolly'. 1971 W. Hillen *Blackwater River* ii. 17 Water running, little hope for tracking-snow. 1981 *Northeast Woods & Waters* Jan. 12/1 Coos County showed the largest increase (87%) where sportsmen had tracking snow for the entire season. 1963 *Ann. Reg. 1962* 445 The tracking station in Great Britain held the signal for only a few moments. 1972 *Daily Tel.* 14 Apr. 4/6 The TV signal is received from the Moon..at the Goldstone tracking station in the Mojave Desert. 1978 G. Greene *Human Factor* IV. i. 197 America maintains a guided missile tracking station and a space tracking station in the Republic. 1962 A. Nisbett *Technique Sound Studio* iv. 87 Recordings played under a tracking weight of about 1½ oz. were reckoned to have a life of about a dozen playings. 1978 *Lancashire Life* Sept. 131/1 The pick-up arm itself should have an adjustment to enable you to set the correct tracking weight for the cartridge.

tracklement (træ·k'lmĕnt). [Origin obscure. Dorothy Hartley claimed to have invented this word. She also claimed that her use of it in this sense was a *spec.* application of an older word, prob. *dial.*, meaning 'appurtenances, impedimenta', but no evidence of such a word has been found.] An article of food, *spec.* a jelly, prepared to accompany meat.
1954 D. Hartley *Food in England* v. 161 (*heading*) Mutton tracklements and condiment. 1959 *Times* 24 Aug. 11/4 A pleasantly astringent, smokily flavoured jelly as a 'tracklement' with mutton. 1971 R. Condon *Vertical Smile* (1973) xxxvii. 259 A saddle of lamb..delicate enough to accept only such a tracklement as rowan jelly. 1978 *Observer* 26 Feb. 35/9 Various salads and tracklements are included in the cold table.

trackster (træ·kstəɹ). *U.S.* [f. Track *sb.* + -ster.] A track athlete.
1974 *Hartsville* (S. Carolina) *Messenger* 22 Apr. 3A/6 The Hartsville High School Cindermen..won a 67-63 victory over the Hillcrest High tracksters. 1979 *Tucson Mag.* Apr. 70/2 The University of Arizona tracksters host two triangular meets at the Stadium.

tract, *sb.*¹ Add: **I. 3. b.** *Tracts for the Times*: also used in *sing.*, with small initials, of any literary work put out to meet a particular need of the times.
1927 A. H. McNeile *Introd. N.T.* 95 The Tübingen conception of the book [*sc.* the Acts of the Apostles] as a tract for the times mediating between the Judaic and the Pauline factions. 1979 E. H. Gombrich *Sense of Order* ii. 41 As a tract for the times the *Seven Lamps* failed to achieve Ruskin's aim of bringing the conditions of the Middle Ages back to industrialized England.
c. *tract society* (earlier example).
1760 Pratt in J. Adams *Wks.* (1850) II. 97, I should be very sorry to have the Tract Society dissolved.

tract, *sb.*³ Add: **3. b.** (*a*) (Earlier and later examples.) Cf. *fibre tract* s.v. *fibre *sb.* 8.
1681 S. Pordage tr. *Willis's Remaining Med. Wks.: Treat. No. 4: Anat. of Brain* xiii. 101 Out of the same tract of the oblong Marrow, lesser paths are carried outwardly, here and there, by particular Nerves, arising from the same, within the Skull. 1803 C. Bell *Anat. Human Body* III. I. iii. 115 It [*sc.* the olfactory nerve] takes its origin by three medullary tracts. 1959 W. Andrew *Textbk. Compar. Histol.* xiv. 566 As one ascends the vertebrate scale, the bundles of white matter or 'tracts' make their appearance running through the gray. 1974 M. C. Gerald *Pharmacol.* x. 189 Ascending nerves transmit sensory impulses up the spinal cord, whereas descending tracts send instructions to effector cells via motor fibers.
c. *U.S.* A plot of land with definite boundaries, esp. one for development; hence, an estate. So *tract home, house.*
1912 *Oregonian* 20 Oct. IV. 6/2 Trading in farm land last week was devoted chiefly to small tracts. There were, however, several large parcels. 1940 S. L. McMichael *Selling Real Estate* (rev. ed.) i. 4 An owner who had a tract of land ripe for development would call in a surveyor. 1954 F. L. Wright *Natural House* I. 108 The plan..was for a housing project on a 100 acre tract near Pittsfield. 1963 D. Hughes *Expendable Man* iii. 72 Raw green tract houses seemed to have taken over the countryside. 1972 Tract home [see *realty² 4]. 1973 *N.Y. Law Jrnl.* 31 Aug. 1/7 The defendants..were developing a tract with cooperative apartments to be constructed on one part. 1977 *New Yorker* 6 June 99/1 Most of his customers live in tract houses that have tiny bedrooms. 1979 *Tucson* (Arizona) *Citizen* 20 Sept. 1B/1 The Lopezes gave up a nice home on the Northwest Side on an acre lot for a nice house on a cramped lot in one of the tracts that have sprung up in south Tempe during the last two years. 1980 *Times Lit. Suppl.* 19 Sept. 1020/4 An amiable, moderately licentious fellow who readily settles down..in a newly built tract home.

traction. Add: **1. e.** *Med.* A sustained pull applied to a part of the body to maintain the positions of fractured bones following reduction of the fracture; the state of being subjected to such a pull; so *in traction.*
1885 *Boston Med. & Surg. Jrnl.* CXII. 545/1 The high pulleys..were used, as before, for oblique traction from the knee bands. 1939 W. C. Campbell *Operative Orthopedics* ii. 97 The majority of apparatus for either suspension or traction of the upper extremity is extremely cumbersome. 1962 *Lancet* 13 Jan. 61/1 The patient had previously been treated by neck traction and by prolonged physiotherapy, without benefit. 1973 'D. Shannon' *Spring of Violence* (1974) iii. 46 They had one leg in traction. 1981 R. S. H. Browne *Basic Facts in Orthopaedics* 95 Traction is used to overcome painful muscle spasm.
3*. *Physical Geogr.* The rolling and bumping of particles along the ground by a stream or the wind.
1914 G. K. Gilbert in *Prof. Papers U.S. Geol. Surv.* No. 86. 15 This second division of current transportation is called by certain French engineers *entrainement* but has received no name in English. Being in need of a succinct title, I translate the French designation..by the word *traction.* 1954 W. D. Thornbury *Princ. Geomorphol.* iii. 48 Traction involves the partial support of the material being transported by the buoyancy of the water or air but consists chiefly of the rolling, pushing, and dragging along of rock particles which are too large to be lifted. 1968 R. W. Fairbridge *Encycl. Geomorphol.* 319/2 Wind carries rock and organic debris by traction, saltation and suspension. 1972 R. J. Small *Study of Landforms* ii. 40 In areas where..chemical weathering is very active, streams may contain much of their load in solution, and traction, saltation and suspension may be correspondingly small.
4. **traction motor,** an electric motor designed for use in traction; **traction splint** (examples).
1900 Parshall & Hobart *Electr. Generators* I. 232 For satisfactory commutation, traction motors are designed with very high magnetisation at full load. 1950 *Times Rev. Industry* Sept. 25/1 The other two [locomotives] will have single-phase a.c. traction motors of special design. 1969 R. W. Smeaton *Motor Applic. & Maintenance Handbk.* viii. 5 Traction motors are very ruggedly built. 1935 *Sun* (Baltimore) 5 Apr. 3/2 The remedy..is use of 'traction splints', devices for automtically pulling ends of broken bones together and holding them. 1976 M. Machlin *Pipeline* xl. 443 He's got a bad break there. You'd better put that leg in a traction splint.

tractlet. (Earlier example.)
1889 E. Dowson *Let.* 27 Oct. (1967) 112, I have still a soul above tractlets.

tractor. Add: **2. c.** Also in mod. use, a rugged, powerful motor vehicle for drawing farm machinery, esp. one with large rear wheels and an elevated driving seat.

1905 *Sci. Amer. Suppl.* 4 Nov. 24948/3 At the recent show of the British Royal Agricultural Society great interest was centered in the Scott motor tractor... The motor in this tractor is a 24-horse-power..standard Aster engine. **1910** *Sci. Amer.* 15 Jan. 51/2 American motor tractors used for plowing and threshing usually develop from 12 to 35 horse-power. **1917** *Isle of Ely & Wisbech Advertiser* 28 Nov., This Tractor will operate on any land. .. It maintains a firm grip without injuring the lightest surfaces. **1932** [see *COMBINE *sb.* c]. **1958** *Economist* 11 Jan. 94/2 Antarctica's native inhabitants, on seeing the tractor marks in the snow, may well..ask themselves what will happen next. **1972** R. ADAMS *Watership Down* xix. 111 Few places are far from human noise—cars, buses, motor-cycles, tractors, lorries.

d. *Aeronaut.* An airscrew mounted at the front of an aircraft so as to exert a pull; an aircraft having this. Usu. *attrib.* (see sense 4 below). Cf. *PUSHER 2 c.

1903 *Work* 18 Apr. 171/1 A screw..working in front and acting as a tractor. **1909** [see *KITE *sb.* 3 c]. **1914** *Sphere* 7 Mar. 302/3 The Short and Sopwith tractors. **1980** H. F. KING *Sopwith Aircraft 1912–20* 20 Concerning the two early Naval Sopwith tractors, it seems worth recording that a demi-official drawing once existed showing just such a machine.

e. The driving section of an articulated lorry.

1926 *Encycl. Brit.* II. 987/2 Another combination for heavy merchandise transportation consisted of a road tractor, which was merely a foreshortened truck chassis, and a semi-trailer. **1951** [see *landing-gear* s.v. *LANDING *vbl. sb.* 8]. **1977** [see *PRIME MOVER 3]. **1982** *New Scientist* 11 Nov. 339/2 A lorry that has been loaded quite legally will tend to become overloaded on the drive axle (the rear of the two axles on the tractor) if it is gradually unloaded from the rear.

f. The mechanism that draws the paper through a printer.

1970 *U.S. Patent 3,511,354* 2 A switch is operated to reduce the speed of the tractor whenever..an excess of forms queued between the tractor and the stacker is such that a forms jam is imminent. **1983** *Austral. Microcomputer Mag.* Sept. 88/3 The TDS-13 daisy wheel printer handles paper up to 15in wide and has variable tractors and a friction platen to accommodate both continuous forms and sheets.

4. *attrib.* and *Comb.*, as (sense *2 c) *tractor-driver, -station*; *tractor-drawn, -mounted* adjs.; (sense *2 d) *tractor aircraft, airscrew, biplane, machine, monoplane, propeller, screw, seaplane*; **tractorman**, one who drives a farm tractor; **tractor-trailer** *U.S.*, an articulated lorry; cf. *trailer-truck* s.v. *TRAILER *sb.* 9.

1969 K. MUNSON *Pioneer Aircraft 1903–14* 110/1 A very early British Breguet bore the legend 'B. 3' on the rudder, the prefix letter indicating a tractor aircraft. **1932** *Rep. & Mem. Aeronaut. Res. Committee* No. 1522. 1 The magnitude of the retardation of air flow..has a mean value of 0·05 for a tractor airscrew in front of a medium body. **1912** S. F. WALKER *Aviation* iv. 28 In the later form of biplane, known as the tractor biplane, the engine and propeller are placed in front. **1969** K. MUNSON *Pioneer Aircraft 1903–14* 98/1 Antoinette III was the alternative title of the Ferber IX, a tractor biplane.. which was abandoned after only a few trial flights in.. 1908. **1943** J. S. HUXLEY *TVA* 43 (*caption*) Terraces, like the one being thrown up by a tractor-drawn grader.. retain from 85 to 90 per cent. of the rainfall. **1971** *Power Farming* Mar. 57/1 The machine is tractor-drawn. **1945** H. J. MASSINGHAM *Wisdom of Fields* x. 208 The tractor-driver despises the hard work of the older countryman and soon..he will need an elevator to lift him on his seat. **1969** R. BLYTHE *Akenfield* 17 Most modern farms need..a good tractor-driver or two, as once they needed good ploughmen. **1928** C. F. S. GAMBLE *Story North Sea Air Station* iv. 67 He was then of the opinion that 'pusher machines' were superior to 'tractor machines'. **1980** H. F. KING *Sopwith Aircraft 1912–20* 20 With a tractor aeroplane that was only a little faster than one of his motor-boats..Sopwith could hardly be content. **1946** J. W. DAY *Harvest Adventure* xiii. 210 Grover, the head tractorman. **1976** *Northumberland Gaz.* 26 Nov., Farmers appreciate the difficulties faced by shepherds and tractormen, he said. **1960** Tractor monoplane [see *MONOPLANE]. **1960** *Farmer & Stockbreeder* 12 Jan. 83/2 Tractor-mounted rotary tiller. **1979** *Internat. Pest Control* Nov./Dec. 139/1 Larger units to fit tractor-mounted spray booms. **1910** R. FERRIS *How it Flies* xx. 473 *Tractor propeller*, a propeller placed in front, so that it pulls the machine through the air, instead of pushing, or thrusting it from behind. **1910** C. C. TURNER *Aerial Navigation of To-day* viii. 127 In many monoplanes a single screw in front is used. It *pulls* the machine, and is often called a 'tractor' screw. **1969** K. MUNSON *Pioneer Aircraft 1903–14* 12 A Sopwith tractor seaplane. **1958** *New Statesman* 5 Apr. 423/2 The state is going to sell..some 20 billion roubles worth of farm machinery at present in the garages and parking lots of some 8,000 machine and tractor stations. **1965** M. MICHAEL tr. *J. Myrdal's Report from Chinese Village* (1967) I. 44 We hire one [tractor] from the tractor station at Yenan. **1949** Tractor-trailer [see *JACK-KNIFE *v.*]. **1977** D. E. WESTLAKE *Nobody's Perfect* xi. 143 A large tractor-trailer was..trying to back into position.

Hence **tra·ctored** *ppl. a.*, ploughed or cultivated by tractors; **tra·ctoring** *vbl. sb.*, activity involving a farm tractor. Also **tra·ctorcade** [*-CADE], a procession of tractors.

1949 E. COXHEAD *Wind in West* iii. 69 One [man] got out the tractor... When the tractoring was finished Les..got himself a fork from the byre. **1966** AUDEN *About House* 37 A house backed by orderly woods, Facing a tractored sugar-beet country. **1977** *Detroit Free Press* 11 Dec. 13-A/3 State and local police said there were no reports of traffic problems or arrests as a result of the so-called 'tractorcade'. **1981** *Observer* 22 Nov. 11/1 Towns throughout Northern Ireland will be choked with 'tractorcades' and marches.

tractotomy (træktǫ·tǒmi). *Surg.* [f. TRACT *sb.*^{3} + -O + -TOMY.] An operation in which certain nerve tracts in the brain are severed or destroyed.

1938 O. SJÖQVIST in *Acta Psychiatrica* (Copenhagen) Suppl. 17. 95 Section of the tract could be performed without accessory lesions or other disadvantages to the patients... The term *trigeminal tractotomy* is proposed for this operation. **1974** R. M. KIRK et al. *Surgery* iv. 61/2 Patients with intractable pain may be helped by division of the posterior nerve roots, spinothalamic tractotomy, and prefrontal leucotomy. **1980** *Daily Tel.* 27 Nov. 3/2 The operations being carried out on the four patients.. was a stereo tractotomy, which involved the destruction of brain tissue at points the size of a pinhead by implanting radioactive particles.

trad (træd), *sb.* and *a. colloq.* **A.** *sb.* **1.** Short for *traditional jazz*. Also *attrib.*

1956 *Melody Maker* 12 May 8 (*heading*) The great trad battle. **1957** *Observer* 13 Oct. 3/5 Lyttelton, who found 'trad' (traditional jazz) became play-acting and a dead-end, has travelled far from the revival days. **1963** *Times* 4 May 13/3 In an ideal world all 'trad' bands would have a good gritty surface noise built in to give an air of authenticity to their records. **1965** G. MELLY *Owning-Up* xv. 194 In 1960 the Trad boom at its height, and a riot at Beaulieu. **1973** G. BEARE *Snake on Grave* viii. 41 The French like good trad and this was real old stuff out of..old Chicago. It was black men's music. **1979** *Globe & Mail* (Toronto) 14 June 13/6 Molson's Brewery is holding a Canadian Jazz Festival July 21 and 22. Fourteen trad bands will take the stand. **1985** *Church Times* 8 Feb. 6/3 A fresh cause for barriers to be raised between the trads and the trendies.

2. Abbrev. of TRADITIONALIST.

1956 *Melody Maker* 12 May 8/4 The 'trads' belong to a strange sort of exclusive society. **1957** S. TRAILL *Concerning Jazz* 11 Those devoted to the older form of the music are referred to by the monstrous and unlovely nickname 'trads', (short for traditionalists). **1960** *Guardian* 11 Jan. 5/2 The ski world is divided to-day between 'trads' and 'modernists'.

B. *adj.* Abbrev. of TRADITIONAL *a.*

1958 *Spectator* 25 July 133/2 A raucous trad-jazz group struck up 'Basin Street Blues'. **1964** *Harper's Bazaar* Nov. 110/2 For people who like to furnish as trad as possible. **1967** E. GRIERSON *Crime of one's Own* xiii. 108 What's going to happen to dear Dave when even his.. dirty books seem tradder than *The Times'* fourth leader? **1974** C. O. BUCHANAN in R. C. D. Jasper *Eucharist Today* ii. 24 The Commission remained resolutely 'trad' until the Liturgical Conference of February 1966. **1981** *N.Y. Times Sunday Mag.* 4 Apr. 90/2 The British fashion press reports London's new look is that of the 'trad English gentleman'.

tradable, *a.* Add: (Later examples.)

1960 *Farmer & Stockbreeder* 5 Jan. 4/1 For good milling qualities around 25s per cwt, delivered, was the tradeable basis. **1975** *Accountancy* Sept. 32/2 It [*sc.* government spending] signifies a transfer of resources from the tradeable to the non-tradeable sector, which would literally kill the productive section of the economy.

Hence **trad(e)abi·lity.**

1979 B. BROWN *Money Hard & Soft on Internat. Currency Markets* 2 A *sine qua non* of currency hardness is free tradability. Residents of the country of issue are permitted to buy and sell a freely tradable currency for foreign exchange. **1983** *Times* 16 Aug. 14/1 The BNOC argument is that Brent crude has a higher 'tradability'.

trade, *sb.* Add: **I. 5. a.** *in trade*, following a mercantile occupation, *spec.* that of a shopkeeper.

1813 [in Dict.]. **1816** JANE AUSTEN *Emma* II. vii. 118 On the other hand, they were of low origin, in trade, and only moderately genteel. **1865** TROLLOPE *Can you forgive Her?* II. xv. 113 There was a little prejudice, because of his being in trade. **1932** LADY DUFF GORDON *Discretions & Indiscretions* v. 60, I could never be presented at Court, because I was in 'trade'. **1953** M. SHARP *Gipsy in Parlour* xii. 125 His father was in trade, and Frederick snubbed him. **1974** 'W. HAGGARD' *Kinsmen* x. 98 When he'd made a great fortune Duncan Gregg had gone up the ladder a little. But not very much, he was still in trade. **1979** A. MCCOWEN *Young Gemini* 53 Living in the Royal Borough of Tunbridge Wells, my father was made to feel over-conscious of being 'in trade'.

6. a. (Earlier examples of the sense 'those engaged in the liquor trade'.)

1846 H. BRETT *Let.* 17 Oct. in *Licensed Victuallers' Guide & Almanack* (1848) 2, I enclose a copy of the Permit..suggesting that a reprint of it in your Almanack would be highly appreciated by the Trade. **1868** *Era Almanack* (Advt. Suppl.), Licensed Victuallers' Protection Society, instituted October, 1833, for the protection of the person and property of the licensed victualler, and for the promotion of the best interests of the trade.

c. Prostitution. *slang.* [Cf. TRADER 1 b.]

1680 OLDHAM in Rochester *Poems* 122 He Heav'n, one large Seraglio, made, Each Goddess, turn'd a glorious Punk, o' th' Trade, And all that sacred place, Was filled with Bastard Gods, of his own Race! **1937** PARTRIDGE *Dict. Slang* 905/2 *The trade* is prostitution: late C. 18–19. **1962** K. A. PORTER *Ship of Fools* 33 Two inordinately dressed-up young Cuban women, frankly ladies of trade, had been playing cards together in the bar for an hour before the ship sailed.

d. The Submarine Service of the Royal Navy. *slang.*

1916 KIPLING *Sea Warfare* 97 No one knows how the title of 'The Trade' came to be applied to the Submarine Service. **1942** G. HACKFORTH-JONES *One-One-One* xviii. 169, I remember in 1919 listening to and looking at the young submarine captains, most of whom had served their four years of war in the 'Trade'. **1982** A. MELVILLE-ROSS *Trigger* xv. 161 It had been tacitly established in 'The Trade' that you did not mourn friends... The Submarine Service referred to itself as 'The Trade'.

e. The Secret Service. *slang.*

1966 'A. HALL' *9th Directive* xviii. 170 'How long,' I asked her, 'have you been in the trade?'..'Three years, on active ops.' **1977** J. GARDNER *Werewolf Trace* x. 87 Heather had that smart plummy voice which spoke of a cut-glass background. The kind of girl the trade enjoyed using: the kind they called a *lady*. **1977** *3rd Rep. R.Comm. Intelligence & Security: Abridged Findings* (Austral.) 4 In the trade, people talk of the 'intelligence cycle'.

9. *spec.* (*N. Amer.*), an exchange of players between two sports clubs or teams.

1913 *Outing* XLII. 133/1 My first big trade was a success. **1968** *Globe & Mail* (Toronto) 10 July 27/5 Riders made another trade, sending Larry De Graw and Bill Cline to Regina Roughriders for Tom Beynon. **1976** *National Observer* (U.S.) 12 June 14/3 The Yankees, who had been spying on Randolph for a year, picked him up last winter in a trade that sent Doc Medich to Pittsburgh for Randolph, Dock Ellis and Ken Brett.

11. b. A prostitute or pick-up used by a homosexual; a homosexual partner; also, such people collectively. *slang.*

1935, etc. [see *rough trade* s.v. *ROUGH *a.* 21 a]. **1941** G. LEGMAN in O. W. Henry *Sex Variants* II. 1177 Trade, generic for male prostitutes to homosexuals, or for heterosexuals to whom homosexuals prostitute themselves. **1968** *Globe & Mail Mag.* (Toronto) 13 Jan. 7/4 If a hustler is not himself homosexual, or maintains the belief that he is not, he is called 'trade'. **1969** *Jeremy* I. iii. 23/1 These are men who because they are too old, or unattractive, cannot pick up free 'trade'. **1975** *Daily Tel.* 24 July 3/6 Many of the boys became male prostitutes... They became known as 'rent boys' and were also referred to as 'trade'.

13*. A trade paper or magazine of the entertainment world. orig. and chiefly *U.S.*

1960 G. MARX *Let.* 21 Mar. in *Groucho Lett.* (1967) 270, I assume the trades are shoved under your door each morning. **1969** [see *PLANT *sb.*^{1} 7 b]. **1978** *Guardian* 11 Dec. 7/2 In Hollywood the two newspapers which report the entertainment industry are known simply as 'the trades'.

14. a. (sense 5) 'of or pertaining to a trade or calling', as *trade journal, magazine, paper, press*; 'caused by or arising out of one's trade', as *trade disease* (example); (sense 8) *trade agreement, association, attaché, balance, boom, delegation, depression, fair, figure* (usu. *pl.*), *relation* (earlier example), *ship* (later example), *token* [TOKEN *sb.* 11].

1934 WEBSTER, Trade agreement. **1940** *Economist* 23 Mar. 514/2 Three agreements between the British and Spanish Governments, a Loan Agreement, a Payments Agreement and a Trade Agreement, were signed in Madrid. **1977** *Whitaker's Almanack 1978* 929/1 Portugal has signed a Trade Agreement with EEC. **1909** WEBSTER, Trade association. **1928** *Britain's Industr. Future* (Liberal Industr. Inquiry) II. viii. §3. 98 Trade Associations are Associations of Traders, Producers, or Employers. **1984** *Economist* 18 Feb. 22 The BBC, IBA and the ITV companies' trade association are all off to Dublin to discuss such an Anglo-Irish deal. **1970** 'D. HALLIDAY' *Dolly & Cookie Bird* vi. 79 The trade attaché moved a bit nearer. **1980** A. COPPEL *Hastings Conspiracy* xxxvii. 227 He was a trade attaché... He would know with whom to speak to provide a friend of the Soviet Union with friendship. **1909** WEBSTER Trade balance. **1928** *Britain's Industr. Future* (Liberal Industr. Inquiry) I. iii. §4. 26 The increased volume of imports, together with a diminished volume of exports, has made the visible trade balance much less favourable. **1984** *Times* 28 Nov. 13/7 Our trade balance sags under imports of consumer goods. **1925** *Scribner's Mag.* July 59 Nothing in the nature of a 'trade boom' could be discovered. **1928** Trade boom [see *trade cycle*, sense 15 below]. **1961** 'J. LE CARRÉ' *Call for Dead* ix. 93 If Blondie was a carrier, it is exceptional..that he should use a trade delegation as a staging post. **1978** R. V. JONES *Most Secret War* liii. 520 He had somehow made contact with the Russian Trade Delegation. **1928** Trade depression [see *trade cycle*, sense 15 below]. **1908** W. JAMES *Mem. & Stud.* (1911) xiii. 322 Priggishness is just like painter's colic or any other trade-disease. **1970** 'D. HALLIDAY' *Dolly & Cookie Bird* iv. 45 One was the commercial attaché..and the three others were straight from Moscow on a trade-fair excursion. **1927** *New Republic* 12 Oct. 194/2 The Washington government is alarmed at the growing hostility toward us in Central and South America, which is beginning to be reflected in our trade figures. **1975** in R. Crossman *Diaries* I. 40 The First Secretary and the Chancellor continued to grapple with the trade figures and to lament the unhappy state of the pound. **1878** *Brooklyn Monthly* Apr. 118/2 The editor of a certain trade journal in New York. **1910** H. G. WELLS *Hist. Mr. Polly* vii. 218 Every issue of every trade journal has its four or five columns of abridged bankruptcy proceedings. **1981** P. VAN GREENAWAY *'Cassandra' Bell* ii. 27 I'm a special features writer for a trade journal—cosmetics. **1907** *Electr. World* XLIX. 674/1 And in other cases space in the trade magazines has been used as before. **1973** E. MCGIRR *Bardel's Murder* iv. 90 Forrest's desk was..bare except for a trade magazine. **1903** E. L. SHUMAN *Practical Journalism* 100 Sometimes the easiest line of approach to these coveted posts is through the avenue of the trade paper or technical journal. **1918** A. BENNETT *Roll-Call* I. ix. 208 It was a chap from the *Builder*, or I wouldn't have seen him. Can't trifle with a trade paper, you know. **1971** *Guardian* 1 July 11/2 The trade papers

try to introduce retailers to modern marketing. **1907** *Electr. World* XLIX. 674/1 In some cases the house organ has taken the place of advertising in the trade press. **1973** 'D. HALLIDAY' *Dolly & Starry Bird* x. 147 She could announce it in the trade press. **1888** E. BELLAMY *Looking Backwards* xiii. 198 A basis of agreement as to what staples shall be accepted..for settlement of accounts, being a preliminary to trade relations. **1935** *Discovery* Feb. 61/2 Last January there were only 49 deep-sea square-rigged trade ships in the world. **1889** G. C. WILLIAMSON (*title*) Trade tokens issued in the seventeenth century. **1933** J. O. MANTON (*title*) Buckinghamshire trade tokens issued in the seventeenth century. **1971** J. R. S. WHITING *Trade Tokens* 11 Broadly speaking I have kept to the strict definition of the term *trade token* (tokens issued for trading purposes, ie as currency).

15. trade allowance (earlier example); **trade binding** (see quot. 1971); **trade book**, a book published by a commercial publisher and intended for general readership; **trade card**, a tradesman's card bearing his name, the designation of his trade, and place of business; **trade counter**, an area in a shop or business where sales are made only to members of the trade; **trade cycle**, a recurring alternation of a period of increased economic activity with one of reduced activity; **trade discount**, a discount allowed by one trader to another, usually one in the same kind of occupation; also *fig.*; **trade dispute**, any dispute between employers and workers, or between different groups of workers, that is connected with the employment or non-employment of any person, with the terms or conditions of employment, or with certain related matters; **trade edition**, (*b*) an edition of a book intended for general sale through bookshops, in contrast to special editions or those sold through book clubs or specialist suppliers; **trade effluent**, effluent produced in the course of a trade or industry; any effluent other than domestic sewage; **trade gap**, the extent by which a country's imports exceed its exports; cf. *trade surplus* below; **trade language**: delete † and add later examples; **trade-last** *U.S.*, a compliment offered in exchange for one that is directed towards the speaker; also, in weakened sense, a compliment, whether reciprocal or not; **trade mission**, a mission sent to another country to promote trade with it; **trade plate**, a temporary number-plate for an unlicensed vehicle; usu. *pl.*; **trade price** (earlier example); **trade-rat**, a pack rat (lit. and fig.); **trade reduction** = *trade-discount* above; **trade-route** (earlier example); **trade-sale** (earlier example); **trade secret**, a device or technique used in a particular trade or (*transf.*) occupation and giving an advantage because not generally known; **trade show** *Cinemat.*, a private showing of a new film to the trade, before release; so **trade-show** *v. trans.*; **trade surplus**, the extent by which a country's exports exceed its imports; cf. *trade gap* above; **trade term**, an expression largely confined to a particular trade; **trade test** (see quot.); hence **trade-test** *v.*, to subject to or carry out a trade test; **trade war**, a situation in which governments act aggressively in international markets to promote their own countries' trading interests; **trade waste** = *trade effluent* above; **trade-weighted** *a.*, esp. of exchange rates, weighted in relation to the importance of the trade conducted with the various countries included.

1837 DICKENS *Let.* ? 21 Apr. (1965) I. 250, I want at the usual trade allowance..a complete set of the Standard Novels up to this time. **1952** J. B. OLDHAM *English Blind-Stamped Bindings* 3 Copies already bound in what are usually called 'trade bindings'. **1971** L. M. HARROD *Librarians' Gloss.* (ed. 3) 645 *Trade binding.* 1. The binding in which a publisher issues a book... 2. Plain calf or sheep bindings which were used in England by publishers from the fifteenth–eighteenth centuries... Until the nineteenth century, purchasers usually bought books unbound or enclosed in wrappers. **1962** Y. MALKIEL in Householder & Saporta *Probl. Lexicography* 9 Other short word-lists..include,..on the tradebook market, glossaries accompanying contemporary novels and short stories. **1977** *Globe & Mail* (Toronto) 17 May 16/7 Last year 39 percent of the company's sales came from elementary school texts, 7 per cent from university texts and 19 per cent from trade books for the general public. **1927** B. C. LANDAUER (*title*) Early American trade cards. **1979** *Early Music* Oct. 475/1 The instrument was certainly rebuilt by Taskin in 1783–4, as attested by his two inscriptions, his trade card glued inside the bentside, and his characteristic workmanship throughout. **1977** *Wandsworth Borough News* 7 Oct. 21/3 (Advt.), Young person required by Builders' Merchants in Battersea to assist on trade counter and learn trade. **1979** P. WAY *Sunrise* vi. 61 David Marriott entered the newspaper building, via the trade-counter door. **1928** *Britain's Industr. Future* (Liberal Industr. Inquiry) v. xxviii. §2. 411 The trade booms..and trade depressions..which were so prominent a feature of the pre-war 'trade cycle'. **1976** *Scotsman* 20 Nov. 3/3 The industry was notoriously vulnerable to the trade cycle, and, at present, world supply of shipbuilding capacity was about twice the level of world demand. **1898** G. VAN DE LINDE *Bookkeeping & Other Papers* 414 *Trade Discount*, see 'Cash Discount'. **1901** *Windsor Mag.* Dec. 199/2 Barclay is simply a surly brute, I never liked him, so you can take the usual trade discount off my estimate. **1977** C. RUNDLE *Accountancy for Everyone* vi. 54 The same rate of trade discount must be taken off the return value as was allowed off the original price. **1875** *Act* 38 & 39 *Vict.* c. 86 §3 An agreement or combination by two or more persons to do or procure to be done any act in contemplation or furtherance of a trade dispute between employers and workmen shall not be indictable as a conspiracy if such act committed by one person would not be punishable as a crime. **1926** *Brit. Gaz.* 12 May 2/1 No trade dispute has been alleged or shown to exist in any of the unions affected except in the miners' case. **1980** *Illustr. London News* Mar. 19/2 In that case it was held that the interpretation of section 13 was subjective, and that a person was protected provided he honestly thought that his action might help one of the parties to a trade dispute to achieve their objectives, and did it for that reason. **1930** A. HUXLEY *Let.* 8 Mar. (1969) 332 With regard to subsequent unlimited trade editions, I imagine you wouldn't have the organization. **1949** R. CHANDLER *Let.* 3 May (1981) 174 Houghton Mifflin..want to publish a trade edition collection of my old stories. **1959** L. M. HARROD *Librarians' Gloss.* (ed. 2) 160 *Large paper copy*, or *edition*, an impression of a book printed on larger and better quality paper than the usual trade edition. **1930** *Engineering* 11 July 47/2 (*heading*) Trade effluents and sewers. **1976** *Eastern Even. News* (Norwich) 9 Dec. 16/6 (Advt.), The post will involve the routine implementation of trade effluent control. **1961** *Listener* 16 Nov. 816/1 Britain's 'trade gap' narrows during October. **1977** P. JOHNSON *Enemies of Society* v. 55 These fresh supplies of bullion bridged the trade-gap between West and East until western industry was sufficiently developed to mass-produce textiles for export. **1840** Trade language [see CHINOOK]. **1907**, etc. [see *MOBILIAN *sb.*]. **1937** M. COVARRUBIAS *Island of Bali* p. xxiii, Malay was the trade language between Balinese and foreigners. **1968** W. J. SAMARIN in J. A. Fishman *Readings Sociol. of Lang.* 661 *Trade language* (*langue de traité*) is usually used for some language not included among the world's majority languages and which is used by some people as a second language in commercial situations. **1891** KIPLING in *Author* July 42 Some day they'll be a Public—not a girl's school swapping Trade-lasts. **1895** *Inlander* Nov. 61 *Tradelast*, n., compliment. **1920** F. SCOTT FITZGERALD *This Side of Paradise* I. 12 It was based upon some 'trade-lasts' gleaned at dancing-school, to the effect that he was 'awful good-looking'. **1935** H. DAVIS *Honey in Horn* iv. 41 The compliment was pointed enough. Uncle Preston looked pleased but not overwhelmed. He was used to trade-lasts from the ignorant. **1949** M. MEAD *Male & Female* 456, I set myself to collect varieties of little-known folk-customs like trade-lasts. **1975** I. SHAW *Nightwork* v. 57 'While on the subject,' she said, 'let me give you a t.l.' 'What's a t.l.?' '..T.l. stands for trade last. A compliment. You gave almost the best performance of anyone I've slept with in this town.' **1964** S. BELLOW *Herzog* 172 Through a Japanese trade mission she also met Mr. Nasser and Mr. Sukarno. **1973** *Times* 24 Apr. (São Paulo Suppl.) p. viii/4 The Japanese presence is very real in São Paulo, and on average two trade missions visit the city every week. **1953** J. WAIN *Hurry on Down* v. 99 The other [formality] was to unscrew the trade plates which the cars carried in place of the regulation number plates they would have when licensed. **1978** J. FLEMING *Day of Donkey Derby* 110 I've got two sets of number plates, and just for luck, two lots of trade plates. **1805** SCOTT *Let.* 29 Mar. (1932) I. 244 He will of course expect what every author is entitled to—half profits upon the trade price when an edition shall be disposed of. **1912** R. A. WASON *Friar Tuck* xxiv. 239 Either the pack-rat reformed into a trade-rat, or else he sold out his claim to a trade-rat. **1948** F. BLAKE *Johnny Christmas* II. 79 Johnny slept that night..disturbed neither by Gitt's snoring nor the scuttle of secretive trade-rats over the packed earth floor. **1970** R. SYMONS *Broken Snare* xxiii. 157 He knew pack rats—trade rats some people called them. They would always make a trade for anything they took. **1852** MRS. GASKELL *Let.* 22 Nov. (1966) 213 Your Uncle Langshaw is to have the trade reduction of price. **1873** T. T. COOPER *Mishmee Hills* 33 Calcutta Chamber of Commerce convened a meeting 'for the purpose of discussing the subject of overland trade-routes with China'. **1791** J. LACKINGTON *Memoirs* xxxi. 230, I purchased very large numbers..at trade-sales of all sorts, as bankrupt sales, sales of such as had retired from business, [etc.]. **1895** *Atlantic Reporter* XXX. 521/1 (*heading*) Injunction —use of trade secrets. **1928** R. B. MCKERROW *Introd. Bibliogr.* II. x. 235 Some of the best [facsimiles] are, I believe, produced by a process of lithography, but the details are probably a 'trade secret'. **1942** *R.A.F. Jrnl.* 13 June 36 The chemicals used to produce this foam are a trade secret, so I am not able to tell you what they are. **1978** G. GREENE *Human Factor* II. iii. 82 They have secrets too—trade secrets. **1919** *Biogram* 8 Mar. 3 (Advt.), If you want a film that will pack your house nightly, book The Bride's Awakening... Trade show will be announced shortly. **1919** *Honey Pot* I. iv. 44 The picture will be trade shown during next February. **1927** G. B. SHAW in *Illustr. London News* 3 Dec. 1004/1 If you have ever been to what is called a 'trade show' and seen all the exhibitors there, [etc.]. **1946** R. CHANDLER *Let.* 30 May (1981) 75 The picture has not even been trade-shown. **1962** N. STREATFEILD *Apple Bough* xv. 214 I'll look out for you at the trade show of the picture. **1984** *Listener* 15 Mar. 8/1 For the other six nights, orchestral works give way to pop, dance.., conferences, trade shows and sport. **1977** *Time* 12 Dec. 18/2 Riding the crest of a gigantic trade surplus, which last week led to a Japanese Cabinet shake-up.., the yen has risen 22% against the dollar so far this year. **1946** O. JESPERSEN *Mankind, Nation & Individual* 164 Some of these trade-terms may have originally sprung up as slang. **1977** *Drive* Sept.–Oct. 113/2 He may remember to avoid such obvious trade terms as *hole in the roof* for sunshine roof. **1934** WEBSTER, *Trade test*, a test of proficiency in a given trade, such as plumbing. **1946** *R.A.F. Jrnl.* May 147 Those who were A.C.2s or A.C.1s will be trade-tested immediately on remustering to their trade. **1960** I. JEFFERIES *Dignity & Purity* vi. 122 Once it had become apparent that I wasn't trade-testing..they [*sc.* the workmen]..did me the favour of answering the questions I..put. **1909** *Cambr. Mod. Hist.* VI. ii. 49 The tariff-war was often the precursor of the trade-war. **1975** J. DE BRES tr. *Mandel's Late Capitalism* xiv. 472 The use of currency manipulations to gain short-term export advantages threatens to turn into a general trade war. **1902** *Encycl. Brit.* XXXII. 525/1 We may..enumerate some of the principal trade wastes; these are from dye-works, print works, [etc.]. **1976** *Financial Times* 11 Feb., Sterling fell to its lowest level ever against major currencies yesterday, with the Bank of England calculation for its trade-weighted average depreciation widening to 30·4 per cent from 30·3 per cent. **1984** *Times* 31 Mar. 21/8 Sterling's trade-weighted value against a basket of currencies fell to the lowest for a year yesterday.

trade, *v.* Add: **6. e.** *to trade down*: to buy or sell cheaper goods, usually in larger quantities; to sell something and buy a cheaper replacement; similarly *to trade up*. Also *fig.*

1942 *Sun* (Baltimore) 22 July 3/2 Catering to the masses, the fur trade is 'trading down', Green said, offering practical furs..in economical designs. **1959** *Wall St. Jrnl.* 14 Apr. 18/3 Americans have followed the traditional pattern of 'trading up' in foods as well as in other goods and services. **1963** *Guardian* 8 May 7/2 It pays to trade up rather than cater for the masses. **1975** *Times* 14 Mar. (Small Car Suppl.) p. vi/7 The phenomenon of trading up from small to bigger cars is well known... In 1974..as many buyers..traded down as traded up. **1977** *Times* 22 June 23/7 Graduates are being given first crack at the jobs in preference to school-leavers. In other words, companies are trading up. **1982** *Nat. Westminster Bank Q. Rev.* Feb. 3 People..may well 'trade up' at various times by increasing their mortgage in order to move into better property.

9. a. (Examples of sense 'to exchange' other than in commerce and barter.) *to trade off*: (earlier example); also *fig.*, to give up in exchange for something else, esp. as a compromise.

1793 in *Mass. Hist. Soc. Coll.* (1810) III. 1 Good crops of corn and rye, which they trade off for spirituous liquors. **1917** *Dial. Notes* IV. 402 *Trade*,..also 'to exchange' in general sense... 'Trade places with me.' **1949** F. FERGUSSON *Idea of Theater* I. 21 In the next part of the fight the opponents trade blow for blow. **1951** *N.Y. Herald Tribune* (Paris) 29 Nov. 3 (*heading*) Insults traded in Commons in 20-hour 20-minute session. **1956** S. SEELY *Radio Electronics* xv. 440 Pulse-duration modulation and pulse-position modulation trade bandwidth for an improvement in signal/noise ratio. **1958** L. URIS *Exodus* I. ix. 55 No American Jew would trade places with a Negro or a Mexican. **1972** *Sci. Amer.* June 22/3 Warheads can be traded off for either ABM penetration aids or increased range. **1974** *Times Lit. Suppl.* 8 Mar. 242/5 The bourgeoisie has 'traded off' some of its control to the armed forces acting in their interests. **1978** M. HESSE in Hookway & Pettit *Action & Interpretation* 6 The pragmatic criterion trades these difficulties for others. **1983** *Times* 29 Apr. 8/7 Punches and insults were traded at a rally addressed by..the South African Prime minister. **1984** *Times* 26 Mar. 2/1 Another skinhead leaned from a window and traded insults with seven youths in the street.

b. *spec.* in *N. Amer. Sport.* Of a club or team: to exchange (one of its players) *for* one or more from another club. Also, to exchange (players) between clubs or teams.

1899 *N.Y. Times* 5 Mar. 8/5 There was very little trading of players during the meeting. **1955** *Sports Illustr.* 7 Mar. 38/3 Branch Rickey..traded Southpaw Paul La Palme to the St. Louis Cardinals for Ben Wade, a relief pitcher. **1972** 'E. LATHEN' *Murder without Icing* vi. 62 Nashville wouldn't be forever. I'd be traded sooner or later. And as long as I shoot those goals in, I can get what I want. **1982** *Philadelphia Enquirer* 13 May 1-c/1 The former UCLA star was traded by the Knicks to New Orleans for Jim Barnett and Neal Wala in 1975.

10. *to trade in*, to give (a used car, etc.) in part payment or exchange *for* a new one. Also *transf.* and *fig.* orig. *U.S.*

1926 G. HUNTING *Vicarion* I. 22 'Don't say you're trying to guard my young innocence, dear,' murmured Carol. 'I traded it in long ago for the new model.' **1955** W. GADDIS *Recognitions* III. ii. 752 You trade in your goddam car, you trade in your goddam wife, and the minute you get used to the goddam thing some bastard puts out a new model. **1973** R. TRAVERS *Murder in Blue Mountains* x. 95 Butler traded in his old black hat for the new one. **1975** D. LODGE *Changing Places* iii. 105 Shall I trade it in for a new one while it's still working? **1977** C. MCCULLOUGH *Thorn Birds* II. vi. 106 Bluey Williams traded in his lovely draft horses and his massive dray for a truck.

11. *intr. Comm.* Of a share: to be bought and sold (*at* a price, etc.).

1976 *Honolulu Star-Bull.* 21 Dec. c-7/1 A 125,000 share block of the stock traded at 17¼. **1981** *Times* 23 May 19/8 Its shares will start trading on June 1. *Ibid.* 8 June 16/1 The huge discount to net assets at which insurance shares have been trading.

tradeable, var. TRADABLE *a.* in Dict. and Suppl.

tradecraft. Add: **b.** *spec.* Skill in espionage and intelligence work; cf. *TRADE *sb.* 6 e.

1961 'J. LE CARRÉ' *Call for Dead* v. 56 He was suddenly alert... Was it the latent skill of his own tradecraft

which informed him? **1979** *Observer* 30 Dec. 7/7 At every juncture of the break-in he made decisions that proved catastrophic, applied 'trade craft' that was ludicrous, and misled his accomplices about matters that were either incriminating to himself or strategic to the break-in's failure.

traded, *ppl. a.* Add: **I. 3*.** *traded option*, an option on a stock exchange (see OPTION 4) which can itself be bought and sold.

[**1973** *Business Week* 16 June 78/1 The Chicago Board of Trade's new Options Exchange is drumming up interest in trading options to buy and sell stock—puts and calls.] **1978** *Daily Mail* 11 Mar. 39 In a month, we shall see the opening in Amsterdam and London of markets in traded options, an innovation which took Chicago by storm in April 1973. **1984** *Daily Tel.* 24 Apr. 16/2 Privately Stock Exchange officials are furious that..a major development in the traded options market has been marred by such uncertainty.

trade-in (trē^i^·d,in). [f. vbl. phr. *to trade in* s.v. *TRADE *v.* 10.] **1. a.** A transaction in which something is traded in; a part exchange. **b.** An item traded in, esp. a used car; also *fig.* **c.** A sum allowed in return for a trade-in.

1917 *Horseless Carriage* 1 Aug. 28 A used car in a trade-in. **1934** J. O'HARA *Appointment in Samarra* ii. 39 That Studebaker sedan, the black one. The one we took on a trade-in from Doc Lurie. **1945** *Word Study* Dec. 3/1 So useful and practical is English becoming that the time may not be far off when many small nations..may consider some sort of 'trade-in' with their national language for the English language. **1954** P. HIGHSMITH *Blunderer* (1956) xxii. 140 Since he hadn't the money for a brand-new car, Kimmel preferred to keep his ancient one rather than acquire something slightly newer on a trade-in. **1960** V. PACKARD *Waste Makers* xiii. 137 The high trade-in proved to be enormously effective in luring prospects. **1969** F. SARGESON *Joy of Worm* ii. 33 He would sell his machine to Jeremy cheap, and..accept as part of the transaction the trade-in of two push bikes in reasonably good condition. **1970** D. G. ALEXANDER *Retailing in England during Industr. Revol.* v. 143 Refurbished footwear which had been received as 'trade-ins' on new footwear. **1972** J. BELFRAGE in G. W. Turner *Good Austral. Eng.* vi. 115 Second-hand car dealers who..beg you to take as-new late models off their hands..for ultra-generous trade-ins on your old bombs. **1980** G. HAMMOND *Reward Game* iv. 46 There's a new sports car that I took as a trade-in.

2. *attrib.*

1927 *Ladies' Home Jrnl.* Dec. 65/1 Buyers..took prompt advantage of the liberal trade-in allowance on their old equipment. **1929** *Collier's* 12 Jan. 9/2 If more than one third of his..transactions..is represented by trade-in cars. **1946** *Sun* (Baltimore) 27 Jan. 22 (Advt.), No more 'tired power' or expensive engine repair. Extra value at trade-in time. **1958** *Economist* 8 Nov. 535/2 A general move in this direction would shave dealers' profit margins and might affect the trade-in values they could offer. **1974** P. FLOWER *Odd Job* iv. 29 They could get new trade-in jobs on HP... He wouldn't take any more old fridges. **1977** *Cork Examiner* 8 June 14/5 (Advt.), Typewriters all makes, new and secondhand, good trade-in allowance. **1980** *Sunday Express* 24 Aug. 22/1 (Advt.), You could get an exceptional trade-in price for your old car.

trade-mark. Substitute for the parenthesis: (secured by legal registration or, in some countries, established by use).

b. (Earlier example.) Also *attrib.* as *adj.*

1869 'MARK TWAIN' *Innocents Abroad* xxiii. 238 We see other monks looking tranquilly up to heaven, but having no trade-mark. **1977** *South China Morning Post* (Hong Kong) 13 Apr. 7/2 Jimmy Shtoow'd, to phonetically adopt his trademark drawl, has an unnerving 'look' for interviewers. **1983** *Daily Tel.* 18 Mar. 17/2 Fans need not worry: her trademark French rayon jersey made up at least half the collection.

tra·de-marked *ppl. a.*

1936 E. B. WHITE *Let.* 24 Dec. (1976) 146 Your public approval of a trademarked product and your influence can be bought at a price. **1983** P. DEVLIN *All of us There* xi. 133 The old [pub]..with its great copper-banded barrels and old trade-marked mirrors.

trade-off (trē^i^·dǫf). [f. vbl. phr. *to trade off* s.v. TRADE *v.* 9.] A balance achieved between two desirable but incompatible features; a sacrifice made in one area to obtain benefits in another; a bargain, a compromise.

1961 *Hovering Craft & Hydrofoil* Oct. 32/2 Propulsion system integration allowing trade-offs between the requirements of lift and forward thrust can be achieved in a variety of ways. **1968** *Economist* 21 Sept. 38/1 It may be that the old argument of the 'trade-off' between inflation and unemployment will be superseded. **1970** A. TOFFLER *Future Shock* xx. 425 To provide data on the social and economic costs of various goals, and to show the costs and benefits of proposed trade-offs. **1972** *Sci. Amer.* June 22/3 Although its nominal range of about 2,500 nautical miles is the same as that of the *A-3*, a trade-off between range and payload is always possible. **1975** *New Yorker* 7 Apr. 55/3 Whether it is prudent, let alone safe, for Congress to try for a trade-off between these two priorities—the environment and the economy—remains a question. **1976** *Nature* 27 May 279/2 The alternative strategy of increasing the protein content of cereals would be a difficult task due to the yield/protein content 'trade-off'. **1976** P. R. WHITE *Planning for Public Transport* ii. 48 The household location may be a compromise rather than a trade-off related solely to a place of employment in the central area. **1978** *Jrnl. R. Soc. Arts* CXXVI. 255/1 A quite significant proportion of judgements in life are a trade-off between safety and cost. **1983** *Times* 3 Sept. 7/2 In the long run, there is no trade-off between inflation and unemployment. **1983** *Listener* 8 Dec. 23/3 The trade-off between housing and other objectives of policy has changed.

tradesman. Add: **2. b.** *tradesmen's entrance* (or *door*): a minor or side entrance to a property for use by tradesmen or workmen.

1892 A. W. PINERO *Magistrate* III. i. 113 We're in the scullery, Guv; let's try and find the tradesmen's door. **1904** E. NESBIT *Phoenix & Carpet* vii. 131 At the side of the house..there is a green gate labelled 'Tradesmen's Entrance'. **1946** 'J. TEY' *Miss Pym Disposes* xiii. 143 'Shouldn't you be going in by the other door?'..'I do not take well to tradesmen's entrances.' **1982** M. HINXMAN *Telephone Never Tells* iii. 21 A well-trodden path that wound round to the rear suggested..that the tradesmen's door was a more familiar mode of entry and not only for tradesmen.

c. *tradesman's token* = *trade token* s.v. *TRADE *sb.* 14 a. Usu. in *pl.*, *tradesmen's tokens*. Cf. TRADE *sb.* 11.

1660 *CSP Dom.* 6 Oct. (1860) 307 Proposition by Sir Wm. Parkhurst..to meet the necessity for small money, and obviate the inconvenience of tradesmen's tokens. **1757** *Gentl. Mag.* XXVII. IX. 498/2 The best account of the money called Tradesmen's Tokens..is to be drawn from..Mr Leake's hist. account of English money, London 1745. **1849** J. Y. AKERMAN *Tradesmen's Tokens* p. i, Notwithstanding the dictum of Pinkerton, many persons are yet found who collect Tradesmen's Tokens. **1892** J. ATKINS (*title*) The tradesmen's tokens of the eighteenth century.

trade-union, trades-union. Add: Now usu. written as two words (without hyphen) except when used *attrib.* **b.** *attrib.*; **trade(s) union congress**, (*a*) a national delegate conference of British trade unions, held annually since 1868; (*b*) (with capital initials) the national confederation of British trade unions, originally formed to organize the annual congress.

1831 LADY E. BELGRAVE *Let.* Feb. in G. Huxley *Lady Elizabeth & Grosvenors* (1965) iv. 97 The tremendous Trade Union Club there [in Manchester]... I wish it could be put down and that someone would shoot O'Connell and Cobbett. **1868** F. HARRISON *Let.* 11 Nov. in *Geo. Eliot Lett.* (1956) IV. 483 Since July I have been quite immersed in my Trades-Union work. [**1878** *Chambers's Encycl.* X. 757/1 The Trades' Congress..holds an annual conference in the different leading towns.] **1888** *Encycl. Brit.* XXIII. 501/2 An annual trades union congress is held in some great centre of industry and population.. at which delegates from almost all the trade unions in the realm are present. **1895** *Nat. Review* XXVI. 163 The Trade-Union Congress..has made itself a really representative body by adopting the principle of one vote one value. **1911** C. E. PERSONS et al. *Labor Laws & their Enforcement* 115 One of its members, Edward H. Rogers, a trade union leader, made some half-hearted recommendations. **1920** *Times* 11 June 17/2 The majority of trade union leaders..have deserted to the camp of the capitalists. **1926** *Brit. Gaz.* 12 May 1/4 Every man who does his duty by the country and remains at work or returns to work during the present crisis will be protected by the State from loss of trade union benefits. **1926** A. CONAN DOYLE *Hist. Spiritualism* I. xiii. 299 The conduct of conjurers [towards mediums] seems to have been usually determined by a sort of trade union jealousy, as if the results of the medium were some sort of breach of a monopoly. **1926** *Law Rep. Chancery Div.* 540 No trade dispute does or can exist between the Trades Union Congress on the one hand and the Government and the nation on the other. **1927** CARR-SAUNDERS & JONES *Social Struct. England & Wales* 51 Trade-union officials. *Ibid.* 77 Trade-union membership advances in waves. **1936** G. B. SHAW *Millionairess* III. 174 You might as well ask me to pay trade union wages as do all that the inspector wants: I should be out of business in a week. **1941** E. WILSON in *Atlantic Monthly* Apr. 480/2 The trade-union leadership is represented only..by an unscrupulous spellbinder. **1964** T. B. BOTTOMORE *Elites & Society* i. 9 Representatives of new social interests or classes (e.g. trade union leaders). a **1974** R. CROSSMAN *Diaries* (1975) I. 335, I am pretty used to Conference now, and being hardened I just opt out of the evening entertainments and the endless trade-union dinners. **1974** P. DICKINSON *Poison Oracle* iv. 100 The hoarse bellowings of an old-style trades union agitator trying to whip an apathetic strike meeting into action. **1975** *Economist* 4 Jan. 75 Mr Gill has been cutting quite a dash since his election as the only communist on the general council of the Trades Union Congress. **1976** *Daily Tel.* 20 July 2/5 When will the British public, 81·5 per cent of whom have no trade union affiliation, realize how rapidly our freedoms are being eroded.

Hence **trade-unionese** *colloq.*, the style of language supposed to be characteristic of public statements by trade-union officials; **trade(s)-unionism** (earlier example); **trade(s)-unionist** (earlier example); **trade-unionize** *v. trans.*, to enrol in a trade union, to form a trade union from among; **trade-unionized** *ppl. a.*

1834 *Times* 10 May 5/4 Is it not somewhat unreflecting on the part of trades' unionists to imagine that they can ..set up an opposition monopoly? **1867** *Blackw. Mag.* June 726/2 When Socialism and Communism in all their various forms died out, Trades-unionism..took their place. **1927** A. P. HERBERT in *Times* 12 Jan. 13/5 There should be prizes for Essays in the Socialist Language, and polysyllabic Resolutions in Trade Unionese. **1960** *Times* 9 Feb. 14/2 (*heading*) Should artists be trade unionized? **1969** H. E. BATES *Vanished World* xii. 156 Nowadays.. we are near-suffocated by tradeunionese, councilese and Americanese, the new extensions of stodge-pudding language that have joined Johnsonese, journalese and politicalese. **1976** *Carn* Feb. 6/1 Public attention has been diverted from the Bretons to the jailed conscripts who tried to trade-unionise the French army. **1982** *Economist* 17 Apr. 51/3 Buy-outs by employees of bust and heavily trade-unionised businesses.

trading, *vbl. sb.* Add: **a.** *trading down, up* (see quot. 1963 and *TRADE *v.* 6 e).

1963 *Gloss. Managem. Terms* (Brit. Inst. Managem.) (Typescript), *Trading down*, a seller's practice of handling cheap or low-grade products in order to secure higher volume sales (usually at a low rate of profit with a high stock turn). *Ibid.*, *Trading up*, a seller's practice of handling expensive or high-grade products in order to gain prestige and secure a better class of trade (usually at a high rate of profit with a low stock turn). **1971** *Daily Tel.* 13 Mar. 20/5 This second house is almost certain to be more expensive... Any capital gain on the first home will almost certainly be more than devoured by the additional cost of this 'trading up'. **1976** *Times* 6 Nov. 18/8 When prices increase..there is a strong tendency to..buying in smaller quantities or items of poorer quality. This is known in advertising jargon as 'trading down'.

b. *trading-boat, goods* (example), *-scow, vessel* (example), *voyage* (earlier example); **trading account**, an account showing the revenue from sales during a period, the cost of those sales, the stock at the beginning and end of the period, and the resulting gross profit or loss; **trading estate**, an area of land specially developed to accommodate light industry; **trading floor**, the area in a stock exchange where the dealing is done; **trading-house** (earlier and later examples); **trading-place**, (*b*) (examples); **trading profit**, profit as shown in a trading account; gross profit; **trading-rat** = *trade rat* s.v. *TRADE *sb.* 15; **trading stamp** orig. *U.S.*, an adhesive stamp given by a retailer to a customer when he buys goods of a certain value and exchangeable in quantity for goods from the company issuing the stamp.

1920 Trading account [see *cost account* s.v. *COST *sb.*^{2} 6]. **1978** J. KELLOCK *Elements of Accounting* x. 174 In the final form, accounts are divided into two sections referred to as the trading account and the profit and loss account. **1738** W. STEPHENS *Jrnl.* 15 June in A. D. Candler *Colonial Rec. State of Georgia* (1906) IV. 156 An Indian Trading Boat arrived. **1867** J. N. EDWARDS *Shelby* xx. 364 Marmaduke..hoped to capture a trading-boat, and thus put an immediate quietus on the cotton trade. **1923** 'R. DALY' *Enchanted Island* xv. 154 I'm trying to figure out our chances of being picked up if we stay here. It's not on the track of any regular trading-boats. **1937** *Ann. Reg. 1936* 8 The Commissioner for the Special Areas of England and Wales..issued a report... His chief object now, he said, was to establish in the Special Areas what he called 'trading estates' for the purpose of attracting to these areas fresh industries, particularly of the lighter type. **1981** B. HINES *Looks & Smiles* 31 They..caught a bus out to the Ring Road where a Trading Estate was being developed to attract new industries to the city. **1947** *Encycl. Brit.* XXI. 422/1 As a market place, the trading floor of the New York Stock exchange affords exactly the same fundamental facilities that a public market does for the housewife. **1971** *Sunday Australian* 8 Aug. 13/9 He also suggested..greater use of electronic equipment, culminating in the elimination of the current trading floor. **1981** *Times* 19 July 22/1 The trading floor appeared unusually empty. **1984** *Christian Science Monitor* 2 Mar. 10/3 They have to figure out what stocks to buy when the bulls return to the trading floor. **1796** *Saskatchewan Jrnls.* (Hudson's Bay Rec. Soc.) (1967) 73 Sent four men with ten horses to Buckingham House for trading goods. **1637** in *Mass. Hist. Soc. Coll.* (1863) VI. 215 They say he came from a trading house which Plymouth men have at Qunnihticut. **1726** S. PENHALLOW in *New Hampsh. Hist. Soc. Coll.* (1824) I. 21 Trading-houses in several places were hereupon engaged. **1899** H. B. CUSHMAN *Hist. Indians* 478 A trading house for the accommodation of the Chickasaws has been established at the Bluffs. **1755** L. EVANS *Geogr. Ess.* 10 The situation of Indian Villages, trading Places, the Creeks [etc.]. **1883** W. E. HOWE *Country Town* xv. 84, I had never been to Twin Mounds, as there was a post-office and a small trading place several miles nearer. **1940** *Economist* 25 May 936/1 A double record of trading profits and also of true net profits has been kept. **1966** *Daily Tel.* 30 June 1/5 Sears Engineering made trading profits of £2,344,000. **1895** *St. Nicholas* Apr. 501/2, I would like to write an entire paper on the droll ways of certain distinguished members of the Wood-rat, Pack-rat, Trading-rat, or Bush-rat genus. **1875** 'MARK TWAIN' in *Atlantic Monthly* Feb. 219/2 He ran over the steering-oar of a trading-scow. **1897** *Catal. Title Entries of Bks.* (Office Reg. Copyright, Libr. of Congr.) No. 326. 21 Trading stamp book issued by Washington Trading Stamp Company. (Received Sept. 30, 1897.) **1901** *Daily Colonist* (Victoria, B.C.) 30 Oct. 2/4 The city council..have passed a by-law prohibiting the use of trading stamps or coupons within the municipality. **1933** *Parl. Papers 1932–3* XII. 387 (Cmd. 4385) 12 A trading stamp company sells collecting books and stamps to retailers. **1964** S. BELLOW *Herzog* 121 Postage stamps and trading stamps soaking on the formica counter. **1977** *Times* 10 May 15/1 The Tesco supermarket chain..has just decided to withdraw from the trading stamp business. **1895** C. M. YONGE *Long Vacation* xviii. 181 He..set her up at Rockquay with the tobacco-shop. She had chosen that place on account of American trading-vessels putting in there. **1745** *Roxana* 427 He.. told me, he could help me to a Share in two Ships, one

was going a trading Voyage to the Coast of Africa, and the other a Privateering.

trading post. orig. *U.S.* [f. TRADING *vbl. sb.* + POST *sb.*^{3}] **1.** A place occupied for purposes of trade, esp. in a region not fully developed.

1796 in *Coll. Georgia Hist. Soc.* (1916) IX. 15 The land.. has been recommended..as proper for a trading post. **1837** [in Dict. s.v. TRADING *vbl. sb.* b]. **1936** D. MCCOWAN *Animals Canad. Rockies* xxi. 190 The fur trading posts of the far North. **1976** *Sat. Rev.* (U.S.) 30 Oct. 23/1 Lamu—a tiny coral island off the Kenya coast..once a prosperous trading post of Omani Arabs.

2. One of the posts or positions on the trading floor of a stock exchange, where stocks assigned to that location are bought and sold.

1951 G. L. LEFFLER *Stock Market* xii. 178 There are now 18 active trading posts... The present posts are horse-shoe or U-shaped stations, occupying 100 square feet... A total of 12 clerks can work inside. **1970** *Toronto Daily Star* 24 Sept. 13/1 A bald man in a red jacket leans heavily against a quiet trading post. His eyes are closed. Another sits on a small seat, one of many which can be pulled down from the sides of the trading posts. **1971** *Reader's Digest* (U.S.) Oct. 58/2 Hundreds of brokers were swarming around 17 horseshoe-shaped trading posts to execute the overnight accumulation of orders to sell, sell, sell.

tradish (trē^i^·diʃ), *a. nonce-wd.* [f. TRAD(E *sb.* + -ISH^{1}.] Of or suggestive of trade or tradesmen.

1803 D. WORDSWORTH *Jrnl.* (1941) I. 243 The houses.. have a tradish look, as if they might have been off-sets from Glasgow.

tradition, *sb.* Add: **7. tradition-directed** *a.*, applied to persons whose behaviour and goals are largely directed by social conventions; cf. *inner-directed* adj. s.v. *INNER *a.* (*sb.*^{2}) I n, *other-directed* adj. s.v. *OTHER *adj. pron.* (*sb.*) D. 2.

1950 D. RIESMAN *Lonely Crowd* i. 9 The society of high growth potential develops in its typical members a social character whose conformity is insured by their tendency to follow tradition: these I shall term tradition-directed people and the society in which they live a society dependent on tradition-direction. **1959** *Times* 3 Sept. 13/5 He insists warmly on the importance of establishing three main categories of social character among writers... There are the inner-directed..the other [*printed* outer]-directed..and the tradition-directed. **1970** E. FLORES in I. L. HOROWITZ *Masses in Lat. Amer.* ix. 333 In countries ruled by tradition-directed, ignorant landlords,..it is impossible to apply sophisticated redistributive policies.

traditional, *a.* Add: **1. c.** Applied to a style of post-war jazz inspired chiefly by the bands of the earliest period of jazz, as opposed to *modern jazz* s.v. *MODERN *a.* 3 a. Cf. *TRAD *sb.* 1.

1950 *Downbeat* 28 July 10/1 This..has been the particular gripe of the traditional jazz adherents. **1980** J. WAINWRIGHT *Man of Law* i. 7 A mutual fanaticism for traditional jazz... The small-group combinations beloved of three decades ago.

traditionalist. Add: **b.** One who plays, appreciates, or supports traditional jazz (see *TRADITIONAL *a.* 1 c). Also *attrib.* or as *adj.*

1951 *Jazz Jrnl.* Sept. 15/1 First, the bands... The stars of the Traditionalist show were 'The Saints'. *Ibid.* 15/2 Why did he have to spoil what was otherwise an excellent job of compering by making his usual crack at the traditionalists? **1962** [see *MODERNIST 6]. **1983** *New Oxf. Compan. Music* I. 990/1 Parker..laid his influence on virtually everything and everyone except the dedicated 'traditionalists'.

traditionalize, *v.* (in Dict. s.v. TRADITIONAL *a.* (*sb.*).) Add: Also, to imbue with or constrain by tradition. Chiefly as **tradi·tionalized** *ppl. a.*, **tradi·tionalizing** *ppl. a.* and *vbl. sb.*

1951 R. FIRTH *Elements of Social Organization* iv. 134 The price system..may be..of a highly traditionalized type, with relative inflexibility in rates over long periods, and considerable resistance on the part of producers and consumers to variation in these rates. **1960** C. GEERTZ *Relig. Java* i. 11 The more traditionalized peasants and their proletarianized comrades in the towns. **1976** *World Politics* XXVIII. 250 There may indeed be a traditionalizing role for military rulers in Africa. **1978** *Econ. Devel. & Cultural Change* XXVI. 763 The presence of the petroleum-extraction industry in the desert region surrounding Augila was having the simultaneous effects of modernizing and 'traditionalizing' oasis life. **1982** *Dædalus* Winter 101 What these revolts appear to have in common..is their class basis and their traditionalizing, but nontraditional, ideologies.

Hence **tradi:tionaliza·tion**, the process of making or becoming traditional; adherence to tradition.

1966 A. R. WILLNER *Neotraditional Accommodation to Political Independence* 3 At the microscopic level, the process of traditionalization, as it is described here for Indonesia, involves the increasing influence of indigenous and particularistic rather than modern, rational criteria on the way in which public officials fulfil their prescribed roles. **1977** *Social Problems* XXV. 135 In the Yom-Kippur War, traditionalization emphasized the centrality of the feminine family roles. **1981** R. & M. M. LAROSSA *Transition to Parenthood* i. 24 The practical implications of traditionalization following birth should also not be ignored. **1981** *Stud. in Compar. Internat. Devel.* Fall–Winter 65 The absorption of h.s. immigrants can be described as entailing not modernization, but rather 'traditionalization'.

traditioned, *a.* For *rare*$^{-1}$ read *rare* and after 'of a kind' add 'or to a degree'. (Further example.)

1940 W. DE LA MARE *Pleasures & Speculations* 14 One of the most ancient and richly traditioned cities of Denmark.

traductor. Restrict † to sense in Dict. (s.v. TRADUCT *v.*) and add: **2.** A device on the side of a railway carriage that picks up and deposits mail bags while the train is in motion.

1959 C. J. ALLEN *Mod. Railways* xv. 187 The sacks [of mail]..are then suspended from hinged traductor arms at the van side. **1970** *Railway Mag.* Oct. 545/2 Post Office Sorting van; has letter sorting racks and some vehicles also have nets and traductor arms. **1978** O. S. NOCK *Gt. Western in Colour* 147/1 Most of these trains had vans fitted with the traductor apparatus for picking up and setting down mails at speed.

Trafalgar. Add: **b.** Special Comb.: **Trafalgar chair** (see quots. 1934, 1969); **Trafalgar Day**, 21 October, the anniversary of the battle of Trafalgar; **Trafalgar Square** *v. trans. joc.*, to harangue (from the practice of 'soap-box' oratory in such public places).

1822 R. BROWN *Rudiments of Drawing Cabinet & Upholstery Furnit.* (ed. 2) p. xii, Many cabinet-makers, for the sake of notoriety, ridiculously give names to furniture quite inconsistent, such as Trafalgar chairs, Waterloo feet, &c. **1934** M. JOURDAIN *Regency Furnit. 1795–1830* 48/2 The well-known Trafalgar chair..was designed as a light 'parlour' chair, and made normally of beech..with caned seat and loose squab cushion. **1969** J. GLOAG *Short Dict. Furnit.* (rev. ed.) 674 *Trafalgar chair.* Various types of single and elbow chairs, with nautical symbols such as anchors and coiled ropes in the backs, were made after the Battle of Trafalgar in 1805... Apart from the incorporation of such ornament these chairs were usually of the graceful Regency type with sabre legs. **1979** E. CAVE *Blood Bond* I. ii. 22 A Trafalgar chair with a tapestry seat. **1918** A. HUXLEY *Let.* 30 Oct. (1969) 167 On Trafalgar Day this year Cobby made precisely the same remarks about Lord Nelson and Lady Hamilton as last year. **1977** *Navy News* Sept. 24/1 The Exeter Flotilla's Trafalgar Day service will be held in Exeter Cathedral on Sunday, October 23. **1895** A. PINERO *Notorious Mrs Ebbsmith* II. 103, I assure you, dear fellow, I was within three feet of her when she deliberately Trafalgar Squared me. **1944** J. AGATE *Ego 6* 35 But not until I had taken the floor and Trafalgar Square'd K as blisteringly as I could.

traffic, *sb.* Add: **2. e.** In phr. (*as much as*) *the traffic will bear* or *stand* and varr.: (as much as) the trade or market will tolerate, as much as is economically viable. Also *fig.*

1931 L. STEFFENS *Autobiogr.* v. ii. 853 His wage-earners had their rents raised to all the traffic would bear. **1936** L. C. DOUGLAS *White Banners* vii. 155 We've had all the worry about you that the traffic will stand. **1964** L. DEIGHTON *Funeral in Berlin* xlvii. 294 How much? That's difficult. What do you think the traffic will bear? **1972** *Guardian* 22 May 14/5 The landlord demands a deposit of anywhere from £3 to £150: this is based on what he thinks the traffic will stand. **1976** *Billings* (Montana) *Gaz.* 4 July 2-E/2 The invariable custom is to hire as many deputy superintendents, associate superintendents, assistant superintendents and administrative assistants as the traffic will bear. **1982** T. FITZGIBBON *With Love* I. v. 32 The small neighbourhood shops were..willing to give credit up to a pound, but no more. They knew to a penny what the traffic would bear.

5. a. Also with reference to air travel (usu. prefixed by *air*, and in *Comb.*: see sense 6 below); also *concr.*, the vehicles, etc., collectively.

1911 E. H. HODGKINSON *Tyranny of Speed* viii. 107 He should hear the traffic coming behind him. **1920** H. B. PRATT *Commercial Airships* vii. 72 Terminal Stations will comprise a landing ground or aerodrome, and will be provided with housing sheds, mooring towers.., with traffic offices, gas generating and storage plant, [etc.]. **1930** W. S. MAUGHAM *Cakes & Ale* ii. 34 There was none of the congested traffic of Jermyn Street. **1935** C. G. BURGE *Compl. Bk. Aviation* 136/1 The cheaper way at intermediate towns or where traffic is light, is to assemble the passengers at a central point, and deal with them at the airport. *Ibid.* 402/2 They were convinced that air traffic would only succeed as a new means of transport if.. international co-operation existed. **1938** [see *traffic control*, sense 6 below]. **1951** R. CAMPBELL *Light on Dark Horse* xvi. 223 The traffic used to collect [in London], like lumber on a river. **1974** J. A. FOSTER in G. P. Howard *Airport Econ. Planning* 5 The data to be collected should not only cover the physical facilities of the airport, but should also indicate the degree of utilization, the volume and composition of traffic, [etc.]. **1982** S. SPENDER *China Diary* 104 There was only one line of traffic in each direction.

d. *Telecommunications.* The messages, signals, etc., transmitted through a communication system; the flow or volume of such business.

1878 *Telegraphic Jrnl.* VI. 15/2 A monopoly of cable traffic to America. **1889** *Telephone* 15 Oct. 476/2 A Pacific Cable... The line..would find traffic enough to pay a fair interest on the investment. **1922** W. F. FRIEDMAN in *Bull. U.S. Signal Office Signal Corps* 1 Oct. 15 In modern military operations, a considerable volume of traffic is available for interception by the enemy. **1935** *Times* 26 Oct. 18/7 These are the busiest hours of the day, during which more than two-thirds of the daily traffic is handled. **1947**, etc. [see *ERLANG 2]. **1977** *Times Lit. Suppl.* 25 Feb. 222/5 British Intelligence could decipher German high-grade cipher traffic. **1978** T. ALLBEURY *Lantern Network* iii. 39 He's got his own W/T operator, but sometimes you may have to take his traffic.

6. (sense 5) *traffic area, block* (earlier example), *congestion, consciousness, flow, noise, stack, stream*; *traffic-conscious, -free* adjs.; instrumental, as *traffic-choked, -crammed* adjs.; **traffic analysis** *U.S.*, in Cryptography, the obtaining of information through analysis of patterns of communication without the decipherment of individual messages; hence **traffic analyst**; **traffic artery** orig. *U.S.*, a main or arterial road; **traffic circle** orig. and chiefly *U.S.*, a traffic roundabout; **traffic control**, the regulation of traffic movement through the use of signals or direct commands from authorized persons; a service with this responsibility; in *Aeronaut.*, also as *air traffic control* (see *AIR *sb.*^{1} III. 1); hence **traffic controller**; **traffic cop** *colloq.* (orig. *U.S.*) = *traffic policeman* below; **traffic court** orig. *U.S.*, a court of law with jurisdiction over motoring offences; **traffic engineer** orig. *U.S.*, one who deals with the design and planning of roads and the control of traffic; hence **traffic engineering**; **traffic island**, a raised or marked area in a road to direct traffic and provide refuge for pedestrians crossing the road; **traffic jam** orig. *U.S.*, a condition in which road traffic cannot proceed freely and comes to a standstill; a stoppage of traffic caused by this, or the vehicles caught in it; also *fig.*; hence **traffic-jammed** *a.*; **traffic lane**, a road carriageway for a single line of moving vehicles; an air or sea route designated as a set course for traffic in order to avoid collisions; **traffic light**, † (*a*) a light used for the guidance of aircraft (*obs. rare*); (*b*) freq. in *pl.*, (one of) a set of lights (usu. red, amber, and green) used for automatic control of road traffic, esp. at junctions; **traffic offence**, an infringement of the law by the driver of a motor vehicle; **traffic officer** = *traffic policeman* below; **traffic pattern**, (*a*) *Aeronaut.* (see quot. 1956); (*b*) the characteristic distribution of traffic on a route; also *fig*; **traffic police**, that branch of the police force concerned with road traffic control; hence **traffic policeman**; **traffic-proof** *a.*, of a horse or pony: that can be ridden safely in traffic; so **traffic-proof** *v. trans.*; **traffic sign**, a roadside sign conveying information, warnings, etc., to drivers of motor vehicles; **traffic signal(s** = *traffic light* (*b*) above; **traffic snarl**: see *SNARL *sb.*^{1} 2 b; **traffic ticket** *U.S.*, an official notification of a traffic offence, issued by a traffic warden or the police; **traffic warden**, a person employed to enforce regulations about the parking of motor vehicles and the use of parking meters.

1937 *Hist. Communications Intelligence in U.S.* (Naval Cryptologic Veterans Assoc.) (1982) 29 Information is obtained from communications by..methods short of cryptanalysis, i.e., traffic analysis. **1979** W. J. HOLMES *Double-Edged Secrets* iii. 18 But until the way was cleared by Rochefort we never discussed the subject of traffic analysis. *Ibid.*, Traffic intelligence summaries were produced each day by two traffic analysts. **1933** *Act* 23 & 24 *Geo. V* c. 53 §27 It is expedient that the existing traffic areas under the Road Traffic Act, 1930,..should be varied. **1977** *Chicago Tribune* 2 Oct. XI. B (Advt. Suppl.) 7/4 Washable—great for traffic areas. **1927** *New Masses* June 7/4 In a side street, close to one of the main traffic arteries of the city of the Angels. **1969** *New Scientist* 13 Mar. 560/1 These additional costs of assimilating a traffic artery into an existing urban area are themselves a massive community burden. **1977** E. V. CUNNINGHAM *One-Penny Orange* (1978) xi. 180 In the twenties..Sunset Boulevard was a quiet carriage road and not the major traffic artery it is today. **1896** *Chambers's Jrnl.* 26 Dec. 822/2 The slow speed and traffic blocks in crowded streets are atp to tell considerably against electricity. **1977** *Listener* 23 June 809/3 An uneventful march from Speakers' Corner to a rally in Trafalgar Square. No one who took part that day was ever out of sight of one or more of the scores of so-called traffic cameras that blanket central London and, on these occasions, are connected to the Operations Room [at Scotland Yard]. **1971** *New Scientist* 1 July 5/1 In traffic-choked cities the slender bike is the fastest means of getting from A to B. **1942** *Policy on Rotary Intersections* (Amer. Assoc. State Highway Officials) 1 The name 'traffic circle' is commonly applied to any intersection design based on the one-way movement of vehicles around a central area. **1970** *Rand Daily Mail* 28 Feb. 7/5 When South Africans say 'traffic circle' for the Englishman's 'roundabout', they give precision to the language. **1934** *Punch* 6 June 634/3 Sir William..urged traffic-congestion which six-line bridge would bring about in Strand, and reminded his adversaries that old bridge was as much national monument as Cenotaph. **1968** E. A. Powdrill *Vocab. Land Planning* iii. 55 The case for urban renewal might be based on one or more of the following factors:..traffic congestion, [etc.]. **1948** C. Greatrex in B. Vesey-Fitzgerald *Bk. of Dog* I. 116 The problem of making a dog traffic-conscious is exceedingly difficult to solve... Far more dogs now possess an inborn instinct of some practical dimension, and as a result show caution and traffic consciousness. **1979** G. N. Knight *Indexing* xiii. 176 An entry such as 'London, its happiness before the invention of Coaches and Chairs' induces a wry smile in our traffic-conscious age. **1936** *Spectator* 10 Jan. 56 The British vice-consulate at Varna, Bulgaria, protests that 'traffic-consciousness' is a horrid word. **1931** *Hansard* (Commons) 6 May 385 Installations of automatic traffic control signals. **1935** *Discovery* Jan. 2/2 Col. O'Gorman appeals to Science for aid in assessing the improvement due to any change of traffic control, of road layout, or of road code, etc. **1938** *Encycl. Brit. Bk. of Year* 33/1 Increasing volume of traffic forced the development of airport traffic control systems... Experiments in traffic control began almost 20 years ago (using signal lights). **1962** E. Snow *Other Side of River* (1963) lxxi. 548 How many policemen do you have in New York? Most of ours are for traffic control. **1971** *Flying* Apr. 13/2 With 10 airplanes on the hook in a traffic-control sector that usually has a couple. **1978** L. Deighton *SS-GB* xxiv. 225 The six-ton mobile command-centre—on hand for traffic control during the funeral procession. **1930** *Engineering* 25 July 106/3 Highfield automatic traffic controller... At present the three-light system appears to be most widely favoured. **1938** *Encycl. Brit. Bk. of Year* 33/1 A glass-enclosed tower is usually provided..to give the traffic controller a full view of the airport and its surroundings. **1973** *Times* 20 Mar. 8/4 The statement also referred to the month-long strike by French air traffic controllers. **1963** *Times* 20 Feb. 14/6 As a contrast to her decorative elegance there was the traffic-controller—I hesitate to say policewoman. **1977** G. Markstein *Chance Awakening* lxvii. 209 'Some big shot?' asked the RAF squadron leader... 'Guess so,' said the traffic controller. **1908** S. Ford *Side-Stepping with Shorty* xiv. 227 It'll be some time before Langdon'll be pestered anymore by the traffic cops. *a* **1930** D. H. Lawrence *Phoenix II* (1968) 425 One God is relative to another God until he gets into a machine; and then it's a case for the traffic cop! **1961** B. Crump *Hang on a Minute, Mate* 23 They..returned to the truck to find a traffic cop standing by it. **1975** R. Butler *Where all Girls are Sweeter* vi. 63 There was the usual jam on Putney Bridge... A traffic cop sorted us out. **1919** *Evening Star* (Washington, D.C.) 11 Mar. 24/1 Approval was given last night..to the plan..to establish a traffic court in this city. **1972** R. Hood *Sentencing Motoring Offender* v. 105 The remainder [*sc.* motoring offences] would be dealt with at Traffic Courts. **1973** *Times* 17 Oct. 20/3 A traffic court is where a motorist pays to give harmless vent to his frustrations with his car, with traffic jams and wardens, with parking regulations and with authority in general. **1959** P. Bull *I know Face* ix. 164, I arrived at dusk in Casablanca and was driven at breakneck speed to Marrakesh along a traffic-crammed road, stiff with the results of accidents. **1916** *Proc. 8th Nat. Conf. City Planning* (U.S.) 69 The Traffic Engineer's duties are along two definite lines. **1959** *Daily Tel.* 8 May 23/3 All the committees were acting independently or their work overlapped. None was able to use the traffic engineers who were being used so effectively in the United States... There seems to be lacking in London the kind of adequately-staffed traffic engineering department which has proved essential in our large cities. **1978** *Jrnl. R. Soc. Arts* CXXVI. 425/1 Increasing the throughput of vehicles on an existing highway..is a field in which traffic engineers have notched up considerable successes as the demand has increased. **1931** *Roads & Streets* Dec. 506/1 A new..occupation has been created in the last few years—that of traffic engineering. **1959** Traffic engineering [see *traffic engineer* above]. **1970** P. Laurie *Scotland Yard* ii. 54 They are initiated into the mysteries of traffic engineering. **1940** R. S. Lambert *Ariel & all his Quality* vi. 146 Broadcasting House..is responsible for a big inward and outward traffic flow; yet there is nowhere to park a car. **1978** *Dumfries Courier* 20 Oct. 10/3 The traffic operation and parking areas have been designed to ensure maximum traffic flow at all times. **1968** R. K. Cox *Retail Site Assessment* xii. 140 The unfortunate word *pedestrianization* means the stopping-up of existing streets ..and their conversion into traffic-free areas for use by pedestrians. **1979** R. Perry *Bishop's Pawn* viii. 137 Dieter was taken..towards Zurich along..relatively traffic-free roads. **1931** E. E. Cummings *Let.* 7 Jan. (1969) 119 I've just returned from the place de la Concorde.. where waited on a traffic-island for 2½ hours. **1935** L. MacNeice *Eclogue for Christmas* in *Poems* 16 On all the traffic-islands stand white globes like moons. **1982** 'C. Aird' *Last Respects* x. 101 The driver negotiated the traffic islands with impatience. **1917** I. Crump *Boys' Bk. Policemen* iii. 55 He plunged into the traffic jam at the next street. **1926** *Sunset* Mar. 38 (*heading*) Traffic jams: how Western cities are trying to reduce congestion on down-town streets. **1939** W. Plomer *Dorking Thigh* (1945) 20 With a traffic-jam outside (for they turned up in scores). **1957** *Economist* 21 Dec. 1038/1 The Home Universities conference could hardly have chosen a more important subject for discussion last week than the traffic jam of students which piles up every summer when school leavers put in their bids for university places. **1976** *Liverpool Echo* 7 Dec. 1/1 A Buckingham Palace spokesman said she had travelled from Sandhurst and there had been heavy traffic jams. **1964** *Economist* 1 Aug. 476/1 They sit traffic-jammed in the intractable streets. **1983** *Listener* 1 Sept. 16/1, I was still sitting in my traffic-jammed car five minutes later. **1905** Kipling *Actions & Reactions* (1909) 150 You could not hoist the necessary N.U.C. lights on approaching a traffic-lane because your electrics had short-circuited. **1937** *Times* 13 Apr. p. x/1 An elaborate Memorandum on the Construction and Layout of Roads, covering many subjects, including safety, standard widths, curves, gradients, traffic lanes. **1948** *Jrnl. R. Aeronaut. Soc.* LII. 90 The American omnidirectional range which defines..traffic lanes..between zones of air traffic control at the airports of departure and arrival. **1972** *Daily Tel.* 5 May 3 The Government is taking powers to prosecute masters of British ships caught travelling the wrong way in traffic lanes through the Straits of Dover. **1912** Traffic-light [see *landing-tower* s.v. *Landing *vbl. sb.* 8]. [**1920** W. P. Eno *Sci. Highway Traffic Regulation* viii. 19 (*heading*) Traffic regulation lights.] **1929** *Sat. Even. Post* 16 Nov. 145/1 T is for Traffic Light, bane of all motorists. **1934** *N. & Q.* 3 Nov. 314/1 These signals somewhat resembled our current traffic lights. **1958** J. Cannan *And be a Villain* i. 5 Pulling up for the traffic lights at the turning into the High Street. **1973** D. Barnes *See Woman* I. 111 Stryker travelled northbound, cursing the traffic engineer who had planned the traffic-light sequence. **1978** S. Wilson *Dealer's Move* vii. 116 There were road works ahead, with temporary traffic lights. **1971** 'G. Black' *Time for Pirates* vii. 112, I..paused to listen, hearing nothing but traffic noises. **1960** *Daily Tel.* 29 Jan. 23/4 A 'ticket' system of optional fixed fines for minor traffic offences. **1981** C. Dexter *Dead of Jericho* xxxii. 176 Cheque for £6, being the penalty fixed for the traffic offence detailed on the ticket. **1915** *Policeman's Monthly* Oct. 5/1 Traffic officers in the center of the street are subjected to many hardships. **1971** *Rand Daily Mail* 4 Sept. 5/2 Johannesburg traffic officers will still attend motor accidents in which there is only minor damage. **1956** W. A. Heflin *U.S. Air Force Dict.* 531/1 *Traffic pattern*, a pattern in the air above or about an airdrome, which is normally followed under visual conditions either by aircraft prior to touchdown or by aircraft after takeoff. **1968** L. O'Donnell *Face of Crime* viii. 106 They landed at Kennedy at 8:32..having been delayed half an hour in the traffic pattern. **1977** *Mod. Railways* Dec. 461/3 These gentlemen cannot both be correct so, for the benefit of those of us who are not familiar with traffic patterns of these routes, would somebody please produce some evidence to prove the point one way or the other? **1977** *Chicago Tribune* 2 Oct. xi. 14/3 But the 'island' arrangement—placing furniture in the middle of a room (if it's large enough), thereby leaving space to walk around it—is a good way to create a traffic pattern. **1978** S. Sheldon *Bloodline* vi. 87 The private Boeing 707-320 was making its final approach to Kennedy Airport, gliding out of the stacked-up traffic pattern. **1906** *Collier's* 20 Jan. 22/3 The effort to find out how it feels to be 'regulated' by the traffic police of New York. **1959** *New Statesman* 3 Jan. 6/3 He also disliked anything that gave traffic police more discretion or wider powers. **1980** E. Leather *Duveen Let.* xi. 134 The Renault..would most certainly receive the attentions of the ever alert Traffic Police. **1917** *Wells Fargo Messenger* Mar. 115/3 The traffic policeman at Norman is still in doubt as to what it was that went by. **1940** Auden *Another Time* 91 Let the traffic policemen wear black cotton gloves. **1980** *Listener* 13 Nov. 653/1 Some of my best friends are traffic policemen. **1971** *Pony* Mar. 347/1 A few months ago I was given the task of traffic-proofing a pony. **1977** *Horse & Hound* 14 Jan. 40/4 (Advt.), Gelding..Viceless, traffic proof. **1915** *Policeman's Monthly* Oct. 5/1 Certain of the cities have given a great deal of thought and attention to traffic signs and signals. **1936** C. Day Lewis *Noah & Waters* 15 Do not be deceived by the two-faced traffic signs. **1973** J. Wainwright *Devil you Don't* 42 McGuire threaded the Jag. through the city streets. He obeyed every traffic sign. **1917** *Harper's Mag.* June 70/2 The Bostonian, supposedly sesquipedalian of speech, has reduced 'a pedestrian who crosses streets in disregard of traffic signals' to the compact *jaywalker*. **1934** *Archit. Rev.* LXXV. 184 Traffic signals did not come to Hyde Park Corner until 1932. **1981** 'J. Ross' *Dark Blue & Dangerous* x. 57 A traffic signal which turned red as he approached it. **1963** L. Deighton *Horse under Water* lii. 219 It's the Seville Traffic Control Zone... If it [*sc.* a plane] gets mixed into that traffic stack I'm not sure that I'll be able to sort it out. *a* **1930** D. H. Lawrence *Last Poems* (1932) 264 The minorities that still see the gleam of life Submit abjectly to the blind mechanical traffic-streams of..The stone-blind bourgeois, and the stone-blind bolshevist. **1981** H. R. F. Keating *Go West, Inspector Ghote* xiii. 155 The monster car, slipping easily from one traffic stream to another. **1950** J. D. MacDonald *Brass Cupcake* iii. 31 Every time you get into that car of yours, you'll get a traffic ticket. **1979** *Tucson Mag.* Apr. 33/2 The typical suburbanite..never goes downtown except for an occasional Community Center event or a traffic ticket to contest. **1984** *Miami Herald* 6 Apr. 18A/1 The judge's bill suggests that, by putting a thumb print on the traffic ticket, the scofflaw might be more easy to trace. **1959** *Punch* 25 Feb. 274/1 Any supposed similarity of function between the police and the newly-proposed traffic wardens vanished with the official statement that the wardens 'would help motorists to find parking space'. **1980** J. McNeil *Spy Game* xx. 197 It was a Ford, parked by the opposite kerb on..double yellow lines. A traffic warden was..preparing to write a ticket.

trafficator (træ·fikēⁱtəɹ). [f. Traffic *sb.* + Indic)ator.] A signal arm attached in former times to either side of a motor vehicle which could be raised and illuminated to indicate the direction in which the vehicle was about to turn; also applied loosely to modern indicators (see *Indicator 3 g).
1933 *Autocar* 13 Oct. 733/1 (*caption*) The Lucas Trafficator is concealed when not in use. **1935** *Times* 17 Oct. 8/4 The least expensive Morris Eight is the two-seater which sells at £118, or £120 10s. with bumpers and trafficators. **1945** *Autocar Handbk.* (ed. 18) x. 196 The Trafficator arm is generally retained by some positive catch arrangement so long as it lies within its sheath, though other patterns leave the arm free so that it may be pulled out without injury. **1976** J. I. M. Stewart *Memorial Service* x. 167 He had actually flicked up a hand—jerkily, like an old-fashioned 'trafficator' on a car—in greeting. **1982** N. J. Crisp *Brink* i. 28 Jenkins signalled a right turn... The trafficators clicked.

‖ **tragédie lyrique** (traʒedi lirik). [Fr., lit. 'lyric tragedy'.] A name given to serious French opera of the seventeenth and eighteenth centuries. Cf. *opéra comique* s.v. Opera 3.
1901 W. F. Apthorp *Opera Past & Present* vi. 117 If the Grand Opera—called *tragédie lyrique* when the libretto conformed to the rules of the classic French *tragédie*—was..a quasi-academic adaptation of the Italian *opera seria* to French taste, the *opéra-comique* may be called the natural growth..of..the Italian *opera buffa*. **1947** A. Einstein *Mus. Romantic Era* xix. 358 In France, there was *tragédie lyrique* and *opéra-comique*. **1976** *Early Music* July 285 In 1966, came a virtually complete recording of Rameau's first and perhaps greatest tragédie-lyrique, *Hippolyte et Aricie*.

tragedize, *v.* Add: **3.** (Earlier example.)
1754 D. Garrick *Let.* 31 July (1831) I. 57 Not like those paltry blasts of art employed in raising storms in a tea-cup, such as tragedizing trivial or even ludicrous situations.
tragedizing *vbl. sb.* (examples).
1780 T. Francklin tr. *Lucian's Works* I. 390 Several of them..fell insensibly into the tragedizing vein. **1813** *Theatrical Inquisitor* II. 182 Ye fair, an *Amateur* before you view, Whose love of tragedising sprung from you.

tragedy. Add: **5.** *tragedy-queen* (earlier examples).
1755 C. Charke *Life* 192 Though it was a valuable Gift, but more proper to ornament the Neck of a Country-Housewife, than a Tragedy-Queen. **1819** [see *Sail *v.*¹ 5 b].

trageremics (trēⁱgərī·miks), *sb. pl.* (const. *sing.*). *Linguistics.* [f. *Trager* (see *Trager–Smith), after *phonemics*.] A mock-technical term for the approach to phonemic analysis characteristic of the American linguist George L. Trager. Hence **tragere·mic** *a.* (see quot. 1967).
1963 R. I. McDavid *Mencken's Amer. Lang.* 249 About 1950 E. Bagby Atwood coined *Trageremics* to designate phonemics according to the specifications of George L. Trager. **1967** —— in *Publ. Amer. Dial. Soc.* XLVII. 4 There does seem to be unnecessary virtuosity in displaying for individual entries the whole range of trageremic differences in the low-front vowel range. *Ibid.* 20 By *trageremic* (a term coined by Atwood) I refer to George Trager's analysis of all English syllable nuclei as composed of nine vowels and three semivowels. **1978** *Amer. Speech* LIII. 171, I mastered Trageremics in the middle 1950s.

Tragerian (trēⁱgiᵊ·riăn), *a.* (*sb.*) *Linguistics.* [f. *Trager* (see next) + -ian.] Of, pertaining to, or characteristic of the approach to linguistic analysis of George L. Trager. Occas. as *sb.*, an adherent of this method.
1962 R. P. Stockwell in *Texas Conf. Probl. Linguistic Analysis in Eng.* 10 The Tragerian pronunciation of *baa* that you wondered about. **1965** *Language* XLI. 168 His phonemics is Tragerian, though he makes tentative use of acoustic distinctive features. **1966** A. A. Hill *Promises & Limitations Newest Type Gram. Analysis* 6 Tragerians always began the description of language with phonemes, and ended it..with syntax. **1974** *Amer. Speech* 1971 XLVI. 126 Since I have usually employed a Tragerian transcription in previous publications on English phonology, it will be adhered to in this paper.

Trager–Smith (trēⁱ·gəɹ smi·þ). *Linguistics.* The names of George L. *Trager* (b. 1906) and Henry L. *Smith* (b. 1913), American linguists, co-authors of *An Outline of English Structure* (1951); used *attrib.* with reference to the method of linguistic analysis and phonemic transcription exemplified in this monograph. See also *Smith–Trager.
1955 *Language* XXXI. 313 The Trager–Smith phonology..is the most nearly successful attempt at a statement of the English phonemic system that has yet been made. **1959** Wimsatt & Beardsley in *PMLA* LXXIV. 586 The relation between meter and Trager–Smith linguistics. **1961** *Dissert. Abstr.* XXII. 1990 From the tapes the words were transcribed phonemically according to the Trager–Smith notation. **1962** *Amer. Speech* XXXVII. 154 A Trager–Smith analysis of the Middle English long mid vowels. **1965** A. H. Roberts *Statistical Linguistic Analysis of Amer. Eng.* v. 35 The system of notation in the count of French, Carter, and Koenig uses only one character to represent a diphthong, the Trager–Smith system uses two.

tragic, *a.* Add: **3. b.** *tragic flaw* = *Hamartia.
1913 L. Cooper *Aristotle on Art of Poetry* ii. 40 For Mary, the tragic flaw of the hero, described as an 'error of judgment', or a 'shortcoming', needs immediate illustration. The single Greek word, *hamartia*, lays the emphasis upon the want of insight within the man, but is elastic enough to mean also the outward fault resulting from it. **1950** W. Farnham *Shakespeare's Tragic Frontier* i. 4 In Brutus then, Shakespeare discovered the noble hero with a tragic flaw. By that discovery he made it possible for English tragedy to reach a greatness hitherto attained only by Greek tragedy. **1970** *English Studies* LI. 235 This flaw in the Hegge Pilate..approximates very closely what is generally meant in dramatic criticism as 'tragic flaw', and the Hegge Pilate may be the first tragic hero in English drama.

tragico-. Add: **tragico-farcical** *a.*, combining tragic and farcical elements; **tragico-historical**

a., combining tragic and historical elements.
1913 E. F. Benson *Thorley Weir* vi. 206 Tragico-farcical situations. **1919** T. S. Eliot in *Times Lit. Suppl.* 13 Nov. 637/3 In *Catiline* Jonson conforms, or attempts to conform, to conventions..of tragico-historical drama.

‖ **trahison des clercs** (tra,izoñ de klẽr). [Fr.] The title of Julien Benda's work *La Trahison des Clercs* (1927), used to denote a compromise of intellectual integrity by writers, artists, and thinkers. Cf. *treason of the clerks* s.v. *Treason *sb.* 1 b.
1935 *New Statesman* 26 Oct. 598/1 Is it one more proof that the intellectual should always advise but never govern? Is it one more case of the *trahison des clercs*? **1952** E. Hyams *Soil & Civilization* 127 True, there has been some reaction away from this state of mind among a few intellectuals, a kind of uneasy, if wholesome, *trahison des clercs*. **1968** J. M. Ziman *Public Knowledge* vi. 123 The aim of Science is understanding, not the accumulation of data and formulae... To fail to construct the building, to leave all the bricks lying round in untidy piles, is the *trahison des clercs* of today. **1978** *Listener* 4 May 559/3 Look, they say, terrorism is a phenomenon of our times. Let us..acknowledge that a diplomat..is fair game... I find this *trahison des clercs*.

trail, *sb.*¹ Add: **II. 9.** (Further examples.) Also *N.Z.* and *Austral.* Cf. *nature trail* s.v. *Nature *sb.* 15 a.
1860 J. Burnett *Let.* 15 Mar. in H. F. von Haast *Life & Times Sir Julius von Haast* (1948) viii. 85 Crossed the Alexander stream and struck Mackay's last year's trail. **1939** *WPA Guide to Florida* (1984) I. 117 In Hillsboro River State Park..are overnight cabins, trails, roads, and a museum. **1958** *Tararua* XII. 25 A trail seems to be something narrow and perhaps rather hard to follow—a way marked only by blazes or worn by animals, usually deer. A track seems to be something broader, cut or formed by man. **1968** Mrs. L. B. Johnson *White House Diary* 2 Oct. (1970) 714 A system of urban and rural trails, including the Pacific Coast Trail from Mexico to Canada. **1968** K. Weatherly *Roo Shooter* 47 The ancestors of the roos used this path. It was miles wide and invisible... When this country is closely settled and these trails are sealed, the red kangaroos will die out. **1977** *Times* 23 Apr. 12/5 There are well marked trails for independent hikers. *Ibid.* 12/7 The benefit of camping and picnic grounds, walking and hiking trails. **1982** G. M. Fraser *Flashman & Redskins* 161 From Santa Fe to Algodones on the river the trail was dotted that night with emigrant camp-fires.
10*. *Radio* and *Television.* A piece of advance publicity (often an excerpt) broadcast prior to the transmission of a programme. Cf. *Trail *v.*¹ 3*; *Trailer *sb.* 4 b.
1973 *Listener* 6 Dec. 798/1 Accidentally switching on early..on Radio 3..I heard..off-putting trails. **1980** *Broadcast* 7 July 24/3 The TV Presentation Department..make hundreds of commercials every year in the form of programme trails.
V. 16. *trail-blazer* (later *fig.* examples), *-blazing* vbl. sb. and ppl. adj., *-breaking* vbl. sb. and ppl. adj., *-cutter, -herd, -herder, -man*; **trail bike** orig. *U.S.*, a motor-cycle designed for use on country tracks rather than on roads; **trail boss** *U.S.*, a foreman in charge of a cattle-drive; **trail head** *N. Amer.*, the beginning of a trail for walkers (occas. also for skiers); an organizational centre at such a place; **trail-hound**, (*a*) [*Hound *sb.*¹ 4 e] = *trail-blazer* above; (*b*) a small hound bred for the sport of hound trailing; **Trail of Tears** *U.S.* (see quot. 1930); **trail-riding**, motor-cycling with a trail bike; **trail-rope**, (*a*) substitute for def.: *U.S.*, a long rope used for tethering animals loosely; (earlier example); **trailway** *N. Amer.*, a route through rough country cleared and maintained for recreational walking.
1969 *Time* 12 Sept. 17 Anyone hoping to escape the..cities for the quiet beauty of our woods, mountains or deserts is in for a rude shock. He is greeted by the rattling snarl of trail bikes, dune-buggies and the like. **1972** *Fairbanks* (Alaska) *Daily News-Miner* 3 Nov. 23/5 (Advt.), Extended bumper on rear for snowmobile, trail bikes. **1976** *New Motorcycling Monthly* Oct. 4/4 Yamaha, of course, have reincarnated the good old 500cc four-stroke single, but in trail-bike trim. **1937** *Discovery* July p. lix/1 Trail blazers of science. **1972** *Daily Colonist* (Victoria, B.C.) 3 Feb. 8/1 Agriculture Minister Shelford lauded his party as being trailblazers in humanity through fiscal astuteness during the throne speech. **1934** Webster, Trail blazing. **1957** V. Packard *Hidden Persuaders* xxi. 233 *Tide*, the merchandisers' journal, admonished America's merchandisers to pay attention to this trail-blazing development as it might be 'tomorrow's marketing target.' **1968** *Globe & Mail* (Toronto) 3 Feb. 43/3 (Advt.), Acres of wonderful wood for trail blazing and riding. **1971** *Advocate-News* (Barbados) 17 Sept. (Guyana Suppl.) p. iv/3 Volunteers..cleared the last few feet of bush for their historic meeting on top of a hill called Point Jason (after a trail-blazing pioneer who supervises the project). **1973** C. Bonington *Next Horizon* xxi. 279 We all agreed that it [*sc.* a climb] was as hard as anything we had ever done, with very little to show for each day's trail-blazing. **1890** *Stock Grower & Farmer* 21 Jan. 6/3 Trail bosses bronzed from exposure..are familiar sights. **1921** [see *Pile *v.*² 3 c]. **1977** *Daily Mirror* 15 Mar. 24/1 What was the name of the actor who played the trail boss in the TV Western series 'Rawhide'? **1912** L. J. Vance *Destroying Angel* xviii. 232 'Must I make talk, then?' she demanded. 'If we must, I suppose—you'll have to show the way. My mind's hardly equal to trail-breaking to-day.' **1965** T. A. Sebeok in *Language* XLI. 80 In this trailbreaking paper, he [*sc.* Trubetzkoy] reduced the supposed multiplicity of vowel patterns to a small number of symmetrical models. **1971** C. Bonington *Annapurna South Face* xiii. 171 Even in descent it took him an hour, and without his trail-breaking from above Martin and Mike Thompson would have had an exhausting time forcing the route from below. **1858** *Brit. Colonist* (Victoria, B.C.) 11 Dec. 2/4 The majority of the Lillooet trail cutters would have remained had it not been grossly mismanaged. **1903** A. Adams *Log of Cowboy* vii. 88 Four..strange men..representing themselves as trail cutters. **1958** *Edmonton* (Alberta) *Jrnl.* 24 June 46/1 The trail-cutters work a four-month season in the winter. **1971** *Islander* (Victoria, B.C.) 30 May 12/4 Another satisfactory water supply can be found at the southern end of the beach, just past the trail head. **1976** *Stillwater* (Montana) *News* 1 July 12/3 Backpackers, fishermen, day hikers, or anyone else using outdoor trails, should sign in on the log book at the trailhead where these are available. **1981** *Nordic Skiing* Jan. 48/2 The Warming Hut on Butternut Lake serves as the trailhead where a skier can..arrange for instruction, rentals, accessories, [etc.]. **1885** *Weekly New Mexican Rev.* 18 June 1/3 The trail herds in Colfax county must go forward or turn back at once. **1962** G. MacEwan *Blazing Old Cattle Trail* i. 1 Ever since the Patriarch, Abraham,..stockmen have been driving trail herds to far places. **1890** *Stock Grower & Farmer* 19 Apr. 3/3 Cattle inspectors of New Mexico were holding up trail herders for one and one-half cents per head. **1931** *Times Lit. Suppl.* 29 Oct. 839/2 The writer is what in her mountaineering vernacular might be called 'a trail hound'. **1972** *Shooting Times & Country Mag.* 27 May 27/1 From all this evolved the trail hound, a smaller, lighter type altogether than his near relative, the Fell foxhound. **1978** R. Hill *Pinch of Snuff* vii. 73 She [*sc.* a cat] was born on a Cumberland farm and reckons she's a trail-hound. **1858** *Brit. Colonist* (Victoria, B.C.) 27 Dec. 3/2 The man..was no trail-man but a stranger. **1891** *Harper's Mag.* Nov. 886/2 The trail-men are sent out to cut what in general parlance would be called a path. **1930** E. Ferber *Cimarron* 40 Tears came to his own eyes when he spoke of that blot on southern civilization, the Trail of Tears, in which the Cherokees, a peaceful and home-loving Indian tribe, were torn [1838–9] from the land which a government had given them by sworn treaty, to be sent far away on a march which, from cold, hunger, exposure, and heartbreak, was marked by bleaching bones from Georgia to Oklahoma. **1978** *Peace News* 6 Oct. 7/2 It was named after the many Long Walks since the Andrew Jackson presidency, including walks like the Trail of Tears in which the Indian people were forced to trek vast distances overland as an expansionist government laid claim to their traditional homelands. **1984** *Miami Herald* 6 Apr. 6A/2 Tribal leaders are calling the reunion the most important event for the Cherokee Nation since the Indians were driven from their southern lands in the 1838 'Trail of Tears'. **1931** C. Aldin in Hunloke & Aldin *Riding* vi. 105 Trail riding..gives us a day's riding *with* a picnic, and teaches us where the side tracks and bypaths on a place like Exmoor lead to. **1979** *Daily Tel.* 13 Jan. 8 These tracks..provide great scope for that non-competitive and gentle form of motor-cycling known as trail-riding. **1826** G. C. Sibley *Diary* 15 Mar. in A. B. Hulbert *Southwest on Turquoise Trail* (1933) 162, I have paid away the following sums, since I left Sta. Fee ..14 trail ropes, 14·00. **1939** *Appalachian Trailway News* July 6/2 The matter of foremost importance was..to obtain the state recognition and interest in the Trailway project. **1940** *Ibid.* Jan. 20/1 The Appalachian Trail or Trailway is entirely a voluntary amateur project. **1972** E. Wigginton *Foxfire Bk.* 276 We'd gone walkin' along th' trailway. **1978** *Globe & Mail* (Toronto) 15 Feb. 1/7 About one-third of the route is already owned by the province and the draft plan calls for the gradual acquisition of a 60-foot-wide 'trailway' from present landowners.

trail, *v.*¹ Add: **I. 1. d.** Phr. *to trail one's coat*, to seek to pick a quarrel; to be provocative in one's conduct. Cf. *to drag his coat-tails, so that some one may tread on them* s.v. Coat-tail.
[**1864**: see Trailer 1.] **1877** C. M. Yonge *Womankind* xxv. 216 Party spirit is equally ready to give offence and to watch for it. It will trail its coat like the Irishman in the fair. **1923** *Daily Mail* 7 Feb. 6 This risk [of war] is greatly increased by the presence of British troops at Constantinople and Chanak. Why should we thus be 'trailing our coats' before the Turks? **1950** D. Divine *King of Fassarai* xxxii. 291, I wouldn't put it in a report!.. I don't trail my coat. **1974** *Times* 4 Nov. 15/1 Nobody trails his coat for another election... There is to be no Commons division on the crisis in agriculture. **1980** J. Ditton *Copley's Hunch* I. ii. 35, I was trailing my coat... Trying to get the Luftwaffe to come up and fight.
3. a. Also, to cause (a person) to accompany or follow one, esp. reluctantly.
1914 W. Owen *Let.* 24 May (1967) 253 Tofield..is married, and trails a French wife about with him, from Berlitz School to Berlitz School. **1977** 'D. Rutherford' *Return Load* i. 21 Sally..trailing a reluctant Josie, was heading for the exit.
3*. To give advance notice of (a radio or television programme). Also *transf.* Cf. *Trail *sb.*¹ 10*, *Trailer *sb.* 4 b.
1941 *B.B.C. Gloss. Broadcasting Terms* 33 *Trail* (*v. trans.*), to draw the attention of listeners to a forthcoming programme or other event of broadcasting importance by means of announcements, recorded excerpts, or other methods calculated to make it widely known. **1942** 'G. Orwell' *Diary* 14 Aug. in *Coll. Essays* (1968) II. 443 Horrabin was broadcasting today... This had been extensively trailed and advertised beforehand. **1960** *Guardian* 8 Nov. 7/2 It remains to me an object of mystery..why the BBC trailed this programme..as unsuitable for young people. **1976** *Daily Tel.* 20 Dec. 8 Powell blamed newspapers for having ignored his embargo—journalists usually receive copies of his speeches a day or two beforehand—but for years his speeches have been 'trailed' without complaint. **1978** *Times* 7 Aug. 12/5 At least by trailing their message on the envelope the senders have..reduced wear on my paper knife. **1980** *Musicians Only* 26 Apr. 11/5 Released to trail a three album blockbuster.
II. 6. c. Also with *in*. *U.S.*
1875 *Fur, Fin & Feather* (ed. 3) 112 Light and drink; drop off and trail in. **1907** S. E. White *Arizona Nights* xvi. 234 With exultant cackles of joy they'd trail in, reachin' out like quarter-horses.
7. b. (Later examples.)
1967 W. Styron *Confessions of Nat Turner* I. 32, I heard Hark's voice trail off in something like a stifled laugh, a gurgle of satisfaction. **1982** *Times* 16 June 17/1 The export expansion should trail off substantially this year.
III. 9. a. Also in *gen.* use, to follow.
1915 H. L. Wilson *Ruggles of Red Gap* (1917) iv. 79 Think of those two poor fellows trailing you over Paris yesterday trying to save you from yourself. **1925** H. L. Foster *Trop. Tramp Tourists* 70 We trailed the other steamer. We trailed her through the *Boca Chica*... We trailed her past the little forts. **1945** B. Macdonald *Egg & I* (1946) xxiii. 228 Sport and the puppy trailed me everywhere, whining and begging me to explain the smoke and excitement. **1957** 'R. Farre' *Seal Morning* ii. 16 No sooner was she past infancy than Lora [*sc.* a seal] started to waddle after me round the croft and trail me over to the byre.
b. To lag behind (someone or something), in a contest, comparison, etc. Also *intr.*
1957 *Times* 6 Sept. 13/2 Hansen's best work came after he had trailed for the first four rounds. **1961** *Wall St. Jrnl.* 24 Mar. 1/1 The value of contracts for residential building awarded last month trailed February, 1960, by 12%. **1972** *Guardian* 10 Aug. 2/3 The Harris Poll today shows that Senator McGovern now trails President Nixon by 23 points. **1979** *Sci. Amer.* Nov. 56/1 Diabetes mellitus and its complications are now thought to be the third leading cause of death in the U.S., trailing only cardiovascular disease and cancer. **1980** *Times* 3 Nov. 2/1 Most MPs seem to expect Mr Silkin to come third with between 30 to 40 votes and Mr Shore to trail with between 20 and 30. **1983** *Times* 19 Feb. 8/4 A few months ago..she was trailing Mr Daley.
IV. 15. *trans. Bowls.* To force (the jack) further up the green with one's bowl.
1908 J. M. Pretsell *Game of Bowls* xi. 194 If a bowl trail the jack through between, and past the line square to the back of, the stationary bowls, it shall score 3. **1923** J. A. Manson *Bowling* 84 The Bowler is required to trail the jack, his own bowl accompanying or 'hugging' it, between the stationary bowls over both of the horizontal lines. **1975** *Oxf. Compan. Sports & Games* 97/2 Occasionally a bowler delivers a bowl which runs on to the jack and stays with it while pushing it a foot or so farther up the green. Basically this is a draw shot delivered with a marginal increase of strength with the object of trailing the jack to a more advantageous position.

trai·lable, *a.* *U.S.* and *Austral.* [f. Trail *v.*¹ + -able.] Of a boat: that may be towed on a trailer behind a motor vehicle; = *Trailerable *a.*
1976 *N.Y. Times* 5 Sept. v. 11 As slip and mooring space becomes more difficult to find, the trailable boat represents an alternative—it can be dry sailed from a boatyard or moored at home. **1977** *Herald* (Melbourne) 17 Jan. 19/2 On the Saturday afternoon trailable yachts, Flying Fifteens, Fireballs and International Cadets will compete.

trailed, *ppl. a.* Add: **2. b.** *trailed slip* (Ceramics), a slip used for decorating pottery by applying it through a nozzle or spout.
1957 Mankowitz & Haggar *Conc. Encycl. Eng. Pott. & Porc.* 15/1 Staffordshire trailed slip posset-pots. *Ibid.* 223/1 They are decorated with 'trailed slip'. **1971** *Country Life* 27 May 1303/1 Trailed slip is the method of pouring out from a cone or can exactly as good cooks decorate birthday cakes with icing sugar. **1974** Savage & Newman *Illustr. Dict. Ceramics* 294 Lead-glazed ware decorated with trailed slip.
3. Publicized in advance. Usu. prefixed by advbs. Cf. *Trail *v.*¹ 3*.
1958 *Spectator* 22 Aug. 247/2 A much-trailed BBC investigation. **1981** *Economist* 20 June 14/1 Ministers at this week's well-trailed cabinet meeting listened to her lectures on the need to cut their spending.

trailer, *sb.* Add: **4. a.** Also *fig.*
1898 J. London *Let.* 6 Dec. (1966) 8 Sent out in this mail, 'trailers' after articles I mailed last September. **1941** H. G. Wells *You can't be too Careful* III. i. 114, I join with Mrs Richard Tewler in deploring the inaudibility of Mrs Humbelay. If only we could have heard those lost trailers of hers, we might have benefited greatly from her ..wisdom. **1952** G. Raverat *Period Piece* vi. 112 This remark was..a sort of trailer, which she hoped might lead to more information.
b. orig. in *Cinematogr.* An excerpt of a film, broadcast, etc., used as advance publicity. Also *transf.* and *fig.*
1928 *N.Y. Times* 11 Mar. VIII. 6 *A trailer*, a few hundred feet of film announcing a forthcoming picture. **1930** *Dancing Times* July 360/2 In a 'trailer' advertising the film, it is announced that the producers have aimed more at entertainment than historical accuracy. **1941** *B.B.C. Gloss. Broadcasting Terms* 33 *Trailer*, microphone announcement or short descriptive broadcast, designed to advertise a forthcoming programme or other event of broadcasting importance. **1942** *Punch* 4 Mar. 169/1 The war..has only just started. The trailer will have given you an idea of what it will be like. **1959** *New Statesman* 21 Mar. 403/2 A quasi-newscaster, giving verbal trailers

of coming attractions, does not fulfil this function. **1959** *Washington Post* 26 Dec. A19/2 Then some wisenheimer from the agency decided we needed a trailer. **1966** *Listener* 19 May 737/1 Whether there is much point in playing a section of a work, except as a trailer, is another matter. **1971** *Daily Tel.* 28 Jan. 1/1 Mr Barber, Chancellor of the Exchequer, used a lunch-time speech yesterday as a 'trailer' for the White Paper on Public Expenditure.. which is to be published this afternoon. **1977** J. Aiken *Last Movement* i. 9 The evening's sunshine was only a trailer for spring. **1978** *Radio Times* 28 Jan.–3 Feb. 70/1 Once upon a time, in the cinema of my youth, there were trailers—unblushingly commercial attempts to lure in the customers by juxtaposing all the dangerous and sexy bits of the film and overlaying them with the most blatant sales pitch of plots and star appeal.
6. a. Now usu. an unpowered vehicle towed behind a car or truck, etc.; *spec.* (chiefly *U.S.*) = Caravan 4 in Dict. and Suppl.
1926 *Kansas City* (Missouri) *Star* 11 June, On the Victory highway most any day now one may see the migratory harvesters—a few walking,..more with their families in cars, and a trailer behind, carrying tents, bedding, and cooking utensils. **1931** J. H. Stone *Caravanning & Camping-Out* x. 64 These trailer-caravans are made so exceedingly light..that a low-power motor can draw them with ease. A trailer runs on two wheels or four. **1951** W. Faulkner *Requiem for Nun* 246 Living now (with now a wife..and..after that a wife and children) in automobile trailers or G.I. barracks on the outskirts of liberal arts colleges. **1966** *Listener* 9 June 839/3 The first motor-diesel vessel to carry caravans—or, as the Americans call them, trailers—across the ocean is now being designed here in San Francisco. **1976** *New Yorker* 17 May 31/1 The miniature control room of a big white mobile broadcast trailer. **1977** P. Way *Super-Celeste* I. 44 Family men from Sydney..on a fishing weekend—their wives.. in the trailers in the park.
c. *trailer-on-flatcar*, used to denote a system of freight transport whereby trailers (and other unaccompanied road freight vehicles) are carried on railway cars. Cf. *Pick-a-back *adv. phr.* (*a.*, *sb.*) b (*b*). orig. *U.S.*
1954 [see *Pick-a-back *adv. phr.* (*a.*, *sb.*) b (*b*)]. **1964** [see *TOFC* s.v. *T* 6 a]. **1979** *Railway Gaz. Internat.* Aug. 719/2 Piggyback, otherwise known as trailer-on-flatcar (TOFC), has proved to be an area of strong.. growth.
9. *attrib.* and *Comb.*, as (sense 4 b) *trailer film*; (sense 6 a) *trailer caravan, hitch* [Hitch *sb.* 6], *-wagon*; **trailer camp** *U.S.*, an area where caravans may be temporarily or permanently parked; similarly **trailer court, park**; cf. *caravan park, site* s.v. *Caravan *sb.* 4; **trailer home** *U.S.* = *mobile home* s.v. *Mobile *a.* 1 h; also **trailer house**; **trailer tent**, a tent which is attached to and erected on a trailer; **trailer-truck** *U.S.*, an articulated lorry; cf. *tractor-trailer* s.v. *Tractor 4.
1921 *Outing* Apr. 39/2 (Advt.), Union Trailer Camp... Whether a week-end jaunt or a vacation tour, this outdoor palace makes it a real one... Your car can draw it easily. **1980** M. Gordon *Company of Women* (1981) III. 254 Mothers bringing their children up in trailer camps. **1930** *Motor Body Building* LI. 98/1 There are, of course, already a considerable number of trailer caravans standardised by body builders specialising in this type of vehicle. **1931** [see sense 6 a above]. **1979** W. H. Canaway *Solid Gold Buddha* xi. 77 The trailer..served him as living quarters and producer's office... He woke to find the trailer-caravan in darkness. **1939** E. S. Gardner *D.A. draws Circle* (1941) xiv. 273 How about taking these folks down to the trailer court, Bill? **1979** *Arizona Daily Star* 5 Aug. (Advt. Section) 14/5, 56 space Travel Trailer Court on Wetmore near new shopping center. Can be converted to regular trailer court. **1941** *Electronic Engin.* XIV. 412 The 'trailer' film, interspersed with regular features, has been recognised by a number of manufacturers as a valuable advertising medium. **1953** C. Armstrong *Catch-as-catch-Can* viii. 69 Do you understand how a trailer hitch operates? **1972** D. E. Westlake *Bank Shot* viii. 58 It has a trailer hitch... It doesn't have any wheels. **1940** H. G. Wells *New World Order* 180 In such large open countries as the United States there has been a considerable development of the mobile home in recent years. People haul a trailer-home behind their cars and become seasonal nomads. **1979** T. Gifford *Hollywood Gothic* xxx. 303 He followed her into the neat little trailer home. **1954**, **1969** Trailer house [see *Mobile *a.* 1 h]. **1975** *Budget* (Sugarcreek, Ohio) 20 Mar. 16/2 They move into a trailer house located on his son-in-law, Ben Peachys farm. **1947** *Daily Oklahoman* (Oklahoma City) 21 Sept. D-6/1 The verdict may well point to a bust in the boom enjoyed this year by the nation's tourist camps, hotels, motels, trailer parks,..and restaurants. **1979** *Arizona Daily Star* 5 Aug. B1/6 A talented cook and editor of her trailer-park newsletter, Rosaaen said she knows why so many women like her are working today. **1971** *Rand Daily Mail* 4 Sept. 10/4 The largest selling trailer tent in Scandinavia... Independent suspension,..easy towing and light weight make it easy on your car. **1981** *West Lancs. Even. Gaz.* 25 Apr. 9 (Advt.), Trailer tent. **1958** A. Budrys *Edge of Sea* in Aldiss & Harrison *Decade the 1950s* (1976) 54 A long-haul trailer-truck driver. **1976** Trailer-truck [see *Rig *sb.*⁴ 3 b]. **1904** Trailer-wagon [in Dict., sense 6 a].

trai·ler, *v.* [f. the sb.] **1.** *trans.* **a.** To advertise or publicize in advance, esp. by the use of excerpts.
1965 *Observer* 5 Sept. 36/1 French 1964 vintages..are already trailered in some American liquor stores as 'the most heralded vintage of the century'. **1977** *Time Out* 28 Jan.–3 Feb. 5/4 Originally scheduled for broadcast last April and 'trailered' on the air, it has never in fact been run. **1979** *Internat. Jrnl. Sociol. of Law* Feb. 99 The book is reminiscent of a Hollywood film, in the sense of its having been extensively trailered. **1983** *Daily Tel.* 27 Aug. 29/2 *The Winds of War Preview*, excerpts from the series which does not start for another fortnight and has already been heavily trailered.
b. Chiefly *N. Amer.* and *Austral.* To transport on a trailer.
1971 *Islander* (Victoria, B.C.) 31 Jan. 7/2 For most of us, snowmobiling requires that we frequently trailer our machines from where we live to more suitable operating locales. **1976–7** *Sea Spray* (N.Z.) Dec./Jan. 55/3 Cover for a vessel while being trailered may be overlooked. **1977** *Austral. Sailing* Jan. 48/1 (Advt.), A little ship like the Marieholm is easy to trailer. **1984** *Gainesville* (Florida) *Sun* 27 Mar. 6B/1 An end to being trailered, to wearing halters, bridles, bits and saddles.
2. *intr.* **a.** To travel or live in a trailer. **b.** To give advance publicity.
1974 *Listener* 17 Jan. 93/3 [They] give up trailering around and settle back near the old folks. **1975** *Time Out* 22 Aug. 3/3 Trailering with stuff like 'The Leisure Press At Bay'.
Hence **trai·lered** *ppl. a.*, towing a trailer, having a trailer (sense *6 a) attached.
1965 'W. Haggard' *Hard Sell* ix. 97 The oil came up from Genoa in convoys of trailered lorries.

trai·lerable, *a.* Chiefly *N. Amer.* and *Austral.* [f. Trailer *sb.* 6 + -able.] Of a boat: that may be transported on a trailer attached to a motor vehicle; = *Trailable *a.*
1971 *Austral. Sailing* Jan. 6/1 (Advt.), The new standard for small yachts. Eyecatching new trailerable mini ocean racer. **1976–7** *Sea Spray* (N.Z.) Dec./Jan. 112/1 (Advt.), 7m Flybridge trailerable Cruiser. **1980** *Outdoor Life* (U.S.) (Northeast ed.) Oct. 152/3 (Advt.), Boat kits—22 trailerable models.

trai·lering, *vbl. sb.* *N. Amer.* [f. *Trailer *v.* + -ing¹.] The act or practice of travelling with or living in a caravan.
1938 *Amer. Speech* XIII. 196 Trailering..Galahading.. newspapering. **1967** (*title*) Woodall's trailering parks and campgrounds. *Ibid.* 33/1 Along Highway 190 from Oaxaca lies the..colonial village of San Cristobal... A bit of offbeat trailering, well worth the trip. **1973** *Daily Colonist* (Victoria, B.C.) 12 Oct. 25/1 A friend of theirs who had retired to a life of trailering told them of his experiences. **1978** *Sunday Sun-Times* (Chicago) 1 Jan. 122/2 Proper weight distribution is essential to insure safe trailering and should be checked before pulling out of the driveway.

trailerite (trēⁱ·lərəit). *N. Amer.* [f. Trailer *sb.* + -ite¹.] One who lives in or travels by caravan.
1940 *Capital* (Topeka, Kansas) 28 Jan. 16B/5 Many trailerites, caught unprepared by Florida's worst cold spell. **1977** *Globe & Mail* (Toronto) 7 Mar. 8/7 Your average trailerite has to be a..jack-of-other-trades to get along while prowling the highways and byways.

trailing, *vbl. sb.* **1. c.** Delete quots. 1873 and substitute earlier example. Now *Obs.*
1775 *Ann. Gaming* viii. 105 What now gives the peculiar advantage to the mace over the cue, is what has been artfully introduced by professed players, under the name of trailing, which is following the ball with the mace to such a convenient distance from the other ball as to make it an easy hazard.
d. A form of bowling played on Scottish greens, the object being to trail or carry the jack into a semicircle drawn beyond two bowls placed three feet apart.
1902 *Encycl. Brit.* XXVI. 329/2 In trailing, two bowls are laid on the turf..and a jack is then deposited equidistant from each bowl... A semicircle is then drawn behind the bowls with a radius of nine feet from the jack. **1923** J. A. Manson *Bowling* 84 Trailing is the section of the Points game which is most worthy of attention.
e. *Ceramics.* A method of decorating pottery by applying slip or glaze through a nozzle or spout. (See also quots. 1960, 1968.)
1940 B. Leach *Potter's Bk.* vi. 145 Glazes are applied by dipping, double dipping..dripping, splashing and trailing. **1960** C. Winick *Dict. Anthropol.* 543/2 *Trailing*, a technique of making broad incised lines in pottery. **1968** J. Arnold *Shell Bk. Country Crafts* xix. 241 Glazes are applied in a liquid state, either by immersion or by brush-work; this is called trailing. *a* **1977** *Harrison Mayer Ltd. Catal.* 18/2 Slip decoration: trailing, feathering. Slip can be applied by all the usual painting, pouring, trailing and dipping methods.
f. The advance broadcasting of excerpts of films, programmes, etc., as a form of publicity. Cf. *Trail *v.*¹ 3*.
1961 *Listener* 17 Aug. 254/3 The trailing of future programmes by announcement or sampling..now seems to be overdone—especially those repeated alluring snippets of coming films which could equally well be false starts of the next programme. **1978** *Broadcast* 20 Nov. 19/3 Intensive trailing on radio and TV.

trailing, *ppl. a.* Add: **1. b.** Also in the names of plants with a trailing habit; *trailing arbutus* = *New England mayflower* s.v. *New England b; also *fig.*
1784 *Mem. Amer. Acad. Arts & Sci.* I. 413 *Cuscuta*... Trailing Cockspur... Borders of brooks and ditches. **1785** H. Marshall *Arbustrum Americanum* 42 Trailing Arbutus..grows naturally upon northern hills, or mountains. **1813** H. Muhlenberg *Catalogus Plantarum Amer. Septentrionalis* 91 *Salix prostrata*. Trailing willow. *Ibid.* 93 *Juniperus prostrata*. Trailing juniper. **1855** Trailing arbutus [see *New England mayflower* s.v. *New England b]. **1861** *Trans. Illinois Agric. Soc.* IV. 462 We have on the lake shore a beautiful trailing evergreen—the Trailing Juniper. **1878** R. T. Cooke *Happy Dodd* 347 A profusion of trailing pine had been stored away in the barn cellar, before frost came. **1899** M. Going *Field, Forest, & Wayside Flowers* 251 The lycopodiums..under the name of..'club-moss', or 'trailing-evergreen', are familiar to almost every one who has summered in New England. **1939** Wodehouse *Uncle Fred in Springtime* i. 18 The male, Barny, was calling me a trailing arbutus..and The Subject was talking about horsewhips. **1979** *United States 1980/81* (Penguin Travel Guides) 633 Among the many woodland trails are no less than 600 varieties of flowering plants—including the trailing yew, unique to this island.
2. c. *trailing vortex* (see quot. 1969).
1929 *Proc. R. Soc.* A. CXXIII. 440 The flow behind the screw is the same as if the screw surface formed by the trailing vortices was rigid. **1949** O. G. Sutton *Sci. of Flight* iv. 112 The trailing vortices actually spring from two wing-tip vortices which, in flight, form just inside the wing tips. **1969** *Gloss. Aeronaut. & Astronaut. Terms* (*B.S.I.*) iv. 15 *Trailing vortex*, a vortex extending downstream from the surface of a body.

trailing edge. [Trailing *ppl. a.*] **1.** The rear edge of a moving body; *spec.* in *Aeronaut.*, that of a wing or other part of an aircraft.
1909 *Flight* 3 July 390/1 These ribs overlap the rear spar and form a flexible trailing edge. **1934** *Flight* 8 Feb. 123/1 The split trailing edge flaps effect a big reduction in landing speed and length of glide. **1962** [see *Riblet]. **1969** *Daily Tel.* 12 Nov. 14/6 Warning red lights mounted in the interior trailing-edge of the doors come on when a door is opened. **1971** I. G. Gass et al. *Understanding Earth* xix. 263/1 The newly generated crust and its upper mantle is effectively welded to the plate's trailing edge.
2. *Electronics.* The part of a pulse in which the amplitude diminishes.
1945 *Nature* 15 Sept. 319/2 The end or 'trailing edge' [of the pulse] marks the beginning of an invaluable clear period in which the radar echoes..can be received free from the overlaying and interfering effect of the primary signal. **1967** *Electronics* 6 Mar. 160/1 Nearly all of the earlier logic systems were based on trailing-edge triggering, so many engineers consider it normal. **1973** *Sci. Amer.* May 109/2 The trailing edge of the next pulse, the third in the series, causes flip-flop A to change state.

trailside, *a.*² (and *sb.*) [f. Trail *sb.*¹ + Side *sb.*¹] Situated at the side of a man-made trail. Occas. as *sb.*, the side of a trail.
1943 F. Gardner *Philippine Indic Stud.* v. 43 A bamboo letter is fastened in a cleft stick and placed by the trailside. **1980** *Beautiful Brit. Columbia* Summer 33 By now the crowd of hikers has thinned out into an ambling string, leisurely pacing themselves with time to..lie back on a trailside carpet of heather and dwarf spirea. **1981** *Nordic Skiing* Jan. 4/3 These are generally outlined on a board at the touring center, on trail maps, or on trailside signs.

train, *sb.*¹ Add: **III. 12. b.** Freq. in *train of thought.* (Further, including *techn.*, examples.)
1651 Hobbes *Leviathan* I. iii. 8 By Consequence, or Trayne of Thoughts, I understand that succession of one Thought to another, which is called (to distinguish it from Discourse in words) Mentall Discourse. **1770** G. White *Let.* 19 Feb. in *Selborne* (1789) II. iii. 125 Your observation..struck me so forcibly, that I naturally fell into a train of thought that led me to consider whether the fact was so. **1899** W. James *Talks* xv. 190 Our habitual associations of ideas, trains of thought, and sequences of action, might thus be consequences of the succession of currents in our nervous systems. **1912** *Proc. R. Soc.* LXXXVII. 93 The electric disturbance produced by a lightning discharge..is probably either a solitary wave or a very short train of waves. **1948** *Proc. IRE* XXXVI. 1457/1 A train of reset pulses is applied to the shift register. *a* **1953** E. O'Neill *Hughie* (1959) 34 [*stage direction*] His train of thought interrupted, irritably. **1955** F. O'Connor *Wise Blood* i. 13 Mrs. Hitchcock lost her train of talk. 'I guess you're on your way to visit somebody?' she asked. **1959** F. Astaire *Steps in Time* (1960) i. 6 It is the easiest thing in the world to become discouraged by a well-meant suggestion which may throw you off your original train of thought. **1967** *Electronics* 6 Mar. 130/2 The circuit will oscillate, simultaneously generating a sawtooth wave and pulse train. **1978** *Nature* 4 May 57/1 When the current with one channel present was recorded for a longer time, characteristic trains of short impulses were observed, separated by relatively long intervals when the channel did not conduct.
16. a. *armoured train*: see *Armoured *ppl. a.* 2 a; *train de luxe*: see Luxe 2.
c. *to pull a train*: see *Pull *v.* 11 g.
IV. 20. b. (Earlier example.)
1783 *Quebec Gaz.* 22 May 2/1 No person shall come with traines, carts or other carriages, loaded with hay, straw or wood within the limits of the market-place.
22. b. *train crash, crew, fare, hostess, ride, station* (U.S.), *-time* (earlier example), *travel, whistle, -wrecker* (earlier example), *-yard*; **train-boy**, (*b*) (earlier example); **train call** *Theatr.*, (a notice of) the time for touring performers to catch a train to the next tour stop; † **train caller**, a railway official who announces the destinations of departing trains (see quot. 1921) (*obs.*); **train-jumper**

orig. *U.S.*, one who travels by train without paying the fare; so **train-jumping** *vbl. sb.*; **train master, trainmaster** *U.S.*, a person in charge of a train or trains; *spec.* a railway official responsible for the movement of all trains over a certain stretch of line; † **train porter**, a railway official in charge of a train over a single-line section of railway (*obs.*); also *attrib.*; **train set**, (*a*) a set of trains, tracks, etc., required for a model railway; (*b*) a set of wagons or carriages, sometimes with an engine, coupled together; **train-sick** *a.* (earlier example); **trainside** *a. U.S.*, at the side of or near a train, taking place next to a train; **train smash** *Naut. slang*, cooked tinned tomatoes, usu. with bacon; **train-staff**, (*a*) (earlier example); **train-stop**, (*b*) the state of a train's being at a stop; a place at which a train stops; **train ticket**, (*a*) a ticket delivered to an engine-driver as authority to travel over a single-line section of railway; cf. *train staff* (a), *staff and ticket* (*system*) s.v. *Staff *sb.*¹ 26; (*b*) a ticket enabling a passenger to travel on a train.
1869 *Atlantic Monthly* July 73/2 [He] prevailed upon me to be his train-boy. **1912** C. Mackenzie *Carnival* (ed. 5) iv. 41 A pal wouldn't let you sleep over the train-call on a Sunday morning. **1933** P. Godfrey *Back-Stage* xvi. 206 He packs his dress-basket, notes down the time of the train-call from the notice-board by the stage door. **1921** *Dict. Occupational Terms* (1927) §706 *Train caller*, a porter whose only duty is to call out destination of a departing train..; also calls out name of own station on arrival of trains. **1939** *WPA Guide to Florida* (1984) I. 5 Seminole names..were even more plentiful before the railroads interceded in behalf of train callers. **1957** D. du Maurier *Scapegoat* x. 133 A train-crash north of Lyons. **1979** P. Theroux *Old Patagonian Express* xix. 293, I don't want to be in a train crash. But I have a very bad feeling about this train. **1904** *McClure's Mag.* Apr. 617/1 As for the train crew, we never had any more trouble with them than if they had been so many sheep. **1976** P. R. White *Planning for Public Transport* viii. 162 If a train is stopped because of a derailment, blockage, etc. between signal boxes..it is necessary for the train crew to protect the trains with emergency lamps. **1905** J. Joyce *Let.* 15 Oct. (1966) II. 122, I will send you 100 crowns to pay your trainfare. **1983** *U.S.A. Today* 19 Apr. 3A/1 In New York, trainfare and bagels were free as 90,000 suburban commuters got their trains back after a six week strike. **1963** *Times* 2 May 16/4 Why don't we have train hostesses like air hostesses..who visit every carriage and see if we are comfortable and happy? **1971** *N.Z. News* 10 Mar. 5 Train hostesses who serve the needs of passengers on New Zealand Railways' 'Southerner' express trains between Christchurch and Invercargill pose in their distinctive transit red uniforms. **1909** Webster, Train jumper. **1930** *Times Lit. Suppl.* 5 June 482/3 Setting out on his trek across the continent,..as hobo, 'train jumper',..and cook in a Great Lakes freighter. **1965** H. P. Tritton *Time means Tucker* i. 18, I..silently cursed myself for being fool enough to take on train-jumping. **1978** *Sunday Mail Color Mag.* (Brisbane) 22 Oct. 6/2 Now, in the early years of this century, train-jumping was an unknown art in Australia. **1880** *News & Press* (Cimarron, New Mexico) 9 Sept. 3/2 Mr. Frank Fulton, train master on this Division,..gave the following information concerning the damage. **1907** J. W. Schultz *My Life as Indian* xviii. 210 Berry declared that he would do no more freighting to the mines with his bull train; he would either sell it or employ some one as a train-master. **1983** *Mod. Railroads* Apr. 18/1 Along with the yardmaster and trainmaster we observed the handling. **1859** *Rep. Accidents on Railways 1858* 17 in *Parl. Papers* XXV. 601 The system..of working by means of a train porter. **1873** *Returns Railways Companies Connections* 11 in *Ibid.* LVII. 765 Single Lines of Railway..Worked under the Train Porter System. **1932** W. Faulkner *Light in August* vi. 132 Perhaps he remembered suddenly the train ride and the food. **1980** *Listener* 4 Sept. 296/2 The train ride down the Peninsula is boring: rubber, rubber, grey rubber trees. **1939–40** *Army & Navy Stores Catal.* 826/1 This excellent train set..comprises a No. 1 Special Locomotive..two No. 1 Pullman Coaches..and rails. **1959** G. F. Allen *Brit. Railways Today & Tomorrow* vii. 133 The Rosters usually indicate the preceding and succeeding use to which each coach of a train set is to be put. **1980** J. Cartwright *Horse of Darius* xi. 158 He was playing with his train set... He..passed his days in a world of trains and model airplanes. **1982** Whitehouse & Allen *E. Treacy—Railway Photographer* 42 The all-maroon train-set of the northbound 'Flying Scotsman' leaving Copenhagen Tunnel. **1905** E. M. Forster *Where Angels fear to Tread* vi. 163 They crossed the Apennines with a train-sick child. **1932** *Sun* (Baltimore) 21 Sept. 1/6 He [*sc.* F. D. Roosevelt]..jollying trainside crowds with localized pleasantries. **1940** *Ibid.* 24 July 1/5 In a series of train-side and platform talks..he [*sc.* W. L. Wilkie] lashed out against leading Democratic party machines by name. **1941** *Weekly Tel.* (Sheffield) 13 Sept. 16/1 The boys of the Navy have a lot of slang... They have given the name 'train smash' to a tomato and bacon breakfast. **1978** *Daily Mirror* 19 June 9/1 Train smash is our nickname for tinned tomatoes with bacon. **1859** *Rep. Accidents on Railways 1858* 17 in *Parl. Papers* XXV. 601 To make the train staff the means of opening the train-ticket box. **1955** W. Gaddis *Recognitions* II. iii. 422 Go to a train station yourself..or a bus station. **1981** *N.Y. Times Mag.* 21 June 10/3 When was the last time you heard a young, rich-affluent-wealthy type use the phrase *railroad station*? Upper-class use is now *train station*. *a* **1963** S. Plath *Ariel* (1965) 43 It is a trainstop, the nurses Undergoing the faucet water. **1977** H. Fast *Immigrants* 10 The food, brought to them at train stops, was a dismal, unchanging diet of cold sausage and stale bread. **1859** *Rep. Accidents on Railways 1858* 17 in *Parl. Papers* XXV. 601 If another engine or train is intended to follow in succession, a train ticket, stating 'staff following', will be given to the person in charge of the leading train, the staff itself being given to the last. **1941** B. Schulberg *What makes Sammy Run?* x. 256 A letter arrived with a train ticket and travelling expenses. **1977** *Lancashire Life* Dec. 60/1 He..showed early business acumen by taking orders for Wakes Week train tickets, sleeping-out on the doorstep of Thomas Cook's the night before the ticket sale to save others' queueing. **1877** F. M. A. Roe *Army Lett.* (1909) 163 From then on to train time, Hal was patted and petted and given dainties. **1979** P. Theroux *Old Patagonian Express* xiii. 200 The difficulties of train travel in Latin America. **1927** R. Lehmann *Dusty Answer* III. i. 128 A far train-whistle roused her. **1981** V. Mehta *Vedi* (1982) i. 3, I remember the train whistle. It blew with a rush of steam. **1883** *Manch. Examiner* 28 Nov. 4/6 It is supposed that the would-be train wreckers were plotting against the Premier's life. **1930** J. Dos Passos *42nd Parallel* I. 16 Dumping grounds, trainyards. **1973** T. Pynchon *Gravity's Rainbow* I. 171 Out the windows.. are a row of bare Army-colored poplars, a canal, a snowy trainyard.

train, *v.*¹ Add: **III. 6. d.** (Earlier example.)
1832 S. Austin tr. *Pückler-Muskau's Tour of German Prince* III. iv. 74, I kept race-horses myself, and had a Newmarket jockey for a time in my service... It amused me greatly to see this fellow 'training' himself.
7. b. *train on* (earlier *fig.* and later examples).
1767 G. Selwyn *Let.* 29 Dec. in *15th Rep. R. Comm. Hist. Manuscripts* App. vi. 225 in *Parl. Papers* 1897 (C. 8551) LI. 1, Lord Beauchamp trains on well, as they say, but *il n' a pas le moyen de plaire*. **1789** *Loiterer* 6 June 7 He *trained on famously* well, and would soon be a very *dashing man*. **1937** *Daily Tel.* 15 Oct. 23/3 He..trained on into a first-rate College oar. **1976** *Horse & Hound* 10 Dec. 48/2 (Advt.), A good sire of fast 2-year-olds that train on.
IV. 9. c. To associate, ally, or co-operate *with*, (*dial.* also *along of*). *N. Amer. colloq.*
1871 J. Hay *Pike County Ballads* 22 It gravels me like the devil to train Along o' sich fools as you. **1889** *Cent. Dict.* s.v. *train*¹, I don't *train* with that crowd. **1892** 'Mark Twain' *Amer. Claimant* i. 5 Have you been training with that ass again? **1907** *Methodist Rev.* Nov. 984 He does not train with the extreme radical theologians. **1935** H. Davis *Honey in Horn* i. 4 A couple of bad-acting sons who got drunk, fought and trained around with thieving half-breeds. **1945** 'L. Ford' *Philadelphia Murder Story* ix. 146 She knew as well as I know now—and I don't train with lawyers—that Malone wasn't going to search the house.
11. b. (Earlier example.)
1856 Ld. Granville *Let.* 12 Feb. in Ld. Fitzmaurice *Life Granville* (1905) I. vii. 163 After acting as godfather, I trained up to town for the Committee of Privileges.

trainability (trēⁱnăbi·lĭti). [f. Trainable *a.* + -ility.] Aptness or capacity for being trained.
1955 G. A. N. Lowndes *Brit. Educ. System* ii. 40 High trainability is a delicate plant. **1960** *Times* 20 Jan. 9/7 The fine quality of the Northern Ireland labour force, its trainability. **1980** *Observer* 4 May 26/6 If a degree indicates intelligence or trainability.

‖ **trainante** (trɛnɑ̃t), *a.* Now *rare.* Also **traînante**. [a. F. *traînante* (fem.), f. pres. pple. of *traîner* to drag, as in *voix traînante*, etc.] Of vocal or musical sounds: dragging, drawling.
1818 C. Morgan *Let.* 16 Sept. in Lady Morgan *Autobiogr.* (1859) 104 The music..possesses here and there fine strains of melody: *per contra*, it is *trainante* often too scientific and old-fashioned in its phrases. **1865** 'Ouida' *Strathmore* I. x. 170 The tranquil *trainante* tones in which he always spoke his rudest things. **1899** *Westm. Gaz.* 5 Oct. 3/1 Talking in their pleasant, *traînante* tones to and of each other.

trainee. Add: Of a person, not now usu. correlative to *trainer*. (Earlier and later examples.) Also *attrib.* or as *adj.*
1841 *Fistiana* 112 An early, light, and nutritive supper would greatly benefit the trainee. **1927** *Daily Tel.* 8 Feb. 13/7 The number of new trainees is now so small that a workshop smaller than that which we are now using will suffice. **1932** *Sun* (Baltimore) 19 July 15/6 (*heading*) Trainees at Meade using machine guns. **1942** *Yank* 4 Nov. 11 Basic trainees beat the softball pitching. **1951** *Good Housek. Home Encycl.* 37/1 The majority of nursery training schools do not make any hard-and-fast rule regarding the work which their trainees should take over. **1953** *Manch. Guardian Weekly* 6 Aug. 7/3 A rice paddy.. through which trainee squads are sent, as on a Korean patrol. **1956** *B.B.C. Handbk.* 1957 138 Sixteen trainee broadcasters. **1962** *Lancet* 1 Dec. 1156/2 The trainee assistant scheme is on the right lines, but there have been complaints that the scheme has occasionally been abused by trainers who used their trainees as ordinary assistants. **1977** *Grimsby Even. Tel.* 31 May 5/7 Miss Sally Dixon, a trainee nurse, was announced as the hospital's Jubilee representative following a ballot. **1978** *West Lancs. Even. Gaz.* 23 Feb. 17 (Advt.), Vacancies for able young persons as trainees.
Hence **trainee·ship**, the position of a trainee, a post as a trainee.
1961 *Listener* 14 Dec. 1021/2 Some management traineeship schemes in industry and commerce. **1964** *New Statesman* 13 Nov. 747/1 Bourgeois and ambitious parents.. thrust their sons into low-paid traineeships with prospects of eventual power and wealth. **1966** *Rep. Comm. Inquiry Univ. Oxf.* II. 193 The category 'other' includes general traineeships and postgraduate apprenticeships. **1980** *Nature* 14 Feb. 610/1 The universities hope that the administration can be persuaded to contribute more than $150 million..to provide a national support programme of grants, traineeships and fellowships.

trainer. Add: **1. a.** Also *attrib.*
1973 *Philadelphia Inquirer* 7 Oct. 13 (Advt.), Convertible or girls sidewalk bike. Trainer wheels included. **1976** *Milton Keynes Express* 2 July 25/1 (Advt.), Baby walker, babycare bath mat and toddler trainer seat £2, or will sell separately. **1977** *Austral. Sailing* Jan. 46/1 For the very young, the best way into sailing is through a sailing club in a trainer class like the Sabot or Manly Junior.
c. An aircraft used in training pilots or other aircrew.
1932 Auden *Orators* II. 59, 1 Moth trainer fully equipped for advanced training. **1950** *Hansard Commons* 21 Mar. 1771 The introduction of the Prentice as the basic trainer has enabled instrument flying instruction to be improved considerably. **1977** 'J. le Carré' *Honourable Schoolboy* xvi. 368 A row of single-engined military trainers.
d. A soft running shoe without spikes; a training shoe.
1978 *Guardian* 22 Dec. 9/3 The Poynton Jemmers, a women's morris dance side..are in mufti tonight... But no trainers or pumps: it's clogs or nothing. **1982** *New Society* 4 Mar. 344/2 Skinny teenage boys in the ubiquitous parkas, jeans and trainers. **1983** *Listener* 28 July 19/1 Dr. Garrow is welcome to don trainers and join the ladies I run with on Saturdays.

training, *vbl. sb.* Add: **2. c.** (Earlier example.)
1786 W. Cowper *Let.* 1 May (1981) II. 531 When you come, I shall talk you into training, as the jockeys say, I doubt not that I shall make a nimble and good walker of you in a short time.
5. *training camp, centre, department, -ground* (earlier example), *prison, programme, session, shoe*; **training-school**, *spec.* (*b*) *N. Amer.*, a vocational institution for juvenile delinquents.
1894 T. B. Aldrich *Two Bites at Cherry* 216, I don't fancy he heard a gun fired, unless it went off by accident in some training-camp for recruits. **1980** *Washington Star* 10 Dec. c6, I felt like a rookie again in training camp. **1926** *Encycl. Brit.* Suppl. III. 819/2 (*heading*) War training centers. **1962** E. Snow *Other Side of River* (1963) lxxi. 549 Shanghai and Manchuria have been the biggest training centers for China's modernization. **1906** *Rep. Brit. Assoc. Adv. Sci.* 271 To take a course of training in some existing training college or training department. **1644** in *Early Rec. Dedham, Mass.* (1892) III. 102 From the Trayning ground to the Cart Bridge. **1950** Training prison [see *Open *a.* 2 d]. **1977** *Herald* (Melbourne) 17 Jan. 2/6 The prisoners were transferred after a security review at country jails following four escapes from Geelong Training Prison in a week last month. **1971** L. B. Johnson *Vantage Point* (1972) iv. 81 Our manpower training programs focused on preparing unskilled men and women for jobs. **1977** *Offshore Engineer* June 36/1 Pilots are selected from employees who show diving aptitude with a full mechanical understanding, and are put through a rigorous training programme. **1905** *First Ann. Rep. N.Y. State Training School for Girls* 4 The Penal Code was amended..so as to authorize the commitment of delinquent girls under the age of sixteen..to the New York State Training School for Girls. **1978** *Globe & Mail* (Toronto) 11 Jan. 7/3 If training schools are closed, then some group homes will have to have a custodial aspect. **1951** *Sport* 6–12 Apr. 12/2 Most clubs have selected their spot-kicker after careful tests during training sessions. **1977** J. M. Johnson in Douglas & Johnson *Existential Sociol.* viii. 242 A worker in Unit One said she had decided to record her time spent in training sessions on line C of the report. **1973** *People's Jrnl.* (Inverness) 4 Aug. 20/2 (Advt.), Training shoes. **1984** *Nutshell* (Gainesville, Florida) Spring 61/1 (Advt.), A remarkable new training shoe designed to take all the wear and tear high-mileage runners can give it.

trainman, train man. **2.** For *U.S.* read 'orig. *U.S.*' and add earlier and later examples; *spec.* (see quot. 1982).
1877 J. D. McCabe *Hist. Great Riots* 48 These motives have been misunderstood, and..produced the present troubles amongst our trainmen. **1960** [see *Shopman 3]. **1964** [see *secondman* s.v. *Second *a.* 7 a]. **1967** *Guardian* 4 Dec. 1/3 British Rail's intention is to replace them [*sc.* firemen and guards] by a new type of railwayman—a 'trainman'. **1982** *Daily Tel.* 19 June 2/7 The six productivity concessions..include..the introduction of 'train men' to combine the duties of guards and travelling ticket inspectors.

trai·n-spotter. [Train *sb.*¹ 16.] One (esp. a small boy) whose hobby is observing trains and recording railway locomotive numbers. Hence **trai·n-spotting** *vbl. sb.*; **trai·n-spot** *v. intr.*
1958 *Spectator* 11 July 70/3 A prize of six guineas is offered for a Train Spotters' Anthem. *Ibid.* 1 Aug. 179/1 Many of them must simply spend their time train-spotting. **1959** *Manch. Guardian* 22 July 2/6 British Railways.. announced a ban on train-spotting..on about twelve main-line stations. **1969** *Times* 1 Nov. (Saturday Rev. Suppl.) p. iii/4 A train-spotting recluse, primed with exact information on lines that have rusted into the ground. **1974** P. McCutchan *Coach North* ii. 15 'I reckon you're all past train spotting.'..The old man didn't look as though he'd ever train spotted. **1978** D. Williams *Treasure up in Smoke* xiv. 126 The Governor appeared to have graduated from train-spotter to Lord Protector in one short morning.

traist, *a.* Restrict †*Obs.* to other senses and add: **3.** (Later *poet. arch.* examples.)

1919 E. Pound *Quia Pauper Amavi* 11 She has her lover till morn, Till the traist man cry out to warn Them. **1955** —— *Classic Anthol.* i. 22 Shall no one be traist? Mother of Heaven, Shall no one be traist?

trait. Add: **6. a.** (Further examples of a culture or social group.) Also *attrib.*

1926 *Amer. Jrnl. Sociol.* Mar. 656 In maize culture as practiced by American farmers we have a fine example of a borrowed culture trait. *Ibid.* 659 The colonists took over all the essential parts of the trait-complex. **1936** R. Linton *Study of Man* xvi. 280 During this [trait] period both the new trait and the old trait or traits with which it is competing become Alternatives within the total culture complex. *Ibid.* xxii. 397 Every trait is intimately associated with some other trait or traits to form a larger functional unit commonly known as a *trait complex*. **1947** G. Murphy *Personality* xxi. 506 Most of the trait names that are used represent general action tendencies; and as soon as they are applied to oneself, or..to others, they stimulate a trait psychology in their user. **1976** A. Haley *Roots* vii. 21 Kunta would always turn and walk away, thus displaying the dignity and self-command that his mother had taught him were the proudest traits of the Mandinka tribe. **1977** R. Holland *Self & Social Context* v. 163 Trait models of professions attempt to list the characteristics of professional activity..as though some essential quality will be revealed by describing and comparing the many examples.

‖ **trait d'union** (trɛ dünyɔ̃). [Fr., lit. 'hyphen'.] A connection between or amongst otherwise unattached characteristics or parties.

1912 J. Buchan *Moon Endureth* iv. 127 He had established no *trait d'union* between the intellect..and the senses. **1934** A. Toynbee *Study of Hist.* I. ii. 190 In the British Empire..this medieval incarnation of political unity [*sc.* the Crown] has latterly acquired a new and unforeseen institutional value as the *trait d'union* between the States Members of the British Commonwealth of Nations. **1959** *Times* 7 Jan. 11/7 M. Mokhtar Ould Daddah likes to think of his country [*sc.* Mauritania] as a *trait-d'union*, or link, between the Maghreb and West Africa.

Trajanic (trēɪdʒæ·nik), *a.* [f. L. *Trāiān-us* + -ic.] Of or pertaining to the Roman emperor Trajan (A.D. 53–117), esp. to the style of triumphal art associated with him.

1906 *Athenaeum* 27 Jan. 113/2 The *extispicium* scene probably represents the..ceremony before Trajan set out on his Dacian campaign, and is Trajanic in style. **1935** *Times Lit. Suppl.* 9 Nov. 780/3 He finds no sign in either the Trajanic or the Aurelian column of a Court art. **1978** *Antiquaries Jrnl.* LVIII. 82 The Dacian victory is perhaps the most significant of all the triumphal themes in Trajanic art.

trajectory, *sb.* Add: Also with pronunc. (træ·dʒéktori).

Trakehner (traké¹·nəɪ). In sense 2 also **trakehner, trakena, trakener.** [a. Ger., f. the name of the Trakehnen stud.] **1.** A saddle horse belonging to a breed first developed at the Trakehnen stud in east Prussia.

[**1900** W. Ridgeway *Origin & Influence Thoroughbred Horse* iv. 472 The extraordinary tractability of the Prussian Trakehnen breed..is a well known feature.] **1926** A. Topham *Chron. Prussian Court* i. 16 The carriages..had been waiting in the shade, each with its four splendid Trakehner horses. **1975** *Islander* (Victoria, B.C.) 13 July 5/2 The Trakehner breed of horses is fairly new in North America, but is famous and long prized in Europe where the breed originated in the areas of Eastern Prussia and Poland. **1980** G. Henschel *Illustr. Guide Horses & Ponies* 155/3 Trakehners often make top-grade show jumpers.

2. A type of fence in the cross-country section of an equestrian three-day event, etc. (see quot. 1973).

1958 S. Willcox *Three Days Running* xvii. 174 We made our way up the incline towards a big trakena fence. **1959** E. Schenderkamp *Encycl. for Horsemen* (rev. ed.) 306/2 *Trakener*, a type of fence used in cross-country... It is a ditch spanned by rails in the centre. **1973** Stratton *Encycl. Show Jumping & Combined Training* 271/1 A trakener is a cross-country obstacle made of a 'knife rest' construction, built on a ditch, with a second rail... **1977** *Horse & Hound* 10 June 25/1 Fences 5 (a trakehner over a dry ditch) and 6 (a sloping palisade followed by a bank and ditch) caused most of the trouble across country.

tralucent, *a.* Delete †*Obs.* and add later examples.

1914 C. Mackenzie *Sinister St.* II. iv. iv. 932 The benign trees that hung down with tralucent green sprays in the lamplight. **1951** —— *Carnival* (rev. ed.) viii. 81 Her hands were long and white; her lips very crimson and tralucent.

tram, *sb.*² Add: **V. 8.** *tram-beam* (*fig.* example), *-fare* (also *transf.*), *-horn*, *horse*, *-refuge*, *ride*, *stop*, *-top*, *-track*, *-train*, *-wagon* (earlier example).

1879 G. M. Hopkins *Poems* (1967) 81 Or to-fro tender trambeams truckle at the eye. **1909** J. R. Ware *Passing Eng.* 249/2 *Tram-fare* (*London Streets*', 1882), twopence. **1922** Joyce *Ulysses* 696 Debit... Tramfare [£.s.d.] 0.0.1. **1978** M. de Larrabeiti *Rose beyond Thames* 87 He bought ..me a second-hand bike so that I could cycle to school and save the tram fare. **1922** *Blackw. Mag.* Apr. 447/1 The blowing of tram-horns. **1891** J. L. Kipling *Beast & Man in India* viii. 206 (*caption*) Bombay tram-horse wearing horse-cap. **1938** *All England Law Rep. Annotated* I. 339 An illuminated bollard at one end of a tram refuge had been damaged in an accident. **1919** R. Fry *Let.* 3 Nov. (1972) II. 465 Marseilles is only one and a half hours tram ride. **1977** *Lancashire Life* Dec. 57/2 One summer's day he changed the routine and took us a tram ride into the country. **1930** R. Lehmann *Note in Music* i. 19 They arrived at the tram-stop to find a solid wedge of humanity struggling to get aboard. **1980** P. Harcourt *Tomorrow's Treason* i. iv. 58 It was a long walk to the nearest tram stop. **1895** G. B. Shaw *Let.* 23 Mar. (1965) I. 504 We..went to her sister's..by tramtop. **1916** Joyce *Portrait of Artist* ii. 70 He heard the mare's hoofs clattering along the tramtrack on the Rock Road. **1911** R. Fry *Let.* 15 Apr. (1972) I. 347 A two hours' journey by a tram-train to the slopes of Mt Olympus. **1824** F. Witts *Diary* 6 May (1978) 38 The tram waggons now may be made to travel without horses by steam.

Traminer (trami·nəɪ). [a. Ger., f. *Tramin* (It. *Termeno*), the name of a village in N. Italy.] The name of any of several varieties of vine and grape widely grown in Germany, Alsace, and elsewhere; the white wine with perfumed bouquet produced from this grape. Also *attrib.*

1851 C. Redding *Hist. & Descr. Mod. Wines* (ed. 3) ii. 48 Then there is the *Traminer* with a small berry, sweet, and fond of a marly soil. **1872** Thudichum & Dupré *Treatise on Origin, Nature, & Varieties of Wine* ix. 281 German wines—... Riesling... Forster Traminer. *Ibid.* xvi. 541 There are also some so-called Franconia grapes, which we will describe as white Traminer. **1948** 'J. Tey' *Franchise Affair* xiii. 144 There is a Traminer for dinner. **1960** *Spectator* 13 May 714 Yugoslav traminer came on to the British market during 1959. **1972** *Guardian* 4 May 13/5 The local variety of the Traminer grape called the Savagnin.

tram-line. Add: Also **tramline.** **2.** *pl.* Either pair of parallel lines bordering the side of a lawn-tennis court, the inner of each pair marking the boundary of the court for singles and the outer for doubles. Occas. *sing.*, one of this pair of lines. *colloq.*

1937 Partridge *Dict. Slang* 906/1 *Tram-lines*, the 4½ ft. wide area on each side of a (doubles) lawn-tennis court: sporting: from ca. 1929. Esp. *down the tram-lines*, i.e. more or less straight along this strip of the court. **1959** *Times* 31 Aug. 12/1 It passed an inch outside the tram-line. **1978** G. Forbes *Handful of Summers* ix. 180 Cliff..hit a two-hander down Rodney's tramlines.

3. *transf.* *fig.* Chiefly *pl.*, inflexible, predetermined or restrictive courses of action, principles, etc.

1948 D. Welch *Brave & Cruel* 68, I ran it [*sc.* a nail-brush] up and down his back until I'd made harsh red tramlines. **1955** H. Spring *These Lovers fled Away* iv. 113 One [obstacle] was..his addiction to social tram-lines. As things had been, they should, in a well-organised world, remain. **1967** *Times Rev. Industry* Oct. 60/2 Tramline thinking..delays improvement in the quality of the trade union movement itself. **1976** *Listener* 22 July 85/3 The government..is forced..more and more on to orthodox, predictable tramlines, and protestations that imagination, vision and radical policy must await more prosperous days. **1979** J. Wainwright *Duty Elsewhere* vii. 28 Most chief constables tend to run on tram-lines... They feel safer.

Hence **tra·m-lined** *a.*, furnished with tramlines.

1924 W. J. Locke *Coming of Amos* xiii. 171 He stepped into the car and..drove at break-neck speed down the tram-lined hill. **1932** Auden *Orators* iii. 89 Tudor from the tram-lined town.

trammie, trammy (træ·mi). *colloq.* (esp. *Austral.* and *N.Z.*). [f. Tram *sb.*² + -y⁶, -ie.] A tram-conductor or tram-driver.

c **1926** 'Mixer' *Transport Workers' Song Bk.* 23 Then he slathers up the 'trammies', As the conductor goes through. **1934** L. G. Gibbon *Grey Granite* iii. 192 The conductor had seen Alick and caught his arm... The trammie held fast, a squat, buirdly bird with a face like a badly-made barn door. *Ibid.*, That shortened the run, in a minute the Docks, the trammie slowed down at a bend. **1945** F. Ryland in *Coast to Coast 1944* 166 The soldier sprang wildly to his feet and knocked the trammie in the mouth. **1972** *Guardian* 6 Dec. 13/5 Trammies, truckies and wharfies (Australian for tram and bus conductors and drivers, truck-drivers and dockers). **1979** *Jrnl. Lancs. Dial. Soc.* Jan. 5 At heart I will always be a trammie, for somehow the outlandish humour of those days was born in trams and synonymous with them.

tra·mming, *vbl. sb.* [f. Tram *v.*¹ 2 + -ing¹.] Conveyance (of coal, ore, etc.) by a tram or trams.

1875 *Stock Exch. Observer* 9 Feb. 3/1 So much for breaking, tramming and carriage by rail. **1983** *New Scientist* 21 Apr. 162 The next operation is 'tramming', or the actual removal of the ore to the surface using wagons.

trammy: see *trammie.

tramp, *sb.*¹ Add: **1. b.** More fully *axle tramp*. Alternate bouncing of wheels on the same axle.

1935 *Story of Knee Action* (General Motors Corp.) 3 Such erratic wheel movements as 'shimmy' and 'tramp' should be eliminated. **1959** *Motor Manual* (ed. 36) v. 121 Independent rear suspension..offers the advantages of reduction of unsprung weight and the elimination of axle tramp and patter. **1977** *Drive* Sept.–Oct. 120/2 Banging and jumping from the rear axle..on fast take-offs from rest or in mid-corner (a condition called 'axle tramp'). **1982** *Motor* 3 July 39/4 Brisk starts are noticeable for their lack of tramp.

3. (N.Z. examples.)

1966 *Weekly News* (N.Z.) 3 Aug. 7/4 Two-day tramps from the Milford Hotel up to the Sutherland Falls. **1984** *N.Z. Woman's Weekly* 30 Apr. 121/2 Day tramps are popular.

4. b. *slang* (orig. *U.S.*). A sexually promiscuous woman.

1922 E. O'Neill *Anna Christie* i. 119 Sure—and another tramp with him. **1936** D. Powell *Turn, Magic Wheel* i. 60 A wayward, double-crossing, lying little tramp. **1959** 'J. Welcome' *Stop at Nothing* iii. 28 You can usually tell..the nice girls from the tramps. **1971** *Sunday Nation* (Nairobi) 11 Apr. 29/2 Even in these permissive times the girl of your age who can't say 'no' can pretty soon earn the title of tramp among her contemporaries. **1979** R. Jaffe *Class Reunion* i. v. 49 Who could blame Richard, so young at prep school, for fooling around with the local tramp?

5. c. An aircraft plying commercially according to demand. Also *attrib.*

1905 Kipling *Actions & Reactions* (1909) 141 These heavy freighters fly down to Halifax direct... They are the biggest tramps afloat. **1948** *Shell Aviation News* No. 115 8/3 At present the majority of freight charters are on a direct 'out-and-home' basis, but the time is coming when, with the parallel development of the Baltic and heavy duty tramp aeroplanes, 'time charter' will be as commonplace as it is with shipping. **1952** 'J. Tey' *Singing Sands* ix. 190 Most of us fly scheduled routes, but some fly tramps. Take anything anywhere.

7. (see also *5 c); (in sense 4) *tramp-woman*.

1902 Hardy *Time's Laughingstocks* (1909) 11 (*poem title*) A trampwoman's tragedy.

trampdom (earlier example).

1891 *Contemp. Rev.* Aug. 257 The tramp also finds it convenient to use the highways, but this is not..common..for it is on the railroads that Trampdom thrives as an institution.

tramp, *v.*¹ Add: **4. a.** *N.Z. spec.* to walk for long distances in rough country.

1935, etc. [implied in *tramping *vbl. sb.*]. **1984** *N.Z. Listener* 28 Apr. 62/1 One of my correspondents tramping with her husband, referred to the 'benched out' track they were following up the hillside.

b. (Earlier example.)

1846 *Swell's Night Guide* 134/2 *Tramp*, to wander as a beggar.

6. b. To transport goods by road to varying destinations as the load requires.

1959 [implied in *tramping *vbl. sb.*]. **1968** in P. G. Hollowell *Lorry Driver* vi. 132 She [*sc.* a lorry driver's wife] didn't like it when I was on tramping... When you're tramping you never know where you're going and when you're coming back.

tramping *vbl. sb.* (further examples).

1935 J. Guthrie *Little Country* xxi. 319 The members of the Tem Tramping Club. **1959** A. McLintock *Descr. Atlas N.Z.* 74 New Zealanders are a people who take full advantage of these open spaces for all manner of recreational activity, including..shooting and fishing..and tramping and mountaineering. **1959** *Times Rev. Industry* June 45/3 ... **1963** *N.Z. Woman's Weekly* 17 June 18/3 [At Jackson Bay] we have the best deer shooting, tramping country and scenery in New Zealand. **1968** P. G. Hollowell *Lorry Driver* ii. 21 Tramping, in contrast to the monotony of trunking, is mainly liked for the variety it affords, both in the types of loads carried, the different parts of the country which are seen, [etc.]. **1972** *N.Z. News* 5 July 3/1 Proper equipment enabled a six-strong tramping party..to survive near freezing temperatures in the rugged Pouakai Ranges. **1984** *N.Z. Woman's Weekly* 30 Apr. 121/2 Tramping, mountaineering and trout fishing are other attractions in the Nelson Lakes National Park.

tramper. Add: **2.** *N.Z. spec.* a person who walks long distances in rough country for recreation.

1960 B. Crump *Good Keen Man* 60, I sent my men, who were more likely to lose themselves than find lost trampers, up and down a branch of the stream looking for boot-prints. **1977** G. Scott *Hot Pursuit* x. 87 One of the temporary shelters for trampers that are dotted about the New Zealand bush. **1984** *N.Z. Field & Stream* Apr./May 46/3 Wool is excellent for New Zealand conditions, and is widely used by trampers and climbers.

trampolin, -ine, *sb.* Add: Now usu. with spelling -ine and pronunc. (-īn). Substitute for def.: A base of elastic material used as a springboard and landing area in acrobatic exercises and displays; now *spec.* a sheet of canvas, nylon mesh, or the like, held in a frame by springs. Also *attrib.* and *fig.* (Further examples.)

1799 *Times* 1 June 3/4 He positively leaps over a large tilted waggon..and does not make use of a spring board or trampoline. **1928** *Daily Express* 13 June 13 A trampoline act, a wire act, in fact, ..a complete vaudeville programme. **1930** *Observer* 1 June 21 They perform on the apparatus resembling a spring mattress, which has been dubbed the 'trampoline'. **1938** N. Streatfeild *Circus is Coming* vii. 109 The things they did when on the trampoline were breath-taking.., shooting up.., then coming down in amazing twists and somersaults. **1960** *Times Lit. Suppl.* 1 July 418/4 [D. H.] Lawrence has been bounced up and down on the trampoline of literary fashion more than most writers. **1961** *U.S. Patents Quarterly* CXXIX. 210/2 As applied to tumbling devices, 'Tram- poline' is completely generic. **1975** *Oxf. Compan. Sports & Games* 1044/2 The trampoline has a steel frame, braced in such a way as to ensure that a landing in the bouncing area..may be made without fear of hitting any structural members.

trampolin(e *v.*, substitute for def.: to perform on a trampoline; also *fig.*; (later examples); **tra·mpolining** *vbl. sb.*, the practice or sport of performing on a trampoline; **tra·mpolinist**, a performer on the trampoline.

1843 in M. W. Disher *Greatest Show on Earth* (1937) 196 (*in illustration*) Mr. W. T. Twist the Trampolinist... Astley's Royal Amphitheatre 1843. **1953** *People & Places* (Chicago) Mar. 6/1 (*caption*) When performing the back turn-over or cannonball, a trampolinist attains a tuck or pike position on takeoff. *Ibid.* 6/2 Descended from a centuries-old European circus art, trampolining is one of the newest..sports to be adopted by the nation's colleges, high schools and private athletic clubs. **1972** *Guardian* 11 Apr. 15/4 Children fenced, trampolined, played volley-ball and football. **1973** *Times* 1 Feb. 16/6 It proved..difficult..to trampoline at all except at a speed dictated by gravity. **1978** R. Hill *Pinch of Snuff* xiv. 142 Isn't she..a trifle overdressed for trampolining? *Ibid.* xiv. 148 Estelle, the teenage trampolinist.

trance, *sb.*¹ Add: **4.** *trance faculty*, *-mediumship*, *music*, *-personality*, *speaker*, *-state* (examples), *-subject*, *-utterance*, *-writing*; *trance-bound* (example), *-eyed* adjs.

1860 J. G. Whittier *Home Ballads* 90 Shine on us with the light which glowed Upon the trance-bound shepherd's way. **1957** C. Day Lewis *Pegasus* 15 A bright bewildered April, a trance-eyed summer. **1909** W. James *Mem. & Stud.* (1911) viii. 190 All the resources of the automatist, including his or her trance-faculty of telepathy. **1870** *Spiritualist* 14 Jan. 37/3 One feature running through the whole range of trance-mediumship, is the fact that the media..feel symptoms of the death pains of the communicating spirits. **1961** Trance-mediumship [see *control *sb.* 4 b]. **1970** *Guardian* 5 June 9/4 The records cover the whole range of Ethiopian music..through cow milking songs to Moslem trance music. **1890** W. James *Princ. Psychol.* I. viii. 211 The poor passive trance-personality had stuck for weeks in the stagnant dream. **1920** Trance speaker [see *automatist 2]. **1890** W. James *Princ. Psychol.* II. xxvii. 601 The suggestion-theory may therefore be approved as correct, provided we grant the trance-state as its prerequisite. **1978** *Amer. Speech* LIII. 59 Felicitas Goodman describes behavior in trance states accompanying glossolalia in congregations mostly in Mexico. **1880** Trance-subject [see *associate *sb.* 7]. **1890** W. James *Princ. Psychol.* I. x. 394 One curious thing about trance-utterances is their generic similarity in different individuals. **1980** 'S. Woods' *Weep for Her* 51 So many things are involved..telepathy, clairvoyance, trance utterance, [etc.]. **1911** W. F. Barrett *Psychical Res.* xv. 218 The group of controls..manifested themselves also in the trance-writings.

tranche. Delete ‖ and 'Now only as a loan-word from French' and add: Now also semi-anglicized as (trɑnʃ). **2.** *transf.* and *fig.*, esp. in *Econ.*, *spec.* an instalment of a loan, a quota, a block of bonds or (esp. government) stock.

1930 *Economist* 10 May (Suppl.) 10/1 The first business of the bank will be the arrangement of a loan to raise $300 million... The first tranche of the combined loan is expected to be offered about the end of May. **1953** *Ibid.* 15 Aug. 470/2 The gas stock (and a few other tranches, too) remains to be sold. **1962** C. A. R. Crosland *Conservative Enemy* v. 83 It is true that, as in Soviet Russia, or any industrial society, the top managerial executives belong to the highest social tranche in terms both of income and prestige. **1963** *Ann. Reg. 1962* 477 On 9 May a further tranche, amounting to £300 million, of 5 per cent Treasury Stock 1986–89 was issued at £84½ per cent. **1964** S. Brittan *Treasury under Tories* iv. 123 The 'gold tranche' represents the portion of British credit facilities automatically available without conditions. **1966** *Listener* 11 Aug. 206/1 It laid down political conditions which Egypt would have to fulfil in order to get the second tranche of the loan. **1973** *Times* 19 Dec. 17/6 To meet the load growth in the meantime, the board says, a 'large tranche of reliable nuclear plant' is required. **1974** *Times* 6 Mar. 15/2 Names have rarely been published except in the cases of the biggest tranches of aid [*sc.* Government aid to private industry]. **1977** *Observer* 3 Apr. 10/2 His second tranche of income-tax cuts. **1978** 'L. Black' *Foursome* v. 40 Later, the next tranche of old houses would be knocked down so that the Brickyard Lane estate could be doubled in size. **1980** *Times* 18 Jan. 19/4 Why not approach the institutional or wholesale money market for funds in large tranches rather than continue with the costly exercise of collecting money from a multitude of smaller savers?

3. ‖ **tranche de vie** [lit. 'slice of life'], a representation of quotidian existence, *spec.* in literature or painting; also *attrib.*

1934 in Webster. **1957** N. Frye *Anat. Criticism* 285 This idolatrous form of mimesis is rare, but the thin line of its tradition can be traced from Classical mime writers like Herodas to their *tranche-de-vie* descendants in recent times. **1958** *Spectator* 11 July 52/1 This *tranche de vie* came into my mind while I was talking. **1959** *Times* 1 Sept. 11/7 The static, informal *tranches-de-vie* landscapes of his Antwerp period. **1970** *New Yorker* 12 Dec. 193/1 It is the hand-held camera..that is designed to deal with the matter of obtaining *tranches de vie*.

‖ **tranchet** (trɑ̃ʃe). [Fr., f. *trancher* to cut.] **1.** (See quot.)

1858 P. L. Simmonds *Commercial Dict. Trade Products* 387/2 *Tranchet* (French), a shoemaker's heel knife.

2. *Archæol.* A flint with a chisel-shaped end, found in some mesolithic and neolithic cultures.

1899 R. Munro *Prehistoric Scotl.* ix. 332 Among the stone relics are small axes (tranchets), precisely similar to those found in the Danish Kjökkenmöddings, knife-flakes, scrapers, &c., but no polished objects. **1926** *Guide to Antiquities of Stone Age* (British Museum) (ed. 3) 150 One of the main flint forms..is the tranchet, or Shell-mound type of axe-head. **1977** G. Clark *World Prehist.* (ed. 3) v. 214 New forms such as tranchets or miniaturized cleavers.

b. *attrib.* and *Comb.*, designating chisel-shaped implements; *tranchet blow*, (a mark made by) a hard stroke at right angles to the main axis of an implement.

1931 *Antiquity* V. 579 An explanation of the tranchet axe. **1949** *Proc. Prehistoric Soc.* XV. 25 Near the base were also found two rather thick roughly made oval hand-axes with a tranchet blow at one end. **1957** V. G. Childe *Dawn Europ. Civilization* (ed. 6) i. 11 Stone tools (including tranchet celts) from high strands on the Norwegian coasts. **1975** J. G. Evans *Environment Early Man Brit. Isles* v. 103 The chisel-ended, or tranchet, arrowhead, specifically designed for immobilizing birds and other small game.

trank² (træŋk). Also **tranq.** Slang abbrev. of Tranquillizer (in Dict. and Suppl.).

1967 *Wentworth & Flexner's Dict. Amer. Slang Suppl.* 708/2 *Trank*,..a tranquilizer, a pill or capsule containing a tranquilizer. **1973** M. Amis *Rachel Papers* 196 As late as four fifteen, mother flaked out: either the party had aggravated her sense of intraspecific alienation, or her tranqs, all day neutralized by adrenalin, had hit her together in one clammy punch. **1976** B. Bova *Multiple Man* xvii. 196 'Y'all got a buzzful of trank in yew, boy.'..I [was] feeling like my head was numb with Novocain. **1980** A. Skinner *Mind's Eye* iv. 34 We'll have to go back to slipping tranks into his coffee.

Hence **tranked** (træŋkt) *a.*, drugged by tranquillizers.

1972 M. Crichton *Terminal Man* iv. i. 135 He was tranked out of his skull with thorazine. **1974** *Observer* 17 Mar. 34/7 Lulling drugs are prescribed; tots shamble eerily about, tranked.

tranny, trannie (træ·ni). Colloq. abbrev. of *transistor radio s.v. *transistor *sb.* 3. Also *attrib.*

1969 *Courier-Mail* (Brisbane) 28 Oct. 14/4 Salt and sand have..been eroding the trannie set rather badly. **1969** *Nova* Nov. 144/3 'How do you feel about the Love Generation now?'..'Sick to my stomach,' he replied, tuning into *The Archers* on his tranny to get back in touch with decent values. **1976** *Listener* 5 Aug. 152/2 The Controller surely had her tranny in the shed with her. **1978** D. Murphy *Place Apart* viii. 158 A solitary, yawning customs officer sitting beside a hideously wailing 'trannie'.

tranq, var. *trank².

tranquillityite (træŋkwi·lïti,əit). *Min.* [f. the name of the Sea of *Tranquillity* on the Moon, where the mineral was first collected: see -ite¹.] A silicate of ferrous iron, titanium, zirconium, and yttrium, $Fe_8(Zr,Y)_2Ti_3Si_3O_{24}$, occurring as dark red laths with a hexagonal crystal structure and found only on the moon.

[**1970** *Sci. Jrnl.* May 23/3 'Tranquilite' is the name Professors Paul Ramdohr and Josef Zaeringer..have given a new mineral they discovered in Moon rock samples collected by Apollo 11 astronauts.] **1971** J. F. Lovering et al. in *Proc. 2nd Lunar Sci. Conf.* (*Geochim. & Cosmochim. Acta* Suppl. II) I. 39 Tranquillityite.., a completely new silicate mineral, was first recognized in Apollo 11 basalt rocks as a possible new phase 'A'. **1981** P. H. Cadogan *Moon* iii. 181 Of all lunar minerals, tranquillityite is perhaps the most important carrier of the naturally radiogenic elements, uranium and thorium.

tranquillizer. Add: Also (*U.S.*) -ilizer. (Earlier and further examples.) Now *spec.* any of a large class of drugs in widespread use since the 1950s for the reduction of tension or anxiety and the treatment of psychotic states; also *fig.*

1800 F. Burney *Jrnl.* 7 Mar. (1973) IV. 402, I find, however, useful employment the best tranquiliser, &..I have less of the violent emotions which have hitherto torn me. **1824** E. Sutleffe *Med. & Surg. Cases* I. 2, I invited the attention of the medical world by introducing this herbaceous tranquillizer [*sc.* ground ivy] to their notice. **1956** A. Huxley *Let.* 21 Jan. (1969) 787 Our impressions of the cutting short of the mescalin experience by this new tranquillizer. **1956, 1957** [see *ataractic *a.* b]. **1958** G. Greene *Our Man in Havana* iv. ii. 176 He found himself taking to truth like a tranquilliser. **1974** *Encycl. Brit. Macropædia* XVIII. 595/1 The antipsychotic tranquillizers act solely on the brain itself, whereas anti-anxiety agents act on the spinal cord as well. **1975** *N.Y. Times* 28 Nov. 37/2 Proposed reforms are only tranquilizers offered in place of the fiscal surgery needed.

b. *attrib.*

1958 *Times* 1 Aug. 7/7 Tranquillizer drugs have cured many animals of the zoo of 'emotional strain and anti-social behaviour'. **1961** *Daily Tel.* 13 Sept. 19/4 Inside the needle is a tranquilliser solution which, when aimed accurately into an unruly dog.., makes the animal docile. **1979** *Globe & Mail* (Toronto) 16 Apr. 9/3 Humane society inspectors arrived and captured the animal with the help of a tranquillizer gun.

tranquillizing, *vbl. sb.* and *ppl. a.* Add: Also (*U.S.*) -ilizing. (Further examples.)

1827 E. Sutleffe *Ess. Insanity* 13 Extract. Glecomæ constituted my principal remedial source. Its tranquillising effects became immediately operative. **1954** *Proc. N.Y. Acad. Sci.* LIX. 41 Reserpine, like *Rauwolfia*, acts well in combination with other hypotensive drugs. Perhaps because of its peculiar tranquilizing, sedative effect, it smooths the course of the hypotensive response of such drugs. **1955** *Sci. Amer.* Oct. 80/1 The new tranquilizing drugs have introduced a new regime in the management of patients in mental hospitals. **1962** R. Carson *Silent Spring* ii. 13 When the public protests..it is fed little tranquilizing pills of half truth. **1974** K. Clark *Another Part of Wood* vi. 223 'Never let the Tories get ye, Kenneth' he [*sc.* Ramsay Macdonald] would say, as Lady Londonderry offered him a Tranquillising pill.

tranquillo, *adv.* Add: Also *adj.* and *sb.*, (a movement or section) played in a tranquil style or tempo.

1939 J. Harrison *Brahms & his Four Symphonies* x. 204 Brahms changes the..shape of the phrases in a *Tranquillo* section. **1976** *Gramophone* May 1766/3 The 'Rondes printanières' in the first part, together with *tranquillo* woodwind melody that frames them, seems to me to be taken..rather too slow. **1980** *Daily Tel.* 17 Mar. 11/2 The 'Serenade'..by..Gerard Schurmann..tested the acoustics of the hall, the 'presto' and 'tranquillo' being at times almost inaudible from the middle of the hall.

tranquilly, *adv.* (Earlier example.)

1756 [see *emanate *v.*].

trans (trɑnz). *colloq.* Also **trans.** (with point). [Abbrev.] **1.** = Translation 2. *U.S.*

c **1877** E. Dickinson in *Poems* (1955) III. 1055 I dare not write until I hear—Intro without my Trans. **1922** H. Crane *Let.* c 18 June (1965) 91 *Physique L'Amour*, which I am lately reading in trans. **1955** E. Pound *Section: Rock-Drill* xciv. 97 No full trans Till 1811.

2. = *trans-continental *sb. Austral.*

1937 E. Hill *Great Australian Loneliness* xxviii. 225 The 'Trans' and its people are a little world sufficient to themselves, a remarkable colony of government servants living in progress and contentment in the desert. **1976** K. Thackeray *Crownbird* v. 85 The next day they got to Kalgoorlie..and changed onto the 'Trans', a long, streamlined diesel.

3. = Transmission d.

1954 *Amer. Speech* XXIX. 103 *Trans*,..transmission. **1976** *Billings* (Montana) *Gaz.* 30 June 9-D/5 (Advt.), Plymouth Fury II, 440 high performance, (highway patrol), trans out, $300 firm. **1981** *Pop. Hot Rodding* Feb. 20/3, I own a '68 Datsun pickup and would like to install a 327 Chevy small-block engine with a 350 turbo trans.

trans-, *prefix.* Add: **5.** *transabdominal*, *-antral* [Antrum], *-capillary*, *-cervical*, *-cutaneous*, *-duodenal*, *-epithelial*, *-glottal*, *-granular*, *-ovarial*, *-ovarian*, *-placental*, *-pyloric*.

The examples given below are cited here despite the note in Dict. after sense 6, since they did not call for the elaborate treatment of a main entry.

1956 *Nature* 18 Feb. 330/2 Although transabdominal puncture of the uterus has been carried out often for therapeutic and experimental reasons without accidents, mere curiosity does not justify the procedure. **1957** *Laryngoscope* LXVII. 566 The degree of orbital tension present in the severe forms of exophthalmos is correlated with the degree of recession obtained by the transantral decompression. **1974** *Nature* 31 May 495/3 Transcapillary and transepithelial water transport. **1963** *Lancet* 19 Jan. 165/2 Transcervical fractures. **1977** *Ibid.* 7 May 983/1 We believe that local warming of the skin and of the capillary blood under the electrode increases P_{CO_2}; the transcutaneous P_{CO_2} values are therefore higher than those found in arterial blood. **1908** Transduodenal [see *retroduodenal* s.v. *retro- 3. b.]. **1975** H. J. Burhenne in Najarian & Delaney *Surg. Liver, Pancreas & Biliary Tract* 104 Percutaneous cholangiography will probably be supplanted by transduodenal cannulization. **1974** Transepithelial [see *transcapillary* above]. **1964** J. C. Catford in D. Abercrombie et al. *Daniel Jones* 31 The trans-glottal air-jets of voice superimpose a periodic fluctuation on the mean air pressure behind the articulatory stricture, resulting in..hiss. **1970** *Language* XLVI. 313 The human..larynx is so constructed that the fundamental frequency of phonation is a function of both the transglottal air pressure drop and the tensions of the laryngeal muscles. **1962** *Science Survey* III. 329 Occasionally the cracks are transcrystalline (trans-granular). **1946** E. A. Steinhaus *Insect Microbiol.* viii. 439 Transovarial transmission of the virus takes place by the viruses penetrating the walls of the ovary and thence entering the developing ovum. **1971** P. C. C. Garnham *Progr. Parasitol.* iii. 34 The tick..may carry over into subsequent generations by transovarial passage. **1954** *Jrnl. Infectious Dis.* XCV. 178/2 The distribution of antibodies in different small age groups of wild birds is further evidence of the transovarian passage of neutralizing antibodies. **1980** *Nature* 7 Feb. 568/2 Another event occurring during oogenesis is trans-ovarian transmission of symbiotic bacteroids..from females to the oocytes. **1902** *Brit. Med. Jrnl.* 17 May 1198/1 Alterations in the transplacental interchanges. **1977** *Lancet* 9 Apr. 795/1 The transplacental leak of fetal red blood-cells. **1905** C. Addison *Ellis's Demonstrations of Anat.* (ed. 12) vi. 298 This plane, from its traversing the pyloric end of the stomach, is called the transpyloric. **1977** *Lancet* 28 May 1157/2 The gastric distension..can be relieved by passing an open-ended nasogastric tube, and adequate nutrition maintained by intravenous or continuous transpyloric feeding.

b. In derived advbs.: *transabdominally*, *-duodenally*, *-ovarially*, *-placentally*.

1962 *Lancet* 8 Dec. 1208/2 A needle passed transabdominally into the liver. **1955** *Radiology* LXIV. 325 When the sphincter of Oddi is sectioned transduodenally, a plastic tube can be inserted into the main pancreatic duct. **1954** *Jrnl. Infectious Dis.* XCV. 168/1 Neutralizing antibodies to western equine encephalitis (WEE) and St. Louis encephalitis (SLE) viruses may be transmitted transovarially. **1979** *Amer. Jrnl. Trop. Med. & Hygiene* XXVIII. 1064 Spores developing in transovarially infected mosquitoes. **1965** *Dorland's Med. Dict.* (ed. 24) s.v. *Listeria*, A septicemic disease which may be transmitted transplacentally in pregnant women.

7. a. *trans-Alleghanian* (earlier example), *-Allegheny*, *-Mississippian* (example), *-Pyrenean*.

1814 *Deb. Congress U.S.* 14 Feb. 1422 Even then the trans-Alleganean wilderness was rustling with the preparation of the savage. **1825** C. D. Colden *Mem.* 93 Why should the trans-Allegany States have remained united with those on the Atlantic? **1831** A. Wilson *Amer. Ornithology* IV. 31 In the trans-Mississippian territories of the United States, the burrowing owl resides. **1875** *Harper's Mag.* Mar. 572/2 The subdivision..into the Trans-Alleghany, Valley, Middle, and Tide-water districts. **1934** A. Toynbee *Study of Hist.* II. 203 Musa had completed the Arab conquest..by occupying the Transpyrenaean province of Septimania along the Gallic coast between the Pyrenees and the Rhône.

b. in substantives with the sense 'the region beyond' or 'one dwelling beyond or on the other side of', as *trans-Alleghanian*, *-Mississippi*, *-Mississippian*.

1774 J. Adams *Diary* 23 Oct. in *Wks.* (1850) II. 401, I went to the Baptist Church and heard a trans-Alleghanian, a preacher from the back parts of Virginia. **1883** *Century Mag.* Nov. 142/1 If the President was to attempt to reach the Trans-Mississippi at all,..he should move on at once. **1898** *Ibid.* Oct. 844/2 The trans-Mississippians have entered upon no line of rural industry with a more intelligent determination to make it a great success than upon dairying. **1949** Beebe & Clegg *U.S. West* 10 Anyone approaching the *matière* of the trans-Mississippi in the nineteenth century as an exploiter of new material is either deluded or an imposter.

8. *trans-Antarctic*, *-Canada*, *-Pyrenean* (earlier example), *-Saharan* (examples), *-Sierran*, *-Tasman*.

1846 R. Ford *Gatherings from Spain* iv. 31 Newfangled transpyrenean reforms, innovations, and botherations. **1888** *Times* 20 Sept. 3/6 Denham, Clapperton, Barth, and other trans-Saharan travellers. **1908** *Busy Man's Mag.* Apr. 95 The Proposed Route of the Trans-Canada Railway. **1916** R. K. Wood *Tourist's N.W.* 315 A campaign for the improvement and construction of roads which.. shall in combination form a trans-Canada motor route. **1933** *Geogr. Jrnl.* LXXXII. 470 Mr. Lincoln Ellsworth.. will then make the Transantarctic flight, which is the sole object of the expedition. **1935** E. B. Buckbee *Saga of Old Tuolumne* 385 The road survived until the State of California came to look with favor upon its possibilities as a tran-Sierran [*sic*] road. **1938** *Times* 17 Feb. 13/4 The flying-boat Centaurus showed her unsuitability for trans-Tasman traffic. **1950** *Pacific Discovery* Mar.–Apr. 4/1 Did you know that a trans-Sierran highway is now being built in Madera County? **1963** P. Drackett *Motor Rallying* iv. 62 The Trans-Canada and Canadian Winter Rallies have not yet reached the lofty eminence of the Safari. **1965** E. McCourt *Road across Canada* 199 The Trans-Canada Highway is an engineering, communications, and scenic marvel. **1966** N. Marsh *Black Beech & Honeydew* viii. 175 In..1928 the trans-Tasman steamer sailed..into Cook Strait. **1978** *Times Lit. Suppl.* 25 Aug. 957/4 Shackleton's unsuccessful transantarctic expedition. **1978** J. Updike *Coup* (1979) i. 4 The capital is Istiqlal, renamed in 1960, upon independence, and on prior maps called Cailliéville, in honor of the trans-Saharan traveller of 1828.

9. a. *Chem.* (Also without hyphen as a quasi-adj. Usu. printed in italic.) Designating a compound in which two atoms or groups are situated on opposite sides of some plane passing through the molecule; hence (of a bond or a reaction), characterized by such a relationship.

[**1888**: see *cis- 3.] **1892** *Jrnl. Chem. Soc.* LXII. 1213 The anhydride of the cis modification invariably melts at a lower temperature than that of the trans form. **1937** *Nature* 3 July 25/1 The trans form of ethylene bromide is considered to be the more 'stable' (preferred) form, even at high temperatures. **1951** C. R. Noller *Chem. Carbon Compounds* xvii. 316 It has been proved that *cis*-2-butene is the isomer boiling at 3·73° and *trans*-2-butene is that boiling at 0·96°. **1956** D. J. Cram in M. S. Newman *Steric Effects in Org. Chem.* vi. 306 The terms *cis* elimination will be used whenever the leaving groups depart from the same side of the incipient double bond, and *trans* elimination when they leave from the opposite side of the incipient double bond. **1972** R. A. Jackson *Mechanism* i. 7 Bromine adds across an olefinic double bond in a *trans* manner. **1976** *Sci. Amer.* Jan. 124/2 Nearly all peptide bonds are trans and planar, meaning that hydrogen and carbonyl oxygen (CO) are on opposite sides of the bond.

b. *transf.* in Genetics, with reference to the location on different chromosomes of dominant alleles of two or more genes or cistrons.

1941 J. B. S. Haldane *New Paths in Genetics* i. 17 There are two geometrically isomeric types of rabbit (to use a chemical analogy) heterozygous for recessive white *c* and recessive yellowfat *y*... The trans-rabbit $+y/c+$ is derived from the crossing of a white-fatted coloured rabbit and a yellow-fatted white. **1957** [see *cistron]. **1973** R. G. Krueger et al. *Introd. Microbiol.* xiii. 385/1 In the diploid the two mutants [*sc.* mutant genes] are said to be trans to one another because they are on different chromosomes.

10. *Biochem.* and *Biol.* In sbs. with the sense 'transfer', as *transacetylase*, *-amination*, *-genosis*, *-methylation*, *-peptidation* (see as main entries).

11. *Physics.* In adjs. and sbs. with the sense 'having a higher atomic number than; beyond (in the periodic table)'.

1952 *Chem. & Engin. News* 21 Jan. 237/2 The trans-californium elements. **1969** *Nature* 26 Apr. 323/1 Dr Glen Seaborg..was able to proclaim that element 104 is the first of the 'trans-actinide' elements. **1973** *Q. Jrnl. R. Astron. Soc.* XIV. 121 The existence of transbismuth elements in nature.

transacetylase (trɑnzæse·tilē¹z). *Biochem.* [f. *trans- 10 + *acetyl + *-ase.] Any enzyme that catalyses the transfer of an acetyl group from one molecule to another.

1950 Stadtman & Barker in *Jrnl. Biol. Chem.* CLXXXIV. 788 The arsenolytic decomposition of acetyl phosphate by the cell-free extracts of *C*[*lostridium*] *kluyveri* can be represented in an analogous way..if it is assumed that the reaction is catalyzed by an acetyl transferring enzyme or trans-acetylase. **1966** *Jrnl. Molecular Biol.* XIX. 576 (*heading*) Transposition of the *lac* region of *Escherichia coli*. II. On the role of thiogalactoside transacetylase in lactose metabolism. **1976** *Ann. Rev. Microbiol.* XXX. 329 To illustrate genetic transfer, Hedges..cites the gene for chloramphenicol transacetylase.

transact, *v.* **2.** (Further examples.)

1883 R. L. Stevenson *Silverado Squatters* 122 With so strong a helper, the business was speedily transacted. **1950** R. Macaulay *World my Wilderness* xxii. 158 What careful, crafty affairs had they transacted in the Hall of the Haberdashers? **1970** D. Jacobson *Rape of Tamar* ii. 14 Gathering together around the king, while he transacts the business of the state.

transactional, *a.* Add: **c.** *Psychol.* Relating to or involving interpersonal or social communication viewed as transactions of attitude between the participants; *spec.* in *transactional analysis*, psychotherapeutic analysis based on the attitudes revealed in such transactions (esp. those of parent, adult, and child); abbrev. T.A. (see *T 6 a); hence *transactional analyst*.

1961 E. Berne (*title*) Transactional analysis in psychotherapy. *Ibid.* 12 Transactional analysis, the social aspect of structural analysis, reveals several different types of 'crossed transactions'. **1969** T. A. Harris *I'm OK—You're OK* i. 13 Transactional Analysis is the method of examining this one transaction..and determining which part of the multiple-natured individual is 'coming on'. *Ibid.* 15 Transactional Analysts claim to have found some of these regularities [of language]. **1976** *Listener* 5 Feb. 141/3 Discussing Basil Fawlty in terms of transactional analysis: as to whether, at certain times, Basil was behaving as a parent, or as a child.

Hence **transa·ctionalist,** one who believes in a theory of social transactions; also *attrib.*

1972 R. E. Ornstein *Psychol. of Consciousness* ii. 32 (*heading*) The transactionalists. *Ibid.* 37 Some of the most relevant psychological experiments have been performed by Jerome Bruner and by the transactionalist group. **1977** P. Johnson *Enemies of Soc.* xv. 202 Then there is a series of schools, associated with the Americans H. S. Becker and Edwin Lemert, known variously as social-control theorists, social-reaction theorists, transactionalists or labelling theorists.

transalpine, *a.* Add: **2. b.** That crosses the Alps, built across the Alps.

1908 *Chambers's Jrnl.* Sept. 647/1 The Simplon is the least steeply graded..of any transalpine railway.

transa·minase. *Biochem.* [f. next: see quot. 1940 and *-ase.] Any enzyme that brings about transamination.

1940 P. P. Cohen in *Jrnl. Biol. Chem.* CXXXVI. 566 Since the original term *Umaminierung*..has been accepted with the English (and French) equivalent of *transamination*..it is suggested here that the enzyme (or possibly enzymes) catalyzing the transfer of amino nitrogen be termed *transaminase*. **1970** R. W. McGilvery *Biochem.* xvii. 354 Much of the nitrogen of the amino acids sooner or later appears in the form of glutamate because of the action of transaminases.

transamination (trɑ:nzæminē¹·ʃən). *Biochem.* [a. F. *transamination* (Schaeffer & Le Breton *L'Action Spécifique des Protides* (1938) 143): see *trans- 10, *amino-, -ation.] The transfer of an amino group from one organic molecule to another, esp. from an amino-acid to a keto-acid.

1939 *Nature* 8 Apr. 609/1 The reversible transfer of amino groups between aminodicarboxylic acids and α-keto-acids, or 'trans-amination'..is of almost universal occurrence in biological objects. **1946** *Ibid.* 12 Oct. 515/2 Not until it reaches the root cells does the transformation of aspartic acid take place (through deamination, transamination, etc.). **1959** A. White et al. *Princ. Biochem.* (ed. 2) xx. 514 In addition to transamination reactions involving L-α-amino acids, transamination of..aldehydes ..has also been observed. **1982** T. I. Diamondstone in T. M. Devlin *Textbk. Biochem.* xi. 546 Virtually every protein amino acid undergoes transamination at some point in its metabolic breakdown.

Hence (as back-formations) **transa·minate** *v. intr.*, to undergo transamination; *trans.*, to deprive of an amino group; to change by transamination; **transa·minated** *ppl. a.*, **transa·minating** *ppl. a.* and *vbl. sb.*

1940 *Ann. Rev. Biochem.* IX. 284 The experiments..disclose the possible operation of the transaminating enzymes within the animal body. **1940** *Jrnl. Biol. Chem.* CXXXVI. 573 Braunstein reports that these amino acids are transaminated to the extent of 17 to 21 per cent. **1955** A. Meister in McElroy & Glass *Amino Acid Metabolism* 1. 14 L-Alloisoleucine and its keto analogue..transaminate at somewhat slower rates than L-isoleucine. **1959** A. White et al. *Princ. Biochem.* (ed. 2) xx. 517 The discovery of the transaminating mechanism. **1970** R. W. McGilvery *Biochem.* xvii. 355 Any other amino acid that will transaminate with α-ketoglutarate may be deaminated in the same way. **1972** *Nature* 8 Sept. 101/2 The group..included amino-acids..that are not transaminated or oxidized in myocardium and those..that can be transaminated and used for energy production. **1974** I. B. R. Bowman in K. Elliott et al. *Trypanosomiasis & Leishmaniasis* 268 It is likely that glutamate is transaminated to α-oxoglutarate.

trans,a·nnular, *a.* *Chem.* [f. Trans- 3 + Annular *a.*] Situated, existing, or occurring between non-adjacent atoms forming a ring.

[**1926** de Barry Barnett & Matthews in *Ber. Deut. Chem. Ges.* LIX. 1429 Es dürfte deshalb angezeigt sein, diese besondere Art tautomerer Umwandlungen unter dem allgemeinen Begriff 'Transannular-Tautomerie' zusammenzufassen.] **1926** *Chem. Abstr.* XX. 3003 The name 'trans-annular tautomerism' is suggested for those cases..where the H migrates across the ring. **1941** *Nature* 21 June 776/1 As the carbon atoms concerned..are components of a cyclic system.., the resulting tautomeric change would be a transannular one and involve the formation of a bridge linkage from C_2 to C_5. **1978** *Further Perspectives Org. Chem.* (CIBA Symposium) 119 This reaction undoubtedly involves fission of the transannular bond.

transat (tra·nzæt). *colloq.* [Abbrev. of Transatlantic *a.*, *sb.*] A type of large deck-chair (see quots.).

1968 *Daily Tel.* (Colour Suppl.) 29 Nov. 76 (Advt.), A few days flat-out on a 'transat'—a long couch like those on the decks of ocean liners—at the Tahiti beach did the trick. **1978** *N.Y. Times* 30 Mar. c8/1 Along with the regular-sized deck chairs,..there's..a transat..—a chair with an attached foot rest that slides neatly under the seat when you don't need it.

transaxle (tra·nzæks'l). orig. *U.S.* [f. Trans(mission + Axle².] In a motor vehicle: an integral driving axle and differential gear.

1958 *N.Y. Times* 1 June x. 21/4 The engineers call the new system a 'transaxle' or a 'traxle'. **1967** P. H. Smith in L. Holmes *Odhams New Motor Man.* vi. 160/1 When access is required to the clutch or gearbox on cars having unitary construction of engine and transaxle,..the complete unit must be removed from the vehicle. **1983** *Fortune* 7 Mar. 110/2 Some of the..contracts for high-margin items like engines, transaxles, and sophisticated electronics are going abroad.

trans-ba·y, *a.* *U.S.* [f. Trans- 3 + Bay *sb.*²] That crosses a bay, *spec.* San Francisco Bay.

1965 *Newsweek* 19 July 71/1 The most challenging and vital element of all, the Trans-Bay Tube, will be begun next year when a 30-foot-deep trench will be dug across San Francisco Bay. **1966** T. Pynchon *Crying of Lot 49* v. 130 She tailed him..back..to the trans-bay bus terminal, where he bought a ticket for Oakland. **1975** *New Yorker* 10 Jan. 29/3 Meanwhile, 1,200 commuters are stranded in a BART train in the collapsed trans-bay tube.

trans-border, *a.* Add: Also, on or from the other side of the border; crossing a border or borders.

1976 *Globe & Mail* (Toronto) 16 Feb. 9/1 It was detonated..where the DC-8 came to rest after the eight-hour, trans-border ordeal. **1976** *Southern Even. Echo* (Southampton) 1 Nov. 1/2 'We have undertaken hot pursuit operations as a result of trans-border aggression by terrorists,' the communique said. **1980** *Times* 26 Mar. 23/4 Restrictions are imposed by various countries on what is becoming known as 'transborder data flow'.

transbus (tra·nzbɒs). *U.S. temporary.* [f. Trans- (or Trans(it *sb.*) + Bus *sb.*²] A bus with special design features, including lower floors to give the elderly and disabled easier access.

1973 *Britannica Yearbk. Sci. & Future 1974* 335/1 The three proposals for the so-called Transbus would have several common features, in addition to length... Lower profile, plus further automatic lowering at stops. **1975** *General Motors Transbus* (Gen. Motors Truck & Coach Div.) I. 24 The use of the pantograph entrance on the Transbus allows use of the wheelchair elevator. **1977** *Time* 5 Dec. 34 He expects that the high cost of the transbuses (as much as $50,000 more than a regular bus) will halt the expansion of the special van service.

transceiver (transi·vəɪ). [f. Trans(mitter + Re)ceiver¹.] An instrument combining a radio transmitter and a radio receiver.

1934 *Electronics* Sept. 273/1 The increasing sales of so-called 'transceivers',..usually one or two-tube affairs operating from batteries, is an indication that the early days of radio may be re-enacted. **1952** *Times* 4 Dec. 7/4 Nowhere except on the transceiver network operated from Alice Springs is it possible for the scholars to play an active part by question and answer to their distant teacher. **1977** *Time* 10 Jan. 44/3 Sales of Citizens Band transceivers were boosted dramatically by gutted prices on current 23-channel gear.

transcendency. b. (Earlier example.)

1902 W. James *Let.* 14 Dec. (1920) II. 179, I believe that the 'transcendency' of the object will not recover from your treatment.

transcendental, *a.* **3.** Add: **e.** *transcendental meditation*: a method of relaxation and meditation based on the theory and practice of yoga popularized in the West by the Maharishi Mahesh Yogi; abbrev. TM (see *T 6 a); hence *transcendental meditator*.

A proprietary term in the U.S.

1966 C. F. Lutes in M. M. Yogi *Sci. of Being & Art of Living* 13 The system on which Maharishi's teaching is based—a simple method of transcendental meditation.. —is indeed systematic and produces measurable and predictable results and is therefore scientific. **1973** *Times* 30 June 14/3 Transcendental meditation is becoming popular as a way of coping with the stress of modern life. *Ibid.* 14/5 Transcendental meditators do not like to publicize the possible dangers inherent in mind-bending techniques. **1975** *Physics Bull.* Sept. 397/2 The aim was to measure the breathing rate and lung ventilation of 15 transcendental meditators before, during and after meditation and compare the values with those obtained for 15 non-meditators. **1976** *Early Music* Oct. 467/1 The place of meditation and mantra made familiar to the West by the practitioners of Transcendental Meditation and similar Yoga techniques. **1976** *Official Gaz.* (U.S. Patent Office) 3 Aug. TM77/2 Class 41—Education and Entertainment. 1,045,673. World Plan Executive Council—United States, Los Angeles, Calif.... Transcendental Meditation. **1980** *Times* 27 May 1/8 Transcendental meditation, as taught by the Maharishi's World Government of the Age of Enlightenment..involves learning the techniques of meditating.

transcendible, *a.* Delete †*Obs.* and add later example.

1953 G. M. Young *Victorian Eng.* xx. 118 Such a body of permitted belief as makes the barriers between England and Rome transcendible.

trans,co·de, *v.* [f. Trans- 2 + Code *v.*] *trans.* and *intr.* To convert from one form of coded representation to another; to change the code of. Hence **trans,co·ding** *vbl. sb.*, the action or result of such conversion.

1962 *Language & Speech* V. 18 A common feature of linguistic and biological information is the ease with which it can be repeatedly transcoded resulting in a multiplicity of codes. *Ibid.*, Print, translation, shorthand and Braille are among many other examples of transcoding. **1965** *Guardian* 24 Mar. 5/6 It is technically possible to 'transcode' from one [television] colour system to another. **1972** *Computers & Humanities* VI. 150 In using either punched-card or paper-tape equipment one has to delegate the transcoding of the special input codes to the computer. **1984** *What Video?* Aug. 5/4 The only course open to you is to have the existing NTSC—standard tape transcoded to UJ—standard PAL 625.

transconductance (trɑns,kǫndɒ·ktāns, -nz-). *Electronics.* [f. Trans(fer *sb.* + *conductance.] The ratio of the (change in) current at one electrode or terminal of an active device, esp. the output, to the (change in) voltage at another, esp. the input; *spec.* = *mutual conductance* s.v. *mutual *a.* 1 f.

1933 K. Henney *Radio Engin. Handbk.* viii. 203 The coefficient showing the effect of plate voltage on the grid current has been termed inverse mutual conductance, or the plate-grid transconductance. **1944** *Electronic Engin.* XVI. 318 The transconductance between these two grids is negative. **1950** *Electronics* Feb. 107/1 In push-pull amplifiers, triodes using this type of neutralization compare favorably with pentodes of the same transconductance as wideband amplifiers. **1981** J. C. Sprott *Introd. Mod. Electronics* vii. 159 At the operating point [of a field effect transistor] the value of the forward transconductance,..and the output resistance,..can be determined.

trans,confe·ssional, *a.* [f. Trans- 3 + Confessional *a.*] Extending across religious denominations; interdenominational.

1975 R. R. Williams in *Critique of Eucharistic Agreement* i. 11 The agreements we have here are by no means the only trans-confessional agreements of modern times. **1983** *Times Lit. Suppl.* 7 Oct. 1103/1 Edward Schillebeeckx..has been much read beyond the bounds of the Roman Catholic community to which he belongs. He has accepted the title of 'transconfessional' theologian.

trans,conforma·tion. *Chem.* [f. Trans- 1 + Conformation.] A change in the conformation of a molecule, esp. a protein; freq. *attrib.*

1954 Lumry & Eyring in *Jrnl. Physical Chem.* LVIII. 110/1 The bulk of denaturation changes consist of changes in secondary bonds... Such rearrangements of the secondary-bonded structure are defined as changes in conformation... We shall thus speak of transconformation reactions rather than denaturation reactions. **1979** *Nature* 18 Jan. 244/1 Besides the folding of the isolated polypeptide chains, additional transconformation reactions occur after the association of the native quaternary structure has taken place.

transconjugant (trɑns,kǫ·ndʒŭgănt). *Biol.* [f. Trans- 1 + Conjug(ate *v.* + -ant¹.] A plasmid or a bacterial cell which has received genetic material by conjugation with another bacterium.

1974 *Jrnl. Bacteriol.* CXX. 1187/2 Transfer frequencies were calculated from the number of transconjugants formed per donor cell in the mating mixture... (Transconjugants are cells of the recipient strain that have received genetic material by conjugation.) **1977** *Nature* 10 Feb. 560/2 Plasmid transconjugants which combine octopine utilisation of the donor and antibiotic resistance of the recipient grow well on this medium.

trans-continental, *a.* (*sb.*) Add: Now freq. as one word without hyphen. (Earlier example.)

1853 *Harper's Mag.* Feb. 550/2 A company..to construct a trans-continental railroad.

B. *ellipt.* as *sb.* A trans-continental railway, a trans-continental train. Chiefly *Canad.*

1907 *Eye Opener* (Calgary) 18 July 3/3 Dealing with the immense expenditure on the G.T.P. transcontinental. **1920** J. M. Gibbon *Conquering Hero* 147 At Winnipeg the transcontinental was boarded by..a rancher on his way back to the foothills. **1955** *Standard* (Shawinigan Falls, Que.) 12 Jan. 2/2 A rate [for carrying grain] fixed..shortly after the first trans-continental opened for business. **1964** J. Carroll *Sky Photographer* 6 You heard the one about the squaw on the transcontinental?

transcribe, *v.* Add: **1. c.** *Biol.* To synthesize a nucleic acid (usu. RNA) using an existing nucleic acid (usu. DNA) as a template, so that the genetic information in the latter is copied. Const. *into* (with the template as obj.), *from*, *off* (with the new acid as obj.).

1962 *Proc. Nat. Acad. Sci.* XLVIII. 544 Only one strand of the DNA is transcribed as functional messenger-RNA. **1973** *Sci. Amer.* Apr. 34/3 Their chromosomes are in a greatly enlarged and uncoiled 'lampbrush' stage where we might be able to see structural details of DNA being transcribed into messenger RNA. **1979** D. R. Hofstadter *Gödel, Escher, Bach* xvi. 517 When mRNA is transcribed off of DNA, the transcription process operates via the usual base-pairing. **1981** L. L. Mays *Genetics* ii. 65 Once RNA is transcribed from DNA, it is cut to its final size, modified in specific ways, and sent to its site of action.

2. b. Also *intr.* for *pass.*

1976 *Gramophone* June 61/1 Vocal ensemble music should transcribe well for brass.

6. a. To make a copy of (a gramophone recording) from a secondary source, not the master recording.

1931 *Gramophone* Dec. 264/1 The Philadelphians have recorded the Fifth Symphony of Beethoven complete on a single 12-in. disc and..thirty-two other discs..have been announced on which existing works..have been 'transcribed'.

b. *Broadcasting.* To record for subsequent reproduction; to broadcast in this form.

1941 W. Abbot *Handbk. Broadcasting* 245 These are inserted into transcribed programs or into a live program.

transcribed *ppl. a.* (further examples).

1961 J. Updike in *New Yorker* 17 June 31/1 A transistor radio somewhere in the sand releases in a thin, apologetic gust the closing peal of a transcribed service. **1981** L. L. Mays *Genetics* ii. 57 Sometimes the transcribed RNA is the final product.

transcriber. Add: **2.** An item of office equipment on which dictation cannot be recorded, but which replays dictated matter for transcription by an (audio) typist.

1931 *Times* 16 Mar. 1/6 All-electric Dictaphone with Shaver and Transcriber: good condition. **1976** *Southern Even. Echo* (Southampton) 3 Nov. (Advt.), Duties will include shorthand or use of transcriber, typing, correspondence, filing, [etc.].

transcript, *sb.* Add: **3.** *Biol.* A length of RNA or DNA which has been transcribed from a DNA or RNA template (respectively).

1961 Jacob & Monod in *Jrnl. Molecular Biol.* III. 352 The molecular structure of proteins is determined by specific elements, the structural genes. These act by forming a cytoplasmic 'transcript' of themselves, the structural messenger. **1972** *Sci. Amer.* Jan. 33/3 We have uncovered evidence that cancer-causing RNA viruses can produce a DNA transcript of the viral RNA. **1982** *Nature* 13 May 130/2 Transcripts initiated further upstream than position −675 will yield a protected fragment of ~1,375 nucleotides.

transcriptase (trɑn,skri·ptē¹z). *Biochem.* [f. Transcript(ion + *-ase.] The polymerase responsible for transcription, which catalyses the formation of RNA from a DNA template; *reverse transcriptase*, the polymerase responsible for reverse transcription, which catalyses the formation of DNA from an RNA template.

1963 Spiegelman & Hayashi in *Cold Spring Harbor Symp. Quantitative Biol.* XXVIII. 162/1 The transcribing DNA-dependent-RNA-polymerase which produces RNA complements will be referred to as a transcriptase. **1965** *New Scientist* 18 Mar. 713/2 An enzyme exists in cells (called 'RNA polymerase' or sometimes 'transcriptase') which is capable of reading the genic code. **1970** *Nature* 26 Dec. 1255/1 Research..on reverse transcriptases in animal cancer viruses and human cancer cells. **1976** *Ann. Rev. Microbiol.* XXX. 23 Kates & McAuslan..initially demonstrated transcriptase in poxvirus cores isolated from infected cells.

transcription. Add: **5. a.** A gramophone record made from a secondary source, not the master recording.

1931 *Gramophone* Dec. 264/2 'Transcriptions', too, for which our unkind readers used to prefer the phrase 'faked records', are not very popular over here. **1968** *Jazz Monthly* Feb. 4/1 Numerous 'pirate' labels also issuing EPs and LPs,..tend increasingly to concentrate on air shots and transcriptions from a variety of sources.

b. *Broadcasting.* The recording of a broadcast for subsequent reproduction; a record or broadcast so made. Also *attrib.*

1932 *B.B.C. Year-bk. 1933* 290 The relaying of the Empire station by overseas transmitters cannot for various reasons be assumed to be possible as a regular practice and therefore the recording of programmes on gramophone discs becomes an important subsidiary method of programme circulation... American programmes are already circulated by this method, which is termed 'electrical transcription'. **1936** *Communication* Mar. 5 (*heading*) The growing importance of transcription broadcasting. *Ibid.* 6/2 The transcriptions in every-day use in broadcast stations include both lateral-cut and vertical-cut recordings. **1943** *B.B.C. Year-bk.* 23 An important broadcasting activity little known in this country is the projection overseas, by means of recordings, of the culture and wartime life of Britain... Known collectively as the London Transcription Service, the activity has been undertaken by the BBC for the Empire since the beginning of the war, and for foreign countries for nearly two years. **1956** *B.B.C. Handbk. 1957* 42 English by Radio lessons..reach an audience of several millions by direct transmission from London, by relays, and by transcription recordings. *Ibid.* 133 Among other transcriptions, the special service for Colonial schools.. proved successful. **1978** *A–Z of BBC* (ed. 2) 225/2 Transcription services are the BBC's channel for selling Radio Programmes to stations overseas.

c. Used *attrib.* to designate equipment used in professional recording or broadcasting transcription, or *gen.* of a standard or type so used.

1936 *Communication* Mar. 8/2 Noise in the output of a transcription equipment is often caused by pickup of the motor vibration. **1943** *Proc. I.R.E.* Feb. 52 (*heading*) The measurement of transcription-turntable speed variation. **1957** *Long Playing Record Library Catal. & Handbk.* 10 The only answer..is the use of a 'transcription motor', implying a high standard of design and finish and individual care in manufacture. **1962** A. Nisbett *Technique Sound Studio* 271 The large transcription tape decks are normally equipped to play either type. **1965** *Wireless World* Aug. 6 (Advt.), The Goldring-Lenco GL 70 transcription unit with its integrally mounted transcription arm continues to be the first choice of discriminating record lovers with custom-built equipment. **1978** *Lancashire Life* Nov. 110/1 (Advt.), All the illustrated units are complete music centres with Dolby cassette deck-belt driven transcription unit.

6. *Biol.* The process by which genetic information represented by the sequence of nucleotides in the DNA of a cell or virus is copied into molecules of RNA, which are synthesized with the DNA serving as a template; *reverse transcription*, the reverse process, occurring in some RNA viruses, by which DNA is synthesized from an RNA template.

1961 Jacob & Monod in *Cold Spring Harbor Symp. Quantitative Biol.* XXVI. 193/1 The second process, which we shall call transcription, allows the gene to perform its physiological function. **1970** *Nature* 27 June 1198/1 For the past twenty years the cardinal tenet of molecular biology has been that the flow or transcription of genetic information from DNA to messenger RNA and then its translation to protein is strictly one way. **1971** [see *promoter 1 g]. **1973** *Sci. Amer.* Apr. 34/2 In prokaryotes, which include the many species of bacteria, transcription and translation of messenger RNA occur at the same time and place. **1977** *Nature* 8 Sept. 122/1 Until recently.. most groups studying reverse transcription *in vitro* found the DNA products to be small relative to the size of the RNA templates.

transcriptional *a.* (examples in *Biol.*); **transcriptionally**: also, in *Biol.*, in a transcriptional way; (further examples).

1970 *Nature* 29 Aug. 916/1 A similar mechanism controls gene expression at the transcriptional level during bacterial sporulation. **1975** *Ibid.* 5 June 462/2 This transcriptionally active DNA represents r-protein genes. **1981** L. L. Mays *Genetics* ix. 416 The prokaryotic systems that operate via transcriptional control often utilize different control systems. **1983** *Nature* 23 June 677/1 Transcriptionally active chromatin.

transcriptionist (trɑn,skri·pʃənist). *U.S.* [f. Transcription + -ist.] An audio typist.

1977 *Chicago Tribune* 2 Oct. xii. 67/6 (Advt.), 60 wpm typing, plus 1 year experience as a medical transcriptionist, or training in this field qualifies you. **1978** *Detroit Free Press* 16 Apr. F 5/5 (Advt.), Medical transcriptionist. For out-patient clinic in university setting. Accurate typing & Medical Terminology necessary.

transcriptor. Restrict †*Obs.* to sense in Dict. and add: **b.** = *transcriber 2.

1957 *Practical Wireless* XXXIII. 532/2 Tape transcriptors are also supplied by Collaro. 1958 *Punch* 17 Sept. 366/3 Anita types. You scarcely see The neat transcriptor at her ear. The tape turns confidentially, And no one else can overhear.

trans,cry·stalline, *a.* [f. TRANS- 5 + CRYSTALLINE *a.*] Of a fracture: passing through individual crystals of a metal rather than following grain boundaries.

1916 H. M. HOWE *Metallogr. Steel & Cast Iron* xxviii. 489 (*caption*) Intergranular and trans-crystalline rupture. 1956 M. C. SMITH *Princ. Physical Metall.* x. 379 The usual fracture of a metal is transcrystalline. 1978 E. LEPA tr. *S. Kocańda's Fatigue Failure Metals* v. 271 The brittle crack propagates..along cleavage planes, yielding a cleavage transcrystalline fracture, or along the grain boundaries, forming an..intercrystalline fracture.

trans,cu·ltural, *a.* [f. TRANS- 3, 4 + CULTURAL *a.*] Transcending the limitations or crossing the boundaries of cultures; applicable to more than one culture; cross-cultural; spec. *transcultural psychiatry*, psychiatry applied to disorders due to migration from one cultural environment to another.

1958 *Internat. Jrnl. Social Psychiatry* III. 245 (*heading*) Some problems of transcultural psychiatry. 1964 I. L. HOROWITZ *New Sociology* 37 Social science is interdisciplinary because social problems are transcultural. 1973 *Observer* (Colour Suppl.) 28 Oct. 34/5 Sailing ships are gone but the sea shanty is still sung: the function is altered but the song remains: the stuff is transcultural. 1976 *Daily Tel.* 27 Apr. 2/8 The congress, on transcultural psychiatry, is being organised by the university and the World Federation for Mental Health. 1978 *Church Times* 3 Feb. 6/5 The resolution of the problem suggested depends on a distinction between meanings which are universal and transcultural and the understanding and expression of those meanings which are limited historically and culturally.

trans,cultura·tion. [f. TRANS- 3 + CULTURE *sb.* + -ATION.] = *ACCULTURATION.

1941 B. MALINOWSKI *Sci. Theory Culture* (1944) ii. 14 Practical problems—such as..the difficulties of culture contact, and transculturation—problems that legitimately belong to anthropology. 1949 *Psychiatry* XII. 184 This paper..has shown that the process of transculturation is not really a process of adaptation to a culture but to a political situation. 1970 R. STAVENHAGEN in I. L. Horowitz *Masses in Lat. Amer.* vii. 287 We use the terms 'transculturation' and 'acculturation' interchangeably.

transcurrence. Restrict † *Obs. rare* to sense in Dict. and add: **2.** *Geol.* The phenomenon of transcurrent faulting.

1971 *Geol. Mag.* CVIII. 40 His [sc. S. W. Carey's] concept of megashears is essentially that of transcurrence as defined here. 1979 *Nature* 1 Mar. 12/2 Such transcurrence juxtaposed features whose dissimilarity had instead suggested to many an oceanic gap.

transcurrent, *a.* Add: **3.** *Geol.* Designating or pertaining to a fault which is primarily due to horizontal displacement; *esp.* one of large dimensions and with a nearly vertical inclination; usu. in *transcurrent fault* (see quots. 1942[2], 1971[1]).

1942 E. M. ANDERSON *Dynamics of Faulting* i. 1 The term 'Blatt' [used by Suess] will be translated as transcurrent fault. *Ibid.* v. 54 The name Transcurrent Faults has been adopted in this memoir for members of the class which are distinguished by lateral movement, with inclinations which, according to theory, are nearly vertical. 1965 G. J. WILLIAMS *Econ. Geol. N.Z.* i. 2/1 Movement along the Alpine Fault..is still going on in the form of major clockwise transcurrent movement with a 300 mile shift. 1971 *Geol. Mag.* CVIII. 33 Transcurrent faults are large-scale strike-slip faults which operate between lithosphere plates. 1977 [see *tear-fault* s.v. *TEAR *sb.*[1] 4].

trans,denomina·tional, *a.* [f. TRANS- 3 + DENOMINATIONAL *a.*] = *TRANSCONFESSIONAL *a.*

1972 *Clergy Rev.* Sept. 658 The first transdenominational church union to take place in Britain in modern times. 1976 *Christian* III. 168 There is a growing hunger for transdenominational eucharistic celebration.

trans,deriva·tional, *a.* *Transformational Gram.* [f. TRANS- 3 + DERIVATIONAL *a.*] Relating to or involving more than one derivation (see *DERIVATION[1] 6 c).

1977 *Language* LIII. 137 One way in which the notion 'superficially optional/obligatory element' could be handled would be by means of a transderivational rule, i.e. a rule referring to another derivation in which the optional element does not occur either in semantic or in superficial representation. 1979 *Trans. Philol. Soc.* 12 In French, by contrast, reflexive deletion is subject to a further transderivational condition.

trans,determina·tion. *Biol.* [f. TRANS- 1 + DETERMINATION.] An alteration of the course of development of an imaginal disc during the culture of *Drosophila* tissue so that it gives rise to a structure that normally develops from a different disc.

1965 E. HADORN in *Genetic Control of Differentiation* (Brookhaven Conf. Rep. BNL-C-44) 157 Since the allotypic organs appear in the offspring of cells which have been first autotypically determined, a change in determination must be postulated. We call this event transdetermination. 1978 *Nature* 2 Feb. 403/2 In general their state of determination (leg, wing, genital, and so on) is conserved during culture, but occasionally a so-called transdetermination occurs whereby a certain type of disk changes to another state of determination.

trans,discipli·nary, *a.* [f. TRANS- 3 + DISCIPLINARY *a.*] Of or pertaining to more than one discipline or branch of learning; interdisciplinary. So **trans,disciplina·rity.**

1972 E. JANTSCH in *OECD: Interdisciplinarity* II. i. 105 The ultimate degree of co-ordination in the education/innovation system,..which may be called *transdisciplinarity*, would..depend on a common axiomatics... The whole education..system would be co-ordinated as a *multi-level, multi-goal* system, embracing a multitude of..interdisciplinary two-level systems, which..will be modified in the transdisciplinary framework. 1977 R. HOLLAND *Self & Social Context* ix. 267 It is therefore possible to assert three promising criteria for new work: reflexivity, transdisciplinarity and the subversiveness of discovery. 1979 *Nature* 3 May 1/2 A recent workshop jointly sponsored by the American Association for the Advancement of Science and the US Department of Energy has been attempting to lay transdisciplinary foundations for a federally supported research programme on the impact of increasing atmospheric carbon dioxide content.

transduce (trɑnsdiū·s, -nz-), *v.* [Back-formation from next.] **1.** *trans.* To alter the physical nature or medium of (a signal); to convert variations in (a medium) into corresponding variations in another medium.

1949 L. L. BERANEK *Acoustic Measurements* xiv. 637 The dynamic range of a microphone is the range of levels of input signals which can usefully be transduced by the instrument. It is..limited at high levels by the amount of distortion which can be tolerated in the transduced signal. 1971 *Nature* 19 Feb. 530/2 Rhodopsin, a visual pigment found in vertebrate retinal rods, transduces light into a neural message. 1973 *Sci. Amer.* Sept. 87/1 (Advt.), It transduced a heart's faint signals into a squiggly line that the doctor could compare with those produced by other hearts. 1977 *Listener* 22–29 Dec. 857/4 Carbon granules..to do their delicate work of transducing sound waves into varying electric currents.

2. *Microbiology.* Of a virus: to transfer (genetic material) from one bacterium to another; also used with the first bacterium as obj. Also, to transfer (a genetic characteristic) from one bacterium to another using a virus.

1952 *Jrnl. Bacteriol.* LXIV. 687 *FA* [sc. filterable agents] from each of the three LT-7 auxotrophs could transduce the other two. 1965 *Virology* XXVII. 290 (*heading*) Variation in composition of chromosome fragments transduced by phage P22. 1973 R. G. KRUEGER *Introd. Microbiol.* xiv. 404/2 In contrast to the specialized transducing phages, the viruses P1 and P22 can transduce a variety of genes in the bacterial genome. 1977 *Lancet* 9 July 94/2, 47 *gen-r* (ex 8799) was a resistant transductant produced by transducing the gentamicin resistance from strain 75/8799 to propagating strain (PS) 47.

Hence **transdu·ced, transdu·cing** *ppl. adjs.*; **transdu·cible** *a.*, susceptible to transduction.

1952 *Jrnl. Bacteriol.* LXIV. 686 All of the transduced cells were still streptomycin resistant. *Ibid.* 687 Several galactose-negative mutants were transducible to galactose-positive by *FA* [sc. filterable agent] from their parental wild type. 1952 *Physiol. Rev.* XXXII. 414 The chemical composition of transducing agents has important implications, if they can be regarded as purified genes. 1962 R. GOLDMAN *Ultrasonic Technol.* iii. 67 A transducing crystal must also be mounted in some sort of holder. 1971 G. S. STENT *Molecular Genetics* xiv. 416 Upon closer examination of the clones of transduced bacteria or transductants..it was found..that all the Gal+ transductants were either actively lysogenetic.., or were at least immune to infection. 1973 Transducing [see sense 2 above].

transducer (trɑns,diū·sər, -nz-). [f. L. *transdūcere* to lead across, transfer, f. *trans* TRANS- + *dūcere* to lead: see -ER[1].] Any device by which variations in one physical quantity (e.g. pressure, brightness) are quantitatively converted into variations in another (e.g. voltage, position).

1924 K. S. JOHNSON *Transmission Circuits for Telephonic Communic.* vii. 46 A structure which, when inserted in a circuit, enables this maximum possible power to be absorbed in the receiving circuit is called an ideal transducer. 1948 [see *QUANTIZER]. 1957 *New Scientist* 12 Dec. 30/2 The excitation of sensory [nerve] endings, physiological 'transducers' of mechanical changes into electrical signals. 1965 *Wireless World* July 31 (Advt.), Racal Magnetic Transducer..converts rotational movement to a pulse output without physical connection to the shaft under test. 1973 M. WOODHOUSE *Blue Bone* ii. 11 We designed and sold a capacitative transducer head for somebody else's flowmeter. 1976 A. HOPE *Hi-Fi Handbk.* 88 Although we have so far considered only moving coil-moving cone loudspeakers, there are other types of 'transducer' on the market which convert electrical energy into sound in different ways. 1980 *Jrnl. R. Soc. Arts* May 348/2 Huxley and Simmons use extremely fast-acting motors and transducers to record force changes in small fractions of a millisecond.

transductant (trɑns,dʌ·ktănt, -nz-). *Microbiology.* [f. TRANSDUCT(ION + -ANT[1].] A cell into which genetic material has been transduced.

1963 *Exper. Cell Res.* XXX. 252 When the histidine-requiring mutants..are crossed by transduction, 1/10[6] recipient cells become stable *his+* transductants. 1978 *Nature* 1 June 355/2 RNA polymerase was therefore isolated from two transductants, differing only at the *alt* locus.

transduction. Restrict *rare* to sense in Dict. and add: **2.** The action or process of transducing a signal.

1947 *Jrnl. Acoustical Soc. Amer.* XIX. 307/1 It is rather interesting..that the direct method of electronic transduction, instead of the indirect method of employing a conventional transducer and then amplifying the output with a vacuum tube, has not been developed. 1970 J. EARL *Tuners & Amplifiers* iv. 87 Low impedance pickup cartridges..using the moving-coil principle of transduction. 1975 *Nature* 17 Apr. 625/1 The transduction of light energy into neural signals is mediated in all known visual systems by a common type of visual pigment.

3. *Microbiology.* The transfer of genetic material from one cell to another by a virus or virus-like particle.

1952 ZINDER & LEDERBERG in *Jrnl. Bacteriol.* LXIV. 681 To help the further exposition of our experiments, we shall use the term transduction for genetically unilateral transfer in contrast to the union of equivalent elements in fertilization. 1960 [see *F III. 1 l]. 1971 *Nature* 18 June 466/1 It has been suggested that transduction of genes by viruses was an important mechanism in evolution for spreading useful mutations between organisms not formally related. 1977 *Lancet* 9 July 94/2 These were derived by selection of sensitive variants from gentamicin-resistant strains or by transduction of this resistance to sensitive strains.

Hence **transdu·ctional** *a.*, of or pertaining to (genetic) transduction.

1956 *Genetics* XLI. 845 (*heading*) Linear inheritance in transductional clones. 1980 *Jrnl. Gen. Microbiol.* CXIX. 51 Transductional analysis revealed that one of the four mutations carried by strain T-693 was responsible for constitutive synthesis of both isoleucine and threonine biosynthetic enzymes.

transductor. Restrict ‖ to sense in Dict. and add: **2.** *Electr.* [See quot. 1939.] A reactor (sense *2 a) having a d.c. winding to control the saturation of a core and an a.c. winding whose impedance is thereby changed, so that a small change in direct current produces a large change in alternating current.

1939 U. LAMM in *ASEA Jrnl.* XVI. 71/2 We should like to introduce some explanation of the word 'transductor'... The purpose of this word is actually to express the idea that as regards the mutual action between its two windings the apparatus resembles a transformer whilst, however, at the same time functioning in the A.C. circuit as a reactor or inductor, *i.e.* a conductor. 1975 *IEEE Trans. Nuclear Sci.* XXII. 1277/1 At Fermilab, transductors are used as current measuring devices in the main accelerator quadrupole power supply feedback loop.

trans,ea·rth, *a.* *Astronautics.* [f. TRANS- 2 + EARTH *sb.*[1]] Of or pertaining to spaceflight or a trajectory towards the earth from the moon or another planet.

1965 *New Scientist* 1 July 12/3 The navigation of the injection into a transearth orbit and the transearth coast are similar to the outward parts of the voyage. 1970 N. ARMSTRONG et al. *First on Moon* xiii. 320 'Now we have only one more thing to worry about.' (She meant transearth injection—the big burn that would let the world know 'Here we come!')

transect (trɑ·nsekt), *sb.* [f. the vb.] **a.** A line or a belt of land along which a survey is made of the plant or animal life or some other feature; a survey of this kind.

1905 F. E. CLEMENTS *Research Methods Ecol.* iv. 176 When longer transects are desired, as in the case of forest formations, tapes of 500 or 1,000 meters should be used with eyelets a meter apart. 1939 [see *QUADRAT 3]. 1974 *Environmental Conservation* I. 57/2 A second transect in an area of high-centred polygons..exhibited less than 2% differences in total thaw by season's end. 1979 *Rescue News* Dec. 4/2 A transect was plotted from the highest corner of the field to the source of the Gypsey Race. Along this over 20 one-metre square trial holes were dug.

b. *attrib.*, as *transect count, line, strip.*

1971 *Country Life* 23 Sept. 739/1 The puffin study, involving transect counts of occupied burrows on the island's vertiginous slopes, had been hazardous. 1973 J. J. MCKELVEY *Man against Tsetse* iii. 175 Certain rounds were arranged to follow transect lines according to compass bearings. 1953 SCOTT & FISHER *Thousand Geese* 220 It was not possible to maintain a constant width of transect-strip.

transempirical, *a.* (Earlier example.)

1904 W. JAMES in *Mind* Oct. 465 Whether there be a trans-empirical reality or not.

transepted (trɑ·nseptėd), *a.* *Archæol.* [f. TRANSEPT + -ED[2].] Having chambers resembling transepts; *spec.* designating a type of gallery grave (see quot. 1956).

1939 G. E. DANIEL in *Proc. Prehist. Soc.* V. 143 The present writer defined a type of megalithic burial chamber which was there called the transepted gallery grave. 1956 R. J. C. ATKINSON *Stonehenge* v. 149 The tomb has..two pairs of chambers opening off its sides, in the manner of transepts in a church; for this reason the type is known technically as a transepted gallery-grave. 1963 [see *gallery grave* s.v. *GALLERY *sb.* 12 b].

transeunce, etc.: see *TRANSIENCE, etc.; **transexual,** etc.: see *TRANSSEXUAL *a.* and *sb.*, etc.

transfection (trɑnsfe·kʃən, -nz-). *Microbiology.* [f. TRANS- or TRANS(FER *sb.* + IN)FECTION.] The introduction of free viral nucleic acid into a cell.

1964 FÖLDES & TRAUTNER in *Zeitschr. für Vererbungslehre* XCV. 61 Infection of cells by the isolated nucleic acid from a virus, resulting in the production of a complete virus will be termed Transfection. 1975 *Nature* 5 June 446/2 The rescue of proviral DNA of Rous sarcoma virus in mammalian cells by 'transfection' to permissive chicken cells. 1980 OLD & PRIMROSE *Princ. Gene Manipulation* i. 7 Transformation of a cell with DNA from a virus is sometimes referred to as transfection.

So **transfe·ct** *v. trans.*, to infect (a cell) with free viral nucleic acid; **transfe·cted** *ppl. a.* (said of the cell and of the acid).

1964 *Zeitschr. für Vererbungslehre* XCV. 61 For the plating of transfected cells it is essential to use tryptone plates. 1966 *Ann. Rev. Microbiol.* XX. 394 It appears that the genetic basis for the ability to be transformed or transfected is similar. 1974 *Nature* 22 Nov. 319/2 We tried to transfect ΦX-DNA to various bacteria belonging to the *Enterobacteriaceae*. 1983 *Sci. Amer.* Jan. 58/2 Investigators have 'transfected' cultured cells by exposing them to naked foreign DNA.

transfer, *sb.* Add: **2. c.** *Psychol.* (More fully *transfer of practice, training.*) The carrying over of the effects of training or practice from the learning of one function to the learning of another. Cf. *negative transfer* s.v. *NEGATIVE *a.* 8 c; *positive transfer* s.v. *POSITIVE *a.* 8 d.

1901 THORNDIKE & WOODWORTH in *Psychol. Rev.* VIII. 386 There is no inner necessity for improvement of one function to improve others closely similar to it, due to a subtle transfer of practice effect. 1924 *Psychol. Rev.* XXXI. 157 There is no evidence for such transfer of training among rats. 1948 E. R. HILGARD *Theories of Learning* ii. 29 The theory proposes that transfer depends upon the presence of identical elements in the original learning and in the new learning which it facilitates. 1970 HALSTEAD & RUCKER in W. Byrne *Molecular Approaches to Learning & Memory* 6 (*heading*) Behavioral modification, née transfer of training.

d. The transference of a worker or player from one location, sphere, sports club, etc., to another; a change of place of employment within an organization.

1895 *Football News* (Nottingham) 2 Nov. 1/6 It is stated that the Forest have offered £70 for Bruce's transfer. 1923 J. D. HACKETT *Labor Terms in Management Engineering* May, *Transfer*, the shifting of a worker from one occupation to another. 1937 [see *DOTTED *ppl. a.* 1 c]. 1970 *Times* 13 Oct. 15/3 Trevor Gould..has been given a free transfer by Coventry City. 1973 J. THOMSON *Death Cap* xiii. 176 'I've been thinking again about putting in for a transfer,' Holbrook said... Finch..wished now that he could have satisfied the Sergeant over the question of his transfer... A good local policeman meant a lot to a small community.

3. For '(*rarely*, a person)' read 'or a person' and add further examples of people.

1929 *Daily Express* 7 Nov. 12/4 The prizes will hardly be glittering enough to attract the best 'transfers'. 1970 *Globe & Mail* (Toronto) 26 Sept. 35/6 The Mustangs have six experienced transfers who must be fitted in. 1979 *Arizona Daily Star* 1 Apr. c 9/1 The return of 33 lettermen—five senior-college transfers, most of whom will likely start—and 10 junior-college transfers.

4. b. (Examples.)

1883 I. M. RITTENHOUSE *Maud* (1939) 187 The sun was just coming up as we crossed the river on the transfer. 1892 S. HALE *Let.* 28 Apr. (1919) 269, I mounted a cable, took a transfer, and went..out into the suburbs. 1903 A. B. HART *Actual Govt.* 207 In most cities there is a system of free transfers, so that, starting from one suburb, one may often travel for a single fare 5, 10, or 15 miles to another suburb.

5. *transfer agent, -boat* (earlier example), *-company* (earlier example), *-deed* (example), *list, market, office* (later example), *payment, price, pricing*; **transfer case,** (*a*) a case in(to) which materials are transferred; (*b*) *Mech.*, (the housing of) a mechanism for dividing the power between a number of axles in a motor vehicle with two or more driving axles; **transfer chamber,** the chamber in which the material is initially heated in transfer moulding; **transfer effect(s),** the result(s) of transfer of training (see sense 2 c above); **transfer factor** *Immunol.*, a substance released by antigen-sensitized lymphocytes and capable of transferring the response of delayed hypersensitivity to a non-sensitized cell or individual into which it is introduced; **transfer fee,** (*a*) (example); (*b*) *Football*, a sum of money paid by one club to another for the transfer of the services of a professional player; **transfer function,** a mathematical function relating the output or response of anything to the input or stimulus; also *transf.*; **transfer line** *Engin.*, a line of work-stations along which a part is automatically conveyed to be subjected to a sequence of automatic machining operations; **transfer machine** *Engin.*, a composite machine that performs a series of operations without the intervention of the operator; **transfer mould** *sb.*, the mould cavity in transfer moulding; also (with hyphen) as *v. trans.*, to make by means of transfer moulding; **transfer moulding,** a moulding process used chiefly for thermosetting plastics in which the material is softened in a heated chamber and then forced by a plunger into an adjacent closed, heated mould cavity where it sets; **transfer orbit** *Astronautics*, an orbit that touches two given orbits and therefore provides a trajectory by which a spacecraft can pass from one of them to the other; **transfer-paper** (earlier example); **transfer pot** = *transfer chamber* above; **transfer-printed** *a.* (examples); **transfer RNA** *Molecular Biol.*, RNA that collects particular amino-acids in the cytoplasm of a cell and conveys them to a ribosome, where they are assembled to form part of a polypeptide or protein molecule; **transfer station**: delete (*U.S.*) and add example.

1869 *Bradshaw's Railway Man.* XXI. 430 All certificates [shall] be signed by both the transfer agent and register. 1978 A. MALING *Lucky Devil* xix. 103 You are the transfer agent for Lucky Devil Minerals. 1882 *Uncle Rufus & Ma* 52 We ferried over in a transfer boat. 1923 H. A. MADDOX *Dict. Stationery* 77 *Transfer Case*, a Binding Case or file for receiving the matter transferred periodically from the live file or loose leaf book. 1949 I. FRAZEE et al. *Automotive Fundamentals* vi. 378 The transfer case is located behind the transmission. 1970 *Southerly* XXX. 216 Gear box and transfer case and differentials growled a deep throated work song with power to spare. 1983 *Judge of Election Handbk.* (*Board of Election Comm., Chicago*) 11/1 Open the transfer case and check that the official ballot cards are for the proper precinct. 1946 DUBOIS & PRIBBLE *Plastics Mold Engin.* 353 It is common practice to standardize on the size and design of the transfer chamber and plunger used with hand molds. 1977 *Times Educ. Suppl.* 21 Oct. 29/5 There has been at least one case of a transfer chamber exploding. 1879 H. T. WILLIAMS *Pacific Tourist* 262/1 The Transfer Company will carry baggage alone for 50 cents. 1869 *Bradshaw's Railway Man.* XXI. 63 Certificates must accompany transfer deeds. 1931 R. PINTER *Educ. Psychol.* xii. 268, I feel sure that actual measurements of the transfer effects would be very disappointing to me as a teacher. 1955 T. H. PEAR *Eng. Soc. Differences* ix. 194 For a time there was a general tendency to be sceptical concerning any claim of transfer-effects. 1963 *Rep. Comm. Inquiry Decimal Curr.* xi. 109 in *Parl. Papers 1962–3* (Cmnd. 2145) XI. 195 Psychologists have advised us that giving an old name to a new currency unit could result in serious 'transfer effect' difficulties. 1977 *Language* LIII. 340 'Transfer effects' may reverse the natural order of the acquisition of a sound or structure. 1956 *Jrnl. Exper. Med.* CIV. 328 The above experiments suggest that release of the transfer factor from sensitive cells may occur under relatively mild circumstances. 1978 J. A. BELLANTI *Immunology II* xiii. 333 Attempts to detect antibody in the transfer factor have always been negative, and since its small molecular weight became known, this possibility has been excluded. 1983 *Oxf. Textbk. Med.* I. v. 10/2 Therapy with transfer factor may..find an accepted place in the management of a limited number of clinical situations. 1869 *Bradshaw's Railway Man.* XXI. 63 Transfer fee, 2s. 6d. each. 1901 *Football News* (Nottingham) 9 Mar. 6/3 New Brighton offered Everton £135 for my transfer, and Burnley were prepared to pay £200, but neither offer came to anything. Consequently I had to come South out of the reach of transfer fees. 1911 A. BENNETT *Card* xii. 296 How are you going to get new blood, with transfer fees as high as they are now? You can't get even an average good player for less than £200. 1976 *Western Mail* (Cardiff) 27 Nov., The three Welsh clubs..could also release experienced players to Somerton Park on a loan basis, should Newport be able to obtain transfer fees for any of their own professionals. 1948 BROWN & CAMPBELL *Princ. Servomechanisms* i. 18 The system behaviour may be represented in terms of transfer functions *KG* (*jω*), which are complex functions of the frequency variable *ω*. 1963 R. W. DITCHBURN *Light* (ed. 2) viii. 304 The transfer function constitutes a better evaluation of the performance of the optical system than a statement of the resolving power. 1971 *Nature* 20 Aug. 564/1 McFarland and Budgell found a transfer function for key pecking by thirsty birds in response to modulations of ambient temperature. 1977 *Gramophone* June 122/1 The speaker converts an electric signal into an acoustic wave and..may be regarded as a 'black box', with a particular transfer function representing its overall response in terms of sound waves. 1981 D. J. FISK *Thermal Control of Buildings* i. 9 The transfer function of a room..is not unique but depends on where heat is input..and where its output temperature is measured. 1956 E. MOLLOY *Automobile Engineer's Ref. Bk.* III. 210 Only one head is used at the next station. This is at the right of the transfer line, and is a plunge-cut horizontal milling head for milling the bearing-retaining slots. 1975 *Sci. Amer.* Feb. 25/1 Under mass-production conditions..the engine block is conveyed automatically along a transfer line. 1951 *People* 3 June 6/7 Still on Spurs' transfer list at a fee is Cyril Toulouse. 1976 *Star* (Sheffield) 29 Oct. 28/6 Terry Eccles..is on the transfer list at his own request. 1951 *Treat. on Milling & Milling Machines* (ed. 4) xvii. 723 In order to reduce handling time, it is often advantageous to perform a number of different machining operations..with a multiple-station automatic machine... Such machines are also known as transfer machines. 1977 *Sci. Amer.* May 89/1 (Advt.), Transfer machines finish rough castings into complex pieces such as engine blocks. 1970 *Globe & Mail* (Toronto) 25 Sept. B 7/4 Both see what is called the transfer market—the hundreds of thousands of Canadians who move from one community to another each year—as increasingly important. 1976 *Evening Times* (Glasgow) 1 Dec. 32/2 Partick Thistle could be involved at both ends of the transfer market before the weekend. 1933 *U.S. Patent 1,919,534* 5/1 A device of the class described for transfer molding an infusibly thermosetting resinous material. 1942 J. SASSO *Plastics for Industr. Use* iii. 36 Figure 13 shows an example of a transfer mold with a transfer plunger entering the transfer well. 1963 H. R. CLAUSER *Encycl. Engin. Materials* 165/1 Radio and television cabinets weighing up to four pounds have been transfer-molded from phenolic materials. 1971 E. W. DUCK *Plastics & Rubbers* iv. 56 Transfer moulds..can be regarded as very crude injection moulds, since they first pre-heat the plastic in the transfer cavity. 1940 J. DELMONTE *Plastics in Engin.* xi. 319 The term, transfer molding, has been used to designate specifically the injection of thermosetting materials. 1963 H. R. CLAUSER *Encycl. Engin. Materials & Processes* 164/1 In the transfer-molding process..the mold is completely closed and under clamping pressure before the material is injected into the mold cavity. This results in little or no flash and accurate control of dimensions. 1869 *Bradshaw's Railway Man.* XXI. 430 All the stock registered in New York at the transfer office now kept by Duncan, Sherman, and Co., shall also be registered at another office. 1961 W. T. THOMSON *Introd. Space Dynamics* iv. 66 Transfer between coplanar circular orbits can be effected by an elliptic orbit with perigee and apogee distances equal to the radii of the respective circles... The cotangential ellipse is known as the Hohmann transfer orbit. 1964 *Listener* 7 May 748/1 Once the transfer orbit has been entered, the probe will be moving in free fall, and no further thrust need be applied. 1841 *Brit. Pat.* 9002 18 Obtaining several copies of marks by the use of surfaces of transfer paper. 1964 S. M. MILLER in I. C. Horowitz *New Sociology* 301 In our country, the redistribution of income takes place to a large extent in transfer payments of welfare and social assistance. 1973 *Times* 9 June 19/2 A transfer payment is a payment made by the scheme of a former employee to the scheme of the employee's present employer in consideration of which the new employer's scheme takes over the responsibility for benefits in respect of the service with the former employer. 1976 *Hansard* (Canada) 17 Mar. 11881/1 The balance of $22 billion was in the form of transfer payments to persons, provinces and corporations for subsidies, and so on. 1963 H. R. CLAUSER *Encycl. Engin. Materials & Processes* 164/2 When the mold is opened, the small amount of material remaining in the transfer pot..is removed. 1971 B. SCHARF *Engin. & its Lang.* vi. 38 The mould (die) consists of two sections, an upper transfer chamber (transfer pot)..and a lower mould. 1969 J. ARGENTI *Managem. Techniques* 76 As the complexity and size of the company increases..the task of preparing a budget becomes highly intricate due to such problems as transfer prices and allocation of central overheads. 1974 *Terminol. Managem. & Financial Accountancy* (Inst. Cost & Managem. Accountants) 17 *Transfer price*, a price related to goods or other benefits transferred from one process or department to another or from one member of a group to another. 1971 D. C. HAGUE *Managerial Economics* III. x. 220 The problem of how to price a product as it leaves one department for another—the problem of transfer pricing—is one that has troubled accountants and managers for many years. 1979 *Abacus* (Sydney) XV. 3 International transfer pricing is concerned with the pricing of goods and services transferred between a company's domestic divisions and foreign subsidiaries or among those foreign subsidiaries themselves. 1869 LADY C. SCHREIBER *Jrnl.* 15 Sept. (1911) I. 37 We bought a couple of transfer-printed Wedgwood plates. *Ibid.* 17 Sept. 39, I coveted a small transfer-printed leaf. 1938 *Burlington Mag.* May p. xvii/2 Some rare Liverpool transfer-printed mugs. 1976 *Times* 24 July 9/3 Wheildon plates..and transfer-printed creamwares. 1961 *Ann. Reg. 1960* 402 It was thought a further kind of RNA, called 'Transfer' RNA, might be necessary to link the amino acids into a protein molecule. 1977 *Time* 4 Apr. 39/2 Aaron Klug..first determined the crystalline structure of transfer RNA (tRNA), the molecule that brings amino acids to the ribosome for assembly into protein. 1869 *Bradshaw's Railway Man.* XXI. 198 The negotiations with the South Eastern in reference to a new transfer station at Waterloo.

b. Chiefly *Electronics.* Used *attrib.* to designate a ratio of two quantities measured simultaneously at two different points of a circuit or device.

1933 [see *MUTUAL *a.* 1 f]. 1943 F. E. TERMAN *Radio Engineers' Handbk.* iii. 200 The transfer impedance is defined as the ratio of the voltage E_1 applied in mesh 1 to the resulting current I_2 of mesh 2. 1966 *McGraw-Hill Encycl. Sci. & Technol.* III. 429/2 The transfer characteristic *H* of the controller mathematically relates the controller M_1 to its input *E*: $M_1 = HE$. 1975 HAVILL & WALTON *Elem. Electronics* iv. 69 For the particular case of the transistor, the output and transfer characteristics are almost linear over substantial regions of the normal operating range. 1980 J. R. O'MALLEY *Circuit Analysis* xix. 467 The most popular of these transfer functions are the transfer impedance.., the transfer admittance.., the transfer voltage ratio or voltage gain.., and the transfer current ratio.

transferable, *a.* Add: *transferable vote*, (in systems of proportional representation) a vote that is transferred to a second or further competing candidate if the candidate for whom it is first cast is eliminated in one of the succession of counts or has more votes than are needed for election; esp. as *single transferable vote* (abbrev. S.T.V. s.v. *S 4 a).

1885 W. E. SMITH *Fair Representation* vi. 32 (*heading*) The single transferable vote. 1909 [in Dict.] 1954 B. & R. NORTH tr. *Duverger's Pol. Parties* II. iii. 358 We must distinguish between the result of proportional representation and the consequences of list-voting, which generally coincides with it (except in the transferable Irish vote). 1972 *Guardian* 19 June 10/4 Mr Whitelaw's reintroduction of the single transferable vote in Ulster local government is likely to please minority groups.

transferase (trɑ·nsfĕrēɪz, -nz-). *Biochem.* [f. TRANSFER *v.* + *-ASE.] Any enzyme that catalyses the transfer of some particular group or molecule from one molecule to another.

1948 *Jrnl. Biol. Chem.* CLXXII. 12 It may be that similar 'transferase' phenomena will be found to be of general occurrence with other types of hydrolases. 1974 [see *OXIDOREDUCTASE]. 1981 *Sci. Amer.* June 70/3 In *Penicillium chrysogenum* an acyl transferase then catalyzes the replacement of this side chain by one derived from a utilizable precursor, such as phenylacetic acid.

transference. Add: **1. b.** *Psychoanal.* [tr. G. *übertragung*.] The transfer to the analyst by the patient of re-awakened and powerful emotions previously (in childhood) directed at some other person or thing and since repressed or forgotten; the process or state of such a transfer; *loosely*, the emotional aspect of a patient's relationship to the analyst; also *transf.* *negative transference*: see *NEGATIVE *a.* 8 c; *positive transference*: see *POSITIVE *a.* 8 d.

[1895 S. FREUD in Breuer & Freud *Studien über Hysterie* iv. 266 Die Uebertragung auf den Arzt geschieht durch falsche Verknüpfung. 1910 tr. *Freud's Orig. & Devel. Psychoanal.* in *Amer. Jrnl. Psychol.* XXI. 215 Every time that we treat a neurotic psychoanalytically, there occurs in him the so-called phenomenon of *transfer* (Uebertragung), that is, he applies to the person of the physician a great amount of tender emotion, often mixed with enmity.] 1911 *Amer. Jrnl. Psychol.* XXII. 434 The reason why the physician is so often the object toward which the transference is made is that the Œdipus complex is almost invariably present in the patient. 1916 C. E. LONG tr. *Jung's Analyt. Psychol.* 245 What has disgusted you in hypnotism is at bottom nothing but the so-called 'transference' to the doctor. 1920 E. JONES *Treatm. Neuroses* 40 He is..reacting not toward the physician, but rather toward the other person who has been brought together ('identified') with the latter in his mind, an occurrence technically known as 'transference'. 1937 A. S. NEILL *That Dreadful School* xi. 155 If you tell a child any vital truth, or if it confides its troubles to you, he or she gets a transference, that is you get all the child's emotions showered on you. 1973 A. JANOV *Primal Scream* xiv. 246 Since I believe that the transference is the neurosis, I think that doing anything else with the patient other than helping him to feel his Pain is to render him a disservice.

3. *attrib.* and *Comb.*, as (sense *1 b) *transference feeling, situation*; **transference neurosis** *Psychoanal.*, a neurotic stage during transference frequently encountered during analysis and considered beneficial to the therapy; **transference number** *Physical Chem.* (chiefly *U.S.*) = *transport number* s.v. *TRANSPORT *sb.* 6.

1964 GOULD & KOLB *Dict. Soc. Sci.* 557/1 Freud's aim in treatment gradually changed..to the interpretation and modification of 'transference feelings' and their underlying unconscious conflicts. 1977 R. HOLLAND *Self & Social Context* iv. 76 The Freudian analyst is prepared to hold the transference feelings, and possible acting out behaviours, long enough for the client to re-experience and go beyond them. 1916 A. A. BRILL tr. *Freud's Three Contributions to Theory of Sex* (ed. 2) iii. 77 Psychoanalysis of the so-called transference neuroses (hysteria and compulsion neurosis) offers us here a reliable insight. 1968 H. RACKER *Transference & Countertransference* i. 15 In the transference neurosis,..the return of the relations to the parents implies the return of the neurotic conflicts with them. 1898 *Jrnl. Chem. Soc.* LXXIV. II. 553 The values were found to be almost independent of the concentration.., the transference numbers for the anions being given by the expressions [etc.]. 1909 [see *HITTORF]. 1966 *McGraw-Hill Encycl. Sci. & Technol.* XIV. 18b/2 The two procedures for determining transference numbers by which most of the available data in the literature have been obtained are the Hittorf method and the moving boundary method. 1933 W. GALT *Phyloanalysis* 72 Only this intense transference-situation as it exists socially, however disguised, can account for the extremes of emotional stimulation and response constantly elicited in the process of phyloanalysis upon the slightest, most trivial occasion. 1977 C. STORR *Tales from Psychiatrist's Couch* 41 A probing of his feelings towards me as a mother figure, an exploration of the transference situation.

transferral. (Earlier example.)

a 1790 J. H. BEATTIE *Ess. & Fragments* (1794) III. 295 Instead of reference, preference, commitment..say *referral, preferral*..and the *transferral* of property, instead of the transferring of property.

transferred, *ppl. a.* (In Dict. s.v. TRANSFER *v.*) Add: *transferred epithet*, an epithet grammatically qualifying a noun other than (though contextually associated with) the noun to which it literally applies.

1866 A. BAIN *Eng. Comp. & Rhet.* 24 The *Transferred Epithet* is a common figure in Poetry. The shifting of an epithet from its proper subject to some allied subject..is seen in... 'Hence to his *idle bed*.' 1947 C. BROOKS *Well Wrought Urn* ix. 159 One can..justify the adjective as a transferred epithet on the model of Vergil's *maestum timorem*. 1958 C. BROOKE-ROSE *Gram. Metaphor* iii. 57, I have not found this transferred epithet in the later texts.

transferrin (trɑnsfe·rin, -nz-). *Biochem.* [f. TRANS- + L. *ferr-um* iron + -IN[1].] Any of several beta globulins found in blood serum which bind and transport iron; = *siderophilin* s.v. *SIDERO-[1] 2.

1947 HOLMBERG & LAURELL in *Acta Chemica Scandinavica* I. 950 We suggest that the new metal-combining protein (iron-binding component) in serum be called transferrin. 1962 H. HEATH in A. Pirie *Lens Metabolism Rel. Cataract* 364 Ascorbic acid has been shown to be necessary for the transfer of plasma-bound iron, transferrin, into the liver and its incorporation into ferritin. 1971 [see *siderophilin* s.v. *SIDERO- 2]. 1977 *Jrnl. R. Soc. Arts* CXXV. 699/1 These transferrin genes show much genetic diversity within herds.

trans,fi:naliza·tion. *Theol.* [f. TRANS- 3 + *FINALIZATION.] The change in purpose or function undergone by bread and wine at the Eucharist through transubstantiation, expressed in terms of finality or teleology. Cf. *TRANSIGNIFICATION.

1965 *Pope Speaks* X. 311 It is not permissible..to discuss the mystery of transubstantiation without mentioning what the Council of Trent had to say about the marvelous conversion of the whole substance of the bread into the Body and the whole substance of the wine into the Blood of Christ, as if they involve nothing more than 'transignification' or 'transfinalization' as they call it. 1966 *Worship* XL. 337 In the eucharist we ought to be concerned with an interpersonal relationship between Christ and us,..in which Christ gives himself to man by means of bread and wine which, by this very gift, have undergone a transfinalization and an ontological and therefore radical transignification. 1975 E. L. MASCALL in *Critique Eucharistic Agreement* v. 73 The writers who introduced the notions of transfinalization and transignification were worried that the term 'transubstantiation' as commonly understood suggested a notion both insufficiently dynamic and insufficiently human.

transfixture. For *rare*⁻¹ read *rare* and add later example.

1955 W. DE LA MARE *Beginning* 215 In this transfixture, a single commonplace word came sallying nonchalantly up out of his memory.

transfluence (trɑ·nsfluĕns). *Geomorphol.* [ad. G. *transfluenz* (A. Penck *Die Alpen im Eiszeitalter* (1909) III. 804), f. L. *-fluentia* flow: see TRANS-.] The flow of glacial ice in quantity across a preglacial watershed with consequent severe erosion.

1949 *Scottish Geogr. Mag.* LXV. 123 The ice-flow thus set up across the divides is termed diffluence or transfluence. 1955 *Sci. News* XXXVIII. 55 At the present time transfluence is occurring at the edge of the Greenland ice-cap. 1970 R. J. SMALL *Study of Landforms* xi. 374 Glacial transfluence is found where the impeded ice cuts out of a valley system not by a lateral distributary but at the very head of the valley.

transfluent, *a.* Restrict *rare* to sense in Dict. and add: **b.** *spec.* in *Geomorphol.*, applied to glacial ice undergoing transfluence.

1951 *Trans. Inst. Brit. Geographers* Pub. No. 15, 2 Like Penck he clearly recognised that where a pass was crossed by diffluent or transfluent ice it was markedly eroded by it. 1968 R. W. FAIRBRIDGE *Encycl. Geomorphol.* 429/2 Many of the breaches made by transfluent ice involve considerable arrangement of the drainage pattern.

transfluxor (trɑnsflʌ·ksəɪ, -nz-). *Electronics.* [f. *TRANS- 10 + FLUX *sb.* + -OR.] (See quots.)

1955 RAJCHMAN & LO in *R.C.A. Rev.* XVI. 303 A novel device which is based on the fact that completely new switching and storing functions become possible when two or more apertures are made in the rectangular hysteresis loop cores, thereby creating a number of distinct legs and flux paths in the magnetic circuit. The new device operates by the controlled transfer of flux from leg to leg in the magnetic circuit and was consequently named 'Transfluxor'. 1963 [see *LADDIC]. 1972 G. J. KLIR *Introd. Methodol. Switching Circuits* v. 185 Interesting properties are exhibited by a magnetic core which has several holes with windings. This element is called a transfluxor.

transform, *sb.* Add: **2.** *Transformational Gram.* A syntactic structure derived by the application of a transformation.

1955 N. CHOMSKY *Transformational Analysis* (Ph.D. Dissertation, Univ. Pennsylvania) v. 26 It seems most natural to characterize these sentences in terms of some notation of grammatical transformation, regarding these sentences as transforms of certain sentences which are derived on the level *P* and which do have *P* markers. 1957 —— *Syntactic Structures* (1962) viii. 88 This sentence [*sc.* I found the boy studying in the library] was a transform..of 'I—found studying in the library—the boy'. 1964 R. H. ROBINS *Gen. Linguistics* vi. 243 The collocational compatibilities..between the particular words of any sentence and its transform are likely to be substantially the same. 1972 R. D. EAGLESON in G. W. Turner *Good Austral. Eng.* v. 98 We can see them [*sc.* two versions of a sentence]..as related to the same underlying structures and as different transforms of it. 1976 *Word 1971* XXVII. 253 There is an embedded sentence corresponding to sentence 7 b, a transform derived from sentence 8 b by a movement of the Aux *haben* ('have').

3. *Geol.* = *transform fault*, sense 4 below.

[1965 J. T. WILSON in *Nature* 24 July 343/1 A junction where one feature [marking a mobile belt] changes into another is here called a transform.] 1971 *Geol. Mag.* CVIII. 27 It appears that there is a mechanical advantage in overcoming overall transtension by stepped transforms. 1978 *Nature* 16 Feb. 617/1 (*caption*) The three transforms are shown by dashed lines and their possible extensions by dotted lines.

4. *Comb.*: **transform fault** *Geol.*, a transcurrent fault terminating abruptly at both ends, *esp.* one that connects two segments of an oceanic ridge; also, any transcurrent fault associated with two lithospheric plates sliding past one another; hence **transform faulting.**

1965 J. T. WILSON in *Nature* 24 July 343/2 It is proposed that a separate class of horizontal shear faults exists which terminate abruptly at both ends, but which nevertheless may show great displacements... The name transform fault is proposed for the class. 1973 *Ibid.* 10 Aug. 341/2 The boundaries between the rigid plates which make up the Earth's crust are..of three distinct kinds: ridges..; trenches..; transform faults, along which plates may slip relative to each other. 1976 *Ibid.* 4 Mar. 14/1 There are several transcurrent faults in solid continental rocks which, because they happen to terminate at structural features which 'absorb' their motion, must also be termed 'transform faults' according to Wilson's strict definition. 1980 *Guardian* 20 Nov. 13/4 There are fault boundaries, known as transform faults, along which the plates merely slide past each other. 1971 *Geol. Mag.* CVIII. 27 The sinuous zone of fission as it extends the length of the north and south Atlantic Ocean is necessarily in many places oblique to the direction of spreading. This situation is generally..accommodated by transform faulting.

transform, *v.* Add: **1. c.** More widely, to subject (any mathematical entity) to a transformation (TRANSFORMATION 2 c in Dict. and Suppl.). Also *absol.*

1972 M. KLINE *Math. Thought* xix. 427 Finding it difficult to evaluate in rectangular coordinates, he transformed to spherical coordinates. 1982 D. M. SCHNEIDER et al. *Linear Algebra* v. 181 The function *f* defined by the equation $f(x) = x^2$. This function transforms a real number into a real number, namely its square.

f. *Molecular Biol.* To change (a bacterial cell) into a genetically distinct kind by the introduction into it of DNA from another cell of the same or a closely related species.

1928 *Jrnl. Hygiene* XXVII. 150 An R strain is most readily transformed into the S variety when the killed culture used is of the same serological type as that from which the R strain was derived. 1947 *Jrnl. Exper. Med.* LXXXVI. 449 Repeated attempts both *in vitro* and *in vivo* to transform D39/Int53 to pneumococcus Type III were unsuccessful. 1981 L. L. MAYS *Genetics* vi. 274 Pieces of DNA of molecular weight less than 1.5×10^5 daltons cannot transform *Haemophilus influenzae*.

g. *Cytology.* To cause (a eukaryotic cell) to undergo transformation (*TRANSFORMATION 3 i).

1959 *Jrnl. Nat. Cancer Inst.* XXIII. 1035 (*heading*) Clonal analysis of variant cell lines transformed to malignant cells in tissue culture. 1982 *Sci. Amer.* Mar. 72/1 Analysis of the DNA of the Rous sarcoma virus has revealed a single gene capable of transforming cells.

2. Delete 'Now *rare*' and add later examples.

1893 HARKNESS & MORLEY *Treat. Theory of Functions* i. 14 If $w = u + iv$ be a one-valued monogenic function of $x + iy$, the systems of orthogonal straight lines $x = a$, $y = b$ transform into systems of orthogonal curves in the *w*-plane. 1970 PASSMORE & ROBSON *Compan. Med. Stud.* II. xxii. 9/2 The function of antigen at the surface of the lymphocyte is to induce it to transform and proliferate into active antibody-producing cells. 1971 *Nature* 26 Nov. 187/1 The larva then transforms to a pupa. 1982 *Suppl. to O.E.D.* III, *Pseudoscalar* sb., a quantity that transforms as a scalar under rotation but changes sign under reflection.

transformant (trɑnsfǭ·ɹmănt, -z-). *Biol.* [f. TRANSFORM *v.* + -ANT[1].] **a.** A transformed bacterium (see *TRANSFORM *v.* 1 f). **b.** = *transformed cell* s.v. *TRANSFORMED *ppl. a.* b.

1957 *Biochimica & Biophysica Acta* XXVI. 71 The yield of transformants is determined by the prior state of competence of a pneumococcal culture. 1959 *Jrnl. Exper. Med.* CIX. 437 Capsulated transformants producing type II polysaccharide have been obtained by reactions between a number of R mutants of type II pneumococcus. 1971 *New Scientist* 8 Apr. 83/2 Cells selected from a population of transformants by virtue of their resistance to agglutination by concanavalin A, grow to a density expected of untransformed rather than transformed cells. 1980 *Amer. Jrnl. Trop. Med. & Hygiene* XXIX. 1045/1 (*caption*) Transformant clones were plated in duplicate.

transformation. Add: **2. c.** (Further examples.) Also, a change of any mathematical entity in accordance with some definite rule or rules; the rules themselves; *spec.* = *MAPPING *vbl. sb.* 2.

1908 [see *LORENTZ]. 1909 *Proc. Section Sci. K. Akad. van Wetenschappen te Amsterdam* XI. 798 A continuous one-one transformation in itself of a singly connected, onesided, closed surface leaves at least one point invariant. 1941 BIRKHOFF & MACLANE *Surv. Mod. Algebra* vi. 128 The 'similarity' transformations of space—those one-one transformations which multiply all distances by a constant factor. 1949 S. LEFSCHETZ *Introd. Topol.* i. 29 If *f* is one to one and bicontinuous (both *f* and its inverse f^{-1} continuous), *f* is said to be a topological transformation or a homeomorphism. 1952 E. T. BELL *Mathematics* vi. 354 The numerical value of f(t) is unaltered when we replace the variable *t* by the linear expression *t* + 1... Thus, the value of the function is invariant under a particular linear transformation. 1958, 1964 [see *MAPPING *vbl. sb.* 2]. 1964 [see *FOURIER]. 1966 S. BEER *Decision & Control* vi. 109 It is possible to specify a transformation that will map the infinite set of natural numbers on to this other finite set. 1982 D. M. SCHNEIDER et al. *Linear Algebra* v. 181 If *V* and *W* are vector spaces, a function or transformation *T* from *V* into *W* is a rule that associates with every vector *x* in *V* a unique vector in *W*.

3. g. *Physics.* Change of one element into another, whether artificially induced or by spontaneous decay. Cf. *TRANSMUTATION 3 a.

1902 RUTHERFORD in *Phil. Mag.* IV. 395 These changes must be occurring within the atom, and the radioactive elements must be undergoing spontaneous transformation. 1926 R. W. LAWSON tr. *Hevesy & Paneth's Man. Radioactivity* xxi. 150 Similar attempts to influence the velocity of transformation of uranium and radium D, by subjecting them to the action of radiation, have also led to a negative result. 1958 O. R. FRISCH *Nucl. Handbk.* IV. 4 In each unit of time a certain definite fraction of the total number of the atoms present will disintegrate but there is nothing to indicate the moment at which a given atom will undergo the radioactive transformation. 1969 *Times* 12 Mar. 4/8 It is to be assumed that the uranium and thorium in the galaxy were created by the nuclear transformation within densely packed matter at high temperatures.

h. *Molecular Biol.* The genetic alteration of a bacterial cell by the introduction or absorption of extraneous DNA (see *TRANSFORM *v.* 1 f).

1928 F. GRIFFITH in *Jrnl. Hygiene* XXVII. 154 Experiments with culture heated at temperatures higher than 60 °C. have rarely been successful in causing transformation of type. 1960 *New Biol.* XXXI. 72 The first clear demonstration of transformation was made in 1928 by Griffith, who discovered that an avirulent and normally harmless strain of pneumococcus was changed into a virulent strain when injected into mice together with some virulent pneumococci that had been thoroughly killed by heating. 1970 AMBROSE & EASTY *Cell Biol.* x. 346 Transformation is a very inefficient process but has proved useful for gene mapping in bacteria where a suitable transducing phage is not known. 1980 *Sci. Amer.* Feb. 36/2 In another process, known as transformation, DNA released by cell death or other natural processes simply enters a new cell from the environment by penetrating the cell wall and membrane.

i. *Cytology.* The modification of a eukaryotic (nucleated) cell so that it comes to possess some or all of the characteristics of a cancer cell.

1943 *Jrnl. Nat. Cancer Inst.* IV. 202/1 The cell transformations appeared after a definite latent interval of several weeks following initial exposure to the carcinogen. 1967 *Nature* 8 July 171/2 The concept of contact inhibition has attracted particular interest since the advent of tissue culture investigations on neoplastic transformation. 1982 *Sci. Amer.* Mar. 71/1 Transformation..is due to the action of a gene, which must be expressed continuously to maintain the cancerous state.

j. *Linguistics.* An operation by which one syntactic structure is converted into another by the application of specific rules; a rule converting deep structure to surface structure (see *deep structure* s.v. *DEEP *a.* IV c, *surface structure* s.v. *SURFACE *sb.* 6 d); the process by which surface structures are generated.

1955 N. CHOMSKY *Transformational Analysis* (Ph.D. Dissertation, Univ. Pennsylvania). 27 A sentence *X* is related to a sentence *Y* if, under some transformation set up for the language, *X* is a transform of *Y* or *Y* is a transform of *X*. 1957 —— *Syntactic Structures* (1962) v. 44 Let us call each such rule a 'grammatical transformation'. 1957 Z. S. HARRIS in *Language* XXXIII. 283 We can proceed to define transformation..based on two structures having the same set of individual co-occurrences. This relation yields unique analyses of certain structures and distinctions which could not be analyzed in ordinary linguistic terms. 1964 *Word* XX. 429 Transformations..may be thought of as manipulations—reordering, combination, addition, deletion—performed on fully formed sentences. 1964 R. H. ROBINS *Gen. Linguistics* vi. 242 Transformation is a method of stating how the structures of many sentences in languages can be generated or explained formally as the result of specific transformations applied to certain basic sentence structures. 1967 D. G. HAYS *Introd. Computational Linguistics* viii. 153 Such a transformation can break down the structure of one sentence and insert all or part of it at a specified place in the structure of the other. 1977 *Canad. Jrnl. Linguistics 1976* XXI. 156 The traditional assumption that transformations do not change meanings.

5. transformation (playing) card, a playing card on which the suit signs are incorporated into a design or picture.

[1848 W. A. CHATTO *Facts & Speculations on Origin & Hist. Playing Cards* iv. 260 In 1811 two different packs of caricature cards, imitated..from the picture-cards in Cotta's Almanack, appeared in England... On the wrapper of both packs the inscription is the same: 'Metastasis. Transformation of Playing-cards.'] 1931 H. T. MORLEY *Old & Curious Playing Cards* 152 Transformation cards, 1828. A pack of 52 cards,..printed from wood blocks. 1960 H. HAYWARD *Antique Coll.* 286/2 *Transformation playing cards*, first issued in London 1808... Making transformation cards from ordinary packs became a fashionable pastime, pen and ink converting cards into designs of topical or personal association. 1966 S. MANN *Collecting Playing Cards* viii. 164 Transformation cards are a rather different case... Their aim is to 'transform' an ordinary pip card into a picture by means of incorporating the pips in their standard positions in a larger overall design.

transformationist, (*b*) = *TRANSFORMATIONALIST *sb.*

1962 J. SLEDD in Householder & Saporta *Probl. Lexicogr.* 145 The transformationists..have little interest in pedagogic problems. 1965 *Language* XLI. 124 A Czech study on the structure of German sentences..is contrasted with a study on the same subject by an American transformationist.

transformational, *a.* (In Dict. s.v. TRANSFORMATION.) Add: *spec.* in *Linguistics*, of or pertaining to a transformation or transformations (sense 3 j); (more fully *transformational-generative* adj.) designating, of, or pertaining to a linguistic model or method of analysis based on the generation of surface structures from underlying structures by transformations; cf. *GENERATIVE *a.* 2 b and TG s.v. *T 6 a.

1955 N. CHOMSKY (*title of Ph.D. Dissertation, Univ. Pennsylvania*) Transformational analysis. 1957 —— *Syntactic Structures* (1962) v. 46 Corresponding to the level of phrase structure, a grammar has a sequence of rules of the form $X \rightarrow Y$, and corresponding to lower levels it has a sequence of morphophonemic rules of the same basic form. Linking these two sequences, it has a sequence of transformational rules. *Ibid.* ix. 101 The similarity between active-passive, negation, declarative-interrogative, and other transformational relations would not have come to light if the active-passive relation had been investigated exclusively in terms of such notions as synonymity. 1961 P. H. MATTHEWS in *Archivum Linguisticum* XIII. 196 (*heading*) Transformational grammar. 1963 F. G. LOUNSBURY in J. A. Fishman *Readings Sociol. of Lang.* (1968) 48 One result of transformational analysis..is to lead linguistics a step closer to a general model for the syntax of language. 1964 R. H. ROBINS *Gen. Linguistics* p. xviii, One of the most striking developments in linguistics in the last few years has been the increasing recognition of the transformational model of linguistic statement. *Ibid.* p. xix, Some transformationalists, as the adherents of transformational-generative grammar are called. 1964 E. A. NIDA *Toward Sci. Transl.* iv. 60 There are some psychological confirmations of transformational theory which seem to have special significance. 1965 *N.Y. Times* 29 Dec. 32 Transformational grammar grew in part from M.I.T. computer experiments to produce mechanical translations of foreign languages. 1966 T. F. MITCHELL in C. E. Bazell *In Memory of J. R. Firth* 354 One cannot help but be puzzled by the refusal of American transformational-generative grammarians to incorporate in their valuable work collocational study of the kind envisaged here. 1966 Y. BAR-HILLEL in *Automatic Transl. of Lang.* (NATO Summer School, Venice, 1962) 7 Transformational grammars seem to have a much better chance of being both adequate and practical. *Ibid.* 12 The fifth aspect of syntactic complexity is, then, transformational history. 1967 [see *GENERATIVIST]. 1968 CHOMSKY & HALLE *Sound Pattern Eng.* ii. 15 These observations suggest a general principle for the application of rules of the phonological component, namely, what we shall call the principle of the 'transformational cycle'. 1972 *Language* XLVIII. 442 Arens fails to show..the essential ideas and assumptions underlying transformational-generative theory. 1973 *Amer. Speech 1969* XLIV. 220 The discussion of Middle English grammar and phonology is based on a transformational-generative view of language. 1973 *Archivum Linguisticum* IV. 35 Presumably, in transformational terms, transitive, intransitive, perfective, imperfective..are here surface structure categories. 1976 J. HOOPER *Introd. Natural Generative Phonol.* i. 4 The phonological rules..apply in sequential order... Each rule may apply only once, and is assigned a particular place in the sequence or is said to be *ordered* with respect to other rules. The theory with this general form will be referred to as transformational generative phonology. 1977 *Dædalus* Fall 119 The proper division of theoretical labor between rewriting and transformational rules (in particular, how to limit the power of the transformational rules in intuitively reasonable ways) has been a central concern of those who have worked with this conception of grammar. 1980 *Word 1979* XXX. 132 Some degree of exposure to transformational-generative grammar, case grammar and other theoretical approaches is evident in several [Vietnamese] works. 1980 *Canad. Jrnl. Linguistics* XXV. I. 1 The standard theory referred to as transformational generative phonology. 1980 *English World-Wide* I. 1. 133 Transformational grammar has shown itself unable to handle the problem of the description of varieties which are as apparent in the 16th and 17th centuries as at the present day.

Hence **transforma·tionalism** *Linguistics*, transformational theory; **transforma·tionalist** *sb. Linguistics*, an adherent of transformational theory; also *attrib.* or as *adj.*; **transforma·tionally** *adv. Linguistics*, by means of transformation(s), according to transformational rules.

1963 *Amer. Speech* XXXVIII. 240 These attributives can be related transformationally to predications. 1964 E. BACH *Introd. Transformational Gram.* viii. 187 The transformationalists have provided elegant and powerful tools for the description of particular languages. 1969 *Neuphilologische Mitteilungen* LXX. 221 Chomskyan transformationalism rejects a scientific approach for an anti-scientific one. 1973 *Amer. Speech 1970* XLV. 125 A question that challenged the structuralist linguistic theory..and that seems no more amenable to a transformationalist solution. 1976 *Archivum Linguisticum* VII. 155 It is interestingly novel, however, within the framework of contemporary transformationalism in America. 1977 *Dædalus* Fall 119 If the interrogative sentence 'Are the men here?' is derived transformationally from the phrase structure underlying the declarative sentence 'The men are here', it would seem to imply that a speaker first thinks of the declarative sentence and then transforms it into the interrogative form. 1978 *Language* LIV. 174 Raimo Anttila's 'Revelation as linguistic revolution'..reads like an extemporaneous sermon on the wickedness of the transformationalists. 1980 *English World-Wide* I. 268 The neglect of the results of earlier scholarship on the part of young linguists, especially of the transformationalist school.

transformative, *a.* Add: **b.** *Linguistics.* = *TRANSFORMATIONAL *a.*, *spec.* in *transformative-generative* adj.

1962 P. STREVENS *Papers in Lang.* (1965) v. 73 The three major modern linguistic theories (*i.e.* phoneme-morpheme grammar, transformative-generative grammar, and system-structure grammar). 1965 *Language* XLI. 213 Recent interest taken by transformative-generative linguists in unidirectional transformations and unique derivations.

transformed, *ppl. a.* Add: **b.** *transformed cell* (Cytology), a eukaryotic cell which has undergone transformation (*TRANSFORMATION 3 i).

1956 *Science* 23 Mar. 503/2 The malignant nature of the transformed cells was demonstrated by the production of sarcomas when the cultures were inoculated into animals of the same strain. 1979 ARMS & CAMP *Biology* xvi. 247 Transformed cells often undergo drastic changes in morphology and metabolism such that they become unresponsive to the normal controls over cell division.

transformer. Add: **2. a.** (Further examples.) In mod. use, a static apparatus in which an alternating or pulsating current in one winding induces an alternating current in a second winding, usu. with different values of voltage and current.

1911 *Encycl. Brit.* XXVII. 173 A continuous current transformer is an appliance which effects a similar transformation for continuous currents, with the difference that some part of the machine must revolve. 1947 R. LEE *Electronic Transformers & Circuits* vii. 187 Line impedance changes abruptly, and transformers may be necessary for good power transfer. 1955 *Sci. News Let.* 15 Oct. 248/3 A transformer is a device very widely used to increase or lower the voltage of an electric current.

b. transformer oil, a high-grade oil with a low sludge content used to cool and insulate transformers and other electrical equipment.

1904 *Electric Club Jrnl.* I. 228 By transformer oil is meant an oil in which the transformer is completely immersed, forming a homogeneous insulation. 1957 E. B. JONES *Instrument Technol.* III. I. 19 The variable resistance at the transmitter end is immersed in transformer oil which keeps it free from dirt. 1979 *Electr. Rev. Internat.* 7 Dec. 47/3 Transformer oil becomes highly viscous at low temperature.

transfo·rmerless, *a.* [-LESS.] That does not have a transformer; also, produced without the use of a transformer.

1949 B. GROB *Basic Television* xxii. 498 (*heading*) Transformerless television receiver. 1975 *Physics Bull.* Feb. 81/2 This has been achieved by using a transformerless supply and an aluminium casting for the frame.

transforming, *ppl. a.* Add: **b.** *transforming principle* (Biol.), a substance that genetically transforms bacterial cells (*TRANSFORM *v.* 1 f).

1944 *Jrnl. Exper. Med.* LXXIX. 155 If the results of the present study on the chemical nature of the transforming principle are confirmed, then nucleic acids must be regarded as possessing biological specificity the chemical nature of which is as yet undetermined. 1965 PEACOCKE & DRYSDALE *Molecular Basis Heredity* iii. 13 The transforming principle has two properties characteristic of genes, namely, determination of a specific inheritable property and self-reproduction.

transfusion. Add: Hence **transfu·sional** *a.*, occurring as a result of or by means of transfusion.

1965 *Endocrinology* LXXVII. 954 (*heading*) Effect of natural and synthetic corticoids on transfusional shock in the rat. 1974 R. ZELEDÓN in K. Elliott et al. *Trypanosomiasis & Leishmaniasis* 70 Besides the normal mechanism of transmission by the insect, two other modes, the transfusional and transplacental, may be epidemiologically important in the maintenance of the disease. 1982 *Jrnl. Pharm. & Pharmacol.* XXXIV. 730 (*heading*) Pyridoxal complexes as potential chelating agents for oral therapy in transfusional iron overload.

transgenosis (trɑnsdʒėnō·sis, -nz-). *Genetics.* [f. *TRANS- 10 + GENE + -OSIS.] The transfer of genes to an unrelated organism and their subsequent expression.

1973 C. H. DOY et al. in *Proc. Nat. Acad. Sci.* LXX. 723/1 The overall phenomenon of transfer and subsequent expression has been termed transgenosis. **1979** I. H. Herskowitz *Elem. Genetics* xiii. 192 Such transgenosis experiments usually involve either the uptake of naked bacterial DNA or the injection of phage DNA into eukaryotic cells.

trans,glo·bal, *a.* [f. Trans- 3 + Global *a.*] That travels across or round the world.

1953 [see *hostess 2 c]. **1981** *TV Times* 25–31 July 29/3 Prince Charles talks about The Transglobal Expedition.

transgress, *v.* Add: **2. a.** (*b*) *spec.* in *Geol.* Of the sea: to spread over (the land). Cf. *transgression 2.

1909 *Bull. Geol. Soc. Amer.* XX. 479 There are periodic recurrences of extensive emergences of the continents and..each one is later invaded or transgressed by continental seas. **1978** *Nature* 13 July 131/1 The down faulted and transgressed blocks on Fig. 1 have been numbered to show the sequence and time when the block was first transgressed.

transgressed, *ppl. a.* (Later example in *Geol.*)

1978 [see *transgress *v.* 2 a (*b*)].

transgressible, *a.* For *rare*$^{-0}$ read *rare* and add example.

1851 H. L. Mansel *Proleg. Logica* 100, I..consider the results of my experience as contingent only and transgressible.

transgression. 2. Substitute for def: The spread of the sea over the land, as evidenced by the deposition of unconformable marine sediments. (Later examples.)

1908, etc. [see *regression 7]. **1975** J. G. Evans *Environment Early Man Brit. Isles* iii. 67 Minor changes of sea level and coastal configuration have continued well beyond the main period of marine transgression.

transhisto·rical, *a.* [f. Trans- 4 + Historical *a.*] (Having significance) that transcends the historical; universal or eternal.

1909 W. R. Inge in *Q. Rev.* Apr. 602 It is not the province of faith to flout scientific knowledge, nor to contaminate the material on which science works by intercalating what M. Le Roy calls 'transhistorical symbols'—myths in fact—which do not become true by being recognised as false, as the new apologetic seems to suggest. **1963** J. A. T. Robinson *Honest to God* i. 24 In order to express the 'trans-historical' character of the historical event of Jesus of Nazareth, the New Testament writers used the 'mythological' language of pre-existence, incarnation, [etc.]. **1976** T. Eagleton *Crit. & Ideology* v. 178 Even where literary science would deem a work to have 'justly' survived, there is no call for materialist embarrassment about the 'metaphysical' quality of such transhistorical status.

transhuman, *a.* Delete *rare* and add further examples.

1936 E. Underhill *Worship* xii. 251 Gazing on the Saints in their manifest humanity, their heroic virtue and 'spiritual persuasiveness', he shares their trans-human experience. **1957** *Economist* 9 Nov. (Suppl.) 12/1 This intensification of life—reaching towards a 'transhuman' level. **1968** S. Rosen in *PN Rev.* (1979) No. 10. 15/2 We cannot return..to Greek, Jewish, Christian, or any other trans-human gods, whose meaning has been effectively destroyed by the decay of the values they represented.

transhumance (trɑnshiū·mɑ̃ns, -nz). [a. Fr., f. *transhumer*, ad. Sp. *trashumar* (f. L. *trans* across, over + *humus* ground, soil).] The transfer of grazing animals to other summer pastures and back, often over substantial distances.

1911 M. I. Newbiggin *Mod. Geogr.* vii. 179 The summer drought makes it difficult for even these hardy animals to obtain food, and necessitates in many regions a curious form of nomadism, to which the name of transhumance is given. Transhumance, still well developed in Spain, is the periodic and alternating displacement of flocks and herds between two regions of different climate. **1931** C. F. Jones *South America* 366 Government concessions to permanent ranchers, who do not desire the migrating flocks,..are reducing transhumance. **1954** M. Beresford *Lost Villages* vi. 204 Sheep which knew transhumance were less averse to being shepherded a score of miles over to a new pasture. **1975** J. G. Evans *Environment Early Man Brit. Isles* vi. 133 We do not know to what extent these Bronze Age people were nomadic, or were practising transhumance, or were settled farmers.

Hence **transhu·mant** *a.*, migrating between regions with differing climates; of or pertaining to transhumance.

1932 E. H. Carrier *Water & Grass* 78 The transhumant flocks. **1967** *Listener* 30 Mar. 426/3 The Sarakatsani—transhumant pastoralists of the Balkans. **1976** *Times Lit. Suppl.* 26 Mar. 353/2 The longer transhumant routes from the Pyrenees southwards to Catalonia.

trans,hydro·genase. *Biochem.* [f. *trans- 10 + *hydrogenase.] Any enzyme which catalyses the transfer of hydrogen from one organic substrate to another. Cf. *dehydrogenase.

1952 S. P. Colowick et al. in *Jrnl. Biol. Chem.* CXCV. 95 It is shown that the enzyme catalyzes a transfer of electrons (or hydrogen)... The enzyme will therefore be referred to here as 'pyridine nucleotide transhydrogenase'. **1978** *Jrnl. R. Soc. Med.* LXXI. 171 About 60% of insulin entering the liver is inactivated by liver enzymes, such as glutathione insulin transhydrogenase.

transience. Add: Also (in sense 2) **transeunce.**

1906 S. S. Laurie *Synthetica* I. I. i. 6 The difficulties that arise in connection with the transeunce. **1914** C. D. Broad *Perception* ii. 105 Leibniz and Lotze would have overlooked the immanence in the whole system,..and fastened on the transeunce within it with respect to its various elements.

transiency. Add: Also (in sense 2 of Transient *a.*) **transeuncy.**

1942 *Mind* LI. 137 Spinoza's central causal theory refers to the world of adequate knowledge as it is directed to *entia in se*, and its application to transeuncy must be governed by derivation therefrom.

transient, *a.* (*sb.*) Add: **1. a.** *spec.* in *Electr.* (cf. sense B. 3 below).

1853 *Proc. Philos. Soc. Glasgow* III. 285 (*heading*) On transient electric currents. **1870** *Phil. Mag.* XXXIX. 428 The galvanometer takes account of the induced transient current as a whole. **1962** *Newnes Conc. Encycl. Electr. Engin.* 821/2 The transient current consequent upon the switching-on of a filament lamp. **1969** J. J. Sparkes *Transistor Switching* v. 124 With the pulse steering circuits added..a transient current may flow.

b. *transient equilibrium* (Nuclear Sci.), the condition in which the half-life of a parent isotope is greater than that of the daughter but comparable to the period of observation, so that after an initial increase the total radioactivity decays with the parent's half-life and the ratio of parent atoms to daughter atoms remains constant.

1912 Makower & Geiger *Practical Measurement in Radioactivity* viii. 111 The name transient equilibrium has been given to this state of apparent equilibrium which exists whenever the life of a product is not negligibly short compared with that of the preceding substance which controls the decay. **1961** G. R. Choppin *Exper. Nucl. Chem.* vi. 84 For a parent with a 1 month half life, observation over a few days will seem to be secular equilibrium, whereas observation over a 3 month period will show transient equilibrium.

2. (Further examples of *transeunt.*)

1903 F. C. S. Schiller *Humanism* iv. 64 The impossibility of explaining such transeunt causation compels to the inference that things are not really separate and independent. **1933** *Mind* XLII. 155 The more responsive P_n is to P_m the more transeunt action there is between the two. **1942** R. G. Collingwood *New Leviathan* xx. 140 It [*sc.* the process of ruling] is *transeunt* when that which rules rules something other than itself. **1949** H. W. B. Joseph *Lect. Philos. Leibniz* iii. 107 That is the difficulty of transeunt causation—an effect produced in one thing by what is just another.

4. (Further *spec.* examples.) Also *transf.*, for transient guests, short-stay.

1818 H. B. Fearon *Sk. Amer.* 44 Boarding..is 8 dollars a week, for what is termed 'a transient man'. **1879** F. R. Stockton *Rudder Grange* xi. 121 We had no accommodations for them, neither had we any desire for even transient visitors. **1891** *Fur, Fin & Feather* Mar. 185 The transient rate for travelers at the Hilsabeck Hotel in Springfield is $1 a day. **1903** *N.Y. Even. Post* 19 Oct. 3 A 12-story transient hotel. **1906** *Springfield* (Mass.) *Weekly Republican* 9 Aug. 16 They will then rent apartments with or without board to transient and permanent guests. **1942** E. Paul *Narrow St.* xxii. 175 Would that..some Turk would..rush me to a transient hotel. I am past that age, and never enjoyed a clandestine situation. **1976** *Times* 29 May 1/8 Was placed in transient barracks, a form of solitary confinement. **1981** *Sci. Amer.* Nov. 37/3 More intensive canvassing of places such as pool halls and transient hotels was done in an attempt to include a greater proportion of people who have no permanent address.

6. *U.S.* (Esp. of printed matter) occasional, isolated, individual.

1831 *Boston Transcript* 18 Apr. 2/3 We shall use all patrons alike, whether they are annual or transient advertisers. **1841** *Lowell* (Mass.) *Offering* I. 245 The clerk asked her if it was a transient paper. **1857** *Harper's Mag.* Feb. 403/1 The prepayment of postage on transient printed matter has been made compulsory. **1857** *Lawrence* (Kansas) *Republican* 28 May 1 All transient advertisements must be paid for in advance. **1904** *Philadelphia Friends' Intelligencer* 15 Oct. p. ii, For transient advertisements, 5 cents per line.

B. *sb.* **2.** For *U.S. colloq.* read '*colloq.* (orig. *U.S.*)' and add: Also, a traveller, a tramp, a migrant worker. (Further examples.)

1941 H. G. Wells *You can't be too Careful* II. xv. 104 Whenever Doober's had rooms to spare a card was put into the ground floor window, and there would be transients for three or four days. **1946** W. S. Maugham *Then & Now* vi. 33 Piero and the courier were to share a straw mattress in a corridor along with a number of transients only too glad to have a roof over their heads. **1959** M. Renault *Charioteer* vi. 114 A respectable tenement full of transients in a time of flux. **1963** C. D. Simak *They walked like Men* iii. 17 He was snoring gently and he looked..like a transient who might have wandered in to find a place to sleep. **1978** *Beautiful British Columbia* Winter 17 Transients pile in each winter to work the oil patch as soon as the muskeg freezes.

3. *Physical Sci.* A transient variation in current or voltage, or in any waveform; a transient condition.

1911 C. P. Steinmetz *Elem. Lect. Electr. Discharges* i. 2 The transient..appears as intermediate between two permanent conditions. **1911** —— in *Jrnl. Franklin Inst.* CLXXII. 41 Transients are not a specifically electrical phenomenon, but occur in any system of forces, where energy storage occurs. **1936** *Physical Rev.* XL. 522/1 Thus *I''* is of importance only in determining the initial transients but not the steady rate of absorption. **1947** R. Lee *Electronic Transformers & Circuits* iv. 102 Transients occur when the load is applied..or removed..causing respectively a momentary drop or rise in plate voltage. **1970** V. M. Albers *World of Sound* ix. 143 The transients from percussion instruments and a piano are quite severe. **1972** *Nature* 21 Apr. 384/1 Total surface fields of over 100 gammas have been observed when large solar field transients pass the Moon. **1979** *Guardian* 25 Oct. 20/1 Accidental abrupt changes of conditions in reactors—these are called transients.

attrib. **1962** A. Nisbett *Technique Sound Studio* 248 In such [microphone] designs..there are minimal inertial effects and therefore a very good transient response. **1975** G. J. King *Audio Handbk.* v. 117 A useful signal for transient appraisal is the square wave provided its rise time is significantly smaller than that of the amplifier.

transignification (trɑːnsignifikēɪ·ʃən). *Theol.* Also **trans-signification.** [f. Trans- 3 + Signification.] The change in the significance of bread and wine at the Eucharist through transubstantiation, expressed in terms of sacramental symbolism. Cf. *transfinalization.

1965, etc. [see *transfinalization]. **1968** J. M. Powers *Eucharistic Theology* iv. 170 It is sometimes proposed that the idea of transsignification is presented as an alternative to the traditional theological idea of transsubstantiation. **1983** M. F. Wiles *Faith & Mystery of God* iii. 38 The approach in terms of trans-signification or establishment of new meaning can make the same point in a more constructive way.

transillu·minator. [f. Transilluminate *v.* + -or.] **a.** An instrument for examining the conjunctiva and the sclerotic of the eyeball by shining light through them. **b.** An instrument for making visible spots on chromatography plates and electrophoresis gels by shining ultraviolet light through them.

1906 *Ophthalmic Record* XV. 209 (*heading*) Transillumination of the eye in the differential diagnosis of intraocular tumors, with the description of an ocular transilluminator. **1925** B. Lang *Routine Examination of Eye* III. 150 The most important use of the transilluminator is to examine the interior of the globe, particularly in cases of detachment of the retina. **1954** S. Duke-Elder *Parsons' Dis. Eye* (ed. 12) vi. 87 For this purpose, special transilluminators may be employed. **1973** *Nature* 16 Feb. 473/2 Both the short-wave and long-wave transilluminator ultraviolet lamps were used for detection. **1978** *Ibid.* 17 Aug. 715/2 (*caption*) Gels were stained with ethidium bromide solution and visualised on an ultraviolet transilluminator.

trans,indivi·dual, *a.* [f. Trans- 4 + Individual *a.*] Not confined to any particular thing or person, more than individual. Cf. *transpersonal *a.*

1936 *Psychol. Monogr.* XLVII. 1. 8 In Scholastic philosophy these realities are regarded as subsisting in the realm of ideas; they are trans-individual. **1938** *Mind* XLVII. 482 The right answer to the question 'What are numerical propositions?' is that they predicate a peculiar kind of trans-individual quality applicable only to groups. **1973** S. Heath in *Screen* Spring/Summer 105 The transindividual system or code..of elements and rules underlying and assuring individual messages. **1977** A. Sheridan tr. *Lacan's Écrits* iii. 49 Its domain is that of concrete discourse, in so far as this is the field of the transindividual reality of the subject.

transistor (trɑnsi·stəɹ, -nz-), *sb.* [Blend of Transfer *v.* and *resistor.] **1.** A semiconductor device, usu. having three terminals and two junctions, in which the load current can be made to be proportional to a small input current, so that it is functionally equivalent to a valve but is much smaller and more robust, operates at lower voltages, and consumes less power and produces less heat.

1948 *N.Y. Times* 1 July 46/1 A device called a transistor, which has several applications in radio where a vacuum tube ordinarily is employed, was demonstrated for the first time yesterday. **1948** Bardeen & Brattain in *Physical Rev.* 15 July 230/1 (*heading*) The transistor, a semi-conductor triode. **1949,** etc. [see *junction transistor* s.v. *junction *sb.* 4]. **1952** *Electronic Engin.* XXIV. 42 Although it is unlikely that the transistor will ultimately displace the electronic valve, there is no doubt that for many electronic applications the transistor..will be preferred because of its robust and compact form. **1953,** etc. [see *field-effect transistor* s.v. *field *sb.* 21]. **1957** *Observer* 1 Sept. 9/7 A novelty now gaining respectability is small-scale radio, with tiny medium-wave sets using printed circuits, and transistors instead of valves. **1962** [see *point contact* s.v. *point *sb.*[1] B. 14]. **1970** J. Earl *Tuners & Amplifiers* iv. 78 Transistors, particularly the bigger power devices, are regarded as being current-operated... Valves, on the other hand, are often regarded as voltage-operated devices. **1973** *Sci. Amer.* Aug. 48/1 The MOS technology produces transistors of the unipolar type in contradistinction to earlier junction transistors, which are bipolar.

2. *ellipt.* = *transistor radio*, sense 3 below.

1961 *Daily Tel.* 13 July 12/2 Few seaside authorities are likely to risk driving away transistor-addicted visitors to other resorts. **1966** J. Betjeman *High & Low* 50 The endless anonymous croak of a cheap transistor Intensifies the loneliness I feel. **1977** *Rep. Comm. Future of Broadcasting* (Cmnd. 6753) iii. 23 Car radios and portable transistors have made radio one of the nation's major daytime diversions.

3. *Comb.*, as (sense 2) *transistor-addicted* adj.; **transistor radio,** a small portable radio having transistors and other solid-state devices in place of valves; **transistor set,** a radio or television set having transistors instead of valves; **transistor-transistor logic,** logic in which transistors take the place of many of the coupling resistors; abbrev. TTL s.v. *T 6 a.

1961 Transistor-addicted [see sense 2 above]. **1958** *Spectator* 1 Aug. 167/2 The new miniature transistor radios. **1965** Auden *About House* (1966) 15 Nobody I know would like to be buried with..A transistor radio. **1957** *New Yorker* 13 July 19/1 The kid got one of those compact transistor sets for his birthday, and is absolutely fascinated by it. **1961** *Ann. Reg. 1960* 451 Television became more than ever an intrinsic part of Japanese life, and transistor sets of compact size made its penetration possible everywhere. **1963** Transistor-transistor logic [see TTL s.v. *T 6 a]. **1977** *McGraw-Hill Yearbk. Sci. & Technol.* 282/2 I²L is transistor-transistor logic (TTL) compatible.

transistorize (trɑnsi·stəraiz, -nz-), *v.* [f. prec. + -ize.] *trans.* To design or make with transistors (rather than valves). Chiefly as **transi·storized** *ppl. a.*, employing transistors; also *fig.*

1953 *Sci. News Let.* 7 Feb. 86/1 Because of their ruggedness, 'transistorized' amplifiers should eliminate many an electronic headache under the restless sea. **1959** *Listener* 28 May 930/1 The compactness of transistorized equipment and their small power requirements make them suitable for use in telephone exchanges. **1963** T. D. Towers *Transistor T.V. Receivers* viii. 125/1 The line output stage has proved to be the most difficult part of the line time base to transistorise. **1963** *Daily Tel.* 5 Oct. 8/2 What Mr. Wilson has done is to lift the whole nationalisation issue..out of its traditional context and give it a streamlined, transistorized modern setting. **1972** *Ibid.* 31 July 3/1 Both are powered by a new six cylinder, 2·8 litre twin-cam engine with transistorised ignition and electronic fuel injection. **1976** A. Hope *Hi-fi Handbk.* 27 When the craze for transistorising everything got under way, in the 1960s, very few true hi-fi enthusiasts would touch a transistor amplifier with a barge pole. **1980** J. Gardner *Garden of Weapons* II. ii. 132 The screech-boxes looked like..radio sets. They were in fact transistorised transmitters.

Hence **transi:storiza·tion,** the use of transistors in electronic apparatus.

1957 *Trans. Soc. Instrument Technol.* IX. 41/1 One feels that 'transistorization' is bound to be introduced rapidly, thus reducing cost, weight and power consumption. **1969** *Radio Times* 4 Dec. 68 (Advt.), The Ferguson Colourstar is brilliantly designed for reliability with full transistorisation and modular construction.

transit, *sb.* Add: **1. c.** *spec.* Public passenger transport; freq. *attrib.* Chiefly *N. Amer.*

1873, etc. [see *rapid-transit* s.v. *rapid *a.* 6]. **1967** *Boston Sunday Globe* 23 Apr. 8/2 Legislature to launch a 10-year, $300 million urban transit program. **1971** *Rand Daily Mail* 27 Mar. 11/1 White bus drivers employed by a transit authority in an unnamed South African city. **1979** *Tucson* (Arizona) *Citizen* 20 Sept. 1 c/3 The Canadian city has good mass transit.

5. (see also sense *1 c) *transit-line, -time* (examples in *Electronics* and *Med.*); **transit camp,** a camp for the temporary accommodation of servicemen awaiting posting, refugees, prisoners-of-war, etc.; **transit lounge,** a waiting-room for transit passengers at an airport; **transit man** *N. Amer.*, a surveyor who uses a transit-compass; **transit passenger,** a passenger making a brief stop at an airport in transit to another destination; **transit visa,** a visa permitting the holder to pass through a country but not to stay there.

1943 G. Greene *Ministry of Fear* II. ii. 144 The place was as comfortless as a transit camp. **1946** E. Linklater *Private Angelo* xiii. 151 [He] made his escape..from a transit camp for prisoners of war near Bari. **1956** Wallis & Blair *Thunder Above* (1959) xi. 113 There were 25,000 East German refugees in West Berlin, living in transit camps built to accommodate 13,000. **1980** D. Lodge *How Far can you Go?* i. 7 Purgatory was a kind of penitential transit camp on the way to the gates of Heaven. **1894** Hardy *Life's Little Ironies* 179 For South Wessex, the year [*sc.* 1851] formed in many ways an extraordinary chronological frontier or transit-line, at which there occurred what one might call a precipice in Time. **1962** J. Fleming *When I grow Rich* iii. 43 After the arrival of the jet, the transit lounge had filled up with people. **1983** *Jetaway* (Air New Zealand) Sept.–Oct. 28/2 Transit lounge for 500 passengers who are transferring flights or have no need to go through customs. **1873** Transit man [*packer[1] 3 c]. **1971** *Islander* (Victoria, B.C.) 8 May 12/1 Harry, a young transit man, had his bed roll next to the Bella Coola trapper. **1955** E. Bowen *World of Love* xi. 223 They let the transit passengers off first. **1972** J. Potter *Going West* 17 Transit passengers were encouraged to alight for the stopover. **1948** Martin & Hynes *Clin. Endocrinol.* iv. 70 The intestinal transit-time is decreased with the production of loose stools or frank diarrhœa. **1962** Simpson & Richards *Physical Princ. Junction Transistors* vii. 117 When operation at higher frequencies is considered these complexities increase many-fold due chiefly to transit-time effects in the flow of minority carriers. **1974** *Brit. Med. Jrnl.* 19 Jan. 108/2 Constipation is best thought of not in terms of transit-time through the gut..but rather the type of faeces produced. **1975** D. G. Fink *Electronics Engineers' Handbk.* VII. 27 Transit time is a large factor in considering the upper frequency limitation of electron tubes. **1925** C. Connolly *Let.* May in *Romantic Friendship* (1975) 81, I..had got as far as sending my passport up for some transit visas. **1979** W. H. Canaway *Solid Gold Buddha* xx. 134 Miller went to the Burmese Embassy..and got his transit visa.

transitation. Delete † *Obs.*, for '(In quots. *humorous.*)' read '(In early quots. *humorous.*)', and add later example.

1915 D. H. Lawrence *Rainbow* vii. 187 Containing birth and death, potential with all the noise and transitation of life, the cathedral remained hushed.

transition. Add: **1. b.** *Physics.* A change of an atomic nucleus or an orbital electron from one quantized state to another, with the emission or absorption of radiation of a characteristic wavelength.

1913 *Phil. Mag.* XXVI. 18 We consequently observe an absorption of radiation which is not accompanied by a complete transition between two different stationary states. **1922** tr. *Bohr's Theory of Spectra* III. iv. 118 Emission lines of the X-ray spectra due to transitions between the stationary states corresponding to these energy levels. **1930** *Science* 10 Oct. 376/2 Very little is known about nuclear properties of atoms because of the difficulties inherent in excitation of nuclear transitions in the laboratory. **1962** [see *level *sb.* 3 e]. **1977** *Whitaker's Almanack 1978* 1035/2 By varying the frequency, a detailed analysis of a particular x-ray transition was made.

4*. *Molecular Biol.* The occurrence in a nucleic acid of one purine in place of another, or of one pyramidine in place of another. Cf. *transversion[1] 3.

1959 E. Freese in *Proc. Nat. Acad. Sci.* XLV. 630 Each base analogue can induce the transitions..in both directions (from A–T into G–H and vice versa). **1981** Pai & Marcus-Roberts *Genetics* xv. 477 Substitutions of bases are further classified as transitions or transversions.

5. a. *transition area, belt, period, point, region, state* (examples), *zone.*

1831 Carlyle in *Fraser's Mag.* Mar. 144/2 *Don Karlos*, a work of what may be called his transition-period, the turning-point between his earlier and his later period. **1841** J. S. Buckingham *America* I. 461 Baltimore..appears from the very first to have been peopled by a race that never had this transition-state to pass through. **1907** E. A. Mearns *Mammals Mexican Boundary* 135 This station lies in the Transition Zone, the highest peaks extending well into the Canadian or lowest section of the Boreal Zone. **1940** *Chambers's Tech. Dict.* 860/1 *Transition region*, the portion of the axis of a young plant in which the change from root structure to shoot structure occurs. **1950** *Texas Studies in Eng.* XXIX. 254 The transition area (where both [s] and [z] are used), is relatively narrow to the west of Philadelphia. **1957** *Publ. Amer. Dial. Soc.* XXVII. 5 The net effect being to create a transition belt. **1969** J. J. Sparkes *Transistor Switching* i. 15 This results in a narrowing of the collector transition region and a widening of the base region. **1972** H. Kurath *Studies in Area Linguistics* iii. 44 The transition area between the North and the Midland reflects partly the complicated history of the settlement. **1977** *Nature* 14 Apr. 662/1 It is becoming increasingly apparent that FMO interactions are even more important in transition states than they seem to be on inspection of isolated reactant orbitals. **1977** A. Hallam *Planet Earth* 11/3 Based on its density distribution, the mantle has been divided into three parts: the upper mantle, which extends to a depth of 400km..; the transition zone, which extends from 400 to about 700km..; and the lower mantle. **1978** D. Joy *Railways in Lancashire* 81/1 This view shows the transition period with the new station completed but the old one not yet demolished. **1982** *Amer. Speech* LVII. 293 A discussion of isoglosses and transition belts..underpins Chambers and Trudgill's theory of urban dialects.

b. Special Combs.: **transition element** *Chem.*, † (*a*) any of the nine metallic elements forming group VIII of the periodic table (see quot. 1922); cf. *transitional element* s.v. *transitional *a.* c; (*b*) any of a large class of metallic elements making up groups IIIA–VIIA, VIII, and IB of the periodic table (groups 3–11 in the new notation), which are characterized by partly filled *d* orbitals and commonly exhibit variable valency and an ability to form coloured complexes; also extended to include elements having partly filled *f* orbitals (see quots. 1962); **transition fit** *Engin.*, a fit between two mating parts such that, within the specified tolerances, there may be either interference or clearance between them; **transition flow** (see quot. 1969); **transition metal** *Chem.* = *transition element* (*b*) above; **transition probability** *Physics*, the probability of a transition between two given states of a system, *spec.* an atom; **transition temperature** *Physics*, the temperature at which a substance acquires or loses some distinctive property, esp. superconductivity.

1922 E. J. Holmyard *Inorganic Chem.* XXX. 530 (*heading*) The 'transition elements'.—Iron, cobalt, nickel; ruthenium, rhodium, palladium; osmium, iridium, platinum. **1953** de Barry Barnett & Wilson *Inorganic Chem.* i. 6 Transition Elements. This term was originally used by Mendeleef for the three triads of elements forming Group VIII of his periodic table.., but this meaning of the term has long since been abandoned. **1962** Cotton & Wilkinson *Adv. Inorganic Chem.* xxiv. 493 The transition elements may be strictly defined as those which, *as elements*, have partly filled *d* or *f* shells. Here we shall adopt a slightly broader definition and include also elements which have partly filled *d* or *f* shells in any of their commonly occurring oxidation states. *Ibid.* 494 The large number of transition elements is subdivided into three main groups: (*a*) the main transition elements or *d*-block elements, (*b*) the lanthanide elements, and (*c*) the actinide elements. **1965** Phillips & Williams *Inorganic Chem.* I. ii. 52 The valences of the transition and lanthanide elements. **1974** *Encycl. Brit. Macropædia* XVIII. 600/2 The so-called transition elements constitute a group of 56 (the majority of the 105 known elements). **1919** Transition fit [see *interference fit* s.v. *interference 5]. **1971** B. Scharf *Engin. & its Lang.* xi. 111 Transition fits are so designed that either clearance or interference may occur when the mating parts are assembled. Push fits may be tolerance fits. **1969** *Gloss. Aeronaut. & Astronaut. Terms* (*B.S.I.*) IV. 11 *Transition flow*, the flow of gases under conditions intermediate between laminar viscous flow and molecular flow. **1942** *Ann. Reg. 1941* 356 Potter..investigated the resistivity..of pure samples of the transition metals. **1970** *New Scientist* 3 Sept. 457/2 The chemistry of the transition metals..is dominated by their complex compounds. **1926** E. Condon in *Physical Rev.* XXVIII. 1185 The theory of transition probability which will now be developed is an outgrowth of a picture proposed by Franck. **1942,** etc. [see *Markov]. **1978** Pasachoff & Kutner *University Astron.* xxiv. 598 Even though this transition probability is so very low, there are so many hydrogen atoms in space that enough 21-cm radiation is given off to be detected. **1930** *Sci. Abstr.* XXXIII. 697 In all these cases, at temperatures below the transition temperature, increasing the intensity of the magnetic field to a certain value..causes the superconductivity to disappear. **1966** C. R. Tottle *Sci. Engin. Materials* vi. 127 For the compound Nb_3Sn the transition temperature is as high as 18·2°K. **1982** *Sci. Amer.* May 30/1 Ferroelectric and ferromagnetic transition temperatures.

transitional, *a.* Add: **c.** Chem. *transitional element* = *transition element* (*a*) s.v. *transition 5 b.

1887 C. M. Tidy *Handbk. Mod. Chem.* (ed. 2) ii. 54 In Group VIII. three elements occur. These are termed by Mendeleef 'transitional elements', that is, elements occurring between the even and the odd series of a long period. **1921** J. R. Partington *Text-bk. Inorganic Chem.* xlviii. 972 The transitional elements.—The eighth group of the Periodic System comprises three sub-groups, with three elements in each, forming the termination of the even series 4, 6, and 8, and connecting the elements of these series with those of the odd series following. For this reason they were called by Mendeléeff the transitional elements.

transitive, *a.* Add: **1.** (Later U.S. example.)

1906 *Springfield* (Mass.) *Weekly Republican* 8 Mar. 6 At present he is in a transitive state.

6. (Earlier and later examples.) [ad. G. *transitiv* (S. Lie *Theorie d. Transformationsgruppen* (1888) I. 212).]

1888 *Amer. Jrnl. Math.* X. 297 If.. a G_r in *xy* can transform every ordinary point of the plane to every other ordinary point of the plane, the G_r is said to be transitive. **1968** [see *primitive *a.* 5 d]. **1971** L. Dornhoff *Group Representation Theory* A. xxxvi. 215 Nontrivial normal subgroups of primitive permutation groups are transitive.

7. *Math.* and *Logic.* Of a relation: such that if it holds between a first and second item, and also between the second and a third, it necessarily holds between the first and the third.

1856 A. de Morgan in *Trans. Cambr. Philos. Soc.* IX. 104 The first is what I shall call *transitiveness*, symbolized in $X—Y—Z = X—Z$; meaning that if *X* stand in the relation denoted by — to *Y*, and *Y* to *Z*, *X* therefore stands in that relation to *Z*. Very many copulæ exist in which this transitive relation is seen. **1870,** etc. [see *intransitive *a.* 3]. **1903,** etc. [see *reflexive *a.* 7]. **1936** A. J. Ayer *Lang., Truth & Logic* iii. 79 As each of these relations is symmetrical..and also transitive..it follows that the groups of visual and tactual sense-contents which are constituted by means of these relations cannot have any members in common. **1956** E. H. Hutten *Lang. Mod. Physics* iv. 143 We take as axiom the statement that the thermal equilibrium is a transitive relation, i.e. if two bodies in thermal equilibrium are in equilibrium with a third body, all three bodies have a common property, that is, the same temperature. **1976** *Nature* 29 Apr. 773/1 Whether or not the young child can make transitive inferences..is still a controversial issue. **1979** Georgacarakos & Smith *Elem. Formal Logic* ix. 330 Many comparative adjectives provide transitive relations, such as *it is taller than*.

transitively, *adv.* Add: **c.** *Math.* and *Logic.* (See senses 6 and *7 of Transitive *a.*)

1889 [see *intransitive *a.* 4]. **1971** C. W. Curtis in Powell & Higman *Finite Simple Groups* iii. 139 Any two bases of a root system *Δ* are conjugate by an element of the Weyl group (i.e., *W*(*Δ*) acts transitively on the set of bases).

transitiveness. (Further examples, corresp. to *transitive *a.* 7.)

1850 A. de Morgan in *Trans. Cambr. Philos. Soc.* (1856) IX. I. 104 The first [copular condition] is what I shall call *transitiveness*. **1903** B. Russell *Princ. Math.* xxvi. 218 Relations may be divided into four classes, according as they do or do not possess either of two attributes, transitiveness and symmetry. **1955** A. N. Prior *Formal Logic* III. i. 220 It is as if one presented a study of transitiveness under the guise of a 'logic' of the relation of ancestorhood.

transitivism (trɑ·nsitiviz'm). *Psychiatry.* [f. Transitive *a.* (*sb.*) + -ism.] A mental state or condition in which a patient attributes to others his own experiences and sensations.

1924 A. A. Brill tr. *Bleuler's Textbk. Psychiatry* ii. 38 The splitting off of parts of a personality in *transitivism* proceeds in a different manner; here the patient's own experiences become detached from him, and are ascribed to another person... Transitivism is an almost common occurrence in schizophrenia. **1960** R. F. C. Hull tr. *Jung's Coll. Wks.* III. 134 The representation of one's own complexes by strange actors in dreams is well known.. in psychopathology we know it in the form of 'transitivism'. **1971** *Internat. Jrnl. Psycho-Anal.* LII. 237 The schizophrenic delusion of transitivism, which relates to the loss of ego boundaries, represents a regression of ego development.

transitivity. Add: Also in senses 2 and *7 of Transitive *a.* (Examples.)

1897 *Monist* Jan. 211 Not only is the relative of correspondence transitive, but it also possesses what may be called antithetic transitivity. **1928** *Mod. Lang. Rev.* Apr. 144 The following chapters [of Jespersen's *Modern English Grammar*] are concerned with..transitivity and predicatives. **1942** J. C. Cooley *Primer Formal Logic* viii. 339 The generalized postulate for the transitivity of the relation, *older than*, could not be expressed without variables. **1969** J. C. Wells *E.U.P. Concise Esperanto & Eng. Dict.* 17 Many English verbs are of varying transitivity, since they can be used either transitively or intransitively. **1980** A. J. Jones *Game Theory* iv. 194 To prove the theorem it is only necessary to verify the three defining properties of an equivalence relation... Reflexivity... Symmetry... Transitivity.

transitivize (trɑ·nsitivəiz, -nz-), *v. Gram.* [f. Transitive *a.* + -ize.] *trans.* To make (a verb) transitive. So **tra:nsitivi·zing** *ppl. a.*; **tra:nsitivi·zer,** an affix that makes a verb transitive.

1964 *Language* XL. 76 With the transitivizing stem formative—cʔi the resulting stem..means 'he sets it down'. **1972** J. L. Dillard *Black English* iii. 98 Pidgin characteristics, like the *-um* (*-em*) transitivizer. **1978** *Language* LIV. 125 An appropriate transitivizing suffix. **1979** *Trans. Philol. Soc.* 128 The causative *hanjāmaya-* would transitivize a 'joining' that was expressed by *han-gam-*, and thereby allow the past participle *hangmata-* to be used for English 'joined' in both its intransitive and passive sense.

transitron (trɑ·nsitrɒn). *Electronics.* [f. *trans(conductance + *-i-* + *-tron.] A pentode in which the suppressor grid is used as the control grid so that the valve exhibits negative transconductance.

1939 C. Brunetti in *Proc. IRE* XXVII. 88/2 For the sake of brevity it has been found desirable to provide a name for the retarding-field negative-transconductance device... The name 'Transitron' is suggested. **1945** *Electronic Engin.* XVII. 383 Those devices such as.. transitrons, whose negative resistance characteristic can be measured by d.c. tests. **1957** *Practical Wireless* XXXIII. 523/1 The timebase employed in the oscilloscope is of the transitron type.

Transjordanian (trɑːnsdʒɔɹdēɪ·niăn, -nz-), *sb.* and *a.* Also (in non-*spec.* senses) **trans-Jordanian.** [f. *Trans-Jordan* (see Trans- 7), *Transjordan(ia)* (see below) + -ian, -an: see *Jordanian *a.* and *sb.*] **A.** *sb.* A person from beyond the river Jordan (see *Jordan[2]); also *transf.*; *spec.* (now *Hist.*) a native or inhabitant of Transjordan (Transjordania), a territory east of the Jordan, now part of the Hashemite kingdom of the Jordan. **B.** *adj.* Of or pertaining to the land beyond the Jordan; *spec.* (now *Hist.*) of or pertaining to Transjordan. Cf. *trans-Jordan, -ic* s.v. Trans- 7.

The emirate of *Transjordan(ia)* was established under British mandate in 1921. The name was retained for the short-lived kingdom (1946–9), now Eastern Jordan.

1920 G. Saintsbury *Notes on Cellar-bk.* iv. 52 Gentiles, as it were, or at least trans-Jordanians to the pure Israel of Medoc. **1922** G. Bell *Let.* 18 Dec. (1927) II. xxii. 660 The conquest of Hayil..will bring Ibn Saud into the theatre of trans-Jordanian politics. **1965** M. Spark *Mandelbaum Gate* iv. 97 He disliked the Lebanese and wished all the Arabs were Palestinian or Transjordanian. **1979** R. Thomas *Eighth Dwarf* xxvi. 251 The Jews are going to have to fight..the Syrians and the Egyptians and..the Transjordanians.

Transkei, *sb.* (*a.*) Add to def.: From 1910 the Cape Colony formed part of the Union of South Africa, which became the Republic of South Africa in 1961; in 1976 the Transkei was the first Black homeland to be given a measure of independence. **Transkeian** *a.* (further examples); (*b*) *sb.* a native or inhabitant of the Transkei.

1974 *Standard Encycl. S. Afr.* X. 564/1 All Transkeian taxpayers over the age of 18 and all other Transkeian citizens over 21..to whom certain disqualifications do not apply are entitled to register as voters. **1976** *Times* 17 Feb. 7/6 The Transkeians want to make independence a success. **1976** *Times* 25 Oct. 13 Tonight the South African flag will be lowered in Umtata and the Transkeian flag will be raised. **1979** J. Drummond *Patriots* ii. 15 The Transkeian labour problem. *Ibid.* iii. 23 Nearly all those who lost their jobs were Transkeians.

translate, *v.* Add: **I. 1. e.** *intr.* To undergo translational motion.

1964 *Amer. Jrnl. Physics* XXXII 261/1 If frame *β*, thus translates rigidly with velocity *u* as measured in *α* then frame *α* translates rigidly with velocity − *v* as measured in *β*. **1979** *Sci. Amer.* Jan. 76/2 One is therefore forced to conclude that these deep structures do indeed constitute the lower portions of the continental plates and that they have been translating coherently with the crust for hundreds or even thousands of millions of years.

II. 2. d. *Biol.* To use (genetic information in messenger RNA) to determine the amino-acid sequence of a protein during its synthesis; also with the RNA as obj.

1961 *Cold Spring Harbor Symp. Quantitative Biol.* XXVI. 101/2 This finding implied that the information encoded in DNA must somehow be transmitted to the ribosomes where it is translated into the amino acid sequence of a polypeptide chain. **1971** *Nature* 24 Sept. 234/2 Messenger RNAs transcribed in the nuclei of eukaryotic cells have to be transported to the cytoplasm to be translated. **1972** *Sci. Amer.* Jan. 25/2 A length of RNA representing a gene is then translated into a particular protein, a molecule constructed with a 20-letter alphabet, the 20 amino acids. **1977** D. E. Metzler *Biochemistry* xv. 936/2 The ribosome faithfully translates the genetic message, adding amino acids to the peptide chain until a stop codon is reached.

IV. 7. *intr.* Const. *into.* To result in, to be converted into, to manifest itself as.

1975 *Lamp* (Exxon Corporation) Winter 11/2 Any delays in bringing fields into production could quickly translate into lower government revenues and an adverse impact on the balance of payments. **1976** *Sci. Amer.* June 69/1 For maneuvers executed early in a mission this uncertainty translates into an error at the target planet on the order of one kilometer. **1977** *Time* 8 Aug. 42/2 The price of raw coffee could gradually decline to about $1 per lb. on the New York market, which would translate into a retail price somewhere in the $2 range.

translatese (trɑns,lēɪtī·z, -nz-). [f. Translat(ion + -ese.] = *translationese. Cf. *translatorese.

1967 *Listener* 8 June 762/1 He..has couched it in the luke-warm translatese of one of his own more unurgent renderings. **1977** *Times Lit. Suppl.* 25 Feb. 202/1 Paralysing woodenness ('I am concerned to determine'), the dull thud of translatese ('Here is the place to mention Pirandello finally'). **1979** *Studies in Eng. Lit.: Eng. Number* (Tokyo) 228 To the very last his Japanese did not get rid of a 'translatese' completely.

translation. Add: **I. 2. c.** *Biol.* The process by which genetic information represented by the sequence of nucleotides in messenger RNA gives rise to a definite sequence of amino-acids in the protein or polypeptide that is synthesized.

1963 *Cold Spring Harbor Symp. Quantitative Biol.* XXVIII. 352/1 Polarity mutations affect the RNA to protein translation. **1968** H. Harris *Nucleus & Cytoplasm* iv. 83 In higher cells translation and transcription are not closely coupled. **1970, 1973** [see *transcription 6]. **1977** P. B. & J. S. Medawar *Life Science* xii. 95 This translation of genetic into structural information is irreversible, so there is no known..method by which germinal DNA could be imprinted with information acquired in an organism's own lifetime.

IV. 7. *translation-equivalent, process, theory*; **translation loan(-word)** = *loan-translation* s.v. *loan *sb.*[1] 5.

1963 J. Lyons *Structural Semantics* iv. 70 It may be impossible to find even a 'roughly equivalent' term in another language.., even though we can find satisfactory translation-equivalents for most..of its hyponyms. **1977** *Language* LIII. 295 Thai *khon* and its (near) translation-equivalents in many languages denote 'people'. **1954** Koestler *Invisible Writing* xi. 132 At which stage of the translation-process all these blessings had slipped in, we could not tell. **1967** M. Argyle *Psychol. Interpersonal Behaviour* v. 90 Social interaction..depends on the existence of a learnt store of central translation processes. **1900** E. Björkman *Scand. Loan-Words in M.E.* I. 12 What I should like to call 'translation loan-words'... Thus..*wæpenȝetæc* 'vote of consent expressed by touching weapons; district governed by such authority'..distinctively English in form, although..of Scandinavian introduction..*wæpen-* having been put instead of the Scand. *vápn*. **1922** O. Jespersen *Language* xi. 215 (*heading*) Translation-loans. *Ibid.*, Besides direct borrowings we have also indirect borrowings or 'translation loan-words,' words modelled more or less clearly on foreign ones, though consisting of native speech-material. **1958** Translation-loan [see *calque]. **1974** R. Quirk *Linguist & Eng. Lang.* vi. 101 We should add here the use of *bower* which is clearly a translation-loan. **1936** J. R. Kantor *Objective Psychol. of Gram.* v. 59 No doubt in the translation theory it is these social and cultural factors that have been unnecessarily converted into psychic guides of bodily action. **1978** C. Hookway in Hookway & Pettit *Action & Interpretation* 27 Given the underdetermination of translation theory by possible observations, we are invited to conclude that in the field of translation, there is no objective fact of the matter. **1980** *Times Lit. Suppl.* 12 Sept. 992/3 An academic researcher in translation-theory,..one of the very few people in the world..working in this field—had undertaken a questionnaire on the subject and now revealed some of its findings.

translational, *a.* Add: Hence **transla·tionally** *adv.*, (*a*) as regards language translation; (*b*) as regards, or by means of, translational motion.

1916 M. A. S. Riach *Air-Screws* i. 8 The depth of the cylinder will then represent the distance advanced through translationally by the point, and therefore by the whole air-screw, at each revolution. **1923** C. D. Broad *Sci. Thought* xi. 433 A rotating physical object which is translationally at rest. **1959** *Archivum Linguisticum* XI. 152 It is..no use trying..to measure range of content as the 'number of meanings'..of translationally equivalent words. **1978** J. Dunn in Hookway & Pettit *Action & Interpretation* 172 A translationally adequate science of human meanings must in principle be pragmatically accessible. **1981** C. H. Llewellyn-Smith in J. H. Mulvey *Nature of Matter* iii. 55 If momentum is conserved then the underlying laws must necessarily be translationally invariant.

translationese (trɑns,lēɪ:ʃənī·z, -nz-). [f. Translation + -ese.] The style of language supposed to be characteristic of (bad) translations; unidiomatic language in a translation; = *translatese, *translatorese.

1957 R. W. Zandvoort *Handbk. Eng. Gram.* IX. ii. 313 The suffix [*-ese*] may be added in a derogatory sense..to denote a mannered diction or style: ..*journalese, translationese*. **1961** *Times Lit. Suppl.* 17 Feb. p. iv/5 Translationese in a version from Hebrew is not always easy to detect. **1964** *Listener* 3 Dec. 911/2 The kind of unthinking 'translationese' which has so often..imparted to translated Russian literature a distinctive, somehow 'doughy' style. **1972** C. Day-Lewis in G. W. Knight *Jackson Knight* 11 When I had lapsed into translationese, he never failed to point it out. **1984** *Times* 21 Mar. 17/5 A useless mixture of mushy colloquialisms..and translationese.

translator. Add: **1. c.** *Computers.* A program that translates from one (esp. programming) language into another.

1958 A. J. Perlis et al. (*title*) Internal translator (IT) a compiler for the 650. **1959** [see *language *sb.* 1 d]. **1972** *Computer Jrnl.* XV. 229/1 General parsing algorithms..used in actual translators are characterised by the use of classes of grammars which are subsets of the class of context free grammars. **1981** Pohl & Shaw *Nature of Computation* vi. 193 Language processors, such as translators and interpreters, are called systems programs or systems software because they are normally part of the total computer system presented to a user.

4. b. A relay set or station which receives television signals and retransmits them without demodulating them.

1958 *New Scientist* 25 Sept. 898/2 Instead of receiving signals, demodulating them to visual or aural form and then re-transmitting, it just passes on the original signals. The BBC calls it a translator. **1960** *Practical Wireless* XXXVI. 330/2 Tentative plans are being considered for a considerable number of additional TV 'translator' stations of very low power. **1966** *Daily Tel.* 17 Oct. 9/5 TIE may have solved the problem with a television translator powered by sunlight... The translator set on a high point would receive and re-transmit signals to isolated communities.

translatorese (trɑns,lēɪ:tərī·z, -nz-). [f. Translator + -ese.] = *translationese. Cf. *translatese.

1915 *Morning Post* 15 Apr. 2/4 The worst 'journalese' is more English than schoolmasters' 'translatorese'. **1967** *Times Lit. Suppl.* 11 May 399/1 There is even a recognizable variant of pidgin English known as 'translatorese'. **1982** I. Hamilton *Robert Lowell* (1983) xvi. 292 Critics might more damagingly have quoted the limp translatorese that crops up throughout *Imitations*.

translocate, *v.* Delete *rare* and add: **a.** Esp. in reference to the transfer of wild animals. (Later examples.)

1971 *Nature* 6 Aug. 374/2 As long ago as 1928 the possibility of translocating rhino from Zululand to other parts of their former range was suggested by the naturalist Herbert Lang. **1980** *Cape Times* 19 Aug. 5/3 The present objective is to translocate as many black rhino as possible to as many of their original areas.

b. (Examples.)

1931 E. C. Miller *Plant Physiol.* xii. 696 It appears that the mineral nutrients absorbed by the roots on one side of a plant are, in a large measure, translocated to and used by the trunk, limbs, and leaves directly above them. **1959** *New Scientist* 12 Nov. 931/3 Substances may be translocated through the fungal threads and be accumulated into the fungal tissue which surrounds the root. **1976** *Sci. Amer.* Sept. 124/3 (Advt.), When the proper rate is applied to the weeds' well-grown foliage, the compound is translocated throughout the plant.

2. *Genetics.* To move (a portion of a chromosome) *to* a new position, esp. on a non-homologous chromosome.

1936 *Discovery* Sept. 269/1 In one of the new races the left ends were removed from both the second chromosomes and translocated to a third chromosome. **1949** Darlington & Mather *Elem. Genetics* v. 102 A piece may be taken out of a chromosome and inserted or translocated wherever else the same or another chromosome may be broken. **1975** *Nature* 3 Jan. 13/3 Genes translocated to positions close to heterochromatin seem in many species to become inactivated.

3. *intr.* To move, change location.

1977 *Lancet* 2 July 15/2 Others..had demonstrable abnormalities of receptor function, such as failure of the receptor hormone complex to translocate to the nucleus.

Hence **transloca·ted** *ppl. a.*; **transloca·table** *a.*, able to be translocated.

1930 *Jrnl. Genetics* XXII. 313 Offspring..will then have an excess (triple amount) of genes of the sort present in the proximal portion of the translocated piece. **1975** *Nature* 4 Dec. 384/1 Discrete translocatable elements (transposons) may constitute relatively common genetic units in bacteria. **1976** *Ann. Rev. Microbiol.* XXX. 517 One must also be aware of the prevalence of insertion sequences..and the newly discovered translocatable sequences..discovered in enteric microorganisms. **1980** R. P. Wagner et al. *Introd. Mod. Genetics* x. 289/2 The two normal chromosomes move to one pole while the two translocated chromosomes go to the other.

translocation. Add: **1. a.** Esp. in reference to the removal of wild animals. (Later examples.)

1962 *Oryx* VI. 215 By translocation is meant the transfer of wild animals from one area to another. **1969** J. Fisher et al. *Red Bk.* 123/2 Several countries have already used the drug and translocation method to introduce or re-introduce rhinos.

b. (Examples.)

1887 H. M. Ward tr. *J. von Sachs' Lect. Physiol. Plants* xxi. 347 For starch also is found at places in the tissue where it has neither been originally produced nor is employed, and thus in a condition of translocation towards the places where it is made use of. **1951** *Chambers's Jrnl.* Aug. 457/1 In the case of these perennial weeds with persistent root-stocks the chlorate acts by the method of translocation, working downward, cell by cell, from the sprayed tops to the lowest root-tip. **1976** Noggle & Fritz *Introd. Plant Physiol.* xii. 357 Translocation of sugars in the phloem is through living sieve elements.

2. *Genetics.* A transposition (sense *7), esp. to a position on a non-homologous chromosome; also, a portion of chromosome that is translocated.

1923 *Anat. Rec.* XXIV. 426 (*heading*) The translocation of a section of chromosome-II upon chromosome-III in Drosophila. **1924** [see *transposition 7]. **1937** C. D. Darlington *Rec. Adv. Cytol.* (ed. 2) vii. 265 They may also arise with translocation of a segment from one chromosome to another or from one arm of a chromosome to another. **1956** [see *aneuploidy]. **1962** *Lancet* 8 Dec. 1229/2 In all cases diagnosed confidently as having Down's syndrome we have found an excess of material of chromosome no. 21, either as an additional chromosome in the regular trisomic type or as a translocation. **1977** *Nature* 3 Nov. 10/3 It is thought that interaction between the two inverted sequences plays an important part in this translocation.

Hence **transloca·tional** *a.*

1930 *Jrnl. Genetics* XXII. 312 (*caption*) Translocational parent. **1965** *Jrnl. Cellular & Compar. Physiol.* LXV. 280/2 There was virtually no translocational movement in the partially rounded individuals.

translunar, *a.* (In Dict. s.v. Translunary *a.*) Add: **1.** (Examples.)

1927 W. B. Yeats in *Monthly Criterion* June 292 Being dead, we rise, Dream and so create Translunar Paradise. **1962** F. I. Ordway et al. *Basic Astronautics* v. 201 Mechta was instrumented to gather data on cosmic and solar radiations in cis- and translunar space.

2. *Astronautics.* Of or pertaining to space-flight or a trajectory from the earth or another planet towards the moon.

1965 *New Scientist* 1 July 12/2 Corrections will be made several times during the translunar phase..based on measurements made by Earth stations..and..by astronauts. **1969** *Daily Mail* 14 Jan. 1/2 Everybody had built up that burn, and of course it was critical. But so was translunar injection, so was re-entry. **1972** *Daily Tel.* 8 Dec. 1 The correction was the result of a manoeuvre which put Apollo on a trans-lunar trajectory 15 minutes earlier than planned.

trans,ma·rginal, *a.* [f. Trans- 4 + Marginal *a.*] Beyond the margin of normal consciousness; subliminal.

1902 W. James *Var. Relig. Exper.* xx. 511 The exploration of the transmarginal field has hardly yet been seriously undertaken. **1915** J. H. Tuckwell *Relig. & Reality* xi. 220 In the larger transmarginal or subliminal region of consciousness..nothing is ever really forgotten. **1968** W. E. Broen *Schizophrenia* viii. 201 The level of stimulation necessary to produce the transmarginal inhibition is said to be lower in individuals with weaker.. nervous systems. **1973** T. Pynchon *Gravity's Rainbow* I. 48 Send them over into one of the transmarginal phases, past borders of their waking selves.

trans,me·mbrane, *a.* [Trans- 3.] Existing or occurring across a cell membrane.

1961 *Amer. Jrnl. Physiol.* CC. 1252/1 Transmembrane resistance measurements were made by putting a rectangular voltage pulse in series with the microelectrodes. **1974** *Ann. N.Y. Acad. Sci.* CCXXVII. 98 (*heading*) The separation of electrons and protons during electron transfer: the distinction between membrane potentials and transmembrane gradients.

transmental, *a.* (Earlier example.)

1905–6 W. James in R. B. Perry *Thought & Char. W. James* (1935) II. lxxvii. 446 That an idea represents an 'object' may mean that it represents something..Transmental altogether, as when it is said to be altogether 'unknowable'.

transmethylation (trɑːns,meþilēɪ·ʃən, -nz-). *Chem.* [f. *trans- 10 + Methyl + -ation.] The transfer of a methyl group from a molecule of one compound to one of another.

1940 *Jrnl. Biol. Chem.* CXXXIV. 787 It was realized that direct proof of the transfer of the methyl group (transmethylation) was highly desirable. **1976** Smythies & Corbett *Psychiatry* v. 45 Schizophrenia might result from some disorder of transmethylation.

transmission. Add: **b.** *spec.* in *Radio* and *Television* (see *transmit *v.* 3 b); also, a series of electric signals or electromagnetic waves transmitted, a broadcast.

1907 *Rep. Brit. Assoc. Adv. Sci.* 731 To determine how many oscillations..take place in a certain wireless transmission. **1921** *Wireless World* IX. 52/2 In Surrey and Kent the transmissions were easily read. **1923** *Radio Times* 5 Oct. 15/2 Transmission from London of Dance Music by Savoy Orpheans. **1929** [see *television transmission* s.v. *television 3 a]. **1959** *Viewpoint* July 10 A few days before transmission the final camera script will be typed. **1962** A. Nisbett *Technique Sound Studio* v. 99 We have.. wide-range hi-fi, demanding not only high-quality transmissions but also a wide dynamic range. **1966** *Listener* 11 Aug. 204/2 *The Black and White Minstrel Show*.. has just completed eighty-five transmissions. **1977** *Rep. Comm. Future of Broadcasting* (Cmnd. 6753) ii. 8 The Government..has to regulate the strength of the transmissions to prevent interference with other stations.

e. transmission electron microscope, an electron microscope in which the electrons are detected after they pass through the specimen; *spec.* one in which all parts of the image are formed at the same time; so **transmission electron microscopy; transmission line,** a conductor or set of conductors designed to carry electricity (esp. on a large scale) or electromagnetic waves with minimum loss and distortion; also *transf.*; **transmission loss,** dissipation of electrical or acoustic power during its passage from one point to another; **transmission print** (see quot. 1960).

1969 *Jrnl. Ultrastructure Res.* XXVII. 403 The resolution of the scanning instrument is an order of magnitude less than that of present transmission electron microscopes. **1971** *Sci. Amer.* Apr. 26 The transmission electron microscope is analogous to a conventional light microscope. **1972** [see *scanning electron microscope* s.v. *scanning *vbl. sb.* 5]. **1968** *Jrnl. Electron Microscopy* XVII. 164/1 Experimental procedures to determine the direction and the sign of the Burgers vector..of a dislocation by means of transmission electron microscopy are described. **1974** *Encycl. Brit. Macropædia* XII. 137/2 Recently, scanning instruments have been used for transmission electron microscopy, with the advantage over the conventional transmission instrument that very low magnification..may be used. **1906** Transmission line [in Dict.]. **1922** Glazebrook *Dict. Appl. Physics* II. 862/1 The problem of finding the transmission loss due to a piece of apparatus inserted in..a telephone line is the problem of finding out how the current entering the receiving side is altered by the inserted or bridged apparatus. **1934** *Discovery* Dec. 348/1 The transmission loss in steel framing is only one decibel per 1,000 feet, a loss which is acoustically negligible. **1946** *Jrnl. Inst. Electr. Engineers* XCIII. IIIA. 33/1 It has become the common..practice to restrict the term 'wave guide' to devices employing a single hollow conductor, and to reserve the term 'transmission line' to devices employing two conductors. **1960** O. Skilbeck *ABC of Film & TV* 137 *Transmission print*, positive copy of a film intended for T.V. showings. **1970** *New Scientist* 15 Oct. (Suppl.) 13/1 The basic elements in microwave circuits are always some form of transmission lines. **1971** *Sci. Amer.* Sept. 235/1 (Advt.), This..guide is ..suited for use as an optical transmission line, carrying laser beams in any direction it is bent. **1975** D. G. Fink *Electronics Engineers' Handbk.* XVIII. 67 The transmission loss [of a radio circuit] is usually expressed in decibels. **1976** *Broadcast* 29 Nov. 18/2 Once 'shot' the film has to be 'processed'... A transmission print is achieved several generations later.

Hence **transmi·ssional** *a.*; **transmissionist** (earlier example).

1899 J. A. Thomson *Sci. Life* xvi. 226 The Lamarckians and Buffonians..believe in the transmission of acquired characters or modifications. They are sometimes, though not elegantly, called 'transmissionists'. **1930** *Observer* 4 May 26 Sometimes the music sounded ghostly. There were a few transmissional hiccoughs too.

transmissivity (trɑns,misi·vĭti, -nz-). *Physics.* [f. Transmissive *a.* + -ity.] The degree to which a medium allows electromagnetic radiation to pass through it.

1913 *Electr. World* LXII. 426/2 The connection between cloudiness and transmissivity. **1928** *Proc. R. Soc.* A. CXXII. 314 The transmissivity of the telescope could then be calculated. **1946** *Nature* 21 Sept. 422/2 The optical efficiency, that is, the sum of the reflectivity and transmissivity, expressed as a percentage of the incident light intensity, is practically 100 for the interference films, owing to the negligible absorption. **1971** *Sci. Amer.* Sept. 89/1 The flux of solar radiation received at the ground is highly variable..because of the variable transmissivity of the atmosphere and the changing degree of cloudiness.

transmissometer (trɑns,misɒ·mitəɹ, -nz-). [f. L. *transmiss-*, ppl. stem of *transmittere* to Transmit *v.* + -ometer.] An instrument for measuring the degree to which light is transmitted through a medium without absorption.

1955 *Sci. News Let.* 24 Sept. 197/1 Other equipment the Weather Bureau plans to purchase includes:... Ceilometers and transmissometers, instruments that tell cloud ceiling height and visibility. **1966** *McGraw-Hill Encycl. Sci. & Technol.* X. 172/1 Reflectometer. This instrument, also called a transmissometer, combines integrating spheres and barrier-layer cells... The transmittance can be measured by placing a sample of the material in the opening between the two spheres. **1975** *Nature* 3 Apr. 414/2 A volumetric concentration [in the sea] on the order of 0·01–1%, obtained from the acoustic data, enabled us to carry out a first-cut check of acoustic and transmissometer readings.

transmit, *v.* Add: **3. b.** To send out electric signals or electromagnetic waves corresponding to (an image, a programme, etc.).

1877 [see *facsimile 2 a]. **1923** *Radio Times* 28 Sept. 2/1 The first occasion..on which the voice of a public man had been transmitted simultaneously through six wireless stations hundreds of miles apart. **1946** *B.B.C. Year Bk.* 93 A fourth European programme network..allowed four different languages to be transmitted simultaneously. **1969** [see *facsimile 2 a]. **1979** *Financial Times* 18 Sept. (Telecommunications Suppl.) p. vii/2 A related system, generically called teletext, uses the same basic format, but the information is transmitted on spare capacity of the normal television broadcast channels.

II. 4. *Radio.* The infin. used, freq. *attrib.*, in the sense 'transmission'; so *on transmit*, of a transceiver: in the state of being able to transmit radio signals, with the transmitter switched on; **transmit button, switch,** the button or switch used to activate the transmitter; also *ellipt.*

1968 J. Sangster *Touchfeather* xiv. 146 The radio suddenly crackled into life... Marvin flipped the transmit switch. **1973** 'A. Hall' *Tango Briefing* xii. 155, I hit the transmit. *Tango.* She answered straight away. **1976** L. Henderson *Major Enquiry* xiv. 89 Keep your personal radio on transmit, don't try to talk but keep the channel open. **1976** K. Thackeray *Crownbird* vi. 122 He pressed the transmit button. 'Listen carefully.' **1980** *Basildon Recorder* 12 Sept. 1/4 Leaving his personal radio on 'transmit', so officers below could hear what was going on.

transmittance. Restrict *rare* to sense in Dict. and add: **2.** *Physics.* The ratio of the transmitted luminous flux to the incident luminous flux.

1919 *Technologic Papers U.S. Bureau of Standards* No. 119. 10 The transmittance T is defined as the fraction of radiant power transmitted by the first surface which is incident on the second surface. **1960** [see *reflectance]. **1980** *Nature* 29 May 313/1 We had to use a shorter path-length (970 m at Riverside and 750 m at Claremont) because of the poor optical transmittance of the polluted atmosphere at our observation sites.

Hence **transmi·ttancy,** the ratio of the transmittance of a solution to that of a similar body of solvent.

1925 *Jrnl. Optical Soc. Amer.* X. 177 (*table*) T_{sol}/T_{sov} = transmittancy. **1936** *Plant Physiol.* XI. 229 The transmittancies were obtained with a Bausch and Lomb visual spectrophotometer. **1959** H. Barnes *Apparatus & Methods Oceanogr.* i. 26 A calibration can then be set up relating transmittancy and concentration from which the concentrations of unknowns may be read when their transmittancies have been determined.

transmitter. Add: **b.** (Earlier and further examples.) Now *esp.* an apparatus for transmitting radio or television signals.

1844 *Brit. Patent* 10,257 2, I have an instrument which I denominate a transmitter... If the operator turns a handle fixed to the wheel any number of discharges..may be passed from London to Liverpool. **1859** [see *printer 2 a]. **1898** *Ludgate* Nov. 79 The Marconi Transmitter. **1912** *Chambers's Jrnl.* Jan. 60/2, I had got our receiver into 'tune' with the transmitter on board a steamer. **1913** E. C. Bentley *Trent's Last Case* ii. 18 Sir James looked at the telephone..and took up the receiver... Presently, as he listened, he..spoke quickly to Mr. Silver over the top of the transmitter. **1934** J. H. Reyner *Television* xvii. 191 It is conceivable that the introduction of some new scanning system may reduce the width of the band needed by a television transmitter providing plenty of detail. **1955** *Radio Times* 22 Apr. 3/2 Unless the listener lives very close to the nearest F.M. transmitter he will need to install a V.H.F. aerial. **1974** G. Markstein *Cooler* lv. 196 They've all been trained to rig up a wireless transmitter secretly, and send messages under the nose of the Gestapo. **1978** *Sci. Amer.* Mar. 58/2 A typical instrument consists of a handset, containing a transmitter and a receiver, that is connected to the base with an extendable cord.

d. *Physiol.* = *neurotransmitter.

1930 *Jrnl. Physiol.* LXX. 142 The only hint as to the chemical nature of the transmitter is provided by the physiological similarity to the action of adrenaline. **1937** J. C. Eccles in *Physiol. Rev.* XVII. 539 The actual transmission across the intercellular gap of the junctional region must depend upon one or more special factors—which will henceforth be called the synaptic or neuromuscular transmitter. **1954** Martin & Hynes *Clinical Endocrinol.* (ed. 2) vii. 168 It has now been shown..that *nor*adrenaline is the sympathetic transmitter-substance. **1970** *Nature* 5 Sept. 1006/2 Enzymes which synthesize transmitter. **1974** *Ibid.* 6 Sept. 14/3 L-Glutamate serves as the neuromuscular transmitter substance of arthropods, taking the place of acetylcholine in vertebrates in this respect.

2. Special Comb.: transmitter-receiver = *TRANSCEIVER.

1950 *Encycl. Radio & Television* 638/2 *Transmitter-receiver*, assembly comprising a sender and a receiver with provision for changing from send to receive, usually by means of a switching system. 1964 *Discovery* Oct. 23 (Advt.), The BBC 81 mobile VHF transmitter-receiver is so compact that the whole unit can be fitted under the dashboard of almost any vehicle. 1978 R. V. JONES *Most Secret War* xxxvi. 313 He had been parachuted back into France with six radio transmitter-receiver sets for distribution to his sub-agents.

transmitting, *vbl. sb.* and *ppl. a.* (In Dict. s.v. TRANSMIT *v.*) Add: *transmitting station*, a building or establishment from which radio or television signals are transmitted.

1923 *Radio Times* 28 Sept. 26/3 The 2LO transmitting station. 1977 *Whitaker's Almanack 1978* 759/1 There are two shortwave transmitting and receiving stations in Freetown.

trans,mo·rtal, *a.* Chiefly *poet.* [f. TRANS- 4 + MORTAL *a.*] Beyond what is mortal, immortal.

1932 BLUNDEN *Halfway House* 67 That much I saw without transmortal talk. *a* 1963 C. S. LEWIS *Poems* (1964) 97 Thou art Lord of the unbreathable transmortal air Where mortal thinking fails.

transmural, *a.* Add: 2. *Med.* Existing or occurring across the (entire) wall of an organ or blood vessel.

1951 BURTON & YAMADA in *Jrnl. Appl. Physiol.* IV. 330 It will be convenient..to call this difference of pressure the transmural pressure. 1968 *Amer. Jrnl. Obstetr. & Gynecol.* CII. 29/2 Hyalinization of vein walls became more patchy and occasionally transmural. 1977 *Lancet* 24/31 Dec. 1331/1 Transmural electrical stimulation of intestinal structures.

transmutation. Add: 3. a. Also in *Physics*, the (actual) change of one element into another, esp. by irradiation or bombardment (as opposed to spontaneous decay). Cf. *TRANSFORMATION 3 g.

1897 *Electrician* 10 Dec. 214/1 Theoretically, if the modern doctrine as to the ultimate constitution of matter be accepted, the transmutation of the elements is a scientific possibility. The fierce atomic bombardments inside a Crookes tube or an electric furnace would seem the most probable of known conditions whereby the operation might be carried on. 1915 K. TORNBERG tr. *Rasch's Electric Arc Phenomena* viii. 184 Since the electrons are ultra-atomic..the electric arc provides a means for the splitting up of matter, which perhaps makes the synthesis or transmutation of chemical elements not entirely beyond possibility. 1926 R. W. LAWSON tr. *Hevesy & Paneth's Man. Radioactivity* xxi. 147 Up to the present we only know of one method that permits us to resolve the nuclei of the atoms artificially, and thus to achieve..the transmutation of the elements. 1969 BENNISON & WRIGHT *Geol. Hist. Brit. Isles* i. 5 The rate of 'decay' or transmutation of radioactive minerals. 1974 *Physics Bull.* Dec. 585/2 The process of transmutation ('neutron burning'), as applied for several years now to radioactive waste, where certain isotopes are transmuted to isotopes with shorter half lives or even to stable ones.

4. transmutation glaze (example).

1904 [see *FLAMBÉ *a.* 1].

transmute, *v.* Add: 1. b. (Further examples.) Also in *Physics*, to change (one element or isotope) *into* another by irradiation or bombardment; to change (one sub-atomic particle) *into* another.

1897 *Electrician* 10 Dec. 214/1 Only the other day we were told how many centuries it would require to 'transmute' a milligram or two of one element into another, in a Crookes tube. 1926 R. W. LAWSON tr. *Hevesy & Paneth's Man. Radioactivity* xxi. 148 This method of transmuting elements is still far from being of any practical importance. 1956 A. H. COMPTON *Atomic Quest* i. 52 The naturally occurring atoms of U-238 and Th-232 can be transmuted into sources of atomic fuel. 1977 *Mod. Railways* Dec. 474/1 Vinyl chloride monomer..leaves by rail for Barry, where it is transmuted into PVC and heavy-duty plastic items. 1979 *McGraw-Hill Yearbk. Sci. & Technol.* 267/2 The pions can produce only a single-charge exchange by transmuting a proton into a neutron. 1981 *Sci. Amer.* Feb. 85/1 Cosmic rays transmute nitrogen 14 in the upper atmosphere into the radioactive isotope carbon 14. 1981 C. H. L. SMITH in J. H. Mulvey *Nature of Matter* iii. 72 In this case, quarks would very occasionally be transmuted into electrons.

c. Delete † *Obs. rare*$^{-1}$ and add later examples.

1962 D. G. COGAN in A. Pirie *Lens Metabolism Rel. Cataract* 294 The morgagnian globules..finally transmute into homogeneous milky fluid. 1970 A. ROMER *Radiochem. & Discovery of Isotopes* 127 The radiothorium, transmuting into thorium X.., would gradually disappear. 1978 *Nature* 29 June 707/1 In the weak interaction of radioactivity it has been known for many years that the neutrino turns into an electron or that an up quark transmutes into a down. 1983 *Penthouse* Sept. 169 Do you think capitalism is an eternal system, or will it transmute into something else?

trans,na·tional, *a.* (*sb.*) [f. TRANS- 3, 4 + NATIONAL *a.*] Extending or having interests extending beyond national bounds or frontiers; multinational. Also *ellipt.* as *sb.*, a transnational company.

1921 N. ANGELL *Fruits of Victory* ii. 63 Much of Europe lives by virtue of an international, or more correctly, a trans-national economy. 1941 J. S. HUXLEY *On living in Revolution* (1944) 144 The outstanding case of what we may call a transnational natural region—an industrial area cutting right across national boundaries—is the great concentration of industry in North-Western Europe. 1941 J. MACMURRAY *Challenge to Churches* 50 The Christian religion is the only possible force which can conceivably create the conditions of a transnational, multi-racial democratic polity. 1956 P. C. JESSUP *Transnational Law* i. 2, I shall use, instead of 'international law', the term 'transnational law' to include all law which regulates actions or events that transcend national frontiers. 1968 *Economist* 13 July 65/2 To these three [*sc.* the ethnocentric, the polycentric, and the geocentric types of multi-national company], Professor Galbraith has added a fourth type, the transnational company, with international stock ownership. 1973 *Reader's Digest* Apr. 167/1 Terrorism..is 'transnational' in scope—that is, there is a kind of global brotherhood of terrorists who share basic beliefs and techniques. 1977 *Irish Democrat* Mar. 3/5 The Brussels dictators would probably just tolerate a secession provided the transnationals continued to rule the roost economically. 1980 *Telegraph* (Brisbane) 5 Sept. 2/2 Now that multinational has become a dirty word,..multinationals are known as transnationals. 1983 *Church Times* 6 May 11/4 It will fall to your lot to assess the transnational corporations..and see whether in fact they do promote Third World development.

Hence trans,na·tionally *adv.*; trans,na·tionalism.

1921 N. ANGELL *Fruits of Victory* i. 14 Transport and credit, operating trans-nationally. *Ibid.* 190 The old individualist 'trans-nationalism'. 1973 *Listener* 20 Dec. 845 The reality of transnationalism, at any rate in the non-Communist world, since cannot be denied. 1976 *Times* 13 Feb. 15/6 Surely, a peaceful world needs respect for the rule of law, not only nationally but transnationally.

transnormal, *a.* (Earlier example.)

1853 MAX MÜLLER in C. Bunsen *Outl. Philos. Univ. Hist.* (1854) I. 282 Pott adds a fourth class, which he calls *transnormal* or *incorporative*, i.e. the polysynthetic American dialects.

trans,o·cean, *a.* [f. TRANS- 3 + OCEAN *sb.*] = TRANS-OCEANIC *a.* 2.

1901 *Daily Record & Mail* 31 Aug. 3 Besides a free trans-ocean passage, passes are issued from Quebec or Montreal to eastern points of the States. 1950 *Travel Topics* Dec. 11 Trans-ocean trips. 1978 J. A. MICHENER *Chesapeake* 102 Steed found it difficult to clarify the difficulties of transocean trading.

transoid (tra·nzoid). *Chem.* [f. TRANS- + -OID.] Designating a compound, group, or structure in which two unlike atoms or groups are situated in the *trans* position relative to one another.

1959 I. L. FINAR *Org. Chem.* (ed. 3) I. xvii. 387 (*caption*) Staggered (transoid). 1965 *Tetrahedron* XXI. 3121 The optical rotatory dispersion of seven other transoid dienes has been determined. 1970 *Nature* 3 Jan. 36/1 Acetylcholine..in a 'transoid' conformation. 1978 *Ibid.* 29 June 785/1 Bathorhodopsin is a common intermediate between pigments based on two different *cis* isomers and thus must itself contain a 'transoid' chromophore.

transom. Add: 3. e. (Examples.)

1907, 1927, 1955 [see *NOSE *sb.* 14 e].

g. (Earlier examples, in the context of a ship.)

1847 H. MELVILLE *Omoo* lxxxii. 373, I would find the ship's articles on the cabin transom. 1851 —— *Moby Dick* I. xvi. 91 Seated on the transom was what seemed to me a most uncommon and surprising figure.

transonic (trunsǫ·nik). *a.* Also trans-sonic. [f. TRANS- 3 + *SONIC *a.*, after *supersonic*, *ultrasonic*.] Pertaining to, involving, capable of, or designating speeds close to that of sound, at which some of the flow round a body is supersonic and some subsonic and there are characteristic changes in the behaviour of an aircraft.

1946 *Britannica Bk. of Year* (U.S.) 833/2 *Trans-sonic*, speeds ranging from 550 to 760 m.p.h. 1946 *Jrnl. R. Aeronaut. Soc.* L. 436/1, I propose to say a little about 'supersonic' aerodynamics first, and then to return to the so-called 'transonic' region, which presents the most difficult problem of all. 1948 'N. SHUTE' *No Highway* xii. 294 The forces on the structure [of an aircraft] were still very much a matter of guesswork in the trans-sonic range. 1966 *New Scientist* 20 Jan. 135/1 A critical phase is reached as the aircraft goes 'transonic'; that is, as it accelerates through the speed band from just below to just above the speed of sound. 1977 *Jrnl. R. Soc. Arts* CXXV. 347/1 Notable examples [of development] since World War II have been jet propulsion, transonic and supersonic aerodynamics, [etc.].

transparency. Add: 1. b. *Linguistics.* The state or quality of being transparent (sense *2 c). With reference to a phonological rule, opp. *OPACITY 3 c.

1971 [see *OPAQUE *a.* 3 c]. 1975 *Canad. Jrnl. Linguistics* XX. 175 Rules tend to be ordered so as to become maximally transparent, where *transparency* is the opposite of *opacity*. 1981 *Ibid.* XXVI. 73 Transparency of the base word is an important factor in determining speakers' choice of neologism.

c. Of reproduced sound: the state or quality of being transparent (sense *3).

1982 R. GREENFIELD et al. *New Penguin Stereo Record & Cassette Guide* p. ix/2, Disc reproduction continues to offer a marginally greater range of sound and a more subtle inner transparency of detail than the equivalent tape. 1983 *What Hi-Fi?* Dec. 102/3 This £119 speaker offers a high quality of finish and a transparency and accuracy of sound that can compete with models twice the price.

d. The state or quality of transmitting or allowing the passage of sound waves without distortion (see *TRANSPARENT *a.* 1 b).

1983 *What Hi-Fi?* Dec. 113/3 The most impressive [feature] was the sheer transparency of the design. 1984 *Gramophone* Mar. 1086/1 The effect (emphasized by the transparency of the CD medium) is of sitting in a small room, very close to the cello and with the lid of the piano wide open.

2. b. (Earlier examples.)

1785 J. WOODFORDE *Diary* 21 June (1926) II. 196 Went to Bums Gardens.. There was tolerable good Music, indifferent singing, some pretty transparencies and tolerable fire Works. 1802 *Sporting Mag.* XIX. 49/1 A transparency, in which a gardener at Hammersmith exhibited a Flower-pot, during the illuminations.

c. (Earlier and later examples.)

1866 *Brit. Jrnl. Photogr.* 1 June 264/1 Many of these lantern transparencies had been enlarged from spots of three originals. 1956 MRS. L. B. JOHNSON *White House Diary* 9 Feb. (1970) 29 The transparency I have..shown a real John Singleton Copley, of the best early American period. 1969 'E. LATHEN' *When in Greece* v. 49 He had brought slides..a veritable library of transparencies. 1973 'H. HOWARD' *Highway to Murder* iv. 46 Like transparencies flashed on a screen I saw pictures of long ago.

transparent, *a.* Add: 1. b. More widely, allowing the passage of any specified kind of radiation. Also, allowing the passage of sound waves without distortion.

1947 *Physical Rev.* LXXII. 1114/1 At sufficiently high energies the nucleus begins to be transparent to the bombarding particles. 1962 *Listener* 3 May 770/1 The needles..would be effective only in centimetre wavelengths—to which..the ionosphere is virtually transparent. 1962 A. NISBETT *Technique Sound Studio* ii. 41 Note that rack (as well as table surface) should be made of acoustically transparent mesh. *Ibid.* 269 The rack itself should be transparent to sound. 1974 *Nature* 13 Dec. 613/2 Their seismic data are said to 'reveal a distinct basement ridge' blanketed by acoustically transparent sediments which 'thin subtly near the crest of the buried ridge'. 1976 *Gramophone* Apr. 1682/2 The latter consists of a thin framework carrying two stand-off ribs to give a three-faced appearance when the black, acoustically transparent, stockinette cloth is stretched over it. 1978 *Nature* 21 Sept. 175/1 A container of superfluid ^{4}He whose walls are transparent to neutrons with velocity of 450 ms^{-1}. 1980 *Sci. Amer.* July 56/1 The interstellar medium is highly transparent to gamma rays. 1982 *Nature* 9 Sept. 122/1 The ocean is transparent to sound.

2. c. *Linguistics.* Obvious in structure or meaning; that can be extrapolated from surface structure; of a phonological rule: that can be extrapolated from every occurrence of the phenomenon, in which every context implies the rule (opp. *OPAQUE *a.* 3 c).

1974 [see *OPAQUE *a.* 3 c]. 1975 [see *TRANSPARENCY 1 b]. 1977 *Language* LIII. 820 A compound may be highly transparent semantically when it is coined. 1980 *Amer. Speech* LV. 38 The hydronymic element *-kill* and names formed from it are semantically opaque, whereas *creek* and *brook* are transparent. 1981 *Canad. Jrnl. Linguistics* XXVI. 73 They..show no preferences either way if both word boundary and formative boundary derivations are transparent.

3. Of reproduced sound: clear, without tonal distortion, not blurred, with each element distinct.

1950 *Audio Engin.* Sept. 33/2 The sound is then said to be *transparent*..; no *acoustic fog veils* the ensemble, and each instrument stands out with *clarity*.

transpecific, var. *TRANSSPECIFIC *a.*

transpeptidation (tra:nspeptidē̆ɪ·ʃən, -nz-). *Biochem.* [f. *TRANS- 10 + *PEPTIDE + -ATION.] Any reaction in which a peptide bond is broken and the free carboxyl (or amino) group joined to the amino (or carboxyl) group of another molecule so as to form a new peptide bond.

1950 J. S. FRUTON in *Yale Jrnl. Biol. & Med.* XXII. 264 Reactions of this type may be termed 'transpeptidation' or 'transamidation' processes. 1975 *Nature* 10 Apr. 482/2 Various cell-free transpeptidation systems are now known to be sensitive to penicillin.

So trans,pe·ptidase [*-ASE], any enzyme that catalyses transpeptidation.

1952 C. S. HANES et al. in *Biochem. Jrnl.* LI. 25/1 The enzyme will be referred to as γ-glutamyl transpeptidase. 1981 *Sci. Amer.* June 68/1 The vulnerable enzymes are either transpeptidases (which cross-link one peptide chain of new peptidoglycan to another by displacing a terminal D-alanine) or D-carboxypeptidases (which remove a terminal D-alanine by hydrolysis without cross-linking taking place).

trans,pe·rsonal, *a.* [f. TRANS- 4 + PERSONAL *a.*] That transcends the personal, trans-individual; *spec.* designating a form of psychology or psychotherapy which seeks to combine elements from many esoteric and religious traditions with modern ideas and techniques.

1905–6 W. JAMES in R. B. Perry *Thought & Char. W. James* (1935) II. 445 That an idea represents an 'object' may mean that it represents something either—1. *Trans-personal*—as when my object is also your object, [etc.]. 1955 *Bull. Atomic Sci.* Apr. 109/1 Science provides the model of a free society of reasonable men co-ordinating themselves voluntarily in the light of a transpersonal standard. 1968 *Jrnl. Humanistic Psychol.* VIII. 77 Transpersonal (or Fourth Force) Psychology is the title given to an emerging force in the psychology field by a group of psychologists and professional men and women..who are interested in those *ultimate* human capacities and potentialities [etc.]. 1973 *Science* 16 June 1103/1 Experiences of ecstasy, mystical union,..and transpersonal knowledge..are simply not treated adequately in conventional scientific approaches. 1980 R. HERINK *Psychotherapy Handbk.* 684 Transpersonal Psychotherapy can be said to have evolved..as the inner or esoteric teachings of all the great spiritual traditions.

transphasor (transfē̆ɪ·zəɪ, -nz-). *Electronics.* [f. *TRANS(ISTOR *sb.* with PHAS(E inserted.] A semiconductor device in which one light beam can be modulated by another.

1979 MILLER & SMITH in *Optics Communications* XXVIII. 101/2 We report the first realization of an optically bistable device in a semiconductor crystal as well as observation of differential gain both in one beam and, via the modulation of the transmission of the laser beam by a second, in a two beam system. This latter device is analogous to the three terminal transistor and, operating by transferred phase thickness, we term it a 'transphasor'. 1984 *Sunday Times* 18 Mar. 80/1 It is thought that a small electrical current can be used to switch an electrical voltage between a high and a low value..The transphasor does exactly the same thing with laser beams.

trans,pheno·menal, *a.* *Philos.* [f. TRANS- 4 + PHENOMENAL *a.* (*sb.*).] That transcends the phenomenal; that goes beyond the phenomenal. Hence trans,phenomena·lity.

1897 W. M. URBAN *Hist. Princ. Sufficient Reason* vi. 87 The validity of this logical postulate is a trans-phenomenal usage. 1904 [see *GENERALIZE *v.* 1]. 1954 W. DESAN *Tragic Finale* ix. 178 Sartre claims that the existence of the external transphenomenal being is the condition of my phenomenal perception. 1963 H. E. BARNES tr. *Sartre's Being & Nothingness* p. l, The phenomenon of being..is an appeal to being; it requires as phenomenon, a foundation which is transphenomenal. The phenomenon of being requires the transphenomenality of being.

trans,pho:sphoryla·tion. *Biochem.* [f. *TRANS- 10 + *PHOSPHORYLATION.] The transfer of a phosphate group from a molecule of one compound to one of another.

1943 *Jrnl. Biol. Chem.* CXLVIII. 119 Transphosphorylation from adenosine diphosphate to glucose also results in the liberation of 1 acid equivalent. 1970 *McGraw-Hill Yearbk. Sci. & Technol.* 62/2 A transphosphorylation reaction..transfers the phosphate from the hydroxyl of the adjacent nucleotide to the hydroxyl of the pyrimidine sugar, giving a cyclic phosphate.

transpire, *v.* Add: 4. Also *impers.*

1903 G. B. SHAW *Man & Superman* 209 We had hardly recovered from the fruitless irritation of this discovery when it transpired that the officers' mess of our most select regiment included a flogging club presided over by the senior subaltern. 1922 W. GERHARDI *Futility* iii. ix. 13 It transpired that four regiments composing the division had gone over to the enemy. 1966 D. J. ENRIGHT *Conspirators & Poets* i. 16 But then, to our surprise, it transpires that he doesn't think much of our critics, either. 1982 T. HAMILTON *Robert Lowell* (1983) x. 141 Yaddo, it transpired, had been under FBI surveillance for some time.

b. Delete sentence beginning 'App. began in U.S. about 1800' and quot. 1802 and add earlier example.

1775 A. ADAMS *Let.* 31 July in J. & A. Adams *Familiar Lett. Revolution* (1876) 91 There is nothing new transpired since I wrote you last.

transplant, *sb.* Add: 1. a. *spec.* in *Surg.*, etc., an organ, tissue, etc., which has been transplanted into another person or animal.

1913 *Arch. f. Entwicklungsmech. d. Organismen* XXXVII. 254 The homeoplastic transplants in guinea pigs 5 and 8 were removed for microscopic examination. 1952, 1963 [see *heart transplant* s.v. *HEART *sb.* 36]. 1977 J. GILLIS *Killers of Sight* ii. 21/2 Look a little closer. This hair hat of mine is a transplant.. I'm a lot older than Trevor.

b. One who is not native to his place of permanent residence. *U.S.*

1900 M. BEAGLE *These Ruins are Inhabited* (1961) v. 73 If she's a good English wife, she doesn't tell him. If she's an American transplant, she does. 1973 T. TOBIN *Let. G. Ade* 2 The bustling city [of Chicago], which was comprised of numerous rural transplants too busy with the business of living to establish traditions. 1979 *Tucson* (Arizona) *Citizen* 20 Sept. 1B/3 Traffic and the heat are two things the transplants mentioned.

3. *Surg.* An operation in which an organ, tissue, etc., is transplanted from one person or animal to another.

1951 *Sun* (Baltimore) 14 Mar. 34/1 He decided to try a transplant [of a tooth]. 1971 *Daily Tel.* 3 May 1/4 Prof. Christian Barnard..is standing by to carry out his first transplant for two years. 1982 *Sunday Times* 2 Sept. 16/2 It is a type of pneumonia usually found only where there are known causes for a breakdown in the immune system, as in cancer chemotherapy or transplants.

4. *attrib.*

1963 *Guardian* 22 Mar. 1/3 Surgeons at St Bartholomew's Hospital, London, this week carried out the hospital's first kidney transplant operation. 1967 *New Scientist* 7 Dec. 584/3 This [*sc.* the immune reaction] is the problem that bedevils the whole of transplant surgery. 1968 *Observer* 7 Jan. 1/1 After that we can think about further selection of transplant patients. 1977 P. B. & J. S. MEDAWAR *Life Science* i. 8 The anxious question of whether or when a potential transplant donor..can be regarded as dead.

transplantable, *a.*, transplantability. (Later examples of each in *Med.*)

1913 *Jrnl. Exper. Med.* XVIII. 482 (*heading*) Transplantability of tissues to the embryo of foreign species. *Ibid.*, Mall has been added to our knowledge of..tissue grafting in warm blooded animals through the study of the transplantable tumors. 1973 *Nature* 18 May 163/1 These tumours are transplantable and have been transplanted for up to six consecutive passages. 1977 *Ibid.* 6 Jan. 56/1 Increased malignancy of lymphoma cells [was] expressed by serial transplantability and capacity for widespread metastases.

transplanter. Add: 1. b. *spec.* in *Surg.*, a surgeon who carries out transplant operations.

1970 *Daily Tel.* 15 June 13/2 For Prof. Roy Calne,..pioneer liver transplanter, there is only one problem: not enough organs are being transplanted. 1971 *Nature* 24 Dec. 449/3 To the transplanter it may well be that if a..antigen can be elicited..certain typing mismatches may be of no consequence. 1977 *Time* 7 Mar. 43 (*caption*) Kidney transplanter Stubbenbeck with patient Jose Serrano after surgery. 1984 A. SMITH *Mind* v. xv. 299 The needs of the transplanters formed a prime motive for redefining death.

transplendency. (Later example.)

1915 C. MACKENZIE *Guy & Pauline* 163 The clouds..suddenly melted in a wild transplendency of blue.

trans,pluto·nium, *a.* *Physics.* [f. *TRANS- 11 + *PLUTONIUM 2 b.] Of a chemical element: having a higher atomic number than plutonium (i.e. 95 or over).

1955 *Physical Rev.* XCIX. 1048/2 It [*sc.* the alpha radiation] was due to a transplutonium element. 1977 *Sci. Amer.* Mar. 29/2 In the absence of fast-neutron reactors..it would be impossible to completely burn the plutonium and its transplutonium derivatives produced by the slow-neutron plants.

transponder (transpǫ·ndəɪ, -nz-). [f. TRANS(MIT *v.* + RES)POND *v.* + -ER^{1}.] = *RESPONDER 2 b.

1945 *Army & Navy Jrnl.* 18 Aug. 1534/4 *Transponder*. The unit in the IFF system which receives the challenge and automatically transmits the reply. 1954 *Electronic Engin.* XVII. 735/3 It became clear that such a 'transponder'..could be set up at any convenient place and would act as a beacon. 1957 [see *RESPONDER 2 b]. 1967 R. J. SERLING *President's Plane is Missing* (1968) iii. 55 Henderson pushed the transponder button... The new identifying transponders were hooked to the altimeter. 1974 *Nature* 16 Aug. 558/2 The position of the Archimède was known within 10–200 m, depending on her distance from the nearest of the acoustic transponders. 1983 *Listener* 8 Sept. 6/1 The transponder code ought to have immediately alerted ground controllers.

transpontine, *a.* Add: [f. L. *pontus* sea.] That is across the sea; on or from the other side of the ocean, *spec.* the Atlantic, i.e. North American.

1891 R. L. STEVENSON *Let.* Oct. (1923) XXII. 414 The last four chapters of *The Wrecker!*..Ours is such rude, transpontine business. 1920 *Times Lit. Suppl.* 15 Apr. 232/2 She [*sc.* an American writer] has investigated her subject with typical transpontine enthusiasm. 1922 JOYCE *Ulysses* 416 Thou sawest thy America, thy lifetask, and didst charge to cover like the transpontine bison.

transport, *sb.* Add: 3. (Later examples with *a* and *pl.*)

1920 D. H. LAWRENCE *Women in Love* i. 12 She experienced a keen paroxysm, a transport, as if she had made some incredible discovery. 1953 E. JONES *Sigmund Freud* I. xi. 275 A female patient suddenly flinging her arms round his neck in a transport of affection. 1978 A. S. BYATT *Virgin in Garden* xviii. 285 Almost she expected him to rouse himself and roar out transports of self-reproach or self-enforcing ecstasy, either of which would have embarrassed her profoundly. 1981 A. SCHLEE *Rhine Journey* ix. 111 She had gone on to describe..the gentle transports she would expect her sensitive nature to endure, the blameless pious raptures.

4. Cf. *tape transport* s.v. *TAPE *sb.*^{1} 4. Also = *transport plane* below.

1940 *Times* (Weekly ed.) 7 Aug. 23 German tactics in attacking aerodromes, from the first attempt to divert the fighters to the arrival of low bombers and the landing of parachutists and troop transports are now well known. 1943 C. MILBURN *Diary* 11 Apr. (1979) 174 Forty transports going to supply the Axis were brought down. 1970 *Nature* 15 Aug. 655/2 The project to build a supersonic transport has run into renewed complaints from the environmentalists. 1974 C. RYAN *Bridge too Far* ii. i. 79 Transports to carry paratroops and tow-planes to pull the gliders must be diverted from their normal task of supplying the advancing armies and grounded.

6. *transport officer*, *-wagon* (examples); transport café, a roadside café for lorry-drivers; Transport House, the name of the headquarters of the Labour Party, used as a synonym for the Labour Party leadership; transport number *Physical Chem.*, the proportion of a current flowing through a particular electrolytic solution which can be attributed to the movement of any given ion species; = *transference number* s.v. *TRANSFERENCE 3; transport plane *Mil.*, an aircraft used for transporting troops, supplies, etc.

1938 'J. CURTIS' *They drive by Night* i. 15 At St. Albans a lorry-driver was sitting hunched up at the counter of a transport café. 1959 *Manch. Guardian* 29 July 14/3 On the London-Birmingham motorway..there will be..two transport cafés, but only one restaurant. 1978 C. A. BEENY *Gentleman of Road* 131 I was a welcome stay at the transport café..on the strength of two cups of tea. 1937 *New Statesman & Nation* 11 Sept. 364/2 This new orientation fits in exactly with that of the political wing of Transport House. 1958 *Spectator* 31 Jan. 124/1 On the Labour side, it does not appear as though Transport House is ever thinking of coming to grips with reality. 1974 R. CROSSMAN *Diaries* (1975) I. 72 If we as a Cabinet have neglected our relationship with the Parliamentary Party we have equally neglected our relations with Transport House and the Party outside. 1897 M. M. P. MUIR tr. *Löpke's Elem. Electro-Chem.* i. iii. 43 A Cu ion will pass over two of six spaces, and a SO_4 ion will pass over four in the same time. The quotients $\frac{2}{6}$=0.33 and $\frac{4}{6}$=0.66 are called by Hittorf the transport-numbers (die Überführungszahlen) for the kation Cu and the anion SO_4, respectively. 1978 P. W. ATKINS *Physical Chem.* xv. 843 The following are brief summaries of the three methods used to measure transport numbers of ions and, through them, individual ion conductivities and mobilities. 1917 W. OWEN *Let.* 4 Feb. (1967) 431 It doesn't necessarily mean a job as Transport Officer straight away. 1945 R. J. OAKES in *Coast to Coast 1944* 99 He had six wounded men to load into the transport plane. 1967 M. SOKOLSKI tr. *Mehr's Victory Factor* xv. 397 The jeep..led us straight into the maw of a transport plane that swallowed the truck. 1866 [J. LEYLAND *Adventures Far Interior S. Afr.* ii.] I travelled by a Dutch transport waggon. 1936 in N. Rouillard *Matabele Thompson* iii. 46, I found myself at no great distance from a transport waggon, laden with goods intended for a trader in the north.

transportable, *a.* Add: 1. a. *spec.* of a computer (see quots.).

1982 *Byte* Nov. 6 Portable..refers to a small, transportable computer on which you can touch-type. 1983 *Times* 31 May 20/4 The new breed of portable computers makes..look immovable by comparison... Once known as portable computers, these machines are now dubbed transportable computers, to distinguish them from smaller machines that are more easily moved. 1985 *Pract. Computing* Jan. 70/1 Although called the Portable, the machine is..what is more normally termed 'transportable'. That is, it is a mains-powered unit which..is too heavy to be carried around all the time.

b. Of software: that can be used on more than one kind of machine.

1979 *Rep. Nat. Res. Council Canada* No. 17659 (ERIC Rep. No. 160126) 4 Facilities for producing 'transportable' computer based course materials. 1977 *Proc. Internat. Symp. Computer Aided Seismic Analysis* 121/1 PAL, through its structured logic and easily transportable software, provides a framework which will readily accept modification and expansion. 1984 *Mini-Micro Systems* Feb. 71/3 DEC is encouraging software producers in Europe to make software transportable between countries.

B. *sb.* A transportable television set, computer, etc. *spec.* one which is heavier than a 'portable' appliance.

1959 *Daily Tel.* 27 Aug. 11/3 The transportable [*sc.* a television] is something comparatively new. It has a carrying handle, yet is too heavy to be classed as a lightweight portable. 1971 *Radio Times* 25 Nov. 35 (Advt.), The Decavision Executive 17" transportable—a black-and-white TV that gives big, bold performance wherever there's a mains point. 1983 *Observer* 19 June 21/1 Although perception of portability differs radically, three categories can be discerned... These are the handhelds, the true portables, and the transportables. 1983 *Austral. Microcomputer Mag.* Nov. 76/1 Despite the provision of a high level of computing power in such recent hand-held portables.., there always should be room for larger transportables.

transportation. Add: 3. (Earlier example.)

1853 J. L. MCCONNEL *Western Characters* 163 He furnished his own 'transportation', and selected his own encampment.

5. *transportation agent* (example), *-wagon*.

1819 J. A. QUITMAN *Let.* in J. F. H. Claiborne *Life J. A. Quitman* (1860) I. 36, I went to the agent of a train of transportation-wagons. 1825 in T. L. McKenney *Memoirs* (1846) I. 299, I was appointed transportation agent for the United States at St. Louis. 1866 'MARK TWAIN' *Lett. from Hawaii* (1967) 274 Her transportation wagons will be the freight cars of the Pacific Railroad.

transporter. Add: 2. Also, a vehicle used to transport other vehicles or large pieces of machinery, etc., by road.

a 1944 K. DOUGLAS *Alamein to Zem Zem* (1946) xi. 68 The tanks were to move by transporter. 1976 L. DEIGHTON *Twinkle, twinkle, Little Spy* xxi. 206 In the Sahara there were only Peugeots, and Landrovers, and the smart little cars that came in by transporter.

transposable, *a.* Add: (Earlier example.) Also transposible.

1835 J. B. ROBERTSON tr. *Schlegel's Philos. of Hist.* I. iii. 95 Instead of the regular art of printing with transposeable letters..this people [*sc.* the Chinese] make use of a species of lithography, which, to all essential purposes is the same. 1847 A. DE MORGAN *Formal Logic* vi. 122 The whole term is always transposible.

transpose, *sb.* Restrict † *Obs. rare* to sense in Dict. and add: 2. *Math.* A matrix got from a given matrix by interchanging each row and the corresponding column.

1937 [see *SIMILAR *a.* 3 c]. 1939 A. C. AITKEN *Determinants & Matrices* i. 15 The resulting matrix is called the transpose of A and is denoted by A'. (In the less recent literature the word *conjugate* is used.) 1961 [see *HERMITIAN *a.*]. 1978 *Nature* 13 Apr. 605/2 The transpose of the eigenvector matrix of climatic data.

transposed, *ppl. a.* (in Dict. s.v. TRANSPOSE *v.*). Add: *spec.* in *Math.*, applied to the transpose of a given matrix.

1858 *Phil. Trans. R. Soc.* CXLVIII. 32 A matrix compounded with the transposed matrix gives rise to a symmetrical matrix. 1907 M. BÔCHER *Introd. Higher Algebra* ii. 21 Two square matrices..of which either is obtained from the other by interchanging rows and columns are called conjugate to each other. [*Note*] Sometimes also transposed. 1972 M. KLINE *Math. Thought* xxxiii. 807 The transverse († transposed or conjugate) matrix is defined as the one in which rows and columns are interchanged.

transposition. Add: 2. a. (Examples in *Math.*)

1965 J. J. ROTMAN *Theory of Groups* iii. 35 Of all the permutations, surely the transposition, which merely interchanges two points, is the simplest. 1972 A. G. HOWSON *Handbk. Terms Algebra & Anal.* x. 51 Any permutation can be expressed as a product of transpositions. 1981 *Sci. Amer.* Mar. 19/2 If you are willing to let cycles share members, however, any cycle can be further broken up into 2-cycles (called transpositions, or sometimes swaps).

b. *Math.* The interchange of each row of a matrix with the corresponding column.

1858 A. CAYLEY in *Phil. Trans. R. Soc.* CXLVIII. 31 Two matrices such as $\begin{vmatrix} a, b \\ c, d \end{vmatrix}$, $\begin{vmatrix} a, c \\ b, d \end{vmatrix}$, are said to be formed one from the other by transposition. 1955 *Geophysics* XX. 300 The transposition of the ξ_{ij}' matrix was necessary to reveal the trend in the u..direction.

6. *Electr.* An alteration of the relative positions of power lines or telephone lines at intervals along their length, in order to minimize effects of mutual inductance and capacitance.

1911 A. B. SMITH *Mod. Amer. Telephony* xxii. 616 Transpositions are unnecessary when there is but one circuit on a pole line. 1959 K. HENNEY *Radio Engin. Handbk.* (ed. 5) xxviii. 26 In the case of pairs used for high frequencies, the locations of the transpositions must be quite precise. 1975 R. L. FREEMAN *Telecommunication Transmission Handbk.* iv. 164 The dipole type of LP antenna is fed by a balanced transmission line with 180° transpositions at alternate dipole elements.

7. *Genetics.* The transfer of a chromosomal segment to a new position on the same or another chromosome.

1924 *Hereditas* V. 174 In his case of transposition (or translocation) a piece of one of the second chromosomes has been fastened to one of the third chromosomes. 1954 B. P. KAUFMANN in A. Hollaender *Radiation Biol.* I. II. ix. 650 The viable types of intrachromosomal rearrangement include inversions and transpositions. 1978 *Nature* 20 July 211/3 The subsequent transposition of tetracycline resistance from this engineered plasmid must therefore be due to the transposition of the toxin gene.

transpositively, *adv.* (Example.)

1946 *Mind* LV. 323 If we accept the premises..that *fa* and ∼*fa* severally imply (∃*x*)..then we must accept the consequences..that they do not have contradictories; or, transpositively, if we want our universal propositions to be non-existential..we must repudiate the proposition that *fa* implies (∃*x*)*fx*.

transposon (trunspǭ·zǫn, -nz-). *Genetics.* [f. TRANSPOS(ITION + *-ON^{1}.] A chromosomal segment that can undergo transposition (sense *7); *spec.* a segment of bacterial DNA that can be translocated *en bloc* between chromosomal, phage, and plasmid DNA in the absence of a complementary sequence in the host DNA.

1974 HEDGES & JACOB in *Molecular & Gen. Genetics* CXXXII. 38 We designate DNA sequences with transposition potential as transposons (units of transposition). 1978 *Nature* 11 May 171/3 The determinants of resistance to at least seven different antibiotics form part of independent structures several kilobase-pairs long, called 'transposons', which can jump between the genomes of bacteria, plasmids and phages of widely different base compositions. 1980 *Sci. Amer.* Sept. 87/3 Some transposons carry genes for traits such as drug resistance that are clearly advantageous to a bacterial host under some circumstances. 1982 *Nature* 21 Oct. 676/1 Transposons, familiar enough in bacteria.., have been identified in.. eukaryotic genomes principally because their DNA sequences can crop up at different places.

trans,provi·ncial, *a.* [f. TRANS- 3 + PROVINCIAL *a.*] That crosses a province.

1916 R. K. WOOD *Tourist's N.W.* 316 A trans-provincial highway is building through southern British Columbia from the mountains..to Hope in the Fraser River Valley. 1952 D. F. PUTNAM *Canadian Regions* 338/1 They have not..been able to get the Ontario Government to construct adequate transprovincial highways.

transpulmonary, *a.* Add: *transpulmonary pressure*, the difference between the pressure in the lungs and that in the intrapleural cavity.

1957 *Jrnl. Appl. Physiol.* X. 191/1 Measurements of the mechanical properties of the lungs depend on the separation of transpulmonary pressure into separate elastic and flow-resistive components. 1977 *Lancet* 3 Sept. 510/1 When airway collapse does occur, transpulmonary pressure rapidly increases with only a small resultant increase in inspiratory gasflow.

transputer (trunspiū·təɪ, -nz-). *Electronics.* [f. *TRANS(ISTOR *sb.* + COM)PUTER.] A chip that incorporates all the functions of a microprocessor, including memory.

1978 I. M. BARRON in D. Aspinall *Microprocessor* vi. 343 The word 'transputer' has been coined to describe the computer on a chip. The word is derived from..'computer' and 'transistor'. 1983 *Times* 2 Nov. 3 Alice [*sc.* Applicative Language Idealized Computer Engine] will incorporate 64 transputers running in parallel, each in itself a 'computer on a chip'. 1984 *Times* 17 Feb. 10/4 The 'transputer' which Inmos, Britain's state-backed chip company, hopes to begin manufacturing in about a year's time,..includes processor, memory and communications on a conventional-sized chip.

trans,ra·cial, *a.* [f. TRANS- 3 + RACIAL *a.*] Across or crossing racial boundaries.

1971 *Time* 16 Aug. 46 The Merediths' decision is part of a growing phenomenon known in sociologists' jargon as transracial adoption. 1976 *Indian Jrnl. Social Work* XXXVII. 152/1 The trend..is now moving towards transracial adoptions.

transreceiver (tra:ns,rīsī·vəɪ, -nz-). [f. TRANS(MITTER + RECEIVER^{1}.] = *TRANSCEIVER.

1942 *Electronic Engin.* XV. 239/2 Inside the Transreceiver there is a small moving coil mechanism. 1956 'N. SHUTE' *Beyond Black Stump* ii. 46 David had a radio transreceiver for use with the Flying Doctor service.

trans,se·xual, *a.* and *sb.* Also (A. 1, B.) transexual, (esp. A. 2) trans-sexual. [f. TRANS- 3 + SEXUAL *a.*] A. *adj.* 1. Of or pertaining to transsexualism; having physical characteristics of one sex and psychological characteristics of the other.

1957 *Amer. Jrnl. Psychotherapy* XI. 85 Other kinds of secondary experiences might give rise to transsexual tendencies. 1963 A. HERON *Towards Quaker View of Sex* 68 In men..the trans-sexual urge occasionally reaches the point of completely assuming the female role. 1969 *Nature* 2 Aug. 448/2 Cases of transsexual children brought up in unusually close contact with their mothers, who were the dominant members of families in which the fathers were ineffectual characters. 1970 [see *sex change* s.v. *SEX *sb.* 5]. 1977 *Proc. R. Soc. Med.* LXX. 792/1 He may sustain sexual relations..by transexual fantasy.

2. Of or pertaining to both sexes. Also, intersexual.

1977 G. MELLY *Rum, Bum & Concertina* viii. 105, I came into her... We made love.., but crossing the transsexual barrier didn't convert me overnight. I continued for some years to prefer boys. 1978 *Logophile* VIII. 8/1 The word 'clerk' *sounds* acceptably undiscriminatory and trans-sexual.

B. *sb.* A transsexual person. Also, one whose sex has been changed by surgery.

1957 *Amer. Jrnl. Psychotherapy* XI. 84 Marriage, motherhood, and a husband are a woman's life and he wants that as a fulfillment of his femininity. These persons form the group aptly termed 'transsexuals'. 1966 *New Scientist* 8 Dec. 577/1 Transsexuals..are a..group of individuals 'who have a burning and overwhelming desire to be of the sex opposite to what they are anatomically'. 1971 *Daily Tel.* (Colour Suppl.) 10 Dec. 21/3 Adult transexuals are so resistant that psychological treatment and even aversion therapy is virtually hopeless so far as they are concerned. 1976 *Washington Post* 26 Jan. A2/8 Britain's state-run health service has started a night school for transsexuals where men who have changed their sex can learn to face the world as women. 1977 *Times* 18 Aug. 6/2 Dr Renée Richards, aged 42, a transsexual who was a man until 1975.

trans,se·xualism. Also trans-sexualism, transexualism. [f. TRANS- 3 + SEXUALISM.] The state or condition of being transsexual (see *TRANSSEXUAL *a.* 1), manifested in an overwhelming desire to belong to the opposite sex.

1953 H. BENJAMIN in *Internat. Jrnl. Sexology* Aug. 12/1 Transvestitism..is the desire of a certain group of men to dress as women, or of women to dress as men. It can be..overwhelming, even to the point of wanting to belong to the other sex and correct nature's anatomical 'error'. For such cases the term Transsexualism seems appropriate. 1954 *Amer. Jrnl. Psychotherapy* VIII. 220 Transsexualism..denotes the intense and often obsessive desire to change the entire sexual status including the anatomical structure. 1963 K. WALKER in G. Turtle *Over Sex Border* 11 Transexualism was only defined with any clarity in 1954. 1969 *New Scientist* 24 July 183/1 The very difficult clinical problem of trans-sexualism. 1974 *Adolescence* Spring 71 The increasing rate of transsexualism. 1977 E. J. TRIMMER et al. *Visual Dict. Sex* (1978) iii. 46 It is impossible for an ordinary person to be propelled into the condition of transsexualism, as it has nothing to do with free will or voluntary choice.

So trans,se·xualist *sb.* and *a.* = *TRANSSEXUAL *sb.*, *a.* 1.

1954 *Amer. Jrnl. Psychotherapy* VIII. 220 The transsexualist is always a transvestite but not vice-versa.... The transsexualist only lives for the day when his hated sex organs can be removed. 1966 *Punch* 28 Dec. 968/1 The transexualist..told us, in a pleasant baritone, how happily married 'she' was—after surgery. 1976 SMYTHIES & CORBETT *Psychiatry* xi. 211 Transexualists believe that a 'mistake' has been made and they really wish to actually become the opposite sex. 1977 A. SHERIDAN tr. *Lacan's Écrits* vi. 209 A trans-sexualist practice, in no way unworthy of being compared with 'perversion'.

trans,sexua·lity. Also trans-sexuality, transexuality. [f. TRANS- 3 + SEXUALITY.] = *TRANSSEXUALISM; *loosely*, bisexuality.

1941 O. LEGMAN in G. W. Henry *Sex Variants* II. 1149 *Homosexuality*..is more widely used than any other of the many terms that have been proposed, such as..*transsexuality*. 1955 J. F. OLIVEN *Sexual Hygiene & Pathol.* xxi. 397 Primary transvestism. *Synonyms*: Idiopathic transvestism... Transsexuality or transsexualism. 1957 *Amer. Jrnl. Psychotherapy* XI. 85 Homosexuality, transsexuality and transvestism. 1973 J. MARKS *Mick Jagger* 142 Jagger's trans-sexuality is just one of the things which makes him qualify as our great hero. 1977 *Lancet* 6 Aug. 261/1 We have investigated 60 patients treated with high doses of methyltestosterone for transsexuality or impotence.

trans-Sibe·rian, *a.* (and *sb.*) [f. TRANS- 8 + SIBERIAN *a.*] That crosses Siberia. Also *ellipt.* as *sb.*, the trans-Siberian railway or express.

1896 [in Dict. s.v. TRANS- 8]. 1911 *Daily Colonist* (Victoria, B.C.) 1 Apr. 14/1 Return from or go to the Coronation via the great Trans-Siberian. 1939 C. ISHERWOOD *Goodbye to Berlin* 278 We'd stop a few days in Warsaw... Then on to Moscow, and take the trans-Siberian. 1964 S. BELLOW *Herzog* 168 She..was repatriated via the Trans-Siberian Railroad. 1974 *Times* 4 Oct. 10/7 The present Trans-Siberian runs for thousands of miles..almost flush with the Chinese frontier. 1977 *Belfast Tel.* 17 Jan. 8/4 Two days out of Moscow on its seven-day journey to Vladivostock, the Trans-Siberian express was two minutes early.

transsignification, var. *TRANSIGNIFICATION. trans-sonic, var. *TRANSONIC *a.*

trans-speci·fic, *a.* *Biol.* Also transpecific. [f. TRANS- 3 + SPECIFIC *a.*] Passing from one species to another.

1963 E. MAYR *Animal Species & Evolution* xix. 586 The proponents of the synthetic theory maintain that all evolution is due to the accumulation of small genetic changes,..and that transpecific evolution..is nothing but an extrapolation and magnification of the events that take place within populations and species. 1964 *Discovery* Oct. 29/2 Communication may be trans-specific. This is particularly true of alarm signals.

trans-syna·ptic, *a.* *Physiol.* Also transynaptic. [f. TRANS- 5 + SYNAPTIC *a.*] Involving transmission of a nerve impulse across a synapse. Hence trans-syna·ptically *adv.*

1954 PENFIELD & JASPER *Epilepsy & Functional Anat. Human Brain* v. 211 Under deep barbiturate anesthesia.., transynaptic conduction is sufficiently impaired so that spikes recorded beyond a synapse are prolonged by temporal dispersion into slower waves. *Ibid.*, Under light anesthesia..the temporal dispersion of transynaptically conducted spikes may be scarcely detectable. 1974 *Nature* 11 Jan. 112/1 These findings suggest that the rise in T-OH activity elicited by reserpine is mediated trans-synaptically through an increase in ganglionic transmission. 1977 *Proc. R. Soc. Med.* LXX. 671/2 Trans-synaptic regulation and adaptation (both orthograde and retrograde) in the brain.

transtage (tra·nstē̆ɪdʒ, -nz-). *Astronaut.* [f. TRANS- + STAGE *sb.*] A final stage of a multi-stage rocket that can be restarted in order to change the flight path or orbit.

1965 *Daily Progress* (Charlottesville, Va.) 6 May 20 After one 90-minute circuit of the globe in the initial orbit, the transtage was to re-start its engines and swing into an elliptical path. 1975 *Daily Colonist* (Victoria, B.C.) 21 May 1/7 Air force officials said the 'transtage' of the Titan III-C rocket and the two 1,200-pound satellites connected to it began rolling and tumbling out of control.

Transtainer (truns,tē̆ɪ·nəɪ, -nz-). *U.S.* [f. TRANS- + *CON)TAINER.] A proprietary name for a mobile gantry crane.

1964 *Official Gaz.* (U.S. Patent Office) 21 July TM 115/1 Pacific Coast Engineering Company, Alameda, Calif... Transtainer. For mobile gantry cranes. First use June 15, 1960. 1969 *Jane's Freight Containers 1968–69* 51 This Transtainer has a 35 foot span and straddles two rows of containers plus a truck roadway. 1977 *Hongkong Standard* 12 Apr. (Business Suppl.) 4/4 This includes a quay crane costing $102 million, four new transtainers valued at $33 million [etc.].

trans,thoracic, *a.* *Med.* [f. TRANS- 5 + THORACIC *a.* (*sb.*).] Occurring or carried out through the wall of the thorax or across the thoracic cavity.

1905 J. F. BINNIE *Manual Operative Surg.* III. xii. 429 Close the wound in the diaphragm with sutures introduced by the transthoracic route. 1965 *Dis. Chest* XLVIII. 297/1 Suction aspiration of the lung (transthoracic lung puncture). 1973 *Sci. Amer.* Feb. 11/1 (Advt.), A plug-in module monitors respiration from a set of skin electrodes that detect changes in transthoracic impedance due to respiration.

transuranic (truns,iuræ·nik, -nz-), *a.* *Physics.* [f. *TRANS- 11 + URAN(IUM + -IC 1 b.] Of a chemical element: having a higher atomic number than uranium (i.e. 93 or over), the highest-numbered element to occur naturally in any but trace amounts.

1935 *Proc. R. Soc.* A. CXLIX. 553 Our hypothesis that the 13-minute and 100-minute induced activities of uranium are due to transuranic elements seems to receive further support. 1958 *Times* 29 Aug. 6/3 The..isolation of the transuranic elements. 1976 *Jrnl. R. Soc. Arts* CXXIV. 586/1 Seventy-two of the ninety-two naturally occurring elements are metals. If one includes the man-made transuranic elements the proportion remains about the same.

Also trans,ura·nian (*rare*), -ura·nium *adjs.*, in the same sense; also *fig.*

1938 R. W. LAWSON tr. *Hevesy & Paneth's Man. Radioactivity* (ed. 2) xxvi. 289 'Trans-uranium' elements were first prepared by Fermi. 1946 [see *AMERICIUM]. 1947 CROWTHER & WHIDDINGTON *Science at War* iii. 129 He [*sc.* Fermi] concluded that he had made new atoms..and supposed that these must be 'trans-uranian' atoms. 1965 C. BROOKE-ROSE in J. Turner *Fourth Ghost Bk.* 181 The rumour spreads..through live material like a transuranian element decaying..into lead. 1971 *Nature* 6 Aug. 366/1 Seaborg..expressed the desire to return to California to resume his research on transuranium elements. 1978 N. FREELING *Night Lords* xxiv. 112 Night is what? A trans-uranium element, perhaps. 1979 *Monitor* (McAllen, Texas) 10 June 11-H/3 A person of almost transuranium energy.

trans,ure·thral, *a.* *Surg.* [f. TRANS- 5 + URETHRAL *a.*] Carried out via the urethra. Hence trans,ure·thrally *adv.*

1933 R. HOWARD *Practice of Surgery* (ed. 4) xxxvi. 1266 (*heading*) Transurethral endothermy. 1934 [see *RESECTIONIST]. 1951 [see *RESECTOSCOPE]. 1972 J. P. MITCHELL *Princ. Transurethral Resection & Haemostasis* xxvii. 227 It is difficult to..photograph the act of resecting transurethrally. 1980 *Brit. Med. Jrnl.* 29 Mar. 937/3 A very reasoned argument in favour of transurethral resection for the majority of benign enlarged prostates.

transvalue, *v.* (Earlier example.)

1899 G. B. SHAW in *Sat. Rev.* 13 May (Suppl.) p. iii/1 Nietzsche..'transvalued' our moral valuations.

transverse, *a.* Add: 1. d. transverse alliteration, in the early verse of some Germanic languages, alliteration of the patterns *abab* or *baab* (the usage of scholars varies somewhat); transverse flute: see FLUTE *sb.*^{1} 1; (now the usual name when specification is required; see also *cross-flute* s.v. *CROSS B, *German flute* s.v. GERMAN *a.*^{2} 4, *QUERFLÖTE 1, *TRAVERSO, etc.); transverse Mercator, the name of a map projection obtained like the Mercator but with the globe turned through 90 degrees relative to the cylinder, so that the great circle where they meet is a pair of meridians rather than the equator; transverse myelitis *Path.*, myelitis which extends across a section of the spinal cord.

1900 O. F. EMERSON in *Jrnl. Eng. & Gmc. Philol.* III. 127 (*heading*) Transverse alliteration in Teutonic poetry. 1920 R. J. MENNER *Purity* p. lvi, Transverse alliteration—abab—appears in 515 For I se wel þat hit is sothe þat alle mannez wyttez. 1942 J. C. POPE *Rhythm of Beowulf* 154 Transverse alliteration occurs once with the whole-line pattern bx/ax/ax/bx. 1949 P. F. BAUM in *Mod. Philol.* XLVI. 146 The most interesting of these minor variations is the crossed or transverse alliteration *ab ab*. 1879 GROVE *Dict. Mus.* I. 536/1 It is held obliquely towards the right side of the player, like the modern transverse flute. 1959, 1976 [see *QUERFLÖTE 1]. 1980 *Early Music* July 313/1 Byzantine musical culture decisively influenced much of the corpus of musical instruments, including..the organ, transverse flute and the bowed instruments. 1921 DEETZ & ADAMS *Elem. Map Projection* 104 In latitudes above 60° where the meridional parts of a Mercator projection increase rather rapidly, charts covering considerable area may be constructed..on..a transverse Mercator, if the locality has predominating north-and-south dimensions. 1969 [see *NATIONAL GRID 2]. 1974 *Encycl. Brit. Macropædia* XI. 476/1 Among the variations of cylindrical projections is the Transverse Mercator, in which the axis of the cylinder is parallel to the Equator, a treatment which has advantages in drawing maps that are long in the north-south direction. 1879 *N.Y. Med. Jrnl.* XXX. 387 (*heading*) Transverse myelitis. 1903 TUBBY & JONES *Mod. Methods in Surg. Paralysis* i. 33 Transverse myelitis in adults is more likely to be confused with the rare condition of poliomyelitis. 1964 J. J. WALSH *Understanding Paraplegia* iii. 16 Approximately 30% of the patients at the National Spinal Injuries Centre develop paraplegia from causes other than injury. Of these a proportion result from infections of various kinds, and are usually grouped under the heading of 'myelitis' or 'transverse myelitis'.

transversion^{1}. Restrict *rare* to sense in Dict. and add: 2. *Logic.* (See quot. 1933.) *Obs.* or *rare*.

1890 E. E. C. JONES *Elements Logic* xix. 143 These may be called *Mixed Eductions*, or *Transversions*. *Ibid.* 148 In Transversion, the most interesting points are that all Inferentials and Alternatives may..be fully and accurately expressed in Categorical form. 1933 C. A. MACE *Princ. Logic* vii. 121 A transversion may be defined as an immediate inference from a proposition of one logical form to another proposition which contains the same 'material content' but is of a different logical form.

3. *Molecular Biol.* The occurrence in a nucleic acid of a purine in place of a pyrimidine or vice versa. Cf. *TRANSITION 4*.

1959 E. FREESE in *Proc. Nat. Acad. Sci.* XLV. 631 The mutagenic effect of the second kind involves the 'transversion' of a nucleotide pair, in which a purine is replaced by a pyrimidine. 1980 *Nature* 8 May 82/2 All the point mutations are transversions in the second position of each of four codons.

Transverter (trunsvɜ̄·ɪtəɪ, -nz-). *Electr. Engin.* Also transverter. [f. TRANS(FORMER + CON)VERTER.] An apparatus for converting alternating current into high-voltage direct current, and vice versa. (A proprietary name in the U.S.)

1916 *Official Gaz.* (U.S. Patent Office) 16 May 1070/1 *Transverter*... Electric motors, electric generators, and electric-motor generators. 1924 *Glasgow Herald* 17 June 9 The machine..is called a transverter... It both transforms, by raising the pressure and converts into direct current. 1975 *Oxford Times* 5 Dec. 15/4 Mr Geoff Day..spoke..about a 384MHz amplifier for use with a 23cm transverter.

trans,ve·sical, *a.* *Med.* [f. TRANS- 5 + VESICAL *a.*] Passing or performed through the bladder.

1926 YOUNG & DAVIS *Young's Practice of Urol.* I. ii. 52 It is most important to empty the bladder at the end of the test period, to determine from this transvesical urine how much, if any, phthalein has escaped outside the catheters into the bladder. 1976 *Lancet* 4 Dec. 1221/2 An uneventful transvesical prostatectomy was done.

transvestic (trunsve·stik, -nz-), *a.* [f. *TRANSVESTISM or *TRANSVESTITE *sb.* and *a.*: see -IC.] Of or pertaining to transvestism; transvestite.

1961 in WEBSTER s.v., Patients with transvestic tendencies. 1966 *Internat. Jrnl. Psycho-Anal.* XLVII. 384 (*heading*) The mother's contribution to infantile transvestic behaviour. 1976 *National Observer* (U.S.) 16 Oct. 10/3 Its basic elements are meticulous screening, designed to exclude applicants who are primarily homosexual, transvestic or psychotic.

So transve·sticism (-isiz'm) (now *rare*) = *TRANSVESTISM.

1937 *Times Lit. Suppl.* 27 Feb. 145/2 By his 'transvesticism'..the mother's brother identifies himself with the mother and all she stands for. 1959 *Times* 31 Oct. 8/7 A group of four from the choir are not difficult to name in spite of temporary transvesticism.

transvestism (trunsve·stiz'm, -nz-). [ad. G. *transvestismus*, f. L. *trans* TRANS- + *vestīre* to clothe: see TRANSVEST *v.*, -ISM.] The action of dressing in the clothes of the opposite sex; the condition of having an abnormal desire to dress in the clothes of the opposite sex.

1928 [see *EONISM]. 1938 *Spectator* 2 Dec. 962/1 So unimportant is the sexual element that transvestism is common in many dance-forms [in Bali] and produces no feeling of embarrassment. 1959 *Listener* 2 July 31/2 The transvestism which is part of the witch/fairy tradition. 1977 E. J. TRIMMER et al. *Visual Dict. Sex* (1978) xix. 199 The most common cases of transvestism are heterosexual men leading otherwise conventional sex lives.

Hence transve·stist *sb.* and *a.* = *TRANSVESTITE *sb.* and *a.*

1944 'M. INNES' *Weight of Evidence* viii. 88 One of those portraits of her in page's costume. She must have been what they call a transvestist nowadays. 1959 P. BULL *I know Face* xi. 197, I played the part of Miss Eve Ashley's mother, and the gentleman dressers at Lime Grove were not unpleased by this transvestist carry-on. 1961 *Guardian* 27 Jan. 8/7, I tired of all those transvestist photographs. 1972 *Daily Tel.* 23 Mar. 8/1 He was a clothes fetishist and eventually a transvestist.

transvestite (trunsve·stait, -nz-), *sb.* and *a.* [ad. G. *transvestit*, f. L. *trans* TRANS- + *vestīre* to clothe: see -ITE^{1}.]

A. *sb.* A person with an abnormal desire to wear the clothes of the opposite sex.

[1910 M. HIRSCHFELD (*title*) Die Transvestiten.] 1922 J. VAN TESLAAR tr. *Stekel's Bi-sexual Love* ii. 69 Among the transvestites (personifiers) we find the most pronounced examples of marked homosexuality and stressed bi-sexuality. 1937 *Human Biol.* IX. 501 The transvestite must attempt to duplicate the behavior-pattern of his adopted sex. 1964 in W. H. Goodenough *Explorations Cultural Anthropol.* 490 The *bate*, male transvestites,..excelled women in butchering, tanning, and other domestic tasks. 1976 SMYTHIES & CORBETT *Psychiatry* xi. 211 Transvestites wear the clothing of the opposite sex to obtain sexual gratification.

B. *adj.* Of or pertaining to transvestism or transvestites; of a person: who wears the clothes of the opposite sex.

1925 A. L. KROEBER *Handbk. Indians California* xxxiii. 497 The transvestite sexual perverts recognized by all North American tribes. 1937 *Human Biol.* IX. 501 An almost senile singer, said to be the last person to know the transvestite initiation songs. 1957 *Observer* 29 Dec. 9/1 According to scholarship, the principal boy in our transvestite pantomimes is a Saturnalian hangover. 1971 *Times* 16 Sept. 12/6 He moves as a suburban schoolmaster between the male aggression of his classroom and the frou-frou of his transvestite boudoir.

transvestitism (trunsve·stitiz'm, -nz-). [ad. G. *transvestitismus*, f. *transvestit*: see prec., -ISM.] = *TRANSVESTISM.

[1913 M. HIRSCHFELD in *Neurol. Zentralblatt* XXXII. 946 (*heading*) Ein Fall von Transvestitismus bei musikalischem Genie.] 1934 in WEBSTER. 1949 [see *MOHAVE]. 1976 *Jrnl. Amer. Psychoanal. Assoc.* XXIV. III. 545 The analysis of a patient with the presenting symptom of transvestitism revealed a prominent set of fantasies of being 'initiated' and taught by a seductive and glamorous woman..in the art of dressing and make-up. 1983 *Financial Times* 4 Oct. 13 If there is a true sub-text to this farcical evening, devoted to voluntary or involuntary transvestitism, it is the satirical reference to psychiatrists' belief that simple things are all the consequence of serious, probably sexual, stimuli.

So transvesti·tic *a.* = *TRANSVESTITE *a.*; transve·stitist *rare* = *TRANSVESTITE *sb.*

1936 A. HUXLEY *Eyeless in Gaza* xviii. 233 Beppo giggled. 'Yes, those transvestitists!' he had to admit rapturously. 1977 *Gay News* 24 Mar. 13/1 Thus heterosexual, homosexual..transvestitic, and dozens of other special relationships hold the potential of working well. 1980 *Jrnl. Sex & Marital Therapy* Summer 124 Individuals rarely seek treatment for..transvestitic problems.

trans-wo·rld, *a.* [f. TRANS- 3 + WORLD *sb.*] That travels across the world; world-wide.

1955 R. J. SCHWARTZ *Compl. Dict. Abbrev.* 178/2 *TWA*,..Trans World Airlines, Inc. 1959 I. & P. OPIE *Lore & Lang. Schoolch.* i. 7 It seems that the schoolchild underground also employs trans-world couriers. 1962 *B.B.C. Handbk.* 50 The BBC Television Service is now looking forward to..transatlantic and even to trans-world television broadcasting.

Transylvanian (tran,silvē̆ɪ·niăn), *sb.* and *a.* [f. *Transylvania* (see below) + -AN.] A. *sb.* A native or inhabitant of Transylvania, formerly the eastern portion of Austria-Hungary, now part of Romania.

1644 MILTON *Areopagitica* 30 Nor is it for nothing that the grave and frugal Transilvanian sends out yearly..their stay'd men, to learn our language and our *theologic* arts. 1757 A. BUTLER *Lives Saints* III. 867/1 The vizir with one hundred and fifty thousand Turks (besides Hungarians, Transylvanians and Tartars) sat down before Vienna. 1927 [see *CARTEL *sb.* 3 d]. 1976 P. HENISSART *Winter Quarry* xvi. 157 The only smart people in this country [*sc.* Hungary] are the Jews and the Transylvanians.

B. *adj.* Of or pertaining to Transylvania; *spec.* applied to a kind of deep red Turkish rug or carpet frequently found in Transylvanian churches.

1843 *Penny Cycl.* XXV. 165/1 The German universities are frequented by a considerable number of Transylvanian students, especially for divinity. 1875 *Encycl. Brit.* III. 116/1 The Transylvanian Highlands extend over Transylvania..into Moldavia and Wallachia. 1915 *Guide to Collection of Carpets* (Victoria & Albert Museum) ii. 45 It seems probable that these 'Transylvanian' carpets were made in Anatolia. 1929 E. G. METCALFE *Metcalfe Coll. Oriental Rugs* vii. 31 This is an excellent specimen of the finest Transylvanian prayer rugs. 1957 *Encycl. Brit.* XIX. 624/1 (*caption*) So-called 'Transylvanian' rug, Asia Minor, 17th century. 1978 R. WESTALL *Devil on Road* x. 64 The [dog's] collar..had as many studs as a Transylvanian torture-implement.

transynaptic, var. *TRANS-SYNAPTIC *a.*

tranylcypromine (tra:nǝil,sǝi·prǒmīn). *Pharm.* [f. *trans*-2-phenyl*cyclo*propyl*amine*, the systematic name, f. *TRANS- 9 a + PHENYL + *CYCLO- 2 + PROPYL + AMINE.] A monoamine oxidase inhibitor, $C_9H_{11}N$, used in the treatment of severe depression (usu. in the form of the sulphate).

1959 *Proc. Soc. Exper. Biol. & Med.* CII. 380 (*heading*) Some pharmacological observations on tranylcypromine (SKF *trans*-385). 1964, 1965 [see *PARNATE]. 1976 *Lancet* 27 Nov. 1164/1 Six patients with neurogenic orthostatic hypotension were treated with a chemical preparation of tyramine and tranylcypromine.

trap, *sb.*^{1} Add: 1. d. A device which allows a pigeon to enter but not to escape from a pigeon loft.

1876 R. FULTON *Illustr. Bk. Pigeons* 274 Let the board, upon which the birds alight (when the trap is closed and admission only to be obtained through the bolting wires) be carried upon a pair of hinges. **1912** W. E. Barker *Pigeon Racing* i. 5 Others..swear by a steeply sloping roof..to compel the birds to drop upon a trap or alighting board. **1961** H. Blunt *Tackle Pigeon Racing this Way* iii. 30 The trap can be made of stout galvanized wire,.. curved to facilitate use by the birds without injury.

2. a. (Earlier and later examples in *Theatr.* use.)

1800 in S. Rosenfeld *Temples of Thespis* (1978) x. 149 Theatre traps and cutting out bricks. **1977** S. Brett *Star Trap* xiii. 142 The stage..had been equipped with the full complement of trap doors... Downstage were the corner traps, small openings used for the appearance or disappearance of one actor... Then there was the Grave Trap centre stage..always used for the Gravediggers' scene in Hamlet.

b. The mouth, esp. in phrr.: *shut your trap!* be quiet!; *to keep one's trap shut*, to remain silent. *slang.*

1776 E. Gibbon *Let.* 19 Dec. (1896) I. 298 You may say in general in the family (if any should bark) that you are satisfied with my conduct, and order them to shut their trap. [**1785**, **1860**: see *potato-trap* s.v. Potato *sb.* 7.] **1866** J. T. Staton *Rays fro' Loominary* 90 Shut thy trap, fayther. **1899** *Star of Hope* 12 Aug. 1/1 Why in h— don't those recruits..keep their traps shut? **1939** A. Huxley *After Many a Summer* II. i. 187 If only the rest were silence! But that's the trouble with poets... They will not keep their traps shut, as we say in the Western hemisphere. **1959** J. Braine *Vodi* xxiv. 255 'Shut your bloody trap,' Dick said. **1981** M. Duffy *Gor Saga* II. 48 If Emily should open her great trap and spill the lot she could find herself deep in trouble.

4. b. In greyhound-racing, the compartment from which a dog is released at the start of a race.

1928 A. R. D. Cardew *Greyhound Racing* 13 The owner ..is invited..to attend the draw for trap places 1 to 6, No 1 being the inside trap. *Ibid.* 18 (*caption*) Leaving the starting trap. **1954** R. Dahl *Someone like You* 251 The traps went up and the dogs flew out. **1977** *Listener* 30 June 847/3 The dogs barked in their traps. Then the hare was running..and the dogs were out.

5. *to be up to trap* (U.S. examples).

1837 W. Irving *Capt. Bonneville* II. i. 15 The beaver now being completely 'up to trap', approaches them cautiously. **1877** J. Habberton *Jericho Road* xix. 170 You needn't come any of your..moral tricks on me. I'm up to trap.

6. (Further Austral. and S. Afr. examples.) Now only *Austral. slang.*

1812 J. H. Vaux *Vocab. Flash Lang.* in *Mem.* (1964) 275 *Traps*, police officers, or runners, are properly so called; but it is common to include constables of any description under this title. **1882** T. Sheffield *Story of Settlement* 85 Amusing tales are told of how the excisemen or 'traps' sent to try to obtain evidence of illicit dealing were hoodwinked. **1935** L. Mann *Human Drift* xxviii. 185 Suddenly there came a cry 'The traps, look out, the — traps. The Joes, Joes.' He turned and saw a cordon of mounted and foot-police endeavouring to surround the crowd which scattered away from them. **1945** [see *demon[2] 1]. **1970** *Sunday Mail Mag.* (Brisbane) 18 Jan. 14/6 After the arrests Derrincourt and Wilson were kept apart. Little did the former know that the latter was telling the traps how William Derrincourt had engineered the whole business.

8. c. *Geol.* An underground rock formation in which an accumulation of oil or gas is trapped; so *oil trap*.

1920 *Econ. Geol.* XV. 249 Trap structures contain the majority of the important accumulations of oil. **1938** D. Hager *Pract. Oil Geol.* (ed. 5) iii. 56 Areas of heavily metamorphosed rocks are unfavorable for gas and oil accumulations, unless the oil has migrated to traps in such rocks. **1946** *Nature* 28 Dec. 931/2 In considering the distribution of structures which might act as oil-traps, the field of inquiry may be limited to those geological formations which provide some indication of the presence of oil. **1969** Bennison & Wright *Geol. Hist. Brit. Isles* xii. 282 Salt-dome structures..commonly give rise to oil traps, salt being capped by gypsum. **1977** *Offshore Engineer* May 38/1 Seismic evidence suggests thick deeply buried deposits onlapping old ridges which may form hydrocarbon traps.

d. *Radio.* A resonant circuit used as a rejector or acceptor circuit to block or divert signals of a specific frequency, esp. to reduce interference in a receiver tuned to a nearby frequency; = *wave trap* s.v. *wave *sb.* 10.

1927 B. F. Dashiell *Pop. Guide to Radio* xii. 236 A properly designed trap should not affect the tuning of the set to any great degree. **1957** *Practical Wireless* XXXIII. 570/1 Should there be any I.F. break-through traps, they should be done [*sc.* aligned] next. **1974** Harvey & Bohlman *Stereo F.M. Radio Handbk.* v. 113 The low-pass filter is followed by a 38 kHz trap..to remove any residual 38 kHz component.

9. d. *Golf.* = *sand-trap 2. Cf. Bunker 4 a in Dict. and Suppl. Chiefly *U.S.*

1890 H. G. Hutchinson *Golf* xiii. 313 That little round trap of a Strath's bunker not three yards nearer you. **1903** J. L. Low *Concerning Golf* ix. 173 Bunkers..refuse to be disregarded, and insist on asserting themselves... That little bunker on the seventeenth green at St. Andrews..that 'trap' bunker at the third green at North Berwick. **1933** F. Ouimet *Game of Golf* xvi. 236 Billy waded into the sand and blasted his ball out of the trap. **1952** B. Cerf *Good for Laugh* 173 How many shots did you have in that trap? **1971** 'D. Halliday' *Dolly & Doctor Bird* xv. 215 Wallace Brady..landed in the long, pale trap in front of the green and stayed there doing explosive shots with a sand-wedge.

10*. *Physics.* A site in a crystal lattice which is capable of temporarily immobilizing a moving electron or hole.

1945 *Proc. R. Soc.* A. CLXXXIV. 366 Thermoluminescence and long-period phosphorescence arise from the release of electrons from metastable levels or traps. **1971** *Physics Bull.* Oct. 579/1 Laboratory studies use uv, x ray or corpuscular..excitation to fill the traps, which are then emptied on heating. **1980** *Cambr. Encycl. Archaeol.* 426/2 In the structure of the crystal lattice of most minerals there are defects or imperfections known as traps.

11. **trap-bat** (earlier example); **trap boat** *N. Amer.*, a boat used for fishing with trap-nets; **trap-drum, drummer**: see *trap *sb.*[7] 3; **trap-gun**, (*a*) (see quot. 1964); (*b*) a shotgun used in trap-shooting; **trap-house**, a shelter from which clay pigeons are released for trap-shooting; **trap-line**, (*b*) *N. Amer.*, a series of hunter's traps; **trap-nest** orig. *U.S.*, a nesting-box which a hen can enter but cannot leave until released; also as *v. trans.*; hence **trap-nesting** *vbl. sb.*; **trap-net** (earlier example); **trap-point** (earlier example); **trapshoot** *N. Amer.*, a trap-shooting contest or event; **trap-shy** *a.*, of an animal: reluctant to approach a trap; hence **trap-shyness**; **trap skiff** *N. Amer.* = *trap boat* above; **trap-yard**, an enclosure into which animals such as horses, sheep, etc. are driven and confined.

1842 Dickens *Let.* 15 July (1974) III. 271 [I play] some most riotous game at trap-bat and ball in the Garden with the children. **1894** *Rudder* Aug.–Sept. 201/1 She was invited to sail over a course with a fleet of Richibucto's famous trap boats. **1974** F. Mowat *Boat who wouldn't Float* iv. 38, I was to join the four-man crew of a trap boat. She was a big, broad-beamed skiff powered by a five-horsepower, 'jump-spark' single-cylinder engine. **1801** J. Barrow *Travels Interior S. Afr.* I. vi. 360 A *stell-roar* or trap-gun, set by a Hottentot. **1947** J. Stevenson-Hamilton *Wild Life S. Afr.* xxxv. 307 Crocodiles are sometimes shot by trap guns. **1964** H. L. Peterson *Encycl. Firearms* 323/1 *Trap gun, trip gun.* 'Trap' as here used, has nothing to do with the sport of trapshooting; it refers to devices designed to shoot automatically at men or animals that come into the line of fire. Many of these trap guns are set off by the tripping of a wire or cord. **1976** *Shooting Times & Country Mag.* 16–22 Dec. 47/2 (Advt.), Nikko 5,000-II o/u trapgun. **1980** *Outdoor Life* (U.S.) (Northeast ed.) Oct. 90/2 None of the major arms makers offered specialized trap guns in 16-gauge, but ammo makers did make a special 2¾-dram-equivalent 1⅛oz. 16-gauge load for trap shooting. **1942** *Tee Emm* (Air Ministry) II. 75 We hear that 50 yards is not the best distance between the two traphouses, if the 'Plus' trap is being used. **1979** G. Hammond *Dead Game* xvi. 206 The local club have put up a tower for high birds [*sc.* clay pigeons]..and built some trap-houses. **1926** *Daily Colonist* (Victoria, B.C.) 7 July 3/5 The disappearance of Charles Olson from his trap line on the headwaters of the Parsnip River in Northern British Columbia remains as much a mystery as when it was first reported. **1954** W. Faulkner *Fable* 82 He had no more doubt of where he was than would the old wolf or lynx when he was near a trap-line. **1970** *Islander* (Victoria, B.C.) 10 May 6/1 The men were absent on the trapline and only the women and children were home. **1901** G. M. Cowell in *Rep. Marine Agric. Exper. Station* 1900 XVI. 97 It was a prime necessity to ascertain the exact record of the eggs produced by each individual. This led to devising the trap nest. **1908** *Ann. Rep. N.Y. Agric. Exper. Station, Ithaca, 1907* 261 For the poultry-man..who..will trap-nest conscientiously..we think there is a large reward. **1910** W. W. Broomhead *Poultry & Profit* iii. 33, I asked Mr. Tapley if he had his fowls trap-nested. **1960** *Farmer & Stockbreeder* 29 Mar. (Suppl.) 3/1 Three hundred pullets are trapnested ..each year. **1973** *Animal Behaviour* XXI. 98/1 When given access to trapnests the domestic hen has a characteristic behaviour associated with oviposition. **1906** *Reliable Poultry Jrnl.* XIII. 353/1, I read an editorial dealing with this problem of trap-nesting. **1956** Wilson & Card *Farm Poultry Production* iii. 60 The best way to get complete records is by trap nesting. **1865** Trap net [see *pound net* s.v. *pound *sb.*[2] 6]. **1885** E. B. Ivatts *Railway Managem. at Stations* 555 It is common to speak of a *catch siding* and *catch points*, of a *trap siding* and *trap points*. **1926** *Daily Colonist* (Victoria, B.C.) 17 July 12/2 Big Trapshoot at Colwood on Sunday. One of the biggest shoots of years is to be held tomorrow. **1976** *Billings* (Montana) *Gaz.* 18 June 4-D/1 (*heading*) Trapshoot set at Huntley. **1922** *Contemp. Rev.* July 90 After a time the rat grows poison-shy and trap-shy. **1947** *New Biol.* II. 19 A population which is thus immune to trapping is sometimes called 'trap-shy'. We do not know what trap-shyness is in terms of rat behaviour, but it is a very real phenomenon. **1968** K. Weatherly *Roo Shooter* 39 The rabbits had become trap-shy. **1934** *Geogr. Jrnl.* LXXXIV. 44 There were some trap-skiffs jigging for cod on the shoal. **1969** Trap skiff [see *jack *sb.*[1] 25]. **1906** *Chambers's Jrnl.* 12 May 380/1 There are..two means of capturing these horses... The first is to build a strong trapyard on their line of retreat, and endeavour to run them into it. **1936** Trap yard [see *crush *sb.* 4 c].

trap, *sb.*[5] **b.** Add: *trap-dike* (earlier example).

1839 [see *intrusion 1 b].

trap, *sb.*[7] Add: **2. a.** (See *traps *sb. pl.* 2.)

b. *Comb.* **trap-drum**, (*a*) a drum forming part of a set of traps (*traps *sb. pl.* 2) as opposed to a drum used in an orchestra or military band; (*b*) *pl.* = *traps *sb. pl.* 2; hence **trap-drummer**, a musician who plays the trap-drums.

1924 see *plastered *ppl. a.* 2]. **1929** T. Wolfe *Look Homeward, Angel* xiv. 180 Mr. Buckner..officiating at the trap drum and tambourine. **1959** W. F. Nolan *Dark Encounter* in H. Q. Masur *Murder Most Foul* (1973) 125 The sharp, sweet cry of horns could be heard above the rolling trap drums. **1978** *West Africa* 16 Oct. 2042/2 Trap-drums, double-bass, guitar. **1903** *Med. Rec.* (N.Y.) 14 Feb. 268/1 Trap-drummer's neurosis: a hitherto undescribed occupation-disease... The man's occupation was to beat a drum by the operation of a pedal which is manipulated with the right foot, while with his hands he plays the other drums, triangle, and the various traps. **1926** H. V. Morton *Nights of London* 168 We entered [the night club]. A trap drummer crouched over his instruments. **1977** *New Yorker* 20 June 93/1 When its complete personnel finally arrived onstage, the Ensemble consisted of two trap drummers.., a reedman who doubled on percussion.., a conga drummer.., a bassist.., and a singing pianist.

trap, *v.*[1] Add: **1. b.** (Later examples with inanimate obj.)

1936 *Discovery* Nov. 349/1 Its dust-particle content is.. trapped in the volume of liquid. **1952** E. R. Janes *Flower Garden* 49 Cloches were used in conjunction with hotbeds, but their efficacy for forwarding purposes depended chiefly upon their ability to trap solar heat. **1970** L. Deighton *Bomber* xxiii. 342 This cold still air trapped smoke from the furnaces and factories and held it like a grey woollen blanket.

2. (Earlier and later examples.)

1831 J. O. Pattie *Personal Narr.* 142 We set 40 traps, and..caught 36 beavers... We concluded..to travel slowly, and in hunters phrase, trap the river clear; that is, take all that could be allured to come to the bait. **1940** *Sun* (Baltimore) 2 Feb. 14/7 We plan to trap only a small part of our marshes. **1969** I. Kemp *Brit. G.I. in Vietnam* v. 105, I would usually go flat on the ground in case the door was trapped with a grenade or claymore mine.

4*. a. *Baseball.* (*a*) To catch (the ball) just after it has hit the ground; (*b*) to hem (a runner) between two fielders.

1892 *Chicago Herald* 16 May 2 Meehan trapped grounders by wholesale. **1912** C. Mathewson *Pitching in a Pinch* viii. 181 A Boston batter tapped one to Merkle which I thought he trapped, but Johnstone, the umpire, said he caught it on a fly. **1939** D. E. Jessee *Baseball* iv. 41 The second baseman will have many opportunities to participate in 'run-down' plays in which a base runner has been 'trapped' between first and second or second and third. **1959** E. Allen *Baseball Play & Strategy* v. 105 As a general rule all fly balls are caught... When a runner retreats toward a base with less than two outs, some fly balls may be purposely trapped and two players retired. **1967** R. Merkle *Concentrated Baseball* 105 On a rundown play between second and third, the trapped runner should be allowed to advance about half the distance toward third base.

b. *Cricket.* To cause (a batsman) to be dismissed leg before wicket.

1919 *Times* 4 July 8/6 The wicket..was nothing like so difficult as made out when once..you had gauged the bowler's spin without being trapped by Trumble's straight one. **1969** *Wisden's Cricketers' Almanack* 679 The use of pads instead of the bat was prevalent with six batsmen trapped leg before in the first innings. **1977** *Sunday Times* 27 Feb. 28/6 Lever broke through in his third over when Sri Lanka batted, trapping Fernando lbw at 16.

c. *Assoc. Football.* To receive and control (the ball), esp. between the foot and the ground.

1950 N. Cardus *Second Innings* 146 When it [*sc.* the football] was passed to him and it fell at his feet he would 'trap' it and lever it to an inch of where he wanted it. **1976** *Times* 2 Dec. 12/1 The ball was cleared from the United penalty area, Dobson trapped it with his left thigh, and hit it with his right foot, and the ball bounced just in front of Stepney and into the net.

trap-door. Add: **c*.** *Computers.* A method of surreptitiously gaining unauthorized access to data belonging to other users of a computer.

[**1976** D. B. Parker *Crime by Computer* xii. 112 The Trojan horse had been rolled into the fortified city and fully accepted. In the unsuspecting environment a trap-door in its belly opened, and out popped the soldiers.] **1977** *New Yorker* 29 Aug. 61/1 The nature of a trapdoor is that, while it is known to and usable by a penetrator, it is unrecognized by and unknown to other users of the system —even to the audit-trail mechanism. **1981** *Courier-Mail* (Brisbane) 27 July 5/5 'Trapdoors' allow people sharing a computer to slip into the confidential data streams of other users. **1982** S. F. X. Dean *Such Pretty Toys* xiv. 191 He just entered the girl's name into the computer..as some sort of routing key or trapdoor to cut off any trace.

c.** *Cryptography.* A piece of secret information that makes it easy to solve an otherwise very difficult code. Freq. *attrib.*

1978 *Communications Assoc. Computing Machinery* XXI. 128 They are called 'trap-door' functions since the inverse functions are in fact easy to compute once certain private 'trap-door' information is known. **1982** Beker & Piper *Cipher Syst.* 376 The general name given to this type of function (i.e. for which there does not appear to be a polynomial time algorithm but for which there is one so long as the method of application is known) is a trapdoor function. In our situation it is intended that the genuine receiver should be the only person who knows how to 'open' the trapdoor. **1984** *IEEE Trans. Information Theory* XXX. 595/1 It is this presence of trapdoors that makes some of the attacks on the additive knapsack cryptosystems feasible.

trapes, traipse, *v.* Add: (The usual spelling now is *traipse*.) **1. a.** Also in gen. use, to tramp or trudge, to go about.

1926 A. Huxley *Let.* 4 Mar. (1969) 268, I don't want to spend unnecessarily on traipsing round the continent. **1968** V. S. Pritchett *Cab at Door* xii. 238, I traipsed for a year from one paint shop to the next round Paris, selling glue, shellac and, for a hungry period, ostrich feathers and theatre tickets. **1976** *National Observer* (U.S.) 9 Oct. 5/4, I spent one day traipsing after Thomson, but his limousine disappeared at very high speed over a hill and by the time my rented sedan got to the other side he had disappeared. **1978** R. V. Jones *Most Secret War* xix. 159 For days we had to traipse for water down six flights of stairs and hundreds of yards to a stand pipe in the road.

2. c. Causatively: to carry or take about in a trailing way.

1814 H. Capel *Let.* July (1955) 53 St. Francis, the tutelary Saint of Brussels who had been previously trapsed round the town with the most astonishing pomp & splendour. **a 1974** R. Crossman *Diaries* (1976) II. 399 Suddenly I saw a picture of the tiny little woman looking upwards and seeing the soles of the feet of the statues above her as she was traipsed miles and miles around on the red carpet.

trapeze. Add: **1. b.** *Sailing.* (See quot. 1961.)

1961 F. H. Burgess *Dict. Sailing* 211 *Trapeze*, in sailing dinghies, a sliding support used by the crew for outboard balancing when they lay up to windward. **1969** *Daily Tel.* 25 Oct. 7/7 The work covers the origins of the trapeze (a means of crew support to help in holding a light dinghy level in a breeze). **1977** *Modern Boating* (Austral.) Jan. 43/3 Try looking at the race with your head upside down sometime when you are..flat out on the trapeze.

2. b. (See quot. 1968); = *trapeze-line*, sense 3 below.

1958 *Spectator* 13 June 761/2 Miss Lee is the only lady member to have adopted the new short skirt..and none of them has so far ventured upon the trapeze. **1968** J. Ironside *Fashion Alphabet* 30 The trapeze was a wide, stiff full-skirted tent shape stopping at the knees and moulding the figure to a high bust in front while falling free from the shoulders at the back. Trapeze was short for trapezium..but somehow a circus trapeze seemed to describe it more visually and the fashion magazines showed dresses on, under or in front of them.

3. *Comb.*, as **trapeze artist**, one who performs acrobatics on a trapeze; **trapeze harness** *Sailing* (see quot. 1981); **trapeze-line** [*line *sb.*[2] 14 b], a fashion style in which the outline of the garment resembles that of a trapezium (cf. sense 2 b above).

1938 *Amer. Mag.* CXXV. 90 Mother Millette started as a trapeze artist. **1981** W. Strathern *Don't look for Me—I'm Dead* iii. 52 The lean and sinewy strength of a trapeze artist. **1946** *Yachts & Yachting* 20 Aug. 383/2 (Advt.), All G.R.P., well equipped boat Needlespar, trolley, cover, certificate, lifting rudder, trapeze harness, ready to race. **1981** B. Webb *Schult's Sailing Dict.* 297/2 *Trapeze*, gear fitted to fast racing dinghies and some keelboats to enable the crew to put all his weight outboard to windward. The crew wears a trapeze harness or belt with a hook, which he slips into a ring on the lower end of the trapeze wire. **1958** *Vogue* Mar. 119 His [*sc.* Yves St Laurent's] wedge-shaped silhouette—called..the Trapeze Line—is flared from narrow shoulders to a smooth wide hemline. **1975** 'M. Fonteyn' *Autobiogr.* II. iv. 173 A 'sack' dress,.. a development of Christian Dior's 'Trapeze' line which had been..worn by so many wrong-shaped ladies.

trapezist. (Earlier example.)

1875 T. Frost *Circus Life* x. 179 The first female trapezist appeared.

trapezium. Add: **1. c.** (This sense is the one that is standard in the U.S., but in practice *quadrilateral* is used rather than *trapezium*.) (Further examples.)

1901 T. F. Holgate *Elem. Geom.* i. 74 If only two sides of a quadrilateral are parallel, the figure is called a trapezoid. If no two sides of a quadrilateral are parallel, the figure is called a trapezium. **1959** G. & R. C. James *Math. Dict.* 400/2 *Trapezium*, a quadrilateral, none of whose sides are parallel.

trapezoid, *sb.* Add: **1. b.** For *? Obs.* read 'Now *U.S.*' (Further examples.)

1901 [see *trapezium 1 c]. **1925** F. E. Seymour *Plane Geom.* II. 103 If the non-parallel sides of a trapezoid are equal, the trapezoid is called an isosceles trapezoid. **1972** *Whitaker's Almanack* 1973 902/2 Flag [of Kuwait].— Three horizontal stripes of green, white and red, with black trapezoid next to staff. **1975** *Sci. Amer.* Jan. 110/3 The matrix is stretched to a trapezoid, then the artist copies the picture by filling in the trapezoidal cells. **1977** *Monitor* (McAllen, Texas) 6 Jan. E-2/4 The Incas apparently adopted many things from the Mollos, of which the trapezoid is only one.

trapezoidal, *a.* Add: **a.** (Further examples.) The sense 'trapeziform' is standard in the U.S.

1955 *Sci. Amer.* Mar. 101/2 The temple has a trapezoidal central pyramid. **1973** *Nature* 21 Sept. 160/1 Trace brightness was modulated by a periodic trapezoidal waveform.

Trapezuntine (træpɪzu·ntəin), *sb.* and *a.* [f. L. *Trapezunt-*, *Trapezus*, Gr. τραπεζόντ-, τραπεζοῦς Trebizond + -ine[1].] **A.** *sb.* A native or inhabitant of the city of Trebizond or Trabzon in north-eastern Turkey.

1900 'Odysseus' *Turkey in Europe* vi. 269 After the conquest of Trebizond in 1461, many noble families migrated from that town to Constantinople, and naturally aspired to the Patriarchate..thus producing another element of discord among the unhappy Greeks, who were now rent by the factions of Trapezuntines and Constantinopolitans. **1956** R. Macaulay *Towers of Trebizond* viii. 79 We went out to explore, and the first thing we saw in the street was the B.B.C. van taking records, and round it stood a crowd of Trapezuntines staring.

B. *adj.* Of or pertaining to Trebizond, its natives, or inhabitants.

1926 W. Miller *Trebizond* i. 11 A conspicuous landmark of Trapezuntine history. **1969** A. C. Bandy in *Ibid.* (new ed.) p. ix, The Trapezuntine Empire..continues to interest scholars and students of Greek civilization. **1980** *Times Lit. Suppl.* 4 July 763/5 Trapezuntine princesses married into the Byzantine imperial house and into other Orthodox ruling families.

trapped, *ppl. a.*[1] **2.** (Earlier example.)

1872 F. Rogers *Specifications Pract. Archit.* II. ii. 106 To put to the water-closet in the basement-story a strong cast-iron trapped basin.

Trappist, *sb.*[1] (*a*.) **1. b.** (Earlier examples.)

1836 [in Dict., sense 1 a]. **1837** J. Binns *Miseries & Beauties of Ireland* II. xiii. 298 The superintendant of the Trappist Settlement at Mount Melleray.

Trappistine. 1. For '*attrib.*' read '*attrib.* and as *adj.*' and add earlier example.

1869 P. G. Hamerton *Wenderholme* III. xi. 96 He observed a Trappistine silence during the repast.

2. (Examples.)

1877 E. S. Dallas *Kettner's Bk. of Table* 279 Trappistine, distilled by the good fathers of the Grace of God in the Doubs. **1920** G. Saintsbury *Notes on Cellar-Book* ix. 138 Benedictine; Trappistine; a certain 'Père Kermann' ..and others will occur.

trappy, *a.*[1] (Earlier example.)

1882 *Daily Tel.* 13 Nov. 2/5 The fences might have been increased in size, however, without being made 'trappy'.

traps, *sb. pl.* Add: **2.** *colloq.* (orig. *U.S.*). In a jazz or dance band, percussion instruments or devices (e.g. wood-blocks, whistles) used to produce a variety of special effects; these together with the standard jazz or dance band drum-kit. Cf. *trap-drum* s.v. *trap *sb.*[7] 2 b.

1903 [see *trap-drummer* s.v. *trap *sb.*[7] 3]. **1925** F. Scott Fitzgerald *Great Gatsby* iii. 56 A great number of single girls dancing individualistically or relieving the orchestra for a moment of the burden of the banjo or the traps. **1938** *Oxf. Compan. Music* 948/2 *Traps*,.. The origin of the word may be from the nineteenth-century colloquial 'traps' meaning baggage of which the individual in question [*sc.* the trap-drummer] has necessarily a good deal. **1947** J. Steinbeck *Wayward Bus* xvii. 278 It's a rubber drum that you beat with a sponge. It's for the drunks that want to play traps in the orchestra. **1967** *Crescendo* May 26/2 When one packed up after a gig, one simply stuffed the snare drum, stand, pedal and traps—all the bits and pieces were known as 'traps' in those days— inside the open side of the B.D. [*sc.* bass drum]. **1982** B. Fantoni *Stickman* ii. 19 Dance-band drummers, beats me why dames go goofy on them. I played the traps a little myself once.

trapunto (trăpu·ntə). Also **Trapunto.** [a. It., = quilting.] A kind of quilting in which the design alone is padded (see quot. 1967).

1929 *Sewing for Profit* (Woman's Inst. Domestic Arts & Sci., U.S.) 73 Besides the usual kind of quilting..there is the trapunto, or Italian, quilting. **1936** *Home Arts Needle-craft* Jan. 5 (*heading*) Quilting in the Trapunto manner. **1967** E. Short *Embroidery & Fabric Collage* ii. 47 In the case of Trapunto, areas of the design are padded where required by splitting the backing and inserting wadding, after which the backing is sewn up again. **1972** *N.Y. Times* 3 Nov. 7/2 (Advt.), Leather-look..coat..with trapunto embroidery. **1979** *Guardian* 25 Apr. 11/8 The needlework designs..used in the country from which the garment hails: embroidery, applique, quilting, trapunto, smocking.

‖ **trascinando** (traʃinæ·ndo). *Mus.* [It., pres. pple. of *trascinare* (see next).] (See quots.); = Rallentando.

1876 Stainer & Barrett *Dict. Mus. Terms* 438/2 *Trascinando* (It.), dragging, delaying the time. **1947** E. Blom *Everyman's Dict. Music* 632/2 *Trascinando* (It.), dragging. **1983** *New Oxf. Compan. Music* 1842/1 *Trascinando* (It.), 'dragging', i.e. holding back, *rallentando*.

trascine (trasī·n), *v.* *nonce-wd.* [ad. It. *trascinare* to drag, pull.] *trans.* To carry, to drag.

1922 Joyce *Ulysses* 48 She trudges,..drags, trascines her load.

‖ **trasformismo** (trazfǫrmi·zmo). Also **Trasformismo.** [It.; cf. Transformism.] In Italy, a system of shifting political alliances, or of changes of allegiance, to form a stable administration or a workable policy. Cf. Opportunism.

1925 A. Solmi *Making of Mod. Italy* vii. 144 The parliamentary phenomena..called *trasformismo*. **1943** C. J. S. Sprigge *Devel. Mod. Italy* ii. 57 A struggle between *Trasformismo* and groups which endeavoured to gain power in the name of definite principles. **1957** M. Carlyle *Mod. Italy* ii. 55 Depretis..with his policy of shuffling and reshuffling of cabinets (*trasformismo*) created the dictatorship of the Cabinet. **1964** *Economist* 27 June 1462/2 This change of heart looked like the wildest sort of *trasformismo*. **1973** P. A. Allum *Politics & Society in Post-War Naples* iv. 116 The Piedmonte cobbler was able to fit fascism into his life experience, the classic *trasformismo* of southern boss politics. Before the March on Rome, everyone was socialist, afterwards, they were all fascists.

trash, *sb.*[1] Add: **1. a.** *spec.* in the U.S., domestic refuse, garbage.

1906 H. de B. Parsons *Disposal of Municipal Refuse* iii. 21 Rubbish is discarded trash, composed principally of all kinds of paper, wood, rags, mattresses, bedding, boxes, ..tin cans,..bottles,..and the like. **1925** *Amer. City* Jan. 54/2 The collection of garbage and trash may be made by the city with its own organization. **1931** W. G. McAdoo *Crowded Years* i. 12 The abandoned..building... Its steps were littered with trash, and many..windows were broken. **1962** A. Lurie *Love & Friendship* vii. 127 Mother used to get up at five in the morning..to sweep the front porch and carry the trash out. **1977** *New Yorker* 24 Oct. 128/3 Truckloads of trash were taken to the prison dump.

4. (Later examples applied to individuals.) *white trash* (earlier and later examples); now also used outside the Southern States of America and in *attrib.* use. Cf. *white *a.* 4 a.

1831 H. J. Finn *Amer. Comic Ann.* 88 'You be right dere,' observed Sambo, '..else what fur he go more 'mong niggers den de white *trash*?' **1833** F. Kemble *Jrnl.* 6 Jan. (1835) II. 112 The slaves themselves entertain the very highest contempt for white servants, whom they designate as 'poor white trash'. **1863** 'E. Kirke' *My Southern Friends* 55 The poor trash..scratched a bare subsistence from a sorry patch of beans and collards. **1901** W. Churchill *Crisis* I. i. 7, I..put a bullet past his ear, just to let the trash know the sound of it. *Ibid.* II. x. 211 It was not even a wild dream that white-trash Lincoln would be elected. **1932** W. Faulkner *Light in August* xvi. 363 Who told you I am a nigger, you little white trash bastard? **1942** B. Robertson *Red Hills & Cotton* viii. 189 If that was the sort of good-for-nothing trash she was, then she could just leave. **1945** E. Waugh *Brideshead Revisited* I. viii. 180 Who are all this white trash, anyway? **1973** *Sunday Times* 10 June (Colour Suppl.) 51/4 He said that all the Australians were white trash. **1977** J. Didion *Bk. of Common Prayer* IV. i. 158 'Lower that white-trash voice,' Warren said.

5. a. (See sense 1 a above) *trash basket, -bin, can, collection, collector, compactor, container, pickup*; **trashbag** (*a*) (further example); (*b*) chiefly *U.S.*, a rubbish bag; **trashman** *N. Amer.* = Dustman 1.

1792 S. Burdy *Life of Late Rev. Philip Skelton* 161 He had a trash bag, as they call it, in which he kept needles, thread, and such like articles, to put a few stitches, if necessary, in his clothes. **1934** Webster, Trash bag. **1960** *Guardian* 2 Feb. 5/1 The provision..of brown paper 'trash bags' for free issue to picnic parties. **1978** *Sci. Amer.* Feb. 158/3 Some household materials are also suitable: wrapping paper, brown paper bags and plastic trash bags (the kind used to line the inside of garbage cans). **1895** *Dialect Notes* I. 395 *Trash-basket*, waste-paper basket. N.Y. City. **1959** *Listener* 5 Mar. 411/? The trash basket in his mother's bedroom. **1972** *New Yorker* 26 Aug. 21/3 Trash baskets stood in ranks on the empty sand, like sentinels. **1955** W. Gaddis *Recognitions* I. i. 38 Janet came in a few minutes later to find him sifting through the kitchen trashbin. **1966** *Punch* 9 Mar. 331/2 How many mushroom enterprises leave the customer with equipment fit only for the trash-bin or the attic? **1976** *Columbus* (Montana) *News* 27 May 1/1 Members of the Columbus High School..have 'bicentennialized' the fire hydrants and trash bins. **1929** *Sci. Amer.* May 445/3 A prominent member of Washington society last winter rolled 25,000 dollars worth of diamonds in a chamois bag and carelessly left the bag on a table where her little son found it and carried it to the trash can. **1936** W. Stevens *New Caravan* 74 A trash can at the end of the world. **1960** *Times* 14 Sept. 12/7 Our dustbins— sorry, trash cans. **1981** 'P. Mallory' *Killing Matter* xvii. 173 A shaggy dog, working the trash cans. **1967** Mrs. L. B. Johnson *White House Diary* 26 Jan. (1970) 480 One of the Mayors said that his budget for trash collection and cleaning had tripled in the last two years. **1979** *Arizona Daily Star* 5 Aug. 1. 12/3 (Advt.), A continuous city-wide program of trash collection was implemented in 1974. **1967** *Boston Sunday Herald* 26 Mar. I. 34/1, I have a bone to pick with the Dedham trash collectors. **1973** *Washington Post* 13 Jan. E 16/3 (Advt.), Optional trash compactors and tub enclosures. **1977** *Chicago Tribune* 2 Oct. XII. 11/7 (Advt.), Kitchen complete with all appliances including disposal and trash compactor. **1968** Mrs. L. B. Johnson *White House Diary* 17 Jan. (1970) 617 We came out loud and strong for one more trash pickup a week..and for more trash containers. **1965** *Amer. Psychologist* Dec. 1014/2 A good idea was translated into banalities about..the friendly postman and trashman. **1971** *Daily Colonist* (Victoria, B.C.) 9 July 26/7 We used to organize regular expeditions..early on the mornings before the trashman came. **1968** Trash pickup [see *trash container* above]. **1978** *Detroit Free Press* 16 Apr. (Parade Suppl.) 31/1 You should see Beverley Hills on trash pickup day.

b. *attrib.* or as quasi-*adj.*, designating that which is worthless or of poor quality. Chiefly *U.S.*

In quot. 1843 perh. short for Trashy *a.* 1.

1843 Dickens *Let.* 7 Aug. (1974) III. 537 We were obliged at the last moment to alter an excellent bill; and the entertainments were very trash. **1940** *Sun* (Baltimore) 23 Jan. 5/1 Flocks of ducks—mostly 'trash' ducks like Black Bay coots and shelducks. **1944** *National Geogr. Mag.* Jan. 27/1 Trash fish and tons of discarded shrimp offal are now valuable in making fertilizer. **1966** 'H. MacDiarmid' *Company I've Kept* xiii. 270 That availability to Yankee trash-culture which has developed apace. **1967** G. Steiner *Lang. & Silence* 103 The serious novel has had to choose topics formerly exploited by trash-fiction. **1971** *Times* 27 Nov. 3/3 The so-called 'trash mail' service, the delivery of unaddressed circulars [by postmen]. **1973** *Daily Colonist* (Victoria, B.C.) 4 Oct. 31/3 Two inventors say they have a great new food product, but the name leaves something to be desired. It's 'trash fish' sausage. **1977** *Amer. N. & Q.* XV. 108/2 The image of Latin America in some German 'trash' novels of the twentieth century. **1979** R. Barnard *Posthumous Papers* xiii. 127 The offspring of shopkeepers, who bribed him with trash food. **1983** *New Yorker* 5 Dec. 160/2 About seven tons of trash fish eventually turned belly-up— Sacramento suckers, hardheads.

trash, *v.*[3] Add: **3. a.** To vandalize (property or goods), esp. as a means of protest. Occas. *intr.*, to perform such acts of destruction. Also *fig.* *colloq.* (chiefly *U.S.*).

1970 *Guardian* 14 May 2 On Sunday night a small gang went out to trash but a sudden rainstorm stopped the attack. **1971** *Time* 22 Mar. 26 Backstage at *Comes a Day* he got drunk and trashed his dressing room. **1974** H. L. Foster *Ribbin', Jivin', & Playin' Dozens* vi. 266 Students or unauthorized visitors who are physically attacking someone, 'ripping off' school equipment, 'trashing', attempting to burn or blow up a building, or otherwise interfering with instruction or threatening a student or worker with physical harm. **1975** 'S. Marlowe' *Cawthorn Jrnls.* (1976) xix. 174 The room..had been trashed..by either the patrons..or by the police. **1976** *Globe & Mail* (Toronto) 24 Aug. 29/8 Mine was one of a group of offices trashed at Trinity College in early June. Trashed, not burglarized. Nothing stolen. Art works smashed. Manuscripts and notes left alone, but books soaked in wine or worse. Furniture ripped, ceiling tiles torn. **1984** *New Yorker* 20 Feb. 43/1 They've trashed the laws.

b. To injure seriously, destroy or kill (someone or something). *U.S. colloq.*

1973 W. McGivern *Reprisal* 196 Don't be squeamish... Remember that Jules Levy, a Jew, trashed the pusher who murdered your son. **1977** C. McFadden *Serial* (1978) l. 107/1 Harvey threatened Spenser with grievous bodily harm... 'Whaddaya wanna trash me for?'

c. To reduce or impair the quality of (a work of art, etc.); to expose the worthless nature of (something), to deprecate. *colloq.* (chiefly *U.S.*).

1975 *New Yorker* 12 May 114/2 In Hollywood, the writer is an underling whose work is trashed, or, at best, he's a respected collaborator without final control over how his work is used. **1976** *Time* 5 Apr. 42/2 The presentation is ignorant, cluttered and coarse, and it trashes the sculpture. Works that need to be walked around..can only be seen frontally. **1977** *Saturday Night* (Toronto) May 72/1 With *Ghost Fox*, I thought, Houston would trash all the melodramatic comic-book stuff about Indians with one neat blow. **1981** *London Rev. Bks.* 2–15 July 12/1 She writes..yet another trashing of radical chic. This might be more gripping had she herself not trashed radical chic already.

trashed (træʃt), *ppl. a.* *colloq.* (chiefly *U.S.*). [f. *trash *v.*[3] 3 + -ed[1].] Bungled, spoiled; ill-treated or injured; run-down. Freq. with advbs.

1926 *Spectator* 22 May 857/1 We don't want any trashed-up peace. **1977** *Sounds* 9 July 30/2 Guthrie was real folk, born (1912) and raised in Oklahoma,..a trashed-on Okie farmer's son when that didn't mean you droned on about being proud of your bigoted ignorance. **1979** *Tucson* (Arizona) *Citizen* 20 Sept. 7B/6 'I've sat through this movie three times.'..'In this trashed-out theater? The picture's that good?' 'It's a lousy picture! I can't get my feet unstuck from the floor!!' **1980** *Dirt Bike* Oct. 14/1 Track-N-Trail has just come up with a solution to the age-old problems of mud, cold, rain, and trashed knuckles for you offroaders.

trasher. (In Dict. s.v. Trash *v.*[3]) Add: **2.** A vandal or wrecker. Cf. *trash *v.*[3] 3 a. *colloq.*

1970 *Guardian* 19 Sept. 9/5 The trasher mixes with the non-violent student. **1975** *Time* (Canada ed.) 9 June 8/2 'As a lifelong Tory and ranking member of the legal establishment,' reports Time Canada's Toronto Bureau Chief Robert Lewis, 'Maloney also is not about to pursue his office like a trasher.'

trashing, *vbl. sb.* (In Dict. s.v. Trash *v.*[3]) Add: **2.** The action of *trash *v.*[3] 3 a; vandalism or an instance of this. *colloq.* (chiefly *U.S.*).

1970 *N.Y. Times* 29 Apr. 40/1 The new breed of campus revolutionaries..are now turning to what they call 'trashing'—the setting of fires, hurling of rocks, smashing of windows. **1971** *Publishers' Weekly* 4 Oct. 24/1 Student riots and trashings of college stores. **1973** D. Martin *Tracts against Times* v. 47 The 'trashing' activities which damaged all property within a half-mile radius of the University of Wisconsin. **1979** *Listener* 3 May 614/2 'Trashing', the destruction of buildings and property, became commonplace.

trashy, *a.* Add: **3.** Of people: worthless, disreputable. *colloq.*

1862 'E. Kirke' *Among Pines* vii. 167 He regarded the white man as altogether too 'trashy' to be treated with much ceremony. **1898** B. Kirkby *Lakeland Words* 148 Of bad habits sairy man. Dick, Ah's flait he's nobbut trashy. **1931** H. Nicolson *Diary* 31 Dec. (1966) 100, I am thought trashy and a little mad. I have been reckless and arrogant. **1935** [see *piney-woods cracker* s.v. *piney wood]. **1977** *Detroit Free Press* 11 Dec. 13-c/1, I honestly couldn't believe such things happened except in low class trashy families.

trat(t (træt), colloq. abbrev. of next.

1969 R. Airth *Snatch!* xv. 149 We stopped off at a little trat Bruno knew of..and had a last one. **1969** *Queen* 17–30 Sept. 18/3 Luigi..who served you dinner in last week's trendy tratt. **1970** *Guardian* 6 Mar. 9/5, I have been thinking about trattorias. Mostly I mean the white-tiled tratts of SW 1, 3, and 7.

trattoria (tratorī·a, tratǫ·riă). Pl. **-ias**, ‖ **-ie.** [a. It., f. *trattore* host, f. *trattare* to treat.] **a.** In Italy, an eating house, a takeaway, a restaurant. **b.** Outside Italy, a restaurant serving Italian food.

1832 W. Gell *Pompeiana* I. iv. 49 A trattoria and coffee-house. **1873** 'Ouida' *Pascarel* I. II. iv. 169 Then he would pass methodically across the piazza to his favourite trattoriâ. **1922** [see *fettuccine]. **1955** E. Pound *Section: Rock-Drill* xci. 74 'Dodici Apostoli' (trattoria). **1966** *Punch* 20 July 112/1 The High Street sees a new Restaurant, a new Trattoria or a new Bistro opening every week. **1971** 'A. Burgess' *MF* ix. 101 The Great Giro..was piling cold plates together like a *trattoria* waiter. **1973** *Times* 25 Aug. 12/7 An insipid stew (*veau Berichonne*), with tough meat, that one would get in lesser known London trattorie. **1975** Felton & Fowler *Best, Worst & most Unusual* 216 Pizzas of all shapes and varieties are served in restaurants and trattorias throughout Italy.

trauma. Delete ‖ and add: Now also with pronunc. (trau·mă). Pl. **traumas, traumata.**

2. a. *Psychoanal.* and *Psychiatry.* A psychic injury, esp. one caused by emotional shock the memory of which is repressed and remains unhealed; an internal injury, esp. to the brain, which may result in a behavioural disorder of organic origin. Also, the state or condition so caused.

1894 W. James in *Psychol. Rev.* I. 199 Certain reminiscences of the shock fall into the subliminal consciousness, where they can only be discovered in 'hypnoid' states. If left there, they act as permanent 'psychic *traumata*', thorns in the spirit, so to speak. **1895** [in Dict.]. **1896** *Brain* XIX. 402 The author thought that the painful area on the thigh increased in size with the accession of fresh 'psychical traumas'. **1927** Bryan & Strachey tr. *Abraham's Sel. Papers* i. 47 He [*sc.* Freud] assigns a secondary rôle to sexual traumas in youth and assumes the presence of an abnormal psychosexual constitution. **1927** Henderson & Gillespie *Textbk. Psychiatry* iii. 56 Trauma may produce mental symptoms in one of two ways. Either it causes structural injury to the brain, or it causes emotional disturbances... In the first instance the mental reaction is of the organic type..in the second the result is usually a psychoneurosis. **1941** S. H. Kraines *Therapy of Neuroses & Psychoses* xv. 343 Simple trauma to the brain does not produce psychotic symptoms. **1950** *Brit. Jrnl. Psychol.* June 235 Complexes can be regarded as 'dissociated partial psyches'..the dissociation being produced by traumata, moral conflicts, etc. **1954** W. Mayer-Gross et al. *Clin. Psychiatry* x. 398 Mental disease due to brain trauma is only rarely a cause of admission to mental hospital. **1967** H. P. Laughlin *Neuroses* iv. 184 The specific trauma or event is not the *raison d'etre*, the cause of the depression. **1980** *Daily Tel.* 6 Dec. 12/3 Among the many shops and houses which advertise 'Counselling' (i.e., fortune-telling), I came across a hand-written notice which said 'Traumas Treated'.

b. In general and *fig.* use.

1977 H. Greene *FSO-1* x. 96 We know the trauma you suffered..but you've gone about it all wrong. **1977** *Mod. Railways* Dec. 486/2 Because of the trauma in the American rapid transit vehicle business—caused largely by the dilution of the market by the aerospace companies and the absence of long runs of standard designs—only two firms had put in tenders. **1978** S. Brill *Teamsters* vi. 248 Much of that trauma had come not from the real damage Dorfman had done but from the way the press had misstated or exaggerated it. **1981** *Daily Tel.* 30 Oct. 1/1 This is a sensible deal. It will mean that we do not have to go through the trauma of possible strike action.

traumatic, *a.* Add: Now also with pronunc. (traumæ·tik). **2. a.** *Psychoanal.* and *Psychiatry.* Of, pertaining to, or caused by a psychic wound or emotional shock, esp. leading to or causing behavioural disturbance.

1889 T. Savill tr. *Charcot's Clin. Lect. Dis. Nervous Syst.* III. 388 The existence of traumatic psychosis [*psychose traumatique*] adds still more to the gravity of the prognosis. **1909** A. A. Brill tr. *Freud's Sel. Papers Hysteria* i. 1 It is quite evident that in 'traumatic' hysteria it is the accident which evokes the syndrome. **1929** *Times Lit. Suppl.* 4 July 528/2 The Freudian traumatic theory. **1941** S. H. Kraines *Therapy of Neuroses & Psychoses* xv. 352 In the traumatic psychoses, damaged brain tissue is irreparable. **1979** A. Storr *Art of Psychotherapy* x. 94 Women who react to traumatic events with depression are generally contending with a variety of difficulties.

b. In general use: distressing, emotionally disturbing.

1962 A. Huxley *Let.* 17 June (1969) 935 Memories of traumatic events in childhood. **1965** M. Naylor *Your Money* ix. 54 You will only have sacrificed one-seventh of your original capital. This is depressing enough, of course, but it is a good deal less traumatic than losing half. **1973** *Howard Jrnl.* XIII. 287 It was felt by some that this would..make leaving prison at the end of the sentence less traumatic. **1977** E. Heath *Travels* viii. 167 Whatever the outcome, the impact on the United States of the decade of war in Vietnam was traumatic. **1984** *Daily Tel.* 2 Feb. 2/3 We see manufacturing industry only feebly recovering from the traumatic experience of the last five years.

3. *Bot.* *traumatic acid*, a plant hormone that is found in damaged tissue in some plants and promotes its repair; dodec-2-ene-1,12-dioic acid, $HOOC(CH_2)_8CH{:}CHCOOH$.

1939 J. English et al. in *Science* 6 Oct. 329/2 It would seem appropriate and convenient to refer to this substance as 'traumatic acid'. **1966** R. M. Devlin *Plant Physiol.* xvii. 427 The effect of traumatic acid on inducing cells to divide does not appear to be general. **1971** J. Z. Young *Introd. Study Man* xii. 160 In plants..injured cells release a..traumatic acid.

Hence (in sense 2 b) **trauma·tically** *adv.*

1972 D. Bloodworth *Any Number can Play* viii. 56 His patriotic and anti-communist postures are almost traumatically compulsive. **1976** *Listener* 11 Nov. 615/1 A view of life, of being, that would lead me ineluctably into false hopes and traumatically real disappointments.

traumatism. Add: **2.** *Psychol.* and *Psychiatry.* A morbid condition of the psyche resulting from repression to the unconscious of emotional wounds or shock which are unacceptable to the conscious mind. Also *loosely* in general use, a shock or unpleasantly startling experience.

1898 H. Ellis in *Alienist & Neurologist* XIX. 610 They may be said to have shown conclusively, what has already been more or less hesitantly suggested by others, that the most typical hysteria is really a 'psychic traumatism'. **1926** J. I. Suttie tr. *Ferenczi's Theory & Technique Psycho-Anal.* 77 A certain amount of infantile sexual experience (that is to say, a little 'sexual traumatism'). **1973** D. Matias tr. C. Metz in *Screen* Spring/Summer 51 The actor gave the impression of reciting a sparse and laconic text which cut across the preceding silence producing a minor aesthetic traumatism for the spectator.

traumatize, *v.* (In Dict. s.v. Traumatism.) Add: **1.** (Further examples.)

1929 *Jrnl. Amer. Med. Assoc.* 13 July 116/2 The white bands disappeared promptly when care was taken not to traumatize the fold of skin with the orangewood stick. **1954** S. Duke-Elder *Parson's Dis. Eye* (ed. 12) xxii. 361 In this way they may be severely traumatized; and at the same time the wave of pressure, striking the retina and choroid, may do considerable damage.

2. To inflict an emotional wound or shock upon; to impair or damage psychologically. Also *fig.*

1949 M. Mead *Male & Female* xvi. 336 Two bitter little rivals may otherwise spend hours quarrelling and traumatizing each other. **1958** *Spectator* 28 Feb. 255/1 A Roman Catholic lad who traumatised me by telling me that God was always about. **1965** *New Statesman* 17 Dec. 960/2 In the intervening period, 34 people were killed..1,032 were injured... The event has traumatised California. **1970** A. Toffler *Future Shock* x. 194 The year 2000 is closer to us in time than the great depression, yet the world's economists, traumatized by that historic disaster, remain frozen in the attitudes of the past. **1974** *Sci. Amer.* Aug. 56/2 For children who come from environments in which the capacity of the family to function has been most severely traumatized by such destructive forces as poverty, ill health and discrimination, the consequences for the child are seen [etc.]. **1979** P. Theroux *Old Patagonian Express* (1980) xiii. 264 The passengers were either asleep or sitting silently, traumatized by the heat.

Hence **traumatiza·tion**; **trau·matized, trau·matizing** *ppl. adjs.*

1935 *Proc. Soc. Exper. Biol. & Med.* XXXII. 1249 If profound and fatal shock is to be obtained in the intact dog by traumatization, the severity of tissue injury must be very much greater than is required to induce shock symptoms in the equally healthy and vigorous animal lacking adrenal glands. **1935** *Amer. Jrnl. Physiol.* CXI. 430 Such traumatized animals lacking adrenals die within 24 hours or less. **1949** M. Mead *Male & Female* v. 118 Two inexperienced adolescents had a first sex-affair..and became traumatized by their own clumsiness. **1950** E. H. Erikson *Childhood & Society* (1951) i. 37 The condition started with such damage, or at least with momentary traumatization. **1966** *Lancet* 31 Dec. 1464/1 Fat from traumatised adipose tissue can on occasion enter the circulation and produce fat-embolisation. **1971** K. Millett *Sexual Politics* (1972) II. iv. 180 We perceive that the traumatizing circumstance of being born black in a white racist society invests skin color with symbolic value. **1977** M. Sokolinsky tr. *Merle's Virility Factor* xvi. 330 There is an excess of love..an instinct that, in women, is no longer stifled by the traumatizing sense of her social inferiority. **1979** *Daily Tel.* 31 Oct. 15/2 Thousands of traumatised survivors of the Pol Pot horrors were starving and dying. **1979** *Nature* 13 Dec. 727/1 It was important to ascertain whether our surgical procedure led to any transient denervation or traumatisation of synapses made by the soleus nerve.

traumato-. Add: **traumatology** (examples); in mod. use, the branch of medicine concerned with the treatment of wounds and serious injuries and with the disabilities they cause; hence **traumato·logist**, a specialist in this.

1935 *Dorland's Med. Dict.* (ed. 17) 1427/2 Traumatologist. **1948** *Excerpta Medica Section IX* II. 944 Traumatologists must devote themselves not only to special treatment, but also to the prevention of injuries. **1981** *Sci. Digest* Aug. 52 'Half a million out of eighty million accident cases a year need Shock Trauma's services,' says one veteran traumatologist. **1948** *Excerpta Medica Section IX* II. 944 The necessity for special traumatology is caused by the progress of civilization. **1964** *Acta Universitatis Carolinae Medica Suppl.* XIX. 296 Modern scientific experiences and results of research of many teams offer more and more effective and better means to traumatology and raise the standard of its work. **1977** *Lancet* 30 Apr. 961/2 A national institute of traumatology in the United Kingdom cannot be put off any longer. **1980** R. Owen et al. (*title*) Scientific foundations of orthopaedics and traumatology.

traumatropism. Add: Also **traumatotropism.**

1965 Bell & Coombe tr. *Strasburger's Textbk. Bot.* 380 Galvanotropism, traumatotropism and thermotropism.. are possibly only special forms of chemotropism.

trautonium (trautōᵘ·niəm). [f. the name of Friedrich *Trautwein* (1888–1956), German scientist and inventor of the instrument, after Euphonium.] An electronic musical instrument, capable of producing notes of any pitch.

1931 *Electronics* July 18/2 The 'Trautonium' is a recent development at the Radio Research section of the Berlin Academy of Music... Paul Hindemith..is himself an excellent Trautonium player and has written music specially for it. **1936** *Discovery* July 224/2 The Trautonium, due to Dr. Trautwein and developed at the Charlottenburg Music High School (Conservatoire), where it so interested Hindemith that he not only wrote for it but even himself learnt to play it. **1959** *Chambers's Encycl.* V. 131/2 Instruments such as the theremin and the trautonium derive music directly from the tuned circuits in oscillation and have no mechanical vibrator except the diaphragm in the loud-speaker. **1978** P. Griffiths *Conc. Hist. Mod. Music* viii. 111 Hindemith wrote a concertino for trautonium and orchestra in 1931.

travail, -aille, *sb.*[3] (Earlier example.)

1801 A. Henry *Jrnl.* 13 Oct. in E. Coues *New Light on Greater Northwest* (1897) I. iv. 190 Chamanau arrived from the hills, bringing his deceased wife on a travaille to be buried here.

‖ **travaux préparatoires** (travo preparătwār), *sb. pl.* *Law.* [Fr., lit. 'preparatory works'.] Drafts, records of discussions, etc., pertaining to legislation or a treaty under consideration (see quot. 1980).

1935 *Harvard Law Rev.* Feb. 562 French courts exhibited the tendency to limit recourse to *travaux préparatoires.* **1957** H. F. Jolowicz *Roman Foundations Mod. Law* ii. 16 No text appears to deal with *travaux préparatoires*, but it is in accordance with the civilian tradition that resort to drafts, speeches in Parliament, and similar evidence is usually allowed more widely on the Continent than in England. **1962** *Listener* 15 Mar. 456/1, I am thinking about our rule which excludes the so-called *travaux preparatoires*; that is to say, statements made in negotiation, prior to the treaty, which might throw light on its meaning. **1980** *Oxf. Compan. Law* 1231/1 *Travaux préparatoires*, materials used in the preparation of, and having formative effect on, the ultimately adopted form of an agreement, or legislation, or an international treaty. Such materials include, in the domestic sphere, reports, proposals and technical advice, in the legislative sphere, Select Committee or Royal Commission or other reports, academic studies, Green Papers, White Papers, and the like, and in the international sphere reports of expert committees, discussions and proposals, drafts, and the like.

travel, *sb.* Add: **5.** *travel bag, -book* (earlier example), *film, literature, permit, poster, ticket, time, voucher, warrant*; *travel-minded* adj.; objective, as *travel editor, -writer* (later examples); instrumental, as *travel-wearied* adj.; **travel agency**, a firm which makes arrangements for the transport, accommodation, etc., of travellers, and which acts as an agent for tour-operators (see *tour *sb.* 12); **travel agent**, one who owns or works for a travel agency; **travel allowance**, (*a*) the amount of money given to a traveller to cover the expenses of a journey; (*b*) under the Exchange Control Bill, the maximum amount of money travellers were allowed to take out of the U.K. during the period 1946–80; **travel brochure**, a booklet advertising travel and describing the features and amenities of holiday resorts or other places of travel; **travel bug** *colloq.*, a strong urge to travel (cf. *bug *sb.*[2] 4 a); **travel bureau** = *travel agency* above; **travel document**, a document required for travel; *spec.* a document allowing foreign travel, held by one not entitled to a passport; **travel folder** = *travel brochure* above; **travel sickness**, nausea induced by the motion of a vehicle; carsickness; hence **travel-sick** *a.* affected by travel sickness; **travel trailer** *U.S.*, a variety of caravan.

1927 *World Travel* Oct. 39/2 (Advt.), Imperial Airways. Daily Services Between London Cologne Brussels Paris... Book through any Travel Agency or direct with the Company. **1975** B. Bainbridge *Sweet William* iii. 79 He was going to get brochures from a travel agency. He thought they should all go to Spain. [**1902** *Encycl. Brit.* XXVII. 227/2 Cook, Thomas (1808–1892), travelling agent, was born at Melbourne in Derbyshire.] **1925** *Times* 1 May 2/2 (Advt.), write to-day for your copy of 'Economy Tours to America', to..leading Travel agents. **1980** S. Brett *Dead Side of Mike* vi. 61 Toby Root played a travel agent. **1937** Travel allowance [see *second *a.* 6 b]. **1951** *Ann. Reg. 1950* IV. 453 The basic annual travel allowances were increased in December to £100 for adults. **1978** A. Waugh *Best Wine Last* xviii. 237 In 1947..the meagre travel allowance was again reduced. **1939–40** *Army & Navy Stores Catal.* 875/2 Popular travel bag, with.. passport pocket. **1968** L. Deighton *Only when I Larf* i. 9 Umbrella in one hand, travel bag in the other, he marched off. **1843** Dickens *Let.* 2 Nov. (1974) III. 591 The travel-book, if to be done at all, would cost me very little trouble. **1953** P. Scott *Alien Sky* I. vi. 75 A plan I have to issue highly coloured travel brochures with a photo of myself on an elephant. **1972** F. Warner *Maquettes* 42 A copy-writer for the travel brochures. **1976** P. Cave *High Flying Birds* i. 13 The travel bug. Ants in your suitcase. **1930** E. Waugh *Labels* iii. 51 Their speech is rich with the words of the travel bureau's advertising manager. **1966** A. K. Train *Spoken like Frenchman* 94 (*heading*) At a travel bureau. **1963** *Listener* 14 Feb. 281/2 Soblen, provided with an Israeli travel document valid for travel to any country except Israel, left by air. **1980** E. Behr *Getting Even* xiv. 166 He was not to volunteer information about the man's travel documents. **1910** *Bradshaw's Railway Guide* Apr. 1054/2 'Clonsilla' En Pension... Recommended by Travel Editor of 'Queen'. **1977** *Chicago Tribune* 2 Oct. IV. 19/5 Travel Editor Holt was born and reared in West Virginia. **1922** Travel film [see *scenic *sb.* 2]. **1978** A. Waugh *Best Wine Last* ix. 107 Marrakesh has been a subject of many articles and travel films. **1955** W. Gaddis *Recognitions* II. v. 488 A tour from a travel folder. **1980** D. Bloodworth *Trapdoor* xii. 68 The secluded Kahala Hilton with its sun-swept beach ..could have been torn straight out of a travel folder. **1934** Travel literature [see *courtesy card* s.v. *courtesy *sb.* 12]. **1955** E. Blunden *Addresses on General Subjects* 285 It is an example of the prolific travel-literature of England. **1932** Travel-minded [see *minded *ppl. a.* 4 c]. **1962** *John o' London's* 1 Mar. 202/1 Everyone these days is travel-minded. **1942** M. Cable *Gobi Desert* 245 It was no longer he who issued the travel permits and received official visits. **1978** T. Willis *Buckingham Palace Connection* v. 96 The British Vice-Consul..had promised to get her the necessary travel permit. **1958** *Times Lit. Suppl.* 10 Jan. 22/5 This is no excuse for a travel-poster jacket and flamboyant title. **1979** R. Jeffries *Murder begets Murder* xv. 91 Sun from dawn to dusk just like the travel posters had promised. **1959** *Times* 13 July 9/1 Some of the children will be travel-sick. **1978** *Times* 30 Dec. 4/4, I felt travel sick as we were driving along the lane. **1900** Dorland *Med. Dict.* 710/2 *Travel-sickness*... Same as *Car-sickness.* **1941** W. Graham *Night Journey* xx. 246, I was talking to your husband on the problem of travel sickness. **1979** R. Perry *Bishop's Pawn* I. 15, I was going to swallow a handful of travel sickness pills. **1949** Dylan Thomas *Let.* 1 Dec. (1966) 341, I must..hurry everything up, as visas, travel-tickets, etc., cannot be too easy to procure. **1980** *Daily Tel.* 26 Jan. 17/3 Auckland..is not the place to buy travel tickets, as there is a 10 per cent tax on them there. **1946** Travel time [see *flying time* s.v. *flying *vbl. sb.* 3]. **1976** P. R. White *Planning for Public Transport* viii. 160 Over routes of about 200 to 250 m.. total travel times by air and rail are similar. **1961** *Mobile Home Jrnl.* Dec. 21/1 Harold Martin..is now the proud owner of a twenty-seven foot Yellowstone travel trailer... The twenty-seven foot model is the largest in the Yellowstone line of travel trailers. **1978** *Sunday Sun-Times* (Chicago) 1 Jan. 122/1 Travel trailers are of two types: The conventional, rectangular-shaped unit constructed of aluminum or molded fiberglass over wall studs, and the fifth-wheel trailer. **1964** L. Deighton *Funeral in Berlin* iii. 21 The travel vouchers and tickets are ordered. **1978** P. Bryers *Cat Trapper* xxviii. 180 His travel vouchers were made out for the motel at Kishinev. **1952** 'R. West' *Meaning of Treason* (ed. 2) I. vii. 156 A travel warrant issued by the Ministry of Labour. **1919** W. de la Mare *Flora* 40 Noonday to night the enigma of thine eyes Frets with desire their travel-wearied brain. **1927** W. B. Yeats *October Blast* 22 Cease to remember the delights of youth, travel-wearied aged man. **1949** C. Graves *Ireland Revisited* viii. 125 Every travel-writer and poet who has visited the Lakes of Killarney has made some attempt to do justice to their loveliness. **1972** W. Garner *Ditto, Brother Rat!* xi. 80 A guest! A famous travel writer.

travel, *v.* Add: **2. d.** (Earlier example const. *in.*) Also, const. *for* the concern for which a commercial traveller works (later examples).

1841 Thackeray in *Fraser's Mag.* Sept. 330/2 I've got a place—a tip-top place..to travel in the West of England in oil and spermaceti. **1872** Geo. Eliot *Middlem.* III. vi. lx .336, I travelled for 'em, sir, in a gentlemanly way—at a high salary. **1922** Joyce *Ulysses* 111, I travelled for cork lino. **1964** 'J. Melville' *Murderers' Houses* ii. 46 He travels for Associated Boxes. It's the big firm on the London Road.

f. *to travel light*: to travel with little luggage. Also *fig.*

1921 E. O'Neill *Emperor Jones* i. 166, I travels light when I wants to move fast. **1931** 'Grey Owl' *Men of Last Frontier* 13 As he has also to break his own trail, he travels light, taking only a sheet of canvas for a windbreak and one blanket. **1954** I. Murdoch *Under Net* xviii. 252, I just couldn't help making money, and I don't want that. I want to travel light. **1977** *Time* 19 Dec. 18/2 West German terrorists are especially difficult to fathom because ideologically they travel light.

3. e. (Later examples.)

1911 G. Stratton-Porter *Harvester* v. 74 Betsey.. wants to meander along the road with a loaded wagon... Betsey, you must travel! **1970** M. Kenyon *100,000 Welcomes* xxi. 178 Mercy, the lorry's travelling. Foot down.

5. b. *Theatr.* To take (costumes, equipment, etc.) with one from place to place.

1930 E. Wallace *Lady of Ascot* i. 15 She had sapphire rings and clips..of an incredible value, and she 'travelled' them, as they say in theatrical circles. **1966** 'J. Hackston' *Father clears Out* 123 The taller of these two guests travelled a broken concertina with him.

c. *Publishing.* To take (books) from place to place in order to promote and sell them. Cf. sense 2 d in Dict. and Suppl.

1937 V. Woolf *Let.* 10 Feb. (1980) VI. 106 We're taking Tuesday off at Rodmell to travel our books in Sussex. **1977** B. Colloms *Victorian Country Parsons* xi. 219 [George Routledge] liked to travel his own books in the north country so that he could keep in touch with booksellers.

travel(l)ator: see *Travolator.

travelled, traveled, *ppl. a.* Add: **1.** (Later examples.)

1966 *Listener* 11 Aug. 210/1 Synge, already a travelled man when Yeats suggested to him that he'd find the Aran Islands more to his liking than Paris was a foreigner in his own country. **1978** G. Greene *Human Factor* II. i. 51 We need travelled gentlemen like you to deal with foreign affairs.

3. (Earlier examples.)

1845 J. C. Frémont *Rep. Exploring Expedition* 163 [To Fort Hall] along the travelled road from the town of Westport..is 1,323 miles. **1869** *Bradshaw's Railway Man.* XXI. 433 The travelled route through this country crosses formidable ranges of mountains.

4. *travelled blood* (see quots.).

1962 'J. le Carré' *Murder of Quality* iii. 41 There's a lot of what we call travelled blood..that's to say, blood spurted from an open artery. **1981** *Event* 9–15 Oct. 29/4 *Travelled blood*... A pathologist's description of blood that has spurted from a severed artery.

traveller, traveler. Add: **1. b.** (Earlier and later Austral. examples.) Also, a gypsy.

1868 M. Clarke in *Australasian* 5 Sept. 305/3, I remember at one station, situated on the main road for 'travellers', that the unhappy cook was 'put on the fire' by a crowd of these gentry. **1891** 'F. W. Carew' *No. 747* ii. 18 A little commercial transaction—known among 'travellers' as 'trucking'. **1945** [see *passage *sb.* 1 e]. **1967** *New Scientist* 7 Dec. 582/3 The question is whether or not gypsies (who call themselves Travellers) are members 'of the Romany race'. **1971** *New Society* 1 July 18/2 Scotland's 2,000-odd itinerant tinkers—or, as they prefer to be called 'travellers'... Some claim descent from the roving Irish tinsmiths or Scottish outlaws, others from the true romany gypsies.

3. (Earlier example.)

1790 J. Wedgwood *Let.* 13 July (1965) 328 Such distinguished favours cannot but make a deep impression on my mind. Nor will they be forgotten by the travellers.

5. a. (Earlier U.S. example.)

1828 A. Sherburne *Mem.* iii. 61 He and other officers contrived to haul the men ashore... He fixed a traveller on the rope, by which he first went ashore, so that he could not wash off.

f. A craftsman's tool used for measuring circumferences, esp. of wheels (see quots.).

1879–81 G. F. Jackson *Shropshire Word-Bk.* 454 *Trindle*, a disc used by blacksmiths for measuring the circumference of wheels—a 'traveller'. **1923** G. Sturt *Wheelwright's Shop* xxiii. 122 The new tyre..had to be measured, as also had the wheel it was meant for. Blacksmiths kept a special implement for this purpose —a 'traveller' or 'tyre-runner'. The traveller was a thin circular disk of iron, six or seven inches across, which the smith would hold out, waist-high, at right angles to himself, and run round wheel and tyre in turn. **1969** G. E. Evans *Farm & Village* xiv. 148, I cut the band to the exact circumference of the stone. I find this with a device I made... It's called a *traveller*... It's a metal wheel. I roll this round the stone and count the revolutions... (This device works on the same principle as the measuring wheel used by the old road surveyors—a trundle wheel or way-wiser.) **1976** *National Observer* (U.S.) 10 July 9/1 'This here is called a traveler,' he says..displaying a round, flat device used to measure the perimeter of a wagon wheel... 'You won't find too many of them left.'

5*. A suitcase, trunk, or travelling bag. Chiefly *U.S.*

1895 *Montgomery Ward Catal.* Spring & Summer 564/3 Canvas Traveler... Large square box made of the best basswood, covered with extra heavy duck... A most handsome ladies' trunk. **1965** *Harper's Bazaar* Dec. 89/1 Cosmetic traveller lined with silk for girls who are on the go all day. **1983** *Country Life* 1 Dec. 1677 (Advt.), Travel-bag—this great, waterproof traveller holds three leakproof bottles.

6. b. **traveller's cheque** orig. *U.S.*, a cheque for one of several specified amounts of money, which can be cashed at a bank in most countries, or used in payment for goods, on the holder's endorsement against his original signature; also written *travellers(') cheque.*

1891 (*title no. 24775 in Library of Congress copyright registration bk.*) American Express Company, Travelers Cheque. Ten Dollars. **1894** *N.Y. Tribune* 11 July 5/2 The American Express Company's travellers' checks are of great assistance to tourists abroad. **1907** M. Rollins *Money & Investments* 218 Express companies have made a speciality of issuing 'travellers' cheques'. **1922** F. Scott Fitzgerald *Beautiful & Damned* II. i. 150 On his dressing-table were spread a number of articles..their tickets to California, the book of traveller's checks. **1957** D. du Maurier *Scapegoat* iii. 32 Twenty-five pounds of travellers' cheques still uncashed. **1969** *Times* 15 Nov. 7/7 The counterfeiting of travellers cheques. **1981** 'E. Lathen' *Going for Gold* iii. 29, I assume these foreign traveler's checks work like American Express?

Hence **travellership** *rare*, in various *nonce* usages.

1920 Joyce *Let.* 20 Aug. (1966) III. 17 Giorgio has been offered a position here in an American Trust Agency which would develop into a secretaryship and travellership for same. **1961** *Times* 24 June 9/6 We left early..accompanied in somewhat distant travellership by an austerely demeanoured delegation from Communist China.

travelling, traveling, *vbl. sb.* Add: **b.** *travelling expenses* (earlier and later examples); *travelling bag* (earlier U.S. example), *cap* (earlier U.S. example), *case, cloak, clock* (earlier example), *dress* (earlier example), *rug, trunk* (later examples); **travelling scholarship** (examples).

1838 C. Gilman *Recoll. Southern Matron* ii. 18 He..called the little boy who held the travelling bag a 'black-faced nigger'. 1790 *Pennsylvania Packet* 2 Jan. 4/2 (Advt.), Trimmings..for Gentlemen's Travelling Caps. 1744 H. Purefoy *Let.* 26 Feb. (1931) I. 119, I desire you will send me an allarum to pull up together with a Travelling Case. 1895 *Montgomery Ward Catal.* Spring & Summer 257/1 Ladies' and men's toilet and traveling cases. 1984 A. Price *Sion Crossing* iv. 48 He walked meekly to the nearest bench..tucking his travelling case under him. 1854 E. B. Browning *Plea for Ragged Schools of London* I If she shakes a travelling cloak, Down our Appian roll the scrolls. 1944 A. Clarke *Coll. Plays* (1963) 245 A figure appears, tall and handsome, in travelling cloak with tricornered hat. 1860 C. M. Yonge *Hopes & Fears* II. xii. 236 Phoebe was strongly tempted to answer, but the little travelling clock struck. 1815 J. Mayne *Jrnl.* Feb. (1909) xii. 282 We ran off, the instant we arrived, in our travelling dresses, were in the theatre at eight o'clock. 1793 J. Woodforde *Diary* 23 Oct. (1929) IV. 75 Travelling Expenses..amounted in the whole to..78.19.7. 1907 G. B. Shaw *John Bull's Other Island* I. 13 An advance on his salary—for travelling expenses. 1977 D. Francis *Risk* xvii. 230 He'd invented travelling expenses to the races for horses which..had never left the yard. 1911 *Daily Colonist* (Victoria, B.C.) 23 Apr. 18/5 (Advt.), Lost—Travelling rug and overcoat, between Empress Hotel and the outer wharf. 1977 C. McCullough *Thorn Birds* iii. 66 The big tartan traveling rugs all the suitcases bore on their outsides. 1911 R. Brooke *Let.* 1 Mar. (1968) 281 Benians, of John's, is staying in Munich a week on his way round the world on one of these Kivet Kahn Travelling scholarships. 1967 E. Lemarchand *Death of Old Girl* i. 13 I'm certainly going to mention the art department's successes, especially Miss Cartmell's travelling scholarship. 1854 *Rep. Trans. Pennsylvania State Agric. Soc.* 97 Their very handsome riding saddle and russet travelling trunk. 1948 *Christian Sci. Monitor* (Midwestern ed.) 12 Feb. 10/4 A 'traveling trunk' program, in which the institute sends out to schools actual trunks of touchable Texas gear.

travelling, traveling, *ppl. a.* Add: **1. b.** (Earlier and later U.S. examples.)

1775 F. Asbury *Jrnl.* 6 Nov. (1821) I. 124 At this meeting we admitted F.P.T.F. and J.H—y as travelling preachers. 1874 E. Eggleston *Circuit Rider* xxvii. 252 The incessant activity of a travelling preacher's life.

c. *Physics. travelling wave*, a wave in which the nodes and antinodes travel (cf. *standing wave* s.v. *standing *ppl. a.* 11 e); freq. *attrib.*, as *travelling wave tube*, an electron tube in which a guided electromagnetic wave is amplified by interaction with a beam of electrons travelling at about the same velocity.

1908 C. P. Steinmetz *Gen. Lect. Electr. Engin.* 273 Where a travelling wave is reflected, the combination of the reflected wave and the incoming wave produces a standing wave or oscillation, that is, a wave in which the voltage maxima and the zero points or nodes have fixed positions on the line. 1946 *Wireless World* Nov. 371/3 The travelling-wave tube..consists..of a long and straight helix of wire supported in an evacuated glass envelope containing also an electron gun for producing an electron beam and a collector. 1963 G. Troup *Masers & Lasers* (ed. 2) v. 66 For the travelling wave maser, the gain coefficient..is [etc.]. 1967 *Oceanogr. & Marine Biol.* V. 25 A travelling wave is an essential feature of surge phenomena in the North Sea. 1974 Travelling-wave tube [see *Karsa 5*].

2. Special collocations. *travelling exhibition*; **travelling circus,** (*a*) a circus which travels from place to place giving performances; (*b*) *Mil. slang*: in the war of 1914–18, a mobile military unit; a squadron of aeroplanes (cf. *circus 2 d); also *fig.*; **travelling library,** a library which is transported from place to place and serves remote rural communities, hospitals, etc.; a mobile library; **travelling salesman** = Traveller, traveler 3; **travelling salesman problem** *Math.*, the problem of determining the shortest route that passes through each of a set of given points once only and returns to the starting point; **travelling stock** *Austral.* and *N.Z.*, livestock which is driven from place to place; freq. in Comb., as *travelling stock road, route*; **travelling stock reserve,** land decreed as stock-routes; cf. *stock-route* s.v. stock *sb.*[1] 63.

1883 *Harper's Mag.* June 137/1 The travelling circus..had journeyed on and left her. 1917 [see *circus 2 d]. 1919 F. A. McKenna *Battery A—1914 Field Artillery in France* 36 Field Marshal Von Hindenburg's troops, nicknamed by the Yanks, 'The Travelling Circus', composed of the famous Prussian Guards and picked Turkish storm troops, were reported in the sector opposite. 1946 *Happy Landings* (Air Ministry) July 1/1 Among the latest exhibits added to the Directorate's 'travelling circus' are components from a Sabre engine. 1974 G. Mitchell *Javelin for Jonah* xi. 137 He had been with a travelling circus for some time, but they dismissed him. 1937 *Discovery* Aug. 236/1 A travelling exhibition which will penetrate into the remotest country districts. 1977 J. R. L. Anderson *Death in City* vi. 87 Bringing 'Pictures to the People' in the form of travelling exhibitions. 1910 A. E. Bostwick *Amer. Public Library* 108 Traveling libraries are simply collections of books sent to communities, associations, or individuals for circulation. 1960 *Library Assoc. Rec.* Aug. 265/2 Travelling Library, a vehicle of small size..shelved or otherwise equipped to provide a rural service to villages and isolated farms and houses. 1982 H. Innes *Black Tide* v. i. 220 The *Mabinogion*..She'll get it from the travelling library. 1898 *South Florida Sentinel* (Orlando) 1 July 3/3 The popular traveling salesman..will leave in a few days. 1964 *Jrnl. Operations Res. Soc. Amer.* II. 393 Little is known about the travelling-salesman problem. 1960 G. A. W. Boehm *New World Math.* 114 Equally expanding is the travelling-salesman problem, with which a good many mathematicians have wrestled unsuccessfully for more than twenty years. 1978 I. B. Singer *Shosha* v. 105 Baskets and boxes accumulated from the time Zelig was a traveling salesman. 1979 Page & Wilson *Introd. Computational Combinatorics* ix. 79 These are very interesting paths and circuits as they can be generated into the well-known 'travelling salesman' problem. 1891 R. Wallace *Rural Econ. Austral. & N.Z.* xxiv. 364 Should the land [of a sheep station] be mountainous or a travelling-stock road pass through it, the numbers [of] occupancy must be increased. 1930 L. G. D. Acland *Early Canterbury Runs* 1st Ser. viii. 196 There was a travelling stock reserve there, and it was in the hut belonging to it that the man was murdered. 1977 C. McCullough *Thorn Birds* vi. 120 There was an official Travelling Stock Route or TSR winding its way near the Barwon River.

travelogue (træˈvelɒg). *orig. U.S.* [f. Travel *sb.* + -ologue, after Monologue *sb.*] An (illustrated) lecture about places and experiences encountered in the course of travel; hence a film, broadcast, book, etc., about travel; a travel documentary.

1903 *Daily Chron.* 16 Apr. 6/7 Mr. Burton Holmes, an American entertainer new to London, delivered last evening the first of a series of 'Travelogues'. 1921 *Glasgow Herald* 7 Nov. 10 The..Travelogue film, 'With Allenby in Palestine and Lawrence in Arabia'. 1927 H. E. Fosdick *Pilgrimage to Palestine* p. vii, Some [books] are simply travelogues in which the successive experiences of the traveller furnish the strand for the narrative. 1931 [see *cinema 4 a]. 1936 [see *documentary a. 4]. 1935 R. Macaulay *Personal Pleasures* 67 Dido and her court feasted Æneas and his warriors, and after supper listened to his mournful travelogue. 1956 S. Ertz *Charmed Circle* 78 Reality lay wholly within the boundaries of the United States, and..what she saw in Europe was a sort of illustrated travelogue. 1976 A. Davis *Television* 119 There were 100 programmes and series from the travelogues of Armand and Michaela Denis. 1983 *Listener* 20 Jan. 27/2 It is part travelogue, part autobiography.

traversa: see *traverso.

traversal (trævəˈsəl, træˈvəːsəl). [f. Traverse *v.* + -al¹.] = Traverse *sb.* 2.

1909 in Webster. 1936 [see *line frequency* s.v. *line *sb.*² 32]. 1955 L. A. Wheeler in W. Pauli *Niels Bohr* 171 The increase of a in the traversal of the fission barrier will be of the order of unity. 1982 *Times* 22 Oct. 10/5 As he began on Wednesday the traversal of all Beethoven's sonatas..it was piercingly evident that here was someone who..more than ever than they play the piano. 1984 *Observer* 15 Jan. 49/8 Speed—the traversal of the maximum space in the minimum time—is the essence of modernity.

traverse, *sb.* Add: **3.** (Earlier example.)

1804 M. Lewis in *Orig. Jrnls. Lewis & Clark Exped.* (1905) VI. 232 A Circumferentor..has also been employed in taking the traverse of the river.

16. Also *spec.* a pair of right-angled bends in a trench for protection against enfilading fire. [Later examples.]

1767 Sterne *Tr. Shandy* IX. xxvi. 115 Uncle Toby..got his wound before the gate of St. Nicolas, in one of the traverses of the trench. 1802 C. James *New Mil. Dict.* s.v. *Trenches*, On the angles or sides of the trench, there are lodgements or epaulements, by means of traverses, in order to hinder the sallies of the garrison. 1917 E. A. Mackintosh *War, the Liberator* (1918) 136 As MacTaggart turned back at the corner of the traverse he felt strangely comforted by the sight of MacRae. 1957 P. Kemp *Mine were of Trouble* iv. 76 It struck me that the trenches were very badly constructed, being..dug almost in a straight line instead of with traverses. 1971 S. Hill *Strange Meeting* ii. 139 The men had been getting their mid-day meal into the traverse when the bomb had landed in the middle of them.

traverse, *v.* Add: In British (but not American) English the vb. is now often stressed on the 2nd syllable, and this has influenced the pronunciation of its derivative forms. The sb. is still normally stressed on the 1st syllable.

traverso (trævəˈsəʊ). *Mus.* Also **traversa.** [a. It.] = *transverse flute* s.v. *transverse *a.* 1 d.

[1776 J. Hawkins *Gen. Hist. Sc. & Pract. Mus.* II. iv. iv. 452 It seems that the invention of the traverse flute is not to be attributed either to the Germans or the Helvetians.] 1801 Busby *Dict. Mus.* s.v. *Traversa*, (Ital.) a German flute. 1884 Grove *Dict. Mus.* IV. 163/1 *Traverso*, (Ger. *Querflöte*), the present form of flute, held *square* or *across* (*à travers*) the performer... In Bach's scores it is called Flauto traverso, Traverso, and Traversiere. 1932 C. S. Terry *Bach's Orchestra* iv. 79 Bach's treatment of the traverso has peculiar interest. 1938 *Oxf. Compan. Mus.* 949/1 *Traversa, traverso* (It.); traversière (Fr.); Traversflöte (Ger.). 1979 *Early Music* July 357/1 Open and forked fingerings..are basically the same on oboe and traverso.

travesty, *sb.* **2.** Chiefly *Theatr.* **a.** Delete *rare* and add later examples. *Spec.* (dressing) in the attire of the opposite sex. Freq. (*en*) *travesti*.

The phr. *en travestie*, which is not recorded in Fr., represents a misinterpretation of the F. pa. pple. as a sb.

1850 Thackeray *Pendennis* II. xi. 102 He went into the pit, and saw..that inimitable artist, Mr. Tom Horseman, dressed as a woman. Horseman's travestie seemed to him a horrid and hideous degradation. 1927 G. B. L. Wilson *Dict. Ballet* 212 *Petipa*, Marie S... Her husband created for her a dance, *The Little Hunch*, in which she appeared travesti. 1959 *Times* 3 Nov. 15/5 Defrance's troupe leaders and the girls in travestie receiving their last-minute counsels remain unaffectedly convincing. 1978 *New Yorker* 11 May 131/1 Nero, Otho, Tamerlane, and Julius Caesar will still have to be played by women *en travesti* or by countertenors. 1980 *Daily Tel.* 15 Dec. 10/2 Shapely young women have been showing off their legs *en travesti* since they were allowed on the English stage.

b. *Comb.* **travesty role,** a role designed to be played by a performer of the sex opposite to that of the character represented.

1958 *Listener* 5 June 955/3 Michel Sénéchal handled the travesty role of Plataée with tact and sang the difficult music in an accomplished style. 1978 *Times* 23 Aug. 11/4 At the Coliseum Dennis Wicks makes the most of his travesty role.

travisher (træˈvɪʃəɹ). [Origin uncertain: *travish* is recorded as a dial. form of *traverse* (*Eng. Dial. Dict.*).] A carpenter's shave used for the final smoothing of chair seats.

1926 *Architects' Jrnl.* LXIX. 128/1 The travisher was invented—a spokeshave with a curved cutting iron, to be used, after the adze, in making the seat more comfortable. 1953 A. Johnson *Household & Country Crafts* xi. 151 For the final smoothing tools, the top two are travishers, for finishing off the elm seats after adzing. 1960 J. Arnold *Shell Bk. Country Crafts* 133 Smoothing off follows with a succession of shavers:—the travisher, the smoothing-off iron and a devil.

travois, -voise, *sb.* Add: Also † **travoi, try-voy,** etc. (which represent an older form). (Earlier and later examples.)

1847 K. Carson in W. F. Cody *Wild West* (1888) 349 The Tlamaths..prevented his body from falling into our hands by throwing it upon a travoi. 1879 J. G. Bourke *Diary* VIII. 847 (MS.), He] says he will stick with the column if he has to be carried on a travois. 1926 C. S. Walgamott *Reminisc. Early Days* 26/2 George..found..five head of horses with two sides the other three horses dragging either lodge poles or try-voy, possibly both. 1959 E. Trebing *Indian Sats* 80 These Indians trained their dogs to help with the transport job by carrying packs or by dragging tent poles and small travois on which burdens could be lashed. 1974 J. A. Michener *Centennial* iv. 149 The travois, that primitive but functional invention for hauling goods, was constructed always from two poles used otherwise to support the tipi.

Travolator (træˈvəʊleɪtəɹ). *orig. U.S.* Also with small initial, and **travelator, travellator.** [f. Travel *v.*, after *escalator.] The proprietary name of a moving pavement designed for use at railway stations, airports, shopping centres, etc. (see quot. 1955).

1955 *N.Y. Herald Tribune* 28 Oct. 4/3 A moving sidewalk with cleated escalator treads has been developed by the Otis Elevator Company... Designers foresee use of the moving platforms, called 'Trav-o-lator', for such congested areas as airports, subway stations, [etc.]. 1957 *Country Life* 13 June 1186/3 Travolators..represent an improvement on the moving-staircase principle, since they will possess a flat surface along which the passengers can walk while they are moving. 1957 *Economist* 28 Sept. 1041/1 Two 300 foot 'travelators'..tubular escalators..that carry pedestrians along slight gradients—are expected to be in operation at the Bank underground station in London towards the end of next year. 1958 *Trade Marks Jrnl.* 26 Feb. 196/1 *Trav-o-lator*..Passenger-carrying conveyors (machines). Otis Elevator Company..New York, Eleventh Avenue, New York, United States of America; Merchants and manufacturers. 1959 *Official Gaz.* (U.S. Patent Office) 23 May 1996 Otis Elevator Company, New York, N.Y. Filed Nov. 28, 1958. *Trav-o-lator*. For Endless Conveyors. First use July 28, 1953. 1959 *Archit. Rev.* CXXVI. 321 Victorian Street, which is a rather dead shopping area, receives wedges of new shops, with rear servicing, on both sides of the travelator system. 1969 W. Haggard *Conspirators* iv. 40 He bought a ticket from the coin-machine, going down to the train on the travolator. He didn't walk but let it carry him. 1969 *New Scientist* 10 Oct. 180/3 Whether all the oceans are continually in movement, like giant travolators. 1977 B. Cocks *Mid-Victorian Masterpiece* xx. 195 Members found that even without a travelator they could reach the Chamber within the six minutes allowed before the lobby doors are locked for a division. 1980 [see *terminal *sb.* 5 a].

trawl, *sb.* **I. 2.** Restrict † *Obs.* to sense in Dict. and add: **b.** *fig.* An act of 'trawling' in order to find a person or persons (esp. a new employee) from among a larger population.

1971 *Daily Tel.* 2 Mar. 3/3 A 'trawl' is being made among civil servants to find a suitable man and an appointment is expected within two months. 1980 C. Moorhead *Fortune's Hostages* v. 97 The generals..rounded up 4,000 suspected leftists. They did very well in the trawl. 1984 *Times* 5 Apr. 1/8, I am going to make a serious trawl through the profession and see if I can find circuit judges whom I can safely appoint.

III. 4. *trawl-line* (earlier example).

1867 G. E. Clark *Seven Years of Sailor's Life* 308 The old mother fish, full of spawn, are snaked on, to their miles of trawl line.

trawl, *v.* Add: **1. c.** (Later *fig.* examples.)

1979 'J. le Carré' *Smiley's People* (1980) xix. 234 Kirov dutifully trawls the émigrés, but without result. 1984 *Observer* 8 Apr. 52/6 We trawled Britain, the United States, Australia and South Africa for a chief executive.

trawler. 3. trawler-man: delete † and add later examples.

1934 (see *Baltimore) 8 Oct. 7/5 Trawlermen of Hampton, Phoebus, Portsmouth and nearby harbors are planning a union. 1958 *Times* 3 Oct. 11/6 The distant water trawlermen..are doing work that for its hazards, long hours, and gruelling conditions would be difficult to compare with any job ashore. 1976 'W. Trevor' *Children of Dynmouth* i. 9 A few boys became trawler-men, but life was easier and richer at the fish-packing station. 1984 *Times* 20 Feb. 14/4 As the son of a trawlerman..he wanted to see the very highest level of safety of people crewing vessels of all kinds.

Traxcavator (træˈkskəveɪtəɹ). *orig. U.S.* Also **traxcavator.** [Blend of Track *sb.*, Tractor, and Excavator.] The proprietary name of a type of mechanical excavator which moves on endless steel bands or tracks (see quot. 1940).

1940 *Official Gaz.* (U.S. Patent Office) 30 Apr. 1035/2 *Traxcavator*. For excavating, grading and loading machinery—namely high shovels, tractor shovels, tractor loaders, bull-dozers, graders, and the like. Claims use since Jan. 22, 1940. 1953 *Law Rep. Queen's Bench Div.* II. 60 A traxcavator, a tracked vehicle with a speed of 2½ miles an hour turned this corner and stopped. 1956 *Trade Marks Jrnl.* 22 Feb. 183/2 Traxcavator..Machines for mechanical handling. Caterpillar Tractor Co.; 800, Davis Street, San Leandro, State of California, United States of America; manufacturers. 1968 *Industrial Tribunals Reports* III. 176 The material was excavated by mechanical traxcavator and a traxcavator. 1977 *N.Z. Herald* 5 Jan. 1 2/5 (Advt.), Mortor. Brown Asphalt Ltd., paving Auckland's carparks, drives, etc since 1948. Own traxcavators, graders, rollers and pavers.

tray, *sb.*¹ Add: **1. a.** Also *spec.* a tray of food brought to one not able or not wishing to eat at table; hence (*loosely*), a light snack.

1914 S. Wood *Wire Virginia* xiv. 196 I'll ask them to bring you up a tray. What would you like? A little bread and fish—or chicken? 1976 E. F. Benson *Trouble for Lucia* ix. 199 We might bring her a tray instead of supper. 1951 L. Hellman *Autumn Garden* ii. 88 You have had no dinner? I have had a tray. 1982 J. S. Borthwick *Case of Hook-Billed Kites* (1984) xx. 72 Mrs. Brest and Miss Felkins had had trays in their rooms.

d. = *sand-box 2 c.

1969 F. MacCann *Cats* ii. 33 The tray, or pan, should have a low side all round it; put it in a dark corner—not close to the cat's bed. 1968 M. Sourbeck *Cat Breeding* 176 Once the use of the tray has been taught in the house it is an excellent plan to encourage the kitten to go outdoors. 1969 'A. Gilbert' *Missing from Home* viii. 103 You really can't have a cat in the flat, there's no outlet, and I think..a tray is unhygienic. I don't care how often it's changed.

e. *Austral.* The part of a truck on which goods are carried.

1960 'N. Shute' *Trustee from Toolroom* v. 100 The semi-trailer stood by the aircraft with the sausage-like component on the tray swathed in hessian. 1980 P. Davis *Australians on Road* xiv. 125/2 Ford management conceived the idea of producing a vehicle which could be sold to be an essential part of farm equipment, yet was comfortable enough for the farmer to take his wife out for an evening in town. The result was a passenger-type cab, married to an enclosed load tray; it was called the coupé utility, later corrupted to 'ute'.

4. tray-cloth (examples); **tray lunch(eon),** lunch served on a tray; a light lunch; **tray-mobile** *Austral.* and *N.Z.* = *mobile *sb.*⁴ a small wheeled table or stand on which food, etc., may be transported; a tea-trolley; **tray stand,** a small table on which to rest a tray; **tray supper,** supper served on a tray; a light supper; **tray-top** (*a*) a rimmed table top which can be removed and used separately as a tray; (*b*) *Austral.* a truck with a pick-up body.

1884 C. M. Yonge *Stray Pearls* 1. xiii. 116 I. Lawrence *Collier's Friday Night* (1934) ii. 59 Beatrice Wyld sits in the armchair, and Nellie Lambert on the sofa, the latter doing drawn-thread work on a white tray-cloth. 1971 *Islander* (Victoria, B.C.) 25 Dec. 5/2 The very best cups and saucers..were set on a hemstitched linen traycloth in a large, round wicker tray. 1970 V. Canning *Great Affair* vi. 58 The steward..served our tray lunches. 1936 P. Bottome *Level Crossing* xvi. 193 After what Nellie called a 'tray luncheon', she suggested showing Deidre the house. 1948 V. Palmer *Golconda* xx. 169 Her attention was on the traymobile she pushed and was loaded for tea. She began to pour out the tea. 1965 G. McInnes *Road to Gundagai* v. 78 Against the vacant wall in the 'dumb waiter' or 'traymobile' on which food and crockery come in from the kitchen. 1844 T. Webster *Encycl. Domestic Econ.* 241 A tray-stand..formed of two frames and girth to fold up. 1895 *Army & Navy Co-op. Soc. Price List* 15 Sept. 311 Cairo Tray Stand. 1825 H. Wilson *Mem.* I. 43 Amy gave us merely a tray-supper, in one corner of the drawing-room. 1933 H. Edwards *All Night at Mr. Stanyhurst's* 23 Me and you were going to the play, and coming home to a tray supper. 1962 M. Carleton *Dread Sunset* (1963) ii. 36 Her own tray suppers looked deceptively simple. 1934 Webster, Tray-top table. 1951 *Catal. Exhibits, South Bank Exhib., Festival of Britain* 144/1 Occasional table with removable tray top. 1969 *Northern Territory News* (Darwin) *Focus '69* 97/3 It is a relief for the semi or tray top crews when they reach Alice—and the bitumen of the Stuart Highway. 1979 *Truck & Bus Transportation* (N.S.W.) Sept. 46/1 The general cartage fleet comprises..three Albions with tray-top bodies.

treacle, *sb.* Add: **III. 5.** *treacle pud(ding, toffee, -well* (earlier example); **treacle-posset** (earlier example); **treacle sleep** *colloq.*, a deep, unbroken sleep (later example).

1739 E. Smith *Compl. Housewife* (ed. 9) 219 Those who can't afford mountain-wine, may drink treacle-posset. 1974 J. Murdoch *Sacred & Profane Love Machine* 314 What about some treacle pud, it's awfully good here. 1861 Mrs. Beeton *Bk. Househ. Managem.* 688 Rolled treacle pudding..suet crust..treacle..grated ginger. 1973 'H. Howard' *Highway to Murder* x. 117, I slept a treacle sleep from eleven to seven without the ghost of a dream or anything. 1885 R. Holland *Gloss. Words County of Chester* 367 Treacle toffy, s., sometimes called toffy sticks, a very favourite sweetmeat amongst Cheshire school children. 1924 'R. Crompton' *William—the Fourth* viii. 151 William had..been..enjoying a *treacle toffee*. 1863 'A. T. Ellis' *Other Side of Fire* xx. 133 Think what fun it will be..parkin and treacle toffee. 1865 'L. Carroll' *Alice's Adventures in Wonderland* vii. 107 The Dormouse..said, 'It was a treacle-well.'

treacle, *v.* Add: **2. b.** Also *intr.*

1915 H. G. Wells *Boon* iv. 124 Going round with the lantern when one is treacling for moths. 1941 —— You can't be too Careful iii. xxii. 218 As moth hunters treacle for moths.

3. (Later example.)

1966 J. Betjeman *High & Low* 25 In blest Bethesda's limpid pool Comes treacling out of Sunday School.

treacly, *a.* Add: *fig.* (Examples in musical contexts.)

1930 J. B. Priestley *Angel Pavement* iv. 176 The organ was shaking out cascades of treacly sound. 1947 Auden *Age of Anxiety* (1948) i. 26 A doomed Sodom dancing its heart out To treacly tunes. 1980 S. Brett *Dead Side of Mike* xiii. 143 Treacly, undistinguished music.

tread, *sb.* Add: **III. 10. b.** (Examples with reference to motor transport.) Also *spec.* the thick moulded surface of a pneumatic tyre, which runs in contact with the ground (as opp. to the sidewalls). Cf. *retread *sb.* 1.

1902 G. L. Freeston in A. C. Harmsworth *Motors & Motor-Driving* 237 The Collier..This type is provided with an exceptionally stout tread. 1913 *Sci. Amer.* 11 Jan. 35/1 The tread is made up independently of the tyre by means of layer upon strips of rubber..in such a way that the center of the tread is thicker than the edges. 1929 *Rubber & Tyre Rev.* Jan. 148/1 On pneumatic tyres for buses and goods vehicles a tread..similar to..tractor tyres can be safely assumed. 1964 *Amer. Speech* XXXIX. 275 Tread, a..the outer, patterned surface of a pneumatic tire. It is made up of the cap which contacts the road surface and the sidewall edges. 1982 M. Russell *Rainbow* iv. 38 People come along fast... The treads find they've a little extra to do.

11. c. *Geomorphol.* The approximately horizontal part of each of the step-like parts of a glacial stairway or similar landform.

1900 [see *stairway b]. 1939 F. E. Matthes *Geol. Hist. Yosemite Valley* 95/2 Rock structure, or, more broadly, rock resistance..determines the position and level, of the treads in a given canyon the individual sills and treads shall develop. 1954 *Jrnl. Glaciol.* II. 421 Fig. 2..shows a *roche moutonnée* of step-like form... The contrast between the smoothed upper tread and the irregular 'plucked' riser can be seen. 1968 R. W. Fairbridge *Encycl. Geomorphol.* 467/1 Where the overdeepened treads are undrained, there rock-cut depressions or partly excavated pools... They are known as paternoster lakes.

IV. 12. (sense 4) *tread-mire*; **tread plate,** (*a*) a footplate or runner which forms or protects the step on a vehicle; (*b*) (see quot. 1967); **tread-trap** *Archæol.*, a wooden device for trapping an animal by the foot.

1888 G. M. Hopkins *Poems* (1967) 105 Stanches, starches Squadroned masks and manmarks treadmire toil there Footfretted in it. 1949 *Automobile & Carriage Builders' Jrnl.* CIV. 59 (heading) P-G-P aluminium tread plates. 1967 *Gloss. Sanitation Terms (B.S.I.)* 57 *Tread plate*, 1. Glazed ceramic, or other hard wearing, edging to the floor finish adjacent to the channel of a urinal. 2. Glazed ceramic, or other hard wearing, non-slip footrests on each side of a squatting W.C. pan. 1952 J. G. D. Clark *Prehist. Europe* ii. 51 A type of tread-trap..appearing for the first time in the Late Bronze Age..symbolizes the part still played by trapping in the closing stages of European prehistory.

Hence **tread-ed** *a.* [-ed²], of a tyre: having or furnished with a moulded tread; **treadless** *a.* [-less], having no tread or treads (esp. of tyres).

1906 *Westm. Gaz.* 6 Mar. 4/2 The substitution of single pneumatic tyres for the present double-treaded ones. 1968 A. Diment *Great Spy Race* x. 186 The treadless tyres teetered on tiptoe across the road. 1973 'D. Rutherford' *Kick Start* vi. 139 The bike canting at a steep angle, we were holding on to the road surface by the treaded edge of the competition tyre. 1974 *Observer* 3 Nov. (Colour Suppl.) 27/2 Wearing treadless shoes, [we] were let into a dusty attic with a forged key. Another prisoner followed us, replacing anything we might move.

tread, *v.* Add: **A. 2.** *pa. t.* Also: treaded (only in phr. *treaded water*: see sense 7).

1944 *Stars & Stripes* (London ed.) 1 May 3 While Huie and four crewmen clambered into the liferaft, three others treaded water for three hours before succeeding in blowing up another raft by lung power. 1952 H. E. Bates *Purple Plain* x. 113 He treaded water for a moment or two. 1974 J. Irving *158-Pound Marriage* viii. 193 He said nothing; we treaded water.

B. 3. b. *to tread on air* (earlier examples); cf. *to walk upon air* s.v. *walk *v.*¹ 5 l.

1796 R. M. Roche *Children of Abbey* I. viii. 154 Such were the ideas of the innocent and romantic Amanda; ideas, which made her seem to tread on air. a 1817 Jane Austen *Northanger Abbey* (1818) II. xi. 223 If Wednesday should ever come!..It came—it was fine—and Catherine trod on air.

4. c. *to tread on the gas*: see *gas *sb.*¹

7. *to tread water*: also *fig.*, to withhold oneself from action, to 'mark time'.

1941 J. Lees-Milne *Ancestral Voices* (1975) 16 Although they miscalculated in assuming that the campaign would be over before last winter, they have been treading water since then, and merely keeping the war line. 1967 *Guardian* 24 May 9 In the absence of the Secretary-General, the UN delegates were treading water. 1980 N. Marsh *Photo-Finish* vi. 180, I am really..treading water until the police arrive.

treadle, *sb.* Add: **4. treadle mat,** a mat or casing which activates a mechanism when stepped on or otherwise depressed.

1937 *Times* 13 Apr. p. xvi/4 Pneumatic treadle mats cover the steps so that the doors when released by the driver may be automatically opened by passengers standing on the steps. 1966 *Electronic Weekly* 16 Mar. 3/1 The passage of the wheels over a treadle mat causes the light to change to red.

treadmill, *sb.* Add: trea·dmiller, a person who is 'on the treadmill' (*fig.*), esp. one who follows a dull and arduous working life.

1923 D. H. Lawrence *Kangaroo* vii. 164 Better a 'wicked creature' any day, than a mechanical treadmiller of a careerist. 1956 H. Gold *Man who was not with It* (1965) viii. 66, I even thought of..that shy treadmiller Joy.

treason, *sb.* Add: **1. b.** *treason of the clerks* = *trahison des clercs.

1940 W. Empson *Gathering Storm* 34 Treason of the clerks, boys, curtains that descend, Lights becoming darks, boys, waiting for the end. 1970 C. C. O'Brien *Camus* iii. 61 The proposition that failure to take an anti-communist stand constituted 'the treason of the clerks' of which Benda spoke. 1979 *Guardian* 6 June 14/7 Ex-King's men will be revealed or those who bought liberalism led them to 'the treason of the clerks'.

4. a. *treason trial*.

1930 *Economist* 11 Dec. 1054/2 Treason trials are the grand eclipse of revolutionary régimes; and for years the Soviet Government, their prosecutors and their courts have managed to stage a series of such performances in almost uninterrupted succession. 1979 A. Melville-Ross *Two Faces of Nemesis* viii. 47 Treason trials don't help the national image.

treasure, *sb.* Add: **2.** Also as an affectionate term of address.

1920 'K. Mansfield' *Let.* 31 Oct. (1977) 194 But, my treasure, my life is ours. You know it. 1967 N. Freeling *Strike Out* 40 Next week, treasure, we're going to make a cruise... Go and buy yourself some clothes.

4. *treasure-box* (earlier example), *-hunter* (earlier and later examples), *-seeker*; **treasure-hunt** (examples) (*spec.* a game in which hidden objects are searched for, often by following a trail of clues); **Treasure State** *U.S. slang*, the State of Montana.

1876 Treasure-box [see *powder keg* s.v. *powder *sb.*¹ 8 b]. 1913 J. Vaizey *Collage Hold* viii. 1 This is fitting..cia treasure hunt'..lots of presents, stowed away in odd corners. 1919 H. H. Jones *Road to En-Dor* (1920) xii. 149 A treasure-hunt has a glamour of its own. 1939 T. S. Eliot *Family Reunion* ii. iii. 110 You have played a..giddy journey. Think of it as like a children's treasure hunt. 1977 M. Green *Children of Sun* 30 One of those Twenties' treasure hunts, in which people drove..across all England, in search for some otherwise unprocurable item. 1851 H. Melville *Moby Dick* II. xxxvi. 142 He proceeds very heedfully, like a treasure-hunter in some old house, sounding the walls to find where the gold is masoned in. 1983 S. Vizinczey *Innocent Millionaire* xiii. 111 Maybe I was meant to be a treasure-hunter. 1890 J. G. Frazer *Golden Bough* II. iv. 367 The treasure-seeker places the rod on the ground after sundown, and when it rests directly over treasure, the rod begins to move as if it were alive. 1982 'C. Aird' *Last Respects* viii. 81 There are treasure-seekers.. Inspector, who would..not care that they were destroying an archaeology. 1934 M. H. Weseen *Dict. Amer. Slang* 421 Treasure State, Montana. 1976 *Billings* (Montana) *Gaz.* 20 June 16-C/1 A solid century of mining has left a heavy mark on that state's gold, silver, copper and coal reserves. So the slogan, 'Treasure State', which used to grace Montana license plates, is still appropriate.

treasury, *sb.* Add: **3. c.** *pl.* Treasury bills.

1922 *Daily Tel.* 12 June 2/1 New secondhand Treasuries were dealt in at 2½ per cent. 1930 *Daily Express* 6 Oct. 14/4 The banks bought short-dated Treasuries at 2 per cent.

6. treasury-bench (earlier example); **treasury note,** (*a*) for *U.S.* read chiefly *U.S.*; = *currency note* s.v. *currency 6; (earlier and later examples); (*b*) (in Dict. s.v. *treasury letter*) (earlier example); **treasury tag** = *India tag* s.v. *India 6; formerly consisting of a length of lace with a blunt pin at one end which was secured through a socket at the other.

1775 F. E. Boscawen *Let.* 28 June in C. Aspinall-Oglander *Admiral's Widow* (1942) 64 Our cruel patriots..would willingly wade through blood, provided it led to the Treasury Bench. 1756 in S. M. Hamilton *Lett. to Washington* (1898) I. 202 Who is hereby required to pay the same in Treasury Notes, to be emmitted by Virtue of the said Act of Assembly. 1820 *Kaleidoscope* 25 July 30/1 Or (summoned by a Treasury-note) Night after night to sit and vote. 1923 A. Huxley *Antic Hay* iii. 50 It was with reluctance that Gumbril parted from his Treasury notes. 1974 *Encycl. Brit. Micropædia* X. 103/2 *Treasury note*, government security, usually marketable, with maturity ranging from one to five years. 1912, 1963 Treasury tag [see *India tag* s.v. *India 6]. 1975 'M. Sinclair' *Long Time Sleeping* iv. 48 A little tray of pins, paperclips and little bits of coloured string known as Treasury tags.

treat, *sb.*¹ Add: **5. a.** (Earlier example.)

1770 J. Wedgwood *Let.* 13 Oct. (1965) 98 Your stay here..was a most agreeable treat to us and all your friends in this part of the world.

b. Delete *vulgar* and add earlier and later examples. Also adjectivally.

1898 [see *duke *sb.* 7]. 1942 *R.A.F. Jrnl.* 18 Apr. 10 We..set light to a..dump of Iraqi petrol which went up a fair treat. 1959 A. Sillitoe *Loneliness of Long-Distance Runner* iii. 185 The sports ground looked a treat: with big tea-tents all round and flags flying. 1960 *Guardian* 23 Dec. 7/2 He had a stroke..but he's come on a treat. 1984 *New Yorker* 17 Sept. 56/1, I knew this floor had life left in it... It's come up a treat.

treat, *v.* Add: **7. d.** Colloq. phr. *to treat 'em rough*, to manhandle (people, etc.), to treat harshly or aggressively. As a motto: see quot. 1918. Also (hyphened) as *attrib. phr.* Chiefly *U.S.*

1918 W. H. Allen *Stories of Americans in World War* 162 The men in the tank service have chosen 'Treat 'Em Rough' as their slogan, and a huge black cat as the emblem and mascot. Any cat that looks black enough and fierce enough is apt to be kidnapped and adopted by some tank battalion. 1930 *Amer. Speech* VI. 83, I never have the same girl twice; I take 'em young and treat 'em rough. 1962 *Times* 6 July 15/4 A treat-'em-rough warden of the old school.

treatabi·lity. [f. Treatable *a.*: see *-bility.] The quality of being treatable; responsiveness to medical or psychotherapeutic treatment.

1957 *Rep. Committee on Homosexual Offences* v. 32 in *Parl. Papers 1956–57* (Cmnd. 247) XIV. 85 The question of treatability. 1959 B. Wootton *Social Sci. & Social Pathol.* viii. 241 If treatability is taken as the test of responsibility, it follows that those who cannot or will not be treated must be regarded as liable to punishment. 1976 *Economist* 21 Aug. 18 The importance of treatability as a criterion. 1982 *Daily Tel.* 29 Oct. 18/6 The inclusion of 'treatability' among criteria for admission.

treatable, *a.* **1. b.** Delete † and add: *Obs.* exc. in *Med.*. (Later examples.)

1974 E. Ambler *Dr. Frigo* ii. 141 Muscular dystrophy..is to some extent treatable and controllable. 1978 *Bull. Amer. Acad. Arts & Sci.* Feb. 29 Heart disease and strokes are related to..potentially treatable but undetected hypertension.

treatise, *sb.* Add: **4.** *Comb.*, as **treatise poem,** a didactic poem of the eighteenth century.

1936 C. S. Lewis *Allegory of Love* vi. i. 233 In our Augustan period we find a form which has not yet been named and which is only less dominant than satire. I mean the long Treatise Poem (if I may risk the invention of a name where one is badly needed) as practised by Thompson, Armstrong, Young, Akenside, Cowper, and the like. 1980 *Times Lit. Suppl.* 1 Aug. 863/3 The eighteenth-century treatise-poem of Akenside and the rest.

treatment. Add: **5. b.** *Cinemat.* A preparatory version of a screenplay, including descriptions of sets and of the camerawork required.

1928 L. North *Parasites* i. 33 We always make treatments of our stories—it's a sort of synopsis suggestin' what to use an' what to put in the discard. 1938 A. Huxley *Let.* 18 Nov. (1969) 437 I've done a fair amount of work: a 'treatment', as they call it in the jargon of the films, of the life of Mme Curie for Garbo. 1959 Halas & Manvell *Technique Film Animation* 342 *Treatment*, preliminary stage to writing script. Assembling ideas and situations for the film in hand. 1981 L. Deighton *XPD* xxix. 240 We asked the FO to request a copy of the treatment... They would have got a copy of the script too.

c. *the full treatment*, the most elaborate manner of dealing with a subject, 'the works', esp. in phr. *to give* (or *get*) *the full treatment*. Also (often less emphatically) without *full*. *colloq.*

1950 E. Hemingway *Across River* xxiv. 170 We'll give breakfast the full treatment. 1958 *Sunday Express* 9 Nov. 17/4 In *No Concern of Mine* it gets the full treatment in a first act which is brilliantly contrived. 1959 *Listener* 4 June 999/2 This programme was admirably free from the piety or boost which seems to be unavoidable when some celebrities are given the treatment. 1967 Mrs. L. B. Johnson *White House Diary* 25 June (1970) 539 She seemed impressed with the kitchen when we took her through. Betty gave her the full treatment about the washer and dryer and disposal. 1973 R. Hill *Ruling Passion* i. iii. 26 I'm really getting the treatment, thought Pascoe. What does he expect from me?

8. *attrib.*, as *treatment plant, room*.

1963 A. Lubbock *Austral. Roundabout* 114 These were the trucks which transport the ore from the mine to the treatment plant. 1975 *Petroleum Rev.* XXIX. 315/1 A treatment plant for the removal of impurities. 1961 I. Fleming *Thunderball* iv. 43 It was a white cubicle treatment-room like all the others. 1977 J. A. Kotarba in Douglas & Johnson *Existential Sociol.* ix. 259 These conversations occurred in the waiting room, double occupance treatment rooms..and occasionally over coffee.

treaty, *sb.* **3. a.** For 'b.' read 'b and in phr. *private treaty*: see *private *a.* 7 f'.

6. *treaty-money* (Canad. examples); **treaty Indian** *N. Amer.* (now chiefly Canad.), an Indian whose tribe or band has signed a treaty with the Government, a ward of the Government; **treaty-port** (earlier example); **Treaty stone,** the stone on which the Treaty of Limerick (3 Oct. 1691) was reputedly signed (see quot. 1866).

1876 R. I. Dodge *Black Hills* 139 Every year since the treaty was signed has witnessed more or less pillage, depredation, and murder, by the treaty Indians. 1936 B. Brooker *Think of Earth* i. v. 59 He wore the shoddy black clothes, moccasins and red neckerchief which the Treaty Indians of the neighbourhood had affected years before. 1973 'M. Campbell' *Halfbreed* ii. 18 Grandma Dubuque was a treaty Indian woman. 1933 *Beaver* June 53 Upon the arrival of the Indian agent, the payment of the treaty money is usually first proceeded with. 1956 H. S. M. Kemp *Northern Trader* (1957) iii. 35 Had he been on the books as an Indian, he would have been considered a ward of the Government, drawn his Treaty Money and supplies, [etc.]. 1863 *Times* 24 Nov. 9/6 (*heading*) The Yang-tze-Kiang and the new Treaty Ports. 1842 J. P. Lawson *Gazetteer of Ireland* 602/2 It is said by tradition that this famous document was signed by both parties on a large stone near Thomond Bridge,..which is locally designated the *Treaty Stone*. 1866 M. Lenihan *Limerick* xxxiv. 271 The treaty is said to have been signed..near the Red Gate... Tradition does not admit that it was signed on what has been called the 'Treaty Stone', which has occupied a place on the North side of Thomond Bridge for many years, and which was originally a stone, used by country people for getting on horses when leaving town. 1922 Joyce *Ulysses* 324 Remember Limerick and the broken treatystone. 1977 *Irish Democrat* Mar. 6/3 In the breach of death my Donal fell and he sleeps near the Treaty Stone.

treble, *sb.* Add: **I. 2. f.** (Earlier example.)

1872 *Young Englishwoman* Oct. 555/1 *Crochet rosette*..3rd row:..work 11 treble on each chain scallop.

g. *Racing.* (*a*) A total of three races won by the same horse; (*b*) a bet on three horses to win the respective races in which they are entered (the usual sense).

1924 C. Hawtrey *Truth at Last* xix. 226 Many [*sc.* starting-price bookmakers] have further altered that rule, now limiting the odds to 50 to 1 for the double event, and 100 to 1 for treble. 1931 *Daily Express* 21 Sept. 15/4 Peacock wound up a fine week in Scotland, where Nevett landed a treble for him on Saturday. 1951 [see *accumulator 4]. 1964 A. Wykes *Gambling* viii. 194 (*caption*) The bettor has staked a total of 27s. 6d. on his four chosen horses with 11 separate bets of 2s. 6d.—the bets consisting of six 'doubles', four 'trebles', and an 'accumulator' bet. 1981 P. Inchbald *Tondo for Short* vi. 70 Cigars, liqueurs, brandy—has he won a big treble or something?

h. *Darts.* A throw into the narrow space between the two middle circles on a dartboard, worth three times the single score for the sector in which the dart lands; the space itself.

1936 [see *double *sb.* 3 s]. 1981 R. Lewis *Seek for Justice* ii. 53 The first dart flicked into the treble twenty. It was followed by a second... The microphone boomed: ..'Can he nail a third treble?..He's done it'.

i. *Assoc. Football.* The winning of three national or international competitions by one football club during a season.

1959 *Listener* 19 Feb. 331/1 The treble of League, Cup, and European Cup seemed to be at least a possibility. 1977 *News of World* 17 Apr. 21/2 That would make me unique, the only player who has done the double and treble.

j. A drink of spirits of three times the standard measure.

1968 J. F. Straker *SIN & Johnny Inch* 131 Sinclair poured yet another whisky. Like the others, it came well up the glass. A good treble, Johnny reckoned. 1979 M. Babson *So soon done For* ii. 12 Crispin poured drinks for himself and Kay. Jeremy, he noted, had given himself a treble.

treble, *a.* Add: **1. c.** Also of a drink of spirits: constituting three times the standard measure.

1964 L. Deighton *Funeral in Berlin* xxiii. 126 He..ordered three treble brandies. 1977 'J. D. White' *Salzburg Affair* xvi. 137 Hendryks..raised an urgent hand for the waiter. 'A treble scotch.'

3. *treble agent*, a spy who works for three countries, his superiors in each being informed of his service to the other, but usu. with actual allegiance only to one; *treble chance*, a form of football pool in which various points are awarded for a draw, an away win, and a home win; *treble X*, a brand or strength of beer.

1967 *Punch* 11 Oct. 542/1 A list of our treble agents in Bulgaria. 1978 W. Wingate *Bloodbath* i. 12 He wants out. Reckons he's earned it as a double, or treble, agent. 1951 'M. Innes' *Operation Pax* ii. iv. 64 'Heard what was last week's treble chance?' he asked. 'Ninety-eight thousand.' 1972 M. Jones *Life on Dole* ii. i. 102 Luck..seems as evasive as the treble-chance. 1856 Geo. Eliot in *Westm. Rev.* Jan. 5 German ennui must be something as superlative as Barclay's treble X. 1858 C. M. Yonge *Christmas Mummers* vii. 95 They began to sing at the next house as loud as if they thought they should get..more than three times treble XXX ale! 1880 E. W. Hamilton *Diary* 7 Dec. (1972) I. 85 The Irish 'soup' (as Mr. Gladstone terms it) is 'thickening' and becoming what the brewers would call 'Treble X'.

tree, *sb.* Add: **B. 1. d.** = Christmas-tree.

[1838 H. Martineau *Retrospect* III. 182, I was present at the introduction into the new country of..the German Christmas-tree... The tree was the top of a young fir, planted in a tub.] 1851 E. Ruskin *Let.* 28 Dec. in M. Lutyens *Effie in Venice* (1965) ii. 236 They wanted me to come in the evening when the tree was lighted to see the presents all divided. 1945 N. Mitford *Pursuit of Love* iii. 23 We got back late for the tree... Uncle Matthew..was struggling into his Father Christmas clothes! 1979 M. McCarthy *Cannibals & Missionaries* i. 19 Distribution of presents..and the darned crèche and parish-house tree to set up.

6. b. (*b*) More widely, any branching system of vessels or organs in the body; (*e*) (further examples); (also in *Linguistics*, etc.) a set of items that can be represented by such a diagram; (*f*) *Oil Industry* (see quot. 1954); cf. *Christmas-tree 2, quot. 1930.

1881, etc. [see *root *sb.*¹ 14 c]. 1930 T. S. Eliot tr. *St.-J. Perse's Anabasis* 59 You shall see me for long time unspeaking under the female tree of my veins. 1952 [see *cavitation 2]. 1954 *Time* 11 Jan. 3/2 (Advt.), These *trees of steel*, with their long *metal roots* extending thousands of feet into the earth, are actually assemblies of valves and fittings which control the flow of oil from reservoirs. Oilmen call them 'trees' or 'Christmas trees' because of the many unusual patterns and designs obtained when this wellhead equipment is put together to control wells of various kinds, varying pressures, and unique producing characteristics. 1958 W. H. Burge in *Information & Control* I. 183 The tree used is a hierarchical network with a finite number of points arranged in levels. 1959 *Nuovo Cimento* Suppl. XIII. 499 The restriction on the number of symbols that can be rewritten in a single rule guarantees that given a terminal string..it will be possible to discover the associated tree or trees. 1972 *Computers & Humanities* VII. 5 With the use of a light gun the linguist can select from alternative expansions in phrase structure trees. 1973 C. W. Gear *Introd. Computer Sci.* vii. 294 If it is necessary to trace through a tree in order frequently, it is worth storing the trace path. 1976 *Canad. Jrnl. Linguistics* XXI. 129 The psychological reality of aspects of deep structure and surface structure trees is open to interpretation in several respects. 1976 *Offshore Engineer* Mar. 6/3 Shell Expro is going ahead with subsea completion of a stepout well on the Brent field this summer and will be using one of the most sophisticated underwater trees in the world... The tree is described as 'wet, diverless', and flowline connection can be carried out from a rig or a drillship. 1977 *Lancet* 4 June 1187/1 Angiography has made it possible to assess with reasonable confidence the state of the cerebral vascular tree. 1977 *Ibid.* 6 Aug. 278/1 After cholecystectomy for gallstones, it is not unusual for a stone to be left behind in the biliary tree. 1978 *Nature* 24 Aug. 745/1 The amount of work involved in searching a tree of moves is B^D, where B (the branching factor) is the average number of alternatives throughout the tree, and D is the depth of search.

7. a. *money* (etc.) *does not grow on trees* (orig. *U.S.*): money (etc.) is not easily obtainable; *out of* (*one's*) *tree* (U.S. slang) (see quot. 1971).

1669 'Poor Robin' *Almanack* sig. B8, Minc'd Pyes do not grow upon every tree, But search the Ovens for them, and there they be. 1750 W. Chancellor *Diary* Nov. in *Pennsylvania Mag. Hist. & Biogr.* (1968) XCII. 471 Africa, where tis so falsly said, that Gold grows on the Trees. 1787 *Amer. Museum* II. 383 When the new government is established, 'money will grow upon the trees'. 1833 F. Marryat *Peter Simple* in *Metropolitan Mag.* Aug. 302/2 *Clothes* don't grow upon trees in ould Ireland. 1932 W. McFee *Harbourmaster* xxi. 371 Can I make money? Does it grow on trees out there? 1964 J. Aiken *Black Hearts in Battersea* (1965) iv. 51 You'll be wanting it yourself come dinner-time. Sausages don't grow on trees in London. 1977 'S. Woods' *Thief or Two* 118, I don't imagine these things [*sc.* jewels] grow on trees.

1966 *Current Slang* (Univ. S. Dakota) Fall 6 *Tree*, n., mind, esp. in the expression 'drive one out of one's tree.'..She drives me right out of my tree. 1971 E. E. Landy *Underground Dict.* 143 *Out of one's tree*, expression meaning (1) be thinking, talking or acting in an irrational way—e.g. *You are talking out of your tree* or (2) be in an unfamiliar place. 1976 N. Thornburg *Cutter & Bone* ii. 45 'We is duh [= the] loanees.' 'You're out of your tree.'

b. *tree of life*, (*d*) a schematized representation of a tree or shrub used as an artistic motif, esp. in oriental work; freq. *attrib.*

1880 G. C. M. Birdwood *Industrial Arts India* 336 The tree of life represented on modern Yarkand rugs is always a pomegranate tree. 1931 A. U. Dilley *Oriental Rugs &*

Carpets Pl. 57 (*caption*) Beluchistan Prayer Rug with Rectangular Niche and Tree of Life. 1960 B. Snook *Eng. Historical Embroidery* 81 Hangings worked in polychrome,..with flowing stems or a Persian 'Tree of Life' rising from a ground of grass-grown mole hills. 1972 *Islander* (Victoria, B.C.) 28 May 5/2 A most recently finished piece [of weaving] is done in the universal tree-of-life symbol. 1977 *Times* 25 June 2/3 A Kashan silk Tree of Life rug..made £3,000.

9. a. *tree-belt* (example), *-crop, -fork, -fruit* (later examples), *-growth, -shadow* (example). **b.** *tree-felling* (earlier example), *-planting* (examples); *tree-shadowing* adj. **c.** *tree-arched, -bound, -grown, -hung, -lined* (examples), *-planted* (examples), *-scattered, -screened, -shaded* (examples), *-shadowed, -surrounded, -tangled, -wrapt* adjs.

c 1857 J. R. Lowell *Power of Sound* in *Uncoll. Poems* (1950) 123 A parson's son, through tree-arched country ways, I rode. 1936 W. Faulkner *Absalom, Absalom!* ix. 365 They walked up the rutted tree-arched drive. 1962 E. Snow *Other Side of River* (1963) lxvi. 502 Another tree belt had been more than half planted over a length of 720 miles. 1886 Hardy *Mayor of Casterbridge* I. iv. 49 From the centre of each side of this tree-bound square ran avenues. 1951 L. MacNeice tr. *Goethe's Faust* ii. 202 Through a great plain Peneios freely takes His bush-bound, tree-bound course through quiet lakes. 1943 J. S. Huxley *TVA* 102 An unusual line of TVA research concerns the development of so-called tree-crops. 1958 *Times* 22 Aug. 12/4 Tree-crops, small fruits, grains, seeds, and livestock are the main farming interests. 1759 *Crit. Rev.* Sept. 178 Why, for example, should we be so complaisant to the French, as to use their terms of *carcasse*,..*abbattement*,..and *coup de main*; when we can say *fire-ball*,..*tree-felling*,..and *bold stroke*? 1922 Joyce *Ulysses* 191 A runaway in blighted treeforks from hue and cry. 1946 Dylan Thomas *Deaths & Entrances* 22 A she bird sleeping..Within the nested treefork. 1946 *Nature* 2 Nov. 605/1, I presented the fundamental and elementary culture of the Mediterranean based on a combination of cereal agriculture and tree-fruit crops. 1970 D. Waterfield *Continental Waterboy* i. 8 A family could make a living off ten acres by growing tree-fruits. 1846 J. G. Whittier *Poems* (1849) 321 Ghosts of old Beliefs still flit and moan..O'er tree-grown barrow and gray ring of stone. 1978 *Detroit Free Press* 5 Mar. (Parade Suppl.) 15/3 Tree-grown cherries..demand years of tender care. 1917 *Amer. Forestry* XXIV. 732 (*caption*) Comparison of 43 years of rainfall and tree growth. 1956 *Nature* 21 Jan. 124/1 The variety and abundance of insect life as a whole rapidly fall off beyond the limits of tree-growth. 1927 J. Elder *Thomasina Toddy* xii. 118 Leafy backwaters, sunny fields, and tree-hung banks to suit all tastes. 1981 *Sunday Express* 11 Oct. (Colour Suppl.) 23/2 There is a little, tree-hung, irregular village square with an island of greenery at its hub. 1910 *Bradshaw's Railway Guide* Apr. 1123 (Advt.), The smartest bijou hotel in London... Situate in wide tree-lined thoroughfare. 1978 R. Ludlum *Holcroft Covenant* xi. 130 He..drove rapidly through the peaceful, tree-lined suburban area. 1879 C. M. Yonge *Magnum Bonum* III. xl. 904 The broad tree-planted streets of the old Quaker city. 1962 E. Snow *Other Side of River* (1963) lxx. 539 Strolling down a tree-planted street of new apartment houses I chose one to enter. 1872, 1902 Tree-planting [see *Arbor Day]. a 1974 R. Crossman *Diaries* (1975) I. 67 Today I was busy with Pritchett about tree-planting on the bends of the Cherwell between Upper Prescote and Prescote. 1980 P. Lively *Judgement Day* vii. 90 He..had refused to contribute to the Tree Planting Fund. 1951 Tree-scattered [see *river-winding* s.v. *river *sb.*¹ 4 g]. 1923 Kipling *Irish Guards in Great War* II. 163 A close and blind land of woods, copses, farms, mills and tree-screened roads. 1909 *Westm. Gaz.* 20 Oct. 4/1 Matthew Arnold's tree-shaded grave lies to the south-east of the church. 1958 O. Caroe *Pathans* xvii. 285 A place he loved, covered with green turf, tree-shaded beside the broad stream. 1954 J. R. R. Tolkien *Fellowship of Ring* i. iii. 87 As silent as tree shadows. 1952 S. Spender *Learning Laughter* ix. 117 Ben Shemen is a charming, tree-shadowed place. 1912 E. Pound *Ripostes* 37 In streams and tree-shadowing Forests on hill slopes. 1915 W. B. Yeats *Reveries over Childhood & Youth* ii. 16 Next to Merville where I lived, was another tree-surrounded house. 1925 A. Huxley *Those Barren Leaves* ii. v. 125 Round as a fruit, tree-tangled, shines The moon. 1886 W. B. Yeats *Mosada* 3 Whose dwelling was a tree-wrapt island.

10. a. tree box, any of several larger varieties of the common box, *Buxus sempervirens*; **tree-climber** = Liana, liane; cf. Tree-creeper 2; **tree-daisy** = *olearia; **tree-fuchsia,** a shrub or small tree, *Fuchsia excorticata*, native to New Zealand and bearing pendent reddish-purple flowers with blue pollen; cf. *konini; **tree lucerne** (see quot. 1965); **tree pæony, peony** = *moutan.

1731 P. Miller *Gardeners Dict.* s.v. Buxus, All the Varieties of the Tree or large Box are proper to intermix in Clumps of Ever-greens. 1785 G. Washington *Jrnl.* 13 Apr. (1925) II. 360, 12 Horse Chestnut Trees..and an equal number of cuttings of the Tree Box. 1858 J. A. Warder *Hedges & Evergreens* ii. 240 Where a moderate or low hedge is needed,..nothing can be better than the Tree-box. 1864 J. A. Grant *Walk across Afr.* 339 A tree-climber (*Landolphia florida*?) lay with its trunk winding like a huge snake. 1926 Tree-daisy [see *daisy-tree* s.v. *daisy 7]. 1958 Cockayne & Turner *Trees N.Z.* (ed. 4) 142 Weeping tree daisy..common in Central Otago. 1906 Laing & Blackwell *Plants N.Z.* 294 (*heading*) Fuchsia excorticata (the tree fuchsia). 1910 L. Cockayne *N.Z. Plants* iii. 29 The tree-fuchsia..offers a transition to the scrambling habit. 1970 *S. Afr. Panorama* Feb. 35/3 Below the platform a minute sunbird with iridescent blue plumage hovered before the crimson blooms of a tree-fuchsia. 1933 *Bulletin* (Sydney) 14 June 25/1 Tree lucerne is very hardy and easily grown from seed. 1965 *Austral. Encycl.* V. 383/1 The white-flowered tagasaste of Teneriffe (*Cytisus proliferus*), which is a very large broom, is often known in Australia as tree lucerne, a name strictly applicable to the yellow-flowered bush *Medicago arborea*. 1981 *Southern Horticulture* (N.Z.) Spring 53/1 Weather conditions could influence the situation, as too could the presence of tree lucerne near-by. 1811 W. T. Aiton *Hortus Kewensis* (ed. 2) III. 315 Chinese Tree Pæony. Moutan. Nat[ive] of China. 1880 [see *moutan]. 1962 I. Murdoch *Unofficial Rose* i. 44 The more intense evening light against a long bed of yuccas and tree peonies. 1980 R. Grounds *Private Life Plants* xxiii. 133 As many as 3,000 flowers have been counted on a tree peony.

b. tree-fly (example); **tree-partridge** (earlier example); **tree-runner,** a brightly coloured Australian nuthatch of the genus *Neositta*, esp. *N. chrysoptera*; **tree-spider,** any of many spiders that live on the trunks or branches of trees; **tree squirrel,** an arboreal squirrel, distinguished from a ground squirrel.

1834 *Chambers's Edin. Jrnl.* III. 357/3 Much wood..during warm and summer months, raining down great store of tree-flies. 1864 J. A. Grant *Walk across Afr.* 93 The..tree-partridge resembles the painted one of India, has yellow legs, beautiful plumage, and weighs about a pound. 1901 A. J. Campbell *Nests & Eggs Austral. Birds* I. 337 The true home of the Orange-winged Sittella or Tree Runner is Eastern Australia. 1964 A. L. Thomson *New Dict. Birds* 545/2 The so-called 'treerunners' or 'sitellas' are widely distributed in Australia. 1904 W. H. Hudson *Green Mansions* ii. 33 The shaft reveals a tangle of shining silver threads—the web of some large tree-spider. 1910 W. de la Mare *Three Mulla-Mulgars* iii. 45 They sat and ate..with scorpions and speckled tree-spiders watching them. 1934 A. Russell *Tramp-Royal in Wild Austral.* xxxviii. 249 So strong and thickly-woven are the webs of the Central Australian tree or orchid spider that small birds are often caught in them. 1822 J. Woods *Two Years' Residence Eng. Prairie* 193 Tree-squirrels are of two or more sorts, and are eaten here. 1872 *Routledge's Every Boy's Ann.* 614/1 Dennis climbs like a tree squirrel. 1968 *Ecol. Monogr.* XXXVIII. 31 (*title*) The adaptive nature of social organisation in the genus of tree squirrels *Tamiasciurus*.

c. tree-box, a frame used to protect a young tree; **tree-bridge,** (*b*) (earlier example); **tree-coral,** a branching coral; **tree diagram** = sense 6 b (*e*) in Dict. and Suppl.; **tree doctor** = *tree surgeon* below; **tree farm** orig. and chiefly *U.S.*, an area of forest managed in a way that ensures the regular production of timber; hence **tree farmer, farming; tree-feeder,** an animal that feeds on the foliage of trees or the insects living on leaves or bark; **tree-house,** (*a*) (earlier example); (*b*) a child's playhouse (sense *2 a) built in a tree; **tree-limit,** the line beyond which trees do not grow, with reference to either altitude or latitude; cf. *tree-line* (*a*); **tree-line,** (*a*) (earlier examples); (*b*) a row of trees; the edge of a wood; **tree-nest,** a nest built in a tree, in contrast to one built at ground level; **tree-path,** the track of an arboreal animal; **tree-people,** in fantasy or fiction: (*a*) persons that live in trees; (*b*) animated trees; **tree preservation order,** an order prohibiting the felling or removal of a tree or group of trees; **tree-pruner** (example); so **tree-pruning** (also *transf.*, the removal of branches from a tree diagram); **tree-ring,** an annual growth ring in the trunk of a tree; hence **tree-ring analysis, dating** = *dendrochronology; **tree-road** = *tree-path* above; **tree-rune** (earlier example); **tree search,** a search in which a situation or entity is represented by a tree diagram, e.g. to facilitate efficient searching; **tree stool,** the stump of a fallen tree as preserved in a peat bog; **tree structure,** a structure in which there are successive branchings or subdivisions; cf. *tree diagram* above; **tree surgeon,** a practitioner of tree surgery; **tree surgery,** the pruning, repair, and preservative treatment of ornamental trees, first professionally organized by John Davey (1846–1923), American landscape architect.

1876 'Mark Twain' *Tom Sawyer* ii. 27 [Tom] sat down on the tree-box discouraged. 1896 J. C. Harris *Sister Jane* 157 Whittling away with his pocket-knife on the tree-box, against which he was leaning. 1805 T. E. White *Jrnl.* 20 July (1904) 26, I..cross'd the creek on a tree bridge an came through the woods. 1871 *Harper's Mag.* June 28 On the confines of this channel may be seen in clear water a perfect forest of coral—tree-coral, we call it, on account of its great size. 1915 E. R. Lankester *Diversions Naturalist* 11 Great tree-coral of these waters—the Paragorgia. 1965 N. Chomsky *Aspects of Theory of Syntax* i. 14 A tree-diagram of a sentence. 1978 *Language* LIV. 15 In a tree diagram, only the configuration of nodes matters, not the length of branches and sub-branches. 1776 tr. *Béardé de l'Abbaye's Ess. in Agriculture* vi. 37 There was a person, who assumed the title of Tree doctor. 1908 *Harper's Weekly* 5 Dec. 15/1 The services of the tree doctor are needed. 1976 'M. Albrand' *Taste of Terror* xx. 115 The tree doctor..took a look at the willow. 1941 *N.Y. Times Mag.* 9 Nov. 13/2 Instead of reseeding sketchily over immense areas, the industry is laying out 'tree farms'. 1942 *Jrnl. Forestry* XL. 596/2 Tree farmers..should be eligible for the same..treatment as other farmers. *Ibid.*, Tree farming appears to be off to a good start. 1962 Tree-farming [see *clear-fell* s.v. *clear *a.* D. 3]. 1973 P. A. Whitney *Snowfire* ii. 21 Julian had gone into tree farming. *Ibid.* xii. 231 He told me..about the controlled growth..on a tree farm. 1984 *New Yorker* 23 Jan. 78/3 Tree farms [in China] have also begun to experience problems with theft. 1914 *Chambers's Jrnl.* Jan. 75/1 A species of rhinoceros..was particularly a tree-feeder. 1953 N. Tinbergen *Herring Gull's World* vii. 66 Great Tits..being tree-feeders, they do not peck at the ground. 1867 J. C. Patteson in C. M. Yonge *Life J. C. Patteson* (1874) II. xi. 275, I am high and dry, and have ..a broad ladder—up to my house. The Mahaga lads and I call it my tree-house. 1949 A. Wilson *Wrong Set* 128 Go see if she's in the Tree House... It's a kind of funny old place she and Hamish made when they were kids. 1979 R. Jaffe *Class Reunion* i. iv. 40 She couldn't go up in the tree house anymore. 1934 *Discovery* June 167/1 They extend well above the local tree-limit. 1953 D. Bannerman *Birds Brit. Isles* I. 209 The typical Swedish race..and the west European race..have a very wide distribution on the continent of Europe, where their combined range extends north to the tree-limit. 1893 *Outing* Aug. 346/1 We struck the tree-line again in the immense ravine between them. 1903 Kipling *Five Nations* 53, I camped above the tree-line—drifted snow and naked boulders. 1936 F. Clune *Roaming round Darling* xiv. 123 The trail..wended down the Barwon, branching off at the various blazed tree-lines to the numerous creeks where they settled. 1977 D. Harsent *Dreams of Dead* 23 In single file the women left the treeline, a flicker at the corner of his eye. 1924 J. A. Thomson *Sci. Old & New* x. 55 A..Tineid caterpillar, found in the tree-nest of one of the Termites. 1953 D. A. Bannerman *Birds Brit. Isles* I. 16 These tree-nests [of crows] are often most conspicuous. 1897 J. L. Allen *Choir Invisible* xv. 227 The grass-path or the tree-path of the cougar. 1954 J. R. R. Tolkien *Fellowship of Ring* ii. vi. 355 That was the custom of the Elves of Lórien, to dwell in the trees... Therefore they were called the Galadrim, the Tree-people. 1964 *Listener* 24 Dec. 1003/1, I think an intelligent plant would be large and virtually immobile; the tree-people in Olaf Stapledon's *Star Maker* might just qualify. [1943 *Act* 6 & 7 *Geo. VI* c. 29 §8 If it appears to any interim development authority that it is expedient..to make provision for the preservation of trees or woodlands..they may..make an order (in this section referred to as an 'interim preservation order') with respect to such trees.] 1947 L. Silkin in *Hansard* (Commons) 22 Apr. 779 One hundred and six tree preservation orders have been submitted for my approval. 1976 *Leicester Mercury* 16 July, A tree preservation order has been made by the Harborough District Council to protect trees in and around the grounds of Little Bowden Rectory. 1887 *Illustr. Catal. Garden Furniture* (J. B. Brown & Son) 83 The 'standard' tree pruners. 1933 R. Tuve *Seasons & Months* iv. 160 February-by-the-fire has been crowded out by putting an extra tree-pruning picture into the series. 1966 *Math. Linguistics & Automatic Translation* (Harvard Univ. Computation Lab. Rep. No. NSF-17) iv-1 (*heading*) A proposal rule of tree-pruning. 1976 J. S. Gruber *Lexical Structures in Syntax & Semantics* ii. iii. 365 We will have the following four derived trees... Each of these will undergo tree-pruning. 1919 A. E. Douglass *Climatic Cycles & Tree-growth* iii. 23 The plan of using tree-rings for the general purpose of a check on astronomical and meteorological phenomena was first formulated in 1901. 1946 F. E. Zeuner *Dating Past* i. 6 Tree-ring analysis is based on a well-known structural feature of wood, namely the annual growth-rings. *Ibid.* 11 The scope of tree-ring dating is extending rapidly. 1977 *Times* 20 July 13/4 Tree-ring analysis—or..dendrochronology—can..help to date..old paintings on oak panel. 1982 *Nature* 6 May 28/1 Others have sought an explanation of..variations in tree rings. 1895 Kipling *Second Jungle Bk.* 218 When he tired of ground-going he threw up his hands monkey-fashion to the nearest creeper,..he would follow a tree-road till his mood changed. 1863 J. M. Mitchell *Mesehowe* 49 The six tree Runes form the word *Arrier*. 1970 O. Dopping *Computers & Data Processing* xxii. 362 Tree search and heuristic programming cover a wide field of problems and are in principle well suited for automatic computation. 1980 *Daily Tel.* 26 May 10/6 In a complex game, a computer normally moves after conducting a 'tree search' of all possible moves, a process which if unlimited by time, would take billions of years. 1898 *Geogr. Jrnl.* XI. 431 The deeply submerged peats and tree-stools indicated..that the post-glacial recovery brought the land-level almost to normal pre-glacial conditions. 1975 J. G. Evans *Environment Early Man Brit. Isles* vi. 140 Peat now covers these hills, but..they..were once forested as is shown by the presence of tree stools at the base of the peat. 1965 N. Chomsky *Aspects of Theory of Syntax* i. 12 The most obvious formal property of utterances is their bracketing into constituents of various types, that is, the 'tree structure' associated with them. 1971 *Computers & Humanities* V. 292 Special language facilities in the fields of list processing, string processing, tree structure operations. 1983 *Austral. Microcomputer Mag.* Aug. 51/2 WangNet's cable topology is a duplicated tree structure. 1908 *Harper's Weekly* 5 Dec. 15/1 The attention given by the tree surgeon to the aged and decaying historical trees of the country..is equal to that given a wealthy invalid by his physician. 1978 *Cornish Guardian* 27 Apr. 19/1 (Advt.), Tree Surgeons. Fully qualified and insured for all felling, planting, pruning and repair work. 1902 J. Davey *Tree Doctor* 14 Learn to do your own tree surgery, or direct it personally. 1973 *Country Life* 7 June 1706/3 (Advt.), Southern Tree Surgery Company (Consultants and Tree Surgeons).

treeless, *a.* (Earlier example.)

1794 W. B. Stevens *Jrnl.* 19 Nov. (1965) ii. 205 On your Tree-less Coast indeed You can have no falling leaves to warn you of the approach of Winter.

treen, *a.* Restrict *Obs.* exc. *dial.* to senses in *Dict.* and add: **B.** as *sb.* = Woodware, esp. when regarded as antiques. Const. as *pl.*

1927 H. V. Morton *In Search of England* i. 7 'Before people used pewter for plates and tankards,' he explained, 'wooden trenchers, drinking cups and bowls—called 'treen'—were used by everyone.' 1949 E. H. Pinto *Treen* 3 The small useful woodware of to-day is the treen of to-morrow... The turnery side of treen-making is much more alive than most people realise. 1971 *Canadian Antiques Collector* Apr. 22/1 Articles ranging from ladles to snuff boxes, candlesticks to combs, may be included in the group known as treen. 1980 *Daily Tel.* 26 Feb. 13/3 Hand-turned treen are a joy to look at and a great pleasure to use. 1981 *Rescue News* Mar. 2/6 The site has also yielded a great deal of domestic material, including a rich collection of pottery, pewter and treen.

treescape. Delete *rare* and add later examples.
1950 *Archit. Rev.* CVIII. 53 (*caption*) A lovely, casual treescape at Marsh Lock..another fine treescape at Cookham Dean. 1960 *Guardian* 15 July 7/1 The treescape seen from inside appears to be part of the walls. 1969 B. Patten *Notes to Hurrying Man* 62 Above treescapes the fawn smells black Smoke drifting. 1978 *Times* 5 June 12/8 This particular polished granite..has a very high reflectivity, and will tend to mirror the treescape of Euston Square.

tree-top, tree top. Add: **b.** *attrib.* passing into *adj.* Of or pertaining to tree-tops; in the tops of trees. Also *fig.*
1896 R. L. Stevenson *In South Seas* iii. iii. 246 The folk of the town streamed by us intermittingly... In the first grey of the morning, and again late in the afternoon, these would straggle past about their tree-top business..and vanish from the face of earth. 1945 *New Yorker* 10 Feb. 23/3 Bombers went to critical missions, at tree-top level. 1961 *Sunday Express* 26 Feb. 5/3 A tree-top hotel..in the Aberdare Mountains. 1977 *Daily Mail* 24 Sept. 15/6 One man with a few machines can run a dawn-to-dusk radio station. Britain's few fugitive tree-top pirates do the same on the odd Sunday afternoon. 1979 P. Niesewand *Member of Club* xxi. 165 Colonel Winter heard the jets flash over; just above tree-top height.

‖**tref** (trev). *Wales.* [W., hamlet, home, town.] A social unit that was once traditional in Wales, consisting of a hamlet or homestead or the community occupying it (see quot. 1889).
1841 A. Owen tr. *Anc. Laws & Institutes of Wales* 1004/2 Trev, a vill:—A territorial division of land containing four gavaels or 256 erws. 1889 H. Lewis *Anc. Laws Wales* i. iii. 57 In modern Welsh 'trev' means a village or town, in ancient times it is said to have meant a single house or homestead... In..laws we have a frequent use of trev in a sense having reference to the land of the joint family, that is to their settlement and group of dwellings. 1895 *Wales* July 304/2 There was a good deal of feasting amongst the hillside trefs. 1900 Rhys & Brynmor-Jones *Welsh People* vi. 218 We now turn to consider briefly the law relating to property in or possession of land... The cymwd was thus divided: Four *erwau* in every *tyddyn* (homestead),..four *gafaelion* in every *tref* (vill or township). 1968 [see *gwely].

trefa, trifa. Add: Also **trayf, treff, treife, trifah,** etc. (Earlier example.) Chiefly *attrib.* or as *adj.*, not prepared according to the Law, applied to any food. Also *transf.*
1837 *Brit. & Foreign Rev.* V. 424 The Jews call..treff all that is prohibited to be used as food. 1892 I. Zangwill *Childr. Ghetto* I. 88 Even pious people eat tripha cheese and butter. *Ibid.* III. 39 Their money is kosher; they are tripha. 1907 —— *Ghetto Comedies* 68 The tripha meat cooked in Simon's mess-tin. 1961 A. W. Moss *Valiant Crusade* vi. 81 If more than one cut is necessary..the carcase is rejected as 'Trifah', i.e. unfit for Jewish..food. 1966 H. Kemelman *Saturday the Rabbi went Hungry* (1967) xix. 115 When a utensil becomes tref, the way you cleanse it is to bury it in the earth. 1975 G. Meir *My Life* i. 5 He served in the Russian army for another thirteen years, and never once..did he touch treife (non-kosher) food. 1978 I. B. Singer *Shosha* ix. 168 To me he's as *trayf* as pork.

‖**treffend** (tre·fĕnt), *a.* [Ger.] Apposite, fitting, pertinent.
1850 S. Austin *Let.* 21 Jan. in J. Ross *Three Generations* (1888) I. 244 It is perfectly and thoroughly treffend. 1900 W. James *Let.* 13 Mar. (1920) II. 119 His surface thoughts ..of a scientific order, were extraordinarily *treffend* and clearly expressed.

trefid: see *trifid *a.* b.

trehalase (trīhā·lēɪz). *Biochem.* [ad. F. *tréhalase* (E. Bourquelot 1893, in *Bull. de la Soc. mycologique de France* IX. 194): see Trehala and *-ase.] An enzyme which catalyses the hydrolysis of trehalose to two molecules of glucose.
1893 *Jrnl. Chem. Soc.* LXIV. 451 The author proposes to call the special ferment trehalase. 1926 Waksman & Davison *Enzymes* x. 182 Trehalase has been demonstrated in the intestinal secretions of certain animals (horse, cattle), certain fishes, in malt, in various fungi, in yeasts and in certain bacteria. 1977 *Lancet* 5 Nov. 982/2 Sucrase, trehalase and lactase activities were absent.

treille. b. (Earlier example.)
1865 F. B. Palliser *Hist. Lace* vii. 119 The thick 'treille' and scanty flowers of the old laces [in West Flanders].

trek, *sb.* For 'S. Africa' read 'orig. S. Afr.' **1. a.** Restrict def. to S. Afr. use and add: Now in gen. use elsewhere, a long journey or expedition, esp. one overland involving considerable physical effort.
1941 I. L. Idriess *Great Boomerang* i. 6 Risky treks against hazards different from those of the general Australian bush. 1968 R. M. Patterson *Finlay's River* 168 There they made camp, cached their canoe and load, and sorted out what they wanted to take for their next overland trek—this time a hunting trip. 1972 D. Craig *Double Take* i. 8 The trekking lesson always ended like this... The other members of the trek looked at Brian.

2. trek Boer, (*a*) a Boer who moved his family and grazing stock from place to place; (*b*) = Voortrekker; also, a participant in a later migration of Afrikaners; **trek-bok,** *pl.* **bokke(n),** an antelope, esp. a springbok, in a migrating herd; **trek-cart,** a light cart used by (boy) scouts for transporting stores, etc.; **trek chain** (examples); **trek-farmer** = *trek Boer* (*a*) above; **trek fever,** an insatiable longing for travelling or wandering in the veld; **trek-net** = Seine *sb.*[1]; hence **trek-netter; trek path,** a right of way across the land of another farmer; **trek sheep,** sheep driven or carried a long way for pasturage.
1835 A. Steedman *Wanderings S. Afr.* II. iii. iii. 53 The next day we met a Trek Boor, with his cattle. 1847 in C. Pettman *Africanderisms* (1913) 513 All the most intelligent of the Trek Boers whom I have seen, look forward with dread to the course the Government are pursuing. 1882 C. du Val *With Show through Southern Afr.* I. 106 Abolition of slavery was the primary cause of the movement of these 'trek Boers'. 1929 D. Reitz *Commando* xxiv. 281 We moved north through country thinly occupied by Nomad Boers (Trek Boers), who spend their lives going from one well to another with their flocks. 1941 C. W. de Kiewiet *Hist. S. Afr.* 17 When the Trekboers entered it with their flocks and tented wagons, they left the current of European life. 1981 *Times Lit. Suppl.* 13 Feb. 159/2 The Afrikaner remains, according to Lambley, the atavistic, insular, racially arrogant trekboer. 1824 *S. Afr. Jrnl.* I. 72 On the approach of the *Trek-Bokken* or migrating spring-boks, the grazier makes up his mind to look for pasturage elsewhere. 1827 G. Thompson *Trav. & Adv. S. Afr.* ii. vi. 274 The destructive flocks of trek-bokken, or migratory springboks, pressed by the long droughts, occasionally inundate the northern parts of the Colony. 1896 H. A. Bryden *Tales S. Afr.* ix. 215, I have passed across these plains through a herd of trek-bokken ..three or four miles broad. 1966 E. Palmer *Plains of Camdeboo* ix. 157 The springbuck migrations..are something we shall never see again. Colonists called them 'trekbokke' or 'travellingbuck'. 1928 R. A. Knox *Footsteps at Lock* v. 43 The bigger boys had gone..with the trek-cart to bring our stores over. 1977 *Drive* Jan.-Feb. 15/2 Boy scouts' trek-cart needed. 1878 H. A. Roche *On Trek in Transvaal* 332 Our oxen were free, walking off a yard or two with our tree-disselboom and trek-chain. 1972 *Farmer's Weekly* (S. Afr.) 21 Apr. (Advt.), Chain traces 55c each; Trek chains R1.55. 1912 *East London Daily Dispatch* 1 May 5 (Pettman), The desirability of amending the railway tariff for trek-sheep to enable trek farmers to avail themselves of the railway. 1966 E. Palmer *Plains of Camdeboo* vii. 128 The mountain bush had housed the first trek-farmers in hard and stormy weather. 1980 First & Scott *Olive Schreiner* i. 28 Boer trek-farmers moved away from British control. 1897 J. P. Fitzpatrick *Outspan* 3 When..this instinct awakens,..it becomes a madness, and they call it trek-fever. 1943 D. Reitz *No Outspan* viii. 106 Gauko-Otawi, the 'Rustplaats' or resting place of the Trekkers. Here it was that in 1878 they had built a church, their trek-fever temporarily stilled. 1913 W. W. Thompson *Sea Fisheries Cape Colony* ii. 46 The seine, or 'trek-net', has from the very earliest period of the European occupation of the country been the type of net generally adopted. 1970 *Cape Argus* 24 Dec. 2 They had cast trek nets in the surf. 1956 J. L. B. Smith *Old Fourlegs* i. 9, I..lived with the coastal trek-netters. 1934 Webster, Trekpath. 1936 *Cape Argus* 18 Mar. 13 The trek-path controversy has led many men to fence their farms. 1955 L. G. Green *Karoo* xii. 142 A trek path is a definite route which a farmer is entitled to follow when leading his sheep to new pastures. 1912 Trek sheep [see *trek-farmer* above].

trek, *v.* For 'S. Africa' read 'orig. S. Afr.' **1. a.** Restrict orig. use to S. Afr. and add examples of further *transf.* applications elsewhere (see *trek *sb.* 1 a). Freq. in trivial use.
1911 C. E. W. Bean *'Dreadnought' of Darling* xxxviii. 342 When the police first saw them they were trekking through the scrub. 1912 [in Dict.]. 1943 *Sun* (Baltimore) 11 June 13/2 The hungry pilot trekking over the tundra should beware of the liver of the Polar bear. 1955 [see *trekking vbl. sb.]. 1976 *Oadby & Wigston (Leics.) Advertiser* 26 Nov. 7/6, I was surprised as I trekked from shop to shop how much prices varied. 1977 C. McCullough *Thorn Birds* xvii. 438 It would mean trekking down to the kitchen again, and..no one appreciates the patter of my little feet.

trekker. (In Dict. s.v. Trek *v.*) Add: **1.** (Earlier and later examples.)
1851 R. Gray *Jrnl. Bishop's Visitation Tour through Cape Colony in 1850* 27 Only a few of the latest trekkers have a friendly feeling towards the English government. 1973 *Sunday Times* (Johannesburg) 9 Dec., Many Coloureds..trekked to..the Cape Flats, but unlike other trekkers before them there was no promised land.

2. A person travelling a long distance, esp. on foot; *spec.* a rambler or hiker. Cf. *pony-trekker* s.v. *pony *sb.* 6 b.
1932 *Sun* (Baltimore) 7 Jan. 2/5 The main body of the small army marching from Pennsylvania..is not expected until late tonight... The trekkers..found no difficulty in obtaining admission to the gallery. 1939 R. Godden *Black Narcissus* iv. 39 It isn't even on the route that the trekkers take. 1968 *Punch* 14 Aug. 234/2 The Association has 'on call' about seventy ponies—for sixty trekkers, six experienced guides, and a few spares. 1977 *Times Lit. Suppl.* 13 May 595/2 It will just about go into the side pocket of your rucksack, and no intelligent trekker should go off to Gurung country without one. 1984 *Times* 18 Feb. 14/1 In their hearts trekkers are all pilgrims.

trekkie (tre·ki). [f. Trek *sb.* + -ie.] **1.** *S. Afr.* A small group of trekkers.
1888 J. Bird tr. *D. P. Bezuidenhout's Narrative* in *Annals of Natal* I. 367 Five men were first sent forward to seek a road to the Drakensberg... A small 'trekkie' (party of emigrants) had preceded us. 1953 J. Collin-Smith *Locusts & Wild Honey* i. i. 10 It was a bright autumn morning when we had inspanned the sixteen oxen, and the wagon wheels had turned, and the little trekkie had started away.

2. Also **Trekkie.** An admirer of the U.S. science fiction television programme *Star Trek*; hence, a space-traveller; one interested (trivially) in space travel.
In S. Africa the form *trekker* is also used for this sense.
1976 *New Yorker* 16 Feb. 39 (*caption*) Of course, I didn't know George was a Trekkie when I married him. 1977 *Time* 15 Aug. 50/2 Berry admits that his first trekkies would not know where they might emerge or if they would ever get back. 1978 *Sunday Sun* (Brisbane) 17 Sept. 45/3 Fans—called Trekkies—still number in their tens of thousands. 1981 *Space World* Aug.-Sept. 6/3 Many of the [L-5] society's other members were considered space Trekkies more interested in social experimentation than in technology. 1983 *Oxf. Univ. Press (N.Y.) Spring Catal.* 27 The audience for science fiction now runs the gamut from the high school 'trekkie' to the serious literary scholar.

trekking, *vbl. sb.* and *ppl. a.* (In Dict. s.v. Trek *v.*) Add: (Later examples.) *spec.* = *pony-trekking* s.v. *pony *sb.* 6 b.
1942 Zuckerman & Bernal in S. Zuckerman *From Apes to Warlords* (1978) vii. 143 The situation in Hull has been somewhat obscured..by the occurrence of trekking, which was made possible by the availability of road transport. 1955 *Times* 22 July 9/6 About 35,000 came last year, and more are expected this summer... They come to fish and shoot or to trek in the mountains. 'Only the English like trekking,' one agent said. 1962 *Times* 21 Apr. 11/3 There is also a list of trekking and riding holiday centres which have been awarded a certificate of approval. 1968 *Punch* 14 Aug. 234/3 While some of the ponies probably would be kept in any case..the majority are now kept principally for the revenue from the trekking. 1972 [see *trek sb. 1 a]. 1976 *Horse & Hound* 3 Dec. 42/4 His book is well illustrated and there are some useful appendices, though that claiming to list a brief selection of trekking centres in fact is so brief as to be almost comic. 1984 *Times* 18 Feb. 14/1 Trekking has an aura which is irresistible to the romantic.

trek-tow. Add: Also **tracktoe, trektou(w).** (Earlier and further examples.)
1822 W. J. Burchell *Trav. Interior S. Afr.* I. 151 The trektouw (draw rope or trace), is a long rope made of twisted thongs of raw hide, made fast by a hook to the staple at the end of the pole, and having iron rings attached to it at proper distances, into which rings the yokes are hooked. 1835 A. Smith *Diary* 8 June (1940) II. 60 In this river the crocodile abounds; one carried off and swallowed a tracktoe belonging to a trader. [*Note*] Trek-touw, or draught rope. 1939 S. Cloete *Watch for Dawn* ii. 26 As the wagon topped the rise they trotted down with slack trek-tous. 1972 A. Scholefield *Wild Dog Running* 214, I didn't know how many had been in the pride but two bullocks had been killed where they stood. A third had been dragged some yards from the *trektou*, the strong leather thong around his neck snapped like cotton.

trellis, *sb.*[2] Add: **1. e.** Short for *trellis stitch*: see *3.
1912 L. F. Pesel *Stitches from Old Eng. Embroideries* 19 (*caption*) Trellis with cross-stitch couching. 1921 A. G. I. Christie *Samplers & Stitches* v. 57 Trellis is used for solid fillings.

3. trellis stitch, in embroidery or knitting, an arrangement of stitches between parallel lines to give a lattice effect.
1921 A. G. I. Christie *Samplers & Stitches* v. 57 The thread for working Trellis stitch should be untwisted for the best effect to be gained. 1974 *Guardian* 26 Jan. 15/2 Trellis, moss and blackberry stitch by which..Aran mothers recognise their drowned sons.

trellised, *ppl. a.* Add: **2. c.** *Physical Geogr.* Of a drainage pattern: resembling the pattern of a vine growing on a trellis, with tributaries flowing in a direction approximately at right angles to the stream they join and bends in the main stream being approximately right-angled.
1895 *Monogr. Nat. Geogr. Soc.* No. 1. vi. 186 (*caption*) On the head waters of Bluestone River..the branches are adjusted to tilted hard and soft beds, forming an example of 'trellised' drainage. 1937 Wooldridge & Morgan *Physical Basis Geogr.* xiv. 192 In all such cases, whether the network of major valleys is rectangular or rhomboidal, the drainage plan may be described as 'trellised'. 1970 R. J. Small *Study of Landforms* vii. 227 The weaker strata are gradually eroded to form strike vales, separated by cuestas associated with the more resistant rocks, and are

occupied by tributary streams which, with the passage of time, form an increasingly dominant component of the trellised pattern.

Tremadoc (trɪmæ·dǒk). *Geol.* The name of a village in Gwynedd, N. Wales, used *attrib.* to designate a series of Tremadocian rocks (see next) about 300 metres thick that form the top of the Harlech Dome and comprise mudstones, shales, and slates; also, more widely, = next.
1847 A. Sedgwick in *Q. Jrnl. Geol. Soc.* III. 145 The anomalous position of the Tremadoc rocks,..and the dislocation of the beds above noticed along the Merioneth coast, were probably all produced by the same set of disturbing forces. *Ibid.* 157 The Festiniog or the Tremadoc group. 1882 A. Geikie *Text-bk. Geol.* 655, 12 species of lamellibranchs from the Tremadoc beds of Ramsay Island and St. David's. 1885 C. Lyell *Student's Elem. Geol.* (ed. 4) xxviii. 448 Mr. Callaway has shown that the Shineton shale of Shropshire is of Lower Tremadoc age. *Ibid.*, *Orthoceras sericeum* and *Cyrtoceras præcox* are of the Upper Tremadocs. 1938 A. K. Wells *Outl. Hist. Geol.* iii. 20 The graptolites..had made their first appearance in Tremadoc (Upper Cambrian) times. 1970 R. M. Black *Elements Palaeont.* xiv. 222 Primitive pterobranchs are found in rocks of Tremadoc age. 1974 *Encycl. Brit. Micropædia* X. 108/3 Important Tremadoc sequences are..known from Australia.

Tremadocian (tremădǫ·kiăn), *a.* *Geol.* [f. prec. + -ian.] Of, pertaining to, or designating a stratigraphic series typified by the Tremadoc beds, orig. placed in the Upper Cambrian but now sometimes regarded as Lower Ordovician. Also *absol.*
1910 W. G. Fearnsides in Monckton & Herries *Geol. in Field* xxxii. 795 Across the denuded edges of Tremadocian and Olenidian rocks rest the basal members of the great Arenig series. 1927 *Q. Jrnl. Geol. Soc.* LXXXVIII. 145 The order of that succession is applicable in all its detail to the type-area of the Tremadocian in the western parts of Wales. 1950 David & Brown *Geol. Commonw. Austral.* I. iii. 135 By Tremadocian time graptolites had begun to make their appearance in Victoria. 1974 *Nature* 18 Oct. 575/1 None of the formal papers included any discussion of the boundaries of the Ordovician, most speakers making it clear whether they regarded the Tremadocian as Cambrian (as in English usage) or Ordovician.

trematode, *sb.* (Earlier and further examples.)
1859, 1877 [see *Bilharzia, bilharzia]. 1962 J. D. Smyth *Introd. Animal Parasitol.* xiii. 138 The great majority of digenetic trematodes are inhabitants of the vertebrate alimentary canal or its associated organs.

tremblant (tre·mblănt), *a.* [f. Tremble *v.* + -ant[1].] Of an ornament, jewel, etc.: incorporating springs or fine projecting wires which tremble or vibrate when affected by movement.
1970 *Times* 26 Mar. 12 A very fine diamond tremblant brooch in the shape of a five-petalled flower brought the same price. 1973 *Country Life* 13 Dec. (Suppl.) 326 A late 18th Century diamond head ornament, the sunburst tremblant. 1979 *Ibid.* 3 May (Suppl.) 56/1 A mid Victorian ruby and diamond tremblant spray brooch.

tremblement. Add: ‖**3.** *Mus.* (tranbləman) = Shake *sb.*[1] 5, Trill *sb.*[2] 1 b.
[1883 *Grove Dict. Music* III. 479 Shake or trill (Fr. *trille*, formerly *tremblement*).] 1884 F. Niecks *Conc. Dict. Mus. Terms* s.v., *Tremblement* (Fr.), a shake. 1893 E. Dannreuther *Mus. Ornamentation* i. xiv. 100 In the Méthode the tremblements are generally marked. 1915 A. Dolmetsch *Interpret. of Music of 17th & 18th Centuries* iv. 163 The 'Tremblement et Pincé' is a shake with a Turn as termination. 1946 E. Blom *Everyman's Dict. Music* 482/1 *Tremblement appuyé* = prepared shake. 1978 *Early Music* Oct. 517/2, I have left aside certain details, for example, whether the *tremblement* is to be played slowly at first and increasing in speed, or is to be trilled equally throughout.

trembler. Add: **4.** Also, such a blade used as a make-and-break sensitive to physical disturbance.
1943 N. Balchin *Small Back Room* xv. 185 It was the terminals I had to get wires on to, not the trembler. 1958 A. B. Hartley *Unexploded Bomb* ii. 13 When the bomb struck its target..the shock caused the trembler to function and electrically to ignite the flash-pellet in the base of the fuze. 1973 J. Drummond *Bang! Bang! Bang!* xii. 26 A Banx is an anti-personnel device... Set it for time-fuse, or a trigger-action, even a trembler. 1978 R. Jansson *News Caper* 8, I held the control column rock steady, as if it were a bomb with a trembler fuse.

trembleuse, *a.* or *sb. attrib.* Add: Also as *sb. absol.*
1869 C. Schreiber *Jrnl.* 17 June (1911) I. 13 A pair of Trembleuses and Stands, ruby glass, with white Smalto inside..£7.

trembling, *ppl. a.* Add: **trembling poplar** (earlier example).
1731 P. Miller *Gardeners Dict.* s.v. Populus. The Trembling Poplar, or Aspen-Tree.

tremblingly, *adv.* (Earlier and later examples of the collocation *tremblingly alive.*)
1733 Pope *Ess. Man* i. 11 Or touch, so tremblingly alive all o'er, To smart, and agonize at ev'ry pore? 1818 C. R. Maturin *Women* I. ii. 40 Ignorant of music as a science, but 'tremblingly alive' to its influence, he listened.

tremblor (tre·mblǫɹ). orig. and chiefly *U.S.* [Alteration of Sp. *temblor* shudder, (in Amer. Sp.) earthquake, influenced by Eng. Trembler.] An earthquake or earth tremor.
1913 C. C. Goodwin *As I remember Them* 193 He..received the sobriquet of 'Earthquake' Stewart, because of the theory that he put out, that the tremblors in California were caused..by..electrical disturbances in the air and in the earth near the surface. 1925 *N.Y. Times* 1 July 2/2 Dr. U. S. Grant..does not agree with Prof. Goode that the same tremblors shook Montana and California. 1926 *Daily Colonist* (Victoria, B.C.) 1 July 1/5 Today's shocks occurred in areas more than 100 miles removed from the lower coast line, where tremblors were felt yesterday. 1977 *Chicago Tribune* 2 Oct. i. 10/5 Californians, long accustomed to earthquake jitters, have become concerned over a rapid-fire series of minor tremblors along a 20-miles stretch of the mysterious San Andreas Fault.

tremendous, *a.* Add: Also *dial.* and non-standard **tremenj(i)ous, treminjous,** etc. **2.** (Further examples.) Also as quasi-*adv.*
1835 [see *lathering vbl. sb.]. 1886 R. D. Blackmore *Springhaven* in *Harper's Mag.* Oct. 755/1 Makes us pay tremenjious for 'most everything. 1888 Kipling *Phantom 'Rickshaw* 92 Then ten men with bows and arrows ran down that valley, chasing twenty men with bows and arrows, and the row was tremenjus. 1892 'Q' *I saw Three Ships* v. 97 'Ay, naybours all,' broke in Farmer Tresidder... 'I shudn' wonder if ye was to see me trottin' to Parlyment House in a gilded coach..I be so tremenjous rich.' 1901 M. Franklin *My Brilliant Career* xvii. 150, I thought them straps couldn't break only onder a tremenjous strain. *Ibid.* xix. 162 'How are you enjoying yourself?'.. 'Treminjous intoirely, sor.' 1952 M. Allingham *Tiger in Smoke* ii. 39 'Was it fun?' 'Tremenjous.' 1977 E. W. Hildick *Loop* vii. 37 She had 'tremenjous powers of seeing the future'.

‖**tremendum** (treme·ndǒm). [Shortened from *mysterium tremendum.] The overwhelming awe which can be part of religious experience.
1923 J. W. Harvey tr. *Otto's Idea of Holy* iv. 12 (*heading*) The analysis of 'tremendum'. *Ibid.* 16 The elements which unfold as the 'tremendum' develops. 1930 A. G. Hebert tr. *Brilioth's Eucharistic Faith & Practice* ii. 65 It is an expression for the awfulness of the holy, the *tremendum*, which belongs to all deep religion. 1950 [see *numinosum]. 1960 R. F. C. Hull tr. *Jung's Coll. Wks.* III. 260 The originally chaotic or frightening impression is replaced by the picture, which, as it were, covers it up. The *tremendum* is..made harmless and familiar.

tremis. Substitute for entry:
tremissis (tremi·sis). Also erron. **tremis(s.** Pl. **tremisses.** [late L., gen. sing. of *trēmis*, f. *trēs* three, after *sēmis* half an as: cf. *semis[1].] A late Roman or early Byzantine gold coin, the third part of a solidus; a Merovingian or other imitation of this.
1706, 1756 [in Dict.]. 1952 *Antiquity* XXVI. 77 All the coins from Sutton Hoo are tremisses (thirds of solidi). 1962 H. R. Loyn *Anglo-Saxon England* ii. 74 The *solidus* (or *sou*) was..essentially..the imperial coin reformed by Constantine; more frequent in Gaul was the *tremissis* (or *triens*), the third part of a *sou*. 1976 *Anglo-Saxon England* V. 176 Small Merovingian gold tremisses..had been penetrating the south-east of England. *Ibid.* 177 There is considerable circumstantial evidence for the equation of a Merovingian tremissis with an Anglo-Saxon gold shilling. 1983 *Times* 6 Apr. 16/4 The discovery of a remarkable Visigothic *tremiss* minted in gold in Southern Gaul in AD 455–475, in the final levels of one site excavated in late 1982, shows that some form of 'urban' life continued until nearly the end of the fifth century.

tremolist (tre·mǒlist). [f. Tremolo + -ist.] One who uses the tremolo.
1927 *Proc. Musical Assoc.* 1926–7 18 The soprano tremolist is no longer tolerated. 1952 B. Ulanov *Hist. Jazz in Amer.* (1958) vi. 63 The first boogie-woogie crew, who must have influenced the second wave of C-major tremolists, the famous Chicagoans.

tremor (tre·mǫɹ), *v.* [f. the sb.] *intr.* To be agitated by a tremor or tremors; to shake or tremble.
1921 *Chambers's Jrnl.* XI. 858/1 The ship tremored, vibrated like mad. *Ibid.* 860/2 Her voice had tremor'd.. with urgency. 1926 M. Walsh *Key above Door* x. 113 His strong, big jowl was..tremoring with the chill. 1928 —— *While Rivers Run* vi. 68 His car was purring and tremoring. 1963 A. Smith *Throw out Two Hands* xvi. 162 They [*sc.* zebras] went by in droves, and the earth tremored beneath them.

tremorine (tre·mǒrīn). *Pharm.* [f. Tremor + -ine[5].] A crystalline compound, 1,4-dipyrrolidino-2-butyne, $C_{12}H_{20}N_2$, capable of inducing the symptoms of Parkinsonism and used in research into this disease.
1956 G. M. Everett in *Federation Proc.* XV. 420/2 'Tremorine', when given in doses of 5–20 mg/kg by all routes, produces a marked sustained tremor. 1970, 1972 [see *oxotremorine]. 1976 *Nature* 15 July 221/2 Tremorine, which is oxidised *in vivo* to oxotremorine by mammals, showed only slight activity against ticks.

trench, *sb.* Add: **2. b.** An elongated channel in the sea-bed; *spec.* one of the very long ones, several kilometres deep, that run parallel to the edge of a continent or an island arc and are believed to mark subduction zones.
1936 *Amer. Jrnl. Sci.* CCXXXI. 401 Recent discoveries of many valley-like trenches that interrupt the outer steeper slopes ('continental slopes') of continental shelves are truly startling. 1948 F. P. Shepard *Submarine Geol.* xi. 283 A series of deep trenches skirt the Pacific. 1963 G. L. Pickard *Descriptive Physical Oceanogr.* ii. 10 The greatest depths in the oceans occur in these trenches. 1972, etc. [see *subduction 5]. 1975 *Offshore Engineer* Sept. 60/1 The Norwegian trench is a pitfall that has always tempered Norway's oil future. Crossing it is beyond the current state of pipeline technology. 1977 A. Hallam *Planet Earth* 94 The island arcs and ocean trenches that border the northwest and southeast Pacific.

3. d. (Later examples.)
1970 G. Jackson *Let.* 23 Mar. in *Soledad Brother* (1971) 188 I've been living in the trenches where it's understood that it's us against them, hide and seek. 1977 *Rolling Stone* 24 Mar. 13/2 He has been in the trenches too long not to be a master at mixing sincerity with evasiveness.

8.* *ellipt.* for *trench-coat*, see sense 9 below.
1972 *New Yorker* 14 Oct. 1 (Advt.), Bonwit's velvet rain trench plays matinees and evenings. 1974 [see *midi-]. 1978 *N.Y. Times* 30 Mar. A9/3 (Advt.), Three styles for sizes 8 to 16. Double breasted trench for reg. and petite.

9. *trench kit, life, light, raid, raiding, rifle, strafing, system, war*; *trench-stale, trench-to-trench* adjs.; **trench boot,** (usu. in *pl.*) combined boot and leggings; **trench-coat,** a waterproofed overcoat worn by officers in the trenches; a long, belted raincoat for civilian use; hence **trench-coated** *adj.*; **trench fever,** an epidemic louse-borne rickettsial disease that was common among soldiers in the war of 1914–18, causing splenomegaly and recurrent fever; **trench foot, feet,** a painful condition of the feet caused by prolonged immersion in cold water or mud, marked by swelling, blistering, and some degree of necrosis; **trench-knife,** a knife with a double-edged blade, orig. used in trench raids; **trench mortar,** a small mortar designed to propel bombs from a front trench into enemy trenches; hence as *v. trans.*; **trench mouth,** Vincent's angina of the mouth (see *Vincent[2]); **trenchoscope** = next; **trench-periscope,** a kind of tube-and-mirror apparatus used in trench warfare (see quot. 1918 for *trenchscope* below); cf. prec. and *periscope 3; **trench-rat,** the brown or Norway rat, *Rattus norvegicus*; **trenchscope** = *trench periscope* above; cf. also *trenchoscope* above; **trench warfare,** hostilities carried on by means of or in trenches; also *fig.*, a protracted dispute or conflict in which the parties seek to maintain their entrenched positions while launching persistent attacks upon their opponents; cf. *trench war* above.
1933 J. Buchan *Prince of Captivity* ii. i. 132 He wore a tattered trench waterproof and..ancient trench-boots. 1973 *Country Gentlemen's Mag.* Mar. 184/2 Officer's brown leather calf length trench boots..practically new. 1916 W. Owen *Let.* 16 Aug. (1967) 405 My poor troops were wet to the bone. (But I had my Trench Coat.) 1918 E. Ferber in *Best Short Stories of 1917* 209 Jo Hertz, in one of those pinch-waist belted suits and a trench coat..was a sight for mirth or pity. 1944 M. Laski *Love on Supertax* xi. 107, I got a job in a trench-coat factory in Manchester. 1978 W. F. Buckley *Stained Glass* xi. 109 A young man dressed in an old army trenchcoat walked slowly out, dragging his wooden leg like a ball and chain. 1941 N. Alley *I Witness* xxxv. 296, I compensated my desire with the wishful thought that I'd be running across his familiar trench-coated figure some time later. 1980 *Listener* 29 May 697/3 A trench-coated Micawberish librarian. 1915 *Lancet* 25 Sept. 734/1 The case of a twice-inoculated soldier suffering from trench fever, whose case was diagnosed as pyrexia. 1917 G. S. Gordon *Let.* 22 May (1943) 77 He says I've got what they call vaguely 'Trench' Fever. 1933 J. Buchan *Prince of Captivity* i. iii. 85 Blown-up, buried, dysentery, trench-fever, and most varieties of wounds. 1976 *West Lancs. Even. Gaz.* 13 Dec. 7/1 After convalescing in England from trench fever, he successfully applied for a commission. 1915 *Lancet* 17 Apr. 812/2 The term trench-foot appears to us to be the most suitable for a condition which has practically only been met with in those who have had to remain for long periods in the trenches. 1916 W. S. Churchill *Let.* 6 Jan. in M. Gilbert *Winston S. Churchill* (1972) III. Compan. ii. 1359, I wish you wd write for me a brief description of your 'trench feet routine'. a 1918 W. Owen *Poems* (1920) 23 But never..fever, trench-foot, shock, Untrapped the wretch. And death seemed still withheld. 1940 *War Illustr.* 12 Jan. 603/1 Though the men had plenty of food all were suffering from exposure and trench feet when they landed at a South Coast port. 1969 [see *immersion foot* s.v. *immersion 5]. a 1976 A. Christie *Autobiogr.* (1977) v. ii. 239 Half our patients seemed to be trench feet cases. 1982 *Times* 31 May 5/3 There have been cases of exposure and trench foot. 1914 *War Illustr.* 19 Dec. 416/1 He inspected their trench kit of goatskins and strawbags. 1926 *Scribner's Mag.* Aug. 194/2 A Boche lad I killed with me trench-knife. 1979 R. Blythe *View in Winter* iv. 188, I had a trench knife in one hand and a pistol in the other. 1917 W. Owen *Let.* 15 Aug. (1967) 484

Nothing like his [*sc.* Sassoon's] trench life sketches has ever been written. 1977 A. Wilson *Strange Ride of R. Kipling* vii. 298 What he [*sc.* Kipling] saw of trench life.. horrified him. 1918 G. Frankau *Judgement of Valhalla* 18 Downwards, and on, where trench-lights shone—For we, we might not rest. 1915 D. O. Barnett *Let.* 19 May 143 In the afternoon we had a trench-mortar duel with the Allymans. 1920 *Chambers's Jrnl.* 20 Mar. 254/1 He shelled it; he trench-mortared it, he raided it. 1973 M. Woodhouse *Blue Bone* xii. 129 What looked like a three-foot metal pipe with a rectangular base... 'Five-centimeter trench mortar,' said Yancy. 1918 *Evening Mail* 1 May 3/4 We have trench mouth, just as we have trench feet. Otherwise known as ulcero-membranous stomatitis, or Vincent's disease. 1946 J. Lees-Milne *Diary* 1 Jan. (1983) 3 Went to the dentist who said it is trench mouth that I am suffering from. 1981 G. Priestland *Priestland's Progress* 8 Chris Rees had to take to his bed with a rare attack of trench mouth. 1915 *Morning Post* 11 Feb. 3/5 The Adams trenchoscope is the latest periscope for use in the trenches. 1915 Trench-periscope [see *hyposcope* s.v. *hypo- II]. 1917 A. G. Empey *Over Top* 313 *Trench raid*, several men detailed to go over the top at night and shake hands with the Germans, and, if possible, persuade some of them to be prisoners. 1961 W. Vaughan-Thomas *Anzio* i. 5 A struggle in the mud, complete with duck-boards, trench-raids and patrols in no-man's-land. 1974 A. Price *Other Paths to Glory* ii. viii. 213 It's a sawn-off Lee Enfield... Used for trench raiding... The trench rifle had been Jarras's newest toy. 1916 G. Frankau *Guns* 22 The trench-rats patter And nibble among the rations. 1917 A. G. Empey *Over Top* 308 There are three things in this world that Tommy loves: a slacker, a German, and a trench-rat. 1918 E. S. Farrow *Dict. Mil. Terms* 629 *Trenchscope*, a simple periscope, used in the trenches (permitting a safe view to the front), consisting of parallel mirrors in a long wooden box, both set 45° to the long axis of the box. 1915 Kipling *France at War* v. 55 One understood after a while the nightmare that lays hold of trench-stale men. 1931 *Times Lit. Suppl.* 22 Oct. 822/1 The squadron was required to assist the hard-pressed infantry by..'trench strafing'. 1918 E. S. Farrow *Dict. Mil. Terms* 629 *Trench system*, all the field-works included in a defense zone. 1923 Kipling *Irish Guards in Gt. War* I. 227 Some half-wiped-out German trench-systems. 1975 P. Fussell *Gt. War & Mod. Memory* ii. 36 From the winter of 1914 until the spring of 1918 the trench system was fixed. 1923 Kipling *Irish Guards in Gt. War* II. 143 There are certain analogies between trench-to-trench attack and 'soccer'. *Ibid.* I. 20 The trench-war was solidifying itself. 1918 F. M. Ford *Let.* 6 Jan. (1965) 86, I could do a very good one [novel] about trench warfare. 1973 *Times* 7 Dec. 18/3 In industry we have had continuing trench warfare deriving from low pay, and authoritarian and remote management. 1978 F. Maclean *Take Nine Spies* iv. 125 He had fought at Dixemude.. before settling down later that winter [in 1914] to the horrors of trench warfare. 1980 R. Barnard *Death in Cold Climate* vii. 69 Sterile trench-warfare with colleagues over matters of principle.

trencher[1]. Add: **5.** (Later U.S. example.)
1889 *Harper's Mag.* Jan. 238/2 He next binds his [beaver] trap to a flat stone 'about the size of a teakettle', opens the jaws, and arranges the 'trencher', as the pan is called, pressure on which springs the trap.

7. trencher-bread (later *Hist.* examples); **trencher-salt:** delete † and add later examples; **trencher table,** a table at which members of domestic staff were seated at meal times.
1882 F. Michel *Crit. Inquiry into Scottish Lang.* iii. 54 [Edinburgh] citizens had four different kinds of wheaten bread: the finest called *manchet*, the second *cheat* or *trencher* bread. 1974 Trencher-bread [see *jumbal, jumble]. 1967 *Times* 24 Feb. 14/6 The most unusual objects in a silver sale at Sotheby's yesterday were two miniature trencher salts, 1¼in. diameter and weighing only to 10 dwt. by George Middleton, 1864. 1970 *Canad. Antiques Collector* Mar. 23/1 Salt cellars were made in many shapes, starting with plain, round or oblong, so-called trencher salts in the early 1700's. 1968 *Ibid.* June 9/1 Dinner..was served on two tables. One was at floor level—the 'trencher' table—where the steward, the curate, the governess and members of the staff were seated. 1977 D. E. Westlake *Nobody's Perfect* ii. iii. 110 Ratty pieces of living-room furniture..and an old trencher table.

trenching, *vbl. sb.* **b.** Add: *trenching-spade* (earlier U.S. example), *-tool* (examples).
1866 *Harper's Mag.* Oct. 636/1 Rough-boarded warehouses crammed with..gun-carriages,..trenching spades, and axes. 1779 in *9th Rep. R. Comm. Hist. MSS.* App. iii. 148 in *Parl. Papers* 1910 (Cd. 5038) XXXV. 675 A large quantity of trenching tools, etc., have been put on board. 1809 A. Henry *Trav. & Adventures Canada* xv. 130 The most common way of taking the beaver is that of breaking up its house, which is done with trenching-tools. 1979 R. Blythe *View in Winter* iv. 172 They was now shellin' like anything... I got my trenchin'-tool out.

tre·nchless, *a.* [f. Trench *sb.* + -less.] Applied to pipe-laying and draining machinery, etc., that dispenses with the cutting of a trench.
1969 [see *pipe-layer]. 1971 *Power Farming* Mar. 60/1 This is a dual-purpose trenchless drainer. 1979 *Daily Tel.* 31 Dec. 15/5 There is the so-called trenchless machine which, in one operation, can lay the pipe without any soil being removed.

trench-plough, *sb.* Add: Also used for trench-making in warfare.
1918 E. S. Farrow *Dict. Mil. Terms*, *Trench-plough*, a kind of plough for opening land to a greater depth than that of common furrows; a plough used in trench making.

trend, *sb.* Add: **4. b.** Now freq. with qualifying word and without const. (Later examples.)
1902 G. W. E. Russell *Londoner's Log-Book* xiv. 243 Beyond doubt, Bounderley's local popularity is waning. The 'trend' is pointing in another direction. 1928 *Granta* 3 Feb. 240 Show me a play that's Russian with a psycho-symbolical trend. 1930 M. A. Magee (*title*) Materials for the study of business trends in location of the women's clothing industry. 1967 Singha & Massey *Indian Dances* i. 38 This can be regarded as a healthy trend since it has aroused a consciousness of the dance.

c. *spec.* in *Educ.* (See quots.)
1960 *Where?* iii. 18/1 'Trend', the, jargon for the increasing tendency of pupils to stay at school beyond the compulsory school-leaving age. 1962 A. Sampson *Anatomy of Britain* xii. 185 Since the war more children are staying on voluntarily than had been expected..(the phenomenon known to schoolteachers as 'Trend'). 1969 H. Perkin *Key Profession* v. 208 The 'trend', that is, the growing demand on the part of the young and their parents for higher education expressed in the tendency to stay on at school beyond the statutory leaving age.

5. *Geol.* A geological formation which is a source of oil or gas. Cf. sense 4 a in Dict.
1939 *Bull. Amer. Assoc. Petroleum Geologists* XXIII. 860 The Jackson trend continued to lead in the number of discoveries with 16 new [oil] fields. 1977 *Time* 5 Dec. 59/1 In Louisiana, the 'trend' (main potential gas-producing formation) lies four miles beneath the green bayous and sugar-cane fields.

6. Special Comb.: **trend analysis,** analysis of (esp. statistical) data in order to detect or study any trend represented in them; **trend line,** a line indicating the general course or tendency of something (as a geographical feature or a set of points on a graph); **trend-spotter,** one who observes (or seeks to predict) the changing tide of fashion, in dress, ideas, etc.; **trend surface,** a mathematically defined surface computed as a best fit to the sampled values of some parameter over an area of interest; so *trend surface analysis.*
1934 M. Sasuly *Trend Analysis of Statistics* i. 6 The primary purpose of this book is to derive formulas and computation schedules that will simplify..practical trend analysis. 1964 M. Argyle *Psychol. & Social Probl.* xv. 183 For purposes of trend analysis it is not particularly interesting to know how particular people have changed, since this confuses developmental with historical trends—we want to compare groups of people today with equivalent groups at a previous date. 1971 *Jrnl. Gen. Psychol.* LXXXIV. 107 Trend analysis showed that performance under steady illumination did not vary over wavelength. 1912 *Q. Rev.* Apr. 532 The trend-lines of mountain systems are the results..of something more than a lateral pressure. 1930 M. Ezekiel *Method of Correlation Analysis* xvi. 239 The residuals from the final trend line might be again plotted against the other curves, to see if any further changes were necessary. 1959 *Listener* 2 Apr. 581/2 The almost level trend-line of coal production. 1965 G. J. Williams *Econ. Geol. N.Z.* iii. 30/2 The lateral displacements are as much as 95 ft from the trend-line of the lode. 1976 *National Observer* (U.S.) 6 Nov., If you slow growth, it means that the trend line for the production of automobiles, refrigerators, houses, and so forth will begin to taper down. 1965 *Punch* 21 Apr. 570/2 As trend-spotters will have spotted, the sweet-and-twenty blonde, who was last year selling us Scotch, shirts and motor-cars from the hoardings is increasingly yielding place to little winsome children. 1980 *Times Lit. Suppl.* 21 Nov. 1316/4, I don't deny that ideology can be adopted as a fashionable mode, but ideas, real ideas, do not make their appearance and disappearance merely to satisfy the shaping whims of trend-spotters; and to pretend that they do is to become a trend-spotter yourself. 1956 R. L. Miller in *Jrnl. Geol.* LXIV. 425 The problem of defining and analyzing contemporary environments of sedimentation is approached from the point of view of mapping 'trend surfaces'. 1959 *Jrnl. Geophysical Res.* LXIV. 823 Trend surface analysis is a procedure for separating the relatively large-scale systematic changes in mapped data from essentially non-systematic small-scale variations due to local effects. 1978 B. Chapman *Clarke's Analytical Archaeol.* (ed. 2) x. 455 (*caption*) The location of the cultural assemblages in the Central Plains which have been analysed by trend-surface analysis.

Trendelenburg (trende·lĕnbǝɹg). *Med.* Also (*erron.*) **-berg.** The name of Friedrich *Trendelenburg* (1844–1924), German surgeon, used in the possessive and *attrib.* to designate certain phenomena observed and medical procedures invented by him, as **Trendelenburg('s) position,** an operating position in which the patient lies supine on a tilted table or bed with the pelvis higher than the head; **Trendelenburg('s) test,** (*a*) a test for disorders of the hip joint or gluteus muscles in which the patient stands on one leg and raises the other, a dropping of the pelvis on the unsupported side being a positive sign (*Trendelenburg's sign*); (*b*) a test for varicose veins in which the leg is raised to drain it of blood and then quickly lowered, immediate and rapid distension of the leg veins indicating valve incompetence.
1892 Keen & White *Amer. Text-bk. Surg.* II. iii. viii. 948 The further manipulations within the pelvis will be much facilitated by the elevation of the pelvis.., the position being known as Trendelenburg's position. 1907 *Practitioner* Aug. 288 The abdomen is opened with patient in the extreme Trendelenburg position. 1912 A. H. Tubby *Deformities* (ed. 2) I. iii. ix. 586 Trendelenburg's sign is this: If the patient stands on the affected limb and flexes the hip of the second side..the pelvis slips downwards on the sound side. 1923 Jones & Lovett *Orthopedic Surg.* xxvii. 556 The so-called Trendelenburg sign.. is very characteristic and explains a good deal of the curious gait. 1930 H. Bailey *Demonstrations of Physical Signs in Clin. Surg.* (ed. 2) xxi. 214 Trendelenburg's test is not diagnostic of congenital hip disease. 1967 G. M. Wyburn et al. *Conc. Anat.* vi. 170/2 With paralysis of gluteus medius or minimus the pelvis drops on the unsupported side when standing on the affected limb (Trendelenberg's sign). 1974 J. D. Maynard in R. M. Kirk et al. *Surgery* xi. 239 Trendelenburg test... Confirmation of retrograde flow in the superficial veins is carried out. 1975 *Year Bk. Ear, Nose & Throat* 116 The patient is placed in the 8-degree reversed Trendelenburg position.

trendily (tre·ndili), *adv.* [f. *trendy *a.* + -ly[2].] In a fashionable or 'trendy' manner; modishly. Also in *Comb.* with *ppl. adjs.*
1967 *Guardian* 24 Apr. 4/4 A trendily dressed couple. 1971 *Rolling Stone* 24 June 3/1, I passed a trendily-attired Mendelsohn in front of Bullock's Westwood. 1974 *Listener* 21 Nov. 685/1 His young, trendily radical mistress. 1980 S. Brett *Dead Side of Mike* ix. 101 A trendily-dressed blonde teenager of thirty-seven.

tre·ndiness. [f. *trendy *a.* + -ness.] The uncritical following of fashionable modes of thought, dress, etc.; the quality of being trendy.
1966 *Guardian* 11 Oct. 6/2 Trendiness seemed worse than going to the common hordes. 1968 *Listener* 26 Dec. 871/2 Its breathless combination of trendiness and self-righteousness. Admittedly, *How It Is* is not alone in turning civil rights, say, into a fashion; and fashion.. into a tyranny. 1972 J. McClure *Caterpillar Cop* iii. 42 The homes..were modest bungalows succumbing..to an ill-becoming trendiness; bright colours had been painted over the exterior woodwork. 1974 *Economist* 5 Jan. 13/1 As a result of yesterday's trendinesses, we have now created in the developed world an unfortunate excess of both birth control devices and anti-pollution controls. 1982 S. Radley *Talent for Destruction* x. 70 Indeed he is! A very popular Rector, I put it down to the fact that he's not trendy. Trendiness doesn't do in a parish like Breckham Market.

trending, *ppl. a.* (In Dict. s.v. Trend *v.*) (Later example.)
1968 D. L. Clarke *Analytical Archaeol.* vi. v. 274 Once again we have six trending variables, each with three crude attitudes.

tre·nd-setter. [f. Trend *sb.* + Setter *sb.*[1]] One who or that which establishes trends in dress, thought, etc.
1960 *Guardian* 17 Mar. 4/3 Powerful trend-setters, outstanding personalities. *Ibid.* 7 Dec. 6/3 The magazines.. are the chief trend-setters. 1966 *Listener* 2 June 814/1 The music and aesthetics of modern Italy's musical trend-setter, Casella. 1973 *Times* 24 Apr. 10/2 His big shouldered jackets are important trendsetters. 1979 D. Hurd *End to Promises* ii. 26 My strongest feeling was satisfaction that the experts, the know-alls, and the trend-setters had been confounded.
Hence (as back-formation) **tre·nd-set** *v. intr.*; **tre·nd-setting** *ppl. a.*
1960 *House & Garden* Oct. 172 The big news from.. those trend-setting couturiers. 1965 *Punch* 7 July 34/3 That hairstyle, do you think it will trend-set? 1971 M. Lee *Dying for Fun* v. 42 Pollie Potter, dress designer, had been trendsetting now for two years. 1976 *Times* 26 Aug. 13/2 The trend-setting intellectuals of our day.

trendy (tre·ndi), *a.* and *sb.* [f. Trend *sb.* + -y[1].] **A.** *adj.* Fashionable, up to date, following the latest trend. (Sometimes dismissively.)
1962 *Punch* 7 Nov. 654/3, I saw the headline 'The Trendiest Twin Set'. 1965 *Sun* 20 May 7/6 The BBC's Debussy film..must have been the first to use the screen credit 'Art Nouveau Consultant'... This is trying to be trendy: what's wrong with art-adviser? 1972 *Lancet* 20 May 1104/1 Pathobiology (a trendy name for general pathology) seems to be a fashionable subject in the United States. 1977 B. Pym *Quartet in Autumn* viii. 74 That was how it had always been and how it would go on in spite of trendy clergy trying to introduce so-called up-to-date forms of worship. 1982 *Chicago Sun-Times* 11 Nov. 88 (*heading*) Bargains still available in trendy neighborhoods.
B. *sb.* One who follows the latest trends or fashions.
1968 J. Fleming *Kill or Cure* xi. 143 She was well in with what is now called the Chelsea set.., there are trendies and *personae non gratae* amongst them. 1971 *New Scientist* 26 Aug. 450/1 Amphetamine..to the young trendy..provides a means of staying awake all night. 1974 *Courier-Mail* (Brisbane) 30 Mar. 8/3 Some of us have fought all our lives to be trendies. 1982 *Listener* 16 Dec. 20/2 The 'trendies'' concern for the individual seems to relate more to his place in society than to his soul.

tre·ndyism. [f. prec. + -ism.] = *trendiness.
1977 *Spare Rib* July 4/1 The coy use of Christian names ..reminds me of the clubby, in-group atmosphere of competitive trendyism I found so depressing at the last WLM conference. 1981 *Observer* (Colour Suppl.) 13 Dec. 66/1 An encroaching tide of trendyism.

trepan, *v.*[1] Add: **c.** *Engin.* To cut an annular groove or hole in (something) by means of a crown saw or other tool; to make (a hole) thus, the core being removed as a solid piece.
1909 in *Cent. Dict. Suppl.* 1919 A. G. Robson *Engin. Machine Tools & Processes* ix. 195 With the cutters at hand it was impossible to get a feed greater than 1/200 inch per revolution when trepanning steel from the solid without breaking the cutter. 1953 G. S. Schaller *Engin. Manufacturing Methods* xiii. 221 The solid forging is trepanned instead of being bored in the conventional manner. 1970 I. Bradley *Myford ML10 Lathe Man.* xi. 64 The smaller holes are best bored, but large holes can be trepanned in order to save a useful piece of material.
trepanning *vbl. sb.* (further examples).
1880 M. P. Bale *Woodworking Machinery* xvii. 168 A Mr. Murdock, in 1810, took out a patent for an improved machine for forming wooden or stone pipes. For boring wood he employed a hollow cylinder, fitted at its extremity with a circular trepanning saw. 1949 W. S. Churchill *2nd World War* II. ii. xviii. 319 Trepanning consisted of making a hole in the bomb casing in order to deal with the explosive contents. 1974 *Sci. Amer.* Jan. 36/2 In trepanning a hollow cathode shaped according to the specified pattern lifts parts of complex shape and uniform thickness from a metal slab.

trepa·nger. [f. Trepang + -er[1].] A trepang-fisher.
1912 A. Searcy *By Flood & Field* xlviii. 302 Our confrere then informed us that as soon as the famous hunter gained his liberty he had purchased a ketch, fitted her up with the requirements of a trepanger, and sailed away. 1934 *Bulletin* (Sydney) 14 Mar. 33/1 The trepangers with their long, sharp iron spears will soon be working the Rowley Shoals.

trepanner[1]. Add: **2.** *spec.* in *Coal Mining* (see quot. 1967).
1956 Atkinson & White in D. L. Linton *Sheffield* 269 Intensive efforts are being made to replace the hand-loading of coal..by mechanical power-loaders and..considerable success is being achieved with..Gloster Getters, Trepanners,..and Huwood Slicers. 1967 *Gloss. Mining Terms* (*B.S.I.*) viii. 27 *Trepanner*, a longwall power loader, usually double ended, the main cutting element of which is a trepanning wheel. 1971 [see *plough, plow *sb.*[1] 5 i]. 1974 P. Wright *Lang. Brit. Industry* xiv. 145 He may..hear..shearer and trepanner for types of coal-cutters.

trephine, *sb.* Add: **2.** = *trephination* s.v. Trephine *v.*
1958 F. G. Slaughter *Daybreak* i. ii. 13 A patient should be more than a chart number..even to a neurosurgeon performing a preliminary trephine. 1976 *Lancet* 9 Oct. 769/2 Bone-marrow aspirate and trephine confirmed acute myeloid leukæmia.

trepidant, *a.* (Earlier example.)
1891 [see *abasia].

treponema (treponī·mă). *Biol.* and *Med.* Pl. **-ne·mata.** Also anglicized as **tre·poneme.** [mod.L. (coined in Ger. by F. Schaudinn 1905, in *Deutsch. med. Wochenschr.* 26 Oct. 1728/1), f. Gr. τρέπ-ειν to turn + νῆμα thread.] An anærobic spirochæte of the genus of this name, the members of which are parasitic or pathogenic in man and warm-blooded animals and include those causing syphilis and yaws.
1908 [see *spirochæte* s.v. *spiro-]. 1922 *Nature* 20 May 667/2 Existence of the treponeme in the cytoplasm of the nerve cells of the cerebral cortex. 1949 M. A. Jennings in H. W. Florey et al. *Antibiotics* II. xxxi. 1011 In a patient with concurrent typhoid fever the treponemata disappeared but the typhoid infection ran its usual course. 1970 *New Scientist* 19 Mar. 543/1 Late in the disease, dormant treponemes persist in some instances in lymph nodes and other tissues. 1981 *Brit. Med. Jrnl.* 18 Apr. 1312/1 Walt Disney's serried ranks of gonococci and treponemes.
Hence **trepone·mal** *a.*, of, pertaining to, or caused by treponemes; **treponemato·sis** (pl. **-oses**), infection with, or a disease caused by, treponemes.
1913 Castellani & Chalmers *Man. Trop. Med.* (ed. 2) xliv. 1191 These drugs seem to act in frambœsia more quickly and powerfully than in any other spirochætal and treponemal condition. 1927 *Jrnl. Laboratory & Clin. Med.* XII. 670 In this paper we are using the term 'treponematosis' to include syphilis and the condition called yaws. 1970 *New Scientist* 19 Mar. 543/1 Venereal syphilis of adults is rare in tropical regions where the endemic treponematoses, yaws and pinta, prevail. *Ibid.*, The results of a quarter of a century's experience with penicillin in the treatment of treponemal disease, notably syphilis, are now available. 1980 *Nature* 7 Feb. 573/2 Treponemal infection in irradiated mice.

treponemicidal (tre:pǒnīmisəi·dăl), *a.* [f. prec. + -i- + -cid(e + -al.] Of, pertaining to, or causing the destruction of treponemata.
1933 J. E. Moore *Mod. Treatment of Syphilis* xxvi. 382 In early meningeal neurosyphilis..the desideratum of treatment is treponemicidal effect applied as powerfully as possible. 1970 *New Scientist* 19 Mar. 543/1, 2·4 mega-units of long acting benzathine penicillin can maintain a treponemicidal blood and tissue level for three weeks or more.

‖**très** (trę), *adv. colloq.* [Fr.: see Tres-.] With English adjs.: very. Usu. with reference to a fashionable or modishly superior quality, freq. as *très snob*, very 'posh'.
[1815 F. Burney *Let.* 10 July (1980) VIII. 285 He is *très* what those on the other side the question call *exalté*.] 1819 Keats *Let.* 3 Jan. (1931) I. 292 A full, true, and très particular account of Miss M's ten Suitors. 1939 'N. Blake' *Smiler with Knife* iv. 63 It's a sort of county club. Très snob. Très cad. On the Thames. 1959 A. Sinclair *Breaking of Bumbo* v. 74 The continental millionaires, who thought it *très snob* to bring out their young in the last society in Europe. 1968 A. Diment *Gt. Spy Race* viii. 126 *Très, très, très* sexy. It was fairly obvious..which section of 'trad' spy-craft Mrs Omega was going to test. 1976 *Publishers Weekly* 8 Mar. 67/2 The picture on the cover, a *très* cool modern kid who doesn't look a bit like the novel's hero. 1978 J. Sherwood *Limericks of Lachasse* xii. 140 Students..who were of good family and *très snob*.

Trevira (trəvīə·rä). Also **trevira.** The proprietary name of a type of artificial fibre or the fabric made from it (see quots.).
1959 *Official Gaz.* (U.S. Patent Office) 24 Nov. TM160/1 Trevira. Owner of German Reg...dated Mar. 21, 1956.. For table linen and bed linen; net, lace, woven and mesh fabrics; textile ribbons, textile trimmings; carpets and rugs; mats; curtains; flags; and felt. 1960 *Trade Marks Jrnl.* 21 Dec. 1650/2 *Trevira*... All goods included in Class 24. Farbwerke Hoechst Aktiengesellschaft Vormals Meister Lucius & Brüning..45, Brüningstrasse, Frankfurt-am-Main—Hoechst, Germany; Manufacturers and Merchants. 1964 *Which?* Aug. 253/3 *Polyester*... Brand names: *Dacron* (USA), *Tergal* (France), *Terlenka* (Holland), *Terylene*, *Trevira* (W. Germany). 1967 *Daily Tel.* 30 Oct. 11/3 Now there's another new test-tube fibre name to memorize... Trevira will pop into Britain's fashion shops from December. 1972 *Guardian* 22 Aug. 9/2 Trevira/rayon shirt. 1977 *N.Z. Herald* 5 Jan. 2-18/3 (Advt.), Lincoln Fabrics... Plain and fancy crimps and treviras.

trevorite (tre·vǒrəit). *Min.* [f. the name of T. G. *Trevor* (1865–1958), South African geologist and mining official + -ite[1].] A black, magnetic, isometric oxide of nickel and ferric iron, $NiFe_2O_4$, belonging to the spinel group.
1921 A. F. Crosse in *Jrnl. Chem., Metallurgical & Mining Soc. S. Afr.* XXI. 126/2 One of the most interesting mineralogical discoveries..in the Transvaal..is an extraordinarily rich nickel ore... This ore is as far as I am able to judge a new and undescribed mineral... I should like to call it 'Trevorite' after Major T. G. Trevor, Mining Inspector for the Pretoria District. 1976 *Mineral. Abstr.* XXVII. 252/2 Microprobe analyses of grains of ferroan trevorite from a serpentine massif in south-western China reported Ni 3·4–8·8% but no Ti.

trews, *sb. pl.* Add: **b.** Trousers in general (tartan or otherwise), including close-fitting trousers worn by women.
1847 H. S. Riddell *Poems* 19 When I brought ben your clase, Sae beaten with the weather, And gae the trews a wee bit touch, Out flew goud guinea frae ae pouch. 1883 F. Sutherland *Sunny Mem. Morayland* 57 The soor-moo'd limmer wears the trews. 1917 N. Munro *Poetry* (1931) iii. 51 His body unadorned by Highland raiment, Trammelled, for glorious hours, in Saxon trews. 1931 E. Linklater *Juan in America* ii. iv. 94 His Tyrolean costume made evident that even his knees were comely—strong, round, and rosy under their brief leather trews. 1958 *Woman's Own* 5 Mar. 16/3 They make a handsome pair, when she's wearing a blouse with matching tapered trews in printed wool. 1959 H. Hobson *Mission House Murder* i. 5 She was wearing tartan trews, a black cashmere sweater and a short red duffle-coat.

trey, *sb.* Add: **2.** (Earlier and further examples.) *spec.* in the U.S., a three-dollar packet of a narcotic.
[1859 Hotten *Dict. Slang* 204 *Tray saltee*, threepence.. tre soldi.] 1887 J. W. Horsley *Jottings from Jail* i. 3 And he who 'does a tray' (serves three months' imprisonment) therein, borrows his word from our Gallican neighbours. 1944 D. Burley in A. Dundes *Mother Wit* (1973) 214 A deuce or tray of haircuts ago. 1960 'A. Burgess' *Doctor is Sick* xiii. 98 'I know all about you. You did a tray on the moor.'..'It wasn't a tray..it was only a stretch.' 1967 [see *nickel bag* s.v. *nickel *sb.* 3 a]. 1972 J. Mills *Report to Commissioner* 98 She wants to buy two treys, $3 bags of heroin. He says he has treys, but wants $3.50 for them. 1977 *National Times* (Austral.) 17 Jan. 11/3 Service of the kind just described is as rare these days as finding a trey in the Christmas pudding.
3. trey-bit (freq. **tray-bit**) *Austral.* and *N.Z. slang* (now *Hist.*), a three-penny piece; also *trey-piece.*
1898 *Bulletin* (Sydney) 1 Oct. 14, 3d. a 'traybit'. 1901 *Bulletin Reciter* (Sydney) 181 Den I socked me bit upon 'er—Ev'ry tray-bit I could bring. 1937 F. Sargeson in *Tomorrow* 17 Mar. 310/2, I upend them to collect the tray bits. 1953 A. Upfield *Murder must Wait* xviii. 162 I'll bet my job against a trey bit you're right. 1977 *Sunday Sun* (Brisbane) 1 May 16 When it comes to unique competitions the people of outback Winton reckon they're the full quid—and you can bet your last zac or traybit on it! 1899 *Bulletin* (Sydney) 14 Jan. (Red Page), We have here [*sc.* in Tauranga, N.Z.]..slang words for..3d.—*thrum, half-tiz, tray,* or *tray-piece.*

trez, var. Tray *sb.*[4]

tri-, *prefix.* Add: **I. 1. a. trialle·lic** *Genetics*, having three different alleles of a gene; **tricontine·ntal,** embracing three continents; **trifu·nctional** *Chem.*, having three functional groups in the molecule; hence **trifu·nctionally** *adv.*; **trilaminar:** also in *Cytology* (examples); **trinu·cleotide** *Biochem.*, an oligonucleotide in which the number of nucleotides is three; **tripare·ntal** *a. Microbiology*, involving or resulting from the infection of a bacterium by three different bacteriophages at the same time; **trisensory** (earlier example).
1944 S. S. Atwood in *Proc. Nat. Acad. Sci.* XXX. 70 Because suitable terms to describe multiple alleles in autotetraploids would facilitate the discussion, the following new terminology is suggested and will be used in this paper:..Triallelic. 1975 *Nature* 24 July 310/2 Triallelic plants..cannot be obtained by any normal form of inheritance and their appearance is strong evidence for the occurrence of an unusual genetic transfer process. 1962 M. Hardwick *Sherlock Holmes Compan.* 199 Watson's astonishing statement about his tri-continental experience of women. 1966 *Economist* 22 Jan. 299/1 The tricontinental conference held in Havana..will increase the prestige of Dr Castro. 1929 W. H. Carothers in *Jrnl. Amer. Chem. Soc.* LI. 2550 Among compounds having more than one functional group, those of the type x—R—y may be called bifunctional, $R''x_3$, trifunctional, etc. 1975 *Nature* 10 Apr. 482/2 An essential feature is that some of these amino acids are trifunctional. 1941 *Jrnl. Amer. Chem. Soc.* LXIII. 3085/2 (*caption*) Schematic representation of a trifunctionally branched three-dimensional polymer molecule. 1889 *Cent. Dict.*, Trilaminar. 1971 *New Scientist* 1 Apr. 24/1 (*caption*) Electron micrographs of mitochondrial membranes reveal a trilaminar or railway track appearance. 1977 *Jrnl. Protozool.* XXIV. 18/1 The trilaminar construction of the ciliate cortex. 1918 *Jrnl. Chem. Soc.* CXIV. i. 48 The simultaneous liberation of the trinucleotide, triphosphonucleic acid, and the mononucleotide, uridine-phosphoric acid, indicates that the three constituent mononucleotides in triphosphonucleic acid must be combined in a different manner from the uridine-phosphoric acid in the parent molecule of yeast-nucleic acid. 1974 *Nature* 25 Oct. 734/2 The two RNAs are known to contain an identical trinucleotide at their 5′ terminal. 1951 *Cold Spring Harbor Symp. Quantitative Biol.* XVI. 471/2 Important information about genetic recombination comes from experiments in which the frequency of triparental recombination is measured. 1961 *Genetics* XLVI. 1314 Occurrence of triparental recombinants between two Hfr and one F⁻ has been demonstrated in *E. coli* K-12. 1976 *Ann. Rev. Microbiol.* XXX. 517 Nonconjugative plasmid transfer by such triparental matings may occur under ideal laboratory conditions. 1894 Trisensory [see *bisensory *a.*].
b. *tri-county, state.*
1974 *News & Courier* (Charleston, S. Carolina) 19 Apr. 6-A/1 A municipal scramble for federal recreation funds is under way in the tri-county. 1978 *Detroit Free Press* 5 Mar. 13/2 These centers cater to some 150,000 deaf adults in the tri-county area. 1963 *Times* 15 Jan. 9/6 A tri-state transportation committee is carrying out a survey..of New York, New Jersey, and Connecticut. 1983 *Listener* 22 Sept. 5/2 The attractions of the US market—and the New York tri-state area in particular—have been appreciated for a long time.
II. 4. b. tri-·axle, a trailer or articulated lorry with three (rear) axles; **triplane** (further examples); **trivoltine:** also **trivoltin** (example).
c. tri-car (earlier example); **tri-·mix,** a breathing mixture for deep-sea divers composed of nitrogen, helium, and oxygen.
1971 M. Tak *Truck Talk* 173 *Triaxle*, a semi with three rear axles and consequently a greater weight-carrying allowance. 1978 *Detroit Free Press* 16 Apr. F 8/1 (Advt.), Trailer: 1967 Ravens 29′ dump on 34′ frame, tri-axle with air-lift. 1981 *Daily Tel.* 10 Dec. 9 (*caption*) A..38-tonne tri-axle lorry. 1903 *Motor* 6 May 279/1 (*heading*) The new Rex 'Tricar'. 1976 *Jrnl. Appl. Physiol.* XL. 605/2 Each subject..breathed either air or a mixture of 36% helium, 21% oxygen, and 43% nitrogen (tri-mix) during cycles of immersed work. 1981 *New Scientist* 12 Feb. 390 Last week a team of divers..broke the world record for a simulated dive, experiencing pressures equivalent to those 686 metres beneath the sea surface... Part of the secret of the test's success was that the divers breathed a recently developed gas mixture, called trimix. 1909 Triplane [see *quadruplane]. 1920 *Glasgow Herald* 10 July 5 The Pullman triplanes of the British Company carry 18 people. 1977 J. Cleary *High Road to China* ii. 47 'What did you fly?'..'Albatros D's and Fokker Triplanes. I was with von Richthofen.' 1888 Trivoltin [see *bivoltine *a.*].
III. 5. a. trite·rpane, a terpane with the formula $C_{30}H_{60}$, analogous to the triterpenes; **trite·rpene,** any of the group of terpenes of formula $C_{30}H_{48}$, found in plant gums and resins; also, a triterpenoid; **trite·rpenoid,** a triterpene or a derivative of one.
1902 F. J. Pond tr. *Heusler's Chem. Terpenes* 432 Several well characterized compounds which occur in elemi-resin belong to the class of triterpenes. 1932 *Chem. Abstr.* XXVI. 3244 (*heading*) Contribution to the accurate determination of the empirical formulas of several triterpenes and triterpenoids. 1945 Triterpene [see *isoprenoid *a.* and *sb.*]. 1965 *Proc. Nat. Acad. Sci.* LIV. 1406 We wish to report now the isolation and identification of..a C_{30}-pentacyclic triterpane from the branched-cyclic alkane fraction of the Green River Shale. 1978 Triterpene [see *steviosıde]. 1978 *Nature* 16 Mar. 217/1 Polycyclic triterpenoids are found in petroleum and their presence was at first taken as evidence of non-marine contribution, as they had been detected in the lipid extract of ferns. 1981 *Jrnl. Chromatogr. Sci.* XIX. 156/1 This study deals with the apparent effect that maturation has on the relative concentration of individual triterpanes [*etc.*].
b. triethylene, containing three non-contiguous ethylene radicals in the molecule.
1920 *Jrnl. Chem. Soc.* CXVII. 1090 (*heading*) Triethylene tri- and tetra-sulphides. 1953, etc. [see *thiotepa* s.v. *thio- 1]. 1962 H. Burn *Drugs, Med. & Man* xix. 193

Tri-ethylene melamine..is used in the textile industry... It is now widely used in the treatment of Hodgkin's disease.

c. **tributyl phosphate,** an oily liquid, $(C_4H_9O)_3PO$, that is a solvent used as a plasticizer and in the solvent-extraction of nuclear fuels; **tricre·syl phosphate** [Cresyl], a colourless liquid, $(CH_3C_6H_4O)_3PO$, used as a fuel additive, plasticizer, and fire retardant; **tri:ethano·lamine** [*Ethanolamine], an oily alkaline liquid alcohol, $(HOCH_2CH_2)_3N$, used as a solvent and a stabilizer; **trihydroca·lcite** *Min.* [ad. Russ. *trigidrokal'tsit*" (P. N. Chirvinsky 1906, in *Ezhegodnik" po Geol. i Mineral. Rosii* VIII. 241/1)], a trihydrate of calcium carbonate, $CaCO_3.3H_2O$, the natural occurrence of which is uncertain; **tri:iodo·thy·ronine** *Biochem.* [*Thyronine], a thyroid hormone similar to thyroxine (tetraiodothyronine) but having greater potency; $HO{\cdot}C_6H_2I{\cdot}O{\cdot}C_6H_2I_2{\cdot}CH_2CH(NH_2)COOH$.

1930 *Brit. Patent* 330,228 1/2 The excess of alcohol and later the tributyl phosphate are distilled off in vacuo. 1957 *Financial Times Ann. Rev. Brit. Industry* 69/1 The concentrate is dissolved in nitric acid..where it is extracted with tributyl phosphate, producing uranium. 1882 *Jrnl. Chem. Soc.* XLII. 839 The authors have thus obtained triphenyl, tri-β-naphthyl and tricresyl phosphate from the corresponding phenols. 1959 *Economist* 10 Jan. 153 (Advt.), The Ignition Control Additive based on tricresyl phosphate..was developed..to overcome serious problems of power loss and rough running. 1962 *Punch* 15 Aug. 218/3 Tricresyl phosphate was a chemical similar to triorthocresyl phosphate. 1897 *Jrnl. Chem. Soc.* LXXII. 314 (*heading*) Triethanolamine (trihydroxytriethylamine). 1939 *Jrnl. R. Aeronaut. Soc.* XLIII. 617 Strips of brass sheet were..immersed respectively in pure glycol, commercial glycol and commercial glycol treated with triethanolamine phosphate. 1976 *New Yorker* 8 Mar. 67/1 (Advt.), Its mild 'heavy-molecular' triethanolamine-base formulation is chemically balanced to remove surface dirt and makeup without penetrating and robbing the sub-layers of the skin. 1910 *Mineral. Mag.* XV. 432 Trihydrocalcite... Hydrated calcium carbonate, $CaCO_3.3H_2O$, occurring as a mould-like encrustation on chalk-marl near Nova-Alexandria. 1928, etc. Trihydrocalcite [see *pentahydrocalcite* s.v. *Penta-]. 1952 *Lancet* 1 Mar. 439/1 (*heading*) The identification of 3:5:3'-L-triiodothyronine in human plasma. 1975 *Jrnl. Endocrinol.* LXIV. 573 In the present culture system the thyroid hormones (tri-iodothyronine and thyroxine) inhibited the action of PTH.

triac (trai·æk). *Electronics.* [f. *Tri(ode *a.* and *sb.* + *A.C.* s.v. *A III.] A three-electrode semiconductor device that will conduct in either direction when triggered by a positive or negative signal at the gate electrode.

1964 E. K. Howell in *IEEE Internat. Convention Record* XII. ix. 86/2 To simplify the control of AC power, a completely new silicon semiconductor has been developed. 'Triac' is a generic term coined for this *triode AC* switch. 1975 *Hi-Fi Answers* Feb. 78/2 The light pulses from the led are picked up by a photo-conductive cell, and applied to the..triggering circuit of the triac. 1981 *Computer-Aided Design* Jan. 7/1 During the past two decades the application of triacs and thyristors in the control of power and machines with complex feedback arrangements has rapidly replaced the use of their conventional counterparts in almost all sectors of industry and public services.

triacetate (traiˌæ·sĭtēˈt). *Chem.* [f. Tri- + Acetate.] †**a.** A compound in which an acetate group is combined with three atoms or molecules of a base. *Obs.*

1860 [see Tri- 5 a].

b. A compound containing three acetate groups in the molecule; *spec.* cellulose triacetate, in which acetate groups replace hydrogen atoms in (notionally) all three hydroxyl groups in each constituent glucose molecule; a man-made fibre made from this.

1895 C. F. Cross et al. *Cellulose* i. 35 Heated at 180° in a sealed tube, in the proportion by weight of 1 of cellulose to 6 of the anhydride, the cellulose is converted into the triacetate. 1921 *Jrnl. Soc. Chem. Industry* 31 Jan. 81/2 If the ratio of chlorine to sulphur dioxide be nearer to unity then the product is mainly cellulose triacetate. 1956 W. J. Roff *Fibres, Plastics, & Rubbers* 25 A method of preparing fully acetylated triacetate possessing exceptionally good electrical properties. 1960 *Which?* Jan. 19/1 Triacetate is very similar to ordinary acetate, but not so soft to handle; it will take permanent pleats and creases. 1978 *Lancashire Life* Mar. 113/3 Each item can be bought separately—the triacetate and nylon shirt at about £13.50.

triacid (traiˌæ·sid), *sb.* and *a.* [f. Tri- + Acid *a.* and *sb.*] **A.** *sb.* [partial tr. G. *triacidlösung* 'three-acid solution'.] A biological stain consisting of methyl green, orange G, and acid fuchsin. Also *attrib.*

1896 A. B. Lee *Microtomist's Vade-Mecum* (ed. 4) xviii. 223 Ehrlich's Triacid Mixture.—According to a custom which, I believe, originated with Ehrlich himself, and which would, perhaps, be better honoured in the breach than the observance, the name of Triacid ('Triacidlösung') has been given to a mixture of the same three dyes as in the Ehrlich-Biondi mixture, but in such proportions that the 'acid' colours therein have a larger share assigned to them. 1899 [see *polychromatic a. 2]. 1929 Slider & Downey in C. E. McClung *Handbk. Microsc. Technique* vii. 248 The triacid stain is used for five minutes. *Ibid.*, Triacid contains methyl green, orange G and acid fuchsin. 1976 P. Collard *Devel. of Microbiol.* v. 54 In 1880 he [*sc.* P. Ehrlich] published an account of his famous tri-acid stain for the differential staining of leucocytes.

B. *adj.* Of a triglyceride: containing three different acid radicals in the molecule.

1945 Chen & Daubert in *Jrnl. Amer. Chem. Soc.* LXVII. 1256/1 'Triacid triglycerides' is used in this report to describe those triglycerides consisting of three different fatty acids. 1977 *Appl. Spectrosc.* XXXI. 122/1 Seven diacid and five triacid saturated triglycerides have been investigated by infrared spectroscopy.

triactor (traiˌæ·ktəɹ). *Canad.* [Tri- 4 c.] A form of betting on race-horses (see quot. 1979); freq. *attrib.*

1976 *Telegraph-Jrnl.* (St. John, New Brunswick) 23 Aug. 12/2 *Charlottetown*—Two big triactor payoffs..highlighted a 10-dash harness racing card here Saturday. 1979 *Beautiful Brit. Columbia* Winter 36 The triactor bettor..had to name the first three finishers, in order, in the last race each day and lived in hopes of making up to $1,000 for his $2 bet. 1984 E. Wright *Smoke Detector* iv. 124 It is sometimes nearly as helpful if you know which horse *can't* win a race... It helps with the triactor, anyway.

triad. Add: **2. c.** (Earlier examples.)

[1809 W. Blake in *Compl. Writings* (1972) 560 Three Ancient Britons overthrowing the Army of armed Romans ..From the Welch Triades.] 1838 [see *Afanc].

j. *Path.* A group of three symptoms or signs.

1899 E. Lang in T. L. Stedman *20th Cent. Pract.* XVIII. 267 Since the work of Hutchinson, who looked upon a keratitis parenchymatosa in conjunction with a frequently observed deafness and anomaly of the incisors as an expression of hereditary syphilis (Hutchinson's triad), affections of the cornea are numbered among the most frequently occurring manifestations of syphilis. 1908, etc. [see *Hutchinson]. 1909 G. Dock in Osler & McCrae *Syst. Med.* VI. xvii. 439 The early idea that the disease [*sc.* exophthalmic goitre] was characterized by a 'triad' of symptoms gives way slowly. 1948 [see *Tridione]. 1954 D. Nabarro *Congenital Syphilis* ii. 14 He pointed out that interstitial keratitis and notched incisor teeth were frequently associated with 8th nerve deafness —the three signs being known as the Hutchinsonian triad. 1982 *Sci. Amer.* Aug. 82/2 The classic triad of the inflammatory reaction is redness, warmth and swelling.

3. *Triads*: also, secret societies, freq. of a criminal character, into which the Triad Society has become divided, and which flourish among overseas Chinese. Also in *sing.* Freq. *attrib.*

1960 D. Whitehead *Journey into Crime* (1961) 27 Triad gangs poured into the streets..and fought the police. 1962 M. & G. Gordon *Journey with Stranger* (1963) xxiv. 139 The Triads, those secret criminal societies that sought to monopolize every kind of racket. 1975 D. Bloodworth *Clients of Omega* vii. 64 A rival society buried your body, it seems. The 18-K Triad. 1976 *Spectator* 14 Feb. 3/2 Chinese Triad gangs made their mark in London by kicking to death a man in a Soho gambling club. 1976 *Daily Colonist* (Victoria, B.C.) 2 Mar. 3/1 A heroin smuggling racket that may be linked with the world-wide Chinese secret society known as Triad. 1977 *Hongkong Standard* 14 Apr. 5/1 As far as triad influence in these committees was concerned, Mr Lam admitted his department had received complaints in the past. 1978 *Daily Tel.* 3 Apr. 3/3 Yard chiefs concerned about the increasing violence between rival triads ..have ordered a further big crackdown.

Triadist. Restrict *Welsh Lit.* to sense in Dict. and add: **2.** *pl.* Members of the Triad Society (see Triad 3). *rare.*

1855 *North-China Herald* 17 Mar. 132/2 The Triadists, who so recently held Shanghae.

triage. Add: **2.** Also with Fr. pronunc. (triɑ̃ʒ). **a.** The assignment of degrees of urgency to wounds or illnesses in order to decide the order or suitability of treatment. Freq. *attrib.* Hence (*rare*) as *v. trans.* (see quot. 1977).

1930 F. A. Pottle *Stretchers* 222 A special triage officer at once surveys the patients to determine the urgency of their injuries. 1945 *Jrnl. R. Army Med. Corps* LXXXIV. 125 The word 'triage', literally 'assessment according to quality', has recently been adopted to describe the process. 1973 *Parade* 18 Feb. 8/1 Chicago's Michael Reese Hospital ..has instituted a 'triage', or selection system, whereby incoming patients are evaluated by an RN as to their degree of urgency, and sent to the appropriate area. 1973 N. Meyer *Target Practice* viii. 95 And presto, we were all second-lieutenant navy doctors. For six months I worked in *triage* there, operating sometimes around the clock. 1975 *Observer* 11 May 6/7 Hundreds of 'borderline' cases have been at the mercy all this week of 'triage' panels trying to sort out patients desperately in need of care. 1976 *Lancet* 15 Nov. 1061/1 Triage at an early stage can label the patient with coma as surgical or medical. 1977 M. Herr *Dispatches* 82 He was so bad that the doctor triaged him, passed him over to treat the ones that..could still be saved. 1978 *Tucson Mag.* Dec. 55/1 An extremely efficient triage nurse, an RN, greets new arrivals and determines the severity of the problem.

b. *transf.*

1974 *Time* 11 Nov. 80 In the West, there is increasing talk of triage, a commonsense if callous concept that teaches that when resources are scarce, they must be used where they will do most good. 1975 *Globe & Mail* (Toronto) 3 Nov. 7/2 The concept should now be applied to countries crippled by food shortages, famine and overpopulation. 'The triage discussions now seek to classify nations into those what do not need help, those that are capable of responding to help and those "broken back" states that are in such difficulty that they cannot be helped,' says Mr Hopper. 1979 *Guardian* 18 Oct. 5/8 There is [*sc.* in New York] an unofficial 'triage' system in which teachers and school administrators concentrate their limited resources on helping those students who seem to be capable of succeeding.

triagonal, *a.* (Earlier examples.)

1665 Hooke *Micrographia* 158 Piggs-hair (A) is somewhat *triagonal*, and seems to have neither pith nor pore. 1794 A. Thomas *Newfoundland Jrnl.* (1968) 121 Its shape is Triagonal and it measures Three Hundred and Fifty Miles in length.

trial, *sb.*¹ Add: **1. a.** Also *transf.* in phrr. *trial by television* or *the media*, subjection of a public figure under some cloud to discussion of his case on television or in the media, usu. in such a way as to imply his guilt.

1960 J. Freeman in *New Statesman* 15 Oct. 556/1 A group of Labour MPs had..written to *The Times* complaining that my questions to Mr. Foulkes, in the BBC programme *Panorama*, about specific..allegations of malpractice in his union amounted to public trial by television. 1968 *Punch* 6 Mar. 327/1 Urged to stop 'trial by television', the Postmaster-General..said..he himself had no power over the content of programmes. 1979 *Broadcast* 4 June 10/1 There have been predictable references to 'trial by the media', 'trial by television'.

2. a. Applied *spec.* in *pl.* to a boat's trial run (see sense *13 a below).

1921 *Daily Colonist* (Victoria, B.C.) 25 Mar. 17/2 On Friday the Traveller was taken out for her trials... Slight trouble arose in the main bearings and the vessel returned to the dock for adjustments. 1969 F. Mowat *Boat who would not Float* (1970) vii. 70 Seamen refer to the first tentative voyage of a newly commissioned ship as her trials.

c. A contest designed to test the capabilities of motor-cyclists or (formerly) car-drivers and their vehicles, in which riding or driving takes place over long distances or rough terrain. Freq. in *pl.* Cf. *reliability trial* s.v. *Reliability 3.

1926 [see *scramble *sb.* 2 b]. 1935 *Encycl. Sports* 429/1 Among the more famous English trials that have been run for many years are the London–Edinburgh run, the London–Exeter run, the London–Land's End Trial and the London–Gloucester Run. 1950 *Oxf. Jun. Encycl.* IX. 328/2 The famous Scott Trial in Yorkshire was the first of such rough-riding fixtures [in motor-cycling contests]. 1963 P. Drackett *Motor Rallying* i. 10 The Thousand Miles Trial of 1903 went several steps nearer to the rally as we know it to-day, with road sections and timed hill-climbs. 1976 *Southern Even. Echo* (Southampton) 3 Nov., Trials enthusiasts will be out in force in Hampshire this weekend to watch the Hood Trophies and Perce Simon Trials—two major events in the British motor cycle calendar.

4. a. *trial and error*, (*a*) also in non-mathematical contexts, the process of succeeding by repeated trying with or without improvement of method by learning from failures; (*b*) *spec.* in *Psychol.*, with reference to the theory that a primitive form of learning results, over a series of trials, from erroneous random responses to a problem being replaced by the correct response, rather than from insight. Freq. (with hyphens) *attrib.*

1894 C. L. Morgan *Introd. Compar. Psychol.* xiv. 241 Such a proceeding can be completely explained in terms of sense-experience. The process was throughout one of trial and error. 1898 E. L. Thorndike in *Psychol. Rev. Monogr. Suppl.* II. viii. 105 If the method of trial and error, with accidental success, be the method of acquiring associations among the animals, the slow progress of primitive man..becomes suggestive. 1900 C. L. Morgan *Animal Behaviour* iv. 139 The method of varied trial and error with the utilization of chance success, is a lengthy and somewhat clumsy process; but it suffices. 1940 Hilgard & Marquis *Conditioning & Learning* x. 252/2 The behavior of animals in the trial and error situation yields evidence of more intelligent behavior than is implied in the simple process of stamping in correct responses and stamping out wrong ones. 1951 Parsons & Shils *Toward Gen. Theory of Action* ii. ii. 129 Invention may be..trial-and-error learning. 1957 J. S. Huxley *Relig. without Revelation* ix. 230 Scientific hypotheses.. are better organisations for coping with our experience of physical phenomena than are trial-and-error methods. 1962 *Listener* 15 Nov. 796/1 There had been fitful, trial-and-error attempts to create a National Assembly. 1967 M. Dobb *Capitalism, Devel. & Planning* v. 242 In the Lange trial-and-error process..it was variable prices (accounting prices) that were fixed by the top-level authorities. 1972 *New Yorker* 26 Aug. 32/1 The American psychologist Edward L. Thorndike..is credited with the first rigorous investigation of trial-and-error, or instrumental, learning.

6. b. *Sport.* A match held to select players for a major team; esp. in *Rugby Football.* Cf. *trial match*, sense 13 in Dict.

a 1914 J. E. Raphael *Mod. Rugby Football* (1918) 249 E. W. Baker played for the South and in other important trials. 1921 E. H. D. Sewell *Rugby Football* 337 He was nominated to play on the wing in the third Trial. 1950

[see *signing *vbl. sb.* 1 b]. 1978 *Rugby World* Apr. 5/2 Jeeps was the man who instigated the new system of trials which operated this season, involving games between England's regions and divisions.

c. A contest in which horses, dogs, etc., are put through various tests and assessed on their performance.

1942 R. B. Kelley *Animal Breeding* xvi. 145 The second kind of trial endeavours to set out a course during which the dog is required to perform the tasks of every-day life. 1946 M. C. Self *Horseman's Encycl.* 196 Working hunter trials. Jumps 4 feet to 4 feet 6 inches. On special course. Details of course to be withheld until one hour before class. 1960 J. Stroud *Shorn Lamb* xv. 175, I was..flitting round the suburbs like a sheepdog at a Trial. 1974 *Encycl. Brit. Micropædia* I. 723/3 The park [at Badminton] is also well-known for its horse trials.

13. a. *trial flight, marriage* (examples), *night, pit, separation* [Separation 3]; **trial balance** (examples); † **trial ball** *Cricket*, a practice ball which a bowler was formerly allowed to bowl before beginning his first over (*obs.*); **trial balloon** = *Ballon d'essai; **trial-bred** *a.*, of a dog bred: to compete in trials (*Trial *sb.*¹ 6 c); **trial court** (earlier example); **trial eight** (examples); **trial judge** (example); **trial jury** (example); **trial lawyer** (examples); **trial run,** a preliminary trip given to a new vessel or vehicle, to test its performance; freq. *transf.* or *fig.*; **trial trench** *Archæol.*, an exploratory trench dug on a site; hence *trial-trench* vb. intr.

1838 Trial balance [see *balance-sheet* s.v. *Balance *sb.* 22]. 1910 *Encycl. Brit.* IV. 227/1 A *trial balance* is thus no very adequate safeguard against fraud. 1978 *Detroit Free Press* 5 Mar. c 16/5 (Advt.), Local Co. needs a mature individual for full-charge bookkeeping thru trial balance. 1830 in R. S. Holmes *Hist. Yorks. County Cricket* (1904) 25 It is usual for a bowler, before he commences, to bowl a trial ball at one of his colleagues. 1870 *Times* 1 Sept. 10/4 The former [bowler] preluded his successes..with three trial balls. 1939 *Sun* (Baltimore) 21 Apr. 3/4 Congressional moves to bar alliance members from relief funds are in the 'trial balloon' stage. ?1949 Q. Howe in W. Safire *New Lang. Politics* (1968) 454/1 He [*sc.* Theodore Roosevelt] also originated the 'trial-balloon' technique and gave favored correspondents 'off-the-record' statements that they attributed to 'authoritative sources'. If the statement caught on, Roosevelt would make it his own. If it fell flat, he would drop it. *a* 1974 R. Crossman *Diaries* (1976) II. 590 Since I'd been wanting to launch a trial balloon on this subject, I aired my views at length. 1948 J. A. Reid in B. Vesey-Fitzgerald *Bk. Dog* 749 The 'trial-bred' collie. 1960 *Farmer & Stockbreeder* 9 Feb. 91/3 Many collies lie down almost instinctively and, among trial-bred dogs, getting them to stay on their feet is far more difficult. 1890 *U.S. Rep.* (Supreme Court) CXXXVII. 347 On review in this court, the rulings of the trial court were sustained. 1873 C. C. Knollys *Oxf. Univ. Challenge Races* p. iii, The crews of the Trial and University Eights. 1900 W. E. Sherwood *Oxford Rowing* 63 In 1858..considerable alterations were made..the most important perhaps being the establishment of the Trial Eight Race. 1909 F. Ash *Trip to Mars* xii. 89, I only took my trial flight in it yesterday! 1892 *U.S. Rep.* (Supreme Court) CXLI. 562 The conclusion of the trial judge was that the second claim of the reissue was an enlargement of the single claim of the original patent. 1884 *Pacific Reporter* IV. 255 A trial jury in this territory is a body of 12 men, possessing the requisite qualifications, duly summoned, and sworn to well and truly try the questions of fact submitted to them by the court, and a true verdict render according to the law and the evidence. 1929 R. R. Morton *What Negro Thinks* 146 His practice is seldom that of a trial lawyer, but rather as an adjuster of cases and an adviser in civil processes. 1983 'E. Lathen' *Green grow Dollars* vi. 48 She..had steered Wisconsin Seedsmen into the arms of the right trial lawyer. 1906 E. C. Parsons *Family* vii. 142 Trial-marriage is a variety of time-marriage, it being distinctly agreed that the relationship may be dissolved by either man or woman at any time. 1930 *New Statesman* 27 Dec. 355/2 The Bishops of Miss Dunbar's Church recently gave an episcopal blessing to birth control, one prominent clergyman approving trial marriage. 1977 *Time* 4 Apr. 11/2 For the moment, the Liberals and Labour are only committed to keep their trial marriage going until the end of this parliamentary session in November. 1825 P. Egan *Life of Actor* vii. 253 Our hero received a letter, offering him a *trial night* in Hamlet. 1904 W. D. Adams *Dict. Drama* I. 187/2 At last in 1817 he was granted, through the influence of friends, a trial-night at Covent Garden. 1905 D. MacKenzie *Let.* in *Observer* (1962) 11 Feb. 11/3 The many preliminary trial-pits sunk in the early years of the excavation. 1966 Trial pit [see *soil survey* s.v. *soil *sb.*¹ 10]. 1903 *Trans. Inst. Naval Archit.* XLV. 295 Some twenty trial runs..were made under various conditions as to speed. 1909 *Chambers's Jrnl.* 25 Sept. 675/1 The first great trial-run of a number of motor-cars. 1949 *Sun* (Baltimore) 22 July 1/2 The House rejected the proposal for a 'trial run' of the Brannan farm program for a period of two years. 1962 *Rep. Comm. Broadcasting 1960* 4 in *Parl. Papers 1961–2* (Cmnd. 1753) IX. 259 Two trial runs of local sound broadcasting were arranged for us. 1974 D. Seaman *Bomb that could Lip Read* xviii. 177 The government desperately wants the new [Irish] Assembly to have a peaceful trial run. 1968 *Listener* 4 Oct. 439/2 We are telling our friends that he's not around because we've agreed to a trial separation. 1978 M. Tripp *Wife-Smuggler* ii. 20 When I suggested..a trial separation she said..I didn't have the guts to make a clean break. 1947 *E. African Ann.* 1946–7 67/2 Once trial trenches had been cut, it became obvious that the site was even more important than we had at first believed. 1954 M. Beresford *Lost Villages* App. iii. 416 The position of each house is clearly visible, thus precluding the need to trial-trench to find structures. 1980 *Rescue News* Sept. 2/3 In the first small trial trench was found a rim of Saxo-Norman pottery.

b. *attrib.* in *pl.* (cf. sense 2 c above), as *trials bike, rider, riding*, etc.

1969 *Daily Tel.* 25 Oct. 8/8 Quite a few leading trials riders are farmers... Light-weight two-stroke machines are now favoured for trials work. 1976 Trials enthusiast [see sense *2 c above]. 1976 *Southern Even. Echo* (Southampton) 10 Nov. 22/1 The cream of British trials riders left the native Yorkshire and came south to compete in rain and mud in the National Perce Simon and Hood Trophy Trials. 1976 *Norwich Mercury* 19 Nov. 12/1 Trials riding is the cheapest form of motor cycle sport. 1977 *West Briton* 25 Aug. 10/3 The new trials bike group..have raised half the £180 for their trials machine by a sponsored walk.

trialist (trai·ălist), *sb.*¹ [f. as Trial *sb.*¹ + -ist.] **1.** One who advocates or follows trialism (sense 2); used *spec.* with reference to a proposed German–Magyar–Slav state. Also *attrib.*

1931 J. A. R. Marriott *Europe 1815–1923* 429 On one detail of their programme the 'Trialists', as they began to be called, were not unanimous. 1935 H. A. L. Fisher *Hist. Europe* III. xxviii. 1097 Would it not be possible..to substitute for the dual monarchy, resting on the dominion of the Magyars and the Germans, a trialist state founded on the equal fellowship of Slav, Magyar, and German? 1980 *Times Lit. Suppl.* 4 Jan. 22/2 The Southern Slav Question builds up to a plea for a modified 'trialist' state of Austrian, Hungarian and Slav units.

tri·alist, *sb.*² Also **triallist.** [f. Trial *sb.*¹ + -ist.] **1.** One involved in a judicial trial (Trial *sb.*¹ 1).

1967 *Guardian* 17 Nov. 11/1 Letter from trialists' relatives. The 'Guardian' received..a copy of a letter addressed to the Prime Minister by the relatives of the 50 Greeks whose trial begins in Salonika today. 1979 J. Drummond *Patriots* vii. 48 The Treason Trialists were acquitted.

2. a. One who takes part in a preliminary match or contest, with a view to being selected for a major team. Cf. Trial *sb.*¹ 6 b.

1960 D. Storey *This Sporting Life* ii. 37, I took this to mean he was watching me, though there were four other trialists in the team. 1962 *Listener* 19 Oct. 628/1 The contestants [*sc.* bridge-players] are all international trialists. 1980 *Daily Tel.* 15 Sept. 12 An Olympic trialist, who was training in the same pool. 1977 *Western Morning News* 1 Sept. 10/7 Richmond include..Dave Whibley, their England trialist full-back.

b. One who takes part in a contest or competition, esp. a motor-cycle trial (*Trial *sb.*¹ 2 c).

1961 *Times* 5 Oct. 4/5 The American motor cycle trialist. 1977 *Daily Tel.* 18 Oct. 22/8 Griffiths, this season's leading trialist, covered the [cycle] course in 5hr 54min 14sec. 1980 *Observer* (Colour Suppl.) 14 Oct. 54 (*heading*) Alan Road reports on the bumpy world of the trialist. 1982 *Daily Tel.* 24 Nov. 3 The future of two top Welsh international show jumping trialists hangs in the balance after an allegation that they tried to influence the selection of a competition judge.

c. One who takes part in clinical tests or trials of new drugs, etc.

1977 *Lancet* 17 Sept. 595/2 The conditions of a trial can be adjusted so that only the most suitable patients emerge with the statisticians' stamp of approval, though trials conducted under such harsh discipline would yield neutral results more often than not and trialists would fade away through frustration. 1983 *Glaxo Group News* Mar. 12/1 The programme, attended by nearly 800 doctor-delegates (mainly trialists and specialists in infectious diseases) from more than 30 countries, included 83 presentations.

triallist: see *tri·alist.

triamcinolone (traiămsi·nŏlōn). *Pharm.* [f. *triamcin-* (etym. unkn.) + *prednisolone.] A synthetic glucocorticoid which resembles prednisolone in its effects but is a more potent anti-inflammatory agent; 9α-fluoro-16α-hydroxyprednisolone, $C_{21}H_{27}O_6F$.

1957 *Jrnl. Amer. Med. Assoc.* 7 Dec. 1821/1 The purpose of this clinical study was to evaluate the anti-inflammatory and antipruritic properties of a new steroid compound, Aristocort diacetate (also known as triamcinolone diacetate). 1961 *Lancet* 12 Aug. 347/2 Some children with nephrosis do not respond to high doses of prednisolone, but they do respond to 'equivalent' doses of triamcinolone. 1979 *Nature* 13 Dec. 736/2 The glucocorticoids dexamethasone, triamcinolone,..stimulated interferon production more than 20-fold.

triangle, *sb.* Add: **1. c.** Esp. a love-relationship in which one member of a married couple is involved with a third party; freq. as *eternal triangle.*

1907 [in Dict.]. 1913 Kipling *Diversity of Creatures* (1917) 358 The couples had rearranged themselves or were re-crystallizing in fresh triangles. 1919 G. B. Shaw in F. Harris *Contemp. Portraits* 2nd Ser. 332 For the modern drama, with its eternal triangle and so forth, he claims nothing, but that it proves adultery to be the dullest of subjects. 1938 H. G. Wells *Apropos of Dolores* iv. 162 He was much more substantial than in the days of our romantic triangle. 1955 H. Kurnitz *Invasion of Privacy* (1956) vi. 48 The details of the story, the way the husband and wife first met..the other woman in the triangle. 1963 A. Heron *Towards Quaker View of Sex* iv. 39 Most examples of the 'eternal triangle' are produced by boredom and primitive misconduct. 1979 J. Phillips *Why Murder?* (1980) ii. 91 A husband, a wife, a lover—the classic triangle.

d. *North Atlantic Triangle*: a name given to the tripartite alliance between Great Britain, Canada, and the United States.

1945 J. B. Brebner (*title*) North Atlantic Triangle. 1957 H. Heaton in L. B. Pearson *Where do we go from Here?* 2 Canada was one of the three points in the 'North Atlantic Triangle'—the others were Great Britain and the United States. 1978 J. Hutcheson *Dominance & Dependency* i. 22 The nationalistic position..has been complicated by Canada's location in the North Atlantic Triangle.

2. a. *the Bermuda* or *Devil's Triangle*: a name given to an area of sea between Bermuda and Florida credited with a high number of unexplained disappearances of boats and aircraft; hence used allusively; *the golden triangle*, an area at the meeting-point of Burma, Laos, and Thailand, where much opium is grown.

1964 V. Gaddis in *Argosy* Feb. 28 (*heading*) The deadly Bermuda triangle. *Ibid.* 116/2 Draw a line from Florida to Bermuda, another from Bermuda to Puerto Rico, and a third back to Florida through the Bahamas. Within this area, known as the 'Bermuda Triangle', most of the vanishments have occurred. 1973 *Bangkok Post* 22 Apr. 1 Both the opium and the morphine base almost certainly originated in the so-called 'golden triangle' where the opium poppy grows in abundance. 1975 *Collier's Encycl. Year Bk.* 1976 162 The Atlantic region known as the Bermuda, or Devil's, Triangle. 1983 *Times* 23 Jan. 5/6 The increasing importance of South-East Asia's 'golden triangle' as a source of narcotics. 1979 N. Price *Tomorrow's Ghost* vi. 99 [She] had stepped out for a breath of fresh air..and she hadn't been seen again... She had turned a good piece of Surrey into a Bermuda Triangle. 1983 *Times* 22 Feb. 4/8 Drugs from the Golden Triangle have been reaching the West in quantity since before 1979.

p. A triangular warning sign placed on the road to mark the presence of a broken-down vehicle or vehicles.

1962 H. Wain *Daimler IX* ii. 8 There's your red breakdown triangle... They're obligatory in France. 1971 H. Pacy *Road Accidents* i. 33 Utilize the special warning triangles larger trucks carry for this purpose. 1977 'J. Fraser' *Heart Eats it.* 12 Superintendent Bill Aveyard helped him as he saw the flashing torches and warning triangles at the side of the road.

4. (sense *2 c) *triangle drama*; (sense 2 j) *triangle-player*; **triangle inequality** *Math.*, the statement that the modulus of the sum of two quantities is less than or equal to the sum of their moduli; (so called from the analogy with the distances between the vertices of a triangle).

1931 E. A. Robertson *Four Frightened People* i. 22 Just another triangle drama. 1961 *Times* 13 May 5/2 In an all too successful attempt to turn Shakespeare's tragedy into a conventional triangle-drama, Rossini's librettist, Berio, built up Rodrigo. 1941 Birkhoff & MacLane *Survey Mod. Algebra* vii. 183 In any Euclidean vector space, length has the following properties:.. $|\xi+\eta| \le |\xi|+|\eta|$ (the triangle inequality). 1972 R. Kaye *Math. Thought* xlvi. 1083 Schwarz's inequality and the triangle inequality are proved for the norm. 1966 J. Joyce *Let.* 4 Oct. (1966) II. 170 AE ought now to write some..dreamy thing about a..Triangle-player. 1971 D. E. Westlake *I gave at Office* xx Arnold dropped an occasional word in, like the triangle-player in the Philharmonic.

triangulable (traiæ·ŋgiŭlăb'l), *a.* *Math.* [f. Triangul(ate *v.* + -able.] Of a topological space: capable of undergoing triangulation (sense *2 b).

1940 *Proc. Nat. Acad. Sci.* XXVI. 359 *Triangulable manifold*, a topological manifold which can be subdivided into a complex. 1967 J. Stewart *Concepts Mod. Math.* xii. 184 We know that S is triangulable, so there exists a map on S (with triangular faces).

triangular, *a.* (*sb.*) Add: **3. A.** (Later examples.) Also *spec.* with reference to the 'eternal triangle' (see *Triangle *sb.* 1 c).

1880 Hardy *Fellow-Townsmen* v. 44 The triangular situation—himself, his wife, Lucy Savile—was the one clear thing. 1908 *Times* 9 July 15/5 [The Australians] were unable to take part in the suggested triangular contest in this country next year. 1914 W. L. George *Dramatic Actualities* 39 The difficulties of matrimony, triangular or other, financial or monetary entanglements. 1968 S. Hynes *Edwardian Turn of Mind* vi. 185 The hero [is] involved in a triangular situation that Shaw took from his own amatory experience. 1976 *Southern Even. Echo* (Southampton) 19 Nov. 17 The Mid Hants (Southampton) girls' netball teams had a successful time in a triangular tournament against Derbyshire and Berkshire at Reading. 1977 *Gay News* 24 Mar. 24/4 Husband, wife and female lodger involved in a triangular relationship.

4. c. **triangular trade,** a multilateral system of trading in which a country pays for its imports from one country by its exports to another; *spec.* (Hist.) in the slave trade (see quots.).

1934 C. M. MacInnes *England & Slavery* iii. 39 After the Restoration a great triangular trade developed be-

tween England, the West Coast of Africa and the West Indies or the continental colonies. 1948 T. S. Ashton *Industrial Revolution 1760–1830* ii. 47 Cloth, firearms, hardware, and trinkets were sent to Africa and exchanged for slaves, who were shipped to the West Indies to pay for the luxuries and raw material which constituted the final cargo in this disreputable, triangular trade. 1971 C. & D. Plimmer *Damn'd Master* ii. 28 With the profits from the sale of the slaves in the West Indies they bought sugar..which, back in Europe, they sold for a second profit with which in turn they bought more goods... This became known as the triangular trade.

triangularly, *adv.* **b.** (Earlier example.)

1890 Kipling in *Pioneer Mail* 15 Jan. 92/3 We nodded triangularly in all good will and swore eternal friendship.

triangulated, *ppl. a.* and *a.* **2.** (Lit. example.)

1960 G. C. Dickinson *Maps & Air Photographs* ii. 40 From about 1750 onwards a few of the better cartographers had begun to accept the need, for accuracy's sake, of providing a full triangulated framework for their county maps.

triangulation. Add: **1.** *spec.* by measuring the angles and one side of each triangle (cf. *Trilateration). Freq. *attrib.*, as *triangulation point* (also *fig.*).

1923 *Geogr. Rev.* XIII. 465 The recent remarking of the Meades Ranch Station calls attention to the unique importance of this triangulation point. 1947 A. R. Hinks *Maps & Survey* (ed. 5) ii. 172 The methods of trigonometrical survey..divide themselves into the following sections: 1. Determination of mean sea level... 9. Transference of the triangulation points to the plane-table sheets. 1977 *Times Lit. Suppl.* 22 Apr. 494/1 Dispassionate yet sensitive, his *Lens*..deservedly became the triangulation-point for a whole generation.

2. b. *Math.* (See quot. 1956); also, the result of such a process.

1940 *Proc. Nat. Acad. Sci.* XXVI. 360 This result is but one of the implications of a triangulation. 1956 E. M. Patterson *Topology* v. 89 Triangulation..is the process of dividing up a [topological] space into pieces which are homeomorphic with the interior of a triangle or its analogues in other dimensions. 1974 *Greek Female Yearbk. Sci. & Technol.* 412/2 These developments [in topology] were highlighted by the solutions in 1969 of the annulus conjecture..and of the triangulation problem for manifolds. 1975 *Sci. Amer.* Oct. 113/1 A triangulation that represents..a minimal five-chromatic map cannot have any vertices with fewer than five neighbors.

triangulator. Add: (Example.) Also, an instrument used in triangulation.

1932 *Geogr. Jrnl.* LXXXIII. 444 Norway, Poland, Romania, and the United States are the most active triangulators. 1938 *Ibid.* XCII. 434 In this exhibition there were also..the Bianchi triangulator.

triathlon (traiˌæ·θlǫn). [f. Gr. τρι- Tri- + ἆθλον contest, after *decathlon*, etc.] An athletic or sporting contest composed of three different events.

1973 *Daily Tel.* 21 July 14 A new event..called the 'Triathlon'. In this all four members of a team have to demonstrate their prowess in clay pigeon shooting, fly fishing and riding a handy hunter-course over jumps. 1981 *Austin* (Texas) *Amer.-Statesman* 28 Mar. (Time Out section) 8 The Triathlon consists of a 2·5 mile surf swim, followed by a 112-mile bike race, followed by a full marathon. 1983 *Times* 26 Feb. 22/1 The word 'triathlon' is the latest that has been attached to the so-called 'Iron Man contests' which..basically consist of a swim in the open sea, a long cycle ride of up to 70 miles and a marathon run.

triatomid (traiˌæ·tǫmid), *a.* (*sb.*) [f. mod.L. family name *Triatomidæ*, f. generic name *Triatoma* (F. L. Laporte 1832, in *Mag. Zool.* II. ix. 11), f. Tri- + Gr. τόμος to cut, in allusion to the antennæ of the insects: see -id³.] Of or pertaining to the family Triatomidæ, now usually included in the family Reduviidæ. Also as *sb.*, a blood-sucking assassin bug of the family Triatomidæ, which includes several vectors of disease.

1955 *Sci. News Let.* 29 Oct. 280/1 The blood-suckers are triatomid bugs and are known to carry the germs from animals, such as opossums, to man. 1961 E. R. & G. A. Noble *Parasitol.* xv. 726 Some triatomids..obtain meals by tapping the blood-engorged bodies of other arthropods. 1971 P. C. C. Garnham *Progress in Parasitol.* ii. 30 In the State of São Paulo..the writer saw typical dwellings in which the incidence of the infection was 19 per cent in the dogs, 16 per cent in the cats, and nearly 100 per cent in the triatomid bugs. 1974 F. Pifano in K. Elliott et al. *Trypanosomiasis & Leishmaniasis* 77 The source of food of the triatomids captured in the palm trees was investigated with the precipitin test.

triatomine (traiˌæ·tǫmīn), *a.* (*sb.*) [f. mod.L. subfamily name *Triatominæ* (see prec. and -ine¹).] = *Triatomid *a.* (*sb.*), when the group is considered as a subfamily of the family Reduviidæ.

1962 Gordon & Lavoipierre *Entomol. for Students of Med.* xxxix. 237 Triatomine bugs are the vectors of Chagas's disease. 1974 *Nature* 30 Nov. 392/2 The triatomine bugs..are medically important as vectors of Chagas' disease in the Americas. 1978 *Ibid.* 27 Apr. 820/2 The second type (type II) was found in acute and chronic cases of Chagas' disease, cats, house mice, rats and guinea pigs, from houses infested by the domiciliated triatomine *Panstrongylus megistus*. 1979 *Pharmacol. & Therapeutics* VII. 86 *T. rangeli* is a non-pathogenic trypanosome with a wide range of hosts which is also transmitted by triatomine vectors.

triaxial, *a.* (In Dict. s.v. Triaxal *a.*) Add: (Further examples.) Also, occurring or responding in three mutually perpendicular directions.

This, rather than Triaxal *a.*, is the usual term.

1924 J. G. A. Skerl tr. *A. Wegener's Origin of Continents & Oceans* xxi. 202 In his latest work Helmert infers, from the distribution of the force of gravity on the earth's surface, that the earth is a triaxial ellipsoid. 1962 *Engineering* 14 Dec. 756/2 A state of triaxial stress, according to its severity, suppresses deformation by shear and makes cleavage more probable. 1975 *Nature* 31 Jan. 327/2 A single range triaxial fluxgate magnetometer capable of measuring fields up to 10 gauss along each orthogonal axis. 1978 *Sci. Amer.* Jan. 48/2 The criterion of triaxial motion meant that a ball-and-socket arrangement would be the best means of achieving a mechanical interlock between components.

Hence **triaxia·lity**, triaxial nature; **tri-axially** *adv.*

1970 *Nature* 14 Mar. 1008/2 Because of triaxiality the Moon theoretically has three free oscillations with periods of about 1, 40 and 800 months in the directions of its axes. 1972 *Physics Bull.* Nov. 669/1 There are clearly no sharp boundaries..but the fracture is dependent on triaxiality of stress and very sensitive to temperature or strain rate. 1982 *Sci. Amer.* May 116/2 (Advt.), By means of a triaxially stabilized attitude control system, [it] directs the solar panels towards the sun.

tri-axle: see *tri- 4 b.

tribadism. Add to def.: *spec.* in modern use (see quots.).

1965 *New Statesman* 26 Mar. 492/3 The first [technique of lesbian intercourse], known as 'tribadism', consists in one woman lying on top of the other and simulating the movements of heterosexual intercourse in such a way as to stimulate the clitoris of each. 1970 G. Greer *Female Eunuch* 263 The prevalence of tribadism as the principal lesbian mode of lovemaking argues the relative unimportance of the masculine fantasy.

tribal, *a.* Add: **1. b.** Characterized by the tendency to form groups or by strong group loyalty.

1951 H. Arendt *Burden of Our Time* ii. viii. 227 Tribal nationalism always insists that its own people is surrounded by 'a world of enemies', 'one against all'. 1970 G. Greer *Female Eunuch* 182 In the tribal teenage situation there are some boys with whom one does not go out. 1977 *Time* 10 Oct. 11/3 A Jew in this rural, tribal and fiercely Christian heartland [*sc.* Savannah, Georgia] is a wanderer indeed.

c. As *sb.*, a member of a tribal community (usu. in *pl.*). Chiefly *Indian English.*

1958 *New India: Progress through Democracy* iii. vi. 378 Illiteracy is almost universal among tribal peoples. 1966 *Economist* 18 Apr. 261/1 More are arriving daily, among them Christian and Buddhist tribals. 1971 *Illustr. Weekly India* 25 Apr. 42/2 An elaborate welfare scheme..reached both the settlers and the tribals. 1973 *Country Life* 14 June 1713/2 ..a ceremonial bag of five tigers slain by tribals in Nagaland in 1969. 1979 *South China Morning Post* 28 Dec. 3/1 Teams of mountain tribals are to be brought in to search for the Singapore Air Force Skyhawks which disappeared over the northern Philippines eight days ago.

2. Special collocations: **tribal council**, the organ of internal self-government of a recognized U.S. Indian tribe, since the Indian Reorganization Act of 1934; **tribal mark**, a scar resulting from the traditional or ritual cutting of the face practised among certain African tribes, hence *tribal marking*; also *tribal scar, scarring*; **tribal territory**, an administrative division of the Northwest Frontier Province of Pakistan (formerly of India).

1948 *Salt Lake Tribune* 17 Dec. 16/5 Other tribal council members, speaking in Navajo..agreed with Mr. Akeah. 1976 *Billings* (Montana) *Gaz.* 5 July 1-a/4 The Northern Cheyenne Tribal Council has launched a drive for redesignation of the air quality level on the southeastern Montana reservation. 1980 *New Age* (U.S.) Oct. 3/1 Through CERT and the tribal councils, thousands of acres of traditional tribal land have been turned over to mining companies searching for uranium and coal. 1897 M. H. Kingsley *Trav. in W. Afr.* xxiii. 530 The cicatrices are sometimes tribal marks, but sometimes decorative. 1965 W. Soyinka *Road* 73 Are you sure you know what I mean? Sort of tall but a little on the short side. Tribal marks, but beginning to wear off. 1925 W. D. Hambly *Hist. Tattooing & its Significance* iii. 176 Are there among the accounts of tribal markings any genuine references to ancient legends, which reveal that these marks have..an important social significance? 1973 'A. Hall' *Tango Briefing* xx. 251 The carved teakwood statuette..was obviously a god, wide nosed and with tribal markings on the forehead. 1977 P. Raymond *Matter of Assassination* iii. 23 ..expressionless Negro face marked with tribal scars across the cheeks. 1962 D. Wilder *Reading Guest* ii. 33 The tribal scars..had been worked..into her cheeks with a needle... The tribal scarring had been outlawed for thirty years. 1908 *Imperial Gazetteer India* XIX. 166 In tribal territory, beyond the districts of British India..alluded to, the Torwāl and Garhwī reside in the higher ranges of Swāt. 1951 Khan & Stark *Young Pakistan* xxiv. 199 In the far north, an observer standing on the high mountains can in fact look down on five different lands—Afghanistan..and Pakistan (Tribal territory) in the south west. 1974 *Encycl. Brit. Macropædia* XIII. 256/2 Between the settled districts to the south and the Afghan border is the tribal territory, whose inhabitants enjoy a large measure of independence.

tribalism. Add: **b.** Loyalty to a particular tribe or group of which one is a member.

1955 *Times* 30 Aug. 9/7 If a stable parliamentary democracy is to be introduced one stable political party, avoiding if possible the extremes of tribalism and anti-tribalism, would seem necessary. 1969 *Busara* (Nairobi) II. ii. 56 According to Reinhold Niebuhr, 'the chief source of man's inhumanity to man seems to be the tribal limits of his sense of obligation to the other man'. In *tribalism* Niebuhr includes race, language, religion, class and culture as important traits. 1976 *Drum* (E. Afr. ed.) Apr. 3/1 Tribalism? Isn't it true that in some firms about 75 per cent of the employees come from one tribe—depending, of course, on which tribe the bosses come from. 1978 *Times Lit. Suppl.* 1 Dec. 1390/5 His call for a fusion or integration of cultures is one that commits him to the course of liberal tolerance, and sets him against closed systems of thought such as Marxism and tribalism.

tribalist. Restrict *rare* to sense in Dict. and add: **b.** An advocate or practitioner of tribalism (sense *b). Also *attrib.* or as *adj.* (cf. next).

1960 *Times* 29 Sept. (Nigeria Suppl.) p. iv/4 Opponents have accused him of being a tribalist. 1970 J. D. Caute *Fanon* vi. 70 The Nationalist Parties drove the political opposition into retrograde, tribalist forms.

tribalistic (trəibăli·stik), *a.* [f. Tribal *a.* + -istic.] = Tribal *a.* 1 a, b in Dict. and Suppl. Also, characterized by tribalism. Cf. *Tribalist b.

1961 in Webster. 1976 *Daily Sketch* (Ibadan) 26 Oct. 1/2 They..recommended that Nigeria's future President should not be arrogant, tribalistic but should be consistent. 1980 in S. Terkel *Amer. Dreams* 320 It [*sc.* a cocktail party] was a tribalistic ritual. I was the meat on the altar.

tribalize (trəi·bălaiz), *v.* [f. as prec. + -ize.] *trans.* To render tribal, to unite on a tribal basis; to imbue with tribal loyalty. So **tri·balized** *ppl. a.*; **tri:baliza·tion.**

1927 *Other Lands* Jan. 59/2 The United Free Church worked more among the tribalized communities. 1959 *Times Lit. Suppl.* 6 Mar. 127/5 There are still many who share Mary Kingsley's predilection for 'tribalized' peoples. 1964 M. McLuhan *Understanding Media* (1967) xxiv. 249 The war had fraternalized and tribalized us. 1984 *Times* 16 Oct. 14/6 The main effect [in Poland] has been the tribalization of politics..the bike boys, the punks..the hippies..the tough and the meek.

tribe, *sb.* Add: **4. b.** A gang of criminals or delinquents. Also, in recent use, a group of hippies or other drop-outs.

1914 Jackson & Hellyer *Vocab. Criminal Slang* 85 *Tribe*, used principally by yeggs and begging bums, though current, too, amongst grafters who operate in cliques. A gang; a class. 1955 D. W. Maurer in *Publ. Amer. Dial. Soc.* xxiv. 83 In general, [pickpocketing] mobs are also known as *tribes*. 1968 *Guardian* 29 Apr. 7/5 The fifteen hundred 'Tribes' of San Francisco..beg their food and sleep where they can. 1973 R. C. Dennis *Sweat of Fear* vii. 44 A room with wall-to-wall mattresses. Sprawled about were a half-dozen members of the tribe.

tribo- (trəi·bo, tri·bo), comb. form repr. Gr. τρίβος rubbing, as **tri:boelectri·city** [ad. F. *triboélectricité* (O. D. Chwolson *Traité de Physique* (1913) IV. i. ii. 268)], electricity generated by friction; so **triboele·ctric** *a.*, **-ele·ctrically** *adv.*; **tri:boelectrifica·tion**, the production of triboelectricity; **tri:bonuclea·tion** (see quot. 1967); **tri·bophysics**, the physics of friction; hence **tribophy·sical** *a.*; **tribosphenic** (-sfī·nik) *a.* *Palæont.* [Gr. σφήν wedge] (see quots. 1936, 1975²); **tribo-the:rmolumine·scence**, thermoluminescence produced in a material as a result of friction.

1917 *Nature* 27 Dec. 337/1 The research provides an explanation of the well-known readiness with which materials change their tribo-electric character. 1967 W. R. Harper *Contact & Frictional Electrification* v. 77 The notorious difficulty of performing triboelectric experiments in damp weather proclaims the importance of adsorbed water. *Ibid.*, Glass cleaned with alkalis became triboelectrically ⊕ve with respect to metals. 1972 *Sci. Amer.* Mar. 54/3 Yarns, plastic films, paper and kindred materials moving at high speeds inevitably lead to triboelectric troubles. 1979 *Adv. Colloid & Interface Sci.* XI. 64 The roles of tribo-electricity in everyday life and in industry are numerous both as a nuisance and hazard and as a phenomenon to be exploited. 1938 G. P. Harnwell *Princ. Electr. & Magn.* i. 1 The production of electrification by the frictional process of rubbing is known as triboelectrification. 1967 W. R. Harper *Contact & Frictional Electrification* v. 85 The loosely bound ions are available to diffuse to another surface brought into contact with the one under consideration, and might be important..for triboelectrification. 1967 *Brit. Jrnl. Appl. Physics* XVIII. 641 If a solid body such as a stirrer rod

is rubbed against the inner wall of a vessel containing either a supersaturated solution of a salt or a supercooled liquid, the nucleation of crystals is likely to occur at the point of contact. This phenomenon is known as tribonucleation ('nucleation by rubbing'). 1974 *Nature* 20 Dec. 696/1 One mechanism for the production of micronuclei in man, tribonucleation, has been suggested by Ikels. 1977 *Jrnl. Catalysis* L. 542/1 The catalyst produced by the tribophysical procedure is more active longer than the conventional one. 1950 *Engineering* 17 Mar. 303/3 The new Division of Tribophysics has developed from a section which, during the war, evolved some interesting theories regarding the fundamental nature of friction and lubrication. 1976 *Sci. Amer.* Apr. 30/2 Sanders, who is chief research scientist in the Tribophysics Division, heads a group that is engaged in studying the reactivity of metallic surfaces, relying mainly on electron microscopy. 1936 G. G. Simpson in *Dental Cosmos* LXXVIII. 797/1 At this stage the upper molar is implanted by two external and one internal root, the latter generally largest, and the lower molars are implanted by two sub-equal roots, one beneath the trigonid and one beneath the talonid... In order to have a single word..by which this type of molar dentition can be unequivocably..designated, I propose to call it 'tribosphenic'. 1975 *Nature* 31 July 402/1 Therian mammals with tribosphenic molars were probably in existence at the beginning of the Cretaceous. *Ibid.*, Tribosphenic molars have an additional cusp on the upper molars (protocone) which fits into a basin (talonid) on the matching lower molar. 1930 Nyswander & Cohn in *Physical Rev.* XXXVI. 1257 The term tribothermoluminescence has been given to the phenomenon under consideration which involves the process of grinding followed by application of heat. 1971 *Nature* 23 July 257/2 Thermoluminescence (TL) of fossil bones and of various kinds of recent biological material has been reported by Jasińska and Niewiadomski, who suggest that such materials could be used for dating purposes, but draw attention to difficulties which arise due to tribothermoluminescence (TTL, which is thermoluminescence derived from the mechanical energy of grinding).

tribology (trəibǫ·lŏdʒi). [f. *Tribo- + -ology.] The branch of science and technology concerned with interacting surfaces in relative motion and with associated matters (as friction, wear, lubrication, and the design of bearings).

Mr. H. P. Jost (see quot. 1968), chairman of a working group of 'lubrication engineers', corresponded with me in 1965 about the need for a new term and accepted *tribology*, a word suggested by Mr. C. G. Hardie of Magdalen College.—Ed.

1966 (Feb.) *Lubrication (Tribology)* (Dept. Educ. & Sci.) 4 Tribology is defined as follows: Tribology is the science and technology of interacting surfaces in relative motion and of the practices related thereto. 1968 H. P. Jost in *New Scientist* 8 Feb. 292 After consultation with the English Dictionary Department of the Oxford University Press—the publishers of the *Oxford Dictionary*—we chose the term 'tribology'. 1969 *Sunday Times* 30 Nov. 30/2 Two chairs of tribology have been established, one at Leeds and another at Salford. 1975 *Sci. Amer.* July 50/3 This field of study is known, more commonly in Britain than in the U.S., as tribology. 1980 *Canada Weekly* 19 Nov. 4/1 As Canada's main centre of expertise on tribology..the..laboratory examines the problem of wear and failure of rails and wheels.

Hence **tribolo·gical** *a.*, of or pertaining to tribology; **tribo·logist**, an expert or specialist in tribology.

1966 *New Scientist* 19 May 423/1 They suggested that the Institution of Mechanical Engineers should organize 'professional institution activities in the tribological sphere'. 1969 *Lubrication Engin.* Feb. 88/3 (*heading*) Dr. Finkin named chief tribologist. 1973 *Nature* 6 Apr. 361/3 The tribological problems the centre has tackled have ranged from making gas bearings to designing window hinges for high rise flats. 1974 *Globe & Mail* (Toronto) 18 Apr. 45/2 Tribologists believe they can save Canadians several hundred million dollars a year. 1977 *Engin. Materials & Design* Aug. 35/2 This compound layer has excellent tribological properties under dry running as well as lubricated conditions.

tribrachic, *a.* (Earlier example.)

1852 J. S. Blackie *Pronunciation of Greek* 43 The tendency to the ineffective tribrachic and even proceleusmatic accent in the termination of our polysyllables.

tribunal, *sb.* Add: **2. c.** Any of various local boards of officials empowered to settle disputes, esp. between an individual and a government department, to adjudicate on fair rents, exemption from military service, etc.; *industrial tribunal*, a board arbitrating in disputes arising out of employment, such as complaints of unfair dismissal. Also (in full *tribunal of inquiry*), a board set up by the government to investigate some matter of public concern.

1916 *Act 5 & 6 Geo. V* c. 104 § 2 An application may be made at any time before the appointed date to the Local Tribunal established under this Act..for a certificate of exemption from the provisions of this Act. 1921 (*title of Act*) Tribunals of inquiry (evidence) act. 1932 L. Golding *Magnolia Street* ii. iv. 321 The Baritone, at least, managed to convince tribunal after tribunal that it would be a mistake to put him into khaki. 1945 [see *rent *sb.*¹ 4 c]. 1949 *Britannica Bk. of Year* 309/2 Public interest was focused upon the judicial tribunal set up, under the chairmanship of Mr. Justice Lynskey, to inquire into allegations of corrupt practices involving several ministers and public servants. 1962 L. Golding *Dict. Local Govt.* 158 A Committee..was set up to consider..the constitution and working of administrative tribunals, particularly those dealing with the compulsory purchase of land. 1971 *Mod. Law Rev.* XXXIV. 657 The industrial tribunal may only recommend reinstatement. *Ibid.*, The tribunals may award..compensation. 1974 M. Meacher *Scrounging on Welfare* iii. 38 In 1971 the Supplementary Benefits Commission reported 153 cases to a Tribunal under Section 12/1.

tribune, *sb.*¹ Add: **2. b.** (with capital initial.) The title of a British weekly journal, founded in 1937, advocating radical left-wing policies: used *attrib.* with reference to this type of socialism, esp. as *Tribune group*, a group of Labour MPs sharing these views.

1952 *Ann. Reg. 1951* i. ii. 33 *One Way Only*, a *Tribune* pamphlet published on 9 July. 1968 *Times* 9 July 9/6 The trade unionists who are moving against the Government are not, as the *Tribune* group supposes, moving to the left. *a* 1974 R. Crossman *Diaries* (1975) I. 562 They were really completely answered by Raymond Fletcher, a left-winger from the Tribune group. 1975 *Times* 16 July 14/7 Those who regarded themselves as the keepers of the socialist conscience—the Tribune group. 1976 A. Price *War Game* i. viii. 165 Not even the Tribune Group will be able to complain about the high cost of security. 1976 [see *Tribunism]. 1977 *Times Lit. Suppl.* 3 June 685/1 Orwell..was a Tribune-style socialist (egalitarian, libertarian).

Hence **Tri·bunism**, socialist policies of the type advocated by the *Tribune*; **Tri·bunite**, a member of the Tribune group (freq. *attrib.*).

1970 *Times* 18 Nov. 12 Frank Allaun, the Tribunite M.P. *Ibid.* 7 Dec. 2 The second day took in a debate led by..Mr. Stan Orme, the redoubtable Tribunite, and Mr. Neil McBride. 1976 *Times* 15 Mar. 13/2 The influx of new MPs..has probably weakened Tribunism rather than reinforced it. But a new generation of Tribune men is beginning to assert its influence. 1977 *Listener* 28 July 125/2, I do not believe that the Tribunite view..goes by default. 1981 *Times* 22 June 2/7 Neither have Tribunites north of the border made any open moves to become involved.

tributary, *a.* and *sb.* **A.** *adj.* **2.** (Earlier example with reference to a stream.)

1823 C. B. Vignoles *Observations upon Floridas* 56 The Choctawhatchie river, and all its tributary streams discharge into the eastern end of this bay.

B. *sb.* **2. a.** (Earlier example.)

1822 W. H. Simmons *Notices of East Florida* iii. 29 [The] appearance [of bonnet leaf]..indicates from a distance, the influx of some tributary of the main stream.

tribute, *sb.* Add: **2. b.** A praiseworthy thing attributable *to*, a testimony *to*.

Often (as by Fowler in quot. 1926) regarded as an incorrect use.

1926 in H. W. Fowler *Dict. Mod. Eng. Usage* 662/2 The debate on the whole was a tribute to the good taste and good form of the House of Commons. 1937 W. H. S. Smith *Let.* 3 May in *Young Man's Country* (1977) ii. 69 Are the Guildford's using their old Coronation Robes? If so, it's a tribute to Lady G.'s figure. 1961 *New Statesman* 8 Feb. 198/1 Perhaps..the gusty vigour of the heroes of the period from Grant to McKinley is a tribute to the American character.

4. **tribute rice** *Chinese Hist.*, a grain tax paid in kind.

1853 *North-China Herald* 26 Mar. 134/4 (*heading*) Destruction of the Che Hien's House and tribute rice. 1959 P. Fleming *Siege at Peking* v. 61 Tungchow, whose prosperity depended on its status as an entrepot for tribute-rice brought by canal from the interior, had economic as well as ideological motives for disliking the railway.

tributyl phosphate: see *tri- 5 c.

tricarballylic (trəi:kāɹbăli·lik), *a.* *Chem.* [f. Tri- + *carballylic acid* s.v. Carb-.] *tricarballylic acid*: a crystalline tribasic acid found in immature beets and produced synthetically; propane-1,2,3-tricarboxylic acid, $HOOC{\cdot}CH(CH_2COOH)_2$.

1865 M. Simpson in *Jrnl. Chem. Soc.* XVIII. 334 This body has been named by Kekule carballylic acid...I propose..in order to avoid confusion, to call it *tri*carballylic acid. *Ibid.* 335, I have prepared and analysed several of the salts and ethers of tri-carballylic acid. 1950 I. L. Finar *Org. Chem.* xvii. 348 Tricarballylic esters have been used as plasticisers. 1980 *Polish Jrnl. Chem.* LIV. 1681 In the course of biological studies on new potential anticonvulsants, a number of substituted tricarballylic acids..were synthesized.

tricarboxylic (trəi:kāɹbǫksi·lik), *a.* *Biochem.* [f. Tri- + Carboxyl + -ic.] *tricarboxylic acid*: any acid with three carboxyl groups in each molecule; *tricarboxylic acid cycle*, the Krebs cycle (see *Krebs).

1894 Perkin & Kipping *Org. Chem.* I. xiii. 246 Citric acid is a tricarboxylic acid, and..forms three classes of salts. 1938 G. H. Richter *Textbk. Org. Chem.* xvi. 302 Aconitic acid is an example of an unsaturated tricarboxylic acid. 1945 *Jrnl. Biol. Chem.* CLXI. 413 (*heading*) Fatty acid oxidation and the Krebs tricarboxylic acid cycle. 1947 *Ibid.* CLXXI. 446 Tricarboxylic acids may be formed in plants by a process involving the addition of CO_2 to an α-ketoglutarate. 1950 [see *Krebs]. 1974 B. A. Newton in K. Elliott et al. *Trypanosomiasis & Leishmaniasis* (Ciba Symp. No. 20) 299 Some of the changes which occur during the development of T. cruzi, such as the activity of the tricarboxylic acid cycle enzymes.., may only be quantitative changes. 1982 M. S. Olson in T. M. Devlin *Textbk. Biochem.* vi. 278 Various investigators defined many of the enzymes and di- and tricarboxylic acid intermediates in this pathway, but it was Krebs who pieced together these components in his formulation of the 'Krebs cycle'.

Tricel (trəi·sel). Also **tricel.** [f. Tri- (in *triacetate) + Cel(lulose *sb.*] A proprietary name for a man-made fibre made from cellulose triacetate, and for material made from this.

1954 *Trade Marks Jrnl.* 22 Dec. 1300/2 *Tricel*... Raw or partly prepared artificial fibrous textile materials not being yarns or threads. British Celanese Limited, Celanese House.., London W.1; manufacturers. 1956 *Official Gaz.* (U.S. Patent Office) 20 Nov. 88/1 *Tricel*... For staple fibres made wholly or partially of cellulose derivatives. 1965 *Guardian* 31 Mar. 13/2 The two fabrics looked identical, but 'Tricel' had got something that the other hadn't got—easy-care properties. 1969 W. J. Burley *Death in Willow Pattern* i. 13 Susan..put on a simple short-sleeved Tricel dress. 1977 *Lancashire Life* Feb. 22/2 A top cup of cotton embroidered nylon tricel.

tricesimo-secundo: see *-mo.

trichinelliasis (trikinelai·ăsis). *Med.* [f. mod.L. *Trichinella*, generic name superseding Trichina (f. L. *-ella*: see -el²) + *-iasis.] = Trichinosis.

1907 *Allbutt's Syst. Med.* (ed. 2) II. ii. 914 The girl had evidently died, not of enteric fever as was supposed, but of trichinelliasis. 1930 E. C. Faust *Human Helminthol.* xxiii. 327 The disease trichinelliasis or, more familiarly, trichinosis, may be divided into three stages. 1977 Schmidt & Roberts *Foundations of Parasitol.* xxiii. 424/2 *Trichinella spiralis* is the only species in the family ..and is responsible for the disease variously known as trichinosis, trichiniasis, or trichinelliasis.

Also **trichinello·sis** [-osis], in the same sense.

1958 *Excerpta Medica* IV. 479, 209 wild rats..and 21 domestic rats..were examined for trichinellosis. 1965 *Jrnl. Amer. Med. Assoc.* 13 Sept. 182/1 So-called meat inspection..provides no mechanism for detection of trichinellosis. 1971 P. C. C. Garnham *Progress in Parasitol.* v. 64 In the Mau Mau emergency the youth went wild on Mount Kenya, ignoring the ancient tribal taboo regarding the consumption of the flesh of certain animals and devoured uncooked wild pig: a severe outbreak of trichinellosis, with many deaths, was the result.

Trichinopoli. Add: *Trichinopoli chain.*

1879 *Encycl. Brit.* IX. 163/2 Round plaited gold chains of fine wire, such as are still made by the filigree workers of India, and known as Trichinopoly chains.

trichlor-, trichloro-. Add: **trichlo:ro-, tri:chloraceta·ldehyde** = Chloral.

1921 J. S. Chamberlain *Textbk. Org. Chem.* 226 The tri-chlor acet-aldehyde then reacts with an alkali present yielding chloroform. 1964 N. G. Clark *Mod. Org. Chem.* x. 189 Acetaldehyde..may be chlorinated to chloral (trichloroacetaldehyde..), an important intermediate in the manufacture of 'D.D.T.'.

trichloroethylene (traiklōᵊ:roe·þilīn). *Chem.* Also **trichlorethylene.** [f. Tri- + Chloro-² + *Ethylene* s.v. Ethyl.] A liquid organochlorine compound, C_2HCl_3, used as a solvent, analgesic, and anæsthetic. Cf. *Trilene.

1889 G. M'Gowan tr. *Bernthsen's Text-bk. Org. Chem.* 60 (*table*) C_2HCl_3, Tri-chloro-ethylene. 1930 *Engineering* 26 Dec. 814/2 The use of chemical cleaners, such as petrol ..and trichlorethylene, is more effective. 1955 *Ann. Reg. 1954* 411 Midwives should be allowed..to use a new pain-relieving vapour called trichlorethylene B.P. 1976 *National Observer* (U.S.) 17 July 2/3 The Food and Drug Administration said it will ban use of a chemical once used in decaffeinated coffees and some beers. The chemical, trichloroethylene, has been found to cause liver cancer in mice.

tricho-¹. Add: **trichoma·niac** *nonce-wd.*, a hair fetishist; **trichomo·nal** *a.*, of, pertaining to, or caused by trichomonads; **tri:chotillo·ma·nia** [ad. F. *trichotillomanie* (H. Hallopeau 1889, in *Ann. de Dermatol. et Syphilol.* X. 441), f. Gr. τίλλεσθαι to pull out (hair)], a compulsive desire to pull out one's hair; hence **tri:chotilloma·niac**, a person with this.

1949 R. Graves *Common Asphodel* 303 From descriptions in his poems it is clear that the first thing that he [*sc.* Milton] saw in a woman was not her bright love-darting eye (as it was to practically all his contemporaries), but her hair. He was, in fact, a trichomaniac. 1948 *Jrnl. Amer. Med. Assoc.* 18 Sept. 231/2 The diagnosis and treatment of trichomonal vaginitis may not always be as simple as depicted. 1970 G. Greer *Female Eunuch* 259 Cases of incurable trichomonal infection are all due to a combination of fear, superstition and doctors' sloppiness. 1905 *Rep. Soc. for Study of Dis. in Children* V. 28 (*heading*) A case of trichotillomania. 1980 *Brit. Med. Jrnl* 29 Mar. 881/2 The loss is patchy and must be distinguished from.. trichotillomania, in which the child pulls out his or her hair and may eat it. 1962 *Woman* 26 May 9 (*heading*) Don't be a trichotillomaniac! That is, in simple terms, someone who tears out hair!

trichome. Add: **2.** *Ent.* In myrmecophilous insects, a tuft of hairs near a gland producing a secretion attractive to ants.

1911 E. Jacobson in *Tijdschr. Ent.* LIV. 177 The bug possesses a very curious tuft of yellow hair (a trichome) which apparently secretes some substance with a flavour agreeable to the ants. 1923 W. M. Wheeler *Social Life among Insects* v. 227 The trichomes surround the openings of singular glands, the aromatic, volatile secretions of which..are licked off by the ants. 1971 E. O. Wilson *Insect Societies* (1972) xx. 390/2 The *Cremastocheilus* adults are furnished with tufts of golden hairs ('trichomes') at the anterior and posterior corners of the thorax.

trichomoniasis (tri:komǫ̆nai·ăsis). *Med.* and *Vet. Sci.* [f. mod.L. *Trichomonas*, generic name (coined in Fr. as *Trico-monas* by A. Donné 1836, in *Compt. Rend.* III. 386), f. Tricho-¹ + *-monas: see *-iasis.] Infection with trichomonads, which in man is often symptomless; *esp.* (*a*) a venereal disease of women caused by *Trichomonas vaginalis*, in which there are vaginal irritation and a discharge; (*b*) a venereal disease of cattle caused by *T. fœtus*, characterized by abortion, pyometra, and sometimes sterility.

1915 *N.Y. Med. Jrnl.* CI. 886/2 (*heading*) Clinical and experimental trichomoniasis of the intestine. 1972 *Daily Colonist* (Victoria, B.C.) 19 May 2/2 Another common cause of persistent or recurrent discharge is a parasitic infection called trichomoniasis. 1980 *Trop. Med. & Hygiene News* Dec. 25 Investigations on trichinellosis, hydatid disease and trichomoniasis.

trichoplax (tri·koplæks). [mod.L. (F. E. Schulze 1883, in *Zool. Anzeiger* VI. 92), f. Tricho-² + Gr. πλαξ plate.] A minute marine animal with a body formed of three layers of cells, formerly included in the genus of this name but now usually considered to be a modified form of a hydrozoan planula.

1897 Parker & Haswell *Text-bk. Zool.* I. iv. 220 Trichoplax is a compressed plate-like body of irregular and extremely variable shape, but circular in the resting condition. 1940 L. H. Hyman *Invertebrates* iv. 243 *Trichoplax* and *Tetroplax* were found actually to be modified planulae of Hydroidea.

trichothecin (triko-, traikopī·kin). *Biochem.* [f. mod.L. *Trichothec-ium*, name of a genus of fungi (H. F. Link 1809, in *Mag. d. Ges. Naturforschender Freunde zu Berlin* III. 18), f. Tricho-¹ + Thecium: see -in¹.] A crystalline trichothecene, $C_{15}H_{19}O_3{\cdot}O{\cdot}CO{\cdot}C_3H_5$, that is an ester of butenoic acid produced by the fungus *Trichothecium roseum* and is toxic to some other fungi.

1948 Freeman & Morrison in *Nature* 3 July 30/1 Antagonistic activity to other fungi by *Trichothecium roseum* Link..has been reported... The name 'trichothecin' is suggested for the active substance. 1981 Cole & Cox *Handbk. Toxic Fungal Metabolites* v. 152 Examples of trichothecenes that do not contain a carbonyl function at C-8 are T-2 toxin,..and trichothecin.

Hence **trichothe·cene**, any of a class of sesquiterpenoids based on a tetracyclic ring system $C_{15}H_{22}O_2$.

1971 *Analytical Biochem.* XLIII. 327 The trichothecenes are a family of closely related tetracyclic sesquiterpenoid metabolites of various strains of *Fusarium*, *Trichoderma*, *Trichothecium*, *Myrothecium*, and other species of imperfect fungi commonly found in soil, and on grains and other foods and feeds. 1978 *Experientia* XXXIV. 1333/1 Samples from 2 lots of corn in France suspected of causing infertility, hyperestrogenic signs and feed refusal in swine were analyzed for zearalenone and trichothecenes. 1982 *Daily Tel.* 24 Mar. 18 Trichothecene toxins should be seen as the murderous successors to mustard gas.

trichromasy (traikrōᵘ·măsi). *Ophthalm.* Also **-chromacy**. [f. Tri- + Gr. χρῶμα colour: see -y³.] Colour vision in which three pure colours, in different combinations, are required to match all the colours that can be perceived (as in normal vision).

1911 *Amer. Jrnl. Psychol.* XXII. 371 Guttmann..identifies color-weakness with anomalous trichromacy. 1923 *Proc. R. Soc.* CII. 359 Trichromasy, in my experience, seems to approach monochromasy directly without passing through dichromasy as an intermediate stage. 1973 (see *protanomaly). 1980 *Nature* 27 Mar. 306/1 Our trichromacy has a 'blue' channel that is about 100 times less sensitive than the 'red' and 'green' channels.

trichromat (traikrōᵘ·mæt). *Ophthalm.* Also †**-ate**. [f. Trichromat(ic *a.*] An individual with trichromasy, esp. an anomalous form of it.

1906 *Arch. Ophthalm. & Otol.* XXXV. 27 A thorough study was made of only two normal and two abnormal trichromates. 1925 M. Collins *Colour-Blindness* i. 14 It is customary to divide these anomalous trichromates into..deuteranomalous trichromates..and protanomalous trichromates. 1940 *Nature* 17 Aug. 226/2 It might happen that a building matched its background for the normal person, yet, for the anomalous trichromat, the two would be distinct. 1978 J. Parr *Introd. Ophthalm.* iii. 69/2 The normal individual requires a minimum of three primaries and is said to be a trichromat.

trichromatism. Add: (*c*) = *trichromasy.

1910 M. Greenwood *Physiol. Special Senses* 239/1 (Index), Trichromatism. 1925 M. Collins *Colour-Blindness* i. 13 Nagel rejects the term colour-weakness as being too wide, and prefers the term anomalous trichromatism. 1946 [see *protanomaly]. 1956 *Jrnl. Optical Soc. Amer.* XLVI. 1075/1 The commonest form of aberrant color vision, namely anomalous trichromatism.

trichrome (trai·krōᵘm), *a.* [f. Tri- + Gr. χρῶμα colour.] = Trichromatic *a.*; *spec.* applied to a stain and method of staining in which different kinds of tissue are stained in one or other of three different colours.

1918 H. Croy *How Motion Pictures are Made* xiii. 289 To show the trichrome pictures a special projecting-machine was needed. 1929 *Jrnl. Technical Methods* XII. 83 Sections intended for the trichrome stain—Hæmalum, Erythrosin–Saffron, should previously undergo a collodionisation. 1931 *Museum Jrnl.* XXII. 213 The artist had at his disposal monochrome, bichrome and trichrome decorations of pleasing though simple geometric designs. 1938 *Amer. Jrnl. Path.* XIV. 237 Trichrome methods are rapidly replacing the ancient hematoxylin-eosin technique so largely used in pathology. 1939 *Times* 10 Mar. 17/6 A number of trichrome jars of varying shapes. 1975 *Nature* 6 Nov. 71/2 Material for histological examination was fixed in neutral formal saline, embedded in wax and sections were stained with haemalum and eosin or Masson's trichrome stain.

Trichuris (trikiū·ris). *Zool.* and *Med.* Also **trichuris**. Pl. **Trichurides** (-idiz). [mod.L., f. Gr. τριχ-, θριξ hair + οὐρά tail.] A nematode worm of the genus of this name, which comprises filamentous worms (whipworms) several centimetres long that are intestinal parasites of man and higher animals.

[1799 *Mem. Med. Soc. London* V. 226 It is now five years since I first discovered a new species of worm..but it was not till lately that I found that the same worm was first mentioned by Roederer, in the year 1760... By him it was first called Trichuris, and under that name I have described it.] 1807 *Edin. Med. & Physical Dict.* II, s.v. *Worms*, Dr. Hooper has seen upwards of twenty trichurides in some fæces of a child of six years old. 1929 E. R. Stitt *Diagnostics & Treatm. Tropical Dis.* (ed. 5) xlii. 517 Of the cosmopolitan round worms *Ascaris* and *Trichuris* are the most common. 1951 P. Kourí in R. B. H. Gradwohl et al. *Clin. Tropical Med.* xxxvii. 842 A great number of Trichuris were found attached to the prolapsed and congested [rectal] mucosa. 1962 [see *ascaris]. 1977 Schmidt & Roberts *Foundations Parasitol.* xxiii. 421/1 (*caption*) A male *Trichuris*. Note the slender anterior end and the stout posterior end.

Hence **trichuri·asis** [*-iasis], infection with such worms, which in man may be inapparent but can cause diarrhœa and other symptoms.

1921 *Amer. Jrnl. Tropical Med.* I. 375 (*heading*) The treatment of trichuriasis with leche de higueron. 1970 Passmore & Robson *Compan. Med. Stud.* II. xix. 29/2 All these species [of worms] are cosmopolitan, though ascariasis and trichuriasis are probably less common in Britain than they used to be. 1981 M. Cheesbrough *Med. Lab. Man. Tropical Countries* I. xxi. 266/1 The laboratory diagnosis of trichuriasis is by finding the characteristic eggs in stool specimens.

trick, *sb.* Add: **I. 1.** *trick of* (or *o'*) (*the*) *loop*, a cheating game; = Fast and loose a, *strap-game* s.v. Strap *sb.* 17. Also *fig.* *Anglo-Irish.*

1886 M. B. Buckley *Diary of Tour in Amer.* 16 The thimble-rigger and trick-o'-loop man are nowhere to be found. 1907 J. M. Synge *Playboy of Western World* iii. 57 And he after bringing bankrupt ruin on the roulette man, and the trick-o'-the-loop man. 1922 Joyce *Ulysses* 318 Norman W. Tupper bouncing in with his peashooter just in time to be late after she [*sc.* his wife] doing the trick of the loop with officer Taylor. 1974 *Listener* 21 Feb. 239/1 Their cities were..crowded..with pilgrims, curiosity-mongers, refugees from justice and trick-of-the-loop men [in medieval Ireland].

2. a. *trick or treat*, a traditional formula used at Hallowe'en by children who call on houses threatening to play a trick unless given a treat or present; also as *sb.*, this practice. Hence *trick-or-treating* vbl. sb. and ppl. adj. orig. and chiefly *U.S.*

1947 *Amer. Home* Oct. 150/2 The household larder needs to be well stocked on October 31, because, from dusk on, the doorbell rings, bright eyes peer through crazy-looking masks, and childish voices in ghostlike tones squeal, croak, or whisper, 'Trick or Treat!' 1950 *Sun* (Baltimore) 31 Oct. 12/1 So let the kids go out tonight and have a grand time with their masquerading and trick-or-treating. 1954 *Ibid.* 22 Oct. 18/4 Now that the 'Trick or Treat' season is upon us, let us hope that thoughtful parents will discourage the practice. 1968 Mrs. L. B. Johnson *White House Diary* 31 Oct. (1970) 731 He and his mother had stopped by the office on their way to 'trick or treating' at some friends' houses. 1973 M. Crowell *Greener Pastures* 64 Trick-or-treat begins at Grandma Latimer's down the road in the little green house. 1974 *New Yorker* 25 Feb. 112/2 Like a horde of trick-or-treating children who have suddenly been turned middle-aged and paunchy by a wicked witch. 1982 *Daily Tel.* 29 Oct. 3/1 A tradition of allowing children out on Halloween 'trick-or-treat' expeditions.

3. *bag of tricks*: see *bag *sb.* 17.

5. b. A robbery, theft; chiefly in phr. *to turn a trick*, to commit a successful robbery or theft. *U.S. slang.*

1865 *Leaves from Diary of Celebrated Burglar & Pick-pocket* xxvii. 94/2 Directly he had done the trick he 'namased' with his booty. 1904 'No. 1500' *Life in Sing Sing* 254/1 *Trick*, a theft. *Ibid.* 258/1 *Turning a trick*, accomplishing a theft. 1904 H. Hapgood *Autobiogr. of Thief* v. 104, I am hounded for the old trick; and the detectives are looking everywhere for these negotiable bonds. 1926 *Flynn's* 30 Jan. 843/1 Ewing was a thief, who..had settled in Chicago... He did not ply his trade here, but after 'turning a trick' outside of the city, would return to Chicago to plan the next excursion into the country. 1935 *Jrnl. Abnormal Psychol.* XXX. 365 *Trick, go on a*, to commit a robbery. 1956 [see *score *sb.* 15 c]. 1979 D. MacKenzie *Raven settles Score* 70 Campbell's claim was that he hadn't turned a trick in a year but the money had to be coming in from somewhere.

c. In Negro folk-magic or hoodoo: a spell cast on a person; an object used to 'conjure' a person or put him under a spell. Cf. *trick-doctor* s.v. Trick *sb.* 13.

1893 [see *door-stone* s.v. *door 8]. 1895 A. M. Bacon in A. Dundes *Mother Wit* (1973) 367/1 Either after or before the cure of the patient is well under way, the doctor will make an effort to find the 'trick' or 'conjure' and to identify the miscreant who has caused the trouble. 1962 [see *root *sb.*¹ 3 c]. 1966 D. J. Crowley *I could talk Old-Story Good* ii. 17 Stories about the return of spirits, murder or curing through obeah, love-potions, 'tricks' (aggressive magic) and 'guards' (protective charms) are all traditional in theme. 1977 J. Dillard *Lexicon Black Eng.* vii. 119 The conjure doctor..is involved in the performance of *tricks*. To *trick* the victim is the frequently recorded phrase.

6. c. Applied playfully to a small or amusing person, animal, or child. *U.S.* and *Austral. colloq.*

1887 *Century Mag.* May 113/1 We uns played tergether w'en we wuz little tricks. 1890 *Stock Grower & Farmer* (Las Vegas) 29 Mar. 7/1 Down in the Panhandle..I used to ride a little trick named Dandy. 1907 H. B. Wright *Shepherd of Hills* iv. 39 She ain't had no mother since she was a little trick. 1941 Baker *Dict. Austral. Slang* 78 *Trick*, an amusing person or child, esp. the latter. 1945 S. Lewis *Cass Timberlane* xxiv. 156 'What kind of a girl he marry?' 'Cute little trick, bright's dollar.' 1951 H. Giles *Harbin's Ridge* ii. 7 She was a little trick of a person. 1963 *Sunday Mail* (Brisbane) 10 Mar. 19/1 My wife was mystified when somebody in Brisbane described our daughter Sally, who is nearly five, as a 'trick'.

II. 7. *at* or (now more usually) *up to one's* (*old*) *tricks*, misbehaving or plotting mischief in one's characteristic way. Cf. quot. 1581 in Dict.

1823 Scott *St. Ronan's Well* II. i. 15 Aweel, I trust he is not at his auld tricks again, goodwife? 1863 H. E. P. Spofford *Amber Gods* 206 'You are at your old tricks again!' said he. 1898 G. B. Shaw *Man of Destiny* 181 What do you mean? Eh? Are you at your tricks again? Do you think I dont know what these papers contain? 1935 *Time* 7 Jan. 55/1 She and Dill are soon up to their old tricks.

9. b. *U.S. slang.* A term of service on a ship. Also, a term in prison.

1933 *Amer. Speech* VIII. iii. 32/2 *Trick*, a prison term. 1939 *Sun* (Baltimore) 28 Jan. 20/6 After serving a few tricks in the penitentiary they might turn State's evidence. 1942 *Ibid.* 19 Mar. 15/1 He reenlisted as a corporal, a rank he held at the end of his former trick. 1975 J. Gores *Hammett* i. 16 He got caught..and did a little trick at Quentin.

9*. a. An instance of the sexual act or any of its variations; usu. *spec.* a prostitute's session with a client. Esp. *to turn a trick*, to perform a sexual act with a casual partner, usu. for money. *slang* (orig. and chiefly *U.S.*).

In quot. 1926 the context concerns repeated sexual acts.

1926 C. Van Vechten *Nigger Heaven* 252, I said, Now, daddy, do you know any more tricks? 1946 Mezzrow & Wolfe *Really the Blues* 30 'Turning a trick' was how they described one session with a john. 1956 B. Holiday *Lady sings Blues* (1973) iii. 30, I had decided I was through turning tricks as a call girl. 1962 A. Lurie *Love & Friendship* xv. 300 Twenty-four dollars a time. That's pretty cheap for a girl like you... In New York, with the right connections, I bet you could get at least a hundred a trick. 1974 *Telegraph* (Brisbane) 16 July 14/2 She said in June: 'I'm lucky if I turn five to 10 tricks a week now.' 1975 J. F. Burke *Death Trick* ii. 20 It was a true lovers' tryst, not a trick. 1977 *Time* 28 Nov. 45/1 Some of the young prostitutes live at home and turn tricks merely for pocket money.

b. A casual sexual partner; usu. *spec.* a prostitute's client. *slang* (orig. and chiefly *U.S.*).

1925 in Odum & Johnson *Negro & his Songs* 189 Lawd, I went to my woman's do' Jes' lak I been doin' befo'; She says, 'I got my all-night trick, baby, An' you can't git in.' 1931 B. L. Reitman *Second Oldest Profession* viii. 118 Lillian has four children. Billy, her man, is a fourth-rate taxi-driver pimp. Billy goes out and gets 'tricks', and she takes care of them in the home where her children are. 1968 B. Turner *Sex Trap* xv. 148, I doubt there's one trick in twenty who isn't a married man. 1973 [see *John 1 f]. 1979 *Globe & Mail* (Toronto) 2 July 10/1 Young male prostitutes vie for tricks, the street name for a client.

V. 12. b. Also (chiefly *U.S.*) *to turn the trick*. **c.** *to miss a trick*, to fail to take advantage of an opportunity or notice something important; esp. *he* (or *she*) *never misses* (*does not miss*, etc.) *a trick* (see *miss *v.*¹ 5 d; earlier examples). *colloq.* (orig. *U.S.*). **d.** *how's* (less freq. *how are*) *tricks?* how are things? how are you getting on? *colloq.* (orig. *U.S.*).

b. [1872 Trollope *Eustace Diamonds* (1873) II. xxxvii. 134 Then the boy was done with and was carried away. She had played that card and had turned her trick.] 1933 *Sun* (Baltimore) 20 Apr. 10/3 It is our hope and prayer that Mr. Farley may turn the trick. We should be glad to see any administrator make a go of the postal service. 1942 *R.A.F. Jrnl.* 3 Oct. 24 But it was the Old Man who really turned the trick. 1960 I. Wallach *Absence of Cello* 230 It takes many years to live without a deliberate confusion about anyone's wants, including our own. Some people never turn the trick. 1976 *Springfield* (Mass.) *Daily News* 22 Apr. 39/2 A couple of American college products turned the trick for the Whalers. North Dakota graduate Alan Hangsleben and New Hampshire alumnus Cap Raeder shared the hero's role in the triumph.

c. [*a* 1916 H. James *Sense of Past* (1917) iv. iii. 266 It was..for him to have kept it as..she preferred it. He had begun so..and how..came it therefore that he now repeatedly missed that trick?] 1922 S. Lewis *Babbitt* xix. 241 'I'll bet..you were a bad old egg when you were a kid!' 'Well, I wasn't so slow!' 'I bet you weren't! I'll bet you didn't miss many tricks!' 1943 N. Coward *Middle East Diary* (1944) 11 He is a highly intelligent man and doesn't miss a trick... He had clear, alert views on the most diverse subjects. 1957 'J. Wyndham' *Midwich Cuckoos* xii. 99 He went on, with a puzzled frown on his brow as he realised that somewhere he had missed a trick; something had been kept from him. 1965 *Weekly News* (Auckland, N.Z.) 10 Mar. 49/1 The fact that the Wellington [boxing] association could match three Auckland fledgling professionals with three unknown Australians and make a profit points to someone else missing a trick.

d. 1915 J. London *Jacket* xiii. 149 'How's tricks?' I asked finally. 1924 W. Holtby *Crowded Street* i. 21 'Well, Mrs. H., how's tricks?' His wife flushed slightly at the vulgarity of his phrase. 1934 D. Runyon in *Collier's* 24 Nov. 8/4 Meyer Marmalade and I are glad to see her looking so well, and we ask her how are tricks. 1959 'A. Fraser' *High Tension* ix. 91 'Well,' he greeted me, 'how's tricks?' 1980 N. Marsh *Photo-Finish* ii. 36 'Gidday,' said Les Smith. 'How's tricks, then, Bert?'

13. (sense 5) *trick-flying*, *-rider*; *trick-leap* vb.; (sense 11) *trick-score*, *-taking*; **trick-cycling**, (*b*) *humorously*, psychiatry (cf. *trick-cyclist* (*b*) below); **trick-cyclist**, (*a*) (earlier example); (*b*) humorous alteration of Psychiatrist; **trick-film**, a film using trick photography; **trick photography**, photography using montage and other technical devices to create visual illusions; **trick picture** = *trick-film* above; **trick question**, a question designed to elicit more information than it appears to on the surface, or to trick the respondent into giving a wrong answer; **trick shot**, (*a*) in golf, etc.: a particularly clever or devious shot; (*b*) a camera shot made by means of trick photography; **trick wheel**, an auxiliary steering wheel on a ship.

1951 G. Frankau *Oliver Trenton* xvi. 139, I picked it up from one of our surgeons, who's rather keen on trick-cycling. 1966 G. B. Mair *Kisses from Satan* vii. 79 Don't try and sell that stuff about trickcycling to someone with rheumatism and gall stones. 1897 *Nat. Police Gaz.* 26 May 14/4 That noted trick cyclist, Lee Richardson, left London for America on Saturday. 1930 H. Wolfe *Uncelestial City* iii. 112 A trick-cyclist gravely reassembling The features of the ectoplasmic dead. 1971 P. Scott *Towers of Silence* v. iii. 345 That's why the trick-cyclist wallah insists on coming. 1977 *Listener* 31 Mar. 414/3 Is neurotic, inadequate, unhappy..is up in Harley Street being sorted out by a trick cyclist. 1912 F. A. Talbot *Moving Pictures* xix. 207 The achievements of Méliès and Paul set a very high standard of excellence in trick pictures. Their popularity precipitated a 'trick film' fever. The market became flooded with so-called magic pictures. 1914 *Chambers's Jrnl.* 6 June 429/2 A certain number of these craft [*sc.* aeroplanes] are kept..for..trick-flying, &c. 1932 E. Wallace *When Gangs came to London* xxii. 198 You used to do trick flying. 1922 Joyce *Ulysses* 428 Bloom trickleaps to the curbstone. 1913 *Technical World* XIX. 464 It merely accomplishes what is known as 'trick photography'. 1928 R. Knox *Footsteps at Lock* xxiv. 238 His cousin was fond of trick photography. 1984 *Listener* 2 Feb. 37/3 He achieved the crucial transformation scenes without the help of trick photography on cutaways. 1912 Trick picture [see *trick film* above]. 1939 R. Stout *Red Threads* xv, in *Mystery Bk.* 516, I could easily ask you some trick questions that would put sweat on your brow. 1954 N. Tomalin in J. Philip et al. *Best of Granta* (1967) ii. 139 He plugs away at trick questions..like: 'You *did* say you were giving the money to Dr Barnardo's didn't you sir?' 1978 P. Niesewand *Underground Connection* 86 The journalists..filed out... He had not expected any trick questions, and none came. 1902 *Billboard* 31 May 18/1 (Advt.), The motor wonders Arthur Stone and Joe Judge pace Frank Armstrong (who knows no fear as a trick rider). 1976 *National Observer* (U.S.) 24 July 18/4 Keitel calmly saying, 'Hi, Joy,' to a trick rider flashing past slung from the side of her horse. 1929 Trick-score [see *overtrick]. 1938 J. Culbertson *Contract Bridge for Beginners* vii. 71 A three-bid in no-trumps will produce a trick-score of 100 points. 1924 C. J. Tolley *Mod. Golfer* 229 He tried to recover by the aid of a trick shot. 1926 *Daily Colonist* (Victoria, B.C.) 20 Jan. 11/3 He told me that Joe Kirkwood, the marvelous trick-shot golfer had just given an exhibition. 1933 *Jrnl. Soc. Motion Picture Engineers* XX. 319 If the subject should contain more than the usual number of so-called 'trick' shots..the shooting time will easily run from 125 to 150 hours. 1981 *Sunday Times* 23 Aug. 54 At an exhibition they expect you to play about 10 frames [of snooker] and then to do some trick shots like hitting a ball into someone's handkerchief. 1983 J. Gardner *Elephants in Attic* xii. 110 My one experience of the film world had been devising trick shots for Alexander Korda. 1936 E. Culbertson *Contract Bridge Complete* xxxvii. 398 The low cards in long and short suits have their own trick-taking power. 1977 *Jrnl. Playing-Card Soc.* May 23 Reversis is historically important as the earliest known negative complex trick-taking game. 1942 *Sun* (Baltimore) 18 June 8/4 Men were stationed at the 'trick wheel'—an auxiliary wheel situated deep inside the ship, where it was operated by hand. 1972 L. M. Harris *Introd. Deepwater Floating Drilling Operations* 244 The emergency steering wheel and trick wheel.

trick, *v.* Add: **I. 1. e.** To put a spell on (a person), 'conjure'. Cf. *trick *sb.* 5 c. *U.S. dial.* (esp. in the speech of *U.S. Blacks*).

1829 *Virginia Lit. Museum* 25 Nov. 384 And, amongst the degraded and ignorant part of our own population, the notion of '*tricking*' or bewitching is universally..received. 1895 L. Herron in A. Dundes *Mother Wit* (1973) 360/2 The conjure doctor's business was of two kinds: to conjure, or 'trick', a person, and to cure people already conjured. 1970 H. M. Hyatt *Hoodoo-Conjuration-Witchcraft-Rootwork* I. 688 Well, if a man is got de skill upon 'im tuh make a man do 'jes whut he want 'im to do without *hurtin'* 'im, dat's *trickin'* a man. 1977 [see *trick *sb.* 5 c].

3. b. To have casual sexual intercourse, esp. for money; chiefly const. *with*. Cf. *trick *sb.* 9* a, b. *U.S. slang.*

1965 C. Brown *Manchild in Promised Land* (1966) vi. 163 Since her mother was laying so many cats, why shouldn't she be tricking. 1967 C. Himes *Black on Black* (1973) 133 He was trying to get his old lady, Tang, to go down into Central Park and trick with some white man so they could eat. 1973 J. Wambaugh *Blue Knight* xii. 207 He tricked with a whore the night before in the Orchid Hotel. 1978 A. Maling *Lucky Devil* xxix. 154 'You *know* him?' I asked. 'We've tricked,' he replied.

IV. 8. *trick and tie* (*trick* app. = to take one's turn at something; cf. Trick *sb.* 9), to be equal or even *with* someone or something; chiefly as *adj.* (sometimes hyphened); occas. as *sb.*

1825 C. M. Westmacott *Eng. Spy* I. 241 Trick and tie you know is fair play. 1829 P. Egan *Boxiana* 2nd Ser. II. 132 The Grecian, in order to make 'trick and tie' with his opponent, put his best foot foremost, and pursued him. 1883 W. H. Cope *Gloss. Hampshire Words & Phr.* 96 *Trick-and-tie*.., equal to each other. 1890 Barrère & Leland *Dict. Slang* II. 374/1 *Trick and tie* (sport). To be *trick and tie*, or touch and go, is to be equal in a race, or other athletic performance. 1905 *Eng. Dial. Dict.* VI. 234/1 [Wiltshire] 'I'll keep trick-and-tie wi' un', will keep even or level with him, in mowing or standing pots of beer or anything else.

trick, *a.*² *U.S. colloq.* [f. the sb.] = Tricky *a.* 2; liable to give way unexpectedly, defective, unreliable.

1961 in Webster. 1968 *Punch* 21 Feb. 256/2 He would have been out there himself, he said, only he had this trick knee, had it since he was a kid, gave him hell. 1977 *Time* 8 Aug. 14/1 Private Citizen Henry Kissinger has a trick back like millions of other Americans. 1977 *Hot Car* Oct. 42/1 Best upholstery went to Sandy Ventriglin for her work on a very trick Viva.

trickeration (trikərēⁱ·ʃən). *U.S. Blacks.* [f. Tricker(y + -ation.] A trick or stratagem (see also quots. 1940, 1970).

1940 *Music Makers* May 37/3 *Trickeration*, struttin' your stuff, muggin' lightly and politely. 1946 Mezzrow & Wolfe *Really the Blues* ix. 138 History was laying some trickeration on us. 1951 L. Hughes *Montage of Dream Deferred* 18, I believe my old lady's pregnant again! Fate must have some kind of trickeration to populate the cullud nation! 1970 C. Major *Dict. Afro-Amer. Slang* 116 *Trickeration*, to show off.

trickle, *sb.*¹ Add: **2.** Special Comb.: **trickle charger** *Electr.*, a device for charging a storage battery at a low rate over a long period; hence **trickle-charge** *sb.* (in quot., *attrib.*) and *v. trans.*; **trickle-charged** *ppl. a.*; **trickle-charging** *vbl. sb.*; **trickle irrigation** (see quot. 1969); hence (as back-formation) **trickle-irrigate** *v. trans.*

1959 *Times* 11 Sept. 7/4 It is ideal for trickle-charge operation. 1974 *Undercurrents* July–Aug. 3/3 This..produced a few hundred milliamps at about 10 volts in a fair breeze—enough to trickle-charge a small battery. 1938 *Proc. Physical Soc.* L. 422 The driving, demultiplying, and amplifying circuits were operated by trickle-charged batteries. 1927 *Observer* 24 July 4/5 A fool-proof set..is provided with a 'trickle charger' for keeping the filament battery up to strength. 1977 *Film & Television Technician* Mar. 11/3 Complete with its internal rechargeable batteries and a built-in trickle charger, the D34 weighs in at only 12 lbs. 1960 E. L. Delmar-Morgan *Cruising Yacht Equipment & Navigation* xvi. 175 This 'trickle charging' is..harmful for the type of batteries used in yachts. 1971 *World Bk. Sci. Ann.* 1972 255 Vast areas in the Negev Desert of Israel are now trickle-irrigated. 1969 *Gloss. for Landscape Work* (*B.S.I.*) v. 13 *Trickle irrigation*, a method of supplying water by means of a restricted, controlled flow to the surface of a growing medium (usually at discrete points, one to each plant). 1975 *N.Z. Jrnl. Agric.* Sept. 43/1 The research workers..are hopeful that trickle irrigation, by maintaining even soil moisture, will give some help.

trickle, *v.* Add: **1. c.** Also used facetiously for 'to make one's way, go'. Cf. *ooze *v.*¹ 2 c.

1912 R. Brooke *Let.* Feb. (1968) 357, *I* wrote to *her*, about her perhaps trickling down to Rugby. 1936 Wodehouse *Laughing Gas* iii. 32 In these circs, it seemed to me that the best way of passing the time would be to trickle over to the table where the drinks were and brace myself with one or two. 1961 —— *Service with Smile* (1962) vii. 100 He headed for the lake. I trickled after him. 1983 *Country Life* 5 May 1194/3 Trickling through morning traffic..it [*sc.* a car] showed its docility by contentedly running in fifth.

3. c. *Sport.* To cause (a ball) to travel slowly over the ground, esp. in golf. Also *to trickle a putt*. Also *absol.*

1902 *Daily Chron.* 20 Aug. 7/3 He can..trickle the ball away to fine-leg with a delicate turn of the wrist. 1903 *Westm. Gaz.* 6 Feb. 3/2 If you bolt at the hole, you will not need to make the same allowance for incline as if you trickled. 1927 *Daily Tel.* 14 Mar. 13/1 'I will trickle the..putt up to the hole.'..It certainly was a trickle..for the ball stopped five feet short of the hole.

4. *Comb.*: **trickle-down** *a.*, of or based on the theory that economic benefits to particular groups will inevitably be passed on to those less well off; also *transf.* as *sb.*, a filtering down (of money or ideas). orig. and chiefly *U.S.*

[1931 W. Rogers in *Tulsa Daily World* 12 July iv. 7/3 What about the old Boys here on the home grounds? Well maybe this thing will eventually reach him in some beneficial way. Lord knows what way it may trickle down to him some day.] 1944 *Antioch Rev.* Summer 192 In agriculture, as in business, they are devotees of the trickle-down philosophy. 1949 H. S. Truman in *Sun* (Baltimore) 6 Jan. 6/1 We have rejected the discredited theory that the fortunes of the nation should be in the hands of a privileged few. We have abandoned the 'trickle-down' concept of national prosperity. 1954 *Sun* (Baltimore) 13 Feb. (B ed.) 2/1 The Administration has already offered us a trickle-down tax program. Now, we are presented with a trickle-down housing program. 1962 C. Walsh *From Utopia to Nightmare* i. 18 There has been a trickle-down [of ideas] and permeation. 1971 *Publishers' Weekly* 6 Dec. 17/2 It is to be hoped that textbook writers come in contact with frontier thinkers or their writings and translate some of the results into educational materials. This is known as the 'trickle-down' process. 1977 *Time* 16 May 38/2 It's classic trickle-down economics. 1980 *Jrnl. R. Soc. Arts* July 508/2 The theory that if you build the industrial capability and increase the GNP of the country there will be a trickle-down to people at subsistence level. 1981 Plate & Darvi *Secret Police* ii. 42 If the sovereign himself is corrupt or family and relatives are corrupt, there can be a trickle-down effect into the ranks. 1984 *New Yorker* 16 Apr. 82/2 To Fink this often sounded suspiciously like Republican trickle-down economics.

trickless (tri·klės), *a.* [f. Trick *sb.* + -less.] Possessing no tricks (*spec.* in *Cards*).

1927 *Observer* 29 May 25 If he finds his partner trickless, he must go down enormously. 1977 *Times* 17 June 12/1 Anyone who regularly makes preemptive opening bids on hands which are trickless in defence has only himself to blame.

trickling, *vbl. sb.* Add: **b. trickling filter** = *percolating filter* s.v. *percolating *vbl. sb.*

1903 W. J. Dibdin *Purification of Sewage & Water* (ed. 3) p. xv, I think it is quite practicable to deal with Leeds sewage on trickling filters without antecedent septic treatment. 1976 *Ann. Rev. Microbiol.* XXX. 266 Bacterial and viral aerosols are generated..and bacteria have been detected at least 0·8 mile from a trickling filter.

tricky, *a.* **2.** (Earlier example.)

1868 C. L. Eastlake *Hints Househ. Taste* vii. 176 Chromo-lithography..accustoms the eye to easily rendered and therefore *tricky* effects of colour which falsify rather than illustrate nature.

tricolour, tricolor, *a.* and *sb.* **A.** *adj.* **1.** Delete '(in form **tricolor**)'. Esp. in reference to a black, white, and tan dog. (Later examples.)

1922 R. Leighton *Compl. Bk. Dog* viii. 113 At Cruft's in 1913 an especially nice tricolour puppy named Jason was placed high above Lerwick Jarl. 1971 F. Hamilton *World Encycl. Dogs* 103 The coat is smooth and short and the colors are tricolor, black, tan and white.

3. a. Employing or pertaining to the use of the three primary colours. Cf. *three-colour *a.*

1898 A. C. Austin (*title*) Practical half-tone and tri-color engraving. 1909 *Chambers's Jrnl.* 27 Mar. 268/1 A remarkable new development..promises to revolutionise the art of printing in photogravure. The process is three-colour; and..the results..are far in advance of anything that has yet been attempted in tricolour printing. 1937 J. S. Mertle *Photolithography & Offset Printing* 101 Three-color photography proper is based on the photographic separation of the colored original into three primary color-record (separation) negatives, which are obtained by exposing three separate color-sensitive plates through what are termed 'tri-color' filters. *Ibid.* 103 Tri-color images. 1968 *Gloss. Terms Offset Lithogr. Printing* (*B.S.I.*) 17 *Tricolour filter*, a colour filter that transmits a primary band of the spectrum, i.e., red, green or blue.

b. *gen.* Three-coloured.

1979 *Washington Post* 20 Oct. F7/2 Even the large sculptures are hard to find among the visual chaos—amid blaring tricolor trashcans, telephone booths, [etc.]. 1983 *Christian Sci. Monitor* 14 Nov. B12/1 Almost anything Chinese—whether Tang Dynasty tricolor jars, Ming molded gourds, or cheap jade trinkets—can be bought in Hong Kong.

B. *sb.* **2. a.** (Earlier example.)

1797 *Political Censor* Mar. 84 To strut and hector over the poor fallen *Tricolor*, and to call on your readers to 'unite under their *own flag*'..entitles you to but very little praise.

b. The green, white, and orange Irish Republican flag.

1969 *Listener* 28 Aug. 268/2 The tricolour had been raised in Bogside once before. 1977 *Belfast Tel.* 22 Feb. 5/5 No Irish tricolours were displayed.

3. A black, white, and tan dog.

1905 J. Watson *Dog Bk.* lxvi. 724 The tricolour has been neglected in the fashion for black and tans. 1971 F. Hamilton *World Encycl. Dogs* 97 Two tricolors can produce a pure golden-sable.

4. A combination of the three primary colours.

1978 *SLR Camera* Dec. 88/3 In theory, the yellow, magenta-red, and cyan-blue tricolour forming the image of the slide should each perfectly reflect two-thirds of the spectrum of white light.

tricoloured, -colored, *a.* (Later example.)

1913 F. T. Barton *Toy Dogs* ix. 41 The Ruby and Tri-coloured, or Prince Charles Spaniels, have a separate classification.

tricontinental: see *tri- 1 a.

tricot². Add: Also, a pair of close-fitting knitted tights. (Earlier and later examples.)

1859 L. Wraxall tr. *Robert-Houdin's Mem.* II. ix. 259 And the whole [false stomach] being concealed beneath a flesh-colour *tricot*, appeared to form part of his body. 1893 [in Dict.]. 1926 *Spectator* 6 Feb. 219/1 Let us hope that the odious 'tricot', or tights, which create, by emphasizing, indecency have disappeared for ever—at any rate in Paris [music-halls].

‖**tricoteuse** (trīkotöz). Also **Tricoteuse**. [Fr.] **1.** A woman who knits; applied *spec.* to women who, during the French Revolution, sat and knitted at meetings of the Convention or at guillotinings. Also *transf.*

1830 Hazlitt *Life of Napoleon Buonaparte* I. vi. 284 It was this [popular fury] that inspired the Furies of the Guillotine, and sat and smiled in the galleries of the Convention with the *tricoteuses* of Robespierre! *a* 1886 M. B. Chesnut in C. V. Woodward *M. Chesnut's Civil War* (1981) viii. 183 Jenny Barron, Jenny Cooper and Mary Hammy have gone to have their photographs taken as 'Tricoteuses', each armed with their knitting. 1905 Baroness Orczy *Scarlet Pimpernel* i. 8 The old hags, 'tricotteuses' [*sic*] as they were called, who sat there and knitted, while head after head fell beneath the knife. 1940 M. Dickens *Mariana* viii. 309 The eyes of the *tricoteuses* nearly came out of their heads, and the less hardened of them dropped a stitch. 1961 *Guardian* 19 Apr. 10/6 Is it fair that the Conservative Women's Conference should..be thought of as..a collection of elegant Tricoteuses? 1972 R. Quilty *Tenth Session* i. 66 The inner circle around the victim, the eager *tricoteuses* who always collected at any drama on the streets. 1973 *Listener* 22 Nov. 727 The wife of the production manager..sits sourly knitting on set like a *tricoteuse* at the guillotine.

2. *Antiques.* (See quot. 1960.)

1960 H. Hayward *Antique Coll.* 287/2 *Tricoteuse* (Fr.), a term probably of 19th-cent. origin, applied to a small work-table surrounded by a gallery, part of which can be lowered to contain sewing materials. 1973 *Country Life* 30 Aug. (Suppl.) 72/2 Mahogany Tricoteuse with black line inlay. English, circa 1810.

Tricotine (tri·kǫtīn). Also **tricotine**. [f. Tricot² + -ine⁴.] A worsted fabric with a double twill.

A proprietary term in the U.S.

1914 *Official Gaz.* (U.S. Patent Office) 27 Oct. 1236/2 'Tricotine'...Woolen, Worsted, Silk, Mohair and Cotton Piece Goods and Piece Goods Made of a Combination of Two or More of Those Fibers. 1920 G. Bell *Let.* 10 Oct. (1927) II. xviii. 502 I've got bills for tricotine and things which Elsa kindly bought for me. 1921 *Daily Colonist* (Victoria, B.C.) 5 Oct. 8/6 Smartly fashioned dresses..of Tricotine and Serge, comprise this special offering. 1930 *Daily Express* 6 Oct. 6/1 (Advt.), Smart in every sense of the word in a tricotine finished faced-cloth.

Tricouni (traikū·ni). Also **tricouni**. [Swiss trade name, app. f. Tri- + Co- + Uni-.] The proprietary name for a kind of climbing-boot nail with a serrated edge. (Now *disused.*)

1914 O. Eckstein in *Climbers' Club Jrnl.* Mar. 77 (*title*) The Tricouni Nail. *Ibid.*, A new nail, bearing the above name, has lately been invented by a Swiss climber, Mr. F. Genecand; it is intended for nailing boots used on mountains. 1920 G. W. Young *Mountain Craft* ii. 82 An even better nail appears to be the lately invented Tricouni nail. 1934 *Trade Marks Jrnl.* 3 Oct. 1276/1 Tricouni... Boot and shoe nails; and cleats and calks; all of ordinary metal. Tricouni S.A..., Geneva, Switzerland; manufacturers. 1946 J. E. Q. Barford *Climbing in Britain* ii. 20 Tricounis are better on snow and ice because the sharp edges bite in well. 1957 J. Masters *Far, Far the Mountain Peak* 143 They had spent those eight hours hanging by fingertips and tricounis over the Bowl. 1972 M. Woodhouse *Mama Doll* ii. 4 Boots nailed with what..are..called tricounis.

tri-county, tricresyl phosphate: see *tri- 1 b, 5 c.

tricycle, *sb.* Add: **2. b.** A three-wheeled motorized invalid carriage.

1974 *Guardian* 26 Mar. 32/4 Disabled drivers' tricycles should be replaced with modified production cars. 1976 *Times* 24 July 1/7 Invalid tricycles are to be phased out... The existing 21,265 tricycle drivers will be able to keep their vehicles until they wear out.

c. An aeroplane with a three-wheeled undercarriage.

1942 *Tee Emm* (Air Ministry) II. 95 The pilot..strikes the sea at the normal three-point landing attitude (slow landing attitude for tricycles).

3. *Comb.* **tricycle undercarriage** *Aeronaut.*, a three-wheeled undercarriage; also **tricycle landing-gear**.

1938 *Sun* (Baltimore) 26 Sept. 13/5 The 'tri-safety' tricycle landing gear, with a nose wheel under the pilot's cockpit. 1960 *Economist* 31 Dec. 1382/2 Development of the tricycle landing gear has been the greatest single factor in making a flier out of a businessman. 1937 *Flight* 18 Nov. 483/1 (*heading*) Simplified control system for all-metal pusher with tricycle undercarriage. 1966 D. Francis *Flying Finish* x. 129 The bump from the tricycle undercarriage as we touched down. 1976 *Official Programme Farnborough Internat. Exhib.* 28 The aircraft has a fixed tricycle undercarriage.

tricyclic, *a.* Add: **2. b.** *spec.* in *Pharm.*, designating or pertaining to a group of anti-depressant drugs based upon a molecular structure of three fused rings; also as *sb.*

1966 *Internat. Jrnl. Neuropharmacol.* V. 299 (*heading*) Association between biochemical and behavioral actions of tricyclic antidepressants. 1972 *Daily Tel.* 4 Mar. 32/3 He said pregnant women should steer clear of all drugs if possible, particularly drugs known as tricyclic antidepressants. 1973 *Sci. Amer.* Sept. 121/2 The chemical structures of these two classes of antidepressants—the monamine oxidase inhibitors and the tricyclics—are quite different. 1980 J. Wainwright *Man of Law* i. 18, I suggested..a good tricyclic antidepressant.

tridentate, *a.* Add: **2.** *Chem.* Of a ligand: forming three separate bonds (usu. but not necessarily with the same central atom). Of a molecule or complex: formed by such a ligand.

1925 *Jrnl. Chem. Soc.* CXXVII. 2030 Werner also identified groups capable of treble attachment to metallic atoms, such tridentate residues containing three fused chelate groups. 1966 Phillips & Williams *Inorg. Chem.* II. xxvii. 347 The cyclopentadienyl and benzene ligands are best regarded as tridentate and their transition-metal complexes are strongly stabilized in the sandwich symmetry only.

tridimensional, *a.* Add: (Earlier example.)

1858 *Q. Jrnl. Math.* II. 182 An analogous method of transformation is applicable to tridimensional space.

Hence **tridime·nsionally** *adv.*

1919 R. T. Browne *Mystery of Space* vii. 240 Trace out the biologic development of each mental faculty..and it will be found that..the nature of each of these has been to express itself tridimensionally. 1956 H. Read *Art of Sculpture* v. 102 One of the most striking developments in our time has been in the direction of linear sculpture, another apparent contradiction of terms that can be justified in so far as the lines are used tridimensionally to indicate volume.

Tridione (traidai·ōᵘn). *Pharm.* [f. Tri- + *-dione.] A proprietary name for an analgesic agent (also called *troxidone*).

1944 Richards & Everett in *Federation Proc.* III. 39/2 Preliminary clinical investigations have shown that Tridione is an effective analgesic. 1948 *Official Gaz.* (U.S. Patent Office) 24 Feb. 607/1 *Tridione*... For anticonvulsant intended for use in treatment of the petit mal triad and other convulsive disorders. 1948 *Trade Marks Jrnl.* 25 Feb. 141/2 *Tridione*... Therapeutic compounds being alkyl derivatives of diketo oxazolidine. Abbott Laboratories.., City of North Chicago,..manufacturers. 1974 M. C. Gerald *Pharmacol.* xi. 212 Trimethadione (Tridione) was originally synthesized as an aspirin substitute; however, its lack of dependable analgesia caused it to be rapidly discarded for this purpose. 1979 R. Jaffe *Class Reunion* i. ix. 95 Time to get to the pharmacy.., fill her prescriptions for Dilantin and Tridione.

tri-dominium. Add: Also applied to the former rule of Great Britain, Greece, and Turkey in Cyprus.

1955 *Times* 25 July 7/7 The second scheme is, broadly, that the island should be turned into some form of 'tridominium' with, perhaps, a British Governor and one Greek and one Turkish deputy Governor, and with all three national flags flying over Government House. 1958 *Times* 20 June 11/2 The next stage may be independence, condominium, or even the perpetuation of tridominium, but it can scarcely be absorption by either Greece or Turkey.

triduum. b. (Earlier example.)

1873 G. M. Hopkins *Jrnl.* 9 Feb. (1959) 230 Began a triduum.

tried, *ppl. a.* Add: **3. b.** Phr. *tried and true*, proved reliable by experience.

1954 W. Faulkner *Fable* 352 His enslavement to the demonic progeny of his own mechanical curiosity, from which he will emancipate himself by that one ancient tried-and-true method by which slaves have always freed themselves: by inculcating their masters with the slaves' own vices. 1967 *Listener* 6 Apr. 474/3 Miss Aukin had had the good sense to use the tried and true concealment gambit by which eventually two young officers, bent on cuckolding a greengrocer, were compelled to hide in the same grandfather clock. 1979 *Tucson Mag.* Apr. 47/1 A beautifully made 'period' movie, written and directed by tried-and-true Michael Crichton.

triene (trai·īn). *Chem.* [f. Tri- + -ene.] Any organic compound containing three double bonds between carbon atoms.

1917 *Chem. Abstr.* XI. 3031 These hydrocarbons give exaltations greater than those of the aromatic series: with the trienes the difference is proportionally greater than with the dienes. 1936 [see *diene value* s.v. *diene b]. 1976 *Nature* 22 Apr. 726/2 We recorded the laser Raman spectrum of vitamin D_3 (in CCl_4), which has a conjugated triene structure.

trier. 10. (Further examples in gen. use.)

1927 C. A. Wilson *Empire's Junior Partner* 227 There are many opportunities to acquire land on easy terms, very many farmers' trading concerns which are ready to back a 'trier'. 1942 *Tee Emm* (Air Ministry) II. 133 One day his Instructor said 'That chap's a trier' and sent him to a Squadron. 1955 *Times* 10 June 4/7 Pritchett has always been a real trier, and his game has looked more mature since his return from the United States.

Triestine (tri,e·stīn), *sb.* and *a.* [f. *Trieste*, name of a city and province in north-eastern Italy + -ine¹.] **A.** *sb.* A native or inhabitant of Trieste. **B.** *adj.* Of Trieste or its inhabitants.

1905 J. Joyce *Let.* 12 July (1966) II. 93 Nora can speak about thirty words of Triestine dialect. *Ibid.* 24 Sept. (1966) II. 112 Two little Triestines went into a hallway and laughed till they beat each other. 1921 *Contemp. Rev.* July 46 Sig. Scaramanga, was a Triestine, and therefore an ex-Austrian subject. 1975 'D. Rutherford' *Mystery Tour* iv. 69 Maria, his homely and chubby Triestine wife.

triethanolamine, triethylene: see *tri- 5 c, b.

trifecta (traife·ktă). *N. Amer.*, *Austral.*, and *N.Z.* [f. Tri- + *per)fecta.] (See quot. 1977.) Also *fig.*

1974 *Daily News* (N.Y.) 6 Feb. 46 The money that is bet on gimmick races like the trifecta is placed in a separate pool from the straight win, place, and show wagering. 1975 *Cleveland* (Ohio) *Plain Dealer* 6 Apr. 13-c/1 Before acting on the request for trifecta wagering..the ORC must advertise in affected counties 30 days before the hearing. 1977 *Daily Colonist* (Victoria, B.C.) 26 Nov. 32/1 To win the trifecta, a bettor must select the first, second and third place horses in order. 1979 *Sunday Mail* (Brisbane) 19 Aug. 6/5 A new bride on Thursday, a $50,000 win in the New South Wales lotteries on Friday and a winning day at the races on Saturday. That's last week's trifecta for Sydney's biggest bookmaker. 1984 *N.Z. Truth* 23 May 6 The T.A.B. is accepting trifecta and quinella on this feature event which is also the 1st leg of the Double and Treble.

triffid (tri·fid). Also **Triffid**. [f. Tri-, prob. after Trifid *a.*, as the plant was supported on 'three bluntly-tapered projections extending from the lower part' of the body.] In the science-fiction novel *The Day of the Triffids*, by John Wyndham (1903–69), one of a race of menacing plants, possessed of locomotor ability and a poisonous sting, which threaten to overrun the world. Hence used allusively of vigorous plants, or *transf.* of anything invasive or rapid in development.

1951 'J. Wyndham' *Day of Triffids* ii. 46 A catchy little name originating in some newspaper office as a handy label for an oddity—but destined one day to be associated with pain, fear and misery—*triffid*. *Ibid.* 54 He had also established that the infertility rate of triffid seeds was something like ninety-five per cent. 1965 *New Scientist* 11 Mar. 619/3 Ninety per cent of British households have television..and neither bindweed, triffids, nor dragon's teeth grew more rapidly than the angular aerial. 1972 S. Hughes in M. Bygrave et al. *Time Out's Bk. of London* ix. 90/1 The south is sprouting with tall dark buildings like triffids. 1977 *Times* 10 Feb. 5/2 Roots and suckers started appearing all over our gardens... They were like 'Triffids'. 1978 R. H. Lewis *Antiquarian Bks.* iv. 86 Books 'taking over' in Triffid style is a common experience.

Hence **triffi·dian** *a.*; **tri·ffid-like** *a.* and quasi-*adv.*

1951 'J. Wyndham' *Day of Triffids* ii. 49 It was assumed..that their characteristic of suddenly losing their immobility and rattling a rapid tattoo against the main stem was some strange form of triffidian amatory exuberance. 1971 *Daily Tel.* 16 Jan. 10/6 This cactus had run wild and, Triffid-like, had taken over thousands of square miles of good agricultural land. 1971 *New Scientist* 9 Sept. 589/3 Even the departure of four million..inhabitants..did not appear to check the spread of the Triffid-like condition.

trifid, *a.* Add: **c.** Also **tref(f)id**. Designating a type of antique spoon (see quot. 1977). Also *absol.* as *sb.*

1892 C. J. Jackson in *Archaeologia* LIII. i. 138 The stem of this [Puritan] spoon is as wide as that of the trifid-ended form which immediately succeeded it. 1911 —— *Hist. Eng. Plate* II. xvii. 521 The earliest *Trifid Spoon* known to the author is one bearing the Dublin hallmarks for 1663. 1927 N. Gask *Old Silver Spoons of England* ix. 96 (*heading*) Lobed-ends, variously called Trifids, Trefoils, Split-ends or Pieds-de-Biche. 1932 *Antique Collector* Nov. 394/1 The Trifid type of handle, introduced with the Restoration, appears to have been evolved in turn from the Puritan. 1952 G. E. P. How *Eng. & Scottish Silver Spoons* I. ii. xiv. 325 The Trefid Spoon, in its fully-developed form, was apparently introduced to London from the Continent at the Restoration of the Monarchy. 1956 G. Taylor *Silver* v. 111 The so-called Puritan spoon seems to be the starting-point for more decorative developments. The plain square end was hammered out into a thin leaf-shaped blade which is most commonly notched on either side of the pointed end to form a simple trefoil... Such spoons are therefore described as 'trefid' (or 'trifid')... Larger trefid spoons with the same characteristics, but a much longer stem, are found. 1974 [see *rat-tail 3 b]. 1977 Fleming & Honour *Penguin Dict. Decorative Arts* 802/1 *Trifid* or *Trefid spoon*, a C17 English type of spoon with a flat handle widening towards the end which is divided into three parts by deep notches.

trifle, *sb.* **6. b.** (Earlier example.)

1755 H. Glasse *Art of Cookery* (ed. 5) xvi. 285 Trifle. Cover..your Dish..with Naples Biscuits..Mackeroons..and Ratafia Cakes..wet them..with Sack, then make a good boiled Custard..pour over it..then put a Syllabub over that.

trifluoperazine (traiflūo,pe·răzīn). *Pharm.* [f. Tri- 5 + Fluo- + *pi)perazine.] A phenothiazine derivative used as a tranquillizer and anti-emetic (usu. administered as the hydrochloride); 10-[3-(4-methylpiperazin-1-yl)propyl]-2-trifluoromethylphenothiazine, $C_{21}H_{24}F_3N_3S$.

1958 *Amer. Jrnl. Psychiatry* CXIV. 747 (*heading*) Triflupromazine and trifluoperazine: two new tranquilizers. *Ibid.* 747/2 The chlorine atom of the phenothiazine nucleus has been replaced by carbon trifluoride... This differentiates..trifluoperazine from prochlorperazine. 1965 [see *Stelazine]. 1977 *Lancet* 15 Oct. 816/2 A 65-year-old woman had been taking 'Parstelin' tablets (tranylcypromine 10 mg, trifluoperazine 1 mg), one three times daily for a depressive illness for 4 years. 1984 *Brit. Med. Jrnl.* 8 Sept. 612/1 He was diagnosed as having an anxiety neurosis and was prescribed trifluoperazine together with orphenadrine.

trifocal (traifōᵘ·kăl), *a.* and *sb.* [f. Tri- + Focal *a.*] **A.** *adj.* † **a.** (See quot.) *Obs.*

1826 J. I. Hawkins in *Repertory Patent Inventions* III. 386 When three pairs [of glasses] are worn in one frame..which I denominate trifocal spectacles, the opening should be a circle of one inch diameter.

b. Of a lens: having three parts with different focal lengths. Of spectacles: having such lenses.

1921 *Amer. Jrnl. Ophthalmol.* IV. 406/2 The writer does not claim that every presbyope should have a pair of trifocal spectacles. 1928 W. S. Duke-Elder *Pract. Refraction* xxiii. 343 For some purposes tri-focal lenses are advocated. 1969 K. Vonnegut *Slaughterhouse-Five* v. 93 She was wearing tri-focal lenses in harlequin frames.

B. *sb.* A trifocal lens; usu. *pl.*, trifocal spectacles.

1899 J. Thorington *Refraction* xi. 283 Trifocals.—Occasionally, a patient is not content with bifocals, but will demand a focal point somewhere between infinity and his working distance. 1946 Berens & Zuckerman *Diagnostic Examination of Eye* i. xii. 319 (*caption*) Double-segment panoptik trifocal for sculptors. 1962 L. S. Sasieni *Optical Dispensing* vii. 172 Where the gardener does not need a distance correction, the best combination may be reading and weak intermediate. A trifocal is the obvious alternative. 1973 R. Hayes *Hungarian Game* xxiv. 147 He was a pallid man with eyes that were weirdly distorted by trifocals.

trifunctional, -ally: see *tri- 1 a.

trifurcate, *a.* Add: **tri·furcating** *vbl. sb.*, in *trifurcating box* (see quot. 1940).

1922 *B. I. Handbk.* (Brit. Insulated & Helsby Cables Ltd.) (ed. 3) 447 (Index), Trifurcating boxes. 1940 *Chambers's Techn. Dict.* 864/2 *Trifurcating box*, a cable dividing box for enclosing the joints between a three-arc or triple concentric cable and three single-core cables or conductor terminals.

trig, *sb.*² Add: *to toe the trig*: see *toe *v.* 2.

trig, *a.*² and *sb.*³ Colloq. abbrev. of Trigonometrical *a.*, Trigonometry. Freq. as *trig point*, *station*.

1862 *McLean Papers* (MS.) XX. 87 Mr Swainson informs me that the Trig Station pulled down was not one erected by him. 1895 W. C. Gore in *Inlander* Nov. 65 *Trig.* n., trigonometry. 1924 P. Marks *Plastic Age* 36 Kane announced the textbook, and when Hugh caught the word 'trigonometry' he actually thrilled with joy. He had had trig in high school. 1926 J. Devanny *Lenore Divine* xx. 184 They reached the trig station at the top, two thousand five hundred feet above sea-level. 1936 H. S. L. Winterbotham *Key to Maps* iii. 28 On the ordnance maps and plans you will find..little triangles with dots inside... Each one represents a place fixed by careful instrumental measurement, and the sum total represents the skeleton..upon which all our maps depend... 'Trig. Points', represented by the triangles, control the map in plan, 'Bench-marks' in height. 1959 J. Braine *Vodi* vi. 92 I've not done those trig. problems. 1968 G. R. Crone *Maps & their Makers* xi. 143 Transference of the trig. points to the sheets issued to the plane tablers. 1976 J. Lee *Ninth Man* i. 48 Andy flunked trig for the second time (damn higher mathematics, anyway). 1981 *Times Lit. Suppl.* 22 May 577/1 Surveyors had to be trained [for Ordnance Survey mapmaking]; trig-points and bench-marks established.

trigenic, *a.* Add: **2.** *Genetics.* [cf. *genic *a.*] Involving or controlled by three genes.

1941 [see *polygenic *a.* 3]. 1979 *Experientia* XXXV. 172/2 The trigenic ratios have been reported here for the first time for the above-mentioned characters.

trigger, *sb.*¹ Add: **3.** (Earlier examples of *quick on the trigger* and further *fig.* examples.)

1808 M. L. Weems *Let.* 22 Apr. in E. E. F. Skeel *M. L. Weems: Works & Ways* (1929) II. 377, I trust that all your Aids will be quick on the trigger. 1842 C. M. Kirkland *Forest Life* II. xlvii. 223 'Pretty quick on the trigger!' muttered Uncle William. 1887 G. H. Darwin in *Leisure Hour* May 354/2 The attraction of the moon or the variation in atmospheric pressure pulls the trigger. 1946 *Lancet* 19 Jan. 97/1 A theory of the nervous initiation of contraction—the trigger without which voluntary muscle remains inert. 1961 M. Laski *Ecstasy* ii. 16 Of the circumstances in which they found themselves when ecstasy took place, they identified certain objects, events, and ideas as standing in some kind of a causal relationship to their ecstatic experiences. These objects, events, and ideas I am calling *triggers*. 1977 J. L. Harper *Population Biol. of Plants* 64 Triggers to development which predict a changing environment will generally be more efficient than those that are themselves the changed conditions.

3*. *Electronics.* **a.** A trigger circuit or trigger tube.

1945 *Electronic Engin.* XVII. 329/1 The charging circuit producing the used forward stroke operates continuously even through the flyback period when the trigger is conducting. 1946 [see *flip-flop *sb.* 6 e]. 1962, etc. [see *Schmitt]. 1969 J. J. Sparkes *Transistor Switching* v. 126 The reasons should be understood for always using a negative-going trigger to drive *npn* transistors off rather than a positive one to drive them into the conducting state. 1981 J. C. Sprott *Introd. Mod. Electronics* x. 239 The Schmitt trigger is useful..for generating square waves from a sinusoidal input.

b. A momentary signal or change in signal level that causes a change of state in a trigger tube or other device.

1948 *Gloss. Computer Terms* (Mass. Inst. Technol. Servomechanisms Lab. Rep. R-138) 11 *Trigger.* See *trigger pulse*. 1953 *Electronic Engin.* XXV. 143/1 A trigger derived from the phantastron is used to initiate the multivibrator circuit. 1979 M. M. Mano *Digital Logic & Computer Design* vi. 210 Asynchronous flip-flops..require an input trigger defined by a change of signal level... Clocked flip-flops are triggered by pulses.

3.** A fission bomb built into a fusion bomb in order to initiate the fusion reaction.

1955 *Times* 13 Aug. 5/4 It was a question of arranging the proper conditions, and there was no reason why fusion energy should not be obtained without the use of a fission bomb as 'the trigger'. 1969 *Listener* 5 June 773/3 It's necessary to have a trigger made of an ordinary fission bomb, and there is good evidence that this must consist of fissile uranium or uranium-235 and not fissile plutonium. 1982 *New Scientist* 2 Sept. 642/1 The X-rays produced by the triggers are absorbed and re-emitted by an ellipsoidal casing of ^{238}U, and the fraction which is re-emitted inwards goes on to strike the main bulk of the thermonuclear fuel.

4. *trigger-action*, *effect*, *question*, *switch*, *word*; *trigger-pulling* adj.; **trigger circuit** *Electronics*, a circuit that behaves like a trigger tube; also, a circuit for producing a trigger pulse; **trigger-fish**: for 'the genus *Balistes*' substitute 'the family Balistidæ'; (earlier example); **trigger-happy** *a. colloq.* [*-happy], over-ready to shoot at anything at any time or on slight provocation; also *transf.* and *fig.*; hence **trigger-happiness**; **trigger man** *slang* (chiefly *U.S.*), a gunman; a hired thug or bodyguard; also *fig.*; **trigger-point**, (*b*) *U.S.*, a price level at which price controls are imposed or re-imposed; **trigger price** *U.S.*, a minimum selling price for steel imported into the U.S., such that any steel imports below that price incur investigation to ensure that dumping is not taking place; **trigger pulse** *Electronics*, a pulse that acts as a trigger (sense 3* b above); **trigger tube** *Electronics*, a vacuum tube that has two operating states and changes rapidly from one to the other in response to a momentary application of, or change in, a signal.

1915 W. M. Bayliss *Princ. Gen. Physiol.* x. 304 The difference between what is sometimes called 'trigger action' and catalysis. *Ibid.*, Supersaturated solutions are cases of 'trigger action'. They remain indefinitely as such until infected with a crystal, and then the rate of crystallisation is independent of the amount of crystals added. 1950 A. Huxley *Lett.* (1969) 623 Trivial acts of selfishness and wantonness may release, as though by a kind of trigger action, a huge avalanche of tragic destiny. 1938 *Rev. Sci. Instruments* IX. 223/1 Another trigger circuit which has inherent possibilities as a counting circuit is shown in Fig. 2. 1951, 1962 [see *bistable *a.*]. 1974 A. Van der Ziel *Introd. Electronics* xi. 262 (*caption*) Transistor monostable circuit with trigger circuit. 1931 *Prof. Papers Inst. Post Office Electr. Engineers* No. 136. 19 The adjustments were such as to avoid definitely the 'trigger' effect. 1949 M. Mead *Male & Female* x. 218 In the..patterning of a woman's sexual receptivity now one part of the body, now another,..may be sensitive enough to develop a trigger effect. 1849 H. Melville *Mardi* I. xlviii. 131 The rank and file of the Trigger-fish—so called from their quaint dorsal fins being set in their backs with a conical curve, as if at half-cock. 1945 C. Burney *Dungeon Democracy* iii. 82 There was much trigger-happiness, men shooting each other, shooting themselves and shooting into thin air. 1970 *Daily Tel.* 23 Mar. 16 The trigger-happiness with which workers are now ready and eager to enforce their fantastically increased claims by industrial action..leaves Britain with the choice between being ruined by runaway inflation or by a series of disastrous strikes. 1978 N. Freeling *Night Lords* xxi. 95 Suddenly she said 'Have you killed people?'..I thought it the usual accusation of trigger-happiness. 1943 F. J. Bell *Condition Red* 190 Yes, they missed us, and the G hereby absolves whoever it was along that section of coast that got a little trigger-happy early one December morning. 1946 *Archit. Rev.* CI. 47/1 On the Acropolis itself a group of trigger-happy gendarmerie lounged with an assumed nonchalance by the lower entrance. 1957 *Time* 2 Sept. 19/1 Some trigger-happy U.S. radio commentators..helped confuse it further by proclaiming that Syria was already Russia's newest satellite. 1971 H. Wilson *Labour Govt.* xxxvi. 937 It was fairly clear that the main issues now were relief and the avoidance of atrocities, which, if they occurred, would be most likely to be caused by trigger-happy young soldiers. 1974 F. Warner *Meeting Ends* i. ii. 7 If only you knew how trigger-happy he is when he gets a dialling tone. 1984 *Miami Herald* 6 Apr. 12 A/2 We have a president who is trigger-happy and who commits troops for impossible missions. 1930 *Amer. Mercury* Dec. 458/2 *Trigger man*, an assassin; a body guard. 'He's trigger man for Big Tony.' 1934 *Sun* (Baltimore) 22 Aug. 13/1, I was the triggerman in both hold-ups. 1954 'N. Blake' *Whisper in Gloom* iii. 42 A graceful, self-possessed, cat-like walk..the tread of the trigger-man. 1974 *Times* 2 May 6/5 P[resident]... You feel that really the trigger man was really Colson on this thing? D[ean] No... He was just in the chain. 1977 *Hongkong Standard* 12 Apr. 9/3 He was said to be a senior triggerman—an overseer of 'hit men'—for reputed mob boss Anthony 'Big Tuna' Accardo. 1981 W. Safire in *N.Y. Times Mag.* 1 Mar. 9/3 The triggerman of this slim but explosive volume is described on the cover as 'formerly professor of Romance languages and literatures at Harvard University, where he is now emeritus professor'. 1952 *N.Y. Times* (Late City Ed.) 12 Aug. 32/6 The federation estimates that at the beginning of suspension the average of current prices was 69.39 per cent of their 1951 highs, which meant that they could rise a little more than 15 per cent before reaching the so-called 'trigger-point'. 1979 H. Kissinger *White House Years* ix. 330, I had to learn an entire vocabulary of international trade, such as 'export subsidy techniques' as well as the arcane complexity of 'trigger points' (at which restraints would go into effect). 1978 *Business Week* 23 Jan. 25/2 (*heading*) Steel trigger-prices start sowing discord. *Ibid.* 26/3 Jack Meyer, assistant director of the Council on Wage & Price Stability, which devised the trigger-price system. 1981 *Economist* 24 Jan. 88/1 The renewal of trigger prices last October coincided with a recovery in demand for, eg, tubes for the oil and gas industry, plates and girders for process plant and construction. 1924 J. A. Thomson *Sci. Old & New* xix. 105 A current of air is necessary as the trigger-pulling stimulus. 1946 *Radar: Summary Rep. & Harp Project* (U.S. Nat. Defense Comm., Div. 14) 144/2 *Trigger pulse*, a pulse which starts a cycle of operations. 1956 *IRE Trans. Electronic Computers* V. 124/1 For trigger pulses of a few mμsec duration there will be little or no interference between the trigger pulse and the change of voltage at the cathodes of the EFP-60's. 1981 J. D. Lenk *Handbk. Digital Electronics* ii. 84 The circuit changes state only when both the input pulse and a clock pulse are present simultaneously. (The clock pulse is also known as a gate pulse or trigger pulse.) 1927 J. Adams *Errors in School* 213 In external written examinations, where..the clever pupil..is led astray by expecting a question, and then treating one of the questions actually set as the one he expected. These 'trigger questions', as they may be called, set off the candidate on the wrong track. 1973 T. Pynchon *Gravity's Rainbow* (1975) i. 147 Hyperkinetically, waiting only the right trigger-question to start blithering 200 words a minute about their special, terrible endowments. 1952 'J. Wyndham' in *Galaxy Sci. Fiction* July 72/1 He brought the cutter up, and pressed the trigger-switch. 1894 *Electrician* 15 June 188/1 Zehnder's trigger tube. 1939 H. J. Reich *Theory & Applications of Electron Tubes* vii. 208 A single pentode may also be used as a trigger tube. 1978 R. V. Jones *Most Secret War* viii. 69 My first step was to take the electronic trigger tube down to my former colleagues at the Admiralty Research Laboratory, to get them to evaluate its performance. 1975 *Listener* 17 July 74/1 Those of us who work in radio..rely on trigger words, Pavlovian clichés which become a kind of bogus mental shorthand.

trigger (tri·gəɹ), *v.* [f. Trigger *sb.*¹] **1.** *trans.* To act as a 'trigger' (sense 3) for, causing another event (esp. a chain reaction) to occur; to stimulate or 'set off'; to activate, to bring about; to spark *off* (an idea, etc.). Also *lit.*, to pull (depress, etc.) the trigger of (a gun or other device).

1930 R. Campbell *Adamastor* 94 When life is triggered by a hair And stands upon the peak of death. 1938 *Sun* (Baltimore) 18 Apr. 8/5 Denmark, whose people thrive on thrift, milk, bacon and eggs, and never need 'triggering' into activity by shot-in-the-arm spendings of borrowed billions. 1948 *Sat. Even. Post* 20 Mar. 39/3 There is certainly no lack of evidence that the typical glaucoma patient has a nervous temperament and that emotional episodes will increase the pressure within the eyeball and even trigger off acute attacks. 1949 *Sun* (Baltimore) 26 Sept. 4/1 The strike..was triggered by two rival AFL unions. 1950 *Ibid.* 21 July 14/3 A system by which the powers are prepared but lie dormant until triggered into action by specific congressional action is the ideal. **1958** W. J. H. Sprott *Human Groups* 163 May it not be that a crowd is 'triggered off' by people whose 'thresholds of mob-involvement'..are low? **1958** *Spectator* 19 Sept. 379/1 But as a space-veteran who once triggered a ray-gun with Flash Gordon, let me advise you to read on. **1958** *Listener* 23 Oct. 648/2 Artists like Joan Mitchell, Al Leslie..have all been triggered by de Kooning's example. **1959** *Ibid.* 18 June 1083/2 Sir Faithful Fortescue..whose loyalties were so finely triggered that he rode across from Parliament side to Royalist. **1968** J. D. McCawley in Bach & Harms *Universals in Linguistic Theory* 168 Chomsky..in effect asserts that all lexical insertion takes place in the base component unless triggered by other transformations. **1972** *Amateur Photographer* 12 Jan. 42/2 (*caption*) Recently there have been a number of flash meters on the market which, when placed at the subject position read out the correct f/stop to use when the flash is triggered. **1973** C. Bonington *Next Horizon* xiii. 190 We tiptoed up the snow, hardly daring to talk, as if the resonance of our voices might trigger off an avalanche. **1977** *New Yorker* 5 Sept. 80/3 Before I improvise, I just listen, and that triggers me. **1978** S. Sheldon *Bloodline* xlii. 361 His eyes were fixed on the ribbon that the girl was wearing around her neck. It triggered a memory. **1978** *New York* 3 Apr. 10/1 The Israeli invasion of Lebanon and the mass murder that triggered it have nearly obliterated from the public consciousness the killing of Egyptian editor Yousef el-Sebai. **1981** *Times* 5 May 15/7 The fact that no danger signals were triggered during the growth of Norton Warburg has alarmed the City.

2. *Electronics.* **a.** To initiate a change of state or a cycle of behaviour in (a device).

1937 *Proc. Cambr. Philos. Soc.* XXXIII. 551 In order that it may be used in a scale-of-two counter, it is necessary to provide some simple means of triggering the circuit, that is to say changing from one stable state to the other. **1945** *Electronic Engin.* XVII. 473 A differential circuit and pulse generator which triggers a thyratron. **1967** *Electronics* 6 Mar. 160/1 A d-c flip-flop is triggered by the leading edge and clamped until the pulse is removed. **1974** A. Van der Ziel *Introd. Electronics* xi. 262 A monostable multivibrator is a circuit that goes through a complete wave form when triggered.

b. *intr.* Of an electronic device: to change state in response to a momentarily applied signal.

1933 *P.O. Electr. Engineers' Jrnl.* XXVI. 63/2 A tube is now manufactured capable of 'triggering' both 'on' and 'off'. **1967** [see *Schmitt].

Hence **tri·ggered** *ppl. a.*, furnished with or activated by a trigger; **tri·ggering** *vbl. sb.* (freq. *attrib.*) and *ppl. a.*

1944 *Electronic Engin.* XVI. 380 The harmonics generated in the circuit provide standard frequencies..for..high speed triggering etc. **1945** *Electronic Industries* Sept. 226 *Triggered spark gap*, a fixed spark gap in which the discharge passes between two electrodes and is struck (started) by a subsidiary electrode, the trigger, to which low power pulses are applied at regular intervals from a pulse amplifier, thus closing the switch. **1958** K. Amis *I like it Here* xii. 152 The sight of it at this moment must have had some triggering effect on Bowen's alimentary canal. **1962** *Listener* 3 May 770/2 This would involve installing a 'triggering' device in the vehicle. **1967** *Ibid.* 30 Nov. 694/1 The triggered responses which might deter me..no longer apply. **1972** *Language* XLVIII. 299 Identity-of-reference deletions involve two coreferential NP's, a vanishing NP and a triggering NP. I propose that the rule of deletion..superimpose the vanishing NP over the triggering NP, keeping both NP nodes. **1977** J. L. Harper *Population Biol. Plants* xviii. 520 Most of these annuals have a precisely triggered transition from the vegetative to reproductive phase depending on photoperiod. **1977** Savage & Rumbaugh in D. M. Rumbaugh *Language Learning by Chimpanzee* xvi. 289 Hockett..defined communication as an act by which one individual 'triggered' the behavior of another ('triggering' in this sense implies that the energy expended during the output of a communicative pattern is unrelated to the energy of the response).

triggerable (tri·gərăb'l), *a.* [f. *trigger *v.* + -able.] Susceptible to triggering.

1964 *Anesthesiology* XXV. 200/1 You will need both triggerable and automatic respirators. **1973** *New England Jrnl. Med.* 4 Oct. 735/1 If such triggerable foci, in which spontaneous activity can be either initiated or terminated by the appropriately timed arrival of a propagated impulse, exist in depressed areas of the heart they might well serve to initiate or sustain fibrillation.

trigonal, *a.* (*sb.*) Add: **1. c.** *Chem.* Characterized by three orbitals lying in a plane and directed to the corners of an equilateral triangle.

1939 L. Pauling *Nature Chem. Bond* 429/2 (Index), Trigonal bond orbitals. **1961** A. Streitwieser *Molecular Orbital Theory* i. 20 Linear combinations of *s*- and two of the *p*-orbitals result in three trigonal sp^2-orbitals. **1979** B. M. Gimarc *Molecular Struct. & Bonding* iv. 79 These complexes are mainly trigonal bipyramidal..in structure with the bonds to the axial ligands..being slightly longer than those to the equatorial ligands.

trigonally *adv.*, restrict *rare*$^{-0}$ to sense in Dict. and add: (*b*) *Chem.*, in a trigonal manner.

1962 P. J. & B. Durrant *Introd. Adv. Inorganic Chem.* xviii. 590 (*heading*) Compounds in which the carbon atom is trigonally hybridised. **1982** *Nature* 25 Feb. 658/1 The borates, where trigonally and tetrahedrally bonded atoms may occur.

trigonitis (tri-, trəigonəi·tis). *Med.* [f. Trigon(e + -itis.] Inflammation of the trigone of the bladder.

1900 in Dorland *Med. Dict.* **1912** R. Guiteras *Urology* II. lxiii. 508 Gonorrhea of the neck of the bladder, trigonitis, is not uncommon. **1974** J. D. Maynard in R. M. Kirk et al. *Surgery* viii. 178/1 In young women, urethral trauma from intercourse commonly leads to recurrent attacks of urethritis and trigonitis.

trigonometrical, *a.* Add: *trigonometrical point, station*, a reference point on high ground, usu. marked by a small pyramidal structure, used in triangulation.

1860 Trigonometrical station [in Dict.]. **1886** T. P. White *Ordnance Survey of U.K.* iv. 68 In very many cases the old trigonometrical points were again observed to or from. **1949** T. W. Birch *Maps* vi. 56 The *trigonometrical points* or *stations* determined by theodolite are accurately plotted on the plane-table sheet.

trigram. Add: **a.** *spec.* in *Psychol.*, a (nonsense) word of three letters used in the study of learning or memory (see esp. quot. 1960).

1960 E. J. Archer in *Psychol. Monogr.* LXXIV. x. 1/2 This study is a re-evaluation of the meaningfulness of all possible three-letter combinations of the Roman alphabet of the form consonant-vowel-consonant... It is herewith proposed to call these combinations 'trigrams', and in view of their form to further specify them as 'CVC trigrams'. **1970** *Jrnl. General Psychol.* LXXXIII. 214 The study..assumes that an object elicits a given number of mediating responses with which a CVC trigram can be associated. **1973** *Jrnl. Genetic Psychol.* CXXIII. 15 Correct recognitions of trigrams on test trials.

b. Also *spec.*, a figure traditionally used in the Chinese philosophy of the I Ching for the purposes of divination.

1877 J. Legge in *Encycl. Brit.* VI. 263/2 The *Yih King*..the rudiments of which are assigned to Fuh-hsi in the 30th century B.C. Those rudiments, however, are merely the 8 trigrams and 64 diagrams, composed of a whole and a broken line. **1923** J. P. Bruce *Chu Hsi & his Masters* vi. 127 The Grand Terminus..produced the two elementary Forms. These two Forms produced the four Symbols, which, again, produced the eight Trigrams. The eight Trigrams served to determine the good and evil issues of events. **1957, 1965** [see *I Ching]. **1974** *Sci. Amer.* Jan. 108/2 There are two ancient ways of displaying the eight trigrams in a circle.

trigrammatic, *a.* For quot. 1834 substitute:

1824 T. Young in *Suppl. Encycl. Brit.* IV. 43/1 The 'trilinguar' or rather trigrammatic Stone of Rosetta.

trihybrid (trəihəi·brid), *a.* [f. Tri- + Hybrid *sb.* and *a.*] **1.** *Genetics.* Of or pertaining to a hybrid that is heterozygous with respect to three independent genes.

1918 *Genetics* III. 591 The undepleted litters in our trihybrid ratio show a suggestive order or arrangement of frequencies.

2. Involving or pertaining to a descent from three different races or types that interbred.

1941 Tindale & Birdsell in *Records S. Austral. Museum* VII. 1. 6 The Australian aboriginal no longer may be considered as a pure race of unusual homogeneity, but a well-blended group of at least dihybrid and probably trihybrid origin. **1969** A. A. Abbie *Original Australians* xi. 215 My own observation on Aborigines in many parts of Australia have not disclosed the physical differences the 'trihybrid theory' demands. **1979** *Nature* 15 Nov. 298/2 In spite of an earlier model postulating a trihybrid origin, a theory favouring a homogeneous colonising stock had gained some acceptance.

trihydrocalcite: see *tri- 5 c.

trihydrol (trəihəi·drǫl). *Chem. Obs.* exc. *Hist.* [f. Tri- + *hydrol.] A supposed trimer of water, $(H_2O)_3$, formerly thought to be present especially in ice.

1900 [see *hydrol 2]. **1932** *Proc. Nat. Acad. Sci.* XVIII. 136 (*heading*) The physiological effect of trihydrol in water. **1957** [see *hydrol 2]. **1972** H. S. Frank in F. Franks *Water* I. xiv. 531 Many of the detailed models..have been overtaken by events and are now of hardly more than historical interest. Perhaps the classic example of such obsolescence is found in the theory put forward in 1900..representing water as a mixture of monohydrol (monomeric, steamlike), dihydrol (dimeric, intrinsically liquid), and trihydrol (trimeric, icelike) molecules.

triiodothyronine: see *tri- 5 c.

tri-jet (trəi·jet). Also **trijet.** [f. Tri- + Jet *sb.*[2]] An aircraft powered by three jet engines. Freq. *attrib.*, as *tri-jet airbus, transport.*

1968 *Economist* 20 Apr. 24/3 Last month Eastern Air Lines and Trans World Airlines jumped in with orders to the Lockheed Aircraft Corporation for 94 of the new-look tri-jet transports. **1969** *New Scientist* 13 Mar. 554/2 The tri-jet air bus to carry 350 is on the way. **1970** *Time* 19 Jan. 44 Originally designed for shorter-range routes than the 747, the trijets are now being offered in stretched intercontinental versions. **1979** *Daily Tel.* 17 Apr. 8/8 Boeing..will be extremely busy introducing two new twin-jet airliners and possibly a tri-jet as well.

trike. Colloq. abbrev. of Tricycle *sb.* and *v.* or (*rarely*) of Tricyclist. Hence **tri·ker**; **tri·king** *vbl. sb.*

1883 *Sporting Life* 22 Apr. 6/5 Londoners call bicyclists 'bikes' and tricyclists 'trikes'. **1884** *Wheel World* Apr. 495/2 The makers of his new trike. **1885** *N. & Q.* 14 Nov. 386/1 Do you *bike* or *trike*? **1890** Barrère & Leland *Dict. Slang* II. 374/1 *Trike*, (common) a tricycle; *triking*, cycling. Do you bike or trike? do you ride a bicycle or a tricycle? **1901** *Pall Mall Gaz.* 15 May 1, I was further gratified with the intimation that the peccant triker had been discovered. **1912** A. S. M. Hutchinson *Happy Warrior* iii. v. 148 I'm all mixed up in this awful trike, you know. **1939** X. Herbert *Capricornia* xiii. 185 There is no fixed time for those who wish to come by horse or trike. Just roll up when you like. **1939–40** *Army & Navy Stores Catal.* 827/1 *Trike*..for ages 3 to 7 years. **1972** *Oxford Times* 4 Feb. 17 An Oxford doctor has condemned the three-wheeled invalid trikes as 'tin coffins'. **1977** *Times* 9 Aug. 5/7 Many disabled people are now turning from the 'to trike or not to trike' argument. **1984** *N.Z. Farmer* 23 Feb. 31/2 (*caption*) Honda farm bikes and trikes remain one of the strongest sellers in Motor Holdings' lineup.

b. A kind of ultralight aircraft.

1981 M. A. Markowski *Ultralight Aircraft* 45 The trike appears to be a viable approach to ultralight flight and allows thousands of hang gliders to be powered, as desired. **1982** *Sci. Amer.* July 63/1 A trike (from tricycle) is a pyramidal tubular frame that holds the engine, the pilot's seat and the landing gear.

Trike, var. *Trique *sb.* and *a.*

trikini (trəikī·ni). Also **tri-kini, Tri-Kini.** [f. Tri-, after *bikini b, with reference to Bi-[2].] Any of various designs of ladies' swimsuit which consist of three main areas of fabric (as pants and a separate covering for each breast).

1967 *Scottish Daily Mail* 7 June 12 Some ingenious fellow has just come up with a Tri-Kini, best described as a handkerchief and two small saucers. The saucers, say the manufacturers, stick on with Velcro, the stuff which fastens at a touch. **1968** *Sunday Truth* (Brisbane) 8 Sept. 1/2 She is wearing the new trikini swimsuit which is to make its debut in Australia next summer. **1970** *Telegraph* (Brisbane) 20 Feb. 15/5 When is a bikini a trikini? When it's a suntanner of a swimsuit like this..deeply decolette [*sic*] and, to put it briefly, even briefer.

trilateral, *a.* Add: **2.** Pertaining to or concerning three countries, parties, etc., esp. with reference to the relations between Europe, the United States, and Japan; *the Trilateral Commission* (see quots. 1973, 1981).

1965 *Economist* 20 Feb. 758/1 Ethiopia. A trilateral agreement is being negotiated with the object of harnessing German and American help in forming an Ethiopian frontier defence force. **1973** *N.Y. Times* 2 Mar. 35/1 A group of distinguished private citizens in the United States, Japan and Europe are now organizing what they call 'the trilateral commission' to study the common problems of these three power centers. **1977** *Times* 28 Apr. 7/4 Mr Jimmy Carter..really believes in what has been called the trilateral system of America, Europe and Japan. **1981** *Washington Post* 18 Mar. A1/2 For the record, the Trilateral Commission is a New York-based policy group formed by Rockefeller in 1973 in a reaction to the nationalistic foreign economic policy of President Nixon, which disturbed some traditional U.S. allies. **1984** *Times* 5 Mar. 12/2 It was thanks to the complicated trilateral arrangement worked out in secret..that he was taken to Pakistan.

Hence **trila·teralism**; **trila·teralist.**

1976 R. H. Ullman in *Foreign Affairs* Oct. 1 'Trilateralism'..is the latest attempt both to describe and to prescribe for the relationship between the United States and the other principal democratic, industrialized, market-economy states. **1977** *Studia Diplomatica* XXX. 231 Dr Kissinger can hardly be said to have been unaware of the 'Trilateralists'. **1977** *Foreign Policy* No. 29. 90 The original trilateralism, which links prominent citizens of the United States, Western Europe, and Japan in an international 'commission', arises from the shared ideological commitment of its members to liberal values and politics. The goals of Arab trilateralism, however, are not expressed in ideological terms. **1981** *Washington Post* 18 Mar. A1/2 Remember those dreaded, three-sided Trilateralists, the international conspirators headed by David Rockefeller who were going to take over the world? *Ibid.* A5/6 But old fears die hard, and the far right is ever-vigilant for signs of creeping Trilateralism within the citadel.

trilateration (trəilætĕrē¹·ʃən). [f. Tri- + L. *later-, latus* side + -ation.] A method of surveying analogous to triangulation in which each triangle is determined by the measurement of all three sides.

1948 *Bull. Géodésique* No. 10. 342 A network of triangles may be computed and adjusted from observed lengths of sides. The development of error in trilateration is different from that in triangulation..as the side error is not accumulative. **1952** G. Bomford *Geodesy* i. 34 It might be possible to use radar trilateration between ground points instead of ordinary geodetic triangulation. **1960** *Times* 4 Mar. 3/2 Experienced in first order triangulation, trilateration and astro-fixes. **1974** *Physics Bull.* Apr. 136/3 A geodetic network was established to an accuracy of 2 mm over the surface of the Laboratory II site using triangulation and trilateration.

trilby (tri·lbi). Also **Trilby.** Pl. **trilbies** or **trilbys.** [The title of a novel by George du Maurier published in 1894, and the name of its heroine.] **1.** *colloq.* **a.** A jocular name for the foot (with reference to Trilby's feet, which were objects of admiration). ? *Obs.*

1895 *People* 7 July, An American paper has spent its energy of psychological investigation on the foot (I beg

pardon, the trilby). **1907** H. E. Dudeney *Canterbury Puzzles* 114 'Two feet—' he murmured. 'Somebody's Trilbys?' I inquired. **1932** U. Sinclair *Candid Reminiscences* i. v. 29 There was a book by the name of 'Trilby', which the ladies blushed to hear spoken of... I knew it had something to do with feet, because thereafter my father always called them 'trilbies'.

† **b.** A particular type of shoe. (Formerly a proprietary name in the U.S.) *Obs.*

1895 *Official Gaz.* (U.S. Patent Office) 16 Apr. 447/1 Boots, shoes and lasts. S. Weil & Co., New York... 'Trilby'. Essential feature—the word 'Trilby'. Used since October 1, 1894. **1895** *Montgomery Ward Catal.* Spring & Summer 509/3 The Trilby... The very latest in ladies' footwear. **1897** *Sears, Roebuck Catal.* 192/1 Our New Trilby... The accompanying cut is an exact reproduction of our new Trilby Shoe.

2. In full *trilby hat*: a soft felt hat, esp. one of the Homburg type with a narrow brim and indented crown; any hat of a similar shape.

[**1895** *Bradford Daily Argus* 12 Nov. 1/8, I have been puzzling my head to account for the reason of so many soft hats being worn at present, and at last I have hit it. It is another phase of the 'Trilby' complaint. In one of the illustrations of the book Little Billee is 'discovered' wearing a hat of this description, so it has been seized upon by those worshippers at the shrine of Trilby, whom nature will not assist in the cultivation of a Svenjali [*sic*] beard or Taffy whiskers.] **1897** *Daily News* 6 Feb. 6/5 In the struggle Mr. Bennett lost his hat, a black 'Trilby'... Mr. Carr..was also wearing a black 'Trilby' hat. **1915** E. Wallace *Man who bought London* iii. 33 He turned sharply to the young man in the trilby hat. **1927** W. Deeping *Kitty* v. 63 The window-sill being a low one Mrs Sarah had a good view of Vernor Street, and the passing of the khaki caps, bowlers, 'Trilbys' and cloth caps. **1958** *Observer* 9 Feb. 11/5 Three men in trilbies and a woman with straggly hair. **1978** *Lancashire Life* Apr. 67/1 Fred was a dapper dresser, and I like to remember him in Summer, wearing a medium-grey lightweight suit, black patent shoes, a beautifully-arranged lilac-coloured pure silk cravat with a massive diamond stick-pin, and a natural straw Trilby. **1979** 'J. le Carré' *Smiley's People* (1980) xxi. 258 Toby..handed him a trilby hat.

Hence **tri·lbied, tri·lby-hatted** *adjs.*, wearing a trilby hat.

1966 *Punch* 12 Jan. 45/2 'Tell him about Indonesia, Dave,' cried a small, trilbied man. **1975** M. Simpson *Chrome Connection* ii. 25 The trilby-hatted husband..gave her a curious look.

Trilene (trəi·līn). *Pharm.* Also **trilene.** [f. *tri(chloroethy)lene.] A proprietary name for a medicinal grade of trichloroethylene, used as an analgesic and light anæsthetic.

1935 *Trade Marks Jrnl.* 13 Feb. 184/1 *Trilene*... Medicated preparations of trichlorethylene for human use. Imperial Chemical Industries Limited,..Millbank, London. *Ibid.* 20 Feb. 214/1 *Trilene*... Trichlorethylene for veterinary and sanitary purposes. Imperial Chemical Industries Limited. **1941** *Brit. Med. Jrnl.* 21 June 925/1 We..found that this firm was already manufacturing, under the trade name of 'trilene', a purified form of the substance specially prepared for medical purposes, though with no reference to inhalation or general anaesthesia. **1944** *Official Gaz.* (U.S. Patent Office) 18 Jan. 349/1 *Trilene* for anaesthetic for cleansing wounds and for the alleviation of pain. **1966** *New Statesman* 8 July 47/3 Where we had practised my wife was completely able to use the breathing for her own purposes, even under the sleepy drug. She never used her Trilene machine at all. **1971** H. Pacy *Road Accidents* ii. 51 Trilene inhalers..provide mild analgesia and are used by the casualties themselves. **1979** G. Bourne *Pregnancy* (rev. ed.) xxiv. 365 The indications and the technique for Trilene analgesia are essentially similar to those described for nitrous oxide and oxygen analgesia.

tri-level, *a.* and *sb.* *N. Amer.* Also **trilevel.** [f. Tri- + Level *sb.*] **A.** *adj.* Having or consisting of three levels; (of a house, etc.) having three storeys or floors on three levels. Cf. *split-level *a.* **B.** *sb.* A tri-level house.

1960 V. Packard *Waste Makers* i. 8 Workingmen today..want their tri-level house in the suburbs. **1965** P. Wylie *They both were Naked* i. ii. 55 Coming down the staircase in Will's suburban home, a tri-level, if there's such a term. **1973** 'D. Shannon' *No Holiday for Crime* iii. 34 Pershing Square with the tri-level parking lot under it. **1974** *Car & Locomotive Cycl.* (ed. 3) S1-25/1 *Tri-level car*, a flat car designed with integral superstructure of posts, bracings and decking to permit triple level loading of automobiles. **1976** *Casper* (Wyoming) *Star-Tribune* 29 June 16/5 (Advt.), Stretch out in over 2000 sq. ft. of tri-level located on 1.8 acres. **1979** *Arizona Daily Star* 5 Aug. (Advt. Section) 16/7 This tri-level showplace home is ideal for entertaining and family living. **1979** *United States 1980/81* (Penguin Travel Guides) 432 Today this tri-level arcade features chic boutiques and restaurants.

trilingualism (trəili·ŋgwăliz'm). [f. Trilingual *a.* + -ism.] The ability to speak three languages; the use of three languages.

1934 H. V. Morton *In Steps of Master* i. 5 And as you go into Jerusalem, glancing at the trilingualism everywhere, the words of St. John come into the mind. **1956** *Nature* 18 Feb. 343/2 If we take into consideration languages read as well as written, no language group can to-day avoid even trilingualism if it wishes to attain the highest standards of culture and scholarship. **1976** *Word 1971* XXVII. 388 The scope of this article constrains us from adding still a third dimension, where more than a language pair is involved, such as the Yaqui-English-Spanish trilingualism of Pascua, near Tucson, Arizona.

trilinguar, *a.* **b.** (Earlier example.)

1824 [see *trigrammatic *a.*]

triliterality. (Earlier example.)

1864 *Ann. Rep. Smithsonian Inst. 1863* 109 Their [*sc.* the Semitic languages'] most fundamental peculiarity is the triliterality of their roots, every Semitic verbal root containing just three consonants.

trillet. (Earlier *lit.* example.)

1867 G. Meredith *Vittoria* I. ii. 25 We require to be refreshed with quavers and crescendos and trillets.

trillion. Add: The sense 'a thousand "billions", or 10^{12}' is now standard in the U.S. and is increasingly common in British usage. (Further examples.)

1919 *Evening Star* (Washington, D.C.) 12 Mar. 2/7 The consideration of reparations has introduced the word 'trillion' in recognizing money... In estimating the war losses of all the powers the first figures of one of the great powers aggregated a trillion francs and those of another power were slightly above a half trillion francs, namely, six hundred billion francs. **1971** *Daily Tel.* 30 Jan. 13 For the current calendar year the 'magic trillion' will be with us. The [American] budget assumed a 9 p.c. growth in GNP this year to $1,065,000 million. **1975** *New Yorker* 21 Apr. 48/1 About two trillion dollars' worth of insurance alone is currently in force. **1980** *Guardian Weekly* 30 Nov. 8/1 Plans to devote a trillion dollars to US military spending over the next five years. **1982** *Nature* 1 July 9/1 Compositions of many important trace species are in the parts per trillion (10^{-12}) by volume range.

trilobal, *a.* Delete *rare*$^{-1}$ and add: *spec.* applied to (man-made fibres having) a cross-section of this form.

1961 *Times Rev. Industry* Mar. 33/3 Bri-nylon 61..owes its softness of handle and subdued, silky lustre to the trilobal cross-section of the individual filaments. **1965** *Guardian* 31 Mar. 13/4 In America textured yarns of heavier denier tri-lobal..nylon have made a big hit in dress jersey. **1979** *Men's Wear* 24 May 34/1 Tri-lobal polyester and viscose do much the same.

trilobe (trəi·lōᵘb), *a. rare.* [f. Tri- + Lobe.] Having three lobes, trilobate.

1931 *Antiquity* V. 272 One is reminded of the trilobe cranial amulet found in a La Tène context. **1950** A. L. Rowse *Expansion Elizabethan England* i. 36 Pendennis and St. Mawes castles,..the latter a perfect..specimen of a Henrician fort of trilobe design, both planned by a German military engineer.

trim, *sb.* Add: **I.** For 'Nautical' read 'Nautical and Aeronautical'. **2. f.** The position of a submarine with respect to the angle between its longitudinal axis and the horizontal.

1917 *Chambers's Jrnl.* Aug. 557/2 When his boat was diving he had to be careful how he changed his position; otherwise the 'trim' was in danger of being upset. **1935** *Sun* (Baltimore) 14 Feb. 8/2 All ballast and emergency fuel dump tanks in the after part of the ship were dropped, and gas was valved from the forward gas cells in an effort to regain the trim. **1942** *Gen* 1 Aug. 3/1 Once out of the harbour the submarine dives for trim and, having caught the trim, she surfaces again. **1974** M. Hebden *Pride of Dolphins* ii. ix. 186 Navigation..is what you will be chiefly responsible for. Navigation and trim. *Ibid.* iii. ii. 224 Addams did a trim dive before he left.

2*. The position of an aircraft with respect to the angle between its longitudinal axis and the horizontal; the condition of static balance of the aerodynamic forces on an aircraft in straight flight; a device or action used to maintain such balance. Freq. *attrib.*

1919 W. B. Faraday *Gloss. Aeronaut. Terms* 17 *Trim*, the inclination to the horizontal of the longitudinal axis when the aerostat is floating freely at rest. **1935** C. G. Burge *Compl. Bk. Aviation* 242/2 A trim of this kind may also be useful on a very powerful fighting aeroplane of small span. **1944** *Times* 3 Apr. 2/4 With the elevator trims gone, the Lancaster was tending to climb all the time. **1962** J. Glenn in *Into Orbit* 12 The enemy shell had knocked out part of my trim controls. **1968** M. Woodhouse *Rock Baby* xxvii. 250 Yancy corrected trim. We flew another mile. **1977** D. Beaty *Excellency* vi. 78 He studied the load and trim sheet. *Ibid.* 81 He took his right hand off the stick, grabbed the trim wheel to push it forward... He let go of the trim, brought his right hand back. *Ibid.* 82 With the aircraft so badly out of trim, the autopilot refused to cope. **1982** J. Savarin *Water Hole* 173 He found the trim to his liking, switched on the auto pilot and relaxed.

II. 3. d. (Later examples.)

1931 G. A. Foan *Art & Craft of Hairdressing* iii. 134/1 It must be realized, however, that this is a shingle *trim*, that the work must be done more lightly, and that..less hair will be removed. **1955** H. D. Steiner *Crowning Glory* iii. 30 The main business of the hair dresser is to see that both the trim and set accord with the natural convolutions of the hair. **1977** D. Bennett *Jigsaw Man* viii. 141 He left with two wigs and an appointment to come back for a trim. **1983** *Chicago Sun-Times* 15 Nov. 41/1 (*heading*) Too few trims in new budget.

e. Delete † *Obs.* (Later examples.)

1948 H. Pepin *Fund. Apparel Design* vi. 137/2 This asymmetrically balanced motif proved suitable for border trim on sleeve. **1964** [see *beading *vbl. sb.* 2 b]. **1982** W. Boyd *Ice-Cream War* i. i. 7 The women all wore white dresses with lacy trims and carried parasols.

g. *U.S.* A shop-window display.

1899 *Harman's Jrnl.* Feb. 7/1 The design for a [men's] furnishing trim..by Harry Harold of Milwaukee, Wis., a window trimmer, is a very clever arrangement. **1926** *Publishers' Weekly* 30 Jan. 328/1 A large red ribbon rosette, from which radiated white satin ribbons to a number of stands at each side of the trim. *Ibid.* 10 July 119/2 When Stone's trim was removed, Wheatly did his 'stuff'. **1945** J. Bradford *Retail Merchandiser's Handbk.* xi. 62 Try to liven up a window as much as possible by using new, bright, and flashy display trims.

h. Ornamental additions or finishings to a vehicle, piece of furniture, or other article; *spec.* the upholstery or interior lining of a motor car. Cf. *trimming *vbl. sb.* 2 c. orig. *U.S.*

1922 *Automobile Trimmer & Painter* Aug. 50/1 The Franklin trim is designed to give a maximum degree of resiliency and durability in order to match these qualities in the rest of the car. **1936** C. W. Seager *Upholstered Furniture* vii. 54 Trim serves the double purpose of concealing the raw edges and seams and supplying a decorative note. **1950** Pomeroy & Walkerley *Motor Year Bk.* 23 Throughout the range is an attractive style of trim incorporating contrasting piping round the edges of the seats. **1957** *Practical Wireless* XXXIII. 532/1 This model costs 98 guineas, and the bow-fronted cabinet is veneered in walnut with gilt trim. **1961** *B.S.I. News* Aug. 26/2 We have not had a single exhaust trim or exhaustor returned with a plating fault. **1962** *Which? Car Suppl.* Oct. 144/1 Leather upholstery and walnut trim. **1969** *Sears Catal.* Spring/Summer 6/2 Matching double handle handbag... Contrasting bar lightly touched with gleaming gold-color metal trim. **1971** 'D. Halliday' *Dolly & Doctor Bird* vi. 85 The sofa..was one of a facing pair in oatmeal with hide trim. **1977** *Time* 4 July 6/3 But the new Soviet President let it be known that he was not pleased with the color of the trim on the wagon's seats.

i. *Cinemat.* A piece of film cut out during editing; *spec.* a very short piece cut out during the final stage of editing. orig. *U.S.*

1934 in Webster. **1948** R. Spottiswoode *Basic Film Techniques* iii. 31/2 If he has to unwind rolls of *trims* (or cut-out sections of shots), he will let them hang on the bins. **1959** W. S. Sharps *Dict. Cinematogr.* 116/1 *Outs*, otherwise *Trims*, the material that is not included in the final edited version of a film. **1964** *Listener* 28 May 899/3 Theseus-Adonis and the Minotaur were separated by a film about Montreal made completely out of 'trims' from taped interviews and very rapid shots exemplifying the culture of cities. **1976** *Broadcast* 23 Aug. 6/2 The shop committee..will consider releasing the trims once it has seen all the documentation. *Ibid.* 6/3 Thames..could also satisfy the union curiosity about the sheer quantity of trims—60 cans.

6. *U.S. slang.* A woman; sexual intercourse with a woman.

1955 *Amer. Speech* XXX. 302 *Chick, crazy freak, local talent, neat job, snatch, talent, trim, unfair sex, n.*, girl, usually pretty. Often used to refer to a woman of loose morals. [**1961** Rigney & Smith *Real Bohemia* p. xvii, *Trim*, cunnilingus.] **1962** E. Lacy *Freeloaders* vi. 125 The broad isn't worth it, no trim is. **1974** H. L. Foster *Ribbin', Jivin', & Playin' Dozens* v. 191 Female student: 'Somebody always askin for some trim and haven't even got anything.'

7. Special Comb.: **trim tab**, (*a*) *Aeronaut.* = *trimming tab* s.v. *trimming *vbl. sb.* 7 b; (*b*) *Naut.*, a hinged tab fitted to the trailing edge of a keel or rudder to facilitate steering.

1944 H. F. Gregory *Anything a Horse can Do* 47 Elevator trim tab on an airplane—trimmed the craft so that when the hands were off the stick, the aircraft had no tendency to nose down or up. **1958** 'Castle' & 'Hailey' *Flight into Danger* vii. 96 The speed slowly dropped. At 160 George adjusted the trim tabs. **1977** *Encycl. Aviation* 188/3 Most airplanes have hinged trim tabs whose incidence is controlled from the cockpit. **1978** *Detroit Free Press* 5 Mar. C21/5 (Advt.), 1977 Sea Ray 24' 233 Merc cruiser..trim tabs—hydraulic.

trim, *v.* Add: **II. 9. b.** *fig.* or in fig. context. To cheat (a person) out of money; to 'fleece'. *slang.*

1600 [in Dict., sense 9 *fig.*]. **1604** Dekker *Newes from Graves-End* Ep. Ded. sig. C, Thou wouldst neuer haue gone to any Barbers in London..but haue bin trimd only there, for they are the true shauers, they haue the right Neapolitan polling. **1917** D. G. Phillips *Susan Lenox* I. vi. 95 Guileful women, bent on trimming him for anything from a piece of plated jewelry to a saucer of ice cream. **1926** *Flynn's* 16 Jan. 639/1, I had simply trimmed a sucker for a few kale seeds. **1940** Wodehouse *Eggs, Beans & Crumpets* 155 Hearing her elders discuss..some burgeoning scheme for trimming the investors. **1955** *Publ. Amer. Dial. Soc.* xxiv. 94 Some of the big circuses carried their own *whiz mobs* to trim the crowds along the way. **1962** J. Ludwig in R. Weaver *Canad. Short Stories* (1968) 2nd Ser. 254 Didn't she know he was going to get trimmed? But what did she care about money by that time?

c. *fig.* To reduce the size, amount, or number of; to eliminate (wasteful expenditure); to reduce the profits of. Also *absol.* orig. *U.S.*

1966 *Wall St. Jrnl.* 21 Nov. 2/2 General Motors Corp. will reduce previously scheduled overtime next month and trim its daily car production pace 3.7%. **1970** *Globe & Mail* (Toronto) 26 Sept. 33/2 If you really want to trim expenses, you can rent a camper. **1976** *National Observer* 13 Nov. 12/2 There is the argument over dollar levels of U.S. spending and what waste might be trimmed from the budget as well as what new items should be included. **1979** *Daily Tel.* 8 Nov. 21/1 Readicut trimmed. Readicut International reveals a 32·5 p.c. drop in interim pre-tax profits. **1981** *Times* 17 Apr. 1/4 British forces on the Continent are likely to be trimmed. **1982** *Times* 9 Jan. 17 (*heading*) American Telephone trims to compete.

10. Also, to defeat.

1927 S. Lewis *Elmer Gantry* iii. 40 No, gee, Judson, I guess you got me trimmed! **1950** Wodehouse *Nothing Serious* 152 Surely..Rodney can trim a man with hay fever? **1977** *Chicago Tribune* 2 Oct. iii. 16/9 Grand Valley of Michigan piled up 324 yards rushing and 90 yards passing to trim Northeastern Illinois 34–12.

11. b. (Earlier example.)

1783 G. Cartwright *Jrnl.* 8 Oct. (1792) III. 29 As those birds [*sc.* eider-ducks] trim the shore along in the flight-times.

13*. *Aeronaut.* **a.** To maintain or adjust the trim (sense *2*) or inclination of (an aircraft or spacecraft, or part of one). Also *absol.*

1909 *Aero Manual* 40 Some..device is necessary to damp any oscillations that may take place in the line of flight..but hand operation of the steering devices must also be used to 'trim' the machine occasionally. **1921** *Discovery* Apr. 95/2 When trimmed up by the bow, the airship will be found to ride satisfactorily. **1924** *Flight* 13 Mar. 149/2 The pilot trims the tail by operating a hand pump..to increase or decrease its angle of incidence. **1942** *Tee Emm* (Air Ministry) II. 85 If he trims it to fly at a certain speed and power, it should stay at that speed despite bumps. **1958** 'Castle' & 'Hailey' *Flight into Danger* vii. 103 Put full flap on, bring your airspeed back to 110 knots and trim to hold you steady. **1976** B. Lecomber *Dead Weight* i. 12, I ease the throttles back a fraction and trim the nose a touch lower. **1978** *Nature* 5 Oct. 415/1 The spacecraft gyros were trimmed to the Ganymede celestial motion and Io was subsequently manoeuvered into the large slot.

b. *intr.* (for *refl.* or *pass.*) of an aircraft.

1921 *Techn. Rep. Advisory Comm. Aeronautics 1917–18* III. 1023 It should be noted that the ability to trim at high speeds is the one essential point of difference between a seaplane and a racing motor boat. **1923** *Flight* 31 May 295/2 In order..that the machine would trim correctly..the pilot was placed ahead of the wing, in which position he balances the rest of the machine.

14. c. Freq. in phr. *to trim one's sails to the wind*, to adapt oneself to circumstances.

1928 L. Strachey *Elizabeth & Essex* viii. 112 Burghley, trimming his sails to the changing wind, thought it advisable..to take the side of Essex in the matter of the Spanish ransoms. **1934** J. E. Neale *Queen Elizabeth I* xiv. 229 She preferred to trim the country's sails to the winds when and how they blew, rather than set them at once for a storm that might not come. **1940** F. L. Allen *Since Yesterday* x. 275 *Fortune*..trimmed its sails so skillfully to the winds of conservatism that it not only became a mine of factual material for future historians but subtly broadened reactionary minds.

trimaran (trəi·mărӕn). [f. Tri- + Cata)maran *sb.*] A boat with a central hull and a float on each side.

1949 *Sun* (Baltimore) 27 Sept. 22/3 A trimaran showed running speed but little drive to windward. **1959** *Engineering* 16 Jan. 86/2 The trimaran configuration, that is a main hull and two outriggers. **1966** T. Pynchon *Crying of Lot 49* iii. 57 Metzger handed Oedipa aboard the about-to-be-hijacked vessel, a 17-foot aluminum trimaran. **1973** *Guardian* 18 May 18/1 [He] sailed out from England to New Zealand three years ago in a 22-foot trimaran, built and designed by himself.

trimeprazine (trəime·prăzīn). *Pharm.* [prob. f. Trime(thyl + Pr(opyl + *pheno-thi)azine.] A phenothiazine derivative that is used for its sedative and antihistaminic properties (usu. in the form of the tartrate); dimethyl(2-methyl-3-phenothiazin-10-ylpropyl)amine, $C_{18}H_{22}N_2S$.

1959 *Canad. Med. Assoc. Jrnl.* LXXX. 125/1 (*heading*) A new drug for control of itching—trimeprazine. **1977** *Proc. R. Soc. Med.* LXX. 627/1 Mild sedation, e.g. trimeprazine tartrate, is usually required for a few days until the child becomes used to the apparatus.

trimer (trəi·məɹ). *Chem.* [f. Tri- + *-mer.] A compound whose molecule is composed of three molecules of monomer.

[**1929** *Chem. Abstr.* XXIII. 3213 The supernatant liquid..yielded 180, 20, 11, and 4g. of the di-, tri-, tetra- and pentamer, resp.] **1939** [see *tetramer]. **1952** *Jrnl. Physical Chem.* LVI. 85 A reaction between 3 molecules of insulin of molecular weight 12,000..to give a 'trimer' of molecular weight 36,000..is postulated. **1969** W. R. R. Park *Plastics Film Technol.* viii. 195 Even polystyrene films..can cause problems if they contain even relatively low percentages of residual monomers, dimers, or trimers. **1974** M. C. Gerald *Pharmacol.* xi. 198 This compound [*sc.* paraldehyde] is a trimer of acetaldehyde.., that is, three molecules linked in a ring.

Hence **trime·ric** *a.*; **tri:meriza·tion**, the formation of a trimer from smaller molecules.

1938 *Jrnl. Chem. Soc.* 11 Polymerisation of βγ-dimethylbutadiene by sulphuric acid dissolved in acetic acid gives dimeric, trimeric and other low-molecular products. **1954** *Adv. Protein Chem.* IX. 419 A similar free energy change might be expected on dimerization and trimerization. **1972** A. A. R. Sayigh et al. in Stille & Campbell *Condensation Monomers* v. 382 The value of these catalysts is limited because the same substances can act as catalysts for the trimerization of isocyanates. **1975** *Nature* 18 Dec. 576/2 The functional significance of the trimeric nature of the molecule is obscure.

trimester. Add: *spec.* each of three such periods into which human gestation is divided.

1916 G. P. Shears *Obstetrics* ii. 27 Most writers divide pregnancy into three periods of three months each, three trimesters, as they are often called. **1938** A. L. Mudaliar *Clin. Obstetrics* vi. 33 The signs and symptoms vary with the different periods of pregnancy, and we shall classify them into..the first, second and third trimesters of pregnancy. **1980** Gates & Meckel *Newborn Beauty* (1981) iii. 79 During the first trimester you may feel nauseated, fatigued, and lose your appetite.

trimethoprim (trəime·þoprim). *Pharm.* [f. Trimeth(yl + O(xy- + *p(y)rim(idine, constituent parts of the systematic name (see quot. 1962).] An antibiotic, $C_{14}H_{18}N_4O_3$, that is usually given in conjunction with a sulphonamide, esp. in the treatment of malaria and of respiratory and urinary infections.

1962 *Jrnl. Medicinal & Pharmaceutical Chem.* V. 1107 Compound LXV, 2,4-diamino-5-(trimethoxybenzyl)-pyrimidine (B.W. 56-72, trimethoprim), has been selected for detailed study and clinical trial. **1968** *New Scientist* 17 Oct. 147/1 A tablet of Septrin actually contains two drugs, one old—a sulphonamide—and one 'new'—trimethoprim. **1974** R. M. Kirk et al. *Surgery* ii. 29/1 Trimethoprim acts synergistically with sulphonamides, since they interfere sequentially at two stages in bacterial synthesis of folic acid. **1977** *Lancet* 2 July 4/1 Much of the shigellosis could be successfully treated with ampicillin trihydrate and closely related antibiotics, or with co-trimoxazole (trimethoprim and sulphamethoxazole).

trimetrical, *a.* **b.** (Earlier examples.)

1835 *Chinese Repository* July 105 The Trimetrical Classic. **1853** *North-China Herald* 21 May 168/1 (*heading*) Pamphlets published by the Insurgents. The Trimetrical Classic. [*Note*] Each line containing three words, and each verse four lines.

Trimetrogen (trəime·trogǫn). Also **tri-metrogon.** [f. Tri- + *Metrogon* (see quot. 1944).] Used *attrib.* with reference to a technique in which aerial photographs are taken simultaneously by a camera pointing vertically downwards and two pointing obliquely in opposite directions.

1944 P. G. McCurdy et al. *Man. Photogrammetry* (Amer. Soc. Photogrammetry) xiii. 648 In the United States the tri-lens system of map compilation is now usually designated as the 'Trimetrogon' method. The word 'Metrogon' is the commercial name of the kind of lens generally used, but the method of compilation is not affected by the kind of lens. *Ibid.*, Either the K-17 or K-3B type cameras are used in the trimetrogon assembly. **1952** *Chambers's Jrnl.* Feb. 84/1 The planes..carried the marvellous trimetrogon aerial cameras, developed during the War, which make possible a thousandfold increase of speed and show the landscape in stereoscopic detail. **1969** G. C. Dickinson *Maps & Air Photographs* xv. 244 Something of the best of both worlds is obtained with photographs taken by multiple camera installations, the chief aim here being to reduce cost or time by increasing the area photographed during each run; well-known types are..the American Trimetrogon type where one vertical camera is flanked by two high oblique cameras giving continuous cover from horizon to horizon at each exposure.

trimmed, *ppl. a.* Add: **a.** *trimmed joist*, a joist which is tenoned into a trimmer (sense 4); cf. *trimming-joist* s.v. Trimming *vbl. sb.* 7.

1876 *Encycl. Brit.* IV. Plate facing p. 482, Trimmed Joists. **1945** H. C. Bissill in N. W. Kay *Pract. Carpenter & Joiner* xvii. 361 (*caption*) Pictorial view of floor of a first-floor, with..bridging joists, ..trimmer, trimming joist, and..trimmed joist. **1966** A. T. Collins *Newnes Complete Pract. Woodworking* 96/1 A further use for the stopped housing is to join trimmed floor joists to a trimmer.

b. In Comb. with advbs. and preps. forming adjs., as *trimmed-down, -up.*

1897 Kipling *Capt. Cour.* ix. 198 It's great to have a trimmed-up meal again. **1971** *Wall St. Jrnl.* 22 July W1/3 Last month the White House vetoed a public works measure, but an administration spokesman said there are signs the President might accept a trimmed-down plan.

trimmer. Add: **4.** (Later examples.)

1897 F. C. Moore *How to build Home* iii. 32 'Trimmer'-beams enter the wall on each side of the chimney. *Ibid.* viii. 113 All hearths shall be constructed with trimmer-arches extending 20 inches from the chimney-breast to a 'skew back'. **1953** *Archit. Rev.* CXIV. 364/1 With the exception of the assembly hall and gymnasia, which are on solid load-bearing walls, the structure is carried on reinforced raft foundations with ground trimmer beams at the edges.

6. Also (chiefly *Austral.* and *N.Z.*) a good or impressive thing or person, a 'smasher'.

1878 'Ironbark' *Southerly Busters* 11, I thought thee a regular 'trimmer', I thought thee a generous man. **1943** N. Marsh *Colour Scheme* v. 92 'Running well, isn't she [*sc.* a car]?'..'She's a trimmer.' **1955** [see *lair *v.*]. **1962** [see *sheila]. **1970** *Private Eye* 13 Mar. 16 Jeez Valda Clissold, cripes she was a real little trimmer. **1970** *N.Z. Listener* 12 Oct. 13/5 Dave, you're a trimmer. You'll go places. Take a top seat.

10. *Electronics.* A small adjustable capacitor usu. used for the fine adjustment of a larger capacitor to which it is connected. Also *trimmer capacitor, condenser.*

1930 Moyer & Wostrel *Radio Construction & Repairing* (ed. 2) xviii. 331 'Trimmer' condensers which are out of adjustment..will impair the quality of sound reproduction. **1939**, etc. [see *padder *sb.*[2] 3]. **1944** *Electronic Engin.* XVI. 362/2 This error can easily be compensated by the use of a two gang trimmer condenser across the two tuning condensers. **1965** *Wireless World* Sept. 422/2 The appropriate trimmer is first adjusted to give a low intermediate frequency. **1971** A. Margolis *Mod. Radio Repair Techniques* (1974) vi. 83 Usually, padder capacitors are found only in oscillator circuits. Trimmer capacitors are found in the RF circuits, mixer circuits and oscillator circuits.

11. *Aeronaut.* = *trimming tab* s.v. *trimming *vbl. sb.* 7 b.

1935 *Aircraft Engineering* Dec. 305/1 The effect of the trimmer on the drag of the whole tail must be kept in mind. **1947** A. C. Douglas *Gliding & Advanced Soaring* ii. 48 If the quick release is the one used for winching, usually below the thrust line, the machine may tend to ride high, necessitating a continual forward pressure on the stick. This may be tiring on long tows unless elevator trimmers are fitted. **1964** G. Lyall *Most Dangerous Game* ix. 62 Watching my hands wander round the cockpit, checking and setting... Trimmers to takeoff, throttle nut tensioned.

trimming, *vbl. sb.* Add: **2. b.** Freq. in phr. *all the trimmings.* Also *fig.*

1927 D. Parker in *New Yorker* 29 Oct. 93/1 He was praised, adored, analyzed, bestsold, argued about, and banned in Boston; all the trimmings were accorded him. **1946** M. Dickens *Happy Prisoner* ix. 195 If he..had to start his illness all over again, with transfusions and penicillin and all the trimmings, it would be their fault. **1949** R. Harvey *Curtain Time* 148 Turkey and all the trimmings. **1973** *Times* 25 Apr. 8/2 Those rich romantics or fantasists who still wish to enter matrimony with a gracious splash can hire for a day the ancestors, the family retainers, their own private stately home and all the trimmings, in the centre of London. **1981** 'W. Haggard' *Money Men* iii. 35 He had a passion for silverside, all the trimmings—carrots and onions and small sweet dumplings. **1984** *Times* 6 Apr. 21/2 Next week he unveils his..personal computer which is said to have all the trimmings but with a no-frills price.

c. The upholstery or interior lining of a motor vehicle. Cf. *trim *sb.* 3 h.

1938 *Motor* 8 Nov. 2/1 (Advt.), 1937, colour scheme green, suitable leather trimming, comprehensively equipped. **1948** *Ibid.* 3 Nov. 385/1 The windscreen is slightly vee'd. Interior trimming is grey leather and the price is £3,775.

7. b. trimming flap = *trimming tab* below; **trimming gear** *Aeronaut.*, apparatus for altering the angle of the tailplane of an aircraft; **trimming plane** *Aeronaut.*, a control surface used to trim an aircraft; **trimming tab** *Aeronaut.*, an adjustable tab or aerofoil attached to a control surface, and used to trim the aircraft; *spec.* one which can be adjusted by the pilot in flight; **trimming wheel** *Aeronaut.*, a control wheel used to trim an aircraft by its action on the tailplane.

1935 *Jrnl. R. Aeronaut. Soc.* XXXIX. 1038 The ailerons and elevators require trimming flaps sufficient to enable the airplane to be flown 'hands off'. **1922** *Flight* XIV. 416/1 The tail plane, for which a very substantial worm trimming gear is provided, has spars of straight taper, spindled out to an I section. **1933** *Aeroplane* 5 Apr. 606/1 There was some mechanical stiffness in the ailerons and no tail trimming gear had yet been rigged. **1921** *Flight* XIII. 325/2 A trimming plane is installed just ahead of the tail plane, which can be adjusted in flight from the pilot cabin by a wheel. **1935** *Aircraft Engineering* Dec. 297 The trimming tab is quite revolutionary in its simplicity. **1942** W. Simpson *One of our Pilots is Safe* ii. 42 To twist the trimming tab controls for rudder and elevator. **1950** *Engineering* 29 Sept. 269/3 All the control surfaces are provided with trimming tabs. **1941** D. Masters *So Few* 99 The trimming wheel is a device which acts upon the tailplane in order to take the physical strain off the pilot. **1956** D. Barnham *One Man's Window* 153, I start turning the trimming wheel to ease me out of this dive.

trimodal (trəimōᵘ·dăl), *a.* [f. Tri- + Mode *sb.* + -al.] Of a frequency curve or distribution: having three modes (*mode *sb.* 7 c). Of a phenomenon or property: described by such a distribution.

1927 *Jrnl. du Conseil* II. 354 Though the significance of the bimodal or trimodal selection in drift caught fish is open to argument, the fact of its existence is not. **1962** *Lancet* 5 May 968/1 The distribution of the values for 'relative incidence' is not a normal one but bimodal (or possibly even trimodal). **1978** *Nature* 3 Aug. 485/2 The results in Fig. 2 show a trimodal distribution when viability was determined after exposure to a standard dose of γ-radiation.

trimonthly, *a.* **a.** (Earlier example.)

1856 *Trans. Mich. Agric. Soc.* VII. 329 The Steamer Superior..made tri-monthly trips from Buffalo to Detroit.

tri·mo:tor. Also **tri-motor.** [f. Tri- + Motor *sb.*] An aeroplane fitted with three engines. Cf. *tri-jet.

1923 *Glasgow Herald* 1 Jan. 8 The aeroplane which it is proposed to use will be a tri-motor. **1931** *Technol. Rev.* Nov. 66/2 The Ford trimotors are *tin geese*, an amphibian is often a *duck*. **1936** *New Yorker* 14 Mar. 25/1 His favorite machines are a tri-motor Junker and a silver-and black steel Immelmann. **1965** C. D. Eby *Siege of Alcázar* (1966) 13 The pueblo of Getafe..in 1936 boasts the busiest airport in Spain, but there are only a few silent trimotors on the airstrips. **1977** H. Fast *Immigrants* iv. 226 He lay awake most of the night, reliving his experience on the big trimotor plane. **1981** R. Thomas *Mordida Man* xviii. 158 The old gentleman said he had flown in everything from Ford trimotors to 747s.

tri-·mo:tored, *a.* Also **trimotored.** [f. Tri- + Motor *sb.* + -ed[2].] Of an aeroplane: fitted with three engines.

1927 *Times* (Weekly ed.) 9 June 632/3 Tri-motored aeroplanes capable of landing on the water..were essential. **1929** *Sat. Even. Post* 7 Dec. 63/1 The giant tri-motored Ford planes of this line. **1968** Miller & Sawers *Technical Devel. Mod. Aviation* ii. 33 The totals..may suggest the general level of operating costs of the tri-motored aircraft in the United States in 1932–33. **1980** R. Butler *Blood-Red Sun at Noon* (1981) i. ii. 28 One of the big tri-motored Junkers.

Trimphone (tri·mfōᵘn). Also **trimphone.** [f. Trim *a.* + Phone *sb.*[2]] The proprietary name of a type of lightweight telephone with a high-pitched quavering (or 'warbling') ringing tone.

1965 *Times* 11 May 6/7 The new instrument is called a Trimphone and, in the words of the Post Office, it does not ring, it warbles. **1969** *Daily Tel.* 18 Apr. 23/2 The £1 charged is also removed from the cost of installing a Trimphone—the one that 'warbles'. **1973** *Trade Marks Jrnl.* 26 Sept. 1852/2 *Trimphone*..B 986,992. Telephones; telephone receivers; telephone transmitters; telephone dialling apparatus; and parts and fittings..for all the aforesaid..The Post Office. **1973** J. Wainwright *Pride of Pigs* 166 Quince hooked his fingers through the carrying handle of the Trimphone, telephone handset. **1977** J. Gardner *Werewolf Trace* i. 18 The trimphone whistled and Greg picked it up, nodding at the voice in his ear.

‖ **Trimurti** (trimū·rti). Also **Tri-murti** and with small initial. [Skr., f. *tri* three + *-mūrti* consisting or formed of.] **1.** In Hinduism, the gods Brahma, Vishnu, and Shiva, conceived as aspects of one ultimate reality. Also *transf.* Cf. Triad 2 b.

1836 J. F. Davis *Chinese* II. xiii. 107 Their own *Trimurti*, or Triad of *Brahma, Vishnu*, and *Siva.* **1877** Monier Williams *Hinduism* vii. 87 These three gods are the first and highest manifestations of the Eternal Essence, and..constitute the well-known *Tri-mūrti*, or Triad of divine forms which characterises Hindūism. **1895** E. W. Hopkins *Relig. of India* xvi. 447 The votaries of these subsects worshipped some, the rising sun, some the setting sun, while some again worshipped the noonday sun, and others, all three as a *tri-mūrti.* **1921** *Encycl. Relig. & Ethics* XII. 457/2 Serving as it does to reconcile rival monotheisms with one another and with the philosophic doctrine of the absolute, the theory of the Trimūrti presents no such close similarity to the Christian doctrine of the Trinity as to render derivation from Christian influences either necessary or probable. **1954** D. S. Sarma in K. W. Morgan *Relig. of Hindus* 34 The three most important functions of that Spirit, the creation, preservation, and destruction of the world, were associated with the great gods Brahmā, Vishnu, and Siva. Thus arose the doctrine of the Hindu Trinity, the Trimūrti. **1980** M. Gilbert *Death of Favourite Girl* ii. 23 Brahma, the creator, with Vishnu, the preserver and Siva, the destroyer..form the Trimurti—that is, the great Hindu Triad.

2. A statue with three faces representing Brahma, Vishnu, and Shiva.

1877 Monier Williams *Hinduism* vii. 87 There is a well-known Tri-mūrti sculptured out of the rock in the caves of Elephanta, at Bombay. **1920** E. B. Havell *Handbk. Indian Art* iii. i. 203 The Bodhisattva holds in his right hand Vishnu's blue lotus, and his tiara bears the three jewelled sun-disks like the Vishnu in the Trimūrti sculpture of Elephanta. **1953** B. Rowland *Art & Architecture of India* xvii. 188 The treatment of the subjects, ranging from the *Trimurti*..to a panel like the Betrothal of Śiva and Pārvatī..reveal the extraordinary ability and scope of the Dravidian sculptor. **1981** G. Priestland *Priestland's Progress* viii. 123 A statue with three faces, like the sublime Hindu *trimurti* in the cave temples of Elephanta.

Trinidadian (trinidæ·diăn, -dē¹·diăn), *a.* and *sb.* [f. *Trinidad* + -ian; see Trinidado.] **A.** *sb.* A native or inhabitant of Trinidad.

1910 L. O. Inniss (*title*) Trinidad and Trinidadians. **1948** *Trinidad Guardian* 24 June 1/2 (*heading*) Trinidadian faces fraud indictment. **1957** [see the adj. below]. **1966** *Listener* 23 June 923/3 White men in Trinidad had only taught the Trinidadians how to play cricket because there were not enough of their own sort to make a team. **1978** 'A. York' *Tallant for Disaster* ii. 25 A Trinidadian by birth, he was tall and thin.

B. *adj.* Of or pertaining to Trinidad or its inhabitants.

1934 A. Huxley *Beyond Mexique Bay* 20 This appeal for Trinidadian autarchy was warmly applauded. **1957** *Amer. Anthropologist* LIX. 817 Trinidad's social structure..shows these racial and national groups to be arranged in a hierarchy by Trinidadians... A consensus of Trinidadian opinion might arrange these groups as follows. **1971** *Advocate-News* (Barbados) 17 Sept. (Guyana Suppl.) p. iii/1 A Trinidadian sportsman has also been ordered to serve two six-month jail sentences. **1980** *Amer. Speech* LV. 31 Examples are Barbadian../basabasa/ 'fussy' and /kǫŋki/ 'boiled corn pudding', both also Trinidadian and Guyanese.

Trinil (trī·nil). The name of a village in Java, used *attrib.* to designate the fossil remains found there by Eugène Dubois in 1891, esp. those of a hominid, *Homo erectus* or Java man. Cf. *Java *a.*, *Pithecanthropus 2.

1896 *Jrnl. Anthropol. Inst.* XXV. 245 The Trinil skull in form and size very closely resembles the type of the anthropoid apes. **1898** E. Dubois in *Sci. Trans. R. Dublin Soc.* VI. 9 The human form of the Trinil femur is not sufficient to prove that it did not belong to the same individual as the skull-cap. *Ibid.* 12 The Trinil individual, if a human being, ought to have been a microcephalic idiot... The Trinil cranium..very much approaches the type of Anthropoid Apes. **1933** A. S. Romer *Man & Vertebrates* xi. 237 The teeth are essentially human in character but do not definitely settle the status of this Trinil man. **1960** C. Winick *Dict. Anthropol.* 210/1 The Trinil man's teeth are large and his dental arch narrow. **1975** G. H. R. von Koenigswald in R. H. Tuttle *Paleoanthropology* 304 The famous hominid of the Trinil fauna is the classic *Pithecanthropus* or *Homo erectus*. *Ibid.* 305 It [*sc.* a skull] is not directly comparable to the Trinil skull cap.

trinitrotoluene (trəinəi:trotǫ·liu,īn). [f. Trinitro- + *toluene* s.v. Tolu-.] Any of three isomeric nitro derivatives, $CH_3C_6H_2(NO_2)_3$, of toluene, esp. the 2,4,6-isomer, used as a high explosive that is relatively insensitive to shock and can be conveniently melted. Abbrev. *T.N.T.* s.v. *T 6 a. Also † **tri-ni:troto-luol** [*toluol* s.v. Tolu-], in the same sense.

1908 *Chem. Abstr.* II. 459 The trinitrotoluene is preferable because it leaves no poisonous gases after explosion..and can be added to safety explosives. **1910** Trinitrotoluol [see *trotyl]. **1915** Kipling *Fringes of Fleet* 21 We lay doggo in twelve-fathom water With tri-nitro-toluol hogging our run. **1916** *Yorkshire Post* 27 Mar. 5/1 Contact with tri-nitrotoluol may result in the occurrence of troublesome skin affections. **1917** *Causation & Prevention Tri-nitro-toluene (T.N.T.) Poisoning* (Nat. Health Insurance) 9 The material known commercially as T.N.T. is mainly composed of one of the three isomeric tri-nitro-toluenes. **1964** N. G. Clark *Mod. Org. Chem.* xx. 402 All the existing substituents direct to one position in the nucleus, and the product is 2,4,6-trinitrotoluene, the well-known explosive TNT. **1973** 'A. Hall' *Tango Briefing* xviii. 227 It has the equivalent of one hundred tons of trinitrotoluene.

trinity. Add: **2. c.** (Later examples.)

1911 [see *Trinity Sunday*, sense 6 in Dict.]. **1981** Barton & Halliburton in *Believing in Church* iv. 90 From Trinity to Advent a series of *historiae* is appointed [to be read].

5. (Later examples.)

1874 K. H. Digby *Temple of Memory* iii. 39 To Cambridge he is gone..with comrades three, Bound like himself for Trinity. **1933** D. L. Sayers *Murder must Advertise* i. 12 'He was at Trinity. Your Trinity, I mean, not ours.' (Mr. Hankin was a Cambridge man.) **1982** M. Young *Elmhirsts of Dartington* ii. 20 He became a 'Trinity man' expected to assume a definite character..to distinguish him from King's men or Christ's men or John's men.

6. Trinity House (later examples); now chiefly concerned with the licensing of pilots and the erection and maintenance of lighthouses, buoys, and other aids to navigation, on the coasts of England and Wales.

1842 Queen Victoria *Jrnl.* 29 Apr. (1980) 17 The Trinity-House steamer goes with us. **1979** *Bull. Yorks. Dial. Soc.* Summer 9 The station at the time was controlled by the Hull Trinity House.

trinket, *sb.*[1] Add: **4.** *trinket-box* (earlier example).

1814 Jane Austen *Mansfield Park* II. viii. 186 You would be..welcome to any other in my trinket-box.

trinket, *v.*[2] Add: Hence **tri·nketed** *ppl. a.* (*rare*).

1922 Joyce *Ulysses* 433 Her hands passing slowly over her trinketed stomacher.

‖ **Trinkhalle** (tri·ŋkhalə). Also **trinkhalle.** [Ger., lit. 'drinking-hall'.] A place at a spa where medicinal water is dispensed for drinking; a pump-room. Also, an establishment at which (alcoholic) drink is served; a refreshment stall.

1873 G. H. Lewes *Diary* 16 Aug. in *Geo. Eliot Lett.* (1956) V. 427 Drinking seltserwasser at a Trinkhalle. **1971** D. MacKenzie *Sleep is for Rich* vi. 189 Chalice was waiting in the *trinkhalle* in an aura of stale tobacco smoke and beer.

trinocular (trəi-, trinǫ·kiŭlăɹ), *a.* [After Binocular *a.* and *sb.*: see Tri-.] Of a microscope or its body tube: having provision for a camera in addition to eyepieces for both eyes.

1960 *McGraw-Hill Encycl. Sci. & Technol.* VIII. 392/1 Trinocular bodies are binocular bodies with a third tube for a camera. **1970** R. P. Loveland *Photomicrography* I. iii. 121 The trinocular body tube can now be obtained from all of the microscope companies. **1972** *Sci. Amer.* Mar. 65/1 (Advt.), The ED-10 fits virtually any microscope. Whether monocular, binocular or trinocular.

trinomial, *a.* Add: **3.** Of the names of married women (esp. in the U.S.): consisting of three elements, the given, maiden, and husband's surname; also applied to those known by this style, whereby the maiden name is in some measure retained.

1966 *Listener* 2 June 806/3 A vast cast of trinominal cuckoos (Harriot Stanton Blatch, Carrie Chapman Catt, Elizabeth Gurley Flynn, Lucretia Coffin Mott, Rev. Anna Howard Shaw *et aliae*) flap across the stage, but hardly one is characterized by any human identifiable characteristic. **1981** *Economist* 28 Nov. 107/2 Mary Boykin Chesnut—and the reader cannot begin too soon to accustom himself to trinomial appellations.

trinucleotide: see *tri- 1 a.

trio. Add: **3.** *Comb.*, as **trio-sonata** [cf. It. *sonata a tre*], a sonata written in three parts, and often performed on four instruments.

1884 Bell & Fuller-Maitland tr. *Spitta's Johann Sebastian Bach* II. iv. iii. 106 The accompaniment..is arranged by one of the master's best pupils..who did the same thing in a trio-sonata of Bach's. **1934** (*title*) J.-M. Leclair: Trio-Sonata in B flat major for 2 violins, violoncello (ad lib.) & Piano. **1958** *Listener* 11 Dec. 1010/2 Purcell wrote two splendid sets of trio-sonatas. **1978** *Early Music* Oct. 561/1, I have chosen the trio sonata in C major for recorder, flute and basso continuo.

triode (trəi·ōᵘd), *a.* and *sb.* [f. Tri- + *-ode[2].] † **A.** *adj.* *Telegr.* Permitting or involving the transmission of three signals simultaneously. *Obs.*

1886 [see *hexode *a.*].

B. *sb.* *Electronics.* **a.** A thermionic valve having three electrodes (also *triode valve*); also, an analogous semiconductor device with three terminals.

1919 W. H. Eccles in *Electrician* 18 Apr. 475/2 It seems very natural to call a vacuous space containing three electrodes a 'tri-electrode', or, for short, a triode. **1919** *Nature* 30 Oct. 178/2 The internal action of a triode valve. **1923** *Mod. Wireless* June 304/2 The amplifier used consisted of a single Western Electric triode. **1943** C. L. Boltz *Basic Radio* x. 155 The thing which is most striking about the triode is the effect of grid voltage on the anode current. **1948** *Physical Rev.* LXXIV. 230/1 (*heading*) The transistor, a semi-conductor triode. **1962** D. F. Shaw *Introd. Electronics* xi. 220 One of the most important uses of the triode valve is for the amplification of a voltage. **1962** Simpson & Richards *Physical Princ. Junction Transistors* viii. 191 The second type of avalanching triode..is obtained by connecting a base lead to either the n_1 or p_2 sections of the diode in Fig. 8.22. **1974** A. Van der Ziel *Introd. Electronics* ix. 216 The vacuum triode is hardly ever used as a small-signal amplifier except at high temperatures, but it is still the only way to develop very large powers.

f. Comb.: **triode-hexode**, a valve containing a triode and a hexode in a single envelope, with separate anodes but a common cathode; similarly **triode-pentode.**

1937 A. T. Witts *Thermionic Valves Mod. Radio Receivers* viii. 164 (*caption*) Circuit for the triode-hexode. **1952** E. Armitage *Wireless Fundamentals* xviii. 325 Sometimes the beat oscillator valve and the mixer valve are both inside a single glass envelope as in the triode-hexode. **1936** Moyer & Wostrel *Radio Receiving & Television Tubes* (ed. 3) iv. 156 (*caption*) Characteristic curves of triode section of triode-pentode tube.

triose. Add: **2.** Special Comb.: **triose phosphate**, any compound in which a hydrogen atom in the hydroxyl group of a triose is replaced by a phosphate group (PO_3^{2-}).

1934 *Chem. Abstr.* XXVIII. 3746 Triosephosphate was sepd. by fractional pptn. with EtOH and Me_2CO. **1982** R. A. Harris in T. M. Devlin *Textbk. Biochem.* vii. 336 Triose phosphate isomerase then catalyzes the reversible interconversion of dihydroxyacetone phosphate and glyceraldehyde 3-phosphate to complete the splitting stage of glycolysis.

trioxan (trəi,ǫ·ksæn). *Chem.* Also **-ane** (-ē¹n). [f. Tri- + *ox(a- + -ane.] A cyclic trimer of formaldehyde, $(—CH_2·O—)_3$, obtained as colourless, pliant crystals that are combustible and very volatile at room temperature.

1915 P. E. Spielman tr. *V. von Richter's Org. Chem.* I. 205 Hexachlorodimethyl Trioxan. **1919** *Decennial Index Chem. Abstr. 1907–16* 4643/2 *s*-Trioxane. **1964** [see *trioxymethylene* s.v. *trioxy-]. **1978** P. S. & C. A. Bailey *Org. Chem.* x. 268 It [*sc.* formaldehyde] is usually sold and transported as low-molecular-weight polymers, such as 1,3,5-trioxane, formalin, and paraformaldehyde.

trioxy-. Add: **trioxyme·thylene** = *trioxan.

1880 *Jrnl. Chem. Soc.* XXXVIII. 25 The distillate left a white amorphous residue..which proved to be identical with trioxymethylene, $C_3H_6O_3$. **1964** N. G. Clark *Mod. Org. Chem.* x. 188 On distilling aqueous formaldehyde with a trace of sulphuric acid, trioxan ('trioxymethylene') is obtained.

trip, *sb.*[1] Add: **I. 3. g.** *colloq.* *this* (or *that*) *trip*: on this (or that) particular time; on the occasion specified.

1746 C. Knowles *Let.* 19 Sept. in J. S. McLennan *Louisbourg* (1918) x. 174 M. le Duc with all his force shan't have Louisbourg this Trip. **1902** S. E. White *Blazed Trail* 188, I guess I'll let you off this trip. **1906** E. Dyson *Fact'ry 'Ands* ii. 18 Copped out that trip, didn't yeh?

h. *Racing.* The distance from start to finish of a race.

1959 *Times* 1 June 16/7 Dan Cupid is a stocky..colt with..little on public form to prove he can get the trip. **1969** *Australian* 24 May 34/5 Koranui: eighth to Deep Court over 14½f here last Sat. Looks tested at this trip from 24 yd. 1977 *Cork Examiner* 6 June 8/9 The highly fancied Pharly, who beat Crystal Palace three weeks ago in the Prix Lupin at Longchamp over a slightly shorter trip.

4*. *slang* (orig. and chiefly *U.S.*). **a.** A hallucinatory experience induced by a drug, esp. LSD. Cf. *down trip* s.v. *DOWN *a.* 1 e.

1959 N. Mailer *Advts. for Myself* III. 245, I took some mescaline... At the end of a long and private trip which no quick remark should try to describe, the book of *The Deer Park* floated into mind. 1960 J. Gelber *Connection* I. 23 All right, junkies. During our trip we will incorporate an allied art—the motion picture. 1966 *Daily Tel.* 10 Aug. 13/3 The tape-recorder picked up the horrifying moans and shrieks of one man who had made 33 pleasurable 'trips' with LSD and was encountering his first 'freakout,' or bad LSD experience. 1971 *Sci. Amer.* Sept. 240/3 One of the volunteers had a bad trip, entering a panicky and nearly psychotic state. 1975 I. Murdoch *Word Child* 301 You were under the influence. He tried to talk to you... I said you were on a trip.

b. *transf.* and *fig.* An experience, esp. a stimulating one.

1966 *Time* 1 July 50 The Jefferson Airplane flies on weekends at a discothèque in Fillmore Auditorium, where projectors flash quivering, amoeba-like patterns on the walls to induce the dancers 'to take a 'trip'..without drugs'. 1968 L. W. Robinson *Assassin* (1969) xii. 128 Their passion was a long one.., as though they hated to come back, ever, from the rocking, tossing, sweet trip. 1970 *Time* 3 Aug. 32 Part of the message is in the drug argot that he [*sc.* Arthur Blessitt] raps out to his street audiences: 'You don't need no pills. Jes' drop a little Matthew, Mark, Luke, and John. Christ is the ultimate, eternal trip.' 1974 *Melody Maker* 13 July 3/7 The drums are bright shiny cab yellow by the way. It's a trip. 1978 G. Vidal *Kalki* iv. 91 On the other hand, the shop itself was not only exotic, it was a trip, as the addicts say.

c. An activity, attitude, or state of mind, esp. one that is delusory or self-indulgent. Cf. *ego-trip* s.v. *EGO 5.

1967 Wentworth & Flexner *Dict. Amer. Slang* Suppl. 708/2 *Trip*,..any activity, outing, period of time, or way of life. Some beat and student use since c 1965. 1969 [see *BIT *sb.*² 4 i]. 1970 J. Popenoe *Inside Summerhill* 104 It [*sc.* a gang] was a great power trip for the 14-year-old boy who was the leader. 1972 V. Ferdinand in A. Chapman *New Black Voices* 470 It's an unbelievable trip to think that the absence of quality is the cause for the exclusion of Black writers when there is so much garbage being dumped on the heads of people by white publishers. 1974 K. Millet *Flying* III. 282 Hoping is a trip, and it's hopeless anyway. 1977 *New Musical Express* 12 Feb. 8/2 Transcribed onto paper his words may sound like a speech by a musician with delusions of grandeur, but Piazzo ain't into that trip. 1979 R. L. Simon *Peking Duck* xvi. 117, I shouldn't bother—politics was a sixties trip.

III. 8. b. *Nuclear Sci.* (See quots.)

1962 *Gloss. Terms Nucl. Sci.* (*B.S.I.*) 122 *Trip*, a reduction in reactor power initiated by any of the safety circuits of the reactor. 1978 *Times* 1 Feb. 4/7 Reactors suffer occasional unscheduled shut-downs or 'trips' from a wide range of causes, such as fail-safe faults on protective equipment, operator errors and faults in conventional non-nuclear equipment.

IV. 9. trip-bucket, a bucket used for raising water from wells in Arabia, operated by a tripping device and pulled by animals; **tripcock**, a device on a train which applies the brakes when engaged by a projection on the track, if the train is passing a signal set at danger; **tripmeter**, an instrument which may be set to record the distance travelled by a vehicle during each trip; similarly *trip* (*mileage*) *counter*, *trip* (*distance*, *mileage*) *recorder*; **trip slip** (examples); **trip switch** *Electr. Engin.* (see quot. 1924).

1926 T. E. Lawrence *Seven Pillars* (1935) xxxix. 229 He told me of the wheel over the well, with its machinery of leathern trip-buckets, raised by oxen upon an inclined path of hard-trodden earth. 1959 W. Thesiger *Arabian Sands* x. 190 Villagers in the Hadhramaut use camels and oxen to raise the trip-buckets from which they water their cultivation. 1906 *Railway Mag.* Apr. 341/2 These automatic signals have a trigger, which, when the signal is at danger, should engage with a trip cock on the vehicles of the train. 1968 O. S. Nock *Railway Enthusiasts' Encycl.* VII. 274 Co-acting with each stop signal is a train stop, mounted beside the track... When the signal is at danger, the arm is raised. If a train overran a signal..a trip cock lever would strike the raised train stop arm. 1959 *Motor Manual* (ed. 36) vii. 195 Another difference with the trip counter is that it may be set to zero at any time. 1977 *Westworld* (Vancouver, B.C.) May–June 34/2 This gradual phasing-in of the changes means that by 1978 all instrumentation, speedometers, odometers and tripcounters will be entirely converted [to metric]. 1955 *Motor* 7 Dec. 765/1 (*caption*) Trip distance recorder re-setting (twist). 1966 T. Wisdom *High-Performance Driving* xi. 114 A driver..may..exchange his present speedo for a similar one with a trip meter. 1972 O. Sela *Bearer Plot* xxi. 134 Elmer hunched over the bicycling machine... The tripmeter read 5.2 kilometres. 1959 *Motor Manual* (ed. 36) vii. 195 When a trip mileage counter is also fitted this will be constructed on similar lines to the main counter. 1962 *Times* 8 May 16/5 The car is well equipped, and noteworthy points include..a trip mileage recorder, brake servo warning light, twin screenwashers. 1970 *Motoring Which?* Apr. 55/1 All six cars had a speedometer, mileage recorder, trip mileage recorder, fuel gauge. 1966 T. Wisdom *High-Performance Driving* xi. 114 Few speedometers these days are fitted with trip recorders. 1977 *Daily Tel.* 14 Dec. 12/6 Instruments include a rev. counter, speedometer trip recorder, and large clock. 1876 *Scribner's Monthly* Apr. 910/2 The conductor, when he receives a fare, will immediately punch in the presence of the passenger, A Blue trip slip for an 8 cent fare. *a* 1884 Trip-slip [see *bell-punch* s.v. *BELL *sb.*¹ 12]. 1924 S. R. Roget *Dict. Electr. Terms* 275/2 *Trip Switch*, a switch for closing the tripping circuit of a circuit breaker. 1977 *Times* 15 July 8/5 The British system includes a series of trip switches, making it a simple matter for the engineer watching the various loads at any time to isolate a power failure.

Trip (trip), *sb.*⁵ Colloq. abbrev. of TRIPOS 2 d.

1909 R. Brooke *Let.* 18 May (1968) 170, I am a prisoner ..in a room where a hundred and eight damned fools are writing Greek verses for the classical Trip. 1925 W. Deeping *Sorrell & Son* xxiv. 229, I want a first in the Science Trip. 1927 R. Lehmann *Dusty Answer* III. i. 126 Trips. Labs. Lectures. Dons. Vacs. Chaperons. The voices gabbled on.

trip, *v.* Add: **I. 5. b.** *slang* (orig. *U.S.*). To experience hallucinations induced by a drug, esp. LSD. Also with *out*. Also *transf.* Cf. *TRIP *sb.*¹ 4* a.

1966 [see *PSYCHEDELIC *sb.* 2]. 1968 *Globe & Mail* (Toronto) 3 Feb. 25/1 'Trip with us,' coaxes Duke Edwards in a sandy voice... 'Trip with us—without the aid of LSD.' 1969 *Daily Tel.* 4 Sept. 23/2 He was asked if he took LSD, and answered: 'I have been tripping for three weeks.' 1971 [see *MANDY, MANDY]. 1976 H. Ferguson *Confessions of Long Distance Acid Head* 13 The bunch with whom I used to trip out and smoke pot with were form-mates of my brother. 1980 J. Scott *Gospel Lamb* iii. 45 Some of the people here were tripping already. Seemed a pity not to bust 'em.

II. 6. b. Also *spec.* in U.S. Sport, to defeat.

1974 *State* (Columbia, S. Carolina) 15 Feb. 6-B/2 The Generals got goals from Mike Gaines and Eddie Hewbrank in the second overtime to trip Airport, 2-0. 1979 *Honolulu Advertiser* 8 Jan. C-2/4 In Rural AJA games, Wahiawa tripped Pearl Ridge 7-5.

III. 14. a. (Later examples.) Also more widely, to cause to operate or respond; *spec.* in *Electronics*, to cause (a bistable device) to change from one stable state to the other; *to trip out*, to render electrically disconnected, esp. as an automatic action.

1936 *Sun* (Baltimore) 25 Jan. 5/8 It was eleven minutes after the electrical apparatus operating the gas generating equipment was tripped before physicians pronounced Foster dead. 1937 *Rev. Sci. Instruments* VIII. 414/2 It is necessary that at each incident pulse the circuit shall be tripped from one equilibrium state to the other. 1950 *Engineering* 20 Jan. 79/3 The gap was in the form of an expulsion tube..this arrangement helping to extinguish quickly the power-follow current so that the transformer was not tripped out. 1953 C. A. Lindbergh *Spirit of St. Louis* II. vi. 307, I tripped my safety-belt buckle..and rolled out over the high rim of the cockpit. 1961 *Ann. Reg.* 1960 396 This light in turn tripped more atoms until none were left in the excited state. 1967 *Electronics* 6 Mar. 126/2 When the critical temperature is reached, the resistance of the thermistor changes to allow the proper value of current to flow, and this trips a relay. 1972 P. Cleife *Slick & Dead* xxviii. 233 Tripping the quick-release of my harness, I leapt from my seat. 1977 *Daily Tel.* 25 Oct. 2/1 'The damn thing didn't even trip our noise meters,' he was quoted as having said after last week's three days of test landings and take-offs at Kennedy Airport. 1978 *Sci. Amer.* Mar. 146/3 When I tripped the switch S_1, the outputs from Q and $\bar{Q}$ of IC 1A changed states: the Q then produced a logical-1 signal. 1981 *New Scientist* 29 Oct. 295/2 Another tree..in East Sussex caused a similar fault, tripping out another 400 kV supergrid line feeding the south coast.

b. *intr.* Of a mechanism or the like: to undergo a sudden change of state; to operate or (also *trip out*) cease to operate.

1940 *Jrnl. Marine Res.* III. 73 When each water bottle trips there will be a sufficient jar..so that the recording stylus..will make a noticeable mark on the temperature depth trace. 1950 *Engineering* 20 Jan. 79/3 In the 14 years under review, sub-station transformers tripped out 140 times. 1977 *Times* 16 July 5/8 Three main power lines..were hit by lightning... This caused four more lines to trip out as the safety devices to stop them overloading came into action. 1980 *Sci. Amer.* Mar. 36/1 The main feedwater pumps in the lower level of the turbine building tripped, interrupting the removal of heat from the primary system. 1981 *New Scientist* 29 Oct. 295/2 As other parts of the grid tripped out, power stations in the South and South-West struggled to meet what demand they could.

15. *Bot. trans.* To operate the pollination mechanism of (certain flowers) by disturbing the keel so that the anthers and style spring out of it.

1909 *Bull. Bureau Plant Industry, U.S. Dept. Agric.* No. 24. 9 Only a slight pressure on the keel is necessary to trip the flower. 1930 *Jrnl. Amer. Soc. Agronomy* XXII. 782 The flowers were left exposed and not tripped artificially. 1978 *Nature* 7 Sept. 54/1 Most inbred lines show poor seed set unless their flowers are visited by bees or artificially manipulated (tripped) and are therefore called auto-sterile.

tripack (trəi·pæk). *Photogr.* Also **tri-pack**, † **-pak.** [f. TRI- + PACK *sb.*¹] A set of three superimposed plates or films with different colour sensitivities and kept in contact, so that three separation negatives can be obtained at one exposure; also (*integral tripack*), a film having three such emulsions on the one base.

1911 *Brit. Jrnl. Photogr.* 3 Mar. (Suppl.) 19 (*heading*) The Ives Tripak system of colour photography. 1924 *Ibid.* Sept. (Suppl.) 36/1 It is fundamentally impossible to obtain satisfactory results by the tri-pack system. 1931 [see *separation negative* s.v. *SEPARATION 14]. 1953 [see *INTEGRAL *a.* 2]. 1957 R. W. G. Hunt *Reproduction of Colour* v. 47 (*caption*) Sensitization of the layers in Kodachrome, Ektachrome, Kodacolor and other similar integral tripacks. 1973 D. A. Spencer *Focal Dict. Photogr. Technol.* 606 The coloured positive images may be produced..*in-situ* by using integral tripack material.

triparental: see *TRI- 1 a.

tripa·rtisa·n, *a.* [f. TRI- + PARTISAN *a.*] Of, representing, or composed of members of, three (political or other) parties.

1959 *Economist* 14 Mar. 950/2 Candidates deliver brief speeches of equal length to an invited bi-partisan or tri-partisan audience. 1965 *Ibid.* 20 Feb. 731/1 The cross-currents of party strife over what should be a tripartisan policy. 1972 D. Dakin *Unification of Greece* v. 76 Otho accepted a tripartisan cabinet of seven who had been recommended by Makriyannis.

tripartism (trəipā·rtiz'm). [f. TRIPART(ITE *a.* + -ISM.] **1.** Division into three political parties or other groups. Also *transf.*

1954 B. & R. North tr. *M. Duverger's Pol. Parties* II. i. 222 A tripartism analogous to that in England at the same period where the Socialist party was taking up a position alongside the two 'bourgeois' parties. 1964 *Listener* 1 Oct. 505/1 If different types of secondary school really were 'separate but equal', the attacks upon tripartism would lose much force.

2. A system under which representatives of three groups engage in consultation, negotiation, or joint action; *spec.* a system of economic planning by representatives of government, employers, and trade unions.

1961 *Guardian* 14 June 18/6 The Soviet insistence on tripartism and triunity. 1974 *Government & Opposition* IX. 405 The British flirtation with planning throughout the 1960s could be characterized as toothless tripartism. 1980 *Times* 17 May 14/3 Meetings with Ministers are not the 'tripartism' that the TUC wants.

tripe¹. Add: **1. c.** ‖ *tripe(s) à la mode de Caen*, a traditional French dish of tripe cooked with carrots, onions, and cow heels in cider or white wine. Also *tripe à la mode* (*du pays*).

1859 Thackeray *Virginians* II. iii. 26 'And an orchard ..and a dish of tripe *à la mode du pays*!'..Museau would ..return to the subject of Normandy, and cyder, and '*trippes* [sic] *à la mode de Caen*.' 1936 E. Waugh *Waugh in Abyssinia* ii. 71 From time to time she would placard the town with news of some special delicacy—*Grand Souper. Tripes à la mode de Caen.* 1968 E. Hyams *Mischief Makers* ii. 20 An improbable garden-restaurant where we sat eating *tripe à la mode*. 1970 Simon & Howe *Dict. Gastronomy* 376/2 The classical way of preparing tripe in France is *Tripe à la mode de Caen*.

3. Now applied esp. to artistic work, opinions, conversation, or the like: worthless stuff, rubbish.

1892, 1895 [in Dict.]. 1902 'T. Le Breton' *Mod. Christian* viii. 80 She puts in six or seven pages of her own tripe. 1927 C. Connolly *Let.* Aug. in *Romantic Friendship* (1975) 313 Ordinary talk is such ghastly tripe once voice and gesture are removed. 1935 I. Miller *School Tie* xiv. 277 'I've tried hard, sir; really I have.' 'Tripe! You've tried to get out of work.' 1952 W. Stevens *Let.* 24 Oct. (1967) 763 Non-objective art without an aesthetic basis seems to be an especially unpleasant kettle of tripe. 1963 [see *CODSWALLOP]. 1973 W. H. Canaway *Harry doing Good* I. ii. 22 The group of girls who were watching some tripe on television.

4. *tripe sausage*; **tripe-hound** *slang*, (*a*) an unpleasant or contemptible person; also *spec.* a newspaper reporter or an informant; (*b*) a contemptuous term for a dog; *spec.* in Austral. and N.Z., a sheep-dog.

1923 J. Manchon *Le Slang* 320 *Tripe-hound*,..un sale cabot, un clebs à poubelles. 1928 D. L. Sayers *Unpleasantness at Bellona Club* xv. 176 If you'll call off your tripe-hounds, we'll let you have an interview and a set of photographs. 1933 L. G. D. Acland in *Press* (Christchurch, N.Z.) 23 Dec. 15/7 *Tripe-hound*, slang for sheep dog. This was common on South Canterbury stations in the 'nineties, and I always thought it a New Zealand word until I came across it in an English novel the other day, 'Early Closing', I think. In the novel the word was applied to a spaniel! 1935 J. Buchan *House of Four Winds* iv. 98 If your tripe-hounds had been worth their keep they would have seen me meet him. 1937 N. Marsh *Vintage Murder* viii. 87 You damned little tripe-hound. 1946 B. Marshall *George Brown's Schooldays* 123 Draw, you sorry tripehound, draw. 1966 'L. Lane' *ABZ of Scouse* 111 *Tripe-hound*, a mongrel dog. Also applied to a racing greyhound that persists in putting up a disappointing performance. 1966 P. V. Price *France* II. 219 *Andouilles limousines*. Tripe sausages, usually served grilled. 1981 'M. Hebden' *Pel is Puzzled* viii. 73 Andouillettes, the tripe sausage of the region.

triped (trəi·ped). [ad. L. *tripēs*, *-ped-is*, three-footed, three-legged, f. *tri-* TRI- + *pēs* foot.] A three-legged animal.

1916 *Daily Colonist* (Victoria, B.C.) 2 July 4/5 A three-legged chicken is the pride of a brood of a dozen hatched last week... The triped is quite lively and is putting lots of joy into the life of the Simpson barnyard. 1954 W. K. Hancock *Country & Calling* 14 Peter [*sc.* a horse] went permanently lame after a week or two of work; he was a triped, William said, and no use to a bush minister in his professional capacity. 1971 I. Brown *Old & Young* xiv. 220 The small child is a quadruped... Then comes the erect and active biped. At last the codger limps along using a stick for an extra limb. As one who has reached the triped stage I am denied the pleasure of..exercise.

tripelennamine (trəipele·nămīn). *Pharm.* [f. TRI- + P(YRIDYL + E(THY)LEN(E + AMINE.] An antihistamine drug, $C_{16}H_{21}N_3$, given orally as the crystalline hydrochloride or citrate.

1947 *Jrnl. Amer. Med. Assoc.* 31 May 454/2 Granulocytopenia is believed not to have been described hitherto among the untoward side effects of tripelennamine hydrochloride, 'pyribenzamine' (N.N.R.) (N'-pyridyl-N'-benzyl-N-dimethylethylenediamine hydrochloride), therapy. 1962, 1974 [see *PYRIBENZAMINE].

tripey (trəi·pi), *a. colloq.* Also **tripy.** [f. TRIPE¹ + -Y¹.] Inferior, trashy, rubbishy, worthless.

1955 E. Blishen *Roaring Boys* IV. 239 How you can tell them to paint in the same tripey way as me, I don't know. 1962 L. Davidson *Rose of Tibet* x. 177 'Don't you like any of our books?..What do you think of this one?' 'It looks a bit tripy to me.' 1968 R. Jeffries *Traitor's Crime* iv. 49 It's a case of the more tripey, the more relaxation. 1971 'M. Underwood' *Trout in Milk* xiv. 145 The jury'll never convict on the tripey evidence the prosecution have brought.

trip-hammer. (Earlier examples.)

1781 S. Peters *Gen. Hist. Connecticut* 265 Anchor-making is done by water and trip-hammers. *a* 1817 T. Dwight *Trav. New-Eng.* (1821) II. 15 Here he built a shop; and set up the first trip-hammer in this part of the country.

triphibian (trəifi·biăn). [Irreg. f. TRI- + AM)PHIBIAN *sb.*] One who or that which is capable of existing or operating in three different spheres, esp. on land, on water, and in the air. (An occasional word.)

1935 *Sun* (Baltimore) 26 Oct. 1/2 Constantinos Vlachos' invention, which he had intended should travel on the ground, in the air and on water, collapsed upon him in flames today as he was demonstrating it on the lawn of the Library of Congress. The device, which Vlachos, a resident of Washington, called a triphibian, has never been off the ground, his wife said. 1943 W. S. Churchill in *Amer. Speech* (1944) XIX. 14/2 He [*sc.* Lord Mountbatten] is what I venture to call a complete triphibian. 1943 *Punch* 8 Sept. 197 (*caption*) Looking Eastward. Mr. Punch's portrait of a 'Triphibian'. 1977 H. Osmond in A. Huxley *Moksha* viii. 39 Soon my dear friend [*sc.* A. Huxley], the wise and gentle triphibian, for that was his own definition of man, was no more.

triphibious (trəifi·biəs), *a.* [Irreg. f. TRI- + AM)PHIBIOUS *a.*] Capable of living or operating on land, on water, and in the air; *spec.* of or pertaining to military operations involving land, sea, and air forces. Hence **triphi·biously** *adv.* (*rare*).

1941 L. Hore-Belisha in *Times* 4 Nov. 2/3 Whether or not amphibious—or rather triphibious—raids at unexpected points along the extensive coastline..would at this stage seriously distract the enemy was doubtful. 1950 A. de Seversky *Air Power* (1952) ii. 25 Either we shall continue to divide our national potential three ways to support an outlived triphibious method of war-making, or we shall concentrate it. 1964 D. Macarthur *Reminiscences* VI. 166 Ground, air, and sea operations were thoroughly co-ordinated. It was a new type of campaign —three-dimensional warfare—the triphibious concept. 1978 *Sci. Amer.* Dec. 29/2 In southeastern Asia, however, some 50 species of frogs have lengthened toes, arrayed in a still wider web, and swim the air—triphibiously. 1982 *Contemp. Rev.* Jan. 51 Their service is universal and triphibious.

tripho:sphopy·ridine nu·cleotide. *Biochem.* [f. TRI- + PHOSPHO- + PYRIDINE + *NUCLEOTIDE.] = *nicotinamide-adenine dinucleotide* s.v. *NICOTINAMIDE b.

1937 *Chem. Abstr.* XXXI. 718 (*heading*) Oxidation of Robison's ester with the aid of triphosphopyridine nucleotide. 1951 *Sci. News* XXII. 80 This reaction, which uses di- or tri-phosphopyridine nucleotide as a hydrogen carrier may thus be a key reaction in photosynthesis. 1962 [see *DIPHOSPHOPYRIDINE NUCLEOTIDE].

triping (trəi·piŋ). *Sc.* [Perh. f. TRIPE¹ + -ING¹.] Coal as brought to the pit-head, not yet cleaned or graded.

1886 J. Barrowman *Mining Terms* 68 *Triping*, a kind of drossy coal. 1921 *Glasgow Herald* 21 Apr. 5 A 'triping' ton..represents about 55 per cent. of round coal, 40 per cent. of dross, and 5 per cent. of dirt. 1923 *Ibid.* 22 Dec. 5 The miner produces what is known in the trade as 'triping'. 1924 *Ibid.* 6 Oct. 8 The coals are filled in the hutches at the coal face as 'triping'—large and small mixed.

tripla. Delete † *Obs.* and add later examples.

1944 W. Apel *Harvard Dict. Mus.* 608/2 Nominally, this proportion was *proportio tripla* (another name for such a *Nachtanz* was *Tripla*). 1980 *Early Music* Jan. 120/2 Whilst I applaud the attempt to maintain a definite relationship with the main pulse, the required effect might sometimes be a stately *sesquialtera* rather than a hectic *tripla*.

tripla: pl. of *TRIPLUM.

triple, *sb.* **1. b.** Delete † *Obs. rare* and add later examples.

1966 D. Bennett *Stranger in his Grave* viii. 66, I wanted to..lie naked in an air-conditioned room with a triple of aspirins inside me. 1981 *Northeast Woods & Waters* Jan. 11/2 We finally settled down and started to pick individual targets and stopped shooting at triples or whole flocks.

2. d. *Chem.* A group of three atoms or ions.

1952 *Jrnl. Chem. Physics* XX. 685/2 I_{111} is an integral involving a triple of atoms arranged as in the figure. 1977 *Sci. Amer.* July 95/2 (*caption*) A sodium cation (Na^+) might attract to its vicinity two independent solvated electrons, forming an 'ion triple'.

e. *Horseracing.* = *TRIFECTA.

1972 *Britannica Bk. of Year* 733/3 *Triple*, *specif.*, a system of betting on races in which the bettor must pick the first, second, and third horses in this sequence in a specified race in order to win. 1976 *N.Y. Times* 21 Aug. 22 They had hit on a triple (picking the first three horses in the right order), and it was the young woman's turn to collect.

6. *Baseball.* A hit which enables the batter to reach third base.

1880 *Chicago Inter-Ocean* 21 June 8/4 In the fifth Farrell's two-baser, Ward's triple hit, Bradley's triple hit, and Walker's fumbled grounder gave the Champions two earned runs. 1887 *Chicago Tribune* 3 May 3/1 He made in succession a single, double, triple, and home run. 1926 [see *HOME *a.* 4]. 1949 *Milwaukie* (Oregon) *Rev.* 4 Aug. 4/4 The hard working first sacker collected his first triple of the year. 1974 *Birmingham* (Alabama) *Post-Herald* 29 June A 14/2 With two out, Bill North singled and scored on Campaneris' triple.

7. = *TREBLE *sb.* 2 j.

1981 W. H. Hallahan *Trade* v. 153 He poured himself another drink—a triple. 1981 G. Markstein *Ultimate Issue* 174 Welk poured them all triples. 'Salut,' he toasted.

triple, *a.* (*adv.*) Add: **A.** *adj.* **5.** *triple agent* = *treble agent* s.v. *TREBLE *a.* 3; *triple alliance*: also, an alliance of trade unions representing miners, railwaymen, and transport workers; *triple bond* (Chem.) [*BOND *sb.*¹ 13 e], a bond in which the two atoms 'share' three pairs of electrons rather than one pair; hence *triple-bonded* adj.; *triple concerto*, a concerto with three solo parts; *triple cross*, the act of betraying one party in a transaction by pretending to betray the other, or of betraying a person who has betrayed another; so *triple-crossing*; *triple crown*, (*b*) also applied to several other instances of winning three victories in of the Kentucky Derby, the Preakness Stakes, and the Belmont Stakes, and, in Rugby Union and hockey, the winning by England, Ireland, Scotland, or Wales of victory over each of the other three in the same season; *triple entente* (earlier example); *triple jump*, an athletic long jump event, also known as the hop, step, and jump (cf. HOP *sb.*² 3 a in Dict. and Suppl.); *triple junction* (Geol.), a region at which the boundaries of three lithospheric plates meet; (Petrogr.), a point where three grain boundaries meet at angles of approximately 120 degrees; *triple mirror* = *three-way mirror* s.v. *THREE-WAY *a.* a; *triple play* (earlier example); *triple point*, (*b*) (Physics), the temperature and pressure at which the solid, liquid, and vapour phases of a pure substance can coexist in equilibrium; the point representing this state in a phase diagram (marked by the junction of three lines); more widely, an analogous state or point for any three phases of a substance; (*c*) (Petrogr.), an invariant point involving three phases, e.g. where kyanite, andalusite, and sillimanite are stable in the system Al_2SiO_5; *triple spacing*, the spacing of typewritten or other text so that two blank lines separate adjacent lines of text (see *triple-spaced*, sense C. 1 a below); *triple time*: also *transf.* in prosody; *triple tonguing*, the use of the tongue to achieve rapid articulation in groups of three notes on the flute and brass instruments; hence (as back-formation) *triple-tongue* vb. intr.; *triple vaccine* (Med.), (*a*) a vaccine containing three species of the *Salmonella* bacteria, used as a prophylactic against typhoid and paratyphoid; (*b*) a vaccine containing diphtheria and tetanus toxoids and the killed whooping cough organism, administered in early childhood as a prophylactic against conditions caused by these.

1968 J. Wainwright *Web of Silence* 94 Jackson figures he's created a double-agent. We don't agree. The way we see it, Schneller's maybe a triple-agent. 1982 T. Heald *Masterstroke* viii. 159 Something in Intelligence. Our Intelligence. Theirs too... A triple agent at least. 1915 *Times* 9 Oct. 3/1 The conference of the Miners' Federation of Great Britain..adopted unanimously the scheme of a triple alliance between railwaymen, transport workers, and miners. 1974 *Times* 8 Jan. (Europe Suppl.) p. ii/2 As Christmas approached, a state of affairs was fast developing reminiscent of that which occurred when the 'triple alliance' of miners, dockers and railwaymen sprang into being during the run-up to the General Strike of 1926. 1889 G. M'Gowan tr. *Bernthsen's Organic Chem.* i. 55 The constitutional formula for acetylene, C_2H_2, is assumed to be CH≡CH, according to which the carbon atoms are joined together by a triple bond. 1971 *Nature* 10 Dec. 333/2 The CO molecule has a very stable triple bond which is difficult to dissociate. 1937 *Jrnl. Amer. Chem. Soc.* LIX. 2091/2 The carbon–oxygen bond [in BH_3CO] presumably involves resonance between double- and triple-bonded structures. 1965 Phillips & Williams *Inorg. Chem.* I. xiii. 484 In the most extreme cases there is some reason for saying that oxygen is triple-bonded. 1879 Grove *Dict. Mus.* I. 389/1 Sometimes concertos are written for more than one solo instrument, and are then known as double, triple, etc., concertos as the case may be. 1932 *Daily Tel.* 8 Oct. 8/3 In March a triple concerto for flute, clarinet, bassoon and orchestra by J. R. Heath will be introduced. 1985 *Daily Tel.* 22 Jan. 9 The London Sinfonietta..had played the Triple Concerto on the first night of the Tippett birthday celebration week. 1971 J. Aiken *Nightly Deadshade* vii. 76, I..resolve to set a few subliminal tests for my fellow-conspirators.. in case they are working the triple-cross. 1978 *Times* 25 Jan. 11/5 Anthony Price is a master of the double (or even triple) cross. 1922 S. Lewis *Babbitt* xix. 236 'Kind of double-crossing.' 'It ain't. It's triple-crossing. It's the public that gets double-crossed.' 1899 *Whitaker's Almanack 1900* 648/2 In their last match at Cardiff against Wales, Ireland won by a try to nothing, securing the triple crown with three straight victories as in 1894. 1946 M. C. Self *Horseman's Encycl.* 413 Race horses which win the Kentucky Derby, the Preakness and the Belmont are said to win the Triple Crown. 1953 *Times* 10 Apr. 4/5 A victory for England [over Scotland at hockey] will mean that the meeting of Ireland and England in Dublin on April 18 will be a fight for the triple crown. 1974 *Sunday Tel.* 7 Apr. 36/7 Hockey history was made on the Cardiff University ground at Llanrumney where Wales won the Championship and took the Triple Crown for the first time since the quadrangular tourney began in 1903. 1976 *National Observer* (U.S.) 22 May 6/4 She may never win a Triple Crown race herself—no woman ever has—but she says she'll 'definitely ride again; it's in my blood'. 1978 *World of Tennis* (BP Yearbk.) 179 The Virginia Slims Championship is the first event in the Triple Crown (Wimbledon and Forest Hills are the others) which has replaced the Grand Slam at the pinnacle of achievement in women's tennis. 1979 M. Boyce *I was There!* 47/2 Twelve..the number of times Wales have won the Triple Crown. 1979 *Harvard Gaz.* 23 Feb. 2 Zoo Captain Bill Renke is the only person to have won the tiddlywinks Triple Crown—the Singles, Pairs, and Team Championships. 1910 *Encycl. Brit.* IX. 949/1 (*heading*) The Triple entente and the Triple Alliance. 1964 M. Watman *Encycl. Athletics* 150/1 Basically, the rules for the triple jump (formerly known as the hop, step and jump) are identical with those governing the long jump. 1972 *N.Y. Times* 4 June 4/7 The runner-up in the voting was John Craft, who set an American triple-jump record of 55·5. 1969 *Nature* 11 Oct. 125/2 Evolution of such triple junctions can produce many of the changes which would otherwise appear to have been caused by a change in the direction or magnitude of the relative motion between plates. 1976 R. H. Vernon *Metamorphic Processes* v. 137 Three grains meet at a point ('triple junction'), the interfacial angles closely approximating 120°. 1979 Mallory & Cargo *Physical Geol.* xvii. 437 Examination of these triple junctions shows that the three branches of the junction may involve convergence, divergence, or translational (sideways) movement. 1982 Aramaki & Ui in R. S. Thorpe *Andesites* iii. 260/1 Japan and the surrounding islands form three chains of island arcs which meet at a trench-trench-type triple junction located at *c.* 34° N and 142° E. 1907 *Yesterday's Shopping* (1969) 110/2 Triple folding mirrors..Size of glass, 9 by 9 in.—20/3. 1920 S. Lewis *Main Street* xiv. 164 A real dressing-table with a triple mirror. 1967 'K. O'Hara' *Unknown Man* iv. 29 She adjusted the triple mirror and took up the eye-liner. 1869 *De Witt's Official Base Ball Guide* 42 Remarks concerning double and triple plays will apply to the third baseman as much as to either of the other base-players. 1872 J. Thomson in *Rep. Brit. Assoc. Adv. Sci. 1871* II. 32 We must suppose three curves (namely, the line between gas and liquid, the line between liquid and solid, and the line between gas and solid) to meet in one point... This point of pressure and temperature for any substance may then be called the triple point for that substance. 1879 [see *ice line* s.v. *ICE *sb.* 8]. 1955 *Sci. Amer.* Mar. 52/3 In another change, the absolute temperature scale was redefined in terms of a single fixed reference point—the triple point of water. 1964 J. H. Brophy et al. *Thermodynamics of Structure* ii. 20 If the specified temperature and pressure are the coordinates of point *D*, all three phases coexist in equilibrium with one another. Point *D* is called the *triple point*. 1966 [see *ice-point* s.v. *ICE *sb.* 8]. 1978 *Nature* 14 Dec. 696/2 One possible reason..is that a solid–solid–liquid triple point could exist near 10^7 Pa..in the proposed phase diagram [of carbon]. 1979 K. G. Cox et al. *Interpretation of Igneous Rocks* iii. 47 Figure 3.1 shows a one-component phase diagram for a substance such as H_2O. It consists of three fields in each of which a single phase exists. Each pair of fields meets in a curve along which two phases co-exist, and the three fields meet in a point U (so-called triple point) at which all three phases co-exist. 1983 D. S. Barker *Igneous Rocks* iii. 28 The 'triple point' where liquid, vapor, and solid coexist in equilibrium has yet to be experimentally located for any silicate. 1957 A. C. Lloyd et al. *Gregg Typewriting for Colleges* 2/1 In triple spacing, typing appears on every third line with 2 blank lines between the typed ones. 1978 *Brit. Med. Jrnl.* 16 Dec. 1724/3 Typing should be on one side of the paper, with double or triple spacing between the lines. 1880 G. M. Hopkins *Let.* 5 Sept. (1935) 107 So far as I know triple time is in English verse a shy and late thing. 1951 W. Morum *Gabriel* I. i. 7

He triple-tongued up the scale to high C. 1967 *Crescendo* Feb. 23/3 The band stopped for several bars, and Charles *triple-tongued*—at breakneck tempo—and never faltered or slowed down once. 1879 Grove *Dict. Mus.* I. 459/2 Triple tongueing is also possible. 1951 W. Morum *Gabriel* II. vi. 211 You could rattle out triple-tonguing stuff when you was a kid. 1961 C. W. Monk in A. Baines *Mus. Instruments* xi. 280 Rapid passages are managed by alternating 't' and 'k' in 'double-tonguing' and fast triplets by 'triple tonguing' 'ttk' (or 'tkt, ktk'). 1917 *Jrnl. Amer. Med. Assoc.* 22 Sept. 100/2 The reaction following the inoculation of the triple vaccine was no different than when the typhoid vaccine was alone used. 1947 *Ann. Rev. Microbiol.* I. 327 A reinvestigation of the antigenicity of the strains routinely used for the preparation of typhoid and 'triple' vaccine..resulted in the introduction of strains rich in somatic antigen. 1970 Passmore & Robson *Compan. Med. Stud.* II. xxii. 22/2 The injection of the triple vaccine may be combined with three oral doses of poliomyelitis vaccine. 1982 *Times* 8 Sept. 1/4 Kate is to have her third triple vaccine against whooping cough, diphtheria and tetanus today.

C. 1. a. *triple-barrelled*, *-hummocked*, *mirrored*, *-tiered*; **triple-spaced,** typed or formatted so that two blank lines separate adjacent lines of text; also as quasi-*adv.*; (see *triple spacing*, sense A 5 above).

1934 W. S. Churchill *Marlborough* II. xxv. 561 Between their squadrons appeared the triple-barrelled guns, which opened a remarkably rapid fire. 1977 *Navy News* Sept. 2/3 Up, up, and away goes H.M.S. Salisbury's Squid triple-barrelled mortar, the last firing mounting in the Royal Navy. 1876 G. M. Hopkins *Poems* (1967) 64 The triple-hummocked Giant's Stool. 1939 R. Chandler *Big Sleep* vii. 45 There was perfume on the triple-mirrored dressing table. 1970 R. Rendell *Guilty Thing Surprised* iii. 38 Between the two mirrors stood a triple-mirrored dressing table. 1946 R. Chandler *Let.* 6 Oct. (1981) 80 The silly little triple-spaced half pages I type on. 1966 F. Stewart *Deadly Nightcap* iii. 40 There were margins of twenty degrees on either side of the typescript, and it was triple-spaced. 1978 M. H. Clark *Stranger is Watching* xix. 81, I typed them triple-spaced. 1807 J. Barlow *Columbiad* VII. 267 Flames, triple tier'd, and tides of smoke, arise, And fulminations rock the seas and skies. 1962 E. Snow *Other Side of River* (1963) ii. 23 A balcony overlooking the Outer City gave a view of the distant, triple-tiered, blue-glazed round roofs of the Temple of Heaven. 1972 M. J. Bosse *Incident at Naha* i. 11 He.. went to his triple-tiered pipe rack.

b. *triple-action*, *-digit*, *-expansion* (earlier example), *-threat*.

1934 Webster 725/2 A *triple-action die*, when in operation, has a movement, produced by springs or a triple-action press, of two punches, two matrices, or a punch and a matrix, within either the upper or lower half of the die. 1960 *Farmer & Stockbreeder* 29 Mar. (Suppl.) 11/1 From Steiner comes news of two brand new hair aids. The first is Pearl Foam, a triple-action shampoo which cleanses, adds lustre and protects the hair from the weather. 1974 M. Taylor tr. *Metz's Film Lang.* ix. 233 This triple-action construction gives the ending of the film..its true meaning. 1976 *N.Y. Times* 7 Mar. III. 15/6 The foot-dragging is caused by Argentina's triple-digit inflation. 1979 *Time* 28 May 12 The cost of living for April had jumped a shocking 8·7%, more than 100% if projected over the entire year. The admission provoked howls of alarm that the country could be heading toward uncontrollable triple-digit inflation. 1882 *Engineering* 12 May 474/1, I may mention that within the last few weeks there has been a steamer completed to work at a pressure of 150 lb. per square inch with triple expansion engines very similar to those fitted in the Aberdeen. 1939 W. H. Baumer *Sports as taught & played at West Point* 40 Any backs who possess the three qualifications of being a good runner, passer and kicker to a marked degree are triple-threat men. 1972 J. Mosedale *Football* v. 67 Football no longer requires the triple threat back—the player who can run and kick as well as punt.

2. a. *triple-locked*, *-quick*.

1974 Hawkey & Bingham *Wild Card* xiv. 123 The triple-locked door to his apartment. 1951 L. MacNeice tr. *Goethe's Faust* 226 How triple-quick we spirits fly!

c. with vbs., as *triple-lock*.

1876 'Mark Twain' *Tom Sawyer* xxxii. 251, I had its big door sheathed with boiler iron..and triple-locked. 1976 B. Bova *Multiple Man* xvi. 177 The first thing I did was..to make certain I was alone..after triple-locking the front door.

triple, *v.* Add: **4.** *Baseball.* To hit a triple (see *TRIPLE *sb.* 6).

1908 *Sporting News* 17 Sept. 4/5 The very next day he tripled with the bases full. 1972 *N.Y. Times* 4 June 3/1 Danny Thompson tripled, and consecutive singles by Rod Carew, Harmon Killebrew, Steve Braun and Bobby Darwin gave the Twins a 2-0 lead.

tri·ple-decker *a.* [f. TRIPLE *a.*, after DOUBLE-DECKER.] **1.** Of sandwiches: consisting of three layers of bread and two layers of filling.

1946 *New Yorker* 16 Mar. 21/2 We are the land of the between-meals snack, of the triple-decker sandwich. 1973 *Cookery Year* (Reader's Digest Assoc.) 133/5 *Club sandwiches*, one of the best inventions of the American kitchen is this triple-decker sandwich.

2. Of bunk beds: arranged one above another in threes.

1979 *Arizona Daily Star* 5 Aug. A 8/2 Rosenthal walked into a tin-roofed quonset hut jammed with triple-decker bunk beds. 1980 *Nat. Geographic* June 866 Those who need rest..retire..to homemade triple-decker triple beds.

triple-hea·der. *U.S.* Also **tripleheader.** [f. TRIPLE *a.*, after DOUBLE-HEADER.] **1.** In baseball, etc., a sporting event at which three consecutive matches are staged. Also *transf.* and *fig.*

1961 in Webster. 1970 *New Yorker* 3 Oct. 34/2 Anybody who can't straighten out a plain old two-league, six-division distribution of twenty-six teams, each of which plays games both inside and outside its own division..isn't in shape for a single Sunday TV triple-header. 1979 *Honolulu Advertiser* 8 Jan. C-2/3 Sheridan Midas Mufflers..annihilated Manoa 15-5 in the opener of a Honolulu AJA Senior Baseball tripleheader yesterday.

2. A situation, occurrence, etc., having three aspects or involving three participants.

1976 *New Yorker* 29 Mar. 16/1 In his first attempt at a tripleheader (writer-director-star) he gets bogged down in an overelaborate production. 1977 *Ibid.* 4 July 23/1 The Cooper-Hewitt Museum, which this year has been throwing new exhibitions at us almost faster than we can catch them, has most recently unveiled a triple-header. 1977 *Washington Post* 13 Jan. C 1/3 Sens. Donald Riegle (D-Mich.), Lowell Weicker (R-Conn.) and Herman Talmadge (D-Ga.) made it a triple-header with their divorce announcements—getting it done before the new congressional session began and, of course, a few months after election time.

tripler (tri·plər). *Electronics.* [f. TRIPLE *v.* + -ER¹.] Any device for producing an output whose frequency or whose voltage is three times that of the input.

1924 S. R. Roget *Dict. Electr. Terms* 244/2 *Static frequency-changer*... Also called doublers and triplers in the cases of twice and three times the original frequency. 1947 L. B. Young in C. G. Montgomery *Technique of Microwave Measurements* vi. 371 In this frequency range the push-pull tripler of Fig. 6.22 is used. 1973 G. J. King *Newnes Colour Television Servicing Man.* I. i. 29/2 The final 24 kV potential is developed only when the tripler e.h.t. output is connected to the tube final anode.

triplet. Add: **2. j.** *Poker.* (See quot. 1864.)

1864 W. B. Dick *Amer. Hoyle* 164 *Triplets* are three cards of the same denomination, and rank higher than two pairs. For example;—three Deuces beat a pair of Aces and Kings. 1887 J. W. Keller *Game of Draw Poker* 14 Full Hand—(Triplets accompanied by a pair)... A full hand beats a flush. 1950 [see *SANDBAGGER 2].

k. (i) *Physics* and *Chem.* A multiplet (sense *a) composed of three lines or energy levels. Freq. *attrib.*, esp. designating an atom with two unpaired electrons and $S = 1$.

1879 *Proc. R. Soc.* XXX. 29 The flame spectrum of magnesium was examined, a green triplet was observed. 1922 [see *SINGLET 3 a]. 1934, 1937 [see *INTERCOMBINATION]. 1950 G. Herzberg *Spectra of Diatomic Molecules* (ed. 2) v. 216 Molecules with an even number of electrons have odd multiplicities (singlets, triplets, ...) since *S* is integral. 1977 *Nature* 3 Nov. 15/1 The Earth has an atmosphere containing diatomic triplet oxygen..essential for life.

(ii) *Particle Physics.* A multiplet (sense *b) of three sub-atomic particles.

1937 [see *SINGLET 3 b]. 1961 M. Gell-Mann in Gell-Mann & Ne'eman *Eightfold Way* (1964) 12 We have a triplet ρ of vector mesons coupled to the isotopic spin current and a singlet vector meson ω^0 coupled to the hypercharge current. 1968 [see *OCTET, OCTETTE 3 c]. 1975 *Sci. Amer.* Oct. 40/1 The pion is a triplet with an average mass of ·137 GeV and three charge states: + 1, 0 and − 1.

3. triplet code *Genetics*, the accepted version of the genetic code in which amino-acids are specified by three successive nucleotides in a nucleic acid molecule.

1957 *Proc. Nat. Acad. Sci.* XLIII. 687 Such triplet codes..have an excess of information, since there are sixty-four different triplets for the twenty amino acids. 1976 P. Collard *Devel. Microbiol.* viii. 107 The correctness of the triplet code was soon verified by the elegant experiments of Nirenberg.

triplex, *a.* (*sb.*) Add: **1. b.** *triplex board*, a type of cardboard consisting of three layers felted together by pressure without the use of adhesive.

1921 H. A. Maddox *Dict. Stationery* 78 *Triplex board*, a cheap class of pasteboard or ticket board which derives its name from the fact that it comprises three layers. 1962 F. T. Day *Introd. to Paper* iv. 46 In the case of Triplex boards one grade of pulp is used for the middle while the two outside sheets are made of a thinner substance to make up the finished board.

2. *Genetics.* Of a polyploid individual: having the dominant allele of any particular gene represented three times.

1921, etc. [see *NULLIPLEX *a.*]. 1929 *Jrnl. Genetics* XXI. 138 If we assume 14/16 to be triplex for Y the unexpected ivory plant could then have arisen by (1) non-disjunction.., or (2) irregularity of disjunction in the equational division.

3. (Usu. with capital initial.) The proprietary name of a type of toughened or laminated glass, orig. *spec.* consisting of two layers of glass and a layer of celluloid between them.

1923 *Trade Marks Jrnl.* 28 Nov. 2495 Triplex. Use claimed from 2nd August 1912... Safety Glass in sheets. The Triplex Safety Glass Co. Ltd. 1927 M. Arlen *Young Men in Love* II. 128 'I live in such a glass house, Peter!' 'Well, so does everyone else.' 'Oh, no, other people's are made of Triplex!' 1930 *Times Educ. Suppl.* 27 Dec. (Home & Classroom Suppl.) p. i/3 The car was fully equipped with three speeds,.. Triplex glass screen, windscreen wiper,..side curtains, and hood bag. 1935 L. MacNeice *Poems* 40 Chromium dogs on the bonnet, faces behind the triplex screens. 1970 P. Dickinson *Seals* i. 26 Even through that grimy and half-opaque triplex, the harbour had seemed awkwardly placed.

4. a. An apartment or other residence on three floors. Chiefly *U.S.*

1932 E. Ferber in *Hearst's Internat.* Mar. 18/1 Photographs in the magazines showed her glamorous apartment—triplex, with balcony overhanging the East River. 1978 *New York* 3 Apr. 91 (Advt.), 40' converted mansion, elevator, magnificent owner's triplex plus high income apartments. 1981 *Times* 3 Aug. 10/5 A 21-room 'triplex' (ie, on three floors) going for 'only' $9 million.

b. A building containing three self-contained residences or suites of rooms; also, one of the dwellings in such a building. Cf. *DUPLEX *sb.* 1. Chiefly *N. Amer.*

1962 *Maclean's Mag.* (Toronto) 10 Mar. 37/1 They wanted to build three triplexes. 'Definitely not,' Reeve Fred Hall told Norman. 1971 A. Blaisdell *Practice to Deceive* i. 2 She lived in one unit of a triplex. 1971 *Rand Daily Mail* (Home Owner) 27 Mar. 7/2 City dwellers are gravitating towards high density living (flat complexes.. duplex and triplex..and so on). 1976 *Billings* (Montana) *Gaz.* 30 June 7-D/1 (Advt.), Accelerated depreciation is still available on this new brick tri-plex. This unit features a fantastic view. 1979 *Arizona Daily Star* 5 Aug. B 1/6 She observes construction workers working on a triplex where her display company's warehouses once stood.

Triplice (tri·plitʃe). Also with small initial. [a. It. *triplice* triple; cf. TRIPLEX *a.* (*sb.*).] The Triple Alliance (see TRIPLE *a.* 5) of Germany, Austria-Hungary, and Italy, formed in 1882 against Russia and France. Also *transf.*

1896 *Daily News* 15 Dec. 5/3 The same demand has been made to the other members of the Triplice. 1897 *Ibid.* 6 Sept. 5/2 The Triplice desires peace at any cost. 1901 *Speaker* 20 Apr. 74/2 It would be misleading to call the *Triplice* a League of peace. 1908 *Daily Chron.* 30 July 3/6 A suggestion for a new alliance, a triplice of Britain, France, and Germany, in the interests of peace, was made by Mr. Prust, of Launceston. 1911 Djavid Bey *Let.* 28 Oct. in R. S. Churchill *Winston S. Churchill* (1969) II. Compan. II. xvii. 1368 The attack of one of the triple alliance powers on our territory has turned the public opinion greatly against the triplice. 1947 *Hist. of The Times* III. x. 250 The conclusion that the Triplice.. was weak as an Alliance, was forced to the forefront of Wallace's mind. 1979 E. Ingram *Beginning of Great Game* ix. 263 The tsar forestalled Palmerston's attempt to erect a near-eastern triplice [of Britain, Austria and France].

Hence **Tri·plicist**, a supporter of the Triplice; also *attrib.* or as *adj.*

1923 J. Buchan *Nations of To-day: Italy* 172 It was able to give a decidedly 'Triplicist' aspect to the enterprise. 1924 *Glasgow Herald* 15 Jan. 7 It is all to the credit of the Triplicists..and very little to that of the Allies.

triploid, *sb.* Restrict *rare.* ? *Obs.* to sense in Dict. and add: **B.** *adj. Genetics.* [Cf. *-PLOID.] (Made up of somatic cells) containing three sets of chromosomes. Also as *sb.*, a triploid organism.

1911 *Ann. Bot.* XXV. 933 There is not, so far as I am aware, a single case of a species whose sporophyte has the triploid number of chromosomes. 1927 *Jrnl. Genetics* XVIII. 183 These five are all the configurations which have been seen among several hundred trivalents of triploids..in *Datura* and *Canna*. 1930 *Times Lit. Suppl.* 10 July 578/3 Triploid apples (e.g., Bramley's Seedling) fruit well. 1957 R. A. Beatty *Parthenogenesis & Polyploidy in Mammalian Devel.* iv. 59 Three of the young [rabbits] lived nearly a year, being the two 'triploids' and a presumed diploid male. 1971 [see *DIPLOID *a.*]. 1973 [see *OCTOPLOID *a.* (*sb.*)].

Hence **tri·ploidy**, the state of being triploid.

1916 *Genetics* I. 237 It seems probable that we have in these plants examples of triploidy (21 chromosomes). 1961 *Lancet* 5 Aug. 318/1 A malformation syndrome associated with triploidy. 1976 *Cytogenetics & Cell Genetics* XVII. 144 Triploidy or diploid/triploid chimerism has been reported in man. 1982 *Annals Human Genetics* XLVI. 223 Human triploidy is a common condition.

‖ **triplum** (tri·plŭm). *Mus.* Pl. **tripla.** [med. L., neut. of *triplus* TRIPLE *a.*; cf. TREBLE *sb.* 4.] In thirteenth- and fourteenth-century polyphonic vocal music, the third voice part, next but one above the tenor.

1782, 1884 [see TREBLE *sb.* 4]. 1944 W. Apel *Harvard Dict. Mus.* 223/1 *Triplum*, *quadruplum* are other parts above the tenor, frequently of the same range as the duplum. 1954 *New Oxf. Hist. Music* II. xi. 354 The motet 'Salve virgo virginum/Est il donc ainsi/Aptatur' has 202 notes in the *triplum* (highest voice), 173 in the *motetus* (middle voice), and only 103 in the tenor. 1977 *Early Music* Apr. 185 A glance through the tripla and dupla of 13th-century motets is enough to show that rhythmic styles could be tolerated in music which made a nonsense of poetic scansion.

tripod, *a.* **1.** (Later examples.)

1939 *Oxoniensia* IV. 101 Thirty-seven base and body fragments of tripod pitcher..buff ware, uniform in colour throughout, fairly well fired and hard. 1963 E. M. Jope in Foster & Alcock *Culture & Environment* xiii. 342 Regional styles are still discernible among these glazed jugs, such as the tripod-pitchers of the twelfth century.

Tripoli. Add: **2.** A large, mild onion; also, the plant producing a bulb of this kind. Also *attrib.*

1822 J. C. Loudon *Encycl. Gardening* III. i. 715 Tripoli, the largest onion grown; oval, light-red, tinged with green and brown, soft and mild. 1873 *Young Englishwoman* Sept. 446/2 The best varieties [of onion] for autumn sowing are the Tripoli, Giant Madeira, [etc.]. 1932 *Times Educ. Suppl.* 27 Feb. (Home & School Suppl.) p. iv/1 The following crops should be started as soon as possible:.. tripoli onions, shallots, [etc.]. 1951 *Dict. Gardening* (R. Hort. Soc.) III. 1425/2 Such onions as the Roccas, White Lisbon, and Tripoli..are of no value for storing.

Tripoline, *a.* Add: **2.** [f. *Tripoli*, the name of a city and port in North Africa.] Of or pertaining to Tripoli, now the capital of Libya. Also as *sb.*, a native or inhabitant of Tripoli. Cf. next.

1819 A. Salamé *Narr. Exped. Algiers* 6 We found she was a Tripoline polacca, (I am sorry that she was not an Algerine). 1843 *Penny Cycl.* XXV. 256/1 The Tripoline cruisers seldom allowed a ship at sea to escape them if they thought they could make a prize of her with impunity. *Ibid.* In 1832, Yussuf, the last basha of the Caramanli family,..having lost the affection of the Tripolines, after a reign of forty years,..was obliged to abdicate. 1909 [see *TRIPOLITAN *a.*].

Tripolitan (tripǫ·lităn), *a.* (and *sb.*) [ad. It. *tripolitano* (in sense a); in sense b, f. *Tripolitania* (see next).] **a.** = TRIPOLINE *a.* 2. **b.** = TRIPOLITANIAN *a.* Occas. as *sb.*

1783 Miss Tully *Let.* 3 July in *Narrative Ten Years' Residence at Tripoli* (1817) 5 The Tripolitan dresses, almost covered with gold and silver,..make a most superb appearance. *Ibid.* 3 Sept. 24 The Tripolitans carry the right hand from the breast to the forehead..where they mean to be respectful. 1888 *Encycl. Brit.* XXIII. 575/2 The Tripolitan pirates soon became the terror and scourge of the Mediterranean. 1896 *Geogr. Jrnl.* VII. 150 The Tripolitan range of hills. 1909 G. W. Furlong *Gateway to Sahara* i. 8 'Tripolitans' signifies the people of the territory, 'Tripoline' a dweller in the town of Tripoli. 1913 R. G. Usher *Pan-Germanism* xii. 184 From the ports on the Tripolitan coast..a flank attack could be directed upon the English communications with Suez. 1928 V. G. Childe *Most Anc. East* ii. 25 In the very heart of the Sahara at In-Ezzan, just south of the Tripolitan borders.

Tripolitanian (tripǫlitēⁱ·niăn), *sb.* and *a.* [f. *Tripolitania*: cf. -IAN.] **A.** *sb.* A native or inhabitant of Tripolitania, the region surrounding Tripoli in North Africa. **B.** *adj.* Of or pertaining to Tripolitania. Cf. prec., sense b.

1942 [see *CYRENAICAN *a.*]. 1943 G. Casserly *Tripolitania* 5 When this book was begun no enemy stood on Tripolitanian soil. 1963 M. Khadduri *Modern Libya* ii. 48 Native Tripolitanians were soon to enter the service and their number steadily increased. 1969 J. Wright *Libya* v. 55 The Roman empire has been called 'a federation of municipalities' and the Tripolitanian townsmen derived more benefit from Roman rule than the countryman. 1978 A. Melville-Ross *Blindfold* x. 60 The Tripolitanians are very much afraid of the nomads. 1981 T. Barling *Bikini Red North* i. 29 Do what you can with your Tripolitanian argot.

Tripolye (tripǫ·lyĕ). Also **Tripolje.** The name of a village near Kiev in Russia, used *attrib.* to denote a neolithic culture typified by remains found there, which flourished in the western Ukraine and in eastern Romania during the late fourth and third millennia B.C.

1913 [see *PLOSHCHADKA]. 1935 *Jrnl. R. Anthrop. Inst.* LXV. 113 In successful search for painted pottery of the Tripolye type. 1937 *Discovery* Jan. 27/1 The Tripolje culture, so called from the site in Southern Russia on which it was first found, has as its most characteristic feature painted pottery. 1957 G. Clark *Archæol. & Society* (ed. 3) vi. 190 Significant differences have been noted between the composition of animal remains from sites of the Tripolje culture marked by painted and grooved pottery and from those of the neighbouring Ousatovo culture. 1970 Bray & Trump *Dict. Archaeol.* 237/2 Tripolye culture came to an end with the expansion westwards of steppe cultures of kurgan or Single-Grave type.

‖ **tripot** (trī·po). [Fr.] A gaming-house, a gambling-den.

1864 W. H. Ainsworth *John Law* I. II. v. 290 The person before us..is a suspected sharper, and a constant frequenter of tripots. 1883 'Ouida' *Wanda* I. vii. 264 A winner at a *tripot*, what a hero for you, mother mine. 1909 R. Nevill *Light come, Light Go* viii. 236 The gaming-resorts of old Paris were filled with people whose reputations for probity were generally a good deal more than doubtful. In one of the best of these *tripots* a gentleman..delayed the game by insisting on searching for a few pieces of gold which he had dropped on the floor. 1930 A. Bennett *Imperial Palace* xlv. 319 The loss of sixty pounds odd..in a Paris *tripot*.

‖ **tripotage** (trīpotā·ʒ). [Fr.] **a.** Underhand dealings, intrigue. Also *fig.* **b.** *rare.* Pawing, handling, fingering.

In quot. 1853 perh. used mistakenly for *tripotée* large quantity.

1779 Dr. Warner *Let.* Mar. in J. H. Jesse *George Selwyn & his Contemporaries* (1844) IV. 38 An infinite deal of lying, on all sides, and *tripotage*. *Ibid.* 42 But it is all of a piece, such a cursed *tripotage*! 1853 C. Brontë *Villette* III. xxxvi. 131 At last I got through my list. The patterns for the slippers, the bell-ropes, the cabas were selected—the slides and tassels for the purses chosen— the whole 'tripotage', in short, was off my mind. 1895 *Nineteenth Cent.* Oct. 548 The recent exposures of political *tripotage*. 1932 *Times Lit. Suppl.* 12 May 343/3 Beau Nash, though dabbling deeply in the *tripotage*, deprecated sensational ruins and suicides. 1958 J. Lodwick *Bid Soldiers Shoot* ii. 41 The freshly shaved Hirsch..cornered the two women and subjected their not unwilling flesh to expert *tripotage*.

trippage (tri·pėdʒ). [f. TRIP *sb.*¹ + -AGE.] The act or process of making a series of short journeys over the same route; the number of such journeys made.

1941 *Sun* (Baltimore) 16 Oct. 7/2 Dairy officials pointed out that the new system would eliminate 'back trippage of drivers' and make for 'straight trippage'. Under the old system, most drivers made early morning deliveries and then, on collection days, repeated the routes to collect the money. The second trip will be eliminated under the daylight delivery system. 1972 *Daily Tel.* 5 Apr. 19/4 If milk distribution costs are to be stabilised, some means must be found to push up the trippage rate [of milk bottles]. 1979 *Ibid.* 6 Aug. 2/6 The average milk bottle makes 23 trips to and from the dairy and customer before it is lost or broken... Bottles are still used because their high trippage more than offsets other costs.

tripped (tript), *ppl. a.* [f. TRIP *v.* + -ED¹.] **1.** *Bot.* Of a flower whose pollinating mechanism has been activated by tripping.

1914 *Bull. U.S. Dept. Agric.* No. 75. 5 In artificially tripped flowers..12 out of 34 tripped set seed. 1956 *Nature* 18 Feb. 334/2 The number of tripped florets was recorded and calyces marked each day. 1980 *Sabrao Jrnl.* XII. 104 The seed set of tripped flowers indicates the level of self-fertility.

2. Operated or caused to respond by contact with a projection or other object.

1921 *Conquest* Jan. 130/1 A roller which reacts the 'tripped' brushes or wipers. 1977 R. Ludlum *Chancellor Manuscript* xxx. 320 Don't worry, there's a set of electronically tripped windows..with..bulletproof glass.

3. *tripped-out*: under the influence of a hallucinogenic drug, esp. LSD. *slang.*

1973 *Listener* 15 Nov. 680/3 The tripped-out nudes in the penthouse flat. 1976 H. Ferguson *Confessions of Long Distance Acid Head* 11 Everyone was gathered round talking about the arrangements they would make for their 'excursion' the following day. They cared so little for my tripped-out state that they turned out the light and left me in the darkened room.

tripper, *sb.* Add: **5.** (Examples.)

1882 J. D. McCabe *New York by Sunlight & Gaslight* 244 The 'trippers', as those men are called who only run three-quarters of a day, get $1·50. 1950 *Reading* (Pennsylvania) *Times* 28 Feb., They had refused to operate 'tripper' or extra, runs because five members of the union had been furloughed. The company said this situation resulted in the failure of nine 'tripper' runs to be made.

7. *slang* (orig. *U.S.*). One who experiences hallucinations induced by a drug, esp. LSD.

1966 T. Leary in *Playboy* Sept. 110/2 These episodes can be dealt with easily by an experienced guide who recognizes where the LSD tripper is caught. 1968 *New Scientist* 3 Oct. 38/3 LSD 'trippers'..need no sleep. 1972 *Village Voice* (N.Y.) 1 June 78/4 When I returned several days later, Wheeler's was in an uproar over the discovery of a dead tripper. 1979 B. Malamud *Dubin's Lives* i. 29 One of the swamis there, a secret acid tripper, got on my nerves.

tripper (tri·pəɹ), *v. colloq. rare.* [f. the sb.] *intr.* To behave like a tripper (sense 4); to take trips or excursions.

1959 G. Jenkins *Twist of Sand* ii. 41 Trippering up and down the coast. 1974 'S. Harvester' *Forgotten Road* iii. 37 They trippered around Istanbul for some days.

tripperish (tri·pəriʃ), *a. colloq.* [f. TRIPPER *sb.* + -ISH¹.] = *TRIPPERY *a.* So **tri·pperishness.**

1898 M. Sadler *Let.* 26 Sept. in M. Sadleir *Michael Ernest Sadler* (1949) x. 166 Stromness is prim little grey stone place of about 2,000 people—quite unspoiled... No obtrusive tripperishness. 1931 E. Waugh *Remote People* iv. 89 Do you think..it would be very vulgar and tripperish to make them scramble for them [*sc.* coins]? 1934 M. Allingham *Death of Ghost* xxi. 245 Nothin' tripperish about them. 1960 'J. & E. Bonett' *No Grave for Lady* xi. 170 The atmosphere was rather tripperish. 1975 S. Lauder *Killing Time on Corvo* iii. 26 She was very sharp.. and tripperish.

trippery (tri·pəri), *a. colloq.* [f. TRIPPER *sb.* + -Y¹.] Of, pertaining to, or characterized by the presence of trippers (sense 4); touristy (somewhat *derog.*).

1924 C. Connolly *Let.* 22 Dec. in *Romantic Friendship* (1975) 41 Italy is so trippery in Spring. 1926 W. J. Locke *Old Bridge* ii. 18 'But let us see all we can to-night.'.. 'That wouldn't be fair to Florence. It's a bit trippery, isn't it?' 1928 *Daily Express* 11 Apr. 9/4 Venice,..the Queen of the Adriatic in its most trippery and least attractive garb. 1969 M. Pugh *Last Place Left* xix. 145 Wordsworth had complained of trippery overcrowding in this belt, more than a hundred years before. 1970 *Nature* 4 July 11/1 The trippery appurtenances and atmosphere which have been established in the neighbourhood [of the Niagara Falls]. 1971 *Country Life* 20 May 1234/3 All that is missing are the monkeys: even though Iguazu is very definitely far from trippery, they have taken themselves off.

tripping, *vbl. sb.* Add: **1.** (Examples in *Bot.*)

1909 *Bull. Bureau Plant Industry, U.S. Dept. Agric.* No. 24. 8 If fertile seed is to be produced in any quantity it is necessary that a certain explosive mechanism within the flower be released. The release of this mechanism, whether it be accomplished by insects or otherwise, is popularly called 'tripping'. 1930 *Jrnl. Amer. Soc. Agronomy* XXII. 782 When the flowers were left exposed and not tripped artificially the gain was 1:1·7 in favor of artificial tripping. 1978 *Nature* 7 Sept. 54/2 In artificial field bean pollination, manual tripping of open flowers is a recommended practice for increasing seed set in auto-sterile lines.

2. (Examples in sense *5 b of the verb.)

1968 L. W. Robinson *Assassin* xii. 128 Their passion was a long one..as though they hated to come back..from the rocking, tossing, sweet trip... But no, the sweet tripping was not over. 1970 K. Platt *Pushbutton Butterfly* vi. 59 The girls weren't wearing brassieres... The skinnier ones just looked flat. Tripping didn't solve everything. 1980 *Times Lit. Suppl.* 24 Oct. 1203/4 When Christiane F. was thirteen years old, she began to frequent a youth club run by the Protestant Church in an overcrowded district of West Berlin. There she started smoking hashish, taking 'uppers and downers' and 'tripping' on LSD.

trippkeite (tri·pki,əit). *Min.* [ad. G. *trippkeit* (G. vom Rath 1880, in *Sitzungsber. d. Niederrheinische Ges. f. Natur- und Heilkunde* 209), f. the name of P. *Trippke* (1851–80), Polish or German mineralogist, its discoverer: see -ITE¹.] An oxide of copper and arsenic, $CuAs_2O_4$, found as soft greenish blue tetragonal crystals.

1881 *Jrnl. Chem. Soc.* XL. 551 Minerals from the Veins of Copper-ore near Copiapo, in Chili.—*a.* Trippkeite. A cupric arsenite..occurring in small bluish-green crystals, in druses of a thick bed of red copper-ore mixed with malachite and copper pyrites. 1976 *Mineral. Abstr.* XXVII. 300/1 The crystal structure of synthetic trippkeite..was refined to R = 0·059.

trippy (tri·pi), *a. colloq.* (chiefly *U.S.*). [f. TRIP *sb.*¹ + -Y¹.] Of, pertaining to, or resembling a hallucinatory experience induced by a drug. Hence **tri·ppiness.**

1969 Fabian & Byrne *Groupie* (1970) xiv. 101 Joe asks if it's trippy, and I tell him that it is a bit. 1975 *Harper's Mag.* June 9 In my trippy daze, dope was the filter for the movie camera in my mind, the regulator of my psychic jets. 1976 *New Yorker* 19 Jan. 48 Robert Wise directed with tame, impersonal good taste; there's none of the blissful trippiness of being carried in the belly of a zeppelin, and none of the carnival vulgarity of the recent disaster thrillers. 1976 Logan & Woffinden *Illustr. New Musical Express Encycl. Rock* 96/1 In 1966 they [*sc.* The Grateful Dead] chose their name; in keeping with the band's image, it was a decision made under the influence of various drugs... The band thought it seemed vaguely appropriate, and certainly it had trippy connotations. 1978 *Maclean's Mag.* 13 Nov. 78 The trippy optimism of the '60s lent importance to such things as creativity and communication, which in the '70s have given way to matters of a homelier urgency. 1980 *New Age* (U.S.) Oct. 54 (Advt.), Trippy music for meditation, massage, free-form movement, tantric loving, and a relaxing environment.

triptane (tri·ptēⁱn). [f. TRI- + -*p*- + *BU)TANE.] A liquid branched paraffin used as a high-octane aviation fuel; 2,2,3-trimethylbutane, $CH_3C(CH_3)_2CH(CH_3)CH_3$.

1943 *Chem. & Engin. News* 25 Sept. 1561/1 Triptane is the most powerful hydrocarbon known for use in internal combustion engines. 1970 M. Smith *Aviation Fuels* x. 65 Some of the best hydrocarbons, such as triptane, do not occur naturally in significant amounts.

tripton (tri·ptǫn). *Biol.* and *Oceanogr.* [ad. G. *tripton* (J. Wilhelmi 1917, in *Arch. für Hydrobiol.* XI. 115), f. Gr. τριπτόν, neut. of τριπτός that which is rubbed or pounded, f. τρίβειν to rub or pound.] The non-living part of the fine particulate matter suspended in water. Cf. *SESTON.

1931 R. N. Chapman *Animal Ecol.* xvi. 325 The dry organic matter has been selected as the measure because it eliminates a large amount of detritus ('Tripton' of Wilhelmi, 1917). 1957 G. E. Hutchinson *Treat. Limnol.* I. xvii. 897 The phytoplankton, or possibly bacteria associated with organic tripton, produce a good supply of thiamin. 1978 *Nature* 21 Sept. 194/1 As spring and summer pass, decaying organic matter (tripton) from the overlying water accumulates on or near the chemocline where its continued decomposition removed any dissolved oxygen from the overlying water mass.

triptych. Add: **3.** *transf.* **a.** A set of three operas or pieces of music intended to be performed together.

1925 R. A. Streatfeild *Opera* (ed. 5) xiii. 304 Puccini's last work is a so-called 'triptych', consisting of three one-act operas. 1928 *Grove's Dict. Mus.* (ed. 3) IV. 283/2 In this triptych the composer's technique is more elaborate than in 'Butterfly'. 1959 *Listener* 31 Dec. 1176/3 Any music that makes a strong visual suggestion like the Debussy orchestral triptychs. 1976 *New Yorker* 1 Mar. 90/1 This season, the triptych has been reassembled: a new 'Tabarro' and 'Suor Angelica' join the 1974 staging of 'Gianni Schicchi'. 1976 *Gramophone* Sept. 424/2 Ormandy's version (which offers 24 minutes' extra music in the form of *Feste romane*, the still more luridly coloured third leaf of the triptych) will do very nicely.

b. *Cinemat.* A sequence of film designed to be shown on a triple screen, using linked projectors.
1976 *Oxf. Compan. Film* 494/2 After the first presentation it [sc. *Napoléon*] was released in a truncated version from which the triptych sequences had been removed: Gance, disappointed by the poor reception, destroyed much of the original footage, including some of the triptych. 1980 *Times* 5 Dec. 11/5 The great triptych—Gance called it Polyvision—in no respect falls short... From the breath-catching moment when the screen is suddenly multiplied to reveal a great panorama of the Grand Army on the Alps, Gance's use of the triptych is light years in advance of anything three-projector Cinerama ever achieved. *Ibid.*, Sometimes the triptych image is a continuous panorama; sometimes it is split into different images. There are superimpositions and mirror images, the whole orchestrated with passion.

‖ **tripudium** (tripiū·diŏm). *Rom. Antiq.* [L.; see Tripudiate v.] A ritual dance (see quots.). Also *transf.* and *fig.*
1909 in Webster. 1922 W. R. Halliday *Lect. Hist. Roman Relig.* iii. 46 A feature of this procession was the dancing of the armed priests... Their leaping dance, the *tripudium* or three step, was accompanied by the clashing of rods or spears against the shields. 1922 Joyce *Ulysses* 50 The foot that beat the ground in tripudium. *Ibid.* 559 He runs to the piano and takes his ashplant, beating his foot in tripudium. 1923 L. Pullan *Relig. since Reformation* viii. 239 The Tübingen school attempted..at the same time 'to force Christian history into the Hegelian *tripudium* of thesis, antithesis, and synthesis'. 1938 B. Schönberg tr. *C. Sachs' World Hist. Dance* vii. 246 The weapon dances of the warriors and the priests of Mars who were grouped together under the name of *Salii*, which is equivalent to *saltantes* or dancers... The *Salii* stamped..in repetends of three beats... From this tripedal character their dance takes the name of *tripudium*. As a choral dance..it had a dance leader whose movements were answered by the two choruses of older and younger men as they walked around in a circle to the rhythmical beating of the shields. 1949 *Oxf. Classical Dict.* 789/1 At certain spots they [sc. the Salii] halted and performed elaborate ritual dances (*tripudium*, cf. Plut. *Num.* 13), beating their shields with their staves.

tripuhyite (tripū·i,əit). *Min.* [f. *Tripuhy*, name of a locality near Ouro Prêto, Minas Gerais, Brazil, where the first specimen was found: see -ite¹.] An oxide of ferrous iron and antimony, $FeSb_2O_6$, found as aggregates of translucent, yellowish to dark brown, tetragonal crystals.
1897 Hussak & Prior in *Mineral. Mag.* XI. 302 From these schists, doubtless, is also derived the tripuhyite, although as yet this new mineral has only been found in fragments loose in the gravel. 1968 [see *ordoñezite].

tripus. Restrict ‖ *Obs. rare* to sense in Dict. and add: **2.** *Zool.* A bone in the Weberian ossicles of cypriniform fishes, linking the ear and the swim-bladder.
1893 T. W. Bridge in *Phil. Trans. R. Soc.* B. CLXXXIV. 83 The horizontal process moves backwards or forwards with the lateral motion of the tripus. 1962, 1970 [see *intercalarium 2].

tri·p-wire. Also trip wire, tripwire. [f. Trip sb.¹ or v. + Wire sb.] **1. a.** A wire stretched near the ground in order to trip up enemies, trespassers, etc. Hence, a wire placed so that contact with it operates a weapon, flash-light, or other device.
1916 A. Knebworth *Let.* 24 Feb. in Ld. Lytton *Antony* (1935) i. 21 He walks forward, he has found his landmark. He thinks he knows where the Huns are. He is coming to the Hun trip wire. He has cut the German trip wire. 1928 *Daily Mail* 3 Aug. 8/3 Trip-wires to ensnare the enemy. 1928 *Daily Tel.* 16 Oct. 18 A flash-light operated by means of a 'trip wire'. 1941 *Illustr. London News* 22 Feb. 233/1 (*caption*) The mine can be fired by various methods such as electric contact or time fuse—trip wire or impact. 1947 D. M. Davin *Gorse blooms Pale* 124 They had time..to lace the stumps with barbed trip-wires. 1960 C. Day Lewis *Buried Day* v. 100 The window-cleaner's tricycle was built up to represent a German tank, which was laagering in a dell I had privily surrounded with trip-wires. 1974 *Times* 21 Jan. 12/5 There's a series of trip wires which set off rockets and flares if they are touched. 1978 'F. Parrish' *Sting of Honeybee* vi. 83 Dan wondered about dogs, electric fences, trip-wires, gin-traps.
b. *transf.* and *fig.*
1971 P. O'Donnell *Impossible Virgin* vi. 117 He was operating on more than one level. He may have meant his offer, but he was laying trip-wires at the same time. 1976 Ld. Home *Way Wind Blows* xiv. 195 A Prime Minister..is well-advised to search every question for the trip-wire which is usually well-concealed, but almost sure to be there, and to think up the riposte which will turn the tables on the Opposition. 1979 P. Niesewand *Member of Club* ix. 63 One other type of sensor..sets up an invisible light beam... If someone walks across it, they interrupt the beam. It's a kind of optical tripwire.
2. *fig.* A comparatively weak military force employed as a first line of defence, whose involvement in hostilities will trigger the intervention of stronger forces. Freq. *attrib.* orig. *U.S.*
1957 *Observer* 1 Sept. 8/3 The German electorate are baffled as to whether Nato is meant to defend their soil, or provide the tripwire for a Soviet–American suicide pact. 1960 *Washington Post* 4 Apr. A 19 Stans suggested that a switch be made to the 'trip-wire' defense theory which would require but one American division. 1966 Schwarz & Hadik *Strategic Terminology* 115 Advocates of this modification ridicule the simple tripwire concept by saying that to all intents and purposes a single U.S. soldier could act as tripwire. 1969 *New Statesman* 11 Apr. 500/3 He [sc. King Hussein] is anxious to make a separate peace with the Israelis on the basis of a demilitarised West Bank, with an Israeli military tripwire on the Jordan. 1976 Ld. Home *Way Wind Blows* xii. 167 There was, however, a running argument among the professionals as to whether the line between the Warsaw Pact and the NATO forces should be thinly held (by a trip-wire) or more strongly manned. 1979 *Jrnl. R. Soc. Arts* CXXVII. 550/2 From the mid 1950s to the late '60s, the West relied on the so-called 'tripwire' strategy. Stated simply, this meant that any aggressive adventure on the part of the Soviet Union would be met by an overwhelming nuclear response. 1980 *Times* 24 May 15/2 What is profoundly discouraging is to find our work impeded by the old discredited trip-wires of the Cold War.

Trique (trī·ke), *sb.* and *a.* Also Trike, Triqui. [Native name.] **A.** *sb.* **a.** An Indian people of Oaxaca, Mexico. **b.** The Mixtecan language spoken by this people. **B.** *adj.* Of or pertaining to this people or their language.
1891 D. G. Brinton *Amer. Race* iii. 148, I do not doubt that Orozcoy Berra was right in placing the Triquis in the same [Tequistlatecan] family. 1900 F. Starr in *Proc. Davenport Acad. Sci. 1899–1900* (1901) VIII. 142 Belmar gives in his *Essays* a brief sketch of the grammar, a list of phrases in Spanish and Triqui and a Spanish-Triqui vocabulary. *Ibid.*, The towns he mentions are none of them Triqui. *Ibid.*, San Andres Chicahuastla is the Triqui town where our work was done. 1911 *Bull. U.S. Bureau Amer. Ethnol.* No. 44. 52 Trike. This language, which belongs to the Zapotecan family, is spoken by a small tribe residing in the central part of the Mixtec area. *Ibid.* 53 Professor Starr..says none of the towns mentioned by Orozco y Berra are Trike..and that the real district of the Trike is situated in the high mountains of the districts of Tlaxiaco and Juxtlahuaca... They form a little island of Trike speech in the midst of the Mixtec area. 1952 J. R. Swanton *Indian Tribes N. Amer.* 639 Trique, a tribe entered by Mason and Johnson as a substock of their Otomanguean stock. 1957 [see *Mazatec sb. and a.]. 1965 *Language* XLI. 67 The identifier tagmeme in Trique noun phrases. 1974 *Encycl. Brit. Micropædia* X. 130/1 *Trique*, Indians of Oaxaca, Mex., speaking a Mixtecan language... Some Trique men work outside the community as labourers... Trique religion includes both ancient Indian and Roman Catholic rites. 1977 T. A. Sebeok *Native Languages Americas* II. 370 Trique. Mixtecan; 4,000 in Oaxaca (1952).

triquetral, *a.* Add: *triquetral bone*, also, an approximately pyramidal bone of the wrist that articulates with the pisiform bone; = *cuneiform bone* (*a*) s.v. Cuneiform a. 1; also *ellipt.* as *triquetral*.
1913 *Cunningham's Text-bk. Anat.* (ed. 4) 223 An exceptional case..in which the centres for the capitate and triquetral bones were already present [at birth]. 1961 S. Zuckerman *New Syst. Anat.* i. iv. 84 The triquetral bone forms a conspicuous prominence distal to the head of the ulna on the medial border of the dorsum of the hand. 1980 *Gray's Anat.* (ed. 36) 371/2 The palmar and dorsal surfaces of the carpal bones, apart from the triquetral and pisiform, are rough for the attachment of ligaments.

triradius (traire¹·diŏs). Pl. -radii. [f. Tri- + Radius.] In dermatoglyphics, a point from which the dermal ridges radiate in three directions at angles of approx. 120 degrees.
1960 *New Scientist* 14 July 129 (*caption*) A finger-print on which a white line has been drawn joining the core of the pattern..to the associated triradius. 1965 *Ibid.* 11 Feb. 345/2 There are discontinuities [in the fingerprint pattern], known technically as 'triradius points', where three ridges meet at a single junction. 1970 [see *loop sb.¹ 4 h]. 1971 J. Z. Young *Introd. Study Man* xxxix. 570 For genetic analysis counts are made from the triradius to the centre of the pattern. 1977 *Sci. Amer.* Dec. 141/1 The resulting patterns are known to the dermatologist respectively as loops, triradii and whorls.

tris (tris), *sb.* *Chem.* Also Tris. [f. Tris-.] **1.** [f. *tris*hydroxymethylaminomethane.] The crystalline compound $(HOCH_2)_3CNH_2$, 2-amino-2-(hydroxymethyl)propane-1,3-diol, used as a buffering agent. Also *tris buffer*.
1959 *Science* 20 Mar. 783/1 This compound is commonly known as 'trishydroxymethylaminomethane' or 'tris buffer'. 1964 *Biochim. & Biophysica Acta* XCII. 133 A uniform suspension of ghosts in 0·5 mM Tris. 1979 *Sci. Amer.* Mar. 102/2 The sodium current in a squid axon is abolished by applying a sizable dose of tetrodotoxin (and also replacing the sodium in the bathing medium with 'Tris' buffer).
2. [f. *tris*-2,3-dibromopropylphosphate.] The organophosphorus compound $(Br_2C_3H_5)_3PO_4$, which is used as a flame retardant.
1976 *St. Louis* (Missouri) *Globe-Democrat* 17 Sept. 6 B/1 A chemical nicknamed 'tris' that clothing manufacturers use as a flame retardant in children's pajamas causes mutations in the genes of bacteria. 1981 M. C. Gerald *Pharmacology* (ed. 2) xxx. 578 To date there is no evidence that Tris causes cancer in humans.

trisazo (trisæ·zo), *a.* *Chem.* [f. Tris- + Azo-.] Containing three azo groups in the molecule.
1904 *Jrnl. Chem. Soc.* LXXXVI. i. 700 Black trisazo-dyes are obtained by diazotising acetyl-*p*-phenylenediamine, [etc.]. 1948 Kirk & Othmer *Encycl. Chem. Technol.* II. 247 Naphthogene Blue 4R.., a trisazo dye. 1966 [see *disazo-]. 1970 R. L. M. Allen *Colour Chem.* v. 62 Three of the various types of trisazo structure are of principal importance.

Triscuit (tri·skit). Also triscuit. [f. Tri-, irregularly after *biscuit*.] The proprietary name of a savoury cracker or biscuit.
1906 *Official Gaz.* (U.S. Patent Office) 27 Mar. 1324/2 Biscuit or crackers. The Natural Food Company..The word 'Triscuit'. 1919 'G. Cumberland' *Set down in Malice* xiv. 174 They have..studios..where one has triscuits for breakfast. 1932 *Trade Marks Jrnl.* 23 Mar. 383/1 *Triscuit*..Solid food products. The Shredded Wheat Company, Limited, Welwyn Garden City..Manufacturers. 1937 G. Frankau *More of Us* viii. 93 Stern strong business-men Whom Shredded Wheat or Triscuit made that hearty They rarely failed to catch the ten-past-ten. 1980 *Times* 22 Dec. 12/7 Ketchup in a bottle, salt in a shaker, triscuits (a savory snack biscuit) in the box.

tri-service (trəisə̄·ɹvis), *a.* Also tri-Service, Tri-Service. [f. Tri- + Service¹ 5 b.] Of, pertaining to, or representing the three armed forces, Army, Navy, and Air Force.
1959 *Times* 18 June (Queen in Canada Suppl.) p. vii/7 A unique experiment in military education, a tri-service college. 1977 *Globe & Mail* (Toronto) 1 Feb. 7/6 Much has been made of the damage supposedly done by unification. The opponents of it, both in the services, and in groups on the outside such as TRIO (Tri-Service Identities Organization), concentrated on this one issue. 1979 *Navy News* Feb. 1/5 The survey, which will be tri-Service, will cover all personnel, including Servicewomen. 1982 *Daily Tel.* 15 Dec. 24/4 The White Paper confirmed the maintenance of a sizeable tri-service garrison in the South Atlantic.

trishaw (trəi·ʃǭ). Also trisha, tri-sha, tri-shaw. [f. Tri- + Rick)shaw.] In the Far East, a light three-wheeled vehicle propelled by pedalling, freq. used as a taxi.
1946 *Sun* (Baltimore) 9 July 5/6 (*heading*) Trishaws may replace Singapore Rickshaws. *Ibid.*, The Rickshaw Association has asked permission for 2,000 additional trishaw licences to provide employment for former rickshawmen. 1955 [see *pedal v. b]. 1961 *Guardian* 30 May 7/3 The Chinese trishaman..who failed to move out of his way. 1971 *Carry Singapore in your Pocket* (Singapore Tourist Promotion Board) (ed. 3) 71 *Trishaw*, a pedal bicycle with side-car attached. This typically Oriental mode of transport is fast dying out in Singapore but a short 'trishaw ride' is certainly worthwhile. 1972 *Daily Colonist* (Victoria, B.C.) 26 Mar. 49/3 The Chinese introduced a souped-up version of the jinricksha in the form of the tri-sha. These bicycle-powered two-seaters weave their way through George Town traffic, as common as taxis are to the streets of other cities. 1977 P. Theroux *Consul's File* 79 The cycling noises approached... It was a trishaw, cruising for fares. 1979 K. Cassilis *Arrow of God* iv. vi. 119 The tri-shaw man came over to help him.

trishtubh, var. *tristubh.

triskaidekaphobia (tri:skəidekăfōᵘ·biă). Also triske-, -decaphobia. [f. Gr. τρεισκαίδεκα thirteen + -phobia.] Fear of the number thirteen.
1911 I. H. Coriat *Abnormal Psychol.* ii. vi. 287 Fear of the number thirteen (triskaidekaphobia). 1953 *N.Y. Times* 8 Nov. E2 A discussion in the U.N. last week on the number of members on a committee raised the question of triskaidekaphobia. 1967 *Daily Tel.* 14 Jan. 18/8 Thirteen people, pledged to eliminate triskaidecaphobia, fear of the number 13, today tried to reassure American sufferers by renting a 13ft plot of land in Brooklyn for 13 cents (10½d) a month. 1976 *Sunday Mail Color Mag.* (Brisbane) 1 Aug. 7/1 Mrs. Ratcliffe suffers from triskedekaphobia..the name psychologists have given for an inexplicable dread of a Friday falling on the 13th of the month. 1979 *Guardian* 13 July 11/6 I'm tempted to diagnose triskaidekaphobia or allergy to 13.

triskele. Add: (Later examples of form *triskelion*.)
1973 T. Pynchon *Gravity's Rainbow* (1975) i. 150 Pins, brooches, opalescent scorpions (her birth sign) inside gold mountings in triskelion. 1977 *Sci. Amer.* Dec. 168 (*caption*) Below, at the left, is another fragment of side-link mold with a decorative whorl and, at right, a triskelion decoration for a chariot linchpin.

trisome (trəi·sōᵘm). *Cytology.* [f. Tri- + *-some³.] A chromosome which is represented three times in a chromosomal complement; also, a trisomic individual.
1921 [see *disome]. 1926 L. W. Sharp *Introd. Cytol.* (ed. 2) xvii. 388 Such 2n+1 forms are called 'simple trisomic' mutants; they have 11 disomes (normal pairs) and one trisome. 1944 [see *monosome 1]. 1979 *Nature* 27 Sept. 280/2 The failure to form trisomes indicates.. that on intact RNA3 stable ribosome binding does not occur simultaneously at both the site near base 29 and the site near base 70.

trisomic (trəisōᵘ·mik), *a.* (*sb.*) *Cytology.* [f. as prec. + -ic.] Of or pertaining to a trisome. Also as *sb.*, a trisomic chromosome, cell, or individual.
1921 A. F. Blakeslee in *Amer. Naturalist* LV. 259 If the Globe and Poinsettia [*sc.* 'mutant' forms of *Datura stramonium*] could be combined to form a mutant with 3 chromosomes each in two of the 12 sets, such a mutant

would be called a double trisomic mutant. 1924 *Genetics* IX. 194 Super-enlarged in general bears the same relation to enlarged as round-leaf globe to globe; i.e., the tetrasomic is an accentuated expression of the trisomic. 1939 *Jrnl. Genetics* XXXVIII. 382 Trisomics are of frequent occurrence and are known in almost every genus employed in genetic work. 1957 R. A. Beatty *Parthenogenesis & Polyploidy in Mammalian Devel.* v. 79 Since the only abnormality would be the presence of one extra chromosome, the trisomic might have greater viability than the full triploid. *Ibid.* vi. 102 In triploid or trisomic *Drosophila*, the position of the centromere has been assessed in this way. 1974 *Nature* 1 Mar. 54/2 Aneuploid embryos found in this study included uniform monosomics and trisomics.
So **tri·somy**, trisomic state; freq. with following numeral denoting the chromosome concerned, as *trisomy-21* (associated with Down's syndrome).
1930 *Bibliographia Genetica* VI. 15 Further observations..disclosed a high degree of intraspecific variation resulting in polyploidy, trisomy, fragmentation of chromosomes [etc.]. 1961 G. Allen et al. in *Lancet* 8 Apr. 775/2 Several others [of the undersigned] believe that this is an appropriate time to introduce the term 'trisomy 21 anomaly' which would include cases of simple trisomy as well as translocations. 1963 [see *dermatoglyphics]. 1965 [see *Down's syndrome]. 1970 Passmore & Robson *Compan. Med. Stud.* II. xxxi. 19/1 The relative frequency of trisomy 13–15 in abortuses indicates that it is less compatible with normal intra-uterine development than either trisomy 17–18 or mongolism. 1977 *Nature* 6 Jan. 65/2 Case 2..showed an excess of α-globin synthesis smaller than expected for functional trisomy of the α-globin genetic region. 1979 *Ibid.* 4 Jan. 57/1 Although the patient had no features of the trisomy 21 syndrome, two first-degree cousins of his father had given birth to children with Down's syndrome.

Tristanesque (tristăne·sk), *a.* [f. the name *Tristan*, the hero of Wagner's opera *Tristan und Isolde* (1865), + -esque.] Resembling the music of *Tristan und Isolde*; *spec.* characterized by tonal ambiguity and chromaticism.
1942 *Scrutiny* XI. 5 Similarly the diatonic system decayed into the deliquescence of feeling and tonal instability which marks the Tristanesque music of the late nineteenth century. 1948 [see *impressionist 2]. 1957 W. Mellers *Man & his Music: Romanticism & 20th Cent.* i. ii. 46 (Consider the Tristanesque opening of the second piece). Liszt the romantic lover here recollects in tranquillity. 1962 *Times* 26 Jan. 16/5 The debt to Wagner was incurred a little later, in the thoroughly Tristanesque love duet towards the end of *A Village Romeo and Juliet*.

tri-state: see *tri- 1 b.

tristful. Add: Hence **tri·stfulness.**
1909 Hardy *Time's Laughingstocks* 152 In the bearing of each a passive tristfulness. 1914 C. Mackenzie *Sinister Street* II. iv. ii. 877 For a whim of tristfulness, for the luxury of consummating the ineffable depression the house created in him.

tristimulus (trəisti·miŭlŏs). [f. Tri- + Stimulus.] Each of three reference colours (as red, green, and blue) which can be combined additively in specified proportions (*tristimulus values*) to produce any given colour. Usu. *attrib.*
1933 *Jrnl. Optical Soc. Amer.* XXIII. 359/2 The specification [for matching any given colour stimulus] consists of giving the amounts of each primary stimulus required for the match. This is known as the tristimulus system of color specification. 1937 G. S. Monk *Light* xvii. 336 The tristimulus values of the recommended standard source..for wave-length 4800 angstroms are given by the ordinates at that wave-length of the three curves. 1962 H. C. Weston *Sight, Light & Work* (ed. 2) vii. 212 The vertical scale labelled Y refers to one of the tristimuli and the horizontal scale labelled X refers to another. 1976 C. Reynolds *Photoguide to Filters* 227 The CIE system is the tristimulus specification most widely used, and is ideally suited to defining the colour of filters.

tristubh (tri·ʃtub). Also trishtubh. [Skr. *triṣṭubha*.] A Vedic metre of eleven syllables (see also quot. 1939).
1869, etc. [see *jagati]. 1939 *Jrnl. Amer. Oriental Soc.* LIX. 159 Ordinarily, the Hindu metricians mechanically define any metrical pāda of eleven syllables as 'triṣṭubh', and any of twelve syllables as 'jagatī'. This cannot be accepted... What we shall call a triṣṭubh may have anywhere from ten to thirteen syllables, a jagatī from eleven to at least thirteen (possibly more). The distinction between the two is solely based on the cadence; a triṣṭubh always ends ⏑ – x..in western terms..in a catalectic diiambus. 1965 *Language* XLI. 11 Three instances are accounted for by the pāda..two by the triṣṭubh cadence. 1971 *Ibid.* XLVII. 64 The last pada of the stanza in question is incontrovertibly triṣṭubh.

tritanopia (trəităṅōᵘ·piă). *Ophthalm.* [mod. L., ad. G. *tritanopie* (J. von Kries 1911, in *Helmholtz's Handbuch der Physiol. Optik* (ed. 3) II. 341), f. *trit-*, var. Trito- + An- 10 + *-opia.] A form of dichromatic colour-blindness marked by reduced sensitivity and discrimination in the blue and green parts of the spectrum. Hence **tri·tanope**, one who has tritanopia; **tritano·pic** *a.*
1915 J. H. Parsons *Introd. Study Colour Vision* ii. i. 159, I shall adopt his [sc. v. Kries's] terms, viz., protanopes, deuteranopes, and tritanopes, corresponding respectively with v. Helmholtz' red-, green-, and blue-blind. *Ibid.* ii. 180 Cases of tritanopia or so-called blue-blindness are rare and mostly due to disease. *Ibid.* 304/1 (Index), Tritanopic vision. 1937 [see *deuteranope]. 1955 *Jrnl. Optical Soc. Amer.* XLV. 614/1 The tritanopic luminosity function..does not indicate any major shift in excitation peaks of the receptor substances. 1959 [see *dichromat, dichromate sb.²]. 1965 *New Scientist* 14 Oct. 134/3 Tritanopic defects are described as autosomal dominant while tritanomaly is considered to be more or less recessive sex-linked. 1974 *Jrnl. Optical Soc. Amer.* LXIV. 1246/1 The tritanope's targets were formed of checks of pale violet and yellow-green and were compared to a blank field of violet. 1978 *Nature* 23 Nov. 390/1 When the eye has been adapted to a bright yellow light, a marked loss of sensitivity to short-wavelength stimuli may be recorded immediately after the extinction of the adapting field. This phenomenon..was termed transient tritanopia by Mollon and Polden.

triterpane, -pene, -penoid: see *tri- 5 a.

tritiated (tri·tiē¹tĕd, -ʃiē¹tĕd), *a.* *Chem.* [f. *triti(um + -ate³ + -ed².] Containing tritium; having had an atom of ordinary hydrogen replaced by tritium. So **tritia·tion**, the introduction of tritium into a compound or molecule in place of ordinary hydrogen.
[1947 M. D. Kamen *Radioactive Tracers in Biol.* vi. 127 A few drops of triterated water may be recovered.] 1956 *Nature* 25 Feb. 379/2 Analyses for tritium were carried out by combustion of the organic compound and conversion of the tritiated water to tritium. 1961 G. R. Choppin *Exper. Nuclear Chem.* xi. 180 The Wilzbach method of tritiation involves the exposure of the unlabeled compound to a multicurie atmosphere of tritium gas for periods of time as long as two weeks. 1978 *Bull. Amer. Acad. Arts & Sci.* Feb. 11 He further developed novel methods for the accurate determination of tritiated compounds in tissues.

triticale (tritikē¹·li). Also Triticale. [f. mod.L. generic names *Tritic*(*um* wheat + *Se*)*cale* rye.] A hybrid cereal grass, of the genus × *Triticosecale*, produced by crossing various species or varieties of wheat and rye.
1952 *New Biol.* XIII. 44 A cross between wheat and rye..is known as Triticale. 1968 *New Scientist* 24 Oct. 181/2 (*caption*) Triticale, a hybrid of wheat and rye developed at the University of Manitoba. 1974 A. J. Huxley *Plant & Planet* xxvii. 309 New races of the great seed plants..have been created in the last few years, including a promising wheat/rye hybrid called triticale. 1979 *McGraw-Hill Yearbk. Sci. & Technol.* 414/1 Although triticale plants were first described in 1876, only in the mid-1960s did they begin to receive great interest from scientists and farmers. 1983 *New Scientist* 13 Jan. 98/1 From a slow start in the 1950s..triticale is now growing on more than half a million hectares, in the USSR, Europe, the United States and South America.

tritish, *a.* Delete *rare*⁻¹ and add: Also triteish.
1980 *Times* 5 June 11/3 A good, if triteish, situation.

tritium (tri·tiŏm). *Chem.* [mod.L., f. Gr. τρίτ-ος third + -ium.] **1.** A radioactive heavy isotope of hydrogen, having two neutrons as well as a proton in the nucleus, which constitutes one part in 10^{18} of the naturally occurring element and is produced for use in fusion reactors and as an isotopic label. Symbols ^3H (H^3), T. Cf. *deuterium, *protium.
1933 Urey & Murphy in *Jrnl. Chem. Physics* I. 513/2 If the H^3 isotope is discovered, we would recommend to the discoverer the consideration of the name tritium for it. 1959 *Sci. News* LI. 12, ^3He is obtained from the decay of tritium which in turn is the product of nuclear reactions involving neutrons from a reactor. 1962 [see *superheavy a. b]. 1972 *Sci. Amer.* Oct. 104/3 Large amounts of tritium (hydrogen 3)..are stockpiled for nuclear weapons and for research in fusion power. 1976 *Nature* 9 Sept. 103/1 Tritium ($t_{½}$ = 12·3 yr) is present on the Earth's surface mainly as HTO.
2. *attrib.* and *Comb.*
1934 *Sun* (Baltimore) 6 Oct. 20/6 Tritium water could be formed by the addition of two more atoms of hydrogen to the heavy water. 1953 J. Blish in E. Crispin *Best SF* (1955) 345 The U.N.'s police would be glad to know that they could have access to a virtually inexhaustible stock of tritium bombs. 1966 *McGraw-Hill Encycl. Sci. & Technol.* XIV. 113/1 Tritium oxide, T_2O, has been prepared by oxidation of tritium gas with hot copper oxide. *Ibid.* 113/2 Tritium-labeled compounds may be prepared by ordinary synthetic chemical methods. 1979 *McGraw-Hill Yearbk. Sci. & Technol.* 153/2 Tritium-helium dating will eventually prove very useful in studying those physical processes which transport or redistribute substances in oceans.

trito-. Add: **tritano·maly** *Ophthalm.* [An-omaly], a rare form of anomalous trichromatism marked by a reduced sensitivity to blue; hence **tritano·malous** *a.*, having tritanomaly.
1943 *Jrnl. Optical Soc. Amer.* XXXIII. 572/1 These [subjects] are diagnosed by most other tests as tritanomalous or blue-yellow blind. *Ibid.* 574/2 These plates are invalid for the detection of tritanomaly. 1946 W. D. Wright *Res. on Normal & Defective Colour Vision* xxiv. 297 The tritanomalous observer has poor hue discrimination in the blue-green wavelengths. 1956 *Jrnl. Optical Soc. Amer.* XLVI. 1075/1 Tritanomaly, discovered by Engelking, seems to be very rare. 1965 Tritanomaly [see *tritanopia].

tritoma. Add to etym. after mod.L. (J. B. Ker 1804, in *Curtis's Bot. Mag.* XX. 744). Add to def.: = *kniphofia. (Earlier examples.)
1804 *Curtis's Bot. Mag.* XX. 744 (*heading*) Glaucous-Leaved Tritoma. 1854 [see *kniphofia]. 1871 J. C. Patteson *Let.* 8 Mar. in C. M. Yonge *Life J. C. Patteson* (1874) II. xii. 514, I like both the red and the yellow tritoma.

tritomite (tri·tŏməit). *Min.* [ad. G. *tritomit* (P. H. Weibye 1850, in *Ann. d. Physik u. Chem.* LXXIX. 299), f. Gr. τρίτομος thrice-cut, in allusion to the shape of the cavities left by the crystals: see -ite¹.] **a.** A borosilicate, fluoride, and hydroxide of cerium earths, calcium, and thorium found as brown trigonal crystals. **b.** *tritomite-Y*: an analogue of this in which yttrium largely replaces cerium.
1856 *Edin. New Philos. Jrnl.* III. 60 The tritomite here analysed was sent me by M. Wiborg of Brevig, labelled as thorite, which it also very closely resembled in appearance. 1962 *Amer. Mineralogist* XLVII. 9 Tritomite, a rare, metamict boro-silicate..is known only from the nepheline syenite pegmatites of Låven, Brevik, and Barkevik, in..southern Norway. 1966 *Ibid.* LI. 156 The terms..spencite, [etc.]. should be dropped... Respectively, these minerals would be known as..tritomite-(Y), [etc.]... The Commission [on New Minerals and Mineral Names of the International Mineralogical Association] has approved this nomenclature.

triton² (trəi·tǫn). *Physics.* [f. *trit(ium + *-on¹.] A sub-atomic particle composed of one proton and two neutrons, the nucleus of the tritium atom.
1942 *Physical Rev.* LXII. 115/1 The nucleus Li^7 is pictured on the alpha-particle model as built up from an alpha-particle group and a triton (H^3) group. 1965 *Wireless World* Sept. 446/2 Semiconductor devices can be employed for counting other types of ionizing particles such as protons, deuterons, tritons, fission fragments, etc. 1976 *Nature* 29 Apr. 749/3 There have been rather few studies of the interactions of tritons with nuclei, mainly because tritons are highly radioactive, with a half-life of about 12 years.

tritonality (trəitonæ·lĭti). *Mus.* [f. Tri- + Tonality.] The simultaneous use of three keys in a musical composition. Hence **trito·nal** *a.*
1931 *Music & Lett.* Oct. 323 Atonalities, bitonalities, tritonalities and their peers are best heard and not seen. 1944 *Scrutiny* XII. 121 The Lydian [mode]..is harmonically treacherous owing to its imperfect tritonal fourth. 1963 *Listener* 17 Jan. 141/2 The third movement is at the opposite tritonal pole, F sharp.

Tritoness. (Later example.)
1956 K. Clark *Nude* vii. 271 A small tritoness,..recently emerged from the excavations in Ostia.

trit-trot. (Earlier example.)
1818 M. Edgeworth *Let.* Sept. (1971) 89 The drollest trit-trot little walk she has.

triumphal, *a.* (*sb.*) Add: Hence **triu·mphally** *adv.* Cf. Triumphantly *adv.*
1897 F. Thompson *New Poems* 109 Thou dost thy dying so triumphally. 1984 *Miami Herald* 6 Apr. 2B/1 Mike Zeck returns triumphally as..the local kid who actually did break into the business.

triumphalism (trəi,ʌ·mfăliz'm). [f. Triumphal *a.* + -ism.] The sense of pride (often linked with ostentation) in the rightness and achievements of one's Church (used *pejoratively*). Also in extended sense.
1964 R. McA. Brown *Observer in Rome* 27, I am greatly impressed by the recognition of human failings in this prayer and by its exclusion of the 'triumphalism' that has often seemed to characterize the church. 1968 *N.Y. Times* 12 Jan. 25 Wayne H. Cowan, managing editor of the liberal Protestant journal, Christianity and Crisis, said the pastoral 'mutes the triumphalism of the past, but still places great emphasis on the mystery and infallibility of the church'. 1972 *Catholic Herald* 9 June 4 Nostalgia for the pre-Conciliar years of exclusivity and triumphalism. 1975 *New Yorker* 10 Mar. 83/1 This contrast is understandable, given what critics of the regime have labelled 'triumphalism'—something that goes way beyond mere ostentation on a colossal scale. 1977 P. Johnson *Enemies of Society* iv. 47 The loss of interest and confidence in the human mind and spirit is, to some extent, concealed by the gigantic triumphalism of late imperial architecture. 1981 G. Priestland *Priestland's Progress* i. 17 John V. Taylor, Bishop of Winchester,..is right when he seeks to turn us from shallow triumphalism or the reshuffling of old dogmas. 1983 *Times* 31 May 13/2 There would probably be an initial outbreak of Tory triumphalism, which would be distasteful and unnecessary.
Hence **triu·mphalist** *a.* and *sb.*, **triumphali·stic** *a.*
1967 H. Chadwick *Early Church* 285 Towards such triumphalist assumptions a twentieth-century Christian is likely to be cool and reserved. 1967 *Times* 22 Apr. 12/5 The anxieties of the lingering triumphalists are increased. 1967 R. McA. Brown *Ecumenical Revolution* vi. 115 It must be acknowledged that later Protestants themselves became as triumphalistic about their own confessions and traditions and denominations as they ever accused the Roman Catholic Church of being. 1970 *Daily Tel.* 2 Dec. 12/7 Elgar's unashamedly triumphalist setting of the National Anthem sounded a defiantly anachronistic note. 1973 *Listener* 19 Apr. 512/1 The busy, businesslike, triumphalist, materially successful France of today. 1980 *Focus* Summer 24/1 The triumphalist tends to interpret what God has done as his own achievement. 1982 *Sunday Tel.* 30 May 9/2 Churches have been stripped of baroque or Italianate furnishings, altars have been heaved forward, 'triumphalist' pictures and symbols stashed away. 1983 *Observer* 28 Nov. 8/3 The journalists ..fed readers and viewers a diet of triumphalistic pap.

trivalent, *a.* Add: **2.** *Cytology*. That is (part of) a trivalent.
1921 *Proc. Nat. Acad. Sci.* VII. 200 In the 8 remaining prophase or metaphase figures, not all the trivalent chromosomes could be distinguished from the bivalents or univalents into which they had divided, or from which they were composed. 1929 [see *quadrivalent a. 2]. 1976 *Genetical Res.* XXVIII. 55 We present data pertaining to..chromosome XVII trisomics of *S*[*accharomyces*] *cerevisiae* which demonstrate that trivalent meiotic association occurs with a high frequency.
3. *Immunol.* Of a vaccine: giving immunity to three forms of a disease.
1959 *New Scientist* 19 Feb. 395 (*caption*) Transfer of single strain vaccine pools to tanks to form final trivalent vaccine. 1961 *Guardian* 25 Oct. 3/5 A 'trivalent' form of vaccine will be used. This contains polio virus of all three types.
B. *sb.* *Cytology*. With pronunc. (tri·vālĕnt). A multivalent consisting of three chromosomes.
1922 *Amer. Naturalist* LVI. 341 There are 12 sets of three united chromosomes each, and these trivalents can be arranged according to the size formula. 1936 *Hereditas* XXI. 305 In triploid hybrids..there was a high frequency of trivalents. 1975 [see *quadrivalent sb. 2].
Hence also **triva·lency**.
1888 *Jrnl. Chem. Soc.* LIV. 1071 The formula for benzene [was]..afterwards given up owing to the difficulty of explaining the trivalency of carbon which it involved. 1927 N. V. Sidgwick *Electronic Theory of Valency* xv. 271 Trivalency [of carbon]..only arises under extreme compulsion, and is excessively unstable.

trivia (tri·viă), *sb. pl.* [mod.L., pl. of L. *trivium* (see Trivium), infl. in sense by Trivial a. 6.] Trivialities, trifles, things of little consequence.
1902 L. P. Smith (*title*) Trivia. 1920 *Glasgow Herald* 21 July 8 His [sc. Mr. Bennett's] method suggests the amount of human interest and knowledge that may lurk in the trivia of holiday experience. 1929 E. Linklater *Poet's Pub* xv. 175 He packed an attaché case with a few shirts,..some toilet trivia. 1947 Auden *Age of Anxiety* 1. 20 Farouche they appear,..loitering through the..Nocturnal trivia. 1961 B. Pym *No Fond Return of Love* xix. 191 The rooms were furnished in a luxuriantly Victorian style, and filled with such nostalgic trivia as waxed fruit under glass, paper-weights, shell and seaweed pictures, and stuffed birds. 1978 *Sunday Times* 26 Feb. 33/7 Besides, trivia has its importance too. Or to put it another way, trivia have their importance too.

trivial, *a.* Add: **6. b.** *Math.* Of no consequence or interest, e.g. because equal to zero; satisfying a given relation on a set with every member of the set; *spec.* applied to a subgroup of a given group that either contains only the identity element or is identical with the given group.
1915 R. D. Carmichael *Diophantine Analysis* ii. 28 We have thus established the fact that Eq. (2) has at least one integral solution which is not trivial. 1941 Birkhoff & MacLane *Surv. Mod. Algebra* vi. 135 The reflexive property is trivial (every group is isomorphic to itself by the identity transformation). 1949 S. Kravetz tr. *H. Zassenhaus' Theory of Groups* i. 10 𝔊 and *e* are trivial subgroups of 𝔊. 1953 [see *proper a. 5 c (ii)]. 1957 L. Fox *Two-Point Boundary Probl.* vii. 192 If *y* vanishes at *x* = 1 then all the derivatives, if finite, are also zero, giving the trivial solution. 1971 G. Glauberman in Powell & Higman *Finite Simple Groups* i. 35 These subgroups..will therefore be non-trivial when *P* is not trivial. 1979 *Proc. London Math. Soc.* XXXVIII. 508 Strong spectrality is trivial since *A*(*K*) = *A*.
7. c. *Chem.* Of the name of a chemical species: not systematic; often used in preference to the systematic name for reasons of convenience or tradition, as *neohexane* (systematic name *2,2,dimethylbutane*) or *formaldehyde* (systematic name *methanal*). Cf. *systematic a 7.
1892 *Nature* 19 May 58/1 The extent to which familiar trivial names shall be retained in the official system [of chemical nomenclature] is therefore a matter of great importance. 1951 *Chem. & Engin. News* 23 July 3036/2 The alchemists used fanciful names; we would class them as 'trivial' names today. 1979 Clark & McKervey in Barton & Ollis *Comprehensive Org. Chem.* I. ii. 40 Several of these trivial names are still universally accepted... However, trivial names for alkanes containing multiple branching can become cumbersome.

trivialization (triviălaizē¹·ʃən). [f. Trivialize v. + -ation.] The act or process of trivializing.
a 1866 J. Grote in *Jrnl. Philol.* (1874) V. 153 A still more important law..is that of *evaporation* or *trivialization*; by which I mean the gradual blunting of the force of a word. 1927 H. G. Wells in *Sunday Express* 26 June 12 The greater danger of promiscuity and the trivialisation of the sexual life lies in a delayed marriage. 1949 Koestler *Insight & Outlook* xxviii. 380 By this process of trivialization and smug understatement, the universe itself becomes a silly thought. 1981 *Times* 21 May 4/7 The growing under-use of human abilities through trivialization of work and through unemployment is damaging to those who suffer from them.

trivializer (tri·viălaizəɹ). [f. Trivialize v. + -er¹.] One who trivializes.
1960 F. Raphael *Limits of Love* iii. vi. 369 There was..nothing the happy man thought about less than death, but nor was there anything the trivialiser was more eager to forget. 1980 *Times Lit. Suppl.* 12 Dec. 1405/3 Taylor has been a popularizer without being a trivializer.

trivializing (tri·viălaiziŋ), *ppl. a.* [f. as next + -ing².] That trivializes.
1961 *New Left Rev.* Mar./Apr. 13/2 Berger would..see all 'abstract' art as the product of a trivialising despair. 1966 *Economist* 28 May 970/3 He has serious doubts about the power of literary studies, as now practised, to act as a counterweight to the trivialising forces of our society. 1970 *Radio Times* 8 Oct. 66 It's true that television is endemically a trivialising medium, but it doesn't follow that it *has* to be it.

tri·vializing, *vbl. sb.* [f. Trivialize v. + -ing¹.] The action of Trivialize v.
1963 A. Heron *Towards Quaker View of Sex* v. 44 There is an almost overwhelming urge throughout society towards the trivializing of sexual actions. 1979 *Listener* 16 Aug. 220/3 Any such trivialising of the sacred backfires. The gods will not be mocked.

trivially, *adv.* Add: **2. b.** Chiefly *Math.* In an inconsequential or uninteresting way.
1941 Birkhoff & MacLane *Survey Mod. Algebra* vi. 148 The conclusion is trivially true. 1956 E. M. Patterson *Topology* ii. 35 Conditions (T.1) and (T.3) are satisfied trivially. 1977 *Language* LIII. 353 But it is trivially true that *all* features characteristic of creole speech will be removed if decreolization is carried far enough.

tri-weekly, *a.* and *adv.* Add: **A.** *adj.* **b.** Also *absol.* as *sb.*, a tri-weekly journal. orig. *U.S.*
1851 C. Cist *Sk. Cincinnati in 1851* 74 These are all dailies, tri-weeklies, and weekly reissues of dailies. 1884 *U.S. Census* VIII. 111 Three months only the *Spy* ran as a tri-weekly, and but three months longer as a semi-weekly. 1978 D. Daiches *Edinburgh* vi. 110 The *Caledonian Mercury* (a tri-weekly founded in 1720).
B. *adv.* **b.** (Earlier example.)
1837 J. M. Peck *Gazetteer Illinois* (ed. 2) iii. 180 The mail..arrives here tri-weekly.

trizonal (trəi·zōᵘnăl), *a.* *temporary.* [f. Tri- + Zonal a.] Of, pertaining to, or consisting of three zones; *spec.* with reference to the British, French, and American zones of occupation in West Germany at the end of the war of 1939–45. So **trizo·nia**, an area of three zones; *spec.* (with capital initial) West Germany as occupied by the Allies.
1947 *Sun* (Baltimore) 21 Nov. 2/8 Advance indications were that a stalemate at London would bring efforts to expand the present American–British occupation area in Western Germany into a 'trizonia' with France. 1947 *Richmond* (Va.) *Times-Dispatch* 28 Dec. ii. 2-B/1 It [sc. the German Socialist party] views the proposed establishment of 'Trizonia' with many misgivings. 1948 *Times* 10 Jan. 4/4 Pending steps towards tri-zonal fusion, administrative reorganization must go on in the bizone to secure greater efficiency. 1948 *Sunday Times* 7 Mar. 1 Establishment of a trizonal régime in Western Germany. 1948 *News Chron.* 28 Aug. 1 He said the West German political situation, particularly the question of the formation of a trizonal area was discussed.

trizzie (tri·zi). *Austral. slang.* Also trizzy. [Orig. uncertain: perh. f. Trey sb. + -ie.] A threepenny piece.
1941 Baker *Dict. Austral. Slang* 78 *Trizzie*, a 3d. piece. 1959 G. Hamilton *Summer Glare* 31 My greatest pal now was John 'Trizzie' Peele... He always had a threepenny bit in his pocket which gave him his nickname. 1966 *Sunday Truth* (Brisbane) 23 Dec. 22/1 When you peppered the Christmas pud. with trey-bits this year we hope you remembered they will be scarcer next Yuletide and unless you hoard some there will be no trizzies at all for..the 1968 plum-duff... A trey-bit or a trizzy is Aussie slang for a three-penny-bit.

Troadic (trōᵘ,æ·dik), *a.* [(f. *Troad*, name of the region about Troy,) f. L. *Trōad-*, *Trōas*, a. Gr. Τρωάδ-, Τρωάς (contr. of Τρωϊάς) Trojan + -ic.] Of or pertaining to ancient Troy and its surrounding regions.
1932 *Antiquity* VI. 77 In the second half of the third millennium, the culture which we may call Troadic spread over a large part of Asia Minor. 1977 G. Clark *World Prehistory* (ed. 3) iv. 157 That Troadic forms reaching Greece were transmitted by way of the Cyclades is confirmed by the occurrence on the mainland of..the ceramic fictiles of 'hour-glass' form.

Trobriand (trōᵘ·bri,ănd). Used *attrib.*: Of, pertaining to, native to, or produced in the *Trobriand* Islands, a small group of coral islands in the Solomon Sea, now forming part of Papua New Guinea; **Trobriand Islander** = *Trobriander.
1922 B. Malinowski *Argonauts W. Pacific* ix. 221 The big bay of Gatu, where once the crews of a whole fleet of Trobriand canoes were killed and eaten. 1935 —— *Coral Gardens & their Magic* II. vi. v. 232 The Trobriand phenomenon of a language of magic..fits into our theory of language. 1937 R. H. Lowie *Hist. Ethnol. Theory* xiii. 231 The reader becomes saturated with the Trobriand atmosphere. 1951 E. E. Evans-Pritchard *Social Anthropol.* iv. 74 Malinowski came to know the Trobriand Islanders well. 1956 R. Redfield *Peasant Society & Culture* 35 The many interrelations of custom, institution, and human need in Trobriand life. 1974 *Country Life* 11 Apr. 838/1 The bronze..can be appreciated..by anyone—atheist, Hottentot, Eskimo, Trobriand Islander.

Trobriander (trōᵘ·bri,ændəɹ). [f. *Trobriand* (see prec.) + -er¹.] A native or inhabitant of the Trobriand Islands.
1922 B. Malinowski *Argonauts W. Pacific* ix. 220 The Trobrianders will sail deep, shaded bays. 1969 *Times* 22 Jan. 12/8 The difficulty of disentangling the Trobrianders' original beliefs from what they have been taught by Europeans.

trochaical, *a.* (Later *fig.* example.)
1930 R. Campbell *Poems* 10 Jack Squire through his own teardrops sploshes In his great flat trochaical goloshes.

trochanter. Add: **trocha·nteral** *a.*
1961 in Webster. 1967 J. H. Sudd *Introd. Behaviour Ants* ii. 15 The coxal and trochanteral joints.

trochotron (trǫ·ko,trǫn). *Electronics.* [f. Trocho(idal a. + *-tron.] A type of magnetron in which there are a number of anodes at different angular positions around the central cathode, with the electron beam able to be switched from one anode to another.
1947 Alfvén & Romanus in *Nature* 1 Nov. 614/1 Research on valves with cycloidal or trochoidal electronic motion has been carried out... The development of the valves, which are called 'trochotrons', has been carried out by G. Hambraeus and..T. Wallmark. 1962 *Jrnl. Brit. Inst. Electr. Engineers* XXIII. 99/1 The Trochotron is a hot-cathode multi-electrode tube containing a number of open box electrodes and operating in a constant magnetic field. 1980 *Jrnl. Nuclear Materials* XCIII–XCIV. A. 352/2 The property of stigmatic focusing used in the trochotron mass spectrometer..is conserved in this device.

‖ **trockenbeerenauslese** (trǫ·kənbērənɑu:s-lēzĕ). Pl. -lesen. [Ger., f. *trocken* dry + *beeren* pl. of *beere* Berry sb. grape + *auslese* selection, choice (wine).] (A) sweet German white wine of superior quality, made from individually selected grapes affected by noble rot.
1963 *Times* 8 Feb. 12/5 Beerenauslesen and Trockenbeerenauslesen. These latter terms mean, respectively, wines made from selected over-ripe single grapes and those made from over-ripe single grapes which are in effect sundried raisins, in a state of *pourriture noble*. 1964 *Harper's Bazaar* Nov. 146/3, I do not think a white wine, except a Trockenbeerenauslese..should be expected to last so long. 1972 W. Garner *Ditto, Brother Rat!* i. 5, I drank so many good wines my tastebuds stopped performing for anything less than a *Trockenbeerenauslese*.

troepie: see *troopie.

trog (trǫg). *slang.* [Abbrev. of Troglodyte.]
1. A speleologist. *rare.*
1955 *People* (Austral.) 7 Sept. 23/3 These are the trogs, as they cosily call themselves,..members of the Sydney Speleological Society, the Sydney University Speleological Society, [etc.].
2. One of a despised social group; a lout, a boor, a hooligan, an obnoxious person.
1956 L. McIntosh *Oxford Folly* 15 This charm school would have been rather a brilliant thing to do... After all, these trogs lead such dreary lives. 1957 J. I. M. Stewart *Use of Riches* i. ii. 23 You've been listening to some disgusting trog being beastly about Rupert, and now you're parroting him. 1960 D. Potter *Glittering Coffin* vi. 89 Trinity..infrequently admits a 'trog' (in other words a grammar school boy). 1961 M. Dickens *Heart of London* iii. 277 Nobody mixes, I mean really *mixes* with the trogs. 1962 J. Fleming *When I grow Rich* xv. 173 One of the trogs appointed himself foreman. 1967 *Guardian* 30 May 2/4, I am thoroughly disgusted. Yesterday I saw two long-haired trogs, one with a ribbon in his hair, wearing the red frock-coats of the Chelsea Pensioners. 1981 M. Duffy *Gor Saga* 87 He'd given her the morning after pill and the little trog had just kept it... She would stick him with a paternity order. 1983 *Granta* vii. 17 The scowling vandals, bus-stop boogies, and soccer trogs malevolently lining the streets.
3. (See quot. 1958.) *N.Z.*
1958 *Tararua* XII. 28 For shelter, hillmen may seek a *trog*, a large overhanging boulder or bluff giving shelter like a cave. 1971 *N.Z. Listener* 19 Apr. 56/5 They found a possie in a bit of a trog and boiled-up.
4. A teenager who camps out or lives in caves. *temporary.*
1965 *Sun* 8 June 7/7 For Mods and Rockers you can now read Trogs and Thunderbirds... A teenager I know explained it to me yesterday: 'Mods do a lot of sleeping out, camping.' 1966 *Daily Tel.* 14 Apr. 23/3 The young people, who called themselves 'trogs' after the word troglodyte,

cave dweller, could be found in the caves at weekends with about 50 permanent 'residents'.

troglo- (trǫ·glo), combining form of Gr. τρώγλη hole, used in the names of various groups of organisms found in caves, as **troglobi·on(t)**, **tro·globite** [a. G. *troglobie* (J. B. Schiner in A. Schmidl *Grötten und Höhlen von Adelsberg* (1854) 240), f. Gr. βιῶν living], an animal living entirely in the dark parts of caves; hence **troglobi(o)·tic** *a.*; **tro·glophil(e)** [a. Ger. (J. B. Schiner *loc. cit.*: see -phil, -phile], a cave-dwelling animal that does not live entirely in the dark; **tro·gloxene** [a. Fr. *trogloxène* (E. G. Racovitza 1907, in *Arch. Zool. Expér. & Gén.* 4th Ser. VI. 437), f. Gr. ξένος guest], an animal that spends occasional short periods in dark caves.
1924 *Glasgow Herald* 13 Dec. 4 Permanent cave-dwellers (the troglobions), like the Dalmatian Proteus newt. 1927 *Ibid.* 2 July 4 The three groups have received various names, such as troglobionts, troglophils, and trogloxenes. 1982 Moyle & Cech *Fishes* xxxi. 444 Most cave waters containing troglobiotic fishes have at least intermittent connections to the outside. 1953 Hazelton & Glennie in C. H. D. Cullingford *Brit. Caving* ix. 268 The family Dendrocoelidae includes a large number of troglobites. *Ibid.* 270 The genus *Candona*..includes many troglobite species. 1971 Webster Add., Troglobitic. 1924 *Glasgow Herald* 13 Dec. 4 We include many of the bats as troglophils. 1947 *Sci. News* V. 52 The troglophili actively seek out and prefer the underground dark. 1953 Hazelton & Glennie in C. H. D. Cullingford *Brit. Caving* ix. 268 Only those species which find the cave temperature suitable can become troglophiles. 1965 B. E. Freeman tr. *Vandel's Biospeleology* vii. 66 Its grey colour and the presence of eyes indicate that this planarian is a troglophile. 1927 Trogloxene [see *troglobiont* above]. 1965 B. E. Freeman tr. *Vandel's Biospeleology* ix. 142 Until quite recently all crabs which had been found in caves were trogloxenes or at most troglophiles.

troika. Add: **2.** A group or set of three persons (*rarely* things) or categories of people associated in power; a three-person commission or administrative council. Also *attrib.*
1945 [see N.K.V.D. s.v. *N II. 1]. 1954 C. P. Snow *New Men* xl. 286 Faith, hope, and hate: that was the troika which rushed him on. 1957 *Times Lit. Suppl.* 15 Nov. 682/1 The so-called *troika*, or commission of three, which authorized summary executions. 1961 *Guardian* 6 June 1/2 Experience of the United Nations action in the Congo..had convinced the Soviet Government of the need for the troika principle to be applied to all international action. 1961 *New Statesman* 9 June 901/2 Krushchev's central doctrine of the 'Troika', the principle of triple-harness—Communist, western and uncommitted—in the administrative, as well as in the policy-making, organs of the UN. 1969 A. Arent *Laying on of Hands* (1971) ix. 91 The landed gentry. Part of the troika who, with the army and the Church, run Spain. 1971 *Nature* 26 Feb. 585/1 Every chemistry department, after all, is now at least a troika of inorganic, physical and organic chemistry. 1974 T. P. Whitney tr. *Solzhenitsyn's Gulag Archipelago* I. i. vii. 281 Real scope entered the picture with the twenties, when *permanently* operating *Troikas*—panels of three, operating behind closed doors—were created to bypass the courts permanently. 1976 *Church Times* 23 Jan. 9/1 The editorship is now a troika consisting of David Jenkins, John Drury and James Mark. 1976 M. J. Lasky *Utopia & Revolution* (1977) ii. 92 Ideas, images, and ideology never quite manage to be harnessed into a controllable troika.

troilism (troi·liz'm). [Perh. f. F. *trois* three: see -ism.] Sexual activity in which three persons take part simultaneously. Also *transf.* Hence **troi·list** *a.*
[1941 *Dorland's Med. Dict.* (ed. 19) 1544/2 *Troilism*,..a psychotic sexual manifestation in which the patient desires the sexual partner of the person for whom he has homosexual yearnings.] 1951 *Ibid.* (ed. 22) 1622/2 *Troilism*,..paraphilia practised by three persons, by two women and a man, or by two men and a woman. 1973 *Sunday Times* 14 Oct. 18/4 Wife-swapping and troilism are mostly in the writers' imaginations, not part of their life. 1976 *Times Lit. Suppl.* 20 Feb. 191/1 The troilist encounter in the film *Performance* is breathlessly mentioned as taking place in 'a great big bed'. 1983 V. Glendinning *Vita* 197 Emotional troilism always attracted her.

Trojan, *a.* and *sb.* Add: **1. b.** *Trojan horse*: according to epic tradition, the hollow wooden horse in which Greeks were concealed to enter Troy; *fig.* a person, device, etc., insinuated to bring about an enemy's downfall; a person or thing that undermines from within; also *attrib.*
1574 [in Dict.]. 1837 S. S. Prentiss in G. L. Prentiss *Memoir of S. S. Prentiss* (1858) i. viii. 188 He cannot so easily introduce his Trojan horse within these walls [seating of contested members in Mississipi House]. I, for one, will hurl a spear against its hollow sides. 1940 *Sun* (Baltimore) 13 May 1/4 Alarmed by the success of Germany's 'Trojan horse' and parachute-troop tactics. 1963 *Listener* 17 Jan. 112/2 The strengthening of the 'special links' between London and Washington made Britain's possible entry into the Common Market 'more than ever likely to be that of a Trojan horse'. 1974 *Datamation* Jan. 57/1 A 'Trojan Horse' technique was used to compromise the security of a campus time-sharing computer system... A computer operator used it..erasing all trace of the illicit Trojan Horse code. 1979 A. Boyle *Climate of Treason* iii. 96 This ambitious Trojan Horse strategy called for the recruitment and indoctrination of compliant intellectuals. 1981 *Courier-Mail* (Brisbane) 27 July 5/4 The 'Trojan horse' technique involves smuggling into a computer system an illegal set of instructions.
2. *Astr.* The epithet of two groups of asteroids which are at the same distance as Jupiter from the sun and approximately 60 degrees ahead of it and 60 degrees behind it in their orbit, so that with the sun and Jupiter they occupy positions of stability at the corners of two equilateral triangles. [So called because the first ones to be discovered were named after heroes of the Trojan War.]
1913 *Jrnl. Brit. Astron. Assoc.* XXIII. 214 Masculine names are reserved for [minor] planets of peculiar interest, viz., those that pass very near the Earth..or have the same period as Jupiter (the Trojan group). 1918 *Mem. R. Astron. Soc.* LXII. 79 The four asteroids—Achilles, Patroclus, Hector, and Nestor—are the Trojan planets. 1979 *Daily Tel.* 17 Apr. 8/6 (*caption*) Troy is one of the two groups of so-called Trojan satellites which always form equilateral triangles between themselves, the sun and Jupiter.
B. *sb.* **4.** *Astr.* A Trojan asteroid.
1918 *Mem. R. Astron. Soc.* LXII. 80 The inclinations of the orbits of the Trojans and Jupiter vary through a quite considerable range. 1954 C. Payne-Gaposchkin *Introd. Astron.* (1956) ix. 234 The theory of the 'Trojans' is a beautiful special application of the dynamical 'Problem of Three Bodies'. 1979 *Icarus* XL. 341/1 There seem to be three times more Trojans at the leading Trojan point.
5. *U.S.* The proprietary name of a make of contraceptive sheath.
First registered as a proprietary term in the U.S. on 26 Apr. 1927.
1951 *Official Gaz.* (U.S. Patent Office) 17 Apr. 757/1 Trojan... For Prophylactic Membranous Articles for the Prevention of Contagious Diseases. 1962 A. Lurie *Love & Friendship* xiv. 264 'Why "Trojans"?' she asked, picking up a small box... 'They lost the war, after all.' 1973 M. Amis *Rachel Papers* 202 After some neck-ricking soixante-neuf and a short period inside her unsheathed, I clawed at the little pink holder and took its final trojan.

troland (trōᵘ·lănd). [The name of L. T. *Troland* (1889–1932), U.S. psychologist and physicist.] A unit of retinal illumination, being the illumination produced by a surface with a luminance of one candela per square metre when the pupil has an area of one square millimetre; orig. called a *photon* (*photon¹ 1).
1944 *Jrnl. Optical Soc. Amer.* XXXIV. 254/2 [Report by the Committee on Colorimetry.] A special unit of retinal illuminance, the photon, was introduced by Troland in 1917 and has been employed to some extent, but its use is circumscribed by the danger of confusion with the elementary quantum of radiant energy which is generally called the photon. The renaming of the unit of retinal illumination as the troland, in honor of Dr. L. T. Troland, its originator and chairman of the Committee..in 1921–22, is recommended. 1976 *Nature* 19 Feb. 570/1 We have..measured sensitivity to short wavelengths after adaptation to retinal illuminances that varied over the range 1·1–6·0 log trolands.

troll, *sb.*² Add: **b.** Also Comb. *troll-like* adj.
1954 J. R. R. Tolkien *Two Towers* iv. 66 A large Man-like, almost Troll-like, figure. 1978 *Trans. Yorks. Dial. Soc.* lxxviii. 18 Joseph is a troll-like figure, the foil to Heathcliff's gigantic, elemental being.

troll, *v.* Delete '*arch.* and *dial.*' and add: **I. 1.** Delete † *Obs.* and add later examples; *spec.* (*slang*) of a homosexual: to walk the streets, or 'cruise', in search of a sexual encounter; cf. sense 13.
1942 E. Langley *Pea Pickers* i. iii. 41 Past rows of hawthorn hedges in leaf, but lacking flowers, we trolled. 1967 A. Wilson *No Laughing Matter* iii. 201 At first..I just got myself picked up... But later I started trolling. 1967 *Listener* 21 Dec. 814/3 They all come trolling on in form-hugging black and do evocative things with chairs and ladders and planks of wood. 1981 R. Barnard *Sheer Torture* xi. 120, I trolled off quite happily and entered the house.
III. 6. (Later example.)
1819 Scott *Ivanhoe* II. vi. 88 Come, trowl the brown bowl to me.
IV. 10. (Further examples.) Cf. Trolly-lolly *int.*
1881 R. L. Stevenson *Virginibus Puerisque* 283 But let him feign never so carefully, there is not a man but has his pulses shaken when Pan trolls out a stave of ecstasy and sets the world a-singing. 1933 H. Allen *Anthony Adverse* III. ix. lxiv. 1190 At Anthony's suggestion they left off the doleful ballads which at first engrossed them and took to trolling more cheerful lays. 1951 N. M. Gunn *Well at World's End* xiv. 99 He felt like a voyageur..trolled a note or two and lifted his tweed hat as if it were a sombrero. 1977 *Rolling Stone* 16 June 69/2 When the Diamonds trolled 'Them Never Love Poor Marcus', I was moved.
12. (Later examples.)
1948 J. Berryman *Dispossessed* 77 Now Tell me. Troll me the sources of that Song—Assigned last week—by Blake. 1971 K. Millett *Sexual Politics* (1972) ii. iii. 137 The old scholar chuckles while trolling the more rakish passages of Catullus.
V. 13. (Later *fig.* examples.)
1966 E. Lindall *Time too Soon* iv. 51 Kamindo had rebuffed him when he had trolled for information. 1984 *Monitor* (McAllen, Texas) 1 May 6A/3 It will troll the Earth's upper atmosphere for magnetospheric, atmospheric and gravitational data.

trolley, trolly, *sb.* Add: **2. b.** In fig. phr. *off one's trolley*, crazy. Cf. Rocker¹ 2 c. *slang.*
1896 Ade *Artie* x. 92 Any one that's got his head full o' the girl proposition's liable to go off his trolley at the first curve. 1903 A. H. Lewis *Boss* xix. 264 She's off her trolley. She toins sick; an' in a week she croaks. 1949 N. R. Nash *Young & Fair* ii. ii. 66 If you suspect Patty, you're off your trolley! 1976 *National Observer* (U.S.) 4 Sept. 13/3 Anybody who buys a luxury liner for use as a floating hotel is off his trolley. 1983 *Times* 5 Feb. 3/1 The London college gym mistress who is suing her former lover for libel in the High Court, heard a lawyer say yesterday that she had 'gone off her trolley' about the affair.
c. A small table or stand on wheels or castors for use in serving food, transporting light objects, luggage, etc. Freq. as the second element in Combs. *supermarket trolley*: see *supermarket 2 b; also *ellipt.*
1937, etc. [see *tea-trolley* s.v. *tea sb. 9 c]. 1944 D. Welch *In Youth is Pleasure* ii. 39 Some [waiters] carried trays poised high in the air; others trundled chromium trolleys with glass shelves, on which brilliant little cakes were piled. a 1948 —— *Voice through Cloud* (1950) iv. 39 When the porters..started to wheel him out of the ward, he sat up on the trolley, so that the red blankets fell off him. 1949 S. Smith *Holiday* xiii. 173 The semolina pudding came in on the trolly. 1963 [see *hostess trolley* s.v. *hostess 4]. 1973 J. Wilson *Truth or Dare* ii. 24, I couldn't help noticing what she'd got in her trolly: sausages, luncheon meat *and* a tub of marge. 1977 *Lancs. Life* Nov. 107/1 (*caption*) A good trolley is an invaluable aid to efficient hostessmanship. 1982 *Amer. Speech* LVII. 154 *Trolley*, surgical dressing cart (British).
4. trolley-bus († trolli-bus), a trackless passenger vehicle that gets its power from an overhead cable by means of a pole and trolley (see 3); also *attrib.*; **trolley shop**, a trolley (see *2 c) from which goods are sold, esp. in a hospital.
1921 *Daily Colonist* (Victoria, B.C.) 13 Oct. 11/7 The trolley-buses..are of the single-deck type, with seats for twenty-four passengers. 1927 *Daily Express* 20 July 9/6 The 'trolli-bus' traffic receipts last week..were £400 more than the receipts for the corresponding week last year. 1939 J. B. Priestley *Let People Sing* i. 13 He climbed into a trolley-bus, which after a mile or so stopped at a large gateway. 1961 C. Willock *Death in Covert* xii. 224 A clang and ring that reminded him incongruously of vibrating trolley-bus wires. 1978 *Country Life* 13 Apr. 956/2 The nearly extinct tram and trolley bus will be revived. 1958 *Times* 7 July (Suppl.) p. xix/1 Volunteer nurses..still give invaluable aid in hospital wards; volunteers take round trolley-shops. 1974 *Country Life* 4 Apr. 797/1 We [sc. the Red Cross] organize out-patients' canteens. Trolley shops. Libraries.
trolley *v.* (further *trans.* examples); **trolleyful** (earlier example).
1889 Kipling *From Sea to Sea* (1899) II. xxvii. 34 The cans..were..slidden along by the trolleyful. 1936 'M. Innes' *Death at President's Lodging* xiii. 234 Ah, well, my picture of Haveland trolleying his own bones plus corpse up the garden path was no doubt, as you suggested, a bit steep. 1978 P. McCutchan *Blackmail North* vi. 69 Stocks being trolleyed in from freighter aircraft.

trolling, *vbl. sb.* Add: **4. trolling motor** *U.S.*, a motor suitable for a boat used in trolling (see Troll v. 13 c); **trolling pole** *N. Amer.*, a horizontal pole rigged on each side of a fishing boat in order to keep the lines clear of the propeller.
1964 M. Weeks *Compl. Boating Encycl.* 530/1 *Trolling motor*, a low-powered, slow-speed motor used for trolling. 1980 *Outdoor Life* (U.S.) (Northeast ed.) Oct. 26/1 The latest high-thrust, 12-volt electric trolling motor, Thruster Plus from Mercury Marine, is the most efficient that I've tested to date. 1960 M. Sharcott *Place of Many Winds* v. 97 Fishermen cut their trolling poles from the forest... The poles must be between thirty and fifty feet long.

trollius (trǫ·li,ŏs). Also Trollius. [mod.L. (J. C. Buxbaum *Plantarum Minus Cognitarum Centuria I* (1728) 15): see Troll flower.] A perennial herb of the genus of this name, belonging to the family Ranunculaceæ, native to Europe, Asia, and North America, and bearing yellow or orange globe-shaped flowers. Also *attrib.* Cf. *globe-flower* s.v. Globe sb. 10 b.
1899 in E. T. Cook *Century Bk. Gardening* 142/1 It must be a sorry garden that fails to satisfy the Trollius. 1952 Bates & Lowther *Breeding Birds Kashmir* 191 We found one [sc. a wagtail's nest] in a tuft of *trollius* leaves. 1962 R. Page *Education of Gardener* viii. 236 We planted its [sc. a stream's] banks..with..clumps of different varieties of yellow trollius. 1978 R. Gorer *Growing Plants from Seed* vi. 76 Plants with hard, shiny seeds, such as..trollius, tiarella, and saxifraga all do best if sown outside in autumn.

trollop, *v.* (In Dict. s.v. Trollop sb.) Restrict *Sc.* to sense (a) and add: (*b*) *spec.* to walk in a slovenly way, to slouch; (earlier and later examples).

1854 M. Dods *Early Lett.* (1910) 63, I felt deeply moved for her, thinking she would trollop away home. **1870** 'Ouida' *Puck* I. vii. 113 There's alas a lot of..bad wimmin a trolloping about. **1925** *Chambers's Jrnl.* 23 May 397/2 We'll go very slow and he can trollop behind.

Trollopian (trǫlōᵘ·piǎn), *a.* and *sb.* Also **-ean.** [f. the name *Trollope* (see below) + -IAN.] **A.** *adj.* **1.** Of, pertaining to, or characteristic of the English novelist Frances Trollope (1780–1863) (mother of Anthony) or her writings.

1847 W. Howitt in *Howitt's Jrnl.* 9 Jan. 18/1 Mrs. Trollope was introduced to the court circles—everything was shown to her, and the urbane minister was so particularly polite, that, instead of a Trollopean laughter, there was nothing but laudation.

2. Of, pertaining to, or characteristic of the English novelist Anthony Trollope (1815–82) or his writings.

1903 G. Gissing *Private Papers Henry Ryecroft* 213 Any Trollopean work that lay upon the counter? **1907** Hardy *Let.* 19 Sept. in *One Rare Fair Woman* (1972) 134 *Our Fatal Shadows*..shows I think a great advance upon your previous novels... It is quite Trollopian. **1939** C. S. Lewis *Let.* 24 Nov. (1966) 171 The Cole and the whole cathedral surroundings in Tours are almost Trollopian. **1957** *Times Lit. Suppl.* 25 Oct. 637/4 A Trollopian setting and situation, a cathedral town and a Dean's daughter about to be married. **1980** *Daily Tel.* 11 July 9/2 The 'Strangers and Brothers' sequence..was a conscious effort to write a social history in novel form on a Trollopian scale. **1983** M. Duggan *Family Life* 173 The paper's Trollopian report next day—'Good Heavens!' cried the chaplain.

B. *sb.* A student or admirer of Anthony Trollope or his writings.

1920 A. O. Godley *Reliquiae* (1926) II. 316 In Trollope one remembers the characters, but I never met more than one Trollopian who knew the plots apart. **1946** *N. & Q.* 23 Feb. 67/1 We receive..from the University of California Press the first number of 'The Trollopian: A Semi-annual Journal Devoted to Studies in Anthony Trollope and His Contemporaries in Victorian Fiction'. **1969** J. Gross *Rise & Fall Man of Lett.* ix. 14 Michael Sadleir..well known as biographer, bibliographer, Trollopean, and author of *Fanny by Gaslight*.

trolly. Add: **2.** [Perh. a different word: cf. *trolleywags* trousers (Barrère & Leland *Dict. Slang*).] *pl.* Ladies' drawers or knickers. *dial.* and (*schoolgirls'*) *slang.*

1891 J. Baron *Blegburn Dickshonary* (rev. ed.) 68 *Trollys* (female underclothing). That's as near as th' paysun'll come to th' meanin' o' this word; aw co'nna wimmen's treawsis, an' he co'n't; 'ca' em drawers, which is wheere they're kept. **1934** B. Pym *Diary* 8 Jan. in Holt & Pym *Very Private Eye* (1984) I. 35, I bought a peach coloured vest and trollies to match. **1971** M. Woman *Eng. Girls' Boarding Schools* vi. 148 Items of clothing carried names, thus 'trolleys' for underwear, 'B squared' for brassieres.

trolly (trǫ·li), *sb.*⁴ *dial.* [Alteration of Trollop *sb.*: cf. also Trull.] = Trollop *sb.* 1. Also *comb.* in **trolly-mog, trollimog** [cf. Moggy 2] in the same sense.

1851 E. Sternberg *Dial. & Folk-Lore Northants.* 117 *Trolly-mog*, a dirty, slovenly female. **1854** A. E. Baker *Gloss. Northants. Words & Phrases* II. 357 Oh! what a trolly she is! **1876** J. Hartley *Yorksher Puddin* 164 He's nobbut a trollop. **1901** F. E. Taylor *Folk-Speech S. Lancs.* s.v. *Troll*, *Trolly*, a loose woman, a trull. **1975** M. de la Mare *Broomsticks* 130 That old trollimog that what lives in Hoggets Bottom. **1974** T. Wright *Lang. Brit. Industry* xvii. 163 Untidy housewives abound, judging by all the so-called *slatterns, trolly-mogs, slovens* and *trollops*.

‖**tromba marina** (trǫ·mba mări·nă). [It., = marine trumpet.] = *trumpet marine, marine trumpet* s.v. Trumpet *sb.* 2 b.

The identification in quot. 1776 is error.

1776 [see Sea-trumpet 2]. **1838** [see Trumpet *sb.* 2 b]. **1948** G. B. Shaw *How to become Mus. Critic* (1960) 325, I should rather like to hear the tromba marina. **1976** D. Munrow *Instruments Middle Ages & Renaissance* 30/2 A French sculpture of the twelfth century gives us our first glimpse of the tromba marina: the three-sided body is about four feet long and tapers towards the pegbox... By the fifteenth century it had acquired two strings of unequal length.

Trombe (trǫmb, ‖trōnb). *Archit.* The name of Felix *Trombe*, 20th-c. Frenchman, used *attrib.* and in conjunction with that of his collaborator J. *Michel* to designate a masonry wall of a kind designed by him, having glass sheeting fixed a small distance in front of it so as to absorb solar radiation, and usu. ventilated internally to release the heat into the building.

1978 *Washington Post* 4 Nov. E33/2 One of its simple forms is the Trombe wall. This consists of a masonry wall just inside a large south-facing glass wall or window. **1980** *Family Handyman* Sept. 83/1 Since these shutters can be motor-controlled from a remote position, you can install them in front of a trombe wall or other passive solar heat collector. **1980** J. J. Greenland in H. J. Cowan *Solar Energy Applications in Design of Buildings* v. 130 A recent development which makes use of solar energy and thermal inertia without excessive loss of heat through glass is the Trombe–Michel solar wall. **1984** *Christian Sci. Monitor* 14 Feb. 21/3 A 'Trombe wall' solar-heating system..helps warm the big structure's lab.

trombiculid (trǫmbi·kiulid), *a.* (and *sb.*) [a. mod.L. family name *Trombiculidæ*, f. generic name *Trombicula* (A. Berlese *Acari Nuovi* (1905) IV. 155), f. *Trombi-dium* (see Trombidid *a.* and *sb.*) + -culus: see -id³.] Of or pertaining to a mite of the family Trombiculidæ, which includes several species having parasitic larvæ which cause or transmit disease in man and other mammals. Also as *sb.*, a mite of this kind.

1950 R. Matheson *Med. Entomol.* (ed. 2) iv. 133 Trombiculid mites and disease. The attacks of various species of mites..usually result, in man, in a marked dermatitis accompanied by intense itching. **1957** *New Scientist* 14 Nov. 33/1 A trombiculid does not feed on a mammal more than once in its lifetime. **1962** Gordon & Lavoipierre *Entomol. for Students of Med.* xiv. 267 The association of trombiculid mites with scrub typhus had been suspected well over a century ago by the Japanese physician Ohtomo. **1978** *Jrnl. R. Soc. Med.* LXXI. 507 The causative organism..is transmitted by the larvae (or chiggers) of several species of trombiculid mites.

trombone. Add: **3.** A green or yellow pear-shaped pumpkin belonging to the Australian variety of this name.

1926 *Jrnl. Agric.* (S. Austral.) Jan. 275 The trombone is not such a good cropper unless it can be watered in hot weather. **1969** *Ibid.* Jan. 208 By far the most popular pumpkin variety in South Australia is the Trombone. It is typically pear-shaped with a curved neck. **1978** *Guardian* 10 Nov. 21/8 According to one leader, a trombone is a non-spherical pumpkin much used in chutneys and pickles. Another, equally well versed in Australian horticulture, tells me that it is a long-necked marrow with a bulbous end (hence the name) which is cooked and tastes exactly like vegetable marrow.

trombone, *v.* Restrict *rare* to sense 1 in Dict. and add: **2.** (Earlier and later examples.) So **trombo·ning** *vbl. sb.*

1864 J. A. Grant *Walk across Africa* ix. 196 When standing here, the hoarse tromboning of the hippopotamus, wishing to come out to graze, echoed from out these rushes. **1866** J. Macgregor *Thousand Miles in Rob Roy Canoe* (ed. 2) iii. 48 Crowds of gaping peasants..jostled against bands drumming and tromboning..., and marching in a somewhat rickety manner over the undoubtedly rough pavement. **1958** R. Harris in P. Gammond *Decca Bk. Jazz* iii. 44 There was one man..who created a legend of tail-gate tromboning—the one and only Kid Ory. **1960** *New Oxf. Hist. Music* III. xii. 426 'Die pusauner pusaunoten mit schallen..die schalmier mit dreyen stymmen, als man sunst gewonlichen singet' ('The trombonists tromboned together in three parts as one is otherwise accustomed to sing'). **1967** *Listener* 26 Jan. 144/3 A contentious fugal start and impervious tromboning behind the story-telling with a piquant sense of expectation.

tromino (trǫ·mino). [f. Tr(i- + D)omino by deliberately false analogy (see quot. 1961).] Any planar shape formed by joining three identical squares by their edges.

1954 S. W. Golomb in *Amer. Math. Monthly* LXI. 676 It is impossible to cover the checker board with 21 straight trominoes, and a monomino in the upper left-hand corner of the board. **1961** [see *pentomino]. **1979** *Sci. Amer.* Apr. 19/1 The games played with the two 3-cell animals (the trominoes) are slightly more difficult to analyze.

tromp (trǫmp), *v.* Var. (orig. and chiefly *U.S.*) of Tramp *v.*¹ Hence **tro·mping** *vbl. sb.*

1891 *Dial. Notes* I. 234 Tromp = tramp. **1895** S. Crane *Red Badge of Courage* x. 103 Yeh was tromped on. **1902** *Dial. Notes* II. 148 He tromped my toe. **1929** W. Faulkner *Sanctuary* (1931) vi. 59 You'll tromp on a loose board and find yourself... **1946** J. Steinbeck *Wayward Bus* ... **1952** E. Ferber *Giant* xx. 334 You want to look out, Bick, she don't get tromped the way they're milling around this year. **1972** Tolkien *Return of King*... Some bad trompings and the sheep..., and they tromped, jumping and grunting rhythmically. **1962** *Ibid.*, And while the singing and the tromping was going on on the river bank, the initiates for baptism came down one by one from a high rock. **1962** J. Steinbeck *Travels with Charlie* I. 12 About that time hurricane Donna was reported tromping her way out of the Caribbean. **1968** J. Crist *Private Eye, Cowboy & Very Naked Girl* 193 Beatniks tromping grapes in the buff. **1974** J. Irving *158-Pound Marriage* v. 117 Edith heard Frau Reiner and the Chetniks whispering and tromping about in the living room. **1975** *New Musical Express* 24 May 20/1 Heat and noise and darkness and a steady, muffled tromping that you can feel through your feet. **1976** M. Machlis *Pipeline* xlvii. 491 He wouldn't care who got tromped to get there either. **1979** *United States 1980/81* (Penguin Travel Guides) 301 Tromping through the ice plants is a botanical education in itself.

‖**trompe l'œil** (trōnp lœ̄y). Also **trompe-l'œil,** *erron.* **d'œil.** [Fr., lit. 'deceives the eye'.] Deception of the eye, an illusion, *spec.* in *Art* with regard to the material reality of the object(s) represented, a (usu. still-life) painting, plaster ornament, etc., intended to give an illusion of reality. Also *fig.* and *attrib.* passing into *adj.*

1889 C. H. Stranahan *Hist. French Painting* vii. 457 The public of connoisseurs who care not for any tricks of 'trompe l'œil', but for art. **1926** A. Huxley *Ess. New & Old* 171 Their taste ran to *trompe l'œil* pictures of fighting giants. **1927** E. Bowen *Hotel* x. 118 The hill..by some *trompe-l'œil* of twilight seemed to topple. **1928** *Observer* 19 Feb. 5/1 The nearest approach to realistic treatment is in Mr. Cedric Morris's picture of a luxurious flowery meadow in North Africa, and this, far from being a *trompe-l'œil*, is a decorative 'pattern'. **1934** R. Fry *Last Lect.* (1939) 107 The carefully exposed reflection of the fallen soldier in the retina of his shield, which is very much in agreement with the Greek *trompe-l'œil*—like that of the 'Grapes of Zeuxis'—which were the stock in trade of art critics like Pliny. **1936** A. Huxley *Eyeless in Gaza* xviii. 231 And the Museum of Sexology: such photographs and wax models—almost too *trompe-l'œil*. **1957** *Listener* 14 Oct. 658/1 Details may be solid, *trompe l'œil*, or flat. **1961** E. Taylor *In Summer Season* i. 11 Facing her, as she turned the stairs, was a *trompe-l'œil* panel, designed to lengthen the passage into an endless arcade. **1964** S. Sontag in *Evergreen Rev.* Dec. 76 Plato's view that all art is an elaborate *trompe l'œil*, and therefore a lie. **1968** *Ideal Home* Nov. 35 The dining-room has *trompe l'œil* marbling. **1970** *New Scientist* 11 June 530/1 *Trompe l'œil* techniques such as false perspectives painted on walls are common. **1974** G. Butler *Coffin for Pandora* viii. 147 Olivia had told her own story and had told it badly... She had led those who listened to her up to a blank wall and confronted them with a *trompe l'œil*. **1978** R. Barnard *Unruly Son* xv. 155 Shelves and books had been painted on the wall, making a perfect *trompe l'œil*.

-tron (trǫn), *suffix. Physics.* The ending of *electron¹ (but cf. Gr. -τρον instrumental suffix), used: **a.** in the names of some kinds of thermionic valves and other electron tubes, as *kenotron* (1915), *pliotron* (1915), *ignitron* (1933); **b.** in the names of a few sub-atomic particles, as *positron* (1933), *negatron* (1934), *mesotron* (1938): cf. *-on¹; **c.** in the names of devices used to accelerate particles, as *cyclotron* (1935), *betatron* (1941), *phantastron* (1943), *levitron* (1960).

1940 *Nature* 6 Apr. 603/1 -*tron* should be restricted to signify either an instrument or a particle, but not both... -*on* taken by itself..seems the most natural ending for a particle. **1940** *Ibid.* 4 May 713/1 The Amsterdam conference [of the International Union of Physics] attempted to dispose of the curious modern theory that the ancient Greek termination for a fundamental particle was '-tron'. It deemed the word 'electron' as against its rivals.

tronc (trǫŋk, ‖trōn). [a. Fr., = collecting box.] In hotels and restaurants, a common fund into which tips and service charges are paid for distribution to the staff. Also *attrib.*

1938 *Observer* 13 Jan. 6/1 The staff are paid from the tronc, or pooling system, whereby all tips are divided. **1960** *Times* 13 Jan. 9/3 The value of a point was set each week according to the amount in the tronc. **1964** *Observer* (Colour Suppl.) 13 Dec. 33/1 In all tronc (pooled tips, which the tronc head waiter distributes according to the number of points each waiter has) they declare an average figure to the Income Tax. **1976** *Rhyl Jrnl. & Advertiser* 9 Dec. 10 (Advt.), Chef—Royal Lido, Prestatyn. Applications are invited for this permanent post at a salary of..plus an annual supplement of £315 and a percentage of tronc. **1981** *Times* 11 June 14/6 The money was managed by management and distributed to the staff..or paid into an independent tronc fund.

trondhjemite (trǫ·ndhjĕmait). *Petrogr.* [ad. G. *trondhjemit* (V. M. Goldschmidt 1916, in *Skrifter udgivna af Vidensk. i Christiania* (Matem.-Naturv. Klasse) No. 2. 75), f. *Trondhjem* (now *Trondheim*), name of a city in western Norway: see -ite¹.] Any leucocratic tonalite, esp. one in which the plagioclase is oligoclase.

1922 *Jrnl. Geol.* XXX. 406 The acid phase is represented by 'trondhjemite' (new name) which Kjerulf calls an oligoclase-granite, and which might as well be characterized as a granodiorite. **1978** *Nature* 12 Oct. 538/2 The metabasaltic sequences may be of considerable thickness and in many areas are associated with intrusive tonalitic gabbros, trondhjemites and minor developments of serpentinized peridotites. **1979** F. Barker *Trondhjemites, Dacites, & Related Rocks* i. 1 The author suggests that the IUGS definition of trondhjemite as leucotonalite be followed, except that andesine-bearing leucotonalite be termed calcic trondhjemite, and that albite-bearing leucotonalite, as well as..

troop, *sb.* Add: **1. c.** *esp.* A group of apes or monkeys.

1929 R. M. & A. W. Yerkes *Great Apes* vii. 71/1 Some observers assert that troops [of gibbons]..will abandon a wounded comrade. **1951** R. Campbell *Light on Dark Horse* v. 84 I blundered into the middle of a huge troop of baboons. **1965** *Listener* 10 June 863/1 Another of the characteristic features of the primates is the size of the group or 'troop' as it is called.

2. b. *pl.* The members of a mob or gang collectively. *U.S. slang.*

1932 *Daily Progress* (Charlottesville, Va.) 7 Apr. 4/4 *Troops*, mob or gang. **1963** *Listener* 4 Apr. 585/2 On the trip back he was met by 'the troops'. With quick dispatch they placed four bullets in the back of his head, disposed of the weapon in a nearby alley, and were gone.

3. c. A company of Scouts comprising not less than three patrols of six Scouts each.

1908, etc. [see *Patrol *sb.* 3 b]. **1959** E. H. Clements *High Tension* x. 167, I ran a Scout Troop in that town and he was one of my troop. **1980** W. Maxwell *So Long, See You Tomorrow* iii. 28 Were we in the same Boy Scout troop?

5. troop-carrier, a large aircraft or armoured vehicle for transporting troops; hence **troop-carrying** *a.*

1943 *Daily Mail* 23 June 3 Among landplanes there are huge new troop-carriers capable of carrying 25 fully equipped soldiers. **1958** *Ibid.* 18 July 1/1 Israel's permission to ferry troop-carriers through her air-space was not..fully cleared. **1964** L. Deighton *Funeral in Berlin* vi. 38 A Volkspolizei troop carrier was parked at the roadside. **1976** A. White *Long Silence* ix. 84 We could hear..the engines of motor-cycles, lorries and troop-carriers. **1927** B. H. L. Hart *Europe in Arms* iii. 32 A force of 1200 men together with 150 machine-guns and 18 light field-guns was carried 100 miles in troop-carrying aircraft. **1977** M. Gilbert *Winston S. Churchill* IV. *Compan.* i. 654 The first use of troop-carrying aeroplanes during a military campaign took place on 21 February 1923.

trooper. Add: **1. b.** (Earlier example.)

1739–40 Richardson *Pamela* (1740) I. 239 She curses and storms at me like a Trooper.

c. A brave or stalwart person. *colloq.*

1948 R. Campbell *Light on Dark Horse* 230 Nina Hamnett (she was a fine Trooper).

3. a. (Earlier example.)

1830 *Hist. Rec. Austral.* (1922) 1st Ser. XV. 770 The Mounted Police, which at present consists of about 68 Troopers.

b. *U.S.* A mobile state policeman. More fully *state trooper.*

1911 *Ann. Rep. 1910* (Pennsylvania Dept. State Police) 16 On arrival of the detail, the mob of foreigners proceeded to stone and shoot at Troopers... Three of the Troopers..fired at their assailants, wounding two Italians. **1921** [see *state *sb.* 38 e]. **1977** *New Yorker* 3 Oct. 43/1, I slowed the car and pulled over, and a state trooper pulled up behind us... The trooper said, Do you need any help.

c. A paratrooper. *orig. U.S.*

1943 *Sun* (Baltimore) 16 Feb. 1/2 English neighbors tasted this clannishness soon after the troopers settled in their midst. **1974** C. Ryan *Bridge Too Far* III. iii. 156 Of the sixteen paratroopers, pilot and co-pilot, only Johnson and two other troopers got out.

4. (Later examples.)

1942 *R.A.F. Jrnl.* 16 May 11, I saw the empty trooper, scrubbed and waiting. **1981** J. Barnett *Firing Squad* II. 114 First objective, a decent cabin on the trooper.

Hence **troopere·ss** (*rare*), a female trooper.

1924 Galsworthy *White Monkey* II. iv. 152 Miss Wilmot was..lying to them like a trooperess. **1927** *Daily Express* 2 Sept. 3 The stories related of the coarse, swearing 'trooperess' are astounding.

troopie (trū·pi). *colloq.* Also (*S. Afr.*) **troepie.** [f. Troop *sb.* + -ie.] In South Africa and Rhodesia (Zimbabwe): a private soldier, esp. a national serviceman without rank.

1974 *Eng. Usage in S. Afr.* May, Some of the troopies..., **1979** *Cape Times* 25 July (*request programme s.v.* *request *sb.*¹ 11]. **1980** *Observer* 3 Aug. 10/1 The FN rifle carried by troopies in the bush.

tropane (trōᵘ·pĕⁱn). *Chem.* Also † **tropan.** [ad. G. *tropan*, f. *tropein* Tropeine: see -ane 2 b.] A tropane, a bicyclic tertiary amine which is a basic liquid obtained from various plants and the parent of a series of compounds which includes atropine, cocaine, and related alkaloids; 8-methyl-8-azabicyclo-[3.2.1]octane, $C_8H_{15}N$.

1909 *Chem. Abstr.* III. 392 The residual strong solidifies in a few days... **1923** *Ibid.* XVII. 1643 (*heading*) The spectrochemistry of derivatives of tropane. **1941** A. Goodman *Pharmacol. & Therapeutics* i. 34 The alkaloids of the atropine group of drugs may be considered as derived from a combination of a piperidine and pyrrolidine ring designated as tropane. **1981** *Phytochemistry* XX. 2409 (*heading*) Tropane alkaloids are known to occur in *Anthocercis littorea* [etc.].

trope, *v.* [f. Trope *sb.* 3.] *trans.* To introduce (a trope) as an embellishment; to embellish with a trope or tropes; to add as a trope to. Hence **troped** *ppl. a.*

1894 W. H. Frere *Winchester Troper* p. xv, The Winchester Tropers..originally contained only a long jubilum or *pneuma*, but later in MS. CC the words were added and the trope troped. **1922** Madan & Craster *Summary Catal. Western Manuscripts Bodleian Library* II. i. 124 These flyleaves I understand from Mr. Bannister to come from a non-monastic breviary, and, as he does not find a troped office for St. K. in English breviaries, I should have supposed them French. **1959** *Listener* 11 Dec. 1134/3 The final word 'portum', set to a long melisma, is troped 'portum in ultimo, da nobis iudicio'. **1977** *Gramophone* Sept. 469/3 The choir missed a golden opportunity by not singing *O come, O come, Emmanuel* in its original fifteenth-century French Franciscan version, as a two-part troped litany.

troph-, var. *tropho- before a vowel.

trophallaxis (trǫfălæ·ksis). *Ent.* [f. *troph- + Gr. ἄλλαξις exchange.] The mutual exchange of food material by adult insects and larvæ. Hence **trophalla·ctic** *a.*

1918 W. M. Wheeler in *Proc. Amer. Philos. Soc.* LVII. 322 In *Belonogaster* the feeding of adults and larvæ is reciprocal... As the relationship is clearly cooperative or mutualistic, I suggest the term *trophallaxis*. **1929** W. Osler *Old Humanities* ii. 13 The nursing function..is really trophallactic... The larva is provided with..an amber-fluid greedily lapped up by the nurse. **1931** W. C. Allee *Animal Aggregations* xii. 358 'Trophallaxis' is the bond that unites parent and offspring in the social insects. **1940** *Jrnl. Compar. Psychol.* XXIX. 456 This process..may be represented as the outcome of a simple conditioned response based upon trophallactic relations in the bee-swarm. **1978** R. J. Elzinga *Fund. Entomol.* vii. 186 Trophallaxis plays a greater role in ants than it does in the social wasps.

trophic, *a.* (*sb.*) Add: **1. b.** *Ecol.* Of or pertaining to the feeding habits of, and the food relationship between, different types of organisms in the food-cycle; so *trophic level*, any of a hierarchy of levels of an ecosystem, each consisting of organisms sharing the same function in the food-web, and the same relationship to the primary producers.

1942 *Ecology* XXIII. 407/1 Food-cycles rarely have more than five trophic levels. **1957** *Ecol. Monogr.* XXVII. 55 (*heading*) Trophic structure and productivity of Silver Springs, Florida. **1974** R. H. Britton in R. Goodier *Natural Environment of Shetland* 123 The nutrient poor categories (dystrophic and oligotrophic) are by far the most numerous and..eutrophic and brackish lochs are fewer... The lochs within each trophic category can be further subdivided according to their superficial area. **1976** *Nature* 22 July 284/1 The trophic base of the arthropod fauna is wind-blown detritus. **1980** *Jrnl. R. Soc. Arts* Feb. 149/2 While they are trophic levels—top carnivores, herbivores, plants, microorganisms—none is dominant.

3. Of a hormone: stimulating the production of another hormone from a specific gland; = *tropic *a.* 4 b.

1945 I. S. Kleiner *Human Biochem.* xxiii. 503 Another hormone of this gland which produces its effect by influencing a different structure, i.e., a 'trophic' hormone, is the adrenotrophic factor. **1971** E. Frieden *Animal Hormones* ii. 19 The site of production of the trophic hormone controlling the adrenal cortex is still in doubt. **1975** *Nature* 27 Nov. 340/2 The existence of a human cell line showing severalfold stimulation by androgens in a defined system, free from the effects of other trophic hormones, should provide a useful reagent for the further study of the mechanism of androgen action.

-trophic (trōᵘ·fik, trǫ·fik), *suffix.* [See Trophic *a.*] **1.** Forming adjs.: (*a*) with the senses 'characterized by nutrition (of a certain kind)', 'finding nourishment in', as in *autotrophic* adj. s.v. *auto-¹, *heterotrophic, *lecithotrophic *adjs.*, *psychrotrophic* adj. s.v. *psychro-; also 'controlled by', as in *neurotrophic *a.*; (*b*) with the sense 'maintaining or supporting (a gland, tissue, etc.)' and hence 'regulating', esp. in the epithets of hormones, as *gonadotrophic, *neurotrophic *adjs.* Cf. *-tropic.

1943 *Endocrinology* XXXIII. 407 The use of -tropic as in *gonadotropic*..reverses and confuses a clear, practical pre-established usage in the broad field of biology... -Trophic, even if not perfectly apt, is close enough in meaning and is not new. **1950** *Lancet* 2 Dec. 708/2 It is becoming current practice to speak of the action of a hormone in controlling an endocrine gland as trophic (e.g., thyrotrophic hormone...). This is surely a misuse of words, from a confusion between trophic action..which is concerned with nutrition, and tropic action..related to movement. Is it too late to revert to thyrotropic..and the like? **1957** *Ibid.* 25 Sept. 701/2 For good or ill, -trophic is all but universal [in the names of hormones], though I saw corticotropic recently in a new edition of a student's biochemistry book from the States.

trophied, *a.* **2.** (Earlier example.)

1830 Sotheby *Madoc* ii. 197 He sits upon a throne of trophied skulls.

tropho-. Add: Also, before a vowel, **troph-. tropho-ctoderm** *Embryol.* = *trophoblast*; hence **trophecto·dermal** *a.*; **trophobiont** *Ent.* [f. Gr. βιοῦντ-, βιοῦν living], an insect which produces a secretion used as food by another; † **tropho·chro·matin** *Cytology* [f. *chromatin: cf. G. *trophochromatin* (W. Lubosch 1902, in *Ergebnisse Anat. und Entwicklungsgeschichte* XI. 783)], chromatin which was thought to be concerned only with the regulation of the metabolism and growth of the cell, and not with its reproduction (*obs.*); **tropholy·tic** *a.* [*-lytic] (of part of a lake) characterized by the decomposition of organic matter, opp. *trophogenic *a.* 2; **trophonu·cleus** *Zool.*, a large nucleus present in some flagellated protozoa, esp. trypanosomes, which regulates the metabolism and growth of the cell; **trophothy·lax** *Ent.* [f. Gr. θύλακ-ος pouch] (see quot. 1971).

1932 M. T. Harman *Textbk. Embryol.* vii. 134 Supposedly the troph-ectoderm produces an enzyme which digests the maternal tissue until the embryo is entirely imbedded. **1980** *Nature* 10 Apr. 550/1 It was recently found that the inner cell mass of the 3.5 day blastocyst is also totipotent and can form trophectoderm when isolated by immunosurgery. **1978** *Ibid.* 7 Sept. 10/3 The strange distribution of this determinant does not fit in with any preconceived notions of trophectodermal formation or differentiation. **1913** E. Wasmann in *Ann. Rep. Smithsonian Inst. 1912* 464 We distinguish..trophobionts or food-producing animals of the ant. **1978** R. J. Elzinga *Fund. Entomol.* vi. 173 These trophobionts are protected by their hosts and are analogous to domestic cows, for they yield food sugar solutions..upon request. **1909** *Q. Jrnl. Microsc. Sci.* LIII. 282 Mesnil..uses a terminology which also has a physiological foundation..; trophochromatin, for chromidial structures of a vegetative function; idiochromidia, for chromidia which enter into the formation of gametes. [*Note*] Cf. Lubosch's ('02) term, 'trophochromatin' and 'idiochromatin'. **1917** *Amer. Nat. Rev. Microbiol.* I. 2 In amoeboid forms special interest attaches to the structure and division of the nucleus. A distinction may be made between the 'trophochromatin' which stains intensely with iron haematoxylin, but takes no part in the formation of the chromosomes, and the 'idiochromatin' out of which the chromosomes are formed. **1957** Tropholytic [see *trophogenic *a.* 2]. **1975** C. A. Cole *Textbk. Limnol.* ii. 10/1 Below the trophogenic layer is a darker tropholytic region, where respiration and decomposition predominate. **1906** H. M. Woodcock in *Q. Jrnl. Microsc. Sci.* L. 181 This is revealed..in the sharp resolution of the nuclear material into trophic and kinetic constituents, which are practically separate and independent, at any rate, during the trypanosome phase. The fertilisation spindle or division nucleus is to be regarded as representing the trophic portion, and it will be convenient, therefore, to distinguish it as the trophonucleus. **1964** M. Hynes *Med. Bacteriol.* (ed. 8) xxviii. 436 Stained preparations [of *Trypanosoma gambiense*] show two nuclear structures, the one, large and centrally placed, is known as the macronucleus or trophonucleus and the other, smaller, placed at the posterior end, is known as the micronucleus or kinetoplast. **1920** Wheeler & Bailey in *Trans. Amer. Philos. Soc.* XXII. 256 The sternal portion of the first abdominal segment is transversely elliptical..and furnished with a food-pouch, the trophothylax. **1971** E. O. Wilson *Insect Societies* iv. 55 The nurse worker first pushes the fragment deep within the trophothylax, the special food pouch located on the lower surface of the thorax just behind the head (and found only in pseudomyrmecine [ant] larvae).

trophogenic (trǫfo-, trōᵘfodʒe·nik), *a.* [f. Tropho- + *-genic.] **1.** *Ent.* Arising from an insect's feeding habits or diet.

1928 W. M. Wheeler *Social Insects* vii. 193 The castes may be blastogenic..in some groups of social insects, and trophogenic..in others. **1980** *Insectes Sociaux* XXVII. 80 The autogenic determination which separates queen and worker castes is succeeded by the larval trophogenic determination which is also at the origin of the various castes.

2. Of part of a lake: characterized by the photosynthetic production of oxygen and organic matter. Opp. *tropholytic* adj. s.v. *tropho-. [tr. G. *trophogen* (E. Naumann *Limnologische Terminol.* (1931) 696).]

1957 G. E. Hutchinson *Treat. Limnol.* I. ix. 583 Oxygen, produced by photosynthesis in the trophogenic layers of the lake, and methane, produced by anaerobic decomposition in the tropholytic layers. **1979** *Ecol. Modelling* VI. 1 (*heading*) The modelling of ³²P kinetics within the trophogenic zone of a small lake.

So **tropho·geny,** the determination of an insect's development by its feeding habits.

1943 W. M. Wheeler *Social Life among Insects* vi. 253 It was formerly supposed that all termite eggs were alike and therefore produced young larvae which..took on the various caste characters as a result of differences in feeding (trophogeny). **1938** *Times Lit. Suppl.* 5 Mar. 19/1 Are the castes of ants determined by the diet of the larvae, that is, trophogeny, or in the egg, that is blastogeny?

tropic, *a.*¹ Add: **II. 4.** In this sense usu. pronounced (trǫ·pik).

b. Of a hormone: = *trophic *a.* 3.

1955 R. M. De Coursey *Human Organism* xviii. 433 The hypophysis or pituitary gland has often been called the master gland of all the endocrines because through its tropic hormones it exerts a regulatory effect over the activity of other endocrine glands. **1966** Lee & Knowles *Animal Hormones* ii. 19 The hormones secreted by the adenohypophysis may be divided into those which control the secretion of other endocrine glands and are named trophic (or tropic) hormones, the remainder act without the mediation of another endocrine gland. **1982** *Jrnl. Clinical Endocrinol. & Metabolism* LIV. 507/1 Rates of [¹⁴C]acetate incorporation into steroids are increased by tropic hormones.

-tropic (trǫ·pik, trōᵘ·pik), *suffix.* [f. Gr. τροπ-ή turning (sb.) + -ic.] Forming adjs. with the senses: (*a*) 'turning or attracted towards', as in Geotropic, Heliotropic, *sympathicotropic *adjs.*; (*b*) 'affecting', as in *neurotropic, *psychotropic *adjs.*; (*c*) in epithets of hormones used interchangeably with *-trophic, q.v.).

tropical, *a.* Add: **2. b.** (Further examples.)

1934 *Discovery* July 207/1 The practical certainty of infection from tropical anaemia or 'hookworm'. **1944** *Living off Land* v. 101 The most insignificant scratch or wound should be tended carefully—it might cause immediate infection and numerous large tropical sores. **1969** Edington & Gilles *Path. in Tropics* x. 435 Tropical splenomegaly syndrome may in some patients end up in lymphosarcoma. **1974** Tropical bubo [see *lymphogranuloma]. **1980** F. A. Nwako *Textbk. Paediatric Surg. in Tropics* lxv. 368/1 The acute tropical ulcer may be complicated by tetanus and gas-gangrene.

d. Of clothing, fabric, etc.: suitable for wearing or using in hot climates; lightweight and porous; also more fully *tropical weight*. Of a (weight of) fabric freq. in *attrib.* use.

1792 W. Bligh *Voy. to South Sea* ii. 26, I gave orders for their light tropical clothing to be put by, and made them dress in a manner more suited to a cold climate. **1920** W. J. Locke *House of Baltazar* x. 121 The tropical drill material which had clothed the troops in Hong Kong. **1924** R. Macaulay *Orphan Island* ii. 31 [We] can be ready in a fortnight. It will take us about that to get our tropical outfit. **1925** in C. Allen *Tales from Dark Continent* (1979) iii. 43 Tropical weight dress coat. **1931** E. Waugh *Remote People* 150 They dried themselves, combed their hair, put on smart tropical suits, and called for dinner. **1938** —— *Scoop* I. iii. 57 William had acquired..six suits of tropical linen. **1942** R. Chandler *High Window* iii. 24 A tropical worsted suit. **1945** E. Waugh *Brideshead Revisited* II. i. 200, I discarded the experiences of those two years with my tropical kit. **1966** P. O'Donnell *Sabre-Tooth* i. 8 He wore the tropical uniform of the United States Army. **1971** 'J. Mayo' *Asking for It* v. 18, I unpacked a cream shirt, my tropical-weight brown suit and put them on. **1972** 'W. Haggard' *Protectors* viii. 93 He wore a tropical suit and a spotted bow tie. **1978** L. Block *Burglar in Closet* i. 4 My suit was a tropical worsted. **1981** G. MacBeth *Kind of Treason* xviii. 178 Mountbatten..was..in immaculate tropical kit.

B. *sb. pl.* Tropical clothes (see sense A. 2 d above).

1934 G. B. Shaw *Too True to be Good* III. 85 Aubrey, in white tropicals, comes strolling along the beach. **1980** J. Hone *Flowers of Forest* III. i. 221 Dressed in immaculate linen tropicals and some kind of old boy's tie.

tropicalize, *v.* Add: **2.** To make suitable for use under tropical conditions. Chiefly in *pass.* or as *pa. pple.*

1941 W. S. Churchill *Second World War* (1950) III. 760 The German tanks recently captured by the Tobruk garrison... Are they tropicalised, desert-worthy, and fitted for use in the very hot weather? **1947** *Short Wave Mag.* V. 234/1 All components are tropicalised. **1972** *Daily Tel.* (Colour Suppl.) 20 Oct. 50 All the woodwork in a Rolls..is 'tropicalised' by three coats of lacquer making it impervious to changes in temperature and humidity. **1978** *Gramophone* Jan. 1335/3 The massive mains transformer is fully tropicalized and the primary winding is adjustable for input voltages of 110–120 and 220–240V AC. **1981** *Hi-Fi Answers* Jan. 7/2 The manufacturer of the British amplifier made a 'Point' of claiming that his equipment had been tropicalised.

Hence also **tro:picaliza·tion.**

1944 *Sun* (Baltimore) 24 Sept. 5/4 For a year past our modern battleships have been undergoing a further measure of modernization and tropicalization to meet the rapid wartime changes in technical apparatus. **1980** *Nature* 7 Feb. p. xii/1 Virtually any required type of impregnation, tropicalisation, potting or moulding (in epoxy, polyester or polyurethane) can be supplied.

tropicana (trǫpikā·nă), *sb. pl.* [f. Tropic *sb.* and *a.*¹ + Ana *suff.*] Things associated with or characteristic of tropical regions; objects from the tropics.

1960 *Spectator* 1 July 12/2 Cockfighting is part of the exotic tropicana Haiti offers to American tourists. **1969** 'G. Black' *Cold Jungle* xii. 165 An astonishing display of tropicana, palms, bamboo, a couple of flame trees. **1976** *Listener* 29 Jan. 121/3 'That slice of the mind which is pure tropicana'—a mixture of Kipling and Maugham.

tropism. Add: Also with the pronunc. (trōᵘ·piz'm).

tropo (trōᵘ·po), colloq. abbrev. of *tropo-scatter.

1966 [see *bill-board, billboard 2]. **1976** *Offshore Engineer* July 5/1 There is little hope of a similar development in the UK sector, since the Post Office has provided the tropo facility and would be reluctant to offer an alternative.

tropo-. Add: **tropoco·llagen** (trōᵘ-) *Biochem.* [so called from its being able to turn into collagen], the molecular constituent of collagen fibrils, formed of three supercoiled polypeptide chains; **tropota·xis** (trōᵘ-) *Biol.* [mod.L., coined in Ger. (A. Kühn *Die Orientierung der Tiere im Raum* (1919) 60)], a taxis in which an animal's movement is in response to the difference in stimulation of two symmetrically placed receptors; hence **tropota·ctic** *a.*, **-ta·ctically** *adv.*

1954 J. Gross et al. in *Proc. Nat. Acad. Sci.* XL. 679 We adopted the term 'tropocollagen' to denote the thin, long particles. *Ibid.*, In the present paper is summarized the evidence..that the various morphologically distinguishable forms of collagen are mutually interconvertible and that the unit of collagen structure involved in these changes of aggregation is the tropocollagen particle. **1971** *Nature* 16 Apr. 437/1 It is usually accepted that collagen is composed of tropocollagen molecules 2900 Å long, 15 Å wide. **1982** J. F. Van Pilsum in T. M. Devlin *Textbk. Biochem.* xxi. 1050 The above classifications of collagens are based on the amino acid sequence and composition of the peptide chains in the tropocollagen molecule. **1940** Fraenkel & Gunn *Orientation of Animals* vii. 89 Arthropods..are, above all, the animals which have eyes suitable for tropo-tactic behaviour. **1979** *Experientia* XXXV. 1457/1 Widely separated nares..would appear to be an adaptation for tropotactic olfactory perception. **1940** Fraenkel & Gunn *Orientation of Animals* vii. 78 The paired receptors of an animal which behaves tropotactically have been compared with the paired reins of a horse. **1934** Tropotaxis [see *telotaxis]. **1979** *Experientia* XXXV. 1457/2 Tubenosed fruit bats locate ripe fruit from among unripe fruit by olfaction, and since detection..has to be made while on the wing, it seems likely that tropotaxis is beneficial.

tropolone (trǫ·pŏlōᵘn). *Chem.* [f. Trop(-ilidine + -ol + -one.] A water-soluble, colourless crystalline compound, $C_7H_6O_2$, which is an enolic ketone whose molecule is formed by a ring of seven carbon atoms and which is the parent of a series of compounds derived from various plants (as colchicine and the thujaplicins); also, any derivative of this.

1945 M. J. S. Dewar in *Nature* 13 Jan. 51/1 If stipitatic acid actually has the resonating structure (I) or (II), it represents a new type of aromatic system; the parent *cyclo*heptatrienolone might be termed 'tropolone'. **1964** *New Scientist* 4 June 613/1 One group of compounds extracted [from timber], the tropolones, was found to be comparable in fungicidal effect to the synthetic fungicides. **1981** *Tetrahedron* XXXVII. Suppl. No. 9. 426/1 The 1,3,2-benzodioxaarsole ether of tropolone..illustrates the possibility of modelling extremely rapid rearrangements.

tropomyosin (trǫpo-, trōᵘpomai·ōsin). *Biochem.* [f. Tropo- + Myosin.] Any of a group of crystallizable proteins related to myosin; *esp.* one found together with troponin in the thin filaments of myofibrils which is instrumental in the mechanism of muscle contraction.

1946 K. Bailey in *Nature* 23 Mar. 369/1 The exact relation of tropomyosin to myosin itself is equally obscure, but the analytical and structural similarities indicate that it is a species of myosin differing mainly in the length of the polypeptide chain. In proposing the present name, we have deemed it desirable to retain the word 'myosin' and to add a prefix which suggests this specific relationship. **1973** *Nature* 5 Oct. 235/3 The finding of non-muscle tropomyosins, of course, raises more questions than it answers. **1979** *Sci. Amer.* May 94 (*caption*) In muscle cells tropomyosin regulates the ability of actin filaments to form cross bridges with adjacent myosin filaments. **1982** J. F. Van Pilsum in T. M. Devlin *Textbk. Biochem.* xxi. 1006 Tropomyosin is a rod-shaped molecule found associated with the actin filaments.

tropone (trōᵘ·pōᵘn). *Chem.* [Trop(ilidine + -one.] A viscous hygroscopic oil, C_7H_6O, which is a cyclic ketone of aromatic character and of which tropolone is the hydroxylated derivative.

1951 *Jrnl. Amer. Chem. Soc.* LXXIII. 876/1 We wish to report the synthesis of 2,4,6-cycloheptatrien-1-one (tropone)..and evidence bearing on its aromatic character. **1981** *Jrnl. Org. Chem.* XLVI. 3575/1 The cycloadducts from 8-azaheptafulvenes and sulfenes, as well as those from tropone and arylsulfenes, readily undergo metalation.

troponin (trōᵘ·pǫnin, trǫ-). *Biochem.* [f. *tropo(myosin + -n- + -in¹.] A globular protein complex consisting of three subunits (*troponin C, I,* and *T*) which is related to and occurs with tropomyosin in the thin filaments of muscle tissue and is important in the mechanism of muscle contraction.

1966 Ebashi & Kodama in *Jrnl. Biochem.* (Tokyo) LIX. 425/1 Tropomyosin-like protein, or 'native' tropomyosin..has been shown to consist of two different proteins, *viz.*, tropomyosin of Bailey type and a globular protein promoting the aggregation of tropomyosin, named troponin. **1982** *Sci. Amer.* June 52/3 Troponin *c* is probably found only in cardiac and skeletal muscle; the troponinlike protein in other tissues is calmodulin.

tropopause (trǫ·pŏ-, trōᵘ·popǭz). *Meteorol.* [f. *tropo(sphere + Pause *sb.*] The upper limit of the troposphere, separating it from the stratosphere, at which the lapse rate falls to zero.

1919 *Geophysical Mem.* (Meteorol. Office) No. 13. 59 The terms 'tropopause' and 'lapse-limit' have been suggested to denote the plane of cessation of the vertical temperature gradient. **1922** *Encycl. Brit.* XXXI. 930 In a cyclone the tropopause is low, in an anticyclone high. **1947, 1971** [see *jet stream a]. **1982** *New Scientist* 17 June 788/1 Venus's tropopause..is much higher than that of the Earth,..some 70 kilometres above the Venusian surface.

troposcatter (trǫ·po-, trōᵘ·poskætəɹ). [f. *tropo(spheric *a.* + Scatter *sb.*] The scattering of radio waves by clouds and local variations in the troposphere so as to extend the range of radio communication.

1959 *Britannica Bk. of Year* 547/1 Tropo-scatter. **1962** *Aeroplane* 7 June 19/2 For VHF ground-air improvements the team suggested the investigation of mountain-top sites and tropo-scatter methods to provide better coverage. **1967** *Technology Week* 23 Jan. 45/1 (Advt.), The Pacific Scatter System..uses ionoscatter, troposcatter, microwave, and vhf. **1976** *Times* 26 May 16/8 High levels of production of oil from the North Sea are dependent on advances in the use of troposcatter.

troposphere (trǫ·pŏ-, trōᵘ·posfīᵊɹ). *Meteorol.* [f. Tropo- + Sphere *sb.*] The lowest region of the atmosphere, extending to a height of 8 to 18 km. and marked by convection and a general decrease of temperature with height.

1914 *Q. Jrnl. R. Meteorol. Soc.* XL. 108 M. Teisserenc de Bort discovered that the atmosphere is divided into two parts, the troposphere, which extends from the surface to about 7 miles, and the stratosphere, which lies above. **1922** *Nature* 2 Feb. 141/1 In the lower layer, called the troposphere, the atmospheric gases are kept well mixed up by winds and convection. **1951** [see *exosphere* s.v. *exo-]. **1982** *New Scientist* 21 Jan. 151/1 Dust in the troposphere, the lowest layer of the atmosphere, is soon washed out by rain.

Hence **troposphe·ric** *a.*, of, pertaining to, or involving the troposphere; *tropospheric scatter* = *troposcatter.

1939 *Proc. IRE* XXVII. 634 It is believed that most of the tropospheric reflections occur at air-mass boundaries or other similar discontinuities. **1955** *Ibid.* XLIII. 1336 (*heading*) Some tropospheric scatter propagation measurements near the radio horizon. **1966** *Electronics* 17 Oct. 137 The system..is a wideband tropospheric scatter and microwave network consisting of 100 sites. It hops across the Mediterranean from the center of Spain to eastern Turkey, covering 6,000 miles and five nations. **1973** *Guardian* 18 May 18/5 With tropospheric fallout there is a steep rise in the level of iodine 131 in milk.

troppo (trǫ·po), *a. Austral. slang.* [f. Tropic *sb.* and *a.*¹ + *-o².] Mentally ill through spending too much time (orig. on war service) in the tropics; (hence simply) crazy, mad. Esp. in phr. *to go troppo*.

1943 G. Johnston *New Guinea Diary* 222 'A man must be going troppo,' he remarks quietly. **1958** R. Stow *To Islands* i. 19 'Terry thinks he's going troppo.' 'Troppo?' 'It's not in the Nurse's Encyclopaedia. Means going queer from being too long in the tropics.' **1968** S. L. Elliott in E. Hanger *Three Austral. Plays* I. i. 33 Only fourteen months here and he's troppo... Know what he does this morning?.. Gets out of bed at reveille and goes off to shave with his spoon and fork. **1975** *Sun-Herald* (Sydney) 30 Nov. 131 Aunty Jack..could bring a badly-needed fresh burst of local comedy (in a troppo way) to our screens.

tropylium (trǫpi·liŏm). *Chem.* [f. *trop(-olone + -yl + -ium.] The cation $C_7H_7^+$ consisting of a ring of seven =CH– groups. Usu. *attrib.*

1954 *Jrnl. Amer. Chem. Soc.* LXXVI. 3204/2 The value..indicates that the tropylium ion is about as strongly acidic as acetic acid when water is the reference base. **1982** T. W. G. Solomons *Fund. Org. Chem.* xi. 375 Tropylium bromide, C_7H_7Br, is insoluble in nonpolar solvents but dissolves readily in water.

trot, *sb.*¹ Add: **I. 1. d.** *on the trot*, (*b*) in uninterrupted sequence, in succession; (*c*) on the run, escaping from confinement, etc.

1952 M. Tripp *Faith is Windsock* vii. 106 Two kites on the trot with crook engines. **1956** *People* 13 May 13/5, I want to be between those posts again when Manchester City reach Wembley next year for the third time on the trot! **1958** M. Pugh *Wilderness of Monkeys* 176, I eloped with one of the boys and we went on the trot from the approved school. Then it came time for his National Service and he went on the trot from the Army. **1973** *Times* 12 Apr. 12/6 Bookmakers lost money for five weeks on the trot. **1982** G. Lyall *Conduct Major Maxim* ii. 16 'He's on the trot,' Maxim guessed. 'Oh Christ, Jim, you can get a district court for that, aiding a deserter.'

e. *the trots* († *trot*), diarrhœa; also *fig. colloq.* Cf. *run *sb.*¹ 14 f.

1808 E. Weeton *Let.* 10 June in *Jrnl. of Governess* (1969) I. 94, I should perhaps be running over to Mr. Ridyard's so *very* often, that ten to one my brother would be..asking what was the matter with me that I was so *often hastily taken*; saying he was sure I was ill of the trot. **1904** in P. Fleming *Bayonets to Lhasa* (1961) xv. 205 He suffers continually from the trots (diarrhoea) which have completely shattered his nerves. **1936** J. G. Cozzens *Men & Brethren* II. 181, I often used to have to hot-foot it over to chapel—a kind of spiritual trots—and pray fervently. **1977** C. McCullough *Thorn Birds* xi. 249 'Go easy on the water at first,' he advised. 'Beer won't give you the trots.'

2. Delete *rare* and add earlier and later examples. In *pl.* (*colloq.*, orig. *Austral.* and *N.Z.*), a series of trotting-races held at a fixed time on a regular course.

1856 *Porter's Spirit of Times* 25 Oct. 128/2 Nothing would have given the lovers of the trotting turf more pleasure than to witness a trot of three miles. **1899** *Bulletin* (Sydney) 21 Jan. 24/2 At the recent big M.L. trots horses well-known this side carried off their full share of prize-money. **1905** A. C. Rice *Sandy* 215 Nelson wants the fellow to drive for him at the fall trots. **1934** T. Wood *Cobbers* ii. 19 We're proud of the Trots in Perth. It's the best course in the world. **1959** *N.Z. Listener* 16 Jan. 14/4 An oddball like myself, wholly uninterested in racing, even night trots. **1968** *Globe & Mail* (Toronto) 17 Feb. 43 Sixth race—Trot, mile. Purse $3,000. **1976** *National Observer* (U.S.) 2 Oct. 7/1 He won the Empire Trot at Syracuse two weeks ago. **1977** *New Yorker* 19 Sept. 131/1 No doubt remembering the crush of more than forty thousand when the trots opened at Meadowlands a year ago, many people stayed away.

5. (Examples.)

1924 P. Marks *Plastic Age* 299 I'm talking about the copying of math problems and the using of trots. **1975** B. Meggs *Matter of Paradise* (1976) vii. v. 173 Somebody suggested..that he get a trot. An absolutely forbidden interlinear translation. The Latin on one line; the English right below it. **1984** *Times Lit. Suppl.* 27 Apr. 44/2 The translations are rarely better than lame trots.

5*. *Austral. colloq.* A sequence, a succession, esp. in a game of chance; a run of luck of a specified kind. See 1 d above.

1911 L. Stone *Jonah* 216 A trot or succession of seven tails followed, and the kip changed hands rapidly. **1919** W. H. Downing *Digger Dialects* 51 *Trot*, an experience (e.g. 'a rough trot'; 'a bad time'). **1937** J. A. Lee *Civilian into Soldier* 99 Sometimes a man would succeed daringly, doubling up and breaking the ring with a long run of heads, 'throwing a trot'. **1949** L. Glassop *Lucky Palmer* 177 He was 'Lucky' Palmer, having a bad trot at the moment, admittedly, but still 'Lucky' Palmer. **1966** P. Mathers *Trot* 90 He..had had a tough trot, humped the bluey, been through it all. **1974** D. R. Stuart *Prince of my Country* v. 33 He's had a damn good trot, old Marney.

II. 6. b. *Naut.* (See quot. 1976.)

1923 *Man. Seamanship* (Admiralty) II. 107 When several targets are secured in line to a trot, only the ends of the trot need be marked by lights. **1950** G. Hackforth-Jones *Worst Enemy* iii. 202 The old ship parted her moorings and drifted down on to a destroyer trot. I had to let go two Admiralty-pattern anchors that were last used in the Crimean War. **1976** *Oxf. Compan. Ships & Sea* 893/1 *Trot*, a multiple mooring for small boats or yachts. The base mooring is laid in a straight line and from it individual moorings rise at intervals spaced to allow the boats room to swing with the tide.

III. 7. trot-boat (see quot. 1955); **trot-line** (earlier example).

1945 'N. Shute' *Most Secret* vi. 124, I can get the trot boat down each evening. **1955** *N. & Q.* Sept. 402/2 A 'trot-boat' is a boat of any size which makes routine visits to discharge or embark passengers, stores, etc., at ships secured to the buoys. **1826** 'Nonius Nondescript' *The* —— 18 Feb. 10 As full of noozes and strings as a fisherman's trot line.

Trot (trǫt), *sb.*³ and *a.* Colloq. abbrev. of *Trotskyist *sb.* and *a.*, *Trotskyite *a.* and *sb.*

1962 D. Lessing *Golden Notebk.* iv. 451, I was a hundred per cent party member, and there was Harry, a dirty Trot, so there were high words and we parted for ever. **1970** G. Greer *Female Eunuch* 22 The most telling criticisms will come from my sisters of the left, the Maoists, the Trots. **1976** *Times* 29 Dec. 8/8 A true *Trot* ought to believe in worldwide revolution. **1983** 'J. le Carré' *Little Drummer Girl* iv. 80 Some kind of loony Trot splinter group.

trot, *v.* Add: **2.** Freq. with specifying adv. or advb. phr.; *absol.* also (contextually) to depart, to leave. (Further examples.)

1847 C. Brontë *Jane Eyre* II. viii. 203 In case I married Miss Ingram, both you and little Adèle had better trot forthwith. **1862** in N. Longmate *Hungry Mills* (1978) viii. 108, I..trot down to a butcher in a better neighbourhood. **1899** O. Wilde *Importance of being Earnest* II. 74 *Chasuble*... At what hour would you wish the ceremony performed? *Jack*. Oh, I might trot round about five if that would suit you. **1954** J. R. R. Tolkien *Fellowship of Ring* I. ix. 166 No time for talking. I must be trotting. **1960** *Cambr. Rev.* 7 May 506/2 It is not true to suppose that the setting-up of machinery for psychiatric consultation merely encourages the 'neurotic' to trot along to the psychiatrist at the least excuse. **1984** *Your Computer* May 25/2 You can..trot up to the Registry with that reference number and get a copy of his death certificate very quickly.

4. d. *to trot out*: also (*N.Z.*) simply *trot*.

1946 F. Sargeson *That Summer* 33 I've got a job in a grocer's shop and I'm trotting a sheila. **1964** B. Crump in *Weekly News* (Auckland) 21 Oct. 46/6, I didn't know she was going steady with you... If I'd known you were trotting her [etc.].

f. To bid against at an auction in order to force up the price; to make or accept a spurious bid for (an item at auction) in order to force up the price. Also with *up*. *slang*.

1864 Hotten *Slang Dict.* 262 *Trot*, to 'run up', to oppose, to bid against at an auction. *Ibid.*, 'We *trotted* him up nicely, didn't we?' *i.e.*, we made him (the private buyer) pay dearly for what he bought. **1955** W. Mankowitz *Make me an Offer* viii. 64 'But it's no good to you?'.. 'Only if it goes reasonable—not if the reserve is high. And not..if it's trotted.' 'We don't do that sort of thing in the country, you know.'

g. *to be able to trot a mouse on it* and varr.: said of particularly strong or thick liquid food or drink. *dial.*

1936 'N. Blake' *Thou Shell of Death* xiii. 229 A cup of tea, sir, after your journey... It's nice strong tea, so it is. Ye could trot a mouse on it. **1970** H. McLeave *Question of Negligence* i. 3 That's the way they make it [*sc.* tea] in Scotland... Sweet as a sheep's eye and strong enough to trot a mouse on. **1975** *Times* 17 May 10/8 A bowl of parsnip soup—'so thick you could trot a mouse on it', as the country saying goes.

Trotskyism (trǫ·tski,iz'm). Also † **Trotskysm,** † **Trotzkyism.** [f. Leon *Trotsky*, the assumed name of Lev Davidovich Bronstein (1879–1940), Russian revolutionary and politician + -ism.] The political or economic principles of Trotsky; a form of Marxism urging world-wide revolution, as advocated by Trotsky.

1925 tr. Trotsky in *Times* 30 Jan. 11/4, I cannot..accept the accusation of trying to pursue my own line of policy (Trotskyism). **1925** —— in M. Eastman *Since Lenin Died* 155, I can nowise accept, however, the accusation that I have pursued a special line (Trotskysm) and tried to revise Leninism. **1930** W. H. Chamberlin *Soviet Russia* iv. 76 There would seem to be little political future in Russia for Trotzkyism. **1942** E. Waugh *Put out More Flags* 48 She believed in a People's Total War; an uncompromising girl whom none of them liked; a suspect of Trotskyism. **1963** [see *capitulationism]. **1977** *Belfast Tel.* 24 Jan. 9/8 Trotskyism is installed as official doctrine and the pro-Soviet sympathies of some of our trade union leaders go unrebuked.

Trotskyist (trǫ·tski,ist), *sb.* and *a.* Also **Trotskist,** † **Trotzkyist.** [f. *Trotsky* (see prec.) + -ist.] **A.** *sb.* A follower or supporter of Trotsky or Trotskyism. **B.** *adj.* Of, pertaining to, or characteristic of Trotsky or Trotskyism.

1927 *Daily Tel.* 22 Nov. 9/2 The adoption of the word 'Russia' by the Trotskyists. *Ibid.* 6 Dec. 11 The struggle between the Trotskists and the Stalinites. **1930** W. H. Chamberlin *Soviet Russia* iv. 74 The Party Congress..laid down the rule that adherence to the views of the Trotzkyist opposition was inconsistent with membership in the Communist Party. **1937** Koestler *Spanish Testament* ix. 178 The P.O.U.M.—the Trotskyist Party—was even more unrestrained in its agitation. **1949** [see *diversionist]. **1959** *Daily Tel.* 15 Apr. 17/5 The recently formed Trotskyist organisation known as the Socialist Labour League. **1973** 'I. Drummond' *Jaws of Watchdog* x. 136 Trotskyists, anarchists and the lunatic fringe of the Underground. **1980** *Washington Post* 6 July C1/2 On the left there are well over 40 different organizations—Leninist, Stalinist, Trotskyist.

Trotskyite (trǫ·tski,ait), *a.* and *sb.* Also **Trotskiite.** [f. *Trotsky* (see *Trotskyism) + -ite¹.] = *Trotskyist *sb.* and *a.*

1919 *Mr. Punch's Hist. Great War* 210 Which am I..Pro-German or Pro-Trotskyite? **1920** [see *Leninite *a.* and *sb.*]. **1953** *Encounter* Nov. 30/2 Young X. starts as a Communist, is soon disillusioned, founds a Trotskyite opposition group of ten people [etc.]. **1957** V. Nabokov *Pnin* vii. 184 Russian emigration was made to mean by astute communist propaganda a vague and perfectly fictitious mass of so-called Trotskiites (whatever these are), ruined reactionaries, [etc.]. **1977** M. Walker *National Front* viii. 210 The Trotskyite parties..are too small and too dependent upon the intelligentsia to present any credible revolutionary threat. **1978** *Jrnl. R. Soc. Arts* CXXVI. 671/1 Their activities were compared to those of the Trotskyites in 1923.

trottie, *sb.* (In Dict. s.v. Trot *sb.*¹ 4.) (Example.)

1924 'L. Malet' *Dogs of Want* vii. § 6 Darling girls, from the time when they were the tiniest trotties till now.

trottie, var. *Trotty *a.*

trotting, *vbl. sb.* Add: (Example corresp. to *trot *v.* 4 f.)

1969 *Daily Tel.* 5 May 19/2 The most common is 'trotting'—taking non-existent bids to force up the price.

b. *trotting-match* (earlier example.)

1822 *Sunday Times* 20 Oct. 3/2 The roan trotting match for 500 sovereigns.

trottoir. (Earlier example.)

1792 A. Young *Trav. France* I. 150 The streets..are wide, and very well paved, with the addition, uncommon in France, of *trotoirs*.

trotty (trǫ·ti), *a. colloq.* Also **trottie.** [f. Trot *sb.*¹ + -y¹.] **a.** Of daintily small proportions. **b.** Of a person: agreeable, amenable.

1891 'L. Malet' *Wages of Sin* II. v. i. 165 Some of the little silk shifties and night-i-gowns were simply too trottie for words. *a* **1913** F. Rolfe *Desire & Pursuit of Whole* (1934) xi. 109 'I can't tell you how trotty we think it of you!'..bibbled the woman. **1928** Galsworthy *Swan Song* I. i. 4 Trotty little ladies with dresses tight blown about trotty little figures.

trotyl (trōᵘ·til). [f. *trini)trot(oluene + -yl.] = *trinitrotoluene.

1910 *United Service Mag.* Feb. 554 A new explosive for shells, mines and torpedoes is being manufactured... It is known commercially as 'Trotyl', its full name being Trinitrotoluol. **1922** *Encycl. Brit.* XXXI. 50 Trinitrotoluene (TNT) which is known officially as Trotyl is..very similar in its action to picric acid and had been discovered by Wilbrand in 1863. **1981** *Chem. Abstr.* 7 Sept. 121/2 (*heading*) Initial phase of the initiation of the detonation process in pressed trotyl.

trouble, *sb.* Add: **1. e.** *my troubles*, a dismissive exclamation: 'don't worry about me': 'I don't care'. *Austral. colloq.*

1895 C. Crowe *Austral. Slang Dict.* 89 *My troubles*, what do I care. **1905** N. Spielvogel *Gumsucker on Tramp* 90 Off again; round Leeuwin Cape; rough seas; My Troubles! I'm coming home. **1947** G. Casey *Wits are Out* 44 'You better lay off Kitty while the old man's about, or there'll be one more out-of-work motor salesman kicking round the city,' Syd suggested. 'My troubles!' Jerry jeered.

f. Usu. with qualifying noun: faulty working of apparatus or machinery, esp. on a motor vehicle; a problem caused by this (*engine trouble*, etc.). Also applied *transf.* to personal relations, as *wife trouble*.

1902 *Trans. Inst. Naval Archit.* XLIV. 213 Although it seems to fit the water tube troubles, it does not answer so well with the furnace troubles. **1909** *Westm. Gaz.* 26 Oct. 2/1 The only other serious difficulty [with the Wright biplane] seems to be what is known, generically, as 'engine trouble'... The forms that this 'engine trouble' takes are various, as every motorist knows. **1981** P. Audemars *Gone to her Death* iii. 61 The local garagist..has wife trouble, because she has the money he needs.

g. *trouble and strife*, rhyming slang for: (*a*) 'life' (*rare*); (*b*) 'wife'.

1908 'Doss Chiderdoss' in *Sporting Times* 11 July 1/3, I shouted, 'Your "bees", or your "trouble and strife"!' Like the hero in 'Highwayman Harry'. **1929** J. B. Priestley *Good Companions* III. ii. 611 The old trouble-and-strife, eh? **1949** A. Wilson *Wrong Set* 62 'Thanks for looking after my old trouble and strife' said Bruce. **1959** J. Osborne *Paul Slickey* II. x. 86 My posh trouble-and-strife, I'll be hers. **1977** G. Fisher *Villain of Piece* i. 7 It's the old trouble and strife—wife. I want to see her all right.

2. b. *the troubles, the Troubles.* Any of various rebellions, civil wars, and unrest in Ireland, *spec.* in 1919–23 and (in Northern Ireland) since the early 1970s.

1880 W. H. Patterson *Gloss. Words Antrim & Down* 109 *Troubles, the*, the Irish rebellion of 1641. **1922** Joyce *Ulysses* 237 Times of the troubles... Somewhere here Lord Edward Fitzgerald escaped from major Sirr. *Ibid.* 613 He vividly recollected when the occurrence alluded to took place..in the days of the land troubles..early in the eighties. **1923** *Times Lit. Suppl.* 11 Oct. 661/3 A weak Government..., a new wave of nationalist exaltation, an untrained army of youths brought up on war rations..—these factors were sufficient to account for the troubles of 1919–21. **1942** E. Waugh *Put out More Flags* iii. 235 The ruins of a police barracks, built to command the road through the valley, burnt in the troubles,..were one green with the grass. **1949** C. Graves *Ireland Revisited* vi. 57 'This was where Michael Dwyer was in keeping during the Troubles,' Mackey vouchsafed. ('In keeping' means being on the run.) **1959** *Listener* 2 July 32/1 The complicated political and personal passions inspired by 'the troubles'. **1968** M. Collis *Somerville & Ross* xxv. 258 As the Troubles were over more than ten years before [1936], how came it that Admiral Boyle, living in quiet retirement and much liked by high and low, was singled out? **1981** M. Kenyon *Zigzag* i. 6 Before the new Troubles..he had fallen in love with romantic Ireland.

3. *to take the trouble* (earlier example).

1830 R. J. Raymond *Oh! Men what Silly Things You Are* (song) 3 She marks you down, fly where you will.. Can wing you, feather you or kill, Just as she takes the trouble.

5. a. Also *to ask for trouble*: see *ask *v.* 16 b. Similarly, *to look for* (or *seek*) *trouble*.

1901 Merwin & Webster *Calumet 'K'* 134 We've got to build the belt gallery—and we'll have no end of a time doing it if the C. & S.C. is still looking for trouble. **1905** *N.Y. Even. Post* 29 Aug. 2 In the possible chance of rounding up all who might be seeking trouble, the police temporarily sequestered and questioned 140 Chinamen. **1912** 'Aurora' *Jock Scott, Midshipman* xiv. 165 But if you are artful you don't often get 'bowled out', unless one of the 'crushers' has a 'down' on you, and is 'looking for trouble'. **1922** E. O'Neill *Anna Christie* (1923) I. 25, I ain't looking for trouble. **1947** W. Motley *Knock on Any Door* 152 Swollen out in their own importance they walked along West Madison looking for trouble.

7. *trouble-maker* (examples), *-shirker*; *trouble-making* ppl. adj. and vbl. sb.; **trouble-hunter** (examples); *spec.* = *trouble-shooter 1; also **trouble hunting** *vbl. sb.*; **trouble lamp, light** *N. Amer.*, a portable lamp (esp. one carried on a motor vehicle), by the light of which roadside repairs, etc., can be done; **trouble man** *U.S.* = *trouble-shooter 1; **trouble spot,** a place where difficulties frequently occur; a scene of (impending) conflict.

1910 Trouble-hunter [see *hell-bent *a.* and *adv.*]. **1924** *New Eng. Telephone Topics* XVIII. 288 Repairmen, the 'trouble hunters', are at work constantly. **1882** T. D. Lockwood *Pract. Information for Telephonists* 135 Every movement made for an accurate preliminary test frequently saves an hour of happy-go-lucky trouble hunting. **1916** *Daily Colonist* (Victoria, B.C.) 9 July 12/4 If a car is not equipped with an extension trouble lamp, it is well to provide among the accessories a pocket flash lamp. **1927** W. Faulkner *Sartoris* III. 196 He was doing something to the engine of it [*sc.* a car] while the house-yard-stable-boy held a patent trouble-lamp. **1952** *Sun* (Baltimore) (B ed.) 5 Jan. 7/4 Their headlights went out... A door slammed shut and cut the wire on the trouble light. **1979** *Arizona Daily Star* 1 Apr. H3/1 He just happened to have a siphon hose; also a trouble light with a cord that seemed long enough to reach back to his home in Mexico City. **1923** *Time* 28 May 1/2 (*heading*) Chief trouble maker. **1931** Kipling *Limits & Renewals* (1932) 191, I took stock o' them, to spot the funny-men an' trouble-makers. **1955** 'A. Gilbert' *Is She Dead Too?* ii. 40 A snooper, or trouble-maker, that was Margaret Reeve. **1981** W. Ebersohn *Divide Night* xiii. 175 A more disciplined age where trouble-makers who went against the government would be dealt with firmly. **1920** S. Lewis *Main St.* xvi. 202, I certainly hope you don't class yourself with a lot of trouble-making labor-leaders! *a* **1974** R. Crossman *Diaries* (1975) I. 77 Manny wouldn't allow it, for fear—as he put it—that the questions raised would be used for trouble-making. **1889** *Cassell's Family Mag.* June 410/1 A special band of what the Americans call 'Trouble-men', who are prepared to attend at once to sudden calls for assistance. **1953** *Herald* (Belle Glade,

Florida) 13 Feb. 1/1 According to Florida Power & Light district manager C. A. Chase, FPL's 'Troubleman' J. J. McCarley located the difficulty, and repair crews worked until 2 am Wednesday repairing broken circuits and restoring service. **1908** A. S. M. Hutchinson *Once aboard Lugger* v. vii. 268 These light-hearts, these trouble-shirkers. **1956** M. E. W. Goss in R. K. Merton *Student-Physician* iv. 258 The regular duties..included the unwritten obligation to assist in..assessing the 'trouble-spots' and suggesting possible solutions. **1963** *Listener* 7 Feb. 260/2 Sir David Eccles wants £200,000,000 a year pumped into the trouble-spots [*sc.* areas of heavy unemployment]. **1981** T. Barling *Bikini Red North* ii. 41 It should be quiet enough, being so far from Montmartre and the other trouble spots.

trou·ble-shooter. orig. *U.S.* [f. Trouble *sb.* + Shooter.] **1.** A person who traces and corrects faults in machinery and equipment (orig. *spec.* on a telegraph or telephone line). Cf. *trouble-hunter, man* s.v. *Trouble *sb.* 7.

1905 *Strand Mag.* Mar. 268/1 A good looking young 'trouble-shooter'—as a mender of telephone lines is called—had..asked her to marry him. **1913** *Red Cross Mag.* Jan. 34/1 Among them are..the 'trouble shooters', highly trained men who are responsible for the repairing of any breaks in the plant or equipment. **1931** B. Starke *Touch & Go* xv. 248 A trouble-shooter for the telephone lines. **1945** H. D. Smyth *Gen. Acct. Devel. Atomic Energy Mil. Purposes* xi. 121 Particularly in the early stages of operation the Berkeley men stationed at Clinton were invaluable as 'trouble shooters' and in instructing operators. **1951** *Engineering* 2 Feb. 133/2 Manufacture..by fully-automatic machine shops..with only a few skilled men as 'trouble-shooters'. **1959** H. Hobson *Mission House Murder* xxi. 140 A post office electronics expert and trouble-shooter.

2. One who specializes in removing or solving difficulties; esp. a mediator in diplomatic or industrial affairs.

1927 *Sat. Even. Post* 15 Jan. 153/3 With the 'Trouble-Shooters' of the North Atlantic Icebergs... Locating and destroying them is the perilous and never-ending duty of the United States Coast Guard cutters. **1933** R. C. Mayer *How to do Publicity* xi. 134 The 'trouble-shooters' in publicity deal mostly with such emergencies. **1940** R. S. Lambert *Ariel & all his Quality* iii. 77 The light had gone out of the Talks.... His successors were chosen to be what Americans call 'trouble shooters'. **1953** W. R. Burnett *Vanity Row* iii. 21 He needed an expert trouble-shooter, untainted by police politics. **1962** R. Buckminster Fuller *Epic Poem on Industrialization* 24 Self-helpless old fashioned business War forced to call in ..Professional Trouble-shooters. **1971** H. Wilson *Labour Govt.* ix. 136 The appointment of an industrial relations 'trouble-shooter' for the industry.

Also **trou·ble-shooting** *vbl. sb.* and *ppl. a.*; hence (back-formation) **trouble-shoot** *v. trans.* and *intr.*, to solve (a problem), to repair; to mediate.

1918 V. W. Pagé *Aviation Engines* 9 Special attention has been paid to instructions on tool equipment, use of tools, trouble 'shooting' and engine repairs. **1938** E. B. White *Let.* 18 Nov. (1976) 186 This place teems with trouble of one sort and another. I am up every morning at twenty past six, trouble shooting. **1941** *Daily Progress* (Charlottesville, Virginia) 19 Aug. 1 (*heading*) Judge Rosenman..is now in capital, trouble-shooting the bottle-neck. **1957** V. Packard *Hidden Persuaders* xviii. 208 One firm that provides psychological bug-hunting services to industry cited the service it performed in trouble-shooting an employee problem in Ohio. **1964** S. Brittan *Treasury under Tories* ii. 53 The Cabinet's *Economic Policy Committee*.., over which the Chancellor presides, is mainly a trouble-shooting body. **1969** *Daily Colonist* (Victoria, B.C.) 8 June 2/7 (Advt.), Analyst—reporting to the President—required for troubleshooting all facets of the Victoria winery operation. **1977** P. Dickinson *Walking Dead* iv. ii. 255 His official status in the Company was a string of vague general nouns, but his job was trouble-shooting. **1978** R. Lewis *Inevitable Fatality* i. 19 I'm a business consultant... My forte is to troubleshoot, to get in and out again.

trough, *sb.* Add: **1. d.** *fig.* A place where food is provided, *spec.* a dining-table; hence, a meal. *colloq.*

1901 'H. McHugh' *John Henry* 95 We left the mob just as all hands were paddling off to the ice-cream trough. **1915** F. M. Hueffer *Good Soldier* i. iii. 38 Why shouldn't we all eat out of the same trough—that's a nasty New York saying. **1930** Wodehouse *Very Good, Jeeves!* iv. 96 The Bellinger..had sung us a few songs before digging in at the trough. **1965** *New Statesman* 14 May 777/1 Things are a bit different at the old trough these days. **1981** 'M. Innes' *Lord Mullion's Secret* viii. 68 If he didn't stir his stumps he would be late for the trough.

e. In various *fig.* phrases applied to a ready source of income, esp. one shared by unscrupulous persons. *colloq.*

1906 J. London *Let.* 20 Oct. (1966) 212 All I can tell you is, that you've got your feet in the trough. **1971** P. Tamony *Americanisms* (typescript) No. 28. 2 Local pimps and fast-buck boys who had hustled to the troughs for fat-staff salaries. **1974** L. Deighton *Spy Story* iv. 47 I'm going to find out what it's costing. We can't go on eating our heads off at the public trough. **1981** J. D. MacDonald *Free Fall in Crimson* xvi. 186 The money would come..to Josie, and you would be able to stay in the trough.

6. b. Also *absol.* (Later *fig.* examples.)

1942 C. S. Lewis *Let.* 20 Jan. (1966) 199 Sorry you're in a trough. I'm just emerging..from a long one myself. **1958** *Sunday Times* 9 Nov. 15/3 E. Nesbit..has therefore been 'in the trough'—widely read, ardently admired, but neglected as a subject for critical appraisals. **1977** *Listener* 28 July 123/3 At the moment his [*sc.* E. M. Forster's] reputation is in the trough; it is said that he is a slight talent, overpraised for extraneous reasons.

d. *Econ.* The lowest level of economic activity or prosperity reached during a recession.

1916 G. B. Shaw *Androcles & Lion* Pref. p. lxvi, Basing..our whole industrial system on successive competitive waves of overwork with their ensuing troughs of unemployment. **1930** *Economist* 29 Mar. 691/1 We are, in fact, in the trough of a depression. **1960** *Ibid.* 8 Oct. 161/2 Even if the recession does not reach its trough until well into the spring. **1981** *Daily Tel.* 9 July 1/6 There is now firmer evidence that the trough in the recession has been reached, said the Treasury yesterday.

e. Hence, the lowest point in a period of any varying quantity; the time when this occurs. Also, the representation of this state on a graph; a point in a wave-form at which the varying quantity is a minimum. Cf. *crest *sb.*[1] 7 e, *peak *sb.*[2] 5 e.

1938 *British Birds* XXXII. 214 This is followed by a more or less distinct trough, after which numbers rapidly increase to a higher autumn peak by mid September. **1958** *Listener* 16 Oct. 605/1 Absence of distortion and the avoidance of marked peaks and troughs in the amplitude-frequency characteristic. **1971** *Physics Bull.* Aug. 462/2 Such currents tend to pile electrons in the potential troughs of the wave and denude the crests. **1976** *Daily Tel.* 22 Mar. 7/1 Chromatography splits the sample into its volatile chemical constituents, and draws an alpine graph with heady peaks and troughs to represent the chemicals coming through.

7. trough garden, a miniature garden comprising a group of small plants, often alpine ones, grown in a trough-like container of real or imitation stone; cf. *sink garden* s.v. *sink *sb.*[1] 14.

[**1923** *Times* 31 May 10/7 Mr. Clarence Elliott's attempt at providing miniature alpine gardens in old stone troughs ..would undoubtedly provide much interest where space was too limited for real gardening.] **1935** C. Elliott *Rock Garden Plants* 288 An invaluable small thing for spilling about and filling up odd sunny corners, and perfect on the trough rock garden. **1950** W. E. Shewell-Cooper *Home, Window & Roof Gardening* viii. 70 The Trough Garden..can be a great joy to the rock garden lover. **1979** M. Soames *Clementine Churchill* xxviii. 471 Here Clementine made a 'trough' garden.

trounce, *v.*[1] Add: **3. c.** To defeat heavily at a sport. *colloq.*

1942 Berrey & Van den Bark *Amer. Thes. Slang* § 649/5 *Defeat decisively*,..trample (on), trounce, walk on [etc.]. **1951** *Sport* 27 Jan.–2 Feb. 3/1 He was omitted from the side trounced 4–0 at Reading. **1972** J. Mosedale *Football* v. 64 Green Bay teams had trounced AFL entries the previous two years.

trouncer. Add: Also *spec.* an assistant to a carman, drayman, or lorry-driver. (Later examples.)

1913 M. S. Reeves *Round about Pound a Week* i. 2 Some of the more enviable and settled inhabitants of this part of the world [*sc.* Kennington]..generally are somebody's labourer, mate, or handyman. Painters' labourers ..trouncers for carmen, are common amongst them. **1923** *Weekly Dispatch* 30 Sept. 3 It was stated that a 'trouncer' was a coal carman's assistant. **1953** *Word for Word* (Whitbread & Co.) 35/2 *Trouncer*, the drayman's mate; so-called because, before the improvement of roads under Telford and MacAdam, he had to 'trounce', i.e., push and manhandle the dray over the innumerable potholes and hazards.

trouper (trū·pəɹ). Also trooper. [f. Troupe + -er[1].] **1.** An actor or performer belonging to a troupe.

1890 B. Hall *Turnover Club* 160 As the 'troupers' come into the station where I sat, they were a sorry-looking lot. **1912** L. J. Vance *Destroying Angel* (1913) vi. 77 I'm as superstitious as any trooper in the profession. **1946** *Boston Transcript* Sept. 6/2 A little knot of interested troupers were looking on as Joe and I met. **1973** 'D. Rutherford' *Kick Start* i. 8 A good trouper can still shimmy in her fifties.

2. *transf.* A reliable, uncomplaining person; a staunch supporter or colleague. Freq. with qualifying adj., as *good trouper*. *colloq.*

1959 P. Bull *I know Face* xi. 194 The phrase 'she's a trouper' now has an old-fashioned and faintly derogatory air and is usually bandied about when someone continues to play with a high temperature or a shattering bereavement. **1961** *Times* 6 July 5/5 Chapman is a good trouper, and..he has a fine record of consistency in match and stroke play. **1976** D. Francis *In Frame* iv. 70, I don't think you're selfish at all. In fact, Maisie, I think you're a proper trouper.

trousers, *sb. pl.* Add: **2. c.** (Earlier example.)

1820 M. Wilmot *Let.* 3 May (1935) 57 Catharine..has not one frock, or pair of trousers fit to wear, now that summer is coming.

d. Applied to the hair on the hind legs of certain dogs, esp. those of long-coated breeds.

1948 C. L. B. Hubbard *Dogs in Britain* 464 Trousers.—The hair on the hindquarters of any dog. **1949** L. E. Naylor *Poodles* v. 44 They [*sc.* poodles] enjoy having any kind of smart clip, because they dislike their trousers being muddy. **1962** R. H. Smythe *Anat. Dog Breeding* vii. 133 The loose trousers wrinkled round the knees and pasterns of the Dachshund.

2*. In various colloq. phrases. **a.** *anything in trousers*, etc.: any man, whether eligible, suitable, or not.

1887 *Lantern* 14 May 3/1 They go crazy over everything that wears trousers. **1979** A. Price *Tomorrow's Ghost* i. 9 Anything in trousers was as much Target for Tonight to Marilyn Francis as Marilyn Francis was for anything in trousers.

b. *not in these trousers*: certainly not.

1920 P. Gibbs *Realities of War* vi. 189 'Come up and have a look, Jack,' he said to one of the blue-jackets. 'Not in these trousers, old man!' said that young man. **1929** R. C. Sherriff *Journey's End* iii. ii. 73 She said, 'Not in these trousers'—(in French). **1939** J. Cary *Mister Johnson* 147 You think perhaps we leave the money in the till and you tief 'em. Not in these trousers, Mister Poldedoodle.

c. *to wear the trousers*: to be the dominant member of a household. Also *transf.* Cf. Breech *sb.* 2, *pants *sb. pl.* 1 e.

1931 R. Campbell *Georgiad* i. 11 It is you must 'wear the trousers' now. **1959** J. L. Austin *Sense & Sensibilia* (1962) ii. 15 It is essential to realize that here the notion of perceiving indirectly wears the trousers. **1963** A. Heron *Towards Quaker View of Sex* iii. 33 A married couple where the woman 'wears the trousers'.

d. (*to catch*) *with one's trousers down*: in a state of embarrassing unpreparedness. Cf. *pants *sb. pl.* 1 c. *slang.*

1966 *Guardian* 31 Mar. 14/8 Catch them with their trousers down. **1967** 'F. Clifford' *All Men are Lonely Now* ii. 254 By that time the shooting will seem to be as haphazard as can possibly be, as if we'd been caught with our trousers down. **1980** J. Gardner *Garden of Weapons* iii. xix. 186 A job... Took us by surprise: with the trousers down.

4. *trouser-fly, -pocket* (earlier example), *-seat*; **trouser-bottoms** *pl.*, the lower parts of the legs of a pair of trousers; **trouser-clips** *pl.*, clips (of various kinds) used by cyclists to confine the trousers round the ankles; **trouser-cuff**, the turn-up on a trouser-leg; **trouser suit**, a woman's suit consisting of matching jacket and trousers; hence **trouser-suited** *a.*; **trouser zip**, a zip used as a fastening, usu. at the front of a pair of trousers.

1920 D. H. Lawrence *Women in Love* xiv. 188 He..pulled on his boots. They were sodden, as were his socks and trouser-bottoms. **1973** A. Ross *Dunfermline Affair* 116 My trouser-bottoms were wet. **1895** *Army & Navy Co-op. Soc. Price List* [illegible] Trouser Clips. Per pair 0/3. **1908** H. G. Wells *War in Air* ii. 33 A small, dissolute-looking shop in the High Street, adorned with..a display of bells, trouser-clips, oil cans, and other accessories. [illegible] Drood adjusted his trouser clips and [illegible] off. **1931**, etc. trouser-cuff [see *cuff *sb.*[1] 4 d]. **1964** C. Barrett *On Wallaby* v. 90, I had fifteen-inch trouser cuffs, [illegible] **1982** H. Engel *Murder on Location* (1983) xxiii. 207, I squeezed [illegible] **1922** Joyce *Ulysses* [illegible] **1852** J. S. Coyne *Box & Cox Married & Settled* 9, I demand your card, sir! You'll find it in my left hand trouser pocket. **1923** D. H. Lawrence *Ladybird* [illegible] **1960** Wodehouse *Jeeves in Offing* viii. 90 Gives a woman a start, naturally, to come into her son's bedroom and observe an alien trouser-seat sticking out from under the dressing table. **1939** *Vogue* Dec. (Advt., verso front cover), Digby Morton, famous tailleur, created this trouser suit in a bright tartan 'Viyella'. **1975** D. Lodge *Changing Places* v. 174 She was waiting for him..in a cream-coloured trouser-suit. **1973** 'D. Halliday' *Dolly & Starry Bird* iv. 47 The girls [illegible] with Pat Nixon hairdos. **1966** [illegible] *Price List* 24 Lightning Trouser Zipps. **1976** P. Dickinson *King & Joker* vi. 70 'Shall I show you something, little girl?' he whispered. [illegible] trouser-zip.

trouserettes, (*b*) short trousers; **trouserless** *a.* (earlier example).

1924 W. de la Mare *Ding Dong Bell* 30 A little boy in velveteen trouserettes. **1961** *New Statesman* 28 Apr. 658/2 The huge arthritic waiters stared..at [illegible] strange check pantaloons. A party of secure business people were eating lobsters, and one of them [illegible] these [illegible] trouserettes. **1848** A. H. Clough *Bothie of Toper-na-Fuosich* v. 33 A heavy pea-coat his trouserless trunk enwrapping.

trousies (trau·ziz), *sb. pl.* Also trowsies. Dial. or colloq. var. of Trousers *sb. pl.* 2 a.

1886 H. Baumann *Londinismen* [illegible] Trousies, trousers. **1913** H. Kephart *Our Southern Highlanders* xiii. 285 [illegible] (horses) [illegible] trousies [illegible] tives). [illegible] Mind now, or you'll [illegible] A. Hunter [illegible] my nice [illegible] v. 146 [illegible] slops of shoes.

Trousseau² (truso). *Med.* [The name of A. *Trousseau* (1801–67), French physician, who described the sign in 1862 (*Clin. Méd. de l'Hôtel-Dieu de Paris* II. xliv. 113).] *Trousseau's sign*: spasm of a muscle evoked by pressure on the nerve supplying it, as seen in cases of tetany.

1887 Vickery & Knapp tr. *A. Strümpell's Text-bk. Med.* 748 Another very characteristic symptom [of tetany] was discovered by Trousseau—'Trousseau's sign'.

It is..this: a fresh paroxysm can at any time be artificially excited by pressure upon the larger arteries and nerves of the arm. **1981** *Brit. Med. Jrnl.* 14 Nov. 1315/2 Trousseau's sign, which depends on the production of tetany, is very painful for the patient.

trout, *sb.*[1] Add: **3. lake trout** (earlier U.S. example); **yellow trout**, a name used in Scotland for the brown trout.

1830 *Cabinet Nat. Hist.* I. 147 In the outlet..from the lake, none of the lake trout were ever found. **1794** *Statist. Acct. Scotl.* [illegible] **1884** *Sat. Rev.* 12 July 61/1 Mr. Thomson caught one sixteen-pounder, which seized a yellow trout he was playing.

4. b. *old trout*: a derogatory term for an old woman. Cf. *old *a.* [illegible]

1897 'S. Grand' *Beth Book* (1898) xxxix. 364 They said.. 'they were blessed if they'd go near the old trout again.' **1914** D. Beatty *Let.* 16 Feb. in W. S. Chalmers *Life & Lett. David, Earl Beatty* (1951) vi. 127 There were some funny old trouts and some sprightly young ones [illegible] **1952** A. Gilbert *And Death came Too* ii. 33 She and her husband always went south to stay with her mother-in-law, an old trout called Lady Dingle. **1972** V. Canning *Rainbird Pattern* iii. 50 She wasn't such a bad old trout. For all her money and position, life hadn't been all good to her.

5. a. *trout-hole* (example), *-hook* (earlier example), *-pond*, *-rod* (earlier example), *-stream* (earlier examples); *trout-fisher* (examples).

1751 Fielding *Amelia* I. iii. xii. 271 It is placed among Meadows washed by a clear Trout Stream. **1770** S. Foote *Lame Lover* i. 15 Oh! clear as a trout-stream. **1839** *Spirit of Times* 15 June [illegible] **1840** *Ibid.* 5 Sept. 319/1 Get a couple of dozen of trout hooks of assorted sizes. [illegible]

b. *trout-lily U.S.*, the yellow dog's-tooth violet, *Erythronium americanum*; cf. *Erythronium; **trout-line**, (*a*) a line used in trout fishing; (*b*) *U.S.* = *trot-line* (Trot *sb.*[1] 7).

1900 *Cent. Dict. Suppl.* 729/1 *Trout-lily*, the yellow dog-tooth violet. **1943** R. Peattie *Great Smokies* 275 The spring beauties and trout lilies..herald the blooming season. [illegible] **1789** J. Woodforde *Diary* 13 July (1927) III. 121 Busy in making up some new Trout lines and for Eels. [illegible] **1934** *Sun* (Baltimore) 4 July 11/5 Crabs are reported to be so scarce that trout-line crabbers are able to catch only two barrels daily.

Trouton (trau·tən). *Physics.* [The name of F. T. *Trouton* (1863–1922), Irish physicist, who published the observation in 1884 (*Phil. Mag.* XVIII. 54).] *Trouton's law* or *rule*: the observation that for many substances the latent heat of vaporization of one mole, divided by the absolute temperature of the boiling point, is a constant (*Trouton's constant*) equal to approximately 88 joules per kelvin.

1895 J. Walker *Introd. Physical Chem.* xii. 114 If the substance is in the state of vapour, the heat of vaporization must be added to the thermochemical data for the liquid. This quantity of heat is approximately proportional to [illegible] the boiling point of the substance on the absolute scale (Trouton's rule). **1901** *Jrnl. Chem. Soc.* LXXX. ii. [illegible] **1922** Glazebrook *Dict. Appl. Physics* I. 561/2 It was shown by Despretz in 1823 that some relationship of the form given above by Trouton held, and later Pictet (1876), Ramsay (1877), rediscovered it independently, but it is now generally referred to as Trouton's rule. **1966** [illegible] **1982** [illegible] Enthalpies of vaporization of unassociated substances follow Trouton's rule.

trove. Add: Hence, a source of treasure, a reserve or repository of valuable things.

1976 *Publishers Weekly* 13 Sept. 97/1 Reaching back to the fifth century and up to today, the authors find a trove of artists whose work merits acknowledgment. **1982** *Sci. Amer.* Aug. 30A/1 Kingdon has himself visited 105 of those areas. He returns with his trove of image and understanding.

trowel, *sb.* **1. d.** (Earlier example.)

1845 G. Dodd *Brit. Manuf.* 4th Ser. v. 128 The workman holds in his right hand a kind of trowel, consisting of a long narrow blade, about a foot in length, decreasing in width towards one end, and having at the other a handle which bends back over the blade. With this trowel..the workman draws the paint over the canvas, smoothing it repeatedly.

troxidone (trǫksidōᵘn). *Pharm.* [f. Tr(i- + Ox- + *-id(ine + -one, elements of the systematic name (see quot. 1952).] An anticonvulsant drug, $C_6H_9NO_3$, used chiefly in treating petit mal epilepsy. Cf. *Tridione.

1952 *Brit. Pharmaceutical Codex 1949* Suppl. 78 Troxidone... *Synonym*: Trimethadone. Troxidone is 3:5:5-trimethyloxazolidine-2:4-dione. **1970** Passmore & Robson *Compan. Med. Stud.* II. v. 60/1 In patients undergoing electroconvulsion treatment the seizure patterns are not modified by troxidone as they are by phenobarbitone and phenytoin.

truant, *sb.* (*a.*) Add: **C.** *Comb.* **truant officer** *U.S.* = *truant-inspector*; **truant-school** (earlier example).

1872 C. L. Brace *Dangerous Classes N.Y.* 348 The Massachusetts system of 'Truant-schools'—that is, Schools to which truant officers could send children habitually truant—does not seem so applicable to New York. **1911** G. F. Ward in S. M. Kingsbury *Labor Laws* 181 The truant officer finds that no certificate has been issued from the central office. **1972** T. Kochman *Rappin' & stylin' Out* 249 What do you do when you ditch school and..a truant officer walks up.

Trubenized (trū·bĕnəizd). [f. *Tru-*, of unknown origin + *-ben-*, said to be f. the name of *Benjamin* Liebowitz, inventor of the process + -ize + -ed[1].] A proprietary name for clothing, esp. shirt collars, made durably stiff by a special process in manufacture.

No longer a proprietary name in the U.S.

1933 *Official Gaz.* (U.S. Patent Office) 26 Dec. 879/2 S. Liebovitz & Sons, Inc., New York... *Trubenized* for dress and negligee shirts and collars. **1939** *Trade Marks Jrnl.* 26 July 1051 *Trubenised*... All goods included in Class 25. Trubenising Limited,..High Holborn, London. **1955** *Radio Times* 22 Apr. 27/3 Shirt..25/6d. with soft collar (or 26/6d. with 'Trubenised' collar). **1969** A. J. Hall *Stand. Handbk. Textiles* (ed. 7) i. 45 Actually this solubility is not a serious disadvantage..for it enables acetate fibres to be employed in the 'Trubenised' process.

Trubetzkoyan (trūbetskoi·ăn), *a.* Also **Trubetskoyan.** [f. the name of Nikolai Sergeevich *Trubetzkoy* (1890–1938), Russian linguist + -an.] Of or pertaining to Trubetzkoy or his theory and methodology.

1940 *Language* XVI. 248 The first part is very largely Trubetzkoyan. **1951** *Ibid.* XXVII. 333 Martinet presents an outline of fairly orthodox Trubetskoyan phonology. **1964** R. H. Robins *Gen. Linguistics* iv. 177 Without the use of further Trubetzkoyan concepts like neutralization. **1977** *Language* LIII. 427 The Trubetzkoyan era, however, was to be closed in a most dignified manner by his virtually finishing the manuscript of his famous encyclopaedic *Grundzüge der Phonologie*.

truce, *sb.* Add: **1. g.** A temporary pause or respite during a game. Hence, used to demand such a truce (cf. sense 2 b in Dict.).

1870 [see *fain *v.*[2]]. **1959** I. & P. Opie *Lore & Lang. Schoolch.* viii. 142 Children were sensitive to the difference between making a truce and surrendering.

4. b. *truce-breaker* (later example).

1949 Koestler *Promise & Fulfilment* ii. v. 265 The Egyptians having officially been branded as truce-breakers.

trucial (trū·ʃiăl), *a.* [f. Truce *sb.* + -ial.] Of, pertaining to, or bound by a truce; used only with reference to the maritime truce made in 1835 between the British Government and certain Arab sheikhs of the Oman Peninsula.

The truce was renewed several times. In 1853 it was succeeded by a Treaty of Perpetual Peace, but the territories to which it had applied continued to be known as the *Trucial States* until 1971, when they became the United Arab Emirates.

1876 *Aitchison's Coll. Treaties* (ed. 2) VII. 44 The possessions of the so-called trucial Chiefs of the maritime tribes of the Persian Gulf. **1891** G. N. Curzon *Persia* II. 452 Adjoining the Trucial states upon the West is the rugged promontory of El Katr. **1911** L. Fraser *India under Curzon* 82 We..bound them by a truce..so that to this day they are known as the Trucial Chiefs of Oman. **1927** P. Cox in *Lett. Gertrude Bell* II. 506 We had treaties of old standing..with the Sheikhs of the Pirate (now the Trucial) Coast of Oman. **1930** A. Rihani *Around Coasts of Arabia* 354 In Trucial Oman also the five independent Sheikhs agree not to enter into correspondence or agreement with any power other than the British Government. **1957** *Times* 24 Aug. 5/4 The Government are prepared to prop up even the ailing and backward sheikhs of the Trucial coast. **1971** *Daily Colonist* (Victoria, B.C.) 15 Aug. 1/4 The announcement of independence had been expected for some time—ever since six of the seven trucial coast emirates agreed to federate without Bahrain when the British military forces depart at the end of this year.

truck, *sb.*[1] Add: **3. a.** (Later examples.) Now usu. in negative contexts: *to have no truck with* (a person or thing), etc.

1899 R. Whiteing *No. 5 John Street* xxvi. 259 Fust time in 'er life..she's ever 'ad any truck with any of them sort. **1929** H. S. Walpole *Hans Frost* iii. iii. 333, I don't want to have any truck with the world at all. **1938** M. K. Rawlings *Yearling* xi. 112 Mebbe your Ma's right. Mebbe you hadn't ought to have no truck with the Forresters. **1948** *Mind* LVII. 17 Others will have no truck with images, and declare that when we remember we are directly apprehending the past occurrence. **1952** J. L. Waten *Alien Son* 97 She would have no truck with so-called midwives who practised spells and incantations. **1960** 'J. & E. Bonett' *No Grave for Lady* i. 13 Wasn't there a story going about..that Lotte Liselotte was having truck with him? **1975** S. Heaney *North* ii. 58 We tremble near the flames but want no truck With the actual firing.

5. *truck-farm, -farmer, -farming*: later examples; **truck crop** *U.S.* = sense 4 c.

1895 *U.S. Dept. Agric. Yearbk. 1894* 133 Soils having over 10 or 12 per cent of clay are too heavy and too retentive of moisture for the early truck crops. **1937** *Sun* (Baltimore) 17 Feb. 7/8 Payments for truck-crop growers have been increased. **1972** R. G. Kazmann *Mod. Hydrol.* (ed. 2) iv. 109 Higher-valued crops such as truck crops are valuable enough to justify far more than this expenditure. **1969** *Observer* (Colour Suppl.) 19 Jan. 8/3 It was a truck farm, which you call a market-garden, don't you, or a nursery? **1976** *New Yorker* 17 May 34/1 Sold their..duplex and moved to a coöperative truck farm. **1979** *Amer. Poetry Rev.* Mar./Apr. 24/1 A few children of Japanese truck farmers and some of us from Youngstown and White Center helped preserve what I snobbishly prefer to think of as peasant vitality. **1973** *Publ. Amer. Dial. Soc.* lx. 10 Raised and lives in Williston, an unincorporated community..mainly devoted to fishing, tugboating, and truck farming.

truck, *sb.*[2] Add: **3. g.** A motor vehicle for carrying goods, troops, etc., by road. Cf. *lorry *sb.* 1 b. orig. *U.S.*

1916, etc. [see *motor-truck* s.v. *motor *sb.* 5 a]. **1930** [see *lorry *sb.* 1 b]. **1932** G. Greene *Stamboul Train* i. i. 3 The passengers cross the quay..round..abandoned trucks. **1950** *Times* 27 Apr. 6/7 Many soldiers in the last war will remember that 'gas' might or might not be petrol and a 'truck' might or might not be a lorry. So it is to-day with 'gearbox', 'transmission', and many others. **1961** L. van der Post *Heart of Hunter* vii. 112 One of the foremost ranchers near Gemsbok Pan was in the Union and coming back by truck across the desert. **1976** *Economist* 16 Oct. 92/1 Of European sales nearly 200,000 were glorified cars, about 400,000 medium-sized vans (up to 3.5 tonnes) and some 240,000 trucks (over 3.5 tonnes). **1984** *Times* 8 Feb. (Energy Suppl.) p. iv./2 This step by the manufacturers to put electric vans and trucks into serious production..has vindicated the enthusiasm of the.. Electric Vehicles Association. **1984** *N.Z. Truth* 23 May 33/3 The new ERF M16, ERF's bid for sales in the mid-size truck market, is on its way to becoming a big seller in Europe.

h. An axle unit of a skateboard to which the wheels are attached.

1976 A. Cassorla *Skateboarder's Bible* 12 The average board came with trucks featuring one cushion rather than two. **1977** *Montgomery Ward Catal.* Spring–Summer 509/1 Heavy duty double-action die-cast trucks. **1978** *Globe & Mail* (Toronto) 15 Aug. 5/2 The wheels are lined with fibrous or plastic material to reduce noise, and steerable axles called trucks are supposed to eliminate squealing on curves and costly wear on rails.

3*. *U.S.* A popular dance (see quots.). Cf. *truck *v.*[2] 5, *trucking *vbl. sb.*[2] 3.

1935 *Sun* (Baltimore) 15 Nov. 14/6 The truck, or truckin', that jerky yet rhythmic dance which combines a bend of the body, a tightening of the hand muscles and a slight strut with the legs, hit the theaters, sidewalks, gin taverns and dance floors of Harlem last summer. **1937** *N.Y. Amsterdam News* 4 Sept. 12/2 Add a bit of the Shag, the new dance sensation that has pushed the 'Truck' out of the limelight, throw in a bit of the Suzi-Q for a spice and then top it all off with the 'Truck'.

4. *truck-driver* (examples in sense *3 g), *-horse* (earlier example); **truck-end** *slang rare*[−1], the buttocks; **truck frame** *U.S.*, the frame of a bogie; **trucklot** *N. Amer.*, a quantity of goods sufficient to fill a truck and sold at a lower rate than a smaller quantity; **truck mixer** (see quot. 1954); **truck stop** chiefly *U.S.*, an establishment which provides refreshments for truck-drivers and fuel and servicing for their trucks.

1931 E. Wilson *Axel's Castle* viii. 278 At Marseilles, he manages to live by unloading cargo and helping truck-drivers. **1973** 'R. MacLeod' *Burial in Portugal* iii. 71 A truck driver leaned out of his cab and shouted a greeting. **1913** D. H. Lawrence *Sons & Lovers* iii. 53 But six shillin' wearin' his truck-end out on a stool's better than ten shillin' i' th' pit. **1850** *Rep. Comm. Patents 1849* (U.S.) I. 164, I claim the shapes and combinations of the truck frame pieces. **1942** W. Faulkner *Go down, Moses* 145 He nudged the log to the edge of the truckframe. **1839** *Spirit of Times* 27 July 246/1 Indeed many of their truck horses are equal to those used on the road. **1943** *Sun* (Baltimore) 15 July 12/3 The regulation covers sales of berries at country shipping points, effective July 13, and carlot and trucklot sales at any receiving point. **1970** *Toronto Daily Star* 24 Sept. 16/3 (Advt.), Attention! all Trucklot Buyers!!.. Prices in this section apply on normal mixed truckloads. **1954** *Gloss. Highway Engin. Terms* (*B.S.I.*) 49 *Truck mixer*, a concrete mixer mounted on a self-propelled chassis, capable of mixing materials during transit from a batching plant to the point of placing. **1976** *Milton Keynes Express* 9 July 16/1 (Advt.), Experienced class II H.G.V. driver required to operate a truck mixer in the Milton Keynes area. **1961** *Amer. Speech* XXXVI. 271 Somewhere on your *run* you will spend some time at a *truck stop*..while the *hasher* serves your *diesel* or *Joe* or

Java. **1973** P. Berton *Drifting Home* vi. 91 Now it is a hodge-podge: tavern, gas station, motel, snack bar, grocery store. Carmacks has become a truck stop.

truckful (earlier example).

1836 Dickens *Pickw.* xxi. 213 He had moved in all his furniture—it wasn't quite a truck-full.

truck, *v.*[2] Add: **1.** (Later examples.) Now usu. with reference to *truck *sb.*[2] 3 g.

1935 *Motion Picture* Nov. 80/3 The heavy electrical equipment was trucked in. **1943** *Sun* (Baltimore) 28 Apr. 8/3 The fighters are trucked in crated from cargo ships. **1954** [see *fat *sb.*[1] 1 b]. **1969** A. Lurie *Real People* 108 He has trucked all his equipment..up here at considerable expense. **1982** L. Kallen *C. B. Greenfield* iv. 47 The produce, trucked in daily from their own upstate farm, was fresh.

2. (Example with reference to *truck *sb.*[2] 3 g.)

1976 M. Machlin *Pipeline* xi. 135 If he stayed with the private contractors who were trucking for Denali, it was improbable he would be recognized.

3. a. *U.S. slang.* Of a vehicle: to proceed. Hence of a person: to go (by truck or otherwise); to move or stroll.

1925 C. R. Cooper *Lions 'n' Tigers* v. 109 One by one the big wagons were trucking toward the first smoking torch at a corner of the grounds. **1938** *Better English* Nov. 51 *Truck, truck on down*, to go somewhere. **1941** Steinbeck & Ricketts *Sea of Cortez* xxiv. 237 We said good-by to Tiburon and trucked on down toward Guaymas. **1970** T. Wolfe *Radical Chic & Mau-Mauing Flak Catchers* 131 They would truck around in the pimp style, too. **1979** *United States 1980/81* (Penguin Travel Guides) 148 You'll still find plenty of people trucking through the streets in flannel shirts, blue jeans, cowboy hats, and boots.

b. Slang phr. *to keep on trucking*, to persevere: a phrase of encouragement.

1972 *Sat. Rev.* (U.S.) 28 Oct. 12 One poster..shows the famous R. Crumb cartoon characters and bears the caption: 'Let's Keep on Truckin'.' **1976** *Southern Even. Echo* (Southampton) 6 Nov. (Advt. Suppl.) 5/4 'To keep in business he's just got to keep on truckin'.' For Karl, and his kind, that can mean upwards of 200,000 miles a year. **1977** *New Yorker* 27 June 79/1 Feels like I frosted the ends of my toes a bit, but they're far from my heart, so I'll keep on truckin.

4. *Cinematogr.* = *track *v.*[1] 3 e.

1929 [see *dolly *sb.*[1] 4 h]. **1942** *Amer. Cinematographer* June 283/2 The camera would start at a long shot, and truck rapidly down toward the background. **1948** A. Huxley *Ape & Essence* (1949) 151 Dr. Poole and Loola enter the shot, and the Camera trucks with them as they come striding down the slope. **1961** G. Millerson *Technique Television Production* iii. 25 (*caption*) Two methods of trucking sideways are shown.

5. To dance the truck. *U.S. slang.*

1937 *Amer. Speech* XII. 183/1 Only negroes can really truck. **1966** M. & J. Stearns in A. Dundes *Mother Wit* (1973) 614/2 Sweetie May trucks provocatively onstage. **1972** W. M. Estes *Streetful of People* vii. 238 Toward the end of the number, the girls turned rosy red and started truckin'.

Hence **trucked** *ppl. a.*; *trucked-in* adj., brought by truck.

1940 *Sun* (Baltimore) 16 Apr. 15/5 Offerings during forenoon rounds in the sheep pens consisted of a few lots of trucked-in native spring lambs. **1966** T. Pynchon *Crying of Lot 49* iii. 56 The trucked-in white sand. **1977** *Time* 7 Mar. 54/3 Lack of green grazing land and hay is also forcing cattlemen either to sell off their thin animals at low prices or fatten them on expensive trucked-in feed.

truckage². (Earlier example.)

1830 W. S. Moorsom *Lett. from Nova Scotia* iii. 80 Away scamper a dozen proud nags waiting for truckage.

trucker². Add: **1.** (Earlier example.)

1853 Dickens *Down with Tide* in *Househ. Words* 5 Feb. 484/2 The Truckers..whose business it was to land more considerable parcels of goods than the Lumpers could manage.

2. A (long-distance) lorry-driver. orig. and chiefly *U.S.*

1961 *Amer. Speech* XXXVI. 273 A fast-driving trucker. **1963** *Times* 2 Feb. 9/6 'Truckers, the real night drivers, are a race apart and 99 per cent honest,' said Taffy. **1966** B. H. Deal *Fancy's Knell* (1967) vi. 84 There was the barbecue and good coffee. Truckers often stopped here. **1971** *Maclean's Mag.* Sept. 72/3 Most truckers..name their rigs after their girl friends. **1978** S. Brill *Teamsters* i. 14 The special nature of their work gives these truckers and warehousemen a stranglehold on the nation's economic life. **1984** *Gainesville* (Florida) *Sun* 30 Mar. 11A/1 The chase started when truckers on I-75 radioed troopers about a late-model Chevrolet station wagon that was weaving on the highway. **1984** *More* (Auckland) May 81/3 And then there are some hardcore longterm truckers who have seen several seasons come and go.

truckie (trɒ·ki). *Austral.* and *N.Z. colloq.* [f. Truck *sb.*[2] + -ie.] = *trucker[2] 2.

1958 *Coast to Coast 1957–8* 201 The truckie looked upwards. 'Whaddya want, mate?' **1968** *Sunday Mail* (Brisbane) 9 June 5/3 Since April last year, more than 50 truckies had appeared in the courts in Queensland on 'pep' pill charges. **1970** P. Carlon *Souvenir* vi. 56 In the truckies' café..the waitress remembered them. **1970** *N.Z. Listener* 21 Dec. 8/4 Another time you'll thumb a truck—truckies will generally stop for you. **1976** *Telegraph* (Brisbane) 24 June 20/2 Detectives have begun a blitz on interstate truckies in a bid to catch color television thieves. **1978** O. White *Silent Reach* viii. 83 We are evacuating this camp... Regular truckies stand to their vehicles.

trucking, *vbl. sb.*[2] (In Dict. s.v. Truck *v.*[2]) Add: **2. a.** Lorry-driving; *spec.* the conveyance of goods by means of a lorry or other motor vehicle. orig. and chiefly *U.S.*

1947 [implied at *parlay *v.* 2]. **1955** *Amer. Speech* XXX. 91 In trucking, *twin screws* mean that there are two sets of rear wheels that are powered. **1961** *Ibid.* XXXVI. 271 Both the [lumbering] industry and the comparative isolation encourage trucking. **1968** *Globe & Mail* (Toronto) 3 Feb. B5/1 Parts one and two of the act, introduced last year, have been implemented and deal mainly with railways. Part three, dealing with trucking, has not yet become law. **1978** *New York* 3 Apr. 88/1 (Advt.), Quick (but careful) Trucking—Immediate Service.

b. *attrib.*

1947 Trucking firm [see *parlay *v.* 2]. **1955** *Amer. Speech* XXX. 91 Some of the words and terms of the trucking industry have been carried over from earlier days of teamsters and steamboat men. **1962** 'E. McBain' *Heckler* iv. 35, I..work for the trucking company. **1969** G. Lyall *Venus with Pistol* ix. 50 I'm in the trucking business. **1978** S. Brill *Teamsters* i. 14 He had negotiated a national trucking contract.

3. The action of dancing the truck. *slang.*

1935 [see *truck *sb.*[2] 3*]. **1938**, etc. [see *Susie-Q]. **1944** C. Calloway *Hepsters Dict.* in *Of Minnie the Moocher* (1976) 260 *Trucking*, a dance introduced at the Cotton Club in 1933. **1959** 'F. Newton' *Jazz Scene* iv. 60 The Black Bottom, Charleston, Lindy Hop, Big Apple, Truckin' and the rest..have been mainly temporary crazes. **1971** E. Bullins in W. King *Black Short Story Anthol.* (1972) 70 The dancers upon the concrete were blocks away, souls in time to the trotting and trucking of the savage song of the threshing floor.

4. *Cinematogr.* (In sense *4 of the vb.) **trucking shot** = *tracking shot* s.v. *tracking *vbl. sb.* 5.

1948 A. Huxley *Ape & Essence* (1949) 46 We..enjoy a trucking shot of mortuary gazebos. **1959** W. S. Sharps *Dict. Cinematogr.* 91/2 *Dollying*, otherwise *Tracking*, *Travelling* or *Trucking*, the moving of a camera on a *dolly* or *camera truck* during *shooting*.

truckster. (Earlier and later examples.)

1843 *Knickerbocker* XXII. 38 All relics of a former age.. exposed for sale in the windows of the trucksters. **1916** H. J. Laski in *Holmes–Laski Lett.* (1953) I. 40 It gets very nauseating to have the measure of progress taken as a commercial truckster might measure it. **1931** A. Uttley *Country Child* v. 69 He had a pack on his back containing ..all the odds and ends of the truckster.

Trudeaumania (trūdōᵘmēⁱ·niă). [f. the name of Pierre Elliott *Trudeau* (b. 1919), former Prime Minister of Canada + -mania.] Enthusiastic or exaggerated admiration for Trudeau.

1968 *Listener* 4 July 5/1 With the phenomenal climb to power of Mr Trudeau a tremendous cult has developed among younger Canadians. It's known as Trudeaumania or Trudolatry. **1971** *Maclean's Mag.* Oct. 28/1 Far more accomplished than Pierre Trudeau, who merely observed the phenomena named Trudeaumania and was shrewd enough to take advantage of it. **1976** *Time* 22 Mar. 32/1 Trudeaumania has long since faded away in Canada.

true, *a.* (*sb.*, *adv.*) Add: **A.** *adj.* **3. e.** Phr. *true for you* [after Ir. *is fíor duit*]: an expression of assent to something said by another. (Stressed on *for*.) *Anglo-Ir.*

1835 R. M. Bird *Hawks of Hawk-Hollow* I. xix. 247 'You are Tapes, the pedler.'.. 'True for you, captain Gilbert!' cried the other, with a stare. **1901** J. Barlow *From Land of Shamrock* 63 They would not, thrue for you. **1980** J. O'Faolain *No Country for Young Men* xv. 329 You're right there... True for you.

f. Purporting to be true. Freq. in collocations used *attrib.* to designate popular magazines which contain (remarkable) stories which purport to be true, as *true confessions, story*; also *true-life story*, etc.

1926 A. Huxley *Jesting Pilate* iv. 260 He walked up and down the train..peddling..True Story Magazines. **1937** [see *pulp magazine* s.v. *pulp *sb.* 5 c]. **1957** C. MacInnes *City of Spades* ii. x. 170 Barbara was..reading a 'true story' magazine. **1958** *Times Lit. Suppl.* 7 Feb. 72/3 She writes well, and—somewhat unusually for a social worker—quotes poetry. She seasons her facts with many 'true-life' stories. **1965** M. Spark *Mandelbaum Gate* vii. 303 Love, love-affairs, men and women and true-life stories formed the daily entertainment and talk of their week's travelling. **1967** 'T. Wells' *What should you know of Dying?* ii. 30 [She] was reading a true confessions magazine. I didn't think they even printed them any more.

g. Colloq. phr. *so* (..) *it isn't true* and varr.: to an almost incredible extent.

1963 *Daily Herald* 25 Apr. 7/6 The Princess was so calm it wasn't true. She was so relaxed. **1964** 'A. Garve' *Ashes of Loda* i. 14 He's..so incompetent about ordinary day-to-day living it's just not true. **1970** R. Rendell *Guilty Thing Surprised* ix. 103 You may be only thirty-six but you're so dead old-fashioned it isn't true. **1982** Barr & York *Official Sloane Ranger Handbk.* 8/1 Sloane Britain is so heavily weighted towards the South and the West it's not true.

4. e. *true to life* (earlier and later examples). Also *true to type*.

1835 *Athenæum* 16 May 372/1 Another character—true to life—is Mrs. Hollis, the fruiterer. **1872** J. M. Langford *Let.* 11 Mar. in *Geo. Eliot Lett.* (1956) V. 254 One feels them all to be true to life. **1929** *Oxford Poetry* 10 Say he died true to type: and then erect A cenotaph; he liked to be select. **1960** *Farmer & Stockbreeder* 1 Mar. 80/2 This was indeed a true-to-type Devon: a good, compact animal with nice fleshing and conformation. **1980** K. Follett *Key to Rebecca* xvii. 193 His preference for 'true-to-life' murders, as opposed to implausible country-house killings.

i. Of bearings: measured relative to true North.

1834 [see Azimuth 2 a]. **1912** [see *projection *sb.* 7 b]. **1969** G. C. Dickinson *Maps & Air Photographs* viii. 125 Bearings measured relative to true north are called true bearings.

j. Of the ground or other surface prepared for ball games: free from unevenness, level and smooth.

1851 in W. G. Grace *W. G.'s Little Bk.* (1909) i. 5 A man is but half a player who is only prepared for true grounds. **1895** H. G. Hutchinson *Golf* (ed. 5) xii. 309 The putting-greens are very good and true. **1934** W. J. Lewis *Lang. Cricket* 297 It [*sc.* the wicket] is said, with regard to its condition, to be *hard* when firm..*plumb* or *true* when it is perfectly level and the ball behaves normally. **1965** L. R. Benaud *Young Cricketer* 86/1 Pitches of today seem to have changed from those of Bulli soil days..when..one played on a true, black, shiny strip as hard as concrete.

5. d. *true left* (or *right*): the side which is on the left (or right) as one looks down from a hill or mountain, or downstream.

1910 J. Buchan *Prester John* x. 177 We followed a narrow shelf on its left side (or 'true right', as mountaineers would call it). **1929** —— *Courts of Morning* iii. iv. 344 Six men were perched high up among the rocks on the right side (what mountaineers would call the 'true left') of the couloir. **1971** *N.Z. Listener* 19 Apr. 55/2 An acquaintance asked..what the reporter had meant by the 'true left' bank of the river. I explained that it was the one on the lefthand side as you looked downstream.

B. *sb.* **3.** (Earlier and later examples.) Now usu. without *the*. Hence *out-of-true* sb., the extent to which a part is out of exact alignment. Cf. Truth *sb.* 6 in Dict. and Suppl.

1876 J. Rose *Compl. Pract. Machinist* vi. 86 If the face plate of the lathe is a trifle out of true, the eccentric will only be out to an equal amount. **1895** J. T. Usher *Mod. Machinist* xxi. 199 The eccentric is..held on the arbor while it is being turned in precisely the same way as it is held on the crank-shaft or axle of the engine, thereby avoiding the tendency to spring it out of true after it is turned, which often happens when it is held for turning by other means. **1970** K. Ball *Fiat 600, 600D Autobook* vi. 59/1 The out-of-true at bearing seats must not exceed ·0008 inch.

C. *adv.* **3. b.** (Later examples.)

1912 *Chambers's Jrnl.* Dec. 810/2 Each variety breeds 'true' in breeders' parlance. **1967** *Listener* 3 Aug. 142/1 But there are two regularities which are equally impressive: organisms breed true, and their structures are orderly.

D. Combinations. **1. d.** Appositively: **true-false** *a. Educ.* and *Psychol.*, denoting a type of test question constructed so that only the words 'true' or 'false' (or another pair of opposites) are acceptable responses; characterizing a test that uses this technique.

1923 P. B. Ballard *New Examiner* vii. 80 The new examination comprised three tests, the first of which was of the True–False type. **1957** D. L. Bolinger in *Publ. Amer. Dial. Soc.* xxviii. 24 Yes–no Qs are essentially true–false Qs. **1965** N. E. Gronlund *Measurement & Eval. in Teaching* viii. 127 Some of the variations..deviate considerably from the simple true–false pattern. **1974** in H. G. Macintosh *Techniques & Probl. Assessment* iii. 25 Other word pairs relating to the statement such as 'greater than–less than'..'faster–slower' and so on. It is the possibilities offered by these other pairs which make the true/false form a particularly useful one.

trueish (trū·iʃ), *a. colloq.* Also **true-ish.** [f. True *a.* + -ish[1].] Partly true, almost true.

1980 I. Murdoch *Nuns & Soldiers* v. 326 'So you think it's true?' 'I think it may be trueish. There's something.' **1981** T. Heald *Murder at Moose Jaw* iv. 44 'True,' said Bognor. 'Or at least true-ish.'

truffle. Add: **1. b. truffle-dog** (earlier example); **truffle hound** = *truffle dog*.

1855 S. Whiting *Heliondé* ii. 47 'If they should see you digging into the surface of the Sun' (he might have added 'like a truffle dog') 'they will certainly think you are demented.' **1975** J. Grigson *Mushroom Feast* 134 There is the whole business of truffle pigs and truffle hounds.

2. A type of confectionery made of a mixture of chocolate and cream, freq. flavoured with rum, shaped into a ball and covered with powdered chocolate.

1926–7 *Army & Navy Stores Catal.* 54/2 Chocolates.. Truffles—lb. 4/3. *c* **1938** *Fortnum & Mason Price List* 8/1 Truffles, with fresh cream (perishable)..Truffles, Rum Flavoured. **1944** D. Welch *In Youth is Pleasure* iv. 61 He imagined the aromatic acrid dust..sticking to [the heart] and coating it..as bright-coloured bitter cocoa powder clings to the rich dark truffle. **1951** *Good Housek. Home Encycl.* 407/1 (*heading*) Chocolate truffles. *Ibid.* 675/1 Turkish delight and marshmallows, chocolates and truffles. **1974** J. Stubbs *Painted Face* xiii. 168 [His] only acquaintance with truffles had been the chocolate variety.

‖ **truite au bleu** (trwīt o blö). *Cookery.* [Fr., lit. 'trout in the blue'.] Trout cooked with vinegar, which turns it blue.

[**1861** Mrs. Beeton *Bk. Househ. Managem.* 44 *Au-bleu*, fish dressed in such a manner as to have a *bluish* appearance.] **1935** M. Morphy *Recipes of All Nations* 37 *Truite au bleu*... One of the most popular ways of cooking them is 'au bleu'—the boiling vinegar in which they are plunged turning them a vivid blue. **1966** B. Glemser *Dear Hungarian Friend* iv. 76 He offered her blue trout, *truite au bleu Duna*. **1978** *Chicago* June 236/2 The Chicago branch of the Paris house offers wonderful dishes like truite au bleu (trout are killed immediately before poaching and sauced with hollandaise at the table).

‖ **truite bleue** (trwīt blö). Also *erron.* **truite bleu.** [Fr., lit. 'blue trout'.] = prec.

1907 E. Glyn *Three Weeks* i. 18 She ate a delicate *truite bleu*. **1948** Wodehouse *Uncle Dynamite* iv. 56 It was a relief when the waiter, arriving with *truite bleue*, broke a tension which had begun to be uncomfortable. **1980** 'M. Harris' *Treasure of Sainte Foy* ix. 123 The prix-fixe is Truite Bleue, then Cassoulet..salad, fruit, and cheese.

‖ **trullo** (tru·lo). Pl. **trulli.** [It.] In Apulia, a small round house built of stone, with a conical roof.

1909 in Webster. **1925** L. V. Bertarelli *Southern Italy* iii. 380 *Alberobello* is a curious village of 'trulli'... These are round huts of stone without mortar, with conical roofs. **1932** *Antiquity* VI. 408 Away in the distance another trullo is visible... No stables or barns are these. We have arrived in 'trulli land'. These trulli are human habitations. **1958** P. Kemp *No Colours or Crest* xii. 252 We lived in a *trullo*, one of the beehive-shaped dwellings that are typical of the Apulian hill country. **1968** S. Jay *Sleepers can Kill* xiii. 134 Those round Trulli houses they have in Southern Italy. **1981** *Italy* (Michelin Tourist Guide) 243 The *Trulli*.—In these..can be seen traces of prehistoric Saracen and Christian civilisations, similar to the *nuraghi* of Sardinia.

truly, *adv.* (*sb.*) Add: **4. b.** *well and truly* (examples); now also for colloq. emphasis: decisively, 'good and proper' (*good *adv.* d).

1458 J. Jernyngan *Let.* 1 June in N. Davis *Paston Lett.* (1976) II. 341 There was not so gret a batayle vpon þe se þis xl wyntyr; and for sothe we were wele and trewly bette. **1477** in *N. & Q.* (1975) July 290/2 The forme of this present lettre whiche ben wel and truly correct. **1521** B. Anslay tr. *C. de Pisani's Bk. Cyte of Ladyes* i. xiii, He shall rewarde them that well and truely maynteyneth hym. **1760** Sterne *Tristram Shandy* I. xv. 82 The said intended marriage..to be well and truly solemnized and consummated. **1895** *Funk's Standard Dict.* s.v. *well* adv., *Well and truly* (Law), conformably to duty; heedfully; used in oaths and affirmations. **1935** *Discovery* Oct. 314/1 The great principle is laid well and truly down—not to attempt to hurry. **1948** G. V. Galwey *Lift & Drop* iv. 72, I am—well and truly married. **1971** D. Potter *Brit. Eliz. Stamps* iii. 41 As soon as the cup was well and truly won by England. **1973** *Guardian* 16 Feb. 13/8 British Brussels is well and truly split on the issue of the TUC's participation in Europe.

Truman (trū·măn). The name of Harry S. *Truman* (1884–1972), U.S. President 1945–53, used *attrib.* in **Truman Doctrine**, the principle first enunciated by Truman in March 1947 that the United States should 'support free peoples who are resisting attempted subjugation'.

1947 *Sun* (Baltimore) 2 Apr. 18/1 At some time, at some point, the Soviet may not take kindly to our extension of the 'Truman Doctrine' to areas bordering upon her claimed sphere of influence. **1948** *Labour Monthly* Aug. 237 The Truman Doctrine aims at making Europe safe against socialism. **1958** *Polit. Sci. Quarterly* LXXIII. 321 The broad terms of the Truman doctrine merely claimed a broad right to intervene, even though there was no intention to exercise this right in full. **1964** Mrs. L. B. Johnson *White House Diary* 10 Mar. (1970) 82 President Truman..said, 'It [*sc.* the Truman Doctrine] was all the Marshall Plan... It's just that the Greeks like the Jews and the Irish can holler louder than anybody else, if you have heard more about the Truman Doctrine than you have the others.' **1977** *Time* 6 June 14/1 By early 1947 Soviet adventurism had inspired the Truman Doctrine, with its pledge of military help to any free people threatened by Communist aggression.

trumeau (trū·mo). Pl. **trumeaux.** [Fr., lit. 'calf of the leg'.] **1.** Also *trumeau mirror*. A pier-glass.

1883 J. W. Mollett *Illustr. Dict. Art & Archæol.* 327/1 *Trumeau*, a pier looking-glass. **1941** *Amer. Speech* XVI. 27 'Über dem Kamin..ist ein Trumeau-Spiegel.'..Cf. 'Trumeau' in Cent. Dict., also in Webster's: not in NED. Nowhere do I find any reference to 'Trumeau Mirror'. **1969** *Canad. Antiques Collector* May 11/1 The trumeau at the left is one of a pair flanking the dining room entrance. **1974** S. Sheldon *Other Side of Midnight* xiv. 278 From mantel top to ceiling rose a heavily carved trumeau mirror.

2. *Archit.* A stone pillar supporting the middle of the tympanum of a doorway, esp. in a church.

1890 C. H. Moore *Gothic Archit.* vii. 262 After the eleventh century the principal portals of great monastic and cathedral churches were commonly divided into two openings by *trumeaux*, or pillars of stone, affording place for sculpture, which consisted usually of a statue with more or less subordinate carving. **1936** A. W. Clapham *Romanesque Archit.* iv. 100 The 'trumeau' at Moissac, also with its superimposed pairs of lions, stands parent to the extraordinary creation at Souillac. **1968** M. Jay tr. *Bazin's Hist. World Sculpture* (1970) vii. 266 (*caption*) *Trumeau* of the old portal, church of Souillac. **1977** *New York Rev. Bks.* 12 May 8/4 The only other Romanesque church with two portals, which originally included trumeaux, is St. Lazare at Autun.

trump, *sb.*[2] Add: **2. a.** *to turn up trumps* (earlier example).

1819 M. Wilmot *Let.* 3 Sept. (1935) 17 A little converted Jew..who received us into his house at a moderate rate and has turned up such *trumps* that I must introduce him to you.

3. b. *Austral.* and *N.Z. slang.* A person in authority.

1937 Partridge *Dict. Slang* 912/2 *Trump of the dump, the*, anyone in authority: New Zealanders': in G[reat] W[ar]. **1941** Baker *Dict. Austral. Slang* 78 *Trump*, a commanding officer. Diggers' slang. **1950** *Landfall* (N.Z.) IV. 126 The hoops are on the last cask by 11.45, and the trump calls out all hands to load the railway wagon. *Ibid.* 127, The trump comes in and calls us gentlemen and wishes us the very best. **1974** D. Stuart *Prince of my Country* xiv. 142 There's a blackfeller or two knows which is which, shafters, and pin, and body, and leaders, the trump says.

4. trump signal (example); also at Bridge.

1901 C. J. Melrose *Bridge Whist* 41 His partner must.. be on the alert to lead trumps through the opponent's strength, and to look out for a trump signal from his partner. **1964** Frey & Truscott *Official Encycl. Bridge* 634/1 Some players use the trump signal whenever they hold three trumps.

† **trumpa** (trɒ·mpă). *Obs.* [ad. F. *trumpeau, trumpo.*] = Sperm whale.

1625 Purchas *Pilgrimes* III. 471 The third sort of Whale is called Trumpa, being as long as the first, but thicker forwards. **1851** H. Melville *Moby Dick* I. xxxi. 215 This [sperm] whale, among the English of old vaguely known as the Trumpa whale.

trumped *ppl. a.* (Earlier examples.)

1728 Fielding *Love in Several Masques* v. xii. 76, I know my Title to be secure, it must be some trumped-up Cheat. **1777** *Sixteenth Ode of Third Bk. Horace Imit.* 21 A Pamphlet fill'd with trump'd-up stories.

trumpet, *sb.* Add: **2. e.** = *horn *sb.* 14* a. Now *Hist.*

1899 *Strand Mag.* Dec. (Advt.), p. xxxv/1 The Gramophone. Berliner's Patent... Length of Trumpet 16 inches. **1904** *Science Siftings* 26 Mar. 353/1 These are again transferred into sound..and transmitted to the audience through a huge trumpet. **1922** S. A. Maycock *Handbk. Gramophone* iii. 18 The hornless models certainly look neater than the instruments which are fitted with trumpets. **1947** F. W. Gaisberg *Music on Record* vi. 81 For the first time they heard sibilants emerge from the trumpet, loud and hissing!

3. *to blow one's own trumpet* (examples).

1854 Mayne Reid *Young Voyageurs* v. 71 They may live to 'blow their own trumpet' a long while yet. **1887** W. S. Gilbert *Ruddigore* i. 12 You must stir it and stump it, And blow your own trumpet. **1952** A. Buckeridge *Jennings & Darbishire* ii. 27, I vote we're not allowed to vote for ourselves because my father says it's swanking to blow your own trumpet. **1983** P. Roberts *Tender Prey* xiv. 165, I was not averse to blowing my own trumpet. Modesty is a fool's game.

6. f. *Metallurgy.* A vertical tube with a bell mouth and a refractory lining, through which metal is poured into runners in uphill casting.

1923 Harbord & Hall *Metallurgy of Steel* (ed. 7) I. i. 37 At one time it was generally considered that sounder ingots could be obtained by bottom casting, but opinions are now much divided as with bottom pouring there is.. some danger of the refractory lining of the trumpet.. being carried into the steel. **1929** W. Lister *Pract. Steelmaking* xxxviii. 370 In this trumpet no wet clay or ramming is used and no weights or clamps are required. **1973** *Times* 12 Feb. (Anchor Project Suppl.) p. ii/6 Mould preparation will be done in a separate bay which is well designed for mould cooling and equipped for..preparation of trumpets and runners for up-run teeming.

7. b. *trumpet-twisted* adj.; **c. trumpet creeper** (earlier example); **trumpet-gall** (examples); **trumpet honeysuckle** (earlier example); **trumpet-leaf** (earlier example); **trumpet-lug** *Archæol.*, a type of tubular handle with expanded ends, found on British neolithic pottery; **trumpet medium**, a spiritualistic medium in whose seances a trumpet megaphone is used; **trumpet pattern**, in medieval art: a shape resembling that of a horn; **trumpet seance**, a spiritualistic seance in which a trumpet megaphone is used; **trumpet-snail** = *ram's horn 6; **trumpet spiral** (see quot. 1959); cf. *trumpet pattern* above; **trumpet style** *Jazz*, a style of piano-playing imitative of a trumpet; **trumpet-vine** (later examples); **trumpet-weed**, (*b*) (earlier example).

1818 W. P. C. Barton *Compendium Floræ Philadelphicæ* II. 43 Trumpet Creeper... Flowers red and orange. **1879** Trumpet-gall [see *nail-gall* s.v. *nail *sb.* 14 a]. **1908** V. L. Kellogg *Amer. Insects* 470 Trumpet-galls on leaves of California white oak. **1731** P. Miller *Gardeners Dict.* s.v. *Periclymenum*, Trumpet Honeysuckle... We have but one species of this Plant at present,..Virginian Scarlet Honeysuckle. **1861** A. Wood *Class-Bk. Bot.* (ed. 10) 222 *S. Gronovii*. Trumpet-leaf..in swampy pine woods. [**1932** S. Piggott in *Archaeol. Jrnl.* LXXXVIII. 76 The horizontally perforated lug..exhibits a 'trumpet-ended' variety at Windmill Hill and Hembury.] **1937** —— in *Antiquity* XI. 450 More important was the occurrence of a type of lug or tubular perforated handle with expanded ends, which the writer distinguished as a 'trumpet-lug' in 1932... At Hembury it was present as a recurrent feature. **1972** L. Alcock *By South Cadbury* v. 109 These suspension tubes—trumpet lugs to give them their technical name—are seen again on pottery from sites like Windmill Hill in Wiltshire. **1912** *Nash's Mag.* July 553/2 Last year the wonderful trumpet medium, Mrs. Wreidt, spent some time at 'Julia's Bureau'. **1968** B. Steiger *Voices from Beyond* iii. 58 Trumpet mediums always seem to be popular at Spiritualist camps. **1937** *Burlington Mag.* Feb. 99/1 We find that admirable curling trumpet-patterns..in the brilliant manuscripts of the early Church in Ireland and Northumbria. **1954** M. Rickert *Painting in Brit.: Middle Ages* 232 *Trumpet pattern*, two whorls.. joined across the open side by a curved line. **1965** L. N. Valentine *Ornament in Medieval Manuscripts* 51 '*French horn*', a trumpet pattern combined with a helix shape. **1912** *Nash's Mag.* July 544/1 The sitting took place at 'Julia's Bureau'. It was a Trumpet Séance, and Mrs. Wreidt..was the medium. **1931** *Daily Express* 15 Oct. 7/3, I am aware you are giving trumpet seances. **1968** B. Steiger *Voices from Beyond* iii. 58 At trumpet seances—almost invariably conducted in the dark—the horn rises, ostensibly lifted by spirit hands. **1901** E. Step *Shell Life* 320 The Ram's Horn or Trumpet-snail, so frequently introduced in fresh-water aquaria. **1965** tr. *H. Janus Young Specialist looks at Land & Freshwater Molluscs* iv 70 Family Planorbidae (Ram's-horn or Trumpet Snails) **1936** A. W. Clapham *Romanesque Archit.* i. 9 Certain Celtic motives such as the trumpet-spiral. **1959** E. A. Fisher *Anglo-Saxon Archit. & Sculpt.* 73 Both single and double spirals were common in Celto-British art... Sometimes the connecting C-line would be double and wider apart in the middle resembling two trumpets joined at their wide ends—hence the term *trumpet spiral*. **1946** R. Blesh *Shining Trumpets* xiii. 320 Hines's trumpet style..was based on Louis Armstrong's trumpet phrasing. **1959** 'F. Newton' *Jazz Scene* vii. 130 Players attempted the feat of adapting the piano to the vocalising style of the other instruments (the so-called 'trumpet style'). **1977** *New Yorker* 6 June 120/1 The so-called trumpet style of jazz piano playing, which Earl Hines originated in the late twenties, consists of hornlike single-note melodic lines in the right hand and on-and-off-the-beat chords, single notes, and countermelodic lines in the left hand. **1895** W. B. Yeats *Poems* 12 Many a trumpet-twisted shell. **1883** *Peterson Ladies' Nat. Mag.* June 460/2 The great porch in front..[was] destitute of railing or ornament, but the creeping trumpet vine. **1978** *Detroit Free Press* 16 Apr. (Gardening Guide) 14/2 Trumpet vine is another woody vine that bears striking flowers. **1830** *Huntingdon* (Pa.) *Courier* 15 Sept. 4/5 American Remedies Wanted..Gravel Wort or Trumpet Weed.

trumph (trɒmf). Sc. and north. var. of Trump *sb.*[2] Phrases: † *to play trumph about* (obs.): to vie in achievements *with*; *what's trumph?* what is happening? what is the news?

1777 R. Forbes *Ulysses' Answer* 29 Achilles played na' trumph about Wi' him, he says; but judge ye. **1819** J. Burness *Plays, Poems* 286 Again Will deals the cartes about Says he, Lads, hearts is trumph. **1895** J. Nicholson *Kilwuddie* (ed. 3) 173 A lass that has that wi' the lads should be trumph. **1908** E. M. Sneyd-Kynnersley *H.M.I.* iii. 30 Corners denoted the 'Jack of trumph'. **1955** W. P. Milne *Eppie Elrick* xxx. 272 Fat's lickly tae be trumph noo? **1969** J. T. R. Ritchie *Golden City* 76 Ye turn the tope card over to see what's to be trumph.

truncate, *v.* Add: **c.** *Math.* To cut short or approximate (a series, etc.) by ignoring all the terms beyond a chosen term. Also *absol.*

1955 M. Loève *Probability Theory* xvi. 233 We truncate X [*sc.* a random variable] at $c > 0$..when we replace X by $X^c = X$ or 0 according as $|X| < c$ or $|X| \geq c$. **1966** J. H. Cadwell *Topics Recr. Math.* xiv. 157 Because of the steadily decreasing terms, when the series is truncated the error incurred is of smaller magnitude than the first term omitted. **1968** P. A. P. Moran *Introd. Probability Theory* ii. 78 It is also sometimes useful to exclude more than one of the possible values of *i*, and in this case it is more often necessary to truncate at the other end. **1981** *Nature* 5 Nov. 14/3 The series in equation (1) should extend to $m = n = \infty$, but in practice the series is truncated at a maximum value..of *m* and *n*, usually in the range 8 to 15.

truncated, *ppl. a.* Add: **2. e.** *Statistics.* Of a frequency distribution or sample: obtained by disregarding values of the variate greater than or less than some chosen value. Of a variate: treated in this way.

1931 R. A. Fisher *Truncated Normal Distribution in Math. Tables* (Brit. Assoc. Adv. Sci.) I. p. xxxiii, The frequency of the truncated distribution in the range *dx*. **1952** A. Hald *Statistical Theory with Engin. Applications* vi. 146 The cumulative distribution functions of three truncated distributions..with degrees of truncation of 10, 30 and 50% respectively. **1971** C. R. Heathcote *Probability* v. 240 Truncated variables are often easier to handle than the original ones, and..under wide conditions, the two sequences have the same asymptotic behaviour.

f. Of soil: having lost the upper horizon(s) as a result of rapid erosion.

1938 A. B. Yolland tr. *A. de Sigmond's Princ. Soil Sci.* xiv. 206 These truncated forest soils possess a peculiar dynamics of their own. **1941** H. Jenny *Factors of Soil Formation* v. 100 Owing to differences in color of the *A* and *B* horizons, truncated profiles are often readily discernible on freshly plowed slopes. **1976** L. F. Curtis et al. *Soils Brit. Isles* 315 Truncated podzols. Here the surface soil has the characters of a B horizon.

truncation. Add: **2. c.** *Statistics.* The cutting off of a frequency distribution at a certain value of the variate.

1937 Yule & Kendall *Introd. Theory of Statistics* (ed. 11) vi. 103 We can picture it as a slightly skew distribution which has been cut off on the left owing to the inadmissibility of negative values of the variate. Discontinuous variates not infrequently give rise to this effect of truncation. **1952** [see *truncated *ppl. a.* 2 e].

d. The loss or removal of the upper horizon(s) of a soil by erosion.

1941 H. Jenny *Factors of Soil Formation* v. 100 An example of widespread truncation is provided by the Cecil series of the Piedmont Plateau. **1972** J. G. Cruickshank *Soil Geogr.* iv. 133 Loss of surface horizons by erosion (truncation) is more common in Oxisols [than in Spodosols].

e. *Math.* The cutting short of a numerical computation or expression before its natural end (if any). Usu. *attrib.* in *truncation error*.

1952 D. R. Hartree *Numerical Analysis* x. 223 The solution of the set of equations..is not, of course, the solution of the partial differential equation on account of the truncation error of the approximation. **1968** Fox & Mayer *Computing Methods for Scientists & Engineers* x. 205 For the fourth-order Runge–Kutta method the truncation error is the factor h^5 multiplying a rather complicated expression involving derivatives of $f(x, y)$. **1973** [see *rounding *vbl. sb.*[1] 1 c].

trundle, *v.* Add: **1. c.** (Earlier and later examples.)

1849 *Punch* 14 July 12/1 In those Days..they did moderately trundle the Ball under-hand; but now they fling it over-handed from the Elbow. **1861** *Baily's Mag.* July 140 Such bowling as was trundled by Mr. Lyttelton and Mr. Salter in this innings is rarely witnessed in a University match. **1870** *Ibid.* Dec. 213 Six out of the eleven have trundled the ball. **1898** G. Giffen *With Bat & Ball* iii. 47 The bowlers, too, trundled with that specially placed on-field in their mind's eye. **1959** *Punch* 3 June 747/2 Four of the team bowl leg-spinners.., and Gupte is reported to be the best of his type now trundling.

trundling *vbl. sb.* (earlier examples in sense 1 c of the vb.).

1861 W. J. Prowse in *Bell's Life* 10 Nov. 6/3 But however good their trundling—pitch or pace, or break, or spin—Still the monarch of all bowlers, to my mind, was Alfred Mynn! **1862** *Baily's Mag.* Apr. 260 The Eleven then commenced batting to the trundling of Moore and Conway.

trundler. b. (Earlier example.)

1871 [see *snick *v.*[2] 2 b].

trunk, *sb.* Add: **I. 1. d.** (see quot. 1950). Cf. *trunk dial*, sense 18 in Dict.

1899 F. J. Britten *Old Clocks and Watches* 316, I am able to give an engraving of a very early specimen of a long-case clock... This case is of oak and panelled. The head is fixed on the trunk, and will not take off. **1950** D. de Carle *Watchmakers' & Clockmakers' Encyclopædic Dict.* 188/1 *Trunk*, refers to the body or main part of the case of a long-case clock. The case is formed of three parts, the trunk, the Hood..and the Plinth. **1978** *Times* 17 June 9/5 A dial clock is basically a clock with a round dial which hangs on the wall, or a round dial with a trunk underneath, which is called a trunk dial or drop dial.

4. c. *Teleph.* (*a*) A telephone line connecting two exchanges a long way apart or in different telephone areas; also (*U.S.*), a line connecting exchanges within the same area (cf. *toll call* s.v. *toll *sb.*[1] 3); (*b*) a line connecting selectors or the like of different rank within an exchange.

1889 Preece & Maier *Telephone* 249 This switchboard is required to distribute the trunks between the different offices [*sc.* exchanges], and also to enable the testing of all the trunk and subscribers' lines to be carried on from one central point. **1908** *Jrnl. Inst. Electr. Engineers* XLI. 120 In America the *local* term remains the same, the *junction* is called 'trunk', and our *trunk* is called a 'toll' or 'long-distance' line. **1921** W. Aitken *Autom. Telephone Syst.* I. 4 The designation of the lines interconnecting apparatus at different switching stages..is somewhat confusing. The American generally speaks of these as trunks, which practically is equivalent to our junctions... For example, the circuits between first and second preselectors,..second and third selectors, and third selectors and connectors, are all trunks or junctions... Very commonly the lines between exchanges in this country [*sc.* Britain], and for the use of which an extra charge is made.., are known as trunks. For these the American terms toll line or long-distance line is used. **1925** Wright & Puchstein *Telephone Communication* ii. 33 It is necessary in central-energy systems to serve districts of more than ten thousand lines by means of a number of central offices connected by trunk circuits or trunks. **1966** *McGraw-Hill Encycl. Sci. & Technol.* XIII. 433/1 When the extension user dials the first digit, the wipers of the first selector step upward to the level corresponding to the digit dialed, and automatically step around in a horizontal arc..until they find an idle trunk. **1978** *Sci. Amer.* June 90/2 A modern telephone exchange is connected not only to its own subscribers but also, through special lines called trunks, to other exchanges serving other subscribers. **1978** P. H. Smale *Telecommunication Syst. I* vii. 60 There are various grades of exchange in order of importance... There are also various grades of interconnecting line, for example subscribers lines, junctions and trunks, and again the number gets fewer as length, importance and cost increase.

(ii) *pl.* The operators who deal with trunk calls. *colloq.*

[**1889** Preece & Maier *Telephone* xx. 353 As the trunk operators had too much to do..a special plan was arranged by which a subscriber requiring a connection to another town mentions the word 'trunk' to the ordinary operator.] **1947** N. Cardus *Autobiogr.* iii. 233 He rang me up at my Manchester house; he was speaking from Harrogate. When 'trunks' gave me notice of his call I expected something urgent. **1977** C. McCullough *Thorn Birds* ix. 203 Give me trunks, please, switch... I want to put an urgent call through.

II. 7. b. *N. Amer.* The luggage compartment of a motor vehicle; = *boot *sb.*[3] 4 c.

[**1929** *Hearst's Internat. Mag.* Nov. 210 (Advt.), Six wire wheels and trunk rack standard equipment. **1930** *Automobile Topics* 6 Dec. 359/3 Rear-end trunks were larger and more prevalent. In one line of cars they were designed into the rear of the body itself.] **1931** *Amer. Home* Apr. 197 The luggage belongs in the trunk on the rear of the car. **1937** *Sat. Even. Post* 2 Oct. 28/3 The enlarged trunks will hold enough luggage to carry you around the world. **1951** J. W. Vale *Mod. Auto Body & Fender Repair* xiii. 162 The deck compartment, sometimes referred to as the rear trunk compartment or the turtle back..may be repaired in the same way. **1960** *Times* 14 Sept. 12/6 When we hired a car in California we found that a car..bristles with surprises. You..find the spare wheel in the trunk. **1964** Mrs. L. B. Johnson *White House Diary* 24 Apr. (1970) 119 Lyndon had transferred to an open convertible. Along the way he made three unscheduled stops, standing on the trunk of his car. **1968** *Globe & Mail* (Toronto) 17 Feb. 49/9 (Advt.), Extra lighting inside, in hood, in trunk. **1975** N. Luard *Robespierre Serial* v. 34 The Belgian unlocked the trunk, stood by the porter while he lifted out two suitcases.

10. i. *U.S.* A floodgate or sluice controlling the flow of water into and out of rice-fields.

1856 in *Documentary Hist. Amer. Industr. Society* (1910) I. 120 Trunk-minders undertake the whole care of the trunks. **1903** 'P. Pennington' *Woman Rice Planter* (1913) i. 8 Each field has a very small flood-gate (called a trunk), which opens and closes to let the water in and out. **1939** *Sat. Even. Post* 10 June 37/2, I opened all my rice-field trunks, so that the flowed field inside would equalize the pressure from the flood outside, thus saving my banks.

IV. 17. c. For *U.S.* read 'orig. *U.S.*' and add later examples. *swimming trunks*: see *swimming *vbl. sb.* 6.

1941 Baker *Dict. Austral. Slang* 78 *Trunks*, swimming shorts. **1956** P. Scott *Male Child* i. vi. 86 Except for a pair of swimming trunks he was naked. **1964** L. Deighton *Funeral in Berlin* xviii. 107 A blond man in very small knitted swimming-trunks. **1982** S. B. Flexner *Listening to Amer.* 61 Men's *trunks* had been in use by professional swimmers and athletes since the 1880s.

d. Knickers; underpants with short legs.

[**1926–7** *Army & Navy Stores Catal.* 705/1 Gent's underwear..Trunk drawers—18/6.] **1936** [see *ankle-sock* s.v. *ankle *sb.* 3]. **1970** *Kay & Co.* (Worcester) *Catal.* 1970–71 Autumn–Winter 452 *Meridian trunks.* New style with shorter leg and continental front.

V. 18 (sense 4 a) *trunk route, -wire* (earlier example); (sense *4 c) *trunk circuit*; (senses 6, 7) *trunk room, -seller, strap*; (sense 15) *trunk-nosed* adj. (earlier example); **trunk-cabin** (example); **trunk-deck** (example); **trunk dialling** (see quot. 1971); **trunk murder,** a murder after which the body is hidden in a trunk; also **trunk murderer; trunk-road** (further examples); *spec.* in *Grand* (also *Great*) *Trunk Road*, the great highway between Calcutta and Amritsar constructed during the British Raj; **trunk-work** (later example in allusion to quot. 1611). See also *trunk line.

1878 F. O. Davenport *On Man-of-War* 197 The captain had a small trunk cabin, a little higher and abaft ours. **1896** Trunk circuit [see *trunking *vbl. sb.*[2] 2]. **1921** *Jrnl. Inst. Electr. Engineers* LIX. 390/2 (*caption*) Trunk circuits radiating from London to provincial towns. **1896** *Nautical Mag.* LXV. 1076 Oscar II..a trunk deck vessel of the type invented by Mr. W. Hök. **1952,** etc. Trunk dialling [see *subscriber trunk dialling* s.v. *subscriber 3]. **1959** *Ann. Reg. 1958* 505 Trunk dialling from Bristol began in December. **1971** *Gloss. Electrotechnical, Power Terms* (*B.S.I.*) iii. ii. 20 *Trunk dialling*, control of an exchange's automatic switching equipment from an exchange in another multi-exchange area over trunk or toll circuits. **1976** *Times* 20 Dec. (Istanbul Suppl.) p. iv/3 Although international trunk-dialling is promised..it can take anything up to 15 minutes to get a dialling tone..in Istanbul. **1905** *Daily Mail* 15 Apr. 5/4 (*heading*) The trunk murder. How the bodies were found. **1936** G. Greene *Journey without Maps* i. i. 11 Another clue in a trunk murder case. **1976** S. Hynes *Auden Generation* v. 136 The crimes..are actual-sounding crimes: a trunk murder at Paddington station, a girl killed on Streatham Common. **1925** P. Selver tr. *K. Capek's Lett. from Eng.* 54 At Madame Tussaud's..in the catalogue I found..Arthur Devereux, hanged 1905, known as the 'trunk murderer', because he hid the corpses of his victims in trunks. **1962** G. Butler *Coffin in Oxford* xiv. 176 Discovered your trunk murderer yet? **1887** Kipling *From Sea to Sea* (1899) I. 114 The Englishman..took off his hat to the tun-bellied, trunk-nosed God of Good-Luck. **1851** *Ret. Public Works India* 146 in *Parl. Papers* XLI. 513 Documents..report the progress of the works on the Great Trunk Road. **1848** J. Bourne *Let.* 24 Apr. in *Railways in India* (ed. 2) 19 The grand trunk road, connecting Calcutta with the north west provinces ..is already a railway all but the rails. **1888** Kipling *Departmental Ditties* (1890) 19 All those hairy gentlemen ..Swaggered down the Grand Trunk Road into Bow Bazar. **1931** J. W. Gregory *Story of Road* xviii. 274 In 1839 it was decided to construct a metalled road, the Grand Trunk Road, from Calcutta to Delhi... By 1849 about £300,000 had been spent on it. **1937** *Archit. Rev.* LXXXI. 155 The Trunk Roads Act comes into operation on April 1st. **1974** *Listener* 2 May 574/3 Bentinck's Governor-Generalship..was not a complete failure, as the Great Trunk Road shows. **1698** J. Verney *Let.* 16 June in M. M. Verney *Verney Lett.* (1930) I. iii. 31 The little long room that is under the Trunk room of the Purple Chamber. **a 1752** Ld. Verney *Will* in *Ibid.* II. xxxiii. 246, I give to her..all the money & Jewels in the cabinet in the Trynk Room. **1860** J. G. Holland *Miss Gilbert's Career* (1866) 293 Cheek was..led to the trunk-room of the lodging-hall. **1952** E. Wilson *Lilly's Story* i, in *Equations of Love* 133 A trunk-room full of the trunks which accompany a large English family in migration. **1970** *Times* 2 June (Container Suppl.) p. i/5 Container ships,..capable of carrying as much general cargo in a year on a trunk-route shuttle service as an entire fleet of traditional break-bulk cargo liners. **1855** Trollope *Warden* xvi. 264 He remembered the shop distinctly; it was next door to a trunk-seller's. **1887** Kipling *From Sea to Sea* (1899) I. 40 Jey Singh..would have hanged those Globe-trotters in their trunk-straps. **1970** *Country Life* 31 Dec. 1296/2 The hood, when up, was secured to the front mudguards by two stout trunk straps. **1885** *List of Subscribers, Classified* (United Telephone Co.) (ed. 6) 8 The very great cost of running and maintaining the Trunk wires between the different Exchanges. **1920** 'K. Mansfield' *Let.* 25 Sept. (1928) 46, I heard again from Methuen to-day. They now say they'd like 2 books for next spring. I think there must have been some trunk work, some back stair work in this on your part.

trunker (trʌ·ŋkəɹ). [f. Trunk *sb.* or *v.*[2] + -er[1].] **1.** (See quot. 1921). Cf. Trunk *v.*[2] 2.

1881 *Instructions to Census Clerks* (1885) 84 Copper miner... Trunker. **1921** *Dict. Occup. Terms* (1927) § 056 *Trunker*, separates slimes from ore by running mixture into and out of a long box, launder or trunk.

2. a. A long-distance lorry-driver; *spec.* one who drives at night, and is not responsible for loading or unloading his vehicle.

1954 in P. G. Hollowell *Lorry Driver* (1968) vii. 192 *Trunker*, night driver. **1958** *Times* 12 Apr. 7/6 The long-distance night lorry driver, or trunker. **1959** *Manch. Guardian* 20 July 2/3 We night trunkers..pull out of the depot at seven or eight in the evening. **1968** P. G. Hollowell *Lorry Driver* vii. 176 Although the trunker is a long distance driver and a night worker he is for the most part able to organize his leisure life as he has regular runs. He is not as able to do this as is the shunter who has the greatest rate of participation in associational groupings.

b. A lorry used for long journeys along trunk roads.

1965 M. Russell *No Through Road* 20, I could see the..cars on the A.1 highway and hear the scream of the all-night trunkers. **1969** *Jane's Freight Containers 1968–69* 10 (Advt.), By freight-liner any number of Abel Air containers are carried by rail to terminals where they are readily transferred to trunker or local delivery units. **1979** *Navy News* Sept. 34/1 (Advt.), Driving artics., trunkers or vans or sorting and loading back at base.

trunking, *vbl. sb.*[2] Add: **1. b.** *spec.* a system of ducts or trunks (sense 10), esp. for cables or purposes of ventilation.

1923 *Man. Seamanship* (Admiralty) II. 284 The quantity of air supplied by a fan depends greatly on..the size and tortuosity of the trunking. **1950** *Engineering* 3 Feb. 123/1 Forced draught fans..arranged to draw air..and discharge it to the boilers through suitable trunking. **1963** *Times* 11 May 8/1, 60 electricians were dismissed for refusing to work on the trunking of electric cables on the frigate Mohawk. **1977** *Timber Trades Jrnl.* 17 Dec. 36/2 To blow the by-products from the hopper into a storage silo via trunking is..becoming a more common practice.

d. The driving of lorries on long journeys along trunk roads.

1968 P. G. Hollowell *Lorry Driver* iii. 64 Trunking, being a night driving job, is heavy on the eyes, and also requires quick response to meet crises particularly on runs where the schedules are tight. **1974** P. Wright *Lang. Brit. Industry* ii. 31 *Trunking* used to mean only night driving, but now, because of improved roads and speeds enabling travel to and from a far unloading point in one day, it includes *day-trunking*.

2. *Teleph.* [f. Trunk *sb.*] The use or arrangement of trunks (sense *4 c). Freq. *attrib.*

1896 *Jrnl. Inst. Electr. Engineers* XXV. 639 Facilities for through trunking—*i.e.*, connecting two or more trunk circuits at intermediate offices to provide communication between towns not directly connected. **1933** *Discovery* Apr. 132/2 When all the direct junctions to an exchange are engaged, subsequent calls to that exchange are passed (or 'routed') through a central exchange. This feature of the system is termed 'alternative trunking'. **1947** *Electronic Engin.* XIX. 66 Automatic telephony is dealt with almost entirely by means of trunking diagrams and trunking calculations. **1979** Danielson & Walker *Telecommunications Systems for Technicians I* ix. 79 A trunking diagram showing the connection of a subscriber to a desired number in a four-digit (up to 9999) exchange.

trunk line. 1. (In Dict. s.v. Trunk *sb.* 18.) (Earlier examples.)

1843 R. Wilson *Let.* 13 Jan. in W. F. Cooke *Electric Telegraph* (1857) i. 219 Mr. Cooke's first object..would be to get the Telegraph laid down on the Croydon or some other trunk line. **1851** Dickens *Our School* in *Househ. Words* 11 Oct. 81/1 The Railway had cut it up... A great trunk-line had swallowed the playground.

2. *Teleph.* = *trunk *sb.* 4 c (*a*).

1883 *Pall Mall Gaz.* 6 Dec. 12/2 The cost of laying a trunk line overhead averages about £25 a mile. **1893** Preece & Stubbs *Man. Telephony* xxi. 336 The main trunk lines—those connecting [London] with the various provincial exchanges—are invariably metallic. **1926** T. E. Lawrence *Seven Pillars* cxi. 590 Young and I cut the telegraph, here an important network of trunk and local lines. **1964** M. McLuhan *Understanding Media* xxx. 306 Radio's use of the telephone in a glorified form of the old trunk-line wire-tapping. **1980** *Wall St. Jrnl.* 26 Nov. 10/2 The Federal Communications Commission voted to authorize American Telephone & Telegraph Co. to install an optical-fiber telephone trunk line between Washington and N.Y.

3. A large or main pipeline for oil or gas, esp. one from a production field to a refinery or terminal.

1896 Redwood & Holloway *Petroleum* II. vii. 475 As soon as the well is found to yield, the tank is connected by a 2-inch pipe with the trunk line. **1925** A. B. Thompson *Oil-Field Exploration & Devel.* II. xix. 1092 Long-distance trunk lines entail nearly as much work in surveys and construction as a railway. **1975** *Offshore* Sept. 195/2 The compressor will push the gas shoreward through a 67 mile, 42 inch diameter trunkline to a platform in West Cameron block 167 offshore Louisiana.

trunnel. Add: **trunnel-head** *U.S.* (earlier examples).

1819 E. Evans *Pedestrious Tour* 270 The cogs, wallower, the trunnel-head and the stones [of a grist-mill]. **1839** *Knickerbocker* XIII. 345 Mill-wheels, and trunnel-heads.

truscottite (trʌ·skǫtəit). *Min.* [ad. Du. *truscottiet* (P. Hövig 1914, in *Jaarboek van het Mijnwezen in Nederlandsch Oost-Indië* XLI. 202), f. the name of S. J. *Truscott* (1870–1950), English mining engineer: see -ite[1].] A hydrated basic silicate of calcium and manganese occurring as crystals of the hexagonal system that are typically pearly white.

1925 *Mineral. Mag.* XX. 466 Truscottite... Pearly white scales from the Lebong Donok gold-silver mine, Benkulen, Sumatra. **1979** *Ibid.* XLIII. 333/1 Truscottite and reyerite are of interest not only as natural minerals, but also because of their potential formation in cements hydrated at elevated temperatures and pressures, as in the casings of deep geothermal wells.

truss, *sb.* Add: **8. truss-block** (example).

1883 *Man. Seamanship for Boys' Training Ships R. Navy* (Admiralty) (1886) 26 Trestletrees are two pieces of hard wood, standing fore and aft... On their after ends an eye-bolt is driven from the lower side, for attaching the truss blocks to.

trusser. 2. c. (Earlier example.)

1857 A. Mathews *Tea-Table Talk* II. 96 It [*sc.* a chicken] had apparently made a vigorous struggle for continued existence, which struggle had evidently distorted its form out of the power of the trusser to regulate.

trust, *sb.* Add: **5. e.** *on trust*: (of a dog) obeying the command to trust (see *trust *v.* 1 b). Also *to play 'Trust'*.

1932 C. Morgan *Fountain* i. 4 In Lewis's compartment were two former sergeants of marine, Lapham and Shordey, upright in opposite corners like dogs on trust. **1939** C. Day Lewis *Child of Misfortune* 196 Eve was trying to teach her puppy to play 'Trust' with a piece of biscuit. **1970** [see *paid *ppl. a.* 4 a].

7. b. (Earlier example.)

1877 J. Wanamaker in J. H. Appel *Business Biogr. John Wanamaker* (1930) x. 137 Why should not individual ownership be permitted to grow peaceably and equally with industries that are bunched into trusts?

8. a. *trust-fund* (earlier examples).

1780 J. Fife *Let.* 29 Feb. in A. & H. Taylor *Lord Fife & his Factor* (1925) v. 121 After his death, you know how I was involved in a trust-fund. **1862** *Harper's Mag.* Aug. 337/1 Mr. Pennington has a considerable practice as a lawyer and a handsome private estate. He has, as a consequence, many trust funds.

b. trust-buster *colloq.* (orig. *U.S.*), one who works for the dissolution of trusts (sense 7 b); *spec.* a government official responsible for the enforcement of legislation against trusts; hence **trust-busting** *vbl. sb.* and *ppl. a.*; **trust corporation** *Law*, a corporation empowered to act as a trustee; **trust deed** (earlier examples); **Trust House,** a hotel owned by a company called Trust Houses, which was founded in 1903 to restore the traditional high standards of the best of the old coaching inns, and merged with Forte Holdings Ltd. in 1970 to form Trusthouse Forte; **trust officer** *N. Amer.*, an officer of a trust company or similar institution who has direct responsibility for the institution's activities as a trustee; **Trust Territory,** a territory administered by a nation acting on behalf of the United Nations Organization; cf. *trusteeship 2 b.

1903 *Chicago Chron.* 11 Apr. 2 Mr. Knox is surprising everybody by his zeal as a trust-buster. **1949** *Time* 9 May 34/3 U.S. trustbusters were still locked in stalemate with the Zaibatsu. **1979** *N.Y. Times Mag.* 30 Sept. 60/2 'Fighting the oil and gas lobby'..was soon to become more fashionable among liberals than it had been since the turn of the century, when the muckrakers and the trust-busters were riding high. **1911** *Daily Colonist* (Victoria, B.C.) 22 Apr. 13/5 Clark McClercher..has been appointed special assistant to the attorney-general with 'trust-busting' duties. **1944** *Chicago Daily News* 8 May 10/1 So we have a faint revival of 'trust busting' to give color to the stuff. **1973** *Business Week* 13 Jan. 32/1 The Administrative Council for Economic Defense (ACED)..claims to be the only trustbusting agency in Latin America. **1925** *Act 15 Geo. V* c. 18 § 30. 422 Where there is a sole personal representative, not being a trust corporation, it shall be obligatory on him to appoint an additional trustee. *Ibid.* § 117. 491 'Trust corporation' means the Public Trustee or a corporation either appointed by the court..or entitled by rules made under..the Public Trustee Act, 1906, to act as custodian trustee. **1967** E. Rudinger *Wills & Probate* 12 The Public Trustee is a government department which can be appointed to be your executor, as can some trust corporations. **a 1754** P. Grant *Decisions of Court of Session* (1813) II. 490 The Lords *nem. con.* found action competent upon the trust-deed. **1812** *Dramatic Censor 1811* 400 The creditors on the trust deed, consisting of authors, performers, tradespeople, and others, had due to them 52,611 *l.* [**1900** Earl Grey in *Econ. Rev.* (1901) XI. 101 Arrangements have already been made for the formation of a Public-house Trust Company (Limited) for the county of Northumberland.] **1902** Earl of Carlisle in *Monthly Rev.* Feb. 36 This decision..may affect the reformed trust houses. **1915** *Nineteenth Cent.* Jan. 68 The whole atmosphere of these Trust Houses..is essentially different to that of the average Trade house. **1928** *Evening News* 18 Aug. 11/7 No attempt was made to define 'disinterested management', but one gathered..that it is supposed to be found in 'trust houses'. **1942** E. Waugh *Put out More Flags* i. 22 Basil will be covered with medals while your silly old yeomanry are still messing in a Trust House. **1972** 'M. Delving' *Shadow of Himself* i. 17, I..looked up hotels. There was a Trust House there, the White Swan. **1905** Kirkbride & Sterrett *Mod. Trust Company* iii. 41 The trust officer must have full authority over his department. **1965** H. Hood in R. Weaver *Canad. Short Stories* (1968) 2nd Ser. 221 You're doing splendidly. You have all your friends in the office and inside of two years you'll be a trust officer. **1976** *Globe & Mail* (Toronto) 16 Feb. 15/6 One prominent trust officer calculated that the increase would cost his company an additional $84,000 a year. **1945** *U.N. Charter* xii. § 75, in *Times* 27 June 8/5 The United Nations shall establish under its authority an International Trusteeship System for the administration and supervision of..Trust Territories. **1970** *Internat. & Compar. Law Q.* XIX. 218 New Guinea was a Mandated Territory... Its change of status in 1946 to become a Trust Territory [etc.]..have not affected section 16.

c. Passing into *adj.* *Soc. Psychol.* Applied to activities in which an individual is required to display trust or confidence, esp. as *trust game*. Also with reference to techniques aimed at measuring or achieving trust.

1967 J. B. Rotter in *Jrnl. Personality* XXXV. 653 (*heading*) Construction of the interpersonal trust scale. **1972** *Psychol. Report* XXX. 850 The authors recognize that it is inappropriate to call this particular activity a trust exercise. **1975** *Psychol. Abstr.* LIII. 1181/1 Behavior ..was related to choice in the 'competition' game but not in the 'trust' game. **1978** *Peace News* 25 Aug. 10/2 As I see it, trust games (for example) don't produce trusting groups any more than brooms sweep floors.

trust, *v.* Add: **1. b.** Imperative: an instruction given to a dog, requiring it to wait for a reward, usu. in a begging position with a titbit placed on its nose. Cf. *trust *sb.* 5 e.

1854, etc. [see *paid *ppl. a.* 4 a]. **1921** W. de la Mare *Mem. Midget* xlix. 331 Finger and thumb outstretched above the cringing little dog.. 'Trust, Plum, trust!' **1930** M. Allingham *Mystery Mile* xiv. 132, I put a bit of sugar on his nose and said 'Trust'. **1974** G. Butler *Coffin for Canary* ix. 104 David was throwing him biscuits as to an old dog and saying 'trust'.

trusteeship. Add: **2. a.** The function of a colonial power or other dominant people as protectors of a subject people.

1936 *Internat. Labour Rev.* June 856 Something must be done at once to remedy a state of affairs which..constitutes a flagrant breach of that ideal of trusteeship of Native races not yet able to stand by themselves under the strenuous conditions of the modern world. **1943** *Ann. Reg. 1942* 88 The old ideas of exploitation [of colonial possessions] had of recent years given place to the new doctrine of trusteeship. **1944** J. C. Smuts in H. G. Wells *'42 to '44* ii. 71, I remember Cecil Rhodes used to say that the proper relation between Whites and Blacks in this country [*sc.* S. Africa] was the relation between guardian and ward. This is the basis of trusteeship. Much later, this principle of trusteeship was put into the Covenant of the League of Nations. **1946** *Ann. Reg. 1945* 9 The whole House was committed to the doctrine of trusteeship, and beyond that of partnership.

b. The administration of a territory by a nation acting on behalf of the United Nations Organization. Freq. *attrib.* Cf. *Trust Territory* s.v. *trust *sb.* 8 b.

1945 *U.N. Charter* xii. § 79, in *Times* 27 June 8/5 The terms of trusteeship for each territory..shall be agreed upon by the States directly concerned. **1946,** etc. [see *mandate *sb.* 4 b]. **1952** *Times* 5 Aug. 3/7 The corporation is a Government-sponsored body set up by the Nigerian Government to develop 250,000 acres of plantations, formerly owned by Germans before the war, in the Cameroons territory, which was formerly under British mandate and is now under United Kingdom trusteeship. **1959** *Listener* 19 Nov. 880/1 The Belgian trusteeship territory of Ruanda-Urundi, Central Africa. **1962** *Observer* 14 Oct. 40/1 Sir Hugh Foot, Britain's representative in the United Nations on colonial and trusteeship questions, resigned.

trustify, *v.* Add: **trustified** *ppl. a.* (later examples); **trustification** (later examples).

1920 A. C. Pigou *Econ. of Welfare* ii. vii. 182 The marginal social net product of activities devoted to bringing about any widespread 'trustification' of industry is likely to be smaller than the marginal trade net product. **1922** *Encycl. Brit.* XXXII. 506/2 The large-scale and 'trustified' American capitalist system. **1938** L. Hogben *Science for Citizen* xiii. 653 Edison's later inventions sponsored the far-flung trustification of American industry. **1965** B. Pearce tr. *Preobrazhensky's New Economics* 152 When there is trustification or syndication of important branches of production within a certain country, prices systematically..deviate from value in the upward direction. **1969** P. Worsley in Ionescu & Gellner *Populism* 223 Petty farmers..wanted a freer, more competitive, less trustified, market economy.

trusty, *a.* (*sb.*) Add: **A.** *adj.* **2. a.** Also *spec.* (orig. *U.S.*) designating a well-conducted convict to whom special privileges are granted.

1856 *Democratic State Jrnl.* (Sacramento, Calif.) 28 Oct. 2/3 The 'trusty guards', (commanded by Pete, Scotty acting as first lieutenant,) have recovered from the effects of their stolen debauch of Saturday. **1926** J. Black *You can't Win* iv. 40 A trusty prisoner appeared at my side. 'Come on, you.' **1968** *Listener* 15 Feb. 209/2 At this prison, the 200 guards are all trusty prisoners.

B. *sb.* **a.** (Earlier and later examples of sense 'a well-conducted convict to whom special privileges are granted'.) For 'in *U.S.*' read 'orig. in *U.S.*'

1855 *San Francisco Citizen* 2 Oct. 2/3 Two 'trusties' named Scottie and Greene, escaped in a whale boat from the State Prison grounds on Sunday night. **1912** in J. Sandilands *Western Canad. Dict. & Phrase-Bk.* **1926** J. Black *You can't Win* iv. 38 He told me to stay there till he could get a 'trusty' to take me upstairs. **1958** *People* 4 May 6/4 He was a trustie working in our records office. **1963** T. & P. Morris *Pentonville* ii. 27 The 'outside' men who go beyond the prison walls are selected 'trusties'. **1969** *Telegraph* (Brisbane) 29 May 2/7 (*heading*) Trusties plan no trouble. **1973** R. Travers *Murder in Blue Mountains* x. 100 The Chief of Police banned all general visitors to Butler's cell and a trusty was put in with him to guard against any attempt at suicide.

truth, *sb.* Add: **II. 6.** So *out-of-truth* sb. Cf. True *sb.* 3 in Dict. and Suppl.

1967 L. Holmes *Odhams New Motor Man.* viii. 189/1 Out-of-truth produces irregular tyre wear.

7*. *Particle Physics.* = *top *sb.*[1] 17*. [An arbitrary choice of name.]

1977 *Sci. Amer.* Oct. 74/2 If the two new quarks do exist, there must also be two new properties of matter, which some physicists have taken to calling 'truth' and 'beauty'. **1978** [see *top *sb.*[1] 17*]. **1979** *Nature* 6 Dec. 546/2 They have included evidence for the 'gluon' (the photon of the quark-quark force), and excited states of the upsilon (which contains a beauty quark and its antiparticle), but 'truth' (the quark beyond and pairing with beauty) remains to be found.

III. 8. c. (Also with capital initial.) A game in which players have to answer truthfully questions put by the others or, in some forms of the game (called *truth, dare, promise*, etc., according to the rules), fulfil an alternative requirement.

1868 L. M. Alcott *Little Women* I. xii. 191 Do you know 'Truth'?..The person who draws at the number has to answer truly any questions put by the rest. **1928** *Sat. Even. Post* (U.S.) 29 Sept. 7/1 The ancient game of truth had begun. 'What's your favorite color, Bill?' **1959** I. & P. Opie *Lore & Lang. Schoolch.* xviii. 377 In 'Truth, Dare, and Promise' each player has to agree either to tell the truth, accept a dare, or promise to do as he is told. **1969** —— *Children's Games* ix. 265 In Kidderminster the game is occasionally called 'Truth, Dare, Promise, or Kiss'; in Peterborough and Swansea, 'Do, Dare, Kiss, or Promise'; in Gloucester, 'Truth, Dare, Warning, Love, Kiss, or Marriage';..in Aberdeen, 'Truth, Dare, Force, or Promise'... At Spennymoor,..the game is called 'Truth, Dare, Will, Force, and Command'. **1970** *Times* 8 July 2/7 On one occasion she said, Carole Hanson..stripped to her panties during a game of 'truth, dare and promise'. **1980** G. M. Fraser *Mr American* xiii. 270 Playing truth or consequences in..Sir Charles's drawing-room.

V. 14. a. *truth-claim, -frequency, -relation.* **c.** *truth-searcher*; *truth-bending, -compelling, -loving* (earlier example), *-seeking* (earlier and later examples) sbs. and adjs.

1977 P. Johnson *Enemies of Society* xvii. 232 The title [*sc. Travesties*] reveals the fact that the whole play is an elaborate exercise in truth-bending and symmetrical confusion. **1979** *Tucson* (Arizona) *Citizen* 20 Sept. 2c/6 Tucson is..about to have a Liar's Contest to call its very own... Big Jim Griffith, banjo plucker, anthropologist, impresario, folklorist, Ph.D and himself no slouch in the truth-bending dept., phoned with the news. **1909** W. James *Meaning of Truth* xiv. 273 Good consequences..are proposed rather as the lurking *motive* inside of every truth-claim. **1977** A. Giddens *Stud. in Social & Polit. Theory* i. 78 An 'empirical intersection' subject to disputation in respect of truth claims. **1925** Joyce *Let.* 27 Sept. (1957) I. 233, I am still under the influence of the 'truthcompelling' drug, scopolamine. **1936** *Mind* XLV. 501 The most complete and ingenious defence of the 'truth-frequency' interpretation of probability statements. **1949** A. Pap *Elem. Analytic Philos.* ix. 169 The truth-frequency of a form of influence is the limit approached by the ratio of the number of cases making both premisses and conclusion true to the number of cases making the premisses true. **1828** Carlyle in *Foreign Rev.* II. 439 He has every feature also of a just, quiet,

truth-loving man. **1947** H. Reichenbach *Elem. Symbolic Logic* iv. § 33.178 The truth-preserving property of derivational processes. **1969** *Aristotelian Soc. Suppl. Vol.* XLIII. 77 It is tempting to generalize and actually define logical rules as truth-preserving grammatical rules. **1907** W. James *Meaning of Truth* (1909) vii. 165 It is *between* the idea and the object that the truth-relation is to be sought and it involves both terms. **1952** *Mind* LXI. 193 The truth relation is repeatedly defined by Peirce in straightforward old-fashioned correspondence terms. **1928** A. Huxley *Point Counter Point* xxvi. 443 Truth-Searchers become just as silly..as the boozers. **1828** Truth-seeking [see *open-minded* adj. s.v. *open *a.* 22 c]. **1896** W. James in *New World* V. 345, I..cannot see my way to accepting the agnostic rules for truth seeking.

d. Special Combs.: **truth-condition** (see quot. 1937); **truth drug,** any substance that is administered to a person in the supposition that it will prevent him from lying; **truth-function** (see quots.); hence **truth-functional** *a.*, **truth-functionality; truth-functionally;** *adv.*; **truth game** = sense 8 c above; also *transf.*; **truth serum,** a truth drug in the form of an injection; **truth set** (see quot. 1966); **truth squad** *U.S. Politics*, a group of people with the task of questioning the truth of statements made by members of an opposing party; **truth-table,** a tabular representation of the truth or falsity of a complex proposition as determined by the possible combinations of truth-values of its components; also *transf.*, esp. in *Computer Sci.*; **truth-value** [ad. G. *wahrheitswert(h)*], the value of truth (or falsehood) assigned to propositions, esp. those of two-valued logic; also *transf.* and *attrib.*

1922 tr. *Wittgenstein's Tractatus* 95 The proposition is the expression of its truth-conditions. **1937** *Mind* XLVI. 191 Propositional complexes which are definable by reference to truth conditions *i.e.*, to propositions whose truth-values are logically *determined* by the truth-values of their arguments. **1978** P. Pettit in Hookway & Pettit *Action & Interpretation* 48 Incompatible sentences have truth conditions which we cannot conceive of as being simultaneously fulfilled. **1931** 'A. Abbot' *Murder of Geraldine Foster* xiii. 179 Neither the lie detector nor the truth drug have ever been officially adopted by the Police Department. **1947** *Lancet* 4 Jan. 39/1 If an attempt were made to use a 'truth drug' against an accused person in an English criminal court, the judge would reject its results as an involuntary confession. **1969, 1973** [see *Pentothal]. **1976** T. Sharpe *Wilt* xiii. 140 You can put me on a lie detector. You can pump me full of truth drugs. **1909** W. James *Meaning of Truth* i. 41 The reader will easily see how much of the account of the truth-function developed later in *Pragmatism* was already explicit in this earlier article. **1910** Whitehead & Russell *Principia Math.* I. i. 8 We may call a function $f(p)$ a 'truth-function' when its argument p is a proposition, and the truth-value of $f(p)$ depends only upon the truth value of p. **1967** *Encycl. Philos.* V. 76/2 *Truth-function*, a function whose arguments and values are truth-values. **1981** *Sci. Amer.* Oct. 155/1 A complete account of how the truth value of a compound sentence is determined by its constituent sentences is called a truth function, and logicians customarily display the evaluation of the truth function in an array called a truth table. **1947** *Mind* LVI. 237 A modern extensional logic containing the ordinary truth-functional modes of statement-construction. **1968** Chomsky & Halle *Sound Pattern Eng.* 387 We note there that certain truth-functional conditions are required for the phonological rules. **1950** W. V. Quine *Methods of Logic* (1952) § 2.8 The property of truth-functionality..is thus enjoyed by negation, conjunction, and alternation. cases, to a semantic component based on truth-functionality. **1950** W. V. Quine *Methods of Logic* (1952) § 7.37 in which it is used in linguistics—unrelated, in many cases, to a semantic component based on truth-functionality. **1950** W. V. Quine *Methods of Logic* (1952) § 7.37 The one statement implies the other truth-functionally. **1935** F. Scott Fitzgerald *Taps at Reveille* 66, I thought you might want to know... I thought maybe you thought I liked somebody else. The truth game didn't get around to me the other night. **1941** 'N. Blake' *Case of Abominable Snowman* xviii. 198 Since we seem to be playing the truth game..did you kill Elizabeth Restorick? **1980** G. Mitchell *Whispering Knights* ix. 42 'Why don't we play the Truth Game?' 'Because nobody will tell the truth.' **1925** F. J. Reynolds *Marvels of 1924* 44 Dr. House believes truth serum should be a part of every court and prison equipment. **1940** [see *scopolamine]. **1977** J. Crosby *Company of Friends* vi. 43 Acetol is the strongest truth serum there is. Do you remember talking, talking, talking? **1963** Truth set [see *solution set* s.v. *solution *sb.* 12]. **1966** *Britannica Bk. of Year* 807/2 *Truth set*, a mathematical or logical set containing all the elements that may be substituted in a given statement of relationships without altering the truth of the statement (the equation $x + 7 = 10$ has as its *truth set* the single number 3). **1952** *Tuscaloosa* (Alabama) *News* 3 Nov. 12/5 The Republican 'Truth Squad' after trailing President Truman across the country on his campaign trips, passed down its final verdict today that the President was 'guilty of over 100 lies, half-truths and distortions.' **1980** *Washington Star* 31 Oct. B1/1 GOP leaders formed what they called a 'truth squad' to campaign this week because they said Democrats have avoided the issues. **1921** E. Post in *Amer. Jrnl. Math.* XLIII. 166 The general notion of truth-table is not introduced [in *Principia Mathematica*]. *Ibid.* 167 So corresponding to each of the 2^n possible truth-configurations of the p's a definite truth-value of f is determined. The relation thus effected we shall call the truth-table of f. **1937** *Mind* XLVI. 191 Truth-tables which analytically exhibit the logical conditions under which a given truth-function..would be true, together with those under which it would be false. **1962** Simpson & Richards *Physical Princ. Junction Transistors* xvi. 401 The above results may be most easily summarized for two transistors with two inputs *A* and *B* in the form of so-called truth tables, two of which are shown below. **1965** P. Caws *Philos. of Sci.* xvi. 122 A decision procedure..for the sentential calculus..consists of constructing a truth-table for the *wff* in question. **1970** O. Dopping *Computers & Data Processing* i. 26 From the truth table we can then derive a logical equation, which can be used as a basis for the program. **1977** *Sci. Amer.* Sept. 84/2 The first step in the design of the circuit is the construction of a truth table that gives the desired output for every possible combination of inputs. [**1891** G. Frege in *Funktion, Begriff, Bedeutung* (1975) 26 Ich sage nun: 'der Wert unserer Funktion ist ein Wahrheitswert' und unterscheide den Wahrheitswert des Wahren von dem des Falschen.] **1903** B. Russell *Princ. Math.* 502 There are, we are told.., three elements in judgment: (1) the recognition of truth, (2) the Gedanke, (3) the truth-value (*Wahrheitswerth*). **1932** Lewis & Langford *Symbolic Logic* vii. 215 In this calculus, it is important to distinguish the truth-*values*, 1, ?, and 0, from the truth-functions, *p*, *Mp*, and *Np*. **1936** Wirth & Shils tr. *Mannheim's Ideology & Utopia* 13 The truth-value of human knowledge in general. **1966** S. Beer *Decision & Control* viii. 165 Then this formulation is subjected to a truth-value analysis. **1978** J. McDowell in Hookway & Pettit *Action & Interpretation* 129 Dummett's realist is attempting, with the truth-value-link manoeuvre, to respect that principle. **1981** A. Born tr. F. Lasson in *Dinesen's Lett. from Afr.* p. xiv, The presumed documentary truth-value of the letters.

try, *sb.* Add: **III. 7.** *attrib.* and *Comb.*, as (sense 4 b) *try-getter, -getting, -scorer, -scoring.*

1954 J. B. G. Thomas *On Tour* iv. 43 The two wings, were the try-getters. **1977** *Western Mail* (Cardiff) 5 Mar. 3/3 England could well feel the backlash of these frustrated Welsh try-getters. **1954** J. B. G. Thomas *On Tour* 12 Without complete forward supremacy try-getting was like needle-hunting in the proverbial haystack. **1930** *Daily Express* 8 Sept. 10/5 Try-scorers for the tourists in the first half were Rew and Jones-Davies. **1976** *Scotsman* 27 Dec. 11/1 Earlier try-scorers for Gala were Gordon Dickson and George Telfer. **1974** *Times* 4 Feb. 7/2 With try-scoring so difficult, Wales missed a golden opportunity of taking a decisive lead midway through the first half.

try, *v.* Add: **5. d.** *to try out*: to test the advantages, possibilities, or qualities of (a material or immaterial thing); also, to test (a person). orig. *U.S.*

1888 *Judge* (N.Y.) 29 Dec. 190/1 Tried Out By Fire. **1899** *N.Y. Jrnl.* 30 July 34/6 Britain will try out heavy motor wagons. **1906** *N.Y. Evening Post* 26 Oct. 1 The new rules have been but partially tried out. **1974** A. Price *Other Paths to Glory* iii. 254 He was being tried out on someone else's problem.

6. c. To submit (a case) for the judgment of a court of law. *U.S.*

1905 S. W. Mitchell *Constance Trescot* 166 'Do you still feel that all chance of settlement is out of the question?' 'Yes; I am instructed to try the case.' **1931** *N. Amer. Rev.* Jan. 22 This is one of his jokes; he knows I can't afford to try criminal cases. It's been fifteen years since I've been in a criminal court.

7. e. To put (a person) to the test to ascertain the truth of what is asserted or believed of him or her. Freq. in imp. *try me.*

1970 V. Canning *Great Affair* xi. 193 'You'll not like it.' 'Try me.' **1971** *Scope* (S. Afr.) 19 Mar. 124/2 'Miss Blandish..ahem..I take it that you can keep a secret?' 'Try me, Mr Stone.' Her voice was like a soft caress. Her suggestion to 'try her' scared J.B. **1984** A. Price *Sion Crossing* vi. 106 'I think maybe you won't like it, Oliver.'..'Try me.'

11. e. To test the effect of (a thing) *on* (a person, thing, etc.). *to try it on the* (or *a*) *dog*: to test the effectiveness of something on someone regarded as being of lesser consequence than those for whom it is ultimately intended; *Theatr.*, to test the possibilities of a play, etc., by performing it as a matinée or before a provincial audience. *colloq.* (orig. *U.S.*).

1890 in Barrère & Leland *Dict. Slang* II. 377/1 'Bootle's Baby' will on the 7th of May be produced somewhere in the provinces. This is what the Americans call trying it on a dog. **1897** *Daily Tel.* 4 Feb. 9/1 If any enterprising person desires to make money from a play or a composition of music he does not boldly attempt the experiment upon the public. His shrewd suspicion that they would avenge the torture induces him to adopt the preliminary precaution of 'trying it on the dog'—a creature of delicate susceptibilities, and very amenable to the influences of Teutonic bands and street-corner cornet solos. **1922** H. Crane *Let.* 29 Sept. (1965) 101, I want to try *Dial* [a literary review] on 'F and H', before it goes anywhere else. **1941** G. Heyer *Envious Casca* xiv. 261 Mathilda had never felt less inclined to listen to a dissertation on the benefits of experience to an actress, and she very rudely told Paula to try it on the dog.

f. *to try in* (Dentistry): to place (a denture or prosthesis) in the patient's mouth to test the fit. [From the prepositional use illustrated in quot. 1896.]

[**1896** C. J. Essig *Amer. Textbk. Prosthetic Dentistry* xi. 408 After the teeth of a full upper denture have been tried in the mouth and found to be correct, a protective rim is to be made.] **1921** D. Gabell *Prosthetic Dentistry* viii. 199 The dentures should be placed in tepid water with some pleasant antiseptic,..and then each separately tried in. **1968** Neill & Nairn *Compl. Denture Prosthetics* 101/2 When trying in the waxed-up dentures initially the opportunity is taken to..check the jaw relationships.

g. *try anything once*: a cliché indicating (often somewhat unexpected) willingness on the part of a speaker to attempt or experience something new.

1921 *Ladies' Home Jrnl.* July 20/1 This slogan runs, 'Try anything once'. **1959** N. Mailer *Advts. for Myself* (1961) 160 'I am sort of curious about the film. I've never seen one [*sc.* a pornographic film], you know.' 'Try anything once, is that it?'

h. *try-your-strength, try-your-weight*: used *attrib.* to designate an apparatus at a fair or the like which tests or measures a person's strength or weight.

1929 J. B. Priestley *Good Companions* i. iv. 135 One o' these try-your-strength things..—down with the 'ammer and up she goes and rings the bell. **1930** R. Lehmann *Note in Music* v. 214 A try-your-strength machine that gave him his money back. **1932** *Radio Times* 29 July 241/1 Everything to make them feel at home..try-your-weight machines, 'diddlers', peeps-at-Paris. **1963** Wodehouse *Stiff Upper Lip, Jeeves* iii. 24 There was plenty and to spare of the Rev. H. P. Pinker. Even as a boy..he must have burst seams and broken try-your-weight machines. **1977** 'E. Crispin' *Glimpses of Moon* vi. 97, I want to try the Try-Your-Strength machine.

try-. Add: **1. try-pot** (earlier example); **try-works** (earlier example).

1795 R. Murry *Jrnl.* 12 Oct. in R. McNab *Hist. Rec. N.Z.* (1914) II. 523 The try pot and steam were as they were left. **1792** Z. Macy *Jrnl.* 15 May in *Mass. Hist. Soc. Coll.* (1810) III. 157 The oil [was] boiled out in the try works at sea.

2. try-in *Dentistry* [*try *v.* 11 f], the experimental trial of a denture or prosthesis in a patient's mouth as a preliminary to any further work; also, the prosthesis itself; **try-on,** (*a*) (earlier example); **try-out,** for '(*U.S. slang* or *colloq.*)' read '*colloq.* (orig. *U.S.*)' and add: (further examples); also, an experimental trial, a test of performance, a trial run or period, *spec.* of a play, etc., in a provincial theatre, etc.; also *attrib.*

1939 R. O. Schlosser *Compl. Denture Prosthesis* xvi. 244 One of the dentures..may be completely processed, and a final try-in made while the opposing denture is still in the wax model state. **1963** C. R. Cowell et al. *Inlays, Crowns, & Bridges* iii. 26 Immediately before the try-in the temporary dressing must be removed. **1970** J. M. Buchanan *Atlas Compl. Denture Prosthesis* xix. 100/2 Replace the characterized mandibular try-in in the patient's mouth. **1977** M. M. Hudis *Dental Lab. Prosthodontics* iv. 99 The teeth are arranged on the occlusion rims so that they may be returned to the dentist for a try-in. **1823** 'J. Bee' *Slang* 181 *Try-on*,..an essay or endeavour to do a thing. **1903** *Sci. Amer.* 30 May 414/1 Cup challengers in their try-outs in British waters. **1915** *Literary Digest* (N.Y.) 21 Aug. 361/2 The new Grinnell Sprinkler equipment..has already had its initial tryout. **1923** *N.Y. Times* 7 Oct. ix. 2/6 *Try out*, an experimental hearing of a new act, usually far away from Broadway. **1928** *Evening News* 18 Aug. 9/3 The play will not be given the provincial 'try-out'. **1933** P. Godfrey *Back-Stage* vi. 76 Had this play not first been produced at a 'try-out' theatre..its author might still be sending it round. **1941** B. Schulberg *What Makes Sammy Run?* ii. 4 Sammy was getting a three-week tryout. **1963** A. Ross *Australia 63* viii. 157 Davidson's final try-out in the nets was watched over..by two selectors, Bradman and Seddon. **1970** A. Glyn *Blood of Britishman* xxi. 275 Brief try-outs of plays before they arrive in London's West End. **1976** A. Davis *Television* 89 It is as difficult to predict which comedy series will succeed as to forecast which records will top the hit parade charts, and that is one of the reasons for the development of try-out series. **1983** *Listener* 17 Mar. 29/3 *Gilgamesh* was chosen to give the new technology a try-out.

trying, *vbl. sb.* **b. trying-pot** (earlier example).

1843 E. Dieffenbach *Trav. N.Z.* I. ii. 51 The blubber is..immediately put into the trying-pots.

trypaflavine (tripăflēⁱ·vīn). *Pharm.* [f. Trypa(nosoma (see quot. 1954) + L. *flāv-us* yellow: see -ine[5].] = *acriflavine.

1913 *Chem. Abstr.* VII. 2944 Trypaflavine..and KCN..give 3,6-diamino-N-methyl-ms-cyanoacridane,..partially converted back into trypaflavine by heating with conc. H_2SO_4. **1954** *Thorpe's Dict. Appl. Chem.* (ed. 4) XI. 381/2 Amongst purely organic substances, Acriflavine..was originally introduced as a trypanocide under the name *Trypaflavin*. **1975** *Biol. Abstr.* LX. 50/1 (*heading*) Investigation of the interaction between trypaflavine and nucleic acid components.

trypan. Add: **trypan blue** [rendering G. *trypanblau*], a diazo dye used as a vital stain and in the treatment of trypanosomiasis and other protozoan infections.

[**1909** *Parasitology* II. 187 Trypanblau and Trypanrot are highly efficient remedies in the treatment of canine piroplasmosis.] **1911** *Chem. Abstr.* V. 3495 (*heading*) Trypanosan, trypan red, trypan blue and parafuchsin in immunization against rabies. **1980** *Nature* 8 May 110 (*caption*) Viable cells (trypan blue exclusion) at 16 h after infection normally contained approximately 10% bi- and multinucleated heterokaryons.

trypanocidal (tri:pănosəi·dăl), *a.* [f. Trypano(soma + -cide 1 in Dict. and Suppl. + -al.] That is fatal to trypanosomes.

1909 *Practitioner* Feb. 248 Trypanocidal substances. **1946** A. A. Morton *Chem. Heterocyclic Compounds* xii. 339 Crude dichloro ortho-para fuchsin proved less toxic but more trypanocidal than para fuchsin, methyl violet, and like compounds. **1976** P. Collard *Devel. Microbiol.* v. 55 Atoxyl is a very effective trypanocidal agent.

So **trypa·nocide,** a trypanocidal agent; = *trypanosomacide* s.v. Trypanosoma.

1917 in Stedman *Med. Dict.* (ed. 4) 1012/2. **1956** *Nature* 31 Mar. 604/2 Contributions on trypanocides, antimalarials, amœbicides and on the chemotherapy of virus infections. **1977** *Lancet* 8 Oct. 769/1 On its own, S.H.A.M. is ineffective as a trypanocide in vivo, but when combined with glycerol it temporarily clears bloodstream trypanosomes in rodents infected with *T. brucei*.

trypanolysis (tripănǫ·lisis). [f. as prec. + *-lysis.] Destruction of trypanosomes; **trypanolytic** *a.* (in Dict. as main entry) (further example).

1905 *Index Medicus* 2nd Ser. III. 200/1 (Index), Trypanolysis. **1936** *Ann. Trop. Med. & Parasitol.* XXX. 377 Trypanolysis by mouse immune serum is very rapidly effective in the presence of added haemolytic complement. **1979** *Infection & Immunity* XXIV. 691/1 Eleven anti-*T*[*rypanosoma*] *brucei* and five anti-*T. congolense* IgG preparations were tested in the trypanolysis assay for the presence of trypanolytic factors.

trypanosome (in Dict. s.v. Trypanosoma). Add: **b.** Used *attrib.* to designate trypanosomes and related hæmoflagellates at a stage in their life cycle when they have an elongated body with the flagellum arising from the posterior end. Cf. *trypomastigote.

1924 Hegner & Taliaferro *Human Protozool.* v. 151 Once these crithidial forms [of *Trypanosoma* spp.] are established in the salivary glands they probably form a permanent source for the production of the infective trypanosome forms. **1949** C. A. Hoare *Handbk. Med. Protozool.* xi. 171 The developmental stages through which trypanosomes pass in the course of their life-cycle..comprise the trypanosome, crithidial and leishmanial forms, and occasionally also the leptomonad form. **1967** A. W. Jones *Introd. Parasitol.* xxix. 412 *Trypanosoma*..may include all forms, and it parasitizes insect and vertebrate hosts. In the latter host, only the trypanosome form occurs.

trypanosomal *a.* (examples of the sense 'trypanosomid'); **trypanoso·mid** *sb.* and *a.* [ad. mod. L. *Trypanosomidæ* (F. Doflein *Die Protozoen als Parasiten und Krankheitserreger* (1901) 57): see -id[3]], (pertaining to or designating) a member of the family Trypanosomidæ, which comprises trypanosomes and related species of flagellate protozoans which at some stage in their life cycle have an elongated body with one nucleus and a single flagellum arising from a kinetosome; also **trypanoso·matid** *sb.* and *a.* (corresponding to the synonymous family name Trypanosomatidæ).

1942 D. L. Belding *Textbk. Clin. Parasitol.* xi. 144 The species..of the genus *Herpetomonas* have four stages in their life cycle, appearing as leishmanian, leptomonad, crithidial and trypanosomal forms. **1956** *Nature* 11 Feb. 279/2 (*heading*) A synthetic growth medium for the trypanosomid flagellate *Strigomonas* (*Herpetomonas*) *oncopelti*. *Ibid.*, This trypanosomid..is parasitic in the digestive tract of Hemiptera. **1962** *Jrnl. Protozool.* IX. 53/1 Because certain species of horseflies..have been incriminated in the transmission of trypanosomes of animals, a study of their own trypanosomatid parasites has a particular interest for students of insect flagellates. **1963** F. G. Wallace in J. Ludvík et al. *Progr. in Protozool.* 70 The criteria that have been used most in the definition of genera among the trypanosomatids of insects have been morphological. **1971** Markell & Voge *Med. Parasitol.* (ed. 3) vii. 115 In both the Gambian and Rhodesian forms of African trypanosomiasis, the parasites occur in the blood stream and central nervous system in the trypanosomal form. **1975** *Nature* 8 May 157/1 Calcium may be responsible for the control of flagellar activity in a trypanosomid flagellate, *Crithidia oncopelti*. **1977** C. D. Becker in J. P. Kreier *Parasitic Protozoa* I. x. 370 Trypanosomes of fish appear to be similar to related trypanosomatids from the blood of other vertebrates.

tryparsamide (tripā·ɹsămәid). *Pharm.* [f. Tryp(anosoma + Ars(enic *sb.*[1] + Amide.] An arsenical organic compound, $C_8H_{10}AsN_2NaO_4.\frac{1}{2}H_2O$, used to treat trypanosomiasis and syphilis of the central nervous system.

1921 *Jrnl. Exper. Med.* XXXIII. 193 (*heading*) Therapeutic action of N-phenylglycineamide-*p*-arsonic acid (tryparsamide) upon experimental infections of *Trypanosoma rhodesiense*. **1935** *Lancet* 26 Jan. 193/2 After one injection of reduced tryparsamide the trypanocidal titre of the urine rapidly rose to an enormous level. **1974** *Trypanosomiasis & Leishmaniasis* (Ciba Foundation) 310 The problem of drug resistance has become of less importance since the discontinuance of tryparsamide for the treatment of established cases, and its replacement by melaminyl derivatives of arsenic such as melarsoprol and melarsonyl potassium.

trypomastigote (tripomæ·stigōᵘt). *Zool.* [f. Gr. τρυπ-ᾶν to bore + -o + Gr. μαστιγ-, μάστιξ whip + -ώτης (see -ot[2], -ote).] A stage in the life cycle of trypanosomes (see quot. 1966). Cf. *trypanosome b.

1966 Hoare & Wallace in *Nature* 17 Dec. 1386/1 We have devised the following new terms...(3) Opisthomastigote.., for former 'trypanosome'..stage, represented..by forms with postnuclear kinetoplast; flagellum arising near it and emerging from its anterior end...(5) Trypomastigote.., for the true 'trypanosome' stage, represented by forms with postnuclear kinetoplast; flagellum arising near it and emerging from the side of the body to run along a long undulating membrane. **1980** J. N. Farmer *Protozoa* vi. 225/1 Trypomastigotes reproduce in the vertebrate host by means of longitudinal binary fission except in the case of *Trypanosoma* (*S.*) *cruzi*.

trypsinize (tri·psinәiz), *v.* *Biochem.* [f. Trypsin + -ize.] *trans.* To treat with trypsin. So **try·psinized** *ppl. a.*, **try·psinizing** *vbl. sb.*

1952 *Jrnl. Immunol.* LXIX. 688 The cultures prepared with trypsinized tissue were inoculated on the 8th day of incubation. **1959** *Virology* VIII. 396 Because these cultures sometimes contained tissue lumps which interfered with plaque production, they were usually trypsinized. **1971** *Nature* 30 July 312/2 After 10–14 days, wells containing single colonies were trypsinized. **1974** F. Warner *Meeting Ends* i. vi. 25 Oh my head! I'm trypsinized, pepsinized, falling to bits. **1979** *Experientia* XXXV. 244/2 When the cells reached confluency standard trypsinizing procedures were used to remove them from the flask surface.

Also **trypsiniza·tion,** treatment with trypsin.

1959 *New Scientist* 19 Feb. 396/1 (*caption*) Trypsinization of monkey kidney for tissue-culture. **1980** *Parasitology* LXXX. 374 They were removed from the T-75 flask by trypsinization.

tryptamine (tri·ptămīn). *Biochem.* [f. *Trypt(ophan + Amine.] An amine, $C_8H_6NCH_2CH_2NH_2$, related to tryptophan, from which it is produced by decarboxylation and which itself is oxidized to indoleacetic acid.

1929 R. A. Gortner *Outl. Biochem.* xx. 445 Tyramine, histamine, and tryptamine, all have a powerful physiological action. **1963** *Lancet* 19 Jan. 127/1 Indole and tryptamine are formed by the action of bacteria on unabsorbed tryptophan in the colonic lumen. **1980** G. Guroff *Molecular Neurobiol.* ix. 127 An overload of tryptophan and/or an amine oxidase inhibitor is required to produce pharmacological evidence for tryptamine formation [in brain tissue].

tryptophan (in Dict. s.v. Tryptic *a.*). Also **tryptophane.** Add: **a.** An amino-acid essential in the diet of vertebrates; β-3-indolylalanine, $(C_8H_6N)CH_2CH(NH_2)COOH$. (Earlier and later examples.) [a. G. *tryptophan* (R. Neumeister 1890, in *Zeitschr. f. Biol.* XXVI. 3).]

1890 *Jrnl. Chem. Soc.* LVIII. 804 The term tryptophan is suggested for the substance which is formed during pancreatic digestion from proteids, and which gives a reddish-violet coloration with bromine. **1922** *Sci. Amer.* July 42/3 Certain of the amino acids, of which about 19 have been found in proteins, are absolutely essential for growth and maintenance, among which are lysine, cystine and tryptophane. **1945, 1949** [see *co-factor 2]. **1956** *Nature* 3 Mar. 422/2 Supplementation of the [rats'] diet with either nicotinic acid or tryptophan had no effect on this depressed growth-rate. **1978** *Detroit Free Press* 5 Mar. (Parade Suppl.) 13 (Advt.), Pay less for more protein potency! A whole and complete protein derived from soybean. 0% carbohydrates, 325 mg tryptophane, 26 grams protein with lecithin and papain. **1979** *Time* 2 Apr. 42/2 Only a small dose of tryptophan—which is found in many foods, notably milk—seems to ease the insomniac to sleep.

b. Special Comb.: **tryptophan synthetase,** a bacterial enzyme that synthesizes tryptophan.

1955 C. Yanofsky in Colowick & Kaplan *Methods in Enzymol.* II. i. 233 Tryptophan synthetase from *Neurospora*... This enzyme has also been called tryptophan desmolase... The name tryptophan synthetase was suggested by the editors of this volume. **1976** *Ann. Rev. Microbiol.* XXX. 413 Where this has been specifically tested—with tryptophan synthetase subunits in *Escherichia coli*..homologies have not been found.

Hence **try·ptophanase** [a. Jap. *tryptophanase* (K. Kurono et al. 1932, in *Jrnl. Agric. Chem. Soc. Japan* VIII. 82): see *-ase], an enzyme catalysing the breakdown of tryptophan into ammonia, indole, and pyruvic acid.

1932 *Chem. Abstr.* XXVI. 5107 (*heading*) Tryptophan-decomposing enzyme, tryptophanase. **1963** *Lancet* 19 Jan. 128/1 Glucose reduces indole formation by inhibiting tryptophanase synthesis.., but does not affect the activity of tryptophanase already present. **1976** *Ann. Rev. Microbiol.* XXX. 421 Tryptophanase from *Escherichia coli* can catalyze the reversal of its normal function, forming L-tryptophan from indole, pyruvate, and ammonia.

Tsaconian, var. *Tsakonian *sb.* and *a.*

tsaddik (tsa·dik). *Judaism.* Also **tsadik, tzaddik, tzaddiq, zaddik,** etc. Pl. **-kim, -ks.** [a. Heb. *ṣaddīq* just, righteous.] A man of exemplary righteousness; a Hasidic spiritual leader or sage.

1873 *New Era* (N.Y.) III. 75 These zadiks, or leaders, have no fixed salary. **1881** [see *rebbe]. **1904** *Jewish Encycl.* VI. 253/2 The Hasidim were, however, particularly noted for the exalted worship of their 'holy' zaddikim. **1907** I. Zangwill *Ghetto Comedies* 409 A *Tsaddik* (wonder-rabbi) was killed in the last *pogrom*. **1933** S. Birnbaum *Life & Sayings Baal Shem* p. iii, Chassidism..is characterized..by a new psychic and material organization of comprehensive scope, at whose center of crystallization stand the *tzaddikim*, the 'righteous ones', the masters of the souls of men. **1941** G. G. Scholem *Major Trends in Jewish Mysticism* ix. 338 The existence of the Zaddik or saint as the actual proof of the possibility of living up to the ideal. **1964** M. Wohlgelernter *I. Zangwill* vii. 111 The numerous sects that mushroomed all over Europe, each with its own *Zaddik* performing untold 'miracles'. **1968** *Observer* 10 Nov. 26/5 He was bound to end up..a great chassidic *Tsadik*. **1974** *Encycl. Brit. Micropædia* X. 225/3 In the..movement known as Hasidism, the Jewish religious leader (*tsaddiq*) was viewed as a mediator between man and God. Since the *tsaddiq*'s life was expected to be a living expression of the Torah, his behaviour was even more important than his doctrine. **1982** *Times* 18 June 13/3 The world has much need of such *tsaddiks* (righteous men).

Tsakonian (tsăkōᵘ·niăn), *sb.* and *a.* Also **-c-, Tz-.** [f. *Tsakon*, a region in the eastern Peloponnese, Greece + -ian.] (Of or pertaining to) a modern Greek dialect spoken in an area of the south-eastern Peloponnese, containing ancient elements not derived from the koine.

1902 *Encycl. Brit.* XXIX. 102/2 Certain peculiar dialects, such as the Tzakonian in the south-eastern Morea. **1925** P. Radin tr. *Vendryes's Language* ii. 46 For the present Tsaconian still seems to use it [*sc.* digamma]. **1933** C. D. Buck *Compar. Gram. Gk. & Lat.* 20 The present Tsaconian dialect, spoken in a small portion of Laconia,..is in part the offspring of ancient Laconian. **1939** A. J. Toynbee *Study of Hist.* VI. 71 This 'Tsakonian' *patois* is the only surviving form of spoken Greek that is not derived from the Attic κοινή. **1972** W. B. Lockwood *Panorama Indo-Europ. Lang.* 7 Tsakonian is the outlandish dialect of perhaps as many as 10,000 speakers in an area difficult of access along the forbidding coast of the Peloponnese between the Parnon Range and the Gulf of Argolis.

tsamba. Add: (Earlier example.) Also **tsampa, tsumpa.**

1852 *Dublin Rev.* XXXIII. 12 As tsamba is not a very toothsome affair, we used to make three or four balls of it, with our tea. **1937** H. W. Tilman *Ascent Nanda Devi* iv. 31 The Tibetan tsumpa—barley, or wheat, which has been parched and then ground into meal. **1952** Morin & Smith tr. *Herzog's Annapurna* v. 78 Khangsar was very poor: not a pound of tsampa to spare. **1979** P. Matthiessen *Snow Leopard* ii. 86 The two outcasts dip up tsampa, the roasted maize or barley meal, ground to powder and cooked as porridge or in tea.

tsamma (tsæ·mă). Also **(t)sama.** [a. Hottentot *t'sama*.] A wild water-melon, *Citrullus lanatus*, native to parts of southern Africa. Also *attrib.*

1886 G. A. Farini *Through Kalahari Desert* vii. 106 We came across the first sama I had seen... It is the 'wild water-melon', resembling the cultivated variety in appearance, both internally and externally. **1933** *Times Lit. Suppl.* 16 Mar. 175/2 Ground melons (tsamas) provide the only water supply for the wandering tribes of bushmen. **1937** A. J. H. Goodwin in I. Schapera *Bantu-Speaking Tribes S. Afr.* ii. 39 The tsama melon..supplies liquid to men and beasts. **1948** L. G. Green *To River's End* i. 16 The dunes topped by their t'samma melons..only form a background. **1958** L. van der Post *Lost World Kalahari* ix. 212 A middle-aged woman sat diligently pounding the seeds of the tsamma. **1974** *Stand. Encycl. S. Afr.* X. 640/2 The name 'tsamma' is used mainly in the Kalahari, and is derived from a Bushman name for the plant.

tsampa, var. *tsamba.

‖ **tsantsa** (tsa·ntsă). Also **tzantza.** [Jivaro.] A human head shrunk as a trophy by the Jivaros of Ecuador.

1923 R. Karsten *Blood Revenge* 14 That the blood feuds which take place within the tribe have an entirely different character from the wars of extermination waged against foreign tribes also appears from the fact that only in the latter case..the victors make trophies (*tsantsas*) of the heads of their slain enemies. **1957** *Encycl. Brit.* XIII. 70/1 The finished 'tsantsa' is about the size of the head of a small monkey, and preserves strikingly the human expression. **1966** J. M. Woodman in *Fodor's Guide to S. Amer.* 177 The famed Jivaro shrunken heads or *tzantzas* are extremely rare. Many good imitations are for sale in almost every gift shop in Ecuador, and a good goat-hide imitation *tzantza* usually costs around $2.60. **1977** *New Yorker* 25 July 81/2 Another is the stealthy sale by dubious chaps of shrunken heads (tsantsa), war trophies of the Jivaro head-hunters.

‖ **ts'ao shu** (tsau ʃu). Also **cao shu, tsaou shoo.** [Chinese *cǎoshū*, f. *cǎo* hasty + *shū* writing.] In Chinese calligraphy, a cursive script developed during the Han dynasty (206 B.C.–A.D. 220) from the 'official' script.

1876 *Encycl. Brit.* V. 655/2 The *Tsaou shoo* or 'grass character'. **1910** *Ibid.* VI. 220/2 Out of the 'official script' two other forms were soon developed, namely the..*ts'ao shu*, or 'grass character', [etc.]. **1974** J.-F. L. Chang in T. C. Lai *Chinese Calligraphy* (1975) p. xiii, Ts'ao Shu or Cursive Script..was originally developed as a quicker version of the Official Script during the second century

A.D. **1978** *Nagel's Encycl.-Guide: China* 326 *Cao shu* is a style in which the characters are abbreviated, giving free reign to the movement. The economy of style is often such that it is hard to read for all but the initiated.

tsatske (tsǫ·tskə), **tchotchke** (tʃǫ·tʃkə). *U.S. colloq.* [Yiddish, f. Slavonic (cf. Russ. *tsatska*).] A trinket or gewgaw; *transf.*, a pretty girl. Also **tsa·tskeleh** [Yiddish *-le* dim. suff.], an affectionate diminutive of *tsatske*.
1964 W. Markfield *To Early Grave* (1965) v. 101 He was no *tsatskeleh*, Leslie, he was in certain respects far from being trustworthy. **1968** L. Rosten *Joys of Yiddish* 408 *Tsatske* and *tchotchke* are used interchangeably... At one time..West End Avenue in New York had an inordinately high proportion of *tchotchkies*. **1970** S. Ellin *Man from Nowhere* (1971) xix. 94 He looked Elinor over appraisingly... 'A real *tsatskeh*,' he said with approval. **1972** M. Glenny tr. *Solzhenitsyn's August 1914* (1974) viii. 77 True, she never did anything to cross him, never even put on her expensive clothes and her *tsatski* (diamonds) at home because he disapproved of it. **1974** *N.Y. Times* 12 July 31 'Décor doesn't add to the glamour of a suit,' an owner pointed out. 'You're not buying the rugs or the lamps or the tsatskes.' **1977** *New Yorker* 1 Aug. 14/1 A..boutique, to the left of the entrance, stocked with a careful selection of New York's best tchotchkes. These include thirteen-inch-long matchbooks.

Tsaubwa, var. *Sawbwa. **Tschaikowskian**, var. *Tchaikovskian. **tschee**, var. Stchi in Dict. and Suppl.

Tschermak (tʃɜ̄·ɹmæk). *Min.* The name of Gustav *Tschermak* (1836–1927), Austrian mineralogist, used *attrib.* and in the possessive to designate the synthetic pyroxene CaAl(AlSi)O_6 as a hypothetical component of natural pyroxenes, or the part Al(AlSi)O_6 of this.
1943 *Amer. Mineralogist* XXVIII. 73 The substitution Al_2/MgSi is that by which the 'Tschermak molecule' is derived from diopside. **1962** *Jrnl. Petrol.* III. 355 The Cs may also be traced to the Tschermak's molecule in the high-CaO titaniferous clinopyroxene. **1970** *Nature* 26 Sept. 1337/1 Omphacite (diopsidic clinopyroxene in which jadeite predominates over Tschermak's molecule). **1974** *Encycl. Brit. Macropædia* XV. 322/1 Ferric diopsides contain ferri-Tschermak's molecule ($CaFe_2SiO_6$).
Hence **tsche·rmakite**, an amphibole end-member of the hornblende group, $Ca_2Mg_3Al_4Si_6O_{22}(OH)_2$, that is rich in aluminium; also, any member of the series this forms with ferrotschermakite (the other end-member). Hence **tschermaki·tic** *a.*
1945 A. N. Winchell in *Amer. Mineralogist* XXX. 29 Hallimond notes that the second of these formulas is often called the Tschermak molecule; the writer would suggest that it be called tschermakite. **1963** W. A. Deer et al. *Rock-Forming Minerals* III. 272 Tschermakitic hornblendes. **1966** —— *Introd. Rock-Forming Minerals* II. 168 Compositions approaching those of the edenite and tschermakite end-members are rare, and the compositions of the majority of the hornblendic amphiboles are intermediate between the two end-member series, tremolite–ferroactinolite and pargasite–ferrohastingsite. **1976** *Nature* 22 Apr. 673/2 The generally tschermakitic-pargasitic nature of all the amphiboles is like that of amphiboles produced synthetically from hydrous basaltic melts.

tschernozem, var. *chernozem.

tsetse. Add: **2.** *attrib.* and *Comb.*, as *tsetse-bitten, -conveyed, -free, -infested, -poisoned* adjs.; **tsetse country, district**, an area infested by the tsetse fly; **tsetse-fly disease** = *nagana.
1906 Rider Haggard *Benita* x. 138 Meyer..had already decreed the death of the tsetse-bitten cattle. **1917** *Nature* 18 Oct. 127/2 The tsetse-conveyed sleeping sickness is being got well in hand. **1853** D. Livingstone *Jrnl.* 23 May (1960) iv. 139 There is only one small strip of Tsetse country to hinder one going westward. **1877** T. Baines *Gold Regions S.E. Afr.* 89 Persons travelling must seek for the *latest* information..when approaching the borders of a Tsetse country. **1851** D. Livingstone *Jrnl.* 29 Aug. (1960) ii. 64 Several Englishmen have lost all their cattle horses & dogs by being led into a Tsetse district. **1895** Tsetse fly disease [see *nagana]. **1932** C. Fuller *Louis Trigardt's Trek* ii. 28 They..were compelled to desist..on account of tsetse-fly disease among their cattle. **1948** T. A. M. Nash *Tsetse Flies in Brit. W. Afr.* i. 7/1 Inside this huge area there are only three districts which are tsetse-free. **1977** *Times* 22 Apr. 10/2 The ever-spreading erosion and desertification wreaked by over-stocking of domestic cattle in tsetse-free areas. **1948** T. A. M. Nash *Tsetse Flies in Brit. W. Afr.* i. 13/1 Should..the river flow from tsetse-infested country northwards, it would be wiser not to plant up the banks. **1977** *Times* 22 Apr. 10/4 Vast tsetse-infested tracts of Nigeria. **1906** Rider Haggard *Benita* x. 138 The time was to come when she would swallow that hard, tsetse-poisoned flesh.

Tshi: see *Twi.

T-shirt (tī·ʃɜɹt). orig. *U.S.* Also **tee-shirt**. [f. the letter T (see below) + Shirt *sb.*] A simple kind of garment, orig. a man's undershirt, typically short-sleeved, round-necked, buttonless and made from knitted cotton fabric, and forming the shape of a letter T when spread out flat; now a similar garment of various designs, widely worn as a shirt by men, women, and children for sport or as casual wear.
1920 F. Scott Fitzgerald *This Side of Paradise* i. i. 25 Amory, provided with 'six suits summer underwear..one sweater or T shirt..' set out for New England, the land of schools. **1944** *Survey Graphic* Aug. 368/3 We have been slow to realize that the high school crowd needs to sit in with us, with all their jive talk, their 'T' shirts and 'sloppy joes'. **1948** *Sun* (Baltimore) 2 Aug. 6/2 A ragged looking pair of trousers and as dark a shirt [as] possible or an old 'tee' shirt. **1957** J. Braine *Room at Top* xxv. 203 Roy..was wearing blue suède shoes, blue linen slacks, an orange T-shirt, and white sunglasses. **1958** *Daily Express* 4 Mar. 3/7 Lord M—..wore a striped tee-shirt. **1973** C. Bonington *Next Horizon* xiv. 202 Sebastian,..presented a sharp contrast to myself, already scruffy in T-shirt and jeans. **1980** *Times* 7 Feb. 13 There will be pop music and phone-ins, tee-shirts and car stickers.
Hence **T-shirted, tee-shirted** *adjs.*, wearing or clothed in a T-shirt.
1957 J. Kerouac *On Road* (1958) ii. vi. 134 His muscular neck, T-shirted in the winter night. **1959** *Times* 21 Oct. 11/3 T-shirted trippers perched on tubular steel chairs. **1973** *Philadelphia Inquirer* (Today Suppl.) 14 Oct. 27 (*caption*) A funeral parlor..is guarded by tee-shirted gang members outside. **1979** *Guardian* 3 Sept. 11/8 The jeaned and tee-shirted young.

tsimmes, -is, varr. *tzimmes.

Tsimshian (tʃi·mʃən), *sb.* and *a.* Also Tsimpshian, etc. [Tsimshian self-designation *čamsián*, lit. inside of the Skeena River.] **A.** *sb.* **a.** An Indian people of the north Pacific coast of N. America; also, a member of this people. **b.** Their language. **B.** *adj.* Of or pertaining to this people or their language.
1836 D. Finlayson *Let.* 29 Sept. in E. E. Rich *McLoughlin's Fort Vancouver Lett.* 1st Ser. (1941) 323 The Pearl Harbour, and Skeenah Indians called the Chimmesyan tribe. **1888** [see *copper *sb.*¹ 4 b]. **1890** *Rep. Brit. Assoc. Adv. Sci. 1889* 853 The Olala is confined to the southern Tsimshian tribes. **1911, 1934** [see *Nass]. **1966** [see *Chilkat]. **1972** *Language* XLVIII. 390 Geographically or genetically contiguous languages which share many features of the proposed analysis, e.g. Squamish and Tsimshian.

tsine. For *sondaicus* substitute *banteng*. Also, = *banteng. (Earlier example.)
1880 [see *banteng, banting].

‖ **tsipouro** (tsi·puro). Also **tsippouro, tsipuro**. [mod.Gr., prob. a. Turk.] A rough and local kind of Greek spirituous drink.
1947 J. Mulgan *Report on Experience* viii. 98 We cheered each other in Fourna, passing *tsippouro*—a lower-class cousin of *ouzo*—across the fire. **1953** X. Fielding *Stronghold* i. 4 [*Tsidoukhia*,] The local name for raki, similar to the *tsipouro* of Epirus, which is distilled from what is left of the grape after it has been pressed for wine. **1969** [see *raki]. **1981** J. Bowman *Crete* (ed. 5) 46 On Crete the men are distinguished from the boys by drinking *raki*, a stronger unflavoured version [of ouzo] (and known in Crete as *tsipouro* or *tzikoudhia*).

tsitsith (tsi·tsis, -it), *sb. pl.* and *collect. sing.* *Judaism.* Also **tzitzit(h), zizith**. [a. Heb. *ṣiṣiṯ*.] The tassels of twisted and knotted cord worn by orthodox Jewish males on the corners of certain garments, esp. the tallith; cf. Fringe *sb.* 1; also, a small rectangular garment, with a large hole in the middle and with tassels attached to each corner, worn over the vest but under the shirt.
Translated as 'fringes' in A.V. and R.V. and as 'tassels' in N.E.B. (see Numbers xv. 38, Deut. xxii. 12).
1675 L. Addison *Present State of Jews* xiii. 99 The wearing of the Zizith or Fringe, they collect from Exodus. **1738** 'Gamaliel Ben Pedahzur' *Bk. Relig. of Jews* 4 The Fringe on each Corner is of eight worsted Threads double twisted [*marg.* Called Zizith]. **1816** J. Allen *Mod. Judaism* xvii. 305 Every male is required to have a quadrangular vestment, which they call *Talleth*... Its principal denomination, Tsitsith,..it receives from the fringes upon which all its sanctity is supposed to depend. **1854** *Asmonean* X. 198/1 In every corner is a string of plaited thread, called *zizis*, or *ziziths*. **1881** *Living Age* CIL. 418/1 The *tzitsis*, or fringes..of the boy's garment have been neglected. **1891** M. Goldsmith *Rabbi & Priest* 29 Shall he earn a few paltry kopecks in making tzitzith (fringes for the praying scarfs)? **1907** I. Zangwill *Ghetto Comedies* 389 These fellows..wear fuses under their waistcoats instead of *Tsitsith* (ritual fringes). **1920** J. A. Robertson *Hidden Romance N.T.* viii. 173, I see you are a Jew, like myself... But you don't wear the Zizith any more than I do. **1962** *New Jewish Encycl.* 475/1 The Tallit is rectangular in shape,..with *Tzitzit* (fringes) at each of its four corners. The Tzitzit is the important part of the Tallit. **1962** *Encounter* Sept. 21/2 They didn't buy ice-cream from the carts in the streets because it wasn't kosher; they never took their caps off in class..they wore *tzitzith* under their vests... They were heroic, absurd, and maddening. **1967** C. Potok *Chosen* i. 16 They all wore the traditional undergarment beneath their shirts, and the tzitzit, the long fringes appended to the four corners of the garment, came out above their belts. **1968** L. Rosten *Joys of Yiddish* 418 *Tzitzit* are meant as reminders of one's duty to the laws of Judaism. **1973** *Times* 16 June 9/4 The distinctive features of Jewish dress—the *tsitsith* (tassels) [etc.].

tsk, *int.* Also **tsck**, and redupl. **tsk-tsk**. [Alveolar click formed by suction: cf. Tchick *sb.*, Tck *int.*] A sound expressing commiseration; an exclamation of disapproval or irritation. So **tsk-tsk** *v. intr.*, to make this sound or utter this exclamation; also *trans.*, to say disapprovingly.
1947 K. L. Pike *Phonemics* ii. 41/1 Do you get..a sound resembling the noise of commiseration which is sometimes written in literature as 'tsk-tsk', or 'tut-tut'? **1952** S. Kauffmann *Philanderer* (1953) xvii. 277 'When she doesn't like a writer or a composer, it sort of dampens my..activity about him. I'm henpecked.' 'Tsk, tsk,' said Ed extra clearly. **1958** L. Durrell *Mountolive* xvi. 307 Balthazar..walked slowly..making the little clucking noise he always made with his tongue..*Tsck, tsck*. **1962** A. Lurie *Love & Friendship* xiv. 273 'Poor Carolyn Hastings is officially engaged to that dreadful Cowie boy.' 'Tsk.' **1966** M. Avallone *Fat Death* 47 'Mr. Noon,' he *tsk-tsked* reprovingly. 'You really pretend not to know?' **1968** *Punch* 20 Mar. 428/3 No amount of tsk-tsking indignation and bitter wit will turn a buy into a bargain. **1976** *New Yorker* 26 Apr. 102/2 She tsk-tsks over Momma. **1983** M. Gee *Sole Survivor* iv. 32 The constable went tsk, tsk and looked at us sadly. There was nothing he could do so he sent us home.

Tsonga (tsǫ·ŋgä), *sb.* and *a.* Also **Thonga**. [Native name: cf. *Tonga⁴.] **A.** *sb.* An African tribal group chiefly inhabiting the Transvaal area of the Republic of South Africa and parts of southern Mozambique; the Bantu language spoken by this people. **B.** *adj.* Of, pertaining to, or designating this people or their language.
1905 *Jrnl. Afr. Soc.* V. 12 There are three main Native languages in South Africa: Zulu, Suto and Thonga. **1907** H. A. Junod in C. W. Chatelain *Pocket Dict. Thonga (Shangaan)* (1909) 3 The Shangaan language, which would be more scientifically called the Thonga language. **1912** —— *Life S. Afr. Tribe* I. 13 The Thonga tribe is composed of a group of Bantu peoples settled on the Eastern coast of South Africa. **1933** *Bantu Stud.* VII. 4 The languages of the South-eastern [*sc.* Bantu] zone fall into four clusters, the 'Nguni', the 'Sotho', the 'Venda' and the 'Thonga'... In the Thonga cluster [are] three main groups, Ronga, Thonga and Tswa. **1937** R. H. Lowie *Hist. Ethnological Theory* xii. 225 The Northern Thonga of Portuguese East Africa use a single word for the maternal uncle, the maternal uncle's son, and the mother's father. **1940** A. A. Jaques in *Bantu Stud.* XIV. 259 It should be noted that it is not correct to spell the name of this people [*sc.* the Shangana-Tsonga] *Tonga* or *Thonga*... Since the recent decisions taken by the Tsonga Language Board concerning orthography, there should be no difficulty in adopting the form *Tsonga*. **1955** M. Gluckman *Custom & Conflict in Afr.* iii. 63 Among the Tsonga of Mozambique..witchcraft passes..from women to their children, so that men can carry it, but not transmit it. *Ibid.* v. 116 Take this description of the Tsonga ritual which organizes the moving of a village. **1970** *Standard Encycl. S. Afr.* II. 106/2 The Tsonga group is often also called the Shangana group. These languages are spoken by the descendants of the Zulu chief Soshangana. *Ibid.* 306/1 The first portions of Scripture in Tsonga..appeared in 1883. **1973** 'S. Harvester' *Corner of Playground* iii. ii. 181 A Swazi king in eighteen-fiftyfour raided the Tsonga to find boys..to sell to the Boers. **1974** *Encycl. Brit. Micropædia* IX. 967/3 Although many Thonga are Christian, a substantial proportion adhere to their own traditional, highly ritualistic religion. **1979** *Jrnl. Imperial & Commonwealth Hist.* VII. 255 The provision of food, lodging and security to Tsonga migrants proved beneficial to Natal.

‖ **Tsongdu** (tsǫ·ŋ,du). Also **Tsong-du**. [ad. Tibetan *t'sogs ḏu*, lit. 'an assembly meets'.] The Tibetan National Council or Assembly.
1905 *Spectator* 18 Feb. 243/2 When the expedition started we had two parties in Lhasa against us,—the Dalai Lama, with his adviser Dorjieff, who urged a breach in the traditional exclusiveness of the State in favour of Russia; and the Tsong-du, or Council, who were for the maintenance of the old system against Russia and Britain alike. **1970** R. D. Taring *Daughter of Tibet* vi. 68 A meeting of the *Tsongdu* (the National Assembly).

‖ **tsores** (tsǫ·rəs), *sb. pl.* (sometimes const. as *sing.*) *U.S. colloq.* Also **tsouris, tsuris, -us, (t)z-**, etc. [Yiddish *tsores*, pl. of *tsore* trouble, woe, a. Heb. *ṣārāh*.] Trouble(s), worries.
1901 M. Wolfenstein *Idyls of Gass* viii. 143 I have other *Zores* (troubles). **1905** W. Wittigschlager *Minna* I. i. 15 We have no money..nor do we propose ever to experience such *zorus*, (trouble). **1929** J. D. Rosenberg *Kosher Americans* xvii. 165 Ve had plenty *tso-res* (troubles) before. **1941** B. Schulberg *What makes Sammy Run?* iv. 70, I had a notion that a little of Billie and Sammy Glick might not be such a bad idea, if only to get my mind off my own tsurus. **1950** *Commentary* X. 67/2 He had enough *tzores* without politics. **1956** 'T. Betts' *Across Board* 103 During a long streak of tzuris (misery) with cards and horses. **1968** L. Rosten *Joys of Yiddish* 412 Oh, have I got *tsuris*! **1972** J. Caine *Hamlet, my Boy* vi. 71 It's no *tsooras* for a child. **1975** *N.Y. Times Bk. Rev.* 16 Mar. 31 She has a bad back..and miscellaneous medical tsouris.

tsotsi (tsǫ·tsi). *S. Afr.* [Origin uncertain: freq. said to be a corruption of *zoot suit*.] An African street thug or hoodlum, usu. from the Black townships and distinctively dressed in narrow trousers or in garments of exaggerated

cut. Also in extended sense. Also *attrib.* Hence **tso·tsi-ism**.
1949 *Cape Argus* 20 July 8/9 (*heading*) Tsotsi gangs who hate Bantu students. **1949** *Cape Times* 10 Sept. 8/6 The 'Tsotsi' may be distinguished by his exceedingly narrow trousers which hardly reach his shoes, or else by his 'zoot suit'. **1952** B. Davidson *Rep. Southern Afr.* ii. v. 121 The conditions out of which have grown such strange and horrible manifestations of maladjustment as *tsotsi*-ism. **1956** H. Bloom *Episode* xv. 273 One could tell they were *tsotsis* by their fancy clothes, by the way they took command... They led the mobs. **1971** *Sunday Times* (Johannesburg) *News Mag.* 28 Mar. 9/7 We do not want agitators to give hooligans and tsotsis the chance to plunder shops and businesses. **1979** A. Brink *Dry White Season* i. vii. 70 'You may be a *lanie*'—his red tongue caressed the syllables of the taunting *tsotsi* word—'but you've got it right here.'

tsouris, var. *tsores.

‖ **tsu** (tsū). *Anthrop.* [Chinese.] A patrilineal kinship group in pre-revolutionary China (see quot. 1939).
1939 H.-T. Fei *Peasant Life in China* v. 84 According to accepted principle, all the patrilineal descendents and their wives that can be traced to a common ancestor within five kinship grades consider themselves as belonging to a kinship group called Tsu. **1957** in K.-C. Chang *Archaeol. Anc. China* (1963) vi. 168 *P'i* (grandmothers and female ancestors) was partnered to *tsu* (grandfathers and male ancestors). **1966** M. Freedman *Chinese Lineage & Society* i. 5 Almost all *tsu* (lineage) relatives in the village are related through a grandfather.

‖ **tsuba** (tsū·bä). Pl. **tsuba, tsubas**. [Jap., shortened f. *tsumiha, -ba*, f. *tsumi* to stop + *ha* (enemy's) blade, sword.] A Japanese sword-guard.
1889 J. J. Rein *Industries Japan* iii. vii. 432 The sword-shell, or guard, Tsuba, is as old as the sword. It is an oval metal plate..with an opening in the middle to admit the blade of the sword. **1909** J. Masefield *Multitude & Solitude* ii. 35 Melyard collected tsuba, and fenced archæologically at the Foil Club. **1970** *Times* 11 Mar. 12/5 Sotheby's held a similar sale, mainly devoted to Japanese sword guards or tsubas. **1976** *Times Lit. Suppl.* 16 Jan. 48/4 (*caption*) Examples of the *tsuba*, or hand-guard, of Japanese swords. 'On them..may be found illustrated the whole of the mythology, customs, legends, folklore, famous scenes, characteristics and celebrated personages and events of the history of Japan.'

‖ **tsubo** (tsū·bo). Pl. **tsubo, tsubos**. [Jap.] A Japanese unit of area, equivalent to approximately 3.95 square yards (3.31 square metres).
1727 J. G. Scheuchzer tr. *Kæmpfer's Hist. Japan* I. iv. iii. 292 Woods and forests pay a ..Ground-rent, which differs according to the number of *Tsubo's*, and the goodness and fruitfulness of the soil. **1902** *Encycl. Brit.* XXXIII. 810/1 Tsubo... Japan... 3·0306 [? *an error for* 3·306] sq. metres. **1972** P. M. Bartz *South Korea* viii. 123/2 In crowded Korea, urban land is sold by the *pyong* (35·6 square feet), the equivalent of the Japanese *tsubo* (3·3 square metres).

tsuica (tsū·ikä). Also **tuica, țuica, tzuica**. [a. Rum. *țuică*.] A Romanian plum brandy.
1927 *Sunday Express* 26 June 17/5 Will you come into the house and have a talk over a glass of tuica? **1938** *Times* 1 Jan. 11/1 The improvement of the facilities for the making of *țsuica* (plum brandy, which is a national drink). **1960** O. Manning *Great Fortune* i. 52, I make my own red wine, white wine, *țuică*, and martini. **1965** A. Sichel *Penguin Bk. Wines* iii. 221 Various fruit cordials, such as plum brandy, are also exported from Rumania; there are two types, called Tzuica and Slibovitza. **1971** T. E. B. Clarke *Wrong Turning* xv. 105 She was drinking red wine with sodawater, he was having beer: there was laughter..at Jean's insistence in sticking to tuica. **1977** 'A. Stuart' *Snap Judgement* 51 She was choking—with emotion? With țuica? Both she and the room smelt strongly of plum brandy. **1981** B. de Breffny *My First Naked Lady* i. 29 Grandfather made me sip a Rumanian liqueur called Tsuica.

‖ **tsukemono** (tsŭkimō^u·no). Also **tsukimono**. [Jap., f. *tsukeru* to pickle + *mono* something.] (See quots.)
1885 *Trans. Asiatic Soc. Japan* XIII. 8 *Tsukemono*, the preserved roots or leaves of certain vegetables. **1920** *Japan Advertiser* 22 Aug. 5/3 Most Japanese meals..are accompanied by vegetable dishes... Tsuki-mono stands for the pickle that accompanies the rice at the end of the meal and is made from vegetables in season, such as cucumbers, eggplant, melon, etc. **1968** P. S. Buck *People of Japan* xiv. 167 *Tsukemono*, the pickled vegetables served with any meal.

‖ **tsukuri** (tsukŭ^ə·ri). [Jap.] In Judo, preparatory action taken to facilitate the breaking of one's opponent's balance. Cf. *kuzushi.
1941 M. Feldenkrais *Judo* 25 Thus you have performed the 'fitting movement' (*Tsukuri*). **1968** [see *kuzushi].

tsumebite (tsū·mĕbəit). *Min.* [ad. G. *tsumebit* (K. Busz 1912, in *Festschr. gewidmet den Teilnehmern der 84. Versammlung Deutsch. Naturforscher und Ärzte in Münster von der Med.-Naturwiss. Ges. in Münster* 182), f. *Tsumeb*, name of a town in Namibia: see -ite¹.] A basic phosphate and sulphate of lead and copper, $Pb_2Cu(PO_4)(SO_4)OH$, found as crusts of transparent green monoclinic crystals.
1913 *Mineral. Mag.* XVI. 373 Tsumebite... Hydrated basic phosphate of lead and copper..found as small, emerald-green, monoclinic crystals on white calamine at Tsumeb, Otavi, German South-West Africa. **1966** *Amer. Mineralogist* LI. 267 Due to the extreme tendency of tsumebite to occur as intimately intergrown crystals, considerable difficulty was encountered in obtaining a suitable, single crystal.

tsumpa, var. *tsamba.

‖ **tsun** (tsun). [Chinese.] A style of Chinese vessel, either wide-mouthed or animal-shaped (see quot. 1974). Also *attrib.*
1958 W. Willetts *Chinese Art* I. iii. 144 The Sung cataloguers proceeded to identify as *tsun* various animal-shaped vessels known to them. **1974** *Encycl. Brit. Micropædia* X. 162/3 *Tsun*,..generic term ('sacrificial vessel') for a wide range of Chinese vessel shapes, generally of the Shang (*c.* 1766–*c.* 1122 BC) and early Chou (*c.* 1122–*c.* 900 BC) dynasties, all of which have an ample interior volume probably meant for containing wine. There are two essential varieties of *tsun*: one is shaped like a much enlarged *ku*; the other consists of various animal shapes, often densely embellished with animal decoration. **1977** Kwang-Chih Chang *Archaeol. Anc. China* (ed. 3) vi. 222 Pottery *li* tripods and large-mouthed *tsun* beakers began to appear in significant numbers.

tsunami (tsunā·mi). Also (repr. a strict transliteration of the Jap. form) **tunami**. [a. Jap. *tsunami, tunami*, f. *tsu* harbour + *nami* waves.] A brief series of long, high undulations on the surface of the sea caused by an earthquake or similar underwater disturbance. These travel at great speed and often with sufficient force to inundate the land; freq. misnamed a *tidal wave* (see Tidal *a.* 1 b). Also *fig.* and *attrib.*
1897 L. Hearn *Gleanings in Buddha-Fields* i. 24 '*Tsunami*!' shrieked the people; and then all shrieks and all sounds and all power to hear sounds were annihilated by a nameless shock..as the colossal swell smote the shore with a weight that sent a shudder through the hills. **1904** *Publ. Earthquake Investigation Comm. Foreign Lang.* (Japan) xix. 6 Records and reports of earthquakes and '*tsunamis*'. **1938** *Nature* 12 Nov. 881/2 The authenticity of the reports of earthquakes mentioned in these catalogues is weighed..with records of tunamis. **1956** *Jrnl. Earth Sci. Nagoya Univ.* IV. 2 The tunamis associated with strong earthquakes are frequent in Japan. **1967** *Technology Week* 23 Jan. 34/1 This system would predict ..tsunami run-up floods. **1970** *Daily Tel.* 27 Nov. 11/7 A *tsunami* generated off Chile by the 1960 earthquake crashed into Japan on the other side at 400 m.p.h. **1972** *Science* 11 Aug. 502/1 The Food and Drug Administration ..is currently swimming through a tsunami of comments generated by its announced intention to alter the regulations concerning the dispensation of methadone. **1981** *Monitor* (McAllen, Texas) 30 Jan. 3A/3 The National Weather Service..issued a *tsunami* warning. **1984** W. Golding *Paper Men* viii. 89 It seemed to me that I could feel the indifferent threat of the earth through the soles of my feet, the volcanoes, earthquakes, tsunamis, terrors of nature's fact.

tsuris, -us, etc., varr. *tsores.

‖ **tsutsugamushi** (tsū:tsugāmu·ʃī). *Path.* [Jap., f. *tsutsuga* trouble, illness + *mushi* insect.] = *scrub typhus* s.v. *scrub sb.¹ 6. Usu. *attrib.* in *tsutsugamushi disease*.
1906 *Index Medicus* IV. Index 216/1 Tsutsugamushi disease. **1908** *Philippine Jrnl. Sci.* B. III. 1 (*heading*) A comparative study of *tsutsugamushi* disease and spotted or tick fever of Montana. **1929**, etc. [see *scrub typhus* s.v. *scrub *sb.*¹ 6]. **1937** *Med. Jrnl. Australia* 21 Aug. 300/1 The absence of a local intradermal lesion speaks against the virus being of the spotted fever or *tsutsugamushi* types of Rickettsia. **1978** *Jrnl. R. Soc. Med.* LXXI. 507 The Japanese synonym, tsutsugamushi disease, clearly had precedence, but the name, scrub typhus, became fully established through widespread use in World War II.

‖ **tsutsumu** (tsŭtsŭ·mŭ). [Jap., to wrap.] The Japanese art of wrapping or packaging items in an attractive and appropriate way.
1975 *N.Y. Times Mag.* 9 Feb. 56 Each of the 300 packages in the show (called 'Tsutsumu, The Art of the Japanese Package') was purchased in 1974 in Japan, where an object's wrapping can be as important as the object itself.

Tswana (tswā·nä), *sb.* and *a.* Also **Chwana**. [Native name.] **A.** *sb.* **a.** = *Bechuana (for which *Tswana* is now the usual term). **b.** The Bantu language of this people (cf. *Sechuana). **B.** *adj.* Of or pertaining to this people or their language.
1930 *Bantu Studies* IV. 211 It is therefore probable that they may be still more widespread, at least among the Sotho-Tswana tribes. **1932** D. Jones *Outl. Eng. Phonetics* xxix. 227 Strong stress without accompanying loudness is a common feature of the Chwana language of South Africa. **1937** *Bantu Studies* XI. 137 The orthography set out in the present pamphlet represents the latest result of a prolonged series of efforts made..by the Education Departments..to arrive at a uniform orthography of Tswana. **1948** M. Guthrie *Classification Bantu Lang.* iv. 68 In Tswana the verbal has the extra suffix -ŋ. **1949** I. Schapera in M. Fortes *Social Structure* 104 In contrast to most other Bantu-speaking peoples of southern Africa, the Tswana tribes of the Bechuanaland Protectorate have remarkably few marriage prohibitions. **1957** [see *Sotho]. **1973** J. J. McKelvey *Man against Tsetse* i. 16 Cumming went..down to the hunting grounds of the Limpopo against the advice of the local people, the Tswanas. **1975** 'D. Jordan' *Black Account* xviii. 96 A girl came down the steps..she was blonde, and slim and tall... 'I thought you were all Tswana here,' I said and she laughed... She explained she worked in Bantu Affairs. **1980** First & Scott *Olive Schreiner* i. 37 Basotho and Tswana refugees.

t test (tī·test). *Statistics.* [f. *t* chosen arbitrarily as a symbol.] = *Student's (t) test* s.v. *Student².
[**1924** R. A. Fisher in *Proc. Internat. Math. Congr.* (1928) II. 498 If for the accurate standard deviation we substitute an estimate based on *n* degrees of freedom we have $t = z\sqrt{n}/\sqrt{S(x^2)}$. **1925** 'Student' in *Metron* V. 105 The present tables have..at Mr. Fisher's suggestion been constructed with argument $t = z\sqrt{n}$.] **1932** R. A. Fisher *Statistical Methods for Research Workers* (ed. 4) v. 116 The validity of the *t*-test, as a test of this hypothesis, is therefore absolute. **1958** M. Argyle *Relig. Behaviour* iii. 16 The 't' test..is generally used when equal measurement units are involved, otherwise various non-parametric methods are available. **1968** *Brit. Med. Bull.* XXIV. 220/2 A selection of commonly required significance tests is available, including paired and unpaired *t* tests, correlation coefficients..and analysis of variance. **1974** *Florida FL Reporter* XIII. 73/1 Correlated one-tailed t-tests were run for both the differences between the two groups.

‖ **tuak** (tŭ·æk). [Malay.] A locally-distilled Malaysian or Indonesian palm- or rice-wine.
[**1850** *Jrnl. Indian Archipelago* IV. 179 Liquid sugar..is obtained by boiling the juice extracted from the *loutar* or *tuak* palm.] **1852** *Ibid.* VI. 317, I could not learn that they prepared any fermented liquor, not even sago-weer or tuak. **1920** C. Lumholtz *Through Central Borneo* II. xxxi. 335 On the Upper Samba the custom still prevails of drinking tuak from human skulls. **1961** P. Kemp *Alms for Oblivion* vii. 119 *Tuak* drinking is a favourite pastime of the elderly men. *Tuak*, or toddy, is brewed from the juice of the sugar palm; it looks and tastes somewhat like farm cider,..though not so intoxicating. **1966** *Festival Malaysia 1966: Calendar of Events* 10/1 Tuak (locally distilled rice-wine) is the drink of the occasion [*sc.* a Dyak Gawai]. **1977** *Borneo Bull.* 7 May 4/5 The main cultural show was at an open night for all participants at the Rest House which would have been even more fun if the tuak had not run out early.

tuan (tuā·n). Also **Tuan**. [Malay.] A master, a lord, formerly esp. a European as spoken to or of by Malays; freq. used as a title of respect or form of address, = 'sir', 'mister'.
1779 T. Forrest *Voy. New Guinea* i. ii. 27, I found Tuan Hadjee in high spirits. **1864** J. T. Thomson *Some Glimpses Life Far East* xxi. 106 'Why,' said he, '*tuan*, these two rows belong to the East India Company's chief official.' **1885** E. Innes *Chersonese* I. ii. 21 Before the Tuan's (master's) marriage, all his mending is cheerfully done for him by his 'boy'. **1900** Conrad *Lord Jim* i. 3 They called him Tuan Jim: as one might say—Lord Jim. **1937** G. Frankau *More of Us* xvi. 166 There is nothing frets your gent, your sahib, your tuan,.. Like having some old donah near his new one. **1958** J. Slimming *Temiar Jungle* ii. 25 'Greetings, Tuan!' His voice is quiet and gentle. **1978** L. Heren *Growing up on The Times* iv. 136 British conscript soldiers were expected to keep the enemy at bay in Malaya while the *tuans* made their piles.

Tuareg (twā·reg), *sb.* and *a.* Also 9 **Tawarek, Tuari(c)k**. [Native name.] **A.** *sb.* (A member of) a nomadic people of the western and central Sahara. Also, the Berber language of this people. **B.** *adj.* Of or pertaining to this people or their language.
1821 G. F. Lyon *Narr. Trav. N. Afr.* iii. 112 The nearest Tuarick to Fezzan are at Ghraat. **1826** Denham & Clapperton *Narr. Trav. & Disc. N. & Central Afr.* p. l, The greater number of Tuaricks follow the nomadic life. *Ibid.* p. lxv, They..laughed heartily at our blundering out a few Tuarick words. *Ibid.* 83 He was dressed in a light blue cotton tobe, with a white muslin turban, the shawl of which he wore over the nose and mouth, in the Tuarick fashion. *Ibid.* App. 160 Under the government of the Tawarék. **1882** [see *Kabyle]. **1933** L. Bloomfield *Language* iv. 67 The *Berber* branch of Semitic-Hamitic..is represented today by various languages, such as *Tuareg* and *Kabyle*. **1955** *Sci. News Let.* 4 June 3 Called the 'People of the Veil', the Tuaregs are a nomadic people who live in the central part of the Sahara desert. **1973** *Times* 19 Mar. 6/3 The tracks..did lead to some Tuareg nomads, and to water. **1977** *Time* 3 Oct. 2/2 The peregrinations of the Tuareg in Niger, Mali and Upper Volta and the nomadic Masai in Kenya and Tanzania frighten their respective governments.

‖ **Tuatha Dé Danann** (tū^ə·hă dē da·năn). Also (*erron.*) **Tuatha de Danaan**, etc. [Ir., f. *tuatha*, pl. of *tuath* Tuath + *dé* + *Danann*, name of the mother of the gods (app. formerly gen. sing. of *Danu*, but later used as nom).] In Irish mythology, a people who inhabited prehistoric Ireland.
1682 P. Walsh *Prospect State of Ireland* f. b2, He has not the least mention of Tuatha-De-Danainn, though a powerful People. **1876** [see *Nemedian]. **1893** W. B. Yeats in *Bookman* May 43/1 The berries were the food of

the *Tuatha de Danaan*, or faeries. **1980** J. O'Faolain *No Country for Young Men* ii. 41 Ever hear of the prehistoric rulers of Ireland and how they fought their wars? The Tuatha De Danaan? They wrapped themselves in a cloud and withdrew into the hills..and turned into fairies.

tub, *sb.* Add: **1. c.** (Earlier example.)
1853 E. Clacy *Lady's Visit Gold Diggings Austral.* 116 Great wooden tubs are filled with the dirt and fresh water.
g. A tub-shaped carton, *spec.* one containing a portion of ice-cream; the contents of a carton.
1939 'G. Orwell' *Coming up for Air* iv. iii. 248, I watched the floats rocking up and down among the ice-cream tubs and the paper bags. **1955** 'C. Brown' *Lost Girls* xii. 139 You bought these ices from a fat woman holding out a podgy hand and saying, 'Two tubs, ducks?' **1981** P. Vansittart *Death of Robin Hood* iv. i. 183 The sallow ground was strewn with..ice-cream tubs, empty tins and ruined shoes.
2. (Earlier example.)
1776 H. Newdigate *Let.* in A. E. Newdigate-Newdegate *Cheverels* (1898) i. 11 To-night I can use my warm Bath..which I cannot at present do conveniently at home having neither Tub nor dressers.
6. d. A fire-engine. *U.S. slang.*
1864 *Student & Schoolmate* Jan. 3 The rope was only half manned and wishing to make myself useful..I joined the party in charge of the 'tub'. **1906** J. D. Lovett *Old Boston Boys* vii. 67 A boy without a 'tub', as they were called in the vernacular, was like a man without a country.
e. A bus; *to work the tubs*, to pick pockets on buses or at bus-stops. *slang* (chiefly *Underworld*).
1929 G. Dilnot *Triumphs of Detection* iv. 52 Snatches of their conversation..told that they were on their way to 'work the tubs'—in other words, to pick pockets at omnibus stopping-places. **1933** C. E. Leach *On Top of Underworld* x. 141 *Tub*, omnibus. **1974** P. Wright *Lang. Brit. Industry* x. 85 Inland transport comes along with *tubs* for buses.
10. tub-bass, a bass string instrument made from a tub; **tub garden**, an area containing plants grown in tubs; **tub-oarsman** (example); **tub-size** *v.*: hence **tub-sized** *ppl. a.*; **tub-wheel** (earlier example).
1958 P. Oliver in P. Gammond *Decca Bk. Jazz* i. 23 Parallels have been drawn between the tub-bass and the African earth-bow, but that the former is a folk memory of the other seems unlikely. **1974** N. Marsh *Black as he's Painted* i. 62 She spent a good deal of time in the tub garden at the back of the house. **1845** E. J. Wakefield *Adventure N.Z.* I. xi. 318 The common men have nothing to do but to ply their oars according to orders; except one, called the *tub oarsman*, who sits next to the tub containing the whale-line, and has to see that no entanglement takes place. *a* **1912** Tub-sized [see A.T.S. s.v. *A III]. **1967** E. Chambers *Photolitho Offset* xvi. 250 Surfaced-sized or tub-sized paper is made by passing the moist paper (15–18% moisture) through a trough containing starch solution. **1815–16** *Niles' Weekly Register* IX. Suppl. 182/2 Many mill owners have laid aside their tub wheels.

tuba¹. Delete ‖ and add: **2. b.** (Earlier example.)
1858 J. A. Symonds *Let.* Nov. (1967) I. 174 The stops of the organ were unusually fine—Tuba, [etc.].

tuba². Add: **1.** For *Dalbergia* substitute *Derris*. (Earlier examples.)
1839 T. J. Newbold *Straits of Malacca* II. xii. 189 The Malays sometimes resort to unfair means of securing the finny tribe, by inserting at low water the roots of the Tuba..into the holes and fissures of the coral reefs. **1890** [see *derrid, derride].
2. (Earlier example.)
1704 in A. & J. Churchill *Coll. Voy.* IV. 447/2 Their Wine or Liquor is drawn from the Palm, or Coco-Tree... The Poor put into it some Bark of Trees which give it a Colour, and a hotter Tast, and then it is call'd Tuba.

tubal, *a.* Add: **2.** *tubal ligation*, ligation of the Fallopian tubes, esp. as a method of sterilization.
1948 H. S. & R. J. Crossen *Operative Gynecol.* (ed. 6) 991/2 (Index), Tubal ligation. **1961** *Biol. Abstr.* XXXVI. 5391/2 (*heading*) Histological studies on the change of the adrenal cortex by means of the oophorectomy and tubal ligation in rabbits. **1975** *New Yorker* 29 Sept. 84/3 Mrs. Santana had intended to get a tubal ligation after Gabriel's birth.

tubbable (tʊ·bāb'l), *a.* [f. Tub *v.* + -able.] Of garments, etc.: washable in a tub; suitable for laundering in a tub or (domestic) washing-machine.
1929 *Chicago Daily Tribune* 20 May (Advt.), Tubbable Wearable Suitable Cotton! **1960** *News Chron.* Mar. 229 Synthetic fibres..permanently pleated and tubbable. **1982** *Sunday Express* 25 July 17/1 Children's clothes this week—cute, sharp and tubbable they come in easy-care cotton.

tubbily (tʊ·bili), *adv.* [f. Tubby *a.* + -ly².] In a tubby manner; with an appearance of tubbiness.
1924 D. H. Lawrence in M.M. *Mem. Foreign Legion* 12 He stuck his front out tubbily, like a bird. **1935** H. Nicolson *Let.* 4 Dec. (1966) 229 Winston rose tubbily and stretched out great arms. **1957** M. Spark *Comforters* ix. 220 He watched the movements of a young fat woman on a houseboat moored nearby... But she did emerge again, with a cup of tea. She drank it propped tubbily on the tiny bridge of the boat.

tubby, *a.* Add: **2.** Also, of general acoustic quality.
1940 *Chambers's Techn. Dict.* 869/2 *Tubby*.., characterised by reverberant booming for frequencies which are familiar when barrels are struck. **1962** A. Nisbett *Technique Sound Studio* 243 *Boomy*, subjective description of a sound quality which has resonances in the low frequencies, or a broad band of bass lift. Expressions with similar shades of meaning are tubby or, simply, bassy. **1981** *Popular Hi-Fi* Mar. 81/3 Drum sound was tubby on both decks.

tube, *sb.* Add: **I. 2. g.** *Electronics.* A sealed container, evacuated or gas-filled, containing two or more electrodes between which an electric current can be made to flow; *spec.* (*a*) a cathode-ray tube; (*b*) (chiefly *U.S.*) a thermionic valve. Freq. in *Comb.* with preceding sb., as *discharge, electron, picture, vacuum tube*, qq.v.
1859, etc. [see *vacuum tube 2]. **1898**, etc. [see *discharge *sb.* 3 b]. **1905** *Electrician* 16 June 335/1 The phosphorescent spot on the screen of the tube follows strictly any changes which occur in the strength of the field. **1915** *Ibid.* 21 May 241/2 In the X-ray tube..the space charge effects are very much exaggerated. **1922** C. W. Taussig *Bk. of Radio* ix. 111 The tubes used are 5 watt transmitting tubes. **1940** H. M. Watson et al. *Understanding Radio* v. 223 As you experiment with this one-tube set, you will hear many stations faintly. **1947** R. Lee *Electronic Transformers & Circuits* i. 3 The limitations which inhere in transformers often influence the choice of amplifier tubes. **1973** G. J. King *Newnes Colour Television Servicing Man.* I. i. 29/2 The output direct from the tripler is too high an impedance to accommodate the normal beam current swings of the tube without serious voltage fluctuations. **1981** Nashelsky & Boylestad *Devices* iv. 128 Production rose from about 1 million tubes in 1922 to about 100 million in 1937.
h. *inner tube*: see *inner *a.* 1 i. Also *ellipt.*
1894 Albemarle & Hillier *Cycling* 471 The outer arch is removed, the inner tube carefully examined, the hole discovered—if necessary, by inflating the tube and immersing it in water. **1904** A. B. F. Young *Complete Motorist* (ed. 2) xi. 246 When the tube and cover are both in place..the air chamber is inflated by means of a pump. **1979** *United States 1980/81* (Penguin Travel Guides) 367 You can buy tubes..at gas stations and stands along the route.
i. A telephone. Cf. sense 7 a and Speaking-tube. *colloq.* or *slang.*
[**1873** C. M. Yonge *Pillars of House* II. xiii. 38 Mr. Underwood breathed through a mysterious tube, and Edgar appeared.] *c* **1899** C. H. Chambers in M. R. Booth *Eng. Plays of 19th Cent.* (1973) III. 401 (*Rings off, and hangs up tube.*) That is another mistake—that telephone. **1959** *Esquire* Nov. 70 *Tube*, can be television, but usually telephone. Example: Buzz me on the tube. Call me up.
j. A type of skate (see quot.).
1923 E. Jessup *Snow & Ice Sports* 220 The 'tubes' are a comparatively recent departure in skate design... The blade..is set in a long hollow tube. Similar but wider tubes support the heel and front plates.
k. *the tube*, television, a television set; also, *the boob tube* [see *boob *sb.* 3]; cf. *the box* s.v. *box *sb.*² 3 j. *colloq.* (orig. and chiefly *U.S.*).
1959 [see sense 2 i above]. **1965** *Sunday News* (N.Y.) 4 Oct. 2 She..is making a name for herself as a singer on the tube. **1966** *Current Slang* (Univ. S. Dakota) Fall 1 Let's catch the late show on the boob tube. **1972** *Observer* 31 Dec. 24/1 Turning to the tube in order to redress the balance with a spot of the old festive vulgarity. **1977** M. French *Women's Room* (1978) ii. 115, I sit and watch the stupid boob tube. **1979** *Radio Times* 11–17 Aug. 19/1 'I see you on the tube a lot,' an American said to me recently in a pub. 'Oh really,' I replied, 'the Piccadilly line?' 'No,' he said, 'the *tube*, the dream machine.'
l. *down the tube(s)*, lost, finished, in trouble; freq. in *to go down the tube(s)* = *to go down the drain* s.v. *drain *sb.* 1 e. *slang* (orig. *U.S.*).
1963 *Amer. Speech* XXXVIII. 168 To fail to pass an examination:..*go down the tubes*. **1975** *New Yorker* 5 May 32/1 It would be ludicrous to end on a note of Chris going down the tube. **1977** J. D. MacDonald *Condominium* xii. 122 We've got too many goodies tucked into the Marliss Corporation to take a chance of it going down the tube. **1982** *Listener* 16 Dec. 35/3 The smile on Sir Freddie's face the week before it was revealed that he was down the tubes to the extent of something over £270 million was the smile of a consummate actor.
m. A bottle or can of beer. *Austral. colloq.*
1969 *Listener* 24 Apr. 588/2 This..extrovert chunders..his way through the kangaroo valley of Earl's Court..buoyed up by innumerable tubes (bottles) of Foster's Beer. **1980** R. Hill *Killing Kindness* xx. 187 'What do you want to do?'..Mow my lawn and then cool off with a tube of lager, thought Pascoe.
4. b. A cigarette. *slang.*
1946 P. Larkin *Jill* 16 Christopher, extending his silver cigarette[-case], said with an uneasy smile: 'Tube for anyone?' **1975** *High Times* Dec. 11/2 (Advt.), Filter tipped tubes give a smoother smoke to the very end.
5. a. (Later example.)
1897 Kipling in *Times* 17 July 13/6 Heathen heart that puts her trust In reeking tube and iron shard.
7. b. Also, any tunnel or tubular bridge for a railway. (Earlier example.)
1847 Queen Victoria *Jrnl.* 15 Aug. (1868) 72 We passed the famous *Swilly Rocks*, and saw the works they are making for the tube for the railroad.
II. 9. c. The penis. *slang.*
1922 Joyce *Ulysses* 750, I suppose the people gave him that nickname [*sc.* Mr de Kock] going about with his tube from one woman to another.
11. b. *Surfing.* The hollow curve of a breaking wave.
1962 *Austral. Women's Weekly* 24 Oct. (Suppl.) 3/4 *Tube*, the area of a dumping wave between the breaking crest and the trough. **1968** *Surfer Mag.* Jan. 89/1 You get back inside the tube and the whole tunnel is glowing. **1979** *Nat. Geographic Mag.* Feb. 235 (*caption*) Shootin' the tube, a surfer threads the eye of a breaker.
12. b. Tube Alloys, the code name of a section of the Department of Scientific and Industrial Research formed in 1940 and concerned with research into the production of an atomic bomb; **tube counter** *Physics*, the now usual form of Geiger counter, as contrasted with the point-counter; **tube curare**, curare kept or transported in bamboo tubes; **tube-fed** *a.*, fed, sometimes forcibly, by passing nourishment through a tube into the stomach; so **tube-feed, tube-feeding** *vbl. sb.*; **tube journey**, a journey in a tube, *spec.* a journey by underground railway; **tube-lift**, a lift for the conveyance of passengers from street-level to an underground railway or vice versa; **tube map**, a map of an underground-railway system; **tube shelter**, an underground tube station used as an air-raid shelter; also *attrib.*; **tube skate** = sense 2 j above; **tube sock**, an elasticized sock with no shaping for the heel; **tube steak** *slang*, a hot dog, a frankfurter; **tube top**, a women's close-fitting elasticated top reaching from the waist to under-arm level; **tube-worm** (later examples).
1942 J. Anderson in M. Gowing *Britain & Atomic Energy 1939–45* (1964) App. iii. 437 When you asked me to take over the supervision of work on the project known as 'Tube Alloys', it was contemplated that..a full scale production would be expected in this country. **1945** W. S. Churchill *Victory* (1946) 221 Imperial Chemicals Industries Limited agreed to release Mr. W. A. Akers to take charge of this directorate, which we called, for purposes of secrecy, the Directorate of 'Tube Alloys'. **1978** R. V. Jones *Most Secret War* xxxv. 309 The British 'Tube Alloys' project, as our own nuclear bomb effort was called. **1930** Tube-counter [see *coincidence 7 a]. **1938** R. W. Lawson tr. *Hevesy & Paneth's Man. Radioactivity* (ed. 2) i. 17 We shall only discuss the two [counters] that are most important.., viz. the point-counter and the tube-counter, both of which were introduced by Geiger. **1898** *Jrnl. Chem. Soc.* LXXIV. i. 284 Paracurara, or tube curara, is imported in bamboo tubes, and is the variety now usually met with in commerce. **1974** *Encycl. Brit. Micropædia* III. 300/3 Preparations have been classified according to the containers used for them: pot curare in earthenware jars; tube curare in bamboo; and calabash curare in gourds. **1909** *Westm. Gaz.* 23 Oct. 3/2 Tube-fed Suffragettes. **1964** *Lancet* 26 Dec. 1349/2 Most babies were getting their first tube-feed within 2 hours of birth. *Ibid.* 1351/1 Tube-feeding is a very much simpler procedure. **1974** *Brit. Med. Jrnl.* 19 Jan. 108/1 The ethical problems of prolonged tube-feeding. **1980** *Ibid.* 21 June 1493/1 More work is needed to assess the relative merits of these proprietary diets compared with the tube feeds prepared in hospitals. **1866** Geo. Eliot *Felix Holt* I. 2 Posterity may be shot, like a bullet through a tube, by atmospheric pressure from Winchester to Newcastle... The tube-journey can never lend much to picture and narrative. **1972** C. Fremlin *Appointment with Yesterday* i. 5 No one could guess..that there is one..that has left its identity behind not just for the duration of the tube journey, but for ever. **1915** E. Wallace *Man who bought London* ii. 19 The 'tube' lift was crowded. **1935** E. Farjeon *Nursery in Nineties* 428 Once she had ventured into a tube-lift—'But never again, my dear Eleanor!' **1962** J. Braine *Life at Top* xxiii. 256, I used to have a Tube map on my bedroom wall when I was at College? **1977** *Times* 15 Nov. 17/8 The Bakerloo line on the tube map. **1942** N. Balchin *Darkness falls from Air* xi. 196 We went..by tube... I wanted to see how the tube shelter business was working out. **1943** C. E. Milburn *Diary* 4 Mar. (1979) 170 There was a terrible accident at a tube shelter last night after the sirens had sounded in London. **1962** *Times* 23 Jan. 13/4 Henry Moore's Tube-Shelter drawings. **1923** E. Jessup *Snow & Ice Sports* 230 'Tube' skates. **1975** *Kingston* (Ontario) *Whig-Standard* 19 Dec. 12/3 As a reporter who has covered various classifications of professional hockey since the invention of tube skates, it is my considered opinion that Robert Earle Clarke is one of the most adept ankle-tappers in the history of the game. **1976** *N.Y. Times Mag.* 18 Jan. 4/2 Monday morning I bought a striped blue pair of training shoes,..tube socks, a sweatband and a book called 'On the Road to Self Improvement: The Joy of Jogging'. **1963** *Amer. Speech* XXXVIII. 272 Frankfurters are *tube steaks*. **1978** *Boston Globe* 15 Aug. 1/1 The food isn't bad which is mainly tube steaks (hot dogs). **1974** *News & Reporter* (Chester, S. Carolina) 24 Apr. 4-c (Advt.), Calico-print elasticized tube tops! **1984** *New Yorker* 23 Apr. 42/1 She was wearing khaki shorts and a lime-green tube top. **1928** Russell & Yonge *Seas* viii. 194 The case of the concealed animals, such as the Piddock or the Tube-worm,..presents almost equal difficulties. **1981** *Sci. Amer.* May 90/3 Occasionally a crab would climb the stalk of a tube worm, presumably to attack its plume.

tube, *v.* Add: **4.** *trans.* and *intr.* To fail, to perform poorly (in). *U.S. slang.*

1966 *Current Slang* (Univ. S. Dakota) Summer 5 *Tube*, to fail; to do a poor job. College students, both sexes. Midwest. 'He tubed every test last week.' **1979** *N.Y. Times Mag.* 30 Sept. 10/3 In time, surfers used the verb 'to tube' to mean 'to do poorly'.

tubectomy (tiŭbe·ktŏmi). *Surg.* [f. Tube *sb.* + *-ectomy.] = *salpingectomy* s.v. *salpingo-.
1925 Stedman *Med. Dict.* (ed. 8) 1057/2 *Tubectomy*, salpingectomy. **1975** *Lancet* 27 Sept. 567/1 (*heading*) Tubectomy by paraprofessional surgeons in rural Bangladesh. **1977** *Time* 7 Feb. 45/3 The government is pushing birth control programmes: 6 million people underwent vasectomies or tubectomies in six months last year.

tubed, *ppl. a.* Add: **2.** Of a race-horse: having a metallic tube inserted in the air-passage.
1925 W. & A. J. Day *Racehorse in Training* 13 Tubed horses are rather a nuisance, as the tube should be taken out and disinfected occasionally. **1955** *Times* 13 May 3/3 In the absence of Defender, the tubed steeplechaser Remy was made favourite for the long-distance race, against four opponents.

tubercular, *a.* Add: **B.** *sb.* A person having tuberculosis.
1952 *Sun* (Baltimore) 2 July 14/3 Maryland has three State institutions for the chronically sick, four hospitals for tuberculars, and four for the mentally ill. **1980** I. Hunter *Malcolm Muggeridge* x. 177 In 1949 Orwell left the Isle of Jura to enter a convalescent home for tuberculars at Cranham in Gloucestershire.

tuberculin. Add: **b.** *attrib.* and *Comb.*, as *tuberculin reaction*, † *treatment*; **tuberculin test**, the injection of tuberculin, usu. intradermally, as a test for the past or present existence of tubercle bacilli in the individual; also as *vb. trans.*; hence **tuberculin-tested** *ppl. a.* (see quot. 1950).
1906 *Review of Reviews* Sept. 366 [It] showed no tuberculine reaction. **1955** *Sci. News Let.* 1 Oct. 221/1 The tuberculin test is a skin test. A tuberculin reaction means that tuberculosis germs have invaded the person's body and sensitized it to proteins of the TB germ. **1900** Dorland *Med. Dict.* s.v. *Tests*, Tuberculin-test. **1908** *Med. Annual* 662 Tuberculin Test.—This test, prepared by the Pasteur Institute of Lille, claims to be diagnostic of tuberculosis in man. **1950** J. G. Davis *Dict. Dairying* 721 Three forms of the tuberculin test have been evolved —the ophthalmic, the subcutaneous and the intradermal. **1966** Tuberculin test [see *intradermic* adj. s.v. *intra- 1]. **1937** *Amer. Rev. Tuberculosis* XXXV. 598 A classification of the 56,688 persons tuberculin-tested according to age disclosed the fact that boys and girls between ten and nineteen years of age comprised almost two-thirds of the groups reported. **1936** Tuberculin-tested [see *certified *ppl. a.* b]. **1950** J. G. Davis *Dict. Dairying* 721 *Tuberculin-tested milk*, milk which is derived solely from a herd of tuberculin-tested cows under licence and kept pure and unmixed with other non-tuberculin-tested milk. **1982** M. Young *Elmhirsts of Dartington* xi. 277 Nielsen admitted to disposing of a small quantity of milk as Tuberculin Tested when it was not. **1908** *Med. Annual* 47 Roemisch thinks that tuberculin treatment gives good results in a class of chronic phthisical patients, in whom the ordinary treatment with fresh air and rest at first gives marked improvement.., but after a time no further improvement can be obtained. **1912** *Nature* 12 Dec. 427/2 The mortality of the phthisical under sanatorium and tuberculin treatments.

tuberculo-. Add: **tube:rculo-pro·tein**, protein from the tubercle bacillus, *Mycobacterium tuberculosis*; **tube:rculosta·tic** *a.* [*-static], inhibiting the multiplication of the tubercle bacillus, *Mycobacterium tuberculosis*.
1912 Hamman & Wolman *Tuberculin in Diagnosis & Treatm.* iii. 231 Tuberculocidin represents the attempts by Klebs to purify tuberculin by alcohol and bismuth precipitation... He also produced tuberculo-protein and tuberculo-sozin. **1954** S. Duke-Elder *Parsons' Dis. Eye* (ed. 12) xvi. 218 In many cases the presence of sensitivity can be demonstrated by skin-reactions to tuberculo-protein or streptococci. **1945** *Jrnl. Immunol.* L. 159 Sera, either from rabbits sensitized with a purified tuberculin protein preparation or from tuberculous rabbits, possessed tuberculostatic activities in the chick membrane. **1971** *Nature* 4 June 301/1 We recently proposed structure (I) for the tuberculostatic antibiotic viomycin.

tuberculoid, *a.* Add: **a.** Also *Path.* (Example.)
1974 *Trypanosomiasis & Leishmaniasis* (Ciba Foundation Symp.) 163 In localized forms we often saw epithelioid or tuberculoid nodules, surrounded by a thick infiltration of lymphoid cells.
b. *Path.* Designating one of the two principal forms of leprosy, characterized by a few well-defined lesions similar to those of tuberculosis, often with anæsthesia.
1938 *Leprosy Rev.* IX. 20 No progress can be made.. unless it is frankly admitted that at the Manila Conference [in 1931] the significance and extent of the tuberculoid phases of leprosy were not fully appreciated. *Ibid.*, Anaesthesia is by no means the sole preserve of tuberculoid leprosy. **1948** E. Muir *Man. of Leprosy* ix. 49 The term 'tuberculoid' is given on account of the histological resemblance of this type to chronic lesions in tuberculosis. It is a somewhat unfortunate term, as it is often confused with 'tubercular', the word formy used for the severeerl form, now called lepromatous. **1971** [see *lepromatous* adj. s.v. *leproma].

tuberculome. Add: More commonly as mod.L. **tuberculo·ma** (pl. **-omas, -omata**).
1897 *Lippincott's Med. Dict.* 1076/1 Tuberculoma. **1908** E. A. Peters in T. N. Kelynack *Tuberculosis in Infancy & Childhood* vii. 55 Seward has collected a hundred cases of other forms of tuberculosis of the nose—*e.g.*, tuberculoma and ulcer—from a great many sources. **1954** S. Duke-Elder *Parsons' Dis. Eye* (ed. 12) xxx. 508 Intracranial tumours (including neoplasms and such lesions as tuberculomata) produce two sets of symptoms. **1974** J. D. Maynard in R. M. Kirk et al. *Surgery* x. 215/1 A tuberculoma, a round focus over 1 cm in diameter, which has failed to disappear after adequate chemotherapy.

tubero-. Add: **tu:bero-hypophyseal, -ial** *a. Anat.* = next; **tu:bero-infundi·bular** *a. Anat.*, relating to the tuber cinereum and the infundibulum; *spec.* applied to a tract of nerve fibres that includes those running from the infundibular and related nuclei of the hypothalamus to the infundibulum.
1962 Tubero-hypophyseal [see *tubero-infundibular* below]. **1969** Truex & Carpenter *Human Neuroanat.* (ed. 6) xix. 494/2 The connections of the hypothalamus with the posterior lobe of the hypophysis are well established for man... A smaller bundle [of nerve fibres], the tuberohypophysial tract, is contributed by the medial cells of the tuber cinereum. **1962** J. Szentágothai in J. Szentágothai et al. *Hypothalamic Control of Anterior Pituitary* ii. 45 This system corresponds to the tubero-hypophyseal tract of Spatz (1951) and Nowakowski (1951). Since it terminates in the median eminence and in the proximal part of the stalk, it would be better to term it tubero-infundibular tract. **1980** *Gray's Anat.* (ed. 36) 966/1 The tuber cinereum around the base of the infundibulum is raised to form a median eminence which ..is superficially marked by a shallow tubero-infundibular sulcus.

tuberose, *a.* Add: *tuberose sclerosis* = *tuberous sclerosis* s.v. *tuberous *a.* 2.
1933 W. R. Brain *Dis. Nervous Syst.* ix. 487 Beyond the fact that tuberose sclerosis is due to a congenital abnormality..little is known about its aetiology. **1963** *Lancet* 12 Jan. 67/2 When in the Royal Air Force during the war I saw a man with tuberose sclerosis it was not unexpected to find in his son evidence of the same, genetically determined, disorder.

tuberous, *a.* Add: **2.** *tuberous sclerosis*, a rare, hereditary, usu. fatal disease in which there are hard swellings on the brain and elsewhere, with symptoms including mental deficiency and epilepsy.
1898 C. K. Mills *Nervous Syst. & its Dis.* v. 503 Tuberous Sclerosis. An interesting form of sclerotic pseudotumor is sometimes found in the brain, especially among the idiotic..and epileptic. **1954** S. Duke-Elder *Parsons' Dis. Eye* (ed. 12) xviii. 313 Tuberous Sclerosis (Bourneville's Disease), occurring in young individuals, is associated with nodular lesions in the central nervous system and the skin, particularly on the face (adenoma sebaceum). **1975** Swaiman & Wright *Practice of Pediatric Neurol.* II. xxxii. 739/1 Tuberous sclerosis is best identified by cutaneous signs associated with neurologic deterioration and myoclonic seizures.

tubi-. Add: **tu·bifex** [mod.L. (J. B. P. A. de M. de Lamarck *Hist. Nat. Animaux sans Vertèbres* (1816) III. 224)], a red oligochæte worm of the genus of this name, found in mud at the bottom of rivers or lakes and used as live food for aquarium animals; also *attrib.*; **tubi·ficid** [mod.L. family name *Tubificidæ*: see prec.], an aquatic oligochæte worm of the family Tubificidæ; also as *adj.*
1952 J. Clegg *Freshwater Life Brit. Isles* xi. 156 *Tubifex* worms lay their eggs in oval capsules and breed very rapidly. **1972** *Sci. Amer.* Oct. 115/1 Among the more commonly viewed organisms are tubifex worms. **1976** *Daily Tel.* 3 Dec. 18 The river is now almost too clean. It has led to a decrease in the tubifex worm population, on which waterfowl feed. **1950** M. Gordon in E. J. Farris et al. *Care & Breeding Lab. Animals* xiv. 382 Tubificids are usually most plentiful about one-half mile downstream from the spot where raw pollution is dumped. **1971** *Nature* 26 Feb. 596/1 Tubificid worms..are typical bottom dwelling animals of all fresh-water lakes and reservoirs. **1971** *Oxf. Bk. Invertebrates* 114/1 Tubificids will emerge, tail upwards, to undulate their bodies gently in the water for respiratory exchange. **1978** *Nature* 17 Aug. 644/2 Chironomid midge larvae and tubificid worms have replaced a previously rich fauna.

tubify (tiū·bifəi), *v. nonce-wd.* [f. Tube *sb.* + -ify.] *trans.* ? To give a tubular form to.
1928 D. H. Lawrence *Lady Chatterley's Lover* xviii. 347 'I don't think Vulcan has a figure that interests me.' 'Not even if it was tubified and tittivated up?'

tubing, *vbl. sb.* Add: **b.** (See quot. 1976.) *U.S.*
1975 *Newsweek* 3 Feb. 69 But the big rage of the ski year—and the most painful—is a pastime called 'tubing'. For experts, the idea is to take a running start and then execute a belly-flop onto an ordinary inflated inner tube. **1976** *Webster's Sports Dict.* 464/1 *Tubing*, the sport or pastime of riding down a river or of sliding down a snow-covered slope on an inflated automobile inner tube. **1979** *United States 1980/81* (Penguin Travel Guides) 367 *Tubing*—Arizona's most popular summer sport. On any given weekend, as many as 20,000 residents strap beer-filled ice chests and their behinds to old inner tubes and float down the five or ten miles of free-flowing Salt River below Saguaro Lake.

Tubism (tiū·biz'm). [f. Tube *sb.* + -ism, after *Cubism.] A style of painting characterized by cylindrical and other mechanistic forms, *spec.* that developed by the French artist Fernand Léger (1881–1955). So **Tu·bist** *a.*
1923 *Weekly Dispatch* 11 Feb. 9 He [*sc.* Gregory Brown] first took up applied art, working in metal. Turning his attention to poster work, he..became one of the pioneers of what was jocularly known as the Tubist School. **1955** *Times* 18 Aug. 13/1 'Tubism' serves well to describe the brand of Cubism that Léger made peculiarly his own. Out of the several possibilities indicated by Cézanne's famous remark about the geometrical basis of natural forms he selected the cylinder, and though he admitted other forms, a tubular tendency runs through most of his compositions. **1960** *Twentieth Cent.* Dec. 526 Léger's 'tubist' portraits. **1978** N. Gosling *Paris 1900–1914* 165 Fernand Léger.. attracted by the Cubist experiments..developed a variety of his own, based on interlocking cylinders—a style which was nicknamed 'Tubism'.

tubocurarine (tiūbo,kiū^ə·rārīn). *Pharm.* [ad. G. *tubocurarin*, f. *tubocurare* tube curare (see Tubo-, Curare): see -in¹.] An isoquinoline alkaloid that is the active ingredient of tube curare and whose chloride, $C_{37}H_{42}Cl_2N_2O_6$, is used as a muscle relaxant.
1898 *Jrnl. Chem. Soc.* LXXIV. i. 284 Paracurarine (tubocurarine). **1935** *Nature* 23 Mar. 470/1 The quaternary alkaloid tubocurarine..has now been crystallised for the first time. **1973** *Reader's Digest* Apr. 206/1 They paralysed him with 70 milligrams of tubocurarine chloride. **1977** *Lancet* 19 Mar. 650/2 Three drugs used in clinical anæsthesia consistently cause arterial hypotension in patients with essential hypertension—halothane, *d*-tubocurarine, and verapamil.

tuboplasty (tiū·boplæsti). *Surg.* [f. Tubo- + -plasty.] The surgical repair of one or both Fallopian tubes.
1961 *Obstetr. & Gynecol.* XVII. 504/2 The combination of tuboplasty and metroplasty was indicated as an attempt to overcome sterility factors and minimize the possibility of subsequent miscarriage. **1976** *Daily Colonist* (Victoria, B.C.) 14 Apr. 2/3 A few pregnancies have resulted from tuboplasty procedures—as for unblocking a tube —but generally the results here, too, have been disappointing. **1977** *Lancet* 5 Feb. 284/2 Regular ovulation and a positive postcoital test were confirmed in each patient before tuboplasty.

tub-thumper. Add: So also (as a back-formation) **tu·b-thump** *v. intr.*; **tu·b-thu:mpery**.
1920 'Sapper' *Bulldog Drummond* iv. 117 The sort of type one sees tub-thumping in Hyde Park. **1927** *Observer* 28 Aug. 11/4 Fanny Hawthorne's refusal to be made an honest woman..may still be regarded as advanced in ethics, even if its expression is not without a suspicion of tub-thumpery. **1933** Dylan Thomas *Let.* (1966) 75 And thank you for..heeding my dogmatic tub-thumpery. **1934** E. Pound *ABC of Reading* 113 A bad reader of fourteeners is almost certain to tub-thump.

tubular, *a.* Add: **1.** *tubular bells*, a series of tuned metal tubes of graded length vertically suspended and struck by hammers; *tubular goods* (Oil Industry) (see quot. 1922); *tubular steel*, steel tubing, esp. as used in the manufacture of furniture; also *attrib.*
[**1884** J. Harrington *Provisional Spec.* 14,270 (Patent Office) (1885) 1 Instead of employing the costly gongs or bells hitherto employed, I employ metallic tubes which I suspend vertically by means of catgut, cord or other suitable material... I arrange the hammers so as to strike the same at their upper parts above the point of suspension thereof.] **1919** A. T. Bassett *S. Barnabas', Oxford* iii. 25 The original bell of S. Barnabas'..did service until 1890, when a set of tubular bells was hung in the tower, and a chiming machine added. **1922** D. T. Day *Handbk. Petroleum Industry* I. 300 The term 'tubular goods' generally covers all classes of pipe, casing and tubing used in drilling or operating oil or gas wells, and comprises the following distinct types: casing, tubing, drive pipe, line pipe and rotary drill pipe. **1930** *Melody Maker* Jan. 69/1 Tubular bells are often considered 'a bit of a bore' by drummers. **1933** *Archit. Rev.* LXXIV. 78/3 Four double tubular-steel legs shaped like hair-pins. **1946** R. Graves *Poems 1938–45* 35 Among box-files and tubular steel chairs. **1957** *Times* 2 July (Agric. Suppl.) p. viii/3 The gate and posts are made of tubular steel. **1962** *B.S.I. News* May 19/2 Glockenspiel, xylophone, tubular bells, celesta. **1981** *Times* 3 July 14/7 Marcel Breuer, the Hungarian-born architect..created the first tubular steel furniture..and in 1926 he equipped the new Bauhaus buildings with furniture of this type.
4. c. *Ophthalm.* Applied to a visual field which is restricted to a small area surrounding the fixation point; *tubular vision* = *tunnel vision* s.v. *tunnel *sb.* 5.
1903 Hansell & Sweet *Text-bk. Dis. of Eye* xviii. 488 When his sight had improved the visual fields were still contracted.., showing the so-called 'tubular' field of hysteria. **1927** H. M. Traquair *Introd. Clinical Perimetry* v. 49 A special form of concentric contraction is the 'tubular' field. The depression is severe and involves the whole field with the exception of an area surrounding the fixation point, producing a great and often extreme

contraction... Such fields are obviously of functional origin. *Ibid.* xii. 222 Hysteria. The typical field change.. is concentric contraction... The field is..tubular in type, a form which is necessarily of subjective origin. **1934** *Amer. Jrnl. Ophthalmol.* XVII. 384/2 (*heading*) Transient tubular vision in postencephalitic Parkinson's disease. **1956** *New Gould Med. Dict.* (ed. 2) 1319/2 *Tubular v*[*ision*], a hysterical phenomenon in which the constricted visual field defies the laws of physical projection and maintains a uniform small size..: popularly called gun-barrel v[ision]; tunnel v[ision].

B. *ellipt.* as *sb.* **1.** = *tubular bridge.*

1861 A. J. Symonds *Let.* 9 Aug. (1967) I. 303 We took a nice walk..to Bangor. We saw 2 trains go through the Tubular—one each way.

2. *pl.* = *tubular goods.*

1975 *North Sea Background Notes* (Brit. Petroleum Co.) 27 Most of each top deck will be occupied by a skid-mounted, electrically-driven drilling rig with appropriate storage for tubulars. **1979** *Shell Trade in Eastern Europe* (Shell Internat. Petroleum Co.) 5 Shell companies' purchases from countries in Eastern Europe include oil, chemicals and some metals and materials and equipment such as tubulars, rotary drilling hose, and items for sale in Shell filling-stations.

tubularia (tiūbiŭlē°·riă). [See Tubularian *a.* and *sb.*] = Tubularian *sb.*

1912–13 A. S. Pringle-Pattison *Idea of God* (1917) iv. 72 The Tubularia, a kind of sea-anemone, re-grows its flower-like head. **1924** *Glasgow Herald* 19 June 258/4 The tubularia and the sea-urchin. **1971** *Oxf. Bk. Invertebrates* 8/2 Tubularia, richly-coloured, with long, drooping polyps, is common locally and found at low tide under rocky overhangs.

tubularian, *a.* (Earlier example.)

1856 Geo. Eliot *Jrnl.* 8 May–26 June in *Lett.* (1954) II. 243 G...brought home several varieties of Polyps..—Tubularian, Plumularian and Sertularian.

tubulin (tiū·biŭlin). *Biochem.* [f. Tubule + -in[1].] Either or both of two similar proteins that are the main constituent of microtubules.

1968 H. Mohri in *Nature* 16 Mar. 1054/1 We believe that the microtubule constituent is a different protein, for which we propose the name 'tubulin'. **1977** *Jrnl. Protozool.* XXIV. 4/1 Microtubules are composed of 2 subunits, termed α and β tubulins, which form dimers and then polymers that are essentially infinitely long. **1978** *Bio Systems* X. 93/1 Tubulin..can undergo self-assembly in the absence of other macromolecules to form a microtubule. **1982** J. F. Van Pilsum in T. M. Devlin *Textbk. Biochem.* xxi. 1042 Tubulin..comprises about 14% of the total protein found in mammalian brain.

T.U.C. (tī,yū,sī). Also **TUC.** Abbrev. of *trade(s) union congress* s.v. *Trade-union, trades-union b.

1910 W. J. Davis *Brit. Trades Unions Congress* I. 105 The Chairman of the Parliamentary Committee presented to the Secretary..an ornate address... The monogram 'T.U.C.'..with the motto..were..beautifully inscribed. **1926** *Manch. Guardian* 4 May 12/4 His Majesty's Government..before it can continue negotiations, must require from the T.U.C...an immediate and unconditional withdrawal of the instructions for a general strike. **1947** *Radio Times* 2 May 16/3 George Woodcock, Assistant General Secretary of the T.U.C. **1955** *Times* 2 May 20/1 It is extraordinary that the T.U.C. should actually be advocating an increase in the tax on distributed profits. **1976** F. Zweig *New Acquisitive Society* ii. i. 81 Three decades of incomes policies have enormously enhanced the status of the TUC and unions at large.

tuchas, var. *tochus.

‖ **tuchun** (dudʒūn, tū·tʃun). *Hist.* Also **Tuchun, Tu Chün.** [Chinese *dūjūn*, f. *dū* govern + *jūn* military.] In China at the time of the Three Kingdoms, the title of a military leader; later, in the early years of the Republic of China, the highest military leader in a province; a warlord.

1917 S. Couling *Encycl. Sinica* 213/2 During the disturbances of the last few years it has frequently happened that the post of Governor has been combined with that of 'Tu Chün'..or Military Governor. This latter is the principal military official of the province. **1920** *Nineteenth Cent.* Dec. 942 Nowadays, a *tuchun* or military governor, ..will address the foreigner in a patchwork of stilted literary phrases jumbled together. **1943** J. T. Pratt *War & Politics in China* xii. 194 After the death of Yuan Shih Kai China for several years presented a sorry spectacle of politicians quarrelling and Tuchuns waging civil war. **1977** J. Cleary *High Road to China* i. 15 He was a Confucian, but he had a more immediate master, the *tuchun*, the war lord, in Hunan.

Hence **tu·chunate,** the rule of tuchuns, the office of tuchun.

1923 *Times Lit. Suppl.* 23 Aug. 558/1 At present the Tuchuns control their respective armies and are giving endless trouble. It would not, however, be sufficient to abolish the Tuchunate. A national army is necessary.

tuchus, var. *tochus.

tuck, *sb.*[1] Add: **1.** Also *fig.*

1878 'Mark Twain' in *Atlantic* Jan. 17/2 We had an iron-clad chicken... He ought to have been put through a quartz mill until the 'tuck' was taken out of him. **1910** *N.Y. Evening Post* 10 Nov. 1 The sight of a wounded man lying on the pavement seemed to take the tuck out of the mob

7*. In diving, gymnastics, etc., (the adoption of) a tuck position (see sense 8 below). Also, in downhill skiing, a squatting position (see quot. 1976).

1951 *Swimming* (Eng. Schools Swimming Assoc.) v. 81 The seat is drawn up and the head dropped slightly forward on the tuck, causing the body to spin. **1956** Kunzle & Thomas *Freestanding* vi. 81 The tuck and open out, as in the backward somersault, should be sharp and distinct movements. **1964** *Trampolining* ('Know the Game' Series) 31/1 It is better to learn the action slowly and then the tuck can be added later for effect and for faster rotation. **1976** *Webster's Sports Dict.* 464/1 *Tuck*.., a position in which the skier squats forward and holds his ski poles under his arms and parallel to the ground that is usually used to minimize wind resistance in downhill racing. **1981** 'E. Lathen' *Going for Gold* xvii. 186 There was..no discontinuity between being earthborne and airborne, no jerking resolution of the hunched-over tuck into the aerial float high over the heads of the spectators. *Ibid.* xxi. 232 Tilly..hunched into the tightest tuck that Dick had ever seen, increased her speed to flat-out downhill velocity.

8. tuck box, a box for storing eatables etc., esp. at a boarding-school (see sense 6); **tuck-comb** *U.S.* = *tucking-comb* s.v. Tucking *vbl. sb.*[1] 5; **tuck position,** in diving, gymnastics, etc., a position in which the thighs are pulled close to the chest, the knees bent, and the hands clasped round the shins; **tuck-stitch,** a stitch used in making a tuck; also *attrib.*; so **tuck-stitched** *a.*

1934 I. W. Hutchison *North to Rime-Ringed Sun* xviii. 207 Tuck-boxes were then opened and supper cooked and demolished. **1978** G. Greene *Human Factor* ii. ii. 70, I used to steal out at night from my dormitory and take him tins of sardines from my tuck-box. **1824** Tuck comb [see *side-comb]. **1870** E. Eggleston *Queer Stories* viii. 63 Sukey's way of doing up her hair in a great knot, behind, with an old-fashioned tuck comb. **1931** *Morning Post* 7 Aug. 14/2 All you have to do is hang on to your 'tuck', or 'balled-up' position a little longer. **1964** *Trampolining* ('Know the Game' Series) 32/2 Allow knees to bend to give a loose tuck position here and this helps to speed rotation. **1974** *Encycl. Brit. Macropædia* XVII. 864/1 In the tuck position, the body is gathered tightly into a ball with the hands grasping the shins firmly. **1926** J. Chamberlain *Hosiery, Yarns & Fabrics* vi. 121 The tuck-stitch is a defect in the plain fabric, but if produced systematically, forms many classes of designs. **1971** *Guardian* 7 Sept. 9/1 Tuck-stitch slipover vest in lambswool. **1922** Joyce *Ulysses* 171 In tuckstitched shirt sleeves.

tuck, *v.*[1] Add: **8.** Freq. with *away*; also *fig.*; also, to hit (a ball) to the desired place.

1912 [in Dict.]. **1936** J. Buchan *Island of Sheep* v. 99 My first business must be to tuck him away comfortably somewhere out of the road. **1958** *Observer* 6 July 24/4 There was greater punch in Miss Gibson's game once she had the first set safely tucked away. **1959** *Times* 29 May 4/6 His low forehand, as he tucks the ball away, is a special weapon of execution. **1966** *Listener* 12 May 702/3 What a pity that it should be tucked away into that most unlikely of all listening hours, the end of a Saturday evening. **1977** *Times* 7 Feb. 7/3 Tueart..outpaced a scattered defence and efficiently tucked away a rebound after Shilton had superbly blocked his first attempt.

tu·ck-away, *a.* [f. the vbl. phr. *to tuck away*: see Tuck *v.*[1] 8 in Dict. and Suppl.] That may be tucked away.

1935 *Sun* (Baltimore) 9 Nov. 1/3 The clipper's spick-and-span tuckaway galley producing hot bouillon, fried chicken and fruit. **1968** *Harrods Christmas Catal.* 5/3 Satin evening pochette with gilt tuck-away chain handle. **1979** *Arizona Daily Star* 1 Apr. (Advt. Section) 9/3, 1968 International 18' high cube van trucks with tuck away tailgates.

tucked, *ppl. a.* **1.** (Example with *in.*)

1963 R. N. Frye *Heritage of Persia* v. 198 The nomadic background of the Parthians may be seen in some equestrian features of dress, such as leggings *cum* boots with tucked-in trousers.

tucker, *sb.* Add: **6.** Also *N.Z.*

1864 J. C. Richmond *Let.* 12 May in *Richmond-Atkinson Papers* (1960) II. 111 It is very hard work humping your blankets and tucker. **1870** *Append. Jrnls. House Representatives N.Z.* D. xl. 4 Tucker Flat..has been looked upon as containing worse than 'tucker' ground. **1911** W. H. Koebel *In Maoriland Bush* xxi. 275 If they had obtained no wages for the first six months or so, they would have obtained their 'tucker' free. **1972** M. Shadbolt *Strangers & Journeys* iii. 43 Later Ned got the tucker cooking. It was stew and spud, like most nights.

Hence **tu·ckerless** *a.* (*Austral.* and *N.Z.*), without food.

1937 E. Hill *Great Austral. Loneliness* x. 82 The rind of the pods..makes an acrid but nourishing food..that tides over the tuckerless white man to the next out-camp. **1946** A. P. Harper *Mem. Mountains & Men* xvi. 162 We were left almost 'tuckerless' on Christmas Day.

tucker (tɒ·kəɹ), *v.*[2] *colloq.* (orig. and chiefly *Austral.* and *N.Z.*). [f. Tucker *sb.* 6.] **1.** *trans.* To supply with food. Also *refl.*

1899 *Bulletin* (Sydney) 21 Jan. 14/3 An oldish widower with three sons..goes out to work with Son No. 1, leaving the other two mites at home to mind the 's'lection' and tucker themselves. Old man comes home every month or so. **1920** B. Cronin *Timber Wolves* 40, I got a friend hereabouts that tuckers me when I'm along this way. **1940** E. I. Lord *Old Westland* xi. 137 He 'tuckered' many a down and out digger. **1964** B. Wannan *Fair Go, Spinner* (1965) iv. 126 In those days, the shearers had to provide their own food supplies—'to tucker themselves', as they put it.

2. *intr.* To eat, to have a meal. Also with *up.*

1903 H. B. King *Bill's Philosophy* 24 I'm sick of starving, when a cove can tucker free. **1940** F. D. Davison *Woman at Mill* 143 We were counting on it [*sc.* a money order] to tucker up with in Bairnsdale. **1959** H. P. Tritton *Time means Tucker* (1965) v. 64 We tuckered at the house and Mrs. Craig fed us till we couldn't eat another thing. **1963** *Weekly News* (Auckland, N.Z.) 5 June 37/2 The cowboy was tuckering at the cookshop on his own.

tu·ck-in, *a.* [f. the vbl. phr. *to tuck in*: see Tuck *v.*[1] 9.] That may be tucked in; *spec.* of a woman's blouse, etc., designed to have its lower edge tucked into the skirt.

1929 *Daily Express* 7 Nov. 5/2 Two [blouses] are 'tuck-in', and the other comes over the skirt. **1965** *Harper's Bazaar* June 65 Slashed tuck-in top.

tucking, *vbl. sb.* **5. tucking-comb** (earlier example).

1822 in *Dict. Amer. Eng.* (1944) IV. 2369/1 Mr. Pettigrew Bot of D McDowell one tucking Comb at $4.50.

tu·ck-up. [f. the vbl. phr. *to tuck up*: see Tuck *v.*[1] 6, 9.] †**1.** A fold or plait of hair. *Obs.*

1749 J. Cleland *Mem. Woman Pleasure* I. 186 His hair, which was of a perfect shining black, play'd to his face in natural side-curls, and was set out with a smart tuck-up behind.

2. A boat of a particular construction (see quot. 1889).

1887 *Forest & Stream* 24 Feb. 94/1 The tuckup could swing 300 sq. ft. if desired. **1889** W. P. Stephens *Canoe & Boat Building* (ed. 4) 239 The peculiar name 'tuckup' is derived from the fact that in building, the flat keel is not carried out straight from the stem to sternpost..but it..'tucks up'..to the height of the waterline.

3. The action or an act of tucking someone up in bed.

1915 H. L. Wilson *Ruggles of Red Gap* iv. 81, I was strangely a little warmed at thinking I might not have seen the last of Cousin Egbert, whom I had just given a tuckup.

Tudeh (tū·de). [a. Pers., lit. 'mass'.] In full *Tudeh party*. The Communist Party of Iran. Also *attrib.*

1946 *Civil & Milit. Gaz.* 16 Mar. 2/2 The Tudeh (Proletarian) Party in Azerbaijan, or what is left of it in Persia. **1966** S. Zabih *Communist Movement in Iran* iii. 73 In its first phase the Communist movement assumed the characteristics of a democratic front... Its organizational expression was the Tudeh party of Iran, formed in early October, 1941. **1979** *Economist* 8 Sept. 60/3 There are about a dozen leftist groups (three Chinese, two Trotskyists), all in opposition except for Tudeh, the official Communist party. **1980** J. Cartwright *Horse of Darius* x. 145 They have eliminated two hundred Tudeh Communists and Russian agents in the last few months. *Ibid.* 153 Let us concentrate on the Tudeh. Does our man in Iran know what they are up to?

Tudesco (tude·sko). Also **Tedesco.** Fem. **Tudesca.** [Ladino, = Sp., Pg. *tudesco* German.] A colloq. term among Sephardic Jews for an Ashkenazic Jew.

1897 I. Zangwill *King of Schnorrers* i. 12 You are a Tedesco. *Ibid.* v. 116 A Sephardi marry a Tedesco! Shameful. **1932** C. Roth *Hist. of Marranos* viii. 234 They accentuated the superficial differences..between them and the *tudescos*, from Germany and Poland, or even the *italianos* and *berberiscos* whose antecedents were nearer to their own. *Ibid.* xii. 316 Jacob Israel Bernal..desired to marry a *tudesca*—a member of the despised German and Polish community. **1949** *Spectator* 4 Nov. 595/1 Not so long ago English Sephardic families used to sit in mourning if one of their members married, not a Goy (or non-Jew), but a Tudesco (or Ashkenazi Jew). **1964** M. Wohlgelernter *Israel Zangwill* ii. vi. 85 The intended union of Menasseh's daughter with a Polish Jew excites..horror in..the *Mahamad*. A Tedesco did not pronounce Hebrew as they did, hence he was inferior.

Tudor, *a.* Add: **2.** (Further examples of buildings not belonging to the Tudor period, = mock Tudor.) Also of interior decoration. Also *Tudor-style* adj.

1902 G. E. Mitton *Hampstead & Marylebone* 23 There is the police-station..and adjacent an interesting Tudor house, which, though not old, is well built. **1928** Kipling *Bk. of Words* 267, I have only to leave the Tudor grill-room, take the electric lift upstairs. **1953** R. Lehmann *Echoing Grove* 177 A chintzy Tudor-style hotel. **1955** M. Gilbert *Sky High* viii. 116 They had a glass of sherry in the Tudor Bar, followed by a meal in the Jacobean Dining-Room. **1970** *Globe & Mail* (Toronto) 26 Sept. 42/6 (Advt.), Charming Tudor bungalow with pretty garden. **1978** J. Pudney *Thank Goodness for Cake* 82 Their Tudor-style manor house. **1979** R. Jaffe *Class Reunion* ii. iv. 156 Would they like this Tudor house, or that Spanish one?

B. *sb.* **a.** Mock-Tudor style. **b.** *N. Amer.* A house in mock-Tudor style.

1939, etc. [see *Stockbroker's Tudor* s.v. *stock-broker, stockbroker b]. **1961** 'J. le Carré' *Call for Dead* iv. 37 The Fountain Café..was all Tudor and horse brasses. **1969** E. Sandon *View into Village* 93 At the Somerton corner is the Boxted and Hartest Club erected..in 1888, in red brick Tudor, with two gigantic oriel windows. **1969** P. Zelver *Honey Bunch* vii. 35 The Swopes lived on one side of the McKittricks in an English Tudor. **1980** *News & Observer* (Raleigh, N. Carolina) 28 Oct. wa-5/5 This tudor is located on a circular street with many trees.

Tudorbethan (tiū:dɒɹbī·păn), *a.* (and *sb.*). [Blend of Tudor *a.* and Elizabethan *a.*] Mock Tudor; imitative of Tudor and Elizabethan styles. Also *ellipt.* as *sb.*

1933 Leavis & Thompson *Culture & Environment* 32 The outbreak of 'Tudorbethan' villas. **1958** *Spectator* 4 July 13/1 The 'mediæval character' of their 1930-Tudorbethan neighbours. **1960** *House & Garden* June 66/2 Proper treatment can make bearable the most vulgar stockbroker's 'Tudorbethan'. **1975** *Times* 13 Aug. 10/4 [Liberty's] store was rebuilt in 1924 to the Tudorbethan designs of E. T. and E. S. Hall. **1977** *Guardian Weekly* 29 July 19/2 A flash of black-and-white Tudorbethan pastiche.

Tudory (tiū·dɒri), *sb.* [f. Tudor *a.* + -y[3].] Mock-Tudor architecture or decoration.

1959 P. Bull *I know Face* viii. 144 The atmosphere of old Tudory and brass ornaments brings my bile to boiling-point. **1973** R. Hill *Ruling Passion* i. i. 11 Above the thatched roof a flock of television aerials..sang their triumph over charm and Tudory.

Tudory (tiū·dɒri), *a.* [f. Tudor *a.* + -y[1].] Imitative or suggestive of Tudor style.

1970 A. Fowles *Dupe Negative* xi. 141 The Tudory dining room. **1974** R. Ingham *Yoris* xiii. 41 The May Pole..was beamy and Tudory and phony.

Tudric (tiū·drik). [App. f. Tud(or *a.* + Cym)ric *a.*] The proprietary designation of a type of pewter (see quots.).

1902 (*title*) 'Cymric' gold and silverwork jewellery and 'Tudric' pewter (Liberty & Co.). **1904** *Jrnl. Soc. Arts* 10 June 638/1 For pewter..only modifications of Celtic forms were used, and these were soon supplemented by floral and plant motives to which the distinguishing name of 'Tudric' was given. **1922** *Trade Marks Jrnl.* 7 Feb. 232 *Tudric*... Pewter Ware. Liberty & Co. (Cymric) Limited, 16, Hylton Street, Birmingham; manufacturers of pewter wares. **1963** *Archit. Rev.* CXXXIII. 108/2 Two or three years afterwards, he was to add the range of 'Tudric' pewter to his enterprises. **1977** Fleming & Honour *Penguin Dict. Decorative Arts* 468/2 From 1894 he [*sc.* Sir Arthur Liberty] produced silver in an Art Nouveau version of the Celtic style which he termed 'Cymric' and from 1903 a new type of pewter, with a high proportion of silver in the alloy, which he called 'Tudric'.

tuff-tuff (tɒf,tɒf). Anglicized f. *teuf-teuf.

1902 E. Glyn *Refl. Ambrosine* v. 62 The tuff-tuff-tuff of a motor car was heard, and it drew up at our gate. **1903** *Daily Chron.* 1 July 3/2 'When one has steered one's "tuff-tuff" all day,' said a Parisian, 'or been driven through the clouds in a balloon.' **1982** N. Freeling *Wolfnight* 98 He managed..with a tufftuff to Kehl..and walked across the Europa Bridge.

tufted, *a.* Add: **1. b.** *spec.* Of a carpet, carpeting, etc. (see quot. 1960). Also *ellipt.* as *sb.*

1960 *Textile Terms & Definitions* (Textile Inst.) (ed. 4) 152 Tufted Carpet consists essentially of a pile yarn of tufts or loops which is inserted into a pre-woven backing and secured by means of a bonding material. **1963** *Which?* Mar. 70/2 Tufted carpets..are easier and cheaper to make than Wiltons or Axminsters. *Ibid.*, Tufteds may have cut or loop pile. **1965** *Guardian* 31 Mar. 14/1 Poor performance fibres..tended to give tufteds such an unwelcome reputation in the 1950s. **1970** *Which?* Sept. 265/2 There is a..labelling code..which requires that the label gives..the type of construction (eg Axminster, Wilton or tufted). **1981** *Times* 10 Aug. 16/1 The company..played the pioneering role in introducing the cheaper-to-make tufted carpet to a British market dominated by traditionally woven Axminster or Wilton.

tufting, *vbl. sb.* Add: **1. a.** (Later examples.)

1967 E. Short *Embroidery & Fabric Collage* iii. 68 There is no reason why good designs should not be carried out in candlewick cotton, using either tufting or couching and surface stitchery. **1976** *Daily Tel.* 21 Oct. 17/2 Knitting, crochet,..tufting and soft-toy making.

c. The process of making tufted carpeting, etc. Also *attrib.*

1965 *Guardian* 31 Mar. 14/1 In inserting the pile vertically rather than horizontally as in weaving, tufting has been limited as to the range of effects which can be produced. *Ibid.*, Tufting machines are exclusively American and British developments. **1970** *Encycl. Americana* XXIII. 763 In tufting, a preconstructed backing is used for the basic carpet structure... As the backing fabric moves through the machine, the pile yarns are stitched through it by a long bank of needles working simultaneously. **1974** J. Gray *Canvas Work* 121 (*heading*) Turkey (Single Knot Tufting)... This stitch, like all tufting stitches, is best worked from bottom to top of the area. **1976** J. Messent *Designing for Embroidery from Anc. & Prim. Sources* i. 28/1 Stumpwork, raised work, high padding and quilting are among some methods of creating a protruding area of interest from an otherwise flat surface, as is a highly textured area such as a shaggy pile or tufting.

tug, *sb.* Add: **3. c.** *tug of love*, a conflict of affections; *spec.* a contest for custody of a minor; also (with hyphens) *attrib.*

Perh. infl. by the title of a comedy 'The Tug of Love' by I. Zangwill (1907).

1973 *Times* 9 Nov. 20/7 The Houghton committee was set up after some highly-publicized 'tug of love' cases, and recommended making it easier for long-term foster-parents..to adopt. **1977** *Daily Mirror* 21 Mar. 13/1 Back home in the arms of her mother, a tiny tug-of-love girl sleeps peacefully. The girl..had been taken to California after being snatched by her father. **1984** *Times* 12 Oct. 2/2 'Tug of love' cases where a child is seized by one parent from another.

4. g. A rope. *U.S.*

1805 M. Lewis *Jrnl.* in *Lewis & Clark Exped.* (1904) I. 369 The white perogue..[was] refitted in a few minutes with some tugs of raw hides and nales. **1852** H. C. Watson *Nights in Block-House* 445 They took a strong tug, made from the raw hide of the buffalo or elk. **1910** W. M. Raine *B. O'Connor* xiv. 216 He stopped as if to fasten a tug.

6. b. Any other towing craft or vehicle, *spec.* (*a*) = *tug aircraft* below; (*b*) a tractor used to tow aircraft on the ground or unpowered road vehicles.

1942 *Jrnl. R. Aeronaut. Soc.* XLVI. 7 Aircraft towing as a method for launching high-performance gliders is a relatively recent development. Up till now, no specially designed aircraft 'tug' has become available. **1945** *Amer. Speech* XX. 227/2 *Tug*, a four- or six-wheeled tractor used for towing planes on the ground or for towing warehouse trailers. **1960** *Times Rev. Industry* Nov. 20/3 A..tractor can be a tug for two..vans. **1981** *Times* 14 Dec. 22/8 Tugs could not move the big jets because of ice.

7*. [Perh. a different word.] *Public School slang.* At Eton College, a student on the foundation, a colleger as distinguished from an oppidan. In wider use, a studious or academic type, a swot.

1864 *Eton School Days* ii. 21 That building on the right is Tuggery, where the Tug-Muttons live; you'll hate the Tugs like anything: all the Oppidans hate the Tugs. **1922** S. Leslie *Oppidan* iv. 48 *Tugs* or Scholars were separated from Oppidans by the same gulf that lay between Professionals and Gentlemen in the world of sport. **1976** R. Pound *A. P. Herbert* i. 23 In Wykehamist parlance, he was a 'tug', a clever chap, whose achievement was held worthier than any playing-field victory. **1977** A. J. Ayer *Part of My Life* ii. 34 Traditionally, the Oppidans despised the Collegers, who tended to come from a lower social stratum, and spoke of them as Tugs, because they were believed to engage in tugs of war for the few pieces of mutton which was all that they were given to eat. **1982** Barr & York *Official Sloane Ranger Handbk.* 71/1 Swots are weeds (at Eton: 'tugs don't wash').

8. tug aircraft, a powered aircraft used to tow a glider or train of gliders; **tug-boating** *U.S.*, working on a tug-boat; † **tug-mutton** = sense 7* above; **tug pilot,** the pilot of a tug aircraft; **tug-rope**: for † read '*obs.* exc. *U.S.*' (later examples).

1931 *Flight* 26 June 578/2 The tug aircraft, as it will probably be called. **1962** [see *parachute aircraft* s.v. *parachute *sb.* 5]. **1976** J. Colville *Footprints in Time* xxxiii. 185 Soon there were fleets of gliders too. As each was released over the river, its tug-aircraft turned steeply away for home. **1941** E. P. O'Donnell *Great Big Doorstep* xxi. 310 If it wasn't for rain, I wooden have a job to hole down. You'd see me tugboatin on the river or some kinda ordinary work. **1973** *Publ. Amer. Dial. Soc.* lx. 1 The coastal fringes are ideally suited to those who make their living from the sea—fishing, whaling.., boat-building, tugboating. **1864** Tug-mutton [see sense 7* above]. **1948** Partridge *Dict. Forces' Slang* 197 *Tug pilot*, the pilot of an aeroplane towing a glider. (Colloquial.) **1978** A. Welch *Bk. of Airsports* iii. 48/2 When experienced as a tug pilot, you will probably be given the occasional cross-country retrieve from a field or private airstrip. **1852** J. Reynolds *Pioneer Hist. Illinois* 236 They often pack their meat..by running a tug rope through each piece. **1891** *Century Mag.* Mar. 774/2 We began by eating the rawhide tug ropes and parfleches.

Hence **Tu·ggery** *Eton College slang*, the collegers' boarding-house; the position or status of a colleger.

1864 [see tug *sb.* 7*]. **1883** J. Brinsley-Richards *Seven Yrs. at Eton* xii. 112 [A boy] who had come from Aberdeen 'to try for Tuggery'—that is, to try and pass on to the foundation as a King's scholar.

tug (tɒg), *a.* *Public School slang.* [Origin uncertain: cf. tug *sb.* 7*.] (Esp. at Winchester College) ordinary, commonplace.

1890 Barrère & Leland *Dict. Slang* II. 378/2 *Tug* (Winchester College), usual, ordinary, common, stale, as tug-clothes, every-day clothes. **1907** *Wykehamist* Mar. 387/1 Accounts of events, 'tug' to the average reader, but recorded in print for the sake of the past and the future. **1951** C. P. Snow *Masters* vi. 53 No one on earth could call Jago tug... He's the least commonplace of men.

tug, *v.* Add: **7. b.** To tow (a glider) by means of a powered aircraft.

1942 W. S. Churchill *Second World War* (1951) IV. 800 The Whitley aircraft..is unsuitable for tugging gliders.

tugged (tɒgd), *ppl. a.* [f. Tug *v.* + -ed[1].] *tugged-at*: pulled at.

1930 Auden *Poems* 25 The tugged-at teat. **1962** I. Murdoch *Unofficial Rose* xxxiv. 320 Tugged-at leaves and whirling branches knew that summer was defeated and departing.

tughra, tuğra, varr. *toughra.

tugrik (tū·grik). Also **tukhrik.** [Mongolian.] A monetary unit of Mongolia, equal to one hundred mongos.

1935, etc. [see *mongo[2]]. **1978** *Financial Times* 10 Jan. 21/2 The Mongolian tugrik has lately been reported to stand at an official commercial rate of 0·225 Russian roubles.

tugtupite (tɒ·gtɒpəit). *Min.* [f. the name of *Tugtup* agtakôrfia in southern Greenland, where it was first found + -ite[1].] A tetragonal aluminosilicate and chloride of sodium and beryllium, $Na_4AlBeSi_4O_{12}Cl$.

1962 H. Sørensen in *Meddelelser om Grønland* CLXVII. i. 219 At the oral presentation of the mineral, it was suggested that it should be termed tugtupite, because it clearly differs from sodalite. **1977** *Mineral. Mag.* XLI. 130 Tugtupite..has a distorted sodalite type of structure.

tuhseel, var. *tahsil.

tui. (Earlier example.)

1832 A. Earle *Narr. Residence in N.Z.* 174 The only sounds which broke the calm were the wild notes of the tooe (or New Zeland blackbird).

tuica, ţuica, varr. *tsuica.

Tuileries (twīlĕri). *Hist.* [Fr., so called because built on the site of an ancient tile-works: see Tuille, Tuile.] A palace in Paris begun by Catherine de Medici in 1564 and destroyed by fire in 1871. It stood on the site between the Champs Elysées and the Louvre, now occupied by the Jardin des Tuileries, and was a residence of the court in royal and imperial France: hence, the royal or imperial family, the court, the administration.

1814 M. Birkbeck *Journey through France* 81 Every paragraph in the public journals is modelled and pared down to suit the temper of the Tuilleries. **1863** [see Politesse]. **1885** H. James *Little Tour in France* xi. 80 The gardens..are the promenade—the Tuileries—of the town [*sc.* Bourges]. **1967** *Listener* 25 May 678/1 By the end of the next decade, [Victor] Hugo..had become ..the tame poet of the Tuileries. **1972** T. Aronson *Queen Victoria & Bonapartes* x. 126 The coolness between Windsor and the Tuileries.

tuille, tuile. Add: ‖ **2.** As **tuile.** A thin curved biscuit, usu. made with almonds. Also *attrib.* and with defining addition.

1943 A. L. Simon *Conc. Encycl. Gastron.* iv. 133/1 *Tuiles d'oranges.* Cream 1 oz. butter, add 1½ oz. castor sugar and cream together. Add 1½ oz. chopped blanched almonds, 1½ oz. chopped candied peel, dessertspoonful flour, dessertspoonful milk. **1966** *Observer* (Colour Suppl.) 3 Apr. 41 In France the biscuits shown below are called 'tuiles'—which is French for roof tiles. These thin, crisp biscuits contain chips of almond. **1972** P. V. Price *Eating & Drinking in France Today* ii. 276 *Tuiles* from Amiens (thin chocolate and orange biscuits). **1976** *Times* 2 Oct. 10/3 The coffee ice was served in a tuile basket. **1979** *Harper's & Queen* Apr. 42/1 Sorbets..in a fragile basket of crisp, sweet, *tuiles*, made without almonds.

Tuinal (tiū·inăl). *Pharm.* [f. *tuin-*, of unkn. origin + *-al* in *Amytal, *Seconal.] A proprietary name for a combination of the two barbiturates quinalbarbitone and amylobarbitone, used as a sedative-hypnotic.

1949 *Trade Marks Jrnl.* 25 May 453/2 *Tuinal*... Medicinal preparations composed of sodium propyl-methyl-carbinyl allyl barbiturate and sodium isoamyl ethyl barbiturate. Eli Lilly and company,.. Indianapolis,.. United States of America. **1952** *Martindale's Extra Pharmacopœia* (ed. 23) I. 259 Tuinal (*Lilly*). 1½ or 3 gr. capsules containing equal parts of Seconal Sodium and Sodium Amytal. **1980** *Daily Tel.* 5 Nov. 3/1 She was drunk and they went on to a party at a friend's house. He saw her take two Tuinal tablets during the night.

tuition. Add: **2. c.** = *tuition-fee. U.S.*

1828 Webster, s.v., In our colleges, the tuition is from thirty to forty dollars a year. **1940** W. Faulkner *Hamlet* ii. i. 101 Your tuition will be paid. **1979** *Arizona Daily Sun* 19 Apr. 1/6 Wettaw..was given the award for his opposition to increases in university tuitions.

tuitionary *a.* (earlier example).

1816 J. B. Gilchrist (*title*) The orienti-occidental tuitionary pioneer to literary pursuits [etc.].

tukal, var. *tukul. **tukhrik,** var. *tugrik.

tukul (tu·kŭl). Also **tukal, tukl.** [Native name.] In Ethiopia and some adjacent regions: a dwelling shaped like a beehive, and constructed with a thatched roof.

1901 H. Vivian *Abyssinia* vii. 174 The capital is rather a camp than a town, and there is no particular trouble in rooting up a tukul and planting it elsewhere. **1920** *Blackw. Mag.* Nov. 675/2 The tukls were strongly built of rough stone. **1936** E. Waugh *Waugh in Abyssinia* v. 188 The office was a small, lightless *tukal* a hundred yards or so off the road. **1958** *Spectator* 18 July 111/1, I would not like my daughter to become the mistress of a *kraal* or a *tukl*. **1971** *Daily Tel.* 12 June 11/1 Rough compounds of mushroom-like *tukals*—round, mud-walled thatched peasant huts. **1981** E. North *Dames* i. 5 Thatched huts known as *tukuls*, some now with corrugated roofs.

tuku-tuku (tu·ku,tu·ku). *N.Z.* Also (as one word) **tukutuku, tuku tuku.** [Maori.] (See quots. 1946 and 1958.)

1936 [see *kakaho]. **1946** *Jrnl. Polynesian Soc.* June 160 *Tukutuku*, ornamental lattice-work of toetoe-reeds (kakaho) cross-laced in various patterns with narrow strips of flax and pingao: a panel of tukutuku was placed between every pair of poupou on the whare walls. **1950** *N.Z. Jrnl. Agric.* Aug. 187/3 The tuku tuku (decorative reed panels) [on the meeting house] are admirable examples of Maori art. **1958** *Listener* 20 Nov. 825/2 The tuku-tuku, a weaving of a wall-covering with split, or rather halved, bamboo as the warp. **1977** *N.Z. Herald* 5 Jan. 1-4/7 The old tukutuku or woven panel walls will in most cases be restored and cleaned.

tuladi (tu̇lădi·). Also **toledi, touladi.** [a. Canad. Fr., f. Algonquin, Micmac.] = Namaycush.

1846 C. L. Hatheway *Hist. New Brunswick* 61 The Toledi..weighs from five to twenty-five pounds,..and is a very voracious fish. **1856** C. Lanman *Adv. Wilds U.S.* ii. 79 The principal fish which it yields are the common trout, tuladi or great gray trout, and a small species of white fish. **1896** *Trans. R. Soc. Canada* II. 135 The Mackinaw trout of the great lakes [is]..the touladi of the country of the Micmacs. **1957** *Field & Stream* (N.Y.) May 90/2 A fourth kind of trout called touladi..the French-Canadians say, is a natural cross between the lake trout and the brook trout.

tularæmia (tiūlărī·miă). *Path.* Also (chiefly *U.S.*) **-emia.** [f. mod.L. *tular-ensis*, specific epithet of the causative organism (f. the name of *Tulare* Co., California) + Gr. αἷμα blood + -ia[1].] An acute infectious febrile disease of man and domestic animals caused by the bacterium *Francisella tularensis*, endemic among wild rodents in N. America and elsewhere, and transmitted to man by biting insects and by other means, producing very variable symptoms.

1921 E. Francis in *Public Health Rep.* (U.S.) 29 July 1731 The names thus far used for this disease are strictly vernacular and do not lend themselves to international usage as easily as a name in Latin form. Accordingly, the name tularæmia is proposed as a technical international name. **1925**, etc. [see *rabbit fever* s.v. *rabbit *sb.*[1] 4]. **1949** A. Huxley *Ape & Essence* 30 On the pressure-tanks of one army are painted the words *super tularemia*; on those of their opponents, *improved glanders*. **1961** R. D. Baker *Essent. Path.* ix. 183 Tularemia is acquired, usually, from the handling of infected rabbits.

tule. Add: Also **toolie, tulé, tuley. a.** (Earlier and further examples.)

1837 P. L. Edwards *Jrnl.* 20 July (1932) 26 Driving her along the margin of a bulrush or Tule pond she turned about. **1845** J. C. Frémont *Rep. Exploring Expedition* 252 They..live principally on acorns and roots of the tulé, of which also their huts are made. **1850** W. R. Ryan *Personal Adv. Upper & Lower Calif.* I. 298 The Indians of the party were despatched to hunt up the banks of the river for toolies. **1892** *Outing* Jan. 329/2 Arriving at a small patch of tuleys about the middle of the lake.

b. tule fog *U.S.*, fog over low-lying ground.

1899 *Monthly Weather Rev.* Dec. 536/2 Connected with this pressure distribution was the prevalence of tule fog in the great valleys of California... This ground or tule fog was so dense as to seriously inconvenience farming operations. **1934** S. E. White *Folded Hills* 340 A *tule* fog lay thick over the bottomlands. **1980** M. G. Eberhart *Casa Madrone* vi. 81 The evening breeze freshened, yet was still moist... Scott said..'The tule fog is coming in.'

tulgey (tɒ·ldʒi), *a.* Also (*erron.*) **tulgy.** A factitious word introduced by 'Lewis Carroll' applied to a wood; (usu. interpreted as) thick, dense, and dark; also *fig.*

1871 'L. Carroll' *Through Looking-Glass* i. 22 The Jabberwock, with eyes of flame, Came whiffling through the tulgey wood. **1936** J. R. R. Tolkien in *Proc. Brit. Acad.* XXII. 250 The jabberwocks of historical and antiquarian research burble in the tulgy wood of conjecture, flitting from one tum-tum tree to another. **1949** E. Taylor *Wreath of Roses* xiv. 216 You came out of that tulgey wood? **1972** K. Bonfiglioli *Don't point that Thing at Me* viii. 71, I battled..with Professor Aschloch's tulgey prose—only German poets have ever written lucid German prose. **1976** *Times Lit. Suppl.* 23 Jan. 71/5 A suitable backcloth to the dark, thick gothic forest of his own tulgey forebodings. **1982** *Ibid.* 10 Dec. 1352/4 The tulgey wood of semiotics.

tulip. Add: **2. b.** (Examples.)

1847 *Punch* 16 Oct. 148/1 This, my tulip, is a *salle de danse*. **1895** 'G. Mortimer' *Tales from Western Moors* iii. 67 'Cos for this, my tulip,..work and me fell out a long time back.

3. b. An explosive charge used to destroy a length of railway track. Now *Hist.*

1918 T. E. Lawrence in *Lett.* (1938) 250 A gang of four men can lay twenty 'tulips' in an hour on easy ballast, and for each two slabs (and single fuse) you ruin a sleeper, a yard of bank and two rails. **1920** *Blackw. Mag.* May 599/2 J. and I tried our prentice hands at the new game of 'planting tulips'. **1956** *Railway Mag.* Mar. 167/1 'Tulips', so called because of the appearance of the track after they had 'flowered', were the most effective means of derailing a train.

5. *tulip-bed* (examples), *-field*, *-glass* (later examples), *-time*; **tulip break(ing),** the variegated colouring of certain tulip flowers, caused by a virus infection; **tulip fire,** a fungus disease of tulips, caused by *Botrytis tulipæ* and producing speckled, discoloured leaves and flowers.

1822 T. G. Wainewright in *London Mag.* June 552/2 A delicate Schiavone, various as a tulip bed with rich broken tints. **1939** Joyce *Finnegans Wake* 526 Or tulipbeds of Rush below. **1958** *Manch. Guardian* 28 May 6/3 These changes of colour are symptoms of the virus disease known as tulip break. **1929** *U.S. Dept. Agric. Yearbk. 1928* 596 (*heading*) Tulip 'breaking' is proved to be caused by mosaic infection. **1969** G. Lyall *Venus with Pistol* vii. 38 Amsterdam was cold... Carlos told me..not [to] stop off to look at any tulip fields. **1931** *Pamphl. Seale-Hayne Agric. Coll.* XXXVI. 27 Several suggestions have been made as possible methods for the control of Tulip Fire. **1976** *Homes & Gardens* June 131/2 Tulips are particularly susceptible to tulip fire; this produces malformed leaves and shoots, which wither and stop growing. **1952** M. Allingham *Tiger in Smoke* ii. 49 He was carrying two large tulip glasses which he had overfilled. **1976** D. Francis *In Frame* v. 83 When Jik opened the champagne he poured it into shining tulip glasses. **1954** L. MacNeice *Autumn Sequel* 41 Its tuliptime and playtime.

tulle. Add: **b.** *tulle gras* (gra) [F. *gras* fatty], a gauze dressing for the skin impregnated with petroleum jelly.

1933 *Jrnl. R. Army Medical Corps* XL. 353 The 'Tulle Gras' dressing..was first placed on the market..by a French firm some years ago. It consists..of a fairly large-mesh gauze net, impregnated with vaseline containing 1 per cent. of balsam of Peru, supplied in sections of about four inches by five inches in size. **1974** R. M. Kirk et al. *Surgery* v. 75 A partial thickness burn, or one that is of doubtful nature, is treated by removing debris, pricking and emptying the blisters, and covering the area with tulle gras..which is non-adherent.

tullibee. (Earlier examples.)

1789 A. Shaw *Let.* 16 Dec. in L. F. R. Masson *Bourgeois de la Compagnie du Nord-Ouest* (1889) I. 33, I take a walk to my traps, return to the house, eat *Tollibees* about nine. **1823** J. Franklin *Narr. Journey to Shores of Polar Sea* II. 711 The Cree name of this fish, ottonneebees, has been corrupted by the traders into tullibee.

tulp (tŭlp). *S. Afr.* Also **tulpe.** [a. Du. *tulp* tulip.] Any of several bulbous plants of the genus *Homeria* or *Moræa* of the family Iridaceæ, esp. *Homeria breyniana* or *Moræa polystachya*, which are native to southern Africa, bear yellow, pink, or blue flowers, and are extremely poisonous to cattle.

[**1795**: see *suikerbos.] **1835** T. H. Bowker *Jrnl.* 10 May (MS.), Lots of Bullocks sick with eating Tulp. **1844** J. Backhouse *Narr. Visit Mauritius & S. Afr.* xviii. 276 Among the grass, on the south side, there was an abundance of the species of *Moræa*, known in the country by the name of Tulip or Tulpe, which is very destructive to cattle. **1871** T. Baines *Diary* 28 May (1946) III. 601 Tulp, a plant poisonous to cattle, grew at Blauwe Krantz. **1896** R. Wallace *Farming Industries Cape Colony* iv. 96 Cattle bred on the ground do not eat tulp unless they are very hungry. **1958** *Cape Times* 23 Aug. 11/6 Poisonous plants, such as tulp,..led to illness among animals. **1973** *Grocott's Mail* (Grahamstown, S. Afr.) 19 June 3 An abundance of the highly poisonous tulp plant..could mean stock losses for unsuspecting farmers.

tulwar. (Earlier example.)

c **1810** W. Hickey *Mem.* (1960) xv. 252 In about half an hour after the first grand attack had thus been made, an alarm was given that the mob armed with *tulwars* (scimitars) had forced the sentries.

tum (tɒm), *sb.*[2] *joc.* Short for *tummy. Cf. *tum-tum *sb.*[4]

1869 W. S. Gilbert *Bab Ballads* 121 They can reduce a bulging tum To measures fair By taking air And exercise in plenty. **1890** A. James *Diary* 18 July (1965) 129 That dissipated organ known in the family as 'Alice's tum'. **1937** G. Frankau *More of Us* xiv. 153 Chilled all those suns for which we oiled the raw tum Or boracized the blistering shoulder blade. **1977** *Time* 14 Feb. 33/2 To re-establish old wisdom and simple certitudes: hot chestnuts in the hand, calories in the tum.

tuman (tumā·n). [Var. Toman[1].] A tribe of Baluchis or Pathans.

[**1816** H. Pottinger *Trav. Beloochistan* iv. 62 An assemblage of these Ghedans [*sc.* tents] constitute a Toomun, or village, and the inhabitants of it a Kheil, or society.] **1907** *Baluchistan District Gazetteer Ser.* III. 372 A tribe is not responsible for the actions of any person who takes up his abode temporarily with it for purposes of cultivation or for grazing. In this case his heirs and his own *tuman* is responsible for his acts. **1950** *R. Central Asian Jrnl.* XXXVII. 284 There are nine Baloch tumans (tribes) of Dera Ghazi Khan. **1974** *Encycl. Brit. Micropædia* I. 773/3 Each tribe (*tuman*) consists of several clans and acknowledges one chief, even though in some *tuman* there are clans in habitual opposition to the chief.

So **tumanda·r,** a Baluchi or Pathan chief.

1907 *Baluchistan District Gazetteer Ser.* III. i. 78 Among the Baloch..a chief or *tumandār* may invite contributions on the occasion of a marriage. **1932** *Cambr. Hist. India* VI. xxv. 455 Casting all fear on one side, he boldly advanced into their mountain retreats and made friends with the tribal chiefs or tumandars. **1979** *Indo-British Rev.* (Madras) VIII. i–ii. 52/2 The Tumandars, or chiefs of the unadministered territory, had come in for their annual reunion with..the political Agent.

tumângong, var. *temenggong. **tumbaga,** var. Tombac in Dict. and Suppl.

tumble, *sb.* Add: **2. d.** *to take a tumble* (*to oneself*): to realize the facts of one's situation; to wake up *to* something, to tumble. *slang* (orig. *U.S.*).

1877 [see *on to, onto *prep.* 2]. **1928** F. Hurst *President is Born* xiv. 182 An iron negro boy, with a hitching ring in his fist, stood..at the curb... Once, some town-wag..had hung a pasteboard tag about his neck, 'Take a tumble to yourself, Joe.' **1944** *Living off Land* v. 106 At one goldfield where malaria broke out virulently no one took a tumble why for a long time. **1949** J. R. Cole *It was so Late* 65 The woman, taking a tumble to our set up, gave me the come on. **1959** M. Gee in C. K. Stead *N.Z. Short Stories* (1966) 267 After a while I give up, and I take a tumble to what's happening. I'm getting the bum's rush. **1973** 'J. Patrick' *Glasgow Gang Observed* viii. 79 Ma wee brother will learn sense; he'll take a tumble tae hissel'.

e. A sign of recognition or acknowledgement, a response; chiefly in phr. *to give a tumble*. *U.S. slang.*

1921 H. C. Witwer *Leather Pushers* xi. 282 Neither of 'em give him a tumble. **1934** J. O'Hara *Appointment in Samarra* (1935) vii. 208, I went in his office and started kidding around... I noticed I wasn't getting a tumble from him, so I finally broke down and asked him, I said what was the matter. **1935** D. Runyon in *Cosmopolitan* Jan. 160/3 He never lets on he knows me, and naturally I do not give Mr. Labez any tumble whatever. **1953** *N.Y. Times Book Rev.* 8 Feb. 17 If the right boy won't give you [*sc.* a girl] a tumble, you've got a problem. **1976** *Washington Post* 19 Apr. c3/6 Der Bingle took a subway ride in New York over the weekend and not a soul gave him a tumble. Bing Crosby said he knew what it meant to be just another straphanger.

4. *slang.* An act of sexual intercourse; a woman giving opportunity for this; chiefly in phr. *to give* (or *get*) *a tumble*.

1903 Farmer & Henley *Slang* VII. 224/2 *To do a tumble* (of women) = to lie down to a man. **1934** H. Miller *Tropic of Cancer* 297 She's a big, healthy bitch... I wouldn't mind giving her a tumble. **1954** J. Trench *Dishonoured Bones* iii. 110 He was..giving la Vitrey a tumble somewhere. **1970** 'J. & E. Bonett' *Sound of Murder* xiii. 172 Most men think that a woman who has been loved by a married man is an easy tumble. **1976** P. Cave *High Flying Birds* iii. 45 'Back-pay', he said, 'plus an advance on a quick tumble tomorrow night.'

tumble, *v.* Add: **II. 3. e.** Of laundry: to be tossed about in the revolving drum of a tumble-drier (or washing-machine).

1970 *Which?* Aug. 240/1 Too much foam will certainly stop your clothes from tumbling freely and so getting clean. **1975** C. Weston *Susannah Screaming* (1976) i. 9 Rees..watched his laundry tumbling inside the barrel of the dryer.

7. b. *to tumble up*: to make haste, orig. (*Naut.*) from below deck. *slang.*

1826 W. N. Glascock *Naval Sketch-Bk.* I. 8 The command was repeated by the boatswain and his mates, who were piping and roaring down the hatchways—'Tumble up, tumble up from below.' **1832** [see sense 7 a in Dict.]. **1838** Dickens *Nicholas Nickleby* (1839) viii. 65 'Now, Nickleby, come; tumble up, will you?' Nicholas..'tumbled up' at once, and proceeded to dress himself. **1842** J. F. Cooper *Wing-and-Wing* I. viii. 125 This sight produced a great commotion in the ship, even the watch below 'tumbling up', to get another sight of a craft so renowned. **1858** Trollope *Three Clerks* II. ii. 40 'Mr. Tudor to attend in the board-room immediately,' said a fat messenger... 'All right,' said Charley—'I'll tumble up and be with them in ten seconds.'

9. b. To have sexual intercourse with. *slang.*

1602, 1698 [see sense 9 a in Dict.]. **1772** T. Bridges *Burlesque Transl. Homer* i. 4 What priest beside thyself e'er grumbl'd To have his daughter tightly tumbl'd? **1922** Joyce *Ulysses* 502 Beware of the flapper and bogus mournful... Tumble her. **1971** 'R. Macdonald' *Underground Man* xxxii. 225 He had tumbled the prettiest girl, and got her with child, and Albert and Fritz had taken the rap for it. **1973** *Guardian* 21 June 10/1 A hip young girl who tumbles him when his wife is away. **1976** R. Lewis *Witness my Death* v. 166 Tommy Elias had tumbled the schoolgirl in the ferns.

10. b. (Earlier example.) Also const. *that*. Also *trans.*, to detect, see through.

1846 *Swell's Night Guide* 58, I..officed Bet, she tumbled to the fake, and stalled off to the dossery. **1901** 'J. Flynt' *World of Graft* iii. 104 Women..tumble more guns 'n all the coppers in existence. **1926** *Variety* 29 Dec. 7/4 The pincher would never tumble that 'nice people' meant an act that kicked in more than the usual vaudeville agent's legitimate commission. **1936** G. Ingram *Muffled Man* iii. 49 You can't go on for ever at any game, and not get tumbled some time or other. **1938** F. D. Sharpe *Sharpe of Flying Squad* xxvi. 262 We thought you wouldn't tumble us, Guv'nor. **1962** *New Statesman* 21 Dec. 899/1 By the time you tumble that your drum has been turned over, we're miles away. **1974** *Times* 7 Feb. 3/7 John Rodger..heard the radio and said: 'They have tumbled us.' **1981** J. Barnett *Firing Squad* vi. 57 Have to have words with Simonson, in case he has tumbled the tattoo.

tumble-. Add: **1. tumble-action,** the tumbling action of a tumble-drier; **tumble-bug** (earlier example).

1958 *Sunday Times* 9 Mar. 22/6 Here [in an electric washing machine] tumble-action replaces a wringer. **1976** *National Observer* (U.S.) 11 Dec. 9/1 Do not use tumble-action dryer. **1805** R. Parkinson *Tour Amer.* 362 A kind of beetle, called a tumble-bug,..in the summer forms a cave in the earth.

2. tumble home: also *transf.* of a motor vehicle; **tumble-up,** substitute for def.: a tumbler designed to be placed upside down on the neck of a carafe.

1924 *Motor* 7 Oct. 450/2 The body is of particularly pleasing lines, with a V windscreen and tumble-home stern. **1968** *Motor Industry Res. Rep.* ix. 25/1 Decreases due to increasing the canopy tumble-home.

tumble-dri·er. Also **-dryer.** [f. Tumble *v.* + Drier, dryer.] A machine for drying washing in a heated drum that turns about a horizontal axis. So **tumble-dry** *v. trans.* (also *absol.*), **tumble-drie·d, -dry·ing** *ppl. adjs.*

[**1952** *Gas Age* 8 May 62/1 The tumbler-type of clothes dryer.] **1962** J. T. Marsh *Self-Smoothing Fabrics* xiv. 228 Different results may be obtained according to whether the fabric is dried on a line or on one of the tumble-drying machines. *Ibid.* xx. 347 Fabrics with wet recovery only cannot be tumble-dried with satisfaction but need line-drying. **1969** *Which?* Nov. 372/2 (*caption*) Line dry or Tumble dry. **1972** J. McClure *Caterpillar Cop* vi. 98 The lady next door has a tumble drier so you put your clothes outside and I'll take them round. **1976** *Ilkeston Advertiser* 10 Dec. 12/1 (Advt.), Don't have wet clothes hanging around, buy a Tumbledryer. **1977** 'M. Yorke' *Cost of Silence* v. 36 Mrs Costello hung her tumble-dried washing round the kitchen. **1982** A. Price *Old Vengeful* vi. 86 We can do your things... Clarkie can wash them, and tumble dry them, and iron them.

tumbler. Add: **5. b.** An inexperienced window-cleaner.

1960 'A. Burgess' *Doctor is Sick* i. 8 Me, I clean windows... I've seen these young ones just starting—'tumblers' we call them—get froze stuck up there on a ladder. **1972** *Times* 20 Sept. 3/2 Forty years ago..novice window cleaners starting at the bottom of the ladder were known in the cold-blooded jargon of the trade as tumblers.

6. c. (Earlier example.)

1850 Dickens in *Househ. Words* 21 Dec. 289/2 The Tumbler with his hands in his pockets, who wouldn't lie down.

13. j. *ellipt.* for *tumbler-drier* below.

1947 W. L. Carmichael et al. *Callaway Textile Dict.* 360/1 *Tumbler*, a clothes-drying device consisting of a revolving cage in which hot air is circulated. **1961** *Listener* 30 Nov. 951/2 The most trouble-free and quick-to-use kind of dryer, the electric 'tumbler'.

14. *tumbler-glass* (earlier example), *-screw* (earlier example); **tumbler-drier, -dryer** = *tumble-drier; **tumbler-punch** (earlier example).

1956 *Good Housek. Home Encycl.* 13/2 Electric Tumbler Dryers dry by means of a revolving drum in a heated cabinet. **1969** *Which?* Nov. 352/2 Generally all the tumbler driers were easy to use. **1795** J. Woodforde *Diary* 14 Oct. (1929) IV. 234 The third [remarkable fact] was, of a Man drinking half Pint Tumbler Glass of Beer and eat the Glass after it. **1843** *Act relating Militia State of Vermont 1842* 80 Each squad of ten men, a wire and tumbler punch. *Ibid.*, Order in which the lock is taken apart... 9. The tumbler screw.

tumbling-. Add: *tumbling girl* (earlier example).

1854 Dickens *Hard Times* vii. 52 Tom Gradgrind's whim, ma'am, of bringing up the tumbling-girl.

‖ **tumbok lada** (tu·mbɒk lā·dă). [Malay, lit. 'pepper-crusher', f. *tumbok* to thump, pound + *lada* pepper.] A small Malayan dagger.

1839 T. J. Newbold *Straits of Malacca* II. xii. 212 The Battas of Sumatra, wear..knives called tombak lada. **1911** *Encycl. Brit.* XVII. 477/1 The Malays use..short stabbing daggers called *tumbok lada*. **1936** G. B. Gardner *Keris* viii. 113 A nasty ripping knife is sometimes made by fixing an old razor blade into a *tumbok lada* hilt. **1967** J. Cleary *Long Pursuit* i. 25 He..took from his belt the *tumbok lada*, the small Malayan knife.

tumbu fly (tu·mbu flai). [f. Bantu name of the insect + Fly *sb.*[1]] A yellow fly with grey markings, *Cordylobia anthropophaga*, found in sub-Saharan Africa, where its larva is a parasite of man and other animals.

1911 A. Alcock *Entomol. for Med. Officers* x. 155 The notorious species is..the Tumbu-fly, the larva of which is a subcutaneous parasite of man and other animals. **1930** *Discovery* Aug. 265/1 Wild beasts..and tumbu flies abound. **1979** C. Allen *Tales from Dark Continent* i. 5 There was the *tumbo* [sic] fly..whose worm manifested itself as a large boil.

tumeler, var. *tummler.

tumesce (tiume·s), *v.* [Back-formation from Tumescence or f. L. *tumēscere* (see Tumescent *a.*).] *intr.* = Tumefy *v.* 2. Also *fig.* Hence **tume·scing** *ppl. a.*

1966 *New Statesman* 10 June 853/3 Forming open alliances which were secret before, exposing sexual and ethical weaknesses as they tumesce under the threat of death. **1976** *Theriogenology* V. 261 Those [monkeys] that tumesced before they were 3 years old tended to be heavier than their peers, but were not necessarily heavier than older, nonswelling females. **1976** *New Scientist* 9 Sept. 528/2 Other aids to prediction [of eruptions]... As a volcano tumesces the distance between fixed points increases, and these distances can be measured accurately with optical instruments using laser beams. **1980** *Sci. Amer.* Feb. 86/1 In a tumescing system such as the Yellowstone one the ring fractures could propagate downward, eventually penetrating the main magma chamber.

tumescence. Add: *spec.* the swelling of a volcano as a result of increasing pressure of magma inside it.

1943 *Amer. Jrnl. Sci.* CCXLI. 243 Easterly tilt has long been known to indicate tumescence of Mauna Loa accompanying the rise of magma pressure preceding eruption. **1976** P. Francis *Volcanoes* x. 310 The tumescence of the Hawaiian volcanoes, however, is slight, only one metre or so at the summit.

tummied (tɒ·mid), *a.* [f. *tummy + -ed[2].] In parasynthetic combs., having a stomach (of a specified kind).

1975 T. Heald *Deadline* iv. 72 He..wondered if she had to wear a corset. She was remarkably flat tummied. **1975** *Times* 5 July 10/5 A carved, fat tummied elephant.

tummler (tu·mləɹ). Also **toomler, tumeler.** *colloq.* (orig. and chiefly *U.S.*). [Yiddish, f. G. *tummeln* stir.] Someone who acts the clown, a prankster; *spec.* a professional maker of amusement and jollity at a hotel or the like.

1966 Adams & Tobias *Borscht Belt* iv. 41 If Mrs. Rappaport complained about not getting her third portion of blueberries or Mrs. Davidoff was scrounging fruitlessly for a dancing partner..it was the Toomler to the rescue. **1968** L. Rosten *Joys of Yiddish* 413 It is the *tummler*'s job to guarantee, to the blasé (but insatiable) patrons of a summer resort, that most dubious of vacation boons: 'Never a dull moment!' **1970** L. M. Feinsilver *Taste of Yiddish* iii. 345 Danny Kaye and other entertainers got their starts as tumelers in the Catskills. **1977** *New Yorker* 12 Sept. 86/3 A summer job as part-time social director and tummler at a hotel in Lake Hopatcong, New Jersey. **1984** *Times* 20 Mar. 12/3 Why should she not believe the agents and promoters, the spivs and *tummlers*?

tummy (tɒ·mi). *colloq.* [Repr. a childish alteration of *stomach*: see -y[6].] The stomach or intestine.

1869 W. S. Gilbert *Bab Ballads* 200 Why should I hesitate to own That pain was in his little tummy? **1884** Kipling *Let.* ? Sept. in Ld. Birkenhead *Rudyard Kipling* (1978) vi. 74, I felt the cramps in my legs dying out and my tummy more settled. **1922** Joyce *Ulysses* 356 Cissy poked him..out of fun in his wee fat tummy. **1936** N. Streatfeild *Ballet Shoes* iii. 40 Laughing so much that they fell on the floor, and their tummies ached. **1962** V. Nabokov *Pale Fire* 191 That was Dad's tummy, I think —not a spook. **1979** *Beautiful British Columbia* Fall 14 The bucks..lie napping with heads tucked against tummies like so many dogs.

2. An abdominal pain or complaint. Freq. with preceding place-name: diarrhœa suffered by visitors there.

1888 Kipling *Story of Gadsbys* (1889) 21 He has nothing more than a wet weather tummy. **1937** F. Stark *Let.* 25 Oct. in *Coast of Incense* (1953) 177, I am spending these five days rather tiresomely in hospital with a *tummy*. **1939** R. Godden *Black Narcissus* vi. 77 'Sister Briony..thinks it must be some local infection, as we all have it.'..'Darjeeling tummy,' said Mr. Dean. **1943**, etc. [see *gippy 1 c]. **1959** L. Durrell *Spirit of Place* (1969) 423 The Mediterranean affliction of high summer known to us all as 'tummy' (Egyptian, Greek or Naples tummy). **1970** N. Marsh *When in Rome* iv. 97 A sudden onslaught of the affliction known to tourists as Roman Tummy..necessitated an immediate withdrawal. **1979** A. V. Badgley *Rembrandt Decisions* (1980) ix. 119 He's sick... Says he's got 'Bombay Tummy'.

3. *attrib.* as *tummy muscle, pain, rumble, trouble, upset*; **tummy ache,** an abdominal pain; **tummy bug,** (a germ causing) a disorder of the stomach; **tummy-button** = Navel *sb.* 1.

1926 Galsworthy *Silver Spoon* iii. viii. 284 Kit had tummyache this morning. **1979** L. & J. Brown *Our Miracle called Louise* iii. 35 'I've got a tummy ache,' I told her. 'It will be gone in the morning,' she said. **1969** M. Pugh *Last Place Left* xvii. 121 We're rather below strength... Some sort of tummy bug. **1945** A. Huxley *Time must have a Stop* ii. 13 Everybody's tummy-button grew inwards like that. **1974** G. Butler *Coffin for Canary* x. 133 My waist and tummy muscles are really taut. **1924** J. Buchan *Three Hostages* xvi. 238 He really has had a bad tummy pain. **1947** G. Greene *Nineteen Stories* 192 It was the doctors who called his complaint Borborygmi: in England we usually call it just 'tummy rumbles'. **1982** N. Marsh *Light Thickens* iii. 91 His tummy rumbles are positively deafening. **1937** W. H. S. Smith *Let.* 12 Jan. in *Young Man's Country* (1977) ii. 51 My tummy trouble persisted for several days after my return to Madaripur. **1926** A. Huxley *Let.* 21 Oct. (1969) 274 Matthew meanwhile flourishes..in spite of a tummy upset.

‖ **tu-mo** (du mo). Also **Tu Mu.** [Chinese *dū mài*.] In Chinese medical theory, the chief passage through which the vital energy circulates, located within the spine; *spec.* in acupuncture.

1972 Da Liu *T'ai Chi Ch'uan & I Ching* (1974) i. 9 In meditation one learns to focus and direct energies which are usually squandered in the mundane perceptions of the five senses. In Taoist meditation these energies are directed through two main channels: *Tu Mu*, a channel along the spinal column..and *Jen Mu*, a channel which passes down the front of the body to the genital region. **1974** *Barefoot Doctor's Man.* iv. 105 The governing 'tu-mo' meridian.

tumorigenic (tiūmŏɹidʒe·nik), *a.* *Med.* Also **tumo(u)r(o)-.** [f. Tumour, tumor + -i-, -o + *-genic.] Capable of causing tumours.

1948 *Cancer Res.* VIII. 410/1 It is thought that they [*sc.* these observations] afford an example of the tumorigenic action of..follicle-stimulating hormone. **1965** *Dissertation Abstr.* XXVI. 1300/2 (*heading*) The salivary gland chromosomes of a tumorgenic strain of *Drosophila melanogaster*. **1971** *New Scientist* 24 June 732/2 Tumourgenic hybrids had 80 chromosomes instead of the expected 116. **1979** *Nature* 11 Oct. 486/1 The significance of EBV [*sc.* a virus] as a tumorigenic agent in humans could be finally established if it were possible to prevent tumours by vaccination. **1980** *Ibid.* 21 Feb. 777/1 The present experiments show that human diploid cells can be transformed *in vitro* into tumorogenic cells by X-ray irradiation.

So **tu·morigen** (and varr.), a tumorigenic agent; **tumorige·nesis,** the production or formation of a tumour; **tu:morigeni·city,** tumorigenic property.

1948 *Cancer Res.* VIII. 397 (*heading*) Hormonal imbalances in tumorigenesis. **1952** J. E. Gregory *Pathogenesis of Cancer* (ed. 2) xii. 147 These elements [*sc.* coal tar, etc.] are called carcinogens. We now know that if the virus was present cancer would develop, but if it was not present only benign tumors and hyperplasia would develop as a result of the irritation. This latter fact makes it appear that a better name for these substances might have been tumorgens, instead of carcinogens. **1967** *New Scientist* 25 May 478/2 More than 50 per cent of the tumorigenicity of 24-hour condensate..was due to stable, non-volatile carcinogens. **1970** *McGraw-Hill Yearbk. Sci. & Technol.* 217/2 Other studies have shown..the dominance of tumorogenicity, and the characteristics of polyoma transformation in hybrids between transformed and nontransformed cells. **1971** *Nature* 17 Sept. 195/1 Irradiated hamster foetal cells protect male hamsters against SV40 tumorigenesis. **1978** *Brit. Med. Jrnl.* 11 Mar. 649/2 It would be a disservice to the public to allow marketing of a compound which is a demonstrated tumorigen. **1980** *European Jrnl. Cell Biol.* XXII. 491 The tumorgenicities of RBCF-1 cells.

tumour, tumor. Add: **5. tumour virus,** a virus that causes tumours.

[**1934** *Lancet* 21 July 117/2 Neutralising antibodies can be shown to be formed against fowl-tumour viruses.] **1950** *Amer. Jrnl. Med.* VIII. 495/2 There is no proof that tumor viruses are of a different nature from other viruses. **1982** *Sci. Amer.* Mar. 69/3 Some tumor viruses are oncogenic (that is, they induce tumors) only in animals that are not their host in nature, whereas other tumor viruses are oncogenic in their natural host.

tump, *sb.* Restrict *local* to senses in Dict. and add: **4.** *fig.* Trivial writing, bad prose.

1917 Kipling *Divers. Creatures* 172 It's the most vital, arresting and dynamic bit of tump I've done up to date. **1933** D. L. Murray *Eng. Family Robinson* ii. 36 Did you ever read such tump as our parish magazine?

tump-tump (tɒmp,tɒmp). [Echoic.] A short sound as of water slopping without splashing, or a large ball being kicked.

1917 [see *punt-about]. **1983** *Listener* 20 Oct. 26/3 The one creation that is seen in innumerable moods is the Fenland water—rising to the tump-tump of the water-pumps.

tum-tum (tɒ·m,tɒm), *sb.*[4] *joc.* [Redupl. *tum *sb.*[2]] = *tummy.

1864 G. Meredith *Let.* 1 Mar. (1970) I. 245, I hope hope your tum-tum is stronger, old boy? **1894** G. du Maurier *Trilby* II. iv. 6 Many other unaccustomed good things, so bad for their little French tumtums. **1930** [see *icky-boo* adj. s.v. *icky *a.* and *sb.*]. **1981** P. Mallory *Killing Matter* ii. 22 Make some toast and coffee. My tum-tum's empty.

tuna[2]. Add: Also with pronunc. (tiū·nă). Substitute for def.: Any of several large marine food and game fishes of the family Scombridæ, belonging to the genus *Thunnus*, *Euthynnus*, *Katsuwonus*, or a closely related genus and found in Atlantic, Pacific, and Mediterranean seas. Cf. *skipjack *sb.*[4], Ton[4], and Tunny. (Earlier and later examples.)

1881 *Proc. U.S. Nat. Mus.* IV. 45 Another *Orcynus*, known as the 'tuna', exists about Santa Cruz Island. **1911** *Chambers's Jrnl.* Jan. 63/2 My Californian friends might well possess their disappointed souls in patience if there were any promise of such compensation..for the loss of the tuna. **1949** *Manch. Guardian Weekly* 24 Mar. 11/2 Water you can't swim in, and good tuna and jacks you can't eat. **1970** *Nature* 19 Dec. 1141/2 Tuna is a fisheries term for the larger tunny species.

b. *attrib.*, as *tuna bait, boat, fisherman, fishing, fleet, meat, packer, packing, salad, sandwich, school*; **tuna clipper** *U.S.*, a powered fishing boat with facilities for catching and storing small fish; **tuna fish,** the flesh of the tunny as food.

1901 *McClure's Mag.* Feb. 370/1 Vincente..is just taking the tuna bait from his gill-nets. **1968** *National Fisherman* July 18-c/2 Much squid was used for tuna bait. **1903** C. F. Holder *Big Game Fishes* iv. 59 The tuna boats of Catalina Island are designed for the purpose. **1956** *Fishery Bull.* LVII. 195/1 (*heading*) Grounds fished by tuna boats operating in the Inner South Seas. **1929** *Pacific Fisherman* Sept. 39/1 The 100-ft tuna clipper 'Enterprise' was driven ashore. **1970** *National Fisherman* Sept. 1-b/2 A tuna clipper is a vessel equipped to carry small fish alive for use as bait; capable of being trimmed in such a way as to bring her stern rail as low as possible in the water; and fitted with a system which permits her to hold her catch for long periods of time. **1917** M. Green *Better Meals for Less Money* xvi. 130 (*heading*) Tuna fish salad. **1922** *Guardian* 27 June 6/2 Tinned goods..are the most suitable for storing... 1 tin 'tuna' fish or crayfish. **1978** *Amer. Poetry Rev.* Nov./Dec. 16/3 To a tunafish gray on a bruised, greenish blue. **1982** J. Gardner *For Special Services* viii. 64 Tuna fish sandwiches were hardly Bond's style. **1919** *Pacific Fisherman Yearbk.* 64/2 The tuna fishermen return to port each evening. **1901** *McClure's Mag.* Feb. 370/1 The rods and reels in tuna fishing are of the finest description. **1926** G. Frankau *My Unsentimental Journey* xvii. 222 His hobby—he told me at once—is tuna-fishing. **1980** *Times* 4 Nov. 18/4 [In Peru] tuna fishing..has now virtually ended. **1932** M. Miller *I cover Waterfront* xvii. 97 The only people who know of him are the two men who live on the island and the crews and skippers of the tuna fleet. **1968** *National Fisherman* Apr. 2-A/2 Commander, a steel-hulled, 20-year-old veteran of the Southern California tuna fleet, sank off Mexico. **1923** A. Ward *Encycl. Food* 537 Very little tuna meat is sold fresh in our markets. **1951** Tressler & Lemon *Marine Products of Commerce* (ed. 2) xx. 448 A machine for molding blocks of tuna meat the exact size for the can is in operation in some canneries. **1922** *Pacific Fisherman* May 43/1 Prominent tuna packers. *Ibid.*, Mr. Ambrose was one of the organizers of the first tuna packing plant in California. **1967** *Commercial Fisheries Rev.* Nov. 64/2 Japanese tuna packers.. switched from fruit canning to tuna packing in September. **1953** *Special Sci. Rep.* (U.S. Fish & Wildlife Service, Fisheries) No. 104. 32 [Chunk tuna] was especially suited to the preparation of tuna salads. **1977** C. McFadden *Serial* (1978) xiii. 32/2 He left Fred's, absent-mindedly leaving her to pay for his tuna salad special. **1957** Heseltine & Dow *New Basic Cook Bk.* 718 (*heading*) Tuna sandwiches. **1977** C. McCarry *Secret Lovers* iii. 42 The passenger compartment of the BMW smelled..of bay rum, Brylcreem, tuna sandwiches. **1930** *Pacific Fisherman* Apr. 41/1 In search of distant tuna schools. **1967** *Trans. Amer. Fisheries Soc.* XCVI. 127/1 Purse seines capture only about half the number of tuna schools upon which they are set.

tuna[3]**.** Substitute for def.: Either of two freshwater eels, *Anguilla dieffenbachii* or *A. australis schmidtii*, found in New Zealand. (Earlier and later examples.)

1843 E. Dieffenbach *Trav. N.Z.* II. iii. 389/2 *Tuna*, eel. **1851** J. C. Richmond *Let.* 25 Mar. in *Richmond-Atkinson Papers* (1960) I. ii. 79 They set off..to catch crawfish & tuna, the eels of the country. **1966** *Encycl. N.Z.* I. 565/1 These are the common freshwater eels called tuna by the Maoris.

tunability (tiūnăbi·lĭti). [f. next: see *-bility.] The capability of being varied in frequency and wavelength.

1969 *Sci. Jrnl.* Apr. 57/3 Spectroscopy is one of the most obvious fields for exploitation of laser tunability. **1977** *Jrnl. R. Soc. Arts* CXXV. 766/2 Tunability and the time-compression of laser energy into very short pulses.. represent considerable technological achievements.

tunable, tuneable, *a.* **2.** Delete *rare*⁻¹ and add: *spec.* capable of having its operating frequency and wavelength varied.

1934 E. Little *Mod. Rhythmic Drumming* (rev. ed.) 26 No outfit is complete without at least one tom-tom. The 'tuneable' models are the best, because any dampness in the atmosphere can be counteracted by the use of the tensioning handles. **1943** C. L. Boltz *Basic Radio* xii. 195 It is seen that the aerial is not 'tunable'. **1957** *Proc. IRE* XLV. 1467/2 If they possess [a maser is] 'tunable', i.e., the useful frequency can be adjusted. **1969** R. B. Fuller *Operating Man. Spaceship Earth* v. 67 The macro-cosmic irrelevancies are all the events too large and too infrequent to be synchronously tunable in any possible way with our consideration. **1969** *Sci. Jrnl.* Apr. 57/3 Lasers capable of producing megawatts of power and tunable right through the visible spectrum. **1973** *Daily Tel.* (Colour Suppl.) 21 Oct. 57 (Advt.), The VHF radio has pushbutton programme selection and each programme is separately tunable. **1979** *Jrnl. R. Soc. Arts* 126/1 In conjunction with a tunable x-ray monochromator, anomalous scattering experiments can be performed close to absorption edges.

Tunbridge (tʌ·nbrɪdʒ). **a.** Used *attrib.* to designate water from the chalybeate spring at Royal Tunbridge Wells in Kent.

a 1661 Fuller *Worthies* (1662) Kent 62 Tunbridge-waters.. Good for Splenitick distempers. **1678** T. Browne *Let.* 1 May (1946) 92 He may hopefully drinke Tunbridge waters, if they passe well. **c 1702** C. Fiennes *Journeys* (1947) 125 Tunbridge waters whose property is to retrieve lost limbs that are benumbed. **1967** E. S. Turner *Taking Cure* iii. 46 Nothing was to be gained from the Tunbridge waters unless the person who drank them was facetious, merry..and jovial.

b. Also **Tonbridge.** Used *attrib.* (rarely *absol.*) to designate wooden articles with a characteristic mosaic decoration, made in and about Royal Tunbridge Wells and nearby Tonbridge by slicing cross-sections from a bundle of thin strips of differently coloured wood glued together, to obtain identical copies of the pattern for sticking on the article to be decorated; chiefly in *Tunbridge ware.*

1773 R. Graves *Spiritual Quixote* I. ii. xiii. 101 His Tunbridge-ware tobacco-dish. **c 1795** *Advt. C. Fellows's Circul. Library, Salisbury,* With various Articles in Tunbridge, Ivory, and Morocco. **1816** Jane Austen *Emma* III. iv. 50 A pretty little Tunbridge-ware box. **1842** Marryat *Perc. Keene* I. ix. 98 In the front windows ..were..prints, caricatures, and Tonbridge ware. **1888** *Encycl. Brit.* XXIII. 607/1 Tunbridge ware..includes work tables, boxes, toys, &c. **1901** J. Black *Illustr. Carpenter & Builder Ser.: Home Handicrafts* 61 Developments of the art of what may be termed 'wood mosaic', and amongst these may be reckoned 'Tarsia work' and 'Tonbridge work'. **1934** N. Marsh *Man lay Dead* xiii. 237, I suddenly remembered..a funny Victorian casket made out of inlaid wood... Antique dealers call those caskets.. Tunbridge boxes. **1973** *Times* 25 Aug. 11/5 It has a Tunbridge-ware change-tray and an Edward vii crown set into the base.

tund, *v.* **1.** (Earlier example of the vbl. sb.)

1866 *School Life at Winchester Coll.* iii. 38 When I was a big Inferior, I have more than once received..a 'Tunding' (thrashing with a stick).

tun-dish, tundish. Restrict 'Now *local*' to sense in Dict. and add: In mod. use, a broad, open container with one or more holes in the bottom, used in various industrial processes, e.g. to feed molten metal into an ingot mould so as to avoid splashing and give a smoother flow.

1926 *Jrnl. Iron & Steel Inst.* CXIV. 74 This tun-dish is provided with one large hole or several smaller holes, dividing the stream of metal into several smaller streams. **1957** *Technology* Aug. 223/1 Plugged into the base of a fireclay tundish..fed with liquid metal from an electric furnace, is a vertical die of..graphite. **1965** *Economist* 25 Dec. 1437/2 In the new [spray steelmaking] process, the liquid iron flows from the bottom of a tundish (a container used to keep a constant head of metal). **1975** *Petroleum Rev.* XXIX. 118/1 The tank vent pipes have now been fitted with tundishes to collect condensate.

tundrite (tʊ·ndrəit). *Min.* [ad. Russ. *tundrit* (E. I. Semenov *Mineralogiya Redkikh Zemel'* (1963) 209), f. Russ. *tundra* tundra (from its being first found on the Lovozero tundra near Murmansk): see -ite[1].] A silicate and carbonate (essentially) of cerium (normal tundrite, tundrite-(Ce)) or neodymium (tundrite-(Nd)), sodium, and titanium found as triclinic brownish- or greenish-yellow crystals.

1965 *Amer. Mineralogist* L. 2098 Tundrite occurs in 3 nepheline syenite pegmatites of Mt. Nepkha, Lovozero tundra, Kola Peninsula. **1974** *Ibid.* LIX. 633/2 Infra-red study of tundrite from a new locality in the Khibina massif showed bands of carbonate; this was confirmed by spectra of the Greenland mineral.

tune, *sb.* Add: **2.** Hence (from the proverb) *to call the tune,* to hold the initiative, to have control of events.

1928 A. Huxley *Point Counter Point* xii. 211 Lucy insisted, when she was with men, on doing as much of the paying as possible. Paying, she was independent, she could call her own tune. **1948** W. S. Churchill *Second World War* I. xi. 92 A match throughout which, despite frequent fluctuation, Hitler called the tune. **1961** *Listener* 25 May 913/1 ... **1978** *Lancashire Life* Nov. 70/1 The Listers had called the local tune ever since the reign of Henry IV, although it was not until 1797 that the head of the family was ennobled.

c. (Earlier example.) Also *the tune the cat died of.*

1820 M. Wilmot *Let.* 4 May (1935) 60, I am made laugh heartily three times a week..when they cut capers in the air with solemn faces, to the tune which the old Cow died of. **1943** H. C. Bailey *Mr. Fortune finds Pig* xxvi. 140 What are they singing?..It sounds like the tune the cat died of.

tune, *v.* Add: **I. c.** *spec.* To make (a radio or television) sensitive to a chosen signal frequency or wavelength; to adjust (any device or component) by varying its operational frequency. Also *absol.*

1899 *Notices Proc. R. Inst.* XV. 475 It is easy to transmit many messages in any direction at the same time. It is only necessary to tune the transmitters and receivers to the same frequency or 'note'.. Tuning is very easy. **1900, 1904** [in Dict.]. **1923** W. H. Eccles *Wireless Teleg.* 309 The primary..is therefore tuned to the incoming waves. **1943** C. L. Boltz *Basic Radio* xii. 192 At the transmitter..the LC circuit is tuned to produce free oscillations at a chosen frequency. **1972** *Daily Tel.* (Colour Suppl.) 3 Mar. 17 For several months as 85 foot radio telescope, tuned to 21 centimetres, was aimed at the stars *Tau Ceti* and *Epsilon Eridani*. **1974** *Guardian* 23 Mar. 10/1 The kind of programme listeners tune to, broadcasters respect, and disc critics certainly rarely miss. **1975** D. G. Fink *Electronics Engineers' Handbk.* xiii. 74 Solid-state microwave masers can be tuned over a wide range of frequencies by adjustment of an external magnetic field.

d. Restrict *local* to sense in Dict. In mod. use, to adjust (an engine or part) to improve its efficiency or some other attribute; also with the vehicle or craft as obj.

1916 R. T. Nicholson *Bk. of Ford* 151 You know now how to 'tune' your carburetter for the best results. **1931** T. E. Lawrence *Let.* 10 June (1938) 724 The R.A.F. detached me to Hythe on special duty, to test and tune their new-type speed-boats for the Schneider Cup. **1955** *Times* 23 Aug. 7/6 The Rootes Group have chosen..an engine..and have tuned it for economy and longevity rather than maximum efficiency. **1978** R. Westall *Devil on Road* i. 3, I can strip her (*sc.* a motor-bike) in a day and put it all back. Tuned her like Yehudi Menuhin's violin.

5*. *intr.* Of a radio, etc.: to be capable of being tuned.

1922 *Wireless World* 1 July 435/1 Will the Reinartz Tuner tune to any wavelength by means of external coils? **1930** J. H. Reyner *Testing Radio Sets* iv. 93 It may be found that the circuit does not tune correctly when the aerial is connected to the proper terminal. **1970** J. Earl *Tuners & Amplifiers* i. 23 The f.m. section will..tune over Band II, usually from about 88 to 108MHz.

6. tune in. **b.** *trans.* and *intr.* To tune a radio or television to (a particular station or transmission, or a particular frequency). Freq. *to tune in on* or *to.* Also *transf.*

1913 *Wireless World* Apr. p. xxviii/1 It is possible to tune out one ship or station and tune in others. **1919** *Ibid.* May 105/2 Operators at the various Government wireless stations in and about towns, who promptly 'tuned in' and listened. **1922** *Westm. Gaz.* 12 Dec., While listening-in on a Lincoln wireless company's apparatus..Mr. Mawer was successful in tuning in to an American broadcasting station. **1929** S. Chesky *Theatre* xxiv. 556 A million or so people may..tell in their parlors and 'tune in' on a song by Al Jolson or a scene from *Twelfth Night*. **1938** S. Lewis *It can't happen Here* 52 He tuned in on a program of old songs. **1936** Auden & Isherwood *Ascent of F6* (1937) II. iii. 100 Turn off the wireless. Tune in to another station. **1940** Auden & Isherwood *Journey to a War* ... **1957** A. C. Clarke *Deep Range* xii. 104 This was an important part of his training. **1970** Hessing & Johnson ... **1964** Mrs. L. B. Johnson *White House Diary* 12 Jan. (1970) 25 We tuned in on my TV appearance in 'The Week That Was'. **1976** *Shooting Times & Country Mag.* 16–22 Dec. 30/3 The Indians, who were in the stand of tundra, but their quarry, as they believe that the hunted can 'tune in' on their thoughts. **1977** 'J. Fraser' *Heart's Ease in Death* vii. 61 You make it sound like a radio serial. Tune in next week for the latest episode.

c. *fig.* To become mentally receptive to, or aware of; to comprehend. Const. as prec. Also *sense.*

1926 *Variety* 29 Dec. 5/4 Those fortunate individuals who can tune in on the conversation of a flock of cannons and follow it without the aid of a central office clerk or an interpreter. **1961** A. Miller *Misfits* ii. 18 His mind is constantly trying to tune in on the world, but the message is never clear. **1977** R. Gadney *Champagne Marxist* xii. 84 He'd been so slow in tuning in to the presence of an intruder.

d. *fig.* To harmonize *with.*

1958 M. MacNeice *I crossed Minch* ii. 25, I had a passion for the wild.. This tuned in with my other passions for the antique, the fantastic, and the Irish.

7. tune off. **b.** *trans.* = *tune out* (sense 7* below); also *intr.* Also *fig.*

1926 Maines & Grant *Wise-Crack Dict.* 14/1 *Tune off that station,* change the conversation. **1931** F. A. Arnold *Broadcast Advertising* 140 The public has its own method of self-protection. The listener may..tune off and find some program less offensive. **1957** *Practical Wireless* XXXIII. 722/1 When the input signal decreases, as one tunes off the station, the valve will conduct.

7*. tune out. *trans.* To eliminate reception of (a radio signal of a particular frequency) by tuning.

1908 *Rep. Brit. Assoc. Adv. Sci.* 1907 621 It is easy to hear the ships in the Channel, but it is also easy to tune everything out and listen to the signals from Cornwall. **1913** [see sense 6 b above]. **1957** *Practical Wireless* XXXIII. 721/1 When..the signal is tuned out, the voltage at point A drops. **1970** J. Earl *Tuners & Amplifiers* ii. 47 The filter tunes out the 19kHz pilot tone. **1977** C. MacNeice *Kind of Treason* xiii. 114 Strand was knowing at the radio. He found the station and tuned the static out.

b. *fig.* To disregard; to cease listening to.

1938 T. E. Lawrence *Let.* 1 May (1938) 599 In the East, you hear everything that's happening, and a great deal more. The selective ear tunes out the false news. **1969** *Sat. Rev.* (U.S.) 5 July 28 If you don't like what I say, you can tune me out. **1973** *Times* 18 Dec. 18/3 Rogers *Children & Language* (1975) xv. 322 He learns, as a result of failure and frustration in school, either to tune out the school or to adopt a defiant, rebellious attitude towards it. **1978** G. A. Sheehan *Running & Being* iii. 34, I have the ability to tune out what is going on around me.

8. tune up. **b.** Now usu. *absol.*

1869 W. S. Gilbert *Bab Ballads* 182 He requested them to tune up and begin. **1920** W. Faulkner *Sound & Fury* 31 'Now you got to tune up.' Dilsey said. **1981** A. Schlee *Rhine Journey* xii. 154 Already the orchestra was tuning up for the next waltz.

tuned, *ppl. a.* Add: **2. a.** *Electronics.* Adjusted so as to resonate at a particular frequency; forming part of a circuit so adjusted. Also *transf.*

1899 J. J. Fahie *Hist. Wireless Telegr.* 182 The circuits are said to be in resonance, or to be electrically tuned. **1928,** etc. [see *Litz]. **1936** *Discovery* June 197/2 This is of importance in short-wave technique, seeing that condensation is constantly varying the capacity of the tuned circuit. **1962** Simpson & Richards *Physical Princ. Junction Transistors* xiv. 328 Two types of high-frequency amplifiers are in general use, tuned amplifiers and untuned wide-band or video amplifiers. **1971** *Nature* 3 Dec.

256/1 In *Diptera,* the antennae are the principal organs of hearing and are tuned and directional. **1980** K. J. Borlman *Colour & Mono Television* I. viii. 97 At Band 4 and 5 frequencies, conventional *L, C* tuning circuits are not practical... Fortunately..it is possible to use tuned transmission lines.

b. Having one's radio or television tuned (*in*) to a particular station; esp. in imp. phr. *stay tuned,* go on listening to this station.

1929 *Amer. Speech* XXXI. 258 Stay tuned for Roy Neal and his program. **1970** *Wall St. Jrnl.* 15 June 1/1, 13 million American homes are tuned in via ABC-TV's.. a previously taped episode of Mr Wolk's 15-year-old show. **1972** *Guardian* 8 Jan. 8/3 Our own Keith Dewhurst ..had stayed tuned for this unique blend. **1977** 'E. Crispin' *Glimpses of Moon* viii. 152 She fiddled with volume control in order to get him satisfactorily tuned in.

c. *tuned in* (fig.): (*a*) in rapport with, in harmony with; const. *to, on;* (*b*) (*slang*) = *switched on* s.v. *switched *a.* and *ppl. a.* 3 b.

1958 *Orf. Mag.* 1 May 398/1 It took us some time to get tuned in... Everyone was saying how much better the level was than the first... Had we just begun to get the wave-length?... A minor triumph in the difficult art of getting acquainted. **1963** N. Marsh *Dead Water* (1964) ii. 39 I like..the way people think. All the same, it's fun trying to get tuned-in. **1968** P. Brook *Empty Space* iii. 85 It is perhaps for this reason that..the pop tradition in England has such wide appeal: non-political, unaligned, it is none the less tuned in on a fragmented world in which bombs, drugs, God, parents, sex, and private anxieties..are all of the same order. **1972** D. Hanson *In High Places* xii. 143 I'd walk alone, slowly shaking off the delights of civilization and getting completely tuned in once again to a mountain environment. **1976** *National Observer* (U.S.) 14 Aug. 12/2 This is a regional museum tuned in to the story of man's relationship to the wilderness. **1977** *Hudson Rev.* Spring 69, I thought he'd be some kind of creep, instead I meet this really sweet cat.. He's really tuned in.

tune-in (tiū·n,ɪn). *U.S.* [f. vbl. phr. *to tune in:* see *Tune *v.* 6 b.] **a.** The state of being tuned to a particular station or channel. **b.** The size of the audience for a station or channel.

1928 in F. A. Arnold *Fourth Dimension* 141, I hit the high spots between 7:00 and 10:00 p.m. four evenings a week with a tune-in on the biggest features that come occasionally. **1951** *Sun* (Baltimore) (B ed.) 20 Mar. 1/7 No figures were available for the afternoon telecast, but the average tune-in from 1:00 to 6 P.M. here is about twelve per cent. **1959** *T.V. Guide* (N.Y. Metro ed.) 8 Aug. 41/1 The network is abandoning..long familiar Sabbath half hours..in favor of occasional hour-long specials to be scheduled late Sunday afternoon when tune-in is higher than it is at midday.

tunellite (tʊne·lait). *Min.* [f. the name of George *Tunell* (b. 1900), U.S. geologist + -ite[1].] A hydrated borate of strontium, $SrB_6O_{10}.4H_2O$, found as colourless or white monoclinic crystals.

1961 R. C. Erd et al. in *Prof. Papers U.S. Geol. Survey* No. 424-C 294/1 Tunellite was first noted..in some samples collected in 1957 from a working of the Jenifer mine, Kramer borate district, California. **1964**, **1968** [see *nobleite]. **1978** *Mexican Geologist* XI. 26 Tunellite ..has been found only in the lower part of the borate zone in the Emet deposits in the clay layers in the Kirka deposits (in Turkey).

tuner. Add: **3.** Any device for varying the frequency to which a radio or television is tuned; *spec.* a separate unit for detecting and preamplifying the programme signal and supplying it to an audio amplifier. Also *tuner unit.*

1909 J. Erskine-Murray *Handbk. Wireless Teleg.* 261 Variometer used as tuner. **1937** *Radio Times* 5 Mar. (Suppl.) p. iv/2 With it, no additional tuners are required. **1946** *Wireless World* LII. 296 All receivers..should be provided with..dial settings to ensure that every station..can be tuned in a tuned receiving station. **1959** W. Greenwood *Text-bk. Wireless Teleg. & Teleph.* vi. 122 (*heading*) The tuner portion of a receiving circuit. *Ibid.* 124 The various tuning inductances and condensers are often assembled in a separate box called the 'tuner'. **1957** E. T. Canby *Home Music Syst.* iii. 37 There are separate units that combine numerous functions: AM-FM radio tuners; ..separate AM tuners and FM tuners; [etc.]. **1959** *Listener* 26 Mar. 542/1 There is no simple tuner for a high-quality FM tuner unit. **1967** *Ibid.* 30 Mar. 424/2 Some parts of the [television] receiver, such as the tuner, can be quite cheap, but will cost no more than in a black-and-white receiver. **1970** *Which?* Apr. 113/1 Most of the tuners had a device to show when you were exactly tuned to a station. **1976** K. Thackeray *Crownbird* vi. 127 Priest rotated the tuner so the static-punctuated voice said that Cedric..

4. Special Comb.: **tuner-amplifier,** a combined radio tuner and amplifier; cf. *receiver[1] 7 c; also abbrev. **tuner-amp.**

1970 J. Earl *Tuners & Amplifiers* i. 21 The tuner-amplifier is a new breed of hi-fi equipment. **1975** *Hi-Fi Answers* Feb. 49/1 Your choice of possible tuner-amps is limited by your requirement for an MW section. **1979** J. Gardner *Nostradamus Traitor* xi. 37 A big reel-to-reel tape machine..was linked to a tuner/amplifier set to a pre-selected clear channel.

Tunesin(e), obs. var. *Tunisine *sb.* and *a.*

tunesmith (tiū·n,smɪþ). *colloq.* (orig. *U.S.*). [f. Tune *sb.* + Smith *sb.*] A composer of popular music or songs; *derog.*, a composer of unoriginal or trifling music.

1926 Whiteman & McBride *Jazz* viii. 171 Jazz..is the hardest of all to write, the tunesmiths say. **1959** E. S. Turner *Court of St. James's* xxiii. 256 A modern lyric writer and tunesmith in Tin Pan Alley. **1962** *Times* 24 Aug. 21/6 How else does it happen that Bellini, dismissed for years as a vapid tunesmith, has become admired in our own day as a master of poised classical music-drama. **1976** *Gramophone* Nov. 843/3 How dare he dismiss Duke Ellington as a mere 'tunesmith'!

tunesome, *a.* For *rare*⁻¹ read *rare* and add later example.

1930 W. de la Mare *Crossings* iii. 82 As neat a brace of nightingales as ever I heard. Shy, but tunesome.

tunester (tiū·nstəɹ). Chiefly *U.S.* [f. Tune *sb.* + -ster.] A song-writer or singer; a musician, composer.

1903 H. Hughes *Love Affairs of Great Musicians* I. xii. 126 (*heading*) A few tunesters of France and Italy—Peri, Monteverdi, et al. **1935** *Amer. Speech* X. 154/2 Singers are now tunesters to advertisement writers for vaudeville and other entertainments. **1936** *N.Y. World-Telegram* 22 Aug. 14/4 At that studio, too, are Mack Gordon and Harry Revel, best known of the Hollywood tunesters. **1983** *Washington Post* 9 May B11/1 Luigi Proietti as a tour guide is pretty lively and alert, like a merry tunester at a funeral.

tune-up (tiū·n,ʌp). orig. *U.S.* [f. vbl. phr. *to tune up*: see Tune *v.* 8 in Dict. and Suppl.]

1. The action or an act, of tuning up (*lit.* and *fig.*).

1924 *Automotive Electrician* Nov. 16 (*heading*) Analyzing the need for winter tune-up. **1959** [see *ray rod]. **1962** *Daily Tel.* 27 Jan. 20/5 Parties begin on election tune-up. ..Preliminary moves to 'tune-up' the Conservative and Labour party machines for the next general election have already begun. **1968** 'E. Lathen' *Stitch in Time* viii. 71 He always does the tune-ups on Dr Weverson's sports cars. **1977** *Rolling Stone* 16 June 12/1 They still take ten-minute tuneups between songs.

2. *Sport* (chiefly *U.S.*). An event that serves as a practice for a subsequent one.

1934 *Collier's* 11 Aug. 48/3 Webb..intended starting Black Gold in a few tune-up races. **1940** *Sun* (Baltimore) 19 Aug. 20/5 The rains..spoiled several attempts to stage an informal flying exhibit as a tuneup for the annual meet. **1946** *Richmond* (Va.) *Times-Dispatch* 10 Apr. 16/3 A couple of Purple Heart fighters..went along fine in easy tuneups. **1949** *Ibid.* 1 Feb. 20/7 Vaughan..who must have benefited from his thorough tune-up against Hughes. **1979** E. Newman *Sunday Punch* xii. 181 'When's Turner?' 'In about four months. I have a tune-up in Pittsburgh first.'

tuney, tune-y, var. Tuny *a.*

tung (tʌŋ). [a. Chinese *tóng*]. **1. a.** Any of three trees of the genus *Aleurites* or *Vernicia* (family Euphorbiaceae), *A. fordii, A. cordata,* and *A. montana,* which are native to China and Japan and are cultivated there and elsewhere for the oil from their seeds. So *tung tree.*

1880 *Cent. Dict.*, Tung-tree. **1914** N. Shaw *Chinese Forest Trees* 11 The Tung..tree or tung oil trees, and tung grow wild in these hills. **1928** *Outward Bound* Jan. 42/1 The tung trees were in blossom. **1950** E. Nicholls *Thrones* xxix. 52 Phoenix to Tung tree A mirror to flowers, as water is to the moon. **1965** J. Ch'en *Mao & Chinese Revolution* (1967) ix. 138 The mountains are covered..with tung-trees. **1979** *Tropical Agriculture* LI. 10 This series of trials has shown that nitrogen is the most important nutrient for tung in Malawi.

b. The oil extracted from seeds of the tung tree, used chiefly in the manufacture of inks, paints, and varnishes; = Wood-oil (c).

1911 *Encycl. Brit.* XX. 46/1 (table) Name of oil... Tung. **1950** *Caribbean Q.* I. iii. 42 A trial should be made with a settlement of 100 families, beginning with pigs and poultry and proceeding to tree crops (coffee, cocoa, citrus, tung). **1962** H. G. Champion *Streets's Exotic Forest Trees Brit. Commonw.* II. 173 The oil [of *Aleurites montana*] is known commercially as Tung.

2. *attrib.*, as *tung nut*; **tung oil,** a drying oil obtained from the seeds of tung trees and used in varnishes, paints, and inks; so *tung oil tree;* cf. Wood-oil.

1937 R. Frost *Let.* 5 Jan. (1964) 287 We will probably end our days growing Tung nuts in northern Florida. **1881** Tung-oil [see *Wood-oil]. **1913** E. H. Wilson *Naturalist in W. China* II. vii. 64 'Tung Oil tree'.. produces this valuable oil. **1937** *A.F. Hill Econ. Bot.* ix. 213 The United States uses so much tung oil that trees of the latter species have been introduced. **1951** R. Mayer *Artist's Handbk.* iii. 112 Tung Oil..is highly valued as an ingredient of industrial paints and varnishes. **1972** *Guardian* 16 Feb. 15/3 American businessmen..are confronted with increasing difficulties in obtaining soya beans, bristles, tung-oil and egg products from China. **1973** *Times* 21 Mar. (China Trade Suppl.) p. xiv/5 Huge tracts of hillside had only recently been terraced for planting with..tung oil trees.

Also **tung-yu** [Chinese *yóu* oil], tung oil.

1788 tr. *Grosier's Gen. Descr. China* I. iv. vi. 449 The work is..daubed over with a kind of oil, which the Chinese call *tong-yeou.* **1913** E. H. Wilson *Naturalist in W. China* II. vii. 66 'T'ung-yu' is the chief paint oil throughout the Chinese Empire. **1973** T. R. Tregear *Chinese* iv. 89 The Yangtze valley is noted for its production of..*tung yu* (wood oil), which forms the basis of paints and varnish.

Tungan (tʊ·-, tu·ŋgan; dʊ-, du·ŋgan), *sb.* and *a.* Also † **Tungani** *sing.* and *pl.*; **Dungan.** [ad. Jagatai *Döngan.*] **A.** *sb.* A member of a Muslim people in China and Russian Central Asia of Chinese descent. **B.** *adj.* Of or pertaining to the Tungans.

1875 Bellew & Chapman in T. D. Forsyth *Report Mission Yarkund* ii. 81 The Kara Khitay, the Khitay, and the Tungani. **1908** Lady Macartney *Diary* 12 July in *Eng. Lady in Chinese Turkestan* (1931) xi. 166 Tungans and Kirghiz, noticeable by reason of their big fur hats. **1927** Cable & French *Through Jade Gate* xxxii. 221 A stranger on first seeing a Tungan youth would probably say: 'What a handsome Arab boy.' *Ibid.*, Whereas the Tungan speaks Chinese, the Turki has his own tongue. **1965** K. P. S. Menon *Many Worlds* xiv. 201 A young Tungan General. **1974** *Encycl. Brit. Micropædia* V. 190/3 Some Chinese Muslims, called Dungans, have settled in the Soviet Union in the villages and towns of the Uzbek Soviet Socialist Republic, and the Alma-Ata *oblast* (region) of the Kazakh S.S.R.

Tungar (tʊ·ŋgɑɹ). *Electronics.* [f. Tung(sten + *ar(gon.] A type of low-voltage discharge tube filled with argon and having a heated cathode of thoriated tungsten, used as a rectifier for currents of a few amperes. (A proprietary name in the U.S.)

1917 *Official Gaz.* (U.S. Patent Office) 4 Dec. 270/1 General Electric Company... Tungar... Rectifying apparatus. Claims use since October, 1916. **1935** Nilson & Hornung *Practical Radio Communication* xxi. 710 The Tungar battery charger..is a device for charging storage batteries from an alternating-current line. **1966** McGraw-Hill *Encycl. Sci. & Technol.* VI. 61/1 Hot-cathode gas tubes... three representative types may be distinguished, (1) the Tungar..; (2) the phanotron; and (3) the thyratron.

Tungkingese, -quinese, obs. varr. *Tonkinese *sb.* and *a.*

tungsten. Add: **3.** *tungsten filament*; **tungsten carbide,** either of two compounds of tungsten and carbon, WC and W_2C, that are very hard and are used for cutting tools and abrasives.

1899 *Jrnl. Chem. Soc.* LXXVI. ii. 104 On heating a mixture of tungstic anhydride..., iron, and petroleum coke..an iron tungsten carbide..is obtained. **1930** *Engineering* 14 Nov. 624/1 The enhanced cutting properties of the newer cutting steels, such as the tungsten-carbide tools. **1963** C. R. Cowell et al. *Inlays, Crowns, & Bridges* iii. 12 Penetrate to just within the dentine, using a small round tungsten-carbide bur. **1973** *Sci. Amer.* July 42/1 Tungsten carbide, a very hard ceramic..has been used as a cutting tool. **1924** Glazebrook *Dict. Appl. Physics* II. 379/2 From 1911..It became obvious that the future of the incandescent lamp for some time to come would be with the tungsten filament lamp. **1962** V. Nabokov *Pale Fire* 192 The dead, the gentle dead—who knows?—In tungsten filaments abide.

tungstenite. Add: **3.** A tungsten sulphide, probably WS_2, that occurs as both hexagonal and rhombohedral polytypes in dark grey scaly aggregates that mark the fingers.

1917 Wells & Butler in *Jrnl. Washington Acad. Sci.* VII. 596 It is..a pleasure to announce the discovery of tungsten sulphide..from..Utah... On account of the apparent resemblance to molybdenite in form and some of its properties the new mineral has been named tungstenite. **1970** *Canad. Mineralogist* X. 731 It is virtually impossible to distinguish the x-ray powder pattern of a given polytype of molybdenite from the same polytype of tungstenite... Consequently it is quite possible that tungstenite is more widespread than the literature indicates.

Tungus (tu·ŋus, tuŋū·s). Forms: 6 **Tingus,** 6–7 **Tongu(e)se,** 8 **Toongus.** [Yakut name of a people called by themselves Evenki.] **a.** (A member of) a people of eastern Siberia. Also *attrib.*

1625 Purchas *Pilgrimes* III. iii. vii. 527 The people.. signified..that they were called *Tingoesi*, and that they dwelt along the great Riuer..the great Riuer Ieniscé. *Ibid.* x. 543 These *Tingusses* report, that there is another great Riuer. **1698** tr. *A. Brand's Embassy Muscovy into China* 48 The Tongueses..have of late years been conquered by the Victorious Arms of the Czars of Muscovy, unto whom they pay a yearly Tribute. **1698** [see Shaman *sb.*]. **1763** J. Bell *Trav. from St. Petersburg* I. iii. 231 The Tongusy..so called from the name of the river, who live along its banks, are the posterity of the ancient inhabitants of Siberia. *Ibid.* 229 When a Tonguse kills an elk or deer, he never moves from the place, till he has eat it up. **1799** W. Tooke *View Russian Empire* II. 98 That the Tunguses originally composed one people with the Mandshee, is apparent not only from the resemblance of features..but also chiefly from the agreement of their languages. **1841** *Penny Cycl.* XXI. 459/2 The whole region..has been abandoned to the Toonguses, who get their subsistence by the chace. **1882** A. H. Keane *Asia* 478 Conterminous on the north with the Buriats are the Tunguses. *Ibid.* 479 The Tungus race. **1914** M. A. Czaplicka *Aborig. Siberia* 52 The Nomadic Tungus are cattle-breeders. **1931** M. Burr *Bolshevik Siberia* 154 When a Tungus hits off the spoor of one of these grand beasts. **1948** A. L. Kroeber *Anthropol.* (rev. ed.) x. 429

The Manchu, the Mongols, and the still earlier Tungus conquerors of China lost their own cultures there. **1974** J. R. Baker *Race* x. 174 The Tungus and Kalmuks (Tungid subrace of Mongolids) are said to be devoid of axillary smell.

b. An Altaic language or group of languages related to Manchu, spoken in parts of Siberia, and since 1931 set down in an alphabet based on the Russian alphabet. Also *attrib.* Cf. *Manchu-Tungus.

1822 tr. *Malte-Brun's Universal Geogr.* I. xxiii. 571 The Tunguse is a dialect of the Mantchou. **1888** *Encycl. Brit.* XXIV. 1/2 An exuberance..of verbal forms, which in Osmanli, Finnish, Magyar, Tungus, and Mordvinian may be said to run riot. **1961** L. F. Brosnahan *Sounds of Language* viii. 177 The area of the simple stress accent.. includes..the Mongolian, Tungus and Paleosiberian languages of eastern Asia. **1977** C. F. & F. M. Voegelin *Classification & Index World's Languages* 335 The Tungus languages are bifurcated into two groups... Northern Tungus..= Evenki... Southern Tungus.

Hence **Tungu·sian** *a.*, of or pertaining to this people or language; *sb.*, a Tungus; also, the Tungus language; **Tungu·sic** *a.*, Tungusian; *sb.*, the Tungus language.

1706 tr. *E. Y. Ides's Trav.* 27 Subject to the Jurisdiction of this City are several Tunguzian Heathens. **1706** [see Shaman *sb.*]. **1763** J. Bell *Trav. from St. Petersburg* I. iii. 231 From all the accounts I have heard..of the natives of Canada..they..resemble..the Tongusians. **1799** W. Tooke *View Russian Empire* II. 100 At that time [*sc.* 1607] many tungusian stems owned the paramount supremacy of the Buriats who had shortly before been expelled from Mongolia. **1839** *Jrnl. R. Geogr. Soc.* IX. 198 The Tungusian, though confined to the eastern extreme of the ancient continent, contains some words common to it with languages spoken in Europe. **1854** Max Müller in C. Bunsen *Christianity & Mankind* III. 277 Some of them are certainly widely distant; as, for instance, the dialects of the Finnic nations in the west, and of the Mongolic and Tungusic tribes in the east. **1855** H. D. Seymour *Russia on Black Sea & Sea of Azof* v. 49 The Tungusians extend on the east from the Yenisei to the Sea of Okhotsk. **1864** *Ann. Rep. Smithsonian Inst. 1863* 111 The rank of the Scythian languages..is but an inferior one... They diminish in value eastward, the Tungusic being the poorest of all. **1885** J. Byrne *Struct. Lang.* I. 391 The Tungusian dialect. *Ibid.* 398 The verb *bi*, which in Tungusian takes *hi* in the present. **1888** *Encycl. Brit.* XXIV. 2/2 Turkic lies much closer to Mongolic than it does to Samoyedic and Tungusic. **1890** [see *Gilyak]. **1914** [see *Palæo-Siberian *sb.*]. **1951** W. K. Matthews *Languages U.S.S.R.* iv. 54 Of the three branches of Altaic the easternmost and least significant numerically is the Manchurian or Tungusic. This comprises two subdivisions, viz. the declining Manchu of Northern Manchuria (Manchukuo), with the related languages..and the more primitive and vigorous Tungus (Evenki) and its cognates. **1956** J. Whatmough *Language* ii. 32 The great belt of Mongolian and Tungusian..is connected with the Turkic languages further west, and with Yakut to the North, reaching through Siberia to the Arctic. **1977** *N.Y. Rev. Books* 14 Apr. 3/1 Founded in 947 as a capital of the Khitan Mongols' Liao dynasty, it [*sc.* Peking] had been used similarly by the Tungusic Chin dynasty 1122–1234.

tunic. Add: **5.** Also **tunic shirt,** a long loose-fitting shirt worn outside the trousers; cf. Caftan, *kaftan.

1918 G. Frankau *One of Them* xxx. 234 Smart bosom itch in horsehair tunic-shirts. **1930** *Daily Express* 6 Oct. 5/1 (Advt.), Men's tunic shirts made of the balloon fabric. **1971** 'D. Halliday' *Dolly & Doctor Bird* viii. 104 Mr Tiko, in a blue tunic shirt and blue trousers.

tunica[1] (tiū·nikă). *Anat.* [L.: see Tunic.] = Tunic 4 a in various mod.L. collocations, as *tunica adventitia* [see Adventitious *a.*], an outer sheath, esp. of a blood-vessel; *tunica albuginea* [L. *albūgin-is* white spot], a white fibrous layer, esp. of the penis or testes; *tunica vaginalis,* a serous membrane covering much of the testis.

1698 W. Cowper *Anat. Humane Bodies* sig. Aa, The tunica albuginea, or proper membrane of the testes. **1828** J. Quain *Elem. Anat.* 530 The tunica vaginalis, or serous covering derived from the peritonæum. **1890** *Gray's Anat.* (ed. 12) 48 The arteries are composed of three coats: internal or endothelial coat (tunica intima of Kölliker); middle muscular coat (tunica media); and external cellular coat (tunica adventitia). **1963** *Lancet* 5 Jan. 19/2 The two layers of the tunica vaginalis, which normally invest the testis alone, extend upwards to cover the whole epididymis and the cord, sometimes as high as the inguinal canal. **1977** *Proc. R. Soc. Med.* LXX. 645/1 They observed degenerative changes mainly in the progressive thickening of the tunica adventitia. **1980** *Gray's Anat.* (ed. 36) 1152/1 The fibrous layer of the eyeball..consists of an opaque, posterior part, the tunica sclera, and a transparent, anterior part, the tunica cornea.

Tunica[2] (tiū·nikă). Also [† **Tonica, Tonika.** [ad. F. *Tonika, Tounika,* perh. ad. Tunica *tóniku* the man.] (A member of) an American Indian people of the lower Mississippi valley; their language, now extinct. Hence **Tu·nican,** † **To·nikan,** a postulated linguistic family of which Tunica was the chief member.

1806 J. Sibley in *Message from President of U.S., communicating Discoveries made in exploring Missouri by Capts. Lewis & Clark* 83 Tunicas. These people lived formerly on the Bayan Tunica..on the Mississippi, east side... Their native language is peculiar to themselves. **1891** D. G. Brinton *Amer. Race* i. v. 91 The Tonicas are frequently mentioned in early French accounts of the colony of Louisiana. **1891** J. W. Powell *Indian Linguistic Families* 125 The distinctness of the Tonika language, has long been suspected, and was indeed distinctly stated by Dr. Sibley in 1806. *Ibid.*, The Tonika are known to have occupied three localities. **1902** *Encycl. Brit.* XXV. 374/1 [Linguistic families of North America] Tonikan, Miss. **1911** J. R. Swanton *Indian Tribes Lower Miss. Valley* 19 The method of distinguishing masculine and feminine pronominal forms is also decidedly unlike, Taënsa employing a suffix while Tunica uses entirely distinct forms. **1947** *Romance Philol.* I. 145 (*title*) Some French loan-words in Tunica. **1965** *Canad. Jrnl. Linguistics* Spring 100 Tunican (comprising Tunica, Atakapa, and Chitimacha).

tuning, *vbl. sb.* Add: **1. c.** Also, the process of making adjustments to the engine of a motor vehicle so as to improve its performance.

1916 R. T. Nicholson *Bk. Ford* 151 With proper 'tuning', you ought..to get from 25 to 30 miles per gallon on give-and-take roads. **1939** W. Hassan in Earl Howe et al. *Motor Racing* (Lonsdale Libr. XXVII) xv. 181 One of the most important items in the tuning of a racing car is the correct interpretation and application of the rules of the race for which it is being prepared. **1971** C. Williams *Car Conversions for Power & Speed* v. 127 The most advanced tuning of all is found on racing engines, where the average small capacity unit may be producing more than twice the power of an equivalent engine in a road car.

d. The adjustment of a transmitter or receiver to a particular signal frequency or wavelength; variation of the resonant frequency of an oscillatory circuit. Also *tuning in,* the action of adjusting a radio set to a desired frequency; the selection (of a frequency) by this process; also *transf.*; *tuning out,* the cutting out (of a radio transmission) by tuning.

1899 [see *tune *v.* 1 c]. **1908** *Rep. Brit. Assoc. Adv. Sci. 1907* 622 The various self-inductions and other arrangements for effecting tuning are similarly wound. **1927** W. E. Collinson *Contemp. Eng.* 113 If they have heard through a friend's set they..will have some inkling of the mysteries of tuning in and tuning out. **1929** *Radio Times* 8 Nov. 386/1 Only three knobs..one for tuning, one for volume and one for wavelengths. **1934** H. Jackson *Maxims Bks. & Reading* 9 Reading is nothing but tuning oneself in to a book in a spirit of reverential subjection. **1940** *Amer. Speech* XV. 247 He allows nobody else to have anything to do with the *tuning-in* and the *tuning-out* (or the *turning-off*) of the radio programs. **1970** J. Earl *Tuners & Amplifiers* iii. 73 Very accurate tuning is essential for good stereo reception. **1975** D. G. Fink *Electronics Engineers' Handbk.* xxi. 11 The transmitter is designed for a minimum of tuning adjustment, and..all tuning can be performed from the front panel using only two controls. **1977** *Listener* 17 Mar. 344/1 A furtive tuning-in to Radio 3.

4. (sense *1 d) *tuning circuit, coil, condenser, indicator, inductance, knob, meter*; **tuning-slide** (examples).

1943 C. L. Boltz *Basic Radio* xii. 195 We then put a coil in the aerial to earth circuit, and couple this inductively to the coil of the tuning circuit. **1923** *Popular Wireless* 13 Oct. (Suppl.) 1 Many wireless amateurs experience considerable difficulty in estimating the maximum wavelengths of their tuning coils. **1978** F. Maclean *Take Nine Spies* iv. 148 The copper wire needed for the tuning coils he managed..to buy in Tokyo. **1913** *Wireless World* Apr. p. xxix, If the aerial tuning condenser was set to its previous value and the tuning-switch (not the aerial tuning inductance) put to the second stop, the maximum signals were again obtained. **1937** F. E. Terman *Radio Engin.* (ed. 2) xiii. 559 A more recent development in tuning indicators is a special miniature cathode-ray tube. **1913** Tuning inductance [see *tuning condenser* above]. **1981** *Sunday Express* (Colour Suppl.) 12 July 33/4 For a monthly subscription fee the tuning knob of a British domestic TV set could then offer a dozen or more channels. **1978** *N.Y. Times* 30 Mar. B11/1 (Advt.), Model STA-52.. includes..tuning meter and a cabinet that's made of genuine walnut veneer. **1885** Tuning-slide [see *shank *sb.* 5 w]. **1961** A. Baines *Mus. Instruments* 358 *Tuning slide,* in wind instruments, a part of the tubing that is made extensible for the purposes of tuning.

Tunisian (tiūni·ziăn), *sb.* and *a.* [f. *Tunis* + -ian, or *Tunisia* + -an (see below): cf. the earlier *Tunisine *sb.* and *a.*] **A.** *sb.* **a.** A native or inhabitant of the country of Tunisia in North Africa (or of its capital Tunis), or of the former Barbary state of Tunis which preceded it. **b.** The demotic speech of the Tunisians. **B.** *adj.* Of, pertaining to, or belonging to Tunisia, or Tunis.

1825 J. C. Loudon *Encycl. Agric.* i. vi. 175 The Tunisians are much more agriculturists than their neighbors either of Tripoli or Algiers. **1843** *Penny Cycl.* XXV. 361/1 Grain is frequently imported into the Tunisian territory. **1891** O. Wilde *Pict. Dorian Gray* xi. 199 Yellow-shawled Tunisians plucked at the strained strings of monstrous lutes. **1902** *Encycl. Brit.* XXXIII. 483/2 No doubt in vulgar Tunisian a good many Berber words remain. **1926** A. Huxley *Let.* 31 Dec. (1969) 279, I hope the children's Tunisian dates will arrive fairly soon. **1958** *Ann. Reg. 1957* 325 The Franco-Tunisian customs union. **1973** 'A. Hall' *Tango Briefing* xvii. 217 There was a police guard ..a young Tunisian with a peaked cap.

Tunisine (tiū·nisīn), *sb.* and *a.* ? *Obs.* Also Tunesin(e), Tunis(s)een. [f. *Tunis* + -ine[1]: see prec.] **A.** *sb.* A native of Tunis, a city and former Barbary state in North Africa; esp. a pirate from Tunis. **B.** *adj.* Of or belonging to Tunis.

c 1670 J. Verney *Let.* in M. M. Verney *Mem.* (1899) IV. v. 159 [The ship was taken] by the Tunisseens. **1738** T. Shaw *Trav. Barbary & Levant* 155 The Tuniseens, are the most civilized Nation of Barbary. **1764** A. Anderson *Origin Commerce* I. iii. 126 The Genoese grew very uneasy, lest the Tunesins..should seize on all their effects. **1843** *Penny Cycl.* XXV. 360/1 Susa..is..one of the wealthy cities of the Tunisine state. *Ibid.* 364/2 The Tunisines in general, like the Algerines, are a mixed race of Turks, Moors and Jews. *Ibid.* 366/2 The Tunesine corsairs continued their excursions at sea until 1655.

tunket (tʊ·ŋkĕt). *U.S. dial.* or *colloq.* Sometimes with initial capital. Also **tunkett.** [Origin doubtful.] Euphem. for *hell*; chiefly *who* (*what, why,* etc.) *in tunket.*

1871 *Scribner's Monthly* II. 630 What in tunket are you making such a to-do about it for? **1894** *Life* 4 Jan. 13/2 What in the name o' Tunkett makes all boys so crazy to leave the old farm? **1905** G. S. Wasson *Green Shay* iii. 37 'Who in tunket is it backs up the old creetur', anyways?' asked Master Fairway. **1922** Joyce *Ulysses* 420 Golly, whatten tunket's yon guy in the mackintosh? **1951** E. Graham *My Window looks down East* vii. 59 'And why not, in tunket?' she says. **1971** H. A. Smith *View from Chivo* xix. 192, I cannot forego the use of harsh language when I think of him; he makes me madder'n tunket.

tunku (tu·ŋku). Also ‖ **tengku** (te·ŋku). [Malay.] A title of rank in certain states of Western Malaysia; = 'prince'.

1879 in C. W. Harrison *Council Minutes, Perak, 1877–79* (1907) 63 Toh Puan, the chief wife of the former Tengku Mentri. **1897** D. C. Boulger *Life Sir Stamford Raffles* x. 314 He also goes on to say that Tunku Long, a native rajah, arrived from Rhio, and that Raffles thereupon acknowledged him as Sultan of Johore. **1911** R. J. Wilkinson *Papers Malay Subjects: Malay Hist.* v. xi. 34 His son, Tengku Antah, on claiming the throne, was opposed by a Sumatran prince. **1956** *Britannica Bk. of Year* 283/2 Tengku Abdul Rahman was to head a delegation to London early in 1956. *Ibid.* 284/1 On Dec. 28–29, less than a week before he left for the London talks, the Tengku..met Chin Peng, secretary of the Malayan Communist Party. **1961** *Listener* 23 Nov. 866/2 The Tunku is in this country to discuss with the British Government proposals for a Federation of Malaysia. **1977** P. Theroux *Consul's File* vi. 48 A boy vaguely related to the Sultan..known locally (but inaccurately) as 'Tunku', The Prince. **1977** *Times* 31 Aug. (Malaysia Suppl.) p. viii/8 When someone is a Tengku or Tunku..his blood is very blue indeed, or very white, as the Malays put it; unless..he comes from Perak where a Raja is usually higher.

tunnel, *sb.* Add: **1. b.** (Example.)

1873 [see *crib *sb.* 10 b].

4. a. (Earlier example.)

1765 T. Lowndes *Let.* 1 July in *Hist. Inland Navigations* (1766) i. 41 Mr. Brindley..is driving a large tunnel through the center of this hill.

g. Applied fig. to a prolonged period of difficulty, suffering, etc. Freq. in phr. *light at the end of the tunnel* and the like: a long-awaited sign that a period of hardship or adversity is nearing an end. *colloq.*

1879 Geo. Eliot *Let.* 7 July (1956) VII. 178 Though I am getting out of the tunnel into daylight, this renewal of weakness..makes it seem as if we should be wiser to defer the visit. **1899** H. James *Awkward Age* x. xxxvii. 437 We've worked through the dark tunnel of artificial reserves. **1922** J. M. Murry *Let.* in A. Alpers *Life K. Mansfield* (1980) xx. 359, I begin to feel that the horror may move away and that there is a big round spot of real daylight at the end of the tunnel. **1943** J. B. Priestley *Daylight on Saturday* xxxv. 283 The work..seemed to him a long way off,..seen at the end of a tunnel. It had retreated from him. **1971** *Guardian* 6 Sept. 2/5 The world has reached a crucial point in its drive to reduce illiteracy, UNESCO reports today. There is now 'light at the end of the tunnel'. **1975** Ld. Robbins *Against Inflation* (1979) xviii. 89, I confess I do not understand the suggestion.. that there is any strong light at the end of the tunnel, the way we are going now.

h. *Aeronaut.* A wind tunnel (*wind *sb.*[1] 31).

1911 A. P. Thurston *Elem. Aeronautics* viii. 84 The *wind tunnel* consists of a tube, passage or tunnel, through which air may be forced or drawn by means of rotating fans, steam jets, or the like. The tunnel may be vertical or horizontal. Sir Hiram Maxim used a horizontal tunnel... Dr Stanton used a wind tunnel..in which the current was vertical and downwards. **1930** Nayler & Ower *Aviation* 116 Essentially, the tunnel consists of a large tube..along which the air is drawn by means of a motor driving a fan. **1972** *Nature* 18 Aug. 379/2 A low density tunnel for simulating supersonic and hypersonic flight at altitudes of 20 to 70 miles.

i. *Sport.* A subway or covered passage by which players pass to or from the field of play.

1950 *Sport* 24–30 Mar. 3/3 He..made for the tunnel under the impression that the game was over. **1976** *S. Wales Echo* 22 Nov., He..threw it towards the players' tunnel where the police were escorting the referee.

5. *tunnel-making, worker* (earlier example); *tunnel-like* (earlier example); **tunnel-back** *local,* the rear extension of a house, containing the scullery and other functional rooms; a

house built in this style; **tunnel diode** *Electronics,* a two-terminal semiconductor device, consisting of a heavily doped *p–n* junction, which has negative resistance at low voltage due to quantum-mechanical tunnelling and is principally used as a high-speed switching device; **tunnel effect** *Physics* = *tunnelling *vbl. sb.* 2*; **tunnel-kiln** (examples); **tunnel of love,** a fairground amusement involving a train- or boat-ride through a darkened tunnel, intended for courting couples; **tunnel vision,** a condition in which there is a major loss of peripheral vision; also, one in which anything away from the centre of one's field of view escapes attention; also *fig.*, inability to see more than a single or limited point of view; hence **tunnel-visioned** *a.*

1957 R. Hoggart *Uses of Literacy* i. 20 They have, almost city by city, their own recognizable styles of housing—back-to-backs here or tunnel-backs there. **1981** C. Dexter *Dead of Jericho* vi. 52 No tunnel-backs to the houses, and so the bicycles had to be left outside. **1959** *Proc. IRE* XLVII. 1204/1 The tunnel diode has a very high admittance. **1982** J. E. Uffenbeck *Introd. Electronics* i. 24 This switching property of the tunnel diode makes it suitable for digital applications. **1932** J. Frenkel *Wave Mech.* iii. 111 (*heading*) Transition through a potential energy mountain (tunnel effect). **1974** G. Reece tr. *Hund's Hist. Quantum Theory* xiv. 187 A barrier is not completely impenetrable. In fact it allows..the 'tunnel effect'. **1901** Tunnel kiln [see *continuous kiln* s.v. *continuous 3]. **1961** M. Kelly *Spoilt Kill* i. 11 We have gas-fired tunnel kilns now... There's very little coal firing left in the [pottery] industry. **1880** 'Mark Twain' *Tramp Abr.* xlvi. 530 One of the shows of the place was a tunnel-like cavern, which had been hewn in the glacier. **1910** Tunnel-making [see *road-building* s.v. *road *sb.* 11 a]. **1954** J. R. R. Tolkien *Fellowship of Ring* 31, I know of no tunnel-making. **1954** *New Yorker* 8 May 100/2 'And the lights!.. There are thirty-eight hundred on that ride alone. Why, even the World's Fair in its heyday—' he cried, and then for a moment, words failed him. 'And yet it's only a Tunnel of Love!' **1968** [see *loop-the-loop* sb. s.v. *loop *v.*[1] 6]. **1976** 'W. Trevor' *Children of Dynmouth* i. 13 The Hall of a Million Mirrors and the Tunnel of Love and Alfonso's and Annabella's Wall of Death were in the process of erection. **1949** Snygg & Combs *Individ. Behav.* vi. 110 It has often been observed that in emotional experiences there exists a very high degree of attention sometimes referred to as 'tunnel vision'. *Ibid.* vii. 125 This narrowing of the field is particularly likely to occur when the individual feels he is threatened. The effect has sometimes been called 'tunnel vision'. **1962** *Times* 3 Apr. 17/2 One of the dangers of 'tunnel vision' in driving was brought home to a motorist who recently took the test of the Institute of Advanced Motorists. **1967** *Freedomways* VII. 137 The confused black college graduate, thrust out into a hostile racist society and handicapped by tunnel vision and a self-negating perspective. **1968** *New Scientist* 29 Aug. 449/3 The alternative theory, that of 'Tunnel Vision'. The idea here is that a high level of arousal causes the brain to select very narrowly from among the signals reaching the eyes. **1979** *Daily Tel.* 7 Apr. 3/2 He was now registered as blind. He had tunnel vision, but even this was imperfect... There was some brain damage. **1980** T. Barling *Goodbye Piccadilly* viii. 169 Prebble had the ghetto mind and the tunnel-vision of a committed social climber. **1985** *Observer* 10 Mar. 5/1 Only someone with Tony Benn's tunnel vision could see the strike as 'a turning point in the battle against monetarism'. **1968** J. Lock *Lady Policeman* vi. 50 What happened to the juvenile after the Court's decision was not really in our province but we would have been tunnel-visioned indeed if we had never felt any concern. **1903** Tunnel worker [see *sand-hog* s.v. *sand *sb.*[1] 10 a].

tunnel, *v.* Add: **4. d.** *intr.* Physics. Of a sub-atomic particle: to pass *through* a potential barrier by tunnelling (*tunnelling *vbl. sb.* 2*).

1938 S. Dushman *Elem. Quantum Mech.* iii. 66 The probability that a particle coming up to the boundary at $x=0$ shall 'tunnel' through the barrier. **1966** D. G. Brandon *Mod. Techniques Metallogr.* iv. 181 Electrons may be able to 'tunnel' through to the far side. **1978** P. W. Atkins *Physical Chem.* xiii. 402 An electron is able to tunnel through even quite high potential barriers (for example, they can escape from the powerful forces inside nuclei, and emerge as β-rays).

tunnelling, -eling, *vbl. sb.* Add: **I. 2*.** *Physics.* The quantum-mechanical process whereby a particle has a non-zero probability of penetrating a finite potential barrier even if it has less energy than the height of the barrier.

1938 S. Dushman *Elem. Quantum Mech.* iii. 68 The 'tunneling effect'..is one of the most important deductions contributed by the new quantum mechanics. **1970** *New Scientist* 1 Oct. 38/2 Tunnelling is a phenomenon particular to quantum mechanics. **1978** P. W. Atkins *Physical Chem.* xiii. 402 The particle might be found on the outside of a container, even though according to classical physics it has insufficient energy to escape. This passage through classically forbidden zones is called tunnelling.

tunny. Add: **b.** *tunny boat, fishing* (hence *-fish* vb. intr.); **tunnyman,** a boat engaged in tunny fishing.

1934 *Yachting Monthly* LVII. 24/1 An ever-interesting panorama is provided by the sardiniers and tunny boats [at Concarneau]. **1974** 'J. Graham' *Bloody Passage* xiii. 173 There are a hell of a lot of tunny boats scattered around. **1977** C. Watson *One Man's Meat* viii. 78 He was with me, tunny-fishing off Scarborough. **1971** 'D. Halliday' *Dolly & Doctor Bird* v. 66 Between its [*sc.* a bridge's] arches tunny-fishing boats were constantly sprinting. **1930** *Sea Breezes* Dec. 94 Whilst the motor is making rapid headway amongst the sardine luggers and crabbers, at present the tunnymen are unaffected. **1961** A. J. R. Fraser Taylor *Diary* 13 Aug. in *Roving Commissions 1962* (1963) 123 The following day a slow passage to Bermeo, tying up late in the evening alongside a tunnyman.

tup, *v.* Add: **1. a.** *transf.* Also of a man: to copulate with (a woman). *coarse slang.*

1970 B. W. Aldiss *Hand-Reared Boy* 96 In Derbyshire's dull dorms... When lesser souls abused themselves, outclassed, Our Dancer, saint and patron, he upped and tupped the matron. **1976** R. Jeffries *Two-Faced Death* xviii. 210 You wouldn't tup her?.. Neither of us cut out for adultery.

tupaiid (tupai·id). [a. mod.L. family name *Tupaiidæ,* f. Tupaia + -id[3].] A tree-shrew of the family Tupaiidæ.

1885 T. Gill in J. S. Kingsley *Riverside Nat. Hist.* (1888) V. 141 The proper diet of the Tupaiids is small insects. **1972** T. A. Vaughan *Mammalogy* vi. 70/1 Paleontological evidence indicates that tupaiids were derived from the Insectivora and not from the Primates.

tupaioid (tupai·oid), *a.* (*sb.*) [a. mod.L. superfamily name *Tupaioidea,* f. Tupaia + -oid.] Of or pertaining to the superfamily Tupaioidea, or resembling a member of this group. Also as *sb.*, an arboreal mammal of the superfamily Tupaioidea.

1912 *Rep. Brit. Assoc. Adv. Sci.* 583 Its Tupaioid ancestor took to an arboreal life. **1972** W. C. O. Hill *Evolutionary Biol. Primates* ii. 18 The order is remarkable for the persistence..of living forms that illustrate several successive evolutionary steps..from tupaioids to man.

Tupamaro (tupămā·ro). [f. the names of the Inca leaders *Tupac Amaru* I (d. 1571) and *Tupac Amaru* II (d. 1781).] A member of a left-wing guerrilla organization in Uruguay. Also *attrib.*

1969 *N.Y. Times* 23 Jan. 12/2 The Tupamaros represent a new approach to guerrilla warfare in Latin America. *Ibid.* 12/4 Tupamaro intelligence sources. **1970** *Guardian* 11 Aug. 11/3 The Tupamaros..would like to see a Cuban-style revolution in Uruguay. **1973** G. Jackson *People's Prison* xxiv. 194 Tupamaro hideouts were even then notoriously subterranean. **1977** *Time* 31 Jan. 54/3 Urban guerrilla movements, such as the extinct Tupamaros of Uruguay, may have seen their day.

‖ **tupan** (tu·pæn). [Chinese.] The civil governor of a Chinese province under the Republican regime.

1925 *Glasgow Herald* 31 Aug. 9 Another mandate appoints General Feng's associate, Sun Yueh, Tupan of Shensi. **1928** T. F. Millard *China* 28 That process grew the crop of tuchuns, tupans, field marshals, and what not, so much heard of in these times. As originally used, the word 'tuchun' was distinctly a military term, and 'tupan' meant an officer who exercised civil authority; but in late years the terms are used indiscriminately. **1936** P. Fleming *News from Tartary* vi. ii. 250 Chin Shu-Jen was succeeded by the present *tupan,* General Sheng Shih-tsai. **1949** F. Maclean *Eastern Approaches* (1951) i. v. 98 The intention was that I should go to Urumchi, the capital of Sinkiang, to contact the Tupan, or Provincial Governor.

Tupi. Add: Also pronounced (tupī·). **a.** Also, the group of tribes speaking this language; a person belonging to one of these tribes. Also *attrib.*

1842 [see The *dem. adj.* and *pron.* B. 3 b]. **1863** [see *jacitara]. **1911** *Encycl. Brit.* XXVII. 410/2 Latham makes the Tupis members of the Guarani stock. **1950** C. Lévi-Strauss in J. H. Steward *Handbk. S. Amer. Indians* VI. 475 The fruit of a Bignoniaceae..was used as a comb by the *Tupi* and other tribes. **1950** C. O. Sauer in *Ibid.* 499 The peanut was..important in Tupí economy.

b. *Tupi-Guarani* (gwărānī·), also unhyphened: a South American linguistic and ethnic stock of which Tupi and Guarani are the most prominent members; a person belonging to this ethnic stock. Also *attrib.* Cf. *Guarani i.

1850 R. G. Latham *Nat. Hist. Varieties of Man* 443 The Guarani. Synonyms.—Tupi, Brazilian, Guarani-Brazilian, Tupi-Guarani. **1876** *Encycl. Brit.* IV. 235/1 They [*sc.* the tribes of Brazil] belong..to one original stock, called by ethnographers, the Tupi-Guarani. **1901** *O.E.D.* s.v. *Jaguar,* According to writers on Tupi-Guarani, *jaguara* or *jagua* is orig. a class-name for all carnivorous beasts. **1933** L. Bloomfield *Language* iv. 73 In South America, we note..the Tupi-Guarani [family of languages], stretched along the coast of Brazil. **1956** E. Haugen *Bilingualism in Americas* ii. 15 Tupí-Guaraní is taught in some Brazilian schools. **1968** M. Gilbert *Cork in Bottle* in *Ellery Queen's Christmas Hamper* (1975) 231 The three mestizos..were Tupi-Guaranis, half Indian, half Spanish. **1977** G. Clark *World Prehistory* (ed. 3) x. 449 Polychrome pottery closely resembling the Tupi-guarani ware of East Brazil.

Tupian *sb.* and *a.* (examples).

1902 *Encycl. Brit.* XXV. 374/1 [Linguistic families of America] Tupian, Amazon R. **1948** A. L. Kroeber *Anthropology* (rev. ed.) xviii. 833 A Tupian tribe, the Chiriguano, having conquered an Arawakan one, the Chané, pushed on westward. **1974** *Encycl. Brit. Micropædia* X. 187/1 *Tupians,* South American Indians who speak languages of the Tupian linguistic group. Tupian-speaking peoples were widespread south of the Amazon.

tupik. Add: Also 9 **toupik, tupic; tubik.** (Earlier and later examples.)

1864 C. F. Hall *Life with Esquimaux* I. ix. 176 On my way,..just outside the angeko's tupic, I noticed an oar of a kia[k] stuck upright in a drift of frozen snow. **1878** C. Hallock *Sportsman's Gazetteer* (ed. 4) 700 *Toupik,* an Esquimaux summer lodge of poles covered with seal-skins. **1895** Kipling *Second Jungle Bk.* 152 One of their hunters came across a *tupik,* a skin-tent. **1920** W. T. Grenfell *Labrador Doctor* vi. 129 Wooden huts had largely replaced the former 'tubiks', or skin tents.

Tupinamba (tupinæ·mbă, *prop.* tupinambā·). Also † **Tupinambo.** [Native name; cf. F. *Topinambou* (1578 in form *Tououpinambaoults*) and Tupi.] A group of extinct tribes on the coast of Brazil; a person belonging to one of these tribes. Also *attrib.*

1810 Southey *Hist. Brazil* I. viii. 229 Nobrega learnt from the Tupinambas that two persons..taught them the use of the mandioc. **1819** Shelley *Let.* 3 Nov. (1964) II. 140 An Otaheitan or a Tupinambo. **1863** H. W. Bates *Naturalist on River Amazons* I. vii. 285 The old historians relate that the island of Tupinambarána was colonised by a portion of the great Tupi or Tupinámba nation, who were driven from the sea-coast near Pernambuco, by the early Portuguese settlers in the 16th century. **1949** Wagley & Galvão *Tenetehara Indians of Brazil* i. 5 The Portuguese..found the Island of Maranhão inhabited by the Tupi-Guaraní Tupinambá. **1974** *Encycl. Brit. Macropædia* XVII. 124/1 The Tupinamba shaman fumigates his rattle with tobacco.

-tuple (tiūp'l), *a.* and *sb.* Chiefly *Math.* [The ending of Quintuple *a.* and *sb.*, etc.] With preceding algebraic symbol: (an entity or set) consisting of as many parts or elements as indicated by the symbol.

1863 *Phil. Trans. R. Soc.* CLIII. 457 The curve *m* is a $(m-1)n$-tuple line on the scroll $S(m^2, n)$. **1910** *Encycl. Brit.* I. 615/2 We may regard it as a (2^n-1)-tuple linear algebra. **1938** *Jrnl. Symbolic Logic* III. 151 Each function is defined over a subset..of the *n*-tuples of natural numbers. **1963** J. Lyons *Structural Semantics* ii. 12 In so far as words can be segmented into morphemes, the lexeme can be defined extensionally as the set of all the ordered *n*-tuples ($n \geq 1$) of morphemes (each *n*-tuple being a word) which are grouped together in setting up the paradigm. **1972** *Computer Jrnl.* XV. 232/1 A structure of a string *A* is an ordered *n*-tuple *T.*

Tupperism. (Earlier example.)

1870 J. R. Lowell *Among my Books* 114 The gradual degeneration of a poetic faith into the ritual of unimaginative Tupperism.

Tupperware (tʊ·pəɹwɛ̄ᵊɹ). [Trade name, f. Earl S. *Tupper,* President of the Tupper Corporation + Ware *sb.*[2]] The proprietary name of a range of plastic vessels, containers, etc., sold exclusively at 'parties' in private homes to which potential purchasers are invited. Freq. *attrib.*; also in allusive use.

1956 *Official Gaz.* (U.S. Patent Office) 12 June tm 53 Tupperware... For Molded Plastic Tumblers, Canisters, Pitchers, Dispensers; Empty Condiment Holders... Empty Soap, Hair Massage and Tooth Brush Boxes [etc.]. First use Mar. 3, 1950. **1961** *Trade Marks Jrnl.* 23 Aug. 1162/1 Tupperware... Small domestic utensils and containers..; combs and sponges; brushes.., Rexall Drug and Chemical Company.., City of Los Angeles, State of California, United States of America; manufacturers. **1965** *Which?* Dec. 373/1 Are more expensive brands [of food container], in particular Tupperware, better than cheaper brands at storing foods? *Ibid.*, You can't buy Tupperware in the shops. It is sold through local 'dealers'... The dealer gets a housewife to be the hostess at a party. **1966** T. Pynchon *Crying of Lot 49* i. 9 Mrs Oedipa Maas came home from a Tupperware party whose hostess had perhaps put too much kirsch in the fondue. **1971** N. Stacey *Who Cares?* viii. 131 We were not training a group of people to give Tupperware parties or sell cosmetics to housewives. **1979** T. Barling *Olympic Sleeper* i. 15 There were too many cowboys on the Thames nowadays, playing sailors in their Tupperware boats.

tur (tūᵊɹ). [a. Russ.] A greyish-brown wild goat, *Capra caucasica,* native to south-eastern Russia.

1894 C. Phillipps-Wolley *Big Game Shooting* II. iii. 51 The tûr is the mountain beast, *par excellence,* of the Caucasus. **1894** R. Lydekker *Royal Nat. Hist.* II. 235 There occur in the Caucasus range..wild goats, known locally as tur. **1925** G. Burrard *Big Game Hunting* 87 The East Caucasian tur..is an undoubted goat. **1965** D. Morris *Mammals* 428 There are several other species which are also called Ibex. These include the Tur, or Caucasian Ibex.

turaco: now the usual spelling of Touraco (in Dict. and Suppl.).

Turanian, *a.* Add: **1.** (Earlier examples.)

1788 *Asiatick Researches* I. 7 A Turanian pronunciation. **1841** J. C. Prichard *Res. Physical Hist. Mankind* (ed. 3)

III. i. 16 A great number of roots are thus to be traced in several of the Turanian languages.

2. (Earlier example.)

1836 J. C. PRICHARD *Res. Physical Hist. Mankind* (ed. 3) I. iv. 267 The skulls of the Esquimaux..bring them into the same class of human races with the Kalmuk and other Turanian nations.

Hence **Tura·nianism**, the principle of uniting speakers of Turanian or Ural-Altaic languages (esp. Turkish); cf. *pan-Turanianism* s.v. *PAN- I.

1922 *19th Cent.* Nov. 835 The seeming paradox of the Bolshevist *régime* cementing Islamism and Turanianism in a widespread brotherhood.

turanose (tiū°·rănōᵘz). *Chem.* [ad. Russ. *turanoza* (A. Alekhina 1889, in *Zhurnal Russkago fiziko-khim. Obshchestva* XXI. 418), after Pers. *Turān* Turkistan, place of origin of the manna used to prepare this: see -OSE².] The reducing disaccharide sugar $C_{12}H_{22}O_{11}$, formed by partial hydrolysis of melezitose; 3-α-D-glucopyranosyl-D-fructose.

1890 *Jrnl. Chem. Soc.* LVIII. 733 Melezitose, on inversion with dilute mineral acids, yields at first turanose and dextrose; the former is a new saccharose of the formula $C_{12}H_{22}O_{11}$. **1927** M. BODANSKY *Introd. Physiol. Chem.* ii. 43 Turanose (fructose+glucose) is obtained by hydrolyzing the trisaccharide melicitose. **1975** *Nature* 10 July 128/1 Maltose was slightly more effective, and sucrose, turanose, kojibiose, trehalose and melezitose all inhibited binding at significantly lower concentrations than glucose.

‖ **turba** (tū°·rbā). *Mus.* [L., = crowd.] A name given to the chorus in Passions and other religious oratorios in which crowds participate in the action. (See also quot. 1889.)

1876 STAINER & BARRETT *Dict. Mus. Terms* 443/1 *Turbæ* (*Lat.*), the chorus part or voice of the multitude in a Passion-music. **1889** *Cent. Dict.* s.v., *Turba*, the chorus in mediaeval passion-plays, representing the Jewish populace. **1947** A. EINSTEIN *Music in Romantic Era* xiii. 173 It is a work with fanatic *turbae*, the 'crowds', as in the Passions. **1962** *Listener* 15 Feb. 317/3 The Roman Church's dramatic Passions now come into view with Victoria's and Byrd's settings of the *turba*.

turban, *sb.* Add: **8. turban squash** (examples); **turban tumour** *Path.*, a rare benign tumour, probably of sweat glands, that spreads over the scalp or thorax in grape-like clusters.

1902 L. H. BAILEY *Cycl. Amer. Hort.* IV. 1713/1 The Turban Squashes..have a 'Squash within a Squash'. **1949** *Nat. Geogr. Mag.* Aug. 162/2 Several years ago a North Dakota horticulturist bred a small variety of turban squash. **1981** *Farmstead Mag.* Winter 38/3 The last of the six types to be mentioned is the turban squash. **1903** H. RADCLIFFE-CROCKER *Dis. of Skin* (ed. 3) II. viii. 961 Sarcoma Capitis, or Endothelioma Capitis (Turban tumours). A peculiar form of tumour, in rare instances attacks, and is limited to the hairy scalp; in extreme cases, covering the whole scalp like a wig. **1974** J. D. MAYNARD in R. M. Kirk et al. *Surgery* ix. 196 Turban tumour... These rare tumours are sub-epithelial basal cell carcinomas, usually of sweat-gland origin which grow steadily and slowly, without ulceration or metastasis, on the scalp, face and thorax.

turban(n)y *a.*, resembling or suggestive of a turban.

1912 A. HUXLEY *Let.* 13 May (1969) 42 The banner bearers..wear marvellous uniforms—usually consisting of a..sort of turbanny object or a cocked hat..white breeches and highly polished top-boots. **1924** E. BOWEN in *Spectator* 5 July 11/1 Yes, but haven't you got *any* goldish sort of turbany thing?

turban, *v.* Add: **b.** To wind in the form of a turban.

a **1861** T. WINTHROP *John Brent* (1883) xvi. 151 A strip of old white blanket..was turbaned askew about his head. **1969** *Daily Tel.* 20 Jan. 11 A long white and cream silk scarf turbanned round the head and floating free.

turbaned, *a.* Add: Also **turbanned. a.** (Later examples.)

1968 T. STOPPARD *Real Inspector Hound* (1970) 12 Mrs. Drudge is the char, middle-aged, turbanned. **1976** G. S. Cox in M. Drabble *Genius of Thomas Hardy* III. 172 A turbanned Indian.

b. Arranged to form a turban. (In quot., *transf.*) *poet. rare.*

1924 E. SITWELL *Sleeping Beauty* vi. 28 The..shore Where curled and turbanned waves sigh 'Nevermore'.

turbary. Add: **3. b.** [tr. G. *torf-*.] Applied to kinds of domesticated sheep and pig of prehistoric times that were first found in turbaries in Swiss lake-dwellings.

1908 R. PUMPELLY *Explorations Turkestan* I. I. v. 67 The turbary sheep (Torfschaf) and..the turbary pig (Torfschwein)..appear towards the end of the neolithic period. **1912** R. LYDEKKER *Sheep & its Cousins* vii. 150 An apparently pure-blooded breed of small sheep inhabiting Crete..is identified by Dr. Keller with the turbary sheep. **1920** J. RITCHIE *Animal Life Scotl.* 40 Even in Neolithic times the Turbary or Peat Sheep..was widely distributed in Scotland. **1936** *Antiquity* X. 203 We find a small sheep with erect horns,..the so-called 'goat-horned' or turbary sheep. **1963** F. E. ZEUNER *Hist. Domesticated Animals* x. 257 In the earlier group of Neolithic lake-dwellings the small turbary pig (*Sus palustris* Rütimeyer) occurs beside the ordinary European wild pig. There is no doubt that this turbary pig was introduced into Switzerland by Neolithic man from the East. *Ibid.* 258 According to Kuhn, the turbary pig has survived to the present day in some of the Alpine valleys. **1972** *Science* 12 May 656/2 With regard to sheep and pig, he believes that the well-known turbary type..is a natural product of malnutrition and poor care.

turbidimeter (tɜ̄ɹbidi·mĭtəɹ). *Chem.* and *Biol.* [f. TURBID *a.* + -I- + -METER.] An instrument for determining the turbidity of a liquid from the decrease in the intensity of a beam of light passing through it.

1905 *Water Supply & Irrigation Papers* (U.S. Geol. Survey) No. 151. 26 The needs of the Survey were found to be met in a satisfactory manner by the use of a turbidimeter devised by Mr. Daniel D. Jackson. **1920** *Jrnl. Biol. Chem.* XLII. 191 Turbidimeters and nephelometers are instruments designed for practically the same purpose. **1973** *Sci. Amer.* June 112/1 Extinction turbidimeters are doubtless the simplest instruments that have been devised for measuring the concentration of solids in suspension. They are based on the principle that turbidity is inversely proportional to the minimum length that a column of fluid must have in order to extinguish at one end of the column a source of light at the other end.

Hence (all also **turbido-**) **turbidime·tric** *a.*, obtained with or employing a turbidimeter; **turbidime·trically** *adv.*; **turbidi·metry**, the use of a turbidimeter, esp. for the quantitative analysis of turbid solutions.

1911 *Jrnl. Industr. & Engin. Chem.* III. 554/1 The discrepancies between the turbidimetric and gravimetric results..were..thought to be caused by the presence of nitrates in the solutions examined. **1918** *Jrnl. Biol. Chem.* XXXVI. 33 This reagent gives quite good results used either turbidometrically or nephelometrically. **1920** *Ibid.* XLII. 196 With several of the substances which have already been standardized for turbidimetry we can multiply the accuracy. **1943** *Jrnl. Bacteriol.* XLVI. 377 Applications of turbidimetry to the study of *in vitro* penicillin effects. **1971** Turbidimetric [see *nephelometric* adj. s.v. *NEPHELO-]. **1975** D. H. BURRIN in Williams & Wilson *Biologist's Guide to Princ. & Techniques Pract. Biochem.* v. 145 Very dilute suspensions may be assayed by turbidometry. **1981** *Jrnl. Protozool.* XXVIII. 371/2 Growth was determined turbidimetrically.

turbidite (tɜ̄·ɹbidəit). *Geol.* [f. TURBID(ITY + -ITE¹.] A sediment or rock deposited, or presumed to have been deposited, by a turbidity current. Hence **turbidi·tic** *a.*

1957 P. H. KUENEN in *Jrnl. Geol.* LXV. 231/1 The term 'turbidite' for all deposits of turbidity currents is more appropriate, and the writer accepts this verbal suggestion of C. P. M. Frijlinck. **1973** *Nature* 9 Feb. 389/2 The lowest (Unit I)..has at the base a phyllitic formation..followed by turbiditic sandstones. **1977** A. HALLAM *Planet Earth* 204/1 The thick clastics and turbidites of the upper Devonian of the Yukon seem to be derived from an oceanic zone of uplift.

turbidity. Add: **2.** Special Comb.: **turbidity current**, an underwater current flowing swiftly downslope owing to the weight of sediment it carries.

1939 D. W. JOHNSON *Origin of Submarine Canyons* iii. 27 By analogy those [currents] due to turbidity will here be called *turbidity currents*. **1950** *Jrnl. Geol.* LVIII. 91/1 The most important types of graded bedding appear to have been produced by the action of turbidity currents of high density on the sea floor. **1977** A. HALLAM *Planet Earth* 54/2 An unusual but geologically very important type of suspension deposit is that produced by a turbidity current.

turbidometric, etc., varr. *TURBIDIMETRIC *a.*, etc.

turbine. Add: Also with pronunc. (-əin). **1. a.** (Earlier example.)

1838 *Railway Mag.* IV. 51 Turbine.—An instrument under this name has lately been invented by M. Fourneyron, worked by water-pressure, which is said to have made a great sensation in Germany. *Ibid.*, It is said that a turbine, only thirteen inches diameter,..under a vertical pressure of water of 118 yards, revolved 2,300 times in a minute, and..realized a power, which estimated in steam, would be equal to that of sixty horses.

d. = *gas-turbine* s.v. *GAS *sb.*¹ 7.

1904 *Proc. Inst. Mech. Engineers* Oct. 1078 Some or all of the available heat energy of the gas can be converted into kinetic energy before causing it to act on the turbine. **1940** A. W. JUDGE *Aircraft Engines* I. vii. 231 An alternative method of driving the supercharger is to couple it directly to an exhaust turbine of the de Laval type. **1950** E. T. VINCENT *Theory & Design Gas Turbines & Jet Engines* vi. 161 The power-plant turbine can be divided into the following units: (1) compressor, (2) combustion chamber, (3) turbine, and (4) regenerator. **1971** B. SCHARF *Engin. & its Lang.* xv. 215 After expansion in the turbine, the combustion gases escape at high velocity through the jet pipe, thus providing the forward thrust for the aeroplane.

2. *turbine blade.*

1911 *Encycl. Brit.* XXV. 843/2 The general arrangement of the steam nozzle and turbine blades is illustrated. **1977** *R.A.F. News* 11–24 May 18/2 A pair of turbine blades from an Orpheus jet engine.

turble, var. *TURRIBLE.

turbo. Add: **3. b.** = *TURBOCHARGER; also, a motor vehicle equipped with this device. Also *attrib.*

1957 *Motor* 6 Mar. 168/1 General Motors have built four turbo vehicles. The 370 b.h.p. Firebird I..had single-stage centrifugal compressor and two-stage turbine layout. **1978** *Country Life* 17 Aug. 460/1 While turbo is a term associated with power, the Saab 99 Turbo has no shortcomings in terms of comfort. **1980** *Daily Tel.* 23 Jan. 14/4 Driving it in Portugal last week, I found it impressively smooth and quiet, with a quicker throttle response from low speeds than usual with a turbo.

turbo-. Add: *turboblower, -compressor*; **tu·rbocar**, a motor car powered by a gas-turbine engine; **turbo-co·mpound** *a.*, applied to a piston engine in which the exhaust gases drive a turbine coupled to the crankshaft; hence **turbo-compou·nded** *a.*, **-compou·nding** *vbl. sb.*; **tu·rbodrill** *Oil Industry*, a drill in which the drilling bit is rotated by a turbine situated next to it in the drilling string and driven by the upflow of mud; also as *v. trans.*; hence **tu·rbodrilled** *ppl. a.*, **tu·rbodrilling** *vbl. sb.*; **turbo-ele·ctric** *a. Engin.*, involving or employing electricity generated by means of a turbine; **turbomole·cular** *a. Physics*, applied to a type of high-vacuum pump in which momentum is imparted to molecules by a high-speed rotor inside a stator, both of which possess inclined slots or blades designed so as to cause the molecules to move axially towards the outlet; **tu·rbopump**, a pump that incorporates a small turbine to provide the necessary mechanical power, used esp. in aircraft and rockets; **turbora·mjet** *Aeronaut.*, any of a class of jet engines combining the operations of a turbojet and a ramjet, either as a turbojet with provision for afterburning, or as a ramjet containing a turbojet which is shut down at high velocities; **tu·rboshaft** *Engin.*, used *attrib.* and *absol.* to designate a gas turbine engine in which the turbine drives a shaft other than a propeller shaft; **tu:rbosu·percharger** *Engin.* = *TURBOCHARGER; hence **tu:rbosu·percharged** *ppl. a.*, **-su·percharging** *vbl. sb.*; **tu·rbotrain**, a train powered by a gas-turbine engine.

1911 *Trans. Inst. Mining Engineers* XL. 580 (*heading*) Turbo-blowers and turbo-compressors. **1947** *Jrnl. R. Aeronaut. Soc.* LI. 95/1 For high altitude cruising the application of turbo-blowers has received much consideration. **1979** *Truck & Bus Transportation* (Austral.) Feb. 41/2 With a turboblower, you literally shovel the air in. [**1950** *Motor* 15 Mar. 183/1 (*caption*) The Rover turbine car easily reached 90 m.p.h.] **1956** *Times* 3 July 4/6 The Rover turbocar..was timed at 152 m.p.h. **1974** D. NYE *Motor Racing Mavericks* xix. 190 This turbine car sparked off a terrific controversy... Wallis was taken on..to build two cars similar to his 'STP Turbocar'. **1954** *Economist* 11 Sept. 11/3 Turbo-compound; piston compound. A combination of gas turbine and reciprocating engine. **1955** C. E. CHAPEL et al. *Aircraft Power Plants* (ed. 2) xvii. 339/2 Aircraft powered by the Wright turbo-compound engine are the Douglas DC-7, the Lockheed Super Constellation, [etc.]. **1983** *Truck & Bus Transportation* (Austral.) July 60/2 Whilst the turbocompound diesel has progressed to the operational stage..the benefits of this concept can be more fully realised and cost justified when used in conjunction with an adiabatic or insulated engine. **1978** *Automotive Engin.* Aug. 85/1 In a turbocompounded engine..the exhaust gases are expanded in a turbine and the power generated is transmitted back to the crankshaft. Turbocompounding can be incorporated in naturally aspirated, as well as turbocharged engines. **1911** Turbo-compressor [see *turboblower* above]. **1922** *Daily Mail Year Bk.* 1923 75/1 By the development of a mechanism known as a 'turbo-compressor', he has enabled aero-engines to maintain their power in the thin air of upper altitudes. **1979** A. L. LYDERSEN *Fluid Flow & Heat Transfer* xi. 327 Turbo-compressors are used for vapour recompression of large vapour volumes..while steam ejectors are used in many smaller installations. **1948** *Oil & Gas Jrnl.* 3 June 58/3 During initial field tests conducted in May.., the new Edco Turbodrill penetrated 950 ft. of shallow section in a wildcat test drilling. *Ibid.* 61/1 Photoclinometer and hole-section surveys showed a total drift of 3 ft. or 10½ minutes from vertical through the section turbo-drilled. **1981** 'D. RUTHERFORD' *Porcupine Basin* iv. 66 We're developing a new turbo-drill on a flexible string which can be reeled out on a drum. **1949** *World Oil* 1 July 88/1 Electric logs..were run after the turbodrilled section was completed. **1955** *World Petroleum* XXVI. 84/3 The advantages of turbodrilling stem from the fact that only the bit is actually involved in the rotating effort. **1977** *Offshore Engineer* May 20/1 (Advt.), Our turbodrilling and directional drilling engineers and equipment are at your service. **1904** Turbo-electric [in Dict.]. **1930** *Engineering* 18 Apr. 513/3 Turbo-electric propulsion must exhibit an overwhelming superiority in reliability and maintenance cost to overcome the disadvantages. **1974** *Encycl. Brit. Macropædia* XVII. 752/2 On most nuclear submarines reduction gears are used between the turbines and the propeller shaft; however, a few incorporate turbo-electric drive. **1969** *Gloss. Terms Vacuum Technol.* (*B.S.I.*) 17 *Turbo-molecular pump*, a molecular drag pump in which the rotor has inclined slots or blades moving between corresponding slots or blades in a stator. **1976** *Physics Bull.* Nov. 499/2 The vacuum system is kept at 10⁻⁷ Torr by about 650 sputter ion pumps and 80 turbomolecular pumps. **1903** Turbo-pump [in Dict.]. **1947** *Aircraft Engin.* Aug. 254/2 The turbo-pump assembly consists of a shaft carrying a single stage impulse steam turbine, on each side of which is a centrifugal pump. **1962** F. I. ORDWAY *Basic Astronautics* x. 411 The turbopump proves to be the best means of pressurizing the propellants for large liquid rocket engines. The bi-propellant turbopump consists of two centrifugal pumps and a gas turbine that supplies the driving power for the pumps. **1979** *Nature* 11 Jan. 84/1 A fire in the high pressure turbo-pump that feeds oxygen into the combustion chamber caused the engine to explode. **1948** *Aviation Week* 23 Feb. 36/2 Turboramjet—A conventional turbojet engine with provision for reheating the gas between the turbine discharge and the exhaust nozzle. **1971** P. J. MCMAHON *Aircraft Propulsion* iii. 119 At the moment the most promising fields of use for the turboramjet would seem to be those in air-breathing boosters for the launching of space vehicles. **1958** P. H. WILKINSON *Aircraft Engines of World 1958/59* 31 Bristol-Siddeley Engines Ltd... is specializing in advanced turbojets, turboprops, turboshafts, [etc.]. **1967** *Jane's Surface Skimmer Systems 1967–68* 130/2 The GE LM1500 turboshaft engine is the result of a company investment in a programme to adapt the J79 jet engine to a free power turbine for commercial use. **1977** I. M. CAMPBELL *Energy & Atmosphere* vi. 127 The gas turbine engine used in aircraft under the names of turbojet, turboprop or turbofan, or in industrial or marine settings as a turboshaft engine. **1944** P. H. WILKINSON *Aircraft Engines of World 1944* 34 The Boeing Flying Fortress B-17 (powered with a turbo-supercharged engine). **1978** *Financial Times* 20 Dec. 21/5 At the moment, Mercedes is unique in marketing a turbo-supercharged diesel car. **1938** A. SWAN *Handbk. Aeronaut.* (ed. 3) II. iv. 206 The inherent advantages of the turbo supercharger are mainly centred around the fact that it possesses remarkable flexibility of speed control. **1971** P. J. MCMAHON *Aircraft Propulsion* xi. 319 From the turbo-supercharger, it is a logical step to develop the fully compound engine. **1979** *Financial Rev.* 14 June 29/1 Turbosupercharging is very much in as far as motoring is concerned these days. **1966** *Time* 27 May 52/3 Canadian National Railways..has now ordered five of the turbotrains developed by the U.S.'s United Aircraft Corp... These light-weight, low-slung, turbojet-powered whiz-bangs should be able to clip nearly an hour off the present five-hour Montreal–Toronto run. **1978** *Times* 9 June 1/8 The turbo-train between Strasbourg and Lyons.

turbocharger (tɜ̄·ɹbo,tʃɑ̄ɹdʒəɹ). *Engin.* Also **turbo charger.** [f. *turbo(super)charger* s.v. *TURBO-.] A supercharger driven by a turbine powered by the engine's exhaust gases.

1934 *Jrnl. R. Aeronaut. Soc.* XXXVIII. 182 After explosion, the first exhaust valve which is connected to the turbo charger, opens. **1961** *Engineering* 9 June 786/1 The advantages obtained by applying turbochargers to diesel engines are well known. **1980** 'D. RUTHERFORD' *Turbo* viii. 107, I gently depressed the accelerator. The turbocharger came in with its characteristic whine and the Saab swept quickly to 120 m.p.h.

Hence (as a back-formation) **tu·rbocharge** *v. trans.*, to equip with a turbocharger; **tu·rbocharged** *ppl. a.*, **tu·rbocharging** *vbl. sb.*

1961 *Engineering* 10 Mar. 370/1 The cycle for the Ford 704 engine of 300 hp rating may be described as a turbo-charged gas turbine. *Ibid.* 9 June 786/1 Turbocharging has brought with it the reduced weight and space requirements. **1970** *Commercial Motor* 25 Sept. 97 Ford offers a turbocharged version of its 360 cu. in. engine,.. while Perkins has turbocharged the 6·354. **1971** *Farmer & Stockbreeder* 23 Feb. 21/3 The combination of air cooling and turbocharging is said to reduce engine noise and smoke emission. **1981** *Sci. Amer.* May 41/2 Turbocharging automotive diesels has made possible fuel-economy advances of between 10 and 15 percent. **1981** D. BOGGIS *Time to Betray* x. 54 The Skylane..was turbocharged, with retractable gear.

turbofan (tɜ̄·ɹbofæn). *Aeronaut.* Also with hyphen. [f. TURBO- + FAN *sb.*¹] **a.** A fan connected to or driven by a turbine. **b.** Used *attrib.* and *absol.* to designate a jet engine employing such a fan for additional thrust; = *fan-jet* (*engine*) s.v. *FAN *sb.*¹ 11.

1911 *Trans. Inst. Mining Engineers* XL. 580 The smaller efficiency of the turbo-fan. **1949** *Gloss. Aeronaut. Terms* (*B.S.I.*) II. 17 *Ducted-fan turbine engine, turbo-fan*, a gas turbine engine in which a portion of the net energy is used to drive a ducted fan. **1959** *Wall St. Jrnl.* 7 July 10/1 The turbofan engine is the latest in the family of jet power plants. **1961** [see *double-flow* s.v. *DOUBLE *a.* C. 2 a]. **1970** *Daily Tel.* (Colour Suppl.) 13 Nov. 14/4 Large turbofan engines of the Spey type tend to expel ingested objects along the cold fan duct, so that they do not enter the engine core. **1980** R. L. DUNCAN *Brimstone* x. 261 A cruise missile..propelled by a jet turbofan.

turbojet (tɜ̄·ɹbodʒet). *Aeronaut.* Also with hyphen. [f. TURBO- + *JET *sb.*³ 5 c.] Used *attrib.* and *absol.* to designate (an aircraft having) a type of jet engine in which the jet gases also power a turbine-driven compressor for compressing the air drawn into the engine.

1945 *Aeronautics* Mar. 28/2 The turbo-jet shows definite advantages in comparatively light aircraft designed to operate at high speeds. **1950** *Ann. Reg.* 1949 424 Propelled by..turbo-jet engines,..the Comet flew at an altitude of 35,000 feet. **1958** *Times Rev. Industry* Dec. 50/2 Within five years it is probable that over 1,300 turbo-jet and turbo-prop aircraft will be in service. **1962** E. SNOW *Other Side of River* (1963) i. 17, I had flown halfway around the world and landed at Peking in a Soviet turbojet manned by a Chinese crew. **1977** *Sci. Amer.* Feb. 23/2 A turbojet engine exhausts its gases at 1,450 degrees, whereas a turbofan engine, because of turbulent mixing at the outlet, exhausts them at 600 degrees.

turbopause (tɜ̄·ɹbopǭz). *Meteorol.* [f. TURB(ULENCE + -O + PAUSE *sb.*] The outer limit of a turbosphere, where the distribution of constituents is due equally to diffusion and turbulent mixing.

1951 S. CHAPMAN in *Jrnl. Atmospheric & Terrestrial Physics* I. 201 The diffusive tendency becomes increasingly effective at greater heights, and it may be (but it is not certain) that at some level diffusion becomes dominant over the turbulent mixing. The name turbopause is suggested for this level. **1967** R. W. FAIRBRIDGE *Encycl. Atmospheric Sci. & Astrogeol.* 5/1 Above the turbopause, which [on earth] lies between 100 and 130 km, diffusion is more important than turbulent mixing in distributing the constituents. **1979** *Nature* 17 May 221/1 The δ-bands, present in the Earth's twilight and night time airglow spectra,..indicate a significant population of atomic nitrogen at or below the Venus turbopause.

turboprop (tɜ̄·ɹboprǫp). *Aeronaut.* Also **turbo-prop.** [f. TURBO- + *PROP *sb.*⁶] Used *attrib.* and *absol.* to designate (an aircraft having) a jet engine in which a turbine is used as in a turbojet and also to drive a propeller; = *prop-jet* s.v. *PROP *sb.*⁶ 2 b. Also **tu:rbopropel·ler.**

1945 H. H. ARNOLD *Third Rep. Commanding General U.S. Army Air Forces* in *War Reports* (1947) 464 This war has evolved six distinct methods of utilizing atmospheric oxygen for propulsion, such as:..*turboprop*—a gas turbine plus propeller. **1946** *Shell Aviation News* No. 102. 5/2 The following..types of planes will be..powered by turbo props. **1947** *Aircraft Engin.* Mar. 79/1 The question can turbo-jets or turbo-propellers be used in transport is timely. **1957** *Times* 17 Dec. 13/2 There has been an increasing demand for equipment for civil aircraft, particularly those employing turbo-propeller engines. **1958** *Daily Mail* 3 Mar. 5/2 The more economical turbo-props could be used to force down the price of air travel. **1958** [see *TURBOJET]. **1968** MILLER & SAWERS *Technical Development Mod. Aviation* vi. 205 A turbopropeller-driven airliner will have an advantage over a jet if they both seat significantly fewer than 50 passengers on stage lengths exceeding 200 miles. **1972** *Physics Bull.* Oct. 580/3 The four turboprops which will power the biggest ship at speeds up to 120 mph are located on the tail spars. **1978** *N.Y. Times* 30 March B19/1 (Advt.), Mechanics... Exp on..air research turboprop engines.

turbosphere (tɜ̄·ɹbosfīᵊɹ). *Meteorol.* [f. TURB(ULENCE + -O + SPHERE *sb.*] A region of a planetary atmosphere in which mixing occurs predominantly through turbulence.

1951 S. CHAPMAN in *Jrnl. Atmospheric & Terrestrial Physics* I. 201 Throughout the lower atmosphere, and up to at least 60 km height, turbulence suffices to mix the permanent constituents effectively, overcoming the diffusive tendency for each to distribute itself independently of the others, according to its molecular weight... It is suggested that this whole region be called the turbosphere. **1979** J. K. HARGREAVES *Upper Atmosphere & Solar-Terrestrial Relations* iv. 52 The well mixed part of the atmosphere, in which composition does not change with height, can be called the turbosphere.

turbostratic (tɜ̄ɹbostræ·tik), *a. Physics.* [f. L. *turb-ātus*, pa. pple of *turbāre* to disturb, disorder + -O + STRAT(UM + -IC.] Of or pertaining to a material (esp. one allotrope of carbon) having a structure intermediate between amorphous and crystalline, consisting of stacked disordered layers.

1942 BISCOE & WARREN in *Jrnl. Appl. Physics* XIII. 370/1 Carbon black is a simple and definite example of an intermediate form of matter, which is distinctly different from both the crystalline and the amorphous states. The term 'turbostratic' (unordered layers) is suggested as a name of this particular class of mesomorphic solids. **1966** *R. Inst. Chem. Lect. Ser.* V. 12 The so-called 'turbostratic' boron nitride is still basically the hexagonal form. **1975** *Physics Bull.* June 260/1 This model is distinctly different from the traditional, 'turbostratic' model..in which crystallites comprise flat, graphite layer planes in disordered stacking.

turbulence. Add: **c.** Of fluid flow (see *TURBULENT *a.* 2 c).

1907 F. W. LANCHESTER *Aerodynamics* ii. 53 When a certain critical velocity is exceeded the continuity is broken and the phenomenon of turbulence manifests itself. **1922** GLAZEBROOK *Dict. Appl. Physics* I. 299/2 The important part played by turbulence in reducing the time of explosion in actual engines, and thus rendering high revolution speeds practicable. **1928** N. SHAW *Man. Meteorol.* II. vi. 284 The variation of wind with height is now treated as the effect of turbulence or eddy-motion in the moving air. **1935** *Discovery* Oct. 390/2 This is remedied either by small fans or, automatically, by injecting some cold air from the dehumidifier with the heated air, causing a form of turbulence in the enclosure. **1980** R. P. BENEDICT *Fund. Pipe Flow* iv. 141 A stability theory has been given that explains the origin of turbulence in terms of arbitrary small disturbances which initiate instabilities in the laminar boundary layer.

turbulent, *a.* Add: **2. c.** Of, pertaining to, or designating flow of a fluid in which the velocity at any point fluctuates irregularly and there is continual mixing rather than a steady flow pattern.

Turbulent flow was earlier called *sinuous* or *eddying flow*.

1895 H. LAMB *Hydrodynamics* xi. 574 The resistance, in the case of turbulent flow, is found to be sensibly independent..of the viscosity of the fluid. **1907** F. W. LANCHESTER *Aerodynamics* ii. 53 When this critical velocity is reached the parallel flow breaks up, and is replaced by an irregular turbulent motion. **1926** H. R. RICARDO *Engines of High Output* 63 In the actual engine cylinder..the mixture..is being whirled about very rapidly; it is, in fact, in a highly turbulent condition. **1930** *Engineering* 7 Mar. 319/3 Hence, when viscous flow changes to turbulent flow the dissipation of energy is increased. **1947** [see *Reynolds stress* s.v. *REYNOLDS]. **1956** A. A. TOWNSEND *Struct. Turbulent Shear Flow* i. 3 In turbulent motion..the motion at any point influences the motion at other distant points. **1968** [see *LAMINAR *a.* 2 a]. **1982** *Sci. Amer.* July 99/1 When the bore of a tube flares gradually (as the aortic lumen does in an incipient aneurysm), the fluid near the wall slows down, generating turbulent flow.

Turcification: see TURKIFICATION in Dict. and Suppl.

Turco-, Turko-. Add: **a.** *Turco-Bulgarian* (example), *-German, -Persian, -Russian, -Tartar, -Tatar, -Tataric.*

1854 MAX MÜLLER in C. Bunsen *Christianity & Mankind* III. 279 The still undivided 'Turko-Tataric' speech. **1854** J. H. NEWMAN *Lect. Hist. Turks* 62 The ancient Turco-Tartar empire..extended to the Caspian and towards the Indus. **1880** A. H. SAYCE *Introd. Sci. Lang.* II. viii. 190 The whole Turanian family..may be divided into five branches, the Finno-Ugric, the Turko-Tatar, the Samoyedic, the Mongolian, and the Tungusian. **1909** G. DRAGE in *Camb. Mod. Hist.* (1921) XI. ix. 276 The settlement of the Turco-Persian frontier question. **1915** W. S. CHURCHILL in M. Gilbert *Winston S. Churchill* (1972) III. Compan. I. 421 The Turco-German fleet. **1923** G. BUCHANAN *My Mission to Russia* I. vi. 64 The Turco-Bulgarian agreement. **1948** D. DIRINGER *Alphabet* 567 Turco-Tatar and Caucasian languages. **1969** V. N. DATTA *Jallianwala Bagh* 5 The Turco-German organisation at Berne. **1982** M. BINGHAM *Princess Lieven* xiii. 189 The Turco-Russian conflict rumbled on.

b. Turcoce·ntric *a.*, centred round Turkey or the Turks; hence **Tu:rcocentri·city; Turcologist**: delete *nonce-wd.* and add later examples; also **Turcolo·gical** *a.*, **Turco·logy; Turcophi·lia** = *Turcophilism.*

1964 *Jrnl. R. Central Asian Soc.* LI. 72 A Turco-centric history. **1969** *Middle Eastern Studies* May 173 The passage is worth noting for Gökalp's turcocentricity: 'Europeans committed an historical sin by translating Attila's title..as "Scourge of God".' **1952** *Round Table* Dec. 25 The well of Turcological study. **1951** W. K. MATTHEWS *Languages U.S.S.R.* iv. 63 Gyula Németh, the Hungarian turcologist. **1976** *Times Lit. Suppl.* 3 Sept. 1077/3 The Russians..began to produce excellent Persianists, Turcologists, Mongolists. **1918** *Q. Rev.* Apr. 513 The transformation of the *Medresses*..into 'National' schools for the teaching of Turkology. **1951** W. K. MATTHEWS *Languages U.S.S.R.* iv. 64 This uses phonetic as well as morphological data, and has received the sanction of Soviet turcology. **1908** J. MORLEY *Recoll.* (1917) II. 245 Am I quite wrong in suspecting a degree of Turcophilia in you? **1967** C. SETON-WATSON *Italy from Liberalism to Fascism* x. 376 William II was impressed. His Turcophilia of the previous autumn had by now quite vanished.

Turcoman. Add: **1. b.** (Later examples.) Cf. *TURKMEN, TÜRKMEN.

1888 *Encycl. Brit.* XXIV. 1/1 Ural-Altaic Languages.. *Turkoman* (*Turkmenian*), west Turkestan, north Persia, and Asia Minor. **1908** [see *JAGATAI]. **1954** PEI & GAYNOR *Dict. Ling.* 222 *Turkoman*, a Near-Eastern language; it belongs to the Southern Turkic group of the Altaic sub-family of the Ural-Altaic family of languages.

3. (Earlier example.) Also, a Turcoman carpet or rug.

1881 C. C. HARRISON *Woman's Handiwork in Mod. Homes* I. 57 Turcoman, when ravelled, produces a superb fringe, like soft chenille. **1962** C. W. JACOBSEN *Oriental Rugs* I. vi. 63 Up until 1940, all rugs known as Turkomans or Bokharas..came from Turkestan. **1975** P. SOMERVILLE-LARGE *Couch of Earth* iv. 54 On the floor..were good carpets, mostly small Turcomans.

4. Turcoman carpet (example), **rug** (examples).

1901 J. K. MUMFORD *Oriental Rugs* xii. 235 One division of these Turkoman carpets, which avoids on the one hand close adherence to the Bokhara device, and on the other the latch-hook style of the Yomuds, is called Beshir. **1922** H. CLARK (*title*) Bokkara, Turkoman and Afghan rugs. **1975** *Oxf. Compan. Decorative Arts* 791/1 A design peculiar to Turcoman rugs is that known as katchli.

turd. Add: **1. a.** (Later examples.)

1922 JOYCE *Ulysses* 649 The horse..added his quota by letting fall on the floor..three smoking globes of turds. **1928** in A. W. Read *Class. Amer. Graffiti* (1935) 81 Now and then a fart is heard Mingling with a dropping turd. **1968** *Listener* 1 Aug. 152/2 His protest at the killing in Vietnam is at least original: he parcels up a rich turd and mails it to the White House. **1981** N. GORDIMER *July's People* 35 It was true that it was difficult to get the children to remember to bury the paper along with the turd.

c. (Later examples as a term of execration or contempt.) Cf. SHIT *sb.* 1 b.

1936 A. HUXLEY *Eyeless in Gaza* xxii. 315 'But why not?' the poor old turd kept asking. **1944** D. WELCH *In Youth is Pleasure* viii. 152 Somebody called Woods a dirty old sod, another called him a great turd. **1965** HOWARD & WEST *Making of Prime Minister* xiii. 204 A purple-faced steward walked up to a scrawny, pale heckler and yelled, 'Shut up, you ignorant turd!' **1978** B. FREEMANTLE *Clap Hands* ix. 66 The man..could make everyone else feel a turd.

2. *turd-coloured* (later example), *-eating* adjs.

1978 in R. Quirk *Style & Communication in Eng. Lang.* (1982) ii. 33 Lofty structures of turd-coloured brick.. seemed to be deserted. **1969** Turdeating [see *cock-sucking* ppl. adj. s.v. *COCK *sb.*¹ 23]. **1978** J. KRANTZ *Scruples* x. 295 Those turd-eating Mexican border guards'll put you away.

Hence **tu·rdish** *a.*, characteristic of a 'turd' or contemptible person.

1936 A. HUXLEY *Eyeless in Gaza* xxii. 315 'Turds to the core,' he said. 'So they can't think anything but turdish thoughts.' **1966** *Punch* 12 Jan. 64/2 An aristocratic, even Byronic work which finds strident anti-communism turdish.

turdion. Delete † *Obs. rare* and add later examples in revived use, in Fr. forms *tordion, tourdion.*

1895 L. GROVE *Dancing* viii. 244 The 'Danse basse' was very grave... It was performed to the accompaniment of psalms. It consisted of three parts—(1) the *Danse basse* proper, (2) the *Retour*, (3) the *Tourdion*. **1914** T. & M. W. KINNEY *Dance* iii. 54 The *Tordion* is another dance of lively origin. Sometimes it was made a vehicle for the grotesque. **1924** SHARP & OPPÉ *Dance* 15 The Tordion was danced with the same steps as the Galliard but more quietly, without spring. **1957** G. B. L. WILSON *Penguin Dict. Ballet* 267 *Tordion*, third section of the Basse Danse. **1974** *Early Music* July 164/2 The tordion, a restrained form of galliard.

turf, *sb.*¹ Add: **4. b.** *transf.* The road or street as the milieu of prostitutes, tramps, etc.; esp. *on the turf*, engaged in prostitution. *slang.*

1860 HOTTEN *Dict. Slang* (ed. 2) 241 *On the turf*, one who occupies himself with race course business; said also of a street-walker. **1899** 'J. FLYNT' *Tramping* I. ii. 28 The road proper, or 'the turf', as the people who toil along its stretches sometimes prefer to call it, is low life in general. **1936** H. ASBURY *French Quarter* xii. 369 During [Kate Townsend's] early years 'on the turf', as the saying went, she was..thrifty and ambitious. **1962** PARKER & ALLERTON *Courage of his Convictions* v. 179, I wouldn't let her go out on the turf, because of this thing I've got about not poncing. **1984** J. O'DONOGHUE *Sergeant Horn's Murder Trap* vi. 41 'I might have been one of Ma Dolma's brasses for all you know.'..'Come off it. You've never been on the turf.'

4*. Usu. with substantive in possessive case or with possessive adj. orig. and chiefly *U.S.* **a.** The streets controlled by a juvenile street-gang and regarded by them as their territory.

1953 CRAMER & KARR *Teen-Age Gangs* i. 4 He had looked forward to drifting pleasantly through the Emerald turf—the term currently used in Brooklyn instead of territory. *Ibid.* 6 No War Hawk was safe if caught on the turf of the Emeralds. And no Emerald was safe on the turf of the War Hawks. **1959** H. SALISBURY *Shook-Up Generation* i. 19 These blocks constituted the 'turf' of a well-known street-gang. **1964** [see *CHIPPY *sb.* 4]. **1973** 'J. PATRICK' *Glasgow Gang Observed* xx. 189 Like most American adolescent gangs..the Young Team attached enormous importance to territory and used the same word 'turf' for it.

b. The part of a city or other area within which a criminal, detective, etc., operates. Cf. *PATCH *sb.*¹ 3 e.

1962 *Sat. Even. Post* 28 Apr. 30/2 Her [*sc.* a social worker's] turf: the lower Bronx. **1966** 'J. ASHFORD' *Consider Evidence* iii. 23 She [*sc.* a prostitute] claimed she could make a hundred quid a week on her turf. **1971** *N.Y. Times* 10 Jan. xx-1/1, I came to Beverly Hills..to see the stars' home turf. **1976** D. BENNETT *Jigsaw Man* (1977) viii. 153 Special Branch would not want to be involved in a killing so far from their own turf. **1978** S. BRILL *Teamsters* ii. 48 As both men sat in prison, they were dividing up Teamsters turf.

c. *transf.* and *fig.* A person's sphere of influence or activity.

1970 *Sat. Rev.* (U.S.) 17 Oct. 67/3 The lives of all our children and the very mindedness of society itself cannot be made whole as long as educators are obsessed by indecent needs to defend their own turfs. **1973** *Family Circle* Apr. 120/1 Male occupations are a turf from which women are excluded. **1977** J. F. FIXX *Compl. Bk. Running* xiii. 157 Dogs, he explained, are assiduous defenders of turf. **1982** 'E. LATHEN' *Green grow Dollars* vii. 55 They think that, on their own turf, they can overawe Ackerman and Werzel.

5. a. *turf-barge, -house* (later examples), *-wall* (earlier example); *turf-coloured* adj.; **c.** *turf-grown* (earlier example).

1922 JOYCE *Ulysses* 218 Father Conmee saw a turf-barge... Father Conmee reflected on the providence of the Creator who had made turf to be in bogs where men might dig it out and bring it to town. **1916** —— *Portrait of Artist* (1969) i. 54 He had skin the same colour as the turf-coloured bogwater in the..bath. **1867** J. G. WHITTIER *Tent Beach* 10 Above..turf-grown wall They saw the fort flag rise and fall. **1789** J. WESLEY *Jrnl.* 26 May (1916) VII. 502 Part of them [*sc.* his congregation] were sheltered by a spacious turf-house, and the rest little regarded the rain. **1967** H. HARRISON *Technicolor Time Machine* (1968) v. 50 Smoke still drifted down from the chimney hole of the squat, turf house. **1849** THOREAU *Week Concord Riv.* 168 But as it were, by a turf wall this valley was concealed.

d. turf-knife (example); **turf-line**, a line formed from turf; *spec.* in an archæological excavation, a layer of soil representing former grassland.

1841 THOREAU *Jrnl.* 20 Apr. in *Coll. Wks.* (1906) VII. 251 The ditching spade and turf knife may be engraved on the coat-of-arms of his posterity. **1935** E. H. W. MEYERSTEIN *Verse Lett. to Five Friends* (1954) 11 One joy.. To take the turf-line of the Pilgrim Road. **1936** *Proc. Prehist. Soc.* II. 214 Well marked turf lines isolated these ditches from the Iron Age above them. **1957** V. G. CHILDE *Dawn Europ. Civilization* (ed. 6) i. 3 Fossil turf-lines of Atlantic age. **1975** J. G. EVANS *Environment Early Man Brit. Isles* vi. 119 At the surface of the buried soil is a thin stone-free horizon or turf line... This is caused by earthworm sorting.

turf, *v.*¹ Add: **4.** *trans.* To throw or kick (a person, etc.) forcibly *out* (occas. *off*); also *transf. colloq.* Without const. (*Public School slang*), to kick.

1888 KIPLING *Only Subaltern* in *Under Deodars* 97 The Colonel will turf you out of that in double quick time. **1905** H. A. VACHELL *Hill* ii. 32 Sorry I turfed that little ass so hard... [*Note*] To 'turf', *i.e.* to kick—*Harroviana*. **1925** WODEHOUSE *Carry on, Jeeves!* 90 The old boy turfed me out, Bertie, because he said I was a brainless nincompoop. **1930** J. B. PRIESTLEY *Angel Pavement* viii. 410 She'd bought hundreds of them [*sc.* magazines]. I've just had them turfed out. **1957** C. MACINNES *City of Spades* II. iv. 128 The guv'nor tried turfing them all out at first..but he's given up the struggle. **1976** J. I. M. STEWART *Memorial Service* iv. 58 These people have become my colleagues. If you use that sort of language about them I'll have to turf you out myself. **1977** 'O. JACKS' *Autumn Heroes* iv. 60 The plane's loaded... I can't turf off passengers.

Turfanian (tɜ̄ɹfēⁱ·niăn). [f. *Turfan* in Chinese Turkestan + -IAN.] A name given to the western dialect of Tocharian, otherwise known as Tocharian A. Also **Tu·rfan, Turfane·se.**

1939 Turfanese [see *KUCHAEAN, KUCHEAN]. **1958** PRIEBSCH & COLLINSON *German Lang.* (ed. 4) I. i. 6 Another extinct group of I.E. dialects is..Turfan (Turfanian, W. Tocharian or Toch. B). **1972** W. B. LOCKWOOD *Panorama Indo-Europ. Lang.* 254 Tocharian A was spoken at Karashahr and Turfan; it is sometimes called Turfanian.

turfite. (Earlier example.)

1836 [see *BOOK-MAKING 3].

Turinese, var. *TORINESE.

Turing machine (tiū°·riŋ). [Named after A. M. *Turing* (1912–54), English mathematician, who described such a machine in 1936.] A notional computing machine for performing simple reading, writing, and shifting operations in accordance with a prescribed set of rules, invoked in theories of computability and automata.

It is represented as a scanner that has a number of internal states and moves left or right along a tape on which is a sequence of symbols. The symbol read and the state of the scanner determine (in accordance with the rules) what replacement symbol is written, what new state the scanner enters, and what move it makes along the tape before the cycle is repeated.

1937 A. CHURCH in *Jrnl. Symbolic Logic* II. 43 [Abstract of Turing's paper.] Certain further restrictions are imposed on the character of the machine, but these are of such a nature as obviously to cause no loss of generality—in particular, a human calculator, provided with pencil and paper and explicit instructions, can be regarded as a kind of Turing machine. **1955** *Sci. Amer.* Apr. 62/1 To understand a Turing machine we need only know its table of commands. **1961** *Proc. Symposium Appl. Math.* XII. 39 A Turing machine plus random elements is a reasonable model for the human brain. **1969** P. B. JORDAIN *Condensed Computer Encycl.* 550 No Turing machine has ever been physically constructed or realized in hardware as a device for its own sake, but general-purpose digital computers have been programmed to simulate Turing machines. **1984** *Sci. Amer.* May 70/1 Beginning with the intuitive idea that a method is an algorithm—a procedure that can be mechanically carried out without creative intervention—he [*sc.* A. M. Turing] showed how the idea can be refined into a detailed model of the process of computation in which any algorithm is broken down into a sequence of simple, atomic steps. The resulting model of computation is the logical construct called a Turing machine.

‖ **turismo** (tū°ri·zmo). [Sp., It.] Tourism as an industry or dedicated pursuit (in the Latin countries of Europe).

1926 R. MACAULAY *Crewe Train* I. v. 29 Poor Humphrey is looking quite sick and wan, either with turismo, the motion of the Andorran diligences, or *l'amour*. **1959** *Times Lit. Suppl.* 20 Nov. 679/4 His descriptions of the countries he collected on his speedometer seem more often to be drawn from the handbooks of *Turismo* than from fresh observation. **1977** J. I. M. STEWART *Madonna of Astrolabe* xii. 179 Ravello hadn't changed much... *Turismo* hadn't gained its hoped-for grip.

‖ **turista** (tū°ri·stā). [Sp., lit. = tourist.] A name for a form of traveller's diarrhœa affecting visitors to Mexico.

1970 *New Scientist* 8 Jan. 47/1 An intestinal attack known as gyppy tummy..in the Middle East;..Montezuma's revenge..and turista in Mexico. **1976** M. MILLAR *Ask for me Tomorrow* v. 40 *Turista* is bad enough..but infectious hepatitis is worse. **1980** *Jrnl. R. Soc. Arts* Jan. 90/1 Even avoiding water does not entirely protect me against turista.

Turk, *sb.*¹ Add: **2. e.** Also *transf.* (sometimes with small initials): any group of young or relatively young men full of new ideas and impatient for change; esp. a radical or 'progressive' element in a political party. Occas. *sing.*

c **1929** in W. Safire *New Lang. Politics* (1968) 496/2 These new Republican warriors were called the Young Turks, a band of about 20 who had mutinied against the feeble leadership of the Old Guard. For Senators they were young men (average age: 56). **1953** W. S. CHURCHILL in *Ibid.* 497/2 You're just like the Young Turks in my government. **1963** D. OGILVY *Confessions Advert. Man* (1964) ii. 24 In hiring, the emphasis will be on *youth*. We are looking for young turks. **1971** A. MIZENER *Saddest Story* 331 E. E. Cummings and Pound..were writers little calculated to attract the Young Turks, for whom they would seem elder statesmen of the modern movement. **1981** J. DUNNING *Deadline* (1982) xvii. 160 Malcolm Dawes had been a career man. He was a young turk, graduating from the FBI Academy in 1952.

4. b. A person of Irish birth or descent. *slang* (usu. *depreciatory*). Chiefly *U.S.*

In this sense perh. really a derivative of Ir. *torc* boar, hog, as suggested by W. A. McLaughlin (*Dialect Notes* (1914) IV. 147–8); but cf. *TURKEY *sb.*² 5* b.

1914 in *Dialect Notes* IV. 148 You Italians have the votes, but it takes us Turks to run the government. **1945** MENCKEN *Amer. Lang.* Suppl. I. 603 *Turk* is used among Roman Catholic priests in the United States to designate a colleague of Irish birth. **1959** *Observer* 1 Mar. 10/1 Their backs are to the wall in a desperate tyre-chain feudal war to protect the integrity of their declining manor against the invasion of 'bubbles and squeaks' (Greeks and Cypriots), 'turks' (Irish) and 'spades' (coloureds). **1971** S. HOUGHTON *Current Prison Slang* (MS.) 17 *Turk*, Paddy, Irishman.

6. c. A Turkish cigarette.

1926 'SAPPER' *Final Count* iii. 65 Why the devil don't you smoke a Corona Corona, you fool! Put out that Turk. **1935** N. MARSH *Enter Murderer* vi. 71 Cigarette? These are Turks. **1965** 'R. ERSKINE' *Passion Flowers in Business* v. 60 Fat, oval Turks in a Wedgwood box.

Turkana (tɜ̄ɹkā·nā). [Native name.] (A member of) an East African tribe living between Lake Rudolph and the Nile; their language. Also *attrib.*

1902 H. JOHNSTON *Uganda Protectorate* II. xx. 887 Turkana has a few more words in it betraying Hamitic.. affinities. **1911** [see *KARAMOJO]. **1930** C. G. SELIGMAN *Races of Africa* vii. 161 The Masai and Turkana are nomadic herdsmen. **1959** A. MOOREHEAD *No Room in Ark* vii. 137 The local tribesmen—people like the Karamojong and the Turkana—have not taken to western civilization in the way nearly all other Africans are doing. **1963** *Times* 7 June 12/3 Thirty-two Turkana tribesmen and women have been killed in inter-tribal fighting in Kenya's Eastern Region. **1977** H. INNES *Big Footprints* I. i. 26 He's a Turkana. He was born up there and he's been back to Lake Rudolph many times. *Ibid.* 27 'Is that Turkana you're talking?'..'No. A mixture of Samburu and Swahili.'

Turkey¹. Add: **3. c. Turkey rug** = *Turkish rug* s.v. TURKISH *a.* 2 b; **Turkey sponge,** a superior grade of commercial sponge from the Mediterranean and Adriatic.

1881 C. C. HARRISON *Woman's Handiwork in Mod. Homes* III. 143 The hardwood floors are stained dark, with Turkey rugs. **1920** F. G. ELLERTON *Let.* 25 Feb. in J. Bailey *Lett. & Diaries* (1935) 201 Now stamp up and down your Turkey rugs. **1902** D. SALOMONS in A. C. Harmsworth *Motors* vi. 94 A large Turkey sponge is best for cleaning the body and wheels of the car. **1968** *Canad. Antiques Collector* Nov. 21/1 A fine piece of Turkey sponge was soaked in this mixture [*sc.* aromatic vinegar] and enclosed in a small container called a vinegar box or vinaigrette.

turkey². Add: **2. d.** *to talk turkey* (earlier example of sense 'to say pleasant things'); now usu. (in this sense also *to talk cold turkey*) to speak frankly and without reserve; to talk hard facts, get down to business. (No longer restricted to N. Amer.)

1824 *Little Bit of Tid-Re-I* II. 109 So that, all things considered, I hope neither the Indian, whom the Yankey could not cheat in the division of their game (a turkey and a buzzard,)..will accuse me of *not talking Turkey* to them in this article. **1903** *Dialect Notes* II. 333 *Talk turkey*, v.phr., to talk plainly: 'I'm going to *talk turkey* with him and see if I can't get him to mend his ways.' **1919** E. HOUGH *Sagebrusher* xiv. 125 Do you know when he got rattled he began to talk Dutch to me? Well, I talked turkey to him. **1928** *Daily Express* 4 Jan. 11/5 She talked cold turkey about sex. 'Cold turkey' means plain truth in America. **1939** A. HUXLEY *After Many a Summer* II. x. 279 'I'll make it worth your while,' he said. 'You can have anything you care to ask for.'..'Ah,' said Dr. Obispo, 'now you're talking turkey.' **1946** E. HODGINS *Mr. Blandings builds his Dream House* (1947) xv. 196 The boss painter..wanted to talk turkey about..the final colours. **1967** A. CHRISTIE *Endless Night* ix. 67 Send for a high powered lawyer and tell him you're willing to talk turkey. Then he fixes..the amount of alimony. **1982** T. BERGER *Reinhart's Women* xix. 270 Maybe I'll be in a position to talk turkey about an arrangement that would work out for us both.

e. *cold turkey*: a method of treating drug addicts by sudden and complete withdrawal of the drug, instead of by a gradual process. Also *attrib.* and as *advb. phr.*; also *transf.* Hence *cold-turkey* vb. trans., to cure of drug addiction by 'cold turkey' treatment. *slang* (orig. *N. Amer.*).

1921 *Daily Colonist* (Victoria, B.C.) 13 Oct. 15/6 Perhaps the most pitiful figures who have appeared before Dr. Carleton Simon..are those who voluntarily surrender themselves. When they go before him, they [*sc.* drug addicts] are given what is called the 'cold turkey' treatment. **1936** *Amer. Speech* XI. 120/1 *Cold turkey*, treatment of addicts in institutions where they are taken off drugs suddenly without the 'tapering off' which the addict always desires. **1941** W. C. HANDY *Father of Blues* xviii. 243 We went on 'cold turkey' that morning without a rehearsal. **1951** *N.Y. Times* 27 June 19 She tried it 'cold turkey' once, which she explained meant merely stopping completely, without any attendant medication. **1960** *Times Lit. Suppl.* 16 Sept. 589/4 *The Scene* is written by a junkie..who was snitched by a flip, busted by the nailers and after a stretch in the pen cold turkeyed..that is to say, was once a dope addict; he was arrested and.. managed to break himself of his addiction. **1962** 'K. ORVIS' *Damned & Destroyed* xi. 72, I made one cold-turkey cure and it near killed me. **1976** S. GEORGE *Fatal Shadows* 154 She took a cold turkey, no methedrine, no sedatives, nothing, just off.

3. b. *plain turkey, scrub turkey*: humorous names for swagmen who haunt, respectively, the Australian plains and the bush [perhaps with partial allusion to the 'turkey' (sense 5 below) which they carry]. *Austral. slang.*

1955 A. MARSHALL *I can jump Puddles* 152 Father.. was familiar with the ways of swagmen... The bearded men who kept to the bush he called 'Scrub Turkeys' and those who came down from the plains he called 'Plain Turkeys'. **1973** F. HUELIN *Keep Moving* 178 *Scrub Turkey*, bagman who has gone Bush. Usually slightly mental or eccentric.

5. Applied more generally to bundles or hold-alls carried by other itinerant workers, vagrants, etc. Also *Canad.* and *Austral.*

1909 *Outlook* 2 Jan. 19/1 A [Colorado] desert miner calls his valise a 'turkey'. **1912** G. H. GIBSON *Ironbark Splinters* 6 So you 'pack' your bloomin' turkey, and you take the northern train. **1931** 'D. STIFF' *Milk & Honey Route* 216 *Turkey*, a bundle, a suitcase or a canvas bag. **1945** BAKER *Austral. Lang.* v. 104 Expressions to describe being on the tramp..[*to*] *coil one's turkey* (strictly this applies to the rolling of a swag) [etc.]. **1963** R. SYMONS *Many Trails* v. 54 The cowboys' 'turkeys'—as they called their bedrolls, in which were wrapped their personal possessions such as tobacco—when the outfit was on the move. **1973** in B. Broadfoot *Ten Lost Years* ii. 19, I took to the road with..my poor clothes tied together with two leather thongs, and that is what they call a turkey.

5*. *U.S. slang.* **a.** An inferior or unsuccessful cinematographic or theatrical production, a flop; hence, anything disappointing or of little value.

1927 *Vanity Fair* (N.Y.) XXIX. 132/3 'A turkey' is a third rate production. **1939** G. MARX *Let.* 27 Oct. (1967) 21 The boys at the studio have lined up another turkey for us... I saw the present one the other day and didn't care much for it. **1941** J. M. CAIN *Mildred Pierce* 176 The beach..was studded with rocks and was therefore unsuitable to swimming. For all ordinary purposes it was simply a turkey. **1962** *Movie* June 18/1 With *The Four Horsemen*, Minnelli was once more landed with a turkey, an old one, too. **1977** H. FAST *Immigrants* III. 201 'Have you ever thought of selling the place?' Jake asked... 'Oh? And who the hell would buy this turkey?'

b. = *TURK *sb.*¹ 4 b; *spec.* an Irish immigrant in the U.S.

1932 J. T. FARRELL *Young Lonigan* i. 31 Dooley was one comical turkey, funnier than anything you'd find in real life. **1966** [see *SALT WATER B. c].

c. A stupid, slow, inept, or otherwise worthless person.

1951 in Wentworth & Flexner *Dict. Amer. Slang* (1960) 556/2 So, if you got a collector [of internal revenue] through the civil service system who was a real turkey, you'd be stuck with that turkey practically until he died. **1969** C. BURKE *God is Beautiful, Man* (1970) 105 You better get real strong like so that you won't be a turkey. **1978** *Time* 3 July 13/3 'Come on, you turkeys! Let's speed this show up!' cries an irreverent observer. **1984** *Tampa* (Florida) *Tribune* 5 Apr. 4D/3, I decided I had had enough of that turkey.

6. *turkey dinner, farm, farmer, -gobbler* (earlier example), *meat, -tail, -wing* (earlier example); **turkey-bush**, an evergreen shrub, *Myoporum deserti*, native to Australia and bearing white flowers followed by purple berries; **turkey-foot,** (*b*) applied to other things resembling a turkey's foot (in quot. 1957, a weaving pattern); **turkey-shoot** (examples); **turkey-trot** (earlier example); also, a fast jogging trot like that of a turkey; hence **turkey-trotting** *a.*; **turkey vulture** (earlier example).

1911 W. R. GUILFOYLE *Austral. Plants* 265 'Turkey-bush'..evergreen shrub..reputed poisonous and injurious to stock. **1936** F. CLUNE *Roaming round Darling* xvii. 163 Shrubs..mingled with turkey-bush flower—white and small, having five petals to each bloom, like an English daisy. **1965** *Austral. Encycl.* IX. 59/2 Turkey-bush, one of several names applied to the inland shrub *Myoporum deserti* because wild turkeys or native bustards have been observed to eat the berried fruits. **1953** G. W. BRACE *Spire* xiii. 125 'Turkey dinner, eh?' 'With gravy,' Sylvia said. **1977** 'W. TREVOR' in D. Marcus *Best Irish Stories* II. 78 The Bulrush Café has a turkey dinner advertised. **1952** *Turkeys* Nov. p. ix. (Advt.), Beale's turkey farm. **1959** *Ibid.* July–Aug. 29 (Advt.), One of Norfolk's leading turkey farmers. **1972** K. BONFIGLIOLI *Don't point that Thing at Me* iv. 40 A rich turkey-farmer in Suffolk. **1932** SIMPSON & WEIR *Weaver's Craft* xii. 126 *Turkey Foot* (24 threads).—Each pedal should be used six times in the following order: 4, 2, 3, 1; repeat. **1935** M. MOORE *Sel. Poems* 71 The firs stand in a procession, each with an emerald turkey-foot at the top. **1836** W. T. PORTER in *Spirit of Times* 9 July 162/1 They seemed to me about the size of a big Christmas turkey gobbler. **1901** A. H. RICE *Mrs. Wiggs of Cabbage Patch* ii. 24 If you ain't never et turkey meat you don't know how good it is. **1972** *Country Life* 30 Nov. 1507/1 The promoters of turkey-meat sales. **1845** S. JUDD *Margaret* I. 62 Its succedanea..were a turkey shoot the next day, and a ball. **1898** H. FREDERIC *Deserter* 81 The farther of the two was now so far away that he seemed a mere dark speck, like the object seen from the gun-line of a turkey-shoot. **1980** *Outdoor Life* (U.S.) (Northeast ed.) Oct. 92/2 Saturday afternoon turkey shoots used to be popular in my community. **1851** J. J. HOOPER *Widow Rugby's Husband* 84 Betsy dodged behind the wild turkey-tail which she carried by way of a fan. **1936** M. MITCHELL *Gone with Wind* xviii. 318 Girls who laughed from behind turkey-tail fans. **1839** *Southern Lit. Messenger* V. 337/1 May-be I didn't set up a high turkey-trot, and peeled it like thunder. **1895** F. REMINGTON *Pony Tracks* 187 He would run me off the plantation at a turkey trot if I did shoot. **1908** W. G. DAVENPORT *Butte & Montana beneath X-Ray* 42 The light fantastic, the turkey trot and the pazamala were indulged in by all to a late hour. **1859** H. E. TALIAFERRO *Fisher's River Scenes & Characters* 36 You're a purty set uv ill-begotten, turkey-trottin' pukes, to raise a quarrel with a peaceuble man, and then run like a gang uv geese. **1823** E. JAMES *Acct. Exped. Rocky Mts.* I. 4 At evening we..saw..several turkey vultures. **1872** MRS. STOWE *Sam Lawson's Oldtown Fireside Stories* 4 'I'll sweep up the coals now,' he added, vigorously applying a turkey-wing to the purpose.

Turkey-carpeted, *a.* (Earlier example.)

1831 M. EDGEWORTH *Let.* 29 Mar. (1971) 505 The library at Eton College is the most *comfortable* I ever was in—turkey carpetted.

Turkey red. Add: **c. Turkey red oil** (also **Turkey-red, turkey Red Oil**), sulphonated castor oil, principally used with alizarin to produce the colour Turkey red.

1879 *Jrnl. Chem. Soc.* XXXVI. 187 This 'Turkey-red oil' is a mixture of sulpho-ricinoleate and sulpho-pyroterebate of sodium. **1903** [see *oil tanning* s.v. *OIL *sb.*¹ 6 c]. **1980** K. G. PONTING *Dict. Dyes & Dyeing* (1981) 172/2 John Mercer did a great deal of research on the production of the necessary turkey red oil produced by the action of sulphuric acid on castor oil.

Turkicize (tɜ̄·ɹkisəiz), *v.* Also **turkicize.** [f. TURKIC *a.* + -IZE.] *trans.* To render Turkic or Turkish. Hence **Tu:rkiciza·tion; Turkicized** *ppl. a.*

1939 C. S. COON *Races of Europe* vii. 233 The Magyars were Ugrians from the region between the Volga and the Urals, who had been partially Turkicized by the Petchenegs and others. **1951** W. K. MATTHEWS *Languages U.S.S.R.* iii. 17 A tiny remnant of a mostly turkicised tribe living in the Turkic-speaking country north of the Altai range. **1964** *Language* XL. 301 It was long believed ..that Chuvash was a Turkicized Finnic language. **1964** G. WHEELER *Mod. Hist. Sov. Central Asia* ii. 24 The Turkicization of the Mongol rulers appears to have started very early after the Mongol state was established.

Turkification. Add: (Later examples.) Also **Turcification.**

1922 BUXTON & CONWIL-EVANS *Oppressed Peoples & League of Nations* v. 132 The Armenians..resisted all attempts at 'turcification'. **1949** J. PARKES *Hist. Palestine* xii. 241 Unhappily at this point the new policy of Turcification caused the government to maintain the deposition [of the Patriarch Damianus], as they had no intention of supporting what appeared—and was—an Arab demonstration. **1976** *Economist* 4 Sept. 45/1 The demographic Turkification of the north will soon be complete.

Turkish, *a.* (*sb.*) Add: **A.** *adj.* **2. b.** *Turkish cigarette* (or *cig*), *tobacco*; **Turkish carpet** = TURKEY CARPET; **Turkish coffee**, the strong (usu. sweet) black coffee commonly drunk in the East, in which the ground beans are boiled thrice over and the liquid is served with the grounds; a cup of such coffee; **Turkish crescent** *Mus.* = *Chinese pavilion* s.v. CHINESE *a.* 2; cf. *jingling Johnny* (*a*) s.v. *JINGLING *ppl. a.*; **Turkish delight**, substitute for def.: a sweetmeat consisting of gelatine boiled, cubed, and dusted with sugar, *RAHAT LOKUM; (earlier examples); cf. *DELIGHT *sb.* 4; **Turkish slipper**, a soft heelless slipper with turned-up toe, a babouche; **Turkish trousers**, baggy oriental pantaloons.

1886 S. W. MITCHELL *Roland Blake* v. 42 The room.. was luxuriously comfortable with a heavy-piled Turkish carpet and easy-chairs. **1977** FLEMING & HONOUR *Penguin Dict. Decorative Arts* 805/2 Turkish carpets.. differ from Persian carpets..stylistically in that their design seems invariably to have been created by weavers rather than painters. **1897** KIPLING *Captains Courageous* i. 6 Any gen'elman got a real Turkish cig on him? **1903** A. BENNETT *Truth about Author* xiii. 172 The aroma of coffee, the odour of Turkish cigarettes. **1950** G. GREENE *Third Man* vii. 56 The Turkish cigarettes that Harry always smoked. **1982** T. KENEALLY *Schindler's Ark* ix. 102 Oskar Schindler—in his coat with the fur lapels.. reaching for another Turkish cigarette. **1854** F. DUBERLY *Let.* 17 Aug. in E. E. P. Tisdall *Mrs. Duberly's Campaigns* (1963) ii. 54 Rum and water till bedtime, or the very strongest Turkish coffee. **1898** G. B. SHAW *Arms & Man* II. 22 (*stage direction*), A small table..is laid for breakfast with Turkish coffee pot, cups, rolls, etc. **1958** L. DURRELL *Mountolive* xiv. 267 She was sitting in the lounge of Shepheards Hotel under the clock with an untouched Turkish coffee before her. **1978** H. KAPLAN *Damascus Cover* xv. 155 Ari ordered a second cup of thick Turkish coffee. **1891** C. R. DAY *Descr. Catal. Mus. Instruments R. Military Exhib., London, 1890* XII. 233 There were specimens of the old Turkish Crescent (*chapeau chinois*), once a favourite adjunct to military bands, and known in former years by the familiar nick-name of 'Jingling Johnny'. They consisted of brass hoops, hung with little bells.., and ornamented with gilded crescents and long streaming tails of horse-hair. They were carried upon poles. **1938** *Oxf. Compan. Mus.* 965/1 *Turkish crescent* or *Turkish jingle*.., a noise maker introduced into military bands..at a time when there was a craze for 'Turkish music'. **1961** J. BLADES in A. Baines *Mus. Instruments through Ages* xiv. 338 The middle of the eighteenth century when European military bands were being increased to include percussion instruments on the lines of the Turkish military music—bass drum, cymbals, triangle, and sometimes also tambourine and the Jingling Johnny or Turkish Crescent. **1877** *Porcupine* 31 Mar. 843/2 Arabs and Greeks vended Turkish delight (horrible compound). **1888** *Boy's Own Paper* Summer 38/1 Instead of the usual boxes of..Turkish delight,..there was little to tempt the youthful appetite but piles of fruit. **1865** DICKENS *Our Mutual Friend* II. III. i. 3 Mr. Fledgeby's appearing erect at the foot of the bed in Turkish slippers, rose-coloured Turkish trousers..and a gown and cap to correspond. **1971** 'D. HALLIDAY' *Dolly & Doctor Bird* v. 65 He [*sc.* a Turkish dancer] wore a cinnamon tunic and trousers with gold Turkish slippers. **1827** M. WILMOT *Let.* 10 Oct. (1935) 305 If you add a *pipe* and some turkish tobacco..you will have the portrait of most of our beaux. **1981** A. MACKAY *Death on Eno* 68 The Turkish tobacco is added for aroma. [**1612** W. STRACHEY *Trav. Virginia* (1849) v. 66 A kynd of leather breeches and stockings, all fastened together..which they tye and wrappe about the loynes after the fashion of the Turkes or Irish Trouses.] **1821** BYRON *Don Juan* III. lxxii. 39 Her orange silk full Turkish trowsers furl'd About the prettiest ankle in the world. **1865** [see *Turkish slipper* above]. **1928** V. WOOLF *Orlando* iv. 140 The Turkish trousers which she had hitherto worn. **1973** *Guardian* 10 Apr. 13/2 Twinsets.. allied to jersey wrapover skirts, loose lopped-off pants, Turkish trousers.

B. *sb.* **2.** (Earlier example of sense 'Turkish tobacco'.)

1859 A. J. MUNBY *Diary* 2 Mar. in D. Hudson *Munby* (1972) 24 Long clays also, & a tall jar..of real Turkish.

turkle (tɜ̄·ɹk'l), U.S. dial. var. TURTLE *sb.*¹ and *sb.*² Hence **tu·rkling** = TURTLING.

1861 O. W. HOLMES *Elsie Venner* 384 Don' wan' no snappin'-turkles in my stable. **1893** H. A. SHANDS *Some Peculiarities of Speech in Mississippi* 65 *Turkle*, Negro for *turtle*. *Turkle-dove* is the common name given by negroes to the *turtle-dove*. **1929** W. FAULKNER *Sartoris* ii. 64 It wuz jes' like shootin' turkles in a slough. **1941** J. STILL *Proud Walkers* in *Sat. Even. Post* 10 May 112/3 The hours crawled turkle-slow. **1978** J. A. MICHENER *Chesapeake* p. ix, *Turkling*: State Senator Frederick C. Malkus, the region's premier turtle trapper, took me turkling, as that sport is called.

Turkmen, Türkmen (pl. *-mens*), varr. TURCOMAN.

1927 W. M. RAMSAY *Asianic Elements Gk. Civilization* viii. 86 The Turkish term *ova* and the Turkmen *oba*... The distinction between the Turks and the Turkmens or Nomads. **1953** O. CAROE *Soviet Empire* xiv. 235 Yet Uzbeks, Kazaks and Turkmens persist in abstract praises of their 'motherlands'. One Turkmen even had the insolence to write a poem 'My Turkmenistan'. **1962** A. TIETZE in Householder & Saporta *Probl. Lexicogr.* 267 Other languages of the Southern Turkic or Oghuz subgroup, as Azerbaijani or Türkmen. **1978** *Times* 18 Oct. 16/8 [In] Uzbekistan..many..speak Tadzhik, Turkmen and other Asian languages.

Turk's cap. Add: **3.** Also, any of several other members of the genus *Cactus.* (Earlier and later examples.)

1731 P. MILLER *Gardeners Dict.* s.v. *Melocactus*, The common or large Melon-Thistle, commonly call'd Turk-cap or Pope's-Head in the West-Indies. **1926** FAWCETT & RENDLE *Flora of Jamaica* V. 283 Turk's Head, Turk's Cap, Pope's Head, Melon Thistle. **1951** *Dict. Gardening* (R. Hort. Soc.) I. 345/2 The Turk's Cap. The species [of *Cactus*] are characterized by the formation of a woolly cap or cephalium when the plants reach flowering size.

turmoil, *v.* **2.** Delete † *Obs.* and add later examples (in quot. 1900 in humorous mock-solemn use). Now *rare.*

1900 W. SEWELL in W. Tuckwell *Reminisc. Oxford* xiii. 237 Garlic, deadlier without question E'en than hemlock: oh digestion... What is this, that still so deep here, Keeps turmoiling in my chest? **1981** T. HOLME *Funeral of Gondolas* v. 216 The noise of the storm receded. Outside, where it still turmoiled, was a long way away.

turn, *sb.* Add: **I. 2. d.** *turn of the screw*: an additional twist to tighten up the hold; an extra twist given to a thumbscrew by way of increasing the torture (in quots. *fig.*).

1796 [see SCREW *sb.*¹ 2 a]. **1853** DICKENS *Bleak House* xxiv. 331 (*heading*) A turn of the screw. **1898** H. JAMES *Turn of Screw* 4 If the child [in a ghost story] gives the effect another turn of the screw, what do you say to *two* children? **1940** *Manch. Guardian Weekly* 1 Mar. 175 Even more far-reaching schemes of increasing direct taxation.. are certain to be realised..whenever the psychological ground is favourable for this further turn of the screw. **1973** *Listener* 14 June 785/2 The first turns of the screw on the car commuter are already being prepared. The GLC wants to put up parking fines from £4 to £20.

6. *Naut.* A twist of rope round a mast, etc.

1886 R. BROWN *Spunyarn & Spindrift* vii. 91, I.. jumped to let go of the main-sheet. But Lord! we was in the white water almost before I could cast the turns off. **1930** *Sea Breezes* 74 Brushing off the snow and hammering gasket turns warmed my hands.

7. c. (Later examples.)

1932 G. F.-H. BERKELEY *Italy in Making* I. xviii. 268 At suitable points were inset 'turns', similar to those used in convents, so as to enable the servants to hand in food for both the cardinals and their attendants, without themselves entering the isolated wing. **1966** M. C. LORANG *Footloose Scientist in Mayan America* 90 Food was passed into the dining room through a '*turn*'..—a hollow roller set into the wall so that when one side is open, the other side is closed.

II. 8. d. *Cards.* The dealing or inversion of two cards in faro; hence *to call the turn*, to guess the order of the last three cards in the pack. Also *fig.*

1864 W. B. DICK *Amer. Hoyle* 207 The two cards drawn from the dealer's box—one for the bank and the other for the player..constitute a turn. **1889** *Cent. Dict.* s.v. *faro*, The showing of two cards constitutes a 'turn'. After each turn new bets are made for another, down to the last three cards of the pack; the only betting allowed after this is on 'calling the turn', or guessing which will show first. **1901** H. JAMES *Sacred Fount* 44 The face of Guy Brissenden, as recognizable at a distance as the numbered card of a 'turn'. **1908** *Sat. Even. Post* 5 Dec. 18/2 Ye-e-s, but this Wallingford person called the turn. **1940** D. W. MAURER *Big Con* viii. 257 The odds are always greater on the last turn, and anyone who wins on that play may make a young fortune. **1964** A. WYKES *Gambling* vii. 169 When three [cards] are left (the 'last turn'), players bet on the order of their appearance.

e. *Cricket.* A deviation of the ball's course after pitching; = BREAK *sb.*¹ 5.

1900 P. F. WARNER *Cricket in Many Climes* 190 The Newlands ground is the most difficult to make runs on in the whole of South Africa, the bowlers always being able to get considerable turn on the ball. **1977** *Observer* 30 Jan. 23/8 A deficit of even 50 runs..could pose problems for England because the spinners have already begun to extract a lot of turn.

11. d. *Golf.* The point in the course (after the ninth hole) at which the players begin the return journey.

1899 *Golf Illustrated* 1 Sept. 336/2 The hole was halved, as were also the eighth and ninth... The match..was all square at the turn. **1930** *Cambridge Daily News* 24 Sept. 7/3 Compston, who went out in 37 and was five up at the turn, won by seven and five.

III. 18. c. The point at which one named period of time gives way to the next; the beginning or end *of* a named period of time, regarded in relation to the transition point between it and the preceding or following period; *spec.* (*a*) *turn of the century*, the beginning or end of the century under consideration; also (usu. with hyphens) *attrib.* or as *adj.*; (*b*) *turn of the year*, the end of winter and the beginning of spring; also, the beginning of the calendar year.

1853 R. S. SURTEES *Mr Sponge's Sporting Tour* lxxi. 395 Who doesn't know the chilling feel of an English spring, or rather of a day at the turn of the year before there is any spring? **1859** [in Dict., sense 18 a]. **1926** O. BARFIELD *Hist. in Eng. Words* xi. 195 Just before the turn of the century there burst..upon England that strange explosion..the Romantic Movement. **1934** J. C. POWYS *Autobiogr.* vii. 300 How well I remember watching out the turn of the centuries—the nineteenth becoming the twentieth—in the little dining-room at Court House. **1935** *Discovery* Oct. 310/2 It is interesting to compare Dr Burr's notes, dating back to the turn of the century, with present conditions. **1947** R. CHURCH in M. Balcon et al. *Eng. Lang. & Lit.* xii. 200/2 Blatant imperialism shouted with a loud voice round the turn of the last Victorian years. **1947** W. H. LEWIS in *Ess. presented to C. Williams* 140 If they reached the port at the turn of the year,..the galleys, stripped to their hulls, would be emerging from their winter hibernation. **1952** G. SARTON *Hist. Sci.* I. xx. 512 We know that Autolycos was the teacher of..Arcesilaos of Pitane (315–240)... This suggests that he resided in Pitane and fixes the date approximately, the turn of the century. **1955** E. BLISHEN *Roaring Boys* iv. 251 The school lavatories..were a product of turn-of-the-century parsimony. **1955** I. & P. OPIE *Oxford Dict. Nursery Rhymes* 4 Romantic lyrics of a decidedly free nature..which were carefully rewritten to suit the new discrimination at the turn of the last century. **1955** *Time & Tide* 19 Nov. 1503/1 Can those who were young in the nineteen-twenties..remember the favourites as vividly as their elders recollect the tunes of the century's turn? **1961** I. MURDOCH *Severed Head* xiii. 120, I brought to mind that it was New Year's eve. Some nearer bells took up the peal... Soon it would be the turn of the year. **1961** *Times* 29 Dec. 11/7 Mr. William Brodie's sets, vaguely turn-of-the-century. **1965** C. E. POCKNEE *Parson's Handbk.* (ed. 13) p. xiv, At the turn of this century there were many who were in that state. **1968** A. M. FARRER *Interpretation & Belief* (1976) 190 It was as a doctrine of free will that Neo-Platonism was embraced by St Augustine at the turn of the fourth to the fifth century. **1970** *New Scientist* 17 Sept. 563/1 The latest estimates suggest that the area will be short of..1 270 000 cu.m. a day in 1981, and more than three million cu.m. a day by the turn of the century. **1970** H. BRAUN *Parish Churches* i. 22 At the turn of the millennium the monastic churches were quite enormous. **1970** *Daily Tel.* 26 Feb. 6/2 It was inevitable that at the turn of the decade there should appear yet more 'condition of Britain' analyses. **1971** *Ibid.* 23 Aug. 8 At the turn of the year, Kuala Lumpur officials talked confidently of reduced Communist terrorism along the Thai-Malaysia border. **1976** *National Observer* (U.S.) 29 May 9/1 Her modest, turn-of-the-century home on a quiet street in Indianapolis. **1976** *Church Times* 9 July 6/2 He begins with a splendid assembly of Church of England men all earnestly proclaiming, at the turn of the eighteenth and nineteenth centuries, doctrines then trendy. **1977** K. M. E. MURRAY *Caught in Web of Words* xv. 282 In 1897 James had set himself the target of completing half the letters of the alphabet by the turn of the century. **1979** *Sci. Amer.* Dec. 96/1 The evolutionary significance of the original Neanderthal discovery and of other human remains uncovered at Paleolithic sites was not apparent until the turn of the 20th century. **1981** 'W. HAGGARD' *Money Men* xi. 122 What was still called the parlour..was vintage turn of the century.

V. 28. b. (*e*) *out of* (*one's*) *turn*: out of one's due order or place in a series; *to talk* or *speak out of* (*one's*) *turn*: to say more than one ought to say, to speak inadvisedly or tactlessly.

1888 *Rules of Golf* 5 Playing out of turn. **1930** 'SAPPER' *Finger of Fate* 186 Well, old boy, our Lady Carrington was talking a little out of her turn. I don't blame her—it's a bit disconcerting to lose a thing like that. **1939** I. BAIRD *Waste Heritage* vi. 79 Easy, Eddy, I shouldn't have spoken out of turn there. **1945** J. B. PRIESTLEY *Three Men in New Suits* i. 7 I'm talking out of my turn, I expect—as usual. **1969** P. ROTH *Portnoy's Complaint* 14, I voluntarily and out of my turn set the table. **1978** P. VAN GREENAWAY *Man called Scavener* ix. 123 I'm going to talk out of turn and you'll be welcome to tell me I should mind my own business.

29. b. (Earlier examples.)

1715 D. RYDER *Diary* 19 Sept. (1939) 101 There was rope dancing and tumbling... There were now and then some good humorous turns came in that made us laugh with a just pleasure. **1861** E. COWELL *Diary* 16 Apr. in M. W. Disher *Cowells in America* (1934) 293 Mr. Ogden, not appreciated and evidently uncomfortable, would not sing a second 'turn'. **1889** G. B. SHAW *London Music in 1888–89* (1937) 234 Five out of six of the 'turns' are of the deadliest dulness.

VII. 37. b. (Later examples.)

1862 M. D. COLT *Went to Kansas* 99 Have just been to the spring for my turn of water. **1981** *Publ. Amer. Dial. Soc.* LXVIII. 48 *Turn of water*.., a container in each hand.

c. (Earlier example.)

1792 G. CARTWRIGHT *Jrnl. Coast of Labrador* I. p. xvi, *Turn of timber*, So much as a man can carry on his shoulders.

39. (Earlier examples.)

1857 *Sat. Rev.* 18 Apr. 348/2 Nobody understands the turn of the market better than Tomkins. **1870** J. K. MEDBERY *Men & Mysteries Wall Street* 78 This neat profit is called a 'turn'.

VIII. 41. turns (*rarely* **turn**) **ratio** *Electr.*, the ratio of the number of turns on the primary of a transformer to the number on the secondary, or vice versa.

1927 R. E. BROWN *Alternating-Current Machinery* iii. 66 The turn ratio of a transformer may be obtained from the designer or may be approximately measured by determining the ratio of the indications of two voltmeters. **1965** *Wireless World* Sept. 431/2 The number of turns on each coil, the turns ratio and the inductances are in no sense critical. **1976** RYDER & THOMSON *Electronic Circuits & Systems* xii. 286 A loudspeaker of $4\ \Omega$ can be made to appear as $400\ \Omega$ on the primary side if we use a transformer with the turns ratio $a = \sqrt{(400/4)} = ..10$.

turn, *v.* Add: **II. 4. a.** (Earlier *absol.* example.)

1796 JANE AUSTEN *Let.* 1 Sept. (1952) 8 Frank..enjoys himself here very much, for he has just learnt to turn.

5. d. *to turn an honest penny*: see HONEST *a.* 4 b; *to turn a profit* (U.S.): to earn or make a profit.

1969 *Time* 21 Jan. 44 Partly because of the competition from IBM it is unlikely to turn a profit before 1970. **1976** *National Observer* (U.S.) 14 Aug. 4/1 The iron rule of business dictates staying open anytime there's a decent chance to turn a profit.

IV. 13. c. *Cricket.* Of the bowler: to cause the ball to 'break' (BREAK *v.* 32 b). Also *intr.* of the ball: to break or turn in its course after pitching.

1898 G. GIFFEN *With Bat & Ball* iii. 47 There are very few men bowling at Mac's pace who can turn the ball on the Adelaide Oval. **1909** W. G. GRACE *W.G.'s Little Book* iii. 33, I don't know the moment he delivers the ball which way it will turn on pitching. **1928** *Morning Post* 2 July 15/1 Garland-Wells is slow right, and can turn the ball both ways. **1930** *Ibid.* 16 July 11/6 The bowlers were making the ball turn more than before luncheon. **1955** [see *FLIGHT *v.* 7].

17. c. In *Assoc. Football*, etc., to get round (an opponent at close quarters) by making it necessary for him to change direction.

1976 E. DUNPHY *Only a Game?* (1977) iv. 121 He turns full backs, he does unusual things on the ball, he creates unusual situations. **1980** *Times* 3 Apr. 13/2 Francis.. turned Buchan and sent in a stinging shot.

V. 29. Restrict †*Obs.* to senses in Dict. and add: **c.** To induce or persuade (a person) to act against his country, former associates, etc., esp. as a spy.

1971 C. EGLETON *Last Post for Partisan* xvi. 162 'How did they turn you?' 'I was shopped... They said I could save my neck if I helped them, and so I agreed.' **1973** T. ALLBEURY *Choice of Enemies* vii. 32 It's my assessment that this officer gave no information to the Russians and was not turned by them. **1979** A. BOYLE *Climate of Treason* ix. 309 The process of cornering and 'turning' the Fifth Man made such limited sharing of the secret almost inevitable. **1982** *Times* 27 Aug. 2/4 Several have been 'turned' only after being shown evidence from another 'supergrass'.

30. c. Also of a criminal, to become an informer, to 'grass'.

1977 *Chicago Tribune* 2 Oct. I. 24/3 Like many other informants, Bompensiero 'turned' in order to avoid jail. **1982** *Times* 27 Aug. 2 (*heading*) Pressures that lead a man to 'turn'.

VI. 39. b. Freq. as *pa. pple.* modifying a sb.

1945, etc. [see *POACHER¹ 1 b]. **1964** *Eng. Studies* XLV. 382 Their Scandinavian conquerors-turned-neighbors. **1973** E. SCHUMACHER *Small is Beautiful* I. iii. 44 The economist-turned-econometrician is unwilling..to face the question. **1982** *Times Lit. Suppl.* 10 Sept. 968/3 Jerome's father was a Nonconformist preacher, turned architect, turned mine-owner.

VII. Phrases. (For *to turn the* (*other*) *cheek* see *CHEEK *sb.* 2 b.)

51. turn one's coat (later examples).

1946 G. MILLAR *Horned Pigeon* ix. 130 Like good policemen all over the world, they were only too willing to turn their coats (to keep law and order, of course). **1981** *Times Lit. Suppl.* 2 Jan. 7/5 What prompted Sohō, Japanese intellectuals have endlessly debated, to 'turn his coat' in the 1890s?

c. *trans.* and *intr.* Also *turn aloose.* To let go (of), to leave hold (of). *U.S. dial.*

1906 *Dialect Notes* III. 162 *Turn loose*,..to get rid of, let go. **1910** *Ibid.* 457 *Turn it loose*,..let go of it. **1929** W. FAULKNER *Sound & Fury* 199 He turned my hands loose. **1934** C. CARMER *Stars fell on Alabama* 139, I jus' got to hold you now. I *can't* turn you aloose. **1935** T. WOLFE *Of Time & River* III. xlii. 376 You git his other hand, Jim, an' try to make him turn a-loose. *a* **1938** —— *Web & Rock* (1947) 34 'You turn loose of me,' the captive panted, 'I'll show you who's the cry-baby!' **1966** R. PRICE *Generous Man* (1967) i. 63 'Safe! We're in awful danger. Turn loose, old fool!' He turned loose and lay flat, small on the ground.

VIII. In combination with adverbs. **66*. turn around.** (See simple senses and AROUND *adv.*) = *turn round*, sense 78 in Dict. and Suppl. orig. *U.S.*

1880 'MARK TWAIN' *Tramp Abroad* xiii. 119, I could see the dim blur of the windows, but in my turned-around condition they were exactly where they ought not to be. **1919** E. O'NEILL *Moon of Caribbees* 114 Smitty does not turn around. **1925** F. SCOTT FITZGERALD *Great Gatsby* i. 9 Turning me around by one arm. *Ibid.* ix. 208 He opened it at the back cover and turned it around for me to see. **1932** W. FAULKNER *Light in August* xiv. 315 When he sat up he found that the sun..now shone upon him from the opposite direction. At first he believed that he was merely turned around. Then he realised that it was now evening. **1945** DYLAN THOMAS in *Horizon* Feb. 83 The weather turned around. **1963** J. JOESTEN *They call it Intelligence* I. iv. 45 A spy..caught..usually is given a chance to switch sides..such a helpless foreign agent is being 'turned around'. **1967** V. C. WELBURN *Johnny So Long* II. ii. 63 He shakes his hands high at the crowd, then turns around and does it to Judy. **1971** *Sci. News* 13 Feb. 108 Children..with severe behavioral problems.. are treated in a 'family' situation... An innovative school program helps 'turn youngsters around'. **1971** *Black World* Mar. 54/1, I felt so sorry for them and they wuz so turned around that one day over to Tony's crib I got high wid em. **1972** B. MOORE *Catholics* i. 21 Order them to turn that boat around and send it back for him. **1976** *National Observer* (U.S.) 29 May 3/4 This thing is so close, so fragile, that anything could happen tomorrow to turn it around. **1977** J. D. MACDONALD *Condominium* xxxiii. 322 We're past our marker... More than a mile. We can turn around. **1978** *Globe & Mail* (Toronto) 16 May 5/1 The Ontario Government has announced a..campaign to turn around the image Ontarians have gained as boorish and unfriendly hosts to foreign visitors.

69. turn back. b. For † *Obs.* read '*Obs.* exc. *U.S.*' and add later example.

1927 *Publishers' Weekly* 12 Feb. 610 We felt that the only course open to us in view of the authors' feelings in the matter was to offer to turn back the book to them, subject to their disposal.

71. turn down. c. For ? *Obs.* read '*Obs.* exc. *U.S.*' and add later examples.

1876 'MARK TWAIN' *Tom Sawyer* vi. 71 He took his place...in the spelling class, and got 'turned down', by a succession of mere baby words. **1946** G. WILSON *Fidelity Folks* 136 We had regular places in the line and turned down those who could not spell a word.

d. For *U.S. slang* read 'orig. *U.S. slang*' and add further examples.

1915 J. CHURCHILL *Let.* 11 Aug. in M. Gilbert *Winston S. Churchill* (1972) III. Compan. II. 1128 Everybody seemed to have 'turned it down'. **1927** A. CONAN DOYLE *Case-Bk. Sherlock Holmes* x. 261 A quarter's rent..in advance and no arguing about terms. In these times a poor woman like me can't afford to turn down a chance like that. **1951** *Sport* 7–13 Jan. 17/1 Many of our suggestions are turned down. **1956** A. H. COMPTON *Atomic Quest* iii. 202, I had been approached with regard to college and university presidencies... I had..turned them down. **1958** P. GIBBS *Curtains of Yesterday* xxvii. 214 Many nations..put forward plans for a gradual process of disarmament, and each plan was turned down by the other delegations. **1979** R. JAFFE *Class Reunion* I. vii. 70 A lot of attractive, eligible men kept asking Annabel for dates, and she didn't like having to turn them down.

g. Also, to lower the temperature of (an electrical appliance, heating system, etc., and *transf.*, that which it heats or cooks), orig. by turning a knob or switch; to reduce the volume of sound from (a radio, record-player, etc.), usu. by turning a knob or switch; to turn (a knob or switch) in order to reduce the temperature, volume of sound, etc. Cf. sense 80 m below.

1941 N. MARSH *Death & Dancing Footman* (1942) xv. 277 'To get back to the wireless.'..'I turned it down.'.. 'You turned it *down*... Not off. Down.'..'I turned it down, and five minutes later somebody turned it up.' **1950** B. PYM *Some Tame Gazelle* x. 111 The beef..would be roasted to a cinder by now, unless Emily had had the sense to turn down the oven. **1961** J. STROUD *Touch & Go* v. 48 'Excuse me if I just turn my liver down?'..She hustled back into the kitchen. **1966** P. WILLMOTT *Adolescent Boys* ix. 170 He was alone, playing records... He said, 'Just a minute, I'll turn this down.' **1969** 'D. RUTHERFORD' *Gilt-Edged Cockpit* vi. 93 Could you turn that transistor down a bit? **1970** 'A. GILBERT' *Death wears Mask* ix. 138 Miss Buxton's client..began to fidget under her drier and call out something about it being too hot. 'Turn it down, dear,' said Miss Buxton crisply. **1970** J. PORTER *Dover strikes Again* ii. 30 Old Mr Revel.. switched on the television set. Miss Kettering..turned the volume control right down. **1975** *Guardian* 21 Jan. 5/1 Turn down your heating a couple of degrees. **1975** 'M. SINCLAIR' *Long Time Sleeping* xii. 143 He stopped to turn down the potatoes.

g*. To let down with a winch or the like.

1929 [see sense 80 r* below].

j. Of business or economic activity or fortune: to decline, worsen.

1960 *Economist* 8 Oct. 261/1 Wall Street has shown an impressive record of moving ahead of business activity. In the recession of 1957–58..it turned down in July, one month before the index of industrial production; and turned up again in December, four months before production did. **1980** *Daily Tel.* 23 July 19 Unemployment is rising fast in all the major economies, partly because the economic cycle is turning down.

72. turn in. a. (Further examples.) Also, to hand in or over; *spec.* to betray or surrender to the police; to trade in; to give up, to stop (freq. with *it*). Also, to register, to produce (a result or performance, etc., of a specified kind).

1830 R. DAWSON *Present State of Australia* v. 201, I asked him..the reason of his having been 'turned in', as they call it, to government. **1912** *Technical World Mag.* June 403/1 When they discover any part of the track.. which they cannot readily repair they turn in a signal that summons the roadmaster. **1919** in F. A. Pottle *Stretchers* (1930) 359 Tomorrow we will turn in what few articles of equipment we have not left at Merritt. **1926** J. BLACK *You can't Win* vii. 85 If either of you gets grabbed ..and thinks he can get a light jolt by turning me in, he's wrong. **1931** H. CRANE *Let.* 15 July (1965) 376 I'm very glad that you spoke about the check... I couldn't figure it out any other way than that you had waited until the following month before turning it in. **1938** F. A. POTTLE *Boswell & Girl from Botany Bay* 17 Bligh published a book..and Edwards turned in a report to the Admiralty. **1947** A. HUXLEY *Let.* 27 July (1969) 573 Jessica Tandy.. is a first-rate actress and seems to be likely to turn in a performance which will make most of the more celebrated Hollywood stars look merely silly. **1948** C. DAY LEWIS *Otterbury Incident* iii. 23 'I'll kill myself, then.' 'Turn it in, Nick!' Ted punched him in a friendly way. **1952** L. DURRELL *Let.* 4 Nov. in *Spirit of Place* (1969) 114 I've turned in my resignation and we are clearing off in December. **1958** *Listener* 23 Oct. 632/2 This company, in common with many others, turned in a sizeable loss. *Ibid.* 13 Nov. 777/1 They didn't have to pay anything really [*sc.* for a new car];..they've turned in their old one..and that only left a £50 balance. **1968** *Globe & Mail* (Toronto) 13 Jan. 39/5 Doug Acomb and Frank Hamill scored two goals each as their line turned in one of its best performances of the season. **1971** *Sci. Amer.* July 5 It turns in the kind of performance that delivers up to 25 miles per gallon. **1973** L. MEYNELL *Thirteen Trumpeters* iv. 66 'Turn it in, Hooky,' he advised himself. 'Go while the going's good.' **1977** M. SOKOLINSKY tr. *Merle's Virility Factor* xii. 236 If she'd gone to bed with you, she would have enjoyed it—and then she'd have turned you in. **1978** 'M. YORKE' *Point of Murder* iii. 35 His Ford Escort..was being turned in for an older car with a higher mileage on the clock. **1979** *SLR Camera* Mar. 53/3 At full aperture the 75mm f2.8 optic turned in a surprisingly good performance. **1982** *Sunday Tel.* 1 Aug. 5/3 It [*sc.* the year 1981–82] has seen Sotheby's turn in its first loss in over 20 years.

73. turn off. d. (Later example.)

1888 'R. BOLDREWOOD' *Robbery under Arms* I. i. 8 You can have..anything you like..you unfortunate young beggar, until you're turned off.

h. Also *transf.* of an electrical appliance, a recording or broadcast, etc. Also *fig.*, esp. (*colloq.*) to put (a person) off, to repel, to disillusion, to cause to lose interest.

1965 *Harper's Bazaar* Apr. 173 *Turned off*: Humperdinck turns me off. **1966** P. WILLMOTT *Adolescent Boys* iii. 51 You can always get a bit if you want it, with the girls with the big mouths... But that sort of thing turns you off after a while. **1967** B. PATTEN *Little Johnny's Confession* 33 Until death comes and turns me off. **1968** J. HUDSON *Case of Need* II. vi. 128, I was just turned off, I wasn't paying attention. **1969** J. GASKELL *Sweet Sweet Summer* 70 Or for the water and electric to stay on all the time, instead of being turned off after midnight. **1971** *Radio Times* 18 Nov. 80, I wonder how many viewers turned off the play, as we did. **1972** *Daily Tel.* 5 Feb. 14 [He] is kinky for short-back-and-sides and turned off by long-haired television performers. **1973** *Sci. Amer.* Dec. 14/2 Having become 'turned off' by economics, I was not sure what I wanted to be. **1975** *Nature* 20 Nov. 228/1 The male [cichlid fish] has bright colour patterns which he can turn on and off quickly. **1977** I. SHAW *Beggarman, Thief* I. viii. 101 School was a big part of his life and he couldn't just turn it off because it would be unimportant to grown-ups at this time. **1979** *Financial Rev.* 24 Oct. 10/1 Many voters were turned off by a strike in the last moment of the campaign. **1982** *Times* 21 Oct. 3, I had three frigates badly turned off in terms of capability. We were running out of steam.

74. turn on. a. (Earlier and later *fig.* examples.) *to turn the tap(s) on*, to start weeping; *to turn it on*, to make a particular effort, esp. to be charming. Also *transf.* of an electrical appliance, a recording or broadcast, etc.; *to turn on the heat*: see *HEAT *sb.* 12 b.

In quot. 1877 of piped music.

1866 'MARK TWAIN' *Screamers* (1871) xxix. 149 There was a good deal of honest snickering turned on this time. **1877** *Punch's Almanack for 1878* 14 Dec. 3/1 Now, recollect, Robert, at a quarter to nine turn on 'Voi che sapete' from Covent Garden. **1883** *Daily Tel.* 8 Feb. 3/2 When she had finished her song she fell a crying... She can turn the taps on at a moment's notice. **1930** A. P. HERBERT *Water Gipsies* iii. 23 They had the wireless, which they turned on often for the 'jazz' bands. **1930** R. MACAULAY *Staying with Relations* iii. 43 They turned on a gramophone and danced. **1948** M. LASKI *Tory Heaven* i. 7 'Don't turn it on till the eleven o'clock news,' Janice called out. **1966** *Listener* 24 Nov. 779/1 Thomas could turn it on and brilliantly, when he wanted to. **1976** E. DUNPHY *Only a Game?* (1977) ii. 52 But Preston, who knows? They could turn it on under the floodlights. **1981** T. HEALD *Murder at Moose Jaw* xii. 144 She used to be some looker... And she could turn it on. But not any more.

c. To excite, interest, fill with enthusiasm; to intoxicate with drugs, to introduce to drugs; to arouse sexually. Also const. *to* the object of interest, etc. *slang* (orig. *U.S.*).

1903 H. JAMES *Ambassadors* xxii. 291 One of his sisters ..had observed her somewhere with me. She had spoken to her brother—turned him on. **1953** W. BURROUGHS *Junkie* ii. 31 We kept the weed in Marian's apartment, turned her on for all she could use, and gave her a 50 per cent commission on sales. **1965** *Harper's Bazaar* Apr. 173 Bach really turns me on. **1966** *Current Slang* (Univ. S. Dakota) Winter 8 *Turn on*,..to excite sexually. **1966** *Guardian* 18 Apr. 13/4 Police in New York said that they had seized enough of the drug LSD to 'turn on' the entire population of New York if it was put in the water supply. **1967** J. HAYES *Deep End* 16 The excitement in her eyes deepened. 'You turn me on, man.' **1967** *Melody Maker* 29 July 10/6 There is a compulsive beat so maybe even the nation's half-wits may be turned on to Lloyd. **1972** J. BROWN *Chancer* iv. 53 It must be about this time he turns her on too—onto heroin. **1975** J. I. M. STEWART *Gaudy* ix. 173 It's a funny thing..how quite sure I was she wasn't going to turn me on. **1976** *News of World* 14 Mar. 5/3 Dinner jacket, wing collar, and bow tie may not sound the sort of gear to turn on a teeny bopper. **1976** *National Observer* (U.S.) 10 Apr. 14/5 My work is important. When I can turn on a student or write a good paper, I'm really happy, I'm elated. **1979** S. WILSON *Glad Hand* I. v. 30 Well..it's probably God's way of saying He takes a rather dim view of what turns you on.

d. *intr.* To become intoxicated; to take drugs. Also with *to*, to become interested in. *slang* (orig. *U.S.*).

1955 [see *POD *sb.*² 1 c]. **1967** *Sunday Truth* (Brisbane) 2 Apr. 63/2 According to Dr. Timothy Leary, the avowed leader of the LSD set, you can turn-on without using drugs. **1969** *Gandalf's Garden* IV. 9/1 It was about this time I had turned on to Zen. **1970** *New Scientist* 12 Nov. 314/1 Young people who turn on by sniffing the vapour of airplane glue..sometimes..drop dead. **1971** *Nature* 12 Feb. 462/2 Increasingly scientists are 'turning on' to the human environment. **1976** *Maclean's Mag.* 17 May 22/3 More and more teen-agers..are turning on with alcohol. **1979** R. JAFFE *Class Reunion* III. iii. 242 She walked in while I was turning on so I offered her some [marijuana].

75. turn out. h. (Earlier example.)

1813 J. TAYLOR *Arator* 117 The phrase 'the land is killed and must be turned out', has become common over a great portion of the United States.

q. *spec.* (*Austral. slang*) to become a bushranger.

1862 *Western Post* 24 Sept. 2/2 He was immediately told by the robber they ought to turn out. **1888** [in Dict.]. **1910** J. CAMERON *Spell of Bush* 131 [The bush] had been his home; for even before he had 'turned out', four walls had never held Michael Moran for long.

77. turn over. h. (Further examples.)

1925 W. L. CROSS *Life Sterne* I. 175 Robert Dodsley had just turned over the management of his business to his brother. **1930** *Publishers' Weekly* 5 Apr. 1896 By retiring and turning the business over to his son. *Ibid.*, 1917 Columbia University has just come into possession of the famous collection of works on economics which Professor E. R. A. Seligman..has turned over..for a price of one-half, or less. **1930** *Harper's Mag.* July 196 My German instinct to care for my own child kept me from turning her over to someone else.

i. Also in extended sense.

1971 *Nature* 24 Dec. 483/2 *Sminthopsis crassicaudata* turns over water at about three times the rate of *Dasycercus cristicauda*. **1976** *Ibid.* 22 July 280/1 This indicates that the bulk of dentinal protein is not turned over.

i*. To search; to ransack (usu. in order to commit robbery). *Criminals' slang.*

1859 HOTTEN *Dict. Slang, Turned over*, to be stopped and searched by the police. **1925** [see *DRUM *v.* 9 b]. **1960** *Observer* 25 Dec. 7/6 The drummers, those squalid daytime operators who turn over empty semi-detached villas while the housewives are out shopping. **1971** [see *GOON 2]. **1981** L. MEYNELL *Hooky goes to Blazes* vi. 83 What about that girl's bedroom that got turned over?

i.** To distress, upset, affect with nausea. Cf. sense 80 k in Dict. *colloq.*

1865 DICKENS in *All Year Round* 7 Dec. 47/2 The discovery turned me over. **1962** N. STREATFEILD *Apple Bough* ix. 126 Proper turned me over, you did. I don't want to lose my old man yet. **1972** *New Society* 11 May 302/1 Escalope I had, though what they do to those calves turns me over.

i*.** To cause (an engine, propeller, etc.) to revolve.

1913 *Autocar Handbk.* (ed. 5) vii. 143 The motor should be able to turn the engine over at not less than 150 r.p.m. **1927** C. A. LINDBERGH *We* v. 75 Learning how to turn the propellers over in starting the engine. **1976** P. ALEXANDER *Death of Thin-Skinned Animal* xx. 209 He.. pulled the wires away from the ignition switch and tied them together. This turned the [car] engine over but she wouldn't start without the choke. She was..cold.

i**.** *Printing.* To carry over (a letter, part of a word, etc.) to the next line.

1925 *Hart's Rules for Compositors & Readers* (ed. 27) 64 In most divisions it is the consonantal letter that should be turned over. **1981** I. A. GORDON in *N.Z. Listener* 14–20 Feb., The fragmented word is 'turned over' into the following line.

i***.** *U.S. Sport.* To lose possession of (the ball) to the opposing team.

1971 *Tuscaloosa* (Ala.) *News* 29 Jan. 8/5 We knew we could force Kentucky to turn the ball over and we did. **1979** *Honolulu Advertiser* 8 Jan. C-4/1 We turned over the ball and we just didn't score.

l. Of a body or part of a body or any collective whole: to replace or renew its constituent parts, to renew itself or be replaced.

1956 *Planning* XXII. 155 The entire research staff has turned over on an average about every three years as the various studies have been completed. **1971** J. Z. YOUNG *Introd. Study Man* xii. 151 The body..has some parts that turn over very little..while others turn over so efficiently that we do not normally think of them as suffering wear. **1973** *Times* 17 Apr. 4/7 Nearly three-quarters of the men..have stayed. The other 27 per cent have 'turned over' several times.

m. Of an engine, etc.: to revolve.

1978 T. GIFFORD *Glendower Legacy* (1979) 281 He.. heard the engine turning over;..the boat quivered.

78. turn round. b. (Examples of *fig.* phrases.)

1802 F. BURDETT *Let.* 18 Aug. in H. Maxwell *Creevey Papers* (1903) I. i. 3, I have scarcely time to turn round, but will not defer sending a line in answer to your very kind letter. **1911** A. BENNETT *Hilda Lessways* (ed. 2) III. ii. 226 He simply walked out of the office!..Didn't give me time to turn round. **1969** *Listener* 14 Aug. 217/2 Before my parents could turn round and ask what I was going to do for a living, I went back and announced I'd already got a job as a ship's musician.

c. (Earlier example with *on*.)

1846 DICKENS *Dombey & Son* (1848) v. 38 You're a good little thing..; and yet you turn round on me, because there's nobody else.

g. In *pa. pple.* Confused, disorientated. *U.S. dial.*

1877 R. I. DODGE *Plains Great West* 46 To me, Detroit is always in Canada, and New Orleans always on the right bank of the Mississippi, because I happened to be 'turned round' when I first arrived in those cities.

h. To prepare (a ship, aircraft, etc.) for its return journey. Cf. *TURN-ROUND 1.

1942 *R.A.F. Jrnl.* 16 May 13 Cleanliness also shortens the time it takes to turn a ship round. **1972** *Nature* 21 Apr. 363/1 Is there..a chance that supersonic aircraft can be turned round at international airports with the speed that will be necessary if operators are to make.. the fullest use of their capital investment?

i. = sense 29 c above.

1966 *New Statesman* 6 May 657/2 SOE's intrigues included..'turning round' captured agents. **1974** 'J. LE CARRÉ' *Tinker, Tailor* xxi. 180 All right. The Russians have turned Tarr round... What sort of plant can he be when we don't believe a word he says?

80. turn up. d. Also, to shorten (a garment or part of one) by increasing the width of the hem or by making a hem; to increase the width of (a hem).

1918 E. & M. WALLBANK *Dress-Cutting & Making* xiv. 89 To turn up the skirt to the required length, a skirt gauge..may be used. *Ibid.* xxi. 126 Turn up the hem, fell down the pleats. **1958** M. JOHNSON *Sewing the Easy Way* (1960) 127 Turn up the hem, matching seam upon seam. **1976** J. TATE tr. *Bodelsen's Operation Cobra* xi. 54 Margrethe had borrowed a skirt from her mother and was busy turning it up. **1979** R. RENDELL *Make Death love Me* iii. 29 Pam turned up the hem of an evening skirt.

g. *to turn up trumps*: see TRUMP *sb.*² 2.

k. (*fig.* examples.)

1932 S. GIBBONS *Cold Comfort Farm* xii. 178 Turns you up, don't it, seein' ter-day's dinner come in 'anging round someone's neck? **1968** M. WOODHOUSE *Rock Baby* i. 11 You don't like the rules. Well, well, Giles. Do you know, sometimes you science boys turn me up.

m. Also, to raise the temperature of (an electrical appliance, heating system, etc., and *transf.*, that which it heats or cooks), orig. by turning a knob or switch; to increase the volume of sound from (a radio, record-player, etc.), usu. by turning a knob or switch; to turn (a knob or switch) in order to increase the temperature, volume of sound, etc. Cf. sense 71 g above.

1941 [see sense 71 g above]. **1962** A. NISBETT *Technique Sound Studio* 259 The volume can be turned up louder. **1967** 'E. LINDALL' *Time too Soon* v. 57 Put on your hi-fi... Just turn it up a bit more. **1971** M. LEE *Dying for Fun* ix. 53 The producer of the radio magazine programme turned up his loudspeaker. **1976** W. CORLETT *Dark Side of Moon* II. 85 It is colder... I must turn up the central heating. **1978** *Sci. Amer.* Apr. 67/1 The microscope [operating] is then turned up to high magnification (25 or 40 diameters).

p. Also *imp.* with *it*, stop it!

1945 J. B. PRIESTLEY *Three Men in New Suits* i. 6 Turn it up, will you... You're arguing with yourself. **1948** M. ALLINGHAM *More Work for Undertaker* (1949) x. 127 Turn it up... Keep it for your reminiscences. **1961** J. B. PRIESTLEY *Saturn over Water* xvii. 240 'Are you sure you can trust her?' 'Yes, Joe. So turn it up.'

r. (Earlier example.)

1782 in *Ann. Reg. 1783* (1785) 122/2 The main-sail a-back; all hands turned up; the main-clue garnets manned.

r*. To draw up with a winch or the like.

1911 D. H. LAWRENCE *Prussian Officer* (1914) 282 The winding-engine rapped out its little spasms. The miners were being turned up. **1929** —— *Pansies* 82 My father was a working man And a collier was he, At six in the morning they turned him down And they turned him up for tea.

s. Also *fig.*

1974 *Nature* 22 Feb. 514/2 Lillie presented preliminary evidence that the zodiacal light spectrum turns up below 2,500 Å. **1977** *Evening Post* (Nottingham) 27 Jan. 4/5 The rate of inflation has turned up again and we must expect to see some further deterioration in the next few months.

v. Also, to arrive or present oneself (with no connotation of unexpectedness or casualness); freq. *neg.*, to fail to arrive when expected. *colloq.*

1903 SOMERVILLE & 'ROSS' *All on Irish Shore* i. 18 And if you'll believe me, the two chaps there had never turned up at all. **1939** G. B. SHAW *Geneva* III. 52 The judge himself hasnt turned up. **1977** *Arab Times* 14 Dec. 2/5 Let women everywhere from this day on encourage men to have the courage not to turn up for war. **1979** J. COOPER *Class* iv. 84 At prep schools they insist you turn up [at sports days] and then ignore you. **1981** R. HAYMAN *K* iv. 38 They..turned up in large numbers to the general meetings.

turn-. Add: **turn-and-bank, turn-and-slip** *Aeronaut.*, used *attrib.* and *absol.* to designate an indicator which shows the pilot his rate of turn and correctness or error in banking; **turn-furrow** (earlier example); **turn indicator**, (*a*) *Aeronaut.* (see quot. 1930); (*b*) = *INDICATOR 3 g; **turn-plough** (examples); **turn-row** (earlier example); **turn-screw** (earlier example); **turn signal** *U.S.* = *INDICATOR 3 g; **turn-turtle** *a.* (*nonce-wd.*), in which one turns turtle and dies.

1933 *Jrnl. R. Aeronaut. Soc.* XXXVII. 930 If operating in daylight, they must have a complete set of instruments including duplicate turn and bank indicators, compass, air speed indicator, [etc.]. **1981** *Pilot* Jan. 45/1 A Motorola low-frequency transceiver with..altimeter, turn-and-bank and a clock. **1955** LIPTROT & WOODS *Rotorcraft* iii. 25 Further instruments include..turn-and-slip indicator. **1978** A. WELCH *Bk. of Airsports* ii. 36/2 The instruments used to cope with cloud flying are either a turn-and-slip indicator or an artificial horizon, or both. **1810** in Thirsk & Imray *Suffolk Farming 19th Cent.* (1958) ii. 77 Various sorts of mould boards, turn furrows or breast plates. **1919** *Nature* CIV. 183/2 A trustworthy turn-indicator and improved compass made accurate navigation through clouds possible. **1930** P. M. HENSHAW *Air Questions & Answers* 180 A Turn Indicator is an instrument that warns a pilot when the machine is turning right or left in circumstances (at night or in a cloud) when he would not be aware of it. **1953** C. A. LINDBERGH *Spirit of St. Louis* II. vi. 303, I glance at the turn indicator, kicking rudder slightly as I do so. The needle jumps over to the side. Yes, it's working properly. **1959** *Motor Manual* (ed. 36) vi. 183 Flashing turn indicators are operated either by a switch..mounted on the steering column, or by an independent switch mounted within easy reach of the driver. **1970** *Motoring Which?* July 88/2 All had warning lights for ignition, turn indicators and headlamp main beam. **1854** G. N. JONES *Florida Plantation Rec.* (1927) 104, I think you will nead about 10 turnplowes. **1907** T. F. HUNT *Forage & Fiber Crops* 352 The land having been plowed with an ordinary mold board or turn plow. **1885** 'C. E. CRADDOCK' *Prophet Gt. Smoky Mts.* 3 A young man..came to a meditative halt in the turn-row. **1778** J. WOODFORDE *Diary* 4 June (1924) I. 226 For a turn screw and picker for a gun pd 0. 1. 0. **1949** *N.Y. Times* 20 Mar. II. 19/5 Mechanical turn signals must be standard equipment on all new motor vehicles sold in Minnesota after July 1. **1977** J. WAMBAUGH *Black Marble* (1978) vi. 79 He started the Plymouth, flicked on his turn signal,..then pulled into traffic. **1951** DYLAN THOMAS in *World Rev.* Oct. 66 Dolphins dive in their turnturtle dust.

turnable, *a.* (Later examples.)

1935 E. R. EDDISON *Mistress* xiii. 262 What's good in Lessingham is right sense..and a wit so turnable for all things alike. **1972** C. MUDIE *Motor Boats & Boating* 37 This grill in some units is turnable to direct the jet in any required direction.

turnabout. h. (Earlier example.)

1789 J. BYNG *Torrington Diaries* 1 June (1938) IV. 109 There was (today) a little Fair, and a Stall, and a Turnabout to make the children sick after their Gingerbread.

turnaround (tɜ̄·ɹn,ǝraʊnd). Also **turn-around.** orig. and chiefly *U.S.* [f. the verbal phr. *turn around* (*TURN *v.* 66*).] **1.** = *TURN-ROUND 1.

1936 *Sun* (Baltimore) 20 Aug. 11/6 The dirigible Hindenburg landed..at the Naval Air Station tonight for what was expected to be the quickest turn-around in transatlantic airship service. **1946** *Ibid.* 9 Apr. 17/6 Another feature of the day's market was the turn-around of two loads of 100-pound Iowa wooled lambs which brought $18.55. **1952** *Ibid.* 9 Oct. 5/4 At present the two queens have a 48-hour turnaround at Southampton. **1971** D. BAGLEY *Freedom Trap* ix. 202 The skipper was returning in the tender. It seemed as though they intended a faster turnaround than Gibraltar. **1974** T. P. WHITNEY tr. *Solzhenitsyn's Gulag Archipelago* I. II. i. 497 So that the prisoner shouldn't attempt to escape during the moment he was in the toilet, and also for a faster turnaround, the door to the toilet was not closed. **1977** *Sci. Amer.* Jan. 127/1 Ships designed to carry a unitized cargo, for ease of transshipment, quick turnaround, low damage and low longshore labor costs.

2. = *TURN-ROUND 2.

1941 *Sun* (Baltimore) 30 Dec. 15/1 Many [railroad bonds] had been depressed to the lowest level since 1938 prior to the turn-around late last week. **1961** *N.Y. World-Telegram* 15 Mar. 31/2 Yet this would mean a turn-around in the trend of residential construction. **1972** M. WILLIAMS *Inside Number 10* iv. 82 He was found an office and in the general turnaround the Private Secretary..was shifted from the study..to a large room on the first floor. **1975** *N.Y. Times* 19 May 18/4 The migration to the Arkansas countryside is part of a national turnaround in population. **1984** *Times* 13 Jan. 13/1 Associated Newspapers..yesterday disclosed a £5m turnaround from loss to profit.

3. A space for vehicles to turn round in, often at the end of a drive or cul-de-sac.

1954 C. ARMSTRONG *Better to eat You* vi. 63 He..ran across the turn-around and the parking apron. **1969** 'F. RICHARDS' *Risky Way to Kill* (1970) vi. 68 She swung the Volks in the graveled turnaround, so that it headed back the way it had come. **1975** *New Yorker* 31 Mar. 29/3 Then they..got into the truck, circled the turnaround, and drove straight off. **1984** *Ibid.* 7 May 57/3, I turned up my driveway and parked in the turnaround.

4. The action or fact of turning round; a point in a team-game at which the teams change ends to play in the opposite directions.

1959 I. & P. OPIE *Lore & Lang. Schoolch.* vii. 116, I can do the splits, I can do the turn-arounds, I can do the kicks. **1960** T. MCLEAN *Kings of Rugby* 212 Then, at the turnaround, when the Lions now had the northerly wind at their backs, the change which came over the game was remarkable. **1962** A. SHEPARD in *Into Orbit* 108, I could feel the capsule begin its slow, lazy turnaround to get into position for the rest of the flight.

5. *attrib.*

1944 *Sun* (Baltimore) 15 Aug. 12/5 The 'Big Inch' line..has done the work of 23,000 tank cars operating on an 18-day turn-around schedule. **1946** *Ibid.* 27 Dec. 6/1 Ease of maintenance and servicing will cut costs and turn-around time. **1950** *Ibid.* 14 Aug. 10/6 The harbor to be built at Lake Washington will consist of a turn-around area 1,000 feet square. **1959** *Wall St. Jrnl.* 29 July 6/3 He described the current period as a 'turn-around year' leading to higher sales and earnings next year. **1976** *National Observer* (U.S.) 17 July 7/3 There ought to be some turnaround time to get into a system different than the one that has existed for nearly 40 years.

turnaway (tɜ̄·ɹn,ǎwēɪ), *sb.* and *a.* Also **turn-away.** [f. the verbal phr. *turn away* (TURN *v.* 68).] **A.** *sb.* The action or an act of turning away or deviating (from a course, etc.).

1922 *Encycl. Brit.* XXXI. 666/1 The British battle fleet turned away two points to port... This was the 'turn-away' which has given rise to considerable controversy. **1976** *Nature* 27 May 278/2 The complete turn-away from crop uniformity by employing mixed varieties or multilines, which are heterogeneous for disease resistance and other characters, may be of advantage only where there is severe and continuous disease present.

B. *adj.* Of a crowd: so large that part of it has to be turned away. Also *transf.*, of business, trade, etc.

1943 *Life* 1 Nov. 76 Since the beginning of her radio show she has enlarged her audience 34%, business at the Persian room is turnaway. **1950** *Richmond* (Va.) *News-Leader* 27 Oct. 11/2 In death, as in life, Al Jolson drew a turnaway crowd. **1968** *Punch* 18 Sept. 395/2 Sunday evensong—a service that always did turn-away business. **1977** *Oxford Times* 22 July 16/5 A French restaurant.. opened at the end of June, and has been enjoying a turn-away trade ever since.

turnback, *sb.* and *a.* Add: Also, an act of turning back. **A.** *sb.* (*b*) (Earlier example.) (*c*) The return of something borrowed or rented. (*d*) A reduction in price; a reversal.

1843 *Vermont Militia Act 1842* 61 *Coat*—dark blue, double breasted;..the skirts to reach to the bend of the knee, with buff kerseymere turnbacks. **1943** *Daily Progress* (Charlottesville, Va.) 19 Nov. 4/1 (*heading*) 13 billion dollar turnback... The War Department is returning to the Budget Bureau more than 13 billion dollars..from its 1943 appropriation. **1949** *Sun* (Baltimore) 18 Mar. 1/2 (*heading*) Second auto turnback. **1967** *Electronics* 6 Mar. 351/2 A rise in turnbacks of rented machines in recent years has Japanese computer makers uneasy. **1977** *Sunday Times Mag.* (Perth, Austral.) 16 Jan. 6/5 When the turn back came on the last three days of trading, it was the farm leaders in particular that held their gains.

turncoat, *sb.* and *a.* Add: **A.** *sb.* **c.** Also, a reversible coat.

1958 *Vogue* Oct. 163 Givenchy's barrelled and narrowing turn-coat, in pearl grey wool reversing to off-white.

turncoating *vbl. sb.* (later example).

1965 *National Observer* (U.S.) 11 Jan. 2 He told Mr. Watson he didn't 'think much of turncoating' when Mr. Watson announced for the House in 1962.

turn-down, *a.* and *sb.* Add: Also **turndown. B.** *sb.* **2. b.** *U.S.* A person who is 'turned down' or rejected, esp. as unfit for military service.

1945 *Daily Progress* (Charlottesville, Va.) 30 Jan. 4/1 The high proportion of turndowns among Negro registrants. **1977** *Sat. Rev.* (U.S.) 23 July 14/1 We've gotten a lot of turndowns, yeah; but we've gotten guys who're willing, too.

3. = *DOWN-TURN, DOWNTURN *sb.*

1957 *Economist* 26 Oct. 283 This slow-down in expansion is not the same thing as a turn-down in activity. **1960** V. PACKARD *Waste Makers* (1961) xiv. 153 How far Americans could continue running up debts..during an economic turndown was not clear. **1979** *Daily Tel.* 5 May 23/2 Shares in EMI were sent reeling yesterday after a warning by the company of likely..losses..following an unforeseen and severe turndown in its music division. **1980** *Jrnl. R. Soc. Arts* Mar. 208/2 The industry..is fearful of a turn down in the volume of work.

turned, *ppl. a.* Add: **3. b.** *U.S. colloq.* Of a person: disposed, natured. Cf. TURN *sb.* 34 a.

1931 *Amer. Speech* VII. 94 *Quar-turned*, droll-natured. **1949** H. HORNSBY *Lonesome Valley* 334 She's the best turned girl I ever talked to. She's as friendly turned as anybody! **1951** L. CRAIG *Singing Hills* 70 A nicer-turned man you never saw when he ain't in liquor. **1972** J. S. HALL *Sayings from Old Smoky* 138 Turned, etc. Having a disposition (of a certain kind)... 'She's a mild-turned girl.' She has a mild disposition.

8. *turned-around, -off, -on, -out* (earlier example).

1880 Turned-around [see TURN *v.* 66*]. **1958** T. STANWELL-FLETCHER *Clear Lands* 156 The two missionaries, each in black clerical garb and turned-around collars. **1966** T. PYNCHON *Crying of Lot 49* v. 117 She thought she saw a turned-around collar but took no chances. **1911** 'O. ONIONS' *Widdershins* iv. 35 The dripping of water from an imperfectly turned-off tap. **1972** D. E. WESTLAKE *Bank Shot* xiii. 97 [He] stared moodily at the turned-off television set. **1976** *Washington Post* 19 Apr. A3/2 Camejo hopes to capture votes of turned-off blacks and blue-collar workers. **1967** *Wall St. Jrnl.* 9 Feb. 1/4 The turned-on generation..will beat a path to your door. **1971** *Guardian* 25 May 8/1 Is it merely a difference in style between Huxley's experience and that of the turned-on teenager? **1977** E. J. TRIMMER et al. *Visual Dict. Sex* (1978) vi. 62 The most obvious change in the turned on male is erection of the penis. **1722** DEFOE *Plague* 272 Some of the Dissenting turn'd out Ministers staid, and their Courage is to be commended.

turner¹. Add: **III. 8.** Also, a member of a *TURNVEREIN.

1854 *Calif. Chron.* 16 May 7/3 We..paid a hasty visit to Russ' Gardens, where the Turners and their compatriots had resumed the sports of the previous day. **1913** *N.Y. Times* 13 Oct. 12/4 There was a big gathering of Turners and guests at the North German Lloyd Line Pier.

Tu·rner³ (tɜ̄·ɹnǝɹ). The name of James *Turner*, eighteenth-century London colour-maker, used *attrib.* and in the possessive to designate a yellow pigment patented by him in 1781; = *mineral yellow* s.v. MINERAL *a.* 5 b.

[**1792** *Act 32 Geo. III.* c. 73 The yellow colour invented by the said James Turner, and which is composed of British materials only, has been found to be far superior to the foreign.] **1835** [see *Montpellier yellow* s.v. *MONTPELLIER]. **1886** H. C. STANDAGE *Artists' Man. Pigments* iv. 46 Turner Yellow, Cassel Yellow,..Mineral Yellow. **1951** R. MAYER *Artist's Handbk. Materials & Techniques* ii. 63 *Turner's yellow*, lead oxychloride... Patented by James Turner, England, 1781. **1970** R. D. HARLEY *Artists' Pigments* viii. 92 Accounts of the legal proceedings give no indication of the place where Turner's yellow was made, although, according to one authority the pigment was manufactured at..Walker-upon-Tyne where it was sold as Turner's Patent Yellow.

Turner⁴ (tɜ̄·ɹnǝɹ). *Med.* The name of Henry Hubert *Turner* (b. 1892) U.S. physician, used in the possessive (less commonly *attrib.*) to designate a syndrome he described in 1938 which affects females and is characterized by developmental abnormalities including an absence of ovaries, underdeveloped breasts and womb, and shortness of stature, and is usu. caused by a missing X chromosome in normally XX cells.

1942 *Amer. Jrnl. Med. Sci.* CCIV. 641 Dr. Lawson Wilkins..showed one of the authors a patient who..has both coarctation of the aorta and webbing of the neck. He called the authors' attention to 'Turner's syndrome'. **1961** *Lancet* 23 Sept. 711/2 It had taken more than 50 years from the recognition of the sex chromosome to the discovery that children with Turner's syndrome had just one X chromosome, and Klinefelter's syndrome 2Xs and a Y. **1970** [see *KLINEFELTER]. **1977** *Lancet* 19 Mar. 649/2 There have been conflicting reports about the level of amniotic-fluid α-fetoprotein (A.F.T.) when a fetus has Turner syndrome.

Turneresque, *a.* Add: (Earlier example.)

1846 LADY TREVELYAN *Let.* 7 Oct. in J. Brown *Lett.* (1912) 417 It is in his boundless prodigality of thought that Turner differs from other painters, and that the more Turneresque he was..the more full of meaning every bit of his work became.

Turnerian *a.* (earlier example); **Turnerism** (earlier example).

1851 Turnerism [see *RAPHAELITE]. **1857** GEO. ELIOT *Scenes Clerical Life* (1858) II. 87 Her cheeks..loomed through a Turnerian haze of net-work.

turn-in (tɜ̄·ɹn,ɪn) *sb.* (*a.*) [f. the verbal phr. *turn in* (TURN *v.* 72).] **1.** An edge of material that is folded inwards, as at a seam; *spec.* in *Bookbinding* (see quot. 1952).

1873 *Young Englishwoman* Mar. 147/2 Pin the edge.. allowing an inch and a half for the turn in. **1901** [see *JOINT *sb.* 4 c]. **1931** *N. & Q.* 28 Feb. 146/1 Showing the price on the 'spine' of the jacket instead of at the foot of the front turn-in. **1933** J. E. LIBERTY *Practical Tailoring* v. 52 It should be remembered that the first turn in is not as an ordinary *turned-in* edge, the turn being on the top and not underneath. **1952** J. B. OLDHAM *English Blind-Stamped Bindings* 66 *Turn-in*, the portion of leather that shows along the edges on the inside of the covers.

2. An entrance, a way by which one may turn in, a road or passage leading off another road.

1959 *Cape Times* 7 July 9/2 What could be more pleasing than the old 'leafy lane' road..to the left of the High Constantia gateway? One wonders what this new and awkward turn-in is going to cost. *Ibid.* 26 Oct. 6/4 The accident..occurred on the Port Elizabeth–Uitenhage road at the Despatch turn-in. **1973** E. LEMARCHAND *Let or Hindrance* xiii. 157 'Look, there's a turn-in at the side of the office.'..Toye negotiated the narrow entry.

3. *attrib.* or as *adj.*

1955 J. E. LIBERTY *Practical Tailoring* (ed. 2) v. 66 Baste the silesia edge to the turn-in edge across the top, and sew it. **1973** S. JENNETT *Making of Books* (ed. 5) xii. 195 It [*sc.* the leather] must be pared thin on all turn-in edges.

turning, *vbl. sb.* Add: **6.** (Earlier example of sense 'a part of something folded over'.)

1886 *Girl's Own Paper* 25 Dec. 202/3 All paper patterns ..are of medium size... No turnings are allowed.

12. *turning off, round* (later example).

1940 *Amer. Speech* XV. 247 He allows nobody else to have anything to do with the..*tuning out* (or the *turning-off*) of the radio programs. **1973** *Black Panther* 5 May 2/1 It is hard to understand how any Oakland residents could have missed some exposure to the..campaign... But turning off can be a total thing. **1976** 'W. TREVOR' *Children of Dynmouth* iv. 84 The abrupt turning-off of the kitchen radio, and the bang of the door. **1966** M. R. D. FOOT *SOE in France* viii. 190 Among [Pierre] de Vomécourt's achievements, this successful turning round again of Mme Carré stands second only to his indispensable contribution towards getting organized resistance going at all.

13. turning circle, the smallest circle within which a ship, motor vehicle, etc., can be turned round completely; **turning-machine** (earlier example); **turning-mill** (earlier example); **turning radius**, the radius of a turning circle.

1903 KIPLING *Traffics & Discoveries* (1904) 39 The endurance, armament, turning-circle, and inner gear of every ship in the British Navy. **1928** *Motor Man.* (ed. 27) 219 *Turning circle*, the minimum diameter of circle within which a car can be turned round completely. **1959** [see *LOCK *sb.*² 15 b]. **1963** *Listener* 21 Feb. 339/2, I cannot myself..accept the suggestion..that Admiral Tryon confused radius and diameter of the ship's turning circle ...Diameter is so inseparable in the mind of a seaman with 'turning circle' that the phrase 'turning circle' is commonly and loosely used to mean its diameter. **1980** *Jrnl. R. Soc. Arts* July 513/2 The dimensions of a ship's turning circle vary approximately in proportion to ship's length. **1983** *Sunday Tel.* (Colour Suppl.) 20 Mar. 18 Although over three feet longer than the Golf, the Volvo's turning circle is seven inches smaller. **1849** E. CHAMBERLAIN *Indiana Gazetteer* 429 There are..one foundry and several turning and carding machines, all driven by water. **1844** *Knickerbocker* XXIV. 184 The uplifted arm of Labor.. meets his eye everywhere, in the paper-mill and grist-mill, and..turning-mill. **1967** *Jane's Surface Skimmer Systems 1967–68* 79/1 Characteristics... Turning radius at cruising speed 1,640 ft (500 m). **1973** T. PYNCHON *Gravity's Rainbow* I. 105 She's fed back who knows how many reams' worth of Most Secret flimsies.., squadron numbers, fueling stops, spin-recovery techniques and turning radii.

turning, *ppl. a.* Add: **7. turning pitch** *Cricket*, a pitch on which the ball turns or deviates on delivery; **turning plough** (earlier example).

1956 N. CARDUS *Close of Play* 31 Parkin had no superior at off-breaks on a turning pitch. **1959** *Listener* 19 Mar. 516/1 Slow and turning pitches. **1850** in J. A. Turner *Cotton Planter's Man.* (1865) 118 Many planters here say they scrape with the turning-plough as well.

turning-point. Add: **1.** (Later examples.)

1956 *Railway Mag.* Mar. 165/2 Katrane ..is the only turning point (a triangle) for engines between Amman and Ma'an. **1977** *Wandsworth Borough News* 16 Sept. 9/3 London Transport should be asked to stop using Medfield-street as a turning-point for their buses.

2. (Earlier example.)

1836 J. KEBLE *Wks. R. Hooker* I. p. li, In the annals of the Church,..we may from time to time mark out what may be called *turning points*.

turnip, *sb.* Add: **4. a.** *turnip-head* (*lit.* and *fig.* examples); *turnip-faced* (cf. sense 3 b).

1939 F. THOMPSON *Lark Rise* iii. 49 The old turnip-faced watches which descended from father to son. **1869** D. G. ROSSETTI *Let.* 1 Mar. (1965) II. 689 The turnip-head falls off the broomstick. **1931** S. KAYE-SMITH *Susan Spray* III. 281 He..saw her standing there..fooling all those turnip-heads, who wanted to be fooled. **1962** *Spectator* 2 Nov. 684 Pop..has become the divisive symbol between the turnip-heads and the giant intellects.

b. turnip greens (earlier example, in *sing.*).

1796 J. WOODFORDE *Diary* 9 Feb. (1929) IV. 262 Dinner to day, fryed Pork & Turnip Green.

turnkey. Add: **2. b.** (Earlier example.)

1855 P. T. BARNUM *Life* vi. 91 The pseudo-dentist went to work, and by dint of hammer, pincers, and 'turnkeys', he extracted the twenty teeth.

3. Used *attrib.* to designate a contract, system, etc., whereby the contractor undertakes to supply or install a complete product or service that is ready for immediate use.

1934 WEBSTER, Turn-key job. **1958** *Times* 16 Dec. 6/5 The station, which is to supply the colony's rapidly growing needs, is being built on a 'turnkey' contract by the English Electric Company, which is supplying all the electrical plant. **1964** *Times Rev. Industry* Feb. 48/2 It now has a reputation for successful completion of contracts on a 'turnkey' basis, starting from a survey of the geology of the country concerned and ending with the handing over of a complete factory with trained staff. **1966** *Economist* 5 Mar. 925/1 The so-called 'turnkey' factories, bought ready to go into production and wholly on credit. **1979** *Personal Computer World* Nov. 32 (Advt.), We offer a variety of turnkey systems. **1980** *Nature* 24 Apr. 657/2 A standard PWR reactor of the type sold by Westinghouse all over the world under turn-key contract. **1984** *Christian Science Monitor* 2 Mar. 17/1 One likely institutional change is the construction of 'turnkey' nuclear power plants.

turn-off (tɜ̄·ɹn,ǫf). Also **turnoff.** [f. the verbal phr. *turn off* (TURN *v.* 73 in Dict. and Suppl.).] **1.** A turning off a main road; a side road; a junction where a track or road branches off a main road.

1881 'R. BOLDREWOOD' *Robbery under Arms* (1888) III. xvii. 255 It's the wrong turn-off that makes a man lose his way. **1894** J. WINSOR *Cartier to Frontenac* 151 The turn off at Lake Athabasca..would have conducted him to the northern tributaries of the Columbia. **1949** F. SARGESON *I saw it in my Dream* xv. 237 The boss managed to keep the sheep nicely bunched together until he'd pushed them past the turn-off. **1955** E. BOWEN *World of Love* xi. 221 The road due soon to go on without them to Galway, for soon would be coming the Turn Off. **1977** *Times of Zambia* 7 Sept. 5/5 The premises situate at Lukashya turn-off, Mungwi Road. **1980** *Beautiful British Columbia* Summer 20 Access to Golden Ears Park is from a turnoff at Haney.

2. Disposal (of cattle) at market; the number or quantity marketed.

1960 *Times* 1 Oct. 7/7 The Territory's annual turn-off of 150,000 cattle. **1961** in *Webster* s.v. *turnoff*, Average annual turnoff of fat bullocks. **1969** *Northern Territory News* (Darwin) *Focus* '69 30/3 Last year the turn-off from the Alice Springs pastoral district was more than 20,500 head, earning more than $2,750,000.

3. The action or an instance of turning (something) off, stopping, or causing to cease functioning.

1967 *Technology Week* 23 Jan. 52/1 (Advt.), Each satellite was cycled through turn-off, cold soak, and re-start during approximately 53 eclipses of the sun. **1970** *Globe & Mail* (Toronto) 25 Sept. 1/1 'It's inconvenient, but it's bearable,' she said of the electricity turnoffs. **1974** R. S. BRAY in *Ciba Symposium* No. 20. 97 Another possible mechanism for the 'turn off' of the cellular immune system is a viraemia. **1978** S. BRILL *Teamsters* iii. 104 Hicks attributed Kleindienst's unusual turnoff of the investigation to 'the love affair between Fitzsimmons and Nixon'.

4. Something that repels, disgusts, or 'turns one off' (see *TURN *v.* 73 h). *colloq.*

1975 *N.Y. Times* 1 Nov. 18/1 Patrons dined on cervelle Grenobloise. 'Sounds better in French,' said the chef... 'Brains is a turn-off.' **1976** *National Observer* (U.S.) 13 Mar. 6/6 Should it become unpleasant or prove a turn-off to either, they stop the game. **1982** *Listener* 23 & 30 Dec. 48/1 At first impression, this uneasy blend of piano quintet and violin sonata..was a gigantic, four-square turn-off.

turn-on (tɜ̄·ɹn,ǫn). [f. the verbal phr. *turn on* (TURN *v.* 74 in Dict. and Suppl.).] **1.** The action or an instance of turning something on; activation.

1962 SIMPSON & RICHARDS *Physical Princ. Junction Transistors* vii. 145 It thus produces a further delay in the turn-on of the transistor. **1967** *Technology Week* 23 Jan. 12/2 Availability of the eight new satellites in orbit, assuming proper operation after turn-on..should provide the U.S. with an estimated 95% certainty of continuous service for shorter path lengths and about 89% for longer-haul traffic. **1969** A. M. CAMPBELL *Episomes* ii. 19 Gene *O* is not required for DNA synthesis, but is necessary for normal turn-on of late phage genes.

2. The action or an instance of turning somebody on; a drug-taker's 'trip'; something which or someone who arouses interest, enthusiasm, or sexual response. *slang.*

1969 FABIAN & BYRNE *Groupie* (1970) xxi. 138 There is enough in one bottle for two turn-ons. **1969** *Telegraph* (Brisbane) 3 June 12/2, I think I'm more of a turn-on now than I ever was when I was trying to conform to that curvy image. **1969** *Sunday Mail Mag.* (Brisbane) 22 June 11/5 Other turn-ons are music.., qualities of appearance or character.., and straight-out sex. **1975** *N.Y. Times* 1 Nov. 15/3 A museum spokesman said nearly 50,000 people visited the galleries during the first 25 days. 'It's been a real turn-on,' she said. **1978** J. KRANTZ *Scruples* iii. 81 Masturbation isn't a great big turn-on in my life. **1982** D. HOCKNEY in S. Spender *China Diary* 189 A medieval city is unstimulating to me, whereas to others it might be a great turn-on.

3. *attrib.*

1967 *Electronics* 6 Mar. 133/2 A desirable feature of this circuit is that the turn-on time of the relay is very sharp. **1972** G. S. HOLT in T. Kochman *Rappin' & Stylin' Out* 204 This 'turn-on' ability in terms of communication style.

turn-out, *sb.* (*a.*) Add: Also **turnout. 3.** *spec.* (The number of) those who turn out to vote in an election.

1970 *Guardian* 20 June 15/5 Only one of the five polls gave full weighting..to differential turnout—the question of how many Labour or Conservative supporters will in fact bother to vote. **1976** *New Yorker* 15 Nov. 204/2 Only eighty per cent of the Democratic turnout voted for Carter. **1976** *Honolulu Star-Bull.* 21 Dec. A-2/4 He called his committeemen into party headquarters on the day after elections and made them account for the turnout in their wards.

turn-over, *sb.* and *a.* Add: Also **turnover. A.** *sb.* **1. b.** The point at which it is necessary to turn over a gramophone record; a break in play at the end of a side of a record.

1931 *Times Educ. Suppl.* 12 Dec. 1/3 With almost incredible perversity the engineers have made the turn-over not at the beginning of the Scherzo, but at the *più mosso*. **1976** *Gramophone* Apr. 1603/3 Now that DG have put the whole work on to one disc,..there is a turn-over in the 'Gretchen' movement (it comes at the beginning of the fourth bar after letter G).

3. d. *Printing.* (See quots.)

1938 L. M. HARROD *Librarians' Gloss.* 652 *Turn-over*, an extension of printed matter, beyond the space allotted. **1956** *Bookman's Conc. Dict.* 44/1 *Break line*, the last line of a paragraph not spaced full out to the measure; also known as..Turn Over. **1981** I. A. GORDON in *N.Z. Listener* 14–20 Feb., I am well aware that newspaper columns are narrow and that words must consequently be broken up into two bits, joined by a hyphen at the end of the first line. This necessity is known by printers as the 'turn-over'.

6. a. Also in extended sense, the amount or number of anything (or of persons) dealt with, processed, etc.; the throughput; *turnover tax*, a tax on the turnover of a business.

1911 G. B. SHAW *Doctor's Dilemma* Pref. p. xxvi, The sixpenny doctor, with his low prices and quick turnover of patients,..makes much more than you do. **1920** *Manch. Guardian* 28 Dec. 11/5 A turnover tax operates whether the transaction is a profit or a loss. **1938** *Sun* (Baltimore) 13 Sept. 3/1 Under the old paper-ballot system the turnover was many times faster. **1944** *Ibid.* 15 June 20/8 It is only by speeding up the rate of turnover that, with our depleted staff, we are able to deliver practically as many babies as in [a] normal period. **1973** E. OSERS tr. *Waldheim's Austrian Example* ii. 30 Economic recovery required..the application of a severe austerity programme which involved..the introduction of a turnover tax. **1976** *Howard Jrnl.* XV. I. 43 The subject-matter is the frequency of remands for medical reports.. in relation to the total turnover of magistrates' courts.

b. The simultaneous synthesis and degradation of a substance in a living organism; *turnover rate, time* (see quot. 1943).

1943 *Jrnl. Gen. Physiol.* XXVI. 326 *Turnover*.—This term refers to the process of renewal of a given substance. ..*Turnover rate*..is the amount of the substance that is turned over by that tissue per unit of time... *Turnover time*..is the time required for the appearance or disappearance of an amount of that substance equal to the amount of that substance present in the tissue. **1961** *Times* 3 Feb. 19/7 In health the myelin sheath is a stable tissue element with little or no evidence of metabolic turnover. **1967** M. E. HALE *Biol. Lichens* iv. 58 Slow rates of protein turnover might well be a characteristic of all lichens. **1977** P. B. & J. S. MEDAWAR *Life Sci.* x. 84 The turnover rate of bodily constituents varies widely from tissue to tissue. **1982** S. G. CHANEY in T. M. Devlin *Textbk. Biochem.* xxv. 1180 The turnover of body protein is a normal process.

c. The number of employees leaving a workforce and being replaced, change of staff.

1955 *Times* 7 June 7/3 The plan was also expected to reduce labour turnover since a qualifying period would be—or should be—needed to secure the guarantee. **1956** W. H. WHYTE *Organization Man* I. v. 58 Employees like it and absence and turnover are low. **1963** E. P. THOMPSON *Making of Eng. Working Class* viii. 246 The labour turnover in the early engineering workshops was prodigious; Galloway, who employed eighty or ninety men in 1824, claimed to have had between 1,000 and 1,500 men pass through his works in the previous twelve years; that is more than a total turnover of the labour force *per annum*. **1977** W. B. Eberhard in Bond & McLeod *Newslett. to Newspapers* ii. 149 Postal rates soared..and personnel turnovers were unusually high.

7. *U.S. Sport.* The (unintended) loss of possession of the ball to the opposing team.

1969 *Eugene (Oregon) Register-Guard* 5 Dec. 1 D/3 Not often does a team commit 27 turnovers and win, but South Eugene did just that. **1975** *New Yorker* 7 Apr. 108/2 Similarly, in their other defensive ploys the Knicks' object was to harass their opponents into committing turnovers—that is, losing the ball by making wayward passes or committing technical infractions. **1979** *Tucson (Arizona) Citizen* 20 Sept. 11D/2 Four turnovers took the Ducks out of the contest.

B. *adj.* (Earlier and further examples.) **turn-over collar** (earlier example); **turnover article** = sense 3 c; **turnover board** *Founding*, a flat board on which a flat-bottomed pattern or half-pattern may be stood for sand to be rammed round it; **turnover cartridge**, a gramophone cartridge with a pivoted mounting for two styluses for use at different speeds.

1605 P. Erondelle *French Garden* i. sig. D 8v, Send me..a Cambrick ruffe..Also bring me these turnover shoes, for they be too high. **1747** H. Glasse *Art of Cookery* ix. 93 Close the two Ends of your Paper as you do a Turnover Tart. **1836** Dickens *Sk. Boz* 1st Ser. I. 238 Soiled buff boots with turnover red tops. **a 1861** T. Winthrop *Life in Open Air* (1863) 318 In jacket and turn-over collar. **1888** *Lockwood's Dict. Mech. Engin.* 391 *Turn over board*, a board used for ramming a pattern upon. **1928** W. Rawlinson *Mod. Foundry Operations & Equipment* xiii. 168 A method adopted in certain instances of repetition work..is that of a 'turnover board', also termed 'bottom board' or 'joint board'. **1944** *Penguin New Writing* XXII. 142 *Moby Dick* marks the turnover point where balance was perhaps precariously achieved. **1952** H. Herd *March of Journalism* vi. 82 The third column had a turnover article giving a retrospect of political events since the beginning of the year. **1958** *Spectator* 20 June 795/1, I was glad to see *The Times* coming out so strongly, in Monday's 'turnover' article, against recent abuses by the courts of their Contempt powers. **1958** *Practical Wireless* XXXIV. 57/2 (Advt.), Latest type lightweight crystal pick-ups with turn-over cartridge. **1962** A. Nisbett *Technique Sound Studio* 264 The pick-up head may consist of a turnover cartridge having styli for coarse and fine groove records on the two sides. **1964** S. Crawford *Basic Engin. Processes* (1969) x. 238 The flat face of the pattern is placed on a turnover board and a suitable size moulding box..is placed over it. **1978** *Rugby World* Apr. 59/2 (Advt.), Best quality stretch nylon socks in plain colours and turn-over tops.

turnpike, *sb.* Add: **I. 5. a.** Now *Hist.* exc. *U.S.* (Later examples.)

1950 *Sun* (Baltimore) 2 June 10/5 One route will be recommended for the turnpike. **1965** *New Statesman* 5 Nov. 713/1 The good, fast, safe roads are toll roads, called in New England by the old name of turnpikes. **1977** *New Yorker* 3 Oct. 43/1 Halfway up the Connecticut Turnpike, I slowed the car.

III. 9. turnpike gate, (*b*) (earlier example); **turnpike sailor** (earlier example).

1793 J. Woodforde *Diary* 21 July (1929) IV. 45 We got to Bruton Turnpike Gate. **1839** H. Brandon *Poverty, Mendicity & Crime* 165/1 Turnpike sailors.

turnpike *v.* (earlier example).

1791 J. Hiltzheimer *Diary* 17 Sept. (1893) 172, I took Mr. Francis..to view the road, from Vine Street to Vanderen's Mill, six miles, which it is proposed to turn-pike.

turn-round (tǝ̄·ɹn,raund). [f. the verbal phr. *turn round* (Turn *v.* 78 in Dict. and Suppl.).]

1. The arrival, unloading, and preparation for the return journey of a ship, aircraft, goods vehicle, etc.; the time taken for this. Also *gen.*, the course of receiving, processing, and sending out again; progress through a system.

1913 A. Bennett *Regent* x. 291 She's going to do the quickest turn-round that any ship ever did... She'll leave at noon to-morrow. **1929** *Evening News* 18 Nov. 5/5 In these cabin ships the engines do not take up nearly as much space..nor is it necessary to provide for such a rapid turn-round at the terminal ports. **1951** *Engineering* 1 June 658/1 An exceptional amount of sickness among the key and other operative staff, resulting in a slower turn-round of wagons. **1958** *Daily Mail* 3 July 4/3 An efficiency service that is..capable of..achieving a quicker turn-round in the hospitals. **1963** *Times* 9 Feb. 9/4 The campaign for quicker 'turnround'..is being conducted jointly by the Road Haulage Association and British Road Services. **1972** L. Lamb *Picture Frame* xiv. 125 The expertness of the waiters in getting a quick turn-round of occupants at their tables. **1974** *Physics Bull.* Apr. 142/2 The total time which the material spends in the office including editing time, turn-round of proofs and any waiting time is usually not more than a month. **1976** *Southern Even. Echo* (Southampton) 12 Nov. 3/4 Heavy demands were made on her [sc. QE2], particularly with the short turn-rounds which were the pattern today. **1976** P. R. White *Planning for Public Transport* viii. 156 On the railways,..increases in average speed..coupled with much quicker turnround of stock, have improved utilization of rolling stock on all-year service, by about 100 per cent. **1979** *Dan-Air In-flight Mag.* Winter 15 His [sc. the captain's] is the responsibility for the proper turn-round of an aircraft at its outward destination.

2. The reversal of a trend, a change to an opposite direction, opinion, etc.

1963 *Times* 29 Jan. 14/5 There is a growing feeling in market quarters that..the end of the upswing is in sight and a turnround may be near. **1976** *Listener* 22 Jan. 86/4 Critics in Britain are much more often attacked these days for their indiscriminate praise of everything new than for their obscurantism. To be fair, there is some perception of this turn-round. **1981** *Times* 22 May 19/2 Associated Engineering's £12.2m turnround from first half profits of £10.5m last year to a loss of £1.7m this time shows how bad conditions are in engineering. **1984** *Daily Tel.* 1 Feb. 18/5 There was a notable turn-round in leading engineers. GKN, down to 201p at the opening, were finally a shade better at 208p.

3. *attrib.*

1920 *Glasgow Herald* 16 Aug. 8 The strike has arisen from a dispute about special payment for quick turn-round voyages. **1969** *Times* 5 Nov. 23/6 Turnround time on orders has been cut from 15 to six days. **1977** *Modern Railways* Dec. 462/1 A turnround time of about 5min would permit the running of a half-hourly service with only the barest of margins.

turnstile. Add: **b.** (Examples with sense 'resembling or shaped like a turnstile'.)

1952 D. G. Fink *Television Engin.* (ed. 2) vii. 367 Decoupling in the antenna itself is accomplished by separating the elements of the turnstile antenna..into two orthogonal groups and connecting these groups in a bridge circuit. *Ibid.* 371 The appearance of two horizontal dipoles at right angles has given rise to the descriptive term *turnstile radiator*. **1959** K. Henney *Radio Engin. Handbk.* (ed. 5) xx. 76 Many forms of vertical arrays have already been evolved using magnetic dipoles, turnstile elements, or current sheets. **1971** *Gloss. Electrotechnical Power Terms (B.S.I.)* iii. vii. 16 *Turnstile aerial*, aerial consisting of one or more tiers, each tier being a combination of two horizontal dipoles arranged in the form of a right angled cross. **1978** *Nature* 5 Oct. 375/2 The VHF system consists of duplicate transponders and a four-element turnstile antenna.

turn-table. Add: Now usu. **turntable. 2.** (*e*) The rotating plate on which a gramophone record is placed to be played; the unit housing this plate.

1908 *Sears, Roebuck Catal.* 195/2 The Type FH Harvard Disc Talking Machine... The turntable, of a special composition metal, is 10 inches in diameter. **1921** P. A. Scholes *Learning to Listen by Means of Gramophone* 157 The motor should be wound up fully for each record played, in order that the turntable can rotate at its normal and even speed. **1960** *Practical Wireless* XXXVI. 421/2 Wedge the turntable so that it does not move. **1962** *Times* 5 July 15/7 The disc had been standing on the turntable for a few minutes before playing. **1979** L. Kallen *Introducing C. B. Greenfield* xiii. 166 An old armoire..held his recordings and the turntable.

turn-up, *sb.* and *a.* Add: **A.** *sb.* **2.** *spec.* The turned-up cuff of a trouser-leg.

1925 *Minister's Rep. of Fashion for Gentlemen* Feb. 8/2 Permanent turn-ups are still worn for outdoor wear. **1933** J. E. Liberty *Practical Tailoring* vi. 77 For turned-up bottoms, called permanent turn-ups, go down on the seam from the mark the width of turn-up required..and make a straight line across the bottom. **a 1944** K. Douglas *Alamein to Zem Zem* (1946) xiii. 81 He had..beautifully cut narrow trousers of fawn cavalry twill, without turn-ups. **1969** B. Malamud *Pictures of Fidelman* (U.K. ed.) i. 25 His blue gabardine suit—a one-button jacket affair and the trousers a little frayed at the turn-ups. **1972** G. Durrell *Catch me a Colobus* vii. 142 She..nosed round our legs eagerly, searching in our turn-ups to see whether she could find anything to eat.

3. a. *fig.* Also, an unexpected occurrence, a surprise. Cf. sense 3 b in Dict. and Suppl.

1942 Berrey & Van den Bark *Amer. Thes. Slang* §178/1 Surprise..bob-up, springer, turn-up. **1961** R. Park *Hole in Hill* (1962) x. 79 'Well, this is a turn-up,' said Dunk in disgust. 'How do we get out?' **1972** *Jazz & Blues* Oct. 26/3 What a turn-up then to find there's another version.

b. Freq. in phr. *a turn-up for the book(s)*; also in *gen.* use (*colloq.*), an unexpected turn of fortune, a surprise.

1948 'J. Tey' *Franchise Affair* xviii. 209 We won by a length and a half on a tight rein; and was that a turn up for the book! **1951** *People* 3 June 2/2 What a Derby Day it was! And what a turn up for the books! **1959** P. Bull *I know Face* ii. 35, I reported my findings to Mr Huth, who said..perhaps I would like to write the script. Now this was quite a turn-up for the book, as very few people..are allowed to say what they write. **1968** 'C. Franklin' *Escape* viii. 104 This was indeed a turn up for the book. 'Penelope!' he exclaimed. **1978** J. Wainwright *Jury People* v. 16 A bit of a turn up for the book, isn't it? Murder, I mean. **1983** *Daily Tel.* 13 Oct. 8/7 Even..the Labour group's spokesman..could not hide his surprise... 'This is a real turnup for the books. I am quite amazed by it..,' he said.

5. The curve of the projecting lower jaw of a bull-dog.

1905 [see *lay-back 1]. **1922** R. Leighton *Compl. Bk. Dog* v. 64 It [sc. a type of bulldog] has certain well-defined characteristics, notably the extreme width and turn-up of underjaw. **1973** J. F. Gordon *Bulldog* (rev. ed.) iv. 53 (*caption*) Nice head and skull, good 'turn-up'.

turnverein (tǝ̄·ɹn,vǝrain). [Ger., f. *turnen* to do gymnastic exercises + *verein* society, club.] In the United States, a gymnastic society, orig. for German immigrants, on the model of those instituted by Jahn (see Turner[1] 8). Also *attrib.*

1852 *San Francisco Herald* 1 Nov. 2/2 The Turnverein Society held another of their fetes yesterday. **1949** *Minnesota Hist.* Mar. 26 The Cincinnati Turnverein built the first Turner Hall in the United States in 1850. **1959** R. Condon *Manchurian Candidate* (1960) xxvii. 267 Petitions and documentations were submitted to the Resolutions Committee by farm lobbies, labour unions,..and national manufacturers' Turnvereins. **1974** *Encycl. Brit. Micropædia* X. 202/1 Turnvereins continue to foster citizenship and cultural programs together with health and physical-education activities, particularly gymnastics.

turpentine, *sb.* Add: **4.** *turpentine wood* (earlier example); **turpentine moth**, substitute for first part of def.: any of several leaf-roller moths of the family Tortricidæ; (example); **Turpentine State** (earlier example); **turpentine still** (examples); **turpentine weed**: also, any of several other herbs containing an aromatic sap; (earlier and later examples).

1842 T. W. Harris *Insects Injurious to Vegetation* 350 Turpentine-moths..injure pines and firs. **1850** Mayne Reid *Rifle Rangers* I. v. 46 The danger is, we may stick in the Turpentine State. **1799** *Wilmington (N. Carolina) Gaz.* 12 Dec. 2/1 Will be sold..at Public Sale... Two turpentine stills. **1935** Z. N. Hurston *Mules & Men* I. iv. 86 One woman had killed five [men] when I left that turpentine still where she lived. **1819** *Western Rev.* I. 95 Among the most remarkable and singular [plants of Kentucky is]..*Silphium therebinthaceum*, Turpentine weed. **1913** W. C. Barnes *Western Grazing Grounds* 236 There is a little green weed (*Gutierrezia*) known locally as snakeweed, fireweed, turpentine weed. **1931** G. H. Vansell *Nectar & Pollen Plants Calif.* 14 Turpentine weed..is also visited freely by bees for nectar. **1890** *Philadelphia Inquirer* 1 June 1/4 He would find there every interest and every occupation of the period fully depicted, from the forests of Maine to the turpentine woods of North Carolina.

turpentining, *vbl. sb.* (Examples.)

1910 C. Van Hise *Conservation Natural Resources of U.S.* iii. 229 (*heading*) Reduction of loss in turpentining. **1971** *Forest Products Jrnl.* Feb. 53/2 The Southeastern Forest Experiment Station conducted a ..study to determine the effect of turpentining on..yields of butt peeler blocks.

turps. Add: **2.** *Austral. slang.* Intoxicating liquor, esp. beer.

1945 Baker *Austral. English* 168 Australians have a fair selection of terms to describe drinking and drinking bouts, such as a *beer-up*,..and *to bash the turps*. **1962** Marshall & Drysdale *Journey among Men* 84 The Sergeant alleged that Ah Fong was a notorious drunkard, forever on the 'turps'. **1973** J. O'Grady *Survival in Doghouse* 57 He's humping a dozen cans with him. Ice cold. And he gets a great welcome. Not only because of the turps, but because with him there we can have a four-handed game.

turquoise, *sb.* (*a.*) Add: Now freq. with pronunc. (tǝ̄·ɹkwoiz). **II. 5. c.** *turquoise-coloured* (earlier example), *turquoise-gemmed*, *-studded* (earlier example).

1823 C. J. Mathews *Jrnl.* 13 Nov. in Dickens *Life Charles J. Mathews* (1879) I. iv. 93 The beautiful turquoise-coloured bay. **1862** G. M. Hopkins *Poems* (1967) 9 One bound o'er dripping gold a turquoise-gemm'd Circlet of astral flowerets. **1901** Kipling *Kim* xiv. 361 A fair-coloured woman with turquoise-studded headgears.

turr (tǝ̄ɹ), *sb. Newfoundland.* Also † **tuir**, **turrh.** [Prob. imit.] = Murre.

1794 A. Thomas *Newfoundland Jrnl.* (1968) x. 144 Here are..Penguins, Hegdowns, Muirs and Tuirs, Ice Birds..and a number of other Sea Fowl. **1853** *Trans. Lit. & Hist. Soc. Quebec* IV. 334 Among the sea birds are Mernettes, Moyocks, Gulls,..Turrhs, Paraquets, Penguins, and divers others. **1960** L. M. Tuck *Murres* 34 A common vernacular name for the murre in Newfoundland is 'turr'. **1974** *Nat. Geographic* Jan. 122/2 The seabirds here called stearin, turr, and tickle-ace are the birds known elsewhere as tern, murre, and kittiwake.

turret, *sb.*[1] Add: **2. b.** Also, a similar structure on a tank, armoured car, or aircraft.

1914 E. A. Powell *Fighting in Flanders* iii. 70 The earlier armoured cars used by the Belgians..consisted of a circular turret, high enough so that only the head and shoulders of the man operating the machine-gun were exposed, covered with half-inch steel plates and mounted on an ordinary chassis. **1933** *Gloss. Aeronaut. Terms (B.S.I.)* iv. 27 *Turret*, a form of cockpit primarily intended for the use of a gunner. **1942** *Tee Emm* (Air Ministry) II. 140 Give yourself a few minutes each day in the training turrets so that your turret manipulation is absolutely one hundred per cent. *Ibid.* 141 For most of the trip your hands will be on the turret controls. **1969** G. MacBeth *War Quartet* 26, I stretched Across my turret, thinking. **1978** J. Irving *World according to Garp* i. 15 This ball turret was a metal sphere with a glass porthole; it was set into the fuselage of a B-17 like a distended navel—like a nipple on the bomber's belly.

4. a. (Examples.)

1898 H. S. Wilson *Pract. Tool-Maker & Designer* vi. 58 Knurling fixtures for both the slide and turret. **1963** [see *semi-automatic *sb.* 1]. **1975** Bram & Downs *Manuf. Technol.* vii. 200 The turret is then indexed to perform a number of drilling, reaming, tapping and counter-sinking operations.

b. *Cinematogr.* and *Television.* = *lens turret* s.v. *lens *sb.* 3.

1951 R. Spottiswoode *Film & its Techniques* iii. 64 On almost all cameras..the lenses are mounted in clusters of three or four on a turret, a revolving device which serves to bring the wanted lens in front of the aperture. **1960** O. Skilbeck *ABC of Film & TV* 138 *Turret*, a circular mounting of several lenses held in readiness for use on the front of a camera. **1961** G. Millerson *Telev. Production* iii. 34 The internal complexity of a zoom lens makes it bulkier than a turret assembly. **1965** J. Von Sternberg *Fun in Chinese Laundry* (1966) vii. 184 We illuminated every possible retreat he might find, and more cameras, turrets, and various lenses were employed. **1976** A. Davis *Television* 27 The turret camera with a revolving disc offering the choice of several lenses of different focal lengths, and the single zoom lens of variable focal length ..were still to come.

5. (sense 1) *turret-room* (earlier example); (senses 2 and 4 b) **turret-mounted** adj.; (sense 4 b) **turret-mounting**; **turret-lathe** (examples); **turret-light**, a light on top of a police car, ambulance, etc., which flashes to signal an emergency.

1898 H. S. Wilson *Pract. Tool-Maker & Designer* vi. 58 The variety of work that may be executed on a screw machine or turret lathe. **1939** *Daily Tel.* 18 Dec. 12/3 (Advt.), Experience should include the setting up and tooling of automatic and turret lathes. **1975** Bram & Downs *Manuf. Technol.* v. 129 The vertical turret-lathe is a chucking machine only. **1972** 'G. North' *Sergeant Cluff rings True* i. 14 A small van..carried a turret-light on its roof and had, 'Police,' lettered on its sidepanels. **1961** *Observer* 21 May 5/1 (Advt.), Twin lenses, standard and telephoto, turret mounted to whisk you smoothly from close-up to long shot. **1963** *Times Lit. Suppl.* 31 May 394/5 The loss of the *Captain*..marked the end of the attempt to combine a full set of sails with steam propulsion and turret-mounted guns. **1923** F. A. Talbot *Moving Pictures* 86 Behind this turret-plate is a second and fixed disk or 'turret-mounting' of identical diameter, but having only two openings, corresponding to the photographing and focussing apertures respectively. **1966** 'A. Hall' *9th Directive* ix. 83, I set it [sc. a camera] up on a tripod with a turret-mounting that was rigid enough for the weight. **1803** *Lett. Miss Riversdale* III. 368, I have been removed from the turret room I occupied, to a bed room on the ground floor.

turrible, turble, *dial.* (chiefly *U.S.*) varr. Terrible *a.*, esp. in sense 3.

1893 H. A. Shands *Some Peculiarities of Speech in Mississippi* 65 *Turrible*... A pronunciation of *terrible* very common among the illiterate, and sometimes heard in the conversation of the educated. **1897** Kipling *Captains Courageous* vii. 148 Jason was tur'ble praoud of his boy. **1903** K. D. Wiggin *Rebecca of Sunnybrook Farm* xxviii. 305 It's a turrible risk splittin' up families. **1912** J. Masefield *Everlasting Mercy* 20 You must be turble strong. **1929** H. W. Odum in A. Dundes *Mother Wit* (1973) 183 Had some turrible times in France. **1966** J. Aiken *Trouble with Product X* vii. 123 'Tis a long pull up from the village and turble weather. **1971** *Black World* Apr. 67 You musta done somethin turble to aggravate me like that.

turriform, *a.* For *rare*$^{-1}$ read *rare* and add further examples.

1959 E. A. Fisher *Anglo-Saxon Archit. & Sculpture* 57 Turriform churches. These were not uncommon and are so-called as they were built in the form of towers. **1970** H. Braun *Parish Churches* iii. 33 The structural nucleus of every Byzantine church is a central turriform structure with four..arches rising from four tall and massive piers.

‖ **turron** (turǭ·n). Pl. **turrones, turrons.** [Sp. *turrón*.] A Spanish sweetmeat resembling nougat, made from almonds and honey; a piece of this.

1918 *Chambers's Jrnl.* Jan. 33 The best hams of Montanchez, the finest turron from Jijona. **1950** E. David *Bk. Mediterranean Food* 153 In the winter there are..the sugar-plums of Nice; Spanish nougats and *turrons*; *pralines* from Aix-en-Provence. **1976** E. P. Benson *Bulls of Ronda* xvi. 100 Their desserts were to his taste... 'You can bring me some *turrones*.' He selected two of the creamiest.

turtle, *sb.*[1] Add: **1. c.** *Rhyming slang.* = *turtle-dove 3. (Usu. in pl.)

1893 P. H. Emerson *Signor Lippo* xiv. 55 A long sleeve cadi on his napper, and a pair of turtles on his martins finished him. **1936** 'J. Curtis' *Gilt Kid* 24 Got any turtles? The Gilt Kid, having no gloves, answered: 'No, but I'll buy a pair.' **1962** *John o' London's* 25 Jan. 82/1 Of course he [sc. the criminal] takes the precaution of wearing *turtles* (short for *turtle-doves*, rhyming slang for gloves).

turtle, *sb.*[2] Add: **2. b.** (Earlier examples of *to turn the turtle*.) (See also quot. 1818.)

1818 'A. Burton' *Johnny Newcome* II. 69 John..in the next week..would take Twice calling, to be once awake; They turned the turtle, cut him down. *Ibid.* 254 *Turn the turtle*, to get under a hammock, and lift it up in the middle, thus pitching the sleeper out on one side of it. **1830** *United Service Jrnl.* June 709 The chance on some equally squally night of 'turning the turtle', as Jack facetiously calls upsetting.

5. turtleburger, a kind of hamburger made from turtle; **turtle-deck**: (*a*) also applied to a similar structure on an aircraft; (*b*) = *turtle-back 1 b; **turtle-frolic** (earlier example).

1946 *Amer. Speech* XXI. 67/1 While in Florida Keys in the service in the Spring of 1940 I came across a road stand selling turtleburgers. **1979** *Daily Tel.* 29 Nov. 18 It must be admitted that the meat can be tasteless and fibrous, but its strength is its versatility: 'turtleburgers', for example, are delicious. [**1912** *Flight* 26 Oct. 966/2 The fabric is..stretched over light formers above the girder so as to provide a kind of turtle-back deck.] **1913** *Flight* 31 May 586/2 The comfort of the pilot has been carefully studied and he is well sheltered behind an aluminium turtle deck. **1937** *Jrnl. R. Aeronaut. Soc.* XLI. 7 It was customary for the pilot to carry the parachute in the cockpit with the cable laid along the turtle deck and fastened securely with adhesive tape. **1954** *Amer. Speech* XXIX. 103 *Turtledeck*, the trunk or turtleback of a roadster. **1967** A. Shennan *Sopwith Snipe Described* 17 To this basic structure was affixed a turtledeck structure of plywood formers and stringers. **1750** F. Goelet *Jrnl.* 2 Oct. in *New-England Hist. & Geneal. Reg.* (1870) XXIV. 53 Had an Invitation to day to Go to a Turtle Frolick.

turtle, *v.*[2] Add: **2.** (Examples.)

1838 in G. C. Anderson *Laws of Bahamas* (1843) 119 If any person..be found turtling, or fishing, against the provisions of this Act. **1952** E. Hemingway *Old Man & Sea* 15 He never went turtle-ing. That is what kills the eyes.

3. Also *intr.*

1920 [see *careen *v.* 4 b].

4. b. *intr.* To act in a manner characteristic of a turtle (perh. *spec.* to bridle or show indignation). Also with *up*. ? *nonce-wd.*

1914 D. H. Lawrence *Widowing of Mrs. Holroyd* I. ii. 28 *Clara*: Turning-out time, Laura. *Laura* (*turtling*): I'm sorry, I'm sure. **1920** —— *Lost Girl* v. 64 It was most curious to see Miss Pinnegar turtle up at the mention of this scheme... She blurted, bridling and ducking her head..like an indignant turkey.

turtle-back. Add: **1. b.** A rounded projecting boot on a motor vehicle; the lid of this. *N. Amer.*

1941 'A. A. Fair' *Double or Quits* iii. 33 He raised up the turtleback in the car. **1971** D. Conover *One Man's Island* 23 The right fender fell off and rolled into a ditch. I stuck it disgustedly in the turtleback and in the village asked Lloyd at the garage to put it back on.

3. (Earlier example.)

1898 G. B. Shaw *Let.* 18 Oct. (1972) II. 68 They all..betrayed gross ignorance on the points they were most cocksure about, such as riding on turtleback and the like follies.

3*. A land form likened to the shell of a turtle.

1913 *Geogr. Jrnl.* XLII. 149 Thick forests..alternate with tall grassland and bare and rocky turtlebacks. **1928** *Amer. Speech* IV. 126 'Hog-back' and 'turtle-back' are common names for hills or ridges suggesting those forms. **1938** *Bull. Geol. Soc. Amer.* XLIX. 1875 Three turtleback areas have been recognized in the Black Mountains [of Death Valley, California]. They prove to be structural as well as topographic features.

turtle-dove. Add: **3.** *Rhyming slang.* A glove. (Usu. in pl.) Cf. *turtle *sb.*[1] 1 c.

1857 'Ducange Anglicus' *Vulgar Tongue* 23 *Turtle doves*,..gloves. **1935** A. J. Pollock *Underworld Speaks* 127/1 *Turtle doves*, a pair of gloves. **1959** I. & P. Opie *Lore & Lang. Schoolch.* xiv. 320 'Turtle doves' for gloves, and so on, normally associated with cockneys, is neither confined to the metropolis, nor to the shift-for-a-living class. **1972** *Lebende Sprachen* XVII. 8/4 *Turtle dove*, glove.

Hence as *v. trans.*, ?to show affection for (another), like a turtle-dove for its mate. *nonce-use.*

1922 Joyce *Ulysses* 196 Take her for me... Jove, a cool ruttime send them. Yea, turtledove her.

tu·rtle-neck, *sb.* (and *a.*) orig. *U.S.* [Turtle *sb.*[2]] **1. a.** A close-fitting roll or band collar, now usu. one intermediate in height between a crew-neck and a polo-neck; formerly also = *polo-neck* (*a*) s.v. *polo[1] 3. **b.** A shirt or jersey with such a collar.

1897 *Sears, Roebuck Catal.* 217/3 Turtle Neck Sweater. Extra heavy knit, all wool, turtle neck. **1939** M. B. Picken *Lang. Fashion* 102/3 (*caption*) Turtle Neck. **1957** *New Yorker* 5 Oct. 112/2 The roll at the round neck suggests a turtleneck, and little inverted tucks do nice things for the bosom. **1960** *Daily Express* 30 Aug. 5/3 Back buttoning turtle necks of matching fabric. **1970** E. Tidyman *Shaft* xiii. 170 Tall, young men in..turtlenecks, leatherjackets. **1977** *New Yorker* 3 Oct. 40/2 Dr. Sayles..asks me to take off my shirt, a turtleneck. **1982** J. Gardner *For Special Services* xvi. 174 Bond, clad now in dark slacks, a black turtle-neck and short jacket.

2. *attrib.* or as *adj.*, esp. as *turtle-neck sweater*.

1895 *Montgomery Ward Catal.* Spring & Summer 483/1 The Turtle Neck Shirt or Sweater, double from waist up, one of the most desirable garments ever invented for cold-weather shooting. **1896** F. D. Roosevelt *Let.* 14 Oct. (1947) 47, I should very much like a red turtle neck sweater for skating and coasting. **1905** *Outing* Mar. 743/1 Shape the coat like a turtle-neck sweater. **1926** *Daily Colonist* (Victoria, B.C.) 7 Jan. 18/4 (Advt.), Balbriggan Turtle-Neck Pull-Overs of fine quality, finished with ribbed band at the waist. **1938** J. Steinbeck *Long Valley* 95 The younger wore a blue turtle-neck sweater. **1946** Wodehouse *Joy in Morning* vi. 42 He dresses like a tramp-cyclist, affecting turtle-neck sweaters and grey flannel bags. **1952** M. McCarthy *Groves of Academe* (1953) xiii. 266 He..signalled to a tall blonde girl in a tight turtle-neck sweater. **1964** *McCall's Sewing* xi. 182/2 Turtle-neck collar. A wide true bias band that stands up from the neck edge and then rolls back over the neckline seam. **1978** G. Greene *Human Factor* II. i. 53 She wore brown trousers and a turtle-neck sweater.

Hence **tu·rtle-necked** *a.*

1931 G. S. Chappell *Gardener's Friend* 7 When long-haired athletes in turtle-necked sweaters were the idols of the hour. **1954** *Encounter* Feb. 47/1 Long-haired men in turtle-necked sweaters marched into the room. **1978** J. Wainwright *Jury People* iii. 13 He wore a cheap wind-cheater over a turtle-necked sweater.

Tuscan, *a.* and *sb.* Add: **A.** *adj.* **d.** Also *Tuscan grass*, *hat*. (Earlier examples.)

1830 in A. Adburgham *Shops & Shopping* (1964) iv. 38 Chip and Tuscan Hats. **1833** in *Ibid.* 40 A new bonnet, composed of Tuscan Grass and prepared whalebone.

B. *sb.* **b.** (Later examples.)

1857 J. F. Maguire *Rome* xxi. 245 Tuscans only, or their descendants to the third generation, are received into the society. **1901** M. Carmichael *In Tuscany* i. 9 Certainly the Tuscan has some real love of the Englishman.

d. The golden-yellow colour of Tuscan straw.

1887 [see sense d of adj. in Dict.]. **1912** *T. Eaton & Co. Catal.* Spring & Summer 3/3 Draped Toque..Colors Black, Navy or Tuscan, with corded silk in colors to harmonize. **1923** *Daily Mail* 16 Apr. 1 Colours:..Old Gold, Tuscan, Mastic.

C. *Comb.* **Tuscan lamb**, a variety of processed lambskin, used mainly to make headwear.

1956, **1962** [see *Lucca]. **1970** *Guardian* 24 Nov. 9/4 A Tuscan Lamb Hat (in black or white) complete with ear muffs.

Tuscarora (tʊskărǭ·ră). [Iroquois, = hemp-gatherer.] An Iroquoian tribe, originally inhabiting Carolina, which, after moving to upper New York State, joined the Iroquois Confederacy of North American Indians, commonly called the Five Nations (thereafter Six Nations), in 1722; a member of this tribe; their language. Also *attrib.*

1650 E. Bland *Jrnl.* 28 Aug. in *Discovery New Britaine* (1651) 3 An Englishman, a Cockarous hard by Captaine Floods, gave this Indian Bells..to lay downe to the Tuskarood King. **1713** in *N. Carolina Colonial Rec.* (1886) II. 2 An order from ye Government of New Yorke to Caution ye Tuscaroras against going to warr with ye English here. **1785** T. Jefferson *Notes Virginia* 390 The Monacans or Tuscaroras..were taken into the confederacy.., making the sixth [nation]. **1878** *Jrnl. R. Soc. Arts* 10 May 537/2 He wants one of my improved phonographs to preserve the accents of the..Tuscaroras, who are dying out. **1910** F. W. Hodge et al. *Handbk. Amer. Indians N. of Mexico* II. 842 The Tuscarora..possessed in early times the 'country lying between the sea shores and the mountains, which divide the Atlantic States'. **1915** J. Buchan *Salute to Adventurers* ix. 137 All this land ..is Sioux country... But cheek by jowl is a long strip held by the Tuscaroras. **1933** L. Bloomfield *Language* iv. 72 The Iroquoian family was spoken in a district surrounded by Algonquian; it includes..the languages of the Iroquois type (Mohawk, Oneida,..Tuscarora). **1976** W. L. Chafe in T. A. Sebeok *Native Languages of Americas* I. 532 The Northern Iroquoian languages which are still spoken are six in number. They include the very closely related languages of the original Five Nations of the Iroquois..plus Tuscarora, now the Sixth Nation, a language somewhat more divergent. **1979** *United States 1980/81* (Penguin Travel Guides) 654 The Tuscarora Indian Reservation is nearby.

tusche (tu·ʃǝ). orig. *U.S.* Also **tushe.** [Ger., back-formation from *tuschen*, f. F. *toucher* to touch.] A greasy black liquid composition used in lithography and other printing techniques; lithographic drawing ink. Also *attrib.*

1885 *Lithographer & Printer* 24 Jan. 114/1 By the term autography we designate..everything that is written with lithographic tushe on transfer paper in order to enable it to be transferred to the lithographic stone. We use the word 'tushe', for though specially coined, it has been adopted by the entire craft of lithographers in this country. **1912** W. C. Brown *Practical Textbk. Lithography* iii. 26 The most useful of all inks for writing or drawing on stone come under the name tusche. **1940** R. Mayer *Artist's Hand-bk.* xii. 440 Lithographic crayons and tusche are always black, regardless of the color in which the final proof is to be printed. *Ibid.* 460 (*heading*) Tusche-wash-out method. **1965** Zigrosser & Gaehde *Guide Coll. Orig. Prints* iv. 49 The crayons used [in lithography]—and their liquid equivalent, tusche or lithographic ink—are made of a mixture of grease, wax, soap, and lampblack... Many technical manipulations are possible:..tusche effects with pen, brush, drybrush or spatterwork, [etc.]. **1967** V. Strauss *Printing Industry* v. 274/2 If lithographic tusche is used as a painting medium [in serigraphy], glue can serve as blocking medium as it will be repelled by the fatty lithographic tusche which is, of course, water-repellent.

tush (tuʃ), *sb.*[4] *slang* (chiefly *N. Amer.*). Also **tushie, -y.** [Abbrev. or dim. of *tochus.] = Backside 3.

1962 *Amer. Speech* XXXVII. 205 Another bilingual children's diminutive, *tushie*—from Yiddish *toches* or *tuches* 'rump'—has appeared in phrases like *tushie slide* 'a slide down a slope on one's bottom', the delights of which a group of Midwestern Jewish children have, I am told, expressed to their Gentile social workers. **1969** P. Roth *Portnoy's Complaint* 47 You'd think I was a

twenty-one-year-old girl; you'd think I hadn't wiped your backside and kissed your little tushy for you all those years. **1970** *Pix* (Austral.) 26 Dec. 11/4 Pretty young girls who walk around with..their tushes out there asking for it. **1973** *N.Y. Times* 10 June II. 1/3, I felt a fork hanging from the seat of my pants. I threw it off, just like Stanley would, and the audience went wild. I mean, there were 1,100 people there, looking at me with a fork up my tush! **1977** *Detroit Free Press* 11 Dec. 23-A/1 Eight hundred guests danced their tushies off on the world's largest discotheque floor. **1981** G. V. Higgins *Rat on Fire* xxviii. 170 Her tush is tight and she's got great boobs. **1984** *Miami* (Florida) *Herald* 6 Apr. 4B/6 (*caption*) So what's a damp tush between good friends?

tush (tuʃ, tʊʃ), *v.*[2] orig. *dial.* [Origin unknown.] *trans.* To pull or drag (a heavy object, esp. a log) along the ground.

1841 C. H. Hartshorne *Salopia Antiqua* 602 *Tush*,..to draw a heavy weight, as *tushing timber*. **1879** G. F. Jackson *Shropshire Word-bk.* 458 'Can yo' carry them faggits to the 'ŏŏd-pil?' 'I dunna know, but if I canna carry 'em, be'appen I can *tush* 'em alung.' **1953** H. L. Edlin *Forester's Handbk.* xv. 238 Felled logs are *tushed*, or drawn over the ground, butt-end foremost, by hauling chains attached to a tractor, a horse-team or a winch, until the nearest hard road is reached. **1963** *Times* 12 Feb. 12/7 Dolgelly's eight oak pillars had originally been snaked or tushed by oxen 10 miles over the mountains from Dinas Mawddwy, where they had been grown.

tushery. Add: Now usu. *gen.*, sentimental or romanticizing writing.

1921 H. S. Walpole *Young Enchanted* I. ii. 42 In literature her great period had been during the Romantic Tushery of 1895 to 1905. **1932** *Times Lit. Suppl.* 21 Apr. 292/4 Unlike many novelists who set their scene in Japan, Mr. John Paris indulges in no sentimental tushery about the Japanese. **1967** *Guardian* 16 May 7/5 What a wonderful vanished world of tushery is brought back by 'The Desert Song'. **1981** *Times* 14 Oct. 13/4 The *Idylls* wound their endless way... Such Arthurian tushery seems far removed from..*In Memoriam*.

tusk, *sb.*[3] Now the usual spelling of Torsk. (Later examples.)

1925 J. T. Jenkins *Fishes Brit. Isles* 164 The Torsk or Tusk..is moderately elongate and covered with very small scales. **1926** *Glasgow Herald* 19 Oct. 3 The inhabitants fit out boats for the..tusk fishing. **1935** *Fisheries Notice* (Min. Agric. & Fisheries) XXIII. 6 Suggested Trade Name. Tusk. **1977** *Grimsby Even. Tel.* 26 May 18/5 Principal sorts were..monk 28, tusk 20.

tussie-mussie: see Tuzzy-muzzy in Dict. and Suppl.

tussock, *sb.* Add: **5. tussock land** *Austral.* and *N.Z.*, uncultivated grassland used for sheep-grazing.

1881 W. Bateman *Colonists* x. 186 The tussock land abounds in the Middle Island.. Prior to breaking up the tussock land the native grass is first burnt. **1928** 'Brent of Bin Bin' *Up Country* ii. 17 The journey started across tussock land alive with purling streams. **1941** Baker *N.Z. Slang* v. 41 The tussock lands are a peculiar feature of this country.

tussock-grass. Add: **1.** (*c*) (Earlier example.) Freq. in *Comb.*, as *tussock-grassland*.

1848 J. White *Jrnl.* 13 Mar. 266 (MS.), The top [of the house] is of the nikau leaf and tussock grass. **1867** J. T. Thomson *Rambles with Philosopher* v. 25 The natural tussock-grass lands..stretched beyond the narrow precincts of incipient colonization. **1959** A. H. McLintock *Descr. Atlas N.Z.* p. xiv, Some pockets of beech forest still survive..along with a depleted tussock grassland.

tut, *v.* Add: **2.** *trans.* To express disapproval of by the exclamation 'tut'; to say disapprovingly.

1972 *Times* 10 Nov. 7/2 He [sc. President Nixon] felt sure some of his ideas would be 'tut-tutted' by 'the Georgetown cocktail set'. **1975** *Nature* 3 Jan. 1/2 The authors never address the problem, instead tut-tutting that university geology courses are unsuited to the demands of petroleum geology. **1984** A. Carter *Nights at Circus* III. vii. 239 But when he embarrassedly confessed there'd been no bang nor damage because the dynamite was damp, I'd 'tut-tutted' his inefficiency.

Hence **tu·tting** *vbl. sb.* and *ppl. a.*

1929 J. B. Priestley *Good Companions* I. i. 25 Ted..was shaking his head and..making a loud tut-tutting noise. **1947** *Manch. Guardian Weekly* 30 Oct. 8/3 Great and glossy cars rolled up in smooth procession. Into this a taxi-cab had strayed, to be hurriedly diverted with much tutt-tutting by police officers into the unimportant wastes of Millbank. **1962** *John o' London's* 19 Apr. 386/2, I simply could not see what all the tutting was about. **1976** T. Heald *Let Sleeping Dogs Die* ix. 183 She sucked her teeth and made little tutting noises. **1984** *Times* 11 June 6/6 There was much tut-tutting and an agreement that something should be done.

tute (tiūt). Colloq. abbrev. of Tutor *sb.* and *v.* or *tutorial *sb.*

1895 W. C. Gore in *Inlander* Nov. 65 *Tute*, tutor. **1934** Webster, *Tute*, *v.t.* & *i.* **1942** Berrey & Van den Bark *Amer. Thes. Slang* §197/9 *Tute*, tutor, to give private instruction. **1955** J. I. M. Stewart *Guardians* III. ii. 217 Jones is my tutor, and this happened at my first tute this term. **1957** D. Balsdon *Oxford Life* 169 But the College tutor's public lecture is an interruption in a week otherwise devoted to teaching pupils in his rooms, listening to their essays and talking about them. These are 'private hours'—'tutes', as undergraduates call them, or tutorials. **1982** T. Heald *Masterstroke* iii. 64 We did political theory tutes together.

tutee (tiutī·). orig. *U.S.* [f. Tut(or *v.* + -ee[1].] A university student (in relation to his tutor); a pupil of a private tutor.

1927 *Amer. Speech* II. 214/1 *Tutee*, English 'pupil'. I met this queer coinage in two academic publications. **1937** *Life* 7 June 58/2 Wolff at work is a two-hour torrent of words covering the high spots of a whole college course. Tutees take hasty notes, try to remember what he says overnight. **1952** M. McCarthy *Groves of Academe* (1953) i. 6 He was feeling more than half tempted to take the letter over to the main hall and post it on the faculty bulletin board, before the arrival of the eleven-o'clock tutee. **1975** D. Lodge *Changing Places* i. 19 His girl tutees suddenly began to dress like prostitutes.

tutor, *sb.* Add: **4. a.** Also used in other British universities and other further education establishments. Also, in Cambridge and some other universities and colleges, a member of the teaching staff assigned responsibility for the general well-being of a student (cf. *moral tutor* s.v. *moral *a.* 3 d).

1933 *Times Lit. Suppl.* 14 Dec. 889/3 He [sc. Sir John Sandys] was for long Senior Tutor of his college, a different thing in Cambridge from Oxford. **1980** L. P. Wilkinson *Century of King's* p. xiv, *Tutor*, a Fellow responsible for a student's general welfare. Every student has one.

6. (Later examples.) Now chiefly applied to books of instruction in playing a musical instrument.

1918–19 *T. Eaton & Co. Catal.* Fall & Winter 383/6 Bellak Piano Tutor..one of the best tutors in use. **1956** F. Reidy in S. Traill *Play that Music* 108 Any tutor I have ever read says that the tip of the reed should be struck with the tip of the tongue. **1981** Ld. Harewood *Tongs & Bones* iii. 60, I wanted to learn the clarinet... A beautiful Boehm arrived together with an English 'tutor'.

tutor, *v.* Add: **5.** *intr.* To study under a tutor. *U.S.*

1900 C. C. Munn *Uncle Terry* 55, I tutored some, read law, and was admitted to the bar. **1920** [see *flunk *v.* 1b].

tutordom. (Further example.)

1957 G. Avery *Warden's Niece* xi. 214 There will be a storm to-night, a very suitable end to my tutordom.

tutorial, *a.* Add: **B.** *sb.* **a.** A period of individual instruction given by a college or university tutor to pupils, either singly or in small groups.

1923 G. Saintsbury *Second Scrap Bk.* 27 For 'Mods' Logic one went to Professor Wall's University lectures and Mr. Sidgwick's tutorials. **1927** W. E. Collinson *Contemp. Eng.* 124 In regard to teaching within the University the only terms worthy of notice are the use of *Oral* (where some universities use *Viva*..) and *tutorials* (practical classes for recapitulating the formal lectures). **1932** C. Bailey in *Handbk. Univ. Oxford* 128 During term each man will attend a 'tutorial', as post-war Oxford has agreed to name it, at least once a week. **1953** A. Moorehead *Rum Jungle* i. 19 Old Jack or Geoff with whom I had attended tutorials at the university. **1966** *Rep. Comm. Inquiry Univ. Oxf.* II. 450 A 'tutorial' is to be taken to mean teaching of not more than three people at one time. **1979** *Washington Post* 26 Oct. B4/1 Windt did not show up for her 11:15 math tutorial that day.

b. Any period of tuition or training; a printed account or explanation of a subject intended for private study.

1978 *Sci. Amer.* Feb. 99/1 (Advt.), You'll find our tutorials on hardware and software invaluable reading, also our reports on home applications and evaluative reviews based on experiences with home computer products. **1980** *Amer. Banker* 15 Jan. 6/1 The 1980 conference will feature state-of-the-art developments to help assure the integrity of information systems. The schedule includes 10 tutorials, three general sessions, [etc.]. **1980** *Washington Post* 5 July B2/1 Sonny Stitt, that world-traveling, one man be-bop workshop, is holding tutorials at the One Step Down through tonight.

tutorship. Add: **2. b.** A post as a tutor, *spec.* in a university.

1925 C. Connolly *Let.* 9 Mar. in *Romantic Friendship* (1975) 62, I don't expect your tutorship will be till April. **1929** S. Leslie *Anglo-Catholic* xviii. 262 Colley had taken a travelling tutorship in Switzerland. **1980** *Daily Tel.* 19 Feb. 14/2 The following elections have been made: Magdalen: Official Tutorship as Tutor in Law, [etc.].

‖ **tutoyant** (tütwayaṅ), *a. rare.* [a. F. *tutoyant*, pres. pple. of *tutoyer* Tutoyer *v.*] Intimate, affectionate; suggesting a degree of familiarity sufficient to 'tutoyer'.

1899 'A. Leslie' *Some Players* 110 He turns facile and covert scorn upon the pretty tutoyant affection of happily wedded woman and man. **1975** *Times Lit. Suppl.* 14 Mar. 274/4 Many of her [sc. Lady Tennyson's] letters to her fiancé and husband have been destroyed, but there are enough affectionately *tutoyant* letters to her 'Ally' to show the nature of their correspondence.

Tutsi (tu·tsi). [Native name.] = *Watusi 1.

1950 Huntingford & Bell *E. Afr. Background* (ed. 2) 123/2 (Index), Tutsi. **1965**, etc. [see *Hutu]. **1972** *Times* 8 May 13/2 The Tutsi feudal regime was destroyed in massacres in Rwanda. **1976** *Daily Times* (Lagos) 3 Nov. 1/2 At the heart of the conflict and Burundi's general political malaise lay the fact that the Tutsi, tall cattle herders who make up only 14 per cent of the population, politically and economically dominate the Hutu who number 85 per cent of the population. **1979** 'D. Grant' *Olympic 5000* i. 27 Ochengwe was a Tutsi... In 1964..the Hutu massacred several thousand Tutsis, the traditional ruling caste of Rwanda.

tutti. (Earlier example of sense 'a passage or movement'.)

1816 G. F. Graham *Acct. First Edin. Festival* 105 The flowing and cantabile style of the subject is well contrasted with the spirited *tutti* beginning at the 28th bar.

tutti-frutti (tu·ti,fru·ti). [It., = all fruits.] **a.** A confection of mixed fruits; *spec.* a mixture of chopped preserved fruits, nuts, etc., used to flavour ice-cream; ice-cream so flavoured. **b.** (Tutti Frutti, Tutti-frutti.) A proprietary name for a chewing-gum with a mixed fruit flavouring. **c.** *attrib.* and *transf.*

1834 *Knickerbocker* IV. Sept. 232 *Tutti Frutti*, (all fruits) is the cognomen of an Italian ice, composed of, or rather flavored with, various fruits. **1876** M. J. Henderson *Cooking* 313 *Tutti Frutti.* When a rich vanilla cream is partly frozen, candied cherries, English currants, chopped raisins,..or any other candied fruits chopped rather fine, are added. **1885** *Official Gaz.* (U.S. Patent Office) 30 June 1564/1 Chewing-gum.—Adams and Sons, New York, N.Y... The words 'Tutti Frutti'. **1888** A. Randall-Diehl *Two Thousand Words* 210 *Tutti-frutti*, a compound of many kinds of sliced fruits mixed with sugar and alcohol, also the name of a chewing gum. **1898** A. M. Binstead *Pink 'Un & Pelican* v. 125 Most of my readers will very well remember old Bob Bignell, who kept the Argyll rooms, and, after the powers put a stop to the dancing and tutti-frutti business, turned the place into the cheery old Trocadero. **1916** *Daily Colonist* (Victoria, B.C.) 4 July 9/1 The pleasing tang of its mellow fruit flavors alone was always sufficient for you to prefer Adams Tutti frutti gum. **1924** *Official Gaz.* (U.S. Patent Office) 21 Oct. 478/1 American Chicle Company, Long Island City, N.Y... Tutti-frutti. Particular description of goods.—Chewing gum. Claims use since October 1882. **1951** *Good Housek. Home Encycl.* 516/1 *Tutti Frutti*: Add chopped pistachio nuts, cherries and angelica. **1974** D. Chantler *Man who followed in Front* ii. 60 'And look at them fuckin' ice cream machines,' he said. 'Tell the waiters..the Tutti-Frutti is off.' **1976** *Listener* 29 Jan. 121/3 Felix Mendelssohn and his Hawaiian Serenaders, and all those kitsch tutti-fruttis which drop like coconuts from a set of plastic palm trees.

‖ **tutti quanti** (tu·ti kwa·nti). [It.] Everyone, everything, all (of this, that, kind).

[**1671** Mme de Sévigné *Let.* 1 Apr. (1862) II. 135 M. et Mme de Duras, à qui j'ai fait vos compliments, MM. de Charost et de Montausier, et *tutti quanti*, vous les rendent au centuple. **1676** —— *Ibid.* 29 July IV. 545 Enfin *tutti quanti*: vous savez ce que c'est que de recevoir un mot de tout ce qu'on trouve en chemin.] **1772** Ld. Chesterfield *Let.* 10 Sept. (1932) VI. 2937, I hope you and *tutti quanti* are in a better plight. **1814** *Edin. Rev.* Jan. 403 All the heretical sects are active partisans of Passion, and furiously inimical to Reason, and to all his adherents, *tutti quanti*. **1864** J. A. Symonds *Let.* 6 July (1967) I. 488 If you come with Green & me you will have all the published Dramatists, tutti quanti, to read at pleasure. **1934** H. James *Art of Novel* xv. 274 Whether 'Daisy Miller',..'Julia Bride' and *tutti quanti* do in fact conform to any such admonition would be an issue by itself. **1948** J. Flanner in *New Yorker* 8 May 43/1 The idea..was to bring over from New York..Barrymore, Bankhead, Hayes, *e tutti quanti*.

tutu[1]. Add: **a.** (Earlier examples.)

1845 E. Meurant *Diary* 4 Oct. (typescript, Alex. Turnbull Library, Wellington) 24 Bullocks..had been eating the tutu bush wich [*sic*] is poisonous to cattle. **1849** [see *slash *v.* 1 c].

b. Phr. *to eat* (*one's*) *tutu* or *toot*, to become acclimatized, *spec.* to colonial life in New Zealand (see quots.). *N.Z. slang* (now *Obs.* exc. *Hist.*).

1857 R. B. Paul *Lett. from Canterbury* ii. 26 [The newly arrived settlers] passed..through the crisis of unreasonableness, false pride, and grumbling, which old settlers call 'eating their tutu'... The tutu, or 'toot',..is a native shrub the leaves of which may be eaten with safety by cattle gradually accustomed to its use, but are often fatal to newly-landed animals. **1889** [in Dict.]. **1941** Baker *N.Z. Slang* iii. 27 *To eat toot* was the pioneer way of describing the period during which new immigrants settled down to the cold facts of New Zealand life. More correctly the expression was *to eat tutu*..the poisonous plant. **1966** G. W. Turner *Eng. Lang. in Austral. & N.Z.* viii. 165 The early colonial phrase 'to eat one's tutu' meaning 'to become acclimatized to colonial life'.

Hence **tu·tued** *a.*, poisoned by eating tutu.

1874 A. Bathgate *Colonial Experiences* xv. 211 Flock-owners have sometimes to contend with a poisonous plant called the tutu (*Coriaria ruscifolia*), commonly pronounced toot... Those [sheep] feeding amongst it..are apt to be affected by it, or be, as the phrase is, 'tutued'. **1878** E. S. Elwell *Boy Colonists* 34 When they [sc. bullocks] were 'tutu'd' the only cures were either to bleed them or to put ammonia on the tip of the tongue.

tutu[2] (tu·tu). Also **tu-tu.** [a. F. *tutu*, childish alteration of *cucu*, dim. of *cul* Cul.] A ballet skirt made up of layers of stiff frills, reaching halfway between the knee and the ankle (*romantic tutu*) or very short and standing out from the legs (*classic tutu*). Also *attrib.*

1910 E. F. Spence *Our Stage & its Critics* ix. 196 She wished to exhibit what in technical slang is called *le tutu*, a term descriptive of the abbreviated costume and possessed also of a secondary meaning. **1913** A. E. Johnson *Russian Ballet* 56 Columbine..attired in a scanty *tu-tu*. **1934** A. L. Haskell *Balletomania* 26 An old-fashioned ballet for this old-fashioned tragedy of naked footlights and a dancer's *tutus*. **1947** N. Nicolaeva-Legat *Ballet Educ.* III. 49 To make a tutu skirt, the basque should first be cut. **1949** Chujoy & Manchester *Dance Encycl.* 486/1 The classic tutu reaches to a little above the knee, the romantic to the ankle. **1958** L. Gibbs *Gowns & Satyr's Legs* xii. 82 Four miniature ballet-girls, each poised gracefully on one toe and wearing a diminutive *tutu*. **1970** B. Cartland *We danced All Night* vii. 196 A snow-white figure in a fluffy *tutu*. **1980** 'M. Fonteyn' *Magic of Dance* 239 The soft, full ballet skirt Marie Taglioni had introduced climbed to just below the knee, then to mid-thigh. As it was shortened, it was made fuller and stood out more and more stiffly until it became the modern tutu.

Tuvaluan (tuvalū·ăn, tuvā·lūăn), *sb.* and *a.* [f. *Tuvalu* (see def.) + -an.] **A.** *sb.* A native or inhabitant of the Commonwealth State of Tuvalu, formerly the Ellice Islands, in the south Pacific. **B.** *adj.* Of or pertaining to Tuvalu.

1975 *Pacific Islands Monthly* Dec. 15/2 The Tuvaluans, known before October 1 as the Ellice Islanders. **1978** *Daily Mirror* 20 Feb. 6/4 In 1975 the Polynesian Ellice Islanders voted to part peacefully from the Micronesians on the Gilberts... Today the Tuvaluans rely on coconuts for food and exports. **1978** *Daily Tel.* 30 Sept. 17/6 On Sunday, Tuvalu will become a constitutional monarchy, with the Queen as Head of State represented by a Tuvaluan Governor-General.

tux (tɒks). Also **Tux.** *U.S.* colloq. abbrev. of *tuxedo.

1922 S. Lewis *Babbitt* i. 11 Everyone knows I can put on as expensive a Tux. as anybody else. **1951** W. C. Williams *Autobiogr.* xii. 61, I agreed to take her to a dance and showed up in a tux. **1974** K. Millett *Flying* (1975) III. 322 Daddy doing his tux. First the black tie. Next the studs.

tuxedo (tɒksī·do). orig. and chiefly *U.S.* Also **Tuxedo.** [Named from *Tuxedo* Park, N.Y., where the jacket was first introduced at the country club in 1886.] **1.** In full *tuxedo coat*, *jacket*. A short jacket without tails, for formal wear; a dinner-jacket.

1889 *Sartorial Art Jrnl.* Aug. 18/1 The low-roll, silk-faced sack, variously called the 'Cowes' coat, the 'Tuxedo' coat, and the Dress Sack, is undoubtedly popular. *Ibid.* Dec. 97/1 The 'Tuxedo' or dress sack is steadily growing in favor for dress negligee purposes. **1900** Ade *Fables in Slang* 130 A jimmy little tuxedo. **1925** H. L. Foster *Trop. Tramp with Tourists* 333 Dress coats and tuxedo jackets were removed. **1931** *Times Lit. Suppl.* 12 Nov. 888/4 Bert..is to pose in public as a successful lover,..in the traditional Tuxedo, with the ribbon of the Garter pinned across his bosom as a Right Honourable. **a 1944** K. Douglas *Alamein to Zem Zem* (1946) vi. 47 Beneath the oil stains their white tuxedos and seductive dresses shone. **1950** J. Vedey *Band Leaders* p. xiii, The budding maestro must make up his mind whether he will present his band in orthodox tuxedos or in some distinctive style of uniform. **1971** 'A. Burgess' *MF* viii. 93 He was in a white tuxedo with black floppy tie.

2. Special combinations. † **tuxedo net**, a kind of net veiling (*obs.*); **tuxedo sofa**, a sofa of a style having back and arms the same height; also *tuxedo-style sofa*; **tuxedo (trade) unionism** (see quot. 1965).

1895 *Montgomery Ward Catal.* Spring & Summer 79/1 Black silk tuxedo net veiling, 13 inches wide. **c 1900** in *Amer. Mail Order Fashions* (1961) 19, 28 in. tuxedo net. Entirely new pattern in black, white and navy. **1961** Webster, tuxedo sofa. **1965** *Economist* 6 Feb. 537/2 Critics accuse him of having become aloof from the rank and file of his union and of practising what they call 'tuxedo trade unionism'... As a 'labour statesman' he associates freely with leading officials in both the government and the steel companies. **1965** *Amer. Home* Mar. 43/1 (*caption*) Curved Tuxedo sofa..illustrates the graceful, softer look for 1965 Modern. **1972** *N.Y. Times* 29 Oct. 85/5 (Advt.), Tuxedo-style sofa. **1977** *Time* 17 Jan. 48/2 Sadlowski calls McBride's style 'tuxedo unionism'. **1977** *Chicago Tribune* 2 Oct. I. 45 (Advt.), An elegant and timeless tuxedo sofa with loose pillows and arm bolsters in a rich champagne beige velvet.

Hence **tuxe·doed** *a.*, wearing a tuxedo.

1934 J. T. Farrell *Young Manhood Studs Lonigan* xviii. 284 They..passed the tuxedoed orchestra, which was playing wildly on a dais. **1973** E. McGirr *Bardel's Murder* ii. 45 Near him was Jack Prat, tuxedoed, urbane.

tuzzy-muzzy, *sb.* (*a.*) Restrict *Obs.* exc. *dial.* to other senses and add: **a.** Revived in 20th. cent., usu. in form *tussie-mussie*.

1958 J. G. Conway *Encycl. Flower Arrangement* vi. 104 Tuzzy-muzzy, a small, neat arrangement with flowers in repeated circular lines, edged with foliage or lace. **1960** V. Williams *Walk Egypt* 180 He gave her a hand bouquet of lemon-colored roses which smelled like lemon too, a sour-sweetness that matched the morning. He said, 'A tussie-mussie... I figured you belonged to have something, a day like this.' **1968** *Herb Grower Mag.* XXI. I. 1 From a little Tussie-mussie to a formal garden at the White House, the warmth of herbs reaches across the miles. **1973** *New Yorker* 7 Apr. 126/3 (Advt.), Tussie-mussies... Delightful bouquets of fragrance herbs, cooking herbs,..and medicinal herbs. A full-color limited edition portfolio.

TV (tīvī·). Abbrev. of *television.

1948 *Fortune* May 82/1 It is not where TV has gone,..but the pace at which it is going that causes all the excitement. **1948** *Time* 25 Oct. 82/2 TV is not only colorblind; its eye is astigmatic. **1957** *Times Lit. Suppl.* 29 Nov. 722/3 He knows 'most of the tricks of popular journalism, films, radio, TV'. **1962** V. Nabokov *Pale Fire* 35 The gauzy mockingbird..Returning to her perch —the new TV. **1964** M. Argyle *Psychol. & Social Probl.* iv. 47 Is the effect of showing violence in films or on TV to arouse or to satiate aggressive feelings? **1977** *R.A.F. News* 22 June–5 July 2/3 (*heading*) TVs for hospital.

b. Freq. *attrib.* and *Comb.* **TV dinner**, a prepared frozen meal that needs only to be heated and is suitable for eating while watching television (formerly a proprietary name in the U.S.).

1948 *Fortune* May 83/1 The average capital investment for a TV station is about $375,000. **1950** 'A. Gilbert' *Is She Dead Too?* vii. 135 That's the best of livin' in a village. It's like livin' on a TV screen. **1953** A. Huxley *Let.* 25 Jan. (1969) 663 One must read 100% in order to be able to leave out 99%, as has to be done in this medium and for a TV audience. **1954** TV aerial [see *show-off *sb.* (*a.*) d]. **1954** *Official Gaz.* (U.S. Patent Office) 27 July 735/2 C. A. Swanson & Sons, Omaha, Nebr... TV Dinner. .. For frozen Turkey Dinner, Including Turkey, Dressing, Giblet Gravy, Sweet Potatoes, and Green Peas. **1955** *Radio Times* 22 Apr. 39/3 Gilbert Harding's outspoken autobiography..the frank, personal story of the popular Radio and T.V. personality. **1955** *Times* 9 Aug. 4/5 Mr. Leon Goodman, chairman of T.V. Commercials, Ltd. **1955** H. Kurnitz *Invasion of Privacy* (1956) iii. 26 He's been in two flops on Broadway and in five T.V. shows. **1958** *New Statesman* 5 July 1/2 The argument about the third TV network. **1961** P. Frankau *Pen to Paper* 21 '*That's it!*' as the T.V. announcer snaps at the winning point of the tennis match. **1961** *Times* 18 May 9/7 One TV antenna. **1962** L. Deighton *Ipcress File* xvi. 92 A girl featured weekly in a badly made TV series. **1962** N. Streatfeild *Apple Bough* xv. 216 They went all the way to the T/V studio on a bus. **1963** H. A. Hargreaves in *New Worlds Science Fiction* Dec. 62 It would have to be a TV satellite. **1964** L. Deighton *Funeral in Berlin* vii. 51 Spectacles swung the TV receiver around so that I couldn't see it. *Ibid.* x. 66 The very same people who made the great little TV film..paid for by the TV company, N.B.C. **1964** J. Masters *Trial at Monomoy* x. 292 TV dinners, and everything quick frozen, dehydrated and prepacked. **1964** M. McLuhan *Understanding Media* (1967) I. i. 19 If the TV tube fires the right ammunition at the right people it is good. *Ibid.* II. xiii. 140 The combined radio and TV channels in the United States. *Ibid.* II. xvii. 176 The cartoon is clue to understanding the TV image. *Ibid.* II. xxxi. 334 The TV camera does not have a built-in angle of vision like the movie camera. **1969** TV licence [see *radio licence* s.v. *radio *sb.* 7]. **1970** Koenig & Dixon *Children are Watching* (1971) i. 6 Inside the stove glinted four TV dinners..nested in silver-aluminum trays,..identical portions of gray pot roast, dinner roll, wrinkled peas and rigid mashed potatoes under foil. *Ibid.* 7 These dinners..she wheeled across the hall into the TV room. **1974** *Guardian* 23 Mar. 10/6 It was dense, thick, primeval onion soup. But it was not a TV play. **1974** *Broadcast* 11 Nov. 16/3 The bulk of the population..needs effortless entertainment... The theatre mogul, the film-maker, and now the TV programme producer, have found the truth of this platitude. **1975** D. Lodge *Changing Places* ii. 49 He filled the micro-refrigerator with TV dinners. **1976** B. Bova *Multiple Man* xix. 209 TV crews were rolling their cameras in. **1976** *Liverpool Echo* 7 Dec. 3/1 The Wirral TV mast saga. For more than two years the B.B.C. have been seeking to erect a 200 ft. mast on Wirral.

tvorog (tvǭ·răk). [a. Russ. *tvórog*.] A soft Russian cheese similar to cottage or curd cheese.

1918 Doane & Lawson *Varieties of Cheese* (U.S. Dept. Agric. Bull. No. 608) 59 Tworog. This is a sour-milk cheese made in Russia..on a large scale by Russian farmers. **1960** K. Davydova *Good Food from Russia* 107 The following recipes have as their main ingredient, tvorog, the Russian version of cottage cheese. *Ibid.* 109 Tvorog Pudding with candied peel. **1973** S. Skipwith *Eat Russian* i. 16 Cottage cheese and curd cheese are the nearest equivalent to the Russian Tvorog which is dry but not 'cheesy'. **1981** J. Trenhaile *Kyril* xxiv. 180 He ate only a biscuit smothered with *tvorog*, the stodgy cottage cheese of the peasants. **1982** L. Chamberlain *Food & Cooking of Russia* (1983) 245 By Good Friday several pounds of *tvorog* (curd cheese) would have been sitting for at least 24 hours under a wooden press to extract..the last drops of whey.

twa(a)-gras(s) (twā·grɑs). Also **dwa-, toa-grass.** [f. *umTwa* name used by Bushmen + Grass *sb.*[1]] Any of several species of the genus *Aristida*, which includes several tall feathery grasses native to southern Africa.

1857 A. Wyley *Rep. Min. Struct. Namaqualand* 44 Twa-grass..grows from two to three feet in height, from a small bushy base... When it is green, oxen, horses, and sheep all thrive upon it. **1896** R. Wallace *Farming Industries Cape Colony* v. 100 'Twa-gras' is the most abundant grass..in the Kalahari region. **1897** Edmonds & Marloth *Elem. Bot. S. Afr.* xvii. 185 The Dwa-grass or Toa-grass. **1920** F. C. Cornell *Glamour of Prospecting* xv. 246 We lit fires of dry *toa* grass. **1946** L. G. Green *So Few are Free* I. x. 137, I can hear the 'twaa grass' crackling in the wind. **1949** K. L. Simms *Sun-Drenched Veld* i. 10 Because they [sc. plains] are thickly peppered with drought-resisting bushes and tufts of toa grass, they constitute wonderful ranching country. **1974** *Stand. Encycl. S. Afr.* X. 367/1 In the times of occasional rain Bushman grass and twa grass..will spring up.

Twaddell. (Earlier example.)

1853 Ure *Dict. Arts* (ed. 4) II. 828 The patentee employs diluted sulphuric acid, at 105° Twaddle.

twaddlesome, *a.* (in Dict. s.v. Twaddle *sb.* (*a.*)). Delete *nonce-wd.* and add later examples.

1892 G. Meredith *Let.* 25 Apr. (1970) II. 1080 *Dorothy Penrose*, an enormous bulk, is uninterruptedly twaddlesome. **1966** K. S. Sorabji in 'H. MacDiarmid' *Company I've Kept* ii. 65 The twaddlesome sentimentalities about trusting the ultimate judgement and good sense of the public.

twain, *numeral a.* Add: † **2. c.** *U.S. Naut.* Two fathoms. Esp. in phr. *mark twain*, the two fathom mark on a sounding-line. Cf. Mark *sb.*[1] 12 b. *Obs.*

1799 J. W. Russell in R. D. Paine *Romance Old Time Shipmaster* (1907) iii. 43 The man in the chains suddenly sung out 'quarter less twain', and we instantly struck. **1863** 'Mark Twain' in A. B. Paine *Mark Twain* (1912) I. xl. 221, I want to sign my articles..'Mark Twain'. It is an old river term, a leads-man's call, signifying two fathoms—twelve feet. **1947** E. M. Mack *Mark Twain in Nevada* xv. 228 How many times when he was on the River had he heard the leadsman..call out, 'By the mark, twain!'

Twainian (twē[1]·niăn), *a.* [f. the name of 'Mark *Twain*' (S. L. Clemens), American writer (1835–1910) + -ian.] Of, pertaining to, or characteristic of 'Mark Twain' or his work.

1938 *Times Lit. Suppl.* 27 Aug. 551/4 When stimulated by the potentialities of human sacrifice or the vapourings of a politician, it [*sc.* the pen] moves to a Twainian climax. **1968** *Word Study* Feb. 2/1 Here is such a Twainian reference, appearing in the late novel *The American Claimant*. **1977** *Time* 28 Feb. 53/1 They contain their share of Twainian 'stretchers', or exaggerations.

Twana (twā·nă). [ad. Twana *tuwáduxq*, in an earlier pronunciation that had *n* for *d*.] **1.** A Salishan people of western Washington; a member of this people. Also *attrib.* Cf. *Salish, *Salishan.

1838 J. Douglas *Let.* 18 Oct. in E. E. Rich *McLoughlin's Fort Vancouver Lett. 1825–38* (1941) 262 [*in a list*] Too a nook. **1889** M. Eells in *Ann. Rep. Board of Regents Smithsonian Inst. 1887* 605 The name of the Twanas is spelled Too-au-hooch, in their treaty. The Klallams pronounce it Tu-an'-hu. The Twanas say Tu-ád-hu... These various pronunciations have been shortened into Twana, now used in all governmental reports. It is said to mean a portage, and to be derived from the portage between the head of Hood's Canal and the main waters of the Sound, where the Indian, by carrying his canoe 3 miles, avoids rowing around a peninsula 50 miles long. **1960** W. W. Elmendorf in *Research Stud.* (Washington State Univ.) Sept. Monogr. Suppl. No. 2. 281 Few non-Twana cared to learn the Twana language. **1978** *Amer. Poetry Rev.* Sept./Oct. 19/1 The Twana of Puget Sound maintained that if a woman in seclusion touched her head with her fingers, it would at once rot away.

2. The language of this people. Also *attrib.*

1886 M. Eells *Ten Yrs. Missionary Work* vi. 34 The Twana language..is said to be so difficult to learn that no intelligent Indian advised me to learn it. *Ibid.* 37 We have often sung in English, Chinook jargon, Twana, and Nisgually, on the same Sabbath. **1960** W. W. Elmendorf in *Research Stud.* (Washington State Univ.) Sept. Monogr. Suppl. No. 2. 280 Twana seems a unique case among all Coast Salishan tongues in its features of areally simple distribution restricted to a single inlet and its fresh-water drainage, and very slight heterogeneity of dialects. **1971** *Language* XLVII. 844 Twana (Salish, Puget Sound area) uses duplication of the root vowel.

† **twang** (twæŋ), *sb.*[4] *Austral. slang. Obs.* [perh. back-formation from Twankay.] Opium.

1898 *Bulletin* (Sydney) 1 Oct. 14/3 A few more W.Q. [*sc.* West Queensland] slang words... Opium 'twang', a Chinaman a 'canary' [etc.]. **1910** O'Brien & Stephens *Material for Dict. Austral. Slang 1900–10* (typescript), *Twang*, opium. **1945** T. Ronan *Strangers on Ophir* (1966) 68 The honest Chinese limits himself to his one pipe of 'Twang' per night. **1966** Baker *Austral. Lang.* (ed. 2) 157 Opium was once known as *twang* in bush slang (the later use of *treacle* was noted earlier).

twanka-pang, twank-a-pank. An imitation of the sound of a banjo or guitar.

1929 J. B. Priestley *Good Companions* I. vi. 201 This banjo..was being *played*. The night retreated hastily before its impudent twanka-pang, twanka-pang. **1980** M. Gilbert *Death of Favourite Girl* i. 6 Twank-a-pank... The guitar quickened to a livelier tempo.

twat. Delete † *Obs.* and add pronunc. (twǫt). Also **twot(t. 1.** (Later examples.)

1919 E. E. Cummings *Let.* 18 Aug. (1969) 61 On Tuesday an Uhlan To her twat put his tool in. **1934** H. Miller *Tropic of Cancer* 55 A man with something between his legs that could..make her grab that bushy twat of hers with both hands and rub it joyfully. **1959** N. Mailer *Advts. for Myself* (1961) 101 The clothes off, the guards are driving them into the other room, and smack their hands on skinny flesh and bony flesh, it's bag a tittie and snatch a twot. **1970** G. Greer *Female Eunuch* 39 No woman wants to find out that she has a twat like a horse-collar. **1973** P. White *Eye of Storm* iii. 137 This young thing with the swinging hair and partially revealed twat.

2. A term of vulgar abuse. Cf. *TWIT *sb.*[1] 2 b and *CUNT 2.

1929 F. MANNING *Middle Parts of Fortune* II. xv. 383 Yes, they let a bloody twat like 'im off. **1933** M. LOWRY *Ultramarine* i. 16 He can't help it if you're just a bloody, senseless twat. **1958** H. WILLIAMSON *Love & Loveless* i. 27 Looked a proper twott to me. **1969** P. ROTH *Portnoy's Complaint* 211 Here comes another dumb and stupid remark out of that brainless twat. **1978** J. UPDIKE *Coup* (1979) iii. 123 Divorce me and you'll have a slot for this new twat, what's her name. **1979** R. FIENNES *Hell on Ice* ix. 134 Sterns not prows, you twot.

3. *U.S. dial.* The buttocks.

1950 *Publ. Amer. Dial. Soc.* XIII. 20 *Twat*,..the buttocks. **1964** M. KELLY *March to Gallows* xii. 132, I could tell her what to do with her twat if she's frightened to sit on it.

tweak, *v.* Add: **4.** *Cricket. colloq.* Of a bowler: to impart spin to (the ball).

1958 D. BRADMAN *Art of Cricket* 94/1 My pal.. R. W. V. Robins, tweaked his leg breaks so hard that he left the ground altogether with both feet.

5. To make fine adjustments to (a mechanism).

1966 *Punch* 16 Feb. 233/1 He has been running a Morris 1100 'tweaked so it'll do nearly 100'. **1971** *Daily Tel.* 13 Oct. 11/1 The three-litre V6 engines..have been 'tweaked' to produce eight per cent. more power. **1978** *Gramophone* May 1960/1 It was possible to improve its performance very considerably by 'tweaking' the internal pre-set controls.

tweaking *vbl. sb.*: also in senses *4 and *5 of the vb.

1949 E. M. WELLINGS in *Boys' Bk. of Cricket* 78/1 Most right-handed off-break bowlers do their spinning largely with the forefinger. Personally I do not even have that finger resting on the ball when bowling an off-break. My 'tweaking' is done by the middle finger. **1975** *Drive* Nov.–Dec. 90/3 These engine hiccups are the result of carburettor tweaking that has been necessary to bring cars in line with current exhaust emission regulations. **1983** *Australian Personal Computer* Sept. 123/1 Most parallel [daisy-wheel] printers should work with a little tweaking.

tweaker. Add: **2.** *Cricket. colloq.* **a.** A bowler who spins or 'tweaks' the ball, esp. a left-arm leg-spinner. **b.** A ball bowled with spin.

1948 C. SLY *How to bowl them Out* ii. 17 The slow bowler should have great patience... There should also be an element of precision in his brain—a slapdash fellow will never make a good 'tweaker'. **1956** R. ALSTON *Test Commentary* 12 Johnson's insidious 'floaters' were likely to be more penetrative than Ring's 'tweakers'. **1961** *Times* 12 June 3/6 Lancing's leg tweaker came on too late. **1976** J. SNOW *Cricket Rebel* 40 When I came into the side Ron Bell, a left-arm spinner, was the only genuine tweaker of the ball on the staff. **1982** *Guardian* 26 July 19/3 The fragile left-arm tweakers of Steele.

twee (twī), *a.* (and *sb.*) *colloq.* [f. *tweet*, an infantile pronunciation of *sweet*.] **1.** Originally: 'sweet', dainty, chic. Now only in depreciatory use: affectedly dainty or quaint; over-nice, over-refined, precious, mawkish.

1905 *Punch* 8 Mar. 178/1, 'I call him perfectly twee!' persisted Phyllis. **1917** M. T. HAINSSELIN *Grand Fleet Days* xv. 91 *Girl*: Oh, here's another little gun; isn't it a darling! Isn't it just too twee for words! **1947** E. HYAMS *William Medium* viii. 164 'Isn't he *twee*!' said Mary, and pinched his cheek. **1956** G. DURRELL *Drunken Forest* x. 193 'What twee individuals?' 'Those knowledgeable sentimentalists who are forever telling me that it's cruel to lock up the poor wild creatures in little wooden boxes.' **1962** *Observer* 25 Mar. 25/3 She has a small and, it must be said, pretty twee cottage. **1967** E. SHORT *Embroidery & Fabric Collage* iv. 102 The best of our designers who have abandoned the rather 'twee' decorative type of embroidered picture. **1973** G. ROBYNS *Wimbledon* xxix. 192 There is..a twee Arcadian outdoor studio complete with white trellis and plastic flowers. **1983** *Listener* 21 July 33/1 Mike Nichols's thriller-fantasy about dolphins should be as nauseatingly twee as the worst Disney—but it isn't.

2. *ellipt.* as *sb.*

1957 *Daily Mail* 29 Oct. 12/8, I cannot understand why television's handling of fashion in evening programmes has never got past the twee.

Hence **twee·ly** *adv.*, in a twee manner; **twee·ness.**

1958 *Spectator* 2 May 565/3 He manages..to resist the temptation to play up the tweeness and tell the English what they expect to hear. **1962** *Guardian* 12 July 7/1 The..highly commendable idea of importing bulk grains ..and passing them, tweely packaged, to cage-bird fanciers. **1973** *Observer* 18 Nov. 36/2 'And no doubt, if the bride is awake and has peeped out through the curtains..,' he speculated tweely. **1981** *Radio Times* 7–13 Nov. 21/2 The word 'herbs'..seems to have become associated with tweeness.

tweed. Add: **b.** *tweed mill* (example); *tweed-jacketed*, *-skirted* adjs.

1928 A. HUXLEY *Point Counter Point* iii. 46 This huge bent old man, pipe-smoking and tweed-jacketed. **1949** C. GRAVES *Ireland Revisited* x. 156 The only really interesting person..was Robert Miller, to whose tweed mill he directed us. **1957** C. MACINNES *City of Spades* II. v. 138 The recalcitrant bowler-hatted or tweed-skirted natives.

tweeded (twī·dėd), *a.* [f. TWEED + -ED[2].] Clad in tweed.

1921 G. B. SHAW *Back to Methuselah* II. 38 *The tweeded gentleman.* (Coming in very slowly.) I have something on my mind. **1949** N. MITFORD *Love in Cold Climate* II. v. 241 Why are the English roads always so covered with these tweeded stumpers? **1979** M. RUSSELL *Touchdown* i. 14 Her companion..was nursing on a tweeded lap..a Thermos flask.

‖ **Tweede Nuwejaar** (twiᵊ·də nǖvəyā·ɪ). Also **Tweede Nuwe Jaar, Tweedenuwejaar.** [Afrikaans, lit. = second New Year.] The second of January, a public holiday in Cape Province, celebrated especially by the Black population.

1947 *Cape Times* 30 Dec. 14 The Coons are ready for the New Year—and tweede nuwejaar—that extra holiday which is taken only in the Cape. **1953** *Ibid.* 1 Jan. 1/6 There will be no issue of the *Cape Times* to-morrow, January 2 (*Tweedenuwejaar*). **1959** *Ibid.* 3 Jan. 7/3 The coons have enjoyed fine weather for New Year and the Tweede Nuwe Jaar and we have enjoyed them too. **1978** *Argus* (Cape Town) 29 Dec. 3 Normal editions of The Argus will be published on Tuesday January 2, Tweede Nuwejaar.

tweedle (twī·d'l), *sb.* *Criminals' slang.* [See *TWEEDLE *v.*[2]] A counterfeit ring; hence, a swindle (involving counterfeit goods); a 'fiddle', 'racket'.

1890 BARRÈRE & LELAND *Dict. Slang* II. 383/2 *Tweedle* (thieves), a spurious ring, used to swindle jewellers and pawn brokers. **1938** F. D. SHARPE *Sharpe of Flying Squad* xxvii. 275 One of the oldest methods of crime is the Tweedle... The Tweedler spots a ring worth a lot of money in a jeweller's shop and goes..to have an exact.. replica made. He goes in..and when the assistant isn't looking very carefully substitutes the fake for the real thing. **1959** J. GOSLING *Ghost Squad* ix. 122 A bloke's tried to pull a tweedle on me with a load of jargoons. **1963** H. SLESAR *Bridge of Lions* (1964) 2 'What's 'is tweedle?' the youth asked suspiciously. **1982** *New Society* 2 Dec. 382/3 Then it was back to the shop for the 'tweedle'—for the switch.

tweedle, *v.*[2] *Criminals' slang.* [prob. f. *tweedle*, var. TWIDDLE *v.*[1], in sense 2 b of the latter.] *trans.* To counterfeit, swindle, practise a confidence trick on. Hence **twee·dler,** one who tweedles; **twee·dling** *vbl. sb.*

1925 E. JERVIS *Twenty-Five Years in Six Prisons* i. 17 Some of the boys go 'tweedling'. I am afraid that the knowing author of the 'cracking-a-crib' book would be flummoxed by 'tweedling'. *Ibid.* 18 The chain is handed over in tissue-paper, and the 'tweedler' departs. **1959** J. GOSLING *Ghost Squad* ix. 114 The tweedler will flog you sawdust cigarettes or dummy diamond rings. **1975** P. G. WINSLOW *Death of Angel* iv. 94 'Tweedling'—small con jobs, mostly against the old and weak. *Ibid.* ix. 195 He was always blubbing to Joss, a tweedler like that has to make himself big to a woman. **1980** 'D. KAVANAGH' *Duffy* ii. 30 Big bad villain. Girls, smokes, bit of smack, mossing, tweedling.

tweedy (twī·di), *a.* [f. TWEED + -Y[1].] **a.** Consisting of or relating to tweed cloth. **b.** Characterized by or given to wearing tweeds. **c.** *fig.* Characteristic of those (e.g. the country gentry) who wear tweeds; heartily informal, exclusively clannish, etc.

1912 R. BROUGHTON *Between Two Stools* xiv. 107 Iris stood before them in tweedy brevity of skirt and pertness of tam-o-shanter. **1928** *Sunday Dispatch* 5 Aug. 17/3 In Scottish country houses you rarely get away from the tweedy atmosphere until the afternoon has worn on. **1930** J. B. PRIESTLEY *Good Companions* II. vii. 445 An angel of a woman, very erect, y'know, and tweedy, and straight out of the Old Moated Grange. **1946** G. D. KLINGOPULOS in *Scrutiny* XIV. II. 144 The only excuse for noticing this tweedy sequel [*sc.* V. Sackville-West's *The Garden*]..is that we need to be reminded..that, in Courses of English, 'The Land' is still too often the substitute for modern poetry. **1949** L. P. HARTLEY *Boat* 80 She nodded very perceptibly in the direction of the tweedy group who were talking to each other as members of the same family do. **1978** I. MURDOCH *Sea* 164 He is a big stout man, always dressed..in tweedy suits with waistcoats. **1980** *Daily Tel.* 20 Mar. 14/5 Miss Foster, who seems to have been a perfectly splendid, large and tweedy lady.

Hence **twee·dily** *adv.*, **twee·diness.**

1964 C. P. SNOW *Corridors of Power* xxiv. 195 Roger.. lolloped tweedily along between them. **1965** *Listener* 27 May 788/1 The exotic tweediness of little Phyllis Benton, so despised by Mr Amis. **1978** R. HILL *Pinch of Snuff* x. 100 He would have classified her as genuine English county with a good seat but not erring on the side of tweediness. **1980** 'A. SKINNER' *Mind's Eye* i. 10 An old James Bond movie relayed rather tweedily from the mainland.

tweek (twīk). (*Radio.* [Echoic.] A type of whistler which is heard as a short, high-pitched chirruping noise.

1933 BURTON & BOARDMAN in *Proc. IRE* XXI. 1479 Two varieties of distinct musical atmospherics have been observed and given the onomatopœic names 'swish' and 'tweek'. **1953** *Phil. Trans. R. Soc.* A. CCXLVI. 114 The 'short whistler', 'tweek', or 'chink'..is the short (about 20 ms) musical tone produced by repeated reflexion between the earth and the ionosphere of the waves from a distant lightning flash. **1981** *Jrnl. Atmospheric & Terrestrial Physics* XLIII. 1271 The appearance of tweeks on whistler sonograms has been discussed in terms of VLF wave propagation through the land-sea and ionospheric waveguide.

tweely, tweeness: see *TWEE *a.*

tweeter (twī·təɪ). [f. *tweet* vb. s.v. TWEET *sb.* and *int.* + -ER[1].] A small loudspeaker designed to reproduce accurately high-frequency sounds whilst being relatively unresponsive to those of lower frequency. Cf. *SQUAWKER, *WOOFER.

1934 *Nature* 25 Aug. 294/1 The extension of the range of acoustic fidelity permitted by the 'tweeter'—the high frequency auxiliary speaker—is a boon. **1952** *Electronic Engin.* XXIV. 583/3 The..Tweeter Unit is of the moving coil pressure type and is similar to that used in the 10in. and 12in. concentric Duplex loudspeakers. **1957** *Times* 3 May 13/4 Bearded young men talk of Beethoven in terms of baffles and tweeters. **1959, 1975** [see *SQUAWKER 3]. **1982** *Hi-Fi Answers* Oct. 77/1 These drive units are pretty unusual beasts, and the NS1000M sports two: a 30mm dome tweeter and a 8·8cm dome midrange unit.

tweeze (twīz), *v.* [Back-formation from TWEEZERS *sb. pl.*] *trans.* To pull out (hair) with tweezers. Also, to pull as with tweezers. Hence **twee·zing** *vbl. sb.*

1932 V. WOOLF *Common Reader* 2nd Ser. 153 He.. tweezed out hairs with a silver tweezer. **1956** *Sun* (Baltimore) 13 June B 12/4 Before tweezing, apply to brow area a soft cloth wrung out in hot water. **1968** B. HINES *Kestrel for Knave* 29 So he tweezed the lashes between his finger and thumb and drew the lid down. **1979** M. MCMULLEN *But Nellie was so Nice* I. i. 9 The mouth was scarlet, the brows tweezed to an almost invisible line. **1982** *Chicago Sun-Times* 20 Nov. 19 When Gordon finished, makeup artist Gloria Percival tweezed Schaefer's brows.

twelfth, *a.* Add: **1. c.** (Earlier example.)

1816 SCOTT *Antiquary* III. ix. 196 'I must be prepared for Lord Glenallan's moors on the twelfth, sir,' said M'Intyre. 'Ah, Hector! Thy great *chasse*, as the French call it.'

d. *ellipt.* The 12th of July, celebrated by Protestants in Northern Ireland as the anniversary of the Battle of the Boyne (1 July (Old Style) 1690) at which William III defeated James II.

1896 M. HAMILTON *Across an Ulster Bog* vi. 57 The greatest excitements of her life—next always to 'the twelfth'—had been occasional Methodist or Plymouth Brethren meetings. **1936** *Ann. Reg. 1935* 115 The Orange celebrations of 'The Twelfth' were on a bigger scale than ever, but the atmosphere was highly charged, and in the evening there were serious riots. **1957** *Belfast News-Letter* 2 July 6 (*heading*) Record 'Twelfth' parades expected. **1978** D. MURPHY *Place Apart* xiii. 269 The Twelfth festivities commemorate the victory at the Boyne.

4. *twelfth man* (Cricket), a twelfth player selected as reserve to the team of eleven.

1876 *Haygarth's Scores* VI. 20 Mr. E. Arkwright..was first choice out of the [Harrow] Eleven, or 'twelfth man' in 1858. **1928** J. BUCHAN *Runagates Club* vii. 195, I saw you play at Lord's. I was twelfth man for Harrow that year. **1976** J. SNOW *Cricket Rebel* 26 There were just a couple of minutes to the off when the two twelfth men appeared on the field, the horses lined up as we gathered round the radio.

twell (twel), *prep.* and *conj.* Also **twel.** U.S. dial. and Black English var. of TILL *prep.*, *conj.*, *adv.*

1837 A. SHERWOOD *Gazetteer Georgia* (ed. 3) 72 *Twell*, for till;—twell night—twell next week. **1893** H. A. SHANDS *Some Peculiarities of Speech in Mississippi* 65 *Twel* (twel). Negro for *till*. **1901** W. CHURCHILL *Crisis* I. iv. 38 Dis ole woman'll wuk fo' you twell de flesh drops off'n her fingers, suh. **1938** W. FAULKNER *Unvanquished* vi. 229 Uncle Cash that druv the Benbow carriage twell he run off with the Yankees two years ago.

twelve, *numeral a.* and *sb.* Add: **B. II.** *sb.* **3. c.** A flower-pot eleven inches in width, of which there were twelve in a cast.

1802, etc. [see *SIXTEEN *sb.* 5]. **1852** G. W. JOHNSON *Cottage Gardeners' Dict.* 392/2 Eleven-inch..[Old name] 12s.

III. c. twelve-note, -tone *attrib. Mus.*, (of the technique of musical composition developed by Arnold Schœnberg (1874–1951) using the twelve notes of the chromatic scale so that none is dominant, as opposed to basing composition on the seven notes of the diatonic scale; cf. *note-row*, *-series* s.v. *NOTE *sb.*[2] 21, *SERIAL *a.* h, *tone-row* s.v. *TONE *sb.* 11; hence **twelve-toner,** a composer employing the twelve-tone technique.

1928 C. GRAY *Hist. Music* vi. 96 A reaching out towards the chromatic or twelve-note scale of to-day. **1959** *Times Lit. Suppl.* 16 Oct. 588/5 The most interesting letters are those in which Schoenberg speaks about his own work and his theory of composition. There is one to the Viennese composer Josef Hauer, in December, 1923, in which are discussed the rival claims of the two composers to have invented the twelve-note system. **1975** *Gramophone* Jan. 1329/3 Moses is an uncompromisingly twelve-note composition. [**1923** A. SCHOENBERG *Let.* 1 Dec. in *Briefe* (1958) 108 Mir handelt es sich ausgesprochen dabei um gar keine anderen Theorien, als um die Methoden der 'Komposition mit 12 Tönen', wie ich das—nach vielen Irrtümern und Abschweifungen—heute (hoffentlich endgültig) nenne.] **1926** *Mod. Music* Mar.–Apr. 6 He (*sc.* Schoenberg), too, is convinced that no tone of the twelve tone system should dominate and that the new structural elements should be sought in sequence of twelve

tones. **1956** AUDEN & KALLMAN *Magic Flute* (1957) 58 A *Geist* whose music was composed from *Angst*, at International Festivals enjoys An equal status with the Twelve-Tone Boys. **1980** *Times* 4 Sept. 12/5 There was an almost missionary zeal in the Schoenberg circle to spread the Twelve-Tone gospel of the master. **1966** N. ROREM *Paris Diary* II. 215, I despair of twelve-toners: they have lost the need for pleasure. **1977** Y. MENUHIN *Unfinished Journey* viii. 165 Bartók pours them [*sc.* chromatic sequences] out with a lavishness of invention which the twelve-toner, working away with his slide rule, will never know.

Twelver[2] (twe·lvəɪ). *Islam.* [f. TWELVE + -ER[1].] A member of the larger of the two Shiah sects (the 'Twelvers' and the 'Seveners'), a follower of the twelve Imams or prophets (cf. IMAM 2 b, SHIAH).

1876 R. D. OSBORN *Islam under Arabs* II. i. 167 The Ismailiens, like the Twelvers, make profession of an exclusive attachment to Ali and his descendants. **1934** R. STROTHMANN in *Encycl. Islām* IV. 353/2 We shall consider the main branch somewhat more fully, the Imāmīs or 'Twelvers'. **1979** *Sunday Tel.* (Colour Suppl.) 27 May 25/4 In Iran the Shias are divided essentially into two sects, the Twelvers and the Seveners.

twentieth, *a.* Add: **1. d.** (Earlier example.)

1888 E. BELLAMY *Looking Backwards* xxvii. 432 My thoughts made better music than even twentieth century orchestras discourse.

e. Special Comb.: **twentieth century cut** *Diamond-cutting* (see quots.); **Twentieth Century (Limited)** the name of an express train running between Chicago and New York from 1902 to 1967.

1903 W. R. CATTELLE *Precious Stones* 63 The 'twentieth century' is a new form of cutting lately introduced. The number of facets is greater than in the brilliant-cut and they are differently shaped and arranged. **1925** KRAUS & HOLDEN *Gems & Gem Materials* vii. 77 The '*twentieth century*' *cut* has eighty or eighty-eight facets. **1970** E. BRUTON *Diamonds* x. 164 Jubilee or twentieth century cut (40+40=80 facets). **1902** *N.Y. World* 16 June 7/2 The 'Twentieth Century Limited', the new fast train,.. yesterday afternoon at 2.45 o'clock pulled out of the Grand Central Station. **1913** E. WHARTON *Custom of Country* III. xxii. 332 The Twentieth Century's generally considered the best route to Dakota. **1980** M. G. EBERHART *Casa Madrone* iii. 39 The Twentieth Century made a record run of eighteen hours from New York to Chicago. .. It is a very good train.

twenty, *numeral a.* and *sb.* Add: **A.** *adj.* **1. e.** Phr. *twenty-four hours a day*, all the time, incessantly.

1914 G. B. SHAW *Misalliance* p. xl, If we were habitually underworked and overfed, our notion of heaven would be a place where everybody worked strenuously for twenty four hours a day and never got anything to eat. **1942** *R.A.F. Jrnl.* 13 June 15 The least useful man..is the ..type, who is belligerent 24 hours a day. **1951** W. FAULKNER *Requiem for Nun* II. i. 143 Shut up in that room twenty-four hours a day. **1980** J. BARNETT *Palmprint* xiv. 149 There are American military planes over the Caribbean twenty-four hours a day.

2. c. Also *twenty-first* with ellipsis of *birthday*; cf. *TWENTY-FIRSTER.

1873 C. M. YONGE *Pillars of House* I. xi. 229 Here was his twenty-first not very far off. **1937** 'M. INNES' *Hamlet, Revenge!* I. i. 23 Celebrating a daughter's twenty-first by dressing her in white satin. **1975** 'J. LYMINGTON' *Spider in Bath* vii. 121 My daughter's twenty-first tomorrow. I should have collected a watch from the jeweller.

B. *sb.* **2. b.** (Earlier and later examples.)

1839 *Spirit of Times* 8 June 162/2 We had the gratification of seeing it [*sc.* his jockeyship] rewarded by more presents of odd fifties and twenties than probably Daniel ever saw in his lifetime. **1977** *Transatlantic Rev.* LX. 140 'God knows neither of you can call me ungrateful.' He put a twenty in front of each of them.

4. (Examples with reference to the third decade of the twentieth century; also *attrib.*)

1898 KIPLING in W. Nicholson *Almanac of Twelve Sports* July, The child of the Nineties..in pursuit of a girl whom The Twenties will dub a 'last-century heirloom'. **1930,** etc. [see *roaring twenties* s.v. *ROARING *ppl. a.* 4]. **1956** A. S. C. ROSS in M. Black *Importance of Lang.* (1962) 97 At Oxford in the late twenties the use of the surname..was a known *gaucherie*. **1969** *Listener* 26 June 903/1 A scandalous title, of course, for a book which was, in the Twenties manner, meant to scandalise. **1976** J. GRENFELL *Joyce Grenfell requests Pleasure* (1977) i. 21, I see us now,..our peculiar Twenties figures forced flat by bust-bodices.

C. a. *twenty-minute* (fig. example). **b.** *twenty-five-pounder*. **d.** Special Comb.: **twenty-first century** *attrib.* or as *adj.*, living in the twenty-first century; characteristic of the imagined conditions of the twenty-first century; **twenty-four carat** *a. colloq.*, (*a*) thoroughgoing, unalloyed, out-and-out; (*b*) genuine, flawless, trustworthy; **twenty-four-hour** *attrib.*, (*a*) lasting twenty-four hours; (*b*) of or pertaining to a system of reckoning the time whereby the hours of the day are numbered from one to twenty-four; (*c*) operating all day and all night, round-the-clock; **twenty questions,** a parlour game in which one party is allowed twenty questions (answered by either 'yes' or 'no') to discover the object of the other's thoughts; *spec.* the name of a popular radio panel game; **Twenty-six Counties,** the counties which by the Irish peace agreement of 1921–2 formed the Irish Free State, now the Republic of Ireland; cf. *Six Counties* s.v. *SIX *a.* 1 d; **twenty-twenty** (also **20/20**) *Ophthalm.*, the Snellen fraction for normal visual acuity, expressed in feet; *colloq.* used to denote good eyesight; also *fig.*; **twenty-two carat** *a. colloq.* = *twenty-four carat* (*b*) above; also *ellipt.* as *twenty-two*.

1964 D. FRANCIS *Nerve* iii. 37 He was what I pictured twenty-first century man should be—intensely alive, curiously innocent. **1979** D. BRIERLEY *Cold War* v. 49 The computer...it's very big, very expensive, very twenty-first century. **1980** *Jrnl. R. Soc. Arts* July 467/2 Everyone in the country must adapt to twentyfirst-century living and working patterns. *a* **1944** K. DOUGLAS *Alamein to Zem Zem* (1946) 10, 25-pounders and quads, Bofors guns in pits with their crews lying beside them. **1983** J. MASTERS *Man of War* xx. 264 Now came the first shell from the 25-pounders. **1900** Twenty-four carat [in Dict.]. **1965** D. FRANCIS *Odds Against* iii. 40 It is you.. who is the dyed-in-the-wool, twenty-four carat, unmitigated bastard. **1968** *Times* 21 Dec. 2/3 The legs in thigh-length boots are still lissom, 24 carat. **1974** G. JENKINS *Bridge of Magpies* v. 71 I'd accepted her story as 24-carat. **1984** *Times* 7 Mar. 28 The rest..had to work up real 24-carat grins. **1908** Twenty-four hour [in Dict.]. **1919** in *Cook's Continental Time-Table* (1973) Mar. p. vi/2 Cook's Continental Time-Table...Based on the 24-Hour System. **1947** CROWTHER & WHITTINGTON *Science at War* 7 A continuous twenty-four-hour watch for strange aircraft started. **1975** C. EGLETON *Skirmish* xvii. 169 He had located a twenty-four hour service station and had had the tyre repaired. **1978** H. KEMELMAN *Thursday the Rabbi walked Out* (1979) xxix. 141, I got sick. It was this twenty-four-hour bug. **1978** G. VAUGHAN *Belgrade Drop* i. 11 A clipboard..contained dates, 24-hour clock times, and a short, neatly typed entry against each. **1939** WODEHOUSE *Uncle Fred in Springtime* xi. 150 No matter how suave her manner for the nonce, she is at heart a twenty-minute egg. **1786** H. MORE *Lett.* (1925) 107 Mrs Fielding and I..diverted ourselves with teaching Sir Joshua and Lord Palmerston the play of twenty questions. **1846** R. BELL *Life of Rt. Hon. George Canning* x. 255 Canning proposed that they should play at 'Twenty Questions'. They had never heard of this game. **1930** 'HAY' & WODEHOUSE *Baa, Baa, Black Sheep* I. ii. 21 All right, a vicarage garden. What are we playing at? Twenty Questions, or something? **1979** E. H. GOMBRICH *Sense of Order* iv. 104 This can be done by a simple series of yes or no answers which allows a questioner to locate an item on a given grid as in the game of Twenty Questions. **1922** *Times* 6 June 16/2 The requirement that Ulster shall deliberately 'contract out' of the arrangements made between Great Britain and the twenty-six Southern counties. *Ibid.*, If Mr. Griffith is right in claiming all but two per cent. of the population in the twenty-six counties as supporters of the Treaty. **1949** C. GRAVES *Ireland Revisited* viii. 102 The Bantrymen have always provided a strong contingent in the Government of the Twenty-six Counties ever since 1922. **1978** [see *SOUTH *sb.* 2 a]. **1979** W. NELSON *Minstrel Code* vi. 45 The garage..belonged to a brother-in-law of one of the Sinn Fein leaders in the Twenty-Six Counties... O'Hagan explained that, within the IRA, no one talked of the 'Irish Republic' by any term other than this. **1875** T. LONGMORE *Man. Instructions Army testing Vision* (ed. 2) iii. 46 The 20-feet types are read at 20 feet, the 30-feet types at 30 feet; then V.=20/20 or 30/30, and the acuteness of vision is normal. **1945** L. SHELLY *Jive Talk Dict.* 35 *Twenty twenty*, excellent. **1951** E. F. TAIT *Textbk. Refraction* ii. 16 The visual acuity of healthy corrected eyes may be much better or considerably worse than that represented by the 6/6 or 20/20 standard. **1956** 'E. MCBAIN' *Cop Hater* (1958) iii. 21 Having 20/20 vision without glasses..he..had been appointed a patrolman. **1962** *Flight International* LXXXI. 426/2 Perfect eyesight is denoted as 20/20 vision, and the newest expression in the US air transport business—ruefully coined, we believe, by somebody in Convair—is 20/20 hindsight. Hindsight, of the 20/20 kind, abounds in plenty. But the aviation business is more interested in 20/20 foresight, because it can prevent people from losing a lot of money. **1977** H. GREENE *FSO-1* ix. 83 We're looking back with twenty-twenty hindsight, now. **1981** P. TURNBULL *Deep & Crisp & Even* ix. 162 He had 20:20 vision... He glimpsed a black shape. **1962** R. COOK *Crust on its Uppers* iv. 45 You come out twenty-two carat. *Ibid.* 47, I tell you they're absolutely twenty-two. **1974** W. J. BURLEY *Death in Stanley Street* ii. 46 I've got a twenty-two carat alibi. **1981** J. BARNETT *Firing Squad* vi. 61 Lady Lowderton was no nutter and her title was twenty-two carat.

twenty-firster (-fɜ̄·ɪstəɪ). *slang.* [f. *twenty-first* (sc. *birthday*) + *-ER[6].] A twenty-first birthday party (until 1970 in the U.K. celebrating the coming-of-age), or one who celebrates this. Also, a twenty-first birthday present.

1912 *Isis* 17 Feb. 204/1 There always are, or seem to be, celebrations of a sort on a twenty-firster. **1930** R. LEHMANN *Note in Music* IV. 129 'What a lovely case!' 'Yes, it was a twenty-firster.' **1964** C. MACKENZIE *My Life & Times* III. 166 The conventional Twenty-Firster at one of the Oxford hotels or clubs. *Ibid.* 214 Our rendez-vous with the rest of the Twenty-firsters was for supper. **1975** J. I. M. STEWART *Young Pattullo* iii. 69 The contraption itself was familiar to me, Ninian once having brought one home from the twenty-firster of an older friend.

twenty-five. Add: **1.** Also in Hockey.

1895 [see *BULLY *sb.*[2] 2]. **1930** *Times* 14 Mar. 7/4 With 12 minutes to go, Home worked his way to the German '25' and then suddenly sent in another long shot.

2. (Earlier example.)

1870 'CAVENDISH' (*title*) The Pocket Guide to Spoil-Five, Twenty-Five and Forty-Five.

twentyish (twe·nti,iʃ), *a.* [f. TWENTY + -ISH[1].] **a.** Of a person, (looking) approximately twenty years old. **b.** Characteristic of the 1920s.

1928 *Daily Express* 20 July 4 On the other hand, her blazer-clad, twentyish escort was impressed. **1940** GRAVES & HODGE *Long Week-End* viii. 125 The discovery ..of the unrifled tomb of the Pharaoh Tutankhamen was given typical Twentyish publicity. Ancient Egypt suddenly became the vogue. **1975** L. DICKSON *Radclyffe Hall at Well of Loneliness* 9, I try to show her as the unique personality she was, part of a 'twentyish scene now vanished in the clearer if harsher light of our times. **1981** 'J. ROSS' *Dark Blue & Dangerous* viii. 43 The woman.. wasn't twenty-ish..or startlingly beautiful.

twenty-one. [TWENTY A. 1 b.] = VINGT-ET-UN, VINGT-UN. Cf. *BLACK JACK, BLACK-JACK 10; *PONTOON *sb.*[2]

1790 E. WYNNE *Diary* 11 Mar. (1952) 30 My aunt winned 80 livers in playing at twentyone. **1917** [see *PONTOON *sb.*[2] a]. **1977** *Time* 21 Nov. 44/2 New croupiers are taught the 'theory of craps', while Twenty-One dealers are told to slap their hands and hold them upward when they leave their posts to show that they are not concealing any chips.

twenty-three. [TWENTY A. 1 b.] Ellipt. for *twenty-three skidoo*: see *SKIDOO *v.* 2 b.

1930 J. DOS PASSOS *42nd Parallel* 72 We want to take a look at this burg an then twentythree. **1933** L. BLOOMFIELD *Language* xxiv. 443 One suspects that the queer slang use, a quarter of a century ago, of *twenty-three* for 'get out' arose in a chance situation of sportsmanship, gambling, crime, or some other rakish environment.

twenty-two. [TWENTY A. 1 b.] A twenty-two calibre rifle; = **two-two* (a) s.v. *TWO IV. 2. Also *attrib.*, as *twenty-two rifle*.

a **1930** D. H. LAWRENCE *Phoenix II* (1968) 464, I went back to the house, and got the little twenty-two rifle. **1937** J. STEINBECK *Red Pony* 13 Jody took his twenty-two rifle up to the cold spring at the brush line. **1958** 'E. MCBAIN' *Killer's Payoff* (1960) xiii. 134 I've got three guns. A shotgun, a twenty-two, and a big-game rifle. **1967** K. S. PRICHARD in *Coast to Coast 1965–66* 182 Quick as light, Mac swung his twenty-two.

twerp (twɜ̄ɪp). *slang.* Also **twirp.** [Of uncertain origin. See quots. 1944, 1957; T. W. Earp of Exeter College, Oxford, matriculated in Michaelmas Term, 1911.] A despicable or objectionable person; an insignificant person, a nobody; a nincompoop.

1925 FRASER & GIBBONS *Soldier & Sailor Words* 292 *Twerp*, an unpleasant person. **1934** J. O'HARA *Appointment in Samarra* iv. 87 'And what a husband.' 'Exactly!.. That little twirp.' **1936** WODEHOUSE *Laughing Gas* xxv. 265 You're simply a lot of low twerps who kidnapped me in order to cash in. **1944** J. R. R. TOLKIEN *Let.* 6 Oct. (1981) 94 He lived in O[xford] at the time when we lived in Pusey Street (rooming with Walton, the composer, and going about with T. W. Earp, the original twerp). **1945** [see *RAT *sb.*[1] 3 a]. **1955** E. POUND *Section: Rock-Drill* xcv. 105 Among all these twerps and Pullizer sponges no voice for the Constitution, No objection to the historic blackout. **1957** R. CAMPBELL *Portugal* 87 T. W. Earp (who gave the English language the word twirp, really twearp, because of the Goering-like wrath he kindled in the hearts of the rugger-playing stalwarts at Oxford, when he was president of the Union, by being the last, most charming, and wittiest of the 'decadents'). **1960** S. BARSTOW *Kind of Loving* I. iv. 91 If she turns me down I'll look more of a twerp than ever. **1973** B. BROADFOOT *Ten Lost Years* xxvii. 309 The R. B. Bennetts of Canada and that despicable little twerp Mac. **1980** *National Times* (Austral.) 21 Dec. 30/3 Kendig's former boss..is a twerp. His offices contain a gallery of framed photographs of [himself]: there he is with John Wayne, with Nixon, [etc.].

Twi (twī, tʃwī). Also **Tshi** (tʃī). **a.** The chief language spoken in Ghana, consisting of several mutually intelligible dialects. **b.** The speakers of Twi. Also *attrib.*

1874 J. G. CHRISTALLER *Dict. Eng., Tshi-Asánte-Akra* p. iv, Tshi is the prevailing language of the Gold Coast. *Ibid.* p. ix, In the publications of the Basel missionaries, the language is called Twi, pronounced as *Chwee* would be in English. **1887** A. B. ELLIS (*title*) The Tshi-speaking peoples of the Gold Coast. **1920** A. W. CARDINALL *Natives N. Territories Gold Coast* 113 There are in the Gold Coast and its dependencies..two great languages: Twi,..and a language which in its simplest form is spoken by the Moshi. **1931** *Times Lit. Suppl.* 17 Sept. 692/3 The author's argument that the Twi people originally came from the grass country. **1955** [see *FANTI *sb.* and *a.*]. **1972** *Bk. Thousand Tongues* (rev. ed.) 441/1 The administrative term 'Akan' is used to denote all forms of Twi (Fante, Asante, and Akuapem)... In the 1850's, when Johannes Christaller translated the Bible into 'Twi', he chose..a form of the language which fell phonetically between the linguistic extremes of Fante and Asante-Akyem... Thus Scriptures were finally provided for the Twi. **1978** *Language* LIV. 458, I would like to enumerate the contributions in terms of the languages treated:..one paper each on Bambara (David Dwyer), Twi (Lynette Nyaggah), Nupe (Isaac George), [etc.].

twi-, twy-, *prefix.* Add: **a.** *twi-natured* (later

example). **c.** *twi-nature.* **e.** **twi-minded.** Delete '?' and add later examples.

1886 E. DOWDEN *Let.* Jan. in *Fragments from Old Lett.* (1914) I. 176 You raise..a difficult general question concerning the destruction of old letters. As usual, *I* am twi-minded. **1932** V. WOOLF *Common Reader* 2nd Ser. 174 From the first he [*sc.* Hazlitt] was a twy-minded man. **1897** W. B. YEATS *Secret Rose* 178 Then the twy-nature faded. **1916** G. SAINTSBURY *Peace of Augustans* viii. 287 The cat was a nymph and the nymph was a cat; the two lines fit the twynatured creature in both its natures.

twice, *adv.* (*sb.*, *a.*) Add: (Examples of *dial.* forms *twicet*, *twict*.)

1789, etc. [see *ONCE *adv.* A. γ]. **1895** *Dialect Notes* I. 375 Some other words reported individually are...onct, twict. **1922** JOYCE *Ulysses* 453 And he interfered twict with my clothing. **1926** R. HUGHES in *Hearst's Internat.* Feb. 44/2 Wha'd' you say, Kid, if I'd 'a' matched Coily up wit' some old vet'ran twicet his weight wit' twicet his ring-gener'lship. **1940** H. G. WELLS *Babes in Darkling Wood* II. i. 137 'Seems like Hitler's thinking twicet,' said a gentleman behind him. **1958** E. BIRNEY *Turvey* v. 48, I los all my good time for climbin the fence twicet.

1. e. (Earlier examples.)

1623 W. PAINTER in *Oxf. Dict. Eng. Proverbs* (1970) 263/1 Thinke twise, then speak, the old Prouerbe doth say. **1640** R. BRATHWAIT *Ar't Asleepe Husband?* vii. 277 You thinke twice before you speake, and may be demanded twice before you answer. **1818** SCOTT *Heart Midl.* in *Tales my Landlord* 2nd Ser. IV. viii. 180 If a fule may gie a wise man a counsel, I wad hae him think twice or he mells wi' Knockdunder. **1853** C. BRONTË *Villette* I. viii. 147, I saw in her countenance a something that made me think twice ere I decided.

5. b. In combination with advbs., forming compound advbs. or adjs. (and sbs.), as *twice-nightly*, *-weekly*, *-yearly*.

1929 D. H. LAWRENCE *Pansies* 133 And let the nodding tempests of verbosity Weekly or twice-weekly whistle round your bottles. **1949** M. STEEN *Twilight on Floods* IV. vi. 611 There was English's.., the first of the twice-nightlies, down in Hackney. **1976** 'W. TREVOR' *Children of Dynmouth* v. 102 The cartoonist responsible..was now, in the sunset of his life, himself the recipient of twice-weekly Meals on Wheels. **1980** M. BABSON *Queue here for Murder* iv. 30 The twice-yearly Sales.

twice-laid, *a.* **b.** (Earlier and later examples.)

1777 P. THICKNESSE *Year's Journey* II. xlvi. 110 My entertainment..was half a second-hand roasted turkey, or, what the sailors call, a *twice-laid* dish, i.e. one which is *done over* a second time. **1937** G. P. LOW *Sea made Men* ii. 35 Bread that had been on one long voyage and had been baked over again. It was called by sailors 'twice laid bread', and poor at that.

twicer. Add: **3.** A crook, liar, cheat; a deceitful or cunning individual.

1924 *Truth* (Sydney) 27 Apr. 6 *Twicer*, a deceitful fellow. **1925** FRASER & GIBBONS *Soldier & Sailor Words & Phrases* 292 *Twicer*, a cunning fellow. **1949** E. WINGFIELD-STRATFORD *King Charles & King Pym* IV. ix. 262 The recent dismissal..of that elderly twicer, Sir Harry Vane.

twiddle, *sb.* (Later *Mus.* examples.)

1908 [see *DYNAMICS 3]. **1975** *New Yorker* 21 Apr. 115/1 She adds to Rossini's tense exclamations a cascade of pretty twiddles.

twiddle-twaddle. Add: Hence as *v. intr.*, to chatter foolishly (*nonce-wd.*).

a **1930** D. H. LAWRENCE *Phoenix II* (1968) 480 All you can do now is to twiddle-twaddle about golden boughs.

twiddly, *a.* (In Dict. s.v. TWIDDLE *v.*[1]) Add: Also **twiddley.** (Later examples.) Freq. in Comb. *twiddly bit*, a fancy or intricate embellishment; a detail.

1912 C. MACKENZIE *Carnival* viii. 83 That's no tune to dance to. You want something to show off the twiddly-bits. **1922** H. JENKINS *Mrs Bindle* ii. 52, I like them little twiddley bits wot you been puttin' into that 'ymn. **1935** 'G. ORWELL' *Let.* Sept. in *Coll. Ess.* (1968) I. 152, I have three more chapters and an epilogue to do, and then I shall spend about two months putting on the twiddly bits. **1964** V. S. NAIPAUL *Area of Darkness* x. 255 He was tall and thin,..and had a brisk, twiddly walk. **1974** *Guardian* 23 Mar. 14/4, I was..enchanted by the dining car, all wood panelling and twiddly bits. **1982** BARR & YORK *Official Sloane Ranger Handbk.* 148/1 Sloane windows need curtains with a capital C: with pelmets, twiddly bits, bands, tassels, tie-backs, edging.

twig, *sb.*[1] Add: **5. c. twig girdler** (examples); **twig-pruner,** substitute for def.: = *oak pruner* s.v. *OAK 9 (examples).

1874 *Rep. U.S. Dept. Agric. 1873* 153 The twig girdler, *Oncideres cingulatus*... The insects girdle the twig before depositing their eggs. **1972** SWAN & PAPP *Common Insects N. Amer.* xx. 452 The Twig Girdler..is rough-surfaced, grayish brown or yellowish brown. **1974** *Ridge Citizen* (Johnston, S. Carolina) 18 Apr. 6/3 The twig girdlers deposit their eggs in the portions of the branches that fall to the ground. **1928** METCALF & FLINT *Destructive & Useful Insects* xix. 664 Maple and oak twig pruner. **1972** SWAN & PAPP *Common Insects N. Amer.* xx. 446 Twig Pruner..Brown, clothed with irregular patches of grayish yellow pubescence.

twiglet (*b*), (with capital initial) the proprietary name of a crisp, savoury snack in the shape of a twig.

1932 *Trade Marks Jrnl.* 15 June 766 *Twiglets*... Biscuits. Peek, Frean & Company, Limited,..London, S.E. 16; manufacturers. **1962** M. FRAYN *Day of Dog* 57 A cheese twiglet slips from my fingers. **1980** J. MELVILLE *Chrysanthemum Chain* 103 Walker set out a..drinks tray ..and opened a tin of imported Twiglets.

twig, *v.*[1] Add: Hence **twi·gging** *vbl. sb.*

1916 JOYCE *Portrait of Artist* iv. 174 To flee from noise which caused him painful nervous irritation such as the sharpening of knives..and the twigging of the carpet.

twig(g)age (twi·gėdʒ). *Literary. rare.* [f. TWIG *sb.*[1] + -AGE.] Twigs collectively.

1923 *Glasgow Herald* 11 Dec. 6/8 Even the umbered purple loom of the birch twigage and its stem washes of amethystine-cream were but enhanced. **1964** *Listener* 16 Apr. 633/3 The dry stuff's best To start with, twiggage and haulm and herby tops.

twiggery (twi·gəri). [f. TWIG *sb.*[1] + -ERY.] Twigs collectively. Also *fig.*

1909 in WEBSTER. **1922** *Contemp. Rev.* Feb. 255 Something that was not merely discursive wooden twiggery. **1931** *Observer* 11 Oct. 26 In the winter of 1929 it was only the twiggery that was killed. The main stems broke out into bud... The numerous end-twigs do habitually suffer in winter. **1980** *Washington Post* 6 July K2/3, I would walk into the apartment..warm myself at the fire, admire my twiggery, [etc.].

twiggy, *a.* Add: Hence **twi·gginess,** the condition or quality of being twiggy.

1927 *Smallholder* 26 Mar. 105/1 Pea sticks ought to last through two seasons. Of course, in the second year they lose most of their twigginess. **1981** *Country Life* 1 Jan. 39/2 *Potentilla fruticosa*... Its winter-brown twigginess should match the tawny beech.

twilight, *sb.* Add: **3. a.** *twilight of the gods* (later examples); also *transf.* Cf. *GÖTTERDÄMMERUNG, *RAGNARÖK.

1888 R. GARNETT (*title*) The twilight of the gods and other tales. **1944** *Sun* (Baltimore) 22 July 2/1 The German nation is split wide open... The twilight of the gods has begun. **1979** A. R. PEACOCKE *Creation & World of Science* ii. 55 Under the pressure of experimental facts and the bold and convincing analyses of Planck and Einstein, there was, as Karl Heim puts it, a 'twilight of the gods' of absolute space, time, object, and determinism.

4. a. *twilight glow*, a diffuse glow in the sky at twilight; *spec.* in *Meteorol.*, that caused by spectroscopic emission in the upper atmosphere from atoms excited by solar radiation; *twilight vision*, vision in which colours are hardly perceptible owing to the dimness of the light; scotopic vision.

1819 BYRON *Don Juan* II. clxxxviii. 213 The twilight glow, which momently grew less. **1955** *Sci. Amer.* Sept. 150/3 There is also a twilight glow, about 100 times as intense as the nightglow but not detectable by the eye because of the brighter sky. **1972** *Ibid.* Jan. 80/3 The spectrum of the twilightglow differs from the nightglow spectrum in that certain features disappear shortly after the end of twilight and others are markedly stronger in twilight than they are during the night. **1980** F. H. LUDLAM *Clouds & Storms* iv. 77/1 The twilight glow continues to fade and its upper border to descend more rapidly than the sun, but it does not disappear below the horizon until the sun's depression exceeds about 16°, and astronomical twilight ends. **1921** Twilight vision [see *rod vision* s.v. *ROD *sb.*[1] 11 c]. **1924,** etc. [see *SCOTOPIC *a.*]. **1950** *Sci. News* XV. 17 It has been suspected for many years that the coloured pigment 'visual purple', found in the retinas of such animals as frogs, is associated with twilight vision. This supposition has recently become a certainty.

e. Special Combs.: **twilight area** = *twilight zone* (*a*) below; **twilight home,** (*a*) a home (see HOME *sb.* 8) for old people or animals; (*b*) = *twilight house* below; **twilight house,** a house in a twilight zone (see *twilight zone* (*a*) below); hence **twilight housing; twilight night** *Baseball* = *TWI-NIGHT; **twilight shift,** a shift worked between the day shift and the night shift; **twilight sleep** [tr. G. *dämmerschlaf* (C. J. Gauss, *c* 1905)], a state of amnesia and partial analgesia induced by the administration of morphine and scopolamine (hyoscine), esp. to lessen the pains of childbirth; **twilight world,** (*a*) a shadowy region; (*b*) a world characterized by uncertainty, obscurity, or decline; (*c*) the world which comes to life after sunset, characterized by merry-making or criminal activities; **twilight zone,** (*a*) *spec.*, an urban area in which housing is becoming decrepit; (*b*) *gen.*, an indistinct boundary area combining some of the characteristics of the two areas between which it falls (cf. sense 4 b in Dict.); (*c*) occas., a dimly illuminated region.

1960 *Daily Tel.* 18 June 8/3 Where debate begins and should be encouraged is over the question whether redevelopment of what Sir Keith Joseph called the 'twilight areas' must wait entirely on these other two housing operations. *a* **1974** R. CROSSMAN *Diaries* (1975) I. 44 A Labour Minister should impose central leadership, large-scale state intervention, in these blighted areas of cities, the twilight areas, which were once genteelly respectable and are now rotting away. **1934** WEBSTER, *Twilight home*, a charitable institution providing a home for aged people. *Colloq., Australia.* **1966** 'K. A. SADDLER' *Gilt Edge* v. 74 Twilight homes for retired beach donkeys. **1968** *Guardian* 5 Apr. 1/6 A plan to modernise Britain's four million twilight homes has been agreed by the Cabinet. **1978** I. MURDOCH *Sea* 493 [I] arranged for her mother to be packed off to a comfortable and expensive 'twilight home'. **1971** *New Society* 1 July 20/2 There were 600,000 'slums' and about two million 'twilight' houses. *Ibid.*, A current comparison of slum and twilight housing. **1971** *Mod. Law Rev.* XXXI. VI. 698 He has sections on ..houses in disrepair, on planning blight and on twilight housing areas. **1949** P. CUMMINGS *Dict. Sports* 478/1 *Twilight-night. Baseball.* A double-header, the first game played late in the afternoon, the second in the evening under lights. **1953** *Sun* (Baltimore) 28 Oct. (B ed.) 21/2 There can be none of those frisky twilight-night double headers. **1970** 'C. AIRD' *Late Phoenix* x. 115 He didn't come home last night after the twilight shift at his factory. **1977** *Wandsworth Borough News* 7 Oct. 18/2 (Advt.), Laundry workers evening shift, 5.30–9.30 p.m. We require a number of part-time workers for clean and simple work on our twilight shift, Monday–Friday. **1912** F. W. HEWITT *Anæsthetics & Administration* (ed. 4) ix. 278 As a matter of actual experience in hospital practice by no means all patients achieve the state of *dammerschlaf*, or 'twilight sleep', which foreign authors advocate. **1922** JOYCE *Ulysses* 159 Twilightsleep idea: queen Victoria was given that. **1971** D. D. MOIR *Pain Relief in Labour* i. 5 Twilight sleep is seldom used today because it causes respiratory depression in the new-born and tends to cause delirium and restlessness in the mother. **1981** J. GARDNER *License Renewed* xiv. 161 A nice mix—Scopolamine with morphine: twilight sleep, like having a baby. **1887** Twilight world [see sense 4 c in Dict.]. **1954** KOESTLER *Invisible Writing* xxvi. 281, I mention this episode as one example of the ambiguities of the twilight world in which we lived. **1963** *Times* 8 May 6/7 But in this unhappy twilight world in which we live in a state of truce—neither peace nor war. **1970** C. MAJOR *Dict. Afro-Amer. Slang* 117 *Twilight world*, the world of all-night parties. **1977** D. SEAMAN *Committee* 116 The twilight world of the mentally ill. **1977** 'J. D. WHITE' *Salzburg Affair* v. 45 The twilight world that exists in every city.. the doctor who will tend a bullet wound, the hotel that will provide accommodation without papers. **1909** *Arena* XLI. Mar. 273/2 Such organization will leave no 'twilight zone', no 'no man's land', for railway corporation dodgers. **1918** *Policeman's Monthly* June 30/1 There still remain twilight zones in most centers of population. **1920** J. G. FREDERICK *Great Game of Business* iii. 23 Be aware that the test of real 'honesty' comes in the 'twilight zone' between what is quite clearly honest and dishonest. **1938** *Jrnl. Royal Aeronaut. Soc.* XLII. 492 The twilight zone extends to about 20° either side of the equisignal zone centre. **1960** *Daily Tel.* 20 June 17/6 There are many towns with 'twilight zones' of shabby and outdated houses. **1969** *Times* 29 Jan. 10/7 It lives between 300 and 500 metres below the surface of the ocean, in the region to which light penetrates with such difficulty that it may be considered as a kind of twilight zone. **1981** *Washington Post* 26 Apr. A1/1 Several key officials charged with formulating foreign policy remain in a bureaucratic twilight zone almost 100 days after Reagan's inauguration.

Hence **twilight** *v. trans.* (earlier example).

1819 KEATS *Song of Four Fairies* in R. M. Milnes *Life, Lett. & Lit. Remains J. Keats* (1848) II. 275 And the beams of still Vesper..Are shed thro' the rain..And twilight your floating bowers.

twilly, *a.* and *sb.*[1] Add: **b.** *sb.* (*b*) (see quot. 1948); so *twill(e)y hole*, a hole left in the centre of a wattle hurdle for the insertion of a pole on which several hurdles may be carried simultaneously.

1893 A. KENNARD *Diogenes' Sandals* vi. 90 A 'twilley' hole, is left in the centre of each hurdle for the insertion of the..pole, on which the shepherds carry them. **1948** E. J. STOWE *Crafts of Countryside* iii. 24 About two-thirds of the way up from the bottom of the hurdle there are two important rods. They are known as 'twillies' (a country name for twisted rods), and are twisted about each other and around the upright sails one by one across the hurdle. *Ibid.* 128 *Twillies*, two twisted rods woven across wattle hurdles just above the twilly hole. **1959** *Times* 2 June 12/7 The split rods are twisted until the complete hurdle is ready with the twilly hole in the middle. **1971** *Country Life* 25 Feb. 424/1 These hurdles differ slightly from sheep hurdles which had a gap or twilly hole in their middle through which a shepherd put his stick to carry a load to the next enclosure.

twin, *a.* and *sb.* Add: **B.** *sb.* **3. d.** *ellipt.* for *twin aerial, bed, -city, -cylinder car, -cylinder engine, -engined aeroplane, -town*, etc. (see C. in Dict. and Suppl.).

1928 [see *sun arc* s.v. *SUN *sb.* 13 a]. **1930** *Times* 14 Mar. 12/3 As soon as the B.B.C. 'twins' came into use and two wave-lengths were available. **1938** O. NASH *I'm a Stranger here Myself* 85 A double bed or twins. **1944** G. L. NUTE *Lake Superior* III. xii. 277 Superior had hoped to be the terminus of the first railroad to the head of the lake, but when one came, in 1870, its terminal was Duluth rather than the other twin. **1955, 1963** [see *SINGLE *sb.* 3 c]. **1975** B. MEGGS *Matter of Paradise* VI. ii. 167 Nice..room, double bed... Should have made a point to ask for twins. **1976** R. BARNARD *Little Local Murder* i. 11 They've done one before—for a town in Essex with a twin in Canada. **1976** B. LECOMBER *Dead Weight* iii. 45 If I was going into the charter business I had to have a twin—and she was.. the only twin on the whole airport that I could afford.

C. *Comb.* **a.** *twin-burner*, *-cylinder* (earlier example), *-engine*, *-float*, *-fuselage*, *-track*. **b.** *twin-engined*, *-tyred*, *-walled*. **c. twin aerial** (*temporary*), a twin-wire aerial; **twin bed,** one of a pair of matching single beds; hence **twin-bedded** *a.* (*a*) tucked up in a twin-bed (*nonce*

use); (*b*) furnished with twin beds; **twin bedstead,** one of a pair of matching single bedsteads; **twin-bill** *Baseball* = *DOUBLE-HEADER c; **twin carburettor,** one of a pair of carburettors in the same engine; so **twin carb.**; **twin city,** (*a*) *N. Amer.*, either of two cities that are very close neighbours; *spec.* in *pl.* (*U.S.*) St. Paul and Minneapolis, (*Canad.*) Fort William and Port Arthur; (*b*) occas. used of a city in the sense of *twin town* below; **twin double,** a system of betting (on horse-races, etc.) in which the winners of four successive races must be selected (i.e. two *doubles* in sequence); **twin floats,** a pair of floats (*FLOAT *sb.* 8 e) on a seaplane; **twin-jet** *a. Aeronaut.*, having two jet engines; also *ellipt.* as *sb.*, a twin-jet aircraft; **twin lamb disease,** a pregnancy toxæmia in sheep, apparently caused by malnutrition; **twin-lens** *a.*, designating a camera with two identical sets of lenses, either for taking stereoscopic pictures, or (more commonly) with one forming an image for viewing and the other an image to be photographed; **twin paradox** *Physics*, in relativity theory, the conclusion that if one of a pair of twins makes a long journey at high speed and then returns, he will have aged less than the twin who remains behind; **twin plate** *Glass Manufacture*, plate glass which is ground and polished on both sides at once; also *attrib.*; **twin prime** *Math.*, each of a pair of prime numbers whose difference is 2; **twin set** (also with hyphen and as one word), a woman's matching jumper and cardigan; **twin soul,** a kindred spirit; also as *attrib. phr.*; **twin species** *Biol.* [tr. F. *espèce jumelle* (L. Cuenot 1929, in *Reunion Plénière de la Soc. de Biol. et de ses Filiales, 17–18 Mai* 85], two species which are morphologically identical but which are separated by reproductive isolation; cf. *sibling species* s.v. *SIBLING 3; **twin town,** one of a pair of towns (usu. in different countries) that have established official links; **twin-tub** *a.*, (of a washing machine) having two separate top-loading drums, one for washing and the other for spin-drying; also *ellipt.* as *sb.*

1913 *Wireless World* June 211/2 The ordinary 'twin' aerial used by the Marconi Company on most of their ship-stations. **1928** J. FROST *Wireless Man.* iv. 17 A twin aerial, or aerial of two wires. **1919** G. B. SHAW *Heartbreak House* p. xxxviii, If the twin flats and twin beds produce a guinea more than Shakespear, out goes Shakespear. **1940** GRAVES & HODGE *Long Week-End* xi. 181 Twin-beds replaced the old..double-bed for married couples. **1973** E.-J. BAHR *Nice Neighbourhood* i. 8, I pictured her.. slipping into bed beside her husband... Of course, they may have had twin beds. **1937** G. FRANKAU *More of Us* xiii. 138 O happy nests (nest's best!) where Bob and Bill Sleep (sleep?) twin-bedded by their spawn-glad spouses. **1960** *News Chron.* 27 July 4/3 The writer asked for the best twin-bedded room, with private bathroom. **1981** 'E. LATHEN' *Going for Gold* iii. 29 The twin-bedded cubicle. **1900** *Heal & Son Catal.: Guest's Room: Paris Exhib.* 4 Twin bedsteads. Each 3 feet by 6 feet 6 inches. **1930** *Heal & Son Catal.: Matter of Taste in Furnit.* 17 Twin bedsteads in limed oak. **1939** in E. J. NICHOLS *Hist. Dict. Baseball Terminol.* (Ph.D. thesis, Pennsylvania State Coll.) 81 Twin bill. **1974** *Anderson* (S. Carolina) *Independent* 19 Apr. 5B/1 Virginia's Cavaliers invade Clemson Friday afternoon for a 1:30 Atlantic Coast Conference twin-bill. **1907** *Yesterday's Shopping* (1969) 219/2 A twin-burner Stove, strongly recommended as a boiler and heater. **1974** *Country Life* 5 Dec. 1772/1 Refrigerator, twin-burner stove..stainless-steel sink. **1967** *Autocar* 5 Oct. 47/1 New car called TC (for Twin Carb.), retaining all the special equipment of the basic model. **1973** 'R. MACLEOD' *Burial in Portugal* v. 93 The Lancia had..a high compression alloy engine which sucked fuel through twin carburettors. **1856** *Rock Island* (Illinois) *Argus* 23 Apr., The church bells of the twin-cities [*sc.* Rock Island, Ill. and Davenport, Iowa] rang out their joyous notes in honor of the achievement [*sc.* bridging the Mississippi]. **1883** *Harper's Mag.* June 73/2 The twin cities [*sc.* St. Paul and Minneapolis, Minn.] ..emulate each other in metropolitan airs. **1912** J. SANDILANDS *Western Canad. Dict. & Phrase Bk.*, *Twin Cities*, when spoken of in Canada, usually refer to Port Arthur and Fort William, neighboring cities and ports in Ontario. **1949** *St. Paul Pioneer-Press* 12 Aug. 1/3 Fog and clouds gave the Twin Cities respite from the hot weather for a few hours Thursday morning. **1968** A. HAILEY *Airport* (1969) III. iii. 319 Detroit and Windsor, the twin cities straddling the [U.S.–Canada] border. **1973** *Guardian* 13 Apr. 10/5 Manchester ..is what they, laughingly I trust, call the 'twin city' of Leningrad. **1980** *Quilt World* Sept./Oct. 28/3 A learning-packed three-day seminar..will be held at a camp on the shores of Silver Lake in the Twin Cities. **1846** T. CRADDOCK *Chemistry of Steam-Engine* 91 The loss..as that shown by the expansive curve, induced me to devise the arrangement I have designated the Twin-Cylinder Engine. **1960** *N.Y. Times* 25 Oct. 43/7 The new method, called the twin double, requires a fan to pick four consecutive winners from the sixth through the ninth races. **1979** *Internat. Herald Tribune* 31 Oct. 23/4 There were seven more races, four more swingers, two daily doubles and a jackpot, or twin double, still to come. Plenty of chances to get well. **1931** *19th Cent.* Feb. 155 Twin-engine Farman 'Goliath' seaplanes. **1916** *War Illustr.* 1 Jan. 474/1 The twin-engined Caudron biplane. **1942** *R.A.F. Jrnl.* 27 June 1 The aircraft employed..were..twin-engined Handley-Page bombers. **1974** E. AMBLER *Dr. Frigo* III. 190 There was a small twin-engined plane waiting. **1913** *Flight* 19 Apr. 436 (*caption*) One of the floats on the twin-float Breguet. **1942** E. SARGENT *Every Boy's Bk. Aircraft* viii. 39 A good example of a military seaplane ..is the Fairey Seafox... It has twin floats. **1977** G. R. DUVAL *World Float Planes* (*caption*), The Fairey Flycatcher was the standard Fleet Air Arm fighter from 1923 to 1934... The land undercarriage was readily interchangeable with twin floats. **1980** P. LEWIS *Brit. Bomber since 1914* v. 204 During 1933 another Fairey twin-float biplane made its appearance as the Fox Mk. IVM. **1931** *19th Cent.* Feb. 159 The twin-fuselage Blériot 125 mono-plane. **1980** *Jane's Encycl. Aviation* II. 305/1 *Blackburn T.B. Twin* (*UK*), large twin-fuselage twin-engined seaplane designed to attack Zeppelin airships with incendiary steel darts. **1946** *Jrnl. R. Aeronaut. Soc.* L. 348/1 It may therefore be possible to make more advanced explorations into the transonic region with similar twin-jet installations. **1953** *Ann. Reg. 1952* 405 An English Electric Canberra B5 twin-jet bomber made history on 26 August by completing a double crossing of the Atlantic in a single day. **1961** E. BROWN *Wings on my Sleeve* 85 We had a standing date to go back to Grove and pick up the first of the Arado 234 B twin jets. **1945** J. F. H. THOMAS *Sheep* v. 93 When in-lamb ewes have a diet which is protein-adequate,..twin lamb disease is never a serious cause of loss. **1974** *Country Life* 28 Mar. 740/1 Twin-lamb disease..is often fatal. **1894** *Country Gentlemen's Catal.* 158/3 Hand-Cameras... The 5×4 Twin Lens Artist Magazine or dark slides—£15 15s. od. **1911** *Encycl. Brit.* XXI. 505/1 (*heading*) Twin-lens and reflex cameras. *Ibid.* 505/2 Stereoscopic cameras are another form of twin-lens cameras. **1977** J. HEDGECOE *Photographer's Handbk.* 19 The twin lens reflex design is much older than the SLR and was one of the most popular 'advanced' types of camera prior to World War II. **1957** *Nature* 5 Jan. 35/2 The 'twin paradox' is not even qualitatively discernible in any experiment that does not involve relative accelerations. **1982** W. R. RINDLER *Introd. Special Relativity* iii. 51 If the twins A and B, in the twin-paradox 'experiment'.., visually observe the regular ticking of each other's standard clocks, describe quantitatively what each sees as B travels to a distant point Q and back. **1939** *Archit. Rev.* LXXXV. 104 Twin-plate has arrived to supersede ordinary plate glass. **1962** *Gloss. Terms Glass Industry* (*B.S.I.*) 28 *Twin-plate process*, a process for making polished plate glass in which rolling, annealing and grinding are carried out on a continuously produced ribbon of glass without first cutting it into sections and in which top and bottom surfaces are ground simultaneously. **1930** T. DANTZIG *Number* iii. 49 It has been shown that the so-called twin-primes, such as (3,5), (5,7),..(41,43), etc., become rarer and rarer as the numbers increase. **1981** *Sci. Amer.* Feb. 19/2 The largest pair of twin primes given in the December column has now been surpassed by an even larger pair discovered in 1980. **1937** *New Yorker* 9 Jan. 62 Here you will find sweater classics—twin sets of the conventional type. **1944** M. LASKI *Love on Supertax* iii. 35 I've got a Worth frock.. I swopped..for my cashmere twin-set. **1970** *Listener* 27 Aug. 289/2 Sophia Loren wandering in a tidy twinset across the USSR. [**1868** Twin soul: in Dict., sense A. 5.] **1927** WODEHOUSE *Meet Mr Mulliner* vi. 172 It seemed to him so plain a proof that they were twin souls that he decided to offer him his hand and heart without delay. *a* **1930** D. H. LAWRENCE *Mod. Lover* (1934) v. 37 You know, love isn't the twin-soul business. **1931** *Archivio Zool.* XV. 289 During his studies upon evolution the A. has been led to propose some neologisms that seemed to him useful and he reunites in this study, defining precisely the sense of them. Those neologisms are: statistic adaptation, homochromy, preadaptation, differentative death, differentative fecundity, twin species. **1971** *Biol. Abstr.* LII. 11353/2 (*heading*) Study of 2 twin species of parasitic copepods. **1955** *Harrogate Advertiser* 18 June 8/3 In the afternoon they met in the Council Chamber to discuss Le Monde Bilingue Twin Town Scheme. **1976** *Southern Even. Echo* (Southampton) 2 Nov. 2/2 Wickham entertained 11 French visitors from their proposed twin town of Villers-Sur-Mer at the Kings Head. **1960** Twin-track [see *Speedwalk* s.v. *SPEED *sb.* 11 c]. **1983** *Listener* 13 Oct. 3/2 When the 'twin-track' approach to European nuclear weapons was devised by NATO (make preparations to deploy, but hold arms control talks at the same time) no one seems to have remembered Euclid's principle that parallel lines can never meet. **1962** *Which?* May 139/1 We carried out washing tests..in a twin-tub washing machine. **1970** *New Scientist* 15 Oct. 134/1 An overnight soak and a wash (in a twin-tub), did remove some..stains. **1916** *Chambers's Jrnl.* Dec. 829/1 The twin-tired commercial vehicle. **1913** D. H. LAWRENCE *Love Poems & Others* 26 As if..the twin-walled darkness had bled In one great spasm of birth.

twin, *v.*[2] Add: **2. a.** Also *spec.* to cause (towns) to be twinned (chiefly in *pass.*): see *TWINNED *ppl. a.* 2 c.

1957 *Harrogate Advertiser* 16 Mar. 13/5 Harrogate was the first town in the country to be twinned with a French town—Luchon. **1983** *Guardian Weekly* 6 Feb. 13/5 One thousand towns and villages from each country have been twinned.

b. *spec.* Of a town or city: to become twinned *with* (another).

1973 *Daily Tel.* 1 Feb. 1/4 Liverpool is to go ahead with a plan to 'twin' with the port of Haiphong in North Vietnam. **1977** *Cornish Times* 19 Aug. 15/5 Pleyber-Christ, the Breton town with which it is proposed that Lostwithiel should 'twin'.

twindle (twi·nd'l), *v.*[2] *nonce-wd. intr.* Used by G. M. Hopkins: prob. a blend of TWIST *v.* and DWINDLE *v.*

1881 G. M. HOPKINS *Poems* (1967) 89 A windpuff-bonnet of fáwn-fróth Turns and twindles over the broth Of a pool so pitchblack, féll-frówning.

twing twang. (Later examples.)

1922 JOYCE *Ulysses* 436 Their paler smaller negroid hands jingle the twingtwang wires. **1953** *John o' London's* 23 Jan. 75/3 Central in the 'orchestra' was a guitar making a lazy twing-twang of sound.

twingy (twi·ndʒi), *a. rare.* Also **twingey.** [f. TWINGE *sb.* or *v.*[1] + -Y[1].] Experiencing twinges of pain.

1865 N. HOGG *Poet. Lett. tu es Brither Jan* (ed. 5) 52 And then tha litt'l pigs wid zook, an twinjy in tha jaw wis took. **1915** V. WOOLF *Voyage Out* ix. 120 Aunt E. cheerful, though twingy, she says. **1974** D. SEARS *Lark in Clear Air* (1976) xv. 187 Hearing that song made me feel twingey and kind of lonesome.

twi-night (twəɪ·nəɪt). *Baseball.* [Blend of TWILIGHT *sb.* and NIGHT *sb.*] (See quot. 1955.) Freq. *attrib.*, as *twi-night double-header* (cf. *DOUBLE-HEADER c). Hence **twi-·nighter,** in same sense.

1939 *Amer. Speech* XIV. 5 Blending..has proved practical in the formation of useful compounds where a combination of ideas is desired; as in..'brunch', 'twinight', 'Anglamerican'. **1953** *Sun* (Baltimore) 28 Oct. (B ed.) 21/2 (*heading*) Twi-nighters? **1955** M. REIFER *Dict. New Words* 214/2 *Twinight, n. Baseball*, a double-header game which begins in the afternoon and continues into the night with artificial illumination. **1975** *New Yorker* 17 Nov. 148/2 The pitching left something to be desired, but the next afternoon, in the opener of a twi-night doubleheader, Frank Tanana struck out seventeen Texas batters. **1980** *N.Y. Post* 4 Aug. 56 The Mets are still— again—two games under .500 and twi-night double-header here tonight..opens another five-game series against the division-leading Expos.

twinkle, *sb.* Add: **4.** A ballroom dance (step), danced to slow Blues music. Also *twinkle step*.

1920 A. E. W. MASON *Summons* xxi. 220 'Do you know the fox-trot?' 'A little.' 'The twinkle step?' 'Not at all.' **1936** A. MOORE *Ballroom Dancing* v. 214 A Twinkle is a figure of three steps. The feet are closed..on the 2nd step and the weight is changed, and the 1st and 3rd steps are both taken in a forward direction or both in a backward direction. **1962** L. K. ENGEL *Fred Astaire Dance Bk.* xv. 47 The Open Twinkle is a slight variation of the basic One Step. **1975** G. HOWELL *In Vogue* 9/2 We got syncopated music and what to do to it—the Baleta, the Maxina, the Twinkle, the Jog Trot, the Vampire, [etc.].

5. *Comb.*, as **twinkle-dress** *poet. nonce-wd.*, a sparkling party dress; **twinkle roll** *Aeronautics*, an aerobatic stunt (see quot. 1962); **twinkle-toed** *a.*, light-footed, nimble; (of a dance) quick, requiring agility.

1960 S. PLATH *Colossus* 59 When on tiptoe the schoolgirls danced, Blinking flashlights like fireflies And singing the glowworm song, I could Not lift a foot in the twinkle-dress. **1962** *Flight International* LXXXII. 269/2 Highlights of an outstanding presentation by the Lightnings were the 'twinkle roll' in which the two wingmen of a three-aircraft formation rolled individually on either side of their leader as they passed low and fast in front of the crowd. **1978** R. JANSSON *News Caper* 7 The fighter..slid over our port wing and did a twinkle roll in front of our nose. **1960** *Farmer & Stockbreeder* 29 Mar. (Suppl.) 10/2 Hand-in-hand with about six other youngsters she was scampering through a twinkle-toed dance which she later informed me is called 'the shuffle'. **1961** *Sunday Express* 7 May 14/3 Abandoned, twinkle-toed dancers leaping about. **1978** *Lancashire Life* Nov. 129/1 John Travolta doesn't have the monopoly of twinkle toed addicts.

twinkle, *v.*[1] **4.** *intr. Dancing.* To perform the twinkle step. *temporary.*

1920 *Punch* 10 Nov. 366/2 *Chassée* to the left, two steps forward, two steps back, twinkle each way. *Ibid.*, I quite enjoyed that twinkling business. **1928** *B.B.C. Handbk. 1929* 201 Wireless dance music is often heard from houses where no one has ever 'twinkled' or 'hesitated' or 'glided' or 'dragged'.

twinkly, *a.* Add: **2.** *Comb.*, as *twinkly-eyed* adj.

1926 G. FRANKAU *My Unsentimental Journey* x. 135 A very self-possessed if somewhat twinkly-eyed young woman. **1974** P. GZOWSKI *Bk. about this Country* 11/2 Professor Hiebert..made twinkly-eyed, flirtatious remarks to Edith.

twinly (twi·nli), *adv. rare.* [f. TWIN *a.* and *sb.* + -LY[2].] To an equal extent, doubly; in an identical degree.

1913 A. O'CONNOR *Poems* 50 A spot for man, bearing his cross, to seek, Where night and day are healers twinly bright. **1981** C. DEXTER *Dead of Jericho* xxii. 123 Even if, in his boyhood, Sergeant Lewis's parents had been twinly blessed with privilege and wealth, it seems unlikely that their son would have won a scholarship to Winchester.

twinned, *ppl. a.* Add: **2. c.** Of a city, town, etc.: linked *with* another (in a different country) for the purpose of friendship and cultural exchange. Usu. *predic.* Cf. *twin town* s.v. *TWIN *a.* and *sb.* C e.

[**1923** E. CONYBEARE *Highways & Byways Cambr. & Ely* iii. 51 William of Wykeham..first conceived the idea of twinned colleges, in the provinces and at the University.] **1957** *Harrogate Advertiser* 16 Mar. 13/5 The congress is to talk about problems facing twinned towns, and report on the progress of the twinning. **1960** *Guardian* 25 Feb. 5/5 About 68 towns in Britain are now 'twinned' with towns abroad. **1976** *Times* 5 July 13/6 Leicester city is twinned with Krefeld and a most encouraging friendship has grown up between the two cities. **1980** *Times* 11 July 14/8 Delegates pay visits to twinned cities.

twinning, *vbl. sb.* Add: **2. b.** The linking (of two towns or of one town *with* another) for the purpose of friendship and cultural exchange. Cf. *TWINNED *ppl. a.* 2 c.

Occas. (as in quot. 1975) used of similar links between institutions such as schools.

1956 *Harrogate Advertiser* 9 June 8/3 French week celebrates the town's pioneer contribution to Le Monde Bilingue in its 'twinning' with Luchon, the spa town in the Pyrenees. **1962** *Guardian* 10 Mar. 16/3 The British Bi-Lingual Association..exists to promote the 'twinning' of towns in Britain with towns abroad. **1973** *New Society* 8 Feb. 284/2 Twinning [of towns]—a translation of the French term *jumelage*—became popular after the second world war, and reached a zenith of municipal goodwill in the late 1950s. **1975** *Globe & Mail* (Toronto) 12 Sept. 5/2 Students in Ontario and nine Caribbean countries and the Bahamas will have a chance to learn more about one another's lives through a new twinning program launched by the Ontario Ministry of Education. **1983** *Listener* 6 Jan. 4/1 There is the twinning of cities.

twirl, *sb.* Add: **c.** *Criminals' slang.* A skeleton key. Cf. *TWIRLER b.

1879 *Autobiogr. of Thief* in *Macmillan's Mag.* Oct. 502/2 It was now that I got acquainted with the use of twirls (skeleton-keys). **1923** J. C. GOODWIN *Sidelights* xxvi. 165 In the room Bill cracked with his twirls we piped a pater. **1980** P. KINSLEY *Vatchman Switch* x. 82 She scarcely heard him open the old lock..with the set of 'twirls'.

d. *slang.* A prison warder.

1891 J. BENT *Criminal Life* 272 Will you go and tell Dutch Doll to come up to try and get me a right twirl (good warder)... There is a twirl here from another stir. **1933** G. INGRAM *Stir* xi. 160 I'm standing orderly on this landing and the twirl'll do anything for me. **1962** *John o' London's* 25 Jan. 82/2 Prison officers..are sometimes referred to as *twirls*.

e. A cake in the shape of a twirl.

1973 E.-J. BAHR *Nice Neighbourhood* ii. 20 My Viennese aunt's recipe for butter twirls. **1979** M. INGATE *Tomb of Flowers* i. 8 All kinds of rolls and buns..twirls that went round and round with currants in between.

twirler. Add: **a.** Also *spec.* (*N. Amer.*), one who leads a marching band; a drum-major or drum-majorette. Cf. *TWIRLING *vbl. sb.*

1949 R. L. LEE *Baton* 2 A twirler must 'sell' himself to the audience..having..a big natural smile. **1965** *Daily Progress* (Charlottesville, Va.) 21 May 9/2 Twirlers from Maryland, Ohio, Pennsylvania, [etc.] will compete in the individual events [in a baton-twirling competition]. **1980** *Times* 7 Aug. 12/7 British twirlers tend to call themselves Majorettes.

b. *Criminals' slang.* = *TWIRL *sb.* c.

1921 J. C. GOODWIN in *Chambers's Jrnl.* 24 Sept. 680/1 Skeleton-keys, or 'twirlers', as the thieving fraternity call them. **1935** R. T. HOPKINS *Life & Death at Old Bailey* ii. 64 In a burglar's kit of tools will be found a jemmy, a bunch of skeleton keys, known as 'twirlers',..wedges, glass-workers' diamond, and a 'treacle plaster'. **1974** 'J. ASHFORD' *Colour of Violence* iv. 34 Weir, who was an expert with the twirlers, forced the lock in six seconds.

twirligig. (Earlier and later examples.)

1902 H. BELLOC *Path to Rome* 58 Investigating the twirligigs of the brain to find out where the soul is. **1942** R. KNOX *In Soft Garments* ii. 11 Take those twirligigs in our brains which are concomitant..of our thoughts. **1980** *Washington Post Bk. World* 2 Nov. 9/4 The dress with an orange twirligig pattern worn by Mrs Umphelby.

twirling, *vbl. sb.* (In Dict. s.v. TWIRL *v.*[1]) (Examples in sense of manipulating a baton as the leader of a marching band.)

1938 BENNER & PAINTER *Art of Baton Spinning* I. ii. 4/1 All actions of baton spinning may be divided into four categories: Looping, Twirling, Spinning and Throwing. **1945** F. H. RODGERS *Keokuk High School Marching Band Student's Handbk.* II. 11 It is not necessary that the drum major do any twirling. **1974** *Cleveland* (Ohio) *Plain Dealer* 26 Oct. 6-D/3 They got fancier with their twirling and began to do tricks. **1980** *Times* 7 Aug. 12/7 The United States Twirling Association..[wants] to have twirling recognized as a full Olympic sport.

twirp, var. *TWERP.

twisel, twissel, *sb.* (*a.*) **3.** Delete † and add later *poet.* example.

1956 [see *PENELOPIZE *v.*].

twist, *sb.*[1] Add: **II. 10.** (Earlier example.)

1830 G. COLMAN *Random Rec.* II. iii. 78 But plague upon their *bapps*..a doughy sort of something, between a roll and a *twist*.

11*. A curled piece *of* lemon (or other citrus) peel used to flavour a drink. Also (colloq.) *ellipt.*

1958 A. L. SIMON *Dict. Wines* 58/1 *Merry Widow*, 50 per cent Byrrh Wine; 50 per cent Dry Gin. Fill glass with ice; stir and strain in cocktail glass; twist of orange peel and serve. **1968** *Spirits* ('Know the Drink' Series) 36/1 *Cuba Libre*, 2 oz. light rum, 1 tablespoon unsweetened lime juice. Pour over ice in glass, top up with Coca-Cola, add a twist of lime or lemon rind. **1971** G. V. HIGGINS *Friends of Eddie Coyle* (1972) viii. 55 He ordered a vodka martini on the rocks with a twist. **1973** [see *LILLET]. **1981** W. SAFIRE in *N.Y. Times Mag.* 2 May 18/3 A *twist* is of course a twist of lemon skin.

III. 12. c. A dance in which the body is twisted from side to side; *spec.* a dance of this kind popular in the early 1960s. Also, music for such a dance.

1894 in *Sunday Times* (1962) 11 Mar. 42/5 They're ready an' willin', An' fair at Kadrillin', But my little Flo does the twist. **1898** J. D. BRAYSHAW *Slum Silhouettes* 239 An' there's no kid abaht it, they can both on 'em darnce. Kitty took fust prize..at the contest at that there 'all in Bow. You orter see 'er do the twist. **1928** *Daily Tel.* 11 May 11/1 'The Twist', created by M. Camille de Rhynal..is designed to cultivate gliding and swaying movements. **1961** *Guardian* 4 Nov. 6/3, I have read recently that a new dance has been introduced in America called 'The Twist'. **1965** M. SPARK *Mandelbaum Gate* iv. 116 My mother makes a party for the girls to do the Twist. **1966** *Crescendo* Nov. 6/1 'Manchild'..is an exciting, driving twist. **1978** S. NAIPAUL *North of South* I. iv. 102 Modishly dressed African men and women dancing what I assumed to be the twist.

15. a. Also, a spiral ornament in the stem of a wine-glass. Usu. with defining word, as *air-twist*, *colour twist* (see under the first elements), *enamel twist*, and freq. *attrib.*; also *transf.*, a glass with this kind of stem.

1897 A. HARTSHORNE *Old Eng. Glasses* 275 The stems are of opaque-white twists of many threads. **1923** H. J. POWELL *Glass-Making in Eng.* iv. 61 A goblet with enamel-twist stem. **1927** W. A. THORPE *English & Irish Glass* 18 Enamel-twists in white or coloured enamels. **1930** T. ROHAN *Old Glass Beautiful* 72 A Norwich twist glass. **1961** C. M. ELVILLE *Collector's Dict. Glass* 81 Those glasses in white monochrome included ales and glasses with straight-sided and bucket-shaped bowls, most of which had enamel-twist stems. **1965** P. M. HUBBARD *Hive of Glass* iv. 42 Have you anything in the way of drinking glasses?.. A twist for choice. **1973** *Guardian* 17 Mar. 18/6 A wine glass with an opaque twist might be worth £25, but with a blue spiral as well £200. A goblet with coloured twists would be worth upwards of £500. **1979** *Radio Times* 7–13 Apr. 25/2 We don't normally touch chipped [glass] items—though we did have a very fine colour twist with a slight chip which went for £800... You can still get a little opaque twist of the 1750s for £20–£30.

c. (Earlier example.)

1880 E. W. HAMILTON *Diary* 30 Nov. (1972) I. 83 The Irish land question evidently weighs heavily on Mr. G... He is afraid of Forster 'getting a twist'. Forster is evidently in favor of very strong measures.

d. *the twist*: cheating, dishonesty; treachery; also in phrs. *on, at the twist. Criminals' slang.*

1933 C. E. LEACH *On Top of Underworld* x. 141 *Twist, at the*, double-crossing. **1938** F. D. SHARPE *Sharpe of Flying Squad* 334 *Twist* (*the*), to change something written or said from right to wrong. Sometimes called 'the Oliver Twist'... A dishonest bookmaker..would say: 'Put the Oliver on it', instead of..'Put the Twist on it'—which might be understood by the 'Mug'. **1977** J. WAINWRIGHT *Day of Peppercorn Kill* 29 Silver-smiths,..one of 'em on the twist. **1979** —— *Duty Elsewhere* x. 36 If I'd wanted you picked up—if I'd wanted to work a twist—would I be here, now? *Ibid.* xxv. 67 Who the hell's poor? *Really* poor? Poor enough to merit going on the twist?

e. Slang phr. *to get one's knickers in a twist*: to become unduly agitated or angry (*joc.*).

1971 *Morning Star* 26 June 2/1 Britain's Foreign Office mandarins have had their knickers in a twist for the past fortnight. **1982** *Brand New York* (Lit. Rev. special issue) 118/3 There is no reason to get one's knickers in a twist and believe the revolution is nigh.

20. c. An unexpected development of events, esp. in a work of fiction; a change from usual procedure.

1941 B. SCHULBERG *What makes Sammy Run?* ii. 31 It's a comedy with a helluva twist in it... *she* kidnaps *him*. **1943** B. SMITH *Tree grows in Brooklyn* (1947) xxvi. 145 She did not report happenings truthfully, but gave them colour, excitement and dramatic twists. **1962** [see *SNAPPER *sb.*[1] 2 e]. **1974** 'E. LATHEN' *Sweet & Low* xii. 125 Well, there's a new twist for you... I wonder how much it's costing Dreyer to go on network TV and remind us all that they specialize in murder, as well as chocolate. **1978** *Navy News* Oct. 3/1 Portraying a sailor came almost naturally to Peter O'Toole when he played Robinson Crusoe in 'Man Friday', which provides a new twist to the Daniel Defoe classic. **1982** M. YOUNG *Elmhirsts of Dartington* ix. 227 The fact that he was nephew to..a staunch opponent of theirs was a twist that..appealed to them both. He got the job.

d. *round the twist* = *round the bend* s.v. *BEND *sb.*[4] 10 c. *slang.*

1960 D. ABSE *House of Cowards* in *Plays of Year 1960–61* XXIII. 190, I knew he was barmy. I knew that man was round the twist, sayin' things like that. **1971** 'F. CLIFFORD' *Blind Side* IV. iii. 178, I ask you. Enough to send you round the twist. **1977** D. BAGLEY *Enemy* v. 38, I swear Ogilvie thought I was going round the twist.

IV. 22. twist grip, a control operated manually by twisting, *spec.* one which serves as a hand-grip, and alters the throttle on a motor cycle or scooter, or the gears on a bicycle; **twist-lock,** a locking device for securing large containers to the trailers on which they are transported; **twist-off,** (*a*) *Oil Industry* (see quot. 1932); (*b*) *attrib.*, that may be removed manually by twisting.

1954 J. MASTERS *Bhowani Junction* i. 13, I was bending over the handlebars, turning the twist-grip throttle. **1962** *Engineering* 2 Nov. 584 Travel controls consist of a twist-grip (the amount of twist governing the speed of travel) and a steering wheel. **1975** *Which?* May 143/4 Twistgrip gear change. **1980** *Outdoor Life* (U.S.) (Northeast ed.) Oct. 26/1 Several times I've bumped that twist grip accidentally, turning the motor on. **1969** *Jane's Freight Containers 1968–69* 137/2 'Tie-down' devices..are designed to mate with the bottom corners of the containers, which are fitted with twist-locks. **1977** *Grimsby Even. Tel.* 26 May 4/8 (Advt.), One new Crane Freuhauf 40 ft PSK twistlock trailer, available for hire, £35 per week. **1932** *Amer. Speech* VII. 271 *Twist-off*.., a breaking off of the rotary drill pipe in the hole by torsional stress. **1964** *Supermarket & Self-Service* May/June 19/2 The new twist-off cap. **1970** W. SMITH *Gold Mine* xiii. 38 The whole rig was seconds away from a twist-off. **1974** P. L. MOORE et al. *Drilling Practices Manual* ii. 14 Other limitations have to be considered such as..pipe wear and the danger of twist-offs. **1981** A. LOPEZ *Compl. Course in Canning* (ed. 11) I. viii. 183 (*caption*) 'Twist-off' or Lug cap.

twist (twist), *sb.*[3] *slang* (chiefly *U.S.*). [short for *twist-and-twirl* (also used), rhyming slang for *girl*.] A girl, a young woman (freq. depreciatory).

1924 *Truth* (Sydney) 27 Apr. 6 *Twist and twirl*, a girl. **1926** *Clues* Nov. 162/2 *Twist*, a girl. **1927** *Dialect Notes* V. 466 *Twist*, n., a loose woman. **1932** [see *DOG-HOUSE b]. **1953** 'R. MACDONALD' in H. Waugh *Merchants of Menace* (1969) 93, I hate to see it happen to a pretty little twist like Fern. **1956** H. GOLD *Man who was not with It* (1965) xvii. 154 I'm just as good as any of those Pittsburgh twist-and-twirls. **1979** [see *PROPOSITION *sb.* 7 b].

twist, *v.* Add: **III. 9. d.** *to twist the tail* (of a person): to annoy, to coerce (someone). *to twist the lion's tail* (U.S.): to provoke the resentment of British people.

1895 *Lit. Digest* 25 May 112/2 Papers in the U.S. take to shouting 'Hands off!' to England... Twisting the lion's tail is a regular electioneering maneuver. **1909** 'O. HENRY' *Roads of Destiny* xvi. 259 [He] twisted the tail of a Connecticut insurance company that was trying to do business contrary to the edicts of the great Lone Star State. **1926** E. L. ABBEY *Twist of Lion's Tail* 9 John Bull takes the lion for his emblem... Twist the lion's tail and how he hollers! **1935** 'N. BLAKE' *Question of Proof* v. 91 Revenge seems to me least likely. Grown men don't kill boys just because they've had their tails twisted by them. **1956** A. WILSON *Anglo-Saxon Attitudes* I. iv. 173 I get a good deal of amusement twisting both their 'advanced' tails, particularly the egregious parson's. **1965** P. O'DONNELL *Modesty Blaise* vi. 72 So they were going to twist his tail for a while. Well,..the side-effects would have to be accepted stoically. **1979** E. NEWMAN *Sunday Punch* vii. 58, I took a silent vow never again to twist the lion's tail editorially.

e. *to twist* (someone's) *arm*: to force or persuade someone to do something. Also used *joc.* when no coercion is needed, esp. with reference to drinking.

1953 *Word for Word* (Whitbread & Co.) 36/2 *Twist one's arm*, to 'persuade' one to have a drink, when no persuasion is needed. **1953** BERREY & VAN DEN BARK *Amer. Thes. Slang* (1954) §221/2 *Force; compel*... twist one's arm. *Ibid.* §223/5 *Induce; persuade*..twist one's arm. **1968** C. COOPER *Thunder & Lightning Man* iv. 65 The National Trust, in their genteel fashion, are beginning to twist my arm. The property must be made to pay its way. **1972** G. BELL *Villains Galore* v. 57 'That looks a very nice little pub over there.'..'All right—you've twisted my arm enough,' admitted Boote. **1977** G. SCOTT *Hot Pursuit* xii. 108 If you'd twisted my arm I would have had to admit that it was even important enough to justify the risks. **1982** H. ENGEL *Ransom Game* xxv. 154, I let him twist my arm into taking a Scotch with water.

12. d. *trans.* To cheat, to defraud. *slang.*

1914 JACKSON & HELLYER *Vocab. Criminal Slang* 95 They had to learn awareness in the school of cold, hard facts, having been..'twisted'..times innumerable. **1956** *People* 13 May 2/3 Don't imagine that all the boys in the trade are out to twist you. **1967** P. RYAN *How I became Yorkshireman* xv. 95 He were..content to be twisted daft wi'out mekking a mouse-squeak after value for his brass.

e. *trans.* and *intr. Insurance.* To induce someone to change a policy from one company to another. Cf. *TWISTING *vbl. sb.* 5.

1924 WEBSTER Add., *Twist, v.t. Life Insurance.* To induce (a person) to drop a policy already in force in a company other than that of the twisting agent for one in the agent's company.—*twister, n.* **1936** *Sun* (Baltimore) 29 July 16/2 He expressed the hope that any agent found 'twisting' or attempting to discourage policyholders in the Pacific Mutual not to retain their insurance would be reported to the Insurance Division.

IV. 14. b. (Earlier example.)

1816 W. LAMBERT *Instructions & Rules for playing Cricket* 20 The Ball may be twisted by the usual mode of under-armed Bowling.

c. *intr.* To dance the twist (*TWIST *sb.*[1] 12 c).

1961 *Guardian* 30 Dec. 5/3 It is a week with only one new film, a small loud monstrosity called 'Hey, Let's Twist'. **1968** J. UPDIKE *Couples* ii. 166 Frank was grotesquely Twisting..opposite Carol Constantine.

15. b. In Vingt-et-un, to receive a card dealt face upwards; also, to deal a card in this manner. Occas. *trans.* and as *imp.* Also *fig.* Cf. *STICK *v.*[1] 6 e.

1921 P. ALSTON *Card Games* 121 If it is not desired to buy, the usual expression is to say 'Twist'. *Ibid.* 122

Having bought a player can then twist; but once having twisted, a card cannot be bought. **1939** PHILLIPS & WESTALL *Compl. Bk. Card Games* 194 The player can either buy cards or can have them 'twisted'..; a card twisted is turned face upwards. *Ibid.*, B has a 9 and a 5; he says 'twist me one'. **1963** G. F. HERVEY *Handbk. Card Games* 285 He can twist: that is to say he elects to receive a card face upwards. **1972** *Guardian* 12 Oct. 1/3 Every pontoon player will understand the dilemma of the Tory chiefs. They are undecided whether to stick or twist on a relatively modest hand of cards. **1976** J. ARCHER *Not a Penny More, Not a Penny Less* xii. 136 The young man on Harvey's left also drew a ten and asked the dealer to twist again.

16. (Earlier example.)

1800 T. BOXALL *Rules & Instructions for Playing Cricket* 18 When the ball goes out of a bowler's hand he must endeavour to make it twist a little.

twisted, *ppl. a.* Add: **2. c.** Of the stem of a wine-glass: having a spiral ornament inside. Cf. *TWIST *sb.*[1] 15 a.

1897 [see *air-twisted* s.v. *AIR *sb.*[1] II]. **1897** A. HARTSHORNE *Old Eng. Glasses* 61 The glasses which we know generically as those with 'twisted stems', that is, with thin air-threads, and opaque white spiral lines in their standards. **1929** W. A. THORPE *Hist. English & Irish Glass* I. 199 Between 1714 and 1745..twisted stems are rare. **1970** G. GROS-GALLINER *Glass* iv. 108 A twisted stem decoration could be achieved by rib-moulding and rib-twisting.

3. b. Of a person: neurotic, emotionally unbalanced; perverted. Also *transf.* and with *up*.

1900 *Dialect Notes* II. 68 [College words.] *Twisted*, pp. as adj. 1. Wrong. 2. Crazy. 3. Confused. **1956** R. M. LESTER *Towards Hereafter* xiv. 165 Those who had held to the twisted idea that all psychical phenomena and spirit communication was 'the work of the devil' began to think again. **1963** *Times Lit. Suppl.* 25 Jan. 64/3 Alan..who is as extroverted as Paul is twisted up. **1963** A. HERON *Towards Quaker View of Sex* iii. 34 The emotional strains ..produce the twisted embittered woman, only too familiar to psychiatrists. **1971** A. MORICE *Death of Gay Dog* v. 72 Old Roger..does so enjoy having lots of money; not like that twisted-up Nancy. **1978** S. SHELDON *Bloodline* xlii. 360 Snuff films..would have been made to be shown privately to wealthy individuals who got their pleasure in twisted, sadistic ways.

4. twisted pair *Teleph.*, a pair of insulated conductors twisted about each other, as in a flex or by alternating their positions on successive telegraph poles.

1923 T. E. HERBERT *Telephony* xxiii. 651 The disturbing effect of a twisted pair in good condition is relatively minute when compared with a single wire. **1979** *Sci. Amer.* Aug. 9/2 (Advt.), Each link can be up to five miles long, and uses a single, shielded, twisted pair cable.

twister, *sb.* Add: **4. b.** Also, a dishonest person, a crook. *slang.*

1863 *Once a Week* IX. 568/2 One swags all that the palmer purchases, and stays outside to render the 'twister' any assistance he may need. **1915** *Film Flashes* 4 Dec. 1 'Twisters'..endeavour to put German films in the picture houses, under the pretext that they were made in a neutral, Continental country. **1930** J. B. PRIESTLEY *Angel Pavement* vii. 367 If you ask me, he looks a rotten twister—bit of a crook or something. **1937** [see *SPIV *sb.*]. **1940** E. POUND *Cantos* lv. 53 And Liu-hoei said Ngan was a twister. **1966** [see *KNUCK 3]. **1976** *Milton Keynes Express* 23 July 7/3 He was said to have called two women teachers 'cheats and twisters' and had refused to apologise for his remarks.

c. (Earlier example.)

1832 P. EGAN *Bk. Sports* I. 348/2 The batsman now his weapon rais'd To meet a puzzling twister.

d. Delete 'In the Mississippi region'. (Further examples.)

1903 G. S. WASSON *Cap'n Simeon's Store* vi. 108 He see in his paper where the English ship Falls of Ettrick was plunked on the Diamond Shoal and had went to pieces in that ole twister of a breeze there was a spell ago. **1930** NEFF & HENRY *Folk-Say Regional Misc.* 48, I never did see so many of them little twisters all a-goin at one and the same time. **1955** *Sci. News Let.* 18 June 388/2 A Weather Bureau meteorologist is making miniature tornadoes in a small box in the hope of learning more about what causes 'twisters'. **1967** *Boston Sunday Globe* 23 Apr. 1/2 The most vicious twisters in the history of the Midwest, striking heavily-populated sections of Northern Illinois and Western Michigan, left 53 dead. **1974** V. NABOKOV *Look at Harlequins* (1975) IV. ii. 162 A group of fifteen schoolchildren..were safe in the sudden darkness of that sturdy building when the twister struck. **1977** J. CLEARY *Vortex* iv. 93 You hear the twister warnings, too?

f. A grossly exaggerated tale; a lie. *Naut. slang.* ? *Obs.*

1834 W. N. GLASCOCK *Naval Sketch-Bk.* 2nd Ser. I. 235 I'm an even-minded man..that's providin' I wasn't provok'd by lying lip,—but if the best man in the sarvus was to come up to me,..to tell *me* such a thund'ring twister..why I'd just..floor the feller as flat as a flounder. **1850** H. MELVILLE *White Jacket* II. xxviii. 184 Among innumerable 'yarns and twisters' reeled off in our maintop during our pleasant run to the north, none could match those of Jack Chase. **1873** [in Dict., sense 6].

g. A type of handcuff (see quot. 1939).

1910 [see *NIPPER *sb.*[1] 4 c]. **1939** *Fortune* July 104/1 A style of handcuff, sometimes called 'twisters', used by the New York police instead of the old bracelet type. It consists of a short length of chain with a T-bar at each end. The policeman wraps it around the prisoner's wrist, twists the two T's like a tourniquet as tightly as necessary to make the prisoners come along like a lamb.

h. A key. *twister to the slammer*: (see quot. 1940). *U.S. slang.*

1940 *Music Makers* May 37/3 *Twister to the slammer*, the key to the door. **1941** J. SMILEY *Hash House Lingo* 55 *Twister*, key. **1944** D. BURLEY in A. Dundes *Mother Wit* (1973) 208 Give the jivers a break and substitute the phrase, 'twister to the slammer', for the word 'key'. **1970** C. MAJOR *Dict. Afro-Amer. Slang* 117 *Twister*, doorkey.

i. = TWITCH *sb.*[1] 3 b (*spec.* sense). *U.S.*

1940 W. FAULKNER *Hamlet* III. ii. 223 He..reached down from its nail in the wall a short, smooth white-oak stick eyed at the end with a loop of hemp rope—a twister which Houston had used with his stallion. **1948** *Richmond* (Va.) *Times-Dispatch* 23 Aug. 4/4 A mean horse may take up to eight hours to shoe, using rope harnesses to tie up the leg being worked on, or even a 'twister' for the horse's nose. **1968** R. F. ADAMS *Western Words* (ed. 2) 334/1 *Twitch*, a small loop of cord with a stick through it used to punish a held horse. The loop is placed vertically around the animal's upper lip and then tightened by twisting the stick. Also called *twister*.

j. One who dances the twist (*TWIST *sb.*[1] 12 c).

[**1942** BERREY & VAN DEN BARK *Amer. Thes. Slang* §583/27 *Hip dancer*...twister, wiggle dancer, wiggler.] **1966** 'K. NICHOLSON' *Hook, Line & Sinker* viii. 97 'I just go on Twist nights, don't I, Di?' 'She's a jolly good twister too.' **1977** J. WILSON *Making Hate* ii. 21 He'd been the runner-up in the Champion Twister competition at the Palais.

6. (Earlier examples.) Also (*dial.*), a blow which makes the victim twist or writhe; also *fig.* in U.S. colloq. phr. *to knock* one *the* (or *a*) *twister*.

1835 in *Amer. Speech* (1965) XL. 133 So, low each pill was a twister. I swallow'd about three Doctor's shops. **1843** J. R. PLANCHÉ *Fortunio* I. ii. 9 Ha, ha! I think that was a twister! **1886** F. T. ELWORTHY *W. Somerset Word-bk.* 783 *Twister*.., a blow with a whip or other instrument. **1896** G. ADE *Artie* vi. 55 That's what knocked me the twister. I thought this fellow was all right. **1908** G. SANGER *70 Yrs. a Showman* xvii. 59, I got a twister well home under his ribs that sent him grunting and staggering. **1934** G. ADE *Let.* 15 Mar. (1973) 181, I was, to use an old slang phrase, 'knocked a twister' when I received your letter [etc.].

8. *Insurance.* An insurance salesman or agent who unscrupulously induces a holder to switch his policy from one company to another. orig. *U.S.*

1924 [see *TWIST *v.* 12 e]. **1979** *Telegraph* (Brisbane) 24 Sept. 24/2 The industry calls it twisting. Presumably its practitioners are called twisters. The industry says that life insurance consumers are being ripped off by its practice.

9. *U.S. slang.* In various senses with reference to the taking of drugs (see quots.).

1936 [see *marijuana addict* s.v. *MARIJUANA 3]. **1936** *Amer. Speech* XI. 127/1 *Twister*. 1. A feigned spasm. **1938** *Ibid.* XIII. 192/1 *Twister*. 3. A speed-ball or whiz-bang [vein-shot of mixed drugs]. 4. A bit of violent retching or vomiting of blood or mucus during withdrawal distress. 5. A ration of narcotics. **1951** *Even. Sun* (Baltimore) 27 Mar. 4/1 A powerful combination of 'bernice snorting' and heroin 'shooting' was called 'blowing speed balls' or 'twisters' or 'whiz bangs'. **1959** J. E. SCHMIDT *Narcotics* 185 *Twister*, an intravenous injection of a potent narcotic taken by a drug addict, esp. a dose composed of heroin or morphine and cocaine.

10. Var. *TWISTOR 1.

11. Comb. **twister's cramp** *Path.*, pain in the hands or fingers produced by twisting or wringing.

1923 E. W. HOPE *Industr. Hygiene & Med.* viii. 516 This process of knotting [the warp threads] is done by a peculiar rolling motion of the fingers. The constant repetition of the movement..gives rise in certain operatives to a peculiar trade affection known as 'twister's cramp', the symptoms of which are pain, usually referred to the base of the thumb, tenderness of the muscles, and sometimes swelling at the base of the thumb. **1967** *Punch* 29 Mar. 458/3 Twister's Cramp can still be acquired by any housewife who is eccentric enough to wring clothes by hand.

twisteroo (twiːstĕrūˑ). *colloq.* [f. *TWIST *sb.*[1] 20 c; cf. *-EROO.] (A narrative with) an unexpected twist.

1963 *Punch* 13 Feb. 237/1 The story was a twisteroo, a variation, of the legend of King Midas. **1965** J. P. CARSTAIRS *Concrete Kimono* xxx. 251, I think when you hear me out the twisteroo will appeal to you. **1970** M. PEI *Words in Sheep's Clothing* iii. 23 There is in the field [of 'weasel' words used on the screen] one suffix which may be described as derogatory, and that is the -eroo of 'twisteroo'.

twistical, *a.* (Earlier U.S. example.)

1805 T. G. FESSENDEN *Democracy Unveiled* v. 158 Certain sages, learn'd and *twistical*... Have prov'd what's wonderful.

twistify (twiˑstifəi), *v. U.S. dial.* and *colloq.* ? *Obs.* [f. TWISTY *a.* + -FY. Cf. TWISTIFICATION.] *trans.* To make twisty (*lit.* and *fig.*); to twist. Hence **twi·stified, twi·stifying** *ppl. adjs.*

1835 R. M. BIRD *Hawks of Hawk Hollow* I. xix. 254 The path is astonishing twistified, and not fit for horse. **1843** 'J. SLICK' *High Life in N.Y.* I. 148 There was the fat nigger a twistifying his whip-lash round the horses' heads. **1845** in C. Cist *Cincinnati Misc.* I. 167, I knew..an individual..who possessed this twistifying talent in high perfection... Many amusing stories of his faculty of shifting have been told me. **1872** *Newton Kansan* 17 Oct. 4/3 [The Republicans] repudiate his twistified explanations now.

twisting, *vbl. sb.* Add: **5.** *Insurance.* (See quots.) Cf. *TWISTER *sb.* 8. orig. *U.S.*

1906 *Even. Post* (N.Y.) 20 Jan. (Financial Section) 7/1 By 'twisting' is meant the persuading of policyholders in one company to transfer their insurance to another. **1962** L. E. DAVIDS *Dict. Insurance* 205/2 *Twisting*, practice of inducing any policyholder to lapse or cancel a policy for the purpose of replacing such policy with another to the detriment of the policyholder. The practice is considered to be unethical as well as illegal. **1979** *Telegraph* (Brisbane) 24 Sept. 24/2 In simple terms twisting works like this. A life insurance agent sells you a policy when he is working for company X... Three years later he switches allegiance to company Y. To make life easier he retraces his steps..to tell you that the policy from company X is outdated.

twistor (twiˑstəɹ). [f. TWIST *v.* + -OR.] **1.** *Computers.* Also **-er.** A non-volatile memory element consisting of an insulated copper wire wound helically round with a wire of readily magnetized material. Freq. *attrib.*

1957 A. H. BOBECK in *Bell Syst. Technical Jrnl.* XXXVI. 1319 Three methods have been developed for storing information in a coincident-current manner on magnetic wire. The resulting memory cells have been collectively named the 'twistor'. **1962** *Flight Internat.* LXXXII. 170/1 A permanent twistor memory and a ferrite-core scratch-pad memory are vital to reliability, assuring continuous operation until an attacking missile is intercepted and destroyed. **1975** D. G. FINK *Electronics Engineers' Handbk.* XXII. 19 A twister memory is used as the semipermanent memory constituting a major portion of the program store in central control.

2. *Physics.* A type of spinor used in some descriptions of space-time (see quots. 1967, 1973).

1967 R. PENROSE in *Jrnl. Math. Physics* VIII. 346 Twistors are..the 'spinors' which are relevant to the six-dimensional space whose (pseudo-) rotation group is isomorphic with the conformal group of ordinary Minkowski space-time. The simplest (non-scalar) twistors constitute a four-dimensional, four-valued representation of the restricted conformal group... The general twistor is then a many-index quantity constructible from the above basic twistors by means of the usual 'tensor type' rules. **1973** —— & MACCALLUM in *Physics Rep.* VI. 244 Twistors (that is to say, the original flat-space twistors about which these notes are mainly concerned) are actually the reduced spinors for the proper pseudo-orthogonal group SO(2,4) which is locally isomorphic with, and 2-1 homomorphic with, the restricted conformal group of flat space-time. *Ibid.*, A twistor (of the simplest type) can be pictured 'classically' as effectively a zero rest-mass particle in free motion, where the particle may possess an intrinsic spin. **1979** *New Scientist* 31 May 737 Twistor theory is incomplete, but it offers considerable hope for a quite new approach to the basic problems of theoretical physics.

twisty, *a.* (*sb.*) Add: **c.** = *TWISTED *ppl. a.* 2 c.

1929 L. P. HARTLEY in *Mercury Story Bk.* 206 The glasses with twisty stems were there.

twit, *sb.*[1] Add: **2. b.** A fool; a stupid or ineffectual person. *slang.*

1934 E. LINKLATER *Magnus Merriman* xvi. 178 He was..a false hero who flaunted himself in fine colours when he was drunk and dwindled to a shabby twit when sober. **1960** F. RAPHAEL *Limits of Love* I. iii. 34 Don't be a twit, Sid. **1964** *Spectator* 10 Apr. 493/1 By making his psychologists a dim bunch of twits he weakens his statements. **1970** N. FLEMING *Czech Point* i. 20 No one but a prize twit or Captain Oates would have ventured out in this weather. **1977** C. MCCULLOUGH *Thorn Birds* xviii. 467 There's no need to get so worked up about it, you twit. **1984** *Observer* 4 Mar. 7/7, I hear and read such phrases as 'geriatric old twit': an expression which would hardly have sprung to the lips of the pious Aeneas.

twit, *int.* and *sb.*[3] **2.** (Later example.)

1922 T. S. ELIOT *Waste Land* (1923) iii. 15 Twit twit twit Jug jug jug jug jug jug So rudely forc'd. Tereu.

twitch, *v.*[1] Add: **3.** (Later example with *off*.)

1934 T. S. ELIOT *Rock* i. 28 As he names them they twitch off their caps and kneel.

8. Also said of a smile.

1930 'SAPPER' *Finger of Fate* 31 The faintest suspicion of a smile would twitch round his lips.

twitched, *ppl. a.* (In Dict. s.v. TWITCH *v.*[1]) Add: Also *spec.*, twitchy, irritable, 'rattled'. *slang.*

1959 P. TOWNEND *Died o' Wednesday* xii. 216 He or she is likely to be..worried..'twitched to the eyebrows' as the flying boys have it. **1981** S. JACKMAN *Game of Soldiers* I. i. 16 The C.O.'s in there and he's a bit twitched.

twitcher. Add: **4.** A bird-watcher whose main aim is to collect sightings of rare birds. So **twi·tching** *vbl. sb.*

1977 *Birds* Summer 59/3 Twitchers are difficult to identify because they are polymorphic. Best clues are behavioural including carrying Zeiss binoculars and *Where to Watch Birds*... Known to have nested in Wandsworth and possess a sense of humour. **1978** *Sunday*

Times 19 Feb. 5/1 (*caption*) The people in the picture are 'twitchers'—birdwatchers. **1980** [see *TICKER[3] 2]. **1980** L. BROWN in Howard & Moore *Compl. Checklist Birds of World* p. vii, Certainly, knowing what a bird is and where it occurs is the first step towards the wider interest that may stem from merely ticking a bird off on a list—twitching, as the vulgar parlance has it. **1982** *Times* 15 June 10/4 Twitchers are only interested in spotting rarities to claim they have seen them. Ornithologists are serious students, who despise and distrust twitchers. **1982** *Brit. Birds* LXXV. 537 The word 'twitcher' has, over the last decade, gained widespread use, meaning a person who 'chases' rare birds... I first heard the term from Bob Emmett at Beachy Head in 1968, when it was used to mean anyone who got 'twitchy' when southeast winds blew and headlands bristled with Pied Flycatchers and Redstarts.

twitchety (twiˑtʃĕti), *a.* Also **twitchetty.** [f. TWITCH *v.*[1] or *sb.*[1] + *-et* + -Y[1], perh. after *crotchety, fidgety*, etc.] Twitchy, nervous; of things, moving back and forth.

1859 HOTTEN *Dict. Slang*, *Twitchetty*, nervous, fidgetty. **1936** W. GREENE *Death in Deep South* III. 214 When I heard the buzzer, I wuz' mos' asleep and I wait two three minutes... It didn't ring again, but I git up anyways and look at the elevator and see where Number One gone twitchety. **1973** *Nature* 28 Sept. 225/3 Variables are, apparently, pulsating, eruptive, symbiotic, eclipsing or twitchety.

twi·tchily, *adv.* [f. TWITCHY *a.*[1] + -LY[2].] Nervously, in a twitchy manner; displaying nervous energy.

1934 in WEBSTER. **1964** J. BRAINE *Jealous God* iii. 70 A girl..who was pretty in an acceptable way, not tartish, not twitchily refined. **1978** J. UPDIKE *Coup* (1979) iii. 108 She no longer moved like the great-granddaughter of a leopard, but more electrically, twitchily, like a modern woman connected to a variety of energy-sources.

twitchy, *a.*[1] Add: **1.** Also said of a smile.

1924 W. M. RAINE *Troubled Waters* xx. 221 When she said goodbye to him it was with a wan twitchy little smile on her face.

Hence **twi·tchiness,** the state or condition of being twitchy; nervousness, fidgetiness, irritability.

1933 J. THURBER *My Life & Hard Times* p. xv, This type of writing is..the manifestation of a twitchiness at once cosmic and mundane. **1984** *Times* 23 Mar. 12/6 (*heading*) Twitchiness in the president's camp.

twitteration. For *nonce-wd.* read *Obs. rare* and add earlier example.

1820 G. COLMAN *X.Y.Z.* I. iii. 26 A female hand! bless me, I'm all over in a twitteration!

twittering, *ppl. a.* **2.** (Later example.)

1936 T. S. ELIOT *Coll. Poems 1909–35* 189 Not here Not here the darkness, in this twittering world.

twizzle, *v.* **2.** (Earlier examples.)

1788 E. PICKEN *Poems & Epistles* Gloss. 248/1 Twisle, to twist, fold. *c* **1840** LADY WILTON *Art of Needle-Work* xvi. 255 In vain she cut and screwed the thread, she burnt it in the candle..she twizled it between her finger and thumb..but enter the eye of the needle it would not.

two, *a.* and *sb.* Add: **B. I.** *adj.* **1. a.** *two cents' worth* (U.S.): = *TWO PENNYWORTH (fig.); cf. *two-cent* adj. in sense IV. 2 below; *no two ways about it*: see WAY *sb.*[1] 14 j.

1942 *Short Guide Gt. Brit.* (U.S. War Dept.) 18 You will hear..Britons openly criticizing their government... That isn't an occasion for you to put in your two-cents worth. **1954** *Sun* (Baltimore) 20 Dec. (B ed.) 14/7 The discussion concerning writers about old age in your column..impels me to add my 2 cents' worth.

2. a. Phr. *that makes two of us*, colloq. formula of agreement: the same is true of me, I am in the same position, I agree.

1956 H. KURNITZ *Invasion of Privacy* x. 71 'He was an amnesia victim.' 'That makes two of us,' said Zorn bitterly. **1974** P. DICKINSON *Poison Oracle* ii. 37 She wanted reassurance. That makes two of us, he thought. **1980** A. E. FISHER *Midnight Men* ix. 116 'She barely understands anything that is g-going on.' 'That makes two of us,' he said.

3. c. *two-thirds*: *spec.* in *Fashion*, applied to a garment that is shorter than the standard full length by about a third.

1963 *Harper's Bazaar* Oct. 56/2 This dreamy two-thirds coat. **1980** *Washington Post* 4 Dec. D3 The writers are really more interested in who's being reassigned to the Ottawa bureau than in whether two-thirds stockings are a trend or not.

5. b. *two-three*: for *dial.* read 'chiefly *dial.* and *U.S.*' (Further examples.)

1930 W. FAULKNER *As I lay Dying* 83 It'd taken them two-three days to get her to town in the wagon. **1949** L. HUGHES *One-Way Ticket* 40, I knock on your door About two-three A.M. **1962** J. F. STRAKER *Coil of Rope* ii. 13 You'll have to wait two-three days if you don't go now. **1976** A. PRICE *War Game* I. viii. 159 The last two-three years he's been working on a post-graduate thesis.

c. (Later examples.)

1818 SCOTT *Heart Midl.* III. ix. 229, I will explain to you in two words the connection betwixt this young woman and me. *a* **1845** S. SMITH *Sk. Moral Philos.* (1850) Lect. xix. 280, I never could find any man who could think for two minutes together. **1956** N. MARSH *Off with his Head* (1957) v. 92, I wonder if I may have two words with Dame Alice Mardian?

II. *sb.* **1. a.** (Earlier example of *to put two and two together*.)

1849 LYTTON *Caxtons* II. XII. i. 254 If they saw that, in proportion to their civility to me, they were depopulated by you, they would put two and two together and renounce my acquaintance.

2. g. Also *in two ups* (Austral. colloq.) = *in two shakes* s.v. SHAKE *sb.*[1] 2 h in Dict. and Suppl.

1934 T. WOOD *Cobbers* iii. 25 He said we'd be there in two ups. **1941** BAKER *Dict. Austral. Slang* 79 *Two ups, in*, in a brief space of time. **1967** J. MORRISON in *Coast to Coast 1965–6* 133 Too close to dark now, Mister, but we'll have you out of that in two ups in the morning.

IV. 1. a. *two-bar, -base, -bearing, -beat* (also *ellipt.* as *sb.*), *-bed, -berth, -blade, -car, -centre, -channel, -colour* (also *fig.*), *-column, -component, -cultures* (cf. *CULTURE *sb.* 5 c), *-deck* (later example), *-digit, -dollar* (also *fig.*), *-door, -drift* (*DRIFT *sb.* 2 f), *-electrode, -front, -hour* (earlier example), *-income, -lane* (*LANE *sb.* 2 d), *-level, -light* (earlier example), *-member, -pack, -part, -pedal, -person, -piano, -pin, -place, -point, -position, -reel, -seat, -sex, -shilling* (earlier example), *-stage, -story* (earlier example; also *-storey*), *-stripe, -term, -tier* (usu. *fig.*), *-topsail, -track* (also *fig.*), *-volume, -wheel* (earlier example), *-wire, -word.* **b.** *two-armed, -banked, -bedded* (earlier examples), *-columned, -decked, -engined, -fingered, -forked, -framed, -fronted, -gunned, -horsed, -named* (later example), *-pronged* (also *fig.*), *-seated, -storied* (earlier example; also *-storeyed*), *-stranded, -sworded, -termed, -tiered, -toothed* (earlier example), *-winged* (later examples). **c.** *two-hitter* (U.S.), *-mover* (earlier example), *-parter, -reeler, -striper* (see *STRIPER 1). **e.** *two-and-a-half striper* (see *STRIPER 1), *two-and-two* (cf. sense I. 2 d in Dict.), *two-in-oneness, two-plus-two, two-to-one.*

1861 'R. HARRINGTON' *Swimming* p. iii, They [*sc.* school children] often passed a river when out for a miserable two-and-two walk. **1957** J. S. BRUNER *Beyond Information Given* (1974) i. 35 The experiment is done on a conventional two-armed bandit, the subject having the task of betting on whether a light will appear on the left or on the right. **1935** L. MACNEICE *Poems* 25 Poetry is not only the bridging of two-banked rivers. **1967** 'M. HUNTER' *Cambridgeshire Disaster* x. 65 He warmed his hands briefly at the two-bar fire. **1880** N. BROOKS *Fairport Nine* 184 Ned made a fine two base hit which brought Watson home amidst great excitement. **1974** J. H. SUBAK-SHARPE et al. in Carlile & Skehol *Evolution in Microbial World* 132 Correlations exist between three-base sequences (that is codons) in messenger RNA and two-base sequences in the DNA coding for that Messenger RNA. **1922** *Times* 20 June 8/5 A short two-bearing auxiliary shaft. **1960** E. L. DELMAR-MORGAN *Cruising Yacht Equipment & Navigation* i. 20 The principle.., as with all two-bearing fixes, attains its best accuracy when the position lines are 90 deg. to each other. **1938** *Swing* July 18/2 *Two-beat swing*.., swing in which the accent is on the second or fourth beats. **1978** *Detroit Free Press* 16 Apr. 11 c/2 Chet Bogan plays two-beat, Dixiebelle sing same. **1792** H. NEWDIGATE *Let.* Feb. in A. E. Newdigate-Newdegate *Cheverels* (1898) viii. 110 She..shew'd me a neat 2 bed Garret where I dare say we shall sleep as well as in a Palace. **1962** E. SNOW *Other Side of River* (1963) lx. 461 Bachelor quarters varied from small two-bed to larger four-bed and six-bed rooms. **1784** H. NEWDIGATE *Let.* in A. E. Newdigate-Newdegate *Cheverels* (1898) iv. 59 We have obtained to Sleep to-night in a tollerable two bedded Room. **1788** J. WOODFORDE *Diary* 19 May (1927) III. 26, I slept at the Kings Head, in a two bedded Room. **1969** M. PUGH *Last Place Left* xxi. 164 He made out sleeper tickets for us..and we shared a two-berth compartment. **1967** *Jane's Surface Skimmer Systems 1967–68* 9/1 Propulsion is provided by two two-blade metal variable pitch propellers. **1983** *Flight Internat.* 10 Sept. 703/1 A pair of two-blade teetering rotors are mounted 50° apart. **1927** *Sat. Even. Post* 9 Apr. 89 This is a two-car country. **1961** M. BEADLE *These Ruins are Inhabited* (1963) viii. 105 Two-car families are rare in England. **1964** J. W. LINNETT *Electronic Struct. Molecules* ix. 152 Sovers showed that a very good description of the excited states could also be achieved by employing two-centre bonding and anti-bonding orbitals only. **1939** *Florida* (Federal Writers' Project) I. 75 Two-channel highways divided by a parkway to reduce the menaces of bright lights and head-on collisions. **1973** G. TALBOT *Ten Seconds from Now* (1974) xvi. 203 A blaring juke-box and two-channel television. **1979** H. KISSINGER *White House Years* xx. 827 The two-channel system and its significance. **1907** *T. Eaton & Co. Catal.* Spring & Summer 43/1 Men's fancy Sweaters,..medium weight, two color effect. **1925** I. A. RICHARDS *Princ. Lit. Crit.* xxvii. 211 *Macbeth*..is a highly successful, easily apprehended, two-colour melodrama. **1967** E. CHAMBERS *Photolitho-Offset* xv. 236 The wide use of lithography for colour work has resulted in the development of two-colour machines. **1916** E. WALLACE *Let.* 13 Nov. in M. Gilbert *Winston S. Churchill* (1972) III. Compan. II. 1583 Do a two column story. **1922** JOYCE *Ulysses* 465 Tom Rochford..jumps from his two-columned machine. **1956** *Nature* 18 Feb. 328/2 Only two-component solvent systems were investigated. **1966** *Listener* 1 Sept. 297/1 As Margaret Mead puts it: 'The recent two-cultures discussion is essentially a lament about..lack of communication.' **1960** *Farmer & Stockbreeder* 29 Mar. 93/1 Sending the farm two-deck lorry to the south-west and back. **1883** *Man. Seamanship for Boys' Training Ships* (Admiralty) (1886) 2 A two-decked ship. So named from having two gun decks below the upper deck. **1963** *Rep. Comm. Inquiry Decimal Currency* 9 in *Parl. Papers 1962–3* (Cmnd. 2145) XI. 195 People find considerably greater difficulty in remembering and manipulating three-digit numbers than two-digit numbers. **1793** *Deb. Congr. U.S.* 3 Jan. (1849) 788 The miserable two-dollar men who were raised for a six months' service. **1873** T. B. HAZARD *Nailer Tom's Diary* (1930) 414/2 Went to Thomas Rodmans Store to get a two dollar bill changed. **1929** *Century Mag.* Autumn 68 He hated what he called 'two-dollar words' and 'high hat' manners. **1908** *Sears, Roebuck Catal.* 61/1 The two-door automatic Model 'A' desk cabinet. **1982** *Sunday Tel.* 1 Aug. 6/2 You can get a 1.8 litre, two-door, four-seat sports saloon for £5,950. **1926** H. MACPHERSON *Mod. Astron.* 156 The results of this study of stars of very large proper motion, scattered all over the sky, were strongly in support of the two-drift hypothesis. **1921** *Wireless World* IX. 187/1 The two-electrode Fleming valve. **1931** *19th Cent.* Feb. 154 The standard two-engined type. **1966** D. FRANCIS *Flying Finish* v. 64 Small two-engined job..cost nearly thirty-five pounds per flying hour to hire. **1910** W. DE LA MARE *Three Mulla-Mulgars* 47 A behemothian bull-Ephelanto..wound his long, two-fingered trunk round Nod's belly. **1978** *Church Times* 27 Jan. 5/3 What she was doing was making a kind of 'I'm as good as you, see if I'm not' two-fingered gesture of defiance at society. **1981** H. CARPENTER *W. H. Auden* (1983) II. iv. 341 He was an adequate two-fingered typist. **1934** DYLAN THOMAS in *Criterion* Oct. 28 The two framed globe that spun into a score. **1946** *New Yorker* 23 Mar. 74/2 One of those two-front wars which 'Mein Kampf' had sensibly argued couldn't be won. **1856** J. G. WHITTIER *Panorama* 12 The two-fronted Future... To-day, your servant, subject to your will; To-morrow, master, or for good or ill. **1936** DYLAN THOMAS *Twenty-Five Poems* 44 And from the windy West came two-gunned Gabriel. **1949** *Time* 10 Oct. 47/2 Accompanied by a grim, 200-lb., two-gunned Big Spring sheriff. **1974** *State* (Columbia, S. Carolina) 26 Apr. 4-B/3 Paul Splittorff pitched a two-hitter for seven innings. **1939** W. B. YEATS *On Boiler* 31 There in a two horsed carriage..Great bladdered Emer sat. **1880** 'MARK TWAIN' *Tramp Abr.* xix. 171 A two-hour pedestrian excursion. **1969** *Guardian* 31 Mar. 2/1 The two-income family spends 15 per cent more on alcohol. **1895** HARDY *Jude* VI. ii. 404 O my comrade, our perfect union—our two-in-oneness—is now stained with blood! **1957** J. KEROUAC *On Road* (1958) II. viii. 156 On a two-lane highway to Baton Rouge in purple darkness. **1957** H. WHITEHALL in N. Frye *Sound & Poetry* II. 142 Metrical patterns..based on the two-level contrast of stressed versus unstressed syllables. **1981** *Beautiful Brit. Columbia* Summer 23/1 A glassed-in two-level public seating area provides a view of busy False Creek. **1845** S. R. GLYNNE *Notes Churches of Cheshire* (Chetham Soc.) (1894) 56 There are large gargoyles at the angles..; in the second stage a two-light window. **1924** O. JESPERSEN *Philos. Gram.* xxii. 306 Every sentence is said to be composed of two parts, Subject and Predicate... Besides such two-member sentences..we may have one-member sentences. **1967** R. S. CHURCHILL *Winston S. Churchill* II. viii. 261 The city of Dundee in 1908..was, like Oldham, a two-member, primarily working-class, constituency. **1868** *Westm. Chess Club Papers* I. 47 A two-mover now knocks me down. **1931** R. GRAVES *To Whom Else?* 17 Two-named one, how shall I call you without duplicity? **1952** V. WILKINS *King Reluctant* I. ix. 131 The involved two-pack patience game known as 'Maréchal Saxe'. **1977** *Listener* 10 Mar. 295/2 He is a two-pack smoker, drinks on occasion to keep going. **1928** *Daily Tel.* 11 Dec. 17/4 'Two-part tariffs' will be introduced by the County of London Electric Supply Co. Ltd. from Jan. 1 for domestic and business consumers. **1939** F. SCOTT FITZGERALD *Let.* 8 June (1964) 106, I have..over half-finished what will be a two-parter for *The Saturday Evening Post*. **1984** J. WAIN in *Listener* 28 June 30/2 They are making two plays out of it, or, perhaps more exactly, a three-hour two-parter. **1961** *New Scientist* 5 Jan. 50/3 Mr. Saunders's two-pedal system..interested me, so I tried it on my car. **1920** D. H. LAWRENCE *Lost Girl* vii. 134 Madame and the German did a screaming two-person farce. **1977** *New Yorker* 17 Oct. 93/1 A new playwright..made a pleasing début on Broadway..with a harsh little two-person comedy. **1952** S. KAUFFMANN *Philanderer* xv. 242 Turning the phonograph very low, he played the Mozart two-piano concerto. **1978** LD. DROGHEDA *Double Harness* x. 92 During the two years that she had been in America, Joan had become a very active pianist, giving a considerable number of two-piano concerts with her partner Harold Triggs. **1894** D. SALOMONS *Electr. Light Installations* (ed. 7) II. vi. 233 The plug to be used with this connector is of the two-pin type. **1962** *B.S.I. News* Feb. 24/1 Three proposals were considered for a standard two-pin plug for use with all-insulated and double-insulated appliances. **1948** H. REICHENBACH *Elem. Symbolic Logic* §17. 83 Both functions are two-place functions; i.e. they possess two arguments. **1963** W. V. QUINE *Set Theory* §1. 13 Suppose the only primitive predicate of some theory is a two-place predicate. **1966** *Publ. Amer. Dial. Soc.* 1964 XLII. 10 *Two-plus-two*,..two-seater with capacity to hold two additional passengers in back. **1977** *Lancashire Life* Mar. 118/3 The 104ZS is cramped in the back and is more of a two-plus-two than a proper four seater. **1960** *Farmer & Stockbreeder* 1 Mar. 131/1 Experience has proved this two-point system. **1972** J. POTTER *Going West* 180 The plane made a jerky two-point landing. **1951** Two-position [see *BISTABLE *a.*]. **1960** *Farmer & Stockbreeder* 8 Mar. 108/2 Two-position Hitch. Easy for field or transport. **1919** W. S. CHURCHILL in M. Gilbert *Winston S. Churchill* (1977) IV. Compan. I. 535 With Mr Balfour's approval, I made the following two-pronged proposals. **1958** O. CAROE *Pathans* xxiii. 375 The success of the two-pronged advance by the Khaibar and the Kurram. **1929** WODEHOUSE *Mr. Mulliner Speaking* ix. 301 Come on, let's beef in or we'll be missing the educational two-reel comic. **1978** *Radio Times* 18–24 Mar. 15/1 Within the next couple of years he made over 20 two-reel Westerns. **1928** *Sunday Express* 3 June 4 The British Screen two-reeler 'Homes of Our King'. **1979** *Guardian* 4 Aug. 9/2 He had made a few modest pictures, two-reelers mostly. **1895** *Montgomery Ward Catal.* Spring & Summer 580/2 Two-seat business wagons. **1880** W. WHITMAN *Daybks. &*

Notebks. (1978) III. 642 Heavy two-seated covered voitures. **1936** G. B. SHAW *Six of Calais* 89 Between them, near the King's pavilion, is a two-seated chair of state for public audiences. **1933** O. JESPERSEN *Essent. Eng. Gram.* xix. 192 When a special indication of sex is wanted with one of the two-sex words, this can always be done by the addition of the adjectives *male* and *female*, respectively: *a male reader, a female cousin*,..etc. **1973** TALAMINI & PAGE *Sport & Society* v. 271 To note [sport's] emergence as a two-sex activity..should not obscure the persistence in sport of male domination, male prejudice, and discrimination against girls and women. **1789** J. WOODFORDE *Diary* 11 June (1927) III. 112 Briton also went into the 2 Shilling Gallery. **1944** R. V. JONES *Most Secret War* (1978) xlv. 460 A two-stage rocket of about 150 tons starting weight could deliver a 1 ton warhead to nearly 3,000 miles range. **1963** *Ann. Reg. 1962* 183 The Skybolt was a two-stage solid fuel missile. **1979** *Dædalus* Summer 50 'Hypocrisy displayed', then, is a two-stage process, a masking followed by an unmasking. **1854** M. S. CUMMINS *Lamplighter* iii. 13 A decent two-storied house. **1874** W. BLACK *Princess of Thule* (ed. 2) I. i. 12 It was a square, two-storeyed substantial building of stone. **1977** P. G. WINSLOW *Witch Hill Murder* II. xvi. 212 The lovely, two-storeyed drawing-room. **1796** *Aurora* (Philadelphia) 15 Apr., That certain one-story Frame shop in front, and Two Story Frame messuage. **1866** A. D. RICHARDSON *Secret Service* ii. 38 We took a two-story car of the Baronne street railway. **1977** *Times* 9 Sept. 2/3 Spacious late-nineteenth-century two-storey terrace houses. **1982** W. BOYD *Ice-Cream War* (1983) II. ii. 120 He passed the ..two-storey building. **1851** H. MELVILLE *Moby Dick* I. ix. 65 Shipmates, it is a two-stranded lesson. **1976** P. COLLARD *Development Microbiol.* v. 64 Synthetic two-stranded RNA polyinosinic acid-polycytidylic acid. **1918** *Jrnl. R. Naval Med. Service* IV. 317 A 'two-stripe doctor'. **1860** W. WHITMAN in *N.Y. Times* 27 June 2/1 The Princes of Asia, swart-cheek'd princes, First comers, guests, two-sworded princes. **1933** A. N. WHITEHEAD *Adventures of Ideas* xv. 230 Also we may well ask whether there are not subtle variations of meaning stretching far beyond the competence of the two-term vocabulary—Judgment, Proposition. **1964** R. H. ROBINS *Gen. Linguistics* I. 26 A two-term relation between the word and the referent. **1968** FOX & MAYERS *Computing Methods for Scientists & Engineers* i. 4 A better approach is to observe that, with a single integration by parts, we can find the two-term recurrence relation. **1981** *Times* 6 Aug. 8/7 A programme which can be implemented only by a two-term Government. **1933** *Mind* LXII. 45 If we consider *Russia is happy, England fears France, Germany prefers England to France, Italy believes that Germany prefers England to France*, we see that these facts form a series in that the first is one-termed, the second two-termed, the third three-termed and the fourth four-termed. **1964** E. BACH *Introd. Transformational Gram.* vii. 157 Prepositional phrase in English is usually a two-termed relation. **1932** *Times* 26 Nov. 12/3 If their system were a two-tier system, the element of uncertainty was reduced to a minimum. **1933** [see *STACK *sb.* 1 d]. **1937** *Burlington Mag.* Oct. 194/1 Ionic impost capitals,..two-tier capitals. **1969** *Punch* 1 Jan. 26/3 It is now working a good deal harder and will work even harder still when the reforms with their two-tier system of voting and non-voting but speaking Peers come into force. **1975** J. P. MORGAN *House of Lords & Labour Govt.* ii. 75 The House can be persistent when an issue catches its imagination—favourites of the 1966–70 Parliament were decimal currency, the two-tier postal service, and the anti-Stansted campaign. **1969** *Listener* 23 Jan. 103/2 The debate is two-tiered: the violence of Chicago..and the violence of the TV shows. **1979** *Arizona Daily Star* 5 Aug. D2/2 The Natural Gas Policy Act of 1978..wiped out the two-tiered pricing structure. **1910** *Motor Man.* (ed. 12) iii. 75 The layshaft is driven by a chain instead of the usual two-to-one gear wheels. **1975** *New Yorker* 17 Nov. 117/1 The Court of Appeals for the District of Columbia Circuit, in a two-to-one vote, upheld the claim. **1742** W. ELLIS *Mod. Husbandman* Sept. xxvi. 124 These are the right profitable sort for Fatting, and not the two-toothed sheep. **1811** *Boston Patriot* 23 Jan. 3/2 A two top-sail sch[ooner] was at H Hole on Saturday. **1944** J. MASEFIELD *New Chum* 137 There used to be that kind of schooner. She was called a 'two-topsail schooner' or a 'maintopsail schooner'. **1934** *New Statesman* 27 Oct. 602/2 It is useless approaching these circles with two-track class-war propaganda. **1961** G. A. BRIGGS *A to Z in Audio* 200 Regular supplies of high quality 7½"/sec pre-recorded stereo tapes will be available, giving twice the playing time of previous two-track issues. **1977** *N.Y. Rev. Bks.* 13 Oct. 18/1 Had President Kennedy been pursuing a 'two-track' policy of offering Castro friendship while plotting his murder? **1925** I. A. RICHARDS *Princ. Lit. Criticism* 1 In a pamphlet or in a two-volume work. **1978** *Early Music* Oct. 544/1 Silvestro Ganassi's two-volume treatise on the viol..is perhaps the most interesting and significant instrumental tutor to have survived from the 16th century. **1744** W. ELLIS *Mod. Husbandman* Jan. i. 16 He plowed up the Surface.. with a two-wheel pecked Share-Plough. **1918** D. H. LAWRENCE *New Poems* 41 Who then sees the two-winged Boat down there? **1949** E. POUND *Pisan Cantos* lxxvi. 45 Benecomata dea Under the two-winged cloud. **1974** *Encycl. Brit. Macropædia* V. 819/1 Diptera, the two-winged, or 'true', flies. **1930** *Engineering* 28 Feb. 278/3 Power distribution is at 230 volts in the two-wire system. **1961** W. F. LEOPOLD in Saporta & Bastian *Psycholinguistics* (1961) 357/1 Two-word verbs of the type 'wake up'. **1977** N. SAHGAL *Situation in New Delhi* xi. 115 They had labelled him with destructive little one and two-word flourishes that could smear an image for millions of readers.

2. two-address *a. Computers*, having two addresses (see quot. 1953); **two and eight** *Rhyming slang*, a state (of agitation); **two-backed beast** = *the beast with two backs* s.v. *BEAST *sb.* 4 b; **two-bagger** *N. Amer. Baseball*, a hit that enables the batter to reach second base safely; **two-bit** *a. U.S.* [*BIT *sb.*[2] 8 b], (*a*) of the value of a quarter of a dollar; (*b*) *fig.*, cheap, petty, worthless (*slang*); **two-body** *a. Physics*, involving or pertaining to two objects or particles; **two-by-four** orig. *U.S.*, a post or batten measuring 2 inches by 4 in cross-section; also (*U.S.*) *fig.* in *attrib.* use: small, insignificant; **two-cent** *a. U.S.*, (*a*) of the value of two cents; (*b*) *fig.* = *two-bit* (*b*) above (*slang*); **two cheers,** catch phr. expressing qualified enthusiasm *for* something, as opposed to the traditional three cheers (see *THREE *a.* 1 g); **two-China(s)** *a. U.S. Pol.*, designating a proposal or policy for admitting to the United Nations representatives of both Communist China and Taiwan; **two cultures**: see *CULTURE *sb.* 5 c; **two-egg** *a.* = *DIZYGOTIC *a.*; **two-eyed stance** *Cricket* (see quot. 1924); **two fingers,** two fingers made into a V-sign as a coarse gesture of contempt; **two-fisted** *a.*, (*b*) *U.S. colloq.*, tough, aggressive; vigorous; (*c*) *Tennis*, of a backhand stroke: played with both hands on the racket; **two-holer** *N. Amer.*, a privy or lavatory accommodating two people; **two-horse** *a.* (earlier example); (*b*) (of a race or other contest) in which only two of the contestants are likely winners; **two-line** *a.*, (*b*) occupying two lines, being two lines long; (*c*) underlined twice; *spec.* applied to a notice of forthcoming parliamentary business in which the attendance of members is requested with two underlines, indicating less urgency than a three-line whip (now the strongest one); **2LO** (tūˌelˌōᵘˑ) [*Lo*(*ndon*)], the call-sign of a radio station established in London in 1922 and taken over the same year by the newly-formed British Broadcasting Company, which used it as call-sign until 1924 and as a name of a programme service until 1930; the station itself; **two-meal,** (*a*) (earlier example); **two(-)minute(s') silence** = *SILENCE *sb.* 2 e; **two nation(s),** used, chiefly *attrib.*, with reference to two irreconcilable groups or factions within a nation; **two-one,** a place in the upper division of the second class in a degree class-list; (a graduate having) an upper-second-class degree; also represented as 2(1), etc.; **two-piece** *a.* and *sb.*, (a suit) consisting of two garments (as a jacket and trousers, or coat and dress) matching or meant to be worn together; (a swimming costume) consisting now of a brassière and briefs, a bikini; also (*U.S.*) **two-piecer; two-place** *a.*, applied to an aeroplane with seats for two people; so **two-placer; two-power standard** (see quot. 1910); **two-revolution** *a. Printing*, applied to a cylinder press in which the impression cylinder rotates continuously, alternately printing and delivering sheets as the forme moves to and fro; abbrev. **two-rev** (also *absol.*); **twos and threes,** a children's chasing game for six or more players; **two-seater,** something (as a vehicle or settee) that has two seats; freq. *attrib.*; **two shoes** (earlier example); **two-shot,** a cinema or television shot of two people together; **two-spot** *U.S.*, (*a*) a playing card with two pips, a deuce; (*b*) a two-dollar banknote; (*c*) *transf.*, an insignificant or worthless person; a two-year prison sentence; **two-star** *a.*, (*a*) given two stars in a grading system in which more stars indicate higher quality (cf. *STAR *sb.*[1] 10 c); (*b*) having or being a military rank that has two stars on the shoulder-piece of the uniform, e.g. major-general or rear-admiral in the U.S. (cf. *STAR *sb.*[1] 6 c); **two-state** *a.*, capable of existing in either of two states or conditions; **two-step**: (*a*) also as *v. intr.*, to dance a two-step; (*b*) *a.*, having or consisting of two successive actions; two-stage; **two-sticker** *Canad. colloq.*, a two-masted boat; **two-suiter,** (*a*) *Bridge* (see quot. 1923); (*b*) orig. *U.S.*, a suitcase large enough to hold two suits and accessories; **two-tailed** *a.*, (*a*) having two tails; (*b*) *Statistics*, applied to a test that tests for deviation from the null hypothesis in both directions; cf. *one-tailed* adj. s.v. *ONE *numeral a., pron.*, etc. B. 33; **two-tone** *a.*, (*a*) in two colours or two shades of the same colour; (*b*) being or producing two notes, usu. alternately at intervals; also *fig.*; so **two-toned** *a.*, in the same sense; **two-tooth** *a.* (later example); *sb.* (examples); **two-two,** (*a*) (usu. represented as .22), used *attrib.* and *absol.* to designate (ammunition for) a gun with a calibre of 0.22 inch; (*b*) (also represented as 2(2), etc.), a place in the lower division of the second class in a degree class-list; see also sense II. 2 g; **two-week** *dial.*, a fortnight; **two-year** *a.*, (*b*) lasting or valid for two years; also *transf.*

1948 Two-address [see *three-address* adj. s.v. THREE *a.* and *sb.* III. 2]. **1953** *Computers & Automation* Dec. 12 *Two-address*, in programming, a system of instructions whereby each complete instruction includes an operation and specifies the location of two registers, usually one containing an operand and the other the result of the operation. **1961** [see *SEQUENCING *vbl. sb.*]. **1982** G. LEE *Hardware to Software* xx. 339 With 3-address instructions (as for 2-address), it is not necessary to have an accumulator. **1938** 'J. CURTIS' *They drive by Night* ix. 103 Give us a hand out, will you? I'm in a right two and eight. **1967** M. CECIL *Something in Common* 129 Poor old Clinker! Bet she's in a proper two-and-eight! *a* **1693** Two-backed beast [see *BEAST *sb.* 4 b]. **1925** G. GREENE *Babbling April* 5 The two-backed beast went trotting in my brain. **1939** DYLAN THOMAS *Map of Love* 63 Here dwells... Sam Rib, the two-backed beasts. He pointed to his Map of Love. **1973** L. SNELLING *Heresy* II. i. 60 Hubby got on to us. Came within an ace of catching us making the two-backed beast in his car. **1880** *Globe* (San Francisco) 16 May 1/4 Willigrod, Smith and J. Whitney led at the bat, the two former getting in each a two-bagger. **1946** *N.Y. Herald Tribune* 24 Mar. VIII. 19/1, I hit a two-bagger with the bases full. **1802** J. DRAYTON *View South-Carolina* 215 Hence the origin of this society; which, from the contributions, being a sum of money called *two bitts*, became known by the appellation of the *two bitt* club. **1880** *Harper's Mag.* May 799 Thompson's Two-bit House, Front St. **1928** S. LEWIS *Man who knew Coolidge* v. 51 There's a man..always got a good story and a two-bit cigar for you. **1932** E. CALDWELL *Tobacco Road* xvii. 183 Tom said she used to be a two-bit slut. **1978** T. WILLIS *Buckingham Palace Connection* viii. 155 Some other two-bit General will try shooting us up. **1956** *Nature* 11 Feb. 268/2 A principal aim of the theory is to show that, with only such two-body forces, one can achieve nuclear saturation. **1978** PASACHOFF & KUTNER *University Astron.* xxvi. 648 The effects that the planets have on each other are much less than the effect that the sun has on each, so the mutual interactions of the planets are treated as small deviations..from the situation that would be present if only the sun-planet two-body problem had to be solved. **1884** E. W. NYE *Baled Hay* 23 The managing editor of the mill lays out the log in his mind, and works it into dimension stuff, shingle bolts, slabs, edgings, two by fours. **1897** 'O. THANET' *Missionary Sheriff* 13 'That how he makes a living?' 'Yes—little two-by-four bakery.' **1906** 'B. M. BOWER' *Phantom Herd* v. 77 Houses..bald board fences save where two-by-fours braced them from falling. **1939** J. B. PRIESTLEY *Let People Sing* i. 5 A little two-by-four provincial agency..giving itself airs now! **1978** *Maledicta* II. 7 You hit him over his ol' skull with a two-by-four. **1979** J. VAN DE WETERING *Maine Massacre* xix. 226 De Gier saw a tableau made out of barn boards, framed neatly by weathered two-by-fours. **1859** L. A. WILMER *Our Press Gang* 42 The Express is a two-cent cash paper. **1899** 'J. FLYNT' *Tramping* I. iv. 86 The next higher type of the town tramp is the 'two-cent dosser'—the man who lives in stale-beer shops. **1902** [in Dict., sense IV. 1 a]. **1908** Two-cent [see *GU[illegible]]. **1951** E. M. FORSTER (*title*) Two cheers for democracy. **1977** *Times* 5 Sept. 6/7 Two cheers for..the National Theatre's debut in community drama. **1962** E. SNOW *Other Side of River* (1963) xx. 149 Mao and other Chinese continued to respect Chiang Kai-shek for one thing, however: he had declined to support the 'two-Chinas' plot aimed at removing Taiwan from the sovereignty of China. **1979** H. KISSINGER *White House Yrs.* xviii. 719 This came close to the two-China solution always vehemently rejected by both Taipei and Peking. **1959** *Listener* 29 Oct. 728/2 Two-egg twins..are derived from the separate eggs fertilized by two different sperms. **1971** Two-egg [see *multiple birth* s.v. *MULTIPLE *a.* 4 c]. **1924** A. C. MACLAREN *Cricket Old & New* viii. 73 What is called the 'two-eyed stance' or the turn of the head to enable the batsman to see the ball with as full a face as possible, but without taking the left shoulder off the line of the [illegible]. **1977** *Sunday Times* 3 July 28/2 Hughie Trumble condemned the two-eyed batting stance. **1971** G. EWART *Gavin Ewart Show* III. 50 Meanwhile on the roof of the Playboy Club..one Bunny Flag. Two ears, like some man's two-fingers-up to Culture. **1977** *Zigzag* June 24/1 There's nothing more the kids want to see than the Pistols at the top of the chart, two fingers pointed at the TV, radio, printing firms and council officials who've tried to stifle them. **1774** P. V. FITHIAN *Jrnl.* (1900) 123 He was Director, and appointed a sturdy two-fisted Gentleman to open the Ball with Mrs. Tayloe. **1903** [see *CLEAN *v.* 4 c]. **1925** E. E. CUMMINGS *Let.* 3 Sept. (1969) 108 [I] was not made to match wits with twofisted gogetters. **1960** *Times* 16 June 18/6 The two-fisted backhand of Australia's Howe. *Ibid.* 16 Sept. 52 Hamdingers being, as the menu says, 'for two-fisted appetites'. **1978** *Times* 4 July 19/4 Miss Kruger, a steady and tenacious competitor with a two-fisted backhand. **1979** *Tucson Mag.* Mar. 18/3 Bonanno..is going to have a two-fisted fight on his hands. The U.S. Criminal Justice system is also bearing down on Bonanno. **1971** *Islander* (Victoria, B.C.) 3 Jan. 12/2 Gone were the woodshed and the two-holer discreetly removed from the house. **1974** S. ALSOP *Stay of Execution* II. 224 The house..is a shooting lodge, built in 1929... There is one bathroom for the ladies, and one for the men, each with a two-holer. *c* **1780** 'J. H. ST. J. DE CRÈVECŒUR' *Sk. 18th-Cent. Amer.* (1925) 138 You have often admired our two-horse waggons. **1976** *Newmarket Jrnl.* 16 Dec., [Darts] With White Lion beating Wellington 7-2 it seems to be a two-horse race at the top. **1977** *Evening Gaz.* (Middlesbrough) 11 Jan. 14/1 [Association football] The two-horse race in Division I looks like continuing. **1984** *Times* 22 Feb. 1/5 If, as is expected, he comes second in New Hampshire, it could turn into a two-horse race. **1901** G. B. SHAW *Admirable Bashville & Cashel Byron's Profession* (rev. ed.) Pref. 288, I like the melodious sing-song, the clear simple one-line and two-line sayings, and the occasional rhymed tags. 1958 WILDING & LAUNDY *Encycl. Parl.* 603 Sir Wilfrid Lawson explained pithily that a one-line whip meant 'you ought to attend'; a two-line whip 'you should attend'; a three-line whip 'you must attend'. 1962 W. NOWOTTNY *Lang. Poets Use* v. 110 One usually expects of couplets that the sense of the passage will fall into two-line chunks. 1976 H. WILSON *Governance of Britain* iii. 46 The chief whip then indicates the kind of whip he will issue for each day—one-line, two-line or three. 1978 W. WHITE in W. Whitman *Daybks. & Notebks.* I. 121 'Roaming in Thought' is a two-line poem, published in the 1881 *Leaves of Grass.* 1923 *Radio Times* 28 Sept. 3/1 (*heading*) A recent talk broadcast from 2LO. 1924 A. R. BURROWS *Story of Broadcasting* viii. 59 A station known as 2LO, a 100-watt set contained in a small teak cabinet, and housed in the cinema theatre on the top floor of Marconi House, London. 1961 E. WILLIAMS *George* xxvi. 415, I would sit for an hour while the faint dream-sounds of 2LO echoed in my head. 1969 *Listener* 17 Apr. 514/3, 30 years after the introduction of broadcasting from 2LO, radio was the dominant or only means of electronic communication. 1981 S. BRIGGS *Those Radio Times* 9/2 The pre-Savoy Hill 2LO studio at Marconi House. 1741 W. ELLIS *Mod. Husbandman* May vii. 112 It is generally made with half skim, and half new; or what is more properly called Two-meal Cheese. 1919, etc. Two(-)minute(s') [see *SILENCE *sb.* 2 e]. [1711 ADDISON in *Spectator* 24 July, There cannot a greater Judgment befall a Country than such a dreadful Spirit of Division as rends a Government into two distinct People, and makes them greater Strangers and more averse to one another, than if they were actually two different Nations.] 1845 DISRAELI *Sybil* I. II. v. 149 Two nations; between whom there is no intercourse and no sympathy; who are as ignorant of each other's habits, thoughts, and feelings, as if they were..inhabitants of different planets... The rich and the poor. 1913 *Times* 8 May 10/3 Rightly or wrongly, the Protestants of Ulster hold to the 'two nations' theory. 1958 O. CAROE *Pathans* xxvi. 434 Only when the British Government's move to bring British authority to an end acquired a momentum so unmistakable that even the unbelieving were compelled to read the signs..was the two-nation theory translated into practical politics. 1976 *Equals* Oct./Nov. 1/5 In contemporary Britain the Two Nations are not so obviously the rich and the poor or the blacks and the whites or the townspeople and the country people, as the urban deprived and the rest of us. 1937 'M. INNES' *Hamlet, Revenge!* I. ii. 34 Gott with his tutor's instinct was placing this young lady's mind provisionally among the good Two-ones. 1963 *Times* 10 May 6/4 The survey showed that 10 per cent of the firsts and 25 per cent of the 2.i (upper second) were earning less than £1,250 a year. 1964 *Guardian* 23 Oct. 14/2 It is argued..that examiners who can award 'two ones' are less ready to give firsts. 1969 M. KELLY *Write on Both Sides of Paper* 17 He was..a two-one graduate. 1976 *Times* 6 Sept. 10/5 For quite a number of years, New Hall got more firsts and 2/1s combined than any other college. 1984 *Oxf. Univ. Gaz.: Suppl.* 16 Feb. 506/1 It [*sc.* division of the second class] will..greatly help the II(1) in competition with officially recognized II(1)s from other universities. *c* 1880 in *Amer. Mail Order Fashions* (1961) 11 A very neat little two-piece suit of calico. 1895 *Montgomery Ward Catal.* Spring & Summer 501/2 Bathing Suits... Two-piece suits. Consisting of quarter sleeve shirts and knee pants. 1933 AUDEN *Dance of Death* 8 Revealing handsome two-piece bathing suits. 1956 *Times* 28 May 13/3 The two-piece, which has now become the generic term for a dress with a matching jacket, has to a great extent replaced the suit for special occasions. 1978 L. DEIGHTON *SS-GB* iii. 29 She was dressed in a tailored two-piece of pink wool. 1963 *New Yorker* 1 June 110 Soak up the sun in our beautifully brief, terry two-piecer. 1978 *Detroit Free Press* 5 Mar. (Spring Fashion Suppl.) 17 (Advt.), The Pierrot collared two-piecer has a gently shirred skirt, with side pockets. 1916 HALL & NILES *One Man's War* (1929) 160 He was flying a two-place fighter. *Ibid.* 190 It wasn't long until Luf spied a two-placer. 1971 *Flying* Apr. 35/2 The jet works the same as a two-place trainer. 1901 *To-Day* 26 Sept. 280 In men and ships..the British Navy is distinctly above the 'two-Power standard'. 1910 *Encycl. Brit.* IV. 613/2 It has..been accepted as a fundamental axiom of defence that the British navy should exceed in strength any reasonable combination of foreign navies which could be brought against it, the accepted formula being the 'two-power standard', *i.e.* a 10% margin over the joint strength of the next two powers. 1914 G. B. SHAW *Fanny's First Play* III. 223 Your honest and sensible statesmen demand for England a two-power standard. 1980 Two-rev [see *stop cylinder* s.v. *STOP *sb.*[2] 29]. 1902 *Encycl. Brit.* XXXII. 5/2 The two-revolution cylinder..makes two revolutions to each impression and can be made to produce 1500 or more impressions on a large sheet in one hour. 1967 *Elsevier's Dict. Printing* 328/2 The cylinder on a two-revolution press constantly rotates, printing a sheet on its first revolution and delivering the sheet on its second. 1896 E. TURNER *Little Larrikin* xxiii. 279 The frantic rushes of the game of 'twos and threes'. 1935 N. MITCHISON *We have been Warned* III. 236 After dinner there were games with the crew... They had fox-and-geese and twos-and-threes. 1983 G. MITCHELL *Greenstone Griffins* ii. 14 Some traditional party games were played... A game of Twos and Threes was in progress. 1891 H. C. BUNNER *Zadoc Pine* 172, I climbed into his 'two-seater', and sat behind talking to Mrs. Tom. 1906 Two-seater car [see SEATER 2]. 1918 T. E. LAWRENCE *Lett.* (1938) 248 Our Bristol Fighter the same day brought down a German two-seater in flames. 1931 *Daily Express* 31 Jan. 7/4 The two-seater airplane which crashed in Bushy Park. 1973 R. LUDLUM *Matlock Paper* xviii. 150 He sat in the Early American two-seater in the outer office. 1977 P. HILL *Fanatics* 145 The helicopter..was a two-seater. 1979 R. JAFFE *Class Reunion* (1980) III. i. 305 Emily..drove to her analyst..in the little two-seater Mercedes. 1858 GEO. ELIOT *Scenes Clerical Life* I. 166 Little Bessie Parrot, a flaxen-headed 'two-shoes', very white and fat as to her neck. 1949 N. STREATFEILD *Painted Garden* xxi. 228 Movies are made with a long shot, a two-shot and a close-up of each person, each taken separately. 1978 *Broadcast* 27 Mar. 8/3 Two grey-suited figures, held in two-shot, recapped results interminably. 1885 *Narragansett Hist. Reg.* III. 213 We were shown a play-card, the two-spot of clubs. 1896 ADE *Artie* vi. 50 You're nothin' but a two-spot. 1901 'J. FLYNT' *World of Graft* 184 They convicted me at last and I got a two-spot. 1909 'O. HENRY' *Roads of Destiny* xviii. 305 We get the heelers out with the crackly two-spots. 1936 H. HAGEDORN *Brookings* v. 73 He knew..when to be discreet and when to bluff on a two-spot. 1951 *Observer* 27 May 4/5 He cannot do better than buy the Michelin Guide to France and tour the two- or even three-star restaurants. 1960 *John o' London's* 7 Apr. 403/1 The unknown two-star general..accumulated power for himself in the name of France. 1976 *Daily Mail* (Hull) 30 Sept. 11/3 The Galant..will do over 35 miles to the gallon on two-star petrol. 1982 'E. LATHEN' *Green grow Dollars* ii. 16 Two-star generals are less accustomed to being balked than most men. 1959 G. TROUP *Masers* iii. 35 The two-state molecules under consideration. 1971 J. H. SMITH *Digital Logic* i. 1 Although most digital devices are 'two state', the essential requirement of a digital device is that it should change from one discrete state to another and not settle into any intermediate state or position. 1910 *Punch* 30 Mar. 223/2 But when the maid my signal sees She 'two-steps' by like winking. 1929 S. LEWIS *Dodsworth* i. 9 The aristocracy of Zenith were dancing... They two-stepped on the wide porch. 1940 *Chambers's Techn. Dict.* 874/1 Two-step relay. 1948 A. C. CLARKE *Across Sea of Stars* (1959) 13 The A. 20 was a two-step rocket. 1978 *Bull. Amer. Acad. Arts & Sci.* Feb. 13 The two-step mechanism of translocation of estradiol finds a precise counterpart in the action of other steroid hormones. 1895 Two-sticker [see *JACK *sb.*[1] 25]. 1931 *Canad. Geogr. Jrnl.* II. 391/1 Our 'two-sticker with a kicker' lies in Sandy Bay harbour, Nova Scotia. 1923 M. C. WORK *Auction Bridge of 1924* vi. 60 A hand containing two suits both strong enough to bid is called a 'two-suiter'. 1958 *Which?* Winter 25/1 The boot accommodated one two-suiter and one week end case, or three week-end cases. 1979 *Arizona Daily Star* 5 Aug. I. 12/1 North had to have a monstrously powerful major two-suiter for his repeated cue-bids, and they invariably took a preference to six spades. 1981 L. DEIGHTON *XPD* xxvii. 223 Stuart lifted a Samsonite two-suiter onto the bed. 1904 *Oxford Mag.* 16 Mar. 275/2 When imported Labour Yellow should eliminate the Fellow, And the Head of every College be a two-tailed Mandarin! 1922 JOYCE *Ulysses* 491 His twotailed black braces dangling at heels. 1931 *Times Lit. Suppl.* 1 Oct. 747/2 They [*sc.* Norman clerks] used a two-tailed capital F. 1945 *Biometrics Bull.* I. 70 If querist has other alternatives to be considered, he may be interested in the two-tailed test discussed in following answers. 1976 *Lancet* 30 Oct. 922/1 Data were analysed by Student's two-tailed *t* test. 1906 GOODCHILD & TWENEY *Technol. & Sci. Dict.* 800/2 *Two tone*,..a term applied to lace composed of cotton of the natural colour (ecru) interspersed with objects of white. 1940 R. CHANDLER *Farewell, My Lovely* xviii. 115 A couple of very nice two-tone Buicks. 1963 *Listener* 21 Feb. 349/2 There are signs, in the two-tone sketchiness of some of the scenes..that the book was written at speed. 1965 G. MCINNES *Road to Gundagai* x. 167 The sad two-tone cadence of the mopoke or Australian owl. 1966 *Statutory Instruments 1966* III. 3501 'Two-tone horn' means an instrument or apparatus which, when operated, automatically produces a sound which alternates at regular intervals between two fixed notes. 1977 *Harpers & Queen* Sept. 51/2 The usual luxury touches like electric windows, two-tone horns. 1981 *Country Life* 22 Jan. 226/3 Deep burgundy *bouclé* jacket..two-tone belt. 1897 *Sears, Roebuck Catal.* 238/1 Striped Undershirts... Neat two toned shades, silver gray predominating. 1951 T. STERLING *House without Door* v. 47 A two-toned gong rang... He opened the door. 1982 R. LUDLUM *Parsifal Mosaic* xviii. 271 He climbed out of the two-toned coupé. 1945 N. MARSH *Died in Wool* viii. 180 A couple more dirty two-tooths for the herd to shear. 1962 *Coast to Coast 1961–62* 26 Kangaroos, goannas and two-tooth ewes constitute the characteristic fauna of Coorabin. 1972 P. NEWTON *Sheep Thief* iii. 26 As two-tooths they all go out to Cow Creek with the ewes. 1980 D. HART-DAVIS *Heights of Rimring* xviii. 212 'What the hell's a two-tooth ewe?' 'A hogget—a second-year ewe. The best mutton there is.' 1895 *Army & Navy Co-op. Soc. Price List* 921 The Remington single shot rifle. Shooting the ·22 rim fire American Cartridge. 1944 'N. SHUTE' *Pastoral* i. 11 A little gun, his own. Two-two. 1973 'M. HEBDEN' *Dark Side of Island* xvi. 134 'What sort was it? Shotgun?' 'No. Two-two.' 1976 J. H. SPENCER *Surgenor Campaign* i. 11 His own Tripos degree class in history, a Two-Two in place of the expected First, had been a disappointment. 1981 LD. HAREWOOD *Tongs & Bones* v. 90, I..at least got a II: ii..; in those days the only respectable thing was either a First or a Third. 1900 H. SUTCLIFFE *Shameless Wayne* 119 It's a two-week come yesterday sin' they fought i' th' kirk-yard. 1927 A. H. MCNEILE *Introd. Study N.T.* v. 141 St. Paul returned [to Ephesus] for his two-year visit. 1962 E. SNOW *Other Side of River* (1963) i. 19 There were six of them..on their way to serve two-year contracts as technical advisers in China. 1975 *Whitaker's Almanack 1976* 802 Institutions of higher education include universities, colleges, professional schools, and two-year colleges.

two-dime·nsional (*also* tū-), *a.* **1.** Having or appearing to have length and breadth but no depth.

1883 [see DIMENSIONAL *a.* 2]. 1898 [in Dict. s.v. Two *a.* and *sb.* IV. 1 b]. 1954 M. RICKERT *Painting in Britain: Middle Ages* viii. 201 Two-dimensional compositions for glass panels. 1967 E. SHORT *Embroidery & Fabric Collage* i. 8 Without realising what effect they will have when translated into a two-dimensional design.

2. *fig.* Lacking depth or substance; shallow, superficial.

1934 C. LAMBERT *Music Ho!* v. 310 It is possible to detach Stravinsky's methods from their contents and apply Stravinsky scoring to any piece of music. Like everything else in his music, it is two-dimensional. 1959 *Times Lit. Suppl.* 9 Oct. 573/4 By comparison *Heart to Heart* seems both shallow and hollow, a two-dimensional manufactured thing for an obvious mass market. 1977 *Broadcast* 7 Nov. 13/1 'Hard Times'..is less a novel than..a strip cartoon with two-dimensional characters.

Hence **two-dimensionality**, the property of being two-dimensional; **two-dimensionally** *adv.*, in, or in terms of, two dimensions; **two-dimensionalness** (*rare*).

1926 H. READ *Eng. Stained Glass* i. 11/1 Two-dimensionality. No attempt is made to put the scene in strict spatial perspective. 1956 *Essays in Criticism* VI. 219 The two-dimensionality of the canvases. 1958 S. SPENDER *Engaged in Writing* 116 Her general two-dimensionalness permitted her to appear different on each side. 1961 WEBSTER, Two-dimensionally. 1968 *Jrnl. Physical Soc. Japan* XXV. 934/1 The computations were made two-dimensionally. 1975 I. STEWART *Concepts Mod. Math.* ii. 22 It is a consequence of the two-dimensionality of the plane that any rigid motion is uniquely specified by what it does to a (non-degenerate) triangle. 1978 *Spectator* (New Canaan High School, Connecticut) 65 The world has become two-dimensionally black and white. 1979 *Jrnl. R. Soc. Arts* July 509/1 The criticisms..that Ruskin saw architecture only two-dimensionally, and that he never seems to have looked at a building structurally, are refuted with ample quotations.

twoer. (Earlier and further examples.)

1887 W. S. CHURCHILL *Let.* 3 May in R. S. Churchill *Winston S. Churchill* (1967) I. Compan. I. iv. 131 We had a game of Cricket this afternoon, I hit a *twoer*, as the expression goes, my first runs this year. 1899 C. ROOK *Hooligan Nights* iv. 64, I..went froo the till. It wasn't much of a 'aul... I don't fink there was more'n free twoers worf to be nicked. 1970 G. F. NEWMAN *Sir, you Bastard* iii. 98 Sneed flipped the notes without undoing the band. 'A twoer there,' Doleman said.

twofer (tū·fəɹ). *U.S. colloq.* Also too-, -fah, -for, -fur. [f. Two *sb.* + (representation of) FOR *prep.*] **1.** A cigar sold at two for a quarter; hence, any cheap cigar.

a 1911 D. G. PHILLIPS *Susan Lenox* (1917) I. xx. 351 He smoked five-cent cigars instead of 'two-furs'. 1922 S. LEWIS *Babbitt* v. 63 'I do like decent cigars—not those Flor de Cabagos you're smoking—' 'That's all right now! That's a good two-for.' 1923 WODEHOUSE *Inimitable Jeeves* xiii. 143, I found him..lying on the bed with his feet on the rail, smoking a toofah. 1943 *Sun* (Baltimore) 3 Feb. 24/6 The latest items on the list of wartime shortages are the 'toofer', the 'nickel cigar' and the '10-center'.

2. a. A coupon that entitles a person to buy two tickets for a specified theatre show for the price of one.

1948 *N.Y. Herald Tribune* 9 Aug. 8/1 Twofers, as they are called in the slang of the trade, entitle the holder to two seats for a given show for the price of one. 1959 P. BULL *I know Face* iii. 56 It was a constant embarrassment to me to watch a nice customer..buy two expensive stalls for 27*s.*, only to be followed by a 'twofer' lady who got the same thing for 13*s.* 6*d.* 1971 *It* 2–16 June 11/3 Me and the missus got twofers for the show. 1982 *Eastern Province* (S. Afr.) *Herald* 5 Aug. 17/2 Twofers are usually available only for near-to-flop shows or those looking tatty after a long run.

b. *transf.*; also *spec.* a Black woman appointed to a post, the appointment being seen as evidence of both racial and sexual equality of opportunity.

1969 *Britannica Bk. of Year* (U.S.) 801/1 *Twofer*, *specif.*, a blazer with matching trousers. 1978 *N.Y. Times* 29 Mar. A11/2 (Advt.), Another terrific two-for. This sun dress..becomes city-smart when you float the boxy jacket over it. 1977 *Time* 3 Jan. 36/1 By appointing her, Carter got a kind of 'twofer': as a black and as a woman, she is proof that the President-elect is trying to open his Cabinet to both groups. 1979 *Daily Tel.* 2 June 10 Personnel departments [in the United States] are told always to try and hire a 'toofer'. 1982 *Record Business* 11 Jan. 2/1 Ian Miles..had best-sellers with..a Wout Steenhuis twofer and a Harry Secombe–Moira Anderson album of duets.

two-footed, *a.* Add: **2.** Of a footballer: able to kick equally well with either foot.

1948 *Sporting Mirror* 19 Nov. 6/2 He's a fine two-footed defender, very speedy in recovery and cool under pressure. 1969 *Listener* 3 Apr. 473/3 Arsenal's Jimmy Robertson, naturally two-footed. 1978 *Sunday Express* 19 Mar. 29/5 Robson drafted the inexperienced but talented, two-footed Russell Osman into his depleted defence.

twofor, var. *TWOFER.

two-forked, *a.* (Later example.)

1923 D. H. LAWRENCE *Birds, Beasts & Flowers* 113 He ..flickered his two-forked tongue from his lips.

twofur, var. *TWOFER.

two-handed, *a.* Add: **6.** *U.S. colloq.* Generous, open-handed.

1929 D. RUNYON in *Cosmopolitan* July 57/1 Miss Missouri Martin..puts the blast on her plenty for chasing a two-handed spender..out of the joint. 1933 G. ADE *Let.* 12 Sept. (1973) 173 He was..a two-handed drinker who could not carry his rum because he was too frail and intellectual.

Hence **two-handedly** *adv.*, with or in both hands; **two-hander,** (*b*) *Theatr.*, a play with a cast of two.

1927 KIPLING *Verse 1885–1926* 730 Two-handedly tossing me jewels. 1976 *Listener* 20 May 648/3 The play was..a two-hander, finely acted by Maurice Denham and Colette O'Neil. 1981 H. BALDRY *Case for Arts* 9 The ever-diminishing casts of the plays which our theatre directors can afford to present... This is the time of the three-hander and the two-hander. 1981 M. KENYON *Zigzag* xxi. 143 Peckover, pickaxe two-handedly poised..stepped towards him.

two-legged, *a.* Add: Also *fig.*

1930 W. K. HANCOCK *Australia* i. 28 In the very best wheat country the farmer will keep a flock of sheep, not only to make his economy 'two-legged', but also because sheep are an essential agricultural implement. 1976 *Milton Keynes Express* 16 July 39 Whereas last season the winners and runners-up of each section met in the semi finals, the winner and runner-up of the one league will meet in a two-legged final.

two natures, *sb. pl. Theol.* The divine and human natures united in the person of Christ. Also (hyphened) *attrib.*

1600, 1651 [see HYPOSTASIS 5]. 1797 *Encycl. Brit.* VII. 43/1 He [*sc.* Eutyches] appeared to allow of two natures, even before the union. 1874 J. H. BLUNT *Dict. Sects, Heresies* 332/2 The Monophysites held that the two Natures were so united, that although the 'One Christ' was partly Human and partly Divine, His two Natures became by their union only one Nature. 1946 E. L. MASCALL *Christ, Christian & Church* iii. 49 In no case is the divine nature seen acting in separation from the human. The two natures are distinct and their union is unimaginably intimate. 1977 D. CUPITT in J. Hick *Myth of God Incarnate* vii. 136 Liddon..did not see in the full two-natures doctrine any threat to the unity of Christ's person. 1982 *Church Times* 24 Dec. 8/3 The doctrine of the 'two natures' in the one person of Christ..does not mean that Jesus was in any sense schizophrenic, acting at one moment as man and the next as God.

twopence. Add: Since the introduction of decimal coinage in Britain in 1971 also pronounced (tū pens) when a sum of money is meant (as in senses 1, 2, and 6).

3. a. Also *for twopence*, very easily, with the smallest encouragement.

1934 R. H. MOTTRAM *Bumphrey's* I. 37 I'm all heavy with that stuff. I could go to sleep for tuppence! *a* 1960 E. M. FORSTER *Maurice* (1973) xxvi. 132 I'd jump out of the window for twopence.

c. *twopence coloured* adj. phr.: orig. with reference to prints of characters for toy theatres that were sold in the early nineteenth century at one penny (1d) for black and white ones and two pennies (2d) for coloured ones; hence, excessively theatrical; cheap and gaudy.

1859, etc. [see *penny plain* adj. s.v. *PENNY 12 c]. 1879 R. L. STEVENSON *Trav. with Donkey* 82 If landscapes were sold, like the sheets of characters of my boyhood, one penny plain and twopence coloured, I should go to the length of twopence every day. 1887 *Art Jrnl.* Apr. 105/1 Picturesque melodrama..has given place to every-day drama..unsusceptible of twopence coloured treatment. 1908 E. TERRY *Story of my Life* x. 243, I never cared much for Henry [Irving]'s Mephistopheles—a two-pence coloured part, any way. 1911 G. K. CHESTERTON *Innocence of Father Brown* xi. 304 It was a twopence coloured sort of incident. 1926 W. DE LA MARE *Connoisseur* 311 My American adventure..is of the 'twopence coloured' variety, rather than the 'penny plain'. 1932 *N. & Q.* 27 Feb. 151/2, I have two very unusual specimens of the twopence coloured. They..have very little brush-work, being decorated in every possible part with foil of different colours. 1948 *Eng. Stud.* I. 95 The Rape of Heraclide is a magnificent specimen of the 'twopence coloured' style. 1966 A. L. COBURN *Autobiogr.* iv. 48 The toy theatre of Stevenson's childhood and mine. A penny plain and twopence coloured. 1968 *Economist* 9 Nov. p. v/1 Almost too much painful detail, which is of no intrinsic interest except by double proof of the horror of the whole; and perhaps slightly two twopence-coloured; but brilliantly readable. 1978 *Guardian Weekly* 24 Sept. 20/3 A penny-dreadful dramatisation..that has been bumped up into a tuppence-coloured theatrical event by sheer production values.

6. (Further examples. Cf. *TWOPENNY *a.* 1 a.)

1972 J. WILSON *Hide & Seek* ii. 28 Alice found she had a two-pence piece in the pocket of her jeans. 1976 DEAKIN & WILLIS *Johnny go Home* ii. 50 Annie called him from a coin box..using a two-pence piece that had been given her. 1979 M. PAGE *Pilate Plot* vi. 94 Dick Goddard pushed a two-pence coin into the slot.

two-penneth, -pennorth, -penn'orth, colloq. contractions of *TWO PENNYWORTH.

twopenny, *a.* and *sb.* Add: **A.** *adj.* **1. a.** *twopenny piece*, any coin of the value of twopence (cf. TWOPENCE 2); in mod. use (freq. pronounced (tū·peni)), a bronze coin first issued in Britain with the introduction of decimal currency in 1971.

1607 E. TOPSELL *Foure-Footed Beasts* 358 A round peece of leather, as broad as a two penny peece. 1907 [in Dict.]. 1972 A. ROSS *London Assignment* 65 The old witch outside began to rap on the glass door [of the telephone kiosk] with her tuppenny pieces. 1974 [see *PHONE *v.* a]. 1976 *Daily Tel.* 5 Nov. 13/6 There were plastic bags of tuppenny pieces from current juniors who had held raffles in 'break'. 1979 J. SCOTT *Clutch of Vipers* viii. 141 The chunk as the Twopenny piece fell.

d. Also *ellipt.* for *twopenny post* or *letter*.

1818 KEATS *Let.* Oct. (1931) I. 263 Haslam..has taken all the Letters except this sheet, which I shall send him by the Twopenny. 1832 DICKENS *Let.* ?Aug. (1965) I. 9 The place has no other name, but a twopenny directed as above will no doubt find us.

†**f.** *twopenny rope*: a cheap lodging-house (see quot. 1836). *Obs.*

1836 [see *ROPE *sb.*[1] 2 f]. 1850 [see *lie-down* s.v. *LIE *sb.*[2] 6].

g. *twopenny library* (Hist.): a lending library, usu. operated from a shop, from which a book could be borrowed for twopence a week.

1935 W. G. TAYLOR in J. Hampden *Bk. World* 85 That the 'reading public' is growing rapidly seems evident from the growth of the 'twopenny library'. 1942 J. B. PRIESTLEY *Black-Out in Gretley* iii. 23 Twopenny libraries bright with book jackets. 1963 'R. FINDLATER' *What are Writers Worth?* 7 The 'twopenny libraries' which spread rapidly throughout the country before the war, and have vanished nearly as quickly in the past decade. 1978 S. HODGES *Gollancz* v. 115 The twopenny libraries..had come into their own..in the mid-thirties.

2. a. *esp.* in phr. (*not*) *to care* (or *give*) *a twopenny damn* (or *hang*); also *ellipt.*

1897 W. S. MAUGHAM *Liza of Lambeth* xi. 215, I don't care a twopenny 'ang for all them blokes. 1898 R. FRY *Let.* 5 Sept. (1972) I. 172 The date for Bissolo's death, for which I don't care just now a tuppenny damn. 1924 LAWRENCE & SKINNER *Boy in Bush* 58, I don't believe she cares a tuppenny for 'em. 1980 D. K. CAMERON *Willie Gavin* ix. 86 It was rich in its lairds, men who.. gave not a tuppeny damn..for anybody.

b. *twopenny upright*, a prostitute. *slang.*

1958 L. DURRELL *Balthazar* ii. 41 A little old tart, button-eyed and razor-nosed—a tart of the Waterloo Bridge epoch, a veritable Tuppeny Upright. 1978 *Maledicta* II. 258 At the turn of the century, an Iowa woman was awarded $200 for being called a 'whore', while in England, at about the same time, a woman was denied any award for being called a 'two-penny upright'.

B. *sb.* **4.** *slang.* The head. [See quot. 1931.]

1859 in HOTTEN *Dict. Slang.* 1889 W. S. GILBERT *Gondoliers* II. 34 A Lord High Archbishop might tell a Lord High Chancellor to tuck in his tuppenny, but certainly not a cook. 1906 GALSWORTHY *Man of Property* III. iv. 321 Fast after him walked George. If the fellow meant to put his 'twopenny' under a bus, he would stop it if he could! 1928 C. E. MONTAGUE *Action* 150 'Into it, Jemmy,' I yelled. 'Into the sewer and tuck in your tuppenny.' 1931 'G. ORWELL' *Coll. Essays* (1968) I. 71 The hop-pickers..also used the abbreviated rhyming slang, e.g. 'Use your twopenny' for 'Use your head'. This is arrived at like this: head, loaf of bread, loaf, two-penny loaf, twopenny.

two pennyworth (tū:pe·niwɒ̄ɹþ), contr. **two-penneth, -penn'orth, -pennorth** (-pe·nəɹþ). As much as is worth or costs twopence. Freq. *fig.*, a small or contemptible amount. See PENNYWORTH.

1851 H. MAYHEW *London Labour* I. 75/2 Two penn'orth for a whet. 1865 'L. CARROLL' *Alice's Adv. Wonderland* x. 160 Who would not give all else for two p (rhyme with 'soup') ennyworth only of beautiful Soup? 1870 *Punch* 5 Nov. 194/1, I walked down the street with just two-penn'orth of swagger on. 1896 G. B. SHAW *Let.* 5 Dec. in *Ellen Terry & Shaw* (1931) 134 You thought two pennorth of flattery all that the occasion demanded. 1924 WODEHOUSE *Leave it to Psmith* vi. 90 Within reason—and if undetected—I see no objection to two-pennorth of crime. 1965 G. MELLY *Owning Up* ix. 105 After it [*sc.* the band] broke up I used to go along every other Tuesday to 'The Three Brewers'..and put in what Mick would call 'my two penn'orth'. 1979 *SLR Camera* Mar. 16/1 No meter, only years of experience and twopenneth of glass.

two-pi·pe, *a.* **a.** Applied to a system of hot-water heating in which a flow pipe supplies the radiators and a separate return pipe receives water from them.

1897 F. DYE *Hood's Pract. Treat. Warming Buildings* (ed. 3) xviii. 328 The arrangement of pipes is on the ordinary two-pipe principle. A flow and return is carried in different directions almost from a central point. 1970 [see *ONE-PIPE *a.* a].

b. Applied to a system of plumbing in which separate waste-pipes and soil-pipes are employed.

1946 E. MOLLOY *Plumbing & Gas-Fitting* x. 203 The 'dual' or 'two-pipe' system, in which the removal of excretal matters..is effected by a pipe connected direct to the drain, and the removal of waste water..effected by a second pipe..discharging..over or into a properly trapped gully. 1972 T. A. TOMPSON *Guide Sanitary Engin. Services* vii. 226 The two-pipe system of sanitation is rarely used today and has been largely replaced by a more economic design.

twos (tuz), *v. U.S. colloq.* [f. Two sb.] *intr.* To keep company with a person of the opposite sex. Hence **two·sing** *vbl. sb.*

1920 S. LEWIS *Main Street* 2 She played tennis,..went 'twosing', and joined half a dozen societies for the practice of the arts. 1924 *Dialect Notes* V. 291 *Twosing*,..going in couples. 1940 W. FAULKNER *Hamlet* II. ii. 119 They chose one another monotonously in the twosing games.

twosome, *sb.* and *a.* For 'Chiefly *Sc.*' read 'orig. *Sc.*' and add: **A.** *sb.* (Later examples.) Also, a pair of lovers.

1926 [see *FOURSOME B. *sb.* 2]. 1945 L. SHELLY *Jive Talk Dict.* 19/2 *Twosome*,..lovers. 1959 *Manch. Guardian* 12 Aug. 6/2 For the leaders of the nations..the rule is two by two. What is one to think of these twosomes? 1972 *Guardian* 4 Dec. 12/1 Government in Australia and New Zealand has become a socialist twosome. 1973 L. SNELLING *Heresy* I. i. 8 Shockley and..Rosamund became a definite twosome in Rome. 1978 J. IRVING *World according to Garp* xii. 232 Along came another Land Rover, as if they were a separated twosome from a column of an army on the move. 1982 H. ENGEL *Ransom Game* xvii. 101 Eddie Milano..used to spend a lot on her. They didn't keep house..—there's a Mrs. Milano out in Fort Erie, I think—but they were known as a twosome in the joints.

2. A game, dance, etc. for two persons.

1911 in *Conc. Oxf. Dict.* 1977 *Washington Post* 3 Sept. D7/4 Irwin and Weiskopf..had to roll in 25-foot birdie putts at the 18th in the last twosome of the day. 1983 *Ibid.* 6 Nov. C2/1 The [golf] competition..also includes two some matches.

two-·stroke, *a.* (and *sb.*) [STROKE *sb.*[1] 12 c.] Working mechanically by means of a succession of reciprocal movements, in and out or up and down; applied *esp.* to (vehicles having) internal-combustion engines in which the power cycle is completed in one upward and one downward stroke of a piston. Also as *sb.*, a two-stroke engine or vehicle.

1855, 1900 [in Dict. s.v. Two IV. 1 a]. 1902 P. N. HANSLUCK *Automobile* vi. 113 Motors of the four-stroke cycle consume less fuel than motors of the two-stroke cycle. 1936 *Economist* 2 May 234/1 In a two-stroke engine, each cylinder gives an impulse at every revolution of the crankshaft. 1963 L. DEIGHTON *Horse under Water* xiv. 59 Fernie propped his two-stroke against the baker's shop. 1972 J. MCCLURE *Caterpillar Cop* ii. 19 The lights pulsed to the quick beat of the two-stroke. 1975 D. MALOUF *Johnno* x. 109 All one scorching summer I blazed from suburb to suburb on a Fanny Barnet two-stroke. 1982 *Sunday Tel.* 1 Aug. 6/1 It [*sc.* a free-wheel device] was..also used by Saab, mainly on its two-stroke models.

twot: see TWAT in Dict. and Suppl.

two-·time, *a. slang* (orig. *U.S.*). **1.** [f. *two times* adv. phr. s.v. Two B. 1 d.] Characterized by something that has happened or been done twice; *two-time loser*: see *LOSER 4.

1897 R. M. JOHNSTON *Middle Georgia* 113 Is a widder, even a two-time widder, got nothin' else to do but..go about grievin' for them that's gone? 1960 *Guardian* 14 July 9/1 A threat to unseat one man..for his two-time allegiance to Eisenhower. 1972 *Publishers' Weekly* 3 Jan. 9 This book by a two-time ABC Master Tournament winner. 1979 A. V. BADGLEY *Rembrandt Decisions* (1980) xv. 212 Maybe they won't go [*sc.* be stolen] any more... Maybe it was just a two-time snatch.

2. [f. the vb.] Double-crossing, two-timing.

1937 E. S. GARDNER *Case of Dangerous Dowager* xii. 231 You think I'm a two-time, chiseling crook.

two-·time, *v. slang* (orig. *U.S.*). *trans.* To deceive (esp. a person to whom one owes loyalty); to be unfaithful to (a spouse or lover); to double-cross. Also *absol.* or *intr.*

1924 J. EDWARDS in P. Oliver *Screening Blues* (1968) ii. 64 She'll two-time you like she double-crossed me. 1926 E. HEMINGWAY *Sun also Rises* I. vii. 66 The drummer shouted: 'You can't two time—'. 1937 WODEHOUSE *Summer Moonshine* (1938) xxiv. 287, I can tell you it pretty near broke me up when I found you were two-timing me that way. Letting another guy send you jewellery. 1959 'M. M. KAYE' *House of Shade* v. 63 You can't go two-timing the police and skipping out of the country on a stolen passport. 1959 M. RENAULT *Charioteer* vi. 117 In return you've done nothing but two-time me. 1975 *Time Out* 10 Jan. 65/2 Do all attractive slim girls think it clever to lie and two-time? 1981 *Sunday Times* 8 Mar. 8/2 Judith Exner..two-timed the late President John Kennedy with a leader of organised crime.

Hence **two-timing** *vbl. sb.* and *ppl. a.*; also **two-timer,** one who double-crosses or is unfaithful.

1927 K. NICHOLSON *Barker* I. ii. 53 You dirty two-timer, you can't get away with that. 1927 'C. WOOLRICH' *Children of Ritz* viii. 149 Two-timing was what they called that, when a wife went out in other company. 1927 *Columbia Record Catal.* 205/2 My two-timing papa. 1935 *Discovery* Nov. 346/2 His criticism of the hypocritical quakers and 'two-timing' politicians..is too strong. 1942 *Daily Mirror* (N.Y.) 7 Aug. 5/4 The girl had a gun..and told McNaughton: 'You're a two-timer and no good... I'm going to kill you.' 1959 N. MARSH *False Scent* (1960) i. 36 You little, two-timing, double-crossing, dirty rat. 1974 G. JENKINS *Bridge of Magpies* xii. 182 I'd written him off as a two-timer who'd run away to save his own skin. 1982 N. PAINTING *Reluctant Archer* iii. 36 Goneril's two-timing of her worthy but dull husband. 1984 *Listener* 26 Jan. 31/3 Rogers plays a chorus girl out to make a show business reputation by confessing to the murder of her two-timing lover.

two-·up, *sb.* and *adv.* (*a.*) [UP *adv.*[2]] **A.** *sb.* **1.** *Austral.* and *N.Z.* A gambling game played by tossing two coins, bets being laid on the showing of two heads or two tails.

1898 *Bulletin* (Sydney) 3 Sept. 32 At 'loo he'd lately scooped the pool; He'd simply smashed the two-up school. 1898 [see *KIP *sb.*[5]]. 1911 L. STONE *Jonah* vi. 213 He marked pak-a-pu tickets, took the kip at two-up, and staked his last shilling more readily than the first. 1916 G. THORNTON *Wowser* viii. 108 Forty young men.. were..playing that favourite but not intellectual game of chance, 'two-up'. 1936 'R. HYDE' *Passport to Hell* viii. 137 Two dozen mounted police charged the two-up schools, involving a heavy loss of stakes as the men scattered. 1948 V. PALMER *Golconda* iv. 27 A fellow of some power, a fellow who had been used to handling the rough crowds of two-up rings. 1952 J. CLEARY *Sundowners* 229 Who's for a game of two-up? Got just the place out the back? 1960 P. WILSON in C. K. Stead *N.Z. Short Stories* (1966) 129 Some had a two-up school going on in the corner. 1965 B. WANNAN *Fair Go, Spinner* IV. 192 The landlord of the pub, who ran the two-up school, roared out to them, 'Stay where you are, around the ring, but put your dough outa sight!' 1980 *Courier-Mail* (Brisbane) 19 June 3/4 Two-up is banned in Queensland. 1982 G. GREER in *Observer* 8 Aug. 21/3 Two-up is Australia's very own way of parting a fool and his money.

2. two-up (and) two-down, a house with two reception rooms downstairs and two bedrooms upstairs. Also *attrib.* or as *adj. phr.* So *two-up-and-two-downer.*

1958 *Listener* 7 Aug. 212/1 The microphone popped in and out of the two-up-and-two-downers. 1962 *Radio Times* 2 Aug. 37/2 Recalling such local institutions [in Liverpool] as 'the two-up, two-downs'. 1970 J. SANGSTER *Touchfeather, Too* v. 125 A smart little two up and two down somewhere in the suburbs. 1973 *Times* 18 May 4/5 Long, straight terraces of miners' cottages, typically two up, two down. 1978 *Times* 18 May 18/4, I am a poor cotton town boy who grew up in a two-up two-down. 1978 M. KENYON *Deep Pocket* vi. 71 The two-up, two-down brick and dinginess of Ensign Terrace.

B. *adv.* Two at a time, two together.

1926 [see *DOUBLE-BANK *v.* 2]. 1967 COX & GROSE *Organization & Handling Bibl. Rec. by Computer* II. 25 Master output for photo-litho reproduction 'two-up'.

two-va·lued, *a.* Chiefly *Logic.* Able to take one or other of only two values; characterized by the usual two truth-values, i.e. truth and falsity.

1918 C. I. LEWIS *Survey Symbolic Logic* iv. 222 The first procedure..interprets the elements of this system as propositions, and adds to it a postulate which holds for propositions but not for logical classes. The result is what has been called the 'Two-valued Algebra'. 1933 *Mind* XLII. 269, I attempt to show that there are fourteen different meanings of implication in a two-valued logic. 1946 *Nature* 14 Sept. 357/1 Ordinary symbolic logic..represents two-valued logic. 1965 N. CHOMSKY *Aspects Theory Syntax* 232 We can regard..number [in German grammar] as a two-valued dimension. 1970 O. DOPPING *Computers & Data Processing* i. 26 All logical connections between two-valued variables can be expressed by means of the three functions NOT, AND, and OR.

two-way, *a.* and *sb.* Add: **1. b.** Of a plug or adaptor: able to accommodate two plugs at the same time.

1923 *T. Eaton & Co. Catal.* Spring & Summer 357 Two-way socket plug. 1977 *Times* 24 Sept. 22/2 You can use two or three-way adaptors..but do be careful not to overload your current supply.

c. Of a loudspeaker: having two separate drive units for different frequency ranges.

1950 *Audio Engin.* Aug. 15/2 Many loudspeakers are two-way: that is, the frequency range is divided and each portion is handled by separate radiating systems. 1960 [see *THREE-WAY *a.* b]. 1978 *Detroit Free Press* 16 Apr. 18A/1 (Advt.), Two-way bass reflex speaker system. 8-in. woofer, 3-in. tweeter.

2. Delete *Math.* *two-way stretch* attrib. phr., designed to stretch in both length and width; also *ellipt.* such an elastic corset.

1932 [see *GIRDLE *sb.*[1] 1 d]. 1959 *News Chron.* 19 Aug. 3/8 A girdle of Ellen Terry's—'I mean a sash, not a two-way stretch.' 1964 *Observer* 22 Mar. 3/1 Mr Clutson invented the two-way stretch elastic. 1977 *Summit* (Austin Reed, Ltd.) Autumn 44 Men's two-way stretch riding breeches from £18.

3. *Electr.* Of a switch, wiring, etc.: that enables a light or other device to be switched on or off at either of two points.

1893 W. P. MAYCOCK *Electric Lighting & Power Distribution* III. xv. 382 Fig. 240 shows a double-pole, 2-way switch, which would be connected in a parallel circuit in the manner illustrated. 1903 [see sense 1 in Dict.]. 1933 D. L. SAYERS *Murder must Advertise* vii. 113 The hall-light..was fitted with two-way wiring. 1960 *Practical Wireless* XXXVI. 395/1 A two-way switch is provided so that C9 may be connected to either.

4. a. Involving or permitting movement or communication in each of two opposite directions; (of a radio) capable of both transmitting and receiving.

1922 *Encycl. Brit.* XXXII. 1027/2 A problem of practical importance is that of two-way radiotelephony enabling two communicators to speak and hear simultaneously. 1927 *Glasgow Herald* 16 Apr. 7 The purpose.. is to strengthen the direct connection between Great Britain and the Middle Western group of American States by increasing the two-way tonnage exchange clearing through the Virginian ports. 1938 *Radio Times* 1 July 6/3 A two-way broadcast arranged by the BBC and the Columbia Broadcasting System of America. 1957 L. F. R. WILLIAMS *State of Israel* viii. 144 There is a growing two-way traffic in literature between Israel and other countries. 1971 A. PRICE *Alamut Ambush* x. 127 He was fixing up the two-way speaker in the porch..so that I could answer the door from here. 1976 L. DEIGHTON *Twinkle, twinkle, Little Spy* xxi. 207 Has he got a two-way radio in the car? 1967 'R. STARK' *Damsel* (1968) II. 47 Eight lanes of two-way traffic were flanked by broad swaths of green grass. 1981 R. HAYMAN *K* xii. 156 A solitude which was unbroken by the two-way traffic in fantasies that went on in his correspondence.

b. Occurring or existing in two directions; reciprocal; *two-way street* (fig.), a situation of mutual action; something that works both ways.

1950 J. JENKS *From Ground Up* ii. 17 The agri-cultural relationship is, like all vital relationships, two-way. 1951 *Sun* (Baltimore) 14 Dec. 5/6 'An amendment..provided that if any company..gave so much as a cigar to a Government employé, the contract should be cancelled... Do you think that should be made stronger?'..'Yes, it is a two-way street.' 1959 *Times* 19 Sept. 6/3 In sharp contrast with other days this week the market was more 'two way'—that is, there was more buying to offset the selling. 1968 H. WAUGH *Con Game* xi. 101 'Our aim in this racket is to use you.' 'I prefer to think of it as a two-way street.' 1969 T. F. TORRANCE *Theol. Sci.* ii. 67 The relation between God and the creature is a two-way relation. 1975 *Language for Life* (Dept. Educ. & Sci.) xiv. 214 Some will have had very little sustained two-way conversation in the home. 1975 *Times Lit. Suppl.* 21 Nov. 1392/5 Tolerance..was a two-way street: if the Germans were to learn to live with the Jews, so too must the Jews learn to live with the Germans. 1982 *Times* 12 Jan. 12/3 The jobbers..encountered a fair amount of two-way business.

c. *two-way mirror*, a mirror which lets through enough light for an observer at the back to see through it, without being seen from the front.

1967 J. GARDNER *Madrigal* i. 3 They directed total concentration through the sighting side of a two-way mirror. 1974 *Times* 23 Jan. 2/5 Everyone laughed at a show business party when looking through a two-way mirror at couples having sexual intercourse. 1982 R. HILL *Who guards a Prince* I. viii. 57 Locking doors was an empty gesture in these days of two-way mirrors.

B. *sb.* A two-way radio.

1963 'W. HAGGARD' *High Wire* v. 54 There's an operator with a two-way somewhere near the junction. I'm picking up his message. 1974 V. BROME *Day of Destruction* v. 51 We've just received a message over the two-way to say he's stuck..in the middle of a swamp.

two-year-old, *sb.* Add: **b.** As the type of a youthful and energetic person.

1912 *Punch* 19 June 470/1 Feeling as he did like a two-year-old, he was convinced that an immeasurable advantage to the country would be gained by placing the ballot in the hands of babies. 1928 GALSWORTHY *Swan Song* III. xvi. 342 Mr. Forsyte was a proper wonder—went at it like a two-year old. 1936 [see *STINKER 6 c].

Tyburnia (taibɜ̄·rniă). [mod.L., f. TYBURN + -IA[1].] A former literary name for the residential district built in the nineteenth century and extending along the Bayswater Road from Marble Arch to Lancaster Gate and northwards.

1848, *a* 1852 [see *BELGRAVIA]. 1878 A. J. C. HARE *Walks in London* II. 104 Tyburn still gives a name to the white streets and squares of *Tyburnia*, which are wholly devoid of interest or beauty. 1902 G. W. E. RUSSELL *For Better? For Worse?* iv. 58 The most gorgeous mansion in Cromwell Road or Tyburnia could never for a moment be quoted as supplying the place of the Hall or the Manor. 1973 *Country Life* 15 Nov. 1528/1 By the mid-1830s in Tyburnia stucco was replacing brick-facing.

Hence **Tybu·rnian** *a.*

1850 THACKERAY *Pendennis* II. xi. 104 The great lawyers are giving grand dinner parties at their houses in the Belgravian or Tyburnian districts. 1860 *Once a Week* 28 July 124/2 The gallows, called 'Tyburn Tree'..had been for years a standing fixture on..the Edgware Road, ..but this was..the second Tyburnian location of the gallows. 1973 *Country Life* 15 Nov. 1528/1 The circular turrets..contemporary with the earliest Tyburnian example..appear in Victoria Square.

tychism (tai·kiz'm). [f. Gr. τύχη chance + -ISM.] The doctrine that objective account must be taken of the element of chance in (philosophical, cosmological, etc.) reasoning.

1892 C. S. PEIRCE in *Monist* II. 533, I endeavored to show what ideas ought to form the warp of a system of philosophy, and particularly emphasised that of absolute chance..which it will be convenient to christen *tychism* (from τυχη, chance). 1926 J. LAIRD *Study in Moral Theory* viii. 173 The theory that every event..is necessary (or *must* occur precisely as it does occur) I shall call *determinism*... The opposite theory I shall call *tychism*, such tychism being either general or restricted. 1978 *Sci. Amer.* July 18/1 At a time when determinism dominated physics Peirce's doctrine of 'tychism' maintained that pure chance—events undetermined by prior causes—are basic to the universe.

tycoon. Delete ‖ and add: **1.** (Earlier examples.)

1857 T. HARRIS *Diary* 28 Oct. (1930) 406 Today, I am told *Ziogoon* is not the proper appellation of their ruler, but that it is *Tykoon*. *Ziogoon* is literally 'Generalissimo' while *Tykoon* means 'Great Ruler'. 1858 *Times* 9 Nov. 7/1 This treaty, in the first place, engages that there shall be perpetual peace and friendship between Her British Majesty and the Tycoon of Japan.

2. An important or dominant person, esp. in business or politics; a magnate. Also *attrib.* orig. *U.S.* (as a nickname of Abraham Lincoln).

1861 J. HAY *Diary* 25 Apr. in *Lincoln & Civil War* (1939) 12 Gen. Butler has sent an imploring request to the President to be allowed to bag the whole nest of traitorous Maryland Legislators. This the Tycoon..forbade. 1886 *Outing* (U.S.) IX. 164/1 The tycoon of the baggage car objected to handling the boat. 1926 *Time* 14 June 32/3 *Married*. Fred W. Fitch, 56, rich hair-tonic tycoon. 1947 AUDEN *Age of Anxiety* (1948) 36 With diamonds to offer, A cleaned tycoon in a cooled office, I smiled at a siren. 1952 *Manch. Guardian Weekly* 3 July 7/3 Warren has.. been the preferred choice of..oil and aviation tycoons with delusions of grandeur. 1958 [see *TYCOONISH *a.*]. 1960 R. W. MARKS *Dymaxion World of B. Fuller* 62/1 Pictures of his latest projects appear regularly on the front cover of the magazines which symbolize the tycoon press. 1982 M. RUSSELL *Rainblast* iii. 21 She has a thing going with Marcus Hicks, the stores tycoon.

Hence **tycoo·nery** [-ERY], the behaviour or status of a tycoon or tycoons; a group of businessmen; **tycoone·ss,** a female tycoon (sense *2); **tycoo·nish** *a.*, characteristic of a tycoon (sense *2); **tycoo·nship,** the status or position of a tycoon (sense *2); the fact of being a tycoon.

1956 *Time* 24 Dec. 47/2 Instead of making a budget, Falk decided to indulge in a bit of extracurricular tycoonery. 1958 *Times* 3 Dec. 6/4 Tycoons are not quite as tycoonish as they were before. 1959 *Times* 19 Nov. 15/5 This immensely long..novel gives us a new Tom Sawyer and takes him up to tycoonery. 1960 *Guardian* 28 Oct. 8/4 A high-powered tycooness must have sharp claws within the velvet paw. 1962 *Punch* 26 Sept. 443/3 The *Express* group stands alone among the major press tycooneries. 1964 R. WARD *Penguin Bk. Austral. Ballads* 15 Rugged individualists separately thrusting their ways.. to an industrial tycoon-ship. 1965 'R. ERSKINE' *Passion Flowers in Business* iv. 48, I..thought it tycoonish in an exciting way. 1970 'D. HALLIDAY' *Dolly & Cookie Bird* viii. 119 Janey's father..had..several irritating habits of tycoonery. 1976 'M. INNES' *Gay Phoenix* iii. 43 Business affairs... A high degree of continuity in their direction was..a *sine qua non* of successful tycoonship. 1983 *Listener* 27 Oct. 34/3 He was busy trying to set up a rival consortium to buy the Sunday Times, competing with (and losing to) Murdoch in tycoonery.

tyee, tyhee. Add: **1.** For *U.S.* read *N. Amer.* and add earlier and later examples.

1792 in E. HEALEY *Hist. Alert Bay* (1958) 15 The Ty-ee, or chief, of the village paid us an early visit. 1866 J. K. LORD *Naturalist in Brit. Columbia* I. 161, I was presented to the chiefs as a Hyas tyee (great chief), one of 'King George's' men. 1877 *Puget Sound* (Washington) *Argus* 23 Nov., With the coming of the military among us came a big church 'tyhee', who told us that the soldiers were come to protect us. 1880 D. M. GORDON *Mountain & Prairie* 117 We were surprised to find, at the head of Stewart Lake, a well-stocked farm, owned and worked by the Indian 'tyee'. 1927 [see *HIGH-MUCK-A-MUCK]. 1963 *Brit. Columbia Digest* Oct. 54/2 A British officer..may have told the Chief that he was now the great Tyee of the country. The Indian took the words literally and elevated himself to top position. 1966 H. MARRIOTT *Cariboo Cowboy* v. 54 The agricultural tyees in both Canada and the United States have taken a wise view.

tyen, obs. var. *T'IEN.

tyke. Add: **2.** Also (unreprovingly), a child, esp. a small boy; occas., a young animal. *U.S.*

1902 *Dialect Notes* II. 248 *Tyke, n.*, a child. 1930 W. FAULKNER *As I lay Dying* 30 'That poor boy,' Cora says. 'The poor little tyke.' 1942 H. K. SMITH *Last Train from Berlin* v. 174 If you think the present Gestapo is brutal, just wait until these little tykes..grow up and become the rulers of Victorious Germany. 1979 *Tucson* (Arizona) *Citizen* 20 Sept. 1E/4 The stripes are nature's way of protecting the tapir tyke until it's old enough to fend for itself. 1981 *Verbatim* VII. III. 23/2 What hours of chair-side fun await the tyke supplied with 'three 6-oz. cans of modeling compound'!

3*. [Assimilated or altered f. TEAGUE.] A Roman Catholic. *Austral.* and *N.Z. slang.*

1941 BAKER *Dict. Austral. Slang* 76 *Tike, tyke*, a Roman Catholic. 1948 R. PARK *Harp in South* xxi. 268 I'll do what I like when I like without the interference of any bone-headed tike. 1961 P. WHITE *Riders in Chariot* viii. 232, I would never ever of suspected you Rosetrees of being tykes. Only the civil servants are Roman Catholics here, and the politicians, if they are anything at all. 1977 D. AITKIN *Second Chair* xviii. 172 Baxter's a tyke (you wouldn't think it from his name..) and he goes to mass each Sunday. 1981 M. GEE *Meg* xvii. 189 Once it fell to me to..explain to a pair of Jehovah's Witnesses that we were all good Catholics in this house... 'At least the tikes have got some style... Shall I nail a crucifix on the door?'

Hence **ty·kish** *a.*, characteristic of a tyke; **tykishness** (further example).

1888 G. M. HOPKINS *Let.* 20 May (1956) 392 There is an old Adam of barbarism, boyishness, wildness, rawness, rankness, the disreputable, the unrefined in the refined and educated. It is that I meant by tykishness (a tyke is a stray sly unowned dog)... Ancient Pistol is the typical tyke,..and the tykish element undergoing dilution in Falstaff and Prince Hall [*sic*] appears to vanish.. in Henry V as king.

tylectomy (tai·lektŏmi). *Surg.* [f. TYLO- + *-ECTOMY.] The excision of a lump or swelling, esp. a cancerous one.

1972 H. ATKINS et al. *Brit. Med. Jrnl.* 20 May 423/2 We are indebted to Dr. W. J. Mann, who has proposed the term 'tylectomy'..for excision of a lump and we use the term 'extended tylectomy' instead of 'wide excision' throughout the paper. 1978 *Jrnl. R. Soc. Med.* LXXI. 341 Stage I cases treated by tylectomy and HVT in the Guy's Hospital trial..had local recurrence rates of 15% at three years and 20% at five years.

tylosin (tai·losin). *Vet. Sci.* [Etym. unkn.; see -IN[1].] A macrolide antibiotic, $C_{46}H_{77}NO_{17}$, produced by *Streptomyces fradiæ* and used esp. to treat respiratory infections in animals.

1961 J. M. MCGUIRE et al. in *Antibiotics & Chemotherapy* (N.Y.) XI. 326 Tylosin, a new antibiotic substance, has been obtained from soil isolates tentatively identified as strains of *Streptomyces fradiae* (Waksman and Curtis) Waksman and Henrici. 1968 [see *MACROLIDE]. 1981 *Antimicrobial Agents & Chemotherapy* XX. 214/1 Tylosin is a complex macrolide antibiotic which is produced commercially by a strain of *Streptomyces fradiae*.., it is also produced by strains of *Streptomyces rimosus*.. and *Streptomyces hygroscopicus*.

Tymba, var. *TEMBU. **tymbal,** var. TIMBAL in Dict. and Suppl.

tympani, var. *TIMPANI.

1876 in STAINER & BARRETT *Dict. Mus. Terms.* 1917 *Lit. Digest* 25 Aug. 28/2 The thirty-seconds scored for the tympani in some of the modern Russian music. 1927 *Melody Maker* May 483/1 There has been much controversy concerning the practical use of tympani in the dance band, and many dance drummers have, owing to the infrequency of tympani passages in dance arrangements, regarded them as instruments unnecessary for dance work altogether. 1965 G. MCINNES *Road to Gundagai* xi. 200 The man with the bass drum and tympani was responsible for the heavy artillery.

tympanist. Add: Now *spec.*, one who plays upon a kettledrum.

1930 *Melody Maker* Jan. 72/3 All that need trouble the tympanist in regard to the minor chords is [etc.]. 1955 L. FEATHER *Encycl. Jazz* 195/1 He studied with Saul Goodman, tympanist with the N.Y. Philharmonic.

tympano-. Add: **ty·mpanogram,** a graphical record of pressure changes obtained in tympanometry; so **tympano·graphy; tympano·metry,** the measurement, for diagnostic purposes, of changes in the compliance of the tympanic membrane as the air pressure is altered in the passage of the external ear; hence **ty:mpanome·tric** *a.*; **ty·mpanopla:sty** *Surg.* [-PLASTY], an operation to repair the middle ear; hence **ty:mpanopla·stic** *a.*; **ty:mpanosclero·sis** *Med.* [SCLEROSIS], thickening of the tympanic membrane, and of the connective tissue in the tympanic cavity; hence **ty:mpanosclero·tic** *a.*

1969 *Arch. Otolaryngol.* LXXXIX. 217/2 The tympanogram has a normal configuration. 1979 J. J. KNIGHT in H. A. Beagley *Auditory Investigation* viii. 158 A flat tympanogram with a negative middle ear pressure is suggestive of serous otitis media. 1977 *Proc. R. Soc. Med.* LXX. 824/1 Surgery should not be undertaken until after tympanography has been done and tomograms taken to exclude abnormalities of cochlear anatomy. 1984 *Lancet* 19 May 1112/2 A sound case can be made for including tympanography in the routine screening tests at school entry. 1970 *Arch. Otolaryngol.* XCII. 255/2 Tympanometric characteristics provide information in analysis of the conductive pathological abnormality. 1984 *Jrnl. Speech & Hearing Res.* XXVII. 257 The influence that repeated tympanometric trials have on the aural-acoustic admittance characteristics of the middle-ear transmission system was studied in 24 young adults. 1956 H. ANDERSON et al. in *Acta Oto-Laryngol.* XLVI. 384 This method—for which we propose the name tympanometry—may..yield very important additional information for the diagnosis of hearing impairments. 1980 *Ibid.* LXXXIX. 480/1 The patients were regularly checked with audiometry and tympanometry. 1955 *Jrnl. Laryngol.* LXIX. 654 Of 110 tympanoplastic cases..60 per cent. have retained or regained 'social' hearing. 1977 *Lancet* 16 July 119/2 Tympanoplastic surgery is a logical development of the modified radical techniques and aims to avoid their main disadvantages—deafness and an open cavity in the mastoid—by creating a functioning middle ear in a healthy closed system. 1955 *Acta Oto-Laryngol.* XLV. 457 It seems advisable in some cases..to operate in two stages: first a conservative radical operation..and, secondly, tympanoplasty. 1960 J. GRANT *Come again, Nurse* xvi. 97 Now a tympanoplasty is different again. That is a plastic repair to the middle ear with a skin graft, to help the sound waves. 1977 *Lancet* 16 July 119/1 Grafting the drum constitutes the simplest and most successful form of tympanoplasty. 1961 *Brit. Med. Dict.* 1491/1 Tympanosclerosis. 1965 I. B. THORBURN in W. G. Scott-Brown et al. *Dis. Ear, Nose & Throat* (ed. 2) II. xx. 454 Tympanosclerosis. Zöllner (1955) drew fresh attention to this condition which was first described as 'Paukensclerose' by von Tröltsch in 1873. *Ibid.*, On otomicroscopic examination a keratinizing appearance distinguishes cholesteatoma, whereas the white tympanosclerotic focus is covered by an intact healthy mucous membrane. 1978 *Jrnl. R. Soc. Med.* LXXI. 354 Scars and tympanosclerosis were only present in the ears with grommets. 1981 *Brit. Med. Jrnl.* 14 Feb. 501/1 Tympanosclerotic plaques are commonly seen in the drum after the use of grommets.

tympany. 1. Delete 'now *rare* or *arch.*' and add later examples.

1923 G. H. WOOLDRIDGE *Encycl. Vet. Med., Surg. & Obstetr.* II. 1023/2 Tympany is..a common accompaniment of rumenitis. 1970 W. H. PARKER *Health & Dis. in Farm Animals* xiv. 186 The swelling in cases of tympany is primarily on the left.

Tyndall (ti·ndăl). The name of John *Tyndall* (see TYNDALLIZATION), used *attrib.* with reference to the scattering of a beam of light by small particles and the blue colour that the scattered light often has (described by Tyndall in 1869).

1910 H. H. HODGSON tr. *Pöschl's Introd. Chem. of Colloids* ii. 12 The Tyndall effect is the basis of ultra-microscopy. 1915 M. H. FISCHER et al. tr. *Ostwald's Handbk. Colloid-Chem.* 8 Liquids which show no definite Tyndall light-cone..are molecular-disperse solutions. 1925 G. BARGER tr. *Freundlich's Elem. Colloidal Chem.* 125 All colloidal solutions with colourless particles show a blue Tyndall light. 1939 H. B. WEISER *Colloid Chem.* xii. 160 There is no pigment in blue eyes; the color is a Tyndall blue which is more intense the finer the particles that give rise to the blue color. 1964 *Oceanogr. & Marine Biol.* II. 366 With the searchlights [of the bathyscaph] on, it is possible to discern very minute particles..by the Tyndall effect. 1966 *McGraw-Hill Encycl. Sci. & Technol.* XIV. 169/1 In aqueous gold sols..strong Tyndall cones are observed. 1971 *Nature* 19 Feb. 573/1 Drops smaller than 1 μm form a 'blue haze' due to selective Tyndall scattering of shorter light wavelengths.

Hence **Ty·ndallmeter** (also as two words), **Tyndallo·meter,** an instrument that makes use of the Tyndall effect for measuring aerosols and suspensions.

1919 TOLMAN & VLIET in *Jrnl. Amer. Chem. Soc.* XLI. 297 (*heading*) A Tyndallmeter for the examination of disperse systems. 1937 *Nature* 25 Sept. 553/1 The hygienic importance of the reduction of dust lends interest to means for its measurement, and the Tyndallometer shown by E. Leitz (London) provides a new and rapid means of its measurement. 1957 H. C. VAN DE HULST *Light Scattering by Small Particles* xix. 405 This problem, of obvious importance in the testing of gas masks, was studied by Gucker and by O'Konski. A Tyndall meter is insufficient for the concentrations of 10^{-8}g/liter and lower that have to be investigated.

Tyneside (tai·nsaid). [SIDE *sb.*[1] 7 a.] The area adjacent to the banks of the river Tyne in England, *spec.* the city of Newcastle-on-Tyne, used *attrib.* to designate things, esp. speech, characteristic of this area.

1824 R. GILCHRIST *Coll. Original Songs* 9 Hail, Tyneside lads! in collier fleets. 1844 R. NICHOLS in M. A. Richardson *Local Historian's Table Bk.* II. 370 (*heading*) The Tyneside angler. 1872 T. & G. ALLAN (*title*) Tyneside songs. 1896 R. O. HESLOP *Bibliogr. List Wks. Illustr. Dialect Northumberland* 6 Two tales of sixty years sin seyne, as related by the late Thomas Bewick, of Newcastle, in the Tyneside dialect. 1923 A. HERBERT *Northumberland* viii. 127 In the Hancock Museum of Natural History may be seen the matchless collection of birds set up by the Tyneside naturalist, John Hancock. 1949 H. L. HONEYMAN *Northumberland* II. ii. 219 If there is anything that can be called a Tyneside type it is perhaps the rather small, wiry kind of man. 1955 P. STREVENS *Papers in Lang. & Lang. Teaching* (1965) ix. 114 A uvular fricative *r*, similar to that used in Tyneside pronunciation of English or in Parisian French. 1978 *Early Music Gaz.* Oct. 3/3 The senior minstrel, aged about 12, gave his mates a good Tyne-side dressing-down for being late and playing wrong notes the night before!

Tynesider (tai·nsaidəɹ). [f. prec. + -ER[1].] A native or inhabitant of Tyneside. Cf. *CLYDESIDER, *MERSEYSIDER.

1895 *Tyneside* May 146/1 The Life of Sir Charles Palmer has been frequently written and is familiar enough to every Tynesider. 1923 A. HERBERT *Northumberland* v. 69 The word 'keel' on Tyneside..was the name for an entire boat... It is the word for which all Tynesiders have most regard. 1950 S. MIDDLEBROOK *Newcastle upon Tyne* xviii. 222 In the 'forties the practice began of holding an annual skiff race..either between two local champions, or between a Tynesider and some challenger from the Thames. 1978 M. STOPPARD in D. Abse *My Medical School* 173 Tynesiders were proud of their College.

type, *sb.*[1] Add: **6.** Also preceding a sb. with ellipsis of *of*, = type of. Cf. *-TYPE 2. *U.S. colloq.*

1966 *Word Study* Dec. 2/2 He could not pick out things like a bridge from 'this type distance'. 1979 *Nature* 22 Nov. p. xvii/1 The 110C systems may be used with virtually any type projector.

7. c. A person of a certain (specified or implicit) character; *one's type*, the sort of person to whom one is attracted (usu. in neg. or interrog. contexts). Also simply, a person; as a gallicism (also with pronunc. (‖tīp)) derogatory. *colloq.*

1922 [see *ABANDON *sb.*[2]]. 1930 KIPLING *Limits & Renewals* (1932) 327, I played piquet with our schoolmaster... *That* was a type upon whom our War had done bad work. 1931 K. BOYLE *Plagued by Nightingale* xv. 123 Luc could fish with Nicholas... It's exactly what the poor *type* would prefer anyway! 1933 'G. ORWELL' *Down & Out in Paris & London* xxix. 216 'Low types,' said the old Etonian, 'very low types.' 1934 F. B. CUTHRELL *Innocent Bystander* vii. 130 Richardson did not interest her, he was not her 'type'. 1942 T. RATTIGAN *Flare Path* II. i. 127 You're the actor type, aren't you? 1948 'N. SHUTE' *No Highway* ii. 41 It didn't do Fisher any good with the R.A.F. types. 1951 J. G. FENNESSY *Sonnet in Bottle* I. v. 28 'Oh, by the way, do you know these types?' and he introduced the two men with him. 1956 'A. BRIDGE' *Lighthearted Quest* 199, I went to look for Colin in that red-haired type's house. 1962 A. LURIE *Love & Friendship* xiii. 250 You wouldn't like him... He's not your type. He's a little fat man. 1965 R. & D. MORRIS *Men & Snakes* i. 16 Although an intrepid explorer type, he made a hasty exit. 1968 M. JONES *Survivor* iii. 51 'I'm asking you if you think she's at all his type.' Stuart shrugged. 'I shouldn't care to say what Martin's type is. Come to think of it, I'd say he has no type.' 1971 D. E. WESTLAKE *I gave at Office* 136, I was not alone in the room. Three army types were there..tall, fat, khaki-uniformed. 1974 J. AIKEN *Midnight is Place* iv. 142 That *type*... He is a brigand! 1979 R. JAFFE *Class Reunion* II. ii. 145 I've always thought you were very beautiful, Annabel. You always were just my type. 1979 A. FRASER *King Charles II* vii. 102 These were scarcely the types to risk life and limb. 1981 'M. HEBDEN' *Pel is Puzzled* xi. 113 'Type over here... He recognises it.' The 'type over here' was a man about thirty-five with long blond hair.

8. b. Add to def.: *esp.* the specimen on which the first published description of a species is based. (Further examples.)

1893 O. THOMAS in *Proc. Zool. Soc.* 242 The following are..the definitions now suggested for the different terms: A Type is a single specimen either unaccompanied by others at the time of description, or else deliberately selected as such by the author out of a series. 1951 G. H. M. LAWRENCE *Taxonomy of Vascular Plants* ix. 205 The term type, used alone and unqualified, generally refers to the holotype. 1964 *Internat. Code Zool. Nomencl.* xiii. 59 The 'type' affords the standard of reference that determines the application of a scientific name. 1970 *Watsonia* VIII. 156 A herbarium sheet stated..to be the 'type'.. is, however, quite different.

e. *Semiotics*, etc. A sign representing a category or set of instances, as opposed to the individual tokens by which the category is instantiated. Cf. *TOKEN *sb.* 1 f.

1908 [see *TOKEN *sb.* 1 f]. 1966 *Publ. Amer. Dial. Soc.* XLVI. 13 The incidence of tokens is..equal to the sum of all forms... The incidence of types..is equal to the number of different forms. 1976 *Biometrika* LXIII. 435 Shakespeare's known works comprise 884 647 total words, of which 14 376 are types appearing just one time, 4343 are types appearing twice, etc.

10. *type category, -character, description, index, -name, -tragedy*; in sense 9, *type-foundry* (earlier example), *size*; *type-founder* (earlier example); *type-creating* adj. (example); *type-marked* adj.; **type approval** (see quot. 1979); **type area,** (*a*) the part of a page covered by print; (*b*) the location of a type-specimen or an area taken as typical of a particular group; *Geol.* = *type site* below; **type-ball,** a spherical ball on certain kinds of electric typewriter on which all the type is mounted; = *golf ball* (b) s.v. *GOLF *sb.* b; **type basket,** the assembly of type-bars in a typewriter; **type-copy** *sb.* (*arch.*), a typewritten copy; so **type-copy** *v. trans.* (*arch.*); **type face,** a set of printing type of a particular design; cf. FACE *sb.* 22, FOUNT[2]; **type facsimile,** a copy of a piece of printing which is either a page-for-page copy using type as close as possible to the original or an exact photographic reproduction; **type-fallacy** *Logic*, the fallacy or mistake of including amongst the members of a type or category something belonging to another type or category (see quot. 1908); **type height** = *HEIGHT *sb.* 1 c; **type locality** = *type-area* (b) above, *type site* below; **type-psychology,** psychological study or theory based on the classification of people or phenomena by type; **type-script** *sb.*, *a.*: hence **type-script** *v. trans.*, to record in type-script; **type-scripted** *ppl. a.*; **type-scripting** *vbl. sb.*; **type-setting** (earlier examples); **type site** *Archæol.*, *Geol.*, etc., a site the features of which are used to define, or are paradigmatic of, a culture, stratigraphic level, etc.; **type-specimen,** (*b*) a printed sheet or booklet showing the variety of type-faces a printer or founder has available; **type test** *sb.* (esp. *Aeronaut.*), a test conducted to determine whether a new piece of equipment meets its specifications; also *attrib.*; hence **type-test** *v. trans.*; **type–token** *attrib.*, in *Semiotics*, etc., pertaining to types and tokens, involving the relationship of type to tokens (see *TOKEN *sb.* 1 f, sense 8 e above).

1967 *Economist* 15 Apr. 270/1 The Europeans are used to the 'type approval' on which most Continental governments insist before [motor] models can be sold. 1979 *Gloss. Terms Quality Assurance* (*B.S.I.*) 11/2 *Type approval*, the status given to a design that has been shown by type tests to meet all the requirements of the product specification and which is suitable for a specific application. 1916 W. H. HAZELL et al. *Estimating for Printers* 19 Before an estimate..can be worked out, the following points must be decided: Number of words in bookwork..; size of type,..and type area of page. 1937 *Burlington Mag.* June 309/2 Consideration of primitive work as craftsmanship..is no more essential for an æsthetic evaluation than the geographical location of type-areas. 1969 *Proc. Geol. Soc.* Aug. 146 Customary stratigraphical usage should be maintained by placing the marker-points as near to the stratigraphical correspondence with traditional boundaries as possible, although not necessarily in the traditional type-area. **1973** S. Jennett *Making of Books* (ed. 5) xvi. 338 There is a theory that the type area should be about 50 per cent of the page area. **1975** Type area [see *type site* below]. **1971** *Computers & Humanities* VI. 43 The character set..is limited to the Selectric type-balls specified by the scanning service. **1977** *Daily Tel.* 3 Aug. 3/5 There has been a real search for a type ball from one of the IBM electric typewriters that were in the office. **1931** M. Crooks *Bk. of Underwood Typewriter* ii. 10 Above and behind the keyboard, occupying practically the centre of the framework, is the type—this part of the machine is known as the Type-Basket. **1968** *Typing* ('Know the Craft' Series) 4/1 Every machine has a type basket and a carriage. **1947** *Amer. Jrnl. Sociol.* LII. 293 Kinship, its relationships and institutions, are the type categories of experience and the familial group is the unit of action. **1931** *Times Lit. Suppl.* 5 Nov. 865/3 Mr. Strauss hits off the foibles of his type-characters with wit and acumen. **1890** H. James *Let.* 10 Nov. (1981) III. 307 A shorter story..which I am just sending off to be typecopied. **1893** *Ibid.* 2 July 416, I have determined to dispatch by the same post as this note, in another cover, a fresh type-copy of the said first act. **1854** Geo. Eliot tr. *Feuerbach's Essence Christianity* vii. 75 Mind presenting itself as at once type-creating, emotional, and sensuous, is the imagination. **1905** Type-description [see *protologue]. **1962** J. A. Ford *Quantitative Method for deriving Cultural Chronol.* iii. 16 Other workers have sought to achieve greater precision by dealing not with types but with various elements, attributes, or 'modes'..that are usually hidden away in the type description. **1967** J. Deetz *Invitation to Archaeol.* 51 An artifact type description is..a statement of a set of somewhat variable attributes which can be observed to occur together in the majority of cases. **1887** T. B. Reed *Hist. Old Eng. Letter Foundries* i. 40 It now remains to trace briefly the origin and development of the leading type-faces used in English Typography. **1923** S. Morison *On Type Faces* p. v, The choice of type face is always a matter of immediate and insistent importance. **1980** B. Crutchley *To be a Printer* v. 65 Bodoni..produced in his lifetime over four hundred type faces. **1900** (*title*) Type Facsimile Society. Publications of the Society for the year 1900. **1966** *Eng. Studies* XLVII. 298 It [*sc.* old spelling] does no harm, provided the reader is not misled into using the book as a type-facsimile. [**1908** B. Russell *Logic & Knowl.* (1956) 75 The division of objects into types is necessitated by the reflexive fallacies which otherwise arise. These fallacies..are to be avoided by what may be called the 'vicious-circle principle'; i.e., 'no totality can contain members defined in terms of itself'.] **1935** *Mind* XLIV. 150 Now, of course, the word 'about' is very ambiguous; but, in one sense of it, to say that a proposition is about itself is to commit the simplest of type-fallacies. **1952** *Mind* LXI. 130 The type-fallacy that only moral goodness itself is good. **1967** *Philos.* XLII. 3 A transition from one to the other would then become tantamount to a category-mistake or type-fallacy. **1797** M. L. Weems *Let.* 13 July in *Works & Ways* (1929) II. 84 A letter was written..containing an order on Mr. Baine the Type Founder for some money. **1809** T. Jefferson *Let.* 28 June in *Writings* (1904) XII. 295 The foundation of printing..is the type-foundry. **1905** C. T. Jacobi *Printers' Handbk.* (ed. 3) 22 There is some uncertainty as to what is type-height, and therefore all those engaged in supplying blocks..to the printer should remember that type-height is ·9175 inch. **1931** R. R. Karch *Printing & Allied Trades* iii. 9 Type heights differ in foreign countries. In England the height is ·917; France, Germany and Spain ·928. **1973** S. Jennett *Making of Books* (ed. 5) ii. 40 Type height, or height to paper, is not the same thing as height of face. **1943** *Mind* LII. 271 In *Principia* the need to avoid a small number of objectionable trains of argument is made the occasion for wholesale elaboration of symbolism (the introduction of type-indices). **1972, 1973** Type index [see *motif-index* s.v. *motif 4]. **1934** Webster, Type-locality. **1937** *Brit. Birds* XXXI. 10 Dr. Ticehurst has given as type-locality Lincolnshire. **1940** *Chambers's Techn. Dict.* 874/2 *Type locality*, the locality from which a rock, formation, etc., has been named and described, usually because of its characteristic occurrence there. **1962** Gordon & Lavoipierre *Entomol. for Students of Med.* liii. 325 The locality from which the holotype was collected is known as the 'type locality'. **1969** *Proc. Geol. Soc.* Aug. 159 Donovan..stated Watchet to be the type-locality of the index species of the zone, but gave no type-locality for the zone itself. **1866** G. M. Hopkins *Jrnl.* 16 May (1959) 137 Hawthorn especially when thrown up with may is very clearly type-marked. **1928** L. P. Smith *Words & Idioms* 40 Some of these type-names give evidence of the impression made on foreigners by the travelling Englishmen of rank. **1974** *Encycl. Brit. Micropædia* X. 219/2 *Type names*,..those names given by the dramatist to characters in his play so that their personalities may be instantly ascertained. **1932** *Brit. Jrnl. Psychol.* July 77 The considerable growth of 'type' psychologies has been a leading characteristic of recent psychological and clinical study and speculation. **1952** H. Read *Philos. Mod. Art* iv. 83 The science of typology—or type-psychology as it is more often called—is comparatively modern. **1976** *Amer. Speech 1974* XLIX. 19 H. Rex Wilson..was probably the first linguistic geographer to propose typescripting entire interviews. **1979** *Amer. N. & Q.* Mar. 106/1 Nor did Fisher ever mention to Crane or to me his *Plowshare* publications of Greenberg, which included 'The Charming Maiden' (June 1918) 'Serenade in Grey', 'Regret At Parting', and 'Where Sweepest Thou' (January 1920), all later typescripted by Hart Crane. **1980** *Amer. Speech 1976* LI. 204 A typescripted record represents several weeks of tedious work. **1981** *Ibid.* LVI. 258 The typescripting or computer-taping and indexing of LAGS field records are now considered supplemental descriptive components of the atlas. **1846** S. F. Smith *Theatrical Apprenticeship* ii. 30 She..would then dismiss us to our type-setting. **1848** *De Bow's Rev.* VI. 52 But the printer is too important..to have his usefulness set aside by the multiplication of type-setting and press-working machines. **1935** *Proc. Prehistoric Soc.* I. 6 The Late Bronze Age assemblage to which the type site offers no significant parallel. **1959** J. D. Clark *Prehist. Southern Africa* vi. 159 Bambata is the type site for the Stillbay Culture in Rhodesia. **1969** *Proc. Geol. Soc.* Aug. 157 Table 1..gives the sequence of stages so far defined for the Quaternary of the British Isles. The stage names are based either on type-sites or type-areas. **1975** J. G. Evans *Environment Early Man Brit. Isles* i. 8 Each interglacial is named after a type site or area where deposits of that stage occur. **1981** P. Salway *Roman Britain* 7 A 'type-site' is, in archaeological jargon, the site after which a culture is named, often the site at which it was first discovered or recognized as distinct. **1922** D. B. Updike *Printing Types* I. ii. 32 It [*sc.* the point system] placed type sizes upon a basis comprehensible to the meanest intelligence. **1978** *Early Music* Oct. 597/3 The physical presentation of *Clementi* is also gratifying. Type sizes are unusually ample. **1875** W. Blades *Some Early Type Specimen Bks.* 3 When printers were their own type-founders their works were their own type-specimen. **1922** D. B. Updike *Printing Types* I. xi. 133 A few 'type specimens' were issued by founders, and some by printers. **1922** *Flight* XIV. 267/1 The Bristol 'Lucifer' engine..has successfully passed its type-tests in accordance with British Air Ministry Type-Test Schedule of May, 1920. **1946** *Sun* (Baltimore) 17 May 6 (Advt.), For sale: Stinson cabin monoplanes... Have been type-tested and declared eligible for certification by CAA. **1978** *Proc. Internat. Conf. Noise Control Engin., San Francisco* 743 Because of the complexity and cost of conducting certification type tests, a study was undertaken to determine the feasibility of using an alternative scheme to obtain approach and takeoff noise levels. **1979** *Gloss. Terms Quality Assurance (B.S.I.)* 11/2 *Type test*, a test or series of tests directed towards approval of a design, conducted to determine whether an item is capable of meeting the requirements of the product specification. **1960** G. Herdan (*title*) Type-token mathematics: a textbook of mathematical linguistics. **1971** *Computers & Humanities* V. 133 A type/token ratio is computed for each text, where type is the number of different words occurring in the text and token is the total number of occurrences of all words in the text. **1979** *Sci. Amer.* Feb. 61/2 The type–token ratio, a parameter that reflects the size of the vocabulary employed by the author, was determined for each text. **1931** S. Beckett *Proust* 7 The tragedy of the Marcel-Albertine liaison is the type-tragedy of the human relationship whose failure is preordained.

type, *v.* Add: **4.** Also *trans.* with *out, up.*
1948 A. Keith *Three came Home* xv. 255 The news that came over the radio was typed out. **1961** 'E. Lathen' *Banking on Death* xvii. 135, I want you to..type up a copy of the Hoffman contract. **1981** C. Dexter *Dead of Jericho* xxxvi. 202 We've got to..get it down in writing, then typed up, and signed.

5. a. To assign to a particular type; to classify; esp. in *Biol.* and *Med.*, to determine the type to which (blood, tissue, etc.) belongs.
1900, etc. [see *typing *vbl. sb.* 1]. **1929** L. H. Snyder *Blood Grouping in Relation to Clin. & Legal Med.* iii. 13 They called attention to the fact that the methods used for typing an unknown blood are based upon the assumption that there are only four iso-agglutination groups and that the blood of every person belongs to one of the four. **1939** *Jrnl. Bacteriol.* XXXVII. 136 Sixty-seven of the double-zone strains have been serologically 'typed' by Lancefield or Plummer. **1946** Gerth & Mills *From Max Weber* (1947) iii. 56 Less 'rational' actions are typed by Weber in terms of the pursuit of 'absolute ends', as flowing from affectual sentiments, or as 'traditional'. **1959** *Times Lit. Suppl.* 3 Apr. 194/4 The simple character, long ago 'typed' and pigeon-holed, often turns out to be much more complex in the light of his correspondence. **1964** *McCall's Sewing* iii. 41/1 Zippers are typed according to their purpose. **1967** A. S. Byatt *Game* xi. 154 Her clothes..typed her: grey pleated skirt, cable-stitch sweater, brogue shoes. **1968**, etc. [see *tissue-type* vb. s.v. *tissue *sb.* 9 b]. **1969** *New Yorker* 12 Apr. 85/1 The exobiologists have insisted that..the astronauts' microflora..have all been typed and catalogued for comparison later. **1977** *Time* 7 Mar. 43/2 The kidney was..then 'typed' so that doctors could choose a patient whose body tissue matched it. **1977** *Jrnl. R. Soc. Arts* CXXV. 87/1 We like to label periods in our history as we like to 'type' people.

b. = *type-cast *v.*
1933 *Sat. Even. Post* 17 June 14/3 The danger of being 'typed' by the producers, which, in turn, fixes you irrevocably in the public eye, is one of those haunting fears that an actor must meet and conquer. **1939** J. Gielgud *Early Stages* ix. 151 Refusing to be typed in 'silly society' parts, she [*sc.* Edith Evans]..achieved her greatest triumph as Millamant. **1959** *Times Lit. Suppl.* 27 Feb. 118/1 There is an inevitable tendency in novels about soldiers and sailors for characters to be typed; but though some of Mr. Armstrong's ship's company conform, others are distinct individuals. **1959** *Listener* 9 July 72/1 This was good documentary in that the characters were not in the least typed. **1960** *Guardian* 20 Oct. 8/7 They were in revolt against the whole Broadway system of typing actors and thus limiting their development.

-type, *suffix.* Add: **2.** [Type *sb.*[1]] Appended to adjs. and sbs. or sb. phrs. forming adjs. with the sense 'of the specified type; typical or characteristic of (..), reminiscent or imitative of (..)'. Cf. *type *sb.*[1] 6.
1887 [see New *a.* 10 b]. **1907** *Yesterday's Shopping* (1969) 97/2 *Australian*... Claret type..Burgundy type. **1922**, etc. [see *shell-type* adj. s.v. *shell *sb.* 39]. **1924** *Sat. Even. Post* 13 Sept. 80 (Advt.), Has the car you have in mind a European type, high compression motor? **1936** *Discovery* Aug. 237/2 Eight astronomical-type stainless steel mirrors. **1939** W. S. Churchill in *War Illustr.* 29 Dec. 522/2 The submarine had not long been in her patrol area before she sighted a big-type U-boat. **1940** *Illustr. London News* 28 Dec. 821 The outbreak of the war found the Fleet Air Arm using for the most part adapted land-type machines. **1949** *Here & Now* (N.Z.) Oct. 23/3 Following the old Spanish custom still prevalent at California-type barbecues, the meat can be served with some version of hot, tangy tomato sauce. **1959** *Chambers's Encycl.* IV. 693/2 The indulines and nigrosines are important azine type dyes. **1960** *Farmer & Stockbreeder* 8 Mar. 77/1 He showed slides of some of his farm-made, 'home-grown' Dutch-barn-type buildings which he believed in putting up. **1970** *Daily Tel.* 18 Aug. 11 Ocean Spirit is the biggest-type glass fibre boat ever made in this country. **1973** *New Statesman* 19 Oct. 557/3 Fifties-type social realist films. **1976** *Derbyshire Times* (Peak ed.) 3 Sept. 15/1 (Advt.), A very spacious older type house.

type-cast (tɔi·p,kɑst), *a.* Also as one word. [f. Type *sb.*[1] + Cast *ppl. a.*] **1.** Formed into type for printing.
1876 [in Dict. s.v. Type *sb.*[1] 10].

2. Of an actor, etc.: that has been type-cast (see *type-cast *v.*); identified with a particular kind of part. Also *transf.* and in extended use.
1946 S. H. Adams *Alexander Woollcott* 314 He is type-cast if ever a man was. **1952** *Time* 2 June 92/2 Hollywood noticed Jules in 1938, changed his name to John Garfield, and launched him on a type-cast screen career of playing himself—the narrow-eyed, rock-hard underdog. **1964** L. Deighton *Funeral in Berlin* xxiv. 135 'So I am type-cast as the loser?' 'It's a one-horse race,' said Samantha. **1965** *Spectator* 29 Jan. 137/2 He is in danger of becoming typecast. **1971** *Daily Tel.* 9 June 13/5 David Shepherd is type-cast as the man who paints elephants and this bothers him. He wants to be known as 'a painter'. **1976** M. Gilbert *Night of Twelfth* xvi. 149 Naval Officers are deceptive people. They're usually type-cast as bluff simple extroverts... Actually, their training, and the lives they have to lead, are calculated to produce the most complex introverts. **1982** M. Babson *Death beside Seaside* v. 41 One gets the feeling she's typecast in her show. **1984** *Listener* 19 Jan. 37/2 It is only in New York, in my experience, that truly typecast events occur.

type-cast (tɔi·p,kɑst), *v.* Also as one word. [Back-formation from *type-casting *vbl. sb.* 2.] *trans.* To cast (an actor) in a role or roles for which he appears to be physically or temperamentally suited or of a kind in which he has been successful; to allocate continually to the same type of part. Also *transf.* and *fig.*, and in extended use, to represent or regard as a stereotype.
1946, etc. [implied in *type-cast *a.* 2]. **1952** Granville *Dict. Theatr. Terms* 192 *Type cast*, to assign parts to artistes who approximate the type drawn in the play. **1958** R. Hoggart in N. Mackenzie *Conviction* 136 We..will no longer type-cast regional accents by class or comedy. **1958** *Wall St. Jrnl.* 6 Nov. 6/2 Right-to-work has been type-cast as a villain in the public mind. **1959** *Times* 9 Nov. 6/4 Sir Thomas Beecham's gifts tend to typecast him as an interpreter of even-numbered Beethoven. **1971** *Country Life* 20 May 1252/1 But it is by using plants in this category as ornamentals in their own right instead of type-casting them purely as herbs for the herb garden that they become most useful in the garden. **1981** *N. & Q.* Oct. 448/2 Hardy's contemporaries, having type-cast him as a novelist, were dubious if not contemptuous of his poetry. **1982** *Daily Tel.* 16 July 15/4 People are not as likely to be typecast by their accents as they once were.

type-casting (tɔi·p,kɑstiŋ), *vbl. sb.* Also as one word. [f. Type *sb.*[1] + Casting *vbl. sb.*] **1.** The forming of metal, wood, etc., into type for printing. Also *attrib.*
1864 T. L. Nichols *Forty Years Amer. Life* I. 381 By the use of type-casting machines a workman can cast ninety brevier types a minute. **1875, 1897** [in Dict. s.v. Type *sb.*[1] 10]. **1967** V. Strauss *Printing Industry* ii. 66/1 Hot-metal machines are type-casting equipment.

2. The casting of an actor in a role or roles for which he appears to be physically or temperamentally suited or of a kind in which he has been successful; the fact of being so cast. Also *transf.* and in extended use, representation as a stereotype or stereotypes.
1927 *Observer* 25 Sept. 13/4 Please do not conclude that I believe in 'type-casting'; an actor should in his time play many sorts of part. **1947** A. Menen *Prevalence of Witches* xi. 190 He says..why pick on *him* to work a miracle? He says it's his beard..; he says its just type-casting, like they do in pictures. **1960** *Twentieth Cent.* Dec. 588 These Dickensian names suggest the general feeling of type-casting. **1977** 'E. Crispin' *Glimpses of Moon* iii. 38 Film-music composers are just as liable to type-casting as actors and actresses. **1979** D. Arkell *Looking for Laforgue* iii. 78 Laforgue was invariably given the part of Colline, the philosopher... It would appear to have been excellent type-casting.

typed, *ppl. a.* Add: **3.** Also with *out, up.*
1971 D. E. Westlake *I gave at Office* 200 They had this typed-out paper, and it would be everything you said... So you'd read it and sign it. **1977** A. Hunter *Gently Instrumental* viii. 116 Gently found Leyston glooming over the policewoman's typed-up notes.

typewriter. Add: **1.** (Earlier example.)
1868 C. L. Sholes et al. *U.S. Patent* 79,265 23 June 4 Thus made, the type-writer is the simplest, most perfectly adapted to its work.

2*. A machine-gun or sub-machine-gun. *slang.*
1915 A. D. Gillespie *Let.* 3 Mar. in *Lett. from Flanders* (1916) 31 The only typewriter here is the machine-gun—the men's nickname for it. **1930** *Sun* (Baltimore) 25 Oct. 10/3 When the 'typewriters', as machine guns are called,

rattle, it is natural for the police to suspect these 'independents' even of shooting each other. **1959** 'J. Christopher' *Scent of White Poppies* vii. 101 'You associate typewriters with more than one guy.'..'Typewriters?' Bella asked. 'Submachine gun. Browning, maybe..firing automatic.' **1973** P. Evans *Bodyguard Man* iii. 25 Al Capone['s]..torpedoes..were mean with a Thompson 'type-writer'.

3. *typewriter cover, ribbon.*
1903 *Christendom* 13 June 393 Please furnish..12 typewriter ribbons. **1934** Webster, Typewriter cover. **1966** D. Francis *Flying Finish* i. 8 Maggie's..typewriter cover askew. **1975** 'D. Jordan' *Black Account* xiv. 74 She was busy..with her typewriter ribbon.

typewriting, *vbl. sb.* (In Dict. s.v. Type-write *v.*) (Earlier example.)
1867 *Sci. Amer.* 6 July 3/1 The subject of type writing is one of the interesting aspects of the near future.

typey (tɔi·pi), *a.* Also **typy.** [f. Type *sb.*[1] + -y[1].] Of a domestic animal: exhibiting the distinctive characteristics of the breed; being a perfect specimen of the breed.
1931 *19th Century* Feb. 210 It is not registered bulls that are needed, nor 'typey' bulls, but proved bulls, whether registered or not. **1951** *Sun* (Baltimore) 9 Oct. (B ed.) 11/3 One bull might be as 'typy' (Angus argot for conforming to perfect physical standards) as another bull. **1976** *Horse & Hound* 10 Dec. 67/4 (Advt.), Pure Arabian filly... Very typey with large eyes and one of the best movers we have ever seen. **1980** *Hunting Ann. 1981* 107 (Advt.), Typey pups sired by our top studs and from matings within our own pack.

typhlo-. Add: **typhlograph,** also = *nyctograph; (earlier example).
1891 [see *nyctograph].

typho-. Add: **typhomalarial** *a.* (earlier example).
1863 J. J. Woodward *Outl. Chief Camp Diseases U.S. Armies* iii. 87 In all the cases of typho-malarial fever,..three several trains of phenomena are to be noted: the malarial, the typhoid, and the scorbutic.

typhoid, *a.* Add: **3.** *Typhoid Mary*, nickname of Mary Mallon (d. 1938), Irish-born cook who transmitted typhoid fever in the U.S.A. Also *fig.*, a transmitter of undesirable opinions, sentiments, etc.
1909 *N.Y. Times* 17 July 3/5 Mary Mallon, known to fame as 'Typhoid Mary',..must remain at Riverside Hospital. **1961** S. P. Hayes in *Webster* s.v., Authoritarianism..is carried by Typhoid Marys, unwitting sources of infection. **1971** 'L. Egan' *Malicious Mischief* (1972) vii. 118, I went to the library..and asked. And of course they looked at me as if I was—was Typhoid Mary. **1976** *Ann. Rev. Microbiol.* XXX. 438 During the next 15 years Typhoid Mary infected well over 200 persons. **1979** R. Jaffe *Class Reunion* ii. vi. 180 Alexander's own misery was the most important thing in Alexander's life, and whatever he gave of it to others was unimportant to him; he was simply unaware. Alexander was the Typhoid Mary of angst.

typhoon. Add: Hence **typhoon** *v. trans.* (*nonce-word*), to batter with the force of a typhoon.
1953 Dylan Thomas *Under Milk Wood* (1954) 50 But with blue hazy eyes the fishermen gaze at that milkmaid whispering water with no ruck or ripple as though it blew great guns and serpents and typhooned the town.

typhus. Add: **c. typhus-louse,** a louse of the kind responsible for transmitting typhus.
1939 Auden & Isherwood *Journey to War* 136 Their upholstery often contained typhus-lice.

typing, *vbl. sb.* Add: **1.** (Examples corresp. to *type *v.* 5 a, b.)
1900 W. Myers tr. *Ehrlich & Lazarus' Histol. of Blood* 32 The simple 'typing' of several hundred cells. **1927** H. W. Jones in *Osler's Mod. Med.* (ed. 3) V. 191 The institution of blood typing has made the operation of blood transfusion reasonably safe. **1928**, etc. [see *blood-typing* vbl. sb. s.v. *blood *sb.* 19]. **1960** *Guardian* 3 Nov. 8/4 The rigid typing of her as a dumb blonde... Hollywood has destroyed..talent by typing. **1962** *Brit. Med. Bull.* XVIII. 64/1 (*heading*) The genetic basis of bacteriophage typing. **1965**, etc. [see *tissue typing* s.v. *tissue *sb.* 9 b]. **1971** *Nature* 9 July 141/2 Each centre coordinates the typing and distribution of cadaveric kidneys to and from a group of collaborating hospitals.

2. *attrib.* and *Comb.*, as *typing agency, bureau, course, error, paper, pool* (see *pool *sb.*[3] 5 c), *purposes, school, speed.*
1973 W. M. Duncan *Big Timer* xi. 75, I should like..a list of the typing agencies within say, roughly, a half-mile. **1935** D. L. Sayers *Gaudy Night* v. 97 [She] ran what was ostensibly a Typing Bureau. **1974** R. Rendell *Face of Trespass* i. 17 My friend Molly that I used to have my typing bureau with. **1952** Leslie & Pepe *Methods Teaching Typing Simplified* (1954) 85 A slightly condensed verbatim report of an actual teaching period within the first week of the typing course. **1981** C. Storr *Vicky* xviii. 128 She'd taken a typing course and she had a job. **1936** A. Dvorak et al. *Typewriting Behavior* xiii. 395 Typing errors are signs of interference, due partly to drills on isolated letter strokes. **1979** M. Page *Pilate Plot* i. 21 He looked at it three times to make sure it was not a typing error. **1944** Mrs. Belloc Lowndes *Let.* 4 June (1971) 249, I got a packet of *pre-war* typing paper—finest quality! **1889** Typing purposes [in Dict.]. **1966** 'G. Black' *You want to die, Johnny?* xii. 223 There are certainly duller jobs waiting for a graduate of typing school than being the secretary to a Regent. **1936** A. Dvorak et al. *Typewriting Behavior* xi. 286 Typing speed grows, but fast motions must also be present from the start. **1976** 'A. York' *Dark Passage* xiv. 182 Her typing speeds and general ability as a secretary were excellent.

typiste (tɔipī·st). [Alteration of Typist, with Fr. termination as in *modiste*, misinterpreted as fem.] A female typist.
1923 *Daily Mail* 26 Feb. 8 An actress who played the part of a typiste. **1969** *West Australian* 5 July 65/1 (Advt.), Shorthand typiste.

typo, *sb.* (*a.*) Add: **a*.** A typographical error.
1892 I. Zangwill *Childr. Ghetto* III. ii. iii. 24 My men..don't like to pass anything till it's free from typos. **1945** E. B. White *Let.* 10 July (1976) 266 As for the 'her'–'hen' typo, I guessed that it was a typo and that it would be caught. **1963** C. D. Simak *They walked like Men* vi. 36, I went through the story again and caught a couple of typos and fixed up another place or two to make language better. **1978** *Times Lit. Suppl.* 15 Sept. 1031/5 Since few proof-readers are perfect, a typo here and there is easily forgiven.

typo-. Add: **typophil** (examples of *typophile*).
1958 *Times Lit. Suppl.* 3 Jan. 12/3 It is only sad that the Typophile Chap Books are not more widely available. **1975** *Dalhousie Rev.* Summer 206 Berenson..was the recognized leader of an energetic group which included..Bertram Goodhue, architect, book-designer and typophile. **1982** *Amer. N. & Q.* Jan./Feb. 89/2 It is good reading, at least for typophiles.

typographica (tɔipǫgræ·fikă), *sb. pl.* [see Typographic *a.*] Examples of fine printing; in quot. 1949 used as the title of a journal dealing with typography.
[**1858** S. L. Sotheby (*title*) Principia typographica.] **1931** *Times Lit. Suppl.* 25 June (Salon Int. du Livre d'Art Suppl.) p. i/3 The former [publishing] house has issued a series of *feuillets d'imprimerie* which..are coveted by collectors of typographica. **1949** *Typographica* i. 3 The first purpose of Typographica will be to present serious analyses of various aspects..of contemporary typography. **1959** *Times Lit. Suppl.* 15 May 289/4 Your account of the George A. Poole collection of typographica..contains one inaccuracy.

typographist. (Earlier example.)
1851 *Tait's Edin. Mag.* Oct. 636/1 The public had a thousand or so of quarto pages for which they paid a folio price, nearly three-fourths of which price was a clear gain to the colluding typographists.

typological, *a.* Add: **3.** Also, pertaining to the study of the use of types in other disciplines or fields of study: see *typology 3.
1913 E. T. Leeds *Anglo-Saxon Settlements* 28 A large amount of information can be obtained from the purely typological method. **1929** V. G. Childe *Danube in Prehist.* 246 By correlating the several stages in the evolution of celts, daggers,..&c...it is possible to divide the Bronze Age up into several typological phases. **1930** *Psyche* X. iii. 80 (*heading*) Typological methods in experimental psychology. **1942** *Antiquity* XVI. 61 What typological evidence there is..confirms the standard view of the invasion in its secondary stages. **1964** R. H. Robins *General Linguistics* viii. 325 It is..possible to compare languages..simply by reference to any significant general features of form or structural organization that they share at any level of analysis... Comparison of languages on this basis is usually distinguished..by the title of *typological comparison*. **1968** *Internat. Encycl. Soc. Sci.* XVI. 178/1 Typological classification, as a subdivision of taxonomy, has characterized a considerable part of the culture of the social sciences. **1971** J. Z. Young *Introd. Study Man* xxvii. 386 We are bound therefore to use arbitrary typological groupings, that is to say to put together those we think seem most alike in appearance or some other character. **1972** *Jrnl. Social Psychol.* LXXXVI. 55 The first definition refers to..the characteristics rather than the persons being used for typological identification. **1980** *Nature* 27 Mar. 341/1 Isaac has stated that it is likely that Olorgesailie may date to before 400 kyr BP and it is suggested on typological grounds that Namib IV is of similar antiquity.

Hence **typolo·gically** *adv.*, by means of, in terms of, or according to typology.
1895 in *Funk's Standard Dict.* **1921** M. C. Burkitt *Prehistory* vii. 94 If a series be made out they are found to grade on the one hand into what typologically are Mousterian points, and on the other into a sort of pseudo-Solutrean. **1943** C. L. Wrenn *Word & Symbol* (1967) 134 Collingwood typologically demonstrates on historical artistic grounds that the inscription belongs to the close of the eighth century. **1964** R. H. Robins *General Linguistics* viii. 326 Languages are typologically classified according to the similarities of form they exhibit with other languages at any level or levels. **1976** G. W. H. Lampe in M. F. Wiles et al. *Christian Believing* (C. of E. Doctrine Cmn.) 55 The Church continued to read it [*sc.* the Old Testament] in its supposed 'spiritual' sense, or typologically. **1980** *Nature* 27 Mar. 340/2 Typologically..Namib IV may closely be paralleled by East African industries from Olorgesailie, Kilombe and Olduvai IV.

typologize (tɔipǫ·lŏdʒəiz), *v.* [f. Typology + -ize.] *trans.* and *intr.* To interpret or classify typologically. Hence **typo·logizing** *vbl. sb.*
1895 *Funk's Stand. Dict.* II. 1955/2 *Typologize*,..To interpret by types; treat typologically... To deal in or with types or figures. **1959** A. Farrer in *Proc. Oxf. Soc. Hist. Theol.* 8 Questions, which..theologians and exegetes are not going to let alone, whether they typologise or not:—for example, why St Mark put several paragraphs in the order in which he placed them. *Ibid.* 9 Here..is an example of the prefiguration scheme which the greatest enemy of typologising can scarcely deny. **1964** *Harvard Educ. Rev.* XXXIV. ii. 356 Then in a sort of neo-Jespersenian typologizing, we find that there are three sorts of monemes. **1969** P. Worsley in Ionescu & Gellner *Populism* 218 When actors see themselves as part of a shared tradition or organized movement..we are plainly dealing with a quite different kind of typologizing, based upon self-identification. **1976** *Times Lit. Suppl.* 12 Mar. 288/2 One might even say they [*sc.* the anthropologists Benedict and Radcliffe-Brown] shared the impulse to typologize. **1978** *Ibid.* 17 Feb. 217/1 It is possible to typologize military regimes.

typology. Add: **3.** The study of classes with common characteristics; classification, esp. of human products, behaviour, characteristics, etc., according to type; the comparative analysis of structural or other characteristics; a classification or analysis of this kind.
1886 *Academy* 8 May 332/1 In his former publications the learned writer gave too high a place to typology. **1929** V. G. Childe *Danube in Prehist.* p. vii, Where stratigraphical or geological evidence is lacking, we must have recourse to typology. This depends on the assumption that types evolved (or degenerated) regularly. **1930** S. Casson *Archæol.* 43 Even after Schliemann archæologists are to be found who will still prefer a typology which is established by *a priori* methods. **1930** *Psyche* X. iii. 82 Scheler..tries to construct a typology on a purely philosophical and phenomenological basis. **1937** J. R. Firth *Tongues of Men* i. 17 Language is the typology of the common elements in your speech and mine, yesterday, to-day, and to-morrow. **1950** T. W. Adorno *Authoritarian Personality* xix. 744 Hardly any concept in contemporary American psychology has been so thoroughly criticized as that of typology. **1953** C. E. Bazell *Linguistic Form* 80 There will therefore be different typologies of language according to the system which is taken as starting-point. **1959** J. J. Michaels in S. Arieti *Amer. Handbk. Psychiatry* I. xix. 358/1 The absence of a psychoanalytic typology may also be explained by the absence of a complete theory of character structure. **1962** [see *a-historical *a.*]. **1964** M. Argyle *Psychol. & Social Probl.* i. 17 One typology in common use is that of mental disorders, largely derived from a classification due to Kraepelin. **1971** *New Scientist* 27 May 534/3 A sort of 'typology' of unsafe drivers, dividing them into groups. **1974** P. H. Matthews *Morphology* 17 In some [languages], grammarians speak of a 'word' without internal grammatical structure: according to the first typologies these were 'isolating' languages... This typology has..been criticised and elaborated. **1977** H. C. Triandis *Interpersonal Behav.* i. 21 The..most important function of the model is to guide investigations of typologies of behaviors, settings and people. **1977** *Dædalus* Summer 89 One direction in which the comparative analogical approach can lead is toward what might be called a typology of cultures and of process.

typothetae (tɔipǫ·pĕtī, tɔi:pǫþī·tī), *sb. pl.* [mod.L., f. Gr. τύπος Typo- + θετός, f. τιθέναι to set, place.] Master printers collectively; *spec.* the members of a N. Amer. association of master printers.
1825 T. C. Hansard *Typographia* i. ii. 43, I cannot find that Gutenberg was encouraged in his labours by the smiles of royal influence. This is the more remarkable, as the then reigning sovereign of Germany, Frederic III, was a monarch 'deeply versed in the learning of the times'...The..emperor..permitted printers to wear gold and silver, and granted coat-armour to the *Typothetae* and *Typographi*, to perpetuate the honour of the discovery. This armorial bearing is still claimed by the professors of the art in Germany. **1865** *N.Y. Times* 22 Mar. 5/2 (*heading*) Inauguration of the Typothetæ. *Ibid.*, At a recent general meeting of master printers of this city, it was decided to organize a permanent association for the general benefit of the the trade and the improvement of the typographic art. A constitution was adopted and the name of 'Typothetæ' was given to the new association. **1888** (*title*) Banquet given by the Typothetae of New York to the delegates of the United Typothetae of America, September 20, 1888. **1915** E. G. Gress in *Fifty Years Typothetae City of N.Y.* 6 The reign of Frederick III., who recognized the German Typothetae, was from 1440 to 1493. **1921** *Daily Colonist* (Victoria, B.C.) 30 Mar. 1/4 The Boston typothetae today announced a reduction of $4 a week in the pay of journeymen printers, and $3 a week in the pay of journeywomen printers.

typy, var. *typey *a.*

tyramine (tɔi^ə·rămīn). *Biochem.* Also **tyramin.** [f. *tyr(osine + Amine.] A crystalline sympathomimetic amine derived from tyrosine and occurring naturally in cheese and other foods, which can cause dangerously high blood pressure in people taking a monoamine oxidase inhibitor; 2-(*p*-hydroxyphenyl)ethylamine, $HO{\cdot}C_6H_4{\cdot}CH_2{\cdot}CH_2{\cdot}NH_2$.
1910 *Biochem. Jrnl.* V. 236 Tyramine,..an organic base which can be produced from tyrosine by action of certain bacteria. **1939** Tyramin [see *hypertensin]. **1974** Passmore & Robson *Compan. Med. Stud.* III. lxii. 22/2 Cheese, red wine, yeast extracts and pickled herrings may contain large amounts of tyramine, and this is normally metabolized by monoamine oxidase... In patients receiving MAO [*sc.* monoamine oxidase] inhibitors, tyra-

mine is absorbed intact and releases the large amounts of stored noradrenaline causing a hypertensive crisis.

tyrannicidal, *a.* (Earlier example.)
1801 W. Dupré *Lexicographia-Neologica Gallica* 282 Projet *tyrannicide*—A tyrannicidal scheme.

tyrannis (tiræ·nis). *Gr. Hist.* [L., a. Gr. τυραννίς rule of a despot.] = Tyranny *sb.* 1. Also *transf.*
1878 T. D. Woolsey *Political Science* I. ix. 406 His [*sc.* Plato's] forms of polity..are..aristocracy, timocracy, oligarchy, democracy, and the tyrannis. Oligarchy arises from overgrown wealth, tyrannis from overgrown liberty. **1910** *Encycl. Brit.* XII. 446/2 Between the Roman principate and the Greek tyrannis there are two essential differences. **1920** *Glasgow Herald* 6 Apr. 6 D'Annunzio..proceeded to establish a tyrannis quite on the classical model.

tyrannosaurus (tirænosǭ·rǝs). Also **tyrannosaur.** [mod.L. (H. F. Osborn 1905, in *Bull. Amer. Mus. Nat. Hist.* XXI. 259), f. Tyranno- + Gr. σαῦρος lizard.] A large bipedal dinosaur of the genus of the same name, known from fossil remains found in North America. Also *fig.*
1905 H. F. Osborn in *Bull. Amer. Mus. Nat. Hist.* XXI. 261 (*heading*) Tyrannosaurus, Upper Cretaceous carnivorous dinosaur. **1927** Haldane & Huxley *Animal Biol.* xi. 240 The Tyrannosaur (was adapted) for preying on large animals. *Ibid.* xiii. 314 The Tyrannosaurus..stood over twenty feet high. **1934** [see *ankylosaurus]. **1957** L. Eiseley *Immense Journey* 64 Tyrannosaurs, enormous bipedal caricatures of men, would stalk mindlessly across the sites of future cities. **1964** J. Wain in *Listener* 13 Aug. 371 Peter Ustinov and Alec McCowen play two 19th-century American railroad barons—Cornelius Vanderbilt and Jay Gould... The two tyrannosaurs fought to a standstill.

2. *Tyrannosaurus rex*, the only species of the genus *Tyrannosaurus*; also *fig.*
1906 *Bull. Amer. Mus. Nat. Hist.* XXII. 284 Tyrannosaurus rex... The complete skeleton of this animal is restored. **1972** D. Bloodworth *Any Number can Play* viii. 56 A stupid, feudal autocrat..who believes in the divine right of Kings, starting with tyrannosaurus rex. **1976** 'J. Ross' *I know what it's like to Die* xxiv. 150 A crane, fancifully recognisable to him as an orange-painted *Tyrannosaurus Rex*, towered above a metal-crushing plant.

tyre, *sb.*[3] Add: **2. b.** *spare tyre*: see *spare *a.* and *adv.* 1 a (a).

3. *tyre-burst, cast, lever, mark, pressure, track, tread*; *tyre chain*, a chain fastened to a tyre to prevent wheel-skid, esp. in snow.
1935 'R. West' *Harsh Voice* ii. 91 A tyreburst made him turn towards the road. **1971** J. Wainwright *Dig Grave* 67 We're busy taking tyre-casts... Taking plaster of Paris casts of tyre-marks. **1958** L. Durrell *Mountolive* iii. 80 His ears had caught the slither and scrape of tyre-chains on the frosty drive outside. **1927** *Cycling Man.* (ed. 7) (Advt.), Tyre levers. The famous 'Jiffy' has no rival. **1959** I. Jefferies *Thirteen Days* vi. 92 Once he found a receptive bitch he wouldn't leave her..until he was prised off with tyre-levers. **1936** 'N. Blake' *Thou Shell of Death* xiv. 258 They had stopped at a fork. Blount was out, scanning the road surface for tyre marks. **1971** Tyre mark [see *tyre cast* above]. **1931** *Man. Morris Minor Car* (Morris Motors, Ltd.) 50 Gauges for testing balloon tyre pressures can be bought from all reputable motor dealers. **1959** *Motor Manual* (ed. 36) v. 130 A word or two should be said on the subject of tyre pressures. The basic fact to be remembered is that it is the air that carries the load. **1931** D. L. Sayers *Five Red Herrings* iii. 38 A fresh set of tyre-tracks in the dust showing where the car had been taken out. **1971** M. Kelly 'Twenty-Fifth Hour' i. 23 There were tyre tracks, wide ones, in the shingle out at each edge of the road. **1925** *Motor Cycling* 30 Dec. (1931) 9 (*heading*) Oil grooves and tyre-tread grooves compared. **1978** R. Westall *Devil on Road* xi. 73 There were hoofprints in the dried-up mud... Wheeltracks too, but no tyre-treads.

Tyrian, *a.* and *sb.* Add: **C.** *Tyrian-dyed* adj. (earlier example).
1847 C. Brontë *Jane Eyre* I. xi. 194 A wide arch..hung..with a Tyrian-dyed curtain.

ty-ring, *vbl. sb.* [f. *tyre* vb. s.v. Tyre *sb.*[3] + -ing[1].] The action of furnishing with a tyre or tyres (= Tiring *vbl. sb.*[3]). Freq. *attrib.*
1923 G. Sturt *Wheelwright's Shop* xxiii. 117 A great business was this tyring—if possible deferred..until a prospect of bright weather set in, and then, it being 'hotted up' in one fire. *Ibid.* xxv. 125 Armed with tyring dogs, the three of us watched, expectant. **1936** *Automobile & Carriage Jrnl.* Dec. 153/2 The exhibits will include the tyring platform and tyre-bending machine. **1945** *Daily Herald* 31 Aug. 4/2 (Advt.), Much of the work had lain in its adaptation to the tyring of tank bogie wheels. **1968** J. Arnold *Shell Bk. Country Crafts* 159 The construction of a wheel, before the tyring stage was reached, required long experience.

tyrocidine (tɔirosai·din, -dīn). *Pharm.* Also **tyrocidin.** [f. mod.L. *Tyro-thrix* (see *tyrothricin) + -cid(e + -ine[5], -in[1], after *gramicidin.] (Any of) a group of crystalline monocyclic decapeptide antibiotics which along with the gramicidins are the active components of tyrothricin, as *tyrocidine A* ($C_{66}H_{87}O_{13}N_{13}$), *tyrocidine B* ($C_{68}H_{88}O_{13}N_{14}$), *tyrocidine C* ($C_{70}H_{89}O_{13}N_{15}$).
1940 Hotchkiss & Dubos in *Jrnl. Biol. Chem.* CXXXVI. 803 This substance was described earlier under the name gramicidic acid by the present writers who recognized that it..will be referred to hereafter as *tyrocidine hydrochloride*. **1952** Battersby & Craig in *Jrnl. Amer. Chem. Soc.* LXXIV. 4019/1 Crystalline tyrocidine hydrochloride is in fact a family of polypeptides... The major peptide has been called tyrocidine A, tyrocidine B, tyrocidine C discovered by countercurrent partition in the solvent system. **1969** Tyrocidin [see *gramicidin]. **1977** J. S. Glasby *Encycl. Antibiotics* (1979) 360 One of the bacteriological examination has been carried out on the mixture of tyrocidins.

Tyrode (tai·rǭᵈd). *Med.* Also **tyrode.** The name of Maurice Vejux Tyrode (1878–1930), American pharmacologist, used *attrib.*, in the possessive, and *ellipt.* as *Tyrode's*, to designate a type of physiological saline solution used to irrigate tissue and in laboratory work.
1910 *Arch. Internat. de Pharmacodynamie* XX. 208 Composition of..Tyrode. **1923** *Jrnl. Pharmacol. & Exper. Therapeutics* XXI. 218 Lowering of pH values of the blood did not occur regularly after injection of..Tyrode's solution. **1962** Harris & Gruber in *A. Pirie Lens Metabolism Rel. Cataract* 374 The cation transport can be measured visually..in a fairly simple medium. Originally, we used Tyrode's solution. **1964** G. W. Smith *Allergy & Tissue Metabolism* iii. 56 These experiments..consisted simply of replacing the oxygenated tyrode solution in which the tissue was bathed with tyrode solution containing either 48/80, stilbamidine protamine sulphate or toluidine blue. **1975** *Nature* 11 Jan. 153/1 The sperm were recovered from the albumin by centrifugation at 2,500g and washed once in Tyrode's.

Tyrolean, *a.* and *sb.* Add: Now usu. with pronunc. (tirolī·ǝn). **a.** *adj.* (Earlier example.) *Tyrolean hat*, a soft felt hat with a brim turned up at the sides and usu. a feather cockade; (later example). **b.** *sb.* Also = *Tyrolean hat* above.
1809 *Times* 7 Nov. 2/3 The greatest part of those prisoners taken..belong to..the Tyrolian Chasseurs. **1957** M. B. Picken *Fashion Dict.* 361/1 *Tyrolean hat*, soft brimmed felt hat. **1973** T. Pynchon *Gravity's Rainbow* iii. 495 Von Göll tips an invisible Tyrolean to old ladies in black who've come out in pairs to get some sun.

Tyroler: also, the dialect of German spoken in the Tyrol; also as *adj.*, = *Tyrolean a.*
1889 M. Howitt *Let.* 11 June in *Autobiogr.* (1889) II. iv. 345 Father Paul..has been ill..yet being able to say Mass this morning, he, a Tyroler Benedictine, remembered our Queen's Jubilee, and made it his intention. **1923** E. Hemingway *Three Stories & Ten Poems* 18 Part of the time he talked in D'Annunzio dialect and sometimes in German Tyroler dialect... Pedazzi decided to talk altogether in Tyroler. **1963** I. Fleming *On H.M. Secret Service* (U.S. ed.) xxvii. 295 A dark-grey Tyroler outfit with the traditional dark-green trimmings and stag's-horn buttons.

tyrosin. Add: Now always **tyrosine** (-īn). An amino-acid that is the precursor of several hormones, including adrenalin; 3-(*p*-hydroxyphenyl)alanine. (Further examples.)
1926 [see *dopamine]. **1955** *Sci. Amer.* July 75/2 The six virus strains were analyzed for their content of three amino acids: tyrosine, tryptophan and phenylalanine. **1965** Lee & Knowles *Animal Hormones* v. 91 Basically, thyroid hormone consists of iodine conjugated with tyrosine. **1972** [see *tyrosinase].

tyrosinæmia (tairǒsinī·miă). *Med.* Also (chiefly *U.S.*) **-emia.** [f. prec. + Gr. αἷμα blood + -ia[1].] Any of several conditions marked by the presence in the blood and urine of abnormally high amounts of tyrosine.
1965 J. Gentz et al. in *Jrnl. Pediatrics* LXVI. 670/1 Tyrosinemia as defined in the present report is a disorder characterized biochemically by a continuous high urinary excretion of tyrosine and tyrosyl compounds and a high plasma tyrosine level, while the concentrations of other amino acids are normal or rather low. **1982** T. I. Diamondstone in T. M. Devlin *Textbk. Biochem.* xii. 589 Transient tyrosinemia of the newborn appears to be due to delayed development of tyrosine transaminase and/or *p*-hydroxyphenylpyruvic oxidase... Persistent tyrosinemia without liver or kidney disease has been reported in six patients, all of whom are retarded. **1983** *Oxf. Textbk. Med.* I. ix. 94/1 A single metabolic lesion to explain all the findings has not yet been identified, and it has been argued that Type I tyrosinæmia is not in fact primarily a disorder of tyrosine metabolism.

tyrosinase. Delete entry in Dict. and substitute:

tyrosinase (tai·rǒs-, tairǒ·sinēᵗz). *Biochem.* [a. F. *tyrosinase* (G. Bertrand 1896, in *Compt. Rend.* CXXII. 1216): see Tyrosin and *-ase.] A copper-containing oxygenase found in many plants and animals which catalyses the formation of quinones from phenols and polyphenols (e.g. melanin from tyrosine) by the addition and then the oxidation of hydroxyl groups.
1896 *Jrnl. R. Microsc. Soc.* 430 In the root of the dahlia and beet, as well as in several fungi..M. G. Bertrand finds a hitherto undescribed diastase, to which he gives the name *tyrosinase*, from its connection with tyrosin. **1904** *Proc. R. Soc.* LXXIV. 313 An extract can be made from the skins of certain pigmented animals..which will act upon tyrosin and produce a pigmented substance. This action suggests the presence of a tyrosinase. **1931** [see *phenolase]. **1974** [see *polyphenol oxidase* s.v. *poly- 2]. **1982** T. I. Diamondstone in T. M. Devlin *Textbk. Biochem.* xii. 614/2 Melanin is formed from tyrosine via L-dopa as intermediate; however, the enzyme that catalyzes the formation of L-dopa in this pathway is not tyrosine hydroxylase, but a copper-containing oxygenase known as tyrosinase.

tyrosine: see Tyrosin in Dict. and Suppl.

tyrosinosis (tairǒsinōᵘ·sis). *Med.* [f. *tyrosine + -osis.] A rare condition of unknown ætiology in which there is increased excretion of the early metabolites of tyrosine but no liver or kidney damage; also (now *rare*) = *tyrosinæmia.
1932 G. Medes in *Biochem. Jrnl.* XXVI. 938 The essential feature in our patient..consists in a complete stoppage of the oxidation of tyrosine at the stage of *p*-hydroxyphenylpyruvic acid... Thus results an excretion not only of *p*-hydroxyphenylpyruvic acid and *l*-*p*-hydroxyphenyllactic acid, but also of tyrosine and *l*-3:4-dihydroxyphenylalanine. For this condition, the name tyrosinosis is proposed. **1942** [see *L-dopa]. **1969** [see *homocystinuria* s.v. *homocystine]. **1974** Passmore & Robson *Compan. Med. Stud.* III. xxxvi. 8/1 Formerly hypertyrosinæmia or tyrosinosis was considered to be a single, exceedingly rare abnormality. **1982** T. I. Diamondstone in T. M. Devlin *Textbk. Biochem.* xii. 589/1 Tyrosinosis and the tyrosinemias have certain biochemical features in common.

tyrothricin (tairǒθrai·sin, -þri·sin). *Pharm.* [f. *Tyrothrix* (see quot. 1940), f. Gr. τυρό-ς cheese + θρίξ, τριχ- hair: see -in[1].] A preparation of gramicidin and tyrocidine which has antibiotic properties, esp. against Gram-positive bacteria, and has been used externally to treat local infections.
1940 Hotchkiss & Dubos in *Jrnl. Biol. Chem.* CXXXVI. 803 Gramicidin will continue to be used as the name of the crystalline neutral substance described before. It is proposed here to apply the name tyrothricin to the bactericidal agent prepared as an alcohol-soluble, water-insoluble material and containing both gramicidin and tyrocidine. This name is derived from the word *Tyrothrix*, a generic name first used by Duclaux to designate sporulating aerobic bacterial species, several of which have since been found to exhibit antagonistic activity toward other microorganisms. **1959** *Sunday Graphic* 25 Jan. 4 (Advt.), Antibiotic throat lozenges contain the highly effective antibiotic tyrothricin. **1974** M. C. Gerald *Pharmacol.* xxvi. 453 Tyrothricin proved too toxic for systemic use in man.

tystie, now the usual spelling of Teistie in Dict. and Suppl.

tyuyamunite (tiǝyāmū·nait). *Min.* [ad. Russ. *tyuyamunit* (K. A. Nenadkevich 1913, in *Izvestiya imper. Akad. Nauk* VI. 945), f. *Tyuya Muyun*, name of a village near Osh, Kirgiziya, U.S.S.R.: see -ite[1].] A hydrous uranyl vanadate of calcium, $Ca(UO_2)_2(VO_4)_2{\cdot}5\text{–}8H_2O$, occurring as soft, yellowish orthorhombic crystals and mined for its uranium content.
1913 *Amer. Jrnl. Sci.* CLXXXV. 440 The calcium carnotite experimented with is probably identical with the tuyamunite of Nenadkevich. **1957** *Financial Times Ann. Rev. Brit. Industry* 30/1 Important secondary ore minerals are the calcium-uranium vanadate tyuyamunite..and torbernite. **1979** *Mineral. Abstr.* XXX. 362/1 The San Rafael mining area is situated along the east flank of the San Rafael Swell in east-central Utah... Coffinite is the most abundant primary ore mineral, and tyuyamunite the most abundant secondary one.

tzaddik, -iq, varr. *tsaddik. **Tzakonian,** var. *Tsakonian *sb.* and *a.* **tzantza,** var. *tsantsa.

‖ **tzedaka(h)** (tsedǫ·ka). [Heb. *ṣĕdāqāh* righteousness.] Charity, the obligation to help one's fellow Jews.
1959 D. D. Runes *Conc. Dict. Judaism* 225/2 *Tzedakah*, righteousness; charity. **1962** *New Jewish Encycl.* 87/1 The Jewish concept of 'charity' has..been refined to mean more than the mere giving of alms; it has been considered as 'Tzedakah', an act of justice and righteousness. **1968** L. Rosten *Joys of Yiddish* 417 Maimonides set down variously rated forms of tzedaka. **1973** *Synagogue Light* Sept. 30/1 To all who have helped our efforts to increase..in the year 5733, let me say that there is no finer form of Tzedakah. **1978** *Jewish Chron.* 6 Oct. 23/3 Acceptance of the mitzvot, the ritual acts of prayer, *tzedaka* and so on.

Tzeltal (tselta·l, tse·ltal; s-), *sb.* (and *a.*) Also **Tzendal, Tzental.** Pl. -al, -ales (-a·lĕs), -als. [a. Sp., earlier also *Tzendal, Sendal*, name of one of the three regions of Chiapas as divided by the Spaniards; ulterior origin uncertain.] (A member of) an Indian people inhabiting parts of southern Mexico; the

Mayan language of this people. Also *attrib.* or as *adj.*
1868 C. H. Berendt in *Ann. Rep. Board of Regents Smithsonian Inst. 1867* 426, I visited the ruins of Palenque, and..was enabled..to collect vocabularies of the Putum and Tzental languages, both spoken in Chiapas. **1871** L. H. Morgan *Systems of Consanguinity* ii. vi. 263 The Chontal language is allied to the Maya of Yucatan. It also affiliates with the Chol and Tzental of Chiapa. **1875** H. H. Bancroft *Native Races Pacific States N. Amer.* I. vi. 645 South of them [*sc.* the Chontales] in Chiapas are the *Choles, Tzendales, Zotziles*, [etc.]. *Ibid.* III. x. 760 Most..are related to..the Maya, of which a dialect called the Tzendal is said to be the oldest language spoken in any of these countries. **1927** O. la Farge in *Tribes & Temples* II. xv. 326 The language itself has been variously named Cendal, Tzendal, Cendales, Tzental, and Tzeltal. I have taken the latter form, as it most nearly represents the Indians' own form, *ts'eltal*, varied in careless speech to *tceltal*. *Ibid.* 325 They speak the Tzeltal language, belonging to that branch of the great Mayance stock more closely related to Maya. **1939** Redfield & Villa in *Contrib. Amer. Anthropol. & Hist.* V. xxviii. 107 Mr. Villa..does not speak Tzeltal. *Ibid.* 110 Most Tzeltal communities are without beehives. **1949** [see *intransitivize *v.*]. **1964** E. A. Nida *Toward Sci. Transl.* x. 239 Among the Tzeltals in Southern Mexico the coming of missionaries..resulted in the conversion of many hundreds of persons, who responded, in their eagerness to describe their spiritual experiences, by creating many new phrases. *Ibid.*, At the same time, the Tzeltal people were undergoing certain adjustments of experience to new elements of language. **1979** L. Campbell in Campbell & Mithun *Lang. Native Amer.* 902 Examination of numerals shows every other one to be Tzeltal alternating with Tojolabal.

tzimmes (tsi·mǝs). Also **tsimmes, -is; tzimmas, -is, -us; (t)zimes.** Pl. same. [a. Yiddish *tsimes* of obscure origin.] A stew or casserole of sweetened vegetables or vegetables and fruit, sometimes with meat. Also *fig.*, a fuss, a confused affair.
1892 I. Zangwill *Childr. Ghetto* I. v. 136 Esther sometimes compounded *Tzimmus*, a dainty blend of carrots, pudding, and potatoes. **1893** *Pall Mall Mag.* Oct. 870 It was a good and simple meal of fish, fowl, and fruits, no 'borsch', no 'tabeches', no 'tzimmas'. **1903** *Jewish Encycl.* IV. 256/1 *Zimes*, or compote, consists generally of cooked fruits..or of vegetables, well spiced. **1923** A. Yezierska *Children of Loneliness* 142 *Tzimmes—blintzes*—a golden-roasted goose swimming in its own fat ravished the senses. **1925** S. Raphaelson *Jazz Singer* 125 Brother Levy, I am the chairman of the Executive Committee. Make yourself a *tsimmes* from it... All right—*all* right. **1925** A. Yezierska *Bread Givers* vi. 102 You ought to taste her *gefülte* fish! Her *tzimes*! **1946** D. Runyon *Short Takes* 199 Why is he making a big tsimmis out of a grunt? **1958** J. Grossinger *Jewish Cookery* p. xi, Rosh Hashonah..one of the many different styles of *tzimmes* is a regular part of the holiday's menu. **1968** L. Rosten *Joys of Yiddish* 411 Since making *tsimmes* took time and various mixings, the word came to mean:..A prolonged procedure; an involved business; a mix-up. **1970** S. Ellin *Bind* xxxviii. 186 I'll be owing money before this *tzimmis* is over. **1970** [see *kishke]. **1974** R. L. Simon *Wild Turkey* (1976) vii. 39 Why are you making such a *tsimmis*? Hecht is dead. **1975** G. Meir *My Life* iii. 37 The Sabbath meal—chicken soup, *gefilte* fish, and meat braised with potatoes and onions, with a carrot-and-prune *tzimmes* on the side. **1978** *Times* 29 July 9/2 On a recent visit, latkes were greasy and tzimmis boring.

tzitzit(h), varr. *tsitsith.

tzolkin (tsǭ·lkin, s-). [Mayan.] The cycle of two hundred and sixty days constituting the sacred calendar of the Maya.
1931 Gann & Thompson *Hist. Maya* viii. 208 After giving the position of a day in their sacred calendar or Tzolkin, as it was probably called, they proceeded to give the position in their 365-day year. **1948** A. L. Kroeber *Anthropol.* (rev. ed.) 549 The 260-day tzolkin..had heavy religious and astrological import. **1977** G. Clark *World Prehistory* (ed. 3) viii. 374 The Maya maintained two distinct counts. For religious purposes they used a sacred year (*tzolkin*) made up of thirteen twenty-day units.

tzores, var. *tsores.

Tzotzil (tsǒᵘ·tsil, tsǒᵘtsi·l; s-), *sb.* (and *a.*) Also **Zotzil.** Pl. **-il, -iles** (-i·lĕs), **-ils.** [a. Sp., ad. Tzotzil *soćil* bat people.] (A member of) an Indian people inhabiting parts of southern Mexico; the Mayan language of this people. Also *attrib.* or as *adj.*
1875 [see *Tzeltal]. **1927** O. La Farge in *Tribes & Temples* II. xv. 326 Their [the Bachajons'] immediate neighbors of the Highlands are the Zotzil. [They] speak a language differentiated from Tzeltal only by a shift of *a, e,* and *b* to Zotzil *o, i,* and *m*. **1939** Redfield & Villa in *Contrib. Amer. Anthropol. & Hist.* V. xxvii. 110 Wheat is..grown in the Tzotzil towns of Chamula and Huistán. **1940** F. Johnson in *Maya & their Neighbors* vi. 108 The Chañabal, and Tzotzil Languages. **1964** E. A. Nida *Toward Sci. Transl.* ix. 218 The same period may be spoken of in very different ways in diverse languages; e.g. the Biblical 'tenth hour' (or 4 P.M.) is called..'the sun is astride the mountain' (Tzotzil) and 'time for untying the oxen' (Bolivian Quechua). **1977** *Language* LIII. 464 Tzotzil is a Mayan language with more than 120,000 speakers, all living in the State of Chiapas, Mexico.

Tz'u Chou (tsu dʒu). Also **Cizhou.** [Chinese *Tz'u Chou* (Wade–Giles), *Cizhou* (Pinyin), place-name in northern China.] Pottery made at Tz'u Chou, or in similar styles elsewhere, from the Sui dynasty onwards.
1910 S. W. Bushell *Chinese Art* II. 22 The Ting-chou and Tz'u-chou porcelains. **1915** R. L. Hobson *Chinese Pott. & Porc.* I. viii. 103 By far the largest group of the Tz'ŭ Chou family consists of the painted wares. Like the rest of the Tz'ŭ Chou pottery..these have a greyish buff body of porcellanous stoneware..covered with a transparent glaze... On this glaze..the painters executed rapid, bold..designs in shades of brown, varying from black to a soft sepia colour. **1934** *Burlington Mag.* May 214/2 The most characteristic Tz'ŭ Chou product is a buff or grey stoneware dressed with white slip. **1972** *Guardian* 15 Nov. 7/1 A firm of New York dealers..yesterday paid £50,000 for a vase of Tz'ŭ Chou stoneware of the Sung dynasty at Sotheby's. **1980** *Sotheby's Preview Calendar of Sales* Apr.–May p. xxiii (*caption*) A Cizhou (Tz'u Chou) pillow... Jin (Chin) dynasty.

tzuica, var. *tsuica. **tzuris,** var. *tsores.

U

2. a. *U-shaped* adj. (further examples); *spec.* designating or pertaining to a valley having such a cross-section, esp. as a result of glacial erosion.

1894 *Jrnl. Geol.* II. 350 Glacial cañons are..U shaped rather than V shaped in cross-profile. **1909** *Bull. Geol. Soc. Amer.* XX. 409 (*heading*) Striations and U-shaped valleys produced by other than glacial action. **1970** R. J. SMALL *Study of Landforms* xi. 367 Virtually all U-shaped valleys in glaciated areas have been produced by the glacial modification of pre-existing river valleys.

b. *U-bend, cross profile, -frame, -section* (hence *-sectioned* adj.), *-tube* (earlier example), *-turn* (also *fig.* and as *v. intr.* and *trans.*), *-valley* (see prec. sense).

1819 M. EDGEWORTH *Let.* 4 Mar. (1971) 177 Excellent lecture room well lighted with *gas* spouting from U tubes at the bottom of the gallery. **1884** W. J. MCGEE in *Proc. Amer. Assoc. Adv. Sci.* XXXII. 238 The effect of temporary occupancy of a typical water-cut cañon by glacier ice will be to..change the original V to a U cross profile. **1909** WEBSTER, *U section*, or *U-section*. **1928** BLUNDEN *Undertones of War* viii. 80 A support line was being made, and very nice and proper it looked..with its clean U-frames..and symmetrical wire anchorages. **1937** *Sun* (Baltimore) 27 July 6/3 He will enforce new ordinances barring U turns in the public square and regulating parking. **1947** AUDEN *Age of Anxiety* (1948) iii. 62 Let down From U-valleys like yarn, Waterfalls..Quietly encourage me on. **1950** 'S. RANSOME' *Deadly Miss Ashley* xi. 145 The driver U-turned into the wake of the police car. **1954** 'N. BLAKE' *Whisper in Gloom* xvi. 232 The armoured car ..made a sudden sweeping u-turn. **1958** R. STOUT *Champagne for One* vii. 82, I do the driving, and I wanted one [*sc.* a car] I could U-turn when the occasion arose. **1959** *Sunday Express* 17 May 6/1 His wife offered him a cup of tea... He U-turned again. **1959** *Times* 30 Sept. 13/6 (Advt.), We don't care *what* you call us so long as you *do* call us when you are in need of cold drawn precision tubes... Not to mention..'U' bends. **1961** WEBSTER, *U-turn*.., something held to resemble a U-turn (as a reversal of policy) <stated that the Administration is making an economic *U-turn*—T. R. Ybarra>. **1963** A. ROSS *Australia 63* ii. 49 At weekends motor boats churn up the water and skiers make U-turns past the koala farm. **1969** E. H. PINTO *Treen* 26 Early folding umbrellas were heavy and clumsy... The most important patents were that of Henry Holland..and that of Samuel Fox & Co. of Stockbridge, Sheffield, for the lighter U-section steel rib in 1852. **1969** *Jane's Freight Containers 1968–69* 578/1 The Conjack is a U-frame container lifter and transporter. **1972** A. ROTH *Heath & Heathmen* i. 16 One of the things which surprised even close observers was the ease with which Heath made a complete U-turn between his first rightwing, or Mark I phase of 1970–1, and his second, leftwing or Mark II phase of 1971–2. **1974** 'A. GARVE' *File on Lester* xx. 93 His wife had once lost a ring down a wash basin waste pipe and it had stuck in a U-bend. **1974** H. MACINNES *Climb to Lost World* ix. 132 Don was carrying up more climbing gear, pitons and bongs—large U-sectioned pieces of alloy or steel..which, when hammered into cracks as climbing aids, emit a resounding 'bong-bong'. **1975** J. G. EVANS *Environment Early Man Brit. Isles* iii. 69 We cannot see..the U-valleys of Lakeland with their sheer rock walls..without feeling the drama and immediacy of the Pleistocene ice. **1977** *Western Morning News* 30 Aug. 2/3 Reg. Cann, F.S.V.A., will sell by Auction,..steel U-sections. **1981** R. LEWIS *Seek for Justice* iii. 81 The car ahead slowed, then turned in a wide U-turn.

c. (Earlier example.)

1873 J. H. BEADLE *Undevel. West* xxv. 521 Four miles from Wingate the valley makes a great U to the northward.

d. = *U-turn*, sense 2 b above.

1971 'G. BLACK' *Time for Pirates* iv. 80, I was gambling on the lane doing a U and coming down again beyond the building. **1973** J. GORES *Final Notice* xv. 95 Ballard went by, U-ed, came back into the parking area.

3*. Symbolic uses: *U* is a coefficient representing the rate at which heat is lost through a structure, in B.Th.U. per hour per square foot per degree difference in temperature between the two sides (or the metric equivalent). Also *U factor, value*.

1928 *Jrnl. Amer. Soc. Heating & Ventilating Engineers* XXXIV. 63 Let U = coefficient of heat transmission. **1958** *Times Rev. Industry* Sept. 24/1 To specify the standard the Ministry [of Power] has approved the U value of thermal transmittance coefficient. **1975** R. H. WARING *All about Home Heating* v. 40 For the same actual thickness of brick, an 11 inch cavity wall has a U factor of about 0·33, compared with a value of 0·43 for a solid brick wall. **1982** *New Scientist* 10 June 692/3 All the external walls are insulated... This should cut the 'U' value from 1·7 to 0·5.

II. 4. a. (All abbreviations given here with a full stop are frequently used without it; those without can also be used with.) *u* (Physics) [f. G. *ungerade* odd], used to designate functions, esp. wave functions, which change sign on inversion through the origin (cf. *ODD *a.* 4*), and atomic states, etc., represented by such functions; U, universal, designating (a certificate given to) films suitable for exhibition to audiences of all age-groups; also *ellipt.* and *fig.*; *u* (Physics), up, a quark flavour; U (= uranium): also with following or preceding numeral indicating the mass number of the isotope symbolized; U.A.B., Unemployment Assistance Board; U.A.E., United Arab Emirates; U.A.P., United Australia Party; U.A.R., United Arab Republic; U.A.W.(A.) (*U.S.*), United Automobile Workers (of America); UCCA (*ʊ*·kă), Universities Central Council on Admissions; U.C.D., University College, Dublin; U.C.L., University College London; U.C.L.A., University of California at Los Angeles; U.D.A., Ulster Defence Association; U.D.C., universal decimal classification; U.D.C., Urban District Council; U.D.I., unilateral declaration of independence (orig. of Rhodesia from the U.K.); UDN (*Vet. Sci.*), ulcerative dermal necrosis, a disease of fish; U.D.R., Ulster Defence Regiment; U.D.S.R. [F. *Union Démocratique et Socialiste de la Résistance*], Democratic and Socialist Union of the Resistance, a left-wing French party under the Fourth Republic; UEFA (yū̦ēⁱ·fă), Union of European Football Associations; U.F., United Free (Church, of Scotland); also, a member of this; U.F.F., Ulster Freedom Fighters; U.G.C., University Grants Committee; UHF, uhf, ultra-high frequency; UHT, ultra heat treated; ultra-high temperature; U.I.L., United Irish League; UJ, universal joint; U.K.A.E.A., United Kingdom Atomic Energy Authority; ULCC, ultra-large crude carrier; ULMS, underwater long-range missile system; UMNO, Umno (*ʊ*·mno), United Malay(s) (*later* Malaysia) National Organization; U.N., United Nations; UNCTAD, Unctad (*ʊ*·ŋktæd), United Nations Conference on Trade and Development; UNDP, United Nations Development Programme; U.N.E.F., United Nations Emergency Force; UNEP, United Nations Environment Programme; UNESCO, Unesco (yūnes·ko), United Nations Educational, Scientific, and Cultural Organization; UNHCR, United Nations High Commissioner for Refugees; U.N.I.A. (*U.S.*), Universal Negro Improvement Association; UNICEF, Unicef (yū·nisef), United Nations International Children's Emergency Fund (now officially the United Nations Children's Fund); UNIDO, United Nations Industrial Development Organization; UNIDROIT, Unidroit [F. *Institut international pour l'unification du droit privé*], International Institute for the Unification of Private Law; UNIP, Unip (yū·nip), United National Independence Party (of Zambia); UNITA, Unita (yūnī·tă) [Pg. *União Nacional por Independência Total de Angola*], National Union for the Total Independence of Angola; UNO, Uno (yū·no), United Nations Organization; U.N.P., United National Party (of Sri Lanka (Ceylon)); UNREF, Unref (*ʊ*·nref), United Nations Refugee Emergency Fund; UNRRA, Unrra (*ʊ*·nră), United Nations Relief and Rehabilitation Administration; UNRWA(PRNE), United Nations Relief and Works Agency (*orig.* for Palestine Refugees in the Near East); U.P., United Party (*spec.* of South Africa); U.P., United Press; U.P., United Provinces (of Agra and Oudh, India), now, Uttar Pradesh; U.P.C., Uganda People's Congress; UPC, universal product code; U.P.I., United Press International; U.P.N.I., Unionist Party of Northern Ireland; u/s, U/S, U.S., u.s., unserviceable; U.S.A.A.F., United States Army Air Forces; U.S.A.F., United States Air Force; U.S.C., Ulster Special Constabulary; U.S.I.A., United States Information Agency; USM, unlisted securities market; U.S.N., United States Navy; U.S.O. (*U.S.*), United Service Organization; U.S. of A. (*colloq.*), United States of America; U.S.P., United States Pharmacopœia; U.S.S., United States Ship; U.S.S.R., Union of Soviet Socialist Republics (Russ. *Soyuz sovetskikh sotsialisticheskikh respublik*); U.T., Universal Time; U.U.U.C., United Ulster Unionist Council; U.V., ultraviolet; U.V.F., Ulster Volunteer Force; U.W.C., Ulster Workers' Council; UXB, unexploded bomb. See also *UFO.

1930, 1962 *u* (= *ungerade*) [see g s.v. *G III. f]. **1965** J. N. MURRELL et al. *Valence Theory* x. 143 Only if there is an odd number of electrons in *u* orbitals can a state of *u* symmetry (e.g. He_2^+) exist. **1922** *Times* 11 Mar. 8/2 An attempt may be made to secure a rather broader interpretation of the division of films into 'U' and 'A' classes. The former are passed for universal exhibition... The 'U' certificate is very difficult to obtain. **1929** M. ALLINGHAM *Mystery Mile* xxiv. 224 They collected the evidence: Datchett collected the blackmail. Not a certificate 'U' production. **1935** [see A, adult s.v. *A III]. **1958** *Spectator* 15 Aug. 221/3 *A Cry from the Streets* (..'U' certificate) is about children. **1968** *Guardian* 26 Sept. 15/2 About the only British medium-budget U showing in London's West End is 'Yellow Submarine'. **1972** D. BLOODWORTH *Any Number can Play* xviii. 182 This particular prophecy was to whom it may concern. U-certificate. Kitay, Americans, British, it didn't matter. **1977** *Punch* 21 Jan. 80 'Hell!' cried the priest. 'What kind of penitent *are* you? Did you at least make love to your wife, see a U film, buy *Reader's Digest*?' 'No,' said the confessant, 'none of those.' **1964**, etc. *u* (Particle Physics) [see *S 15]. **1975** *Sci. Amer.* Oct. 43/1 Just two of the quarks, the *u* and the *d*, suffice to explain the structure of all the hadrons encountered in ordinary matter. **1982** *McGraw-Hill Yearbk. Sci. & Technol.* 386/1 Neutron decay corresponds to a transition $d \rightarrow u$ and Λ^0 decay to a transition $s \rightarrow u$. **1938** R. W. LAWSON tr. *Hevesy & Paneth's Man. Radioactivity* (ed. 2) xxiii. 221 Uranium is a mixture of three isotopes, uranium I (^{238}U), uranium II (^{234}U), and ^{235}U. **1940** *Manch. Guardian Weekly* 10 May 367 The discovery of a natural substance whose chief characteristic is the release of great energy is announced... It is known as U-235. **1968** *Listener* 22 Aug. 230/3 But an analysis of the radioactive fallout associated with the bombs showed that they were made from U235 and not plutonium as had generally been expected. **1979** A. G. MADDOCK in Harbottle & Maddock *Chem. Effects Nucl. Transformations in Inorg. Systems* xxi. 388 In uranyl sulphate solution most of the ^{238}U appears as uranyl ion. **1937** 'G. ORWELL' *Road to Wigan Pier* v. 77 When a man's stamps are exhausted.. he receives twenty-six weeks' 'transitional benefit' from the U.A.B. (Unemployment Assistance Board). **1972** *Whitaker's Almanack 1973* 958/1 The United Arab Emirates (formerly the Trucial States) is composed of seven Emirates... The approximate area of the U.A.E. is 32,000 square miles. **1979** *Financial Times* 12 Jan. 3/4 The Iranian community based on the UAE, which, in the northern Emirate of Dubai, runs into thousands. **1936** *Age* (Melbourne) 4 May 8/5 The attitude likely to be adopted by the full-blooded protectionist members of the U.A.P. towards the Government's proposal for abolition of duty on cement from Great Britain. **1964** J. JUPP *Austral. Party Politics* vi. 130 The U.A.P. leader, R. G. Menzies, called a conference at Canberra in October 1944. **1958** *Spectator* 20 June 793/1 There is no more reason to think that they [*sc.* the Lebanese] want to join the UAR in the immediate future than there is that the UAR wants them to join it. **1966** M. WOODHOUSE *Tree Frog* xxv. 190 The U.A.R. were supposed to be developing a delta-wing of their own. **1981** *Economist* 24 Jan. 43/1 In

1963, when a second attempt was made at a UAR, this time a tripartite one, the Syrians and Iraqis spent weeks in Cairo arguing against Nasser's federal ideas. **1946** W. WEINSTONE *Case against David Dubinsky* iv. 35 They now have in Walter P. Reuther and the presidency of the U.A.W. a strong base from which to operate more effectively. **1970** *Toronto Daily Star* 24 Sept. 35/1 There is some possibility that the UAW strike at General Motors will reduce the number of UAW delegates. **1936** *Business Week* 5 Dec. 16/2 The U.A.W.A. has a remarkable story to tell of its growth. **1963** F. D. FAWCETT *Cycl. Initials & Abbrev.* 149/1 *UCCA*, Universities Central Council for Admissions. **1971** *Where* Dec. 357/3 The particular sections on the UCCA form which need careful filling in. **1976** *Times Higher Educ. Suppl.* 6 Aug. 8/2 A Nigerian universities joint matriculation scheme, which includes a combined school leaving and university entrance examination, and a form of UCCA selection. **1955** R. J. SCHWARTZ *Compl. Dict. Abbrev.* 179/2 *UCD*, University College, Dublin. **1962** B. INGLIS *West Briton* vii. 123 The 'National', as U.C.D. was known in Trinity. **1979** U.C.D. [see T.C.D. s.v. *T 6 a]. *a* **1912** W. T. ROGERS *Dict. Abbrev.* (1913) 192/2 *U.C.L.*,..University College, London. **1979** W. H. CANAWAY *Solid Gold Buddha* xiv. 101 Das had a first class degree in law..and had studied full-time for it at UCL. **1941** B. SCHULBERG *What makes Sammy Run?* viii. 185 We drove past Westwood Village, the home of UCLA..Hollywood's version of campus life. **1977** H. FAST *Immigrants* iv. 290 He asked how far it was to the U.C.L.A. campus at Melrose and Vermont. **1972** *Times* 15 May 2/6 One UDA man said today [etc.]. **1978** D. MURPHY *Place Apart* vii. 134 The main loyalist paramilitary force, the UDA, is not illegal. **1935** *ASLIB Information* Dec. 4/2 (*heading*) English edition of the U.D.C. **1965** *Rev. Internat. Documentation* XXXII. 19/1 For many years the problems of indexing and retrieving information have involved the use of tree-structured classification schemes such as the UDC system. **1905** F. H. COLLINS *Author & Printer* 383/1 *U.D.C.*, Urban District Council. *a* **1974** R. CROSSMAN *Diaries* (1977) III. 576 There were the C.C.A., the A.M.C., the London boroughs, the G.L.C., the U.D.C.s and R.D.C.s. **1965** *Economist* 8 May 634/1 They shrug off..the threat of Britain imposing sanctions if Mr Ian Smith's government made a unilateral declaration of independence ('UDI'). **1966** *Listener* 17 Feb. 251/3 The British often 'lean over backwards' in their determination to do justice to George Washington and the other U.D.I. heroes. **1972** A. PRICE *Col. Butler's Wolf* ix. 99 He lived in Rhodesia...left shortly after UDI. **1980** *Times* 30 July 13/2 Both he and his wife Avis are from Yorkshire—not quite the kind that think Yorkshire should declare UDI. **1968** CARBERY & STRICKLAND in *Irish Veterinary Jrnl.* XXII. 171/2 This paper describes the symptoms, gross and microscopic lesions, epizootiology and bacteriology of UDN. **1972** *Trout & Salmon* Feb. 9/1 The effects on spawning stocks of UDN..are dangers just as significant. **1969** *Times* 20 Nov. 5/5 The Bill provided the necessary legal framework for the establishment of the U.D.R. **1978** D. MURPHY *Place Apart* iii. 53 No UDR, no British army patrols, nothing to flaw the impression of a traditional rural community. **1945** *Times* 25 Oct. 3/3 Joint Socialist and U.D.S.R. (*Union Democratique et Socialiste de la Résistance*) lists, 961, 704. **1965** F. R. WILLIS *France, Germany, & New Europe* vi. 138 The Pleven cabinet, a coalition of the four most 'European' parties in the Assembly (Socialists, MRP, Radicals and UDSR), hoped in this way to avoid a national German army and advance European integration at the same time. **1963** I. WILKES *British Initials & Abbrev.* 102/2 *UEFA*, Union of European Football Associations, P.O. Box 16, Berne, 15, Switzerland. **1972** *Rothman's Football Yearbk. 1972–73* 808 Last season, the U.E.F.A. took over the running of this colourful competition, re-named it the U.E.F.A. Cup and clearly stated that only clubs which finished high in their respective domestic leagues will be accepted. **1982** *Daily Tel.* 23 Dec. 16/1 UEFA have joined FIFA in their condemnation of the referee clampdown. **1913** *Northern Whig* 3 Dec. 8/5 The Rev. John C. Ingles, minister of the North U.F. Church, Crieff. **1922** 'R. WEST' *Judge* I. i. 17 The snatching of the Church funds from the U.F.'s by the Wee Frees. **1973** *Times* 13 Nov. 2/6 The U.F.F...have claimed responsibility for at least 10 murders. **1947** *Universities Q.* Aug. 326 There is..another field of major importance which the UGC is not touching; the whole issue of the aims and methods of university teaching. **1979** *Jrnl. R. Soc. Arts* Oct. 707/2 In 1964..the U.G.C. lost its distinctive rôle of a body advising the Treasury directly. **1937** *RCA Rev.* II. 30 For low-power u-h-f transmitters and receivers, special tubes having low internal capacities..are available. **1939** *Amat. Radio Handbk.* 161/1 The devices of Fig. 16 are satisfactory for matching U.H.F. aerials to open wire lines. **1955** *Times* 27 July 9/6 Its task is confused by the haphazard growth of early v.h.f. (very high frequency) stations and the consequent need to resort to the u.h.f. (ultra-high frequency) band for many of the new licences. **1978** *Jrnl. R. Soc. Arts* CXXVI. 437/1 Chief Officer Rowe..remained with the lifeboats and maintained communication with the rescue vessels using portable UHF radio. **1968** *Sunday Truth* (Brisbane) 27 Oct. 32/3 Local sale of UHT milk. **1981** *Economist* 24 Jan. 49/3 The UHT process—sterilisation by heating to 132 degrees centigrade—produces milk that keeps for months without refrigeration, but tastes peculiar. **1983** *Times* 7 Feb. 10/2 A repellent whitish fluid ..known as ultra heat treated (UHT) milk. [**1963** J. F. HOGERTON *Atomic Energy Handbk.* 569/1 (*heading*) Ultra high-temperature reactor experiment (UHTREX).] **1968** E. PUGH *Dict. Acronyms & Abbrev.* 174 *UHT*, ultra high temperature. **1901** *Notes from Ireland* 1 Dec. Suppl., A U.I.L. meeting in Longhrea. **1976** J. V. O'BRIEN *William O'Brien* viii. 194 UIL circles in America. **1970** K. BALL *Fiat 600, 600D Autobook* 165/2 UJ Universal joint. A coupling between shafts which permits angular movement. **1959** *Times Rev. Industry* Aug. 17/1 U.K.A.E.A. research section. **1983** *Listener* 24 Nov. 5/2 In 1955 a new quango, the UKAEA, was created from various research and production units of the Ministry of Supply. **1973** *Sunday Post-Herald* (Hong Kong) 20 May (Business sect.) 5/3 The European yards..worry that their..share of new orders for VLCCs and ULCCs will remain small. **1981** E. CORLETT *Revolution Merchant Shipping* 26/1 The world was about to be presented with new initials, VLCC and ULCC—the very large and the ultra large crude carriers were at hand. **1970** B. WEDERTZ *Dict. Naval Abbrev.* 234/1 *ULMS*, Underwater Long-range Missile System. **1973** *New Scientist* 30 Aug. 483 The ULMS will have a greater range than either Polaris or Poseidon. **1946** *Malay Mail* 14 May 4/5 Delegates from the newly formed United Malays National Organization will..attend. A two-day conference..has just ended at which delegates from all over Malaya and Singapore passed and ratified a charter incorporating the U.M.N.O., as the new organization will in future be known. **1979** *Straits Times* 27 Nov. 1. 12/8 Umno Youth leader Haji Suhaimi bin Datuk Haji Kamaruddin. **1946** *N.Y. Times* 11 Apr. 1/6 If the U.N. remained at Hunter or went to Lake Success, it would have to erect an auditorium. **1946** J. S. HUXLEY *Unesco* I. iii. 14 Other U.N. agencies such as the F.A.O. and the World Health Organisation. **1978** L. HEREN *Growing up on The Times* iii. 84 The assassination of the UN mediator had frightened most Israelis. The UN had been the indispensable midwife at the birth of their country. **1964** *Newsweek* 6 Apr. 31 UNCTAD..might very well lead to a major realignment of the world, with the world's poor nations in solid opposition to the rich. **1967** *Economist* 19 Aug. 670/1 Unctad, the United Nations body which tries to help the less developed countries. **1977** *Arab Times* 14 Dec. 6/6 A five-day session of an UNCTAD committee. **1966** *U.N. Monthly Chron.* Feb. 21/1 (*heading*) New Projects Approved by UNDP Governing Council. **1956** *N.Y. Times* 29 Nov. 12/6 The United Nations Emergency Force has been in Egypt.. two weeks... It may be..the UNEF, in terms of potential effectiveness.., must be rated as equivalent to a substantially larger military body. **1964** *Ann. Reg. 1963* 141 Owing to the continued failure of a number of Member States to pay their assessed contributions to the costs of the U.N. Emergency Force in the Middle East (U.N.E.F.) ..the financial situation remained serious. **1974** P. GORE-BOOTH *With Great Truth & Respect* 364 The Israelis had not helped their case by refusing to allow the UNEF to operate on their side of the frontier. **1973** *U.N. Monthly Chron.* Aug.–Sept. 84/1 The Council adopted..a resolution noting the report of the Governing Council of UNEP on its first session. **1980** *Africa* Jan. 43/2 UNEP's plans for the Red Sea. **1946** J. S. HUXLEY (*title*) UNESCO: its purpose and its philosophy. **1946** A. HUXLEY *Let.* 27 Oct. (1969) 551, I was delighted..to learn that all goes well with you and the infant Unesco. **1948** *Hansard Commons* 11 Mar. 1504 Someone was being brought from U.N.E.S.C.O. in order to help formulate an education policy. **1955** *Times* 17 June 5/2 The Universal Copyright Convention, which has been promoted by Unesco, will at last come into force on September 16. **1983** *Whitaker's Almanack 1984* 964 (*heading*) United Nations Educational, Scientific and Cultural Organization (UNESCO). *Ibid.*, In most Member States National Commissions serve as a link with Unesco. **1953** *U.N. Yearbk. 1952* 492/2 The report of the United Nations High Commissioner for Refugees (UNHCR)..dealt with the activities of the High Commissioner's Office. **1984** *Listener* 8 Mar. 12/2 The UNHCR includes Laotian and Kampuchean exiles in its total of 891,967 resettled up to the end of 1983. **1921** *Nation* (N.Y.) 28 Dec. 750/1 Any phenomenon among the colored population, like the U.N.I.A., white persons at first..regard as a huge joke. **1977** *Western Political Q.* XXX. 171 Thus we arrive at an understanding of the enhanced emotionality/reduced intellectuality characteristic of groups like the UNIA. **1948** *U.N. Bull.* 1 Mar. 184/2 The word UNICEF, pronounced 'Uni-sef', is now known far and wide in Europe. **1958** *Listener* 5 June 952/3 Danny Kaye's touching film.. about the part played by Unicef in conquering disease among children all over the world. **1982** *Whitaker's Almanack 1983* 962/1 United Nations Children's Fund (UNICEF). **1967** *Ann. Rep. Sec. Gen.* [*U.N.*] *16 June 1966–15 June 1967* xi. 145/1 The General Assembly decided that the United Nations Industrial Development Organization (UNIDO) should function as an autonomous organization. **1980** *Jrnl. R. Soc. Arts* June 440/2 At this moment, UNIDO III—the third conference of the UN International Development Organization—is in its final hours in Delhi. **1959** *Uniform Law Cases* (UNIDROIT) p. iii, One of the activities collateral to the unification of law, which the International Institute for the Unification of Private Law (UNIDROIT) has taken over, has been the study of the most appropriate measures to ensure uniformity in the interpretation of texts of uniform law. **1974** A. WATSON *Legal Transplants* xv. 91 The work of the legal committees of the Council of Europe and of Unidroit is regarded as important as an indication of European trends. **1959** *Central African Post* 26 Oct. 1 (*heading*) Chona's Congress merges with U.N.I.P. **1961** *Times* 2 Nov. 9/2 Youth branches of Unip in the Western Province. **1977** *Times of Zambia* 7 Sept. 6/5, I see UNIP everywhere and I see it as one of the best organised parties in the world. **1967** *N.Y. Times* 17 Sept. 1. 17/3 Two Angolan nationalist organizations, the Popular Movement for the Liberation of Angola and the Unita, have been sending small guerrilla bands into eastern Angola from Zambia. **1972** J. BIGGS-DAVISON *Africa—Hope Deferred* i. 9 U.N.I.T.A. of Angola and C.O.R.E.M.O. of Mozambique. **1978** *Guardian Weekly* 2 July 6 Secretary of State Cyrus Vance..defeated a move by President Carter's national security advisor, Zbigniew Brzezinski, proposing the covert sending of arms to UNITA guerrillas fighting in Angola against the Marxist Government of President Agostinho Neto. **1945** *Tuscaloosa* (Alabama) *News* 15 Nov. 4/4 The ideas range like this..let the United Nations Organization (UNO) handle it [etc.]. **1946** H. H. HENSON *Let.* 14 May (1950) 181 The degraded haggling of the international market to which the cosmic idealism of U.N.O. has now degenerated. **1953** P. SCOTT *Alien Sky* I. iii. 32 An official observer from Uno or one of those dreary organisations. *a* **1974** R. CROSSMAN *Diaries* (1976) II. 353 The arrival of the U.N.O. troops who are there to police the peace settlement between the Greeks and Turks. **1946** *Ceylon Daily News* 9 Sept. 4/2 The speech of Mr. D. S. Senanyaka, the leader of the State Council, at a meeting of the Executive Committee of the United National Party ..is..remarkable for its fairness and vigour... The leader delivered his speech under the auspices of the U.N.P. **1979** *Round Table* Jan. 49 No defeat in the annals of Sri

4. b. *Ugly* (or *ugly*) *American* (in allusion to the title of the book: see quot. 1958), an American who behaves offensively abroad.
1958 Lederer & Burdick (*title*) The ugly American. 1965 *Atlantic Monthly* May 152 A host of odd and funny foreigners: bogus Russian counts, semi-aristocratic Slavic ladies, German officers, and an early type of the ugly American abroad. 1968 *Sat. Rev.* (U.S.) 9 Mar. 76 I don't think we were Ugly Americans; perhaps just unaware, or Unlettered. 1980 D. Williams *Murder for Treasure* x. 100 That awful man..thinks you're swinging the deal and he needs Edgar to blow it by acting the Ugly American.
C. *sb.* **3.** *the uglies* (slang), depression, bad temper; (see also quots. 1903 and 1974).
1846 *Swell's Night Guide* 77, I know as how I've got the uglies. 1903 Farmer & Henley *Slang* VII. 251/1 *Ugly*,.. In *pl.* = delirium tremens; the horrors. 1939 N. Last *Diary* 18 Oct. in *N. Last's War* (1983) 20 A gloom seems over us all. I've shaken off my fit of the uglies, but I felt I'd just like to crawl into a hole. 1974 *Petroleum Rev.* XXVIII. 672/1 Nitrogen narcosis, popularly called 'raptures of the deep' but perhaps more accurately described as 'the uglies', is the malady caused by nitrogen under pressure, interfering with the normal function of the nervous system.
ugly *v.* (later example); also with *up*.
1946 *Sun* (Baltimore) 5 Feb. 8/7 Hands uglied by winter weather? 1965 *New Statesman* 26 Nov. 850/2 He uglies up the very places where one expects an opposite treatment. 1979 *Listener* 23 Aug. 248/2 Ever since *Grease* uglied up the Fifties..the nostalgia industry has taken a curiously tough turn.

Ugrian, *a.* (Earlier example.)
1838 *Jrnl. R. Geogr. Soc.* VIII. 390 He will investigate in that region the primitive as well as the present abodes of the nations belonging to the Ugrian race.

Ugric, *a.* Add: (Earlier example.) Also as *sb.* = Ugrian *sb.* 2.
1854 Max Müller in C. Bunsen *Christianity & Mankind* III. 445 If we compare the Ugric and Tamulic Numerals. *Ibid.* 453 In Mandshu..we find juan for 10, the same root we met before in Ugric. 1964 J. Atkinson tr. *Vuorela's Finno-Ugric Peoples* 2 Ugric comprising Vogul, Ostyak and Hungarian.

Ugro-. Add: *Ugro-Finn*; *Ugro-Tartarian* (earlier example).
1848 J. C. Prichard in *Rep. Brit. Assoc. 1847* 241 The Turanian, or as I shall term them, Ugro-Tartarian languages, or the languages of High-Asia and other regions... Ugro-Tartarian nations. 1862 *Temple Bar* Nov. 549 The Ugro-Finns, whom they have driven northwards. 1880 A. H. Sayce *Introd. Sci. of Lang.* II. viii. 190 It is more than doubtful whether we can class the Mongols physically with the Turkish-Tartars or the Ugro-Finns.

uguisu (ugwī·zu). [Jap.] A bush warbler of delicate olive-green plumage, *Cettia diphone*, native to Japan.
1871 A. B. Mitford *Tales of Old Japan* I. 37 The *uguisu*, by some enthusiasts called the Japanese nightingale. 1941 N. Takatukasa *Jap. Birds* 40 Pride of place among the native songbirds is therefore given to the *uguisu*. 1974 K. Rexroth *New Poems* 33 Maple leaves, an uguisu Sings as if in spring.

uh, *interj.* Add: **b.** *U.S.* Expressing hesitation: = *er.
1962 J. D. MacDonald *Only Girl in Game* vi. 85 'Are you cashing cheques?'..The man hesitated. 'Uh...Yes, we are.' 1973 *National Observer* (U.S.) 3 Feb., He wanted most awfully to see the one in the advertisement about being, uh, well, you know. 1977 *N.Y. Rev. Bks.* 4 Aug. 32/4 'Perhaps we should, uh, wait,' I said.
c. = *eh *int.* 3.
1977 'E. Trevor' *Theta Syndrome* ii. 28 'Was it okay, Doc?' 'Uh?' 'The tube.' 'Oh, sure.' 1978 G. Greene *Human Factor* v. iii. 272 'Weren't those my very words?' 'Uh,' Mr Barker said.

uh (ɒ), repr. the indefinite article in the speech of U.S. Blacks.
1893 H. A. Shands *Some Peculiarities of Speech in Mississippi* 65 *Uh*, the common negro form for the indefinite article *a*. This is generally written *er* by dialect writers, but no sound of *r* is ever apparent in the negro pronunciation. 1933 *Publ. Texas Folklore Soc.* XI. 105 Dey's jes' ez good uh fish in de creek ez evah been caught. 1973 *Black World* Oct. 74 Locking up folks lives For stealing less than uh hundred dollars.

uh-huh (ɒ·hɒ; see quot. 1982), *adv. colloq.* (orig. *U.S.*). [Imitative.] A spoken affirmative or non-committal response to a question or remark; 'yes', 'oh yes?'
1924 *Dialect Notes* V. 278 *Uh-hŭh*, yes. 1925 *Ladies' Home Jrnl.* May 22/3 The policeman behind the desk said to Buckbarrow: 'Here's something! A runaway kid.' ..'Uh-huh,' commented Buckbarrow. 1941 E. Carr *Klee Wyck* 111 'Uh huh,' he nodded. 1947 'N. Blake' *Minute for Murder* v. 111 'You ought to go on the films, Blount.' 'Uh-huh?' 1969 D. Dalby in A. Dundes *Mother-Wit* (1973) 139/1 African usage can also explain the frequent use by Americans of the interjections uh-huh, for 'yes', and uh-uh for 'no'. Similar forms, especially for 'yes', occur in scattered parts of the world, but nowhere as frequently and as regularly as in Africa. 1978 J. A. Michener *Chesapeake* iii. 123 'You're to..mind your manners. This is your last chance.' 'Uh-huh,' Timothy grunted, staring with contempt at the wretched spot to which he was being taken. 1982 J. C. Wells *Accents of English* III. vi. 556 There are also the grunts sometimes spelt *uh-huh* and *uh-uh* respectively. The first, 'yes', is phonetically [ɦ̃ə̃, ʔhʌ̃, ˈmmm], hence nasal or nasalized; it usually has a rising tone pattern... The second, 'no', is [ˈʔəʔə, ʔʌ̃ʔʌ̃, ˈʔmʔm]..; it is not necessarily nasal, and has an accented final syllable, with an obligatorily falling tone pattern.

uh-uh (ɒˌɒ·; see quot. 1982 in prec.), *adv. colloq.* (chiefly *U.S.*). [Imitative.] A spoken negative response to a question or remark; 'no'.
1924 *Dialect Notes* V. 278 *Ŭh-uh*, negation. 1930 D. Hammett *Maltese Falcon* xvi. 194 'Do you know who he is?'..'Uh-uh,' he said, 'but I'd guess he was Captain Jacobi.' 1970 H. Waugh *Finish me Off* (1971) 133 She shook her head. 'Uh-uh. He's my daddy.' 1977 *Sounds* 9 July 19/1 Uh-*uh*. I thought not.

Uhuru (uhū·rū). Also **uhuru.** [a. Swahili, = freedom.] National independence of an African country, *spec.* Kenya.
1961 *Times* 4 Aug. 11/5 The British base in Kenya will be liquidated, all British troops evacuated, and the police totally Africanized. These are the basic conditions of *uhuru*. 1976 *Drum* (E. Afr. ed.) June 4/2 Kenya became independent in 1963. Much has been done to defend and demonstrate the hard-won uhuru. 1980 *Oxford Diocesan Mag.* Jan. 18/1 Any Christian struggle for *uhuru* in South Africa must involve the church as well as the state. 1982 D. Bagley *Windfall* xx. 192 The coming of Uhuru [to Kenya] must have been painful for the [club] membership who had to adapt to a determinedly multiracial society. 1984 *Listener* 24 May 11/1 An entire continent has seemed hell-bent on self-destruction, despite uhuru, despite the bright hopes of the many thousands who died seeking it.

Uigur, *sb.* and *a.* Add: **Uighur** is now the usual form. **A.** *sb.* **1.** (Further examples.)
1966 D. Wilson *Quarter of Mankind* vii. 89 Unlike the Tibetans, Mongols and Uighurs, these are relatively primitive people. 1969 *Listener* 3 July 6/2 There are millions of people living in the border areas, who speak neither Russian nor Chinese. They're Khazaks, Uzbeks, Uighurs, Tartars, many of them still leading nomadic lives, herdsmen born to the saddle. 1977 *Times* 4 July 5/4 The eight to ten million inhabitants of Sinkiang, the majority of whom are Uighurs and Kazakhs.
B. *adj.* (Earlier example.)
1747 *Astley's Gen. Coll. Voy.* IV. ii. iv. 457/2 He says, they are the same with the Wigûr, Oygûr, or Jugur Characters.

uillean pipes (i·lyin), *sb. pl.* Also **-ann.** [See Union pipes *sb. pl.*] = Union pipes *sb. pl.* *Rarely* in *sing.*
The etymological relationship between the two terms is uncertain. This name is now the usual one.
[1876: see Woollen *a.* 1 ¶.] 1906 *Grove's Dict. Mus.* (ed. 2) II. 509/1 The later bagpipe, the Uillean or Union pipe, blown with a bellows, became popular in Ireland. 1962 *Times* 16 Nov. 16/1 Ceilidh bands of the venerable Uilleann pipes. 1974 *Irish Democrat* Nov. 2/4 Irish concert in..Acton Town Hall... Tommy McCarthy (Uillean pipes and concertina). 1980 *Observer* 10 Feb. 14/8 The ensemble consists of two fiddles, flute, harp, a single-headed drum called the bodhran and the uillean pipes. 1984 *Listener* 3 May 28/1 Making extensive use of the most beautiful sound on earth, that of uilleann pipes in skilled hands, it was first and foremost a portrait of a contented man.

uintaite (yu̯i·ntăˌəit). *Geol.* Also **uintahite.** [f. the name of the *Uinta* Mountains, Utah + -ite[1].] = *gilsonite.
1888 [see *gilsonite]. 1904 G. P. Merrill *Non-Metallic Minerals* xiii. 373 The principal use of uintaite thus far has been in the manufacture of varnishes for ironwork and baking japans. 1951 E. N. Tiratsoo *Petroleum Geol.* xi. 291 The following are the principal bitumens found:—(a) Uintaite.., also called Gilsonite. 1974 *Encycl. Brit. Micropædia* IV. 546/2 *Gilsonite*, also called uintaite, lustrous, jet-black, natural bitumen.., an asphaltite.

|| **ujamaa** (udʒamā·). [Swahili, = consanguinity, brotherhood, f. *jamaa* family, a. Arab. *jamā'a* group (of people), community.] The name given by President Nyerere of Tanzania to a kind of socialism he introduced in that country in the 1960s, in which village co-operatives were established based on equality of opportunity and self-help; so *ujamaa village*.
1962 J. K. Nyerere *Ujamaa: Basis of African Socialism* in *Ujamaa: Ess. Socialism* (1968) 12 'Ujamaa', then, or 'Familyhood', describes our socialism. 1962 *Economist* 8 Sept. 892/2 Mr Nyerere's thesis that Ujamaa (familyhood) is the basis of African socialism. 1969 *Reporter* (Nairobi) 13 June 21/1 *Ujamaa* villages will be encouraged, to spread more wealth across the rural areas. 1970 *Drum* (E. Afr. ed.) Feb. 19/2 Tanzania is now well under way with its second Five-Year Development Plan, a £400 million investment programme tailor-made for the strengthening of ujamaa throughout the land. 1971 *Standard* (Dar es Salaam) 7 Apr. 1/8 A sum of 15,422/20 is to be spent by the government on drilling ten boreholes in various villages of Hanang district, including two ujamaa villages. 1979 *New African* Mar. 57/2 Out of Tanzania's total 16 m. population, more than 13 m. are living in about 8,000 *ujamaa* villages. 1983 *N.Y. Times* 15 Aug. C14/6 Most of the cast participate in the theater arts and black studies programs of Ujamaa.

|| **uji** (ū·dʒi). [Jap.] In feudal Japan, a name indicating which ancestral noble family the bearer belonged to; a patriarchal lineage group comprising all those with the same *uji*.
1876 W. E. Griffis *Mikado's Empire* (1877) I. xii. 117 The family name (*uji*) precedes the personal..name. 1896 F. Brinkley *Japanese-Eng. Dict.* 1555/1 Although the offshoots from these noble families took various names.., yet these were not, strictly speaking, the family names or *uji* in the now-accepted sense. 1931 G. B. Sansom *Japan* I. ii. 36 The society consisted of patriarchal units called *uji*, which were communities formed of a number of households of the same ancestry. 1970 J. W. Hall *Japan* iv. 29 Being of the upper class the *uji* possessed surnames and bore titles of respect. 1974 *Encycl. Brit. Micropædia* X. 238/1 The *uji* members..were supported by the labour of common workers, who were organized into subunits of the *uji*... Imperial rule over the various autonomous *uji* remained weak until the adoption of centralized government in the early 8th century.

|| **ujigami** (ūdʒigā·mi). [Jap., f. *uji* *uji + *kami* good.] In feudal Japan, the ancestral deity of an *uji*; later, the tutelary deity of a particular village or area.
1897 L. Hearn *Gleanings in Buddha-Fields* i. 18 The peasants were going to celebrate their harvest by a dance in the court of the *ujigami*. 1931 G. B. Sansom *Japan* I. ii. 36 The members of a clan all worshipped a guardian god, the *uji-gami*, or clan god. 1970 J. W. Hall *Japan* iv. 32 The *uji-gami* venerated by members of certain *uji* were human or totemistic progenitors.

ukata, var. *yukata.

uke[1] (yūk), colloq. abbrev. of *ukulele.
[1915 *N.Y. World* 18 July 2m/4 Those who pretend to know say that the euk did more than anything else to put the lulu in Honolulu.] 1921 *Quill* July 21 The music store of Gino Polluce on Blocker Street was raided by Inspector Bullem but no ukes were captured. 1949 L. Feather *Inside Be-Bop* i. 7, I fooled around with the uke. 1960 N. Hilliard *Maori Girl* 133 A big grey hair. It looks just like a uke-string, it's that big. 1976 W. Goldman *Magic* iii. xiv. 227, I won't sing no more 'Cause I ain't got my uke.

|| **uke[2]** (ū·ke). [Jap., f. *ukeru* to receive, be passive, defend.] In Judo, the passive partner, the one who is acted upon.
1956 [see *reap *v.*[1] 2 e]. 1961 *New Statesman* 22 Sept. 402/2 The thrower's body is turned facing the same way as the Uke's... He gets both feet inside the Uke's. 1972 *Judo Illustrated* Sept. 48/2 Tori and uke should be friends and know each other's attitudes well. 1984 *Coaching Award Scheme* (Brit. Judo Assoc.) 6/2 If nage-no-Kata is mastered it vastly widens tori's repertoire, but just as vital, it will eliminate uke's fears of falling—right or left side.

ukeke (ūkē·kē). [a. Hawaiian *'ūkēkē*.] A Hawaiian stringed instrument consisting of a strip of wood with two or three strings that are played with the fingers and mouth.
1891 W. D. Alexander *Brief Hist. Hawaiian People* xiv. 91 The *ukeke* was a strip of flexible wood or bamboo, mounted with two or three strings of *olona* or of cocoanut fiber. 1970 *Western Folklore* XXIX. 239 The guitar was introduced [into Polynesia]..in 1879 by Portuguese immigrants, the only native stringed instrument being the two- or three-stringed *ukeke*, which was never played with a slider. 1980 *New Grove Dict. Mus.* XIX. 323/2 There is no string instrument native to Hawaii other than the *ukeke*, a musical bow.

ukelele, var. *ukulele.

|| **ukemi** (ū·kemi). [Jap., f. *uke* *uke[2] + *mi* body.] In Judo, the art of falling safely.
[1942 C. Yerkow *Mod. Judo* ii. 26 *Ukemi* is the Japanese word and means 'to fall away'. The aim is to break the fall *before* the body reaches the mat or ground, so that no jar or shock is felt. Thus the equivalent for 'ukemi' might be *break*-falling.] 1956 K. Tomiki *Judo* iii. 54 It is essential to learn the art of falling. This is called *ukemi* (breakfall). 1969 G. R. Gleeson *Anatomy of Judo* iv. 66 In spite of the much vaunted effectiveness of traditional ukemi..very few indeed considered they were capable of taking the falls in the last two sets of nage-no-kata. 1984 *Coaching Award Scheme* (Brit. Judo Assoc.) 29/2 No ukemi is asked as such, but in the first section, the candidate is required to act as Uke and Tori.

|| **ukiyo-e** (ūkīyoyē·). Also **ukiyo-we, -ye,** and without hyphen. [Jap., f. *ukiyo* fleeting world (f. *uku* to float, go by fleetingly + *yo* world) + *e* picture.] A Japanese art-form consisting of wood-block prints or paintings of scenes from everyday life simply treated; a picture belonging to this art-form. Usu. *attrib.*
1879 *Trans. Asiatic Soc. Japan* VII. 358 Its founder is still celebrated as the author of the *Ukiyo-we* or popular style. 1898 L. Hearn *Gleanings in Buddha-Fields* v. 115 The Ukiyo-yé artist drew actualities, but not repellent or meaningless actualities... He looked for dominant laws..for the order of the beautiful as it was and is. 1915 *Dial* LIX. 375/1 In the forty years or thereabout since the color prints by the Ukiyoe masters first came to the attention of art lovers..the circle of their ardent admirers has steadily widened. 1955 *N.Y. Times* 24 Apr. ii–1. 9/1 Japanese tend to see things flatly. This is..most noticeable in the ukiyoe prints. 1957 *Encycl. Brit.* XII. 966/2 Ukiyo-e grew to be almost exclusively the art of the populace of Edo. 1971 *Times Lit. Suppl.* 20 Aug. 997/3 The 'Pictures of the Passing World', the *ukiyo-e*, which from the late seventeenth century constituted the latest and most inventive phase of the Popular School. 1983 *Sunday Tel.* (Colour Suppl.) 26 June 31/2 The tradition of the *Ukiyo-E* print, which had been bought by a Japanese public uninfluenced by Western work, was effectively dead by the 1880s.

ukkers, ukky, varr. *uckers, *ucky.

Ukrainian, *a.* (Earlier example.)
1804 M. Wilmot *Jrnl.* 12 Aug. in Londonderry & Hyde *Russian Jrnls.* (1934) I. 124 His dress a jacket lin'd with Ukrainian sheeps' skin.

ukulele (yūkəlē·li). Also **ukelele.** [a. Hawaiian *'ukulele*, f. *'uku* flea + *lele* jumping: see quot. 1957.] A small four-stringed Hawaiian guitar that is a development of a Portuguese instrument introduced to the island *c* 1879.
1896 *Hawaiian* Feb.–Mar. 789 Then comes the twang of the *ukelele*, the soft, melodious cadence of the hula song. 1900 *Century Mag.* June 164/2 Kolomono..holds the ukulele, a stringed instrument which may or may not be indigenous to the island. 1913 R. Brooke *Coll. Poems* (1918) 28 Somewhere an eukaleli thrills and cries. 1919 Wodehouse *Damsel in Distress* vi. 93 You see the handsome sophomore from Yale sitting beside her on the porch playing the ukalele. 1932 D. L. Sayers *Have his Carcase* xii. 146 Campers..brought gramophones or concertinas or ukeleles. 1950 'D. Divine' *King of Fassarai* xviii. 144 Queer the notions you get about South Sea islands,..all..hula-hula skirts..and ukeleles. 1957 *Amer. Speech* XXXII. 309 The machete was heard one day by the vice-chamberlain of King Kalakaua's court, who..asked to be taught to play it... This vice-chamberlain was a British army officer named Edward Purvis; but the Hawaiians..called him *ukulele* because his lively playing and antics and his small build suggested a leaping flea. The new instrument became a great success,..and someone started calling them *ukuleles*. 1978 L. Thomas *Ormerod's Landing* iv. 67 The simple boy produced a dramatic strumming chord on the ukelele.

ulalu, var. Ululu in Dict. and Suppl.

ulcerogenic (ɒlsĕroˌdʒe·nik), *a. Med.* [f. Ulcer + -o- + *-genic.] = Ulcerative *a.* 1.
1959 *Amer. Jrnl. Digestive Dis.* IV. 903 (*heading*) Ulcerogenic drugs. 1971 *Nature* 15 Oct. 498/1 These results strengthen the hypothesis that nicotine sensitizes the duodenum to the ulcerogenic property of acid flowing from the stomach.

ule, var. *hule.

|| **ulendo** (ule·ndo). [Nyanja.] In central Africa: a trek, a safari.
1921 *United Free Church Miss. Rec.* June 189/1 The settlements near the missions may not be so interesting to visit as those one sees on ulendo or trek in the more backward districts. 1927 *Other Lands* Oct. 14/2 Two months on ulendo through Nyasaland would be an ideal holiday. 1971 *Countryman* Autumn 88 A month-long *ulendo* on foot into the western side of the Luangwa Valley in the wet season was one of the moments of my life. 1979 C. Allen *Tales from Dark Continent* vi. 79 Touring through the district, a process..known as a safari in East Africa, as *ulendo* in Central Africa and as 'going to bush' or 'going on trek' in West Africa.

uliginose, *a.* **2.** (Example.)
1967 V. Nabokov *Speak, Memory* vi. 138 Pretty Cordigera, a gemlike moth, buzzed all over its uliginose food plant.

ulli, var. *hule.

'ullo (ɒlo·), *int.* Also **ullo.** Colloq. or joc. pronunc. of *hullo *int.*
1895 A. W. Pinero *Second Mrs. Tanqueray* iii. 112 'Ullo, 'ullo! Whisky and potass! 1955 M. Allingham *Beckoning Lady* ii. 31 'Ullo, now what? 1968 P. Dickinson *Skin Deep* iii. 36 Ullo, ullo! The other fellah caught in the nest, eh? 1980 *Daily Mail* 17 Jan. 4/1 [*Cartoon: policeman speaking*] 'Ullo, 'ullo, 'ullo..and where do you think *you're* going?

-ulose (-iūlōuz), *suffix*[2]. *Biochem.* [f. (Læv)ulose.] Used (in place of -ose[2]) to form the systematic names of ketoses from the names of the corresponding aldoses, esp. ketoses having the carbonyl group at the second carbon atom; as *hexulose*, *ribulose, *sedoheptulose.

ulotrichous, *a.* Add: Hence **ulo·trichy,** woolly-hairedness.
1924 A. C. Haddon *Races of Man* (ed. 2) i. 5 *Ulotrichy*, or woolly hair. 1936 [see *cymotrichy].

|| **ulpan** (ulpa·n). Pl. **ulpanim** (-i·m). [mod. Heb. *ulpān*.] An intensive course in the Hebrew language, orig. for immigrants to the modern state of Israel; a centre providing such a course; also in extended use.
1950 *Israel Govt. Yearbk.* 119/1 The intensive Hebrew language courses (Ulpanim) are attended by lawyers, engineers, teachers, newspapermen..and senior officials. 1973 *Jewish Chron.* 19 Jan. 7/1 The erection of temporary classrooms and religious ulpanim. 1975 *Times* 27 May 2/3 The courses, run by the extramural department of the University of Wales, are a development of the Ulpan method of language teaching. 1977 *New Yorker* 9 May 42/3 You should go downtown..where they got regular language factories... They even got an *ulpan*, like in Israel.

Ulster. Add: **3.** (Earlier examples.)
1876 L. Troubridge *Jrnl.* 31 Dec. in J. Hope-Nicholson *Life amongst Troubridges* (1966) 149 [I] came swaggering into Dulcie's bedroom..in an ulster, revealing the immortal check trousers. 1877 Trollope *Amer. Senator* I. xx. 209 She once offered to lay an Ulster to a sealskin jacket.
4. b. Also *Ulsterwoman*.
1971 *Guardian* 18 Sept. 9/6 A lady assures us that she is proud to be an Ulsterwoman. 1981 A. T. Q. Stewart *Edward Carson* v. 87 An Ulsterwoman who was the wife of a high official.
c. *Ulster fry* (see quots.).
1941 J. D. Carr *Case of Constant Suicides* 43 That stuff that looks like slices of boloney is called Ulster Fry. 1978 J. Galway *Autobiogr.* (1979) v. 57, I remember eating liver and sausages, roast beef for dinner on Sunday and, of course, an Ulster fry for breakfast—bacon, eggs, potato and soda bread. 1979 *Guardian* 22 June 19/5 The notorious 'Ulster fry'—most things you can think of cooked in a pan.
Ulsterette (earlier example); hence also **U·lsterite** *temporary*, a native or inhabitant of Ulster; **Ulsteriza·tion**, the policy of replacing British security forces in Northern Ireland by Northern Irish ones; also in extended use.
1881 J. W. Buel *Border Outlaws* 187 Each wearing a long linen ulsterette over a heavy fall coat. 1920 *Glasgow Herald* 11 Nov. 9 The amendment was criticised by a number of members in addition to the Ulsterites. 1921 Lloyd George *Let.* 24 Nov. (1973) viii. 195 The Irish negotiations have taken a turn for the worse... This time it is the Sinn Feiners. Last week it was the Ulsterites. 1925 J. O'Connor *Hist. Ireland 1798–1924* II. xx. 260 Southerners hope and think, and the Ulsterites fear and affect to think that the four counties could not stand by themselves. 1977 *Belfast Tel.* 28 Feb. 1/6 Attempts by the Government to move towards an 'Ulsterisation'..of the security forces. 1979 *An Phoblacht* 29 Sept. 1/3 Given Britain's Ulsterisation policy, then that increased repression is likely to be led by the RUC. 1980 *Times* 15 Sept. 12/3 One could be speaking about the ultimate 'Ulsterization' of the West Bank and Gaza.

ult. (Later examples.)
1935, 1962 [see *prox.].

ultima (ɒ·ltimă). [L. *ultima* (sc. *syllaba*), fem. of *ultimus* last.] The last syllable of a word.
1913 [see *affection *sb.* 1 b]. 1964 *Language* XL. 24 PItW [*sc.* Proto-Italo-Western] posttonic syllables survived into Rhaeto-Romance (1) in paroxytones if the ultima was /a/. 1977 *Amer. Speech* 1975 L. 46 It also occurs in unstressed position, for example, in the ultima of *always* and the penult of *annihilated*.

|| **ultima ratio** (ɒ·ltimă rē·ʃio). [L.] Final sanction.
1848 Mill *Polit. Econ.* I. ii. ix. 375 The Irish cottier..protects himself by the *ultima ratio* of a defensive civil war. *a* 1902 S. Butler *Way of All Flesh* (1903) lxv. 292 There can be no doubt about faith and not reason being the *ultima ratio*. 1910 G. K. Chesterton *George Bernard Shaw* 26 The very logic of the Irishman makes him regard war or revolution as extra-logical, an *ultima ratio* which is beyond reason. 1933 *Times Lit. Suppl.* 29 June 436/4 The rod is still the *ultima ratio* of so many teachers. 1969 P. Anderson in Cockburn & Blackburn *Student Power* 226 The cultural limitations of bourgeois reason in England were thus politically rational: the *ultima ratio* of the economy founded both.

ultimate, *a.* Add: **2. e.** Applied to the values of a mechanical property corresponding to fracture or breakage of the object concerned.
1858 W. J. M. Rankine *Man. Appl. Mech.* ii. iii. 273 The Ultimate Strength of a solid is the stress required to produce fracture in some specified way. 1869 —— *Man. Machinery & Millwork* 485 The column headed 'Ultimate Extension' gives the ratio of the elongation of the piece, at the instant of breaking, to its original length. 1876 [see *stress *sb.* 5 c]. 1922 Glazebrook *Dict. Appl. Physics* I. 156/2 The ultimate stress was in some cases greater when the extension was fast than when it was slow. 1962 Read & Watson *Introd. Geol.* viii. 446 The properties which control the reactions of rocks are..the values of the elastic limit and the ultimate strength.
f. *ultimate Frisbee* U.S., a form of the game of Frisbee (see *Frisbee, frisbee).
1972 in *Amer. Speech* 1974 (1976) XLIX. 301 In *ultimate Frisbee*, two seven-man teams play on a field 60 yards by 40 yards each game lasting 48 minutes. 1980 *Boston Globe* 7 Mar. 32 The game of Ultimate Frisbee now is included on the list of activities accepted for academic credit at prestigious Worcester Polytechnic Institute, one of the nation's top five suppliers of engineers. 1984 *New Yorker* 9 Apr. 36/1 Ultimate Frisbee..bears only a slight resemblance to the mellow..Frisbee played on beaches.
B. *sb.* **1.** *the ultimate*, the best that can be achieved or imagined; the 'last word'; const. *in*.
1958 S. J. Perelman *Most of S. J. Perelman* 343 The Central Hotel,..advertised as the ultimate in gaiety and chic. 1971 *Hi-Fi Sound* Feb. 67/1 In the medium-cost category our ideal is a system in which the most attractive 'top' features are recognisable while the ultimate in power handling or 'monitoring' capability has been lost. 1981 P. Davies *Edge of Infinity* viii. 168 The ultimate in elaborate organization is the human body and mind. 1981 *Monitor* (McAllen, Texas) 8 July iii. 1/1 Shop for your man on the first floor and yourself on the second. They have *the ultimate* in fashions.

ultimatory (ɒltimē·tŏri), *a.* [f. Ultimate *a.* + -ory[2].] Having the character of an ultimatum.
1928 E. Blunden *Undertones of War* xi. 127 The company..was..exhausted, and its commander appealed..for relief in ultimatory terms.

u:ltimobra·nchial, *a. Anat.* [f. L. *ultim-us* last + -o + Branchial *a.*; so called because the gland develops from the most caudal pharyngeal pouches.] Applied to a gland in the neck which in many lower vertebrates regulates the calcium level in the body but in man and several higher vertebrates is absorbed into the thyroid during embryonic life.
1913 *Gray's Anat.* (ed. 18) 123 A pair of diverticula arise from the fifth branchial pouch and form what are termed the ultimobranchial bodies..: these fuse with the thyreoid gland, but probably contribute no true thyreoid tissue. 1968 *New Scientist* 15 Feb. 375/2 The ultimobranchial glands are small and situated in the neck region in birds and reptiles, but are generally absent in mammals. 1976 *Path. Ann.* XI. 222 Parafollicular C cells are not restricted to the thyroid, but can also be found in other structures embryologically connected with the ultimobranchial body, ie, parathyroid IV and thymus IV.

ultion. (Later example.)
1901 G. B. Shaw *Admirable Bashville* iii. 323 My mission here Is to wreak ultion for the broken law.

Ultisol (ɒ·ltisǫl). *Soil Sci.* [f. Ulti(mate *a.* + *-sol.] A type of highly weathered, leached, red-yellow or red acid soil marked by a clay-rich B horizon and found in warm, humid climates.
1960 *Soil Classification: 7th Approximation* (U.S. Dept. Agric.) xv. 226/1 The Ultisols include most soils that have been called Red-Yellow Podzolic soils, Reddish-Brown Lateritic soils, and Rubrozems in the United States. 1972 J. G. Cruickshank *Soil Geogr.* iv. 132 In their geographical distribution, the Ultisols—otherwise known as red-yellow podsolics or ferruginous soils—merge into grassland soils on their arid margins and into Alfisols and Inceptisols (brown earths) on their cool humid margins (eg, in North America).

ultra, *sb.* **2.** (Earlier example.)
1823 H. More *Let.* in R. B. Johnson *Lett. Hannah More* (1925) 201 My friend F. is an ultra of the first magnitude. The poor must not only read English, but ancient history, and even the sciences are to be laid open to them.

ultra, *prep.* Add: **2.** (Earlier example.)
1876 *Law Times Rep.* XXXIV. 697/2 Here she was asked to do something which was no part of her service..it being something *ultra* her service to go to the kitchen.
3. In the L. phr. *ultra crepidam* (kre·pidæm) [see Ultra-crepidarian *a.* and *sb.*], on matters beyond one's knowledge.
1883 G. M. Hopkins *Let.* 7 Nov. (1956) 332 Pope was the great master of metre of his day,..but..he was nothing *ultra crepidam*. 1895 *Econ. Jrnl.* V. 589 Meeker than the ancient master, he has refrained from breaking out against the criticisms which have been *ultra crepidam*.

ultra-, *prefix.* Add: **1. b.** *ultra-stellar*.
1906 [see *night-web* s.v. *night *sb.* 13 a].
2. c. With sbs. in this sense, as *ultrafiltration*, *-microscope*, etc. (qq.v. as main entries).
3. a. *ultra-beloved, -bourgeois, -Catholic, -cautious, -civilized, -clean, -clear, -clerical, -critical, -fast, -feminine, -generous, -German* (earlier example), *-intellectual, -left, -leftist, -militant, -miniature, -miniaturized, -nationalist, -nationalistic, -patriotic, -professional, -pure, -rapid, -rationalist, -respectable, -revolutionary* (later example), *-rightist, -romantic, -sentimental, -smart, -tropical* (earlier example), *-violent*.
1923 D. H. Lawrence *Birds, Beasts & Flowers* 183 Everybody so dear, and yourself so ultra-beloved. 1934 G. B. Shaw *On Rocks* Pref. 161 Even Lenin and his colleagues, all ultra-bourgeois (otherwise they would never have so absurdly overestimated the intellectual resources of the proletariat [etc.]). 1830 J. S. Mill *Let.* 20 Aug. in *Victorian Stud.* (1957) I. 143 Three were printers of an ultra-Catholic and royalist journal. 1930 G. B. Shaw *John Bull's Other Island* Pref., in *Works* XI. 71 The ubiquitous sodalities of that new ultra-Catholic Church called Socialism. 1946 *Mind* LV. 141 To suppose that no improvement can be effected by so doing is an ultracautious and conservative position. 1842 E. A. Poe in *Graham's Mag.* Feb. 125/2 Hugh has grown to a stalwart man—the type of man *the animal*, as his father is of man the ultra-civilized. 1931 E. Blunden *Votive Tablets* 170 Byron..with his ultra-civilised fraternisation with the gentlemen. 1970 Ultraclean [see *sonication* s.v. *sonicate *sb.* and *v.*]. 1907 W. James in *Jrnl. Philos., Psychol. & Sci. Methods* IV. 295 Your letter is so ultraclear. 1938 *New Statesman* 23 July 162/1 Maria-Cristina's ultra-clerical court. 1907 W. James *Pragmatism* v. 192 Common science or corpuscular philosophy, ultra-critical science..or idealistic philosophy, all seem insufficiently true. 1939 *Sun* (Baltimore) 28 Jan. 2/6 Ultrafast American two-engined light bombers, capable of being used for combat after they have dropped their bomb loads. 1977 J. Hedgecoe *Photographer's Handbk.* 41 Ultra-fast films are most useful for working in dim existing-light conditions. 1964 Ultra-feminine [see *slimline, slim-line *a.* a]. 1976 *Time* 27 Sept. 94/2 Melba Till Allen owes much of her success at the polls to her charming, ultrafeminine manner. 1903 W. James *Let.* 8 Apr. in R. B. Perry *Thought & Char. W. James* (1935) II. 375 Your review..was as usual ultra-generous, and I thank you for it. 1837 E. A. Poe in *Amer. Monthly Mag.* June 566 Young men..alive with an exaggerated sense of honour. They abound in the most ultra German opinions respecting the *duello*. 1927 T. S. Eliot in *Seneca's Tenne Tragedies* p. xiii, It is assumed..that Shakespeare had acquired some extra- or ultra-intellectual knowledge superior to a philosophy. 1954 Koestler *Invisible Writing* xv. 186 He had discovered both its weaknesses as a play, and its ultra-Left tendency. 1977 P. Johnson *Enemies of Society* xviii. 240 Ulrika Meinhof..became the ideological leader of the German Baader-Meinhof gang of ultra-Left terrorists. 1947 *Partisan Rev.* XIV. 398 Gorky's became an 'opposition school' for 'ultra-leftist' Bolsheviks. 1970 *Times* 11 Mar. 6 A new, ultra-militant shopkeepers' organization, known as the 'Tour du Pin Movement'. 1968 *Sci. News* 7 Dec. 573 An ultra-miniature vacuum pump..will be worn directly under the astronaut's chin. 1968 *World Book Sci. Ann.* 1968 215 Using a multitude of ICs, ultraminiaturized, working in parallel, together with functional circuits, he may build electronic devices tomorrow that seem like impossible dreams even today. 1927 H. Dobbs in *Lett. Gertrude Bell* II. 555 It had indeed alarmed the ultra-nationalist party to find a section of the British press averse from the extension of the alliance. 1977 *Time* 27 June 22/3 Some are concerned about Israeli intransigence and afraid that the new ultranationalist Premier-designate, Menachem Begin..may make a settlement all the harder to achieve. 1974 *Daily Colonist* (Victoria, B.C.) 1 Oct. 5/1 The Chinese call the Amur the Black Dragon River, and the Japanese named one of their most ultranationalistic societies after it in the days when they were itching to go to war with Russia. 1801 W. Dupré *Lexicographia-Neologica Gallica* 283 Some ultra-patriotic writers and journalists have likened Bonaparte to the conquerors who plundered Greece. 1926 J. S. Huxley *Ess. Pop. Sci.* 159 Proud and ultra-patriotic Gods. 1928 E. O'Neill *Strange Interlude* iv. 147 In his ultra-professional manner—like an automaton of a doctor. 1961 Ultra-pure [see *conductivity 2]. 1973 *Nature* 13 Apr. 482/3 The preparation, handling containment and analysis of ultrapure materials. 1921 *Sci. Amer.* 9 Apr. 288 Diagram of the multiple prism ring used for the ultra-rapid camera. 1969 *Jane's Freight Containers 1968–69* 445/1 The Boeing has three baggage/cargo compartments. Two, fully containerised for ultra-rapid ground handling. 1907 W. James *Pragmatism* iii. 123 A mere mess of anarchy and confusion..will pragmatism often seem to ultra-rationalist minds in philosophy. 1925 T. Dreiser *Amer. Trag.* (1926) ii. i. 152 Older and more conservative families who constituted the ultra-respectable element of the city. 1958 *People* 4 May 2/3 The workers have captured even the ultra-respectable Royal Academy this year. 1974 tr. *Wertheim's Evolution & Revolution* i. 47 The ultra-revolutionary Red Guards in turbulent China,..the dissatisfied students in France or the rebellious young Negroes in the United States—all of them are..rebelling against the *status quo*. 1974 *Ann. Reg. 1973* 313 Whereas previously most criticism had accused him of being ultra-leftist, the media now argued that most of his faults were in fact ultra-rightist. 1846 E. A. Poe in *Godey's Mag. & Lady's Bk.* Mar. 135/1 As a drama, we find 'Elfrida' faulty in the extreme. Its situations are ultra-romantic, or improbable. 1931 *Times Lit. Suppl.* 14 May 385/1 Young Browning in his period of ultra-romantic chaos. 1926 E. O'Neill *Great God Brown* Prologue 11 The sound of the school quartet rendering 'Sweet Adeline' with many ultra-sentimental barber-shop quavers. 1904 *N.Y. World Mag.* 1 May 6/1 She is likewise one of the prettiest girls in society, and ultra-smart. 1939 O. Lancaster *Homes Sweet Homes* 74 Ultra-smart householders who reacted instantaneously to every change of fashion. 1848 E. A. Poe *Eureka* 83 On the Melville islands..we find traces of *ultra-tropical* vegetation. 1972 *Daily Colonist* (Victoria, B.C.) 3 Mar. 5/3 Another standard argument for the ultraviolent film is that in real life, violence is not the gunshot flash and quick dissolve, [etc.].

4. a. *ultra-Catholic, -feminist, -leftist, -modernist, -nationalist, -patriot, -rationalist, -rightist, -royalist* (earlier example); also with adj. used as collective sb., as *ultra-feminine, -left, -right*.
1837 J. S. Mill in *Westm. Rev.* XXVIII. 71 The ultra-Catholics with their bigotry and pretensions to priestly domination. 1907 *Cambr. Mod. Hist.* X. xvi. 530 Belgium..was divided.., as it is still, into two irreconcilable parties, the ultra-Catholics or Clericals, and the Liberals. 1926 W. de la Mare *Connoisseur, & Other Stories* 184 Mr. Thripp indeed was no lover of the ultra-feminine. 1979 *Maledicta* III. 17 Even though many ultra-feminists have attacked you.., you still insist that women are superior to men. 1971 *New Scientist* 30 Sept. 740/2 The ultra-left found its voice by protesting against the new Tokyo international airport. 1971 *Time* 11 Oct. 13 As an ultra-leftist, of course, Wu would hardly expect a warm welcome from as revisionist a country as the Soviet Union. 1926 H. W. Fowler *Mod. Eng. Usage* 709/1 Let us not be ultra-modernists & assume that *whence* & *whither*..are dead & buried. 1936 G. B. Shaw *Millionairess* Pref. 121 Only the stupidest or craziest ultra-Nationalists believe that people corralled within the same political frontier are all exactly alike. 1938 Ultra-patriot [see *samurai-minded* adj. s.v. *samurai 2]. 1976 *Times Lit. Suppl.* 21 May 602/4 De Mille and the ultra-patriots who had been smearing the guild's president..for opposing the introduction of a compulsory loyalty oath. 1909 W. James *Pluralistic Universe* 352 Mr. Bradley, for instance, is an ultra-rationalist. 1977 *Time* 21 Nov. 39/3 She's willingly making herself a stalking horse for the ultra-right. 1976 *New Yorker* 8 Mar. 28/1 In 1973, he [*sc.* Lin Piao] became an 'ultra-rightist'. 1817 *Ann. Reg. 1816* 109/2 The latter were accordingly eminently monarchical in their principles, and were invidiously branded with the title of *ultra-royalists*.
b. *ultra-leftism, -liberalism* (earlier example), *-marathon, -nationalism, -Protestantism* (earlier example), *-radicalism* (later example), *-rationalism, -speed, -supernaturalism, violence*.
1949 *Horizon* May 315 'Trotskyism' (i.e., ultra-leftism). 1835 J. E. Alexander *Sk. Portugal* viii. 192 The priest..seized a musket by the barrel and hit him a crack with the butt over the head to show his ultra-liberalism, (priests being generally suspected of being anti-constitutional). 1977 J. F. Fixx *Compl. Bk. Running* xxiv. 268 Corbitt knows as much about long races—ultramarathons, runners call them—as anyone alive. 1949 *Horizon* June 398 Men who have grown old in the ways of self-interest and ultranationalism. 1839 J. S. Mill *Let.* 27 Dec. in *Wks.* (1963) XIII. 415 The principal peculiarity of this school is hostility to what they call ultra-Protestantism. 1949 I. Deutscher *Stalin* 405 The Comintern was now indeed engaged in a mock fight. Its ultra-radicalism was so unreal. 1899 W. S. Churchill *Let.* 19 Jan. in R. S. Churchill *Winston S. Churchill* (1969) II. Compan. I. p. xxvii, Who should draw the line where the maximum of human pain and doubt may be allayed and the minimum of ultra-rationalism be incurred? 1946 *Mind* LV. 109 This ultra-rationalism, as I may call it, put forward by Leibniz. 1963 C. R. Cowell et al. *Inlays, Crowns, & Bridges* ix. 103 Some operators..then undertake most of the rest of the preparation at ultra-speed. 1950 R. A. Knox *Enthusiasm* i. 2 If I could have been certain of the reader's goodwill, I would have called my tendency 'ultrasupernaturalism'. For that is the real character of the enthusiast; he expects more evident results from the grace of God than we others. 1972 *Times* 6 May 9/2 It will serve as a textbook study of how to photograph acts of ultraviolence without giving offence. 1977 *Time* 24 Jan. 14/2 It was like an orgy of 'ultra-violence' from Stanley Kubrick's *A Clockwork Orange*... About 200 masked youths rioted last week in the industrial town of Mestre near Venice.

ultrabasic (ɒltrăbē·sik), *a.* (*sb.*) *Petrol.* [f. Ultra- 3 + Basic *a.*] Of an igneous rock: having a silica content that is lower than that of the basic rocks (cf. Basic *a.* 2 b), *spec.* less than 45 per cent by weight. Also as *sb.*
1881 J. W. Judd *Volcanoes* (ed. 2) 317 There are some rarer materials..that present a most wonderful resemblance to the stony portions of meteorites. These materials we may call 'ultra-basic rocks'. 1885 —— in *Q. Jrnl. Geol. Soc.* XLI. 354 Those rocks which contain an excessive proportion of the bases, especially magnesia and ferrous oxide, and are therefore composed largely of unisilicates, may be conveniently classed as ultra-basic rocks. 1893, 1898 [in Dict. s.v. Ultra- 3 b]. 1938 R. W. Lawson tr. *Hevesy & Paneth's Man. Radioactivity* (ed. 2) xxv. 274 Stone meteorites are still more strongly basic than the ultrabasic terrestrial rocks. 1964 *Economic Geol.* LIX. 804 In the low-lying regions the ultrabasics are largely masked by soil. 1970 *Nature* 28 Mar. 1227/2 Most of the volcanics were basic, but during the upper Cretaceous ultrabasics..were intruded. 1977 *Sci. Amer.* Mar. 96/3 In a typical greenstone belt early volcanic activity has produced ultrabasic and basic lavas.

ultracentrifuge (ɒltrăse·ntrifiūdʒ), *sb.* [f. *ultra- 2 c + Centrifuge *sb.*] A very high-speed centrifuge, now usu. generating over 100,000*g*, used to separate small particles and large molecules in a liquid and to determine their sedimentation rate (and hence their size).
1924 Svedberg & Rinde in *Jrnl. Amer. Chem. Soc.* XLVI. 2678 In analogy with the naming of the ultra-microscope and ultra-filtration apparatus we propose the name *ultra-centrifuge* for this apparatus. 1939 *Ann. Reg. 1938* 374 The ultracentrifuge showed the presence in plant juices of carbohydrates of well-defined molecular weight. 1948 *New Biol.* V. 33 Tiny air-driven turbine ultra-centrifuges which rotate at speeds up to 5,000 revolutions per second, giving forces equal to half a million times that of gravity, will stratify many of the finest cell inclusions. 1968 *New Scientist* 3 Oct. 6/3 A pilot plant for enriching uranium by the ultracentrifuge method. 1978 H. McLeave *Borderline Case* (1979) xiii. 132 The laboratory..had everything, including..several ultracentrifuges.
Hence **u:ltracentrifugal** (stress variable) *a.*, involving or employing an ultracentrifuge; **u:ltracentrifugally** (stress variable) *adv.*, by means of an ultracentrifuge; with respect to the ultracentrifuge.
1930 *Jrnl. Amer. Chem. Soc.* LII. 2904 The ultracentrifugal study of gelatin solutions has been fruitful in revealing the heterogeneous nature of the sols. 1943 *Jrnl. Exper. Med.* LXXVII. 460 Gratia and Paillot had reported polyhedral bodies and ultracentrifugally purified material from jaundiced blood to be serologically unrelated to ultracentrifugally purified material from the blood of normal silkworms. 1953 *Jrnl. Amer. Chem. Soc.* LXXV. 67/2 Recently they succeeded in preparing an electrophoretically and ultracentrifugally homogeneous mucoprotein. 1962 M. Rabaey in A. Pirie *Lens Metabolism Rel. Cataract* 310 Ultracentrifugal analysis permitted us to classify this protein as a member of the so-called group of β-crystallins. 1977 *Lancet* 8 Oct. 742/1 A typical ultracentrifugal scan and the distribution of IgE in the corresponding fractions is shown.

ultrace·ntrifuge, *v.* [f. prec. sb.: cf. *centrifuge *v.*] *trans.* To spin in an ultracentrifuge. Chiefly as **ultrace·ntrifuged** *ppl. a.*, **-ce·ntrifuging** *vbl. sb.* Also **u:ltracentrifuga·tion**, the action or process of ultracentrifuging.
1934 *Proc. Soc. Exper. Biol. & Med.* XXXI. 707 (*heading*) Effect of ultracentrifuging on Paramecium. 1936 *Trans. Faraday Soc.* XXXII. 301 In this case again, analogous to those of viscometry and ultracentrifugation, the dilutest solutions yield the clearest result. 1946 *Nature* 5 Oct. 488/2 Virus can be separated by differential ultracentrifuging into fractions with widely different properties. 1947 *Ann. Rev. Microbiol.* I. 364 Ultracentrifuged concentrates of virus. 1971 *Nature* 18 June 447/1 The extract was spun at 25,000*g* for 20 min to remove large particles before being subjected to ultracentrifugation at 105,000*g* for 90 min to sediment out the microsomes and membrane fractions. 1974 *Ibid.* 13 Dec. 605/2 The injection of adult mice with ultracentrifuged heterologous γ globulin induces immunological unresponsiveness to a subsequent injection of the globulin. 1977 *Lancet* 8 Oct. 742/1 The distribution of radioactivity in a radiolabelled (^{125}I) myeloma IgE preparation ultracentrifuged under identical conditions.

u:ltraco·ld, *sb.* and *a.* [f. Ultra- 3 + Cold *a.*, *sb.*] **A.** *sb.* Extreme coldness.
1967 *Britannica Bk. of Year* (U.S.) 804/3 *Ultracold*, excessive coldness especially near absolute zero. 1977 A. Hallam *Planet Earth* 15/1 The Moon's atmosphere..provides little protection from the direct heat of the Sun, and little insulation from the ultracold of space.
B. *adj. Nucl. Physics.* Of a neutron: having an energy of the order of 10^{-7} eV or less.
1969 *JETP Lett.* IX. 26 Ultracold neutrons are produced and propagate in accordance with the theoretical expectations. 1978 *Nature* 9 Mar. 127/1 Ultracold neutrons are slow neutrons with velocities of less than $7\ m\ s^{-1}$. 1979 *Sci. Amer.* June 106/1 A striking property of ultracold neutrons is their total reflection from solid surfaces.

u:ltracytoche·mistry. *Biol.* [f. *ultra(*structural* adj. s.v. *ultrastructure + *cytochemistry* s.v. *cyto-.] Ultrastructural cytochemistry. So **u:ltracytoche·mical** *a.*; **u:ltracytoche·mically** *adv.*
1963 *Proc. Amer. Assoc. Cancer Res.* IV. 16/1 Ultracytochemical studies of the thymus of birds with myeloblastosis..revealed adenosinetriphosphatase activity. 1965 *Virology* XXV. 162/1 The technique for ultracytochemistry..was proved to be an effective method for investigating intracellularly developing viruses. 1981 *Physiol. Plant Path.* XVIII. 339 (*heading*) Phenolic ultracytochemistry of tobacco cells undergoing the hypersensitive reaction to *Pseudomonas solanacearum*. *Ibid.*, The objective of this study was to ultracytochemically determine if any phenolic deposition and PPO [*sc.* polyphenoloxidase] might be detected. *Ibid.* 343 With the ultracytochemical method, particulate as well as soluble activity can be demonstrated.

ultradian (ɒltrē·diăn), *a.* [f. Ultra- + L. *di-ēs* day + -an.] Designating cycles of physiological activity which recur with a period shorter than one day but longer than one hour. Cf. *circadian *a.*
1961 F. Halberg in *39th Rep. Ross Conf. Pediatric Res.* 15 Observations have been made on frequencies higher than circadian, yet lower than the respiratory or cardiac cycles (ultradian), and frequencies lower than circadian, but higher than the menstrual cycle (infradian). 1978 *Nature* 3 Aug. 490/1 In humans, REM sleep..occurs every 90–110 min and is presumed to be under the control of an underlying ultradian oscillatory mechanism.

ultrafiche (ɒ·ltrăfīʃ). [f. *ultra- 2 c + *fiche.] A microfiche in which the linear reduction of the image size is of the order of 100 or more; documentary material of this kind.
1971 *Collier's 1971 Year Book* 322/1 (*caption*) On one card, over 3,000 pages are reduced to microscopic proportions... The cards, called ultrafiches or microfiches, are magnified in the reader. 1972 *Physics Bull.* Sept. 522/1 Flat sheets of film, known as ultrafiche and measuring 6 in × 4 in can contain more than 3000 pages of normal text. 1972 *Times* 1 Dec. (Suppl.) p. ii/6 Ultrafiches of the photochronic micro image process store up to 4,000 pages on a 4 in by 6 in film. 1975 *New Yorker* 22 Dec. 30/1 They brought out the Bible on an ultrafiche... The whole book on a piece of film two inches square. 1976 P. Harcourt *Dance for Diplomats* xii. 135 The really vital thing is a bit of ultrafiche—that's some form of microfilm. 1981 Jacobstein & Mersky *Fund. Legal Res.* (ed. 2) 447 In some instances up to 1700 pages can be placed on a single sheet of ultrafiche.

ultrafi·ltrate. *Biol.* [f. next, after *filtration*, *filtrate*.] Liquid that has passed through an ultrafilter.
1928 *Biochem. Jrnl.* XXII. 633 There was slightly less inorganic phosphorus in the ultrafiltrate than in either the original or the residual plasma. 1957 G. E. Hutchinson *Treat. Limnol.* I. xv. 811 Two determinations of the copper content of ultra filtrates indicate that about 80 per cent of the so-called organic copper is associated with colloidal material. 1977 *Lancet* 18 June 1294/1 Ultrafiltrates of pepsin digests of the mucosa.

u:ltrafiltra·tion. *Biol.* [f. *ultra- 2 c + Filtration.] The action or process of filtration through a medium sufficiently fine to retain colloidal particles, certain viruses, or large molecules.

1908 *Chem. Abstr.* II. 1155 Ultrafiltration... Filtration of this type is accomplished by forcing under pressure the solutions through colloid membranes such as collodium, glacial acetic acid-collodium and hardened gelatin. 1946 *Nature* 9 Nov. 665/2 The blood pressure reduction cannot have augmented the osmotic pressure by more than..about 3·2/1000 of the total osmotic pressure. This change is so small that ultra-filtration cannot play any considerable part in the formation of aqueous humor. 1959 *New Scientist* 15 Jan. 135/3 Most marine animals possess some type of ampullary excretory organ... It is probable ..that these organs produce an excretory fluid by a process of ultrafiltration. 1964 [see *gradocol]. 1978 *Sci. Amer.* July 88/2 The pores of an ultrafiltration membrane are on the scale of 10^{-9} meter, or one nanometer; average pore diameters range from less than a nanometer to about 10 nanometers.
So **u·ltrafilter** *sb.*, a medium or membrane used for ultrafiltration; also as *v. trans.*, to subject to ultrafiltration; **ultrafi·lterable** *a.*, capable of passing through an ultrafilter.
1908 *Chem. Abstr.* II. 3183 (*heading*) Porosity of ultra filters. 1911 *Ibid.* V. 2 Ultrafiltration... A discussion of the merits of the Bechhold and Burian ultrafilters and of the author's modifications. 1920 *Biochem. Jrnl.* XIV. 539 The oxygenase remains on the ultrafilter. 1928 *Ibid.* XXII. 633 A large proportion of the inorganic phosphate of the plasma is ultrafilterable through cellophane. 1946 *Physiol. Rev.* XXVI. 586 The first indication that the adrenotrophic factor could be ultrafiltered was presented ..in 1934. 1961 *Lancet* 5 Aug. 284/2 The plasma was ultrafiltered. 1974 *Nature* 8 Nov. 176/1 The haemagglutinating activity in galactose..was found in the retentate and could be concentrated quantitatively and washed free of monosaccharide by ultrafiltration over an XM-50 ultrafilter (Amicon) which nominally retains proteins of molecular weight greater than 50,000. 1976 *Ibid.* 8 Apr. 487/1 The rise in brain tryptophan..correlates..with the fraction of tryptophan which is not protein bound (free or ultrafilterable). 1978 *Ibid.* 2 Mar. 65/2 The supernatant was ultrafiltered twice..to separate the proteins into three fractions with molecular weights..above 50,000, between 10,000 and 50,000 and below 10,000.

ultra-high (ɒltrăhəi·), *a.* [f. Ultra- 3 + High *a.*] **1.** *Radio.* Of a radio-frequency: in the range 300 to 3,000 megahertz. Abbrev. UHF s.v. *U 4 a.
1932 *Proc. IRE* XX. 95 Among the significant and comparatively recent developments in electrical engineering is the efficient and dependable production of ultra-high-frequency oscillating circuits. 1935 *Discovery* Sept. 278/2 Special valves are being developed suitable for direct signal amplification at the ultra-high frequencies. 1940 *Jrnl. R. Aeronaut. Soc.* XLIV. 216 Another application of ultra-high frequencies is to provide a system of instrument landing. 1951 *Electronic Engin.* XXIII. 452 The use of ultra high frequencies for communication purposes is a relatively new and rapidly developing branch of radio. 1978 *Dædalus* Spring 212 There is already a good deal of listening in the ultrahigh frequency radio bands.
2. *gen.* Exceedingly high; freq. in attrib. phr. in comb. with following sb.
1936 *Discovery* Mar. 88/1 The G.W.R. and the L.N.E.R. have been even more enterprising in introducing some ultra-high-speed expresses. 1962 F. I. Ordway et al. *Basic Astronautics* vii. 327 Ultrahigh-risk equipment. 1966 D. G. Brandon *Mod. Techniques Metallogr.* iv. 184 Methods for obtaining ultra-high vacua (UHV) are now highly developed. 1972 D. Halberstam *Best & Brightest* 558 At West Point he again had ultrahigh visibility and he impressed everyone. 1977 P. Johnson *Enemies of Society* i. 4 The extended family also dissociated procreation from the responsibility to maintain offspring, and so produced ultra-high birthrates. 1981 *Economist* 24 Jan. 49/3 How long can Britain go on protecting its farmers from competition from imported UHT (Ultra-High Temperature, or long-life) milk?

ultralight, *a.* and *sb.* [f. Ultra- 3 + Light *a.*[1]] **A.** *adj.* (Stress variable.) Extremely lightweight; *spec.* applied to a small, low-speed, inexpensive, usually one-seater aircraft that has a small engine but can also soar and whose fuselage is an open framework without an enclosed cockpit. **B.** *sb.* (Stressed *u·ltralight*.) An ultralight aircraft.
1974 *Soaring* Jan. 44/1 (Advt. heading), Ultralights. 1974 *Flying* Mar. 30/1 Hang gliders have been in the news. ..Sometimes called 'ultralight gliders', they represent the minimum machine capable of controllable sustained flight. 1974 *Nat. Geographic* Dec. 771/2 His ultralight 30-foot canoe was a perfectly balanced sailing machine. 1982 *Sci. Amer.* July 60/1 Only within the past few years has the conjunction of the hang glider and the small engine (Go-Kart or snowmobile) brought the long-sought objective into being as the ultralight airplane. 1982 *N.Y. Times* 22 Aug. i. 1/4 Swarms of these new craft—known to enthusiasts as ultralights—are filling the air. 1983 *Economist* 27 Aug. 47 First on the runway is likely to be Eipper Aircraft of California, the biggest maker of ultralights (almost the same as microlights in Europe).

ultramafic (ɒltrămæ·fik), *a.* (and *sb.*) *Petrol.* [f. Ultra- 3 + *mafic *a.*] Of an igneous rock: composed chiefly of mafic minerals. Also as *sb.*
1933 R. A. Daly *Igneous Rocks & Depths of the Earth* ii. 31 The more important groups [of igneous rocks] are the granite clan.., the ultramafic clans, and the 'feldspathoidal' clans. *Ibid.* xxii. 565 Each of the ultramafic and ultrabasic species..treated in this chapter has its own unsolved problems. 1948 *Bull. Oregon State Dept. Geol. & Mineral.* XXXIX. 8 The ultramafics include bodies of dunite, pyroxenite, and gabbro which have been extensively altered to hornblendite and serpentine. 1965 G. J. Williams *Econ. Geol. N.Z.* x. 143/1 The useful non-metallic minerals are more abundant in those ultramafics which have been subjected either to metamorphic or metasomatic changes. 1967 P. J. Wyllie *Ultramafic & Related Rocks* 1/1 Ultramafic monomineralic rocks composed of the following mafic minerals would certainly be ultrabasic..forsterite..fayalite..hornblende..biotite. 1971 I. G. Gass et al. *Understanding Earth* i. 23/1 Only in the ultramafic rocks..does Al_2O_3 fall to significantly lower values than those quoted.

ultramicro- (ɒltrămai·kro), *prefix* and quasi-*adj. Chem.* [f. Ultra- 2 + Micro- 2 a, *8 b.] Formative element denoting chemical analysis or research which involves very minute quantities (of the order of a few microgrammes or less), as in *ultramicroanalysis*, *-analytical* adj., *-chemistry*, *-chemical* adj. Also used without a hyphen as an independent word.
1937 *Ann. Rev. Biochem.* VI. 90 The methods of capillary colorimetry are necessarily more difficult than the other ultramicro methods which have been discussed. 1940 *Ibid.* IX. 599 Drop analysis.—Ultramicrochemical methods have been advanced..in regard to the number of constituents determinable. 1946 *Chem. Abstr.* XL. 6017 (*heading*) Ultramicroanalysis. III. A method for enriching copper by selective adsorption and for destroying organic matter in the determination of copper. 1946 *Chem. & Engin. News* XXIV. 1195/2 The field which embraces the chemical study of material on this minute scale of operation has been given the name 'ultramicrochemistry' by P. L. Kirk, a pioneer investigator in the field of quantitative chemistry on the microgram scale. 1946 *Nature* 31 Aug. 313/1 (*heading*) Ultra-micro methods in nuclear chemistry. 1962 H. Heath in A. Pirie *Lens Metabolism Rel. Cataract* 365 Ultra-microanalytical methods have to be used to detect the last traces of ascorbic acid. 1971 *Analytical Biochem.* XLIV. 503 (*heading*) Quantitative ultramicroanalysis of amino acids in the form of their DNS-derivatives. 1973 *Biol. Abstr.* LVI. 1717/1 (*heading*) Ultramicrochemical methods for the analysis of tissues and cells of the inner ear. 1979 Ma & Rittner *Mod. Organic Elemental Analysis* xi. 366 Since it is now possible to carry out organic synthesis at the microgram level, the applications of ultramicro analytical methods become apparent.

ultrami·crofiche. [Ultra- 3.] = *ultrafiche.
1967 in Webster Add. 1974 *Encycl. Brit. Macropædia* IX. 569/1 Film has not been used widely as a medium of primary distribution. One example is the Library of American Civilization, a retrospective collection on ultra-microfiche of the full original text of about 20,000 titles.

ultrami·croscope. [f. *ultra- 2 c + Microscope *sb.*, or a back-formation from next.] An optical microscope used to detect particles smaller than a wavelength of light by illuminating them at an angle, so that the light scattered by the particles (Tyndall scattering) can be observed against a dark background.
1906 *Jrnl. R. Microsc. Soc.* 366 The ultramicroscope is adapted to the determination of the identity and purity of oils. 1927 [see *polydispersity* s.v. *polydisperse *a.*]. 1936 *Discovery* Nov. 347/2 The ultramicroscope, which by specially illuminating fine particles so that they themselves shine with reflected light, makes them perceptible through an ordinary microscope when they would not otherwise be so. 1974 *Sci. Amer.* May 88/2 We examined samples of oil in an ultramicroscope.
Hence **ultramicro·scopy**, the art or practice of using an ultramicroscope.
1906 *Jrnl. R. Microsc. Soc.* 366 (*heading*) Ultramicroscopy of oleosole. 1910 [see *Tyndall]. 1961 R. D. Baker *Essential Path.* i. 7 Advances in histochemistry and ultramicroscopy are increasing the scope of pathologic anatomy as an investigative science.

u:ltramicrosco·pic, *a.* Formerly also with hyphen. [f. Ultra- 2 + Microscopic *a.*] Of such minute size as to be invisible or indeterminate under the ordinary light microscope; of or pertaining to the use of the ultramicroscope.
1870, 1905 [in Dict. s.v. Ultra- 2]. 1932 *Nature* 2 Jan. 21/1 Optical ultra-microscopic examination of..films. 1946 F. Schneider *Qualitative Organic Microanalysis* iv. 103 An instrument for estimating the refractive index of particles of ultramicroscopic size. 1958 *Times Lit. Suppl.* 17 Jan. 33/2 A description of this extraordinary population of all shapes and sizes from rabbits to ultramicroscopic organisms. 1978 *Nature* 22 June 610/1 Proteolytic enzymes..are emitted through ultramicroscopic pores in the eggshell.
Hence **u:ltramicrosco·pical** *a.*, in the same sense; **u:ltramicrosco·pically** *adv.*
1904 *Jrnl. R. Microsc. Soc.* 711 There is a very disturbing adsorption effect of the glass planes on the ultramicroscopical particles. 1906 *Ibid.* 366 Fats and ethereal oils were treated with chlorides of the metals, and the products of reaction ultramicroscopically studied. 1929 A. S. C. Lawrence *Soap Films* v. 88 Microscopically, we see the actual particles; ultramicroscopically, we only see the light diffracted by particles themselves invisible. 1976 *Jrnl. Compar. Path.* LXXXVI. 516 Light and ultramicroscopical features of spontaneous glomerular lipoidosis are described. 1981 *Jrnl. Protozool.* XXVIII. 308 Thin sections of treated cells, examined ultramicroscopically, exhibited vacuolations..and severe mitochondrial damage.

ultrami·crotome. [f. *ultra- 2 c + Microtome.] A microtome for cutting sections thin enough for electron microscopy (typically about 300 nanometres thick).
1953 *Jrnl. Electron Microsc.* I. 44 (*heading*) A simplified thermal expansion type ultramicrotome. 1966 D. G. Brandon *Mod. Techniques Metallogr.* 92 A plate glass or diamond knife ultramicrotome has proved to be the most successful method available for producing thin sections of biological specimens for electron microscope examination, and serious attempts have been made to adapt the technique to metallurgical specimens. 1980 *Amer. Jrnl. Tropical Med. & Hygiene* XXIX. 775/1 Ultra-thin sections..were cut with either glass knives or diamond knives on an ultramicrotome.
Hence **ultrami·crotomed** *ppl. a.*, sectioned with an ultramicrotome; **ultramicro·tomy**, the practice or technique of using the ultramicrotome.
1949 *Science* 15 July 66/2 A promising new development in ultramicrotomy is presented. 1976 *Nature* 8 Apr. 513/1 Detailed examinations of a Georgia kaolinite using high resolution electron microscopy of ultramicrotomed sections of the kaolinite mineral revealed not only the expected 7 Å (001) spacing, but also occasional 10 and 14 Å spacings. 1978 *Ibid.* 30 Mar. 433/1 Ultramicrotomy and ion beam thinning of films..have provided greater information on film morphology and composition. 1982 *Electrochimica Acta* XXVII. 245 Electron microscopy.. of stripped anodic films and ultramicrotomed sections of the aluminium substrate and the anodic film have been used to examine directly the general film growth over the macroscopic metal surface.

ultramontane, *sb.* **1. b.** (Earlier examples.)
1829 *Dublin Even. Post* 3 Oct. 3/1 Supposed that the Catholic Clergy in Ireland were *Ultramontanes* of the same class. 1845 J. H. Newman *Ess. Devel. Christian Doctrine* ii. 129 An unbeliever, as Gibbon, assumes one hypothesis, and an Ultra-montane, as Baronius, adopts another. 1865 E. B. Pusey *Eirenicon* 326 The present Ultramontanes have apparently changed the old Ultramontane doctrine of the inerrancy of the Pope.

ultra-sho·rt, *a.* Also **ultrashort.** [Ultra- 3.] **a.** *Radio.* Applied to radio waves significantly shorter than the usual 'short waves': in mod. use, shorter than 10 metres, corresponding to a frequency greater than 30 MHz (i.e. in the VHF range).
1926 E. V. Appleton in *Proc. Cambridge Philos. Soc.* XXIII. 155 (*heading*) On the diurnal variation of ultra-short wave wireless transmission. [*Note*] The term 'ultra-short' is here applied to wave-lengths less than the critical band indicated by the magneto-ionic theory (i.e. about 200 meters). It seems desirable, for historical reasons, to retain the term 'short waves' for the broadcasting wave-lengths 200–600 metres. 1935 *Discovery* Sept. 278/1 A difficulty is..the limited range of these ultra-short wave-lengths, as we are dependent upon the direct waves propagated more or less parallel to the earth's surface. 1947 *Nature* 4 Jan. 16/1 The electrical interference from motor-car ignition systems..is a serious factor in the reception of ultra-short waves. 1974 *Radio Engin. & Electronic Physics* Dec. 20 (*heading*) Multiple diffractor amplification of ultrashort waves.
b. *gen.* Extremely short (in length or duration).
1962 R. H. Smythe *Anat. Dog Breeding* 29 An ultra-short loin. 1975 *Bio Systems* VII. 45 Accordingly, we may introduce an ultrashort time scale. 1977 *Jrnl. R. Soc. Arts* CXXV. 766/1 The same lasers can be used to generate ultra-short light pulses.

ultrasonic (ɒltrăsǫ·nik), *a.* [f. Ultra- 2 + *sonic *a.*] **1. a.** The more usual synonym of *supersonic *a.* 1.
1923 *Proc. & Trans. R. Soc. Canada* XVII. iii. 145 The wave-lengths of ultra-sonic waves are very convenient for experiment. 1926 *Encycl. Brit.* III. 592/1 The utilisation of ultrasonic waves by Professor Langevin..in connection with deep-sea sounding. 1928 *Observer* 30 Dec. 3/6 This was the first occasion in this country on which the 'ultra-sonic' waves have been shown to an audience. 1938, 1948 [see *hypersonic *a.* 1]. 1957 *New Scientist* 9 May 26/3 Owing to the restricted frequency response of the human ear, those organisms which produce ultra-sonic sounds, as do many insects and bats, for example, have in particular been thought of as 'dumb'. 1976 R. Dawkins *Selfish Gene* iv. 67 The..song of the humpback whale, with..its frequencies spanning the whole of human hearing from subsonic rumblings to ultrasonic squeaks. 1985 *Sunday Times* 27 Jan. 80/8 An ultrasonic beam passed through liquid generates bubbles which act as a cleansing agent.
b. Employing or operated by sound waves or vibrations having a frequency above the range of human hearing (i.e. greater than 15–20 kHz); used esp. with reference to devices and techniques which make use of the reflected echo of an ultrasound pulse.
1923 *Proc. & Trans. R. Soc. Canada* XVII. iii. 142 Two ultra-sonic generators..were placed facing one another, 60 cms. apart, in a large tank of water. 1935 [see *reflectometer]. 1949 A. R. Weyl *Guided Missiles* 25 Photo-electric, magnetic, acoustic (ultra-sonic) and other methods have been successfully used for proximity fuses. 1961 *Lancet* 30 Sept. 750/1 Ultrasonic methods have been described for the detection of intra-abdominal masses, breast tumours, [etc.]. 1976 *Offshore Platforms & Pipelining* 174/1 Ultrasonic inspection followed the completion of the welds.
2. Designating speeds above that of sound; supersonic. *rare*$^{-1}$.

1942 [see *infrasonic *a.* 2].

Hence **ultraso·nically** *adv.*, by means of ultrasound.

1955 *Sci. News Let.* 26 Mar. 206/3 Ultrasonically treated honey..showed no signs of crystallizing. 1968 *Amer. Mineralogist* LIII. 1558 Samples of the ash were cleaned ultrasonically. 1974 *Nature* 2 Aug. 410/2 Electrodeposition in an ultrasonically agitated solution.

ultrasonication (ɒ:ltrăsǫnikēɪ·ʃən). [f. *ultra- 2 c + *sonication.] = *sonication* s.v. *sonicate *sb.* and *v.*

1965 *Dissertation Abstr.* XXV. 4366/1 Ultrasonication causes the complete shedding of the embryonic periderm leaving a residual layer of originally basal cells. 1979 *Chem. & Physics of Lipids* XXIV. 257 The solubility of the three steroid hormones..was measured after shaking and ultrasonication.

So **ultraso·nicate** *v. trans.* = *sonicate *v.*; **ultraso·nicated** *ppl. a.*

1969 *Bot. Mag.* (Tokyo) LXXXII. 162 (*heading*) Electron microscopy of replicas from ultrasonicated *Chara* oosporangial walls. 1974 P. Pulido in K. Elliott et al. *Trypanosomiasis & Leishmaniasis* (Ciba Symp. No. 20) 280 Lyophilized *T. cruzi* was resuspended in buffer and ultrasonicated. 1980 *Exper. Parasitol.* L. 360/1 Two milliliters of the suspension was ultrasonicated on ice.

ultrasonics (ɒltrăsǫ·niks), *sb. pl.* [f. *ultrasonic *a.*: see -ic 2.] **a.** Ultrasonic waves; ultrasound.

1924 *Ann. Rep. Canadian Nat. Research Council* 23 Ultra Sonics are sound vibrations of a pitch higher than those which are audible. 1931 *Nature* 21 Feb. 284/2 Prof. F. L. Hopwood discussed ultra-sonics or inaudible sounds. 1957 A. C. Clarke *Deep Range* i. 10 Sub 5 was still close enough to the mother ship for radio to work, but before long he'd have to switch to the ultrasonics. 1960 *New Biol.* XXXI. 32 The bacteria which have been treated by ultrasonics..cannot multiply any more. 1972 R. E. Ornstein *Psychol. of Consciousness* (1975) ii. 19 We normally consider that our senses are the 'windows' to the world... But such a view..is not entirely true, for a primary function of sensory systems taken as a whole is to discard 'irrelevant' information, such as X-rays, infrared radiation, or ultrasonics.

b. The branch of science and technology concerned with the study and use of ultrasonic waves. Const. as *sing.*

1940 [see *Asdic]. 1959 *Sunday Times* 21 June 3/2 Ultrasonics has..been established for a number of years in submarine detection, underwater signalling, echo-sounding, fish location, flaw detection in metals and in many fields of industrial measurement and control. 1977 *Time* 28 Nov. 56/2 Since the original development of the technique for cardiology in the 1950s, ultrasonics has been used to explore other areas of the body, notably, the developing fetus in the mother's uterus.

ultrasonography (ɒltrăsǫnǫ·grăfi). *Med.* [f. Ultra- (in *ultrasound*) + *sono- + -graphy.] A technique which makes use of echoes of ultrasound pulses to delineate objects or areas of different density within the body, esp. for diagnostic purposes.

1960 *Arch. Ophthalmol.* LXIV. 180/1 Ultrasonography is an invaluable aid in the diagnosis and management of orbital disease because it can visualize and localize the position of orbital lesions when all other tests are negative. 1967 *New Scientist* 26 Jan. 195/2 Ultrasonography..is capable of detecting pregnancy even before a woman's urine test becomes positive. 1980 *Brit. Med. Jrnl.* 29 Mar. 940/2 Serum luteinising hormone concentrations and serial ovarian ultrasonography are helpful screening procedures.

So **ultraso·nogram**, an image obtained by ultrasonography; **ultraso·nograph**, an apparatus for producing ultrasonograms. Also **ultrasono·grapher**, one who specializes in ultrasonography.

1958 *Arch. Ophthalmol.* LX. 266/2 The ultrasonogram represents a horizontal section, or planigram, through a level of the eye. 1975 *Proc. 2nd European Congr. Ultrasonics in Medicine* 129 Ultrasonographs employed to visualize internal eye structures use focusing systems which concentrate the ultrasonic beam into a small area. 1975 *Daily Tel.* (Colour Suppl.) 7 Feb. 21/1 Now the second generation of ultra-sonographs—some capable of using a computer to 'freeze' the movement of a single heartbeat—are extending the boundaries of medical knowledge. 1979 *Brit. Med. Jrnl.* 13 Oct. 934/2 They are not in conflict with the ultrasonographers' longitudinal data. 1981 *Daily Tel.* 24 Apr. 3/3 Hydrocephalus occurs in about two of every 1,000 babies. With the use of ultrasonograms..doctors can usually detect the condition while the foetus is developing.

ultrasound (ɒ·ltrăsaund). [f. *ultra- 2 c + Sound *sb.*³] Sound waves or vibrations with frequencies greater than those audible to the human ear, or greater than 20,000 Hz; also, ultrasonic techniques.

1923 *Proc. & Trans. R. Soc. Canada* XVII. III. 143 When the electrical exciting sources were alternating in potential, the tonic trains of ultra-sound so produced—120 pulses per second—could be heard in the stethoscope. 1936 *Jrnl. R. Aeronaut. Soc.* XL. 25 Ultra-sound waves are generated by a quartz oscillator arc propagated through a surrounding liquid medium. 1953 J. Y. Cousteau *Silent World* 104 Are porpoises equipped with sonic or ultra-sound apparatus by which their squeaks give them the feel of unseen bottom topography? 1958 *Oxford Mail* 30 June 4/4 Some firms now use ultrasound to detect flaws in metal. 1968 *New Scientist* 16 May 347/1 The moth's sensitivity to ultrasound may have other uses. 1984 *Times* 16 Nov. 12/4 Any woman who is worried about malformations would be subject to a detailed ultrasound scan.

ultrastable (ɒltrăstēɪ·b'l), *a.* [f. Ultra- 3 + Stable *a.*] Stable against all subsequent disturbances, even those not taken into account in the design of the system. Hence **ultrastabi·lity.**

1952 W. R. Ashby *Design for Brain* viii. 91 This process is most clearly shown in what I shall call an ultrastable system: one that is absolute and contains step-functions in a sufficiently large number for us to be able to ignore the finiteness of the number. *Ibid.*, The principle of ultrastability will be stated formally: an ultrastable system acts selectively towards the fields of the main variables, rejecting those that lead the representative point to a critical state but retaining those that do not. 1967 R. Whitehead in Wills & Yearsley *Handbk. Management Technol.* 66 Cybernetics seeks to produce an ultra-stable solution between an organism and its environment. *Ibid.*, If self-regulation and ultra-stability are desirable things..then we must be prepared to learn from the systems that exhibit them. 1970 G. Ordish tr. *R. Chauvin's World of Ants* 203 What do we mean here by the word 'adaptation'? It should mean a series of successive changes of tactics which only stop, as Ashby says, when the system reaches an 'ultrastable' state within the limits set by the physiology of the individuals in question. 1976 *Listener* 24 June 821/4 Such thought..would itself become an historical system—in the modern jargon, an 'ultra-stable' system, one which could take its own corrective action when things went wrong.

u·ltrastru:cture. [f. Ultra- 2 + Structure *sb.*] Structure of biological material that is visible only under greater magnification than can be obtained with optical microscopy.

1939 F. O. Schmitt in *Physiol. Rev.* XIX. 270 The term ultrastructure as used in this review denotes the submicroscopic organization of cellular and tissue components. 1948 *New Biol.* V. 34 Experiments with the centrifuge and micromanipulator show..that organised development can be interfered with by breaking up this ultrastructure, but as we have seen, it often reforms itself after such a breakdown. 1970 *Jrnl. Neurosurg.* XXXII. 142/2 The ultrastructure of the capillaries in human malignant brain tumors has been studied. 1977 J. L. Harper *Population Biol. Plants* v. 131 A seed that is going to germinate may begin to show marked changes in the ultrastructure of its cells within 30 min of wetting.

Hence **ultrastru·ctural** *a.*, of or pertaining to ultrastructure; **ultrastru·cturally** *adv.*, with regard to ultrastructure.

1939 *Physiol. Rev.* XIX. 270 To learn the details of the ultrastructural organization of such systems it is necessary to know the chemical composition and the physical and chemical properties of the molecules which compose the system. 1971 *Nature* 2 Apr. 334/2 Ultrastructurally, as well as by light microscopy, the human parafollicular cells were similar to those of other mammals. 1975 *Ibid.* 7 Aug. 459/3 Ultrastructurally, the size (diameter 600±150 Å), but not the number, of nuclear pores is remarkably constant in all eukaryotic cells. 1977 *Proc. R. Soc. Med.* LXX. 670/2 The anatomical, histological, ultrastructural and physiological aspects of normal peripheral nerves.

Ultrasuede (ɒ·ltrăswēɪd). *U.S.* Also **ultra-suede.** [f. Ultra- + Suède.] A proprietary name for a synthetic non-woven fabric resembling suede. Also as *sb.*, a garment made of this.

1973 *Official Gaz.* (U.S. Patent Office) 13 Mar. TM 96/2 Spring Mills, Inc., Fort Mill, S.C...*Ultrasuede*. For non-woven suede-like fabrics... First use Nov. 24, 1971. 1974 [see *man-made *a.*]. 1976 *National Observer* (U.S.) 2 Oct. 16/1 Rich chocolate-brown carpet, white ultrasuede couch, and..two pink ultrasuede armchairs. 1977 *Detroit Free Press* 11 Dec. 22-A/2 She dressed in Von Furstenberg shirtwaists and Halston ultrasuedes. 1981 A. Cross *Death in Faculty* i. 8 Her suit, ultra-suede, was worn over a turtleneck knit.

ultrathi·n, *a.* Also **ultra-thin.** [f. Ultra- 3 + Thin *a.*] Extremely thin; *spec.* in *Biol.*, applied to a section cut with an ultramicrotome.

1949 *Science* 15 July 68/2 The new technique provides an inexpensive, practical method for producing ultrathin sections of tissue in almost a routine fashion. 1961 *Technology* Feb. 52/4 Fabrics treated with an ultrathin coating of a polyamide..are said to retain the soft texture of wool, and yet may be washed repeatedly without shrinking. 1962 *Sci. Survey* III. 164 These ultrathin sections are today about 250 times thinner than the conventional sections used in light microscopy. 1969 *New Scientist* 16 Oct. 131/2 The new ultrathin transistor is particularly important for uhf applications. 1976 *Ann. Rev. Microbiol.* XXX. 116 (*caption*) Ultrathin section of cell from a diseased mushroom.

ultraviolation (ɒ:ltrăvaiŏlēɪ·ʃən). *slang.* [Humorous blend of *ultraviolet *a.* and *sb.* and Violation.] Irradiation with ultraviolet light. So **ultravi·olate** *v. trans.*

1978 *Oxf. Univ. Gaz.* 23 Oct. Suppl. 1. 167 (*heading*) Why does yeast die from ultra-violation? 1979 *Amer. Speech* LIV. 119 Ultraviolation. William J. Payne tells us that, when cells are subjected to ultraviolet radiation (for mutagenesis), they are said to have been ultraviolated. Editors..strike the term from papers submitted for publication, but scientists frequently use it in conversation. 1984 *New Scientist* 3 May 46/1 At Stanford University.. medical scientists no longer expose cells to UV light... They prefer to ultraviolate them.

ultra-violet, *a.* Substitute for entry s.v. Ultra- 1 c:

ultravi·olet, *a.* and *sb.* Also with hyphen. [f. Ultra- 1 c + Violet *sb.*¹] **A.** *adj.* **1.** Lying beyond the violet end of the visible spectrum: the epithet of electromagnetic radiation (and of the part of the spectrum containing it) which has a wavelength shorter than that of violet light (about 420 nm.) and (in modern use) no shorter than that of the longest X-rays (of the order of 4 to 40 nm.).

1840 [see *lavender *a.* 2]. 1870 J. Tyndall *Nine Lect. Light* 36 As regards the ultra-violet rays; when they are permitted to fall upon certain substances..they render the substance luminous. 1875, 1887 [in Dict. s.v. Ultra- 1 c]. 1904 *Sci. Siftings* 12 Mar. 320/2 These ultra-violet rays..are most effective in the treatment of such diseases as tuberculosis of the skin, *i.e.*, lupus. 1928 Galsworthy *Swan Song* I. xi. 84 They talk about these ultra-violet rays. Plain sunshine used to be good enough. 1935 *Discovery* Aug. 225/1 Those physical agents which have been found of service in the treatment of various diseases—such agents as ultra-violet and infra-red radiation. 1947 *Sci. News* IV. 43 Ultra-violet light is a strong bleacher. 1955 *Times* 2 July 7/7 Ultra-violet radiation is useful, both to the naked eye and by photography, to determine whether varnish has perished. 1979 T. B. Akrill et al. *Physics* xviii. 248/1 The mercury vapour spectrum also includes a considerable amount of ultra-violet radiation.

2. a. Involving, producing, or pertaining to ultraviolet radiation or its use; *ultraviolet catastrophe*, an indefinite increase that the Rayleigh–Jeans law predicts should occur in the radiation emitted by a black body at successively shorter wavelengths (where the law in fact becomes invalid).

1922 [see *lavender *a.* 2]. 1934 *Discovery* May 138/1 In the detection of forgery, the deciphering of illegible or faded documents, over-writing and the like..ultraviolet photography is the most useful process to employ. 1935 J. Dougall tr. *Born's Atomic Physics* vii. 189 We have here what is called the 'ultra-violet catastrophe'. 1958 *New Biol.* XXVII. 12 The conclusion reached on a chemical basis regarding the presence of lignin has been confirmed using the ultra-violet microscope for spectro-micrographic purposes. 1966 M. Jammer *Conceptual Devel. Quantum Mech.* i. 17 A situation which was later, following Ehrenfest, referred to as the 'ultraviolet catastrophe'. [*Note*] P. Ehrenfest,..*Annalen der Physik* 36, 91–118 (1911). The fourth chapter of this paper is entitled 'Die Vermeidung der Rayleigh-Jeans-Katastrophe im Ultravioletten', where the term 'ultraviolet catastrophe' appeared for the first time. 1974 *Daily Tel.* 17 Jan. 17/6 An ultra-violet photograph taken from an Aerobee rocket..has revealed the development of a huge ball of hydrogen. 1976 *Progress in Sci. Culture* (E. Majorana Centre) Spring 20 The paradox is the ultra-violet catastrophe of the theory of black-body radiation, which was predicted on the basis of classical statistical mechanics and Maxwell's theory of radiation.

b. Sensitive to ultraviolet radiation.

1940 *Chambers's Techn. Dict.* 875/2 Ultra-violet cell. 1969 *Rev. Sci. Instruments* XL. 311 (*heading*) A secondary standard vacuum ultraviolet detector. 1975 J. Taylor *Superminds* iii. 52 At this point the primitive ultra-violet detector was used.

B. *ellipt.* as *sb.* The ultraviolet part of the spectrum; *near, far ultraviolet*, the part close to, or far from, the visible spectrum.

1887 [in Dict. s.v. Ultra- 1 c]. 1931 *Discovery* Mar. 86/2 The possibility is discussed of extending the study of the spectrum of stars into the regions of the ultra-violet. 1954 *Jrnl. Amer. Chem. Soc.* LXXVI. 3847/1 Samples of barium and calcium titanates were sealed in quartz tubes containing phosphorus pentoxide for 72 hours and then exposed to ultraviolet. 1962 L. S. Sasieni *Optical Dispensing* xiii. 326 In snow glare protection is required against the ultra-violet. 1974 *Nature* 4 Jan. 44/1 At the molecular level, the effects of far-ultraviolet (< 300nm) and near-ultraviolet (300 to 420nm) on biological systems are quite different.

‖ **ulu**¹ (ū·lū). Also **oo(d)loo.** [Eskimo.] An Eskimo woman's knife having a crescent-shaped blade.

1864 C. F. Hall *Life with Esquimaux* I. 291 One of the Innuit women slit them down with her oodloo till they did fit. 1940 *Beaver* (Winnipeg) Mar. 24 The women then take their semicircular knives or 'ooloos', and flench the hide from the blubber. 1958 W. Willetts *Chinese Art* I. ii. 77 They [*sc.* knives] are especially common in the North-western Eskimo area, where they occur as the 'woman's knife' or *ulu*, mounted in wood or ivory handles. 1964 *Nat. Geogr. Mag.* May 718/2 Some Eskimo women want their favourite ulu to be buried with them. 1972 S. Burnford *One Woman's Arctic* (1973) ii. 35 Tabitha.. hacked off some bannock with her round, razor-sharp ooloo.

‖ **ulu**² (ū·lū). Also **Ulu.** [Malay (*h*)*ulu* head, upper part.] The upstream, interior part of Malaya; the Malayan jungle.

1878 F. McNair *Perak & Malays* xxxv. 424 Future surveys will confirm this description of the Ulu—interior, or up-stream. 1936 R. H. Bruce Lockhart *Return to Malaya* II. iii. 108 In my time every healthy young man preferred the 'ulu' to the town, not merely because he found the life more attractive in itself, but also because it offered greater chances for promotion. 1963 J. Kirkup *Tropic Temper* 106 In the ulu or deep jungle country, no Chinese would ever dare to look at a Malay girl. 1977 *Borneo Bull.* 7 May 4/3 The ulu people may now be less willing to flock into a town which can cope only primitively with big numbers of visitors.

ululance (ɒ·liulăns). *rare*⁻¹. [f. as Ululant *a.*: see -ance.] Ululation.

1951 W. Faulkner *Requiem for Nun* III. 244 At last, the last of silence too: the county's hollow inverted air one resonant boom and ululance of radio.

ululu. Add: Also **ulalu, ul-ul-loo, ulululu.**

1834 *Knickerbocker* IV. 15 Well may they raise the ul-ul-loo. 1889 W. B. Yeats *Wanderings of Oisin* 80, I must away by wood and sea And lift an ululu forlorn And fling my laughter to the sun. *a* 1955 W. Stevens *Opus Posthumous* (1957) 26 'Olu' the eunuchs cried. 'Ululalu.'

‖ **ulus** (ū·lŭs). Also †**Oolooss.** [Turk.] In Afghanistan, a tribe.

1815 M. Elphinstone *Caubul* II. ii. 159 The name of Oolooss is applied either to a whole tribe, or to one of these independent branches. The word seems to mean a clanish commonwealth. 1902 *East India (N.-W. Frontier) Mahsud-Waziri Operations* 125 in *Parl. Papers* (Cd. 1177) LXXI. 649 In return for the allowances, both Maliks and Levies are expected to serve Government, to control the body of the clansmen, the 'ulus', and to arrange that individual offenders are surrendered for trial. 1953 O. Caroe *Soviet Empire* iii. 35 The Mongol *ulus* or hordes.

ulvöspinel (ɒ:lvospine·l). *Min.* Also **ulvo-.** [ad. Sw. *ulvöspinell* (F. Mogensen 1943, in *Blad för Bergshandt. Vänner* XXVI. 135), f. the name of the *Ulvö* islands, Sweden: see Spinel.] A mineral of the spinel group, Fe_2TiO_4, frequently found as lamellæ in magnetite.

1947 *Mineral. Abstr.* X. 6 X-ray examination shows two spinel phases with cell edges 8·40 and 8·47Å., close to the values for magnetite (8·37) and ferrous orthotitanate Fe_2TiO_4 or titanspinel (8·50). It is assumed that the ore contains a large amount (100,000 tons) of the latter which is named Ulvöspinel. 1963 D. W. & E. E. Humphries tr. *Termier's Erosion & Sedimentation* i. 6 The Curie point is 575°C. for pure magnetite and 675°C. for hematite; it alters in the solid solutions of magnetite (Fe_3O_4) and ulvo-spinel. 1970 *Sci. Jrnl.* May 32/3 If the lava lake was sufficiently enriched in titanium dioxide, its base would tend to collect concentrations of ilmenite, ulvospinel and pseudobrookite. 1971 *Nature* 3 Dec. 264/1 The rock contains relatively small amounts of the phases that we have found in other Apollo basalts (nickeliferous iron, troilite ulvöspinel, ilmenite, [etc.]).

ulys (yū·lis). *nonce-wd.* [Coined by W. de la Mare.] An imaginary mountain flower.

1912 W. de la Mare *Listeners* 58 The icy hills far off from me With frosty ulys overgrown.

Ulysses (yūli·sīz, yū·lisīz). [See Ulyssean *a.*] Used as the type of a traveller or adventurer; occas. also, of a crafty and clever schemer.

1611 Coryat *Crudities* 160 Famous Sir Iohn Mandeuil our English Vlysses. 1876 Geo. Eliot *Dan. Der.* II. iii. xxii. 68 Klesmer was as versatile and fascinating as a young Ulysses. 1915 J. Buchan *Salute to Adventurers* iv. 68, I had been dreaming of foreign parts..and here on a Glasgow stairhead I had found Ulysses himself. 1959 T. H. White *Godstone & Blackymor* 172 That Ulysses, that circumnavigator, that..cheated husband. 1981 P. Vansittart *Death of Robin Hood* II. vii. 84 Richard [Lionheart]..was invisibly present, a resentful Ulysses marking his victims.

um, *int.* Add: **1.** (Later examples.)

1933 P. MacDonald *Mystery of Dead Police* viii. 63 'I don't think,' she said, 'that you two know each other.. Mr. Revel—Captain Um-ha'. 1974 C. Hampton *Savages* v. 39, I publish, I mean I have had published, a few what we used to call slim volumes of verse, um, poetry, you know.

3. Used to indicate assent.

1913 J. Vaizey *College Girl* vii. 98 'Shall I tell you?'.. 'Um!' 'Very well, then.' 1964 L. Deighton *Funeral in Berlin* xlii. 271 'You knew I was working for the Israeli Intelligence.' 'Is that who you work for?' I said.. 'Um,' she said. 1974 N. Freeling *Dressing of Diamond* 49 'Um,' said Vera vaguely... The 'um' was only a symbol.

Hence as *v. intr.*, to make an utterance indicative of hesitation; also as *sb.*

1913 Kipling *Songs from Books* 74 Sometimes in a smoking-room, through clouds of 'Ers' and 'Ums'. 1962 P. Purser *Peregrination 22* vii. 35, I ummed in agreement. 1980 J. O'Faolain *No Country for Young Men* iii. 67, I drink to doubt..and stutters and ums and ahs. Beware of the smooth-speaker.

Uma, var. *huma.

umangite (yū·mäŋgəit). *Min.* [ad. G. *umangit* (F. Klockmann 1891, in *Zeitschr. f. Krist. und Min.* XIX. 265), f. the name of Sierra de *Umango* in Argentina: see -ite¹.] A copper selenide, Cu_3Se_2, found as dark red, violet, or black tetragonal crystals.

1891 *Jrnl. Chem. Soc.* LX. 1435 The author has found three rare minerals of great interest... 2. Umangite.—Associated with eukairite, there occurs a mineral..which proved on analysis to be a variety of copper selenide... The new mineral occurs in finely granular masses. 1978 *Mineral. Abstr.* XXIX. 198/2 Umangite (Cu_3Se_2), berzelianite ($Cu_{2-x}Se$), and athabascaite (Cu_5Se_4) are found in fracture fillings in a carbonatized mafic intrusive rock from Christopher Island, Baker Lake, North West Territories.

Umayyad (uma·yæd), *sb.* and *a.* Also α **Umeiyad**; β **Om(m)ay(y)ad, Omeyyad, Ommiyan,** etc. [f. the name of *Umayya*, ancestor of Muhammad.] **A.** *sb.* A member of a Muslim dynasty which ruled the Empire of the Caliphate from 660 (or 661) to 750 and founded an emirate in Spain in 756. **B.** *adj.* Of or pertaining to this dynasty.

1758 tr. *de Marigny's Hist. Arabians* II. 331 The descendants of Abbas had always refused to acknowledge the Ommiyans as lawful Caliphs. *Ibid.* III. 5 Abdollah.. vented his rage on the tombs of the Ommiyan Caliphs. 1788, 1872 [see *Abbasid, -ide *a.* and *sb.*]. 1907 D. S. Margoliouth tr. *Zaydân's Umayyads & 'Abbâsids* ii. 63 The pivot on which the policy of the Umayyads turned.. was the recovery of the sovereignty which they had enjoyed in pagan days. *Ibid.*, The Umayyad desire for exclusive sovereignty..led them to commit many acts which blacken their memories. 1924 W. Muir *Caliphate* (ed. 2) lx. 432 In passing from the Umeiyad to the 'Abbâsid Caliphate, we reach..a fresh departure... The first new feature is, that while the Umeiyad Caliphate.. was co-ordinate with the limits of Islâm, this is no longer true of the 'Abbâsid. *Ibid.* 433 Of the Umeiyads, the Syrians remained the last support. 1927 D. Hogarth in *Lett. Gertrude Bell* I. xiii. 353 Her journey was a pioneer venture which..cast much new light on the history of the Syrian desert frontiers under Roman, Palmyrene, and Ummayad domination. 1950 B. Lewis *Arabs in Hist.* iv. 68 Umayyad society was based on the domination of the Arabs. *Ibid.* 78 The last of the Umayyads, Marwân II.., was a clever and capable ruler, but he had come too late to save the dynasty. 1975 F. Heer *Charlemagne & his World* viii. 106 In 750 the Ommayyad dynasty was overthrown by the Abbassids, descendants of Mohammed's uncle... One Ommayyad..escaped the massacre of his dynasty in Baghdad..and made his way to Spain, where he became emir. 1981 *Economist* 24 Jan. 43/1 It is taken for granted that present day Damascus and Baghdad automatically oppose each other because of the hostility between the Ummayad and Abbasid dynasties 1,200 years ago.

umbershoot (ɒ·mbəɪʃūt), *nonce-wd.* [Perh. fanciful formation f. Umbrella and Shoot *sb.* Cf. Umber *sb.*¹] (A word of obscure meaning.)

1922 Joyce *Ulysses* 189 Crosslegged under an umbrel umbershoot he thrones an Aztec logos.

umbilical, *a.* and *sb.* Add: Also with pronunc. (ɒmbiləi·kăl). **A.** *adj.* **2.** *umbilical cord.* **c.** *transf.* (*a*) *Astronaut.* A cable or other linking device supplying essential liquid or electrical services; *spec.* the connection between a guided missile and its launching equipment, or that joining a space-walking astronaut to his craft. Similarly *umbilical connection, pipe, tower,* etc.

1948 *Gloss. Guided Missile Terms* (U.S. Research & Development Board) 69 *Umbilical cord*, a cable fitted with a quick disconnect plug at the missile end, through which missile equipment is controlled and tested while missile is still attached to launching equipment or parent plane. 1958 *Times* 1 Mar. 6/3 The 'umbilical pipes' through which liquid oxygen was being pumped to top up its [*sc.* the missile's] fuel tanks. 1959 *Manch. Guardian* 3 Jan. 5 The 'umbilical cord' is a widely detachable cable through which the missile is powered and controlled while still on its 'ivory tower'. 1962 J. Glenn in *Into Orbit* 188 A special countdown started for dropping the umbilical cord which had been providing external power and cooling for the capsule until now. 1963 *Times* 31 May 19/4 The Apollo spacecraft and the three-stage Saturn MK.5 launch vehicle will be erected in the vertical position, together with the umbilical tower on a fabricated base. 1966 *Daily Tel.* 12 Oct. 21/5 The value of the umbilical tether employed on all space walks so far is being questioned. 1967 *Economist* 23 Dec. 1237/2 (*caption*) That grip of gold: umbilical cord linking astronaut to his space ship. 1970 N. Armstrong et al. *First on Moon* vii. 147 We can see the LM umbilical connection quite well.

(*b*) A cable or pipe providing a deep-sea diver with essential electrical and similar supplies. Similarly *umbilical cable, link, pipe,* etc.

1968 *New Scientist* 17 Oct. 127/2 Helium-distorted speech, picked up through the microphone, is transmitted by shielded cables in the umbilical pipe. 1969 *Ibid.* 2 Oct. 11/2 Life support requirements have been provided by way of an umbilical link to a surface station. 1970 R. Barton *Oceanology Today* v. 123 It submerges and then travels along the seabed on four large hydraulically driven wheels... Power is provided through an umbilical cable. 1975 *Offshore* Sept. 115/2 Moreover, the Globule has..an emergency life support of 96 hours and batteries which can be recharged when in operation through the umbilical cable and the induction cable. 1979 *Daily Tel.* 11 Aug. 2/5 The divers asked for their chamber to be lowered to the sea-bed by means of their life-support umbilical cord. 1981 *Times* 23 May 3/3 He found the umbilical lifeline to the [diving] bell..in tatters.

(*c*) In other misc. uses.

1962 *New Scientist* 9 Aug. 285/1 Steering was done from another car travelling behind and a little to one side of the crash car by an attachment which is referred to by the research workers as an umbilical cord. 1968 *Listener* 15 Aug. 200/1 Tiny portable cameras carried by stalwart chaps loaded with power packs and aerials and umbilical cords like a spaceman operating outside his spaceship. 1977 *Rolling Stone* 30 June 117/3 Their giant electrostatic Model One..was the first electrostatic speaker system I know of that got rid of that cumbersome 'umbilical cord' (the line cord you had to connect to a wall outlet to power its high-voltage polarizing supply).

B. *sb.* **2.** *transf.* and *fig.* (also in sing.).

1936 W. Faulkner *Absalom, Absalom!* 259 That River which runs not only through the physical land of which it is the geologic umbilical. 1960 *Times* 18 Oct. 13/6 The umbilicals are..expendable power lines which nurture the bird (missile) while it is on the ground and vanish when it takes flight. 1966 J. A. Chamier *Cannonball* iv. 34 He picked up a microphone with its spring-spiral umbilical from its recess. 1974 *Petroleum Rev.* XXVIII. 674/3 Underwater vehicles may be..tethered (powered by an umbilical from a surface vehicle) or untethered (free swimming). 1977 *Times* 17 Oct. 14/7 On the humid roof sat batteries of television crews... Wires and umbilicals led..to a generator outside. 1982 D. Hart-Davis *Level Five* i. 9 Newman held the end of the hot-water umbilical... The steady flow began, cold at first, then warm, then hot... He..felt the warm jets course through the [diving] suit.

umbilically, *adv.* Add: Also *transf.* and *fig.*

1936 D. Gascoyne *Man's Life is this Meat* 19 Umbilically detached, of sorrowful mien and at the same time tricked out in cobwebs,—these vanquished ones, whose breathings propagate violence and fear. 1951 M. McLuhan *Mech. Bride* (1967) 51/1 The self-consciousness and uneasiness of those still umbilically attached to such guides. 1963 *Adv. Space Sci. & Technol.* V. 241 Mass transfer (gas, liquid, solid cargo, and personnel) can be done umbilically or by docking. 1979 G. Macdonald *Camera* xiii. 187/2 After being linked umbilically to the visual arts for half of the last century, the cultural influence of the contemporary photograph is now slight.

umbiliciform, *a.* (Earlier example.)

1867 [see *after-shaft].

umbracious, *a.* For *rare*⁻¹ read *rare*, and add later example.

1983 P. Levi *Flutes of Autumn* i. 12 That scruffy umbracious margin where the town was just beginning to dissolve the countryside in its dark acid.

umbrell. Add: Also **umberell.** (Earlier and later U.S. examples.)

1816 U. Brown *Jrnl.* 18 Aug. in *Maryland Hist. Mag.* (1916) XI. 151 Never was as wet in my Clothing, through Great Coat umbrell & all. 1910 G. B. McCutcheon *Rose in Ring* I. i. 5 The drizzling rain..blew softly into the faces of the few who enjoyed the luxury of 'umberells'.

umbrella. Add: **5. c.** A screen of fighter aircraft or a curtain of fire put up as protection against enemy aircraft.

1941 [see *air umbrella* s.v. *air *sb.*¹ III. 2]. 1942 *Hutchinson's Pict. Hist. War* 18 Mar.–9 June 102 The task of a fleet working in confined waters is a most difficult one. It is eased if long-range fighters or aircraft-carriers..are available to provide a protective umbrella. 1945 *Sun* (Baltimore) 17 Mar. 2-0/5 The giant bombers of the United States 15th Air Force, continuing their methodical pounding of Germany's vital fuel sources, were escorted by an umbrella of fast American fighters. 1946 *Ibid.* 26 June 8/3 Gun crews pumped deadly umbrellas of anti-aircraft fire above the harbor. 1967 *Electronics* 6 Mar. 73 (Advt.), The Army's new Missile Mentors.. now provide major U.S. cities with air defense umbrellas.

6. d. *U.S. Mil. slang.* A parachute.

1933 C. K. Stewart *Speech Amer. Airman* 99 *Umbrella*, parachute. 1980 J. Ditton *Copley's Hunch* II. i. 117 It takes ages to come down on an umbrella... Then you have to get rid of the chute.

8*. Authority, protection, means of defence; controlling or unifying agency. Freq. in phr. *under the umbrella* and var.

1948 *Hansard Commons* 22 Jan. 388 Giving the smaller Powers a chance to evolve, under the umbrella of the Four Powers. 1949 *Ibid.* 24 May 1213, I am not taking away from the occupants of these flats the umbrella which the law intends to give them. 1952 *Sun* (Baltimore) 15 Feb. (B. ed.) 2/6 The big broad budgetary umbrella under which all manner of wasteful sin is committed. 1958 *Economist* 25 Oct. 297/2 The political division of Germany may become once more a looming *casus belli*, to be contested by the conventional forces that can regain their freedom of action under the atomic umbrella. 1962 H. E. Beecheno *Introd. Business Stud.* xi. 93 What they have done is to bring all the operations, or most of them, under the umbrella of one firm and cut out various 'middlemen' as separate concerns. 1965 *Listener* 1 July 7/2 Europe seems unenthusiastic to exchange the American nuclear umbrella for a French one. 1973 E. Bullins *Theme is Blackness* 7 Theater workers and institutions that presently work from under the Black umbrella. 1976 *Howard Jrnl.* XV. i. 55 Many of the former approved schools continue in their former practices, albeit under a new umbrella. 1983 *Listener* 12 May 5/3 The harsh truth is that Sweden is not under the NATO umbrella.

9. a. *umbrella-stand* (earlier example); (sense *5 c) *umbrella barrage*; (sense *8*) *umbrella basis, policy.*

1944 T. H. Wisdom *Triumph over Tunisia* 134 There was..an intense umbrella barrage over the two Tunisian ports. 1961 *Wall St. Jrnl.* 30 Nov. 19/3 Companies with assembly plants, warehouses and other properties in 15 countries, for example, are realizing it is to their advantage to write insurance on an 'umbrella' basis. 1963 *Daily Tel.* 23 Sept. 21/6 Miss Hawkes said that CND's 'umbrella policy' of accepting any group simply because it had pacifist aims, was not acceptable to her. 1837 Dickens *Pickw.* xxxiv. 378 A mahogany umbrella stand.

b. *umbrella fern.*

1882 T. H. Potts *Out in Open* 53 There Cunningham's *Gleichenia* grows marvellously robust, its stiff many-branched fronds rise, tier above tier, in curved fan-like form—which habit, doubtless, induced settlers to call this species the 'umbrella fern'. 1959 A. McLintock *Descr. Atlas N.Z.* 30 Where the forest has been cleared and burnt, the resulting cover is all too often low scrub, rushes, and umbrella fern.

10. **umbrella bridge**, a temporary raised traffic lane with ramp approaches, in use while building work is conducted below; **umbrella defence,** in Amer. football, an alignment of the backs resembling the shape of an open umbrella; **umbrella field** *Cricket*, an arrangement of close fieldsmen (esp. in the slips) spread in a cordon about the batsman; **umbrella organization,** an organization which represents and protects separate member bodies; **umbrella type,** used (freq., with hyphen, *attrib.*) to denote any structure which resembles (part of) an umbrella in shape.

1962 *Daily Tel.* 14 Sept. 15/7 Supports will be sunk to take a steel 'umbrella' road bridge while the Oxford Circus underground station is rebuilt. 1973 *Times* 24 Mar. 2/8 A temporary 'umbrella' bridge is to be erected in the Charing Cross railway station forecourt. 1950 *Sun* (Baltimore) 23 Oct. 17/3 An umbrella backfield defense.. had no special name. It was..a formation to provide width and depth for pass defence. It has the general shape of an umbrella. 1972 J. Mosedale *Football* x. 145 Owen installed what was called 'the umbrella defense', so-called because the alignment of the defensive backfield resembled an open umbrella. 1954 Miller & Whitington *Gods or Flannelled Fools?* vi. 225 Hassett had been loath to set the..'Umbrella' field..with eight men stationed in an inner arc behind the batsman from backward point to square-leg. 1963 *Courier-Mail* (Brisbane) 21 Nov. 17/1 The conversation..consisted of snippets like..'in an umbrella field'. 1950 *Times* 8 May 2/7 The domestic poultry-keepers could also win independence, and it is doubtful whether either group needs an 'umbrella' organization set up at the Ministry of Agriculture. 1983 *Out of Town* Dec. 17/1 They [*sc.* naturalist rectors] showed an equally native disposition to flourish best as part of an amiable, protective institution. Today the BBC has replaced the Church as the umbrella organisation. 1913 *Wireless World* June 210/1 M.V...asks..whether the 'umbrella type' of aerial would be most suitable, or whether the use of another mast, 10 feet high,..would improve matters. 1940 *Chambers's Techn. Dict.* 876/1 *Umbrella-type alternator*.., a vertical-shaft alternator,..in which the field system is overhung and revolves around the stationary armature. 1962 *Daily Tel.* 4 Dec. 15/3 An open market..has been replaced by a modern covered market. It has an attractive umbrella-type roof. 1963 *Guardian* 27 Feb. 5/2 Why on a semisports car should the handbrake be of the 'umbrella' type? 1971 *Jamaican Weekly Gleaner* 17 Nov. 9/1 The promenade..would contain umbrella type shops and stalls to accommodate the present proliferation of peddlers.

11. *attrib.* passing into *adj.* Of words, names, etc.: covering a number of meanings or associated terms; general, catch-all.

1949 G. Ryle *Concept of Mind* vi. 198 The range of higher order acts and attitudes, which are apt to be inadequately covered by the umbrella-title 'self-consciousness'. 1957 *Listener* 8 Aug. 201/1 Cancer is one of the umbrella words. It covers a number of disease conditions. 1974 *Country Life* 5 Dec. 1723/1 The numerous [Ilex] hybrids which are gathered beneath the umbrella name of *I.* × *altaclarensis.* 1977 *Times Lit. Suppl.* 29 Apr. 530/3 Any one of half a dozen umbrella titles would equally well match the variety of the contents of this military miscellany, *War and Society.*

Umbrian. Add: **A.** *sb.* **2.** (Earlier example.)

1854 C. C. J. Bunsen *Christianity & Mankind* III. 91 In Umbrian, the *D* between two vowels passes into a specific *R*, expressed in the national alphabet by a peculiar letter, in Latin by *RS*.

B. *adj.* **2.** (Earlier example.)

1836 *Dublin Rev.* July 443 The continuous efforts of the Umbrian School, even after..the death of Perugino.

Umbro-. Add: *Umbro-Samnite* (earlier example).

1858 *Bibliotheca Sacra* XV. 99 The Latin stands related to all this Umbro-Samnite class of special dialects, as, in Greek, the Ionic to the Doric dialect.

umfaan (u·mfān). *S.Afr.* Also † **oomfaan, umfane.** [Afrikaans, ad. Zulu *umFana* small boy.] A young African boy, esp. one employed in domestic service.

1852 C. Barter *Dorp & Veld* xiv. 213 The Kaffir *umfane* (boy) when he becomes an *indola* (man), shaves his head, and sews into the scalp a circular coronet of reeds. 1878 H. A. Roche *On Trek in Transvaal* iii. 39 Your wife..if she be so lucky as to have floors at all, will make..that provoking '*Oomfan*' clean them for her. 1907 P. FitzPatrick *Jock of Bushveld* 194 Jim had fought at 'Sandhlwana, and could tell of an umfaan sent out to herd some cattle within sight of the British camp to draw the troops out raiding. 1932 F. W. Fitzsimons *Snakes* xiv. 173 No! Snakes do not suck milk from cows; but Kafir umfaans and pigs do. 1964 S. Milne *False Witness* xv. 167 'You have a servant on the premises?' 'Yes, an *umfaan*. He is sixteen years old.' 1977 J. McClure *Sunday Hangman* ii. 13 He'd strung himself up on a thorn tree... Some umfaans made a report.

umfundisi (umfu·ndisi, umfundi·zi) *S.Afr.* Also 9 **fundis, umfundis(e; mfundisi.** [a. Nguni *umFundisi, Mfundisi* teacher.] A teacher, a minister, a missionary; also used as a respectful form of address.

1825 W. Shaw *Diary* in C. Sadler *Never Young Man* (1967) 64 The Caffres knew me to be a 'Fundis' (teacher). 1833 S. Kay *Trav. Caffraria* iii. 73 Never have we been safe; but the *Umfundis* shall be our bush. *Ibid.* xii. 317 But if we neglect her, the *Umfundis* (Missionary) will be angry. 1837 F. Owen *Diary* 12 Oct. (1926) 44 We don't know what we have learned this morning, for the Umfundis (teacher) sent us away so soon. 1863 J. S. Dobie *S. Afr. Jrnl.* (1945) 78 Called at a kraal and got enlightened on road to the umfundisi (missionary). 1905 G. Calloway *Sk. Kafir Life* 9 Oh! Mfundisi; I am weak, I cannot work to-day. 1923 *Other Lands* Oct. 19/1 Their dear old *Umfundise* was the Moderator-elect. 1948 A. Paton *Cry, Beloved Country* I. ii. 14, I bring a letter, umfundisi. *Ibid.* iii. 21 They saw his clerical collar, and moved up to make room for the umfundisi. 1973 *Eastern Province Herald* 6 Aug. 6/7 Job's comforters are..the umfundisi and a cheerful scavenger from the municipal rubbish dump.

‖ **Umgangssprache** (u·mgaŋsʃprāχə). [Ger., = colloquial speech.] The vernacular language between standard and dialect speech customarily used as a means of communication within a linguistic community.

1934 Priebsch & Collinson *German Lang.* viii. 350 The schools..are playing a considerable part in removing the *Umgangssprache*..from local influences by basing the language of instruction more strictly upon the literary language. 1961 R. E. Keller *German Dialects* i. 8 Wherever the ideal of the standard has led to the abandonment of the native dialect a third form arises between the two: the *Umgangssprache*. 1976 *Amer. Speech* 1973 XLVIII. 215 We assume that the Swiss immigrants on their arrival in Wisconsin used Swiss German as their 'Umgangssprache' and had at least some knowledge of standard German.

Umklapp (u·mklæp). *Physics.* Also **umklapp.** [tr. G. *umklappprozess* (R. Peierls 1929, in *Ann. d. Physik* III. 1073), f. *umklappen* to turn down or over.] Used *attrib.* to designate interactions in a crystal lattice in which their total momentum is not conserved, and the momentum of the initial excitations is reversed. Abbrev. *U-* or *u-process.*

[1937 *Physical Rev.* LII. 690/1 These authors neglect transitions due to the '*Umklappprozesse*' of Peierls.] 1951 *Proc. R. Soc.* A. CCVIII. 90 Further measurements made on a corundum crystal confirm the importance of the 'Umklapp' processes, postulated by Peierls, in causing thermal resistance. 1960 J. M. Ziman *Electrons & Phonons* iii. 133 We shall refer to *U*-processes, and, where wave vector is conserved,..*N*-processes. 1974 H. E. Hall *Solid State Physics* viii. 207 An Umklapp process can be thought of as one in which a phonon is Bragg reflected simultaneously with absorbing or emitting another phonon. 1975 H. M. Rosenberg *Solid State* vi. 99 In a continuous medium u-processes cannot occur. 1976 F. Cap tr. *Busch & Schade's Lect. Solid State Physics* 67 At low temperatures the umklapp processes are essential for the establishment of thermal equilibrium in the crystal.

umlaut. Delete ‖ and add: **b.** The diacritical sign (¨) placed over a vowel to indicate that such a change has taken place.

1938 H. Faulk *Common-Sense German Course* 3 The so-called modified vowels are distinguished by the modification mark or umlaut(¨) on the vowel. 1952 M. Pei *Story of Lang.* I. ix. 93 English makes use of no subsidiary characters, save for the apostrophe. Many other languages use accent-marks, umlauts, cedillas. 1970 [see *computer 2 b].

Hence as *v. trans.*, to make such a change; **u·mlauting** *vbl. sb.* and *ppl. a.*

1938 W. F. Twaddell in *Monatschr. f. Deutschen Unterricht* XXX. 177 The *i, ī,* or *j* which 'caused' the umlauting was no longer present in MHG. 1943 E. A. Nida *Handbk. Descriptive Linguistics* II. v. 84 What changes the stems so materially is the umlauting produced by the *e* vowel of *-et*. 1976 *Language* LII. 154 Moulton.. assumes that all instances of [a] were umlauted simultaneously to [ä]. 1977 *Ibid.* LIII. 18 As a result of this second rule, back vowels are created on the surface which, unlike all the other back vowels, fail to umlaut in umlauting environments. 1983 *Word* XXXIV. 120 The color of prothetic vowels, unless and until umlauted by the next syllable, was that of the laryngeal which was vocalized.

Umlimo, var. *Mlimo.

‖ **umma** (u·ma) Also **Umma, ummah.** [Arab. *'umma* people, community, nation.] **1.** The Islamic community, founded by Muhammad at Medina, comprising individuals bound to one another by religious ties on a tribal model.

1885 T. P. Hughes *Dict. Islam* 654/2 *Ummah*,..a people, a nation, a sect. The word occurs about forty times in the Qur'an. *Ummatu Ibrāhīm*, the people of Abraham ..*Ummatu Muhammad*, the people of Muhammad. 1919 H. U. W. Stanton *Teaching of Qur'an* vi. 71 The..term.. *ummah*, i.e. religious community. Of this it is said that mankind were originally one *ummah*, and that Allāh, had He pleased, could have kept them so. 1934 *Encycl. Islam* IV. 1015/2 Muhammad frequently discusses the question why mankind consists of a plurality of ummas and has not remained a unit. 1974 B. Lewis in Schacht & Bosworth *Legacy of Islam* (ed. 2) iv. 157 From the start, the Islamic *umma* had a dual character. On the one hand it was a political society..; on the other it was a religious community, founded by a prophet and ruled by his deputy. 1976 *Jrnl. R. Soc. Arts* CXXIV. 613/1 The flexibility of government in Islam goes back—doesn't it?—to the concept of 'Umma' in Islam, the idea that Islam came actually to build up an Umma, a community, rather than impose a doctrine. 1979 *Economist* 5 May 82/2 The governance of the Moslem community, the *umma*.

2. (With capital initial.) The name of a nationalist political party founded in the Sudan in 1945.

1946 *Economist* 9 Mar. 369/1 The western Sudanese.. speak through the Umma or nationalist party, which wants 'a union in which the two partners enjoy internal and external autonomy'. 1946 *Times* 30 Sept. 4/5 The Umma Party of Sudan. 1958 *Listener* 21 Aug. 256/2 Both these ministers are members of the Umma, the Mahdist party. 1965 K. D. D. Henderson *Sudan Republic* vii. 89 There emerges in March 1945 a new political party calling itself the *Umma*, the Community Party with the slogan of 'the Sudan for the Sudanese'. 1979 M. Deeb *Party Politics in Egypt* ii. 40 The defunct Umma Party. 1981 *Economist* 24 Jan. 43/2 There are parochial or communal parties which do not favour or are fearful of, absorption into larger units: these include the Christian Phalange in Lebanon, the Umma in Sudan, the Neo-Destour in Tunisia.

umohoite (yūmohōᵘ·əit). *Min.* [See quot. 1953.] A hydrous uranyl molybdate, $UO_2MoO_4.4H_2O$, found as monoclinic and orthorhombic, usu. dark-coloured, crystals.

1953 P. F. Kerr in *Rocks & Minerals* XXVIII. 480/1 The name 'umohoite' is given to the mineral by combining the chemical symbols U, Mo, H, and O with the mineral suffix 'ite'. 1980 *Mineral. Abstr.* XXXI. 82/2 Umohoite from Shinkolobwe, Shaba, Zaire.., is regarded as a magnesian variety: dark green to black, 0·65% Mg, orthorhombic... Two other varieties are described.

ump (ɒmp), *slang* (chiefly *U.S.*). Abbrev. of Umpire *sb.*, *spec.* in baseball. Also **umps** (cf. *-s²).

1915 'High Jinks, Jr.' *Choice Slang* v. 74 Every time that Umps starts talking his tongue gets twisted around his eye tooth and he can't see what he's saying. 1942 *Sun* (Baltimore) 30 June 12/4 (*heading*) Durocher fined..for quarrel with umps. 1952 B. Malamud *Natural* 46 I'd've thrown him out too if I was the ump. 1975 *New Yorker* 17 Feb. 25/3 That's why Nick Colosi, National League ump, was a featured attraction at the Auto Show last week. 1979 *Arizona Daily Star* 8 Apr. c1/6 A few bad calls by the rookie umps will no doubt be cause for more outcries from the baseball world.

umpah, var. *oompah. **umph,** var. *oomph.

umpiring, *vbl. sb.* (Earlier example.)

1851 W. Clark in W. Bolland *Cricket Notes* vii. 149 Umpiring is a very arduous and often unthankful office.

umpteen (ɒ·m(p)tīn, ɒm(p)tī·n), *a.* and *sb.* *colloq.* Also **umteen.** [f. *ump(ty + Teen *sb.*² after *thirteen*, etc.] **A.** *adj.* An indefinite number, used in the sense 'many, several', etc. **B.** *sb.* Such a number in the abstract.

1918 *Blackw. Mag.* Mar. 290/1 Men from five continents and umpteen colonies. 1919 *Athenæum* 1 Aug. 695/1 As 'umpty' means 'dash', it is fairly evident that..'umpteen' (or 'umteen'), which means 'any number of times', comes from this source. 1922 *Public Opinion* 11 Jan. 48/2, I entered into it and prepared to drop umteen floors. 1923 [see *umpty *sb.*] 1930 J. B. Priestley *Angel Pavement* ii. 82 I've got umpteen things for him to sign. 1973 K. Giles *File on Death* ii. 39, I leave business to the Estate managers, six of 'em with umpteen clerks and typists. 1976 A. Price *War Game* I. viii. 141 A potential offender against section umpteen of the Road Traffic Act.

Hence **umpteenth** (stress variable) *a.*

1918 E. A. Mackintosh *War, the Liberator* 99 That's the umpteenth Bosche that I've killed today. 1921 *Blackw. Mag.* Apr. 475/1 It was our umteenth breakdown. 1952 M. Laski *Village* v. 89 'I always did say I'd go back if.. she needed me,' she said for the umpteenth time. 1980 P. Harcourt *Tomorrow's Treason* 19 For the umpteenth time I'm telling you. I haven't done anything.

umpty (ɒ·mpti), *sb.* and *a.* [A fanciful verbal repr. of the dash (—) in Morse code. Cf. *iddy-umpty.] **A.** *sb.* An indefinite number, usu. fairly large. (Often used on an analogy with *twenty*, etc.) *Mil. slang.*

1905 *Outing* July 389/2 The undergraduates..whisper to their guests, who stroked the crew in umpty-seven, the year we won by twenty lengths. 1919 W. Lang *Sea Lawyer's Log* 70 Umpteen or 'umpty', it should be explained, is to the Navy what x is to Euclid—the symbol of an unknown or unmentionable quantity. 1923 *Daily Mail* 3 Sept. 1 (Advt.), Umpteen to umpty Fahrenheit, Wolsey keeps you comfy, quite. 1924 Kipling in *Hearst's Internat.* July 16/2 The bettin' was even on my drawin' a V.C. or getting Number Umpty Rest-Camp.

B. *adj.* **1.** An indefinite number of this kind, in adjectival use.

1917 P. MacGill *Gt. Push* xii. 250 When I go back to blighty I'll go to bed and I'll not get up for umpty-eleven

months. 1939 J. D. Carr *Black Spectacles* xx. 289 Again, once more, and for the umpty-umph time, we had been hocussed by still *another* of Chesney's ingenious tricks. 1959 W. Faulkner *Mansion* xv. 354 'I never got to Heidelberg,' Charles said. 'All I had was Harvard and Stalag umpty-nine.' 1974 R. L. Simon *Peking Duck* xi. 84 A drill press of umpty-ump kilotons' capacity.

2. *fig.* Of a person, place, or circumstance: unpleasant.

[1925 Fraser & Gibbons *Soldier & Sailor Words & Phrases* 294 *Umpty iddy, to feel*, so so. Not very well. All upside down.] 1948 M. Allingham *More Work for Undertaker* xiii. 158 Things a bit umpty at home, I rather suspect. 1970 A. Draper *Swansong for Rare Bird* vii. 54, I was worried in case the guvnor was umpty about the night before. The last thing I wanted was a slanging match. 1974 N. Marsh *Black as he's Painted* iii. 76 Very umpty little dump. 1980 C. Fremlin *With no Crying* xix. 117 This rather umpty friend of his.

Hence **u·mptieth** *a.*

1917 'Contact' *Airman's Outings* 216 The umptieth squadron then had the only machines of this type in France. 1984 *Business Rev. Weekly* 25 Feb.–2 Mar. 92/2 Two hours later, after being wiped out for the umptieth time, frustration rears its ugly head.

umquhile, umwhile, *adv.* and *a.* Add: **A.** *adv.* **2.** (Further example.)

1832 F. Trollope *Dom. Manners Amer.* I. vii. 93 A drawing..representing Hebe and the bird, umquhile sacred to Jupiter.

B. *adj.* **a.** (Later examples.)

1934 J. Buchan *Free Fishers* ii. 40 What do you think of your umquhile pupil, Professor? 1976 *Times Lit. Suppl.* 21 May 606/3 The reference to the *ci-devant* Lord Stansgate is notoriously getting shorter and shorter;.. with further reduction en by en the umquhile peer will disappear completely.

‖ **umu** (u·mu). *N.Z.* [Maori.] = *hangi; also, the food prepared in this oven.

Also used in other Polynesian areas in the Pacific.

1845 E. J. Wakefield *Adventure in N.Z.* I. iv. 75 The tangi had terminated; the *umu* or 'cooking holes' were smoking away for the feast. 1889 S. P. Smith in *Trans. N.Z. Inst.* XXII. 98 An oven of stones, exactly like a Maori *umu* or *hangi*. 1950 *Landfall* IV. 85 The passage describes in detail the well-remembered lighting of the *umu* fires. 1972 M. Shadbolt *Strangers & Journeys* xxii. 474 Mother, grandmother, and children scattered around in vague blue smoke for their *umu*, or earth oven. 1974 T. Heyerdahl *Fatu-Hiva* i. 46 No banquet..can better regale guests than such a juicy, fresh..Polynesian *umu* served without cost..in the open air of a tropical night.

umutsha, var. *moocha.

‖ **Umwelt** (u·mvelt). Pl. **Umwelten.** [Ger., = environment.] The outer world, or reality, as it affects the organisms inhabiting it.

1964 M. King *Heidegger's Philos.* II. ii. 96 One suggestion which Heidegger undoubtedly intends to convey with *Umwelt* is of a world that is closest and most familiar to man... We shall paraphrase *Umwelt* by 'the first and nearest world'. 1966 J. S. Bruner *Beyond Information Given* (1974) xviii. 318 Modern ethnological conceptions are centrally concerned with representation in such mechanisms as releasers and imprinting, much of it deriving from the originating idea of the *Umwelt* first proposed by von Uexküll. 1971 E. O. Wilson *Insect Societies* (1972) xi. 209/2 The various species of ants are generally similar to the honeybee in their *Umwelten*. 1977 A. Sheridan tr. *J. Lacan's Écrits* i. 4 To break out of the circle of the *Innenwelt* into the *Umwelt* generates the inexhaustible quadrature of the ego's verifications.

umzimbeet (umzimbī·t). *S.Afr.* Also **umzimbit(i).** [a. Xhosa *umSimbithi* ironwood.] A South African tree, *Millettia grandis* or *M. caffra* (family Leguminosæ), which bears clusters of pink or purple flowers, may be evergreen or deciduous, and has very heavy, hard wood; the wood itself.

[1851 R. Gray *Jrnl. Bishop's Visitation Tour Cape Colony* II. 99 There are several kinds of valuable wood unknown in the colony. Two of the hardest and most useful are called by the natives 'Unizimbeti' and 'Umnebelala'.] 1870 C. Hamilton *Life & Sport S.-E. Afr.* 6 The wheels are made of the famous Natal wood called 'umsimbiti' or ironwood, from its strength and durability. 1902 G. S. Boulger *Wood* II. 335 Umzimbit..Known also as 'White Ironwood'. 1907 T. R. Sim *Forests & Forest Flora Cape of Good Hope* 203 Umzimbeet is a light-demanding tree. 1950 [see *flat-crown* s.v. *flat *a.* 15]. 1955 W. Gaddis *Recognitions* II. i. 293 Bird-of-paradise flowers..among the native white pear..and umzimbiti. 1972 Palmer & Pitman *Trees S. Afr.* II. 923 The umzimbeet is a medium to large tree of the coastal forests. *Ibid.* 925 The wood of the umzimbeet is extremely heavy, hard, and strong.

un-¹. Add: **7. a.** (This and subsequent sections contain examples illustrating the recent formations that are best attested in the O.E.D. files.) *unadult, un-African, unairworthy, unambivalent, unarcadian, unarchæological, un-Australian, unbitchy, unblameworthy, unblasé, unbureaucratic, un-Byronic, uncerebral, uncharismatic, unchic* (also *absol.*), *un-Chinese, unchipper, unchoosy, un-Christmassy, uncomfy, uncomposite, uncomradely, unconscient, uncool* [esp. *cool *a.* 4 e] (also *absol.*), *uncooperative, uncosy, uncranky, uncreditworthy, uncuddlesome, uncuddly, un-Darwinian, undeducible, un-Dickensian, undimensional, undisastrous, undoctrinaire, undynamic, unecological, uneconomic, unecstatic, unecumenical, unegoistic, unegotistic, unegotistical, unerotic, unetymological, unexotic, unfaery* (poet.), *unfeline, unfeminist, unflamboyant, unflashy, unfond, unfresh, unfurtive, un-Gaelic, ungay, ungimmicky, unglamorous, ungroovy, unhep* (also *absol.*), *unhip* (also *absol.*), *unhors(e)y, unideological, unintrospective, unirksome, unironic, unironical, un-Islamic, un-Italian, un-Jamesian, un-Japanese, unjingoistic, unkeen* (also *absol.*), *unkosher, unlegendary, unlocal, unmarital, un-Marxist, unmawkish, unmeritocratic, unneurotic, unodoriferous* (earlier example), *unopen, unpacifist, unphonemic, unphon(e)y* (also *absol.*), *unphotogenic, unplatonic* (earlier examples), *unpolemical, unpolicemanly, unpositive* (earlier example), *unprestigious, unpriggish, unprivate, un-Proustian, unpugnacious, unradiogenic, unresilient, unrevolutionary, unrisky, unroadworthy, un-Russian, unscenic, unseductive, unselective, unsemantic, unsexy, un-Shelleyan, unsnobbish, unsorry, un-Spanish, unspecial, unspectacular, unstarchy, unsterile, unstiff, unstuffy, unsycophantic, unsymmetric, untendentious, untense, untherapeutic, unthistly, unticklish, untogether, untouristy* (also *absol.*), *untraditional, untrendy, unurgent, unutilitarian, unviable, un-Victorian, unviolent, unvisual, un-Western, un-with-it, unworthwhile, unyoung.*

1944 A. L. Rowse *Eng. Spirit* xxxiii. 229 There was something curiously unadult, ungrown-up about him. 1976 R. B. Parker *Promised Land* xxiv. 152 She's not a fool, but she's misled, maybe unadult. 1923 D. H. Lawrence *Birds, Beasts & Flowers* 31 An ultimate desperateness, un-African. 1979 *Guardian* 6 Sept. 11/7 Parliament [in Kenya] has just thrown out as 'unAfrican' a Marriage Bill which would have outlawed wife-beating. 1907 *Cornh. Mag.* May 617 For such a motion the machine would be longitudinally unstable, and, shall we say, 'unairworthy'—to coin an analogue for the word 'unseaworthy', as applied to ships. 1979 *Daily Tel.* 8 Nov. 3/5 A fine..has been imposed on Braniff International Airways for conducting hundreds of flights with aircraft that were allegedly in an 'unairworthy condition'. 1977 *Lancet* 2 July 36/2 We felt that obstetricians and midwives should remain unambivalent in their efforts to advise smoking mothers to give up the tobacco habit during pregnancy. 1981 *London Rev. Bks.* 5–18 Feb. 9/1 No such vision is likely to be utterly unambivalent, regret being as intrinsic to the human condition as is hope. 1962 Auden *Dyer's Hand* (1963) 350 If the landscape of New England is unarcadian, so is its social life. 1927 *Observer* 11 Dec. 16/5 The workmanship and designs of ancient Peruvian pottery..seemed, to my unarchaeological eye, so like the Etruscan as to be almost identical. 1965 J. A. Michener *Source* (1966) 890 Swinging his pick with unarchaeological vigor he felt its point bite through a thin layer of semi-rock and then leap forward into nothingness. 1965 G. McInnes *Road to Gundagai* v. 82 The larder was notable for two very un-Australian..gadgets. 1963 S. Farrar *Death in Wrong Bed* x. 148 You're the unbitchiest woman I know. 1973 *Country Life* 18 Oct. 1210/3 This most readable and unbitchy of biographies. 1966 S. Smith *Frog Prince* 78 Touch, where the feeling is most vulnerable, Unblameworthy. 1860 Queen Victoria *Let.* 6 June in R. Fulford *Dearest Child* (1964) 258, I like them extremely, so nice, natural, sensible, quiet and so unblasé. 1977 C. Wood *James Bond, Spy who loved Me* vii. 58 Only the most skilful and unblasé eye would be able to detect an unfamiliar outline. 1970 *Daily Tel.* (Colour Suppl.) 16 Oct. 68/1 The first stages..were remarkably easy, unbureaucratic, and free from the famous red tape. 1936 F. R. Leavis *Revaluation* vii. 270 The aesthete who achieved so un-Byronic and so un-Shelleyan a note in the contemplation of human suffering. 1959 *Times Lit. Suppl.* 23 Jan. 45/4 Even his forged confessions have an unbyronic slime about them. 1934 F. Scott Fitzgerald *Tender is Night* II. i. 153 His criterion of uncerebral phrase-making was that it was American. 1971 P. Ziegler *King William IV* ii. 29 The Hanoverians had brought their own style of home-spun and singularly uncharismatic monarchy to England. 1960 *Guardian* 6 May 10/6 Anything thicker is totally unchic. 1975 *51 Newsmagazine* 12 Sept. 2/2 You'll read about New York's politics, events, labor, the chic and unchic. 1934 Webster, Un-Chinese. 1974 Dawa Norbu *Red Star over Tibet* xi. 176 Norzin-la occasionally uttered the words '*San fan, shuang jian*' in her un-Chinese accent before the uneducated masses. 1969 J. Fowles *French Lieutenant's Woman* lvii. 403 The evening that saw him so unchipper in his place of refreshment. 1948 M. Allingham *More Work for Undertaker* xiv. 177 The Fuller gang..made quite a name for themselves as being remarkably unchoosy. 1927 D. H. Lawrence *Let.* 12 Dec. (1962) II. 1026 The post is so tiresome here, and altogether one feels so unchristmassy. 1982 *Daily Tel.* 24 Dec. 1/6 The Leader of the Opposition ..was in a most unChristmassy temper. 1888 Kipling *Wee Willie Winkie* (1889) 51 I'm *so* uncomfy! Come and tuck me up. 1925 A. Huxley *Those Barren Leaves* III. vi. 222 Such an uncomfy house! 1981 'A. Hall' *Pekin Target* xvii. 181 Put it under my head..no need to be uncomfy. 1920 W. B. Yeats *Michael Robartes & Dancer* 3 It follows from this Latin text..that all beautiful women may Live in uncomposite blessedness. 1968 *Listener* 23 May 654/3 The foreign editor charges his chief with suppressing news vital to Party comrades: his chief has rebuked him to the extent of three columns for 'the anarchy of emotions and passions that give rise to uncomradely and irresponsible actions'. 1929 R. Bridges *Testament of Beauty* IV. 184 Like as in unconscient things whence conscience came, ther is also thru'out conscient life. 1953 W. Burroughs *Junkie* (1972) vi. 61 'It's better to meet alone like this.' His smile was ambiguously sexual. 'Nick is a very un-cool guy.' *Ibid.* xiv. 145, I learned the new hipster vocabulary..'cool', an all-purpose word indicating anything you like or any situation that is not hot with the law. Conversely, anything you don't like is 'uncool'. 1958 G. Lea *Somewhere there's Music* xx. 175 Like, buy my forthcoming book on what's uncool in American education. 1960 [see *daddy-o]. 1961 R. Russell *Sound* v. 101, I dunno, old man, to the average colored person the average gray acts like he's in a sweat most of the time. Hungup. Uncool. 1966 *Punch* 30 Nov. 824 The uncool in the audience clapped heartily. The hip young watched in stony contempt. 1968 *It* 1–14 Nov. 16/1 The whole place [*sc.* Turkey]..is very very uncool. The Turks seem to be ready to turn with a malicious vengeance on young Europeans for the least (often no) provocation. 1979 *Guardian* 5 July 9/3 Those men who keep their cool are dragged, willy nilly, into violence not of their making and are then tarred with the same brush as their uncool brethren. 1934 Webster, Uncooperative. 1944 J. S. Huxley *On Living in Revolution* xi. 114 The neighbours had at first been wholly unco-operative. 1970 *Morning Star* 11 May 4/4 The kindergarten teacher told me that she deals in the same way with unco-operative children. 1893 M. Beerbohm *Let.* 3 Dec. (1964) 82, I thought the streets would be rather un-cosy. 1976 *Times Lit. Suppl.* 26 Mar. 344/2 Paris, where they knew nobody, and where their *collègues de carrière* were, at the outset at least, singularly uncosy. 1935 E. Bowen *House in Paris* II. vii. 159 Mrs. Michaelis joined two more un-cranky committees. 1941 *Scrutiny* IX. 376 Yet, if it is his weakness to seem unaware of the amount of uncreditworthy coinage he puts into circulation, at least he knows the problems which must beset anyone who writes poetry to-day. 1960 *Times* 29 Feb. 14/7 Cases which the banks would consider uncredit-worthy. 1946 M. Dickens *Happy Prisoner* vii. 122 Elizabeth..began to give Heather more help with the children, which she did on modern hospital lines, uncuddlesome, but extremely efficient. 1963 *Punch* 13 Nov. 693/1 Huge animals all around; gross, uncuddly. 1899 A. H. Japp *Cuckoo* III. 162, I regard this phrase as in itself very unhappy—and, in fact, un-Darwinian. 1955 H. Hodgkinson *Doubletalk* 43 'The modern theory of mutations' is regarded as un-Darwinian. 1977 D. Cory *Bennett* iv. 121 If he..had evolved from a mere *something*..to a sentient personality, then Bennett in a most un-Darwinian way gave an opposite impression. 1854 Geo. Eliot tr. *Feuerbach's Essence Christianity* viii. 84 The *differentia specifica*..is always in the ordinary sense inexplicable, undeducible. 1952 *Mind* LXI. 267 What is unexpectable—or unpredictable or undeducible. 1948 F. R. Leavis *Great Tradition* iii. 133 The un-Dickensian subtlety—the penetrating analysis. 1982 *Times* 30 June 8/5 Damn un-Dickensian sentimentality. 1940 W. Faulkner *Hamlet* III. i. 173 She owns no dimension against the lambent and undimensional grass. 1931 V. Sackville-West *All Passion Spent* III. 218 He managed to be an undisastrous Prime Minister of England during five..difficult years. 1963 *Economist* 28 Dec. 1317/2 Moving..towards undisastrous answers. 1962 *Times* 4 Apr. 6/5 Her sensibility inclines her to an undoctrinaire approach. 1976 *Daily Tel.* 2 Dec. 18 He was sensible, undoctrinaire, moderate—a consensus man. 1960 C. Day Lewis *Burial Day* vii. 140 E., despairing..of so undynamic a lover..married a don. *a* 1974 R. Crossman *Diaries* (1975) I. 33 Their proposals..struck me as extremely undynamic and dull. 1976 N. Postman *Crazy Talk* 7 What is wrong with Orwell's advice is that it is unecological. It places language outside of any context in which it is used. 1909 Webster, Uneconomic. 1953 *Maori Affairs Act* (N.Z.) 71 For the purposes of this Part of the Act, the expression 'uneconomic interest' means a beneficial freehold interest the value of which..does not exceed the sum of twenty-five pounds. 1971 'G. Black' *Time for Pirates* ii. 40 Vicious, totally uneconomic price cutting. 1858 Queen Victoria *Let.* 28 July in R. Fulford *Dearest Child* (1964) 125 With your..poetical, romantic mind how can you be so unecstatic? 1953 *Essays in Criticism* III. 95 A universe..forever beyond.. tomorrow's unecstatic nights at home. 1970 *Irish Jurist* V. 97, I have given these individuals the benefit of the doubt (if the phrase is not unecumenical). 1934 Webster, Unegoistic. 1942 E. Bowen *Bowen's Court* x. 270 Henry was unegoistic. 1939 D. Cecil *Young Melbourne* i. 25 She cared for few people; but these she loved with a strong, unegotistic affection. 1977 *Gramophone* Aug. 309/3, I should stress that this is a completely unegotistic performance, devoid of any mannerisms or interpretative point-making. 1932 E. Wharton in *Scribner's Mag.* Feb. 113/1 Like many perfectly unegotistical women Catherine Glenn had no subject of conversation except her own affairs. 1930 V. Sackville-West *Edwardians* vi. 275 Morning is bleak and unerotic. 1962 Auden *Dyer's Hand* (1963) 374 The personal relation was completely unerotic. 1876 *Fortn. Rev.* 1 Apr. 568 What is now called the etymological or historical spelling of words, is, in many cases, utterly unetymological and unhistorical. 1960 P. H. Reaney *Orig. Eng. Place-Names* i. 13 A similar unetymological *-ham-* appears in the early forms of Alphamstone, *Alfelmestune* 1086, *Alfhampston* 1318, 'Ælfhelm's farm'. 1934 Webster, Unexotic. 1951 'J. Wyndham' *Day of Triffids* vi. 116 Her pleasant though unexotic countenance. 1983 *Out of Town* Dec. 16/3 For a haunt of Jungle Jims the BBC at Bristol is distinctly unexotic. 1885 W. B. Yeats in *Dublin Univ. Rev.* May 83/1 Peace, peace, the earth's a-quake. I hear Some barbarous, un-faery thing draw near. 1869 'Mark Twain' *Innocents Abroad* ix. 87 Their unfeline conduct in eating up all the Tetouan cats aroused a hatred toward them in the breasts of the Moors. 1930 *Times Lit. Suppl.* 4 Sept. 698/3 Rachel, sweet and serene, destined to be the loyal and unfeline friend. 1924 *Blackw. Mag.* Jan. 73/2 It is only my retrograde feminine mind that *will* jump to these unfeminist conclusions. 1980 M. Drabble *Middle*

Ground 2 'I know my limitations,' said Kate. 'That's a very unfeminist remark,' said Hugo provocatively. **1934** Webster, Unflamboyant. **1962** *Times* 12 Nov. 4/1 Camberabero..had a neat and unflamboyant partner in Laforgue. **1981** J. Johnston *Christmas Tree* 6 The older ladies would wear hats, neat, unflamboyant hats. **1967** *Punch* 29 Nov. p. x/2, Unflashy surroundings designed to relax diners... Very good French cooking. **1979** *Guardian* 9 Oct. 19/8 King is a quiet, unflashy man. **1804** D. O'Connell *Let.* 20 Aug. (1972) I. 115, I know you are unfond of that expedition—But if I do go I solemnly pledge myself to you that I will not put my foot in other than Capt. O'Sullivan's own boat. **1964** *Punch* 11 Nov. 732/2 A medley of mutually unfond nationalities. **1980** *Country Life* 3 July 69/4 She was not un-fond of her children, but they stood in the way of..a full working life. **1854** C. M. Yonge *Heartsease* II. iii. iv. 164 In spite of clinging unfresh muslin and shrinking figure, with the unmistakable air of high breeding. **1976** J. Colville *Footprints in Time* xi. 63 The unfresh air of the Central War Room. **1967** J. Porter *Chinks in Curtain* vii. 68, I spun round and assumed as unfurtive an air as possible. **1949** St. J. Ervine *Craigavon* II. lix. 278 His singularly un-Gaelic name. **1977** A. T. Q. Stewart *Narrow Ground* I. i. 28 The markedly un-Gaelic physical characteristics of the people now living in the Irish-speaking areas of the south and west. **1936** M. Franklin *All that Swagger* xxxv. 273 Laura and Humphrey, both big and ungay, would make a fine buggy pair. **1977** *Logophile* IV. 9/2, I have known some very, very un-gay homosexuals, and many gay heterosexuals. **1963** *Punch* 22 May 753/3 An ungimmicky, original writer. **1977** *Times* 15 Feb. 8/6 It is the goblet that captures me. I love the ungimmicky, familiar and simple lines. **1934** Webster, Unglamorous. **1960** *Times* 15 Jan. 16/4 Mr. David Tindle's unglamorous but sympathetically conceived 'Girl dressing'. **1967** P. Welles *Babyhip* (1968) viii. 73 They're so ungroovy but we have to do it this way. **1944** C. Calloway *Hepsters Dict.*, *Unhep*, not wise to the jive, said of an icky, a Jeff, a square. **1961** L. Hughes *Ask your Mama* xii. 84 (*heading*) Liner notes for the poetically unhep. **1940** *Music Makers* May 37/3 *Unhip*, not wise to the jive, an icky..a square. **1959** C. MacInnes *Absolute Beginners* 131, I climbed in the rear seat, with a fine view of their..un-hip Jermyn street hair-dos. **1968** *Esquire* Apr. 88/2 There is nothing more detrimental to anything hip than to have it fall into the square hands of the hopelessly unhip. **1974** H. L. Foster *Ribbin', jivin', & playin' Dozens* v. 180 Because the teacher may be unhip and square middle class, he has no idea that the game is being run. **1931** Ld. Wodehouse in 'Marco' *Introd. Polo* p. v, Living in the 'unhorsey' atmosphere of ships. **1955** H. Smith *Horseman through Six Reigns* xvi. 163 A very unhorsy and rather ungainly lady. **1935** N. Mitchison *We have been Warned* III. 298 Tom refused to admit that there is anything unideological about cafés. **1981** *Daily Tel.* 28 Nov. 12/5 To live as an ordinary, totally unideological private person. **1913** Webster, Unintrospective. **1931** *Times Lit. Suppl.* 26 Mar. 241/2 Such a straightforward lie was possible to so unintrospective an artist as Milton, but not to Donne. **1929** W. Faulkner *Sartoris* III. viii. 259 And so there was another bond between them, but unirksome. **1938** G. Greene in *Spectator* 4 Nov. 782/2 Miss Richardson's unironic and undetached method. **1979** L. Lerner *Love & Marriage* i. 18 Poetry like this is essentially unironic. **1934** Webster, Unironical. **1942** *Scrutiny* X. 345 Shakespeare's power to present acceptably and movingly the unironical vision (for us given in Miranda and Ferdinand) goes with his power to contemplate the irony at the same time. **1958** O. Caroe *Pathans* i. 13 The un-Islamic flavour of the names of the two sons of Sarbanr..will not escape notice. **1984** *Times* 31 Mar. 4/8 Artificial methods of birth control are unislamic. **1934** Webster, Un-Italian. **1939** *Burlington Mag.* May. 227/2 Both of them have an un-Italian flavour. **1973** P. Evans *Bodyguard Man* iv. 36 The mainlanders..think of the people from the islands as being un-Italian. **1931** T. H. Pear *Voice & Personality* iv. 35 The workings of the mind are..designated in single stark letters. Their messages are un-Jamesian and un-Proustian, but concise. **1979** F. Kermode *Genesis of Secrecy* iv. 81 This model,..a useful way of thinking about the relation of character to narrative structure..is a very un-Jamesian way. **1934** Webster, Un-Japanese. **1974** *Country Life* 28 Mar. 716/1 The more convincing the make-up, the more disconcerting are un-Japanese physiques. **1962** *Economist* 15 Sept. 990/2 A majority of the unjingoistic people in the country. **1978** Cadogan & Craig *Women & Children First* viii. 185 An honest picture of the practical, unjingoistic attitudes which characterized most RAF personnel. **1966** A. E. Lindop *I start Counting* iv. 67 Grandad was on edge until he'd settled his mice... Aunt Lucy had been a bit unkeen to have them. **1971** *Daily Express* 7 Dec. 1/5 Mrs. Indira Gandhi and Mrs. Golda Meir do make one feel tremendously unkeen on Women's Lib. **1983** Barr & York *Official Sloane Ranger Diary* 4/2 The Keen have to put up with the Unkeen. **1924** G. B. Stern *Tents of Israel* i. 20 For nearly fourteen years she must have been ..cooking for them unkosher food. **1965** *Guardian* 4 Aug. 6/4 The Israeli Embassy..had almost been faced with an Israeli parliamentary inquiry into its allegedly sinful un-kosher way of life. **1939** S. Spender *Still Centre* 97 In you The Caesars tamed by dying, fired again Their lives in the unlegendary sky. **1854** Geo. Eliot tr. *Feuerbach's Essence Christianity* xxii. 215 God sees..all locality in an unlocal manner. **1977** Unlocal [see *uncentral *a.*]. **1876** Trollope *Prime Minister* III. iv. 65 That coarse un-marital and yet marital roughness. **1956** M. McCarthy *Sights & Spectacles* p. xii, The notion that abstract reasoning can crush a fact.., a wholly un-Marxist notion, was nonetheless the principle on which most of our criticism was practiced. **1966** *Punch* 13 Apr. 524/2 His exquisite black-and-white short *Shorts*.., with its touching yet unmawkish dwelling upon the hemline. **1966** *Guardian* 10 Oct. 16/8 The present pattern is in many ways strikingly unmeritocratic. **1938** D. Jones *Let.* 15 Feb. in R. Hague *Dai Greatcoat* (1980) ii. 85 The *only* time of day when I feel more or less un-neurotic..is after dinner. **1980** S. Brett *Dead Side of Mike* x. 113 A very balanced and unneurotic personality. **1856** Bagehot in *National Rev.* Oct. 365 If Miss Westbrook had married.. a gentleman, suppose, in the tallow line..her society would have been a gentle relief from unodoriferous pursuits. **1823** R. H. Froude *Let.* 12 Aug. in G. Battiscombe *John Keble* (1963) iv. 75 Keble..is neither positive on this subject nor un-open to conviction on any. **1981** 'A. Cross' *Death in Faculty* v. 47 The few older men..were so stuffy, so unopen to any views but their own. **1916** J. Buchan *Greenmantle* iii. 30 He got a brussels-sprout in the eye, at which..he swore in a very unpacifist style. **1976** S. Hynes *Auden Generation* vii. 195 The book is pacificist propaganda.., but the position he takes is an oddly belligerent and un-pacifist one. **1965** W. S. Allen *Vox Latina* i. 15 Such a complication..is 'unphonemic'. **1974** D. G. Scragg *Hist. Eng. Spelling* iv. 61 Mulcaster..was frequently forced to compromise his endeavour to be guided by sound because what he calls custom favoured a flagrantly unphonemic spelling. **1959** *News Chron.* 19 Aug. 4/3 'I'm no intellectual,' he says with unphoney modesty. **1973** *Publishers' Weekly* 26 Mar. 61/3 Individually these British lads (and lass) are enjoyably unphony. **1982** *Listener* 11 Feb. 27/3 Let our final words honour the exceptionally unphoney—the first performance of Edward Cowie's..Concerto for orchestra. **1934** C. Lambert *Music Ho!* iv. 244 The present vogue for mechanical realism..is bound to disappear..as the Turneresque steam engine gives way to the unphotogenic electric train. **1977** R. Lacey *Majesty* 14 She is actually quite small.. and..really rather unphotogenic. **1749** J. Cleland *Mem. Woman Pleasure* II. 219 Our acquaintance..innocent, at first..changed nature, and ran into unplatonic lengths. **1866** J. S. Mill in *Edin. Rev.* CXXIII. 340 The answer of the Xenophontic Sokrates to the question of Hippias is very un-Platonic. **1934** Webster, Unpolemical. **1936** *Mind* XLV. 397 Philosophy, according to Jaspers, is 'unpolemical'. **1965** *Times Lit. Suppl.* 25 Nov. 1079/2 There is evidence of good, unpolemical work being done. **1936** 'M. Innes' *Death at President's Lodging* ix. 164 Appleby had brought out his notebook—not without a certain diffidence over the remains of the Dean's elegant and unpolicemanly luncheon. **1980** P. G. Winslow *Counsellor Heart* xi. 148 Manning..sitting quietly at his desk, might have been an accountant. For once his unpolicemanly appearance did not amuse Capricorn. **1865** Mill *Comte* 80 Political economy..he deems unscientific, unpositive, and a mere branch of metaphysics. **1968** 'D. Halliday' *Photogenic Soprano* xiv. 160 All the sordid and unprestigious details of your warm friendship with Kenneth. **1980** *Word 1979* XXX. 197 Speakers of unprestigious urban dialects were sharply disadvantaged in education and employment. **1933** H. Walpole *Vanessa* III. 484 She learnt now to be patient, tolerant and unpriggish. **1969** J. Clarke *Foxon's Hole* xxii. 133 Roger really was feeling rebellious, and Gawaine..found it much easier to tell this new, unpriggish Roger..about Mick. **1974** P. Gzowski *Bk. about this Country* 190 I'd..have to become what I call an unprivate person in relation to social issues. **1931** Un-Proustian [see *un-Jamesian* above]. **1981** *Times* 2 Mar. 6/2 Scott Moncrieff's unProustian style. **1934** Webster, Unpugnacious. **1962** Auden *Dyer's Hand* (1963) 305 As creatures go, he [*sc.* the ant-eater] is unpugnacious. **1981** *Daily Tel.* 18 July 28/3 He fought an uncomplicated, unpugnacious and honest campaign. **1968** *Earth & Planetary Sci. Lett.* V. 220/1 Hamilton invoked the melting of Lewisian rocks to account for the unradiogenic leads in the Skye acid rocks. **1949** St. J. Ervine *Craigavon* II. lxvi. 296 Whose unresilient and unimaginative mind. **1963** *Times* 17 May 7/2 (Advt.), They took an ordinary piece of hard, unresilient plastic and 'let the air in'. **1976** *Gramophone* Mar. 1499/1 The music is steamrollered into a kind of impersonal uniformity, its lines unphrased, its rhythms unresilient, and the subtle interaction of words and notes quite lost. **1974** Moore & Parry *Twentieth-Cent. Russ. Lit.* ii. 22 His unrevolutionary deed in surrendering to despair and killing himself. **1909** 'Mark Twain' *Is Shakes. Dead?* 13 When an unrisky opportunity offered..and he was feeling good, I showed it to him. **1934** Webster, Unroadworthy. **1955** *Times* 20 Aug. 5/1 The cars hired by correspondents to take them to the frontier were unroadworthy. **1978** N. Freeling *Night Lords* xxxii. 147 Being suddenly told your home is unhygienic and your car unroadworthy. **1919** tr. *Turgenev's Smoke* x. 73 A dandified air utterly un-Russian. **1976** W. Greatorex *Crossover* 159 She took in the flared tartan trousers... He looked so un-Russian she laughed. **1926** G. Frankau *My Unsentimental Journey* xiii. 178 And so away from Denver by this route which seems not so unscenic. **1978** J. Updike *Coup* (1979) vi. 219 His trip down from the Massif, by the unscenic highway. **1791** E. Inchbald *Simple Story* I. ii. 12 Nor upon that event did he think it necessary..to fly the roof of two such unseductive innocent females as Mrs. Horton and her niece. **1937** M. Borden *Black Virgin* ii. 26 He looked at..her neat unseductive clothes and thanked God she was like that. **1934** Webster, Unselective. **1956** *Nature* 10 Mar. 489/1 Government shooters co-operated in recording results of unselective shooting [of red deer] to make these data available. **1968** *Punch* 24 Jan. 105/2 Flat-rate, unselective welfare was possible when the vast majority of people were earning poor wages and when taxation was relatively light. **1933** L. Bloomfield *Language* x. 158 The situations of the several speakers contain some common features, and..the differences between these situations are irrelevant (unsemantic). **1968** *N.Y. Times* 21 Feb. 56/1 His barrage of blithely unsemantic bombast sweeps you up in such phraseology as 'the political hue that blightens the eye' and the oracular 'protocol takes precedence over procedure'. **1959** *News Chron.* 25 Aug. 6/6 He thought the most unsexy things a woman could wear were trousers, boat necklines, and high-heeled shoes. **1973** M. Amis *Rachel Papers* 193, I would get..a kind of hollow pressure at the back of my throat (..not unsexy) which nevertheless I had to lose, and the only way to lose it was to go on coughing. **1936** Un-Shelleyan [see *un-Byronic* above]. **1896** 'M. Rutherford' *Clara Hopgood* iv. 41 His unsnobbish, deferential behaviour.. showed that he understood who they were and that the little house made no difference to him. **1959** G. D. Painter *Proust* I. vii. 100 She was unmusical, nonpolitical, and in the social sense unsnobbish. **1934** Webster, Unsorry. **1963** J. Fowles *Collector* II. 195 I'm not really sorry. But I'm not absolutely unsorry. **1973** 'G. Butler' *Coffin for Pandora* viii. 163 Alice had departed.., not unsorry, I thought, to get away. **1846** R. Ford *Gatherings from Spain* xxiii. 330 Some muscular.. performer..screams forth his couplets..to the imminent danger of his own trachea, and of all un-Spanish acoustic organs. **1959** H. Thomas *Establishment* 15 Spain, where for four hundred years any idea..which had not existed in the golden sixteenth century has been automatically frowned upon as 'Un-Spanish'. **1970** T. Hilton *Pre-Raphaelites* v. 151 Here is a dull day, unspecial, cold and leaden, and yet a momentous day for those people who are leaving home. **1983** *Woman's Weekly* 8 Jan. 18/1 Anna, feeling very un-special indeed, agreed..that there was no point in looking back. **1926** *Public Opinion* 30 Apr. 433/3 An unspectacular honesty and a certain literary sobriety..mark this novel. **1977** M. V. Brian *Ants* 11 Most of the species have wide temperate Eurasian distributions and, though unspectacular compared with some tropical types, play an important part in terrestrial ecosystems. **1968** R. Marett *Through Back Door* iv. 37 The bluff, nautical and unstarchy Chief Press Censor. **1934** Webster, Unsterile. **1953** R. Lehmann *Echoing Grove* 29 But still the stones seemed rocked, the unsterile mounds, reimpregnated, exhaled dust's fever; a breath, impure, of earthbound anguish. **1977** *Lancet* 26 Mar. 688/1 Unsterile cystoscopes have been in use for years. *c* **1873** V. Montagu *Let.* in G. Battiscombe *Queen Alexandra* (1969) ix. 127 It is very jolly here indeed,.. very unstiff and only a certain amount of etiquette. **1958** J. Pope-Hennessy in *Lonely Business* (1981) III. 248 They said above all she..was 'such a *darling*'—such fun —so unstiff. **1929** D. H. Lawrence *Pansies* 117 Space, of course, is alive. That's why it moves about; and that's what makes it eternally spacious and unstuffy. **1974** K. Clark *Another Part of Wood* vi. 220 To be one of Lady Cunard's regular guests was to have reached somewhere very near the top of unstuffy, new world society. **1933** H. Walpole *Vanessa* II. 414 Will's love for [his master] Adam had been..protective, selfless, and also gay, simple, unsycophantic, man to man. **1964** *Punch* 23 Sept. p. xv, Cockney writers..are refreshingly unsycophantic. **1909** Webster, Unsymmetric. **1957** L. Fox *Numerical Solution Two-Point Boundary Probl.* iv. 103 In this case we would have to use some unsymmetric formula involving only internal points. **1935** J. Laird *Enquiry into Moral Notions* 10 In my view the times are propitious for attempting an untendentious comparison of ethical ideas ..with equity and..patience. **1962** *Times* 5 Dec. 17/6 (*heading*) Untendentious story of the Berlin Wall. **1934** Webster, Untense. **1959** *Punch* 25 Feb. 277/2 To become untense you must relax. **1974** *Times Lit. Suppl.* 8 Nov. 1248/4 A melodrama in a curiously untense key. **1961** *Lancet* 2 Sept. 549/1, I..support your editorial.. questioning the value of euthanasia... Justification for such untherapeutic therapy can rest only on a total lack of understanding of the nature..of Man. **1978** *Sci. Amer.* Feb. 50/2 Untherapeutic though many nursing homes are, living conditions in most of them are at least tolerable. **1906** Kipling *Puck of Pook's Hill* 6 They sat down in the unthistly centre of the Ring. **1953** Auden in R. Humphries *New Poems by Amer. Poets* 7 The water-scorpion finds it quite unticklish. **1969** Fabian & Byrne *Groupie* xxix. 206 As a group they are something else, but off-stage they're pretty untogether, and they need someone like me to get them to the top. **1976** Untogether [see *staggeringly *adv.*]. **1962** *Harper's Bazaar* Aug. 57/3 Basutoland is for the sophisticated traveller seeking the untouristy. **1979** *Homes & Gardens* June 23/2 Today more visitors bring their cars and instead of making an automatic beeline for London they venture into the most remote and 'untouristy' localities. **1934** Webster, Untraditional. **1958** *Times* 16 Dec. 11/1 We await the rest of the season with untraditional excitement. **1971** 'E. Ferrars' *Stranger & Afraid* xii. 187 The understanding that the university should be one of the new ones and as untraditional as possible. **1968** *Guardian* 26 Sept. 8/5 A BBC man, worried that Priestley might be too untrendy for the 'Wednesday Play'. **1978** *Times* 4 Sept. 13/2 My untrendy and doubtless outdated opinions about the high seriousness of political debate. **1934** Webster, Unurgent. **1939** C. Day Lewis *Child of Misfortune* II. iv. 212 There was something intimate yet un-urgent in the touch. **1949** *New Yorker* 18 June 24/2 The mist incorporates the pulse, rapid but unurgent, of a motorboat. **1907** H. Rashdall *Theory of Good & Evil* I. vi. 158 Such an obviously unutilitarian precept as that which condemns cannibalism. **1955** S. Spender *Making of Poem* i. 16 Architecture..expresses the tension of the aesthetic against the useful. At the other extreme, music is completely unutilitarian. **1931** *Times Lit. Suppl.* 7 May 360/3 Courses which are ultimately impolitic though not formally 'unviable'. **1964** G. Wheeler *Mod. Hist. Soviet Central Asia* 234 The steady proliferation of independent although economically unviable states. **1982** M. Wallace *Brit. Govt. Northern Ireland* i. 18 The Irish negotiators..had clearly felt that a commission would so reduce the area of Northern Ireland that it would become unviable. **1916** L. Strachey *Let.* 25 Feb. in *Virginia Woolf & Lytton Strachey* (1956) 56 Oh, it's very, very unvictorian! **1981** F. Inglis *Promise of Happiness* v. 144 Ideal parents..so un-Victorian but also so excellently authoritative. **1934** Webster, Unviolent. **1963** *Listener* 3 Oct. 522/3 Even his rhetorical phrases..came out with a lack of ferocity which has great charm for people enamoured of unviolent politicians. **1980** U. Curtiss *Poisoned Orchard* x. 100 That tranquil, un-violent setting. **1931** V. Woolf *Waves* 168, I may find something unvisual beneath [the mind's eye]. **1943** H. Read *Education through Art* 147 This leaves two of our categories unaccounted for, the expressionist and the decorative, but a little consideration will show that one of these is *a priori* excluded from Bullough's apperceptive types, because it is essentially un-visual. **1903** *Lippincott's Monthly Mag.* Aug. 230 To tell the truth, Barrett's calm philosophy irritated her not a little. It was painfully un-Western. **1981** G. MacBeth *Kind of Treason* xii. 116 The Japanese were traditionally un-Western in their attitude to declarations of war. **1965** K. Amis *James Bond Dossier* ix. 88 These days there are few things more un-with-it than feeling..the slightest anti-American sentiment. **1977** *Grimsby Even. Tel.* 5 May 8/3 There is only one word for them, a word so passé, fuddy-duddy and un-with-it that one blushes to use it. **1959** *Economist* 23 May 716/2 The importation or maintenance of party ardours..must seem

singularly un-worthwhile. **1974** *Nature* 19 Apr. 715/3 Contributions so speculative as to make such an effort unworthwhile. **1925** 'H. H. Richardson' *Way Home* III. v. 281 Emmy with the hard and unyoung look her face assumed when she spoke of her stepmother. **1972** *National Observer* (U.S.) 27 May 12/1 An unyoung 'youth' leader travels about the country urging audiences to 'kill your parents'.

b. (*a*) *unactionable, unarrangeable, unbackable, unbilletable, unbiodegradable, unblinkable, unbroadcastable, unbudgeable, unbuggable, unbuildable, uncapturable, uncashable, unclaimable, unclearable, uncollectable, uncondonable, unconfessable, unconfusable, unconsortable, uncontactable, uncopyrightable, uncounterfeitable, uncrackable, undeflectable, undevelopable* (earlier example), *undeviable* [*deviable *a.*], *undiagnosable, undisseverable, undownable, undreamable* (also *absol.*), *unexorcizable, unexperienceable, unexploitable, unfak(e)able, unfalsifiable, unfaultable, unfilmable* (also *absol.*), *unfulfillable, unidentifiable, unignorable, uninterruptable, unjammable, unkinkable, unkissable* (also *absol.*), *unmanœuvrable, unmeltable, unoccupiable, unpatentable, unplaceable, unplumbable, unpoliceable* (also *absol.*), *unquantifiable* (also *absol.*), *unrecapturable, unregenerable, unrejectable, unreviewable, unruffl(e)able, unsayable* (also *absol.*), *unshar(e)able, unshockable, unskiable, unsmoothable, unsplinterable, unsplittable, unsquashable, unstageable, unstat(e)able* (further examples), *unstressable, unswallowable, untappable, untestable, untippable, untransmittable, unverbalizable, unvintageable, unweighable, unwettable, unwipeable, unwithstandable*; (*b*) *unputdownable, unswitchoffable, unwearoutable*.

(*a*) **1952** *Federal Suppl.* (U.S.) CVI. 1019/1 All this might still be unactionable but for the fact that [etc.]. **1980** *Times* 16 Nov. 3/1 The case was dismissed by the High Court in London in preliminary proceedings as 'unactionable in English law'. **1821** *Blackw. Mag.* X. 525 An unarrangeable mass of contraries and shades of difference. **1982** *Book Collector* Winter 457 Vivaldi was virtually unarrangeable. **1928** *Daily Express* 2 Aug. 13/1 Zahrat.. walked away from three moderate rivals in the Alexandra Handicap. She was..unbackable, odds of nine to two having to be laid on. **1973** *Nation Rev.* (Melbourne) 24–30 Aug. 1418/3 Who's Who was literally unbackable on form and treble figure odds ought to have been obtainable. **1940** *Daily Mirror* 17 Dec. 2/2 My wife was asked to give temporary lodging to some young girls who were described by the authorities as unbilletable. **1977** *Water SA* (S. Afr.) Jan. 18/1 The influent TKN [*sc.* Total Kjeldahl Nitrogen] is split into three fractions (i) an unbiodegradable fraction [etc.]. **1978** D. Bloodworth *Crosstalk* xxxi. 239 An imitation crocodile handbag in some unbiodegradable plastic. **1936** *Times Lit. Suppl.* 31 Oct. 884/4 A stark unblinkable picture of life as it is being lived somewhere to-day. **1966** *Guardian* 28 Dec. 7/2 There is, finally, the unblinkable fact that the [U.S.] Cabinet is weary. **1975** *Listener* 6 Feb. 163/1 The interview was quite unbroadcastable, due..to the Venerable brother's imperfect English. **1936** L. C. Douglas *White Banners* iv. 73 'Now that's where you're wrong, Marcia.' Paul was kind, but unbudgeable. **1981** *Time* 7 Dec. 74/2 It is an almost unbudgeable popular belief..that cats and dogs have an instinctive rivalry. **1979** R. Cassilis *Arrow of God* II. vi. 33 The Service's beautiful, new, and allegedly unbuggable headquarters. **1956** R. A. Heinlein *Door into Summer* v. 82 Take the great Leonardo da Vinci, so far out of his time that his most brilliant concepts were utterly unbuildable. **1851** H. Melville *Moby Dick* II. xxxix. 274 The only spout in sight was that of a Fin-Back, belonging to the species of uncapturable whales, because of its incredible power of swimming. **1925** *Glasgow Herald* 7 Feb. 4 In the translation something elusive and..uncapturable has vanished. **1906** 'Mark Twain' *Autobiogr.* (1924) II. 288, I don't know what Dick got, but it was probably only un-cashable promises. **1978** *Daily Tel.* 28 Oct. 1/4 The £100,000 cheque handed to Korchnoi as defeated finalist in the World Chess Championship is uncashable until he signs a declaration that he has lost the title. **1884** W. James in *Unitarian Rev.* XXII. 205 Whether the world be the better or the worse for having either chances or gifts in it will depend altogether on *what* these uncertain and unclaimable things turn out to be. **1955** *Bull. Atomic Sci.* Apr. 146/3 Most unfortunate is the cleared scientist who marries into an unclearable family. **1984** *Sunday Express Mag.* 26 Feb. 28/1 The RE major.. has abandoned whole fields and beaches as 'unclearable'. **1927** *Glasgow Herald* 18 July 10 They had huge outstanding accounts up country, which were uncollectable. **1981** *Sunday Tel.* 26 July 19/6 Some £12 million that is now officially designated as 'almost uncollectable'. **1935** W. de la Mare *Early One Morning* 136 Hatred of elderly and yet not wholly uncondonable cant. **1919** C. B. Jordan tr. *Ibañez's Four Horsemen of Apoc.* I. v. 143 The infallible remedy for the most unconfessable of diseases. **1967** P. White in *Coast to Coast 1965–6* 232 He lied smiling, ashamed..of his..unconfessable experience. **1934** Webster, Unconfusable. **1935** W. de la Mare *Early One Morning* 533 Each is in accord with its writer's after-work, and unconfusable with the work of any other poet. **1907** Joyce *Chamber Music* xxi, He who hath glory lost, nor hath Found any soul to follow his... That high unconsortable one. **1976** *Daily Tel.* 26 Oct. 1/6 The 3,500-ton Swiftsure class hunter-killer vessel is said to be 'uncontactable' beneath the Arctic ice-pack at the North Pole. **1926** Whiteman & McBride *Jazz* viii. 178 The earliest jazz was found uncopyrightable by certain judges. **1982** *Amer. Speech* LVII. 308 It seems possible that other dictionary makers may well have adopted this simple method of protecting their essentially uncopyrightable labors. **1912** E. Pound in *Poetry Rev.* Feb. 73 A man's rhythm must be interpretative, it will be, therefore, in the end, his own, uncounterfeiting, uncounterfeitable. **1932** *Blue Book Mag.* Sept. 12/1 He owned that uncounterfeitable trait which goes with what we call good birth. **1959** 'B. Mather' *Achilles Affair* I. ii. 21 The code we were using was as near to uncrackable as any code can be. **1954** W. Faulkner *Fable* 263 The forsaken ..betrothed pursuing,..undeflectable. *a* **1866** J. Grote *Moral Ideals* (1876) xv. 371 The other kind of knowledge lies at the base.., insoluble and undevelopable. **1929** S. Leslie *Anglo-Catholic* xiv. 199 This continuity stretching back beyond modern history, undeviating and undeviable. **1951** W. Faulkner *Requiem for Nun* II. 210 Who am I..to set the puny appanage of my office in the balance against that simple undeviable aim? **1926** *Jrnl. Amer. Med. Assoc.* 20 Nov. 1721/2 Some intangible,.. undiagnosable sinus disease. **1978** R. Bannister *Brain's Clinical Neurol.* (ed. 5) p. iii, Disorders either undiagnosable or untreatable or both. **1897** H. James *Spoils of Poynton* xiii. 152 Mona was undisseverable from her prey. **1957** *Sci. Amer.* Apr. 50/2 The undownable question remained: Why were tau and theta [particles] exactly alike in every respect except this one? **1978** J. Updike *Coup* (1979) vi. 224 His motionlessness a mask for his suffocating struggle with the resurging, undownable fact of Sheba's absence. **1906** W. de la Mare *Poems* 81 In its future loomed the undreamable. **1957** L. MacNeice *Visitations* 53 When the undreamable Dream comes clearer. **1922** *Blackw. Mag.* June 800/2 Because of those unexorcisable..Oriental ancestors of his, nothing will ever bring him to regard the game [*sc.* golf]..with the laconic reverence that a Scotsman..expects. **1909** W. James *Meaning of Truth* p. xii, Things of an unexperienceable nature may exist ad libitum, but they form no part of the material for philosophic debate. **1976** *Word 1971* XXVII. 130 If grammar is 'autonomous and independent of meaning', then it is unexperienceable in the surface structure of sentences—therefore unlearnable, and therefore innate. **1923** *Blackw. Mag.* Sept. 398/1 A rich spring of natural oil, as yet unexploited, and perhaps, on account of the climate and neighbours, forever unexploitable. **1978** A. Gilchrist *Cod Wars* vi. 42 Such resources were largely hidden or disregarded or unexploitable until modern techniques could be applied. **1955** P. Larkin *Less Deceived* 41 You cannot always keep That unfakable young surface. **1977** H. Osborne *White Poppy* xxii. 160 That loud wah-wah voice that..was the unfakeable mark of an English gentleman. **1934** Webster, Unfalsifiable. **1963** *Guardian* 25 Jan. 7/6 Any..existential statement which is unrestricted in scope and therefore unfalsifiable. **1977** P. Johnson *Enemies of Society* xi. 153 The proposition was left sufficiently vague to allow for it to be updated..in the light of scientific advances; it is thus a classic example of an unverifiable, or unfalsifiable, statement, and as such scientifically useless. **1965** *Punch* 21 Apr. 572/3 They plump for a return to this hotel, reminding each other of its unfaultable charms. **1928** *Sunday Express* 29 Apr. 4 'Piccadilly', Arnold Bennett's original story.., is said to be not wholly disconnected with his novel 'The Pretty Lady', the great unfilmed (and unfilmable). **1958** *Times Lit. Suppl.* 15 Aug. p. xxviii/4 Film studio executives sat up and took notice of the work of writers whom they had previously dismissed as unfilmable. **1963** E. Humphreys *Gift* II. iv. 238 It didn't matter how unfilmable the book might be, other producers were after it. **1983** *Listener* 6 Jan. 28/1 Le Carré is on record as saying that it is unfilmable. **1934** Webster, Unfulfillable. **1968** *Punch* 4 Sept. 337/1, I ended up spending the night with my friend.., after signing numerous, apparently unfulfillable bonds and promises for the Customs officer. **1982** H. Kissinger *Years of Upheaval* vi. 197 A demand as seemingly reasonable as it was unfulfillable. **1909** Webster, Unidentifiable. **1923** M. Sadleir *Desolate Splendour* ix. 142 The small girl, in charge of her young uncle and the unidentifiable Margery, gambolled..across an uneven field. **1971** S. Hill *Strange Meeting* iii. 179 The hammering noise went on, only slightly muffled and then another, unidentifiable sound, though the voices had ceased. **1955** P. Larkin *Less Deceived* 26 A tense, musty, unignorable silence. **1981** *London Rev. Bks.* 18 June–1 July 7 The speed, wit and range of the book make it unignorable and exhilarating. **1977** *Sociology* XI. 99 We would observe the syntactical momentum employed for an uninterruptable delivery. **1959** *Economist* 11 Apr. 140/2 The [Soviet] government fears that this information has already become unjammable. **1963** *Daily Tel.* 4 Mar. 1/7 Typhon is designed to knock down aircraft and winged missiles by homing on them with its 'unjammable' radar. **1935** *N. & Q.* 4 May 314/1 *Unkinkable*. A new cable-trade term, applied to a fresh and useful flexible electric cord, which is guaranteed not to ravel or ruck up when in use. **1977** *Offshore Engineer* May 16/2 (Advt.), The right rope for the job... H and T Multiplait is:..unkinkable. **1936** G. B. Shaw *Simpleton* II. 63 His poor little secret vice of cigaret smoking?..Faugh! The unkissable... The air poisoner. **1962** M. Drabble *Summer Bird-Cage* vi. 99, I was standing stock still and quite unmanoeuvrable. **1940** W. Faulkner *Hamlet* II. ii. 141 He brought a gross with him in the suitcase, specially made up for him outen asbestos, with unmeltable snaps. **1922** Joyce *Ulysses* 697 A thatched..2 storey dwellinghouse..with agreeable prospect..over unoccupied and unoccupyable interjacent pastures. **1854** *U.S. Reports* (U.S. Supreme Court) LVI. 132 If the inventor of printing had..claimed his art as something distinct from his machinery, the doctrine now advanced, would have declared it unpatentable to its full extent as an art. **1980** *TWA Ambassador* Oct. 21/3 Left unpatentable are elements of nature—a newly discovered wildflower, for instance—and laws of nature such as philosophical theories and mathematical principles. **1935** *Sun* (Baltimore) 10 Jan. 22/4 The increase in the number of unplaceable children taken under care. **1940** Dylan Thomas *Portrait of Artist as Young Dog* 117 We stood in the scooped, windy room of the arch, listening to the noises from the muffled town,..unplaceable sounds,..an engine coughing like a sheep on a hill. **1972** *Human World* Feb. 21 The absence of unplaceable graduates in Britain and the U.S.A. was one of the chief causes of the relative immunity of these countries to Marxism. **1984** *New Yorker* 9 Apr. 39/2 The music was unplaceable. It must have been a Russian idea of American pop. **1921** H. Williamson *Beautiful Years* 118 The man certainly was queer: he no longer thought as he had at first, that there was anything deep in his nature. Either that, or it was so deep as to be unplumbable. **1929** S. Leslie *Anglo-Catholic* x. 122 The pitiless unplumbable spaces filled him with terrors. **1980** D. Moraes *Mrs Gandhi* viii. 116 While she usually answers complex and difficult questions fluently, perfectly innocuous queries seem to drive her back into unplumbable depths of silence. **1971** *Nature* 3 Sept. 2/2 This device let the Vienna agency escape from the invidious position of policing the unpoliceable into which it has been jockeyed. **1981** *Times* 25 July 5/1 Protests in other centres would render the rest of New Zealand unpoliceable. **1890** A. C. Fraser *Locke* 147 The idea of the *unquantifiable* is 'suggested' by the positive ideas of spaces and times which we have had in our sense experience. *Ibid.*, We are *as* remote from the unquantifiable infinite..as we were at the beginning of the process. **1979** *Financial Times* 2 Apr. 27/4 Job security is more generally regarded as unquantifiable. **1925** T. Dreiser *Amer. Tragedy* I. II. xxii. 307 So wild and unrecapturable is the fever of youth. **1964** W. McCord in I. L. Horowitz *New Sociology* 428 We must eschew..the romanticism which longs for the virtues of an unrecapturable past. **1930** *Times Lit. Suppl.* 30 Oct. 884/3 The unregenerate and unregenerable realm of life. **1982** *Observer* 17 Jan. 31/9 It is typical of my unregenerable soul that I can only see this as a marvellous theme for a novel. **1934** Webster, Unrejectable. **1963** M. Drabble *Summer Bird-Cage* viii. 127, I felt rather guilty.., as if I had been reading a diary instead of simply receiving unrejectable impressions. **1934** Webster, Unreviewable. **1955** *Bull. Atomic Sci.* Apr. 129/3 With so much unreviewable power in the executive there is bound to be doubt whether the company is getting the information which it really needs. **1968** *Guardian* 15 Nov. 6/7, I find Mary McCarthy's account of her visit to North Vietnam almost unreviewable. **1960** E. W. Swanton *West Indies Revisited* 226 Worrell, however, is unruffleable. **1981** F. Inglis *Promise of Happiness* ii. 65, I read them for..the unrufflable, wholly impossible calm of big boys. **1870** W. Knight *Colloquia Peripatetica* p. xii, The thought might penetrate into that shadowy region where language almost breaks down in the effort, as he put it, 'to say the unsayable'. **1905** E. F. Benson *Image in Sand* vii. 99 The only things worth saying are just those which are unsayable. **1954** R. Jarrell *Pictures from Institution* IV. 150 Her name was not Rosenbaum on the [gramophone] records but her own real unsayable Russian name. **1979** 'J. le Carré' *Smiley's People* (1980) iv. 49 Strickland incanted the unsayable: 'No coat-trailing..No émigrés. No bugger all.' **1902** W. James *Varieties Relig. Experience* xx. 499 That unshareable feeling which each one of us has of the pinch of his individual destiny. **1976** *Daily Tel.* 19 Oct. 2/1 He said the doctor's belief in independence was fostered by the ultimately unshareable responsibility he accepted for his patient. **1928** C. Bell *Civilization* vi. 173 At this time of day a civilized person, male or female, should be unshockable. **1980** First & Scott *Olive Schreiner* iv. 132 The tenor of the history..was..cool and unshockable. **1950** *Times* 15 Feb. 7/5, I should like to ask Lord Montgomery what he does when the 'soft' snow is practically unskiable. **1978** *Detroit Free Press* 2 Apr. 17D/3 Help came in the form of the snow drought during the 1976–77 ski season that left Western and New England slopes unskiable. **1851** H. Melville *Moby Dick* III. xxvii. 176 'Canst thou smooth this seam?'..'Aye, man, it is unsmoothable.' **1925** 'R. Crompton' *Still—William* xi. 207 William watched him, smoothing back his unsmoothable hair. **1924** *Motor* 21 Oct. 623/3 An M.E. rear screen with Triplex unsplinterable glass. **1962** L. S. Sasieni *Optical Dispensing* xiii. 329 Unsplinterable lenses are provided as protection. **1926–7** *Army & Navy Stores Catal.* 101/3 Shaving brushes..with unsplittable handles. **1945** R. Knox *God & Atom* i. 13 It might be possible..to resolve the molecule into its component parts, but beyond that lay something smaller yet, completely indivisible; it was christened accordingly 'the atom'—the unsplittable thing. **1978** M. Midgley *Beast & Man* (1979) x. 214 Language..seems an unmistakable, unsplittable single thing. **1956** M. Stewart *Wildfire at Midnight* ii. 26 He's little and round and quite, quite sorbo... Unsquashable. **1975** *Daily Tel.* 7 Apr. 9/1 Ghastly statistics about the vast number of totally unstageable scripts that come pouring through the post on to the desks of successful theatre people. **1981** *Times Lit. Suppl.* 13 Feb. 161/5 The work [*sc.* a Russian play] is unstageable today. **1955** P. Heron *Changing Forms of Art* p. xiv, Indeed, as with all good art, the whole truth is unstatable. **1963** *Times Lit. Suppl.* 24 May 376/4 This appeal to unstateable personal and inter-personal experience. **1972** *Language* XLVIII. 83 Of course, one can state rules in global grammar that are unstatable as transformations. **1950** D. Jones *Phoneme* 150 Semivowels, such as j and w, would appear to be essentially unstressable. **1968** *Economist* 14 Dec. 21/1 Shifting to the side of their cheeks some unswallowable differences, they agreed to put up common candidates in 1969's general election. **1910** *Chambers's Jrnl.* 10 Sept. 644/1 The secrecy provided by the use of cryptic combinations of meaningless consonants can be equally assured by the various methods used for rendering the electric waves 'untappable'. **1979** C. McCarry *Better Angels* II. ii. 115 'I found that a computer out there..knew some things that only our computer was supposed to know.'..'But I thought our equipment was untappable.' **1909** Webster, Untestable. **1964** F. Bowers *Bibliogr. & Textual Criticism* III. vii. 93 Generalized opinions that are not only untested ..but are also (usually) untestable. **1950** O. Nash *Family Reunion* (1951) 102 Where hotels and restaurants and service stations are operated by untippable unoffendable machinery. **1968** *Listener* 27 June 851/1 Even in Europe —where tipping started—there are such things as untippable taxis, though one has perhaps to go to Helsinki to find them. **1978** *Broadcast* 28 Aug. 19/2 This commercial ..took place in almost undecipherably muddy twilight. They can't really mean that, I thought..it looks untransmittable. **1953** J. S. Wilkie *Sci. Mind & Brain* iv.

41 Some things are virtually unverbalizable: is it at all credible that one could verbalize the face of a friend? **1978** *Language* LIV. 268 A variant of this position holds that experience, to be authentic, must be unverbalizable ('the idea, once expressed, is a lie'). **1877** O. Wilde in *Irish Monthly* Dec. 746, I stood by the unvintageable sea. **1915** Unvintageable [see *polyphloisbic *a.*]. **1909** Webster, Unweighable. **1938** R. W. Lawson tr. *Hevesy & Paneth's Man. Radioactivity* (ed. 2) xi. 132 A parent substance present in unweighable quantities. **1971** *Nature* 23 July 260/2 By trapping the effluent from gas chromatographic columns we have obtained unweighable samples (<50µg) of the natural pheromone. **1955** *New Biol.* XVIII. 112 Flowers, if floating, must be more or less cup-shaped, and even if lifted above the water surface must, like the leaves, be unwettable. **1977** J. L. Harper *Population Biol. Plants* ii. 46 When light unwettable seed lands on water, it tends to become concentrated at the edges of the water. **1912** G. B. Shaw *Let.* 10 Dec. in *B. Shaw & Mrs. Campbell* (1952) 68 There would be some terrible breach of etiquette some unwipable-out insult, if she gave me a card for you. **1971** *Daily Tel.* (Colour Suppl.) 4 June 39/1 This little hole (felt-lined usually, and utterly unwipeable)..which they actually call a glove-box. **1981** *Times* 4 July 10/1 Tape costing a quarter of the price of these unwipeable tapes. **1931** *Times Lit. Suppl.* 12 Mar. 196/2 The episodes of Susan's life..and her ruthless imposition of herself upon..her third husband, and upon his Poor Christians as a kind of unwithstandable Deborah—are told with admirable spirit. **1970** R. Price *Gt. Roob Revolution* 5 The classic rube withstood the unwithstandable winter at Valley Forge.

(*b*) **1947** R. Chandler *Let.* 5 Jan. in *R. Chandler Speaking* (1966) 66, I found it absolutely..unputdownable. **1982** *Brit. Med. Jrnl.* 15 May 1466/2 The novel is highly readable and quite unputdownable. **1974** *Guardian* 21 Mar. 10/4 The immediacy which makes TV sometimes ..absolutely un-switch offable. **1984** *Listener* 10 May 38/1 Alfred Brendel's recording of Beethoven's Five Piano Concertos, with the Chicago Symphony Orchestra conducted by James Levine..is unswitchoffable. **1968** *Economist* 25 May 89/2 The engine..is a neat—and presumably unwearoutable—air-cooled flat-four unit in the back of the car. **1979** *Field* 3 Oct. 942/1 (Advt.), Almost unwearoutable socks.

c. *unbridegroomlike* (earlier example), *unbutlerlike, uncatlike, unpolicemanlike*.

1830 M. O'Brien *Jrnl.* 15 May (1968) xii. 114 Mary and Fanny exclaimed at this most unbridegroomlike proceeding. **1924** J. Sutherland *Circle of Stars* iii. 24 His manner was unimpeachable, his tone superlatively correct, but somewhere in his eyes was a most unbutler-like gleam. **1955** Auden *Shield of Achilles* i. 22 For an uncatlike Creature who has gone wrong, Five minutes on even the nicest mountain Is awfully long. **1978** D. Williams *Treasure up in Smoke* xx. 186 There was a most unpoliceman-like implication in the Chief Inspector's tone.

8. a. (*a*) *unacclaimed, unaddled, unaffrayed* (arch.), *unassessed, unattributed, unbeglamoured, unbobbed* [*bobbed *a.* b], *unbombed, unbothered, unbuffered, unbugged, unbypassed, uncaked, uncatalysed, uncloned, uncluttered, uncobbled, uncoded, uncopyrighted, uncouponed, undeafened, undepicted, undiffused, undisbursed, undistanced, unearmarked, unenhanced, unenthused, uneroded, unexploded, unfaked, unfattened, unfazed, unfilmed* (later example), *unflustered, unfocus(s)ed, unformatted, unfussed, ungrazed, ungritted, unhipped* [*hip *v.*[5]], *unhousetrained, uniced, uninjected, unkeepered, unladdered* [*ladder *v.* 2], *unlaundered, unlenited, unmonitored, unneutered, unoriented, unpackaged, unpadded, unphotographed, unplanned, unposed, unpowered, unprogrammed, unprovenanced, unrationed, unrattled, unseeded, unshingled* [*shingled *ppl. a.*[1] 3], *unsignposted, unslanted, unsliced, unslipped* [*slipped *ppl. a.*[2]], *unsourced, unspayed, unsponsored, unstaffed, unstreamed* [*streamed *a.* 2], *unstructured, unsupercharged, unsurfaced, untagged, untelevised, untenured, unupholstered, unvetted*; (*b*) *unacculturated, unactivated, unagglutinated, unallocated, uncalibrated, unchlorinated, undemarcated, unencapsulated, unexhilarated, unfractionated, unintegrated, unmotivated* (also *absol.*), *unmyelinated, unpatinated, unrefrigerated, unreplicated, unsegregated, unsublimated, unsyncopated* (later examples), *unvegetated*; (*c*) *unactualized, un-Americanized, unanæsthetized, unanatomized, un-Anglicized, unbowdlerized, uncolonized* (cf. Uncolonize *v.*), *undiphthongized, un-Hellenized, unindustrialized, unlemmatized, unnationalized, unnormalized, unparasitized, unpasteurized, unplasticized, unpoliticized, unpolymerized, unpressurized, unpublicized, unsclerotized, unsocialized, unstabilized, unstandardized* (also *absol.*), *unsterilized, ununionized, unvandalized, unverbalized*.

(*a*) *general*. **1921** W. de la Mare *Mem. Midget* xlvi. 314 That other less professional *début* which poor Mr. Crimble ..had left unacclaimed. **1899** *Leeds Mercury Weekly Suppl.* 5 Aug. (E.D.D.), We should call money that had been acquired without effort, 'unaddled brass'. **1935** W. de la Mare *Early One Morning* xiv. 179 Flattery of the young is so usual that it is remarkable that a child with a 'fine head' manages to keep the brains in it unaddled. **1935** T. S. Eliot *Murder in Cathedral* i. 20 Archbishop, secure and assured of your fate, unaffrayed among the shades. **1922** *Times* 15 Nov. 13/6 That unassessed quantity, the man in the street. **1970** *Jrnl. General Psychol.* LXXXII. 146 Ebbinghaus was faced with a similar problem of unassessed previous experience in learning phenomena. **1972** N. Freeling *Long Silence* II. 233 He had been to look at a picture... The price was high, and might have been a great deal higher but that it was, as he pointed out, quite unattributed. **1901** G. B. Shaw *Three Plays for Puritans* Pref. p. xxix, A thing compared to which Falstaff's unbeglamoured drinking and drabbing is respectable. **1927** *Blackw. Mag.* Nov. 601/1 By her own account the only unbobbed head in Hampstead. **1931** W. Faulkner in *Harper's Mag.* Sept. 394/2 The fine, soft cloud of her unbobbed hair gleamed like the chestnut's flank. **1928** M. Beerbohm *Variety of Things* 128 One likes to think of him there among the unbombed Lakes. **1980** A. Price *Hour of Donkey* iv. 63 One unbombed veteran to another obviously much-bombed one. **1912** W. Deeping *Sincerity* xxviii. 216 Wolfe..seemed to brush Turrell's arms aside as though they were mere sticks. His crisp, sinewy punches landed serenely. He looked unmarked, unbothered. **1965** J. A. Michener *Source* (1966) 794 Bagdadi, apparently unbothered by history, pressed on. **1936** *Nature* 7 Nov. 800/1 The native hæmoglobin of the horse, in unbuffered dilute salt solutions at the isoelectric point. **1966** *Punch* 2 Feb. 169/2 If any Tom, Dick or Harry can meet in a collapsible plastic balloon and chatter away unbugged, there is no knowing what they will say. We shall be back to the old, bad days of secret diplomacy. **1962** Simpson & Richards *Physical Princ. Junction Transistors* xi. 255 A better solution is to leave a portion of the emitter bias resistor un-by-passed, thereby introducing negative-current feedback. **1954** T. Gunn *Fighting Terms* 44, I..felt my body sweet, Uncaked blood in all its channels flowing. **1939** *Jrnl. Organic Chem.* IV. 434 No uncatalyzed addition has yet been observed. **1976** *Nature* 24 June 659/1 After the establishment of the kinetic orders of the catalysed and uncatalysed reactions a mechanism consistent with all the results was deduced. **1971** *Ibid.* 2 Apr. 276/1 If single-cell clones of cultured cells are derived from women heterozygous for *X*-linked genes determining enzyme variants, each clone shows the activity of only one of the two possible genes, whereas uncloned cultures derived from many cells show both. **1980** *Jrnl. Immunol. Methods* XXXIV. 153 Uncloned and cloned populations. **1925** *Glasgow Herald* 6 Nov. 8 The adjective uncluttered is an American term of strong commendation when used to describe a wall, a mantelpiece or even a whole room. **1960** J. Stroud *Shorn Lamb* xxii. 239 New modern houses.. with their plate-glass and uncluttered lines. **1980** *Times* 29 Feb. 26 (Advt.), Spring-time in Greece. Uncluttered beaches, [etc.]. **1922** Joyce *Ulysses* 422 An uncobbled tramsiding set with skeleton tracks. **1918** J. W. Gerard *Face to Face with Kaiserism* xx. 230 The bags were secretly opened and our uncoded despatches and letters read. **1971** B. Patten *Irrelevant Song* 24 A bleak and uncoded message whispers Down all the nerves. **1852** Mrs. Stowe *Let.* 27 Sept. in N. Spain *Mrs Beeton* (1948) i. iv. 51, I do not conceive that I have any *claim* on an English publisher for any of the profits of an uncopyrighted work. **1950** A. Lomax *Mister Jelly Roll* 292 He established *The Tempo-Music Publishing Company* to protect, publish, and push Morton's uncopyrighted and unpublished works. **1935** *Economist* 9 Nov. 930/1 They had actually harvested rather more than that quantity, but the balance, being uncouponed stock, had been carried forward to the current season's account. **1967** *Punch* 13 Sept. 388/1 Unfortunately the police caught him with several bales of uncouponed suit material. **1935** W. B. Yeats *Dramatis Personae* vii. 25 Only at pictures did he look undeafened and unblinded. **1979** *New Scientist* 17 May 537/2 It also sings normally next season, and goes through the usual subsong and plastic song again in the same way as undeafened birds. *c* **1884** E. Dickinson in *Poems* (1955) III. 1120 Bugles call the least of us To undepicted Realms. **1940** W. Faulkner *Hamlet* III. ii. 212 That perfect marriage of will and ability with a single undiffused object. **1964** *Economist* 7 Mar. 917/3 Loans still undisbursed. **1951** W. Faulkner *Requiem for Nun* III. 262 There is the clear undistanced voice as though out of the delicate antenna-skeins of radio. *a* **1868** [see *earmark *v.* 3]. **1971** *Oxf. Univ. Gaz.* 25 Feb. 709/2 That the Curators of the University Chest be authorized to expend from the unearmarked money in the Higher Studies Fund a sum of £10,000. **1934** Webster, Unenhanced. **1942** R. Frost *Witness Tree* 41 We gave ourselves outright..to the land vaguely realizing westward, but still unstoried, artless, unenhanced. **1963** *Punch* 9 Oct. 516/2 It [*sc.* industry] is markedly unenthused about the Tories. **1977** M. Kenyon *Rapist* ix. 109 Keane, unenthused, watched from the door. **1924** *Brit. Weekly* 13 Nov. 147/3, There is no 'basalt and bronze' for a Prime Minister like a still uneroded majority of 210. **1978** J. Updike *Coup* (1979) ii. 48 Its façade is topped with eight marble statues of an unreal whiteness, uneroded in this climate. **1889** *Cent. Dict.*, Unexploded. **1924** A. D. Sedgwick *Little French Girl* II. i. 100 Their mutual secret..that Giles visualized as an unexploded bomb..liable at a touch to.. scatter the family happiness to fragments. **1952** A. Cohen *Phonemes of English* 28 We find that..in *nipped* [the [p] is] unexploded. **1981** B. Langley *Autumn Tiger* ii. 21 *Berlin, February 1945*...a sign in red paint which warned 'Danger! Unexploded Bomb.' **1902** Kipling *Let.* 30 Nov. in C. Carrington *Rudyard Kipling* (1955) xv. 369 A grey stone lichened house..with old oak staircase, and all untouched and unfaked. **1968** A. Diment *Bang Bang Birds* vi. 102 One of the Birds, gazing at them with apparently un-faked adoration. **1895** Hardy *Jude* I. viii. 58 Three young unfattened pigs had escaped from their sty. **1951** T. Capote *Grass Harp* ii. 50 'Of course they are nearer God,' he said, unfazed by the disapproving, sober faces around him. **1977** P. Roth *Professor of Desire* 55 They were all screaming at one another, but he just walked along, unfazed. **1928** Unfilmed [see *unfilmable*, sense 7 b above]. **1913** H. Walpole *Fortitude* II. ii. 165 Beaming, calm, and unflustered as though he had just come from the next street. **1967** T. Stoppard *Rosencrantz & Guildenstern are Dead* II. 58 (*He turns away*. Guil grabs him and spins him back violently.) (*Unflustered*.) Now if you're going to be subtle, we'll miss each other in the dark. **1886** Hardy *Mayor of Casterbridge* II. xxii. 307 His eyes lighting on them with an unfocused gaze. **1939** H. Nicolson *Diary* 4 Feb. (1966) 390 V. and I go round to the Beales where there is a Television Set... Compared with a film, it is a bleary, flickering,..unfocused, interruptible thing. **1979** G. Macdonald *Camera* xii. 165/1 The greatest portraitist of Victorian England was the unfocussed Julia Margaret Cameron. **1967** E. R. Lannon in Cox & Grose *Organiz. & Handling Bibliogr. Rec. by Computer* IV. 82 The second type of data, sometimes called unformatted, takes on special meaning due to its surrounding elements. **1982** *Which Computer?* June 30/3 The M20 has..an on-line unformatted data storage capacity of 320 KB. **1907** T. Hodgkin *Let.* 27 Oct. in L. Creighton *Life & Lett. T. Hodgkin* (1917) xii. 282 When Christ..said that he would have us ἀμερίμνους he meant almost what we mean by 'unfussed'. **1981** N. Freeling *One Damn Thing after Another* xxxii. 236 She phoned home. To hear Arthur's voice, unfussed and matter-of-fact. **1903** Kipling *Five Nations* 155 For the ungrazed upland, the untilled lea Cry, and the fields forlorn. **1972** J. L. Harper *Population Biol. Plants* 450 Within this zone of taller, ungrazed herbage the balance of competitive interactions is changed. **1977** *Belfast Tel.* 19 Jan. 1/2 Environment Department officials were caught napping today by the heavy overnight frost, which left a network of icy Ulster roads ungritted. **1938** *Amer. Speech* XIII. 314/2 *Jump salty*, implies an unexpected change in a person's..knowledge... An unhipped person may become hipped. **1944** D. Burley *Orig. Handbk. Harlem Jive* 90, I hunch the pinball layout, Jack, and it's an unhipped tilt. **1955** Auden *Shield of Achilles* iii. 66 Where should we be but for them? Feral skill, un-housetrained. **1967** A. Wilson *No Laughing Matter* III. 232 He's doing PPE. Whatever that may be. It sounds very un-housetrained. **1864** Geo. Eliot *Let.* 8 Mar. (1956) IV. 138 Iced water is what I always long for ..while water *un*-iced is detestable to me. **1971** 'J. Bell' *Hole in Ground* iv. 53 A plate of scones and an uniced cake. **1934** Webster, Uninjected. **1946** *Nature* 31 Aug. 311/2 There are two natural folds of skin in cattle at..the root of the tail, and the uninjected fold serves as a control. **1977** P. Dickinson *Walking Dead* I. i. 20 He only injected three pairs of rats in a batch... Nor could he run uninjected rats. **1938** J. W. Day *Dog in Sport* xi. 143 There were partridges, naturally bred, more or less unkeepered. **1977** H. Douglas-Home *Birdman* vi. 71 Keepering established a balance which could not have existed in the earlier predator dominated times. Anyone who doubts this need only take a train from Edinburgh to Glasgow, unkeepered country today. **1956** 'C. Blackstock' *Dewey Death* v. 102 She was..her well-groomed self, with..unladdered stockings. **1978** J. Goodman *Last Sentence* i. 28 She and Cecil were entered..at the Streatham Locarno..and she *had* to have unladdered hose for that. **1895** *Montgomery Ward Catal.* Spring & Summer 36/3 Soft or Unlaundered Waists. **1965** T. Capote *In Cold Blood* ii. 125 Some bits of unlaundered laundry. **1978** 'R. Cassilis' *Winding Sheet* III. 199 Politicians..should never be allowed to hear unlaundered secrets. It made them uncomfortable. **1953** K. Jackson *Lang. & Hist. in Early Britain* 554 Original unlenited Brit. *b*. **1977** *Word 1972* XXVIII. 99 Three of the 7 semi-speakers showed switching between lenited and unlenited forms of *daor* 'expensive'. **1966** R. H. Rimmer *Harrad Experiment* (1967) 92 Locating my leopard coat under a heap of fur coats in the unmonitored cloak room. **1976** *Guardian* 23 July 15/6 Now at last it seems that a researcher from Stirling University may begin work, and both staff and inmates see this as a positive development, while regretting that the first years have slipped past largely unmonitored. **1962** G. Butler *Coffin in Oxford* ii. 38 The undying burning hate of one un-neutered tom for another. **1971** *Nat. Geographic* May 735/2 The Seigneur [of Sark]..exercises the right to keep the Island's only 'unneutered' bitch. **1931** V. Woolf *Waves* 205 Here am I marching up and down the terrace alone, unoriented. **1947** [see *oriented *ppl. a.* 2]. **1962** U. Weinreich in Householder & Saporta *Probl. Lexicogr.* 30 Features of a geographic area can be studied even from an unoriented map. **1969** W. R. R. Park *Plastics Film Technol.* ii. 33 Unoriented film is very brittle at below zero temperatures. **1948** A. N. Keith *Three came Home* xi. 199, 100 cigarettes, unpackaged and unlabelled. **1984** *Times Lit. Suppl.* 10 Feb. 148/1 Here is a manual for the unpackaged, mildly enterprising and tolerably well-heeled holidaymakers. **1923** A. Huxley *Antic Hay* x. 152 The wearing exhaustion to which long-protracted sitting on unpadded seats subjects them. **1979** A. Buck *Dress in 18th-C. Eng.* ii. 31 The coat was losing its stiffness, and the pleats of its skirts, now unpadded, were set further back. **1914** G. Bell *Let.* 21 Jan. (1927) I. xiii. 327 And we came at two o'clock to the last of the castles, Bair, as yet unplanned and unphotographed. **1929** 'E. Queen' *Roman Hat Mystery* v. 86 You are by no means an unphotographed young lady—I saw your picture in the paper to-day. **1981** J. Wainwright *All on Summer's Day* 80 An inspection which (literally) left no square inch unsearched or unphotographed. **1909** Webster, Unplanned. **1914** [see *unphotographed* above]. **1942** J. S. Huxley in *Harper's Mag.* Sept. 346/1 The need for entering upon our revolution consciously.., deliberately guiding its course instead of allowing its blind forces to push and buffet our unplanned lives. **1959** *Listener* 4 June 981/1 These people are spontaneous in almost all they do. Children are 'unplanned'. **1961** 'J. le Carré' *Call for Dead* xvii. 186 The obviously unplanned and unproductive selection of files. **1973** *Times* 19 Apr. 19/2 Nearly half of *all* pregnancies are unplanned. **1981** N. Tucker *Child & Book* i. 45 A shoe.. may..house a large and almost certainly unplanned family. **1934** Webster, Unposed. **1960** *Guardian* 13 July 7/2 The best pictures of people are unposed and are taken while they are intent on what they are doing. **1968** E. Hyams *Mischief Makers* i. 12 This unselfconscious, unposed animation was not what I was used to. **1963** *Times* 26 Feb. 9/4 After being boosted to orbital speed, the 45ft. long Dyna-Soar will be capable of manoeuvering freely during reentry into the earth's atmosphere, and then making a horizontal landing. As it will be unpowered at that stage this will call for particular skill and judgment

by the pilot. **1975** S. Johnson *Urbane Guerilla* III. 108 The canal..plunging..through a half-mile tunnel forbidden to unpowered craft. **1941** W. C. Handy *Father of Blues* (1957) xiv. 194 Other New York composer-conductors were introduced here and there and asked to play unprogrammed numbers. **1953** *Sun* (Baltimore) 12 Mar. 2/1 The 'unprogrammed' funds referred to come to the Army from several sources. **1959** *New Biol.* XXVIII. 113 In order to keep such a process going in its most elementary form, some prearrangement, in the form of strings of unprogrammed material to be used, has been found necessary. **1974** *Sci. Amer.* May 61/2 When the machine is off or unprogrammed, the knob moves freely in all directions. **1976** H. Wilson *Governance of Britain* iv. 85 Apart from pre-arranged ministerial meetings with documents circulated and written and sometimes oral briefing, each day sees a number of often unprogrammed meetings with ministers. **1967** *Antiquaries Jrnl.* XLVII. 209 Another animal..is the springing canine with collar on an unprovenanced ?flagon handle also in the British Museum. **1919** A. P. Herbert in *Punch* 22 Jan. 62/1 The free unrationed blotting-pad. **1940** C. E. Milburn *Diary* 10 Mar. (1979) 26 To church..then home for the last unrationed meat dinner. **1982** T. FitzGibbon *With Love* I. v. 31 We would have friends to dinner, and with food rationing becoming stringent, stuff and braise unrationed sheeps' hearts. **1934** Webster, Unrattled. **1953** *Manch. Guardian Weekly* 20 Aug. 7 He was quite unrattled, in good physical shape. **1976** J. Colville *Footprints in Time* xxvii. 144 His second-in-command at the Embassy..was unrattled by the crescendo of disaster to the allied cause. **1952** *N.Y. Times* 8 Sept. 27/5 The third-seeded player, Irene Rawcliffe of Nutley High, was eliminated by unseeded Lynn Anderson of Mount St. Mary's Academy 6-1, 6-1. **1977** *Washington Post* 9 Mar. D4, I think anybody, including the unseeded squads, has a chance to make the final. **1928** *Daily Express* 23 May 8/4 Among the fifty horsewomen only five have unshingled hair. **1933** A. G. Macdonell *England, their England* xiv. 246 A network of narrow lanes, unsignposted. **1981** M. E. Atkins *Palimpsest* ii. 12 Breakdowns *always* occurred in foul weather on unsignposted roads. **1964** F. Bowers *Bibliogr. & Textual Criticism* VI. i. 165 With selected though unslanted evidence, a point should be reached at which our common-sense view of probability rebels at being asked to accept any more coincidence as the result of mere chance. **1974** J. Irving *158-Pound Marriage* vi. 131 It was an old-fashioned, unslanted, glass windshield. **1889** *Cent. Dict.*, Unsliced. **1968** D. E. Allen *Brit. Tastes* iii. 77 Bread, being..made more by small local bakers, tends to be unwrapped (and..unsliced) more frequently than elsewhere. **1978** *Times* 15 Apr. 3/6 There will be bloomers, cobs,..sliced and unsliced, white and brown loaves on the shelves. **1940** R. E. Smith in *Maya & Their Neighbors* xvi. 243 Unslipped jars with flaring necks. **1977** *Sci. Amer.* Mar. 130/2 The pots range from rough, unslipped pieces—'earthenware' in modern terminology—to thinner-walled pieces with smooth and glossy surfaces. **1851** H. Melville *Moby Dick* III. xviii. 128 This ante-mosaic, unsourced existence of the unspeakable terrors of the whale. **1977** *Kuwait Times* 1 Nov. 6/3 In an unsourced report from Bucharest, the Egyptian agency also said that a decision whether to reconvene the Geneva conference would be made within the next two weeks. **1939** J. R. Kinney *How to raise a Dog* vi. 144 During their periods the unspayed bitch is sometimes an aesthetic problem because of the discharge. **1979** 'J. Ross' *Rattling of Old Bones* ix. 83 The woman..[had] the sexual hunger of an unspayed cat in her eyes. **1930** E. Rice *Voyage to Purilia* i. 7 We wanted to travel quietly and unobtrusively... This could be done only if our voyage was unsponsored and unheralded. **1979** *Tucson* (Arizona) *Citizen* 20 Sept. 2D/1 The Patio Pools team amassed 4,882 pins..against Diamond Life and unsponsored Team No. 4 in the first round of action. **1934** Webster, Unstaffed. **1957** [see *remotely *adv.* 2 b]. **1979** V. L. Pandit *Scope of Happiness* xxxvii. 288 The Embassy remained closed and unstaffed. **1962** *Guardian* 30 Mar. 10/4 A large *unstreamed* school in which these distinctions have no place. **1981** *Indexer* XII. 152/2 Six unstreamed classes in a local comprehensive school. **1936** F. & G. M. Heider tr. *Lewin's Princ. Topological Psychol.* xii. 134 If one puts a rat into the maze without food, he gets a chance to 'orient' himself, which means that what is first an unstructured field becomes structured. **1948** M. Sherif *Outl. Soc. Psychol.* vii. 181 Psychological principles which are..applicable in the case of the individual facing an unstructured situation. **1965** *Times Lit. Suppl.* 25 Nov. 1049/3 This was..an anarchically 'unstructured' society. **1973** *Computers & Humanities* VII. 171 The current trend in elementary education is in the direction of open, unstructured classrooms. **1977** *Lancashire Life* Mar. 106/2 Spring coats are lean and unstructured, with their line emphasised with braid or other fabric trimming in contrast colour. **1984** *Guardian* 5 Mar. 11/4 Jackets are loose and unstructured. **1929** V. W. Pagé *Mod. Aviation Engines* I. xii. 343 A supercharged plane requires a somewhat larger propeller than an unsupercharged one. **1974** *Encycl. Brit. Macropædia* VII. 936/1 The loss in power suffered by unsupercharged engines at high altitudes can be largely restored [by supercharging]. **1953** W. Moore *Bring Jubilee* (1955) ii. 13 It was impossible to maintain unsurfaced highways in good condition. **1971** D. Potter *Brit. Eliz. Stamps* vi. 68 A few copies of the 10d on unsurfaced paper..were found in 1970. **1909** Webster, Untagged. **1957** D. L. Bolinger in *Publ. Amer. Dial. Soc.* XXVIII. 112 The evolution of the secondary towards tag-ness has not reached the point of admitting an untagged secondary on the second A of the most conducive intonation. **1967** *Oceanogr. & Marine Biol.* XVI. 440 The basis for this is an assumption that tagged and untagged fish are caught in the same proportions. **1962** *Punch* 11 Apr. 555/2 Shakespeare, untelevised and unrecorded, was never asked to take it all back and substitute some more ethical appeal. **1971** *Guardian* 25 Aug. 8/5 'Polaris—The Secret World' (Yorkshire) was wholly untelevised territory. **1969** *Federal Suppl.* (U.S.) 11 Dec. 1053/1 Person, even though untenured, may not be denied public employment for unconstitutional reason. **1981** 'A. Cross' *Death in Faculty* v. 49 An untenured professor in search of 'contacts'. **1919** G. B. Shaw *Heartbreak House* I. 1 A row of books under the window provides an unupholstered window-seat. **1938** S. Beckett *Murphy* v. 63 Two massive upright unupholstered armchairs. **1962** *Guardian* 15 Feb. 6/6 Unscripted, unvetted discussion programmes. **1979** T. Wiseman *Game of Secrets* viii. 100 The company..in certain cases actually prohibited unsanctioned intercourse with unvetted outsiders.

(*b*) ending in *-ated*. **1969** *Nature* 26 July 419/1 European children, during their first year at school, were like unacculturated African tribal adults in that many failed to interpret the perspective features of a picture as representing depth. **1956** A. Huxley *Adonis & Alphabet* 156 The problem of what we may politely call 'unactivated sludge'. **1950** Race & Sanger *Blood Groups in Man* iii. 34 Such points as..the shaking free of unagglutinated cells enmeshed in agglutinates are skilfully dealt with. **1979** *Jrnl. R. Soc. Med.* LXXII. 450 An unagglutinated suspension of the recipient's blood. **1869** *Bradshaw's Railway Man.* XXI. 183 Deducting all proper expenses and interest on the unallocated debt. *a* **1974** R. Crossman *Diaries* (1977) III. 953 It looks to me as if he might find himself unallocated and put back into the pool. **1965** *Canad. Jrnl. Linguistics* Fall 45 The whole process is wasteful, much as if one were to work with a number of highly sensitive but quite uncalibrated scientific instruments. **1974** *Jrnl. Water Pollution Control* XLVI. 2153 (*heading*) Bacteriology of chlorinated and unchlorinated wastewater effluents. **1902** G. F. Stout in H. Sturt *Personal Idealism* I. 42 All demarcated figure presupposes what we may call undemarcated figure. **1967** M. Ayub Khan *Friends not Masters* x. 161 A similar situation could arise on our own undemarcated borders in the Sinkiang and Baltistan areas. **1965** *Amer. Jrnl. Clin. Path.* XLIII. 112/2 The tumor was..usually unencapsulated and often diffusely infiltrated the adjacent muscle. **1851** H. Melville *Moby Dick* I. xxxiii. 235 He swings himself to the deck and in an even, unexhilarated voice, saying 'Dinner, Mr Starbuck,' disappears into the cabin. **1956** *Nature* 18 Feb. 326/2 The unfractionated polymer..gives crystals which are smaller and more opaque to electrons. **1889** *Cent. Dict.*, Unintegrated. **1920** R. R. Marett *Psychology & Folk-Lore* vi. 121 An unintegrated or imperfectly rational type. *a* **1974** R. Crossman *Diaries* (1977) III. 484 An absolutely self-perpetuating oligarchy of R.H.B.s completely unintegrated with the community services of the local authorities. **1922** J. Riviere tr. *Freud's Introd. Lect. Psycho-Anal.* iv. 48 They [*sc.* acts] also appear to be unmotivated, insignificant and unimportant. **1947** *Mind* LVI. 353 Witness the familiar sort of situation where someone has met with misfortune from a natural, apparently unmotivated, cause. **1953** E. Simon *Past Masters* I. 18 The house..bleak despite a mass of unmotivated Victorian architectural curlicues. **1964** M. Argyle *Psychol. & Social Probl.* v. 70 A man with an unmotivated tendency to attack perambulators and handbags. **1973** *Sci. Amer.* July 22/3 A fairly simple change yielded a considerable response from a group of patients who had previously been characterized..as 'unmotivated' to seek postpartum checkups. **1977** P. Strevens *New Orientations Teaching of English* i. 10 They are unmotivated towards learning English and sometimes even hostile towards doing so. **1978** H. R. F. Keating *Long Walk to Wimbledon* II. 18 His world was at the mercy of the unmotivated. **1915** *Brain* XXXVIII. 384 The unmyelinated fibres of spinal nerves are distributed chiefly to the skin. **1980** *Jrnl. R. Soc. Med.* LXXIII. 268 Using the electronmicroscope, Frank..described unmyelinated nerve fibres in the inner third of fully formed human coronal dentine. **1926** *S. Afr. Jrnl. Sci.* XXIII. 785 The Earlier Stone Age implements are heavily patinated, the later are unpatinated or only slightly so. **1946** F. E. Zeuner *Dating Past* vi. 193 The gravel has yielded..in a fresh, unabraded and unpatinated condition: a middle Levalloisian comparable with that found by Burchell. **1934** Webster, Unrefrigerated. **1964** *Supermarket & Self-Service* (Johannesburg) Feb. 5/2 On the left is a 12 ft. long cabinet with three unrefrigerated shelves. **1968** D. E. Allen *Brit. Tastes* 234 Milk was scarce and the sealed, sterilised variety would stay fresh, though unrefrigerated, for seven days. **1971** *Nature* 25 June 491/3 This fact enabled Hill and Hillova to separate.. the newly replicated double helices..from unreplicated DNA. **1976** *Billings* (Montana) *Gaz.* 20 June 12-A/1 It would be 'almost criminal to look upon this unreplicated study as representative'. **1905** *Rep. Evolution Committee R. Soc.* II. 110 Some gametes are bearing r.p. [*sc.* the characters of both 'rose comb' and 'pea comb' together (in poultry)] unsegregated. **1954** *Harper's Mag.* Sept. 47/2 On the first day of the fall term, Amy Miller..was so apprehensive about entering an unsegregated school that she persuaded her father, a truck driver, to accompany her. **1977** *Listener* 17 Feb. 215/3 One of the motives..was to bring an unsegregated band to perform before an unsegregated audience. **1923** J. S. Huxley *Ess. Biologist* vii. 271 We may perhaps best say that a sublimated instinct has more and higher values attached to its satisfaction than one unsublimated. **1952** Gerth & Martindale tr. *Weber's Anc. Judaism* v. 126 Acculturation is generally productive of entirely new..phenomena given the..compelling need of absorbing a series of as yet unsublimated ideas. **1967** N. Marsh *Death at Dolphin* iv. 94 An actor..was embarrassed rather than released by unsublimated chunks of raw association. **1922** W. J. Locke *Tale of Triona* ix. 97 A company of women groping ..after ideals in unsyncopated time. **1950** Blesh & Janis *They all played Ragtime* (1958) ii. 44 The second and third themes, printed unsyncopated, cry out for syncopation when one plays them. **1953** K. Jackson *Lang. & Hist. in Early Britain* 689 The AS. *Cerdic* seems to have been borrowed..from a syncopated Pr. W. **Car'dig*..side by side with unsyncopated *Ceredig*. **1939** *Geogr. Jrnl.* XCIV. 221 A high pyramid of unvegetated sand was a prominent landmark. **1970** *Nature* 2 May 429/2 The unvegetated mudflats of Sanderson's Gulf.

(*c*) ending in *-ized*. **1938** A. Farrer in T. E. Jessup et al. *Christian Understanding of Man* 211 Christianity asserts ..that there is a true nature of man..actual in the Divine mind and never wholly unactualized in men. **1875** A. J. C. Hare *Days near Rome* I. 1 The real, true, un-Anglicized, un-Americanized country. **1931** *Times Lit. Suppl.* 18 June 481/4 An un-Americanised..honest English England. **1950** O Nash *Family Reunion* (1951) 134 He undergoes the lecturing Like unanesthetized vivisecturing. **1934** Blunden *Mind's Eye* IV. 256 A world as yet unanatomized by scientific accuracy. **1875** Un-Anglicized [see *un-Americanized* above]. **1964** B. Trnka in D. Abercrombie et al. *Daniel Jones* 186 The French nasals in the occasionally un-Anglicized pronunciation of Modern French words. **1873** J. Davies *Hesiod, & Theognis* II. i. 130 The shape in which the poetry of Theognis has come down to us is as unlike the original form and drift as a handbook of maxims from Shakespeare is unlike an undoctored and un-Bowdlerised play. **1957** *Times Lit. Suppl.* 6 Dec. 738/2 A quarter of a century ago the complete, unbowdlerized edition of 1951 might have made a stir. **1862** J. S. Mill in *Westm. Rev.* XXII. 510 The vast uncolonized region of Arkansas, and Texas. **1979** *Jrnl. Brit. Interplanetary Soc.* XXXII. 222/1 The relative availability of uncolonized stars will be rapidly and dramatically restricted. **1921** E. Sapir *Language* 176 The long undiphthongized *u* is still preserved in Lowland Scotch. **1951** W. K. Matthews *Languages U.S.S.R.* iii. 35 Undiphthongised long *e* and *o*. **1907** G. Murray *Rise Greek Epic* ii. 55 It was a hard task for an island in that position to keep itself un-Hellenized. **1934** A. Toynbee *Study of Hist.* II. 79 This region remained unhellenized much longer than many places that were far more distant from the Aegean. **1960** C. S. Lewis *Studies in Words* ii. 24 In defiance of chronology, we begin with some account of the Latin and English words in their un-hellenised condition. **1934** Webster, Unindustrialized. **1962** *Rep. Constitution Commission, Pakistan 1961* iv. 40 The unindustrialized portion of Bengal. **1978** R. Mitchison *Life in Scotland* viii. 154 The Scottish pattern of tall tenements..is to be found in older unindustrialized burghs. **1971** Unlemmatized [see *key-word-in-context* s.v. *key *sb.*[1] 17]. **1968** O. Wynd *Sumatra Seven Zero* ii. 16 She'll have a handsome income for as long as the place stays unnationalized. **1957** L. Fox *Numerical Solution Two-Point Boundary Probl.* vii. 170 This constant is the reciprocal of the corresponding current estimate of λ, obtained as the ratio of successive unnormalized first components. **1889** *Cent. Dict.*, Unparasitized. **1949** *Proc. Soc. Exper. Biol. & Med.* LXX. 580/2 To make such values really comparable with those obtained for unparasitized blood it is necessary to know both the parasite count, and the proportions present of parasites of different sizes. **1909** Webster, Unpasteurized. **1955** J. G. Davis *Dict. Dairying* (ed. 2) 789 Special fluting arrangements to prevent any unpasteurised milk seepages from reaching the outlet connection. **1975** *Listener* 14 Aug. 205/2 The beer here is unpasteurised, unassisted into the glass by CO_2. **1946** *Electronic Engin.* Feb. 54/3 Unplasticised polyvinyl chloride. **1982** Sears & Darby *Technol. of Plasticizers* v. 301 Rigid, unplasticized PVC..essentially obeys Hooke's Law for quite some distance. **1976** D. Davin *Woman-Work* (1978) ii. 57 Communist experience in the towns had been limited to clandestine work in which contact with the mass of unpoliticized women was very difficult. **1879** *Jrnl. Chem. Soc.* XXXV. 758 They [*sc.* terpenes] are ultimately rendered optically inactive, a considerable proportion remaining unpolymerised. **1976** *Jrnl. Molecular Biol.* CV. 527 Is there an appreciable amount of unpolymerized protein free within the cytoplasm, as there probably is for microtubules? **1958** A. Toynbee *East to West* 37 Even this unpressurized plane would fly high enough to make certain of that [*sc.* flying blind]. **1963** *Economist* 2 Feb. 413/2 Some of the best thinking had been done under pressure..far from..optimum conditions... Its case for an unpressurised future might have seemed stronger. **1974** P. Gore-Booth *With Great Truth & Respect* 57 But I did, between Chicago and Minneapolis, take my first aeroplane flight; nobody warned me what an unpressurized flight does to your ears. **1978** S. Radley *Death & Maiden* xiii. 123 The real, unpressurised Adnams Suffolk ale. **1978** *Detroit Free Press* 16 Apr. E2/4 (Advt.), Package of 3 Nassau brand unpressurized tennis balls. **1959** *Times* 28 May 16/6 Equally important is a constant, unpublicized two-way exchange. **1977** M. Edelman *Polit. Lang.* vii. 123 Both the publicized and the unpublicized aspects of policymaking processes have functions to serve. **1967** *Ann. Entomol. Soc. Amer.* LX. 1134/1 The second type of integument is found in the softer unsclerotized regions of the body. **1976** D. J. Horn *Biol. Insects* v. 212 Greater freedom of movement between sclerotized parts is afforded by thin, unsclerotized membranes at joints of moveable body parts. **1934** Webster, Unsocialized. **1948** H. V. Hodson *Twentieth-Cent. Empire* x. 112 The still unstabilized and unsocialized citizens. **1964** P. Meadows in I. L. Horowitz *New Sociology* 451 Locating the sources of human hostility in.. unsocialized, primitive impulse, Freudian social theory provided the same kind of conservative blast at political and economic radicalism. **1948** Unstabilized [see *unsocialized* above]. **1960** *Times* 23 Feb. 12/2 The other [class of satellite] was a smaller unstabilized instrumentation package of 50 lb. **1975** *Petroleum Rev.* XXIX. 89/3 Unstabilised raw crude oil consisting of crude oil and natural gas liquids will be pumped at increasing rates of throughput to Teesside. **1909** Webster, Unstandardized. **1920** S. Alexander *Space, Time, & Deity* II. 262 Error is real only as possessed by the unstandardised believer. **1930** E. Weekley *Adjectives & Other Words* 154, I need hardly say that he was one of the elderly and unstandardized—the sort of man..who says *onct* for *once*. **1909** Webster, Unsterilized. **1945** *Mind* LIV. 327 Pasteur found that putrescible liquids exposed to unsterilised air..developed infusoria. **1979** *United States 1980/81* (Penguin Travel Guides) 333 The atmosphere in this 19th-century structure is distinctly unmodern..and unsterilized. **1973** *Guardian* 23 Feb. 17/2 The American unions seemed to accept a situation in which some 68 million workers..remained ununionized. **1984** *Listener* 1 Mar. 15/2 There is..an inability to see what this feels like from the point of view of people who may be unskilled workers who are still predominantly ununionised. **1968** P. Dickinson *Skin Deep* iii. 42 There was still a telephone kiosk..momentarily unvandalised too. **1909** W. James *Pluralistic Universe* 348 Its purity is only a relative term, meaning the proportional amount of unverbalized sensation which it [*sc.* experience] still embodies. **1977** P. Strevens *New Orientations Teaching Eng.* xii. 149 The internalized, unverbalized rules according to which sentences are created.

c. *un-clued-up*.

1970 D. Francis *Rat Race* ix. 116 The Derrydowns Six had been hired by an un-clued-up trainer. 1982 *Times* 20 Apr. 1/5 Cardinal Hume..said..'I am terribly unclued up on what constitutes a war.'

9. a. *unautumned, unaveraged, unbanked* (also *absol.*), *unbra-ed, unbrassièred, undentisted, ungated, ungoggled, unillusioned, unjacketed, unmooned, unpupilled, unrectored, unscripted, unspeeched* (later example).

1920 J. Masefield *Enslaved* 6 The old un-autumned beauty that never goes away. 1905 Webster, Unaveraged. 1924 G. B. Shaw *St. Joan* p. xli, The unaveraged individual, representing life..never at its merely mathematical average. 1965 *Economist* 19 June p. vii, One way of jerking the clearing banks into providing better facilities for the great unbanked public. 1980 *Daily Tel.* 9 Dec. 15 The banks are well aware that they can best pay their own wages bills by drumming up more business from the great unbanked. 1962 J. P. Carstairs *Pardon my Gun* i. 20 The large cusp-like un-bra-ed bosom ..stood, prow-like. 1965 G. McInnes *Road to Gundagai* ix. 148 He touched his hat to local Mums with shapeless unbrassiered bosoms. 1981 M. Gee *Dying, in Other Words* 74 She had pressed her unbrassièred breasts on his shoulders and said he was gorgeous. *a* 1916 H. James *Sense of Past* (1917) 290 His perfect and soignées teeth..which that undentisted age can't have known the like of. 1954 *Norfolk Mag.* June 57/2 Immediately beyond the farmhouse is the now ungated entrance to the grounds. 1978 C. Tomlinson *Shaft* 21 Did Eden Greet us ungated? 1914 M. Beerbohm in *Eng. Rev.* Dec. 18 From time to time (for I too was ungoggled) I looked round to nod and smile. 1940 G. Frankau *Self-Portrait* lxii. 385, I drove like a fool, the car open and my eyes ungoggled. 1934 Webster, Unillusioned. 1940 W. Faulkner *Hamlet* II. ii. 130 The father, the lean pleasant shrewd unillusioned man..had been betrayed at the last. 1982 *London Rev. Bks.* IV. xxiv. 23/2 We competitors are unillusioned about the lipservice of the walls of leaning bodies and megaphone hands that we sprint between. 1925 *Times* 17 Aug. 6/2 A single unjacketed saucepan of a capacity equal to twice the amount of water to be put into it. 1970 D. Dodge *Hatchetman* iii. 52 They used lead bullets, unjacketed; illegal, under the Hague Convention. 1980 N. Marsh *Photo-Finish* iv. 86 He..came upon a book... It was unjacketed and the spine was rubbed. 1926 S. Leslie *Cantab* (ed. 2) x. 115 The unmooned sky inked out the Universe. 1940 W. Faulkner *Hamlet* II. ii. 127 The long return through night-time roads across the mooned or unmooned sleeping land. 1914 R. Brooke *Let.* Mar. (1968) 568 They come to you by night,..& their eyes—unpupiled balls of white—fall out too, & they stink & shine. 1955 W. Gaddis *Recognitions* I. vii. 231 He did not move, nor did his unpupiled eyes betray any surprise. 1944 Auden *For Time Being* 31 She cannot conceivably tolerate in her presence..the rival..who does not rule but defiantly is the unrectored chaos. 1953 *Ann. Reg.* 1952 449 Previously an unscripted defamatory broadcast was treated as slander. 1966 G. N. Leech *Eng. in Advertising* ix. 89 So far my remarks about the spoken language have applied mainly to unrehearsed and unscripted speech. 1922 Joyce *Ulysses* 49 His mouth moulded issuing breath, unspeeched.

b. *unfootnoted*.

1964 *Economist* 19 Dec. 1356/1 A straightforward, unfootnoted historical narrative.

10. *unalloying, unarresting, uncompelling, unconflicting, unendearing, unenticing* (examples), *unexhilarating, unexpanding, unfascinating* (earlier example), *unflowering, unforfending, unforthcoming, unmatching, unminding* (cf. Unminding *vbl. sb.*), *unmourning, unselfpitying, unselfregarding, unselfseeking, unswinging* [*swinging *ppl. a.* 3 b, c], *unthreatening*.

1812 Shelley *Retrospect* in *Compl. Poet. Wks.* (1904) 970/2 When mountain, meadow, wood and stream With unalloying glory gleam. 1906 G. Higgins in A. Smith *Life of St. Agnes* p. viii, It may be objected..that a book treating of times and persons and occurrences of so ancient a date must be dull and unarresting. 1958 *Times* 25 Nov. 14/4 It was disconcerting to hear..so musically unarresting an account from her of so genial a work. 1920 W. J. Locke *House of Baltazar* xi. 130 We sought for possible imperative objectives, and one so apparently uncompelling as China never occurred to us. 1977 *Gramophone* Jan. 1139/3 How best to write about a set which is noteworthy but ultimately uncompelling? 1899 'J. Oxenham' *Rising Fortunes* xxvii. 182 His was a character of sharp contrasts—of perfect consistency, from his own point of view—of unconflicting opposites. 1938 F. Scott Fitzgerald *Let.* 18 Apr. (1964) 29 We are unconflicting on 90% of things. 1926 W. de la Mare *Connoisseur* 49 The eyes..were now peering vacantly..over a trim but unendearing moustache at the crumbs on his empty plate. 1981 F. Inglis *Promise of Happiness* viii. 202 That is a lapse into avuncularity... In the novel..it is not unendearing. 1915 W. J. Locke *Jaffery* v. 63 Mountains.. were made for goats and cascades and lunatics..; and the more jagged and unenticing they are, the greater is their specious air of stupendousness. 1948 D. Welch *Brave & Cruel* 95, I wondered that Mr. Mellon could show such fondness for Phyllis; she seemed so very unenticing to me. 1978 *Observer* 16 Apr. 38/1 The furniture was unenticing, but some American dealer would take a 40-foot container of assorted items. 1811 Jane Austen *Sense & Sensibility* II. xiii. 253 The nature of her commendation..was..very unexhilarating to Edward. 1909 W. J. Locke *Septimus* x. 114 He walked to the window and looked out into the unexhilarating street. 1968 *Punch* 3 Jan. 32/2 With such an unexhilarating cast-list you'd expect a muted, subfusc drama. But there are surprises in store. 1905 *Spectator* 11 Mar. 363 Countries which exclude coloured labour,.. though possessing a larger area and..advantages of soil and climate, continue to suffer from..unexpanding revenue, heavy taxation, and a stationary population. 1957 'O. Edwards' *Talking of Books* 231 The time came, alas —as it always must in the case of expanding collections of books and unexpanding homes—when something had to go. 1866 Geo. Eliot *Felix Holt* III. xliii. 149 The love of this not unfascinating man. 1942 T. S. Eliot *Little Gidding* iii. 12 Being between two lives—unflowering, between The live and the dead nettle. 1891 Hardy *Tess* III. lvii. 255 Their every idea was temporary and unforefending, like the plans of two children. 1920 'O. Douglas' *Penny Plain* xiv. 141 He might be so shy and unforthcoming that he would put people off. 1934 R. A. Knox *Still Dead* xii. 152 He was certainly a very repressed, unforthcoming sort of boy. 1966 I. Jefferies *House-Surgeon* i. 13 Smiling uncomfortably and looking round for unforthcoming support from Grant. 1981 'E. Lathen' *Going for Gold* xvi. 183 They were guarding a secret... It tells us why Miss Deladier..was so unforthcoming when we spoke to her. 1939 D. Wallace *E. Anglia* ii. 35 This unique church with its two unmatching towers. 1969 C. Armstrong *Seven Seats to Moon* xiii. 119 A bottle and two unmatching glasses. 1945 Dylan Thomas in *Life & Lett.* July 28 Under the unminding skies. 1945 —— in *New Republic* 14 May 675/2 By the unmourning water Of the riding Thames. 1932 W. Faulkner *Light in August* xi. 221 The hard, untearful and unselfpitying..yielding of that surrender. 1983 *N. & Q.* Dec. 576/1 The voice that emerges..is that of a writer who is..generous in spirit, self-aware and unselfpitying. *a* 1945 E. R. Eddison *Mezentian Gate* (1958) ii. 21 An ambiency of beauty that lived in her whole frame and posture, an easefulness and reposefulness of unselfregarding grace. 1982 W. J. Walsh *R. K. Narayan* i. 27 An oblique and as unself-regarding a treatment as anything like an autobiography could possibly be. 1931 W. S. Churchill *World Crisis* VI. iii. 53 Fop, dandy, la-di-da; amiable, polite and curiously un-selfseeking. 1963 *Times* 7 June 7/3 Edward Grey, 'the most completely unselfconscious and unselfseeking politician I have ever known'. 1958 K. Goodwin in P. Gammond *Decca Bk. Jazz* xiii. 153 Rumsey—himself a rather dull, plodding, completely unswinging bassist. 1967 *Punch* 23 Aug. 275/2 It only takes you a day or two in this utterly unswinging country to realise what a frightful crushing bore the legendary London Scene has become. 1969 *Daily Tel.* 29 May 22/7 An unswinging Anglo-Irishman who stands for the common decencies now so acutely unfashionable. 1972 *Jazz & Blues* Feb. 22/1 Marshall is a little unswinging in places but Weston is suitably solid. 1903 E. Wharton *Sanctuary* I. i. 11 The very lifting of the cloud—remote, unthreatening as it had been.

11. *unarguably, unassessably, unboringly, unbridgeably, unconsciently, unhungrily, unidentifiably, unignorably, unprotestingly, unselectively, unurgently*.

1929 *Daily Express* 3 Jan. 8/6 'You know Mr. Umph, don't you?' with an intonation implying that not to know Mr. Umph leaves one unarguably in the certifiable class. 1978 *Nature* 20 Apr. 666/3 By excluding a sizable class of environmental variables, high within group heritability, unarguably does raise the probability of the remaining classes of possible explanations. 1937 J. R. R. Tolkien *Hobbit* xii. 231 Surely, O Smaug the unassessably wealthy, you must realize that your success has made you some bitter enemies? 1921 F. G. Ellerton *Let.* 6 Sept. in *John Bailey* (1935) 214 You do his life most frightfully well and unboringly and your observes on *Paradise Lost* and *Paradise Regained* are most excellent. 1932 *Times Lit. Suppl.* 21 Jan. 33/3 The unique, the unbridgably different. 1979 *Church Times* 30 Nov. (Christmas Bk. Suppl.) p. xi/1, Christ and God are presented as living in a world unbridgeably remote. 1929 R. Bridges *Testament of Beauty* II. 51 So, tho' slowly and unconsciently, he remembereth. 1922 Joyce *Ulysses* 172 His eyes unhungrily saw shelves of tins, sardines, gaudy lobsters' claws. 1965 'R. Macdonald' *Far Side of Dollar* xxiii. 204 Susanna broiled me a steak, and chewed unhungrily on a piece of it. 1934 F. M. Ford *Let.* 27 Sept. (1965) 235, I never guyed you either identifiably or unidentifiably. 1982 A. Price *Old Vengeful* ii. 42 The new voice..was.. unidentifiably classless. 1960 E. Bowen *Time in Rome* ii. 41 No tribute to an assassinated uncle can have been ever more unignorably placed. 1953 R. Lehmann *Echoing Grove* 80 'I didn't realize she was so completely your kept woman.' 'She wasn't,' he said unprotestingly. 1972 P. Cave *Judas Freaks* xv. 104 She followed unprotestingly. 1929 A. N. Whitehead *Process & Reality* 161 The pattern refers unselectively to any eternal objects. 1977 *Nature* 3 Nov. 15/2 Quantitative measurements.. show hydroxyl radicals..to interact unselectively with most biological molecules at rates which are practically diffusion controlled. 1980 V. Cunningham *Spanish Civil War Verse* 45 Jack Lindsay's Declamation now seems unurgently lengthy and even turgid.

12. *unaccentuation, unamaze, unamazement, unbook, unclarity, uncomfiness, uncountry, uncrackability, uncreditworthiness, undeath, undeathliness, undecrease, undeviation, uneducation, unenlightenment, unfreshness, unfulfilment, unglamorousness, unimportancy, unintelligentsia, uninterruptability, uninvolvement, unlight, unmeritocracy, unpriggishness, unrepose, unsurprise, unwettability*.

1879 G. M. Hopkins *Let.* 27 Feb. in Hopkins & Dixon *Corr.* (1955) 22 Wherever there is an accent or stress, there there is also so much unaccentuation, so to speak, or slack. 1936 W. Faulkner *Absalom, Absalom!* i. 8 Then in the long unamaze Quentin seemed to watch them. 1954 —— *Fable* 369 Thinking in a sort of quiet unamazement, Yes, I probably knew from the moment he sent for me what door I should have to emerge from. 1965 *Probl. Communism* July/Aug. 56 (*heading*) Another un-book. 1982 *Underground Grammarian* Nov. 5/1 Reading experts always need tricky new gimmicks to put in their unbooks. 1934 Webster, Unclarity. 1936 *Mind* XLV. 503 Prof. Reichenbach's discussion of material, formal, and tautologous implication..suffers from the same unclarity. 1980 'J. le Carré' *Smiley's People* v. 54 'Vladimir telephoned the Circus at lunch-time today, sir,' Mostyn began, leaving some unclarity as to which 'sir' he was addressing. 1914 L. S. Woolf *Wise Virgins* ii. 39 'I hate being uncomfortable. But then I don't think I often am, except when I..cut my nails.' 'Oh, that sort of uncomfyness!' said Gwen. 1964 W. Golding *Spire* ix. 178 In this uncountry there was blue sky and light, consent and so sin. 1923 *Daily Mail* 10 Jan. 9 (Advt.), The Vitreosil Globe, because of its 'uncrackability', lies close to the mantle. 1966 *Economist* 22 Oct. 413/2 Even the unlucky companies mentioned above are..credit worthy enough: the depths of uncreditworthiness below remain murky indeed. 1933 L. Riding *Poet* v. 125 The sum of the first and second sign Shall be undeath of the moon. 1974 *Globe & Mail* (Toronto) 24 July 13/4 There is, every now and then, a film that escapes this sort of un-death. 1922 Joyce *Ulysses* 379 She prayed to God the Allruthful to have his dear soul in his undeathliness. 1898 Hardy *Wessex Poems* 5 Knowing that, though Love cease, Love's race shows undecrease. 1932 W. Faulkner *Light in August* xix. 435 He was going fast too,..with..the implacable undeviation of Juggernaut or Fate. *a* 1936 G. K. Chesterton *Common Man* (1950) vii. 39 It is not their uneducation but their education that I scoff at. 1950 T. S. Eliot *Cocktail Party* II. 110 While still in a state of unenlightenment, You could always say: 'he could not love any woman.' 1967 T. Keneally *Bring Larks & Heroes* xxii. 178 A creature of dish-eyed unenlightenment. 1897 St. L. Strachey *From Grave to Gay* 288 We can keep the unfreshness of our eggs to ourselves, but not so the unfreshness of our jokes. 1969 J. Cheever *Bullet Park* vi. 79 She..exhaled the faint unfreshness of humanity at the end of the day. 1851 H. Melville *Moby Dick* (U.S. ed.) lxxxi. 397 Oh! that unfulfilments should follow the prophets. 1881 'L. Malet' *Counsel of Perfection* xiv. 323 There is a wonderful compensation in the unfulfilment of prophecy. 1978 F. King *Action* ii. 12 The pathos of his unfulfilment. 1958 *Economist* 26 July 277 He has probably had to contend..also with the poor pay offered and the unglamorousness of the whole proceeding. *a* 1960 E. M. Forster *Maurice* (1971) xlvi. 226 Excuse me if I work at unimportancies for a bit now. 1930 G. B. Shaw *Philanderer* Pref., in *Works* VII. 68 That far more numerous body which may be called the Unintelligentsia was as unconscious of Ibsen as of any other political influence. 1964 R. H. Robins *Gen. Linguistics* v. 211 The unitary behaviour of such forms in sentences and their uninterruptability..are the grounds for ascribing single word status to them. 1966 *Punch* 2 Nov. 682/2 He remains an aloof catalyst. Yet his very uninvolvement has a value of its own. *a* 1973 J. R. R. Tolkien *Silmarillion* (1977) viii. 74 A cloak of darkness she wove about them ..an Unlight, in which things seemed to be no more, and which eyes could not pierce, for it was void. 1970 *New Scientist* 31 Dec. 586/2 Large families are not the prerogative of the feckless poor whether in the Third World or in the urban ghettoes of the unmeritocracy. 1955 S. Spender *Making of Poem* 75 There are standards of sensibility or of 'unpriggishness' which are a development of what Samuel Butler approvingly called 'being a nice person'. 1868 J. H. Newman *Verses on Various Occasions* 312 The prison where they roam in hopeless unrepose. 1954 W. Faulkner *Fable* 343 Thicker and denser than the stars in its concentration of anguish and unrepose. 1932 —— *Light in August* xvii. 371 He seemed to stand aloof,..thinking with a kind of grim unsurprise: 'Byron Bunch borning a baby.' 1955 W. Gaddis *Recognitions* II. iii. 393 A, M, D, G, sequence of unsurprise (Lao-tse's 84-year gestation), right Nicodemus? 1937 *Nature* 26 June 1107/2 The unwettability of natural cotton is generally ascribed to the existence of a film of wax or oil covering its outer surface.

13. *unforgetting*.

1930 Auden *Poems* 8 Unforgetting is not today's forgetting For yesterday.

14. (nonce-wds.) *unbloom, undie, unfulfil, untouch, unvision*.

1898 Hardy *Wessex Poems* 8 And why unblooms the best hope ever sown? 1952 Dylan Thomas *Coll. Poems* p. x, These seathumbed leaves That will fly and fall Like leaves of trees and as soon Crumble and undie. 1922 Hardy *Late Lyrics & Earlier* 37 Though duties due that press to do This whole long day I unfulfil. *Ibid.* 34 And time untouched me with a trace Of soul-smart or despair. 1917 —— *Moments of Vision* 207 I'll not unvision A shape which, somehow, there may be.

16. In telegrams *un-* is sometimes used in nonce formations with any part of speech to avoid a separate negative.

1936 E. Waugh *Waugh in Abyssinia* 161 Cables were soon arriving... '*Require earliest name life story photograph American nurse upblown* [*sc.* bombed] *Adowa*.' We replied '*Nurse unupblown*,' and after a few days she disappeared from the news. 1967 *Observer* 8 Oct. 2/5 Regret expelled by Syrians after twenty-four hours unreason given. 1968 *Punch* 3 Apr. 485/3 All those compound words like 'unproceed'..in press cables.

un-[2]. Add: **3.** *unbake, unban, unbatten, unbore, unclutter* (examples), *undip* [*dip *v.* 6 d], *unendear, uninvent, unjumble, untape, untease*.

1864 G. M. Hopkins *Poems* (1967) 15 Wasteful wide huge-girthèd Nile Unbakes my pores, and streams, and makes all fresh. 1968 *Guardian* 5 Sept. 9/2 Book censorship has eased since the Dail passed a law which un-bans a book after 12 years in purdah. 1983 *Bookseller* 4 June 2072/3 *Lady Chatterley's Lover*: unbanned in South Africa in 1980, it has been selling through Leserskring at a steady 1,000 a quarter. 1788 W. Bligh in R. M. Bowker *Mutiny* (1978) iii. 90 Unbattened all the Hatchways. 1927 R. A. Freeman *Certain Dr. Thorndyke* iv. 62 Osmond..unbattened the doors, and, opening them, slid the wriggling captive down the ladder on to the cabin floor. 1922 M. A. von Arnim *Enchanted April* xiii. 213 Rose felt right down at her very roots that if you have once thoroughly bored somebody it is next to impossible to unbore him. 1930 *Times* 29 Nov. 13/6 Sadler's Wells.. appears to go farthest in uncluttering the conditions for witnessing the performance. 1977 A. Scholefield *Venom* III. 90 He liked to clear everything away, to unclutter the stage. 1966 T. Wisdom *High-Performance Driving* x. 104 Drivers..use their headlights badly. They do not undip them fast enough after meeting other traffic. 1865 G. M. Hopkins *Poems* (1967) 21 Still thou bind'st me to fresh fealty..for nothing here Nor elsewhere can thy sweetness unendear. 1962 *Guardian* 4 Oct. 6/5 It may not be possible to un-invent the motor-car. 1982 *Church Times* 22 Oct. 5/1 Nuclear weapons cannot be uninvented. 1966 J. Derrick *Teaching English to Immigrants* v. 206 Pupils can also be asked to unjumble and write out in correct order a series of sentences that should obviously follow a certain sequence (such as the description of a picture story or of a series of actions). 1979 J. Gardner *Nostradamus Traitor* xliii. 204 He's the bloody expert on Nostradamus. He's even..unjumbled the prophecies. 1968 'R. Macdonald' *Instant Enemy* xxxiv. 217 He untaped my wrists and ankles. 1981 E. Ward *Baltic Emerald* xxi. 175 Henry untaped the envelope and photographs from my back. 1932 Auden in *Rev. Eng. Stud.* (1978) Aug. 287 A piece of paper, neatly Folded in a hexagon shape Which Sampson took and unteased and read.

4. a. *unblouse, undogcollar* (nonce), *unpearl* (nonce).

1922 Joyce *Ulysses* 254 Miss bronze unbloused her neck. 1953 Dylan Thomas *Under Milk Wood* (1954) 74 Esau,..undogcollared because of his little weakness, was scythed to the bone one harvest by mistake. 1936 A. E. Housman *More Poems* 18 For these of old the trader Unpearled the Indian seas.

b. *unzone* (later example).

1907 Joyce *Chamber Music* xi, Begin thou softly to unzone Thy girlish bosom into him.

5. *uncrate*.

1929 W. Faulkner *Sound & Fury* 235, I went on to the back, where old Job was uncrating them [*sc.* cultivators]. 1963 *Ann. Reg. 1962* 520 In addition, jet bombers, capable of carrying nuclear weapons, are now being uncrated and assembled on Cuba.

6. b. *unfather*.

1876 [see *unchild *v.*].

7. *unportal* (nonce).

1922 Joyce *Ulysses* 461 The Timepiece (*Unportalling*.) Cuckoo. Cuckoo. Cuckoo.

9. Another redundant or extended use (= 'peel off') exists in *unpeel*. Cf. Unpeeled *ppl. a.*

1904 W. James *Let.* 16 June in R. B. Perry *Thought & Character W. James* (1935) II. 487 The original 'that' may vanish in the infinitely regressive superposition of human 'whats'—we can't today unpeel them wholly. *a* 1914 in *Penguin Bk. Austral. Ballads* (1964) 122 Then the sheila raced off squealin', And her clothes she was un-peelin'. 1969 *Washington Post* 14 Apr. A27/3 Aides believe that the Kennedy ploy..will ultimately unpeel Republican Senators from the anti-ABM forces.

un, 'un[1]. (Later examples.)

1919 G. B. Shaw *Heartbreak House* III. 109 Right in the gravel pit: I seen it. Serve un right! 1932 D. L. Sayers *Have his Carcase* xi. 134 Soon as I come in zight of 'un, I zee un. 1977 P. Hill *Liars* x. 140 The ladies liked 'un.

un, 'un[2]. (Earlier example.)

c 1810 W. Hickey *Mem.* (1960) iv. 64 The young 'un there wanted to be off.

unabatable, *a.* (Earlier example.)

1788 T. Paine *Amer. Crisis* (1817) 170 The enmity is perpetual, unalterable, and unabateable.

unacceptable, *a.* Add: (Examples in phr. *the unacceptable face of*—.)

1973 E. Heath in *Hansard Commons* 15 May 1243 It is the unpleasant and unacceptable face of capitalism, but one should not suggest that the whole of British industry consists of practices of this kind. 1973 *Times* 5 June 1/6 He called the actions of the *News of the World* 'the unacceptable face of journalism'. 1975 A. Beevor *Violent Brink* vii. 210 Let us look at the unacceptable face of Communism. 1982 *Guardian* 8 Jan. 18/1 The unacceptable face of modern men's tennis was seen again during the Barratt world doubles championship..yesterday when Hank Pfister..swore at the match umpire and then accused him of cheating.

unaccountable, *sb.* **2.** (Later examples.)

1833 R. Dyer *Nine Yrs. Actor's Life* 66 His non-engagement in London is amongst the unaccountables of metropolitan management. 1961 'J. le Carré' *Call for Dead* viii. 83 Of all the unaccountables in the case, that worried him most.

unaccustomedly, *adv.* (See small-type note s.v. Unaccustomed *ppl. a.*) (Examples.)

1963 *Economist* 2 Nov. 472/1 Unaccustomedly emotional language. 1980 H. Curtiss *Poisoned Orchard* iii. 22 She had put her car away, unaccustomedly, because of the sub-zero temperature forecast.

una corda (ū·nă kǭ·ɹdă). *Mus.* [It., = one string.] As a direction in music for the piano, etc.: use the soft pedal (which on grand pianos softens the tone by shifting the hammers so as to strike only one or two strings instead of two or three for each note). Also *attrib.*, designating such a pedal mechanism. Cf. *soft pedal *sb.*

1849 *Hamilton's Celebrated Dict.* s.v., *Una corda* (Italian). Implies that a passage is to be played upon only one string. 1876 Stainer & Barrett *Dict. Mus. Terms* 444/1 *Un, Una, Uno* (It.), one; as *Una corda*, one string. 1909 *Cent. Dict.* Suppl. s.v. *pedal* n., The direction *una corda* signifies simply the use of this [shifting] pedal. 1931 D. F. Tovey *Beethoven Sonatas for Pianoforte* xxix. 140 Beethoven's double shift from *tre corda* through *due corda* to *una corda*. 1961 [see *escapement 3]. 1979 *Sci. Amer.* Jan. 99/2 To prevent this the piano is equipped with the una corda pedal, whose mechanical function is to shift the entire keyboard so that a hammer strikes only two strings of a unison triplet.

unadopted, *ppl. a.* (Examples corresponding to *adopted *ppl. a.* 1 b.)

1938 [see *adopted *ppl. a.* 1 b]. 1963 *Times* 18 May 10/7 It is virtually undistinguished except for the number of roughly surfaced streets compelling a glance at signs proclaiming that they are unadopted. 1970 J. Burke *Four Stars for Danger* ii. 21 That road up to Bryncroeso Hall is unadopted. If the people at the Hall don't want to do anything about it, nobody's going to make them.

unaffectioned, *ppl. a.* (Later example.)

1911 K. Hare *Green Fields* 5 His tuneable unaffectioned voice that loved the matter Has in the grey room conjured up the sunshine.

unaffixed, *ppl. a.* Add: Also as *adj.* [Un- 9 a].

1964 R. H. Robins *Gen. Linguistics* vi. 260 In Japanese the formation of words..referring to more than one entity ..leaves the resultant words syntactically equivalent to unaffixed forms.

unaflow, var. *uniflow *a.*

unali·gned, *ppl. a.* [Un-[1] 8.] **1.** Not physically aligned.

1936 F. M. Ford *Let.* 2 July (1965) 254 This damned new machine—unaligned a's and all which means that it will have to go back to the shop. 1962 Corson & Lorrain *Introd. Electromagn. Fields* iii. 107 An unaligned dipole therefore has greater energy than an aligned one.

2. = *non-aligned *a.* Also *absol.* as *sb.*

1961 *Guardian* 28 Jan. 2/3 India and the 'unaligned nations'. 1962 C. B. Marshall in L. W. Martin *Neutralism & Nonalignment* ii. 13 (*heading*) On understanding the unaligned. 1965 *Listener* 10 June 851/1 Africans feel the need to be wholly African, unaligned, even in their form of government.

unali·ke, *a.* [Un-[1] 7.] Different; not alike.

1934 'L. G. Gibbon' *Grey Granite* 210 If Ewan had been as that other Ewan..and she paused... Was he so unalike? 1944 *Penguin New Writing* XX. 61 Arthur continued to wonder..how these two unalike girls had come to set up together. 1961 B. Vawter *Conscience of Israel* v. 127 If they are two prophets in agreement on fundamentals, they are also severely unalike in numerous ways. 1978 D. Murphy *Place Apart* xii. 250 Cycling is by far the best way to travel around the North where regions only twenty miles apart can be so very unalike.

unalive, *a.* Add: **b.** Lacking in vitality; not living. Also *fig.*

1905 M. Dods *Let.* 14 Apr. (1911) 176 How you can think yourself empty and unalive I don't know. 1931 H. S. Walpole *Judith Paris* II. v. 338 All the members of the Herries family, with whom she had so lately been, seemed unreal and unalive. 1935 E. Bowen *House in Paris* II. x. 209 The street reflected the blind windows and a strip of unalive wet sky. 1954 R. Speaight *George Eliot* iv. 51 The true criticism of Stephen is not that he is unattractive but that he is unalive.

unaltered, *ppl. a.* Add: Of an animal: not castrated.

1946 H. A. Smith *Rhubarb* ii. 28 An unaltered cat is by nature inclined to wildness. 1967 A. Lewin *Unaltered Cat* II. iv. 117 Have you ever tried keeping an unaltered cat in an apartment?

un-American, *a.* Add to def.: Contrary to the ideals and interests of the United States of America. (Further examples.)

1893 H. B. Fuller *Cliff-Dwellers* xviii. 242 Does it seem unreasonable that the State [*sc.* Illinois]..which has done most..to check alien excesses and un-American ideas, should also be the State to give the country the final blend of the American character and its ultimate metropolis? 1917 T. Roosevelt *Let.* 10 Jan. in *Proc. Congr. Constructive Patriotism* 172 Everything is un-American that tends either to government by a plutocracy or government by a mob. 1936 F. R. Marvin (*title*) Fools gold: an exposé of un-American activities and political action in the United States. 1938 *N.Y. Times* 27 May 2/2 The House..today passed a resolution providing for the appointment of a seven-man committee to investigate un-American propaganda in the United States. 1958 E. H. Clements *Uncommon Cold* iii. 97 Visas aren't handed out by Americans to anyone. They are averse to Communists, criminals, all un-American activities. 1980 in S. Terkel *Amer. Dreams* 363 The Red scare was on, the witch hunt. The House Un-American Activities Committee came to town.

un-Americanism (recent examples).

1955 C. L. Sewrey in *Comprehensive Dissert. Index* (1973) XXVIII. 116/3 The alleged 'un-Americanism' of the Church. 1960 *Guardian* 26 Aug. 5/1 To question either now seems tantamount to un-Americanism. 1978 *Amer. Speech* 1975 L. 315 The sin is neither as recent a matter of concern nor as diagnostic of un-Americanism as Dwight Bolinger implies.

unamusing, *ppl. a.* (Earlier example.)

1794 F. Burney *Jrnl.* 2 Mar. (1973) III. 43 He found it very unamusing to have a Walk without any *but*.

unanalysable, *a.* Add: Hence **una:nalysabi·lity.**

1941 *Mind* L. 335 The sort of unanalysability which the P[ure] E[go] theory attributes to class (1) 'someone' sentences. 1973 E. Jong *Fear of Flying* i. 3 There were 117 psychoanalysts on the Pan Am flight to Vienna and I'd been treated by at least six of them... God knows it was a tribute either to the shrinks' ineptitude or my own glorious unanalyzability that I was now..more scared of flying than when I began my analytic adventures some thirteen years earlier.

unanalytic, *a.* (Earlier example.)

1865 *North Brit. Rev.* XLIII. 27 Reflective Realism is only a change in the unanalytic manner of thinking about objects.

Unani, var. *Yunani.

unanimism (yūnæ·nimiz'm). Also ‖ **unanimisme** (ünanimism). [ad. F. *unanimisme*, f. *unanime* unanimous + *-isme* -ism.] A French poetic movement of the early twentieth century which emphasized the submersion of the poet in the group consciousness and which was characterized by simple diction, absence of rhyme, and strongly accented rhythms.

1931 [see *populism b]. 1936 *Times Lit. Suppl.* 7 Mar. 197/3 But Mr. Buchanan is not indulging in the fallacies of *unanimisme*. 1959 *Oxf. Compan. French Lit.* 724/1 *Unanimisme*, a 20th-century poetic movement which owes much to the Whitmanesque doctrine of universal brotherhood as well as to more modern psycho-philosophical theories of group emotion. 1964 *Listener* 27 Aug. 315/2 This man [*sc.* Apollinaire]..under the banner of Unanimism..had loudly and proudly identified himself with everything quick and living. 1971 J. Willett in A. Bullock *Twentieth Century* x. 235/2 Jules Romains, whose faith in the anonymous mass..inspired his short-lived doctrine of Unanimism.

¶ unani·sm, ‖ unanisme.

1919 W. B. Yeats *If I were Four-&-Twenty* (1940) ii. 4 There has been a development in various forms of literature—in French 'unanisme' for instance—towards the expression..of an emotional agreement with some historical or local group. 1958 *Times Lit. Suppl.* 4 July 383/1 A study..of unanism in the plays of Jules Romains.

unanimist (yūnæ·nimist), *sb.* (and *a.*) *Also* ‖ unanimiste (ünanimist). [ad. F. *unanimiste*: see prec., -ist.] An adherent of unanimism. Also *attrib.* or as *adj.*

1915 A. Huxley *Let.* Dec. (1969) 88 This good man [*sc.* D. H. Lawrence]..proposes..to go to the deserts of Florida there..to found a sort of unanimist colony. 1918 [see *surrealist *a.* and *sb.*]. 1929 V. Payen-Payne in M. S. McLaren *Douze Sonnets de Varlet* 9 Varlet..has been classed among the Unanimists because of his friendship with Jules Romains and Georges Duhamel. 1938 *Times Lit. Suppl.* 10 Sept. p. iii/1 From such an incident does the writer, with something like the *unanimiste* technique of Jules Romains..develop a theme of nation-wide, or even world-wide, import. 1959 *Oxf. Compan. French Lit.* 724/2 The *unanimistes* had their own, fairly recognizable, technique of versification. 1974 *Encycl. Brit. Micropædia* X. 253/1 The Unanimiste theories of prosody..resembled those of the American poet Walt Whitman.

unappreciable, *a.* (Earlier example.)

1801 F. Burney *Let.* 29 Oct. (1975) V. 21 My dear Charlottina, whom I regard as an unappreciable consolation & delight to you.

unappreciative, *a.* (Earlier example.)

1840 J. S. Mill *Let.* 23 Dec. in *Wks.* (1963) XIII. 453 A man of Molesworth's sort of limitation has a natural tendency to be intolerant, because unappreciative of ideas & persons unlike him & his ideas.

unapprehended, *ppl. a.* **2.** (Later example.)

1977 *Daily Tel.* 5 Mar. 3/1 It included 341 multi-offenders and concluded with 60 unapprehended men wanted in Dallas this year.

unapropos (ʊnæprǫ̆pō^u·, ʊnæ·prǫ̆pō^u), *a.* and *adv.* [Un-[1] 7, 11 b.] Not apropos; irrelevant(ly); inappropriate(ly).

1840 H. Mozley *Let.* 13 Oct. in D. Mozley *Newman Family Lett.* (1962) III. 92, I have none of the un-a-propos propensities that would stand in the way of engagement with some ladies. 1940 J. Pope-Hennessy *Let.* in J. Lees-Milne *Ancestral Voices* (1975) 267 Christianity.. seems a clear and tranquil stream running parallel to but utterly detached from the turgid river of the war, and ..too unapropos for words. 1956 R. Robinson *Landscape with Dead Dons* xxi. 194, I was chatting to Christelow when it occurred to me quite unapropos that one of the things they had in common was *The Book*.

unarithmetic, *a.* Delete *Obs.*[−1] and add later example.

1953 *Trans. Amer. Inst. Electr. Engineers* LXXII. 597/2 The unarithmetic representation for zero is standard in the 2-out-of-5 code.

unarmed, *ppl. a.* Add: **1.** *transf.* Freq. in phr. *unarmed combat*.

1947 'N. Shute' *Chequer Board* ix. 262 'You went on to a course in unarmed combat. What did they teach you to do there?' 'We was taught how to attack an armed man just with our hands and feet.' 1957 J. Braine *Room at Top* xxx. 252 A hand on my shoulder..started the Unarmed Combat reflexes working. 1973 J. R. L. Anderson *Death on Rocks* ii. 39 If I could get my hands on him—well, I'd been quite good at what the Army calls unarmed combat.

unartificial, *a.* **2.** (Later example.)

1982 *N. & Q.* Aug. 361/2 He demonstrates that Wordsworth considered a good epitaph to be an expression in unartificial language of the deep feelings of the bereaved for the dead.

unary (yū·nări), *a.* [f. L. *ūn-us* one + -ary[1], after Binary, Ternary *adjs.*] **1.** *Chem.* Of a chemical system: consisting of a single component.

1923 A. C. D. Rivett *Phase Rule* i. 25 For systems of one, two, three, four, five (and so on) components, one uses the terms unary, binary, ternary, quaternary, quinary (and so on), respectively. 1980 *Mineral. Abstr.* XXXI. 311/1 Sixteen possible configurations of phase diagrams have been deduced for unary four-phase multi-systems.

2. *Math., Logic,* and *Linguistics.* Of an operator, operation, or transformation: involving or operating on a single element.

1931 *Bull. Amer. Math. Soc.* XXXVII. 487 Let p be the result of an undefined unary operation on a K-element p, and $p+q$ the result of an undefined binary operation on the K-elements p, q. 1940 [see *singulary *a.*]. 1961 *Jrnl. Assoc. Computing Machinery* VIII. 579 The five binary arithmetic operators (↑, ×, /, +, −), the two unary arithmetic operators (+, −). 1965 *Language* XLI. 270 Robert Stockwell has described rules that indicate the colorless patterns for kernel sentences and unary transformations. 1968 J. J. C. Smart *Betw. Sci. & Philos.* ii. 24 'Not' can be thought of as a unary sentence connective. 1973 C. W. Gear *Introd. Computer Sci.* iii. 104 The unary arithmetic operators take precedence over all of the binary operators and must be performed first. 1976 J. S. Gruber *Lexical Structures in Syntax & Semantics* II. i. 266 If the treatment is through a unary transformation that alters structure, the dependency of the transformation on apparently semantic factors becomes a matter of graver theoretical consequences.

3. Composed of a single item or element.

1968 P. M. Postal *Aspects Phonol. Theory* i. 13 Natural languages have structures which are such that the markers on every level can be looked upon as sets (sometimes unary sets) of strings of elements. 1968 *Amer. Documentation* Jan. 73/1 Items are either unary or multiple, depending upon whether they are composed of a single piece of information (which may itself be composed of any number of characters or words) or of two or more separate pieces of information.

unascended, *ppl. a.* (Earlier example.)

1820 Shelley *Prometh. Unb.* III. iv. 203 The loftiest star of unascended heaven.

unashamedly, *adv.* (See small-type note s.v. Unashamed *ppl. a.*) (Examples.)

1905 W. J. Locke *Morals of Marcus Ordeyne* iv. 50 She was wearing a deep red silk peignoir,..unashamedly Parisian, which clung to every salient curve of her figure. 1928 *Manch. Guardian Weekly* 10 Aug. 115/2 They break out..into dance measures of an unashamedly negroid vigour. 1983 M. Gilbert *Black Seraphim* iii. 35 Amanda ..was listening unashamedly.

unassailability. (See small-type note s.v. Unassailable *a.*) (Recent examples.)

1952 *Mind* LXI. 504 Its unassailability has been bought at the price of making no claim about the world. 1957 J. S. Huxley *Relig. without Revelation* i. 27 The unassailability of private property.

unassailableness. (Earlier example.)

1854 Geo. Eliot tr. *Feuerbach's Essence Christianity* xiv. 136 The truth and unassailableness of their subjective feelings.

unate (yūnē[i]·t), *a. Math.* [f. L. *ūn-us* one + -ate[2].] Of a logical function: containing no variable in both negated and unnegated forms. Hence **una·teness.**

1960 *Proc. IRE* XLVIII. 1336/1 The first of these, Theorem 1, gives one of the most easily recognized properties of a setting function, unateness... A unate function is defined as one from which one of each pair of complementary literals..can be eliminated. 1978 S. C. Lee *Mod. Switching Theory* iv. 118 Any threshold function is unate.

unatmosphe·ric, *a.* [Un-[1] 7.] That conveys no suggestion of tone or mood; that fails to evoke associations. Hence **unatmosphe·rically** *adv.*

1913 E. F. Benson *Thorley Weir* iii. 68 Arthur..found this peremptory young savage slightly alarming. For himself he demanded that social intercourse should be conducted in a serene atmosphere of politeness, of manners... He thought he had seldom seen anybody so unatmospheric. *Ibid.* 69 Frank looked at him quite unatmospherically. 1938 *Scrutiny* VII. 179 His [*sc.* Roussel's] instrumentation is unatmospheric. 1963 *Listener* 28 Feb. 393/2 A couple of very stagy and unatmospheric outdoor sets. 1976 *Gramophone* Oct. 584/2 The 'Royal Hunt and Storm' begins unatmospherically, with the strings at once forceful and casual, carrying none of the sense of light slowly dawning on a classical landscape.

unattached, *ppl. a.* Add: Hence **unatta·chedness.**

1936 Wirth & Shils tr. *Mannheim's Ideology & Utopia* III. iv. 140 The 'unattachedness' of the intellectuals. 1981 M. Warner *Joan of Arc* iv. 86 The..unattachedness of the displaced and the poor.

unattributable, *a.* (Later examples.)

1967 *Listener* 9 Nov. 594/2 The 'unattributable' story, which is valid information to be published as long as the source is not identified. 1972 T. Lilley *K Section* xli. 229 The information is unattributable and for your ears only.

unauthentically, *adv.* (Later example.)

1975 *Daily Tel.* 17 Mar. 10/2 John Constable's amusingly, if probably unauthentically, informal continuo-comments.

unavailability. (See small-type note s.v. Unavailableness.) (Examples.)

1855 D. G. Rossetti *Let.* 29 July (1965) I. 265 If Mr. Oakes should..be no longer in the above capacity,.. let me know at once of Mr. O.'s unavailability. 1967 *Listener* 13 July 61/2 In England the young composer is sadly hampered by the unavailability of one possible salvation—electronic equipment. 1980 J. Krantz *Princess Daisy* xvi. 267 *That's* what drives him crazy.. your essential *unavailability*.

unavailingly, *adv.* (Earlier example.)

1748 Richardson *Clarissa* IV. 170 Must..those arms ..be used to repel brutal force; all their strength, unavailingly perhaps, exerted to repel it?

unavertible, var. of Unavertable *a.*

1897 J. L. Allen *Choir Invisible* vi. 78 Their own inescapable tombs, their own unavertible ruins. 1929 *Daily Tel.* 15 Jan. 11/7 The public is seldom concerned with the future until it has become an unavertible disaster.

unavoidability. (See small-type note s.v. Unavoidable *a.*) (Examples.)

1858 D. G. Rossetti *Lett.* (1965) I. 336 Three evenings a week at least are..used up away from work by unavoidabilities. 1938 *Mind* XLVII. 47 Arguments which seek to prove unavoidability by reference to determinism are to be dismissed as ridiculous.

unbated, *ppl. a.* **2.** Delete † *Obs.* and add later example.

1979 A. Williamson *Funeral March* xv. 81 'Do you know how Andersson died?'..'Well—I suppose it was the unbated spear?'

unbeautifully, *adv.* (Later examples.)

1879 Geo. Eliot *Let.* 5 Mar. (1956) VII. 111 The margin seems perilously and unbeautifully narrow. 1922 *Daily Mail* 23 Nov. 8 A Frenchwoman would as soon be seen in shabby shoes as with wisps of hair escaping unbeautifully from her coiffure. 1984 A. Price *Sion Crossing* xi. 222 His lip twisted unbeautifully.

unbecoming, *ppl. a.* Add: **d.** Const. *to*.

1893 Wilde *Lady Windermere's Fan* III. 90 There is nothing in the whole world so unbecoming to a woman as a Nonconformist conscience. 1901 G. B. Shaw *Capt. Brassbound's Conversion* II. 245 Brandyfaced Jack: I name you for conduct and language unbecoming to a gentleman. 1980 *Washington Post* 1 June A17 But to set out to fix blame is..unbecoming to America. 1982 *Financial Times* 21 Aug. 4 Rumours flew about in a manner most unbecoming to the month.

e. *ellipt.* after a sb. in *conduct unbecoming* (*sc.* of a military officer); also *transf.*

1971 B. England (*title*) Conduct unbecoming. 1976 *Time* 5 July 46/3 The Merchant Marine Academy, which classifies copulation in the barracks as conduct unbecoming and worthy of dismissal. 1976 *Times Lit. Suppl.* 6 Aug. 988/4 Conduct unbecoming in a man of letters must be reprimanded.

unbegot, *ppl. a.* Delete † *Obs.* and add later example.

1936 A. E. Housman *More Poems* 14 This is for all ill-treated fellows Unborn and unbegot.

unbeing, *vbl. sb.* Delete † *Obs.* and add later examples.

1935 G. Barker *Janus* 16 The lighter bird of being, obstructing my line of sight, entirely conceals the form of the bird of unbeing. 1936 T. S. Eliot *Coll. Poems 1909–35* 191 Love is itself unmoving... Caught in the form of limitation Between un-being and being. 1944 L. MacNeice *Springboard* 17 From the lubber depths of my unbeing.

unbeknown, *ppl. a.* Add: **1.** (Later examples.)

1947 *Sun* (Baltimore) 31 Oct. 10/3 That this number is insufficient is being quietly demonstrated at the present time, unbeknown to the general public. 1952 J. L. Waten *Alien Son* 121 Unbeknown to our parents we sneaked away from the street one summer day.

ellipt. 1901 J. Barlow *From Land of Shamrock* 215 Openin' it at the breakfast-table, unbeknown, to set people passin' remarks, and risin' a laugh on him.

2. (Later example.)

1976 *Jrnl. Lakeland Dial. Soc.* No. 38. 39 Ah nivver really spock t' dialect proper, but Ah hev a gey lot ev dialect sayings Ah offen let slip oot, unbeknawen like!

unbeknownst, *a.* or *adv.* Add: Also, = Unbeknown *ppl. a.* 1. (Earlier and later examples.)

Now of much wider currency than in the 19th. cent.

1848 Mrs. Gaskell *Let.* 11 Nov. (1966) 61 You don't see me, but I often am sitting in the rocking-chair unbeknownst to you. 1907 J. M. Synge *Playboy of Western World* III. 70 Burying your poor father unbeknownst when..we could have given him a decent burial. 1932 W. Faulkner *Light in August* i. 16 Interfering with his work unbeknownst to him. 1952 A. Christie *They do it with Mirrors* xiii. 122 One of those smart lads may have got out of the College buildings unbeknownst. 1979 *Dædalus* Summer 99 Here, illusion, unbeknownst to those who believed they had overcome it, made its most triumphant reentry. 1982 *London Rev. Bks.* 20 May–2 June 3 A whole other wife and children all unbeknownst to Ackerley until after his father's death.

unbelted, *a.* Add: (Further examples, of a seat belt.) Also *absol.* as *sb.*

1970 *Guardian* 24 Aug. 7/4 An unbelted driver can die (and has died) at speeds as low as 13 miles an hour. 1973 [see *non-trivial *a.*]. 1977 *Times* 5 July 15/5 A higher proportion of belted than unbelted drivers remain in control of their vehicles.

‖ **unberufen** (unbərū·fən), *int.* [Ger., unauthorized, gratuitous.] 'Touch wood!' (Touch *v.* 29 b).

1858 Queen Victoria *Let.* 10 Apr. in R. Fulford *Dearest Child* (1964) 88 Poor Leopold..bruises as much as ever but unberufen 1000 times—is free from any at present. 1911 Beerbohm *Zuleika D.* xv. 225, 'I will choose ..whatever moment within my brief span of life shall seem aptest to me. *Unberufen*,' he added, lightly tapping Mr. Druce's counter. 1960 N. Mitford *Don't tell Alfred* xvi. 180 'My point is that going to Eton will have minimized the danger of such extreme phases in the case of Charlie and Fabrice.' 'Unberufen,' said Alfred.

unbiolo·gical, *a.* [Un-[1] 7 a.] **a.** Not in accord with the findings of biology. **b.** Not such as occurs in the course of nature, as studied in biology.

1934 *Mind* XLIII. 519 Positivism and Kantianism were pre-evolutionary and unbiological. 1950 [see *stereotyped *ppl. a.* c]. 1977 P. Johnson *Enemies of Society* xv. 197 Many of the central ideas of psychoanalysis are profoundly unbiological.

unbirthday. (Further examples.)

1930 W. de la Mare *On the Edge* 272 'Have you one really simple, lovely..trinket suitable for a lady?..An un-birthday present?' 1977 *Lancet* 2 July 31/2 A friend who has come to the front door, unexpectedly bearing a small unbirthday present.

unblinking, *a.* (See small-type note s.v. Unblinkingly.) (Examples.)

1909 in Webster. 1923 *Chambers's Jrnl.* 10 Nov. 791/1 John..could see only the unblinking stars. 1965 G. McInnes *Road to Gundagai* xvi. 283 The unblinking stare of the lizard on the sun-baked rock.

unblissful, *a.* Add: Hence **unbli·ssfully** *adv.*

1849 C. Brontë *Shirley* III. iv. 107 His whole nature seemed serenely alight: he stood..musing not unblissfully.

unblock, *v.* **1.** (Later examples.)

1969 H. Perkin *Key Profession* v. 186 Even that..did not unblock the channels of advance. 1974 J. Mann *Sticking Place* ii. 30 She did not want to un-block drains or carry the dustbins.

unblushingly, *adv.* (Earlier example.)

1752 Richardson *Let.* 18 Apr. (1964) 207 For the word *Love* has enabled some People to talk of a Passion fearlessly, unblushingly.

unbosom, *v.* Add: Hence also **unbo·soming** *vbl. sb.*

1910 *Blackw. Mag.* Jan. 57 And with a voice of growing strength renewed His vague unbosomings. 1935 Dylan Thomas *Poems* (1971) 46 Summer to him Is the unbosoming of the sun.

unbound, *ppl. a.*[1] Add: **4.** Also of a particle: = *free *a.* 14 b.

1971 *Sci. Amer.* June 25/1 Once a few unbound electrons are produced on the surface of the solid hydrogen isotope, these free electrons rapidly pick up energy from the incident oscillating electric field.

unboyish, *a.* (Earlier example.)

1838 S. G. Goodrich *Fireside Educ.* 107 Charles, the eldest boy, with a patience most *unboyish*, was holding a skein of yarn for grandmamma to wind.

unbranning, *vbl. sb.* (Earlier example.)

1848 *Rep. Comm. Patents 1847* (U.S.) 373 In addition to the *unbraning* of the berry, the wheat undergoes an operation [etc.].

unbrave, *a.* Add: (Later and *absol.* examples.)

1896 *Godey's Mag.* Feb. 172/1 She saw before her his old self—strong, not unbrave, not disloyal. 1981 *London Mag.* Oct. 18 The sensible thing, as many unbrave realised at the time, would have been to opt for ground duties with the RAF.

unbreakable, *a.* (Examples corresponding to *break *v.* 16 b.)

1929 T. M. Johnson *Our Secret War* iv. 180 Is there an unbreakable code? 1944 H. McCloy *Panic* 73 There's no such thing as an unbreakable cipher! 1963 R. V. Jones in Brown & Foote *Early Eng. & Norse Studies* 223 An unbreakable W/T code.

unbundle, *v.* Add: **2.** To introduce a system of separate charging for (items previously charged for collectively, esp. computer hardware and software).

1969 *Datamation* XV. 69 (*heading*) IBM 'unbundles' hardware/services charges... Will software be next? 1971 *New Scientist* 15 July 140/1 Programs, courses, computer maintenance and systems engineering would be priced separately (or 'unbundled') from the computers themselves. 1977 *Business Week* 18 Apr. 83/2 Banks may then be forced to unbundle costs and charge an additional 30¢ or so for each check processed. 1983 *Austral. Microcomputer Mag.* Dec. 40/3 The 8086 processor has been unbundled and is now an option.

Hence **unbu·ndling** *vbl. sb.*, (the introduction of) a policy of separate charging for related items.

1969 *Datamation* XV. 69/1 IBM 'expects to make changes in the way it charges for and supports its data processing equipment'. The word for it is 'unbundling'. 1971 E. F. Schoeters in B. de Ferranti *Living with Computer* viii. 72 These arguments are becoming academic in the light of separate pricing of hardware, software, support and staff training, commonly known by the unlovely name of 'unbundling'. 1981 *Economist* 11 July 65/2 Independent Japanese software houses, helped by separate pricing ('unbundling') of software and hardware by computer makers, are starting to write their own packaged programmes.

unbutton, *v.* Add: **1. c.** Also *fig.*

1956 N. Streatfeild *Judith* III. 235 She definitely unbuttoned about her letters. 1978 O. White *Silent Reach* ix. 97 If I cleared you, he'd probably unbutton.

unbuttoned, *ppl. a.* **2.** *fig.* (Earlier and later examples.)

1885 G. Meredith *Diana of Crossways* I. i. 8 He was careless of social opinion, unbuttoned, and a laughter [*sic*]. 1918 'K. Mansfield' *Let.* 6 Feb. (1928) I. 124 On my life, I'd almost rather, like that English lady, not know whether my husband went to the lavatory or not, than be so unbuttoned. *a* 1967 J. R. Ackerley *My Father & Myself* (1968) 23 An unbuttoned stage of mellowness and ease. 1983 *Maledicta 1982* VI. 90 They are not by any means considered polite or even acceptable, except in a mood of broad and unbuttoned humor among close friends.

uncalled, *ppl. a.* **4.** (Earlier example.)

1869 *Bradshaw's Railway Man.* XXI. 199 The financial statement to same date... Discount on shares, 1837—£400,000... Uncalled—£619,525.

uncannily, *adv.* **2.** (Earlier example.)

1822 A. Cunningham *Traditional Tales* II. 267 He skirls sae uncannilie.

uncanopied, *ppl. a.* (Later example.)

1890 R. Bridges *Shorter Poems* III. xv. 61 Uncanopied sleep is flying from field and tree.

uncapped, *ppl. a.* (Examples corresponding to *cap *v.*[1] 1 c.)

1955 *Times* 3 Aug. 4/6 (*caption*) The only uncapped player in the British Isles team to meet South Africa at Johannesburg on Saturday. 1977 *World of Cricket Monthly* June 92/2 The form of uncapped 24-year-old seamer Bill Bourne pierced the gloom like a ray of sunshine.

uncaring, *ppl. a.* Add: Hence also **unca·ringness.**

1930 A. Huxley *Brief Candles* 5 The superficial charm and good humour of the man seemed to overlie a fundamental hardness, an uncaringness, a hostility even. 1955 E. Bowen *World of Love* iv. 78 Oh, how the vice of uncaringness had been hers; she had neither heart nor wish for a living creature.

uncentral, *a.* (Later examples.)

1911 R. Brooke *Let.* Mar. (1968) 292 One's whole personality was there—only, somehow without the point. One was curiously *uncentral*. 1977 D. Jones *My Friend Dylan Thomas* iv. 40 This unlocal, uncentral world where the pubs are bad and the people are sly.

uncertain, *a.* Add: **4. b.** Phr. *in no uncertain terms*, emphatically, very clearly indeed.

1958 L. Durrell *Balthazar* vi. 139 And what's more, I told Abdul so in no uncertain terms. 'Lay a finger on the girl..and I'll get you run in.' 1976 J. Snow *Cricket Rebel* 132 Here we were bowling them out so that they could take advantage of it [*sc.* a green wicket]. They did that in no uncertain terms. 1977 *Time Out* 28 Jan.–3 Feb. 7/1 Five months after the programme was axed, Edmonds was told in no uncertain terms why it didn't fit.

c. Phr. *of uncertain age*.

[1817: see Dict.]. 1877 F. H. Burnett *Theo* v. 137 The blandishments of a single gentleman of uncertain age. 1930 A. Christie *Murder at Vicarage* iv. 33 There is no detective in England equal to a spinster lady of uncertain age with plenty of time on her hands. 1952 J. Cannan *Body in Beck* vii. 152 The fair sex is very credulous, especially in the case of maiden ladies of uncertain age.

uncertainty. Add: **1. a.** Also, the quality of being indeterminate as to magnitude or value; the amount of variation in a numerical result that is consistent with observation.

1853 *Proc. R. Irish Acad.* V. 372 As to the sun and moon, it is more doubtful. In the transit they have larger probable errors than the stars. For the sun I obtained.. the first limb $\pm 0^s{\cdot}116$, the second $\pm 0^s{\cdot}087$;..while stars ..had but $\pm 0^s{\cdot}097$. This greater uncertainty arises from the strong contrast between the bright and dark surfaces whose boundary we take. 1861 G. B. Airy *Theory Errors of Observations* I. 4 Strictly speaking, we ought..to use the word 'uncertainty' instead of 'error'. For we cannot at any time assert positively that our estimate or measure, though fallible, is not perfectly correct; and therefore it may happen that there is no 'error', in the ordinary sense of the word. 1930 Ruark & Urey *Atoms, Molecules & Quanta* xviii. 619 If a coordinate q is measured with an error of the order Δq, the uncertainty, Δp, of the conjugate momentum introduced by our measurement is such that $\Delta q{\cdot}\Delta p \simeq h$. 1943 M. W. White et al. *Practical Physics* i. 12 As applied to the final result of a measurement, the accuracy is expressed by stating the uncertainty of the numerical result, that is, the estimated maximum amount by which the result may differ from the 'true' or accepted value. 1974 G. Reece tr. *Hund's Hist. Quantum Theory* xii. 161 We thus have the relationship $\Delta E \Delta t \simeq h$ between the uncertainty in the determination of energy and the evaluation of a point in time. 1975 *Physics Bull.* Apr. 165/2 The PTB developed a new measuring apparatus capable of accurate measurements of diameter on pistons of 850 mm and cylinders up to 1200 mm in diameter. The uncertainty in Q, dQ/Q, was estimated to be 3×10^{-4}.

4. *Econ.* (The quality of) a business risk which cannot be measured and whose outcome cannot be predicted or insured against (see quots. 1921 and 1964). Cf. *risk *sb.* 2 a.

1921 F. H. Knight (*title*) Risk, uncertainty and profit. *Ibid.* i. 20 A *measurable* uncertainty, or 'risk' proper, as we shall use the term, is so far different from an *unmeasurable* one that it is not in effect an uncertainty at all. We shall accordingly restrict the term 'uncertainty' to cases of the non-quantitative type. It is this 'true' uncertainty..which forms the basis of a valid theory of profit. 1929 G. O'Brien *Notes on Theory of Profit* ii. 17 The assumption of uncertainty is therefore a disutility and must be rewarded. Is uncertainty bearing on this account, entitled to rank as a separate factor of production. 1964 Gould & Kolb *Dict. Social Sci.* 606/1 In its broadest definition the term uncertainty is used by economists to refer to any situation in which a set of alternative outcomes is not fully predictable. 1969 D. C. Hague *Managerial Economics* vii. 137 To conform to established terminology we shall, from now on, use the word uncertainty to mean the same thing as non-insurable risk.

5. (*Heisenberg's*) *uncertainty principle* (Physics), a principle of quantum mechanics implying that certain pairs of observables (e.g. the momentum and position of a particle, the energy and lifetime of a quantum level) cannot both be precisely and simultaneously known, and that as one of any pair is more exactly defined, the other becomes more uncertain. Also *transf.* Cf. *Heisenberg, *principle of indeterminacy* s.v. *indeterminacy b.

The principle is usually stated as an inequality such that the product of the uncertainties of the pair of observables cannot be less than a quantity of the order of Planck's constant.

[1928 *Physical Rev.* XXXII. 570 The principle of uncertainty is particularly clear in this [*sc.* Weyl's] system.] 1929 Condon & Morse *Quantum Mech.* i. 21 (*heading*) The quantum uncertainty principle. 1931 *Times Lit. Suppl.* 5 Nov. 852/4 Perhaps the most remarkable discovery that has been made in connexion with atomic theory is the so-called Uncertainty Principle. 1955 W. Heisenberg in W. Pauli *Niels Bohr* 15 It was now [*sc.* in 1927] assumed in quantum mechanics that real states can always be represented as vectors in Hilbert space (or as 'mixtures' of such vectors). The uncertainty principle was the simple expression for this assumption. 1977 *Time* 14 Mar. 74/1 Even in the age of the Uncertainty Principle and culture fracture, Warren has not lost his sense of life as a sustained drama. 1982 A. M. Lesk *Introd. Physical Chem.* x. 309 What Heisenberg's uncertainty principle asserts is that for *no* state of *any* system can *all* dynamical variables be arbitrarily well-determined.

uncertified, *ppl. a.* Add: **3.** Not certified insane.

1889 G. B. Shaw *London Music in 1888–89* (1937) 101 Brinio..is a patriotic Batavian with two sisters, one of whom is mad and the other sane... Ada, the uncertified one, is beloved by Aquilius. 1938 S. Beckett *Murphy* ix. 161 There were a few such fortunate cases, certified and uncertified, enjoying all the amenities of a mental hospital. 1969 M. Pugh *Last Place Left* xiv. 102 Some uncertified genius had hit upon a new virus mutation method.

unchalked, *ppl. a.* (Later example.)

1938 J. Steinbeck *Long Valley* 113 The house was unscarred, uncarved, unchalked.

uncharge, *v.* **3.** (Later example.)

1902 *Infantry Training* (War Office) I. 119 Charging and Uncharging Magazines in two ranks.

uncharmed, *ppl. a.* Add: **2.** *Particle Physics.* Not possessing the property known as charm.

1972 *Physics Lett.* XXXIX. B. 349 Hadrons containing the fourth type of quark may be as low as 700 MeV above their uncharmed counterparts. 1975 *Physics Bull.* Apr. 181/1 If SU(3) symmetry were perfect, one should clearly distinguish a neutral octet member..from a singlet.., even though they are both neutral, non-strange and of course uncharmed. 1977 *Sci. Amer.* Oct. 68 Charm.. becomes manifest only in hadrons that include a charmed quark or antiquark in combination with uncharmed quarks. 1983 *Canad. Jrnl. Physics* LXI. 124/1 Uncharmed baryons.

unchartered, *ppl. a.* **2.** (Earlier example.)

1812 *Weekly Reg.* II. 19/2 Those planters..who should place confidence in the paper of unchartered banks.

unchild, *v.* Add: Hence **unchi·lding** *ppl. a.*

1876 G. M. Hopkins *Wreck of Deutschland* xiii, in *Poems* (1967) 55 Wiry and white-fiery and whirlwind-swivelled snow Spins to the widow-making unchilding unfathering deeps.

unchildish, *a.* (Later examples.)

1888 F. H. Burnett *Sara Crewe* I. 22 She spoke in a strange, unchildish voice. 1925 V. Woolf *Common Reader* 169 An astonishing and unchildish story, *Love and Friendship*.

unchildlike, *a.* (Earlier contextual example.)

1833 J. S. Mill in *Monthly Repos.* VII. 62 This most grown-up and unchildlike age.

unchinked, *ppl. a.* (Earlier example.)

1819 *Niles' Reg.* XVII. 30/2 A year ago there were only 'five or six unchinked cabins' on the town plot.

unchurchly, *a.* (Earlier example.)

1815 J. Mayne *Jrnl.* 1 Jan. (1909) 237 For although the most wretched, spiritless animals had been purposely selected, yet the novelty of the scene and the shouts of the people would sometimes elicit an unchurchly amble.

uncinate, *a.* Add: **b.** *uncinate gyrus*, the hook-shaped anterior part of the hippocampus involved in the perception of olfactory stimuli.

1883 *Gray's Anat.* (ed. 10) 487 The uncinate gyrus extends from the posterior extremity of the hemisphere to the fissure of Sylvius. 1980 *Ibid.* (ed. 36) 999/1 The tail [of the uncinate gyrus] separates the rest of the inferior surface of the uncus into an anterior uncinate gyrus and a posterior intralimbic gyrus.

c. Involving or affecting the uncinate gyrus: applied to a type of epileptic fit in which hallucinatory sensations of taste and smell are experienced.

1899 J. H. Jackson in *Lancet* 14 Jan. 79/2 There is very often the Dreamy State in cases of this group of epileptic fits—the Uncinate Group. 1948 A. Brodal *Neurol. Anat.* x. 338 An uncinate attack may be followed by an ordinary epileptic fit, in which case the uncinate attack represents an aura. 1974 E. Niedermeyer *Compendium of Epilepsies* v. 109 In human epileptic conditions, Jackson's term of 'uncinate epilepsy' corresponds with amygdaloid insular seizure manifestations.

uncirculated, *ppl. a.* Add: (Examples of money.) Also *absol.* as *sb.*

1938 J. Coffin *Coin Collecting* viii. 100 It is highly doubtful just what price might be given for an 1804 silver dollar in uncirculated condition. 1962 L. Brown *Coin Collecting* i. 17 Owing to modern methods of minting, where coins move along conveyor belts, slide down shutes and are packed into bags, it is almost impossible to obtain a modern coin..in anything like FDC [perfect mint state] condition and so the term 'uncirculated' was introduced. 1978 J. L. Hensley *Killing in Gold* v. 61 I'll stay with crisp uncirculateds..at least for type.

uncircumstanced, *ppl. a.* (Later example.)

1943 S. Sassoon in *Country Life* 25 June 1136/2 Cloud shadows..Dwell and dissolve; uncircumstanced they pause and pass.

uncivic, *a.* Add: Hence **unci·vically** *adv.*

1931 A. Huxley *Music at Night* II. 81 We do not admit ..that there should be citizens treated uncivically.

uncivilization. (Earlier contextual example.)

1880 W. Morris *Hopes & Fears for Art* (1882) 107 The attainment of these very comforts is what makes the difference between civilisation and uncivilisation.

uncivilized, *ppl. a.* Add: Also *absol.*

1900 tr. *J. Deniker's Races of Man* vii. 251 Among the uncivilised, it is not a question of absolute right, of absolute morality; everything is reduced to a very restricted altruism, not extending beyond kin and immediate neighbours.

unclassified, *ppl. a.* Add: (Examples corresponding to *classified *ppl. a.* b, c.) Also *absol.* as *sb.*

1935 *City of Oxford Council Rep.* 1 Mar. 401 It is recommended that the following programme be submitted to the Ministry of Transport..:—*Bridges*... *By-passes*... *Classified Roads*... *Unclassified Roads*. 1958 M. Kelly *Christmas Egg* III. 129 The road was lonely, unclassified... He must walk back to A260. 1972 O. Sela *Bearer Plot* iv. 28 He would swoop off the main roads onto undulating unclassifieds that sometimes were little more than dirt tracks. 1978 *Country Life* 8 June 1634/3 Of all the trackways of these Lincolnshire Wolds the most romantic is the Bluestone Heath Road, now an unclassified lane.

absol. 1952 *Manch. Guardian Weekly* 14 Feb. 13 My work is unclassified, but the fund out of which I am paid comes from a government agency. 1968 'J. le Carré' *Small Town in Germany* v. 72 He's down here collecting the mail... Everything. Classified or Unclassified, it didn't make no difference. 1980 D. Bloodworth *Trapdoor* xxiii. 143 The file was not unclassified... It was marked Top Secret.

uncle, *sb.* Add: **1. f.** In B.B.C. Radio: formerly, a male announcer or story-teller

for children's programmes. Cf. sense *2 b and *auntie b (b).
1923 *Wireless Weekly* 8 Aug. 183/3 The Director of Programmes received me into the actual studio, where he and the other Uncles have so much fun over the Children's Hour. **1981** S. Briggs *Those Radio Times* 12/1 Long before the Corporation was called 'Auntie'..it had dozens of 'aunts' and 'uncles' on its staff. **1985** *Times Lit. Suppl.* 22 Mar. 330/3 Knight began to broadcast..after the war, becoming extremely popular as Uncle Max on the children's programme *Nature Parliament.*
g. *universal uncle*: see *universal *a.* 9 b.
h. A male friend or lover of a child's mother.
1962 *Listener* 31 May 935/2 His mother has never been married, has lived for some years at a time with a series of 'uncles' who have been the fathers of these siblings. **1968** *Ibid.* 1 Aug. 155/2 The play is a simple tale of a boy who, lacking a resident father, grows up under the influence of various temporary 'uncles'.
2. b. (Earlier and later U.S. examples and examples corresponding to senses 1 f, h above.)
1830 S. P. Holbrook *Sketches by Traveller* 111 In many families..the children are taught to address the older servant as *uncle* or *auntee*, and this is sometimes more than a figure of speech. **1876** 'Mark Twain' *Tom Sawyer* xxviii. 216 He let's [*sic*] me, and so does his pap's nigger man, Uncle Jake. **1923** *Radio Times* 28 Sept. 11/2 Children's Stories—Uncle Donald and Auntie Betty. **1937** Partridge *Dict. Slang* 337/1 Her children call him 'Uncle'. **1945** T. Rattigan *Love in Idleness* i. 280 Oh, don't call him sir, Michael. Call him—I know—call him Uncle John. **1962** W. Faulkner *Reivers* ii. 30 His wife.. was Delphine, Grandmother's cook. At that time he was 'Uncle' Ned only to Mother. I mean, she was the one who insisted that all us children..call him Uncle Ned.
c. Also *ellipt.*; also, (the members of) a federal agency.
1849 *Placer Times* (Sacramento, Calif.) 1 Sept., Two Express Lines have been established between our City and San Francisco. Our old Uncle will have to 'stir his stumps' else his 'regular' arrangements will become a *dead letter.* **1950** H. E. Goldin *Dict. Amer. Underworld Lingo* 231/1 *Uncle*... (Plural) G-men; agents of the Federal Bureau of Investigation. **1953** W. Burroughs *Junkie* x. 98 'He belongs to Uncle, now,' said the [police] captain to my wife as they left the house. **1966** T. Pynchon *Crying of Lot 49* i. 17 The well-known portrait of Uncle that appears in front of all our post offices. **1971** G. V. Higgins *Friends of Eddie Coyle* ii. 14 That's not working for uncle, Eddie. You got to put your whole soul into it. **1978** 'P. Mann' *Steal Big* ii. 8 The nerve I had. Uncle had made me prove it time and again.
e. *Uncle Tom Cobleigh* (or *Cobley*) *and all*: a name given to the last of a long list of persons (see quot. *c* 1800 for the ballad alluded to); a whole lot of people.
c **1800** *Widdicombe Fair* in G. Bantock *One Hundred Songs of Eng.* 72 Tom Pearce's old mare doth appear gashly white Wi' Bill Brewer, Jan Stewer, Peter Gurney, Peter Davy, Dan Whiddon, Harry Hawk, old Uncle Tom Cobleigh and all, Old Uncle Tom Cobleigh and all. **1933** E. A. Robertson *Ordinary Families* xiii. 287 When Dru.. repeated to Margaret some gossip about an engagement, Margaret said casually, 'Oh, and to Uncle Tom Cobley an' all, I suppose!' **1941** J. D. Carr *Case of Constant Suicides* iv. 55 They're all here: the Fiscal, and the law-agent..and Uncle Tom Cobleigh and all. **1963** L. Klein *Fabian Tract No. 349* i. 2 We..are exhorted to pant along behind the industrious Germans, Japanese, Russians, Americans and Uncle Tom Cobley. **1966** *Guardian* 10 Sept. 14/8 It seems clear that a compromise, half-way solution had equally been ruled out by Government, Opposition, economists, press, TV, Uncle Tom Cobleigh and all. **1981** D. Boggis *Time to Betray* xxv. 130 Stupid little man, dragging in old Uncle Tom Cobley and all.
f. *Uncle Ned* (Rhyming slang), (*a*) bed (also ellipt. as *uncle*); (*b*) head.
1925 Fraser & Gibbons *Soldier & Sailor Words* 294 *Uncle Ned*, bed. **1955** F. Brown *Martians, go Home!* ii. iii. 68 Hi got to speel or there's no weeping willow for my Uncle Ned. **1964** *Listener* 31 Dec. 1053/1, I have spent an hour fixing the big, loose curls on top of my Uncle Ned. **1974** J. Gardner *Corner Men* xiv. 194 Get out of that Uncle Ned, slide into your threads, and come down the nick with us. **1982** J. Scott *Uprush of Mayhem* x. 105 'You did right, shoving him back in his uncle.'..Uncle. Uncle Ned, Cockney rhyming slang for bed.
3*. *to cry* (*holler, say*, etc.) *uncle*, to acknowledge defeat, to cry for mercy. *N. Amer. colloq.*
1918 *Chicago Herald-Examiner* 1 Oct. 11 Sic him Jenny Jinx—make him say 'Uncle'. **1939** *Amer. Speech* XIV. 267 'He hollered "calf rope"' or 'He hollered "uncle",' are publishments of his defeat. **1941** B. Schulberg *What makes Sammy Run?* vi. 139 Kit was the one who did him some good. 'Okay,' I said. 'I'll cry uncle.' **1962** W. Stegner *Wolf Willow* iii. iii. 237 With good hay land and good range [we can] make this God darned country holler uncle. **1972** D. Delman *Sudden Death* v. 122 'Stop it, darling, please.' 'Say uncle.' 'Uncle.' **1980** *Amer. Speech* 1976 LI. 281 Most American schoolboys are..familiar with the expression *cry uncle* or *holler uncle*, meaning 'give up in a fight, ask for mercy'. *Uncle* in this expression is surely a folk etymology, and the Irish original of the word is *anacol* (*anacal, anacul*) 'act of protecting; deliverance; mercy, quarter, safety', a verbal noun from the Old Irish verb *aingid* 'protects'... My unscientific sampling of English speakers in Britain a few years ago indicated that *cry uncle* is not familiar in England or Scotland.
4. *uncle-figure.*
1959 *Listener* 10 Sept. 375/1 To a majority of Americans, Adlai Stevenson is an uncle-figure—'good old Uncle Adlai'. **1975** *Times* 8 Mar. 7/4 Such an uncle-figure as Johnny Carson..on late night television.

Hence **u·ncle-ish** *a.*
1928 A. Huxley *Point Counter Point* x. 160 An occasional chaste uncle-ish kiss on the forehead. **1981** P. Dickinson *Seventh Raven* xii. 166 He'd get much more mileage out of seeming friendly and uncle-ish.

uncleaned, *ppl. a.* (Earlier example.)
1854 Mrs. Gaskell *North & South* (1855) I. xiii. 157 The uncleaned corners of the room.

Uncle Tom. orig. *U.S.* [Uncle *sb.* 2 b.] The name of the hero of *Uncle Tom's Cabin*, a novel (1851–2) by Harriet Beecher Stowe, used allusively for a Black man who is submissively loyal or servile to White men. Also *transf.* and in extended use.
1922 [see *New Negro, new negro]. **1942, 1960** [see *handkerchief-head* s.v. *handkerchief b]. **1971** *Guardian* 15 July 3/1 Arafat was always attacked by the Marxist-orientated militants as being a Palestinian 'Uncle Tom', neither sufficiently radical or violent. **1972** M. J. Bosse *Incident at Naha* i. 37 Some people consider him an Uncle Tom because he doesn't study Afro-American culture. **1975** M. Bradbury *History Man* v. 84 The girl I'm living with..says I have a slave mentality... She says I'm an Uncle Tom. **1977** *New Yorker* 22 Aug. 66/3 Pryor goes through his part pop-eyed, playing Uncle Tom for Uncle Toms. **1978** *Church Times* 24 Feb. 2/4 Many parishes do have a youngster on the PCC, but... It's only tokenism. The youngsters are 'Uncle Toms', in a way. **1981** *Bull. Amer. Acad. Arts & Sci.* May 41 Uncle Tom's virtues as a worker change when the vices of his condition have to go.
attrib. **1953** Berrey & Van den Bark *Amer. Thes. Slang* (1954) §579/2 *'Straight jazz'*,..schmaltz,..unadulterated corn, Uncle Tom music. **1959** 'F. Newton' *Jazz Scene* v. 88 The savage hostility to 'Uncle Tom' musicians, which for the first time split the community of jazz players. **1960** [see *Jim Crow', Jim-crow, jim crow 3]. **1971** *Black Scholar* Dec. 20 The harshest discrimination that I have encountered *in the political arena* is anti-feminism—from both males and brainwashed 'Uncle Tom' females. **1978** G. Greene *Human Factor* iii. iii. 127 Been to the African University in the Transvaal where Uncle Tom professors always produce dangerous students. **1979** *Guardian* 14 Apr. 8/8 You got an Uncle Tom figure there in your book?..You know, kinda creepy black always trying to ingratiate himself with the white folks.
Hence **Uncle Tom** *v. intr.*, to act in a manner characteristic of an Uncle Tom; also with *it*; **Uncle Tom(m)ery, Uncle Tom(m)ing** *vbl. sb.*, **Uncle Tom(m)ish,** *a.*, **Uncle Tom(m)ism.**
1937 Uncle Tomism [see *nigger *sb.* 1]. **1944** C. Himes *Black on Black* (1973) 198 Here come a big Uncle Tomish lookin' cat in starched overalls. **1947** S. Lewis *Kingsblood Royal* x. 52 Why, you gold-digging, uncle-tomming, old, black he-courtesan! **1950** Patterson & Conrad *Scottsboro Boy* iii. iv. 219 The prisoners clowned for the white folks—the guards, the prison heads, and their families. It looked like a lot of Uncle Tom-ing to me, and I didn't enjoy seeing whites laugh at the coloured guys' pranks. **1960** *New Left Rev.* Nov.–Dec. 49/1 Armstrong's clowning is just Depressing. It isn't that he 'uncle toms' but that the act is so automatic and lifeless. **1961** *Guardian* 1 Dec. 6/5 The Uncle-Tommish innocence of the..Negro. **1967** *Listener* 23 Feb. 264/2 Not all Jews will like it, though. One of my acquaintance finds it patronizing and demeaning, the Jewish equivalent of Uncle Tommism. **1967** *Punch* 9 Aug. 210 An obligation.. applies constantly to all underdog groups, constantly tempted by rewards to uncle-tom, to pull the forelock. **1968** *New Yorker* 17 Aug. 21 The guests will be arriving any minute now. Please, Amanda, try to Uncle Tom it a little just for tonight. **1972** *Guardian* 21 Oct. 9/2 The young black studies' teacher..was striking out against Uncle Tommery. **1976** *Gramophone* Feb. 1321/2 Now today it has almost the opposite effect from what was intended. Many blacks relate it to the era of Uncle Tom-ism. **1979** *N.Y. Times* 12 Feb. B13/1 To tell women that if they just behave nicer, if they shuffle and Uncle Tom a little more, that they will be more successful, is simply not accurate. **1981** *Cape Argus Mag.* 24 Oct. 2/2 With..that substantial brush of Uncle Tommery with which his opponents have tarred him, Dr Cedric [Phatudi] is a walking..paradox... He has none of the stridency of a Chief Buthelezi,..or the bombast of a Matanzima.

unclipped, -clipt, *ppl. a.* Add: **c.** Not fastened with a clip.
1922 Joyce *Ulysses* 23 Buck Mulligan stood on a stone, ..his unclipped tie rippling over his shoulder.

uncock, *v.*² (Earlier example.)
1745 W. Ellis *Mod. Husbandman* VII. ii. 80 Then this Nobleman thought it high time to uncock all the wheat again.

uncognisant, *a.* Add: Also **uncognizant.** So **unco·gnizantly** *adv.*
1843 [see *pigeon-pie* s.v. *pigeon *sb.* 5 a].

uncommendable, *a.* (Later examples.)
1959 I. & P. Opie *Lore & Lang. Schoolch.* xviii. 377 Almost any group of 12-year-olds, asked what are their favourite after-dark games, will name doorbell-ringing, and similar uncommendable activities. **1983** T. de Vere White *Johnnie Cross* i. 12 The public had shown uncommendable restraint in the book shops.

uncommissioned, *ppl. a.* **2.** (Earlier example.)
1822 M. Edgeworth *Let.* 12 June (1971) 407 The Nelson—just finished but uncommissioned a first rate man of war 120 guns.

uncommitted, *ppl. a.* Add: **4.** (Further examples, esp. as between two power groups.) Also *absol.*
1956 Ball & Killough *International Relations* xxiii. 494 The Arabs had rejected association with the West and insisted on maintaining a 'neutral' position. Many Arab leaders regarded the Middle Eastern states as likely to be safer from Soviet attack if they remained uncommitted to the West. **1958** *Listener* 6 Mar. 390/2 It brings in not only countries committed to Western alignment, like the Philippines, Siam, and Pakistan, but also uncommitted countries of the neutral bloc, like India, Burma, and Indonesia. **1958** *Spectator* 22 Aug. 259/2 The key..lies..in a transformation of American relations with the uncommitted world. **1959** *Oxf. Mag.* 11 June 469/2 Islam is likely to gain far more of the uncommitted than is Christianity. **1959** *News Chron.* 10 July 4/2 The uncommitted voter. **1961** *Daily Tel.* 31 Aug. 10 Neutralism, Mr Foster Dulles once declared, is immoral. Nowadays, when the 'uncommitted' constitute a decisive, if not solid, phalanx in the United Nations, Western statesmen eschew such language.

uncomplimentary, *a.* Add: (Earlier example.)
1842 J. S. Mill *Let.* 25 Nov. in *Wks.* (1963) XIII. 558 People call you by various uncomplimentary names indicative of self-conceit.
Hence **u:ncomplime·ntarily** *adv.*
1909 'Mark Twain' *Is Shakes. Dead?* xi. 127 It would grieve me to know that any one could think so injuriously of me, so uncomplimentarily, so unadmiringly of me.

uncompromised, *ppl. a.* Add: Hence **unco·mpromisedness.**
1851 H. Melville *Moby Dick* iii. xxi. 145 This strange uncompromisedness in him.

uncompromisingly, *adv.* (Earlier example.)
1834 J. S. Mill in *Monthly Repos.* VIII. 527 The military tribunals..Mr. Abercromby..steadily and uncompromisingly opposed.

uncomputable, *a.* (Later examples.)
1906 Somerville & 'Ross' *Some Irish Yesterdays* 51 Large statements as to her uncomputable value had not her tail in youth been shut into a stable door and given a double angle like a bayonet. **1979** *Sci. Amer.* Nov. 30/2 Consider a more traditional way of encoding the halting problem in an uncomputable irrational number.

uncondescending, *ppl. a.* (Later example.)
1969 *Guardian* 31 July 2/3 A performance that was obviously pleasing and uncondescending.

unconditional, *a.* (Earlier and later examples of *unconditional surrender.*)
1830 Hazlitt *Life Napoleon Buonaparte* IV. l. 119 We will have more, namely the original stake we played for; unconditional surrender of the right of nations to chuse their own government. **1901** H. Campbell-Bannerman in *Hansard Commons* 14 Feb. 89 Unconditional surrender was our first and last word. **1930** G. B. Shaw *Apple Cart* ii. 69 In plain terms we require from you an unconditional surrender. **1949** *New Statesman* 30 July 115/2 The sharp argument about 'unconditional surrender' in the House of Commons last week revealed much that was of more than historical significance... He [*sc.* Mr. Bevin] had not protested when Mr. Churchill brought back this fatal slogan from Casablanca. **1956** A. Huxley *Adonis & Alphabet* 81 The only completely unconditional surrender will come when everybody—but *everybody*—is a corpse. **1974** *Times* 20 Dec. 1/2 The Kuwait authorities insisted that the [hijackers'] surrender was 'unconditional'.

unconditioned, *ppl. a.* Add: **2. a.** (Earlier example.)
1796 F. A. Nitsch *Gen. View Kant's Princ. concerning Man* 127 Reason..produces the idea of an unconditioned Limitation.
b. *unconditioned reflex*, an inborn, instinctual reflex or reflex action (cf. *conditioned *ppl. a.* 7 b). So *unconditioned stimulus.*
1906, 1927 [see *conditioned *ppl. a.* 7 b]. **1937** *Discovery* Jan. 17/2 Its instincts, or, to use Pavlov's expression, its unconditioned reflexes. **1972** *New Yorker* 26 Aug. 32/3 In the vocabulary that Pavlov adopted to describe his findings, the meat powder was labelled an 'unconditioned stimulus'.
unconditionedness (earlier example).
1854 Geo. Eliot tr. *Feuerbach's Essence Christianity* iv. 54 The metaphysical attributes of eternity, unconditionedness,..and the like abstractions.

unconducive, *a.* (Later example.)
1984 *Times Lit. Suppl.* 25 May 593/4 A number of local CIA men do seem to have taken steps unconducive to the success of Kennedy liberalism.

unconfirmed, *ppl. a.* **1.** (Later examples.)
1920 *Conference of Bishops of Anglican Communion* 30 The irregularity of admitting to Communion the baptized but unconfirmed Communicants of the non-episcopal congregations. **1977** R. L. Wolff *Gains & Losses* ii. ii. 122 Danger besets the unconfirmed girls... They find their own vicar crusty and aloof.

unconformity. **1.** (Later example.)
1982 K. Smidt (*title*) Unconformities in Shakespeare's history plays.

unco·njugated, *a.* *Med.* and *Gram.* [Un-¹ 8 a.] Not conjugated.
1909 in Webster (undefined). **1963** *Jrnl. Clin. Endocrinol.* XXIII. 820/2 Measurement of unconjugated cortisol in the urine affords a reliable index of the biologically active fraction of circulating cortisol. **1964** *Language* XL. 277 The description of pronoun position with verbal constructs..is simple and clear: where the unconjugated form precedes the conjugated, it is equated with the other elements that cause anteposition. **1977** *Proc. R. Soc. Med.* LXX. 598/2 Some patients do have an increase in biliary unconjugated bilirubin.

unconnected, *ppl. a.* **2.** (Earlier example.)
1745 Hume *Let. from Gentleman* (1967) 32 Suppose Mankind, in some primitive unconnected State [etc.].

unconscious, *a.* Add: **2. c.** *Psychol.* Applied to mental or psychic processes of which a person is not aware but which have a powerful effect on his attitudes and behaviour, *spec.* in Freud's psychoanalytic theory, processes activated by desires, fears, or memories which are unacceptable to the conscious mind and so repressed; also designating that part of the mind or psyche in which such processes operate. Freq. *absol.* Cf. *collective unconscious* s.v. *collective *a.* 2 e; *id².
a **1884** M. Pattison *Mem.* (1885) vii. 329, I cannot help observing the remarkable force with which the Unconscious—*das Unbewusste*—vindicated its power. *Ibid.* 330 By whatever name you call it, the Unconscious is found controlling each man's destiny without, or in defiance of, his will. **1912** Freud in *Proc. Soc. Psychical Res.* XXVI. lvi. 315 The term *unconscious*, which was used in the purely descriptive sense before, now comes to imply something more. It designates not only latent ideas in general, but especially ideas with a certain dynamic character, ideas keeping apart from consciousness in spite of their intensity and activity. *Ibid.* 318 The system revealed by the sign that the single acts forming part of it are unconscious we designate by the name 'The Unconscious', for want of a better and less ambiguous term... And this is the third and most significant sense which the term 'unconscious' has acquired in psychoanalysis. **1914** [see *co-conscious *a.* and *sb.*]. **1925** C. E. M. Joad *Mind & Matter* iv. 111 This greater part is known as the unconscious mind, or simply as 'the unconscious'. The theory of the unconscious is based mainly on the work of..Freud. **1946** *Mind* LV. 21 Perhaps further investigation following Wisdom's hint that philosophical views are the vehicles for expressing 'unconscious fantasies', will lead to an understanding of this point. **1956** R. F. C. Hull tr. *Jung's Symbols of Transformation* in *Coll. Wks.* V. ix. 443 The Miller case is a classic example of the unconscious manifestations which precede a serious psychic disorder. **1959** N. Mailer *Advts. for Myself* (1961) 216 To put it crudely, I would think I was dropping people when they were dropping me. And of course my unconscious knew better. **1977** A. Sheridan tr. *J. Lacan's Écrits* iii. 50 The unconscious is that chapter of my history that is marked by a blank or occupied by a falsehood: it is the censored chapter.

unconsiderable, *a.* Delete † *Obs.* and add later examples.
1914 W. J. Locke *Fortunate Youth* i. 19 Sky and grass and trees and white mass of ladies..and unconsiderable men and boys became a shimmering blur. **1976** *Brit. Jrnl. Sociol.* XXVII. 109/2 The rank and file of British Educationalists, sociologists of education and educational psychologists have concentrated their considerable energies and occasionally some not unconsiderable talents on devising arguments to legitimate the equalization of opportunity on moral grounds and to support it on scientific grounds.

unconstitutional, *a.* Add: **1.** (Earlier example.)
1734 *Country Jrnl.* 16 Nov. 1/2 Lest all other Provisions should be ineffectual to keep the Members of the House of Commons out of this unconstitutional Dependency..the Wisdom of our Constitution hath thought fit that the Representatives of the People should not have Time to forget that they are such; that they are empowered to act for the People, not against them.
Hence **unconstitu·tionalism.**
1920 *Glasgow Herald* 11 Dec. 7 It is the first area of dry land which has shown itself after the deluge of unconstitutionalism in that part of the country [*sc.* Ireland]. **1949** St. J. Ervine *Craigavon* ii. lxxvi. 344 He might have made an effective debating retort to the Ulstermen about unconstitutionalism.

uncontrollability. (See small-type note s.v. Uncontrollableness.) (Examples.)
1909 *Chambers's Jrnl.* June 342/2 Wherever horses are there must be unsanitary filth, and sometimes uncontrollability born of nerves or vice. **1980** *New Scientist* 31 Jan. 341/2 The uncontrollability of medical costs.

unconvincible, *a.* (Earlier and later examples.)
a **1747** J. Wesley *Wks.* (1829) V. xxxvii. 476 An unadvisable and unconvincible spirit. **1979** C. McCarry *Better Angels* ii. vii. 119 He made his argument to her, knowing that Patrick..was unconvincible.

uncoped, *ppl. a.* (Later examples.)
1972 J. G. Vermandel *Last Seen in Samarra* xii. 81 Wearing a housecoat, and with her red hair curlier than he'd ever seen it before. Uncoped with. **1981** E. North *Dames* i. 4 They..think back over life uncoped with.

uncorrectable, *a.* (Later examples.)
1970 *New Yorker* 30 May 26 A regularly scheduled airliner..radioed..to report..an undiagnosed and uncorrectable loss of power. **1970** *Time* 21 Sept. 57 For many years, facial paralysis has been uncorrectable.

uncorrectible, *a.* (See small-type note s.v. Uncorrectable *a.*) (Examples.)
1902 A. H. Buck *Ref. Handbk. Med. Sci.* (ed. 2) IV. 528/2 Finally the deformity becomes permanent and uncorrectible. **1952** G. Sarton *Hist. Sci.* I. xxi. 546 Superstition is of necessity more conservative than science, because it is uncorrectible and unprogressive.

uncountable, *a.* Add: **2. a.** *spec.* in *Math.*, infinite and incapable of being put into a one-to-one correspondence with the integers. Opp. *countable *a.* 2 c.
1952 R. L. Wilder *Introd. Foundation Math.* iv. 88 Some mathematicians do not admit the existence of an uncountable set of real numbers as a legitimate consequence of the argument. **1964** T. O. Moore *Elem. Gen. Topology* i. 16 The set *R* of all real numbers is uncountable. **1971** *Sci. Amer.* Dec. 98/1 If the final destination of the bird is not specified, an uncountable infinity of such graphs can start at *C* and end anywhere on the track between *A* and *B*.
4. *Gram.* That cannot be counted; invariable in number; *spec.* of a noun: that cannot form a plural or be used with the indefinite article.
1924 [see B below]. **1948** A. S. Hornby et al. *Learner's Dict. Current Eng.* p. x, The sign [U] is a warning that the noun..stands for a material, quality, etc. that is uncountable. The noun..may not be used in the indefinite article and must not be used in the plural. **1961** R. B. Long *Sentence & its Parts* ix. 225 There is a tendency to assign uncountable-plural status and *mumps are dangerous to adults* is heard alongside the preferred *mumps is dangerous to adults*. **1966** J. Derrick *Teaching Eng. to Immigrants* ii. 71 Foreign learners may misuse these 'uncountable' nouns by analogy with 'countable' ones and say such things as 'I want two milks', 'This is a rice', 'These are moneys', 'This is an ink', etc. **1980** *Chambers Universal Learners' Dict.* p. viii, *nu* This is short for *noun uncountable* and means that a noun (or a particular meaning) labelled in this way may not be used in the plural form.
B. *sb. Gram.* An uncountable noun or its referent.
1924 O. Jespersen *Philos. Gram.* xiv. 188 There is a class of 'things' to which words like one, two are inapplicable; we may call them uncountables, though dictionaries do not recognize this use of the word *uncountable*, which is known to them only in the relative sense 'too numerous to be (easily) counted'. **1965** K. Schibsbye *Mod. Eng. Gram.* ii. 100 Though uncountables are normally in the singular, some of these (nearly) always appear in the plural: *oats, riches.* **1981** *Fremdsprachen* XXV. 236 Modern grammarians often divide nouns according to their capacity to be used with numerical values into: countables and uncountables.
Hence **u:ncountabi·lity,** the property of being uncountable; **uncountably** *adv.* (later examples).
1952 R. L. Wilder *Introd. Foundation Math.* iv. 88 The proof of the uncountability of *R*. **1955** J. L. Kelley *Gen. Topology* iv. 122 The product of uncountably many topological spaces does not generally satisfy the first axiom of countability. **1977** *Sci. Amer.* Jan. 115/3 Conway's proof of the uncountability of Penrose patterns ..can be outlined as follows. **1981** *Ibid.* Nov. 29/1 Their number, however, will be uncountably infinite.

uncounterfeit, *v.* Add: Hence **uncou·nterfeiting** *ppl. a.*
1912 [see *uncounterfeitable* s.v. *un-¹ 7 b (a)].

uncouple, *v.* Add: **3. a.** *Biochem.* To separate the processes of (phosphorylation) *from* those of oxidation.
1948 *Jrnl. Biol. Chem.* CLXXIII. 808 These results indicate that DNP [*sc.* dinitrophenol] reversibly uncouples phosphorylation from oxidation. **1977** D. E. Metzler *Biochemistry* vii. 366/1 Arsenate is said to uncouple phosphorylation from oxidation.
b. *Physics.* To cause to cease to interact; to decouple (sense *2 a).
1980 *Chem. in Brit.* XVI. 456/2 This excited state may return to groundstate or undergo a chemical reaction or may uncouple two electron spins (intersystem crossing) to yield a triplet state.

uncoupled, *ppl. a.*¹ (Examples corresponding to *uncouple *v.* 3 a.)
1954 *Proc. Nat. Acad. Sci.* XL. 919 Mitochondria suspensions taken from thyroxine-treated animals remain uncoupled. **1981** *Plant Physiol.* LXVIII. 1485/1 The light saturated rate of photosystem I-dependent electron transport..was increased by a high concentration of DCMU added to broken and uncoupled chloroplasts.

uncoupled, *ppl. a.*² Add: **b.** *Physics.* Not physically interacting.
1965 W. T. Thomson *Vibration Theory & Applications* vi. 167 The two pendulums behave as if they were uncoupled and independent of each other. **1981** *Sci. Amer.* July 56/1 Wiesel's computer simulations of the evolution of the resonance begin with Io, Europa and Ganymede in uncoupled orbits and with Io driven outward by its tidal interaction with Jupiter.

uncoupler. Add: *spec.* in *Biochem.*, any agent that causes the uncoupling of oxidative phosphorylation.
1956 *Science* 22 June 1107/1 The factor specifically inhibits phosphorylation without affecting the oxidation of β-hydroxybutyrate and thus simulates the action of 2,4-dinitrophenol and other known uncouplers. **1976** *Sci. Amer.* June 44/3 Then we tried uncouplers: agents that allow electron transport to proceed but that in effect disconnect it from phosphorylation and thus from the ATP synthesis it usually accomplishes. **1979** *Nature* 8 Feb. 486/1 We have found that the rapid efflux of Ca^{2+} brought about by collapse of the membrane potential by uncouplers or antimycin A..is not affected by tetracaine.

uncoupling, *vbl. sb.* Add: **a.** (Examples corresponding to *uncouple *v.* 3 a, b.)
1954 *Proc. Nat. Acad. Sci.* XL. 919 As much thyroxine was carried down by rat mitochondria which did not show any appreciable uncoupling. **1983** *Nature* 10 Feb. 512/1 A progressive uncoupling of [bacterial] growth and H_2S production was observed during approach to the stationary state.
b. *attrib.* and *Comb.*, as *uncoupling chain, lever, pole, rod*; **uncoupling agent** *Biochem.*, = *uncoupler.
1956 *Jrnl. Biol. Chem.* CCXXII. 338 Certain uncoupling agents such as thyroxine and Ca^{++} caused rapid swelling of the mitochondria. **1981** *Arch. Microbiol.* CXXIX. 94/1 Inhibition was relieved by low concentrations of uncoupling agent. **1879** *Car-Builder's Dict.* 172/1 *Uncoupling-chain*, a chain by which the uncoupling lever of a Miller-coupler is connected with the coupling-hook or draw-bar. *Ibid.*, *Uncoupling-lever, for Miller-coupler*, a lever attached to the platform of a car, and connected by a chain with a Miller coupling-hook or draw-bar to disengage or uncouple it from the one on the adjoining car. **1895** *Ibid.* (ed. 3) 139/2 *Uncoupling-rod*,..a rod connecting the uncoupling-lever with the lock of an automatic coupler. **1976** A. White *Long Silence* ix. 88 Two trains were being worked... I counted five men walking about the yard with long un-coupling poles.

unco·very. [f. Uncover *v.*, after *discovery, recovery*, etc.: see -ery.] The action of uncovering or bringing to light.
1963 *Listener* 12 Sept. 377/2 When we indulge in.. deduction..the theorem contains the discovery (or, more exactly, the uncovery of something which was there in the axioms and postulates, though it wasn't actually evident). **1977** *Times Lit. Suppl.* 25 Mar. 336/1 Dr Ray's uncovery of a dusty trove of illustrated books in the basement of a London dealer recalls the accidental discovery of the golden bowl.

uncramp, *v.* (Examples of *intr.* use.)
1937 V. Woolf *Diary* 21 Apr. (1984) V. 80 What a mercy to use this page to uncramp in! after squeezing drop by drop into my 17 minute BBC. **1952** E. Hemingway *Old Man & Sea* 70 His left hand was still..tight... It will uncramp though, he thought.

uncredited, *ppl. a.* (Later examples.)
1959 *Times* 17 Aug. 12/6 This version (adapter uncredited) concentrated on the two main conflicts in the book. **1977** *Rolling Stone* 30 June 102/3 The uncredited musicians play sparely but with enough fire to make their presence, and this entire album, memorable.

uncre·olized, *ppl. a.* [Un-¹ 8.] Of a language or dialect: not creolized; that has not undergone creolization.
1980 *English World-Wide* I. i. 50 The greater contact with uncreolized English on the American mainland had altered the identity of this speech. **1982** D. Sutcliffe *Brit. Black Eng.* i. 23 And English—that is, uncreolized English—is also the official language in Jamaica.

uncri·nkle, *v.* [Un-² 7, 3.] **a.** *intr.* To lose crinkles, to become less crinkled. **b.** *trans.* To remove crinkles from.
1904 G. A. B. Dewar *Glamour of Earth* viii. 173 The tiny leaves will be uncrinkling in a day or two about the dark twiggy bole. **1911** W. De Morgan *Likely Story* v. 136 He uncrinkled a result of the shape of that letter-box. **1935** E. Bowen *House in Paris* ii. vii. 165 No one with him to smile and make his face uncrinkle.

uncrossed, *ppl. a.* **4.** (Earlier example.)
1882 [see *open cheque* s.v. *open *a.* 22 c].

uncrowned, *ppl. a.* Add: **2. b.** *uncrowned king* (*queen*), a man (woman) exerting autocratic influence over a specified sphere; a dominant man (woman). Const. *of.*
1917 J. W. Gerard *My Four Years in Germany* ii. 22 Heydebrand, is known as the 'Uncrowned King of Prussia'. *a* **1940** in *Harper's Bazaar* (1969) Oct. 36/3 Lady Dashwood, uncrowned Queen of Diabolo. **1978** 'S. Woods' *Exit Murderer* 154 The uncrowned King of the diamond smugglers. **1981** I. Boland tr. *Ginzburg's Within Whirlwind* ii. iv. 226 Old General Nikishov.. had this handsome lady living with him... the uncrowned Queen of Kolyma.

uncrushable, *a.* (Further examples.)
1929 *Daily Mail* 20 July 8/3 What are our scientists and inventors doing that they have not yet invented uncrushable linen? **1935** C. Ellis *Chem. Synthetic Resins* I. xxx. 641 Natural or artificial textile fibers can be given an uncrushable finish by the use of urea-formaldehyde resins condensed with alkalies. **1954** M. Stewart *Madam, will you Talk?* x. 78, I..shook out the green dress, thanking heaven and the research chemists for uncrush-

able materials. **1983** 'E. Anthony' *Company of Saints* i. 16 She..put on one of the long, uncrushable shifts that are a godsend to travellers.

uncushioned, *ppl. a.* (Earlier contextual example.)
1852 H. W. Dulcken tr. *Pfeiffer's Visit to Holy Land, Egypt, & Italy* i. 23 Uncushioned benches serve for seats by day and for beds by night.

uncustomary, *a.* Add: Hence **u:ncu·stomarily** *adv.*
1909 Webster, Uncustomarily, adv. **1966** *Punch* 2 Mar. p. viii/2 Functional, even austere decor, and uncustomarily fast service don't mark this splendid Chelsea restaurant as unluxurious. **1982** *Nature* 25 Feb. 642/1 Your uncustomarily superficial and misleading article.

uncut, *ppl. a.* Add: **5.** Also, without excisions or omissions.
1946 *Partisan Rev.* Nov.–Dec. 577 O'Neill's new play.. is guaranteed to last two-and-a-half hours longer than any other play, with the exception of the uncut *Hamlet.* **1953** K. Reisz *Technique Film Editing* xii. 193 The documentary or story-film editor's job..requires a subtler understanding and interpretation of the shades of meaning in the uncut shots. **1960** *News Chron.* 15 Mar. 8/3 His film 'India 1958', so far only seen in the uncut version at private showings. **1967** *Guardian* 9 May 5/3 (*heading*) Film director defends uncut 'Ulysses'. *Ibid.*, 'Ulysses' is due to be shown uncut at the Academy Cinema in London from June 1.
6. Undiluted, unadulterated.
1967 C. Drummond *Death at Furlong Post* vii. 104 All six of the Dancer's gaffs have been taken to pieces. Four ounces of uncut heroin. **1978** T. Williamson *Technicians of Death* vi. 43 They can produce very large amounts of uncut heroin.

undared, *ppl. a.* (Later example.)
1936 Auden *Look, Stranger!* 12 And into the undared ocean swung north their prow.

undation. For † *Obs. rare* read: *Obs. rare* exc. in **undation theory** *Geol.* [tr. Du. *undatietheorie* (1932)], the theory that selective internal heating of the earth's mantle causes large wave-like folds to appear in the crust.
1932 R. W. van Bemmelen in *Natuurk. Tijdschr. Nederlandsch. Indië* XCII. 93 The geologic history of the western part of the Sunda arc has been examined according to the Undation Theory. **1950** P. H. Kuenen *Marine Geol.* ii. 146 In his undation theory, van Bemmelen (1939) postulates a primary salsima layer formerly enveloping the whole earth and now forming the floor of the oceans. **1975** *Nature* 3 Apr. 386/1 The..hypothesis bears some resemblance to the minority view of Van Bemmelen.. and others, whose 'undation theory' proposes that selective high radioactive heating in the mantle produces warping of the overlying crust followed by lateral spreading under gravity.

undead, *a.* Add: Also, not quite dead but not fully alive, dead-and-alive. In vampirism, clinically dead but not yet at rest. (Later examples.) Also *absol.* as *sb.*
1897 B. Stoker *Dracula* xxvii. 381 There remain one more victim in the Vampire fold; one more to swell the grim and grisly ranks of the Un-dead. *Ibid.* 382 This then was the Un-Dead home of the King-Vampire. **1920** H. G. Wells *Outl. Hist.* 286/2 Presently by some amazing miracle he would become undead again and return, and set up his throne with much splendour and graciousness in Jerusalem. **1936** Dylan Thomas *Twenty-Five Poems* 4 They suffer the undead water where the turtle nibbles. **1949** D. L. Sayers tr. *Dante's Divine Comedy* I. viii. 118 Why walks this man, Undead, the kingdom of the dead? **1956** C. S. Lewis *Till we have Faces* xiv. 169 Shadow and monster in one, may be, a ghostly, un-dead thing. **1959** *Twentieth Cent.* Dec. 427 The vampire or 'undead' can only move about freely..between sunset and sunrise. **1972** P. H. Kocher *Master of Middle-Earth* (1973) iv. 62 They still inhabit their original bodies, but these have faded and thinned in their component matter until they can no longer be said to exist in the dimension of the living. Their flesh is not alive, not dead, but 'undead'. **1981** J. Sutherland *Bestsellers* v. 59 The good old folkloric remedies for killing the undead.

undeaf, *v.* (Later example.)
1933 W. de la Mare *Fleeting* 53 Fame with trump and drum Cannot undeaf the dumb.

undeca- (ɒnde·kă), before a vowel **undec-**, comb. form of L. *undecim* eleven, as in Undecagon; cf. Hendeca- in Dict. and Suppl. In *Chem.* and *Biochem.* used to form the names of molecules that contain eleven carbon atoms or consist of eleven of the second element, as **unde·cane,** any of a series of isomeric hydrocarbons $C_{11}H_{24}$, *esp.* the liquid unbranched member $CH_3(CH_2)_9CH_3$; **u:ndecape·ptide,** any polypeptide composed of eleven amino-acids; **u:ndecyle·nic acid,** a yellow, water-insoluble carboxylic acid, $CH_2{:}CH(CH_2)_8COOH$, which is used as an antifungal agent.
1899 *Jrnl. Chem. Soc.* LXXVI. i. 816 The same substances were also detected in neutral creosote from paraffin oil; in this, undecane was found. **1971** E. O. Wilson *Insect Societies* xii. 237/1 Undecane and the mandibular gland substances..evoke the alarm response at concentrations of 10^9–10^{12} molecules per cubic centimeter. **1960** *Jrnl. Biol. Chem.* CCXXXV. 3645/2 Imidazole-heme undecapeptide exhibits a more complex behavior. **1979** *Nature* 8 Feb. 480/2 Several lines of evidence suggest that the undecapeptide, substance P, is involved in synaptic transmission in various areas of the central nervous system. **1879** *Jrnl. Chem. Soc.* XXXVI. 306 Undecylic acid, $C_{11}H_{22}O_2$, prepared by heating undecylenic acid with red phosphorus and hydriodic acid.., is a colourless transparent substance. **1952** S. Spender *Learning Laughter* xiii. 179 Castor oil plants can also be broken down into eneanthol and undecylenic acid. **1979** *Jrnl. Pharm. Sci.* LXVIII. 384/1 The antifungal activity of products containing undecylenic acid and its salts was demonstrated some time ago.

undecidable, *a.* Add: **b.** *Logic* and *Math.* Of a proposition, theorem, etc.: incapable of being either proved or disproved.
1937 *Mind* XLVI. 60 Gödel has shown that the particular sentence in question is undecidable, *i.e.*, neither it nor its negation is demonstrable. **1966** S. Beer *Decision & Control* x. 208 The network language which spans the gap between a problem situation and its conceptual model also contains undecidable sentences. **1979** *Sci. Amer.* Feb. 5/1 Logicians have been able to show that even simple and mathematically interesting statements may be undecidable.
Hence **u:ndecidabi·lity,** the property of being undecidable.
1942 *Mind* LI. 260 It therefore raises the issue of undecidability in the arithmetical as well as in the linguistic realm. **1967** S. C. Kleene *Math. Logic* v. 279 By the essential undecidability of S, S_1 is undecidable. **1971** *Sci. Amer.* Aug. 99/1 In contrast, the Platonists, who count among their number even Gödel himself, believe (like Einstein) that the undecidability in mathematics is a statement about the inherent limitations of our present axiomatic mode of investigation and not about the mathematical objects themselves.

undecided, *sb.* Add: **2.** An undecided person, one who has not made up his mind.
1968 *Listener* 31 Oct. 568/2 Who can decide what an undecided is going to do? **1974** *Times* 7 Oct. 4/4 The Labour Party is picking up more support from the undecideds than any other party.

undecision. Delete † *Obs.* and add later example.
1930 W. Faulkner *As I lay Dying* 13, I mislike undecision as much as ere a man.

undecylenic: see *undeca-.

undefeatable, *a.* Add: (Later examples.) Hence **undefea·tably** *adv.*
1938 J. W. Day *Dog in Sport* xvi. 218 Of great heart and wisdom, with courage undefeatable. **1943** —— *Farming Adventure* xvii. 192 All bids were overtopped by the undefeatable broad Norfolk of an aircraftman who bought recklessly. **1980** *Oxf. Diocesan Mag.* Mar. 19/2 A power which will work unceasingly and undefeatably until the kingdoms of the world become the Kingdom of God.

undefecated, *ppl. a.* (Earlier example.)
1812 *Dramatic Censor 1811* 325 We have not met with any thing on the stage, more abounding in pure, unalloyed, undefecated absurdity, than the '*Wood Dæmon*'.

undefe·ndable, *a.* [Un-¹ 7 b.] **a.** Of a place: that cannot be defended.
1931 W. S. Churchill *World Crisis* VI. viii. 126 Belgrade, the capital, stood actually upon the Danube at the frontier and was undefendable.
b. Of a person = Defenceless, defenseless *a.* 1.
1938 E. Bowen *Death of Heart* i. vi. 109 Her tears were like a flag lowered at once: she felt herself to be undefendable. **1977** W. M. Spackman *Armful of Warm Girl* 114, I never felt so undefendable and I didn't even know who you were.

undefinably, *adv.* (Earlier example.)
1796 F. Burney *Camilla* V. x. xi. 470 This reverie, poignantly agitating, yet undefinably soothing.

undelightfully, *adv.* (Earlier example.)
1749 J. Cleland *Mem. Woman Pleasure* I. 86 The extreme whiteness of her skin was not undelightfully contrasted by the smooth glossy brown of her lover's.

undelighting, *ppl. a.* **2.** (Later example.)
1984 *Times Lit. Suppl.* 27 Apr. 449/4 Trakl himself, whose cold, undelighting, unhuman speech, with its small, select and poisoned vocabulary, is like no other in German.

undema·nding, *ppl. a.* [Un-¹ 10.] Not rigorous or exacting; tolerant; easy. So **undema·ndingness.**
1939 'A. Bridge' *Four-Part Setting* xi. 143 'She..is the most unexacting person I ever met... She *asks* less of you than anyone I know.'..Rose Pelham..had not hit on that peculiar undemandingness. **1940** W. Faulkner *Hamlet* iii. ii. 180 [She] was loyal, discreet, undemanding, and thrifty with his money. **1958** *Times* 18 Dec. 14/1 An undemanding public can be blamed in these cases for spoiling art. **1971** Hindell & Simms *Abortion Law Reformed* v. 122 His publishing activities were fairly undemanding.

undenizened, *ppl. a.* (Later example.)
1887 G. M. Hopkins *Poems* (1967) 103 Undenizened, beyond bound Of earth's glory, earth's ease, all.

undenominationalist. (Earlier example.)
1879 T. Legge in A. Peel *Lett. to Victorian Editor* (1929) 323 Somerville Hall will belong to the undenominationalists.

under, *prep.* Add: **I. 4. a.** Also *under the counter*: see *counter *sb.*³ 4 b; *under the table*: see *table *sb.* 5 e and 6 d.
b. (Later example.) Also, of a motor-cycle to the rider.
1942 [see *nip *v.*¹ 4 c]. **1980** *Dirt Bike* Oct. 57/1 If you're a specialist, you must think long and hard about the MAG 3. Especially if one knifes under you on a flat corner.
II. 10. e. (Earlier example.)
1846 S. Magoffin *Diary* 19 Sept. in *Down Santa Fé Trail* (1926) 135 He has been under the Doctor for some time.
f. *Mus.* = Conducted by.
1887 E. Dannreuther tr. *Wagner's On Conducting* 63 Fancies of this sort, however, were not permitted during the strictly classical performance, under the veteran Capellmeister, at the Munich Odeon. **1910** G. B. Shaw *How to become a Musical Critic* (1960) 278 A performance under Manns of a Mozart symphony. **1943** *N.Y. Times* 9 May ii. 5/5 The City Amateur Symphony Orchestra, under Judge Leopold Prince, will give its annual series of Summer concerts. **1962** *Listener* 12 Apr. 661/2 The BBC Northern Orchestra under Jacques-Louis Monod. **1976** Y. Menuhin *Unfinished Journey* (1977) xvi. 342 During Sir Thomas Beecham's sponsorship of it, I performed the Viotti A Minor Concerto with him—the last time I played under Sir Thomas's baton.
13. f. *Math.* With sbs. denoting an operation performed.
1901 L. E. Dickson *Linear Groups* ii. xi. 252 *G* contains..such conjugate cyclic subgroups, all of whose substitutions are conjugate under *G*. **1940** E. T. Bell *Devel. Math.* xx. 394 The constancy of the cross ratio of four collinear points under projection. **1956** E. M. Patterson *Topology* iv. 84 y_a is the image of x_a under the homeomorphism. **1974** *Encycl. Brit. Macropædia* XI. 657/1 A conic has a central projection that is another conic, but some properties are not preserved under projection.
IV. 22. b. Also prefixed to a number and used as a *sb. pl.* (usu. with *the*) to denote the class of persons who have not yet attained that particular age. Cf. *over- 13 b.
1937 E. Garnett *Family from One End Street* v. 98 John was hopping along..in a sack race for 'under tens'. **1939** E. R. Boyce *Infant School Activities* 201 As much provision as possible should be made for the sort of period suggested for the 'under-fives'. **1941** [see *cross-sectional *a.*]. **1953** K. Tennant *Joyful Condemned* xii. 100 He added her to the other two under-sixteens, Else and Violet. **1968** *Catholic Herald* 15 Mar. 12/5 The Cenacle, Grayshott... Retreat for the Under Thirties. **1973** M. Amis *Rachel Papers* 153 But then, you see, we did the sort of lyrically zany thing that the under-twenties do fairly often.

under, *adv.* Add: **3. d.** In a state of unconsciousness; below the level of consciousness; *spec.* under anaesthetic, in a trance. Also, under the influence of alcohol. *colloq.*
a **1936** 'G. Orwell' *Shooting Elephant* (1950) 28 Doctors..thinking it funny to start operating before you were properly 'under'. **1946** K. Tennant *Lost Haven* (1947) xix. 332 Well, one night when I was down at the pub, this cove..is beside me and he's well under. **1960** M. Spark *Bachelors* vii. 98 He attempted to question me while I was under the other night. **1979** D. Anthony *Long Hard Cure* xxvii. 207, I..stretched out on my bed, and let the music take me under.
4. c. *to get out* (*stand*, etc.) *from under*, to escape or get away from a dangerous or awkward situation. *colloq.* (orig. *U.S.*).
[**1857** *Chicago Times* 6 Oct., To enable me to stand from under the present crash, I shall offer my entire stock for the next 30 days at a great sacrifice.] **1861** *Cincinnati* (Ohio) *Commercial* 24 Apr. (*heading*) Stand from under. **1875** *Scribner's Monthly* Nov. 124/2 The system is rotten.. and, if the nation cares for its life, the quicker it gets 'out from under' the better. **1916** *Lit. Digest* (N.Y.) 8 Jan. 88/2 The next to 'get from under'. **1916** 'Taffrail' *Pincher Martin* iii. 35 Shouts of 'stand from under!' and empty bags came from the deck above. **1938** *New Statesman* 20 Aug. 298/2 The extension of anti-Semitic persecution in the business field [in Germany] has probably led to a certain amount of 'getting out from under' sales by Jews. **1951** H. McCloy *Alias Basil Willing* xiv. 178 I'll get out from under by going to the police myself before anyone else. **1966** 'H. Calvin' *Italian Gadget* ix. 143 Maybe you'll come some day... If I ever get from under Count Capucci. **1974** 'J. Ross' *Burning of Billy Toober* xvi. 157 I'll buy you a dinner when I get out from under.

under-, *prefix*¹. Add: **I. 4. c.** *underredden.*
1866 G. M. Hopkins *Jrnls. & Papers* (1959) 138 The meadows yellow with buttercups and under-reddened with sorrel.
5. a. *under-bodice* (earlier example), *-shorts* (*U.S.*), *-slip*, *-waist* (*U.S.*).
1873 *Young Englishwoman* Apr. 194/1 (*heading*) Under-bodice of jaconet, insertion, and lace. **1960** 'E. McBain' *See them Die* (1963) iii. 28 Murchison..tugged at his undershorts, and wondered if it was any cooler upstairs. **1978** W. F. Buckley *Stained Glass* xii. 129 She had opened the door, exhilarated at the prospect of seeing Paul lying there as she so regularly came on him, dressed only in his undershorts. **1922** Joyce *Ulysses* 321 The bride..looked exquisitely charming in a creation carried out in green mercerised silk, moulded on an underslip of gloaming grey. **1968** B. Hines *Kestrel for Knave* 23 His mother was

standing in her underslip, a lipstick poised at her mouth. **1857** in A. V. G. Allen *Life Phillips Brooks* (1900) I. vi. 209 Thick winter underwaists and socks. *a* **1911** D. G. Phillips in *Hearst's Mag.* (1916) Feb. 137/1 She bought a pair of shoes for a dollar,..two underwaists for a quarter.
b. *underchin.*
1906 Galsworthy *Man of Property* i. iii. 44 Between the points of his stand-up collar,..the pale flesh of his underchin remained immovable. **1978** J. A. Michener *Chesapeake* 19 The geese [had]..jet-black head and neck, snow-white under-chin.
c. *underclearance, -deep, -mire, -structure.*
1930 *Engineering* 15 Aug. 197/3 The [U.S.] War Department..imposed the limiting conditions of 100 feet underclearance above the level of mean high water. **1967** *Jane's Surface Skimmer Systems 1967–68* 64/1 Riser bars may be used, depending on load underclearance. *a* **1930** D. H. Lawrence *Last Poems* (1932) 7 As if any Mind could have imagined a lobster dozing the under-deeps. **1913** —— *Love Poems & Others* 27 And even in the watery shells that lie Alive within the oozy under-mire, A grain of this same fire I can descry. **1943** *Mind* LII. 135 These instincts or reflexes are the second point of contact where behaviour science, factually though not methodologically, rests upon its biological understructure. **1980** *Dædalus* Spring 99 Stories about dreams..often deliberately obfuscate the understructure of common sense.
d. *under-dark, -dusk, -gold, -mist, -night, -pattern.*
1916 D. H. Lawrence *Amores* 137 Bright blue crops Surge from the under-dark to their ladder-tops. **1923** —— *Kangaroo* i. 8 It..was like a whole country with towns and bays and darknesses. And all lying mysteriously within the Australian underdark, that peculiar lost, weary aloofness of Australia. **1914** —— in *Eng. Rev.* Feb. 305 And lamps like venturous glow-worms steal among The shadowy stubble of the under-dusk. **1929** —— *Pansies* 17 Twilight thick underdusk..While darkness submerges the stones. **1922** —— in *Eng. Rev.* June 509 Fishes, with their gold-red eyes, and green-pure gleam, and under-gold. **1917** —— *Look! We have come Through!* 48 Over there is Russia—Austria, Switzerland, France, in a circle! I here in the undermist on the Bavarian road. *Ibid.* 120 Where the seed sinks in To the earth of the under-night Where all is silent. **1934** T. S. Eliot *Elizabethan Essays* 190 What distinguishes poetic drama from prosaic drama is a kind of doubleness... The drama has an under-pattern, less manifest than the theatrical one.
II. 6. a. *under-boss, -matron, -waiter.*
1942 M. Harcourt *Parson in Prison* 20 The whole school was assembled before the underbosses. **1964** *Amer. Speech* XXXIX. 305 Over each [Mafia] family presides a boss... Beneath the boss are an underboss, also known as *sotto capo*, and a *consiglieri.* **1972** *N.Y. Times Mag.* 4 June 95 In the restructured family on which Joe Colombo solidified his hold as boss, another tantalizing figure emerged, Charles (Charlie Lemons) Mineo... Mineo has become a unique kind of underboss. **1976** 'W. Trevor' *Children of Dynmouth* ii. 42 The under-matron, Miss Tomm, had come into the dormitory and asked him to come with her to the study. **1880** Under-waiter [see *landlord *sb.* 4]. **1921** E. M. Forster *Let.* 17 May in *Hill of Devi* (1953) 81 He worked like an under-waiter in a Soho restaurant.
9. a. *underself, -taste.*
1890 W. James *Princ. Psychol.* I. viii. 206 Barring a certain common fund of information, like the command of language, etc., what the upper self knows the under self is ignorant of, and *vice versa.* **1914** W. De Morgan *When Ghost meets Ghost* ii. xvi. 662 This underself of hers may have vibrated in response to the strange hints he had thrown out. **1908** H. James *Spoils of Poynton* p. xxiii, An air of comedy comparatively free from sharp undertastes. **1980** R. B. Kitaj *Artist's Eye* (Nat. Gallery) 3 Their lives at the sinister heart of the Baudelairean city, the spell its compelling undertaste cast on them.
IV. 10. a. *under-calculate, -emphasize, -fulfil, -graze, -react, -recover; under-endowed, -equipped, -financed, -fulfilled, -funded, -garrisoned, -grazed, -gunned, -industrialized, -informed, -policed, -powered, -publicized, -researched, -stained, -stressed* ppl. adjs.; *under-funding, -grazing.*
1836 Dickens *Let.* 8 Oct. (1965) I. 181 There really is not time, unless Hansard's people, have greatly undercalculated the quantity sent. **1910** *Practitioner* Feb. 152 Cough..is one of the..most significant symptoms of pulmonary tuberculosis... It is too often undercalculated by the patient and his friends. **1983** *Platt's Oil Marketing Bull.* 15 Aug. 1/1 It has been undercalculating its Windfall Tax payments. **1964** *Amer. Psychologist* XIX. 14/2 If I have seemed to underemphasize the importance of inner capacities..it is because I believe that this part of the story is given by the nature of man's evolution. **1909** H. G. Wells *Ann Veronica* xvii. 344, I remarked that science was disgracefully under-endowed, and confessed I'd had to take to more profitable courses. **1969** N. W. Pirie *Food Resources* viii. 191 They are almost all under-endowed while money is squandered on projects with little bearing on the world's real needs. **1960** *Times Lit. Suppl.* 2 Sept. 553/4 Our teenagers are cast out into the world with boredom as the only memory of their ill-disciplined, under-equipped schooldays. **1923** *Daily Mail* 30 Jan. 4 This Department is under-staffed, under-financed, unprovided with many of the safeguards it has itself demanded. **1977** M. Edelman *Political Lang.* v. 100 An under-financed and uncoordinated reaction to widespread destitution becomes a 'war on poverty'. **1950** A. Lee *Soviet Air Force* 77 Unlike the later Five Year Plans, the target for the first was 'underfulfilled'. **1964** *Ann. Reg. 1963* 230 According to Premier Široký..the plan as a whole had been underfulfilled by 1·2 per cent. **1982** T. J. Binyon *Swan Song* v. 32 The professor..accused me..of consistently underfulfilling the department's norms in teaching, research and administration. **1970** *Nature* 8 Aug. 551/2 It also suggests, perhaps intentionally, that the project is grossly underfunded. **1963** *Economist* 27 Apr. 342/1 Over-funding last year could be compensated by under-funding this year. **1981** *Daily Tel.* 17 Oct. 12/3 The continual underfunding of the Royal Shakespeare Company..was endangering its ability to.. retain its talented staff. **1936** Auden & Isherwood *Ascent of F6* i. ii. 24 We're under-garrisoned and under-policed and..we're in a blue funk that the Ostnians will come over the frontier. **1977** J. L. Harper *Population Biol. of Plants* xiv. 438 Swards were overgrazed in winter and spring and undergrazed in summer and autumn. **1960** *Farmer & Stockbreeder* 8 Mar. 71/1 Those swards which needed improvement were the undergrazed type. **1933** *Jrnl. R. Agric. Soc.* XCIV. 24 This plot was subjected to overstocking in winter and early spring, followed by gross undergrazing during the summer and autumn. **1974** *Times* 7 Jan. 12/3 To tear up large areas at once has led too often to undergrazing, drainage difficulties, [etc.]. **1928** C. F. S. Gamble *Story North Sea Air Station* xiii. 214 As a fighting machine the H.12 was under-gunned for her size. **1944** *Return to Attack* (Army Board, N.Z.) 8/1 The armoured brigades..were equipped with..both types [of tank] fast-moving but under-gunned compared with the German tanks. **1964** I. L. Horowitz *New Sociol.* 33 We cannot examine demography without basing our analysis on some definite correlation of..underindustrialized and overpopulated. **1968** *Punch* 31 July 141/1 The under-informed voter. **1936** Under-policed [see *under-garrisoned* above]. **1978** *N.Y. Times* 30 Mar. A20/5 Underpoliced and unkempt, [the bus terminal]..serves as headquarters for an ominous army of hookers, muggers and pimps. **1905** Kipling *Actions & Reactions* (1909) 128 Under-powered craft..can ascend to the limit of their lift. **1980** 'M. Harris' *Treasure of Sainte Foy* i. 4 The small Renault is underpowered and rather cheaply built. **1971** H. Wilson *Labour Govt.* xxxvi. 739 One of the under-publicised achievements of comprehensive secondary education. **1965** Under-react [see *overreact *v.*] **1982** *Economist* 5 June 37/2 The markets may, on average, have underreacted to the publication of the money figures. **1967** Under-recover [see *over-recovered* s.v. *over- 27 b]. **1942** M. McCarthy *Company she Keeps* v. 239 [He] went back to the public library; perhaps..the material was under-researched. **1982** *Pol. Sci. Q.* XCVII. 474 Rawlings's intervention, so sadly underresearched. **1941** Understained [see *heterochromatin]. **1956** Understained [see *heterochromatic *a.* 2]. **1900** *Phil. Mag.* L. 132 A lower factor of safety might..be used in such cases, where there is a large reserve of understressed material. **1928** *Observer* 17 June 8 It is a curious book. Colloquial and offhand, deliberately understressed in feeling and description, [etc.]. **1969** *Harper's Bazaar* Sept. 27/1 An expensive car must be one hundred per cent reliable, and this..means an under-stressed engine of the simplest possible kind.
b. *under-capacity, -emphasis, -fulfilment, -population, -recovery, -registration.*
1962 E. Snow *Other Side of River* (1963) lxxxvi. 725 The American problem of abundance or overcapacity to produce commodities and undercapacity to consume them. **1916** E. Pound *Let.* 17 Apr. (1971) 76 In 'Impression', I don't think 'dissolved' is just the right word, though I recognize that you may have been aiming at a sort of restraint or under-emphasis which *can* be effective. **1977** M. Edelman *Political Lang.* v. 83 Hess and Torney found a repetitive emphasis in the schools on the values of loyalty, authority, and law, and an underemphasis on citizens' rights. **1962** E. Snow *Other Side of River* (1963) vii. 58, I can't recall visiting any mine or factory where 'underfulfillment' was predicted. **1922** *Daily Mail* 29 Nov. 8/4 The absurd underpopulation of the country parts. **1966** *Times* 28 Mar. (Austral. Suppl.) p. xii/6 Faced with the difficulties of isolation and under-population, managements argue that secondhand top name overseas packages are a cheaper..investment. **1961** *Ann. Reg. 1960* 511 The Sugar Board..revealed a loss..; the deficit brought the Board's total 'under-recovery' to £7·1 million. **1952** C. P. Blacker *Eugenics* 160 Lorimer gives good reasons for thinking that this discrepancy arose from under-registration of deaths, especially of infantile deaths, in the intervening period.
c. *under-luminous.*
1971 *Nature* 23 Apr. 517/1 With this value..Fig. 1 suggests that the primary is grossly underluminous for its mass.

under-, *prefix*.² Add: *under-ice, -strength.*
1966 *Times* 28 Feb. (Canada Suppl.) p. xvi/6, The submarine freighter..must have an under-ice capacity of 800 to 1,000 miles. **1976** *Jrnl. R. Soc. Arts* CXXIV. 638/2 BAS has undertaken surface sledge traverses for major anomalies and to interpret the under-ice rocks and principal structural features such as George VI Sound. **1959** *Encounter* Nov. 17/2, 175 divisions in the Red Army ..as against 14 under-strength divisions in the U.S. army. **1971** R. Petrie *Thorne in Flesh* ix. 117 An understrength police force was at full stretch.

under-achie·ver. *Psychol.* Also as one word without hyphen. [Under-¹ 10 b.] Someone whose actual performance consistently fails to reach the level predicted by intelligence tests or other measures of ability. Cf. *over-achiever.
1953 *Jrnl. Abnormal Psychol.* XLVIII. 533/2 If his grades fell a full rank below prediction he was labelled an 'underachiever'. **1962** 'I. Ross' *Old Students Never Die* viii. 105 Nowadays we have a name for them: the kids with the high potential and the low grades. We call them 'under-achievers'. **1968** D. Lawton *Social Class, Lang. & Educ.* i. 6 For a number of reasons working-class children tend to be under-achievers. **1973** E.-J. Bahr *Nice Neighbourhood* v. 47, I identify with the underachievers of this world. **1975** *Kingston* (Ontario) *Whig-Standard* 6 Sept. 27/6 The survey also found that those not using seat belts also were under-achievers in school.
So **under-achie·vement**; **under-achie·ve** *v. intr.*, **under-achie·ving** *ppl. a.* and *vbl. sb.*
1951 *School Rev.* LIX. 472 (*title*) Factors related to over-achievement and under-achievement in school. **1953** Underachieving *vbl. sb.* [see *overachieving *vbl. sb.*]. **1954** *Jrnl. Educ. Psychol.* Oct. 322 It is virtually impossible for a pupil at or near the..first percentile [on an intelligence test] to 'under-achieve'. **1965** in M. Kornrich *Underachievement* 553 A role for the counselor may be to help the underachieving student. **1972** *Guardian* 21 July 12/5 Parents who want their children to go to a popular school may..encourage them to underachieve. **1982** *Secondary Educ. Jrnl.* XII. iii. 1/2 Underachievement is not confined to pupils in secondary schools. *Ibid.*, Many of these pupils are..not achieving their full potential—in other words they are underachieving.

under-age, *a.* Add: **1*.** Carried on by someone below the legal age (for the activity).
1978 *Morecambe Guardian* 14 Mar. 19/8 He went on, about supervision on the rally site, and the danger of under-age drinking. **1983** *Sun* 8 June 15/3 He persuaded her to pose nude and sing about under-age sex.

u·nder-and-o·ver, *adj.* (and *sb.*) *phr.* Also unhyphened. [Under *adv.* 1.] = *over-and-under *a.* Also *absol.* as *sb.*
1881 W. W. Greener *Gun & its Development* 380 (*caption*) Under and over Wedge-fast Gun. **1911** *Encycl. Brit.* XXIII. 336/1 There is also Greener's 'under and over', the rifle barrel being topmost (usually 16-bore shot-gun barrel and ·450 rifle barrel). **1931** G. Burrard *Modern Shotgun* I. ix. 235 (*heading*) The Woodward 'under and over' gun. **1958** *Spectator* 1 Aug. 163/1 The 'professionals'..tended to shoot with under-and-over guns. **1969** C. Chevenix Trench *Shooter & his Gun* viii. 97 (*caption*) Double-barrelled under and over.

under-arm, *a.* Add: **1. b.** In other sporting contexts. Also as *adv.*
1929 W. E. Collinson *Spoken Eng.* 90 I'll have to serve underarm, I've strained my wrist. **1960** E. W. Swanton *West Indies Revisited* 230 An under-arm throw by Smith. **1974** Mills & Butler *Tackle Badminton* v. 39 Take a good underarm swing, turning your left shoulder towards the net. **1976** *Times* 3 Feb. 9/4 When one recalls that most women of the period [*sc.* 1909] would have added a hat to the recommended [tennis] ensemble one can see why they had to serve under-arm.
3. *Dressmaking.* Of a seam: that edges the lower half of the arm-hole of a garment, or that joins the underside of a sleeve or the side of a bodice.
1908 M. E. Morgan *How to dress Doll* (1973) v. 43 Put the seam of the sleeve a little to the front of the under-arm seam. **1941** L. I. Wilder *Little Town on Prairie* ix. 85 Laura sewed the whalebone stays onto the underarm seams. **1964** *McCall's Sewing* xi. 161/1 Pin the underarm seam of bodice and sleeve. *Ibid.* 162/1 On the pattern there will be a slash marking at the curve of the underarm.
4. Of a bag or case: carried under the arm.
[**1925** *T. Eaton & Co. Catal.* Spring & Summer 279/1 Under-the-arm bag.] **1927** *Glasgow Herald* 21 Jan. 8 Whatever she carries about with her she keeps in an attaché case, an 'under-arm bag', or some other receptacle. **1974** *Harrod's Christmas Catal.* 19 Under-arm document case £27.50.
5. Applied to various items of personal care used on the armpit, as *under-arm deodorant, razor*. Also *under-arm odour.*
1947 H. M. McLuhan in *Horizon* Oct. 132 The means of defeating under-arm odour. **1968** A. Diment *Great Spy Race* viii. 122 At twelve, showered, shaved with her small under-arm razor..we were on the coast road. **1976** J. Wainwright *Walther P. 38* 21 You use under-arm deodorant.

u·nder-bark, *a.* (and *adv.*). [Under-².] Measured or taken without including the bark of a tree trunk. Also as *adv.* Cf. *over-bark.
1911 C. L. Hanson *Forestry for Woodmen* xiv. 192 If it [*sc.* a log] is 16 inches over bark it will be taken as 15 inches under bark. **1927** *Forestry* I. 8 Sample plots..gave the age as 123 years..and the underbark quarter girth volume per acre over 8,000 cubic feet. **1953** H. L. Edlin *Forester's Handbk.* xiv. 215 Under-bark measure is strictly the volume of the log as measured after the bark has been peeled off it; but in practice it is often taken as the over-bark measure less the customary bark allowance. **1967** Scott & Palmer *Hiley's Woodland Managem.* (ed. 2) ix. 130 Prices which are quoted for timber nearly always apply to the under-bark measurement.

underbed. (Further examples.)
1725 in F. Kidder *Exped. Capt. Lovewell* (1865) 93 A feather bed and under bed and bed furniture. **1778** *New Hampsh. Hist. Soc. Coll.* (1889) IX. 108 We have cut up all the sheets, table cloths, underbeds, towels, &c. **1868** G. G. Channing *Recoll. Newport* 254 The bed or under-bed of straw was laid on cords, and the feather bed above. **1978** *Morecambe Guardian* 14 Mar. 6/5 You can buy beds already fitted with underbed drawers. **1982** *Habitat Catal. 1982/83* 122 Sturdy underbed storage on fixed nylon castors.

under-belly. Add: Also **underbelly. 1.** Now, the belly or underside of an animal. Also *transf.*, esp. of motor vehicles. Cf. Under-body 2, 3 b.
1960 *Times* 3 Mar. 8/6 Both propellers were buckled and the front underbelly was severely dented. **1963** *Listener* 10 Jan. 65/2, I creep along on my cat's underbelly Nursing the floor for smells. **1964** S. Bellow *Herzog* 327 The great car got up the hill slowly, scraping its underbelly on rocks.
2. *fig.* **a.** A vulnerable part, esp. in phr. *soft underbelly.*

1942 W. S. Churchill in *Hansard Commons* 11 Nov. 28 We make this wide encircling movement in the Mediterranean..having for its object the exposure of the under-belly of the Axis, especially Italy, to heavy attack. 1949 *Life* 31 Oct. 36/2 An all-out attack on the 'soft underbelly' of socialism. 1959 E. H. Clements *High Tension* vii. 130 The educational organisation..attacked the soft under-belly of the nation: the children. 1976 J. Crosby *Snake* (1977) x. 45 She was..sticking her knife into the disgusting underbelly of male chauvinism, and..into the soft underbelly of capitalism. 1980 *Encounter* May 86/2 The plan was to..punch their way through the soft, unsuspecting underbelly of Iran.

b. The underside or inferior part of something, which is often unnoticed or concealed.
1962 *Listener* 6 Dec. 957/2 It is a picaresque language from the under-belly of a culture, the speech of people on the move..of the wagon, railroad, and camp. *Ibid.* 27 Dec. 1106/3 The many programmes of popular music, the under-belly, so to speak, of the broadcast structure. 1976 *National Observer* (U.S.) 21 Aug. 16/3 What we're seeing is the underbelly of some of the great legends of American history. 1981 'D. Jordan' *Double Red* ix. 97 The seamy underbelly of American capitalism.

underbid, *v.* Add: **4.** *Bridge.* To bid less on (a hand) than its strength warrants. Also *intr.*
1908 R. F. Foster *Auction Bridge* 29 It is a mistake to underbid the hand. 1945 'S. J. Simon' *Why you lose at Bridge* 58 The average player overbids his big hands and underbids his small ones. 1969 A. Truscott *Gt. Bridge Scandal* xii. 159 He had decided to underbid his hand. 1974 *Times* 16 Feb. 13/3 He did not wish to deny diamonds..nor could he afford to underbid by signing off in Three No Trumps.

u·nder-bid, *sb.* *Bridge.* [f. the vb.] **a.** A bid of a number of tricks insufficient to surpass the previous bid. *rare.* **b.** A lower bid than is warranted by the strength of a hand or of a partnership's combined hands.
1923 *Daily Mail* 6 Oct. 6/4 The under-bid of 2 spades is automatically raised to 3 spades by the fact of one of the opponents calling attention to it. 1945 'S. J. Simon' *Why you lose at Bridge* 96 Most of the time I make what seems to me the best bid in the circumstances, whether it is an overbid, an underbid, or even an anti-system bid. 1977 *Times* 10 Dec. 13/6 The response which is outstanding is a gross underbid of One No Trump.

underbidder. Add: **2.** *Bridge.* A player who under-bids.
1923 [see *overbid *v.* 2 c]. 1945 'S. J. Simon' *Why you lose at Bridge* 58 On the cancelled hands, both overbidder and underbidder remain oblivious of their enormities.

u·nderbit. *U.S.* [Under-[1] 5 b.] An earmark to indicate ownership, made on the lower part of the ear of cattle. Cf. Under-bitted *ppl. a.*, Underkeel.
1837 *Knickerbocker* X. 408 The young bridegroom boasted that he had taken an 'under bit out of his left ear'. 1869 *Overland Monthly* III. 126 A red mulley cow, with a crop and an underbit in the right [ear]. 1915 *Dialect Notes* IV. 185 *Under-bit*, a triangular cut from the lower side.

u·nderbite, *sb.* [Under-[1], after *overbite.] The projection of the lower jaw or the lower incisors beyond the upper.
Not a term used in Dentistry.
1976 M. Machlin *Pipeline* xlvi. 484 Coutts was a big lantern jawed man with a pronounced underbite that gave his chin an appearance something like an Arctic ice-breaker. 1982 *Guardian* 26 Oct. 8/6 You stick your jaw way out until it's almost as if you have an underbite.

u·nder-blanket. Also as one word. [Under-[1] 5 b and Under *a.* 2 a.] A blanket laid under the bottom sheet, as opp. to one used as a covering; now often an electric blanket.
[1746], 1819 [see Under *a.* 2 a]. 1930 *Daily Express* 6 Oct. 9/7 (Advt.), Under blankets. 1971 *Guardian* 12 July 3/1 (Advt.), Electric underblanket. Single size approx. 50″ × 26″. 1979 *Daily Tel.* 24 Jan. 14 A simple pin magnet together with a bag of corks placed between the under sheet and the underblanket has made me immune [from cramp].

under-body. Now usu. unhyphened. **1. b.** For *U.S.* read *U.S. dial.* and add example. Also, any undergarment for the body, esp. an under-bodice or under-shirt.
1873 in *Mag. Albemarle County* (Va.) *Hist.* (1963) XX. 57 She gave me some mighty pretty nansook to make an underbody. 1936 E. Glasgow *Vein of Iron* I. i. 12 She hoped the minister couldn't see the top of her red flannel underbody, which would poke up at the neck, though it was sewed to her petticoat.

3. b. (Further examples.) Freq. *attrib.*, with reference to rust protection.
1956 *Autocar* 19 Oct. 596/1 Under-body sealing and sound-deadening compound. 1976 *Sci. Amer.* Feb. 77 (*caption*) Spot-welding robots..are used in assembling the underbodies of Chevrolet Novas, Pontiac Venturas and Buick Skylarks at the General Motors plant. 1976 *Time* 20 Dec. 56/2 (Advt.), The entire underbody is sealed against the elements. 1977 *R.A.F. News* 8–21 June 5/3 (Advt.), Like fitted front seat belts, full underbody seal and servo-assisted front disc brakes.

u·nder-bo:nnet, *a.* [Under-[2].] Pertaining to, situated, or occurring beneath the bonnet of a motor vehicle (i.e. in or of the engine).
1962 *Times* 24 Apr. 16/2 Oil supply, from an under-bonnet tank. 1973 *Times* 3 May 35/2 A few pounds worth of sound damping material would probably work wonders in suppressing underbonnet and underfloor noise. 1977 *Custom Car* Nov. 13/3 Presumably, Ford believe that customers in that price bracket leave under-bonnet checks to a garage.

u·nderbrim. [Under-[1] 5 e.] The underside of the brim of a hat; a trimming or lining attached to this.
1908 *Sears, Roebuck Catal.* 1036/1 This child's hat is..made on wire frame with the entire upper and under brim very closely covered with shirred pink miliners' mull. 1922 Joyce *Ulysses* 344 A hat of wide-leaved nigger straw contrast trimmed with an underbrim of eggblue chenille. 1923 *Daily Mail* 5 Mar. 15 A pretty girl's hat is a mushroom shape of rough biscuit straw with a flower-covered underbrim. 1935 Brand & Mussared *Millinery* viii. 64 Tack the underbrim to the interlining at the headline, work from the centre-front in either direction.

underbrush, *sb.* Add: **a.** (Earlier U.S. example.)
1775 J. Jeffrey *Jrnl.* 3 Apr. in *Essex Inst. Hist. Coll.* (1914) L. 107 The fire ran among the leaves & dry underbrush for upwards of a mile.
b. *fig.*
1888 *Kansas State Hist. Soc. Trans.* (1890) IV. 274 The underbrush of forgetfulness has so grown that but few in Kansas know that Joel K. Goodin ever lived. 1927 *Bulletin* (Glasgow) 6 Apr. 16/1 Taking up a tarnished knife and fork, he pushed aside the underbrush of onions and came face to face with his steak. 1938 [see *pretrial *sb.*]. 1968 *Med. World News* 20 Sept. 86/2 It drags the reader through dense verbal underbrush.
underbrush *v.* (earlier and later examples); underbrushing *vbl. sb.*
1824 *Canad. Mag.* III. 246 Every year we..have a quantity of wood land under-brushed. 1838 *Six Years in Bush* 80 Underbrushing consists in cutting and removing all the young trees and brushwood. 1863 Under-brushing [see *line *sb.*[2] 26 e]. 1933 D. G. Cameron *Twigs from Oak* 124 That process..called 'underbrushing', was continued over a space about twenty yards wide. 1964 E. C. Guillet *Pioneer Days Upper Canada* (ed. 2) 122 After underbrushing the piece of land the workers proceeded with logging.

underbush, *sb.* (Earlier examples.)
1849 S. C. Brees *Panorama N.Z.* 27 Crawling along the ground, among the underbush and trees. 1867 tr. *Guéranger's Life St. Austin* x. 97 Others grafting fruit trees, or thinning the underbush.

underca·pitalize, *v.* [Under-[1] 10 a.] *trans.* To furnish with insufficient capital (to achieve a desired result). Chiefly as **underca·pitalized** *ppl. a.*
1934 Webster, Undercapitalize. 1962 E. Snow *Other Side of River* (1963) lxvii. 509 The vicious paradox of China's undercapitalized economy. 1976 *Broadcast* Dec. 4/2 The External Services of both the BBC and these other broadcasting organisations are under-capitalised. 1982 J. Fox *White Mischief* i. 15 The early settlers..most of them were chronically undercapitalised.

under-carriage. Add: Now usu. unhyphened. **b.** The landing-gear of an aircraft.
1911 Harper & Ferguson *Aerial Locomotion* 92 The under-carriage was formed of wheels alone. 1922 *Aerial Age* Aug. 420/2 A shock-absorbing type of under-carriage. 1935 C. G. Burge *Compl. Bk. Aviation* 160/2 In addition to the undercarriage, alighting gear includes subsidiary items such as tail skids, wing-tip skids, and floats. 1946 *Happy Landings* July 7/1 A characteristic fault of this aircraft made it necessary to operate the undercarriage lever two or three times. 1969 G. Macbeth *War Quartets* 71 One clutched the under-carriage Of a packed hospital-plane. 1977 D. Beaty *Excellency* vi. 80 The three green undercarriage lights, the two white engine lights, the red inverter lights.

u·ndercart. *slang.* [Under-[1] 5 b; cf. Under-carriage.] = *under-carriage b.
1934 *Popular Flying* Sept. 295 (*caption*) The British Klemm 'Eagle'..going great guns, with undercart out of the way. 1939 J. L. Rhys *World owes me Living* xvi. 244 We've got de-icers, variable pitch airscrews, retractable undercart. 1948 'N. Shute' *No Highway* v. 127 Honey had ruined a Reindeer at Gander by pulling up its undercart. 1958 'Castle' & Hailey *Flight into Danger* vii. 103 Look, Janet, I think you'd better work the undercart lever and call off the air-speed as the wheels come down. *a* 1963 J. Lusby in 'B. James' *Austral. Short Stories* (1963) 225 Do remember that release-tit on the undercart-lever.

undercast, *v.* Restrict † *Obs.* to senses in Dict. and add: **4.** [Under-[1] 10 a.] *Theatr.* To allot (a part) to an inadequate player; to cast (a player) in an insufficient role.
1827 J. Boaden *Mem. Mrs. Siddons* II. xiii. 66 The Anna, by Miss Wheeler, was rather under-cast. 1920 G. B. Shaw *Let.* in *Bernard Shaw & Mrs. Patrick Campbell* (1952) 206 Overcasting a part means undercasting an artist. 1957 *Observer* 12 May 15/2 [The part] is admittedly under-written; that is no reason why it should have been undercast. 1970 *New Yorker* 22 Aug. 59 Rossellini deliberately undercasts him, as he does everyone else. Colbert is made to look uninteresting and rather bourgeois. 1977 *Times* 9 July 10 The play was woefully undercast.

undercha·racterize, *v.* [Under-[1] 10 a.] *trans.* To depict or play with insufficient characterization or subtlety. So **undercha·racterized** *ppl. a.*; **u:ndercharacteriza·tion.**
1960 *Times* 13 Jan. 6/6 Minor characters, either under-characterized..or over-characterized, but in both cases quite lifeless. 1963 *Times* 14 Feb. 8/1 The outer movements..sounded mechanical and under-characterized. 1968 *Time* 26 Apr. 50 *Couples* is flawed by overwriting and undercharacterization. 1970 *Times* 16 Dec. 11 In the wry second movement..the players needed more detailed help with phrasing, and the work as a whole was under-characterized. 1977 *Gramophone* Feb. 1270/1 Impressive as his playing is, in the last resort it is a degree under-characterized.

undercharge, *sb.* Add: Also, an adjustment made to correct an account (see quot. 1861).
1861 E. B. Ivatts *Handbk. Railway Station Managem.* 86 The undercharges or amounts that are afterwards added when the proper charge has not been made,—(when a charge has been once made on an invoice, the figures in English accounts are never altered, errors being corrected by undercharge or overcharge). 1920 *Rep. Departmental Comm. Industrial Assurance Companies* 3 If there is under-payment,..it is not due to any undercharge on the assured.

u·nder-chosen, *a.* *Sociology.* [Under-[1] 10 a.] Denoting those who, in a sociometric group study, are chosen by others as companion, fellow-worker, etc., significantly fewer times than average. Cf. *over-chosen *a.*
1943, 1956 [see *over-chosen *a.*]. 1968 Lindzey & Byrne in Lindzey & Aronson *Handbk. Social Psychol.* (ed. 2) II. xiv. 467 Persons who receive significantly more choices than would be expected by chance may be..pooled for comparison with a similarly selected group of under-chosen persons.

u·nderclass. [Under-[1] 5 b.] A subordinate social class; *spec.* [ad. Sw. *underklass*] the lowest social stratum in a country or community, consisting of the poor and the unemployed.
1918 J. Maclean in 'H. MacDiarmid' *Company I've Kept* (1966) iv. 124 The whole history of Society has proved that Society moves forward as a consequence of an under-class overcoming the resistance of a class on top of them. 1963 G. Myrdal *Challenge to Affluence* iii. 40 Less often observed..is the tendency of the changes under way to trap an 'underclass' of unemployed and, gradually, unemployable persons and families at the bottom of a society. 1964 *Observer* 12 Jan. 10/7 The Negro's protest today is but the first rumbling of the 'underclass'. 1966 *New Statesman* 19 Aug. 247/2 The national economic growth has been bought at the expense of industrial workers and the poor (largely Negro) under-class. 1977 D. M. Smith *Human Geogr.* x. 299 While in South Africa the black proletariat is large enough radically to change society on its own, the American under-class is in a minority and has much less potential power. 1981 D. G. Glasgow (*title*) The Black underclass: poverty, unemployment and entrapment in ghetto youth. 1981 *New Statesman* 17 July 14 Many people... have warned of trouble if we went on creating a permanent underclass, and typing it by colour. 1982 *N.Y. Rev. Bks.* 12 Aug. 15/2 The distinctive feature of any underclass is that its members do not conform to the conduct expected of the poor. 1985 *Times* 11 Jan. 5/1 Modern Britain is fostering an underclass of unemployed and unskilled workers.

underclothes. Add: = Underclothing. (Earlier examples.)
1835 R. M. Bird *Hawks* I. 19 Under-clothes of some white summer-stuff. 1855 D. G. Mitchell *Fudge Doings* I. 47 She supplies her cook with cast-off under-clothes.

underclu·b, *v.* *Golf.* [Under-[1] 10 a.] *refl.* To select (for oneself) a club which will not satisfactorily strike the ball the required distance. Also *intr.* and **underclu·bbed** *ppl. a.*, **underclu·bbing** *vbl. sb.*
1900 H. Hilton in *Outing* Sept. 654 A besetting sin of nearly all golfers is to what may be termed 'underclub themselves'. 1923 *Daily Mail* 8 May 12 When in doubt underclub yourself and hit hard. 1931 T. H. Cotton *Golf* vii. 56 There are more shots lost through underclubbing than anything else. 1936 P. A. Vaile *Short Game* 33 At least ninety-five per cent. of those that swing putters are grievously under-clubbed. 1955 *Times* 2 May 4/1 Burgess underclubbed at the 11th. 1961 *Times* 28 Mar. 4/2 Underclubbing at the 16th caused their opponents to take three putts. *Ibid.* 8 Sept. 4/1 Wolstenholme underclubbed himself. 1975 D. Langdon *How to talk Golf* 78 *Under-clubbing*, a drive which falls short of reaching the green at a short hole. 1979 *Tucson Mag.* Jan. 38/3 Everybody underclubs.

undercoat. Add: **4. a.** *Painting* and *Decorating.* A layer or layers of paint applied before the finishing or top coat; the paint used for this. Also *transf.* in electro-plating, cementing, etc.
1873 [see Priming *vbl. sb.*[1] 4]. 1901 *Decorators' & Painters' Mag.* I. i. 45 It is a rather risky experiment to use red lead as an undercoat for vermilion. 1927 W. Deeping *Kitty* xxvi. 329 'The next job will be to paint

'em.' 'Couldn't I do that?' 'Don't see why you shouldn't, sir. Two good coats over an undercoat.' 1930 Field & Weill *Electro-Plating* 90 The copper is soft and easily buffed, and this is required of the copper which forms the undercoat for nickel. 1958 [see *primer *sb.*[2] 3]. 1960 *Farmer & Stockbreeder* 23 Feb. 76/2 Marley Mix is available..in the following mixes; fine concrete, bricklaying mortar, cement mortar, internal rendering, undercoat mix 1:2:4 concrete. 1976 *Handyman Which?* May 47/1 You'll get a better result by using two coats of primer/sealer beneath the gloss instead of one coat of the primer/sealer and one of undercoat.
b. = *underseal *sb.*
1960 *Motor* 13 July 21 The Flintkote under-coat service is available in many countries. 1968 *Globe & Mail* (Toronto) 15 Jan. 19/7 (Advt.), Automatic transmission, power disc brakes, undercoat, [etc.].
Hence as *v. trans.*, (*a*) to apply an undercoat of paint to; (*b*) to coat the underside of (something, esp. a motor vehicle) with paint, sealant, or the like; **u·ndercoated** *ppl. a.*, **u·ndercoating** *vbl. sb.*
1922 *Decorator* Feb. p. xii (Advt.), Dixon's white paint inside glossy and flat undercoating. *Ibid.* Apr. p. xx (Advt.), The new semi-liquid flat white oil paint for undercoating paint or enamel. 1947 *Automobile Topics* Dec. 20/2 The future of the automobile undercoating business largely rests upon..a proper job of spraying. 1953 *Ibid.* Mar. 26/1 More than half of the nation's postwar automobiles have been undercoated. 1960 *Farmer & Stockbreeder* 23 Feb. 117/1 It can be brushed or sprayed—one coat being sufficient on suitably..primed surface—no undercoating being required. 1978 *Detroit Free Press* 5 Mar. c 22/2 (Advt.), BMW 2002, red,..undercoated, 57,000 mi. 1979 P. Wallage *Restoration Post-War Cars* ii. 25/2, I once had to tackle one which had been undercoated and painted so well it was almost a shame to scrape it off.

under-consu·mption. *Econ.* [Under-[1] 10 b.] Insufficient consumption; a demand for goods and services exceeded by supply or insufficient to call forth the full potential supply. Cf. Over-production.
1895 [see Under-[1] 10 b]. 1906 J. A. Hobson *Evolution Mod. Capitalism* xi. 288 When the disease is at its worst, the activity of producer and consumer at its lowest, we have the functional condition of under-production due to the pressure of a quantity of over-supply, and we have a corresponding state of under-consumption. 1931 *Times Lit. Suppl.* 14 May 374/3 They are thus on the way, not only to disentangling the mesh of popular theory connected with the words 'over-production' and 'under-consumption' but to destroying the presupposition..that ..over-production..may..occur. 1961 *Economist* 22 Apr. 348/1 'Under-consumption' at the bottom of the social scale was apparent to a slight extent in dental and ophthalmic care. 1974 B. Pearce tr. *Amin's Accumulation on World Scale* I. i. 112 After a certain level of development has been reached, possibilities of saving become greater than investment needs (governed by the volume of consumption). We have here a general theory of underconsumption.
Hence **underconsu·mptionist,** an advocate of a theory of underconsumption; also *attrib.* or as *adj.*
1936 A. L. Rowse *Mr Keynes & Labour Movement* 41 In sympathy..with the attitude of the under-consumptionists all along. 1948 G. Crowther *Outl. Money* (ed. 2) v. 149 'Under-consumptionist' theories. 1974 M. B. Brown *Economics of Imperialism* ix. 216 Some neo-Marxists..have accepted an underconsumptionist interpretation of Marx and therefore see the increase of surplus value and the search for its disposal as the main source of the continuing expansionism.

undercoo·l, *v.* **1.** [Under-[1] 8 c.] **a.** *trans.* = *supercool *v.* a.
1895 C. S. Palmer tr. *Nernst's Theoret. Chem.* I. iii. 65 Probably the two curves would intersect asymptotically at abs. zero, if one could under-cool a liquid so far. 1962 *Zeitschr. f. Metallkunde* LIII. 600/2 Many molten metals slagged in Pyrex glass can be highly undercooled irrespective of the amount. 1977 *Scripta Metall.* XI. 253 Ingots..were under-cooled and solidified under increasing gravity-levels.
b. *intr.* = *supercool *v.* b.
1969 *Cast Metals Research Jrnl.* V. 137/1 Cooling curves for ductile iron showed no distinct eutectic arrest but undercooled during eutectic solidification. 1977 *Scripta Metall.* XI. 253 The prepared boules were heated to 230°C, and allowed to undercool.
2. [Under-[1] 10 a.] *trans.* To cool insufficiently.
1983 *New Scientist* 13 Jan. 83/2 The continued loss of core-cooling water..meant that the reactor's core was seriously undercooled.
Hence **undercoo·led** *ppl. a.*; **undercoo·ling** *vbl. sb.*
1895 C. S. Palmer tr. *Nernst's Theoret. Chem.* I. iii. 65 The dotted line is the pressure curve of the under-cooled liquid substance, and forms the continuation of the pressure curve of the liquid. 1913 *Engineering* 10 Oct. 510/3 The under-cooled liquid phase can in many cases be retained in the metastable condition at temperatures far below the normal freezing-point by a process of undercooling. 1936 *Forestry* X. 128 The point at which ice formation begins may be still further lowered by undercooling. If water is kept absolutely still, it is possible to lower its temperature below the freezing point without ice formation... This undercooling often takes place in plants. 1980 *Sci. Amer.* Apr. 84/3 Above the glass temperature, in the undercooled liquid phase, the atoms are free to make extensive translational movements.

u·ndercount. *Statistics.* [Under-[1] 10 b.] An incomplete enumeration; *spec.* the amount by which the number enumerated in a census falls short of the actual number in the group.
1955 *Population Studies* (United Nations) No. 23. 11/2 Is there a possibility of an under-count in the 1950 census in the Vega Alta? 1964 *New Statesman* 13 Mar. 390/3 Allowing for a 5 per cent undercount in 1952 and for a 2 per cent increase per annum,..the Northern total was just about what had been expected. 1981 *Sci. Amer.* Apr. 70/1 Preliminary study of the raw census data now suggests that the 1980 census was remarkably complete: the undercount is turning out to be much smaller than had been projected.
Hence as *v. trans.*, to enumerate incompletely; **u·ndercounting** *vbl. sb.*
1955 *Population Studies* (United Nations) No. 23. 11/1 There was serious under-counting in the Southern States at the time of the 1870 census. 1970 *Daily Tel.* 22 Aug. 3/6 Mayor Lindsay is convinced that New York will be seriously undercounted in the 1970 census figures. 1978 *Guardian Weekly* 30 Apr. 17/1 In the United States, the Mexican-American population, which may have been severely undercounted in the 1970 census, is growing steadily.

under-cover, *a.* (and *sb.*). Also **undercover.** [Under-[2].] **A.** *adj.* **1.** That is situated or occurs under cover; sheltered.
1854 [see Under-[2]]. 1959 *News Chron.* 14 July 4/4 A good fun-fair and some under-cover amusements. 1971 *Engineering* Apr. 61/1 The first mentioned can be very useful when under-cover storage is not available.
2. a. Of a person: operating in secret within a community or organization, esp. as a detective or spy. Freq. as *under-cover agent, man.* Also applied to organizations or agencies.
1920 U. B. Sinclair *100%* 266 [Mrs. Godd] had written in a large, bland, girlish hand her opinion of 'under cover' men and those who hired them. 1932 W. Faulkner *Light in August* ii. 41 Whiskey can be bought from Brown almost on sight, and the town is..waiting for him to..produce from his raincoat and offer to sell it to an undercover man. 1933 *B'nai B'rith Mag.* XLVIII. 108/1 German steamship lines were 'in the front ranks of Nazi-ism—active under-cover nests of conspiracy'. 1951 E. Paul *Springtime in Paris* iii. 67 An undercover worker for the Party. 1958 *Spectator* 20 June 809/3 The Providence Island Company..which served as undercover organisation for the opposition to Charles I. 1970 *Toronto Daily Star* 24 Sept. 12/8 The undercover agent is the surest way of handling narcotics problems in factories.
b. Surreptitious, covert. Of an activity: conducted or existing in secret.
1933 *Light* Nov. 3/1 That undercover atheism which disguises itself under other names... The undercover atheism which so often sneaks by in the clothing of religion itself. 1935 R. McKay *'Intelligence' Game* 169 Perhaps they were made careless because they had an undercover 'tip' about the coming of the Austrian. 1953 M. McCarthy *Groves of Academe* iii. 50, I have been useful to them from time to time in various little under-cover jobs. 1960 *Times* 27 Aug. 7/3 The menace of recruitment to under-cover prostitution on a large scale cannot be ignored. 1979 N. Mailer *Executioner's Song* (1980) I. xvii. 285 Years ago, when a patrolman, Nielsen did some under-cover work in narcotics.
B. *sb.* An undercover agent. *slang.*
1962 'K. Orvis' *Damned & Destroyed* vii. 50 The Horsemen have sneaked plants, undercovers, right inside here. 1972 J. Mills *Rep. to Commissioner* p. x, She was a very good detective. She was a narcotics undercover.

undercure (stress variable), *v.* [Under-[1] 10.] *trans.* To cure (plastic or rubber) for less than the optimal period.
1916 [see *overcure *v.*]. 1965 *Encycl. Polymer Sci. & Technol.* II. 53 To increase the bending capacity still more, the laminated sheets are usually slightly undercured.
So **u·ndercured** *ppl. a.*, **undercu·ring** *vbl. sb.* Also **u·ndercure** *sb.*, the process or result of undercuring.
1912 *Jrnl. Soc. Chem. Industry* 16 Dec. 1100/1 Nos. (1) and (2) [sc. specimens of rubber] were distinctly undercured at 35 lb. 1915 *Ibid.* 15 Oct. 990/2 In the case of undercures the curve often breaks off too soon for this to be apparent. 1916 *Ibid.* 31 Aug. 872/2 If the impulse is insufficient..we have the condition of undercuring and the rod falls back, corresponding with a gradual deterioration of the vulcanised product. 1940 J. Osborne *Dental Mech.* xi. 128 Under-cured vulcanite is soft and weak, and permeable to mouth fluids and bacteria. 1952 [see *over-cured ppl. adj.* s.v. *overcure *v.*]. 1965 *Encycl. Polymer Sci. & Technol.* II. 53 This undercure slightly reduces the quality of the surface, but other advantages are gained so that today a substantial fraction of the total production of decorated, laminated sheet stock is plasticised. 1968 *SPE Jrnl.* Nov. 47/1 Undercured resins are inferior..but long cure times are uneconomical. 1969 *Jrnl. Paint Technol.* XLI. 702 (*heading*) Effect of undercuring on durability of siliconized polyester enamels with hexakis-(methoxymethyl) melamine.

undercurrent. 2. (Earlier example.)
1792 [see *reaction 4 a].

undercut, *sb.* Add: **2. b.** *Dentistry.* A horizontal cut at the base of a tooth cavity; *esp.* one made to anchor a filling more securely.
1892 R. Ottolengui *Methods of filling Teeth* i. 6 Many fillings have failed through the well-meant but unwise efforts of the operator to give great retentive strength to his cavity by deep undercuts. 1923 J. B. Parfitt *Operative Dental Surgery* (ed. 2) x. 87 The condition for retention, namely, some kind of 'undercut', has already been fulfilled. 1980 J. R. Grundy *Conservative Dentistry* vii. 41/1 Minor undercuts should be removed during cavity preparation.
c. *Mining.* A long, thin cut made under a vein of ore or a face of coal. Cf. *overcut *sb.* c.
1892 *Trans. Federated Inst. Mining Engineers* I. 130 (*table*) Depth of undercut. 1902 A. S. E. Ackermann *Coal-Cutting by Machinery in Amer.* ii. 31 They hole about 2in. in the clay, and partly because of this and partly because of two dirt bands at the bottom, they got practically the whole of the coal taken out of the undercut. 1939 B. L. Coombes *These Poor Hands* vii. 109 With such a large undercut there was the likelihood of it [sc. the roof] falling any second. 1959 G. D. Mitchell *Sociol.* viii. 136 Gummers..clean out the undercut so that when the shot is fired the coal will have space in which to fall.
d. *gen.* A space formed by the removal or absence of material from the lower part of something.
1914 [see *interference 4 b (i)]. 1964 F. O'Rourke *Mule for Marquesa* iii. 58 Fardan found an undercut in the south wall two miles from the mouth. 1971 *Country Life* 15 July 141/3 Some sickness has made them lie out of sight in one of those undercuts that sheep rub for themselves in an eroding bank of light soil. 1977 *Design Engin.* July 54/2 In load bearing applications, undercuts, knurls, lugs, and dovetails are used to provide the component with surfaces onto which the molten metal can lock as it shrinks.
3. A projection on a pattern corresponding to an undercut portion of the mould.
1909 in *Cent. Dict. Suppl.* 1935 *Die-Casting* (Machinery's Yellow Back Series No. 4) i. 10 To provide the undercut a collapsible core is necessary.

undercut, *v.* Add: **2. c.** Also in *Lawn Tennis*, to impart backspin to (the ball) by slicing down on it below the centre (in quot. *absol.*, to play a stroke which would have this effect).
1926 E. Bowen *Ann Lee's* 86 Mr. Barlow..walked springily about..hacking, slashing, and under-cutting with his racquet at the air.
d. *Mining.* To cut away the under-part of (a vein of ore or a face of coal); to obtain (coal, etc.) in this way.
1883 [in Dict., sense 2 a]. 1892 *Trans. Federated Inst. Mining Engineers* I. 130 The function of all these machines is to undercut the coal in the same way as has hitherto been done by hand labour. 1939 B. L. Coombes *These Poor Hands* vii. 108 It [sc. the coal-cutter] undercut the coal to the depth of the jib. 1945 D. H. Rowlands *Coal* xiii. 172 The very first coal-cutter was patented in the eighteenth century, and since then hundreds of inventors have worked on the problem of undercutting the coal-face. 1982 *Sci. Amer.* Sept. 66/1 By the end of World War II 90 percent of the coal mined in the U.S. was undercut by machine.
3. b. *fig.* To render unstable; to render less firm, to undermine.
1955 W. J. Bate *Achievement Sam. Johnson* ii. 81 In the very activity or process of wishing, there are inherent liabilities that are able to undercut the wish itself. 1976 *National Observer* (U.S.) 13 Nov. 1/3 Many vowed that their children would not grow up with the same sort of expectations and handicaps that had so undercut their own self-reliance. 1977 L. Gordon *Eliot's Early Years* iii. 63 The wry, derisive note..undercuts the posturing of Saint Narcissus. 1981 R. Hayman *K.* xi. 146 He was aware of undercutting all his gestures towards healthy living by starving himself of sleep.

undercut, *ppl. a.* Add: **2.** In *Lawn Tennis*, applied to a stroke which undercuts or imparts backspin to the ball.
1920 [see *chop *sb.*[1] 4 e]. 1977 *New Yorker* 10 Oct. 152/2 He shifted from his usual top-spin backhand to a sliced undercut backhand—a stroke that many of us had seen him use only rarely.
3. *Mountaineering.* Of a handhold: cut from below, and used esp. to maintain balance when climbing.
1950 tr. *Mountaineering Handbk.* (Assoc. Brit. Members Swiss Alpine Club) vi. 46 Hand and footholds make progress possible... They can be horizontal, oblique, vertical or undercut. 1965 A. Blackshaw *Mountaineering* vi. 161 Side-holds and undercut holds are valuable for maintaining balance or for moving 'in opposition'. 1975 W. Unsworth *Encycl. Mountaineering* 120/1 An undercut hold is one that is upside down, but it can be useful.

underda·mped, *a.* [Under-[1] 10 a.] Of a physical system: incompletely damped; damped to the extent of allowing only a few oscillations after a single disturbance. So **underda·mping** *vbl. sb.* Cf. *overdamped *a.*
1934 W. V. Houston *Princ. Math. Physics* iv. 45 Write the expression for the work done by an underdamped oscillator against the damping force. 1951 [see *hunt *v.* 7 b]. 1952 D. E. Christie *Intermediate College Mech.* xii. 297 We shall limit our discussion to what is called underdamping. 1976 *Water Resources Res.* XII. 71/1 In the underdamped case the water level oscillates about the equilibrium level.

u·nderdamper. [Under-[1] 5 b.] Any of the set of dampers in an upright piano of a type

in which the dampers are placed below the hammers. Freq. *attrib.*
1870 E. Brinsmead *Hist. Pianoforte* 74 Square piano action, with under dampers, and a screw in each key. 1909 *Northampton Independent* 30 Oct. 29/2 (Advt.), A magnificent overstrung underdamper, upright iron grand. 1933 R. E. M. Harding *Piano-Forte* iv. 56 The under damper is also found in connection with the Single Action. 1980 *Early Music* Apr. 241 (Advt.), After John Broadwood, 1801, a simple under-damper square fortepiano.

underdete·rmine, *v.* [Under-[1] 10 a.] *trans.* To account for (a theory or phenomenon) with less than the amount of evidence needed for proof or certainty. Freq. as *pa. pple.* Cf. *overdetermine *v.*
1966 J. J. Katz *Philos. of Lang.* ii. 11 As in other sciences, the evidence underdetermines a generalization. 1977 A. Giddens *Stud. in Social & Polit. Theory* 11 Even in the most developed areas of the natural sciences, theories are underdetermined by facts. 1979 A. R. Peacocke *Creation & World of Science* ii. 73 It is salutary to remember that most cosmological theories are underdetermined by the facts. 1984 *Times Lit. Suppl.* 14 Dec. 1453/3 Early exploratory models will be drastically underdetermined by the data.
Also **u:nderdetermina·tion,** the state or quality of being underdetermined; **underdete·rmined** *ppl. a.*
1966 J. J. Katz *Philos. of Lang.* iv. 114 Such underdetermination means that the slack must be taken up by considerations that have to do with the simplicity and generality of the hypotheses. 1977 A. Giddens *Stud. in Social & Polit. Theory* 11 The underdetermination of theories by facts in social science is certainly greater than in natural science. 1978 Hookway & Pettit *Action & Interpretation* p. xi, The argument is that with underdetermined theories as to someone's beliefs and desires, there is no language-independent realm of meanings in regard to which the theories differ.

under-deve·loped, *a.* [Under-[1] 10 a.] **a.** Incompletely developed, in various senses of the vb.
1892 [see Under-[1] 10 a]. 1910 H. G. Wells *Hist. Mr. Polly* vii. 165 Mr. Polly felt himself the faintest under-developed simulacrum of man that had ever hovered on the verge of non-existence. 1944 *Ann. Reg. 1943* 328 A creed of this kind [sc. Fascist] will appeal to certain types of under-developed Europeans. 1975 *Gen. Systems* XX. 107/1 Anthropology is disorganized and crisis-ridden, yet vastly underdeveloped.
b. *spec.* designating a country or other region in which economic and social conditions fail to reach their potential level or an accepted standard. Cf. *developing *ppl. a.* and *Third World.
1949 H. S. Truman in *Congressional Record* 20 Jan. 478/1 Fourth, we must embark on a bold new program for making the benefits of our scientific advances and industrial progress available for the improvement and growth of underdeveloped areas. 1950 *U.S. Dept. State Bull.* 25 Sept. 497 (*title*) Methods of financing economic development of underdeveloped countries. 1956 T. Balogh in A. Pryce-Jones *New Outl. Mod. Knowl.* 519 The typical under-developed country even in the more recent past has only attracted capital for the exploitation of primary products. 1964 Auden in *Listener* 1 Oct. 525/2 It's heartless to forget about the under-developed countries. 1977 P. Johnson *Enemies of Society* xix. 253 Among the so-called underdeveloped countries, the group most likely to carry through industrialization are those which have profited from the 'energy crisis' which has given them truly gigantic quantities of investment capital. 1984 *Guardian* 5 Nov. 12/2 Few people begrudge the aid we send to the starving or merely under-developed regions of the earth.
Also **u:nderdeve·lopment,** incomplete development.
1891, 1892 [see Under-[1] 10 b]. 1927 H. G. Wells in *Sunday Express* 20 Feb. 12/6 The development of Soviet flying is retarded by comparative poverty and the economic underdevelopment of the huge regions concerned. 1958 *Listener* 21 Aug. 274/1 It has for too long been..assumed that inserting a technical skill into a different society will automatically overcome that aspect of underdevelopment to which it is applied. 1969 A. G. Franks *Latin Amer.* (1970) i. 3 Most studies of development and underdevelopment fail to take account of the economic and other relations between the metropolis and its economic colonies.

u:nderdiagno·sis. *Med.* [Under-[1] 10 b.] Failure, on a significant scale, to detect or correctly diagnose all the cases of a disease examined.
1966 *Gastroenterology* LI. 1074 (*heading*) Under-diagnosis of biliary tract disorders? 1971 *Brit. Med. Bull.* XXVII. 14/1 Under-diagnosis has also contributed to the apparently low incidence. 1982 *Brit. Med. Jrnl.* 29 May 1585/1 This low figure for Britain is unlikely to represent underdiagnosis.
So **underdi·agnose** *v. trans.*, **underdi·agnosed** *ppl. a.*
1974 *Med. Jrnl. Austral.* 12 Jan. 49/1 (*heading*) Cœliac disease: an underdiagnosed disorder with important implications. 1975 *Gut* XVI. 81 The condition may have been underdiagnosed in Britain. 1979 *Brit. Med. Jrnl.* 15 Dec. 1582/3 Surely this does not show general practitioners to be disgracefully underdiagnosing or undertreating hypertension, especially with regard to detection, as a great number of these people had not seen their doctor for some considerable time. 1983 *Ibid.* 17 Sept. 827/1 He considers the disorder to be seriously underdiagnosed.

u:nder-differentia·tion. *Linguistics.* [Under-[1] 10 b.] The incomplete differentiation of phonemic elements in a language, esp. in loanwords taken from a language in which certain phonemic distinctions do not correspond to those in the receiving language. Cf. *differentiation 1 b.
1953 U. Weinreich *Languages in Contact* ii. 18 Under-differentiation of phonemes occurs when two sounds of the secondary system whose counterparts are not distinguished in the primary system are confused. 1962 *Canadian Jrnl. Linguistics* VII. 82 These three types of substitution have been variously called..under-differentiation, over-differentiation, and reinterpretation. 1976 *Amer. Speech* 1973 XLVIII. 220 Underdifferentiation is the failure to distinguish two or more sounds of the donor language because of the existence of a single counterpart in the recipient language. 1983 *English World-Wide* IV. 53 The orthography gives near maximum representation of significant sounds but allows for underdifferentiation in spelling.

u:nder-dispe·rsion. [Under-[1] 10 b.] A reduced degree or amount of dispersion; *spec.* in *Ecol.*, a greater evenness in the distribution of individuals than would be expected on purely statistical grounds. Cf. *over-dispersion. Also **u:nder-dispe·rsed** *ppl. a.*
1935 *Ann. Bot.* XLIX. 794 There is significant under-dispersion or aggregation among the individuals of the population. 1940 *Proc. Linnean Soc. N.S.W.* LXV. 135 The distribution of individual species in the areas examined must be either random, or over- or under-dispersed. 1946, 1957 [see *over-dispersion]. 1962 D. R. Cox *Renewal Theory* iii. 42 Expression..is less than one, so that there is apparent under-dispersion relative to the Poisson distribution. 1964 K. A. Kershaw *Quantitative & Dynamic Ecol.* vi. 98 Several species showed either 'overdispersion' or 'underdispersion'. 1982 G. A. F. Seber *Estimation of Animal Abundance* (ed. 2) xii. 479 The classification into overdispersed and underdispersed is not sufficiently fine.

underdo·ctored, *a.* [Under-[1] 10 a.] Chiefly of an area: insufficiently supplied with doctors; short of doctors.
1960 *New Left Rev.* May–June 23/2 There were..more than 3,500 people per doctor..and, in some under-doctored areas, many more. 1976 *Lancet* 9 Oct. 778/1 Too few graduates from British medical schools are willing to work in underdoctored areas. 1980 *Brit. Med. Jrnl.* 29 Mar. 951/2 Its working party's report on underdoctored areas.

underdog. Add: Hence **u·nderdogger,** one who supports the underdog in a contest; **underdo·ggery.**
1938 H. Belloc in *Tablet* 1 Jan. 8/1 Anyhow, the difficulty and injustice of under-doggery is softened in all sorts of ways by the virtues of charity and humility. 1969 D. Thomson *Aims of Hist.* 68 It was no doubt natural, perhaps inevitable, that the approach of early enthusiasts for economic history should be strongly tinged with under-doggery. 1970 *N.Y. Times* 17 Aug. 26/4 'We underdoggers have to try harder,' he [sc. Governor Rockefeller] explained to reporters. 1977 *Time* 3 Oct. 54 After three crushing defeats, Australia's loyal underdoggers were busy recalling all the old familiar whiny excuses. 1978 *Times* 2 Sept. 7 The angel with the perfect smell, the innocent, the do-gooder, the outsider, the perfect stranger. I was a great underdogger. 1981 *London Rev. Bks.* 2–15 July 24/3 He bore a grudge for not getting the Nobel prize... Reviews..have made much of O'Hara's underdoggery.

underdrain, *v.* Add: Hence **underdrai·ning,** *vbl. sb.*
1805 [see Underdrain *v.*]. 1879 *Harper's Mag.* June 135/2 Other minor improvements have been made, such as the under-draining of a low tract. 1969 G. E. Evans *Farm & Village* ii. 27 This is under-draining, sometimes called thorough-draining, the technique by which a hollow drain is made under the soil into which excess water can soak and be taken off.

u·nder-drawing. [Under-[1] 5 b.] A preliminary sketch, subsequently covered by layers of paint.
1934 *Burlington Mag.* Feb. 85/2 In the latter picture the under-drawing is done for effect and lacks entirely the astounding precision of Holbein's portraiture. 1968 *New Scientist* 3 Oct. 37/2 Mediaeval European paintings..were made on wood... Preliminary designs were sketched ..with bone black or carbon black..and this under-drawing was then..covered with paint layers.

underdress, *sb.* **2.** (Earlier example.)
1806 E. Wynne *Diary* 22 July (1940) III. viii. 291 My bridal array consisted of a white satin under dress and a patent net over it, with a long veil.

u·nderdrive. [Under-[1] 10 b.] An auxiliary speed-reducing gear in a motor vehicle which may be brought into operation in addition to the ordinary gears to provide an additional set of gear ratios.
1929 [see *overdrive *sb.* 1 a]. 1941 *Automobile Engineer* XXXI. 48/1 Chrysler's new 'underdrive' transmission, when coupled with the fluid drive and conventional clutch, is one of the outstanding mechanical innovations of the year. 1981 *Sunday Express* (Colour Suppl.) 4 Oct. /4 (Advt.), The only car with underdrive.

under-edge. Add: Also *spec.* in *Cricket*, the inside- or bottom edge of the bat.
1960 E. W. Swanton *West Indies Revisited* 143 He..dragged the ball into his stumps off the under-edge. 1977 *Sunday Times* 30 Jan. 30/4 Willis bowled Kirmani off the under-edge.

under-employ·ed, *a.* [Under-[1] 10 a.] Insufficiently employed, not used to the optimum capacity. Chiefly of persons or machinery.
1908 W. S. Churchill in R. S. Churchill *Winston S. Churchill* (1969) II. Compan. II. xii. 852 We cannot distinguish between the unemployed and the under-employed. 1941 *Economist* 22 Feb. 241/1 The existence of idle or under-employed machinery—and labour—is sheer waste to the war effort. 1962 *Daily Tel.* 15 June 14/2 There is ample evidence of a progressively under-employed industry and of a turn to export business to employ spare capacity. 1980 *Jrnl. R. Soc. Arts* Feb. 150/2 No sensible organization keeps an expensive designer under-employed.

under-employ·ment. [Under-[1] 10 b.] Insufficient use of resources; *spec.* a situation in which the number of the unemployed exceeds the number of job vacancies, producing a labour surplus.
1909 S. & B. Webb *Public Organization of Labour Market* (Minority Rep. Poor Law Commission II) iv. 185 The evil of Under-employment is shown in its most common form in the great class of Casual Labourers. 1926 Galsworthy *Silver Spoon* II. i. 112 To free the country from..under-employment, and over-population. 1944 A. L. Rowse *Eng. Spirit* xxxii. 225 Carlyle [in *Past & Present*]..went on to point to the dilemma of over-production and under-employment which is a recurrent trouble of *laissez-faire* capitalism. 1961 *Times* 1 Dec. 13/6 This kind of underemployment is to be seen in many sectors of the economy. 1979 *Dædalus* Spring 90 Under-employment does exist, though, and..there are strategies to pay certain categories of women to take time off or to leave the labor market early.

u·nderfelt. [Under-[1] 5 b.] Felt used as an underlay, esp. beneath a carpet.
1895 *Harrod's Catal.* 1487 Paper and felt for laying under carpets—..Underfelts..0/10 and 1/1 per yard. 1907 *Army & Navy Stores Catal.* 270/4 Electric Illuminating Table Cloth. Is used as an underfelt and covered with a white table cloth. 1934 *Discovery* June 155/2 Carpets are much more effective as absorbents [of sound] at the lower frequencies when they are used with underfelts. 1975 N. Luard *Robespierre Serial* xiii. 119 Carswell slept..in the hall, on a roll of underfelt which had been accidentally left behind.

u·nderfill, *sb.* *Metallurgy.* [f. Under-[1] 10 b, after *overfill *sb.*] The condition of there being insufficient metal to fill the aperture between rolls, the impression of a die, etc., so that the desired shape is not taken; a bar or the like that is too small for the rolling it is to undergo.
1924 F. W. Dencer *Detailing & fabricating Struct. Steel* xxvii. 355 Underfills.—A defect of this kind may result from cobbling, and generally is cause for rejection of the pieces affected. 1929, 1957 [see *overfill *sb.*]. 1970 *Welding Engineer* July 24/2 With three of four tetra axis welds inclined to the bottom plane, weld underfill would make conventional welding difficult without a special positioner.

under-filling, *vbl. sb.* Add: **II. 2.** *Metallurgy.* (Stressed **underfi·lling** and written without a hyphen.) [Under-[1] 10 a.] The action or result of causing an underfill.
1953 [see *overfilling *vbl. sb.*]. 1968 R. N. Parkins *Mech. Treatm. Metals* ii. 78 In designing a sequence of passes the critical factor is to ensure that each groove in turn is just filled with metal so that its form is accurate... Underfilling will produce an imperfect shape.
So **underfi·ll** *v. trans.*
1968 [see *overfill *v.*].

u·nderfit, *a.* *Physical Geogr.* [f. Under-[1] 10 c, after *misfit.*] Pertaining to or designating a stream which, if its average flow in the past was at present-day levels, would be expected to have eroded a smaller valley than it has done. Cf. *overfit *a.*
1913 W. M. Davis in *Ann. Assoc. Amer. Geographers* III. 3 Underfit Rivers.—The chief object of this paper is to call attention to an explanation recently suggested for the peculiar relation that is frequently observed between the small-curved meanders of a river and the larger-curved meanders of its valley, a relation that has been called 'underfit'. 1939 A. K. Lobeck *Geomorphology* vi. 199 The beheaded stream, having lost much of its volume, acquires mature characteristics. It develops small meanders not suited to the size of the valley. It becomes a misfit, or an underfit, stream. 1964, 1968 [see *misfit *sb.* 2]. 1970 R. J. Small *Study of Landforms* ii. 46 In post-glacial times, the old channels have been choked with alluvium, and the meandering valleys are now followed by 'underfit' streams, possessing insufficient power to effect further erosion.
Hence **underfi·tness,** the property or state of being underfit.
1913 *Ann. Assoc. Amer. Geographers* III. 24 The lower Seine in Normandy shows a slight underfitness.

1964 *Prof. Papers U.S. Geol. Survey* No. 452-A. 29/2 The underfitness of streams in incised meandering valleys can no longer be denied.

u·nderfloor, *a.* [Under-[2].] That occurs or is situated below floor level; also, operating below the surface of the floor, as *underfloor heating* (often by an electrical element embedded in a concrete floor).
1899 [see Under-[2]]. 1934 H. M. Vernon *Princ. Heating & Ventilation* v. 92 The scheme adopted took the form of under-floor heating by means of a pipe running round each room at a distance of about 4 in. from the walls. 1953 D. J. Warburton *ABC Commercial Vehicles* 8 An under-floor-engine passenger chassis. 1977 M. Drabble *Ice Age* i. 61 The flat she'd lived in with Len had had the lot: de luxe washing machine,..underfloor central heating, two bathrooms, shower... The lot.

underflow, *sb.* Add: **2.** The (more or less horizontal) flow of water through the ground, *spec.* underneath a river bed.
1890 *Rep. Secretary of Agric. 1890* (U.S. Dept. Agric.) 472 Papers on underflow and subterranean water. *Ibid.* 475 Mr. [J. W.] Gregory became convinced of the existence at moderate depths below the surface..of an almost continuous drainage supply, which he has termed the under-sheet or under-flow water. 1905 *U.S. Geol. Survey Water-Supply & Irrigation Papers* No. 140. 85 Such wells, several hundred feet in depth, with perforations opposite the best water-bearing material, would utilize a large part of the underflow which now escapes to the sea. 1951 H. E. Thomas *Conservation of Ground Water* iii. 138 Underflow is the only method of movement in certain reaches of some streams, where the surface channel is dry. 1970 W. C. Walton *Groundwater Resource Evaluation* vi. 375 Water leaving basins includes streamflow, evapotranspiration, and subsurface underflow.
3. *Computers.* [After *overflow *sb.* 2 d.] The generation of a number that is too small to be represented in the device meant to store it.
1959 E. M. McCormick *Digital Computer Primer* xi. 155 A number which is too small (yet is not 0) will cause an 'underflow'. 1970 O. Dopping *Computers & Data Processing* vi. 102 In machines with automatic floating-point arithmetic..there may also be indicators for overflow and underflow in the characteristic. 1985 *Personal Computer World* Feb. 182/2 Computer users pay far less attention to underflow than they do to overflow or rounding.

underflow, *v.* Add: **3.** *intr.* *Computers.* [Cf. prec., sense 3.] Of the result of an arithmetical operation: to become too small for the device meant to store it.
1965 *IBM Systems Jrnl.* IV. 32 If the exponent underflows or the result fraction is zero, the result should be treated as 'zero'. 1973 C. W. Gear *Introd. Computer Sci.* ii. 73 Give examples of pairs of numbers A and B in internal form such that..A × B underflows in floating-point arithmetic.

underfo·cus. [Under-[1] 10.] The situation in which the beam of an electron microscope is focused to a point somewhat short of the specimen. So **underfo·cused** *ppl. a.*, **-fo·cusing** *vbl. sb.*
1953 C. E. Hall *Introd. Electron Microsc.* x. 268 The pattern appearing at the screen is an underfocused image. 1971 *Jrnl. Physics* A IV. 806 These large phase shifts can be exactly cancelled by underfocus of the objective lens. *Ibid.*, This underfocusing will produce a large chromatic aberration. 1979 *Nature* 17 May 227/1 The contrast of the images taken from very thin parts of crystal at about 700 Å underfocus is known to reflect approximately the projection of the structure; the dark dots show the sites of Re ions.

underfoot, *a.* **1.** (Examples of *conditions underfoot*, esp. applied to the state of the going in *Horseracing.*)
1976 *Eastern Even. News* (Norwich) 9 Dec., Underfoot conditions at Chelmsford made it difficult for everyone. 1979 *Oxford Times* (City ed.) 5 Jan. 6 They [sc. postmen] begin their round in darkness before underfoot conditions have a chance to improve.

underfoot, *adv.* Add: **5.** Of a person or persons: about one's feet, constantly (and irritatingly) present; 'in the way'. *colloq.*
1891 *Harper's Mag.* June 62/1 He muttered something about children being underfoot and staring at such times. 1922 S. Lewis *Babbitt* xviii. 230 Kenneth Escott and she were always under foot. 1959 M. Scott *White Elephant* i. 3 It has been a trying month for her too, with Deryk always underfoot. 1981 'S. Woods' *Dearest Enemy* i. 38 It's really too much of a nuisance having him always underfoot when I'm trying to prepare my own meals.

u·nderfoot, *sb.* *rare.* [f. the adv.] The surface of the ground at the foot of a tree.
1910 W. De Morgan *Affair of Dishonour* iv. 50 This morning was no time for breakfast under the cedar trees. For all the underfoot, where grass grew, was no better than a sponge. 1959 E. Collier *Three against Wilderness* xxiv. 248 No flame could lick far into the forests so long as their underfoot was moist.

underfoo·ting. [f. Under foot *adv.*: see -ing[1].] **1.** *Ir.* In turf-cutting: (see quots.).
1942 E. E. Evans *Ir. Heritage* xv. 136 In breast-work the digger stands facing the turf bank, and thrusts his spade horizontally, whereas in underfooting he stands on the bank and cuts vertically. 1957 —— *Ir. Folk Ways* xiv. 189 There are two methods of cutting, vertically or 'underfooting' and horizontally or 'breasting', and they tend to be used respectively on thin upland and deep lowland bogs. 1979 *Country Life* 27 Sept. 963/3 In some areas the [peat] turf is cut horizontally, known as breasting, but most turf is cut vertically, which is termed 'underfooting'.
2. The ground under foot, esp. with regard to its condition. Cf. Footing *vbl. sb.* 5.
1948 *Sun* (Baltimore) 29 Apr. 19/1 Shivaree..was timed in 1.12 3–5, which was excellent in the sloppy underfooting. 1961 L. Mumford *City in History* x. 309 Some three centuries before wheeled vehicles became common, the street lost its natural underfooting. 1976 M. & G. Gordon *Ordeal* (1977) xviii. 125 He slowed only when he encountered rough underfooting.

under-frame. Add: **a.** Also, the substructure of a motor vehicle. Cf. *sub-frame* (b) s.v. *sub- 3 a.
1903 *Lanchester Motor & Carriage Descriptive Man.* i. 9 The transmission from the car frame to the front under-frame. 1954 D. J. Warburton *ABC Brit. Lorries & Road Services* 28 The underframe is also available for the construction of other lightweight bodies.
b. *Furnit.* The framework supporting a seat or table-top.
1934 *Burlington Mag.* Nov. 201/2 The armchair[s] of walnut..show in the carving on the top-rail and under-frame..that they are closely connected with the Indian high-back chairs. 1952 Hoard & Marlow *Cabinet-maker's Treasury* 266/2 *Underbracing*, arrangement of stretchers to brace the legs of chairs, tables... Also *Underframe*. 1982 *Habitat Catal.* 1982/83 53/1 The underframe and legs of the tables are in solid timber.
underframing (examples of furniture).
1902 *Carpet & Upholstery Trade Rev.* 1 Sept. 73/1 The chair..is of a different type... The underframing begins to resemble rather closely the style known as 'Louis Quatorze'. 1969 E. H. Pinto *Treen* xv. 218, I think that the underframing was cut down to support the top, which originally was a loose board.

u·nderfug. *Schoolboys' slang.* [Under-[1] 5 a + *fug *sb.*] An undervest; also, underpants.
1924 H. de Sélincourt *Cricket Match* ii. 32 'I do hope you'll be sensible and wear your nice, warm undervest, John.'..'I'm not vain, Maria..But..I heard one of the toffs say..'My God! Look, he's wearin' an underfug.' 1946 B. Marshall *George Brown's Schooldays* iv. 18 The matron kept everybody's spare shirts, underfugs and towels and dished clean ones out once a week. 1967 *Listener* 2 Mar. 299/3 Felsted School calls radiators *fugs* and underpants *under-fugs*.

underfu·nction, *sb.* Chiefly *Med.* [Under-[1] 10.] = *hypofunction* s.v. *hypo-. Hence as *v. intr.*, to exhibit underfunction; to have a diminished capacity for acting and responding; **underfu·nctioning** *vbl. sb.*
1941 *Trans. Assoc. Amer. Physicians* LVI. 51 The adrenal cortex is underfunctioning because of pituitary disease. *Ibid.*, The term 'hypoleydigism' is defined as underfunction of the cells of Leydig of the male gonad. 1964 M. Argyle *Psychol. & Social Probl.* vi. 74 The person suffering from a behaviour disorder is usually ineffective, underfunctioning in his work and social relations. 1979 *Brit. Med. Jrnl.* 15 Dec. 1582/2 It is not uncommon to diagnose pathological underfunctioning in the presence of values which may be above average for the population but which are known..to fall below the norm for that particular individual. 1981 *Pharmacopsychiatria* XIV. 3 An underfunction of nonadrenergic synaptic transmission in depression.

u·nderfur. [Under-[1] 5 d.] An inner layer of shorter fur on an animal.
[1892 H. Poland *Fur-Bearing Animals* 200 The under fur of this Seal is long compared with other species.] 1904 *Ann. & Mag. Nat. Hist.* XIII. 424 Underfur of back about 15 mm. in length. 1968 [see *overfur]. 1971 *Nature* 4 June 331/1 The fur was very spiny with virtually no soft underfur.

undergird, *v.* Add: The *fig.* use is now dominant. (Further examples.)
1973 *Times* 30 July 20/6 It involves an understanding of the variety of sub-systems which undergird community life, health, education, welfare, [etc.]. 1975 *Church Times* 15 Aug. 16/5 Dr. Taylor believes that such smaller groups as these are going to become the essential sub-structure of Church life that is needed to undergird the larger congregation in the parish church. 1981 *Dædalus* Spring 19 'Why shouldn't people have these rights?' was the question that undergirded these laws. 1983 *Salisbury Rev.* Spring 18/1 The same instinct directing a man to preserve his life, undergirds his attachment to the institutions which inform his life as a participant in the historical stream of the national culture.

under-glaze, *a.* and *sb.* Add: **1. a.** Also in other collocations, as *under-glaze design, work*, etc. (Earlier and later examples.)
1879 J. C. L. Sparkes *Pottery Painting* 28 Oil mediums may be used for over-glaze and for under-glaze work. 1925 Heal & Son *Catal.: Table Wares*, A simple brush-work under-glaze design in blue and red. 1967 M. Chandler *Ceramics in Mod. World* ii. 49 (*caption*) Chinese porcelain bowl with blue underglaze design.
2. (Earlier and later examples.)
1879 J. C. L. Sparkes *Pottery Painting* 45 An under-glaze flesh tint may be made of claret brown. 1947 J. C. Rich *Materials & Methods of Sculpture* ii. 49 An under-glaze color is a concentrated pigmenting agent that depends upon the glaze subsequently applied over it for 'life' or gloss and brilliance. 1980 R. Adams *Girl in Swing* i. 11, I took out the Copenhagen plate, with its underglaze blue wave mark.

undergra·duacy. *rare.* [f. Undergradua(te *sb.* + -cy.] The state or position of (being) an undergraduate in a university.
1927 C. Connolly *Let.* 21 Apr. in *Romantic Friendship* (1975) 295 This sense of living in a sequel tends to be the canker gnawing at the heart of undergraduacy. 1964 *Economist* 11 Jan. 89/2 The rather unworldly cloister of undergraduacy.

undergra·duatish, *a.* [f. Undergraduate *sb.* + -ish[1].] Characteristic of an undergraduate; undergraduate-like. Cf. Undergraduate *a.* 3.
1925 [see *floater 5]. 1931 *Times Lit. Suppl.* 10 Dec. 1001/3 He seems, for example, to give way to an undergraduatish melancholy at Cambridge. 1980 W. Ash *Incorporated* iii. 33 You're more interested in undergraduatish cleverness than in pursuing an interest which ..unites us.

undergradue·tte. *colloq.* (now *rare*). [Altered f. Undergraduate *sb.*: see -ette.] A female undergraduate.
1919 *Observer* 23 Nov. 17/4 The audience was chiefly composed of under-graduates and under-graduettes. 1926 *Spectator* 23 Oct. 677/1 There are too many undergraduettes at Oxford. 1934 W. Gerhardie *Resurrection* xcv. 355 A professor, a don, an undergraduette in black cotton stockings, a green-grocer's boy—all move on bicycles. 1940 Graves & Hodge *Long Week-End* viii. 123 The subject of the duel was reported in the London Press to be an 'undergraduette of Somerville'. 1972 J. Potter *Going West* 70 He ostentatiously walked out arm-in-arm with the prettiest undergraduettes. 1980 *Listener* 23 Oct. 549/2 In the Oxford of 1939..I remember a red-headed 'undergraduette' (as we then rather quaintly would have called her).

u·ndergrip. [Under-[1] 5 b.] **a.** *Gymnastics.* A hold in which the hands pass beneath the horizontal bar, with the palms facing the gymnast. **b.** *Mountaineering.* = *underhold.
1920 Naylor & Temple *Mod. Physical Education* 109 The starting position is reached as in Exercise 90, the hands being in 'under-grip'. 1955 J. E. B. Wright *Technique of Mountaineering* iv. 76 (*caption*) An under grip. 1976 T. & J. Cousminer tr. *Kaneko's Olympic Gymnastics* 214 Hang straight from the horizontal bar... If your palms face backwards, you are using the reverse grip (also called the undergrip). 1977 D. Law *Starting Mountaineering & Rock Climbing* vi. 67 In the under grip [hold] the hand is turned over, fingers pointing upwards; it is quite sufficient to maintain balance.

underground, *adv.* Add: **2. b.** Into hiding or surreptitious activity, esp. in phr. *to go underground*: applied chiefly to political organizations and their representatives which continue to operate in secret (and often subversively) after becoming officially unacceptable.
1935 *Ann. Reg. 1934* II. 198 The Socialist leaders..decided that it was best to accept defeat: those leaders who were known made for the frontier, the others 'went underground' and began at once the organising of an illegal party. 1949 'M. Innes' *Journeying Boy* xxiii. 285 There was nothing for it but a quick get-away and a going underground for good. 1960 E. Bowen *Time in Rome* iv. 118 Like Resistance workers in the occupied countries, long-ago Christians, from time to time, strategically 'went underground'. 1977 *Times* 15 June 1/5 The crucial factor was the subsequent move underground by a handful of those people [sc. Soviet sympathizers], and the covers they assumed to mask their new role as agents.

underground, *a.* and *sb.* Add: **A.** *adj.* **1. d.** *underground mutton*, a rabbit; rabbit meat. *Austral. slang.*
1946 A. J. Holt *Wheat Farms of Victoria* viii. 129 'Underground mutton' (rabbit) is almost always available for those who like it. 1965 E. Lambert *Long White Night* xv. 138, I thought a feed of underground mutton would go all right for my tea. 1979 D. R. Stuart *Crank Back on Roller* 151 Maybe a rabbit, though I was never what y' could call over fond of the ole underground mutton meself.
2. c. (*a*) Also in other collocations, as *underground line, service*, etc. (now often *attrib.* uses of the sb.); (*b*) also *fig.*
(*a*) 1883 [see *metropolitan *a.* 2 b]. 1926 *Times* 6 May 3/1 In London most of the tube and underground services were in force or others were expected to be to-day. 1926 *Daily Express* 11 May 1/3 The Underground trains yesterday showed no difference from any ordinary Monday morning. 1975 J. Symons *Three Pipe Problem* xviii. 204 He lost his way to the Underground station. 1982 G. Lyall *Conduct of Major Maxim* vii. 62 With the Underground map in his diary he worked out a route that involved two changes of train.
(*b*) 1868 C. M. Yonge *Chaplet of Pearls* I. xviii. 251 There was what in later times has been termed an underground railway amid the persecuted Calvinists. 1940 *Sun* (Baltimore) 14 Oct. 2/3 A so-called 'underground railway' delivering prominent scholars and writers from Nazi-dominated parts of Europe. 1979 R. Laidlaw

Lion is Rampant xviii. 137, I could probably have got you smuggled to the Border via my 'Underground Railway'.

4. c. Designating (the activities of) a group, organization, or its representatives, working covertly to subvert the aims of a ruling (often occupying) power. Cf. *RESISTANCE 1 c.

1939 [see *RESISTANCE 1 c]. 1939 *War Illustr.* 9 Dec. 392/3 Even in the completely occupied territory there was underground activity. 1944 *Times* 18 July 2/3 An exhibition of the underground press of Europe. 1950 G. Brenan *Face of Spain* v. 118 Until recently people suspected of Underground activity had been arrested and court-martialed. 1965 J. A. Michener *Source* (1966) 784 The night in the Belgian port when another English underground operative had called, 'One more place in the lorry. Look lively.' 1974 J. White tr. *Poulantzas's Fascism & Dictatorship* iv. ii. 186 The KPD's underground apparatus turned out to be non-existent.

d. Of or pertaining to a subculture which seeks to provide radical alternatives to the socially accepted or established mode; *spec.* manifested in its literature, music, press, etc.

1953 *Observer* 13 Sept. 9/3 Its detached picture of barren tragic love..in a furtive fantastic 'underground' sector of London. 1962 *Movie* Dec. 4/2 Fuller is not an 'underground' director whose films actually *do* the opposite of what they overtly *say*. 1969 *Oz* May 36/1 He talked solidly for nearly forty-five minutes—he'd said it all before..to all the underground papers in the States. 1970 A. Toffler *Future Shock* xii. 248 The underground movie..is flourishing even more than the underground press. 1977 *New Yorker* 9 May 126/3 He is attracted to buying *Mainline* because of its history as a radical underground weekly.

e. *underground economy*, the economic sector of private business deals in which tax liability is not reported; the 'black' economy. Cf. *BLACK *a.* II c. *U.S.*

1978 *Business Week* 13 Mar. 74 Neither the government nor labour unions seem to have any control over the underground economy. 1980 *Economic Rev.* (Fed. Reserve Bank of Atlanta) Jan.–Feb. 8/2 It suggests increasing activity in the 'underground economy' (gambling, bartering, and other unreported income). 1982 *Christian Sci. Monitor* 1 July B-4/1 Local custom means heavy traffic in illegal drugs and prostitution, an underground economy that is left alone as long as there is no trouble. 1983 *Chicago Sun-Times* 9 Sept. 38/2 The 30 cents a pound is essentially tax-free, part of the underground (or curbside) economy.

B. *sb.* **3.** Esp. the one in London. Usu. with capital initial. (Earlier examples.)

1866 J. R. Planché *Orpheus in Haymarket* III. 32 For Tartarus I happened to be bound, And was just starting by the Underground. 1875 L. Troubridge *Life amongst Troubridges* (1966) 123 Set off from Victoria Station by the Underground for Shepherd's Bush.

4. *fig.* **a.** A group or movement organized secretly to work against an existing regime, often by violent means; *spec.* an 'underground' resistance movement. Usu. with *the*.

1946 Koestler *Thieves in Night* 277 The Hebrew underground began as a purely political movement. 1946 Auden *Nones* (1952) 61 By night our student Underground At cocktail parties whisper round From ear to ear. 1958 L. Uris *Exodus* I. xii. 72 The Danes, by the middle of 1941, had established a small but determined little underground. 1966 A. Sachs *Jail Diary* iii. 30 With her husband Dennis on trial for his life for supporting the underground, she was hoping for some peace. 1973 T. Pynchon *Gravity's Rainbow* I. 96 Captain must allow for the real chance she's a British spy, or member of the Dutch underground.

b. Any unofficial group or movement which seeks to provide alternatives to the forms of expression and action sanctioned by the society in which it exists.

1959 [see *TRANS-WORLD *a.*]. 1969 *Oz* Apr. 27/2 The underground's a great crowd really. We enjoy them as people..just being what they are. 1970 *Guardian* 28 Feb. 9/2 We are going to try..to establish a definition of the underground... The word applies to a life-style, to a group of people who live..outside the constraints of ordinary society..and who largely..have come to the realisation of this new life-style through the use of cannabis and LSD. 1974 M. C. Gerald *Pharmacol.* i. 13 On the one hand, he has been enticed by 'notes from the underground', exalting the wonders of 'acid';..simultaneously, he has often been subjected to..the lies and half-truths of the well-meaning establishment.

undergrounder (later examples).

1954 L. MacNeice *Autumn Sequel* xv. 94 The hole becomes..a wide Doorway; the undergrounders burst and spout. 1974 K. Millett *Flying* iv. 380 This patron saint of all undergrounders.

u·nderground, *v.* [f. the adv.] *trans.* To lay (electricity or telephone cables) below ground level. So **u·ndergrou:nding** *vbl. sb.*

1889 in *Cent. Dict.* 1961 *Times Rev. Industry* Jan. 40/1 In..transmitting power, there are only two alternatives—the overhead and underground..and it's 16 times as dear to run [cables] underground... I think undergrounding is out of the question. 1964 *Times* 2 Sept. 11/6 We were recently obliged to underground three and a half miles of the Thorpe Marsh–Stalybridge line. 1971 P. Gresswell *Environment* 71 In the case of a line of pylons..he [*sc.* the citizen] needs to be told not simply how much undergrounding would cost..but how this relates to the total capital expenditure. 1977 *New Scientist* 27 Jan. 188/2 The CEGB spent £220,000 undergrounding 2 km..cable..to preserve the view of Oxford as seen from Elsfield.

u·nder-hold. *Mountaineering.* [Under-[1] 5 b.] A hold in which the hand grasps a downturned edge or point from beneath with the palm upwards, used esp. to maintain balance. Cf. *UNDER-GRIP b.

1920 G. W. Young *Mountain Craft* iv. 162 A cling 'under'-hold keeps body and eyes free at the length of our arms, bent or straight according to our convenience. 1948 'Mountaineers' *Mountaineers' Handbk.* ix. 68 In a cling under hold..the hands pull upward and exert a tension through the body against the feet, pressing them into the rock. 1965 A. Blackshaw *Mountaineering* vi. 175 (*caption*) Note the use of the under-hold..to permit a high reach with left hand.

under-insu·rance. [Under-[1] 10 b.] Insurance (of goods, property, etc.) at less than their real value.

1893 *Standard* (Boston) 20 May 518/1 Mr. Goodwin considers the tendency toward under-insurance of safe risks as the cause of the general requirement of the 80 per cent coinsurance clause. 1930 F. E. Wolfe *Princ. Property Insurance* ix. 147 Underinsurance is ignored if the practice is followed by fixing rates from time to time. 1949 R. S. Sharma *Insurance Princ. & Pract.* xxiv. 268 One of the difficulties in arriving at a scientific basis of rating in fire insurance is of under-insurance. 1976 *Business Week* 5 July 24 The hazards of underinsurance ..exploded in Mississippi as a massive liquidity crisis triggered the..'bank holiday'.

under-insu·re, *v.* [Under-[1] 10 a.] *trans.* To insure at less than the real value. Also *intr.* or *absol.* Cf. *OVER-INSURE *v.*

1911 *Rep. Labour & Social Conditions in Germany* (Tariff Reform League) III. 145 If a person..under insured to avoid the rates he would lose in case of fire. 1921 W. N. Bament *Co-Insurance* 3 The natural inclination..is to under-insure and take the chance of not having a total loss. 1982 *Economist* 18 Sept. 25 The problem is that the less well-off might underinsure.

So **under-insu·red** *ppl. a.* (also *absol.* as *sb.*).

1893 *Standard* (Boston) 20 May 518/2 The sufferers of slight and partial losses are able..to draw their full damage without reference to the insurance carried, which, in the case of small insurance in proportion to value carried, enables a person who is thus under-insured to draw more than an equitable portion. 1909 W. J. Graham *Romance of Life Insurance* v. 66 Life, the great creator and conservator of the world's wealth, is grossly underinsured when compared with fire insurance. 1912 E. R. Hardy in H. P. Dunham *Business of Insurance* I. x. 179 In case of important properties where mortgage loans have to be protected..all such, although their property may not be subject to any greater loss than the under insured, are paying for the under insured property. 1947 P. Gordis *How to buy Insurance* i. 8 We have established the inadvisability of being under-insured. 1962 A. Lurie *Love & Friendship* iv. 69 A regular fire-trap the whole house. Oh they're wild about it now... Underinsured, that's what it was. 1981 M. Duffy *Gor Saga* ii. 102 He was underinsured... His capital had been in his stock.

u·nderkill. *colloq.* (orig. *U.S.*). [f. Under-[1] 10 a, after *overkill*.] (The mobilization of) insufficient capacity, esp. of nuclear weapons, to kill and effect destruction to the level of strategic requirements. Freq. *fig.*

1964 *New Scientist* 10 Sept. 647/2 Mr Goldwater has engaged in a press war..over the future strength of United States' strategic retaliatory force—what for lack of a better description might be called the question of underkill. 1971 *Time* 5 July 13 Harvard's Paul A. Freund maintained that 'risk for risk, the law has opted for underkill in duels over publication'. 1984 *Financial Times* 20 Jan. 21/1 There is no limit to the silliness resulting when that director goes over the top and weds visual overkill to narrative and dramatic underkill.

u·nderlap, *sb.* [Under-[1] 5 b, after *overlap*.] †**1.** *Geol.* The fact or state of underlapping. *Obs. rare*[−1].

1883 J. G. Goodchild in *Geol. Mag.* Decade II. X. 226 Overlap, and its correlative Underlap, as it is here suggested to employ them, thus denote the stratigraphical relations of one part of a formation to another where the original extent of the lower part occupied a smaller area than that of the next higher members of the same series.

2. a. In warp knitting, the lateral movements of the guide bar made on the side of the needle remote from the hook.

1926 J. Chamberlain *Hosiery Yarns & Fabrics* vii. 170 The underlap may be at the back of any convenient number of needles. 1952 D. F. Paling *Warp Knitting Technol.* i. 4 The first movement is applied to the guide bars, and consists of a lateral motion in front of the needles ..known as the *underlap*. 1971 D. G. B. Thomas *Introd. to Warp Knitting* iii. 24 In most modern machines, the underlap takes place when the needles are down. The underlap is sometimes called a *shog*.

b. A piece of material that extends beneath another fold or pleat.

1943 I. R. Duncan *Needles & Pins* xiv. 284 Stitch the two pieces together at the top and bottom for a dress.. to form an over and under lap for the placket. 1964 *McCall's Sewing* ii. 33/1 *Underlap*, a part of a garment that extends or laps under another pleat. 1979 *Tucson* (Arizona) *Citizen* 20 Sept. 2B/1 Put the coat on over a skirt or dress and button it closed. Place a pin at the desired new hem length at center front on both the overlap and underlap.

underlay, *sb.* Add: **2. e.** *gen.* A layer which underlies another; a substratum. Also as *fig.* use of next sense.

1876 W. Whitman *Jrnl.* 10 June in *Specimen Days* (1883) 87, I write..here by the creek... For underlay, trees in fulness of tender foliage—liquid reedy, long-drawn notes of birds. 1964 *Listener* 15 Oct. 572/1 The main characteristic of the new popular manner is an aggressive surface over a soft underlay; it is noisy..about minor issues, but takes good care not to tread any..controversial ground. 1977 C. McCullough *Thorn Birds* (1978) II. vii. 174 'I do not intend to make a long eulogy,' he said in his clear, almost Oxford diction with its faint Irish underlay.

f. *spec.* (sheets of) material laid beneath a carpet (usu. felt or paper, or as an integral rubberized backing) or a mattress, for protection and support.

1907 *Army & Navy Stores Catal.* 275/2 Moth proof Cotton underlay to go between wire bottom of bed and mattress..4/8. 1923 *Daily Mail* 21 Feb. 8 (Advt.), Cedar Felt is an improved paper felt underlay for carpets. 1929 W. Deeping *Roper's Row* xxi. 233 They'll do for underlays on the beds. I shan't waste them. 1959 *Spectator* 7 Aug. 165/2 Underlay adds to the life of a carpet, absorbs pressure from furniture and footsteps and acts as a sound absorber and insulator. 1980 D. Adams *Restaurant at End of Universe* xx. 117 The thin tufted nylon floor covering was black, and when they had lifted up a corner of it they had discovered that the foam underlay also was black.

4. *Early Mus.* The placing of text in relation to music. Cf. sense *1 e of the vb.

1969 G. Reaney in Reese & Snow *Essays in Musicology in Honor of Dragan Plamenac* xvi. 250 The most difficult type of underlay consists of a moderately large group of notes with rather fewer syllables. 1974 *Times Lit. Suppl.* 14 June 642/2 When Petrarch's lines are restored to their proper shape and order and deployed in reasonable fashion below the notes of the vocal part, most if not all of the problems of underlay disappear. 1980 *Early Music* Jan. 21/1 Directions for tempo and dynamics, and frequently for underlay also, are absent from manuscript and printed sources alike.

underlay, *v. trans.* Add: **1. e.** *Early Mus.* To place (the text of a song, etc.) in relation to the music.

1934 A. Hughes *Anglo-French Sequelae* 7 Is there any satisfactory reason adducible to explain why the music of the Verba-passages should be underlaid with a text, and not the rest? 1949 W. Apel *Notation of Polyphonic Music 900–1600* (ed. 4) II. ii. 118 This composition also serves to illustrate the problem of text-underlaying in early music... The original frequently leaves considerable room for doubt..as to the 'correct' placing of the words. 1960 D. Stevens *Hist. Song* 88 The text does not fit the ligatures of the tenor part... But it makes good sense when underlaid to the soprano part.

u·nderlayment *U.S.* [f. Underlay *v.* + -MENT.] Carpet underlay; material laid beneath roofing tiles, etc.

1956 *Study of Installation Wood Block Finish Flooring* (U.S. Nat. Res. Council) 17 The sub-floor should be covered with an underlayment of 15 lb. asphalt impregnated felt. 1975 'E. Lathen' *By Hook or by Crook* x. 97 You'll be sure that the men have instructions about the underlayment.

u·nderlead. [Under-[1] 4 d.] *Bridge.* The lead of a low card in a suit of which a higher card is held by the player.

1934 E. Culbertson *Contract Bridge Red Bk. on Play* xli. 531 An Ace underlead not only has the advantage of the surprise effect on the enemy but it retains the control of the suit. 1947 C. H. Goren *Better Bridge* ii. 101 This is one of the rare cases in which the under lead of an Ace is recommended. 1964 *Official Encycl. Bridge* 650/2 Another motive for an underlead is an urgent desire to get a particular lead from partner.

Also **underlea·d** *v. trans.*, to lead with a card of lower value than (another card of the suit held).

1945 Phillips & Reese *How to play Bridge* III. 111 It is dangerous to underlead Aces against suit contracts. 1962 *Listener* 26 July 154/3 The question was whether East would risk underleading his queen of diamonds, to put partner in with the jack for a spade lead. 1973 *Daily Tel.* 13 Jan. 9/5 West should therefore underlead his ♡A (assuming he has it) for a second club ruff.

underleaf. Add: **2.** Also, a lower leaf.

1922 Joyce *Ulysses* 220 The lychgate..showed Father Conmee breadths of cabbages, curtseying to him with ample underleaves. 1969 F. E. Round *Introd. Lower Plants* viii. 99 Yet another feature of this group is the basic conservatism of the organs; they differ in detail but vegetatively there are only three basic growth forms, (*a*).. (*b*) a dorsiventral leafy thallus with two rows of lateral leaves and a third modified row of 'underleaves'.

underline, *sb.* Add: **4.** The caption or text beneath a picture, esp. in a book, newspaper, etc.

1924 N. J. Radder *Newspaper Make-Up & Headlines* vi. 109 Practice varies with respect to the underline below the cut... When the underline is brief it is much like the second deck of a headline and the copyreader must.. avoid repetition of words..in the overline. 1947 A. E. Davies in J. G. Herzberg *Late City Edition* xxii. 206 The editor writes captions to run above the picture, and the more detailed underlines to run below them. 1956 H. Williamson *Methods Bk. Design* p. xiii, In order to

simplify the text pages, the underlines of the illustrations refer only to the specific points which are illustrated. 1971 Baskette & Sissors *Art of Editing* x. 203 Picture texts are known by many names—cutlines, captions, underlines (or overlines), legends.

underline, *v.*[2] **2.** (Earlier examples.)

1825 P. Egan *Life of Actor* ii. 67 A file of old play bills which might do to bind, With only the play for next night—underlined. 1838 *Actors by Daylight* I. 5 A new burletta is underlined for this week.

underlining, *vbl. sb.* (Earlier example.)

1838 L. Hunt *Let.* July in *Lett. C. Dickens* (1965) I. 685, I have marked..my *most* favourite passages... The underlinings imply admiration.

underloaded (stress variable), *a.* [Under-[1] 10 a, after *overloaded*.] Not loaded or burdened to capacity; *spec.* in *Geomorphol.* of a stream: carrying less than the maximum amount of sediment for the given conditions, and eroding its bed.

1898 I. C. Russell *River Development* iv. 68 But if the bottom current is underloaded, the material is carried slowly forward. 1942 O. D. von Engeln *Geomorphol.* xii. 234 A stream may be underloaded in which case it will ordinarily be downcutting, or it may be loaded to capacity. 1964 K. G. Lockyer *Introd. Critical Path Analysis* viii. 70 When too much work is required of a work source, the work source is said to be *Overloaded*, whilst if too little is needed it is said to be *Underloaded*. 1970 R. J. Small *Study of Landforms* ii. 57 At some times streams erode vertically (because they are underloaded), at others they erode laterally (because they are graded), and at others they deposit (when they change from a graded to an overloaded condition).

Hence **u·nderload,** an occurrence or state of being incompletely loaded.

1964 K. G. Lockyer *Introd. Critical Path Analysis* viii. 76 With experience, a great facility is obtained in viewing a histogram and..assessing whether a 'peak' (i.e. an overload) can be toppled into a 'valley' (i.e. an underload) in order to 'smooth out' the loading. 1971 *Cabinet Maker & Retail Furnisher* 24 Sept. 535/1 Unless the home market improves..APV could have..an underload position in some of their factories. 1981 *N.Y. Times Mag.* 8 Feb. 12/2 Conservatives are foolishly worrying about lithosphere underload.

u·nderlord. [Under-[1] 6 a.] A subordinate or lesser lord; one in authority below another (opp. *overlord*).

1929 Belloc *Joan of Arc* ii. 39 The Kingdom was held in fealty from God alone of Whom Charles was Vassal and Under-Lord. 1937 F. L. Packard *Dragon's Jaws* 251 And if the old temple is the secret meeting place ..and Fan Chao-tao and his underlords..are to be there —what then? 1969 *Daily Tel.* 29 Oct. 16 In the sphere of the other main new overlord (Mr Benn) one nominated underlord has already resigned, presumably because he thinks it won't work. 1985 *Financial Times* 1 Feb. 21/5 The appointment of Lord Young..to the Cabinet as what he calls a kind of 'underlord' responsible for training matters.

Hence **u·nderlordship** *rare.*

1937 Belloc *Crusade* ii. 38 The title King meant that he had no overlord. He was the summit in a pyramid of overlordships and underlordships.

underlying, *ppl. a.* Add: **2.** *Linguistics.* Applied to a significant feature of a linguistic unit or structure which needs to be represented in its derivational or syntactic analysis, but which may not be apparent from the standard written or spoken form. Cf. *deep structure* s.v. *DEEP *a.* IV. c.

1933 L. Bloomfield *Language* xiii. 210 By a feature of modulation common to nearly all constructions of English morphology, the underlying form keeps its stress, and the bound form is unstressed. 1958 C. F. Hockett *Course in Mod. Linguistics* xxviii. 241 If the underlying form in a secondary derivative is not a single morpheme, its structure is also covered by the classification. Thus the underlying form of *actress-* is the secondary derivative *actor-*. 1970 J. Lyons *Chomsky* 67 This system of phrase structure rules will generate a large (but finite) number of what we may call *underlying strings*. 1978 *Language* LIV. 387 Actually, rather than 'deep structure' we should think in terms of the 'underlying structure' of Generative Semantics, a highly abstract and universal level that is much more closely akin to the Freudian unconscious than is the 'deep structure' of classical transformational grammar.

Hence **underly·ingly** *adv.*

1973 A. H. Sommerstein *Sound Pattern Anc. Greek* i. 2 There is underlyingly a system of five short..and five corresponding long vowels. 1982 J. C. Wells *Accents of English* i. 76 It is possible to argue..that an occurrence of the phoneme /r/ is present underlyingly in the critical environments, but that it is obligatorily vocalized or deleted by rule so that it is not present phonetically.

underman. Add: **2.** [tr. G. *untermensch.*] *spec.* opp. *OVERMAN *sb.* 4; a person with sub-human attributes.

1910 T. Common tr. *Nietzsche's Joyful Wisdom* III. 179 The inventing of Gods, heroes and supermen of all kinds, as well as co-ordinate men and undermen—dwarfs, fairies,..devils—was the inestimable preliminary to the justification of the selfishness..of the individual. 1936 A. N. Field *All these Things* vi. 206 Planned Economy calls for Supermen to control and Undermen to submit. 1963 L. Trilling in N. Frye *Romanticism Reconsidered* 92 *More life*: perhaps it was this boast of the Underground Man that Nietzsche recalled when he said, 'Dostoevsky's Underman and my Overman are the same person.' 1974 W. Kaufmann tr. *Nietzsche's Gay Science* III. 192 *Untermensch* (underman, for a subhuman man) is not rare.

3. *Railways.* A subordinate member of a gang of platelayers.

1921 *Dict. Occup. Terms* (1927) § 577 *Underman*.., a worker in a permanent way gang supervised by ganger. 1928 *Daily Express* 18 Feb. 7/3 Sidney Coleman, an underman platelayer, said he was one of the gang which was 'packing' sleepers on the down main line. 1931 *Cambridge Daily News* 5 Sept. 5/2 Andrew White, an underman.. said he shouted as hard as he could to warn the men that a train was approaching.

u·ndermass. *Geol.* [Under-[1] 5 c.] An older, deformed body of rock overlain by younger, undeformed strata.

1942 O. D. von Engeln *Geomorphol.* ix. 168 The geomorphic significance of the abrupt, commonly marked in kind, transition between the materials above and below a plane of angular unconformity is indicated by the introduction of the terms cover mass and undermass. 1956 D. L. Linton *Sheffield* ii. 35 We..interpret it as the product of the cycle(s) of erosion which stripped away the presumed chalk cover, removed some hundreds of feet of the undermass, and saw the first group of adjustments of the superimposed streams to the underlying structures. 1966 [see *plate tectonics* s.v. *PLATE *sb.* 20].

undermeaning. (Earlier example.)

1841 Ruskin *Let.* 12 Feb. in *Lett. to College Friend* (1894) vii. 87 Rhyme and rhythm are..thoroughly injurious where there is no mystery, when there is not some undermeaning.

underminingly, *adv.* (Later example.)

1832 F. Burney *Mem. Dr. Burney* III. 176 The accumulation of the whole had, slowly and underminingly, brought him into the state that has been described.

u:ndermodula·tion. *Radio* and *Cinemat.* [Under-[1] 10 b.] Modulation that is less than it need be, so that the available capacity of a transmitter or medium is underused or the signal is too weak. So **undermo·dulate** *v. trans.*, **-mo·dulated** *ppl. a.*

1940 *Chambers's Techn. Dict.* 877/1 Undermodulation. 1962 A. Nisbett *Technique Sound Studio* v. 94 Programme meters... Its uses are:..(ii) to provide indications of levels which would result in under- or over-modulation at the recorder or transmitter. *Ibid.* 100 Reducing the effective service area of the transmitter by undermodulating a whole programme. 1974 *Listener* 2 May 573/3 Why are the transmissions on Radio 3 invariably under-modulated?

undermost, *a.* **1. b.** (Earlier example.)

1822 D. Wordsworth *Jrnl.* 22 Sept. (1941) II. 367 A girl, bare-legged, petticoats kilted to her undermost, of white flannel.

under-nou·rished, *a.* [Under-[1] 10 a.] Having had insufficient nourishment, esp. over a sustained period; in a state of semi-starvation. Also *fig.*

1910 Galsworthy *Justice* iv. 98 He's under-nourished. It's very trying to go without your dinner. 1928 *Manch. Guardian Weekly* 10 Aug. 110/1 Most of the Indian children are badly under-nourished. *a* 1942 B. Malinowski *Sci. Theory of Culture* (1944) viii. 80 When African labour is drafted into European enterprise, whether mines, plantations, or factories, it is usually found that the workers are undernourished with reference to the efforts which they will have to put into their performances. 1962 E. Snow *Other Side of River* (1963) xxxix. 289, I asked one how he explained his fine white teeth if he had been undernourished. 1972 *Computer Jrnl.* XV. 200/1 The element of black magic is it satisfies one of our most undernourished psychological needs.

Also **u:nder-nou·rishment.**

1920 H. G. Wells *Outl. Hist.* 575/2 Scores of millions were suffering and enfeebled by undernourishment and misery. 1944 M. Laski *Love on Supertax* v. 62 Never before had he been brought into direct contact with under-nourishment. 1971 *Sci. Amer.* Oct. 20/1 A child who has suffered undernourishment very early and for an appreciable length of time will never reach normal size for his age.

u:nder-occupa·tion. [Under-[1] 10 b.] The occupation of a dwelling or dwellings by fewer people than had been planned; *spec.* applied to local authority accommodation that is considered to be underused.

1961 *Guardian* 6 Nov. 2/4 There is a good deal of under-occupation. 1966 [see *one-person* s.v. *ONE B. IX. 32 a]. 1977 *Jrnl. R. Soc. Arts* CXXV. 116/2 Parallel with the so-called scandal of empty homes, we also hear of the scandal of under-occupation. It is all wrong, the critics say, for an elderly council tenant to occupy a three bedroom house, when she could be living in a one-room new council flat.

Also **under-o·ccupy** *v. trans.*, **under-o·ccupied** *ppl. a.*

1961 *Guardian* 28 Nov. 8/2 From the point of view of a local authority housing official a house with spare bedrooms is 'under-occupied'. 1966 J. Tunstall *Old & Alone* xiv. 281, 52.3 per cent of old people under-occupying housing were owner-occupiers, against 38.5 per cent who paid rent. 1970 *Times* 23 Mar. 5/1 To find room in this way for the elderly and others needing small homes would release seriously underoccupied large houses for bigger families. *a* 1974 R. Crossman *Diaries* (1975) I. 75 The interesting thing at Leicester..was the admirable way they are trying to deal with the problem of the people who grow elderly on their huge housing estates and then under-occupy their three-room council houses.

under-officered, *a.*[2] (Earlier example.)

1856 *Tracts on Increase of Episcopate in Eng. & Wales* I. 3 In respect to her Bishops, the Church of England is, so to speak, under-officered.

underpaid, *ppl. a.* (Earlier example.)

1817 Ld. Colchester *Diary* 17 Feb. (1861) II. l. 604 This was objectionable, as giving currency to the vulgar opinion that public men..were overpaid, whereas, they were notoriously underpaid.

u·nderpainting. *Art.* [Under-[1] 5 d.] (The application of) a layer of paint subsequently overlaid by another painted surface; a painting underlying a finished work. Cf. *OVER-PAINTING *vbl. sb.*

1866 [see Under-[1] 5 d]. 1905 *Spectator* 25 Feb. 287/2 In much of his painting..Whistler seems to have begun with a cool grey underpainting. 1937 *Discovery* July 212/1 Constable himself did not disdain the use of a warm monochromatic under-painting. 1951 E. Paul *Springtime in Paris* iii. 63 There must have been an underpainting of that tone, because it showed also around the edges of the canvas. 1979 Ld. Clark in *Hist. Today* Nov. 725/1, I had it X-rayed, and there appeared an underpainting that was unquestionably the work of Filippino.

Also **u·nderpaint,** a layer of paint applied before another coat or finish.

1934 *Burlington Mag.* Jan. 19/2 There are indications of a dark grey underpaint. 1976 D. Francis *In Frame* ii. 38 A gloss on the lips..which one couldn't do until the under paint was dry.

u·nderpan. [Under-[1] 5 b.] The protective metal covering fitted beneath the engine, clutch, and transmission of a motor vehicle.

1913 A. C. Clough *Dict. Automobile Terms* 275 *Sod pan*, a protective covering, fitted under the engine..and other parts of an automobile... Syn.: Pan, underpan. 1934 Webster *Under pan*, in automotive vehicles, a protective metal covering fitting under the engine, clutch, and transmission case. 1968 *Jane's Freight Containers 1968–69* 530/1 Hi-tensile pressed steel bottom rail with cross-members pressed into all welded leak proof underpan of great strength. 1977 *Drive* Mar.–Apr. 57/3 Opel's anti-rust measures include..a wax spray in box sections and over the underpan.

u·nderpants, *sb. pl.* [Under-[1] 5 a.] An undergarment covering the lower part of the body (and part of the legs); short knickers, briefs. Cf. Pants *sb. pl.* 1 b.

1931 R. Campbell *Georgiad* I. 15 The living image of a country lover, In woolly underpants, a sort of Faun. 1955 J. Morrison in B. James *Austral. Short Stories* (1963) 144 Without having removed shirt or underpants, he shuffled down between the blankets. 1961 M. Beadle *These Ruins are Inhabited* iii. 37 The Salesman spread out an impressive..array of undershirts, underpants, knee-high wool socks..and red Rugby shirts. 1979 R. Jaffe *Class Reunion* I. xi. 105 Chris yanked off her..slip and bra and underpants.

under-pa·rted, *a. rare.* [Under-[1] 10 a: cf. Underpart *sb.* 2.] Of an actor: cast in an insufficient role. Cf. *UNDERCAST *v.* 4.

1890 G. B. Shaw in *Star* 2 May 2/3 Mr Coffin, being a handsome young man, and considerably under-parted to boot, had an easy time of it. 1898 —— *Our Theatres in Nineties* (1932) III. 290 Miss May Harvey..is almost dangerously underparted. 1984 *Daily Tel.* 14 Aug. 7/8 Only the underparted Andrew Cruickshank as the Abbot, with his hissed cry of 'heresy!' in a superb scene of denunciation, strikes the right balance between malevolence and outrageousness.

u·nderpass. orig. *U.S.* [Under-[1] 5 b.] A (section of) road providing passage beneath another road or a railway; a subway. Cf. *OVERPASS *sb.*

1904 *Springfield* (Mass.) *Weekly Republ.* 16 Sept. 8 The need of an underpass at the union railroad station in this city. 1929 [see *OVERPASS *sb.*]. 1943 *Sun* (Baltimore) 29 Oct. 4/3 The Ponca-Lombard street bypass..is designed to bypass the Eastern underpass. 1959 *Manch. Guardian* 26 June 8/7 A four-lane underpass..would run under Euston Road. 1973 [see *RINGWAY].

underpay, *v.* Add: Also **underpay·ment.**

1848 Mill *Pol. Econ.* II. III. xiii. 87 There was an underpayment to one set of persons, and an overpayment to another. 1920 [see *UNDERCHARGE *sb.*]. 1972 *Accountant* 26 Oct. 507/3 He received eight changes of code number within..five months. He also received a notice alleging £42.67 underpayment of tax.

underperfo·rm, *v.* [Under-[1] 10 a.] **1.** *intr.* To perform in a manner which falls below expectation.

1976 *Business Week* 26 Jan. 71/1 Institutions..underperform over the all-important long term. 1977 *Daily Mail* 17 Mar. 6/1 What's wrong with comprehensive schools..is..not that they fail to teach children properly. .. They just 'underperform'. 1984 *Observer* 26 Feb. 27/2 Our shares have underperformed since 1977. But this time we have performed.

2. *trans.* Of shares, etc.: to perform less well than (the general market).

1975 *Dun's Rev.* June 92/3 A company that fits the criteria..might underperform the market for several years before somebody wants to acquire it. 1979 *Daily Tel.* 18 July 21 The shares of Midland Bank have underperformed in an astonishing way..those of its rivals. 1984 *Amer. Banker* 14 Mar. 13/4 Further doubts could make bank stocks underperform even a downward moving stock market.

Hence **underperfo·rmance; underperfo·rmer.**

1975 *Forbes* 15 Aug. 82/3 Stocks like International Flavors, IBM..have all been underperformers in this bull market. 1976 *Business Week* 29 Mar. 86/2 The Japanese economy shows signs of continued underperformance. 1977 *Daily Mail* 17 Mar. 6/1 What's wrong with comprehensive schools, according to the Ministry inspectors, is 'widespread underperformance'. 1981 *Times* 18 May 18/6 Phillips & Drew has put it [*sc.* a named share] on its list of expected underperformers.

under-petticoat. (Later examples.)

1865 Dickens *Our Mutual Fr.* II. iv. xvi. 295 Why one should go out to dine with one's own daughter..as if one's under-petticoat was a backboard. 1924 R. Macaulay *Orphan Island* xiii. 156 Flora looked with interest at Rosamond's cami-knickers, Rosamond at Flora's under-petticoat of scarlet-dyed cocoa-nut cloth.

underpinning, *vbl. sb.* Add: **2. c.** *spec.* (*U.S. slang*), the legs. Chiefly in *pl.* Cf. Underpinner[2].

1848 E. Bennet *Mike Fink* 9/2 Nothing like long under-pinins fur travel. 1895 *N.Y. Dramatic News* 5 Oct. 6/1 Do cigarette girls at work wear their dresses decollete at the bottom and show their underpinning? 1934 A. Woollcott *While Rome Burns* 48 There he was at last with his underpinnings shot from under him. 1974 R. B. Parker *God save Child* (1975) xiii. 95, I learned Vic's technique for developing 'sinewy and shapely underpinnings'.

under-pitched, *ppl. a.* Add: **2.** *Cricket.* Of a ball: pitched short of a length; = Short *a.* 16.

1927 *Daily Tel.* 14 June 6/1 Allen..began to pull Larwood's slightly under-pitched one—a manoeuvre which showed..that he meant to get close to the ball. 1963 *Times* 14 Feb. 3/1 Knight, it is true, is a disappointing fielder and looks vulnerable to the under-pitched ball.

underplant, *v.* Restrict *rare* to sense a. **b.** For def. read: To plant or cultivate the ground about (a tall plant) *with* smaller ones. (Examples.) Chiefly as **underpla·nted** *pa. pple.* and *ppl. a.*

1891 W. Schlich *Man. Forestry* II. iv. 244 Oak woods should be underplanted when the process of opening out has set in. 1959 *Geogr. Rev.* XLIX. 29 Clumps and spinneys were underplanted with evergreens. 1962 R. Page *Educ. of Gardener* iii. 101 There were a few old trees which Duchêne had underplanted with yews. *Ibid.* xi. 302 The cypresses are underplanted with sheets of the bright gentian blue echium. 1971 *Guardian* 17 Apr. 7/7 We plan..to underplant a bush of *Spiraea arguta*..with the pale lemon tulip. 1980 *Amat. Gardening* 25 Oct. 24/2 Transform a corner into a mini-woodland garden with light foliage trees, such as silver birch, underplanted with small flowered daffodil varieties.

u·nderplanting, *vbl. sb.* [f. prec.] The planting or cultivation of smaller plants in between taller ones; a plant so grown; *spec.* in *Forestry*, the process of growing shade-bearing trees among taller ones which they may eventually replace.

1914 G. Jekyll *Colour in Flower Garden* (ed. 3) ix. 86 [Variegated mint] is one of the prettiest things as an underplanting to anything of white or yellow colour. 1928 R. S. Troup *Silvicultural Systems* xv. 170 Underplanting of the thinned coppice and standards should be carried out with shade-bearers only. 1953 H. L. Edlin *Forester's Handbk.* vii. 108 As a rule the purpose of underplanting is to raise a successor crop, to supersede the present main one, which will gradually be eliminated. 1962 R. Page *Educ. of Gardener* iii. 99 The Shablikine roses share a terrace with an underplanting of agapanthus. 1978 *Country Life* 10 Aug. 367/1 There is an extensive pinetum in which the trees are sufficiently widely spaced to allow generous underplanting.

underplay, *v.* Add: **3.** *Theatr.* To underact (a part); to perform with deliberate restraint. Also *fig.* **a.** *intr.*

1896 G. B. Shaw *Let.* 16 Mar. (1965) I. 612 Waring.. will not exactly fail: he will only underplay, and all the papers will treat him with great politeness. 1962 *Movie* Dec. 10/2 In *Dawn Patrol* we underplayed, dispensing with the emoting and ham-acting. 1975 *New Yorker* 17 Nov. 159/1 Tiant..did not have full command of his breaking stuff and was forced to underplay.

b. *trans.*

1897 [see *ROUTINE *v.*]. 1941 B. Schulberg *What makes Sammy Run?* xi. 208 He came in quietly, underplaying the scene. 1976 *Daily Tel.* 11 Aug. 9/8 Whatever happened..to the amazing gift Mr Williams showed here for underplaying his lines?..In the Hancock world..he let the character..speak.

4. *fig.* To play down the importance of (something); to present less emphatically than usual.

1949 M. Mead *Male & Female* v. 124 So among the Arapesh the six-year-olds are treated gently, their sex underplayed. 1965 *New Statesman* 16 Apr. 599/3 They slowly start approaching the peasants, buying provisions at high prices and..underplaying their communist identity. 1973 C. Bonington *Next Horizon* iii. 60 The others had tried to underplay my narrow escape, but she had heard them talk about it amongst themselves. 1980 *Amer. Speech* LV. 120 Scargill's book underplays, to some extent, the influence of America on Canadian English. 1983 R. Scruton *Aesthetic Imagination* ii. 15 The point of under-playing intention is to insist on the public character of the aesthetic object.

Hence **underplay·ed** *ppl. a.*, **underplay·ing** *vbl. sb.*

1896 G. B. Shaw *Let.* 16 Mar. (1965) I. 612 XYZ's underplaying will not hurt Janet. 1951 T. Sterling *House without Door* iii. 26 It was not much different from any other bar..but it had a kind of underplayed elegance which he liked. 1962 *Listener* 13 Dec. 1025/1 Kenneth More's easy underplaying adapts to the television screen much better than one might have thought. 1973 J. J. McKelvey *Man against Tsetse* ii. 83 Davies' analysis of the controversy stressed the underplayed contributions of Nabarro and Baker.

underpri·vilege. [f. next.] The state of being underprivileged; lack of what are considered the normal amenities of life.

1937 *Nation* 25 Sept. 324/2 The greater part of the book deals with the factors that underlie the spread and control of syphilis, its relation to poverty, underprivilege, [etc.]. 1966 P. Scott *Jewel in Crown* v. 207 He..had so obviously set his heart on lifting himself by his bootstraps from the state of underprivilege into which he had been born. 1976 *Times* 8 Nov. 3/2 The probation service..has a duty to oppose policies which intensify underprivilege.

underpri·vileged, *a.* (and *sb.*) [Under-[1] 10 b.] **1.** Less privileged than others; *spec.* experiencing a standard of living which falls short of an accepted norm, socially disadvantaged. Chiefly applied to persons.

1896 J. Barnes *Princetonian* xxxiii. 380 It was very quiet in the little square that was filled with nurse-maids and children moving about inside the railings—several little underprivileged ones peering in at them from the outside. 1935 Z. N. Hurston *Mules & Men* 18 These people, being usually under-privileged, are the shyest. 1939 C. R. Cooper *Designs in Scarlet* ii. 27 Don't try to say that the problem is one of 'underprivileged districts', and not of such surroundings as yours. 1948 'J. Tey' *Franchise Affair* xiii. 140 He wrote a..letter about her.. pointing out how under-privileged she had been. 1960 M. Spark *Bachelors* viii. 109, I was definitely underprivileged by birth..though not delinquent. 1978 K. Hudson *Jargon of Professions* iv. 96 Writers and speakers on education..refer..to..'underprivileged children'.

2. *absol.* as *sb.* with *the*.

1935 A. P. Herbert *What a Word!* iii. 46 She had spent a long time persuading one of the 'underprivileged' to go to hospital to have an operation. 1945 'L. Lewis' *Birthday Murder* (1951) iii. 36 Not infrequently he got drunk, coming home late from some revel with the underprivileged. 1977 M. Drabble *Ice Age* I. 34 They did not read novels, or go to good films, or read the arts pages of newspapers, or listen to music or discuss the problems of the underprivileged.

under-rea·d, *v.* [Under-[1] 10 a.] **a.** *trans.* (See quot. 1934.) **b.** *intr.* Of a gauge, dial, etc.: to show a reading lower than the true one. **c.** *trans.* Of the reading public: to read (an author, a book, etc.) with less than normal frequency or with less than due appreciation.

1934 Webster, *Underread*,..to take a reading below the correct reading of (a test); to read (a temperature, measurement, weight, etc.) as lower than that actually registered. 1975 *Daily Tel.* 7 May 12/4 Under an EEC directive..it would be illegal for speedometers to under-read at all. 1977 *Lancet* 30 Apr. 952/1 The device over-read the systolic pressure by an average of 2.7 mm Hg (S.D. 5.2) and under-read the diastolic pressure by an average of 2.4 mm Hg (S.D. 9.3). 1982 E. Dipple *Iris Murdoch* p. ix, Although the role of critic as evangelist may seem questionable, it strikes me as necessary, given the current tendency to under-read and underestimate Murdoch's work. 1984 *N. & Q.* June 274/2 Some progress is made towards consolidating that concept through isolating characteristic aspects of fiction by under-read, minor, and 'popular' novelists as well as by those more familiar names later accorded their separate chapters.

u·nderreamer. *Oil Industry.* [Under-[1] 5 b.] A drilling bit that can be used to drill a hole below the casing, of sufficient size for the casing to be lowered further. So **u·nderreaming** *vbl. sb.*

1912 Under-reamer [see *ROTARY *a.* 2 c]. 1922 D. T. Day *Handbk. Petroleum Industry* I. 246 Underreaming is accomplished through the medium of an ingeniously designed bit (termed an 'underreamer,' because it reams under the casing). 1939 D. Hager *Fund. Petroleum Industry* viii. 186 If for any reason the hole has been cased at a certain depth and it is necessary to carry the same diameter of hole deeper without removing the casing, an underreamer is used.

under-reco·rd, *v.* [Under-[1] 10 a.] *trans.* **a.** To make too few recordings of (a work or performer). Chiefly *pass.* **b.** To record using too low a signal, so that the sound is obscured by other sounds or by instrument noise. **c.** To record (data, information, etc.) insufficiently or inadequately. So **under-reco·rded** *ppl. a.*, **under-reco·rding** *vbl. sb.*

1958 P. Gammond *Decca Bk. Jazz* xv. 183 There was Teddy Weatherford, a most able but under-recorded musician. 1962 *John o' London's* 18 Jan. 66/3 Nobody can complain that Britain's favourite oratorio is under-recorded. 1968 P. Oliver *Screening Blues* iii. 122 Though Sunnyland Slim's stentorian voice was under-recorded, his shouting style of blues singing could be heard outlining the vocal line with its exaggerated rise. 1971 H. Wilson *Labour Govt.* (1974) xxxv. 911 It became clear that for four or five years our national export figures had been increasingly under-recorded. *Ibid.* 913 In the autumn of 1970 the statisticians had discovered that the 1970 figures were still subject to substantial under-recording. 1977 *Gramophone* Mar. 1481/2 One can under-record to a peak of a few dB below zero on the VU meter. 1980 *Times Lit. Suppl.* 31 Oct. 1232/4 Such crucial problems as how to compensate for the under-recording of the activities of the poorer inhabitants of Halesowen.

under-rehea·rsal. [Under-[1] 10 b.] Insufficient rehearsal of a play, piece of music, etc., for performance. Hence **under-rehearsed** *a.*

1900 G. B. Shaw *Let.* 30 Dec. (1972) II. 215 The second performance has all the flatness of a second night plus the effects of under rehearsal. 1937 *Discovery* Nov. 331/1 The major difficulty with the B.B.C. transmissions is under-rehearsal. 1960 *Times* 24 Nov. 8/5 The performance itself had an air of under-rehearsal. 1963 *Times* 7 Mar. 15/5 Those under-rehearsed in the intrigues of the narratives can suck them out from his obliquities. 1982 J. Sherwood *Shot in Arm* vi. 51 The show was badly under-rehearsed.

under-repo·rt, *v.* [Under-[1] 10 a.] To fail to report (income, events, information, etc.) fully. So **under-repo·rted** *ppl. a.*, **under-repo·rting** *vbl. sb.*

1959 *Washington Post* 12 Nov. A2/1 Harrington..made public Tuesday night that new steps are needed to deal with the 'shocking' problem of under-reporting of income. *Ibid.*, Americans 'under-reported' their 1957 income by about $24 billion. 1972 *Science* 26 May 857/2 He found that people underreported both their use and knowledge of contraceptives. 1976 A. Grey *Bulgarian Exclusive* v. 35 Bulgaria's the most under-reported, unnoticed East European country of 'em all. 1980 *Sci. Amer.* Oct. 70/2 The actual total was surely higher, not only because of countries not included in the survey but also because of underreporting in the records of those countries that were included.

underri·de, *v.* [Under-[1] 4 b.] **a.** *trans.* To form the basis on which (something) occurs; cf. Override *v.* 3 a. *rare.*

1956 M. Duggan in C. K. Stead *N.Z. Short Stories* (1966) 84 It underrode everything and was to him no more than suspected.

b. *Geol.* Of a mass of rock, water, etc.: to move underneath (another mass).

1970 *Bull. Geol. Soc. Amer.* LXXXI. 1665/1 The seismically active zone of movement..between the moving oceanic crust and upper mantle, and the relatively stationary material that it underrides. 1977 A. Hallam *Planet Earth* 54/2 Sediment-charged rivers spread material out over the lake floor by under-riding the lake water.

under-ru·ff, *v.* *Bridge.* [Under-[1] 10 a, b.] *intr.* To undertrump. Also as *sb.*

1945 'S. J. Simon' *Why You lose at Bridge* 151 He made a gallant attempt to avert the impending squeeze by under-ruffing with the seven of Hearts. 1960 T. Reese *Play Bridge with Reese* 185 The under-ruff to preserve possibilities is uncommon but worth noting. *Ibid.*, I under-ruff with the 5 of spades. 1964 R. L. Frey *Official Encycl. Bridge* 651/1 In the following deal an underruff was necessary at the third trick, because East could not spare any cards in the side-suits. *Ibid.*, East had a discard problem which he solved by underruffing with the diamond deuce. 1979 *Country Life* 4 Oct. 1127/3 The Rabbit ..underruffed with his Eight. *Ibid.*, Without the under-ruff the contract would have been secure.

underrun, *sb.* Add: **2.** [Under-[1] 10 b: see sense *4 of the vb.] (An instance of) underrunning; the extent to which a programme, project, etc., underruns.

1941 *B.B.C. Gloss. Broadcasting Terms* 34 *Underrun*, extent by which a programme falls short of its allotted time. 1967 A. Battersby *Network Analysis* (ed. 2) xiv. 238 An example is shown in Fig.,14.2, which displays the extent to which the project is ahead of schedule or behind it, referred to respectively as 'over-run' and 'under-run'.

3. [Under-[1] 4 d.] The act of running under something, *spec.* (of a vehicle) under the back of the vehicle in front. Used *attrib.* and in *Comb.*, as **underrun bar, bumper,** etc., a guard attached to the back of a large-wheeled vehicle to prevent other vehicles running underneath.

1969 *Jane's Freight Containers 1968–69* 532/1 *Underrun bar*, at the rear of the chassis a massive bumper bar is provided. 1970 *Daily Tel.* 27 May 11/4 What are needed are under-run bumpers or similar protective structures at the rear of lorries, intended to catch a car that might run into them. 1973 *Care on Road* (RoSPA) Feb. 9/1 Referring to under-run accidents on motorways, the report says: 'There were severe and multiple fractures of skull and face with associated brain damage.' 1974 *Ibid.* Oct. 7/2 Some trailer-manufacturers do fit under-run bumpers. 1980 *Daily Tel.* 24 May 30/4 Under-run accidents are one of the most lethal types of collisions, particularly on motorways.

underrun, *v.* Add: **4.** [Under-[1] 10 a.] *intr.* Of a broadcast programme, item, etc.: to run for less than its allotted time.

1941 *B.B.C. Gloss. Broadcasting Terms* 34 *Underrun*, to fall short of the allotted time (of a programme). 1962 [see *OVERRUN *v.* 10 d].

5. Of a car: to run under a larger vehicle in front. Cf. *UNDERRUN *sb.* 3.

1972 *Care on Road* (RoSPA) Sept. 9/4 A carefully designed bumper is essential at the rear of every truck. It must be large enough to prevent the car under-running.

under-runner. Add: **3.** [Under-[1] 4 a.] A subterranean stream. *N.Z.*

1921 H. Guthrie-Smith *Tutira* v. 34 At last there is created a subterranean stream, or, in shepherd's phrase, an 'under-runner'. 1949 F. Sargeson *I saw in my Dream* II. xiv. 211 It had been one of those under-runners, with the opening all grown over with biddy-bid. 1959 *Mother Earth* Jan. 428 On the down (lower) country, the problem is one of water retention rather than erosion, though there are some bad cases of the latter, consisting mainly of 'under-runners'.

undersa·turated, *ppl. a.* [Under-[1] 10 a.] **a.** Not saturated, falling short of being saturated; *spec.* (of a solution) not containing as much solute as is possible in equilibrium conditions.

1832 *Nat. Philos.* (Libr. Useful Knowl.) II. v. ii. 13/1 In a deficiency of fluid, or in matter under-saturated [with electric fluid]. 1957 G. E. Hutchinson *Treat. Limnol.* I. ix. 592 It appears that the surface of the lake is significantly undersaturated throughout the spring months. 1961 L. Martin *Clinical Endocrinol.* (ed. 3) ii. 56 The arterial blood is chronically under-saturated with oxygen. 1973 *Sci. Amer.* Jan. 105/2 When the bases of the clouds are fairly high and there is a thick layer of warm, undersaturated air below them.

b. *Petrol.* Of a rock or magma: containing insufficient free silica (or some other specified oxide) to saturate all the bases present; consisting wholly or partly of undersaturated minerals. Of a mineral: unable to form in the presence of free silica; unsaturated.

1913 S. J. Shand in *Geol. Mag.* Decade V. x. 510 An undersaturated magma (i.e. one which on solidifying would give rise to a partsaturated or unsaturated rock) is capable of entering into chemical combination with the silica of invaded rock masses. 1947, etc. [see *OVERSATURATED *ppl. a.*]. 1965 G. J. Williams *Econ. Geol. N.Z.* xi. 168/2 A very fine-grained phonolite or undersaturated trachyte of high soda content. 1983 D. S. Barker *Igneous Rocks* xi. 259 All xenoliths for which a mantle origin seems likely have been found in silica-undersaturated rocks.

u:ndersatura·tion. [Under-[1] 10 b.] The state of being undersaturated.

1913 S. J. Shand in *Geol. Mag.* Decade V. x. 510 It will also be desirable to distinguish partsaturated rocks.. from wholly unsaturated ones; as a general term to cover both partsaturation and unsaturation we may employ *undersaturation*. 1957 G. E. Hutchinson *Treat. Limnol.* I. ix. 595 Both supersaturations and undersaturations are likely to be recorded at the time of vernal circulation. 1978 *Nature* 19 Oct. 639/1 Magmas with increasing degrees of undersaturation can be produced only by progressively decreased amounts of partial melting of a peridotitic upper mantle if H_2O is the only volatile component present. 1981 *Jrnl. Chem. Physics* LXXIV. 5243/2 The theory predicts evaporation rates at any degree of undersaturation.

underscore, *v.* Add: **b.** *fig.* To point up, to emphasize, to reinforce; = Underline *v.*[2] 1 b.

1891 [in Dict.]. 1939 *Sun* (Baltimore) 17 Apr. 8/2 A look at the gold statistics underscores the fears which are so often expressed on this score. 1952 S. Kauffmann *Philanderer* (1954) iv. 58 He could feel the stupidity of what he was writing being underscored by the stupidity of what was being written in the editorial rooms next door. 1966 D. F. Galouye *Lost Perception* iv. 37 Headquarters would not be caught unawares. And, in the interest of underscoring that point, the International Guard detail had been tripled. 1979 *Nature* 13 Sept. 98/1 The near catastrophe this year at the nuclear plant at Three Mile Island, Pennsylvania, has underscored dangers which could arise even in the absence of war. 1984 *Toronto Star* 28 Mar. A 3/1 The report..underscores the common complaint made by several groups about institutional racism.

underscoring *vbl. sb.* (earlier and later examples).

1751 Richardson *Clarissa* (ed. 3) VII. xcix. 386 You will perhaps, Mr. Walton, wonder at the meaning of the lines drawn under many of the words and sentences (*underscoring* we call it). 1941 [see *PURPLE *a.* 3]. 1983 *N. & Q.* Apr. 100/2 The use of an emboldening effect for titles makes the volume easier on the eye than does the usual method of underscoring.

underscrub. Add: **2.** (Earlier examples.)

1870 S. F. Prentice *Tale of N.Z.* (MS.) 56 To force a passage through lighter underscrub. 1891 [see *FALL *v.* 51 c].

Hence **u·nderscrubbing** *vbl. sb. N.Z.*, the cutting down of underscrub.

1935 N. R. McKenzie *Gael fares Forth* vi. 79 To the younger boys was assigned the task of 'under-scrubbing' or cutting the small trees, shrubs, vines..and other plants which grow so luxuriantly on the forest floor. 1948 D. L. G. Mundy *There's Gold in Hills* ix. 83 The underscrubbing in New Zealand would be done for about 4s. to 5s.

u·nderseal, *sb.* Also **Underseal.** [Under-[1] 4 d.] A waterproofing material for use as a protective coating on the under-parts of motor vehicles; in the *U.S.*, a proprietary term for a particular brand of this.

1948 *Official Gaz.* (U.S. Patent Office) 14 Sept. 307/2 Minnesota Mining & Manufacturing Company... *Underseal* for rust, corrosion and abrasive resistant rubberized coating material. 1964 *Times* 7 Feb. p. iii/4 The wise motorist, for example, protects his new car with Underseal. 1968 Busby & Holtman *Main Line Kill* v. 45 The professionals..bring it in pasted on the underneath of their cars with underseal. Jacks would never think of stripping off underseal looking for junk. 1973 *Sci. Amer.* Dec. 138/3 (Advt.), Underbody, wheel arches, longitudinal members and the underside of the luggage-compartment floor carry a thick coating of underseal.

u·nderseal, *v.* [f. prec.: see Under-[1] 4 a.] *trans.* To coat (the underbody of a vehicle) with waterproof material. Also *absol.* So **u·ndersealed** *ppl. a.*, **u·ndersealing** *vbl. sb.*

1958 *Autocar* 24 Oct. 658/1 Undersealing of cars. 1960 *Farmer & Stockbreeder* 12 Jan. 108/1 Unitary all-steel construction with undersealed bodies gives extra strength and durability. 1961 *Economist* 21 Oct. 252/2 The body is undersealed. 1963 N. Freeling *Gun before Butter* III. 141 He liked to get a car in good condition; he cursed the cheap peeling chrome..the corner-cutting factories that did not underseal. 1976 *Horse & Hound* 3 Dec. 66/1 (Advt.), These ex-Army Land-Rovers are undersealed from new and have heavy duty chassis, springs and suspension. 1980 *Motoring Which?* July 388/2 '*Rustproofed*', '*undersealed*' and 'Ziebarted' were the terms cropping up most frequently.

under-secretary. Add: *parliamentary under-secretary* (*of State*), a member in a department of State, ranking below a minister: see *PARLIAMENTARY *a.* 1; *permanent under-secretary*: see *PERMANENT *a.* 1 d; also used for various other ranks in the administrative civil service, esp. as a title next below *deputy secretary*.

1904 *Rep. War Office Committee* in *Parl. Papers* 9 (Cd. 1932) VIII. 101 The Council should consist of seven members—four military and three civil—with the Permanent Under-Secretary as Secretary. 1917 G. Bell *Let.* 20 July (1927) II. xvi. 420, I wish you would go and see Sir A. Hirtzel, the Permanent Under-Secretary. 1959 C. Dixon *Civil Service* v. 47 The Administrative Class is the top Class of the Civil Service. Direct entrants come in as Assistant Principals and then proceed by promotion to Principal; the next grades are Assistant Secretary and Under-Secretary (or Assistant Under-Secretary in some departments). 1980 Kellner & Crowther-Hunt *Civil Servants* vii. 167 By the time a civil servant is ripe to join the highest ranks of the Service—Under Secretary, Deputy Secretary and Permanent Secretary—he needs to have acquired certain qualities. *Ibid.* 170 Where Assistant Secretaries handle the details of specific issues, Under Secretaries supervise wide areas of policy.

under-se·xed, *a.* [Under-[1] 10 a.] Having a lesser degree of sexual desire than the average; lacking in sexual feelings or desire, uninterested in sexual gratification.

1931 J. S. Huxley *What dare I Think?* ii. 61 Grafts and extracts of the reproductive glands have been used successfully in under-sexed cases. 1947 'N. Blake' *Minute for Murder* ix. 207 Alice simply wouldn't understand the feeling of all-for-love. She's madly civilised. And she's under-sexed. 1965 *Mod. Law Rev.* XXVIII. v. 606 To use the words of the learned judge, 'the husband was undersexed'. 1972 H. J. Eysenck *Psychology is about People* ii. 88 The introvert..is slightly 'under-sexed', but again not unhealthily so.

undershoot, *v.* Add: **1.** *fig.* Also *spec.* of financial performance.

1977 *Daily Tel.* 10 Mar. 21 Evidence that the public sector borrowing requirement is still firmly under control with only a month of the current financial year to go has added to signs that the money supply is also going to undershoot even the lower end of the target range. 1982 *Times* 23 Apr. 21/3 The public sector borrowing requirement in the financial year just ended undershot the Government's original estimate.

2. *trans.* and *intr.* Of an aircraft or pilot: to fail to reach (a designated landing-point) while attempting to land. Cf. *OVERSHOOT *v.* 1 d.

1918 W. G. McMinnies *Pract. Flying* v. 79 If he sees that he is going to undershoot, it is a good plan to make another circuit. 1938 *Sun* (Baltimore) 8 Jan. 3/2 He undershot storm-swept Newark airport. 1947 A. C. Douglas *Gliding & Adv. Soaring* 68 If you are still high, wait a bit before turning round into wind, but it is always better to come in slightly high and then do gentle S turns into wind, than undershoot. 1969 *Gloss. Aeronaut. & Astronaut. Terms* (B.S.I.) II. 4 *Undershoot*, to land, or to follow an approach path which would cause an aircraft to land, short of the intended area.

Hence **undershoo·ting** *vbl. sb.*

1928 B. Studley *Pract. Flight Training* xxiv. 293 Undershooting is always to be avoided. 1982 *Age* (Melbourne) 3/1 The Treasurer..also outlined 'savings' of $361 million made by the Government on earlier appropriation bills..because of normal undershooting of Budget outlay targets.

u·ndershoot, *sb.* [f. the vb.: cf. *OVERSHOOT *sb.*] The action or result of the vb.; *spec.* in *Electronics*, a small variation in signal immediately before, and in the opposite direction to, a sudden (larger) change.

1934 in Webster. 1938 [see *power approach* s.v. *POWER *sb.*[1] 18]. 1956 Amos & Birkinshaw *Television Engin.* II. i. 26 (*caption*) Waveform of a pulse with undershoot of the leading and trailing edges. 1959 *Listener* 26 Feb. 371/1 The score now stands at two undershoots by the United States to one powerful drive by the Russians. 1969 [see *OVERSHOOT *sb.*]. 1978 *Nature* 12 Oct. 550/1 One of them is a progressive rise in time of the overshoot and undershoot (for spikes following the first) amplitudes with nearly constant firing level. 1982 *Times* 23 Apr. 23/7 When one allows for the adverse impact on Government resources of the civil servants' dispute, the undershoot is larger still.

underside. Add: **2.** *Comb.* **underside-couching,** a form of couching (Couch *v.* 4 b) in which the couched thread is drawn through the fabric to the underside by each of the couching stitches (cf. *surface couching* s.v. *SURFACE *sb.* 6 d); also **underside-couched** *a.*

1936 *Burlington Mag.* Oct. 182/1 Underside-couching is used for the metal threads, and also for some of the silk ones. *Ibid.* 187/1 The heraldic shields..are underside-couched throughout. 1964 tr. *A. Geijer's Textile Treasures Uppsala Cathedral* 24 Embroidery of multicoloured silk, in split stitch, and gold in underside couching forming a herringbone pattern.

undersle·pt, *ppl. a.* [Under-[1] 10 a: cf. *undersleep* vb.] Having had insufficient sleep; suffering from or characterized by lack of sleep.

1943 *Our Towns* (Women's Group on Public Welfare) ii. 23 It has become a commonplace to say that the town child is underslept. 1966 K. Amis *Anti-Death League* III. 340 Lucy took an illustrated magazine to Churchill's bedside, but she too was underslept, and in a few minutes she nodded off. 1981 P. Shea *Voices & Sound of Drums* ii. 14 A neglected, underslept appearance.

u·nderslung, *ppl. a.* [Under-[1] 4 a.] **a.** *Mech.* That is slung under (another part, etc.), or from the under part of (something).

1909 in Webster. 1920 *Sci. Amer.* 30 Oct. 453/1 Those who have had to do with the handling of lumber have..discovered that..shapes long and narrow..are frequently best manipulated by the aid of an underslung carriage. 1924 S. R. Roget *Dict. Electr. Terms* 282/1 Underslung Monorail Railway. See Suspended Railway. 1932 S. C. H. Davis *Motor Racing* v. 74 The whole of the front axle had parted from its underslung spring on one side. 1951 *Engineering* 14 Sept. 329/1 A marine Diesel engine has an under-slung crankshaft. 1968 *Gloss. Terms Materials Handling* (*B.S.I.*) iv. 8 *Under-slung crane*, a crane in which the travelling end carriage wheels run on the lower flanges of the gantry beams. 1982 *Daily Tel.* 24 May 5/6 Helicopters..hover to lift underslung loads.

b. In various *transf.* and *fig.* uses.

1931 'D. Stiff' *Milk & Honey Route* 216 *To get underslung*, means to ride under a train and have the *shacks* [brakemen] throw things at you, or to have them drag a piece of iron on a string under the car so that, bounding up and down, it will punish you plenty. 1931 F. L. Allen *Only Yesterday* vi. 130 Charles G. Dawes..entranced the newspaper-reading public with his picturesque language, his underslung pipe, and his broom-waving histrionics. 1940 *Amer. Speech* XV. 248 Though his figure is somewhat *underslung*, he is by way of being a sportsman. 1978 S. Sheldon *Bloodline* xii. 163 She had a tiny potbelly, and her derrière seemed—this was the only word Elizabeth could think of—underslung.

undersow, *v.* Add: **2.** [Under-[1] 4 a.] **a.** To sow (a later-growing crop) on land already seeded with another crop. **b.** To sow land already seeded with (one crop) *with* a second, later-growing crop. Also *absol.*

1950 *N.Z. Jrnl. Agric.* Aug. 145/1 In arable farming districts many spring-sown pastures are established with a cover crop, usually by undersowing the grass mixture on land which is carrying an autumn- or winter-sown cereal. 1951 P. Oyler *Feeding Ourselves* vii. 65 If the wheat has not been under-sown with ryegrass and clover ..the stubble will either be scarified..or it will be shallow ploughed. 1960 *Farmer & Stockbreeder* 5 Jan. 102/3 Also include (or undersow) 15lb per acre Westerwolths ryegrass. 1963 *Times* 10 June 7/2 High Mowthorpe..has grown barley for nine successive years, comparing the effects of giving additional nitrogen, undersowing with trefoil and ploughing in straw. 1981 *Daily Tel.* 21 Sept. 10/8 Partridges..lost a major source of food when the practice of undersowing barley with grass was abandoned from the mid-1960s onwards.

So **undersow·ing** *vbl. sb.*, **undersow·n** *ppl. a.*

1943 H. J. Massingham *Men of Earth* viii. 142 There was hardly a crop in the garden that was not undersown to another. 1960 *Farmer & Stockbreeder* 12 Jan. 107/2 What is your opinion of sowing seeds along with corn when undersowing is practised? *Ibid.* 1 Mar. 106/2 (Advt.), Particularly recommended for undersown crops. 1963 *Times* 10 June 7/1 Undersown cereals can be difficult to deal with in a wet year where the corn is slow to ripen. 1972 *Nature* 21 Jan. 135/2 Field experiments..using various types of undersowing on the levels of pest insects and their enemies.

underspe·nd, *v.* Add: **2.** [Under-[1] 10 a.] To spend less than (an estimate or budget).

hair coat like Daphne's—understated, sophisticated, collegiate.

understatement. Add: **b.** The quality of being understated (sense *b) or modest in design.
1967 *Guardian* 21 July 7/4 The kind of subtly simple dress..that 'Vogue' would call understated... Understatement is the thing. 1977 'J. Fraser' *Hearts Ease in Death* xiii. 142 Her simple Hardy Amies dresses, designed with all the understatement..that had become the Queen's dressmaker's hallmark.

u·nderstater. [f. Understate *v.* + -er[1].] One who understates or avoids overstatement.
1957 *New Yorker* 2 Nov. 114/3 Mr. Sinatra..is probably not a natural-born understater. 1977 P. Ustinov *Dear Me* vii. 96 Britain had already supplied Hollywood with a whole battalion of elegant understaters, immaculate actors of the Du Maurier school, who..were able to play anything..without their assumed characters..being allowed to affect their performances.

understeer, *sb.* [Under-[1] 10 b.] A tendency in a motor vehicle to take too wide a turn when made to deviate from the straight.
1936 [see *oversteer *sb.*]. 1957 S. Moss *In Track of Speed* xii. 149, I..watched Castellotte sorting out his Ferrari, the front wheels on full understeer with the inside one off the ground. 1964 *Which?* Apr. 36/2 The *Bedford Romany* had light steering and very little understeer—it was hardly necessary to turn the steering wheel more when going faster round a bend. 1971 *Daily Tel.* 31 Mar. 11/5 A disadvantage of the bigger engine is the heavier steering and even more marked understeer, made worse by the steep roll angles in fast cornering. 1981 G. G. Hilditch *Further Look at Buses* i. 8/2 You make your approach run perhaps changing down..and then wind on a bit of lock. The bus doesn't notice thanks to a wonderful degree of understeer so you add a few more degrees..and the resulting oversteer runs away with you.

understeer, *v.* Add: **2.** [Under-[1] 10 a.] *intr.* To exhibit understeer. Hence **u·ndersteering** *vbl. sb.* and *ppl. a.*
1936 *Proc. Inst. Automobile Engineers* XXX. 730 If a car under-steers, more rudder is required at the higher speeds. *Ibid.*, The under-steering of the parallel front road springs is..due to the lack of cornering power. 1948 R. Dean-Averns *Automobile Chassis Design* ii. 49 Understeering vehicles,..when put into a turn, endeavour to straighten up and require constant *increase* of lock as the speed on the bend is increased. 1966 J. Miles in T. Wisdom *High-Performance Driving* v. 47 If the wheels are pointing exactly round the line of the circle some cars will actually travel in a wider circle. These are under-steering cars. 1970 *Motoring Which?* Apr. 45/1 Both tended to understeer (run wide of the corners). 1972 *Sci. Amer.* Aug. 20/2 Of the three characteristics understeering is the most desirable.

u·nderstorey. Also -story. [Under-[1] 5 d.] The (layer of) vegetation growing beneath the level of the tallest trees in a forest. Cf. Underwood 1. Also *attrib.*
1945 *Ecology* XXVI. 280/1 The setback in growth for the dominant trees..may occur because of too dense a hardwood understory. 1958 *New Biol.* XXV. 62 The removal of the dense shrub understorey of rhododendron and laurel from another catchment increased the yield of water. 1970 *New Scientist* 30 Apr. 212/1 The total removal of herb and scrub understory removes a habitat diversity which makes young conifer plantations so attractive to wildlife. 1975 W. Condry *Pathway to Wild* xi. 191 A woodland bird reserve needs a wealth of understorey shrubs for birds to nest in. 1983 M. Gee *Sole Survivor* xxi. 224 Although the sun spreads a watery light in the understoreys of the bush the cold stays on.

understudy, *v.* Add: **3.** *intr.* To act as an understudy.
1909 in Webster. 1939 W. Fortescue *There's Rosemary* x. 74 He was very kind and encouraging, but assured me that I should only waste my time in London, understudying and playing small parts. 1962 J. McCabe *Mr. Laurel & Mr. Hardy* i. 26 He went on as a single again..and understudied for *Home from the Honeymoon*.

undertaking, *vbl. sb.* **1. d.** (Earlier example.)
1843 Dickens *Mart. Chuz.* (1844) v. 52 There's other businesses. Undertaking, now. That's gloomy.

underthru·st, *v. Geol.* [Under-[1] 4 b.] *trans.* To be forced underneath. Chiefly as **underthru·st** *ppl. a.*, **u·nderthrusting** *vbl. sb.*
1893 E. A. Smith in *Amer. Jrnl. Sci.* CXLV. 306, I..suggest the following terms, *under fault, faulted under fold, underthrust fault,* as applicable to this type of structure. 1908 H. B. C. Sollas tr. *Suess's Face of Earth* III. ix. 396 Holmquist concludes that, owing to the continuous subsidence of the eastern part (i.e. of the shield or foreland) the overthrusting practically becomes an underthrusting. 1931 *Geol. Mag.* LXVIII. 39 The front of the underthrust mass will be concave towards the direction of underthrusting. 1965 G. J. Williams *Econ. Geol. N.Z.* xix. 350/1 The south-east flank is steep and in places faulted or underthrust. 1971 I. G. Gass et al. *Understanding Earth* xix. 286 The major mountain chains..occur where moving plates collide and one underthrusts the other. 1977 [see *Richter].

underti·p, *v.* [f. Under- 10 a + Tip *v.*[4]] To give an inadequate or insufficient gratuity to (one who has been of service). Hence **underti·pped** *ppl. a.*
1975 'A. Hall' *Mandarin Cypher* iv. 64, I got out and paid and under-tipped..but..he went away happy..with a Hong Kong dollar. 1979 *Guardian* 4 Aug. 19/3 The undertipped taxi-driver who returned his passenger's gratuity. 1984 *New Yorker* 24 Dec. 37/1 My twenty-five-year-old daughter undertipped the airline porter.

undertone, *sb.* **1. a.** (Earlier example.)
1762 D. Garrick *Let.* 24 Jan. (1831) I. 135 It naturally gives him a slow tremulous under-tone of voice.

u·ndertread. [Under- 5 d.] A layer of reinforcement in a rubber tyre.
1968 *Wall St. Jrnl.* (Eastern ed.) 29 Jan. 11 The new tire is made with a rubber impregnated mat of steel wire under the tread. Called a reinforced undertread, the mat contains about 80,000 pieces of wire. 1971 *Flying* Apr. 7/1 (Advt.), A deep undertread adds muscle... Next time you buy tires, ask for Goodyear's Flight Custom. 1982 *Coal Age* Apr. 140/2 An undertread or protective cushion of rubber can also be placed on all three tire designs between the tread..and body ply.

undertreat, *v.* Add: Also, to provide with insufficient (medical, etc.) treatment.
1908 H. James *Portrait of a Lady* I. p. xxi, So early was to begin my tendency to overtreat, rather than undertreat..my subject. 1979 [see *underdiagnose* vb. s.v. *underdiagnosis].

u·ndertrial. *India.* [Under-[2].] A person held in custody awaiting trial.
1966 *Times of India* 5 May 9/4 An infuriated group of about 50 undertrials tore down the walls of the judicial lock-up. 1979 V. L. Pandit *Scope of Happiness* xv. 102 The Allahabad District Jail..was used for under-trials or for those prisoners about to be released after completion of their term in other parts of the province.

u·ndertrick. *Bridge.* [Under-[1] 10 b.] A trick required to make up the number of the bid or contract, but not taken. Also *attrib.*
1908 *Laws of Auction Bridge* § 50 When he fails, his adversaries score, above the line, 50 points for each undertrick, *i.e.*, each trick short of the number declared. 1929 M. C. Work *Compl. Contract Bridge* i. 4 Vulnerable..seems appropriate enough when it increases the losses for undertricks. 1952 J. B. Pick *Phoenix Dict. Games* 266 Undertrick points—that is, penalty points when declarer fails to fulfil his contract—are scored by opponents above the line. 1981 Reese & Dormer *Bridge Player's Alphabetical Handbk.* iii. 220 (*heading*) Penalties for undertricks. *Ibid.*, Not vulnerable, 50 a trick. Doubled, 100 for the first undertrick, 200 for each subsequent one.

under-u·se, *sb.* [Under-[1] 10 b.] Insufficient use (of a facility, etc.); use below the optimum level.
1960 *Guardian* 26 Nov. 2/7 There might, in some places, be a considerable 'under-use' of [parking] meters. 1971 *Country Life* 18 Feb. 348/2 Under-use leads to poor maintenance of roofs, gutters and party walls. 1975 *Physics Bull.* Sept. 411/3 A current wastage of communication in research (due to underuse of existing services) of the order of 9%.

under-u·se, *v.* [Under-[1] 10 a.] *trans.* To make insufficient use of (a facility, etc.); to use (something) below the optimum level. So **under-u·sed** *ppl. a.*
1960 *Farmer & Stockbreeder* 8 Mar. 51/2 Farming can have no overwhelming case to retain all its present land while it is so obvious that some of it is still under-used. 1970 *Guardian* 9 Nov. 10/2 A country can never recover by persistently under-using its resources, as Britain has done for too long. 1976 *Field* 18 Nov. 948/1 The substantial amount of derelict and under-used land which exists in urban areas..indicates that these transfers could be reduced to not more than 29,700 acres per year.

under-u·tilize, *v.* [Under-[1] 10 a.] *trans.* To make use of (equipment, resources, etc.) insufficiently or below the optimum level; to underuse. So **u:nder-utiliza·tion** *sb.*, **under-u·tilized** *ppl. a.*
1954 *Newsweek* 29 Nov. 89/1 Underutilization of capacity in some lines is inevitable for some time. 1958 *Economist* 18 Oct. 204/1 Mr Amory may go on living down to his..nickname of Derick-or-little-by-little this winter (when industrial capacity may be unnecessarily underutilised). 1964 K. G. Lockyer *Introd. Critical Path Anal.* viii. 83 To increase utilization it is common practice to employ the under-utilized resources on another project. 1977 *Lancet* 2 July 26/2 The cause of the lactic acidosis is almost certainly both underutilisation of lactic acid as well as overproduction. 1978 *N.Y. Times* 30 Mar. A 15/1 The small-boat mariners have turned to the 'junk fish', or underutilized species like perch and pollock, while waiting out the 12-day Federal ban on cod.

underway·, *adv.* [The phr. *under way* s.v. Way *sb.*[1] 38 taken as one word.] **a.** *Naut.* Of a vessel: under way; having begun to move through the water.
1934 W. Nelson *Seaplane Design* vi. 67 The depth of the step affects..the moment to change trim when underway near the planing speed. 1940 P. V. H. Weems *Marine Navigation* (1941) v. 82 Before getting underway the navigator should determine by azimuth or by a known true bearing of a prominent object whether any error

exists in the compass reading. 1978 *Sunday Mail* (Brisbane) *Boating Suppl.* 26 Nov. 16/6 Technically a vessel is underway when not moored, at anchor or aground.

b. *transf.* Of a process, project, activity, etc.: having begun, in progress, being done or carried out. Of a person: having started doing something.
1935 *Sci. Amer.* Apr., 198/1 Uncle Sam now has underway one of his greatest construction projects. 1950 *Chem. Engin. Progress* XLVI. 112/1 Construction should be underway by 1952. 1955 [see *program, programme *sb.* 2 b]. 1973 M. Amis *Rachel Papers* 23 Once underway, though, Gloria would have been able to detect few noteworthy points of contrast between sexual arousal and rabies. 1976 *New Scientist* 16 Dec. 660/2 Experiments..with an electro-optically q-switched neodymium-glass laser are underway. 1981 W. Boyd *Good Man in Africa* ii. ix. 124 They walked arm-in-arm into the club where the dance was underway.

underwear. (Earlier examples.)
1872 *Picayune* (New Orleans) 2 Apr. 6/1 (Advt.), Ladies' underwear... Best Merino Undershirts, fine finish;..fine French corsets. 1875 *Cassell's Family Mag.* I. 113/2 It will scarcely be necessary to give the address where the patterns of hygiene 'under-wear' are to be obtained. It would be more to the purpose to tell how the so-called 'under-wear' is made and trimmed in Paris.

underweight, *sb.* Add: **2. a.** The condition or quality of being lighter than average for one's height and build.
1935 H. S. Diehl *Textbk. Healthful Living* x. 143 Both underweight and overweight are handicaps to employment,..for women overweight is a greater handicap than underweight. 1952 R. Macaulay *Let.* 16 May (1961) 313, I suspect that thinness and under-weight may be a physical quality that you and I share.
b. [Absol. use of adj.] An underweight person.
1910 A. Bryce *Laws of Life & Health* vii. 247 It is quite a common thing for underweights to live to eighty, ninety, or even a hundred. 1935 H. Roberts *Everyman in Health & Sickness* ii. iv. 152 The longest lived are found among the underweights.

underwei·ght, *a.* Also under-weight. [Under-[2].] **a.** Not sufficiently heavy, lacking in weight; *spec.* of a person: lighter than average for one's height and build; too thin.
1899 [see *over-weight *a.*]. 1925 C. E. Turner *Personal & Community Health* ii. 53 The underfed person drags himself about—underweight, pale,..and pessimistic. 1971 'D. Halliday' *Dolly & Doctor Bird* i. 5 He was a tall, underweight man in his early thirties. 1979 *Irish Times* 28 Sept. 3/6 Two British boats had been found to be underweight but no action was taken.
b. *fig.*
1960 T. Reese *Play Bridge with Reese* 15 Partner is sufficiently pleased with the result not to notice that I was under-weight for my medium no-trump. 1977 *Time* 26 Sept. 42/1 The plot of *Annie Hall* has the two underweight egos twine together, rose and briar. 1985 *Times* 11 Jan. 17/6 With many institutional investors still underweight, the price is set to move up.

underwhe·lm, *v. joc.* [f. Under-[1] 10 a, after Overwhelm *v.*] *trans.* To leave unimpressed, to arouse little or no interest in. Chiefly as **underwhe·lmed** *pa. pple.* and *ppl. a.*, and **underwhe·lming** *ppl. a.*
1956 T. K. Quinn *Giant Corporations* viii. 61 He wrote..commending the action of one of the giant corporations for a..price reduction at a time when prices were rising. I was underwhelmed, and investigated. 1968 *Punch* 17 July 81/2, I agree that the wretched parents are not to be penalised till 1969 and I'm sure they are under-whelmed with gratitude. 1970 *Daily Colonist* (Victoria, B.C.) 3 July 47/1 Victoria Fair got off to a quite underwhelming start..with a concert in the University of Victoria gym. 1972 *Times Lit. Suppl.* 29 Dec. 1569/4 Both the prose and the play are underwhelming. 1978 *Ottawa Citizen* 2 Mar. 3/3 The Sparks Street post office ran out of applications..but a survey of other post offices ..showed the public was generally underwhelmed. 1984 *Observer* 15 Jan. 9/3 He was..fluent in speech and crashingly dull. If there was an opportunity to be underwhelming, he unfailingly seized it.

underwing. Add: **1. b.** The part of a bird's body concealed when the wings are folded.
1956 G. Durrell *Drunken Forest* vi. 110 Two jacanas bathing, their underwings flashing buttercup yellow. 1977 C. McCullough *Thorn Birds* iv. 75 Big pale-grey parrots with brilliant purplish-pink breasts, underwings and heads.
3. Also, situated or occurring beneath the wing of an aeroplane.
1946 Taylor & Allward *Spitfire* 27 (*caption*) The Spitfire's eye—the underwing camera installation. 1947 *Shell Aviation News* No. 106 23/1 Underwing or integral fuelling of aircraft is receiving a great deal of attention in the United States and Great Britain. 1949 *Gloss. Aeronaut. Terms* (B.S.I.) ii. 14 *Under-wing radiator,* a radiator fitted below a wing. 1979 J. Garnett *Backfire is Hostile* xiii. 143 Aircraft are..carrying underwing stores.

underwooded, *a.* (Earlier example.)
1811 [see *oak barren* s.v. *oak 9].

u·nderwool. [Under-[1] 5 a, d.] **1.** Wool used to make underwear; woollen undergarments. *rare*[−1].
1910 W. de la Mare *Private View* (1953) 223 The typical Englishman..a scorner of great-coats and underwool.
2. A layer of wool lying beneath the outer layer on an animal.
1939 J. S. Huxley in *Geogr. Mag.* Dec. 78/2 They [*sc.* Soay sheep] have..the same general colouration..with dense underwool. 1965, 1968 [see *qiviut].

underwork, *v.* **2. c.** (Earlier example.)
1869 H. Sidgwick *Let.* 8 Mar. in A. & E. M. Sidgwick *Henry Sidgwick* (1906) 191, I am rather underworking.

underworld. Add: **4. a.** (Later *fig.* examples.)
1903 J. London in *Ainslie's Mag.* Oct. 76/1 And without a word, when his *wanderlust* gripped him, he was off and away into that great mysterious underworld he called 'The Road'. 1913 C. J. Hogarth (*title*) Dostoyevsky's Letters from the underworld. 1929 *Amer. Speech* IV. 337 The following word-list..does..record representative words and phrases commonly used by 'knights of the road', 'migratory workers', and denizens of the so-called 'underworld'. 1972 F. Fitzgerald *Fire in Lake* iii. 126 They managed to create an underworld of warlords, secret societies, and bandit groups.
b. *spec.* The world of criminals or of organized crime (usu. with *the*); hence, the inhabitants of this region.
1900 *McClure's Mag.* Aug. 356 (*heading*) True stories from the Underworld. *Ibid.*, Their life amongst them [*sc.* the criminal classes] is not to break laws, but to understand as thoroughly as possible the motives and methods of that great part of the community which they describe as 'The Under-World'. 1903 'J. Flynt' *Rise of R. Cloud* iv. 136 Susan was the accepted Queen of the local Under World. 1926 *Westm. Gaz.* 22 Mar., Four of the most dangerous women in London's underworld began long terms of imprisonment during the weekend. 1956 H. Kurnitz *Invasion of Privacy* xxiii. 146 Remember the code of the underworld and what happens to a squealer. 1977 *Time* 8 Aug. 16/2 He..was presumably executed by the underworld. 1981 M. Moorcock *Byzantium Endures* ix. 235 Through an acquaintance in the Podol underworld, I had two copies of my passport printed, complete with photographs.
c. The slang of the criminal underworld. *rare.*
1927 *Vanity Fair* (N.Y.) Nov. 132 'Taking him for a ride' is underworld for enticing a person to death.
5. *attrib.* and as *adj.* (esp. in sense *4 b).
1929 D. H. Lawrence *Pornography & Obscenity* 12 Genuine pornography is almost always underworld. 1955 D. W. Maurer in *Publ. Amer. Dial. Soc.* xxiv. 12 Perhaps the fact that underworld areas have not been traditionally considered 'respectable' for academic research has discouraged some investigators. 1977 *Time* 4 July 8/2 Widely believed to have underworld as well as high society connections, Riachi was found murdered in his apartment.
Hence **u·nderwo:rldling,** a member of an underworld.
1928 *Tablet* 21 Jan. 89/1 One of the points on which Protestant Underworldlings have agreed to blacken the Church. 1962 N. Marsh *Hand in Glove* v. 157 'What can I do for you, Super?' Moppett asked him with the slight smile of the film underworldling.

underwrite, *v.*[1] Add: **3.** (Further *fig.* examples.) Also, to support or reinforce (an idea, quality, etc.); to lend support to (a party, etc.).
1938 *Sun* (Baltimore) 5 Sept. 6/3 This sectional purpose in the bill was in effect underwritten by the new Administrator. 1962 *Listener* 15 Mar. 480/1 The fact that Hirst was at the Royal College of Art..at the time of that journalistically stimulated style known as 'New Realism', to some extent underwrites this basic quality of factualness in his work. 1965 *Ibid.* 10 June 851/2 We must not always find ourselves underwriting the regimes of yesterday while our opponents support those of tomorrow. 1978 *Jrnl. R. Soc. Arts* CXXVI. 439/1 Their sponsors (usually governments) underwrote the belief that an exhibition's ultimate social and economic benefits would outweigh its immediate costs.
4. c. (Later example.) Also *transf.*, to guarantee by military or other power.
1964 *Ann. Reg. 1963* 224 A free West Berlin could have its social system underwritten by the United Nations, and foreign troops could remain there 'for a certain period' under the U.N. flag. 1979 T. Benn *Arguments for Socialism* ii. 50 Big business..underwrote the cost of the campaign to keep Britain in the Common Market at the time of the 1975 referendum.

underwriting, *vbl. sb.*[2] *rare*[−1]. [f. Underwrite *v.*[2]] Writing with less than acceptable power or fervour, 'low-key' writing.
1938 O. Sitwell *Trio* 69 These days of 'under-writing', as of 'under-acting'.

undescended, *a.* Add: *spec.* in *Med.*, applied to a testis that has not descended into the scrotum from its fœtal position in the abdominal cavity.
1897 in *Lippincott's Med. Dict.* 1901 *Ann. Surg.* XXXIV. 234 The undescended testicle is an extremely interesting organ. 1908, etc. [see *maldescended *a.*]. 1964 [see *maldescent]. 1980 *Amer. Jrnl. Roentgenol.* CXXXV. 211/2 Until recently, localization of the undescended testicle has remained a clinical urologic problem.

2. (Later examples.)
1928 *Scotsman* 6 Apr. 7/2 Their enthusiasms were reached only by tact and wise consideration... The atmosphere unfrozen; and even the hotel people became polite and gentle. **1958** *Times* 11 Nov. 8/1 Members of his [sc. de Gaulle's] staff point out that he never expected educated Muslim opinion to 'unfreeze' all at once. **1968** K. O'Hara *Bird-Cage* xiii. 99 A small Scotch to his hand, a steak quietly unfreezing in the kitchen. **1981** P. Vansittart *Death of Robin Hood* iii. iii. 143 She slowly unfroze, motioning him to the damson ottoman.

un-French, *a.* (Earlier example.)
1803 *Lett. Miss Riversdale* II. 249 Madame de Sainval.. prides herself much upon being so unfrench as to admit it.

unfunny, *a.* Add: Hence **unfu·nnily** *adv.*; **unfu·nniness.**
1927 *Daily Mirror* 10 Dec. 4/1, I saw quite a lot of Mark Twain, and my chief astonishment was his regular unfunniness. **1958** N. Marsh *Singing in Shrouds* (1959) viii. 160 Could he hit quite such an all-time-low for unfunniness, do you suppose? **1963** V. Nabokov *Gift* iv. 221 A student who unfunnily plays the fool. **1973** *Daily Tel.* 16 Mar. 14/4 The amiable progress began..at Harwich, which Mr Cutforth..unfunnily commended for being pleasant and rather empty. **1980** *Times Lit. Suppl.* 21 Nov. 1342/5 For me, it had all the wearisome unfunniness of back numbers of *Punch* perused in the dentist's waiting-room.

unfussy, *a.* Add: (Earlier example.)
1870 H. Froude *Let.* 12 Aug. in G. Battiscombe *John Keble* (1963) iv. 75 The unfussy way in which he [sc. Keble] goes on, and the complete indifference which he seems to feel to the opinions of people.
Hence **unfu·ssily** *adv.*; **unfu·ssiness.**
1960 *Guardian* 29 Oct. 5/6 The film..becomes what, unfussily, it tries to be: a genuine human document. **1968** D. E. Allen *British Tastes* iv. 95 The Welsh unfussiness about social class. **1977** K. Benton *Red Hen Conspiracy* viii. 50 She handled the little car well, unfussily.

ungardened, (*ppl.*) *a.* (Later examples.)
1928 'Brent of Bin Bin' *Up Country* ix. 139 Shy, ungardened, but industrious and dependable, he's worked his sentence out without moral mishap. **1981** *Times Lit. Suppl.* 8 May 520/1 There were too..the un-gardened gardens, the unapologetic messes, too sodden in winter..for any caller to beat a path to their owners' doors.

ungear, *v.* **1.** (N. Amer. examples.)
1846 T. L. McKenney *Mem.* I. vii. 157 Wading into the stream, we ungeared the obstinate animal, and led him out. **1878** J. H. Beadle *Western Wilds* xv. 237 At 2 p.m. we..ungeared the mules, and crawled under the wagon for shade. **1975** E. Wigginton *Foxfire 3* 235 When I come in at night I'd put m'mules up an' ungear 'em.
2. Also *fig.*
1931 Galsworthy *Maid in Waiting* xxii. 188 He'll almost certainly get up against something now he's back. If he does it will ungear him in no time.

ungenially, *adv.* (Earlier example.)
1814 F. Burney *Let.* 4 Sept. (1978) VII. 457 O drive, as fast as you can, this W[illia]m L[ocke] who has broken so ungenially upon your happiness, from your mind.

unglued, *ppl. a.* (Later and *fig.* examples: in a confused or incoherent state or condition.) Normally preceded by *come* or *become*.
1870 Crown Princess of Prussia *Let.* 15 Feb. in R. Fulford *Your Dear Letter* (1971) 266 All the woodwork comes unglued and the tiles is all cracked and bent. **1871** M. A. von Arnim *Enchanted April* xx. 322 Now that Mrs. Fisher too had at last come unglued,—Rose protested at the expression, and Lotty retorted that it was in Keats. **1971** *Wall St. Jrnl.* 13 Aug. 1/5 Already-negotiated agreements may come unglued. **1980** K. Coppel *Hastings Conspiracy* xxviii. 181 What's gone wrong? You wouldn't be here if something weren't coming unglued.

ungood, *a.* (Later examples.)
1949 'G. Orwell' *Nineteen Eighty-Four* i. 53 If you have a word like 'good', what need is there for a word like 'bad'? 'Ungood' will do just as well—better, because it's an exact opposite, which the other is not. **1964** W. Golding *Spire* viii. 150 Jocelin found the capstone ungood to look at.

ungovernable, *a.* (and *sb.*). Add: Hence also **ungo:vernabi·lity.**
1968 *Economist* 18 May 3/1 [Mr Wilson's] going might merely contribute to Britain's drift towards ungovernability. **1977** *Financial Times* 1 Apr. 22/8 Beyond the immediate question of the invasion perhaps the most serious problem is what has been termed Zaire's fundamental ungovernability. **1983** *Guardian Weekly* 16 Jan. 9/3 Now comes the threat the Germans call 'ungovernability'.

ungracious, *a.* Add: **5. b.** *ungracious living* (opp. *Gracious *a.* 2 c).
1976 *Listener* 4 July 13/2 He had gone straight to a bodge-yer-up builder for a slab of ungracious living. **1976** R. Rendell *Demon in my View* i. 11 The houses were warrens... Ungracious living was evinced by a row of doorbells, seven in an eight-roomed house.

ungraded, *ppl. a.* **1.** (Earlier example.)
1845 S. Judd *Margaret* I. vi. 33 The roads rough, ungraded, and divided by parallel lines of green grass.

ungrammatical, *a.* Add: Hence also **u:ngrammatica·lity; ungrammaticalness** (later example).
1961 *Word* XVII. 6 On the other hand, sentences which are not universally rejected do not show ungrammaticality either total, or to a significant degree. **1967** R. W. Langacker *Language & its Structure* ii. iv. 94 It is important not to confuse ungrammaticalness with excessive complexity that makes a sentence difficult or impossible to use. **1969** D. T. Langendoen *Study of Syntax* ii. 8 Ungrammaticality arises not when there is merely internal contradiction within a linguistic object, but when it is felt that the object possesses some gross deformity in comparison with sentences in the language. **1975** *Language* LI. 579 The two sources of ungrammaticality in written texts are slips of the pen..and deliberate breaches of grammar.

ungra·ssed, (*ppl.*) *a.* [Un-[1] 8, 9.] Not sown with grass.
1934 in Webster. **1947** *Sun* (Baltimore) 26 June 10/3 The visitors saw evidence of erosion in corn fields and former timberland, steep embankments along roadsides and ungrassed drainage ditches. **1974** R. Adams *Shardik* xiii. 339 At a little distance were four or five mounds of new-turned earth, ungrassed and strewn with a few flowers.

ungrate, *v.* **2.** *a.* (Later *nonce* example.)
1922 Joyce *Ulysses* 394 The men of the island, seeing no help was toward as the ungrate women were all of one mind.

un-Grecian, *a.* (Earlier example.)
1799 F. Burney *Let.* 14 Nov. (1973) IV. 359 William there may see Noses to his mind—& if difficult already, make himself 10 times more so with every ungrecian one he sees.

ungroomed, *ppl. a.* (Earlier example.)
1829 G. Griffin *Collegians* I. x. 216 Close behind..on that ungroomed, ungroomed creature..rides the crafty Ulysses of the assemblage.

u·ngrown-up, *ppl. a.* and *sb.* [Un-[1] 8 c, 12.] **A.** *adj.* Not grown-up; immature.
1937 *Mind* XLVI. 515 He may feel anxiety in the face of the infinite threats of his ungrown-up super-ego. **1945** A. L. Rowse *Eng. Spirit* xxiii. 229 There was something curiously 'unadult, ungrown-up' about him. **1960** C. Storr *Marianne & Mark* x. 126 She thought this a very ungrown-up thing to do. **1980** J. Lees-Milne *Harold Nicolson* xi. 201 To some extent he..remained ungrown-up in that his code of social behaviour was what he had imbibed from his..parents and schoolmasters.
B. *sb.* An ungrown-up person. *rare.*
1966 J. Lees-Milne *Diary* 21 Feb. (1983) 21 J. just the same sweet ungrown-up he always will be.

unguardedness. (Earlier example.)
1818 T. Brown *Brighton* III. i. 38 So also does he argue with ability, when unguardedness does not break in upon him.

unguent, *v.* (Later example.)
1918 A. Quiller-Couch *Foe-Farrell* xvi. 273 'I understand,' said I, holding up from my business of unguenting the stoker, who was not badly burnt.

u-nguent, *a. rare.* [f. Unguent *sb.* or *v.*] Of a person: emollient in manner, unctuous.
1931 Belloc *Cranmer* ii. 30 He shrank, withdrew, was suave and unguent.

unhappen, *v.* Add: (Later example.) Also *trans.* To cause not to happen or to have happened.
1975 I. Murdoch *Word Child* 135 All these things did happen. Keeping them secret isn't going to unhappen them. **1980** — *Nuns & Soldiers* iii. 187 It's..so unreal. All this can unhappen. You can unhappen it just by saying we won't speak of it again.

unheaded, *a.* (Later examples, corresponding to Headed *a.* and *pa. pple.* 6.)
1968 Spring *Lett. of Travel* (1920) 253 Even unheaded clippings from them [sc. various newspapers] declared their origin. **1980** J. Jeffries *Dead Man's Bluff* xiv. 189 How did you know which wasn't the letter you wrote on this unheaded paper?.. The letter was written on unheaded notepaper.

unhealthy, *a.* Add: **2. b.** *Mil. slang.* Dangerous.
1915 A. D. Gillespie *Lett. from Flanders* (1916) 266 All this place is 'out of bounds' to the troops, for it must be an unhealthy corner when shells are falling. **1930** E. Raymond *Jesting Army* i. vi. 93 The Gully Ravine..was now 'unhealthy': it was being sprinkled with shrapnel bullets.

unhear, *v.* For † *Obs.*[-1] read *rare* and add later examples.
1864 J. Fraser *Offer made to Gentleman of Qualitie* 182 Many of ws hes done weal worse having condemned the kirk the spouse of Jesus Christ unheard and unheared that he accused hir and praised him self. **1953** Dylan Thomas *Let.* 22 June (1966) 408 The woman next to me was shrouded so I spoke to her all the way more..and more wildly.., and she unheard all my delirium with a smile.

unhearable, *a.* Add: Also *absol.* (Later examples.)
1841 T. Carlyle *Let.* 19 July in T. Wemyss Reid *Life Lord Houghton* (1890) I. vi. 267 Drawing..things unspeakable into things unhearable. **1931** R. Campbell *Georgiad* i. 10 With the Unhearable their ears we'll din.

unheedy, *a.* For † *Obs.* read 'Now *rare*' and add later example.
1919 R. Bridges in W. Carõe *Tom Tower* (1923) 112 Nor 'mong them was a single person..so void of scruple and unheedy.

‖unheimlich (u·nhaimliχ), *a.* [Ger.] Uncanny, weird.
*c***1877** W. James in R. B. Perry *Thought & Character of W. James* (1935) I. xxi. 499 To human nature there is something uncanny, *unheimlich*, in the notion of a universe stripped so stark naked [as pure phenomenalism would have it]. **1900** G. Bell *Let.* 9 Apr. (1927) I. v. 81 But it is very strange—'unheimlich', some silly German said and it's not as silly as it sounds at first. **1945** B. Russell *Let.* 20 Feb. in *Autobiogr.* (1969) III. i. 41 The new ways on the Campus make it strange and unheimlich to me.

unhe·lpful, *a.* Add: Hence also **unhe·lpfully** *adv.*
1981 *Cent. Dict.* **1971** *Daily Tel.* 21 Apr. 12/3 The situations are promising, but the play becomes unhelpfully confusing.

unhistorically, *adv.* (Earlier example.)
1846 Geo. Eliot tr. *Strauss's Life of Jesus* I. ii. i. 310 We might..be led to the supposition that the words for *the remission of sins*..was commonly used in relation to Christian baptism, and was thence transferred unhistorically to that of John.

unholy, *a.* and *sb.* **2.** (Earlier example.)
1842 Dickens *Let.* 24 Dec. (1974) III. 401, I am reminded of my promise to see to the Pantomime, and am come off in an unholy hurry.

unhoodwinked, *ppl. a.* (Later example.)
1904 Kipling *Traffics & Discoveries* 38 Let Zeus adjudge your landward kin,.. But ye the unhoodwinked waves shall test.

unhook, *v.* **1.** *fig.* (Later examples, corresponding to *Hook *v.* 5 b.)
1960 *Guardian* 17 June 22/8 Girls who have been in trouble over drugs have been helped to stay 'unhooked'. **1977** D. Garfield *Record* xxiii. 328 We..made a junkie out of her... I'll just get her unhooked.

unhope. (Later *poet.* example.)
1895–6 Hardy *Poems of Past & Present* (1902) 214 But death will not appal One who, past doubtings all, Waits in unhope.

unhuman, *a.* **3.** (Earlier example.)
1841 W. James *Let.* Sept. (1920) I. 38, I have noticed fleeting shades of expression on her face..*unhuman*, ghoul-like, fiendish-cunning.

unh-unh (ʌ·ʔʌ̃, ə̃ʔə̃), *int.* Expressing negation or denial.
1921 J. Cornish *Provincials* 38 'I guess no one'd buy her.' 'Unh-unh.' Gloom flooded back. **1963** M. McCarthy *Group* xi. 255 'I should think he would step in as a doctor,' Polly said mildly. Gus shook his head. 'Unh-unh,' he said. 'That's what they have to watch out for.' **1977** *New Yorker* 20 June 31/2 'Unh-unh, I don't feel like it.'

unhurrying, *ppl. a.* (Later examples.)
1928 *Glasgow Herald* 15 May 5/2 Through all this..blaze of conflict the old Vindictive, still unhurrying, was wailing the lighted waters towards the entrance. **1928** *Daily Express* 25 July 10/2 Others..detect something fine and typical of the national character in the mild and unhurrying dignity of this annual contest. **1972** P. D. James *Unsuitable Job for Woman* iii. 107 Benskin arrived, unhurrying, imperturbable.

unhy·phenated, *ppl. a.* [Un-[1] 8.] **1.** Not joined by a hyphen; not written with a hyphen.
1934 in Webster. **1960** *Amer. Speech* XXXV. 215 The authors seem to be unsure whether some forms occur as nicknames or as generic epithets... Some unhyphenated terms comprising two words have only the first capitalized. **1980** *Washington Post* 25 Apr. (Weekend Section) 4 [Noel Hume's]..double barreled name goes unhyphenated because such things are regulated by law as well as custom in his native Britain.
2. But not employing a hyphenated term (such as *French-Canadian*) in describing one's political or cultural allegiance. Of such allegiance, etc.: not divided. Cf. *Hyphenated *a.* 2. *N. Amer.*
1970 L. B. Pearson *Words & Occasions* 229, I am a Canadian who speaks English. There are millions [of others] who speak French... Others..have an ancestral language which they use. But we are all, or should be, Canadians—and unhyphenated with pride in our nation and its citizenship. **1973** *Saturday Night* (Toronto) Oct. 17/2 Many Canadians..were interpreting his One Canada to be on a par with the call for unhyphenated Canadianism. **1981** *Washington Post* 30 Jan. 415 Unhyphenated Democrats, at least those brave enough to acknowledge their political heritage, have a brand-new patron saint.

unhypocritical, *a.* (Earlier example.)
1854 Geo. Eliot tr. *Feuerbach's Essence Christianity* 310 The unhypocritical, honest acknowledgment of sensual life is the acknowledgment of sensual pleasure.

uni (yū·ni). Also **Uni.** Chiefly *Austral.* and *N.Z.* Colloq. abbrev. of University *sb.* **1.** Also *attrib.* and *Comb.* Cf. *Univ.

1898 *Bulletin* (Sydney) 17 Dec. (Red Page), The only classical idioms I have found..are *rotter*, i.e., an adept in learning anything; and *panem agere*, Sydney Uni. slang for 'doing a loaf'. **1953** *Lincoln* (Nebraska) *Daily News* 28 Feb. 1/7 (*heading*) Uni. men depart to judge debates. **1929** K. S. Prichard *Coonardoo* 165 Stay in Perth, go to the uni, be a lawyer or doctor, or something? **1962** A. Seymour *One Day of Year* 103, I think I might ditch my course. Leave Uni. **1966** 'L. Lane' *ABZ of Scouse* 112 *Uni-type*, a university student. **1975** M. Bradbury *History Man* x. 189 All the girls at the uni, what she calls the uni, in her set talk about separating. **1984** *Metro* (Auckland) Mar. 15/1 The poor fool who sat the exam and got potted has been threatened with having his own accreditation cancelled and may be banned from sitting the exam for another two years—when he planned to go to uni in this year.

uni-. Add: **1. a.** *unicon·sonantal, -segmental*; **unia·lgal** *Bot.* (see quot. 1914); **unicu·spid** *Zool.* = *unicuspidate* s.v. Uni- 1 a; **unifa·cial**, (*b*) *Archæol.*, of a flint tool, etc.: (worked) on one side only; cf. *bifacial *a.*; also *absol.* as *sb.*; **unili·near**, (*b*) of an evolution, a theory, etc.: having a single line of development or progression; **unimo·dular** *Math.*, having a determinant whose value is 1; **univo·cular**, substitute for def.: = *monovular *a.*; (further examples); **unipo·tent** *Med.* and *Biol.*, of a cell: capable of giving rise to only one type of cell or tissue; **univa·llate** *Archæol.*, having a single encircling rampart; cf. *multivallate *a.*; **univa·riant** *Physical Chem.*, of a chemical system: having one degree of freedom (cf. *freedom 10 b); **univa·riate** *Statistics*, involving or having one variate or variable.
1914 G. M. Smith in *Trans. Wisconsin Acad.* XVII. 1173 According to the usage of some authors, a pure culture is one that contains only one algal species; others understand it to be a culture of single algal species that is also free from other organisms... To differentiate between the two I propose the term *unialgal culture* to designate one which contains but a single species of alga, but which may contain other organisms. **1946** E. G. Pringsheim *Pure Cultures of Algae* vi. 79 The separation of the purification process into two stages, the first involving the preparation of unialgal or species-pure cultures, the second that of bacteria-free or absolutely pure cultures. **1979** *Nature* 27 Sept. 290/2 This infective filtrate caused the destruction of cultures of four unialgal strains of *Microcoleus*. **1942** *Amer. Jrnl. Sci.* CCXL. 60 The.. **1967** D. Dirringer *Alphabet* ii. 60 The phonograms were bi-consonantal or uni-consonantal. **1894** *Gould Dict. Med.* 1373/1 Unicuspid. **1948** A. L. Hand *Mammals Eastern Rockies* 54 The skull when viewed from the side appears to hold only 5 unicuspid teeth. **1977** *Lancet* 17 Sept. 610/2 Surgical exploration..revealed a unicuspid aortic valve with a 'horseshoe' appearance. **1951** *S. & Q. Anthropol.* ed. 6) iv. 345 The distinction between tools made on cores and tools made on flakes should be observed, and also that between tools flaked on both faces (so-called bifacials) and those flaked or retouched on one side only (unifacials). **1957** *Jrnl. R. Anthropol. Inst.* Jan. 119 A wide range of choppers and chopping tools, core, flake, bifacial and unifacial, is still in use. **1981** *Science* 4 Sept. 1115/2 Simple unifacial tools. **1939** *Athenæum* 12 Mar. 299 A worldwide unilinear evolution. **1939** *Mind* XLVIII. 369 It is an order in which the thoughts in the chains of reasoning are not linked in a unilinear, but in a 'global' fashion. **1974** tr. *Wertheim's Evolution & Revolution* i. 22 In Soviet Russia [unilinear] evolution..the issue of unilinear evolution also came in for serious discussion. **1953** J. J. Sylvester in *Camb. & Dublin Math. Jrnl.* VII. 57 The linear transformations are supposed to be always taken such that the modulus..is unity; or, as it may be phrased, the transformations are uni-modular. **1973** L. J. Tassie *Physics Elementary Particles* vi. 136 The group SU(3) is the group of all unitary unimodular 3 × 3 matrices. **1948** M. Innes' *Night of Errors* ix. 102 The two men were uncommonly like each other—a most striking family resemblance. But then I suppose they were what are called uniovular twins. **1952** *Ann. Eugen.* XVI. 360 Observations on a group of uniovular twins. **1964** J. Pollitt *Depression & its Treatment* vii. 91 Kallman..showed that 96 per cent of uniovular twins of manic-depressive partners were similarly affected. **1979** G. Bourne *Pregnancy* (rev. ed.) xxi. 448 These babies will be identical, or uniovular twins since they have exactly the same genetic structure and the same chromosomes. **1974** *Brit. Jrnl. Haematol.* XXVI. 604 Stem cells are unipotent..a cell is unipotent and exclusively committed to maturation along the erythroid pathway. **1955** Uni-segmental [see *multi-segmental* s.v. *multi- 1 b]. **1957** *Word* XIII. 83 The data discussed there share with these data the fact of unsegmental modification. **1950** *Archaeol. Jrnl.* CVII. 56 The first univallate enclosure on Eddisbury Hill was preceded by a palisade structure. **1979** L. Laing *Celtic Britain* ii. 36 The simple univallate hillforts were in some cases given further ramparts. **1899** Univariant [see *invariant *a.* b]. **1940** Glasstone *Text-bk. Physical Chem.* vi. 467 When two phases are in equilibrium..the condition must correspond to a point on one of the lines..in Fig. 99; only temperature or pressure need be arbitrarily fixed in order to define the system, and the latter has one degree of freedom, i.e., it is univariant. **1978** P. W. Atkins *Physical Chem.* vii. 181 The system is univariant when two phases are present; there is only one degree of freedom. **1948** *Biometrika* XX. 32 Various writers struggled with the problems that arise when samples are taken from univariate and bivariate populations. **1958** *Brit. Jrnl. Psychol.* XLIX. 451 (*heading*) The influence of univariate selection on factorial analysis of ability. **1973** *Nature* 16 Mar. 210/3 The wealth of mathematical forms with which we can express the frequency or probability distributions of univariate theory.
b. *uniface* (later examples), *-pivot*.
1911 H. M. Hobart *Dict. Electr. Engin.* II. 591/1 (*heading*) Unipivot measuring instrument. *Ibid.*, The chief advantage of the unipivot instrument is that, owing to the fact that the moving system is supported on a single jewel, it may be entirely lifted off when the instrument is out of use. **1940** *Chambers's Techn. Dict.* 878/1 *Unipivot instrument*, an instrument whose moving-coil system is balanced on a single pivot passing through its centre of gravity. **1944** *Antiquity* XVIII. 117 Amid the amazing expressions of the goldsmith's art which Scandanavia..produced between the 14th and mid-5th centuries, perhaps the most interesting is the bracteate, the product unifaced medallions. **1977** *Signature* May/June 34/4 By October 1914 watermarked paper was produced and used. This had a uniface surface, as did the first issue. **1977** *Gramophone* Aug. 377 (Advt.), A unique system of magnetic stabilization on the unipivot bearings.
2. *unidimensionality, -linearism.*
1953 C. E. Bazell *Linguistic Form* 3 Unidimensionality. There is only one dimension of succession. **1975** *Human Relations* XXVIII. 795 Another factor in the unidimensionality of the approach. **1964** P. Worsley in I. L. Horowitz *New Sociol.* 374 It will certainly have to eschew unilinearism and the West-European ethnocentrism of nineteenth-century schemas.

uniaxial, *a.* Add: **3.** Characterized by one axis of alignment or action.
1965 C. R. Atkinson in P. D. Ritchie *Physics of Plastics* v. 250 The cube is subjected to a simple tensile stress normal to the *x* faces,..i.e. a state of simple uniaxial tension. **1969** W. R. R. Park *Plastics Film Technol.* ii. 28 Uniaxial orientation takes place during the drawing of a filament. Here the polymer chains are aligned in one direction. **1982** *Jrnl. de Physique: Lettres* XLIII. 585 The dynamics of an amorphous polystyrene melt is studied..during stress relaxation following a uniaxial deformation.
Hence **unia·xially** *adv.*
1909 in Webster. **1969** *Jrnl. Appl. Physics* XL. 1301 The coercive force for a pair of identical interacting uniaxially anisotropic dipoles of arbitrary bond angle is calculated. **1979** *Nature* 15 Mar. 222/1 The entire assembly would then be hot-pressed, either uniaxially or isostatically, at a temperature of 1,200–1,300°C.

unicameral, *a.* Hence **unica·meralism,** the system of having only one legislative chamber.
1934 *Fortn. Rev.* Oct. 742 With rare unanimity the *civilized world* has rejected the notion of unicameralism, and has decided that two legislative chambers are..necessary to a modern democracy. **1957** A. C. Breckenridge *One House for Two* i. 4 Unicameralism at best simply became a thing of historical interest.

† unicepto·r (yū·ni,septɔɹ). *Immunol. Obs.* [f. Uni- 2 + Receptor.] In Ehrlich's theory of immunization, a receptor having only one combining or haptophoric group of atoms, by which it unites with the immunizing body but not with the complement. Cf. *amboceptor.
1902 *Jrnl. Exper. Med.* VI. 281 According to the manner of action he distinguishes 'unicepters' and 'amboceptors'. **1910** in *Lippincott's New Med. Dict.*

uniclinal: delete entry in Dict. s.v. Uni- 1 a and substitute:
uniclinal (yūniklo·năl), *a. Geol.* [f. Uni- + Gr. κλίν- (see Clino-) + -al.] Characterized by a uniform angle of dip; formed by uniformly dipping strata; = *homoclinal* adj. s.v. *homocline. (In quot. 1884 = *monoclinical *a.* b.)
1846 Dawkin *Geol. Obs. S. Amer.* vii. 197 The masses having these different inclinations, are separated from each other by parallel vertical faults,..often giving rise to separate, parallel, uniclinal ridges. **1884** A. J. Jukes-Browne *Student's Handbk. Physical Geol.* 342 When a set of horizontal beds is suddenly bent up or down into a sharp curve, and then continued horizontally at a higher or lower level, the flexure is called a monoclinal or uniclinal curve. **1932** A. W. Grabau *Textbk. Geol.* I. xxii. 733 Thus a series of uniclinal ridges is produced, the crest line of the original anticline. **1937** Wooldridge & Morgan *Physical Basis Geogr.* xii. 159 This general process of asymmetrical development (of valleys) is called uniclinal shifting. [*Note*] The word 'uniclinal'..is preferable to 'monoclinal', since the asymmetrical development of valleys bears no necessary relation to 'monoclines'. **1970** R. J. Small *Study of Landforms* xi. 402 Glacial streams which happen to be superimposed on to the slopes above a col will undergo 'uniclinal' shifting along the ice-rock contact towards the centre of the col. **1965** *New Oxf. Atlas* 106 Newer sediments..with gentle or moderate uniclinal dips.

unicycle. Add: For *U.S.* read orig. *U.S.* and add later examples. Hence as *v. intr.* and **u·nicycling** *vbl. sb.*
1897 *Wheelist Ann.* 31 It's all been so slow at Slochester Park, That of fun or of frolic I've not seen a spark, Until yesterday week, when we had a new race: That for rocking-horse and unicycle meet. **1978** J. Irving *World according to Garp* vi. 123 There's no *tradition* of bears on unicycles here. *Ibid.* 128 'He unicycled up in the lunatic's left-behind clothes. *Ibid.* 128 Even his unicycle, alas! It was all among unicycling, was irretrievable. **1980** *Radio Times* 23–29 Feb. 10/1 He rejects the 'pure' dumbshow of performers like Marcel Marceau, preferring to mix.. verbal humour, video, the odd spot of unicycling, and mask work.

unideal, *a.* **4.** (Earlier example.)
1838 E. Fitzgerald *Let.* 8 June (1979) 36 The best painter of the unideal Christ is, I think, Rembrandt.

unidentified, *ppl. a.* (Examples of *unidentified flying object*: cf. *UFO.)
1950 *Chambers's Jrnl.* Nov. 668/1 Project Saucer revealed that it had analysed 375 incidents of 'unidentified flying objects'. **1954** 'R. Crane' *Hero's Walk* x. 81 Five unidentified objects were discovered last night..flying..in the direction of Earth. **1966** *New Scientist* 14 Apr. 87/1 Community after community observed unidentified flying objects. There were the inevitable sightings by police. **1976** L. Deighton *Twinkle, twinkle, Little Spy* i. 11 The strictly infra dig. pictures of landings by unidentified flying objects, or what the sci-fi freaks call ufology.

unidire·ctional, *a.* [f. Uni- 1 + Directional *a.*] Having or being motion in one direction; operating or functional in one direction only.
1883, 1894 [in Dict. s.v. Uni- 1]. **1946** *Nature* 9 Nov. 674/2 The simplified apparatus was..actuated by unidirectional air flow. **1955** G. G. Sutton *Sci. of Flight* 36 The effect is to give an additional but unidirectional motion to the molecules of air. **1963** K. Bennett *Technique Sound Studio* iii. 23 Cardioids—sometimes called unidirectional microphones—have a heart-shaped response. **1963** Y. Malkiel in Householder & Saporta *Probl. Lexicogr.* 13 The kernel of Howell's..*Lexicon tetraglotton* (1660) is a unidirectional dictionary providing for the translation of each English entry into French, Italian, and Spanish. **1971** I. G. Cass et al. *Understanding Earth* xiii. 174/2 The uni-directional flow of rivers is complicated by the development of meanders. **1974** *Sci. Amer.* Mar. 26/1 For years psychologists have been concerned with the effectiveness of unidirectional modes of communication such as highway signs, books, lectures and television programs. **1984** *Trans. Philol. Soc.* 6 The impression that [linguistic] evolution is constant, continual and unidirectional is a mirage. **1984** *What Video?* Aug. 15/1 Its electronic viewfinder, unidirectional mike but it is disappointing.
Hence **u:nidirectiona·lity; unidire·ctionally** *adv.*
1958 W. D. Kingery *Ceramic Fabrication Processes* viii. 70/1 A much more suitable method is to apply pressure from all sides instead of unidirectionally. **1960** A. C. Spaulding in Dole & Carneiro *Essays in Science of Culture* 454 Unidirectionality of formal changes through time is simply the idea of cultural evolution. **1968** *Science* 21 June 1365/3 The unidirectionality observation..may be placed in correspondence with certain findings concerning functional neuroanatomy of OKN [sc. optokinetic nystagmus]. **1970** *Proc. I.E.E.E.* Aug. 1233/1 Unidirectionality increases transducer conversion efficiency by 3 dB. **1976** *Nature* 10 June 518/1 Replication..proceeds either unidirectionally or bidirectionally away from this site. **1980** *English World-Wide* I. i. 16 If there was a straightforward conflict at the level of decision-making this lack of unidirectionality can be expected to re-emerge at the implementation stage.

unification. Add: **2.** *Comb.* **Unification Church,** an evangelistic religious and political organization (see quot. 1973) founded in 1954 in Korea by Sun Myung Moon, and subsequently known as a cult religion in the U.S. and elsewhere. Its adherents are sometimes contemptuously called *Moonies*.
1973 *Time* 15 Oct. 129/3 By 1954 he had founded the Holy Spirit Association for the Unification of World Christianity—known more simply as the Unification Church. **1975** *Americana Ann.* 481/1 A..group in Korea, called the Unification Church, claiming Christian origins and suggesting Messianic status for its leader, the Rev. Sun..Moon, engaged in evangelistic work in the United States. **1976** *Daily Colonist* (Victoria, B.C.) 18 Mar. 7/2 The fringe religious groups include..Unification Church, [etc.]. **1979** *Minutes Gen. Synod House of Bishops* 3 May 4 The Archbishop of York made a statement..relating to the Unification Church.
Hence **unifica·tory** *a.*, tending to unify.
*a***1897** W. Wallace *Lect.* (1898) 84 The monistic, if that means the unificatory, instinct is irresistible. **1932** W. T. Stace *Theory of Knowl.* vii. 156 The other chief logical character of unificatory constructions is that they are simply serviceable fictions. **1971** A. Kirk-Greene in J. Spencer *Eng. Lang. W. Afr.* 143 In Nigeria, Hausa..played the same unificatory role until its use was forbidden..during the 1939–45 war.

unified, *ppl. a.* Add: **2.** Special collocation. *unified* (*field*) *theory* (Physics): a field theory that describes two or more of the four interactions (orig. gravitation and electromagnetism) previously described by separate theories.
1935 E. A. Milne *Relativity, Gravitation & World-Structure* i. 13 This view is in violent opposition to the demand for a 'unified field theory', a demand for further geometrical modification of the space used so as to be able to employ these modifications in describing electromagnetic phenomena. **1959** *Listener* 26 Mar. 544/2 The first unified field theory was produced not by Einstein but by Hermann Weyl, in 1918. **1979** *New Scientist* 1 Mar. 663/2 QCD is a particularly favoured theory because it is very similar to the 'field theories' of the other forces in nature, and this would be a great help in formulating 'GUTs', the 'grand unified theories of the strong, weak and electromagnetic forces which currently challenge theorists. **1983** *Nature* 27 Jan. 285/1 Minimal SU(5), one of the simplest of the 'grand unified theories', or GUTs, which links all the non-gravitational forces.

unifilarly (yūnifi·lăli, -fi·lăli), *adv. Biochem.* [f. as Unifilar *a.* + -ly[2].] In a single strand of a DNA duplex.
1974 *Nature* 13 Sept. 156/2 The unifilarly substituted chromatids fluoresced more brightly than the bifilarly substituted sister chromatids. **1980** *European Jrnl. Cell Biol.* XXII. 352 Unifilarly-substituted DNA.

uniflow (yū·ni,flōᵘ), *a.* Also **una-** (now *rare*). [f. Uni- 2 + Flow *sb.*[1]] Involving flow in one direction only; *spec.* applied to: (*a*) a type of reciprocating steam engine in which the steam in the cylinder flows in one direction from inlet(s) to outlet; (*b*) scavenging in an internal-combustion engine in which there is a similar flow of waste gases in the cylinder. Also *ellipt.*, a uniflow steam engine.
1912 tr. *J. Stumpf's Una-Flow Steam-Engine* i. 1 Una-flow engines..may be made with a single expansion stage. **1913** G. F. Gebhardt *Steam Power Plant Engin.* (ed. 4) iv. 379 A 20 × 30, 300 horse-power Nordberg uniflow engine using very wet steam. **1942** L. B. Chapman *Marine Power Plan* (ed. 2) v. 127 The uniflow engine has been widely used for stationary work in both Europe and the United States. **1949** C. W. Chapman *Mod. High-Speed Oil Engines* I. 25 There are three main systems employed: cross scavenging, loop scavenging and uniflow scavenging. **1963** H. W. Dickinson *Short Hist. Steam Engine* (ed. 2) viii. 156 The uniflow engine is a gallant attempt on the part of the reciprocating engine-makers to hold the field against the advance of the turbo-alternator and the electrical drive. **1963** R. R. A. Higham *Handbk. Papermaking* vii. 184 Condensate..This vat is similar in construction to the uniflow type. **1972** G. Watkins *Textile Mill Engine* II. 74 The gearing of the waterwheel was destroyed by grit in a flood, and the uniflow then drove the whole until the mill was closed. **1981** Burghardt & Kingsley *Marine Diesels* iv. 62 Figure 4-8(d), illustrating an opposed piston engine, is another method of uniflow scavenging.

uniform, *sb.* Add: **2. d.** *transf.* The customary dress or mode of appearance characteristic of persons of a certain age, class, or lifestyle.
1930 G. B. Shaw *Apple Cart* i. 23 [*Pointing to his blouse.*] *Boanerges.* The uniform of Labor, your Royal Highness. **1967** *Listener* 17 Aug. 197/3 One day one had one's hair flopping down one's back, short skirts which barely cleared the knee. (Ironically that's the uniform of grown-ups nowadays, isn't it?) **1976** D. Fletcher *Don't whistle 'Macbeth'* 37 The discreet beads, the silver bracelet,..court shoes..were identical with the uniform of thousands of women..of the middle class.
3. a. For † *Obs.*[-1] read *rare* and add later examples.
1900 J. K. Jerome *Three Men on Bummel* ix. 208, I believe there is a heavy fine for joking with any German uniform. **1970** G. Jackson *Let.* 24 Mar. in *Soledad Brother* (1971) 189, If a uniform denied some small request, we would take to the bars.
b. Short for *uniform branch* (see below).
1978 F. Branston *Sergeant Ritchie's Conscience* i. 13 'Spoken to the Chief?' he said.. 'Uniform have done that,' guessed Ritchie.
4. b. *spec.* in the police force, distinguished from the plain-clothes section, esp. in *uniform branch*. (Later examples.)
1938 F. D. Sharpe *Sharpe of Flying Squad* i. 15, I don't wish to detract from the valuable work carried out by the Uniform Section..but..the Flying Squad plays a leading part in this work. **1970** P. Laurie *Scotland Yard* ii. 51 The school also provides instructors to train uniform officers. **1972** *Police Rev.* 17 Nov. 1509/1, I would like to express my gratitude..for the efforts of both the C.I.D. and the uniform branch. **1982** *Observer* 5 Dec. 26/2 P.C. Winslow *Counsellor Heart* ii. 41 Uniform Branch have their complaints of noise.. Late parties.

uniform, *v.* **3.** (Earlier U.S. examples.)
1861 O. W. Norton *Let.* 8 June in *Army Lett.* (1903) 12 We are to be uniformed and equipped immediately. **1888** *Cape Branch* (New Jersey) *News* 7 Apr., In our spirit of imitation do we not go too far when we talk of uniforming the shop girls in the big dry goods store.

uniformed, *a.* Add: *spec.* of police officers. See *uniform *sb.* 4 b. (Later examples.)
1922 *Rep. Tax Cases* 1913–21 VII. 176 Detective officers receive the same rates of pay as uniformed officers. **1945** *Law Rep. King's Bench Div.* 420 When they saw the uniformed officer they hurried to the opposite direction. **1973** 'E. Peters' *City of Gold & Shadows* v. 81 Sergeant Comstock, of the uniformed branch..came from a long line of native fishermen.

uniformize, *v.* Restrict *rare* to sense in Dict. and add: **2.** *Math.* To transform (an equation or expression) so that each variable is expressed as a single-valued function of a new parameter; to parameterize.
1899 *Phil. Trans. R. Soc.* A. CXCII. 1 The only automorphic functions known hitherto which have been applied to uniformise functions whose genus is greater than unity, are [etc.]. **1946** E. T. Bell *Devel. Math.* xxi. 474 The curve $y^2 = x^4 + 1$ is uniformized by $x = \sin t$, $y = \cos t$. **1972** M. Kline *Math. Thought* xxxix. 938 The parametric equations (11) or (12) are said to uniformize the algebraic equation (10).
Hence **u·niformizing** *vbl. sb.* and *ppl. a.*; **u:niformiza·tion.**
1899 *Phil. Trans. R. Soc.* A. CXCII. 3 Comparatively little of the published work on automorphic functions.. has been written in connexion with the uniformisation of algebraic forms. *Ibid.* 2 The analytical connexion between the uniformising variable *t* and the variables *x, z*, of the algebraic form. **1933** E. Sapir in *Encycl. Social Sci.* IX. 160/2 Language acts as a socializing and uniformizing force. **1954** Rosenbloom & Proskauer *Methods Algebraic Geom.* III. xi. 161/2 A necessary and sufficient condition that P be a rational curve is that there exist a set of uniformizing parameters at P. **1972** M. Kline *Math. Thought* xxxix. 938 Clebsch's results on the uniformization of curves of genus 1 by means of elliptic functions of a parameter made it possible to establish for such curves remarkable properties about points of inflection, [etc.].

u-nijunction. *Electronics.* Also **uni-junction.** [Uni- 1 b.] A negative-resistance device consisting of a rectifying *p-n* junction in the middle of a length of semiconducting material that has an ohmic contact at each end, used as a switching element.
1957 *Automatic Control* Feb. 24 The silicon unijunction transistor has characteristics resembling a neon light; it can be used for pulse..and a transistor to yield an oscillator. **1962** Simpson & Richards *Physical Princ. Junction Transistors* viii. 192 A device whose regenerative property is obtained in a somewhat different manner is the uni-junction transistor or double-based diode. **1981** Nashelsky & Boylestad *Devices* viii. 291 Recent interest in the unijunction transistor (UJT) has, like that for the SCR, been increasing at an exponential rate.

unilateral, *a.* Add: **I. 1. c.** Of car-parking: restricted to one side of the street.
1945 *Rep. Watch Comm. Oxf. City Council* 7 June, The Council desired to preserve the effect of the war-time order which prohibited parking in King Edward Street. **1954** *Highway Engin. Terms* (*B.S.I.*) 57 Unilateral waiting (prohibition of), a system under which vehicles are prohibited from waiting on one side of a carriageway. The side may be fixed or alternated. **1959** *Listener* 14 May 841/1 Or, if it is unilateral parking, that you go to the other side next month.
II. 4. d. *unilateral disarmament*, disarmament (in recent use, *spec.* of nuclear weapons) by one state, irrespective of whether others take similar action. Also *unilateral disarmer*.
1929 *Times* 15 Nov. 14/3 Lord Salisbury agreed that unilateral disarmament had probably reached its limits. **1935** C. R. Attlee in *Hansard Commons* 22 May 375, I want to recall to the House what our position is..in regard to the question of defence... We do not stand for unilateral disarmament. **1960** [see *disarmer b]. **1969** Plano & Olton *Internat. Relations Dict.* 237 By demonstrating peaceful intentions rather than merely talking about them, unilateral disarmament theorists believe, one side could put the arms-race cycle into reverse by evoking reciprocation of its disarmament initiatives. **1980** *Observer* 14 Sept. 11 To adopt the unilateral disarmament option would be akin to behaving like a virgin in a brothel. **1980** *Times* 17 Nov. 15/3 Mr Foot..is a unilateral disarmer. **1984** S. Townsend *Growing Pains of A. Mole* 96 Went back to Pandora's and watched the Labour Party Conference vote for unilateral disarmament... If elected the Labour Party would chuck all their nuclear weapons away.

unila·teralism. [-ism.] = *unilaterality* s.v. Unilateral *a.*; *spec.* (*a*) = *unilateral disarmament* s.v. *unilateral *a.* 4 d; (*b*) *U.S.*, the pursuit of a foreign policy without allies or irrespective of their views.
1926 *Public Opinion* 30 Apr. 434/2 We must..surmount national and social unilateralism in the domain of the spirit. **1935** *Punch* 25 Sept. 354/2 Laugh heartily when politicians talk about 'unilateralism'. **1959** *Manch. Guardian* 20 July 6/2 There is..little danger of Labour Party unilateralism. **1962** *Listener* 9 Aug. ..Unilateralism in this sense has come to mean 'no alliances' while unilateralism is more often used to mean an attitude of withdrawal. **1968** N. Nicolson in H. Nicolson *Diaries* III. 385 The policy of unilateral renunciation by Britain of the atomic bomb had been advocated by Bevan... The Labour Party Conference had voted.. for unilateralism. **1979** H. Kissinger *White House Years* xxiv. 1089 From an early hostility to the American alliance with Japan..the Chinese leaders soon came..to view it a guarantee of America's continued interest in the Western Pacific and a rein on Japanese unilateralism. **1984** *New Statesman* 16 Nov. 10/1 After the Labour defeat at the General Election, the installation of Cruise missiles and a seemingly unequivocal Labour commitment to unilateralism, many CND activists are looking at the Labour leadership..to the polls in order to win the political debate on disarmament.

unila·teralist, *sb.* and *a.* [-ist.] **A.** *sb.* One who favours or adopts a policy of unilateral disarmament.
1927 *Daily Tel.* 14 Mar. 9/7 The lack of foresight on the part of the 'Unilateralists'..led..to the..postponement of Germany's entry into the League. **1959** *Guardian* 15 Oct. 10/4 Defeats among unilateralists were matched by defeats among believers in the 'great deterrent'. **1960** [see *multilateralist* adj. and sb. s.v. *multilateral *a.*]. **1980** *Times Lit. Suppl.* 14 Nov. 1277/2 For Bevan was emphatically no fellow-traveller, no isolationist, no unilateralist and certainly no pacifist.
B. *adj.* Of or pertaining to unilateral disarmament, or to unilateralists or their activities.
1959 *Manch. Guardian* 20 July 6/2 There is..little danger that it [sc. the Labour Party conference] will go along the unilateralist road. **1962** *Listener* 6 Sept. 6/1 By insisting that this country should abandon nuclear weapons and any alliance that possesses them, Mr Cousins is taking up a unilateralist position. **1963** *Ann. Reg. 1962* 22 Mr Gaitskell's recapture of Labour Party defence policy from the hands of the Campaign for Nuclear Disarmament..had left him exposed as the chief target for unilateralist retaliation.

u:nilateraliza·tion. *Electronics.* [f. Unilateral *a.* + -ization.] Neutralization (sense *1 d), esp. of resistive as well as reactive feedback. So **unila·teralized** *ppl. a.*
1954 *IRE Trans. Circuit Theory* May 23/1 (*caption*) Lossless unilateralization. *Ibid.*, Suppose..that a number of active devices are to be placed in cascade, each being unilateralized with lossless coupling in order to avoid the problem of wave reflection back through the chain. **1962** [see *neutralization 1 d]. **1974** Harvey & Bohlman *Stereo F.M. Radio Handbk.* v. 99 The winding direction of L₁ is such that signals at the lower end of this coil are in antiphase with the signals at the lower end of L₂. In this way the stage is unilateralized. **1981** E. Oxner in A. D. Evans *Designing with Field-Effect Transistors* iv. 14 The advantages of neutralization and unilateralization are that input and output are effectively isolated from each other and maximum stable gain results.

unili·neal, *a.* [f. Uni- 1 + Lineal *a.*] **1.** *Anthropol.* Of or relating to a kinship system in which group membership, inheritance, etc., are established through either the father's or the mother's lineage. Opp. *multilineal *a.*
1935 A. R. Radcliffe-Brown *Structure & Function Primitive Society* (1952) ii. 36 The problem..of the nature and function of the unilineal transmission of rights. **1947** *Advancem. of Sci.* IV. 219/1 This leads to the emergence of more stable groupings of kin according to a principle of unilineal affiliation. **1957** V. W. Turner *Schism & Continuity in Afr. Soc.* ii. 291 Nor is their polity one consisting of homologous unilineal descent groups. **1963** *Brit. Jrnl. Sociol.* XIV. 44 Both [patrilineal and matrilineal] kinds of lineage system are described as unilineal. **1964** Gould & Kolb *Dict. Social Sci.* 367/1 A dual or double unilineal system, while both parents are severally members of the same group as their child, he has kinsmen..who are not necessarily unilineal with him. **1977** *N. Z. Jrnl. Hist.* XI. 73 The author..stresses the importance of unilineal kinship... **1980** *Times Lit. Suppl.* 8 Aug. 992/1 Positive marriage rules which established alliance relations between unilineal descent groups.
2. Of a theory, progress, etc.: that adheres to one line of development, *spec.* that of uniform stages in the evolution of culture.
1953 G. Clark *Archaeol. & Society* (ed. 3) vi. 173 Those who believed in the unilineal progress of culture. **1962** M. S. Zenkovsky (ed.) D. Chawkins *Readings Sociol. of Lang.* (1968) 298 It can be said that the 100 word list shows no more validity for this type of unilineal study than the 200 word list. **1968** *Encycl. Brit.* IX. 529/2 In the second half of the nineteenth century a belief in unilineal and progressive evolution of mankind through progressive succession and similar stages of development was generally held by anthropologists and was most clearly defined in them. **1980** *Times Lit. Suppl.* 15 Aug. 921/5 Hallpike seems to be a unilineal evolutionist for some kind of cognitive operation (Heal [stages] of developmental psychology), and a permissive relativist in the theory of primitive thought.
Hence **unili·nealism,** adherence to unilineal views; **unili·neally** *adv.*, in a unilineal manner.
1947 *Advancem. of Sci.* IV. 219/2 Factors tending to produce both bilateral and unilineally organized statuses and groups of unilineally related kin. **1957** K. A. Wittfogel *Oriental Despotism* ix. 371 (*caption*) The spread of a 'Marxist-Leninist' neo-unilinealism. **1964** Gould & Kolb *Dict. Social Sci.* 405/2 A belief..in direct lines descent group which has recognized in their case unilineally. **1965** *Times Lit. Suppl.* 16 Mar. 279/1 Soviet Marxism, as it crystallized during the Stalin period, had a clear and sharp outline, with its theory of a single dominant historical highway... This explains the unilinealism.

unilingual, *a.* Add: **2.** *absol.* as *sb.* = Monoglot *sb. rare.*
1968 *Publ. Amer. Dial. Soc.* xxxvi. 9 A monolingual (also called a monoglot or a unilingual) is a person who knows only one language. **1976** *Word* 1971 **1981** *Times* 382 Swedish was usually spoken to unilinguals.

unimer (yū·nimaɹ). *Chem.* [f. Uni- 2 + *-mer.] One of the single molecules (usu. macromolecules) that go to make up a multimeric aggregation. Cf. *multimer.
1967 *Chimia* XXI. 55/1 Methods to determine the association of macromolecules and to calculate the molecular weight of the unimer..are discussed. **1972** [see *multimer]. **1983** *Polymer* XXIV. 1155/1 ..

unimodal (yūnimōᵘ·dăl), *a.* [f. Uni- + Mode *sb.* + -al.] Of a frequency curve or distribution: having one mode (*mode *sb.* 7 c). Of a phenomenon or property: described by such a distribution.
1923 *Biometrika* XIV. 339 The distribution according to size of family..would be represented by one of the unimodal curves of the Pearson types. **1932** J. S. Huxley *Probl. Relative Growth* vii. i. 210 The frequency-curve for female body-length is unimodal. **1975** *Sci. Amer.* Feb. 90/1 The unimodal solar-day rhythms of organisms geared to the 24-hour solar day.
Hence **unimoda·lity,** the property or quality of being unimodal.
1934 in Webster. **1967** *Ann. Math. Statistics* XXXVIII. 1296 (*heading*) A note on the unimodality of distribution functions of class L. **1978** *Nature* 3 Aug. 504/1 The switch from bimodality to unimodality (that is convergence of two peaks into one) is what is expected for X-chromosome inactivation.

u:nimole·cular, *a. Chem.* [f. Uni- + Molecular *a.*] **a.** [ad. F. *unimoléculaire* (J. H. Van 't Hoff *Études Dynam. Chim.* (1884) 8).] In chemical kinetics: having or pertaining to a molecularity of one; involving the fragmentation or internal transformation of a single molecule in the rate-determining step of a reaction (rather than the collision of a pair of molecules); in quot. 1901, first-order (see *order *sb.* 10 f). Cf. *monomolecular *a.* b.
1901 *Jrnl. Chem. Soc.* LXXX. ii. 647 The reaction between ferric salts and metallic iodides is unimolecular for the iron salt and bimolecular for the iodide. **1948** [see *unimolecular s.v. ...]. **1964** R. A. Jackson *Mechanism* i. 5 Unimolecular reactions occur as a result of reorganization of bonds in the starting material, with or without rupture into fragments. **1978** P. W. Atkins *Physical Chem.* xxvi. 863 Most reactions can be broken down into a sequence of steps that involve either a unimolecular reaction, in which a single molecule shakes itself apart or into a new configuration, or a bimolecular reaction.
b. = *monomolecular *a.* c.
1925 *Proc. R. Soc.* A. CIX. 303 Unimolecular films are thus to be anticipated in those cases in which the surface to be adsorbed is not very active, e.g., on diamond. **1945** S. Brunauer *Adsorption of Gases & Vapours* I. i. 6 When a surface can take up only one layer of adsorbed gas, the adsorption is called unimolecular. [*Note*] I. Langmuir (proposing the concept in 1917) used the term 'monomolecular'. However since this is a term of mixed Greek and Latin derivation the use of the term unimolecular is preferable. **1978** K. J. Laidler *Physical Chem. with Biol. Applic.* xi. 462 Ammonia molecules are rather strongly adsorbed, so that a monolayer, or unimolecular layer, may be formed.
Hence **unimole·cularly** *adv.*
1901 *Jrnl. Chem. Soc.* LXXX. ii. 647 Strontium and calcium iodides act unimolecularly. **1938** *Jrnl. Chem. Physics* III. 155 Suppose we have a non-linear molecule of *n* atoms decomposing unimolecularly. **1974** Gill & Willis *Pericyclic Reactions* vi. 203 Both 2,5-dihydrofuran and 1,4-cyclohexadiene are decomposed unimolecularly into hydrogen and respectively furan and benzene.

unimpressively, *adv.* (Earlier example.)
1832 J. S. Mill *Let.* 29 May (1963) I. 30 The opinions I have put forth in these different articles..are expressed so boldly and yet unimpressively that I can scarcely bear to look back upon such poor stuff.

unimprovability. (Earlier example.)
1814 H. C. Robinson *Diary* 28 May (1967) 34 The doctor's favourite opinions of the unimprovability of mankind met with no approbation from the Lambs.

unindicted, *ppl. a.* (Later examples.)
1978 *N.Y. Times* 30 Mar. 213/6 [He] had been named by a grand jury as an unindicted co-conspirator in a series of allegedly illegal break-ins. **1979** *N.Y. Rev. Bks.* 8 Feb. 8/1 The publishers of this chaotically dull..work are unindicted co-conspirators who should be named unindicted co-conspirators in this shabby effort to get those five hundred men off the hook.

unindividualized, *ppl. a.* (Earlier example.)
1844 Poe in *Columbian Lady's & Gentleman's Mag.* Aug. 69/2 Man thus divested would be God—would be unindividualized.

unineme (yū·ni,nīm), *a. Cytology.* [f. Uni- 1 + Gr. νῆμα thread.] Of a chromatid: having (as usual) just one duplex of DNA.
1963 *Proc. Nat. Acad. Sci.* XLIX. 794 Autoradiographic experiments..have yielded results compatible to a polyneme rather than a unineme structure for the chromosome. **1972** *Proc. R. Soc.* B. CLXXXI. 21 Most cytologists refused (and some still refuse) to accept the simplest and most direct deduction from Taylor's classical labelling experiment, namely that a chromatid, prior to replication, is unineme. **1981** *Chromosoma* LXXXIII. i. 4 The unineme concept is supported by genetic data.
Hence **unine·mic** *a.*, in the same sense; **u-ninemy,** the state of having one duplex of DNA per chromatid.
1970 *Cold Spring Harbor Symp. Quantitative Biol.* XXXV. 553/1 The current status of the concept of 'uninemic' is, cytologically, quite incorrect. **1972** *Proc. R. Soc.* B. CLXXXI. 26 Another line of evidence for 'uninemy'..was provided a few years ago by Miller. **1980** *Tsitologiya* XXII. 83 Chromosomes are uninemic, i.e., each chromonema consists of a single DNA molecule (or a single chain of linked DNA molecules) whose ends are located in telomeres. **1981** *Chromosoma* LXXXII. 1 (*heading*) Evidence for the uninemy of eukaryotic chromatids.

uninhi·bited, *ppl. a.* [Un-[1] 8.] Not inhibited; unrestrained.
1880 W. James *Feeling of Effort* 24 The motor idea, uninhibited by remote associations,..discharges by the preappointed mechanism into the right muscles. **1929** B. Russell *Marriage & Morals* x. 111, I think that uninhibited civilised people, whether men or women, are generally polygamous in their instincts. **1949** M. Mead *Male & Female* xii. 255 'Why,' ask the uninhibited American child of 1949, 'does no one ever go to the bathroom in a book?' **1959** P. H. Johnson *Last Resort* xiv. 293 He coughed once or twice, blew his nose with an uninhibited trumpeting. **1971** S. Hill *Strange Meeting* ii. 147 Hilliard stood, pitying them their lack of privacy..yet envying them too, their carefully ordered life and clear uninhibited friendships and enmities. **1980** A. N. Wilson *Healing Art* xvi. 154 In the States..he was being, according to his own brand of uninhibited.
Hence **uninhi·bitedly** *adv.*
1959 *Times* 30 Jan. 7/6 An informal folk concert in which the audience uninhibitedly joins. **1961** L. O Broin *Dublin Castle & 1916 Rising* vi. 47 Birrell was accustomed to express himself most uninhibitedly to Nathan about personalities. **1976** *New Society* 22 Jan. 147/1 If, as a child, things don't go your way and you're miserable, you can make the point by screaming, kicking or flinging yourself uninhibitedly on the floor. Adolescents and adults cannot show their feelings so uninhibitedly.

uninterested, *ppl. a.* Add: **3.** In this sense *disinterested* is increasingly common in informal use, though widely regarded as incorrect: see *disinterested *ppl. a.* 1. (Later examples.)
1906 G. Greene *Dr Fischer* ix. 28, I wouldn't say that he was totally uninterested in his wife. **1972** *London Rev. Bks.* 19 Nov.–2 Dec. 21 Classical historiography was on the whole uninterested in the tragedies of the past. D. Fraser *Alanbrooke* iii. 53 He gave, at that time and later, the impression of being uninterested in people except at an agreeably superficial level, devoted only to practical or professional pursuits.

uninterpreted, *ppl. a.* Add: *spec.* in *Logic*, applied to a calculus, formula, or symbol that has no meaning assigned to the symbol(s) independently of the calculus.
1898 A. N. Whitehead *Universal Algebra* i. 5 When the marks and their rules of arrangement are such as appear likely to receive an interpretation..then the art of manipulating such marks may be called..an uninterpreted calculus. **1967** H. Reichenbach *Elem. Symbolic Logic* 165 The uninterpreted formula, a logical syntax or the usage of symbols equipped with certain structural properties. **1963** H. Wang *Survey Math. Logic* iii. 29 It is more correct to call the sentence-like expressions of Q statement schemata..because the symbols for predicate, functions, and individuals are uninterpreted.

uninterruptible, *a.* (Later example.)
1984 *Sunday Tel.* 22 July 13/8 He no longer discusses. He just gives uninterruptible lectures.

uninvitedly, *adv.* (Earlier example.)
1814 E. W. Hamilton *Diary* 12 May (1972) I. 272 The Government..could not fail to take heed of information tendered to them uninvitedly.

union, *sb.*[1] Add: **I. 1. f.** (Later example.)
1960 C. Winick *Dict. Anthropol.* 554/2 *Ritual union*, sexual intercourse on special occasions, as part of a ceremony.
II. 7. f. *India, Pakistan,* and *Bangladesh.* A local administrative unit comprising several rural villages. Freq. *attrib.*
1885 *Bengal Local Self-Govt. Act* ii. § 38 in *Acts of Lieutenant-Governor of Bengal in Council* (1886) 78 The Lieutenant-Governor may, by order in writing, constitute any village or group of villages into a Union; and may prescribe for such Union the number of members of which the Union Committee shall consist. **1959** *Pakistan Q.* Winter 14/2 Basic Democracy creates institutions at the Union, Tehsil, Divisional and Provincial levels.... At the Union level there will be a Union Council made up of ten elected and five nominated (non-official) members. **1964** R. W. Gable in *Inayatullah District Admin. W. Pakistan* 29 Thus a Basic Democrat [sc. union councillor] will represent a..constituency, the number of voters. **1969** ...union councils for the villages (or a union, or groups of villages); *tehsils* (in West Pakistan) and *thanas* (in East Pakistan); districts; and divisions. In urban areas there are Town Committees and Union Committees in place of Union Councils. **1973** *Bangladesh Times* 19 Jan. 1/8 Elections in 229 unions will be held today (Wednesday) in 18 districts of the country.
g. *Math.* The set that comprises all the elements (and no others) contained in any of two or more given sets; also, the operation of forming such a set.
1941, 1968 [see *intersection 3 b]. **1970** O. Dopping *Computers & Data Processing* i. 28 The formation of the union is analogous to a logical addition. **1973** J. C. Hensey *Mod. Math.* 10 The union of X and Y, written X ∪ Y (also written X+Y and spoken as 'X union Y') comprises all elements which are elements in set X, set Y, or both. **1979** *Computer Jrnl.* XV. 195/1 The advantages of the 'union' operation is that it represents a practically the union of all the characters associated with the previous sets.
V. 11. a. (*b*) *Union Army*.
1866 'F. Kirkland' *Bk. Anecdotes* 376 Colonel Bailey.. [believed] that their capture or destruction would involve the destruction of the Union army. **1931** E. O'Neill *Homecoming* i, in *Mourning becomes Electra* (1932) 27 He wears the uniform of an artillery captain in the Union Army. **1973** A. Dundes *Mother Wit* 362 The Union Army..was ostensibly fighting in part to end slavery.
c. *union baron, -basher, -card, dues, hours, house, negotiator, pay, scale, ticket*; **union-bashing, -busting** (also as *pres. pple.*).
1974 *Socialist Worker* 26 Oct. 11/5 There is a need for links with the other unions in the entertainment industry and beyond, not just Media Conferences where Labour MPs and union barons appear and TV directors nod approvingly. **1977** *Times* 14 Sept. 1/3 Voices in the Conservative Party arguing moderation rather than 'union bashing' in its approach to the closed shop... Sir Keith Joseph had been depicted as an 'enthusiastic union basher'. **1980** *Illustr. London News* Mar. 19/3 It [sc. the Employment Bill] is not revolutionary, it is not union-bashing, but it imposes some legal restraints on secondary strike activity and provides some stimulus to union democracy. **1923** J. London *Valley of Moon* 198 They're all union-bustin' to beat the band. **1947** *Sun* (Baltimore) 26 June 1/7 Union-busting act. **1874** *Rep. Proc. Internat. Typographical Union* N. Amer. 84 The International Typographical Union shall issue..a card, with appropriate design, to be called 'the Union Card'. **1908** H. Haggard *Poison People* iv. 141 There's..an efficient Trade Union... You'll need a Union card. **1977** *Undercurrents* June–July 11/4 Being an anarchist I don't take dole nor can I afford union dues. **1945** Union hours [see *hour *sb.* 4]. **1975** *Socialist Worker* ..Union house [see *scab *sb.* 4 c]. **1977** M. Ford *Let.* 27 Mar. (1965) 257 Doing what I can to persuade any publishers..[to see] that the Press is one of the Union houses. **1964** *Mod. Law Rev.* XXVIII. 111 ..the local union negotiator (shop steward and the like, who is so vital to the operation of collective bargaining) will..usually be an employee. **1924** D. H. Lawrence *Widowing of Mrs Holroyd* iii. 75 Well, if he's badly hurt, there'll be the Union-pay, and sick-pay—we shall manage. **1976** *Honolulu Star-Bull.* 21 Dec. 2-5/6 The artists will be paid union scale, and the Kennedy Center is donating the space, he added. **1891** A. Francis *Oke the Knight* 19, I went to two or three cities, but I couldn't get work, having no union ticket. **1908** Kipling *Let. of Travel* (1920) 167 Is it difficult to get skilled labour in here?.. Then it is hard without the union ticket. [see *lapse *v.* 2.].
12. union catalogue, a catalogue of the combined holdings of several libraries; **union dye,** a dye that will satisfactorily dye the two materials of a union cloth, esp. cotton and wool, at the same time; so **union dyeing** *vbl. sb.*; **union list,** a union catalogue, esp. one giving details of periodical holdings in several libraries; **union purchases,** a method of cargo-handling (see quots.); **union shop** orig. *U.S.*, a shop, factory, trade, etc., in which employees must belong to or join a trade union; a post-entry closed shop (see *post-entry *a.*); **union suit**: restrict † to sense in Dict.; (*b*) chiefly *N. Amer.*, a one-piece under garment reaching to the ankles; = *combination 9; **union-wide** *a.*, that involves or encompasses the whole of a trade union (movement).
1897 *Libr. Jrnl.* Sept. 437 One of the latest examples of co-operative library work is the union catalog of medical literature recently completed in Denver. **1982** *Rogers Dict. Soc. N. Amer.* (1979) 83 Most union catalogs are made up from individual libraries' catalog cards and are created by dispensing with the subject element in the individual library catalogs. **1960** Owen & Standage *Dyeing & Cleaning Textile Fabrics* 36 Union dyes are..of more general adaptation to the renovating of garments than any other class of dyewares. *Ibid.* 93 Full directions are given later for union dyeing. **1954** A. J. Hall *Student's Textbk. Textile Sci.* iv. 194 Union dyeing. **1975** Two variations of piece dyeing are union dyeing and cross dyeing. **1885** *Libr. Jrnl.* X. 130 A union list of periodicals in these libraries. **1968** *Bodl. Libr. Rec.* VIII. 63 Union list of serials in the science area, Oxford. **1978** *Amer. N. & Q.* XVII. 9/1, I am using a union list of 19th century photographically illustrated books in library collections throughout the country. **1938** B. Cunningham *Cargo Handling at Ports* (ed. 2) v. 46 The principle of using the double line with a single hook (on the combined process of lifting and slewing, called in this country [sc. Great Britain] the Union Purchase. **1961** *B.S.I. News* Mar. 13/1 Greater safety for stevedores handling cargo by the union purchase method (the operation of two ships' derricks in tandem). **1904** *McClure's Mag.* Feb. 370/1 Many stores, restaurants, and saloons display placards in their windows advertising the fact that they are 'union shops'. **1937** F. M. Ford *Let.* 27 Mar. (1965) 276, I will..ask the publishers, whether or not the Riverside Press, which prints this book, is or is not a union shop. **1977** *Time* 7 Mar. 28/2 The section permits states to ban the so-called union shop, which requires employees to join unions. **1892** *Ladies' Home Jrnl.* Sept. 29/3 Yet, you will be surprised to know that..you may buy..union suits. **1946** W. Faulkner *Intruder in Dust* vii. 147 The sagging knees, the..union suits..he had already..overalls..and unionsuits. **1967** E. S. Turner *Taking Cure* xii. 187 Smedley..urged the wearing of the union suits for both adults and children. **1937** *Nation* 14 Aug. 165/2 Assuming..union-wide elections. **1981** *N.Y. Times* 24 Mar. 14/4 The operators also gave up a union-wide guarantee.

‖Union Corse (ünion kɔrs). [Fr., lit. 'Corsican union'.] A criminal organization controlled by Corsicans, operating in France and elsewhere. Cf. next.
1963 I. Fleming *On Her Majesty's Secret Service* v. 31 'The Union Corse', more deadly and perhaps even older than the Unione Siciliana, the Mafia.. It controlled most organized crime throughout metropolitan France and her colonies. **1973** *Times* 21 May 14/3 A key figure was a former model..said to have been living with an Union Corse racketeer..before his death in a gun battle with Mexican police. **1978** W. Garner *Möbius Trip* (1979) iii. 99 The drug-running firm featured the Moor's head and eagle of the official Union Corse.

‖Unione Siciliana (ünio·ne sitʃilia·na). Also (*erron.*) Unione Siciliano, etc. [It., lit. 'Sicilian union'.] A criminal organization controlled by Sicilians, operating in Italy and the United States.

Similar but not equivalent to the Mafia. The Union's roots in the U.S. were apparently as a mutual protection society amongst early Sicilian immigrants.
1924 *Chicago Daily Tribune* 13 Nov. 5/5 At [Merlo's] death he was chairman of the board of the Unione Siciliana, with hundreds of thousands of members. **1924** *Chicago Herald Examiner* 13 Nov. 3/1 A guard also was maintained at the wake for Michael Merlo, head of the Unione Siciliano society. **1930** F. D. Pasley *Al Capone* (1931) v. 231 Lombardo fell in his tracks, two dumdum bullets in his brain, the third president of the Unione Sicilione to die by the gun. **1956** C. Cockburn *Discord of Trumpets* xvi. 218 He had recently been elected the President of the Unione Siciliano, a slightly mysterious, partially criminal society, which certainly had its roots in the Mafia. **1970** P. Geddes *November Wind* iv. 35 The Trust..came into existence when the Mafia and *L'Unione Siciliana* were both on the slide.

unionism. Add: **a.** (Further examples relating to trade unions.)
1904 W. T. Mills *Struggle for Existence* (ed. 8) xxxv. 487 The New Unionism of recent years has been continuously enlarging the number of those to be included. **1976** F. Zweig *New Acquisitive Society* i. v. 65 White-collar unionism has grown over the last two or three decades much faster than manual workers' unionism. **1978** S. Brill *Teamsters* vii. 279 The appeal of unionism diminished, especially in the eyes of younger workers.

Unionist, *sb.* and *a.* Add: **A.** *sb.* **1. c.** The name remained the official designation of the alliance of Liberal Unionists and Conservatives until 15 January 1922, when the Irish Free State was established. The title is retained, however, by loyalist parties in N. Ireland (the official *Unionist*, *Ulster*, and *Democratic Unionist* Parties, etc.); also in the full name of the Conservative (and Unionist) Party, and in the Scottish Unionist Association (see quot. 1982 of the adj. *1 b).
1953 *Times* 23 Oct. 8/6 The Unionists were returned to power again in the general election in Northern Ireland. **1964** G. D. M. Block *Source Book of Conservatism* 67 In Ulster and Scotland 'Unionist' has entirely carried the day, but in the Southern Kingdom the 'new term' of Conservative has survived as an official name of the Party. **1974** *Times* 18 Feb. 14/8 Opposed to them are three loyalist groups: the Rev Ian Paisley's Democratic Unionists, Mr William Craig's Vanguard Unionists, and Mr Harry West's Ulster Unionists.
B. *adj.* **1. b.** (Earlier and later examples.)
1885 *Cork Constitution* 1 Dec. 3/2 (*heading*) Unionist meeting. **1955** *Times* 16 May 3/5 The fact that Sinn Fein has put forward candidates for West Belfast, Mid-Ulster, and Fermanagh and South Tyrone may mean that a Unionist gain will be recorded. **1965** *Times* 6 Mar. 6/2 The Scottish Conservatives set the seal on their reorganization plans by approving a new constitution today... The title of the organization is to be changed from the Scottish Unionist Association to the Scottish Conservative and Unionist Association. **1974** *Times* 5 Sept. 2/5 Mr Brian Faulkner..launched his own political group yesterday. It is..called the Unionist Party of Northern Ireland. **1982** R. Rose *Understanding United Kingdom* iii. 68 The Conservative Party officially styles itself the Conservative and Unionist Party. *Ibid.* 69 In Scotland, the party is organized separately. Its modern foundation dates from amalgamation in 1912 of Scottish Conservatives and the Liberal Unionist Association under the name of the Scottish Unionist Association.

unionize, *v.* Add: **2.** *intr.* To become unionized; to join or constitute a trade union.
1969 *Computers & Humanities* IV. 95 It would be especially inappropriate in this context to get into the question of whether scholar/teachers should unionize or professionalize. **1974** J. White tr. *Poulantzas's Fascism & Dictatorship* vi. iv. 292 The agricultural protelariat unionized in massive numbers. **1978** J. L. Hensley *Killing in Gold* viii. 99 There is a new breed of schoolteachers... They unionize, they strike.
unioniza·tion (examples); **u·nionized** *ppl. a.* (examples).
1896 *Proc. Internat. Typogr. Union* 30/2, I am sure there would be little to gain in its unionization. **1900** *Amer. Review of Reviews* XXI. 651 This year [*sc.* 1900] the strikes were notably successful—New England papers reporting that nearly every 'unionized' town in that section has now the eight-hour workday for..building trades workmen. **1918** *World's Work* XXXV. 486 The issue of the strike being once more unionization. **1920** A. C. Pigou *Econ. of Welfare* iii. iv. 388 The facts..do not warrant us in supposing that local non-governmental Boards would fail if tried on the less completely unionised soil of the Continent. **1947** J. Bertram *Shadow of War* i. iii. 26 Under the reorganised National Government of 1946, the voluntary unionisation of labour in China is legalised in the country as a whole for the first time. **1957** W. H. Whyte *Organization Man* 42 The employers..resisted unionization. **1978** S. Brill *Teamsters* v. 159 He was..emphasizing the benefits of unionization. **1980** *Times* 23 Jan. 10/6 We should also ask whether big, bureaucratic, hierarchical, highly unionized, cumbersome and cautious local government machines are the best instruments for delivering social services. **1984** *Guardian* 5 Nov. 6/5 Mr Mondale's day started early yesterday with church in Memphis, just as it began on Saturday among unionised car workers outside Detroit.

unionized (ɒn,ai·ǒnaiz'd), *ppl. a.*² Also un-ionized. [Un-¹ 8.] Not ionized.
1900 B. D. Jackson *Gloss. Bot. Terms* 283/1 *Unionized*, when the molecules are undivided. (J. F. Clark). **1915** *Chem. Abstr.* IX. 1412 Thus HX and YOH (un-ionized) may unite to form HX.YOH. **1962** D. H. Calam in A. Pirie *Lens Metabolism Rel. Cataract* 439 At the low pH employed, only strongly acidic groups remain charged, most of the carboxyl groups are unionized. **1973** *Sci. Amer.* Apr. 57/1 Later neutral, or un-ionized, sodium was also found to be a constituent of the interstellar medium.

unipare·ntal, *a.* *Biol.* [Uni- 1.] Of, pertaining to, or derived from, one parent.
1900 K. Pearson *Grammar of Sci.* (ed. 2) xi. 469 The algebraical discussion of this problem..involves no further assumptions than those already made for uniparental inheritance. **1963** E. Mayr *Animal Species & Evolution* ii. 27 Clandestine sexuality appears to be rather common among so-called asexual organisms. The expression 'uniparental reproduction' is being used increasingly, instead of 'asexual reproduction', to overcome this and other difficulties. **1975** *Nature* 29 May 401/2 This indicates that streptomycin resistance was transmitted in a non-Mendelian, uniparental way [in tobacco hybrids].

unipersonal, *a.* **2.** (Contextual example.)
1843 G. Crane *Princ. Lang.* v. 207 Verbs, which thus express action without a definitely conceived subject, are called *impersonal*, or perhaps more properly *unipersonal*, verbs.

unipod (yū·nipǫd). *Photogr.* [f. Uni-, after Tripod *sb.* 4.] A one-legged support for a camera. Cf. *monopode *sb.*
1935 *Camera* Feb. 361/2 The Unipod. Pocket models, which extend to 4ft. 3 ins. and 5 feet... This real aid to steadiness should meet with a universal sale. **1937** L. C. Peltier in Fraprie & Jordan *Photographic Hints & Gadgets* 40 (*title*) Unipod made from flexible steel tape measure. **1951** G. H. Sewell *Amat. Film-Making* (ed. 2) iii. 34 An alternative to the tripod is the 'unipod'. That is a single stick or rod, with a screw at the top for fixing the camera. **1979** *Mod. Photogr.* (N.Y.) Oct. 60/1 (Advt.), Now that you've steadied your camera on a conventional unipod, what are you going to do about your arms?

unipolar, *a.* (and *sb.*). Add: **1. a.** Also *unipolar induction*: electrical induction in which a continuous direct current is produced in a conductor joining a magnetic pole and equator by the rotation of either the conductor or the magnet. Cf. *homopolar *a.* 2.
1884 S. P. Thompson *Dynamo-Electric Machinery* ix. 176 The same fundamental idea has been worked upon by Messrs. Siemens and Halske, who have produced a so-called 'unipolar' machine. [*Note*] This sounds like a *lucus a non lucendo*, for the machine has two poles. But the name is derived from the term 'unipolar induction', which Continental electricians give to the induction of currents by a process of 'continuous cutting'. **1888** [in Dict.]. **1982** *Astrophysical Jrnl.* CCLXII. 87 A potential drop $\sim 10^{19}$ volts is generated by the unipolar induction of a rotating accretion disk surrounding the black hole.
b. Also, involving, or operating by means of, unipolar induction.
1883, 1884 [in Dict.]. **1884** [see prec. sense]. **1920** *Whittaker's Electr. Engineer's Pocket-bk.* (ed. 4) 139 Unipolar machines have not been able to compete with heteropolar types. **1940, 1973** [see *inductor 3 b]. **1975** *Nature* 6 Feb. 416/2 The satellite Io acts as a unipolar inductor and a source of constant e.m.f. across the Io flux tube as it moves through the Jupiter magnetic field. **1975** A. Shadowitz *Electromagnetic Field* xi. 410 A unipolar generator (or motor) differs from a homopolar generator (or motor) in that the rotating disk or cylinder is also a uniform magnet.
3. *Electronics.* Of a transistor or similar device: involving conduction by charge carriers of a single polarity.
1952 W. Shockley in *Proc. IRE* XL. 1313/1 In order to distinguish between the more conventional transistors and the analog types, we propose to use the words *bipolar* and *unipolar*. *Ibid.* 1365/2 In a 'field-effect' transistor, the current flow is carried by one type of carrier only... For this reason the name 'unipolar transistors' is proposed. **1973** *Sci. Amer.* Aug. 48/2 Most integrated circuits produced in the 1960's were of the bipolar type, but production of the newer unipolar type is growing rapidly. **1981** J. C. Sprott *Introd. Mod. Electronics* viii. 169 The bipolar transistor is so named because current is carried simultaneously by charges of both polarities (electrons and holes) rather than by a single species, as in the FET which is an example of a unipolar device.
4. *Psychol.* Of a psychiatric disorder: characterized by depressive but not manic episodes.
1965 *Acta Psychiatrica Scandinavica* Suppl. No. 180. 87 We..have assumed that 'middle age depressions' according to Stenbäck and 'unipolar depressions' according to Leonhard are merely different terms, which..cover the same nosographic entity. **1982** Donlon & Rockwell *Psychiatric Disorders* v. 76 Unipolar disorders are much more common than bipolar.

unipole (yū·nipōᵘl). *Radio.* [f. Uni- 2 + Pole *sb.*²] = *monopole³ 2. Also *attrib.* or as *adj.*
1950 H. P. Williams *Antenna Theory & Design* II. v. 226 Fig. 5.13 (*b*) shows a folded unipole antenna which is suitable for mounting in an attic. **1958** *Engineering* 31 Jan. 157/3 It works into groundplane folded unipole aerials on top of 80 ft. masts. **1973** *Electronics Lett.* IX. 300/2 The input impedance of a coaxially fed short unipole can be brought to resonance by encasing the antenna in a slab of high-permittivity dielectric material.

unique, *a.* Add: **2. a.** (Later examples.)
1908 K. Grahame *Wind in Willows* viii. 168 'Toad Hall,' said the Toad proudly, 'is an eligible self-contained gentleman's residence, very unique.' **1912** Chesterton *Manalive* i. iii. 86 Diana Duke..began putting away the tea things. But it was not before Inglewood had seen an instantaneous picture so unique that he might well have snapshotted it. **1939** *Country Life* 11 Feb. p. xviii/2 (Advt.), Almost the most unique residential site along the south coast. **1960** [see *diquat]. **1980** *Verbatim* Autumn 15/2 A high-ranking state Alcoholic Beverage Commission official said Friday that Wednesday's retroactive renewal and transfer of the beverage permit of the rural Bloomington Liars' Lodge by the Monroe County Alcoholic Beverage Board was 'unique but not uncommon'.
b. (Later examples.)
1934 G. B. Shaw *On Rocks* ii. 262 You dont appreciate him. He is absolutely unique. **1938** [see *chef d'école].

un-Irish, *a.* (Earlier example.)
1829 G. Griffin *Collegians* III. xli. 225 Suicide is a very un-Irish crime.

unirradiated, *ppl. a.* (Later examples.)
1955 *Radiation Res.* II. 364 When both haploid cells were unirradiated, zygote formation occurred in about 95% of the pairs. **1978** *Jrnl. R. Soc. Med.* LXXI. 670 Patients with grade III astrocytoma who were irradiated had greater 1–5 year survival rates than those unirradiated.

unirrigated, *ppl. a.* (Earlier example.)
1876 'Mark Twain' *Tom Sawyer* iv. 44 When he emerged from the towel, he was not yet satisfactory, for the clean territory stopped short at his chin... Below..there was a dark expanse of unirrigated soil.

uniselector (yū·nisilektəɹ). *Teleph.* and *Electr.* [f. Uni- 2 + Selector.] A selector (sense *c(b)) which has a wiper free to rotate about an axis but not to move along it.
1930 *Gloss. Terms Telegraphs & Telephones* (*B.S.I.*) 27 Uniselector. **1938** C. W. Wilman *Automatic Telephony* (ed. 2) iii. 20 Single-motion switches are commonly termed uniselectors, lineswitches being known as subscribers' uniselectors. **1956** G. A. Montgomerie *Digital Calculating Machines* ix. 181 The other major component of interest to us is the stepping switch or uniselector. **1971** J. H. Smith *Digital Logic* i. 4 An electronic telephone exchange uses static switching, but..a conventional exchange..uses relays and uniselectors consisting entirely of moving parts.

u·nisex, *a.* and *sb.* [f. Uni- + Sex *sb.*] **A.** *adj.* Of, pertaining to, or characterized by a style (of dress, appearance, etc.) that is designed or suitable for either sex; not peculiar to one sex, sexually indeterminate or neutral.
1968 *Life* 21 June 87 With-it young couples..are finding that looking alike is good fashion as well as good fun. The unisex trend was launched by..the teen-agers. **1968** *Manch. Guardian Weekly* 21 Nov. 4 Greenwich Village..has just spawned the world's 'first unisex boutique for men and women from 16 to 25'. **1969** *New Yorker* 5 Apr. 99/1 Unisex metallic trouser suits. **1969** *Daily Tel.* 6 June 17 Unisex fashions have literally gone to children's heads, with look-alike brother and sister haircuts. **1970** P. Carlon *Death by Demonstration* vi. 71 A lot of the men..were friendly with her on a strictly unisex level. I mean there was nothing in the slightest degree like romantic attraction. **1972** *Sat. Rev.* (U.S.) 1 July 47/1 'How clean,' 'how Spartan,' 'how unisex' the Chinese appeared to be. **1976** J. I. M. Stewart *Memorial Service* xv. 262 A sexuality quite as strong as Anna's moved beneath the androgynous or 'unisex' persona she had created for herself. **1980** *Times Lit. Suppl.* 21 Mar. 320/1 A group of student actors, all with cropped hair, sallow cheeks and dressed in unisex denims, are warming up in front of a packed house.
B. *sb.* A condition or phase during which people of both sexes appear to be indistinguishable in dress and outward behaviour.
1969 *Sunday Mail* (Brisbane) 11 May 24/8 It's unisex where men and women have abandoned the old 'vive la difference' school of thought in dressing and added a fillip to the old guessing game of 'is it a he or a she?' **1972** *Nature* 28 Jan. 234/2 It could be pertinent to recall that, at the time these results were obtained, the adolescent trend towards unisex was strongly under way. **1976** T. Sharpe *Wilt* i. 5 Eva Wilt was too easily influenced..to be allowed out with a woman who believed that..unisex was here to stay.

unisexual, *a.* Add: **3.** = *unisex *a.*
1970 *Sunday Times* 29 Nov. 29/2 Adolescents of both genders strode along..with books and long flaxen unisexual hair. **1978** C. Sykes in R. Buckle *U & Non-U Revisited* 52 By the 1960s unisexual umbrellas were commonplace in Germany.

unisexuality. Add: **b.** In sense of *unisexual *a.* 3.
1971 *Daily Tel.* 13 Aug. 9/2 If he meant anything serious at all in his comedy..it can only have been the tedium arising from such coy and quaint treatments of unisexuality. **1973** *Ibid.* 25 Sept. 18 The notion of a blatant unisexuality such as she implies is not the aim of these anti-discrimination laws—or of the majority of feminists.

unit, *sb.* (and *a.*). Add: **1. c.** *spec.* one kilowatt-hour, as the unit used in measuring and charging for mains electricity; also, the unit used for metered telephone calls.
1891 *Minutes of Proc. Inst. Civil Engineers* CVI. 16 Fuel used..has..fallen from 12 lbs. to 7·9 lbs. per unit generated. **1926–7** *Army & Navy Stores Catal.* 344/3 Electric radiators... Two bars..consumes 2 units per hour when full-on. **1961** *Which?* Dec. 334/1 Local and trunk calls are divided into 2*d.* 'units', the amount of the time you get for your 2*d.* depending on the distance. **1972** *Daily Tel.* (Colour Suppl.) 12 May 15/1 Food freezers are inexpensive to run, using about 2 units of electricity for each cubic foot per week.

f. *unit of account*, a monetary unit in which accounts are kept; *spec.* in the European Economic Community (see quots. 1977 and 1982).
1882 R. Bithell *Counting-House Dict.* 311 *Unit of Account*, the unit of value in which accounts are kept. It may, or may not, coincide with any coin in circulation... The Anglo-Saxon unit of account was the shilling.. but no coin called a *shilling* was issued before the reign of Henry VII. **1959** A. H. Robertson *European Institutions* ii. 42 After offsetting these balances (measured in 'units of account' equivalent to the gold value of the U.S. dollar) against each other, Members are left with a credit or debit account *vis-à-vis* the [European Payments] Union. **1973** *Physics Bull.* Apr. 207/1 The four year allocation for direct research work amounts to 157·2 million units of account (ua, equal to the predevaluation US dollar, ie about £65m). **1977** *Times* 6 Dec. (Europa Suppl.) p. iii/6 *Units of account*, embryonic European currency used as a device for calculating the EEC budget, fixing farm prices and in certain transactions with non-Community countries. The value of the *unit of account* in national currencies depends on the purpose for which it is being used. **1982** J. Phillips *Dict. Trading Terms* 67 The unit of account is now equivalent to a group, or 'basket' of fixed amounts of European currencies, and is described as a 'basket unit of account'.
g. A basic measure of educational attainment credited to a student for completing the number of hours of study assigned to one section of an academic course. Cf. *credit *sb.* 13 d. *U.S.*
1894 *Univ. of Chicago Weekly* 4 Oct. 4/1 The system of majors and minors, units and flunks, is harder to understand than any other 'credit' method in operation among educational institutions. **1930** A. Flexner *Universities* ii. 47 When a college catalogue states that fifteen units of high school work are required for matriculation, a unit, as defined by the College Entrance Examination Board, represents one year's study in any one subject in a high school. **1945** C. V. Good *Dict. Educ.* 436/1 *Unit*..(3) a basic measure used in calculating the amount of credit to be assigned to any particular course or the number of graduation credits earned by a pupil or student in completing a course..(*a*) in secondary education, one *unit* equals approximately 120 hours of classroom or laboratory work in a given subject..(*b*) in higher education,..one *unit* may equal 1 hour of class or laboratory work per week during one term, semester, or school year. **1974** *Aiken* (S. Carolina) *Standard* 18 Apr. 4-c/3 Their required 18 units of study. **1981** D. Rowntree *Dict. Educ.* 335 *Unit*... 3 (US) In high school, one hour in class per day of a subject (for five days a week over the academic year) counts as one course unit of that subject.
h. The standard unit of quantity by which bread and petrol were rationed during and immediately after the war of 1939–45; a coupon of this value.
1939 *Punch* 18 Oct. 439 (*caption*), I can't move on—I've used up all my units. **1946** [see B.U. s.v. B III. b]. **1948** *Daily Tel.* 26 Oct. 5 Nine Hundred Petrol Units were stolen from an office at Swanley, Kent. **1963** S. Cooper in Sissons & French *Age of Austerity* ii. 41 The Bread Unit represented seven ounces of bread... A large loaf..would require four Units, one pound of flour three Units.
2. d. A group of buildings, wards, etc., in a hospital; *spec.* one equipped to provide a particular type of health care.
1893 D. Galton *Healthy Hospitals* xiii. 229 Separation of the ward unit has been the principal feature of modern hospital construction in Germany and..the United States. **1911** W. Osler in *Lancet* 28 Jan. 212/1 There might be, as at Berlin and Vienna, two or three medical and the same number of surgical units. **1927** J. E. Stone *Hospital Organization & Management* xiv. 287 The operating theatre unit is a very important part of a hospital. **1955** R. F. Bridgman *Rural Hospital* ii. 70 With the rural hospital as a base, mobile health-units may be organized, through which modern medical techniques can be taken to the villages in sparsely populated countries with poor communications. **1965** *Nursing Times* 5 Feb. p. vi/1 (Advt.), Plastic Surgery and Burns Unit..Regional Thoracic Surgical Unit. **1976** *Amer. Speech 1973* XLVIII. 195 The inhalation therapist, for example, supplies hospital floors, or units, with oxygen masks and in emergencies is summoned to start and operate O_2 tents 'oxygen tents'.
e. A piece of (esp. storage) furniture or equipment which may be fitted with other pieces to form a larger system, or which is itself composed of smaller complementary parts. Freq. *attrib.* Also *transf.* (see also *unit audio, construction*, sense 3 c below).
1912 L. Weaver *House & its Equipment* 44 The unit system of bookcases, by which they are built up of sections of standard size, and are thus capable of indefinite expansion. **1930** *N.Y. Times* 10 Aug. v. 14/4 He [*sc.* Franz Schuster] has developed a kind of 'unit' furniture. He reduces the shapes of chest and cabinet to their fundamental forms and by standardizing their measurements permits the combination of parts by the manufacturer. **1937** [see *kitchen unit* s.v. *kitchen *sb.* 5 c]. **1944** J. van Druten *Voice of Turtle* i. 3 The kitchen has an icebox, stove and sink in a combined unit in the left wall. **1958** *Engineering* 7 Mar. 320/2 The cooking unit ..is mounted on top of a storage cupboard with a sliding serving shelf. **1974** *Gramophone* Nov. 1009 The connoisseur takes pride in choosing separate units. **1978** [see *recliner 2]. **1981** M. E. Atkins *Palimpsest* ix. 92 I'm going to..start with the kitchen. I'll have units all round, a new sink and cooker.
f. An accommodation unit in a larger building or group of buildings, esp. in a block of flats or a motel. *U.S.*, *Austral.*, and *N.Z.*
1932 F. L. Wright *Autobiogr.* ii. 223, I lingered in Los Angeles aided by my son Lloyd working on the new unit-block system. **1937** *Tourist Court Jrnl.* Oct. 6/2 Being separate units each cottage is assured of ample ventilation..through the windows on each side. **1953** *Hotel Monthly* Nov. 27/1 Additional units will be added to the Kahler Ranchotel. **1963** D. B. Hughes *Expendable Man* ii. 46 No one was waiting for him at the motel. No one stopped him at the door of his unit. **1971** 'A. Blaisdell' *Practice to Deceive* i. 2 She lived in one unit of a triplex. **1973** *Sun-Herald* (Sydney) 26 Aug. 103/1 We live in a unit in a delinquency-prone inner area. **1980** 'D. Shannon' *Felony Files* x. 230 It was a pleasant, unpretentious furnished apartment in a six-unit place.
g. = *film unit* s.v. *film *sb.* 7 c.
1959 E. H. Clements *High Tension* i. 13 The hectic urgency of everyone else in the unit. **1962** L. Davidson *Rose of Tibet* i. 26 Location work would have finished in Calcutta and..the unit would have moved up into the foothills of Everest.
3. c. Special Combinations. **unit audio,** a sound reproduction system which comprises separate matching parts; **unit cell** *Cryst.*, the smallest structural unit having the overall symmetry of a crystal, which by repetition in three dimensions gives the entire lattice; **unit character** *Genetics*, a character inherited according to Mendelian laws, esp. one controlled by a single pair of alleles; also, † the alleles themselves (see quot. 1966); **unit construction,** modular construction, esp. of buildings (cf. *modular *a.* 1 b); **unit cost** *Accounting*, the cost of manufacturing or otherwise processing one unit of production; **unit factor** *Genetics* = *gene (cf. *factor *sb.* 7 b); now *hist.*; **unit-holder,** one who holds securities in a unit trust; **unit-linked** *a.*, of a life assurance policy (see quot. 1979); **unit load,** a package of goods arranged for shipment, etc., as a single unit (esp. on a pallet) to facilitate handling; **unit matrix** *Math.* = *identity matrix* s.v. *identity 10; **unit membrane** *Biol.*, any lipoprotein membrane composed of two electron-dense layers enclosing a less dense layer, found enclosing many cells and cell organelles; **unit price,** the price at which a single unit of a commodity is sold; **unit pricing** (see quot. 1970); **unit train** *N. Amer.*, a train allocated to transport a single commodity (i.e. coal or grain) at special rates between two points; **unit trust,** an investment group investing combined contributions from many persons in various securities and paying them dividends in proportion to their holdings.
1966 *Hi-Fi News* Nov. 592/3 'Unit audio' is the name given to a new range of matching equipment... Two loudspeakers, a tuner-amplifier, an amplifier and a tape recorder..are available. **1976** *Gramophone* Nov. 880/1 People without expert knowledge will generally find it easier to buy a 'unit audio' system made up from matching units from the same manufacturer. **1915** W. H. & W. L. Bragg *X-Rays & Crystal Structure* viii. 116 Only calcium and carbon atoms are shown in their places in the unit cell of the structure. **1930** G. P. Thomson *Wave Mech. Free Electrons* iii. 40 Each unit cell of a crystal lattice contains the same amount of matter similarly arranged. **1966** C. R. Tottle *Sci. Engin. Materials* iii. 50 In many cases it is convenient to avoid drawing the complete crystal lattice extended over many unit cells, and merely to draw the unit cell itself. **1977** A. Hallam *Planet Earth* 114 All crystalline substances have lattices built of one of these types of unit cell. **1902** Bateson & Saunders *Rep. Evol. Comm. R. Soc.* I. 126 The purity of the germ-cells, and their inability to transmit both of the antagonistic characters, is the central fact proved by Mendel's work. We thus reach the conception of unit-characters existing in antagonistic pairs. Such characters we propose to call allelomorphs. **1903** *Biometrika* II. 286 Mendel was the first to systematically analyse the differential characters of a race or species into a series of unit-characters, each of which might..be inherited independently of the others. **1915** T. H. Morgan et al. *Mechanism of Mendelian Heredity* ix. 210 So much misunderstanding has arisen amongst geneticists themselves through the careless use of the term 'unit character' that the term deserves the disrepute into which it is falling. **1945** M. F. Glaessner *Princ. Micropalaeontol.* v. 79 The numerical values of morphological features (unit characters) plotted against numbers of specimens in which the progressive values occur, tend to arrange themselves in a regular curve. **1966** E. A. Carlson *Gene* ii. 13 The unit-character [of Bateson] combined the 'differentiating character' used by Mendel with the 'formative element' which he assumed to represent it in the germ cell. **1921** *Conquest* May 291/2 Houses built on the 'Unit' construction system... The concrete blocks are made of a standard size, the dimensions of windows and door openings being made multiples of the block size. **1959** *Motor Manual* (ed. 36) v. 3 The body now forms the main structure of the car... This form of construction is now known by a variety of names, including 'integral construction', 'unit construction' and 'chassisless construction'. **1964** *McCall's Sewing* ii. 33/1 *Unit construction*, organisation of sewing procedure so that an entire garment section is completed before it is joined to another. **1914** E. H. Jones *Unit Construction Costs from New Smelter* 1498 These unit costs..represent delays in material shipments..., delayed plans,..labor troubles, [etc.]. **1962** *Listener* 17 May 835/1 Wages and unit costs rise faster than ever. **1978** J. Kellock *Elements of Accounting* iii. 66 The stock valuation..is calculated by taking a physical stock count of the stock on hand at the end of the period and multiplying each item by the appropriate unit of cost... Unit cost is the cost of purchasing or manufacturing identifiable units of stock. **1911** Unit factor [see *gene]. **1926** J. S. Huxley *Ess. Pop. Sci.* 9 Inheritance takes place by means of separable units, generally known as unit-factors or genes. **1966** E. A. Carlson *Gene* iv. 26 When unit-character was changed to 'unit-factor' or to 'factor' alone, Castle no longer dissociated the transmitting agent from its effect on a character. **1965** M. Naylor *Your Money* 87 One great advantage of unit trusts is that unit-holders can buy or sell at any time, and get the 'true' value of their investment. **1969** *Daily Tel.* 8 Feb. 6/3 A unit trust management pays capital gains tax at 30 p.c. for either long- or short-term gains, irrespective of the tax position of the individual unitholder. **1969** *Times* 30 Apr. 28/3 Nearly every time we open the paper we read of the attractions of Unit Trusts and Unit-linked life assurance policies. **1979** F. E. Perry *Dict. Banking* 258/2 *Unit-linked policy*, a type of life assurance policy where a part of the premium is invested on behalf of the assured in a unit trust. **1939** *Steel* 12 June 54/1 (*caption*) Unit load of four edgewise-wound copper coils on pallet handled by fork truck. **1945** D. L. Beattie *Unit Load Materials Handling* I. 9 Once the unit load is established for a material, there is every reason to expect that this unit load may be standardized. **1970** *Times* 2 June (Container Suppl.) p. i/1 The British Transport Docks Board has a nationwide network of nine well equipped ports, providing some of the most advanced facilities for unit-load handling. **1862** Unit-matrix [see *premultiplication]. **1972** M. Kline *Math. Thought* xxxiii. 807 The product of a matrix and its inverse is the unit matrix, denoted by *I*. **1959** J. D. Robertson in E. M. Crook *Structure & Function Subcellular Components* (Biochem. Soc. Symposium XVI) 33 Perhaps the gap substance or the character of the unit membrane surfaces here is different. **1970** Ambrose & Easty *Cell Biol.* v. 173 The chloroplast, like all plastids, is bounded by a double unit membrane. **1934** Webster, Unit price. **1953** [see *field *sb.* 15 f]. **1977** P. Way *Super-Celeste* i. 45 Whoever sold a plane here would sell it to NATO through the 1990s. Sell here and the unit price would come down. **1970** *Wall St. Jrnl.* 17 June 40/4 Mr. Alldredge..told the meeting that 'unit pricing', or the marking of all packaged commodities with the price per unit weight, 'would be frighteningly expensive'. **1971** *Guardian* 3 June 2/6 Unit pricing, a new system under which large supermarkets have to indicate the cost of food items by measure, completed its first day here [*sc.* in New York]. **1967** *Times Rev. Industry* Apr. 47/2 Unit trains working on the principles now applied in the Great Lakes coal trade would discharge coal through ground hoppers without uncoupling. **1979** *Sci. Amer.* Jan. 29/1 (*caption*) Coal-carrying 'unit' trains have been developed to move coal expeditiously at low cost, usually between one mine and one customer, which is most often an electric utility. The trains shuttle back and forth without being uncoupled, acting much like a conveyor belt. **1936** *Economist* 18 Apr. 135/2 Three new trusts with different degrees of flexibility have recently appeared, which extend the activities of the unit trust movement into new fields. **1958** *Spectator* 18 July 108/1 All unit trusts have this much in common: investments..are deposited with a bank or insurance company, acting as trustee, who issues participation certificates (called units) in exchange. **1980** *Times* 5 Jan. 18 Most of the unit trusts managed not to lose too much money for unit holders last year.

unitard (yū·nitɑ̄ɹd). orig. *U.S.* Also **Unitard, unitards.** [f. Uni- + *(leo)tard.] A tight-fitting one-piece garment of stretchable fabric which covers the body from neck to feet, worn by gymnasts, dancers, and as a fashion garment. (Formerly a proprietary name in the U.S.) Cf. *cat-suit* s.v. *cat *sb.*¹ 18.
1961 *Official Gaz.* (U.S. Patent Office) 5 Dec. TM15/1 Danskin Inc., New York... Unitard. For one piece form-fitting integral neck-to-ankle..garment, with sleeves, for gymnastics, and dance use. **1978** *Newsweek* 13 Feb. 68/3 Danskin now offers 75 styles for men, women and children, including..a one-piece neck-to-toe 'Unitard'. **1978** *Washington Post* 11 Dec. B8/2 Stan Fowler..had cut, dyed and sewn together the pieces of his 'milliskin unitard' to get the Spiderman look of bright blue and red from head to toe. **1979** *N.Y. Times* 2 Feb. C16 One woman, dressed in unitards and a wide gold sash, bounces dull-eyed in the center of the rooms. **1984** *Times* 14 Feb. 9/2 Unitards—one-piece suits with built-in leg interest like stripey panels at the calf.

Unitarian, *sb.* and *a.* Add: **A.** *sb.* **2. c.** (Earlier example.)
1832 *Ann. Reg. 1831 Hist.* xv. 464/2 The Unitarians were dispossessed of the government of that province [*sc.* Entre-rios, Argentina], and the preponderance of Buenos Ayres restored.
d. A critic who ascribes the *Iliad* and the *Odyssey* to the same author. Opp. *separatist *sb.* 1 f. Cf. Chorizontes *sb. pl.*
1959 *Times Lit. Suppl.* 13 Mar. 138/4 Any Unitarian must depend very largely on demonstrating some recognizable pattern or design in the Homeric poems as evidence of single authorship. **1976** [see *separatist *sb.* 1 f].
B. *adj.* **2. d.** Applied to the theory that the *Iliad* and the *Odyssey* are the work of a single person. Cf. *Homeric question* s.v. Homeric *a.*, and prec. sense.
1865 M. Pattison in *North British Rev.* June 277 Even on the more special question of the origin of the Homeric poems,..we may safely say that no scholar will again find himself able to embrace the unitarian hypothesis.

unitary, *a.* Add: **4. d.** *Math.* and *Physics.* Applied to mathematical entities that in some specific way are described by or related to a *unitary matrix*, one which when multiplied by the transpose of its complex conjugate gives the unit matrix; *unitary group*, the group of all square unitary matrices of a given size; *unitary symmetry*, the symmetry of a unimodular unitary group as used to relate the properties of different sub-atomic particles.

1908 H. Hilton *Introd. Theory Groups Finite Order* iii. 16 The substitution *A* is called..unitary if $AA' = 1$. **1935** P. A. M. Dirac *Princ. Quantum Mech.* (ed. 2) v. 111 We can now see that a unitary transformation transforms observables into observables. **1937** *Physical Rev.* LI. 109/2 The representations of the four-dimensional unitary group will characterize the multiplet systems. **1941** Birkhoff & MacLane *Survey Mod. Algebra* ix. 255 One may adopt the properties of linearity, skew-symmetry, and positiveness as the postulates for an inner product..in an abstract vector space over the complex numbers; the space is then called a unitary space. *Ibid.*, A linear transformation *T* of the space is unitary if it preserves lengths $|\xi T| = |\xi|$. **1961** M. Gell-Mann in Gell-Mann & Ne'eman *Eightfold Way* (1964) i. 12 We attempt..to treat the eight known baryons as a supermultiplet, degenerate in the limit of a certain symmetry but split into isotopic spin multiplets by a symmetry-breaking term... The symmetry is called unitary symmetry and corresponds to the 'unitary group' in three dimensions in the same way that charge independence corresponds to the 'unitary group' in two dimensions. **1969** [see *Lie]. **1975** *Physics Bull.* Apr. 176/2 Strong interactions among nucleons are invariant under a group of unitary symmetry transformation which changes protons into neutrons and vice versa—the group SU(2). **1979** Cheng & O'Neill *Elementary Particle Physics* xiii. 275 The mathematical groups on which elementary particle physicists have concentrated most of their attention are..continuous groups described by matrices *U* that are unitary: $U^\dagger U = 1$.
6. Special collocations: *unitary taxation* (U.S.), a system of taxation by which a company or business is taxed on a proportion of its worldwide earnings, and not just on those made within the jurisdiction of the taxation authority (i.e. a State government); also *unitary tax*.
1977 *Washington Post* 14 Aug. A8/4 Brown is moving to alleviate concerns of multinational companies over the state's 'unitary tax'. **1979** *Economist* 7 July 91/2 Unitary taxation (purists call it combined or consolidated income reporting)..hits a company not on profits made in a given state but on a percentage of the parent's total (multi-state or world-wide) income. **1984** *Miami Herald* 6 Apr. 22A/3 When the governor crawfishes out of his ill-advised unitary tax (his great tax-reform promise), the greedy ones might try to increase property taxes again.
Hence **u·nitarily** *adv.*, in a unitary manner; **unita·rity,** the property of being unitary.
1932 *Amer. Jrnl. Math.* LIV. 149 Similar unitary matrices are..always unitarily equivalent. **1959** *Nuovo Cimento* XIII. 354 The unitarity of the *S*-matrix implies that it can be written in terms of a hermitian matrix *A* in the form S=[etc.]. **1969** *Nature* 24 May 720/1 SU2 and SU3..are simple unitary symmetries in which the unitarity ensures that a set of possibilities has unit probability. **1979** J. C. Polkinghorne *Particle Play* vi. 88 S-matrix theory played a valuable role in highlighting certain general properties (unitarity, crossing, analyticity) which are important aspects of relativistic quantum mechanics.

united, *ppl. a.* Add: **4. a.** *United Empire Loyalist*: see Loyalist; *United Front*, a common alliance of political groups; *spec.* in Communism: (*a*) = *Popular Front; (*b*) in Chinese communism, an alliance with the Kuomintang; subsequently, a coalition of several parties in a Communist government; also *transf.*
1934 A. Werth *France in Ferment* xiii. 277 The rank and file of the United Front and of Bergery's Front Commun are merely unhappy and disgruntled people. **1935** B. Russell *Relig. & Sci.* ii. 42 The wars of religion made a 'united front' desirable. **1937** E. Snow *Red Star over China* iii. iii. 101 The building up of a United Front, such as has been advocated by the Communist Party ever since 1932. **1943** J. T. Pratt *War & Politics in China* xv. 264 Ch'iang Kai Shek..sternly rejected the overtures of the Shensi communists who naively proposed to enter the united front as the equal allies of the Kuomintang. **1954** *Round Table* Dec. 46 United Front leaders in East Bengal. **1958** J. Cannan *And be Villain* i. 6 That blasted Primrose will have arrived..and the old witch herself will be there. ..I'll have a united front to cope with. **1971** J. J. Taylor in D. J. Dwyer *China Now* (1974) xxii. 412 Mao's insistence on the development of an armed capability..is a reflection of the history and the experience of the Chinese Communist Party, most notably the failure in 1927 of the First United Front with the KMT and the marked Communist success with the Second United Front between 1937 and 1945. **1980** S. J. Burki *Pakistan under Bhutto* ii. 22 The Muslim League was squarely beaten in Bengal by the United Front.
b. United Kingdom: (after the formation of the Irish Free State in 1921) of Great Britain and Northern Ireland.
d. United Nations: in the war of 1939–45, the Allied nations who united against the Axis powers; hence, an international peace-seeking organization of these and many other States, founded by charter in 1945 (in full, *United Nations Organization*), with a permanent headquarters in New York; abbrev. *U.N.* s.v. *U 4 a; cf. *League of Nations; *Security Council* s.v. *security 12 e; *trusteeship 2 b.
1942 *Daily Tel.* 28 Jan. 3/3 But at any rate it will be long enough for Japan to inflict..losses upon all of the United Nations who have..possessions in the Far East. **1942** H. A. Wallace in *N.Y. Times* 9 Nov. 19/7 The first article in the international law of the future is undoubtedly the United Nations' Charter. **1944** [see *Security Council* s.v. *security 12 e]. **1946** A. Boyd (*title*) The United Nations Organisation handbook. **1953** R. Niebuhr *Christian Realism & Polit. Probl.* (1954) ii. 25 The necessarily minimal constitutional structure which we have embodied in the United Nations. **1958** *Times* 7 Aug. 7/5 It emphasizes..the case for creating a permanent United Nations police force, including an increased number of United Nations observers. **1974** P. Gore-Booth *With Great Truth & Respect* 141 Mrs Eleanor Roosevelt came to propose that the organization be called 'The United Nations'... I put forward a motion to the effect that we accept Mrs Roosevelt's proposal subject to a committee of jurists being satisfied that the term 'United Nations' presented no legal difficulty. **1980** *Jrnl. R. Soc. Arts* July 501/2 The *inter-governmental* component of world conservation is represented by the United Nations Environment Programme (UNEP).

United States. 1. b. (Earlier examples.)
1776 *Jrnls. Continental Congress* (1906) VI. 865 Resolved, that the inhabitants of Canada, captivated by the United States..be released and sent home.
attrib. **1819** G. Flagg *Let.* 12 June in *Trans. Illinois State Hist. Soc. 1910* (1912) XV. 165 They settle on united States land.

Uniterm (yū·nitɜ̄ɹm). *Library Science.* Also † **unit term.** [f. Uni- 2 or Unit *sb.* (and *a.*) + Term *sb.*] The name for a system of indexing whereby each of a series of documents is accessible through an alphabetical index of subject headings; a keyword which forms one of these subject headings.
1952 M. Taube in *Amer. Documentation* III. 213/2 The basic ideas of unit terms as a substitute for standard indexing for subject headings and logical combination and order as a substitute for..alphabetic cross-reference structures. **1953** —— et al. *Studies in Compar. Indexing* I. 5 We have used the name 'coordinate indexing' for this general method and the more specific name 'Uniterm System' to designate a particular manual application of coordinate indexing. **1961** T. Landau *Encycl. Librarianship* (ed. 2) 100/1 Co-ordinate indexing..is based on the conception of the Uniterm... In the Uniterm system each book..or other item is numbered as it is received and..it is possible to gain an approximate idea of the date of any currently-published item from its serial number. The title, salient contents, etc.,..are then analysed into fundamental terms usually of one word each. These constitute the Uniterms. **1976** *Gloss. Documentation Terms* (*B.S.I.*) 70 *Uniterm*, originally a single word selected from, and characterizing a part of the subject matter of, a document, for use in a co-ordinate indexing system. Now loosely used as a synonym for keyword or descriptor.

unitiza·tion. [f. next + -ation.] **1.** The joint development of a petroleum source which straddles territory controlled by several companies.
1930 *Handbk. Unitization of Oil Pools* (Mid-Continent Oil & Gas Assoc.) i. 15 The term 'Unitization' refers to the practice of unifying the ownership and control of an actual or propective oil or gas pool by the issuance or assignment of units or undivided interests in the entire area with provision for development and operation by an agent, trustee or committee representing all holders of undivided interests therein. **1938** D. Hager *Practical Oil Geol.* (ed. 5) ix. 263 The ideal is for one oil concern to own a whole field and to unitize the leases so that property owners are allowed a royalty on all the oil produced in proportion to their property holdings. *Ibid.*, Unitization as an ideal is fine for the oil producer. **1952** *South Western Reporter* (U.S.) CCXLIX. 917/1 Respondents suggest that the conclusion we have reached will discourage the making of unitization agreements. **1977** *Internat. & Compar. Law Q.* XXVI. 353 Discussions are already under way between Norway and the United Kingdom on the unitisation of the Statfjord oil and gas field.
2. The packaging of cargo into unit loads (see *unit *sb.* (and *a.*) 3 c); = *palletization.
1953 *Times* 20 Jan. 2/5 Addresses will be given on.. 'Packaging and air freight', and 'Packaging and unitization'. **1967** D. Wilson *Use of Expendable Pallets* 1 Since most modern handling systems of unit loads are based around the fork-lift truck there has to be a common denominator of any method of unitisation i.e. that it is capable of being lifted and moved by fork-lift truck. **1981** E. Corlett *Revolution Merchant Shipping* 12/1 Militating against the spread of container unitisation was the lack of infra-structure.
3. Conversion of an investment trust to a unit trust (see *unit *sb.* (and *a.*) 3 c).
1974 *Daily Tel.* 10 Aug. 17/4 The cost of unitisation 'would be high'. **1982** *Observer* 7 Feb. 20/1 Fleming..may well find that unitisation cannot be avoided in respect of some of the trusts.

unitize, *v.* Delete *rare* and add: **1.** (Later examples.)
1939 D. Hager *Fund. Petroleum Industry* ix. 201 The new drilling outfits are unitized, i.e., the various sections are mounted on single steel frames which are enclosed in steel cases to protect the workers from the machinery. **1962** *Engineering* 21 Sept. 369 As much as possible of each engine has been 'unitized'.
2. *techn.* In the senses of: **a.** *unitization 1.
1938 D. Hager *Pract. Oil Geol.* (ed. 5) ix. 263 The ideal is for one oil concern to own a whole field and to unitize the leases.
b. *unitization 2.
1962 [see *containerize *v.*]. **1968** *Globe & Mail* (Toronto) 13 Feb. B1/8 The parcel was not drilled, but was unitized into one of the Rainbow pools and is producing revenue pro-rated to the company's share of the pool's reserves. **1973** (*title*) Packing for profit 1: the economic advantages of unitising break-bulk cargo.
c. *unitization 3. Also *intr.* or *absol.*
1978 *Daily Tel.* 4 Feb. 21/1 With the current average discount on investment trust shares running at around 30 p.c. there is considerable pressure from the private shareholder to unitise the trusts, which would mean a payout at the full value of the underlying investments. **1982** *Sunday Times* 7 Feb. 54/1 Robert Fleming..promised to take the question of unitising seriously.
unitized *ppl. a.* (later examples).
1947 *Sun* (Baltimore) 23 Aug. 7 You must see the Nash '600' to realize how far into the future this big car takes you..with..girder-strong unitized body and frame. **1950** *Nucleonics* May 15/2 The fitting of unitized furniture into the laboratory is a critical consideration which must be applied to the dimensioning of all of the rooms. **1961** *Times* 22 Feb. 11/6 In the United States mechanized and 'unitized' general cargo handling is racing ahead. **1964** *Economist* 26 Sept. 1243/1 The roads have..introduced.. unitised trains made up of wagons carrying one product to one customer. **1970** R. P. Loveland *Photomicrography* I. iii. 121 Unitized construction allows change of body tube between binocular and monocular for photomicrography. **1978** *N.Y. Times* 29 Mar. D10/5 The Institute of Mental Health, a unitized psychiatric hospital, requires a clinical psychologist PhD.

Univ (yū·niv), colloq. abbrev. of University *sb.* 1; *spec.* University College (Oxford, etc.) Cf. *uni.
1896 F. W. Maitland *Let.* 6 Dec. (1965) 153 We have several good MSS at the Univ. Lib. **1903** Farmer & Henley *Slang* VII. 256/1 *Univ, subs.* (Oxford University), University College. **1950** M. Marples *University Slang* 131 *Oxford colleges*... *University*: always *Univ* nowadays, but apparently *Varsity* during the second half of the last century.

univalent, *a.* Add: **1.** Now usu. with pronunc. (yūni,vē·lěnt).
2. *Cytology.* (yūni·vālěnt). [ad. mod.L. *univalens* (introduced in Ger. by O. Hertwig 1890, in *Arch. f. mikrosk. Anat.* XXXVI. 6).] Of a chromosome: remaining unpaired during meiosis. Cf. *monovalent *a.* 3.
1898 *Zool. Jahrb.* (Abt. für Anat.) XII. 79 The chromosomes of the 1st reduction division are univalent, when they occur in the normal number. **1916** *Jrnl. Morphol.* XXVII. 226, I believe he [*sc.* Haecker] is wrong in thinking that the 'univalent' chromosomes making up these 'bivalents' are to be considered as members of homologous chromosome pairs. **1971** *Nature* 19 Feb. 570/2 Many authors have reported that the behaviour of univalent chromosomes at meiosis is irregular. **1981** *Jrnl. Cell Biol.* LXXXVIII. 281 During meiosis I in males of the mole cricket *Neocurtilla* (*Gryllotalpa*) *hexadactyla*, the univalent X_1 chromosome and the heteromorphic X_2Y chromosome pair segregate nonrandomly.
3. *Immunol.* = *monovalent *a.* 2 b.
1939 *Bacteriol. Rev.* III. 76 If Hooker and Boyd now believe antibody to be univalent they must abandon the entire basis for these and other of their calculations. **1940** *Jrnl. Exper. Med.* LXXI. 271 Antibody is referred to as low grade, incomplete, imperfect, or 'univalent' in instances in which it is incapable by itself of precipitating with added antigen. **1977** M. W. Steward in Glynn & Steward *Immunochemistry* vii. 245 There are several reports which show the reduction in neutralization of both animal viruses..and bacteriophages..when univalent antibodies have been used.
B. *sb. Cytology.* (yūni·vālěnt). A univalent chromosome.
1912 *Jrnl. Exper. Zool.* XIII. 350 In the first division the X- and Y-chromosomes divide as separate univalents. **1921** *Ann. Bot.* XXXV. 173 These chromosomes were of two types, as revealed during meiosis, fourteen being bivalents and twenty-one univalents. **1971** *Nature* 19 Feb. 570/2 In the triploid species *Leucopogon juniperinus* univalents are close to the poles at metaphase, but..they are not in the same plane as the spindle on which the bivalents are positioned.

universal, *a.* (*adv.*) and *sb.* Add: **A.** *adj.* **9. b.** Also *Universal Aunts*, the name of a company incorporated in 1922 and based in London, which provides domestic assistance to its clients through a staff of professional helpers; hence *Universal Aunt*, a member of the organization; usually *transf.* Cf. *aunt 1 b. Similarly *Universal Uncle*.
1922 *Certificate of Incorporation No. 185,178* (Department of Trade) 20 Oct., I hereby certify that Universal Aunts, Limited is this day incorporated under the Companies Acts..and that the Company is Limited... Registrar of Joint Stock Companies. **1923** *Westm. Gaz.* 10 Jan. 7/6 Associations such as 'Universal Aunts' and 'Useful Women' who supply workers for..social work. **1929** M. Allingham *Mystery Mile* iii. 38 He's really a sort of 'Universal Aunt', isn't he? 'Your adventures undertaken for a small fee.' **1931** —— *Look to Lady* iii. 42, I am..a

sort of universal uncle, a policeman's friend and master-crook's factotum. **1937** A. Thirkell *Summer Half* viii. 229 The universal uncle went down to dinner. **1961** *Listener* 12 Oct. 576/1 His role of cultural Universal Uncle. **1978** M. Dickens *Open Book* iv. 33 There was a domestic agency in Knightsbridge called Universal Aunts—now in Chelsea—which was famous for doing things that real aunts ought to be doing, like meeting small boys from India at the boat train and taking them across London to their train for school.
c. Also *universal decimal classification*, a form of decimal library classification (see *decimal *a.* 1 a).
1930 [see *decimal *a.* 1 a]. **1949** *College & Research Libraries* Oct. 333 The Universal Decimal Classification (U.D.C.). This last was the name given to the 'Brussels expansion' of the *Decimal Classification and Relative Index* of Melvil Dewey. **1958** *B.S.I. News* Sept. 12/1 The B.S.I., with the support of ASLIB and the Library Association, is arranging a series of one-day discussions on universal decimal classification.
14. universal donor *Med.*, a person whose blood is group O (so called because before the discovery of other blood group systems group O blood was thought to be compatible with that of any individual); **Universal Product Code** *N. Amer.* (see quot. 1979); **Universal Provider,** the name of a well-known general store formerly trading in London; freq. with small initials and *transf.*; **universal quantifier** *Logic* [tr. G. *allgemeiner quantificator* (Łukasiewicz & Tarski 1930, in *Sprawozdania z Posiedzeń Towarzystwa naukowego Warszawskiego* (Wydział III) XXIII. 44)], a quantifier referring to all the members of a universe or class; **universal set** *Logic* and *Math.* = Universe 2 d in Dict. and Suppl.; **universal time,** Greenwich time calculated from midnight at the Greenwich meridian (rather than from noon, as formerly).
1922 G. Keynes *Blood Transfusion* iv. 72 Individuals of Groups I and IV have therefore been named 'universal recipients' and 'universal donors' respectively. **1976** Edington & Gilles *Path. in Tropics* (ed. 2) 480 Universal donor group O blood should not therefore be employed for transfusing A or B recipients without prior investigation. **1974** *Consumer Reports* (U.S.) May 364/2 A computer..knows the package code (the grocery industry has agreed on a universal product code). **1979** *Hammond Almanac 1980* 761/2 *Universal Product Code*, a pattern of lines and numbers by which information about a product may be encoded for automatic scanning by a device.. that records its price for charging the consumer as well as its stock numbers, inventory, etc. **1884** *List of Subscribers* (London & Globe Telephone Co.), *Whiteley, W.*, Universal Provider, Westbourne Grove, W. **1903** Beerbohm *Around Theatres* (1924) I. 461 As a curate he has to offer that consolation of which he is universal provider. **1953** *Guardian* 11 Sept. 6/5 To our children we will always be a kind of Universal Provider, vague of face but soft of bosom. **1962** *Sunday Express* 23 Dec. 2/4 The world-famous 'Universal Provider'. Anything from a flea to an elephant..Whiteley boasted he could provide. [**1845** *Encycl. Metrop.* I. 207/1 When the subject of a Proposition is a common Term, the universal signs ('all, no, every') are used.] **1936** *Amer. Jrnl. Math.* LVIII. 353 Then..$(x)P$ and $(\exists x)P$ are propositions of elementary number theory, where (x) and $(\exists x)$ are respectively the universal and existential quantifiers. **1940**, etc. [see *quantifier 1 a]. **1961** J. E. Whitesitt *Boolean Algebra* iii. 60 $\forall_x$ is called the universal quantifier of the variable *x* and is usually read 'for all *x*' or 'for every *x*'. **1980** E. P. Lynch *Applied Symbolic Logic* i. 11 The universal quantifiers are 'all', 'for every', 'for all', and so on. [**1910** Whitehead & Russell *Principia Mathematica* I. i. 30 The class determined by a function which is always true is called the universal class, and is represented by V.] **1959** Universal set [see *solution set* s.v. *solution *sb.* 12]. **1975** I. Stewart *Concepts Mod. Math.* iv. 57 In any particular problem, the sets one is concerned with often lie inside some reasonably small universal set. **1882** *Monthly Notices R. Astron. Soc.* XLII. 205 (*heading*) Universal time and the selection of a prime meridian. *Ibid.*, The American Meteorological Society further considered it desirable that in the future a universal time reckoned from the meridian 180° from that of Greenwich should be generally introduced. **1969** *Times* 24 June 4/7 The object rose above the eastern horizon at 2.49 a.m., universal time.
B. *sb.* **5.** An artificial language invented for universal use by H. Molenaar; also known as *Pan-Roman. Cf. sense 3 c of the adj.
1907, etc. [see *Pan-Roman, Panroman]. **1928** O. Jespersen *International Lang.* 40 Among numerous systems of the same type, but not worked out to the same extent as Neutral, I shall here mention only H. Molenaar's *Universal* (1906). **1947** [see *neutral *sb.* 5].
6. *Linguistics.* Any of the fundamental rules or features proposed as universal attributes of natural languages (see quots.).
1948 B. W. & E. G. Aginsky in *Word* IV. 109 What are the universals of language?.. All languages employ sound sequences in which may be discerned a limited number of recurring types of speech-sound segments. These..are meaningless.., but enter into the meaningful units of form, the morphemes. All languages employ such morphemes in sequences. **1964** Katz & Postal *Integrated Theory of Linguistic Descriptions* 160 A formal universal is a specification of the form of a statement in a linguistic description, while a substantive universal is a concept or set of concepts out of which particular statements in a linguistic description are constructed. **1965** N. Chomsky *Aspects of Theory of Syntax* i. 28 The study of linguistic universals is the study of the properties of any generative grammar for a natural language. *Ibid.*, It is useful to classify linguistic universals as *formal* or *substantive*. A theory of substantive universals claims that items of a particular kind in any language must be drawn from a fixed class of items. **1972** Hartmann & Stork *Dict. Lang. & Linguistics* 245/2 Examples of universals are the conventional character of language.., the duality of transmission and reception, the presence of names and deictic elements. A distinction is sometimes made between *substantive universals*, i.e. features of sound substance such as the phonological elements..and *formal universals* which are made explicit by the linguist in the form of grammatical rules. **1973** *Language* XL. 178 It seems to be a language universal that productive inflectional morphemes are not only very short, but also employ a reduced inventory of phonemes.

universalism, *a.* Add: Also *spec.* = Universalistic *a.* 1. (Earlier examples.)
1837 W. Jenkins *Ohio Gazetteer* 357 It has..three houses for public worship (methodist, presbyterian, and universalian). **1852** J. Reynolds *Hist. Illinois* 327 He is one of the Universalian Baptists.

universalism. Add: **3.** (Earlier example.) Also *spec.* in *Sociol.* and *Econ.*, contrasted with *particularism 5 b and *regionalism 1.
1835 *Leigh Hunt's London Jrnl.* 11 July 221/1 What (if we might take the liberty to coin a word) we would call the universalism of the Homeric poetry. **1939** T. Parsons in *Social Forces* May 462/2 The fact that the central focus of the professional rôle lies in a technical competence gives a very great importance to universalism in the institutional pattern governing it. **1947** —— *Weber's Theory Social & Econ. Organization* 72 Ethical universalism, the insistence on treatment of all men by the same generalized, impersonal standards. **1955** *Bull. Atomic Sci.* Apr. 142 The humanitarian theme of the two preceding centuries certainly persisted, but universalism yielded step by step to national particularism.

universalist, *sb.* Add: **1. b.** *transf.* One who believes in the brotherhood of all men in a manner not subject to national allegiances.
1944 [see *regionalist]. **1952** V. Gollancz *My Dear Timothy* xii. 110 A universalist is a person for whom nations don't exist, only persons. **1955** *Times* 25 Aug. 7/2 It seemed unfair to call him an expatriate, for he was a true universalist. It is a man who loves his fellow men..that Mr. MacDonald has succeeded in being a one-man civilizing factor in this territory.

u:niversali·zable, *a.* Chiefly *Philos.* [f. Universalize *v.* + -able.] That can be made or rendered universal; capable of universal application.
1952 A. E. Duncan-Jones *Butler's Moral Philos.* viii. 171 In order that he shall be said to make a moral judgement, his attitude must be 'universalisable'. **1955** *Proc. Aristotelian Soc.* LV. 170 If too few details are included in the maxims, it will be difficult to find *any* that will be universalisable at all. **1977** P. Baelz *Ethics & Belief* ii. 18 Reasons are impersonal, or inter-personal. They are logically impartial, or universalizable. **1982** *Times Lit. Suppl.* 24 Dec. 1423/4 That a moral imperative should be universalisable seems to be at the heart of what we understand when we understand it as a *moral* imperative.
Also **unive:rsalizabi·lity;** **u:niversali·zably** *adv.*
1952 A. E. Duncan-Jones *Butler's Moral Philos.* viii. 173 What we are calling the universalisability of his attitude consists in his being disposed..to respond in an equivalent way to people and situations of a given kind, should he come to consider them. **1954** W. D. Ross *Kant's Theory of Ethics* 33 But the man who tells the lie may well retort to Kant 'Why should the test of universalizability be applied to my act regarded in this very abstract way, simply as a lie?' **1963** R. M. Hare *Freedom & Reason* vi. 91 This argument would break down if 'ought' were not being used both universalizably and prescriptively. **1982** *Times Lit. Suppl.* 2 July 713/1 'Ought'..commits one to universalizability, in the sense that if I say you ought to do *A*, I implicitly affirm that everyone ought to do *A* in identical circumstances.

universe. Add: **2. d.** *universe of discourse*: also *absol.* For def. read: the totality of entities under consideration; all those that the terms of a proposition may refer to. Also (as *universe*) in *Statistics*, = *population² 2 d. (Earlier and later examples.)
1849 A. De Morgan in *Trans. Cambridge Philos. Soc.* VIII. 380 By not dwelling upon this power of making what we may properly (inventing a new technical name) call the universe of a proposition, or of a name, matter of express definition, all rules remaining the same, writers on logic deprive themselves of much useful illustrations. *Ibid.*, Let the universe in question be 'man': then *Briton* and *alien* are simple contraries. **1898** A. N. Whitehead *Treat. Universal Algebra* I. ii. v. 110 If we extend the Universe of self-evident propositions either by some natural or conventional definition, we may extend the conception of conversion. **1911** G. U. Yule *Introd. Theory Statistics* ii. 17 For actual work on any given subject, no term is required to denote the material to which the work is so confined... But for theoretical purposes some term is almost essential to avoid circumlocution. The expression the *universe of discourse*, or simply the *universe*, used in this sense by writers on logic, may be adopted. **1939** A. E. Treloar *Elements Statistical Reasoning* i. 8 Such a type of selective sampling from this universe is wholly impossible. **1967** G. Wills in Wills & Yearsley *Handbk. Management Technol.* 191 Numbers of calls made by sales representatives is a meaningless item of statistics unless it can be related to..the total universe of outlets which can handle such a product. **1972** *Science* 23 June 1306/2 The universe of discourse is severely restricted in this jargon. **1975** *Brit. Jrnl. Sociol.* XXVI. 37 The universe from which the sample was drawn was all Royal Navy officers stationed in England.

university, *sb.* Add: **I. 1. a.** Const. without article: *at* (or *to*) *university*, etc.
1959 *Listener* 22 Jan. 153/2 Is the son of a miner working-class, suppose he has gone to university. *Ibid.* 5 Mar. 405/2 At school or at university. **1968** *New Society* 22 Aug. 266/2 'He's at university' (very widely used) is certainly non-U.
b. Also in phr. *the university of life*, the experiences of life, considered as a means of instruction. Cf. *the school of hard knocks* s.v. *school *sb.*¹ 4 b.
1959 A. Glyn *I can take it All* i. 12 A revolting cliché like 'educated in the University of Life'. **1972** [see *life *sb.* 12 d]. **1978** P. Hill *Enthusiast* iii. 25 Bob..had the chance to educate him in the real world of people, in the university of life.
c. *University of the Air*: † (*a*) *Aeronaut.* (see quot. 1931); (*b*) an organization which provides a course of (higher) education partly through radio and television broadcasts; *spec.* an early name for the Open University (see *open *a.* 22 c).
1931 *Civil & Milit. Gazette* (Lahore) 4 Nov. 2/3 To meet the demand for trained and competent personnel..the College of Aeronautical Engineering has been formed—the University of the Air. **1959** *Twentieth Cent.* Nov. 369 Any government that was determined could give us a University of the Air tomorrow. **1963** *Glasgow Herald* 9 Sept. 1/1 Mr Wilson..suggested the broadcasting time for the 'university of the air' could be obtained by allocation of the fourth television channel... There could also be appropriate radio facilities. **1969** *Radio Times* 27 Nov. 12 Originally named 'the University of the Air', the Open University offers an exciting new opportunity for adults throughout the country to study for degree qualifications through the media of integrated television, radio and specially-designed correspondence courses. **1984** *Listener* 10 May 7/2 When the idea of a university of the air was first floated, sceptics abounded.
III. 6. a. *university campus, education* (earlier and later examples), *entrance, grant, library* (earlier and later examples), *oath, town* (earlier example); *university-level* adj.
1623 W. L'Isle *Sax. Treat. conc. Old & New Test.* f. e 3, I meane ere long to let the world know what is more remaining; as more I have seene both in our Universitie Libraries, and that of Sir Robert Cotton. **1721** Swift *Let. to Young Gentleman* 26 You cannot but have already observed, what a violent Run there is among too many weak People against University Education. **1823** J. S. Mill in *Morning Chron.* 8 Feb. 3/1 The violation of the University oath, in every case where its observance interferes in the slightest degree with the convenience of the swearer, is a complete proof that the ceremony of swearing affords no security whatever for veracity in any other case. **1857** C. M. Yonge *Dynevor Terrace* I. v. 68 Fitzjocelyn was twenty-one, and had nearly finished his university education. **1872** O. W. Holmes *Poet at Breakfast-Table* 23 The soil of the University town is divided into patches of sandy and of clayey ground. **1919** *Treasury Minute* 14 July in *Rep. University Grants Comm.* 2 in *Parl. Papers* (1921) XI. 362 The Chancellor of the Exchequer states..that he proposes to appoint a Standing Committee..to enquire into the financial needs of University Education... He proposes that the Committee should be known as 'The University Grants Committee'. **1939** University campus [see *campus]. **1957** *Encycl. Brit.* VIII. 941/2 The relationship between the G.C.E. and university entrance qualifications had in the early 1950s reached only a temporary definition. **1962** E. Snow *Other Side of River* (1963) xxx. 226 Among full-term university-level students 283,000 were engineers. **1974** N. Freeling *Dressing of Diamond* 149 A university education served this much purpose..that emotional problems always got placed at the level of intellectual argument. **1982** H. R. Lane tr. *M. Vargas Llosa's Aunt Julia & Scriptwriter* xv. 266, I had a sudden inspiration and headed..for the university campus. **1982** 'J. Ross' *Death's Head* xiii. 75 Mrs Knostig's work is financed by a university grant. **1983** *Chron. Higher Educ.* 19 Oct. 1 More and more university libraries are charging fees for interlibrary loans.
b. *university student*.
1914 G. B. Shaw *Misalliance* p. lxv, An office boy of fifteen is often more of a man than a university student of twenty. **1981** A. Edwards *Sonya* iii. 45 Putting his school in the hands of a young university student who had become his assistant.
c. *university don, professor, staff, teacher*.
1907 G. B. Shaw *Major Barbara* ii. 236 This love of the common people may please an earl's granddaughter and a university professor. *Ibid.* iii. 277 How can you succeed in business when you are willing to pay all that money to a University don who is obviously not worth a junior clerk's wages! **1948** M. Laski *Tory Heaven* iv. 54 The White Paper analysing the origins and opinions of University staffs appalled the whole nation. **1954** M. Beresford *Lost Villages* viii. 266, I went north in April 1948 to become a University teacher. **1971** Halsey & Trow *Brit. Academics* xv. 413 We see the results in their university teacher children. **1977** *Times* 19 Sept. 3/4, 10,000 more students..with a reduction in the unit cost per student..is effectively increasing university dons' teaching loads. **1977** M. Kenyon *Rapist* iv. 42 Miss Hitchcock, a university professor, arrived in Ireland only yesterday.

d. *university-educated* adj.; also with vbl. sbs., as *university teaching*.

1923 R. Macaulay *Told by Idiot* I. 44 No creature was ever more solemn, more earnest,..than the university-educated young female of the eighties. **1962** C. L. Barber in F. Behre *Contrib. Eng. Syntax* 21 Those countries where a great deal of university-teaching is carried out in English. **1981** J. Halkin *Fatal Odds* iii. 49 Well-heeled families, university-educated.

7. university college = College *sb.* 4 d, *spec.* one which is not or was not empowered to grant degrees (see also quot. 1981); **university member**, a member of the House of Commons representing a university or a group of universities (university seats were abolished in Britain in 1948).

1838, etc. [see College *sb.* 4 d]. **1954** *Times* 1 July 9/6 University College of North Staffordshire..is the only university college empowered to grant degrees. **1981** D. Rowntree *Dict. Educ.* 335 *University colleges.* 1. The name formerly given to the UK civic universities which, when first set up, did not have the power to grant their own degrees and usually granted those of London University instead. The last such (Leicester) became autonomous in 1957. 2. In addition, Oxford, Cambridge, Durham and London each has a college named University College, and it is also the name of the independent university at Buckingham. **1867** *Hansard Commons* 18 June 29, I share the opinion of my hon. Friend the Member for Birmingham, that we are already overstocked with University Members. **1949** A. P. Herbert in *Punch* 27 Apr. 453/3 Mr. Haddock, by the way, complains rather bitterly that the University Members are being flung out of Parliament in the sacred name of 'One-Man-One-Vote'. **1979** J. Adam Smith *John Buchan & his World* 80 Buchan was delighted, especially because a university member could sit fairly loose to party.

univoca·lity. [f. Univocal *a.* + -ity.] The state or condition of being univocal (sense 1 b).

1934 *Theology* XXIX. 342 The Scotists, for whom analogy of being gives way to univocality of being. **1959** *Analysis* XX. 7 Calling them by one name because of three features that they all lack (predicative meaning, ascriptive force, and univocality). **1977** E. von Glasersfeld in D. M. Rumbaugh *Language Learning by Chimpanzee* v. 125 It is unlikely that this univocality will be preserved when more correlators and conceptual lexigram classes are added.

unjust, *a.* **1. b.** (Further examples, in *unjust enrichment.*)

1942 *Law Rep.* 4 July 135 It is clear that any civilised system of law is bound to provide remedies for cases of what has been called unjust enrichment or unjust benefit, that is, to prevent a man from retaining the money of.. another which it is against conscience that he should keep. **1962** A. Turner *Law of Trade Secrets* iv. iv. 346 The term 'unjust enrichment' refers to the use to which a disclosure is put. **1973** *N.Y. Law Jrnl.* 24 July 12/3 The defendants ..have counterclaimed for $7,500, in damages for unjust enrichment.

unjustified, *ppl. a.* Add: **3. b.** With type not adjusted to fill up the line or produce an even margin.

1961 Webster, *Unjustified*,..of a line of type: not adjusted properly to fill the measure. **1963** *Times* 6 Mar. 9/6 The *Oklahoma City Times* published its regular editions today with type set entirely by computer... 'Unjustified' perforated tape, such as might be made by a reporter working on a special tape-producing typewriter, is fed into the computer, which cuts a perforated tape. **1971** *Sci. Amer.* Aug. 116/2 This 300-page book, with.. its text in unjustified typescript, indexes the work of the center. **1980** C. Burke *Printing Poetry* iii. 32 Some.. of the problems in a prose setting can be obviated by setting it 'ragged right', that is by leaving the right-hand margin unjustified.

unk (ʌŋk), *sb.* Also **unkie, unky.** Colloq. or nursery abbrevs. of Uncle *sb.*

1907 A. Bennett *Grim Smile of Five Towns* 21 'You carry me down-stairs, unky?' the little nephew suggested. **1957** W. Faulkner *Town* iii. 65 'We need a grindstone,' Gowan said. 'Unk Noon,' Top said. 'We'll take the gun like we're going rabbit hunting,' Gowan said. So they did: As far as Uncle Noon Gatewood's blacksmith shop on the edge of town. **1959** N. Marsh *False Scent* (1960) ii. 50 It's so hard to explain, Unky. **1971** A. Diment *Think Inc.* ii. 19 Didn't you realise you should come home ..and tell Unkie Rupert all about it? **1977** C. McCullough *Thorn Birds* xvii. 408 There's been tons of male influence for your children with the Unks around.

unkingdomed, *ppl. a.* (Later example.)

1917 J. B. Cabell *Cream of Jest* (1923) xix. 109 The Stuarts or the Valois or the Cæsars, or other dynasties long since unkingdomed.

unki·nk, *v.* [Un-[2] 3, 7.] **a.** *trans.* To take the kinks out of, to smooth, to straighten. **b.** *intr.* To lose the kinks, become straight. Also *fig.*

1891 Kipling & Balestier in *Cent. Mag.* Dec. 193/1 Tarvin got himself out of the cart, unfolding his long stiffened legs..and unkinking his muscles one by one. **1947** *Daily Progress* (Charlottesville, Va.) 20 Mar. 8 Designed to un-kink highways. **1947** J. Steinbeck *Wayward Bus* ii. 18 Unkinking the cable behind him on the ground. **1972** C. Short *Blue-Eyed Boy* viii. 82 Gradually my soul began to unkink. **1980** J. McNeil *Spy Game* xiv. 138 The road unkinked and far ahead Corrigan was hopping past a goods van.

unknowing, *vbl. sb.* For † read 'revived in mod. use, esp. in phr. *cloud of unknowing* (after quot. *a* 1400)' and add later examples.

a **1400** (*title*) Þe clowde of vnknowyng. **1629** A. Baker *Commentary on Cloud* (1924) II. iii. 355 This cloud of unknowing..is but the self-same knowledge and sight of God which I and others do usually term the light, sight and knowledge that we have by our faith. **1911** E. Underhill *Mysticism* II. vii. 415 Reason finds itself, in a very actual sense, 'in the dark'—immersed in the Cloud of Unknowing. **1939** T. S. Eliot *Family Reunion* II. ii. 110 Accident is design And design is accident In a cloud of unknowing. **1957** *Oxf. Dict. Christian Church* 402/2 This [progressive deification of man] is to be obtained by a process of 'unknowing', in which the soul leaves behind the perceptions of the senses as well as the reasoning of the intellect. **1976** H. Montefiore in *Christian Believing* 154 Even in granting as much as this to doctrine and dogma, I have to enter into the cloud of unknowing and assert the Church's apophatic tradition. **1984** *Daily Tel.* 9 Feb. 16/5 This is one of those incidents which cloud the mind with mystery, forming, as we brood on it, a positive cloud of unknowing. **1984** *Sunday Tel.* 8 July 18/4 Ideally, the fearful void of unknowing should be filled by the ballast of faith.

unknow·ledgeable, *a.* [Un-[1] 7 b.] **1.** Unknowable. *rare.*

1920 C. M. Grieve in *Northern Numbers* 69 The barrier ..Lifts sheerly..To the unknowledgeable skies. **1926** W. de la Mare *Connoisseur* 303 Such facts were strange, and, as you might say almost unknowledgeable.

2. Not knowledgeable; uninformed.

1969 *Islander* (Victoria, B.C.) 9 Nov. 7/2 Many unknowledgeable people find their way into the mountains. **1974** H. L. Foster *Ribbin', Jivin', & Playin' Dozens* iv. 128 By using the inversion process blacks were able to take advantage of the unknowledgeable white opponents.

unknown, *ppl. a.*[1] and *sb.* Add: **A.** *adj.* **1. b.** (Further examples of *unknown God*: see quot. 1526 in Dict.) *unknown soldier* or *warrior*, an unidentified soldier whose tomb symbolizes that of all those killed in battle.

1607 E. Topsell *Four-footed Beasts* 96 The Arabians sacrifice a camell to the vnknowne god. *c* **1848** M. M. Sherwood *Last Days of Boosy* (ed. 2) 152 The child.. addressed the fearful name..as I had taught him to do to the to him unknown God who made the heavens. **1920** *Times* 11 Nov. 14/1 The Unknown Warrior..was brought to London by night... He lay..awaiting burial today in the Abbey among the greatest of his race. *Ibid.* 12 Nov. 13/1 The body of the Unknown Warrior..was buried in Westminster Abbey yesterday, the King being chief mourner. **1942** E. Waugh *Put out More Flags* i. 36 Rupert Brooke, Old Bill, the Unknown Soldier—thus three fond women saw him. **1947** E. M. Forster in *Harper's Mag.* July 9/2, I add the proviso 'if all goes well' because success lies on the knees of an unknown God. **1970** *Times* 3 June 5/6 Princess Margaret drove today to Mount Avala to lay a wreath of red poppies on the tomb of the unknown soldier. **1980** I. Murdoch *Nuns & Soldiers* i. 42 The soldiers at the Unknown Warrior's grave in Warsaw.

c. *unknown quantity* (examples of *fig.* use).

1865 W. Bagehot in *Fortn. Rev.* 15 May 21 The first election of Mr. Lincoln..was government by an unknown quantity. **1883** [see Quantity 12]. **1951** *Sport* 30 Mar.–5 Apr. 6/3 Mel Ford, the Aberavon prop who gets a 'cap' is an unknown quantity in the North. **1973** *Times* 16 Apr. 14/2 Unless some formula is found for substantial alternative investments for Arab oil money it will continue to be an unknown quantity on the world money market.

B. *sb.* **2. b.** More widely, an unknown thing; an unknown factor (merging with *fig.* use of sense 3).

1878 E. Dickinson *Poems* (1955) III. 849 Let my first Knowing be of thee With morning's warming Light—And my first Fearing, lest Unknowns Engulph thee in the night. **1947** H. S. Shelton in Dewar & Shelton *Is Evolution Proved?* xi. 319 We cannot, therefore, correlate all these unknowns and say what variations result from what changes. **1948** R. Spiller et al. *Literary Hist. U.S.* 1159 Sensitivity to spiritual unknowns. **1964** M. Gowing *Britain & Atomic Energy 1939–1945* i. 36 He acknowledged the unknowns in the situation.

unknownst. (Earlier example.)

1805 E. Cavanagh *Let.* 4 Oct. in Londonderry & Hyde *Russ. Jrnls. M. & C. Wilmot* (1934) II. 185 An Army, God knows, might live *unknownst* in the House!

u·nk-unks, *sb. pl. U.S. slang.* [duplication of the first three letters of the word *unknown* + *pl.* ending *-s*.] A condition resulting from fear of or apprehension about the unknown; unknown factors.

1970 *Time* 9 Mar. 63 Lately the industry has suffered a succession of blows:..As a result, aerospacemen have come down with a severe case of what they call the 'unk-unks'—the 'unknown unknowns'. **1970** *Guardian Weekly* 1 Aug. 24 'Unk Unks'..are the villains favoured by Lockheed to explain the extraordinary series of financial disasters that has led it to the verge of bankruptcy.

unky: see *unk *sb.*

unlabelled, *ppl. a.* Add: Also **unlabeled.** (Examples in *Biol.* and *Chem.*: cf. *labelled *a.* d.)

1935 [see *label *v.* 2]. **1981** *Jrnl. Gen. Microbiol.* CXXVI. 97 Catechol methyltransferase..was localized immunocytochemically in the yeast *Candida tropicalis* by the unlabelled antibody enzyme method.

unladen, *ppl. a.* (Later examples, of the weight of a vehicle.)

1930 *Road Traffic Act 20 Geo. V* 24 in *Parl. Papers 1929–30* IV. 140 The weight unladen of any vehicle shall be taken to be the weight of the vehicle..exclusive of the weight of water, fuel or accumulators used for the purpose of propulsion. **1959** *Motor Manual* (ed. 36) xiii. 270 'Unladen weight' is open to all sorts of interpretation and may bear little relation to what the caravan weighs when it eventually comes to the customer.

unlaid, *ppl. a.* and *sb.* Add: **1. d.** Of a woman with whom no one has had, or a particular person has not had, sexual intercourse. *slang.*

1962 'E. McBain' *Like Love* (1964) iv. 56 What it all meant was: 1. Gaspipe. 2. Sober. 3. Unlaid. **1977** *Sunday Times* 27 Mar. 42/2 A thousand places visited and not absorbed, a thousand paperbacks unread, a thousand unlaid airhostesses.

un-Latin, *a.* Restrict † *Obs.*[−1] to sense in Dict. and add: **b.** Not true to the character of the Latin language or of Latins.

1846 [see Un-Greek *a.*] **1858** J. A. Symonds *Let.* 28 Nov. (1967) I. 173 After the Greek paper came..Latin prose from [Sir Francis] Bacon—a queer and Un Latin piece. **1932** *Times Lit. Suppl.* 21 Jan. 37/3 In one of these [Catalan] poems he writes:..'I know not what thou wouldst say to me..': certainly an un-Latin vagueness.

unlawyer-like, *a.* (Earlier example.)

1869 Taylor & Dubourg *New Men & Old Acres* III. 53 Everything hurried through in the most unlawyer-like manner.

unleached, *ppl. a.* (Earlier example.)

1804 in J. Roberts *Penn. Farmer* 111 Are leeched or unleeched ashes most beneficial as manure?

unleaded, *ppl. a.* Add: **3.** Of petrol, etc.: without added lead. Also *ellipt.*

1965 *Oil & Gas Jrnl.* 20 Dec. 26/1 The industry association will make a study next year of the cost of producing unleaded gasoline. **1970** *Daily Tel.* 14 Oct. 11/8 While a change to unleaded petrol reduces exhaust emission and engine deposits..it necessitates a reduction in the compression ratios to cope with lower octane fuels. **1981** J. D. MacDonald *Free Fall in Crimson* iv. 38 He pulled up to the pump... He took six and four-tenths gallons of unleaded, which came to eight sixty-four.

unleave, *v.* Add: **b.** *intr.* To lose or shed leaves. *rare.*

1880 G. M. Hopkins *Poems* (1967) 88 Margaret, are you grieving Over Goldengrove unleaving?

unless, *prep. phr.*, etc. Add: **2. c.** (Examples in *unless and until.*)

1937 D. Jones in *Le Maître Phonétique* Apr.–June (Suppl.), We should as a rule stick to that pronunciation unless and until we find another native whose speech we have reason to think is more characteristic. **1956** A. Wilson *Anglo-Saxon Attitudes* II. ii. 335 Mother and son had both arrived with the fixed determination of not leaving unless and until either of the two women..should have paid handsomely to secure their departure. **1983** *Times* 1 Feb. 15/1 Unless and until the government also proposes the abolition..of the GLC, the savings in public money are likely to be minimal.

unliberated, *ppl. a.* (Examples corresponding to *liberate *v.* d.)

1970 *Time* 31 Aug. 18 Unliberated honorifics like 'Mrs.' and 'Miss' are replaced by the noncommittal 'Ms.'. **1971** S. Berman *Underground Guide to College of your Choice* 144 Most chicks are still unliberated and the virginity index is still quite high. **1975** *Listener* 9 Oct. 486/4 Ann is..the unliberated woman striving for acceptance in a permissive world. **1981** J. Sutherland *Bestsellers* vii. 85 The unliberated condition of woman—incarcerated, flagellated, degraded, violated—was celebrated time and again.

unlimited, *ppl. a.* Add: **4.** Of a hydroplane: having no limit placed on its engine capacity. Also *absol.* as *sb. U.S.*

[**1953** W. A. Shrader *Fifty Yrs. of Flight* 54 Fifth Annual All-American Air Races are held at Miami, Fla. In the free-for-all unlimited engine displacement race, James Wedell is the winner.] **1956** *Rudder* Apr. 40/2 A dozen open regattas for unlimited inboard hydroplanes have been scheduled. **1959** *Yearbk. Amer. Power Boat Assoc.* 165 Unlimited Hydroplanes shall be designated by the prefix letter 'U'. **1962** *New Yorker* 29 Sept. 102/2 The sleek-shaped unlimiteds, gaily painted like some archaic 'flying circus'. **1972** *Collier's Encycl. Year Bk. 1971* 521 Unlimiteds, those manta-ray shaped thunderboats that have over 200 mph on straightaways. **1976** *World Bk. Year Bk.* 218 The national championship series for unlimited hydroplanes consisted of 10 races with $350,000 in purses.

unlingui·stic, *a.* [f. Un-[1] 7 a + Linguistic *a.*] Not related to linguistics.

1960 [see *stance *sb.*[2] 1 e]. **1962** Y. Olsson in F. Behre *Contrib. Eng. Syntax* 98 Unlinguistic speculations.

unlisted, *ppl. a.* Add: **2.** *spec.* **a.** *Stock Exchange.* Formerly designating securities not dealt in on the Stock Exchange; also (*N. Amer.*), those sold over the counter (see *over the counter *adv.* (*a.*) b); now in *unlisted securities market*, a market for securities in

small companies admitted for trading on the Stock Exchange but not bound to comply with the rules for listed securities.

1905 [in Dict.]. **1921, 1929** [see *over the counter *adv.* (*a.*)]. **1979** *Times* 13 Dec. 17/5 A new Unlisted Securities Market (USM) has been proposed which will provide a half way house between a private company and an official listing. **1983** *Sunday Tel.* 10 Apr. 22/6 The Unlisted Securities Market (U.S.M.) is now an established part of The Stock Exchange London.

b. Of a telephone (number): = *ex-directory *a.* Chiefly *U.S.*

1937 R. Chandler in *Dime Detective Mag.* Nov. 48/2 Willie Peters..did a sideline selling unlisted telephone numbers bribed from maids and chauffeurs. **1942** E. S. Gardner *Case of Careless Kitten* xiv. 137 Mason's unlisted telephone was ringing as he opened the door of his apartment. **1974** 'J. le Carré' *Tinker Tailor Soldier Spy* xi. 84 A belly-ache from Admin..about the misuse of unlisted Circus telephones for private calls. **1980** G. V. Higgins *Kennedy for Defense* v. 48 Gretchen doesn't give out the unlisted number. If I had a call, it was from somebody important.

unliterary, *a.* Add: Hence **unli·terariness.**

1961 C. S. Lewis *Experiment in Criticism* viii. 76 His very unliterariness saves him from confusing the two [*sc.* art and knowledge].

unliterate, *a.* Add: **b.** Unliterary; not interested in reading or literature. Also *absol.*

1950 M. Mead *Male & Female* xiii. 271 The Gesell norms used by the reading mother or the neighbourhood gossip of the unliterate. **1960** *Guardian* 13 June 5/2 The innumerate humanist and the unliterate scientist were equally inadequate.

unliv(e)able, *a.* **2.** (Earlier example.)

1834 M. Edgeworth *Tour in Connemara* (1950) i. 43 The want of window curtains..gave the whole an unfinished unlivable appearance.

unlived, *ppl. a.*[2] Add: **b.** [Un-[1] 8 c.] With *-in*. Appearing to be uninhabited; unused by the inhabitants.

1931 H. Nicolson *Diary* 28 Jan. (1966) 67 The room has an unlived-in appearance. **1943** D. Welch *Jrnl.* 15 June (1952) 74 There is the feeling about the whole place of utter solitude, stealth, the ghostly, unlived-in, fascinating feeling of week-end houses. **1975** *New Yorker* 29 Dec. 22/1 Any room..has an unlived-in corner. **1979** J. Gardner *Nostradamus Traitor* xxiii. 95 The vestibule led to the main room, large and uninviting... The place looked unlived-in.

unliving, *ppl. a.* Add: Hence **unli·vingness.**

1914 D. H. Lawrence *Prussian Officer* 80 The sick man lay as if dead... Miss Louisa was heavy-hearted under the load of unlivingness. **1928** [see *rubber-necked* s.v. *rubber *sb.*[1] 13 c].

unload, *v.* Add: **1. b.** (Later examples.) Also with *on*.

1887 B. Harte *Millionaire & Devil's Ford* 158 He might unload his gossip because Mamie wouldn't have him. **1904** *Minneapolis Times* 28 June 6 Dr. Dowie has landed in New York and unloaded an interview in praise of President Roosevelt. **1976** J. I. M. Stewart *Memorial Service* v. 66 This was probably why he unloaded on me these useless gobbets of information. **1978** 'S. Woods' *Exit Murderer* 160 If we succeed in identifying Mr. X I shall unload the whole thing on them [*sc.* the police].

2. a. (Later *fig.* examples.) *esp.* To confide in someone, to divulge information, etc. Also with *on*.

1904 W. H. Smith *Promoters* i. 8 I'm so full of it that I shall burst if I don't unload. **1972** 'J. Godey' *Three Worlds of Johnny Handsome* (1973) ii. 27 If you get along with your cell partner, you usually unload. I did the same with him. **1978** 'D. Kyle' *Black Camelot* xii. 185 'What's the problem with this German? Why won't he unload?'.. 'He thinks once he comes through..we'll knock him off because he knows too much.' **1984** *Miami Herald* 30 Mar. 3 B/1 Your letter exhibits a great deal of bottled rage. I strongly suggest that you unload on a counselor.

6. a. (Earlier example.)

1870 J. K. Medbery *Men & Mysteries Wall St.* 138 To unload, is to sell out a stock which has been carried for some time.

b. *transf.* To sell or dispose of (anything); to get rid of by sale; *esp.* to dispose profitably of something that is unwanted or that constitutes an embarrassment. Also with *on*.

1884 *Boston Jrnl.* 15 Mar. 2/3 There is a flavor of reviving an excitement in order to unload oil lands. **1901** Merwin & Webster *Calumet 'K'* ii. 30 They're going to make a mighty good try at unloading it on him and making him pay for it. **1929** D. Hammett *Dain Curse* v. 40 That dinge of yours—Rhino Tingley—was picked up in a hock shop last night trying to unload some jewelry. **1946** *Time* 25 Mar. 82/3 Many a landlord..has already unloaded a building, at a fat price, on his tenants. **1976** 'M. Delving' *China Expert* v. 56 Tashjian..had only the day before unloaded an extremely dubious Han tomb figurine on an unsuspecting dealer.

c. To get rid of (a person).

1973 E.-J. Bahr *Nice Neighbourhood* xiii. 139 John wasn't about to be any problem when she had everything set up and wanted to unload him. **1982** C. Watson *Whatever's been going on at Mumblesby?* xvi. 150 Cork-Bradden's main object was to 'unload' the woman on to me, in order to placate his wife.

unlogical, *a.* Add: Also, not involving logic. Hence **unlo·gically** *adv.*

1897 W. M. Urban *Hist. Princ. Suff. Reason* iv. 45 Sufficient Reason he speaks of as a metalogical truth.. which functions unlogically in space, time, causality and motivation. **1922** tr. *Wittgenstein's Tractatus* 43 We cannot think anything unlogical, for otherwise we should have to think unlogically.

unlo·nely, *a.* [Un-[1] 7 a.] Not lonely. Also *absol.*

1952 J. Steinbeck *East of Eden* 396 The poison of loneliness and the gnawing envy of the unlonely. **1967** H. W. Sutherland *Magnie* iv. 63 A man should sleep with his wife in winter. It's warm—unlonely. **1971** P. Scott *Towers of Silence* III. iv. 197, I am not unlonely.

unluck. (Earlier example.)

1795 E. Wynne *Diary* 16 Jan. (1937) II. ii. 12 This feast has great many misfortunes we have every day more unluck.

unmade, *ppl. a.* **3. b.** (Later examples with *up*.)

1936 D. Powell *Turn, Magic Wheel* II. 133 Their decently un-made-up little wives. **1960** *News Chron.* 12 Oct. 3/2 Her unmade-up face smiling. **1971** J. Turner *Stone Dormitory* vi. 63 He went on along the unmade-up road. **1978** R. Hill *Pinch of Snuff* iv. 42 He was roused from the unmade-up spare bed by Ellie pulling his hair.

unmaid, *v.* (Later example.)

1922 Joyce *Ulysses* 386 She was there unmaided.

unmakable, *a.* (Later examples.)

1939 W. Fortescue *There's Rosemary* I. 282 The designer had drawn a lovely but unmakeable model which had only one side. **1974** E. Brawley *Rap* (1975) I. xi. 183 He knew escape was impossible, an unmakable caper. **1979** *Country Life* 4 Oct. 1127/2 Seven No Trumps was unmakeable.

unman, *v.* Add: Hence also **unma·nningly** *adv.*

1947 Dylan Thomas in *Horizon* Dec. 302 For who unmanningly haunts the mountain caverned eaves.

unmanned, *ppl. a.*[1] **1.** (Examples relating to aviation, space travel, etc.)

1906 *Nature* 8 Nov. 35/2 The machines he made and launched were all 'unmanned'. **1907** *Ibid.* 4 Apr. 538/2 During the course of the last few years very rapid strides have been made in investigating the upper air by means of manned and unmanned balloons. **1946** *Congressional Digest* May 154/2 'Drone' aircraft—they are unmanned, radio controlled. **1954** *Economist* 11 Sept. 3/2 The manned fighter is surrendering some of its duties to wingless, almost tailless, unmanned missiles. **1962** F. I. Ordway et al. *Basic Astronautics* xiii. 552 Unmanned satellites and guided missiles. **1969** *Listener* 20 Feb. 232/2 The pictures you brought back from the Moon were not as good as those taken on an unmanned flight. **1972** *Guardian* 22 Feb. 2/4 The unmanned Soviet moon probe Luna 20 made a soft landing on the moon's surface last night. **1977** *R.A.F. News* 11–24 May 3/4 The RAF was watching the developments concerning Cruise missiles and other unmanned systems... If unmanned vehicles could be developed to take on some of the roles of air power this would be welcomed.

unmarked, *ppl. a.* Add: **1. a.** Add to def.: having no distinguishing or identificatory mark. (Further examples.)

1936 M. Mitchell *Gone with Wind* xxx. 502 The thousands in unmarked graves who would never come home. **1960** 'E. McBain' *Give Boys Great Big Hand* (1962) v. 42 The unmarked police sedan pulled to the curb. **1967** L. Hughes *Panther & Lash* 65 They buried Lumumba In an unmarked grave. But he needs no marker—For air is his grave. **1975** *New Yorker* 2 June 101/1 The two patrolmen had driven in their unmarked car to the west end of the Mercer Island Floating Bridge. **1979** C. McCarry *Better Angels* III. iii. 208 He had flown in from Paris..in his own unmarked plane.

b. (Earlier example.)

1791 F. Burney *Jrnl.* Dec. (1972) I. 103 Our visit to Mrs. Montagu turned out very unmarked; I met my good Mrs. & Miss ord, & a little chat with them was all my entertainment.

c. Of a linguistic construction, form, etc.: not marked (see *marked *ppl. a.* 1 c).

1933, etc. [see *marked *ppl. a.* 1 c]. **1964** C. Barber *Present-Day Eng.* iv. 105 *Author*..can be used of both men and women; this is called the *unmarked* member of the pair [*author: authoress*]. **1978** *Sci. Amer.* Nov. 95/1 In the case of 'wide' and 'narrow', 'wide' is the unmarked word: asking 'How wide is the road?' does not suggest that the road is wide, but asking 'How narrow is the road?' does suggest that the road is narrow. **1980** *Amer. Speech* LV. 88 Features of an unmarked register may be imported into a marked one, but not vice versa.

unmarried, *ppl. a.* **1.** (Further examples, of parents.)

1834 *Rep. Poor Laws* 196 in *Parl. Papers* XXVII. 200 An unmarried mother has voluntarily placed herself in the situation of a widow. **1933** D. C. E. Peel *Life's Enchanted Cup* i. 9 People did not look upon unmarried mothers with so lenient an eye as they came to do during the war. **1965** Hall & Howes *Church in Social Work* ii. 50 Unmarried fathers and 'other men in moral difficulty'. **1972** *Guardian* 15 Nov. 9/1 One [letter] suggests that if unmarried mums were only encouraged to keep their babies, this sort of thing couldn't happen. **1983** J. Gardner *Elephants in Attic* iv. 29 In the thirties unmarried mums were not the 'in' thing.

2. (Later examples, of parenthood.)

1930 R. Lehmann *Note in Music* iv. 157 He carried on the splendid tradition of unmarried fatherhood. **1962** *Sunday Express* 30 Dec. 19/5 The problems of unmarried motherhood. **1980** C. Fremlin *With No Crying* i. 6 Friends..passionately defending her right to unmarried motherhood.

unmechanized, *ppl. a.* (Later examples.)

a **1930** D. H. Lawrence *Last Poems* (1932) 180 But oh, men, men still unmechanised,..what are you going to do, entangled among all the engines? **1937** B. H. L. Hart *Europe in Arms* xxii. 296 The enemy is not only taken unaware but is himself unmechanized. **1973** R. Lane Fox *Alexander the Great* iv. 79 An agricultural and unmechanized world.

unmeditative, *a.* (Earlier example.)

1831 *Examiner* 12 June 370/1 The inert, unobserving unmeditative mass.

unmentionability. Add: **b.** The fact of (some things') being unmentionable.

1909 G. B. Shaw in *Three Plays by Brieux* (1911) p. xv, Their imitators assumed that unmentionability was an end in itself. *c* **1925** D. H. Lawrence *Virgin & Gipsy* (1930) i. 12 The children were brought up in this atmosphere of cunning self-sanctification and of unmentionability.

unmentionable, *a.* and *sb.* Add: **b.** (Earlier example.) Also, ·underpants, and (chiefly *joc.*) underwear, esp. women's.

1823 *London Mag.* Oct. 433/2 Liston, in a pair of *unmentionables* coming half-way down his legs. **1910** O. Johnson *Varmint* 221 Each [boy] was required to don upper and lower unmentionables. **1930** *Amer. Speech* V. 497 Silk nighties, panties, and undies in general..these articles were consistently grouped in the common speech as 'unmentionables'. **1972** *Daily News-Miner* (Fairbanks, Alaska) 3 Nov. 1/7 The Russians..'d buy up perfume, wrist watches and ladies' unmentionables. **1974** *Times* 5 Jan. 10/6 Fear of being ambulanced away to a place where nurses will..snigger at your frayed unmentionables.

c. *sb.* A person or thing not to be mentioned (by name). Chiefly *pl.*

1928 *Public Opinion* 8 June 547/3 If you pinched a penny of his pay you passed beyond the pale, you became an unmentionable, you ceased to be a comrade. **1939** F. Thompson *Lark Rise* iii. 51 Those parts of the human body then known as 'the unmentionables'. **1975** *Nature* 17 Jan. 149/3 The lowest grade, dirty, wet mixtures of plastic, paper and other organic unmentionables could at least provide heat, if distilled in a pyrolysis plant set up on a wasteplex site. **1976** *Sat. Rev.* (U.S.) 30 Oct. 24/2 *Feijoada*, a Brazilian mix of rice, beans, and pork unmentionables. **1981** *Times* 6 Mar. 6/5 Already the unmentionables have been mentioned and the unpublishables published in the main..weeklies. **1982** 'P. Loraine' *Sea-Change* III. vii. 168 The whole plan was based upon unmentionables.

unmet, *ppl. a.*[1] (Examples corresponding to Meet *v.* 7.)

1962 *Guardian* 13 July 10/4 Housing needs often make unmet demands. **1969** *Listener* 6 Feb. 172/3 There remains a massive unmet need for advice and aid from legally trained people. **1980** *Times Lit. Suppl.* 6 June 652/2 The particular need of the rapidly growing sectors, unmet by the traditional capital market, was to be of significance.

unmetaphorical, *a.* Add: (Earlier example.)

1746 J. Wesley *Let.* 25 June (1982) II. 199 In as plain and unmetaphorical words as the nature of the thing would bear.

Hence **unmetapho·rically** *adv.*

1829 G. Griffin *Collegians* II. xxiv. 193 Sitting down to her, unmetaphorically, bitter draught with the meekest resignation. **1932** H. H. Price *Perception* i. 13 This assumption when openly and unmetaphorically stated, is.. extraordinary. **1963** R. M. Hare *Freedom & Reason* xi. 210 Prepositions expressed in terms of them [*sc.* metaphors] are, when put unmetaphorically, true.

unme·tered, (*ppl.*) *a.* [Un-[1] 8, 9.] Not measured by means of a meter; not provided with a meter.

1909 in Webster. **1967** *Guardian* 13 Sept. 8/3 It is quite untrue that airport taxis are unmetered... The tariff is clearly displayed inside the cab on the taximeter. **1977** C. Brandreth *Parking Law* 57 Any unmetered spaces are provided for loading and unloading and picking up and setting down passengers. **1980** *Blair & Ketchum's Country Jrnl.* Oct. 111/2 A 1975 report on metropolitan area water usage and 'unmetered water'—quantities leaked, used to flush mains or extinguish fires, or consumed free by public facilities.

unmi·ssable, *a.* [Un-[1] 7 b.] That cannot or should not be missed.

1934 in Webster. **1972** *Guardian* 18 Mar. 10/2 Radio 4's unmissable 'Word in Edgeways'. **1978** *Times* 28 Aug. 8/8 William Randolph Hearst's fairy castle..is an unmissable monument to random acquisitiveness. **1980** *Daily Tel.* 23 Aug. 10 One act in the 'unmissable' category at the Edinburgh festival must surely be the Prvimaj Pirot Workers Culture Art Band from..Yugoslavia. **1983** *Observer* (Colour Suppl.) 13 Mar. 32/3 One unmissable part of her daily routine.

unmistakable, *a.* Add: Hence also **u:nmistakabi·lity.**

1923 J. M. Murry *Pencillings* 13 It matters only if another writer should arise who..will take advantage of some of Henry James's explorations and use them in order

to increase his own unmistakability. **1972** *Daily Tel.* (Colour Suppl.) 19 May 41/3 One thing all [these villages] possess, and Torremolinos-rampant does not, is unmistakeability.

unmix, *v.* Add: **a.** (Later example of *trans.* use.)

1973 *Sci. Amer.* Apr. 115/2 That is why it is easy to mix cream into a cup of coffee but difficult to unmix the two.

b. *intr.*

1968 *Physical Rev. Lett.* XX. 318/1 These particles will phase 'unmix' causing the appearance of a third wave, 'the echo'. **1971** I. G. Gass et al. *Understanding Earth* i. 18/1 A high temperature feldspar..frequently 'unmixes' if cooled slowly to form a crystal composed of discrete blebs or laminae of two compositionally different feldspars. **1980** *Phil. Mag.* A. XLI. 637 For situations in which a true thermodynamic equilibrium can be attained, the equilibrating species unmix, producing a heterogeneous composition.

unmi·xing, *vbl. sb.* [Un-[2] 8.] The process by which the components of a mixture separate.

1929 *Amer. Mineralogist* XIV. 235 Pentlandite is supposed to be one of the components of the 'unmixing' of a solid solution of (Fe, Ni)S. **1934** *Proc. Nat. Acad. Sci.* XX. 452 This is evidently the reason for unmixing of a solid solution (in mineralogical parlance) or the formation of a segregate phase (in metallographic parlance). **1950** [see *solvus]. **1974** *Nature* 9 Aug. 480/1 To avoid any appreciable unmixing occurring on cooling, the sample was rapidly removed from the hot furnace into a water-cooled brass jacket.

unmodish, *a.* Delete † *Obs.* and add later example.

a **1974** R. Crossman *Diaries* (1977) III. 225 Of all the places which are not exactly with-it that dreary part of South London is the worst, brand new and yet unpopular and unmodish.

unmollified, *ppl. a.* (Later examples.)

1934 W. S. Churchill *Marlborough* II. xxiv. 531 Slangenberg, unmollified, objected even to receiving the order of his own Government from the Commander-in-Chief under whom he was to serve. **1968** *Punch* 12 June 853/1 Animals not allowed, said the hatchet manageress. Good God, woman, this is our last night in England. She was unmollified.

unmortal, *a.* Delete † and add later example.

1935 Dylan Thomas in *New Verse* Aug.–Sept. 2 My man of leaves and the bronze root, mortal, unmortal.

unmould, *v.* Add: **1. b.** (Further examples.) Also *absol.*

1971 *Daily Tel.* 15 July 15/2 The no-cooking puddings.. can await your pleasure in the freezer..; but do remember not to unmould them until the moment of service. **1972** *Ibid.* 28 Dec. 11/3 Unmould, and dust with sifted icing sugar and serve with orange sauce. **1977** *Lancashire Life* Mar. 111/2 Pour into a suitable mould and chill or leave to set. Unmould onto lettuce leaves and serve with mayonnaise. **1982** L. Chamberlain *Food & Cooking of Russia* (1983) 265 Refrigerate for a few hours. Unmould on to a serving plate.

unmoving, *ppl. a.* **2.** Delete *rare*[−1] and add later example.

1971 S. Hill *Strange Meeting* 188 We have had a pep talk from the Brigadier, and last week, a pep letter came round to all officers and N.C.O.'s—entirely unmoving.

unmusicality. (See small-type note s.v. Unmusical *a.*) (Examples.)

1946 *Penguin Music Mag.* Dec. 14 Our vaunted choral singing is really the proof of our fundamental unmusicality, because any amateur can sing in a chorus. **1963** *Times* 17 May 18/6 Beethoven (whom to admit disliking is still tantamount to an admission of unmusicality). **1984** *Listener* 12 Jan. 28/3 Their unmusicality demands public exposure.

unmusicalness. (Later examples.)

1873 C. M. Yonge *Pillars of House* I. x. 201 Geraldine resembled Fulbert in unmusicalness. **1922** *Daily Mail* 10 Nov. 7 She had been painfully struck by..'the unmusicalness' of the bells of public clocks.

unnerve, *v.* Add: Hence also **unne·rvingly** *adv.*

1962 I. Murdoch *Unofficial Rose* xxxiii. 313 There was something unfocussed, something a little unnervingly fragmentary, in his present apprehension of Lindsay. **1976** T. Heald *Let Sleeping Dogs Die* i. 10 Parkinson had been unnervingly friendly.

unneutral, *a.* (Later examples.)

1895 T. A. Walker *Man. Public Internat. Law* iv. ii. 165 A neutral Government is in general responsible in respect of the unneutral employment of its territories and territorial waters. **1949** W. S. Churchill *2nd World War* II. II. xx. 358 The transfer to Great Britain of fifty American warships was a decidedly unneutral act by the United States. *a* **1974** R. Crossman *Diaries* (1976) II. 366 Most of the gains we made by the July measures would have been upset by an unneutral policy.

unnoticeably, *adv.* (Earlier example.)

1872 Geo. Eliot *Middlem.* IV. viii. lxxx. 285 She would make as quietly and unnoticeably as possible her second attempt to see and save Rosamond.

unobservable, *a.* Add: **3.** *sb. pl.* Things which cannot be observed.

1944 *Mind* LIII. 227 It is not at all certain that there are not highly respected scientific hypotheses which allege the existence of unobservables. **1980** *Times Lit. Suppl.* 17 Oct. 1181/2 The logical positivists held that for a sentence to be meaningful it must be capable of experiential verification. In consequence they had a central problem with discourse about unobservables... Their standard line of solution was some kind of definitional reduction of unobservables to observables.

Hence **u:nobservabi·lity.**

1944 *Mind* LIII. 224 The positivist principle..does not tell us anything at all about the observability or unobservability of the facts stated in P. **1979** *Amer. Pol. Sci. Rev.* Mar. 164/2 Unobservability is one reason..for the relative unfamiliarity of political things.

unobtainable, *a.* (Examples in *Teleph.*)

1930 [see *number *sb.* 4 f]. **1961** E. Waugh *Unconditional Surrender* II. iv. 96 When they tried to ring him up they were told the number was 'unobtainable'. **1968** P. Geddes *High Game* v. 58 He..retreated to telephone. .. A misconnection to Guildford, two unobtainable signals and nine minutes later he had finally made it. **1976** 'K. Royce' *Bastillo* vi. 82 He..rang Todashi... He got the unobtainable tone.

unoccupied, *a.* Add: **3. c.** *spec.* Designating that part of France not held under German military occupation during the war of 1939–45. Cf. *occupied *ppl. a.*

1940 *New Statesman* 19 Oct. 380 A Jewish friend (who recently returned from unoccupied France to his home in Paris) told me that he liked the English immensely. **1942** Mrs. Belloc Lowndes *Let.* 3 July (1971) 230, I hear dreadful accounts of France—all old people and delicate people, are dying—especially..in the unoccupied districts. **1978** A. Price *'44 Vintage* xviii. 210 A chateau south of the Loire..in the Vichy zone of unoccupied France.

unofficial, *a.* and *sb.* Add: **1.** *unofficial strike*: one not endorsed by the relevant union.

1946 'G. Orwell' in *Partisan Rev.* Summer 321 There is resentment against long hours and bad working conditions, which has shown itself in a series of 'unofficial' strikes. **1955** *Times* 24 June 4/3 The Minister of Labour could not deal with unofficial strikes in the normal way. That was a matter for the union concerned to re-establish its authority over its own members. **1972** *Guardian* 24 Nov. 10/1 Lower-paid hospital workers are resorting to a series of unofficial strikes.

2. a. spec. *unofficial member* = *private member* s.v. Private *a.* 2 c.

1879 *Tablet* 31 May 709/1 An unofficial member. **1893** Erskine May *Law of Parl.* (ed. 10) viii. 245 The relative precedence of government business, and business in charge of unofficial members,..is prescribed by the standing orders. **1970** A. P. Herbert *In Dark* iii. 74 There spoke, besides two long-suffering Ministers, 14 private (or 'unofficial') members—12 against and 2 in favour.

unoperated, *ppl. a.* Add: In *Med.*, not having been operated on.

1932 J. S. Huxley *Probl. Relative Growth* iv. iv. 126 The effect of regeneration on the normal growth of neighbouring unoperated structures. **1975** *Year Bk. Ear, Nose & Throat* 119 Increased uptake in the unoperated ear.

unordered, *ppl. a.*[1] Add: **2. b.** *Linguistics.* Of rules: not requiring to be applied in a particular order.

1968 *Language* XLIV. 696 Unordered disjunctive sets [of rules] are abbreviated by variables. **1970** *Canadian Jrnl. Linguistics* XV. 97 These rules are considered unordered; the initial numbers are inserted merely for reference. **1979** *Trans. Philol. Soc.* 106 The type of phonological theory which allows the maximum simultaneity of application of rules is the so-called 'unordered rule hypothesis'.

unoriginal, *a.* and *sb.* Add: Hence **unori·ginally** *adv.*

1934 in Webster. **1963** V. Nabokov *Gift* v. 297 'And so I'll never see him again,' he told himself, unoriginally. **1964** *Archivum Linguisticum* XVI. 15 Several handbooks discuss our problem either too cursorily or too unoriginally to invite analysis.

unorthodox, *a.* Add: Hence **uno·rthodoxly** *adv.*

1934 in Webster. **1969** *Daily Tel.* 25 Apr. 21/3 Everything seemed to be right, both in this unorthodoxly proportioned work itself and in the expounding of it. **1977** A. Wilson *Strange Ride R. Kipling* vi. 285 The loves in the village world..are enjoyed precariously and unorthodoxly.

unostentatiousness. (Contextual example.)

1901 *Chambers's Jrnl.* Apr. 234/2 A pattern of unostentatiousness.

unpack, *v.* Add: **2. a.** *fig.* (Later examples.)

1966 *Philosophy* XLI. 30 If we unpack what is meant by 'positive morality' we come down to the opinions of individual people. **1979** A. R. Peacocke *Creation & World of Sci.* i. 39 The theological enterprise has always involved much unpacking and elaborating of this image.

5. *Computers.* To convert (an item of stored data) *into* two or more separate items; to retrieve data from (a record). Cf. *pack *v.*[1] 3 g.

1954 *Computers & Automation* Dec. 22/2 *Unpack*, to separate packed items of information each into a separate machine word. **1959**, etc. [see *pack *v.* 3 g]. **1960** Gregory & Van Horn *Automatic Data-Processing Systems* iii. 104 The 13 instructions given above..are required to unpack each meter-reading card and set up the quantity for bill calculation. **1972** *Computer Jrnl.* XV. 199/1 A stream..might be formed by one stream function which unpacks words into bytes.

unpacker. (Earlier example.)

1768 J. Wedgwood *Let.* 13 June (1965) 64 He writes a good Hand, and will be more useful in that Respect than as an unpacker.

unpalatable, *a.* Add: Hence also **u:npalatabi·lity.**

1934 in Webster. **1974** *Nature* 25 Jan. 170/1 The potential unpalatability of a current policy designed to affect the demographic situation in the next century. **1981** *Birds* Summer 37/2 Inedible insects advertise their unpalatability by wearing warning colours—usually black and yellow or black and red.

unpapered, *ppl. a.* (Earlier contextual example.)

1805 F. Burney *Let.* 29 May (1975) VI. 517 It is unfurnished, indeed, unpapered, & every way unfinished.

unpatterned, *ppl. a.* Add: **3.** Not formed or cast into a pattern.

1949 M. Mead *Male & Female* ix. 190 Man, the heir of tradition, provides for women and children. We have no indication that man..unpatterned by social learning, would do anything of the sort. **1960** T. McLean *Kings of Rugby* xi. 192 Too much unpatterned play. **1977** *Sci. Amer.* Jan. 117/1 Consider an analogous isomorphism exhibited by a sequence of unpatterned digits such as pi.

unpeg, *v.* Add: (Later examples.) Also *fig.* Hence **unpe·gging** *vbl. sb.*

1923 J. M. Keynes *Tract on Monetary Reform* 121 The present situation..did not begin until after the 'unpegging' of the leading exchanges in 1919. **1975** 'M. Yorke' *Small Hours* i. 14 A young woman came to unpeg them [*sc.* the clothes]. **1977** *Guardian Weekly* 7 Aug. 10/1 Unpegging sterling from the US dollar is a thoroughly sensible move. **1977** *Times* 29 Oct. 19/4 The Israel pound fell sharply as it was unpegged from its official rate of 10 to the dollar. **1981** *Times* 21 Jan. 11/4 They chat as they unpeg the washing.

u·npeople, *sb.* [Un-[1] 12.] **a.** People excluded from 'the people' as understood politically. **b.** Unpersons.

1962 E. Snow *Other Side of River* (1963) I. 380 There is less mystery in China today than in most Communist countries about what both people and 'unpeople' find unbearable. **1968** *Guardian* 28 Nov. 11/2 (*heading*) Czech unpeople down on the farm. **1970** *Guardian Weekly* 21 Mar. 6 People on foot on a hot road..walking from nowhere to nowhere... Tired people. Unpeople. **1975** P. Lively *Going Back* vii. 88 Who are those people the Greeks knew about—the ones who wander around the Styx? The unburied dead: grey, unpeople. I am one of them. **1975** Ld. Hailsham *Door wherein I Went* vi. 28 Communists teach that there was no such person as Jesus at all... He is the unperson to end all unpeople.

u·nperson. [Un-[1] 12: introduced by 'George Orwell'.] A person who, usu. for political misdemeanour, is deemed not to have existed and whose name is removed from all public records. In extended use, a person whose existence or achievement is officially denied or disregarded; a person of no political or social importance.

1949 'G. Orwell' *Nineteen Eighty-Four* II. 159 Syme was not only dead, he was abolished, an *unperson*. **1954** *Economist* 18 Sept. 883/2 Beria is already an 'unperson', the record of his career 'unfacts'. **1961** *Guardian* 28 Apr. 8/5 The concentration camp was a factory for processing people into un-persons. **1962** *Listener* 15 Feb. 308/1 From the Soviet point of view they are 'unpersons', ignored or slandered in Soviet travesties of literary history. **1969** H. E. Salisbury *Siege of Leningrad* I. iii. 24 Berezhkov omits any mention of Dekanozov's name or of the Dekanozov-Weizsäcker meeting. Because of his execution in 1953 Dekanozov apparently has become an unperson. **1981** P. Dickinson *Seventh Raven* xi. 151 You've got absolutely nothing to do..in hospitals... Places like that tend to turn you into a kind of unperson. **1983** *Listener* 16 June 4/1 He omitted the then Foreign Secretary, Francis Pym, who seemed even then to have become an unperson.

Hence **unpe·rson** *v.* to make into an unperson (usu. in *pa. pple.*); **unpe·rsoning** *vbl. sb.* Also *transf.*

1966 Page & Burg in *Soviet Stud.* XVIII. 96 (*title*) Unpersoned: the fall of Nikita Sergeyevitch Khrushchev. **1973** *Listener* 4 Jan. 8/2 On television... The addition of a face..turns the newsreader into a person, but the job then requires him to unperson himself. **1976** *Times Lit. Suppl.* 13 Feb. 156/3 The unpersoning process [in Czechoslovakia] had gathered momentum and many of the notables of 1968–69 were being rapidly transmogrified into the nobodies of the 1970s. **1977** *Listener* 16 June 790/2 One of them dead and the other efficiently unpersoned and confined to a political asylum. **1983** *Daily Tel.* 12 Mar. 14/2 In 1956, Bob..brought in Hamilton.. as editorial director... But in the new edition it looks as if all the work was done by his successor, Harold Harris. 'It is no trivial matter to be "unpersonned",' says Hamilton.

unperturbed, *ppl. a.* (Examples in *Physics*: cf. *perturbed *ppl. a.* 2.)

1937 [see *perturbed *ppl. a.* 2]. **1967** Margerison & East *Introd. Polymer Chem.* ii. 67 A 'poor' solvent, on the other hand, is one in which the polymer dimensions approach those of the unperturbed configuration. **1974** G. Reece tr. *Hund's Hist. Quantum Theory* xiv. 184 The function ψ^0 is a solution of the 'unperturbed' problem $H^{(0)}$.

unphysiological, *a.* Add: Also **unphysiolo·gic** *a.* (chiefly *U.S.*); **unphysiolo·gically** *adv.*

1934 Webster, *Unphysiologic.* **1948** *Endocrinology* XLIII. 118 The amount of food consumed..was unphysiologically large in most instances. **1969** *Obstetrics & Gynecol.* XXXIII. 419/1 It is unphysiologic for newlywed apareunia. **1970** *Nature* 17 Oct. 210/1 The significance of this effect *in vivo* may be limited, for it takes place *in vitro* only at unphysiologically high levels of glucose.

unpick, *v.* **2.** (Earlier contextual example.)

1808 Jane Austen *Let.* 7 Oct. (1932) I. 217 Your gown shall be unpicked.

unpicturesquely, *adv.* (Earlier example.)

1841 in H. H. Wilson *Trav. W. Moorcroft & G. Trebeck* I. I. i. 39 Bilaspur is not unpicturesquely situated.

unpin, *v.* Add: **5.** *trans. Chess.* To release (a piece that has been pinned). Hence **unpi·nning** *vbl. sb.*

1878 S. Loyd *Chess Strategy* 145 The key unpins the Black Queen. **1906** A. C. White *Tours de Force* p. xxxvii, The more general tasks can be grouped under several heads: checking, sacrificing, pinning and unpinning, [etc.]. **1967** V. Nabokov *Speak, Memory* (rev. ed.) xiv. 289 Themes in chess..are such devices as forelaying, withdrawing, pinning, unpinning and so forth.

u·npin, *sb. Chess.* [f. the vb.] The action or fact of releasing a piece that has been pinned.

1922 Hume & White *Good Compan. Two Mover* 187 There are six unpins, five of which are direct..while the sixth is an Interference Unpin. **1928** [see *half-pin* s.v. *half II. i].

un-pin-dow·nable, *a.* [Un-[1] 7 b (*b*) + *pin *v.*[1] 6 c.] That cannot be pinned down or defined.

1966 *Punch* 14 Feb. 249/2 Of all the great detectives he has the most un-pindownable character. **1973** *Guardian* 28 June 14/6 Harold Pinter is..our most tangible and un-pin-downable playwright.

unpinned, *ppl. a.* Add: **a.** (Later examples.)

1776 F. Burney *Diary* 6 Apr. (1889) II. 139 Two of her curls came quite unpinned, and fell lank, on one of her shoulders. **1821** M. Edgeworth *Let.* 27 Nov. (1971) 281 An end of her frill was unpinned.

b. Of a grenade: having had the pin (*pin *sb.*[1] 1 n) removed.

1974 P. Dickinson *Poison Oracle* vi. 159 The Jap pilot who brought the plane down..on to an inadequate runway with an assassin sitting beside him holding an unpinned grenade.

unpiteous, *a.* Restrict † *Obs.* to sense 1 in Dict. and add: **2.** (Later example.)

1954 W. Faulkner *Fable* 289 In our country, in our hard and unpiteous mountains, people used..children as people in lands savage with dangerous animals used guns and bullets: to defend, preserve themselves.

unplayable, *a.* Add: Hence **unplay·ably** *adv.*

1955 *Times* 6 June 3/1 Before luncheon the occasional ball turned and lifted nastily, and the one that accounted for May did so almost unplayably. **1978** *Daily Tel.* 21 Feb. 13/6 Next the unplayably simple A flat, made to sound telling and enigmatic at the same time, a great success.

unpleasable, *a.* (Later example.)

1951 A. L. Rowse *English Past* 24 The two sons were growing up not much to the liking of their virtuous, unpleasable mother.

unpleasantness. Add: **b.** *the late unpleasantness* (U.S.), the American Civil War; also *transf.*

1868 D. R. Locke *Ekkoes from Kentucky* 23 (Th.), That cuss cum back here, doorin' the late onpleasantniss, kernel of a rigiment. **1872** *Harper's Mag.* Feb. 479/1 During our 'late unpleasantness' a convalescent hospital was established..in Nashville. **1903** *N.Y. Times* 19 Sept. 3 The only soldier to be killed from Orange during the late unpleasantness with the Filipinos. **1930** S. Henry *Conquering Great Amer. Plains* 33 When 'the late unpleasantness' terminated, the Texan drover lacked a satisfactory market. **1948** *Amer. Jrnl. Nursing* Dec. 770/2 Among the stamps of particular interest to nurses to come out of the 'late unpleasantness', are a Hungarian Red Cross series,..and a German child welfare series. **1981** *Times* 25 Aug. 7/1 The British and the Germans.. were all of that older generation which well remembers the late unpleasantness between our two countries.

unpleasure. Add: **a.** (Earlier example.)

1792 F. Burney *Jrnl.* Jan. (1972) I. 111, I told him, very truly, of the pleasure with which I had re-entered his Roof—but I [write] the *un*pleasures that followed!

b. *Psychoanal.* [tr. G. *unlust.*] The sense of inner pain, discomfort, or anxiety which results from the blocking of an instinctual impulse by the ego and is the opposite of the affect of pleasure.

1919 A. Wohlgemuth in *Brit. Jrnl. Psychol.* Monogr. Suppl. VI. I. 1 The affective elements are Pleasure-Unpleasure. **1925** I. A. Richards *Princ. Lit. Crit.* xii. 95 Unpleasure being frustrated, chaotic, mal-successful activity. **1949** Koestler *Insight & Outlook* xv. 208 The nervous processes which determine the pleasure-unpleasure tone of emotional experience. **1974** *Nature* 11 Oct. 500/2 The monkeys' 'interest' overrode—so long as it lasted—their 'unpleasure'.

unplucked, *ppl. a.* (Examples of eyebrows.)

1959 'E. Peters' *Death Mask* i. 8 Her brows were still high, shapely, and unplucked. **1974** *Times* 26 Oct. 8/8 Her chalk-white complexion, emerald, Kohl-rimmed eyes with unplucked eyebrows.

unplug, *v.* (Examples of an electrical appliance: opp. *plug *v.* 1 e.)

1942 *Tee Emm* (Air Ministry) II. 141 Nothing is more annoying than to find..one glove [of an electrically-heated flying suit] unplugging itself every time you move your arm. **1969** A. Glyn *Dragon Variation* vi. 184 Mr. Jackson unplugged the television and turned out the lights.

unpolicied, *ppl. a.* (Later example.)

1947 Auden *Nones* (1952) 64 The unamerican survivor Hears angels drinking fruit-juice with their wives Or making money in an open Unpolicied air.

unpolitic, *a.* (Later example.)

1978 J. Updike *Coup* (1979) i. 20 The unpolitic loyalty of the fearful.

unpolitical, *a.* Add: Hence **unpoli·tically** *adv.*

1930 E. Sitwell *Alexander Pope* v. 63 The young and un-politically-minded Pope. **1962** *Times* 5 Dec. 17/6 It [*sc.* a film] is unpolitically concerned with people.

unpompous, *a.* (Later examples.)

1928 *Daily Express* 8 May 5 This is an unpompous age, and the tendency to poke fun at that ancient spectacle, grand opera,..does not decrease. **1946** D. Welch *Jrnl.* 9 Apr. (1952) 197 He quickly dropped the 'sir' when he was sitting beside me... This I knew was a compliment in some way. I suppose I interpreted it as a tribute to an unpompous or young appearance. **1959** J. Osborne *World of Paul Slickey* I. iii. 29 We can safely say in a not unpompous way, blind Them with words! **1978** *Even. Standard* 29 Sept. 20/5 At yesterday's function Sir Peter read a neat, unpompous little speech.

unpornogra·phic, *a.* [Un-[1] 7 a.] Not pornographic.

1938 *New Statesman & Nation* 23 July 156/2 This admirable and totally unpornographic novel. **1943** D. Welch *Jrnl.* 20 Apr. (1952) 61 As Bernard Shaw said to me when we were all called to give evidence over the *Well of Loneliness* fuss, 'Here I am, asked to say something about this pathetic book, and I don't know what to do, because I know it's serious, unpornographic, but it's so bad as literature.' **1968** *Punch* 5 June 829/1 The story is not only quite unpornographic, it is also very funny.

unpowdered, *ppl. a.* Add: **3.** Not wearing face-powder; without face-powder.

1917 J. F. MacDonald *Two Towns—One City* III. ii. 210 Flushed, and dishevelled, and unpowdered Fifine becomes. **1921** W. de la Mare *Mem. Midget* xli. 276 Her clear, unpowdered skin had the faint sheen of a rose. **1956** 'C. Blackstock' *Dewey Death* iii. 59 The fair hair was falling over her face; her skin was blotched and unpowdered. **1974** J. Mann *Sticking Place* v. 88 Look at her now, shabby, unpowdered.

unpracticalness. (Earlier example.)

1843 J. S. Mill *Let.* in *Wks.* (1963) XIII. 579 The chief fault seems to me that of entire unpracticalness.

unprayable, *a.* Add: **2.** That cannot be uttered as a prayer. *rare.*

1941 T. S. Eliot *Dry Salvages* ii. 9 The prayer of the bone on the beach, the unprayable Prayer at the calamitous annunciation.

unpredictable, *a.* Add: Hence **u:npredictabi·lity.**

1920 S. Alexander *Space, Time, & Deity* II. 324 Unpredictability is not limited to human determinism. **1955** *Bull. Atomic Sci.* Mar. 87/1 Their freedom imparts an unpredictability to historic events. **1977** J. L. Harper *Population Biol. Plants* 769 Organisms in nature live in environments that contain rhythms and unpredictabilities, patterns and noise.

unpressed, *ppl. a.* **1.** (Examples corresponding to *press *v.*[1] 4 a.)

1932 W. Faulkner *Light in Aug.* xix. 419 A tall, loosejointed man with a constant cob pipe,..wearing always loose and unpressed dark gray clothes. **1968** J. Ironside *Fashion Alphabet* 96 Soft pleats with edges rounded and left unpressed. **1974** J. Flint *Cecil Rhodes* (1976) ii. 19 Rhodes was notable for the shabbiness of his unpressed suits. **1977** *Lancashire Life* Nov. 115/3 (*caption*) A dress with an intricately seamed bodice falling into unpressed pleats.

unpretty, *a.* Add: Hence **unpre·ttily** *adv.*

1929 W. Faulkner *Sartoris* II. vi. 157 She mouthed her food unprettily. **1982** R. Barnard *Death & Princess* xiv. 141 The Princess Helena pouted unprettily.

unpriceable, *a.* Add: **b.** That cannot be priced.

1951 P. Larkin *XX Poems*, And to hear how the past is past and the future neuter Might knock my darling off her unpriceable pivot.

unprinceliness. (Earlier example.)

1855 G. H. Lewes *Life & Wks. Goethe* I. iv. i. 310 The princely unprinceliness of selling to the Jews a diamond ring.

unprintable, *a.* and *sb.* Add: Hence **unpri·ntably** *adv.*

1934 in Webster. **1940** E. Hemingway *For whom Bell Tolls* iii. 44 Go then unprintably to the campfire with thy obscene dynamite. **1965** D. Francis *Odds Against* xi. 153 Jones-boy unprintably told Chico where he could find his coffee. **1977** *Gay News* 7–20 Apr. 23/2 It's infuriating trying to write about Anger because so much is unprintably scandalous.

unproblematic, *a.* (Later examples.)

1944 C. G. Myrdal et al. *Amer. Dilemma* II. 1032 The system of mores, conceived of as a homogeneous, unproblematic, fairly static, social entity. **1981** *Word 1980* XXXI. 196 Something that was not possible within the tonetic stress marks theory proves relatively unproblematic for the configurational model. **1984** *Listener* 19 Apr. 16/2 Nor is there any unproblematic external standard with which to compare competing paradigms.

unprocessed, *ppl. a.* (Later examples, corresponding to Process *v.*[1] 3.)

1959 *Times* 6 Mar. 11/7 The United States..want the unprocessed raw materials. **1962** E. Snow *Other Side of River* (1963) lxxix. 607 In Peking I had been assured that I would be permitted to take out unprocessed film. **1974** *Brit. Med. Jrnl.* 19 Jan. 109/2 Unprocessed bran is an ideal additive to the diet in such cases.

unprofessional, *a.* and *sb.* Add: Hence **unprofe·ssionalism.**

1934 in Webster. **1955** A. L. Rowse *Expansion Elizabethan England* 382 The clue to English ill-success was, as usual, amateurishness, unprofessionalism, in the beginning. **1977** *Time* 28 Feb. 22/1 The Carter Administration was still marred in some areas by confusion and unprofessionalism; the staff work was sometimes shoddy and key memos poorly prepared.

unprotestant, *a.* (Earlier example.)

1839 J. S. Mill in *Westm. Rev.* Apr. 504 The unprotestant and unchristian doctrines..which are extensively professed.

unpublishable, *a.* Add: Also as *sb.*

1981 [see *unmentionable *a.* and *sb.* c].

unpublished, *ppl. a.* Add: **1. b.** Of an author: having had no writings published.

1934 R. Macaulay *Milton* 26 A severe judgment on a writer already so fruitful, though as yet unpublished. **1976** *Scotsman* 20 Nov. (Weekend Suppl.) 3/8 (Advt.), This second anthology of new poems..includes work by.. other established poets as well as by very talented but previously unpublished poets.

unpunctuality. (Earlier example.)

1814 Jane Austen *Mansfield Park* II. iv. 84 Their remoteness and unpunctuality, or their exorbitant charges.

unpunctuated, *ppl. a.* (Earlier example.)

1860 W. James *Let.* 12 Aug. in R. B. Perry *Thought & Character W. James* (1935) I. 196 Alice must be locked up alone..to write a letter, unassisted, uncorrected and unpunctuated, to her loving brothers.

unquakerlike, *a.* (Earlier example.)

1832 F. Trollope *Domestic Manners of Americans* II. xxx. 154, I overheard many unquakerlike jokes.

unqualifiedly, *adv.* (Earlier example.)

1861 Mill *Repr. Govt.* iii. 57 On the question of strikes ..it is doubtful if there is so much as one among the leading members of either House, who is not firmly convinced that the reason of the matter is unqualifiedly on the side of the masters.

unquick, *a.* **2.** (Later example.)

1925 D. H. Lawrence *Refl. Death Porcupine* 111 The novel..can't exist without being 'quick'. The ordinary unquick novel, even if it be a best seller, disappears into absolute nothingness.

u·nquote, *v.* [Un-[2] 3.] *intr.* Used as a formula in dictation, etc.: terminate the quotation. See *quote *v.* 4 c.

1935 E. E. Cummings *Let.* Mar. (1969) 139 But he said that if I'd hold up publication of No Thanks for 15 days he'd kill unquote a page of Aiken. **1935**, etc. [see *quote *v.* 4 c]. **1969** *New Yorker* 11 Oct. 48/2 Then Mr. Tanks announced the last downtown stop. He said, 'Madison Square Garden, Penn Station..et cetera, et cetera, et cetera, unquote'.

unquoted, *ppl. a.* (Later examples: cf. Quote *v.* 7.)

1969 *Times* 2 May 31/6 Cranfields is an unquoted public company in flour milling and baking. **1977** Johns & Greenfield *Dymond's Capital Transfer Tax* iii. 54 (*heading*) Condition (4) Sale of unquoted shares or debentures.

unra·ced, *ppl. a.*³ [Un-¹ 8 + Race *v.*¹] That has not taken part in a race.
1955 *Motor* 26 Jan. 57/1, 2-seater sports, one owner, pale green, unraced, £975. 1963 *Times* 21 May 3/6 Lord Derby's brown gelding, Robinson Crusoe, unraced as a two-year-old but winner of two races at Newmarket this season, gave a first rate display of stamina and gameness at Leicester yesterday. 1976 *Horse & Hound* 3 Dec. 9/4 Miles Away was bought by the combination at the Ballsbridge November Sales for 10,000 guineas, a record price in Ireland for an unraced gelding of his age.

unracy, *a.* (Earlier example.)
1847 *Dublin Rev.* Sept. 228 The style..is seldom chargeable with the defects of unracy or unidiomatic phraseology, commonly objected to Johnson's.

unrailed, *ppl. a.* (Earlier contextual examples.)
1842 *Rep. Comm. Children's Employment* 384 in *Parl. Papers* XVI. 1 Women and children employed to carry coal on their backs in unrailed roads. 1891 Kipling & Balestier in *Century Mag.* Nov. 35/1 The unrailed bridge that crossed the irrigating-ditch.

unrea·cted, *ppl. a. Chem.* [Un-¹ 8.] Not having undergone reaction.
1950 *Jrnl. Amer. Chem. Soc.* LXXII. 10/1 If the recovered unreacted bismuth compound is considered, the yield..was 62·4%. 1977 A. Hallam *Planet Earth* 33/3 After neutralizing or evaporating off any unreacted acid, the resulting rock solution is then sprayed into a hot flame.

unrea·ctive, *a. Chem.* [Un-¹ 7 a.] Not reactive (*reactive *a.* 5 a).
1934 in Webster. 1946 *Nature* 28 Sept. 437/1 In this substance..the Cl atoms are unreactive. 1965 Phillips & Williams *Inorg. Chem.* I. x. 348 Molecules with very high dissociation energies such as C_2, N_2, and O_2 would be expected to be unreactive, while Li_2, Be_2, and F_2, with very low dissociation energies, would be reactive. 1978 *Jrnl. Physical Chem.* LXXXII. 2554/1 The zwitterion is relatively unreactive.
Hence **unreacti·vity.**
1946 *Nature* 14 Sept. 383/1 This chemical unreactivity of the fluorine atom of the FCH_2-group is shared by many of the compounds mentioned in this communication. 1978 *Jrnl. Physical Chem.* LXXXII. 2549 The unreactivity of glycine-like, α-amino acid zwitterions does not appear to be due to rate-limiting ring closure.

unreadable, *a.* Add: Hence also **unrea·dably** *adv.*; **unreadableness** (earlier example).
1780 F. Burney *Diary & Lett.* (1842) I. vii. 316 In the evening we had Mrs. Lambert, who brought us a tale, called 'Edwy and Edilda', by the sentimental Mr. W—, and unreadably soft, and tender, and senseless is it. 1838 J. S. Mill in *Westm. Rev.* Aug. 506 He could stop nowhere short of utter unreadableness. *a* 1974 R. Crossman *Diaries* (1975) I. 564 She may be right that if you want to impress people you must make things unreadably long.

unre·alist, *sb.* and *a.* [Un-¹ 12, 7 a.] **A.** *sb.* One who is not a realist. **B.** *adj.* Not realistic or realist.
1934 Webster, Unrealist, *n.* 1936 E. Gill *Let.* 10 Oct. (1947) 367 It is unrealist to write..saying 'art is a way of life..that should govern all we do and make, not to say think'. 1938 A. L. Rowse in *Pol. Q.* Jan.–Mar. 24 The hopeless doctrinaires,..the chronic unrealists, who are the despair of the [Labour] Party. 1958 S. Spender *Engaged in Writing* 67 They, with their fairy stories,..are the unrealists. 1973 *Screen* Spring/Summer 40 Certain types of cinema (modern, realist, 'un-realist', [etc.]). *Ibid.* 48 Films which are deliberately 'un-realist' (legends, fairy stories, films of the *phantastique* genre, etc.).

unrealistic, *a.* Add: Hence **unreali·stically** *adv.*
1961 *Listener* 16 Nov. 797/2 Our critics often use their weapons unrealistically, and unhistorically. 1977 *Times Lit. Suppl.* 20 May 610/3 He kidded him for making American life unrealistically grim.

unrecallable, *a.* (Later example.)
1930 W. de la Mare *On the Edge* 293 Not that she had ever confessed this in so many raw unrecallable words.

unreconstructed, *ppl. a.* Add: *spec.* (orig. *U.S.*) not reconciled to the outcome of the American Civil War; hence *gen.* not reconciled or converted to the current political orthodoxy; unreformed; die-hard. (Earlier and later examples.)
1867 *Harper's Weekly* 9 Nov. 707/2 The Democratic candidates in Maryland are..of the 'unreconstructed' kind. *a* 1936 Kipling *Something of Myself* (1937) vii. 195 There came..with her married daughter the widow of a Confederate Cavalry leader; both of them were what you might call 'unreconstructed' rebels. 1944 *Univ. of Chicago Mag.* May 6 The young are in increasing numbers, not only in Oklahoma but in much of the vast inland region of our republic, moving to the cities, leaving the unreconstructed small towns to their elders and to decay. 1946 J. Flanner in *New Yorker* 9 Mar. 80/1 Nuremberg defense counsel have just offered..an absolutely first-rate demonstration of the still unreconstructed prewar German mind. 1949 B. A. Botkin *Treas. S. Folklore* I. i. 4 As the type and symbol of the 'unreconstructed Southerner', Donald Davidson selects 'Cousin Roderick', an idealized Middle Georgia country gentleman, who combines the 'bearing of an English squire' with the 'frontier heartiness' of A. B. Longstreet's *Georgia Scenes*. What principally distinguishes Cousin Roderick from 'Brother Jonathan' (the Vermont 'unreconstructed Yankee' who resembles the Georgian in so many ways)..is the fact that he does not work with his own hands. 1962 A. Sampson *Anat. of Britain* xxx. 480 He and his board have reacted..to revelations about their monopoly, with the old-fashioned sang-froid of unreconstructed businessmen. 1968 *Economist* 6 Jan. 34/3 The trial had an even greater impact because of the blunt rebukes hurled at the defendants by Judge Harold Cox, once considered one of the more unreconstructed judges in the South. 1972 R. Plant in Cox & Dyson *20th-Cent. Mind* II. iv. 91 The major opposition to Roosevelt's New Deal policies in fact came from these unreconstructed *laissez-faire* liberals who failed to realize that planning was necessary in order to make individual freedom a *real* possibility for the masses. 1973 [see *reconstructed *ppl. a.* b]. 1979 *Guardian* 2 Oct. 10/2 A further entrenchment of unreconstructed union power. 1981 'A. Cross' *Death in Faculty* i. 12 The place is pretty much unreconstructed..an ancient bathroom complete with mahogany bathtub.

unreduced, *ppl. a.* Add: **7.** *Phonetics.* Of a vowel sound or other phonetic element: not reduced (see *reduced *ppl. a.* 6 b).
1953 K. Jackson *Lang. & Hist. Early Brit.* 658 Unreduced forms are almost never found in Welsh. 1964 E. Blancquaert in D. Abercrombie et al. *Daniel Jones* 300 English weakened syllables are pronounced unreduced.

unreflectingly, *adv.* (Earlier examples.)
1696 J. Sergeant *Method to Sci.* II. ii. 131 The Former comes by Experience Unreflectingly; the Later is attain'd by Study and Reflexion. 1854 A. G. Henderson tr. *Cousin's Philos. of Kant* p. lx, Thou wouldst not unreflectingly confer it upon the first comer.

unrelatable, *a.* (Later examples.)
1963 Bruner & Olver in J. S. Bruner *Beyond Information Given* (1974) xx. 361 The lists proceed from near to far items, from easily associated to almost unrelatable elements. 1968 *Listener* 1 Aug. 147/3 The relative pronoun hangs in the air, unrelatable to anything in particular.

unreleased, *ppl. a.* (Later examples.)
1960 *Lang. & Speech* III. 109 (*heading*) On the perception of unreleased voiceless plosives in English. 1982 *Times* 13 Feb. 6/1 This successful, though unreleased, movie.

unremaining, *ppl. a.* (Later example.)
1936 A. E. Housman *More Poems* 63 Here the child comes to found His unremaining mound.

unremember, *v.* (Later example.)
1983 *Spectator* 16 Apr. 22/2 We cannot unremember Auschwitz.

unremunerated, *ppl. a.* (Examples.)
1895 J. Williams *Briefless Ballads* 51 We mourn in unremunerated pain Those brave old days. 1945 A. L. Rowse *West-Country Stories* 3 His services were unremunerated.

unrenewable, *a.* (Later example.)
1977 *Church Times* 25 Feb. 7/2 Its chronic prodigality in the use of unrenewable fossil fuels.

unrepair. (Earlier example.)
1843 Thackeray *Irish Sk.-Bk.* II. xii. 225 A dismal state of unrepair.

unrepresentative, *a.* Add: Hence **unreprese·ntativeness.**
1958 A. Toynbee *East to West* 221 This unrepresentativeness of the capital is one of its generic defects. 1980 Butler & Pinto-Duschinsky in Z. Layton-Henry *Conservative Party Politics* viii. 198 Does Conservative unrepresentativeness harm the party in other ways?

unreprodu·ctive, *a.* [Un-¹ 7.] Not reproductive.
1930 W. K. Hancock *Australia* viii. 161 Informed opinion was ready to welcome the report of the British Economic Mission (January 1929), which warned Australia that unreproductive developmental expenditure was imposing 'a heavy burden on the general community.' 1968 P. Scott *Day of Scorpion* (1973) II. ii. 330 Hers were the menstrual flows of a virgin, sour little seepages such as Barbie Batchelor had presumably sustained for a good thirty years of her unreproductive life.

unrepulsive, *a.* (Later example.)
1946 D. Welch *Jrnl.* 12 Jan. (1952) 185 The person who built this church was unclogged with 'book-learning' and so his church is unrepulsive and almost pretty and good.

unresponsible, *a.* Add: Hence also **u:nresponsibi·lity.**
1935 E. Hemingway *Green Hills of Africa* xii. 235, I..settled, happily, with the darkness into the unresponsibility of victory; only emerging to direct M'Cola in where to cut.

unrested, *ppl. a.*¹ **b.** (Later examples.)
1940 N. Last *Diary* 22 Oct. in *Nella Last's War* (1983) iii. 81, I got up, tired and unrested, after a very broken night. 1981 D. Boggis *Time to Betray* xvii. 94 Nigel lay sleepily, unrested, worried.

unrip, *v.* **3. b.** For † read 'Now *rare*' and add later example.
1928 Barrie *Peter Pan* II, in *Plays* 42 Unrip your plan, Captain.

unrope, *v.* Add: Also *absol.* for *refl.* in *Mountaineering.*
1902 G. Bell *Let.* 3 Aug. (1927) I. vii. 146 About 6 we got to where we could unrope—having been 48 hours on the rope. 1957 J. Masters *Far, Far the Mountain Peak* 67 She began to unrope, and a moment later the second *cordée* had joined them.

unsacerdotal, *a.* (Earlier example.)
1844 J. S. Mill in *Edin. Rev.* LXXIX. 33 The Papacy could..indulge within certain limits their [*sc.* the Franciscans'] most unsacerdotal preference of grace to the law.

unsa·ckable, *a.* [Un-¹ 7 b.] Not sackable.
1980 *Daily Tel.* 14 Feb. 14/8 It is a fair description of the real world of unsackable functionaries whom not God nor Sir Derek Rayner nor TNT will remove. 1984 *Listener* 12 July 17/2 The system introduced in the 1940s..made the men who handled the catches in the registered fishing ports unsackable.

unsaddling, *vbl. sb.* (Further *attrib.* examples.)
1951 *People* 3 June 7/5 Horses shied; off scampered the rabbit to the unsaddling enclosure. 1972 D. Francis *Smokescreen* iv. 51 We walked down from the stands and over towards the unsaddling enclosures.

unsafe, *a.* Add: **2. c.** *unsafe period,* the part of the menstrual cycle during which conception is most likely.
1961 G. Greene *Burnt-Out Case* III. i. 82 He sometimes allowed her to be alone during her monthly or unsafe periods. 1969 *Times* 20 Nov. 3/3 The Roman Catholic Church has revised its ideas about the..'unsafe period' and now makes it four days around the mid-point of the female cycle.

unsalt, *a.* (Later example.)
1935 E. Bowen *House in Paris* II. ix. 143 Here the sea air was washed unsalt by the rain.

unsalubrious, *a.* (Later examples.)
1951 M. McLuhan *Mech. Bride* (1967) 118/2 Nobody can do much about changing the noisy and unsalubrious character of the big cities in which most of us live and work. 1971 *Daily Tel.* 13 Dec. 7/8 There have been demonstrations in several [French] jails recently against unsalubrious conditions.

unsa·turate. *Chem.* [Un-¹ 12.] Any unsaturated compound, esp. an unsaturated fat or fatty acid. Cf. *polyunsaturate* sb. s.v. *polyunsaturated *a.*
1934 *World Petroleum* Apr. 123/2 Whereas straight run fuel contained less than two percent unsaturates, cracked gasoline contained from 10 to 40 percent unsaturated compounds. 1959 Logan & Maggiolo in E. S. Pattison *Industr. Fatty Acids* v. 41 These unsaturates can be cleaved to produce monobasic and dibasic acids and their derivatives. 1974 *Radiation Res.* LIX. 109 When these substances undergo moderately rapid reaction the surface temperature of the unsaturate is considerably higher than 77°K.

unsaturated, *ppl. a.* (Examples corresponding to *saturated *ppl. a.* 3 b.)
1866 *Notices Proc. R. Inst. Gt. Brit.* IV. 417 The two nitrogen arms which are left exposed sufficiently indicate that two attraction units remain unsaturated. 1872 *Phil. Mag.* XLIII. 259 A radical..is a portion of a molecule, a group of atoms, the affinities of which do not wholly saturate one another, the radical being uni-, bi-, tri-, quadri-, &c. valent, according as 1, 2, 3, 4, .. affinities are left unsaturated. 1916 *Jrnl. Amer. Chem. Soc.* XXXVIII. 778 It has been generally assumed that what is known as a bivalent element must be tied by two bonds to another element or elements, or remain with an 'unsaturated valence'. 1951 I. L. Finar *Org. Chem.* iv. 67 The acetylenes are unsaturated hydrocarbons that contain one triple bond. 1982 J. E. Fernandez *Org. Chem.* v. 89 The alkenes are unsaturated: that is they contain fewer hydrogen atoms than the alkanes.

unsaturation. Add: *Chem.* The condition of a compound, esp. an organic one, of having one or more multiple bonds in its molecule. Cf. Saturation 3 a in Dict. and Suppl. (Further examples.)
1932 I. D. Garard *Introd. Org. Chem.* v. 59 The cause, or explanation, of the unsaturation is not the same in all instances. 1964 N. G. Clark *Mod. Org. Chem.* vii. 115 Chemical reagents attack the site of unsaturation but this occurs less readily than in the case of the related olefins. 1975 *Sci. Amer.* Mar. 78/3 The greater the unsaturation in a fat or a wax, that is, the greater the number of double bonds, the more likely it is that the substance will be liquid at low temperatures.

unscalped, *ppl. a.* (Earlier example.)
1726 C. Lyman in S. Penhallow *Hist. Wars New-England* 22 We found seven dead upon the spot: Six of whom we scalpt, and left the other unscalpt.

unscattered, *ppl. a.* (Later examples.)
1941 in M. Gowing *Britain & Atomic Energy 1939–45* (1964) 400 The 'elastically' scattered neutrons are deviated only through small angles by the collisions and to a final approximation may be treated as unscattered. 1966 D. G. Brandon *Mod. Techniques Metallogr.* 48 The thickness of film that will reduce the unscattered transmitted intensity to $1/e$ of its original value is given by $t = 20V/\rho$.

unscented, *a.* (Later examples.)
1884 *Girl's Own Paper* 26 Jan. 271/3 Employ a mild, unscented soap at night. 1914 [see *sapele]. 1979 C. McCarry *Better Angels* I. x. 92 Philindros was odourless; he..used an unscented soap.

unsceptre, *v.* (Later example.)
1897 F. Thompson *New Poems* 110 Unsceptre thee of state and place!

unscra·mble, *v.* [Un-² 3.] **1.** *trans.* To reverse the process of scrambling (eggs). Also in *fig.* contexts.
This would appear to be the earlier sense though pre-1926 printed evidence has not been found.
1926 R. H. Tawney *Relig. & Rise of Capitalism* ii. 88 But the discovery of the sage who observed that it is not possible to unscramble eggs had already been made. 1928 C. Sandburg *Good Morning, America* 18 Can you unscramble eggs?..J. Pierpont Morgan's query as to court decrees dissolving an inevitable industrial combination. 1969 *P.E.N.* IX. 46 She quoted, as an example, 'Mr. Enoch Powell called last night for the denationalisation of all State-owned industries and explained exactly how to unscramble the eggs.' 1980 J. Wainwright *Tainted Man* 131 You demanded retribution..*your* law..unscrambles all eggs.
2. To put into or restore to order; to disentangle; to make sense of (something) confused; to extricate from (or *from*) a state of confusion or muddle; to separate into constituent parts; to 'dismantle' (an organization or system); *spec.* to restore (a signal) by applying the reverse of the process previously used to scramble it; to render intelligible in this way.
1923 Wodehouse *Inimitable Jeeves* x. 104, I collapsed on to the settee and rather lost interest in things for the moment. When I had unscrambled myself I found that Jeeves and the child had retired. 1927 [see *scramble *v.* 3 c]. 1952 *Times* 17 Nov. 6/1 The Government propose to 'unscramble' nationalized road haulage service. 1955 [see *scramble *v.* 3 c]. 1956 *Archivum Linguisticum* VIII. 153 Since Trubetzkoy's untimely death the task of unscrambling them [*sc.* phonology and phonemics] has been pursued with wearisome halts. 1956 Auden & Kallman *Magic Flute* (1957) 116 So english, remodel Our lines as you please, Unscramble the drama and jumble the keys. 1958 *Times* 19 Aug. 9/2 Very broadly the intention seems to be to 'unscramble' from the French legislature those territories which wish to become federated. 1959 *Daily Tel.* 6 Mar. 14/3 By putting money in a box attached to a set the viewer automatically 'unscrambles' the transmission. 1963 *Ibid.* 10 Jan. 1/8 The process of 'unscrambling' Northern Rhodesia..will take very much longer. 1973 'I. Drummond' *Jaws of Watchdog* ii. 31 The message was unscrambled by a radio on a fast, low-profile motor-yacht. 1974 *Listener* 9 May 597/1 When the Conservatives returned to office in 1951 they didn't unscramble the National Health Service. 1978 R. Ludlum *Holcroft Covenant* xxxvi. 420 You should bring him instead a potion to unscramble his doddering brains. 1981 *Sunday Express* (Colour Suppl.) 12 July 33/4 Only those who pay the extra rental are provided with a device to unscramble the film signals. 1983 *Listener* 18 Aug. 8/1 It was often more fruitful than trying to unscramble what he was actually saying.
Hence **unscra·mbled** *ppl. a.*; **unscra·mbling** *vbl. sb.*
1955 J. G. Davis *Dict. Dairying* (ed. 2) 993 (*caption*) Unscrambling machine. 1958 *Listener* 25 Sept. 463/2 The unscrambling attachments to the receiving sets. 1959 *Times Lit. Suppl.* 16 Jan. 30/3 To be brought up short on page 78 by a sentence scrambled by the printer almost beyond unscrambling. 1959 *Daily Tel.* 6 Mar. 14/3 In another method, the 'unscrambling' is done by dialling, as on a telephone. 1959 *Times* 11 June 5/7 The transparency of the texture and the clean, fresh sounds of unscrambled timbres. 1964 L. Deighton *Funeral in Berlin* xxx. 156 There was a din of unscrambled noise before Charlotte Street switched the scrambler into the circuit.

unscra·mbler. [f. prec. + -er¹.] A device for unscrambling scrambled messages or signals.
1968 A. B. Carlson *Communication Systems* v. 218 The system shown..is a simplified speech scrambler... Show that an identical system will suffice as an unscrambler. 1976 'A. Hall' *Kobra Manifesto* vi. 80 We began reading the signals as they came off the integral unscrambler. 1979 *Maclean's Mag.* 26 Mar. 59/1 Methods range from such simple expedients as wrapping antennas with aluminum foil to black-market sales of 'unscramblers' that decode signals.

unscreened, (*ppl.*) *a.* Add: **2. b.** Not investigated or checked for security: see *screen *v.* 4 c.
1970 R. Clapperton *Victims Unknown* xi. 101, I have been severely criticised..for authorising the employment on intelligence work of an unscreened individual. 1979 H. Kissinger *White House Years* xix. 748 The beady eye of Secret Service agents Ready and McLeod, who were not about to leave me to the mercy of unscreened foreigners.

unseasonableness. a. (Later examples.)
1853 C. M. Yonge *Heir of Redclyffe* II. i. 2 Mrs. Ashford put the matter off for the present by the unseasonableness of the weather. 1971 *Daily Tel.* 3 July 9/1 Summer..is the season when unseasonableness becomes most glaring and least sufferable.

unsectarianize, *v.* (Earlier example.)
1832 J. S. Mill *Let.* 17 Sept. in *Wks.* (1963) XII. 118 The editor & his writers..are but Unitarians & liberals, unsectarianized.

unsedentary, *a.* (Later example.)
1915 W. B. Yeats *Tribute to Thomas Davis* (1947) 15 A gallant unsedentary man.

unseeming, *ppl. a.* Restrict † *Obs.* to sense in Dict. and add: **2.** Unapparent. *rare.*
1923 D. H. Lawrence *Birds, Beasts & Flowers* 174 The elephants ponderously, with unseeming swiftness, galloped uphill in the night.

unselfconsciously, *adv.* (See small-type note s.v. Unselfconscious *a.*) (Examples.)
1921 H. Crane *Let.* 26 Nov. (1952) 72, I..carry these encumbrances..deftly and un-selfconsciously. 1958 P. Winch *Idea of Social Sci.* IV. iv. 102 Science..applies its criteria unselfconsciously. 1977 P. Strevens *New Orientations Teaching of Eng.* ii. 17 Young children (say, age 6–13)..tend to learn easily and unselfconsciously.

unse·llable, *a.* [Un-¹ 7 b.] Lacking a buyer; that no one wants to buy.
1975 *Daily Tel.* 22 Oct. 16 The Wardle farm was derelict and regarded as unsellable before the family..bought it two years ago. 1981 *Times Lit. Suppl.* 18 Sept. 1061/1 New ways of getting the johns to spend their money on previously unsellable old tat.

unsensational, *a.* (Earlier example.)
1854 Geo. Eliot tr. *Feuerbach's Essence Christianity* xxii. 213 God sees..all objects of sense in an unsensational manner.

unsentimental, *a.* (Earlier example.)
1752 H. Walpole *Let.* 23 June (1974) XXXVII. 340 He..was even so unsentimental as to talk, of desiring to make her happy.

unseriousness. (Later examples.)
1973 I. Robinson *Survival of English* ii. 62, I am not discussing in this book the gross popular examples of breakdown or coarsening..nor shall I more than mention the increasing unseriousness of the denominational press. 1978 *Detroit Free Press* 5 Mar. B 9/1 Aliette works so hard she brooks no unseriousness.

unsex, *v.* Add: Also *fig.*
1955 Auden *Shield of Achilles* i. 27 A day that meekly takes The potter's cuff, a gravel that as concrete Will unsex any space which it encloses.

unshadow, *v.* (Further examples.)
1895 A. Machen *Three Impostors* 93, I, too, burned with the lust of the chase, not pausing to consider that I knew not what we were to unshadow. 1926 D. H. Lawrence *David* xv. 121 Is my heart fireless..? Kindler! it shall not be so! My heart shall shine to Thee, yea, unshadow itself. 1953 A. Clarke *Moment Next to Nothing* II. 41 All that men think of would be immaterial Could they but watch these women shadowing, Unshadowing themselves.

unshakeable, *a.* Add: Hence also **u:nshak(e)abi·lity.**
1907 W. James in *Philos. Rev.* Jan. 9 The result claimed [from yoga, etc.]..is strength of character, personal power, unshakability of soul. 1952 Gerth & Martindale tr. *Weber's Ancient Judaism* xv. 387 The complete unshakability of its communities by the foreign environment from which they segregated themselves.

unsharp, *a.* Add: Hence **unsha·rpness.**
1961 in Webster. 1967 E. Chambers *Photolitho-Offset* xi. 161 Normal action of the diffusing stop, when all the small apertures are clear, results in a moderate degree of unsharpness. 1977 J. Hedgecoe *Photographer's Handbk.* 314/2 Overall unsharpness can be caused by gross misfocus.

unshi·rted, *a. U.S. slang.* [Un-¹ 9.] In phr. *unshirted hell,* serious trouble; 'a bad time'.
1932 *Sun* (Baltimore) 6 Jan. 10/3 When he proposed certain policies on prohibition..he was given what is known in rural districts hereabouts as 'unshirted hell'. 1954 F. P. Keyes *Royal Box* v. 67 She and his playboy son are going to fall for each other and then there will be unshirted hell. 1979 H. Kissinger *White House Years* xxi. 897 I've been catching unshirted hell every half-hour from the President who says we're not tough enough.

unshook. (Later example.)
1893 Kipling *Many Inventions* p. ix, May I look with heart unshook On blow brought home or missed.

unshrinkable, *a.* Add: Hence **u:nshrinkabi·lity.**
1934 in Webster. 1946 *Nature* 5 Oct. 476/2 A high degree of unshrinkability is obtained with 10 per cent of polymer within the fibres. 1963 A. J. Hall *Textile Sci.* v. 249 This treatment is continued until almost complete unshrinkability is obtained.

unsight, *v.* Add: **3.** *trans.* Of a participant in a game: to deprive (another player or an official) of a clear view. Freq. in *pa. pple.*
1923 *Daily Mail* 15 Jan. 11 The referee..was unsighted and so, apparently, was the linesman to whom he appealed. 1928 *Daily Express* 8 June 17/2 Gibbons..would have been caught by slip if the wicketkeeper had not unsighted that fieldsman. 1951 *Sport* 30 Mar.–5 Apr. 10/2, I felt sorry for Hendon's goalkeeper, Reg Ivey, who was unsighted both times and had hitherto played a very sound game! 1972 *Times* 3 Mar. 7/2 Mr Kelly was travelling some two to three lengths behind Mr Smith and inevitably that unsighted him for a reasonable distance in front of Mr Smith. 1976 *Alyn & Deeside Observer* 10 Dec. 3/7 It was afterwards revealed that the umpire had been unsighted and had not seen the ball hit by a Chester player in the circle.
4. To make unseen. *nonce.*
1914 Hardy *Satires of Circumstance* 17 Nor God nor Daemon can undo the done, Unsight the seen.

unsi·gned, *a. Math.* [Un-¹ 9.] Of a number: without a plus or minus sign, or a bit representing this.
1953 *Electronic Engin.* Jan. 6/1 One advantage possessed by the complement notation for signed numbers is that the processes of addition and subtraction can be carried out by exactly the same methods on signed and unsigned numbers. 1970 O. Dopping *Computers & Data Processing* xvii. 273 The 'compare' instruction can hardly be used for numerical information except when it is certain that all the keys are unsigned, that all keys have a plus sign, or that all keys have a minus sign. 1979 *Sci. Amer.* June 18/2 The problem asks whether for different values of n it is possible to pair each number in one subset with a number in the other so that the $2n$ sums and absolute, or unsigned, differences of the numbers in each pair are all distinct.

unsilenced, *ppl. a.* (Later examples: cf. *silenced *ppl. a.* b.)
1923 *Autocar* 7 Dec. 1151/2 Unsilenced engines. 1971 *Daily Tel.* (Colour Suppl.) 12 Nov. 21/4 A motor-bike of approximately six litres, unsilenced. 1981 M. Kenyon *Zigzag* xxv. 171 Kelly's new unsilenced gun was shaping up..effectively.

unsi·mplify, *v.* [Un-² 3.] *trans.* To make less simple; to state in a more complex form.
1858 F. W. Faber *Foot of Cross* iv. 219 Why should she..unsimplify her worship, by disuniting in thought what God had united. 1960 R. A. Knox *Occas. Sermons* xx. 94 We try to simplify modern politics, by making them all black and white, all heroes and villains; and in doing that we only unsimplify ourselves. 1975 *Sci. Amer.* Oct. 100/1 (Advt.), To understand, then, why there is a controversy, it is necessary to unsimplify the issue.

unska·ted, *ppl. a. rare.* [Un-¹ 8.] Not skated on.
1936 Dylan Thomas *Twenty-Five Poems* 19 December's pools..Lie this fifth month unskated.

unskin, *v.* (Later example.)
1935 T. S. Eliot *Murder in Cathedral* i. 20 And our hearts are torn from us, our brains unskinned like the layers of an onion.

unsmoothed, *ppl. a.* Add: **b.** Of data, etc. (cf. *smooth *v.* 1 e). **c.** Of a voltage: with any ripple left in.
1945 *16th Census: Population: Diff. Fertility: Women by Children under 5* (U.S. Bureau of Census) 2/2 An indication of the effect of inaccuracy in reported ages of women on the fertility statistics is given by a comparison of the unsmoothed figures with the smoothed figures. 1957 *Practical Wireless* XXXIII. 557/1 The frequency of the mains input is 50 c.p.s. and the output 'unsmoothed' H.T. has a 100 c.p.s. ripple. 1981 *Jrnl. Geophysical Res.* LXXXVI. 8002/1 The relative velocities are computed from the unsmoothed data.

unsnarl, *v. intr.* (Earlier example.)
1844 'J. Slick' *High Life N.Y.* II. xxviii. 167 All on 'em seemed kinder tangled up and trying to unsnarl all over the floor.

unsneck, *v.* Add: (Later examples.) Also *intr.*
Apparently now not restricted to *north.* and *Sc.*
1932 L. G. Gibbon *Sunset Song* i. 77 She unsnecked the door of the kirkyard wall, passing through to the Manse. 1948 A. Jobson *This Suffolk* iii. 50 A little wicket gate..snecks and unsnecks, to let one in or out. 1967 'G. North' *Sergeant Cluff & Day of Reckoning* xv. 139 Mole..unsnecked the door of the bedroom.

unsni·b, *v.* [Un-² 3.] *trans.* To unfasten (a catch); to unlatch.
1904 Glegg & Duncan *Law Reparation Scotl.* (ed. 2) ii. 39 When he unsnibs the window to clean it. 1966 J. McClenaghan *Moving Target* xiv. 148 He unsnibbed the chain which held the dogs together, and..they slipped away. 1980 R. Hill *Spy's Wife* xxiii. 178 She snibbed her bedroom door, but after a few seconds in bed got up and unsnibbed it.

unsociably, *adv.* (Later examples.)
1977 *Gramophone* June 117/2 Such indiscretions apart, and the inevitable rash of black sheep playing music at unsociably high volumes..the show had much to offer the keen visitor. 1981 A. Fraser *Splash of Red* xiii. 145 She had once been prepared to toil unsociably for the whole of August.

unsocial, *a.* Add: **b.** *unsocial hours,* socially inconvenient working hours.
1973 *Times* 4 Dec. 1/1 A proposed unsocial hours payment in recognition of the odd times of the day and night that a [train] driver has to report for duty. 1976 *Evening Post* (Nottingham) 17 Dec. 15/7 (Advt.), Waiter-waitress required for Lenton Hall of Residence... Good holidays and unsocial hours payment for week-end work. 1982 *Economist* 13 Nov. 50/1 If the government is to avoid the annual pay squabble with the nurses the new review body should first establish realistic pay scales..taking into account the unsocial hours. 1984 *Brit. Med. Jrnl.* 21 July 145/2 The unsocial hours during which most emergency operating is done has meant that much of it has been unsupervised.

unsocialism. Add: **b.** An absence of socialism. *rare.*
1889 G. B. Shaw *Fabian Ess. Socialism* 4 The gambling spirit urges man to..secure some acres of her [*sc.* Stepmother Earth]... This is Private Property or Unsocialism.

unso·cialist, *sb.* and *a.* [Un-¹ 12, 7.] **A.** *sb.* One who is not a socialist. **B.** *adj.* Not socialist.
1892 G. B. Shaw *Fabian Soc.* 6 Socialist statesmanship must..consist largely of taking advantage of the party dissensions between the Unsocialists. 1893 —— *Impossibilities of Anarchism* 4 It was bad enough to have to contend with the conservative forces of the modern unsocialist State. 1935 N. Mitchison *We have been Warned* IV. 423 She'd been afraid..in strange, unsocialist towns. 1967 *Spectator* 28 July 103/3 To an unsocialist, socialism is as unmoral as it is fatuous. Suez revealed the extent to which the Conservative party has given up the idea of an unsocialist morality. 1979 *Guardian* 9 Oct. 4/8 You can't..let this community be destroyed. It's un-Christian, let alone unsocialist.

unsolvable, *a.* Add: Hence also **u:nsolvabi·lity.**
1947 *Jrnl. Symbolic Logic* XII. 7 They become decision problems of recursively enumerable sets of positive integers of the same degree of unsolvability as the complete set *K*. 1979 *Sci. Amer.* May 131/1 One approach for circumventing unsolvability is to limit the kinds of statements given to the computer.

unsophistically, *adv.* (Later example.)
1890 W. James *Princ. Psychol.* I. xiii. 500 A formulation of the facts which offers itself so naturally and unsophistically.

unspan, *v.* (Later example.)
1914 T. A. Baggs *Back from Front* xxiv. 120 They unspanned in a neighbouring field and invited me to supper.

unspeakableness. (Later example.)
1963 B. Friedan *Feminine Mystique* viii. 182 After the loneliness of war and the unspeakableness of the bomb,..women as well as men sought the comforting reality of home and children.

unspeaking, *ppl. a.* (Later example.)
1935 E. Bowen *House in Paris* II. iv. 129 Karen herself had more than once been the victim of that unspeaking smile.

unspiked, *ppl. a.* (Further examples: cf. *spiked *a.*² 3 a, b.)
1969 *Lithos* II. 138 Unspiked measurements of Sr^{87}/Sr^{86} and Sr^{88}/Sr^{86} were made for all other samples. 1980 *Nature* 31 Jan. 438/2 Completely separate sets of equipment were used for spiked and unspiked sample solutions, thus eliminating the possibility of cross-contamination.

unspoilableness. (Earlier example.)
1873 C. M. Yonge *Pillars of House* I. xi. 232 Geraldine thought it was a great proof of his unspoilableness.

unspoilt, *ppl. a.* (Further examples, of the environment.)
1925 C. Connolly *Let.* 8 Apr. in *Romantic Friendship* (1975) 68, I hope to see some good unspoilt villages. 1939 *Country Life* 11 Feb. p. v (Advt.), 3 miles south of Dorking in beautiful unspoilt country, with lovely views. 1968 T. Wolfe *Electric Kool-Aid Acid Test* v. 59 Kesey wasn't primarily an outdoorsman. He wasn't that crazy about unspoilt Nature.

unspoo·l, *v.* [Un-² 3.] *trans.* To unwind (thread, tape, etc.) from a spool; *spec.* in *Cinematogr.,* to project (a film); also *intr.* of the thread, etc., or the film shown. Also *fig.*
1940 *Amer. Speech* XV. 205/1 *Unspool,* to project a film. 1961 S. Plath in *London Mag.* Aug. 8 The heath grass glitters and the spindling rivulets Unspool and spend themselves. 1962 A. Nisbett *Technique Sound Studio* 271 *Spill,*..to unspool a quantity of tape by accident. 1973 *Listener* 22 Feb. 254/2 The new play..unspools inside Christopher's head. 1980 *Times* 10 Oct. 14/1 A noisy adventure film..opened (or unspooled as local jargon has it) in Delhi.

unsporting, *ppl. a.* Add: Hence **unspo·rtingly** *adv.*, **unspo·rtingness.**
1932 R. Campbell *Taurine Provence* 17 The 'unsportingness' of hunting an animal in an enclosure. *a* 1974 R. Crossman *Diaries* (1976) II. 222, I was now in the dock for unsportingly challenging the rules when I'd lost a round in the parliamentary game. 1978 R. V. Jones *Most Secret War* xxv. 217 The Joint Intelligence Committee decided, very unsportingly, I thought, to hold back Colvin's account while they invited Ewen Montague..to write an officially approved account.

unspringy, *a.* (Later example.)
1936 *Scrutiny* IV. 398 The new verse moves line by line, the characteristic single line having..an evenly distributed weight—a settled, quite unspringy balance.

unspru·ng, *ppl. a.*² [Un-¹ 8 b + *sprung *ppl. a.*²] Not provided with a spring or springs. Of a (dance-)floor: not constructed so as to be resilient.
1928 C. F. S. Gamble *Story N. Sea Air Station* i. 32 The floats of seaplanes were practically unsprung. 1939 M. Allingham *Mr Campion & Others* I. viii. 171 A small unsprung dance-floor. 1973 R. Perry *Ticket to Ride* xii. 162 Both bunks boasted a mattress of sorts, thin and unsprung.

unsqueamish, *a.* Add: Hence **unsquea·mishly** *adv.*; **unsquea·mishness.**
1922 F. L. Lucas *Seneca & Elizabethan Tragedy* iv. 97 With Tudor unsqueamishness the audience then proceeded to watch Tereus dining off his son's flesh. 1959 *Times* 24 Jan. 7/7 The Calvinism that Burns satirized..was unsqueamishly aware of man's carnal nature. 1970 *Daily Tel.* 6 Feb. 17 The nurses are..tireless. All the virtues of the Victorian heroine are there, with unsqueamishness added for good measure. 1976 *National Observer* (U.S.) 27 Mar. 19/2 Vidal's fiction is true to the spirit of 1876, an age that was unsqueamishly exploitive and loved the grand scale.

unstable, *a.* Add: **3. c.** *Physics.* (Cf. *stable *a.* 3 d.)
1904 F. Soddy *Radio-Activity* viii. 123 Our knowledge of unstable atoms is necessarily limited. *Ibid.* 124 Radioactivity..has thus introduced us to a whole series of new unstable elements. 1924 O. Lodge *Atoms & Rays* ii. 33 The possibility of building up still more complex, and probably still more unstable, elements..remains a subject for future discovery. 1981 C. E. Swartz *Phenomenal Physics* xxiii. 715 Most elements exist naturally as a mixture of several isotopes... The diagram shows all of the stable isotopes, and many of the unstable ones.

unstained, *ppl. a.* **1.** (Further examples.)
1918 *Heal & Son Catal.: Cottage Furnit.* 3 This furniture is left quite plain, unstained and unpolished. 1936 'N. Blake' *Thou Shell of Death* ii. 27 A cream-washed bedroom, furnished..with unstained oak.

Unstan (*ʊ*·nstăn). The name of the site of a chambered tomb on Mainland, Orkney, used *attrib.* to denote a type of early neolithic pottery originally found on that site.
1932 *Proc. Prehistoric Soc.* VII. 64 The hatched and shaded Triangle motif occurs on the Unstan bowls. 1954 S. Piggott *Neolithic Cultures* viii. 248 The bulk of the pottery..can best be classed as Unstan ware from the thirty or so vessels found in that tomb. 1978 *Times* 27 May 3/3 One post-hole also yielded shreds of 'Unstan' pottery, characteristic of the early Neolithic period of the Orkney Islands.

unstarred, *ppl. a.* Add: *spec.* of a Parliamentary question: denoting that a written reply is required.
1902 *Hansard Commons* 18 June 958 To ask the First Lord of the Treasury if arrangements can be made by which the answer to an unstarred Question shall be communicated to the Member asking the Question within an hour after the sitting of the House on the day..the Question is asked. 1919 Ld. Curzon in *Hansard Lords* 11 Mar. 633 His alternative, I think, was this—that one is to assume that a Question in future, starred or unstarred, is a Question only. 1978 T. Willis *Buckingham Palace Connection* i. 10 Prayers had been said, and the business of the afternoon began, as usual, with a series of four Unstarred Questions.

unstayable, *a.* (Later examples.)
1940 G. Greene *Power & Glory* II. i. 113 It had been set up in a minute clearing by a small farmer the forest must have driven out,..an unstayable natural force which he couldn't defeat with his machete. 1980 D. K. Cameron *Willie Gavin* xviii. 177 He had risen on the unstayable tide of his master's success.

unstick, *v.* Add: **2.** *Aeronaut. intr.* (occas. *refl.*). To take off (*from* the surface of the ground, water, etc.).
1912 *Aero* May 134/1 It attains a high speed when running awash and 'unsticks' (*decoller*) very easily. 1913 *Flight* 19 Apr. 432/1 A head wind helps them materially to 'unstick' themselves from the water. 1913 *Captain* Sept. 1074/2 It is necessary to design the floats with the greatest care, so that they may 'unstick', or leave the water easily. 1926 *Chambers's Jrnl.* Sept. 581/2 It needs almost as long a run as the ordinary aeroplane to 'unstick'. 1964 G. Lyall *Most Dangerous Game* xix. 152 When she unstuck from the water..we went away low. 1977 J. Gardner *Werewolf Trace* i. 15 The British Airways Trident unstuck from the cold stressed-concrete.
Hence **unsti·cking** *vbl. sb.*
1926 'N. Shute' *Marazan* vi. 206, I took the whole length of the aerodrome to get off. It was some time since I had flown a Thirty-four, and unsticking was never her strong point at the best of times.

u·nstick, *sb. Aeronaut.* [f. the vb.] The moment of take-off. Also *attrib.*
1935 C. G. Burge *Encycl. Aviation* 606 *Unstick,* the moment during the take-off when an aerodyne definitely leaves either the ground (or a solid platform) or water. 1936 *Jrnl. R. Aeronaut. Soc.* XL. 519 Stalling to the extent of losing two-thirds of the static thrust, while it increases the 'unstick' time due to reduced acceleration, has only a small effect on the length of the take-off run. 1966 D. Francis *Flying Finish* xii. 154 Inside the windowless lavatory compartment it was impossible to tell the exact moment of unstick, but the subsequent climb held me close anyway against the wall, as I faced the tail. 1969 K. Munson *Pioneer Aircraft 1903–14* 154/2 The 1904 multiplane had an 'unstick' speed of some 34 m.p.h... and was tested at Streatham, but apparently made no flights of significant length.

unstoppable, *a.* Add: Hence **u:nstoppabi·lity; unsto·ppably** *adv.*
1961 Webster, Unstoppably. 1966 D. F. Jones *Colossus* i. 13 There is no way of walking back. The whole point is the Project's unstoppability. 1975 K. P. Barr in Barr & Line *Ess. Information & Libraries* iii. 40 He had..the entire programme in his head from the start and moved firmly and unstoppably towards its completion. 1980 R. Adams *Girl in Swing* xii. 125 He was generous to the point of embarrassment, having..a kind of baffling unstoppability when it came to paying restaurant bills. 1984 *Times* 11 Aug. 19/8 An audience of adults and children chattered unstoppably.

unstrea·m, *v. Educ.* [Un-² 3 + *stream *v.* 14: cf. *unstreamed* s.v. *un-¹ 8 a (a).] *trans.* In a school, to end the practice of streaming different abilities; to fail to stream in this way. Also *absol.*
1961 *Listener* 12 Oct. 565/3 Primary headmasters who have unstreamed their schools report that they have been able to raise the standard of work of the school as a whole. 1969 *Guardian* 29 Sept. 12/2 As a head, when a majority of the staff wanted it, he unstreamed a school. 1971 *Daily Tel.* (Colour Suppl.) 2 Apr. 18 (*caption*) You can't unstream—as the Swedes have done—and still hang on to classroom discipline.
Hence **unstrea·ming** *vbl. sb.*
1964 *Listener* 3 Dec. 904/1 Teachers..oppose unstreaming because they think it would threaten the interests of the 'A' stream. 1972 *Guardian* 9 June 9/5 Unstreaming, or teaching children in mixed ability groups.

u·nstress, *sb. Phonetics.* [Un-¹ 12.] Absence of stress; the pronunciation of a syllable, etc., without stress.
1945 E. K. Chambers *Eng. Lit. at Close of Middle Ages* i. 61 In this play aureate language..is exaggerated, almost to the point of burlesque, and is accompanied by anapaestic unstress, but not alliteration. 1953 C. E. Bazell *Linguistic Form* iii. 30 In the case of phonemic oppositions, the most striking case of a contradiction between the criteria of freedom of distribution and frequency would be the opposition of stress and unstress. 1970 B. M. H. Strang *Hist. Eng.* vi. 341 Hesitation between the spellings *u* and *o* in unstress does not indicate a sound midway between the two phonemes.

unstressed, *ppl. a.* (Further examples.)
1927 W. Deeping *Kitty* ix. 118 There seemed to be comfort for him in those fields... So peaceful and unstressed. 1946 P. Bottome *Lifeline* ii. 17 The peasants..unstressed, simple people.

unstring, *v.* Add: **3. c.** *intr.* Of the nerves: to be released from tension, to become lax.
1906 Hardy *Dynasts* II. vi. v. 308 My nerves unstring, my friends, my flesh grows weak. 1972 D. Bloodworth *Any Number can Play* xv. 130 He systematically slackened his body and mind..feeling the knots twitch loose, the nerves unstring.

unstu·ck, *ppl. a.* [Un-¹ 8 b.] **a.** *to come unstuck:* see *come *v.* 24 d. **b.** *to get, come,* etc., *unstuck* (Aeronaut.), to get into the air, to take off: cf. *unstick *v.* 2.
1913 A. E. Berriman *Aviation* xvi. 157 It is not easy to acquire a proper flight-speed while trying to rise from the water, and it is only with considerable difficulty that pilots are able to get some machines 'unstuck'. 1920 *Flight* 17 June 639/2 The machine had a very low landing-speed, got 'unstuck' after a very short run, and was very easy to fly. 1934 *Ibid.* 8 Feb. 121/2 No one seeing her for the first time would have expected her to come unstuck so quickly as she did. 1958 'N. Shute' *Rainbow & Rose* viii. 295 We came unstuck at the fourth flare. 1979 *Truck & Bus Transportation* Apr. 16/3 With the motors running at high pitch, the end of the asphalt loomed up very quickly and with a deft flick of the wrist by the 'skipper' the aircraft became 'unstuck' just in time.

unsubstantially, *adv.* (Later examples.)
1890 W. H. Dawson *Unearned Increment* vii. 84 It matters not to the speculator how unsubstantially his houses are built. 1972 P. D. James *Unsuitable Job* iii. 96 The sitting-room was elegantly but unsubstantially furnished.

unsubtle, *a.* (Later examples.)
1942 A. L. Rowse *Cornish Childhood* ii. 29 My father..was a man of simple texture, upright, hard-working..but he was uneducated, unintrospective, unsubtle. 1978 R. Nixon *Mem.* 526 The Soviets moved troops to the Chinese border in an unsubtle attempt to tie up Chinese forces and prevent them from going to the aid of Pakistan.
Hence **unsu·btly** *adv.*
1934 in Webster. 1959 *Times* 24 Oct. 9/1 French vowels fall unsubtly from her lips. 1976 *Daily Tel.* 8 July 16 Anyone who goes around announcing, or unsubtly implying, that he is so terribly tough, aggressive and exciting is unlikely to be any of those things.

unsupportable, *a.* **3.** (Later examples.)
1904 A. H. Sayce *Monument Facts & Higher Critical Fancies* (ed. 2) i. 18 The unsupported and unsupportable assumptions of the modern scholar. 1984 *Times* 6 June 5/3 The present overregulated system of air transport was quite unsupportable.

unsurprising, *ppl. a.* (Later examples.)
1927 *New Republic* 12 Oct. 219/1 The unsurprising result is a sermon of the first class, and a novel of the third. 1975 *Listener* 18 Dec. 819/1 Adaptations of Dickens's works..are meant to make you feel good. This is unsurprising, as..it could be argued that this is exactly how Dickens intends you to feel.
Hence **unsurpri·singly** *adv.*
1961 in Webster. 1972 *Bankers Mag.* (Boston, Mass.) Winter 73/1 Unsurprisingly, no agency has ever approved any proposal that would reduce its own supervisory role. 1979 *Field* 28 Nov. 1417/2 The passage concluded unsurprisingly.

unswayable, *a.* (Later examples.)
a 1945 E. R. Eddison *Mezentian Gate* (1958) i. 5 He was..hard-necked and unswayable in policy. 1979 *United States 1980/81* (Penguin Travel Guides) 84 Bostonians..have the Irish tradition of impassioned and unswayable conviction.

unsweetened, *ppl. a.* Add: Also *ellipt.* as *sb.,* unsweetened gin. ? *Obs.*
1886 Green & Hall *Jack in Box* (Brit. Libr. MS. LCP 53369 Q) f. 37, And if to hint sir, I might be so bold. *Jack.* Unsweetened with a little water cold. 1890 Barrère & Leland *Dict. Slang* II. 390/1 Those who are partial to the *unsweetened* or 'Old Tom'.—*Bird o' Freedom.* 1910 Wodehouse *Psmith in City* xii. 100 The messengers were..endeavouring to restore their nerve with about sixpenn'orth of the beverage known as 'unsweetened'.

unswept, *ppl. a.* Add: **b.** Of the wing of an aircraft: not swept-back (see *swept *ppl. a.* 3), not having sweepback or sweep-forward (see *sweep- 3).
1946 *Jrnl. Brit. Interplanetary Soc.* VI. 95 The application of sweep-back to wing shapes implies a swept V-shaped wing, in which the centre section is effectively unswept. 1977 *R.A.F. News* 11–24 May 6 (Advt.), Wingspan: 63 ft. unswept.

unsyllabic, *a.* (Further examples.)
1932 W. L. Graff *Language & Languages* 56 The sound that possesses the highest degree of sonority in a syllable is called syllabic or sonant, the others are unsyllabic or con-sonant. 1964 H. Kurath in D. Abercrombie et al. *Daniel Jones* 146 There is no difference in the system of consonants, except that unsyllabic /ə/..appears only in dialects that lack postvocalic /r/. 1970 *Publ. Amer. Dial. Soc.* 1968 L. 6 The unsyllabic phone [ə]..will here be treated as the semi-vowel /ə/.

unsympathetic, *a.* (Examples corresponding to *sympathetic *a.* 2 b.)
1937 L. MacNeice in *Ess. & Stud.* XXII. 145, I have recently been to an exhibition of nineteenth-century French painting and was surprised to find it so unsympathetic. 1946 *Ann. Reg. 1945* 438 Dreiser was a most unsympathetic personality, yet during the 1920's he was regarded as probably the most impressive figure in American literature since Walt Whitman. 1968 B. Foster *Changing Eng. Lang.* ii. 75 *Sympathetic* and *unsympathetic* have latterly taken on an extra meaning in imitation of French idiom (or possibly also Spanish), where *sympathique* means 'likable'. 1981 *Daily Tel.* 12 Feb. 16/6 An insidious invasion, unfamiliar, unsympathetic, alien.

unsystematized, *ppl. a.* (Earlier example.)
1832 F. Burney *Mem. Dr. Burney* III. 323 His internal resources were too diffuse and unsystematized.

unta·ck, *v.*² [Un-² 4 + *tack *sb.*⁷ b, *v.*⁵] *trans.* To remove the saddle and bridle from (a horse). Also *absol.*
1962 W. Faulkner *Reivers* viii. 185 We went back to the barn and untacked and Lycurgus brought a bucket and a rag and Lycurgus washed him down..before stabling and feeding him. 1977 F. Parrish *Fire in Barley* iv. 40 Would you like me to untack her?..Take off the saddle and bridle?

untalented, *a.* (Later example.)
1981 *London Rev. Bks.* 19 Nov.–2 Dec. 16 Wittgenstein..was impatient with untalented or lazy pupils.

untalked, *ppl. a.* (Later examples.)
1807 E. Weeton *Let.* 3 Dec. (1969) I. 53 She has for a long time been untalked of. 1926 G. Bell *Let.* 25 Jan. (1927) II. xxvi. 745 We seem to have left such a lot of things untalked about.

untarnishable, *a.* (Earlier example.)
1887 in A. Adburgham *Shops & Shopping* (1964) vii. 77 A thread of untarnishable gold or silver interwoven with the worsted.

untelling, *ppl. a.* Add: Also *U.S. dial.* **1.** Delete † *Obs.* and add further examples.
1834 *Chambers's Edin. Jrnl.* 28 June 170/1 It being untelling the number of lodgers who used to elope without coming to a settlement. 1925 E. C. Smith *Mang Howes* 1 The road was thrang wui droves o nout—aa keinds, untellin. 1951 H. Giles *Harbin's Ridge* iii. 18 It's untelling the times me and Faleecy John have walked the snake-back of that barnyard fence.
2. (Further example.)
1941 *Sat. Even. Post* 10 May 36/2 A man's notions are untelling.

unte·nse, *v.* [Un-² 3, 7.] *trans.* and *intr.* To render or become less tense or rigid; to relax.
1970 N. Fleming *Czech Point* i. 7, I hooked my ski-sticks on to the T-bar and with some difficulty untensed my leg muscles. 1970 E. Tidyman *Shaft* (1971) i. 10 Anderozzi untensed: Shaft wasn't going to run. 1976 A. J. Russell *Pour Hemlock* vii. 72 She untensed and started walking again. 1978 *Times Lit. Suppl.* 1 Dec. 1406/4 The triplets helped Tomlinson to un-tense his diction.

‖ **untergang** (u·ntəɹgaŋ). [Ger., = decline, downfall.] An irreversible decline, esp. leading to the destruction of culture or civilization.
1938 L. MacNeice *I crossed Minch* ix. 133 The Untergang, the collapse of civilisation. 1962 *Listener* 12 July 51/2 There were times when Zarathustra—and, I believe, Nietzsche himself—longed for the *Untergang,* the going down, the descent, among the many, for the many, into death. 1965 *New Statesman* 30 July 166/1 The rooms and houses seemed on a depressingly small scale for a civilisation which it had always pleased me to think of as suffering the disease of gigantism which afflicts societies in full *untergang.*

‖ **untermensch** (u·ntəɹmenʃ). Pl. **-menschen.** [Ger.] Esp. with reference to the Nazi régime (1933–45): a racially inferior person, a subhuman person. Cf. Übermensch.
1964 *Punch* 13 May 723/1 A Negro American..is a benighted *Untermensch.* 1966 *Sat. Rev.* (U.S.) 26 Mar. 34 To the Germans, Lithuanians were *Untermenschen,* a second-class people to be exploited and, when politically expedient, enslaved. 1974 A. Goddard *Vicana Pursuit* II. 60 The Jews were shown to be people beyond the pale—*untermenschen* who had murdered Christ. 1981 R. Barnard *Sheer Torture* xii. 132 Maria-Luisa had been shouting insults... Comes from the gutter. Scum. *Untermensch.*

‖ **unteroffizier** (u·ntərɒfitsiᵊ·ɹ). *Mil.* [Ger.] A German non-commissioned officer.
1917 T. E. Lawrence *Let.* 10 July (1938) 228 We entered Akaba..with 600 prisoners, about 20 officers, and a German unteroffizier. 1942 *Order of Battle of German Army* (U.S. War Dept. General Staff) 33 *Hauptgefreiter,* no equivalent grade... *Unteroffizier,* Corporal. 1980 R. Butler *Blood-Red Sun at Noon* (1981) I. i. 22 *Unteroffizier* Neumann was..shot down.

untha·nking, *ppl. a.* [Un-¹ 10.] Not giving thanks (for something); ungrateful.
1902 Kipling *Five Nations* (1903) 138 Arid, aloof, incurious, unthinking, unthanking, gelt. 1932 Chesterton *Chaucer* i. 36 We may sometimes be unthinking about it; unthinking and especially unthanking.

u·nthink, *sb.* [Un-¹ 12.] Passive acceptance; failure to use logical reasoning.
1958 *Spectator* 14 Nov. 641/2. Mr. Wechsler..got on his feet rampaging against 'latrine prose' and the tedious doctrine of 'un-think'. 1961 *Guardian* 9 June 11/6 The undiscriminating unthink of the new middle class. 1974 P. Gore-Booth *With Great Truth & Respect* 421 And let us not talk of a 'Lost Empire': that is a piece of 'unthink' which implies that, if only we had been cleverer, Australia would still have been governed from London.

unthought, *ppl. a.* Add: **1. b.** Also with *out.* (Cf. Think *v.*² 15.)
1919 M. K. Bradby *Psycho-Anal. & its Place in Life* x. 12 An unthought-out attitude is shown and resulting unhappiness. 1933 W. E. Orchard *From Faith to Faith* vi. 80 It may..reveal to those who cling to a merely traditional and un-thought-out orthodoxy what it is that often inspires such liberalism. 1978 R. Ludlum *Holcroft Covenant* xiv. 163 There were canvases with bold dashes of color and heavy, un-thought-out lines.

unthriftiness. 2. Delete *Obs. rare* and add later examples.
1950 *N.Z. Jrnl. Agric.* Apr. 327/1 At the first sign of unthriftiness calves should be drenched with phenothiazine. 1955 W. W. Denlinger *Complete Boston* 83 Much of the unthriftiness and langor that are charged to worms are caused by other diseases. 1973 *Country Life* 7 June 1660/1 If the calves are under stress..unthriftiness or mortality will follow.

unthrown, *ppl. a.* (Examples corresponding to *throw *v.* 32 d.)
1959 G. Greene *Complaisant Lover* I. i. 4 *Victor* (*quite unthrown*). Take him any way. 1977 C. Wood *James Bond, Spy who loved Me* xiv. 128 Bond tried to appear unthrown.

until, *conj.* For examples of *unless and until* see *unless *prep. phr.,* etc. 4 c.

untipped, *ppl. a.*¹ Add: **2.** *spec.* Of a cigarette: without a filter tip. Also *absol.*
1968 *Times* 15 Nov. 8/8 More men than women smoke heavily, inhale deeply and prefer untipped cigarettes. 1969 J. Elliot *Duel* I. iii. 67 'Smoke?'..We both took untipped and she lit them. 1973 H. Gilbert *Hotels with Empty Rooms* xiii. 111 The smoke from a hundred untipped Gitanes lay low in the air.

u:ntouchabi·lity. [f. Untouchable *a.*: see -ility.] The quality or state of being untouchable; the condition or status of an untouchable; the social practice of having a class or caste of untouchables. Also *transf.*
1919 F. B. Fisher *India's Silent Revolution* v. 110 The depressed classes are now holding Untouchability Conferences at frequent intervals. 1928 *Speeches Maharaja of Baroda* 244 The question of untouchability was particularly Hindu. 1929 *Nineteenth Cent.* Dec. 763 His own system..tends to perpetuate such social abuses as untouchability and child marriage. 1952 A. R. Radcliffe-Brown *Structure & Function in Primitive Society* vii. 138 An extreme sanctity or untouchability attached to a chief born of a brother and sister who were themselves children of a brother and sister. 1958 *Times* 9 Sept. 9/6 The origin of untouchability in Japan is not clear. 1978 L. Heren *Growing up on The Times* ii. 44 Untouchability, surely the greatest injustice in the history of mankind. 1979 J. Wainwright *Tension* xvii. 59 She has a perpetual air of untouchability about her. 1982 *Times* 24 June 3 Among the first generation [of Asians], untouchability is practised..in Britain... The Hindus of Southall have a separate temple for their untouchables.

untouchable, *a.* Add: **3.** (Examples referring to social outcasts.) Also *transf.*
1909 *Indian Spectator* 23 Oct. 843/2 Persons in mourning are..considered to be defiled and untouchable for some days. 1910 *Times* 29 July 5/6 In non-essentials Brahmanism soon found it expedient to relax the rigour of caste obligations, as for instance to..travel even in their own country in railways..without incurring the pollution of bodily contact with the 'untouchable' castes. 1943 G. Muff *Let.* in *Times* 8 July 5/5 There was a gulf between the public school and the elementary school—a caste system; when all the while we knew the child of the worker was neither 'untouchable' nor belonged to a depressed class. 1963 T. & P. Morris *Pentonville* ii. 27 The work of the general work cleaners is of 'untouchable' status and is frequently given to the Maltese and 'blacks' for this reason. 1979 A. Brink *Dry White Season* III. vii. 237 He is untouchable, protected by the entire bulwark of his formidable system.
Hence **untou·chableness,** the state or condition of being untouchable.
1909 *Times of India* (Mail ed.) 23 Oct. 19/3 The Hon. Mr. Ghokale..thought if only the untouchableness went, it would be a comparatively easy matter to help these classes. 1916 *Indian Review* Feb., in M. K. Gandhi *Coll. Works* (1964) XIII. 232 This miserable, wretched, enslaving spirit of 'untouchableness'. 1970 *Daily Tel.* 7 Feb. 9/6 It [*sc.* Verdi's 'Macbeth']..had to overcome our strong proprietorial feelings about the untouchableness of Shakespeare.

untou·chable, *sb.* [f. the adj.] A Hindu of a hereditary low caste, contact with whom was regarded as defiling members of higher castes. Also *transf.* and *fig.* Cf. *Harijan.
Use of the term, and the social restrictions which accompany it, were declared illegal in the constitution adopted by the Constituent Assembly of India in 1949 and of Pakistan in 1953.
1909 *Indian Spectator* 23 Oct. 843/2 Our untouchables were not clean. 1911 *Times* 2 Feb. 5/5 When it is remembered in what manner the lower classes are treated in daily life it may appear strange that the higher castes should be so..alarmed at the prospect of the untouchables ceasing to be regarded as Hindus. 1920 *Asiatic Review* XVI. 172 The term 'untouchable', as a name for the 'depressed classes', or 'outcastes', is a revival of the most ancient designation of these people. 1926 [mentioned in Dict. s.v. Untouchable *a.* 3]. 1928 *Daily Express* 22 May 10/2 Those in Whitehall may go on thinking there is something extremely meritorious in treating Russia as a diplomatic untouchable. 1931, etc. [see *Harijan]. 1960, 1969 [see *Eta, eta²]. 1975 *Guardian* 27 Jan. 5 Five hundred Untouchables—low caste Indians—marched on Downing Street yesterday. 1978 N. J. Crisp *London Deal* vi. 104 'Well, I've become a non-copper.' 'You mean you've been suspended?'..'A modern untouchable.' 1981 G. Priestland *Priestland's Progress* vi. 93 The Indian untouchable who becomes a Christian often has to pay a heavy price for his liberation.

untouched, *ppl. a.* Add: Hence **untou·chedness.**
1889 Kipling *From Sea to Sea* (1899) I. xviii. 145 The utter untouchedness of the town was one-half the charm.

untrilled, *ppl. a.* (Earlier example.)
1866 *Jrnl. Amer. Oriental Soc.* VIII. 342 The production of this untrilled *r* may be carried as far back in the mouth as we choose.

untuned, *ppl. a.* Add: **3.** Of an electronic device: not tuned to any one frequency; able to deal with signals of a wide range of frequencies. Also *transf.*
1905 *Electrician* Mar. 822/1 A forest would have great influence in impeding if not in staying altogether the transmission of signals over its surface, more especially in an untuned system. 1962 Simpson & Richards *Physical Princ. Junction Transistors* vii. 136 Small-signal operation in the region beyond cutoff is usually confined to two types of amplifier: 1. The tuned band-pass type... 2. The untuned wide-band or video amplifier. 1978 G. Sims *Rex Mundi* vii. 47 The disorganisation..extended even to the absence of..canned *bouzouki* music. Instead there was the untuned blare of a radio station.

unty·pable, *a. Med.* and *Biol.* [Un-¹ 7 b.] That cannot be assigned to a specific type.
1950 *Britannica Bk. of Year* 682/2 Untypable. 1961 *Lancet* 22 July 173/2 Both strains [of Staphylococci] were untypable. 1977 *Ibid.* 14 May 1047/1 The isolate derived from the mincer was of serotype 1, and that from the butter beans was untypable.

‖ **unum necessarium** (ū·nŏm nesesē̂·riŏm). [mod.L., ad. Vulgate *unum est necessarium* one thing is necessary (Luke x. 42).] The one, or the only, necessary thing; the essential element.
1931 H. H. Henson *Let.* 21 Oct. (1950) 63 So long as episcopacy is looked upon as the *unum necessarium* of a

Christian Church, I am sure that no reunion with the presbyterian and congregational churches is possible. **1937** *Times Lit. Suppl.* 15 May 379/1 Mr. Wilkins has the *unum necessarium* of a storyteller in this kind. **1938** W. S. Churchill *Marlborough* IV. xxxi. 522 His exclusion was, he said, the *unum necessarium*.

unuseful, *a.* (Later examples.)
1977 *Western Morning News* 1 Sept. 8/1 A correspondent in Dartford, Plymouth, says he finds it is the unnecessary or 'unuseful' noises which annoy him most. **1983** *N. & Q.* Aug. 357/2 The transference of the term to English literature is unuseful. **1984** *Ibid.* June 259/1 The concatenation of Rolle, Langland and Wycliffe is decidedly unuseful.

unutterable, *a.* **2. b.** (Earlier example.)
1849 *Jrnl. Amer. Oriental Soc.* I. 423 The endless, and to a European, unutterable jargons of the other class.

unvenom, *v.* (Later example.)
1906 F. Thompson in *Dublin Rev.* Apr. 381 You did, with thrift of holy gain, Unvenoming the sting of pain, Hive its sharp heather-honey.

unvizard, *v.* Add: Also *absol.* for *refl.* (Later example.)
1911 H. G. Wells *New Machiavelli* iv. i. 388 People who unvizard to talk more easily at a masked ball.

unvocalized, *ppl. a.* (Earlier example.)
1855 C. R. Lepsius *Stand. Alphabet* 39 A belongs, therefore, to the unvocalised strong fricatives.

unwaged, *ppl. a.* Add to def.: not receiving a wage, out of work. Also *absol.* (Later examples.)
1975 Dalla Costa & James in *Power of Women* (1975) 33 (*heading*) The productivity of wage slavery based on unwaged slavery. **1981** *Sunday Mercury* (Birmingham) 30 Apr. 21/2 (Advt.) Women's Theatre Group presents their new play on Nuclear Power, Breaking Through, at Small Heath Community School.. £1·50 waged, £1 unwaged. **1982** *Libr. Assoc. Rec.* (Vacancies Suppl.) 30 Nov. Fee £5 per line for waged persons or £1 per line for those who are unwaged. **1982** *Guardian* 24 Feb. 10/4 They're supposed to do unwaged work for their whole lives.

unwalkable, *a.* Add: **c.** Unfit for walking on.
1976 *N.Y. Times Mag.* 11 Sept. 85/2 The top step is 2 feet high and unwalkable. **1981** *New Scientist* 16 July 141 A nasty mixture of cow dung and raw sewage makes the sand unwalkable. **1982** *London Rev. Bks.* IV. xxiv. 8/3 These proud owners of this awful model house, tripping around on an artistic but farcically unwalkable pattern of paving-stones.. conduct themselves with a sober sense of import and duty.

unwanted, *ppl. a.* Add: Also *absol.* as *sb.*
1932 *New Statesman* 10 Sept. 284/2 We don't take no notice although she did 'ave a little Unwanted. **1949** *Here & Now* (N.Z.) Oct. 28/1 The new freedom has to be weighed against the risk of rearing a generation of unwanteds, filled with hostility. **1976** *Daily Record* (Glasgow) 4 Dec. 31/5 The problem is not peculiar to Saints.. Bigger teams heave out unwanteds, their feet not touching the ground.

unware, *a.* **2. a.** *ellipt.* (Later example.)
1866 Hardy *Wessex Poems* (1898) 10, I lived unware, uncaring all that lay locked in that Universe taciturn and drear.

unwashed, *ppl. a.* **2. b.** (Earlier examples with and without *great*.)
1830 Lytton *Paul Clifford* I. p. xix, He is certainly a man who bathes and 'lives cleanly', (two especial charges preferred against him by Messrs. The Great Unwashed). **1840** *Tipperary Constitution* 21 Aug. 2/3 The learned gentleman then briefly addressed the few *unwashed*, who were attracted to the scene.
unwashedness (earlier example).
1890 W. Booth *In Darkest England* ii. ii. 106 You can have a thorough wash-up at last, after all these days of unwashedness.

unwatering, *vbl. sb.* (Earlier examples.)
1909 R. H. Richards *Text-bk. Ore Dressing* xv. 452 Unwatering devices are used to diminish the water carried by sand, or the sand carried by water. **1935** *Economist* 21 Dec. 1283/2 Their consulting engineer on the spot estimated that the unwatering of the old mine should commence about the beginning of February. **1976** *Northern Miner* (Toronto) 19 Aug. 13/3 Stage I included.. unwatering, sampling and surface and underground drilling.

unwed, *ppl. a.* (Later examples, of parents.)
1967 E. S. Gardner *Case of Queenly Contestant* (1973) viii. 103, I would keep on working as long as I was able. Then I would go to a home for unwed mothers and have my child. **1981** R. McClure *Coram's Children* x. 124 In 1764 forty-nine children were reclaimed by their parents, often unwed mothers.

unweighted, *ppl. a.* (Examples corresponding to Weighted *ppl. a.* 2 c in Dict. and Suppl.)
1927 Bowley & Stamp *National Income 1924* 23 The unweighted average is obtained by adding up the percentages and dividing by the number of them. **1966** *Rep. Comm. Inquiry Univ. Oxf.* II. 260 (*heading*) Student/staff ratios (unweighted) for certain universities. **1977** *Econ. & Social Rev.* (Dublin) VIII. 146 In cases where an unweighted regression applied to all twenty-six Irish counties would yield inefficient estimates, re-estimating the equation with Dublin omitted is in practice equivalent to applying a full correction for heteroscedasticity.

unwelcome, *sb.* Add: **2.** Lack of welcome; a cold reception. *rare.*
1912 D. H. Lawrence *Trespasser* i. 2 A stranger.. was assured of his unwelcome.

unwell, *a.* **b.** (Later examples.)
1934 J. Rhys *Voyage in Dark* i. vi. 78 When I was unwell for the first time it was she who explained to me, so that it seemed quite all right. **1964** E. Bowen *Little Girls* ii. iii. 98 Mrs Kimasin, herself unwell today, had not the slightest intention of going in.

unwhipped, *ppl. a.* Add: **3.** Not directed by the interests of a political party; not subject to a party whip.
1959 *Manch. Guardian* 11 Aug. 6/4 The argument for the independent 'unwhipped' councillor. **1971** Hindell & Simms *Abortion Law Reformed* xi. 233 It might.. turn out that members would cast their ballots to some extent along party lines, even though unwhipped. **1979** *Guardian* 6 July 26/2 The plan is for an unwhipped vote on a motion covering the principle of the return of the death penalty.

unwhitewashed, *ppl. a.* (Earlier and later examples.)
1797 J. C. Davie *Let.* July in *Lett. from Paraguay* (1805) 118, I would have had the whole house.. left unwhitewashed. **1909** M. B. Saunders *Litany Lane* III. xx. 188 It is to be feared that in her suppressed excitement she betrayed the unwhitewashed Hilda.

unwifed, *ppl. a.* For **1840** read **1834**.

unwind, *v.*^{1} Add: II. **5. c.** *fig.* Add: *esp.* To relieve from tension or anxiety, to cause to relax. (Later examples.)
1958 B. Malamud in *Partisan Rev.* Spring 180 He managed to unwind himself and relax. **1975** W. Haggard *Scorpion Tail* ix. 131 They sat down.. the wheel.. wants unwinding.
3. *fig. esp.* To obtain relief from tension or anxiety; to relax. (Later examples.)
1938 D. Baker *Young Man with Horn* iv. vii. 276 He was tired.. If he'd ever unwound and relaxed, it would have been all over, he couldn't have lifted a finger. **1958** *Radio Times* 3 Oct. 34/1 (Advt.) After interviews Edana finds she can 'unwind' with 'Aspro' and a cup of tea. **1982** M. Russell *Backblast* ii. 14 He loved the solitude.. Helped him unwind.
unwinding *vbl. sb.* (further examples).
1933 H. G. Wells *Bulpington of Blup* ix. 398 He began to apply to this dread and mystery the 'unwinding'. **1971** D. Clark *Sick to Death* ii. 23 The unwinding part of the day when most people like to take their ease. **1977** A. Morice *Scared to Death* xiv. 94 There is something anti-climactic about going straight home.. when the curtain comes down and it is quite pleasant to indulge in some gentle unwinding.

unwinnable, *a.* Delete Chiefly *Sc.* and add later examples.
1972 D. Halberstam *Best & Brightest* 493 The war was unwinnable, or at least it was for a civilized government. **1975** F. Beeston *Vows & Law* 18 What would make the case absolutely unwinnable would be if the tenant was also coloured. **1980** *Daily Mail* 24 Dec. 15/5 The fifth year of his unwinnable Vietnam-style war.

unwinter, *v.* **1.** (Later *intr.* example.)
1944 L. Mumford *Condition of Man* iv. 108 Spring was coming to Western Europe: the darkness and cold were almost over. But it did not unwinter suddenly.

unwished, *ppl. a.* **2.** (Later examples.)
1918 *Glasgow Herald* 21 July 4/2 An echo—modified.. by reinforcements of sound at one point or unwished-for diminuendo at another. **1959** 'O. Mills' *Stairway to Murder* v. 49 Her collection of unwished-for guests.

unwishing, *ppl. a.* (Later example.)
1951 S. Spender *World within World* iii. 120 When he even no longer feels desire, he can in an idle, abstract and unwishing kind of way prove to himself.. that the hidden life of forbidden wishes exists.

unwomanliness. (Earlier example.)
1854 H. James *Let.* in R. B. Perry *Thought & Character W. James* (1935) I. 135 The machinery by which it works is lying, theft, fraud and every species of unmanliness and unwomanliness.

unworkable, *a.* Add: Hence also **unworkably** *adv.*
1927 C. Hollis *Amer. Heresy* 164 A plan, unworkably complicated. **1972** W. A. Pantin *Oxford Life* iii. 25 As late as 1850 some people thought that a Congregation which might amount to over 100 would be unworkably large.

unworldliness. (Earlier example.)
1803 D. Wordsworth *Jrnl.* 28 Aug. (1941) I. 286 That visionariness which results from a communion with the unworldliness of nature.

unwrapped, *ppl. a.* Delete *rare*$^{-0}$ and add contextual examples.
1921 *Daily Colonist* (Victoria) 20 Mar. 8/1 Why buy soap in packages when you can buy it cheaper 'unwrapped'? **1968** D. E. Allen *British Tastes* iii. 77 Bread, being brought to the door and made more by small local bakers, tends to be unwrapped (and, of course, unsliced) more frequently than elsewhere. **1978** A. Neave *Nuremberg* vii. 78 He handed over to Andrus and myself odd little packets of food and three bars of unwrapped chocolate for analysis.

unzi·p, *v.* [Un-^{2} 3, 7: cf. next.] **1.** *trans.* To unfasten the zip of; to unfasten (a zip). Also *transf.* and *fig.*
1939 L. MacNeice *Autumn Jrnl.* iii. 18, I shall.. make the world my sofa, Unzip the women and insult the meek. **1951** G. Frankau *Other Tremaine* xvi. 138 Sancha.. unzipped her skirt. **1959** *Encounter* Aug. 25/2 When his lips cracked.. out, even a loving wife can unzip them. **1972** R. Quilty *Tenth Session* i. 140 Shirley.. emerged with a brace of glistening Carlsbergs.. She tossed one over and perched again to unzip her cat. **1976** 'D. Fletcher' *Don't whistle 'Macbeth'* 156 Unzip me, please? **1979** R. Jaffe *Class Reunion* i. xi. 105 She reached for the zipper of her skirt and unzipped it.
2. *intr.* To unfasten by means of a zip. Also *fig.*
1971 *Daily Tel.* 11 Oct. 11 (Advt.), A double bed that simply unzips into two singles when preferred. **1979** D. Attenborough *Life on Earth* i. 19 DNA.. is shaped like two interlocked helices. During cell division, these unzip, splitting the molecule.

unzi·pped, *ppl. a.* [Un-^{1} 8; partly also f. prec.] With the zip unfastened; not zipped up. Also *fig.*
1951 N. Blakiston *Canon James* 69 He could not turn his eyes away from its nakedness, its shameless unzipped modernity. **1971** M. Underwood *Trout in Milk* ix. 105 A pair of unzipped trousers. **1978** J. Gardner *Dancing Dodo* xxiii. 218 The men wore protective silver coveralls, unzipped and with the hoods down. **1980** J. McClure *Blood of Englishman* v. 46 He lay unzipped from pubic arch to jaw bone.

unzi·pper, *v.* [Un-^{2} 3: cf. next.] *trans.* = *Unzip *v.* 1.
1961 in Webster. **1977** J. Cheever *Falconer* 202 He unzippered the sack.

unzi·ppered, *ppl. a.* [Un-^{1} 8; partly also f. prec.] = *Unzipped *ppl. a.*
1953 *Manch. Guardian Weekly* 20 Aug. 7/1 He had on linen slacks, a polo shirt, and an unzippered windcheater. **1961** K. M. Joseph *I'll soldier no More* 291 Reaching inside his unzippered windcheater.

‖ **uomo universale** (ŭwō·mo ūnīvĕrsā·le). Pl. **uomini universali.** [It., = universal man.] A man who excels in the major fields of learning and action. Cf. *Renaissance man* s.v. *Renaissance 1 d.
1963 *Times* 28 May 10/7 The newspaper poses the question who is to replace this *uomo universale*, with his good and bad traits. **1979** *Guardian* 16 May 12/6 Time was when the prestigitator Uomo Universale was considered a pillar of any civilized society. **1983** *Punch* 4 May 28/2 Some others, by comparison, are positive *uomini universali*.

up, *sb.* Add: **2. d.** A state of mental stimulation or excitement. Cf. *high *sb.* 1 h. *U.S. colloq.*
1966 Rote & Winter *Lang. Pro Football* iii. 144/1 Up,.. state of being emotionally prepared for a game. **1979** *N.Y. Times* 1 Apr. 9/2 Young women want to be with it. The shapes in your act are sleek and slim and they come in colors you can get an up from.
e. Phr. *in two ups*: see *two *sb.* 2 g.
3. (Earlier example.)
1849 F. B. Head *Stokers & Pokers* ix. 82 Her daughter.. listens for the rumbling of the 5 a.m. up-train.
4. Colloq. phr. *on the up-and-up*. **a.** Honest(ly), straightforward(ly), 'on the level'. orig. and chiefly *U.S.*
1863 *Chapparal Rep.* (Unionville, Nevada) 4 July 2/1 Now that would be business, on the dead up-and-up. **1930** D. Hammett *Red Harvest* vii. 71 He phoned the old man's residence to find out if the check was on the up-and-up. **1932** Wodehouse *Hot Water* i. 20, I kept telling her the whole thing had been strictly on the up-and-up, but she wouldn't listen. **1952** M. Allingham *Tiger in Smoke* iii. 65 They've got to be on the up-and-up, see? **1974** P. de Vries *Glory of Hummingbird* xiii. 200 Thus I ended.. on the up-and-up. I had recovered some hours to.. a thoroughly shady enterprise.
b. Steadily rising, improving, or increasing; prospering, successful.
1930 *Sun* (Baltimore) 18 Aug. 6/1 From now on, we are led to believe, law and order will be on the up and up, as the current phrase is. **1937** G. Heyer *They found him Dead* xviii. 317 It certainly wasn't on the up-and-up when I knew him. He was picking up a living doing odd jobs for any firm that would use him. **1959** *Encounter* Oct. 25/2 Private travel is on the up and up. **1971** *Farmer & Stockbreeder* 23 Feb. 10 (*heading*) Drainage work on the up and up. Drainage work completed in England and Wales during the coming year is likely to reach an all-time high.
5. In Winchester College Football, a forward.
1869 *Wykehamist* Oct. 6/1 J. W. Barry, a good and persevering 'up'. **1900** R. T. Warner *Winchester* ix. 142 It begins with a 'hot' or scrimmage, in which all the 'ups' take part. **1975** *Oxf. Compan. Sports & Games* 397/2 In the 15-a-side—XV—game, a team is composed of eight 'ups', four 'hotwatchers', and three 'behinds' or 'kicks'.
6. *U.S. slang.* A prospective customer.

1942 Berrey & Van den Bark *Amer. Thes. Slang* §542/21 Forward, front, up, a prospective customer. **1949** *N.Y. Times* 1 May 62 The hottest salesman who ever turned a looker into an up. **1977** *Drive* Sept.–Oct. 112/1 You go to buy a car, offering your Old Faithful in part-exchange.. In New York, you would be the up. 'I'm sitting at my desk. The guy comes through the door, so I gotta get up. See?'
7. = *upper *sb.*^{2} Freq. in *pl.* *slang.*
1969 R. Jaffe *Fame Game* xi. 164 Bonnie had taken a pill, one of the little cache of Ups Bonnie got from the queens in the gay bars. **1972** H. Perkins *Singing Millionaire* iii. 3 'Meth,' he said, 'not meths. Methedrine. Speed. Up. Chrystal. Crank.' **1978** P. G. Winslow *Coppergold* 172 'She did take pills, ups, if you get me.' Capricorn understood her to mean amphetamines.

up, *a.* Add: **3. a.** (Later examples.)
1910 H. G. Wells *Hist. Mr Polly* iv. 107 Mr Johnson, at large: 'Let's the beer up!' It's the 'eated room.' **1934** S. Beckett *More Pricks than Kicks* 12 Their bottled stout was particularly excellent and well up.
4. b. Of a lift, escalator, etc.: ascending, moving upwards, carrying persons to a higher floor. Also applied to the button which operates or summons this.
1948 G. V. Galwey *Lift & Drop* v. 124 Lord Swale.. was usually the sole first occupant of the 'up' lift. **1967** 'M. Carrel' *Dark Edge of Violence* v. 42 The man then slammed the gate and punched the 'Up' button. **1976** J. Wainwright *Bastard* i. 16 He fell down the escalator. The 'up' escalator.
5. In a state of emotional or nervous stimulation, either naturally or as a result of taking drugs; excited, elated; at a peak of performance. Cf. *high *a.* 16 c. *colloq.*
1943 [see *knock *v.* 7]. **1964** H. Selby *Last Exit to Brooklyn* 49 Waiting for the time to fly, as they started to get up on benny. **1972** *Times* 31 May 7/6 She is playing lovely tennis. In the language of the game, she is really up for this one. **1975** W. Safire *Before Fall* ii. v. 55 He's too 'up' to sleep—can you sit around with him until he runs out of gas? **1981** *Gossip* (Holiday Special) 10/2 He was very up about his job (in the CBS studio mailroom) and people in general.
6. *Particle Physics.* Applied to a quark carrying a flavour with a charge of $+2/3$; symbol *u* (*U 4 a). [An arbitrary choice of word, which appeared in print later than the symbol for the up quark.]
1975 *Sci. Amer.* Oct. 41/4 Gell-Mann designated the three quarks *u*, *d* and *s*, for the arbitrary labels 'up', 'down' and 'sideways'. **1977** *McGraw-Hill Yearbk. Sci. & Technol.* 208/1 The common baryons, the proton and the neutron, are composed of only up (*u*) and down (*d*) quarks (proton = uud, neutron = udd). **1978** *Nature* 2 Feb. 406/1 The best known meson is the pion (π) which is a combination of an up and a down quark. **1982** *Sci. Amer.* Nov. 134/1 Charm, like up, down and strange, is a quark flavor.

up, *v.* Add: **I. 4*. a.** *Cards.* To raise (a bid, stake, etc.). Cf. *raise *v.*^{1} 32 a. Also *transf.* Chiefly *U.S.*
1855 *Harper's Mag.* Apr. 488/1 I'd 'a' upped it till the hot place froze over! *Ibid.* 489/1 I'll up that! the old man was saying. **1942** Berrey & Van den Bark *Amer. Thes. Slang* §746/3 *Raise*, go (it) one (or more) better, hike,.. up. **1984** *Listener* 3 May 16/1 Some competitors see it as his way of upping the ante.
b. To increase or raise (prices, production, mechanical power, etc.). *colloq.* (orig. *U.S.*).
1934 *Amer. Speech* IX. 76/1 In Birmingham on September 17, Dr. Sterling J. Foster.. warned his hearers that 'if a certain fatal mistake is made, taxes will be upped on every house in the city'. **1943** *Sun* (Baltimore) 1 Dec. 9/3 (*heading*) Shot of water ups engine power. **1953** W. Burroughs *Junkie* xi. 116 You can only buy P.G. so often, or the druggist gets wise. Then he packs it, or ups the price. **1957** Wodehouse *Over Seventy* viii. 94 These negotiations are better left to one's agent. I have instructed mine to arrange for a flat payment of ten guineas, to be upped, of course, if they want to know what I had for dinner at that amusing château in the wine country. **1969** *Daily Tel.* 16 Apr. 23/2 This 28 per cent increase.. the annual bill of the trade from £36 million to about £47 million. **1974** R. Crossman *Diaries* (1975) I. 208 I'd taken his text with the Dame before lunch and cautiously suggested that we should make our target 135,000 houses.. Harold immediately upped me to 150,000. **1978** G. A. Sheehan *Running & Being* x. 135 Athletes upped their practice time fivefold.
c. To promote in rank. *colloq.*
1945 H. Brown *Artie Greengroin* 182 Someday that mess sergeant is going to fill the Spam full of arsenic and knock off the whole company for a laugh. The day he does that they'll probably up him to tech. **1970** G. F. Newman *Sir, You Bastard* iii. 111 Both the detectives' names and ranks were correct; neither was upped to DCS. **a 1974** R. Crossman *Diaries* (1975) I. 669 Harold Wilson breezed up and said 'Meet your new Lord President.'.. They were astonished that Harold had upped me into the stratosphere.
d. To improve, to 'boost'. *colloq.*
1968 *Globe & Mail* (Toronto) 17 Feb. 28 You can up your morale all so easily. **1976** *Daily Express* 29 June 5/4, I did make a perfunctory attempt to up my image by purchasing chic glasses.
II. 5. b. (*b*) (Later examples.)
1935 E. E. Cummings *Let.* 31 Jan. (1969) 135 And he ups and hands Am [*Eimi*] such a boost as would knock Karl Marx's whiskers out of Benjamin G. Woozeythought's cabinet d'aisance. **1958** 'A. Gilbert' *Death against Clock* 81 So you upped and fled. **1961** O. Nash *Coll. Verse* 33 One of these days not too remote I'll probably up and cut your throat. **1973** *Black World* Jan. 62/1 It did no good. I upped and died. **1979** J. Rathbone *Joseph* I. i. 20 As soon as we could we upped and fled.

up, *adv.*^{1} Add: **III. 26. up to—. e.** Bridge. *to lead up to*: to lead in a manner which allows (a particular card or suit) to be played from the third or fourth hand. Also after the sb.
1911 L. Leigh *Blue Bk. Bridge & Auction* iii. 97 In a trump deal, if the lead has been a low card the suit cannot be more than moderately strong, and the third hand.. may lead up to a weak holding in Dummy's hand. **1927** L. Hattersley *Auction & Contract Bridge Clarified* xxv. 251 The Queen should *never* be led up to the Ace with the vain idea of making a finesse. **1950** G. S. Coffin *Learn Bridge* iv. 26 He must lead a ◇ away from his king up to dummy's ace-queen. **1964** R. L. Frey *Official Encycl. Bridge* 655/2 The old maxim recommending a lead 'up to weakness' is valid but not very helpful. **1973** Reese & Dormer *Compl. Bk. Bridge* xvii. 223 He leads up to and not away from dummy's high cards.
26*. up until—. = *up to—*, sense 26 c (*c*). Cf. *up till—*, sense 25.
1938 *Tablet* 28 May 698/2 Up until the time when *Mit Brennender Sorge* and the associated Encyclicals appeared, there was indeed some reason for believing that the idea of Catholic Action was to be interpreted more or less in such a manner. **1971** *Sci. Amer.* Oct. 118/3 Up until the past few years all the pictures we saw of that world.. seemed less photographic, for all their authenticity, than maplike.
29*. *ellipt.* for *up with—* (sense 30 e). *colloq.*
1937 S. Beckett in A. Chisholm *Nancy Cunard* (1979) xxiii. 241 Up the Republic! **a 1966** 'M. na Gopaleen' *Best of Myles* (1968) 330 'Up the Prince of Wales' or something, I suppose. **1980** M. McMullen *My Cousin Death* (1981) xi. 131 Conor's taken him off.. Up Conor, I say.

up, *adv.*^{2} Add: **I. 3. e.** Of a woman's hair: worn tied or pinned on top of or at the back of the head, not hanging down; *spec.* as an indication of entry into adult society.
1911 Beerbohm *Zuleika D.* xiv. 207 Her hair, tied back at the nape of her neck, would very soon be 'up'. **a 1976** A. Christie *Autobiogr.* (1977) iv. i. 166, I was now ready to 'come out'. My hair was 'up', which at that period meant.. large knots of curls high up on the head.
6. d. *Baseball.* At bat.
[**1862** *N.Y. Sunday Mercury* 13 July 6/1 Crane came up to open the inning.] **1896** *Sun* (N.Y.) 13 May 4/1 At the beginning of the tenth inning the score was a tie. Van Haltren was the first New Yorker up. **1909** R. H. Barber *Double Play* xvii. 208 The fourth man up chose a ball to his liking and sliced it down the first-base line. **1942** P. Gallico *Lou Gehrig* viii. 97 Koenig was up next, a precision machine at getting a man along to second with hit or sacrifice. **1976** E. Blackwell in *Baseball between Lines* 52 They got a man in scoring position with two out and Buddy Kerr up.
7. d. In various colloq. phrases: *up and about, around*, active, moving about, esp. of a person who has been ill, no longer in bed; *up and doing*, busy and active.
1817 H. Granville *Let.* in B. Askwith *Piety & Wit* (1982) vii. 103 We are all much better for her presence—it says 'up and doing', she looks so reviving. **1893** 'Mark Twain' in *Century Mag.* Dec. 234/2 She was up and around the same day. **1901** Up and doing [see sense 19 in Dict.]. **1909** *Dialect Notes* III. 385 *Up and about*,.. used in expressing moderate health. **1927** G. Hunting *Vicarion* xxi. 350 It required another week for him to get up and about. **1946** K. Tennant *Lost Haven* (1947) 3 Steamy rains wash.. from men's minds all desire to be up and doing. **1978** *Lancashire Life* Sept. 88/2 It was not unusual to hear her up and about in the middle of the night, checking on a seriously ill patient.
8. e. Cf. sense 15* below.
f. Of a foxhound or a follower of the hunt: keeping pace with the fox; present at its death.
1839 'Harkaway' *Jrnl.* 4 Jan. in E. A. Pease *Cleveland Hounds* (1887) iii. 63 Sly Reynard ran down the lane a field's length, and then took the fields. This gave the leaders a sob and the second-raters time to get up. **1889** F. Mason *Flowers of Hunt* 199 Ride as they might, the pace was so great that only a select few were on anything like terms with the hounds. 'Only eight of us up!' remarked Tom Chirpington. **1908** *Punch* 8 Apr. 267/1 Biggest ole dog-fox what ever I see!.. Nobody up but the Master an' me! **1972** *Daily Tel.* 21 Nov. 19 On the second occasion the pack accounted for a brace of foxes, but the Princess's party was not up at either kill.
II. 10. c. Also *fig.*
1922 Joyce *Ulysses* 295 Bob's a queer chap when the porter's up in him.
11. d. Delete 'Chiefly with *what*'. Also, amiss, wrong. (Earlier example.)
1838 Mrs. Gaskell *Let.* 19 Aug. (1966) 37, I did not mention a word to Lucy but she must have guessed something was 'up'.
e. Of food, drink, etc.: ready, served; freq. (*tea up!*, etc.) as an indication that something is ready to be served, eaten, or drunk. *colloq.*
1941 J. Smiley *Hash House Lingo* 55 *Up*. This is usually added to another as 'coffee up' 'waitress up' or 'bread up' and designates the want or approach of a person or thing. **1950** 'D. Divine' *King of Fassarai* xxi. 177 They heard her voice, 'Chow up!' **1972** J. Porter *Meddler & her Murder* xi. 138 Grub's up!.. Them as wants forks can fetch 'em! **1981** J. Wainwright *All on Summer's Day* 14 'Tea up.' Wooley.. carrying a steaming pot.
12. f. (Earlier example.)
1829 P. Egan *Boxiana* 2nd Ser. II. 243 When time was called, it was 'all up' with Bob, and Jem was declared the winner.
13. b. Also (in *Computing*), in working condition. Freq. in phr. *up and running*. Cf. *up time.
1978 *Computing* 9 Feb. 1/1 British Steel's giant private packet-switched network is up—and running successfully. **1978** *Nature* 24 Aug. 746/1 The host computer had just broken down, forcing a delay until it could be brought up again. **1983** *Austral. Personal Computer* IV. 106/3 A lot of other facilities need to be available to make a complete up-and-running software package.
III. 15. up against—. (Earlier example.)
1896 Ade *Artie* i. 7, I saw I was up against it.
15*. up at—, attending (a specified college or university). Cf. sense 8 e in Dict.
1873 Trollope *Lady Anna* (1874) I. viii. 96 The grand idea that young Jack Bluestone, who was up at Brasenose, should marry the Lady Anna. **1926** J. Buchan *Dancing Floor* i. i. 11 It's old Milburne. He's up at Magdalen with me.
15. up for—. a.** Liable to, having to submit to, due to receive; under consideration for, having been proposed for.
1918 L. E. Ruggles *Navy Explained* 52 Down for a shoot,.. or up for a shot are all the same. It means that a man has been placed on the report and will have to appear at the mast before the captain. **1921** [in Dict., sense 14 b]. **a 1936** Kipling *Something of Myself* (1937) ii. 29 The Prefects.. were all of the 'Army Class' up for the Sandhurst or Woolwich Preliminary. **1979** A. Scholefield *Point of Honour* 34 She said he was up for a medal... And a few weeks later there was the announcement of the VC.
b. *up for grabs*: see *grab *sb.*^{2} 1 b.
17. up to—. d. (Earlier example.) orig. *U.S.*
1896 Ade *Artie* ii. 11 Up to me—see!
e. In phr. *to be up to* (a master), to be tutored by (him). *Public school colloq.* (chiefly *Eton College*).
1874 C. M. Yonge *Life J. C. Patteson* I. ii. 19 The lower remove of the fourth form.. was then 'up to' the Rev. Charles Old Goodford, *i.e.* that was he who taught the division so called in school. **1910** A. Huxley *Let.* 15 Feb. (1969) 33 This half we are all up to that ignorant creature Heygate. I have successfully proved his ignorance. **1927** H. E. Wortham *Oscar Browning* vii. 99 Curzon was subsequently 'up to him'. **1977** A. J. Ayer *Part of my Life* ii. 36 In the official language of the school.. to be in a master's division was to be up to him.
IV. 19. Also *up-and-at-'em*.
1909 O. H. Ball *Their Oxford Year* 193 It was always the up-and-at-'em aspect of things that appealed to him. **1933** Dylan Thomas *Let.* 25 Dec. (1966) 82 You like the ..'up-&-at 'em'.. shoutings of Mr. Kipling.

up, *prep.*^{2} Add: **I. 1. c.** (Earlier and later examples.) *local.*
1774 P. V. Fithian *Jrnl.* 16 July (1900) 209 She then retired up chamber. **1893** S. Hale *Let.* 22 May (1919) 282 Louise.. is now carrying some new pails up garret.
d. *vulgar.* Of a man: having sexual intercourse with.
1937 Partridge *Dict. Slang* 926/2 *Up*,.. in coïtion with (a woman): low: late (? mid-) C. 19–20. **1973** 'J. Patrick' *Glasgow Gang Observed* xii. 108 We've aw been up her. **1977** C. Watson *One Man's Meat* viii. 74 The younger man said to the older: 'I'd rather be up her than up in Newcastle.'
3. b. *up yours*, an exclamation of contemptuous rejection, often used *imp.* (and accompanied by an impolite gesture) [shortened f. *up your arse* (or a similar expression): cf. *shove *v.* 10 e]. *coarse slang.*
1956 'E. McBain' *Cop Hater* ii. 18 'How's the graft these days?'.. 'Up yours,' Carella answered drily. **1969** B. Malamud *Pictures of Fidelman* vi. 208 Fidelman blew a.. green horse for Beppo.. 'Up yours,' said the glass blower. **1970** A. Toffler *Future Shock* viii. 154 The upraised finger—the 'up yours' gesture—appears to be gaining greater respectability. **1975** J. Symons *Three Pipe Problem* xviii. 208 She made a V sign at the audience, said distinctly 'Up yours'. **1978** J. Hyams *Pool* xi. 168 'Up yours, Richie Lesser,' Freda said without malice. 'I'm smarter than you are any day.'
II. 6. a. (Earlier example of *up the stage*.)
1795 T. Wilkinson *Wandering Patentee* III. 244 She turned quite round up the stage, (though not in character) as much as to say, kiss—.
b. *U.S.* Up in (the), up at. Cf. sense 1 c.
1845 S. Judd *Margaret* II. ix. 344 You will find.. in the bottom of my chest, up garret, five dollars and a quarter. **1862** M. D. Colt *Went to Kansas* 274 My nephew is.. teaching among the Indians up Lake Superior. **1884** 'Mark Twain' *Huck. Finn* xxvi. 237 Up garret was a little cubby with a pallet in it. **1923** R. Frost *Two Witches in New Hampshire* 66 Then we asked was there anything Up attic that we'd ever want again.
c. Up at. *colloq.* and *dial.*
1960 M. Spark *Ballad Peckham Rye* vi. 122 Collie Gould up the Elephant with young Leslie. **1967** J. Burke *Till Death us do Part* xi. 165 'Where you been? That's what I want to know.' 'Up the pictures.' **1975** A. Drummond *Thames Jrnls. Vicesimus Lush* 23 Vicesimus Lush.. was living in a cottage near the Hape mine—'up the Hape' in the local speech.

up-, *prefix.* Add: **I. 2. a.** *upclimb, -cry* (later example), *-curve* (examples), *-draw, -flutter, -haul* (examples), *-reach, -shine.*
1920 *Blackw. Mag.* July 69/2 The lonely halts of the

long upclimb. **1929** O. F. Dudley *Masterful Monk* viii. 88 There would undoubtedly be an upcry from Rome. **1928** *Nation* 27 June p. iii/3 Our circulation is on the up-curve. **1950** Fraser & Thomson *Honest Bread* vii. 64 The upcurve in public drinking. **1912** J. London *Son of Sun* i. ii. 23 Grief, with a quick updraw of his knees to the other's chest, broke the grip and forced him down. **1929** D. H. Lawrence *Pansies* 35 And then the geese scuttled in.. and round the ring they went.. then doubled, and back, with a funny up-flutter of wings. **1981** *Sunday Express Mag.* 14 June 24 (*caption*) Uphaul line with knots, used to pull sail up from water. **1984** *Times* 25 Aug. 11/2 Taking all the weight on my legs I eased the sail out of the water using the uphaul and paused for it to drain. **1926** D. H. Lawrence *David* xiii. 100 So the upreach of his love fails him. **1934** F. Scott Fitzgerald *Tender is Night* ii. iv. 176 The upshine of a street-lamp.
b. *up-road* (examples).
1938 X. Herbert *Capricornia* (1939) xiii. 185 Up-road guests may come by special train.. at excursion rates. **1984** *Times* 18 July 19/2 'Footsie' started on the uproad.
III. 4. b. upspi·n *trans.*; **upbu·bble** *intr.*
1865 E. Caswall *May Pageant* ii. 25 Close to where St. Oswy's ancient well Up-bubbles from its arch'd and mossy cell. **1954** L. MacNeice *Autumn Sequel* xxvi. 160 Wells of words Upbubble. **1925** E. Blunden *English Poems* 27 The darkening room by use well knows Each thread of life that these upspin.
5. (*b*) **upbrushed, -starched, -stiffed.**
1894 W. J. Locke *At Gate of Samaria* (1895) viii. 87 A shapely neck, on which clustered coquettishly a few tiny madcap curls below the smooth, upbrushed, fair hair. **1968** *Guardian* 24 July 7/1 He was responsible for the 'up-brushed' coiffures. **1922** Upstarched [see *Sambo]. **1922** Joyce *Ulysses* 39 In a Greek watercloset he breathed his last:.. stalled upon his throne, with upstiffed omophorion, with clotted hinderparts.
7. (*b*) **up-pouring.**
1918 W. Stevens in *Others* Dec. 9 A deep up-pouring from some saltier well Within me, bursts its watery syllable.

-up, *suffix.* The adverb Up appended to vbs. (sbs., etc.) as a suffix forming substantival or adjectival compounds (usu. derived from a simple vbl. phr.: see Up *adv.*^{1} II), implying an instance or spell of an activity, an abundance or abuse of something, characterized by the action of the specified vb., etc. Such combinations are typically disyllabic and stressed on the first element: e.g. *balls-, beer-, booze-, brush-, foul-, fuck-, mop-, nosh-, show-, sign-up*, see at main element in Suppl.; *blow-, break-, build-, call-, clean-, close-, cock-, cover-, dust-, flare-, hold-, jam-, knees-, lay-, link-, make-, mark-, mock-, muck-, mug-, pick-, pile-, pin-, press-, pull-, punch-, push-, round-, run-, set-, shape-, smash-, speed-, stack-, stand-, stick-, tip-, tune-, walk-, warm-up*, see as main entries in Suppl.

up-along, *adv. dial.* Along in a particular or specified direction; in the world at large; in or towards a larger community outside an isolated region. Freq. in Cornish speech: up North, uphill. Also as *sb.*
a 1552 J. Leland *Itinerary* (1721) IV. 89 The Beauty of Bermingham.. is one Street going up alonge almost from the left Ripe [= bank] of the Brooke. **1905** E. Phillpotts *Secret Woman* iii. ii. 249 'I wish I had the mastery of the Word that man hath.' 'Very fine,' said Mr. Tapp, 'but I hate fog—whether 'tis up-along or in church. The man goeth in a mist, an' his landmarks fail him.' **1908** K. Grahame *Wind in Willows* ix. 199 'Nice little farm,' replied the wayfarer briefly. 'Upalong in that direction'—he nodded northwards. **1913** H. S. Walpole *Fortitude* i. iv. 45 Well, 'ere's the end of yer as yer are... Up along they'll change yer. **1959** *Coast to Coast 1959–60* 136 Wish she'd stop going then. Where we heading? Up-a-long or down-a-long? **1963** C. Berry *Portrait of Cornwall* (1984) iii. 54 'Coming up-along, are 'ee?' asks a West Cornwall man of his friend at the foot of the hilly street. **1966** J. Aiken *Trouble with Product X* vii. 128, I said as how you'm taking Miss Whatname upalong to hospital. **1972** *Even. Telegram* (St John's, Newfoundland) 29 June 3/1 If you have some of your people down from upalong, keep a modest bearing and don't go boasting. They may say, in their curious mainland accents: 'Waal, gee! Will ya look at all those birds. Why, this is incredible!'

‖ **Upanayana** (upanā·yănă). Also **Upanayan, Upanayanam.** [a. Skr. *upanayana*, f. *upa* towards + *nayana* leading, bringing (f. *nī* to lead).] An initiation ceremony, one of the rites undergone by Hindu Brahman boys between the ages of eight and sixteen.
1817 tr. *Dubois' Descr. People of India* ii. i. 92 The children of Brahmans are invested with the Cord when they come to the age of seven or nine years... The ceremony.. is called *Upanayana*, or, *the Introduction to the Sciences. Ibid.*, I had some difficulty in bringing myself to detail the whole of this ceremony of the Upanayana, it is so filled with minute and trifling superstition. **1877** J. N. Wilson *Indian Caste* I. v. 188 The *Upanayana* (or sacrificial endowment with the string) of a Brâhman should take place in his seventh year. **1919** S. Cave *Redemption Hindu & Christian* iv. 97 The three higher castes alone can perform the *upanayana* ceremony, they alone can qualify themselves for the study of the Veda. **1935** L. S. S. O'Malley *Pop. Hinduism* iv. 112 For the twice-born castes the most essential is the *upanayan* or investiture with the sacred thread, which takes place in a boy's eighth, eleventh or twelfth year according to his caste. **1950** M. K. Gandhi *Hindu Dharma* iv. 91 The *upanayanam* ceremony, though I have discarded it myself, it has.. a deep meaning. The sacred thread is a sign of new birth, a regeneration. **1971** *Illustr. Weekly India* 4 Apr. 11/1 Among all the sixteen *sanskars*, the *upanayana* is the most important. It signifies the practice of making over a boy to a learned religious teacher for education.

up-anchor, *v.* (Earlier example.)
1889 'Mark Twain' *Connecticut Yankee* xi. 137, I was all complete and ready to up anchor and get to sea.

up-and-co·ming, *a.* [Up *adv.*^{2} 19.] **a.** *U.S.* Active, alert, wide-awake, energetic.
1889 *Harper's Mag.* Dec. 146/2 Can't you hear just how up an' comin' it was? **1901** [see Up *adv.*^{2} 19]. **1926** F. N. Hart *Bellamy Trial* i. 10 Redfield's pretty up and coming for a place of its size. **1946** R. Lehmann *Gipsy's Baby* 59 They are of course admitting the blond curls of my brother; perhaps the juicy, up and coming appearance of all the four. Jolly-looking family. **1954** Wodehouse *Jeeves & Feudal Spirit* ii. 20 His attitude towards me had been that of an official at Borstal told off to keep an eye on a more than ordinarily up-and-coming juvenile delinquent.
b. Promising, making progress, beginning to achieve success. orig. *U.S.*
1926 R. Hughes in *Hearst's Internat.* Feb. 44/1 The up-and-coming young prize-fighter 'Curly' Boyle. **1950** G. Hackforth-Jones *Worst Enemy* i. 19 Next to Meriton I would have placed Peters (on his reports) as a most up-and-coming youngster. **1959** 'R. Simons' *Houseboat Killings* vi. 67 'He's one of our up-and-coming young men.' 'Oh. So he is a successful artist?' **1977** B. Pym *Quartet in Autumn* xviii. 161 Their fashionable little house in that up-and-coming district by the common.
Hence **up-and-co·mer,** an up-and-coming person; **up-and-co·mingness.**
1890 Up-and-comingness [see Up *adv.*^{2} 19]. **1944** *Gen* 9 Sept. 24/2 I've named only a few of the up-and-comers. **1968** T. Stoppard *Real Inspector Hound* (1970) 40 Ah—yes—well, I like to give young up and comers the benefit of my—er—of course, she lacks technique as yet—. **1982** 'W. Haggard' *Mischief-Makers* vi. 77 He's one of their up-and-comers.

up and down, *adv., prep., a.,* and *sb.* Add: **A.** *adv.* **1. a.** (Later *fig.* example.)
1974 M. Birmingham *You can help Me* iv. 106 Robin had been exceptionally quiet at lunch... He usually showed off in front of visitors. Still, he did go up and down.
c. Delete †*Obs.* and add later example. Now also often in sense 'of varying quality'.
1945 C. S. Lewis *Let.* 28 May (1966) 207 Mrs. Moore is up and down; very liable.. to fits of bad jealousy.
7. (Earlier example.)
1854 'O. Optic' *In Doors & Out* 29, I told her, up and down, that she was not what she used to be when she lived with you.
C. *adj.* **3. b.** Also, of a person: subject to alternating or changing moods.
1960 I. Cross *Backward Sex* 125 She had always been an up-and-down sort of person, depressed one minute and elated the next.
D. *sb.* **II. 6.** A rapid or cursory survey or perusal; the 'once-over'. *rare.*
1923 Wodehouse *Inimitable Jeeves* x. 102 'Read this letter.' He gave it the up-and-down.

up-and-do·wner. *slang.* Also **up-and-a-downer, upper and downer.** An up-and-down fight or argument (Up and down *a.* 5 a); a violent quarrel.
1927 *Daily Tel.* 6 Sept. 7/7 My daughter's young man wants to be king of the castle,.. and the trouble is that the missus thinks more of him than she does of me. We has a bit of an up-and-a-downer last night, and he has the cheek to fetch a policeman. **1932** P. MacDonald *Rope to Spare* xii. 174, I 'appened to hear them in a proper up-and-downer. **1955** M. Gilbert *Sky High* v. 71 Regular upper-and-downer you seemed to be having with Miss Palling. **1978** D. Williams *Treasure up in Smoke* viii. 73 I've just had the most glorious up-and-downer with my brother.

up-and-o·ver, *a.* [Up *adv.*^{1} + Over *adv.*] Denoting a type of garage door which opens by moving horizontally as it is raised.
1959 *Motor Manual* (ed. 36) ix. 221 Up-and-over doors, which are counterbalanced and swing away into the roof space.. are gaining popularity. **1968** P. Marlowe *Hire me a Hearse* xii. 169 The garage up-and-over door.. had a tendency to rattle on its pulleys. **1979** J. Cooper *Class* xi. 197 The Weybridged house has.. an up-and-over garage door.

up and u·nder. *Rugby Football.* [f. Up *adv.*^{1} + Under *adv.*] A high kick intended to give the kicker and some other members of his team time to reach the point where the ball will come down. Also *up-and-under kick*.
[**1949** D. M. Davin *Roads from Home* ii. 28 'Up and under,' their captain cried and Star forwards.. raced ahead to be under it when it fell.] **1960** V. Jenkins *Lions down Under* 208 An early tackle when he followed through after kicking an up-and-under. **1960** *Times* 18 Nov. 22/2 Oxford regained the lead from a high up-and-under kick. **1979** J. P. R. Williams *JPR* i. 32 Up on Lock's Common the wind is at its strongest; this is where dad chose for his barrage of up and unders for us all to catch.

Upanisha·dic, *a.* Also **Upaniṣadic, Upnishadic.** [f. Upanishad + -ic.] Of or pertaining to the Upanishads.
1921 M. A. Buch *Philos. Shankara* i All the main ideas of Shankara's theory are anticipated by the Upnishadic seers. **1927** Belvalkar & Ranade *Hist. Indian Philos.* II. vii. 242 (*heading*) A critical exposition of Upanishadic texts. **1937** *Mind* XLVI. 407 Witness his cogent criticism of Zeller and Burnet on Parmenides and his comments on the Upaniṣadic theories of self-consciousness. **1954** A. L. Basham *Wonder that was India* vii. 332 The great Śaṅkara himself.. maintained the rigid Upaniṣadic doctrine of salvation by knowledge. **1964** *Language* XL. 112 A series of studies of Vedic and Upanishadic problems of interpretation.. preluded his inheritance of Bloomfield's projected Vedic variants.

upa·rch, *v. Geol.* [f. Up- 4.] *trans.* To raise (strata, etc.) to form a broad dome or anticline. Hence **upa·rching** *vbl. sb.*, the raising of strata into an anticlinal form; the structure so produced.
1877 G. K. Gilbert *Rep. Geol. Henry Mountains* iv. 95 So soon as the lava can uparch the strata it does so, and the sheet becomes a laccolite. **1911** *Ann. Assoc. Amer. Geographers* I. 43 The present form and altitude of the range is therefore not due to the monoclinal displacement of the compound mass.. but to the broad and simple up-arching of a much later date. **1954** *Jrnl. Glaciol.* II. 420 In a small quarry.. I again photographed up-arching of granite. **1970** R. J. Small *Study of Landforms* iii. 94 The strong tensional forces arising from uparching produce structural weaknesses.. at the crest of an anticline. **1979** *Prof. Papers U.S. Geol. Survey* No. 1127. 1/1 A regional north-south compression.. uparched several east-trending anticlines in central Washington.

upbeat, *sb.* Add: **2.** (Examples with reference to Old English poetry.)
1942 J. C. Pope *Rhythm of Beowulf* 49 Anacrusis derives its effect.. from being placed in the up-beat or arsis. **1948** S. O. Andrew *Postscript on Beowulf* 120 Anacrusis is an introductory unstressed up-beat.
3. *fig.* An optimistic or positive mood, development, etc.; a pleasant occurrence.
1950 C. McCullers in *Theatre Arts* Apr. 28/1 The publisher says this character must not die and the book should end on an 'up beat'. **1955** *N.Y. Times Mag.* 1 May 28 (*heading*) Upbeat for modern dance. **1969** H. Waugh *Young Prey* xiii. 118 Breakfast with a pretty airlines hostess.. was Frank Sessions' only upbeat of the morning. **1976** *National Observer* (U.S.) 31 July 17/6, I don't think he's going to end tragically—he's in the classic sense a comic character, on the upbeat rather than the down.

upbeat, *a.* [f. the sb.] *colloq.* (orig. and chiefly *U.S.*). Cheerful, happy; hopeful, optimistic, positive; lively, vigorous.
1947 *N.Y. Herald-Tribune* (U.S.) 26 Sept. 16 (*heading*) Dizzy Gillespie, Yardbird Parker, Thelonius Monk get nod in up-beat set. **1952** *Variety* 2 Jan. 5, '51 was 'Upbeat' but '52 looms as 'Challenge Year'. **1961** *John o'London's* 25 May 591/3 Diana Sands as Beneatha brings a much-needed touch of up-beat comedy. **1965** *Punch* 25 Aug. 275/2 Like Queen Victoria I am inordinately cheered up by the delivery of pieces of upbeat information, not merely about my own luck, but about others, even. **1977** C. McFadden *Serial* (1978) xxvi. 58/2 I'm feeling a lot more upbeat about Gregor now. **1984** *Times* 27 Mar. 10/2 (*heading*) Upbeat mood as Hongkong talks start again.

u·pbound, *a.* and *adv. U.S.* [Up *prep.*^{2} 2.] Going upstream.
1884 'Mark Twain' *Huck. Finn* xii. 99 We watched the ..up-bound steamboats fight the big river in the middle. **1976** *Advance News* (Ogdensburg, N.Y.) 18 July 9/1 The Britannia, Royal Yacht of Queen Elizabeth II and Prince Philip of England, is due to transit the Dwight D. Eisenhower Locks upbound on the St. Lawrence Seaway.

upbrushed, -bubble: see *Up- 5, 4 b.

u·p-card. *U.S.* [Up *a.*] In various card games: a card turned face up on the table; a turn-up (Turn-up *sb.* 3 a).
1938 Wood & Goddard *Compl. Bk. Games* 223 The next card, called the up-card, is faced up beside the undealt cards, or stock. **1951** *Amer. Speech* XXVI. 102/1 *Up cards*, cards that are dealt face up in stud [poker]. **1964** *Life* 27 Mar. 80 Charts at top tell player how to play his hand, depending on the dealer's up card. **1974** J. Scarne *New Compl. Guide to Gambling* 375 This example should make it clear to the reader why Black Jack experts attach so much importance to the dealer's upcard.

up-cha·nnel, *adv.* and *a.* [Up *prep.*^{2} 2.] (Moving, leading, etc.) towards the upper end of a channel.
1893 Kipling *Seven Seas* (1896) 4 Go, get you gone up-Channel with the sea-crust on your plates. **1898** [in Dict. s.v. Up *prep.*^{2} 2]. **1925** Belloc *Cruise of 'Nona'* 135 Anyone going up-Channel makes for the lighthouse.. for this shortens an up-Channel course. **1936** E. G. Bowen in H. C. Darby *Hist. Geogr. Eng.* i. 24 The up-Channel shift of the important trade routes. **1954** —— *Settlement Celtic Saints in Wales* ii. 45 Llanilltud.. and Nant Carban ..are.. conveniently well up-Channel to provide an easy crossing.

upchuck (ɒ·ptʃɒk), *v. U.S. slang.* [Up *adv.*^{1}: cf. *to throw up* s.v. Throw *v.*^{1} 48 b.] *intr.* To vomit. Also *trans.*
1960 Wentworth & Flexner *Dict. Amer. Slang* 561/2 *Up-chuck, upchuck*, v.i., v.t. to vomit. Since c 1925... Considered a smart and sophisticated term c 1935, esp. when applied to sickness that had been induced by over-drinking. **1967** 'T. Wells' *What should you know of Dying?* i. 19 Anyway, Natalie had to upchuck, it's that kind of bug. **1975** P. de Vries *Glory of Hummingbird* ii. 21 'Did I ever tell you about the time I got sick in Miss Haley's dancing class?'.. He.. had quite copiously upchucked. **1981** 'A. Cross' *Death in Faculty* i. 9 She upchucked it.. onto the bedroom floor.

u·pcoast, *adv.* [Up *prep.*^{2}] Situated or travelling further up the coast.
1909 J. Masefield *Multitude & Solitude* xii. 296 He lay in his bunk in a cabin.., on his way up-coast. **1931** H. T. Concannon *St. Patrick* 241 Muirchu writes of the Saint as having sailed up-coast. **1974** W. R. Hunt *North of 53* i. 5 St Michael, just upcoast from the Yukon River delta.

upcoming, *ppl. a.* Add: **b.** That is about to happen, etc., forthcoming. Chiefly *U.S.*
1959 A. Huxley *Let.* 3 Sept. (1969) 878, I am working like mad.. on my lectures for the upcoming semester. **1967** Mrs. L. B. Johnson *White House Diary* 5 June (1970) 521 Today it was work on the speeches for the upcoming trip to New England. **1971** *Publishers' Weekly* 1 Nov. 18/1 Booth and Haley did their upcoming books no harm. **1973** *Express* (Trinidad & Tobago) 27 Apr. 31/3 The up-coming contest is open to all karate associations. **1982** T. Barling *Terminate with Prejudice* iv. v. 104 An upcoming visit by a member of the Soviet Politburo. **1984** *Daily Tel.* 25 Feb. 17 No change in the law was required, the existing statute was merely being clarified and there was no need to insert a special clause in the upcoming Finance Bill.

up-conve·rter. Chiefly *Electr.* [Up- 8.] A device for converting a signal to a higher frequency. So **up-conve·rsion;** also **up-conve·rt** *v. trans.*, **up-conve·rted** *ppl. a.*
1958 *Bell Syst. Technical Jrnl.* XXXVII. 989 (*heading*) Gain and noise figure of a variable-capacitance up-converter. *Ibid.*, In up-conversion (modulation) a power gain results, with the power added to the input signal being supplied by the beat oscillator. In down-conversion a power loss results. **1968** *Microwave Jrnl.* Aug. 52/2 The 12·4 GHz signal from the frequency up-converter is first down-converted to 160 MHz.. then upconverted to 12·4 GHz in a single frequency parametric upconverter. **1970** *New Yorker* 11 Apr. 34 A laser beam can be used to upconvert infrared light to visible light. **1972** *Physics Bull.* Jan. 18/2 Since the upconverted radiation now has a wavelength close to both the helium-neon and krypton (0·605 μm) wavelengths, an accurate interferometric comparison is much easier to perform than with the 9·3 μm line directly. **1972** Upconverter [see *parametric *a.*^{1} b]. **1975** D. G. Fink *Electronics Engineers' Handbk.* xiv. 72 There is a distinct advantage to up-conversion; image rejection is easily achievable by a simple low-pass filter. **1976** *Which?* Nov. 262/3 UHF signals received at the aerial are sometimes translated to VHF or even lower frequencies (HF). These VHF (or HF) signals can't be plugged directly into a UHF TV... They need either a box (called an up-converter) to turn them back into UHF signals, or a specially modified TV. **1983** E. Trundle *Beginner's Guide Videocassette Recorders* v. 91 The up-conversion process during replay involves a heterodyne system.

up-country, *sb., a.,* and *adv.* Add: **1.** *sb.* **b.** (*b*) (Earlier examples.)
1817 M. L. Weems *Let.* in *M. L. Weems: Works & Ways* (1929) III. 176, I have a number due in the up country. **1835** J. H. Ingraham *South-West* I. xxiii. 248 In the cabin are the merchants and planters of the 'up country'.
2. *adj.* (Earlier and later examples.) Also, situated in, etc., a part of the country away from any town. (Now rare in the U.K.)
In this sense and as *adv.* sometimes implying 'higher in altitude'.
1810 M. L. Weems *Let.* in *M. L. Weems: Works & Ways* (1929) III. 27 Not thinking the little up-country post offices safe in matters of money [etc.]. **1831** *Canton Miscell.* V. 321 Up-country stations in India. **1859** F. Fuller *Five Years' Residence N.Z.* xi. 233 The up-country tracks were not then marked out. **1911** C. E. W. Bean '*Dreadnought' of Darling* xx. 193 The fact that each of seven station hands *could* agree to put in £70 may throw some light upon the conditions amongst upcountry workers in Australia. **1911** W. H. Koebel *In Maoriland Bush* xxii. 284 Homely little upcountry race meetings. **1923** [see *station *sb.* 13 d]. **1929** K. S. Prichard *Coonardoo* xvi. 162 Poor degraded wretches,.. drifting about the up-country towns and settlements along the coast. **1962** J. Frame *Edge of Alphabet* xviii. 102 How I would have liked to.. live in an upcountry house whose tall windows faced the mountains. **1965** B. Wannan *Fair Go, Spinner* iv. 131 The Drongo once got a job on an up-country farm. **1970** *Kenya Farmer* Feb. 9/1 One of the greatest pleasures that Kenya can offer—a weekend on an 'up country' farm. **1977** *Church Times* 29 July 6/3 A BCMS missionary nurse in upcountry Uganda. **1981** *Times* 2 Feb. 16/7 He was the first Kandyan (upcountry) Sinhalese Buddhist to be appointed Governor-General, his predecessor having been a low country Sinhalese Christian.
3. *adv.* (Later examples.) Also, away from any town, remote.
1866 M. A. Barker *Station Life N.Z.* (1870) vi. 37, I had danced with a young gentleman whose station was a long way 'up country'. **1891** G. Chamier *Philosopher Dick* i. ii. 43 Here, up country, we have mostly a moving population of tramps, topers and outcasts. **1911** E. M. Clowes *On Wallaby* iv. 86 Up-country the sundowner, or *bona fide* worker in search for a job, will find 'tucker' for the asking at any farm or station. **1911** W. H. Koebel *In Maoriland Bush* xiii. 183 Nature is a staunch friend to this obsolete atmosphere. Nowhere, perhaps, does it come to its aid with greater fervour than in New Zealand, up-country. **1962** J. Frame *Edge of Alphabet* i. 14 As if his head were a secret gully somewhere up-country. **1962** *Housewife* (Ceylon) Apr. 10 Children's woollen clothes very moderately priced and ideal for wear, both up-country and abroad. **1971** *Inside Kenya Today* Mar. 48/1 Farmers.. can grow a variety of crops—just as much as their counterparts up-country could. **1980** *Word 1979* XXX. 167 Transportation problems, levels of literacy, as well as economic factors, tend to restrict the diffusion of the newspapers up-country.

u·pcurrent, *sb.* [Up- 2.] A rising current of air.
1909 F. W. Lanchester in *Flight* 22 May 296/1 When a plane falls normally through a vertical path, there is set up around the edges an up-current of air. **1921** *Flight* 8 Sept. 604/1 This map brings out very clearly.. the manner in which Klemperer followed the valleys and took advantage of the gusts, up-currents, etc., which were caused by the nature of the country. **1931** *Air Ann. Brit. Empire* 30 The most elementary form of up-currents are caused by the deflection of winds over hills.. necessary for soaring flight. **1953** N. Tinbergen *Herring Gull's World* iv. 39 The influence of the upcurrent on their flight is seen most impressively when.. one forces the gulls to make a detour out over the beach. **1979** B. L. C. Johnson *Pakistan* iv. 55/1 It limits the scope for upcurrents in the surface monsoon air stream to rise very high.

u·p-cut. *Engin.* [Up- 2.] A cut made by a cutter rotating so that the teeth are moving upwards when cutting; so *up-cut milling*.
1934 Webster, Upcut. **1950** J. A. Oates in A. W. Judge *Machine Tools & Operations* III. ii. 113 Down-cut milling produces a dull surface, in constrast to the bright and polished type resulting from ordinary up-cut milling. **1964** S. Crawford *Basic Engin. Processes* vi. 166 The normal or conventional method of milling is known as up-cut milling. **1973** J. G. Tweeddale *Materials Technol.* II. vi. 146 Figure 6.4 c shows an 'upcut'.

upda·table, *a.* [f. next + -able.] That may be updated.
1976 *Business Week* 5 Apr. 44/2 Eastman Kodak Corp. and Xerox Corp. are experimenting with nonsilver imaging and updatable microfiche. **1977** *Financial Times* 19 May 17/1 GEC have developed a magnetic card which looks much like existing credit cards, but carries updatable information which frees banks from their present large central accounting systems.

upda·te, *v.* orig. *U.S.* [Up- 4; cf. Up to date, up-to-date *adv. phr.*] *trans.* **a.** To bring (information, esp. written material or material recorded in some other form) up to date.
1948 *Time* 11 Oct. 22/1 [The speech] had been corrected and updated after last-minute teletype reports from.. Paris. **1959** *Times Lit. Suppl.* 20 Mar. 155/2 He would streamline the production, add a chorus of dancing girls, update the lyrics, and upjazz the music. **1962** C. Walsh *From Utopia to Nightmare* 12 All current allusions have been 'updated' (as of early 1962). **1979** *Ottawa Law Rev.* XI. 815 The entry for 'Hudson's Bay Company' needs to be updated.
b. To supply (a person) with the most recent information; to bring (a person) up to date.
1952 *N.Y. Herald-Tribune* 14 Mar. 21/2 He updated me on a couple of gimmicks. **1970** J. Earl *Tuners & Amplifiers* 7 The book will have appeal, too, to the older hand, allowing him to become quickly updated without the task of researching into piles of technical literature. **1975** *Sunday Times* 16 Nov. 44/4 No doubt I'll update myself in a few weeks and emerge as a real Seventies mum.
c. Of equipment, processes, etc.
1959 *Listener* 4 June 998/1 This prescience has now been 'up-dated', as they say, by reference to nuclear warfare. **1970** *Islander* (Victoria, B.C.) 8 Feb. 7/2 Despite ..updating the equipment to include new freezing facilities.. this [whaling] venture also folded. **1977** *N.Y. Rev. Bks.* 13 Oct. 45/1 To update the 'dirty test tube' analogy, the effects of the contaminants have been estimated and found to be minor. **1984** *Daily Mail* 1 Dec. 14/1 (Advt.) This way, you can update your equipment whenever you want.
Also **upda·ted** *ppl. a.*; **upda·ting** *vbl. sb.*
Quot. 1910 is an isolated early British example.
1910 Hardy *Satires of Circumstance* (1914) 222 Your up-dated modern page. **1954** *Newsweek* 27 Sept. 58/1 The new research center is an updated transplant of the National Bureau of Standards world-wide radio investigations formerly based in Washington. **1959** *Spectator* 31 July 134/2 Despite the up-dating and jazzy typographical treatment, one fairly quickly perceives an enduring pattern of traditional attitudes. **1974** Adby & Dempster *Introd. Optimization Methods* iv. 96 Round-off errors in the computations of the up-dating process can lead to non-positive definitive S_k and divergence. **1978** *Jrnl. R. Soc. Arts* CXXVI. 672/1 The plan adopted is said to be a revised and updated version of an outline plan for the development of the economy produced in 1975.

u·pdate, *sb.* orig. *U.S.* [f. the vb.] **1.** New information received or supplied; an updated version of something. Also, an account or report of the present state of affairs. *colloq.*
1967 *Britannica Bk. of Year* (U.S.) 804/3 *Update*, current information for use by a computer in order to make the calculations necessary for achieving a particular goal. **1970** N. Armstrong et al. *First on Moon* v. 116 When you're ready to copy.. I've got a couple of small flight plan updates. **1975** B. Garfield *Hopscotch* ix. 92 Thanks for the update. Keep on it. **1976** *Physics Bull.* July 313/2 Quarterly updates are available for each course. **1977** H. Greene *FSO-1* xiii. 119 One of his people will.. give you a full update on related problems. **1983** *Brit. Med. Jrnl.* 6 Aug. 417/1 The editors of *Harrison's Principles of Internal Medicine* have bridged the gap between the 9th and 10th editions with a series of updates. **1984** *Miami Herald* 6 Apr. 11D/1 An embarrassing update of the 1960s comedy filmed on Fort Lauderdale Beach. **1985** *Times* 5 Jan. 17/7 The new album is a high-grade Motown update, funky, sleek and packed with potential hits.
2. The action or result of updating; the supplying of new information, data, etc. Freq. in *Computers.*
1968 *Amer. Documentation* Jan. 75/1 Thus, interrogations, extractions and updates can be based upon items or subitems, but not upon units smaller than these. **1972** *Computer Jrnl.* XV. 191/2 The effect, of course, is to put the clock back to the time of the last dump and all updates made since then must be repeated. **1976** *Gramophone* Aug. 353/3 Such 'updates' as they are called, involve switching individual tracks of the tape from the 'read' to the 'write' mode. **1982** *ICL News* No. 96. 5/1 A pre-Conference session.. was followed by an overall update and presentations on latest versions of data management software.
3. *attrib.*
1969 *Sunday Times* 13 July 13/2 *Update pad*, information on spacecraft attitudes, and other data, transmitted to the crew in standard format. **1972** *Computers & Humanities* VII. 127 Working with this printout, the editor prepares data for the update module in which insignificant variants are deleted and citations added. **1983** *Word Ways* Nov. 197 It.. is supplemented by update lists which the computer automatically generates.

up-dip, *a.* and *adv. Geol.* [f. Up *prep.*^{2} + Dip *sb.*] **A.** *adj.* (Stressed *u·p-dip.*) Situated or occurring in a direction upwards along the dip. **B.** *adv.* (Stressed *up-di·p.*) In an up-dip direction.
1916 F. H. Lahee *Field Geol.* viii. 191 When a fault is approximately parallel to the strike of the strata, the block on the side of the fault toward which the dip of the beds is measured may be termed the down-dip block, and the other one is the up-dip block. **1957** G. E. Hutchinson *Treat. Limnol.* I. i. 156 The bays originated as artesian springs, which tended to migrate inland up-dip. **1965** [see *lenticularity]. **1974** R. D. Grace in P. L. Moore et al. *Drilling Practices Manual* xiii. 329 It predicts up-dip deviation when drilling into softer rock.

up-do (ɒ·pdū). *colloq.* (orig. *U.S.*). Also **updo.** [f. Up *adv.*^{1} + Do *sb.*^{1} 2, after *hair-do.*] A style of dressing women's hair by sweeping it up and securing it away from the face and neck.
1938 *Sun* (Baltimore) 29 Nov. 9/4 The 'up-do' hair style does not click with De Paul University men students... The 'up-do' probably will go back down after movie stars get tired of it. **1966** J. S. Cox *Illustr. Dict. Hairdressing & Wigmaking* 157/1 *Up-do*, a hair style dressed away from the forehead and neckline. Updo: American Spelling. **1984** *Daily Tel.* 3 Sept. 13/5 She has not appeared to be wearing anything in her hair. If Miss Resnik returns to space a second time, Mr Menchaca said he would consider putting in 'braids or an up-do'.

updo·ming, *vbl. sb. Geol.* [Up- 7.] The upward expansion of a rock mass into a dome shape.
1964 *Bull. Vermont Geol. Survey* No. 23. 7 Updoming was the deforming mechanism of the anticlinorium. **1977** A. Hallam *Planet Earth* 78/3 The heating of the rocks beneath the rift zone causes expansion, and this accounts ..for the updoming.

updraught. Restrict † *Obs. rare* to sense in Dict. and add: **2.** Also (chiefly *U.S.*) **updraft.** [Draught *sb.* 24.] An upward-moving current of air. Cf. Down-draught 1. Also *fig.*
1909 *Daily Graphic* 26 July 10/1 While the planes are fighting against the wind their area will be comparatively small, but when floating on an up-draught they will be expanded as a fan expands. **1935** C. Day Lewis *Time to Dance* 31 Some aimed at a small objective but the fierce updraught of their spirit Forced them to the stars. **1953** N. Tinbergen *Herring Gull's World* iv. 28 A north wind caused an updraught, which enabled the birds to hang motionless in the air for many minutes. **1954** R. J. Schwartz *Dict. Business & Industry* 533 Updraft. **a 1963** S. Plath *Crossing Water* (1971) 38 A sheet of newsprint on the fire Levitating a numb minute in the updraft. **1983** D. Boggis *Woman they sent to FIGHT* iv. 28 Brakes off, undercarriage down.. *Slam* through the updraught off the spinney.

updry, *v.* (Later *poet.* example.)
1889 F. Thompson in *Merry England* Sept. 306 Till Time, the hidden root of change, updries, Are Birth and Death inseparable on earth.

up-end, *v.* Add: **1.** Also *fig.*
1970 G. Greer *Female Eunuch* 37 By growing their hair they managed to up-end some strange presupposition about its sexual significance.
2. *spec.* of waterfowl: to dip the head below water and raise the tail into the air, when searching for food in shallow water. (Later examples.)

1927 E. SANDARS *Bird Bk. for Pocket* II. 126 [Brent Geese] sometimes when feeding at high tide, up-end like Ducks. **1954** A. W. P. ROBERTSON *Bird Pageant* i. 20 Avocets..often when water-borne..up-end like ducks. **1957** R. A. H. COOMBES in D. A. Bannerman *Birds Brit. Isles* VI. 266, I never saw them 'upending' in the shallows as the other geese were fond of doing. **1981** *Birds* Summer 47/2 An up-ending mallard is a feeding mallard, but this, our commonest duck, is as adaptable in its feeding technique as it is in its choice of food.

upfield, *adv.* and *a.* **A.** *adv.* (Stressed *upfie·ld.*) [UP *prep.*^{2}] **1.** *Football*, etc. In or towards the end of the field nearest the goal which the team is attacking.

1951 *Sport* 27 Apr.–3 May 8/1 The wee Scot whipped the ball upfield to the late Jack Lambert before Huddersfield had organised their ranks. **1960** *Times* 29 Feb. 3/3 France worked their way upfield again. **1976** *Evening Advertiser* (Swindon) 31 Dec. 19/4 He would not stay upfield wide on the wing and was coming back too deep.

2. *Chem.* and *Physics.* In a direction corresponding to greater field strength.

1965 *Jrnl. Chem. Physics* XLII. 636/2 The $^{205}Tl^+$ resonance is shifted upfield by a very large amount. **1976** *Nature* 29 July 424/3 The statement..makes the dangerous assumption that no aromatic protons are shifted upfield or downfield out of this region by local ring current effects.

B. *adj.* (Stressed *u·pfield.*) **1.** *Football*, etc. Directed into or occurring in a part of the field near(er) the goal attacked.

1960 *Times* 18 Nov. 22/2 Long upfield kicks by stand-off Evans. *Ibid.* 6 Dec. 18/3 Upfield support of forwards by the backs. **1983** *N.Y. Times* 30 Jan. v. 8/6 Manley is the upfield pass rusher.

2. *Chem. & Physics.* Situated or occurring in the direction of greater field strength.

1967 *Inorg. Chem.* VI. 1133/2 Increased metal to ligand π bonding produces an upfield coordination chemical shift. **1979** *Macromolecules* XII. 763/1 These spectra show an upfield peak separated from a smaller peak, 2·36±0·1 downfield.

upfloor. Add: (Later archaizing example in sense 'an upper floor').

1922 JOYCE *Ulysses* 381 Sir Leopold heard on the upfloor cry on high and he wondered what cry that it was whether of child or woman.

upfront, *adv.* and *a. colloq.* (orig. *U.S.*). Also **up-front, up front.** [f. UP *adv.*^{1}, *adv.*^{2} + FRONT *sb.*] **A.** *adv.* **a.** At the front, in front.

1937 [see *LATCH *v.*^{1} 1]. **1954** A. FULLERTON *Bury Past* I. i. 9 In their sealed compartment up front the sound came thinly. **1968** *Listener* 25 July 107/2 'Get your parachutes on.' He added reluctantly: 'We're having a little trouble up front and you may have to jump.' **1976** *Sunday Post* (Glasgow) 26 Dec., Alloa just deserved their success because they were a bit sharper up front.

b. *transf.* Of payments: in advance; initially. Also openly, frankly.

1972 *Britannica Bk. of Year* (U.S.) 733/3 *Up front, adv.*, in advance (actors demanding $1 million *up front*). **1980** *Maledicta* III. II. 238 That is much like the 'frank' people these days who confess their inadequacies or even villainies *up front* and then are shocked when you do not cancel those things out as exculpated by admission. **1982** S. WILSON *Dealer's Wheels* xiv. 135 'How much cash did you have in mind?' 'Five thousand, up front.' 'I beg your pardon?' 'In advance.'

B. *adj.* That is in the forefront; honest, open, frank; (of money) paid in advance. **a.** *attrib.*

1967 *Time* 29 Dec. 24/2 What the up-front struggle really amounts to is an angry, private little war between two people. **1975** *New Yorker* 20 Jan. 31/1, I'd like to suggest that we get most of the people involved below the line, so we won't need much of that scarce up-front bread. **1977** *Bulletin* (Sydney) 22 Jan. 88/1 Lunch with local film people who want to buy film rights... They chat about 'up-front money' and consider whether Robert Altman or Werner Herzog would do the better job. **1980** *Fortune* 24 Mar. 70/2 The up-front problem in the division today is that it does not garner deposits. **1982** *Chicago Sun-Times* 14 Oct. 100/1 'We've never had a budget to market these properties before,' he said. 'This time we have $250,000–$300,000 in upfront money.' **1985** *Broadcast* 11 Jan. 7/1 As well as the up-front money from RAI on these two documentaries RAI Channel 1 will also be taking up to 100 hours of natural history and scientific documentaries from Horizon..and QED.

b. *predic.*

1970 J. LENNON in J. Wenner *Lennon Remembers* (1972) 67 That game of 'Well, I'm going to be up-front because..a few people said she'd got a lousy name in New York.' **1973** D. LANG *Freaks* 119 You might as well live in a totalitarian state that's up front with it, at least. **1978** M. PUZO *Fools Die* xxiii. 258 He would..make them understand in a nice way that he would trade space for a piece of ass. He was that upfront about it. **1980** *Washington Star* 2 July c2 Sirois was totally 'up front', O'Malley said, in his efforts to inform and ask for help from the Caps.

u·pgiven, *ppl. a. poet. nonce-use.* [UP- 5.] Surrendered, resigned. Cf. GHOST *sb.* 1.

1947 DYLAN THOMAS *In Country Sleep* in *Horizon* Dec. 304 The upgiven ghost Of the dingle torn to singing.

u·pglide. *Phonol.* [UP- 2 a: see next and GLIDE *sb.* 4.] An upward glide.

1930 [see *DOWN-GLIDE]. **1957** *Publ. Amer. Dial. Soc.* XXVIII. 115 Normally *or* marks the division between rise and fall, itself occurring at the end of the upglide. **1972** W. LABOV *Language in Inner City* i. 9 In New York and other northern cities,..the back upglide of *ball* and *hawk*, so characteristic of many southern areas, is rarely heard.

u·pgliding, *ppl. a. Phonol.* [UP- 6 b: see prec.] Gliding towards a high(er) vowel.

1933 O. JESPERSEN *Essentials Eng. Gram.* iv. 42 The *r*, or the resulting vowel [ə], also prevents the formation of the up-gliding diphthong [ou] in words like *oar, board*, [etc.]. **1948, 1962** [see *INGLIDING *ppl. a.*]. **1970** *Publ. Amer. Dial. Soc.* 1968 L. 19 An upgliding diphthong occurs sporadically. **1981** *Amer. Speech* 1977 LII. 203 Variants listed as deviating from his constructed general pattern of simple, upgliding, and ingliding vowels.

upgra·dable, *a.* Also **upgradeable.** [f. *UPGRADE *v.* + -ABLE.] Capable of being upgraded or raised to a higher standard; *spec.* in *Computers*, applied to (the storage size of) a system. Hence **upgradabi·lity.**

1974 *Physics Bull.* Dec. 568/1 (Advt.), ND 100 has a completely solid state memory with 1K, 2K, or 4K memory capacity (field upgradeable). **1982** *Which Computer?* June 44/3 The ease of software upgradability certainly makes it attractive. **1983** *Daily Tel.* 14 June 6 (Advt.), The Exxon 8431 System features an advanced central controller with upgradeable storage up to 320 MB. **1985** *Personal Computer World* Feb. 42/2 (Advt.), Works equally well on 5¼", 3½", or 3" disks, with upgradability built-in from 100K to 2.6 Mbyte.

up-grade, *sb.* and *adv.* Add: **1. a.** (Earlier example.)

1873 J. H. BEADLE *Undevel. West* xv. 257 Forty miles of staging over boulder and rocky up-grade.

2. *on the upgrade* (*fig.*) (later examples). Also, improving.

1914 A. BENNETT *Price of Love* vi. 110 'Auntie still on the up-grade?' he inquired... He guessed..that Mrs. Maldon must be still better. **1926** *Ladies' Home Jrnl.* Oct. 143 Monty's been the hardest child we've had to handle,..but I believe he's on the upgrade. **1951** 'J. TEY' *Daughter of Time* iv. 55 Don't you worry. I'm on the up-grade. Even my temper has improved. **1972** *Daily Tel.* 21 Jan. 1/2 Various indices..show that the economy is now firmly on the upgrade.

3. [f. the vb.] An upgraded version. Also, an upgrading to a higher standard (chiefly, in *Computers*, to a larger and more powerful system).

1980 *Nature* 17 Apr. 599/1 The upgrade of this facility, PBFA-II, will deliver 3.5 MJ at 100 TW and will be available for experiments in 1986. **1982** *Which Computer?* June 55/2 They can reach down to the more sophisticated microcomputer user, holding out the possibility of an upgrade to a larger machine. **1983** *What's New in Computing* Jan. 30 (Advt.), PT7/5 powerful local processing..allows simpler planning of mainframe upgrades.

upgra·de, *v.* Also **up-grade.** [f. UP- 4: cf. the sb.] **1.** *trans.* To increase the grade or status of (a job); to raise (an employee) to a higher grade or rank. Also *fig.*

1920 [implied at *UPGRADING *vbl. sb.*]. **1928** *Daily Express* 5 Mar. 3/5 His job..has become so important that the post is being 'up-graded'. **1942** *Tee Emm* (Air Ministry) II. 130 In future they are going to be upgraded [from angels] to archangels. **1955** [see *DOWN-GRADE *v.*]. **1976** *National Observer* (U.S.) 20 Nov. 14/2 The commission agreed to upgrade controllers at 32 facilities, but fewer controllers would be promoted than the union demands. **1985** *Contact* (Pre-School Playgroups Assoc.) Feb. 23/1 Teachers who underestimate the richness of children's homes decide to upgrade the children's language skills.

2. a. To raise (something, esp. equipment or facilities) from one grade to another; to improve or enhance physically.

1935 [implied at *UPGRADED *ppl. a.*]. **1955** M. REIFER *Dict. New Words* 217/2 *Upgrade*, to improve the grade of a product. **1959** *Globe & Mail* (Toronto) 17 July 13/3 If Canada were to maintain and extend its capabilities in this field, it was necessary that its effective weapons be continuously upgraded. **1964** G. L. COHEN *What's Wrong with Hospitals?* vii. 143 These old places are really quite comfortable, you know; they've been up-graded out of all recognition. **1971** *Daily Tel.* 13 Sept. 10/6 The money..would be spent on financing sewerage schemes and upgrading existing effluent treatment facilities. **1976** A. HOPE *Hi-fi Handbk.* 14 If you wish to up-grade a music centre, you can only throw it out and replace it with a better one. **1979** 'J. LE CARRÉ' *Smiley's People* xvii. 211 He booked a first-class seat and said he would upgrade his economy ticket on arrival at the airport.

b. More generally, to raise to a higher level; to improve.

1959 G. D. MITCHELL *Sociol.* 120 Generally it was found that women were more likely than men to up-grade themselves. **1962** M. MCLUHAN *Gutenberg Galaxy* 43 For until men have up-graded the visual component communities know only a tribal structure. **1977** *New Yorker* 24 Oct. 79/2 They have upgraded their criminal expertise by watching such programs as..'Kojak'. **1983** *Times* 29 Oct. 16/4 A lunch in the City..led to one broker upgrading his forecast for the year.

c. *absol.* or *intr.* (for *refl.*).

1950 *Sun* (Baltimore) 2 Oct. 9/1 Movers say that the bulk of families moving are 'upgrading'. By that they mean that they are occupying new apartments renting for much more than they have been paying. **1967** *Boston Sunday Globe* 23 Apr. B41/5 The head of a family..wants to upgrade—and usually does. **1970** J. EARL *Tuners & Amplifiers* iii. 60 This may differ slightly if we already possess some items of equipment and merely wish..to upgrade by changing a particular item..for an equivalent of improved quality.

Hence **upgra·ded** *ppl. a.*; **upgra·ding** *vbl. sb.*

1920 *Glasgow Herald* 19 Aug. 8/3 The government proposals for..the upgrading of the unskilled labourers..were rejected by the operatives. **1935** *Times* 1 Oct. 4/4 The rates on the upgraded stretch and the 1 in 22½ Dashwood Hill were 15 and 50 and 30 and 44 m.p.h. **1938** *Encycl. Brit. Bk. of Year* 214/1 The Burnham Committee..dealt with the up-grading and down-grading of Elementary Schools for the assessment of Head Teachers' salaries. **1959** *Times Lit. Suppl.* 9 Oct. 579/1 The danger of 'escalation'—the upgrading of the size of the weapons in reply to the apparent use by the enemy of bigger types. **1976** P. R. WHITE *Planning for Public Transport* viii. 186 A preliminary study suggests that investment in an upgraded rail link may provide net benefits as high as for a parallel road scheme.

uphaliday. 1. c. See next.

Up-Helly-Aa (ʌp,heli,ā·). Also **-A'.** [See UPHALIDAY.] (A revival of) a traditional midwinter fire-festival held at Lerwick in the Shetland Islands (see quots.).

1872 *Saturday Herald & Shetland Gaz.* (Kirkwall) 10 Feb. 1/3 Monday night was what is here [*sc.* in Lerwick] called 'Up-haly a', and the youths here indulged in their savage sports of tar-barrel burning. **1901** *Shetland News* 5 Jan., The principal Festival of the season to Lerwegians, namely 'Up-helly A',..is now celebrated with..Norse galleys, torch-light processions, and guizing galore. **1934** W. MOFFAT *Shetland* 129 Up-Helly-Aa night was the twenty-fourth night of the Helli or Holy Days, and that period of feasting, drinking, singing and rejoicing concluded with a great flare-up on Up-Helly-Aa night. **1948** C. E. MITCHELL *Up-Helly-Aa* 2 Up-Helly-Aa is the final day of the Jola Bød or Yule Period which commenced on the fifth day of January, old style, and ended twenty-four days later in a climax of revelry. **1955** *Shetland Times* 28 Jan., When a man is elected to the Up-Helly-A' Committee he can expect to become Guizer Jarl according to his seniority. **1972** *Guardian* 15 Jan. 14/5 The fire festival of Up-Helly-A in the Shetlands, which celebrates the passing of midwinter with..torchlight processions..and a lifesize longboat that goes up in flames.

upholding, *ppl. a.* Add: Hence **upho·ldingly** *adv.* (nonce poet. formation).

1934 KIPLING *Non nobis Domine* in *Verse* (1940) 512 O Power by whom we live—Creator, Judge and Friend, Upholdingly forgive.

upholster, *v.* Add: **2. b.** *fig.* Also, to provide with (fine or smart) clothes, to dress. (Earlier example.) Cf. *UPHOLSTERED *ppl. a.*

1873 'MARK TWAIN' *Gilded Age* xxxiii. 300 It had cost something to upholster those women.

upholstered, *ppl. a.* Add: *fig.* and *transf.* Also, of persons (*euphemistically*): (well-)-dressed; plump, stocky. See also *WELL-UPHOLSTERED *ppl. a.*

1924 WODEHOUSE *Leave it to Psmith* iii. 60 This sumptuously upholstered young woman. **1929** D. H. LAWRENCE *Pansies* 31 The upholstered dead that sit in deep arm-chairs. **1958** S. J. PERELMAN *Most of S. J. Perelman* 21 Where are those big, jolly, upholstered girls one used to see? **1976** R. LEHMANN *Sea-Grape Tree* 6 British travellers..stoutly upholstered middle-aged couples from the Midlands.

up Jenkins (ʌp dʒe·ŋkinz). [f. UP *adv.*^{1} + the surname *Jenkins*.] The name of a parlour game resembling tip-it.

1889 K. GREENAWAY *Bk. Games* 25 *Up Jenkins*. The company divide into two parties and sit round a table. One party then puts their hands under the table and a shilling or other small article is placed in one of the palms. The other side then calls out: 'Up Jenkins!' and the players whose side has the shilling concealed must all place their closed hands on the table. The opposite side must then guess in which hand the shilling is concealed. **1916** D. H. LAWRENCE *Twilight in Italy* 193 Afterwards we played 'Up, Jenkins'. **1928** E. WAUGH *Decline & Fall* III. iv. 255 'I must ask you both to put your hands on the table in front of you,' said the warden. 'Like Up Jenkins,' said Margot faintly. **1933** S. JAMESON *Day Off* 83 They played Up Jenkins, in which a button is passed frenziedly from hand to hand. **1946** L. P. HARTLEY *Sixth Heaven* viii. 164 They..pressed their hands on the table as if for 'Up Jenkins'. **1951** S. SPENDER *World within World* iii. 144 Sometimes he would play childish games such as 'Up Jenkins'. **1981** P. THEROUX *Mosquito Coast* xvi. 212 We played Up Jenkins on the Gallery floor.

uplift, *sb.* Add: **3.** The support or lift gained from a garment that raises part of the body, esp. the bust; the (part of the) garment which achieves this.

1929 *Radio Times* 8 Nov. 435/3 A supporting stocking..Its gentle uplift massaging action has a beneficial effect upon the varicose veins. **1934** *Times* 22 June 17/6 The skirt covers neatly fitting trunks and clings closely to the figure; the top has a knitted uplift. **1955** J. P. DONLEAVY *Ginger Man* xxv. 297 Bras the uplift of which will put a new lust into the hearts of these citizens. **1957** *Daily Mail* 25 Oct. 10/4 When the vulgarity of too much uplift, too much emphasis on the female form finally overwhelmed us Dior sensed it before we did. **1959** *Housewife* June 22 The bra that gives a natural uplift.

4. An increase (in prices, wages, etc.).

1949 *Times* 26 Oct. 5/5 The whole conception of uplift..assumes that the manufacturer of consumer goods has..

two prices, one for sales to the wholesaler and one for direct sale to the retailer. **1952** *Sunday Express* 15 Nov., These appliances are given an uplift of 100% between the maker and the public. **1955** *Canadian Tax Jrnl.* III. 99 If the price of the transaction differed [from normal]..an uplift would be applied to the actual sale price to determine price for tax purposes. **1962** E. GODFREY *Retail Selling & Organization* xv. 158 If goods normally purchased through a wholesaler are bought at a lower price direct from a manufacturer, they may be subject to uplift. **1979** *Daily Tel.* 1 Dec. 21 The Ford agreement..consists of a 20·5 p.c. uplift in wages plus an extra 1 p.c. to cover an increase in the supplementary payment.

5. *attrib.* **a.** In sense 2.

1915 *Sphere* 23 Jan. 110/2, I find in an American paper a scornful reference to one of the 'uplift' magazines. **1930** J. BUCHAN *Castle Gay* ii. 32 Thomas was beginning to be much in request by uplift societies. **1940** R. S. LAMBERT *Ariel & all his Quality* ii. 50 The 'uplift' experiment..fell between..the professors of adult education..and the broadcasting experts. **1977** P. JOHNSON *Enemies of Society* ix. 122 Schneider and Dornbusch identified four common characteristics in the religion idealized in these uplift books: activism, optimism, individualism and pragmatism.

b. **uplift bra(ssière),** a brassière that provides uplift (sense 3 above).

1932 *Week-End Rev.* 1 Oct. 373/1 An 'up-lift' brassière would make you look rounder, of course. **1949** M. MEAD *Male & Female* iv. 80 Far enough removed from the uplift bras and the way Grandfather looks when Granddaughter wears one of them. **1957** J. D. MACDONALD *Executioners* (1959) vii. 134 She wore nothing under pants and blouse except an uplift bra. **1977** E. J. TRIMMER et al. *Visual Dict. Sex* (1978) iv. 51 The flavour of that era of uplift bras and pencil-skirts is still nostalgically preserved.

uplift, *v.* Add: **3. b.** More generally, to collect or pick up (something other than money); *spec.* of a bus: to take up (passengers). Chiefly *Sc.*

1961 *Alexander's (Midland) Bus Timetable, Falkirk* 171 Only passengers who are travelling beyond Milngavie Cross will be uplifted between Glasgow (Buchanan Street) and Milngavie Cross. **1967** E. RUDINGER *Wills & Probate* 109 The court is asked to confirm that the executors who have sworn the inventory are the persons entitled to uplift and administer the various items of estate listed in the inventory. **1968** 'S. JAY' *Sleepers can Kill* xvii. 175 Somebody..has left a message for Felson. The objective is to uplift the message without being detected. **1976** *Buses* XXVII. 421 Near-empty SBG buses, none allowed to uplift potential passengers, glide into town. **1982** G. HAMMOND *Game* xii. 129 The letter was waiting at the airport for him. I checked up yesterday, and the letter was uplifted.

4. b. = RAISE *v.*^{1} 32. Cf. sense *4 of the sb.

1962 E. GODFREY *Retail Selling & Organization* xv. 158 The Inland Revenue Department uplifts the price to what the goods would have cost had they been purchased through a wholesaler. **1971** *Timber Trades Jrnl.* 14 Aug. 45/2 Devaluation of sterling..technically should have had the effect of uplifting import prices by 16·7%.

uplifter. Add: **3.** *N. Amer.* One engaged in social reform; a 'do-gooder'.

1923 [see *DO-GOOD]. **1935** S. LEWIS *It can't happen Here* xvii. 191 Social workers, both amateurs and long-trained professional uplifters. **1971** J. H. GRAY *Red Lights on Prairies* iv. 97 As the publisher of the paper explained when uplifters complained..it was no part of the responsibility of a newspaper to police the morals of its tenants.

upli·ftment. Chiefly *Black* and *Indian English.* [f. UPLIFT *v.* + -MENT: see UP- 2.] The action or process of improving or raising to a new standard; *spec.* amelioration of economic or social conditions; the result of this.

1926 *Brit. Weekly* 15 Apr. 46/3 Native women of the educated class might be potent influences in the upliftment and betterment of their people. **1973** *Caribbean Contact* Feb. 11/3 A company of West Indians formed for the economic upliftment of the people of the region. **1976** *Nigerian Chron.* 18 Aug. 12/2 Mr Onette congratulated the people of Osomba..for their initiative towards educational up-liftment in their community. **1979** P. NIHALANI et al. *Indian & Brit. Eng.* I. 187 The upliftment of the rural areas should be a top priority for the government. **1984** *Times* 25 Aug. 5/2 His immediate priorities would be the economic and educational upliftment of his people.

u·plight. [UP- 2.] = next.

1982 *Program* 1982/83 (Erco trade catal.) 56 Up-lights are the free-standing version of downlights. They illuminate the ceiling. **1983** *Homes & Gardens* Mar. 92/2 We decided to have uplights to cast light on to the fabric.

u·plighter. [UP- 2.] A light placed or designed to throw illumination upward.

1969 *Queen* 17–30 Sept. 96/2 Rotaflex uplighters in plain black cans, shining on, say, a picture. **1974** *Habitat Catal.* 116/1 Uplighter..to stand on the floor or a low table to light a strategically placed plant or shapely object. **1978** J. KRANTZ *Scruples* iii. 67 He installed three ten-foot-tall Kentia palms he got wholesale at Kind's, lit them from underneath with uplighters. **1983** *Homes & Gardens* Nov. 137 Best lighting for watching television comes from an uplighter placed behind the set; it cuts down glare and avoids reflections on the screen.

u·plink. [UP- 2.] A communication link for transmissions from the earth to a satellite, weather balloon, etc. Freq. *attrib.*

1968 W. M. GRIGGS *AMSS Prototype Radiosonde* (Rep. AD-680-409) 4 The radiosonde is designed to perform two functions... As a transponder, it must receive the uplink range tones. **1975** *Sci. Amer.* June 127/1 Since its launch in late 1972 it has offered a repeater service open to all, with an uplink at two meters and a downlink at 10. **1982** *New Scientist* 9 Sept. 682/1 These 'uplink' signals are at frequencies between 5·9 and 6·4 gigahertz.

uplook, *sb.* (Earlier example.)

1836 T. POWER *Impressions of Amer.* II. 235 The Virginian, tall of stature,..with an open up-look.

Upmann (ʌ·pmăn). The proprietary name of a make of (Havana) cigar.

[**1878** *Trade Marks Jrnl.* 16 Jan. 59 H. Upmann... Henry Upmann and William Rocholl, trading as H. Upmann and Co., Havana, Cuba; cigar manufacturers. Manufactured tobacco, cigars. **1912** *Official Gaz.* (U.S. Patent Office) 15 Oct. 770/1 H. Upmann & Co., Habana, Cuba... Cigars.] **1969** N. FREELING *Tsing-Boum* xxi. 148 He..had a cigar going, an Upmann that could have been made of tightly-rolled hundred-franc notes. **1979** A. SCHOLEFIELD *Point of Honour* 142 He was dispensing Upmanns from a humidor.

u·pmanship. *colloq.* = *one-upmanship* s.v. *ONE *numeral a., pron.*, etc. 29* c.

1962 *Spectator* 23 Nov. 837/2 (Advt.), Upmanship is the art of being one up on all the others. **1967** *Ibid.* 25 Aug. 212/2 He obviously thought I was trying upmanship on him before that word had been invented. **1976** *Casper* (Wyoming) *Star-Tribune* 29 June 19/1 (*caption*) Hospital upmanship: My Doc' is better'n yours! **1980** *Brit. Med. Jrnl.* 6 Sept. 666/1 In a moment of inspired upmanship this young moonlighting casualty officer had quietly dropped the word that he was a famous specialist in exotic diseases.

u·p-market, *a.* and *adv.* [UP *prep.*^{2}] **A.** *adj.* **1.** Of merchandise, etc.: characteristic of or designed for the more expensive end of the market; superior, expensive, 'quality'.

1972 *Times* 28 June 14/4 Glass has still, apparently, a more up-market image than plastic. **1974** *Daily Tel.* 17 Sept. 6 An 'up-market' £1,950 version of the Austin Allegro, hand-finished by the Vanden Plas coachbuilders. **1976** *Daily Mail* (Hull) 30 Sept. 18/3 Sarila Kitchens are the ultimate in luxury and design. Naturally, they are also up-market in price. **1980** *National Times* (Austral.) 24 Aug. 4/3 Franchised lines, apart from the usual T-shirts, badges, posters and dolls run up to a range of $200 couturier ensembles and an upmarket line of cosmetics.

2. *transf.* and *fig.*

1976 *Listener* 6 May 584/1, I cannot allow other people to have all the most up-market remarks. **1979** *Early Music* Jan. Suppl. 13/3 Trust House Forte's up-market evenings..are presenting several early music groups in 1979. **1981** *Times Lit. Suppl.* 30 Jan. 117/5 Linguistic shamanism is respected, for there have been some distinguished practitioners. Sir Ernest Gowers was an up-market shaman. So was H. W. Fowler.

B. *adv.* Towards the more expensive end of the market. Also *transf.* and *fig.*

1975 *Daily Tel.* 12 Apr. 12/4 Lyons-Tetley go up-market a bit with their Red Label and Orange Label [tea]. **1978** *South China Morning Post* 24 Nov. 14/7 Protectionism will increase in European Common Market countries so that Hongkong will have to diversify upmarket. **1980** *Times Lit. Suppl.* 24 Oct. 1206/2 Even the secessionist New English Art Club moved up-market to become a nursery for the Academy, with its paintings of..the pleasing and the picturesque. **1983** *Daily Tel.* 14 Mar. 11/8 Slightly upmarket of the rest of the channel's evening fare, *Omnibus* (BBC-1) has settled for an Everyman figure..as presenter. **1984** W. GOLDING *Paper Men* xi. 125 To contemplate the nature of predestinate insects or, moving up-market, Lobsters and crabs.

Also as *v. trans.*, to raise the standing of (a product) in the market, esp. by advertisement or actual improvement. Hence **up-ma·rketing** vbl. sb.

1972 *Times* 25 Aug. 7 Mathew Clark wants to upmarket Noilly Dry French, plugging heavily the drink's provenance compared with that of Italian Martini and Cinzano. **1975** *Listener* 4 Dec. 735/3 Leyland..decided the group's future lay in..improving its cars and charging more for them... It would take a lot of up-marketing to keep it [*sc.* Leyland] at even half its present size. **1977** *Daily Mirror* 21 Mar. 12/2 The British Sausage Bureau, in short, is trying to up-market its succulent product. **1980** *Times Lit. Suppl.* 19 Sept. 1030/4 In 1819 Harris, now joined by his son, sensed the way trade was going and boldly up-marketed his nursery books still further.

upon, *adv.* Add: **1. b.** (Later examples after biblical use: see quot. 1611.) Cf. CLOTHE *v.* 9.

1895 L. JOHNSON *Poems* 34 Old ramparts, gray and stern; But comely clothed upon With wealth of moss and fern. **1930** *Month* Mar. 230 Ancient stones, like Ezekiel's dry bones, need to be clothed upon.

u·p-patient. [UP *a.*; Cf. UP *adv.*^{2} 7 a.] An in-patient in hospital no longer confined to bed.

1952 'R. GORDON' *Doctor in House* xi. 119 Two up-patients, dressing-gowned old gentlemen. **1959** *Manch. Guardian* 19 Aug. 5/6 Topham is an up-patient... We started him with a couple of hours up each morning. **1976** J. GRENFELL *Joyce Grenfell requests Pleasure* (1977) xi. 163 Up-patients sat on the grass on grey army blankets.

upper, *sb.* Add: **1. a.** (Earlier example.)

1789 G. PARKER *Life's Painter of Variegated Characters* v. 36 If the top and leg of a jack-boot were joined to a dog-skin upper and a buff sole.

c. (Further examples.) Formerly in fig. phr. *to walk* (etc.) *on one's uppers.* Also *down on one's uppers.*

1886 *Lantern* 8 Sept. 4/3 The Royal Street actors who are walking on their uppers, must mourn..when they..hear of some of the boys spending 200 a week yachting. **1895** ROBERTS & MORTON *Adv. Arthur Roberts* xi. 143, I know two actors who were left, as the term goes, 'on their uppers', in a town in the heart of the Midlands. **1899** 'J. FLYNT' *Tramping with Tramps* I. v. 117 I's been a moocher, an' now I's shatin' on me uppers. **1903** *Judy* 9 Dec. 577/1 'What would you do if you were in my shoes?' 'Eh? Oh, then I should be fairly down on my uppers.' **1918** *Blackw. Mag.* May 602/2 We are pretty well down on our uppers as regards transport. **1985** D. WILLIAMS *Wedding Treasure* ii. 31 My guess is the swine's on his uppers... He's going for the ten thousand a year.

2. b. *ellipt.* for upper deck, storey, berth, etc.

1938 'GIRALDUS' *Merry Malloe Again* 179 Just sit down opposite the hatch and contemplate your new shipmates as, one by one, they troop on the 'upper'. **1955** F. O'CONNOR *Wise Blood* i. 17 The man in the station..had sold him a berth..an upper one... A sign said to get the porter to let you into the uppers. **1968** *Globe & Mail* (Toronto) 13 Feb. 30/6 (Advt.), Well located duplex with extra large upper. **1969** *Down Beat* 17 Apr. 16/1 Two chartered sleeping cars carried the men across the United States like royalty—and nobody ever had to take an upper.

4. a. *Public school slang.* A pupil of the upper school.

1929 J. BUCHAN *Courts of Morning* II. viii. 379 The Eton Beagles in the fields beyond Slough, and himself and Lariarty, both newly become uppers, struggling desperately to keep up with the field. **1937** [see *PI-JAW *sb.*].

b. *colloq.* An upper-class person.

1955 T. H. PEAR *Eng. Social Differences* iii. 90 While many people use 'person' indiscriminately, some 'uppers' employ it chiefly in a derogatory sense. **1967** [see *LOWER *a.* (*sb.*) 3 a]. **1968** *Economist* 27 Apr. p. v/1 The genuine uppers' genuine feeling of superiority.

u·pper, *sb.*^{2} *slang* (orig. *U.S.*). [f. UP *v.* + -ER^1; cf. *UP *a.* 5.] **1.** A drug (esp. an amphetamine) which has a stimulant or euphoric effect, often in the form of a pill.

1968 *Current Slang* (Univ. S. Dakota) Fall 50 *Upper*, type of drug that makes you feel active. Amphetamine is a commonly used stimulant of this kind. **1969** FABIAN & BYRNE *Groupie* (1970) xix. 133, I wasn't used to so many uppers all at once. **1976** 'R. ROSTAND' *D'Artagnan Signature* vii. 45 The trained-down leanness..likely came from epsom salts and uppers. **1981** 'D. SHANNON' *Murder most Strange* ii. 52, I want all your pills, man, all the uppers and downers you got.

2. *transf.* and *fig.*

1973 *Time* 1 Jan. K2 It certainly is a relief to know that State 2 is an upper; but by that time, who cares? **1976** *National Observer* (U.S.) 27 Nov. II. 22/3 Shorty's alchemy with rum and other potions produces something called a Goombay Smash that is a definite upper. **1977** *Time* 18 Apr. 45/2 Singing is a real upper. It makes me feel dizzy and energetic.

upper, *a.* Add: **I. 3. a.** *fig.* Also in phr. (*crushed*, etc.) *between the upper and the nether millstones*, between two irreconcilable opposing forces.

1902 G. B. SHAW *Mrs Warren's Profession* Pref. p. xxi, Thus am I crushed between the upper millstone of Mr Redford, who thinks me a libertine, and the nether popular critic, who thinks me a prude. **1921** L. STRACHEY *Queen Victoria* v. 167 His position, crushed between the upper and the nether millstones, grew positively unbearable. **1948** F. R. COWELL *Cicero & Roman Republic* vii. 131 [A] Consul..was a link between Senate and People, responsible to both, an unenviable situation between an upper and a nether millstone. **1950** A. BRYANT *Age of Elegance* ii. 40 Not only were the French Armies of Portugal and the North..marching to the fortress's relief, but Soult and Joseph with 60,000 men were threatening Madrid from Valencia. The British were in danger of being crushed between the upper and nether millstones.

7. *upper cut* (earlier example); also in *Cricket*, a cut that sends the ball up (*obs.* exc. *humorously*).

1842 *Spirit of Times* 17 Sept. 339/2 Giving him a sharp upper cut as he fell. **1865** *F. Lillywhite's Guide to Cricketers* 128 [He] has..made some good scores this year, his 'upper cut' being particularly effective. **1872** *Baily's Mag.* June 44 Smith's plucky innings was considerably helped by hits known as 'upper cuts'. **1955** A. Ross *Australia 55* 133 Two upper-cuts sent second and third slip on futile chases to the boundary.

II. 8. a. (Further examples, of domestic servants.) Cf. *UPSTAIRS *a.* c.

1776 J. WOODFORDE *Diary* 3 June (1924) I. 182 One of them is to be an upper servant and she lived very lately with Mr. Howes. **1783** *Ibid.* 24 Mar. (1924) II. 65 Betty, my Upper Maid stayed at home being Washing Week. *a* **1817** JANE AUSTEN *Persuasion* (1818) III. vi. 102 Her upper-housemaid and laundry-maid. **1872** [see *QUAVER *v.* 3 b]. **1958** R. WILLIAMS *Culture & Society* III. 329 That kind of confidence which will enable the upper servants to supervise and direct the lower servants.

b. *Comb.* (Predic. examples of *upper class.*) Also used to indicate finer gradations in the social scale, as *upper-lower, -middle, -upper, -working* (class); see also *UPPER-MIDDLE-

CLASS *a.* and *sb.* as main entry. Also *upper bourgeois.*

1907 M. BERENSON *Let.* 6 Mar. (1983) 139 So you aren't a 'County Family'..'not knowing yet' whether you're Upper Middle or only Middle. **1938** E. BOWEN *Death of Heart* III. vi. 420 Oh, shut up, darling... Don't be so upper class. **1940** HARRISSON & MADGE *War begins at Home* v. 103 The response was best from the upper and middle classes; whilst the upper-working or artisan class supplied its quota. **1955** T. H. PEAR *Eng. Social Differences* iii. 88 Radio-copies of lower-middle, upper-lower and middle-lower speech. **1955** *Caribbean Quarterly* IV. II. 109 The 'upper-upper' French Creoles who number under 200 consider all the rest to be of mixed ancestry. **1964** T. B. BOTTOMORE *Elites & Society* iv. 82 The postwar reforms of recruitment to the French higher civil service..have changed the ethos of education for the administrative elite—have made it more 'managerial' and less 'upper class'. **1964** C. BARBER *Ling. Change Present-Day Eng.* ii. 18 The general pattern of usage is as follows: *pudding* (upper and upper-middle), *sweet* (middle), [etc.]. **1967** M. ARGYLE *Psychol. Interpersonal Behaviour* iv. 83 A development in the British class system during the 1950s was the emergence of an upper working class. **1970** P. DICKINSON *Seals* vi. 123 The harryings of ninety sets of upper-bourgeois parents. **1974** H. WAUGH *Parrish for Defence* (1975) xxxix. 182 You're closer to where it's at if you're only upper middle class than if you're upper upper.

absol. (Later example.)

1967 N. FREELING *Strike out where not Applicable* 74 She was too much upper to say cheers.

c. *upper-bracket* adj., belonging to or characteristic of the higher section of a (social) scale; upper-class, wealthy, influential; cf. (*income*) *bracket* s.v. *BRACKET *sb.* 5 c; also *upper-income* adj.

[**1943** Upper bracket: see *BRACKET *sb.* 5 c.] **1945** S. LEWIS *Cass Timberlane* 157, I go there all the time, to shave the upper-bracket crooks when they got too big a hang-over to walk. **1957** O. NASH *You can't get there from Here* 130 And, should upper-bracket dreamers wake, Squab o' Neptune, and Plankton Steak. **1960** N. MITFORD *Don't tell Alfred* ix. 98 He's all set for the upper-income group—tax free of course. **1973** J. ROSSITER *Manipulators* xiv. 138 Luckhurst's overcoat and suit had been expensively tailored, the watch on his wrist an upper-bracket Rolex. **1978** *Detroit Free Press* 16 Apr. 8A/1 The migration of middle and upper-income whites out of the central cities has become a headlong flight.

9. b. With an ordinal number, designating: (*a*) the senior division of a class or form at school, as *Upper Sixth* (*form*), etc.; (*b*) the upper division of a second-class honours degree.

1856 C. M. YONGE *Daisy Chain* I. ix. 90 They are a low, ungentlemanly lot just now, about sixth and upper fifth form. **1858** J. A. SYMONDS *Let.* 6 June (1967) I. 145 Bosanquet is an Upper Sixth Boy and nearly a Monitor. **1905** R. BROOKE *Lett.* (1968) 29, I decided that Swinburne would be a great aesthetic blessing to the starved Upper Sixth. **1935** C. ISHERWOOD *Mr. Norris changes Trains* i. 7 An incident..To do with the upper fourth form classroom. **1981** E. NORTH *Dames* ii. 24 The form previously called Middle Five B would now be known as Upper Fifth. **1982** *Oxford Times* 25 June 11/3 He decided he wanted to see America, after graduating at Oxford with an Upper Second in Geography.

III. 11*. **upper air** = *upper atmosphere*; **upper atmosphere,** the upper part of the earth's atmosphere; now *spec.* that above the troposphere; **upper berth** *U.S.*, the higher of two bunks set one above the other, usu. in a boat, sleeping-car, etc.; **upper circle,** the tier of seats above the dress circle in a theatre; cf. CIRCLE *sb.* 11 b; **upperclassman** *U.S.*, a junior or senior student in school or college (as opp. FRESHMAN).

[**1877** *Q. Jrnl. Meteorol. Soc.* III. 447 It would not..do to imagine that every cloud in the rear of another cloud was cirrus, and yet it might be a distinct upper air current.] **1895** *Ibid.* XXI. 182 The scheme of exploring the upper air by means of small balloons was first proposed by Geheimrat A. Meydenbaur. **1957** *Encycl. Brit.* XXII. 883B/1 In meteorology the exploration of the upper air, using instruments carried by balloons, kites or aircraft, is an important part of observational technique. **1979** G. B. NAVARRA *Atmosphere, Weather & Climate* x. 357 Aircraft provide a highly mobile way of gathering data in the upper air. **1895** *Q. Jrnl. R. Meteorol. Soc.* XXI. 182 The Upper Atmosphere.—On July 7th, 1894, Dr. Assmann set free at Charlottenburg a balloon, equipped with self-registering apparatus for the determination of meteorological data in the higher regions of the atmosphere. **1933** *Discovery* Dec. 359/2 There is a monthly sequence of intense magnetic activity when wireless waves are absorbed and not reflected by the upper atmosphere. **1976** *Physics Bull.* Dec. 554/2 The author looks at the earth's upper atmosphere, taking the tropopause as its lower boundary. **1894** F. M. CRAWFORD *upper berth.* **1917** J. HUSBAND *Story of Pullman Car* 29 The upper berth might be closed in the day time and also serve as a receptacle for bedding. **1829** *Harlequin* 20 June 46 The upper circle and lower gallery were reserved for company to view the entertainment. **1889** *Theatre* May 292 The house consists of four tiers, pit and stalls, dress circle, upper circle, and gallery. **1926** A. CONAN DOYLE in *Liberty* 18 Dec. 9/1 Amberley..had taken two upper-circle seats at the Haymarket Theater. **1871** L. H. BAGG *4 Years at Yale* 70 Only a few upper-class men will be found there. **1897** [in Dict., sense 9]. **1933** F. SCOTT FITZGERALD *Let.* 26 June (1964) 502 Insofar as upperclassmen are concerned I saw a rather depressed runt at the Yale game. **1978** J. IRVING *World according to Garp* ii. 48 He befriended Garp in a very decent fashion while he was an upperclassman at Steering and Garp was just starting out.

III. 12. upper crust. c. (Earlier example of sense 'a hat'.)

1829 P. EGAN *Boxiana* 2nd Ser. II. 461 Ward..threw his upper-crust into the ring.

e. For 'Chiefly *U.S. colloq.* (also *attrib.*)' read: *colloq.* (orig. *U.S.*). Freq. *attrib.* or as *adj.* (Further examples.)

1863 G. DU MAURIER *Let.* May in *Young G. du Maurier* (1951) 204 All Millais' [paintings] are upper crust. **1878** W. S. GILBERT *H.M.S. Pinafore* II. 31 Two tender babes I nussed: One was of low condition, The other, upper crust, A regular patrician. **1957** *New Statesman* 19 Oct. 1/2 Views which are commonplace in upper-crust circles. **1966** D. FRANCIS *Flying Finish* vii. 82 One particular horse from an upper crust stud. **1973** M. AMIS *Rachel Papers* 116, I stalked up to the door and rapped, with an upper-crust rap, on the knocker.

f. Hence **upper-crusty** *a.*, aristocratic, (socially) superior.

1977 *Time* 26 Dec. 1/1 The politicians' wives, looking upper-crusty. **1980** *TWA Ambassador* Oct. 77/3 We lived in Garden City, the upper-crusty part.

13*. **upper dog.** [After UNDERDOG.] The victorious party in a contest; the one who has the upper hand or a position of superiority. Cf. *OVERDOG. *rare.*

1903 G. BOWLES in *Parl. Debates* 18 Feb. 224 If it came to a question of force, we should always be the 'upper dog' in Persia. **1940** 'G. ORWELL' *Inside Whale* 82 One has got to change sides when the underdog becomes an upperdog. **1971** *Daily Progress* (Charlottesville, Va.) 30 Mar. 4/2, I am for the underdog—the achiever, the succeeder.

18. upper ten. a. (*b*) (Earlier example.)

1846 A. J. H. DUGANNE *Daguerreotype Miniature* 20 Major Peyton Florence was held in great reverence by the 'upper ten'.

b. upper-tendom (earlier example).

1848 F. A. BUCK *Yankee Trader in Gold Rush* (1930) 13, I suppose you and those other families left out constitute an *Upper Ten-dom*, in the place.

upper-middlebrow, *sb.* (*a.*) [UPPER *a.* 8 b.] One whose cultural interests lie between those of the middlebrow and those of the highbrow. Also *attrib.* or as *adj.*

1956 R. CONQUEST *New Lines* p. xiv, A sort of upper-middle-brow equivalent of the horror-comic. **1959** *Economist* 28 Feb. 787/1 Magazines for upper middle-brows. **1974** *Country Life* 14 Mar. 588/2 The middle to-upper-middle brow range that Chichester programmes now generally encompass. **1977** *Listener* 11 Aug. 170/1 In *Life* magazine, nearly 30 years ago, an American critic said: What culture and civilized living we have today is provided by the interaction of..the aesthetic radical highbrow and the somewhat more conservative upper-middlebrows.

upper-middle-class, *a.* and *sb.* [UPPER *a.* 8 b.] **A.** *adj.* Of, pertaining to, or characterized by the class of polite society next below the upper class.

1872 *Times* 26 Dec. 5/4 Upper and upper middle-class society is represented in all its grades. **1906** GALSWORTHY *Man of Property* I. i. 3 Those privileged to be present at a family festival of the Forsytes have seen..an upper middle-class family in full plumage. **1949** W. L. WARNER in M. Fortes *Social Structure* 4 Excerpts from the Total Configuration of an upper-middle-class doctor illustrate how this technique is used. **1967** M. ARGYLE *Psychol. Interpersonal Behaviour* iv. 80 English upper-middle-class speech includes considerable understatement; a person who fails to follow this convention is regarded as boastful. **1982** 'W. HAGGARD' *Mischief-Makers* ii. 20 Willy's skin was the colour of coal... He thought of himself as an upper-middle-class Englishman.

B. *sb.* **1.** Usu. unhyphened. This class itself, or its members.

c **1885** M. DARWIN *Let.* in G. Raverat *Period Piece* (1952) vi. 98 The upper middle class think they are acting rightly by over-protecting their daughters. **1891** [see MIDDLE CLASS *sb.*]. **1903** G. GISSING *Private Papers H. Ryecroft* 210 A lad whose education ranks him with the upper middle class. **1923** A. BENNETT *Riceyman Steps* I. ii. 5 He had the air neither of a bookman nor of a member of the upper-middle class. **1955** H. ROTH *Sleeper* xii. 99 Stamford, Connecticut, is a typical transplanting ground of the upper middle class. **1965** R. WILLIAMS in *Tribune* 5 Mar. 1/5 The English upper-middle class.

2. *ellipt.* as *upper-middle*, a member of the upper middle class. Usu. in *pl.*

1955 [see *middle-middle* s.v. *MIDDLE *a.* 6]. **1967** *Economist* 18 Feb. 607/2 In Birmingham, Manchester, Liverpool, the upper-middles do not choose to live in the city. **1978** R. WESTALL *Devil on Road* xv. 111 When my parents quarrel, they..fight in hoarse whispers. But like a lot of upper-middles..Derek and Susan let it all hang out.

Upper Vo·ltan, *a.* and *sb.* [f. *Upper Volta* (see def.) + -AN: cf. UPPER *a.* 1.] **A.** *adj.* Of or pertaining to Upper Volta, an inland republic in West Africa (a former French colony whose independence was recognized in 1960, now called Burkina). **B.** *sb.* A native or inhabitant of Upper Volta. Cf. *VOLTAIC *sb.* and *a.*^{2}

1972 *N.Y. Times* 28 Nov. 2/4 Horsemen from the Upper Voltan fringes of the Sahara called out, 'Pa! Pa! Pompidou!' **1975** *Congress. Q.: Editorial Research Rep.* (Daily Service) 7 May, For many Upper Voltans the real villain of the piece is not Mali but France. **1976** *Facts on File* 17 Apr. 269/3 Upper Voltan trade unions had welcomed the formation of the new government. **1983** *Times* 25 Jan. 6/1 There are more than 500,000 immigrants from Niger, as well as..Togolese, Beninese, Upper Voltans and Chadians.

uppie (ʌ·pi). *slang.* [-Y^6, -IE.] = *UPPER *sb.*^{2}

1966 *Sunday Times* (Colour Suppl.) 13 Feb. 35/4 *Uppies and downies*, pep pills and tranquillisers. **1975** J. F. BURKE *Death Trick* (1976) ii. 22 There's nothing in the box but a few uppies. I haven't got a regular prescription.

u·ppishly, *adv.* [f. UPPISH *a.* + -LY^2.] **a.** *Cricket.* Of the ball, etc.: in a slightly upward direction, esp. so as to give some chance of a catch.

1904 P. F. WARNER *How we recovered Ashes* 109 Just after he had passed his hundred, he sent one uppishly through the slips. **1955** A. ROSS *Australia 55* 137 Hutton..then turned Archer a shade uppishly through the short-legs. **1963** *Times* 11 Jan. 10/6 Guest..drew one false stroke out of Dexter which went uppishly past gully.

b. In an uppish (sense 2 d) manner.

1911 G. B. SHAW *Blanco Posnet* Pref. 307 The majority of the Committee began by taking its work uppishly and carelessly. **1960** M. SPARK *Ballad Peckham Rye* vii. 141 'Miss Merle Coverdale, one of my unofficial helpers,' Dougal said uppishly. **1981** *Times Lit. Suppl.* 26 June 727/2 Rowland Lacy, wooer of the uppishly poised Rose.

uppity (ʌ·piti), *a. colloq.* (orig. and chiefly *U.S.*). [f. UP *adv.*^{1} + -*it*- + -Y^1: cf. *BIGGITY *a.*] Above oneself, self-important, 'jumped-up'; arrogant, haughty, pert, putting on airs. Cf. UPPISH *a.* 2 d. **a.** *attrib.*

1880 J. C. HARRIS *Uncle Remus* 86 Hit wuz wunner deze yer uppity little Jack Sparrers, I speck. **1933** *Times Lit. Suppl.* 9 Nov. 776/2 Grammy is living contentedly enough with an 'uppity' young creature named Penny. **1952** F. L. ALLEN *Big Change* II. viii. 130 The effect of the automobile revolution was especially noticeable in the South, where one began to hear whites complaining about 'uppity niggers' on the highways, where there was no Jim Crow. **1982** B. CHATWIN *On Black Hill* v. 28 He had a head for figures and a method for dealing with 'uppity' tenants.

b. *predic.*

1932 *Sun* (Baltimore) 23 Aug. 6/2 [She] could have plenty o' friends. The trouble with her is she thinks folks too common to bother with unless they're too uppity to bother with her. **1947** 'N. SHUTE' *Chequer Board* 68 They've been here alone too long, and they've got uppity. **1955** F. O'CONNOR *Wise Blood* v. 89, I reckon you ain't as uppity as you was last night. **1966** D. BAGLEY *Wyatt's Hurricane* i. 27 The Navy is trying to build up Cap Sarrat as a substitute for Guantanamo in case Castro gets uppity and takes it from them. **1973** P. WHITE *Eye of Storm* viii. 381, I came prepared to rough it... It's Dorothy who grows uppity if all the cons aren't mod.

Hence **u·ppitiness,** the quality of being 'uppity'; an instance of this.

1935 H. L. DAVIS *Honey in Horn* x. 145 Clay's bravery and uppitiness had done nothing. **1966** *Listener* 27 Oct. 622/1 She had decided that Joyce was 'pretentious' and 'under-bred'... But who was Virginia Woolf to talk (in this purely literary sense) of 'uppitiness'? **1975** *Ibid.* 9 Oct. 479/1 Few delegates seemed versed in *Private Eye* nomenclature and would, anyway, disapprove of such uppityness. **1982** R. BARNARD *Death & Princess* ii. 17 Joe may appreciate my couthness..but he can sniff out uppitiness.

up-pouring: see *UP- 7 (*b*).

u·p-push. *rare.* [UP- 2.] A pushing upwards.

1910 J. W. GREENE *Clin. Course Dental Prosthesis* 46 There is now a little general up-push all along the yielding line. **1940** *Sun* (Baltimore) 5 Apr. 23/1 (*heading*) Rails continue up-push in bonds. **1960** R. W. MARKS *Dymaxion World B. Fuller* 28/1 If such a man took the proper headlong attitude, with respect to air resistance, the amount of up-push required to keep him plummeting forward should be no more than that supplied by his leg muscles to give him his initial altitude.

upraised, *ppl. a.* Add: **2.** Of the eyes: (earlier examples).

1707 E. SETTLE *Siege of Troy* I. ii. 9 Hail, beauteous goddess, all divine, Our up-rais'd eyes and hearts all thine. **1817** COLERIDGE *Biog. Lit.* II. xxiv. 309 The upraised Eye views only the starry Heaven.

upraising, *vbl. sb.* (Later examples.)

1926 *Daily Colonist* (Victoria, B.C.) 19 Jan. 17/4 The upraising going on in the underground working is progressing very satisfactorily. **1936** BELLOC *Battle Ground* xi. 225 Some said he was Elias, come back to the upraising of men.

upra·te, *v.* [UP- 4.] *trans.* To raise to a higher standard, to upgrade; *spec.* to improve the performance of (a mechanical device or process); to increase the value of (a commodity, grant, etc.).

1965 *National Observer* (U.S.) 8 Feb. 15/1 The first stage of the Saturn 1-B is similar..except that its engines have been uprated to produce more power. **1968** *Listener* 1 Aug. 159/1 At a time when petrol prices were rather unstable, Esso uprated their Extra..to super or five-star grade. *a* **1974** R. CROSSMAN *Diaries* (1976) II. 372 Every two years we up-rate pensions in order to see that they

don't lose their purchasing power. **1980** 'D. RUTHERFORD' *Turbo* ii. 29 The [car] springs had been uprated to carry the extra weight. **1984** *Daily Tel.* 25 Jan. 17/4 Government plans to uprate early leavers' pensions by up to 5 p.c.

Hence **upra·ted** *ppl. a.*, **upra·ting** *vbl. sb.* and *ppl. a.*

1967 *Jane's Surface Skimmer Systems* 1967–68 116/2 More recent models are fitted with the uprated..version of this marine diesel. **1968** *Listener* 1 Aug. 159/1 The recent mass switch to five-star petrol may..be partly due to the uprating of Extra. *a* **1974** R. CROSSMAN *Diaries* (1976) II. 372, I can tell you we should now expect a pension up-rating Bill to be published within the next fortnight. **1981** *Sunday Express* (Colour Suppl.) 14 June 21 An uprated sports suspension makes sure that all this extra power is kept under control.

‖ **uprava** (uprā·vă). [Russ., = authority.] In Imperial Russia: the executive board of a municipal council.

1870, 1886 [see *DUMA]. **1918** G. A. BIRKETT in R. Beazley et al. *Russia* III. iii. 432 Thirty-four provinces of European Russia..received self-government under the law of 1864. Each district had its own zémstvo council, which elected an executive board (*upráva*) sitting constantly and reporting to the annual meeting of the council. **1954** G. VERNADSKY *Hist. Russia* (ed. 4) x. 221 The representatives elected a board known as the *uprava* for a term of three years. **1967** H. SETON-WATSON *Russ. Empire 1801–1917* xi. 385 Every city was to have a municipal council..and they elected from their number an executive board (*uprava*).

upre·nd, *v. rare.* [UP- 4.] *trans.* To pull or tear up; to uproot.

1911 CHESTERTON *Ballad White Horse* I. 9 Earthquake swallowing earthquake Uprent the Wessex tree. **1932** *Sphere* 10 Dec. 435/1 (*caption*) A 130+ m.p.h. gale..uprent the expedition's wireless mast.

upright, *sb.* Add: **3. b.** Also *spec.* in *Football*, a goalpost (as opp. to the crossbar).

1910 *Glasgow Herald* 14 Feb. 12/6 Barr..had little difficulty in placing the ball between the uprights. **1927** W. E. COLLINSON *Contemp. Eng.* 24 There were all the 'Kennings' in the sporting columns..Sphere for football, the uprights or sticks for the goalposts. **1951** *Sport* 27 Apr.–3 May 4/1 Bill rapped the upright with a penalty-kick. **1954** J. B. G. THOMAS *On Tour* 72 Morkel hit an upright with his conversion attempt. **1977** *Irish Press* 29 Sept. 18/2 Mick Lawlor's 24th minute left-footed drive..hit the bottom part of the upright and rebounded into play.

f. *Basketry.* A plane used for shaving skeins to a required width.

1842 *Encycl. Brit.* IV. 429/1 In order to bring the split into a shape still more regular, it is passed through another implement called an upright, consisting of a flat piece of steel, each end of which is fashioned into a cutting edge. **1907** *Jrnl. Soc. Arts* 11 Jan. 190/1 For finer work the rod is split into three or more skeins by a cleaver; the splits are then successively drawn through a shave to remove the central pith and through an upright to render them uniform in width. **1929** A. G. KNOCK *Fine Willow Basketry* 37 For most skein work the skeins can be used after being shaved, but for..extra fine skein work of other kinds they are made uniform in width and as narrow as required by being drawn through the upright. **1961** L. G. ALLBON *Basic Basketry* iii. 18 Two specialized planes complete the process. The shave skims away the pith and renders the skein of even thickness; the upright straightens up the sides of the skein to an even width. **1981** T. W. BAGSHAWE *Basket Making in Bedfordshire* 16 (*caption*) Uprights for shaving to uniform width.

g. In a crossword puzzle, one of the clues whose solution is to be entered vertically in the frame.

1917 M. T. HAINSSELIN *Grand Fleet Days* xv. 96 How far have you got? Only as far as the 'uprights'—Belgium and Berlin. **1967** *Sci. Amer.* Sept. 268/2 The first stanza gives clues for two words, called the uprights, that are spelled vertically by the initial and final letters of the words to which clues are given by the numbered stanzas.

up-river, *a.* Add: orig. *U.S.* **1. a.** (Earlier examples.)

1774 I. LITCHFIELD *Jrnl.* 19 Apr. in W. J. Litchfield *Litchfield Family in Amer.* (1906) I. 334, I..ordered them to meet at upriver meeting house by the Sun an hour high Compleat in arms with 4 Days provision. **1857** *Trans. Mich. Agric. Soc.* VIII. 732 Charming villages are also rapidly springing into existence along the up river bank of the Grand River in this country.

b. (Earlier examples.)

1836 *Southern Lit. Messenger* II. 698/1, I had never imagined that any thing half so grand..awaited us on our up-river jaunt. **1857** W. CHANDLESS *Visit to Salt Lake* I. i. 1, I passed a few days there, waiting for an up-river boat.

up-ri·ver, *adv.* [UP *prep.*^{2} 2.] Towards or in the direction of the source of a river.

1848 THOREAU in *Union Mag.* Nov. 220/2 Only a few axe-men have gone 'up river' into the howling wilderness which feeds it. **1887** *Harper's Mag.* Apr. 667/1 Logs were usually cut and hauled in summer-time to the banks of streams, often a long distance 'up-river'. **1929** BELLOC *Joan of Arc* ii. 38 He had been all day up-river in the marshes shooting quail. **1981** M. NABB *Death of Englishman* I. i. 25 Upriver the ghost of the Ponte Vecchio..was straddling nothing.

uproarious, *a.* **2.** (Earlier example.)

1818 LADY GRANVILLE *Let.* 12 Aug. (1894) I. 135 We arrived here to dinner and found Hart in uproarious spirits.

uproll, *sb. rare.* [UP- 2.] A rolling movement upwards.

a **1885** G. M. HOPKINS *Poems* (1967) 192 Moist..With the uproll and the downcarol of day and night delivering Water. **1984** A. PRICE *Sion Crossing* v. 83 David must have been..not quite senior enough to have sighted the gun and pulled the lanyard on the uproll—?

uproo·tedness. [f. UPROOTED *ppl. a.* + -NESS.] The state of being uprooted; a condition of severance from one's natural origins. Chiefly *fig.*

1927 J. LAVRIN *Russ. Lit.* v. 29 The negative features of a romantic type: uprootedness, fear and hatred of reality. **1954** KOESTLER *Invisible Writing* xxii. 247 What a refugee craves most is relief from his permanent feeling of uprootedness. **1974** O. COOK *Eng. Country House* i. 7 The contrast..between our uprootedness and the continuity and stability of the life led in the great mansion.

uprush, *sb.* (Earlier example.)

1871 BROWNING *Balaustion's Adventure* 31 In the fire-flash of the appalling sword, the uprush and the outburst.

upsaddle, *v.* (Earlier example.)

1838 in W. B. Boyce *Notes S. Afr. Affairs* vi. 148 Dingaan..inveigled them within an enclosure..just as they were up-saddling to depart, and massacred..every living soul of them.

‖ **upsara** (ʌ·psɒrā). *Indian Mythol.* Also **apsaras.** [Hindi *apsarā*, f. Skr. *apsarās.*] A celestial nymph, one of the wives of the Gandharvas (heavenly minstrels).

[**1846** J. T. THOMPSON *Dict. Hindee & Eng.* 17/2 *Upsura*..a female dancer or courtezan in *Swurg*, the hindoo paradise.] **1865** [see *Vedaic* adj. s.v. VEDA]. **1879** J. DOWSON *Class. Dict. Hindu Mythol. & Relig.* 20 The Apsarases..are fairylike beings... They are the rewards in Indra's paradise held out to heroes who fall in battle. **1892** KIPLING *Barrack-Room Ballads* 105 Above the dark *Upsaras* flew, beneath us plashed the blood. **1932** F. M. ATKINSON tr. H. de Wilman-Grabowska in J. Hackin et al. *Asiatic Mythol.* 113 The Apsarases are the recognized courtesans of the sky. **1968** B. WALKER *Hindu World* II. 143 In Hindu mythology the celestial nymph or *apsarā* (ap-sara, wet-flow) is a personification of the mists or clouds in the form of a beauteous damsel.

u·pscale, *a. U.S.* [UP *prep.*^{2}] At the higher end of a (social) scale; superior, of a high quality; 'up-market'.

[**1945** *Newsweek* 2 Apr. 68/2 The Mullikin selling method, which he called 'upscaling', aimed to build up the idea that prefabricated houses were rich-looking, comfortable and permanent.] **1966** *One Hundred Basic Media Terms Defined* (Batten, Barton, Durstine & Osgood Inc. Media Dept.) 15 Downscale is a term used to describe a market or audience with above-average representation at the lower end of the socio-economic scale. Its opposite is 'upscale'. **1975** *New Yorker* 1 Dec. 44/1 A lot of advertising people don't think jazz fans have that all-important rage to consume. They don't think the jazz audience is 'up-scale' enough. **1977** P. STREVENS *New Orientations Teaching of English* xiii. 163 The *up-scale* lab has all typical facilities, plus full remote control of each student machine. **1983** *Chicago Sun-Times* 24 Nov. 110 '20-20' generally finishes behind 'Hill Street', whose upscale audience it shares. **1984** *Christian Science Monitor* 2 Mar. 10/4 He cites Chi-Chi's Mexican restaurants as appealing to customers who want a bit more upscale atmosphere.

upset, *sb.* Add: **I. 3. c.** *Basketry.* Usu. **upsett.** The first section of waling, which sets the stakes firmly in place.

1907 T. OKEY in *Jrnl. Soc. Arts* 11 Jan. 191 (*caption*) It will be seen that the bye-stakes are merely inserted in the upsett. **1946** N. WYMER *Eng. Country Crafts* vii. 73 He then lets in the stiff 'uprights' and fastens them securely by working in several willows together to form the 'upset'. **1977** B. MAYNARD *Basketry* 19/2 You may like to start the upsett with one round of 4-rod wale.

upset, *v.* Add: **I. 1. e.** Usu. **upsett.** (Pa. t. and pa. ppl. also **upsetted.**) (*a*) To bend upwards (a stake) plaited into the base of a basket to form part of the frame for the side; (*b*) to form the 'upset' of (a basket); also *absol.*

1875 *Encycl. Brit.* III. 423/1 Osiers..are forced or 'scallumed', between the rods of the bottom from the edge towards the centre, and are turned up, 'upset', in the direction of the sides. **1907** T. OKEY in *Jrnl. Soc. Arts* 11 Jan. 190/2 If a foot is needed it is now put on by inserting the tops cut off from the stakes alongside the upsetted stakes. **1912** —— *Art of Basket-Making* vi. 33 Very coarse work..is upsetted with a pair instead of a wale. **1945** H. J. MASSINGHAM *Wisdom of Fields* v. 94 My basketer..after 'scalluming' the foreign rods..proceeded to wale up the sides, 'upsetting' as it is called, that is to say, building up the frame. **1977** B. MAYNARD *Basketry* 19/2 Stake up with the 40 side stakes, nip them at the edge of the base and tie them together tightly at the top. ..Upsett with six rounds of waling with No. 6 cane keeping the sides quite vertical.

upset, *pa. pple.* and *ppl. a.* Add: **4.** Physically disordered: said esp. of the stomach.

1973 C. LEACH *Send-Off* iii. 26 All you got is a burned back and an upset tummy. **1980** J. GARDNER *Garden of Weapons* III. iii. 249 Hoffer had sent him back because of his upset stomach.

upse·ttable, *a.* [f. UPSET *v.* + -ABLE.] Capable of being upset.

1930 R. C. HUTCHINSON *Thou Last Devil* xii. 167 He was, of course, in an upsettable mood. **1958** J. CANNAN *And be a Villain* vi. 145 'Hugh is..the most upsettable.' 'Couldn't agree more,' said Laura, 'thirteen's too young for funerals.' **1973** J. ELSOM *Erotic Theatre* iii. 53 In the dialogue, he tiptoes around the subject with the clumsy concern of Bertie Wooster, surrounded by upsettable coffee tables.

upsetting, *vbl. sb.* Add: **1. g.** *Basketry.* = *UPSET *sb.* 3 c; the formation of this.

1924 C. CRAMPTON *Canework* 13 *Upsetting*, the three or four rows [of weaving] worked at the commencement of the side..to 'set' the stakes in order. **1937** A. H. CRAMPTON *Raffia Wk. & Basketry* II. ii. 44 The first row of upsetting is worked with four canes, each passing in front of three stakes and behind one. **1953** A. G. KNOCK *Willow Basket-Wk.* 20 The number of rounds of waling forming the upsetting should, in fine basket-work, never be less than two, and seldom less than three. **1964** H. HODGES *Artifacts* x. 146 Waling was frequently used at the base of baskets in order to hold the upright stakes securely..in which case it is..referred to as upsetting.

upse·ttingly, *adv.* [f. UPSETTING *ppl. a.* + -LY^2.] In an upsetting manner; distressingly.

1926 'L. MALET' *Dogs of Want* vi. 155 Barbara Heritage's graceful silhouette had, somehow, become interchangeable with that of the Margate peroxide blonde of ten years back. Most upsettingly, but, also, most unjustifiably so. **1981** *Washington Post* 26 Apr. H9/1 To traditionalists it is upsettingly disheveled chaos.

upse·ttingness. [f. as prec. + -NESS.] The quality of being upsetting or disturbing.

1922 *Glasgow Herald* 23 Feb. 6/8, I have heard an old man speak with asperity of the 'upsettingness' of one who had adopted 'Mac', which his father had not used. **1949** G. RYLE *Concept of Mind* iv. 93 The degrees of upsettingness..are ordinarily characterised as degrees of violence. **1977** 'M. INNES' *Honeybath's Haven* xviii. 167 He had taken the reason to have been simply the general upsettingness of Melissa's turning up on them in that restaurant in Rome.

u·pshift, *sb.* Chiefly *U.S.* [UP- 2.] A movement upwards (esp. in various devices); *spec.* a change to a higher gear in a motor vehicle.

1839 [see UP- 2]. **1951** W. K. TOBOLDT et al. *Automatic Transmissions* iv. 129 The point of upshift from second to third and from third to fourth is excessively high. **1967** D. G. HAYS *Introd. Computational Linguistics* iv. 65 In standard paper-tape machines, a single capital letter is represented by three frames of tape, one for *upshift*, one for the character, and one for *downshift*. **1978** L. PRYOR *Viper* (1979) ix. 175, I missed one upshift and smashed gears.

u·pshift, *v.* Chiefly *U.S.* [UP- 4.] *intr.* To change into higher gear in a motor vehicle. Also used of the gear itself. Cf. *SHIFT *v.* 12 e. Hence **u·pshifting** *vbl. sb.*

1956 W. H. CROUSE *Automotive Mechanics* (ed. 3) xxiv. 513/2 If the brake band is released and at the same time the clutch is applied, the planetary set upshifts or goes into direct drive. *Ibid.* 515/2 Upshifting is controlled by both throttle opening and car speed. **1979** *Truck & Bus Transportation* 31 July 25/2 The flexibility..covers up an out-of-sequence ratio gap between 3rd and 4th gears. It is not at all apparent when upshifting..but it is noticeable when changing down.

upshoot, *sb.* **2.** (Earlier example.)

1850 W. HOWITT *Year-Bk. of Country* iv. 172 His oaks and elms in his park..what are they but the towering upshoots of his prejudices?

upsidaisy, *int.* Add: Also **ups-a-daisy, upsy daisy,** etc. (Later examples.)

1912 J. SANDILANDS *Western Canad. Dict. & Phrase-Bk.*, *Ups-a-daisy*, the tender words of the fond father when engaged in baby-jumping. **1934** D. L. SAYERS *Nine Tailors* I. ii. 61 Hoops-a-daisy, over she goes! **1940** *Horizon* Mar. 204 Come on, sonny, that's the way! Upsy-daisy! **1948** 'P. QUENTIN' *Run to Death* xx. 153 'Upsy daisy.' I picked her up, swung round and dumped her in the bath-tub. **1969** 'I. DRUMMOND' *Man with Tiny Head* vii. 92 'Upsidaisy,' said Jenny. 'No time to waste.' **1984** *New Yorker* 13 Feb. 125/1 The Great Rudner is..given the most labyrinthine acrobatic choreography—a tortuous series of slithers, blind leaps, upsy-daisy lifts, and ass-over-heels floorwork.

Hence *ellipt.* as **u·psa** (and varr.) *int.*

1922 JOYCE *Ulysses* 491 Hoopsa! Don't fall upstairs. **1928** E. M. FORSTER *Life to Come* (1972) 141 'Upsa! Take care!' 'Upsa!' were some drops of brandy, which Conway had spilt.

upside. Add: **4.** *Comm.* An upward movement of share prices, etc.; also = *upside potential.* Also as *adj.*, esp. *upside potential*, the possibility of (gain from) such a rise in value.

1961 in WEBSTER. **1969** *Punch* 5 Feb. 206/1 The share price..needs to improve by only 50 per cent... Likewise, your original upside potential would be considerably higher. **1977** R. E. MEGILL *Introd. Risk Analysis* xvi. 175 Most parameters have up-side and down-side possibilities in addition to a most probable value. **1983** *Sunday Tel.* 27 Feb. 20/7 The share price..is vulnerable to even the most modest of bear rumour and..the downside substantially exceeds the upside. **1983** *Observer* 29 May 19/1

Oct. 8/3 Even without 'Cotton Ed's' upstage behavior, this plea would have been unimpressive. **1947** N. MARSH *Final Curtain* xii. 188 All upstage and county! **1966** 'J. HACKSTON' *Father clears Out* 84 He was a little patronizing to me, upstage in his bearing.

2. That is situated or occurs at or towards the rear of the stage. Also as *sb.*, the back of the stage, furthest from the audience.

1933 P. GODFREY *Back-Stage* i. 14 The up-stage O.P. flood. **1959** [see *DITHER *v.* 1]. **1974** F. WARNER *Meeting Ends* I. viii. 30 Hanging on upstage side of pump..are a brightly coloured towel and a black shawl.

upsta·ge, *v.* [f. the adv.] **1.** *trans. Theatr.* To move upstage of (another actor), forcing him to face away from the audience; to divert attention from (a fellow performer) to oneself, to 'steal the scene' from.

1933 [implied in *UPSTAGER, *UPSTAGING *vbl. sb.*]. **1958** B. NICHOLS *Sweet & Twenties* 200 Miss Tempest always 'upstaged' her—..she slowly pushed her chair to the rear so that..Miss X was obliged to turn away from the audience. **1958** *New Statesman* 6 Sept. 302/2 So if Strether's 'tragedy' seems..rather trivial, that is partly because the hero has been upstaged by the backdrop. **1972** *Islander* (Victoria, B.C.) 18 June 6/3 While Gracie was singing her famous aspidistra song, a stray dog wandered onto the stage and upstaged her. **1976** *Early Music* Oct. 400/1, I would hesitate before *telling* an Aeneas that he is being literally upstaged during his climactic high F.

2. *fig.* To put (a person, etc.) at a disadvantage; to outshine. Also, to treat in a haughty or snobbish manner. *colloq.*

1921 H. C. WITWER *Leather Pushers* x. 268 Nada Nice has upstaged the Kid..at your orders. **1946** 'BRAHMS' & 'SIMON' *Trottie True* v. 103 The Duchess stood there exerting her personality to its most ducal extent... 'Blast,' said Trottie under her breath. 'Upstaged.' **1967** *Daily Tel.* 10 Apr. 10/6 My sister..didn't get on with the other debs—she was more beatnik. They were rude and kept up-staging her. **1974** 'R. TATE' *Birds of Bloodied Feather* iii. 57 It was the only occasion in Edward's life that he had been upstaged by his younger brother.

So **upsta·ger**; **upsta·ging** *vbl. sb.* and *ppl. a.*

1933 P. GODFREY *Back-Stage* iii. 40 With the chronic up-stager the only remedy for the other actors is to withhold their speech until they have deliberately taken up a position favourable to themselves. *Ibid.*, 'Crowding' and 'up-staging' are tricks of the selfish actor...'Up-staging' is to take up a position nearer to the back of the scene than the other players. This forces them to turn three-quarter-back to the audience when speaking to the up-stage actor. **1968** M. WOODHOUSE *Rock Baby* v. 46 I'd been running a bright, upstaging little war with Driver and McKellar and patting myself on the back. **1982** P. LOVESEY *False Inspector Dew* II. iii. 29 With her dramatic training, she knew all about up-staging.

upstairs, *adv.*, *sb.*, and *a.* Add: **A.** *adv.* **1. c.** (*a*) *Parliament.* In phr. *to send* (*a bill*) *upstairs*, to refer (a bill) for its committee stage from the floor of the House to a standing committee. Cf. COMMIT *v.* 4.

1835 *Mirror of Parl.* 17 June 1399/2, I rise to suggest, that if this Bill requires mature deliberation with respect to its details, it should be referred to a Committee up-

e. Mentally, 'in the head'. Chiefly in phrases indicating weak (or abundant, etc.) mental capacity. *slang.*

1932 J. FARRELL *Young Lonigan* i. 44 Aw, she's all vacant upstairs. **1945** A. KOBER *Parm Me* 179 '*Meshugeh* upstairs!' Jennie laughed and drummed on her forehead. **1952** G. W. BRACE *Spire* xxvii. 322 He just ain't right upstairs. **1962** A. HUXLEY *Island* xi. 184 'One may be very stupid upstairs.' She patted the top of her head. **1974** P. DE VRIES *Glory of Hummingbird* (1975) v. 63 Assurances that my progress here would be well lubricated by my having 'plenty upstairs'.

f. *the Man Upstairs*, euphem. and slang for 'God'. Cf. *UP THERE 1. *U.S.*

1961 J. HELLER *Catch-22* xviii. 184 'When you talk to the man upstairs,' he said, 'I want you to tell Him something for me.' **1971** *Guardian* 8 Oct. 9/4 Everything that's happened to me has been very much because of the Man Upstairs. **1976** H. KEMELMAN *Wednesday Rabbi got Wet* xix. 120 I'd just lost my wife, see. I guess the Man Upstairs wanted her.

B. *adj.* **1. c.** *fig.* Of, pertaining to, or characteristic of life 'above stairs' (i.e. in the private rooms of a household, as opp. to the servants' quarters) (chiefly *Hist.*); refined, genteel, privileged. Freq. in phr. *upstairs-downstairs* adj., denoting the social contrast between employer and domestic servant.

1942 'M. INNES' *Daffodil Affair* I. ii. 13 Almost might she be said, in upstairs language, to be receiving. [**1971** *TV Times* 9 Oct. 6/2 A six-part series, *Upstairs, Downstairs* (or Secrets of an Edwardian Household), starting Sunday, lovingly, critically examines life at that time through one family.] **1974** D. FRANCIS *Knock Down* i. 6 To one side of the sale ring, and built to a specification as Upstairs as the wooden circle was Downstairs, was a magnificent turn of the century stable yard. **1975** *New Society* 6 Nov. 305/2 Our system is too monolithic, based as it is on an upstairs-downstairs view of music. Upstairs people choose the classics on Radio Three... Downstairs, people listen to pop. **1978** G. MITCHELL *Mingled with Venom* i. 4 She had been taken on as kitchen-maid... She was..a ready learner of upstairs speech and manners. **1980** E. BEHR *Getting Even* i. 18 The lady in black with white lace cap and apron, a real upstairs-downstairs figure.

upstand, *sb.* Add: *spec.* a turned-up edge of any flat surface or sheeting, esp. in a roof space where it meets the wall; also *attrib.* (see quot. 1963).

1955 *Archit. Rev.* CXVII. 283/3 It [*sc.* a roof tile]..is readily adapted to different plan shapes having an easy repertoire of trimming details at eaves, verges and upstands. **1956** *Ibid.* CXX. 10 The steel box frames forming the turbine-house rooflights make a self-contained rigid frame with sliding joints supported on concrete upstand curbs from the main roof. **1963** *Gloss. Building Terms* (*B.S.I.*) 16 *Upstand beam*, a beam projecting above an adjoining slab. **1974** W. E. KELSEY *Building Construction* viii. 141 It [*sc.* a damp-proof course] is best placed where the upstand or skirting of the roof covering is turned into a mortar joint. **1983** *Ideal Home* June 69 An upstand at the back of the worktop will make it easier to clean and more hygienic—no crumbs or food waste can get trapped in the gap at the back.

also, designating part of a State remote (esp. north) from a large city, as *upstate New York.*

1901, 1904 [see UP *prep.*[2] 6]. **1935** *Amer. Speech* X. 107 Pronunciation in upstate New York... Upstate speech has been studied..by three previous investigators. **1949** *Southern Weekly* 16 Nov. 3/2 This is the figure with complete returns from Greater New York and 19 up-State districts missing. **1969** *Wall St. Jrnl.* 30 Sept. 28/3 Five days at an upstate New York county fair. **1978** J. UPDIKE *Coup* vi. 219 Not a metropolis, but a small industrial city, of the type you call..upstate. **1983** 'W. HAGGARD' *Heirloom* i. 6 He had taken her back to up-State New York.

C. *ellipt.* or as *sb.* An upstate region; a rural area.

1965 *Economist* 23 Jan. 323/1 It was the Democrats from 'upstate' (outside the city) who swung behind Mr Kennedy at once. **1972** *Village Voice* (N.Y.) 1 June 78/2 The Liberal Socialists, who came down every week from upstate for meetings. **1974** *Progress* (Easley, S. Carolina) 24 Apr. II. 8/4 Many of us in the upstate do not appreciate the value of the Tidelands..to our environment.

Hence **u·pstater**, one who lives or comes from upstate.

1944 E. A. HOLTON *Yankees were like This* xvii. 202 The Cape Codder and the Up-Stater are blood brothers with only the difference in the backdrops of sea or mountain. **1965** *Economist* 23 Jan. 323/1 Now, around the figure of Senator Kennedy,..the upstaters have begun to develop some strength. **1975** *New Yorker* 10 Feb. 106/3 In a move to put upstaters' minds at ease about one thing, Carey..assured them, 'The capital of this state is not New York City or elsewhere.'

up stick(s: see STICK *sb.*[1] 7 in Dict. and Suppl., and UP *adv.*[1] 29.

upstiffed: see *UP- 5.

up-stream, *adv.* and *a.* Add: **A.** *adv.* **2.** In the oil and gas industries: at or towards the source of production; *spec.* at a stage in the process of extraction and production before the raw material is ready to be refined.

1973 *Auckland Star* 10 Feb. 18 The most natural way the oil producers can spend their vast wealth..is in developing the industry itself. Huge investment is needed 'upstream'. **1983** *Business Week* 21 Feb. 61/1 KPC has not neglected foreign activities upstream. It wants to expand exploration in the U.S.

B. *adj.* **4.** Relating to the stages in the production of oil and gas before the raw material is ready for refining.

1965 WILLIAMS & MEYERS *Oil & Gas Terms* (ed. 2) 429 Gathering activities are..ended when gas reaches a central point... Facilities used before this point of demarcation are upstream facilities. **1981** *Times* 3 Apr. 17/1 Upstream activities..are overshadowed by the comfortable world supply position while downstream, weak demand is making it more difficult to recoup rising operating costs.

u·pstreet, *adv. colloq.* and *dial.* [UP *prep.*[2] 5.] Up or along the street; in, into, or towards the higher part of a town, etc.

1828, etc. [see *DOWN-STREET *adv.* (*a*.)]. **1933** [see *PIE *sb.*[2] 4 a].

up-stroke. Add: Also **upstroke. 4.** *Physiol.* The part of a nerve impulse when the action potential is becoming more positive.

1974 *Nature* 4 Jan. 69/2 The upstroke of the action potential is considered the result of a Hodgkin cycle. **1979** *Sci. Amer.* Sept. 65/2 The depolarizing upstroke of the action potential is produced mainly by the inflow of sodium ions into the terminal.

u·psurge. [UP- 2.] **1.** A sudden rise or increase of feeling.

1928 *Catholic Times* 11 May 11/5 His books are an upsurge of primitive passion. **1944** D. WELCH *Jrnl.* 20 June (1973) 123, I remember..saying, 'That's a wonderful poem,' and his vital upsurge of agreement. **1958** *Times* 28 June 9/3 There has been a great upsurge of interest in recent years in regal pelargoniums. **1971** *Times* 22 Feb. (Canada Suppl.) p. vii/2 More recently there has been a noisy upsurge of English-Canadian nationalism in Toronto.

2. An uprising, an insurrection.

1930 *Aberdeen Press & Jrnl.* 4 Feb. 7/5 The beginnings of a widespread revolutionary upsurge..are visualised in a proclamation issued by the Red International. **1937** E. LYONS *Assignment in Utopia* (1938) II. ii. 68 The Chinese heritage asserted itself. [Eugene] Chen traveled to the land of his forebears and, without knowing its language, became a leader in the revolutionary upsurge. **1963** *Ann. Reg. 1962* 291 Political machinery for collective leadership to guarantee the revolutionary upsurge.

3. A sharp rise in economic activity, demand, prices, etc.

1935 *Sun* (Baltimore) 22 Apr. 8/2 A sharp upsurge in the postal receipts at Denver. **1955** *Times* 14 July 13/1 It was hardly to be expected that the paint industry could fail to share in the general upsurge of industrial activity over the past 18 months. **1962** *Listener* 7 June 980/2 The speediest possible upsurge in the production of meat and milk. **1974** *Guardian* 25 Jan. 14/1 An upsurge in exports this year has boosted the national output by 6·9 per cent. **1985** *Times* 8 Jan. 17/2 Distillers Co...was another to fail to join the upsurge.

4. A rapid growth in number or size.

1955 A. L. ROWSE *Expansion of Elizabethan Eng.* 161 The sudden upsurge..is witnessed precisely by Hakluyt.

1974 *Sunday Post* (Glasgow) 31 Mar. 9/1 In this year of strife there's been an upsurge of Scots wanting to emigrate.

5. *lit.* A surging upwards.

1969 *Daily Tel.* 5 Feb. 1/8 There was such an upsurge of gas that 14 surplus workmen were evacuated by helicopter. But then the safety valve clicked in.

upsu·rgence. [f. prec. + -ENCE, after *resurgence*, etc.] = *UPSURGE.

1934 in WEBSTER. **1945** *Sun* (Baltimore) 5 Feb. 40/2 Mr. Harry L. Hopkins' view of what is needed to keep the Germans and Japanese from new upsurgences is certainly not a frivolous one. **1958** *Church Times* 22 Aug. 3/4 He emphasized the great dangers of the upsurgence of non-Christian religions in the Eastern world—particularly Buddhism. **1971** I. G. GASS et al. *Understanding Earth* xii. 160/1, I have mentioned the possible effect of plant upsurgence during the Carboniferous.

u·psweep. [UP- 2.] **1.** An upward movement in a long, sweeping curve; a raising or lifting up. Also *fig.*

1898 W. J. LOCKE *Idols* xx. 285 'Who knows?' said Minna, with an insolent upsweep of her lazy lashes. **1954** M. OLIVER *Failing Wine* xxvii. 100 But what are we to say of the upsweep of the moral curve? **1976** *Daily Tel.* 2 Aug. 10 As a result I believe we would see an upsweep in the morale of the whole country.

b. *spec.* An upswept hair-style.

1946 'P. QUENTIN' *Puzzle for Fiends* (1947) xxi. 151 She moved out of the room..absently patting the stray hairs of her upsweep. **1978** J. UPDIKE *Coup* iii. 105 The trim Dacron skirt and jacket, with secretarial upsweep and..dearth of bangles.

upswept, *a.* (In Dict., UP- 5(*b*).) Add: **2.** Applied to (a style of) hair brushed up and fastened at the top of the head; = *swept-up* s.v. *SWEPT *ppl. a.* 1.

1938 *Vogue* 15 June 34 Those toy-size doll hats..on many French heads already—and what a fine solution they are to up-swept hair. **1939** M. B. PICKEN *Lang. Fashion* 160/3 *Up-swept*, term applied to style of hairdress with smooth, high-swept back and small curls on top of head. **1946** 'P. QUENTIN' *Puzzle for Fiends* (1947) xvii. 119 Her rakish upswept hair-do. **1959** *News Chron.* 12 Aug. 3/6 Two oval faces, with smooth up-swept hair. **1976** J. GRENFELL *Joyce Grenfell requests Pleasure* xiii. 182 Lynn showed me how to achieve a fold at the back of my up-swept hair. **1981** A. LURIE *Lang. Clothes* iii. 72/1 Her upswept hairdo, puffed out over pads of wire and horsehair.

3. More generally, having an upward sweep, curved upwards.

1960 *News Chron.* 16 Sept. 1/7 The woman the police are seeking..wears glasses with blue upswept frames. **1961** *Shropshire Mag.* Apr. 43/1 (Advt.), New spectacle frames... Two-tone 17/-; Upswept 20/-. **1961** WEBSTER s.v., Upswept rear fenders. **1984** *Washington Post* 9 July c7/1 There he is, with his striped pantaloons and upswept conquistador hat.

u·pswing. [UP- 2.] **1.** *Golf.* = *back-swing* s.v. *BACK- A. II. *rare.*

1922 WODEHOUSE *Clicking of Cuthbert* vi. 145 His upswing was shaky, and he swayed back perceptibly.

2. a. *Econ.* An upward movement or trend in economic conditions; a (period of) improvement in trading activity. Freq. in advb. phr. *on the upswing.*

1934 *Sun* (Baltimore) 20 Oct. 1/2 Reflecting a sharp upswing in retail trade throughout the country, data on current employment and pay roll trends were made public today. **1946** [see *DOWN-SWING]. **1953** *Times* 31 Oct. 1/11 They do not..reflect the vigorous upswing in the industrial output this year, as revealed by the monthly Treasury indices. **1967** *N.Y. Herald Tribune* (Internat. ed.) 11–12 Feb. 7/7 (*heading*) Market closes on upswing; color TV comes into demand. **1973** 'R. MACLEOD' *Burial in Portugal* iv. 90 When he'd bought in, Consolidated had already been shading at 130 and Maltsters had been on the upswing at 146. **1983** *Daily Tel.* 20 Oct. 21/2 The latest batch of cyclical indicators suggest that the upswing in the economy will continue well into 1984.

b. *transf.* and *fig.*

1947 *Ann. Rev. Microbiol.* I. 351 The seasonal upswings of influenza A or B are much less uniform in time than those of other diseases such as diphtheria, measles, or chickenpox. **1951** *Sunday Pictorial* 21 Jan. 12/5 There is a general upswing in your affairs. **1963** 'E. MCBAIN' *Ten Plus One* (1964) ii. 27 Gang violence..seemed to enjoy an upswing during the summer months. **1976** A. CASSORLA *Skateboarder's Bible* 13 The Sport was once again on the up-swing.

uptake, *sb.* **1.** Delete '(and still chiefly)' and add: *quick* (*slow*, etc.) *in the uptake* (later examples). Also with *at the uptake* (later examples), and now most commonly *on the uptake.*

1911 E. M. CLOWES *On Wallaby* vi. 162 They..are not so 'smart', so quick in the up-take, as themselves. **1927** H. A. VACHELL *Dew of Sea* 259 For a moment the chieftain was puzzled. But he was fairly quick at the up-take, replying after a pause. **1931** D. L. SAYERS *Five Red Herrings* viii. 92 A good girl..but slow in the uptake. **1940** R. S. LAMBERT *Ariel & all his Quality* viii. 190 No one was 'quicker on the uptake'; no one responded quicker to a nod or a wink. **1949** *Here & Now* (N.Z.) Oct. 13/1 An energetic, likeable, cockily pugnacious figure, but slow, almost Neanderthally slow, on the uptake. **1957** H. NICOLSON *Journey to Java* vi. 106 Being quick at the uptake, he then realized that the flat had been visited by of uranium 238, each member resulting from the decay of the previous one.

1941 in M. Gowing *Britain & Atomic Energy 1939–1945* (1964) 394 We have now reached the conclusion that it will be possible to make an effective uranium bomb..which would be equivalent as regards destructive effect to 1,800 tons of T.N.T. **1955** *Times* 14 June 6/5 Information appears to be coming to light here which confirms that the so-called hydrogen bomb exploded at Bikini last year was a uranium bomb involving a triple process of fission-fusion-fission. **1964** M. Gowing *Britain & Atomic Energy 1939–45* i. 38 As war began there was much speculation about Hitler's supposed 'secret weapon' and..the uranium bomb was among the candidates for this title. **1942** Pollard & Davidson *Applied Nucl. Physics* x. 187 The answer lies in the discovery of uranium fission by Hahn and Strassman. **1955** J. Lindhard in W. Pauli *Niels Bohr* 193 Through the discovery of uranium fission it became possible to investigate the penetration of highly charged nuclear fragments. **1956** A. H. Compton *Atomic Quest* 321 Uranium fuel must compete economically with such energy sources as coal. **1899** *Collective Index Trans. & Abstr. Chem. Soc. 1873–1882* 629/1 Uranium hexafluoride. **1941** in M. Gowing *Britain & Atomic Energy 1939–45* (1964) 395 Work on a fairly large scale is needed to develop the chemical side for the production in bulk of uranium hexafluoride, the gaseous compound we propose to use [for the manufacture of ^{235}U]. **1971** Uranium hexafluoride [see *hex]. **1984** *Times* 27 Aug. 1/6 The ship's owners identified the material as uranium hexafluoride, a radioactive gas. **1914** *Phil. Mag.* XXVIII. 825 The equation for the complete disintegration of uranium is U→8He+Radium G (uranium lead). **1955** [see *thorium lead* s.v. *thorium 2]. **1955** *Bull. Geol. Soc. Amer.* LXVI. 1141/1 (*heading*) Uranium-lead age of the granite. **1977** A. Hallam *Planet Earth* 184/2 Isotopic age measurements by the uranium-lead and rubidium-strontium methods on most meteorites have yielded a solidification age close to 4600 million years. **1909** *Chem. News* 26 Mar. 146/2 (*heading*) A new radio-active product of the uranium series. **1973** Uranium series [see *thorium series* s.v. *thorium 2].

Uranus. Add pronunc. (yurē¹nəs). For 'one' read 'two'.

Urartian (urā·rtiăn), *sb.* and *a.* Also **Urartæan, Urartean.** [f. the name of *Urart(u*, an ancient kingdom in eastern Asia Minor + -ian.] **a.** = *Khaldian b. **b.** A native or inhabitant of Urartu.

1934 Webster, Urartaean *a.* and *n.* **1939** [see *Khaldian]. **1939** A. Toynbee *Study of Hist.* VI. 62 Rediscovered language..Sumerian, Akkadian, Elamite, Urartian,..and so on. **1950** H. L. Lorimer *Homer & Monuments* v. 176 Though the early history of the Urartians is obscure, they are believed to have come eastward to Lake Van from Anatolia. **1964** G. Roux *Ancient Iraq* xiv. 192 Their language [*sc.* that of the Hurrians]..is neither Semitic nor Indo-European, but belongs to the vague so-called 'Asianic' group, its nearest relative being Urartian, the language spoken in the country of Urartu (Armenia) in the first millennium BC. **1965** J. Puhvel *Evidence for Laryngeals* 84 The Urartean *Kuḫai* may correspond to the later city of Κολχίς in Armenia. **1972** W. B. Lockwood *Panorama Indo-Europ. Lang.* 175 The I[ndo-E[uropean] immigrants imposed their language on both Hurrians and Urartians and doubtless the non-IE people are ultimately responsible for many of the peculiar and unexplained features of Armenian. *Ibid.* 263, Hurrian is a non-IE language which has affinities with Urartian, the language of the kingdom of Urartu ('Ararat'), centred on the area between the lakes Van, Urmia and Sevan. **1977** *Antiquaries Jrnl.* LVII. 264 The Copenhagen cauldron can be regarded as Greek work of Urartian inspiration in the eighth or seventh centuries B.C.

urban, *a.* and *sb.* Add: **A.** *adj.* **3.** Special collocations: *urban blight*, the gradual unfolding or existence of slum areas, waste land, ghettos, etc., within a city or town (cf. *blight *sb.* 4 b); *urban district*: see District *sb.* 3 b; *urban district council*, the local council of an urban district; abbrev. *U.D.C.*; *urban guerrilla*, a guerrilla operating in cities or towns and involved in kidnapping, etc.; also *attrib.*; *urban guerrilla warfare*, irregular kidnapping, bombing, etc., by urban guerrillas; *urban renewal* (orig. *U.S.*), the clearance and redevelopment of slum areas, waste land, ghettos, etc., within a city or town; also *attrib.* and *fig.*; *urban–rural* adj., of or pertaining to both town and country; *urban sprawl*, the uncontrolled expansion of urban areas; *urban village*, a small self-contained district or community within a city or town (= *village *sb.* 1 e); so *urban villager*, an inhabitant of an urban village.

1938 M. L. Walker (*title*) Urban blight and slums. **1975** M. Bradbury *History Man* i. 14 A real town of urban blight and renewal, social tensions, discrimination. **1894** Urban district council [see sense A. 1b]. **1973** *Whitaker's Almanack 1974* 617/2 Urban district councils are also highway authorities. **1967** G. Jackson *Let.* 31 Jan. in *Soledad Brother* (1971) 106, I have made inroads into political economy..and when I can get hold of them some of the works on urban guerrilla warfare. **1972** *Guardian* 17 Jan. 11/2 The army..is claiming only that it has the measure of the urban guerrilla in Belfast. **1979** G. Seymour *Red Fox* iii. 44 The threat..posed by the rash of urban guerrilla groups. **1955** *Statutes at Large U.S.A.* LXVIII. 1. 622 The heading of title I of the Housing Act of 1949..is hereby amended to read Title I—Slum Clearance and Urban Renewal. *Ibid.*, A fund..known as the 'Urban Renewal Fund', shall be available for advances, loans and capital grants to local agencies for urban renewal projects. **1961** E. A. Powdrill *Vocab. Land Planning* iii. 55 'Urban renewal' can be defined as a generic expression or term which comprises the elements of redevelopment (demolition and rebuilding), rehabilitation (repairing, altering and remodelling), and conservation (the preservation of buildings or groups of buildings). **1966** H. Nielsen *After Midnight* (1967) iv. 61 He heads some extremist group... Sort of an urban renewal project of the Ku Klux Klan. **1977** D. M. Smith *Human Geogr.* v. 104 Some change such as urban renewal or the construction of a new hospital has the capacity to shift the welfare frontier. **1984** *Times* 24 Apr. 13/4 Our delegation visited and discussed urban renewal. **1953** A. E. Smailes *Geogr. of Towns* ii. 33 There is no longer either socially or physically a simple clear-cut dichotomy of town and country; rather it is an urban-rural continuum. **1970** R. Stavenhagen in I. L. Horowitz *Masses in Lat. Amer.* vii. 256 There may be those who see in this situation only an ecologic relation, an 'urban-rural' conflict. **1958** *Listener* 25 Sept. 448/2 The result is a disturbing incidence of crime, juvenile delinquency, road accidents, congestion, urban sprawl, ignorance, and ugliness. **1979** *Monitor* (McAllen, Texas) 8 July 8D/3 Houston's urban sprawl is greater than that of Los Angeles. **1962** H. J. Gans *Urban Villagers* i. 4 The West End was an urban village, located next to Boston's original and once largest skid row area, Scollay Square. **1974** *Times* 15 Apr. 2/4 A device to create an 'urban village' in a predominantly middle-class area. **1962** H. J. Gans (*title*) Urban villagers.

B. *sb.* **2.** = *urban district council* (see sense A. 3 above). *rare.*

1952 M. Laski *Village* ii. 40 'What do you think of this proposal to bring us under Walbridge Urban?'..'Well, I suppose it might mean some street lighting.'

urbanism. (In Dict. s.v. Urban *a.* and *sb.*) Add: Also, town-planning; town-life; urbanization.

1929 *Times* 16 July 17/6 In all the opening speeches..the newly coined word 'urbanism' was prominent. It denotes town-planning, [etc.]. **1934** A. Huxley *Beyond Mexique Bay* 258 Many primitive virtues are obviously incompatible with urbanism and industrialism. **1952** Gerth & Martindale in *Weber's Anc. Judaism* p. xviii, Weber saw the civic society of Palestine as a variation of ancient Mediterranean urbanism. **1977** *Ecologist* VII. 52/3 The dynamic of urbanism as we know it makes inevitable the syndrome of violence, alienation, high crime rates and delinquency that we associate with our cities. **1980** *Times Lit. Suppl.* 26 Sept. 1061/5 The discussion of the Capitol's urbanism and topography.

Urbanist¹. Add: **1.** Also *attrib.*

1926 S. F. Smith *Great Schism of West* 22 To say..that the urbanist succession was valid is by no means the same as [etc.].

u·rbanist². [f. Urban *a.* and *sb.* + -ist.] A specialist in or advocate of town-planning.

1930 *Times Lit. Suppl.* 11 Sept. 708/4 To do so would be to entrust the fate of a city to the technical urbanist. **1964** *Economist* 5 Sept. 917/1 The 'urbanists' fight for more railways to revive the city's heart. **1979** *Jrnl. R. Soc. Arts* Nov. 776/1 The Urbanist's rôle is to design fragments of the city which reflect the culture and ideals of these groups while contrasting them with other orders—pragmatic or architectural.

Hence **urbani·stic** *a.*, **urbani·stically** *adv.*

1959 *Listener* 12 Feb. 289/1 Though he [*sc.* Le Corbusier] has derived so much from the study of the city, his own urbanistic achievements are scarcely to be considered to rank alongside his architectural ones. **1975** *Times Lit. Suppl.* 13 June 660/2 Urbanistically, there is no Middle America. **1983** *Listener* 21 Apr. 7/3 An international competition was held for an urbanistic plan for the Sassi.

urbanite (ɜ̄·ɹbănait). [f. Urban *a.* and *sb.* + -ite¹.] A dweller in a city or town.

1897 *Advance* (Chicago) 29 Apr. 542/1 They will capture streets.., will say to urbanites and suburbanites, 'Stand and deliver.' **1927** *Irish Rosary* July 508 All comes from splendid organizing, bringing peasants, villagers, townsmen and urbanites to meet..and work with bishops, priests, university men. **1960** *Twentieth Cent.* Apr. 347 The urbanite's usual excessive enthusiasm for the beauties of nature. **1979** *Church Times* 29 June 10, I was..only doing what hundred of thousands of other Londoners do every day, as do urbanites in other big cities.

urbanize, *v.* Add: **3.** To accustom to life in a city or town. Chiefly as *pa. pple.* and *ppl. a.*

1948 *Rep. Native Laws Commission 1946–48* (Dept. Native Affairs, S. Afr.) 19/1 There are large numbers of Natives in a transitional state, who are partly urbanised but have not yet broken their ties with the Reserves. **1963** *Daily Tel.* 18 Jan. 15/3 Robins are becoming more 'urbanised'..and will now take crumbs from the hand at the cafeteria. **1969** *Ibid.* 9 Sept. 13/2 The 'good old Mother Earth' myth..can turn the most urbanised of people into horticultural maniacs. **1974** *Times* 22 Apr. 7/3 What about the eight million urbanized blacks living in white cities?

urbanology (ɜ̄ɹbănǫ·lŏdʒi). [f. Urban *a.* and *sb.* + -ology.] The study of cities or towns and their problems. So **urbano·logist.**

1967 *Time* 28 July 11 The 'urbanologist' aspires to be a student of the entire city,..whose concerns go beyond brick and mortar to budgets and laws, souls and sensibilities. Just as the word urbanology is a cross between Latin and Greek, the science..is a mélange of disciplines. **1972** *Daily Colonist* (Victoria, B.C.) 27 Apr. 33/1 The solution as some urbanologists see it is to come up with a new concept in automotive packaging. **1972** *New Yorker* 16 Sept. 124/2 A landscape that transcends all the nickel-and-diming defeats of our urbanology. **1975** M. Bradbury *History Man* iii. 51 The local council, now impressed with Howard's urbanology, have accepted this scheme;..the surviving houses in the terrace will ultimately be restored.

urbicide (ɜ̄·ɹbisaid). [f. L. *urbis, urbs* a city + -cide 2.] The destruction of a city or its character.

1966 *Harper's Mag.* May 94/1 This fearful instrument of urbicide will be not only the tallest, but unquestionably one of the ugliest buildings in the world. **1972** *N.Y. Times* 15 Oct. iv. 14 It does no good to speculate at what point real estate becomes art, or history, or a talisman of place. When it does, it enters the public domain. To destroy it is an act of urbicide.

urbiculture (ɜ̄·ɹbikɒltiŭɹ). [f. L. *urbis, urbs* a city + Culture *sb.*] **a.** The development of cities and towns and urban life; the cultivation of urban interests. **b.** = *rurbanization* s.v. *rurban *a.*

1954 *Daily Progress* (Charlottesville, Va.) 26 July 16/1 A bill introduced by Rep. J. Arthur Younger..to create a U.S. Department of Urbiculture. **1959** [see *rurbanization].

Hence **u:rbicu·lturist.**

1977 *Sat. Rev.* (U.S.) 23 July 48/1 The urbiculturists—give a multimillion-dollar skyscraping enclave or two—move now not to knock down and build anew, but to make over or, to use today's mot juste, to recycle.

‖ **urbi et orbi** (ɜ̄·ɹbi et ǭ°·ɹbi), *adv. phr.* [L.] Of papal proclamations, blessings, etc.: to the city (of Rome) and to the world. Also *transf.*, for general information or acceptance; to everyone.

1867 *Chambers's Encycl.* IX. 671/2 *Urbi et Orbi*.., a form..used in the publication of papal bulls, for the purpose of signifying their formal promulgation to the entire Catholic world, as well as to the city of Rome. **1876** Mrs. Tait tr. *Klaczko's Two Chancellors* viii. 268 These were the expressions contained in an official document of incontrovertible authority, a diplomatic manifesto which announced *urbi et orbi* the lofty thoughts of the Imperial Government of France. **1889** M. S. Van de Velde *Cosmopolitan Recoll.* II. v. 165 The Conclave met; the election of Camerlingo Pecci was foreseen, it was voted without opposition, and at one p.m., March 3rd, announced *urbi et orbi* from the *loggia* of St. Peter. **1907** G. W. E. Russell *Pocketful of Sixpences* 21 On the eve of the General Election of 1880 she [*sc.* Lady Burdett-Coutts] issued a proclamation, *urbi et orbi*, enforcing the need for a 'strong government'. **1924** J. O. Field *Uncensored Recoll.* xiii. 239 The great pose of Morès..was a deep-rooted hatred of England, to which sentiment he was always giving very loud and dramatic utterance, *urbi et orbi*. **1958** *Listener* 27 Nov. 861/1 The familiar figure dressed in white..giving his benediction *urbi et orbi* from the loggia of St. Peter's. **1961** *Times* 7 Jan. 9/3 The voice of the poet addressing himself *urbi et orbi*. **1973** M. Bence-Jones *Palaces of Raj* v. 105 One of the nobles, whose bard would, every two hours of the night, proclaim *urbi et orbi*..his titles and honours. **1980** *Times* 27 Dec. 3/6 The Pope's traditional Christmas homily *Urbi et Orbi* (to the city and the world).

Urbino (ɒɹbī·no). Name of a city in the province of Le Marche, Italy, and of a former duchy, used *attrib.* to designate (items of) majolica made there from the fifteenth to the seventeenth centuries.

1881 C. C. Harrison *Woman's Handiwork* ii. 104 An Urbino drug-pot, or a Delft plaque. **1883** J. W. Mollett *Illustr. Dict. Art & Archæol.* 331/1 *Urbino ware*, made at Urbino, under the patronage of its Duke. **1924** Rackham & Read *Eng. Pottery* iv. 47 Another class of Italian motives is the strange brood of caryatids, sphinxes, winged beasts, dragons, and semi-monstrous birds... The seed of this kindred may have come in some cases from some piece of Urbino maiolica. **1940** *Burlington Mag.* Aug. 64/2 A large Urbino dish in the Musée de Cluny. **1952** B. Rackham *Italian Maiolica* v. 24 A dish at Cambridge..interesting as an example of Urbino ware with lustre enrichment added by Maestro Giorgio at Gubbio—is now considered to be the work of the son. **1973** *Times* 3 Nov. 2/2 The sale also contained an Urbino majolica dish painted with the murder of Amphiarus by Francesco Xanto.

urchin. Add: **8.** **urchin cut**, a short style of haircut for women; also as *ppl. adj.*, and *urchin hair-cut*.

1951 [see *cut *sb.*² 16 a]. **1958** 'E. Dundy' *Dud Avocado* i. vii. 132 A lot of rather gorgeous..girls floating around with urchin hair-cuts. **1958** *Woman's Own* 4 June 9/1 She was long-limbed, slender, her shining hair urchin-cut. **1958** *Woman* 11 Oct. 32/3 She was small and boyish, with urchin-cut hair. **1979** N. Freeling *Widow* v. 26 A small girl with an urchin face and an urchin-cut.

urea. Add: Delete ‖. Also with pronunc. (yurī·ă).

1. b. A urea-formaldehyde plastic or resin.

1935 *Economist* 7 Dec. 1142/2 Phenolic plastics are used for various mouldings, urea is translucent plastic. **1961** H. R. Symonds *Source Bk. New Plastics* II. ii. 32 A new urea especially prepared for the bonding of particle board. **1969** L. K. Arnold *Introd. Plastics* vi. 83 The ureas are somewhat more expensive [than the phenolics] but have the advantage of being available in a wide range of light colors.

2. **urea-formaldehyde,** used *attrib.* and *absol.* to designate plastics, resins, etc., made by condensation of urea with formaldehyde; **urea resin**, a synthetic resin derived from urea; a urea-formaldehyde resin.

1928 *Brit. Patent* 291,473 3/2 When using urea-formaldehyde resins, alkaline catalysts may be used to obtain the intermediate condensation product. **1933** [see *phenol-formaldehyde* s.v. *phenol c]. **1976** *Milton Keynes Express* 9 July 14/4 The majority of homes built in the last thirty years can be insulated by filling the cavity with..a foaming resin called urea formaldehyde. **1984** *Christian Science Monitor* 2 Mar. 23/1 The safety of urea-formaldehyde foam insulation..has been a subject of debate for years. **1937** R. S. Morrell et al. *Synthetic Resins* ii. 52 Tootal, Broadhurst, Lee Company Limited have patented the use of urea resins to prevent the crushing of cellulosic fabrics. **1979** P. Stevens *Polymer Chem.* xii. 329 Decorative interior plywood is normally glued with urea resin because the dark-colored phenolic resins can stain the veneer.

ureaplasma (yū°rī,æplæ·zmă). *Biol.* Pl. -plasmas. [f. Urea + Plasma.] A micro-organism of the genus *Ureaplasma*, formerly included within the genus *Mycoplasma* (cf. *mycoplasma), which is characterized by the ability to metabolize urea.

1974 M. C. Shepard et al. in *Internat. Jrnl. Syst. Bacteriol.* XXIV. 160 It is reasonable to propose establishing a new, separate genus in the family *Mycoplasmataceae* in which to classify the T mycoplasmas isolated from man and lower animals. The name *Ureaplasma* is proposed for this new genus, which at present contains a single human species. **1979** *Jrnl. Med. Microbiol.* VIII. 258 [illegible] **1981** *Annales de Microbiol.* CXXXII B. 172 Ureaplasmas were isolated first from the human genito-urinary tract.

urease (yū°rī,ē¹z). *Biochem.* [f. Ure(a + *-ase.] An enzyme produced in bacteria and certain plants which converts urea into carbon dioxide and ammonia.

1892 *Jrnl. R. Microsc. Soc.* 515 M. P. Miquel describes a new diastase which according to the nomenclature of Duclaux is called urease. **1926** *Jrnl. Biol. Chem.* LXIX. 437 The jack bean appears to contain a very large amount of urease. **1967** M. E. Hale *Biol. Lichens* iv. 58 These lichens often contain urease enzyme systems. **1982** T. I. Diamondstone in T. M. Devlin *Textbk. Biochem.* xii. 564 The source of this ammonia is urea, which, being freely diffusible, finds its way into the large intestine, where it is acted upon by bacterial urease.

Urecholine (yū°rīkō°·lin, -in). *Pharm.* Also **ure-.** [f. Ure(a + *Choline.] A proprietary term in the U.S. for the carbamic acid ester carbaminoyl-β-methyl choline chloride, $C_7H_{17}ClN_2O_2$, a quaternary ammonium compound used as a parasympathomimetic agent to stimulate bowel or bladder muscle activity.

1941 *Official Gaz.* (U.S. Patent Office) 29 Apr. 1104/2 Merck & Co., Inc.... Urecholine for medicinal preparations for the treatment of disorders of the peripheral circulatory system and for stimulation of the para-sympathetic nervous system. **1954** S. Duke-Elder *Parsons' Dis. Eye* (ed. 12) v. 70 Two drugs are sometimes employed which combine the direct and indirect methods of stimulation—doryl (carbaminoylcholine) and urecholine (carbaminoyl-β-methyl-choline chloride). [illegible] Bethanechol (Urecholine) and methacholine (Mecholyl) have many of the actions of acetylcholine on glands. **1981** Gerald & O'Bannon *Nursing Pharmacol. & Therapeutics* vi. 64/2 Bethanechol (Urecholine), possessing relatively specific effects on the smooth muscle of the gastrointestinal tract and urinary bladder, is clinically employed to relieve postoperative abdominal distention and urinary retention.

uredinium (yū°ridi·niəm). *Bot.* Pl. -inia. [f. L. *uredin-, urēdo* Uredo + -ium.] A cluster or sorus of uredospores and the hyphæ which bear them. So **uredi·nial, uredi·noid** *adjs.*

1900 B. D. Jackson *Gloss. Bot. Terms* 283/1 Uredinial. **1905** J. C. Arthur in *Bot. Gaz.* XXXIX. 221 For the sorus of the second spore-stage,..I propose uredinium;..derivatives uredinial, urediniospore [or uredospore], urediniosore, etc. **1916** *Mycologia* VIII. 182 [illegible] **1929** J. C. Arthur et al. *Plant Rusts* i. 13 Less easily recognized as aecia, with spores borne singly on pedicels and usually without peridia or paraphyses, are the stylosporic or uredinoid aecia. [illegible] **1974** G. W. Burns *Plant Kingdom* ix. 168 (*caption*) Diagrammatic section through a uredinium. *Ibid.* 169 The uredinia begin their development beneath the epidermis of the host. **1982** Hiratsuka & Sato in Scott & Chakravorty *Rust Fungi* i. 12 The uredinial state is the most destructive spore state of such rusts as wheat stem rust..and poplar leaf rust.

uredium (yurī·diəm). *Bot.* Pl. **uredia.** [mod.L., f. Uredo.] = *Uredinium. Hence **ure·dial** *a.*

1907 *Nature* 8 May 800/2 Forty-four days after the inoculation was made, one observed small uredia on the upper surface of several of the pustules. **1937** Uredial [see *æcium]. **1949** B. B. Mundkur *Fungi & Plant Dis.* vii. 155 Towards the end of the growing season the formation of uredia ceases and teliospores begin to appear. **1967** *Trans. Brit. Mycol. Soc.* L. 192 Aecia usually result from a fertilization process..but they may also originate from aeciospores and, thus, can be a repeating spore form (uredial aecia as in *Cronartium* and *Coleosporium*). **1970** J. Webster *Introd. Fungi* ii. iv. 369 A single uredium may contain from 50,000 to 400,000 spores.

uredosorus (yuridōsō°·rəs). *Bot.* Pl. **-sori.** [f. Uredo + Sorus¹.] = *Uredinium.

1905 *Bot. Gaz.* XXXIX. 221 The structure which arises from a single, fertile hyphal mass, or hymenium, either with or without a peridium, now usually called spermogonium, aecidium, uredosorus, teleutosorus, or kindred names. **1930** *N.Z. Jrnl. Sci. & Technol.* XII. 126 The term 'uredosorus' is applied to sori in which the spores are borne singly on distinct pedicels. **1973** *Nature* 17 Aug. 463/1 The majority of plants in the plot were infected, some severely (500–2,000 uredosori per plant).

uredospore. Add: Also **uredi·niospore** [cf. *uredinium], **ure·diospore**, in the same sense.

1905 Urediniospore [see *uredinium]. **1937** *Nature* 8 May 801/1 The uredospores germinate normally and readily infect wheat seedlings. **1970** J. Webster *Introd. Fungi* ii. iv. 368 [illegible] **1978** *Bio Systems* X. 32/1 [illegible]

ureilite (yū°rī·ilait). *Geol.* [ad. Russ. *ureilit* (Erofeev & Lachinov 1888, in *Zhurnal Russ. fiziko-khim. Obshchestva pri Imper. St. Petersburgsk. Univ.* XX. 213), f. the name of Novo-Urei, a village near Penza, in the vicinity of which a meteorite belonging to this class fell.] Any of a group of calcium-poor achondrite meteorites that consist mainly of olivine and pigeonite.

[**1889** *Jrnl. Chem. Soc.* LVI. 225 The authors find, after comparison with known classes of aerolites, that the meteorite described above differs from all of them in many respects and propose to name this form ureilite.] **1916** *Mineral. Mag.* XVIII. 38 The ureilites consist of a coarsely crystalline aggregate of olivine and monoclinic pyroxene cemented together by a film of nickel-iron and carbonaceous matter. **1962** B. Mason *Meteorites* iii. 115 The ureilites form a unique group of three meteorites, quite distinct from all other achondrites. **1972** I. G. Gass et al. *Understanding Earth* viii. 116/1 The presence of diamond in the small group of ureilites appears to be due to extraterrestrial shock effects.

ureotelic (yū°rī,ote·lik), *a.* *Biochem.* [ad. It. *ureotelico* (A. Clementi 1916, in *Atti della R. Accad. dei Lincei: Rendiconti* XXV. I. 366), f. *ureo-* Ureo- + Gr. τελικός final.] Of an animal or its metabolism: producing nitrogenous waste chiefly in the form of urea.

1929 [see *uricotelic *a.*]. **1972** M. S. Gardiner *Biol. Invertebrates* xiii. 590 Animals are conveniently classified as ammonotelic, ureotelic, or uricotelic.

uretero-. Add: **ureterocele** [Gr. κήλη tumour, rupture], an outward protrusion of the wall of a ureter; **uretero·graphy**, radiography of the ureter; **ureterosigmoido·stomy** [*-stomy], the operation of implanting the ureters into the sigmoid flexure of the colon.

1913 *Interstate Med. Jrnl.* XX. 346 On close observation of the right ureter orifice..during urinary excretion, a considerable cystic dilation was noticed and the condition was diagnosed as a ureterocele. **1976** *Lancet* 11 Dec. 1301/1, I would have appreciated more information about those boys with radiological abnormalities who did not undergo surgery, the one with a right ureterocele, and the 9-year-old with grade-III reflux. **1921** Stedman *Med. Dict.* (ed. 2) 961/2 Ureterography. **1926** Young & Davis *Young's Pract. Urol.* II. xiii. 241 Ureterography.—The contour of the ureter may be shown by filling it with various shadowgraphic media and taking the x-ray film or plate in the same manner as described for pyelography. **1977** *Proc. R. Soc. Med.* LXX. 182/1 Mr Rose described the methods by which today an accurate and precise diagnosis may be made by early high-dose urography and ureterography. **1934** Webster, Ureterosigmoidostomy. **1940** L. Dodson *Urol. Surg.* xxxii. 600 If this fails, muscle transplants or ureterosigmoidostomy may be tried, depending upon the extent of the deformity. **1977** *Proc. R. Soc. Med.* LXX. 536/2, 2 teenage girls with bladder exstrophy and previous bilateral ureterosigmoidostomy performed in childhood.

urethane. Add: Now usu. with pronunc. (yū°re·þē¹n). [f. as ad. F. *uréthane* (J. B. A. Dumas 1833, in *Ann. d. Chim. et Physique* LIV. 236).] **b.** Any ester of carbamic acid. **c.** By extension, any compound which includes the group —NH·CO·O— in the molecule.

1856 *Jrnl. Chem. Soc.* VIII. 276 Butylic urethane..forms beautiful nacreous scales. **1876** *Encycl. Brit.* V. 578/1 The neutral ethereal salts of carbamic acid are known as urethanes. **1926** *Jrnl. Amer. Chem. Soc.* XLVIII. 2176 It was decided..to prepare the mono- and di-carbethoxy-guanidines... The compounds may be regarded as urethanes. **1945** C. M. Fisher et al. in G. S. Whitby *Synthetic Rubber* xxv. 926 The diisocyanate reacts with the terminal hydroxyls of the polyester, forming urethane linkages. **1968** R. O. C. Norman *Princ. Org. Synthesis* xiv. 447 The isocyanate is converted into an amine or a urethane. **1976** A. L. Ternay *Contemporary Org. Chem.* xxii. 688 Urethanes have high melting points, making them useful as derivatives of phenols.

2. = *polyurethane; also *attrib.* and *fig.*

1956 [see *processable *a.*]. **1969** *Jane's Freight Containers 1968–69* 463/2 Construction: Aluminium extrusion and sheet,..urethane foam, laminated wood. **1971** *Cabinet Maker & Retail Furnisher* 1 Oct. 14/2 Children's stools made of ICI's new solid urethane plastic are being supplied to infant and primary schools. **1976** A. Cassorla *Skateboarder's Bible* 9 The past can be shrouded by the translucent urethane gauze of the present. **1979** *Arizona Daily Star* 8 Apr. A3/3 (Advt.), Madly flattering sandals of soft kid urethane, with flexible, easy going soles.

urethro-. Add: **urethro·graphy**, radiography of the urethra.

1932 R. M. Le Comte *Man. Urol.* i. 38 Urethrography. This is done by filling the urethra with a radiopaque solution and taking an x-ray film. **1962** J. D. Maynard in R. M. Kirk et al. *Surgery* vii. 157/1 Urethrography. Micturating cystograms do not demonstrate the anterior urethra.

Urfirnis (ū°·ɹfɪ̄ənis). [Ger., f. *ur-* Ur- + *firnis* varnish, veneer.] A form of early Greek pottery (see quots.).

1912 Wace & Thompson *Prehist. Thessaly* ii. 21 Urfirnis Ware..was first found..at Orchomenos... The whole vase is covered with a thin semi-lustrous wash which varies in colour from red-brown to..black... The Orchomenos Urfirnis naturally occupies a middle position between the Tirynthian and that from Lianokladhi. **1928** Peake & Fleure *Steppe & Sown* x. 126 Into this mixed population [around Corinth] came the Cycladic traders with their characteristic *Urfirnis*, or primitive glazed pottery. **1939** J. D. S. Pendlebury *Archaeol. Crete* ii. 73 The pottery, peculiar to Vasilike, is a highly specialized form of 'Urfirnis' ware—the pottery covered with a lustrous black or brown wash—which develops in Greece and the Islands, no doubt with an admixture of local origin. **1957** V. G. Childe *Dawn Europ. Civilization* (ed. 6) v. 63 [illegible]

urgency. Add: **I. b.** (Earlier example.)

1881 E. W. Hamilton *Diary* 14 Mar. (1972) I. 116 In consequence of these tactics on the part of the Tories, the Government of course failed this afternoon to get the necessary majority to vote urgency.

III. 9. (Earlier example.)

1881 E. W. Hamilton *Diary* 14 Mar. (1972) I. 116 Another piece of news which took one by surprise was Sir S. Northcote's manifesto deprecating resort to the urgency rules of which Mr. Gladstone had given notice for supply this evening.

urger. Add: **1. b.** A man who obtains money illegally or discreditably, esp. as a tipster at a racecourse. *Austral. slang.*

1919 V. Marshall *World of Living Dead* 69 The truly light-fingered gentry, the racecourse urger (tip slinger), the magsman..never hesitate to express their contempt for the more roughly inclined of the profession. **1934** *Bulletin* (Sydney) 21 Mar. 20/3 He was a tout or an urger, I gathered. 'Mixed up in racecourses,' was the way he put it to me. **1960** A. Kimmins *Luggs O'Leary* v. 74 'An urger,' explained Luggs patiently, 'is a man who looks around for suckers like you and tips each one a different horse. Someone's got to win.'

urgingly, *adv.* (In Dict. s.v. Urging *ppl. a.*) (Earlier example.)

c **1882** 'Mark Twain' *Speeches* (1923) 104, I say it beseechingly, urgingly.

‖ **Urheimat** (ū°·rhəimāt). [Ger., f. *ur-* Ur- + *heimat* home, homeland.] The place of origin of a people or of a language.

1934 Priebsch & Collinson *German Lang.* i. 17 The 'cradle' or primeval home (*Urheimat*) of the Indo-Europeans has been a question of some controversy. **1950** J. R. Firth *Papers in Linguistics 1934–51* (1957) xv. 178 The more romantic theorists who enjoy Indo-European fantasies and from *Ursprache* go on to speculate on the *Urvolk* and the *Urheimat*. **1976** D. R. Snow in W. Cowan *Papers 7th Algonquian Conf.* 339 (*heading*) Archaeological implications of the Proto-Algonquian Urheimat.

Uriah Heep (yuraiă hīp). The name of a character in Dickens's *David Copperfield* (1850) used allusively for a man who is hypocritically humble. Also *attrib.* Hence **Uriah Heepish** *a.*, reminiscent of Uriah Heep.

1876 [see *Ananias]. **1915** J. Webster *Dear Enemy* 77 The Uriah Heepish attitude toward trustees that characterized Mrs. Lippett's manners. **1947** I. Brown *Say the Word* 106 Uriah Heepish creatures. **1974** *Listener* 4 Apr. 438/1 'If I may'—often issues from the lips of the Uriah Heeps. **1976** A. J. Russell *Pair Hemlock* v. 57 A Uriah Heep kind of guy named Logan.

urial, now the usual spelling of Oorial, *Ovis orientalis*.

1903 [mentioned s.v. Oorial]. **1912** R. Lydekker *Sheep & its Cousins* xiii. 263 The range of the urial is very wide. **1950** T. Longstaff *This my Voyage* ix. 180 We were astonished to see a herd of urial. **1977** G. B. Schaller *Mountain Monarchs* ix. 227 Urial ewes, as well as urial lambs, differed from the others in being almost passive.

uricase (yū°·rikē¹z). *Biochem.* [a. F. *uricase* (Battelli & Stern 1909, in *Compt. Rend. des*

Séances & Mem. de la Soc. de Biol. LXVI. 412): see Uric *a.* and *-ase.] An enzyme which promotes the conversion of uric acid into allantoin and is found in certain insects and most mammals other than primates.

1910 *Jrnl. Biol. Chem.* VII. 172 If we limit the use of this term..to the oxidizing enzyme which produces allantoin and CO_2 from uric acid, and appreciate there may possibly exist other enzymes which destroy uric acid.., we find that either the kidney or the liver of every animal investigated, except man..possesses uricase. **1934** *Times Lit. Suppl.* 19 Apr. 286/1 The preparation of soluble uricase from ox kidney. **1956** *Nature* 31 Mar. 622/2 It was also possible to demonstrate the presence of the uricase enzyme in those fungi. **1977** D. E. Metzler *Biochem.* xiv. 883/2 Injection of the uric acid-hydrolyzing enzyme uricase has been tested experimentally.

uricosuric (yū°rikosū°·rik), *a.* *Med.* [f. Uric *a.* + -o + *-s-* + Uric *a.*] Causing or characterized by an increased excretion of uric acid or urate in the urine.

1948 *Amer. Jrnl. Med.* IV. 774/1 This effect is similar to that produced by other drugs..all of which produce a simultaneous increase in minute excretion of urate and a decrease in plasma urate concentration. This pattern of pharmacologic actions has been termed the 'uricosuric effect'. **1961** *Lancet* 19 Aug. 430/2 All phosphate levels were estimated before starting, or at least one month after finishing, uricosuric therapy. **1974** M. C. Gerald *Pharmacol.* xiv. 269 High doses of aspirin, about 5 g, have a uricosuric action. **1980** *Brit. Med. Jrnl.* 17 May 1212 An orally active diuretic with uricosuric properties.

uricotelic (yū°rikote·lik), *a.* *Biochem.* [ad. It. *uricotelico* (A. Clementi 1916, in *Atti della R. Accad. dei Lincei: Rendiconti* XXV. I. 366), f. *urico* Uric *a.* + Gr. τελικός final.] Of an animal or its metabolism: producing nitrogenous waste chiefly in the form of uric acid or urates, as in certain insects, birds, and reptiles, rather than urea.

1924 *Proc. R. Soc.* B. XCVII. 227 Clementi enunciated the doctrine that arginase is present in the livers of all those animals which have what he calls a 'ureotelic' metabolism—a metabolism, that is, in which urea is the final product of protein degradation—and absent from the livers of those in which protein metabolism is 'uricotelic'—ending in uric acid. **1934** *Biochem. Jrnl.* XXVIII. 1390 The uricotelic character of the metabolism of this gastropod is supported by new *in vitro* experiments. **1954** *New Biol.* XVII. 45 The uricotelic nitrogen metabolism of insects..is economical of water since uric acid can be excreted without much water loss. **1979** *Jrnl. Compar. Physiol.* CXXXIII. 211/2 Xeric adapted arboreal frogs of at least two genera..excrete nitrogen predominantly in the form of urate (uric acid), in proportions and amounts similar to uricotelic reptiles.

uridine (yū°·ridīn). *Biochem.* [ad. G. *uridin* (Levene & Jacobs 1910, in *Ber. d. Deut. Chem. Ges.* XLIII. 3152), f. *ur-acil* *uracil: see *-idine.] A pyrimidine nucleoside, $C_5H_9O_4{\cdot}C_4H_3N_2O_2$, in which the base is uracil and the sugar ribose, and which is a constituent of RNA and various intermediates in cell metabolism.

1911 *Jrnl. Chem. Soc.* C. i. 96 Nitrous acid effects the quantitative elimination of the amino-group from cytidine, and uridine is obtained. **1946** [see *nucleoside]. **1970** R. W. McGilvery *Biochemistry* xx. 477 Uridine triphosphate can be used for nucleic acid synthesis or in the formation of metabolic intermediates such as uridine diphosphate glucose. **1976** *Sci. Amer.* Feb. 33/2 RNA..differs from DNA in that all the thymidine units are replaced by uridine.

uridylic (yū°rĭdi·lik), *a.* *Biochem.* [f. prec. + -yl + -ic.] *uridylic acid*: the phosphoric acid ester, $C_5H_8O_3(PH_2O_4){\cdot}C_4H_3N_2O_2$, of uridine, one of the four nucleotides present in RNA.

1933 *Jrnl. Biol. Chem.* CI. 529 Cytidine is readily transformed into uridine, having properties identical with those of the uridine from uridylic acid. **1946** [see *nucleoside]. **1975** *Sci. Amer.* May 25/2 There are four kinds of nucleotide, named for their bases: adenylic acid, uridylic acid, guanylic acid and cytidylic acid, better known as *A, U, G* and *C*.

urinalysis. Delete *U.S.* and add further examples.

1971 *Nature* 12 Mar. 113/1 All subjects had a routine physical check-up, including complete blood counts, urinalysis, chest X-ray, [etc.]. **1980** *Daily Tel.* 15 Oct. 19/6 A post-fight urinalysis showed the presence in Ali of opiates and the drug Phenothiazine.

urine, *sb.*¹ Add: **3.** *urine-sodden* adj.

1912 *Man. Elem. Milit. Hygiene* (War Office) x. 62 The front of the latrine rapidly becomes a urine-sodden quagmire. **1944** *Public Health* LVII. 137/1 Nauseating odours assail one's nostrils on entry, and the source is usually located in some urine-sodden faecal-stained mattress in an upstairs room.

urine·tte. [pseudo-Fr., f. Urin(e *sb.*¹ + -ette.] (See quot. 1967.)

1954 J. Pudney *Smallest Room* 35 The 'urinettes' have always been there. **1967** *Gloss. Sanitation Terms* (*B.S.I.*) 58 *Urinette*, a urinal like an elongated W.C. pan, for female use.

‖ **urinoir** (ürinwār). [Fr.] A public urinal.

1955 G. Greene *Quiet American* ii. iii. 147 The old women gossiped as they had always done, squatting on the floor outside the urinoir. **1962** *Spectator* 16 Nov. 770 A wrought iron urinoir in Holborn.

Urkey (ɜ̄·ɹki). [Origin unknown.] A local name of a children's game (see quot.). Also, the person who is 'it' in this game. As *v. trans.*, to defeat in this game. Cf. *lerky.

1938 E. Thomas *Childhood E. Thomas* iii. 53 The best game was an evening one, called Urkey. One boy who was Urkey stood still by a tin can while the others hid. When a shout told that they had found a hiding place he went in search of them. His object was to see one and run home to the can, crying 'I Urkey Johnny Williams'.

urlar (ū°·ɹlăɹ). [Gael.] (See quots.)

1889 *Cent. Dict.* s.v. *Pibroch*, It consists of a ground-theme or air called the *urlar*, followed by several variations,..the whole concluding with a quick movement called the *creanduich*. **1900** C. S. Thomason *Ceol Mor* 5 The Ground or *Urlar*, which corresponds to the *Thema* or Theme of ordinary modern music. **1925** J. P. Grant *Piobaireachd* 27 All that appears is the Urlar and a Doubling. The first and third lines of the Urlar correspond with Angus Mackay's MS. **1962** A. MacLeod *Eighth Seal* iv. 44 Flora had been working out the *urlar*, the groundwork or theme of her pibroch. **1977** *Meanjin* (Austral.) XXXVI. 1. 80 *Urlar* and *Siubhal*..are respectively the 'ground' and 'variations' in classical pipe music.

urn, *sb.* Add: **7.** **urnfield**, a cemetery of individual cremation graves with remains in pottery urns, *esp.* as used by North European peoples from *c* 1200 B.C. onwards; also *attrib.*, esp. designating peoples using this rite or their cultures.

1889 [see Urn *sb.* 7]. **1928** V. G. Childe in *Antiquity* II. 37 On the continent as in Britain the later phases of the Bronze Age are marked by the spread of large cremation cemeteries generally termed urnfields. One of the several groups of urnfield cultures in Central Europe occupies such a pre-eminent position that it may even claim to be the parent of all the rest. It is known as the Lausitz or Lusatian culture. **1958** T. G. E. Powell *Celts* 38 The dead were generally cremated, and the broken bones placed in an urn for burial in a flat cemetery. Many of these cemeteries..have been called urnfields so that the descriptive labels 'Urnfield Period' and 'Urnfield Culture' have come into use. **1968** A. Powell *Military Philosophers* iv. 177, I walked up the road.., leaving them [*sc.* Welshmen] to move eastward towards the urnfields of their Bronze Age home. **1979** B. Cunliffe *Celtic World* 15/1 This period, generally referred to..as the Urnfield period, is typified by the appearance of large cremation cemeteries, the ashes of the dead interred in urns. The tradition took form in Hungary sometime in the thirteenth century B.C. and was rapidly adopted further west.

urning (ɜ̄·ɹniŋ). Now *rare*. Also **Urning.** [a. Ger., coined by K. H. Ulrichs ('Numa Numantius') in 1864.] A homosexual. Cf. *Uranian *a.*¹ 1 c, *sb.*², *uranism.

1883 *Jrnl. Nervous & Mental Dis.* Apr. 200 For himself and fellow-Urnings there was nothing left but this unnatural love. **1892** C. G. Chaddock tr. *Krafft-Ebing's Psychopathia Sexualis* 235 The urning loves and deifies the male object of his affections. **1896** J. A. Symonds *Probl. Mod. Ethics* vii. 91 Man, Woman, and Urning—the third being either a male or a female in whom we observe a real and inborn, not an acquired or a spurious inversion of appetite. **1909** E. Carpenter *Intermediate Sex* (ed. 2) 135 According to the information of De Joux..the number of Urnings in all Europe is about five millions.

uro-¹. Add: **urodyna·mics**, the branch of medicine concerned with the containment and flow of urine in the body; hence **urodyna·mic** *a.*; **urolithi·asis**, lithiasis in the bladder or urinary tract; **urothe·lium** [Epi)thelium], the epithelium of the urinary tract, esp. the bladder; hence **urothe·lial** *a.*; **urotro·pine** (-pīn), † **-in** [ad. G. *urotropin* (A. Nicolaier 1895, in *Deut. med. Wochenschr.* 22 Aug. 541/1): see *-tropic, -ine⁵] = *hexamethylenetetramine* s.v. *hexa-.

1963 *Jrnl. Urol.* XC. 730/2 Practical application of electromanometric urodynamic studies depends upon additional information. **1981** *Brit. Med. Jrnl.* 23 May 1706/3 The clinical and urodynamic evaluation of incontinence in elderly patients. **1954** D. M. Davis in *Ann. Surg.* CXL. 839 (*heading*) The hydrodynamics of the upper urinary tract (urodynamics). **1977** *Lancet* 13 Aug. 335/2 The past decade has witnessed an enormous step forward in our knowledge and understanding of the mechanisms involved in incontinence in women and consequently of the principles of management. The new science that has evolved is that of urodynamics. **1865** W. Roberts *Pract. Treat. Urinary & Renal Dis.* ii. iii. 209 (*heading*) Gravel and calculus. (Urolithiasis.) **1926** Young & Davis *Young's Pract. Urol.* I. vi. 388 Urolithiasis is one of the four important and serious conditions causing hematuria. **1966** *Lancet* 31 Dec. 1455/2 The prevalence of urolithiasis in Arad was lower than in the Beersheba settlers, and the urinary output was higher. **1977** *Ibid.* 26 Mar. 684/2 Close similarities between Balkan nephropathy and urothelial tumours with respect to geographical clustering, age, and sex. **1954** P. A. Narath in M. Campbell *Urology* I. i. iii. 88 Absorption is accomplished by the urothelium which lines the tract. **1977** *Proc. R. Soc. Med.* LXX. 413/2 In patients with bacterial infections of the lower urinary tract it is possible that small amounts of nitrosamines may be produced which could initiate neoplastic or preneoplastic changes in the urothelium. **1895** *Amer. Practitioner & News* XX. 486 The name urotropin was applied to hexamethylenetetramin owing to the changes which its administration brought about in the urine. **1897** *Lippincott's Med. Dict.* 1097/2 Urotropine. **1898** *Therapist* VIII. 115/1 (*heading*) Administration of urotropine and its effects upon the urine. **1940** [see *hexamethylenetetramine* s.v. *hexa-]. **1967** J. A. Simmons et al. in H. S. Pieser *Crystal Growth* (Jrnl. Physics & Chem. Solids Suppl.) 270/1 In the cases of urotropine and arsenolite the growth forms change when the crystal starts with disturbed growth.

urobilin. Add: [Coined as G. *urobilin* by M. Jaffe 1869, in *Arch. f. Path. Anat. & Physiol.* XLVII. 406.]

urobilinogen (yū°rəbaili·nŏdʒĕn). *Biochem.* [f. prec. + *-ogen.] Any of a group of colourless tetrapyrrole compounds produced by the reduction of bilirubin, esp. by bacterial action in the gut, and forming urobilin upon subsequent oxidation.

1893 A. Eichholz in *Jrnl. Physiol.* XIV. 331 Reserving the name urochromogen for the body isolated from urine, one may perhaps adopt the name urobilinogen for this artificial product. **1947** *Radiology* XLIX. 303/2 This excessive destruction of red cells is also indicated by a maintained increase in excretion of fecal urobilinogen and urinary bilirubin. **1974** R. M. Kirk et al. *Surgery* vi. 97 Jaundice... The kidneys excrete conjugated bilirubin..and urobilinogen..so that the urine is dark.

uroboros (yū°rǫbǫ·rǫs). Also **ouroboros, uroborus.** [ad. Gr. οὐροβόρος, also οὐρη-, devouring its tail (freq. connected with δράκων).] The symbol, usu. in the form of a circle, of a snake (or dragon) eating its tail.

1940 H. G. Baynes *Mythol. of Soul* vi. 221 Thus the *uroborus* symbol represents our psychic continuity with the immemorial past. *Ibid.*, Geber, or Jabir, the most famous of the Arabian alchemists, who lived in Kufa about A.D. 776, used the *uroborus* to represent a closed system or magic ring, denoting the idea of an eternal process. **1953** R. F. C. Hull tr. *Jung's Psychol. & Alchemy* in *Coll. Wks.* XII. iii. v. 357 The alchemical parallel..is the double nature of Mercurius, which shows itself most clearly in the Uroboros, the dragon that devours, fertilizes, begets, and slays itself and brings itself to life again. **1957** N. Frye *Anat. Criticism* 157 Alchemical symbolism takes the ouroborus and the hermaphrodite..in this redemptive context. **1975** Hughes & Brecht *Vicious Circles & Infinity* Fig. 11 The ouroboros, the snake with his tail in his mouth, is the prototype of the vicious circle. ..The 'Endless Snake' depicts an ouroboros who has become one with himself. It has fallen into the mathematical sign for infinity.

urochloralic (yū°:roklōræ·lik), *a.* *Biochem.* [tr. G. *uro-chloralsäure* urochloralic acid (von Mering & Musculus 1875, in *Ber. d. Deut. Chem. Ges.* VIII. 666), f. Uro-¹: see *chloralic* adj. s.v. Chloral.] *urochloralic acid*: a metabolite formed in the body after chloral has been administered (see quot. 1977).

1875 *Jrnl. Chem. Soc.* XXVIII. 1040 The authors propose to give it the provisional name of urochloralic acid. **1882,** etc. [see *glycuronic *a.*]. **1931** *Chem. Abstr.* XXV. 349 (*heading*) The pharmacology of urochloralic acid with special regard to the diuretic action of the sodium salt of this acid. **1977** *Martindale's Extra Pharmacopoeia* (ed. 27) 753/2 Chloral hydrate is..excreted slowly in the urine as trichloroethanol and its glucuronide (urochloralic acid).

urodæum, -eum. (yū°rodī·ɒ̆m). *Zool.* Pl. **-æa, -ea.** [mod.L., irreg. f. Uro-¹ + Gr. ὁδαῖος on or by the way.] The part of the cloaca into which the urinary ducts open.

1888 H. Gadow in *Phil. Trans. R. Soc.* B. CLXXVIII. 28, I propose to designate the typical urino-genital and fæcal chambers the Urodæum and Coprodæum in accordance with Professor E. Ray Lankester's terms Stomodæum and Proctodæum. **1897** Parker & Haswell *Text-bk. Zool.* II. xiii. 368 The cloaca is a large chamber divided into three compartments, the coprodæum.., the urodæum.., and the proctodæum. **1959** W. Andrew *Textbk. Compar. Histol.* xi. 444 Urine formed in the kidney is concentrated in the cloacal chambers (urodeum and coprodeum) where water is absorbed through the walls. **1975** *Nature* 17 Jan. 217/2 At the other end, the proctodeum opens into urodeum leading to the coprodeum.

urogastrone (yū°rogæ·strǒ°n). *Biochem.* [f. Uro-¹ + Gastr(ic *a.* + -one; cf. *enterogastrone* s.v. *entero-.] Any of a number of closely related humoral agents in mammalian urine which retard gastric secretion and motor activity.

1940 J. S. Gray et al. in *Proc. Soc. Exper. Biol. & Med.* XLIII. 228 Until such time as the gastric inhibitory factor is definitely identified, we propose to call it urogastrone. **1946** *Nature* 31 Aug. 305/1 (*heading*) Urease in the gastric mucosa and its increase after a meat diet, soya bean flour diet or urogastrone injections. **1966** Wright & Symmers *Systemic Path.* I. xiv. 494/1 Extracts of normal urine contain a substance, urogastrone, that inhibits the action of pepsin and hydrochloric acid. **1975** *Nature* 25 Sept. 325/1 The two urogastrones were shown to be water-soluble polypeptides of relatively low molecular weight and the difference between them could be shown by various physical techniques including acrylamide gel electrophoresis.

urography (yurǫ·grăfi). *Med.* [f. Uro-¹ + -graphy.] Radiography of the urinary tract.

1925 W. F. Braasch in *Jrnl. Urol.* XIV. 631 The present widespread employment of urography as an aid in the diagnosis of surgical conditions affecting the urinary tract merits careful consideration. **1962** *Lancet* 19 May 1049/1 Excretory urography showed that both kidneys were functioning well. **1977** *Proc. R. Soc. Med.* LXX. 187/1 An accurate and precise diagnosis may be made by early high-dose urography and bulb ureterography.

Hence **urogra·phic** *a.*; also **u·rogram**, a radiogram of the urinary tract.

1925 W. F. Braasch in *Jrnl. Urol.* XIV. 632 The personal equation may be a large factor in interpreting the urogram. *Ibid.* 634 That it is often the cause of obscure abdominal pain..is not borne out by pathologic, urographic or clinical evidence. **1940** P. S. Pelouze *Office Urol.* xii. 520 Every patient in whom a movable kidney is felt or suspected should be subjected to urographic study. **1980** *Nature* 17 Apr. 619/1 The success of transplantation and the cross-sectional area of the kidneys were demonstrated on excretory urograms.

urokinase. (yū°rokəi·nē¹z). *Biochem.* and *Pharm.* [f. Uro-¹ + *kinase.] An enzyme which catalyses the conversion of plasminogen to plasmin and is used in the treatment of blood clots.

1952 G. W. Sobel et al. in *Amer. Jrnl. Physiol.* CLXXI. 768/2 Normal human and dog urine contains..a potent activator of profibrinolysin (plasminogen). The activator ..we have designated urokinase. **1977** *Daily Colonist* (Victoria, B.C.) 23 Oct. 10/5 Urokinase is used in the treatment of blood clots and scientists said it has the potential of preventing 50,000 deaths per year in the United States alone. **1983** Davies & McNicol in *Oxf. Textbk. Med.* II. xix. 132/2 Urokinase is produced naturally by cells in the kidney and is present in human urine, probably to aid maintenance of ureteric patency.

urolagnia (yū°rolæ·gniă). *Psychol.* [f. Uro-¹ + Gr. λαγνεία, act of coition, lust.] Sexual pleasure derived from urination. Hence **urola·gnic** *a.*

1906 H. Ellis *Stud. Psychol. Sex* V. iii. 47 These [functions] are sometimes termed the scatalogical group, with the two subdivisions of urolagnia and coprolagnia. *Ibid.* 62 So far we have been concerned with the urolagnic..variety of scatalogical symbolism. **1940** Hinsie & Shatzky *Psychiatric Dict.* 545/1 *Urolagnia*.., pleasure connected with urine... Some individuals gain gratification in watching others urinate. **1980** *Spectator* 5 July 20 Was it his mother's fault that he suffered from urolagnia?

uronic (yurǫ·nik), *a.* *Chem.* [f. Uro-¹ + -*n*- + -ic 1 b, or f. *glyc)uronic *a.*] *uronic acid*: any derivative of a monosaccharide in which the —CH_2OH group has been oxidized to a —COOH group.

1925 D. R. Nanji et al. in *Jrnl. Soc. Chem. Industry* 253 T/2 Lactose oxidised by bromine in presence of calcium carbonate yields a dicarboxylic acid from which galacturonic acid can be obtained on hydrolysis, whilst sucrose under similar treatment yields glycuronic acid. We shall refer to these compounds as 'uronic' acids. **1946** *Nature* 24 Aug. 271/1 The viscosity of solutions containing the 'recovered' sodium alginate is..higher..than that of solutions of salts of low molecular uronic acids. **1982** N. B. Schwartz in T. M. Devlin *Textbk. Biochem.* viii. 430 The long heteropolysaccharide chains are made up largely of disaccharide repeating units, in which one sugar is a hexosamine and the other a uronic acid.

uropod (yū°·ropǫd). *Zool.* [f. Uro-² + Gr. ποῦς, ποδ- foot.] Orig., any abdominal appendage of a crustacean; now *spec.* each of the sixth and last pair of abdominal appendages of malacostracan crustacea, which together form part of the tail fan in lobsters.

[**1884** A. Sedgwick tr. *Claus's Elem. Text-bk. Zool.* x. 454 The three posterior pairs of abdominal feet (uropoda) are well developed and often much elongated.] **1893** T. R. R. Stebbing *Crustacea* iv. 46 In Amphipoda the fourth and fifth pairs [of abdominal appendages] are more or less adapted for springing, and bear the name of uropods, or tail-feet. This name is also given to the appendages of the twentieth segment whenever they are present. **1933** R. H. Wolcott *Animal Biol.* xliii. 251 The appendages of the last abdominal segment [of the crayfish] are broad and flat and are called uropods. **1974** *Encycl. Brit. Macropædia* V. 547/1 In hermit crabs the uropods are modified into structures that help to hold the abdomen of the crab in the gastropod shell in which it lives.

uroporphyrin (yū°ropǭ·ɹfĭrin). *Biochem.* [ad. G. *urinporphyrin* (H. Fischer 1915, in *Zeitschr. f. physiol. Chem.* XCV. 34): see Uro-¹ and *porphyrin.] Any of a group of porphyrins, occurring esp. in the urine during certain types of porphyria, in which each of the pyrrole rings has one acetate and one propionate side chain.

[**1915** *Chem. Abstr.* IX. 2884 Methyl ester..prepd. by the action of MeOH-HCl upon urinoporphyrin.] **1924** *Ibid.* XVIII. 2720 For uroporphyrin and coproporphyrin ..the spectra are now shown to be quite distinct. **1939,** etc. [see *porphobilinogen]. **1955** [see next]. **1958** *New Biol.* XXV. 75 The African plantain-eaters [*sc.* a family of birds]..are unique in possessing two uroporphyrin pigments, green turacoverdin and the turacin. **1978** A. R. Battersby in Porter & Fitzsimons *Further Perspectives Org. Chem.* (CIBA Symp. No. 53) 37 M. Thompson mixed the protiomethyl ester of uroporphyrin-II with the corresponding deuteriomethyl ester.

uroporphyrinogen (yū°:ropǭɹfĭri·nŏdʒĕn). *Biochem.* [a. G. *uroporphyrinogen* (Fischer & Zerweck 1924, in *Zeitschr. f. physiol. Chem.* CXXXVIII. 242): see prec. and *-ogen.] Any porphyrinogen in which the pyrrole rings have side chains as in a uroporphyrin.

1924 *Chem. Abstr.* XVIII. 3063 (*heading*) Uroporphyrinogen heptamethyl ester and a new conversion of uro- into coproporphyrin. **1955** *Science* 17 June 878/2 Uroporphyrins and very small amounts of coproporphyrins have been recovered after incubation of uroporphyrinogen I or III..with *Chlorella* preparations. **1982** T. M. Devlin *Textbk. Biochem.* xxii. 1077/1 This enzyme acts on the side chains of the uroporphyrinogens to form the coproporphyrinogens.

uropygium. (Earlier example.)

1771 G. White *Let.* in *Selborne* (1789) xxxv. 92 The trains of those magnificent birds [*sc.* peacocks]..growing not from their *uropygium*, but all up their backs.

urs (ū°ɹs). Also **'urs, Urs.** [a. Arab. *'urs*, lit. 'marriage ceremony'.] A ceremony celebrating the anniversary of the death of a Muslim saint. Also *attrib.*

1839 T. J. Newbold *Straits of Malacca* I. v. 252 There is no particular day..on which they congregate to perform the Urs, or pilgrimage. **1885** T. P. Hughes *Dict. Islam* 655/2 *'Urs*.., the ceremonies observed at the anniversary of the death of any celebrated saint or *murshid*. **1974** *Encycl. Brit. Macropædia* IX. 920/1 The Muslim masses also celebrate the death anniversaries of various saints in a ceremony called *'urs* (literally, 'nuptial ceremony'). The saints, far from dying, are believed to reach the zenith of their spiritual life on this occasion. **1975** *Bangladesh Observer* 18 July 8/1 Five hundred Muslims from Bangladesh and 150 from Pakistan were among about one lakh devouts who attended the urs. **1979** *Morning News* (Karachi) 24 May 3/4 An exquisite silver embroidered velvet green 'Chadar'..will be flown..from Lahore for laying at the Mazar of Hazrat Khwaja Moeenuddin Chishti during the Urs celebrations starting at Ajmer Sharif from May 28.

‖ **Urschleim** (ūə·rʃləim). [Ger., f. *ur-* Ur- + *schleim* slime.] In early biology, the original form of life; protoplasm.

1921 G. B. Shaw *Back to Methuselah* p. xxix, Lorenz Oken..defined the original substance from which all forms of life have developed as protoplasm, or, as he called it, primitive slime (*Urschleim*). **1958** *Times Lit. Suppl.* 15 Aug. 454/4 When first it was investigated a belief arose in a plain of *Urschleim* on the ocean bed, a protoplasmic half-living matter, in which the process of creation was still at work.

ursid (ɜ̄·ɹsid), *a.* and *sb.* [a. mod.L. family name *Ursidæ*, f. L. *ŭrsus* bear, adopted as a generic name by Linnæus (*Systema Naturæ* (1735)): see -id³.] **A.** *adj.* Of, pertaining, or belonging to the family Ursidæ. **B.** *sb.* A mammal of this family.

1921, 1973 [see *procyonid *sb.* and *a.*]. **1976** *Nature* 22 Apr. 700/1 They have revealed..the ursid nature of the giant panda. **1979** *Ibid.* 12 July 138/1 Groups thus added to the Sahabi faunal assemblage include primates, ursids, [etc.].

ursinia (ɜ̄ɹsi·niă). [mod.L. (J. Gaertner *De Fructibus et Seminibus Plantarum* (1791) II. 462), f. the name of John *Ursinus* (1608–66), German botanist + -ia¹.] An annual or perennial aromatic herb of the genus of this name, belonging to the family Compositæ, native to South Africa, and bearing orange, yellow, or white flowers.

1928 *Gardeners' Chron.* 7 July 9/1 This Ursinia remains open until dark. **1959** *Listener* 26 Mar. 572/1 For the cool greenhouse ursinias make lovely pot plants. **1976** *Hortus Third* (L. H. Bailey Hortorium) 1140/1 The annual ursinias of the flower garden are grown from seeds planted indoors or directly in the open.

‖ **Ursprache** (ū°·ɹʃprāχə). Also **ursprache.** Pl. **-en.** [Ger., f. *ur-* Ur- + *sprache* speech: see note s.v. Ur-.] = *proto-language.

1908 T. G. Tucker *Introd. Nat. Hist. Lang.* ix. 163 The Semitic Ursprache (or common parent). **1922** O. Jespersen *Language* iii. 85 The basis of the whole was not an artificially constructed nebulous *ursprache*, but the familiar forms and words of an historical language. **1937** *Science & Society* I. 158 The term 'Indo-European school' is..applied..to all scholars using the general methods employed in the study of these languages (for instance, the building of *Ursprachen* on the basis of detailed comparisons in surviving languages). **1950** [see *Urheimat]. **1964** R. H. Robins *Gen. Linguistics* viii. 319 The 'starred forms'..give reasons for assuming the existence of an earlier relatively unitary stage of I[ndo]-E[uropean] (or any other family established in this way), the so-called Ursprache, or parent language; but they do not constitute it, or represent word forms in it or any other language. The evidence for the earlier existence of an Ursprache is not the same as bits of the Ursprache itself. *a* **1975** A. Toynbee *Greeks & their Heritages* (1981) 122 Greek, Latin and Sanskrit..are all close enough in structure to their common progenitor, the *Ursprache*, to have retained the *Ursprache*'s principal characteristic.

‖ **Urtext** (ū°·ɹtekst). Also **urtext.** [Ger., f. *ur-* Ur- + *text* Text *sb.*¹.] An original text; the earliest version. Also *attrib.* or as *adj.*

1932 *Times Lit. Suppl.* 14 July 511/3 In these volumes ..we have the nearest thing possible in Chopin's case to an *Urtext*. **1959** *Cambr. Rev.* 6 June 598/2 Authoritative editions allegedly based on urtexts. **1963** S. Weintraub *Private Shaw & Public Shaw* iv. 119 The earlier version still retains advocates, because of its more complete, ur-text quality, and the comfortable feeling that no Procrustean games were played with its vocabulary and sentence structure. **1974** *Early Music* Oct. 259/1 The edition is *urtext*, with prefatory staves, showing the original clefs and signatures. **1982** *Times* 2 Apr. 14/2 An *urtext* edition of the 21 Schubert piano sonatas. **1983** *London Rev. Bks* 7–20 July 21/4 Elaborate versions often point back to the gospel of Mark as a kind of cryptic *Urtext*.

urucuri. (Earlier example.)

1860 [see *inajá].

Uruguayan (yū°rŭgwai·ăn, ū°r-, -gwē¹-), *a.* and *sb.* [f. *Uruguay*, the name of a republic in eastern S. America + -an.] **A.** *adj.* Of or belonging to Uruguay or its inhabitants; produced in or characteristic of Uruguay. **B.** *sb.* A native or inhabitant of Uruguay.

1869 *Jrnl. R. Geogr. Soc.* XXXIX. 340 A piece of solid gold weighing over 1 lb..has been taken out of the quartz on the surface, and is in the hands of the Uruguayan minister of war. **1889** *Cent. Dict.*, *Uruguayan*..*a.* and *n.* **1912** *20th Cent. Impr. of Uruguay* 225/2 The Uruguayan soil contains the necessary chemical components for the general run of agricultural farming. **1923** R. Seton *Memories Many Years* 113 There I met the Uruguayan idling about. **1972** *Daily Tel.* 11 Apr. 13/8 While duelling is perfectly within Uruguayan law, there are one or two ways in which the authorities seek to discourage it. **1976** *Nature* 22 July 243/2 The Uruguayans in their zeal to stamp out the Tupamaro guerillas seem to have 'processed' an extraordinarily large proportion of the population.

Uruk (u·rŭk). The Sumerian name of an ancient city in southern Iraq (mod. Warka), used *attrib.* in *Uruk period* to designate a phase in Sumerian-Akkadian culture.

1932 *Antiquity* VI. 198 The Jemdet Nasr culture is preceded at Uruk (Warka)..by an older urban civilization in which writing was being developed, that is termed the URUK I culture. *Ibid.* 522 (Index), Uruk culture. **1934** V. G. Childe *New Light Most Anc. East* vi. 146 The next phase is..best known as a result of the excavations so meticulously conducted and so promptly published by the Germans at Erech. It is accordingly designated the Uruk period (Uruk is just the Babylonian form of the Hebrew Erech). **1947** D. Diringer *Alphabet* i. 41 The earliest extant written cuneiform documents, consisting of over one thousand tablets and fragments, discovered mainly at Uruk or Warka, the Biblical Erech, and belonging to the 'Uruk Period' of the Mesopotamian pre-dynastic period, are couched in a crude pictographic script and probably Sumerian language. **1958** L. Cottrell *Anvil of Civilisation* vii. 94 Phase Two, the so-called 'Uruk' period, [was discovered] by the Germans at Warka.

‖ **urushi** (urŭ·ʃi). Also 8 urusi. [Jap.] The Japanese lacquer tree, *Rhus verniciflua*; also, Japan lacquer, used for coating Japan ware.

1727 [see *Japan varnish (tree)*]. **1881** *Encycl. Brit.* XIII. 590/2 The varnish tree is of several kinds, and the *Urushi* tree growing in Japan..supplies, it is said, a finer gum than any other of the same species. **1909** *Cent. Dict.* Suppl., *Urushi*,..Japanese lacquer; varnish. **1965** W. Swaan *Jap. Lantern* x. 120 Lacquer is made from the sap of the *urushi* or lac tree. **1983** *New Scientist* 21 Apr. 149/1 Far Eastern lacquer, or *urushi*, is a poisonous exudate from the stems of the lacquer tree, or varnish sumach.., which can be tapped like those of rubber trees.

urushiol (ŭrŭ·ʃiǫl). *Chem.* [f. Jap. *urushi* (see prec.) + -ol.] An oily phenolic liquid causing skin irritation which is present in various plants and is the main constituent of the lacquer obtained from the Japanese lacquer tree.

1908 *Chem. Abstr.* II. 2307 According to the author [*sc.* K. Miyama] the main constituent of the lac is a polyhydrophenol with unsaturated radicles in the side chains, the name Urushiol is proposed for it. **1945** *Jrnl. Amer. Med. Assoc.* 7 Apr. 920/2 It is highly probable that urushiol is the irritant common to ivy, sumac and the lac trees. **1974** J. E. Underhill *Wild Berries Pacific Northwest* 95 Sumach is closely related to Poison Ivy.. and Poison Sumach... It lacks the poisonous oil, urushiol, that makes them such a hazard.

us, *pers.* and *refl. pron.* Add: **I. 1. c.** *one of us*: see *one *numeral a.*, *pron.*, etc. B. 15 b.

f. Persons like ourselves; ordinary citizens, as opp. to those in authority.

1948 *Observer* 17 Oct. 5/2 The 'whistle stops'..break down the difference between 'them' and 'us'. **1957,** etc. [see *them *pers. pron.* 1 e]. **1962,** etc. [see *them *pers. pron.* 3 c]. **1984** *Times Lit. Suppl.* 26 Oct. 1206/1 The Chinese awareness of 'us and them' is intense, pervasive and hard to break down.

2. b. For † Obs. read '*Obs. exc. dial.*' and add later examples.

1861 O. W. Norton *Army Lett.* (1903) 35 We..then took possession of a stack of wheat and made us good beds and slept well. **1907** S. E. White *Arizona Nights* x. 166 We got us timbers, and made a scow. **1928** 'M. Chapman' *Happy Mountain* 42 We'll make us a heap o' cash money. **1942** *Sat. Even. Post* 22 Aug. 42/3 Le's go and wake us up a preacher.

5. d. Also as *adj.*, suited to or representative of our tastes, personality, etc.; appropriate for us. Usu. *predic.*

1940 M. Dickens *Mariana* viii. 312 'How could you know I'd like something like this?'..'It just looked absolutely us, somehow.'

usage, *sb.* Add: **10.** *attrib.*, as (sense 8) *usage guide, label, labelling.*

1972 R. D. Walshe in G. W. Turner *Good Austral. Eng.* xi. 241 There is a need for up-dating. This is true of most of the 'usage guides' which editorial offices and printeries use in order to impose a consistency of style. **1980** *Amer. Speech* LV. 134 Readers of these usage guides too often take them as gospel truth. **1967** F. Christensen in P. B. Gove *Role of Dict.* 24/2 The new dictionary has dropped the usage label *colloq*...altogether. **1981** *Dictionaries* II.–III. 75 Usage-label treatment is just as important to the composition student as treatment of synonyms. **1977** *Computers & Humanities* XI. 89/2 In this kind of usage labeling, 'label' is not to be taken literally. **1982** *Papers Dict. Soc. N. Amer.* 1977 66 The first of these anxieties gives weight to the need for rather particular status-labelling—and this term (used in Webster's) also seems..preferable to 'usage-labelling'.

Usbeck, Usbeg, etc., varr. *Uzbek.

use, *sb.* Add: **IV. 17.** (Examples of *for the use of* with the obj. of *of* preposed.)

1959 [see *Mallaby-Deeley]. **1971** D. Francis *Bonecrack* iv. 44 There was..an armchair of sorts, visitors for the use of.

21. b. (Earlier example.)

1872 *Harper's Mag.* June 158/2 He was an obstinate fellow..and moreover, he 'had no use for' the defendant any way.

V. 22. use immunity *U.S. Law* (see quot. 1972); **use-life,** useful life.

1972 *New Yorker* 25 Mar. 86/3 The suggested revision, known as 'use immunity', would prevent anyone who was compelled to testify from being prosecuted on the basis of that testimony. **1976** *Ibid.* 19 Apr. 42/3 Under use-immunity law, however, people who were compelled to testify could later be prosecuted as long as the government did not base its case against them directly or indirectly, on their own testimony. **1950** *Chambers's Jrnl.* Mar. 191/2 It is now reported that the magnesium type of dry-cell has a 'use-life' of about thirty hours. **1972** *Computers & Humanities* VII. 87 Interactive systems on today's scale are very recent; for the program designer there are obstacles of rapid change, little standardization, and relatively high development expenses in relation to the probable use-life of the programs.

use, *v.* Add: **II. 11.** (Further examples.) Now *esp.*, to take or consume (an alcoholic drink, a narcotic drug) regularly or habitually.

1902 Kipling *Traffics & Discoveries* (1904) 15, I don't use rum as a rule, but I did then, because I needed it. **1906** W. Churchill *Coniston* I. ix. 104 Unlike Jethro, he 'used' tobacco. **1921** J. Buchan *Path of King* xiv. 276 It's curious that a man who don't use tobacco or whisky should be such mighty good company. **1929** D. Hammett *Dain Curse* xiv. 149 He..picked up the morphine paper... 'What do you suppose this is doing here?' he asked. 'She uses it.' **1942** J. D. Carr *Seat of Scornful* iv. 43 'Not that I've got any objections to 'em,' Mr. Morell assured him, alluding to the tobacco and the spirits. 'Just don't use 'em.' **1962** H. Burn *Drugs, Med. & Man* x. 106 The best known drugs of addiction are morphine, heroin and cocaine. Somewhat less well known are cannabis..and pethidine. These are the main drugs, other than alcohol, which are used by addicts. **1982** J. Wainwright *Anatomy of Riot* 31 Okay, a little hash here an' there. He sold it. Maybe he used it. Okay, he used it.

14*. *to be able to use*, to be in need of, to be in a position to benefit from, to want. *colloq.*

1956 'N. Shute' *Beyond Black Stump* viii. 217 But I could use a river, and the sight of snow on a mountain. **1958** V. Scott *Savage Affair* ix. 152 Listen..they said they might see their way to an advance. Two-fifty apiece. You could use two hundred and fifty dollars, couldn't you? **1961** R. Godden *China Court* 258 'I could use a gin,' said Bella. **1976** *Ulverston* (Cumbria) *News* 3 Dec. 1/6 The tarn is the most beautiful part of the village and The Landings can use some cleaning up. **1977** *Oxford Diocesan Mag.* July 18/1 We could have used more time to explore this avenue.

V. 20. Add to note: and *colloq.* in *did* (*not*) *use* (or *used*) *to*; see also *usen't, *useter. (Further examples.) *used to could*: see *can *v.*[1] A. 7.

1857 C. M. Yonge in *Monthly Packet* Jan. 34 'Things didn't use to be so stupid when Ned was there!' sobbed Gilbert. **1873** —— *Pillars of House* II. xvi. 105 Did Alda use to be nice, or is it love? **1925** S. Lewis *Arrowsmith* xviii. 192 Didn't we used to have fun. **1927** E. Hemingway *Men without Women* 154 He certainly did used to make the fellows he fought hate boxing. **1935** E. Farjeon *Nursery in Nineties* III. i. 124 Mama, did you use to be a flirt? **1963** V. Nabokov *Gift* ii. 117 And now I continually ask myself what did he use to think about in the solitary night. **1974** *Radio Times* 28 Feb. 25, I suppose I did use to be a prophet of doom. **1983** *Listener* 10 Feb. 31 (*heading*) Adrenalin sports are big. TV didn't used to be one.

24. To take drugs. *slang.*

1953 W. Burroughs *Junkie* (1964) x. 104 The reason it is practically impossible to stop using and cure yourself is that the sickness lasts five to eight days. Twelve hours of it would be easy, twenty-four possible, but five to eight days is too long. **1960** C. Cooper *Scene* 15 Why don't you bust a cap with me? It's choice. I used this morning and I'm still nice. **1962** 'K. Orvis' *Damned & Destroyed* xi. 71 Almost twenty-four hours..since I've had a fix... Are you the only one?..You forget I use, too.

used, *ppl. a.* Add: **I. 1. b.** *spec.* (*a*) (esp. of a vehicle) = Second hand, second-hand B. 2; also in hyphenated attrib. phr.; (*b*) of paper currency: not in mint condition.

1931 *Punch* 3 June 589/2 It seems there is little demand nowadays for used cars. People find they can get all the walking exercise they need without buying a second-hand car. **1932** L. N. Wright *Links of Old Family Tree* x. 184 She was tempted to go into the business of dealing in used furniture. **1938** [see *repossession 1]. **1955** F. O'Connor *Wise Blood* iv. 67 By six-thirty, he was down town, looking for used-car lots. **1960** I. Jefferies *Dignity & Purity* i. 12 He gave me five hundred quid in used ones. **1966** J. B. Priestley *Salt is Leaving* i. 12 Albert is doing quite nicely in the used-car business. **1970** 'D. Halliday' *Dolly & Cookie Bird* vii. 109 Nine thousand used dollar bills.

II. 5. (Earlier example.)

1839 H. McLeod *Let.* 18 Jan. in *Papers M. B. Lamar* (1922) II. 423, I will come down in a few days with Genl Rusk, but I am really so 'used up' now, that I cannot undergo the fatigue.

u·sed-to-be, *sb. U.S.* [cf. Use *v.* 20 *Comb.*] A person whose time of popularity or efficiency is past; also = *ex *sb.*[1]

1853 [see *has-been *sb.*]. **1911** [see *never-was* s.v. *never *adv.* 8]. **1942** L. Hughes *Shakespeare in Harlem* 50, I want to tell you 'bout that woman, My used-to-be—She was de meanest woman I ever did see.

useful, *a.* and *sb.* Add: **1. c.** Applied to an odd-job man. *Austral. colloq.*

1866 R. Henning *Let.* 16 May (1966) 219 There are three men employed about the place [*sc.* a logging business]. The bullock-driver, the punt-man and a 'generally useful' man. **1900** H. Lawson *Middleton's Peter* in *Stories* (1964) 1st Ser. 293 There were two rooms..attached to the stables. One was occupied by a man who was 'generally useful'.

2. c. *useful load* (Aeronaut.), the difference between the maximum permitted weight of an aircraft and its weight when empty, including cargo, passengers, crew, fuel and (with some writers) fixed equipment such as radios also; similarly *useful weight.*

1909 A. Berget *Conquest of Air* II. iv. 205 A 'useful weight' in the form of fuel and oil to the extent of 80 kilogrammes. **1914** *Sphere* 7 Mar. 298/2 The most remarkable of the new aeroplanes is the Sikorsky... It can carry a useful load of over a ton. **1978** D. B. Thurston *Design for Flying* ix. 112 If the airplane is intended for the ..private market, thin skin could be used and the weight saved converted to useful load.

d. Of a performer or performance: reasonably effective, fairly successful.

1955 *Amer. Speech* XXX. 23 A horse with a seemingly excellent racing background will be described as good; and a horse the record of which..would seem good will be described as useful. **1959** *Listener* 23 July 129/1 He..played a useful game of cricket..with a local league side. **1971** N. Stacey *Who Cares?* i. 19, I had been a useful school sportsman and got into the first eleven at most sports at Dartmouth.

e. Of a woman's dress; practical, unostentatious; suitable for a variety of occasions.

1963 *Observer* 3 Nov. 33/1 In the dress trade, 'a useful little dress' means one with no distinguishing characteristics; 'romantic' means 'cleft to the waist'. **1968** M. Jones *Survivor* i. 22 The dinner was arranged... She had dressed with restraint in a 'useful' black dress.

B. *sb.* **1.** See sense 2 b in Dict.

2. An odd-job person. *Austral. colloq.*

1898 A. Joyce *Homestead Hist.* (1969) 41 Our friends had met with a trained carpenter in town, whom with his wife they had hired for £20 a year, the man as general useful,..and his wife as cook. **1935** K. Tennant *Tiburon* 37 Roman stepped out of the room next to the laundry of O'Brien's Hotel, where he was barman, yardman and general useful. *a* **1963** J. Fountain in 'B. James' *Austral. Short Stories* (1963) 277 Every boy..knew the..circumstances of..Maggie's affair with the useful from the hotel.

usen't, colloq. shortening of *used not.*

c **1863** T. Taylor in M. R. Booth *Eng. Plays of 19th Cent.* (1969) II. 96, I usen't to mind unkind looks and words much once. **1907** G. B. Shaw *Major Barbara* III. 255 That is a new accomplishment of Andrew's, by the way. He usent to drink. **1929** 'H. H. Richardson' *Ultima Thule* III. v. 279 Usen't Richard to say that it was etiquette in the profession to treat a patient's relatives.. as so many cretins?

user[1]. Add: **1. a.** *spec.* A person or organization that makes use of a computer. Freq. *attrib.* and *Comb.* (cf. sense 5 below).

1967 [see *user-assigned* adj., sense 5 below]. **1973** C. W. Gear *Introd. Computer Sci.* iv. 151 The software makes it possible for a user to prepare procedures and have them executed by the computer with a minimum of effort. **1984** *Which Micro?* Dec. 8/1 Micro breakdown..drives users to despair.

b. A person who takes narcotic, etc., drugs. orig. *U.S.*

1935 A. J. Pollock *Underworld Speaks* 129/2 *User*, a person addicted to any of the poisonous habit forming drugs; a hop-head; dope fiend. **1953** W. Burroughs *Junkie* (1964) vi. 58 The owner knew Tony had been a user and had told him to stay off the stuff or get another job. **1969** *Guardian* 3 Dec. 9/1 She was taking six grains of heroin a day... She had been a user for about six months. **1975** H. White *Raincoast Chron.* (1976) 144/2 There had always been users around in the shadowy back streets. **1983** *Easyriders* Feb. 111/4 Harley man, 29 ..seeks lady 5'7" or under for friend, lover, and partner... No boozers or heavy users.

5. *attrib.* and *Comb.*, as *user benefit, charge, cost, fee, group*; *user-assigned, -supplied* adjs.; *user-processing*; **user-definable** *a. Computers*, having a function or meaning that can be specified and varied by a user; so **user-defined** *a.*; **user-friendly** *a. Computers*, easy to use; designed with the needs of users in mind; also *transf.*; hence **user-friendliness**; **user interface,** the means by which a person is enabled to use a computer; **user-orientated, -oriented** *adjs.*, designed with the user's convenience given priority; **user-programmable** *a. Computers*, capable of being programmed or assigned a function by the user.

1967 Cox & Grose *Organization & Handling Bibl. Rec. by Computer* 84 The valves dcn, date, day, etc. are the user-assigned names for the bibliographic data fields. **1972** *Computers & Humanities* VI. 303 We will study the cost-effectiveness of stripping away various features.. with least prejudice to user-benefit. **1945** *Sun* (Baltimore) 29 Sept. 3/2 John J. Pelley, president of the Association of American Railroads, urged..that 'user charges' be levied against highway, waterway and airway carriers. **1976** *Times* 29 Nov. 12/2 It is possible to raise only one cheer for user charges as a means of avoiding the dilemma of cutting public services or increasing taxes. **1936** J. M. Keynes *Gen. Theory Employment* vi. 70 We have defined the user cost as the reduction in the value of the equipment due to using it as compared with not using it. **1969** D. C. Hague *Managerial Economics* (1971) II. v. 103 The cost of servicing and repairs is part of what economists call 'user cost', the cost of using the car (or any other asset) instead of leaving it idle. **1972** *IBM Technical Disclosure Bull.* XIV. 3553 (*heading*) Representation of tree data structures as matrices suitable for user-definable traversing. **1983** *MicroComputer Printout* Sept. 69/1 A programmable character set (also called user-definable) may let you design your own characters—mathematical symbols, or foreign alphabets. **1985** *Personal Computer World* Feb. 133/1 The number of user-definable characters that you can enter at the Model 4 Keyboard in this way is limited to 20. **1968** *Simulation* XI. 304/2 User-defined functions. **1984** *Which?* July 302/1 In rating graphics..we looked at..the range of graphics features provided with the computer-graphics characters, ..sprites,..user-defined graphics (shapes you can set up yourself) and so on. **1984** *Personal Software* Winter 26/2 The program will allow you to have up to fifty user defined commands. **1980** *Tucson* (Arizona) *Star* 4 Mar., He sees no reason why 'user fees'..should not be charged for libraries. **1984** *Gainesville* (Florida) *Sun* 28 Mar. 12A/5 It is clear that (Proposition 1) fails to allow for increased demand on government-owned utilities and other user-fee services resulting from population increases and other causes. **1979** *Interfaces* May 72 User friendliness is a term coined by Harlan Crowder to represent the inherent ease (or lack of ease) which is encountered when running a computer system. **1982** *Daily Tel.* 7 Dec. 2 (Advt.), The system has been very well received—a credit to the HP 3000's user-friendliness. **1977** Birss & Yeh *Set Theoretic Data Structures* (*STDS*) 31 STDS-I does not provide the user with a sufficiently 'user-friendly' interface to allow noncomputer scientists to easily work with a data base. **1982** *New Scientist* 30 Sept. 931/1 The program it works from is 'user-friendly' insofar as the commands are based on initials such as CV for 'centre vertically' and FD for 'forms design'. **1984** *Which Micro?* Dec. 3 (Advt.), Every computer manufacturer now claims its products are 'user friendly'. **1984** *Listener* 13 Dec. 38/1 No TV show (not even the news) could close without reference to this user-friendly family of dolls. **1972** *Accountant* 26 Oct. 518/2 To satisfy the information needs of entrepreneurs, investors, and other outside user-groups is to guarantee a full and active future for the accountant. **1983** *Lebende Sprachen* XXVIII. 48/1 Language mediators represent one of the largest user groups of terminology. **1968** *Proc. Internat. Fed. Information Processing* (1969) I. II. 570 Three levels of user interface are defined for the information retrieval language. **1983** *Byte* Feb. 36/1 The company's first task was to devise a new user interface—that is, a new and better way for humans to interact with the computer. **1985** *Personal Computer World* Feb. 137/2 The operating system, GSX and GEM Services bind together to form the programmer interface, but they do not provide the user interface directly. **1969** *Computer Aided Design* Winter 5/2 The plea for more languages..which are specific to particular problems, and in this sense 'user-orientated'. **1978** *Hi-Fi News* Sept. 218 (Advt.), There's a low distortion equaliser, tape duplication switch..plus a host of user-orientated features. **1964** *Communications Assoc. Computing Machinery* VII. 290/1 (*heading*) An experiment in a user-oriented computer system. **1979** *Dictionaries* I. 110 User-oriented service programs..could put the student in immediate contact with a wealth of lexicographic information. **1958** *Newnes Compl. Amat. Photogr.* 283 Kodak Ltd. propose eventually to market processing kits..for the user-processing of Kodacolor Film. **1976** *Scientific & Technical Aerospace Rep.* XIV. XXII. 2927/1 To fulfil all desired capabilities, a user programmable communication device is required. **1983** *What's New in Computing* Jan. 60/1 The full alphanumeric and function key keypad..includes..eight user-program-

mable keys. **1969** *Computers & Humanities* III. 130 It contains the ability to perform, *inter alia*, the following tasks:.. Generate reports of extracted data according to user-supplied format.

useter (yū·stəɹ). Also **useta, uster.** Repr. an informal or uneducated pronunc. of *used to* (see Use *v.* 20 in Dict. and Suppl.).

1890 *Dialect Notes* I. 69 When it rains and wets our old rooster, He don't look like he useter. **1898** J. D. Brayshaw *Slum Silhouettes* 215 I've just seed Liz Dukeson,.. her that Cocky uster cart abaht. **1921** E. O'Neill *Emperor Jones* v. 185 Is you sellin' me like dey uster befo' de war? **1935** in Z. N. Hurston *Mules & Men* (1970) I. viii. 172 Naw, dat's Uncle Yistiddy, he's a useter-be! **1937** D. L. Sayers *Busman's Honeymoon* i. 57 You mind the little cottage down by the river..where old Blunt useter live? **1952** M. Steen *Phoenix Rising* vi. 134 Dey say she gone to Harlem... Useta teach school. **1978** L. Block *Burglar in Closet* (1980) i. 8, I useta beg her to keep some of that stuff in a safe-deposit box. **1982** A. Taylor *Caroline Minuscule* x. 79 'He said it'd be all right for me to doss down here... He useter—' 'Liar. He couldn't have done.'

ush, *v.*[2] **2.** (Earlier example.)

1890 *Harper's Mag.* Dec. 160/2 The six gentlemanly cow-boys..swore that whoever should prove to be the lucky man, the others would ush for him at the ceremony.

ushabti (uʃæ·bti). *Egyptology.* Also **ushebte, ushebti.** Pl. **ushabtis, -iu.** [a. Egyptian *wšbty* answerer.] A figurine of a deceased person, made of faience, stone, wood, etc., and placed in an ancient Egyptian tomb to act as a substitute for the dead person in any work he might be called upon to do in the afterlife. Also *ushabti-figure.* See also *shabti, *shawabti.

1885 E. A. W. Budge *Dwellers on Nile* viii. 164 The figures placed with the dead were called *ushabtiu*, and were inscribed with the name of the deceased and the sixth chapter of the Book of the Dead. **1894** H. M. Tirard tr. *Erman's Life in Anc. Egypt* xiii. 317 Besides these models of servants and of sailors which replaced the earthly domestics of the departed, there were many other figures of a different kind... These are the so-called funerary statuettes, or as they are termed in Egyptian the Ushebte, *i.e.* the *answerers.* **1910** *Encycl. Brit.* V. 709/1 This polychrome faience was also now [*sc.* in the Eighteenth Dynasty] used for the *ushabti* figures which were placed in the tombs; hitherto they had been made exclusively of stone or wood..the plain blue and black of the ordinary vases was adopted. The *ushabtis* of King Seti I...are fine specimens of this type. **1925** *Glasgow Herald* 27 July 8 The ushabtiu figures..were deposited either in the coffin itself or near to it. **1933** A. W. Shorter *Everyday Life in Anc. Egypt* viii. 175 The actual meaning of the word *ushabti* is not known for certain, one explanation being that the word is derived from the verb *usheb* 'to answer', and therefore means 'answerer'—i.e. one who comes at the summons of the deceased. *Ibid.* 176 The *ushabti*..represents the deceased himself in mummy form.. The *ushabti*-figures of the Eighteenth Dynasty.. are made of wood or stone... The manufacture of faïence *ushabtiu*, however,..became popular. **1957** L. Durrell *Justine* I. 57 The air was all at once full of..tear-bottles, Ushabti, and Sèvres. **1961** A. H. Gardiner *Egypt of Pharaohs* iii. 32 The tombs of the well-to-do often yield hundreds of small statuettes mostly of faience or wood now generally known by their later name of *Ushabti* figures or 'answerers'..; the earlier writing Shawabti is of doubtful meaning. **1979** H. Evans *Mystery of Pyramids* ii. 62/2 To make this everlasting life as comfortable as possible, the spirit called upon the magical aids and implements provided in its earthly tomb. These facilities even included model figures, *ushabtis*, that could act as substitutes if the gods called for workers in the fields.

Ushak (u·ʃak). [ad. Turk. *Uşak.*] Also **Ouchak, Oushak.** The name of a town in W. Turkey used *attrib.* or *ellipt.* to denote a type of antique rug made there.

1901 J. K. Mumford *Oriental Rugs* iv. 35 For rugs of the heavier quality, such as the ponderous Oushaks and Anatolians, the sheep of the Asia Minor plains produce a wool that is adequate in length, and..soft to the touch. **1905** M. C. Ripley *Oriental Rug Bk.* xv. 115 In Turkish carpets of large size many styles are grouped under the trade name 'Ouchak' in which modern methods are observable. **1922** Kendrick & Tattersall *Hand-Woven Carpets, Oriental & European* I. II. i. 100 The carpets of Asia Minor, of the kind called Ushak, which date from the fifteenth to the seventeenth centuries, are invariably woven with the Ghiordes knot. **1931** A. U. Dilley *Oriental Rugs & Carpets* Pl. 45 (*caption*) Ushak prayer rug, late sixteenth-century. **1952** B. Miall tr. *Jacoby's How to know Oriental Carpets & Rugs* 106 From the very first the Ouchaks differed very perceptibly, in colour and design, from the Persian carpets. **1975** 'E. Lathen' *By Hook or by Crook* xvi. 152 We should list some of our best rugs... Or just mention the Ushak. **1977** *Times* 12 Feb. 14/6 (*heading*) Rare Ushak carpet is sold for £42,000. **1984** *Times* 1 June 12 (*caption*) Charles Sternberg with the large Ushak (or Mahal) for which he paid £48,600.

usher, *v.* Add: **1. d.** Also, to act as an usher in a cinema. *U.S.*

1973 *Publishers Weekly* 27 Aug. 243/1 A 13-year-old boy who ushers in a movie house. **1980** M. Gordon *Company of Women* (1981) I. i. 26 It was teen-agers who flocked to see that kind of movie. Mary Rose had to usher at those movies now.

ushere·tte. [f. Usher *sb.* + -ette.] A female usher in a cinema or theatre.

1925 *College Humor* Aug. 66/2 The obese usherette in the toney movie house who has to wear a different fancy dress costume every week. **1926** *Bulletin* 27 Feb. 2 Thirty beautiful girls..will receive visitors to..the Plaza, at its opening on Tuesday. They will be called 'usherettes'. **1948** *Times* 24 Feb. 7/4 Will the usherette hush the little fellow crying in the stalls. **1960** M. Spark *Ballad Peckham Rye* vii. 136 She's an usherette at the Regal from six-thirty to ten-thirty. **1980** R. Butler *Blood-Red Sun at Noon* (1981) II. i. 129 Nothing in her career as theatre usherette, conjuror's assistant and night-club hostess had prepared her for this kind of life.

ushership. 2. (Earlier example.)

1788 W. Cowper *Let.* 30 Nov. (1982) III. 233, I was under his ushership at Westminster.

usitative, *a.* For *rare*[−1] read *rare* and add later example.

1939 [see *consuetudinal *a.* and *sb.*].

usnea. (Earlier *attrib.* example.)

1854 Thoreau *Walden* 137 The single spruce stands hung with usnea lichens.

u·s-ness. [-ness.] The fact of being or feeling united in mind, feeling, or purpose; the fact of forming a unity.

1927 *Glasgow Herald* 7 Mar. 6 In a magazine story... 'It's Us-ness that makes love,' the heroine of the tale is made to say to the hero, 'it's being two of you together against the world.' **1958** *Listener* 27 Nov. 874/1 One of the dangers of 'The Method' is that it tends to reduce everything to us-ness, if I may so put it. **1966** P. J. Kavanagh *Perfect Stranger* vii. 77 There was a pleasant feeling of 'us-ness' in the unit. Were the prisoners 'us' or had they comfortably become 'them'? **1976** *Christian* III. 167 This has the effect of buttressing and reinforcing the me-ness of me (insistently individual) as part of the intensified us-ness of us (compellingly corporate).

‖ **usque ad nauseam** (ʊ·skwe æd n-ǭziæm). Also **ad nauseum usque.** [L., = 'right up to sickness'.] = *ad nauseam.

1616, 1693 Ad nauseam usque [see *ad nauseam]. **1819** T. Moore *Tom Crib's Memorial* p. xxxi, That person has already been exhibited, perhaps, 'usque *ad nauseam*', before the Public.

ussingite (ʊ·siŋəit). *Min.* [ad. Da. *ussingit* (O. B. Bøggild 1913, in *Meddelelser om Grønland* (1914) LI. 103), f. the name of N. V. *Ussing* (1864–1911), Danish mineralogist: see -ite[1].] A triclinic basic sodium aluminosilicate, $Na_2AlSi_3O_8OH$, found as reddish-violet crystals.

1914 *Chem. Abstr.* VIII. 3170 (*heading*) Ussingite, a new mineral from Kangerdluarsuk, Greenland. **1975** *Jrnl. Geol.* LXXXIII. 763/1 In both localities, ussingite is late-stage in character, occurring as a pegmatitic mineral in Lovozero [in the Kola Peninsula, USSR]..., and as a hydrothermal vein mineral, in the Ilimaussaq intrusion [in Greenland].

ustad (u·stād). Also **Ustad.** [Pers., Urdu.] A master, esp. of music.

1903 M. A. Stein *Sand-Buried Ruins Khotan* viii 125 The few 'ustads' (masters) who cared to attend to my orders. **1967** V. Mehta *Portrait of India* 40 He is Hindustani music's unrivalled *ustad* (Urdu for 'master'). Actually, '*ustad*' is a sort of Muslim counterpart of '*guru*'. **1971** *Femina* (Bombay) 16 Apr. 8/3 She was able to become the pupil of the well-known Ustad Atah Khan of Patiala.

Ustashi (ustā·ʃi), *sb. pl.* (also taken as *sing.* with pl. *-s*). Also **-chi, -ci, -sha, -ša, -še, -si.** [a. Serbo-Croatian *Ustaše* pl., *Ustaša* sing., insurgent rebel.] (Members of) a party and separist movement of Croatians; the soldiers and supporters of the autonomous Croatian régime between 1941 and 1944: as *sing.*, a member or supporter of the Croatian separatist movement. Also *attrib.*

1932 *Times* 8 Oct. 11/6 The Ustasi move in small groups from village to village and organize their adherents, train them in shooting, and disappear as suddenly as they came. **1943** C. Hollingworth *German just behind Me* xiii. 258 There are also the terrorist societies such as the Croat *Ustashi*. **1946** A. M. Hyamson *Dict. Internat. Affairs* 334 *Ustaci, The*, the party of Croatian Fascists who, under Pavelitch and subject to the overriding veto of Germany and Italy, governed Croatia after its conquest in 1941 by Germany. **1949** Koestler *Promise & Fulfilment* II. i. 217 The Arab troops in this region were..reinforced by.. Croat Ustachis. **1949** F. Maclean *Eastern Approaches* III. iii. 334 The independent State of Croatia..was a kingdom, but its King, the Italian Duke of Spoleto, had wisely omitted to take up his appointment and power was in the hands of Ante Pavelić and his Ustaše, supported by the Wehrmacht. *Ibid.*, Orthodox churches were..burned down with the screaming congregation inside them (an Ustaše speciality, this). **1961** E. Waugh *Unconditional Surrender* III. i. 210 There are five or six divisions of Cetnics and Ustachi..the Serb and Croat Quislings. **1972** *Guardian* 28 Jan. 11/2 If Ustashi activists..planted a bomb on board the Yugoslav airliner..it will be their biggest outrage since the war. **1973** *Nation Rev.* (Melbourne) 24–30 Aug. 1405/1 Frequent statements that the Ustasha exists in Australia and that the croatian community..condones terrorist acts. **1976** *New Yorker* 22 Mar. 64/2 Two Croatian nationalists—members of Ustaša, the party that ran an 'autonomous' Croatian state for the Nazis during the Second World War, killed hundreds of thousands of other Yugoslavs, and operates now as a secret terrorist society—tortured and then murdered the Yugoslav ambassador to Sweden. **1976** J. Colville *Footprints in Time* xxxvii. 213 The fanatic Croat Ustasi. **1978** 'G. Vaughan' *Belgrade Drop* vii. 46 The caves where the Ustaše, the Croat fascists whose fanaticism in Hitler's cause had rivalled that of the SS, had butchered the Partisans. **1980** *Listener* 28 Feb. 265/3 The Ustashas were Croat Fascist collaborators.

usual, *a.* Add: **3.** Freq. in *usual channels* [Channel *sb.*[1] 8]; *usual office(s*: see *office *sb.* 9 b.

1905 *Hansard Commons* 16 May 500 In reply to the Question of the hon. Member for Waterford, I have to ask him to communicate with my right hon. friend near me through the usual channels. **1946** *Erskine May's Law of Parl.* (ed. 14) xii. 245 The Government Chief Whip.. together with the Chief Whips of the other parties, constitutes what is known as the 'usual channels', through which communications pass as to business arrangements and other matters which concern the convenience of Members as a whole. **1975** J. P. Morgan *House of Lords & Labour Govt.* viii. 213 The usual channels had collapsed and, in the absence of the customary arrangements, the Government could only hope that they might limit discussion by imposing a guillotine.

uszza, ussza (ʊ·zā), *int.* [App. a mere exclamation, but cf. Huzza *int.*] A shout of anger or effort.

1913 D. H. Lawrence *Sons & Lovers* ii. 30 As he sat on his heels, or kneeled, giving hard blows with his pick, 'Uszza-uszza!' he went. **1914** —— *Widowing of Mrs. Holroyd* II. 36 Let me smash that bloody door in. Come out—come out—ussza!

‖ **uta** (yū·ta). [Jap., poem, song, f. *utau* to sing.] A Japanese poem; *spec.* = *tanka[2].

1855 R. Hildreth *Japan* xxxiv. 335 He found out certain words which he brought together into an *Uta*, or verse. **1897** *Japan Times* 23 Mar. 2/6 The collection of the *uta* of the old school being a riddle to the ordinary reading public. **1911** *Encycl. Brit.* XV. 169/1 Such couplets [which admitted Chinese words] were called *shi* to distinguish them from the pure Japanese *uta* or *tanka.*

Ute (yūt), *sb.*[1] and *a.* Also 9 **Eutaw, Utaw,** etc. [Shortening of *Utah*, a. Sp. *Yuta* an unidentified Indian language.] **A.** *sb.* **a.** A Shoshonean Indian people inhabiting parts of Colorado, Utah, and New Mexico; a member of this people. **b.** The language of this people. **B.** *adj.* Of or pertaining to the Ute or their language.

1826 in D. M. Frost *Notes on General Ashley* (1960) App. B. 136 The Eutaws and Flatheads..express a great wish that the Americans should visit them frequently. **1846** G. R. Gibson *Jrnl.* 13 Oct. in *S.W. Hist. Ser.* (1935) III. 252 About fifty Ute Indians came in and held a council with Colonel Doniphan. **1846** M. B. Edwards *Jrnl.* 2 Nov. in *Ibid.* (1936) IV. 212 Major Gilpin..made a long and arduous journey through the country of the Ute and Navajo. **1885** D. G. Brinton in *Pennsylvania Mag.* Apr. 30 The simple form of the verb may convey three different ideas, as in Ute, where the word for 'he seizes' means also 'the seizer'. **1907** *Univ. Calif. Publ. Amer. Archaeol. & Ethnol.* IV. III. 67 The Ute vocabulary was obtained, also in 1900, among the Uintah Ute. **1933** [see *Paiute *sb.* b]. **1958** L. C. Pritchett *Cabin at Medicine Springs* (1959) xxii. 199 Tha's talk of herdin' all the Utes out of Colorado. **1973** A. H. Whiteford *N. Amer. Indian Arts* 93 Plains lazy-stitch sewing is characteristic of the work of..the Ute of the Plateau. **1979** C. McCarry *Better Angels* III. iii. 180 Woolen socks, hand-knitted by Ute women.

ute (yūt), *sb.*[2] Colloq. (chiefly *Austral.* and *N.Z.*) abbrev. of *utility 4 b.

1943 Hunt & Pringle *Service Slang* 68 *Ute*, short for Utility truck, a light van used by the Army. **1951** E. Lambert *Twenty Thousand Thieves* 178 He gets pissed one night, pinches a ute from the transport lines. **1961** J. Danvers *The Living Come First* i. 17 I'd like you to take the ute and drive in to Alice. The wire we ordered has arrived. **1971** *N.Z. Listener* 8 Nov. 15/5 Wind whipped at the ute and rocked it on the road's sharp corners. **1981** *National Times* (Austral.) 25–31 Jan. 23/4 We used to help load Teddy into the back of the ute. **1984** *NZ Farmer* 12 Apr. 39/1 Now Nissan has followed it with a tough new 4 × 4 ute, known at this stage just as the 720.

utile (yū·tili), *sb.* [a. the specific epithet of the tree's Latin name: see Utile *a.*] The timber of a large West African forest tree, *Entandrophragma utile*, of the family Meliaceæ; also, the tree itself.

1956 *Handbk. of Hardwoods* (Forest Prod. Res. Lab.) 228 Utile reaches a height of 150 ft or more. *Ibid.*, In structure and general appearance utile resembles the closely allied sapele..and is somewhat more open in texture. **1970** *Timber Trades Jrnl.* 21 Mar. 54/1 There are now signs of weaker prices for utile. **1980** *Yachts & Yachting* 29 Feb. 680/2 As nice a piece of red utile as you'll see this side of a stereo cabinet.

Utilidor (yuti·lidōᵊɹ). *Canad.* Also **u-.** [f. Utility *sb.* + *-dor* ad. Gr. δῶρον gift. For the formation cf. Cuspidor, *humidor, and Thermidor.] The proprietary name of a system of enclosed conduits used esp. for carrying water and sewerage in regions of permafrost.

1957 *Maclean's Mag.* 14 Sept. 92/2 At many outposts such as Churchill, Norman Wells and Frobisher ingenious insulated conduit boxes called 'utilidors' are used to convey water, sewage and heating pipes to their destinations. The idea is that the warmth from the heating pipes is just enough to keep the other two pipes from freezing. **1969** *Official Gaz.* (U.S. Patent Office) 15 Apr. TM 112 Ric-Wil, Incorporated, Barberton, Ohio. Filed June 12, 1967. *Utilidor.* For prefabricated conduits used in underground services... First use at least as early as Apr. 30, 1950. **1977** *Islander* (Victoria, B.C.) 21 Aug. 14/3 Each year one expects to pay $5000 for utilities hooked to each house by the famous utilidor.

utility, *sb.* Add: **1. d.** The intrinsic property of anything that leads an individual to choose it rather than something else; in game theory, that which a player seeks to maximize in any situation where there is a choice; the value of this, as (actually or notionally) estimated numerically.

1881 F. Y. Edgeworth *Math. Psychics* p. vi, It is argued from mathematical considerations that the basis of arbitration between contractors is the greatest possible utility of all concerned. **1934** *Economica* Feb. 53 Suppose ..that we have a utility function given; that is to say, we know, for the individual in question, how much utility he would derive from any given set of quantities of the goods on the market. **1944** von Neumann & Morgenstern *Theory of Games* i. 16 We feel..that one part of our assumptions at least—that of treating utilities as numerically measurable quantities—is not quite as radical as is often assumed in the literature. **1948** *Jrnl. Polit. Econ.* LVI. 280/1 Choices among riskless alternatives are explained in terms of maximization of utility: individuals are supposed to choose as they would if they attributed some common quantitative characteristic—designated utility—to various goods and then selected the combination of goods that yielded the largest total amount of this common characteristic. **1960** A. Rapoport *Fights, Games, & Debates* iv. 64 The more he worked, the more he would get and so the more utility would accrue to him on that account. The more he worked, however, the more tired he would get and the more *negative* utility would accrue to him on that account. **1965** *Papers Regional Sci. Assoc.* XV. 162 The individual will tend to locate himself at a place whose characteristics possess or promise a relatively higher level of utility than in other places which are conspicuous to him. **1968** G. Owen *Game Theory* vi. 133 This cannot be determined simply by measuring the increase and decrease of utility which this action causes to the two individuals, for..the units of utility scales are arbitrary and thus cannot be used for interpersonal comparisons. **1973** *Proc. R. Soc.* B. CLXXXIV. 421 In such a calculus [of medicine] it is necessary to assume that it is possible to attach a measure of worth or value to any state of health, and for this measure we use the word utility. **1980** A. J. Jones *Game Theory* iv. 178 If you assign a utility of −50 to the loss of $10, then it will not pay you to engage in this gamble if your utility for a gain of $20 is less than 75. *Ibid.* 180 One alternative possesses a larger utility than another because it is preferred, not the other way round.

3. c. *public utility*: see *public *a.* 2 i. Also simply *utility.* Also, a company providing such a service or supply. In *pl.* also shares in such a company. chiefly *N. Amer.*

1930 *Daily Express* 6 Sept. 2/6 Utilities joined in the forward movement, which gathered considerable momentum in the final dealings, and the closing was strong. **1935** *Economist* 26 Jan. 195/1 Steel shares were merely steady, but utilities were inclined to take fresh heart... Mayor La Guardia indeed turned his attention this week from power utilities to the transit companies. **1936** L. C. Douglas *White Banners* iii. 46 It was a new..sensation to be free of bill-collectors... You would have thought him another person than the..apologizer..who had abased himself before..the credit departments of the stores and the utilities. **1942** *Sun* (Baltimore) 7 Mar. 20 Additional housing..within the limits of existing utilities..is immediately desirable. **1957** [see *Dow-Jones]. **1968** *Globe & Mail* (Toronto) 17 Feb. 5/1 Toronto Hydro's request for a 700-foot smokestack on the..steam heating utility. **1974** *N.Y. Times* 8 Dec. IV. 3/1 American Electric Power Co., the nation's largest privately owned utility.

4. a. Also, the activity of a utility actor, the taking of any small part.

1803 T. Wilkinson *Let.* in A. Mathews *Mem. C. Mathews* (1838) I. xvii. 359, I take it for granted in very full plays you will..not object to pantomime utility, being a lord mayor, a witch, &c. **1846** G. A. à'Beckett *Quizziology of Brit. Drama* 10 The heavy man, who is paid thirty shillings a week for doing the 'respectable utility'.

b. A utility vehicle (see sense 5 c (a) below). chiefly *Austral.* and *N.Z.*

1944 *Coast to Coast 1943* 162 They walked..to where a fairly new utility was parked with one or two other cars and an old truck. **1949** *Automobile & Carriage Builders' Jrnl.* Nov. 34 (*caption*) Two views of an Alvis Fourteen utility. **1960** *Guardian* 28 Dec. 2/3, I bought a 1949 Jowett Bradford utility two years ago. **1961** D. Stuart *Driven* iv. 40 He had come to the well in the utility and found the footmen washing their clothes. **1968** K. Weatherly *Roo Shooter* 101 As he passed the bore he noticed a very battered looking utility parked there. **1971** *N.Z. Listener* 8 Nov. 15/5 He got in the utility by Honi, who waited with the motor running. **1984** *Auckland Star* 2 Mar. A3/7 A 54-year-old Southland man..died in a four-wheel drive utility accident.

c. A utility suit (see sense 5 c (c) below).

1945 J. B. Priestley *Three Men in New Suits* ii. 12 His clothes were rather like yours. He must have been dressed in his best Utility. **1957** J. Braine *Room at Top* xviii. 164 The suit was my demob Utility.

d. *Computers.* Short for *utility program* (see sense 5 a below).

1972 *Computers & Humanities* VI. 277 There are..a number of natural-language utilities available, but often they do not fit exactly the user's research design. **1983** *Daily Tel.* 30 Apr. 25/3 Commodore..has now produced a utility, called 'simons basic' which is both delightful and infuriating to use.

e. Short for *utility room* (see sense 5 a below).

1976 *Field* 18 Nov. 948 (Advt.), Drawing room, dining room, study, cloakroom, kitchen, utility, etc.

5. a. *utility value*; (sense *3 c) *utility bill, company, executive, pole, wire*; **utility area,** a part of a house set aside for general use rather than for habitation; **utility curve,** a graph of a utility function; **utility function,** a mathematical function which ranks alternatives according to their utility to an individual; **utility man,** (*a*) (earlier example); **utility program, routine** *Computers*, one for carrying out routine tasks associated with the use of a computer; **utility room,** a room set aside for domestic appliances such as a boiler for a central heating system, a washing machine, etc., and for the storage of other equipment; **utility routine**: see *utility program* above.

1969 'E. Lathen' *Murder to Go* (1970) xiv. 135 The new Hedstrom house..consisted of one spacious common room and assorted bedrooms, serviced by tiny utility areas. **1978** *Morecambe Guardian* 14 Mar. 24/2 (Advt.), Accommodation comprises: entrance hall, lounge, dining room, kitchen, 2 bedrooms, bathroom/WC. Utility area. **1976** *National Observer* (U.S.) 24 Jan. 13/3 Warmer-than-usual weather has reduced the schools' projected utility bills. **1984** *New Yorker* 30 Jan. 48/1 Whether it's utility bills or telephone bills or taxes or health costs or heating costs, it seems that the cost of the average family raising their family is going up. **1926** *Daily Colonist* (Victoria, B.C.) 1 Jan. 18/3 The progress and prosperity of a community naturally is reflected in the fortunes of the utility companies whose duty it is to supply such essentials as light, water, power, [etc.]. **1980** *Amer. Speech* LV. 54 Presumably, the utility companies..specify certain hours of the day as the peak hours. **1948** *Jrnl. Pol. Econ.* LVI. 297/1 A single prize will be the optimum..if..every chord from the utility curve at the current income to the utility of a higher income is everywhere above the utility curve. **1969** Utility curve [see *indifference 8]. **1931** F. L. Allen *Only Yesterday* iii. 59 Book-censors, Jew-haters, Negro-haters, landlords, manufacturers, utility executives, upholders of every sort of cause,..all wrapped themselves in Old Glory and the mantle of the Founding Fathers. **1979** *Time* 8 Jan. 39/1 Says a utility executive in Iowa: 'People have been had too many times.' **1934** Utility function [see sense 1 d above]. **1977** *Dædalus* Fall 84 Models now start from mathematical representations of personal choice (utility functions) where work and leisure stand in the 'correct' relationship to each other and where mathematics imposes a 'correct' income and substitution effect. **1849** *Theatrical Mirror* 1 Oct. 37/1 He is a great friend of the utility-man's. **1961** in *Webster*, His car sideswiped a utility pole. **1973** *Advocate-News* (Barbados) 15 Feb. 4/2 The car fell into a three-foot gully and came to rest against a utility pole which kept it from overturning. **1964** T. W. McRae *Impact of Computers on Accounting* i. 24 Utility programmes have been devised for the same reason as the autocodes. **1980** S. Hockey *Guide Computer Applications in Humanities* ii. 29 Most computer centres have a utility program which makes copies of magnetic tapes. **1953** C. Armstrong *Catch-as-catch-Can* xviii. 153 They were in a small lumber room, utility room, the Californian equivalent of a cellar. **1959** *Housewife* Oct. 69/1 The utility room is fitted with a sink and drainer and also houses the fuel store and gas boiler. **1964** G. L. Cohen *What's Wrong with Hospitals?* vi. 115 A partial answer comes from North America, where the treatment and utility rooms are increasingly made up of movable partitions. **1978** *Cornish Guardian* 27 Apr. 14/7 (Advt.), The accommodation comprises..rear hallway, wash house/utility room, [etc.]. **1962** Utility routine [see *housekeeping *sb.* 5]. **1980** M. S. Wu *Introd. Computer Data Processing with BASIC* ix. 188 If an installation frequently receives data from various locations on small reels of magnetic tape, a utility routine may be written to combine the data from the short reels onto one large reel of tape prior to processing. **1926** A. E. Taylor *Plato* xvi. 431 Our question is not what 'art' has the.. greatest utility-value, but simply which sets up the most severe standard of truth and accuracy. **1974** B. Pearce tr. *Amin's Accumulation on World Scale* I. i. 64 It is impossible to 'measure' statistically the advantage (or disadvantage) that the underdeveloped countries derive from international exchange, whether one looks at the matter from the angle of labor value or from that of utility value. **1968** *Listener* 22 Aug. 235/1 Ugly poles carrying utility wires. **1973** *Sunday Advocate-News* (Barbados) 25 Feb. 18/6 We experience a lot of damage done to roofs by overhanging trees over-hanging verandahs and low-hanging utility wires.

c. Intended for use; generally useful; (passing into) merely functional; *spec.*: (*a*) of a motor vehicle: designed for carrying both passengers and goods; (*b*) of a player in a team game: able to play competently in more than one position; (*c*) during and in the aftermath of the war of 1939–45, designating clothes and household goods made in a standardized form in accordance with an official allowance of materials (cf. *austerity 4 b).

1895 *Montgomery Ward Catal.* Spring & Summer 585/1 Combination Barrel and Utility Cart... Cart will pick up barrel or body in a moment. **1908** Utility vehicle [see sense 5 b]. **1911** Utility man [see def. s.v. *utility man* (b)]. **1911** *Daily Colonist* (Victoria, B.C.) 25 Apr. 20/2 (Advt.), 1000 Yards Utility Cloth, fast colors..Per yard, 20¢. **1920** *Car* 3 Nov. 225/1 No greater mistake could be made than in the assumption that America's resources are limited to the production of 'utility' motor vehicles. **1942** *Daily Express* 24 Mar. 4/1 (*caption*) Frankly, Meadows, can you see *me* in a utility suit? **1943** *Ann. Reg. 1942* 32 Exempting so-called 'utility' cloth and clothing and boots and shoes from purchase tax. **1943** *Archit. Rev.* XCIII. 3 An exhibition has recently been held of utility furniture and utility pottery. **1944** F. Clune *Red Heart* 11 A utility truck stood at the edge of the landing-ground. **1947** *Jane's Fighting Ships 1946–47* 57 *Savage* and ships of 5 later groups have been described officially as a 'utility' type. **1948** *Observer* 8 Feb. 5/3 The dental profession.. will..refuse to enter a service which..will lower the present high standard of dental treatment to a utility level. **1949** *Automobile & Carriage Builders' Jrnl.* Nov. 49/1 The utility car, estate wagon or shooting brake has been standardized by some motor manufacturers. **1950** *RAC Guide & Handbk. 1950–51* 52 The utility vehicle, shooting brake, or station wagon is not a class of vehicle known as such to the law. **1952** M. Laski *Village* i. 14 Mrs. Wilson made the tea and then they both sat down on their camp-beds and sipped it out of thick white Utility cups. **1959** W. Golding *Free Fall* iv. 100 No, madam, I'm sorry, we don't supply them at that price. This is a utility model. **1963** *Daily Tel.* 13 Nov. 25 (*heading*) 'Utility' state opening without the Queen..Peers and M.P.s in ordinary dress. **1970** *Times* 2 Nov. 14/8 They would be the centres, with Davies (a centre last year) on the right wing and Linnecar, a utility player, on the left. **1971** *E. Afr. Standard* (Nairobi) 10 Apr. 2/2 Kadir, an intelligent utility half-back. **1974** *Country Life* 26 Sept. 828 Utility clothes, furniture and household goods..were the wartime government's solution. **1976** *National Observer* (U.S.) 16 Oct., Pickup trucks and four-wheel-drive utility vehicles have jumped in popularity in the West and may bid for a national status rivaling the van's. **1978** *Early Music* Oct. 587/2 The selection of illustrations ..ranges from fine illuminated manuscripts to simple 'utility' articles. **1984** *NZ Rugby News* 9 May 2/2 Utility Laurie Holmes has shifted to Sydney.

utilization. Add: **2.** *Comb.*, as **utilization factor,** the proportion of a given resource which is being used or is available for use.

1921 A. P. Trotter *Elem. Illumination Engin.* viii. 81 The ratio of the flux of light which reaches the working plane, whether directly or by reflection, to the whole flux of light from a lamp, is called the utilization factor, or coefficient of utilization. **1982** *IEEE Jrnl. Solid-State Circuits* XVII. 671/1 A gate utilization factor of ~95 percent has been achieved in programming the gate array.

‖ **uti possidetis** (yū·tai pǫsidī·tis). [late L., = as you possess, f. *uti* (*ut* conj.) + *possidētis* 2nd pers. pl. of *possidēre* to possess.] (See quot. 1980.)

1681 [see Interdict *sb.* 2 a]. **1763** J. Bell *Trav. from St. Petersburg* I. II. vi. 307 All matters were soon accommodated..on the footing of *uti possidetis*; i.e. each of the parties [*sc.* Russia and China] retaining the people and territories that then belonged to them. **1780** Burke *Let. Affairs Ireland* 20 The six resolutions were to be considered as a sort of *uti possidetis*. **1823** Byron *Don Juan* x. xlv. 75 A sort of treaty or negotiation, Between the British cabinet and Russian,..Something about the Baltic's navigation, Hides, train-oil, tallow, and the rights of Thetis, Which Britons deem their 'uti possidetis'. **1905** *Cambr. Mod. Hist.* (1907) III. xxi. 709 In 1593 a truce had been arranged on the *uti possidetis* basis. **1934** A. Toynbee *Study Hist.* I. 381 He made peace..on a basis of *uti possidetis*. **1980** *Oxf. Compan. Law* 1269/1 *Uti possidetis*, in the Roman law, an interdict whereby the colourable possession of real property by a *bona fide* possessor was continued until the rights of parties were finally determined. The phrase is sometimes referred to as a principle under which property not expressly provided for in a treaty terminating hostilities is to remain in the hands of the party who happened to have possession of it when hostilities ended.

Uto-Aztecan (yū:to,æzte·kăn), *sb.* and *a.* Also **Uto-Aztekan, Utaztekan.** [f. *Ute *sb*[1] and *a.* + -o + *Aztecan* adj. s.v. *Aztec *sb.* and *a.*] **A.** *sb.* A family of languages spoken in central America and western North America. **B.** *adj.* Of or pertaining to this language family.

1891 D. G. Brinton *Amer. Race* 44 Different bands of the same linguistic stock were found, some on the highest, others on the lowest stages of development, as is strikingly exemplified in the Uto-Aztecan family. *Ibid.* III. i. 118 Of all the stocks on the North American Continent, that which I call the *Uto-Aztecan* merits the closest study. **1913** E. Sapir in *Jrnl. de la Société des Américanistes de Paris* X. II. 384 Southern Paiute has developed a number of secondary forms of the original Uto-Aztekan vowels and consonants. *Ibid.* 422 In..Ute-Chemehuevi prenominal elements occur as suffixes..not prefixes, as ordinarily in Uto-Aztekan. **1935** *Amer. Anthropologist* XXXVII. 608 Are there any traits that distinguish 'Shoshonean' from the rest of Uto-Aztecan? If there are, I do not know what they are. **1937** J. R. Firth *Tongues of Men* vii. 93 The Uto-Aztekan group of related languages, spoken by the Indians of Utah, Arizona, California, and Mexico. **1968** *Occas. Papers Idaho State Mus.* XXII, (*title*) Utaztekan prehistory. **1979** I. Davis in Campbell & Mithun *Lang. Native Amer.* 409 The differences between Uto-Aztecan and Kiowa-Tanoan both in phonology ..and grammar are striking.

Also **U·to-Aztec** *a.* and *sb.*

1932 W. L. Graff *Lang.* xi. 430 The Uto-Aztec group comprises the three branches of Shoshonean, Pima-Sonoran, and Nahuatl. **1956** J. Whatmough *Lang.* (Map facing p. 22) Uto-Aztec.

Utopianize, *v.* Add: **2.** *intr.* = *Utopia-ize* vb. s.v. Utopia.

1905 H. G. Wells *Mod. Utopia* ii. 62, I pass..from Utopianising altogether, to ask the question that.. Schopenhauer failed completely to answer. **1911** —— *New Machiavelli* I. iv. 141 We planned, half in earnest and half Utopianizing, a League of Social Service.

utopiate (yutōᵘ·piēⁱt). [Blend of Utopia and Opiate *sb.*] A drug which induces fantasies of a utopian existence; a euphoriant.

1964 R. Blum (*title*) Utopiates: the use and users of LSD 25. *Ibid.* xiv. 292 The movement promises much—a return to paradise, a Utopia of the inner life—and so LSD-25 becomes, if one may be allowed a neologism, a 'Utopiate'. **1983** *Guardian* 11 Aug. 18/3 The typical recruit for drug addiction is the young, weak, inadequate person for whom the drug is a 'utopiate', an escape from the problems of reality.

utriculoplasty. (yutri·kiŭloplæsti). *Surg.* [f. Utricul(us[2] + -o + -plasty.] An operation to reduce the size of the uterus by removing part of the uterine wall.

1910 *Practitioner* June 788 The operation of utriculoplasty as carried out by Howard Kelly. **1911** V. Bonney in *Proc. R. Soc. Med.* IV. (Obstetr. & Gynecol. Sect.) 272, I have ventured to apply to it the term 'utriculoplasty' as best describing its object. **1974** Passmore & Robson *Compan. Med. Stud.* III. xxviii. 68/2 Plastic repair of a bicornuate uterus (utriculoplasty) may be indicated in cases of recurrent abortion.

‖ **ut supra** (ut sᴵū·prā). [L., f. *ut* as + *suprā* Supra *adv.*, (*a.*), *prep.*] As previously, as before (in a book or writing), as above. Also shortened **ut sup.**

c **1450** in J. Stainer *Early Bodl. Music Sacred & Secular Songs* (1901) I. no. lxii, Chorus. vt supra What tydynges. **1520** R. Elyot in T. Elyot *Governour* (1880) I. 310 And the prest vid to synge ut supra. **1526** [see Supra 1]. **1651** T. Ireland *Abridgment Rep. Sir J. Dyer* 202 The going at large *ut supra*, is not an escape. **1668** [see Supra 1]. **1875** Paley & Sandys *Select Private Orations Desmosthenes* I. 164/1 He let the arbitrators give judgment against him by default (compare Mid. *ut sup.*) and then moved for a new trial. **1959** E. Pound *Thrones* xcviii. 43 That the books you read shall be Cheng King Ut supra.

utter, *a.* Add: **4. f.** As a trivial emphasizer.

1898 G. B. Shaw *You never can Tell* IV. 308 Certainly not. It's utter bosh. Nothing can be in better taste. **1914** —— *Misalliance* 33 You are the only really clever..man I know who has given himself away to me by making an utter fool of himself with me. **1930** N. Coward *Private Lives* III. 78 You're talking utter nonsense! **1956** *Times* 3 Jan. 3/6 Professor Richard van der Riet Woolley, the newly appointed Astronomer Royal, said..that the prospect of interplanetary travel was 'utter bilge'.

utter, *v.*[1] **2. c.** For † *Obs.* read 'Now *arch. rare*' and add later example.

1977 'M. Innes' *Honeybath's Haven* iv. 37 He..placed on the seat beside him his hat, his gloves, and the handsomely illustrated brochure uttered by the proprietors of Hanwell Court to their prospective clients (or inmates).

utterance[1]. Add: **I. 5.** Freq. in *Linguistics*, spoken or written words forming the complete expression of a thought. (Used with varying degrees of technicality.)

1932 A. H. Gardiner *Theory of Speech & Lang.* iv. 206 Under the term 'utterance' writing must be included. **1951** Z. S. Harris *Methods in Structural Linguistics* ii. 14 An utterance is any stretch of talk, by one person, before and after which there is silence on the part of the person. The utterance is, in general, not identical with the 'sentence'. **1955** J. L. Austin *How to do Things with Words* (1962) v. 61 Written utterances are not tethered to their origin in the way spoken ones are. **1964** M. A. K. Halliday et al. *Linguistic Sci.* iv. 95 The utterance, the smallest institutional unit of language activity. **1966** Y. Bar-Hillel in *Automatic Transl. of Lang.* (NATO Summer School, Venice, 1962) 17, I have already suggested..to distinguish artificially between them *qua* technical terms and use 'utterance' for observational entities and 'sentence' for theoretical ones.

III. 7. *attrib.* and *Comb.*, as *utterance-type*; *utterance-final, -initial, -interior, -medial* adjs.; *utterance-finally* adv.

1953 C. E. Bazell *Linguistic Form* 5 The most common English utterance-type. **1957** in *Amer. Speech* 1972 (1975) XLVII. 229 Even assuming there may be for some speakers certain contrasts in utterance-interior positions which might require both /č/ and /tš/ in the inventory, surely there are no such contrasts in utterance-initial and utterance-final sequences. **1964** W. Jassem in D. Abercrombie et al. *Daniel Jones* 346 Voiced stops do not occur utterance-finally in Polish. **1970** *Language* XLVI. 80 Even in his speech the laryngeal is lost in utterance-medial positions. *Ibid.* 82 For the nasal consonants, the variant which occurs in utterance-initial syllable onset before nasal vowels is taken to be basic. **1977** *Ibid.* LIII. 318 The optional utterance-final particle *-o* may be added to both imperatives and vocatives. **1978** C. Hookway in Hookway & Pettit *Action & Interpretation* 32 The notion of utterance type is itself translational, as Wallace puts it.

utu. (Earlier example.)

1828 in W. Colenso *Papers* (typescript) III. 18 Until another chief has been killed as an utu or payment.

uvala (ū·vălă). *Physical Geogr.* Also **ouvala.** [a. Serbo-Croat *uvala* hollow, depression.] A depression in the ground surface occurring in karstic regions (see quots.).

1902 *Geogr. Jrnl.* XX. 429 Dr. Cvijić's researches have led him to consider the uvala (Karstmulde) as an intermediate form between doline and polye. The uvala is a large, broad sinking in the karst with uneven floor, formed by the breaking down of the wall between a series of dolines. **1921** *Geogr. Rev.* XI. 600 As time goes on, the divisions between neighboring dolines are broken down; and larger depressions, called 'uvalas' or 'ouvalas', are created. **1922** *Geol. Mag.* XIX. 401 Several of the smaller ..poljes of Southern Herzegovina..appear to result from the collapse of underground watercourses. To such depressions it would perhaps be better to restrict the use of the local Bosnian term 'uvalas'. **1954** W. D. Thornbury *Princ. Geomorphol.* xiii. 323 The Bosnian term uvala is most commonly applied to the larger depressions resulting from the collapse of extensive roof sections over underground watercourses. What have been designated above as compound sinkholes are sometimes called uvalas, but this usage of the term does not seem justified. **1970** R. J. Small *Study of Landforms* iv. 152 In many areas closely adjoining sotchs have amalgamated, through lateral extension, to give larger depressions comparable with the 'uvalas' of the Karst proper.

uveoparotid (yū·viopărǫtid) *a. Med.* [f. Uve(a + -o + Parotid *a.* and *sb.*] Affecting or involving the uvea and the parotid gland. So **u·veoparoti·tis,** inflammation of the uvea and the parotid.

1917 *Brit. Jrnl. Ophthalmol.* I. 619 He came to the conclusion that..uveo-parotid fever is nothing more than atypical mumps. *Ibid.* 620 The case for mumps as the sole cause for uveo-parotitis is as yet, 'not proven'. **1961** *Lancet* 16 Sept. 614/1 In 5 out of 13 patients who have uveoparotitis the pulmonary disease..has led to severe fibrosis. **1967** [see *lymphogranuloma]. **1974** Passmore & Robson *Compan. Med. Stud.* III. xv. 1/1 A generation ago textbooks gave accounts of Hutchinson's disease, Besnier's disease, Boeck's disease, lupus pernio and uveoparotid fever... All these conditions..are different expressions of the disorder now known as sarcoidosis. **1975** *Brit. Med. Jrnl.* 27 Dec. 731/2 Two middle-aged sisters, living separately, developed sarcoidosis with bilateral uveoparotitis and cranial palsies within six months of each other.

uxorilocal (ʊ:ksorilōᵘ·kăl), *a. Anthrop.* [f. L. *uxōri-us* (f. *ūxor* wife) + Local *a.*] Applied to or denoting residence after marriage in the area of the wife's home or community. Cf. *matrilocal *a.*; *residence *sb.*[1] 1 d; *virilocal *a.*

1936 R. Firth *We, the Tikopia* xvi. 596 The greater tendency to uxorilocal settlement at marriage. **1948** L. Adam in *Man* Jan. 12/2 The adjectives 'patrilocal' and 'matrilocal' do not fulfil these requirements [of accuracy and clarity]... I therefore propose adoption of the adjectives 'virilocal' and 'uxorilocal' to indicate whether a married couple shares the domicile with the family of the husband or of the wife. **1958** *Man* Apr. 69/1 Marriages are about equally virilocal and uxorilocal. **1974** A. P. Wolf *Relig. & Ritual in Chinese Soc.* 156 When a family has no sons who survive to marry, they ordinarily must arrange an uxorilocal marriage for a daughter or an adopted daughter.

Hence **u:xorilo·cally** *adv.*

1963 *Brit. Jrnl. Sociol.* XIV. 24 So that now you may read about individual couples 'marrying avunculocally' or 'living uxorilocally'. This seems to me an abuse of terminology. **1974** A. P. Wolf *Relig. & Ritual in Chinese Soc.* 148 From her father's point of view, a daughter is an outsider. She can achieve the right to a place on his altar only by marrying a man who agrees to reside uxorilocally.

Uzbek (u·zbek, ʊ·z-). Also formerly **Usbeck, Usbeg, Uzbeg,** and other varr. [Russ.] One of a Turkic people of central Asia, forming the basic population of the Uzbek SSR (Uzbekistan), and also living in Afghanistan; the language of this people. Also *attrib.* or as *adj.*

1616 T. Roe *Let.* 17 Jan. in *Embassy to Court of Gt. Mogul* (1899) II. 113 The King..intendeth the conquest of the Vzbiques, a Nation between Smarchand and him. **1715** J. Stevens *Hist. Persia* xxii. 221 To make war on the Usbecks. **1788** Gibbon *Decl. & F.* VI. lxiv. 292 A descendant of Zingis, who reigned over the Usbeks of Charasm, or Carizme. **1834** A. Burnes *Bokhara* I. viii. 262 We now found ourselves among the Uzbeks. **1841** J. Wood *Personal Narr. Journey to River Oxus* xiv. 215 Murad Beg, the head of this Uzbek state. **1876** E. Schuyler *Turkistan* I. iii. 106 The Uzbeks are the descendants of the Turkish tribes who..migrated to this part of Asia [*sc.* Tashkent], both before and since the time of Tchinghiz Khan. **1889** G. N. Curzon *Russia in Central Asia* vi. 153 The huge sheepskin bonnets..disappeared in favour of the capacious white turban of the Uzbeg or the Tajik. **1891** A. Constable tr. *Bernier's Trav. in Mogul Empire* 120 There are probably no people more narrow-minded, sordid or uncleanly, than the Usbec Tartars. **1900** 'Odysseus' *Turkey in Europe* iii. 101 Uzbek also is not strictly a linguistic name, but political, and denotes the Turks who are, or were, the ruling faction in the Khanates of Khiva, Bokhara, and Kokand. **1927** *Glasgow Herald* 15 Aug. 13 Agitation has been set on foot among the Uzbek people, on the southern banks of the Oxus, who now demand affiliation with their Soviet kinsmen across the river. **1929** [see *Jagatai]. **1933** L. Bloomfield *Language* iv. 68 The Turkish..family of languages ..Turkish, Tartar, Kirgiz, Uzbeg, Azerbaijani. **1957** H. Bower *Short Guide Soviet Life* 7 Constituent Republics (SSR) are with capitals in parenthesis..Turkmen SSR (Ashkhabad) Uzbek SSR (Tashkent) Tadzhik [etc.]. **1961** [see *Iranize* vb. s.v. *Iranian *sb.* 2]. **1964** G. Wheeler *Mod. Hist. Soviet Central Asia* ix. 212 Firqat (1858–1909) ..did the first translation of a Russian classic—Tolstoy's *What Men Live By*—into Chagatay, or old Uzbek, as it is now called, in 1877. **1974** T. P. Whitney tr. *Solzhenitsyn's Gulag Archipelago* I. I. ii. 59 On the Volga Canal construction site newspapers were published in four national languages: Tatar, Turkish, Uzbek, and Kazakh. **1976** *Times* 3 Nov. 16/5 A dark-haired agronomist from an Uzbek collective farm.

Uzi (ū·zi). Also (*erron.*) **Uzzi.** The name of Uziel Gal, Israeli army officer, used *attrib.* or *absol.* to designate an Israeli type of submachine gun designed by him.

1959 E. O'Ballance *Sinai Campaign* ii. 73 The personal small arms of the man was either the Mauser rifle or the Uzi sub-machine-gun. *Ibid.*, The Uzi is one of the finest guns of this type I have ever handled. **1966** L. Davidson *Long Way to Shiloh* xi. 156 Too many men and too many Uzzis which were much too easy to fire. **1971** *Scope* (S. Afr.) 19 Mar. 17/2 A loaded Uzzi sub-machine-gun..is always within reach. *Ibid.* 22/1 The Uzzi..fires a clip of 9 mm bullets in double-quick time and is accurate up to 100 metres. **1976** G. Seymour *Glory Boys* viii. 94 Two other young men, both of whom had been issued with licences..to carry Uzi sub-machine-guns. **1981** C. R. Lajeunesse *Dead Man Running* lxii. 177 Cyril was cut in two by bullets from Weasel's Uzi.

V

V. Add: **I. 2. a.** (Further examples.) Cf. VEE I.

a 1917 W. De Morgan Old Madhouse (1919) ii. 14 Until you've seen her in a low neck, or at least a V, you really can't tell. 1930 Sat. Even. Post 13 Dec. 11/3 Midway, the range was cleft from summit to base by a V more than 2000 feet deep. 1958 E. Birney Turvey vii. 78 The V of Ashanti spears above the fireplace. 1985 J. Melville Death Ceremony xxi. 156 A pulsing at her throat above the V of her delicate silk under-kimono.

b. V-blouse, body, -formation, -front, girder, hut (earlier example); **V aerial, antenna,** an aerial in which the conductors form a large horizontal V that transmits principally along its axis; **V-belt,** a belt which is V-shaped in cross-section in order to give better traction on a pulley; **V-block,** a metal block with a V-shaped recess cut in it to hold a cylindrical object while it is being worked on; **V-eight,** an internal combustion engine with eight cylinders arranged in two rows of four at an angle to each other, forming a V-shaped cross-section; freq. attrib. and written **V-8;** also, a motor vehicle with such an engine; so **V engine; V-neck,** a garment neckline in the shape of a V; freq. attrib.; also absol., a garment, as a pullover etc., with a V-shaped neckline; **V-thread,** a screw thread which is V-shaped in profile. Cf. *VEE 2.

1961 Amat. Radio Handbk. (ed. 3) xiii. 385/2 The V aerial produces one major beam along its axis, together with a rather complex pattern of minor lobes. [1931 Proc. IRE XIX. 1822 Fig. 41 is a polar diagram showing the power distribution for a V wire, having sides equal to one wave, in the plane of the wires.] 1932 Ibid. XX. 1033 These [curves] were taken during the process of adjustment of a V antenna system in which both antenna and reflector units each consisted of 2 V wires one above the other. 1959 K. Henney Radio Engin. Handbk. (ed. 5) xx. 60 V antennas are arranged to utilize this main lobe from all wires of the system. 1911 C. S. Lake Motor Cyclist's Handbk. viii. 116 The most common form of transmission on a motor cycle is that of a 'V' belt. 1973 A. Parrish Mech. Engineer's Ref. Bk. xiii. 7 One of the problems associated with V belts..is the variation in velocity ratio which occurs from time to time. 1901 Shop & Foundry Pract. (Colliery Engineer Co.) II. §10.45 Cylindrical parts are usually supported on V blocks. 1971 B. Scharf Engin. & its Lang. viii. 54 V-blocks are usually made in pairs so that they can support a long tube. 1915 Contemp. Rev. Sept. 370 Quaker maidens now wear tucks or V-blouses or anything else that is the fashion. 1885 'Mrs. Alexander' Valerie's Fate iv. 69 That black satin and lace costume..with the clear sleeves and a V body. 1930 Autocar 2 May 841/2 It is said..that the V eight..is superior as regards compactness of build. 1936 Motor Man. (ed. 29) ii. 40 The V-eight engine comprises two blocks, each containing four cylinders, which are set at right angles on a single crankcase. 1942 G. R. Gilbert in D. M. Davin N.Z. Short Stories (1953) 252 Lena would giggle and look pleased as though she had Prince Charming waiting in a V8 outside the kitchen door. 1959 I. Jefferies Thirteen Days x. 162 A V-8 pick-up was parked, with a pair of Haganah in the back. 1963 Bird & Hutton-Stott Veteran Motor Car 57 Often claimed to be the first V-eight to be marketed. 1982 Times Lit. Suppl. 5 Mar. 249/4 The comparison between the purist Bugatti engine..and an American V-8 engine of thirty years later. 1924 Motor Man. (ed. 25) i. 19 An eight-cylinder V engine. 1967 Economist 29 July 425/3 Berliet has a range of v-engines due to appear soon. 1949 Koestler Insight & Outlook xi. 163 The V-formation of migrating geese. 1974 'J. Graham' Bloody Passage i. 10 A flight of Brent geese drifted across the sky..in a v formation. 1895 Montgomery Ward Catal. Spring & Summer 92/1 Men's Rutland V front collars. 1919 Brit. Manufacturer Nov. 24/1 The output of 'T', 'V', and lattice girders of all gauges. 1851 Lyttelton (N.Z.) Times 1 Feb. 3 One V hut was blown away. 1910–11 T. Eaton & Co. Catal. Fall & Winter 20/1 Women's coat sweater... The V-neck and fronts have wide, plain knitted border. 1921 [see house-dress s.v. *HOUSE sb.[1] 19]. 1949 O. Nash Versus 62 They lose their rubbers and store their V-necks. 1970 T. Lewis Jack's Return Home 127 He had on a white shirt..., a bottle green V-neck, twill trousers. 1978 Detroit Free Press 5 Mar. 30 (Advt.), T-shirts or V-neck shirts. a 1877 Knight Dict. Mech. III. 2061/2 (caption) V-thread. 1887 [in Dict.]. 1939 [see square thread s.v. *SQUARE a. 14 a]. 1971 B. Scharf Engin. & its Lang. xi. 101 A distinction is also made between V-threads and square threads, according to their cross-section.

c. V-shaped adj. (further examples); spec. designating or pertaining to a valley having such a cross-section, esp. when contrasted with a U-shaped valley; also in Comb. with other adjs., ppl. adjs., and sbs., as V-cut, -fronted, -like, -necked, -type, -winged adjs. Cf. *VEE 2.

1912 S. Ford Shorty McCabe's Odd Numbers 107 Maybe Cornelia will have some plans of her own, thinks I, as I gets into my silk faced dinner jacket and V-cut vest. 1977 H. Kaplan Damascus Cover iv. 36 She wore a V-cut peasant blouse. 1927 Blackw. Mag. Jan. 76/1 They stamp the snow from their V-fronted high-heeled jackboots. 1929 E. Linklater Poet's Pub xxiv. 258 A cattle-track.., branching V-like off the road. 1971 V-necked [see *SIGNATURE sb. 4 b]. 1894 V-shaped [see U-shaped s.v. *U 2 a]. 1907 Bull. Geol. Soc. Amer. XVIII. 355 Downstream from the glacial region the valley ceases to be U-shaped. It becomes narrow and V-shaped, and the terraces die out. 1937 [see *HYPSOGRAPH]. 1970 R. J. Small Study of Landforms i. 6 On the resultant maps..the forms of valley floors (whether V-shaped, rounded, or flat-bottomed)..[are] shown. 1967 Technology Week 20 Feb. 35/1 The 'V' type antenna is used because it can be designed to give good performance even if distorted. 1876 G. M. Hopkins Poems (1967) 178 Through the velvety wind V-winged..To the nest's nook I balance and buoy.

d. as v. intr. Of geese: to fly forming the shape of a horizontal V. Cf. V-formation, sense 2 b above and V 2 a (quot. 1894).

1907 Canad. Mag. XXIX. 21/1 Then across his senses came the nearing doom—the honk, honk of wild geese V-ing their way along the shadow trail of the night sky. 1970 R. Lowell Notebk. 245 The usual autumn Flight of Canada geese V above it, moonborne. 1972 J. Gores Dead Skip ii. 14 Another batch of kids..their cries as full of spring as geese V-ing north.

3. V-agent, any of a group of organophosphorus nerve gases having anticholinesterase activity; VX, a type of V-agent, O-ethyl S-2-diisopropylaminoethyl methylphosphonothiolate.

1964 Kirk & Othmer Encycl. Chem. Technol. (ed. 2) IV. 874 Aerosolized V agents also are quite lethal by inhalation. 1975 Nature 10 Jan. 82/3 The nerve gas codenamed VX is the most toxic of a family of V-agents produced by British chemical warfare scientists since World War Two. 1966 New Statesman 16 Dec. 900/1 The Americans..have also produced less volatile gases called V agents, particularly a liquid called VX. 1980 Sci. Amer. Apr. 36/1 Some 5,000 tons of VX were made between 1961 and 1967.

III. 5. a. vs., versus.

1889 in Cent. Dict. 1949 E. Pound Pisan Cantos lxxix. 74 Kumasaka vs/ vulgarity. 1967 Boston Sunday Herald 26 Mar. vi. 4/7 It has developed a way of seeing Europe that seems the perfect answer to the group vs. non-group argument. 1970 Jrnl. Gen. Psychol. LXXXIII. 133 Conditions..are optimal for divergent vs. convergent tasks.

b. v = very (earlier examples); V, victory, spec. used as the symbol of allied victory in the war of 1939–45 (cf. V-Day below; *VE, *VJ, *V-SIGN); V, v, volt; VA (U.S.), Veterans' Administration; V.A., Vice-Admiral; V.A., visual acuity; V.A.D., (a member of a) Voluntary Aid Detachment; V. and A., Victoria and Albert Museum; VASCAR, Vascar, visual average speed computer and recorder; V-bomber (see quot. 1955); V.C., Vice-Chancellor; V.C., Victoria Cross (earlier example); also, a holder of the Victoria Cross; VC (orig. and chiefly U.S.) = *VIET CONG sb. and a.; V.C.H., Victoria County History (or Histories); V.C.O., Viceroy's Commissioned Officer; VCR, videocassette recorder; V.D., v.d., venereal disease; freq. attrib.; V.D., Volunteer Decoration (formerly awarded in the Territorial Army or the Royal Naval Volunteer Reserve); V-Day, V-day, V Day, Victory Day, used variously with reference to allied victory in the war of 1939–45 (cf. VE-day s.v. *VE; VJ-day s.v. *VJ); also transf. and fig.; VDT, video (or visual) display terminal; VDU, vdu, visual (or video) display unit; V.F. (R.C.Ch.) = vicar foran(e) s.v. VICAR 4 c; V.F.A. (Austral.), Victorian Football Association; V.F.L. (Austral.), Victorian Football League; V.F.R., visual flight rules; VFW (U.S.), Veterans of Foreign Wars; V.G. = VICAR GENERAL (sense 2); VHD, video high density (system); VHF, vhf, very high frequency: applied to radio waves with a frequency between 30 and 300 MHz; VHS, Video Home System; VIP, vasoactive intestinal (poly)peptide; VISTA, Volunteers in Service to America; VLA (Astron.), very large array (system of radio telescopes in the U.S.); VLBI (Astron.), very long baseline interferometry (method of measuring signals from a radio astronomical source); VLCC, very large crude (oil) carrier; VLDL (Biochem.), very low-density lipoprotein; VLF, very low frequency: applied to radio waves with a frequency between 10 and 30 kHz; V.L.R., very long range; VLSI (Electronics), very large scale integration (or integrated); V-Mail (U.S.), victory mail (see quot. 1966); VOA, Voice of America; VOR, VHF omnirange (cf. omnirange s.v. *OMNI-); VP, verb phrase (in Transformational Grammar); V.P., Vice-President (examples); V.P.P. (India), value payable post (see quot. 1975); V.S., veterinary surgeon (example); V.S.O., Voluntary Service Overseas; a member of the organization thus called; V.S.O.P., very special old pale (brandy); V-STOL, V/STOL, VSTOL, vertical and short take-off and landing; VTO(L), vertical take-off (and landing); VTR, video tape recorder; VU, vu (Electronics) [prob. abbrev. of volume unit: cf. quots. for this s.v. *VOLUME 10 c, where it is implied to be so], a unit in which some types of volume indicator are calibrated, a sine wave with a power of 1 mW being assigned the reference value of 0 VU (see quot. 1940 and cf. volume indicator s.v. *VOLUME 10 c); freq. attrib., esp. in VU meter, any volume indicator employing the VU scale. See also (as main entries) *VAT, *VE, *VEEJAY, *V.I.P., *VJ, *V-SIGN.

1863 Q. Rev. Jan. 159 It is said also, that the prisoners have been known to make an example of a warden who was not in their opinion sufficiently liberal with his V.G.'s ('Very Good,' as marked in the accounts). 1891 W. S. Churchill Let. 1 Jan. in R. S. Churchill Winston S. Churchill (1967) I. Compan. I. v. 219 V-happy V. well. 1941 C. Milburn Diary 20 July (1979) 104 The V campaign (. . . —) was launched by Mr Churchill today. V for victory, the opening notes of Beethoven's V (fifth) symphony. 1973 D. Westheimer Going Public iv. 67 She raised her hand in a peace sign... He realized it was not the peace sign at all. To those of the old woman's generation it was V for Victory. 1889 W. P. Maycock Pract. Electr. Notes & Definitions II. 23 Comparison of various E.M.F.s... Continuous dynamo = 10V to 500V. 1943 C. L. Boltz Basic Radio x. 166 Mains valves..can be operated with 200V, 300V, and even 400V on the anode. 1966 Wireless World July (Advt. section) 83 Power pack kits. Fully smoothed output 250 v. 60 mA. H.T. and L.T. 6·3 v. 1·5 amps. 1945 Newsweek 16 Apr. 40/2 Last week..unfavourable publicity hit the VA. 1976 N. Thornburg Cutter & Bone xii. 275 He spent a long time with the shrinks in VA hospitals. 1794 R. F. Greville Diary 23 Aug. (1930) 300 We near'd the Minotaur, on which V.A. McBride took that Opportunity of hoisting out His barge. 1915 W. S. Churchill 23 Jan. in M. Gilbert Winston S. Churchill (1972) III. Compan. I. 444 In the absence of Adl Carden, Adl de Robeck will have a temporary rank of V.A. 1932 Optician LXXXIII. 398/2 No effect on the peripheral V.A. of one half of an eye was noted when the other half was illuminated. 1982 M. Urvoy et al. in François & Maione Paediatric Ophthalmol. 399/2 For a subjective measurement of V.A., we have four groups of tests. 1915 G. Bell Let. 10 Feb. (1927) I. xiv. 359 She is a V.A.D. part of a detachment which is going up as orderlies to the Cross Hospital at G.H.Q. 1916 Lancet 18 Mar. 651/1 Whether he had received any complaints from the V.A.D. hospitals as to the strict nature of the regulations governing the movements of convalescent wounded. 1980 'M. Yorke' Scent of Fear x. 85 In that earlier war she had become a VAD nurse. 1937 Partridge Dict. Slang 929/1 V. and A., the..The Victoria and Albert Museum: museum-world coll.; late C. 19–20. 1958 Listener 28 Aug. 317/1 At the V. and A. both the Morris dining-room and the magnificent Poynter grill room..have gone. 1977 J. Aiken Last Movement vi. 115 He sold them [sc. pictures] all to the V & A. 1966 N.Y. Times 1 Oct. 39/5 Trademarked Vascar, the instrument divides the time a car takes by the distance it travels, and shows the answer in miles an hour. 1967 Traffic Digest & Rev. May 3/2 VASCAR is a device which allows an operator by measuring quantities of distance and time to compute the speed of vehicles on a highway. The chief advantage of VASCAR, according to its inventor, Arthur N. Marshall of Richmond, Virginia, is that the officer in a car equipped with VASCAR can maintain regular patrol and still clock the speed of other vehicles on the road. 1973 Times 13 Aug. 4/3 Essex police are..to introduce a new speed detection device. Known as VASCAR—visual average speed computer and recorder—it has been tested in the country for two years. 1983 Times 24 Feb. 3/8 He had decided to make a Vascar speed check with the equipment fitted in his vehicle, which meant choosing two fixed features on the road—this case a large tree and a bridge. 1955 Britannica Bk. of Year 489/2 Concentration on air-power was reflected in terms like V-Bomber (the initial referring to the types, Victor, Vulcan and Valiant). 1958 Spectator 10 Jan. 59/1 This weapon is the natural armament of the 'V' bombers. 1975 in R. Crossman Diaries I. 57 Mr Wilson proposed to replace the M.L.F. with an Atlantic nuclear force, which would include American Polaris submarines, British V-bombers, 'some kind of mixed-manned, jointly owned elements' and the British Polaris submarine that Labour was not going to scrap after all. 1715 in Bodleian MS. Ballard 49 f. 154, We the V.C. & Heads of Houses & Professors think it incumbent on Us to make this publick Declaration of our Utmost Abhorrence & Detestation of Such Offences. 1866 Law Rep.

(Chancery Appeal Cases) I. 66 V.C. Wood held that the Plaintiff's having come to the nuisance did not disentitle him to equitable relief. 1883 J. A. H. Murray Let. 8 Nov. in K. M. E. Murray Caught in Web of Words (1977) xii. 227 The V.C. had to rush off in a cab. 1953 M. Davidson Medicine in Oxford ii. 26 It seems not unlikely that the latter may have consulted the V.-C. about Francis's migration. 1971 Rayden's Pract. & Law of Divorce (ed. 11) I. iii. 43 Sir Richard Kindersley V.-C., nevertheless said [etc.]. 1859 A. Thackeray Let. in H. Ritchie Lett. A. T. Ritchie (1924) v. 111 Papa gave us a letter to read ..from Edward Thackeray's colonel recommending him for a V.C. 1929 Daily Express 7 Nov. 13/5 Mr. Thomas Dinesen, the Danish V.C., and the only foreigner to win the decoration, arrived at Liverpool-street Station. 1964 N.Y. Times 16 Sept. 4 V.C.—They are of course the Vietcong, the enemy. But when a private displeases his sergeant, he may hear 'you knucklehead V.C.!' 1965 Punch 11 Aug. 214/1 Some of them [sc. GIs] responded to a professional sergeant's claim that they were..eager to get to grips with the 'VCs', as the Viet Cong are now known in the trade. 1968 Listener 23 May 656/3, I felt rather anxious that the patrol might have disappeared and left me in the empty suburb with Mr Van and the VC snipers. 1977 N.Y. Rev. Bks. 23 June 6/3 A nineteen-year-old Marine is discovered cutting the ears off a dead VC. 1931 Times Lit. Suppl. 10 Sept. 683/1 Thornton Riseborough, according to the V.C.H., appears always with a double title after the twelfth century. 1965 Listener 8 Apr. 531/3 The V.C.H., as it is known to all its users, is a great work of reference, but it is unreadable. 1945 C. J. Auchinleck Let. 24 Nov. in Mansergh & Moon Transfer of Power (1976) VI. 531 Officers, V.C.O.s, and I.O.R.s who became officers in the I.N.A. 1977 'D. MacNeil' Wolf in Fold ii. 15 The acting squadron commander's a VCO—a risaldar named Jalala Khan. 1971 New Scientist 26 Aug. 469/1 So that the television does not have to be adapted to take the recorder, the VCR is put between the TV and its aerial. 1983 Listener 12 May 3/1 VCRs whirred away as people took advantage of watching the latest movies. 1984 What Video? Aug. 5/4 The cassette is totally incompatible with British VCRs and TV sets. 1920 Ann. Rep. Chief Med. Officer, Ministry of Health II. iii. (caption facing p. 163), V.D. clinic. Suggested plan of arrangement of a..hut. 1920 F. Fox G.H.Q. vi. 87, I do not know where the idea sprang from that v.d. was very common in the Army. 1962 E. Snow Other Side of River (1963) xxxv. 262, I didn't spend my old man's money learning to become a V.D. quack for a gangster society. 1978 'L. Black' Foursome ii. 15, I don't do it for money—only with men I like the look of. And I haven't got VD. 1901 T. F. Fremantle Bk. of Rifle p. v, The Hon. T. F. Fremantle, V.D. 1946 Jrnl. R. United Service Institution XCI. 129 Captain C. A. R. Shillington, V.D., R.N.V.R. 1941 Newsweek 28 July 22/3 Encouraged by the success [of the V propaganda campaign], Britain proclaimed July 20 as 'V Day'. 1942 Time 16 Mar. 11/1 We at Hercules are eager to learn of any new material, process, or equipment..which can enable us to create more employment after V-Day. 1945 Times 5 Apr. 5/2 To-day the battle still rages with loss and peril in Europe. On V Day it will still go on over great stretches of land and water in the Far East. 1949 Koestler Promise & Fulfilment I. xiii. 146 It was Jewry's V-day—the first since the time of the Maccabeans. 1967 A. Christie Endless Night xxiii. 211 'Well,' said Greta with a deep satisfied sigh, 'we've made it.' 'V-Day all right,' I said. 1975 Nature 16 Oct. 557/1 If reporters can operate typewriters with the accuracy necessary for an OCR reader they can probably operate keyboards producing punched tape for the computer or sophisticated visual display terminals (VDTs) on-line to the computer. 1979 Globe & Mail (Toronto) 20 Feb. 9/1 Mr. Brown described himself as not very mechanically minded, but said he has worked hard to become knowledgeable about the video display terminals—known as VDTs. 1982 A. Clements Microcomputer Design & Construction ii. 236 The main output device..of many microprocessor systems is the video display terminal (VDT). 1968 Brit. Med. Bull. XXIV. 192/2 The data-terminal..may consist of a 'video display unit' (VDU), in effect the combination of a television-like display tube with a keyboard. 1970 Computer Management Nov. 52 (caption) Entering data via the keyboard of the VDU. 1976 Liverpool Echo 24 Nov. (Advt.), Hardware consists of an ICI 1900 mainframe linked to mini-computers with disc storage, local printers and VDUs. 1982 What's New in Computing Nov. 5/3 Because the entire unit is stalk mounted, the vdu angle can be adjusted for best visibility. 1985 Personal Computer World Feb. 195/2 Designs published to date have concentrated on putting the intelligence in the node controller which then allows operation of the system through an ordinary VDU. a 1912 W. T. Rogers Dict. Abbrev. (1913) 197/1 V.F., Vicar Forane. 1922 Joyce Ulysses 312 The rev. John Lavery, V.F. 1936 Age (Melbourne) 1 May 7/8 (heading) V.F.A. Seconds. 1969 Melbourne Truth 12 July 24/1 Dandenong and Preston meet in the most tension packed VFA game of the season. 1936 Age (Melbourne) 1 May 7 (heading) V.F.L. Season opens on Saturday. 1969 Melbourne Truth 12 July 2/6 The new kicking out-of-bounds rule introduced by the VFL this season. 1949 Jrnl. R. Aeronaut. Soc. LIII. 967/2 Under V.F.R. (Visual Flight Rules) it appears that communication takes place between the ground and the aircraft for an aggregate time of about 60 seconds. 1974 L. Deighton Spy Story xv. 146 This aircraft's electronics were primitive. Flying V.F.R. meant he'd have to put it down before dark. 1981 Pilot Jan. 13/1 A special VFR clearance. 1920 Foreign Service Mag. Dec. 12/1 The V.F.W. is an organization for service. That is the purpose of its existence. 1977 C. McFadden Serial (1978) xlviii. 103/2 You're gonna get all those calls again from people who want you to sing 'God Bless America' at VFW conventions. 1871 Tablet 14 Oct. 502/1 Very Rev. Dr. O'Shea, P.P., V.G. 1922 Joyce Ulysses 312 The rt rev. Mgr M'Manus, V.G. 1980 New Scientist 13 Nov. 442 In the US, VHD will face stiff competition from Philips's laser-reading (VLP) system. 1984 What Video? Aug. 11/1 The juke boxes use Thorn EMI VHD disc players and discs (not available for the home). 1932 Admiralty Handbk. Wireless Telegr. 1931 p. ii, The range of frequencies of the æther waves used in wireless communication is now subdivided as follows:.. Above 30,000 kc./s... Very high Frequencies (V.H.F.). 1951 'N. Shute' Round Bend ii. 50 A small V.H.F. radio set. 1955 Times 29 July 5/4 The present system of amplitude modulation in the v.h.f. maritime services should be changed to one of frequency modulation. 1956 B.B.C. Handbk. 1957 134 The introduction of very high frequency transmissions, with frequency modulation (VHF/FM) in several parts of the country, was the major development of the year in sound broadcasting. 1974 Harvey & Bohlman Stereo F.M. Radio Handbk. ii. 9 Also, at v.h.f., there was sufficient bandwidth available for hi-fi quality. 1982 Daily Tel. 30 July 3/5 (Advt.), Simple to use VHS recorder with 10-day timer. 1984 What Video? Aug. 10/2 SKC..is also launching a range of high grade cassettes in standard lengths in VHS and Beta formats. 1972 Bioorganic Chem. II. 30 (heading) Synthesis of the vasoactive intestinal peptide (VIP). Ibid. 87 Information on partial sequences of VIP became available recently. 1983 R. G. Long et al. in Oxf. Textbk. Med. I. xii. 50/1 VIP secretion has been demonstrated after direct neural stimulation. 1964 Amer. Forests Oct. 13/1 The act provides for establishment of the Volunteers In Service To America (VISTA)—a sort of domestic peace corps. 1980 New Age (U.S.) Oct. 42/2 NOFA..sponsored VISTA workers to help set up farmers' markets in New Hampshire and Vermont. 1974 VLA [see *OPTICAL a. 2 a]. 1978 Pasachoff & Kutner University Astron. xxvi. 669 When fully operational..the VLA will make pictures of a field of view a few minutes of arc across, with resolutions comparable to the 1 arc sec of optical observations from large telescopes, in about 10 hours. 1969 Sci. Jrnl. Aug. 63/2 This interferometer system which is called the very long baseline interferometer (VLBI) is unusual in that there is no connection between the receiving elements. 1982 Sci. Amer. May 85/3 The VLBI maps now being made are as good as the maps made with linked telescopes 10 years ago. 1968 Punch 24 Apr. 612/3 The introduction of VLCCs (very large crude carriers, supertankers of up to 200,000 tons) will cut transport costs dramatically. 1974 Nature 19 Jan. 196/1 In December 1969 three VLCCs had serious explosions in one of their centre tanks during tank cleaning. 1975 Times 30 June 16/5 Tanker rates continued to increase..as Very Large Crude Carrier (VLCC) premiums moved up. 1977 Time 21 Nov. 40/1 Among the largest and lightest of these globules are the very-low-density lipoproteins (VLDL). 1938 Admiralty Handbk. Wireless Telegr. 1938 I. (Nomenclature of Waves), On the basis of a recent C.C.I.R. recommendation, promulgated in French; a suitable nomenclature, likely soon to be accepted internationally, may be given in English as follows:— Below 30 kc/s...Very Low Frequencies (V.L./F.). 1962 L. Deighton Ipcress File xx. 136 There was a V.L.F. (very low frequency) radio wave-length and a compass bearing. 1983 New Scientist 13 Jan. 93/1 Scientists in California have discovered that the Earth's magnetic field can act as a giant neutral amplifier for very low-frequency (VLF) radio waves. 1943 W. S. Churchill in Hansard Commons 8 June 566 We took the measures which have thrown the long-range aircraft—the very long-range aircraft—the V.L.R., as they are called—, effectively in to the anti U-boat struggle. 1946 Happy Landings July 7/3 In April, 1945 [he] became A.O.A., Tiger Force (V.L.R.), Bomber Force for the bombing of Japan. 1978 World Book Year Bk. 309/1 In 1977..the semiconductor segment of the industry was virtually on the threshold of a new frontier—very large scale integration (VLSI). 1979 W. J. Caelli Microcomputer Revolution p. xvii, VLSI—Very Large Scale Integration. This term flows on from the LSI designation and refers to component densities of well over 1000 components. 1983 Listener 25 Aug. 25/2 The amount of VLSI (Very Large Scale Integrated) circuitry needed to enable them to sync Autocue clichés with corny visuals is as great as that employed in the space shuttle. 1984 Ann. Rep. Racol Electronics PLC 6/1 The collaborative development of a microelectronic very large scale integrated (VLSI) process. Ibid. 7/1 The design of VLSI chips. 1942 N.Y. Times 13 June 17/6 The new V-Mail for United States overseas forces, patterned after the British microfilm postal system, was started when letters were delivered to President Roosevelt today. 1943 R. Vance They made me a Leatherneck 44 'George never lets up on V mail to that female,' Weber observed. 1966 Sunday Times (Colour Suppl.) 4 Dec. 73/4 [GI jargon.] V-Mail, letters to or from home, reproduced photographically to conserve shipping space. 1949 Bull. U.S. Dept. of State 27 Mar. 396/2 The second part [of a broadcast] originating in the VOA offices in New York, will include news. 1975 New Yorker 26 May 28/3 The basic problem is that V.O.A. has been placed at the intersection of journalism and diplomacy: the practice of one of these disciplines negates the practice of the other. 1955 Times 17 Aug. 6/4 There was controversy earlier this year over how soon—if at all—Tacan should replace VOR/DME..as the standard air navigation system in the United States for civil as well as military aircraft. 1982 T. Beattie Diamonds xviii. 157 'Could you confirm your VOR is monitoring?' 'Freetown roger... The VOR is unserviceable.' 1972 Hartmann & Stork Dict. Lang. & Linguistics 249/1 In transformational-generative grammar, the verb phrase is that constituent of a sentence which contains the predicate (or complement or adjunct). .. The abbreviation VP is used in phrase structure rules. 1976 Word 1971 XXVII. 248 The main verb or an auxiliary verb, if there is one, is placed in the final position of a verb phrase in German deep structure while it occupies the initial position of a VP in surface structure. 1887 Irish Times 30 Nov. 5/3 Sir Andrew Hart, V.P., T.C.D. 1925 C. S. Lewis Let. 14 Aug. (1966) 103 When the V.P. [of Magdalen College, Oxford] laid a red cushion at his feet I realized..that this was going to be a kneeling affair. 1978 M. Puzo Fools Die xxvii. 310 With his bosses, like the VP in charge of production at Wartberg's Tri-Culture International Studios,..he was much more frank, more human. 1888 Kipling Barrack-Room Ballads (1892) 112 How he met with his fate and the V.P.P. 1975 C. Allen Plain Tales from Raj viii. 93 With VPP or Value Payable Post, you paid the postman the value of the goods in the parcel. 1952 A. M. Sullivan Last Serjeant xiii. 139 The ancient claimant to the degree of V.S. was a little more learned but often a little less skilful than the country cow doctor. 1960 Voluntary Service Overseas 4 Volunteers give their service free... This leaves V.S.O. with the task of raising funds to cover the cost of travel and insurance. 1962 Times Lit. Suppl. 28 Dec. 1007/3 Mrs. Dickson's husband was largely responsible for initiating the scheme, known as Voluntary Service Overseas (VSO). 1965 Listener 7 Jan. 21/2 He was the first British V.S.O. to come to Libya. 1967 Guardian 30 May 5/5 The conflict came to a head shortly before my VSO year came to an end. 1980 Jrnl. R. Soc. Arts Jan. 111/1 She has a VSO working with her and six or seven dedicated staff. 1981 E. North Dames xiii. 256 Your children will..work for V.S.O. and Amnesty International. 1907 Yesterday's Shopping (1969) 99/2 J. and F. Martell's..V.S.O.P.—108/0. 1951 R. Postgate Plain Man's Guide to Wine ix. 125 Five Stars should indicate a good brandy; higher-up the various firms have their own indications: X.O., V.S.O.P., Cordon bleu, and so forth. 1982 M. O'Donnell Devil's Prison II. i. 88 The waiters had left them alone with a jug of coffee and a bottle of VSOP. 1960 Aeroplane XCVIII. 234/1 In the United States, NASA applies the STOL appelation to any fixed-wing type capable of operating from a 500-ft. strip surrounded by 50-ft. obstacles. This is sufficiently drastic, however, to eliminate all but a handful of experimental aircraft, which may more accurately be described as V-STOL (very short take-off, etc.) types. 1961 New Scientist 23 Feb. 462/2 Construction of economical V/STOL aircraft..is..a much more urgent and profitable line of development than supersonic aircraft will ever be. 1977 R.A.F. News 30 Mar.–12 Apr. 13/2 The future of military VSTOL would seem to be assured in a maritime context. 1954 Aviation Week 26 Apr. 30/2 New approaches to the problems of developing vertical-rising aircraft are being explored... NACA has been doing basic research in the VTO field for more than a decade. 1955 Sci. Amer. Apr. 106/3 V.T.O. aircraft (vertical take-off) are being developed vigorously in both England and the U.S. 1963 Ann. Reg. 1962 390 Bristol-Siddeley had produced a prototype supersonic VTO fighter. 1955 Wall St. Jrnl. 4 Feb. 3/4 Bell Aircraft Corp. announced it has built and flown the first jet-propelled vertical rising airplane which takes off and lands without needing a runway. The test VTOL (vertical take-off and landing) airplane weighs about 2,000 pounds, is 21 feet long, has a wing span of 26 feet, and carries only the pilot. 1958 Times 1 Mar. 7/3 V.T.O.L. designs are as yet in their infancy. 1979 N. Slater Falcon i. 24 The basic [plane] design owed much to the VTOL Harrier. 1954 Britannica Bk. of Year 638/1 VTR (video tape recorder). 1968 Globe & Mail (Toronto) 17 Feb. 51 (Advt.), Minimum of two years experience preferably in colour telecine and Ampex VTR. 1982 J. Gardner For Special Services xii. 109 Each [cabin] had a large sitting room with television, stereo and VTR. 1940 H. A. Chinn et al. in Proc. IRE XXVIII. 14/2 It was thought..that there would be less confusion in adopting the new standards if a new name were coined for expressing the measurements. The term selected is 'vu', the number of vu being numerically the same as the number of decibels above or below the new reference-volume level. 1944 Ibid. XXXII. 601/1 A key located to the left of the VU meter should be used. 1959 K. Henney Radio Engin. Handbk. (ed. 5) xiii. 19 The A scale emphasizes the VU markings and has an inconspicuous voltage scale. 1976 Canad. Jrnl. Linguistics Spring 70 The speaker made every attempt to maintain equal intensity across syllables by monitoring his output on a VU meter.

d. Of German words: V, Vergeltungswaffe, 'reprisal weapon'; used to denote German missiles of the war of 1939–45, as V-1 = flying bomb s.v. *FLYING vbl. sb. 3; hence V-bomb; V-2, a type of rocket bomb; hence V-weapons; VW, Volkswagen.

1944 Times 30 June 3/5 For two weeks London has now been subjected to ceaseless bombardment by the German weapon V1. 1944 Life 21 Aug. 17/1 It seems probable..that the V-2, successor to the V-1 robot bomb, will be a heavy rocket. 1944 Evening Sun (Baltimore) 13 Sept. 1/6 The Germans, after toning down their 'V' weapons threat for a few days, are now making new threats. 1944 Sun (Baltimore) 20 Nov. 3/4 Lord Vansittart found no substance in objections that German V-bombs debar a Big Three meeting in Britain. 1952 M. Allingham Tiger in Smoke ii. 43 Night. V2 time... Remember V2's?.. Suddenly, no warning, no whistle, wallop! 1962 F. I. Ordway et al. Basic Astronautics ii. 23 The modern space carrier vehicle is a direct descendant of the V2 guided missile developed by the German army during World War II. 1978 D. Kyle Black Camelot x. 156 Hitler remains confident he can win the war... The V-weapons, I suppose. 1982 T. FitzGibbon With Love II. viii. 155 After D-Day, 6 June 1944, a new horror arrived..the pilotless flying bombs, called V1s and known..as 'doodle-bugs'... In September..the first rockets (V2s) reached London. 1958 S. Abbey Bk. of Volkswagen v. 45 It is possible to improve the performance of the VW by a standardized engine tuning process... The VW owner is saved the cost of several accessories which are desirable. 1962 A. Lurie Love & Friendship vii. 135 Maybe it's just the fellow feeling of his Volkswagen for my Renault; he would be equally helpless if the VW broke down. 1977 C. McFadden Serial (1978) iii. 13/1 She..extricated Kat Vonnegut..from the rear of her VW bus.

e. Of French words: V.D.Q.S., vin délimité de qualité supérieure, a wine of superior quality from amongst the wines of a limited area.

1962 Wine Mag. Sept.–Oct. 253/2 Next in order of quality are the V.D.Q.S. wines, or, to give them their full title, the Vins Délimités de Qualité Supérieure. 1966 P. V. Price France: Food & Wine Guide 135 Below the A.C. wines come those marked V.D.Q.S.— vins délimités de qualité supérieure. 1974 Times 2 Dec. (Suppl.) p. ii/1 The VDQS stamp..[is] a stamp of quality awarded by the French Government. VDQS stands for 'Vins Délimités de Qualité Supérieure'.

IV. Symbolic uses.

6. Particle Physics. V is used to designate the heavy unstable particles that produce

characteristic V-shaped tracks when they decay (V-events), now identified as hyperons and kaons. Obs. exc. hist.

1950 P. M. S. Blackett Let. 12 July (MS.), We have been discussing here the question of nomenclature and I would like to ask your views about the following suggestion. This is that we should call the special type of track that you and we have observed v-tracks and the particle or particles which make them v-particles. The advantage of this seems to be that the letter v is reasonably unallocated and that the name has strong mnemonic values as, in fact, the tracks are v shaped. 1951 Nature 31 Mar. 503/1 Six charged V-tracks are due to the decay of new unstable particles. Ibid. 503/2 Two schemes are suggested to explain the photographs: $V^0 \to p^+ + \pi^- \ldots;\ V^0 \to \pi^+ + \pi^-$. 1952 Sci. Amer. Jan. 26/2 The V-particles appear to be somewhat more massive than a proton or neutron, because in some instances a proton is a decay product. 1968 M. S. Livingston Particle Physics v. 98 If an incident neutral particle has an interaction leading to two charged particles, the vertex of the V event shows the location, and the balance of transverse momenta identifies the incoming direction of the neutral particle. 1974 Frauenfelder & Henley Subatomic Physics vii. 170 By about 1952, many V events had been seen, and a mystery had developed: the V particles were produced copiously but decayed very slowly.

Vaad Leumi (vā·ād lə,umī·). [Heb. wa'ad committee + lə'ummī national.] A national committee of Palestine Jews, serving as their official representative during the period of the British Mandate from 1920 to 1948.

The form Vaad Haleumi in the 1926 example includes the definite article.

1926 Zionist Rev. Feb. 127/2 The Second Jewish National Assembly appoints a Vaad Haleumi composed of 38 members. 1932 Palestine Post 4 Dec. 1/2 Mr I. Ben Zevi, a member of the Vaad Leumi Executive. 1941 Contemp. Jewish Rec. IV. 428/2 The threat of Nazi occupation caused the Agency Executive..to ratify an agreement between the Vaad Leumi and the Right bloc. 1949 Koestler Promise & Fulfilment I. xv. 169 On March 1 the Vaad Leumi met in Tel Aviv. 1963 Times 24 Apr. 16/1 He later became one of the founders of the General Council for Palestine Jews (Vaad Leumi), the executive body of the Jewish community, and served as chairman or president from 1931 until it was dissolved with the establishment of the State of Israel. 1971 Encycl. Judaica XVI. 49 The Va'ad Le'ummi represented the yishuv in its relations with the Mandatory government and the Arab leaders and dealt with internal matters (such as the school system) which were delegated to it by the Zionist Executive.

vaalhaai (fā·lhai). S. Afr. Also **Vaalhai.** [Afrikaans, f. Du. vaal pale + haai shark.] A local name for the tope, Galeorhinus galeus; = TOPE sb.[2] Also attrib.

1947 K. H. Barnard Pict. Guide S. Afr. Fishes 10 Tope; Vaal-haai... This medium-sized (6 feet) cosmopolitan shark has recently become of considerable economic importance in South Africa for the extraction of vitamins from its liver-oil. 1949 Cape Times 24 Sept. 1/6 Any person capturing a Vaalhaai shark shall land it in a whole state. 1958 Cape Argus 14 June 13/4 The vaalhaai..grows to about 6 ft. and is harmless to man. 1973 Stand. Encycl. S. Afr. IX. 603/2 In South Africa that [sc. the liver] of the liver-oil shark Galeorhinus (vaalhaai) is especially rich, and from 1940 to 1955 this species was specially hunted for it.

‖ **vaaljapie** (fā·lyāpi). S. Afr. Also **Vaal Japie.** [Afrikaans, lit. 'tawny Jake', f. as prec. + japie dim. of the name Jaap f. Jakob.] Rough young wine, inferior wine.

1945 Cape Times 21 May, What I say is 'Come quick, go quick,' and Vaal Japie is my best friend. 1949 L. G. Green Land of Afternoon 59 Young wine, not matured but about six months old, is known as Vaaljapie.... It takes its name from its tawny colour though some varieties are red. 1958 Cape Times 29 Nov. 1/5 Some woodcutters..made him drunk on vaaljapie and called the police while he was asleep. 1968 D. J. Opperman Spirit of Vine 242 Brandy and vaaljapie have always had an irresistible attraction for these people. 1975 Stand. Encycl. S. Afr. XI. 464/2 The wine ration given by farmers to their labourers..is referred to as 'boys' wine', or 'vaaljapie'.

‖ **Vaalpens** (fā·lpens). S. Afr. Pl. **-pens, pense(n).** Also with small initial in attrib. use. [Afrikaans, f. as prec. + pens paunch, belly.] **a.** A name for a member of the Ba-Kalahari tribe. **b.** colloq. A nickname for a Transvaler.

1871 J. Mackenzie Ten Years North of Orange River iii. 53 Their fellow-countrymen to the south..sometimes call them 'Vaalpensen', which is the Dutch for Bakalahari, the ill-favoured and lean vassals of the Bechuanas. 1899 Eastern Province Herald (Port Elizabeth) 6 Dec. 3/4 A South African Dutchman writes us a somewhat bitter letter... He writes as a Dutch Afrikander, a Vaalpens in fact. 1900 A. H. Keane Boer States iii. 32 Here [in the Bosch Veld] is also the true home of the Vaalpens, most degraded of all the South African aborigines. 1916 Eastern Province Herald (Port Elizabeth) 28 Sept. 3 The Vaalpens reported that one of our oxen had been mauled. .. We saddled up and with three Vaalpens soon found where the lion had caught the ox. 1934 Star (Johannesburg) 1 May 13/1 For the past 50 years and more Free Staters have been known among Dutch-speaking South Africans as Blikore (tin ears) and Transvaalers as Vaalpense, the latter so called after a certain native tribe of that name who lived in the Transvaal. a 1936 E. Marais Soul of Ape (1969) iii. 92 Here in Waterberg..a case of 'homing' in a descendant of the so-called 'vaalpens pygmies' that at one time inhabited the Bushveld of the northern Transvaal. 1970 Personality Competition, The 'Vaalpens' are very scarce in South Africa nowadays.

‖ **va banque** (va bank). Gambling. [Fr., lit. 'go bank'.] In baccarat and chemin-de-fer, a bet against the whole of the banker's stake. (In quots., fig.) Cf. *BANCO int.

1946 A. J. P. Taylor Course of German Hist. ii. 38 Both dynasties desired the defeat of Napoleon; but the Hohenzollerns, having nothing more to lose, were ready to bid va banque—the Habsburgs were not. 1966 Economist 12 Nov. 683/3 Disraeli..was an adventurer who played the great game va banque with a courage and effrontery that commanded, perhaps even deserved, success.

vac², colloq. abbrev. of *VACUUM v.

1942 N. Last Diary 23 July in Nella Last's War (1981) vii. 212, I hurried home to bake my bread, vac the dining-room and dust. 1970 J. Wainwright Prynter's Devil iii. 50 Vac the room first, kiddo. Then start the repaint job. 1971 Guardian 26 Nov. 11/5 Little ladies in nylon overalls were noisily vaccing the deep red carpet. 1981 J. Wainwright Urge for Justice II. i. 100 My cleaning lady..vacs and polishes around.

vac³, colloq. abbrev. of vacuum cleaner s.v. VACUUM 4.

1974 P. Wright Lang. Brit. Industry xvii. 164 The brave new indoor world of vacs, mixers, mincers and all the rest. 1976 Star (Sheffield) 20 Nov. (Advt.), Cash paid for Washers, Vacs and Fridges in good condition. 1979 Arizona Daily Star 1 Apr. (Advt. Suppl.), Wards has jiffy vacs priced low as 22.88.

vac, abbrev. VACANT a.: see sit(s) vac s.v. *SIT sb.²

vacance. Add: **5.** poet. nonce-use. A rendering of Fr. absence in the original.

1930 T. S. Eliot tr. St.-J. Perse's Anabasis viii. 53 To the scale of our hearts was such vacance completed!

vacancy. Add: **II. 6. c.** A vacant room in a hotel, guest-house, etc. Usu. attrib. as vacancy sign, a signboard advertising available accommodation, or in pl.

1953 'R. Macdonald' Gone Girl in Lew Archer, Private Investigator (1977) 24 The first motel I came to..was decorated with a vacancy sign. 1970 R. H. Greenan Nightmare iv. 15 The place..is a bit sleazy... There's a vacancy sign out. 1972 Guardian 17 May 12/3 Hotels are replacing the 'Vacancy' signs with hoardings saying 'For Sale'. 1973 Value Added Tax Tribunals Rep. I. 165 Students have at their choice, provided vacancies are available, three types of accommodation. 1982 M. Babson Death beside Seaside xiii. 111 Most of the trippers will be leaving this afternoon. There'll be plenty of vacancies.

III. 9. c. Cryst. A defect in a crystal lattice consisting of the absence of an atom or ion from a position where there should be one.

1951 Physical Rev. LXXXII. 551/1 The experiments described below seem to be the most direct evidence, to date, that diffusion in close-packed metals occurs, predominantly, through the movement of vacancies. 1958 [see *SCHOTTKY 2]. 1967 A. H. Cottrell Introd. Metallurgy xx. 365 The quench..produces a supersaturated solution of vacancies and..these vacancies agglomerate to form dislocation rings or other defects which harden the metal by acting as obstacles to gliding dislocations. 1971 New Scientist 25 Mar. 664/2 The interstitial migrates by jumping from one site to another, while the vacancy migrates as a result of a neighbouring atom jumping into the vacant hole. 1974 Encycl. Brit. Macropædia V. 334/1 Theoretical considerations require that all crystals have vacancies except at absolute zero temperature.

vacant, a. Add: **7. a.** vacant-eyed, -minded (earlier example), -seeming adjs.

1836 Poe in Southern Lit. Messenger Apr. 339/2 Not a broad, forced, loud vacant-minded joke, but a quiet, pungent, sly, laughter-moving conceit. 1922 D. H. Lawrence Aaron's Rod xiii. 186 It was a large, vacant-seeming, Empire sort of drawing-room. 1936 L. H. Myers Strange Glory II. ix. 150 A boy of about ten.., ill-nourished and vacant-eyed. 1965 J. A. Michener Source (1966) 59 From the shores of Morocco..came frightened, dirty, pathetic Jews, illiterate, often crippled with disease and vacant-eyed.

b. In phr. vacant possession, with reference to premises (esp. those offered for sale): available for occupation by the purchaser, not occupied by the vendor or a tenant or tenants.

The legal interpretation of the term can be modified in certain circumstances by agreement between the vendor and the purchaser.

1825 H. Roscoe Treat. Law of Actions relating to Real Property I. 546 Ejectment cannot be maintained, as on a vacant possession, where there is any thing left by the tenant on the premises, however trifling. 1883 Wharton's Law-Lexicon (ed. 7) 287/1 In case of vacant possession the writ may be served by posting a copy on some conspicuous part of the property. 1927 Daily Tel. 24 May 4/7 Vacant possession at Michaelmas will be given of the Manor Farm, 428 acres, at Oxwick. 1946 Law Rep. (King's Bench) 264 A vendor who leaves chattels of his own on property sold by him to an extent depriving the purchaser of the physical enjoyment of part of the property has failed to give vacant possession. 1973 Country Life 15 Mar. 713/1 The average price of vacant-possession farms in England is £273 an acre. 1976 Morecambe Guardian 7 Dec. 28/1 (Advt.), Three bedroom semi-detached house with vacant possession.

vacate, v. Add: **1. b.** (Later example.)

1895 'Mark Twain' in N. Amer. Rev. July 3 In the 'Deerslayer' tale this rule is vacated.

4. c. (Earlier example.)

1836 Knickerbocker VII. 15 Ned and I were vacating..at his father's charming residence.

vacating (văkēi·tiŋ), ppl. a. [f. VACATE v. + -ING².] That is retiring from office, etc.

1921 Act 11 & 12 Geo. V c. 21 Sched. I. 6 Where the unexpired portion of the term of office of the vacating member is less than one year. 1981 N.Y. Times 18 Jan. XI. 20 Mrs. Kennelly suggested that the leader of the Senate or the House from the party of the vacating member be able to appoint a substitute legislator.

vacation, sb. Add: **III. 9.** (Earlier example.)

1860 Players 14 Apr. 121 Since Mr Kean's vacation of the Princess's, Miss Murray has joined the present company at the St. James's Theatre.

IV. 10. Comb., as **vacation home** U.S., a house used by the owner for holidays or at weekends (cf. holiday home (b) s.v. *HOLIDAY sb. 4 a); **vacation job,** paid employment for a student during vacation from a university, polytechnic, etc.; **vacation-land** U.S., an area attracting holiday-makers.

1969 'E. Lathen' When in Greece xiii. 146 The whole area is honeycombed with vacation homes. 1978 D. Williams Treasure up in Smoke iv. 40 A big-spending visitor who maintained a vacation home. 1971 Guardian 23 July 6/7, 24,400 students registered for vacation jobs. 1977 D. James Spy at Evening x. 68, I had..managed to get a vacation job as reserve stoker in a block of flats. 1927 Scribner's Mag. Apr. 100 The glories of Yellowstone Park take on new meaning this year with our discovery of Gallatin Gateway! It is the great sensation of vacation-land. 1977 Chicago Tribune 2 Oct. IV. 18/1 The region is a vacationland of great variety. There are spas, county fairs, antiques and crafts shows, [etc.].

vacationer, (a) for U.S. read orig. U.S. and add later examples; **vaca·tioning** ppl. a. and vbl. sb. (U.S.).

1926 Scribner's Mag. Aug. 7/2 Advice to vacationing young folks: In the dog days, don't be too Sirius. 1961 Times 28 Nov. 13/7 The organized and the individual vacationers. 1967 Idle Moments (Austral.) Sept. 27/3 The vacationer's body also will have reduced metabolic needs which may take two or three weeks for adjustment. 1976 National Observer (U.S.) 24 Jan. 8/2 (Advt.), Off-the-Beaten Path names the really low cost Florida retirement and vacationing towns. 1978 Detroit Free Press 16 Apr. (Parade Suppl.) 13/1 Increasingly, vacationing Americans are taking to the woods, mountains, deserts and seashores.

vaccinate, v. **1. b.** For 'virus' read 'vaccine'. (Earlier and later examples.)

1882 E. Klein in 11th Ann. Rep. Local Govt. Board in Parl. Papers (C. 3337. I) XXX. II. 509 In Pasteur's case the sheep inoculated with such bacilli..are not killed by anthrax, but 'vaccinated', and protected. 1955 [see *POLIO 1]. 1983 Oxford Times 3 June 18/2 Women are being urged to check that they have been vaccinated against German measles after a serious outbreak of the disease.

vaccination. 1. b. Substitute for def.: The inoculation of an individual with any vaccine in order to induce or increase immunity. (Further examples.) [The use of the term for diseases other than small-pox is due to Pasteur (Trans. 7th Session Internat. Med. Congr. (1881) I. 90).]

1896 Lancet 19 Sept. 809/2 These anti-cholera inoculations have served as a pattern for the typhoid vaccinations. 1955 Sci. Amer. Apr. 44/3 Tests show that antibody persists for an appreciable time after vaccination with the killed-virus vaccine. 1978 T. R. Bowry Immunol. Simplified iv. 14 Cholera vaccination is still required for travel to a few countries.

2. vaccination mark.

1899 A. E. Housman in Univ. College Gaz. 22 Mar. 34/2 Vain his laced boots, and vain his eyebrow dark, And vain, ah! vain, his vaccination mark. 1914 D. H. Lawrence Widowing of Mrs. Holroyd III. 92 And such arms on 'im! Look at the vaccination marks, Lizzie. 1983 N.Y. Times 20 Mar. 13/1 Her face was as powdered and pitted as the vaccination mark on her arm.

vaccine, sb. **2. b.** Substitute for def.: A preparation of the causative organism or substance of a disease (or its products) that has been specially treated for use in vaccination (see also quot. 1983). (Earlier and later examples.)

1882 E. Klein in 11th Ann. Rep. Local Govt. Board in Parl. Papers (C. 3337. I) XXX. II. 509, I have not yet succeeded in discovering the method employed by M. Pasteur..for the production of 'vaccine' protective against anthrax. 1892 Brit. Med. Jrnl. 28 May 1157/2 M. Pasteur has for some years directed his attention to the treatment of epilepsy by antirabic vaccine. 1911 Lancet 16 Sept. 814/1 The treatment of hay fever by hypodermic inoculations of pollen vaccine. 1931 W. T. Vaughan Allergy & Appl. Immunol. xxix. 328 While the use of immune sera has not been successful in the treatment of typhoid fever, the vaccine has more than demonstrated its efficiency as a preventive. 1962 J. H. Burn Drugs, Med. & Man xii. 132 Evidence concerning the value of whooping-cough vaccine to protect children against whooping-cough. 1980 D. J. Rapp Allergies & your Family xiii. 220 Measles vaccine should be avoided by children who are receiving steroids. 1983 Sci. Amer. Feb. 48/2 There has been increasing interest in the preparation of synthetic vaccines, which is to say vaccines containing not intact viruses but merely peptides..that

have been constructed in the laboratory to mimic a very small region of the virus's outer coat.

3. Special Combs.: **vaccine-damaged** a., harmed, esp. seriously, as a result of being vaccinated; also absol.; **vaccine therapy,** (G. B. Shaw) **vaccinetherapy,** treatment of a disease with an appropriate vaccine.

1976 S. Wales Echo 22 Nov. 8/8 As a parent of a vaccine-damaged daughter who is classed as severely subnormal I can say every word Mr. Peter Ellershaw wrote was true. 1980 Times 15 Jan. 4/1 (heading) New campaign to win state help for the vaccine-damaged. 1907 Jrnl. Infectious Dis. IV. 313 (heading) The opsonic index and vaccine therapy of pseudodiphtheric otitis. 1911 G. B. Shaw Doctor's Dilemma p. lxxxviii, The theory and practice of securing immunization from bacterial diseases by the inoculation of 'vaccines' made of their own bacteria: a practice incorrectly called vaccinetherapy (there is nothing vaccine about it). 1974 Mycopathologia et Mycol. Applicata LIII. 25 (heading) Pompholyx of the hands and feet. Its etiology, pathogenesis and specific vaccine therapy.

vaccinee. Delete rare and add later examples.

1970 Nature 24 Oct. 307/1 Some of the vaccinees might unknowingly be pregnant. 1977 Lancet 20 Aug. 400/1 First results were obtained from 20 sera of vaccinees.

‖ **vacherin** (vaʃ'ræñ). [Fr.] **a.** A kind of cheese (see quot. 1936). **b.** A confection of meringue and whipped cream.

1936 A. L. Simon Catechism concerning Cheeses 44 Vacherin. A French Savoy and Swiss cheese made during the winter. 1960 E. David French Provincial Cooking 34 We were then beguiled into eating a sweet called a Vacherin glacé. This turned out to be..ice-cream, glacé fruits, frozen whipped cream, and meringue. 1965 A. Roudybush Season for Death (1966) xxiii. 133 An impromptu supper of..venison baked in wine and great vacherin meringues was served. 1969 R. & D. De Sola Dict. Cooking 231/1 Vacherin, Swedish cheese used as a spread, creamy and aromatic. 1979 N.Y. Times 18 May C14/4 A vacherin of nutted meringue layers with whipped cream and strawberries.

vacky (væ·ki). colloq. abbrev. Also **vaccy, vakky.** [f. *EVACUEE: cf. -Y⁶.] An evacuee, esp. a child evacuated from the city to the country esp. at the beginning of the 1939–45 war.

1940 John o' London's Weekly 3 May 147/1 These little vackies and their country friends. 1940 J. Cary Charley is my Darling ii. 16 There are only eight vackies in the place. 1949 E. Coxhead Wind in West iii. 66, I was a vakky. My home's in Clapham really, but our school was sent to Bedfordshire. 1962 M. Duffy That's how it Was viii. 71 The Wortbridgers distrusted the vaccies with their quick ways and sharp, pinched faces. 1979 Bookseller 25 Aug. 832/2 A 'must' book for all ex-'vaccies, and their parents, and their children. 1982 J. Aiken Whisper in Night 151 The Welsh landscape had changed..since he was a Vakky, sent here for safety from the Birmingham bombs.

vacuolating (væ·kiu,ǒlēⁱtiŋ), ppl. a. Med. [Back-formation from VACUOLATION: see -ING².] vacuolating agent or virus: a papovavirus, orig. obtained from rhesus monkey kidney tissue, which is capable of causing tumours in animals and animal tissue cultures; also called SV 40 (cf. SV s.v. *S 4 a); also, a virus related to this.

1960 Sweet & Hilleman in Proc. Soc. Exper. Biol. & Med. CV. 420/1 This agent has been called the 'vacuolating virus' by us because of the prominent cytoplasmic vacuolation seen in infected cell cultures... The discovery of this new virus, the vacuolating agent, represents the detection..of a hitherto 'non-detectable' simian virus of monkey renal cultures. 1965 Listener 11 Mar. 369/2 A remarkable situation is also presented by..vacuolating virus or simian virus 40. This virus..was accidentally given to millions of human subjects with early batches of polio vaccine. 1977 Jrnl. Virology XXI. 179/1 Papovaviruses of the simian virus 40 (SV 40)-polyoma subgroup ..occur in the mouse.., the rabbit (rabbit kidney vacuolating virus [RKV]), [etc.].

vacuolized (væ·kiu,ǒləizd), ppl. a. [f. VACUOLE + -IZE + -ED¹.] = VACUOLATED ppl. a.

1910 Ann. Bot. XXIV. 733 These bands and masses may be homogeneous or vacuolized. 1961 Lancet 2 Sept. 525/2 The cells were enlarged and their cytoplasm was vacuolised.

vacuome (væ·kiu,ōᵘm). Cytology. [a. F. vacuome, f. L. vacu-us empty: see *-OME.] The vacuoles of a cell collectively (see quot. 1976).

1926 [see *PLASTIDOME]. 1936 Nature 20 May 914/2 The vacuome and chondriome systems can be distinguished by their behaviour to vital stains. 1947 Ann. Rev. Microbiol. I. 9 In the dinoflagellate, Polykrikos schwartzi, ..mitochondria, Golgi material, vacuome, fat globules and glycogen have been described. 1959 W. Andrew Textbk. Compar. Histol. v. 146 As it [sc. the food particle] approaches the vacuome (Golgi substance), particles from this structure are attracted to its surface. 1960 L. Picken Organization of Cells vi. 235 According to a theory elaborated by Parat and Painlevé..all cells contain two systems, unrelated though of profound importance in cell economy: the chondriome—a lipoidal 'phase' manifested as isolated mitochondrial networks; and the vacuome—a dispersed phase of aqueous droplets staining with neutral red, to be equated with the Golgi apparatus. 1976 R. Rieger et al. Gloss. Genetics & Cytogenetics (ed. 4) 564 Vacuome, originally, a system comprising a variety of inclusions of plant cells, from the highly hydrated tonoplast to the dense aleuron granules, all with the common ability to stain vitally with neutral red. Today, all the membrane-bounded spaces of the cell with the exception of the mitochondria and the plastids which make up the chondriome..and the plastome respectively. 1981 Jrnl. Ultrastructure Res. LXXVI. 317 Multivesicular bodies seen in section of the unicellular green alga Scenedesmus obtusiusculus are profiles of a continuous cisternum, the vacuome.

vacuum. Delete ‖ and add: **2. c.** ellipt. for vacuum cleaner, sense 4 below. colloq. (orig. U.S.).

1910 Judge 9 Apr. 8/2 A vacuum was the only thing she could be trusted to handle with safety. 1922 Hotel World 25 Mar. 14/1, I have three vacuums going all day. 1960 Farmer & Stockbreeder 8 Mar. (Suppl.) 10/1 Is it better to have a powerful or a handy cleaner? That has always been a problem when choosing a 'vacuum'. 1977 New Yorker 24 Oct. 88/3 Green Haven's dep [sc. deputy superintendent] of administration was preoccupied with the size of the wet-dry vacuums being used to clean the prison kitchen.

4. vacuum-jacketed adj.; **vacuum activity** (see quot. 1981); **vacuum aspiration,** a method of induced abortion in which the contents of the uterus are removed by suction through a tube passed into it via the vagina; **vacuum bottle** = vacuum flask below; **vacuum chamber,** a chamber designed to be emptied of air; **vacuum cleaner,** an electrical appliance for removing dust (from carpets and other flooring, soft furnishings, etc.) by suction; also transf. and fig.; hence (as a back-formation) **vacuum-clean** v. trans., = *VACUUM v.; so **vacuum-cleaning** vbl. sb.; **vacuum deposition,** deposition of a substance by allowing it to condense from the vapour in what is otherwise a vacuum; so **vacuum-deposit** v. trans., **-deposited** ppl. a.; **vacuum extraction** Obstetr., the application of suction to a baby's head to assist its birth; so **vacuum extractor,** a cup-shaped appliance for achieving this; **vacuum-fitted** a., of a railway car: furnished with a vacuum brake; **vacuum flask,** a vessel with a double wall enclosing a vacuum so that liquid in the inner receptacle retains its temperature (cf. THERMOS in Dict. and Suppl.); also transf. in attrib. use; **vacuum fluctuation** Physics, a fluctuation in field strength in a nominally field-free vacuum, occurring in consequence of the quantization of any radiation field; **vacuum forming,** a type of thermoforming in which a vacuum is used to draw the plastic into the mould (see quot. 1974); **vacuum grease,** a grease which because of its low vapour pressure is suitable for sealing joints in a vacuum apparatus; **vacuum packaging,** (a) = vacuum-packing vbl. sb. s.v. *VACUUM-PACK v.; (b) the vacuumized container used in vacuum-packing; **vacuum polarization** Physics, the spontaneous appearance and disappearance of electron–positron pairs in a vacuum; **vacuum pump,** a pump for evacuating a container of air or other gas; (examples); **vacuum-tight** a. = AIR-TIGHT a.; **vacuum wax** = vacuum grease above.

1953 N. Tinbergen Herring Gull's World iv. 35 It is possible..this was just a kind of 'vacuum activity' due to the accumulation of the urge to paddle. 1981 Oxf. Compan. Anim. Behaviour 579/1 Vacuum activities occur in the apparent absence of the external stimuli that normally elicit the activity. 1967 Obstetrics & Gynecol. XXX. 28 (heading) A critical view of vacuum aspiration: a new method for the termination of pregnancy. 1974 Passmore & Robson Compan. Med. Stud. III. xlii. 7/2 Before the 12th week it is customary and simplest to evacuate the uterus via the vagina by vacuum aspiration. 1978 F. Weldon Praxis ii. 13 If by mistake they fall pregnant, they abort by vacuum aspiration. 1910 Chambers's Jrnl. June 413/2 The vacuum-bottle has entered so extensively into the domestic circle as to become regarded almost as indispensable. 1976 Daily Colonist (Victoria, B.C.) 6 Mar. 19/5 Use lunch buckets or vacuum bottles to keep foods hot or cold. a 1877 Knight Dict. Mech. III. 2687/2 Huffer..claims the use of exhaust steam from an engine to condense in a vacuum-chamber, and so raise water to turn a wheel. 1971 Materials & Technol. II. v. 230 A certain amount of water is evaporated from the clay as it passes through the vacuum chamber. 1912 Sci. Amer. 22 Nov. 442/2 (caption) Vacuum-cleaning an automobile. The exhaust gases of the motor create the suction. 1924 Kipling Debits & Credits (1926) 149 The organ-bench, whose purple velvet cushion was being vacuum-cleaned on the floor below. 1956 Good Housek. Home Encycl. (ed. 4) 316/2 Wickerwork..requires to be brushed or vacuum-cleaned regularly. 1973 Physics Bull. Feb. 110/3 These..could then be scraped and vacuum cleaned from the carpet. 1977 J. R. L. Anderson Death in City i. 10 A team..did the rooms,..vacuum-cleaning floors. 1983 Daily Tel. 8 Apr. 16/3 Soviet spies are vacuum cleaning the West for its industrial and scientific secrets. 1903 Vacuum cleaner [in Dict.]. 1907 Yesterday's Shopping (1969) 113/1 The 'Witch' Dust Extractor is a vacuum cleaner suitable alike for carpets, upholstery, clothing, &c. 1962 A. Lurie Love & Friendship i. 12 She..did not hear the noise of the vacuum cleaner sucking its way up the stair carpeting. 1972 World Bk. Sci. Ann. 1971 342 The vacuum cleaner is actually a hydraulic pump designed to lift as much as 400 tons of material from the sea floor daily. 1976 Times 1 Mar. 13/3 The conservation of fish stocks..in the face of over-developed world fishing fleets and 'vacuum cleaner' techniques of fishing. 1902 Let. 20 Oct. in Goblin Story (Goblin Ltd.) (1969) 9, I have submitted the subject of the Vacuum Cleaning Company's operations to the King. 1916 J. Webster Dear Enemy 71 'Well?' said she, her tone implying that I was a vacuum-cleaning agent. 1939 Country Life 11 Feb. p. vi (Advt.), Vacuum cleaning plant. 1946 Nature 14 Dec. 862/1 Mollwo reported on the density of vacuum-deposited salt layers. 1960 McGraw-Hill Encycl. Sci. & Technol. X. 596/2 Fabrication of printed resistors by vacuum deposition is an expensive process and there has been no general application of this technique except for precision resistive elements. 1982 Jrnl. Colloid & Interface Sci. XC. 335 When certain materials such as Se or Sn are vacuum-deposited onto heated polymers under appropriate conditions, a most unusual structure can form. Ibid. 337/1 The structure of the monolayer assemblies formed by vacuum deposition is almost independent of the deposition rate. Ibid., The vacuum deposited structure is little altered if the system pressure during deposition is allowed to rise. 1961 Lancet 22 July 189/2 They concluded that vacuum extraction does not distress the fœtus. 1975 I. Illich Med. Nemesis vii. 120 The vacuum extraction method has rendered the interruption of pregnancies safe, cheap and simple. 1954 T. Malmström in Acta Obstetrica et Gynecologica Scandinavica XXXIII. Suppl. No. 4. 5 The vacuum extractor ad modum Dr Malmstrom is constructed and manufactured in co-operation with R. Soderberg & Co, Gothenburg, Sweden. 1960 Proc. R. Soc. Med. LIII. 749/1 The vacuum extractor consists of a flattened round metal cup with bulging sides. 1980 S. Kitzinger Pregnancy & Childbirth 260 Sometimes a vacuum extractor, or ventouse, is used instead of forceps. 1937 Vacuum-fitted [see passenger train s.v. *PASSENGER 7 a]. 1968 Listener 15 Aug. 210/1 We..have a brake which is known as the vacuum brake, and our wagons are known as vacuum-fitted. 1917 Harrods Gen. Catal. 946/3 The 'Icy-Hot' Vacuum Flasks, improved designs, will keep liquids hot for 24 hours or cold for 3 days. 1926 Daily Colonist (Victoria, B.C.) 20 Jan. 5/1 This vacuum flask will keep things hot for you indefinitely. 1958 New Statesman 23 Aug. 214/2 Investigation in person will clearly have to wait until travellers are provided with armour-plated vacuum-flask suits. 1978 M. Babson Tightrope for Three vi. 39 She..filled the kettle..and..the vacuum flask as well. 1955 L. Rosenfeld in W. Pauli Niels Bohr 88 The 'vacuum' fluctuations of the field variables. 1973 Nature 14 Dec. 397/1 The laws of physics place no limit on the scale of vacuum fluctuations. The duration is of course subject to the restriction $\Delta E \Delta t \sim h$. 1979 J. C. Polkinghorne Particle Play v. 74 Buffeted by vacuum fluctuations the elementary particle has a pretty rocky ride. This is the origin of the infinities which people found when they first tried to calculate with quantum field theories. 1946 J. Sasso Plastics Handbk. for Product Engineers v. 308 In the vacuum-forming process the plastic sheet is heated to approximately 250F. 1974 Encycl. Brit. Macropædia XIV. 521/1 In simple vacuum forming, the plastic sheet is clamped over an airtight box, from which air is exhausted, rapidly drawing the heated plastic down on a 'former' within the box, thus reproducing its shape. 1946 Jrnl. R. Aeronaut. Soc. L. 393 Connection to the vacuum system is made by a ground metal joint sealed with vacuum grease. 1977 Vacuum XXVII. 431/1 There are three factors involved in choosing a vacuum grease—cost, convenience, and its effect on the vacuum system. 1946 Nature 20 July 105/1 The best way to carry methane on motor-vehicles is as a liquid in vacuum-jacketed tanks. 1970 Sci. Jrnl. Aug. 82/3 The 'lagged storage tank' for LNG would need to be vacuum jacketed. 1954 L. C. Barail Packaging Engineering xx. 249 The cans filled with the whole products are then sealed by means of vacuum machines, a process known as vacuum packaging. 1982 R. Manheim tr. Grass's Headbirths iv. 65 The cut in the vacuum packaging could be patched up. [1935 Physical Rev. XLVIII. 55/2 The existence of [such] an induced charge corresponds to a polarization of the vacuum.] 1951 Ibid. LXXXI. 664/1 We shall illustrate this assertion by applying such a gauge invariant method to treat several aspects of the problem of vacuum polarization by a prescribed electromagnetic field. 1979 Cheng & O'Neill Elementary Particle Physics iii. 82 The two dominant processes that modify the electron's interaction with the proton.., resulting in the Lamb shift, are the vertex correction and vacuum polarization. 1882 Chem. News 27 Oct. 192/1 In the machine in question the refrigeration is effected simply by the evaporation of water under a special vacuum-pump. 1979 J. Matley Fluid Movers v. 267/1 A good seal between the inlet and outlet sides is essential in all mechanical vacuum pumps. 1927 Brit. Jrnl. Radiol. XXIII. 143 The glass is directly sealed to a chromium-iron alloy which is perfectly vacuum-tight. 1946 Nature 23 Nov. 756/2 Vacuum-tight seals through hard glass with bare tungsten wire are difficult to make. 1964 M. Gowing Britain & Atomic Energy 1939–1945 viii. 219 New techniques had to be developed for ensuring that the machinery was..exceptionally vacuum-tight. 1926 J. H. Smith tr. L. Dunoyer's Vacuum Practice iii. 142 Vacuum waxes. Golaz wax has been in use since Regnault's time. 1971 Sci. Amer. Aug. 108/1 The joint was sealed with Apiezon W-100 vacuum wax.

vacuum (væ·kiu,ǒm), v. colloq. (orig. U.S.). [f. the sb.] trans. To clean (a room, carpet, etc.), or to remove (dust, etc.) with a vacuum cleaner. Also fig. and absol. Occas. intr. for pass.

longer can we assume that Greece is the hermetically sealed Olympian miracle, any more than we can consider Israel the vacuum-packed miracle from Sinai. **1960** *Farmer & Stockbreeder* 19 Jan. (Suppl.) 41/3 The final processing involves..draining, swabbing and then vacuum-packing in transparent film bags. **1962** *Spectator* 4 May 602/3, I was at a testing of vacuum-packed bacon. Vacuum-packing is likely to spread, because it is so convenient for the handlers. **1975** *Times* 18 Feb. 13/2 The vacuum-packed fine arts sector. **1984** *Listener* 5 Apr. 10/2 The girl said it came vacuum-packed.

vacuum tube. **1.** An evacuated tube or pipe, esp. one along which vehicles or other objects can be propelled by allowing air to enter behind them.

1784 in *Pall Mall Mag.* (1896) Aug. 586 Grand Vacuum Tube Company. Direct to Bengal. **1846** *Patent Jrnl.* 4 July 90/1 Fig. 1 is a ..section of a railway carriage and traction tube. Fig. 2 is a cross section..showing the communication between the carriage and the piston. The traction or vacuum tube..is the same as generally used. **1920** D. H. Lawrence *Touch & Go* 7 A system of vacuum tubes for whooshing Bradburys about from one to the other. **1972** *Daily Tel.* 30 Dec. 6/5 The explanation, it is believed, is that while it was being scooped up it became mixed with the dark grey soil usually found on the Moon. It could not have changed as a result of exposure to the Earth's atmosphere, because it is stored in vacuum tubes. **1973** *Times* 15 Oct. 6/3 High-speed surface systems should be studied as an alternative to air travel, including advanced systems such as the gravity vacuum tube, which would give high speeds for amazingly low energy consumption. **1974** [see *people mover* s.v. *people *sb.* 8].

2. An evacuated tube (sense *2 g) (orig. a glass cylinder); *spec.* one used as a thermionic valve.

1859 *Phil. Trans. R. Soc.* CXLVIII. 15 The direct discharge is that which is visible when taken from two wires hermetically sealed in a vacuum tube. **1880** *Rep. Brit. Assoc. Adv. Sci. 1880* 260 The band-spectrum of nitrogen... It was first observed by Plücker (1858) in a vacuum tube. **1901** *Engineer* 17 May 507/2 One of the most interesting exhibits was a number of electric vacuum-tube lamps. **1915** W. H. Eccles *Wireless Telegr. & Teleph.* 376 A. Langmuir has applied his vacuum tube with third electrode (the 'pliotron') in a manner different from those just described. **1923** *Electr. Communication* II. 157/2 No single advance has contributed so largely to change our whole picture of art as the advent of the thermionic valve or vacuum tube as it is designated in

conquerors; the phrase as a maxim or utterance.

1612 J. Selden in Drayton *Poly-Olbion* viii. 124 Whence, vpun a murmuring complaint among the Romanes, crying væ victis, came that to be as prouerbe applied to the conquered. **1792** Burke *Let.* 3 Jan. in M. Arnold *Lett., Speeches & Tracts on Irish Affairs* (1881) 259, I cannot say *væ victis*, and then throw the sword into the scale. **1819** Scott *Ivanhoe* I. ii. 39 The *vae victis*, or severities imposed upon the vanquished. **1856** M. Bernard in *Oxford Ess.* II. ii. 90 The stonecutter has laboured to produce a vigorous representation of the *vae victis*—of triumphant pride and abject humiliation. **1904** G. K. Fortescue in *Cambr. Mod. Hist.* (1907) VIII. xvi. 512 *Vae victis* was one of the few unchanging revolutionary maxims. **1936** J. Nehru *Autobiogr.* l. 401 *Vae victis* seems to run like a thread through these utterances. **1944** S. Bellow *Dangling Man* 112 Life is hard. *Vae victis!* The wretched must suffer.

vaesite (vēⁱ·zəit). *Min.* [f. the name of J. F. *Vaes*, 20th-c. mineralogist in the Belgian Congo + -ite¹.] A mineral of the pyrite group, ideally nickel sulphide, NiS_2, found as grey isometric crystals.

1945 P. F. Kerr in *Amer. Mineralogist* XXX. 483 Material approaching NiS_2 in composition and having a pyrite type of lattice has been found in the Kasompi mine of the Belgian Congo. This mineral is tentatively called vaesite. **1962** W. A. Deer et al. *Rock-Forming Minerals* V. 132 They suggest that many of the natural bravoites are actually vaesite solid solutions or two phase mixtures of nickel-rich pyrite and iron-rich vaesite. **1978** *Mineral. Abstr.* XXIX. 80/2 The first occurrence of vaesite in Japan is described from the Kuroko-type deposits of the Kosaka mine... It contains small amounts of Fe, Cu, and Co.

‖ **va-et-vient** (va e vʸæ̃). [Fr., lit. 'goes-and-comes'.] Coming and going, toing and froing; commerce, exchange; bandying (of argument).

1919 R. Fry *Let.* 3 Nov. (1972) II. 465 There is..a considerable *va et vient* of Paris artists. **1933** G. Arthur *Septuagenarian's Scrap Book* 18 For a large percentage of women life, day in day out, is a happy *va-et-vient*, sometimes rich in excellent results of practical value, sometimes only resolving itself into a flutter of delicious nothings. **1959** C. Spry *Favourite Flowers* iii. 28 Looking back over the quite gentle *va-et-vient* of this argument I know that really we are in agreement. **1966** *Guardian* 19 Mar. 7/4 All the well-intentioned wordy *va-et-vient*

with Vagabondia. The latter is a more ancient and less alluring realm of the Old World.

vagabonding, *vbl. sb.* (In Dict. s.v. Vagabond *v.*) (Later example.)

1925 H. L. Foster *Trop. Tramp with Tourists* 148 In years of vagabonding about the far corners of the earth.. I have never seen [etc.].

vagabondish, *a.* **1.** (Earlier example.)

1805 C. Wilmot *Let.* 4 Aug. in Londonderry & Hyde *Russ. Jrnls.* (1934) II. 165 His appearance was rather *Vagabondish* at the first glimpse.

vagabondizing, *vbl. sb.* (Earlier example.)

1755 C. Charke *Life* 223 The Aversion I had conceived for Vagabondizing..and the good Nature of my Friends in Chepstow put it strongly in my Head to settle there.

vagal, *a.* Add: Hence **va·gally** *adv.*, by, or by means of, the vagus.

1951 *Gastroenterology* XIX. 263 In dogs with vagally innervated or vagally denervated gastric pouches intravenous injections of Banthine..and of atropine..almost completely inhibited the secretion of hydrochloric acid. **1961** *Lancet* 26 Aug. 475/1 Unconsciousness can follow irritation of vagally innervated viscera. **1978** *Nature* 28 Sept. 323/2 Vagally induced release of Ach.

vagancy. Delete † *Obs.* and add later example.

1945 A. J. Macdonald *Episcopi Vagantes in Church Hist.* 8 The sentence of deprivation, which was intended as a means of checking episcopal vagancy, in effect only made it worse, by causing the offender to pass from being an occasional vagans into a condition of permanent vagancy.

‖ **vagantes** (vāgæ·ntīz), *sb. pl.* [L. pres. pple. of *vagārī* to wander.] The scholar monks who travelled about Europe in the Middle Ages. Occas. in sing. form *vagans*.

1927 H. Waddell *Wandering Scholars* p. v, The historical interest of the *Vagantes* as one of the earliest disintegrating forces in the mediaeval church has been left on one side. **1945** [see *vagancy]. **1946** *Scrutiny* Dec. 90 Carols, nursery rhymes and the songs of the vagantes. **1973** M. Blackett *Mark of Maker* viii. 79 Helen began serious work on selection and translation of the lyrics of her 'vagantes'. **1982** F. Corrigan in H. Waddell *Songs*

of Wandering Scholars 9 Her research brought her face to face with *vagantes*, wandering scholars, from the very dawn of Christianity.

vagary, *sb.* Add: Now freq. with pronunc. (vēⁱ·gări). **1. c.** An irregular course or distribution.

1923 *World's Work* May 566/1 Huster's attention was arrested by the uniformity in the course and distribution of nerves in comparison with the vagary of blood-vessels.

vagile (væ·-, vēⁱ·dʒəil), *a. Biol.* [ad. G. *vagil*, f. L. *vag-us* wandering, straying.] Of an organism or group of organisms: having the ability to disperse or be dispersed in a given environment. Hence **vagi·lity** (also *fig.*).

1903 *Amer. Geologist* XXXI. 214 Their recent descendants find it..advantageous..to prey upon the much richer fauna of the vagile benthos. **1937** Allee & Schmidt tr. *Hesse's Ecol. Animal Geogr.* vi. 76 The power of dispersal of an animal species may be summed up as its 'vagility'. The less the vagility of a species, the less it is able to overcome barriers..and hence the more numerous the opportunities for independent variation. **1967** *Oceanogr. & Marine Biol.* V. 478 Many small vagile species swim in the water washing the algae. **1979** E. A. & L. D. Bloom *Satire's Persuasive Voice* vi. 269 Eighteenth-century satire registers a widening discontent with the collapse of native culture and, conversely, a search for its own kind of moral vagility. **1981** *Science* 12 June 1291/2 Taxa with small Nₑ [*sc.* effective population size] due to low vagility..are expected to experience rapid..speciation rates.

vagina. Add: **1. c.** *vagina dentata* (Anthrop. and Psychol.). [L. *dentata* adj., having teeth, toothed], the motif or theme of a vagina equipped with teeth which occurs in myth, folklore, and fantasy, and is said to symbolize fear of castration, the dangers of sexual intercourse, of birth or rebirth, etc.

1908 R. H. Lowie in *Jrnl. Amer. Folklore* XXI. 108 Its variants in North America are innumerable; the most important being the crushing entrance to heaven, snapping doors,..and the vagina dentata. **1926** J. I. Suttie tr. *Ferenczi's Further Contributions Psycho-Anal.* xxxii. 279 Anxiety in regard to the mother's vagina (vagina dentata = birth anxiety). **1932** *Jrnl. R. Anthrop. Inst.* LXII. 219 There is in fact one fear, the neurotic dread of sexual intercourse, which is symbolized in the same manner as *vagina dentata* by many people in many countries. **1950** [see *beast *sb.* 4 b]. **1958** W. R. Trask tr. *Eliade's Birth & Rebirth* iii. 51 The return [to the womb] implies the risk of being torn to pieces in the monster's jaws (or in the vagina dentata of Mother Earth). **1976** *Listener* 26 Feb. 249/3 *Vagina dentata*, that prototype male nightmare which..always seems to crop up more often in books than real life. **1983** *Observer* 5 June 32/3 The heroes fight against being fed into the all-consuming mouth..of a nightmarish gigantic vagina dentata.

vaginal, *a.* Add: **2.** *vaginal smear*: see *smear *sb.* 3 b.

Hence **vagi·nally** *adv.*, via the vagina; in the vagina.

1929 *Amer. Jrnl. Obstetr. & Gynecol.* XVIII. 245 We are very slow to examine primiparae vaginally. **1966** Masters & Johnson *Human Sexual Response* ix. 137 A subjective awareness of tachycardia described frequently as feeling the heartbeat vaginally. **1979** *Nature* 6 Sept. 15/1 The new compounds seem to produce fewer gastro-intestinal side effects and can be given intramuscularly, vaginally or orally. **1982** 'D. Shannon' *Motive on Record* iii. 53 She'd been raped vaginally and anally.

vaginismus. Add after 'vagina': in response to physical contact or pressure. (Further examples.)

1936 W. Shaw *Textbk. Gynæcol.* xi. 234 Typical cases of vaginismus always have a psychical basis. **1974** Passmore & Robson *Compan. Med. Stud.* III. xxxv. 35/1 Frigidity..also includes the symptoms of those women in whom spasm of the perineal muscles (vaginismus) make the [sexual] act painful..or impossible. **1984** *Listener* 8 Mar. 34/4 Dr Stanley gives some candid advice on premature ejaculation and vaginismus, the involuntary muscular spasm of the outer third of the vagina which prevents penetration.

vagitus (vădʒəi·tŏs). [L., f. *vāgīre* to utter cries of distress, to wail.] A cry or wail; *spec.* that of a new-born child.

1652 N. Culverwel *White Stone* in *Elegant Discourse Light of Nature* ii. 119 Thou hast not yet the strength a well grown Christian; well, but is there the *vagitus* of an Infant? **1825** R. Hooper *Lexicon Medicum* (ed. 5) 1237/1 *Vagitus*, the cry of young children; also the distressing cry of persons under surgical operations. **1921** *19th Cent.* July 28 The various inspired articles..hardly went beyond the vagitus, the earliest cry of the new-born method. **1938** S. Beckett *Murphy* v. 71 To go back no further than the vagitus, it had not been the proper A of international concert pitch,..but the double flat of this. **1957** V. Nabokov *Pnin* ii. 47 He actually seemed to forehear the babe's vagitus. **1977** A. Sheridan tr. *J. Lacan's Écrits* iii. 31 My speech was to be nothing more than a *vagitus*, an infantile cry.

vago-. Add: **vago·tomy** *Surg.* [-tomy], an operation in which the vagus nerve is cut, either as a research technique or as a means of reducing the rate of gastric secretion; hence **vago·tomized** *ppl. a.*

1906 J. R. Murlin tr. *R. Tigerstedt's Text-bk. Human Physiol.* vi. 190 The hearts of dogs which had survived bilateral vagotomy for several months presented no anatomical changes whatsoever. **1948** *Gastroenterology* X. 415 He then produced ulcers in the vagotomized animals. **1955** W. Gaddis *Recognitions* i. v. 180 You ought to go back to analysis. Or have a vagotomy. **1974** R. M. Kirk et al. *Surgery* vi. 79 Alternatively, the available gastric acid may be reduced by performing partial gastrectomy, or vagotomy and a drainage procedure if fibrous scarring has made the oesophagus too short to restore the cardia to the abdomen. **1975** *Nature* 20 Nov. 227/1 In response to isoprenaline, vagotomised rats also failed to drink normally.

vagotonia (vēⁱgotōᵘ·niă). *Physiol.* Also anglicized as **vagotony** (vēⁱ-, văgǫ·tŏni). [f. Vago- + *-tonia.] The state or condition in which there is increased influence of the autonomic nervous system and increased excitability of the vagus nerve. Also **vagoto·nus**, in the same sense.

1916, etc. [see *sympathicotonia]. **1929** A. Kuntz *Autonomic Nervous System* xviii. 432 The clinical conception of vagotonia and sympatheticotonia was first formulated by Eppinger and Hess in 1909... They assumed that clinical cases exhibiting a functional imbalance between the sympathetic and parasympathetic nerves could be classified as vagotonic or sympatheticotonic. **1977** *Lancet* 12 Nov. 1027/2 Beta-blockers..are probably unable to turn the situation so far towards parasympathicotony (vagotony) that more profuse lachrymation would be resumed. **1981** *Acta Astronautica* VIII. 798 Numerous disorders have been related to vagotonia.

Hence **vagoto·nic** *a.*, displaying or promoting vagotonia.

1929 [see above]. **1977** *Proc. R. Soc. Med.* LXX. 159/1, I consider it dangerous to use a widespread sympathetic block..combined with halothane (with its vagotonic action and myocardial depressant effect).

vagrance. (Later U.S. example.)

1951 W. Faulkner *Requiem for Nun* iii. i. 266 Of course, the other niggers would just be in and out over Saturday and Sunday night for fighting or gambling or vagrance or drunk.

vague, *a.* Add: **3. a.** Used in superl. with ellipsis of *idea, notion*. Cf. *faint *sb.* 5 d.

1968 N. Marsh *Clutch of Constables* viii. 201 'Have you seen this particular photograph, Mr Pollock?'..'Haven't the vaguest.' **1981** D. Uhnak *False Witness* (1982) xi. 94 'Any particular place your customers go on Tuesday nights?' 'I haven't the vaguest.'

7*. *Comb.*, as *vague-looking* adj.

1904 W. H. Hudson *Green Mansions* vi. 82 How different she seemed now; the brilliant face grown so pallid and vague-looking! **1948** Wyndham Lewis *Let.* 25 Oct. (1963) 469 This super-dream..is I imagine too vague-looking to be practical.

‖ **vague** (vāg), *sb.*³ [Fr., lit. 'wave'.] A movement, trend, vogue. Cf. *nouvelle vague.

1962 *John o' London's* 19 Apr. 371/3 Here is one requisite which must be regarded as essential in any new *vague*. It must not, whatever else it is, be vague. **1970** R. Lowell *Notebk.* 219 The *vague*, the vogue, what do they tell the critic? **1974** *Times Lit. Suppl.* 20 Dec. 1439/4 He has beautifully caught the old Hungarian *vague* of British cinema under Alexander Korda... The Hungarian *vague* also swept on to the Denham studio floor such interesting flotsam as Gabriel Pascal.

vague, *v.*¹ **a.** γ. (Later example.)

1874 L. Troubridge *Life amongst Troubridges* (1966) 88 We vagued about until tea-time.

vagulate (vēⁱ·giŭlēⁱt), *v. rare.* [Fanciful formation f. L. *vagul(us* nonce dim. of *vagus* wandering + -ate³; perh. influenced by Undulate *v.*] *intr.* To wander in a vague manner; to waver.

Only in the writings of Virginia Woolf.

1918 V. Woolf *Diary* 3 Nov. (1977) I. 213 Emphie vagulates in & out of the room. **1921** *Ibid.* 6 Mar. (1978) II. 97 All is too soft & emotional. Now for writing or anything I believe you must be able to screw up into a ball & pelt straight in people's faces. They vagulate & dissipate. **1930** —— *Let.* 27 June (1982) IV. 182 Poor dear Angus vagulating like some pale anemone in a cranny.

vagulous (vēⁱ·giŭləs), *a. nonce-wd.* [f. L. *vagulus*: see prec.] Wayward, vague, wavering.

1919 V. Woolf *Diary* 12 July (1977) I. 291, I like Forster very much, though I find him whimsical & vagulous to an extent that frightens me with my own clumsiness and definiteness.

‖ **vahana** (vā·hana). Also **vahan.** [Skr. *vāhana*, lit. 'conveyance', f. *vah-* to carry, transport.] In Indian mythology, the mount or vehicle of a god.

1810 E. Moor *Hindu Pantheon* 16 Lacshmi..is with him [*sc.* Vishnu] on his *vahan*, or vehicle, Garuda. **1879** J. Dowson *Class. Dict. Hindu Mythology* 330 *Vāhana*. 'A vehicle'. Most of the gods are represented as having animals as their vâhanas. **1891** J. L. Kipling *Beast & Man* vi. 122 The bull receives high honours as the vâhan or steed of Shiva. **1971** *Illustr. Weekly India* 11 Apr. 11/1 (*caption*) Return of the procession after lustral rites of a Jina, with the gods and goddesses on their *vahanas*.

vahine (vāhī·ne). Also **vahini.** [Tahitian.] A Tahitian woman or wife. Cf. *wahine.

1950 F. H. Lyon tr. *Heyerdahl's Kon-Tiki Exped.* viii. 217 They had a lively and spirited leading singer in a luxuriantly fat *vahine*. **1960** H. E. Bates *Aspidistra in Babylon* 187 Better take a *vahini*..and settle down. *Ibid.* 227, I will build a house and live like your *vahine*. It cost nothing. I'll be your *vahine* and work for you. I'll work for you and you can love me. **1980** *London Mag.* July 24 The village of Mataiea, where Gauguin..lived..with his beloved *vahine* Teha'amana.

Vai (vəi). Also **Vei, 9 Vy.** [ad. a native name.] (A member of) a people of the southern coasts of Liberia and Sierra Leone; also, their language. Also *attrib.* or as *adj.*

1845 *Encycl. Metrop.* XX. 18/1 The Country near Cape Mount was anciently inhabited by the Vai, or Fai, and Pwi Tribes. *Ibid.* 18/2 The Vai, or Vei, were not reduced without difficulty. **1849** S. W. Koelle (*title*) Narrative of an expedition into the Vy country of West Africa. *Ibid.* 9 The Vei language is very simple, harmonious, with scarcely any grammatical inflexions, rich in vowels. **1854** —— *Outl. Gram. Vei Lang.* p. iv, If we compare the dialects spoken by the Manis and the Veis, we discover a difference which can scarcely have arisen in less than a couple of centuries. **1923** A. L. Kroeber *Anthropol.* ix. 225 All the values of his signs were syllabic. The same holds of the West African Vei writing invented in the nineteenth century by a native. He..fell back on syllabic signs even though this necessitated two hundred different characters. **1932** W. L. Graff *Language & Languages* xi. 434 Of some 435 Sudanese languages..only five or six are written. These are Vai, of the Nigero-Senegalese group,..[etc.]. **1936** G. Greene *Journey without Maps* ii. i. 99 Mr. Reeves was a Vai, a Mahommedan. **1949** L. D. Turner *Africanisms in Gullah Dial.* i. 14 The Vai word for 'fat' is..lit. 'mouth full'. **1957** M. Banton *W. Afr. City* vii. 127 Vai territory in Sierra Leone amounts only to some twenty-five square miles on the coastline next to the Liberian frontier. There are hardly any Vai in the colony. **1965** *Sierra Leone Lang. Rev.* (*Afr. Lang. Rev.*) IV. 81 The derivations listed for these 85 terms are *prima facie* astonishing, for 63 are Mende and 21 are Vai. **1974** J. R. Baker *Race* xxi. 411 The Mende tribe (a subdivision, like the Vai, of the Mandingo group of tribes). **1979** *Africa* XLIX. 92 The Vai script is so famous that it is surprising to realize that the language is spoken by, at most, 40,000 people. **1983** *Word* XXXIV. 136 Reading and writing Vai script was associated with certain specific skills.

Vaisākha, var. *Vesak.

‖ **Vaishnava** (vəi·ʃnavă). [a. Skr. *vaiṣṇavá* relating, belonging, or sacred to Vishnu; a worshipper or follower of Vishnu.] **A.** *sb.* A member of one of the three great divisions of modern Hinduism, exclusively devoted to the worship of the god Vishnu as the Supreme Being. **B.** *adj.* Of or pertaining to this division of Hinduism.

1845 [see *Sakta]. **1876** [see *Saiva *sb.* and *a.*]. **1882** [see Vedism]. **1903** *Times* 2 May 16/2 The 'holy basil' is ..planted before every Vaishnava house, and every Vaishnava wears necklaces, or armlets, and carries a rosary, made up of sections of its stalks or roots. **1944** E. Thompson *Robert Bridges* xi. 95 They both thought it worthless and in bad taste, as Rabindranath Tagore considered Bengali Vaishnava erotic mystical verse. **1968** *Jrnl. Mus. Acad. Madras* XXXIX. 61 To the Vaishnava devotees all over India Sri Krishna has been the Godhead for meditation.

Hence **Vai·s(h)navism** = Vishnuism; **Vai·s(h)navite** *a.*, of or pertaining to Vaishnavas or Vishnuism.

1877 [see *S(h)aivism* s.v. *Saiva *sb.* and *a.*]. **1919** N. Macnicol *Psalms of Maratha Saints* 5 There are one hundred and eight *tulsi* beads on the rosary worn by the Vaisnavite devotee. **1934** *Nature* 5 May 680/2 The ingrained love of life disclosed by the religions of Saktism and Vaisnavism among the Bengalis, comparable to that found among the Aryans, is a racial psychological trait to be associated with the brachycephalic Bengali castes. **1956** [see *S(h)aivism* s.v. *Saiva *sb.* and *a.*]. **1967** Singha & Massey *Indian Dances* i. 35 Of the Vaishnavite temples those at Belur and Halebid have an important significance. **1969** *Indo-Asian Culture* Oct. 70 Both Saivism and Vaishnavism were popular in Srihatta and the neighbouring region during the late Gupta and medieval times.

vaisya. Add: Also **vaish(y)a.** (Later examples.)

1920 H. G. Wells *Outl. Hist.* I. xx. 159/1 The four main castes seem to have been:..The Vaisya—herdsmen, merchants, money lenders, and land-owners. **1956** R. Pieris *Sinhalese Social Organization* v. i. 170 The existence among the Sinhalese of a system akin to the classical Hindu four-fold scheme—priestly *brahmins*, *kshatriyas* or warriors, the pastoral *vaisyas*, and the lowly *sūdras*..is attested in the survival in Ceylon of the place-name 'brahmin village'. **1958** *Spectator* 17 Jan. 70/3 Brahmins, Kshattriyas, Vaishas and Sudras. **1971** *Illustr. Weekly India* 11 Apr. 11/1 Rishabhadeva also classified the people according to their occupations, the kshatriyas who were warriors, the vaisyas who were traders, and the sudras who made their living by manual labour. **1971** K. Kent in C. Bonington *Annapurna South Face* 309 *Vaishya*, common man or merchant.

‖ **vajra** (va·dʒră). [a. Skr. *vájra.*] In Hinduism and Buddhism, a thunderbolt or mythical weapon esp. one wielded by the god Indra.

1788 *Asiatick Researches* I. 241 His weapon, *Vajra*, or the thunderbolt. **1854** A. Cunningham *Ladák* xiii. 371,

valley, obs. var. VLEI in Dict. and Suppl.

Vallombrosan (vælǒmbrǒᵘ·săn), *a.* and *sb.* [f. *Vallombrosa* near Florence, Italy + -AN.] **A.** *adj.* Of or pertaining to Vallombrosa; *spec.* designating a Benedictine congregation established there in the eleventh century by Giovanni Gualberto. **B.** *sb.* A monk of this congregation.

1851 E. B. BROWNING *Casa Guidi Windows* I. xxix. 77 The Vallombrosan brooks were strewn as thick. 1884 A. J. C. HARE *Cities Central Italy* I. xi. 297 The habit of the Vallombrosans was light grey, but the late monks wore a black cloak and a large hat when abroad. 1901 M. CARMICHAEL *In Tuscany* 141 They were succeeded in 1793 by monks of the Vallombrosan Congregation of Benedictines. 1922 JOYCE *Ulysses* 332 The monks of Benedict of Spoleto, Carthusians and Camaldolesi,.. Oratorians and Vallombrosans.

valorization. Delete *U.S.* and add: Also *gen.*, evaluation, giving validity to, making valid. (Later examples.)

1957 *Times* 28 Dec. 10/1 The announcement of a retention and valorization scheme by the Brazilian authorities in the spring touched off the rise in prices [of cocoa]. 1975 J. DE BRES tr. *Mandel's Late Capitalism* 9 The first four chapters..deal..with..the connection between the development of capitalist technology and the valorization of capital itself. 1975 *Times Lit. Suppl.* 22 Aug. 942/1 The new structuralist model, with its valorization of the synchronic system over the older historical, genetic, diachronic modes of understanding. 1980 J. P. FARRELL *Revolution as Tragedy* v. 247 A Victorian protest against the Romantic valorizations of tragedy that announce themselves in the works of Scott, Byron, and Carlyle.

valorize (væ·lǫraiz), *v.* [f. VALOR + -IZE.] *trans.* To raise or stabilize the value of (a commodity, etc.) by a centrally organized scheme; *gen.* to evaluate, to make valid. Hence **va·lorized** *ppl. a.*

1921 *Contemp. Rev.* July 53 It attempted both to regulate the output and to stabilise and to 'valorise' the prices. 1934 C. LAMBERT *Music Ho!* II. 79 Artists..have valorized the dream. 1976 T. EAGLETON *Crit. & Ideology* v. 164 Criticism becomes a mutually supportive dialogue between two highly valorised subjects: the valuable text and the valuable reader. 1976 *Daily Tel.* 1 Nov. 16 Merely 'valorising' specific duties..to recoup inflation since April would bring in £250 million. 1978 *Dædalus* Summer 48 The *Emile* and the *Contrat social* provide the explanation of the positively valorized concepts (virtue, morals, patriotism). 1984 *Christian Science Monitor* 2 Mar. B8/4 He has actually managed to suggest that the mire and blood, the 'refuse' of the embittered heart, is valorized by the poetic artifacts created from it.

valorousness. (In Dict. s.v. VALOROUS *a.*) For *rare* $^{-0}$ read *rare* and add examples.

1920 G. SANTAYANA *Character & Opinion in U.S.A.* vii. 214 Their valorousness and morality consist in their indomitable egotism. 1922 E. R. EDDISON *Worm* xxvii. 346 The Demons..by their strength and valorousness set free the Lord Goldry Bluszco.

valpack (væ·lpæk). *U.S.* Orig. (a proprietary name) **Val-A-Pak.** Also **valpac, valpak,** etc. [perh. f. elements of VALISE, PACK *sb.*[1], etc.] A type of soft zip-up travel bag.

1934 *Official Gaz.* (U.S. Patent Office) 30 Oct. 1005/2 Val-A-Pak. For Hand Bags. 1935 *Esquire* Dec. 212/1 (Advt.), Val-A-Pak hangs flat against the wall of a stateroom berth or closet. *Ibid.*, By the exclusive Val-A-Pak construction, the bag whether flat or folded for carrying, always conforms to its contents. 1946 J. P. MARQUAND *B.F.'s Daughter* ix. 111 Later in his room at the Bachelor Officers' Quarters he opened his valpack and took out a packet of letter paper. 1966 E. WEST *Night is Time for Listening* ii. 45 He hung out his Valpak, removed his overcoat, and lit a cigarette. 1967 *Sat. Rev.* (U.S.) 22 Apr. 52/2 Experienced travelers in 1947 favoured the Valpack to carry clothes. It was like an Army kitbag and zipped up. 1977 E. LEONARD *Hunted* (1978) ix. 86 He threw extra clothes into a valpac.

Valpolicella (vælpǫlitʃe·lă). Also earlier **Val Policella.** The name of a valley in the western Veneto, Italy, used to designate the red or rosé wine made there.

1903 N. NEWNHAM-DAVIS *Gourmet's Guide to Europe* ix. 164 A bottle of Val Policella is exactly suited to this kind of repast. 1935 SCHOONMAKER & MARVEL *Compl. Wine Bk.* v. 130 Of the Veronese wines, the red Valpolicella is decidedly the best. 1950 E. HEMINGWAY *Across River* xii. 103 What about a fiasco of Valpolicella? 1970 A. SILLITOE *Start in Life* vi. 294 Glasses of Valpolicella not yet touched. 1979 C. CURZON *Leaven of Malice* xi. 126 *Tournedos Rossini* with salad and a reasonable Valpolicella.

valproic (vælprǒᵘ·ik), *a.* *Pharm.* [f. VAL(ERIC *a.* + PRO(PYL + -IC.] 2-Propylpentanoic acid, $C_7H_{15}OOH$, a branched-chain fatty acid; also, a salt of this. Hence **valpro·ate,** a salt of this acid, esp. the sodium salt, an anticonvulsant drug given orally in cases of epilepsy.

1972 *Approved Names* (Brit. Pharmacopœia Commission) Suppl. IV. 6 Valproic acid. 1974 *Brit. Med. Jrnl.* 15 June 584/1 The anticonvulsant effects of sodium dipropylacetate (sodium valproate) were first reported in France. 1977 *Lancet* 22 Oct. 860/1 The antiepileptic activity of valproic acid (dipropylacetic acid) was recognized serendipitously when the agent was administered to animals as a vehicle for a series of test compounds. 1983 F. E. DREIFUSS in Pedley & Meldrum *Rec. Adv. in Epilepsy* I. iii. 35 Clinical experience with Valproate extends over nearly 20 years. *Ibid.*, The term Valproate is used here to refer both to valproic acid and its salts (sodium or magnesium).

Valsalva (vælsæ·lvă). *Med.* The name of Antonio M. *Valsalva* (1666–1723), Italian anatomist, used *attrib.*, in the possessive, and *absol.* to designate the action, described by him, in which an attempt is made to exhale air while the nostrils and mouth, or the glottis, are closed, so as to increase pressure in the middle ear and the chest.

1886 *Buck's Handbk. Med. Sci.* II. 577/2 To insure a proper performance of this (Valsalva's) method of inflation, it is merely requisite that the patient shall make a strong expiratory effort while holding both his mouth and his nose firmly closed. 1893 H. RICHARDS in C. H. Burnett *Syst. Dis. Ear, Nose, & Throat* I. 112 The patient is the operator of the Valsalva inflation. 1913 *Jrnl. Physiol.* XLVI. 186 The radial artery was released and the first rise of the Valsalva experiment was resumed. *Ibid.*, In Fig. 5 is shown another tracing of occlusion of the aberrant radial artery during Valsalva. *Ibid.* 187 The first rise in the sphygmographic curve in Valsalva's experiment is due to increased blood-pressure with increase of the calibre of the artery. 1943 *Bull. Hist. Med.* XIV. 313 Only the historian can fully realize the relative unfruitfulness of the vast amount of experimentation and of study devoted to the Valsalva. 1972 *Sci. Amer.* Feb. 85/2 In attempting to stop the heart the yogis usually performed what is called the Valsalva maneuver. 1977 *Lancet* 28 May 1140/1 In shallow diving an over-forceful Valsalva manœuvre may give rise to neuro-sensory hearing loss, with or without vertigo.

Valspeak (væ·lspīk). *U.S.* [f. VAL(LEY *sb.* + *-SPEAK.] = *Valleyspeak* s.v. *VALLEY *sb.* 7.

1982 *People Mag.* 13 Sept. 90/2 On the record, in pure, uncut Valspeak, Moon laments in bubbly staccato that, 'Like my mother like makes me do the dishes. It's like so *gross*.' 1982 *Guardian* 26 Oct. 8/4 Valspeak is the latest teentalk craze to surface in the real world.

valuation. Add: **1.** *attrib.* Also *valuation office, officer.*

1905 G. B. SHAW *Let.* 3 Jan. (1972) II. 503 A high official in the Valuation Office. 1925 *Act* 15 & 16 *Geo. V* c. 90 §55(1) It shall be lawful for rating authorities, assessment committees, and county valuation committees to appoint for the purposes of this Act such rating officers, valuation officers and other officers as they think fit. 1976 *Evening Post* (Nottingham) 14 Dec. 4/7 A spokesman for the Inland Revenue said it was customary for the valuation officers to try to reach agreement with the ratepayers. 1978 *Ryde's Rating Cases* (1979) XXI. 143 He had stated that on evidence now to hand the valuation office erred in some cases in not projecting forward the 1967/1969 rents.

valuative, *a.* Delete †*Obs.* $^{-1}$ and add later examples.

1946 [see *APPRAISIVE *a.*]. 1958 *Listener* 12 June 983/1 Valuative disputes arise both about what is to be passed on [by education] and about the manner of passing it on.

value, *sb.* Add: **I. 1.** In phr. *value for money* (freq. *attrib.*).

[1806: see Dict.]. 1902 W. S. CHURCHILL in R. S. Churchill *Winston S. Churchill* (1969) II. Compan. I. 132 The first limitation is therefore the restriction of a committee on estimates to the 'merit' or 'value for money' aspect of expenditure. 1929 *Radio Times* 8 Nov. 444/3 Lissen are famed for the keen value-for-money of every component. 1951 [see *PROMOTION 1 e]. 1976 *Jrnl. R. Soc. Arts* CXXIV. 509/2 Every natural salesman knows that the true test of marketability is the subtle ratio between specification and price which even after years of anxious thought Consumers' Association can still only call 'value for money' and can still only nominate and not measure. 1977 *Management Today* July 72/2 Both in extroversion and value-for-money terms, the XJS retains the edge. 1984 *Times* 30 Apr. 25/3 The marketing manager will be impressed by an easy-to-handle, value-for-money approach.

2. c. *good value* (colloq.), entertaining, worth keeping company with, worth seeing, etc.

1930 R. LEHMANN *Note in Music* III. 102, I rather like her... She might be good value, given a chance. 1934 R. KNOX *Still Dead* vii. 91 Lisbon..is really pretty good value, coming into the harbour especially. 1937 AUDEN & MACNEICE *Lett. from Iceland* xii. 161 The geyser was better value, it went off. 1961 J. B. PRIESTLEY *Saturn over Water* vii. 104 Her eyes shone with excitement. Either the Garlettas were exceptionally good value or she was longing to get out of this house. 1979 'S. WOODS' *This Fatal Writ* 34 If it weren't for the chap's confounded habit of asking questions, he would be finding his client rather good value..because of the refreshingly casual way in which he treated..a very serious charge.

d. *value added* (Econ.), the amount by which the value of an article is increased at each stage of its production by the firm or firms producing it, exclusive of the cost of materials and bought-in parts and services; also *attrib.*, esp. in *value added tax* [cf. G. *mehrwertsteuer*], a tax levied on the value added to an article or the raw material forming it at each stage during its production or distribution.

1935 *Social Research* II. 161 We may call the value added tax a 'refined sales tax'. *Ibid.*, A tax which chooses as its basis of assessment the sales after deduction of all expenses for raw material and for repair or replacement of equipment, that is, a tax on the 'value added by manufacture'. 1940 *Jrnl. Polit. Econ.* XLVIII. 652 A rather imperfect form of a value-added tax was incorporated by Mr. Arthur L. Johnson in his revised old age pension bill in 1939. 1951 P. A. SAMUELSON *Economics* (ed. 2) xi. 236 If we insist upon decomposing the 10 cents of final product represented by the bread into the contributions of the different stages of production, we can always do so by concentrating on the so-called 'value-added' at each stage of production. 1957 J. F. DUE *Sales Taxation* vii. 125 The basic intent of a value-added tax is to tax each firm on the sum of the value which it has added in the manufacturing and distribution process. 1962 *Economist* 3 Mar. 827/1 The French system of a value-added tax—that is to say, a tax on net sales, or total sales less the cost of bought-in materials and components. 1967 *Ibid.* 21 Jan. 243/2 The aids must..bring in industries with rapid growth rates and a high value-added element. 1977 M. WALKER *National Front* vi. 142 A Government in mid-term (which had moreover introduced the unpopular Value Added Tax the previous month).

6. In *Philos.* and *Social Sciences*, regarded esp. in relation to an individual or group; *gen.* in *pl.*, the principles or standards of a person or society, the personal or societal judgement of what is valuable and important in life.

1902 J. M. BALDWIN *Dict. Philos. & Psychol.* II. 823/2 Since value is a function of desire or judgment, expressing a relation between subject and object. 1918 THOMAS & ZNANIECKI *Polish Peasant* I. 21 By a social value we understand any datum having an empirical content accessible to the members of some social group and a meaning with regard to which it is or may be an object of activity. *Ibid.* 33 Sociology..has this in common with social psychology: that the values which it studies draw all their reality, all their power to influence human life, from the social attitudes which are expressed or supposedly expressed in them. 1921 *Times Lit. Suppl.* 3 Nov. 705/4 In the effort, again, to give his characters and scenes the vivid impression of reality, the novelist, whether voluntarily or not, cannot avoid revealing not merely his powers of mind and imagination, but his spiritual and philosophical bias, his views of society, of religion, his 'values'. 1933 *Economica* XIII. 30 Like all human action social behaviour is determined..in accordance with standards of value or through conscious belief in standards assigning intrinsic value to certain types of behaviour. 1938 E. BOWEN *Death of Heart* III. iv. 394 You've got a completely lunatic set of values. 1950 I. BERLIN in *Foreign Affairs* XXVIII. 382 Crumbling values and the dissolution of the fixed standards and landmarks of our civilization. 1955 *Times* 10 May 8/3 Restoring to Germany the basic values of democratic civilization. 1958 *Listener* 9 Oct. 548/1 The reason..lies, I believe, in the structure of Arab society..and in its economic values. 1964 GOULD & KOLB *Dict. Soc. Sci.* 744/1 Social scientists for the most part..have confined their attention to values..as empirical variables in social life whose *scientific* importance is not so much dependent on their validity and correctness as..upon the fact that they are believed..by those who hold them. 1970 N. CHOMSKY *At War with Asia* vi. 299 By their willingness to die, the Asian hordes..exploit our basic weakness—our Christian values which make us reluctant to bear the burden of genocide, the final conclusion of our strategic logic.

attrib. (Further examples.)

1910 *Mind* XIX. 227 Our æsthetic, ethical and directly sensational judgments are all expressions of the fundamental value-attitude in specifically different relations. 1936 *Mind* XLV. 289 A will to maintain the society in spite of one's dissatisfaction with certain elements in its total value-pattern. 1949 G. BATESON in M. Fortes *Soc. Structure* 49 Of these differences between von Neumannian and human systems, only the differences in value scales..concern us here. *a* 1952 H. READ *Anarchy & Order* (1954) 195, I am going to assume..that 'good' and other ethical value-terms have only an emotive meaning. 1952 *Mind* LXI. 290 Attempts to construct various value-hierarchies. 1952 R. M. HARE *Lang. Morals* i. 8 'Bad' is a value-word, and therefore prescriptive. 1954 W. K. HANCOCK *Country & Calling* vi. 173 Myrdal saw only one remedy for it—to state explicitly the value-premises of his exploration so that writer and reader alike would always be able to distinguish the discussion of *is* from the discussion of *ought*. 1964 A. EDEL in I. L. Horowitz *New Sociol.* 220 Some value-attitudes..are ushered into the inner sanctum of science. 1977 A. GIDDENS *Stud. in Social & Polit. Theory* i. 91 The initial statement conceals a value premise. *Ibid.* 94 Ultimate values..involved in both the value hierarchies of individual actors, and those of overall cultures. 1970 S. L. BARRACLOUGH in I. L. Horowitz *Masses in Lat. Amer.* iv. 123 The principal 'causes' of the agrarian problem are to be found in the population explosion,..and changing value patterns. 1977 BULLOCK & STALLYBRASS *Fontana Dict. Mod. Thought* 47/2 Any field of human discourse in which the general value-terms 'good' and 'ought' figure falls within the range of axiology. 1977 P. HOWARD *New Words for Old* xiii. 117 (*heading*) Value Words... The change from description to evaluation is one of the most potent agents for decaying meanings.

f. The quality of a thing considered in respect of its power and validity for a specified purpose or effect.

1906, etc. [see *news value* s.v. *NEWS *sb.* (*pl.*) 6 b]. 1933, etc. [see *nuisance value* s.v. *NUISANCE 3]. 1935 A. P. HERBERT *What a Word!* 8 To-day, instead of 'fun', we learn to speak of 'entertainment-value'. 1937, etc. [see *ENTERTAINMENT 13]. 1962 C. WINSTON *Hours Together* vi. 129 In a decade..all this would also be a joke, and the names of these neighbourhoods..would have the same comedy value as the names he had grown up with. 1966 H. MOORE *On Sculpture* 167 These small figures, seen so

much bigger, take on an extra importance and impressiveness, and are a proof that size itself has an emotional value. 1978 K. HUDSON *Jargon of Professions* 1 The solidarity-value of such nonsense-language certainly should not be underestimated. 1979 J. RABAN *Arabia through Looking Glass* v. 196 It's a synthetic soil-substitute, better than real soil by far. Its nutritional value's *much* higher.

8. Special Combs.: **value analysis,** the systematic and critical assessment by an organization of design and costs in relation to realized value; also *transf.*; **value analyst,** one who undertakes a value analysis; **value calling** *Bridge*, a system of estimating bids which takes into account the scoring values of the suits; **value engineering,** the modification of designs and systems according to value analysis; **value-free** *a.*, free from criteria imposed by subjective values or standards; purely objective; = *value-neutral*; hence *value-freedom*; **value-judgement** [cf. G. *werturteil*], a judgement predicating merit or demerit of its subject; **value-laden** *ppl. a.* = *value-loaded* ppl. adj.; hence *value-ladenness*; **value-loaded** *ppl. a.*, weighted or biased in favour of certain values; **value-neutral** *a.*, involving no value judgements, neutral with respect to (personal or group) values; **value-orientation,** the direction given to a person's attitudes and thinking by their beliefs or standards; so *value-oriented ppl. a.*; **value-system,** any set of connected or independent values; **value theory,** (*a*) *Pol. Econ.*, the (Marxist) labour theory of value; (*b*) *Philos.*, axiology.

1955 K. E. BOULDING *Econ. Anal.* (ed. 3) xxxiii. 714 We can, therefore, perform something with the tools of economic analysis which might be called 'value analysis', and which should be of use in clarifying the choices involved in economic policy. 1963 *Engineering* 9 Aug. 162/1 When good engineering, manufacturing and purchasing practices are supplemented by value analysis, the cost of a product..can be reduced by up to 25 per cent. 1977 R. HOLLAND *Self & Social Context* i. 12 It will be necessary to apply sociological techniques of value analysis in order to reach a full understanding of the theories. 1969 J. ARGENTI *Managem. Techniques* 265 Some companies establish permanent teams under a Value Analyst. 1927 *Daily Express* 8 Nov. 1/5 We do not consider that there is any general desire for the adoption of majority calling in place of value calling. 1959 *Ship & Boat Builder* Oct. 349 (*heading*) Can value engineering cut costs? 1962 *Engineering* 19 Oct. 515 Value engineering is a well-established technique in the USA... They define it as 'the systematic application of techniques and principles which aim at cutting production costs'. 1973 *Lebende Sprachen* XVIII. 73/1 *Value engineering* is a technique for reducing total cost while maintaining or improving the overall usefulness of the product or service. 1949 J. A. PASSMORE in Feigl & Brodbeck *Readings in Philos. of Sci.* (1953) 674 (*heading*) Can the social sciences be value-free? 1979 *Nature* 19 July 185/1 Science and technology are not neutral or value-free but are instruments of power, and that means political power. 1984 *Times Educ. Suppl.* 30 Nov. 3/2 Europe Singh, a maths teacher..believes maths and the sciences have wrongly been considered to be neutral and value-free. 1959 P. RIEFF *Freud* viii. 299 Scientific energies, by the facile transformation of the objectivity necessary to science into..'value-freedom', are easily enlisted to the aims of society, whatever these may be. 1892 J. ORR in *Thinker* II. 146 Two kinds of knowledge are distinguished by Ritschl—the one, religious knowledge which moves solely in the region of what he calls worth or value-judgments. 1899 [see sense 6 *attrib.*]. 1941 J. S. HUXLEY *Uniqueness of Man* xi. 229 Even in natural science, regarded as pure knowledge, one value-judgment is implicit—*belief in the value of truth*. 1961 *Listener* 30 Nov. 912/1 The decision depends on what may ..be called policy considerations; that is, where the court has to make a value judgment. 1975 *Amer. N. & Q.* XIV. 53/2 Robert Frost's penchant for 'the fact' (as in 'Mowing') provides a useful measuring stick for determining the worth of value judgments about him. 1980 *Times Lit. Suppl.* 3 Oct. 1085/2 The method adopted here is a detailed interpretative analysis of poetic language and structure, liberally sprinkled with value-judgments. 1971 *Ibid.* 13 Aug. 958/4 For them, even the internal content of science is value-laden, and to some extent ideologically determined. 1977 *Jrnl. Politics* XXXIX. 24 The growing acceptance of the thesis that political science is necessarily a value-laden discipline. 1978 M. HESSE in Hookway & Pettit *Action & Interpretation* 8 A distinction between two sorts of 'value-ladenness' in social science. 1951 D. RIESMAN *Individualism Reconsidered* (1955) 33 Obviously, the very term 'masses' is heavily value-loaded. 1974 tr. *Wertheim's Evolution & Revolution* 35 To state that a given situation shows 'progress' or 'evolution'..in relation to another situation implies the use of value-loaded criteria. 1946 GERTH & MILLS tr. M. Weber in *From Max Weber* (1947) ix. 247 Even a pirate genius may exercise a 'charismatic' domination, in the value-neutral sense intended here. 1979 *Dædalus* Winter 55 'Excellence' is not a value-neutral concept. 1951 G. W. ALLPORT in Parsons & Shils *Toward Gen. Theory Action* IV. i. 365 Prejudice is manifestly a value-orientation. 1968 W. E. LAMBERT et al. in J. A. Fishman *Readings Sociol. of Lang.* 488 In general, value orientations do not play an important role in predicting who will or will not do well in French. 1980 N. ABERCROMBIE et al. *Dominant Ideology Thesis* ii. 48 System integration is defined in terms of the processes whereby value-orientation patterns are institutionalised at the social level via the mechanism of social roles with the effect of organising the behaviour of adult members of society. 1962 N. J. SMELSER *Theory Collective Behav.* iii. 49 Behind a vast array of religious and political value-oriented movements lie the same kinds of strain. 1977 *Bull. Amer. Acad. Arts & Sci.* Oct. 16 It is at this point that value-oriented parameters for assessing progress become necessary. 1936 *Mind* XLV. 288 Persons who are not *Buerger* (citizens)..like the Jews in Nazi Germany, or the bulk of the Bantu in the Union of South Africa. For such as these, the relation to the value-system embodied in the state is of the most tenuous and indirect kind. 1969 *Listener* 3 July 3/1 Two American sociologists examined the value system of a small rural town in the American Mid-West. 1980 *Jrnl. R. Soc. Arts* June 416/2 A society in which there are overlapping different value systems which create different structures. 1887 G. B. SHAW *Let.* 17 May (1965) I. 169 Socialism does not stand or fall by the Value Theory. 1941 *Mind* L. 198 The contributions to aesthetics, value-theory, theology and Spinozistic lore have all..been published before. 1966 S. BEER *Decision & Control* x. 221 We are now in the field of value theory, the subject which attempts to bring managerial value judgments within the compass of decision theory. 1979 D. MCLELLAN *Marxism after Marx* xxiii. 309. There has been a vigorous and sophisticated defence of traditional Marxian value theory.

valued, *ppl. a.* **2.** (Later example.)

1913 R. H. GRETTON *Mod. Hist. Eng. People* I. ii. 52 Rents were refused above the 'valued rent' scale.

valuta (vălʹū·tă). Pl. **valute, valuten.** [(G. a.) It *valuta*:– late L. *valuta, use as sb. of fem. pa. pple. of L. *valēre* to be worth.] Foreign currency; a monetary standard, the currency constituting an acceptable medium of exchange, the valuation constituting an acceptable rate of exchange. Also *attrib.*, *transf.*, and *fig.*

1893 R. BITHELL *Counting-House Dict.* (rev. ed.) 314 *Valuta*, Russian paper money. 1921 *Glasgow Herald* 28 Oct. 11 The speculation in valute and devisen has become ..ridiculous, as the great boom on the Berlin exchange proves. 1924 LUCAS & BONAR tr. *Knapp's State Theory of Money* 106 Everywhere there is a valuta or standard money. *Ibid.* 165 If there is more than one kind of money in the country, the value in valuta money is always meant. *c* 1939 E. BENN *Diary* in *Happier Days* (1949) xiv. 171 He doubted if the *valuten* could be found to send the Führer to England. 1954 E. H. CARR *Interregnum* i. 29 A general decree was issued 'On Valuta Operations'... It confined transactions in foreign valuta to the Exchanges. 1964 S. BELLOW *Herzog* 143 A mob broke into his house.. looking for *valuta*. 1967 *Economist* 19 Aug. p. xxix/1 Russia is still the chief importer of Skoda products, but the emphasis is gradually shifting to the West and other areas. A nominal level of *valuten* income is now set. 1972 M. MUGGERIDGE *Chron. Wasted Time* I. v. 252 All were deeply moved;..as, indeed, was I, despite my valuta origins and prospects. 1979 F. E. PERRY *Dict. Banking* 262/1 *Valuta*, rate of exchange, value, currency; entries in bank books which fix charges to be made or interest charges due, such as the determination of opening and closing dates for the relative periods.

valva (væ·lvă). *Ent.* [L.: see VALVE *sb.*[1]] = VALVE *sb.*[1] 2 b, *c.

1802 W. KIRBY *Monographia Apum Angliæ* I. 110 *Valvæ*. These have been frequently noticed. Swammerdam calls them *appendages of the sting*. 1826 [see VALVE *sb.*[1] 2 b]. 1924 *Ann. Entomol. Soc. Amer.* XVII. 276 The term valvæ or valves was used to denote the two lateral outer appendages. 1975 *Entomologist's Gaz.* XXVI. 198 There is in my material a further specimen..showing some variation in the valva.

valvar, *a.* Delete *rare* and add later examples. In *Med.* = VALVULAR *a.* 3.

1895 *Jrnl. R. Microsc. Soc.* 669 Dr. O. Müller proposes a new terminology for diatoms... The plane which passes through the apical and the transapical axes is the valvar plane. 1946 *Nature* 26 Oct. 588/1 Division took place along the longitudinal axis in what would have been the valvar plane. 1955 *Guy's Hospital Rep.* CIV. 372 Kirklin *et al.*..have also drawn attention..to this association between valvar and infundibular stenosis. 1977 *Amer. Heart Jrnl.* XCIII. 461/1 Mitral regurgitation or other mitral valvar abnormality has been reported in four other cases. 1980 *Plant Systematics & Evolution* CXXXV. 266 (*caption*) Region of the intercalary bands seen parallel to the valvar plane.

valve, *sb.*[1] Add: **I. 2. c.** *Ent.* = *VALVULA 2.

1802 W. KIRBY *Monographia Apum Angliæ* I. 110 Linneus, in his character of *Ichneumon*, calls them the valves of the vagina of the aculeus. They are the covers of the genuine vagina. 1919 *Ann. Entomol. Soc. Amer.* XII. 277 The sternal region of segment nine is shifted.., thus bringing the bases of the dorsal and inner valvulæ into the same transverse plane with those of the ventral valves. 1969 R. F. CHAPMAN *Insects* xvii. 325 If the insect oviposits in plant or animal tissue the valves are sclerotised and lanceolate.

d. *Ent.* A clasper of a male butterfly.

1864 *Trans. Linn. Soc.* XXV. 35 There remain therefore only the characters of the perfect insect, the most important of which are the anal valves in the male. These.. are furnished with projecting points or spines..which serve to attach the male more firmly to the female *in copulâ*. 1883 *Ibid.: Zool.* II. 332 *C. Eubule* has a very curious valve, armed as elaborately, and as singularly, as that of many a *Papilio*. 1964 R. M. & J. W. FOX *Introd. Compar. Entomol.* iii. 109 Other orders with periphallic claspers (harpes, harpagones, valves, etc.) form them from the coxae and exite styli of the appendages of at least the ninth segment.

3. c. Each of the two siliceous cell walls of a diatom, similar in shape but slightly different in size, with one overlapping the other.

1852 A. PRITCHARD *Hist. Infusorial Animalcules* (rev. ed.) III. 295 Siliceous valves are deposited *exterior* to a cell-membrane. 1857 [see sense 3 b in Dict.]. 1898 H. C. PORTER tr. *Strasburger's Text-bk. Bot.* II. i. 313 Both valves are so strongly impregnated with silica, that, even when subjected to intense heat, they remain as a siliceous skeleton, retaining the original form and markings of the cell walls. 1973 R. G. KRUEGER et al. *Introd. Microbiol.* iii. 125/1 The newer valve of any diatom is invariably the smaller valve, because it is constructed within the confines of the older valve.

II. 6. a. *fig.* (Later examples.)

1930 AUDEN *Poems* 56 No chattering valves of laughter emphasized..the sessile hush. 1933 E. O'NEILL *Ah, Wilderness!* IV. i. 116 Seizes this as an escape valve—turns and fixes his youngest son with a stern forbidding eye.

b. *Electronics.* = *thermionic valve* s.v. *THERMIONIC *a.*

1905 J. A. FLEMING in *Proc. R. Soc.* LXXIV. 478 We have in this vacuum valve and associated mirror galvanometer a means of detecting feeble alternating electric currents or oscillations. *Ibid.* 479 This arrangement of a differential galvanometer and two valves transforms.. more of the alternating oscillation into direct current than when one valve alone is used. 1924 GIBSON & COLE *Wireless of To-Day* xxiv. 307 Monster valves have now been manufactured absorbing as much as 100 kw. each, and in consequence of the tremendous heat generated, the electrodes are specially constructed permitting water to circulate within for cooling purposes. 1928 *Electr. Communication* VI. 241/2 The high-power amplifying tubes, which over here [*sc.* in Britain] we call 'valves'. 1943 C. L. BOLTZ *Basic Radio* x. 165 All battery-fed radio apparatus utilizes directly-heated valves, which need 2V on the filament. 1968 M. GUYBON tr. *Solzhenitsyn's First Circle* lxxv. 476 He was a radio engineer by training, and hadn't a box containing two valves been found during the search of his flat?

c. *Chess.* (See quot. 1930.)

1930 WHITE & HUME *Valves & Bi-Valves* 7 In chess problem terminology, the designation of a Valve has been given to any move which simultaneously opens one line while it closes another. In a broad sense, Valves include a large domain with many varied combinations of themes. There is a much narrower application of the term: Valve, and that is the particular case where not only is the move made by Black, but both of the lines affected are also Black. 1936 P. W. SERGEANT tr. *Znosko-Borovsky's Art of Chess Combination* II. v. 62 In the interception of lines one closes lines to the adversary at critical points, while here one opens lines to one's own pieces. The skilled hand deals with the valves on this side and on that.

7. a. *valve line, stem*; also used to designate brass instruments whose range is increased by the addition of valves, as *valve horn, trombone* (so *trombonist*), *trumpet*; in sense *6 b, esp. designating apparatus employing valves, as *valve circuit, detector, heater, -holder, oscillator, receiver, set, voltmeter.*

1915 HAWKHEAD & DOWSETT *Handbk. Wireless Telegraphists* 119 If an E.M.F. be applied to the valve circuit a more sensitive condition is obtained. 1934 *Times Rev. 1933* 1 Jan. p. ix/4 The Wireless Exhibition at Olympia illustrated the exceptional advances made during the year in valve and valve-circuit technique. 1915 HAWKHEAD & DOWSETT *Handbk. Wireless Telegraphists* 120 The valve detector is used with various circuits. 1929 *Radio Times* 8 Nov. 443/3 Get the necessary output from a Regentone mains unit for A.C. valve heaters. 1960 *Practical Wireless* XXXVI. 326/2 Commence wiring the valve heaters by taking a tightly twisted pair of insulated wires from V1 to V2. 1922 *Wireless World* 4 Mar. 748/1 The base..carries the valve holder. 1960 *Practical Wireless* XXXVI. 392/2 Positions for the valve holders can be marked out from the measurements indicated in Fig. 2. 1877 Valve horn [see *valve trumpet* below]. 1938 *Oxf. Compan. Mus.* 439/2 The valve horn has the immense advantage of..a chromatic series, for three valves add instantaneously to the air column a length corresponding respectively to a semitone, two semitones, and three semitones. 1959 *Listener* 4 June 1001/3 The more limpid yet penetrating tone of the narrower-bored French valve-horn. 1977 *Early Music* Apr. 221/2 Scores of the valve horn era. 1871 *Leisure Hour* 8 Apr. 222/2 The balloon had been gyrating, and the valve-line become twisted. 1963 [see *rip line* s.v. *RIP *sb.*[4] 5]. 1969 *Gloss. Aeronaut. & Astronaut. Terms* (*B.S.I.*) VII. 8 *Valve line*, a cord for the operation of a valve. 1935 *Discovery* Aug. 226/1 In some forms of valve oscillator, the high-tension supply is such that only half of the wave of the a.c. feed mains is rectified. 1913 *Wireless World* Nov. 478/1 A valve receiver of rather longer range than usual is used. 1929 *Radio Times* 8 Nov. 437/1 This popular Loud Speaker unit..gives..perfect results with any valve Receiver. *Ibid.* 397/1 Her nursery..is wired for broadcasting and..her movements or cries are now heard loud in the sitting-room..where our valve set is placed. 1981 S. BRIGGS *Those Radio Times* i. 28/2 By 1926..the valve set playing through loudspeakers replaced the simple crystal set. 1888 *Lockwood's Dict. Mech. Engin.* 397 *Valve stem*, a valve spindle or rod. 1889 J. M. WHITHAM *Steam-Engine Design* iv. 98 If the valve.. is long, the weight, friction, and diameter of the valve-stem are increased. 1970 K. BALL *Fiat 600, 600D Autobook* i. 13/1 Valve stems, once bent, cannot be straightened satisfactorily. 1883 O. COON *Harmony & Instrumentation* xxvi. 73 The valve Trombone is often substituted for the one with a slide. 1979 *Jazz Jrnl.* XXXII. 11/1 (*caption*) Thad Jones on valve trombone. 1946 R. BLESH *Shining Trumpets* (1949) xii. 263 The valve trombonist Brad Gowans. 1877 E. PROUT *Instrumentation* v. 81 The valve-trumpet..possesses, like the valve-horn, a complete chromatic scale. 1979 *Oxf. Junior Compan. Music* 331/4 The trumpet now in everyday orchestral use is the *valve trumpet*. That is to say, it has extra lengths of tubing coiled alongside its main tube, and these can be brought into action by pressing down a 'valve'. 1925 Valve voltmeter [see *slide-back* s.v. *SLIDE- c].

c. valve head *Mech.*, the part of a lift valve that is lifted off the valve aperture to open the valve; **valve train** *Mech.*, in an internal-combustion engine, the gearing and linkages by which the crankshaft is caused to open and close a valve at the proper time.

1904 A. B. F. YOUNG *Compl. Motorist* (ed. 2) iv. 91 The valve-head is provided with a slot for the insertion of a tool for grinding purposes. 1971 B. SCHARF *Engin. & its Lang.* xii. 176 One also differentiates between ordinary, high lift and full lift safety and relief valves according to the distance by which the valve head is automatically raised. 1955 W. H. CROUSE *Automotive Engines* vii. 222 The L-head engine uses a relatively simple valve train, or valve mechanism. 1981 *Pop. Hot Rodding* Feb. 31/1 A Sig Erson camshaft actuates the mildly worked valve-train, composed of Crane valve-springs, steel retainers, and TRW chrome-moly pushrods.

valve, *v.* For *rare* read '*rare* exc. in ballooning, etc.' and add: **1.** Also *fig.*

1960 R. W. MARKS *Dymaxion World B. Fuller* 11/1 The harnessing factor—the activity which 'valves' the mass-energy of the universe to human advantage—is inventive wisdom born of intuition and experience and put to use in a global industrial complex.

2. (Further examples in ballooning.)

1936 *Nat. Geogr. Mag.* LXIX. 71 Andy valved, and valved again and again! 1963 A. SMITH *Throw out Two Hands* v. 63 You valve a little. You start coming down.

3. *trans.* To discharge gas from (an airship or balloon) by opening a valve; to discharge, or let *off*, (gas) thus.

1925 *Sci. Amer.* Nov. 301/3 She was swept rapidly upwards..Commander Lansdowne valved her freely, pointed her nose down with engines running, and she came down with..rapidity. 1928 *Engineering* 3 Aug. 141/3 As an airship uses up its fuel, it is necessary to reduce the lift, and hitherto this has been done by simply 'valving' off some of the hydrogen used for inflation. 1928 *Daily Tel.* 18 Sept. 9/6 The extra lifting effect of the expanding gas cannot be counteracted by allowing the gas to escape, or in other words by valving the gas. 1936 *Nat. Geogr. Mag.* LXIX. 71 He opened the valves for..another half-minute interval (which is a very long time for a balloon to be valved at low altitude).

valved, *a.* Add: **2.** (Further examples.) Also *fig.*

1856 M. C. CLARKE tr. *Berlioz's Treatise Mod. Instrumentation* (1858) 146/2 *Valved trumpets*,—called so on account of a movable valve similar to that of the trombone, and which is moved by the right hand,—are adapted to produce the truest intervals. 1923 D. H. LAWRENCE *Birds, Beasts & Flowers* 26 To me, all faces are dark, All lips are dusky and valved. 1927 *Bull. U.S. Nat. Museum* No. 136. 51 A valved trumpet marked 'alto B flat'... It has three rotary valves. 1970 J. EARL *Tuners & Amplifiers* 7 This book is concerned essentially with transistored equipment, for the days of the valved amplifiers and tuners have now gone for ever.

valvifer (væ·lvifəɹ). *Ent.* [f. L. *valva* VALVE *sb.*[1] + -I- + L. *-fer* (see -FEROUS).] In some insects, a modified coxa on each of the eighth and ninth abdominal segments that forms a basal plate of the ovipositor and bears a valvula.

1917 G. C. CRAMPTON in *Jrnl. N.Y. Entomol. Soc.* XXV. 236 Basal sclerite of valvula of ovipositor (Valvifer). 1935 R. E. SNODGRASS *Princ. Insect Morphol.* xix. 611 The basal apparatus consists essentially of two pairs of lobes or plates, the first and second valvifers..which support the ovipositor shaft by the bases of the valvulae. 1981 *Animal Behaviour* XXIX. 299/1 The valvifers at its base could be seen to be moving rhythmically back and forth.

valving (væ·lviŋ), *vbl. sb.* [f. VALVE *sb.*[1] + -ING[1].] **a.** A system or arrangement of valves; valves collectively. **b.** Opening and shutting in the manner of a valve; valve-like operation.

1948 *Chem. Industries* Feb. 222/1 (*heading*) Valving CC's. 1957 *Petroleum Engineer* Oct. c. 66 (*heading*) Valving critical in elaborate distribution system. 1968 *Jrnl. Dental Res.* XLVII. 1013/2 Physiologic and temporal features of oral and palatopharyngeal valving were correlated with velocity and volume of oral-nasal flow, and intraoral air pressure. 1973 *Sci. Amer.* Dec. 143/1 [A giraffe's] elevated brain is supplied with oxygen by blood vessels that manage a lot of valving and pumping. 1978 *Ibid.* May 38/1 The valving [of the Newcomen steam engine] was automatic from a very early stage. 1981 *Jrnl. Air Pollution Control Assoc.* XXXI. 377 Thermal mass flowmeters, with microprocessor controlled flow and automatic valving, supply variable levels of concentrated odorant and clean dilution air to a static mixing device.

valvotomy (vælvǫ·tǒmi). *Surg.* [f. VALVE *sb.*[1] + -O + -TOMY.] An operation in which an incision is made into a valve, esp. of the heart.

1903 *N.Y. Med. Jrnl.* LXXVIII. 296/2 The rectitis should be recognized by the general practitioner, that the rectum may be so treated that absorption of the exudate may result rather than the formation of an organized deposit rendering rigid the rectal valve—thus may valvotomy be made unnecessary. 1912 *Proctologist* VII. 155 Valvotomy is a justifiable operation, as it frequently relieves obstipation and constipation. 1962 *Lancet* 8 Dec. 1195/2, 7 days later aortic valvotomy was successfully performed. 1980 *Brit. Med. Jrnl.* 23 Feb. 563/2 We would like to plead for the continued use of closed valvotomy, notably in third world countries where large numbers of patients with mitral stenosis could overwhelm the relatively small number of surgical units equipped for open heart surgery.

valvula. Add: **2.** *Ent.* An elongated blade-like process attached to the coxa on the eighth or ninth abdominal segments of some insects and forming part of the ovipositor; = *VALVE *sb.*[1] 2 c.

1917, 1935 [see *VALVIFER]. 1964 R. M. & J. W. FOX *Introd. Compar. Entomol.* iii. 106 The ovipositor is present in its fully developed form, with three pairs of valvulae, in Phasmida, Grylloblatodea, Dictyoptera, Odonata and Corrodentia. 1978 H. V. DALY et al. *Introd. Insect Biol. & Diversity* ii. 36/2 Eggs issue from the genital opening.., pass down a channel formed by the valvulae, and are deposited in the ovipositional medium.

valvulotomy (vælviūlǫ·tǒmi). *Surg.* [f. VALVUL(E + -O + -TOMY.] = *VALVOTOMY.

1916 J. F. BINNIE *Operative Surg.* (ed. 7) II. xxxi. 332 Such operations may be named internal valvulotomy. 1951 *Lancet* 21 Apr. 896/2 Valvulotomy for mitral stenosis. 1970 *New Yorker* 21 Nov. 68/2 Any two-bit surgeon who can do a valvulotomy can have a patient in this unit.

vamose, vamoose, *v.* Restrict *U.S.* to sense 2 and add: Now usu. *vamoose.* **1.** (Earlier U.S. and further examples.)

[1827 W. CLARKE *Every Night Bk.* 30 They have done more foolish things in their day—but vamos.] 1834 *Knickerbocker* IV. 455 Be off, you good-for-nothing rascals —*vamos*! 1892 STEVENSON & OSBOURNE *Wrecker* xvi. 254 Well, of course he can vamoose with the entire speculation, if he chooses. 1936 F. CLUNE *Roaming round Darling* ix. 82 The river was going downhill, and the country growing more and more similar in appearance to the Lachlan, before it vamoosed in the marshes. 1958 'J. REEVES' *Mulbridge Manor* xii. 155 'See anyone?' asked Winston. 'Not a soul. Whoever it was has vamoosed.'

2. (Earlier example.)

1847 'M'C' *Let.* 2 Apr. in *Rough & Ready Ann.* (1848) 245 On the morning after I wrote the letter to father.. they..stacked their arms and colors, and 'vamossed the ranch'.

Hence **vamo(o)sing** *vbl. sb.*

1862 J. R. LOWELL *Biglow Papers* 2nd ser. v. 75 Or, when the vamosin' come, ever to find [etc.].

vamp (væmp), *sb.*[3] *U.S. slang.* [Origin unknown.] A volunteer fireman.

1877 *Fireman's Jrnl.* I. 15/1 Our old friend..seems to have the run of the old vamps. 1942 BERREY & VAN DEN BARK *Amer. Thes. Slang* §850/2 *Vamp*, a volunteer fireman.

vamp (væmp), *sb.*[4] [Abbrev. of VAMPIRE *sb.*] A woman who intentionally attracts and exploits men; an adventuress; a Jezebel; freq. as a stock character in plays and films.

a 1911 CHESTERTON *Lunacy & Letters* (1958) xxxvi. 178 Thackeray took it for granted that Mary Stuart was a vamp. 1918 *N.Y. Times* 15 July 9 Enid Bennett In a New 'Vamp' Story... 'The Vamp'..is a pleasing light comedy..in which Enid Bennett..appears as Nancy; an ingenuous wardroom girl at a musical comedy theatre where she hears sophisticated chorus girls tell how the female of the species may make the male buy her dinners and diamond bracelets by 'vamping' him... So Nancy, taking a tip from the chorus girls, 'vamps' him—and the wedding is a quick result. 1930 G. B. SHAW *Wks.* VII. 156 Ask yourself whether, if the lot in life therein described were your lot in life, you would not rather be a jewelled Vamp. 1973 *Times* 22 Dec. 9/2 Exotic red flowers like the lips of vamps. 1976 H. R. F. KEATING *Filmi, Filmi, Inspector Ghote* v. 44 She was..playing the Vamp in a film.

vamp (væmp), *v.*[3] [f. *VAMP *sb.*[4]] **1.** *intr.* To behave seductively; to act as a vamp, to be a vamp. *rare.*

1904 ADE *True Bills* 60 Any time that he fills in from eight o'clock to Midnight he certainly has to do some Vamping. 1922 *Observer* 1 Oct. 5/4 Trollope's Signora Neroni certainly vamped.

2. *trans.* To act as a vamp towards; to attract and exploit (a man, occas. a woman).

1918 [see *VAMP *sb.*[4]]. 1927 *Observer* 20 Mar. 15/3 Her friend, Violet Usher..shamelessly vamped Randall, and he felt obliged to marry her. 1939 L. M. MONTGOMERY *Anne of Ingleside* xxv. 166 Don't try to vamp me, woman. I've paid you all the compliments I'm going to. 1973 T. PYNCHON *Gravity's Rainbow* II. 245 Eager young chaps with patent-leather hair rush about trying to vamp the ladies. 1979 D. ANTHONY *Long Hard Cure* xix. 150 Gavin's secret girl, who vamped him on the nights those four women were assaulted.

va·mping, *vbl. sb.*[2] [f. *VAMP *v.*[3] + -ING[1].] Seductive behaviour; acting as a vamp.

1904 [see *VAMP *v.*[3] 1]. 1918 *N.Y. Times* 15 July 9 Nancy tries her 'vamping' tactics on him and he confesses his crime.

vampire, *sb.* Add: **2.** *spec.* = *VAMP *sb.*[4] (Further examples.)

1903 G. B. SHAW *Man & Superman* IV. 170 You lie, you vampire, you lie. 1918 *National Police Gaz.* (U.S.) 20 Apr. 4 (*caption*) Theda Bara... Vampire of the Screen. 1919 *Honey Pot* I. 42 Miss Maitland was a 'vampire' of an entirely new type. 1920 C. D. FOX *Who's Who on Screen* 301 Louise Glaum, who is credited with having given to the screen one of the most perfect vampire characterizations, was born near Baltimore. *a* 1953 E. O'NEILL *Long Day's Journey into Night* (1956) IV. 165 Made whores fascinating vampires instead of poor, stupid, diseased slobs they really are. 1968 *Word Study* Dec. 4/2 A vampire is a woman who uses sex to facilitate the acquisition of money or other signs of wealth. 1978 LD. BIRKENHEAD *Rudyard Kipling* vii. 99 A grim but authentic picture.. of callow subalterns trotting beside the rickshaw wheels of faded provincial vampires.

5. *vampire-winged* adj.; **vampire trap** (earlier example).

1846 S. F. SMITH *Theatr. Apprenticeship* viii. 63 Down I went through the trap-door (it was what actors call a Vampire trap) before any one was aware of my intentions. 1831 POE *Poems* 64 Some tomb, which oft hath flung into black And vampyre-winged pannels back.

Hence **va·mpiredom,** the state of being a vampire (sense 1); the acts of a vampire; **vampiric** *a.* (earlier and later examples); **va·mpirine** *a.* = VAMPIRIC *a.*; **vampirish** *a.* (further examples).

1933 *Times Lit. Suppl.* 28 Sept. 653/3 The more obvious literary possibilities of vampiredom were thoroughly explored and exploited nearly forty years ago. 1972 *Daily Tel.* (Colour Suppl.) 12 May 56 There before the horrified gaze of the living was all the evidence of vampiredom—twisted position, torn shroud and blood. 1853 D. G. ROSSETTI *Let.* 17 Apr. (1965) I. 136 Such are the vampyric notions of reciprocity. 1963 *Listener* 24 Jan. 165/2 She [*sc.* Marilyn Monroe] had all the physical equipment of the vamp, but the spirit of the girl next door... Marilyn was never truly vampiric on the screen, and she was never a 'taker' in life. 1914 in D. McCarthy *Drama* (1940) 129 This is too much for Vanya; he explodes at the old vampirine humbug, and..dashes from the room. 1946 BLUNDEN *Shelley* x. 135 Byron began and dropped a thriller which was becoming vampirine. 1929 *Sunday Dispatch* 13 Jan. 1/2 Among my own friends my reputation is notoriously the reverse of vampirish, money means nothing to me. 1944 R. LEHMANN *Ballad & Source* v. viii. 300 Mother fastened vampirish eyes on her. 1981 N. TUCKER *Child & Book* vii. 198 Religious references.. to the Virgin Mary behaving in a way that is distinctly vampirish have been glossed over.

vampish (væ·mpiʃ), *a.* [f. *VAMP *sb.*[4] + -ISH[1].] Suggestive or characteristic of a vamp. Hence **va·mpishness.**

1922 S. FORD *Trilby May crashes In* iv. 60, I have got something more over the footlights than just my ankles and a few vampish hip motions. 1922 *Observer* 1 Oct. 5/4 'The Vavasour' in 'Strathmore'..out of pure vampishness makes a conquest of the hero. 1928 *Ibid.* 15 July 12/4 Not international 'stars' of vampish wiles, but hundreds of Miss Betty Balfours should be sought and cherished. 1971 *Petticoat* 24 July 24/3 At last a girl who admits that she would like to look vampish and look a 'tramp' or 'tart'. 1972 J. WILSON *Hide & Seek* ii. 36 Her crêpe dress, low-cut and forties style, very vampish and sexy. 1977 *Time* 14 Mar. 32/1 That vampish young lady in black hardly looks like the type to drive a race car or ride a bucking bronco.

vampy (væ·mpi), *a.* [f. *VAMP *sb.*[4] + -Y[1].] That is a vamp; vampish. So **va·mpiness.**

1928 *Sunday Express* 29 July 4 The varnished vampiness of Greto Garbo. 1949 *Richmond* (Va.) *Times-Dispatch* 7 Oct. 10 Barbara Stanwyck has come up with the statement that the vampy, predatory female makes a good wife. 1977 *Spare Rib* July 49/1 You see a lot of stilettoes, tights, revealing things, black vampy make-up.

van, *sb.*[3] Add: **1. a.** Now usu. motorized.

b. Also *police van*: see *POLICE *sb.* 6. Also, a light covered vehicle employed to carry passengers.

1973, etc. [see *van pool*, etc., below]. 1979 *N.Y. Rev. Bks.* 25 Oct. 7 (Advt.), Every morning, 1,441 Gulf employees who used to drive their own cars to work now make the trip in vans.

c. A gypsy caravan or one used by a showman; a holiday-maker's caravan; a camper.

1858 DICKENS *Going into Society* in *Househ. Words* Extra Christmas No., 7 Dec. 22/1 The House was so dismal arterwards, that I giv it up, and took to the Wan again. 1876 C. M. YONGE *Three Brides* II. v. 107 He was born in one of they vans, and hadn't never been to school. 1926 KIPLING *Debits & Credits* 237 You can stand at your door and mock when the gipsy-vans come through. 1952 *Motor Manual* (ed. 34) xiii. 244 The owner who wants to tow his caravan from home to some pleasant site where he can leave the van during the summer months. 1972 *Guardian* 5 Feb. 14 The trailer tent..[is] easier to tow than a full-scale trailer 'van. 1976 *Kingston* (Ontario) *Whig-Standard* 4 Aug. 25/4 In California travelling in his own van. 1979 *Observer* 15 July 2/5 Australians who fly in and buy a van to take them round Britain are finding they have to omit the extremities from their itinerary. 1980 R. HILL *Killing Kindness* vii. 62 Dave Lee had gone off in his van.

2. (Earlier example.)

1854 DICKENS *Hard Times* II. i. 143 Very heavy train and vast quantity of it [*sc.* luggage] in the van.

3. *van-guard*; **van pool** *U.S.*, a group of workers sharing a van provided by their employer to transport them to work; so *van-pooler, -pooling.*

1921 *Dict. Occup. Terms* (1927) §723 *Van guard*.. travels with and guards contents of mail vans. 1931 *Daily Express* 22 Sept. 7/3 A vanguard..was accused of being concerned with another man..in stealing a motor-car. 1973 *Sunday Bull.* (Philadelphia) 7 Oct. (Parade Suppl.) 14/2 (*heading*) Van pooling. *Ibid.*, The 3M Company in St. Paul, Minn., bought six 12-passenger vans and

assigned them to workers to form 'van pools'... Van pool driver-coordinators receive free rides and can use the van during off-duty hours. 1974 *Woman's Day* (N.Y.) Oct. 138/2 Van pools are working so well at Minnesota Mining and Manufacturing's huge complex..that the company now has fifty-seven company-owned vans. 1976 *National Observer* (U.S.) 16 Oct., Van pooling is a new commuting style looking for a permanent place in transportation between private cars and mass-transit facilities. 1983 *Mass Transit* (U.S.) Apr. 24/1 The average vanpooler saves about 400 miles of driving and 25 gallons of gas a month.

van (væn), *sb.*[5] *Lawn Tennis.* Abbrev. of VANTAGE *sb.* 6.

1927 in W. E. Collinson *Contemporary Eng.* 36. 1960 N. HILLIARD *Maori Girl* 47 They counted in lawn-tennis style with vans and loves. 1977 *Fremdsprachen* XXI. 125 Van in, van out: my van, your van.

vanadian (vănēⁱ·diăn), *a.* *Min.* [f. VANAD(IUM + *-IAN.] Of a mineral: having a (usu. small) proportion of a constituent element replaced by vanadium.

1930 W. T. SCHALLER in *Amer. Mineralogist* XV. 572 Vanadium—vanadian. 1951 C. PALACHE et al. *Dana's Syst. Min.* (ed. 7) II. 1084 The so-called eosite from Leadhills, Scotland, apparently is a highly vanadian wulfenite.

vanadiate. (Earlier example.)

1833 *Rep. Brit. Assoc. Adv. Sci. 1832–3* 470 Vanadiate of lead.

vanadic. Add: (Earlier example.] Now *Obs.*

1833 *Rep. Brit. Assoc. Adv. Sci. 1832–3* 469 Vanadic acid..in the state of powder is yellow.

vanadium. (Earlier example.)

1833 *Rep. Brit. Assoc. Adv. Sci. 1832–3* 468 Vanadium ..was discovered by Sefström towards the end of 1830, in the iron from the forges of Eckersholm in Sweden.

vanadous, substitute for entry in Dict.: **vanadous** (væ·nădǒs), *a.* *Chem.* [f. VANAD(IUM + -OUS.] Of or containing vanadium, *spec.* in an oxidation state of +2.

1850 H. WATTS tr. *Gmelin's Hand-bk. Chem.* IV. 90 (*heading*) Bisulphide of vanadium.—Vanadous sulphide. 1858 [in Dict.]. 1887 C. M. TIDY *Handbk. Mod. Chem.* (ed. 2) xvii. 504 Preparation [of Vanadium].—By heating vanadous chloride (VCl_3) in a current of dry hydrogen, air and moisture being excluded. 1927 *Amer. Mineralogist* XII. 236 There may also be assumed to be present..:..a vanadous acmite, $Na_2O{\cdot}V_2O_3{\cdot}4SiO_2$. 1954 *Thorpe's Dict. Appl. Chem.* (ed. 4) XI. 833/1 Vanadous Oxide, VO, has not been prepared pure. 1962 H. L. KERN et al. in A. Pirie *Lens Metabolism Rel. Cataract* 386 The tubes were.. gassed with N_2—CO_2 (95:5) from which the oxygen had been removed by passage through vanadous sulphate. 1980 *Jrnl. Amer. Chem. Soc.* CII. 3035/1 The change from a d^4 (chromous) to a d^3 (vanadous) ion results in a profound change in the stoichiometry.

Van Allen (væn æ·lėn). Also (*erron.*) **van Allen.** The name of James A. *Van Allen* (b. 1914), U.S. physicist, used *attrib.* to designate each of two regions reported by him in 1958, partly surrounding the earth at heights of several thousand kilometres and containing intense radiation and many high-energy charged particles trapped by the earth's magnetic field; also applied to similar regions around other planets.

1959 *Nature* 8 Aug. 439 (*heading*) The upper boundary of the Van Allen radiation belts. 1965 *Wireless World* Sept. 447/1 This [sensitivity to radiation damage] limits the application of junction detectors in reactor instrumentation and in the van Allen belts. 1971 *Nature* 28 May 217/2 The circular polarization of about one per cent which is emitted by the planet Jupiter indicates the presence of a field of 0·4 gauss in its van Allen belts. 1974 *Times Lit. Suppl.* 13 Sept. 968/2 It girds us Transparently like the Van Allen Belt, Simply like the tie-belt of a mac. 1983 *New Scientist* 13 Jan. 93/2 They penetrated the Van Allen radiation belt (which is contained by the Earth's magnetic field) and loosened showers of electrons.

vanaspati (vănɒ·spăti). [Skr. *vánas-páti*, lord of the wood, forest tree.] A vegetable ghee used in India. Also *attrib.*

1949 *Food Manufacture* XXIV. 500/2 On account of the shortage of butterfat in India, the consumption of hydrogenated oils, known locally as *vanaspatis*, has recently been greatly increased. 1969 *National Herald* (New Delhi) 29 July 9/3 Oils and seeds sought lower levels following fears of nationalization of the vanaspati industry. 1973 *Indian Express* (Bombay) 29 Oct. 4/5 Several temples used vanaspati instead of ghee to keep the lamps before the deities burning. 1979 *Eastern Economist* 14 Sept. 534/1 Ensuring price stability as well as availability of vanaspati.

Vanbrughian (vænbrū·iăn), *a.* [f. the name of Sir John *Vanbrugh* (or *Vanburgh*) (see below) + -IAN.] Of, pertaining to, or characteristic of the architecture, landscaping, or plays of Sir John Vanbrugh (1664–1726). Also *ellipt.* as *sb.*, Vanbrughian style.

1947 J. LEES-MILNE *Diary* 3 July (1983) 178 The windows tame Vanbrughian. 1972 *Country Life* 5 Oct. 812 The heavy Doric frieze and Vanbrughian chimneypiece. 1979 *Garden* (Victoria & Albert Mus.) 63/2 Early in the twentieth century there was..a Vanbrughian [garden] mode. 1980 *Country Life* 17 July 243/2 A tale that is.. gorgeous, pyrotechnic, Vanbrughian.

vancomycin (væŋkomǝi·sin). *Pharm.* [f. *vanco-*, of unkn. origin + *-MYCIN.] A glycopeptide antibiotic produced by the actinomycete *Streptomyces orientalis* and active against most Gram-positive bacteria.

1956 M. H. MCCORMICK et al. in *Antibiotics Ann. 1955–6* 606 A new antimicrobial agent, vancomycin, has been isolated..from strains of *Streptomyces orientalis*, n. sp. 1974 R. M. KIRK et al. *Surgery* ii. 29/1 Vancomycin is effective against some penicillin-resistant staphylococci but is too toxic for routine use. 1982 *Brit. Med. Jrnl.* 22 May 1508/2 In the past few years..vancomycin has re-emerged as a valuable antibiotic in three clinical settings: endocarditis, multiresistant staphylococcal infections, and as the first-line treatment of pseudomembranous colitis.

Vandal, *sb.* Add: (With small initial.) **2. b.** *attrib.* and *Comb.*, as *vandal-proof, -resistant* adjs.

1971 H. PACY *Road Accidents* iv. 115 A vandalproof phone, consisting of a loudspeaker and microphone sheltered behind heavy steel grids. 1977 *Linlithgowshire Jrnl. & Gaz.* 15 Apr. 3/5 We have tried to make the hall vandal-proof by introducing a number of safeguards. 1977 C. BRANDRETH *Parking Law* 50 Boston (Mass.)..have vandal-resistant meters, the coins encased in a heavy iron box.

vandalistic, *a.* Add: Hence **vandali·stically** *adv.*

1922 *Weekly Dispatch* 10 Dec. 7 Drew a picture of Spion Kop, of myself,..and others of the party on the tablecloth... One of the party began vandalistically to cut the square out of the cloth. 1970 J. STEER *Conc. Hist. Venetian Painting* v. 134 The *Presentation of the Virgin* [by Titian]..still hangs in the position for which it was designed, although a hole has been vandalistically cut in the left-hand side to make room for an extra door. 1980 *Times* 30 Aug. 10/3 Customers who behave so inconsiderately, not to say vandalistically.

vandalization. Add: Also, an act of vandalism.

1984 *Times* 16 June 18/8 We know one couple who finally gave in to silent local pressure to move out (five break-ins..two vandalizations).

vandalize, *v.* Add: (Later examples.)

1905 *Chambers's Jrnl.* Dec. 17/1 A very remarkable church, spoilt and vandalised by the introduction of galleries and deal pews. 1968 *Wall Street Jrnl.* 19 Feb. 1/1 These militants roamed the campus, vandalizing the bookstore and a dining hall and terrorizing students. 1971 J. WAINWRIGHT *Last Buccaneer* I. 48 He had to run more than a mile before he could find a kiosk which had not been vandalised. 1978 *Daily Tel.* 17 Aug. 2/8 Newcastle..is to incur even heavier expenditure by importing more costly American [parking] meters said to be harder to vandalise.

Hence **va·ndalized** *ppl. a.*, **va·ndalizing** *vbl. sb.*

1971 *Oxford Times* 13 Aug. 5/7 In a two-week campaign they restored 21 vandalised kiosks to working order. 1979 U. CURTISS *Menace Within* xvi. 169 Inquiries..in connection with the vandalizing of a local school. 1984 *Times* 8 Oct. 14/3 In Minster Court, Liverpool, buyers rushed for well landscaped flats in what had been a vandalized estate.

van de Graaff (væn dǝ grɑf). Also **Van de Graaff.** The name of R. J. *van de Graaff* (1901–67), U.S. physicist, used *attrib.* and *absol.* to designate a machine devised by him for generating electrostatic charge by means of a vertical endless belt which collects charge at its lower end from needle points connected to a voltage source and carries it to similar points at the top connected to the inside of a metal dome, whose potential is thereby increased; also a particle accelerator based on this machine.

1934 *Rev. Sci. Instruments* V. 18/1 The paper discusses the design and construction of a Van de Graaff electrostatic generator capable of an output of 500 microamperes and more than a million volts. 1961 G. R. CHOPPIN *Exper. Nuclear Chem.* xii. 193 Positive ions..are obtainable in cyclotrons as well as Van de Graaff and linear accelerators. 1973 *Physics Bull.* Dec. 722/1 Both tandem van de Graaffs and cyclotrons currently accelerate a variety of high quality beams. 1980 J. W. HILL *Intermediate Physics* xviii. 172 Two metal plates about 5–7 cm apart are..connected to..earth and the dome of a Van de Graaff generator.

vandendriesscheite (vændėndrī·ʃ(ǝ)ǝit). *Min.* [a. F. *vandendriesscheite* (J. F. Vaes 1947, in *Ann. de la Soc. géol. de Belgique: Bull.* LXX. 217), f. the name of A. *Vandendriessche* (1914–40), Belgian mineralogist: see -ITE[1].] A hydrated oxide of lead and hexavalent uranium, $PbU_7O_{22}{\cdot}12H_2O$, found as orange orthorhombic crystals.

1947 *Mineral. Abstr.* X. 146 Vandendriesscheite,.. pseudo-hexagonal barrel-shaped crystals, perfect cleavage. 1971 *Ibid.* XXII. 158/2 Among the secondary V- and Th-minerals occurring in the uraninite-bearing pegmatite at Einerkilen [in Norway], the following have been identified..: thorogummite, vandendriesscheite, [etc.].

Vanderbilt (væ·ndǝɹbilt). [Name of Cornelius *Vanderbilt* (1794–1877), U.S. railway magnate, and his descendants.] One who resembles a member of the Vanderbilt family in being exceptionally rich; a millionaire. So **Va·nderbilter.**

1885, 1910 [see *ROTHSCHILD 1]. 1975 M. AMIS *Dead Babies* xv. 73 So then of course some Vanderbilters get along from Nashville and Skip starts to hang out with them.

Van der Hum (væn dǝɹ hɒ·m). Also **Vanderhum, † vanrhum.** [perh. a personal name.] A South African brandy-based liqueur made with nartjies.

1861 *Let.* 14 Oct. in *Cape Monthly Mag.* (1870) Oct. 225 Mrs M...has even promised to show me how to brew liqueurs, and distil 'vanrhum',—the latter a most aromatic and powerful *elixir vitæ*. 1893 KIPLING *Many Inventions* 330 Judson's best Vanderhum, which is Cape brandy ten years in the bottle, flavoured with orange-peel and spices. 1913 D. FAIRBRIDGE *That which hath Been* xxi. 240 If you want a *zoopje* there is Van der Hum on the table yonder and a bottle of old dop. 1949 A. WILSON *Wrong Set* 41 Wait..till he tasted the Van Der Hum after the meal, then he would see what the Union could do. 1966 [see *NARTJIE]. 1978 K. BONFIGLIOLI *All Tea in China* xi. 141 A black, fat-bellied bottle of something called 'Van Der Hums' which he had prudently laid in at the Cape.

van der Waals (væn dǝɹ vā·ls). *Physics.* Also **Van der Waals.** The name of Johannes *van der Waals* (1837–1923), Dutch physicist, used *attrib.* and in the possessive to designate (*a*) an equation of state of a gas (proposed by him in 1873) that allows for intermolecular attraction and finite molecular size, viz. $(p + a/V^2)(V - b) = RT$ (where p is the pressure, V the volume, R the gas constant, T the absolute temperature, and a and b are constants); (*b*) short-range attractive forces between uncharged molecules arising from interaction between (actual or induced) electric dipole moments; so *van der Waals attraction*, etc.

1895 C. S. PALMER tr. *Nernst's Theoret. Chem.* II. ii. 186 The equation assumes the form $(p + a/v^2)(v - b) = RT$. This is van der Waals' equation of condition. 1923 *Jrnl. Chem. Soc.* CXXIII. 3403 At the critical point in particular, Van der Waals's equation predicts that pv will be $2\frac{2}{3}$ times as great as the value given by the simple law of Boyle. 1930 *Engineering* 5 Dec. 700/3 There were, further, different types of cohesion forces. (1) Van der Waals cohesion. 1964 J. W. LINNETT *Electronic Struct. Molecules* ii. 28 No chemical bond is formed and, except for a weak van der Waals (polarization) attraction at large distances, the two atoms repel one another. 1965 P. CAWS *Philos. of Sci.* xxii. 166 The gas law..is hopelessly inadequate in most cases, and has to be replaced by Van der Waals' equation..which is itself only approximately valid. 1979 *Sci. Amer.* Dec. 145/1 If the sauce is kept cool, the colloidal particles are unlikely to touch one another with a frequency that would allow the van der Waals forces to make them coalesce into pools of butter.

Vandyke, *sb.* Add: Also **Van Dyke. 5*.** = *Vandyke beard.*

1909 in WEBSTER. 1933 H. ALLEN *Anthony Adverse* III. VIII. lix. 949 A carefully trained moustache and a vandyke turning white contrived to be at once distinguished and benign. 1951 E. PAUL *Springtime in Paris* xi. 204 The 'Zazous' quarter surpasses even Oberammergau, as far as beards are concerned... The hearties have stiff whiskers, and some of the Central and South American boys sport silken Van Dykes. 1974 P. DE VRIES *Glory of Hummingbird* xvi. 252 A false mustache and Vandyke..completed the transformation. 1982 S. B. FLEXNER *Listening to Amer.* 65 The first group mainly wore short *Imperials* or waxed and pomaded *Van Dykes*.

Hence **Van Dy·ckian,** pertaining to or characteristic of the paintings of Vandyke.

1942 *Burlington Mag.* Feb. 43/2 The stage effects of court painting in the Van Dyckian vein. 1979 *N. & Q.* Feb. 71/2 The pose..is pure convention: a Van Dyckian cliché lifted from Van Dyck's etching of Pieter Brueghel the Younger.

vanette (væne·t). [f. VAN *sb.*[3] + -ETTE.] A small van.

1921 *Glasgow Herald* 23 July 8 A company..were sent out from Jandola, followed..by Lewis guns in vanettes and an armoured motor car. 1957 V. W. TURNER *Schism & Continuity in Afr. Society* v. 147 Once Mukanza was in my vanette he was my responsibility and no longer that of the village members.

Van Gelder (væn ge·ldǝɹ). The name of a Dutch paper-maker, used *attrib.* and *absol.* to designate a fine handmade paper with deckle edges.

1892 in G. Meredith *Jump to Glory Jane* 2 This edition is limited to an issue of 1,000 copies..and a special issue of 100 copies on Van Gelder paper. 1928 S. J. LOOKER *Booklover's Catal.* Jan. 6 Long Ago, First American Edition,..one of 925 copies on Van Gelder paper. 1934 H.

Hiler Notes Technique Painting iii. 224 Some good papers for practical working are cartridge, Whatman,.. Van Gelder, [etc.]. 1960 G. A. Glaister Gloss. Bk. 429/2 Van Gelder, one of the most famous of all hand-made papers, produced for a century or more in Holland and said to contain 100% rag. It is used by artists and publishers of fine editions. 1974 Country Life 17 Jan. 75/1 The..tragic etching of 1904, Le Repas Frugal on van Gelder paper.

vanguard. Add: **1. c.** In *Communism*, the elite party cadre which, according to Lenin, would be used to organize the masses as a revolutionary force and to give effect to communist planning.

1928 E. & C. Paul tr. Stalin's Leninism I. 95 As far back as 1902, foreseeing the special role of our Party, he [sc. Lenin] thought it necessary to point out that: Only a party guided by an advanced theory can act as vanguard in the fight. 1941 tr. Lenin's One Step Forward 85 The stronger our Party organizations..the richer..will be the influence of the Party on the elements of the working class masses..guided by it. After all, the Party, as the vanguard of the working class, must not be confused with the entire class. 1952 P. Selznick Organizational Weapon 10 Only the vanguard is exposed to a full statement of communist aims and methods. 1973 C. D. Kernig Marxism, Communism & Western Soc. VIII. 270/1 The vanguard proves itself both in and after the victorious proletarian revolution by taking over the leadership in all spheres of life.

d. The name of a political party in Northern Ireland, representing a secession from Ulster Unionism. Freq. *attrib.*

1972 Times 14 Feb. 2/4 A crowd of 1,500 attended a rally called by Mr William Craig, MP for Larne, who has founded what he calls the Vanguard Movement. Ibid. 13 Mar. 1/4 When it began three months ago, Vanguard was described by one Government minister as a 'comic opera'. 1973 Times 31 Mar. 1/2 Mr William Craig... chose the steps of Stormont to announce the formation of the Vanguard Unionist Progressive Party. 1975 Butler & Sloman Brit. Political Facts (ed. 4) xvi. 354 On June 28, 1973 the..Assembly was elected; its composition was 22 Ulster Unionist..15 Loyalist Coalition (7 Vanguard.. 8 Democratic Unionist), [etc.]. 1982 M. Wallace Brit. Govt. in Northern Ireland vii. 133 The Vanguard Unionists had decided in November 1977 to return to the official Unionist Party, while remaining a 'movement'.

attrib. (Further examples: cf. *avant-garde 2.)

1958 Times 9 Sept. 5/1 But it was not the pretentious nonsense that so often passes for 'vanguard' culture. 1977 Guardian Weekly 16 Oct. 19/3 One prominent vanguard dealer in London..had only sold one single work so far this year. 1980 Listener 13 Nov. 648/3 Today, vanguard art seems to have lost its 'political' role.

Hence **va·nguardism**, the quality of being in the vanguard of a political, cultural, or artistic movement (cf. *avant-gardism); **va·nguardist**, a person in the vanguard of a political, cultural, or artistic movement; also *attrib.* or as *adj.*; (cf. *avant-gardist(e)).

1934 Webster, Vandguardist. 1952 P. Selznick Organizational Weapon ii. 91 (heading) Vanguardism in semicolonial areas. 1964 Spectator 15 May 664/1 It is not true that this complex personality comes out of the same box as Rauschenberg. That vanguardist exclaims, 'Hell, what a marvellous universe.' 1971 New Yorker 16 Oct. 148/2 The integration of art and non-art is the essence of twentieth-century vanguardism in art. 1976 Ibid. 12 Jan. 22/1 Just being there must make the average armchair nihilist want to argue with his fellow-man over the merits of anarcho-syndicalism, vanguardism, and radical conformism. 1976 Spare Rib Nov. 21/1 After wading through some vanguardist rhetoric, and an informal discussion.. things became even more complicated. 1981 Times Lit. Suppl. 24 Apr. 465 (heading) From vanguardism to fascism.

vanilla. Add: **3. c.** A vanilla ice cream.

1955 T. Sterling Evil of Day viii. 84 You should go to Schrafft's for a plain vanilla with marshmallow sauce. 1970 Guardian 13 July 9/3 They'll have a vanilla, our Ethel, ta.

4. *vanilla ice cream*; *vanilla-flavoured*, *-sweet* adjs.; **vanilla slice**, an oblong pastry containing custard flavoured with vanilla and usu. iced.

1972 Harrods Christmas Catal. 59/4 French marrons in vanilla flavoured syrup. 1974 L. Deighton Spy Story xix. 207 That big vanilla-flavoured ice-cream sundae. 1904 Vanilla ice cream [see *sundae]. 1911 [see *Neapolitan ice]. 1974 Times 6 Apr. 12/5 Serve the oranges very cold with vanilla ice cream. 1930 Radiation Cookery Book (ed. 12) vii. 119 Vanilla Slices. Ingredients—½ lb. puff paste. A few drops vanilla, ½ pint thick custard... When the pastry is cool, spread the custard on one strip and put the second strip over the top; cover with icing, and cut with a sharp knife in strips about 2 inches wide. 1979 This England Winter 19/1 I've brought you one of your favourite vanilla slices as well! 1940 L. MacNeice Plant & Phantom (1941) 59 All her vanilla-sweet forgotten vaudeville nights.

b. Passing into *adj.*: vanilla-coloured, vanilla-flavoured. (Not clearly distinguishable from some of the quots. in sense 4 a.)

1946 A. Christie Come, tell me how you Live viii. 133 A vanilla soufflé, for a wonder, goes right. 1962 Vanilla soda [see *soda[1] 4 b]. 1980 H. Engel Suicide Murders xviii. 121, I ordered a vanilla marshmallow sundae and a vanilla milkshake. 1982 M. Russell All Part of Service iii. 23 A box of vanilla fudge. 1984 Guardian 5 Oct. 17/8 Old-fashioned vanilla sundae with hot fudge sauce

vanillyl (væ·nilil). *Chem.* and *Biochem.* [f. Vanill(ic *a.*, -in + -yl.] The radical, $OCH_3C_6H_3(OH)CH<$, derived from vanillin; *vanillylmandelic acid* [Mandelic *a.*], the end-product of the metabolism of adrenaline and noradrenaline, measured to test for certain tumours; 4-hydroxy-3-methoxymandelic acid, $OCH_3C_6H_3(OH)CH(OH)COOH$.

1876 Jrnl. Chem. Soc. XXIX. 76 This body, which is still under investigation, is vanillyl alcohol. 1932 Jrnl. Pharmacol. & Exper. Therapeutics XLV. 165 The salts of vanillyl-ethylamine and methylamine were found to be quite soluble. [1957 Armstrong & McMillan in Pharmacol. Rev. XI. 395 Because of the cumbersome nature of the proper name for this compound..the trivial name 'vanilmandelic acid' is proposed, and the abbreviation 'VMA'.] 1961 Clin. Chem. VII. 257 The determination of 3-methoxy-4-hydroxymandelic acid, also called vanillylmandelic acid (VMA), in urine has recently come into use as a confirmatory test for pleochromocytoma. 1979 Nature 1 Mar. 41/2 Many oxidative degradations have also been carried out to break coal down into simpler species; however, isolation and identification of p-hydroxyl (IV), vanillyl (V), and syringyl (VI) groups, which are characteristic lignin oxidation products, have not yet been confirmed. 1979 Jrnl. R. Soc. Med. LXXII. 533 Twenty-four hour urinary vanillylmandelic acid (VMA) estimations were within normal limits.

vanish, *sb.* **1.** (Later example.)

1872 'Mark Twain' Roughing It iii. 33 He..left for San Francisco at a speed which can only be described as a flash and a vanish.

vanish, *v.* **4.** Delete 'Now *rare*' and add later examples, chiefly with reference to conjuring.

1934 H. G. Wells Exper. Autobiogr. I. v. 264 Lenin conjured government by mass-democracy out of sight, 'vanished' it as conjurors say, by his reorganization of the Communist Party. 1949 [see *disappear v. 3]. 1981 Daily Tel. 4 Jan. 6/8 Thurston..could make a girl disappear from a cage suspended in mid-air, or vanish a girl playing a piano (and the piano).

vanishing, *vbl. sb.* Add: **2.** *vanishing point*: also *fig.*

1913 Times 7 Aug. 8/2 The danger of operation, gas operation, is retreating to a vanishing point. 1963 Lancet 12 Jan. 96/2 Routine application of this test reduces to vanishing-point the mortality from air embolism during neurosurgery in the sitting position.

3. *vanishing act, trick.*

1923 H. C. Bailey Mr. Fortune's Practice vi. 160 She goes off walking at night with nothing but what she stood up in. If you ask me to believe she meant to do the vanishing act..I can't see how it's likely. 1981 P. Mallory Killing Matter vii. 79 He's chosen to pull a vanishing act just when the painting is lifted. 1973 G. Sims Hunters Point xviii. 174 At the end of the path Jaeckel disappeared and it was Buchanan's turn to stop, momentarily at a loss..mystified by the vanishing trick.

vanishing, *ppl. a.* Add: **3.** *vanishing cream*, a cosmetic cream that is readily absorbed by the skin; also *fig.*

1916 Daily Colonist (Victoria, B.C.) 11 July 14/5 (Advt.), Pond's Vanishing Cream. 1923 W. A. Poucher Perfumes & Cosmetics III. vi. 390 Vanishing Creams.. consist of stearic acid, partially saponified with alkali, the bulk of the fatty acid being emulsified by the soap thus formed. a 1951 in M. McLuhan Mech. Bride (1967) 47/2 It's like your wife's vanishing cream—not greasy or sticky. 1981 E. Ambler Care of Time ix. 149 If his luck holds, he's getting the defectors' vanishing-cream treatment..somewhere in North America.

‖ **vanitas** (væ·nitās). [L., lit. 'vanity, emptiness'.] **1.** *vanitas vanitatum* (from the Vulgate transl. of Eccles. i. 2): vanity of vanities, an exclamation of disillusionment or pessimism; futility.

1565 J. Jewel Replie unto M. Hardinges Answeare iii. 164 This labour may well be called Vanitas vanitatum. 1848 Thackeray Vanity Fair lxvii. 624 Ah! Vanitas Vanitatum! Which of us is happy in this world? Which of us has his desire? or, having it, is satisfied? 1910 Chesterton G. B. Shaw 105 In Shakespeare he saw nothing but profligate pessimism, the vanitas vanitatum of a disappointed voluptuary. 1978 H. Shaw in Islands (N.Z.) Aug. 105 Ah Vanitas! Vanitatum! All had been so totally suppressed, yet..I had sensed certain emanations.

2. (Usu. with capital initial.) Used *attrib.* and *absol.* of a 17th-century Dutch genre of still-life painting that incorporated symbols of mortality or mutability.

1909 M. L. Clarke tr. Bode's Great Masters Dutch & Flem. Painting 241 The artist [sc. J. Davidsz. de Heem] became a follower of the peculiar art movement of the older Leyden still-life painters, as several pictures of a 'Vanitas' show. Ibid. 239 The 'Vanitas' presentments required a monochrome treatment. 1942 Burlington Mag. Dec. 293/1 Paul..was a painter of some consequence, as shown for instance by his Vanitas still-life at Berlin (1636). 1947 Bergström & Taylor tr. Bergström's Dutch Still-Life Painting 17th Cent. iv. 154 On account of its literary symbolism, often rather elaborate, the Vanitas occupies a special place among the forms of Dutch still-life painting. 1963 E. H. Gombrich Meditations on Hobby Horse 104 Is the spread of the genre [sc. still life] due to the religious appeal of the vanitas motif? Ibid., Any painted still life is ipso facto also a vanitas. 1980 Times 13 Aug. 14/3 A painting by Frans Hals, 'Young Man Holding a Skull'..is a vanitas, a reminder of human vanity.

Vanitory (væ·nitŏri). orig. *U.S.* Also **vanitory.** [f. Vanit(y + -ory[1].] A proprietary name for a vanity unit.

1951 Official Gaz. (U.S. Patent Office) 13 Mar. 355/1 Vanitory. For combination dressing table-lavatory unit. 1958 Trade Marks Jrnl. 16 July 720/2 Vanitory. All goods included in class 11... 22nd November 1957. 1960 Spectator 26 Feb. 300 Bathing-dressing-rooms, complete with vanitories. 1961 Ibid. 10 Mar. 345 The appearance of Vanitory units in the ballroom as well as the bedroom for which they were designed. 1972 House & Garden Dec.–Jan. 82/1 One of the bathrooms..with built in vanitory unit. 1984 Sunday Tel. (Colour Suppl.) 29 Apr. 40/4 A roll-out bidet which, when not in use, is neatly concealed in a vanitory unit beneath the basin.

vanitous (væ·nitəs), *a. rare.* [f. Vanity + -ous or ad. F. *vaniteux*.] = Vain *a.* 4.

1900 G. Meredith Let. 12 July (1970) III. 1351 An accurate perception of foibles in those whom we love..is only a vindication of our intellect—the seeing in what way our hero or friend or beloved is a little vanitous and pretentious. 1905 Ibid. 30 July 1536 French criticism.. instructs without wounding any but the vanitous person. 1930 Musical Times 1 Mar. 210/2 It is only the very young, or the very vanitous, who think of making a world of themselves.

Hence **va·nitously** *adv.*

1939 V. Woolf Diary 3 Mar. (1984) V. 207, I was pleased, vanitously, to find that Inez thinks me a poet-novelist.

vanity. Add: **4. b.** (Later example.)

1894 'Mark Twain' in Century Mag. June 236/1 The claim that the knife had been stolen was a vanity and a fraud.

5*. *N. Amer.* **a.** = *vanity table* (see sense 6 below). **b.** = *vanity unit* (sense 6 below).

(a) 1937 'E. Queen' Door Between xiv. 148 She sat down before the vanity to cold-cream her face. 1967 'V. Siller' Biltmore Call 124 Her make-up and perfume bottles and jars were still on a kidney-shaped vanity. (b) 1967 Boston Sunday Herald 26 Mar. (Advt.), Classic elegance for your bathroom is yours with this 30×20-in. vanity... Vitreous china top and bowl. 1977 Chicago Tribune 2 Oct. xii. 8/2 (Advt.), Ceramic tile baths and vanities. 1984 Tampa (Florida) Tribune 5 Apr. (Sears Suppl.) 9/3 Start your bath remodeling with this lovely vanity.

6. *vanity-bait*; **vanity bag** (examples); **vanity basin**, a wash-basin for a vanity unit; **vanity box** (examples); **vanity case** (examples); **vanity mirror**, (*a*) a small make-up mirror, *esp.* as a fitting in a motor vehicle; (*b*) a dressing-table mirror; **vanity number plate**, *U.S.* **vanity plate** (see quot. 1967); **vanity press, publisher** orig. *U.S.*, a publisher who publishes only at the author's expense; so *vanity publishing*; **vanity set**, (*a*) a set of cosmetics or toiletries; (*b*) *U.S.*, a matching bath and vanity unit; **vanity table**, a dressing-table; **vanity unit**, a unit comprising a wash-basin set into a fixed dressing-table.

1907 Yesterday's Shopping (1969) 404/1 The Vanity Bag. Containing on one side mirror, separate pocket for powder..with puff,..gusset pockets..for gold on other side, pocket for cards, papers, &c. 1946 K. Tennant Lost Haven (1947) xviii. 295 The silver-beaded white vanity-bag that carried her money, her lipstick, her handkerchief. 1974 J. Cleary Peter's Pence iii. 78 The note..had said she was spending the night with Fergus... She had taken a small vanity bag and slipped out of the apartment. 1816 Jane Austen Emma II. xiv. 278, I should never have expected you to be lending your sanction to such vanity-baits for poor young ladies. 1972 House & Garden Dec.–Jan. 84/3 'Luxe' vanity basin..£11.25. 1912 G. R. Chester Five Thousand an Hour ii. 10 She carried her own vanity box. 1978 N. Marsh Grave Mistake iv. 136 A vanity box..lay on the table. 1913 S. Story Spirit of Paris 52 The exquisite femme du monde..has a final glance at herself, 'vanity case' in hand. 1957 Practical Wireless XXXIII. 531/1 The Sky Casket by Every Ready ..is of the vanity-case type—no controls or other external components being visible. 1979 M. McCarthy Cannibals & Missionaries xi. 323 He had slipped a folded plan of the house..into Eloise's vanity case, under her powder-puff, camouflaged by a thick layer of face-powder. 1959 Observer 1 Mar. 21/5 Visors with vanity mirror are flush fitting. 1966 T. Pynchon Crying of Lot 49 i. 16 A half hour in front of her vanity mirror. 1971 'D. Halliday' Dolly & Doctor Bird viii. 100 The vanity mirror..was surrounded by fourteen ormulu makeup lamps. 1983 Listener 27 Oct. 25/2 (Advt.), Such thoughtful touches as an illuminated vanity mirror..and seat back map pockets are all standard. 1983 Daily Tel. 10 Oct. 13/4 They will sell you a personalised or 'vanity' number plate for as little as ten dollars. 1967 Britannica Bk. of Year (U.S.) 804/3 Vanity plate, an automobile license plate bearing distinctive letters, numbers, or a combination of these and usually available at extra cost. 1974 'D. Shannon' Crime File (1975) xi. 194 The drivers who wanted to pay extra.. could buy the vanity plates. 1969 C. Armstrong Seven Seats to Moon vi. 61 Are you planning to pay that Vanity Press to publish your father's book? 1976 N.Y. Times Bk. Rev. 7 Mar. 12/2, I read this book with the kind of horrified fascination with which one reads vanity press confessions. 1922 Holliday & Van Rensselaer Business of Writing 138 Numerous devices are employed by the 'vanity publisher' to lead the innocent author on toward becoming famous in his own eyes and those of his friends. 1978 Amer. N. & Q. Nov. 45/2 There are all sorts of literary histories, ranging from the pathetic things by which vanity publishers con local authors into expenditures far beyond their means, up to the CHEL and comparable works. 1984 H. Spurling Secrets of Woman's

Heart 19 Ivy placed Pastors and Masters..with a small firm of 'vanity' publishers called Heath Cranton in Fleet Lane, paying for publication herself. 1960 G. A. Glaister Gloss. Bk. 429/2 Vanity publishing, publishing on behalf of and at the expense of an author who pays for the production and often for the marketing of his book. 1981 V. Glendinning Edith Sitwell iii. 45 She had emerged from vanity publishing to the real thing... 'I have found a publisher.' 1930 A. P. Herbert Water Gipsies xix. 285 She lightly powdered her face. Lily had..given her a 'vanity set'. 1970 Washington Post 30 Sept. b11/6 (Advt.), New tub. Toilet and vanity set. 1979 D. Cook Winter Doves ii. i. 45 He had packed dolls' Vanity Sets into boxes. 1936 L. C. Douglas White Banners i. 9 The mirror of the vanity table. 1954 W. Tucker Wild Talent (1955) xiv. 183 A vanity table likewise revealed occupied drawers. 1980 L. Birnbach et al. Official Preppy Handbk. 191/2 Women's locker rooms often boast vanity tables with combs and face powder. 1973 Vanity unit [see *limed ppl. a. 2 b]. 1983 Sunday Tel. 21 Aug. 31/4 The property has modern conveniences, including..fitted kitchen, bathroom with vanity unit and plenty of power points throughout.

vanner[2]. Add: **2.** *N. Amer.* An owner or operator of a van, *esp.* one who uses the van for recreation.

1973 Hot Rod Nov. 71/3 Hot Rod editor Terry Cook and art director Jervis Hill indulged their..fantasies by forming vans into what may be the largest "Keep on Truckin'" ever created. The vanners were willing to help. 1976 National Observer (U.S.) 10 Apr. 15/1 Vanners.. converge on campgrounds and recreational areas every week end for van fetes all over the country. 1976 Kingston (Ontario) Whig-Standard 4 Aug. 25/3 Coming up for vanners in the area, are a few special events.

Vannetais (va·nte, ‖ vantę), *sb.* and *a.* [Fr., f. *Vannes* (see below).] (Designating) a dialect of Breton spoken in the region of Vannes in Brittany.

1953 K. H. Jackson Lang. & Hist. in Early Britain ii. xiv. 301 He argues that the change to -ec took place also in Vannetais. 1967 —— Historical Phonol. of Breton ii. i. 18 The Vannetais dialect coincides on the whole quite closely on the west with the diocesan boundary in the valley of the Ellé, as far north at least as near le Faouet. 1977 Word 1972 XXVIII. 14 The mixed character of the Vannetais dialect of Breton is explainable by the profound influence of these early Romance dialects. Ibid. 19 The area now occupied by Vannetais had less chance of retaining Gaulish than did the northwestern part of Brittany.

Vannic (væ·nik). [f. *Van*, the name of a town on the site of an ancient centre of Armenian civilization + -ic.] = *Khaldian b. Also as *adj.* Cf. *Urartian *sb.* and *a.*

1882 Jrnl. R. Asiatic Soc. 14 July 380 The Vannic inscriptions were noticed by Sir H. Rawlinson. 1888 A. H. Sayce Hittites vii. 134 The language of Ararat itself, the so-called Vannic, may belong to the same family of speech. 1897 W. H. Ward in H. von Hilprecht Rec. Res. in Bible Lands 172 We judge that the land of the Hittites, invaded by these Vannic kings, stretched along the upper course of the Euphrates. 1908, etc. [see *Khaldian]. 1974 Encycl. Brit. Macropædia I. 832/2 The terms Chaldean and Vannic have also been used as designations for Urartian during earlier stages of research. 1977 C. F. & F. M. Voegelin Classification & Index World's Lang. 162 Vannic Haldean... Formerly spoken in the Mount Ararat region near Lake Van.

vanrhum, obs. var. *Van der Hum.

Vansittartism (vænsi·taːtiz'm). [f. the name of Robert Gilbert *Vansittart* (1881–1957), English diplomat + -ism.] The foreign policy advocated by Sir Robert (later Lord) Vansittart, *spec.* with regard to the demilitarization of Germany. Also *transf.*

1941 Economist 27 Dec. 778/2 It is against this background that the futility of both Vansittartism and appeasement can best be judged. 1944 D. J. Enright in Scrutiny Spring 103 It is a fine thing to see their books published at a time when, among the circles of the learned, a kind of Vansittartism is beginning to pervade the sphere of culture. 1967 R. C. D. Jasper George Bell xiv. 261 Bell..had good cause to suspect that the Government had determined upon a policy of treating all Germans alike whether they were Nazis or not... And it was with 'Vansittartism' in mind that he wrote a letter to The Sunday Times on 20 January 1941. 1978 N. Rose Vansittart xii. 260 A prolonged occupation by the allied forces; the complete destruction of the German army; drastic control of German heavy industry; the total disarmament of Germany; and the re-education of the German people. This was the essence of Vansittartism.

So **Vansi·ttartite**, a supporter of the policy of Vansittart; also *attrib.* or as *adj.*

1942 New Statesman 7 Mar. 166/3 Those who were once appeasers and then Vansittartites. 1973 R. W. Clark Einstein xxi. 567 His..refusal to budge from an almost Vansittartite approach to Germany.

vantage, *sb.* Add: **3.** Also, a vantage-point.

1908 L. M. Montgomery Anne of Green Gables ii. 31 It was already quite dark, but not so dark that Mrs Rachel could not see them from her window vantage. 1969 M. Bragg Hired Man viii. 83 There were halloes every few minutes and the men themselves became hunters, climbing the heights in anticipation of a vantage which would give them a total view and enable them to race down when the kill was near.

6. (Further example.)

1904 J. P. Paret Lawn Tennis 352 Vantage-in (or vantage server). A term used to indicate that the server has won the 'vantage' point (opposite of 'vantage-out'). Vantage-out (or vantage striker) [etc.].

7. (sense 3) *vantage-nook.*

1930 Blunden Summer's Fancy 30 The many vantage-nooks That nature sets about the wooing weald.

van't Hoff (vænt hǫ·f). *Chem.* Also **Van't Höff.** The name of J. H. *van't Hoff* (1852–1911), Dutch chemist, used *attrib.* and in the possessive to designate rules or hypotheses put forward by him with reference to: **a.** Stereo-chemical properties of molecules.

1888 Jrnl. Chem. Soc. LIV. 597 (heading) Investigation of the second Van't Hoff hypothesis. 1979 A. Beknazarov tr. Potapov's Stereochemistry iv. 221 In the early thirties of our century there appeared data contradicting the van't Hoff second postulate—the assumption of free rotation about single bonds.

b. The osmotic pressure of solutes in solution.

1890 Jrnl. Chem. Soc. LVIII. 845 (heading) Deductions from Van't Hoff's theory. Ibid. 846 The author..discusses the determination of the value of i for solids in solution from Van't Hoff's equation. 1926 [see *lyotropic a. 1]. 1978 P. W. Atkins Physical Chem. viii. 223 The equation simplifies into the van't Hoff equation: $\Pi V = n_B RT$. The van't Hoff equation is very suggestive of the ideal gas equation.

c. The thermodynamics of chemical reactions.

1909 H. J. H. Fenton Outl. Chem. I. xiii. 159 Van't Hoff's 'principle of mobile equilibrium' is an application of this rule [of le Chatelier] to the case in which the imposed constraint is an alteration of temperature. 1923 Chem. Abstr. XVII. 2211 This [calculation] is offered as a kinetic explanation of van't Hoff's rule that for a temp. rise of 10° the speed of a reaction is doubled. 1978 P. W. Atkins Physical Chem. ix. 262 This [sc. the Gibbs-Helmholtz equation] immediately implies the van't Hoff equation.

¶ In Du. (rarely in Eng.) written *van 't Hoff.*

vanthoffite (vænthǫ·fəit). *Min.* [ad. G. *vanthoffit* (K. Kubierschky 1902, in *Sitzungsber. d. Preuss. Akad. d. Wissensch. zu Berlin* XXI. 407), f. the name of J. H. *van't Hoff, who first synthesized the compound: see -ite[1].] A sulphate of sodium and magnesium, $Na_6Mg(SO_4)_4$, found as colourless, grey, or pale yellow monoclinic crystals that are transparent with a vitreous lustre and occur esp. in oceanic salt deposits.

1902 Jrnl. Chem. Soc. LXXXII. ii. 406 The following salts are present in the mixture:..$3Na_2SO_4$, $MgSO_4$, to which the name vanthoffite is given. 1966 Prof. Papers U.S. Geol. Survey No. 550-b. 125/1 The halite rock in which the double sulfate minerals occur is medium to coarse grained... Loewite, vanthoffite, bloedite, and leonite are constituents of the ore zones.

Vanuatuan (vænwatū·an), *a.* [f. *Vanuatu* + -an.] Of or pertaining to the Vanuatu Republic (formerly the Condominium of the New Hebrides) or any of its constituent islands.

1980 Sunday Times 24 Aug. 1/1 Two British soldiers of the joint military force withdrawn from the rebel Vanuatuan island of Espiritu Santo on Monday have been charged with looting. 1983 Defense & Foreign Affairs Weekly 28 Mar. 4 Within an hour of hoisting its flag, the Vanuatuan naval ship was being 'peaceably escorted' out of the area.

vaporetto (væpore·to). Pl. **-etti, -ettos.** [It., dim. *vapore* f. L. *vapor* steam.] In Venice, a canal motor-boat (orig. steamboat), used as a form of public transport.

1926 C. Beaton Diary 25 Aug. in Wandering Years (1961) v. 121 We set off blindly, taking the vaporetto. 1941 W. Graham Night Journey v. 62, I can catch a vaporetto from the Quay. 1956 R. Macaulay Towers of Trebizond xxiv. 278, I remember it when there was nothing on the canals but gondolas, none of those horrid steamers and vaporetti. 1963 Economist 6 Apr. 37/1 The tourists.. prefer the speed of the municipally-owned vaporettos to the romance of the older hand-driven craft [sc. gondolas]. 1977 Time 28 Nov. 30/3 Other names..soon became household words at the staid Café Florian, on every vaporetto water bus and in the intellectual salons of the Salute island. 1984 New Yorker 23 Apr. 92/3 As soon as I got off the vaporetto, I felt that this was where I should be.

vapour, *sb.* Add: **1.** (Further examples. Cf. next sense.)

1845 Encycl. Metrop. IV. 246/2 Comparing a given space filled with gas, and another saturated with vapour, at a given temperature; if we suppose that space to be diminished, the gas will be compressed..but the vapour will be partly condensed. 1863 E. Atkinson tr. Ganot's Elem. Treat. Physics iv. i. 93 Heat..converts liquids.. into the aeriform state in which they obey all the laws of gases. This aeriform state of liquids is known by the name of vapour, while gases are bodies which, under ordinary temperature and pressure, remain in the aeriform state.

2. (Further examples.) In mod. scientific use, a fluid that fills a space like a gas but, being below its critical temperature, can be liquefied by pressure alone.

1823 Faraday in Q. Jrnl. Sci., Lit., & Arts XVI. 237 Now that we know the pressure of the vapour of chlorine, there can be no doubt that the following passage describes a true liquefaction of that gas. Ibid. 239 During the condensation of the gas in this manner, a liquid has been observed to deposit from it. It is not, however, a result of the liquefaction of the gas, but the deposition of a vapour (using the terms gas and vapour in their common acceptation) from it, and when taken out of the vessel it remains liquid at common temperatures and pressures. 1823 H. Davy in Phil. Trans. R. Soc. CXIII. 165 The compression [of gases] resulting from their slow generation in close vessels..may be easily assisted by artificial cold in cases where gases approach near to that point of compression and temperature at which they become vapours. 1883 E. Atkinson tr. Ganot's Elem. Treat. Physics (ed. 11) vi. v. 312 A vapour may be defined as being a gas at any temperature below its critical point. Hence a vapour can be converted into a liquid by pressure alone, and can therefore exist in the pressure of its own liquid, while a gas requires cooling as well as pressure to convert it into a liquid. 1979 T. B. Akrill et al. Physics xi. 141/1 It is conventional to use the term vapour to describe a gas which is at a temperature below the critical temperature for that substance, but there is no obvious difference between a vapour just below T_c..and a gas just above T_c.

4. For † *Obs.* read 'Now *rare*'. (Later examples.)

1841 J. Romilly Diary 16 Apr. (1967) 214 Ray.. reminded me I had said I w[d] give a Guinea when the Peterhouse wall was replaced by an iron-rail:—this work is now going on:—I had forgotten this vapour; but produced the Guinea. 1940 W. B. Yeats If I were Four-&-Twenty iv. 8 Men whose lives had been changed by Balzac, perhaps because he cleared them of Utopian vapours.

5. a. *vapour-capacity, -phase, -pipe*; **vapour lock**, an interruption in the flow of a liquid through a pipe as a result of its vaporization; **vapour pressure**, the pressure exerted by a vapour; **vapour-proof** *a.*, impervious to vapour; **vapour tension** = *vapour pressure* above; **vapour trail**, a visible trail of condensed water vapour in the sky, in the wake of an aircraft; also *fig.*

1922 W. G. Kendrew Climates of Continents 215 The sea is then coolest relatively to the land, so that the vapour-capacity of air blowing from the sea is increased over the land. 1930 S.A.E. Jrnl. XXVII. 93/1 The more volatile a fuel, the greater will be the tendency to boil in the fuel-feed system as the engine warms up. If the fuel boils, then interruptions of flow due to vapor lock may be expected. 1951 O. Berthoud tr. P. Clostermann's Big Show 39 My jettison-tank gave out—probably a vapour-lock in the feed pipes. 1974 Times 22 Mar. 15/4 Filter King is said..to prevent carburettor flooding..and to prevent vapour 'lock'. 1946 Nature 19 Oct. 562/1 A process has been developed for the preparation of motor fuel and other petroleum products by a method based on vapour-phase cracking of the vegetable oils contained in seeds. 1964 N. G. Clark Mod. Organic Chem. xxv. 515 The vapour-phase nitration of propane. 1913 V. B. Lewes Oil Fuel 79 A horizontal cylindrical boiler with a dome from which a broad vapour-pipe leads the distilling vapours to the condensers. 1875 Encycl. Brit. III. 385/2 As regards the atmosphere, evaporation [of water] goes on until the maximum vapour pressure for the temperature has been attained, at which point the air is said to be saturated. 1978 P. W. Atkins Physical Chem. vii. 175 The vapour pressure of water at 100°C is 1 atm. 1946 Sun (Baltimore) 17 May 13/3 About a foot of earth was scraped away from the site and a layer of vapor-proof material placed on the ground. 1963 Engineering 16 Aug. 205/1 A heat resisting gasket..is fitted..to render the unit vapourproof and weatherproof. 1981 Oil & Gas Jrnl. 20 Apr. 167/2 The recorder case is made of die cast aluminum, and it's vapor-proof. [1845 Phil. Trans. R. Soc. CXXXV. 169 Cyanogen..yielded on different occasions results of vaporous tension differing much from each other.] 1864 Vapour tension [in Dict.]. 1933 W. Lindgren Mineral Deposits (ed. 4) x. 116 Sudden separation of the gaseous phase will take place..only if the vapor tension of the solutions is greater than the external pressure. 1941 Picture Post 3 May 23/1 The vapour trails are left by the R.A.F. fighters weaving in and out of the German formation. 1948 L. Durrell Let. in Spirit of Place (1969) 98 Vapour-trails of cows on the pampas, desolation. 1977 W. McIlvanney Laidlaw xxvi. 115 The vapour trails left by interrupted conversations. 1979 C. Priest Infinite Summer 18 There, high in the blue, were several curling white vapour-trails, but no other sign of the German bombers.

b. *vapour-turbanned.*

1892 W. B. Yeats Countess Kathleen 125 Under that cold and vapour-turbanned steep.

vapouring, *vbl. sb.* **3.** (Earlier examples.)

1866 'Mark Twain' Lett. from Hawaii (1967) 118 All the good sound sense or point there was in his vaporings could have been boiled down into half a page of foolscap. Ibid. 242 The creature has got no sense, but his vaporings sound strangely plausible sometimes.

vaquero. Add: Also **vacero.** **1.** (Earlier examples.)

1826 in E. C. Barker Austin Papers (1924) 1483 D[n] Luciano..rarely has communication with any one, except his Vaceros Comes to Town to Hug his Frigoña, once a month. 1831 F. W. Beechey Narr. Voy. Pacific II. 40 Intrusting their baggage to the care of two vaqueros (Indian cattle drivers) who were to accompany them.

var (vāɹ). *Canad.* Also **varr.** [dial. var. of Fir.] The balsam fir, *Abies balsamea.*

1793 P. Campbell Trav. N. Amer. 57 The men set to work, and..cut the crops of a species of ever green wood, which they call varr [in New Brunswick]. 1842 R. H. Bonnycastle Newfoundland II. 266 The bark of the.. var..was taken off. 1907 J. G. Millais Newfoundland iv. 99 Trees standing clear against a brown mass of tall 'vars' and spruces. 1959 Atlantic Advocate Jan. 74 In the feel of drawknife in wood, smoothing a shingle out of straight-grained var, there was something that smoothed the mind. 1982 Evening Telegram (St. John's, Newfoundland) 8 Feb. 24 Many an axe is biting into a piece of crooked var.

vara. Add: **2.** *Bullfighting.* A long spiked lance used by a picador.

1932 E. Hemingway Death in Afternoon i. 6 The picador can perform his mission with the spiked pole, or vara. 1934 R. Campbell Broken Record 83 Trident, vara, and lance. 1967 McCormick & Mascareñas Compl. Aficionado i. 20 It [sc. tercio] is used both of the entire lidia ('bull fight'..), which is said to be divided into the tercio of the vara (from the picador's long pole tipped by the pic), of the banderillas; and the faena. 1968 [see *morillo].

varactor (vāræ·ktəɹ). *Electronics.* [f. *var(iable re)actor.*] A reverse-biased *p–n* junction whose capacitance depends on the bias voltage in a definite way. Also called *varactor diode.*

1959 Pfann & Garrett in Proc. IRE XLVII. 2011/1 (heading) Semiconductor varactors using surface space-charge layers. 1965 Wireless World July 62 (Advt.), Elimination of manual tuning by a unique self-tuning system (using servo controlled varactor diodes). 1975 Physics Bull. Sept. 401/1 Varactor diodes are extensively used in voltage-variable tuning applications, such as in modern solid-state radio and television circuits. 1979 A. Bar-Lev Semiconductors & Electronic Devices xvii. 325 Varactors are also used for frequency multiplication or harmonic power generation in the microwave range.

varanus (væ·rānəs). [mod.L. (B. Merrem *Tentamen Systematis Amphibiorum* (1820) 58): see Varan.] = Varan. Cf. *goanna, *iguana 2, Monitor *sb.* 5 in Dict. and Suppl.

1934 A. Russell Tramp-Royal in Wild Austral. viii. 60 What should come scuttling across the track but a big varanus liazrd. 1953 A. Smith Blind White Fish in Persia vi. 99 A Varanus in fact, a type of lizard, but it had the appearance of a Rhinegold dragon.

vardo. Delete † *Obs.* and add: Also **varda.** [Romany.] Now *spec.* a gypsy caravan.

1934 P. Allingham Cheapjack vi. 59 As me brother David pointed out to me in the vardo afterwards, this 'ere finger called me a bastard—and he was right. 1940 [see *Reading 3]. 1977 New Society 29 Sept. 643/2 You know what a varda is. It's one of those painted wagons, drawn by horses, that gypsies donate to folk museums. 1980 R. Hill Killing Kindness xxvi. 263 Don't forget I'm pure bred Romany... To end my days sitting on the vardo steps puffing away at an old pipe. 1982 Times 30 Apr. 10/3 This 'ere Gorse Hill..was always too tiring for an 'orse after pullin' a varda..10 or 12 mles.

vargueño (vāɹgē·nʸo). [ad. Sp. *bargueño*, *vargueño* adj., of Bargas, a village near Toledo, the former place of manufacture.] A kind of cabinet made in Spain in the 16th and 17th centuries, with numerous small compartments and drawers behind a fall front which opens out to form a writing surface.

1911 E. Foley Bk. Decorative Furnit. I. 159 Prominent amid the Hispano-Moresque decorative woodwork probably made by these Moorish craftsmen is the vargueño: a box with a door in front, evolved from the chest or hucha, and mounted on a stand. Ibid. 160 It has been reasonably conjectured that the vargueño cabinet took its title from the small town of Vargas, near Toledo. 1923 G. L. Hunter Decorative Furnit. ix. 265 The most distinctive piece of Spanish furniture is the vargueño. Like the two-story cabinets or highboys of other countries, it was a development from the chest. 1960 H. Hayward Antique Coll. 24/2 Bargueño or vargueño, Spanish cabinet with fall-front enclosing drawers and often mounted on a stand. 1975 Oxf. Compan. Decorative Arts 809/2 The most typical vargueños are in the Mudejar style... In 1568 the guilds of Mexico required a cabinet-maker to be able to make a vargueño..as a condition of qualification.

vari- (vēᵊ·ri), comb. form of Variable *a.* (in some words infl. by Variation or Various *a.*).

variable, *a.* and *sb.* Add: **A.** *adj.* **6. d.** *variable cost* (see quot. 1974).

1953 [see prime cost s.v. *prime a. 9 a]. 1967 A. Battersby Network Analysis (ed. 2) xii. 200 There are other cases which more closely resemble the variable-cost control, as when the number of men allocated to a job is variable and governs duration. 1969 D. C. Hague Managerial Economics i. 15 The remainder of total cost is made up of those costs that do vary with output—with what are therefore known as variable costs. 1974 Terminol. Managem. & Financial Accountancy (Inst. Cost & Managem. Accounts) 24 Variable cost, a cost which, in the aggregate, tends to vary in direct proportion to change in the volume of output or turnover.

9. Special collocations. **a.** In *attrib.* use, as *variable-area, -length, -reluctance, -speed.*

1920 Flight XII. 5/1 Variable area or variable camber wings—or other means for reducing the landing speed of a machine. 1957 Manvell & Huntley Technique Film Music iii. 171 This optical wedge is caused to oscillate over a light slit by the motion of the pendulums, producing a variable-area type of sound-track. 1959 E. M. McCormick Digital Computer Primer viii. 119 These difficulties are avoided in many business computers by organizing the storage to accommodate variable-length records. 1970 Computers & Humanities IV. 327 The data fields of the customer's data base in variable-length capability. 1980 C. S. French Computer Sci. xi. 60 Variable length records mean difficulties for the programmer but better utilisation of storage. 1959 W. S. Sharps Dict. Cinematogr. 138/1 A variable-reluctance microphone is a microphone which depends for its operation on variations in the reluctance of a magnetic circuit. 1875 Variable-speed [in Dict., sense 7 a]. 1978 P. Griffiths Conc. Hist. Mod. Music viii. 116 Cage went further and created..his Imaginary Landscape no. 1, in which musicians have to perform with frequency recordings on variable-speed turntables. 1984 B. Francis AA Car Duffer's Guide 67/2 A variable-speed fan is provided to boost the air-flow throughout the car's interior at low speeds.

b. *variable geometry*, a configuration of component parts that can be varied; *spec.* in *Aeronaut.* = *variable sweep* below; freq. *attrib.*; *variable-mu* adj. (Electronics), (of a valve) having an amplification factor that can be varied; *variable-pitch* adj., of or pertaining to † (*a*) a propeller in which the blades are shaped so that the pitch varies along their length (see quot. 1912) (*obs.*); (*b*) a propeller, fan, etc., in which the angle of the blades with respect to the direction of air flow can be adjusted, esp. while they are in motion; *variable sweep* (Aeronaut.), sweep (sense *14*) that can be varied during flight according to requirements; usu. *attrib.*

1957 Jrnl. Brit. Interplanetary Soc. XVI. 166 The design characteristics of such gliders may include..variable geometry lifting surfaces, to be very thin at high speeds and to have good lift properties at landing. 1963 Times 10 June 10/1 He believed that variable geometry would give a better aircraft. 1969 Sci. Jrnl. Apr. 48/1 The wing surface [of a bat's wing] is an elastic membrane of skin stretched between the four long fingers, with at least 11 movable joints in each wing. This very variable geometry permits a high degree of manoeuvrability and control. 1970 New Scientist 30 July 236 (caption) These drawings show the variable geometry of the air intake on the Olympus 593-3B engine. 1977 P. Way Super-Celeste 39 The rival to the Super-Celeste, the variable geometry wing F-24, was airborne again. 1930 Ballantine & Snow in Proc. IRE XVIII. 2102 (heading) Reduction of distortion and cross-talk in radio receivers by means of variable-mu tetrodes. 1962 [see *mu 2]. 1971 E. N. Lurch Fundamentals of Electronics (ed. 2) v. 143 The tube employing this special grid is termed cutoff, supercontrol, or variable-mu. 1912 Ledeboer & Hubbard tr. Duchêne's Mech. of Aeroplane vii. 197 M. Drzewiecki, with the idea of getting the maximum efficiency out of every part of the propeller, varies the pitch at each point so that the actual angle of incidence is everywhere the optimum angle. Propellers of this kind are known as variable-pitch propellers, and are consequently no longer true screws. 1918 W. H. Berry Aircraft in War & Commerce vii. 91 Worked in combination with a variable pitch propeller the gearbox would allow the pilot a range in power and speed. 1919 H. Shaw Text-bk. Aeronaut. xi. 147 In the variable pitch type, the blade angle is varied continuously from the hub outwards in such a way that the slip column of air is forced backwards with a velocity which is uniform throughout. 1952 [see static thrust s.v. *static a. 7]. 1971 Daily Tel. 4 Nov. 3 (Advt.), Variable pitch engine cooling fan. Large range to fit most popular cars. 1974 Encycl. Brit. Micropædia I. 375/1 Additional features have been engineered into variable-pitch propellers; first, the ability to feather..a propeller on an engine stopped in flight. 1954 Economist 11 Sept. 12/3 Variable sweep. After take-off with wings only partially swept, means are provided to sweep them back more sharply in flight. Strictly experimental. 1965 New Scientist 22 Apr. 217/1 The development of the variable-sweep wing in collaboration with the French might lead to a load-carrier of outstanding performance. 1966 Economist 22 Jan. 330/1 Massive, variable-sweep aircraft whose wings pivot around stanchions as thick as tree-trunks. 1978 D. Küchemann Aerodynamic Design of Aircraft iv. 120 E. von Holst flew models with various arrangements of variable sweep.

B. *sb.* **1. b.** *Computers.* A data item that can take on more than one value during or between programs and is stored in a particular designated area of memory; the area of memory itself; (also *variable name*) the name referring to such an item or location.

1837 C. Babbage in B. Randell Origins Digital Computers (1973) 23 The number of variables which can be contained within the store will depend on the length of the rack. 1843 Scientific Mem. III. 666 We have..written Variable with a capital letter when we use the word to signify a column of the engine, and variable with a small letter when we mean the variable of a formula. 1957 [see *subscript sb. 2 b]. 1967 Beckett & Hurt Numerical Calculations & Algorithms i. 23 Variable names are usually limited in length because they must be coded as numbers and space is required to store these numbers... The number of characters used to define a variable is arbitrary, provided that it does not exceed a given maximum. 1972 Bergman & Bruckner Introd. Computers & Computer Programming ix. 288 [In FORTRAN] the data locations are divided into constants and variables... Variables are locations into which we may read numbers or place (store) numbers at execution time. 1975 H. Katzan Introd. Computer Sci. v. 108 The names hours, rate, and pay are variables... A variable is an identifier that names a data item. 1979 Sci. Amer. Dec. 85/1 In programming languages a variable is not an item of data but a label for a location in the memory of the computer. The value of a variable at any moment is the information currently stored there. 1981 W. S. Davis Computers & Business Information Processing xiii. 234 A FORTRAN variable consists of from one to six letters or digits, the first of which must be a letter. 1982 Cooper & Clancy Oh! Pascal! i. 10 The variables that computers use are like the memory keys of hand calculators—they store values... Variables can hold different types of values, including integers.., characters.., or even logical values.

c. *Logic.* A symbol whose exact meaning or referend is unspecified, though the range of possible meanings usually is.

1910 Whitehead & Russell Principia Mathematica I. i. 4 In mathematical logic, any symbol whose meaning is not determinate is called a variable. Ibid., If a statement is made about 'Mr A and Mr B', 'Mr A' and 'Mr B' are variables whose values are confined to men. 1937, etc. [see *open a. (adv.) 11 j]. 1939 Jrnl. Philos. XXXVI. 702 Whereas the singular existence statement calls the alleged existent by name, e.g., 'Pegasus', the general existence statement does not; the reference is made rather by a variable 'x', the logistical analogue of a pronoun 'which', 'something which'. 1954 A. J. Ayer Philos. Ess. ix. 216 The range of our ontological commitments may..be reduced by our ability to recast some of our existential statements in such a way that variables which take a certain type of value disappear from them. 1969 Feys & Fitch Dict. Symbols Math. Logic 6 In interpreting a formalized language it is usual to specify the respective ranges of the various variables of the language. 1978 S. Haack Philos. Logics iv. 39 In general, prefixing a quantifier binding one of its free variables to an open sentence with n free variables yields an open sentence with n−1 free variables.

Variac (vēᵊ·riæk). *Electr.* [f. Vari(able *a.* + -*ac*, perh. repr. *a.c.*, alternating current.] A proprietary name for a type of autotransformer in which the ratio of the input and output voltages can be varied.

1933 Official Gaz. (U.S. Patent Office) 26 Dec. 874/2 General Radio Company, Cambridge, Mass... Variac for transformers. 1938 Trade Marks Jrnl. 16 Feb. 188/2 Variac... Electrical transformers (not being machines). General Radio Company.., Cambridge, State of Massachusetts. 1943 Electronic Engin. XV. 516/3 The movement of the sliding contact on the Variac is effected by a small reversible motor. 1945 [see *interlock sb. 2 b]. 1983 Sci. Amer. Feb. 118/3, I plugged the immersion heater into a Variac so that I could experiment with the rate at which heat was released in the water.

‖ **varia lectio** (vēᵊ·ria le·ktio). Pl. **variae lectiones.** [L., fem. of *var-ius* Various *a.* + *lectio* reading: cf. Lection.] = *various reading* s.v. Various *a.* 8 d.

1652 N. Culverwel Light of Nature x. 81 'Tis some accurate piece that passes so many Criticks..without any Variæ lectiones. 1871 [see Q=Quarto 2 s.v. *Q II. 2 a]. 1969 R. Renehan Greek Textual Crit. 91 Some MSS have a varia lectio διακωλυόντα. 1980 Times 24 May 14/1 The Book Collector..is the place to read..about incunabula and..variae lectiones.

variance. Add: **I. 3. d.** *U.S. Law.* An official dispensation from a building regulation.

1925 New Hampshire Public Laws xlii. 191 To authorize upon appeal in specific cases such variance from the terms of the ordinance as will not be contrary to the public interest. 1938 Atlantic Reporter CC. 521 A literal enforcement of the ordinance may be disregarded to permit a variance, while conditions for an exception must be found in the ordinance and may not be varied. 1973 N.Y. Law Jrnl. 19 July 13/7 Judgment is granted in favor of the petitioner..directing the issuance of a variance for the installation of a second kitchen in the premises. 1977 Sat. Rev. (U.S.) 3 Sept. 47/3 The only hotel on the island [sc. Bermuda] ever given a variance to be built on a hilltop. The permission took 12 years to get.

e. *Econ.* The difference between actual and expected costs, profits, output, etc., in a statistical analysis.

1964 Times Rev. Industry Apr. 8/2 An expression of the variation between standard cost and actual achieved cost is called 'variance'. In 1951 under manual control the variance was 31.6 per cent. 1967 D. Goch in Wills & Yearsley Handbk. Managem. Technol. 145 The following example will illustrate the application of this method of variance analysis to an imaginary product. 1974 Terminol. Managem. & Financial Accountancy (Inst. Cost & Managem. Accountants) 13 Variance analysis, that part of variance accounting..which relates to the analysis into constituent parts of variances between planned and actual performance. 1978 Accountants' Rec. Dec. 15/1 Causal factors of the sales variance have been established by analysis... Sales variance analysis..provides data enabling management to reconsider its marketing philosophies.

4*. *Statistics.* The mean of the squares of the deviations of a set of quantities; the square of the standard deviation.

1918 R. A. Fisher in Trans. R. Soc. Edin. LII. 399 It is..desirable in analysing the causes of variability to deal with the square of the standard deviation as the measure of variability. We shall term this quantity the Variance. 1948 New Biol. IV. 34 Measurements of the resting blood pressure of a group of fifty young men... We compute the mean blood pressure and the variance of the observed values, the variance being a measure of scatter of values around the mean. 1970 Nature 25 July

376/2 Analysis of variance gave highly significant population and fertilizer effects. 1977 R. E. Megill Introd. Risk Analysis iii. 23 Because all values are squared in the variance the standard deviation is larger than the mean deviation.

variant, *sb.* Add: **1. b.** Also *spec.* a textual variation in two or more copies of a printed work (not necessarily implying reimpression).

1927 R. B. McKerrow Introd. Bibliogr. ii. vi. 208 Besides these added lines, The Devil's Charter exhibits a very large number of striking variants in different copies, some being mere corrections of literal errors, others important alterations in wording. 1953 C. Hinman in Shakespeare Q. IV. 280, I have been able to construct an instrument which..has enabled me to collate well over a hundred folio pages a day for some months... Taking reasonable care, the investigator can hardly fail to note any variant, however minute, in two copies of the page being examined. 1972 P. Gaskell New Introd. Bibliogr. 357 Warner Barnes..machine-collated an average of six copies of each of his author's eighteen primary editions... In ten of them he found possible evidence of concealed.. reimpressions; in another four, variant states of the type not indicating reimpression... Most of the variants shown up by the machine were trivial.

variate (vēᵊ·riăt), *sb. Statistics.* [f. as Variate *a.*] † **a.** The value of an attribute common to a number of individuals in any one instance; an observed value of a variate (sense *b). *Obs.*

1899 C. B. Davenport Statistical Methods i. 1 A variate is a single magnitude-determination of a character... Integral variates are magnitude-determinations of characters which from their nature are expressed in integers. Such magnitudes are determined by counting; e.g., the number of teeth in a porpoise. 1906 R. H. Lock Rec. Progress Study of Variation iv. 90 A variate is one of the separate numerical values from which a curve of variability can be constructed; the biometrician usually deals with some such number as 1,000 variates.

b. A quantity or attribute having a numerical value for each member of a population; *esp.* one for which the values occur according to a frequency distribution.

1909 Biometrika VII. 97 We have, σ_1 and σ_2 being the standard deviations of the two variates and r their correlation, $\bar{p} = r\sigma_1/\sigma_2\bar{q}$. 1925 R. A. Fisher Statistical Methods for Res. Workers i. 5 The variable quantity, such as the number of children, is called the variate, and the frequency distribution specifies how frequently the variate takes each of its possible values. 1952 Sci. News XXIV. 109 The histogram gives the frequency with which the weight, a variate, takes values in a certain range in the sample. 1968 [see *random a. 1 b]. 1971 Brit. Med. Bull. XXVII. 2/1 Epidemiology also has a contribution to make..to the study of attributes (or variates, as some call them) such as intelligence..and blood pressure.

variation. Add: **II. 8. b.** (Further examples.)

1810 R. Woodhouse Treat. Isoperimetrical Probl. ii. 23 If problems involving merely one property, the maximum, require the variation of two, and those involving two properties, the variation of three elements [etc.]. 1847 Cambr. & Dublin Math. Jrnl. I. 264 We have, in this case, by Lagrange's theory of the variation of the arbitrary constants, the formulæ da/dt=[etc.]. 1918 H. C. Plummer Introd. Treat. Dynamical Astron. xii. 134 This is the foundation of Lagrange's method of the variation of arbitrary constants. 1966 H. Pollard Math. Introd. Celestial Mech. iv. 91 We shall begin with the undisturbed system $\dot{x}_1 = -x_2$, $\dot{x}_2 = k^2x_1$, and apply the method of variation of parameters.

e. The difference between the values of a function at either end of a subinterval; the sum of such differences for all the non-overlapping subintervals into which a given interval is divided; the upper bound (if any) of this sum when all possible modes of subdividing the interval are considered.

1905 J. Pierpont Lect. Theory Functions Real Variables I. xii. 349 An important class of limited integrable functions is formed by functions with limited variation. 1911 Q. Jrnl. Pure & Appl. Math. XLII. 57 Although in forming the positive (negative) variation over (a, b) we considered all possible sets of non-overlapping intervals, we may without loss of generality confine ourselves to sets consisting of a finite number of intervals only. 1946 H. & B. S. Jeffreys Methods Math. Physics i. 23 The total variation is of interest since it is related to the condition for existence of a Stieltjes integral..and to the determination of the total length of a curve. 1971 E. R. Phillips Introd. Anal. & Integration Theory ix. 251 Let us assume that the set of nonnegative variations..is bounded from above.

III. 13. b. *variation*, the amount by which some quantity changes in value, or the addition made to the quantity; *esp.* the change in a function when there is a small change in the variables or constituent functions of the function.

In the calculus of variations the function concerned is usu. an integral, and the aim is to find what relation between the variables in the integrand makes the integral a maximum or a minimum.

1810 R. Woodhouse Treat. Isoperimetrical Probl. iii. 45 The general form of these equations is, $P.bg - Q.ci + R.d\theta$, $d\theta$ being a variation of the ordinate similar to the variations bg and ci. 1814 J. Toplis tr. Laplace's Treat. Analytical Mech. ii. 46 Of all the curves along which a moving body, subjected to the forces P, Q, and R, can pass from one given point to another given point, it will describe that in which the variation of the integral $\int v ds$ is nothing, and in which, consequently this integral is a minimum. 1834 [see *least a. 1 e]. 1845 Encycl. Metrop. II. 209 To obtain the variation of a function of y we must write $y + \delta y$ for y, and having expanded the new function according to the powers of δy, subtract from it the original function, and the first term of the difference will be the variation required. 1934 W. V. Houston Princ. Math. Physics v. 56 The variation of the integral is defined as its value when α has the infinitesimal value δα, minus its value when α is zero. 1972 M. Kline Math. Thought xxiv. 578 By equating the variation of the integral to zero and by using a crude limiting process to transform the resulting difference equation, he obtained the differential equation which must be satisfied by the minimizing arc.

14. b. *elegant variation*: in writing, the stylistic fault of studiedly avoiding repetition by using different words for the same thing. Also *transf.* and in ironic use.

1906 H. W. & F. G. Fowler King's English iii. 178 The locking of arms is..only an elegant variation for clinging. 1926 H. W. Fowler Mod. Eng. Usage 131/1 It is the second-rate writers..that are chiefly open to the allurements of elegant variation. 1947 Partridge Usage & Abusage 16/1 Mr Herd then picks on the device known as 'elegant variation'. 'If,' he says, 'the mayor has been mentioned, he makes further appearance as 'the civic chief', 'the leader of our official life'..and so on.' 1981 Guardian Weekly 5 July 21 Most of the costumes are.. elegant variations on dancers' practice dress. 1982 Washington Post 7 May d2/6 ... rarely occurs less than twice per page, and often as many as five, with a 'f---' or two stuck in for elegant variation.

c. *Ballet.* A solo dance.

1912 Dancing Times Aug. 420/2 [Grahn's] career was interrupted by an accident while rehearsing a variation which she was to perform at [a] benefit. 1948 Ballet Ann. II. 49 She attacks the formidable difficulties of the variation and adagio with an ease and confidence. 1980 'M. Fonteyn' Magic of Dance 65 He makes the preparation for his 'variation', or solo, with utmost care and accuracy.

15. variation method *Physics*, a method for finding an approximate solution to Schrödinger's equation by varying the trial solutions to find which gives the lowest value for the energy and is therefore closest to the true solution; **variation order**, an order authorizing a change in an original order or contract (see quots.); **variation principle** *Physics*, the principle (employed in the variation method) that the energy corresponding to an arbitrary wave function cannot be less than the actual lowest energy of the system under consideration.

1935 Pauling & Wilson Introd. Quantum Mech. vii. 182 The variation method is..very frequently used to obtain approximate wave functions as well as approximate energy values. 1960 [see quantum-chemical adj. s.v. *quantum 7 a]. 1974 P. W. Atkins Quanta 96/2 With these approximations in hand the variation method is applied to determine the best linear combination of atomic π-orbitals to describe the structure of the molecule. 1940 Chambers's Techn. Dict. 886 Variation order,..a document giving authority for some alteration in work being done under contract. 1975 E. B. Ceadel in Barr & Line Essays on Information & Libraries iv. 58 Where a change is made in the contract arrangements, it is listed by the architect as a variation order, which is subsequently costed by the quantity surveyors. 1977 Daily Tel. 19 Nov. 3/2 An order was prepared..requiring him to leave by Nov. 17... The Home Secretary..had considered whether, instead of issuing a 'variation order', he should proceed under another provision of the Immigration Act 1971 with a view to deportation. 1923 Variation principle [see quantum condition s.v. *quantum 7 a]. 1975 H. F. Hameka Quantum Theory Chem. Bond i. 28 We shall discuss the two methods of approximation that are most widely used in quantum chemistry, namely the variation principle and perturbation theory.

variational, *a.* Add: *spec.* with reference to Variation 13 b in Dict. and Suppl.

1897 Acta Math. XXI. 99 His substitution of the Variational Curve for the ellipse as the intermediate orbit is..of primary importance in the Lunar Theory. 1937 E. C. Kemble Fund. Princ. Quantum Mech. iv. 130 Let J denote the integral $J[y, \lambda]$... Let δy denote the first variation in $y(x)$... Let it be required to find a function $y(x)$ and a corresponding value for the parameter λ which satisfy the variational equation $\delta J = 0$. 1957 L. Fox Numerical Solution Two-Point Boundary Probl. viii. 259 With a single second-order equation we were able to fix both y_0 and y_1, so that y_n was a function only of λ, and by solving a variational equation for $\partial y/\partial \lambda$ we were able to alter λ systematically until a solution was obtained. 1962 C. S. Ogilvy Tomorrow's Math ix. 125 Questions asking what shape, what path, or what form will yield a minimal or optimal result are known as variational problems. 1971 Physics Bull. Jan. 17/1 We introduce a trial wave-function ψ containing as many variational parameters as we feel to be necessary or sufficient, and then choose the values of these parameters in such a way that the Rayleigh ratio..shall be stationary.

variationist. Add: **1. b.** One who practises variation or introduces variety in anything.

1926 H. W. Fowler Mod. Eng. Usage 132/1 The writers are confirmed variationists. 1928 J. Y. T. Greig Breaking Priscian's Head 66 Mr Fowler..pokes fun at the 'elegant variationist'. 1981 N.Y. Times 25 Jan. ii. 25/4 Johns is by nature a variationist—someone who hates to let a good idea go while there is still something that can be done with it. Ibid. 2 Oct. c28/6 It works wonderfully well in terms of Mr. Judd's skill as a variationist. Every one of the 10 sections of the work has the same basic set of themes.

2. One who studies variations in usage among different speakers of the same language.

1975 Amer. Speech 1973 XLVIII. 37 What variationists have discovered is the extraordinary stylistic and sociological importance of such variation, and the apparently remarkable consistency with which the frequency of a given alternant is patterned throughout society and throughout an individual's stylistic repertoire. 1978 Archivum Linguisticum IX. 48 Committed variationists.. seem to disagree on how variation is to be incorporated into the grammar of a language. 1983 Canadian Jrnl. Linguistics XXVIII. i. 82 It is not, however, a problem for the variationists, but for those who would arbitrarily reduce linguistics to a branch of cognitive psychology.

variceal (værisī·ăl), *a. Med.* [f. L. *varic-*, Varix + -al; *-eal* after *corneal, laryngeal*, etc.] Of, pertaining to, or involving a varix.

[1947 Jrnl. Amer. Med. Assoc. 8 Nov. 630/1 (heading) Intraesophageal venous tamponage. Its use in a case of varical hemorrhage from the esophagus.] 1961 Postlethwait & Sealy Surg. Esophagus xii. 392/1 Others.. later used tamponade for control of variceal hemorrhage. 1981 Brit. Med. Jrnl. 17 Jan. 189/2 Haemorrhage from varices around an artificial intestinal stoma..is well documented... Variceal haemorrhage from a colostomy has been less frequently reported.

varicocelectomy (væːrikosīle·ktŏmi). *Surg.* [f. Varicocel(e + *-ectomy.] An operation to remove a varicocele.

1894 Gould Med. Dict. 1590/1 Varicocelectomy.., excision of a varicocele. 1950 A. I. Dodson Urol. Surg. (ed. 2) xliv. 719 (heading) Inguinal varicocelectomy. 1980 S. J. Silber How to get Pregnant (1981) v 149 The first successful case of restoration of fertility after varicocelectomy.

varicap (vēᵊ·rikæp). *Electronics.* [f. *vari(able cap(acitance.*] = *varactor.

1967 Aegerter & Habian in IEEE Trans. Broadcast & Television Receivers Nov. 103/1 A varicap is a variable capacitance diode. 1975 G. J. King Audio Handbk. iv. 98 Tuner-amplifiers in which varicaps are employed for tuning must also feature supply regulation. 1981 J. C. Sprott Introd. Mod. Electronics vi. 137 A typical varicap has a capacitance that varies over about a factor of 10 in the picofarad range.

variegated, *ppl. a.* **I. c.** (Earlier example.)

1818 M. Edgeworth Let. 4 Oct. (1971) 107 All the varieties of trees and shrubs..he has now revealed to view—The tulip tree and acacia and variegated oak and.. the variegated rhododendron.

varietal, *a.* Add: **1.** Similarly in *Min.* with reference to minerals.

1933 [see *œllacherite]. 1966 [see *caswellite].

2. Of wine: made predominantly from a single variety of grape; also, of or pertaining to the vine or grape of a particular variety. orig. *U.S.*

1941 Schoonmaker & Marvel Amer. Wines xi. 260 The expense of..certification could be borne by a small per-gallon tax levied upon varietal wine. Ibid., Until it [sc. identification of the grape varieties grown in America] has been accomplished, a system of honest varietal labeling is virtually unattainable. 1955 J. Storm Invitation to Wines 72 The success of varietal wines in California has influenced some Eastern vintners. 1968 Amer. Speech 1967 XLII. 80 Varietal types of wines. 1973 Bulletin (Sydney) 25 Aug. 7/2 (Advt.), An aromatic wine..with varietal flavour. 1979 A. Hailey Overload iii. v. 214 Among other things, Nim Goldman was a wine buff. He had a keen nose and palate and especially liked varietal wines from the Napa Valley. 1981 Times 2 Mar. 12/6 Limousin oak casks..contributed another complex dimension to its distinct varietal personality.

B. as *sb.* A wine made from a single variety of grape. orig. *U.S.*

1955 J. Storm Invitation to Wines 72 The wines labeled according to the name of the chief informing grape are called varietals. 1977 H. Fast Immigrants iv. 254 There is no need for you to try to produce a varietal, which simply means a wine produced out of a single variety of grape. 1977 Times 15 Nov. (Italian Wine Suppl.) p. iii/3 Lambrusco, the famous wine of Modena..is a varietal, the Lambrusco being the grape from which it is vinified. 1979 Tucson (Arizona) Citizen 28 Apr. (Weekender Mag.) 7/2 A 1977 Emerald Riesling..displayed a high acidity that may be typical of the varietal.

varietally *adv.* (later example).

1942 M. F. Mabon ABC of America's Wines ii. 13 The majority of varietally named wines happen to be blends.

varietist (vărəi·ĕtist), *sb.* [f. Variet(y + -ist.] One who chooses variety, esp. in the satisfaction of (sexual) desires; also *attrib.* or as *adj.* Hence **vari·etism**, the practice or result of such choices.

1911 G. B. Shaw Getting Married Pref. 153 The reassurance that man is not promiscuous in his fancies must not blind us to the fact that he is (to use the word coined by certain American writers to describe themselves) something of a Varietist. 1925 T. Dreiser Amer. Tragedy I. i. xiii. 101 So captivated was he by this savor of sensuality and varietism that was about her. 1968 Ann. Amer. Acad. Pol. & Social Sci. July 65/2 Some sexual varietists..believe that they need plural sex-love affairs. Ibid., Such individuals manage to have their varietist inclinations fulfilled. Ibid. (heading) Sexual varietism. 1977 N.Y. Rev. Bks. 15 Sept. 26/2 'Free love' never meant libertinism..but quite simply that there should be no coercion in sexual relationships... There were 'exclusivists' and 'varietists'.

variety. Add: **8.** Also, this species of entertainment, including its presentation on radio and television. (Earlier and later examples.)
1904 [see *legit., legit]. **1929** *Illustr. London News* 13 Apr. 609/1 (*caption*) Broadcasting variety from 2LO. **1967** *Stage* 2 Mar. 3/4 Variety makes a comeback to Edinburgh on Monday. **1977** J. Fleming *Every Inch* III. v. 134 He realized..that variety was not..on the way out. It was..very much alive.
9. a. For *U.S.* read *N. Amer.* and add earlier U.S. and later Canad. examples.
1768 *Boston Even.-Post* 21 Nov. 3/3 Just imported in the Bristol Packet..and to be Sold by William Jackson At his Variety Store,..Nails, Brads, and Tacks of all sorts. **1790** *Columbian Sentinel* (Boston) 15 Sept. 4/2 To be sold, at J. Brazer's Variety-store,..Holland Gin, of the best kind, in cases. **1965** H. Hood in R. Weaver *Canad. Short Stories* (1968) 2nd Ser. 218 We proceeded to the general store, grocery store, variety store, butcher shop, what would you call it? **1975** *Weekend Mag.* (Montreal) 1 Nov. 25/1 Variety store owners just grin and bear the hockey card mania.
b. (Earlier and further examples.)
1868 *Oregon State Jrnl.* 17 Oct. 3/1 Variety Troupe.—This troupe gave an entertainment in the Court House. **1878** *Appleton's Jrnl.* XIX. 36/2 A 'music-hall', a place of entertainment like that which we call a 'variety theatre' in America. **1882** [see *advance agent* s.v. *advance *sb.* V]. **1895** Stuart & Park (*title*) The variety stage. **1926** G. B. Shaw *Transl. & Tomfooleries* 232 This is not a serious play: it is what is called a Variety Turn for two musicians. **1967** *Stage* 2 Mar. 3/4 A new variety-revue opens at the Palladium Theatre. **1982** C. Castle *Folies Bergère* ii. 60 Provincial English variety theatres before World War Two.

varifocal (vɛ̄ᵊrifōᵘ·kăl), *a.* [f. *vari- + Focal *a.*] **a.** Having a focal length that can be varied.
1945 *Jrnl. Soc. Motion Picture Engineers* XLV. 466 A new positive Vari-Focal view-finder for motion picture cameras. **1962** *Jrnl. Optical Soc. Amer.* LII. 353/2 The two-lens varifocal system consists of a single movable component placed behind a fixed component. **1967** *Appl. Optics* VI. 1085 (*heading*) Stereoscopic display using rapid varifocal mirror oscillations. **1979** *Ibid.* XVIII. 5/1 We have used a varifocal lens to compensate for changes in refractive index of lens elements with wavelengths from 190 nm to 600 nm.
b. *Ophthalm.* = *omnifocal* adj. s.v. *omni-. Also *ellipt.*
1975 L. S. Sasieni *Princ. & Pract. Optical Dispensing & Fitting* (ed. 3) xii. 373 Varifocal lenses have two areas of specified powers, for distance and reading, and an intermediate region in which the power gradually changes, from one to the other. **1982** *Ophthalmic & Physiol. Optics* II. 75 (*heading*) Theoretical aspects of concentric varifocal lenses. *Ibid.*, Substantial amounts of unwanted astigmatism in the transition zone are intrinsic to the design and to other forms of concentric varifocal.

varihued (vɛ̄ᵊ·rihiūd), *a.* [f. Hued *ppl. a.*, after *varicoloured*.] = Vari-coloured, varicoloured *a.*
1921 *Nat. Geogr. Mag.* Sept. 274 The varihued deposits resemble brilliant mosaics. **1950** *John o'London's Weekly* 7 July 435/1 Queer fish of another sort..make his book an aquarium of vari-hued specimens.

varimax (vɛ̄ᵊ·rimæks). *Statistics.* [f. Vari(ance + Max(imum.] A method of factor analysis in which uncorrelated factors are sought by a rotation (*rotation 1 d) that maximizes the variance of the factor loadings. Usu. *attrib.*
1956 H. F. Kaiser (title of thesis, Univ. of California) The varimax method of factor analysis. **1958** —— in *Psychometrika* XXIII. 187 Dr. John Caffrey suggested the name varimax. **1969** *Language* XLV. 89 The purpose of the varimax rotation was to focus these indices on possible natural groupings of the words according to their positional characteristics. **1974** R. L. Gorsuch *Factor Analysis* x. 193 Varimax is available on virtually every computer which has any programs for factor analysis. **1977** M. Mac Gréil *Prejudice & Tolerance in Ireland* xv. 467 Factor No. 3 'Republican Political'..: The high varimax loadings of the four categories which constitute this cluster is akin to that of the racial cluster in Factor 1.

vario-coupler (vɛ̄ᵊ·riokʌpləɹ). *Electr.* ? *Obs.* Also **variocoupler.** [f. *vari- + -o + Coupler.] A device, consisting of two unconnected coils, one inside the other, whose relative position can be varied to vary the mutual inductance between two circuits.
1922–3 *T. Eaton & Co. Catal.* Fall & Winter 401/4 Variocouplers, built up on black fibre cylinder, with taps, fitted with lugs. **1925** P. J. Risdon *Crystal Receivers & Circuits* ii. 21 A vario-coupler..is a sort of combination of a variometer and an ordinary tapped tuning inductance. **1937** *Discovery* Feb. 54/2 (*in figure*) Vario-coupler.

variometer (vɛ̄ᵊriǫ·mɪtəɹ). [f. *vari- + -ometer.] **1.** An instrument for measuring variations in the intensity of the earth's magnetic field.
1889 in *Cent. Dict.* **1916** *Proc. R. Soc.* A. XCII. 321 There can be no question of the enormous simplification of the field work which such a portable variometer produces. **1972** *Nature Physical Sci.* 25 Dec. 184/2 Telluric recording equipment was operated in conjunction with a magnetic variometer array.
2. *Electr.* (*a*) An inductor (*inductor 3 d) whose total inductance can be varied by altering the relative position of two coaxial coils connected in series, and can thus be used to tune a circuit. (*b*) A device achieving a similar function by means of permeability tuning.
1908 C. C. F. Monckton *Radio-Telegr.* xv. 255 The antenna and the receiving circuit are tuned to the incoming oscillations by means of the variometer. **1921** *Wireless World* IX. 6/1 For amateur use the variometer has the advantage that it is cheap compared with a variable condenser. **1926** 'R. Keverne' *Carteret's Cure* vii. 168 Mr. Pell turned to the listening set and picked up the head-phones while his fingers played deftly with the variometer. **1959** K. Henney *Radio Engin. Handbk.* (ed. 5) iii. 14 Another form [of variable inductor], which is rather bulky..and now little used, is the variometer... Much more common is the variation of inductance through the variation of the permeability of the core. **1977** *Elektor* Mar. 22/1 The variometer is an updated version of the old permeability tuners.
3. An instrument for indicating the rate of climb or descent of an aircraft.
1924 *Sci. Amer.* Feb. 83/2 Air speed indicator... Turn indicator... Variometer. **1930** Startup & Kinnear tr. *Stamer & Lippisch's Gliding & Sail-Planing* x. 111 An instrument called a variometer exists for the purpose of indicating ascent and descent and is much used in balloons. **1969** *Daily Tel.* (Colour Suppl.) 21 Nov. 64/2 A variometer can confuse you; you might be going down in a rising thermal, so it's no good just knowing how you are doing relative to the air you're in. **1973** *Q. Jrnl. Meteorol. Soc.* XCIX. 768 Height changes give a check on the calibration of the variometer.

variorum. Delete ‖ and add: **1. c.** (Earlier examples.) Also applied to a single reading.
1850 *N. & Q.* 19 Oct. 325 A very curious variorum reading. **1873** (*title*) Osorio. A tragedy as originally written in 1797 by Samuel Taylor Coleridge. Now first printed from a copy recently discovered by the publisher with the variorum readings of 'Remorse' and a monograph on the history of the play in its earlier and later form by the author of 'Tennysoniana'.
d. Used, chiefly *attrib.*, to denote an edition, usu. of an author's complete works, containing variant readings from manuscripts or earlier editions.
This use is deplored by some scholars.
1955 *Times* 11 June 9/3 How delighted he [*sc.* Johnson] would have been to have known of the Yale University project for a really complete variorum edition of his works. **1957** (*title*) The variorum edition of the poems of W. B. Yeats. **1964** *Times Lit. Suppl.* 29 Oct. 979/4 The variorum edition of Emily Dickinson..(meaning, it appears, an edition containing variant readings). The..use of the term is incorrect. **1980** *N.Y. Times* 26 July II. 22 The book, a variorum, records every usable fact and opinion about the play—every significant textual variant, every influential interpretation, [etc.]. **1983** P. G. Ruggiers in Beadle & Griffiths *St. John's College Cambridge MS. L.1.* p. xiii, Having then presented these two essential manuscripts, consideration was given by the Variorum editors [*sc.* the editors of the Variorum Chaucer] to providing other materials essential to the understanding of an evolving Chaucer text.
2. (Later examples.) *Sc.*
1876 J. Smith *Archie & Bess* 95 It's a lang time since I heard ye sing't and it'll aye be a variorum. **1901** R. Trotter *Galloway Gossip* 11 They widna let the Paraphrases be sung in the kirk, or tunes wi' variorum about them. **1930** in *Sc. Nat. Dict.* (1974) IX. 520/2 A'm guid at readin' write gin it has nane o' your variorums and whirligigs.

variphone (vɛ̄ᵊ·rɪfōᵘn). *Linguistics.* [f. *vari- + Phone *sb.*¹] One of two or more sounds used interchangeably by the same speaker in the same phonetic context.
1932 D. Jones in E. P. Hamp et al. *Readings in Linguistics II* (1966) 32 Variphones are found in some varieties of German, where *p* and *b*, ʃ and ʒ, and other corresponding voiced and voiceless consonants are apparently used indifferently. **1934** Yuen-Ren Chao in M. Joos *Readings in Linguistics I* (1966) (ed. 4) 39/2 Bloomfield makes no explicit mention of free phonemes or variphones. *Ibid.*, Variphones are also phonemes, except that the choice of the exact shade of the sound used is determined by psychological and physiological factors other than those of phonetic environment. **1950** D. Jones *Phoneme* xxviii. 206 ŋg and ŋ form a variphone in the speech of Midland districts of England.

Variscan (vāri·skăn), *a. Geol.* [f. L. *Varisc-ī* (see quot. 1906¹) + -an, after G. *variscisch* (E. Suess *Antlitz der Erde* (1888) II. iii. ii. 131).] Of, pertaining to, or designating a mountain system that formerly extended from southern Ireland and Britain through central France and Germany to southern Poland, or the orogeny that gave rise to it during late Palæozoic time.
1906 H. B. C. & W. J. Sollas tr. *Suess's Face of Earth* II. iii. ii. 111 It is thus fitting that the name of this range, which includes most of the German horsts, should be borrowed from the land of the Varisci or the Vogtland; and we shall name it therefore the Variscan range, from the Latin name of Hof in Bavaria (*Curia Variscorum*). **1908** *Ibid.* III. iv. viii. 346 The Carnic mountains are a chain..and of Variscan age. **1956** W. Edwards in D. L. Linton *Sheffield* i. 22 Earth-movements of the Variscan orogeny were already in action during deposition of the Lower Carboniferous. **1969** Bennison & Wright *Geol. Hist. Brit. Isles* xi. 255 The building of the Variscan mountain chains profoundly affected the geography of Europe, both in the distribution of land and sea and climatically. **1974** *Encycl. Brit. Macropædia* VI. 14/1 The European geosyncline is called the Variscan.

variscite (væ·risəit, variskəit). *Min.* [ad. G. *variscit* (A. Breithaupt 1837, in *Jrnl. f. prakt. Chem.* X. 506), f. med.L. *Varisc-ia*, name of the Vogtland district of E. Germany where the mineral was first found: see -ite¹.] An orthorhombic hydrated aluminium phosphate, $AlPO_4.2H_2O$, which is dimorphous with metavariscite and occurs usu. as green or colourless, translucent or transparent, fine-grained masses, crusts, and nodules.
1850 D. T. Ansted *Elem. Course Geol., Mineralogy* ix. 187 Variscite is also a phosphate of alumina, of a green colour. **1912** *Proc. U.S. Nat. Museum* XLI. 415 The variscite and variscite-matrix from this locality would yield very beautiful gems for the so-called barbaric jewelry. **1964, 1975** [see *redondite]. **1979** *Arizona Highways* Apr. 40/1, I did a series of shields with variscite, which is a dark stone that has some turquoise in it.

vari-sized (vɛ̄ᵊ·ri,səizd), *a. rare.* [f. Sized *ppl. a.*¹ 1 b, after *varicoloured*, etc.] Of various or different sizes.
1936 F. Clune *Roaming round Darling* xxi. 208 Beneath..were nineteen dogs... All were vari-coloured, vari-bred, vari-sized and -shaped. **1940** W. Faulkner *Hamlet* iv. i. 243 Vari-sized and -coloured tatters torn at random from large billboards.

varistor (vɛ̄ᵊri·stəɹ). *Electr.* [f. Var(iable *a.* + *res)istor.] A semiconductor diode whose resistance varies non-linearly with the applied voltage.
1937 *Bell Telephone Q.* Apr. 114 A varistor is a device, for example a copper oxide rectifier, whose resistance varies with the applied voltage. **1957** *Electronic Engin.* XXIX. 386/1 Varistors exhibit a practical resistance variation of several decades over their normal rated working range of current or voltage. **1978** *Sci. Amer.* Mar. 59/1 The compensation is termed equalization and is achieved by adding two varistors, or variable resistors, to the speech network.

Varityper (vɛ̄ᵊ·ritəipəɹ). orig. and chiefly *U.S.* Also **Vari-typer, varityper.** [f. *vari- + Typer.] The proprietary name of a kind of typewriter that has interchangeable type faces. Also, a kind of type-composing machine with similar operation.
1928 *Official Gaz.* (U.S. Patent Office) 16 Oct. 494/1 *Varityper.* For typewriter machines. **1931** *Times Lit. Suppl.* 1 Oct. 760/4 Latest Hammond (Varityper) Typewriter... Turn knob and you get another type shuttle—italics, smaller letters or accented, or Greek, each with close or open spacing. **1948** *Economist* 13 Nov. 781/1 The varityper..has kept the Chicago newspapers going through a compositors' strike. **1956** *Trade Marks Jrnl.* 11 Jan. 27/1 *Vari-typer*... Type composing machines for office use and parts thereof... Ralph C. Coxhead Corporation,..Newark, State of New Jersey, United States of America; manufacturers. **1959** *Oxf. Univ. Gaz.* 3 Dec. 369/2 New 'Vari-typer' composition service with small-offset-litho reproduction. **1977** *San Antonio* (Texas) *Express News* 30 Oct. 5E/4 She designed the layout, typeset it at home on an old varityper and made the negatives for the printer.
Hence **Va·ri-type** *v. trans.*; **Va·rityping** *vbl. sb.*
1955 H. Quastler *Information Theory in Psychol.* p. iv, 'Information Theory in Psychology' was Vari-typed by Diane Folk, Stenographic Bureau, University of Illinois. **1960** *World's Press News* 30 Dec. 15 Varityping plus photo litho printing results in lower cost and faster delivery for brochures, handbooks, leaflets, catalogues.

varlamoffite (vāɹlāmǫ·fəit). *Min.* [a. F. *varlamoffite* (H. Buttgenbach *Les Minéraux de Belgique et du Congo Belge* (1947) 182), f. the name of N. *Varlamoff*, mining engineer who found the mineral: see -ite¹.] A yellowish cryptocrystalline powdery material, perhaps $(Sn,Fe)(O,OH)_2$, that is related to or perhaps identical with cassiterite.
1948 *Mineral. Abstr.* X. 354 A new mineral varlamoffite, H_2SnO_3, as yellow earthy material mixed with grains of cassiterite from Kalima, Belgian Congo, is here included for the first time. **1970** *Mineral. Mag.* XXXVII. 626 Most of the reported occurrences of varlamoffite show a strong spatial association with stannite, and most workers have concluded that it is a secondary mineral resulting from the attack of meteoric waters on stannite.

varment, varmint, *sb.*¹ and *a.*¹ Add: **1. a, b.** Also **varmit.**
1856 *Porter's Spirit of Times* 20 Sept. 38/2 Why, massa, dat 'tarnal varmit hab fooled me bad. **1974** *Evening Herald* (Rock Hill, S. Carolina) 18 Apr. 13/3 She didn't care if my face was black. She didn't treat me like no varmit or nothing.
Hence **va·rminty** *a. colloq.*, suggestive of or resembling a varmint; sharp, cunning.

1907 A. Conan Doyle *Thr. Magic Door* ii. 34 Louis, thin, ascetic, varminty. **1922** R. Leighton *Compl. Bk. Dog* xvii. 261 The General Appearance of the West Highland White Terrier is that of a small, game, hardy-looking terrier..with a 'varminty' appearance. **1962** R. H. Smythe *Anat. Dog Breeding* 52 The 'varminty' eye of a Fox Terrier.

‖ **varna** (vā·ɹnă, vɒ·rnă). [Skr. *varṇa*, colour, appearance.] Any one of the four early castes or social classes of Hindu society (see Caste 2 a); the system or basis of this division.
1838 *Penny Cycl.* XII. 230/1 The division of Hindus into..castes..existed from the earliest times. In Sanskrit they are called *varnas*, that is, 'colours'. **1876** *Encycl. Brit.* IV. 203/1 The idea of caste is expressed by the Sanskrit term *varna*, originally denoting 'colour', thereby implying differences of complexion between the several classes. **1931** [see *jat²]. **1959** *Times Lit. Suppl.* 1 May 262/5 Viewing the *varna* system of ancient India through rose-tinted spectacles. **1974** *Encycl. Brit. Macropædia* IX. 347/2 The traditional view that *varna* reflected the organization of Indian society has recently been questioned; it has been suggested that..the concept of *jāti* is more central to caste functioning than *varna*, which may be the theoretical rationalization. **1977** M. & J. Stutley *Dict. Hinduism* 323 *Varna.* The word is generally used to denote colour, but its primary meaning, derived from the root *vṛ* is 'screen, veil, covering, external appearance', and hence colour is only one of the many aspects of the term.

varnish, *sb.*¹ Add: **1. f.** A resinous deposit formed in engines by the oxidation of fuel and lubricating oils.
1948 A. P. Fraas *Combustion Engine* viii. 241 The resins may deposit with the sludge or may form thin adherent coatings on engine parts. In the latter case the coatings are called varnish or lacquer because of their appearance. **1967** *Boston Sunday Herald* 26 Mar. 1. 1/2 Don't be fooled by slow cranking because 'varnish' on the pistons will cause so much drag that a hot engine may resist cranking until it cools. **1981** *Pop. Hot Rodding* Feb. 84/1 Oxidation produces new compounds, loosely termed sludge, varnish and acid, which are detrimental to the lubrication system.
5. *varnish remover*; *varnish-like* adj.
1916 *Nature* 25 May 269/2 Boiled linseed oil on exposure to the air is converted by oxidation into a hard varnish-like product. **1965** Finer & Savage *Sel. Lett. J. Wedgwood* 13 The ancient red-figure vases were decorated by reserving the figures on a red pottery ground, surrounding them with a black varnishlike slip. **1965** P. D. Samman *Nails in Disease* ix. 83 Nail varnish and varnish removers and excess manicuring may be of some importance aetiologically. **1973** C. Williams *Man on Leash* ix. 131 That crap they shot into my arm. Battery solution or varnish remover.

varnish, *v.* **1.** (Later *absol.* example.)
a **1817** Jane Austen *Persuasion* (1818) III. xi. 234 He drew, he varnished, he carpentered.

varnishing, *vbl. sb.* Add: **1. b.** = *vernissage. Cf. *varnishing day*, sense 3 in Dict.
1951 N. Mitford *Blessing* ii. iv. 188, I was to go round at six, take her to a varnishing, and then dine with her. **1971** R. A. Carter *Manhattan Primitive* vii. 63 Thatcher preferred the French word for openings; 'varnishings' were what they usually turned out to be.

varoom (vāru·m), var. *vroom.

Varroa (væ·roă). *Ent.* Also **varroa.** [mod.L., coined in Ger. (A. C. Oudemans 1904, in *Entomol. Ber.* (Amsterdam) XVIII. 161), f. the name of Marcus Terentius *Varro* (116–27 BC), Roman scholar.] A small mite, *Varroa jacobsoni*, which is a fatal parasite of the honeybee in the Far East and has spread to other parts in modern times; infection with this mite.
1974 *Jrnl. Washington Acad. Sci.* LXIV. 10/2 '*Varroa* disease' of the honey bee. **1978** *Bee World* LIX. 165 Bees imported from Japan brought *Varroa* to Paraguay. **1979** *Guardian* 26 July 1/4 The barely visible varroa began infesting Russian honey-bees in 1964 and since then it has flourished dramatically. **1984** L. A. Stephens-Potter *Beekeeper's Man.* x. 111/2 In countries where varroa is present it has spread with alarming rapidity and caused enormous losses of bee colonies. **1984** *Times* 11 Oct. 3/2 At the queen examination centre..Ministry of Agriculture scientists make sure she is free of Varroa.

Varsity. Add: **a.** (Earlier and further examples.) Now (in the U.K.) somewhat *joc.* exc. in *varsity match*, a sporting contest, esp. the annual Rugby football match between the universities of Oxford and Cambridge.
1863 *Baily's Mag.* Jan. 360, I had conjured up all the most extravagant and erroneous ideas as to my 'Varsity career'. **1885** A. Edwardes *Girton Girl* III. xvi. 291, I'm darned..if it be'ant my Varsity man, after all! **1908** E. H. W. Meyerstein *Let.* 21 Oct. (1959) 33 It seems to me that the real attractions of varsity life are reserved for the sportsman and the loafer. **1921** *Granta* 30 Nov. 146/1 What would you do if in the 'Varsity match just as you were off, on a clear half-way run down the field somebody ripped your shorts off? **1927** C. Connolly *Let.* 25 Jan. in *Romantic Friendship* (1975) 228 O blessed chastity (rare at the varsity). **1967** F. Sargeson *Hangover* iii. 19 And that damn cleaning man across from the varsity. **1979** *Times* 6 Dec. 4 (*heading*) The varsity match. **1982** *Guardian Weekly* 27 June 23 [He] would become the first university Blue to miss the Varsity match because he will be playing [cricket] for England. *Ibid.* 19 Dec. 23 Cambridge won 20-13, but on the next day Oxford had some revenge in winning the soccer varsity match at Wembley 4-2. **1983** *Metro* (Auckland) Feb. 13/4 Instead he went to varsity with the ambition of taking a science degree.
b. *U.S.* Applied *attrib.* to sporting events, teams, etc., at or of a university or college. Occas. *absol.*, a college team. (See also quot. 1930.)
1891 *Outing* Dec. 241/1 The 'varsity captain whistled a lively air. **1898** *Ibid.* June 305/1 The..excitement generally attendant upon..trips by 'varsity nines. **1902** L. L. Bell *Hope Loring* (1908) 79 Being on the 'Varsity football team. **1928** E. O'Neill *Strange Interlude* viii. 278 It's Gordon's last race, his last appearance on a 'varsity. **1930** *Amer. Speech* V. 243 *Varsity*..is applied always to a student team, as in athletics or debate, not to an institution. **1973** R. Ludlum *Matlock Paper* iv. 32 The coach of varsity soccer. **1978** *Nature* 29 June 794/1 In the intercollegiate ('varsity') sports, which play a large role in American colleges, Conant replaced an autonomous body led by the alumni association, which is the norm in American universities and is the source of athletic scholarships, by a departmental structure administered by a faculty committee. **1978** *New Yorker* 9 Oct. 184/3 The Harvard varsity..came up a winner 10-0, against Massachusetts. **1979** *Tucson* (Arizona) *Citizen* 20 Sept. 1D/4 He went on to stardom at Niles McKinley High School... McKinley had a 21-6-1 record in Sammy's three years of varsity football. **1980** *N.Y. Times* 1 June xxiii. 12/3 Classic Yale-Harvard Regatta in New London... The four-mile varsity race..will be rowed on the Thames River.

Varsovian, *a.* Add: (Earlier example of spelling *Warsovian*.) Also as *sb.*, a native or inhabitant of Warsaw.
1764 H. Grenville *Let.* 10 Mar. in *Desp. & Corr. 2nd Earl of Buckinghamshire 1762–5* (1902) II. 153 The apprehensions of your Warsovian correspondent appear but too well founded. **1959** *Listener* 17 Dec. 1071/2 The Palace of Culture is intensely disliked and resented by the Varsovians. **1978** M. Tripp *Wife-Smuggler* i. 16 One of the best restaurants..where Varsovians go when they wish to treat..distinguished guests.

vartabed (vā·ɹtăbed). Also **vardapet, vartabad, vartapet,** etc. [Armenian.] A member of an order of clergy in the Armenian church (see quot. 1847).
1718 J. Ozell tr. *Pitton de Tournefort's Voy. Levant* II. viii. 303 These Vertabiets, who make such a noise among the Armenians, are not in reality great Doctors. **1841** L. Coleman *Christian Antiq.* xxiii. 470 The vartabeds live not among the people, but in convents. **1847** J. Wilson *Lands Bible* II. 482 The monkish clergy are denominated Vartabads or Doctors, and it is their peculiar office to teach and preach. **1875** *Encycl. Brit.* II. 549/2 The vartabed, or doctor of theology..has frequently charge of a diocese, with episcopal functions. **1923** *Blackw. Mag.* Aug. 252/1 The Patriarch and an Armenian *vartabed*..are pushed inside. **1957** *Oxf. Dict. Christian Church* 87/1 Bishops..are usually chosen from among the vartabeds. **1964** [see *rasophore]. **1982** *Encycl. Brit. Macropædia* VI. 140 Another class of priests is represented by the vartabeds, or doctors, who must remain unmarried.

varus². (Earlier example.)
1756 [see *ephelis].

varve (vāɹv). [ad. Sw. *varv* layer, turn.] A pair of thin layers of clay and silt of contrasting colour and texture which represents the deposit of a single year (summer and winter) in still water at some time in the past (usu. in a lake formed by a retreating ice-sheet); they have been used to establish a chronology of the late glacial and post-glacial period. Also *attrib.* and *Comb.* Hence **varved** (vāɹvd) *a.*, characterized by such layers.
[**1887** *Encycl. Brit.* XXII. 740/1 The glacial clay consists generally of..darker and lighter coloured layers, which give it a striped appearance, for which reason it has often been called *hvarfvig lera* (striped clay).] **1912** G. De Geer in *Compt. Rend. XI Session Congrès Géol. Internat.* 253 The Swedish word *varv*, subst. (old spelling: hvarf), means as well a circle as a periodical iteration of layers. An international term for the last sense being wanted it seems suitable to use the transcription *varve*, pl. -s, in Engl. and Fr. **1929** C. R. Longwell *Pirsson's Textbk. Geol.* (ed. 3) I. v. 126 Lake Deposits.—Remarkably banded clays have been formed in patches within the glaciated regions... The Swedish geologists call them varved clays. **1936** *Times Lit. Suppl.* 9 May 392/2 Varve-variation in its general features is a function of solar radiation. **1948** *Bull. Geol. Soc. Amer.* LIX. 646 In this basin (Fossil, Wyo.) hundreds of thousands of beautifully preserved fish are entombed in the varved sediments. **1953** *Antiquity* XXVII. 35 What we most need now, apart from archaeology, is a study of the post-glacial climatic fluctuations in the Tigris and Euphrates valleys. Here the techniques of Western Europe, such as pollen analysis and varve-counting, seem unlikely to be of any use. **1957** G. E. Hutchinson *Treat. Limnology* I. i. 8 The dating of the events, based primarily on the varve chronology, is in fair accord with the radiocarbon chronology. **1974** *Nature* 15 Nov. 182/1 The thickness of the varves..provide [*sic*] a guide to the year-by-year changes in mean climate; but that is at best a crude measure.

vary, *sb.* (Later example.)
1929 R. Bridges *Testament of Beauty* i. 26 We should not in the field of Reason look to find less vary and veer than elsewhere in the flux of Life.

vas. Add: **1. a.** *vas deferens* (pl. *vasa deferentia*) [L. *dēferens*, pres. pple. of *dēferre* (see Deferent *a.*¹ and *sb.*)], a fibromuscular tube which carries spermatozoa from the epididymis at ejaculation, in man joining the duct from the seminal vesicle at the prostate gland to form the ejaculatory duct.
1578 J. Banister *Hist. Man* vi. 87 (*marginal note*) Why these ij. are called Vasa deferentia. **1713** W. Cheselden *Anat. of Humane Body* iv. i. 163 On the upper part of the Testicles, are hard Bodies call'd Epididymi; which are evidently the beginnings of the Vasa Deferentia. **1849–52** R. B. Todd *Cycl. Anat. & Physiol.* IV. 981/1 The vas deferens is round and indurated,—harder than any other excretory duct in the body, by which character it is easily distinguished, when handled, from the other parts constituting the spermatic cord. **1975** *Nature* 9 Oct. 488/1 For collection of spermatozoa, the vasa deferentia were excised aseptically from adult Wistar strain rats.

VASCAR, Vascar: see *V III. 5 b.

vascular, *a.* Add: **1. e.** *vascular wilt disease*): wilt disease involving the vascular system of a plant; *spec.* = *Panama disease* s.v. *Panama.
1946 *Nature* 16 Nov. 712 The presence of a vascular wilt disease of the oil palm. **1951** *New Biol.* XI. 76 Vascular Wilt Disease..of bananas in the Old World made its appearance in Central America towards the end of the nineteenth century. **1972** [see *Panama disease* s.v. *Panama].

vasculature (væ·skiŭlătiŭɹ). *Anat.* [f. L. *vāscul-um* Vascular *a.* + -ature, after musculature.] The vascular system and its arrangement in the body or a part.
1934 in Webster. **1962** A. Mettler *Neuroanatomy* (ed. 2) Concerning the vasculature of the vertebral canal. **1962** *Lancet* 22 Dec. 1327/1 Burton has shown that a critical opening and a critical closing pressure are peculiar also to the intrarenal vasculature. **1975** *Nature* 18 Dec. 580/2 ..the vasculature of the..

vasculitis (væskiŭləi·tis). *Path.* Pl. **vasculitides** (-ai·tidīz), **vasculitises.** [f. L. *vāscul-um*, dim. of *vās* Vas + -itis.] An inflammatory reaction in a blood vessel; any of various conditions characterized by such reactions.
1900 in Dorland *Med. Dict.* **1945** *Jrnl. Amer. Med. Assoc.* 2 June 337/1 We believe that some of these cases could more properly be grouped with cases of nodular vasculitis. **1976** *Proc. R. Soc. Med.* LXIX. 927/1 The vasculitides and other non-articular rheumatoid disease are distinctly uncommon in this age group. **1983** Byron & Beaulieu in *Oxf. Textbk. Med.* II. xvi. 365/2 For clinical purposes, the vasculitides can be divided into those with skin lesions and those without.
Hence **vasculi·tic** *a.*, of the nature of, or characteristic of, vasculitis.
1971 *Clin. & Exper. Immunol.* IX. 754 Endothelial changes seem to be an important component of a number of vasculitic disorders. **1980** *Brit. Jrnl. R. Soc. Med.* LXXIII. 208 His rheumatoid disease became more active and vasculitic lesions appeared once again on his fingertips. **1983** T. J. Ryan in *Oxf. Textbk. Med.* II. xx. 18/1 There have been a number of more recently described vasculitic syndromes in which urticaria is the only skin manifestation.

vasculotoxic (væskiŭlo,tǫ·ksik), *a. Physiol.* [f. as prec. + -o + Toxic *a.*] Affecting the vessels of the body adversely. Hence **va:sculotoxi·city.**
1957 *Dorland's Med. Dict.* (ed. 23), Vasculotoxic. **1962** *Endocrinology* LXXI. 505/1 The vasculotoxic and nephrotoxic effect of renin. **1976** *Lancet* 20 Mar. 534/2 The vasculotoxicity of renin is also suggested by its presence in excessive amounts in human malignant and renovascular hypertension. **1977** *Lancet* 4 Jan. 83/2 There is as yet little information on the relations of blood-levels to nephrotoxic or vasculotoxic potential.

vasculum. Add: Pl. **-a, -ums. 2.** (Earlier examples.)
1782 J. Lightfoot *Let.* 11 July in W. H. Curtis *William Curtis* (1941) 59, I am extremely obliged to you..for a beautiful Vasculum. **1828** G. Graves *Naturalist's Pocket-Bk.* 99 These [specimens] must be gathered on a dry day, and placed in a common tin vasculum or botanical box. **1839** J. D. Hooker *Let.* in L. Huxley *Life J. D. Hooker* (1918) I. 47 Two Botanising vascula.

vase. Add: **3.** *vase-maker*: quot. 1894 should be cited as '**1770** J. Wedgwood *Let.* 19 May (1965) 92'; **vase carpet,** an oriental (esp. Persian) carpet with a pattern incorporating a stylized vase of flowers.
1953 *Guide to Collection of Carpets* (Victoria & Albert Mus.) 27 Another type of pattern, in which animal and bird life are absent, has given rise in Germany to the name 'vase-carpets'. **1983** *Eastern Carpet* (Hayward Gallery Catal.) 31/2 Another group of Persian carpets with beautiful floral designs are the so-called vase carpets, believed to have been made in the Kerman area.

vasectomized, *ppl. a.* Add: Also, having undergone ligation of one or (usually) both *vasa differentia*. So **vase·ctomize** *v. trans.*
1923 *Physiol. Rev.* III. 342 If..the female rat undergoes a sterile coition with a vasectomized male the corpus luteum persists for a longer period.

1970 *Nature* 11 Apr. 162/2 Spontaneously ovulating females were mated with vasectomized males of strain A, whose sterility had been checked repeatedly. **1977** *Spare Rib* June 31/1 Male, 31, healthy wealthy and dull, seeks girlfriend. Vasectomised. Ideal pet for liberated woman. **1980** *Aggressive Behavior* VI. 218 All six adult males were vasectomized by double ligature of the vas deferens four months before the first copulation was seen in November. **1982** *Times* 6 Dec. 6/8 One man said the number three song on the Thai hit parade 'I'm vasectomized' had inspired him [to have a vasectomy].

vasectomy. Add: Also, ligation of one or (more commonly) both *vasa deferentia*, usu. performed to render the subject infertile.
1897 *Lancet* 11 Sept. 658/1 A case of double vasectomy. **1900** [see Vas 1 a]. **1923** *Physiol. Rev.* III. 339 Pregnancy is prevented through coition being made sterile..by submitting the male to vasectomy. **1970** *Daily Tel.* 2 May 1/4 Vasectomy has recently become much more usual among men as a method of family planning. **1972** [see *semes 2]. **1979** *Toronto Star* 15 May E7/3 Commercial sperm banks in several major U.S. cities have reported growing numbers of donors, particularly..men who choose to store their sperm before undergoing vasectomies. **1985** *Times* 8 Mar. 10/7 A shipyard mechanic who had a vasectomy last autumn said yesterday he was 'lucky' after learning his wife is pregnant with sextuplets.

Vaseline, *sb.* Add: Also **vaseline.** (Further examples.)
Vaseline is still (1984) a proprietary term.
1897, etc. [see *petroleum jelly*]. **1924,** etc. [see *proprietary *a.* 1 a]. **1927** *Trade Marks Jrnl.* 3 Aug. 1383/2 Vaseline 479,707. Petroleum Jelly and Disinfectant Soap for Veterinary use. Chesebrough Manufacturing Company Consolidated... 11th April 1927. **1930** A. Huxley *Brief Candles* 13 Hearts of putty, hearts of vaseline. **1965** [see *soft ground* s.v. *soft *a.* 27]. **1974** *Cleveland* (Ohio) *Plain Dealer* 26 Oct. 4-d/1 Walter Youngblood, one of his trainers, is smearing Vaseline on his face, torso, arms and legs. **1976** Botham & Donnelly *Valentino* xiv. 107 Vaseline, and other hair-care applications, would experience a significant sales-boost. **1978** J. Irving *World according to Garp* xix. 410 The Vaseline made his ear slippery.
b. The greenish-yellow colour of Vaseline as used in the manufacture of glass; glassware of this colour.
1966 J. Laver *Victoriana* 168 The Victorians used a wide variety of colour in their glass—Bristol Blue..and the yellowy-green known as Vaseline. **1973** *Washington Post* 13 Jan. E1/7 (Advt.), Old glassware, Red Mark Prussia, Ruby, Vaseline, [etc.]. **1975** *Daily Colonist* (Victoria, B.C.) 26 Oct. 30/8 Vaseline is a greenish yellow glassware resembling its namesake. It dates from 1870 to the present day. **1980** *Times* 1 Nov. 24/7 Its shade is made of vaseline glass—a turn of the century technique which Christopher Wray has reintroduced.
Hence as *v. trans.*, to treat with Vaseline; **va·selined** *ppl. a.*; **va·selining** *vbl. sb.*
1921 *Librarian* Nov. 74 The combination of vaselining and varnishing is the best thing for the text books. **1934** V. M. Yeates *Winged Victory* II. ix. 262 His neck was still sore with all yesterday's twisting to watch his tail. He ought to have vaselined it last night. **1938** F. Chester *Shot Full* xii. 114 A coloured man, his hair..vaselined to stick up. **1942** R. W. Raven *Surg. Care* xviii. 138 A piece of vaselined-gauze should be placed lightly over the colostomy. **1965** 'Malcolm X' *Autobiogr.* iii. 55 He draped the towel around my shoulders, over my rubber apron, and began again vaselining my hair. **1977** *Rolling Stone* 16 June 39/1 Vaselined hair and a semidignified zoot suit. **1983** *Times* 21 Apr. 14/6 'Have you Vaselined your nipples?' asked a solicitor from Peckham.

vaso-. Add: **vasoconstri·cting, -dila·ting** *ppl. adjs.*, **-dila·tion**; **vasoliga·tion** *Surg.*, ligation of a vessel, esp. of the vasa deferentia; **va·sospasm** a sudden constriction of a blood vessel, resulting in reduced flow; hence **vasospa·stic** *a.*; **vasova·gal** *a.*, involving the vagus nerve and the vascular system: applied to an attack (often the result of emotional stress) in which there is a slowing of the pulse and a fall in blood pressure, causing pallor, fainting, sweating, and nausea; **va:sovaso·stomy** *Surg.* [*-stomy], an operation to reverse a vasectomy by rejoining the cut ends of the *vasa deferentia*.
1925 *Proc. R. Soc.* B. XCVII. 325 The toxic and vasoconstricting action..has not been confronted by us provided we make use of freshly defibrinated blood. **1975** *Daily Colonist* (Victoria, B.C.) 16 July 2/3 One of the vasoconstricting drugs..can be helpful. **1956** *Nature* 18 Feb. 340/1 The well-known vasodilating effect of an increased concentration of carbon dioxide. **1962** in L. Kudrow *Cluster Headache* i. 9 Vasodilating headache: a suggestive classification. **1977** *Lancet* 30 July 231/1 A drug such as isoprenaline with a cardiac stimulant and peripheral vasodilating action can improve tissue perfusion. **1908** *Practitioner* Aug. 348 The tourniquet is then released, the affected members then become bright red, owing to a sudden vasodilation. **1974** M. C. Gerald *Pharmacol.* vi. 111 Beta-receptor activation causes a widening of small blood vessels (vasodilation). **1926** W. N. Berkeley *Princ. & Pract. Endocrine Med.* ix. 299 Vasectomy or vasoligation in old men is said..to cause atrophy of the sperm mechanism of the testis. **1932** C. R. Moore in E. Allen *Sex & Internal Secretions* vii. 314 There is no acceptable evidence that vasectomy or vasoligation has any rejuvenating effect. **1973** *Washington Post* 13 Jan. A-8/4 The experiments involved the use of the vasectomy technique, or a closely related one called 'vasoligation', to close off the two tiny ducts that carry the male spermatozoa. **1902** *Buck's Handbk. Med. Sci.* (rev. ed.) V. 74/1 Sedation of maniacal excitement and relaxation of vasospasm in melancholic stupor are better accomplished by warm than cold baths. **1977** *Lancet* 14 May 1039/2 Workers with vibrating hand tools, such as pneumatic drills and chain saws, are at risk of episodic vasospasm, particularly when the vibration is associated with cold exposure. **1932** *Glasgow Med. Jrnl.* CXVIII. 146 Promising results have been obtained in certain cases of scleroderma in which a vasospastic element is present. **1980** *Brit. Med. Jrnl.* 18 Oct. 1033/2 The results..indicate that prostaglandin E_1 given by central venous infusion is a safe and effective method of treating severe vasospastic disease. **1907** W. R. Gowers *Border-Land of Epilepsy* ii. 18 When the vaso-motor spasm preponderates, the case may seem to differ from the type more than it really does. Such cases may be termed 'vaso-vagal'. **1974** R. M. Kirk et al. *Surgery* iii. 41/2 Vasovagal..shock, and fainting caused by an emotional crisis, produce marked dilation of vessels in the muscles, reducing circulatory blood volume. **1976** *Nature* 27 May 334/2 Some had a frank vasovagal reaction with yawning, bradycardia and pallor and could not continue the test procedure. **1949** *New Gould Med. Dict.* 1119/1 Vasovasostomy. **1957** *Jrnl. Urol.* LXXVIII. 79 For this study vasovasostomy, rather than epididymovasostomy was chosen. **1982** *Jrnl. Andrology* III. 21/2 Seven patients undergoing vasovasostomy for reversal of male sterility secondary to vasectomy.

vaso,a·ctive, *a. Physiol.* [f. prec. + Active *a.*] Affecting the physiological state of blood vessels, esp. their calibre; *vasoactive intestinal (poly)peptide*, a polypeptide 28 amino-acids long which is a neurotransmitter found esp. in the brain and gastrointestinal tract; abbrev. *VIP*.
1958 *Brit. Jrnl. Pharmacol.* XIII. 113 (*heading*) Vasoactive substances in the nasal mucosa. **1962** *Times* 17 May 14/4 New Appointments... M. R. Lee, M. A. (Oxon)..to investigate the control of the generation of vasoactive polypeptides in the blood, at the..Radcliffe Infirmary, Oxford. **1970** *Nature* 28 Feb. 864/2 If the liver is normally the main site for inactivating this vasoactive intestinal peptide, its failure of destruction when liver function is seriously impaired..could account for..peripheral vasodilatation. **1972** [see *VIP* s.v. *V 5 b]. **1976** *Nature* 22 Jan. 224/1 Release..of vasoactive lymphokines, such as prostaglandins,..might not necessarily leave a morphological trace.
Hence **vasoacti·vity,** vasoactive power.
1968 *Experientia* XXIV. 1126/1 These results have been interpreted as evidence that plasma has..intrinsic vasoactivity. **1982** *Amer. Jrnl. Path.* CVII. 289/2 When such lesions were induced in sheep and the lymph plasma draining the stimulated nodes was assayed for vasoactivity, a hyperemia-inducing activity was indeed found to be present.

vasopre·ssin. *Physiol.* [f. as next + -in¹.] A polypeptide hormone present in the neurohypophysis of mammals which controls the retention of water in the kidneys and when given in large quantities raises the blood pressure by vasoconstriction.
1928 [see *oxytocin]. **1951** A. Grollman *Pharmacol. & Therapeutics* xxvi. 553 Vasopressin is also used in the form of a tannate... A single injection of this preparation every 24 to 48 hours may adequately control the polyuria and polydipsia of patients with diabetes insipidus. **1965** Lee & Knowles *Animal Hormones* ii. 29 The basic action of one of the two hormones, vasopressin or an analogous substance, is apparently the same in all terrestrial vertebrates, namely the maintenance of the osmotic tension of the extracellular fluid, and indirectly that of the intracellular fluid. **1976** Smythies & Corbett *Psychiatry* vii. 115 After surgery vasopressin secretion induced by stress makes it impossible to dilute urine. **1981** *Sci. Amer.* Oct. 114/1 Vasopressin, a peptide hormone, turns out to be also a neurotransmitter: nerve cells in the hypothalamus, a part of the brain, rely on vasopressin to signal other nerve cells in the brain.

va·sopressor, *a.* and *sb. Pharm.* [f. Vaso- + Pressor.] **A.** *adj.* Causing the constriction of blood vessels. **B.** *sb.* A drug with this effect.
1928 *Proc. Soc. Exper. Biol. & Med.* XXVI. 243 Recently Kamm [et al.]..separated from pituitary extracts what they believe to be nearly pure vasopressor and oxytocic hormones. **1964** *Brit. Jrnl. Oral Surg.* II. 131 The conditions under which these experiments have been carried out are..far removed from clinical situations since large amounts of vasopressor were used. **1966** Wright & Symmers *Systemic Path.* II. xxix. 1034/1 It has not yet been demonstrated that the vasopressor action of the hormone is of any importance in the normal human. **1975** *Amer. Heart Jrnl.* XC. 233/2 Vasopressors and inotropic drugs were discontinued and cardiopulmonary bypass support was ultimately discontinued. **1977** *Proc. R. Soc. Med.* LXX. 157/1 Vasopressor drugs were very seldom used and then only in an emergency.

vassalled, *ppl. a.* (In Dict. s.v. Vassal *v.*) (Later example.)
1933 W. Faulkner *Green Bough* 63 Thrall to the vassalled garrison that keep Thy soft unguarded breast's white citadel.

vassaldom. For *rare*⁻¹ read *rare* and add later example.
1965 J. A. Michener *Source* (1966) 224 In the end Moab had been reduced to a kind of vassaldom.

vastidity. Delete † *Obs.* and add later examples.
1929 R. Bridges *Testament of Beauty* iv. 190 The spiritual idea of Friendship, the huge vastidity of its essence. **1931** C. Williams *Many Dimensions* xvii. 260 In the foreground of that vastidity, he saw rising the Types of the Stone. **1962** W. Nowottny *Lang. Poets Use* iv. 83 The sense of vastidity of meaning in the sonnet derives from these many transformations of old into new.

‖ **vastrap** (fa·strap). Also † **vast-trap.** [Afrikaans, f. *vas* firm(ly), (Du. *vast*) + *trap* step, stand.] A quick South African folk dance; the music for this dance.
1913 *East London* (S. Afr.) *Dispatch* 3 Jan. 5 The *vast-trap* was performed by a number of nondescript characters who provided much amusement by their antics. **1926** O. Schreiner *From Man to Man* 360 Then she paused and began a reel. 'This is the "vastrap",' she said... 'It's what the Hottentots dance.' **1944** M. de B. Nesbitt *Road to Avalon* xiv. 114 The band is playing a lively 'vastrap'. **1957** *Cape Times* 17 Jan. 7/1 Rock and roll has affiliations with our own vastraps and tiekiedraais to which Coloured bands used to thrum the beat. **1970** L. G. Green *Giant in Hiding* xi. 105 You remember the candle light, the *vastrap* music.

vat, *sb.* Add: **5. vat colour** = *vat dye*; **vat dye, dyestuff,** a water-insoluble dye that is applied in a reducing bath that converts it to a soluble leuco-form with affinity for the fibre, the colour being obtained upon subsequent oxidation; so **vat-dyed** *a.*; **vat dyeing** *vbl. sb.*
1912 L. A. Olney in A. Rogers *Industr. Chem.* xxxviii. 768 The reduction vat colors have come into great prominence during recent years owing to their great resistance to practically all of the color destroying agencies. **1947** Kirk & Othmer *Encycl. Chem. Technol.* I. 968 With respect to all-around tinctorial and fastness properties, vat colors have no peer in any class of dyes. **1903** C. Salter tr. *G. von Georgievics's Chem. Dye-Stuffs* 4 Vat Dyes..have no affinity for textile fibres, and can only be fixed thereon by reduction or subsequent oxidation. **1981** H. Gutjahr in L. W. C. Miles *Textile Printing* v. 159 Vat dyes provide a wide range of colours of good all-round fastness properties, but great care, or specialized equipment, is required for their successful use. **1946** M. R. Fox *Vat Dyestuffs & Vat Dyeing* iv. 55 (*heading*) Fastness tests for vat-dyed wool and silk. **1960** *Farmer & Stockbreeder* 22 Mar. (Suppl.) 11/2 These vat-dyed, colour-fast, 48in wide cotton furnishing materials, are identically patterned on both sides. **1912** L. A. Olney in A. Rogers *Industr. Chem.* xxxviii. 756 The alkaline bath of indigo white is commonly called an indigo vat, and this process of coloring is usually spoken of as vat dyeing. **1946** M. R. Fox (*title*) Vat dyestuffs and vat dyeing. **1914** F. W. Atack tr. *Wahl's Manuf. Organic Dyestuffs* xix. 220 For a long time the vat-dyestuffs were limited to Indigo, its derivatives, and the Indophenols. **1973** *Materials & Technol.* VI. vii. 488 Nylon shows very little affinity for the vat dyestuffs.

VAT (vī,ē¹,tī·, væt), also **vat,** abbrev. of *value added tax* s.v. *value *sb.* 2 d. Cf. TVA s.v. *T 6 a.
1966 *Economist* 29 Oct. 432/2 This may be true of the conventional VAT. **1971** *Guardian* 29 Sept. 5/3 The substitution of VAT (value added tax) for purchase tax and SET in June 1973. **1978** *Hi-Fi News* Sept. 204 (Advt.), All prices include vat at 12½%.
Hence **VA·T-, va·table** *a. colloq.*, liable to VAT; **VA·Tman** *colloq.*, a Customs and Excise officer who deals with VAT.
1973 *Times* 31 Oct. 14/8 A glance at how Alice is faring will show how plays themselves are affected. At present she and her like are VAT-able and non VAT-able according to circumstances. **1976** *Daily Tel.* 4 Nov. 21 The last [*sc.* higher indirect taxes] would put up the cost of living but, to a large extent, the extra cost of vatable items..is voluntary. **1977** T. Heald *Just Desserts* ii. 30 Those absolute fiends from the Inland Revenue or..the dreaded VATmen. **1979** *Accountancy Age* 16 Feb. 7/1 (*heading*) Convincing fanciful VATmen you don't have the goods. **1984** *Listener* 11 Oct. 42/1 The income-tax inspector knows as little about it as the VAT-man.

vaterite (vā·tĕrəit, f-). *Min.* [ad. G. *vaterit* (W. Meigen 1911, in *Verhandl. d. Ges. deutsch. Naturforscher und Ärzte* LXXXII. II. i. 124), f. the name of H. A. *Vater* (1859–1930), German mineralogist and geologist: see -ite¹.] A relatively rare, metastable form of calcium carbonate, $CaCO_3$, that is polymorphous with calcite and aragonite and crystallizes in the hexagonal system.
1913 *Mineral. Mag.* XVI. 374 Vaterite... Vater's third modification of calcium carbonate prepared artificially in the form of minute spherules. **1962** *Acta Chirurg. Scand.* CXXIV. 324/1 X-ray crystallographic studies have shown that the three crystalline forms of calcium carbonate, namely vaterite, aragonite and calcite, can all occur in concretions from the pancreas, biliary- and urinary tract. **1975** *Science* 25 Apr. 363 The spicules of the common tropical ascidian, *Herdmania momus*, are mineralized with vaterite.

‖ **Vaterland** (fā·təɹlant). [Ger., = fatherland.] A German's fatherland.
1852 C. Lever *Daltons* I. xxii. 184 Hansel was the kind, quaint emblem of his own dreamy 'Vaterland'. **1894** G. du Maurier *Trilby* I. ii. 111 The buttercups and daisies of the Vaterland! **1950** *New Yorker* 26 Aug. 70/2 Germans whose every act and thought was directed..toward the enhancement of the *Vaterland*. **1977** N.

Freeling *Gadget* iv. 151 A stone slab had carved upon it the..names of the village boys dead for the Vaterland in '14–'18.

Vatican. Add: **2. a.** *Vatican Council*: now also, the Second Vatican Council (1962–5), also called Vatican II, which is noted for the introduction of the vernacular for the Mass and other reforms.
1961 *Times* 12 Dec. 6/3 On Christmas Day the Pope is to issue his Bull summoning the second Vatican Council. **1963** E. Schillebeeckx in *Life of Spirit* June 504 Is it possible at this stage to surmise what the outcome of Vatican II is going to be? Yes, in principle it is. The majority have expressed themselves in favour of a different approach. **1969** A. Richardson *Dict. Christian Theol.* 354/2 Vatican II..teaches that bishops possess, by virtue of their consecration, 'the fullness of the priesthood', including the threefold role of sanctifying, teaching and governing. **1976** *Jrnl. R. Soc. Arts* Mar. 164/1 The Second Vatican Council may well have an understanding of the Sack of Rome and the Council of Trent which was denied to writers in the earlier part of this century. **1979** D. Gageby in J. J. Lee *Ireland 1945–70* 132 John Horgan, now a member of the Dáil, writing about Vatican II and the ecclesiastical changes. **1983** *Times* 16 May 12/6 The Protestant Ethic..is definitely out of fashion, theologically. Today, they all quote Temple, or Vatican II.
b. Special Combs.: **Vatican City** (or **State**), the temporal state established by the Lateran Treaty of 1929, comprising the area immediately surrounding the Vatican Palace in Rome and headed by the Pope; **Vatican roulette** *colloq.*, the rhythm method of birth control, as permitted by the Roman Catholic Church.
1929 *Times* 8 Feb. 14/1 The new Papal State, which is to be known as the 'Vatican State' or as the 'Vatican City', will be confined to a small extent of territory in the neighbourhood of the Vatican. **1974** *Encycl. Brit. Macropædia* XIX. 36/1 The Vatican City is in all ways independent from Italy. **1962** *Western Folklore* Jan. 34 Vatican roulette—the rhythm system of birth control. **1965** D. Lodge *British Museum is falling Down* 172 That's another thing against the safe method there are so many things that can affect ovulation... No wonder they called it Vatican Roulette. **1977** H. G. Burger in B. Bernardi *Concept & Dynamics of Culture* 458 When Roman Catholicism permitted only the 'rhythm method' of birth control, sometimes satirized for its ineffectiveness as 'Vatican roulette', the Toronto archbishop authorized birth control pills.

Vaticanology (vætikănǫ·lŏdʒi). *colloq.* [f. Vatican + -ology.] The study of the history, or analysis of the policies, of the Vatican. So **Va:ticano·logist.**
1975 *Guardian Weekly* 14 Dec. 27 His publishers describe him as 'a leading Vaticanologist', and he does appear to know more than most people about the Curia. **1976** *Times Lit. Suppl.* 13 Aug. 1018/4 It is..several generations since Chesterton's sallies delighted the devout, and Vaticanology has few students in non-Catholic countries. **1979** M. Craig *Man from Far Country* xv. 173 Father F.X. Murphy, a renowned Vaticanologist. **1982** *Times Lit. Suppl.* 12 Feb. 154/1 It is high time that Vaticanology was recognized as a serious field of historical inquiry. **1984** D. Yallop *In God's Name* 62 If the 111 cardinals were as perplexed as the Vaticanologists then the Church was in for a long, confusing Conclave.

vaticinatory (vătisinē¹·tŏri), *a.* Delete *rare*⁻¹ and add further examples.
1895 *Starting Price* 23 Mar. 3/3 To test the vaticinatory skill of professional and amateur prophets, we offer a prize of £10 for a tipping contest. **1930** Wyndham Lewis *Apes of God* xii. vii. 483 There was, as well as a great vaticinatory verve, a certain boastfulness about this Fascist. **1980** V. Cunningham *Penguin Bk. Spanish Civil War Verse* 38 Cornford is blamed for not being vaticinatory enough about these 'political surrealists' who couldn't wait for the revolution.

Vatinian, *a.* For † *Obs.* read 'Now *rare*' and add later example.
1927 T. Wilder *Bridge of San Luis Rey* ii. 62 The Archbishop of Lima..hated her, with what he called a Vatinian hate and counted the cessation of her visits among the compensations for dying.

vatje (fa·kʸi). *S. Afr.* Also **fadje, fagie, fatje, fikey, vaatjie, vaitje.** [ad. Afrikaans *vaatjie*, dim. of *vat* water-carrier (see Fat *sb.*¹).] A small cask or barrel for carrying water or wine; a soldier's canteen.
1850 R. G. Cumming *Five Years Hunter's Life S. Afr.* i. 7/1 The..general stores which I carried with me were as follows:..2 large 'fagie' or water-casks, [etc.]. **1871** J. McKay *Reminisc. Last Kafir War* ii. 8 The soldier acts as the beast of burden, having been supplied with a large wooden vessel, by soldiers called 'fadje', or keg, capable of holding about half a gallon; and in this he had to carry with him what water he thought necessary. **1871** J. Mackenzie *Ten Yrs. North of Orange River* vii. 115 Khosimore jealously guarded his 'vatjes', or water-vessels. **1891** E. Glanville *Fossicker* xix. 166 One of the three rose up..took a final pull at the water 'fikey', and stretched himself on the bare ground. **1909** *Chambers's Jrnl.* Dec. 28/1 Some Congo brandy and a *vatje* of water. **1951** L. G. Green *Fully Many a Glorious Morning* xviii. 234 We gave up part of a *vaatjie* of red wine and soaked it and then roasted it in front of a huge fire. **1970** in *Voorloper* (1976) 833 The coloured people buy their wine in a vaatjie for celebrations.

VATman: see *VAT.

vauclusian (vǫklū·ziăn), *a. Physical Geogr.* Also **Vauclusian.** [f. the name of the Fontaine de *Vaucluse* in S. France: see -ian.] Applied to a type of spring, often large, occurring in karstic regions, in which the water is forced out under artesian pressure.
1937 Wooldridge & Morgan *Physical Basis Geogr.* xix. 294 Cheddar Gorge..and the spectacular Wookey Hole with its 'Vauclusian spring', show true karstic features. **1977** A. Hallam *Planet Earth* 82/3 There are two main types of karst spring, one where the water issues by means of free flow, the other where the water issues under forced or artesian flow: the latter type is sometimes known as a vauclusian spring. **1980** J. C. Schmid tr. *A. Bögli's Karst Hydrol.* ix. 124 In a genuine vauclusian spring the water course runs upwards through the rock.

vaude (vǫd, vōᵘd). *N. Amer. colloq.* Also **vaud.** [Abbrev. of Vaudeville.] Vaudeville (sense 2); a vaudeville theatre. Also *attrib.*
1933 *Ladies' Home Jrnl.* Mar. 21/3 He was wearing a cap with purple and green in it and the loudest checked suit I have ever seen on anybody outside of a vaud. **1951** Green & Laurie (*title*) Show biz, from vaude to video. **1975** R. Davies *World of Wonders* (1977) i. viii. 105 A rabble of acts..which played for rotten pay in the worst vaude houses.

vaudeville. Delete ‖ and add: Now freq., esp. *U.S.*, with pronunc. (vǫ·dvil, vǫ·dəvil). **2.** (Earlier and further examples.) Now in frequent use in the U.S. to designate variety theatre (Variety 9 b) or music hall.
1827 T. Dibdin *Reminiscences* I. xii. 268, I also had the honour..of being selected by her Royal Highness the Princess Elizabeth to write a sort of *vaudeville* farce. *c* **1831** J. H. Reynolds in J. R. Planché *Recollections & Reflections* (1872) I. xii. 181 And lure for you the light Vaudeville from France. **1911** G. B. Shaw in *Daily Graphic* 2 Dec. 4/3 There are vaudeville theatres in America and variety theatres in England. **1917** *Lit. Digest* 25 Aug. 28/2 The phrase 'Jaz her up' is a common one to-day in vaudeville and on the circus lot. When a vaudeville act needs ginger the cry from the advisers in the wings is 'put in jaz'. **1940** R. Chandler *Farewell, my Lovely* v. 30 You would find them in tank town vaudeville acts..in the cheap burlesque houses. **1967** *Stage* 2 Mar. 4/1 The merging of vaudeville and tombola as a major February attraction at Blackpool last week succeeded. **1976** *New Yorker* 8 Mar. 57/1 The play is a lewd, tragic vaudeville about the life of a bankrupt pursued by creditors. **1976** J. Crosby *Nightfall* xxxiii. 197 You're a fifth-rate vaudeville actor. **1982** *Verbatim* Autumn 23/2 The subject is given light-hearted treatment, as though this jargon were little more than food for variety turns (i.e., vaudeville acts).

vaudevillain (vōᵘ·d(ə)vilăn, vǫ·d-). *U.S. colloq.* ? *Obs.* [f. Vaudeville, after Villain *sb.*] = *vaudevillian *sb.*
1909 *Sat. Even. Post* 5 June 17/2 One thrifty vaudevillain made sixty-five weeks one season in this way. **1916** J. K. Bangs *From Pillar to Post* xvi. 326 The major was a great believer in the value of Author's Readings by what he used to call 'running mates'—teams, as the vaudevillains have it.

vaudevillian (vǫd-, vōᵘd(ə)vi·liăn), *sb.* and *a.* orig. and chiefly *U.S.* Also **-ean.** [f. Vaudeville + -ian.] **A.** *sb.* A performer in vaudeville. **B.** *adj.* Of or pertaining to vaudeville.
1913 *Technical World Mag.* Mar. 19 'That's great!' cried one of the 'vaudevillians', clapping his hands appreciatively. **1924** *Sat. Even. Post* 4 Oct. 70/2 Give them to some small-time vaudevillian to repeat. **1930** *Punch* 8 Oct. 415 The modish songs..serve pleasantly to exercise the vaudevillean gifts of Mr. Coward and Miss Lawrence. **1961** *John o' London's* 5 Oct. 373/1 The elder Glasses were vaudevillians (music-hall artists). **1962** J. D. Salinger *Franny & Zooey* 120 The children's father, a former international vaudevillian. **1975** *Times* 2 Aug. 12/1 (*heading*) Behind the vaudevillian smoke-screen. **1980** *Redbook* Oct. 213/3 Later, years later, he was to win an Academy award playing an old vaudevillian in *The Sunshine Boys*. **1985** *Listener* 23 May 37/1, I don't just mean old vaudevillians like George Burns.

vaudevillist. (Earlier example.)
1839 [see *stenographist].

vaudoux. Add: Also **vaudo.**
1862 J. Le Grand *Jrnl.* 20 Dec. (1911) 57 Heard today of the existence of a negro society here called the 'vaudo' (I believe).

vault, *v.*² Add: **3.** *trans.* To cause to rise *to* or *into* a considerably higher position or situation.
1976 *National Observer* (U.S.) 31 July 1/2 Nadia Comaneci's electrifying gymnastics performances vaulted her from obscurity to world-wide renown. **1977** *Detroit Free Press* 11 Dec. 2-D/2 Severiano Ballesteros of Spain shot a three-under-par 69 Saturday and vaulted his team into a three-stroke lead over Canada after 54 holes of the 25th World Cup Golf Championship Saturday.

vauntingness. For *Obs.*⁻⁰ read *rare* and add example.
1955 E. Bowen *World of Love* v. 83 If she chose to make history out of her vicissitudes, that was really from vauntingness—nothing beat her.

vauntless (vǭ·ntlės), *a. nonce-wd.* [f. Vaunt *sb.* + -less.] Not bragging or boasting.
c **1879** G. M. Hopkins *Poems* (1967) 82 Tongue true, vaunt- and tauntless.

Vauxhall. Add: **b. Vauxhall lamp, light,** an ornamental glass lantern designed to hold a candle and used for outdoor illumination.
1907 *Yesterday's Shopping* (1969) 258/2 Garden or Vauxhall lamps. For illuminations, &c. Diamond moulding, size 2¼ in. by 3¼ in... Wired ready for hanging. **1974** *Country Life* 17 Oct. 1075/3 A pleasure ground..at Tollard Royal..in late Victorian and Edwardian times... The Gardens were illuminated with thousands of Vauxhall lights.
2. Used *attrib.* and *absol.* to designate antique plate glass resembling that made at the Vauxhall Glassworks from *c* 1663 to the end of the 18th cent.
1830 W. Dawson in W. H. Bowles *Hist. Vauxhall & Ratcliff Glass Houses* (1926) 58 Edward Dawson of the Glass House Vauxhall who died Jan. 12 1755 made some great improvements which enabled John Dawson his son to produce those brilliant glasses still distinguished as 'Vauxhall plates'. **1900** *Archit. Rev.* June p. xxii, The mirror has a flat bevel, after the manner of the old Vauxhall plates. **1926** W. H. Bowles *Hist. Vauxhall & Ratcliff Glass Houses* iii. 19 To this day all mirrors made in this country between 1670 and 1750 are styled by dealers 'Vauxhall'. **1972** *Country Life* 27 Jan. (Suppl.) 24/1 (Advt.), Queen Anne walnut bureau-bookcase..retaining original Vauxhall mirror door. **1975** *Oxf. Compan. Decorative Arts* 571/1 In 1663 the second Duke of Buckingham established a plate-glass manufactory at Vauxhall, of which the products have become legendary. The term 'Vauxhall glass' has for long been used to describe any old-looking mirror although no existing example can be traced with certainty to that source. **1977** *Times* 20 Aug. 8/5 English merchants began to export Vauxhall glass..to China.

vauxite (vǭ·ksəit). *Min.* [f. the name of George *Vaux* (1863–1927), U.S. mineral collector + -ite¹.] A secondary mineral that is a hydrated basic phosphate of aluminium and ferrous iron, $Fe^{2+}{}_2Al_2(PO_4)_2(OH)_2.6H_2O$, and occurs as blue, transparent triclinic crystals, usu. in association with paravauxite and wavellite.
1922, 1944 [see *paravauxite* s.v. *para-¹ 2 c]. **1974** *Amer. Mineralogist* LIX. 843 Montgomeryite, $Ca_4Mg(H_2O)_{12}[Al_4(OH)_4(PO_4)_6]$,..possesses chains of corner-linked Al–O octahedra which are topologically and geometrically equivalent to the chains in vauxite.

väyrynenite (vē,irī·nĕnəit). *Min.* [ad. G. *väyrynenit* (Volborth & Stradner 1954, in *Anz. Math.-Naturwiss. Kl. Österr. Akad. Wissensch.* XCI. 21), f. the name of H. *Väyrynen*, 20th-c. Finnish mineralogist: see -ite¹.] A phosphate and fluoride of beryllium and manganese, $BeMnPO_4(OH,F)$, found as red transparent monoclinic crystals.
1954 *Chem. Abstr.* XLVIII. 4380 Optical data of väyrynenite are presented. **1977** *Mineral. Abstr.* XXVIII. 207/2 Crystals acquired in a bazaar in Chitral, Pakistan, have been identified as väyrynenite.

VE (vī,ī·). [f. the initial letters of *V*ictory in *E*urope.] Used *attrib.* and *absol.* to denote the victory of the Allied forces over those of Germany during the second World War; esp. as *VE-day*, designating the date of Germany's surrender, 8 May 1945.
1944 *Washington Post* 10 Sept. 3/1 James F. Byrne, director of War Mobilization, found a new designation for the two victory days... Last night he referred to the date of Germany's impending surrender as V-E (victory in Europe) Day, and the day of Japan's defeat as V-J (Victory over Japan) Day. **1945** *Daily Mirror* 8 May 1/2 Today is VE-Day—the day for which the British people have fought and endured five years, eight months and four days of war. **1956** A. H. Compton *Atomic Quest* iv. 273 At Yalta,..President Roosevelt endeavored..to obtain Stalin's commitment to enter the war against Japan. Eventually Stalin agreed to do so within three months after V-E day. **1974** P. Lively *House in Norham Gardens* x. 130 Mrs Hedges..talked about..D-day and VE night. **1977** *Wandsworth Borough News* 16 Sept. 12/4 The jubilee street parties are reminiscent of those marvellous V.E. and V.J. celebrations, when pianos were pulled on to the streets and everyone—strangers or not—were welcomed. **1978** R. V. Jones *Most Secret War* xxxvi. 312 Some months after V-E Day he was arrested in the American Zone of Germany.

veal, *sb.*¹ Add: **3. a.** *veal and ham pie* (earlier examples), *veal cutlet* (earlier examples).
1728 E. Smith *Compl. Housewife* (ed. 2) 41 (*heading*) To make veal cutlets. **1811** Jane Austen *Sense & Sens.* II. iv. 53 Preferring salmon to cod, or boiled fowls to veal cutlets. **1848** Thackeray *Pendennis* (1849) I. v. 49 That girl, sir, makes the best veal and ham pie in England. **1861** Mrs. Beeton *Bk. Househ. Managem.* 942 *Saturday.* 1. Rump-steaks, broiled, and oyster sauce, mashed potatoes; veal-and-ham pie.
b. veal calf, (*a*) (later examples); **veal**

parmigiana [It. *parmigiano* Parmesan cheese], a dish of small escalopes of veal and cheese; **veal piccata**, a dish of small escalopes of veal.

vealer. (In Dict. s.v. VEAL *sb.*¹) Delete *U.S.* and add examples.

veatchite (vī·tʃait). *Min.* [See quot. 1938 and -ITE¹.] Any of three polymorphs of a hydrous strontium borate, $Sr_2[B_5O_8(OH)]_2 \cdot B(OH)_3 \cdot H_2O$, of which two (*veatchite* and *p-veatchite*) are monoclinic and one (*veatchite-A*) is triclinic.

... the Permian lower evaporite bed of the Eskdale No. 2 boring in Yorkshire... Single-crystal X-ray work now proves that this mineral is *p*-veatchite. **1979** *Amer. Mineralogist* LXIV. 362 A third modification of veatchite occurs in Emet colemanite deposit, Kütahya, Turkey, as white cauliflower-shaped nodules associated with colemanite, hydroboracite, realgar, orpiment and montmorillonite. The new mineral is triclinic... Crystals are transparent, colorless... The new modification is named veatchite-A.

veber, obs. var. WEBER¹ in Dict. and Suppl.

Veblenian (vĕblī·niăn), *a.* and *sb.* [f. the name *Veblen* + -IAN.] **A.** *adj.* Of or pertaining to the work of Thorstein Veblen (1857–1929), U.S. economist and social scientist, esp. the ideas (as of conspicuous consumption) expounded in his book *Theory of the Leisure Class* (1899). **B.** *sb.* One who supports or advocates Veblenian ideas.

1931 *Encycl. Social Sci.* V. 388/2 The Veblenian terminology of American institutionalism. **1953** D. RIESMAN *Thorstein Veblen* viii. 180 Those high-income people..who buy..with a Veblenian ascetic eye on their own motivations. **1968** *Internat. Encycl. Social Sci.* IV. 462/2 Economists working in the Veblenian tradition. **1973** *Hist. Political Econ.* V. 449 (*heading*) Hoxie's economics in retrospect: the making and unmaking of a Veblenian. **1982** *Jrnl. Econ. Issues* XVI. 757 The Veblenian dichotomy is the central analytical tool of institutional economists in the Veblen–Ayres tradition. **1982** J. D. MACDONALD *Cinnamon Skin* xix. 210 Veblen died in 1929 at the age of seventy-two.. I have never been a Veblenian myself.

vecchio (ve·ki͵o). For *Obs.* read *rare* and add later examples.

1938 E. HEMINGWAY *Fifth Column* 273 'Three marsalas,' said the young gentleman to the girl behind the pastry counter. 'Two, you mean?' she asked. 'No,' he said, 'One for a vecchio.' **1944** G. B. SHAW *Let.* 4 Dec. in *To a Young Actress* (1960) 180, I am a vecchio, nearly eightyeight and a half.

Vectian (ve·ktiăn), *a. Geol.* [f. L. *Vect-is*, name of the Isle of Wight + -IAN.] Of or pertaining to the Isle of Wight or the Lower Greensand strata exposed there; *spec.* (see quot. 1961).

[**1845** W. H. FITTON in *Q. Jrnl. Geol. Soc.* I. 189 If hereafter a change [from Lower Green Sand] be thought desirable, he [*sc.* the author] conceives that the new denomination should be taken from the *Isle of Wight*..; and if such a case should arise, he suggests the name of *Vectine* for the strata now called Lower Green Sand, from the ancient name of that island,—*Insula Vectis* of the Romans.] **1885** A. J. JUKES-BROWNE in *Geol. Mag.* Decade III. II. 298 The Lower Greensand can be studied so well in the Isle of Wight..that no name can be more appropriate than *Vectian*; and I regard the introduction of a new name as preferable to the adoption of the French names *Aptien* and *Urgonien*. **1922** [see *SELBORNIAN *a.* 2]. **1961** *Palaeontology* III. 502 The Vectian Province comprises the Isle of Wight and a small part of the Dorset mainland where a strip of Lower Greensand extends westwards from Swanage to Lulworth Cove. **1969** BENNISON & WRIGHT *Geol. Hist. Brit. Isles* xiv. 323 In this Vectian Province the Lower Greensand becomes attenuated westwards at a more rapid rate than the Wealden (Neocomian) Beds.

vector, *sb.* Add: **2. a.** (Earlier and further examples.) *axial vector* = *PSEUDOVECTOR *sb.*; *polar vector*, a vector which changes sign when the signs of all its components are changed.

1846 W. R. HAMILTON in *Phil. Mag.* XXIX. 27 The algebraically imaginary part, being geometrically constructed by a straight line, or radius vector, which has, in general, for each determined quaternion, a determined length and determined direction in space, may be called the vector part, or simply the vector of the quaternion. **1873** J. C. MAXWELL *Treat. Electr. & Magn.* I. 9 A Vector, or Directed Quantity, requires for its definition three numerical specifications, and these may most simply be understood as having reference to the directions of the coordinate axes. **1903** *Nature* 22 Oct. 610/1 ... does not discriminate between 'polar' vectors, e.g. forces and 'axial' vectors, e.g. couples. **1968** M. S. LIVINGSTON *Particle Physics* v. 101 Angular momentum is an axial vector quantity, unlike linear momentum, which is a polar vector.

b. An ordered set of two or more numbers (interpretable as the co-ordinates of a point); a matrix with one row or one column; also, any element of a vector space.

[**1873**: see prec. sense. **1881** J. W. GIBBS *Sci. Papers* (1906) II. 17 The numerical description of a vector requires three numbers.] **1922** J. B. SHAW *Vector Calculus* i. 6 A vector is usually designated by a triple as (*x*, *y*, *z*), and usually such triple is called a vector. **1938** R. A. FRAZER et al. *Elem. Matrices* i. 2 A row of numbers is called..a vector of the first kind..; ... matrix is referred to as a vector of the second kind. **1940** D. E. LITTLEWOOD *Theory Group Characters* ... matrix..of order n^2 may be regarded as ... column vectors. **1961** *Communications Assoc. Computing Machinery* IV. 424/2 The analysis of a program into phases and sequences is accomplished by associating with each sequence a vector of ones and zeros. **1965** PATTERSON & RUTHERFORD *Elem. Abstr. Algebra* v. 155 The polynomial ring *P*(*F*) is a vector space over the field *F*... The polynomials are in this case the vectors of this vector space. **1976** *Biometrika* LXIII. 438 ... vector ($n_1, \ldots, n_{k0}$) will have a multinomial distribution with N_0 trials. **1981** N. RAU *Matrices & Math. Programming* i. 16 Up to this point, vectors have been considered simply as a special case of matrices... In the rest of this book..'vector' will always be taken to mean column vector.

c. *Aeronaut.* A course to be taken by an aircraft, or steered by a pilot.

1941 D. MASTERS *So Few* xxx. 333 'I've got to get a Hun tonight. I'll give you a bottle of champagne if you put me on to one.'..'All right, ..I'll give you a vector.' **1951** O. BERTHOUD tr. *Clostermann's Big Show* ... climbing flat out on vector 095. **1978** R. V. JONES *Most Secret War* xxi. 177 The ground station ordered the bomber to steer a course of 270° (i.e. due west) for 10 miles because it was east of the beam, and this was the vector required to bring it to the right point to start its bombing run.

d. *Computers.* A sequence of consecutive locations in memory; a series of items occupying such a sequence and identified within it by means of one subscript; *spec.* one serving as the address to which a program must jump when interrupted, and supplied by the source of the interruption.

1961 *Communications Assoc. Computing Machinery* IV. 61/1 Since it is often necessary to refer to memory addresses and the contents of memory cells in this discussion, the almost-legitimate device of the 'Memory vector' will be used. This is done by assuming that the memory of the machine in question to constitute a one-dimensional vector, named 'Memory'. **1962** E. W. DIJKSTRA *Primer of ALGOL 60 Programming* 37 The simplest example of such an array is a vector, i.e. a sequence of subscripted variables. **1967** D. G. HAYS *Introd. Computational Linguistics* ii. 26 When the program stops, the J-th column of the matrix B has been copied into the vector G. **1967** P. A. STARK *Digital Computer Programming* xiv. 255 After checking the transfer vector to see that the subroutine name is there, the loader goes back into the mainline program and fixes the linkage so that the mainline program will use the transfer vector. **1975** R. H. ECKHOUSE *Minicomputer Systems* vi. 186 The new contents of the PC [*sc.* program counter] and the PS [*sc.* processor status] are loaded from two preassigned consecutive memory locations called an interrupt vector... The contents of these vectors are determined by the programmer. **1982** *Economist* 3 Apr. 128/3 Individual units of data (ie, binary numbers) are stored in the computer's memory in long lists called vectors. **1982** R. A. SPARKES *Microcomputers & Interfacing* v. 199 A test must be included into the routine to discover if the user wants to return to normal working, in which case ISR [*sc.* interrupt service routine] vector is then returned to 58926 and the extra routine is by-passed. **1984** *Personal Software* Winter 89/3 During loading a message is displayed on the screen and the keyboard and screen vectors are changed to effect automatic program execution.

3. a. *Med.* and *Biol.* A person, animal, or plant which carries a pathogenic agent and acts as a potential source of infection for members of another species. Also *transf.* Cf. *CARRIER 1 l (i).

1922 [see *ARTHROPODA, ARTHROPOD]. **1944** *Nature* 5 Aug. 167/2 Regular transmission of viruses by the egg of the host plant (the insect vector does not become ..., unless it suffers) probably does not occur. **1963** R. CARSON *Silent Spring* xvi. 220 An even more serious problem concerns the vector of yellow fever. **1972** *Lancet* 17 June 1338/2 The stethoscope is yet another vector of pathogenic organisms. **1974** PASSMORE & ROBSON *Compan. Med. Stud.* III. xii. 64/2 The spread of each of the insect-borne diseases depends on a complicated chain of events involving..a reservoir of infection which may be either in man or in other animals,..the insect vector and..a susceptible human population. **1976** *Dumfries & Galloway Standard* 25 Dec. 8/6 A small boy recovered after being bitten by a rabid bat, it being thought that the disease had been modified by passage through this unusual vector.

b. *Genetics.* A bacteriophage which transfers genetic material from one bacterium to another; also, a phage or plasmid used to transfer extraneous DNA into a cell.

1958 *Abstr. 7th Internat. Congr. Microbiol.* 53 The modified phage so produced has been identified as the transducing vector. **1968** W. HAYES *Genetics of Bacteria & their Viruses* (ed. 2) xvii. 478 Some other temperate phages..can act as vectors for the transfer, to recipient bacteria, of virtually any region of the host chromosome. **1976** *Proc. Nat. Acad. Sci.* LXXIII. 3838/1 Gene transfer between two closely related mouse cell lines has been carried out, using as the vector a cell-free preparation of metaphase chromosomes and nuclei. **1982** T. M. DEVLIN *Textbk. Biochem.* xx. 986/2 The methodology involves obtaining the DNA of the desired gene; placing the DNA into a vector or vehicle capable of transporting the gene and maintaining it inside an E. coli cell; and determining whether the gene is functional in *E. coli*. **1983** *Sci. Amer.* Jan. 51/2 Plasmids are routinely used as vectors for introducing foreign DNA into bacteria. **1985** OLD & PRIMROSE *Princ. Gene Manipulation* (ed. 3) 222 This may have the additional property of being a shuttle vector, capable of stable replication in *E. coli* and *A. tumefaciens*.

4. *transf.* and *fig.*

1926 *Spectator* 23 Oct. 735/1 Even if there was a listener-in within the narrow vector of the vibrations, he could not hope to receive the messages at the rate at which they will be sent. **1954** W. FAULKNER *Fable* 82 He identified himself, naming his battalion and its sector. **1957** L. DURRELL *Bitter Lemons* 37 But I was on a different vector, hunting for other qualities which might make residence tolerable, or might isolate me from my fellows. **1970** *Listener* 13 Apr. 466/1 Once, a long vector of geese flew over. **1977** A. HECHT in *Oil. Bk. Contemp. Verse* (1980) 173 The athlete's dancing vector, the spirit's need, And muscle's cleanly diction. **1979** *OCP Studies in English* (Univ. Cape Town) Sept. 19 Antithesis is, of course, the chief trope of the rape genre, not only in respect of the obvious contrasts drawn between the heroine's purity and her shame, but also with regard to the vectors of these states—contrastive images of light and dark.

5. a. In the sense 'of the nature of a (mathematical) vector, representable by a vector'; also as *adj.*

1846 [see sense 2 a above]. **1873** J. C. MAXWELL *Treat. Electricity & Magn.* I. 9 A vector quantity has direction as well as magnitude, and is such that a reversal of its direction reverses its sign. **1880** [in Dict., sense 2]. **1881** J. W. GIBBS *Sci. Papers* (1906) II. 17 The vector operator called ∇ ... the concentration of *u*, whether *u* is scalar or vector. **1962** CORSON & LORRAIN *Introd. Electromagnetic Fields & Waves* i. 1 Wind velocity, gravitational force, and electric field intensity are examples of vector quantities. **1975** *Nature* 18 Sept. 191/1 β-decay within the quarks... **1982** ... there is no nuclear spin to flip.

b. In the sense 'involving (mathematical) vectors', as *vector addition, algebra, analysis, calculus, method*, etc.

1873 KELLAND & TAIT *Introd. Quaternions* iii. 32 (*heading*) Vector multiplication and division. **1881** A. S. HARDY *Elem. Quaternions* i. 3 The operation of vector addition is commutative. **1881** J. W. GIBBS *Sci. Papers* (1906) II. 17 An algebra or analytical method in which a single letter or other expression is used to specify a vector may be called a vector algebra or vector analysis. **1897** [in Dict., sense 2]. **1904** *Rep. Brit. Assoc. Adv. Sci. 1903* 53, I cannot help thinking that he would have used vector methods throughout if he had found ready to hand a vector algebra in harmony with the rest of his notation. **1953** A. HUXLEY *Genius & Goddess* 28 The great man would get bored and quietly fade away, leaving me to solve Timmy's problem by some method a little simpler than vector analysis. **1964** J. W. LINNETT *Electronic Struct. Molecules* i. 7 The resultant angular momentum is, therefore, by vector addition. **1970** *Nature* ... **1968** E. T. COPSON *Metric Spaces* ix. 138 This is ordinary vector algebra without scalar and vector products. **1969** *Jane's Freight Containers 1968–69* 147/2 It has been assumed that the vector sum (the actual value of the lashing tension) will not exceed 90 400 kgf. **1972** M. KLINE *Math. Thought* xxxii. 786 By Maxwell's time a great deal of vector analysis was created by treating the scalar and vector components of the quaternion separately. **1973** H. M. SCHEY *Div, Grad, Curl, & All That* i. 1 Much of vector calculus was invented for use in electromagnetic theory and is ideally suited to it. **1978** *Sci. Amer.* Jan. 117/1 For any operation that can be applied to a single operand (such as the extraction of the square root) there is a corresponding vector operation that consists of applying the same operation to every element in a vector.

c. *Particle Physics.* Used *attrib.* to designate particles with a spin of 1; *vector boson*, esp. any of a group of three heavy bosons (the W⁺ and Z⁰, qq.v. in Suppl.) thought to exist as mediators of the weak interaction. [See quot. 1976.]

1942 *Physical Rev.* LXII. 403 The β-matrices in the vector meson theory can be reduced to the simpler ζ-matrices and Σ matrices (spin matrices). **1949** *Ibid.* LXXVI. 784/1 The exchange of two charged vector mesons. **1959** *Bull. de l'Acad. Polonaise des Sci.: Série des Sci. Math.* VII. 729 Recently a hypothesis of a charged vector boson, which would mediate the weak interactions, was much discussed. **1968** M. S. LIVINGSTON *Particle Physics* xii. 218 The agency of this weak force is presumed to be a vector boson. **1975** *Nature* 3 Apr. 387/2 In a simple SU(4) scheme there is an obvious place for one φ particle, in the same multiplet as the well established vector mesons ρ, ω and φ. **1976** *Sci. Amer.* Jan. 46/1 They are called vector bosons because the quantum-mechanical equation that describes particles with a spin of 1 takes the form of a four-dimensional vector. **1978** *Nature* 12 Oct. 483/1 These fermions interact through vector particles: the photon, charged and neutral massive vector bosons, and gluons which are responsible for the electromagnetic, weak and strong interactions, respectively. **1983** *Ibid.* 27 Jan. 285 Physicists in Geneva have discovered the intermediate vector boson.

d. Special Combs.: **vector address** *Computers*, an address specified by an interrupt vector (see sense 2 d above); **vector-borne** *a.*, (of a disease or pathogen) transmitted or carried by a vector (sense 3 a above); **vector field**, a field defined at each point by a vector quantity; a map from a space to a space of two or more dimensions; **vector function**, a function whose value is a vector quantity; **vector potential**, a potential function that is a vector function (see *POTENTIAL *sb.* 4 a); **vector product**, a vector function of two vectors, (a_1, a_2, a_3) of length *a* and (b_1, b_2, b_3) of length *b*, equal to ($a_2b_3 - a_3b_2$, $a_3b_1 - a_1b_3$, $a_1b_2 - a_2b_1$), representing a vector perpendicular to them both and of magnitude *ab* sin θ (where θ is the angle between them); **vector space**, a group whose elements can be combined with each other and with the elements of a scalar field in the way that vectors can, addition within the group being commutative and associative and multiplication by a scalar being distributive and associative; **vector triple product**, a vector function of three vectors equal to the vector product of one of them with the vector product of the other two, i.e. **a** × (**b** × **c**).

1975 R. H. ECKHOUSE *Minicomputer Systems* vi. 189 The various vector addresses and priority levels for the teletype, high-speed reader/punch, and clock on the PDP-11 are as follows. **1984** R. A. SPARKES *Microcomputers in Sci. Teaching* v. 199 All we have to do is to change the vector address and the ISR [*sc.* interrupt service routine] will start by executing our routine instead. **1956** *Nature* 25 Feb. 367/1 With its further expansion..the designation of the East African Malaria Unit has been changed to the East African Institute of Malaria and Vector-Borne Diseases. **1969** *Lancet* 12 Jan. 109/2 This tumour..might be due to a vector-borne virus. **1971** P. C. C. GARNHAM *Progress in Parasitol.* i. 1 In order to prevent this wide a diffusion of parasitology it is useful to impose some sort of restriction, and this perhaps can best be done by adding to the classical subjects of protozoology and helminthology, only vector-borne infections of other types. **1922** J. B. SHAW *Vector Calculus* iii. 26 A vector field is a system of vectors each associated with a point of space, or of a surface, or a point of a line or curve. **1932** [see *scalar field* s.v. *SCALAR *sb.* 2]. **1976** *Physics Bull.* Sept. 387/1 The vector fields that are particularly relevant to cosmology are those representing the motion of particles and electric charges. **1873** J. C. MAXWELL *Treat. Electr. & Magn.* I. 10 Quantities of this class require three numerical specifications. They are expressed in the language of Quaternions by linear and vector functions of a vector. **1971** [see *POTENTIAL *sb.* 4 a]. **1972** M. KLINE *Math. Thought* xxxii. 786 Maxwell noted..that the curl of a gradient of a scalar function and the divergence of the curl of a vector function are always zero. **1873**, etc. [in Dict.]. **1962** CORSON & LORRAIN *Introd. Electromagn. Fields* v. 186 We shall now show that the magnetic induction **B** is related to a certain quantity **A** through the equation **B** = ∇ × **A**, where the vector **A** is called ... the vector potential. **1962** Vector product [in Dict.]. **1901** [see *scalar product* s.v. *SCALAR *sb.* 2]. **1965** [see *outer product* s.v. *OUTER *a.* 3]. **1972** A. G. HOWSON *Handbk. Terms Algebra & Anal.* xxxiv. 269 Given two vectors, **a**, **b**, we define their vector product (cross product or outer product) denoted by **a** ∧ **b** (or **a** × **b**). **1937** A. A. ALBERT *Mod. Higher Algebra* 319/2 (Index), Vector space; see Linear vector space. **1965** [see *vector field* above]. **1970** *Nature* 19 Dec. 1234/2 A vector space is built up linearly by means of 'scalar' multipliers from a number field. **1972** A. G. HOWSON *Handbk. Terms Algebra & Anal.* vii. 30 Homomorphisms of vector spaces..preserve linear combinations of the type $\lambda_1a_1 + \lambda_2a_2 + \cdots + \lambda_na_n$. **1901** GIBBS & WILSON *Vector Analysis* ii. 72 The vector triple product **a** × (**b** × **c**) is ... vector product must be used to express the component of a vector **B** which is perpendicular to a given vector **A**. **1964** Vector triple product [see *scalar triple product* s.v. *SCALAR *sb.* 2].

vector (ve·ktəɹ), *v.* [f. the sb.] *trans.* **a.** To direct (an aircraft) on its course or towards a target.

1945 *Radar* 34 If the pilot had to ditch, the radar set spotted where he went down, vectoring out to that exact spot the air-sea rescue planes. **1958** *Daily Mail* 24 Oct. 9/2 If the pilot of a military aircraft wants to cross one of the ten-mile-wide airways linking the big airports..he can be 'vectored' across by R.A.F. radar. **1976** B. JACKSON *Flameout* i. 20 He'd been first officer of a DC-8 vectored over Newark when its Constellation had collided with the jet.

b. *gen.* To direct, esp. towards a destination; to change the direction of.

1966 *New Scientist* 27 Jan. 213/3 A flexible cup to hold a rocket's nozzle and so allow it to be vectored or swung for steering purposes. **1978** K. AMIS *Jake's Thing* ii. 20 This time Brenda's tone was warm but the warmth was firmly vectored on her friend. **1979** KRAFT & TOY *Mini/Microcomputer Hardware Design* viii. 291 The address found..vectors the processor to the appropriate device service routine. **1985** *Your Computer* Aug. 21/2 Version 1.2 contains several new features, such as the ability to vector output to one of a number of outputs.

Hence **ve·ctored** *ppl. a.*; *vectored thrust*, thrust that can be varied in direction.

1960 *Aeroplane* XCVIII. 261/1 The Ryan VZ-3RY Vertiplane 'vectored-slipstream' VTOL research aircraft was destroyed early this month during a test flight. **1962** *Flight International* LXXXI. 234/1 The future will undoubtedly see larger, heavier and more complex VTOL aircraft using the principle of 'vectored thrust'. **1973** *Black World* May 6/2 Our scholars and leaders and common people must have vectored minds if we are to prevail. This means we must actively seek stabilizing forces. **1982** *Daily Tel.* 15 June 3/1 He flew at 90 feet, the height at which Harrier had found it necessary to use the special jump-jet tactic of using vectored thrust..to 'stop' in the way of an attacking aircraft.

vectorca·rdiogram. *Med.* [f. VECTOR *sb.* + *cardiogram* s.v. *CARDIO-.] An electrocardiogram (usu. a photograph of an oscilloscope display) that represents the directions as well as the magnitudes of electric currents in the heart.

1938 WILSON & JOHNSTON in *Amer. Heart Jrnl.* XVI. 15 In 1920 Mann..constructed a number of curves of this kind and called them monocardiograms. Granting the priority of this name, we prefer to call them vectorcardiograms in order to emphasize the true nature of the difference..between them and ordinary electrocardiographic curves, which are scalar functions of the time.

Hence **vectorcardio·graphy**, the practice or technique of obtaining and interpreting vectorcardiograms.

1966 *Brit. Heart Jrnl.* VIII. 160 (*heading*) Possible application to vector-cardiography. **1976** *Lancet* 18 Dec. 1330/1 A parallel science of vectorcardiography grew up in specialist centres, in which cardiac currents were studied on oscilloscopes in more than one dimension at a time with both leads and electrodes appropriate to three orthogonal or right-angled axes.

vectored (ve·ktəɹd), *a. Computers.* [f. VECTOR *sb.* + -ED².] Of a facility for interrupting a program: supplying the address to which the program must jump when it is interrupted.

1976 M. HEALEY *Minicomputers & Microprocessors* iv. 136 The obvious advantage of vectored interrupts is the low time that exists between an interrupt being accepted and the specific service routine commencing. **1979** *Sci. Amer.* July 1/2 (Advt.), The new Series Sixteen stands out for state-of-the-art technology as well..too: more than 255 vectored interrupts and battery backup for memory retention during power failures. **1982** A. CLEMENTS *Microcomputer Design & Construction* vi. 170 Motorola have introduced a hardware device giving the 6800 a vectored interrupt facility.

vectorial, *a.* Add: **3.** *Path.* Of or pertaining to the ability to act as a vector of a disease.

1964 *Bull. World Health Organization* XXXI. 71 In malaria eradication the residual insecticide exerts upon the mosquito's vectorial capacity a direct insecticidal impact. **1981** *Trop. Med. & Hygiene News* Feb. 24 The vectorial capacity (i.e., the risk of transmission of the parasite) was found to be about 1000 times the critical value required for the maintenance of endemic malaria.

Hence **vectori·ally** *adv.*, as a vector or vectors.

1895 *Phil. Trans. R. Soc.* A. CLXXXVI. 706 The integral of this rotation taken vectorially throughout a small volume including the initial and final positions of the electron is equal to the strength of the electron multiplied by its linear displacement. **1909** RIEDEL & PIERCE *Direct & Alternating Current Man.* vi. ... must be added vectorially to obtain the resultant current. **1947** *Proc. Physical Soc.* LIX. 24 If the ... admittances ... circle of diameter 1/(S + S₀) is obtained. **1978** *Sci. Amer.* May 126/2 The point will have two components of apparent motion... The components add vectorially, and the point appears to move diagonally rather than either horizontally or vertically.

vectoring (ve·ktəriŋ), *vbl. sb.* [f. VECTOR *sb.* + -ING¹.] **1.** The action of *VECTOR *v.*

1959 W. A. HEFLIN *U.S. Air Force Dict.* 553/1 Vectoring is usually done from the ground, or from a mother aircraft.

2. *Computers.* The provision or use of interrupt vectors.

1977 E. E. KLINGMAN *Microprocessor Systems Design* xii. 352 These [lines] are for 'cascading' several 8214s if more than eight interrupting devices need vectoring. **1979** *Personal Computer World* Nov. 84/4 Software vectoring of interrupts to allow more than one interrupt driven peripheral at a time, and also multi-programming.

vectorscope (ve·ktəɹskōᵘp). *Electronics.* [f. VECTOR *sb.* + -SCOPE.] A type of oscilloscope used to analyse colour television signals (see quot. 1957).

1957 SMITH & MATLEY in *Electronic & Radio Engineer* XXXIV. 198/1 The vectorscope..has been designed to display the chrominance component of the colour television signal as a pattern of vectors. **1979** A. LIVY *Colour & Black & White Television Theory & Servicing* xviii. 652 Some manufacturers recommend the vectorscope for the alignment of the bandpass amplifier tuned circuits.

vedalia (vǝdēⁱ·liă). *Ent.* Also **Vedalia.** [mod.L. (E. Mulsant 1850, in *Ann. de la Soc. d'Agric. de Lyon* II. 901), of unkn. etym.] The ladybird *Rodolia cardinalis* (formerly included in the genus *Vedalia*), which is native to Australia but has been imported into California and elsewhere to control scale insects.

1889 *Insect Life* II. 112 The Vedalia [*sic*] has multiplied in numbers and spread. **1935** H. T. FERNALD *Applied Entomol.* (ed. 3) xxvi. 222 In Australia it [*sc.* the cottony cushion scale] had an enemy known as the Vedalia. **1964** *Discovery* Oct. 62/3 The vedalia ladybird..in less than a year had brought about the virtual elimination of the cottony cushion scale. **1973** P. A. COLINVAUX *Introd. Ecol.* xxix. 410 All stages of the vedalia beetle were carnivorous.

Vedanta. (Earlier example.)

1788 *Asiatick Researches* I. 223 The word *māyā*, or *delusion*, has a more subtile and recondite sense in the *Vēdānta* philosophy.

Veddoid (ve·doid), *sb.* and *a.* *Physical Anthrop.* [f. VEDD(A + -OID.] **A.** *adj.* Belonging or pertaining to a racial group of uncertain status typified by the Veddas of Sri Lanka, characterized by dark skins, short stature, and wavy hair, and occurring chiefly in parts of southern Asia. **B.** *sb.* A member of this group.

1948 A. L. KROEBER *Anthropology* (ed. 2) iv. 139 The Veddoids are almost invariably culturally retarded hill or jungle people who evidently represent an old stratum of population pushed back by Caucasians or Mongoloids, or almost absorbed by them. **1956** *Nature* 7 Jan. 41/2 The sickle cell trait was found to be present in a number of Veddoid communities of southern India. **1965** S. COLE *Races of Man* vii. 85 The Veddoids are short, average stature 157 c.m., skin colour is chocolate brown, hair very wavy or curly. **1974** *Encycl. Brit. Macropædia* XIX. 1081/1 Non-Mediterranean Veddoids (Australoids) are found as minorities [in Southern Yemen]. *Ibid.*, The original islanders [of Socotra in the Arabian Sea] forming about half the population, are Veddoid. **1977** C. CLARK *World Prehist.* (ed. 3) vi. 260 Low-headed people [in the Indian sub-continent] with retreating foreheads, pronounced supraorbital ridges and relatively broad noses with depressed roots appear to be related to a Veddoid-Australoid stock of indigenous character.

veddy (ve·di), repr. a childish, affected, or (*U.S.*) British pronunc. of *very*. Freq. *joc.*

1859 E. EDEN *Semi-Detached House* xix. 257 Charlie would gravely say, 'Yes, veddy true, pooty Rachel.' **1938** *Amer. Speech* XIII. 157/2 For jocularity, newspaper writers..now sometimes write *veddy* for 'very'. **1960** L. KAUFFMANN *Waldo* i. 14 Elegant and veddy sophisticated and all that sort of thing. But Waldo's father was veddy British. **1968** ... **1975** *Publishers Weekly* 24 Feb. 114/1 Note humorous anachronisms as Dylan's fish-girl pours him tea from a bone china teapot—veddy British!

vedette. Add: **2.** More widely, any motor launch. Also *ellipt.*

1977 E. W. MIDDLETON *Lifeboats of World* v. 178, 90 fast vedettes. **1982** H. A. WILLIAMS *Some Day I'll find You* i. viii. 23 If it was that low tide the passengers had to land in vedettes.

3. With pronunc. (vǝde·t). A stage or film star.

1963 *Listener* 28 Mar. 572/2 The 'pop' singers' predecessors were the music-hall vedettes, among them Marie Lloyd and Maurice Chevalier, who..lit up the souls of their audiences with wit and innuendo. **1980** L. ST. CLAIR *Obsessions* viii. 130 At the sight of..the Hollywood *vedette*, the maître d'..escorted his guests to the windows table.

Vedic, *a.* and *sb.* (Earlier examples.)

1848 *Rep. Brit. Assoc. Adv. Sci. 1847* 321 It may be observed that in the Vedic hymns..some dialectic differences and many grammatical discrepancies occur. **1853** *Jrnl. Amer. Oriental Soc.* III. 297 In many of the points in which Vedic and Sanskrit disagree, the former strikingly approaches the [Iranian] as spoken by the westward, the language of the Avesta [etc.].

‖ **veduta** (vedū·ta). Pl. **vedute; vedutas.** [a. It. *veduta* a view, f. *vedere* to see.] A realistic, detailed picture of a town scene with buildings of interest, esp. one belonging to the genre represented by eighteenth-century Italian artists such as Canaletto, Guardi, and Piranesi; *veduta ideata* (pl. *vedute ideate*), a picture in this style but showing an imaginary scene, esp. one by Pannini.

1906 *Studio* (Special Summer No.) A. p. iii, Rudolf [von Alt] was the leader of the Viennese 'Veduta' painting, the true biographer of Vienna. **1934** *Burlington Mag.* Aug. 71/2 Topographical exactitude and the precision of linear perspective—the two essential elements of a *veduta*. **1964** *Ibid.* Sept. 126/1 Little *vedutas* of the Swedish countryside. **1959** *Listener* 25 June 1118/1 Such *vedute ideate* were..a regular branch of Italian landscape painting since the latter part of the seventeenth century. **1961** *Guardian* 16 Nov. 7/7 The boom in Venetian *vedute* by Guardi in 1948. **1967** O. SIMS *Last Bird Friend* xx. 180 There was an eighteenth-century Venetian *veduta* ... painting. **1970** *Oxf. Compan. Art* 199/1 Canaletto..enlarged his repertory to include subjects from the Venetian mainland ..and *vedute ideate* or imaginary landscapes. **1978** *New Yorker* 3 Apr. 64/2 Instead it dwells on the fresh spring light and dark shade falling over ordinary buildings, with a sparseness, accuracy, and austere gaiety that suggest Canaletto's *vedute* of London.

Hence **veduti·sta** (pl. -**ti**, -**e**), a painter of *vedute*.

1962 R. G. HAGGAR *Dict. Art Terms* 355/1 The most notable *vedutisti*..were Pannini and Piranesi, and Canaletto and Guardi. **1967** C. ROUGVIE *When Johnny Died* ii. 55 *Vedute*..are..paintings of Venice..both Canaletto and Guardi were *vedutiste* (sic).

vee. Add: **1.** The name of the letter V, used to denote things having or arranged in this shape. (Examples.)

1933 *Jrnl. R. Aeronaut. Soc.* XXXVII. 845 This..can easily be achieved by making the vee of the hull sufficiently deep. **1939** *War Illustr.* 29 Dec. 538/2 A Squadron of fighters or bombers will fly in a 'vee' with three flights of three machines each. **1950** B. PYM *Some Tame Gazelle* i. 9, I must have something to cover up the neck of my green frock. Perhaps it would have been better if I hadn't tried to alter it to a Vee. **1965** G. MCINNES *Road to Gundagai* xiii. 222 A tall man with..a shock of receding curly hair in a Vee. **1970** *Commercial Motor* 25 Sept. 97 The trend in diesel engine development in this country is now clearly towards vees and/or turbocharging. **1977** 'E. CRISPIN' *Glimpses of Moon* vi. 93 At some stage she stood in the vee with her back to the parapet..and lost her balance and went headlong.

2. *attrib.* and *Comb.*, as *vee aerial, antenna, belt, block, formation, joint, neck(line), thread; vee-necked, -shaped* adjs.; **vee engine**, an engine with two lines of cylinders inclined so as to form a V. Cf. *V 2 b.

1929 *Amat. Radio Handbk.* 158/2 Such an arrangement ..is known as the RCA Vee aerial. **1950** A. L. ALBERT *Electr. Communication* (ed. 3) xii. 481 (*caption*) The nonresonant Vee antenna is composed of two wires several wavelengths long, and connected to ground through terminating resistors. **1928** R. S. ELLIOTT *Antenna Theory & Design* ix. 436 The Vee antenna..is a simplification of the rhombic, with legs 3 and 4 removed. **1931** *Motor Cycle* (R. London Rubber Co. Ltd.) 39/2 A range of twelve Poman New Vee Fan Belts will fit 134 popular cars. **1971** *Power Farming Mar.* 44/1 Machines..lift the root by gripping the leafy top growth of the plant between a pair of..rubber vee-belts. **1893** J. G. HORNER *Princ. Fitting* v. 64 The vee blocks are employed chiefly for supporting shafts and circular spindle-like work generally. **1975** BRAM & DOWNS *Manuf. Technol.* ii. 27 Place the component in a suitable vee-block and slowly rotate it about the diameter to be checked. **1951** J. M. BRUCE *Brit. Aeroplanes 1914–18* 223 It was one of the first aeroplanes to have the then-new Rolls-Royce Condor twelve-cylinder vee engine. **1972** *Practical Motorist* Oct. 70/1 Vee engines have a non-return valve for the crankcase emission control system in the..outside rocker cover. **1960** *Times Rev. Industry* Nov. 18/1 Cylinders in Vee formation. **1964** S. CRAWFORD *Basic Engin. Processes* iii. 93 Fig. [7] (iii) shows the 'single vee-joint suitable for plates up to $\frac{1}{2}$ inch in thickness. **1970** *Islander* (Victoria, B.C.) 15 Feb. 15/4 The original interior finish has been retained. The walls of pine 'Vee' joint; pews, chancel and rails of Douglas fir have darkened and mellowed with age. **1970** *Times* 24 Mar. 9/4 They are all of the same shape, vee neck front. **1973** *Guardian* 10 Apr. 13/2 Checked vee-necked dress with contrast ribbing. **1949** *Sun* (Baltimore) 14 Mar. 8 (Advt.), Elongated vee neckline, especially becoming to the larger woman! **1959** *Listener* 31 Dec. 1157/2 Conditions made it impossible to take a happy turn..but, very unfortunately, they were because perfectly vee-shaped. **1964** S. CRAWFORD *Basic Engin. Processes* i. 36 The vee-shaped grooves are accurately ground parallel and square with the outside faces. **1914** E. PULL *Handbk. Workshop Pract.* xii. 229 (*heading*) Vee threads. **1975** BRAM & DOWNS *Manuf. Technol.* iv. 120 The buttress thread combines the anti-friction advantage of the square thread with the strength of the vee thread.

Hence **veed** *a.*, V-shaped.

1934 *Times* 23 Oct. 7/4 The radiator is in a case which has a 'vee'd' dummy honeycomb front. **1958** *Times* 19 Aug. 8/7 A moveable section—in front of the driver—in a sloping, divided, veed screen. **1972** C. MURRAY *Motor Boats & Boating* 144 Outwardly the racing boats have developed from a heterogeneous collection towards a clean, fine lined, deep Veed fleet.

veeboer (fī·bū³ɹ). *S. Afr.* Also as two words and with capital initial(s). Pl. **-boer, -boere, -boers.** [Afrikaans, f. *vee* cattle, livestock (f. Du. *vee*: see FEE *sb.*¹) + BOER.] A livestock farmer.

1824 *S. Afr. Jrnl.* I. 29 Poor Gert Schepers, a Vee Boer of the Cradock District, was less fortunate in the year 1823. **1902** ... *Aga* [*sic*] The plants were known to the veeboer or schaapboer as the cause of the troubles they produce long before any scientific investigation of their properties had been made. **1944** J. MOCKFORD *Here are S. Africans* iv. 39 Here the free burghers, the vee-Boers, rapidly acquired qualities unknown to the more sedate residents of Table Valley. **1954** H. GIBBS *Background to Bitterness* i. 25 Most free men at the Cape wanted to become a *veeboer*, cattle-farmer. **1965** M. G. ATMORE *Cape Furniture* iii. 56 Thus arose the 'veeboere', each tenant on some 6,000 acres of cattle ranching country.

veejay (vī·dʒēⁱ). Also **vee-jay, VJ.** *slang* (chiefly *U.S.*). [Pronunc. of the initial letters of *video jockey*, after D.J. s.v. *D III. 3; cf. *DEE-JAY, DEE-JAY.] One who presents a programme of (popular) music video recordings, esp. on television.

1982 *N.Y. Times* 4 July 11. 17/1 The image..flashes.. to a loft studio where a video jockey—or 'veejay', a cross between a disc jockey and a TV emcee—announces what has just been played on MTV. **1983** *Amer. Way* June 170/1 VJs, or video jockeys, at MTV's studio cue up as many as 12 hours of video music. **1984** *Sunday Tel.* (Colour Suppl.) 29 Jan. 16 (*caption*) Mike Read and colleagues—today's disc jockeys will be tomorrow's 'veejays'.

veena, var. VINA in Dict. and Suppl.

veep (vīp). *U.S. colloq.* Also **Veep.** [f. the initials *V.P.* (vīpī); cf. *JEEP *sb.*] A vice-president.

1949 *News-Age-Herald* (Birmingham, Alabama) 12 June D-14 (*heading*) 'Veep' Barkley's name now often tied with some eligible widow's. **1953** *N.Y. Times* 19 June 25/1 (*heading*) Woman 'veep' urged. **1961** *Manila Times* 12 Jan. 14/3 Pelaez' presidential bid as vice-pres... **1977** *Time* 17 Oct. 14/1 Kremlinologists have been speculating about who might be named to the newly created post of Vice President. A Veep was needed to take over the fatiguing ceremonial functions of the presidency. **1983** *Fortune* 28 Nov. 152/3 His Makati business club constituents would be happy to nominate E.Z. for veep.

veery. For *Turdus* read *Hylocichla* and add earlier example.

1838 THOREAU *Jrnl.* in *Writings* (1906) VII. 70 Sometimes I hear the veery's silver clarion.

veg (vedʒ). Pl. **veg, veges.** Colloq. abbrev. of *vegetable* (in quot. 1898, of *vegetarian* (restaurant)). Cf. *VEGIE.

1898 A. BENNETT *Man from North* xi. 95 You know Miss Roberts at the veg—red-haired tart. **1918** G. FRANKAU *One of Them* xxviii. 219 Clive of the Court, begone to veg, and joint! **1940** WODEHOUSE *Quick Service* ii. 21 Have a custard apple? It's on the house. The fruit and veg department has just given of its plenty. **1945** BOWEN *Heat of Day* ... 'The spring veg...' **1966** *Listener* ... **1980** *Times* 15 May 9/2 A crudely painted sign that said *Veg Soup Half-Penny Per Cup*. **1983** *Truck & Bus Transportation* June 81/3 We'll heat a can of veges. **1984** *Economist* 3 Nov. 18/1 Treasury officials paying for their meat and two veg are rightly suspicious.

vegan (vī·găn, *U.S.* ve·dʒăn). [f. VEG(ETABLE *sb.* + -AN.]

1. A person who on principle abstains from all food of animal origin; a strict vegetarian.

1944 D. WATSON in *Vegan News* No. 1 'Vegetarian' and 'Fruitarian' are already associated with societies that allow the 'fruits' of cows and fowls, therefore..we must make a new and appropriate word... I have used the title 'The Vegan News'. Should we adopt this, our diet will soon become known as the vegan diet, and we should aspire to the rank of *vegans*. **1945** *Ibid.* Feb. 5 The members have asked how 'Vegan' is pronounced. Vegan, not Vee-jan. **1955** *Fresh Press* 19 Nov. 6/8 A true-blue Vegan, I'm assured..will even exclude from his or her diet, milk ...and honey. **1965** *New Scientist* 20 May 520/1 Vitamin B₁₂..is found almost exclusively in animal foods, so that strict vegetarians (like vegans) can go short unless they take special precautions to ensure a supply. **1977** J. F. FIXX *Compl. Bk. Running* xiv. 170 There are..three times called a vegan; the lacto-vegetarian..; and the lacto-ovo-vegetarian. **1979** J. I. M. STEWART *Our England* 177 Robin had discovered the duty of being a vegetarian. Indeed, he had become a vegan, and that seemed to mean that he could eat virtually nothing at all. **1986** *Times* 1 Feb. 12/2 'Beaumilk: milk that's never even seen a cow' is to vegans, who deplore exploitation of animals and eat nothing derived from them, a highly desirable commodity.

2. *attrib.* or as *adj.*

1944 [see sense 1 above]. **1945** *Vegetarian Messenger* XLII. 163 Following the articles and correspondence regarding the use of dairy products..in *The Vegetarian Messenger* last year, a number of our members who do not use animal products of any kind formed themselves into a group which has since adopted the title of 'The Vegan Society'. **1951** *News Chron.* 13 Dec. 3 A true vegetarian or vegan diet may not be scientifically adequate, said Dr. Hill. **1973** *Listener* 8 Feb. 178/1 The good ecological life: no car, vegan cooking, and a mangle technology in a tumbledown cottage. **1978** *Peace News* 25 Aug. 18/3 A group of people from 21 years ago: vegan farming community in Tennessee..are coming to visit Britain in late September or early October. **1984** *Listener* 9 Aug. 17/2 The facts that CIWF is able to marshal must drive many who read this literature to a vegan diet.

Hence **ve·ganism**, the beliefs or practice of vegans; abstention from all food of animal origin.

1944 *Vegan News* May 1 Veganism is the practice of living on fruits, nuts, vegetables, grains, and other wholesome non-animal products. **1972** *New Scientist* 4 May 297/2 Universal vegetarianism would..tend to disrupt organic farming and the organic cycle—soil, plant, animal, man. It would also, if logically carried on as in Veganism, abolish milk and eggs. **1977** S. R. L. CLARK *Moral Status of Animals* ix. 185 Veganism is a better project than lacto-vegetarianism, though we may in the end be able to take *some* milk from our kin without injustice.

Veganin (ve·dʒănin). Also **veganin.** *Pharm.* A proprietary preparation of aspirin, paracetamol, and codeine phosphate used as an analgesic.

1926 *Trade Marks Jrnl.* 23 June 1442 *Veganin*... Chemical substances prepared for use in medicine and pharmacy. Gödecke & Co., Chemische Fabrik Aktien Gesellschaft..Berlin-Charlottenburg, Germany; manufacturers. **1943** S. ERTZ *Anger in Sky* iv. 98 Take an aspirin, take two, take three. Take veganin. Take anything. **1951** G. GREENE *End of Affair* iii. vi. 141 'I've got a bad headache, that's all'..'I'll get you some veganin.' **1979** G. WATSON *Black Jack* i. 9 There was a stock of Veganin on the hand.

vegeculture (ve·dʒikɒltiūɹ). [f. VEGE(TABLE *sb.* + CULTURE *sb.*] The cultivation of vegetables. Hence **vegecu·ltural** *a.*

1957 H. J. ... *Agric. Origins* ... how to grow vegetables, salads, and herbs. **1963** J. D. CLARK in Braidwood & Willey *Courses toward Urban Life* 23 The distribution of the bored stone in central and south-central Africa suggests that they may also sometimes accompany a vegecultural,..even an incipient agricultural, level of economy during the later stone age times. **1964** S. COLE *Prehist. E. Afr.* x. 275 Only after the introduction of vegeculture do we find evidence of food production... This does not mean that some form of vegeculture did not exist—perhaps mainly on plants with edible roots or tubers. **1977** C. CLARK *World Prehist.* (ed. 3) v. 216 It looks as if there was a greater emphasis on vegetable food and even a form of vegeculture.

vegetable, *sb.* Add: **2*.** *fig.* A person who leads an uneventful or monotonous life, without intellectual or social activity; also, one reduced by illness to little more than a physical body. Cf. VEGETABLE *a.* 5.

1921 G. B. SHAW *Back to Methuselah* I. 16 What use is this thousand years of life to you, you old vegetable? **1933** A. HUXLEY *Let.* 10 Oct. (1969) 373, I was so glad to hear from Norah that you were going on as well as could be expected. It will be a weary business for a bit,..sitting still and being a vegetable. **1953** *Chicago Daily Sun-Times* 28 Nov. 8/1 'But the doctors told us that Rocky is a vegetable, incapable of thinking for himself.' **1963** J. DAWSON *Ha-Ha* iii. 48 I'm going to go on working..Tony says he would hate a wife who was just a vegetable. **1976** SMYTHIES & CORBETT *Psychiatry* vi. 143 Eventually they become bedridden..and become helpless 'vegetables'. **1980** B. CASTLE *Castle Diaries* 24, I hope and pray she will die with dignity and not be reduced by a stroke into a vegetable.

3. a. *vegetable dish* (earlier example), *garden* (earlier example), *juice, oil, patch, rack.*

1820 D. WORDSWORTH *Jrnl.* 20 July (1941) II. 39 A level bottomed oval vessel like that commonly used as a vegetable dishes. **1889** OUIDA *N.* 12/1 Back of its hacienda is a fine orchard and vegetable garden. **1907** *Yesterday's Shopping* (1969) p. 313/1 (Index), Vegetable racks. **1921** *Daily Colonist* (Victoria, B.C.) 20 Mar. 17/1 The born gardener is still looking ahead in the Fall when other people store their tools and never do a hand's turn in the vegetable patch until the Spring urge comes upon them again. **1926** *Ibid.* 16 Jan. 15/3 Listed in the..general cargo..is a large quantity of vegetable oil, both in bulk. **1975** P. G. WINSLOW *Death of Angel* ix. 69 Vegetable juice and soya beans. **1977** *Times* 8 May 13/4 EEC imports of vegetable oils and oil-cakes. **1978** G. MITCHELL *Mingled with Venom* x. 101 The three points were the pear tree, the vegetable patch and the..compost-heap. **1979** J. D. MACDONALD *Green Ripper* (1980) i. 15 One of her vegetable juice cocktails. **1979** A. CLARKE *Poisoned Web* x. 78 Two girls passed the vegetable patch..staring at the runner-beans.

vegetable, *a.* Add: **7.** *vegetable fat, oil*, fat or oil obtained or manufactured from plants; *vegetable lard*, a solid cooking fat prepared from vegetable products; *vegetable spaghetti*, a variety of vegetable marrow bearing fruits whose flesh resembles spaghetti in appearance; also, the fruit itself or its flesh; *vegetable sponge* = *dishcloth gourd* s.v. *DISH-CLOTH 2.

1797 *Encycl. Brit.* XIII. 122/1 Vegetable oils are obtained by expression, infusion, and distillation. **1884** *Ibid.* XVII. 741/1 The ordinary method for separating vegetable oils and fats from the seeds, nuts, &c., of which they form constituent parts is by pressure. **1884** *Cent. Dict.* s.v. *Sponge-gourd*, The netted fiber from the interior of the fruit is used for washing and other purposes, hence called vegetable sponge. **1894** C. R. A. WRIGHT *Animal & Vegetable Fixed Oils* xv. 305 Amongst the Hindoos and others whose religious beliefs preclude the use of animal fats..a large sale now exists for purely vegetable fats of buttery consistence (vegetable lard). **1918** C. A. MITCHELL *Edible Oils & Fats* iii. 33 Coconut oil is treated with alcohol and animal charcoal and the resulting product, which is practically tasteless, is sold as 'vegetable lard' or 'vegetable butter'. **1954** M. H. SMITH *Insect Virus* vii. 107 Whenever it has been shown that viruses of animals or higher plants go through cycles as described for bacteriophages, the terms provirus, vegetative virus and infective virus are appropriate for the corresponding stages. **1973** [see *PROPHAGE]. **1982** FRAENKEL-CONRAT & KIMBALL *Virology* i. 11 The relatively simplicity of extracellular viruses, termed the dormant phase, and the complexity of their interaction with host cell components leading to their replication, termed the vegetative phase, have placed viruses among the most useful tools in the study of all phenomena related to replication, information transfer, mutation, and many other aspects of molecular biology.

1956 *Vegetable fat* [see *hydrogenated* ...]. **1976** *Sci. Amer.* ... **1967** *Guardian* 7 Nov. 22/8 If vegetable spaghetti or custard marrow, ..grow them up a fence... **1973** *Times* 15 Nov. 22/8 If ... vegetable spaghetti ... **1978** J. CROCKETT *Vegetables & Fruits* 164/3 The vegetable spaghetti is part of the trailing type, not bush, so allow room for it to roam. **1984** *Gardening from 'Which?'* Mar. 75/1 The loofah's other common names of vegetable sponge and dishcloth gourd describe its uses.

vegetal, *a.* Add: **4.** *vegetal pole* (Embryology), the lower pole of an ovum or a young embryo, which divides more slowly than the upper (animal) pole and in telolecithal ova contains most of the yolk. Cf. *animal pole* s.v. *ANIMAL *C.* 1.

1914 W. E. KELLICOTT *Textbk. Gen. Embryol.* iii. 92 The vegetal pole is frequently occupied largely by the relatively inert food substances, the materials in general related with the vegetative organs of the developing embryo. **1926** JORDAN & KINDRED *Textbk. Embryol.* v. 32 When the yolk is more abundant,..it tends to segregate at one pole, thus determining a yolk-free pole, the animal pole, and a yolk-laden pole, the vegetal pole. **1947** L. B. AREY *Devel. Anat.* (ed. 5) ii. 31 At the other end of the polar axis is the vegetal pole. Its territory tends to be more sluggish and is concerned with the development of nutrient organs. Cytoplasmic components..are often disposed in a polarized or stratified way. This is well illustrated in telolecithal eggs, where animal pole is more protoplasmic and whose vegetal pole is more yolk-laden. **1982** *Nature* 16 Mar. 255/1 (*caption*) Injection was aimed at the vegetal pole.

vegetarian, *sb.* Add: **1. a.** (Earlier example.)

1839 F. A. KEMBLE *Jrnl. Residence on Georgian Plantation* (1863) 251 If I had had to be my own cook, I should inevitably become a vegetarian.

vegetarianism. (Earlier examples.)

1851 DUNGLISON *Dict. Med. Sci.* (ed. 8) 896/2 *Vegetarianism*,..a modern term, employed to designate the theory that man..ought to subsist on the direct productions of the vegetable kingdom and totally abstain from flesh and blood. **1852** *Punch* 7 Aug. 68/1 Vegetarianism is evidently progressing.

vegetation. Add: **I. 4.** (Earlier example.)

1760 M. MONTAGU *Let.* 25 Oct. (1967) III. 245, I am not surpris'd at the long Vegetation of the D[uche]sse of Argyle.

II. 6. c. Used *attrib.* with reference to the death and regeneration of plant life and the alternation of the seasons as symbolized or represented in religious or cultic beliefs and rituals.

1914 J. G. FRAZER *Golden Bough: Adonis Attis Osiris* (ed. 3) II. iii. viii. 126 Professor Ed. Meyer also formerly regarded Osiris as a sun-god; he now interprets him as a god of vegetation. **1918** *Ibid.* 63 Gray & Moore *Mythol. All Races* XI. 23 Here there seems to be indication of a vegetation cult. *Ibid.* 75 Closely connected with the earth goddesses are their children, the vegetation-deities. **1922** T. S. ELIOT *Waste Land* 53 Anyone who is acquainted with these works [by Frazer and Weston] will immediately recognise in the poem certain references to vegetation ceremonies. **1967** *Listener* 6 Apr. 421/1 In 'sac' the Easter holiday' should have been all outside broadcasts of drunken baroque processions..., villages abusing each other's religious banners..., vegetation ceremonies still describable in pace-egging and mummers' plays.

Hence **vegeta·tional** *a.*

1936 *Spectator* 4 Sept. 354/2 The dark areas observable on the surface of Mars are vegetational regions. **1959** *New Biol.* XXVI. 32 These communities..are prevented, by haymaking and the grazing of sheep, cattle, and rabbits, from entering the normal phases of vegetational succession. **1977** J. L. HARPER *Population Biol. Plants* iv. 95 The buried seed populations of mature climax communities generally contain a living..record of the past vegetational history of the succession.

vegetative, *a.* Add: **1. d.** *vegetative pole* (Embryology) = *vegetal pole* s.v. *VEGETAL *a.* 4.

1892 E. L. MARK tr. *O. Hertwig's Text-bk. Embryol.* i. 11 The dissimilar poles are distinguished:..the under, heavier and richer in yolk, as the vegetative pole. **1909** [in Dict.]. **1941** B. M. PATTEN *Human Embryol.* iv. 60 In mammals, as is the case with surprising uniformity throughout the animal kingdom, the mitotic spindle of the first cleavage division forms at right angles to an imaginary axis passing through the ovum from animal to vegetative pole. **1958** — *Foundations Embryol.* ii. 52 The region opposite the animal pole is called the vegetative, or vegetal, pole because the small, sluggish cells which arise from this region, it remains itself relatively less active.

e. *Biol.* Pertaining to or being a stage in the replication of a virus at which noninfective viral components are synthesized and assembled within the host cell prior to its lysis.

1953 M. DELBRÜCK in *Cold Spring Harbor Symp. Quantitative Biol.* XVIII. 1/1 One new feature is the recognition that the infecting virus undergoes an obscure stage of multiplication in the cell ... which may be called the vegetative phase, by analogy to the use of the word 'vegetative' in the bacteriology of sporulating bacteria. **1954** M. H. SMITH *Insect Virus* vii. 107 ...

6. b. (Further examples.)

1969 *Sci. Jrnl.* Feb. 12/3 Two other patients..had flat EEG readings for prolonged periods, but subsequently recovered, having been clinically in the 'vegetative' state. **1972** *Lancet* 1 Apr. 734/1 Patients with severe brain damage due to trauma..may now survive indefinitely... Such patients are best described as in a persistent vegetative state. **1976** *National Observer* (U.S.) 19 June 2 (*caption*) Her constitutional right to privacy includes the right to live or die, although Karen Ann Quinlan..lives on in 'a vegetative state' for 14 months... The Quinlans last March won court permission to turn off Karen's life-sustaining devices. **1982** *Brit. Med. Jrnl.* 20 Oct. 1022/2 The term 'persistent vegetative state' was suggested in 1972 to describe those patients with irreversible brain damage..who on recovery from deep coma pass into a state of seeming wakefulness and reflex responsiveness but do not return to a cognitive sapient state.

vegie (ve·dʒi). Also **veggie.** Colloq. abbrev. of *vegetable*. Usu. *pl.* Cf. *VEG.

1955 C. BROWN *Lost Girls* iii. 132, I did get a job for myself, selling vegies at a stall in the market. **1966** [see *KINDY]. **1973** *Philadelphia Inquirer* (Today Suppl.) 14 Oct. 8/3 Split-rice (rice and veges wrapped in seaweed). **1976** *New Yorker* 8 Mar. 28/3 They wash and chop veggies and bake bread. **1979** *Chicago Sun-Times* 13 July 17 Last week, ... veggies..have been called... **1983** *Chicago Sun-Times* 27 Aug. (Guide) 83 Also included, a fresh veggie and..a carafe of one of eight wines of your choice. **1984** *Grass Roots* (Australia) Apr. 9/2 The veggie gardens are fenced.

Vegliote (velyo·t). Also **Vegliot, Veliote.** [f. *Vegli*·*a* + -OTE.] An extinct dialect of Dalmatian, formerly spoken on the island of Veglia (now called Krk) off the Dalmatian coast of Yugoslavia.

1910 *Encycl. Brit.* XIV. 891/2 The Vegliote dialect is the last remnant of a language which some long time ago extended..along the Dalmatian coast. **1933** [see *DALMATIAN *a.* and *sb.* 2]. **1960** W. D. ELCOCK *Romance Lang.* i. 185 Lat. *mensa*, attested as *mesa*..is also found in Rheto-Romance..and in Vegliote (*maisa*). **1974** *Encycl. Brit. Macropædia* XV. 1038/1 Romanian, Vegliot, Spanish, and perhaps Rhaetian show similar development of..stressed vowels. **1984** *Trans. Philol. Soc.* 20 Vegliot is a 'dead' language once spoken on the island of Veglia (Krk) off the Jugoslav coast.

vehicle, *sb.* Add: **I. 3. d.** In a metaphor, the literal meaning of the word or words used metaphorically, as distinct from the subject of the metaphor; the image or idea whose associated meaning is the subject of the metaphor. Opp. *TENOR *sb.* 1 d.

1936, etc. [see *TENOR *sb.* 1 d]. **1957** S. ULLMANN *Style in French Novel* vi. 214 It is an essential feature of a metaphor that there must be a certain distance between tenor and vehicle. Their similarity must be accompanied by a feeling of disparity. **1977** *Studies in Eng. Lit.: Eng. Number* (Tokyo) 36 In this metaphor, the tenor and vehicle are not opposed to each other perfectly.

II. 7. a. (Further examples.)

1947 *N.Y. Times* 20 Mar. 3/1 The statement 'there is no defense against a guided missile'..assumes that this bomb is launched by a space vehicle at a velocity well in the supersonic range. *Ibid.*, V-2 type vehicles. **1973** *Sci. Amer.* Feb. 23/3 [They] mounted simulator probes on a vehicle towed horizontally behind a ship. **1982** *Nature* 27 May 243/1 Vehicle capturing the people, whether ships or aircraft, moving in an environment in which radar detection..begins to be technically straightforward, have become exceedingly vulnerable to attack.

b. A space rocket, in relation to its (actual or intended) payload.

1959 *Daily Tel.* 13 Mar. 15/6 Black Knight is entirely a research vehicle and will never be manufactured into a weapon. **1967** *Technology Week* 20 Feb. 36/1 A satellite launched into an eight-hour orbit by an *Atlas-Agena* vehicle. **1983** *Chicago Sun-Times* 13 July 12/5 Last year, the Soviets proposed a sub-ceiling of 1,100 on MIRVed missiles and bombers within the over-all ceiling of 1,800 strategic nuclear delivery vehicles.

8. *fig.* A song, play, film, etc., that is intended or serves to display the actor (or performing artist) to the best advantage.

1863 *Illustrated Times* 15 Aug. 103/3 Then came..Lady Gifford's quasi-comedy of *'Finesse'*, which..simply served as a vehicle for some of Mr. Buckstone's practical directions and preposterous costumes. **1868** in P. Bailey *Leisure & Class in Victorian Society* (1978) vii. 152 A good song must be written, not for its own sake, but for that of the singer.. It must really be a vehicle. **1922** *N.Y. Times* 16 Oct. 20/2 Charles Ray is back on Broadway..in his vehicle being 'A Tailor Made Man', and his stopping place for a week is the Strand Theatre. **1948** Mrs. P. CAMPBELL *Let.* 19 Jan. in B. *Shaw & Mrs. Campbell* (1952) 264, I only saw it as a commonplace, ordinary and good sort of working vehicle, or whatever the right word or expression is for a play that a star might consider for herself. **1967** *Listener* 9 Feb. 195/2 *Times* 16 Aug. 5/4 It might be well to rule out at the beginning the 'vehicle' specially created to exploit the talent and personality of the individual actor. **1966** *Listener* 27 Jan. 132/1 If at the start of his career an actor has to struggle to gain acceptance from the critics, it is possible for him to go ahead given a series of inadequate performances in..vehicle movies and plays designed to show off his personal qualities. **1979** *Rolling Stone* 7 June 16/6 Star names and vast budgets..wedded to a limp vehicle, can destine a feature to oblivion.

9. *Comb.*, as *vehicle-park; vehicle-actuated* adj.; **vehicle-mile**, a distance of one mile travelled by one vehicle, used as a statistical unit; **vehicle mine**, a land-mine designed to destroy vehicles.

1937 *Times* 13 Apr. p. viii/3 With the wide adoption of electromatic vehicle-actuated traffic lights of recent years, complaints of unnecessary delay at signal-controlled crossings have been much dispelled. **1963** *Gloss. Highway Engin. Terms (B.S.I.)* 62 *Vehicle actuated traffic signals*, traffic signalling equipment in which the duration of the red and green signals and the time of the cycle vary in relation to the traffic flow into and through the controlled area. **1975** *Times* 24 Nov. 19 The vehicle-mile..travelled by the whole traffic in each area and through..**1976** *Times Rev. Industry* Feb. 29/2 The statistics of vehicle-miles travelled..have increased faster than road transport in general. **1976** *New Yorker* 16 Feb. 68/2 About eighty per cent of highway use (measured in vehicle-miles) is auto travel. **1944** K. DOUGLAS *Alamein to Zem Zem* (1946) xiv. 82 The verges of gaps in the road were mined with anti-personnel and vehicle mines. **1972** *Times* 27 Dec. 6/2 There had been almost no interference with road building, apart from the odd vehicle mine. **1940** *Vehicle park* [see *gun-park* s.v. *GUN *sb.* 13 a]. **1976** *Cumberland News* 3 Dec. 11/6 (Advt.), Training hall and use of adjacent site as vehicle park at Parkhill Road, Kingstown, Carlisle.

Vei, var. *VAI.

veil, *sb.*¹ **8.** (Example.)

1897 *Quinland* I. xiii. 186 Aunt Hepsa says he was born with a veil over his face, and that was the reason he was so sure that we must not inquire about.

veilleuse. Now usu. a night-light; also, a small, decorative, bedside food-warmer, usu. burning oil in a wick and made of pottery or porcelain so as to let out some light. (Further examples.)

1897 *Private Life of Queen* xv. 122 The Queen betakes herself to bed... The ladies-in-waiting extinguish all the lights but the *veilleuse*. **1959** *Apollo* LXX. 35/1 The *veilleuse* of porcelain is..a hollow pedestal on which sits either a covered warming bowl..or a teapot. **1961** *Times* 1 Sept. 9/6 One was the earthenware bedside food warmer. Some collectors know these comparatively little sets as *veilleuses*. **1966** R. H. PINTO *Treen* 124/2 The oddest material for making veilleuses was probably wood and the fact that so few have survived is doubtless ..because of their inflammability. **1983** *Thames & Hudson Catal.* July–Dec. 31 The *veilleuse*, intended as a food or tea warmer for bedroom use, consists of..a supporting hollow pedestal, a small burner (*godet*) and a covered bowl or..teapot.

vein, *sb.* Add: **IV. 15. c.** vein-gold, gold occurring in a vein or veins.

1848 W. COLTON *Jrnl.* 6 Nov. in *Three Years Calif.* (1850) xxiii. 312 Vein-gold in these hills is as uncertain and capricious as lightning. **1956** G. TAYLOR *Silver* i. 2 The banketts of Witwatersrand, in which vein-gold and alluvial deposits are mixed. **1979** *Econ. Geol.* LXXIV. 1420 (*heading*) Fluid inclusion and geochemical studies of vein gold deposits.

d. In sense 3 b, as **vein-banding**, a symptom of some virus diseases of plants, characterized by a change of colour along the main veins of leaves; freq. *attrib.*

1930 *Bull. Kentucky Agric. Exper. Station* No. 309. 481 Veinbanding is a common disease of tobacco in locations where potatoes have been grown. **1957** *Phytopathology* XLVII. 139 (*heading*) Electron micrographs and physical barriers on the spread of pepper veinbanding mosaic virus. **1979** *Jrnl. Horticultural Sci.* LIV. 23/1 Gooseberry vein banding virus..is aphid transmitted; and..although it is widespread in Europe..., little is known of its economic importance.

veining, *vbl. sb.* **1. b.** (Earlier example.)

1840 [see *WEBBING *vbl. sb.* 3*].

veinlet. Add: **c.** *spec.* in *Geol.* Cf. VEIN *sb.* 7.

1927 [see *TANTERITE]. **1963** [see *RUTHENARSENITE]. **1974** *Encycl. Brit. Macropædia* XIX. 699/1 Sussexite occurs as hydrothermal fibrous veinlets in the U.S. at Franklin, N.J.

veitchberry (vī·tʃbɛri). [f. *Veitch*, surname of a family of nurserymen + BERRY *sb.*¹] A hybrid bramble produced in 1925 by crossing a raspberry and a blackberry; the fruit of this, resembling a large reddish blackberry.

1925 *Daily News* 26 Aug. 3/2 One of the novelties at the Royal Horticultural Society's Show, in Vincent-square, W., held yesterday, is the veitchberry—produced from a blackberry and a raspberry. *Ibid.* (*caption*), The veitchberry compared with the size of a halfpenny. **1956** *Good Housek. Home Encycl.* (ed. 4) 118/1 Plant veitchberries and wineberries 6 feet apart. **1980** *Amat. Gardening* 25 Oct. 20/2 Other hybrid berries..include the boysenberry, with large fruits but darker than a Loganberry, the Veitchberry and the Youngberry.

velar, *a.* Add: Hence **vela·rity**, velar quality.

1952 *Archivum Linguisticum* IV. 72 The apostrophe after *m* and *g* indicates velarity. **1964** [see *ALVEOLAR *a.* and *sb.*].

velaric (vɪlæ·rik), *a. Phonetics.* [f. VELAR *a.* + -IC.] Produced or characterized by a velar articulation, in which there is total or partial closure between the back of the tongue and the velum.

1934 in WEBSTER. **1938** D. M. BEACH *Phonetics of Hottentot Lang.* vi. 74 A..bilabial implosive may be made by closing the lips together, raising the back of the tongue to touch the velum, and then producing a partial vacuum between the lips & the velum by lowering the body of the tongue. This bilabial implosive may be termed a bilabial velaric click. **1959** [see *GLOTTALIC *a.*].

velarization (vīlərɑizē¹·ʃən). *Phonetics.* [f. VELAR *a.* + -IZATION.] A (normally secondary) articulation of a consonant, in which the back of the tongue is raised to or towards the velum; also, in some languages, applied to the articulation of some vowels.

1915 G. Noël-Armfield *Gen. Phonetics* xvii. 102 If it be found necessary to indicate velarisation in a script it may be done by placing a small ~ over the usual symbol. 1936 *Language* XII. 17 A further example is the velarization of the nasal consonant of Spanish *Cinco* [θiŋko], which is still felt to be a member of the n-phoneme. 1962 A. C. Gimson *Introd. Pronunc. Eng.* iv. 30 In the so-called 'dark' [ɫ]..there is an essential raising of the back of the tongue towards the velum (velarization). 1975 J. R. Macpherson *Spanish Phonology* vi. 43 When *a* precedes a velar consonant, the semivowel *w*, or a velar vowel, it is..attracted towards the point of articulation of the following sound and may become slightly velarized. The diacritical sign ~ can be placed under the vowel to indicate a degree of velarization. 1976 *Archivum Linguisticum* VII. 93 These [Rhenish] dialects further have velarization of final [n] in, for example, [wiŋ] *Wein*, [bruŋ] *braun*.

So **ve·larize** *v. trans.* (*rare*), (*a*) to prescribe velarization for; (*b*) to produce by velarization; **ve·larized** *ppl. a.*

1915 G. Noël-Armfield *Gen. Phonetics* xvii. 102 Similarly there are consonants, otherwise normal, in the production of which the back of the tongue is raised to the ~ position. These are known as velarised sounds. *Ibid.*, [Note] Confusion should not be made between velarised and velar consonants. ɫ is quite different from ŋ. The Arabic emphatics are velarised consonants. 1939 L. H. Gray *Found. Lang.* iii. 57 There are..many modifications of consonants, such as palatalised (e.g., Irish te..'hot'..); velarized or pharyngalized (e.g. the final sound of English *little, feel*). 1962 D. De Camp in Le Page & De Camp *Jamaican Creole* II. 137 Postvocal /l/ is so strongly velarized that such pairs as /fuul/ *fool* and /fuo/ *for* are acoustically very similar. 1977 *Word* XXVIII. 173 The rule which velarizes an alveolar consonant must precede the one which reduces pretonic syllables. 1980 A. Alpers *Life K. Mansfield* 12 His sisters..often called him 'Bogie', or 'Bogey' (which perhaps was velarised from 'Boy').

velation. For *rare*⁻⁰ read *rare*⁻¹ and add: **a.** (Example.)

1922 Joyce *Ulysses* 719 The visible signs of postsatisfaction? A silent contemplation: a tentative velation: [illegible]

Velcro (ve·lkro). Also **velcro.** [f. F. *vel(ours) cro(ché)* hooked velvet.] A proprietary name for a fabric made in narrow strips for use as a fastener, one strip having tiny loops and the other hooks so that they can be fastened or unfastened simply by pressing together or pulling apart.

1960 *Trade Marks Jrnl.* 30 Nov. 1521/1 Velcro. Narrow fabrics in imitation of velvet being textile smallwares for use as fasteners or fastenings for clothing. 1961 *Practitioner* July 113 We have been experimenting for some time with the new Bri-nylon fastener, 'velcro', using it particularly for patients who have difficulty in doing up buttons, trousers and belts. 1968 *Guardian* 9 May 9/4 False ringlets on a Velcro band. 1971 'A. Yoxall' [illegible] vii. 95 They had dressed him in a blue [illegible], which..was secured up the front by a strip of velcro. 1979 P. L. G. Bateman *Household Pests* ix. 76 Fly screening of windows..is fairly easy with flexible plastic fly screens which use Velcro to fasten them to the window frames. 1982 *Habitat Catal.* 1982/83 14/1 The roll overarm cushions, filled with polyester, can be held in place with velcro, if required.

Hence **Ve·lcroed, ve·lcroed** *a.* (or *pa. pple.* of *Velcro* vb.), provided with Velcro; attached by Velcro.

1972 O. Sela *Bearer Plot* xxiv. 152 He..pulled the carpet. 'Note the velcroed edge... Just put it back and it sticks.' 1981 J. D. MacDonald *Free Fall in Crimson* xviii. 204 By early June I was walking... I worked with the weight Velcroed around my ankle every evening. 1983 *Washington Post* 13 Feb. c-3/3 [illegible]

veldt, veld. Add: The spelling *veld* is now the only permissible form in S. Africa and the more usual form in other varieties of English.

1. (Earlier examples.)

1785, 1801 [see next sense]. 1835 A. Steedman *Wanderings S. Afr.* I. ii. 92 Here for the first time we bivouacked in what is called the Veld.

2. a. (Earlier and further examples.)

1785 G. Forster tr. *Sparrman's Voy. Cape of Good Hope* II. xiv. 144 The *land-drost* has appointed one of the farmers, with the title of *veld-corporal*, to command in these wars. 1801 J. Barrow *Trav. S. Afr.* I. 378 Louw, the Veld Commandant, readily offered his services. 1946 R. Campbell *Talking Bronco* 66 With veld-flowers as an afterthought. 1959 *Cape Times* 18 July 2/5 The veld flowers should be excellent this year.

b. veld-cornet (earlier example); **veld-craft,** skill in matters pertaining to survival on the veld; **veld-kost:** also **-kos** (later examples).

1802 G. M. Theal *Rec. Cape Colony* (1899) IV. 324, I have the honor to add the original report of the Veld Cornet Nicholaus Johannes Roets, sent to me with those Hottentots. 1905 D. Blackburn *R. Hartley, Prospector* 205 He associated with greedy, scheming directors, who were ignorant of veld-craft. 1910 J. Buchan *Prester John* xiv. 232 The veld-craft I had mastered had taught me a few things. 1980 *Country Life* 9 Oct. 1258/1 Olive Schreiner..possessed considerable veldt-craft and also the power of minute observation. 1948 *Cape Argus* (Mag. Section) 23 Oct. 1/7 The summer rains of 1915 had revived the fountains and veldkos was abundant. In a drought year he could not have survived. 1961 *Africa* XXXI. 231 When !U had a baby, her sister, Di!ai, gathered veldkos for her for five days.

veldt-shoe. Add: Also **vel(d)skoen** (fe·l(t)skun), pl. **-skoen, -skoene, -skoens.** Now, a heavy boot or shoe for outdoor work. (Further examples.)

The most common forms in S. Afr. English are *veld-skoen* and *veldskoen* (pl. *-e* or *-s*) but the anglicized form *veld-shoe* is occasionally used.

1892 *Journal* (Grahamstown) 11 July 3 (Advt.) Colonial-made 'Star' Standard screw veldschoens and boots. 1919 *Manch. Guardian* 10 Dec. 4 (Advt.), I believe this is the latest development of the Lotus welted-veldtschoen. 1939 S. Cloete *Watch for Dawn* xxi. 312 [illegible] 1942 [illegible] 1945 *Cape Times* 29 Oct. 1/3 He was wearing..grey flannel trousers, a white silk shirt and velskoene. 1962 L. Deighton *Ipcress File* xxxii. 207, I put on Irish tweed with Veldtshoen, cotton shirt, and wool tie. 1969 J. Selby *Boer War* i. 34 They wore wide-brimmed felt hats and soft veldschoen on their feet. 1974 *Sunday Times* (Johannesburg) [illegible] 1975 [illegible] 1977 *Listener* 3 Mar. 279/4 (Advt.), Veldtschoen... Virtually waterproof... Durable leather soles...£21.50.

veleta (vĕlī·tă). Also **valeta.** [f. Sp. *veleta* weathervane.] A ballroom round dance for couples in triple time, originating in England in 1900.

1900 A. Morris (*sheet-music title*) Veleta. 1911 D. H. Lawrence *White Peacock* I. viii. 146 She made me take her through a waltz, a minuet, a mazurka, and she danced capitally. 1928 [illegible] 1936 [illegible] 1952 [illegible] 1954 [illegible] the pre-war dances — ..the Maxina, the Valeta. 1966 [see *Gay Gordons* s.v. *gay* a. 9]. 1974 D. Smith *Look back with Love* xii. 113 Most of the evening was taken up with waltzes, barn dances and veletas. 1978 S. Sherlock in D. Abse *My Medical School* 105 We danced the Veleta, the Gay Gordons, the Dashing White Sergeant.

velic (vī·lik), *sb.* and *a. Phonetics.* [f. Vel(um + -ic.] **A.** *sb.* (See quot. 1943.) **B.** *adj.* Pertaining to or involving the velic or its movement.

1943 K. L. Pike *Phonetics* iv. 58 The upper part of the soft palate facing the nasopharynx is the velic (so called in this discussion to distinguish it from the velum, which represents the lower side toward the mouth). The closure of the nasal passage is therefore a velic closure (in contrast to velar closure, where the tongue touches the soft palate). 1971 Chin-Wu Kim in W. O. Dingwall *Survey Linguistic Sci.* 34 At time *a*, the lip rounding for [u] and [w] is superimposed on [k]; at time *b*, velic opening takes place. 1975 *Amer. Speech* 1972 XLVII. 24 *Pin* and *bin* are the same at the outset and in the middle and *partly* the same at the beginning; for both, the lips are closed and the velic is closed. 1981 *Canad. Jrnl. Linguistics* Spring 81 The key features involved are velic closure (for nasalization), apical closure (for stops, fricatives, liquids), and vocal cord vibration (voicing).

Velikovskianism (veliko·vskiăniz'm). [f. the name *Velikovsky* + -ian + -ism.] The (controversial) theories of cosmology and history propounded by Immanuel Velikovsky (1895–1979), Russian-born psychologist, based on the hypothesis that other planets have approached close to the Earth in historical times. Also **Veliko·vskyism,** in the same sense. So **Veliko·vskian** *a.*, of or pertaining to Velikovskianism; **Veliko·vskyite,** an adherent of Velikovskianism.

1972 P. Moore *Can you speak Venusian?* vi. 61 Therefore, say the anti-Velikovskyites, it is rather difficult to see how a comet could change into a planet. *Ibid.* 62 Neither of these gentlemen accepted Velikovskyism lock, stock and sinker. 1974 *Science* 15 Mar. 1061/1 The earliest known examples of writing..contain references to regular appearances of Venus in the sky 1500 years before its Velikovskian brushes with Earth. 1978 *Ibid.* 20 Jan. 288/2 A platform to square off pro- and anti-Velikovskyites. *Ibid.*, In a succinct foreword Isaac Asimov discusses the human psychology of Velikovskianism. 1979 *N.Y. Rev. Bks.* 25 Oct. 55/2 Velikovsky's most recent contributor to Velikovskian publications, is right on one count. *Ibid.* 52/3, I suppose it is too much to expect the Buffalo philosopher, when he teaches his next course on Velikovskianism, to let his students know some of the overwhelming evidence against such hoary balderdash.

vellumy, *a.* (In Dict. s.v. Vellum.) (Example.)

1925 H. A. Maddox *What Stationer & Printer ought to know about Paper* (ed. 3) i. 14 There are smooth vellums which derive their title from..a vellumy thickness and clarity of appearance.

velly (ve·li). A representation of a Chinese pronunc. of 'very'; also used *joc.*

1898 *Applause Reciter* 16 Ole man talkee, 'No can walk, Bimeby lain come, velly dark; Have got water, velly wide!' 1937 D. & H. Teilhet *Feather Cloak Murders* ii. 48 The Chinese woman rushed to him. 'You neahlly killed... Velly neahlly killed.' 1948 L. Durrell *Spirit of Place* (1969) 98 Shakespear him velly fine bigspeak sing song man. 1970 J. Lees-Milne *Another Self* iii. 47 Poor Janie still loves her little cousin velly, velly much. 1974 N. Bentley *Inside Information* xiii. 135 Velly good, sir.' Cheng bowed.

velocimeter. Add: **veloci·metry,** the measurement of speed, esp. speed of flow, by special techniques.

1969 *IEEE Trans. Aerospace & Electronic Systems* V. 687/1 (*heading*) Fundamentals of holographic velocimetry. 1978 *Sci. Amer.* Aug. 30/3 Swedish and Japanese workers have been testing laser velocimetry, a technique that utilizes the reflection of light from small particles in the flow to determine the magnitude and direction of the circulatory patterns. 1980 *Recent Adv. Surg.* X. 84 The first to apply the technique of Doppler ultrasound velocimetry to the detection of pulsatile arterial flow in the digital vessels.

velocipede. 1. (Earlier example.)

1818 W. Sewall *Diary* 19 June (1930) 53/2 Then I went to the circus and rode on the velocipede, which is a new machine.

velocity. Add: **1. c.** In scientific use, speed together with the direction of travel, as a vector quantity.

1847 *Proc. R. Irish Acad.* III. 345 We may always imagine a succession of straight lines, or vectors, to be drawn from some one point, as from a common origin, in such a manner as to represent, by their directions and lengths, the varying directions and degrees (or quantities) of the velocity of the moving point. 1873 J. C. Maxwell *Treat. Electr. & Magn.* I. 9 The velocity of a body, its momentum, ..an electric current, ..are instances of vector quantities. 1883 *Encycl. Brit.* XV. 680/1 We are concerned only with what we may call the 'speed' of the motion. (We purposely avoid the use of the term 'velocity' here, because it properly includes direction as well as speed.) 1965 A. F. Abbott *Ordinary Level Physics* v. 50 In ordinary conversation the word 'velocity' is often used in place of speed. In science, however, it is important to distinguish between these two terms.

d. *Econ.* The rate at which notes and coins change hands; the rate of spending in an economy.

1909 I. Fisher in *Jrnl. R. Statistical Soc.* LXXII. 618 When we know statistically the velocity of circulation of money we shall be in a position to study inductively the 'quality theory' of money. 1911 — *Purchasing Power of Money* ii. 17 Velocity of circulation, or rapidity of turnover, is simply the quotient obtained by dividing the total money payments for goods in the course of a year by the average amount of money in circulation by which these payments are effected. 1930 J. M. Keynes *Treat. Money* II. xxiv. 20 The expression 'velocity (or rapidity) of circulation' first came into use before the development of the cheque system... The 'velocity' measured the average frequency with which a token..was used..to change hands. 1967 *Economist* 19 Oct. 209/1 The note issue is a good indicator because the velocity of notes, unlike that of deposits, is fairly steady. 1982 *Chase Economic Observer* Jan.–Feb. 3/1 Velocity, the rate of turnover of money, is typically measured as the ratio of GNP to the narrowly defined money stock.

3. velocity head [Head *sb.* 17], the velocity pressure of a fluid expressed in terms of the height from which the fluid would have to fall to attain the velocity exerting this pressure; **velocity microphone,** a microphone whose diaphragm is freely exposed to the air on both sides and so responds to the particle velocity within a sound wave rather than its pressure; **velocity potential** [tr. G. *geschwindigkeitspotential* (H. von Helmholtz 1858, in *Jrnl. für die reine u. angewandte Math.* LV. 25)], a scalar function of position such that its space derivatives at any point are the components of the fluid velocity at that point; **velocity pressure,** that part of the total pressure exerted by a fluid which is due to the velocity it possesses.

[1881 *Encycl. Brit.* XII. 462/2, $v^2/2g$ may be termed the head due to the velocity *v*.] 1884 A. Daniell *Text Bk. Princ. Physics* xi. 276 We may say that the velocity-head and the pressure-head are together equal to the total head. 1937 O'Brien & Hickox *Appl. Fluid Mech.* ix. 174 The velocity head may be used in Bernoulli's equation is the average kinetic energy per unit weight of water flowing. 1979 A. L. Lydersen *Fluid Flow & Heat Transfer* i. 5 (*caption*) Pressure head ($p/\rho g$) and velocity head ($V^2/2g$) for frictionless flow from point 1 to point 2. 1931 H. F. Olson in *Jrnl. Soc. Motion Picture Engineers* XVI. 695 The ribbon microphone..can therefore very appropriately be termed a 'velocity microphone'. 1951 A. Smithson *Fund. Radio Communications* xiii. 281 Velocity microphones..may be designed to be unidirectional in their response. 1978 V. Capel *Microphones in Action* ii. 19 The polar diagram of a velocity microphone is different from anything we have discussed so far. 1867 P. G. Tait tr. Helmholtz in *Phil. Mag.* XXXIII. 485 In integrating the hydrodynamical equations, the assumption has been made that the components of the velocity of each element of the fluid in three directions at right angles to each other are the differential coefficients, with reference to the coordinates, of a definite function which we shall call the velocity-potential. 1907 F. W. Lanchester *Aerodynamics* iii. 91 Fluid in irrotational motion has a velocity potential. 1937 [see *stream function* s.v. *stream sb.* 9 c]. 1980 Bober & Kenyon *Fluid Mech.* ix. 417 The velocity potential, Φ, or the stream function, Ψ, are often introduced into fluid-flow problems because they frequently reduce the difficulty in obtaining a solution to a particular problem. 1904 *Proc. Inst. Mech. Engin.* Feb. 298 They used dry, clean air, and therefore it was possible to keep the Pitot tube extremely small, and to measure the static pressure in the close neighbourhood of the point at which the velocity pressure was measured.

1959 N. C. Harris *Mod. Air Conditioning* xv. 293 Velocity pressure is best measured by a Pitot tube combined with a draft gage which reads in inches of water. 1969 *Oceanology* IX. 585 The instrument is based on the measurement of the velocity pressure created by the wind.

velodrome. (Further examples.)

1973 *Trinidad Guardian* 1 Feb. 18/9 Neither the all-weather athletic track nor the cycling velodrome will be trampled upon at Carnival time. 1975 *Daily Colonist* (Victoria, B.C.) 22 Aug. 19/6 Taillibert designed the 70,000-seat stadium, velodrome, an underpass..and the competition swimming pool.

velodyne (ve·lo-, vī·lodain). *Electr.* [f. L. *vēlo-x* swift + Gr. δύν-αμις force.] A device in which the output of a tachogenerator is fed back so as to keep the rotational speed of a shaft proportional to an applied voltage.

1947 Crowther & Whiddington *Science at War* 83 The Velodyne..is an electro-magnetic system that integrates the resultant of several motions. 1952 *Electronic Engin.* XXIV. 382 An electro-mechanical differential analyser, employing velodynes. 1972 *Internat. Jrnl. Control* XVI. 37 The amplified output is fed to the split field of the velodyne motor, the armature of which is supplied from a constant current source.

velometer (velǫ·mitəɹ). [Irreg. f. Velo(city + -meter or by contraction of Velocimeter.] **† 1.** A kind of governor for a marine steam engine. *Obs.*

1878 *Engineer* 13 Sept. 190/1 A marine governor or anti-racer... The 'Velometer'..has..control entirely by force of its own merits.

2. An instrument for measuring the speed of air, or of an aircraft through the air.

1914 *Flight* 24 Jan. 82/2 The velometer..balances a pressure due to the speed through the air against the apparent weight of a column of liquid. 1939 *Jrnl. R. Aeronaut. Soc.* XLIII. 913 The design and application in mines of a spring-controlled vane-type air flow meter known as the Velometer. 1974 *IEEE Trans. Instrumentation & Measurement* XXIII. 203/1 A sonic velometer uses the passage of sound waves through air to determine the air velocity.

velours. 2. a, b. (Earlier examples.)

1794 A. Young *Trav. France* (ed. 2) I. xix. 550 Rouen. ..At present, the velours and *cotton toiles* are the most flourishing. The fabrics spread over all the country... They have also some woollens. 1805 T. Fremantle *Let.* 14 Nov. in Wynne *Diaries* (1940) III. vi. 234 Mistress Tittler with a black Velour pelisse. 1822 M. Edgeworth *Let.* 20 Feb. (1971) 357 My sage-colored French velours simule pelisse and Fanny and Harriets purple ditto are quite the thing for carriage visits.

|| velouté (*velute*). *Cookery.* [Fr., = velvety.] In full *velouté sauce.* A white sauce made with chicken or veal stock.

1830 [see *Ravigote*]. 1835 E. A. Poe in *Southern Lit. Messenger* May 516/2 He mentioned..Muriton of red tongue and Cauliflowers with Velouté sauce. 1868, etc. [see *Allemande sb.* 3]. 1936 Lucas & Hume *Au Petit Cordon Bleu* 71 Have ready ½ pint of *velouté* sauce. 1948, 1965 [see *Suprême sb.* (a.) 1]. 1973 *Times* 6 Dec. 9/6 The vegetable and fish chefs will have made the velouté sauce, lobster sauce and a little bechamel sauce.

veltheimia (velthai·miă). [mod.L. (J. G. Gleditsch 1769, in *Nouveaux Mém. Acad. R. Sci. Berlin* 66), f. the name of August Ferdinand, Graf von Veltheim (1741–1801), German patron of botany + -ia¹.] A bulbous plant of the genus of this name, belonging to the family Liliaceæ, native to South Africa, and bearing thick, oblong leaves and spikes of pink, red, or yellow flowers.

1806 *Curtis's Bot. Mag.* XXIII. 1091 (*heading*) Glaucous-leaved veltheimia. 1946 M. Free *All about House Plants* (1979) 199 (*caption*) Veltheimia, a dependable winter-flowering bulb with glossy green foliage. 1980 *Flower & Garden Mag.* Sept. 24 The remarkable veltheimia deserves more attention as a houseplant.

velvet, *sb.* Add: **I. 1. c.** *to tip the velvet* (example).

1821 P. Egan *Real Life in London* I. ix. 182 And when they were jolly well primed, they got a regular Velvet, in which they would continue..their own life 'Let that go for now... We began to hang their nobs. And tip their velvet...' [illegible]

d. (Earlier example.)

1749 J. Cleland *Mem. Woman Pleasure* II. 114 The deceiving him became so easy, that it was perfect playing upon velvet.

f. Gain, profit, winnings; *to the velvet,* to the good. *slang.*

1901 S. E. White *Westerners* xxiii. 228 They's a good many ton of ore in four hundred foot of shaft.'..'Let that go for now... We can call that 'velvet'.' 1908 K. McGaffey *Sorrows of Show Girl* 240 Before the whistle blew for dinner I was several hundred to the velvet. 1912 F. Irwin *Fine Points Auction Bridge* 56 Do your doubling early in the rubber (so as to pile up 'velvet' for yourself). 1940 Wodehouse *Eggs, Beans & Crumpets* 138 It would be money for jam... Just so much velvet. 1942 *Amer. Speech* XVII. 93/2, I have been taking in plenty of velvet these days working the Fair. 1951 E. Paul *Springtime in Paris* ii. 38 A good French mechanic..would have to work two and one half days to earn 2,430 francs, which on account of taxes..would not be all velvet.

g. A velvet dress.

1851 E. Ruskin *Let.* 6 Nov. in M. Lutyens *Effie in Venice* (1965) II. 212, I had on my black velvet because it was mourning. 1944 R. Lehmann *Ballad & Source* v. v. 222 We stood revealed in our long-sleeved velvets—Jess's sapphire blue, mine claret-coloured. 1963 N. Streatfeild *Vicarage Family* ii. 16 Isobel's velvet was of a pale green with a very full skirt.

II. 4. c. *velvet-blue, -brown, -green, -red.*

1924 'J. Sutherland' *Circle of Stars* iv. 26 She looked up sharply to see herself in the mirror, a rather pale face..and dusky velvet-brown eyes. *Ibid.* ix. 95 Overhead the sky is deep velvet-blue all a-fire with stars. 1952 A. G. L. Hellyer *Sanders' Encycl. Gardening* (ed. 22) 31 *Sanderianus,* large, velvet-green with copper-red veins. 1976 *Flintshire Leader* 10 Dec. 27/1 (Advt.), New Fords for immediate delivery... Granada 2000GL, Velvet Red.

5. *velvet-coated.*

1881 O. Wilde *Poems* 74 After yon velvet-coated deer the virgin maid will ride. 1926 M. Leinster *Dew on Leaf* II. vi. 227 Your pretty face can charm of itself,..as a pale rose standing still lures the velvet-coated bee.

7. velvet carpet, a cut-pile carpet similar to Wilton; **velvet glove,** an appearance of suavity and gentleness of manner, esp. one that masks determination or inflexibility (cf. *iron hand* s.v. *iron a.* 3 c); also *attrib.* (with hyphen); **velvet-painting** (earlier example); **velvet sauce** = *velouté.*

1860 Geo. Eliot *Mill on Floss* II. iv. iii. 188 Good society has its claret and its velvet carpets. 1908 L. M. Montgomery *Anne of Green Gables* xxix. 325 'Velvet carpet,' sighed Anne luxuriously, '*and* silk curtains!' 1979 A. S. Garstein *How-To Handbk. Carpets* ii. 15 At first velvet carpets were woven of solid colors in pile heights ranging from closely woven low pile to longer 'plush' fabrics. 1850, etc. Velvet glove [see *iron a.* 3 c]. 1946 W. S. Maugham *Then & Now* xxii. 125 The velvet glove was off and the mailed fist was bared. 1969 S. Hyland *Top Bloody Secret* II. 202 Concentrated velvet-glove charm. 1973 'M. Innes' *Appleby's Answer* iv. 41 Blackmail..of the very most genteel and velvet-glove sort. 1809 *Charges against H.R.H. Duke of York* 386/1 Did he ever instruct you in velvet-painting? 1893 T. F. Garrett *Encycl. Pract. Cookery* II. 398/2 Velvet sauce (Velouté). 1952 E. White *Good Eng. Food* iv. i. 171 Aspic Cream. Take ½ pint of liquid aspic jelly and mix it with ½ pint of velvet sauce.., ½ pint thick cream and four sheets of gelatine.

b. velvet ant (examples); also, a parasitic wasp of the family Mutillidæ, having a velvety appearance.

1748 M. Catesby *Nat. Hist. Carolina* App. 13 The Velvet Ant... The whole body and head resembled crimson velvet. 1842 T. W. Harris *Insects Injurious to Vegetation* 14 Stinging velvet-ants..are predaceous in their habits. 1932 E. Step *Bees, Wasps, Ants Brit. Isles* 57 The Velvet-ants are not Ants: the name is only a courtesy title given when..the wingless female..was a more familiar object than the winged male. 1968 T. Wolfe *Electric Kool-Aid Acid Test* xxi. 305 He can *feel*..every verruga fly, velvet ant, murine fleas and crabs. 1975 *Sci. Amer.* Dec. 115/1 The natural enemies of *Bembix* include mutillid wasps (sometimes called velvet ants).

velvet, *v.* Add: **2.** Also *fig.*

1959 R. Graves *Coll. Poems* 96 We velveted our love with fantasy Down a long vista-row of Christmas trees.

velveteen. Add: **1. c.** A dress of velveteen.

1873 *Young Englishwoman* Aug. 414/1 Would a black velveteen be suitable for the autumn? 1932 'E. M. Delafield' *Thank Heaven Fasting* II. i. 156 Run upstairs and put on the green velveteen. It suits you.

velvetiness. Add: Also *fig.*

1906 Galsworthy *Man of Property* II. iv. 167 The peculiar exasperation, velvetiness, and mockery, of which Bosinney's manner had been composed.

Venda (ve·ndă), *sb.* and *a.* [An African word; since 1973 the name of a self-governing Bantu homeland in north Transvaal.] **A.** *sb.* **a.** (A member of) a Bantu people inhabiting the north-eastern Transvaal and southern Zimbabwe. **b.** The language of this people. **B.** *adj.* Of or pertaining to this people or their language.

[1905 *Jrnl. Anthropol. Inst.* XXXV. 365 That part of the Bantu race of which I have to treat in this paper calls itself *Bawenda,* that is to say, people of Wenda, or inhabitants of Wenda.] 1908 *Jrnl. Afr. Society* VII. 412 Venda. Spoken by the Bavenda in North Transvaal. 1912 H. A. Junod *Life S. Afr. Tribe* I. i. 18 We find colonies of emigrated Thonga in many parts of the Transvaal. The largest is in the Spelonken district... There they are called *Magwamba* by the Venda and Bvesha who possessed the country before them. *Ibid.* II. vi. i. 327 There is a third tradition relating to the first man, but it has..been borrowed from the Venda or Pedi tribes. 1921 *S. Afr. Jrnl. Sci.* Apr. 208 All these Venda clans (and the Malemba also) speak of their mountain. 1931 H. A. Stayt *BaVenda* p. xi, These boys..were familiar with the whole country and understood the different Venda dialects. 1937 [see *inkosi*]. 1950 I. Schapera in Radcliffe-Brown & Forde *Afr. Syst. Kinship & Marriage* 144 Paternal aunts..do not command anything like the great authority they are said to possess among the Venda. 1977 *Times* 23 Nov. 20/8 We are oppressed..not as Zulus, Xhosas, Vendas or Indians. We are oppressed because we are black. 1979 J. Drummond *I saw him Die* vii. 74 She began talking in some dialect I didn't understand—Venda, perhaps. 1979 *Guardian Weekly* 12 Aug. 6/3 In last year's general election, Chief Mphephu's ruling Venda National Party was decisively defeated by the Opposition Venda Independence Party.

|| vendange (vandaʒ). Also † 8 **vandange.** [Fr.: see Vendage.] In France, the annual grape harvest, the vintage (sense 2).

1766 Smollett *Trav.* I. viii. 133 The mountains of Burgundy are covered with vines... The *vandange* was but just begun, and the people were employed in gathering the grapes. 1852 Mrs. E. Twisleton *Let.* 9 Oct. (1928) iv. 56 The *vendange* for the year is over, but the crop is not good and..they say some of the best wines will not come from it. 1893 Somerville & 'Ross' *In Vine Country* vii. 147 Of course we left vowing to return for the *vendange* next year. 1944 W. Fortescue *Mountain Madness* xi. 83 And when the *jasmin* harvest was over, there was the *vendange,* my grapes to be picked and made into wine. 1969 B. Weil *Dossier IX* xiii. 180 It's a good time to go south. The vendange is just starting. 1972 *Daily Tel.* 8 Apr. 14/7 Last year's *vendange.*

Hence **vendangeur,** a grape-picker.

1893 Somerville & 'Ross' *In Vine Country* v. 92 We had to drive for some distance before we saw the first group of *vendangeurs,* stooping waist-deep in the vines, snipping off the bunches and putting them into square wooden baskets. 1971 *Country Life* 2 Dec. 1501/3 The carver of a misericord in Ely cathedral certainly knew what the job was like when he displayed an Ely *vendangeur* of 1340 or thereabouts.

|| Vendémiaire (vandemiɛr). [Fr., f. L. *vindēmia* vintage.] The first month of the French republican calendar, introduced in 1793, extending from Sept. 22 to Oct. 21.

1799 *Times* 1 June 3/1 In the month of last Vendemiaire, year 7. 1811 Mrs. M. Blackburne [illegible] vi. 267 An important measure of police law..was not presented and voted until the 6th and 7th of *Vendémiaire* of the year IV. 1933 R. Matthews *French Revolution* xx. 293 Bonaparte..with the true adventurer's instinct accepted the command (Vendémiaire 13). 1982 *Encounter* Dec. 32/1 *Vendémiaire* could be joyfully evoked..by those who had not the first idea of the process of wine-making.

|| vendemmia (vende·mia). [It.] The Italian grape-harvest or vintage. Cf. *vendange, *vendimia.

1826 M. Kelly *Reminisc.* I. 179, I..was delighted by the appearance of the elegant villas..belonging to noble Venetians, who, during the *vendemmia* season, pass their *vendemmia* there. 1871 C. M. Yonge *Little Lucy's Wonderful Globe* iii. 21 Ah, ah! 'tis the *vendemmia!* all may eat grapes. 1926 D. H. Lawrence *Let.* 17 Oct. (1962) II. 942 The peasants finished the *vendemmia* two days after we got here—but the wine is still to be made. 1978 J. Pearson *Façades* xxviii. 473 The vineyards had yielded a record crop in the *vendemmia* that year.

vendeuse (vandöz). [Fr.] A saleswoman; *spec.* one employed in a fashion house.

1913 E. Wharton *Custom of Country* xiii. 181 Slender *vendeuses* floating by in a mist of opopanax. 1926 'R. West' *Thinking Reed* ii. 61 This girl had been a *vendeuse* with a great French couturier. 1957 M. Sharp *Eye of Love* iii. 37 In the show-room Miss Molyneux, *vendeuse* and model, and Miss Harris, who fitted, were as usual discussing the private lives of film-stars. 1976 *Times Lit. Suppl.* 3 Sept. 1074/5 Went into a smart milliners and asked the *vendeuse* to 'show me a hat for an ugly middle-aged woman whose husband no longer loves her'.

|| vendimia (vendi·mia). [Sp., = vintage.] The Spanish grape harvest; also, the festival celebrating the end of the vintage. Cf. *vendange, *vendemmia.

1965 *Listener* 30 Sept. 493/3 The *vendimia* or harvest festival at Jerez..is a time of gaiety and laughter. 1975 *Harpers & Queen* May 93/2 The well-known *vendimia* or wine festival. 1978 *Times* 21 Nov. 12/2 The crowning of the queen of the vendimia or wine harvest. 1979 *Country Life* 25 Oct. 1392/1 (*caption*) Last month's ceremony was the 32nd Vendimia to be celebrated in Jerez de la Frontera.

vending, *vbl. sb.* Add: **2.** Special Comb.: **vending machine,** a slot machine from which comestibles or other small goods may be obtained.

1895 *Funk's Stand. Dict.,* *Vending-machine,* a machine having a mechanism, controlled by the dropping of a coin into a slot, for delivering any small article which it has been charged, as a slot machine for selling chewing-gum. 1947 J. C. Rich *Materials & Methods of Sculpture* i. 15 A ridge about..to low relief is used for coins, with a raised rim about the edge of the coin to protect the relief from such wearing factors as vending machines, turnstiles, and stacking. 1980 *Times* 31 May 2/8 The tea trolley is being wheeled back into offices and the anonymous vending machine, the butt of many jokes..is on the way out.

Venedotian, *a.* (Further example.)

1877 J. Rhys *Lect. Welsh Philol.* iv. 145 Here also may be mentioned..the Venedotian versions of the Laws of Wales, which Aneurin Owen found to be in manuscripts of the 12th century.

veneer, *sb.* Add: **4*.** *Dentistry.* = *veneer crown* in sense 5 below.

1930 G. M. Hollenback in I. G. Nicholls *Prosthetic Dentistry* xiii. 653 It..does not provide as good retention, nor as good support for the abutment tooth as a partial veneer. 1965 L. A. Weinberg *Atlas Crown & Bridge Prosthodontics* xii. 232/1 Plastic veneers or porcelain

jacket restorations should be tried in the mouth before final finishing. 1975 D. Stananought *Laboratory Procedures Inlays, Crowns & Bridges* ii. 22 Full veneers may be constructed entirely in metal, or in a combination of metal with either acrylic resin or porcelain on the labial surface.

5. veneer crown *Dentistry,* a crown in which the restoration is placed over the prepared surface of a natural crown.

1927 *Dental Cosmos* LXIX. 951 The next application [of porcelain to dental restoration]..was the so-called porcelain veneer crown which was the progenitor of the present highly perfected porcelain-jacket crown. 1954 J. E. Ewing *Fixed Partial Prosthesis* xi. 61 The three-quarter partial veneer crown is primarily concerned with esthetics, for which function it owes its origin. 1975 D. Stananought *Laboratory Procedures Inlays, Crowns & Bridges* ii. 21 The restoration..may be a full veneer crown (shell crown) with the restoration covering the whole of the crown, or a partial veneer crown (three-quarter crown) with the labial or buccal surface of the crown excluded from the restoration.

venepuncture (ve·ni-, vī·nipɒ:ŋktiŭɹ). *Med.* Also **veni-.** [f. L. *vēna* Vein *sb.* + Puncture *sb.*] Puncture of a vein, esp. with a hypodermic needle to withdraw blood or for intravenous injection.

1923 *Jrnl. Biol. Chem.* LVI. 106 Blood was drawn by venepuncture from a normal man. 1935 Whitby & Britton *Disorders of Blood* xxiv. 475 Blood is obtained by venipuncture without using a tourniquet. 1957 W. Martin in R. K. Merton *Student-Physician* 196 Being able to do a venipuncture without difficulty. 1974 F. Ellis in R. M. Kirk et al. *Surgery* viii. 172/1 A period of six weeks is necessary before the veins are sufficiently dilated to be used for regular venepuncture. 1979 *Arizona Daily Star* 5 Aug. (Advt. section) 3/4 Venipuncture hospital experience preferred.

veneratingly, (ve·nĕrēɪtiŋli), *adv. rare*⁻¹. [f. Venerating *ppl. a.* + -ly².] In a reverential manner.

c 1925 V. Woolf in *Mrs Dalloway's Party* (1973) 68 She and Bertram sat down on deck chairs, she looked at the house veneratingly, enthusiastically.

venereal, *a.* and *sb.* Add: **2. d.** *sb.* A person with venereal disease.

1788 W. Bligh in R. M. Bowker *Mutiny!!* (1978) vi. 266 Three Venereals in the List. 1933 *Sun* (Baltimore) 23 Feb. 8/3 A veneral draws $257.50 a month, or more than a man blinded by an enemy shell.

venereally (vĭni°·riăli), *adv.* [f. Venereal *a.* + -ly².] **a.** By sexual intercourse. **b.** With venereal disease. *rare.*

1945 [see *bejel*]. 1973 *Sci. Amer.* Apr. 119/2 His [*sc.* Linnæus's] chief practice was among the venereally ill young rakes of Stockholm. 1978 *Nature* 28 Sept. 334/2 Genital warts..are benign tumours which are mostly venereally transmitted.

venereology. Add: (Examples.) Hence **venereolo·gical** *a.*; **venereo·logist,** an expert or specialist in venereology.

1894 Gould *Dict. Med.* 1595/2 Venereology. 1934 Webster, Venereologist. 1944 *Jrnl. R. Army Med. Corps* LXXXIII. 102 Non-gonococcal urethritis..is worrying many venereologists nowadays. 1946 *Nature* 17 Aug. 242/2 (*heading*) Adviser in venereology to the War Office. 1961 *Lancet* 29 July 246/2 Their study..included others admitted to non-venereological departments. 1970 *Daily Tel.* 25 Sept. 6/8 Permissive doctors..were condemned yesterday by Dr Ambrose King, consulting venereologist to the London Hospital. 1972 *Lancet* 10 June 1295/1 He could not see why venereology should be considered a specialty in its own right. 1979 D. Barlow *Sexually Transmitted Diseases* iii. 29 This is the only country in the EEC..that recognizes venereology as a distinct specialty. 1982 *Acta Dermato-Venereologica* LXII. 367/2 Every patient with a finding or history of Pediculosis pubis should be offered a venereological examination including test for both gonorrhea and syphilis.

venerous, *a.* Restrict † *Obs.* to senses 1 and 3 and add: **2.** (Later example.) Now *rare.*

1968 W. D. Jordan *White over Black* i. 33 Long before the first English contact with West Africa, the inhabitants of virtually the entire continent stood confirmed in European literature as lustful and venerous.

venesector. Delete *rare*⁻¹ and add further examples. In mod. use, one who takes blood samples as a profession; a phlebotomist.

1976 *Nursing Times* 17 June A53/3 (Advt.), Venesector required to take blood samples from patients for Pathology Department. 1977 *Lancet* 5 Nov. 969/2 The use of a technician rather than a medically qualified venesector is economical and reduces the need to involve the family doctor in the somewhat troublesome, repetitive, and perhaps tedious routine of taking, packaging, and posting, blood-samples.

Venetian, *sb.* and *a.* Add: **A.** *sb.* **1. b.** The dialect of Italian spoken by the inhabitants of Venice.

1598 [see *Lombard sb.*¹ 1 c]. 1642 [see *Roman sb.*¹ 3 c]. 1852 E. Ruskin *Let.* 27 June in M. Lutyens *Effie in Venice* (1965) II. 328 You may imagine how I am put to it sometimes when three or four people question me at once in Venetian. 1901 M. Carmichael *In Tuscany* 99 Had Dante..written in Venetian..there would have been two classical languages in Italy today. 1921 R. L. Gales *Old-World Essays* 140 It is strange to think of these careless people, all talking Venetian in the clear Paduan air four hundred years ago. 1980 *Listener* 23 Oct. 528/2 Catalan..is to Castilian Spanish what Venetian is to Italian.

B. *adj.* **1.** Also, of or pertaining to the ancient Veneti of Gaul or Venetia. *rare.*

1866 W. P. Dickson tr. *Mommsen's Hist. Rome* IV. v. vii. 252 The legions expended their time and strength in the sieges of the Venetian towns. *Ibid.* 253 Caesar caused ..the people of the Venetian canton to the last man to be sold into slavery. 1897 *Archæol. Jrnl.* LIV. 393 Through the Venetian traders the beautiful southern designs..on the golden and bronze ornaments in Ireland might..have been introduced.

2. *Venetian blue,* a turquoise or cobalt blue; *Venetian-Gothic* adj. (earlier example); also *absol.* as *sb.*; *Venetian raised (point) lace,* a point lace in which all outlines are in relief; = *punto a rilievo* s.v. *punto¹ 7; also *Venetian (raised) point* (cf. *Venice point* s.v. Venice 1).

c 1840 Lady Wilton *Art of Needlework* vii. 75 A rich robe of Venetian blue embroidered with golden eagles. 1947 J. H. Bustanoby *Princ. Color* 12/2 *Cobalt blue,* similar brands: Cobalt ultramarine, Hungary blue,..Venetian blue. 1976 *Milton Keynes Express* 18 June 40/4 (Advt.), New Kitten Estate. Venetian blue. 1849 E. Ruskin *Let.* 3 Dec. in M. Lutyens *Effie in Venice* (1965) I. 81 We..looked over several Palaces... The outsides are splendid Venetian Gothic. 1933 J. Betjeman *Ghastly Good Taste* vii. 116 The red brick Flemish revival of Brondesbury and the Venetian Gothic of Kew. 1980 J. Lees-Milne *Harold Nicolson* I. ix. 170 They went up the Canal in a gondola, arguing about Venetian-Gothic. 1872 F. B. Palliser *Notes Hist. Lace* 27 Finest raised Venetian point. 1882 Caulfield & Saward *Dict. Needlework* 513/1 The raised Venetian Points were not worked before 1600. *Ibid.,* The Venetian Raised Points are extremely rich and varied. 1960 H. Hayward *Antique Coll.* 295/1 *Venetian raised point lace,* needlepoint. 1974 [see *punto¹ 7].

Hence **Vene·tianly,** *adv.*

1851 H. Melville *Moby Dick* II. xii. 91 Through all the wide contrasting scenery..flows one continual stream of Venetianly corrupt and often lawless life. 1965 *Guardian* 3 Apr. 7/5 If you want to eat well and Venetianly in Venice..go to the best restaurants.

Venetic, *a.* Add: Also, of or pertaining to the language of the Veneti.

1922 *Jrnl. Anthropol. Inst.* XXV. 214 Indiscriminate combinations of letters from the Venetic alphabet. 1949 M. S. Beeler *Venetic Lang.* 4 That the Venetic alphabet in particular is in fact derived from the Etruscan is now certain.

B. as *sb.* The language of the ancient Veneti.

1904 [see *Eteocretan a.* and *sb.*]. 1932 [see *Messapian sb.* and *a.*]. 1977 [see *Rhaetic sb.*].

Venezuelan, *a.* and *sb.* **a, b.** (Earlier examples.)

1820 *Ann. Reg. 1819* 242/1 On every point the efforts of the Venezuelans were crowned with success. *Ibid.* 242/2 The Venezuelan fleet.

venial, *sb.* For † *Obs.* read 'Now *rare*' and add later example.

1969 J. D. Crichton in J. Fitzsimons *Penance* ii. 32 One thing that the present system inhibits, with..its neat parcels of 'mortals' and 'venials'—the sin-grid, in fact—is the expression of the diffused feeling of sinfulness.

Venice. Add: **1.** *Venice lace, point* (earlier example); *Venice soap* (earlier examples).

1865 F. B. Palliser *Hist. Lace* iv. 45 It is not, however, till the reign of Elizabeth, that Italian cutworks and Venice lace came into general use. 1929 *Oxford Poetry* 3 Every fallen shadow weaves Venice lace for kindly aunts. 1974 *Times-Picayune* (New Orleans) 15 Aug. v. 6/1 A flounce capelet edged in venice lace. 1865 F. B. Palliser *Hist. Lace* iv. 47 This is our Rose (raised) Venice point, the Gros Point de Venise, the Punto a rilievo, so highly prized and so extensively used for albs, collerettes..and costly decoration. 1673 J. Ray *Observations Journey Low-Countries* 202 Venice-Sope..is very like and nothing inferiour to *Castile-Sope.* It is made of the best Oil Olive. 1792 J. Woodforde *Diary* 26 July (1927) III. 364 At D[itt]o. for Venice Soap 1. 0.

venidium (veni·diɒm). [mod.L. (C. F. Lessing 1831, in *Linnæa* VI. 91), f. L. *vēna* vein, in allusion to the ribbed achenes of some species.] An annual or perennial herb of the genus *Venidium,* belonging to the family Compositæ, native to South Africa, and bearing cream or yellow flowers.

1937 R. Hay *Annuals* 230 Venidiums prefer a light soil. 1962 [see *cleome*]. 1976 *Hortus Third* (L. H. Bailey Hortorium) 1147/2 Venidiums are usually treated as annuals in the garden.

venison. Add: **4. b.** *venison ham, pâté, steak.*

1772 B. Romans *Jrnl.* 16 Jan. in *Nat. Hist. Florida* (1775) 331, I purchased some bear, bacon and venison hams of them. 1788 M. Cutler *Jrnl.* 7 Sept. (1888) I. 419 Dined..on venison steak and squirrel pie; very good dinner. 1833 J. Hall *Harp's Head* 214 A little further up were venison steaks, then fried ham. 1860 E. J. Lewis *Jrnl.* 26 May in *Colorado Mag.* (1937) XIV. 219 Bought some venison ham for a dollar. 1975 *Harpers & Queen* May 68/2 The chef makes the most delicious venison paté. 1980 J. Wainwright *Man of Law* xliv. 205 Venison pâté sandwiches and watered-down whisky. 1980 C. & T. Conran *Cook Bk.* II. 248/1 Charcoal grilled venison. 1 venison steak from the leg or loin, weighing 2–3 lb.

Venizelist (venizēɪ·list), *a.* and *sb.* [f. the name of the Greek statesman Eleuthérios *Venizélos* (1864–1936) + -ist.] **A.** *adj.* Of, pertaining to, or supporting Venizélos or his political policies.

1915 *Times* 12 June 5/3 The Venizelist Press, in discussing the Government electoral programme, observes that the promised reforms are almost the same as those brought forward by M. Venizelos. 1916 *Morning Post* 25 Apr. 6/5 (*heading*) Repressing a Venizelist ovation. 1920 *Glasgow Herald* 4 May 7 Even the so-called Venizelist Divisions were nothing more than a collection of supers. 1931 *Times Lit. Suppl.* 19 Mar. 218/2 He confesses that..he was blown about by every wind of doctrine—now Venizelist, now Royalist. 1946 R. Capell *Simiomata* II. 61 There were senior officers of integrity, old Venizelist colonels like Tsamákos and Matsoukos in Thessaly. 1972 D. Dakin *Unification of Greece* xvi. 238 The majority of the government was moderate Venizelist. 1980 J. Lees-Milne *Harold Nicolson* I. vi. 84 Harold..was..put to work on..the supply of stores to the Venizelist forces in Egypt.

B. *sb.* A supporter or adherent of Venizélos or his political policies.

1915 *Times* 15 June 5/3 The processions formed by the Venizelists became so numerous that several attempts were made by the cavalry to disperse them. 1920 V. J. Seligman *Victory of Venizelos* 8 The views..of leading 'Venizelists' and 'Constantinians'. 1946 R. Capell *Simiomata* III. 127 There used to be furious feeling between Venizelists and monarchists. 1981 *Times Lit. Suppl.* 22 May 584/3 Is it not a little unfair to say that there have been no distinguished Cretans apart from King Midas and El Greco? The verdict will not please the Venizelists.

Hence **Venize·lism.**

1931 C. Mackenzie *First Athenian Memories* xv. 380 This result was not gained merely by abstentions in the Islands and the new territories where Venizelism was naturally predominant.

Venn diagram (ven). [Named after its inventor, John *Venn* (1834–1923), English logician.] A group of circles that may or may not intersect according as the logical sets they represent have or have not elements in common.

[1884 J. N. Keynes *Stud. & Exerc. Formal Logic* III. v. 207 The application of Mr Venn's diagrammatic scheme to syllogistic reasonings. 1894 *Ibid.* (ed. 3) III. iv. 298 Syllogisms in *Barbara* and *Camestres* may be taken in order to show how Dr Venn's diagrams can be used.] 1918 C. I. Lewis *Survey Symbolic Logic* i. 77 This method resembles nothing so much as solution by means of the Venn diagrams. 1952 W. V. Quine *Methods of Logic* 70 Whiteness of a region in a Venn diagram means nothing but lack of information. 1970 O. Dopping *Computers & Data Processing* i. 28 By means of Venn diagrams..set theoretic operations such as intersection can be shown in geometrical form. 1977 J. L. Harper *Population Biol. Plants* xxiii. 725 These results are interpreted as a Venn diagram in Fig. 23/11a to show presumed niche relationships between the groups of species.

veno- (vīno), Comb. form of L. *vēna* Vein *sb.*, employed in terms relating to the vascular system, as **venocly·sis** [Gr. κλύσις drenching], the introduction of liquid into the circulation by an intravenous drip; **venoconstri·ction,** constriction of a vein; **veno-occlu·sive** *a.*, characterized by occlusion of veins: applied esp. to a tropical disease in which this is the chief pathological feature; **ve·nospasm,** sudden, transient contraction of a vein; **venosta·sis,** a reduction (induced or spontaneous) in the flow of venous blood from a part of the body.

[1910 M. F. Donahoe *Man. of Nursing* ix. 166 In preparing for a veneclysis the nurse should take as much care as for an abdominal operation.] 1926 *Texas State Jrnl. Med.* XXI. 664/2, I regard the main indications for venoclysis as being any form of grave collapse, grave toxemia, prolonged shock, or serious dehydration from any cause. 1961 *Lancet* 2 Sept. 538/2 Some such device is essential for venoclysis in infancy. 1937 K. J. Franklin *Monograph on Veins* x. 127 Perfusion of the corpus striatum with hot and cold fluids resulted in superficial venodilation and venoconstriction respectively. 1977 *Proc. Soc. Med.* LXX. 691/2 Blood samples were taken without venoconstriction at 13:00 and 14:00. 1954 G. Bras et al. in *Arch. Path.* LVII. 285 (*heading*) Veno-occlusive disease of liver with nonportal type of cirrhosis, occurring in Jamaica. 1969 Edington & Gilles *Path. in Tropics* xi. 489 It is now generally accepted that veno-occlusive disease in the West Indies is due to ingestion of 'bush tea' containing the alkaloids of *C[rotolaria] fulva.* 1950 P. Wood *Dis. Heart & Circulation* i. 13 Venospasm is avoided by proper skin anæsthesia, and by choosing a catheter that is not too large for the vein. 1977 *Lancet* 1 Jan. 29/2 The damage is caused by pulmonary venospasm (occurring as the shock is relieved). 1931 R. J. E. Scott *Gould's Med. Dict.* (ed. 3), Venostasis. 1965 *Thrombosis & Diathesis Hæmorrhagica* XIV. 501 Venostasis was induced by means of a sphygmomanometric cuff (systolic pressure diminished by 20 mmHg). 1976 *Lancet* 11 Dec. 1265/2 Despite the presence of venostasis in the legs, intermittent compression of the arms during and after surgery reduced the incidence of deep venous thrombosis..in the legs to half that in control patients.

venography (vēnǫ·-, vīnǫ·grăfi). *Med.* [f. *veno- + -graphy.] Radiography of a vein after injection of a contrast medium.

1935 *Arch. Surg.* XXXI. 272 Venography..showed an obstruction in the axillary vein distal to the first rib. 1980 *Brit. Med. Jrnl.* 18 Oct. 1039/1 No arteriovenous malformation was detected by venography or arteriography.

Hence **venogra·phic, -ical** *adjs.*; **venogra·phically** *adv.*; also **ve·nogram,** a radiogram of a vein.

1935 *Arch. Surg.* XXXI. 272 A venogram made with a stabilized solution of thorium dioxide on the third day after admission showed a point of obstruction in the axillary vein distal to the first rib. 1940 *Surg., Gynecol. & Obstetr.* LXXI. 701/1 Venographic studies were carried out. 1940 *Acta Chir. Scand.* Suppl. LXI. 22 The venographical method has proved to be a very valuable diagnostic aid in manifest and suspect..thrombosis. 1940 Venographically [see *thromboembolic* adj. s.v. *thrombo-*]. 1972 *Science* 16 June 1236/3 A considerable advantage of these images [produced by ultrasonic Doppler technique] is that they look like present-day arteriograms and venograms. 1978 *Lancet* 12 Aug. 331/2 Limbs with chronic venous insufficiency which had no venographically detected thrombosis. 1980 *Ibid.* 16 Feb. 332/2 We found venographic evidence of compression of popliteal veins by such a cyst in only 1 case.

vent, *sb.*² Add: **I. 2. c.** *full vent,* advb. phr., at full pitch; to the utmost of one's capacity.

1927 D. H. Lawrence *Mornings in Mexico* 11 It is so unlike him, to be whistling full vent, when any of us is around.

II. 10. f. = *port *sb.*³ 4 d.

1940 *Electronics* Mar. 54/2 The vent should be located near the speaker... The vent areas need not necessarily be circular. 1975 G. J. King *Audio Handbk.* vi. 143 (*caption*) Inside of Rectavox Omni Mk II loudspeaker system, showing the bass and treble units, the tube extension from the vent and the frequency divider network at the bottom.

vent, *sb.*⁴ Theatr. slang abbrev. of Ventriloquist.

1893 R. Ganthony *Practical Ventriloquism* III. 89 The Vent: does not suffer provided he makes capital out of unforeseen interruptions. 1945 L. Lane *How to become Comedian* xiv. 116 When an imaginary character answers from the roof the 'vent' looks upwards. 1976 *National Observer* (U.S.) 4 Sept. 6/3 We've got magicians here... We've got jugglers, mentalists, clowns, and vents.

vent, *v.*² **I. 1. a.** Delete † *Obs.* and add later examples. Also, to empty (a confined space) of gas in this manner.

1947 J. C. Rich *Materials & Methods of Sculpture* xi. 355 Molds may be vented to permit the ready escape of air from undercuts. 1969 *Times* 23 May 1/2, To close their hatch, the pressure in the tunnel had to be lowered. 'I am not able to vent the tunnel,' Commander John Young reported at 6.15 p.m...last night. 1978 *Daily Tel.* 18 Aug. 30/6 The balloon's crew were then able to..fall into the most likely airstream by..venting the balloon—letting out the helium and allowing it to drop.

2. a. Delete † *Obs.* and add later examples. Freq. *pass.*

1962 F. I. Ordway et al. *Basic Astronautics* v. 197 After arrival on the moon the fluid is vented. 1969 *Daily Mail* 15 Jan. 5/4 The rocket..vented quite a lot of fuel overboard and the fuel formed millions of ice particles. 1980 *Nature* 29 May 278/3 A total of 10 millicuries of krypton-85 was vented to the atmosphere during the procedure and the engineers received a whole body radiation dose of 10 to 15 mrem. 1983 *Sci. Amer.* Apr. 80/1 The pilot vented the ballast tanks, surrounding *Alvin* with a column of bubbles.

II. 11. Delete 'Now *rare*' and add later examples. Also used of a force causing an outlet to be made.

1966 *Economist* 19 Feb. 686/3 Some of these test explosions 'vent' through the earth's surface and thus contaminate the atmosphere. 1970 *Times* 15 Apr. 1/7 The particles have diminished greatly—almost ceased now, which indicates maybe what was venting has almost stopped. 1980 *Courier-Mail* (Brisbane) 27 Nov. 31/4 Fuel was venting from the tanks. The loss was so great that it was doubted that they would make it to an airfield.

vented, *ppl. a.* **b.** (Later example.)

1977 *Sci. Amer.* Apr. 27/2 Fortunately the vented gas did not ignite.

vented (ve·ntĕd), *a.* [f. Vent *sb.*² + -ed².] Equipped with vents or apertures. Cf. *ported *a.*¹ 2.

1940 *Electronics* Mar. 34/1 The vented speaker enclosure uses a conventional cone type dynamic speaker mounted in a box lined with sound absorbent material. 1957 *IRE Trans. on Audio* V. 38/2 The vented enclosure allows the hf speaker to be mounted with its centroid to lie on the same point as that of the lf speaker. 1975 [see *ported *a.*¹ 2].

venter¹. Add: **I. 3*.** *Bot.* The enlarged, basal part of an archegonium, where the egg cells develop.

1887 Balfour & Garnsey tr. *K. Goebel's Outl. Classification & Special Morphol. Plants* 175 The archegonium when fully formed consists of a thick and rather long stalk, a roundish-ovoid venter resting on the stalk, and above it a long slender neck usually twisted on its axis. 1938 G. M. Smith *Cryptogamic Bot.* II. ii. 17 The mature venter is therefore 12 to 20 cells in perimeter instead of six cells as in the neck. 1978 T. L. Hufford *Botany* vii. 177 The archegonia are frequently long stalked with an only slightly enlarged venter (egg chamber) and an elongated neck.

ventifact (ve·ntifækt). [f. L. *vent-us* Wind *sb.*¹ + -i- + *fact-us,* pa. pple. of *facere* to make.] A faceted stone shaped or altered by wind-blown sand.

1911 J. W. Evans in *Geol. Mag.* Decade V. VIII. 335 If a general expression be required for any wind-shaped stone, we might speak of a 'ventifact', on the analogy of artifact, sometimes spelt 'artefact',..and of 'ventiduct', which has been employed in architecture. 1935 *Times* 28 Jan. 15/4 Dr. Spencer brought back many fine examples of wind-worn rock and ventifacts for further study. 1944 [see *dreikanter*.] 1949 K. P. Oakley *Man the Tool-Maker* ii. 10 Stones splintered by fire, or faceted by sandstorms (ventifacts or dreikanters) are occasionally mistaken for the work of man. 1970 [see *dreikanter*].

ventilate, *v.* Add: **I. 5. a.** Now *Obs.*

c. *trans.* To supply air to (the lungs); to supply air, esp. artificially, to the lungs of; also *transf.*

1919 Flack & Hill *Textbk. Physiol.* xxxii. 288 It is only when the lungs are well ventilated that the parts most remote from these surfaces of direct expansion are brought properly into action. 1946 J. F. Fulton *Howell's Textbk. Physiol.* (ed. 15) xxxix. 871 Under normal rest conditions..5·6 litres of air..are available to ventilate the alveoli. 1971 *Nature* 21 May 181/1 When necessary the lungs were ventilated mechanically with a Palmer respiration pump. 1975 *Ibid.* 23 Oct. 674/1 Animals were artificially ventilated with a mixture of N_2O–O_2–CO_2. 1978 *Sci. Amer.* Aug. 95/2 Heating the hypothalamus caused the fish to ventilate its gills faster. 1979 *Daily Tel.* 6 Nov. 3/6 All they had in fact when the plug was pulled was a corpse being ventilated by a machine.

8. c. *slang.* To shoot (someone or something) with a gun, usu. to kill. Also of a bullet: to make a hole in (something).

1875 C. B. Lewis *Quad's Odds* 473 Some of our folks cleaned up their revolvers..hoping to get a shot at McGrady and to ventilate the mule. 1917 [see *cockpit* 3 c]. 1948 'R. Macdonald' in H. Q. Masur *Murder Most Foul* (1973) 103 'A man was shot in one of his rooms.'..'Who was it got himself ventilated?' 1979 C. Egleton *Backfire* ix. 98 You'd just better pray he doesn't kill somebody..because he's talking about ventilating people.

ventilated, *ppl. a.* Add: Also, supplied with continual fresh air.

1951 D. W. Richards in Cecil & Loeb *Textbk. Med.* (ed. 8) 844/1 The trapping of air in poorly ventilated air spaces, on repeated deep breathing, is shown by a step-wise rise in the spirogram. 1975 *Anaesthesia & Intensive Care* III. 237 (*heading*) A simple clinical method of quantitating the effects of chest physiotherapy in mechanically ventilated patients. 1980 *Daily Tel.* 3 Nov. 16 A substantial sample has..accrued from the innumerable ventilated patients who have not become organ donors.

ventilation. Add: **I. 3. a.** Now *Obs.*

b. The supply of fresh air or oxygen to the lungs (or gills), by the process of breathing or artificially.

1891 A. D. Waller *Introd. Human Physiol.* iv. 136 The bronchus being blocked a portion of the lung is cut off from ventilation. 1919 Flack & Hill *Textbk. Physiol.* xxxii. 284 To facilitate gaseous interchange, the process of breathing or ventilation of the lungs takes place. 1951 E. A. Stead in Cecil & Loeb *Textbk. Med.* (ed. 8) 1052/2 In cardiac failure the increased ventilation..is produced by the reflex stimulation of respiration from the congested lungs and great vessels. 1974 *Nature* 19 Apr. 631/1 Where the patient is being kept 'alive' artificially by ventilation the doctor..must decide when the brain has died and when the ventilator has passed the point at which it will invoke any involuntary response in the patient. 1978 *Sci. Amer.* Aug. 95/2 The fish could..wait until an oxygen deficit occurs and then respond by increasing its gill ventilation.

4. c. *ventilation duct.*

1937 *Discovery* Nov. 345/1 The original roof of the Saxon chapel was almost certainly of wood, probably with ventilation ducts or cowls of a shape which may have suggested to the 16th century builders their idea for the dormer roof lights. 1964 *Listener* 12 Nov. 751/2 The claim to carry a ventilation duct across another's property was just as novel as the claim to protection against the weather.

ventilator. Add: **1. b.** (Earlier example.)

1822 M. Edgeworth *Let.* 9 Mar. (1971) 369 We went one night to the House of Commons—to the Ventilator.

c. *Med.* = *respirator 3.

1961 I. W. B. Grant in D. Dunlop et al. *Textbk. Med. Treatment* (ed. 8) 945 The patient retains fairly powerful respiratory movements and may have difficulty in synchronizing with any ventilator which is not triggered by his own inspiratory efforts. 1976 *Lancet* 13 Nov. 1069/1 It has become commonplace for hospitals to have deeply comatose and unresponsive patients with severe brain damage who are maintained on artificial respiration by means of mechanical ventilators. 1982 *Times* 10 June 7/2 Mr Argov was taken off his ventilator for two hours yesterday, although he remained unconscious.

ventilatory, *a.* Restrict *rare*⁻¹ to sense in Dict. and add: **2.** Of, pertaining to, or serving for ventilation of the lungs.

1946 J. F. Fulton *Howell's Textbk. Physiol.* (ed. 15) xxxix. 871 The maximum voluntary ventilatory rate is a useful index of the capabilities of an individual for doing work. 1963 J. B. Hickam in Beeson & McDermott *Cecil-Loeb Textbk. Med.* (ed. 11) 524/1 Pulmonary tests are awkward to group according to particular kinds of function because they frequently overlap categories. A common and useful procedure is to divide them into tests of ventilatory and respiratory function. 'Ventilatory function' refers to lung volumes and the process of moving gas in and out of the lungs from ambient air to alveolar wall. 'Respiratory function' refers to the transfer of gas between alveoli and blood. 1980 *Daily Tel.* 16 Oct. 20 There is a time in the management of patients with severe brain damage..when the decision as to continue or not with ventilatory support is made.

venting, *vbl. sb.*¹ Add: **I. 1. a.** *spec.* The emission into the atmosphere of radioactive dust and debris from an underground nuclear explosion.

1963 *Wall St. Jrnl.* 27 Sept. 1/1 More unsettling, however, is the possibility that ventings of underground blasts have been adding undetected amounts of radioactive iodine to milk. 1971 *Nature* 30 July 291/3 One fear..is that if massive venting does occur..radioactive fallout could be carried outside United States territory. 1980 *New Scientist* 3 July 4/1 The venting of radioactive krypton-85 from the crippled reactor at Three Mile Island..began at 8 am last Saturday.

venting, *ppl. a.* Restrict † *Obs.*⁻¹ to sense in Dict. and add: **2.** Of gas: that finds escape by the action of venting.

1974 *Physics Bull.* June 253/1 An air filter cartridge for dehydration of the venting gas.

|| Ventôse (vantōᵘz), *sb.*² [Fr., ad. L. *ventōsus*: see Ventose *a.*] The sixth month of the French republican calendar, introduced in 1793, extending from 19 Feb. to 20 Mar.

1802 C. Wilmot *Let.* 13 Mar. in *Irish Peer* (1920) 50 (*heading*) 13 March, 1802. 22 Ventose. 1910 *Encycl. Brit.* XI. 171/1 The winter months were Nivôse, the snowy, Pluviôse, the rainy, and Ventôse, the windy month. 1943 J. M. Thompson *French Revolution* xxiv. 463 The attempt, shown in the *maximum* decree of February 21st [1794] and the so-called 'Laws of Ventôse'..to conciliate the working-class support. 1981 *Encounter* Dec. 41/2 The Year III... Ventôse this time witnessed a sudden and massive thaw.

|| ventre à terre (vantr a tɛ̄r) *adv. phr.* [Fr., lit. 'belly to the ground'.] **a.** In the posture assumed (esp. in sporting prints) by a horse at full gallop; hence at full speed, 'all out'. Also *attrib.*

1847 Thackeray *Van. Fair* (1848) iii. 476, I instantly called for the carriage, and..drove *ventre à terre* to Nathan's. 1867 'Ouida' *Under Two Flags* I. xiii. 302 You know what the Arabs are... They..pick up their sabre from the ground, while their horse is galloping *ventre à terre.* 1918 G. B. Shaw in F. Harris *Oscar Wilde* II. 28 To be called on to gallop ventre à terre to Erith. 1947 J. Stevenson-Hamilton *Wild Life S. Afr.* xxii. 170 He [*sc.* a charging lion] goes not in a series of bounds, but at a *ventre à terre* gallop, and incredibly fast. 1974 *Country Life* 2 May 1059/1 His personal progression from the 19th-century *ventre à terre* animal motion—*à la* Herring and Pollard—to the transverse and rotatory images. 1977 'E. Crispin' *Glimpses of Moon* xi. 223 Man and horse..went on to gallop almost *ventre à terre* in the direction of the hedge.

b. Lying on one's stomach, prone.

1960 *Times* 18 July 15/4 Down at the firing point they formed a line *ventre à terre.* 1968 *Economist* 27 Jan. 4/3 The Cresta, which is run by the St. Moritz Toboggan Club, on which the rider uses a skeleton, or the bob run in which the bobsleigh crew descend either *ventre à terre* or sitting up.

ventriculitis (ventrikiŭlai·tis). *Path.* [f. L. *ventricul-us* Ventricle + -itis.] Inflammation of the lining of the ventricles of the brain.

1926 in R. J. E. Scott *Gould's Med. Dict.* 1946 *Med. Jrnl. Australia* 13 Apr. 513/1 A case of staphylococcal ventriculitis. 1967 *Brit. Med. Jrnl.* 27 May 542/2 Three babies with meningitis and ventriculitis were successfully treated with gentamicin. 1977 *Proc. R. Soc. Med.* LXX. 234/1 A considerable percentage will survive ventriculitis from an infected back.

ventriculo- (ventri·kiŭlo), Comb. form of L. *ventricul-us* Ventricle, occurring in various words in *Anat.,* as **ventri:culoa·trial** *a.*, involving or connecting a ventricle (usu. of the brain) and the atrium of the heart; **ventri:culoperitone·al** *a.*, involving or connecting a ventricle of the brain and the peritoneum.

1959 *Amer. Jrnl. Dis. Children* XCVIII. 467/2 (*heading*) Clinical observations on twenty hydrocephalic children subjected to ventriculoatrial shunt. 1962 *Lancet* 12 May 991/1 The experiments with the rubber slings designed to reduce ventriculoatrial reflux were not consistent. 1977 *Proc. R. Soc. Med.* LXX. 235/2 The ventriculo-atrial shunt is the most physiological, and in spite of its drawbacks I prefer it to ventriculopleural or ventriculoperitoneal drainage except in special circumstances. 1913 *Ann. Surg.* LVII. 468 Ventriculoperitoneal drainage. In 1905, Kausch performed this operation by uniting the lateral ventricle with the peritoneal cavity by means of a small rubber tube, placed subcutaneously. 1977 *Lancet* 30 Apr. 951/1 A ventriculoperitoneal shunt was inserted for relief of the hydrocephalus.

ventriculography (ventrikiŭlǫ·grăfi). *Med.* [f. *ventriculo- + -graphy.] Radiography of the brain with the cerebral fluid of the ventricles replaced by air or some other contrast medium. Hence **ventri:culogra·phic** *a.*; also **ventri·culogram,** a radiogram of the ventricles of the brain.

1918 W. E. Dandy in *Ann. Surg.* LXVIII. 5 (*heading*) Ventriculography following the injection of air into the cerebral ventricles. *Ibid.* 7 Owing to the lighter weight of air, the ventriculogram represents the ventricle farthest from the X-ray plate. **1932** *Amer. Jrnl. Roentgenol.* XXVII. 660/1 The size of the lateral ventricle, and the patency of the foramen of Munro, can be determined readily during this particular part of the ventriculographic procedure. **1936** *Brit. Med. Jrnl.* 28 Mar. 661/2 The usual procedure..was to make a ventriculogram first thing in the morning, and start operating two or three hours later. **1947** *Radiology* XLVIII. 59/1 Ventriculography..showed the posterior portions of both lateral ventricles to be dilated and revealed similar enlargement of the temporal horns. **1974** C. B. T. Adams in R. M. Kirk et al. *Surgery* xiv. 273/2 Often, the brain plugs the fracture and a rise of intranasal pressure by nose blowing forces the air into the brain, an aerocele, and, occasionally, into the ventricle, producing a ventriculogram on X-ray. **1978** *Science* 24 Feb. 854/1 Asymmetries in the size of the left and right lateral ventricles [of the brain] can..be demonstrated..by injecting air into the ventricles during pneumoencephalography and ventriculography.

ventriloquial, *a.* **2.** (Earlier example.)

1818 in A. Mathews *Mem. Charles Mathews* (1838) II. 452 So well did he by his ventriloquial power imitate the voice of a child, without any movement of the natural organs of speech.

ventripotent, *a.* Add: Hence **ventri·potence** *rare.*

1922 [see *ovablastic *a.*].

ventro-. Add: **ventrome·dial** *a.*, both ventral and medial; situated towards the median line and the ventral or anterior surface; hence **ventrome·dially** *adv.*, in a ventromedial direction.

1908 H. E. Santee *Anat. Brain & Spinal Cord* vi. 340 Only the ventro-medial group is present above the sixth cervical segment. **1942** F. A. Mettler *Neuroanatomy* x. 199 Ventromedial to these small cells and encapsulated by them, is a round collection of cells of varying size. **1974** D. & M. Webster *Compar. Vertebr. Morphol.* v. 99 In Crocodilia the girdle is primarily endochondral with a large dorsal scapula and a ventromedial procoracoid. **1960** J. D. Boyd in G. H. Bourne *Struct. & Function Muscle* I. iii. 68 These mesenchymal cells migrate ventromedially round the notochord. **1978** *Nature* 31 Aug. 871/1 The bristles on the wing margin..send similar short, dense projections ventromedially. **1984** *Brit. Med. Jrnl.* 25 Aug. 455/1 Patients with hyperthalamic defects—especially in the ventromedial-arcuate region, which has been implicated in GH regulation—may show raised plasma immunoreactive GRF concentrations.

venture, *sb.* Add: **10. b.** Special Comb.: **venture capital** = *risk capital* s.v. *risk *sb.* 3; hence **venture capitalist**; **Venture Scout,** a male or (since 1976) female member of that section of the Scout Association for those between 16 and 20 years of age (cf. *rover scout* s.v. *rover[1] 3 d); hence **Venture Scouting.**

1943 M. A. Shattuck in *Addresses at Membership Forum* (Nat. Assoc. Investment Companies) 22 Industry during the last decade has not only lacked venture capital for new enterprises; it has also lacked venture capital for established concerns. **1971** *Financial Mail* (Johannesburg) 26 Feb. 681/1 These are some of the..successes which have brought just about every major US institution into the venture capital arena. *Ibid.* 681/2 Some venture capitalists insist on a majority equity stake. **1981** *Sci. Digest* Aug. 118/1 Venture capital, the money that bankrolls people with an innovative product, dried up in 1969. **1966** Venture scout [see *rover[1] 3 d]. **1978** *Broadcast* 27 Mar. 20/3 With many young men of 20 Venture Scouts..'Boy' Scout is hardly an accurate description for a large part of our membership. **1982** *Scouting* Sept. 582/1 This is the year the Venture Scout Section celebrates its fifteenth birthday. *Ibid.* 582/2 (*caption*) Female Venture Scouts have played an important part in the Section since 1976. **1967** *Venture Scouting* (Scout Assoc.) xv. 168 Venture Scouting is not always cheap; it isn't easy to get the right kind of gear. **1983** *Times* 23 Aug. 2/5 Venture Scouting has increased from a membership of 30,000 in 1979 to more than 36,000 this year... Girls..make up about 20 per cent of overall numbers.

Venturi (ventiū°·ri). Also **venturi.** The name of G. B. *Venturi* (1746–1822), Italian physicist, used: **a.** *attrib.* to designate a short constriction in a tube between two longer tapered portions that are usu. of unequal length but terminate with the same diameter, so that there is a drop in pressure in a fluid flowing through the constriction which may be used to determine the rate of flow or used as a source of suction; also devices having this form and the effect involved.

1887 C. Herschel in *Trans. Amer. Soc. Civil Engineers* XVII. 231 Bourdon's anemometer is founded upon the property of a Venturi tube to exercise a sucking action. *Ibid.* 232 Then came the Bourdon anemometer..and now the instrument herein described, the Venturi Water Meter. *Ibid.* 239 The actual operations of the Venturi meter as applied to a water-pipe in ordinary service. **1894** W. K. Burton *Water Supply of Towns* 285 The 'Venturi tube', which has long been used as a means of roughly estimating the velocity of air-currents, has recently been used as a water-meter. **1917** *Jrnl. Agric. Res.* IX. 115 Preliminary experiments on a new type of device, called the 'Venturi flume', for measuring water in open channels. **1930** *Engineering* 24 Jan. 97/1 Each branch has a Venturi section for measurement of flow. **1931** *Handbk. Aeronautics* (R. Aeronaut. Soc.) viii. 515 Venturi anemometer heads use a venturi for producing a suction in order to increase the pressure difference obtainable with a pitot static tube. **1937** *Times* 4 Oct. 23/4 Provision against freezing and against blocking by ice and snow must be made in carburettors, venturi tubes, and the vents of fuel and oil tanks in new passenger aircraft. **1951** *Industr. & Engin. Chem.* June 1325/1 Deposition of a spray on the walls must be considered in the design and operation of several types of industrial equipment, including Venturi scrubbers and atomizers, combustion chambers in which liquid fuels are burned, and spray dryers. **1971** J. W. Ireland *Mechanics of Fluids* ix. 340 A venturi-flume is to be installed in a channel conveying water with the object of raising the level of the water upstream. **1973** *Times* 12 Feb. (Anchor Project Suppl.) p. ii/6 Collection of exhaust gases from the BOS vessels will be by the OG suppressed combustion system, and their cleaning by a two-stage venturi scrubber system. **1974** *Sci. Amer.* Feb. 100/3 The venturi effect in the narrow spaces between buildings can increase the velocity of the wind. **1977** *Lancet* 24 Sept. 641/1 We describe a method of pooling the contents of several packs by means of a simple vacuum box worked from an ordinary water tap and a Venturi pump.

b. *absol.*

1887 C. Herschel in *Trans. Amer. Soc. Civil Engineers* XVII. 239 The water was passing through the Venturi with similar velocities. **1921** *Sci. Amer.* 15 Oct. 275/1 An idea of the shape of the venturi becomes obvious by comparing it with an old-fashioned blunderbuss. **1942** G. Casey in Murdoch & Drake-Brockman *Austral. Short Stories* (1951) 361 Venturis down below [in a mine] and water everywhere to keep the dust down. **1952** *Archit. Rev.* CXI. 138/2 A drip-proof high efficiency burner with fixed venturi, providing flame stability at low consumptions. **1976** K. Thackeray *Crownbird* ii. 36 The wind played strange tricks, the entrance acting as a venturi with gusts swirling in great spirals.

venue. Add: **5. e.** *Theatr.* The site of a theatrical performance, *spec.* one used by touring companies.

1967 *Stage* 2 Mar. 21/4 (Advt.), Clubland Promotions urgently require first-class artists..for venues in South and West Wales. **1980** *Times* 25 Nov. 6/4 Cologne Opera and San Francisco Ballet have both inspected the theatre and are considering it as a venue. **1984** *Times* 8 Mar. 4/2 The largest computerized box office system in Britain, which will link every main venue in the city. Instant ticket printing means that venues can put on performances at short notice.

Venus[1]. Add: **I. 1. e.** *Archæol.* A palæolithic female figurine distinguished by exaggerated breasts, belly, and buttocks. Cf. *steatopyga.

1912 R. Munro *Palæolithic Man* x. 239 Laugerie Basse has supplied a fragment of bone with a pregnant woman and reindeer engraved on it..; the Venus..a headless statuette carved in ivory. **1920** H. C. Bailey *Call Mr. Fortune* 191 'My new palæolithic Venus.' 'You left her in the library... There are not many men..who have a Hottentot Venus to lose.' **1937** Auden & MacNeice *Lett. from Iceland* xvii. 245 We leave to that poor soul A. M. Ludovici the Venus of Willendorf. **1958** *Times Lit. Suppl.* 11 Apr. 198/4 One archaeologist's interpretation of palaeolithic figurines or 'Venuses' as 'the characteristic products of unregenerated male imagination'. **1979** Mills & Mansfield *Genuine Article* ii. 34 The Venus figurine[s]..date from around 25,000 to 22,000 B.C... One of the most celebrated is the 'Venus of Willendorf' from Austria.

Venusberg (vī·nɒsbɛ̄ɹg, ‖vē·nɒsbɛrk). [Ger., lit. 'mountain of Venus'.] The court of Venus, in German legend and esp. in Wagner's *Tristan* found in a grotto on the Hörselberg mountain; hence in *transf.* use, any environment whose primary characteristic is sensual pleasure.

1855 [see *saga[1] 2]. **1890** G. B. Shaw *How to become Mus. Critic* (1960) 192 Not even a visit from the ghost of Sterndale Bennett could have waved him back from the Venusberg then. **1925** H. Crane *Let.* 28 Feb. (1965) 199 A perfect Venusberg of flowers and shrubbery. **1934** H. G. Wells *Exper. Autobiogr.* I. iii. 107 Bowkett had his own secret incidental twilight Venusberg. **1962** *Guardian* 15 Nov. 9/3 We had wonderful plans for Windsor. It was to have been an expense account Venusberg. **1981** M. J. Benkovitz *Aubrey Beardsley* iv. 110 Beardsley depicted..[a] more violent encounter between satyrs and the residents of the Venusberg.

Hence **Venusbe·rgian** *a.*

1896 G. B. Shaw *Let.* 19 Feb. (1965) I. 602 A Venusbergian Asta is too much. **1977** *Gramophone* Dec. 1087/1 Those whining Wagnerian tubas in 'Mars' are matched by the almost Venusbergian sensuousness of 'Venus' here.

Venusian (vėniū·siăn), *a.*[1] *rare.* [f. L. *Venusia* (see below) + -an.] Of or pertaining to Venusia, an ancient town in southern Italy, and birthplace of the poet Horace; hence used allusively.

1616 B. Jonson *Poetaster* in *Workes* 310 And to his steps my Genius inclines, Lucanian, or Apulian, I not whether; For the Venusian colonie plowes either. **1875** F. Arnold *Our Bishops & Deans* I. v. 289 But the Venusian adage is true how the generous wine evermore retains the early flavour which was accidentally imparted to it.

Venusian (vėniū·siăn), *sb.* and *a.*[2] [f. Venus[1] II + -ian.] **A.** *sb.* **1.** *Science Fiction.* A supposed inhabitant of the planet Venus; also, the language spoken by such a being.

1874 A. Blair *Ann. 29th Cent.* III. iii. 56, I suspected from the circumstances the frames of the Venusians were so constituted that sustenance was superfluous. **1897** J. Munro *Trip to Venus* ix. 173 'The good of it?' rejoined the Venusian; 'it is beautiful, and gives us pleasure.' **1927** *Spectator* 24 Dec. 1127/1 It is the Venusians who record the Last Judgment. **1955** W. Girvan *Flying Saucers & Common Sense* i. 15, I do not think it is because of a desire that I should be visited by Martians or Venusians. **1972** P. Moore *Can You Speak Venusian?* xvi. 167 He is fluent in Venusian, Plutonian and Krügerian. **1977** *N.Y. Rev. Bks.* 12 May 36/4 There are several stories which insist on the..general nastiness of human beings, who thus take on the..role which used to be assigned to Martians and Venusians and the like in Fifties science fiction.

2. *Astrol.* A person born under the sign of Venus and thus supposedly subject to the influence of the planet. *rare.*

a **1963** L. MacNeice *Astrol.* (1964) ii. 55 But the sensuality of the *Kalendar* Venusians is not of the violent 'dark god' type.

B. *adj.* **1.** Of, pertaining to, or characteristic of the planet Venus or its supposed inhabitants.

1874 A. Blair *Ann. 29th Cent.* III. iii. 58 At the first glimpse I received of Venusian waters, my admiration was taken by storm. *Ibid.*, I now commenced the acquisition of the Venusian language. **1934** *Jrnl. Brit. Interplanetary Soc.* II. 4 It cannot be over-emphasised that Venusian life cannot be other *than* Venusian. **1956** C. Simak *Time & Again* ix. 42 Earth news..was followed by Martian news, by Venusian news, by the column from the asteroids. **1968** *Times* 12 Oct. 18/6 Further reason for doubting the radius of Venus..has emerged from a computer simulation of the Venusian atmosphere. **1982** *Austin* (Texas) *Amer.-Statesman* 28 Feb. A-17/1 The first [space probe]..is expected to drop through thick Venusian clouds.

2. *Astrol.* Designating or pertaining to the movement or influence of the planet Venus. *rare.*

1913 'Sepharial' *Kabala of Numbers* II. xii. 189 Suppose..that a child is born when the influence of Venus is predominant, and the local mode of etheric vibration is what may be termed Venusian. **1924** C. E. O. Carter *Conc. Encycl. Psychol. Astrol.* 30 Attractiveness is an essentially Venusian quality. *Ibid.* 81 In the horoscopes of poets and painters we see the faculty of artistic imagination denoted by the mixture of Lunar, Mercurial, and Venusian action.

Vepsian (ve·psiăn). Also **Veps, Vepsic, Vesp, Wepsian,** etc. [f. Russ. *Vépsi* + -an, -ian.] **a.** (A member of) a Finnish people dwelling in the region of Lake Onega, now in the north-west of the U.S.S.R. **b.** The Finno-Ugric language spoken by this people. Also *attrib.* or as *adj.*

1859 R. G. Latham *Descr. Ethnol.* I. xix. 409 *Tshud*..is a word which the Slavonians of Novgorod applied to the Non-slavonic populations with which they came in contact. We shall see its import more clearly after a notice of the Vod, and the Vesp. **1863** —— *Nationalities of Europe* I. xvi. 162 Of the Vesp dialect, *eo nomine*, I have seen no specimens. **1877** A. H. Keane tr. *Hovelacque's Sci. of Lang.* iv. 91 With the Suomi are grouped..the *Chudic* [language], situated in a very scattered district south of Lake Onega; the *Wepsic*, which is northern Chudic, [etc.]. **1878** *Encycl. Brit.* VIII. 700/1 Finnic or Ugrian represented by..Finnish proper..Karelian..Tchudic..Vepsic. **1879** *Ibid.* IX. 219 The Chudic, a Slav term..now restricted to the Veps or northern Chud and the Votic or southern Chud, dwelling..round the shores of Lake Onega. **1921** [see *Ludian]. **1933** L. Bloomfield *Language* iv. 68 The other languages of the Baltic branch [of the Finnish-Lapponic languages], Carelian, Olonetsian, Ludian, Vepsian, Livonian, Ingrian, and Votian, are far smaller, and some of them are near extinction. **1933** *Amer. Anthropologist* XXXV. 309 The Leningrad section studied the Finnish tribes, the Vesps, the Izhors, and some purely Russian peoples. **1955,** etc. [see *Ludian]. **1964** *Language* XL. 98 The Veps, Votic, and Estonian cognates. **1977** *Ibid.* LIII. 477 Since the boundaries of the European continent make up the geographical framework of H's analysis, he includes such little-known languages as Votyak, Cheremis, and Vepsian (Uralic), and Bashkir, Karaim, and Kalmyk (Altaic).

‖ **vera causa** (vē·rā kau·zā). *Philos.* [L., lit. 'real cause'.] A true cause which brings about an effect as a minimum independent agency.

[**1687** I. Newton *Philosophiae Naturalis Princ. Math.* iii. 402 Causas rerum naturalium non plures admitti debere, quam quæ & vera [*sic*] sint & earum Phænomenis explicandis sufficiunt.] **1831** J. F. W. Herschel *Prelim. Discourse* ii. vi. 144 To such causes Newton has applied the term *veræ causæ*; that is, causes recognized as having a real existence in nature, and not being mere hypotheses or figments of the mind. **1865** Mill *Exam. Hamilton's Philos.* xxiv. 469 By *verae causae* Newton meant agencies the existence of which was otherwise authenticated. **1890** W. James *Princ. Psychol.* I. ii. 67 Inhibition is a *vera causa*, of that there can be no doubt. **1927** B. Russell *Analysis of Matter* ii. 19 In Newton, 'force' plays a great part, and there seems no doubt that he regarded it as a *vera causa*. **1942** R. G. Collingwood *New Leviathan* xxxiii. 279 Stupidity, I reply, is not a *vera causa*. If he thought that, I want to know why; and to say 'because he was stupid' is not an answer. **1977** *Brit. Jrnl. Hist. Sci.* X. 238 Darwin's commitment to the *vera causa*—or 'true cause'—principle.

‖ **vera copula** (vē·rā kǫ·piŭlā). *Law.* [L., lit. 'true union'.] Sexual intercourse; coitus requiring erection and penetration but not necessarily the power of conception.

1845 J. E. P. Robertson *Rep. Cases Argued & Determined in Ecclesiastical Courts at Doctors' Commons* I. 1. 299 If there be a reasonable probability that the lady can be made capable of a *vera copula*—of the natural sort of coitus, though without power of conception, I cannot pronounce this marriage void. **1945** *Times Law Rep.* LXI. 526/2 It is well established that there must be what Dr. Lushington referred to as *vera copula*. **1961** *Times* 30 June 5/4 The question was whether the fact that the wife would have an artificial vagina would prevent sexual intercourse achieved by such means from constituting, in law, *vera copula*. **1971** *Rayden on Divorce* (ed. 11) I. vi. 156 Sexual intercourse or consummation is sometimes referred to as *vera copula*.

veranda, verandah. Add: **1. c.** (Later N.Z. examples, esp. one built over the pavement outside business premises.)

1940 F. Sargeson *Man & Wife* (1944) 45 The house is a very old house. Once it was a grocer's shop..but..the old man couldn't get another shopkeeper to take it. So he had the verandah roof pulled down, and the front altered. **1959** M. Shadbolt *New Zealanders* 80 An untidy collection of bicycles and boys in the shade of a shop verandah.

2. *veranda-chair.*

1902 *Chambers's Jrnl.* 14 June 437/2 Miss Tresscott..sought to show her disapproval..by turning her veranda-chair with its back to the gay scene. **1973** *Advocate-News* (Barbados) 17 Feb. 12/6 (Advt.), Household furniture... It includes:—Verandah chairs.

verapamil (vĕræ·pămil). *Pharm.* [f. V(al)er(onitr)il(e with inserted *-apam-* (of unkn. origin).] A drug, $C_{27}H_{38}N_2O_4$, that is given orally and intravenously (usu. as the hydrochloride) in the treatment of angina pectoris and cardiac arrhythmias.

1967 *Lancet* 5 Aug. 310/2 The effect of verapamil..on tachycardia was studied in 6 healthy male volunteers. **1983** R. Balcon in *Oxf. Textbk. Med.* II. xiii. 171/1 Nifedipine and verapamil are the two calcium antagonistic agents..widely used in the treatment of angina pectoris.

veratr-. Add: **vera·tridine** [*-idine], an alkaloid ester, $C_{36}H_{51}NO_{11}$, present in veratrum and sabadilla and having anti-hypertensive properties, but now used chiefly for experimental purposes.

1907 *Brit. Pharmaceut. Codex* 242 Cevadine is accompanied in cevadilla seeds by the alkaloids veratridine..and sabatrine. **1935** *Jrnl. Chem. Soc.* 123 On hydrolysis, veratridine gave veratric acid and..cevine. **1954** O. Krayer in V. A. Drill *Pharmacol. in Med.* xxxiii. 8/2 The monoester veratridine is of no clinical use because it has an inadequate therapeutic range and its duration of action is too short. **1976** *Nature* 25 Mar. 337/2 The effects of veratridine, an alkaloid well known for its ability to increase Na^+ permeability in nerve cells, were examined. **1976** [see *sabadine].

verb. Add: **1. b.** *verb-complement, -form, -phrase, -stem* (earlier example), *-tense*; *verb-centered, -like* adjs.; also in combinations (freq. *attrib.*) with the sense 'verb and —', as *verb-adverb, object.*

1887 Southworth & Goddard *Our Lang.* ii. iv. 35 A verb-phrase is a group of words used as a single word. **1888** B. H. Kennedy *Rev. Lat. Primer* 94 Verbs..in which the Verb-Stem was formed by a so-called Thematic vowel added to the root. **1912** A. D. Sheffield *Gram. & Thinking* vi. 92 These classes we can then subdivide, thus... Auxiliaries, forming 'verb phrases' (*is white, may go*) [etc.]. **1924** H. E. Palmer *Gram. Spoken Eng.* 121 The Anomalous Finites are the only verb-forms which may be shifted to front-position. **1933** *Amer. Speech* VIII. iv. 7/1 The verb-adverb combination *make up* was capable of expressing at least fourteen different meanings. **1935** Verb-like [see *noun-like* adj. s.v. *noun 1 c]. **1935** T. Hudson-Williams *Short Introd. Stud. Compar. Gram.* 6 The personal endings of a verb-tense or the case-endings of a noun serve all practical purposes. **1957** *Publ. Amer. Dial. Soc.* xxviii. 132 If the IW appears with no accompanying verb-complement..the contrast..is suspended. **1962** C. L. Barber in F. Behre *Contrib. Eng. Syntax* 21, I shall use traditional terminology..in the discussion of verb-tenses and of subordinate clauses. **1963** F. T. Visser *Hist. Syntax* iv. 356 The verb-object relation..is always easily recognisable when the complement is in an unequivocal genitive case. **1964** R. H. Robins *Gen. Linguistics* 331 Abaza, a Caucasian language, has been cited as an extreme case of a verb-centred language. **1965** N. Chomsky *Aspects of Theory of Syntax* ii. 64 The grammatical relation Verb-Object holds of the pair (*frighten, the boy*). *Ibid.* 105 'Unspecified—subject is working at this job quite seriously', where 'at this job' is a Verb-Complement. **1965** *English Studies* XLVI. 76 The statements about the correlations between verb-forms and contextual factors. **1966** *Ibid.* XLVII. 56 The problem of the verb-adverb combination (e.g. he *takes off* his hat). **1974** *Amer. Speech* 1970 XLV. 266 Not only does he place himself in the mainstream of..the verb-centred approach.., but he also clearly anticipates the drift towards..the verb-centered generative semantics of..Chafe. **1977** *Dædalus* Fall 118 It might have a rule that rewrites sentence as Noun Phrase followed by Verb Phrase. **1978** *Language* LIV. 118 When we consider the range of verb-like elements, the explanation in terms of proximity becomes more interesting.

verbal, *a.* and *sb.* **A.** *adj.* **1. b.** For † *Obs.* read *rare* and add later example. Also, ready to use words; articulate.

1976 *National Observer* (U.S.) 12 June 15/2 It's made a difference... She's more confident and verbal.

2. e. *verbal diarrhœa* (*colloq.*), a tendency to talk too much; extreme verbosity. Cf. Diarrhœa 2.

1823 *London Mag.* Sept. 281/1 All our modern tragedists indulge in..the talking-principle... A verbal diarrhoea is the epidemic disease which afflicts the whole tribe. **1938** N. Marsh *Death in White Tie* xiii. 146 Her chief complaint is..acute verbal diarrhoea. **1981** N. J. Crisp *Festival* vi. 142 This fathead suffers from verbal diarrhoea.

f. *verbal conditioning* (Psychol.), the reinforcing of certain verbal responses with the object of establishing the use of particular words or ways of speaking.

1954 *Jrnl. Exper. Psychol.* XLVIII. 355 (*title*) Influence of awareness of reinforcement on verbal conditioning. **1967** M. Argyle *Psychol. Interpersonal Behaviour* vii. 131 People may embark on all kinds of self-improvement,..including the modification of styles of interaction as in operant verbal conditioning. **1971** *Jrnl. Gen. Psychol.* Apr. 267 Verb impression value..was an important variable influencing response tendencies in verbal-conditioning and sentence-making tasks. **1979** J. P. Houston et al. *Invitation to Psychol.* v. 177 Verbal conditioning is another example of reward training in which a particular response is rewarded if it occurs.

7. Forming compound adjs. with the sense 'verbal and —', as *verbal-acoustic, -metrical, -visual*, etc.

1901 E. B. Titchener *Exper. Psychol.* I. xii. 393 The verbal-motor type stands, in the author's experience, next in order of frequency to the visual. **1911** S. S. Colvin *Learning Processes* vii. 107 If I recall the name as written or printed on a page, I have a verbal-visual image; if I recall the name as spoken by some one, I have a verbal-acoustic image; if I recall the name in terms of the movements of my throat in speaking, I have a verbal-motor image. **1948** L. Spitzer *Linguistics & Lit. Hist.* 201 The verbal-metrical scheme of the strophe. **1972** R. E. Ornstein *Psychol. of Consciousness* ii. 39 The scientist, the writer, the mathematician are examples of the culturally 'dominant' verbal-rational mode. *Ibid.* x. 225 Modern Science is primarily verbal-logical.

B. *sb.* **3.** *Gram.* A word or group of words performing the function of a verb.

1935 [see *particle *sb.* 3]. **1965** *Amer. Speech* XL. 206 Adverbials which serve purely as modifiers of sentences and verbals. **1978** *Language* LIV. 88 Some support for the status of the items listed as verbals is that they are considered predicates in generative-semantic treatments of syntax.

4. *colloq.* A verbal statement, *spec.* a damaging admission, alleged to have been made by a suspected criminal and offered in evidence against him at a trial. Freq. *pl.*

1963 *Times* 22 Feb. 6/5 Who will believe you after I said I wasn't going to make a verbal? **1974** 'M. Underwood' *Pinch of Snuff* xxi. 171 'Have a look through the police evidence.'..'At least, they haven't put in any verbals.' **1980** *Daily Mirror* 24 June 19 Opponents of 'verbals' (alleged admissions of arrested persons to police which are not signed but are admitted in evidence) could see the meeting as a chance to further their campaign.

5. *slang.* Insult or abuse. Esp. in phr. *to give* (someone) *the verbal* and var.

1973 *Time Out* 2–8 Mar. 13/2 We faced them, and gave them a load of verbal across the street. **1977** *Times Educ. Suppl.* 21 Oct. 43/2 Insulting and humiliating rivals—'giving them the verbal'. **1982** *Observer* 10 Oct. 40/7 Each 'ball' consisted of a distinctly lethargic head-high bouncer.., followed by a rousing collection of verbals (money will be paid to lip-reading viewers for translation).

verbal, *v. colloq.* [f. *verbal *sb.* 4.] *trans.* To attribute a damaging statement to (an accused or suspected person). Also const. *up.*

1963 *Times* 22 Feb. 6/5 Those chaps were about and they won't be able to verbal me. **1970** P. Laurie *Scotland Yard* vi. 146 The aggravation of it. He verbals up my villain and then says he'll let him off. **1973** *Courier-Mail* (Brisbane) 17 Oct. 10/4 Finch has claimed that detectives 'verballed' him—fabricated a confession to arrest him. **1981** C. Ross *Scaffold* 145 'He's made no statement yet either.' 'But you verballed him?'..The police officer said nothing.

Hence **ve·rballing** *vbl. sb.*

1973 *Observer* 11 Nov. 15/2 'Verballing'—putting damaging remarks or 'verbals' into suspects' mouths—has existed as long as detectives have been dealing with criminals. **1977** 'C. Aird' *Parting Breath* iv. 47 It wasn't, the policeman consoled himself, really and truly verballing. Verballing was putting words into a man's mouth—and statement.

verbalistic (vɜ̄ɹbăli·stik), *a.* [f. Verbalist + -ic.] Of, pertaining to, or characterized by verbalism.

1934 *Mind* XLIII. 409 (*title*) Verbalistic tendencies. **1953** H. H. Price *Thinking & Exper.* viii. 237 It is no sillier than the purely Verbalistic theory which is at present fashionable. **1975** *Amer. Speech* 1969 XLIV. 128 We live in a highly verbalistic culture.

Hence **verbali·stically** *adv.*

1934 *Mind* XLIII. 411 To do full justice to a speaker's intended meaning..instead of verbalistically insisting that he 'must' mean exactly what he *says*. **1940** *Theology* XL. 426 Pose the question thus verbalistically, and at once one suggests the method of lexicographical analysis.

verbality. Add: **1.** (Later example.)

1877 W. R. Alger *Life Edwin Forrest* II. xxii. 767 He was no starveling fed on verbality and ceremony, no pigmy imitator or empty conformist.

2. (Later examples.)

1899 W. James *Talks to Teachers* 257 We are stuffed with abstract conceptions, and glib with verbalities and verbosities. **1935** B. Malinowski *Coral Gardens & their Magic* II. vi. vi. 246 The fact that the community are aware of the spell and know its wording is the most important clue to the appreciation of the verbalities of magic.

verbalizable (vɜ̄·ɹbăləi·zăb'l), *a.* [f. Verbaliz(e *v.* + -able.] Capable of being expressed in words; able to be verbalized. Hence **ve:rbalizabi·lity.**

1951 C. Kluckhohn et al. in Parsons & Shils *Toward Gen. Theory of Action* iv. ii. 397 'Verbalizable' is not to be equated with 'clearly and habitually verbalized'. The actor's values are often..incompletely..verbalized by him. But implicit values remain 'conceptions'..which can be put into words by the observer... Verbalizability is a necessary test of value. **1981** *N.Y. Times Mag.* 24 May 10/4 The path was cleared for the substitution of the verbalizable 'or' by the unspeakable '/' in the legalistic term 'and/or'.

verbalization. Add: **2.** A verbal expression or statement.

1951 C. Kluckhohn et al. in Parsons & Shils *Toward Gen. Theory of Action* iv. ii. 398 Implicit values will be manifested only in behavior and through verbalizations that do not directly state the pertinent values. **1965** *Times Lit. Suppl.* 16 Sept. 812/1 Many of these potential multiple nouns will founder because..a more efficient verbalization is to hand. **1976** *Howard Jrnl.* XV. 1. 17 The practitioner 'sees through' the client's verbalizations to a reality that lies behind them.

verbalizer (vɜ̄·ɹbăləizəɹ). [f. Verbalize *v.* + -er[1].] One who registers stimuli or thoughts mentally in verbal terms rather than through the senses (cf. Visualizer); one who verbalizes.

1937 *Brit. Jrnl. Psychol.* Jan. 278 The meaning of many a word may be more clear-cut to a visualizer, because accompanied by a visual image, than to a verbalizer. **1952** *Ibid.* May 122 It would..be expected that the better verbalizers would be the better learners. **1975** C. L. Burt *ESP & Psychol.* 7 A verbalizer who (like myself) may be almost devoid of visual or other kinaesthetic imagery, tends rather to think in terms of abstractions and for him material and mechanical structures lose much of their theoretical importance. **1980** *Times Lit. Suppl.* 15 Feb. 173/2 In Ireland what a man said was deemed more important than what he did. The tradition of verbalizer is old and deep..and sprang from the imperative to keep the old cultural matrix alive.

verbicide[2]. Add: Hence **ve·rbicidal** *a.*, tending or liable to destroy the sense or value of a word.

1978 *Forum on Med.* Apr. 85/1 Other verbicidal entertainers who were school dropouts have also received 'honorary' degrees. **1985** *Times* 27 Mar. 12/8 The verbicidal impulses of Pentagonese.

verbid (vɜ̄·ɹbid). [f. Verb + -id[1].] *Gram.* A word, as an infinitive, gerund or participle, which has some verbal characteristics but lacks the power of forming sentences. Also *attrib.* and in *Comb.*

1914 O. Jespersen *Mod. Eng. Gram.* II. 7 We shall..restrict the name of verb to those forms that have the eminently verbal power of forming sentences, and..apply the name of verbid to participles and infinitives. **1940** Bryant & Aiken *Psychol. of Eng.* 23 Such a unit [*sc.* John having called the dog] is called a verbid clause, because *having called* is not a full verb. **1966** R. L. Allen *Verb System Present-Day Amer. Eng.* 136 Of the 4800 verb and verbid-clusters included in the corpus, 1191..are clusters with some form of BE as the verb.

verbigeration. (In Dict. s.v. Verbigerate *v.*) (Earlier example.)

1886 *Jrnl. Nerv. Dis.* XIII. 699 Some emotion, at variance with his 'verbigeration'.

verbo-. (Examples of *verbomania*.)

1923 Ogden & Richards *Meaning of Meaning* ii. 89 A veritable orgy of verbomania. **1977** J. Garvin in D. Ó Muirithe *Eng. Lang. in Ireland* 112 *Finnegans Wake* abounds in Anglo-Irish idiom..but it is subsumed into..'slanguage' which leaves the book adrift in a polyglottic sea of verbomania.

‖ **verboten** (feɹbō·tĕn), *a.* [Ger.] Forbidden; not allowed.

1912 R. Brooke *Old Vicarage, Grantchester* (1916) 8 Meads towards Haslingfield and Coton Where *das Betreten*'s not *verboten*. **1916** J. Buchan *Greenmantle* v. 63, I got very bored, for I had nothing to read and my pipe was *verboten*. **1949** E. Benn *Happier Days* xviii. 217 The unfortunate German, bred and trained from childhood to understand that everything is *verboten* unless specifically permitted. **1968** Finch & Cain in J. Marmor *Mod. Psychoanal.* xvii. 446 The obviously important but almost verboten subject of the implicit structure of rewards. **1979** R. Mutch *Gemstone* viii. 94 She was *verboten*, strictly off limits.

‖ **verbum sap.** Add: Also **verb sap.**

1841 E. Brontë in *Brontës: Lives, Friendships & Corr.* (1932) I. x. 238, I have just made a new regularity paper!

and I mean *verb sap* to do great things. **1889** E. Dowson *Let.* 9 Nov. (1967) 115 Mark that horse & when he runs next plump on him with all your available capital. Verb Sap. This tip I may add doesn't come from Museum Mansions but from a man who spotted Primrose Day for the C'wtch. **1976** J. I. M. Stewart *Young Pattullo* ii. 40 Mind your Ps and Qs, if I were you. *Verb. sap.*—eh?

verd. Restrict † *Obs.* to senses in Dict. and add: **1. b.** *poet.* The colour green. *rare.*

1915 G. C. M. Birdwood *Sva* 59 The whole paradisaical scene shining in the setting sun with the transcendent resplendence of its various verds and shimmering gold. **1929** S. Leslie *Anglo-Catholic* ix. 110 The pavement was tesselated purple and verd. Great tapestries hung between pillars.

‖ **verdelho** (vĕɹde·lʸu). [Pg.] A prolific vine yielding a white grape orig. grown in Madeira and now found in Portugal, Sicily, Australia, and South Africa; also, a medium white Madeira wine made from this grape.

1824 A. Henderson *Hist. Anc. & Mod. Wines* viii. 249 Among the various species of grapes cultivated, those called *verdelho*, *negra*..and *sercial*, yield the best wines. **1883** *Encycl. Brit.* XV. 178/1 The wine usually termed Madeira..is made from a mixture of black and white grapes, which are also made separately into wines called Tinto and Verdelho. **1926** P. M. Shand *Bk. Wine* iii. 41 Two thirds of the vines grown on the island are of the celebrated Verdelho species. **1935** A. L. Simon *Dict. Wine* 249 *Verdelho*, one of the white fortified wines of Madeira, which used to be made with the Verdelho grape, now all but extinct. **1959** W. James *Word-bk. Wine* 195 *Verdelho*, a white grape which gives its name to a sweet dessert type of madeira, though the vine is not grown on the island to any great extent. **1980** *Times* 27 Nov. 20/4 The two drier styles of wine, Sercial and Verdelho.

Verdian (vɛ̄·ɹdiăn), *a.* and *sb.* [f. the name of Giuseppe *Verdi* (1813–1901), Italian composer + -an.] **A.** *adj.* Of, pertaining to, or characteristic of Verdi or his music. **B.** *sb.* An admirer of Verdi; a (skilled) exponent of his music.

1947 A. Einstein *Mus. Romantic Era* xvi. 284 One should not think that *Otello*..became..a drama with music. It remained Italian opera, Verdian opera, which preserves almost completely its connection with the composer's earlier work. **1962** *Listener* 21 June 1091/2 An opera in a broad Verdian style brought up to date. **1976** *Gramophone* Aug. 334/1 Bergonzi is much the more stylistically correct Verdian. **1977** *Times* 20 Dec. 10/5 The Verdians cannot get out of it by saying that Beethoven was the greater genius. Of course he was. **1978** *Times* 10 Oct. 9/4 A noted Verdian and a remarkable Violetta..Miss Sass was perhaps ill-advised to begin her first London recital with Schumann. **1980** *Daily Tel.* 4 June 15/2 [Its] moral and imaginative qualities..have always endeared it to dedicated Verdians. *Ibid.* 15/4 His direction growing in authentically Verdian conviction.

‖ **Verdicchio** (vɛɹdi·kio). [It.] A white grape grown in the Marche region of Italy; also, the dry white wine made from this grape.

1940 H. J. Grossman *Guide to Wines* vi. 80 Not many great wines are grown in this region, but two demand attention: a fine white, light wine—the Verdicchio di Jesi [etc.]. **1954** E. David *Italian Food* 310 The white wines of the Castelli di Iesi, in the Marche, produced from Verdicchio grapes, and called usually by that name, are pleasant and refreshing to drink on the spot. **1967** F. Schoonmaker *Encycl. Wine* 305 *Verdicchio*, superior Italian white wine grape; also its wine. **1978** 'A. Stuart' *Vicious Circles* 109, I ordered an ice-cold bottle of Verdicchio.

verdict, *sb.* Add: **1. c.** *open verdict*, a verdict of a coroner's jury affirming the commission of a crime (but not specifying the criminal), or the occurrence of a suspicious but unexplained death.

1769 [in Dict.]. **1864** Dickens *Mut. Fr.* (1865) I. i. iii. 18 Inquest to-morrow, and no doubt open verdict. **1977** 'M. Innes' *Honeybath's Haven* xiii. 123 His jury..will bring in what is called an open verdict.

verdictive (vɜ̄·ɹdiktiv). *rare.* [f. Verdict *sb.* + -ive.] An utterance which consists in the delivering of a verdict. Also *attrib.* or as *adj.*

1955 J. L. Austin *How to do Things with Words* (1962) xii. 150, I call these classes of utterance..by the following more-or-less rebarbative names:..Verdictives..Exercitives [etc.]. *Ibid.* 152 A verdictive is a judicial as distinct from legislative or executive acts... Verdictives have obvious connexions with truth and falsity. *Ibid.* 160 We may dispute as to whether these [expositives] are not verdictive, exercitive, behabitive, or commissive acts as well. **1975** *Language* LI. 106 Examples of performatives assuming the agent-authority are verdictives..such as *acquit, convict, find* etc. (mostly judicial).

verdin (vɜ̄·ɹdin). Also **Verdin.** [a. F. *verdin* yellow-hammer.] A small grey tit with a yellow head, *Auriparus flaviceps*, found in south-western North America.

1881 *Amer. Naturalist* XV. 217 Another minute species of the titmouse family, the verdin or yellow-headed titmouse. **1903** *Atlantic Monthly* July 103 The same fretful verdin was talking about something with the old emphatic monotony. **1939** G. B. Pickwell *Deserts* 139/2 This is the voice of the Verdin, a true bird of the desert. **1974** L. Gombrich tr. *von Frisch's Animal Archit.* (1975) 211 The verdin barricades its nest with thorns.

verdite[2] (vɜ̄·ɹdəit). [f. *verd-* (as in Verdure) + -ite[1].] A green ornamental rock from South Africa.

1908 *Mineral Industry* XVI. 810 During 1907 there was found on the south bank of the North Kaap river, in South Africa,..another..stone that has the deep green color of..chromiferous syenite... It has a hardness of about three, and is susceptible of a high polish; the color is a rich chrome green and the stone contains a chrome-muscovite and some argillaceous material... The stone is obtained in blocks weighing one ton or more... The name verdite has been suggested for it. **1916** *Chambers's Jrnl.* Apr. 272/2 Articles of jewellery and ornaments are made of verdite. **1935** *Archit. Rev.* LXXVIII. 104/1 In the Midland Headquarters in the City, it is said that each of the Verdite pilasters..cost a Thousand pounds, and there are a considerable number of them. **1979** Hurlbut & Switzer *Gemology* xiii. 143/1 Verdite is a green rock consisting of fuchsite (a chrome mica), and clayey mica.

‖ **verdomde** (fəɹdǫ·mdə), *a.* *S. Afr. slang.* Also **verdoemde, verdomd.** [Afrikaans, ad. Du. *verdoemd.*] Damned, infernal.

1850 R. G. Cumming *Five Years Hunter's Life S. Afr.* ix. 53/1, I overheard him remark to three other gruff-looking Boers who stood beside him that I was 'a verdomd Englishman'. **1878** T. J. Lucas *Camp Life & Sport S. Afr.* 232 Ah, you verdomde beast! **1909** R. Dehan *Dop Doctor* 90 He was no Baas of mine the verdoemde rooinek! **1913** D. Fairbridge *That which hath Been* 125 This verdoemde Governor is obstinate on this point. **1926** E. Lewis *Mantis* i. iv. 80 Why, even with Government helping you not to make a goat of yourself farming's a *verdomde* lottery, believe me. **1964** A. Trew *Smoke Island* vi. 106 Pride wouldn't allow him to claim that achievement for his people now, and this *verdomde* native knew it.

verdure. I. 3. Delete † *Obs.* and add later *Hist.* examples.

1934 *Burlington Mag.* Feb. 65/2 The work of the oldest tapestry factory in England... A 'verdure', a type very popular with Worcestershire weavers. **1977** *Anc. Tapestries from Belgium* (Nat. Arts Centre, Ottawa) 77 Oudenarde was famous for its 'verdures' (tapestries representing trees or foliage).

Verel (vi°·rĕl). orig. *U.S.* The proprietary name of a synthetic acrylic fibre.

1956 *Official Gaz.* (U.S. Patent Office) 2 Oct. TM3 *Verel*, for synthetic staple fiber. **1956** *Trade Marks Jrnl.* 24 Oct. 1050/2 Verel B755,848. Raw or partly prepared synthetic textile fibres. Eastman Kodak Company..343, State Street, Rochester, State of New York, United States..Manufacturers. **1956** *America's Textile Reporter* 22 Mar. 13/2 Processing and ironing temperatures should be held to a maximum 300 degrees Fahrenheit when 100 per cent 'Verel' fabrics are being handled. **1959** A. J. Hall *Stand. Handbk. Textiles* (ed. 5) i. 73 Verel fibres are more hydrophile than usual for synthetic fibres. **1963** —— *Textile Sci.* ii. 87 This [acrylic] group includes all those fibres made from polymers consisting wholly or partly of acrylonitrile, and important members are..Courtelle, Verel, Zefran, etc. **1978** P. G. Tortora *Understanding Textiles* ix. 134 For many years only two modacrylic fibers were made commercially: Dynel..and Verel.

‖ **Verfremdungseffekt** (feɹfre·mduŋs,efe·kt). [Ger.: see *alienation 1 d.] = *alienation effect* s.v. *alienation 1 d. Similarly **Verfremdung,** distancing, alienation (occas. in sense of *alienation 1 c).

1945 *Kenyon Rev.* VII. 470 The distancing ('Verfremdung') which Brecht desires is complemented by his concreteness which makes any escape from the reality demonstrated upon the stage impossible. **1951** *World Rev.* Jan. 67/2 Brecht generally sticks closely and with success to his technique of alienation (*Verfremdungseffekt*, he calls it). **1959** *New Statesman* 28 Mar. 442/1 His famous principle of *Verfremdung*, for example: of the distancing of the actor from his part, of breaking that theatrical illusion which mesmerises the audience into a witless passivity. **1960** J. Bayley *Characters of Love* iii. 145 We are to feel nothing but alienation, a Brechtian *Verfremdungseffekt*. **1980** *Times Lit. Suppl.* 17 Oct. 1172/5 Rylance Waters..provides a brilliant *Verfremdungseffekt*. **1981** F. Inglis *Promise of Happiness* iii. 86 If, with Marx, we are to speak of *Verfremdung*, estrangement or alienation, as the most destructive and wounding experience of labour under capitalism.

verge, *sb.*[1] **I. 1. a.** Delete † *Obs.*[−1] and add later example.

1887 [see *fellatio].

7. c. *verge watch* (later examples).

1963 *Times* 6 Feb. 12/4 An anonymous purchaser gave 290gns. for a seventeenth-century verge watch by John Drake. **1977** *Cleethorpes News* 6 May 22/4 (Advt.), Clocks, silver watches, verge watches.

IV. 16. b. Also, an unpaved strip of land, usu. planted with grass, separating a pedestrian pavement from a road; a (grass-covered) edging to a road. Cf. *grass verge* s.v. *grass *sb.*[1] 13.

1953 H. E. Bates *Nature of Love* 16 She had just time to pull the pram into the verge before Parker went past her and the Ford, bouncing, hit the snake fence thirty yards beyond. **1955** *Times* 9 July 7/5 If there were three lanes the slow traffic would be far more inclined to keep within their lane nearer the verge and allow faster vehicles to pass. **1972** *Human World* Nov. 29 The other passengers..crowded towards the windows. The excitement was caused by a small tribe of gypsies encamped on the verge. **1979** J. Grimond *Memoirs* vii. 108 Before the stupid and expensive cutting of the verges the roadsides were thick with clover and cow-parsley.

verge, *v.*[3] [Back-formation f. Verger[2] 1.] *intr.* To act as a verger; to be a verger. Hence **ve·rging** *vbl. sb.*

1900 W. How *Lighter Moments* 54 He werges up one side of the church and I werges up the other. **1926** *Punch* 13 Oct. 400/2, I verges up the centre aisle; he verges up the sides. **1927** H. V. Morton *In Search of England* i. 14 The profession of verging appears to induce mousey manners. **1976** *Church Times* 29 Oct. 18/4 (Advt.), A Christian couple required by St. Paul's Church, Slough. Lady to do cleaning, gent for verging at week-ends.

verge-board. (Earlier example.)

1827 T. F. Hunt *Designs for Parsonage Houses* 5 Ornamental barge-boards... [*Note*] By some called *verge*-boards.

vergée (vɛ̄·rʒe). [Anglo-Norman, f. F. *terre vergée* measured land.] In the Channel Islands, a superficial measure of land, varying between Jersey and Guernsey (see quot. 1971).

In quot. 1915, in Flanders.

1834 H. D. Inglis *Channel Islands* I. ix. 180 Rent in the neighbourhood of St. Helier is as high as from £3 to £4 per vergée... Two vergées and one quarter, make an English acre. **1858** F. F. Dally *Channel Islands* v. iv. 235 The Guernsey vergée is forty perches, and two vergées and a half are rather more than the English acre. **1915** *Contemp. Rev.* July 107, 400 *vergées* (about one acre and a half), were kept under cultivation. **1971** *Nat. Geogr. Mag.* May 722/2, I do 50 *vergées* of potatoes a year... The vergée, an old Norman measure, is still the legal land unit of the Channel Islands. On Jersey, two-and-a-quarter vergées make one English acre, but Guernsey disagrees, and counts two-and-a-half to the acre. **1977** *Jersey Even. Post* 26 July 11/2 The IDC would like to transfer 1½ vergées of land at the corner of Rouge Bouillon and Roussel Street.

vergence (vɜ̄·ɹdʒĕns). [f. Verge *v.*[2] + -ence.] **1.** *Ophthalm.* The simultaneous movement of both eyes towards or away from one another, as when they focus on a point that is nearer or farther away.

1902 E. C. Ellett in *Jrnl. Amer. Med. Assoc.* 18 Oct. 969/2, I would suggest as a temporary table at least, a sort of composite arrangement, as follows: Movements of the visual axis... Associated disjunctive movements (vergences)..convergence..supervergence. **1940** F. B. Chavasse in Ridley & Sorsby *Mod. Trends Ophthalmol.* xxvi. 280 By growth, extension, and refinement of this reflex is developed the proper vergence (notably convergence) of the eyes upon (say) the prey. **1979** *Sci. Amer.* Jan. 88/1 Abrupt target jumps generate two types of eye movements: slow, smooth vergence eye movements..and sharp, rapid saccadic eye movements.

2. *Geol.* [ad. G. *vergenz* (H. Stille 1930, in *Nachrichten von d. Ges. d. Wissensch. zu Göttingen, Math.-Physik. Klasse* iii. 380).] The direction in which a fold is inclined or overturned.

1960 *Trans. Edin. Geol. Soc.* XVIII. 51 Isoclinal recumbent folds with an average trend ENE-WSW; vergence SSE to NNW. [*Note*] The authors use the term 'vergence'..to indicate the direction of overturning of folds. **1978** *Nature* 7 Dec. 604/1 At the northern apex of the Tyrrhenian Sea, two mountain chains face each other displaying opposite vergence: the Corsican alpine chain and the Northern Apennines.

verglas (vɛ̄·ɹglā). Also (formerly) **ver-glas, ver glas.** [Fr., f. *verre* Verre + *glas* (mod.F. *glace*) ice: see Glace *sb.*[1]] = *silver thaw* s.v. Silver *sb.* and *a.* 21.

1808 H. Gray *Lett. from Canada* (1809) 301 During the thaw, a very extraordinary effect is produced, sometimes, on the trees. The Canadians call it a *ver-glas*. **1863** E. H. Walshe *Cedar Creek* 310 Perhaps you could tell me the cause of the *ver glas*? What makes that thin incrustation of ice over the trunk and every twig? **1886** G. Meredith *Let.* 26 Dec. (1970) II. 844, I dread the many possible things from verglas to bronchitis. **1925** N. E. Odell in E. F. Norton *Fight for Everest: 1924* iii. iii. 310 A considerable snowfall will in spring and summer have evaporated into thin air in a few hours without any visible melting. Consequently, and fortunately, 'verglas' does not exist. **1950** tr. *Mountaineering Handbk.* (Assoc. Brit. Members Swiss Alpine Club) vi. 69 Verglas, which is the result of a frozen trickle of water or rainwater, and which covers the rocks like a sheet of glass, is no less dangerous. **1972** D. Haston *In High Places* i. 12 Verglas can also be very difficult. This is a thin veneer that forms on rock, caused by the thawing and re-freezing of running water or drips.

veridian (vĭri·diăn), var. Viridian *sb.* and *a.*

1902 R. Fry *Let.* 13 Oct. (1972) I. 197 The trees [are] strawy green, the glass almost veridian. **1950** M. Peake *Gormenghast* lxiii. 355 The waters became stained with evergreen from the softest olive to veridian. **1974** R. Rendell *Face of Trespass* viii. 70 A new name..in veridian neon, stuck up above the portico.

veridically, *adv.* (Earlier example.)

1832 F. Burney *Mem. Dr. Burney* II. 179 Next to Shakespeare himself, Pope draws human characters the most veridically, perhaps, of any poetic delineator.

veridi·city. [f. Veri(dical *a.* + -icity.] Veridicality.

1937 *Theology* XXXIV. 284 The veridicity of psychic phenomena. **1942** *Mind* LI. 67 The Pragmatist holds that what works is true. If the arguments of this paper were wrong, that would probably be about the best status one could find for the veridicity of memory. **1977** *N.Y. Rev. Bks.* 29 Sept. 21/3 There are various reasons, particular and general, why this example of tampering with what Mr. Stewart likes to call 'veridicity' is important.

verifiability. Add: **a.** (Earlier example.)

a **1881** A. Barratt *Physical Metempiric* (1883) p. xxv, I cannot believe that the test of sensible verifiability will ever satisfy the higher longings..of humanity.

b. *Philos.* The fact of being capable of verification (sense *3 c). Freq. *attrib.*

1936 A. J. Ayer *Lang., Truth & Logic* i. 31 Every empirical hypothesis must be relevant to some actual, or possible, experience, so that a statement which is not relevant to any experience is not an empirical hypothesis, and accordingly has no factual content. But this is precisely what the principle of verifiability asserts. **1936** M. Schlick *Gesammelte Aufsätze* (1938) xiv. 348 When we speak of verifiability we mean logical possibility of verification. **1939** I. Berlin in *Proc. Aristotelian Soc.* XXXIX. 225 The thesis which I shall try to establish is that the principle of verifiability or verification after playing a decisive rôle in the history of modern philosophy..leads to wholly untenable consequences. **1951** H. Reichenbach in Feigl & Brodbeck *Readings Philos. of Sci.* (1953) 93 The verifiability theory of meaning lays down rules for the construction of meaningful expressions. **1967** *Encycl. Philos.* VIII. 241/1 We shall understand the verifiability principle as claiming that the cognitive meaning..of a sentence is to be determined by reference to the verifiability..of the statement expressed by the sentence.

verification. Add: **3. c.** *Philos.* The action or process of verifying a proposition or sentence through empirical experience (associated esp. with logical positivism). Freq. *attrib.*, esp. as *verification principle.*

1932 M. Schlick *Gesammelte Aufsätze* (1938) viii. 181 The meaning of a proposition is the method of its verification. **1934** C. I. Lewis in *Philos. Rev.* XLIII. 131 Suppose it maintained that no issue is meaningful unless it can be put to the test of decisive verification. **1936** A. J. Ayer *Lang., Truth & Logic* 12, I adopt what may be called a modified verification principle. **1937** *Mind* XLVI. 348, I should have thought the first duty of any advocate of a verification theory of meaning would be to inquire how his theory itself was to be verified. **1956** J. O. Urmson *Philos. Anal.* vii. 107 The verification principle is not essentially a very novel or obscure doctrine except in its traditional formulation. This formulation is that the meaning of a statement is the method of its verification. **1963** W. H. Walsh *Metaphysics* i. 15 Logical Positivists..sought to fashion, in their celebrated Verification Principle of Meaning, a weapon which would destroy metaphysics once and for all. **1977** A. Giddens *Stud. in Social & Polit. Theory* i. 45 What came to be called the 'Verification Principle' went through numerous versions, as the inadequacy of Schlick's original formulation..became very rapidly apparent.

d. *spec.* (see quot. 1972).

1953 *Ann. Reg. 1952* 145 This plan provides for the discussion of the regulation of all armaments..and of their disclosure and verification by two committees. **1962** *Listener* 1 Nov. 720/1 Mr Khruschev says he agrees to order dismantling of missile sites in Cuba under U.N. verification. **1972** *Dict. Milit. Terms* (U.S. Dept. Defense) 316/1 *Verification*.., in arms control, any action, including inspection, detection, and identification, taken to ascertain compliance with agreed measures.

4. (Earlier example.)

1789 A. Young *Jrnl.* 8 June in *Trav. France* (1792) I. 103 If..by the verification of their powers in one chamber, they shall once come together, the popular party hope that there will remain, no power afterwards to separate.

verifica·tionism. [f. Verification + -ism.] The philosophical doctrine or principles associated with verification (sense *3 c).

1939 I. Berlin in *Proc. Aristotelian Soc.* XXXIX. 226 It will tend to show that historical connexion between phenomenalism and 'verificationism' is not a logical one, and that the failure of the latter does not necessarily invalidate the former. **1943** *Mind* LII. 98 Full-blooded verificationism..has now commonly been abandoned in favour of what might be called methodological verificationism. **1965** [see *operationalism].

verifica·tionist, *a.* and *sb.* [f. Verification + -ist.] **A.** *adj.* Of, pertaining to, or supporting verificationism. **B.** *sb.* An adherent or supporter of verificationism.

1940 *Mind* XLIX. 456 He [*sc.* Eddington] is as empirically 'verificationist' as any Logical Positivist. **1941** *Mind* L. 163 Most verificationists contend that realism is meaningless. **1952** R. M. Hare *Lang. Morals* i. 8 The so-called 'verificationist' theory of meaning. **1956** J. O. Urmson *Philos. Anal.* xi. 169 If..the verificationist takes the line that the verification principle is a rule of language for the use of the word 'meaningful'..it is a hard saying. **1980** *Times Lit. Suppl.* 17 Oct. 1181/5 How far such verificationist arguments are sound..depends on questions of meaning.

verifier. Add: **1. b.** That which verifies.

1845 J. S. Mill in *Edin. Rev.* LXXXII. 384 In the philosophy of society,..we look upon history as an indispensable test and verifier of all doctrines and creeds. **1944** *Mind* LIII. 341 If I say 'you are hot', this sentence expresses my belief.., and if you are hot, it indicates your state... Your hotness can be called the 'verifier' of my sentence. **1959** B. Russell *My Philos. Devel.* xv. 185, I call the fact which makes the statement true its 'verifier'.

2. b. A keyboard device for checking whether a card or paper tape is correctly punched by indicating any discrepancy when it is inserted and the data on it keyed a second time.

1940 W. J. Eckert *Punched Card Methods* 9 The mechanical verifier is used to check the punching of the initial data on the cards. **1949** [see *interpreter 5 a]. **1956** G. A. Montgomerie *Digital Calculating Machines* viii. 154 More complicated electrically driven verifiers are available and in some makes they resemble the electrically driven punches. **1978** J. Kellock *Elements of Accounting* xii. 214 After the information has been punched on to the cards, the cards are verified on a machine known as a verifier.

verifying, *vbl. sb.* Add: (Later examples.) Cf. prec. 2 b.

1932 *Monthly Notices R. Astron. Soc.* XCII. 700 In commercial practice punching is usually verified by a special verifying punch. **1954, 1965** [see *keypunching* vbl. sb. sv. *keypunch *v.*]. **1968** *Brit. Med. Bull.* XXIV. 207/1 Four-fifths of the costs of linking a record by computer arise in preparation of the data—i.e., abstraction, classification and coding, card or tape punching and verifying. **1973** F. R. Crawford *Business Systems with Punched Card Data Processing* v. 89 Proper design of documents and card forms is essential for accurate, efficient card punching and verifying.

verisimilous, *a.* For † *Obs.* read 'Now *rare*' and add later example.

1958 *New Scientist* 23 Oct. 1119/2 Narratives which, although verisimilous to a degree, are shot through and through with the credulous and superstitious outlook of the seventeenth century.

‖ **verismo** (veri·zmo). [It.] Realism or naturalism in the arts; esp. with reference to Italian opera of the late 19th century. Freq. *attrib.* Cf. Verism.

1908 R. Dunstan *Cyclopædic Dict. Music* 435/1 *Verismo*, truth, naturalism. **1919** G. Kobbé *Compl. Opera Bk.* 91 What is true of 'Aida', is equally applicable to the whole School of Italian *versimo* [*sic*] that came after Verdi—Mascagni, Leoncavallo, Puccini. **1940** *Scrutiny* IX. 51 If the opera is too realistic it may be boring... The *verismo* of the latterday Wagnerians cannot escape this charge. **1954** *N.Y. Times* 28 Dec. (Late City Ed.) 21/1 Gian-Carlo Menotti's 'The Saint of Bleecker Street'..is Italian 'verismo' opera. **1970** R. Lowell *Notebk.* 224 *Verismo* is no vehicle for Death. **1976** *Times* 23 Jan. 14/8 The movement towards *verismo* in film and theatre. **1978** Ld. Drogheda *Double Harness* xxi. 289 Unfortunately Colin Anderson himself took against what he dismissed somewhat derisorily as Franco's *verismo* method. **1983** *Observer* 29 May 33/2 Kiri Te Kanawa, happily back on Mozartian territory after her unhappy excursion into *verismo*.

veritableness. Delete *rare*[−1] and add later examples.

1890 W. James *Princ. Psychol.* I. vii. 189 But the psychologist must not only *have* his mental states in their absolute veritableness, he must report them and write about them. **1926** R. Clements *Stately Southerner* 142 Whatsoever shape the apparition may have taken, a belief in its veritableness has persisted from of old.

‖ **vérité** (verite). [Fr., = truth.] Realism or naturalism, esp. in cinema, radio, and television; documentary method. Also *attrib.* and in *Comb.* Cf. *cinéma-vérité; Verism.

1966 *Listener* 15 Dec. 880/2 There is an advanced cinema, a cinema of *vérité* which is rightly admired by intellectuals. **1976** *Ibid.* 23 & 30 Dec. 841/1 Jack Gold's early vérité-style account of an industrial dispute. **1978** *Broadcast's Programme Edin. TV Festival* 10/2 The film was heavily criticised..[as] impressionistic, partial..an indication of the problematic nature of verité's claim to be..untainted evidence. *Ibid.*, Verite film brings to the screen the pathetic death of a drowned boy. **1979** *Daily Tel.* 7 Dec. 17/3 The method of 'radio vérité' reporting, with victims telling their own emotional stories and the guilty parties often getting quite annoyed, make [*sic*] compulsive listening. **1984** *Listener* 1 Mar. 32/1 What is over-the-top in a vérité film may be just what is required in sit-com.

verkramp (fɛ̄ɹkra·mp), *predic. a.* *S. Afr.* Also **verkrampte** when used *attrib.* [Afrikaans, lit. 'narrow, cramped'.] Narrow or reactionary in political, religious, social, etc., matters.

1968 *Green Bay Tree* 29 June p. xlv, The most *verkrampte* Nationalist politicians are frightened of liberal foreign propaganda appearing on their screens. **1969** *Sunday Times* (Johannesburg) 24 Aug., On a very strict, narrow interpretation, my guess is that most Nationalists are verkramp. **1971** *Ibid.* 11 Apr. 3 Mr. — is verkramp. He can make what threats he likes, I'll keep on wearing split skirts. **1973** *Deb. Senate S. Afr.* 17 May 2790 A Sunday paper fairly recently referred to certain members of the United Party as a 'verkrampte mafia'. This was not defamatory. **1974** *Argus* (Cape Town) 2 Aug. 3/2 It is believed that conservative resistance to more enlightened policies has been significantly reduced and that verkrampte views will in future meet with little sympathy. **1974** *Rand Daily Mail* 30 Nov. 1/1 This reasoning has now become an excellent excuse for talking verlig and acting verkramp. **1977** *Time* 31 Oct. 19/3 The Cabinet was clearly aware of charges by the *verkrampte* (conservative) wing of the Nationalist Party that the government's power and authority were eroding in the face of protests by urban blacks. **1981** *Observer* 5 Apr. 11/7 Like most black African Muslims, Edvis favours a liberal interpretation of the Koran, being *verligte* rather than *verkrampte* in that regard. Very wisely he likes to drink Pinot Noir from California.

Hence (the *attrib.* form **verkrampte**) as *sb.* A person with rigidly traditional views, esp. as regards apartheid and the preservation of white supremacy. Cf. *verlig *a.*

At first applied only to politicians but now extended to any person holding narrow-minded or ultra-conservative views on social, religious, etc., matters.

1967 *Race Relations News* Sept. 7/4 The verligtes-verkramptes controversy must have raised in many minds the hope of a change of outlook in influential South African circles. **1968** *Economist* 17 Aug. 32/2 The real object was to strengthen Mr Vorster in the ideological dispute that has arisen within his party in recent months between the *verligte* ('enlightened') wing and the *verkramptes* (that is, literally, the cramped ones). **1969** *Times* 18 Sept. 9/1 The Army and its political directors have come in for much criticism..from Unionist *verkramptes*, for apparent irresolution in bringing down the barricades. **1971** *Leader* (Durban) 7 May 6/5 With a so-called outward-looking policy, the Government is beginning to see things in better light. It has shaken itself of some of the old die-hard Calvinistic verkramptes and is able to look at the world at large a little verlig. **1979** *Economist* 8 Sept. 59/3 Now Transvaal *verkramptes* (Conservatives) are saying he exploited Muldergate to destroy their leaders.

Verlainesque (vɛɹlɛ̄ⁱne·sk), *a.* [f. the name *Verlaine* (see below) + -esque.] Of, pertaining to, or characteristic of the French poet Paul Verlaine (1844–96), or his works.

1891 E. Dowson *Let.* 20 Mar. (1967) 189 Its an attempt at mere sound verse..a vague, Verlainesque emotion. **1915** Wyndham Lewis *Lett.* (1963) 68 He told me he had written a lot of filthy sexual verse... He described it as Verlainesque, damn his shifty little eyes.

verlig (fɛ̄ɹli·χ), *predic. a.* *S. Afr.* Also **verligte** when used *attrib.* [Afrikaans, = enlightened.] Regarded as progressive or enlightened, in political, religious, social, etc., matters. Cf. *verkramp *a.*

1968 [see *verkrampte *sb.*]. **1969** *Sunday Times* (Johannesburg) 24 Aug. 25 The odd thing about the whole affair is that the word 'verlig' should become an embarrassment to the Nationalists... There must be few countries where it is an insult to be called 'enlightened'. **1969** *Rand Daily Mail* 25 Sept. 17/3 Verlig long before it was comparatively easy to be so, he condemned arrogant racialism or nationalism wherever he found it. **1970** *Times* 19 Aug. 8/4 It is not just non-aligned countries such as Tanzania, Zambia or Yugoslavia, which are benefiting from the new spirit of goodwill emanating from Peking. China's *verligte* policies have even percolated as far as London. **1971** [see *verkrampte *sb.*]. **1974** *Argus* (Cape Town) 2 Aug. 3/2 The United Party is believed to have irrevocably committed itself to a more boldly verligte political stance. **1974** [see *verkramp *a.*] **1977** *Cape Times* 10 Oct. 3/4 The vast majority of Nationalist voters are a lot more verlig than they think and support the need for change in South Africa. **1981** *Guardian* 14 Dec. 6/5 The Verligte or enlightened faction of his party.

Hence (the *attrib.* form **verligte**) as *sb.* A person regarded as progressive or enlightened in politics, esp. with regard to the future of apartheid. In extended use, one who is enlightened or broadminded in other matters.

1967 [see *verkrampte *sb.*]. **1969** *Guardian Weekly* 24 July 14 If the *verkramptes* of Louis Stoffberg and Dr Herzog win the day, against Mr Vorster's *verligtes* and their good neighbour policy towards black Africa, it will be a useful gain for Peking. **1970** *News/Check* 29 May 9 People who later on were to become known as verligtes sided with Silbersteins; the verkramptes summed up the book as sinful. **1976** *Times* 26 June 5/4 The increasingly vocal Verligtes in the universities. **1980** *Economist* 21 June (S. Afr. Suppl.) 6/1 To the more radical verligtes who make up his praetorian guard, a sort of strategy is clear.

vermiculite. Add: **1. c.** Flakes of this mineral used as a moisture-holding medium for plant growth or as a protective covering for the storage of bulbs or tubers.

1950 *Los Angeles Times Home Mag.* 12 Feb. 38/3 Vermiculite is one of the finest storage materials for bulbs, tubers and corms as it insulates them from sudden temperature changes. **1981** *Farmstead Mag.* Winter 71/1 All that is necessary for this March project is a few waterproof trays, fiber or wooden slats, vermiculite for seed germination, and soilless mix such as 'Promix' for filling the flats. **1983** *Which?* Sept. 398/3 Wash off as much soil as possible, then stand the tubers, stem-down, in a dry, frost-free place..before..storing in boxes of dry peat or vermiculite.

vermigrade (vɜ̄·ɹmigrēⁱd), *a.* *rare*[−1]. [f. Vermi- + *-grade.] Proceeding in a worm-like manner. (In context *fig.*)

1938 S. Beckett *Murphy* viii. 134 Celia was in a state indeed, trembling and ashen. The footsteps overhead had become part and parcel of her afternoon, with the rocking-chair and the vermigrade wane of light. An Ægean nightfall suddenly in Brewery Road could not have upset her more than this failure of the steps.

vermilionize, *v.* For *rare*$^{-1}$ read *rare* and add later *fig.* example.

1924 W. J. Locke *Coming of Amos* iv. 43 If I had found myself at five-and-twenty with fifty thousand pounds lying at the bank, verily, I believe, I should have vermilionized the cosmos.

vermin, *sb.* (and *a.*). Add: **5. b.** *vermin-proof* adj. **c.** *vermin-eaten, -like* (later example) adjs.

1837 Dickens *Let.* 7 Sept. (1965) I. 304 If the *Pickwick* has been the means of putting a few shillings in the vermin-eaten pockets of so miserable a creature, [etc.]. 1931 R. Campbell *Georgiad* i. 16 So many poets..sigh to share their [*sc.* gypsies'] vermin-eaten ways. 1914 Joyce *Dubliners* 86 A horde of grimy children populated the street... He picked his way deftly through all that minute vermin-like life. 1937 *Discovery* Dec. 388/1 Vermin-, rot- and fire-proof. 1982 M. Young *Elmhirsts of Dartington* iv. 81 A vermin-proof rice store.

verminous, *a.* Add: **4. c.** *verminous bronchitis* = *hoose, hooze *sb.*

1925 [see *parasitic *a.* 5]. 1970 W. H. Parker *Health & Dis. in Farm Animals* xx. 268 This is the first sign of the disease known as Husk, or Verminous Bronchitis.

verminicide (vəɹmi·nisəid). [f. Vermin *sb.* + -cide^1.] A preparation for killing vermin.

1925 *Blackw. Mag.* Sept. 313/2 During the summer verminicide became a necessary item in the 'bazar'. 1965 N. H. Johnson *Compl. Bk. Dogs* (1968) xii. 387 Detergents and verminicides, as well as sleeping and reducing pills, are equally dangerous.

Vermont (vəɹmǫ·nt). The name of one of the north-eastern states of the United States of America, used *attrib.* in *Vermont merino*, a sheep belonging to a breed developed there.

1891 R. Wallace *Rural Econ. Austral. & N.Z.* xxvii. 359 The Vermont Merino..is one of the most perfect as regards good covering and density of fleece. 1957 *New Biol.* XXII. 92 Vermont merinos were imported into Australia from the United States in the 1880s. These sheep had wrinkly skins, and many that crossed the Pacific by 1890 presented a concertina-like appearance.

† **Vermonteer.** *Obs.* [f. prec. + -eer.] = next.

1778 *15th Rep. R. Comm. Hist. MSS.* App. 396 in *Parl. Papers 1897* (C. 8551) LI.1 We may hence learn that we are not to flatter ourselves with the hope of winning over the Vermonteers to the Crown side. 1801 *Hist. Rev. & Directory N. Amer.* II. 346 Great numbers of the Vermonteers bring their produce here.

Vermonter (vəɹmǫ·ntəɹ). [f. as prec. + -er^1.] A native or inhabitant of the State of Vermont.

1787 A. Hamilton *Works* (1886) VII. 6 The peace found the Vermonters in a state of actual independence. 1851 H. Melville *Moby Dick* I. vi. 51 There weekly arrive in this town [*sc.* New Bedford] scores of green Vermonters. 1961 W. Vaughan-Thomas *Anzio* v. 67 Its tough, stocky commander, General Ernie Harmon, was a Vermonter. 1979 *United States 1980/81* (Penguin Travel Guides) 662 For Vermonters maple sugaring time is a celebration of spring. 1984 *Sci. Amer.* May 22/3 [So] wrote Vermonter Zerah Colburn in 1833.

Vermontese (vɜ̄ɹmǫntī·z), *sb.* and *a.* Also † **Vermonteze.** [f. as prec. + -ese.] **A.** *sb.* = prec. Also *collect.* **B.** *adj.* Of or belonging to Vermont.

1783 *Polite Traveller* 100 The persons, manners, and customs, of the Vermontese, are nearly the same with those provinces from whom they emigrated. 1798 I. Allen *Nat. & Pol. Hist. Vermont* 280 Our Vermontese house-wives are not a little vain of their knowledge in making home-made wines. 1804 T. G. Fessenden *Orig. Poems* 123 Hamilton was an impudent Vermonteze. 1845 *Knickerbocker* XXVI. 583 We should be pleased to hear these lines applauded by the Vermontese. 1851 *San Francisco Picayune* 20 Sept. 2/4 On a late visit to an ancient Vermontese lady, he was asked, 'Did you see Professor Webster hung?'

Vermoral (vɜ̄·ɹmorăl). Also **-el.** [Said to be the name of V. *Vermorel*, French manufacturer.] A type of sprayer used in the war of 1914–18 to produce a fine spray of water that would absorb residual poison gas.

1916 'H. Rae' *Maple Leaves* 264 'A vermoral sprayer is a new form of frightfulness for spraying the front-line trench.' 'What with? Liquid fire?'..'No, liquid water.' 'For the flies?' 'No, for the gas.' 1929 R. Graves *Good-Bye to all That* xv. 204 Vermorel-sprayers had cleared out most of the gas, but we still had to wear our masks. 1930 Blunden *De Bello Germanico* iii. 30 A 'Vermoral Sprayer' for nullifying gas.

vermouth. Add: **c.** *vermouth-cassis*: see *cassis.

vernaccia (vəɹna·tʃa). [It.: see Vernage.] A wine (usu. white) produced in the San Gimignano area of Italy and in Sardinia; also, the grape from which it is made.

1824 A. Henderson *Hist. Anc. & Mod. Wines* ii. vi. 237 Another source of error arises from the circumstance of several of the best Tuscan wines receiving their appellations from the grapes which yield them, as, for example,.. the *Vernaccia*, &c; and as these names are not confined to Tuscany, but are common to the growths of other parts of Italy, the difficulty of distinguishing them is..further increased. 1833 C. Redding *Hist. Mod. Wines* ix. 241 Their sweet wines the Italians call *Abbocati*... Of the former kind are the Moscatello, Aleatico, and Vernaccia, a white wine, of considerable note among the writers of Italy. 1929 P. M. Shand *Bk. of Other Wines* 85 The vines grown are—*red*: Chiavennasca.., Vernaccia, [etc.]. 1966 C. Ray *Wines of Italy* ix. 95 Vernaccia di Serrapetrona, sweetish sparkling red wine made from the Vernaccia Rossa grape, and interesting to those who are interested in sparkling sweet red wines. 1973 *Times* 31 Jan. (Mediterranean Suppl.) p. ii/4 Pecorino sheep cheese and vernaccia wine have enjoyed steady, if limited, markets outside the island. 1975 *Times* 14 Apr. (Ital. Wines Suppl.) p. iii/4 A..dry golden vernaccia, from..San Gimignano (not to be confused with that potent vernaccia in Sardinia).. is dry enough to accompany fish.

vernacular, *a.* and *sb.* Add: **A.** *adj.* **6.** *spec.* in *vernacular architecture*, architecture concerned with ordinary domestic and functional buildings rather than the essentially monumental.

1857 G. G. Scott *Remarks Secular & Domestic Archit.* p. ix, I want to call attention to the meanness of our vernacular architecture, and to the very partial success which has hitherto attended the attempts at its improvement. 1939 *Country Life* 11 Feb. 154/2 It is as delightful an example as one could find of Georgian vernacular architecture. 1976 [see *spinning *vbl. sb.* 7 c]. 1977 *Dædalus* Summer 3 The studies of so-called vernacular architecture (like barns) no longer seem eccentric in an atmosphere in which architecture can be defined not in terms of monuments but as any changes at all that man makes in his environment.

B. *sb.* **1. a.** Also, the informal, colloquial, or distinctive speech of a people or a group. Cf. sense 1 b below.

1925 F. N. Scott in *McNaught's Monthly Mag.* May 144 (*heading*) English and American vernacular. 1930 G. B. Shaw *Admirable Bashville* 89 With the advent of compulsory education sixty years ago,..newspapers and plays alike soon came to be written by illiterate masters of the vernacular. 1984 *Gainesville* (Florida) *Sun* 28 Mar. 7B/2 Observe feminist vernacular: Call it a 'personhole cover'.

b. (Later example with reference to colloquial speech.) Cf. sense 1 a above.

1975 L. Gillen *Return to Deepwater* iv. 62 In the vernacular,..I couldn't care less what you do.

4. A vernacular style of building. Cf. sense A. 6 above.

1910 *Encycl. Brit.* II. 436/1 The culture of the 'Queen Anne' type of architecture..presented a simple vernacular of construction and detail. 1967 *Listener* 7 Sept. 292/3 What was normally North American about these houses..was their general internal planning..an open-plan vernacular that still works well today. 1977 M. Girouard *Sweetness & Light* ii. 25 They came back to London to design buildings in which Gothic merged into farmhouse vernacular.

verna·cularist. [f. Vernacular *a.* and *sb.* + -ist.] **1.** An advocate of the use of a regional mode of speech; a speaker or writer in a regional or demotic idiom.

1867 [see *Anglicist 1]. 1926 *Glasgow Herald* 27 July 10 There was ample material..in the industrial struggle with which Clydeside was so familiar, which in the hands of a Vernacularist of genius could produce a play so striking as 'Strife'. 1974 *Sat. Rev. World* (U.S.) 4 Dec. 46/2 Creosote bush, which the Spaniards called *hedonillo* but the American vernacularists termed 'little stinker'.

2. *R.C.Ch.* An advocate of the use of the vernacular in the liturgy. Also *attrib.* or as *adj.*

1956 *Catholic Herald* 9 Mar. 2/4 Was St. Thomas a Vernacularist? 1982 *Observer* 25 Apr. 30/3 Mr St John-Stevas's Latin is not good.., but that is a venial sin in these vernacularist days.

vernacularity. 1. (Later examples.)

1904 'O. Henry' in *N.Y. World Mag.* 25 Dec. 2/6 Remsen touched his cap..and took refuge in vernacularity. 1943 Entwistle & Gillet *Lit. of Eng.* iv. 41 He [*sc.* John Lyly] cultivated..unexpected vernacularity amid refined 'conceits'.

vernalization (vɜ̄:ɹnăləizēᶦ·ʃən). [f. Vernal *a.* + -ization, as tr. Russ. *yarovizátsiya*.] The technique of exposing seeds, young plants, etc., to low temperatures in order to hasten subsequent flowering; also, the natural process induced by cold weather which this technique imitates. Also *transf.* and *fig.*

1933, 1934 [see *jarovization]. 1957 V. Nabokov *Pnin* ii. 47 The vernalization of the visas, and the preparations [for going to America]. 1971 E. O. Wilson *Insect Societies* viii. 154/1 The vernalization (chilling) effect that renders *Myrmica* and *Formica* brood queen-potent can be interpreted as a token stimulus. 1974 A. J. Huxley *Plant & Planet* x. 100 In beet..the rosette normally made in the first year..requires winter chilling for flowering the following summer. This winter chilling is called vernalization. *Ibid.* xxvii. 315 Vernalization consists..in giving the seed to be sown in spring a cold period..with a small amount of water.

Hence (by back-formation) **ve·rnalize** *v. trans.*, to treat or affect (seeds, etc.) in this way; **ve·rnalized** *ppl. a.* (in quots. *transf.*), **ve·rnalizing** *vbl. sb.*

1933 Whyte & Hudson in *Bull. Imperial Bureau Plant Genetics* No. 9. 8 (*heading*) Technique for vernalizing long-day plants. 1946 *Nature* 5 Oct. 485/2 The time taken for development to have begun in all the vernalized [sponge] gemmules..is less than in December. 1947 *Ibid.* 4 Jan. 32/1 This [*sc.* premature flowering] may be due.. to a vernalizing effect on the..germinated seed..of naturally experienced low soil..temperatures in early spring. 1971 E. O. Wilson *Insect Societies* viii. 150/1 It is from some of these vernalized larvae that the yearly crop of queens is matured in the spring. 1976 *Sci. Amer.* Sept. 99/3 The crop flowers and produces grain in the spring after being vernalized, or induced to flower, by the low temperatures in winter.

vernation. 2. (Later example of extended use.)

1929 S. Leslie *Anglo-Catholic* x. 121 In the vernation of the year Edward felt the old desire for the earth, a renewed longing to labour and drive the plough.

Verné (vę·rne). Also **Verneh.** [Origin unknown.] **a.** A Caucasian pileless rug or kilim. **b.** An Anatolian brocaded rug. In full *Verné kilim, rug*, etc.

1922 Kendrick & Tattersall *Hand-Woven Carpets* I. iv. 165 Thin and soft Kilims made in the neighbourhood of Shusha are known in the trade by the name of *Verné*. *Ibid.* II. Pl. 143 (*caption*) Verné Kilim carpet. 1931 A. U. Dilley *Oriental Rugs & Carpets* Pl. 53 (*caption*) Verné animal and bird khilim. 1973 *Country Life* 29 Nov. 1823 Verné rug 7′ 1″ × 5′ 6″..Armenia circa 1900. 1975 *Times* 13 June 3/1 An unusual feature of the sale was the presence of two antique Verneh rugs. 1981 I. Bennett *Oriental Rugs* I. 364/2 There are several varieties of so-called bird verneh rugs.

Vernean (vɜ̄·ɹniăn), *a.* Also **Vernian.** [f. the name of Jules *Verne* (1828–1905), French author: see -an, -ian.] Of, pertaining to, or characteristic of the science fiction of Jules Verne. Also **Jules Vernean.**

1960 K. Amis *New Maps of Hell* (1961) i. 37 The book closes with a straightforward Vernean sermon on the dangers of scientific progress considered as an embodiment of human arrogance. 1964 *Listener* 12 Nov. 743/2 The conquest of space no longer strikes us as Wellsian or Jules Vernian. 1980 *Time* 1 Dec. 60/2 Visionaries like Jules Verne were suggesting a better way. A bullet-shaped vehicle..could be propelled far faster by using powerful electromagnetic fields. Now, as a result of lab work in the U.S. and abroad, the Vernean scheme shows promise of becoming a practical reality.

Verner's law: see Law *sb.*^{1} 17 c.

Verneuil (vęrnö·y). The name of A. V. L. *Verneuil* (1856–1913), French chemist, used *attrib.* with reference to a technique invented by him for producing artificial rubies.

1912 *Chem. Abstr.* VI. 55 A description of the Verneuil method of preparing artificial rubies. 1951 Kirk & Othmer *Encycl. Chem. Technol.* VII. 163 A typical Verneuil crystal-growth apparatus is shown schematically in Figure 2. 1979 Mills & Mansfield *Genuine Article* vii. 116 The little bubble which you often see in a Verneuil synthetic.

vernier. Add: **3.** *Astronautics.* Used *attrib.* and *absol.* to designate a small auxiliary rocket engine for effecting minor changes in the velocity or attitude of a spacecraft.

1958 *Observer* 12 Oct. 1/2 Eight small 'vernier', or guidance rockets,..had been fitted to Pioneer. Any or all of these could be fired from the earth to make any necessary correction in the final phase of the flight. 1960 *Aeroplane* XCVIII. 562/1 In its nose would be a third stage employing low-thrust 'vernier' rocket motors. 1968 A. C. Clarke *2001* xxxv. 184 The main thrust died and only the verniers continued to nudge *Discovery* gently into orbit. 1968 *Sci. Jrnl.* Nov. 76/2 The rocket..had no fewer than 20 main thrust chambers and 12 swivelling verniers—small motors for fine control of speed and direction. 1979 J. W. Cornelisse *Rocket Propulsion & Spaceflight Dynamics* x. 212 A third method to ensure ignition after coasting is the use of small (Vernier) rockets, which give the vehicle a sufficiently large acceleration to position the propellant at the sump.

vernis martin (vęrni· mārtæ·ṅ). Also **vernis Martin, Vernis Martin,** and hyphenated. [Fr., f. *vernis* varnish + *Martin* (see below).] A lacquer or varnish used in the eighteenth century by the French brothers Étienne, Guillaume, Julien, and Robert Martin and their contemporaries on a range of furniture, ornaments, etc., to imitate oriental lacquer. Also *fig.*

1877 C. Schreiber *Jrnl.* (1911) II. 18 Some curious boards for the game of Loto [*sic*], done in Vernis Martin. 1883 J. W. Mollett *Illustr. Dict. Art & Archæol.* 337/1 *Vernis-Martin work*, a Japanese style of painting and enamelling on furniture, carriages, and small objects. 1899 R. Whiteing *No. 5 John St.* xvi. 163 That polish of the world which is not exactly vernis-Martin for transparency. 1911 Louisa of Tuscany *My Own Story* vi. 87 It was a magnificent historical vehicle painted in vernis Martin. 1942 *Burlington Mag.* Apr. 104/2 A fine collection of small boxes—vernis martin, ivory, etc.,..will be sold in May. 1963 *Times Lit. Suppl.* 15 Mar. 192/4 A fine example of a 'Vernis Martin' binding of *c.* 1800. 1978 *Country Life* 8 June (Suppl.) 97/3 French vernis martin snuff-box, c. 1780.

‖ **vernissage** (vęrnisā·ʒ). Also **Vernissage.** [Fr.] A day before the exhibition of paintings on which exhibitors may retouch and varnish their pictures already hung. Now usu. denoting a private view of paintings before public exhibition. Cf. *varnishing *vbl. sb.* 1 b.

1912 *Queen* 20 Apr. 643/2 The Salon Nationale des Beaux Arts. The vernissage was on Saturday, and was marked by the usual miscellaneous crowd of French people and foreigners. 1920 R. Fry *Let.* 28 Sept. (1972) II. 491 My landscape is accepted all right at the Autumn Salon... I went to..give it some *vernis à retoucher* at the *Vernissage*. 1930 *Observer* 26 Jan. 10 The Indépendants have once more occupied the Grand Palais... The crowd at the vernissage did not seem to be moving so fast. 1958 *Spectator* 11 July 56/2 Artists were excluded from the *vernissage* of the twenty-ninth *Biennale* at Venice. 1961 *Times* 18 Mar. 3/7 At the height of the season art critics in Paris receive an average of 20 vernissage invitations each week. 1967 *Times* 3 May 9/7, I found myself completely at sea at the 'vernissage' of an exhibition of paintings..at one of London's..art galleries. 1979 M. Soames *Clementine Churchill* ii. 23 Mlle Henri had procured tickets for a *Vernissage*.

vernix. Restrict ‖ and *Obs. rare* to sense in Dict. and add: **2.** *Med.* In full, *vernix caseosa* [mod.L. *caseōsus*, f. L. *caseus* cheese]. A greasy deposit covering the skin of a baby at birth.

1846 Dunglison *Dict. Med. Sci.* (ed. 6) 785/1 Vernix caseosa. 1882 W. T. Lusk *Sci. & Art of Midwifery* iii. 75 In the fifth month the surface of the fetal body is covered by the *vernix caseosa*, a whitish substance composed of.. surface epithelium, down, and the products of the sebacious glands. 1956 *Nature* 18 Feb. 330/1 The specimens [of amniotic fluid] were centrifuged and the vernix and supernatant fluid removed. 1978 *Jrnl. R. Soc. Med.* LXXI. 212 Copious vernix caseosa is often present. 1980 *Brit. Med. Jrnl.* 25 Oct. 1138/1 With difficulty but determination she gave birth to an enormous child coated in so much vernix that it seemed to wear a cream cheese pack.

Veronese, *sb.* Add: **1.** (Examples in *sing.*)

1673 J. Ray *Observations Journey Low-Countries* 220 The Antiquities of Verona written by Torellus Saraina a Veronese. 1967 *Guardian* 10 May 7/2 A Veronese himself, Professor Forlati came to Venice in 1912.

veronica^2. Add: **2.** In *Bullfighting*, a movement typical of the first tercio in which the matador swings the cape in a slow circle round himself in order to persuade the charging bull to follow the movement of the cape. Also *fig.*

1926 E. Hemingway *Sun also Rises* xviii. 217 The bull wanted it again, and Romero's cape filled again... He made four veronicas like that, and finished with a half veronica that turned his back on the bull. 1936 R. Campbell *Mithraic Emblems* 18 Enemy of my inward night..whose arm against the Bull designs The red veronicas of light. 1957 A. MacNab *Bulls of Iberia* vi. 57 In two-handed passes, the cape is held on either side of the collar. The fundamental pass is the *verónica*... The name *verónica* is derived from the attitude of the man holding the cape out in his two hands, which resembles that in which St. Veronica is depicted holding out the towel to Our Lord on the way to Calvary. 1976 *Listener* 29 Apr. 541/1 The oldest of old-time waltzes, the couples dancing a mile apart, the women—executing neat veronicas with the men in swallow-tails—got up like Christmas paper bells.

‖ **veronique** (verǫnik). [Fr.] **1.** *Cookery.* (Usu. in form *Véronique*.) Applied to dishes (esp. of fish or chicken) prepared or garnished with grapes.

1907 G. A. Escoffier *Guide Mod. Cookery* 302 *Filets de soles Véronique.* Raise the fillets of a fine sole. 1940 A. L. Simon *Conc. Encycl. Gastron.* II. 103/2 *Véronique*, stuffed fillets of sole, rolled and poached, garnished with one muscat grape upon each fillet and muscat grapes round the dish. 1958 B. Pym *Glass of Blessings* iii. 38 He was even now preparing them a delicious sole véronique. 1960, etc. [see *sole *sb.*^{2} 1 c]. 1963 A. Simon *Guide Good Food & Wines* 379 Véronique, stuffed fillets of sole, rolled and poached, garnished with one muscat grape upon each fillet and muscat grapes round the dish. 1979 *United States 1980/81* (Penguin Travel Guides) 298 The menu is imaginative,..—trout Véronique (poached and topped with green grapes and hollandaise sauce).

2. = prec.

1931 R. Campbell *Georgiad* ii. 32 My passes brought the colour to his cheeks, The loudest he to cheer my veroniques. 1932 —— *Taurine Provence* iii. 72 A heroic series of repeated veroniques against a bull of *grande vaillance*.

‖ **verre églomisé** (vęr' eglomize). [Fr., f. *verre* (see Verre) + *églomisé *a.* and *sb.*] Glass decorated with a layer of engraved gold (see quots. 1971, 1977).

1907 E. Dillon *Glass* viii. 140 The variety of painted glass known in later times as *verre églomisé*. 1933 *Connoisseur* XCII. 372/2 The term '*verre églomisé*'..has for long been applied to this work, in common with the art of painting under glass. 1941 *Burlington Mag.* Feb. 59/1 The cabinet illustrated..and a pair of looking-glasses, with glass borders ornamented with *verre églomisé* decoration in scarlet and gold..were brought from Spain five or six years ago. 1967 [see *églomisé]. 1971 *Country Life* 15 July 150/2 He was a native of Amsterdam and worked in *verre églomisé*—that is gold and silver leaf laid under glass and engraved with a pointed tool. It is a very ancient technique going back to Roman times, but its customary name is derived from that of a well-known 18th-century picture-frame maker named Glomy who employed the method extensively. 1977 Fleming & Honour *Penguin Dict. Decorative Arts* 826/2 *Verre églomisé*, glass decorated on the back by unfired painting or, usually, by gilding... The painting or gilding is protected by another sheet of glass or by a coat of varnish or a layer of metal foil.

Verrocchiesque (verǫkie·sk), *a.* [f. the name of Andrea del *Verrocchio*, a Florentine painter and sculptor (1435–88 + -esque.] Suggestive of or resembling in subject or style the works of Verrocchio. Also **Verro·cchian** *a.* [-ian].

1902 R. Fry *Let.* 23 Jan. (1972) I. 188 He went on to Verrochian and Botticellian ideas as the *Archangel and Virgin Enthroned* suggest. 1933 *Burlington Mag.* Mar. 140/1 The whole technique is closest to the most Verrocchiesque of all Leonardo's drawings. 1942 *Ibid.* Oct. 243/2 This panel has been attributed to Verrocchio; and ..we must expect to find in it pronouncedly Verrocchiesque features.

verruga. Delete ‖ and substitute for def.: The second, chronic stage of an infection by the bacterium *Bartonella bacilliformis*, characterized by wart-like skin lesions: see *Oroya fever. Also *verruga peruana, peruviana* [Sp. *peruana* Peruvian; *peruviana*, mod.L. rendering of this]. (Earlier and later examples.)

[1825 W. B. Stevenson *Hist. & Descr. Narr. Twenty Years' Residence in S. Amer.* I. xiv. 347 *Berrugas*, warts of a peculiar kind, are common in some of the valleys of the coast [of Peru].] 1873 T. J. Hutchinson *Two Years in Peru* II. xx. 58 He was getting through an attack of that dreadful disease, the verrugas, and appeared but the shadow of a man. 1949, etc. [see *Oroya fever]. 1961 R. D. Baker *Essent. Path.* ix. 190 Bartonellosis is a peculiar disease which occurs in Peru and is characterized by Oroya fever and verruga peruviana. 1968 T. Wolfe *Electric Kool-Aid Acid Test* xxi. 305 He can *feel*.. every verruga fly, velvet ant, murine fleas and crabs.

Versailles (vę̄ɹsai·). The name of a hunting lodge to the south-west of Paris built by Louis XIII and enlarged into a palace by Louis XIV in the 17th century, used to denote: **1.** *transf.* A building of similar style or splendour. Also *fig.*

1749 J. Cleland *Mem. Woman Pleasure* I. 134 But had it been a dungeon..his presence would have made it a little Versailles. 1899 Kipling *From Sea to Sea* I. xvii. 159 Jeypore Palace may be called the Versailles of India. 1959 M. Crosland tr. *J. Rovan's Germany* 140 Potsdam, the Prussian Versailles. 1968 *N.Y. City* (Michelin Tire Corp.) 47 The Hotel Pierre..a sort of '40-story Versailles'. 1977 'R. Player' *Month of Mangled Models* vii. 125 It was Jules Goncourt who had called the whole house the Versailles of Whoredom.

2. The site of the peace conference held there at the conclusion of the 1914–18 war which gave its name to the treaty signed there in 1919. Also *transf.*

1928 J. Buchan *Runagates Club* x. 273 The soldiers.. would have made a cleaner and fairer job of it than the kind of circus that appeared at Versailles. 1936 G. B. Shaw *Millionairess* 117 There remained the clauses of the Versailles Treaty by which Germany was to be kept in a condition of permanent, decisive, and humiliating military inferiority to the other Powers. 1967 *Sat. Rev.* (U.S.) 8 Apr. 16/3 Most German politicians..ominously refer to this treaty as another 'Versailles'. 1971 *Guardian* 5 Aug. 10/4 In 1936..the Guardian was busily condemning the Versailles terms. 1981 J. Wainwright *Urge for Justice* i. ix. 59 Versailles is being repaid...Hitler..is putting greatness back into Germany.

versal, *sb.* Restrict † *Obs. rare* to sense in Dict. and add: **2.** A special style of ornate capital letter used at the beginning of a verse or paragraph, etc., esp. in an illuminated manuscript; in modern calligraphy applied to capitals built up by inking between pen strokes and having serifs in the form of long, thin, straight lines. Freq. *attrib.* as *adj.*

1895 E. F. Strange *Alphabets* ix. 257 (*caption*) Versal letters. *Ibid.* 258 The versal is [a letter]..at the beginning of a chapter or section thereof. 1906 E. Johnston *Writing & Illuminating* vii. 114 The earlier Versals had very simple and beautiful pen shapes... After the fourteenth century they were often..overdone with ornament. *Ibid.* xii. 205 In twelfth-century MSS. long delicate flourishes are commonly found, in red, blue, or green—matching the colours of the Versals. 1912 A. W. Pollard *Fine Books* vi. 84 The small red letters at the beginning of each verse of a psalm, sometimes called versals. 1979 T. Gourdie *Calligraphic Styles* 53 Versal Capitals are a pen-built form of the 'Trajan Roman' and are normally used to begin verses and chapters but may also be used to make impressive panels of lettering. 1981 D. Mahoney *Craft of Calligraphy* 90 A versal letter is built up with a definite number of pen strokes. *Ibid.* 93 The stem width is the same in Versals of varying height.

versatile, *a.* Add: **1. c.** Both heterosexual and homosexual. *slang.*

1959 [see *bent *ppl. a.* 5 c]. 1960 M. Spark *Ballad Peckham Rye* iii. 32 Dougal was probably pansy. 'I don't think so... He's got a girl somewhere.' 'Might be versatile.'

‖ **vers de société** (vęr də sosiete). [Fr., lit. 'verse of society'.] Verse that treats of topics provided by polite society in a light, often witty style.

1796 I. D'Israeli *Miscellanies* 149 This species of poetry can only exist in an age when refinement is introduced into literature, as well as into everything else. We must, therefore, look for it..among a people the most refined among it's [*sic*] neighbours... It has been significantly called '*Vers de Société*'. 1817 G. Crabbe *Jrnl.* 11 July in *Poet Wks.* (1834) I. 248 His poetry [is] far beyond that implied in the character of *Vers de Société*. 1867 F. Locker-Lampson *Lyra Elegantiarum* 9 Smoothly written *vers de société*, where a boudoir decorum is, or ought always to be, preserved; where sentiment never surges into passion, and where humour never overflows into boisterous merriment. 1934 T. S. Eliot *Elizabethan Ess.* 35, I am used to..being informed that something which I meant seriously is *vers de société*. 1976 *Times* 26 Jan. 7/5 Quirky *vers de société*..by Ogden Nash.

‖ **vers d'occasion** (vęr dokazyoṅ). [Fr.] Light verse written for a special occasion (see Occasional *a.* 2 b in Dict. and Suppl.).

1867 F. Locker-Lampson *Lyra Elegantiarum* 11 *Vers de société* and *vers d'occasion* should be short, elegant, refined, and fanciful. 1933 *N. & Q.* 26 Aug. 143/1 Many of his pieces are somewhat unskilful *vers d'occasion*. 1946 R. Lehmann *Gipsy's Baby* 71 She was a great one for *vers d'occasion*. Upon my birthday..I received..a pale pink scalloped gilt-edged card inscribed..in a frame of hand-painted roses. 1960 *20th Cent.* Sept. 276 The *vers d'occasion* includes some limericks and clerihews. 1982 B. Askwith *Piety & Wit* vii. 112 The bons mots, the puns, the vers d'occasion, travel badly in time.

verse, *sb.* Add: **1.** (Examples with reference to Old English poetry.)

1715 E. Elstob *Rudiments Gram. Eng.-Saxon Tongue* 68 The Saxon Verses consist of three, four, five, six, seven, eight, or more syllables. 1883 H. M. Kennedy tr. *Ten Brink's Hist. Eng. Lit.* I. 22 The sentence rarely closes with the ending of the verse. 1938 A. Campbell *Battle of Brunanburh* 16 Sievers showed once and for all the combinations of accentual elements, which might be used to build a verse. 1958 A. J. Bliss *Metre of Beowulf* 1 The term 'verse' is here used instead of the more cumbrous 'half-line' or 'hemistich'.

5. c. That part of a modern popular song which leads into the chorus, or separates one chorus from another. Cf. *chorus *sb.* 6 c.

1927 *Melody Maker* Aug. 782/2 The verse is then taken 'hot' by the trumpet, who gives a fine example of what modern 'hot' playing..should be. 1929, 1935 [see *chorus *sb.* 6 c]. 1966 *Melody Maker* 7 May 13/1 Wonder charges through the verse and builds up into the repetitious chorus.

8. a. *verse-beat, -end, -form, -line, -pair, -rhythm, -unit.*

1943 E. Sitwell *Poet's Notebk.* xxviii. 134 The verse-beat is *not* very strong in this passage. 1930 T. Sasaki *On Lang. R. Bridges' Poetry* i. v. 24 The fully strong stress at the verse-end. 1887 G. M. Hopkins *Let.* 20 Oct. (1956) 381 The style of prose is a positive thing and not the absence of verse-forms. 1906 G. P. Krapp *Andreas* p. xlvi, The distinctively epic verse-form. 1966 *English Studies* XLVII. 97 In a difficult and restrictive verse-form, one might expect the poet to resort to the use of convenient whole-line units more..often. 1927 D. H. Lawrence *Mornings in Mexico* 66 He had written the thing [*sc.* a love-poem] straight ahead, without verse-lines or capitals. 1953 *Speculum* XXVIII. 449 The recurrence of verses and verse-pairs in Anglo-Saxon poetry. 1930 T. Sasaki *On Lang. R. Bridges' Poetry* i. v. 21 Lines in verse.. form units of verse-rhythm intermediate between a 'foot' and a 'stanza'. 1942 J. C. Pope *Rhythm of Beowulf* 22 In no case is it necessary to pass beyond the limits of accentual adjustment that verse-rhythm everywhere allows. 1948 *Mod. Philology* XLVI. 77 When the character of the dipody, or verse unit, is examined, the first impression is one of extreme variation. 1966 *English Studies* XLVII. 96 The verse-unit, the half-line, was quite short.

b. *verse drama, epistle, epitaph, letter, play.*

1925 R. Graves *Welchman's Hose* 31 Then the first draft of a verse-epitaph. 1931 *Times Lit. Suppl.* 5 Nov. 850/2 It might be inferred certainly from the verse-epistles [of Burns] alone: not quite so certainly from the prose letters alone. 1952 T. S. Eliot *Film of 'Murder in the Cathedral'* 7 *Murder in the Cathedral* is, I believe, the first contemporary verse play to be adapted to the screen. 1962 *Times* 14 Aug. 11/1 Jean Cocteau's verse-drama *Renaud et Armide*. 1963 M. H. Abrams in N. Frye *Romanticism Reconsidered* 37 In a verse-letter of 1800 Blane identified the crucial influences in his spiritual history as a series beginning with Milton.

c. *verse-painting, -reading* (later example), *-speaking* ppl. adj. and vbl. sb., *-writing* (earlier example).

1942 Blunden *Romantic Poetry & Fine Arts* 19 A single touch of his originality in the 'Ancient Mariner' holds the secret of his verse-painting. 1938 L. MacNeice *Mod. Poetry* ii. 41 This [*sc.* the *Golden Treasury*] was my chief verse-reading for two years. 1933 *Amer. Speech* VIII. iv. 39/2 Outside of the school there may be a place for verse-speaking choirs. 1980 *Times* 5 Sept. 11/7 His verse-speaking consists of a heavy lurch from beat to beat. 1755 M. Barber in Colman & Thornton *Poems by Eminent Ladies* I. 23 There's nothing I dread, like a verse-writing wife.

verseless, *a.* (Later examples.)

c 1873 G. M. Hopkins *Note-bks. & Papers* (1937) 221 Where verse ends and prose (or verseless composition)

begins. 1975 *Gramophone* May 2020/3 The rapid succession of verseless office antiphons..gave me a slight feeling of dizziness because of their varying pitches.

Versene (vɜ̄·ɹsīn). Also **versene.** A proprietary name for a preparation containing ethylenediamine tetra-acetic acid (q.v. s.v. *ethylene 2) or a similar chelating agent.

1944 *Official Gaz.* (U.S. Patent Office) 29 Aug. 734/2 The Martin-Dennis Company, Newark, N.J... *Versene* for water softening agents in powdered and liquid form with or without detergents. 1951 *Trade Marks Jrnl.* 14 Nov. 1040/2 *Versene*... Chemical products used in industry. Bersworth Chemical Company,..Framingham,.. Massachusetts, United States of America; manufacturers. 1952 *Nature* 19 July 119/2 Although versene protects the enzyme systems from the inactivation caused by the incubation, it does not reverse such inactivation. 1980 *Bull. Environmental Contamination & Toxicol.* XXIV. 543 (*heading*) Acute fish toxicity of the Versene family of chelating agents.

versicler. For *rare*$^{-1}$ read *rare*. (Earlier example.)

1860 G. Meredith *Let.* ? Aug. (1970) I. 61 But do. pray, exclude some of your present versiclers.

versificatory, *a.* For † *Obs.*$^{-0}$ read *rare*. (Example.)

1963 V. Nabokov *Gift* iii. 146 It was then also that my versificatory illness began.

versine (vɜ̄·ɹsəin). [Expansion of Versin after Sine^2.] = *versed sine* a s.v. Versed *a.* 1 a.

1943 R. A. Hamnett *Brit. Railway Track* ix. 273 To obtain the versine of a turnout curve on a chord joining the switch heel and the intersection point of the crossing, (G–H) must be substituted for 'G'. 1958 Clark & Clendinning *Plane & Geodetic Surveying* (ed. 5) I. 634 (*heading*) 'Versine' method of establishing intermediate points on the Euler spiral transition curve. 1976 J. B. Garner et al. *Surveying* 278 Versines are used to compute slope corrections.

version. Add: Also with pronunc. (vəˑɪʒən).

‖ **vers libre** (vęr lībr). Pl. **vers libres.** [Fr., free verse.] Poetic writing in which the traditional rules of prosody, esp. those of metre and rhyme, are disregarded in favour of variable rhythms and line lengths; a composition in this style; = *free verse* s.v. *free *a.* D. 2.

1902 *Encycl. Brit.* XXVIII. 497/2 M. Viele-Griffin..and M. Gustave Kahn..gave us *vers libres* which, but for their typographical arrangement, are indistinguishable from prose. 1912 *Poetry* I. ii. 65 Mr. Richard Aldington is a young English poet, one of the 'Imagistes', a group of ardent Hellenists who are pursuing interesting experiments in *vers libre*. 1920 *Glasgow Herald* 8 July 4/2 Mr. Bunker..is most interesting in his vers-libres; unfortunately they are not poetry, though happily free of the pretentiousness of most vers-librists. 1928 T. S. Eliot in E. Pound *Sel. Poems* p. viii, I remarked some years ago, in speaking of *vers libre*, that 'no *vers* is *libre* for the man who wants to do a good job'. The term, which fifty years ago had an exact meaning, in relation to the French alexandrine, now means too much to mean anything at all. 1937 W. Inge *Modernism in Lit.* 13, I have no doubt that cubism and futurism and most of *vers libre* will soon pass into limbo. 1955 C. Carrington *Rudyard Kipling* xiv. 352 Kipling..experimented in almost every conventional verse-form, and wrote remarkably successful *vers libres*. 1978 *Times Lit. Suppl.* 1 Dec. 1406/3 His was.. a phrased rather than a cadenced poetry. His rhythmic strategy meant that he retained the option of vers libre, the thematic resources of dream, and the technical resources of modernism.

Hence **vers-li-brist(e** (also **-libre-ist** and as one word without hyphen), a writer of *vers libre*.

1916 *Independent* LXXXVIII. 104/3 If the public can be convinced that the vers libristes have something to say worth attention, it will have more sympathy. 1926 *British Weekly* 21 Jan. 395/2 Vers-libre-ists are in poetry very much what the futurists were in art, rebels against the established order. 1957 *Archivum Linguisticum* IX. 142 Certain *vers-libristes* of the Symbolist movement. 1969 J. Gross *Rise & Fall Man of Lett.* viii. 229 By 1911 Ezra Pound was a regular contributor, the verslibrists were arguing their case, Imagism was already in the air. 1981 *N. & Q.* Dec. 571/2 Whether one should see in the experiments of modern vers-libristes a fusion of the two systems..must remain a matter for individual response.

‖ **Versöhnung** (fęɹzö·nuŋ). [Ger., conciliation, propitiation.] A reconciliation of opposites.

1867 J. A. Symonds *Let.* 22 Aug. (1967) I. 750 The truest *Versöhnung* in art I know is to be found in Beethoven's C Minor Symphony. There he first posits all the contradiction of passions, aspirations, and sorrows, then combines them..so transfiguring them that the termination is triumph; the victory and majesty of the soul are wrought out of its defeats and humiliations. 1976 G. Talbot *Permission to Speak* viii. 106 It was a jarring note in an evening of festive international *versöhnung*.

‖ **Verstandesmensch** (fęɹʃta·ndęsmenʃ). Pl. **-menschen.** [Ger.] A matter-of-fact person; a realist.

1879 W. James *Coll. Ess. & Rev.* (1920) 133 Such also is the attitude of all hard-minded analysts and *Verstandesmenschen*. 1938 *Mind* XLVII. 528 He rejects as an unhistorical caricature the view of Kant as a cold *Verstandesmensch*.

‖ **Verstehen** (fęɹʃtē·ən). *Social Sci.* Also **ver-.** [Ger., comprehension.] The use of empathy to understand human action and behaviour, as a method of interpreting historical and sociological phenomena. Also **verste·hende** (fem. pres. pple. of *verstehen* comprehend) *a.*, employing *Verstehen*.

1933 *Economica* XIII. 31 In this sense *verstehende* sociology is rationalistic. 1934 *Psychol. Bull.* XXXI. 298 Ch. Bühler makes use of *Verstehen* as well as statistics in her analysis of life histories. 1937 T. Parsons *Structure Social Action* xvi. 583 This is the first appearance in Weber's methodology of the fundamentally important concept of *Verstehen*. 1948 T. Abel in Feigl & Brodbeck *Readings in Philos. of Sci.* (1953) 682 The characteristic feature of the operation of *Verstehen* is the postulation of an intervening process 'located' inside the human organism, by means of which we recognize an observed.. connection as relevant. 1974 R. D. Jessop *Traditionalism, Conservatism & Brit. Political Culture* i. 18 A pure *verstehende* technique which treats action as the logical concomitant of the actor's ideas, motives, reasons, and beliefs. 1979 *Internat. Jrnl. Sociol. of Law* VII. 327 Problems of *verstehen* sociology—that the analysis should have some regard to the meanings that the actors give to their actions.

vert (vɜ̄ɹt), *a.*^{2} *poet. nonce-wd.* [Cf. Vert *v.*^{1}] Turning.

1947 Auden *Age of Anxiety* (1948) ii. 49 O Primal Age When we danced deisal, our dream-wishes Vert and volant.

vertex. Add: **1. d.** *Math.* A junction of two or more lines in a network or graph (Graph *sb.*^{1} 1 in Dict. and Suppl.); = *node *sb.* 7 b.

1931 *Proc. Nat. Acad. Sci.* XVII. 125 A graph *G* is composed of two sets of symbols: vertices, *a*, *b*, . . ., *f*, and arcs, *α*(*ab*).., *β*(*ac*), . . ., *δ*(*ef*). 1942 G. T. Whyburn *Analytic Topol.* x. 182 Such a decomposition of a graph *A* into vertices and edges is called a subdivision of *A*. 1975 I. Stewart *Concepts Mod. Math.* xi. 160 A network has two main parts: (i) a set *N*, whose elements are called nodes or vertices, (ii) a way of specifying when two vertices are joined together. 1979 *Sci. Amer.* May 98/3 Alpha-actinin was localized primarily at the vertexes of the network and tropomyosin was localized along the short fibers connecting the vertexes.

3. a. *vertex presentation*, a presentation (Presentation 8 b) in which the vertex of the fœtus lies nearest to the cervix as labour begins.

1841 F. H. Ramsbotham *Princ. & Pract. Obstetr. Med.* 135 (*heading*) Comparative frequency of the various modes of vertex presentation. 1899 [in Dict.]. 1974 *Encycl. Brit. Micropædia* VIII. 195/2 In vertex presentations the head of the fetus most commonly faces to the right and slightly to the rear. This position is said to be the most usual one, because the fetus is thus best accommodated to the shape of the uterus.

vertical, *a.* and *sb.* Add: **A.** *adj.* **3. d*.** *Mus.* Involving, pertaining to, or directed at the relationship between notes sounded simultaneously, rather than the pattern of successive notes; harmonic or chordal rather than melodic.

1889 *Cent. Dict.*, *Vertical composition*, musical composition in which the chief attention is put on the harmonic structure of the successive chords. 1928 *Grove's Dict. Mus.* (ed. 3) V. 164/1 Later events have made it almost superfluous to discuss..his [*sc.* R. Strauss's] theories of 'vertical hearing'. 1942, etc. [see *horizontal *a.* (*sb.*) 4]. 1946 A. Bliss in A. L. Bacharach *Brit. Mus. of our Time* xi. 156 As in all his music, one must concentrate on horizontal as well as vertical listening so as to savour the beauty and interest of the inner parts.

e. *vertical breeze* = *breeze *sb.*^{2} 3 b; *vertical cut*, motion of a recording stylus up and down, rather than from side to side; also *attrib.*; cf. *hill and dale* s.v. *hill *sb.* 1 b; opp. *lateral cut* s.v. *lateral *a.* 3 j; *vertical file, filing*; *vertical gust* = *vertical breeze* above; *vertical interval*, the vertical distance between the heights represented by adjacent contours on a map; *vertical man*, a living man, one standing upright (as opposed to a recumbent or dead one); *vertical recording*, magnetic recording in which the direction of magnetization is at right angles to the plane of the recording medium. Also in collocations used *attrib.*, as *vertical-shaft, -spindle, -take-off.*

1925 Fraser & Gibbons *Soldier & Sailor Words* 296 *To suffer from a vertical breeze* (also *vertical gust*), to be nervous. 1934 D. L. Sayers *Nine Tailors* iii. 279 He got a vertical breeze up. 1965 J. R. Hetherington *Selina's Aunt* 59 The term 'vertical breeze' was co-temporary [with 'wind up'], and may have been either the originating phrase or a further refinement. 1935 J. Mills *Fugue in Cycles & Bels* (1936) xi. 145 Vertical-cut phonograph discs of the most recent type can record from 40 to 9000 cycles. 1975 [see *lateral *a.* 3 j]. 1977 *Gramophone* Apr. 1522/1 Every one of these hill-and-dale vertical-cut labels had given place to lateral-cut issues under the same mark by that year [*sc.* 1920]. 1906 *Library Jrnl.* XXXI. 13 A newspaper man..goes to the vertical file, picks out a handful of articles on the subject. 1909 *Independent* (N.Y.) 18 Nov. 1126/1 An assistant..deposits the article in an oblong vertical filing-envelope, ten by eleven inches. 1917 *Daily Mail* 19 July 4/5 Stalled his 'bus and pancaked thirty feet...crashed completely...put a vertical gust up me. 1925 Vertical gust [see *vertical breeze* above]. 1885 G. W. Usill in H. S. Marrett *Pract. Treat. Land & Engin. Surveying* (ed. 4) 320 In this way a table may be calculated showing the horizontal equivalents for the required vertical interval at each degree of slope up to about 30°. 1969 G. C. Dickinson *Maps & Air Photographs* iv. 62 Although contours are widely understood several aspects of their significance are not always fully appreciated. For example their effectiveness in representing terrain is closely controlled by the vertical interval. 1930 Auden *Poems* 2 Let us honour if we can The vertical man Though we value none But the horizontal one. 1961 *Guardian* 16 Feb. 10/5 He was..a 'vertical man', and that in an age when intellectuals have been found flat on their faces. 1975 G. Howell *In Vogue* 61 T. S. Eliot was one vertical man who was honoured..by contemporary writers. 1982 *Sci. Amer.* July 71/3 A number of companies in the U.S., Europe and Japan are working on high-density memory systems based on vertical recording. 1983 *Austral. Microcomputer Mag.* Aug. 67/1 It has announced prototypes of vertical-recording technology disk drives. 1940 *Chambers's Techn. Dict.* 891/2 Vertical shaft alternator. 1967 Vertical-shaft [see *planetary *a.* 1 f]. 1935 *Discovery* May 143/1 Vertical spindle pump. 1964 S. Crawford *Basic Engin. Processes* vii. 190 Vertical-spindle machine employing the face of a cup or segmental wheel. 1935 *Jrnl. R. Aeronaut. Soc.* XXXIX. 1137 So that probably on any day one could actually hover in an autogiro; and they knew also that with the machine which had been illustrated they could achieve vertical take-off as well. 1960 *Daily Tel.* 26 Apr. 1 Britain, France and West Germany are to co-operate in developing a supersonic, vertical-take-off military aircraft. 1972 *Guardian* 28 June 1/2 The fourth of the RAF's vertical take-off Harriers to crash in the past few weeks came down yesterday at Düsseldorf.

7. Of or pertaining to the different levels of a hierarchy or progression. **a.** Extending over or involving successive stages in the production of a particular class of goods. Opp. *horizontal *a.* (*sb.*) 3 b.

1920 *Westm. Gaz.* 2 Dec. 6/1 The vertical Trusts constructed by Stumm, Thyssen and the other raw-material magnates. 1927 *Daily Tel.* 11 Oct. 15/4 He had created what is technically called a vertical combination, embracing every stage of the soap industry. 1959 *Listener* 5 Nov. 768/2 The existing vertical firms have been operating in a market dominated by the factors created by horizontal trading and few indeed have controlled their price policies by vertical statistics and vertical objectives. 1960 [see *agribusiness]. 1962 R. B. Fuller *Epic Poem on Industrialization* 27 A corporation gun nuzzling trick;..precipitating vertical merger. 1967, 1968 [see *horizontal *a.* (*sb.*) 3 b]. 1975 *N.Y. Times Mag.* 3 Oct. 15 Proponents of the effort call it vertical divestiture, by which they mean forcing the largest oil companies to pick one activity—production or refining or transportation/marketing—and sell off the other parts of the action. 1975 J. De Bres tr. *Mandel's Late Capitalism* xii. 384 The process of centralization can only find expression in a growing centralization of capital, among other things, in the form of vertical integration of big companies.

b. Involving differences or changes of level as in social class, income group, or the like.

1927 P. A. Sorokin *Social Mobility* vii. 133 There are two principal types of social mobility, horizontal and vertical. 1931, etc. [see *horizontal *a.* (*sb.*) 3 c]. 1976 F. Zweig *New Acquisitive Society* i. v. 52 The shedding of middle-class values and style of life in the younger generation..is of much deeper significance, transcending the confines of vertical mobility.

c. *vertical union*, a trade union which draws its members from a particular industry without regard to their individual crafts; *vertical market*, one comprising all the potential purchasers in a particular occupation or industry.

1933 *Sun* (Baltimore) 1 Sept. 2/1 This means a vertical union in each industry, free of domination or control either by employers or outside labor leaders. 1937, 1950 Vertical union [see *horizontal *a.* (*sb.*) 3 d]. 1978 *Business Week* (Industr. Ed.) 17 July 36G H-P's role has been primarily as a systems company emphasizing vertical markets needing a wide variety of supporting electronics. 1983 *Austral. Microcomputer Mag.* Aug. 16/3 HiSoft believes there is a big need for vertical market software, in which a common shell is modified to suit individual needs. 1984 *Sydney Morning Herald* 10 Nov. 6/1 (Advt.), They are presently expanding into a new and highly promising vertical market and offer a Sales Management Opportunity. 1985 *Which Computer?* Apr. 45/1 This means that BOS is one of the richest potential sources of vertical market software written in the UK for the UK market.

d. *vertical proliferation* (see quots.).

1966 *Economist* 22 Oct. 350/2 Like other near-nuclear nations, they are unwilling to promise to stay out of the club unless its members will promise to halt what Canada's foreign minister has called their 'vertical proliferation'; that is, promise to stop testing, producing and piling up nuclear arms. 1980 *Sci. Amer.* July 31/2 In the circumstances what can be done to curb both 'vertical' proliferation (the increase in the numbers and kinds of nuclear weapons in the hands of the nuclear-weapons states) and 'horizontal' proliferation (the further spread of nuclear weapons to nations that do not already have them)?

e. *vertical thinking*, deductive reasoning; opp. *lateral thinking* s.v. *lateral *a.* 1 b.

1966, 1967 [see *lateral *a.* 1 b]. 1970 G. Greer *Female Eunuch* 108 The take-over by computers of much vertical thinking has placed more and more emphasis on the creative propensities of human thought.

8. Pertaining to or being an aerial photograph taken looking vertically downwards.

1925 Jones & Griffiths *Aerial Surveying by Rapid Methods* ii. 8 Such a procedure will..be necessary when mapping any large area, whether the work be done by

'vertical' or 'oblique' photographs. 1932 *Jrnl. R. Aeronaut. Soc.* XXXVI. 503 The first field operation is the vertical photography along strips about thirty miles apart. *Ibid.*, As soon as each vertical flight was completed the films were developed. 1974 P. R. Wolf *Elem. Photogrammetry* vi. 117 Relief displacement often causes straight roads, fence lines, etc., on rolling ground to appear crooked on a vertical photograph. 1983 J. C. McCormack *Surveying Fund.* xxii. 404 The oblique view is more easily understood by the public than is the plan view contained in vertical aerial photographs.

B. *sb.* **2. d.** *Austral. Opal-Mining.* (See quots.)

1934 *Geol. Survey, Mineral Resources* (New South Wales Dept. Mines) No. 36. 116 The mineral is found also in vertical or sub-vertical joints and cracks..known locally as 'verticals'. 1967 I. L. Idriess *Opals & Sapphires* 48 A vertical seam cuts in: that is, a seam running downward from the roof..which in general we used to call a 'vertical'.

5. A vertical aerial photograph (see sense A. 8 above).

1925 Jones & Griffiths *Aerial Surveying by Rapid Methods* vi. 69 These [oblique photographs] provide valuable information about the nature of the ground which eventually is to be mapped by verticals. 1954 W. D. Thornbury *Princ. Geomorphol.* xxi. 535 Verticals are more widely used than obliques in geologic field work. 1976 J. B. Garner et al. *Surveying* xiii. 233 If it is required to photograph a long strip of land, many photographs will be required. Each photograph should be a good vertical.

ve·rticalize, *v.* [f. Vertical *a.* + -ize.] *trans.* and *absol.* To render vertical, in any sense. Hence **ve·rticalized, ve·rticalizing** *ppl. adjs.*

1959 *Times Rev. Industry* Dec. 6/3 The..Fair..was one of the first..trade fairs to verticalize by organising.. specialised events during..a year, rather than staging one general fair. 1964 M. Critchley *Developmental Dyslexia* ix. 59 There is a tendency for the child to convert figures, especially those that are verticalised and closed, into a man by drawing a face within the enclosure. 1965 *Economist* 2 Jan. 54/2 The three or four main 'verticalising' companies will have less to feed their appetites and more time to digest what they have. 1982 *Christian Sci. Monitor* 21 Sept. B2/2 As a matter of corporate policy, Hewlett-Packard divisions are 'verticalized', each one functioning as a small company on its own.

Hence **ve:rticaliza·tion,** the action or result of verticalizing.

1962 G. Perle *Serial Composition & Atonality* iii. 42 Verticalization..represents a fundamental concept of atonal composition—that any group of notes which is statable in horizontal succession is also statable as a simultaneity.., a concept cometimes designated by the rather questionable term, 'vertical melody'. 1964 M. Critchley *Developmental Dyslexia* ix. 59 There is some disorientation of the background, usually by rotation of mobile figures, or by 'verticalisation'. 1965 *Economist* 19 June 1439/3 Courtaulds being particularly weak in weaving so that effective verticalisation ends at the yarn stage. 1970 *Daily Tel.* 10 Jan. 7/7 He does renounce.. that festering sore of serialism, the automatic verticalisation of melodies. 1979 *Nature* 15 Mar. 226/1 The late alkaline rocks in the ring-complexes may be related to verticalisation of the Benioff plane accompanied by a diapiric rise of mantle material. 1984 *Christian Sci. Monitor* 7 Feb. 39/2, I praised Masotti's movement of his necktie from his stomach to his pocket and its slight verticalization.

vertically, *adv.* Add: **1. a.** (Examples in *Mus.* Cf. *vertical *a.* 3 d*.)

1934 *Hound & Horn* July–Sept. 596 He..feels that he is tired of exploiting the folk tune, horizontally, vertically, atonally, seriously, or comically. 1954 *Grove's Dict. Mus.* (ed. 5) VIII. 126/2 Pre-war controversy as to the justification of programme music and 'hearing vertically' began to have no more than antiquarian interest. 1969 *Listener* 26 June 905/1 Strauss's counterpoint tends to be the result of thinking 'vertically', against strongly defined chord-progressions.

2. Throughout the different levels of a hierarchical system.

1933 *Sun* (Baltimore) 2 Sept. 2/1 With an industry organized vertically, the logical labor organization is vertical. 1958 *Listener* 9 Oct. 547/2 Differences throughout the Arab sectors run both vertically and horizontally: between religious sects, social strata, settlers and nomads. 1962 *Economist* 27 Oct. 393/1 The industry should..be concentrated into fewer, vertically-integrated firms. 1974 J. White tr. *Poulantzas's Fascism & Dictatorship* iii. ii. 95 The big stores competed for growth as they were vertically integrated into the industrial trusts.

verticillium (vɜ̄ɹtisi·liə̆m). *Bot.* [mod.L. (C. G. Nees von Esenbeck *Das System der Pilze und Schwämme* (1816) 56), f. Verticill(us + -ium.] A hyphomycete fungus of the genus of this name, some species of which cause plant disease; *verticillium wilt*, wilt caused by this fungus, affecting many flowers, vegetables, and fruit bushes.

1916 *Sci. Proc. R. Dublin Soc.* XV. 63 (*heading*) The Verticillium wilt of the potato. 1931, etc. [see *hadromycosis* s.v. *hadrome]. 1951 *Sci. News* XXI. 110 Verticillium wilt of hops..has been a serious disease in Great Britain since 1932. 1971 *Country Life* 18 Feb. 381/1 So serious have they [*sc.* chrysanthemum diseases] become (two caused by viruses and one by a verticillium, in particular) that the future popularity of this florist's flower is said to be in danger. 1978 *Sci. Amer.* Aug. 68/3 In California..plantings [of tomato] are mainly limited to cultivars that resist the wilt caused by fusarium and verticillium fungi.

vertisol (vɜ̄·ɹtisǫl). *Soil Sci.* Also **Vertisol.** [f. Verti(cal *a.* + *-sol.] A clayey soil with little organic matter found in regions having distinct wet and dry seasons, characterized by deep, wide cracks when dry and an uneven surface owing to the swelling and shrinking of the clay.

1960 *Soil Classification: 7th Approximation* (U.S. Dept. Agric.) ix. 124/1 The central concept of Vertisols is one of soils that crack widely, and that often remoisten from water that runs into the cracks rather than from water that percolates through the soil. 1970 *Nature* 2 May 429/1 Dark, cracking tropical clays, that is, vertisols, are characteristic of..the restricted clay plugs and backswamps of the Omo Floodplain [in Ethiopia]. 1981 *Jrnl. Soil Sci. Soc. Amer.* XLV. 668/1 In Vertisols, because of swelling and shrinking, bulk density of the soil is dependent on moisture content.

very, *adv.* Add: **2. a.** (Further examples.) *very difficult, severe*: two of the categories used in classifying rock climbs; also *absol.*; *very high* and *low frequency* (Telecommunications): see VHF, VLF s.v. *V 5 b; freq. *attrib.*

1920 *Radio Rev.* Sept. 579 (*heading*) Circuit for producing very high frequencies. 1938, etc. [see VLF s.v. *V 5 b]. 1951, etc. [see *severe *a.* 9 b]. 1958 *Economist* 26 July 271 With very high frequency radio broadcasting.. providing almost perfect reception, the collector of classical music..can make high fidelity recordings..on a £2 tape. 1967 *Electronics* 6 Mar. 68/2 Navy project officers expect a go-ahead..on construction of a worldwide very-low-frequency Omega navigation system. 1969 'A. Garve' *Ascent of D. 13* ii. 35 I've been climbing ever since I was a kid... I've done more Very Severes than I can remember. 1972 *McGraw-Hill Yearbk. Sci. & Technol.* 229/2 Some of the techniques developed for determining the geodetic coordinates for land use..can be applied at sea... In addition, the very-long-baseline interferometry (VLBI) technique is potentially applicable. 1974 *Encycl. Brit. Macropædia* XV. 425/1 The variation of carrier frequency is known as the frequency deviation, and for very-high-frequency broadcasting it can reach ±75 kilohertz. 1976 *Time* 24 May 64/3 The Japanese government..has declared the development of Very Large Scale Integrations—the technical heart of the next generation of computers—a 'national project'. 1978 Pasachoff & Kutner *University Astron.* xxvi. 667 With this ability, astronomers can make up an interferometer of two or more dishes very far apart, even thousands of kilometers. This technique is called very-long-baseline interferometry. 1982 *Times* 14 Jan. (Information Technol. Suppl.) p. v/4 Very large-scale integration (VLSI)... VLSI puts as many as 100,000 components on a chip.

e. Qualifying a sb. or proper name used adjectively (for emphasis).

1937 [see *predicatival *a.*]. 1968 *Listener* 21 Mar. 389/3 The total effect is very Kirov: it has more in common with the Leningrad *Cinderella*..than with ours. 1978 *Hot Car* July 87/3 *Scallops*, a very fifties paint idea, consisting of a long U-shaped design, the ends of which taper off to points.

4. *veryvery* (as one word).

1969 A. Lurie *Real People* 16 Croquet's become very-very intense this year. 1977 *Transatlantic Rev.* lx. 68 You have a very nice face... And were very very nice to me.

Very^2 (ve·ri, vī^ə·ri). Also (*erron.*) **Verey, Vérey.** The name of Edward W. *Very* (1847–1910), U.S. naval officer, used *attrib.* with reference to a coloured pyrotechnic flare projected from a special pistol for signalling or temporarily illuminating an area; as *Very light, pistol*, etc.

1907 *Jrnl. Mil. Service Inst. U.S.* XLI. 368 In connection with night signaling it may be well here to mention the Very system,..found serviceable in sea-coast signaling. 1915 D. O. Barnett *Let.* 17 Mar. 95 When the 'Very' pistol came, I fired a rocket. 1917 R. Nichols in E. B. Osborn *Nurse in Arms* 49 Before he was aware The 'Verey' light had risen..on the air It hung glistering. 1920 *Blackw. Mag.* June 747/2 Very flares were continually being fired into the air to light up dark corners. 1928 Blunden *Undertones of War* ii. 16 Another officer.. showed me..how to fire a flare... He had with him a cumbrous brass gun, called a duck-gun; from this he fired a Vérey cartridge. 1930 C. R. Samson *Fights & Flights* 177 My sole equipment consisted of an electric torch.. and a Verey-light pistol. 1959 *Chambers's Encycl.* XI. 390/1 The Very pistol, a short-barrelled smooth-bore weapon of 1-in. or 1½in.-calibre, firing a cartridge which is in effect a short roman candle throwing up a single star, is the most generally used pyrotechnic signal. 1976 'A. York' *Dark Passage* xiii. 162 He..found a Very pistol.. and fired. The glowing orange ball arced over his head, and then hung, perhaps a hundred feet above the yacht. 1981 J. B. Hilton *Playground of Death* ii. 15 What we were really playing at was War... A Roman candle..was a Very light.

Vesak (ve·sāk). Also **Wesak; Vai-, Ve-, Visākha.** [Skr. *vaiśākhá*, name of a month, f. *vi-śākhā* branched, (also) the name of a constellation, f. *ví* apart + *śā́khā* branch.] An important Buddhist festival commemorating the birth, enlightenment, and death of the Buddha, observed on the day of the full moon in the month Visākha (April–May).

1927 E. J. Thomas *Life of Buddha* iii. 34 In 1922 the Feast of Wesak (Visākha) in Ceylon was at full moon on May 10. 1961 R. A. Gard *Buddhism* iv. 156 Doctrinally speaking, there is the practice of venerating the Buddha.. especially the Vesākha/Vaiśākha-puja (often called 'Wesak' in South and Southeast Asia). 1965 *Festival Malaysia 1965: Calendar of Events* 9/3 May 15—Wesak Day: Celebration of Lord Buddha's Birthday. 1971 *Carry Singapore in your Pocket* (Singapore Tourist Promotion Board) (ed. 3) 38 May 9. Vesak Day. Celebration of Lord Buddha's birth, death and enlightenment. 1978 C. Humphreys *Both Sides of Circle* x. 109 The Buddhist Society too carried on.. *Wesak*, the annual equivalent of Christmas Day, was held in May as usual. 1978 *Oxford Diocesan Mag.* Aug. 20/3 It is instructive to know what our Jewish neighbours are doing and why at, for example, Yom Kippur, or our Buddihst friends at Vesakha or the Moslems at Meelad al Nabi.

vesicant, *sb.* Add: Also, a vesicant substance for use in warfare.

1923 *Biochem. Jrnl.* XVII. 275 We are trying to find what common factor..underlies the effect of the mustard gas group of vesicants. 1936 A. Huxley *Olive Tree* 87 To go and throw thermite, high explosives and vesicants upon the inhabitants of neighbouring countries. 1936 [see *blister gas* s.v. *blister *sb.* 4]. 1944, 1964 [see B.A.L. s.v. *B III]. 1974 M. C. Gerald *Pharmacol.* vii. 137 Lewisite is an arsenic-containing vesicant capable of producing immediate pain.

vesico-. Add: Also occas. used for Vesiculo-. **ve:sicopu·stular** *a.* = *vesiculo-pustular* adj. s.v. *vesiculo-; **ve:sico-ure·teral, -urete·ric** *adjs.*, of or pertaining to the bladder and the ureters; *spec.* applied to a flow of urine from the former back into the latter.

1902 H. W. Stelwagon *Treat. Dis. Skin* 63 The lesions often do not become strictly purulent, but are of a sero-purulent character, forming vesicopustules, and when such a feature is a predominant one, the eruption is usually designated vesicopustular. 1951 Whitby & Hynes *Med. Bacteriol.* (ed. 5) xxiv. 393 The viruses of this group..characteristically produce a vesico-pustular eruption. 1906 Dorland *Med. Dict.* (ed. 4) 815/1 Vesicoureteral. 1951 M. Campbell *Clin. Pediatric Urol.* i. 67 Vesico-ureteral reflux also results from obstructions of the lower urinary tract. 1982 *Jrnl. Urol.* CXXVIII. 774/1 Vesicoureteral reflux is a common abnormality occurring in children and a major cause of renal failure in early adult life. 1965 *Brit. Jrnl. Urol.* XXXVII. 531 Vesico-ureteric reflux may follow operations on the vesico-ureteric junction. 1977 *Proc. R. Soc. Med.* LXX. 149/2 This finding is not necessarily applicable to children with symptomless infection even though infection in these children is accompanied by vesicoureteric reflux in 23% of cases.

vesicularity (vĭsikiŭlæ·riti). [f. Vesicular *a.* + -ity.] Vesicular condition.

1908 *Mineral. Mag.* XV. 129 The first results..were usually increased vesicularity in the quartz and assumption by much of it of the idiomorphic condition. 1978 *Nature* 20 July 218/1 Vesicularity of seafloor basalt, when plotted against depth of eruption, defines a family of three curves.

vesiculate, *v.* Add: **2.** (Examples.)

1966 *Earth-Sci. Rev.* I. 158 A gas-charged lava flow.. might vesiculate so violently at and near its surface that it would form glass and pumice fragments. 1971 *Nature* 31 Dec. 539/1 Pulsations of ejected molten lava vesiculated and disintegrated passing immediately into the grey ash cloud. 1978 *Jrnl. Histochem. & Cytochem.* XXVI. 1 When a semisynthetic diet containing 1% orotic acid.. is fed to rats, the endoplasmic reticulum..of hepatocytes vesiculates and lipoprotein..droplets accumulate within the vesicles.

Hence **vesi·culating** *ppl. a.*

1966 *Earth-Sci. Rev.* I. 158, 10 years later Iddings (1909) ..omitted the vesiculating flow idea. 1971 *Nature* 31 Dec. 538/2 Blobs and droplets of expanded, molten, vesiculating magma are carried aloft with an expanding vapour. 1979 *Biochim. & Biophys. Acta* DL. 222 Electron microscopy of vesiculating cells shows physical continuity between cell plasma membrane and vesicle membrane. 1983 *Sci. Amer.* Nov. 148/3 Both pumice and ash are frothy glassy materials created by the chilling of vesiculating magma.

vesiculation. Add: Also *Geol.*

1914 R. A. Daly *Igneous Rocks & their Origin* xiii. 264 The basal assumption, that vesiculation occurs at great depth in a volcanic conduit, is..difficult to test. 1971 *Nature* 31 Dec. 538/2 Both articles describe the vesiculation of a magma in terms of the formation, motion and coalescence of bubbles.

vesiculo-. Add: **vesi:culo-bu·llous** *a.*, characterized by or involving both vesicles and bullæ; **vesi:culo-pu·stular** *a.*, characterized by or involving both vesicles and pustules.

1923 R. W. MacKenna *Dis. Skin* xii. 250 In every well-marked case of Dermatitis herpetiformis it is the vesiculo-bullous lesion which predominates. 1980 *Brit. Med. Jrnl.* 18 Oct. 1041/2 Herpes gestationis (HG) is a severely itching, vesiculobullous skin affection. 1911 Stedman *Med. Dict.* 959/1 Vesiculopustular. 1946 *Nature* 27 July 119/1 A fully developed vesiculopustular reaction developed at the site of inoculation. 1973 *Ibid.* 16 Feb. 425/3 There was a prodromal illness of a few days' duration followed by a vesiculopustular rash which was peripheral in distribution.

Vespa (ve·spă). Also **vespa.** Pl. **Vespas,** ‖ **Vespe** (ve·spe). [It., lit. 'wasp, hornet'.] A proprietary name for an Italian make of motor scooter.

1950 *Trade Marks Jrnl.* 22 Mar. 270/2 *Vespa*... Motor Cycles. Piaggio & C. Societa Perazioni,..Genova, Italy:

manufacturers. **1956** A. Thorne *Baby & Battleship* i. 28 Those eternal bantam motor-bicycles, the little *Vespe* that the Italians love. **1958** J. Cannan *And be a Villain* vii. 164 It was one of the first of the motor scooters—before the Vespas. **1960** Auden *Homage to Clio* 80 This could be a reason why they take the silencers off their Vespas. **1965** D. Du Maurier *Flight of Falcon* v. 56 The young were everywhere, pouring out of lecture rooms, laughing, talking, getting on to vespas. **1983** 'D. Rutherford' *Stop at Nothing* iv. 76 A three-wheeler Vespa with a miniature van on the back.

vespasian (vespēⁱ·ʒăn). [Anglicized form of next.] = next.
1938 I. Goldberg *Wonder of Words* 139 Vespasian. **1980** *Times* 27 Feb. 6/8 The City of Paris has decided to erect..three space age vespasians..new 'chalets of necessity'—as they are called.

‖ **vespasienne** (vespazi,e·n). [Fr. (19th cent.), shortening of *colonne vespasienne* Vespasian column, f. the name of Titus Flavius *Vespasianus*, Roman Emperor, 69–79, who introduced a tax on public lavatories.] A public lavatory in France.
1922 E. E. Cummings *Enormous Room* iii. 44 My first request was permission to visit the *vespasienne*. **1962** P. Brickhill *Deadline* x. 130 A pissoir, or, to give it its polite name, a vespasienne, after the emperor who was so solicitous of man's frailty. **1975** *Times* 4 Mar. 6/1 The vicissitudes of a village *vespasienne*.

vesper. Add: **4.** (Later examples.)
1874 O. W. Holmes *Poetical Wks.* (1892) I. 352 How blest to the toiler his hour of release When the vesper is heard With its whisper of peace! **1927** A. Clarke *Son of Learning* 18 *Brother*: I must ring the Vesper.
8. b. *vesper-born.*
1928 Blunden *Retreat* 50 O vesper-born, Stiff-necked I stand like that hewn knotty tree, As if heaven were my halo.

vesperal, *a.* **a.** (Later examples.)
1887 L. Johnson *Incense* in *Ireland* (1897) 60 Pensive and solitary old age finds Calm in the vesperal, mild air. **1951** V. Nabokov *Speak, Memory* iv. 47 The day would take hours to fade, and everything..would be kept in a state of infinite vesperal suspense.

ve·spering, *ppl. a. rare*⁻¹. [f. Vesper + -ing².] Flying westwards (towards sunset).
1914 Hardy *Year's Awakening* in *New Weekly* 21 Mar. 9 O vespering bird, how do you know?

vespertilian, *a.* Delete *rare*⁻¹ and add later examples.
1911 W. J. Locke *Glory of Clementina Wing* xxii. 277 As the studio was rigorously closed to him during the daylight hours his visits were vespertilian. **1955** R. Graves *Crowning Privilege* 222 The fiend..Flaunts vespertilian wing and cloven hoof.

vespertilionid, *a.* Add: Also as *sb.*, a bat of the family Vespertilionidæ.
1965 B. E. Freeman tr. *Vandel's Biospeleology* xxvii. 454 The ears of vespertilionids are immobile. **1976** *Nature* 15 Apr. 628/1 Within our sample alone, species are represented whose primary foods are..insects (the mormoopids, natalids, vespertilionids and molossids).

vessel, *sb.*¹ Add: **4. c.** An airship or hovercraft.
1915 *Sphere* 3 Apr. 22/1 The long covering of the balloon seemed to have been broken. Some people were running beside the vessel. **1916** *Ibid.* 18 Mar. 293/1 As an airship rises it encounters air which has less supporting power, and ultimately..the vessel floats in equilibrium. **1957** I. Asimov *Naked Sun* ix. 93 Baley was in an air-borne vessel again, as he had been on that trip from New York to Washington. **1972** *Daily Tel.* 25 Apr. 1/4 British Rail's hovercraft Princess Anne made an emergency landing on a sandbank yesterday... The vessel was beached at Andressells, eight miles north of Boulogne.

vest, *sb.* Add: **3. b.** Now *N. Amer.* (Further examples.)
1925 F. Scott Fitzgerald *Great Gatsby* ix. 202 While he took off his coat and vest I told him all arrangements had been deferred. **1937** H. G. Wells *Brynhild* vii. 103 He was sitting without jacket or vest, looking neat and healthy in his shirt and black tie. **1968** *Globe & Mail* (Toronto) 17 Feb. 7/3 Hooking his thumbs in his vest, he answered questions in a calm, almost offhand manner. **1978** J. Irving *World according to Garp* ii. 37 Bodger..tucked in his shirt, which was escaping ..from under his tight vest.
d*. A short, sleeveless jacket for a woman. *U.S.*
1909 in Webster. **1974** *Times-Picayune* (New Orleans) 14 Aug. iii. 1 (Advt.), Plaid vest, 18.00. **1978** *Detroit Free Press* 5 Mar. (Spring Fashion Suppl.) 11/1 Vests have never looked quite so fresh and right as they do this spring. They can be the perfect sleeveless jacket, topping all the softness.
d.** A singlet denoting membership of a representative athletics team.
1971 N. Stacey *Who Cares?* ii. 25 It was harder to get a Blue than an international vest.
e. vest-pocket: also *attrib.* as *adj.*: small enough to fit into a vest pocket, very small of its kind; also *fig.*; **vest-slip** = sense 3 b in Dict. and Suppl.
1848 *Sporting Life* 29 July 274/1 This vest pocket companion for cricketers. **1897** 'Mark Twain' *Following Equator* 629 Toy peaks, and a dainty little vest-pocket Matterhorn. **1912** *Brit. Jrnl. Photogr.* 5 July 525 The vest-pocket 'Tenax' camera. **1931** *Times* 16 Mar. 1/3 (Advt.), Unique vest-pocket treatment for catarrh. **1947** *Horizon* Apr. 152 Our provincial hotels with their vest-pocket electric fires. **1983** *Chicago Sun-Times* 9 July 15 Vest-pocket garden-parks provide relaxed places for people to interact with one another. **1984** *Newslet. Amer. Dial. Soc.* Sept. 23/1 He was responsible for..a series of popular vest-pocket dictionaries and reference books. **1920** *Punch* 9 June 456/2 My top-hat was on my head and my vest-slip was all right. **1922** Vest-slip [see *Oxford 1 b].

vest, *v.* **II. 6. b.** (Earlier example.)
1848 C. H. Hartshorne *Eng. Medieval Embroidery* 130 The sides [of the altar] must be covered or vested... Where the table stands away from the wall, the back must be vested likewise.

vested, *ppl. a.* Add: **1. c.** Of a suit: three-piece, having a waistcoat. Chiefly *N. Amer.*
1976 *Daily Colonist* (Victoria, B.C.) 6 Oct. 5/8 (Advt.), Save a great deal on a Tip Top vested suit. **1982** 'W. R. Duncan' *Queen's Messenger* viii. 74 Ross..nattily dressed as usual in his vested dark suit.

vestee. Restrict *rare* to sense in Dict. and add: **1.** Also *transf.*
1937 *Virginia Q. Rev.* XIII. 591 In a department of English, as in any other going business, the proprietary interest becomes vested, and in old and reputable departments the vestees have uniformly been gentlemen who have gone through the historical mill.
2. [f. Vest *sb.*] A vest (sense 3 a or *3 d) or dickey (sense 6). Chiefly *N. Amer.*
1904 *T. Eaton & Co. Catal.* Spring & Summer 87/3 Boys' Vestee Suits, three-piece style, with single-breasted vest. **1916** *Daily Colonist* (Victoria, B.C.) 7 July 14/5 (Advt.), Silk Crepe de Chine Waists..featuring the small button front and vestee effect outlined by French Veining. **1928** *Daily Express* 17 Aug. 4 (*heading*) Vestee and cuff set. **1943** *Sun* (Baltimore) 4 Feb. 17/2 The Chicago Cubs..have worn sleeveless baseball shirts, or vestees, for the last three years. **1963** J. Mitford *Amer. Way of Death* iv. 56 The women's lingerie department of Practical Burial Footwear supplies a de-luxe package..of 'pantee, vestee' and nylon hose.

vestibular, *a.* Add: **a.** *spec.* Of or pertaining to the vestibule of the ear or its function as an organ of equilibrium. (Further examples.)
1902 D. J. Cunningham *Text-bk. Anat.* 482 Its fibres..end in the nucleus of Deiters, the chief vestibular nucleus. **1923** *Physiol. Rev.* III. 209 (*heading*) The function of the vestibular apparatus. **1948** A. Brodal *Neurol. Anat.* vii. 213 The parts of the labyrinth generally assumed to be concerned with vestibular function..are..the saccule and the utricle, and the three semicircular canals. **1956** A. C. Guyton *Textbk. Med. Physiol.* xlvi. 575/1 Equilibrium is controlled especially by the vestibular apparatus. **1962** V. Grissom in *Into Orbit* 80 The doctors call this sensation 'vestibular nystagmus', and it is an uncontrollable movement of the eyeballs which occurs when the balance mechanism in your body gets all messed up from the twisting and tumbling. **1983** *Oxf. Textbk. Med.* II. xxv. 29/2 Either the cochlear or the vestibular fibres of the eighth nerve may be damaged within the brainstem, although vestibular symptoms are appreciably more common from this cause.

vestibulo- (vestiˑbiŭlo), comb. form of L. *vestibul-um* Vestibule *sb.*, used chiefly with reference to the vestibule of the ear, as **vesti:bulo-au·ditory** *a.*, involving the vestibules and hearing; **vesti:bulocerebe·llar** *a.*, applied to nerves running from the vestibular nucleus of the brain to the cerebellum; **vesti:bulo-cochle·ar** *a.*, designating or pertaining to the vestibular and cochlear nerves jointly; **vesti:bulo-o·cular** *a.*, involving both the vestibular and the oculomotor nerves; **vesti:bulospi·nal** *a.*, applied to a tract of nerves in the spinal cord that originate in the vestibular nucleus of the brain. Also Vestibulotomy.
1945 *Arch. Ophthalmol.* XXXIII. 149/1 The vestibulo-auditory symptoms began abruptly with vertigo, tinnitus and deafness. **1980** *Medicine* (Baltimore) LIX. 426/1 Cogan syndrome..is a disease of young adults and consists of acute interstitial keratitis..with vestibuloauditory dysfunction. **1932** *Jrnl. Compar. Neurol.* LIV. 150 Vestibulocerebellar fibers. **1974** D. & M. Webster *Compar. Vertebr. Morphol.* xii. 292 The vestibulocerebellar tract runs from the vestibular nuclei in the brainstem through the restiform body to terminate in the cerebellar cortex. **1962** *Gray's Anat.* (ed. 33) 1141 The vestibulocochlear nerve appears in the groove between the pons and the medulla oblongata. **1974** *Encycl. Brit. Micropædia* III. 413/1 In nerve deafness, some defect in the sensory cells of the inner ear..or in the vestibulocochlear nerve prevents transmission of sound impulses from the inner ear to the auditory centre in the brain. **1982** *Audiology* XXI. 172 Streptomycin..passes into the fetus ..producing vestibulocochlear neural abnormalities. **1921** Tilney & Riley *Form & Function Central Nervous System* xxx. 532 Vestibulo-ocular associated reflex. This reflex is elicited by electrical, thermal and mechanical stimulation of the receptors in the vestibule of the internal ear. **1963** *Jrnl. Theoret. Biol.* IV. 215 The simplest description of this vestibulo-ocular reflex is to say that, if the head is rotated in one direction, the eyes tend to rotate, with respect to the head, in the opposite direction, and then come back to their undeviated position. **1979** *Sci. Amer.* Jan. 88/1 The two other major types of eye movements—the vestibulo-ocular movements, which maintain visual stability during head movements, and the smooth-pursuit eye movements, which follow a moving object such as a flying bird—do not appear..since neither of these types of movements is elicited by target jumps. **1899** L. F. Barker *Nervous System* lviii. 960 It would seem that this uncrossed descending vestibulo-spinal neurone system has been described by various authors under different names. **1948** A. Brodal *Neurological Anat.* v. 118 The most important descending connexion appears to be the vestibulo-spinal tract. **1983** *Brain Res.* CCLIX. 217/1 The vestibulospinal system has a complex organization both in aspects of its afferent connection and of its spinal projection.

vestigial, *a.* Add: **a.** (Earlier example.)
1877 *Gray's Anat.* (ed. 8) p. cxlvi, The only remains of the Wolffian body in the complete condition of the female organs are two rudimentary or vestigial structures.
2. Telecommunications. *vestigial side band*, a side band which is partially attenuated (usu. at the higher frequencies) before transmission; freq. *attrib.*, with reference to a system in which one full side band and one vestigial side band are transmitted (with or without the carrier), esp. to improve the transmission of low-frequency components of the signal.
1940 D. G. Fink *Princ. Television Engin.* vii. 288 At the transmitter, the principal problem of vestigial side-band transmission (as the above-described system is called) is the design of a filter having the pass characteristics shown in Fig. 162. **1966** *McGraw-Hill Encycl. Sci. & Technol.* I. 352/1 In standard television broadcasting in the continental United States..the vestigial sideband has a bandwidth one-sixth that of a full sideband. **1979** A. A. Liff *Colour & Black & White Television Theory & Servicing* ii. 55 The FCC standards require that the lower sideband be transmitted in vestigial sideband form and that the upper sideband be transmitted in its entirety.

vestigium. b. (Later example.)
1970 H. Braun *Parish Churches* ix. 124 These [arches] are not, as might have been supposed, the remains of an earlier clearstory, but simply vestigia of the original windows of a pre-medieval nave.

vesting, *sb.* (Earlier example.)
1813 *Weekly Reg.* IV. 295/1 For the best and handsomest fancy vesting, of cotton..a premium of..forty dollars.

vesting, *vbl. sb.* **1.** (Later *attrib.* examples.)
1922 *Act* 12 & 13 *Geo. V* §188. 238 In relation to settled land, 'vesting deed' or 'vesting order' means the instrument whereby the land is conveyed or vested; 'vesting assent' means the instrument whereby a personal representative, after the death of a tenant for life..vests the land in the successor in title..; 'vesting instrument' means a vesting deed, assent or order. **1948** *Secretary* July 180 (*heading*) Vesting day, 1st April, 1948. **1950** *Engineering* 24 Mar. 318/3 It is difficult to forecast whether or not the iron and steel industry will..pass into State ownership on the vesting date, January 1, 1951. **1964** *Financial Times* 23 Mar. (Defence Suppl.) 14/5 With 'vesting day' only a few days off, staffs are still settling down. **1967** *Economist* 1 Apr. 21/2 For compulsory purchase, the commission may use a procedure called the 'vesting declaration' which virtually cuts out conveyancing, while granting a firm title. **1977** *Belfast Tel.* 19 Jan. 7/2 A vesting order for the land was signed on February 12 last. **1982** *Equity & Law Life Ann. Rep. 1981* 16 Vesting bonus of 30% of the total pension will be allotted to personal pension deferred annuities.

Vestinian (vestiˑniăn), *sb.* and *a.* [f. L. *Vestīni* + -an.] **A.** *sb.* **a.** A member of an ancient Oscan people who lived in the Gran Sasso d'Italia area of Italy. **b.** The language of this people. **B.** *adj.* Of or pertaining to this people or their language.
1578, etc. [see *Marrucinian sb. and a.]. **1933, 1939** [see *Marsian sb. and a.]. **1966** M. S. Beeler in Birnbaum & Puhvel *Anc. Indo-European Dial.* 53 So far as I know no voice has..ever been raised to question the propriety of regarding these languages (and the minor dialects of the Paelignians, Marrucinians, Vestinians, and Volscians) as differentiated forms of a single common ancestor..here termed Oscan-Umbrian.

vestmental, *a.* For *rare*⁻¹ read *rare* and add later example.
1978 *New York* 3 Apr. 92/1 Hermitage prospers now on the commerce of those who pursue chic and never catch it... They occupy every table, tailored flannels and sharkskins seated beside superb little Bendel numbers—vestmental perfections without a stitch of art.

vestock (veˑstǫk). [Blend of Vest *sb.* and Stock *sb.*¹] A clerical stock that extends to the waist.
1975 *Rep. Patent Cases* 473 The..vestock..can be described as a sort of black bib, with a hard upstanding collar round at the front. **1981** *Oxf. Diocesan Mag.* May 24 (Advt.), Fine bespoke tailors and hosiers. Now stock clerical shirts, collars, stocks, vestocks, etc.

vestry¹. Add: **2. b.** In England the sole remaining power of the vestry is the election of churchwardens, its other ecclesiastical powers having been transferred to the parochial church council and the annual parochial church meeting; *select vestry* (*Obs.* exc. *Hist.*) (see quots, 1845, 1906).
1698 *MSS. House of Lords* (1905) III. 261 Dec. 9. Poor Relief (Select Vestries) Bill.—Draft of an Act for preventing the Poor's being cheated. **1845** *Encycl. Metrop.* XXIII. 486/2 In other Parishes,..the rate-payers place the powers intrusted to them in the hands of a small Body, chosen by them from their own number, called sometimes the Select Vestry, and sometimes the Guardians of the Poor. **1906** S. & B. Webb *Parish & County* i. v. 173 The Close or..Select Vestry...consisted of a body of one or two dozen persons, or occasionally more, serving for life and filling vacancies among themselves by co-option. **1921** *Parochial Church Councils (Powers) Measure* 4 From the commencement of this Measure there shall be transferred to the Council of every parish—(i) All powers, duties and liabilities of the Vestry of such parish relating to the affairs of the Church, except as regard the election of churchwardens and sidesmen and as regards the administration of ecclesiastical charities. **1959** Jowitt *Dict. Eng. Law* II. 1610/1 The Select Vestries Bill is the title of a Bill always formally read a first time in the House of Lords at the beginning of a new session before the House proceeds to debate the king's or queen's speech, as an assertion of its independence.

vestryman. Add: Also *fig.*
1910 *Blackw. Mag.* Apr. 585/1 Mr. Shaw is the vestryman of dramatists. His work savours horribly of St. Pancras.

Vesuvius (vĭsiū·viŭs). [The name of a volcano near Naples (see Vesuvian *a.* and *sb.*).] A great explosion of emotion; something or someone liable to sudden outbursts.
1845 Poe in *Broadway Jrnl.* 6 Sept. 136/2 The poetical and critical world of England were, about six years ago, violently agitated (in spots) by the eruption of 'Festus', a vesuvius-cone at least—if not an..Ætna. **1886** G. Meredith *Let.* 9 June (1970) II. 815, I confess that a faint form of decent excuse for your conduct..would have partly appeased my natural indignation. I put it by among other things for the Day of my Vesuvius. **1929** D. H. Lawrence *Pansies* 130 The women are like little volcanoes... It is rather agitating, sleeping with a little Vesuvius.

veszelyite (veˑsĕlyəit, -lĭ,əit). *Min.* [ad. G. *veszelyit* (A. Schrauf 1874, in *Anzeiger d. Akad. d. Wissensch., Wien* XI. 136), f. the name of A. Veszely, Hungarian mining engineer: see -ite¹.] A hydrated basic phosphate of copper and zinc, $(Cu,Zn)_3(PO_4)(OH)_3 \cdot 2H_2O$, occurring as aggregates of bluish monoclinic crystals.
1875 E. S. Dana *Dana's Syst. Min.* (ed. 5) App. ii. 59 Veszelyite... In crystalline crusts on garnet-rock. **1920** *Nature* 1 July 569/2 Rare zinc-copper minerals from..Northern Rhodesia... Still rarer are the copper-zinc minerals aurichalcite and veszelyite; the latter forms minute sky-blue monoclinic crystals.., and differs from the original material from Hungary in its colour and in containing little or no arsenic. **1977** *Mineral. Abstr.* XXVIII. 207/2 The veszelyite appears to have been formed in an oxidation zone by low-*T* hydrothermal processes.

vet, *sb.* Add: **2.** A doctor of medicine. *slang.*
1925 Fraser & Gibbons *Soldier & Sailor Words* 296 *Vet, the*, medical officer. **1938** [see *pill shooter* s.v. *Pill sb.¹ 5]. **1965** M. Spark *Mandelbaum Gate* v. 119 'The vet gone?' Gardnor said... 'You might have a relapse.' **1975** A. Powell *Hearing Distant Harmonies* ii. 83 Saw my vet last week. Said he'd never inspected a fitter man of my age.

vet, *sb.*² N. Amer. abbrev. of Veteran *sb.* in Dict. and Suppl.
1848 *Sporting Life* 17 June 190/2 The same remark may be applied to a much younger man than the above 'vets', whose Spring-like qualities seem to defy old winter. **1866** *Pictorial Bk.* 452/2 Colonel A..took it upon himself to chide the exasperated and unfortunate 'vet' for using such unchristianlike language. **1926** *Amer. Speech* I. 369/1 The [baseball] players are, as in the army, 'vets' or 'rookies' according to the length of time they have served. **1936** *Esquire* Sept. 159/2 'Jesse Lasky's Broadway Booneing' means that the vet producer is scouting plays and talent in N.Y. **1951** M. McLuhan *Mech. Bride* (1967) 144/2 The *Fortune* survey editors..were surprised to find nearly all the vets in favor of getting inside a *big* business. **1968** *Listener* 27 June 841/1 The scene is New York,..the academic 'host' is Columbia University, where a number of young Second World War vets..are making gestures at working for degrees. **1979** *Tucson* (Arizona) *Citizen* 20 Sept. 8c/1 When he talked to vets in other service organizations, he found a lot of them felt they owed something to the Veterans Administration too.

vet, *v.* Add: **3.** To examine carefully and critically for deficiencies or errors; *spec.* to investigate the suitability of (a person) for a post that requires loyalty and trustworthiness.
1904 Kipling *Traffics & Discoveries* 270 These are our crowd... They've been vetted, an' we're putting 'em through their paces. **1924** H. A. Vachell *Quinney's Adventures* 267 Shelagh 'vetted' Mr. Dolan's brogue, and passed it as sound enough for an Irish-American. **1925** E. F. Norton *Fight for Everest: 1924* iii. vi. 338 He should have all equipment..completely ready three or four months before shipment—only thus can everything be properly 'vetted' and criticized. **1938** G. Arthur *Not Worth Reading* viii. 110 The official in Pall Mall..who 'vetted' us..swallowed without a gulp some rather mendacious replies as to one's technical knowledge of the various parts of a Canadian boat. **1947** E. Waugh *Let.* 29 May (1980) 251 The romantic castle was condemned by the architect I sent to vet it, as moribund. **1959** Duke of Bedford *Silver-Plated Spoon* vi. 128 We went through an awful period while Brownie was 'vetted' at a series of interviews with relations, each more embarrassing than the last. **1978** G. Greene *Human Factor* ii. i. 63 HQ had had her vetted.

veteran, *sb.* and *a.* Add: **A.** *sb.* **1. b.** Any ex-serviceman. Chiefly *N. Amer.*
Not always distinguishable from sense 1 a.
1798 W. Dunlap *André* p. iv, The Author has gone near to offend the veterans of the American army who were present on the first night. **1838** *Southern Lit. Messenger* IV. 796 When the revolutionary pension-law was enacted, a majority of the war-worn veterans had travelled..beyond the reach of human reward. **1912** I. S. Cobb *Back Home* 38 Saturday..was also Veterans' Day, when the old soldiers were the guests of honor of the management. **1931** *U.S. Laws & Statutes* XLVI. i. 1016 The President is authorized..to consolidate and coordinate any hospitals and executive and administrative bureaus..into an establishment to be known as the Veterans' Administration. **1946** *Sun* (Baltimore) 7 Oct. 2/5 The War Assets Administration ordered its surplus building materials pushed into the veterans housing program within 60 days. **1954** *Birmingham* (Alabama) *News* 11 Nov. 41/1 The Sheppard murder trial jury took a Veterans Day holiday today. **1974** C. Ryan *Bridge too Far* iv. x. 348 'It was the heaviest volume of fire I ever encountered,' recalls Sergeant Spencer Wurst, a 19-year-old veteran who had been with the 82nd since North Africa. **1979** G. F. Newman *List* viii. 67 He was planning for Veteran's Day weekend. November 11 fell on a Monday this year. **1982** *Times* 13 Dec. 3/6 (*heading*) Falklands veteran killed at 22.
3. A veteran car.
1965 *Guardian* 6 Nov. 3/8 A 'Veteran' is any car made before December 31, 1904. **1968** [see *Edwardian a. 3].
B. *adj.* **3. b.** Applied to cars made before a certain stipulated date (normally a vehicle not less than fifty years old). Also *transf.* (often coupled with *vintage*).
1933 A. G. Macdonell *England, their England* xiii. 222 A three-mile clatter in the veteran car—'never can get the new car,' explained Mr. Fielding. **1958** M. Kelly *Christmas Egg* iii. 107 An apartment..was devoted to the needs of those who..wished to play vintage and veteran records. **1959** *Listener* 29 Jan. 201/2 A veteran car is generally considered to be a pre-1917 model though a few extremely exclusive owners recognise only pre-1905 cars as qualifying for the title. **1961** J. Sheldon (*title*) Veteran and vintage motor cycles. **1965** L. Hunt (*title*) Veteran and vintage aircraft. **1980** *Times* 1 Nov. 14/2 Nineteen-thirty had seen the foundation of the Veteran Car Club of Great Britain.

veto, *sb.* Add: **2. a.** *liberum veto* [L. *liber* free], a power of veto possessed by every member of a legislative body, *spec.* that which existed under the later Polish monarchy. Also *transf.*
1792 W. Coxe *Trav. Poland* I. i. v. 96 In all state-matters of the highest importance no resolution of the diet is valid, unless ratified by the unanimous assent of every nuntio; each of whom is able to suspend all proceedings by his exertion of the *Liberum Veto*. **1831** J. Fletcher *Hist. Poland* iii. 89 It was in this King's reign [*sc.* 1648–68] that the *liberum veto*, or privilege of the deputies to stop all proceedings in the diet, by a simple dissent, first assumed the form of a legal custom. **1931** E. Howell *Mizh* (1979) x. 96 What he had in mind was the *liberum veto* which any young hot head could exercise by the use of his rifle. **1938** H. V. Hodson *Slump & Recovery* viii. 245 Her [*sc.* Bolivia's] *liberum veto* thus stood in the way of any increase of quota allowances. **1952–4** *Proc. Brit. Acad.* XXXVIII. 217 In Jan. 1831 the opposition of a single Cabinet Minister to the proposed grant to Queen Adelaide for an outfit, compelled the Cabinet to abandon it: an example of the *liberum veto*.
3. *veto-free, -proof* adjs.
1959 *Daily Tel.* 27 Apr. 10 The Prime Minister..proposed breaking the existing deadlock over a control system by a 'quota' plan limiting the number of veto-free inspections per year. **1972** *National Observer* (U.S.) 16 Sept. 2/5 The debt-limit extension, veto-proof because of its importance, could become a Christmas tree of Democratic proposals. **1973** *San Francisco Examiner* 20 Sept. 9/2 The American Federation of Teachers..reacted to the vote by calling for the election of a 'veto-proof Congress..no longer intimidated by the big stick of a Nixon veto'.

vetoism. (Earlier example.)
1815 D. O'Connell *Let.* 13 June (1972) II. 47 A priest suspected of Vetoism loses all his respect.

vetoist. (Earlier example.)
1815 *Dublin Evening Post* 19 Jan. 2 The Vetoists..have been routed.

vetter (ve·təɹ). [f. Vet *v.* + -er¹.] One who vets people or things.
1972 *Daily Tel.* (Colour Suppl.) 9 June 7/3 One should be able to assume..that for the BBC's announcers and commentators, as well as for the authors, vetters and readers of scripts, literacy and oral propriety are automatic conditions of employment. **1982** *Listener* 16 Dec. 3/3 The Americans..are privately scathing about the failure of positive vetters to pick up an obvious security risk. **1984** *Daily Tel.* 1 Nov. 2/7 This follows recommendations by the Security Commission into the case of Geoffrey Prime, the spy, whose psychological flaws went undiscovered by security vetters.

ve·tting, *vbl. sb.* [f. Vet *v.* + -ing¹.] The action or process of vetting a person or thing; *esp.* the investigation of a person's background and credentials to determine his loyalty or trustworthiness; *positive vetting*, vetting which includes a search for weaknesses of character or anything else that could render the subject vulnerable to exploitation.
1918 H. A. Vachell *Some Happenings* iv. 42 Doctors were so ridiculously cocksure! All the same, he felt mildly interested in the vetting... Constitutionally he was as sound as a bell. **1927** *Observer* 17 July 13/1 The 'vetting' of applicants for loans would involve the State in an expenditure on itinerant investigators and inspectors. **1955** H. Macmillan in *Hansard Commons* 7 Nov. 1499 At the beginning of 1952, a regular system of positive vetting was introduced. This procedure entails detailed research into the whole background of the officer concerned. **1970** *Canadian Antiques Collector* July–Aug. 5/2, I was pleased to see that the vetting committee had done their work well. **1978** G. Greene *Human Factor* i. iii. 38 Of course, he belongs to the slack vetting days, but I'd say he was clear. **1982** *Daily Tel.* 20 July 3/1 There are some 68,000 government posts currently requiring positive vetting. Civil servants of Under Secretary rank and above are automatic candidates.

‖ **veuve** (vöv). [Fr.] **1.** A widow; also prefixed as a title to the name.
1793 F. Burney *Jrnl.* 13 Apr. (1972) II. 77, I told him I should have the pleasure to present him to another sister—*la veuve!* **1868** C. M. Yonge *Chaplet of Pearls* I. xix. 258 He found her..with her hands in Veuve Laurent's flour. **1894** M. M. Verney *Mem. Verney Family* III. ix. 330 We hope Mrs. Aris *veuve* was treated to some of those lovely phrases for home consumption. **1922** E. E. Cummings *Enormous Room* iii. 42 She must have been very pretty before she put on the black. Her friend is also a *veuve*. **1953** D. Parry *Going up—going Down* v. viii. 252 It's quite a popular sport now, to drop in and cheer the Veuve Tyndale. **1977** T. Heald *Just Desserts* viii. 190 In the hall were two men unmistakably French..policemen..arguing..with la Veuve.
2. Veuve Clicquot (klīko), *erron.* **Cliquot**, the shortened form of a proprietary name for champagne produced by the firm of Veuve Clicquot-Ponsardin in Reims.
1898 W. J. Locke *Idols* xx. 278, I am glad that you prefer champagne *extra sec*... So many women go for Veuve Cliquot, when they can. **1906** Galsworthy *Man of Property* i. ii. 33 His pint of champagne was dry and bitter stuff, not like the Veuve Clicquots of old days. [**1926** *Trade Marks Jrnl.* 9 June 1355 Veuve Clicquot-Ponsardin... Champagne wines. Bertrand de Mun & Cie., successeurs de Veuve Clicquot-Ponsardin.., Reims, France; wine merchants.] **1964** [see *Taittinger]. **1968** *Guardian* 20 Sept. 11/4 A fine pink Veuve Clicquot, say, still remains a fine wine. **1979** P. Alexander *Show me a Hero* xvi. 171 He came in from the other reception room with a bottle of Veuve Cliquot and two glasses on a tray.

‖ **vexata quæstio** (vekˑsā·tă kwĭ·stio). [L.] = *vexed question* s.v. Vexed *ppl. a.* 4.
1813 *Edin. Rev.* Oct. 143 We do not mean to enter upon the *vexata quæstio* of the tones and delivery. **1838** J. S. Mill in *Westm. Rev.* XXXI. 393 The *vexata quaestio* of codification..has now passed into one of Practice. **1889** G. N. Curzon *Russia in Central Asia* ix. 368 This has long been a *vexata quæstio* of Central Asian politics. **1978** *Language* LIV. 425 For the vexata quaestio of the 'gorgia toscana', cf. most recently Izzo 1972.

vexedness. (In Dict. s.v. Vexed *ppl. a.*) For *rare*⁻¹ read *rare* and add later example.
1909 W. J. Locke *Septimus* xi. 160 Zora, regarding the egoist with mingled admiration and vexedness, could only say, 'Oh!'

vexillology (veksilǫ·lŏdʒi). [f. L. *vexill-um* flag + -ology.] The study of flags.
1959 *Arab World* (N.Y.) Oct. 13/1 One of the most interesting phases of vexillology—the study of flags—is the important contribution to our heritage of flags by the Arab World. **1961** *Flag Bull.* Fall 7/2 Editors Grahl and Smith use 'vexillology' and its cognates, vexillologist, vexillological. **1966** *Occasional Newslet. to Librarians* Jan. 4 This unknown specialist has demonstrated his great knowledge of heraldry and vexillology. **1970** W. Smith *Flag Bk. U.S.* i. 3 In 1965 the first International Congress of Vexillology was held in the Netherlands.
Hence **vexillolo·gical** *a.*; **vexillo·logist.**
1961 [see above]. **1963** *Recall* (Boston) Oct. 4 (*heading*) Travel notes and vexillological addendum. **1965** W. Smith *Bibliogr. Flags of Foreign Nations* p. v, Its three principal activities have been the encouragement of contacts and exchanges of information between vexillologists around the world, the coordination of research efforts, and the building up of a library of books and other flag materials. **1971** *Daily Tel.* 19 Nov. 13/7 Between 60 and 70..historians, antiquaries, designers and students of heraldry are expected at an international vexillological congress. **1973** *Smithsonian* Dec. 50/2 Redividing the states would mean redesigning the flag. In vexillologist Whitney Smith's scheme, he retains the symbolism of 13. **1983** *Christian Science Monitor* 8 Apr. 20/1 Father Young is the official community vexillological custodian.

via, *sb.* Add: **2.** *esp.* One in Italy or one of the great Roman roads. Cf. Way *sb.*¹ 1 c. (Earlier and later examples.)
1673 J. Ray *Observations Journey Low-Countries* 369 We departed from Rome and began our journey to Venice; riding along the Via Flaminia..which reaches as far as

Rimini. **1822** M. Wilmot *Jrnl.* 21–25 May in *More Lett.* (1935) 166 Cecilia Metella's Tomb..so well preserved amidst the many ruined tombs with which the old Via Appia abounds. **1851** E. B. Browning *Casa Guidi Windows* i. iv. 8 That winter-hour, in Via Larga, when Thou wert commanded to build up in snow Some marvel of thine art. **1929** Kipling *Limits & Renewals* (1932) 99 What was your best day's march on the Via Sebaste? **1973** G. Sims *Hunters Point* xix. 177 The frontier town of Menton..where the French Route Nationale joins the Italian Via Aurelia. **1984** A. Elliot *On Appian Way* 6 This poem represents a journey from Rome to Brindisi, more or less following the Via Appia.
b. *Via Crucis* (krū·tʃis) = *Way of the Cross* s.v. Way *sb.*¹ 4 g; also *fig.*, an extremely painful experience that has to be borne with fortitude; *Via Dolorosa* (dǫlŏrōᵘ·ză) [L. *dolōrōsus* Dolorous *a.*], the route in Jerusalem that Christ is believed to have followed from Pilate's judgement-hall to Calvary; also *fig.*, = fig. sense of prec.
1844 *Orthodox Jrnl.* 10 Aug. 100 Thirteen small altar pieces surround the arena [of the Colosseum], and Benedict XIV. introduced here the *Via Crucis*, or devotion to the passion, performed by a brotherhood of monks every Friday afternoon. **1878** R. L. Stevenson *Inland Voyage* 186 Fitly enough may the potentate bestride his charger, like a centurion in an old German print of the *Via Dolorosa*; but the toys should be put away in a box among some cotton, until..the children are abroad again. **1901** M. Carmichael *In Tuscany* 236 The loggia is 250 feet in length, on one side of it is a *Via Crucis* in bas-relief, on the other frescoes representing scenes from the life of the Saint. **1904** H. O. Sturgis *Belchamber* ii. 23 Each step in the *via dolorosa* of his existence was fated to be more awful than the last. **1910** C. E. Montague *Hind let Loose* xv. 290 Fay, with his feeble pity for all souls in trouble, winced to see this one..sneaking..down the same *via dolorosa*. **1923** G. M. Trevelyan *Manin & Venetian Revolution* iv. 68 Silvio Pellico, whose narrative of his own martyrdom was the guide-book of Italian patriots on their *via crucis*. **1944** *Horizon* July 71 Every human being makes the *via crucis* from innocence to experience. **1972** *Guardian* 30 Nov. 14/5 'Via Galactica' [*sc.* a play] turns out to be Peter Hall's Via Dolorosa. **1980** O. Manning *Sum of Things* xix. 154 In the Via Dolorosa a procession was advancing slowly over the spacious, creamy flagstones. **1982** 'M. Hebden' *Pel & Staghound* xvii. 191 The Lord never intended the Via Crucis to be travelled with ease, child. **1984** *Observer* 25 Nov. 9/2 Whatever might have happened to other 'wets' on their *via dolorosa* between 1979 and 1983, it was..hardly on the cards that Mrs Thatcher would simply drop Peter Walker.
3. *via media* (earlier and later examples).
1834 J. H. Newman *Via Media* (Tracts for the Times No. 38) sig. A3ᵛ, The glory of the English Church is, that it has taken the Via Media... It lies *between* the (so called) Reformers and the Romanists. **1936** E. Underhill *Worship* xv. 324 The *Via Media* eludes not only the extremes of Catholic and Protestant cultus, but also the heights and deeps of the spiritual life. **1978** *Christian* IV. 355 What the Chalcedonian definition attempted was a compromise, a *via media*, between conflicting interpretations of the union of the human and divine natures of Christ.
b. Theol. *via negativa*, the approach to God in which his nature is held so to transcend man's understanding that no positive statements can be made about it; the way to union with God in which the soul leaves behind the perceptions of the senses and the reasoning of the intellect; also *transf.*, a way of denial; so *via affirmativa*, the approach to God through positive statements about his nature.
1856 R. A. Vaughan *Hours with Mystics* I. iv. ii. 124 These two paths, the *Via Negativa* (or Apophatica) and the *Via Affirmativa* (or Cataphatica) constitute the foundation of his [*sc.* Dionysius's] mysticism. **1899** W. R. Inge *Christian Mysticism* iii. 114 When Luther had the courage to break with ecclesiastical tradition, the *via negativa* rapidly disappeared within the sphere of his influence. **1942** [see *isness]. **1956** G. Macleod *Only One Way Left* viii. 152 The Via Negativa: the way of interior denial. Unfortunately the Via Negativa cuts dead across the Emmaus Road. **1963** J. A. T. Robinson *Honest to God* v. 95 The *via negativa* underlay the whole medieval 'way of perfection'. **1974** *Encycl. Brit. Macropædia* IV. 549/2 He [*sc.* the pseudo-Dionysius] recognizes the partial validity of the positive approach (*via affirmativa*). **1977** *Times Lit. Suppl.* 21 Jan. 88/5 As regards the Revolution, M Meyer's work is an exercise in the *via negativa*.

via, *prep.* Add: **1.** (Examples of the sense 'by way of' referring to a specific person or thing.)
1856 C. M. Yonge *Daisy Chain* ii. xxii. 589 Ethel's misanthropy was happily conducted off *via* the Cocksmoor children. **1899** Kipling *Stalky & Co.* 56 Beetle..overturned a student's lamp, which dripped, *via* King's papers..on to a Persian rug. **1931** R. Fry *Let.* 3 Mar. (1972) II. 654, I was ever so glad to hear, via Helen, of you. **1958** A. Sillitoe *Saturday Night & Sunday Morning* iv. 60 Arthur and his father walked via the scullery into the living-room. **1959** M. Gilbert *Blood & Judgement* xiii. 138 More..had come to the Police via the Lower Deck of the Royal Navy. **1981** G. Household *Summon Bright Water* iii. 149 He led me to talk of my interest in ancient economies and thus, via agriculture in the Forest of Dean, eased the way to my impressions of Broom Lodge.
2. By means of, with the aid of.
1930 in *Amer. Speech* VI. 122 Eastbay youth admits thefts via fish pole. **1972** M. Kaye *Lively Game of Death* (1974) vii. 41 Any deal..would have to be..concluded via contracts, attorneys, the whole *schmeer*. **1977** *Rep. Comm. Future of Broadcasting* iv. 30 It would in theory be possible to provide five more services with national coverage via satellite.

viability¹. Add: In *transf.* use: now *esp.* feasibility; ability to continue or be continued; the state of being financially sustainable. (Further examples.)
1955 *Bull. Atomic Sci.* Mar. 81/1 Considerations of defense, in addition to mobilizing offensive strength, do not in the least imply softness or lack of viability. **1962** *Listener* 11 Oct. 549/1 They are a main factor in giving it [*sc.* the country] such economic viability as it possesses. **1971** *Nature* 19 Feb. 518/2 Mr Stein's apparently innocent bill to limit noise at New York airports..could be..a serious threat to the viability of Concorde. **1977** K. M. E. Murray *Caught in Web of Words* viii. 150 He also had some doubts about the viability of the work financially.

viable, *a.*¹ Add: **c.** In recent use *esp.* workable, practicable, esp. economically or financially. (Further examples.)
1955 *Scottish Jrnl. Theol.* VIII. 93 A viable faith in the twentieth century must be able to take into itself a certain scepticism and relativism with regard to all rational systems. **1958** 'A. Burgess' *Enemy in Blanket* xv. 174 It was time he..planted the seeds of a viable relationship between his wife and himself. **1958** *Economist* 8 Nov. 485/2 The plans must..be such as to make the farm 'viable or more viable'—i.e. capable..of yielding its occupier at least the income of a skilled agricultural worker. **1962** *Listener* 5 Apr. 605/2 The Russian nuclear capacity appears..to be..not capable of destroying anything like enough of the American potential for a Russian first strike to be a viable proposition. **1962** P. Gradon in Davis & Wrenn *Eng. & Medieval Stud.* 66 This simple explanation..is not viable for another group of texts to which I now wish to turn. **1971** H. Macmillan *Riding the Storm* iv. 146 It [*sc.* Jordan] was not in economic terms a viable state without British support. **1977** M. Wiles in J. Hick *Myth of God Incarnate* i. 3 They do not of themselves prove that the concept of a 'Christianity without incarnation' is a viable concept.

viaducted (vəi·ădɒkted), *a.* [f. Viaduct + -ed².] Provided with a viaduct.
Quot. 1963 is pa. pple. of a finite verb *viaduct*.
1963 *Engineering* 18 Jan. 80 The dual carriageway will be viaducted to link up with existing roads. **1981** J. Pearson *Kindness of Dr. Avicenna* xviii. 171 The *autostrada*..a marvel of tunnelled mountains, viaducted gorges.

‖ **viale** (viā·le). Pl. **viali.** [It.] A broad street in an Italian city.
1969 M. Gilbert *Etruscan Net* i. vii. 88 Broke..drove up the Viale, using dipped headlights. **1979** G. Watson *Black Jack* xxviii. 236, I..walked over to the window looking on the broad *viale* below. **1982** J. O'Faolain *Obedient Wife* ii. 39 They'd hang around the warm squares and *viali* in groups.

vialled, *a.* (In Dict. s.v. Vial *sb.*) (Later example.)
1906 Hardy *Dynasts* II. v. i. 188 The riskful blood of my previsioned line..To linger vialled in my veins alone.

viander¹. **4.** (Earlier example.)
1761 *London Gaz.* 17–21 Jan. 2/1 The Vyanders and Principal Inhabitants of the Borough of Newport.

viatic, *a.* For † *Obs.*⁻¹ read: *Obs. rare* exc. for revived *nonce-uses*.
1974 V. Nabokov *Look at Harlequins* (1975) iv. i. 156 Look at that strange fever rash of viatic tabulation in which I persevered. **1976** —— *Details of Sunset* 74, I love the process of settling into viatic quarters—the cool linen of the berth, the slow passage of the station's departing lights.

viatical, *a.* (Earlier and later examples in *Bot.*)
1847 H. C. Watson *Cybele Britannica* I. 66 Viatical. Plants of road-sides, rubbish heaps, and frequented places. **1932** G. C. Druce *Comital Flora Brit. Isles* 349 *A*[*grostis*] *verticillata*... Viatical. Germanic, Waysides, waste places.

vibe (vəib), *sb.* **1.** *pl.* Abbrev. of *vibraphone. *colloq.*
1940 *Swing* July 17 Lastly, some too-formal ensemble riffing with vibes. **1957** L. Feather *Bk. Jazz* xvi. 134 Adrian Rollini, known earlier as a bass saxophonist, had been concentrating on vibes since the early 1930s. **1962** J. Wain *Strike Father Dead* v. 211 He could play just as well with a group led by an alto and including vibes, clarinet..or even flute. **1977** *Rolling Stone* 24 Mar., He fell back on his musical training to support the family, playing trumpet and vibes in a succession of third-rate cabaret bands.
2. Abbrev. of *vibration 3 d. Usu. *pl.* *slang.*
1967 *Sunday Times* 1 Oct. 10 We're not getting the right vibes. **1970** J. Lennon in J. Wenner *Lennon Remembrance* (1973) 67 'You give off bad vibes.' That's what George said to her [*sc.* Yoko Ono] and we both sat through it, and I didn't hit him, I don't know why. **1972** *Daily Tel.* (Colour Suppl.) 1 Dec. 56/1 As I arrive they have finished rehearsing *Hair* in the lobby [of the Chateau Marmont, Los Angeles]... Will this diminish the place's vibes? **1976** *New Mus. Express* 31 July 16/1, I remember when I first saw The Grateful Dead... I didn't like them much but I felt that same vibe—something is happening. **1981** P. Carey *Bliss* iv. 151 Always give out a good vibe, never let them think you hate them. **1983** E. Rossiter *Lemon Garden* vi. 96 The damned thing's got bad vibes... Throw it in the lake.

Hence **vi·bist**, a player on the vibraphone.
1955 *Jazzbook 1955* 20 The musicianly sounds produced by vibist Milt Jackson and pianist Thelonius Mark. **1962** *Melody Maker* 21 July 7/1 The vibist's phrasing shows his early Gospel penchant. **1977** *Listener* 23 June 829/2 Drummer/vibist Cal Tjader.

vibraharp (vəi·brăhāɹp). [f. Harp *sb.*¹, after *vibraphone.] A vibraphone; orig., a proprietary name for a particular U.S. make of vibraphone.
1930 *Official Gaz.* (U.S. Patent Office) 15 Apr. 545/1 J. C. Deagan, Inc., Chicago, Ill. Filed Jan. 22, 1930. *Vibra-harp*... Claims use since Aug. 1, 1928. **1949** L. Feather *Inside Be-Bop* iii. 88 Milton Jackson, vibraharp; b. Detroit, 1923. **1959** [see *electric guitar* s.v. *electric a. 2 b]. **1968** *Jazz Monthly* Jan. 25/2 The programme too is attractively old-fashioned..and the accompaniment, on most tracks by guitar, clarinet, vibraharp, organ and drums, discreet and musicianly. **1977** *Melody Maker* 26 Mar. 34/1 The history of the vibraphone (or vibraharp, as it's sometimes called) in jazz has been strictly delineated by its most famous exponents.
Hence **vi·braharpist**, a player on the vibraharp.
1943 *Metronome* Oct. 32/1 Darned if I can remember hearing ten vibraharpists in my life! **1966** *Melody Maker* 29 Jan. 8/6 Ayers is a vibraharpist. **1978** *Country Life* 5 Oct. 1003/3 Milt Jackson..the great vibraharpist.

Vibram (vəi·brăm). Also **vibram.** The proprietary name of a kind of moulded rubber sole used on climbing boots; also applied to a boot having this sole; so **Vibram-soled** adj. Also *absol.*
1950 tr. *Mountaineering Handbk.* (Assoc. Brit. Members Swiss Alpine Club) ii. 22 Later experience shows that boots with moulded rubber soles—'Vibrams'—are an advantage on rock climbs but are dangerous on greasy or iced rock. **1957** Clark & Pyatt *Mountaineering in Britain* xvi. 237 Vibrams, boots whose soles consisted of rubber moulded into the shape of conventional nails, were just becoming available in Britain when war broke out. **1963** 'G. Carr' *Lewker in Norway* iv. 76 The climbing breeches, thick stockings, and Vibram-soled boots he would wear. **1976** *Trade Marks Jrnl.* 24 Nov. 2470/1 *Vibram*... Soles, heels, heel-tips, cork-soles and inter-soles for shoes. Vibram S.p.A..., Albizzate, Italy; manufacturers and merchants. **1979** C. McNeish *Youth Hostelling* ii. 15 I'm sure you can visualise the effect a hundred pairs of vibram-soles a day can have on a lino floor.

vibrant, *ppl. a.* Add: Hence **vi·brantly** *adv.*
1926 *Record Home & Mission Wk. United Free Church of Scotland* July 316/1 The Christian Church should make its protest vibrantly felt. **1976** *Gramophone* Mar. 1481/1 The warm, vibrantly sympathetic recording acoustic makes no little contribution.

vibraphone (vəi·brăfōᵘn). *Mus.* [f. Vibr-a(tion, Vibra(to, etc. + *-phone 1.] A percussion instrument consisting of a series of metal bars, arranged as in a xylophone, and characterized by the vibrato that can be given to the notes, an effect produced either by electrically rotated vanes in tube resonators under the bars or electronically.
1926 [see *marimbaphone]. **1929** *Melody Maker* Apr. 359/1 Lester takes a Vibraphone solo. **1934** S. R. Nelson *All about Jazz* ii. 51 The leading drummers..are expert on the xylophone and vibraphone. **1957** L. Feather *Bk. Jazz* xvi. 134 The vibraphone, like the xylophone, was an instrument long associated with novelty music. **1966** *Listener* 10 Mar. 364 The programme was completed by Stockhausen's *Refrain*, for piano, celesta, and vibraphone. **1978** P. Griffiths *Conc. Hist. Mod. Music* ix. 137 The vibraphone is used as a substitute for the gamelan while xylorimba and percussion suggest the influence of black African music.
Hence **vi·braphonist**, a player on the vibraphone.
1929 *Melody Maker* Dec. 1119/1 Amateur vibraphonists are requested not to attempt to copy it! **1978** *Detroit Free Press* 16 Apr. (Detroit Suppl.) 29/2 How else could he have known that Milt Jackson would become one of the world's greatest vibraphonists.

vibrated, *ppl. a.* Add: *spec.* Designating or pertaining to concrete which is being or has been compacted with a vibrator (*vibrator 1 b (a)).
1930 *Concrete & Constructional Engin.* XXV. 686/1 The vibrated concrete acts like a liquid and the vibrator rises automatically. **1948** *Archit. Rev.* CIII. 148 The roof-covering is of pre-cast vibrated concrete slabs. **1976** *Derbyshire Times* (Peak ed.) 3 Sept. 10/6 (Advt.), Paving slabs... Remanufactured top quality vibrated plain and coloured paving in a range of sizes.

vibration. Add: **3. d.** Now, an intuitive signal about a person or thing; (*pl.*) atmosphere. Usu. *pl.*
In some instances more or less identical with the *fig.* use of sense 3 b.
1899 O. Wilde *Importance of being Earnest* i. 28 There is very little music in the name Jack, if any at all, indeed. It does not thrill. It produces absolutely no vibrations. **1901** *Chambers's Jrnl.* Apr. 263/1 There is a man in Denver, Thomas J. Shelton who is said to be making his £10,000 a year by selling what he calls 'vibrations'... Mr Shelton's 'vibrations'..he himself explains as being a special force of his inner consciousness, which can be sent through space to purchasers by his mere act of will; and claims for the 'vibrations' so sent a subtle power capable of influencing a man in any direction that may be desired. **1919** Conrad *Arrow of Gold* iv. ii. 157 The Blunt atmosphere, the reinforced Blunt vibration stealing through the walls... Nothing to me, of course—the movements of Mme. Blunt, *mère*. *Ibid.* 164, I listened deferentially to the end yet with every nerve in my body tingling in hostile response to the Blunt vibration, which seemed to have got into my very hair. **1922** Joyce *Ulysses* 497 You can rub shoulders with a Jesus, a Gautama, an Ingersoll. Are you all in this vibration? **1934** M. Allingham *Death of Ghost* i. 23 Other people's pictures in *his* studio—it's sacrilege, isn't it? The vibrations won't be right. **1956** R. M. Lester *Towards Hereafter* ii. 36 When I had a sitting with a medium who was obviously on the same vibration as myself the results were first-class. **1957** J. Kerouac *On Road* ii. vi. 146 Something curiously unsympathetic and cold between them was really a form of humor by which they communicated their own set of subtle vibrations. **1968** T. Wolfe *Electric Kool-Aid Acid Test* ii. 20 Something's getting up tight, there's bad vibrations. **1971** J. Mandelkau *Buttons* v. 62 William showed me upstairs to what was going to be my 'home' for the next few weeks and let me wander around the house bumping into people and picking up on the vibrations. **1977** Miller & Swift *Words & Women* i. 4 Names do seem to give off vibrations of a sort. **1979** N. Mailer *Executioner's Song* ii. ii. 525 Stupor on top of old woe was the sad vibration Dennis was getting from Maximum.
7. Comb. *vibration-proof* adj.; **vibration damper** = *damper 4 c; *spec.* a device for counteracting torsional vibration in a crankshaft.
1932 *Motoring Encycl.* 183/1 Special forms of frictional clutch on the front end of the crankshaft to reduce engine vibration are dealt with under the heading Vibration Damper. **1936** *Gloss. Terms Electr. Engin.* (*B.S.I.*) 87 *Vibration-damper*, of an overhead line. A device attached to a conductor, and designed to suppress vibrations caused by the action of wind. **1952** *Chambers's Jrnl.* Jan. 64/1 Most of the risk of deterioration can be eliminated by using vibration-dampers under machinery. **1961** [see *damper 4 c]. **1968** R. H. Bacon *Car* v. 66 (*caption*) Crankshaft torsional vibration damper, fan belt pulley and camshaft drive. **1917** C. C. Turner *Aircraft of To-Day* vii. 114 The compass..must be carried in a vibration-proof bed.

vibrational, *a.* Add: **c.** *Physics.* Involving or resulting from particular modes of vibration or oscillation of the atoms in a molecule.
1923 [see *isotope effect* s.v. *isotope 2]. **1946** *Nature* 26 Oct. 593/2 From an analysis of a suitable vibrational spectrum, it is usually possible to trace the vibrational-levels in the ground-state from $V = 0$ to $V = c$. 20. **1970** I. E. McCarthy *Nuclear Reactions* i. v. 109 The nuclear probability amplitude for the excitation of vibrational states is obtained from the quantum theory of harmonic vibrations. **1973** C. Sagan *Cosmic Connection* iv. 27 There are vibrational transitions that occur when two atoms in a given molecule oscillate with respect to each other.

vibra·tionally, *adv.* [f. Vibrational *a.* + -ly².] By vibration; as regards vibrational motion.
1961 *New Scientist* 16 Mar. 687/1 Vibrationally assisted forming results in more uniform deformation [of metals]. **1964** [see *de-excite* vb. s.v. *de-excitation]. **1974** *Nature* 30 Aug. 714/1 The object may be a very massive star..which is vibrationally unstable. **1979** *Sci. Amer.* May 104/2 The advent of the laser would seem to be a chemist's dream come true, enabling him in principle to selectively excite reactants—electronically, vibrationally or rotationally, as he wishes.

vibrationless, *a.* (Further examples.)
1925 *Motor* 15 Dec. (Advt. Suppl.) 44, Ansaldo. The beautiful Italian car... Vibrationless. Silent. Fast. **1964** W. L. Goodman *Hist. Woodworking Tools* 157 To make for quicker and vibrationless cutting, the flat gullets vary in size in groups of three or four. **1982** *Sci. Amer.* Nov. 9 (Advt.), The instrument must be of rugged construction and vibrationless, without the aggravating oscillations of long-tubed conventional telescopes.

vibra·toless, *a.* [-less.] Without vibrato.
1975 *Daily Tel.* 12 Aug. 9/7 The pure vibrato-less style of Munrow's singers took on a piercing beauty. **1983** *N.Y. Times* 19 Dec. c13/1 The soloists for the Bach..sang in an even-toned, vibratoless style.

vibrator. Add: **1. b.** *spec.* (*a*) A device for compacting concrete by vibration before it has set. (*b*) A small electrically operated device for producing sexual stimulation.
1930 [see *vibrated ppl. a.]. **1950** *Archit. Rev.* CVII. 331/3 When vibration was first introduced the lack of proper equipment prevented its use except on very large contracts, but many types of vibrator are now on the market. **1981** Mindess & Young *Concrete* x. 289 External vibrators can be clamped to formwork, when the proper use of internal vibrators is not possible.
1953 A. C. Kinsey et al. *Sexual Behaviour in Human Female* v. 163 Douches, streams of running water, vibrators,..and still other methods were occasionally employed [in masturbation]. **1970** in G. Greer *Female Eunuch* 217 As vibrators have been mentioned, may I add that it need not be the penis-shaped battery model. **1979** *Brit. Med. Jrnl.* 13 Oct. 883/2 In a study in which patients used a vibrator to achieve orgasm, all the major tranquillisers were said to interfere with ejaculation. **1983** Cauthery & Stanway *Compl. Bk. Love & Sex* xxxii. 386 The biggest-selling sex aid by far is the vibrator.

vibrio. 2. (Earlier examples.)
1850 *Proc. Acad. Nat. Sci. Philadelphia* IV. 228 With them were also found two species of Vibrio. *Ibid.*, Much higher confervæ..are endowed with inherent power of movement, not very unlike that of the Vibrio.

vibriocidal (vibriōᵘsəi·dăl), *a. Med.* [f. Vibrio + -cide + -al.] Destructive to vibrios.
1962 *Jrnl. Immunol.* LXXXIX. 265/1 Complement is required for the vibriocidal effect. **1978** *Nature* 14 Dec. 709/1 Table 1 indicates a good correlation between protection to challenge and vibriocidal antibodies in the serum.

vibrionic, *a.* (Earlier example.)
1850 *Proc. Acad. Nat. Sci. Philadelphia* IV. 235 All the innumerable objects of living nature..from the vibrionic filament to the noble oak..are the result of a force in connection with an amorphous vesicle, the organic cell-wall, with the contained nucleus.

vibriosis (vibri,ōᵘ·sis). *Vet. Sci.* [f. Vibrio + -osis.] Infection with, or a disease caused by, vibrios.
1951 W. N. Plastridge in *Univ. Pennsylvania Veterinary Extension Q.* 3 Apr. 62 Improved methods of diagnosis and the increase of brucellosis-free herds account for the present increase in the interest in vibriosis. **1951** —— et al. in *Jrnl. Amer. Veterinary Med. Assoc.* June 367 In consideration of the several clinical manifestations of *V. fetus* infection and the probable existence of infection ..in some animals in the absence of clinical evidence, the term 'vibriosis' is proposed to indicate *V. fetus* infection. **1960** R. A. Runnells et al. *Princ. Veterinary Path.* xix. 560/2 In Western Australia vibriosis is reported to be the immediate cause of at least 75 per cent of the infertility in dairy cows. **1980** *Nature* 10 Apr. 566/2 Epizootics of the widespread fish disease vibriosis.

vibro-. Add: **vi·bro-massage,** massage with a special vibrator; **vibrota·ctile** *a.*, of, pertaining to, or involving the perception of vibration through touch.
1923 *Daily Mail* 10 Aug. 5/2 The owner-experts get their features in knots..and may be seen going off for vibro-massages in the evening. **1968** *Listener* 11 July 45/3 People who see nothing either comic or disturbing about our export eastwards of..vibro-massage, canned TV, [etc.]. [**1934** tr. R. H. Gault in *L'Année Psychologique* XXXIV. 2 Le sens tactile (ou comme je préfère..dire, les 'sens vibro-tactiles').] **1934** R. H. Gault in *Jrnl. Acoustical Soc. Amer.* V. 253/2 We may legitimately describe the results of our vibro-tactile experiments as evidences of hearing, even though we are assured that the ear is not involved. **1980** *Pflügers Arch. European Jrnl. Physiol.* CCCLXXXIV. 170/1 Vibrotactile stimuli above the tuning threshold of the nerve fibers can elicit a continuous afferent impulse volley without adaptation during some seconds.

vibrograph. (In Dict. s.v. Vibro-.) Add: *spec.* An instrument for measuring or recording mechanical vibrations.
1904 [in Dict.]. **1946** [see *accelerometer]. **1965** *Economist* 25 Dec. 1434/1 It took them some time to work out the best way of getting the job done without causing too much damaging vibration in the [Abu Simbel] temples—vibrographs were installed to check that the danger level was not exceeded. **1970** *Soviet Jrnl. Optical Technol.* XXXVII. 456 A method is given..for the calculation of the amplitude-frequency and phase-frequency characteristics of a vibrograph for recording small angular vibrations.
Hence **vi·brogram**, a record produced by a vibrograph.
1932 *Jrnl. R. Aeronaut. Soc.* XXXVI. 209 Vibrograms were obtained showing the vibration of different parts of the fuselage at various frequencies. **1950** *Engineering* 21 Apr. 431/2 The principal parameters of a vibrogram from which its effect on a building can be estimated are the maximum amplitude and the average frequency of the vibration. **1975** *Soviet Appl. Mech.* XI. 445 (*heading*) Damping determination according to vibrograms of random oscillations in linear systems.

vibronic (vəibrǫ·nik), *a. Physics.* [f. Vibr(ational *a.* + *electr)onic *a.*] Of or pertaining to electronic energy levels or transitions associated with the vibration of the constituent atoms of a molecule.
1941 R. S. Mulliken in *Physical Rev.* LIX. 880/2 The vibronic wave function ψ_{ev} is a function of the electronic and vibrational coordinates. **1966** G. Herzberg *Molecular Spectra & Molecular Structure* i. 22 For half-integral spin we have two-valued vibronic species. **1976** *Nature* 22 Apr. 675/1 No vibronic structure has been detected in the absorption spectrum of the chromophore of rhodopsin.
Hence **vibro·nically** *adv.*
1966 D. H. Whiffen *Spectroscopy* xii. 160 Transitions which are permitted solely on this account are called vibronic transitions or may be said to be vibronically allowed. **1978** *Nature* 19 Jan. 236/2 The results indicate that the afterglow is due to an emission to S_0 of the monomer or to a vibronically similar ground state.

Vic¹ (vik), colloq. abbrev. of Victoria², used to denote the Royal Victoria Theatre, London (popularly known as the *Old Vic*).
1858 G. A. Sala *Twice round Clock* (1859) 269 There is a transpontine theatre, situated laterally towards the Waterloo Road... This is the Royal Victoria Theatre... It is popularly termed the 'Vic'. **1888** Kipling *Soldiers Three* (1890) 55 'Ave you ever bin in the Pit hentrance o' the Vic. on a thick night? **1915** *Times* 3 Oct. 11/5 At the Royal Victoria Hall ('the Old Vic')..a Shakesperian season opened last night. **1951** *Oxf. Compan. Theatre* 582/1 In 1833 it [*sc.* the Coburg Theatre] was renamed the Victoria, from the fact that Princess (later Queen) Victoria visited it once. It soon became affectionately known as the Old Vic. **1972** W. Garner *Ditto, Brother Rat!* iv. 32 The Vic bars get pretty hectic, especially with only one interval. **1976** *Vogue* 15 Mar. 15/3 So goodbye Old Vic, tiptoeing off into the wings of theatrical history.

vic² (vik). Used for *v* in telephone communications and in the oral transliteration of code messages. Hence also in *Mil.* use, a V-formation of aircraft.
1913 [see *pip sb.⁴]. **1927** W. E. Collinson *Contemp. Eng.* 98 Other artillery terms which spread were..vick for V, also ack emma for a.m. [etc.]. **1940** *Flight* 28 Mar., 295 Spitfires staged some plain and fancy beat-ups..in formation (excellent vic and echelon). **1942** I. Gleed *Arise to Conquer* 12 We flew back as a squadron, four sections in vic. **1952** M. Tripp *Faith is Windsock* iv. 61 The Squadron was leading today's raid, and E-Easy..led the Squadron. To port and slightly behind Easy's vic was C-Charlie, also with a vic.

vicar. Add: **2. a.** Now also in the Church of England a priest who is a member (team vicar) of a team ministry (*team sb. 11) under the leadership of a team rector.
1972 *Daily Tel.* 7 Aug. 10/5 Only the leader of the team, usually called 'Rector', is the beneficed freeholder incumbent inducted by the bishop. His colleagues ('vicars') are licensed by the bishop as members of the team. **1977** [see *rector 3 a]. **1984** *Church Times* 27 Jan. 17/1 (Advt.), Vicar required... N. Birmingham Team Ministry, modern vicarage.
4. c. *vicar apostolical.*
1731 in O. Blundell *Catholic Highlands Scotland* (1917) II. 99 His Holiness..appoints him also Vicar Apostolical with singular powers. **1847** J. A. Manning *Pius XI* I. 168 Differences broke out between the Vicar Apostolical and the Chilian government. **1849** [in Dict.].

vicarage. Add: **6. vicarage tea-party,** used as the type of something mild, innocuous, and uneventful.
1973 *Times* 11 Apr. 1/2 Mr Heath's appearances in the Commons are never vicarage tea parties. **1984** *Daily Tel.* 31 Jan. 12/5 He surveyed the smoking ruins of..a fine old Elizabethan rectory,..and said: 'It makes the Dissolution of the Monasteries look like a vicarage tea-party.' **1984** *Guardian* 6 June 10 Politicians..who fear that events are about to take a nasty turn, frequently say that what they are predicting will make D-Day (or the ravages of Genghis Khan..—every speaker will have his own comparison) look like a vicarage tea party.

vicariad (vikēᵊ·ri,æd). *Ecol.* [f. as next + -ad.] = *vicariant sb.
1944 S. A. Cain *Found. Plant Geogr.* xviii. 266 Most of the studies of vicariads have referred to species. **1960** N. Polunin *Introd. Plant Geogr.* vii. 202 True vicariads (which have arisen from a common stock) should be distinguished from false ones which have not this close genetic relationship... True vicariads..may be classified according to the manner of their separation from one another into (1) horizontal (geographical), (2) altitudinal (physiographic), (3) habitat (ecological), and (4) seasonal (..closely related forms differing in their times of development). **1972** D. M. Moore in D. H. Valentine *Taxonomy, Phytogeogr. & Evolution* viii. 119 The aquatic *Littorella*, absent from North America, is effectively an amphitropical vicariad and certainly results from a much more recent migration.

vicariant (vikēᵊ·riănt), *sb.* and *a. Ecol.* [f. L. *vicāri-us* substitute (see Vicar) + -ant¹, tr. G. *vic-*, *vikarirend*, pres. pple. of *vikarieren* (now *vikariieren*) † to substitute for: cf. *vic-*, *vikarirend spezies* in F. Unger *Ueber den Einfluss des Bodens auf die Vertheilung der Gewächse* (1836) iii. 92; M. Wagner *Die Darwin'sche Theorie* (1868) in *Die Entstehung der Arten durch räumliche Sonderung* (1889) 56.] **A.** *sb.* A vicariant form of a plant or animal.
1952 L. Croizat *Man. Phytogeogr.* iv. 224 Cruciferae and Capparidaceae..could bear being characterized as vicariants. **1957** P. Dansereau *Biogeogr.* ii. 104 On the Nimba mountains of West Africa, Schell..has cited many mountain vicariants in genera of both rain-forest..and savana. **1979** Cardona & Contandriopoulos in D. Bramwell *Plants & Islands* viii. 155 In the *Genista acanthoclada* DC. group..the following subspecies are vicariants: *G. acanthoclada* ssp. *acanthoclada* (Greece and Aegean Islands), ssp. *echinus* (Spach) Vals. (S. Anatolia), [etc.].
B. *adj.* Being or involving varieties, species, communities, or the like that have evolved out of effective contact with one another from a common ancestral stock, esp. in habitats that are similar though separated.
1952 L. Croizat *Man. Phytogeogr.* vi. 355 This form in the end will split up into..vicariant species. **1957** P. Dansereau *Biogeogr.* i. 22 Vicariant varieties, subspecies, species, and genera may differ not only in leaf shape..but also in habitat requirements. **1957** A. Macfadyen *Animal Ecol.* i. 8 The practical advantages include the possibility of using certain 'vicariant' species as 'indicator organisms' for the detection..of certain types of habitat. **1972** D. Bramwell in D. H. Valentine

Taxonomy, Phytogeogr. & Evolution ix. 153 A high degree of vicariant evolution is to be expected in an archipelago such as the Canaries where..sets of ecological conditions are replicated from east to west on a series of islands with similar vegetation zones. **1982** *Sci. Amer.* July 33/2 Whether they are vicariant species, with common parents but evolved in place, or whether they first became distinct and then spread is unclear.

Hence **vica·riance,** the existence of vicariant forms; the separation or subdivision of a population by the appearance of a geographical barrier and subsequent differentiation.

1957 A. Macfadyen *Animal Ecol.* xv. 223 It is also possible to use other criteria..such as frequency, dominance or even 'vicariance'. **1972** D. Bramwell in D. H. Valentine *Taxonomy, Phytogeogr. & Evolution* ix. 151 Adaptive radiation..is a positive process where genetical response to the stimulus of the environment is the main factor; but vicariance, that is divergent evolution in which geographical isolation has been a very important factor, is perhaps a more passive process where the interaction of genetic drift and weak selection result in the establishment of distinctive characters in populations which occupy essentially similar ecological habitats to their parents. **1979** *Nature* 15 Feb. 562/1 It is possible that the presence of hadrosaurs in Laurasia and Gondwanaland represents vicariance rather than dispersal; if so, hadrosaurs would be expected to occur on most continents. **1981** Nelson & Platnick *Systematics & Biogeogr.* i. 45 Speciation, at least in animals, usually involves a process of geographic isolation.., and occurs when a formerly continuous population is divided by the appearance of a barrier (a process called vicariance).

vicariism (vikēᵊ·ri,iz'm). *Ecol.* Also **vicarism.** [f. as *vicariant *sb.* and *a.* + -ism, or ad. G. *vikarismus.*] = *vicariance* s.v. *vicariant *sb.* and *a.*

1939 *Bull. Misc. Information* (R. Botanic Gardens, Kew) v. 229 Vicarious species are particularly instructive to study. These are species inhabiting contiguous but not (or scarcely) overlapping areas. Vicarism may be altitudinal or geographical. **1944** S. A. Cain *Found. Plant Geogr.* xviii. 265 The concept of vicariism is equally applicable to other than the specific category, i.e., to subspecies, sections, etc., and to communities. **1953** R. Good *Geogr. Flowering Plants* (ed. 2) xi. 212 Another particular kind of vicariism, which may be described as climatic, is illustrated..by the two littoral species *Ipomoea pes-caprae* and *Calystegia soldanella*. The former is almost or quite pan-tropical, but the latter is more or less pan-temperate and the two meet at the Kermadecs and elsewhere in similar latitudes. **1969** M. D. F. Udvardy *Dynamic Zoogeogr.* iv. 192 The concept of vicarism is used widely..when comparing certain types of distributional areas. **1974** *Boissiera* XXIII. 296 The distribution patterns of the different chromosome races agree with certain main types of false vicariism.

vicarious, *a.* Add: **4. d.** Experienced imaginatively through another person or agency.

1929 R. S. & H. M. Lynd *Middletown* xvii. 237 To Middletown adults, reading a book means overwhelmingly what story-telling means to primitive man—the vicarious entry into other, imagined kinds of living. **1948** E. Waugh *Loved One* 31 He had lived his twenty-eight years at arm's length from violence, but he came of a generation which enjoys a vicarious intimacy with death. **1976** A. Powell *Infants of Spring* ix. 146 My father, between spasms of grumbling about school bills, and occasional resistance to attitudes of mind inevitably acquired at Eton, had taken a fair amount of vicarious pleasure in my being there.

6. *Ecol.* Of two or more species, etc.: similar to one another and occurring without the other(s) in different areas; usu., = *vicariant *a.*

1932 Fuller & Conard tr. *Braun-Blanquet's Plant Sociol.* vi. 160 Closely related species found upon lime and clay slates he [*sc.* Unger] called substitute or vicarious species. **1937** R. Hesse et al. *Ecol. Animal Geogr.* vi. 77 Transitional variation may be wanting at the boundary between the ranges of vicarious forms which are then considered specifically distinct. **1960** N. Polunin *Introd. Plant Geogr.* vii. 201 With higher groupings—and even families and whole communities may in a sense be vicarious—there is less reason to suppose that their mutual exclusiveness is due to competition. **1981** P. Stott *Hist. Plant Geogr.* viii. 115 Vicarious evolution has been invoked..to explain the distribution [in the Canaries] of endemic species in *Centaurea* sect. *Cheirolophus* subsect. *Flavifiorae.*

vicariously, *adv.* Add: **3.** At second hand, at one remove. Cf. *vicarious *a.* 4 d.

1925 F. Scott Fitzgerald *Gt. Gatsby* vii. 157 Jordan and I tried to go, but Tom and Gatsby insisted..that we remain—as though..it would be a privilege to partake vicariously of their emotions. **1957** L. Durrell *Justine* ii. 127 Those interminable monologues about a life which has long since receded, lost its vital momentum, only to live on vicariously in the labyrinths of memory. **1984** *N.Y. Times* 13 May vi. 70/2, I think the greatest moral pitfall is not that we witness too much bang bang, but that, for the most part, we perceive it vicariously.

vi·carish, *a.* [f. Vicar + -ish¹.] Suitable for or characteristic of a vicar.

1938 *Times Lit. Suppl.* 26 Oct. 734/2 Two maids..an amiable vicar and his very vicarish wife and a certain Captain Carling complete the party. **1976** *New Society* 29 Jan. 206/3, I was also subject to frequent visits by vicars, who popped their heads round my door in a vicarish manner every so often. **1979** K. M. Peyton *Marion's Angels* i. 2 'It doesn't *belong* to her,' they said. 'It belongs to God, and so does Marion.' A typical, vicarish remark, they said.

Vicat (vī·kā). *Engin.* The name of L. J. *Vicat* (1786–1861), French engineer, used *attrib.* with reference to an apparatus for measuring the consistency and setting time of Portland cement and other materials.

1904 C. F. Marsh *Reinforced Concrete* iii. 125 For this test the Vicat needle apparatus may be employed. **1910** *Brit. Standard Spec. for Portland Cement* (Engin. Standards Comm.) 10 The mixture shall be plastic when filled into the Vicat mould. **1956** J. N. Anderson *Appl. Dental Materials* xviii. 211 An alternative method of testing for initial set [of plaster] is to use the Vicat needle. **1979** I. Soroka *Portland Cement Paste & Concrete* ii. 29 (*caption*) Vicat apparatus for determining the standard consistence and setting times of Portland cement.

vice, *sb.*¹ Add: **1. c.** *ellipt.* for *vice squad*, sense 8 b below. *slang.*

1967 C. Drummond *Death at Furlong Post* iv. 42 From his days on Vice Reed remembered the large free-spenders. **1976** *New Society* 4 Mar. 481/2 A woman they know is a junkie... She proceeds to tell them how she got picked up by the 'vice' the night before.

8. a. Also *spec.* with reference to certain crimes, esp. organized prostitution, as *vice den, racket, trade*, etc. orig. *U.S.*

1903 *McClure's Mag.* Nov. 89 In New York, Croker has failed signally to maintain vice-bosses whom he appointed. **1915** *Sci. Amer.* 30 Jan. 98/3 The Puritan conception of life, like that of vice-crusaders, suffragettes, and most crusaders, scorns all trifling with its weighty realities. **1937** Vice society [see *strip-teasing* vbl. sb. s.v. *strip-tease]. **1938** F. D. Sharpe *Sharpe of Flying Squad* vii. 78 Lots of the other men in the vice racket..were pulled in and interviewed. **1952** *Manch. Guardian Weekly* 8 May 3 The relations of one of these with a Chicago vice-syndicate may be merely an unfair reflection on Governor Warren. **1962** *Spectator* 6 July 12/2 The vicelands of Notting Hill. **1971** *It* 2–16 June 1/1 (*heading*) Vice girls of Princedale Road—the shocking truth! **1975** J. Gores *Hammett* (1976) v. 37 In a vice raid..police..trailed a group of three boys..to the house of prostitution..and jailed the inmates of the..vice-den. **1976** *Billings (Montana) Gaz.* 30 June 4-A/1 Suppose that you..prevented vice officers from arresting a drug suspect. **1981** P. O'Donnell *Xanadu Talisman* v. 100 His wife was.. sold into the vice trade.

b. Special Comb.: **vice ring,** a group of people criminally involved in organized prostitution; **vice squad** [*squad *sb.*¹ 4 c] orig. *U.S.*, a police unit concerned with the enforcement of laws relating to prostitution, drug abuse, illegal gambling, etc.

1938 F. D. Sharpe *Sharpe of Flying Squad* xi. 125, I don't think..they were..connected with any vice ring. **1981** C. Scott *Heavenly Witch* vi. 80 Men in charge of vice rings spread rumours that the converts were paid to testify. **1905** *N.Y. Times* 22 June 8/6 Six of Capt. Cottrell's Tenderloin detectives will report to Capt. Egger this noon for duty on the Vice Squad. **1939** *Daily Tel.* 18 Dec. 8/4 Scotland Yard's vice squad..has been instructed to give special attention to small clubs opened in Soho since the outbreak of war. **1978** L. Heren *Growing up on The Times* v. 167 The vice squad might have had a beady eye on me, but I was glad to hold his hand.

vice, *sb.*⁷ Add: Also prefixed by *Mr.* as a form of address to a vice-chairman or vice-president.

1894 G. du Maurier *Trilby* II. 102 A table..at one end of which sits Mr. Chairman..and at the other 'Mr. Vice'. **1916** M. Diver *Desmond's Daughter* ii. iii. 61 The President at the far end of the table had lifted his glass. 'Mr Vice—the Queen,' said he. **1976** T. Jeal *Until Colours Fade* xxxvii. 325 The president of the mess rose..and brought down his silver mallet. 'Mr Vice, the Queen,' that officer said, addressing the vice-president at the opposite end of the table.

vice-. Add: **a.** *vice-chief of staff, -editor, -lieutenant* (later example), *-minister.*

1943 W. S. Churchill *End of Beginning* 69 The Chiefs of Staff Committee are assisted by a Vice-Chiefs of Staff Committee. **1978** R. V. Jones *Most Secret War* xlv. 458 The Vice-Chiefs of Staff..advised that the threat was over. **1976** *National Observer* (U.S.) 12 June 22/2 A fellow with the title of vice editor. **1963** *Times* 4 June 13/5 His native county of Lincolnshire, of which he was Vice-Lieutenant for many years. **1909** E. M. Satow in *Cambr. Mod. Hist.* XI. xxviii. 865 Ōki of Hizen, and Itō, Inouyé and Yamagata of Chōshiu were retained as vice-Ministers. **1976** *Eastern Daily Press* (Norwich) 19 Nov. 1/5 The agreement was signed by Iran's vice-minister of war.

b. *vice-chair* (earlier example).

1839 Dickens *Nickelby* xlviii. 475 A farewell-supper..at which Mr. Snittle Timberry would preside, while the honours of the vice chair would be sustained by the African [Sword-]Swallower.

c. *vice-preside* vb.

1889 G. B. Shaw *London Music in 1888–89* (1937) 94 You are patronized by the Lord Mayor, presided over by the Duke of Westminster, and vice-presided over and councilled by nearly five dozen illustrious persons.

‖ **vice anglais** (vīs ańglę). [Fr., lit. 'English vice' (Vice *sb*¹.).] The vice to which the English are said to be particularly prone (esp. with reference to corporal punishment).

1942 [see *self-medication]. **1947** *Horizon* Nov. 269 No novelist in the last decade of the century could create a character who practised *le vice anglais* without attributing to him some of Swinburne's physical characteristics. **1950** E. H. W. Meyerstein *Let.* 6 Jan. (1959) 369 You must remember that it is much more important socially for a boy to play for his public school in cricket..than to win a prize... It is a very queer trait and a modern psychologist might quite possibly link it up with sadism ('le vice Anglais'), success consisting in *hitting* something, though that 'something' be no more than a ball. **1962** *Times* 21 July 4/5 *Whack-O!*—a programme whose long life testifies, if nothing else, to the continued popularity of *le vice anglais.* **1976** F. Muir *Frank Muir Book* 73 Flogging in public schools..has been given as the reason for so many..English gentlemen enjoying a little corporal punishment later in life, 'le vice Anglais'. **1983** *Daily Tel.* 5 Aug. 12/1 The true *vice anglais* is the hatred felt by a sub-section of the middle class for their own country and their desire to humiliate her.

vi·ce-champion. [f. Vice- + Champion *sb.*¹ 4.] The runner-up in a sporting contest.

1981 *N.Y. Times* 5 July v. 7/1 Only 251 fans showed up..to see Sao Paulo, the vice champion of Brazilian professional teams, play in that city of eight million people. **1984** *Soviet Union* ix. 53 The men's team, vice-champion of Europe, has twice beaten the world champion Pakistan team.

vice-like, *a.*² (Earlier example.)

1835 E. A. Poe in *Southern Lit. Messenger* June 570/1 Clutching with a vice-like grip the long-desired rim.

vice-president. Add: **vice-presidency, -presidential** (earlier examples).

1804 *Guardian of Freedom* (Frankfort, Kentucky) 28 July 2/3 He is charged with having been long intriguing for the vice presidency. **1854** T. H. Benton *30 Years' View* I. 45/1 Mr. Calhoun was the only substantive vice-presidential candidate before the people. **1884** G. W. Cable *Dr. Sevier* xlvi. 341 With a presidential candidate on one side and his vice-presidential man Friday on the other.

viceregal, *a.* (Earlier example.)

1836 Disraeli *Lett. of Runnymede* x. 86 Ascending at last even to the Viceregal throne of India.

vice-regent. Add: Hence **vice-re·gency.**

1930 Belloc *Wolsey* ix. 243 He drafted a form of Vice-regency, a delegation of Papal power to himself.

vicereine. Add: Now usu. with pronunc. (vəi·srēⁱn).

vicey-versey, vicy-versy, (vəi·si vɔ̄·ɹsi), repr. colloq. or joc. pronunc. of Vice-versa *adv. phr.*

1858 J. R. Lowell in *Atlantic Monthly* Aug. 371 How far from these unhappy days When all is vicy-versy! **1979** K. O'Hara *Searchers of Dead* ix. 102 Actors work on directors as well as vicey-versey.

Vichy. Add: **1.** (Earlier *ellipt.* example.)

1882 *Harvard Lampoon* 26 Jan. 87/2 Vichy and Seltzer possessed of limpidity.

2. Used to denote the government of France which operated from Vichy (1940–44) in collaboration with the Germans; freq. *attrib.* and in *Comb.* Also *transf.* and *fig.*

1941 [see *collaboration 2]. **1942** 'G. Orwell' in *Partisan Rev.* Mar.–Apr. 159 Both Vichy and the Germans have found it quite easy to keep a façade of 'French culture' in existence. **1942** 'M. Home' *House of Shade* vi. 96 Vichy French territory had been entered at Bardai. **1949** Koestler *Promise & Fulfilment* i. vii. 76 When, in 1941, British troops invaded Vichy-held Syria, they needed a vanguard of Commandos. **1965** B. Sweet-Escott *Baker Street Irregular* ii. 54 Efforts to start something in Morocco, then occupied by the Vichy French, had been a failure. **1966** G. Greene *Comedians* i. iii. 69 During the war..I served in the Political Intelligence Department of the Foreign Office, supervising the style of our propaganda to Vichy territory. **1970** R. Wingate *Lord Ismay* iv. 59 The German ability to use Vichy-controlled Syria as a route for her aircraft. **1979** P. Way *Sunrise* xiv. 146 The jail..had been convent no. 787 of the Poor Clares under West Africa's Vichy régime. **1980** *Times* 27 May 4/2 He went on to claim there was a 'Vichy mentality' in parts of the Foreign Office. The Home Office was 'stuffed with reactionaries'. **1981** M. Warner *Joan of Arc* xiii. 263 Maurras's adherence to both Vichy and Joan of Arc.

Hence **Vi·chyist,** a supporter of the Vichy government.

1943 *Rev. Foreign Press* (Foreign Office Research Dept.) 23 Aug. 105/1 (*heading*) 'Combat' and' Liberté' demand severe measures against all Vichyists. **1958** *Times* 3 Dec. 8/5 M. Soustelle was in charge of intelligence and espionage, first against occupying Germans and then against Vichyists and Communists.

Vichyite (vi·ʃi,əit), *sb.* and *a.* [f. Vichy + -ite¹.] **A.** *sb.* = *Vichyist* s.v. *Vichy. **B.** *adj.* Supporting the Vichy government (see *Vichy 2). Also *fig.*

1943 *Ann. Reg. 1942* 287 Yesterday a 'Vichy-ite', collaborating with the Nazis..today a French patriot. **1945** *Times* 30 Oct. 3/4 Vichyite officials in Indo-China. **1968** [see *collaborateur 2]. **1978** *Times* 9 Jan. 8/5 Each school of thought has its own pet profile of 'Treasury Man', whether it be the Institute of Economic Affairs seeing in him the symptoms of chronic Keynesianism or the Tribune Group of left wing Labour MPs sniffing for 'Vichyite' collaborators with the International Monetary Fund. **1979** *N.Y. Rev. Bks.* 8 Feb. 14/4 A *collabo*, someone who picked the wrong side, a Theban Vichyite?

vichyssoise (viʃiswā·z). Also with capital initial and ¶ **vichycoise, vichysoisse.** [a. Fr., in full **crème vichyssoise glacée** iced cream (soup) of Vichy.] A soup made with potatoes, leeks, and cream, usu. served chilled; a bowl of this soup.

1939 *Vogue's Cookery Bk.* 7 Creme Vichycoise. This is a special favourite in iced soups. **1941** A. Escoffier *Escoffier Cook Bk.* 262 Vichysoisse, now called Crême Gauloise, is made by adding cream and chilling. **1943** I. S. Rombauer *Joy of Cooking* (ed. 3) 63/1 Vichyssoise (French potato soup)... Now called 'de Gaullesoise' in a New York restaurant. **1950** M. Miller *Sure Thing* 35 Tetrazzini..and two Vichyssoise to start out, large coffees, later. **1959** *News Chron.* 30 Oct. 3/4 It happened..at the Savoy Hotel. The diner tasted his chilled Vichyssoise soup. **1969** *New Yorker* 20 Sept. 163/1 Vichyssoise..is usually served very cold. **1982** J. Aiken *Whisper in Night* 100 The night had grown stiflingly oppressive and humid. 'The air is like Vichyssoise,' said my uncle.

Vici (vəi·səi). *N. Amer.* Also **vici.** [perh. a. L. *vīcī*, pa.t. of *vincere* to conquer.] The name of a chrome-tanned kid leather used for shoes and boots. Freq. *attrib.*, esp. as *Vici kid.*

Formerly a proprietary name in the U.S.

1888 *Shoe & Leather Reporter* 19 July 133/3 Robert H. Foerderer courts the smiles of fortune with a horseshoe enclosing the word 'vici'. **1891** *Official Gaz.* (U.S. Patent Office) 13 Oct. 128/2 Kid, goat, and similar light weight leathers... The word 'Vici'. **1904** 'O. Henry' in *Everybody's Mag.* Feb. 187/1 He was the color of vici kid, and his whiskers was like excelsior made out of mahogany wood. **1906** *Daily Colonist* (Victoria, B.C.) 6 Jan. 12/7 (Advt.), Men's Vici Kid Shoes. **1937** H. H. Kroll *I was Sharecropper* x. 193, I had a pair of vici kid shoes for two dollars and a half. **1946** *Harper's Mag.* Oct. 311/1 There.. would be Pa in his Sunday clothes and vici shoes. **1946** H. R. Quimby *Pacemakers of Progress* vi. 115 Dressing high boots were often made with tops of silk or thin kid..; the lower parts were of patent leather or vici kid in black, bronze, white, or a color.

vicinism (vi·siniz'm). *Biol.* [f. L. *vicīn-us* (see Vicine *a.*) + -ism.] (See quots. 1905, 1959.) Hence **vi·cinist,** a form produced by vicinism.

1905 H. De Vries *Species & Varieties* 188 For this purpose I propose the word vicinism..as indicating the sporting of a variety under the influence of others in its vicinity. *Ibid.* 201 Of two hundred seeds one became a blue atavist, or rather vicinist, while all others remained true to the white type. **1929** *Hereditas* XIII. 188 The percentage of vicinism has been exactly determined for one *Tetrahit* line... The flower colour of this line..is recessive to the red flower colour of the line T–G. **1959** *N.Z. Timber Jrnl.* Feb. 58/2 Vicinism... The tendency to variation caused by natural crossing with related forms growing nearby.

vicious, *a.* Add: **II. 9.** *vicious circle.* **c.** *gen.* A situation in which action and reaction intensify each other; a self-perpetuating process of aggravation. Similarly *vicious spiral*, in which the ill-effects are cumulative. Cf. *spiral *sb.* 2 d.

1839 [in Dict., sense 9 a]. **1892** H. James *Notebks.* (1947) 130 The whole situation works in a kind of inevitable rotary way—in what would be called a vicious circle. **1929** [see *Dosto(y)evskian *a.*]. **1940** M. Nicholson *How Britain's Resources are Mobilized* (Oxf. Pamphlets on World Affairs No. 30) 24 The result, when supplies of goods are short, is to bid up prices, thus raising the cost of living, inspiring demands for increased wages and starting the 'vicious spiral' of inflation. **1958** *Spectator* 11 July 60/2 All stress disorder is subject to this vicious-spiral rule. **1965** *Listener* 11 Nov. 741/2 It is sometimes necessary to enact laws against racism as a first step towards breaking a vicious circle. **1975** *Times* 23 Aug. 1/5 This is a vicious spiral of..mounting prices and declining traffic volume. **1982** *Times* 26 Aug. 3/8 It is a vicious circle. The boats cannot be sure of selling their fish until the processors invest in the new plant to handle it, and the processors cannot risk their money until they are sure that the fleet has guaranteed fishing areas and catch quotas.

10. *vicious abstraction* (Philos.), the abstraction of one quality or term from a thing or concept at the expense of other qualities or terms of which it is also composed; hence *vicious abstractionism.*

1883 F. H. Bradley *Princ. Logic* 511 If we recognize these elements our unit is not solitary; if we ignore them we fall into vicious abstraction. **1909** W. James *Meaning of Truth* xiii. 249 Let me give the name of 'vicious abstractionism' to a way of using concepts which may be thus described. **1932** H. H. Price *Perception* vii. 173 To use the language of the Idealist tradition, they only seem to be mere acceptances through a 'vicious abstraction'.

Vickers (vi·kəɹz). The name of a manufacturing company, orig. *Vickers*, Ltd., used: **1. a.** *attrib.* With reference to any of a variety of products of the firm, esp. armaments, aircraft, etc. Occas. also in the possessive.

1913 *303-Inch Machine Guns* (Ordnance College: Field Artillery & Small Arms Branch) 62 (*heading*) Gun, Vickers', ·303-in., Light. **1917** A. G. Empey *Over Top* 313 Vickers gun. **1919** *Vickers News* 15 Oct. 7/2 Vickers engines are occasionally tested on some such oil as Texas. *Ibid.* 1 Nov. 25 (*caption*) A Vickers airship is also seen leaving Barrow with letters for Sheffield. *a* **1944** K. Douglas *Alamein to Zem Zem* (1946) xiii. 79 Our supporting infantry manned their Vickers M.G.s on the flanks. **1945** *Penguin New Writing* xxiv. 31 In front of me the man carrying the Vickers ammunition. ? **1949** B. W. A. Dixon *Aircraft* i. 14/1 The Vickers 'Vimy', a twin-engined bomber designed originally for the long-distance bombing of Berlin. **1953** C. A. Lindbergh *Spirit of St. Louis* ii. vi. 295 Alcock and Brown..got across the ocean in their twin-engined Vickers bomber. **1974** *Encycl. Brit. Macropædia* XIX. 688/2 The Maxim machine guns, later often known as Vickers weapons, were used throughout the world well into the 1960s.

b. *absol.* One of a series of machine-guns manufactured by the company and used in both World Wars, esp. the ·303 or Vickers Maxim.

1917 G. Frankau *City of Fear* 18 You know what it's like in a listening-post, The Very candles aflare, Their bullets smacking the sand-bags, our Vickers combing your hair. **1918** T. E. Lawrence *Lett.* (1938) 243 A frontal attack of eighteen men, two Vickers, and two large Hotchkiss. **1942** E. Waugh *Put out More Flags* iii. 195 Nigel was full of questions;..what was the difference between a Bren and a Vickers. **1946** R. Capell *Simiomata* i. 36 The caiques scattered with Brens and Vickers raking them.

2. *attrib.* With reference to a method of testing the hardness of a material (esp. metal) by measuring the indentation produced when a small diamond pyramid is applied to the surface under a specified load.

1926 *Automobile Engineer* XVI. 105/1 In designing the Vickers machine..an impression having a diameter equal to three-eighths that of the ball has been taken as ideal. **1930** *Engineering* 12 Sept. 318/2 The hardness of hardened steel was further raised from Vickers number 950 to 1,100 by this treatment. **1973** J. G. Tweeddale *Materials Technol.* I. iv. 77 The most standard form of test is the Vickers hardness system performed by pushing a standard square pyramid diamond indenter..into the surface of the material.

vicomte (vi·koṅt). [Fr.] A viscount. Similarly **vicomtesse** (vikoṅte·s), a viscountess.

c **1786** T. Blaikie *Diary of Scotch Gardener* (1931) 204 The Vicomtesse de Pons..had a Park to Make. **1847** C. Brontë *Jane Eyre* I. xv. 286, I knew him for a young roué of a vicomte—a brainless and vicious youth. **1889** C. M. Yonge *Reputed Changeling* II. xxi. 71 The mother.. was too proud of him to miss any opportunity of exhibiting him to an experienced mother and grandmother like the vicomtesse. **1928** C. A. Nicholson *Hell & Duchess* v. 94 You can scarcely blame the vicomte for suggesting a change. **1968** R. Amberley *Incitement to Murder* iii. 77 His uncle and aunt will be staying in the vicomte's house. **1980** C. Watson *Bishop in Back Seat* (1981) xxviii. 157 There was a sometime film producer,..a French count,..a vicomtesse who wrote best-sellers.

Viconian (vikōᵘ·niăn), *a.* [f. the name of Giovanni Battista *Vico* (1668–1744), Neapolitan philosopher + -ian.] Designating the theories or doctrines of Vico, esp. those concerned with the historical development of culture and its cyclical or repetitive character.

1957 N. Frye *Anat. Crit.* i. 62 Joyce and his Viconian theory of history. **1967** *Listener* 23 Mar. 391/3 Implying that the work should properly be..bound in such a way as to have no first or last page, a very embodiment of Viconian circularity. **1976** T. Eagleton *Crit. & Ideology* iv. 157 Human societies are specific sectors of that cosmic process, determined by its unalterable laws, moving inexorably through their Viconian cycles.

victim, *sb.* Add: **4.** *victim-hero.*

1962 *John o' London's* 1 Mar. 211/3 Young Cordier.. isn't the perfect victim-hero. **1975** P. Fussell *Gt. War & Mod. Memory* i. 8 Air bombardment, which was supposed to shorten the war, prolonged it by inviting those who were its targets to cast themselves in the role of victim-heroes and thus stiffen their resolve. *Ibid.* vi. 220 Guy Crouchback, Waugh's victim-hero.

victimization. Add: **c.** *spec.* With reference to the imposition of penalties on trade union members who go on strike.

1923 *Westm. Gaz.* 23 Apr., The expression 'no victimisation' implied that a striker's place was not to be filled by an outsider. **1925** S. O'Casey *Juno & Paycock* i, in *Two Plays* 8 Why did they sack her? It was a clear case of victimization. **1957** *Listener* 26 Dec. 1057/1 Victimisation, that 'doubletalk' word of today, was then a vivid and frequent reality. **1961** *Daily Tel.* 16 Dec. 7/1 They will receive severance pay from the company and a share of a 'victimisation fund' to tide them over until they find other jobs. **1966** [see *Newspeak]. **1974** *Socialist Worker* 23 Nov. 1/3 From the miners' ballot to the strikes of bakers and journalists, to the victimisation strike at Intex in Manchester, the will to fight is there.

vi·ctimless, *a.* orig. and chiefly *U.S.* [f. Victim *sb.* + -less.] Applied to a crime in which there is no injured party. Hence occas. *transf.* of the offender.

1965 E. M. Schur *Crimes without Victims* 171 A comparison of the three situations analyzed in this book may provide some hints as to the factors determining the expansion of deviance in the victimless crime sphere. **1971** *Wall St. Jrnl.* 29 Mar. 1/1 We have to find ways to clear the courts of the endless stream of 'victimless crimes'... There are more important matters..than minor traffic offenses, loitering and drunkenness. **1976** *Australasian Express* 1 Oct. 4/3 Prostitutes are to be given a say in possible changes to laws covering victimless crimes in N.S.W. **1982** *Times* 1 Sept. 3/2 A large amount of victimless, but technical crime. **1983** *Sunday Tel.* 3 July 10/3 If we sent fewer trivial victimless offenders to prison, the numbers in custody could very easily be halved without detriment to public safety.

victimo·logy. [ad. F. *victimologie* (B. Mendelsohn 1956, in *Rev. Internat. de Criminol. et de Police Technique* Apr.–June 97/1), f. *victime* Victim *sb.*: see -logy.] The study of the victims of crime, esp. of the psychological effects on them of their experience. Hence **victimo·logist.**

1958 *New Statesman* 5 July 6/1 We ought to establish a new science of victimology with chairs at the universities, field workers studying the effects rather than the causes of crime, and a special department assessing the impact of sex-crime on women. **1964** *Economist* 1 Feb. 417/1 'Victimology', an unusual aspect of the sociology of murder. **1971** *Time* 5 July 46 Some victimologists contend that rape victims invite attack. **1978** *Practitioner* Feb. 301/1 Any one of these fields of study—stress, coping, captivity, victimology—is a springboard for analysing the particular plight of the victim of terrorism.

Victor Charlie (vi·ktǝɹ tʃā·ɹli). *U.S. Services' slang.* [f. the communications code-names for the initial letters of *Viet Cong*; cf. VC s.v. *V 5 b.] = *Charley, Charlie 8.

1966 *New Yorker* 18 June 135/2 He..was told about being shot out of a helicopter by Victor Charlie. **1968** *Saturday Night* (Toronto) Aug. 15 [Westmoreland's] men say they have to get them one 'Victor Charlie'. **1982** T. Barling *Terminate with Prejudice* ii. i. 48 Casualties.. was inordinately high. Victor Charlie had too much bushcraft and too much jungle to hide in.

Victoria². Add: **1. b.** A sovereign minted in the reign of Queen Victoria.

1870 E. G. E. Ward *Jrnl.* 9 Nov. in D. P. Carew *Many Years, Many Girls* (1967) i. 35 Let a packet of the bright, solid, sell-milled 'Victorias' reach you, and see if you do not deem them 'golden angels'! **1958** [see *pecorino].

7. a. *Victoria sandwich*, a sponge cake consisting of two layers of sponge sandwiched together with a jam filling; also *Victoria sponge* (*sandwich*). (The ingredients and style of presentation have not always been the same since the mid-19th century.)

1861 Mrs. Beeton *Bk. Househ. Managem.* 751 Victoria sandwiches... Spread one half of the cake with a layer of nice preserve, place over it the other half. **1902** *Little Folks* LV. 448/2 Auntie Kate told her cook to make some cakes,..some little cakes,..and some Victoria sandwich. **1976** *Burnham-on-Sea Gaz.* 20 Apr., Victoria sandwich. **1934** *Woman's Jrnl. Home Cookery* 150 (*heading*) Victoria sponge with grated pineapple. **1951** *Good Housek. Home Encycl.* 701/2 *Victoria sponge sandwich*, a sponge made with fat, which enables it to keep moist longer than the fatless type. **1980** D. Clark *Poacher's Bag* v. 121 He.. was..handing round wedges of Victoria sponge.

b. *Victoria green.*

1890 in Webster. **1934** H. Hiler *Notes Technique Painting* ii. 117 *Victoria green*, a potter's pigment, introduced by William Burton, but unfortunately not current, though it is absolutely permanent. *a* **1977** *Harrison Mayer Ltd. Catal.* 38/1 Body slip and glaze stains..colour..Victoria Green.

Victorian, *a.*² and *sb.* Add: **A.** *adj.* **1.** (Earlier examples.)

1839 *Athenæum* 2 Nov. 825/1 Perhaps the Annean authors, though inferior to the Elizabethans, are, on a general summation of merits, no less superior to the latter-Georgian and Victorian. **1850** E. P. Hood *Age & its Architects* ii. 71 The Victorian Commonwealth is the most wonderful picture on the face of the earth.

2. *fig.* Resembling or typified by the attitudes supposedly characteristic of the Victorian era; prudish, strict; old-fashioned, out-dated.

1934 in Webster. **1950** G. B. Shaw *Farfetched Fables* 72 He was helping the movement against Victorian prudery in a very practical way as a nudist. **1965** M. Spark *Mandelbaum Gate* vi. 157 In an emergency, one can't be Victorian about things, you know. **1977** P. G. Winslow *Witch Hill Murder* ii. xvi. 217 He was becoming rather heavily paternal to Linda. A Victorian parent. **1977** *Time Out* 17–23 June 5/2 Elsewhere in the files is an even worse example of what workers described as 'Victorian industrial relations'.

3. Special collocations: *Victorian Gothic* adj., designating the style of architecture typical of the Gothic Revival (see *Gothic *a.* 1 d); freq. *absol.* as *sb.*; *Victorian-Italianate* adj., designating a style of architecture revived in the nineteenth century in imitation of that of the Italian Renaissance.

1910 H. G. Wells *New Machiavelli* (1911) i. iii. 59 A new church in the Victorian Gothic. **1934** T. E. Tallmadge *Story Eng. Archit.* (1935) viii. 256 This structure [*sc.* the Albert Memorial] typifies to the last degree the Victorian Gothic style. **1961** *Times* 18 May 16/6 This small jewel of Victorian-Gothic architecture. **1980** 'L. Black' *Eve of Wedding* ii. 14 A huge red-brick mansion on three floors, heavy Victorian Gothic, a massive door in the centre of the front façade, leaded windows. **1963** A. Lubbock *Austral. Roundabout* 97 The public buildings are mostly in Victorian-Italianate style, pillared and porticoed; painted white, or in Edinburgh-rock colours. **1982** S. Radley *Talent for Destruction* vi. 40 The Victorian Italianate tower of the town hall.

B. *sb.* **2. b.** *U.S.* A house built during the reign of Queen Victoria.

1959 *House Beautiful* June 100 (*heading*) The virtues of a Victorian. **1978** J. Gores *Gone, no Forwarding* (1979) ix. 56 The house was an old Victorian, a Queen Anne which had been converted into rental units.

Hence **Victo·rianist**; **Victorianize** *v.* (later examples); **Victo·rianized** *ppl. a.*

1940 *Burlington Mag.* Apr. 127/2 The church had been so thoroughly 'Victorianised' that the discovery was all the more unexpected. **1946** J. W. Day *Harvest Adventure* ii. 27 The gatehouse of Butley..owes its renaissance from a Victorianized ruin to a lovely house, full of medieval grace, to Dr Montague Rendall. **1970** *Guardian* 1 Oct. 11/8 Gillian Avery is an eminent Victorianist. She has written..neo-Victorian children's books.., she has edited a series of Victorian revivals. **1974** *Times* 22 Apr. 14/3, I amused myself by guessing which fellow-passengers were members of the Victorian Society. The man opposite ..did not quite fit my vision of a Victorianist. **1976** I. Murdoch *Henry & Cato* i. 47 The tall Victorianized sash windows, which also served as doors, reached down to the ground. **1979** *Guardian* 3 Sept. 2/5 The building is unusually well preserved because it was never Victorianised or modernised. **1982** *UCT Studies in English* (Univ. Cape Town) Oct. 68 Writing as a Classicist and Victorianist, Jenkyns shows the enormous extent to which Hellenism influenced the generations of Victorians between about 1832 and the First World War.

Victorian, *a.*³ Add: **B.** as *sb.* A native or inhabitant of Victoria.

1862 *Temple Bar* Sept. 286 The Victorians went pluckily in for their second innings. **1901** A. W. Jose *Australasia* x. 152 They are men of Melbourne, Brisbane, or Adelaide rather than Victorians or Queenslanders. **1943** K. Tennant *Ride on Stranger* v. 47 All in the carriage were staunch Victorians, and his scathing references to the climate of Sydney were greeted with approval. **1973** *Sun-Herald* (Sydney) 26 Aug. 28/2 It's 41 years since Phar Lap died, but he lives on with a new generation of Victorians.

Victoriana (viktōᵊriā·nă). [f. Victorian *a.*²: see *-iana.] **1.** Matters relating to the Victorian period; attitudes characteristic of that time.

1918 E. Pound in *Future* Oct. 265/1 For most of us, the odour of defunct Victoriana is so unpleasant..that we are content to leave the past where we find it. **1931** *Times Lit. Suppl.* 23 Apr. 326/2 This book—Victoriana. A symposium of Victorian Wisdom. **1960** *Time* 8 Jan. 13/5, I think Shakespearian production is bogged down in a mire of Victoriana in this country. **1961** *Guardian* 21 Apr. 6/3 She provides many interesting domestic details for students of Victoriana. **1977** *S. Wales Echo* 18 Jan., Mr. Litterick added: 'He was talking straight Victoriana. That statement is straight out of Queen Victoria's age—it is arrogance.'

2. Objects, as furniture, ornaments, etc., made in the Victorian period; also, buildings or architecture of that era.

1940 *Illustr. London News* 11 May 640 (*caption*) The latest vogue in 'Victoriana'. Glass paper-weights to be sold at Sotheby's. **1947** N. Marsh *Final Curtain* xiii. 197 The terrifying Victoriana within. **1958** *Listener* 12 June 973/1 Victoriana is now the fashion in furniture and house decoration. **1958, 1959** [see *Betjeman]. **1968** *Canad. Antiques Collector* July 20/1 Whether the taste be for pine, Sheraton mahogany, or Victoriana, a lovely bouquet can be arranged to enhance antiques. **1978** *Lancashire Life* Mar. 71/1 There was one notable omission—namely, that of the dear old Princes Theatre. This was, prior to modernisation in 1932, a perfect specimen of Victoriana.

Victorianism. (In Dict. s.v. Victorian *a.*² and *sb.*) Add: Victorian attitudes or style; (an example of) that which is characteristic of the Victorian era. (Later examples.)

1913 Chesterton *Victorian Age in Lit.* iii. 196 The real revolts that broke up Victorianism at last. **1913** Mrs. H. Ward *Coryston Family* xi. 216 A heavy gold setting, whereof the Early Victorianism cried aloud. **1942** J. Lees-Milne *Jrnl.* 5 Feb. in *Ancestral Voices* (1975) 16 Others took the absurd view that this important house is once again pure Jacobean since the Victorianisms have been purged by the fire, which was the best thing that could have happened. **1974** M. Tippett *Moving into Aquarius* 37 The two ways forward out of Victorianism seem to be equally dispiriting. **1978** *Dædalus* Fall 72 He jettisoned much of what we think of as Victorianism, but not on the whole, the values of his family, school, and university.

Victo·rianly, *adv.* [f. Victorian *a.*² + -ly².] In a Victorian manner or style; also, prudishly, formally.

1917 *Duke of Norfolk Notebk.* 112 In conventional rooms, furnished Victorianly. **1933** C. Williams *Shadows of Ecstasy* i. 22 The clothes were late Victorian; the whole picture was Victorianly idyllic. **1940** E. F. Benson *Final Edition* xii. 249 No doubt this boom in promiscuity was partly due to reaction, for an undue reticence had been Victorianly observed about sexual instincts. **1967** A. Wilson *No Laughing Matter* iii. 250 This..conventionally, almost Victorianly clad huge body.

victorine, *sb.*¹ (Earlier example.)

1848 Geo. Eliot *Let.* 31 May (1954) I. 263 We do not find it too warm, however, for I have even felt the want of my Victorine.

victor ludorum (vi·ktər lūdōᵊ·rŏm). [L., = victor of the games.] The overall champion in a sports contest (usu. at a school or college); the championship itself. Also *transf.* and *fig.*

1901 J. H. Gray in W. B. Thomas *Athletics* v. 103 The sack race was no longer a consolation race, for Mr. Thornton the *victor ludorum* is returned as the winner. **1926** H. M. Abrahams *Athletics* xiii. 114 There is far too much of this *victor ludorum* and champion athlete business at all schools. **1950** [see *running *vbl. sb.* 2 e (*b*)]. **1963** A. Howard in Sissons & French *Age of Austerity* i. 30 Labour, as the electoral *victor ludorum*, was collecting its long-coveted trophies. **1966** *Listener* 28 July 144/1 Last week's *victor ludorum* was undoubtedly 'Rendezvous with Death'..the story of the sinking of the Lusitania. *Ibid.* 27 Oct. 617/1 Ian outshone..generations of Etonians.. being Victor Ludorum two years running, the only boy in living memory to win that honour. **1980** 'T. Hinde' *Sir Henry & Sons* xiii. 104 The school honour-board inscribed with the names of the winners of the Victor Ludorum.

victory, *sb.* Add: **5.** *victory ball, celebration, dance, night, parade*; **victory bond,** a bond issued by the Canadian and British governments during or immediately after the war of 1914–18; **victory garden,** a vegetable garden maintained to provide food in wartime (*spec.* in the war of 1939–45); **victory point** *Bridge*, a point scored in a championship representing a number of international match points in accordance with an agreed scale; **victory roll,** a rotational manœuvre about a longitudinal axis performed by an aircraft as a sign of triumph (cf. *roll *sb.*² 1 d); also *fig.*; **victory sign,** a signal made by holding up the hand with the palm outwards and the first two fingers spread apart to represent the letter V (for *victory*) to indicate triumph (cf. *V-sign 2 a); also = *V-sign 1 b.

1945 Auden *Coll. Poetry* 117 We were To go to a great banquet and a Victory Ball. **1952** M. Laski *Village* ii. 42 To bedeck the village hall on the night of the Victory Ball. **1917** *Canad. Year Bk. 1916–17* 693 On November 12, 1917, preparations were completed for the issue of a fourth Canadian War Loan in the form of five, ten and twenty year 'Victory Bonds' in denominations as low as $50. **1919** *Times* 14 June 19/1 The provision for Victory Bonds being accepted at their face value as cash for payment of death duties. **1977** Johns & Greenfield *Dymond's Capital Transfer Tax* xvii. 346 Victory bonds were accepted only for tax chargeable on death. **1945** J. Reith *Diary* 4 May (1975) vii. 347, I am dreading the victory celebrations and have no sort of heart for them. **1978** Cadogan & Craig *Women & Children First* vi. 131 When the war ends [Susan]..is..caught up in the spurious gaiety of the victory celebrations. **1921** A. Noyes *Sel. Verse* iii. 5 (*title*) A victory dance. **1976** *Billings (Montana) Sunday Gaz.* 27 June 1-D/1 The Sioux and the Arapaho have come to share in this victory dance with us. **1942** Wyndham Lewis *Let.* 10 Sept. (1963) 336 Why doesn't she lie low..and work in her victory-garden? **1978** H. Wouk *War & Remembrance* i. 13 She had started a victory garden and seemed the merrier for it. **1952** M. Laski *Village* ii. 41 The coloured electric lights that Mr. Waters had dug out for Victory night. **1977** *Time* 26 Sept. 17/2 The victory-night hoopla of Norway's Laborites was, however, a bit premature. **1931** F. L. Allen *Only Yesterday* i. 10 Every other city has its victory parade. **1982** Warner & Sandilands *Women beyond Wire* xi. 179 The war in Europe became progressively less of an Axis victory parade. **1962** *Listener* 8 Nov. 786/3 In the second half Britain went further ahead and won the match 6–0 in victory points. **1972** *Daily Tel.* 21 June 14/5 Britain scored 42 victory points out of 60 during Monday's play in the World Bridge Olympiad. **1942** *Tee Emm* (Air Ministry) II. 58 On returning to his aerodrome he began a victory roll, got into a spin and failed to recover. **1971** A. Price *Alamut Ambush* vi. 73 There was no point in doing a victory roll, however. It might even be premature if he failed to handle Havergal with compassion. **1981** T. Barling *Bikini Red North* xi. 241 The F15 Eagle.. passed overhead, turning in a victory roll. **1942** M. Dickens *One Pair of Feet* vii. 132 Her bell rang, not once, but as if she were giving the Victory sign in Morse. **1959** M. Steen *Woman in Back Seat* ii. viii. 310 One gave the 'victory' sign. 'We've beaten 'em.' **1978** L. Thomas *Ormerod's Landing* v. 100 They went quickly, only the idiotic boy turning around and giving the victory sign.

Victrola (viktrōᵘ·lă). Also **victrola.** [f. the name of the *Victor* Talking Machine Co. + *-ola.] The proprietary name of a kind of gramophone.

1905 E. Johnson *Let.* 9 June in *Amer. Speech* (1961) XXXVI. 116 The word Victrola is similar to nothing I have ever heard of and seems to me to have a sound suggestive of music, and would in all probability be the best word to use. **1906** *Official Gaz.* (U.S. Patent Office) 9 Jan. 644/2 Victor Talking Machine Company... Filed Dec. 1, 1905. Victrola. **1906** *Trade Marks Jrnl.* 9 May 628 Victrola 279,292. Philosophical instruments, scientific instruments, and apparatus for useful purposes. **1916** 'J. Webster' *Dear Enemy* 318 The children..had a lot of new records for the victrola. **1937** N. Coward *Present Indicative* vi. 259 A tall Renaissance chair with a red velvet cushion under which lived an electric victrola. **1952** J. Steinbeck *East of Eden* 374 He bought a Victor victrola..and he went regularly to see what new records had come in. **1979** P. Theroux *Old Patagonian Express* xvii. 259, I found an old phonograph. It was literally a Victrola, a 1904 Victor.

vicy-versy: see *vicey-versey, vicy-versy.

vide, aphetic form of Divide *v.* Delete † *Obs.*⁻¹ and add: **1.** (Later example.) Now only U.S. Blacks'.

1935 Z. N. Hurston *Mules & Men* 20 'Way after while when He ketch dat Jew, He's goin' to 'vide things up more ekal'.

2. as *int.* A set parliamentary cry for the division of the house into two groups voting on each side of a question for the purpose of counting. Cf. Divide *v.* 10.

1893 *Harper's Mag.* Dec. 39/2 This is usually done by shouting 'Divide! divide!' or, as the word is generally pronounced, ''Vide! 'vide!' **1908** H. W. Lucy *Mem. Eight Parliaments* vi. 242 Opposite and around him was a crowd of hilarious gentlemen shouting ''Vide! 'vide! 'vide!' **1951** J. Biggs-Davison *George Wyndham* xiv. 206 From the Conservative Back Benches came a shout of ''Vide! 'Vide!'

videnda, *sb. pl.* (Later examples.)

1790 J. Byng *Torrington Diaries* (1954) 312 Thus is one often humbug'd by printed accounts of visionary videnda! **1964** Auden in *Listener* 1 Oct. 525/1 Windows averted from plausible videnda but admitting a light one could mend a watch by.

video (vi·dio), *sb.* [f. L. *vidē-re* to see + -o, after *audio-.] **I.** *attrib.* **1.** Of or pertaining to (*a*) television, or (*b*) the visual element of television broadcasts or the signals representing it.

1935 [see *audio- II]. **1937** *RCA Rev.* July 17 Another important consideration has to do with the difficulty of feeding and switching video circuits. **1944** *Jrnl. Television Soc.* IV. 69/2 Trap circuits are commonly used..to keep the sound carrier out of the video channel. **1945** J. Barzun *Teacher in Amer.* xix. 280 The copywriter.. must be the same who was impelled to call television 'the video art'. **1949** *Proc. IRE* XXXVII. 289/2 Video Techniques. Notable progress was made in the comparatively new field of television recording on film. **1951** Koestler *Age of Longing* 194 An American video commentator. **1959** N. Mailer *Advts. for Myself* (1961) 157 A hippopotamus of a television-radio-and-phonograph cabinet with the blind monstrous snout of the video tube. **1967** *Boston Herald* 1 Apr. 1/1 News commentator Chet Huntley claimed Friday he had the support of his video partner, David Brinkley, and all but three of 40 NBC newscasters in his bid. **1979** *Daily Tel.* 17 Dec. 12/6 Such moments..were rare in Alick Rowe's *Two People* (ITV)... Video-drama remains uncertain about the love story.

2. Of or pertaining to video recording; **video art,** art in which a video recording is the medium; so **video artist.**

1955, etc. [see *video film*, etc., sense 8 below]. **1972** *Harper's Mag.* June 92 Few of these programs are going to win awards as video art. *Ibid.* 90 Russ Connor is now in charge of granting money through the New York State Council to independent video artists. **1973** *Art Internat.* Mar. 42/2 The video sculpture *De La*, which utilizes the machine constructed for shooting *La Région Centrale*, carries on a discourse on perception at the same time that it exists as a fascinating mechanical object. **1975** *N.Y. Times* 14 Apr. 33 Mr. Gillette..is one of a growing breed of video artists, for whom the TV screen has become an esthetic medium... They produce videotapes that take ingenious advantage of the technology. **1975** *New Yorker* 5 May 44/1 The video-art movement, which has been in high gear for more than three years now, can have come as a surprise to practically no one. **1976** *Broadcast* Dec. 20/1 In a video world he has still a film-stock contribution to make to the overall impression of the whole. **1977** C. McFadden *Serial* (1978) iv. 14/2 He said he'd had amazing results just acting out his anger with his patients. He was also big on video feedback.., role-playing..and Japanese hot tubs. **1978** *Chicago* June 14/4 Channel 11 producer Tom Weinberg [is] a video artist and part-owner of the White Sox. **1979** *Daily Tel.* 15 Dec. 15/5 Home video equipment is giving the Mafia a foothold in Britain, with illegal distribution of pornography and pirated films... 'Videocrime' is growing so fast that it will be the Crime of the '80s. **1980** *Times* 25 Jan. (Audio Visual Rep.) p. ii/5 Video programmes are being used by the National Bus Company for training drivers. **1982** *Listener* 23 & 30 Dec. 10/3 If you don't like what broadcast television offers, then you can supplement the diet from the video shop. **1984** *What Video?* Aug. 6/2 The VR-3995 has four video heads for noise-free still frame and slow motion.

II. *absol.* or as *sb.* **3.** That which is displayed or to be displayed on a television screen or other cathode-ray tube; the signal corresponding to this.

1937 *Printers' Ink Monthly* May 45/2 *Video*, the sight channel in television, as opposed to audio, the sound channel. **1940** *Broadcasting* 1 June 32 Video seen 230 miles at sea. Clear steady images picked up during test. **1946** [see *display *sb.* 1 c]. **1949** *Hollywood Q.* Winter 157 And pipe the finished output of these segments, both video and audio, instantaneously and simultaneously to the kinescope recorders. **1951** *Proc. IRE* XXXIX. 8/1 One cycle of video during active horizontal scanning represents one dark and one light picture element on a particular scanning line. **1960** J. L. Bernstein *Video Tape Recording* p. vii, Directors, editors, cameramen, and others..would benefit if they could learn the processes involved in recording video on tape. **1964** *Times* 7 Feb. p. iv/3 Except for its width..video tape looks exactly like sound recording tape. But it records not only sound but a continuous picture—video—as well! **1976** *Aviation Week* 10 May 131/1 An IBM scan converter transforms radar video into a format suitable for presentation on the TV monitors. **1977** *Gramophone* Aug. 361/2, I see it as the precursor of the all-purpose high quality cassette recorder that will record both video and audio. **1979** W. C. Brandenburg *Introd. Television Servicing* ii. 4/2 Both the audio and video can be broadcasted from the same antenna. **1982** G. White *Video Techniques* vi. 134 Sound is as important as the video and often more difficult to edit.

4. Television as a broadcasting medium. *U.S. colloq.*

1941 *Amer. Mercury* Nov. 581/2 *Vidio*,..television. **1946** *Time* 25 Feb. 72 NBC published a 55-page booklet, listing words & phrases commonly used in video. **1954** *Billboard* 13 Nov. 21 Most of the big name spinners have taken a fling in video during the last five years, but their survival-average has been low. **1979** *Boston Globe* 10 Apr. 32 Their play was flashed by video to an adjoining room where experts commented on it before a throng.

5. A video recorder; also, a VDU.

1958 *Observer* 26 Jan. 14/6 The Video is like a combined tape-recorder and cinema camera. It records your television appearance complete with sound track and can be played back at the touch of a switch. **1979** *Television & Home Video* Mar. 7/2 There's not a lot of point in owning a home video and using it to record the rubbish you might otherwise have missed. **1982** *Times* 7 May 17/5 Last year over 900,000 videos were rented or sold in Britain. **1983** *What's New in Computing* Jan. 5/1 The rest of the machine, the discs, the power supplies and the videos are all retained or upgraded and existing software can be run side by side with new software. **1984** S. Townsend *Growing Pains A. Mole* 190 We are the only family in our street who haven't got a video.

6. A video recording; videotape as a recording medium.

1968 *Observer* 14 Jan. 28/4 The days of the disc, in the pop world at least, are numbered. For soon will come the video. We will have the top 20 videos which you plug into your home video-machine. **1978** *Radio Times* 4–10 Mar. 4/2 We've got some video of a man he has already made contact with... We'll just have to cut in with that if necessary. **1981** *Church Times* 7 Aug. 5/3 They..went down to BBC television... Later he popped round to the school and showed them a video of themselves. **1983** *New Scientist* 3 Mar. 569/1 The BBC recognised early on that there was money to be made from selling archive programmes on video. **1984** *Melody Maker* 6 Oct. 3/1 Spandau Ballet have just returned from Hong Kong where they filmed the video for 'Highly Strung'.

7. The production or use of video recordings.

1970 *It* 9–24 Apr. 7 There are also groups of people exploiting video in any way they can think of. **1977** *N.Y. Rev. Bks.* 23 June 25/4 Made images move (cinema) and achieved their simultaneous recording and transmission (video). **1980** *Times* 31 Mar. 24/6 There are enough able practitioners around to demonstrate how effectively video, like any other artistic tool, can be used. **1980** C. MacCabe *Godard* 26 You envisaged a different kind of distribution: film and video as a handcraft industry. **1982** *Listener* 11 Feb. 34/3 The good news is that things in video could be worse. The bad news is that things in video will get worse.

III. 8. Special Combs. (see also *video-).

Most of the following are written as one word or two, with or without a hyphen, so that the distinction between them and words f. *video- is to some extent arbitrary.

video amplifier, an amplifier able to amplify the wide range of frequencies present in video signals and suited to delivering a signal to the picture tube of a television set; **video art, artist** (see sense 2 above); **video camera,** a camera for producing an electrical signal corresponding to a changing scene, suitable for feeding to a video recorder; **video display terminal** or **unit** = *visual display unit* s.v. *visual *a.* 6 e; abbrev. VDU s.v. *V 5 b; **video film,** orig., a cinematographic film of a television broadcast; now, a recording on a videocassette; **video frequency,** a frequency in the range employed for the video signal in television (viz. a few hertz to several million hertz), *esp.* one in the higher part of this range; freq. *attrib.*; **video game,** a game played by electronically manipulating images displayed on a television screen; **video map,** a map produced electronically on a radar screen to assist in navigation; so **video mapping**; **video nasty** *colloq.*, a horror video film; **video piracy,** the illegal production and sale of copies of commercial video films; so **video pirate**; **video-player,** a machine used in conjunction with a television for playing videocassettes; **video signal,** a signal that contains all the information required for producing the picture in television broadcasting; **video terminal** = *video display unit* above.

1937 *Electronics* Aug. 22/1 A video amplifier is one that is responsive to picture signals, and therefore, is an extremely good audio as well as a very wide band radio frequency amplifier. **1975** D. G. Fink *Electronics Engineers' Handbk.* xx. 55 Finally the video amplifier raises the signal to from 50 to 100 V to drive the picture tube. **1978** *Tucson Mag.* Dec. 28/1 Videocameras for home productions range from $300 to $1500. **1983** *New Scientist* 24 Mar. 793/1 It is neither easy nor ethical to perch with notebook or video camera over spontaneous scenes of human mating or aggression. **1976** *New Yorker* 1 Mar. 23/2 A V.D.T.; or Video Display Terminal, is a machine that combines a television screen with a typewriter keyboard and is connected to 'a gigantic network of computers'. **1983** *Boston Globe* 12 July 3/5 Video display terminals, or VDTs, have become commonplace in many offices. **1968** Video display unit [see VDU s.v. *V 5 b]. **1969** *Jane's Freight Containers 1968–69* 114/3 New video display units provide instantaneous information to answer inquiries from customers concerning the location of cars. **1983** *Engin. & Mining Jrnl.* May 99/1 Engineers have at their disposal newly developed hardware and software, with modern distributed control systems using computer interface and video display units. **1944** R. E. Lee *Television* ix. 182 Even with the financial savings of videofilm, television is the most expensive way of selling things that man has ever concocted. **1955** *Amer. Cinematographer* XXXVI. 365/1 Within a matter of minutes, this video film recording can be quick-processed and screened. **1978** *Washington Post* 29 Mar. D-2/5 The flight scheduled to carry the video film of Sunday's game 'did not take off' Monday night. **1983** *Daily Tel.* 10 June 2/3 Management brought in video films which were shown in updated form every few months to enable employees to know how the company was functioning. **1937** *Electronics* Aug. 23/2 There are several ways of designing an amplifier to give satisfactory response for video frequencies. **1956** Amos & Birkinshaw *Television Engin.* II. i. 28 The video-frequency signal generated by scanning such a scene has a uniform level which is interrupted by the sync signals. **1982** *Electronics* 6 Oct. 63 In development.. is a..generator for cathode-ray-tube applications offering an 80-MHz video frequency. **1973** *Business Week* 10 Nov. 212/1 The astonishing ability of the video game to lure quarters from the public and the electronic techniques used in its design are forcing major changes on the coin-operated game business. **1983** *Wall St. Jrnl.* 5 Jan. 1/2 Coleco's 1982 earnings 'appear to be well in excess of five times 1981 levels of $7.7 million,' the third-largest maker of video games said. **1961** *Aeroplane* CI. 74/2 All displays are capable of accepting video map information. **1982** *Radar-82* (IEE Conf. No. 216) 307/1 Producing a video map very much more accurate than the radar data is unnecessarily costly. **1954** *Sun* (Baltimore) 18 Mar. (B ed.) 20/4 Another of the new developments, 'video mapping', electronically reproduces an area map on the radar scope—showing terrain hazards, airport runways and radio flight paths. **1962** *Flight Internat.* LXXXI. 283/2 The range of synthetic displays to be added to the raw radar..will include video mapping. **1983** *Listener* 10 Feb. 4/1 The same children are sometimes known to see video 'nasties' or horror films screened later on. **1984** *Ibid.* 14 June 3/2 Unless one has seen a video nasty..it is difficult to imagine the depths of degradation to which certain producers are willing to sink. **1980** *Times* 2 Aug. 2/4 Audio and video piracy..usually involve the copying of a legitimate product such as videotaped film. **1983** *Daily Tel.* 20 June 17/2 Customs and Excise is claiming the right to see documents seized under a court order by a firm of London solicitors acting for the British Videogram Association in a video piracy case. **1982** *Ibid.* 17 Nov. 2/3 A video pirate was fined £500 yesterday by magistrates in Chepstow, Gwent, in a test case which could start a new clampdown on the illegal copies. **1983** *New Scientist* 3 Mar. 569/1 The deal is a blow to the video-pirates, who have cashed in by selling illegal copies. **1970** *Sci. Amer.* Dec. 37/1 (Advt.), The little cartridge simply drops into the videoplayer you will have attached to your set. **1972** *Listener* 8 June 775/3, I would guess that in about fifteen years most people would have a video-player of one sort or another in their homes. **1983** *N.Y. Times* 23 Apr. 48/4 The man..slipped a cassette into his video player... A comedian..popped into view on the television screen. **1937** J. C. Wilson *Television Engin.* ix. 346 Fig. 210..shows a triode valve having its grid biased approximately at zero volts, so that on receipt of positive video signals it runs into grid current and the video-signal voltage is dropped across the series grid resistance. **1937** [see *fade *v.*¹ 9 b]. **1957** Amos & Birkinshaw *Television Engin.* I. i. 17 The composite signal obtained by combining a picture with a synchronising signal is known as a video signal. **1976** A. Davis *Television* ii. 28 The main problem was the high speed at which the tape needed to run in order to accommodate the wider range of frequencies of video signals compared with audio signals. **1970** *Computer Management* Nov. 52/2 The key point in all video terminal applications, present or future, is the ease of handling and understanding information presented in a familiar form. **1983** *N.Y. Times* 28 Jan. 18/2 I.R.S. operators will still do the talking, but the computers will catalogue and display on video terminals all the taxpayers' facts, figures and prior promises.

video (vi·dio), *v.* [f. the sb.] **1.** *trans.* To make a video recording of.

1971 *Jrnl. Soc. Motion Picture Engineers & Television Engineers* LXXX. 414 *Video*, to dub by rerecording the video signal recovered from the tape being copied. **1984** *What Video?* Aug. 4/2 My work..is videoing anything to do with the emergency services (road accidents, fires etc.).

2. *U.S.* To televise. *rare.*

1973 E. Bullins *Theme is Blackness* 5 Colored people knew they were black, unique, separate and had a future. For paeans of Blackness were videoed throughout Black America.

video- (vi·dio), *video *sb.* used as a formative element (cf. *video 8), as in **videocasse·tte,** a cassette of videotape; freq. *attrib.*; **vi·deo-co:nference,** an arrangement in which television sets linked by telephone lines or the like are used to enable a group of people in different places to see and hear whichever of them is speaking at the time; hence **video-co·nferencing** *vbl. sb.*, telecommunication in the form of a videoconference; **vi·deo-disc,** a disc on which (moving or static) visual images have been recorded in non-representational form for subsequent reproduction on a television screen or the like; **vi·deophone,** a telephone incorporating a television screen on which the other person may be seen speaking; also as *v. trans.*; **vi·deoporn** *colloq.*, pornography on video film; **video-te·lephone** = *videophone* above.

1970 *Times* 26 Sept. 12/1 The video-cassette comprises a pre-recorded, pre-packaged programme of pictures and sound, designed to be 'played' on a television set with a converter. **1976** *Broadcast* 23 Aug. 8/1 The videocassette editor..is a small desk-top box..controlling two Sony video-cassette recorders. **1978** *Lancashire Life* Nov. 112/1 (Advt.), The Philips N1700 uses a 2½ hour video cassette no larger than the average paperback, and like an audio cassette the video cassette recordings can be played back whenever you like and as often as you like. **1982** M. Babson *Death beside Seaside* xv. 132 That's a video cassette recorder! Where did he get that? We couldn't afford that. **1977** *Communications: Technol. for Better Tomorrow* (Proc. 13th Ann. Internat. Conf. Communications) III. 75/1 An experimental videoconference link has been installed in France. **1983** *Times* 25 June 8/5 Whether a face-to-face meeting is in any sense 'better' than a videoconference is a different question. **1977** *Communications: Technol. for Better Tomorrow* (Proc. 13th Ann. Internat. Conf. Communications) III. 75/1 Characteristics concerning typical pictures in videoconferencing. **1983** *New Scientist* 7 Apr. 7/3 British Telecom has been running video conferencing studios for the past 12 years. **1967** *Ibid.* 2 Feb. 285/3 We shall soon have on the market the video-disc, about the size of a gramophone record and costing about 22 shillings. **1975** *Broadcast* 28 July 11/2 Consider the mind-boggling experience of a visit to your local bookstore to purchase a videodisc with perhaps 150 books on it—or maybe 54,000 slides of paintings. **1982** *New Scientist* 21 Oct. 150/1 Video discs are the video equivalent of the long-playing record. **1982** *Jrnl. R. Soc. Arts* Nov. 781/1 The floppy disc can hold the equivalent of 40 pages of the *Concise Oxford Dictionary*, but the video disc the whole of the *Encyclopaedia Britannica*. **1955** C. L. Moore in 'E. Crispin' *Best SF* 129 He..called Maltzer's apartment by videophone. *Ibid.* 131 Maltzer videophoned him. **1971** *New Scientist* 1 July 14/2 The West Germans have recently begun operating a 240-mile experimental videophone link. **1976** *TV Guide* (U.S.) 1 May, To reach other videophone outlets, the user merely dials the telephone. **1979** *Daily Tel.* 15 Dec. 15/5 The programme claimed that a 'videoporn' cassette of Linda Lovelace's 'Deep Throat'..was made in America with Mafia money. **1982** *Punch* 13 Jan. 83/2 Before videoporn and clockwork dildo moved inexorably in. **1964** M. McLuhan *Understanding Media* xxii. 233 The possibility of shopping by two-way TV, or video-telephone. **1978** *Jrnl. R. Soc. Arts* CXXVI. 270/2 The interview will remain the mainstay of any selection procedure: even if it is transferred to the video telephone.

videogram (vi·diogræm). [f. *video- + -gram.] **1.** An apparatus for making or reproducing video recordings. *rare.*

1963 *Times* 17 Aug. 4/2 A young Wolverhampton design consultant said today that he had invented a record player which reproduced vision as well as sound..Mr. Mason's videogram—as he has named it—could be on sale soon. **1972** *Guardian* 11 Apr. 15/2 They have ordered 60 cassette recorders—videograms is the coming word.

2. A prerecorded video recording; a commercial video film or disc.

1972 *Publishers Weekly* 10 Apr. 27/2 Mr. Geranton insisting on the distinction between a video copy of a previous work (say a film) and an original 'videogram'. **1976** [see *phonogram 2]. **1980** *Daily Tel.* 15 Oct. 18 The newly formed British Videogram Association..will put an end to people taping a film on one television channel while they watch football..on the other. **1982** R. Dean *Home Video* vi. 71 Videogram sales and rentals are now dominated by features, with 'adult' subjects.. running close behind. **1983** *Times* 29 Jan. 4/2 Owners and distributors of videograms (pre-recorded video cassettes and discs) are to be asked..to approve cinema-type certifications.

So **video·grapher,** one who makes videograms; **video·graphy,** the process or activity of making a video recording using a video camera; the use of such a camera.

1976 (*title of periodical*) Videography. **1980** *Daily Tel.* 18 June 17/2 Videography is what they are calling it and it will revolutionise home movies. Following hard on the sales of video recorders..comes the video camera. **1983** *N.Y. Times* 24 July ii. 30/6 One more advantage that the cinematographer has over the videographer is in the amount of lighting required. Video requires a lot of light to get a good result. **1984** *What Video?* Aug. 40/1 Colour settings are especially important with underwater videography. **1985** *Audio Visual* Feb. 54/3 Although obviously intended for a primary audience of stills photographers, David Kilpatrick's..volume could just as easily serve as a guide to in-house videographers.

videophile (vi·diofəil), *a.* and *sb.* orig. *U.S.* [f. *video- + -phile.] **A.** *adj.* Of or pertaining to a videophile or videophiles. **B.** *sb.* One who is very keen on watching television or video recordings.

1978 *Washington Post* 2 July M-4/2 In time there will be a videophile subculture: it is evolving now, as members of the baby-boom generation come to realize that television has played the cultural role in their lives that the movies and radio played in their parents' lives. **1980** *Business Week* 7 July 77/1 Pioneer Electronic Corp...is betting that it can turn its large base of audiophiles into videophiles. It..recently set up a new subsidiary..to produce music for videodiscs. **1984** *New Yorker* 20 Feb. 42/1 Kodak's new system had a maximum recording time of ninety minutes—not enough for a football game. The product, he said, would have 'some appeal to the upscale videophile'.

video recording, *vbl. sb.* Also with hyphen and as one word. [f. *video *sb.* + Recording *vbl. sb.*] **1. a.** The process of making a cinematographic film of what appears on a tele-

vision screen. **b.** The process of recording on videotape; videotaping.
1949 *Tele-Tech* May 31/3 The most obvious problem in designing a suitable photographic camera for video recording is that of reconciling the 30 frames per second rate of American television with the 24 frame per second rate which is standard in motion pictures. **1953** *Proc. IRE* XLI. 466/1 Work on video recording has not been restricted to film. Advances were made in the experimental recording of video information on tape. **1962** R. Bretz *Techniques Television Production* (ed. 2) xxiii. 495 A standard sound-record head contacts the tape a half-second after the video recording has been done. **1983** *Nature* 23 June 651/3 These..combine the recovery of [sea-]bed samples with the photography and video-recording of marine life.

2. a. A film of a television broadcast. **b.** A recording on videotape.
1950 *Tele-Tech* Nov. 32 (*heading*) Video recordings improved by the use of continuous moving film. **1968** A. Whitney *Every Man has his Price* vii. 59 She nearly died when we showed that videorecording to her! **1978** P. Marsh et al. *Rules of Disorder* iv. 112 Oxford fans were shown..a video-recording which was made..during their game with Plymouth Argyle. **1984** *Guardian* 5 Oct. 2/3 The court heard a video recording of an interview given by Mr. Scargill to Channel 4.

So **vi·deorecord** *v. trans.*, to make a video recording of; **vi·deorecorded** *ppl. a.*; **video recorder,** an apparatus for making video recordings; *spec.* a form of tape recorder for recording television programmes from a signal inside the set.
1951 *Jrnl. Soc. Motion Picture & Television Engineers* LVI. 222 (*caption*) Video recorder. **1954** *Tele-Tech* May 129 To be..acceptable to television stations, the magnetic video recorder must be capable of handling color signals with similar fidelity. **1960** J. L. Bernstein *Video Tape Recording* i. 1 The video recorder is able to provide immediate playback after a scene has been shot. **1961** *Britannica Bk. of Year* 537/1 Science also provided a term for a relatively new process with *video-record*, to record..in pictures rather than sound. **1972** *Daily Tel.* 22 Jan. 2/5 The disadvantages of having all schools broadcasts from ITV in the mornings (..many teachers may now videorecord them if they have the equipment). **1975** *Language for Life* (Dept. Educ. & Sci.) xxii. 322 Such activities would involve the play-back in school hours of video-recorded evening programmes. **1978** *Lancashire Life* Sept. 131/1 Now the video-recorder has arrived in the home. **1979** *Jrnl. R. Soc. Arts* Apr. 265/2 This teaching sequence is video-recorded by the use of a simple CCTV system. **1979** *Financial Rev.* (Sydney) 5 Oct. 23/1 A US District Court judge has ruled that the non-commercial use of home video recorders to record television broadcasts is lawful. **1983** L. Deighton *Berlin Game* vi. 65 You should have heard George when the au pair dropped his wretched video recorder.

videotape (vi·diotē^i^p), *sb.* Also as two words or hyphenated. [f. *video- + Tape *sb.*[1]] **1. a.** Magnetic tape on which can be recorded moving visual images such as television programmes (as well as sound).
1953 *Wall St. Jrnl.* 2 Dec. 1/4 With further development of video tape techniques, numerous possibilities will open up. Small portable television cameras are already in wide use in industry, in stores, banks and schools. **1965** *Listener* 10 June 857/2 A course of television mathematics lectures is now being recorded. Where these are transmitted from videotape the lecturer will be in the room with the students. **1973** C. Bonington *Next Horizon* viii. 117 We'll record the climb on video-tape the day before it actually goes out on the air. **1980** B. W. Aldiss *Life in West* i. 15 Standing by the bookstall was a white board announcing that the television series had been captured on videotape and would be shown in its entirety over the four evenings of the conference. **1985** *Listener* 24 Jan. 37/1 Videotape soon became a vehicle for prerecorded films and music programmes.

b. A length of videotape, or the recording it carries.
1960 *Guardian* 7 May 1 Video-tapes and film versions of the ceremony were being flown all over the world. **1967** *New Scientist* 2 Feb. 285/3 Americans have been at great pains to discover whether students learn as much from televised lectures and video-tapes as they do from 'live' lectures. **a 1974** R. Crossman *Diaries* (1976) II. 135 They wanted a video-tape to be made of every word said in the House of Commons and the House of Lords. **1977** *Language* LIII. 883 Videotapes of the lectures have been broadcast nationally. **1984** *What Video?* Aug. 14/1 (Advt.), You can easily view a negative or positive film image on a TV screen and also record it on a video tape.

2. *attrib.* **a.**
1958 *Times* 24 July 5/2 The BBC's VERA which tape-records complete television shows, picture and sound combined, and the AMPEX and R.C.A. videotape machines which do the same job for the independent television contractors, will greatly facilitate the provision of such Press shows. **1963** J. N. Harris *Weird World Wes Beattie* (1964) viii. 90 Sometimes she was in New York, holding a can of toilet-bowl cleanser in front of a video-tape camera. **1976** B. Jackson *Flameout* iii. 40 A man hefting a videotape camera on his shoulder brushed past, focusing his camera for another long shot. **1977** *Times of Swaziland* 11 Feb. 5 (Advt.), Kosie Smith v's Victor Galindez—light heavy weight title fight—in our Video-Tape Shows starting 8.30 every night. **1978** *Broadcast* 13 Nov. 23/1 The Circle is presently involved in forming..a video tape library. **1979** *Globe & Mail* (Toronto) 5 May 48/2 The [tennis] club has 24 courts,..videotape play-back, ball machines and individual stroke analysis.

b. videotape recorder, a tape recorder that will record and replay videotape recordings; **videotape recording,** a recording on videotape; also, the making of such a recording.
1954 *Proc. IRE* XLII. 731/2 Video tape recorders were demonstrated. *Ibid.* 756/2 Experimental video tape recording systems were demonstrated. **1959** *Daily Tel.* 9 Mar. 13/2 Television has entered an era of video tape recording that will profoundly influence future development. **1959** *Listener* 19 Mar. 499/2 Some executives think that video-tape recordings will help them to beat the film menace. **1963** *Guardian* 11 Sept. 7/6 Increasingly the television camera and videotape recorder, an unwieldy picture development of the ordinary tape-recorder, are being used where once the cine camera stood alone. **1971** *New Scientist* 16 Sept. 633/1 Videotape recording (VTR) offers them a freedom which is impossible with film. **1978** *Tucson Mag.* Dec. 28/1 Since videotape recorders hit the scene, television addicts don't have to choose between two conflicting programs. **1980** *Brit. Med. Jrnl.* 29 Mar. 889/1 Before and during stimulation the patients were examined clinically; videotape recordings made of their movements; [etc.]. **1984** *Listener* 19 July 34/2 There is..an enormous park of video-tape-recorders in Britain.

Hence **vi·deotape** *v. trans.*, to record on videotape; **vi·deotaped** *ppl. a.*, **vi·deotaping** *vbl. sb.*
1964 *Times* 7 Feb. (Advt. Suppl.) p. iv/3 Even an expert cannot tell the difference between live and videotaped programmes. **1970** R. Parkes *Death-Mask* xv. 186 An agonising interview had been video-taped at the television studios. **1977** J. D. Douglas in Douglas & Johnson *Existential Sociol.* i. 10 This form of linguistic ethnomethodology imposed the further constraint that only those accounts that could be tape-recorded or videotaped could be studied. **1977** *Times* 29 June 7/6 The FBI had set up an antique shop..fitted with wiretaps and hidden videotaping equipment. **1979** *Time* 30 July 12 After the lesson Carter ran through the speech and watched a videotaped replay, then practiced again, until he and Rafshoon were satisfied. **1980** B. W. Aldiss *Life in West* i. 28, I see..that some ingenious person..has videotaped my television series.

videotex (vi·dioteks). Chiefly *U.S.* [f. *video- + Tex(t *sb.*[1]] = next.
1978 *Globe & Mail* (Toronto) 16 Aug. b2/2 The Federal Department of Communications has developed a two-way television system called Videotex that it claims is superior to rival British and French systems. **1980** R. Woolfe *Videotex* i. 4 This book is about two-way, interactive videotex... One-way videotex..is referred to as teletext. Two-way videotex is called plain videotex. **1980** *Economist* 9 Aug. 71/1 America is proceeding equally slowly with a cousin of teletext, known as viewdata or videotex. **1981** *Sci. Amer.* Oct. e5/1 (Advt.), The Prestel videotex service has been operating for two years and is the largest information retrieval network in the world. **1983** *Amer. Banker* 8 June 8/1 The role banks will play in the emerging videotex field—a technology which permits consumers to access a variety of data bases and transaction services with a home terminal—has been much debated.

videotext (vi·diotekst). [f. *video- + Text *sb.*] Any information system in which a television is used to display alphanumeric information selected by the user; viewdata or teletext, esp. the former.
1980 *Times* 25 Jan. (Audio Visual Rep.) p. i/6 There is evidence of companies in Britain becoming increasingly aware of the importance of video communications through the viewdata services such as the Prestel system of the Post Office, and a wide variety of other forms of computer-generated videotexts. **1980** *New Scientist* 3 July 49/4 The British pioneered viewdata (now known as videotext according to international standards) with Prestel, but the Gallic Teletel is running hard to catch up. **1981** *Ibid.* 10 Dec. 721/1 The first big order for a public videotext system—in West Germany—has gone to IBM, not to British Telecom's Prestel. **1983** *Christian Sci. Monitor* 15 Apr. b-12/2 Cable television is just one of several emerging video technologies, which include videodisc, teletext, and videotext.

vidicon (vi·dikɒn). [f. *vid(eo- + *icon(oscope.] A kind of small television camera tube in which the image is formed on a transparent electrode coated with photoconductive material, the video signal being obtained from the variation in the current flowing to or from this as it is scanned by a beam of (usu. low-speed) electrons.
1950 P. K. Weimer et al. in *Electronics* May 72/1 The name 'vidicon' has been coined to distinguish these tubes from the photoemissive tubes. **1961** *Listener* 2 Nov. 726/3 A vidicon tube is also used in a small lightweight 'radio camera'. **1966** *McGraw-Hill Encycl. Sci. & Technol.* XIII. 465/2 Nearly all closed-circuit television cameras utilize a vidicon. **1978** *Nature* 5 Oct. 414/1 The UV vidicons, when properly erased and prepared for use, can be exposed for several hours on very weak planetary emissions. **1981** 'A. Hall' *Pekin Target* xvi. 158 These are hard copies of some stuff we took from high altitude with vidicon cameras.

vie, *sb.*[1] Restrict † *Obs.* to senses in Dict. and add pronunc. (vī). ‖ **3.** Used in a number of mod.Fr. phrases, as **vie de Bohème** (də bo,em), a Bohemian way of life; also *attrib.*; **vie de château** (də ʃato), the way of life of a large country house; aristocratic social life; **vie de luxe** (də lüks), a life of luxury; **vie d'intérieur** (dæ̃nteryör), private or domestic life; **vie en rose** (añ rōz) [app. from a French song by Edith Piaf containing the line '*je vois la vie en rose*'], a life seen through rose-coloured spectacles; **(la) vie intérieure** (æ̃nteryör), one's inner life, the life of the spirit; **(la) vie intime** (æ̃ntīm), the intimate personal life of a person; **Vie Parisienne** (parizyen), Parisian life, the name of a popular French magazine; used *attrib.* to denote a characteristic quality of voluptuous appeal; **vie romancée** (romañse) [see *romancé *ppl. a.*], a fictionalized biography.
1888 Mrs. H. Ward *Robert Elsmere* I. i. vii. 174 That golden *vie de Bohème* which she alone apparently of all artists was destined never to know. **1957** 'P. Quentin' *Suspicious Circumstances* i. 5 Two days later I was installed in a suitably Vie de Bohème apartment looking out on the Luxembourg gardens. **1980** S. T. Haymon *Death & Pregnant Virgin* ii. 12 Who could have prophesied that Paul Falkener, that personification of the *vie de Bohème*, would fall so completely under her spell? **1924** A. D. Sedgwick *Little French Girl* i. vii. 61 Is there a *vie de château* in the neighbourhood? **1979** A. Buchan *Scrap Screen* i. 4 The denigrated sentiment of nostalgia..feeds on reconstructions of the *vie de château*. **1920** 'K. Mansfield' *Let.* Nov. (1928) II. 81 You realize the vie de luxe they are living—the very table—sweets, liqueurs, lilies, pearls. **1929** O. Seaman *Interludes of Editor* 81 And, when in Town you take your meed, I'll mark the *vie de luxe* you lead. **1889** M. H. van de Velde *Cosmopolitan Recoll.* I. vii. 235 Under the roof that shelters them it appears to her [*sc.* the Queen of Italy] that there is no room for that happy *vie d'interieur* of which she is so honestly fond. **1933** G. Arthur *Septuagenarian's Scrap Bk.* 201 The *vie d'intérieur* makes far less appeal. Take a look at the English drawing-room..conspiring to give a delicious sense of the room being lived in, a sense which is wholly foreign to the French *salon*. **1957** O. Nash *You can't get there from Here* 45 He was wafted into a glamorous *vie en rose* of amorous ruses. **1974** M. Cecil *Heroines in Love* vi. 157 So many hopes had tumbled that magazine writers were reluctant to present an unending *vie en rose*. **1912** C. Mackenzie *Carnival* xxvii. 299 Suffragism viewed in retrospect was shoddy embroidery for the *vie intérieure* of Jenny. **1977** A. Fraser in A. Thwaite *My Oxford* 175 If caught, alone and inexplicably loitering, it was conventional to snatch up a book of poetry (Donne was rather smart) and indicate sudden world-weariness, a preference for *la vie intérieure*. **1984** *Listener* 5 Apr. 23/1 A work in which the *vie intérieure* of the American and European bourgeoisie is brought out into the daylight. **1891** S. Webb *Let.* 23 May (1978) I. 272 *La vie intime*—I want to talk to you about very frankly. **1939** *Times Lit. Suppl.* 7 Jan. 4/4 A novel with real characters, who, however, are not likely on that account to prosecute him for libel, despite his plain-spoken exposure of their *vie intime*. **1979** L. Lerner *Love & Marriage* p. x, Those social historians who believe that *la vie intime* has a history which can be studied. **1890** Kipling *Abaft the Funnel* (1909) 287, I replied that all my French was confined to the Vie Parisienne [*sc.* the French weekly *La Vie Parisienne*] and translations of Zola's novels with illustrations. **1936** C. Connolly *Rock Pool* iii. 57 She..gave him long, soft Vie Parisienne glances. **1983** L. MacDonald *Somme* viii. 78 The saucy *Vie Parisienne*, whose cut-out pictures enlivened the decor of almost every dugout on the Western Front. **1941** *New Yorker* 13 Dec. 34/1 'The Beloved Returns', last year's full-length *vie romancée* about Goethe, started as a short sketch. **1976** *Times Lit. Suppl.* 9 Jan. 39/5 In biographical matters we have *vie romancée*: 'if the question flashed through Schiller's mind, he must have dismissed it'.

Vienna. Add: **1. a. Vienna Circle** [tr. G. *Wiener Kreis*], the name given to a group of empiricist philosophers, scientists, and mathematicians active in Vienna from the 1920s to 1938 who were chiefly concerned with methods of verifying statements, the formalization of language, the unification of science, and the elimination of metaphysics (cf. *logical positivism* s.v. *logical *a.* 7); **Vienna coup** *Bridge*, the playing of the highest ranking card of a suit as a preparation for eventually forcing an opponent to discard winning cards; **Vienna cross** (earlier example); **Vienna sausage,** a small frankfurter made of pork, beef, or veal (cf. *wienerwurst* s.v. *wiener *a.* 2); **Vienna Secession** = *secession 3 d; **Vienna steak,** a flat rissole made of minced beef; **Vienna white** (earlier example).
1934 *Philos. Rev.* XLIII. 125 Last..one would mention the logical positivism of the Vienna Circle. **1956** J. O. Urmson *Philos. Anal.* vi. 100 The main attack on metaphysics as such came from philosophers whom we have not previously discussed, the Vienna Circle and their associates. **1973** B. Magee *Popper* i. 11 The fashion prevailing..for his generation there and then was the logical positivism of the Vienna Circle. **1933** *Sunday Times* 5 Feb. 5/1 The coup formerly only known as the 'Vienna Coup', but now, more appropriately, also termed the 'Squeeze'. **1959** Reese & Dormer *Bridge Player's Dict.* 248 In technical terms, the Vienna Coup consists of the play of the top-ranking card of a suit so that a card of lower rank will be correctly positioned as a one-card threat in an automatic squeeze. **1974** *Country Life* 5 Dec. 1811/3 This is the Vienna Coup, essential if a squeeze on East is planned. **1881** C. C. Harrison *Woman's Handiwork* i. 23 Vienna cross stitch..consists of two stitches crossing each other..exactly alike on both sides of the material. **1958** *Catal. County Stores, Taunton* June 4 Vienna Sausages—a tin 4/-. **1977** *New Yorker* 27 June 64/3 A modest but welcome food cache—cornmeal, canned vegetables, Vienna sausage. **1964** Lake & Maillard *Dict. Mod. Painting* (ed. 3)

182/2 Between 1898 and 1903, he [*sc.* Klimt] was president of the Vienna Secession. **1971** *Guardian* 19 Jan. 8/2 There stands in angular rectitude at the Vienna Secession Exhibition at the Royal Academy—a clock..by Adolf Loos. **1900** A. N. Whybrow *Day-by-Day Cookery Bk.* 97 Vienna steaks... Take 1½lbs. each of rather lean beef or veal, trim.., chop very finely, add..onion or shallot,..and 2 well beaten eggs... Form into steaks and fry in butter. **1951** G. Greene *End of Affair* ii. iii. 71 Henry chose a Vienna steak... He..expected something like a Wiener Schnitzel. **1854** Vienna white [see *Krems].

b. (Earlier example.)
1845 E. Acton *Mod. Cookery* xx. 545 This dish is sometimes called in England a Vienna cake; and..also..a *Gateaux de Bordeaux*.

c. Used *attrib.* and *absol.* to designate (items made of) a hard-paste porcelain, often richly decorated, manufactured at Vienna from 1719 to 1864.
1784 H. Walpole *Descr. Strawberry-Hill* 9 A tea-pot and bason, six handle cups and saucers, with bottles in black, of Vienna china. **1875** C. Schreiber *Jrnl.* 13 Feb. (1911) I. 352 A most interesting Vienna tea service. **1885** *Encycl. Brit.* XIX. 640/1 Vienna porcelain..is greyish in tint; its paintings are very poor, and it depends for its effect chiefly on gilt-moulded scroll-work. **1900** F. Litchfield *Pott. & Porc.* vii. 307 A..more tawdry description of Vienna china has been placed on the market... This would seem to have damaged the sale of the better class of modern Vienna. **1976** *Times* 30 Mar. 19/3 £3,850..for a Vienna teabowl and saucer decorated in schwarzlot and gold. **1980** *Radio Times* 27 Sept. 14 (Advt.), Pair of Vienna porcelain vases and covers... Dark blue and cream.

2. as *sb.* A Vienna sausage.
1963 M. Levinson *Taxi!* xi. 132 The modern Jewish taxi-driver..doesn't mind..a plate of ham with his viennas, the traditional Jewish 'hot dog'. **1971** *Sunday Times* (Colour Suppl.) 27 June 50/2 *Frankfurters and Viennas* are the world's best-known boiling sausages. Made from a mixture of pork and veal and well smoked. Reheat for 10 minutes.

Viennese, *sb.* and *a.* Add: **A.** *sb.* **b.** (Examples.)
1849 E. Ruskin *Let.* 22 Dec. in M. Lutyens *Effie in Venice* (1965) i. 94 Hanoverian German I understand pretty well, but Viennese, which these officers speak, is very different. **1981** Ld. Harewood *Tongs & Bones* vi. 108 Marion..talked fluent German (or rather Viennese),..rattling the guttural Viennese 'r' round her throat.

B. *adj.* **b.** Special collocations: **Viennese Secession** = *secession 3 d; **Viennese waltz,** (a piece of music for) the form of waltz originating in Vienna, characterized by a romantic, nostalgic quality (see quots. 1980, 1983).
1906 *Studio* (Special No.) p. a. iv, When, on the 3rd of April, 1897, nineteen young artists came together, to found the Viennese 'Secession', they chose as their leader Rudolf von Alt. **1960** *Burlington Mag.* CII. 395/1 The Viennese Secession movement (of which Klimt was co-founder and..first president). **1915** Gibson & Miles tr. *Gayda's Mod. Austria* xxxi. 293 These worthy citizens..give a light,..almost healthy tone to the urban life of the Empire—like the melodies of Strauss's Viennese waltzes. **1920** G. Grossmith in A. M. Cree *Handbk. Ball-Room Dancing* 12 Later came tight skirts and slow 'Bostons'. Then divided skirts, the 'Viennese Waltz', 'Two-Step',..'Turkey-Trot' [etc.]. **1979** *Listener* 19 July 71/1 'Musique-Dance', the latter..segueing from Viennese waltzes..to 'oom-pah' drinking-songs. **1980** *New Grove Dict. Mus.* XX. 206/1 The Viennese Waltz compositions of the second half of the 19th century, especially when played with the slight anticipation of the second beat of a bar and the subtle use of rubato which are characteristics of the traditional Viennese performance, remain a popular feature of concerts. **1983** *New Oxf. Compan. Mus.* II. 1966/2 The Viennese waltz developed the characteristic of a slight anticipation of the second beat of the bar,..which gives a delightful and distinctive lilt to the playing.

Vierendeel (vī·rəndē^ə^l). *Engin.* The name of A. *Vierendeel*, Belgian engineer and architect, used *attrib.* with reference to a type of girder introduced by him in 1896 in which there are no diagonal members, overall rigidity of the structure being maintained by rigidity at the joints of the vertical and horizontal members.
1930 *Engineering* 20 June 790/1 The types of steel coaches usually dealt with..could be reduced to two, in one of which the sides of the body were used as girders of the Vierendeel type. **1947** *Archit. Rev.* CI. 84/1 A special system was employed, using two contiguous spaces together with the columns and beams of the two upper floors, the whole forming a Vierendeel frame. **1970** *Internat. Jrnl. Solids & Structures* VI. 353 (*heading*) The minimum weight design of Vierendeel frames.

Vierkleur (fī^ə^·ɹklö̂ɹ). [Afrikaans, f. Du. *vier* four + *kleur* colour.] The flag of the nineteenth-century Transvaal Republic (1857–77 and 1881–1900), distinguished by three horizontal stripes of red, white, and blue, and one vertical green stripe on the left-hand side. Also *attrib.*
1900 W. S. Churchill in *Morning Post* 17 July 8/1 Someone..produced a Union Jack (made during imprisonment out of a Vierkleur). **1911** H. H. Fyfe *S. Afr. To-day* 106 Yet in Cape Town shop-windows are exhibited cards lamenting the 'Vierkleur', the four-coloured flag of the Transvaal. **1929** D. Reitz *Commando* iii. 20 On the left of the track stood a large marquee over which floated the vier-kleur flag of the Transvaal, indicating General Joubert's headquarters.

Viet (vyet), *sb.* and *a.* Abbrev. of *Vietnamese *a.* and *sb.*
1958 J. Buttinger *Smaller Dragon* i. 29 These Viets were probably moving into southeast China and toward the Indochinese peninsula at some time between 500 and 300 B.C. **1967** *Freedomways* VII. 128 Viet brothers come give us a hand We fight for freedom We fight for land. **1979** *Daily Tel.* 19 Jan. 4/6 (*heading*) Five Khmer areas fight Viet troops. **1979** *Tucson* (Arizona) *Citizen* 20 Sept. 1e/1 (*heading*) Viets say Nixon plots with China.

‖ **vi et armis** (vī et ā·ɹmis), *adv. phr.* [L., lit. 'with force and arms'.] Violently, forcibly, by compulsion; *spec.* in *Law*, causing direct damage to person or property; also *attrib.* Cf. Force *sb.*[1] 5 c.
1618 W. Fulbecke *Parallel of Civil & Common Laws* I. 80 a This Writ will not lye, for it is *vi et armis*. **1703** in *Eng. Rep. King's Bench* (1909) XCII. 137 So a man shall have an action against another for riding over his ground, though it do him no damage; for it is an invasion of his property, and the other has no right to come there. And in these cases the action is brought vi et armis. **1762** Smollett *Launcelot Greaves* I. iv. 101 He compelled, *vi & armis*, a rich farmer's son to marry the daughter of a cottager. **1819** Keats *Let.* 31 July (1958) II. 134 If I had not put pen to paper since I saw you this would be to me a vi et armis taking up before the Judge. **a 1846** B. R. Haydon *Autobiogr.* (1927) iii. xvii. 338, I looked astounded, but casting a glance round the table easily saw..I was to be set at that evening *vi et armis*. **1873** C. M. Yonge *Pillars of House* II. xix. 181 It was current in the nursery that he was a black man who expelled us *vi et armis*. **1941** W. A. Percy *Lanterns on Levee* xv. 179 The squad kept marching; it marched now like Beatrice Lillie, *vi et armis*, clear through an unyielding detachment planted in front of it. **1957** R. F. V. Heuston *Salmond's Law of Torts* (ed. 12) i. 5 To walk peacefully across another man's land is a forcible injury and a trespass, no less than to break into his house *vi et armis*.

Viet Cong (vyet kǫ·ŋ); *sb.* and *a.* Also **Vietcong.** [Vietnamese, lit. = Vietnamese Communist.] **A.** *sb.* (A member of) the Communist guerilla force(s) active in Vietnam between 1954 and 1976. **B.** *adj.* Of or pertaining to (the members of) this movement.
1957 *Ann. Reg. 1956* 321 Though small groups of dissidents remained in the jungles they ceased to be a danger, and the Communist dissidents known as the Viet Cong were equally ineffective. **1961** *Times* 27 Sept. 13/7 At the base of the Vietcong hierarchy is the village 'cell'. **1964** *Asia Mag.* 7 June 6 The war prevails, and it will, until the Viet Cong is wiped out. **1967** *Observer* 30 Apr. 12 The little cream schoolhouse is essential to the American dream of what we are doing in Vietnam, and it is essential for the soldiers to believe that in *Vietcong* hamlets no schooling is permitted. **1981** B. Granger *Schism* (1982) viii. 78 The village headman was a Viet Cong.

Viet Minh (vyet mi·n). Also **Vietminh.** [f. *Việt*-Nam *Dôc-Lâp Dông-Minh*, Vietnamese Independence League.] A nationalist independence movement (1941–50) in French Indo-China; *loosely*, the movement succeeding this; a member of one of these movements. Also *attrib.*
1945 *Times* 14 Sept. 3/6 There seem to be only two parties of any significance at the moment—the Viet-min and the Communists. **1945** *Times* 16 Nov. 3/3 It is believed some effort may be made to come to a compromise with the Viet Minh 'Government' in Hanoi. **1962** E. Snow *Other Side of River* (1963) lxxxv. 684 The free Vietnamese armed forces..were guerrilla groups led largely by the Vietnamese Independence League—Viet Minh, for short. **1968** R. West *Sk. from Vietnam* i. 21 The Vietminh, precursors of the Vietcong, employed assassination and terror. **1975** *New Yorker* 21 Apr. 133/1 The structure of the three basic committees is reminiscent of the old Vietminh organizations established in the South between 1946 and 1954.

Vietnamese (vyetnămī·z), *a.* and *sb.* Also **Viet Namese.** [f. *Vietnam* (see below) + -ese.] **A.** *adj.* Of or pertaining to the S.E. Asian country of Vietnam (formed in 1945 by the union of the former French colonial provinces of Annam, Tongking, and Cochin-China, and between 1954 and 1976 divided into North and South Vietnam), its inhabitants, or their language.
1947 *Facts on File* Apr. 129 Yesterday French troops claimed that 3 Japanese officers were killed near Hanoi while leading Viet Namese troops. **1951** M. B. Emeneau (*title*) Studies in Vietnamese (Annamese) grammar. **1951** *Vietnam: Old Nation—Young State* (U.N.), As a result the Vietnamese language developed considerably and was able to free itself entirely from Chinese influence. **1962** E. Snow *Other Side of River* (1963) lxxxv. 685 Gracey refused to see Vietnamese leaders who wished to negotiate terms of cooperation for disarming the Japanese. **1969** *Listener* 12 June 814/3 Whenever I see the members of the village council, I say to them: 'I'm not here to teach you what to do. I'm not Vietnamese. I don't look like you. I don't even speak your own language. You've got to make your own decisions.' **1977** *Early Music* July 365/2 (*caption*) Trân Van Khe playing the Vietnamese fiddle. **1983** P. Niesewand *Scimitar* xi. 247 It was an excellent meal, starting with..Vietnamese spring rolls..and finally squirrel fish, a whole baked fish.

B. *sb.* **a.** A native, *collect.* the natives, of Vietnam. **b.** The language of this people, considered by some philologists to belong to the Mon-Khmer family.
1947 H. R. Isaacs *New Cycle in Asia* viii. 157 Matters came to a head in Hanoi on December 19, 1946, when clashes in that city resulted in generalized warfare... The French charged that the Vietnamese were the instigators of the outbreak. **1950** *Times* 25 Apr. 7/2 The new State of Viet Nam has to rely so much on foreign support that many patriotic Vietnamese hold back or join with the Communists. **1953** W. W. Gage (*title*) Verb constructions in Vietnamese. **1965** *New Statesman* 30 Apr. 680/2 As a Vietnamese, I feel I have the right to express my resentment at the US air raids on Vietnam. **1972** M. Sheppard *Taman Indera* 3 The people whom we now know as Chinese, Thais, Khmers, Vietnamese and Malays. **1978** E. Tidyman *Table Stakes* ii. i. 141 He..shouted a string of orders in Vietnamese.

Vietnamize (vye·tnămaiz), *v.* [f. *Vietnam* (see prec.) + -ize.] *trans.* To give a Vietnamese character to; to make Vietnamese; to transfer to Vietnamese (esp. as opposed to American) influence or control. (This now *Hist.* only.)
1957 *Holiday* Aug. 119/1 Cholon's Chinese ebullience has definitely quieted down as a result of Diem's drive to 'Vietnamize' the Chinese. **1969** *Daily Tel.* 25 Oct. 5/3 It was learned..that America had handed over another base to South Vietnam. Two of the United States major port facilities, including the one in Saigon, were next on the list to be 'Vietnamised'. **1972** *Nature* 22 Sept. 185/2 Last week's statement..criticizes the policy of the United States in 'Vietnamizing' the war.

Hence **Vie:tnamiza·tion**; **Vie·tnamizing** *vbl. sb.*
1957 *Holiday* Aug. 119/3 When the new idea of 'Vietnamization' is being fostered in the country..it is not surprising the Vietnamese prefer to approach..foreigners with care and reticence. **1969** *Daily Tel.* 11 Oct. 18/1 The Vietnamisation of the war continued with the transfer of 80 United States Navy river patrol boats to the South Vietnamese Navy. *Ibid.* 18 Nov. 14/4 The price for 'Vietnamising' the war..is clearly a Vietnamising of the country's politics as well. **1972** F. Fitzgerald *Fire in Lake* xvi. 404 'Vietnamization', the ironic name for the slow withdrawal of American ground troops and the buildup of Vietnamese armed forces to fight an American-directed war in their stead. **1980** *Word 1979* XXX. 125 The Committee also identified such new problems as..the Vietnamization of foreign terms.

Vietnik (vye·tnik). orig. *U.S.* [f. *Viet(nam* + *-nik after *beatnik.] (A usu. pejorative term applied to) an active opponent of American military involvement in the war between North and South Vietnam. Also *attrib.* Cf. *peacenik.
1965 *Time* 22 Oct. 25a/1 A ragtag collection of unshaven and unscrubbed—they could be called Vietniks—turned out last weekend to promote the most popular new anti-cause. *Ibid.*, The Vietnik rallies—which also attracted some tweedy faculty members..—seemed to bear out a Senate Internal Security report issued last week. **1966** *Wall St. Jrnl.* 31 Mar. 18/2 If the Communists believe that draft-card-burners and other Vietniks represent majority American opinion, they are simply misinformed. **1969** *T.V. Times* (Brisbane) 4 June 6/1, I am opposed to the Vietniks who want us to get out... I saw too many twisted bodies of women and children, mutilated by the Vietcong. **1977** *Time* 18 Apr. 47/2 Sometimes the argument has sounded like a replay of old Vietnik protests.

‖ **vieux** (vyö), *a.* The Fr. word for 'old', used in various idiomatic phrases, as *vieux jeu* (ʒö) [lit. 'old game'], (something or someone) old-fashioned, hackneyed, outmoded, 'old hat'; *vieux marcheur* (marʃör) ['old campaigner', f. *Le Vieux Marcheur* (1909), a play by Henri Lavedon], an elderly womanizer; also *transf.*; *vieux port* (pǭr) ['old port'], the old harbour area of a modern French seaport; *vieux rose* (rōz) ['old rose'] = *old rose b.
[**1888** H. James *Reverberator* II. iii. 57 His father..didn't think it well painted... 'Poor dear papa, he only understands *le vieux jeu*!' **1891** M. S. van de Velde *French Fiction of Today* I. i. 7 Their contempt for what they call vieux jeu.] **1896** Mrs. H. Ward *Sir George Tressady* ii. xvi. 361 'Did you see the new piece at the Français?' He made a face. 'Not I! One couldn't be caught by such *vieux jeu* as that!' **1900** G. Murray *Let.* 30 July in G. B. Shaw *Coll. Lett.* (1972) II. 181 Cæsar..did not invent the sarcasm of calling the soldiers 'Citizens'. It was vieux jeu. **1918** R. Fry *Let.* 14 Oct. (1972) II. 435 We are *vieux jeu* and incorrigibly nineteenth century. **1955** *Times* 30 June 15/1 Are not dilatory bus drivers and conductors by now somewhat *vieux jeu*? **1972** M. Glenny tr. *Solzhenitsyn's August 1914* lxii. 600 To indulge in arguments with hostile, half-baked youth struck him as *vieux jeu* and boring. **1975** *Harper's & Queen* June 128/3 Henri's parents are very French *vieux jeu*... My mother-in-law talks about Louis XV as though she *knew* him. **1920** G. B. Shaw *Shaw on Theatre* (1958) 133 All the young men are cads and cowards, all the old men *vieux marcheurs*. **1937** *Times Lit. Suppl.* 16 Jan. 39/1 Of all the vieux marcheurs in nature, the cock pheasant has almost the worst reputation. **1940** G. Arthur *Concerning Winston Spencer Churchill* 108 The immortal hero of Khartum—so ran an odious whisper—was addicted to the brandy bottle; Mr. Gladstone, so it was vulgarly

suggested, pursued paths associated with a *vieux marcheur*. **1945** E. Waugh *Brideshead Revisited* i. ii. 44 Those leprous façades in the vieux port at Marseille. **1957** 'P. Quentin' *Suspicious Circumstances* xvii. 198 The lights on all the boats moved in the Vieux Port. **1981** 'D. Rutherford' *Porcupine Basin* i. 11 Be at the Café des Voyageurs in the *vieux port*... Take a table on the terrasse. **1890** *Girl's Own Paper* 11 Oct. 31/2, I should get your vintage dress-maker..to make you a *vieux rose* zephyr dress. **1926** *Daily Colonist* (Victoria, B.C.) 10 Oct. 23/2 The colors include white..old rose, gold..vieux rose, mid-brown, [etc.]. **1941** E. Bowen *Look at Roses* 55 Light blared on the *vieux rose* curtains.

view, *sb.* Add: **III. 16. c.** *on view* (earlier example).
1850 *Punch* 19 Oct. 164/1 The South Western Railway ..keeps a quantity of hissing, smoking, screaming engines always 'on view'.

18. b. *to take the long view*, to have regard for more than the present; to provide for the future.
1924 *Times Trade & Engin. Suppl.* 29 Nov. 247/2 Those who took the long view and ordered more than just to meet current needs are now reaping the benefit of such a policy.

c. *to take a poor view*: see *poor *a.* (*sb.*) 5 f. Also (in same sense) *to take a dim view*.
1941 *Newsweek* 7 July 27/2 Take a dim view—disapprove. **1947** H. Greve *Something in Country* ii. 13 Mr Everard took a dim view of his youngest niece. **1977** *Daily News* (Perth, Austral.) 19 Jan. 6/3 Bukovsky said he took a dim view of the way the West was pursuing detente.

IV. 19. a. *view day* (later example).
1850 *At. Jrnl.* July 224 The rooms were crowded during the 'view' days with visitors.

b. *view-painter*, *-painting*; **view card,** a picture postcard showing a view; **view-finder** (earlier example); **view-phone,** a (proposed) device for enabling telephone users to see each other during a call; **viewsite,** a site (for a house or other building) with a view (sense 8 b).
1901 E. Banner *On Several Occasions, Buying* a view-card and a book. **1973** *Express* (Trinidad & Tobago) 17 Mar. 14/2 Hobbies—art, stamps and collecting viewcards. **1889** View-finder [see Finder 3 d]. **1968** *Listener* 20 June 796/1 Flocks of academic view painters..used to set up their easels round the picturesque little harbour at St. Ives. **1971** *Country Life* 2 Sept. 536/3 Worcester's view-painters may fairly be compared with their Chelsea counterparts. **1861** ..Topographical view-painting was not confined to Worcester. **1982** *Nat. Gallery News* Feb. 1 Tourists visited Venice, wanting to take away reminders of the city they might not see again. The growing market must have encouraged both Canaletto and Guardi to turn to view painting. **1964** F. Pohl *Alternating Currents* (1966) 48, I turned off the viewphone, got up and walked out. **1966** *New Scientist* 24 Nov. 440/3 A 'viewphone' service which could enable telephone users, to see each other during a call. **1978** *Times* 2 Nov. 27/4 The Post Office itself has listed..the main telecommunications services..envisaged for the years 1985 and 2000... By 1985 there will be..radiopaging, confravision..viewphone. **1971** *Sunday Express* (Johannesburg) 28 Mar. 20/2 (Advt.), Each stand a spectacular viewsite! **1977** *Chicago Tribune* 2 Oct. (TV Week Suppl.) 4/1 (Advt.), This economical two-story home turns its back to the street to take beautiful advantage of a viewsite) to create an uncommon sense of privacy even in a busy urban setting.

view, *v.* Add: **1. d.** *absol.* To look over a property to assess its suitability for purchase or rent.
1914 'E. Bramah' *Max Carrados* 29 The place is to be let... We will go on to the agents and get a card to view. **1967** N. Marsh *Death at Dolphin* i. 9 We hold the keys. Were you wanting to view? **1971** C. Sims *Deadhand* ii. iii. 94 Is it the house? No good—you can't view with-out a card.

2. c. To watch (television); to watch on television. Also *absol.*
1935 *Discovery* Sept. 277/2 The comfort and interest with which the [television] pictures may be viewed in a semi-darkened room. **1936** *Times* 3 Nov. 9/4, I should be unwilling to lay heavy odds against a resident in Hindhead viewing the Coronation procession. **1958** *Something Done* 15 A number of people might now be viewing without having bothered about taking out a licence. **1956** *B.B.C. Handbk.* 1957 104 Asking them what broadcasts they listened to or viewed 'yesterday'. **1958** *Listener* 25 Dec. 1090/1 They view on average for thirteen hours a week. **1961** C. Mackenzie *Paper Lives* x. 145, I was anxious for Humphrey Mowart to explain to me the operation of the viewing-box..of a television set so that I shall be in control of the machine when we are what they call viewing.

viewing *ppl. a.* (examples corresponding to *view *v.* 2 c).
1958 *Spectator* 6 June 730/3 It was a great deal better than half a dozen others which the viewing public has taken to its heart. **1975** *Listener* 18 Sept. 368/1 NHK..will continue..to give Japan's viewing public all of what it wants.

view-able, *a.* [f. View *v.* + -able.] **1.** That may be viewed, inspected, or looked over.
1909 in Webster. **1924** E. Pound *Lett.* (1971) 189 No. The Studio is not viewable till I get back. **1959** *Historic Houses & Castles in Gt. Britain & Northern Ireland* 51 Adm. 2/-, Chd. 1/-. Gardens not viewable. **1976** *New Yorker* 1 Mar. 64/3 He said that three hundred people had arrived in Oakland—one, on Monday, of 'viewable' bodies and two, on Tuesday, of 'non-viewable' bodies.

2. *spec.* Of a television programme: capable of being viewed with pleasure or interest, worth watching.
1957 *Observer* 22 Dec. 8/6 B.B.C.'s Friday night documentary... This piece, though viewable, was often wildly unconvincing. **1971** *Daily Tel.* 21 Aug. 7 The secret of the endlessly viewable National Film Theatre series on television..seems to lie in quite simply in its subjects. **1979** *New* | 21–27 Sept. 32/3 The viewable..

viewability.
1958 *Observer* 5 Oct. 19/4 You could argue hard about the precise degree of depth and reality in Granada's *The Liberty Man*, but there was no doubt about its viewability. **1962** *Listener* 2 June 813/1 This fascinating teaser, which I would match against any recent television drama for sheer viewability. **1977** *New Yorker* 16 May 114/1 Nickname..carefully molded his holes for exceptional viewability.

viewdata (viū·dēᵗə). orig. also **Viewdata.** [f. View *v.* + *data*, pl. of Datum.] A system enabling a user's television set to show alphanumeric information selected from a computer database, with a telephone line providing two-way communication with the computer.
1975 S. Fedida in *Communications Networks* 275 An experimental study of a computer-based information distribution and retrieval interactive system, called viewdata, has been carried out. **1975** *Times* 24 Sept. 2/8 The telephone and television set could jointly to provide a significant new type of home information service... Known as Viewdata, the proposed new service was demonstrated publicly for the first time yesterday. **1976** *Nat. Electronics Rev.* Jan.–Feb. 11/1 Unlike these specialist orientated services, Viewdata has been developed as an interactive information service for use by the general public. **1979** *Guardian* 24 Sept. 20/6 The Post Office tried to register the name Viewdata for its product but this was refused on the ground that it was too all-embracing a title. Then it chose Prestel—and led a campaign to get viewdata accepted internationally as the generic term for computerised public information systems. **1979** *Austral. Financial Rev.* 21 Dec. 19/3 A meeting of a viewdata technical standards group in the Netherlands..has provided a draft standard which includes both Prestel and the French Teletel system. **1981** [see *teletext]. **1983** *Computerworld* 23 May 1D-14/1 Telematics..regards its viewdata system as being important because..the rise in such dissemination of information.. **1984** *N.Z. Farmer* 14 Feb. 16/1 Two Viewdata pilot trials are scheduled to start in the North Island this year.

viewer. Add: **2. b.** One who watches television. Cf. *televiewer. Also *attrib.*
1935 *Times* 16 May 9/3 When television sets are as numerous as broadcast receivers it is possible that there may be a demand for some form of record..with which the viewer would be independent of wireless broadcasts. **1936** *Times* 3 Nov. 9/1 At Alexandra Palace yesterday, when the new television service of the B.B.C. was officially opened, the Postmaster-General and others had to address themselves not only to listeners, but to 'viewers'. Within a radius of some 25 miles, 'viewers' saw and heard a ceremony of which the speakers rightly described as historic. **1960** ...There was a speech by Mr. R. C. Norman, chairman of the B.B.C. who was the first to use the word 'viewers' in its new meaning. **1946** *Electronic Engin.* XVIII. 276 The poor light at Wimbledon did not entirely account for television's failure to convey to viewers, comfortably seated at home, the breathless thrill [etc.]. **1955** *Britannica Bk. of Year* 189/1 Viewer Research..an investigation into the relative popularity of programmes shown. **1957** *Listener* 5 Dec. 936/1 In football and boxing the viewer has an immense advantage over the spectator on the spot. **1961** A. Wilson *Old Men at Zoo* iii. 125 The viewer responses so far..is jolly encouraging. **1966** G. N. Leech *Eng. in Advertising* v. 50 The (television advertising) copywriter..does not despise the triviality of real-life conversation. In fact by deviating from it he stands to lose credibility and viewer involvement. **1971** G. Charles *Destiny Waltz* i. 11 'You're probably familiar with some of our feature programmes if you watch a lot.' 'Yes, I'm a viewer all right,' said Jimmy. 'I've so far believe about TV.' **1977** *Rep. Comm. Future of Broadcasting* i. 3 The first decision we took was to ask individual viewers and listeners for their opinions. **1985** *Broadcast* 11 Jan. 5/4 The better the viewer tends to be, it results would reflect badly on broadcasters in the minds of viewers.

4. An optical device for looking at film transparencies or the like.
1935 [see microfilm viewer s.v. *microfilm sb. 2]. **1940** A. M. Sowerby *Wall's Dict. Photogr.* (ed. 15) 23/2 The principle involved is exactly that of the 'viewers' now sold for placing in front of small transparencies. **1958** *Woman's Own* 8 Oct. 58/3 The X-ray on the viewer. **1972** *Sci. Amer.* Jan. 65/1 When viewed through a stereoscopic viewer, the resulting image is truly three-dimensional. **1972** M. J. Bosse *Incident at Naha* ii. 67, I wondered what Virgil would learn from the microfilm. He had taken it to the university library where he could use the viewers.

Hence **viewer-ship,** the viewers of a television programme collectively; the number of such viewers.
1954 *Sun* (Baltimore) 30 Apr. 16/3 The extent of viewership of the subcommittee hearings was far from excessive. **1957** *Economist* 24 Aug. 637/2 More reliable estimates of 'viewership'. **1964** D. Susskind *Let.* 26 Mar. in G. Marx *Groucho Lett.* (1967) 223 Please keep watching 'Open End'... We need viewership very urgently. **1979** *Arizona Daily Star* 1 Apr. (Tucson T.V. Suppl.) 12/3 CBS..restored 'The White Shadow' to its former Monday night time period, where it once again is drawing a big viewership.

viewing, *vbl. sb.* Add: *spec.* (*a*) (*U.S.*) The action of taking a last look at the body of a dead person before the funeral; a time during which visitors may so view a body; (*b*) the activity of watching television; an instance or period of this.
(*a*) **1944** *Casket & Sunnyside* Dec. 31/3 At least during the trying hours of the viewings we had individualized the setting. **1955** Habenstein & Lamers *Hist. Amer. Funeral Directing* xii. 571 The viewing is controlled in some parts of the country by the funeral director to the extent that certain visiting hours are established. **1966** *Washington Post* 20 Feb. 33/7 In addition to the viewing there might be a wake. **1963** J. Mitford *Amer. Way of Death* vi. 95 Demonstrations of being an undertaker..the justifications for universal embalming and 'viewing' may be, they have nonetheless proved very effective and so far have been safe from authoritative contradiction. **1968** H. Waugh '30' *Manhattan East* (1969) 182 The funeral services..will be held at the Morelli Funeral Home..Viewing will be Monica Glazzard's mortal remains were to be displayed was on Lexington Avenue..and the viewing was ..Friday night from 8 p.m.
(*b*) **1956** *20th Cent.* Nov. 335 After a year's viewing the newcomer to television does view rather less than at first. **1960** *Guardian* 22 July 4/6 Much of the output..[is] unsuitable for 'family viewing'. **1974** *Times* 21 Feb. 16/8 The material is largely composed of old favourites, but that is an irony of television..a second, even third, view or viewing.
attrib. (Further examples.)
1937 *Television* 15 May 23/2 A few weeks ago a similar viewing post was opened in Berlin. **1940** *Chamber's Techn. Dict.* 893/1 Viewing room, a small projection theatre in a studio, for viewing rushes and completed films and other pictures. **1949** *Radio Times* 23 Aug 43/1 Automatically raised to viewing position by press button control. **1955** 'J. Christopher' *Year of Comet* i. 5 He knew her viewing habits. **1962** B.B.C. *Handbk.* 1963 104 It is of obvious interest and importance to the BBC to know how those of the television public who have a choice of programmes, divide their viewing time. **1969** *20th Cent.* Nov. 335 American viewing figures..are higher than ours. **1961** G. Millerson *Telev. Production* iii. 32 Clearly, if we use any camera lens-angle other than the audience's viewing angle (around 12° to 20°), distortion will occur. **1969** J. Elliot *Duel* x. 68 Cutting room and a tiny viewing theatre. **1972** 'H. Calvin' *Take Two Popes* iii. 23 Ramon could see himself herded into the viewing area. **1973** *Listener* 23 Aug. 261/3 There's a BBC view that if we have a spare viewing-room ready for researchers, we are bad housekeepers. **1975** *Broadcast* 14 July 6/2 BBC viewing hours have been slight since viewing for programmes has decreased. **1979** *Language for Life* (Dept. Educ. & Sci.) ix. 14 Nothing this Committee can say in isolation will change the viewing pattern of those evening hours which children spend in front of a set. **1975** M. Kenyon *Mr Big* xx. 188 At the terminal we watched him from the viewing platform but..was not even certain he was looking at the correct one.

view-port. [f. View *sb.* + Port *sb.*[3]] **1.** A window in a spacecraft or in the conning tower of an oil rig.
1957 T. Sturgeon *Thunder & Roses* 174 You just don't look through viewports too much. **1978** *Offshore Engineer* Apr. 61/1 The forward 'sphere'..has eight round viewports: two above, one forward, two each side and below and behind.

2. *Computers.* A defined part of a VDU display, such as may be allocated to a particular category of information.
1973 Newman & Sproull *Princ. Interactive Computer Graphics* vii. 126 It is possible to clip several different objects to different viewports and display them simultaneously on the screen. **1982** J. E. Scott *Introd. Interactive Computer Graphics* ix. 183 The graphics package provides a full range of two- and three-dimensional viewing transformations, including window selection, viewport assignment, and clipping.

viff (vif), *sb.* and *v.* Also **Viff** and in capitals. *colloq.* [f. initial letters of *vector(ing) in forward flight*.] **A.** *sb.* 'Viffing'; the ability to 'viff'.
1972 *Flight* 28 Dec. 946 The one distinct and inherent advantage possessed by the Hawker Siddeley aeroplane..is the ability to vector in thrust. The possibility of VIFF (vectoring in forward flight) have been under consideration for some time. **1981** F. K. Mason *Harrier* viii. 118 Initial investigations demonstrated that VIFF considerably enhanced the aircraft's potential in air combat.

B. *v. intr.* Of an aircraft: to change direction abruptly as a result of a change in the direction of thrust of its engines. So **vi·ffing** *vbl. sb.* and *ppl. a.*
1981 B. Gunston *Harrier* vi. 78 No matter how inferior the starting position, the viffing Harrier invariably won the engagement. **1982** *Times* 3 May 3/7 With the exception of the Russian Yakovlev Yak 36MP, it is the only high-speed aircraft in the world which can take off without the advantage of an airfield, fly backwards, and 'viff' (vector in forward flight)—that is 'stop' in mid-air and swerve acutely sideways. *Ibid.* 15 June 4/2 There had been few dogfights in the air and thus few opportunities for the Harriers to demonstrate their unique ability to 'viff'—or vector in flight. **1982** *Nav Limits Suppl.*... Harrier..to accelerate, decelerate or turn with superlative ease. **1982** *New Scientist* 27 May 557/3 US Marine Corps pilots developed a technique known as 'viffing'. **1983** *Times Lit. Suppl.* 14 Oct. 1109/2 The VSTOL Harrier with its swivelable jets and ability to 'viff'.

Vigenere, Vigenère (vīʒənęr). The name of Blaise de *Vigenère* (1523–96), French scholar and student of ciphers, used *absol.* and *attrib.*

with reference to a polyalphabetic cipher described by him (*Traité des Chiffres*, 1586).
1916 P. Hitt *Man. Solution Mil. Ciphers* vii. 51 The cipher disk method is practically the Vigenere cipher with reversed alphabets. **1943** [see *modularly *adv.*]. **1944** H. McCloy *Panic* 73 He tapped the cipher message... 'This is obviously a Vigenère, and all Vigenères can be broken.' **1957** *Encycl. Brit.* V. 929/2 The Confederate army used the Vigenère cipher—which the Federal army cryptanalysts are said to have solved every time a message in it was intercepted. **1961** Shulman & Weintraub *Gloss. Cryptogr.*, *Vigenere cipher*, a periodic substitution cipher originally employed with a 26 × 26 tableau and later converted into a slide with a normal clear alphabet over a normal cipher alphabet. **1963** H. S. M. Coxeter *W. W. R. Ball's Math. Recreations & Ess.* (rev. ed.) xiv. 400 If the different alphabets are related as in a Vigenère square, certain properties of symmetry can be used to advantage.

vigent (vəi·dʒĕnt), *a. rare.* [ad. L. *vigent-*, *vigens*, pres. pple. of *vigēre* to thrive.] Flourishing; vigorous; prosperous.
1590 W. Clever *Flower of Physicke* 4 Forthwith after these painefull defatigations, let naturall sweate and quiet sleepe, consolidat and refresh the body, to become more vigent, and the stomacke more sharpe. **1930** J. Wall *Durham Cathedral* iv. 183 Durham College..after several changes of fortune is now vigent as Trinity College.

viggerish, var. *vigorish.

vigil, *v.* (Later examples.)
1915 G. Frankau *Tid'apa* vii. 40 Two days and two nights has he vigiled—the doctor dozes and blinks. **1975** J. Montague *Slav Dance* 57 We vigil by the dying fire, talk stilled for once.

vigilance. Add: **1.** *committee of vigilance* (U.S.) = *vigilance committee*.
1837 H. Martineau *Society in Amer.* II. 139 He was brought to trial by the Committee of Vigilance; seven elders of the presbyterian church of Nashville being among his judges. **1857** J. W. Gihon *Geary & Kansas* 35 A committee of vigilance..was appointed, whose duty it was to observe and report all such persons [i.e. abolitionists].

vigilante. Add: **1.** For *U.S.* read orig. *U.S.* and add: (Earlier example.) Also *transf.* and *attrib.*
Now normally with pronunc. (vidʒilæ·nti).
1856 C. Nordhoff *Man-of-War Life* xv. 255 The second day after the expiration of their liberty, notice was given the *vigilantes*, ashore, that five dollars reward would be paid for every man of the crew rendered on board. **1890** N. P. Langford *Vigilante Days* i. xiv. 181 In the name of Vigilante justice [some men] committed crimes which..were wholly indefensible. **1918** G. Frankau *One of Them* xxiv. 185 Shall Muses condescend to gutter-trilling; Or Pegasus, whose tail the strap-hitched star tickles, Feed like a cow-hocked yellow Rosinante In Grecian mangers of a Vigilante? **1939** Joyce *Finnegans Wake* 78 What vigilantes and ridings then. **1948** *Daily Tel.* 22 May 3/6 The Advertising Association is asking 24 advertising clubs..to form a 'vigilantes' committee to report cases of advertising..of a doubtful..nature. **1959** *Times* 22 Oct. 8/2 An appeal by Nottingham 'vigilantes'—a group of businessmen—for an inquiry into the city's Labour-controlled administration has been rejected. **1980** *Sunday Times* 24 Aug. 2/3 But the critical officers say that if normal police cover is continually reduced..there is a danger..that could lead to highly undesirable vigilante activity. **1984** S. Townsend *Growing Pains A. Mole* 42, I offered to set up a vigilante group but my father said that anyone who has carried an old mattress 300 yards in the dark is not going to be put off dumping it by a gaggle of spotty schoolboys.

Hence **vigila·ntism** (orig. *U.S.*), the principles or activities of vigilantes or vigilance committees.
1937 *Sun* (Baltimore) 27 Sept. 2/7 A public investigation of 'vigilantism' in strike areas was announced today through the American League Against War and Fascism. **1942** W. Stegner *Mormon Country* 96 Perhaps even those incidents were purely unofficial and spontaneous acts of devout Mormons, the Mormon equivalent of lynch law and vigilantism. **1953** *Economist* 19 Sept. 775/3 In the United States, neither private vigilantism nor the government seems prepared to treat a favourable verdict as final. **1979** *Times* 6 Dec. 3/1 Africa was confronted, he said, with a choice between a system of collective security and a system of international vigilantism. **1985** *Listener* 10 Jan. 9/1 The one genuine, spontaneous popular institution in the West was vigilantism.

vigintennial (vəidʒinte·niăl), *a. rare.* [f. Viginti-, after Biennial *a.* and *sb.*, Triennial *a.* and *sb.*, etc.] Occurring once in twenty years.
1921 *Glasgow Herald* 29 Jan. 13/3 Their [*sc.* the planets'] vigintennial conjunction is due a few months hence.

vigintivirate. Delete † *Obs.* and add later examples.
1904 Tyrrell & Purser *Corr. Cicero* (ed. 3) I. 319/2 Cicero..represents Caesar as having been offended at his refusal to become a member of the vigintivirate. **1976** *Classical Q.* XXVI. 312 He was born in about A.D. 55, and..we might assign the *latus clavus* and a post in the vigintivirate to this reign.

vignette, *sb.* Add: **2. b.** A brief verbal description of a person, place, etc.; a short descriptive or evocative episode in a play, etc.
1880 E. Simcox *Diary* 28 Mar. in K. A. McKenzie *Edith Simcox & George Eliot* (1961) iii. 61, I have thought ..of writing a little book of 'Vignettes'. **1901** [see Thumbnail 2]. **1934** *Punch* 19 Dec. 698/1 Its writer gets and provides what entertaiment she can from them—witness her amusing vignette of the unfortunate *Habibullah*. **1957** *Practical Wireless* XXXIII. 558/1 The play was supposed to evoke the Edwardian era in a series of tiny vignettes interspersed with 'instrumental effects'. **1958** *Times* 12 Aug. 10/3 Miss Maria Lapinska, as his [dancing] partner came nearest to touching the heart in a wartime *vignette* entitled *1940*. **1980** *Jrnl. R. Soc. Arts* Mar. 226/1 Let me quote one vignette.

4. (Earlier example.)
1790 *Loiterer* 2 Jan. 5 Three..volumes in duodecimo; which, with..a handsome vignette frontispiece, will cut a respectable figure.

vignette, *v.* Add: **2.** *Optics.* To modify so as to give rise to vignetting of an image.
1945 *Jrnl. Optical Soc. Amer.* XXXV. 499/1 Otherwise, light rays coming from those points of the light source farthest from the optical axis will not spread out over the entire striation field but will be vignetted by the condenser aperture. **1961** *Jrnl. Sci. Instruments* XXXVIII. 93/1 At the edge of the field of ±3½°, the meridial section of the aperture is vignetted to about 80% of its axial value. Over the vignetted aperture, both the meridian plane and secondary plane sections of the emergent wave front lie [etc.]. **1973** *Optical Engin.* XII. 20/2 At 1 mm diameter pinhole in 1·5 mm lead severely vignettes the field off-axis.

vignetted *ppl. a.* (further example).
1961 [see *vignette *v.* 2].

vignetting, *vbl. sb.* Add: **1.** (Earlier example.)
1842 E. A. Poe in *Graham's Mag.* Apr. 201/1 The peculiarities of the design, of the *vignetting* and of the frame.

2. *Optics.* A dimming or disappearance of an image at its edge as a result of the blocking of some off-axis rays during their passage through the optical system.
1930 L. B. W. Jolley et al. *Theory & Design Illuminating Engin. Equipment* xxix. 449 This vignetting effect may result in a reduction in the illumination towards the edges of the image even in the most carefully designed lenses. **1959** Born & Wolf *Princ. Optics* iv. 187 Designers sometimes rely on vignetting to obliterate undesirable off-axis aberrations. **1974** *Sci. Amer.* Mar. 112/1 The diameter of the mirrors..must be at least as large as the ruled area of the gratings to prevent vignetting..and the scattering of stray light into the image. **1978** *SLR Camera* Aug. 64/1 The negative had good contrast and no discernible vignetting. **1984** *What Video?* Aug. 40/3 An lie [etc.]. **1973** *Optical Engin.* XII. 20/2 A 1 mm diameter pinhole in 1·5 mm lead severely vignettes the field off-axis.

vignoble. Delete † *Obs. rare* and add later examples.
1928 Joyce *Let.* 27 May (1966) III. 178, I found that the oldest vignoble in all Provence is the Clos S. Patrice. **1978** *Country Life* 16 Nov. 1618/1 Jurançon was almost wiped out by the phylloxera... The *vignoble*..was in danger of complete disappearance.

vigogne. 2. (Earlier example.)
1873 *Young Englishwoman* Feb. 75/1 Polonaise..of deep marine blue vigogne, a black silk tunic, trimmed with black Llama lace.

vigonia. 1. b. (Earlier examples.)
1803 T. Jefferson *Let.* 23 Apr. in R. G. Thwaites *Orig. Jrnls. Lewis & Clark Exped.* (1904) VII. 217, I saw a robe of what they called the Peruvian sheep, and I took to be of the Lama or Vigogna. **1818** M. Edgeworth *Let.* 29 Oct. (1971) 130 There is no black *Vigonia*—the Vigonia wool will not take black dye. Tell me which you prefer the Merino or the Queens cloth.

vigorish (vi·gəriʃ). *U.S. slang.* Also **viggerish,** etc. [prob. f. Yiddish, ad. Russ. *vyigrysh* gain, winnings.] The percentage deducted by the organizers of a game from the winnings of a gambler. Also, the rate of interest upon a usurious loan. Also *transf.* and *fig.*
1912 A. H. Lewis *Apaches of N. York* 51 Stuss licks up ..a round full fifth of all the East side earns, and to viggresh should be given the black glory thereof. **1943** *N.Y. Times Mag.* 31 Oct. 2/4, I have heard the word 'viggerish' used for the cut the house takes in dice games. **1959** R. Condon *Manchurian Candidate* xv. 201 Eugénie Rose Cheyney..loved Marco. That fact gave Marco a large edge, tantamount to wiping out the house percentage in banker's craps. No matter what the action, that is a lot of vigorish to have going for anybody. **1959** L. Katcher *Big Bankroll* iv. 51 Today,..with pari-mutuel betting..the various states, the race tracks, the local communities, all scoop their percentage off the top of each bet. The state practices a new, and highly profitable, type of 'vigorish'. **1964** 'E. McBain' *Ax* vi. 119 'Was he taking a house vigorish?' 'Nope.' 'What do you mean? He wasn't taking a cut?'.. Then why'd he risk having the game in his basement?' **1966** 'E. V. Cunningham' *Helen* viii. 98 Being a widower who lived alone in the Cattleman's Club, he also bore areputation for having no use for the vigorish. He was honest when he wanted to be honest, and he was afraid of no one. **1968** *National Observer* (U.S.) 12 Feb., Mr. and Mrs. K were paying interest rates—called 'vigorish' in the loan-shark business—of 25 per cent a week. **1978** *Film Rev. 1978–9* 13/1 The companies are not in any way stealing from the picture-makers. They have to have built-in vigorishes—or else they'd go broke. Who pays for the 21 million dollars lost on *The Sorcerer*? The Studio!

‖ **viguier** (vigye). [S. Fr. var. of *vicaire*: see Vicar.] **a.** *Hist.* A magistrate in pre-Revolutionary southern France. **b.** Each of two government officials in Andorra (see quot. 1983).
1744 M. W. Montagu *Let.* 12 June (1966) II. 331 We have a new Vice Legate... The Magistrate next to him in place is call'd the Viguier, who is chose every year by the Hotel de ville. **1898** H. L. Smith in H. Spender *Through High Pyrenees* iii. iii. 291 The President of the French Republic nominates a viguier with an unlimited tenure of office, and the Bishop of Urgel nominates another, who must be an Andorran, and who holds office for three years. **1922** *Glasgow Herald* 24 July 5 The chief of these is the supreme judicial authority, though the Viguiers—in theory at least—are also the heads of the militia. **1983** *Whitaker's Almanack 1984* 793/1 The sovereignty of Andorra is vested in two 'Co-Princes', the President of the French Republic and the Spanish Bishop of Urgel... They are represented by Permanent Delegates of whom one is the French Prefect of the Pyrenees Orientales Department at Perpignan and the other is the Spanish Vicar-General of the Diocese of Urgel. They are in turn represented in Andorra la Vella by two resident 'Viguiers' .., who have joint responsibility for law and order and overall administration policy, together with judicial powers as members of the Supreme Court.

vihara (vīhā·rā). Also † **vihar, vihare.** [Skr.] In Sri Lanka and India, a Buddhist temple or monastery.
1681 R. Knox *Ceylon* iii. iii. 74 Their Temples are styled Vehars... Many of the Vehars are endowed and have Farms belonging to them. **1875** [see *chaitya]. **1901** Kipling *Kim* i. 8 There were..fragments of statues and slabs crowded with figures that had encrusted the brick walls of the Buddhist *stupas* and *viharas* of the North Country. **1913** L. Woolf *Village in Jungle* i. 12 Years ago..he had gone on a pilgrimage to the vihare at Medamahanuwara. **1930** J. Still *Jungle Tide* ix. 220 In the *vihara*, every Buddhist monk is a member of an order, a celibate by profession. **1956** R. Pieris *Sinhalese Social Organization* i. iv. 24 The *vihara*..villages..were exempt from their jurisdiction. **1978** C. Humphreys *Both Sides Circle* v. 56 In 1928 three monks were sent from Ceylon by the Anagarika to found the first Western *vihara* or monastery.

‖ **vihuela** (vihwē·la). [Sp.] An early Spanish stringed musical instrument; *spec.* one of two early types of guitar (more fully *vihuela de mano*) or viol (more fully *vihuela de arco*).
1832 *Amer. Railroad Jrnl.* I. 15/1 The music..consisted of several vihuèlas (a small kind of guitar). **1870** C. Engel *Descr. Catal. Musical Instruments S. Kensington Museum* 46 The *guitar* is evidently an importation from the East..In Spain it had formerly also the name of *vihuela*..and in England, at the time of Henry VIII. we find it occasionally called 'the Spanish viol'. **1910** *Encycl. Brit.* VII. 514/1 Spanish..vihuela de arco. Guitarra Latina or vihuela de mano. **1961** T. Dart in A. Baines *Musical Instruments* 184 One hybrid..made a great contribution to the development of European music. Its origins lie in Spain, home of the flat-backed *vihuela de mano*, the five- or later six-stringed guitar. This instrument was held and played much like the plectrum guitar of the present day; the cross-fertilization consisted in applying to it a playing technique..associated with..the fiddles... The result was the *vihuela de arco* ('bowed' vihuela), which was to become known..as the *viol*. **1968** *New Oxf. Hist. Music* IV. xi. 560 Ortiz deals with the viol (*vihuela*), for which he describes three kinds of playing. **1975** *Gramophone* Dec. 1002/1 An attractive collection of Spanish Renaissance *vihuela* music..excellently played (on the lute) by Konrad Ragnossnig, would prove enjoyable.

Hence **vihueli·sta,** a player of the vihuela.
1925 J. B. Trend (*title*) Luis Milan and the vihuelistas. **1978** *Early Music* Oct. 623/1 It is the largest book of the vihuelistas and is..the most complete and representative collection drawn from the rich and varied musical repertoire of mid-16th-century Spain.

‖ **vila** (vī·lā). Pl. **vilas, vile.** [Serbo-Croat and Slovenian.] In Slavonic mythology: a fairy, a nymph, a spirit. Cf. *wili, willi.
1827 J. Bowring *Servian Pop. Poetry* p. xxxvii, An omnipresent spirit—airy and fanciful—making its dwelling in solitudes..a being called the *Vila*. **1887** *Folk-Lore Jrnl.* V. 347 Whosoever has compared the northern elves with the Slavonic vilas..will see that they are of modern origin. **1911** *Encycl. Brit.* XXIV. 689/2 The [Serbian] peasants believe in charms and omens, in..ghosts, the evil eye and *vile* or white-robed spirits of the earth, air, stream and mountain, with hoofs like a goat and henna-dyed nails and hair. **1922** D. H. Low *Kraljević's Ballads* iv. 21 In Serbian song Vilas are represented as jealous and capricious beings but on the whole not unfriendly to mankind. **1974** *Encycl. Brit. Macropædia* XVI. 876/1 Particularly feared are maidens who died before marriage and are believed to be addicted to the kidnapping of bridegrooms and babies... They are called ..*vile*..in Serbo-Croatia and Bulgaria.

vile, obs. var. *ville[3].

Vilene (vəi·lin). Also **vilene.** A proprietary name for backing or an interlining for clothing material, etc.
1954 *Trade Marks Jrnl.* 9 Feb. 159/2 *Vilene*... Lining and stiffening materials for clothing, all being textile piece goods;..The Viledon Company Limited, 15, New Street, Bishopsgate, London E.C.1.; merchants. **1960** *News Chron.* 10 Oct. 8/4 This [dress] is made in fine black wool, with a vilene-backed skirt. **1961** *Guardian* 24 Mar. 12/4 Putting a very fine netting of 'Vilene' between two

VILLAED

layers of paper, and the resulting material handles like cloth. **1976** *Woman's Weekly* 6 Nov. 44/2 Cut hymn book from Vilene interlining and sew to right hand.

vi·llaed, *a.* [f. Villa + -ed[2].] Covered with villas.

1791 A. Seward *Let.* 30 July (1811) II. xxix. 95 A pretty little lawn..admits the near hill, so magnificently villaed. **1937** G. Greene *Nineteen Stories* (1947) 51, I had forgotten too the turning to the left up a steep, villaed hill.

villaette. Add: Also **villarette.** (Earlier example.)

1792 F. Burney *Jrnl.* 2 June (1972) I. 184 Mrs. & Miss ord & myself set off for Sudbury, near Harrow, where her very elegant Relation, Mr. Orde, has a villarette. The House is half old, half new, but well fitted up.

villafy, *v.* (Earlier example, in sense (b).)

1865 C. M. Yonge *Clever Woman of Family* I. vii. 156 My sister lives..at Little Worthy, the next parish... It has a railroad in it, and the cockneys have come down on it and 'villafied' it.

Villafranchian (vilăfræ·ŋkiăn), *a. Geol.* [ad. F. *villafranchien* (L. Pareto 1865, in *Bull. Soc. Géol. de France* XXII. 262), f. the name of *Villafranca* d'Asti in N. Italy, in the vicinity of which exposures of this series occur: see -ian.] Of, pertaining to, or designating a stratigraphical stage in Europe variously assigned to the Upper Pliocene and the Lower Pleistocene (see quots.). Freq. *absol.*

1893 A. Geikie *Text-bk. Geol.* (ed. 3) 1016 (*table*) Villafranchian. **1955** G. G. Woodford tr. *M. Gignoux's Stratigraphic Geol.* x. 558 From the viewpoint of evolution of mammalian faunas, the Burdigalian..and Villafranchian mark important changes. **1972** *Gloss. Geol.* (Amer. Geol. Inst.) 778/2 *Villafranchian*, European stage: lower Pleistocene... It is the terrestrial equivalent (in France and Italy) of the marine Calabrian. Before 1948, it was used for the latest division of the Pliocene. **1973** *Nature* 15 June 391/2 Recent data have shown that the continental type Villafranchian is essentially late Pliocene (pre-Calabrian). **1974** *Encycl. Brit. Micropædia* X. 436/3 It is likely that the Villafranchian includes within it the Plio-Pleistocene boundary. *Ibid. Macropædia* XIV. 568/2 The Villafranchian fauna was..greatly modified by extinctions and new additions.

village, *sb.* Add: **1. e.** A small self-contained district or community within a city or town; *spec.* † (*a*) see sense 1 c in Dict.; (*b*) (with capital initial) = *Greenwich Village.

1865 [see sense 1 c]. [**1924**: implied in *Greenwich Villager* s.v. *Greenwich Village.] **1929** E. Wilson *I thought of Daisy* i. 16 Sue Borglum's pleasantry had been in the vein of the Village; Daisy's was in the taste of Broadway. **1931**, etc. [see *Greenwich Village]. **1949** M. Allingham *More Work for Undertaker* xii. 156 London is made up of many villages. **1952** *N.Y. Times* 17 Aug. viii–ix. 1w/5 (*caption*) Sketch of section of..cooperative multi-family for Holliswood, Queens, to be known as Hilltop Village. **1971** A. Thorburn *Planning Villages* iv. 24 The word 'village' has a pleasant and attractive connotation for most of us, sufficiently so for it to be borrowed by many estate agents regardless of the context, and to be applied to self-contained neighbourhoods in towns (e.g. at Banbury and Washington, Co. Durham). **1975** *Harper's & Queen* June 35/1 Hampstead—the loveliest of London's historical 'villages'. **1977** *Guardian Weekly* 25 Sept. 8/3 They call it a village but Skokie is geographically a middle-income suburb of Chicago with a population of 70,000. **1979** M. McMullen *But Nellie was so Nice* i. iv. 23 She had grown up in the Village, on West Ninth Street between Fifth and Sixth.

4. a. (Further examples.)

1636 Massinger *Duke of Florence* ii. iii. sig. E1, 'Tis a plaine Village Girle Sir, but obedient. **1803** G. Colman *John Bull* iv. ii. 46 One of the prettiest little village-churches you ever saw in your life. **1817** Scott *Rob Roy* I. v. 111 The domestic chaplain, the village doctor.. and my uncle. **1818** T. G. Fessenden *Ladies Monitor* 124 Learning should never pose a woman's head,.. Whose wealth and beauty sanction higher aims, Than those of village-school instructing aims. **1824** M. R. Mitford *Our Village* I. 6 The village shop, like other village shops, multifarious as a bazaar; a repository for bread, shoes, tea, cheese..for every thing, in short. **1841** Dickens *Barnaby Rudge* xxv. in *Master Humphrey's Clock* III. 74 They hurried through the village street. **1842** Tennyson *Poems* II. 201 He is but a landscape-painter, And a village maiden she. **1843** *Cumberland Pacquet* 1 Aug. 3/1 His [*sc.* a raven's] masterpiece was his correct repetition of the Lord's prayer; which..would have done no discredit to many a village schoolmaster. **1847** C. Brontë *Jane Eyre* III. v. 121 To be a village-schoolmistress, free and honest. **1852** Dickens *Bleak Ho.* (1853) vii. 61 A dark-eyed, dark-haired, shy, village beauty comes in. **1852** Thackeray *Esmond* I. ii. 58 The village people began to be reconciled presently to their lady. **1853** C. M. Yonge *Heir of Redclyffe* I. vi. 94 A village boy, whom he caught misusing a poor dog. **1854** —— *Heartsease* I. i. 2 A party of village children..gathering cowslips. **1855** Mrs. Gaskell *North & South* I. ii. 27 Mr. Hale..was anxious for the village postman, whose summons to the household was a rap on the back-kitchen window-shutter. *Ibid.* II. xxi. 281 The house fronted the village green. **1859** Geo. Eliot *Adam Bede* I. ii. 28 Mr Rann's leathern apron and subdued griminess can leave no one in any doubt that he is the village shoemaker. *Ibid.* II. xvii. 5 That village wedding..where an awkward bridegroom opens the dance with a high-shouldered, broad-faced bride. *Ibid.* xviii. 36 An experienced eye would have fixed on him at once as the village blacksmith. **1860** C. M. Yonge *Hopes & Fears* I. viii. 316 The apartment was not much behind that at the village inn at Hiltonbury. *Ibid.* II. xviii. 347 It was..interesting to observe his impression of the English village-life at Hiltonbury. **1861** —— *Young Step-Mother* v. 58 The pleasures to which he had been introducing Gilbert, were not merely..the rabbit-shooting and rat-hunting of the farm, nor even the village cricket-match. *c* **1873** C. Rhodes *Let.* in J. Flint *Cecil Rhodes* (1974) ii. 24 Whether I become the village parson.. remains to be proved. **1873** W. D. Howells *Chance Acquaintance* 38 Under the porch of the village store some desolate idlers..had clubbed their miserable leisure. **1890** W. Booth *In Darkest England* ii. iii. 138 Every effort will be made to establish village industries, and I..hope..we may be able to restore some of the domestic occupations which steam has compelled us to confine to the great factories. **1891** J. L. Kipling *Beast & Man in India* viii. 194 The village Elders stand before him with joined hands to learn his Lordship's commands. *Ibid.* xii. 316 The Eastern cat..is used in a frequently-quoted saying about doubtful matters. 'If the Punchayat (village council) says it's a cat, why, cat it is.' **1892** C. M. Yonge *Old Woman's Outlook* 167 The village shopkeeper, the maker of the 'vinosity' bread. **1894** Kipling *Second Jungle Bk.* (1895) 34 As soon as the villagers saw the smoke in the deserted shrine, the village priest climbed up..to welcome the stranger. **1895** C. M. Yonge *Long Vacation* i. 5 An expedition to play the zither and sing at a village fête. **1907** G. B. Shaw *Major Barbara* 148, I myself have had a village idiot exhibited to me as something irresistibly funny. **1912** R. Marsh *Judith Lee* i. 10 Dickson was at my bedside..and Pierce, the village policeman. **1913** Chesterton *Victorian Age in Lit.* ii. 143 Hardy became a sort of village atheist brooding and blaspheming over the village idiot. **1915** J. Buchan *Salute to Adventurers* i. 9 She was presently driven out of the place by..the baillie, and the village dogs. **1920** 'O. Douglas' *Penny Plain* xxv. 296 The village women, with little girls in clean pinafores clinging to their skirts. **1923** M. Kennedy *Ladies of Lyndon* ii. 70 I'm afraid..that Modern Art wouldn't be quite suitable... It's only simple village folk, Mr. Ervine. *Ibid.* 96 The Village Room is quite two miles off. **1924** H. de Sélincourt *Cricket Match* i. 10 Down by the Village Room, where pictures are shown on Friday evenings..and into the village square again. *Ibid.* ii. 27 Best cricket going, village cricket. **1926** L. Elmhirst in M. Yonge *Elmhirsts of Dartington* (1982) vi. 138 Like the village community of earlier times,..the school community..must engage in many practical enterprises. **1929** C. Day Lewis *Transitional Poem* ii. 24 It is high time to renounce This village idiocy. **1930** K. Boyle *Plagued by Nightingale* (1931) xi. 87 Tomorrow was the village *fête*. **1930** A. P. Herbert *Water Gipsies* xxiii. 350 She went many times..to the Chiswick Church in the little old village street beside the river. **1932** L. Golding *Magnolia Street* i. vi. 107 He would put up at the village pub until the moment to pounce was due. **1933** A. Thirkell *High Rising* i. 25 She had formed a habit of ordering groceries..on a gigantic scale, from the village store. **1939** L. Bemelmans *Life Class* iii. v. 246 From the railroad station came crates and boxes with materials from New York and Vienna; the village children's hair was full of excelsior as they helped unpack them. **1939** F. Thompson *Lark Rise* x. 186 'Everybody who was anything'..kept a maid..stud grooms' wives, village schoolmasters' wives. *Ibid.* xii. 225 The position of a village schoolmistress was a trying one socially. **1942** M. Cable *Gobi Desert* 122 To buy the printed likeness of the kitchen god at a village fair. **1943** A. Christie *Moving Finger* vii. 82 Emily Barton..has a mental picture of men as..smoking cigars, and in the intervals dropping out to do a few seductions of village maidens. **1948** F. Thompson *Still glides Stream* ii. 18 At that time village houses had no numbers or names. **1949** D. Smith *I capture Castle* i. 5 He does nothing but read detective novels from the village library. **1950** *New Yorker* 23 Sept. 66/3 A lively, exalted young novice.. was formerly a village schoolteacher. **1951** R. Firth *Elements of Social Organization* iv. 134 Such behaviour is a function of the social structure, with its emphasis on the village community and the kinship group. **1952** M. Laski *Village* i. 12 The village boys and girls still danced sedately. *Ibid.* iii. 52 The village people..normally had their radios on all day. **1953** G. E. & K. R. Fussell *Eng. Countrywoman* iv. 123 Farmhouse and cottage furniture was made by the village carpenter. **1953** S. Bedford *Sudden View* i. x. 94 In France, village *curés*.. exact fees from their parishioners. **1954** A. Seton *Katherine* vii. 119 Celibacy might be asked of monk or friar but hardly from..a village parson. **1957** J. Cary in R. S. Surtees *Mr. Sponge's Sporting Tour* (1958) p. xi, The few hundred who..had been able to..travel farther than the county town or the village fair. **1957** 'M. M. Kaye' *Shadow of Moon* xxi. 309 He called upon the Kotwal—the village headman. **1960** J. R. Ackerley *We think the World of You* 40 The small blank eyes mooned stolidly at me..it was like being gaped at by the village idiot. **1967** A. Cordell *Bright Cantonese* ii. 28 The village elders were waiting for me on the little airstrip at Hoon. **1969** R. Blythe *Akenfield* ii. 60 Texts in glass cases hang outside.. and can be read by passengers in the village bus, which just stops there. *Ibid.* iv. 89 When the village shopkeeper sends things to out-of-the-way cottages..he's going to charge.. something for the service. **1969** *Listener* 12 June 814/1 The village chief himself asked us to a dinner of dried deer and shrimp crackers. **1971** O. Norton *Corpse-Bird Cries* i. 3 Most of the fishing would probably be done in the company of a village bobby. *Ibid.* v. 96 Such a simple life, isn't it, being a village copper? **1973** *Stornoway Gaz.* 13 Jan. 7/2 The annual general meeting of Portree Village Council was held in the Portree Hotel, at which there was a fairly large attendance. *a* **1974** R. Crossman *Diaries* (1976) II. 157 Then we rushed him down to the village hall, which has been built by the energy of Len Edwards, our remarkable local inventor of laundry machinery. **1975** *Country Life* 9 Oct. (Suppl.) 26/1 (Advt.), Family house on village green. **1976** 'H. Carmichael' *False Evidence* ii. 25 The cottage was set in restful countryside, a quarter of a mile from the village inn. **1976** P. R. White *Planning for Public Transport* vii. 142 Local village shops may also substitute for day-to-day needs. **1978** P. Van Greenaway *Man called Scavener* ii. 21 Unhonoured assignations with village beauties. **1978** J. Porter *Dead Easy for Dover* ii. 25 We had the funeral on the Saturday... The village church was absolutely packed. **1978** 'J. Melville' *Axwater* ii. 57 The village girls used to go there at the new moon and wish. Like a wishing well. **1979** 'S. Kemp' *Goodbye Pussy* x. 125 The village doctor..saw her once in a while for measles and mumps. **1980** G. Sims in *Winter's Crimes 12* 149 The village Postmistress, who delighted in gossip. **1981** M. Warner *Joan of Arc* i. 20 Women who..were ducked in village ponds to find out whether or not they were witches. **1981** E. Clark *Send in Lions* xiii. 121 The village postman..had seen Kemp that morning. **1981** A. Edwards *Sonya* xi. 175 On Christmas Eve the village priests came to hold a vespers service. **1981** C. Miller *Childhood in Scotland* 63 We wore heavy black boots made by the village shoemaker. **1983** *Times* (Saturday Suppl.) 29 Jan.–4 Feb. 4/4 (*caption*) Rekindling village life in Chelsea.

c. *village-based, -made* adjs.

1976 P. R. White *Planning for Public Transport* vii. 140 In this situation the village-based independent may score over the larger operator. **1883** R. Jefferies *Nature near London* 221 Each village-made crook had an individuality, that of the blacksmith.

d. village college (see quot. 1981); **village constable,** (*a*) *Hist.*, in Papua, a local man through whom the orders of the Australian administration were transmitted; (*b*) a police constable stationed in a village; **village gossip,** (*a*) the idle talk of a village (see Gossip *sb.* 4); (*b*) a woman who gossips (see Gossip *sb.* 3); **village Hampden,** a person like John Hampden (1594–1643), one without means or influence who opposes a powerful local person or organization (in imitation of quot. 1751); **village pump,** a village's communal water pump; freq. used allusively (cf. *parish pump* s.v. *Parish *sb.* 7 b).

1924 H. Morris (*title*) The village college. **1981** D. Rowntree *Dict. Educ.* 342 *Village colleges*, a UK scheme in community education initiated in rural Cambridgeshire in the 1930s, with a number of colleges each serving a village not only as a secondary school but also as a cultural and recreational centre for old and young alike out of school hours. The scheme was later adopted by several other largely-rural counties. **1924** 'R. Daly' *Outpost* xxvii. 259 'They say they will have no chief but the village-constable,' he said, 'and no sorcery except that of the white man.' **1943** F. Thompson *Candleford Green* vii. 115 The village constable was still regarded by many as a potential enemy. **1965** *Austral. Encycl.* VI. 468/1 In the period before World War II, for the carrying out of policy, Administrations to some extent relied on selected headmen (called luluais) in New Guinea and on village constables in Papua. **1981** B. Knox *Killing in Antiques* viii. 166 The village constable got there in ten minutes flat then ran for his car radio. **1847** C. M. Yonge *Scenes & Characters* xvi. 201 Jane sought for amusement in village gossip. **1952** M. Laski *Village* ix. 146 He'd tell her the village gossip. **1972** P. D. James *Unsuitable Job for Woman* iv. 143 The confiding relish of a village gossip about to relate the latest scandal. **1751** Gray *Elegy* 8 Some Village-Hampden that with dauntless Breast The little Tyrant of his Fields withstood. **1857** C. M. Yonge *Dynevor Terrace* I. i. 3 He stalked along like a village Hampden, muttering, 'The old tyrant shall see whether I'm to be trampled on!' **1957** D. Piper *English Face* iv. 111 Many of Johnson's portraits have now lost their names..yet some of them no doubt were 'village Hampdens'. **1978** A. Sanders *Victorian Historical Novel* i. 23 Reade is not mourning the silence of village Hampdens, for..he is aspiring to a history which is 'familiar rather than heroic'. **1925** V. Woolf *Common Reader* 112 The talk of old women round the village pump. **1953** G. E. & K. R. Fussell *Eng. Countrywoman* vi. Plate 48 (*caption*) The village pump at Fressingfield, Suffolk: still used. **1955** G. Gorer *Exploring Eng. Character* iv. 61 What a 20-year-old Hereford student calls 'the proverbial "village pump" attitude and conflicts' seems to bedevil the life of many. **1978** *Listener* 6 July 27/2 Officially, the village of Montaillou was subject to the bailiff of the local count of Foix... But Professor Ladurie is able to tell us about the exercise..of authority at the level of the village pump.

e. (With capital initial.) Of, pertaining to, or characteristic of Greenwich Village. *U.S.*

1950 T. Sterling *House without Door* iv. 41 She's not a Village artist... She was very wealthy once. **1979** M. McMullen *But Nellie was so Nice* i. i. 7 He wore a sort of Village uniform—corduroys and a turtle-necked dark jersey.

village *v.* (*b*) to visit a village in a pastoral capacity.

1871 F. Kilvert *Diary* 24 Feb. (1977) 130 Villaging about to Mrs Jones at the Infant School, Jo Phillips and Margaret Griffith. **1981** 'M. Innes' *Lord Mullion's Secret* 27 The Vicar of Mullion, an old man given to antique usages, sometimes described himself as having been 'villaging'—by which he meant going round the cottagers and chatting them up.

villagiza·tion. [f. Villag(e *sb.* + -ization.] In Africa and Asia, concentration of population in villages; the transfer of control of land to villagers communally; in Tanzania, *spec.* = *ujamaa. Cf. *Bhoodan.

1963 *Punch* 10 July 42/1 'Villagisation' is the word. **1965** P. Fordham *Geogr. African Affairs* iv. 85 Part of the security measures during the Mau Mau rebellion was the forcible 'villagization' of the Kikuyu and the consolidation of fragmented holdings. **1970** *Guardian Weekly* 22 Aug. 2 The majority of land-owners in more than 140,000 Indian villages have declared themselves in favour of Gramadan (gift of village),..which involves

the principle of villagisation (as distinct from nationalisation) of land. **1980** *Sci. Amer.* Sept. 152/1 A campaign of 'villagization' was begun, and over the next several years a massive transfer of rural population was accomplished.

villain, *sb.* Add: **1. b.** *transf.* (Later example.)

1832 *Q. Rev.* Mar. 234 Perchance one hound in ten may throw his tongue as he goes to inform his comrades, as it were, that the villain is on before them.

d. Esp. in phr. *villain of the piece* (now usu. *transf.*).

1854 A. C. Mowatt *Autobiogr. Actress* vii. 133 Ayesha, *the villain of the piece*,..had received a great wrong. **1867** [in Dict.]. **1928** Wodehouse *Money for Nothing* ix. 200 I'm sure you're on the right track. This bird Twist is the villain of the piece. **1937** *Discovery* May 163/1 Fascism, in its ultra-national aspect, is the villain of the piece. **1978** P. Sutcliffe *Oxf. University Press* v. i. 173 Ernest Barker and others took on Nietzsche and Treitschke, who could be regarded as the ultimate villains of the piece.

e. A professional criminal. *slang.*

1960 [see *bogy[1], bogey[1] 4]. **1963** L. Deighton *Horse under Water* xxxi. 125 This villain is doing a nice Cabinet Minister's home. **1975** *Sunday Tel.* 7 Dec. 1/2 A flying squad officer said: 'As far as we know these are no ordinary villains. We believe they are Irish IRA.' **1977** L. Meynell *Hooky gets Wooden Spoon* xiii. 156 There'll be a getaway car..waiting close to the house with a villain in it... I don't like thieving villains.

villainous, *a.* **7.** *villainous-looking* adj. (earlier example).

1777 J. Woodforde *Diary* 22 July (1924) I. 209 The Hangman was an old Man and a most villainous looking Fellow indeed.

villamaninite (vilămæ·ninait). *Min.* [f. *Villamanín*, name of the locality (north of León in NW. Spain) near which the mineral was found + -ite[1].] A black isometric mineral, $(Cu,Ni,Co,Fe)S_2$, of the pyrite group.

1919 *Nature* 20 Nov. 326/1 Dr. W. R. Schoeller and A. R. Powell: Villamaninite, a new mineral. **1968** *Neues Jahrb. f. Mineral.: Monatshefte* 189 The conditions under which villamaninite formed must have been quite extreme. Investigations of Ni-Cu ores from Katanga, Carrol County in Maryland and So. Wisconsin..where low temperature mineralizations are supposed to have occurred, failed to produce a new villamaninite locality. **1977** *Mineral. Abstr.* XXVIII. 205/1 Villamaninite breaks down to α-NiS, carrollite ($CuCo_2S_4$), and some Ni and Co sulphates at around 400°C.

‖ **villancico** (vilʸanþi·ko). *Mus.* [Sp.] A Spanish and Portuguese musical form (see quots.).

1822 J. B. White *Lett. from Spain* ix. 327 The music.. was..used in a species of dramatic interludes in the vulgar tongue, which were sung, not acted, at certain intervals of the service. These pieces had the name of *Villancicos*, from *Villano*, a Clown, shepherds and shepherdesses being the interlocutors in these pastorals. **1849** G. Ticknor *Hist. Spanish Lit.* I. xxiii. 440 The 'Villancicos' that follow—songs in the old Spanish measure with a refrain and occasionally short verses broken in—are more agreeable, and sometimes are not without merit. **1876** Stainer & Barrett *Dict. Mus. Terms* 446/2 *Villancico* (*Sp.*), a species of song of two or more stanzas, each containing seven lines, belonging to the poetry of the 15th century, which, like the madrigal, is of an epigrammatic form—formerly very popular in Spain... Those motets which are sung during high mass on Christmas-eve are always called Villancicos. **1937** M. N. Hamilton *Music in 18th Cent. Spain* 5 In Eslava there were references to *villancicos*, which during the eighteenth century seemed to be the routine accomplishment of every chapel master. **1959** *Collins' Mus. Encycl.* 701/1 *Villancico*... (1) A type of song..current in Spain in the late 15th and 16th cent. It is characterized by the fact that it begins with a refrain, which is subsequently repeated after each verse... (2) In the 17th and 18th cent. a cantata for soli and chorus with instrumental accompaniment, frequently on the subject of Christmas. (3) In modern Spanish a Christmas carol. **1968** *New Oxf. Hist. Mus.* IV. iv. 135 In earlier times courtly love had been a favourite theme for *villancicos*... In the later fifteenth century a more popular tone invaded *villancicos*... The increasing number of religious *villancicos* found in the sixteenth century points towards the transformation of the form in the seventeenth century into an extended sacred cantata. **1980** *Early Music Gaz.* Jan. 16/1 The three papers are:..Manuel Carlos de Brito on 'A little-known collection of Portuguese Baroque Villancicos and Romances'; [etc.].

Villanova (vilănōᵘ·vă). *Archæol.* The name of a hamlet near Bologna, Italy, where archæological finds were made, used *attrib.* to designate an Italian culture of the early Iron Age.

1901 W. Ridgeway *Early Age Greece* I. ii. 237 After the Terramare came the Early Iron Age, usually termed the Villanova period by the Italian archaeologists, from the discovery of a large number of its characteristic remains at Villanova near Bologna. **1910** *Encycl. Brit.* V. 721/2 The next period..is known as the 'Villanova' period,..or as the period of pit-tombs (*a pozzo*), from the form of the graves in which the pottery has been found.

Villanovan (vilănōᵘ·văn), *sb.* and *a. Archæol.* [f. prec. + -an.] **A.** *sb.* An inhabitant of Italy during the Villanova period. **B.** *adj.* Of, pertaining to, or designating the Villanova period. Also *absol.*

1924 D. Randall-MacIver *Villanovans & Early Etruscans* 3 In the fifth century the Villanovans disappear as a recognizable people, stifled no doubt..by the dominant Etruscans who had settled beside them. **1928** —— *Italy before Romans* v. 73 Villanovan helmets and girdles and Villanovan ossuaries..may be seen side by side with Hallstatt swords. *a* **1930** D. H. Lawrence *Etruscan Places* (1932) ii. 55 It seems as if the primitive inhabitants of this part of Italy always burned their dead, and then put the ashes in a jar, sometimes covering the jar with the dead man's helmet, sometimes with a shallow dish for a lid, and then laid the urn with its ashes in a little round grave like a little well. This is called the Villanovan way of burial, in the well-tomb. **1957** [see *Atestine *a.* and *sb.*]. **1974** *Encycl. Brit. Micropædia* X. 438/3 During the first quarter of the 7th century an Orientalizing civilization..was superimposed on the Villanovan in Tuscany. The northern Villanovans of the Po valley..continued to produce a geometric art..as late as the last quarter of the 6th century. **1980** *Times Lit. Suppl.* 17 Oct. 1179/4 The Iron Age possessors of the material culture (or cultural material) conventionally known as 'Villanovan' ought at least to have been mentioned.

villarette, var. Villaette in Dict. and Suppl.

Villar y Villar (vī·lār *i* vī·lāɹ). [Sp., = Villar (a surname) and Villar.] The proprietary name of a Havana cigar.

1878 *Army & Navy Stores Price List* 60 Cigars and tobacco... Villar y Villar. Reina Victoria... 62/- per 100. **1903** W. Stevens *Let.* 20 Oct. (1967) 68 But to-night I've been polite..have..puffed a Villar y Villar.

ville[3] (vil, vəil). *slang* (now *U.S.*). Also 9 **vile.** [a. F. *ville* town.] A town or village.

1837 L. C. Boynton *Jrnl.* 11 Sept. in *Proc. Amer. Antiquarian Soc.* (1933) XLIII. 336 Amherst [Mass.] is a very pleasant little ville, but still and inactive. **1859** Hotten *Dict. Slang* 114 *Ville*, or *vile*, a town or village,—pronounced *phial*, or *vial*. *French*. **1891** 'F. W. Carew' *No. 747* xxxv. 416 We made a long round back to vile. **1939** Joyce *Finnegans Wake* 130 The brick of the viled ville of Barnehulme. **1977** M. Herr *Dispatches* 10 Once we fanned over a little ville that had just been airstruck.

Ville[4] (vil). *slang.* Also **'Ville, (')ville.** [Abbrev.] *the Ville*, Pentonville Prison in Greater London.

1903 'No. 77' *Mark of Broad Arrow* ii. 35 Before the notorious bank forgers had been in 'The Ville' a single hour they were located in different wings of the prison, thus rendering personal contact with each other impossible. **1962** *John o' London's* 25 Jan. 82/3 Pentonville Prison is *the ville*. **1962** D. Warner *Death of Bogey* iv. iii. 146 A stretch in the Ville or on the Moor. **1936** [see *fly-paper* s.v. *fly *sb.*[1] 11]. **1967** C. Drummond *Death at Furlong Post* xii. 152, I don't want to see your great face up against me in the 'ville for twenty years. **1972** L. Henderson *Cage until Tame* vi. 51 Yeah, that's right, he was in the 'Ville.

-ville (vil), *suffix. colloq.* [ad. F. *ville* town.] A terminal element appended to sbs. (which freq. have a pl. suff.) or adjs. to denote: (*a*) a fictitious place; (*b*) a particular quality suggested by the word to which it is appended.

In *U.S.* usage orig. as — *from* —*ville*.

1567 Harman *Caveat* (ed. 2) sig. G3, Rome vyle London. **1843** G. Cruikshank *Comic Almanack* sig. F4[v], Tripe & Trotter Depôt..Meatville. **1891** *N.Y. Sporting Times* 11 July 3/4 Then he was as frisky as a young colt and a slugger from Sluggersville. **1906** F. H. Burnett *Shuttle* (1907) xxxviii. 384 That girl is a winner from Winnersville. I take off my hat to her. **1932** *Magnet* 17 Sept. 13/3 I'm telling you you're the biggest bonehead from Boneheadville. **1939** [see *route march* s.v. *route *sb.* 4]. **1956**, etc. [see *Squaresville]. **1959**, **1961** [see *Cubesville* s.v. *cube *sb.*[1] 1 c]. **1961** [see *Squaresville]. **1962** P. Mortimer *Pumpkin Eater* xiii. 118 Tiny bit boring, between you and me. Strict secrets of course. English Rose stuff. Deathville, as far as I'm concerned. **1964** [see *relocate *v.* 2]. **1967** J. Aitkin *Young Meteors* i. 35 University? Man, that's just dragsville. **1972** *Publishers' Weekly* 2 Oct. 56/1 There are some who will simply not get the fun of it out there in mass-marketville. **1979** *National Times* (Austral.) 13 Oct. 5/2 But there is a mite more to leadership, even here in Mediaville [*sc.* Washington], than looking pretty and carrying a resonant baritone voice.

‖ **ville lumière** (vil lümyę̄r). [Fr., = town or city of light(s).] A brightly-lit city or town; an exciting modern city or town; *la Ville Lumière*, Paris.

1920 W. J. Locke *House of Baltazar* xiv. 166 London ceased to be a city of dreadful night. In his enthusiastic eyes it had almost become a *ville lumière*. **1922** A. M. Hyamson *Dict. Eng. Phrases* 353/2 *Ville Lumière, La*, (Fr., the light-city; the city of lights): Paris. **1923** W. J. Locke *Moordius & Co.* vii. 94 We are but nursing the lamp of la Ville Lumière till better times. **1955** E. Waugh *Officers & Gentlemen* 137 Cape Town at the extremity of two dark continents was a *ville lumière* such as Trimmer had sought in vain. **1980** *Listener* 25 Sept. 389/1 The Eiffel Tower..proclaimed *la ville lumière* to be the modernist capital.

villiaumite (vi·liōmait). *Min.* [a. F. *villiaumite* (A. Lacroix 1908, in *Compt. Rend.* CXLVI. 215), f. *Villiaume*, name of a French explorer in whose collection the mineral was first identified: see -ite[1].] Native sodium fluoride, NaF, occurring as red, pink, or orange transparent isometric crystals.

The symmetry is erron. described in quot. 1908

1908 *Jrnl. Chem. Soc.* XCIV. ii. 201 This new mineral, called villiaumite, is tetragonal and pseudo-cubic, with three perfect cleavages at right angles to one another. **1970** *Amer. Mineralogist* LV. 126 Thermodynamic calculations show that fluorite is to be expected as the common fluoride mineral in both silica-saturated and undersaturated igneous rocks; but in calcium- and silica-poor assemblages villiaumite is favored, but in calcium-poor quartz-bearing rocks, cryolite will prevail.

villiform, *a.* (Earlier example.)

1846 [see *close-set* s.v.* close C 1 and 2].

‖ **villino** (vilī·no). Pl. **villini.** [It., dim. of *villa* Villa.] A small (rural, suburban, or urban) house in Italy (in quot. 1935, in France).

?**1863** Mrs. Gaskell *Let.* 16 July (1966) 708 The pleasure of our intercourse at the Villino Trollope. **1923** A. Huxley *Let.* 12 Nov. (1969) 223 They have a Città Giardino just outside to the north and the number of horrible little villini..is extraordinary. **1935** E. Bowen *House in Paris* i. ii. 31 Though Mrs. Arbuthnot had not.. been lent the villino at Beaulieu, she had been lent a flat at Mentone. **1958** 'M. Innes' *Long Farewell* i. 14 Then let's go up to the *villino*. The view's just as good from there. **1980** *Times* 11 Oct. 10/7 One restaurant..occupies a former *villino* or small house with a garden.

Villonesque (viyone·sk, vilone·sk), *a.* [f. the name of François *Villon* (1431–1480 or 1489), French poet + -esque.] Characteristic of (the style of) Villon.

1932 E. Rickword tr. *Coulon's Poet Under Saturn* vi. 163 A name evocative of the cells, the Villonesque name of Pierre Duchâtelet. **1937** *Times Lit. Suppl.* 20 Mar. 222/1 The self-portrait..is of a somewhat Villonesque character. **1979** M. A. Screech *Rabelais* iii. 86 His Villonesque ricks against the university and church dignitaries.

‖ **villotta** (vilǫ·tă). Also **villota.** Pl. **villot(t)e.** [It.] A type of villanella, originating in northern Italy.

1876 Stainer & Barrett *Dict. Mus. Terms* 446/2 *Villotte* (*It.*). The name given to the first secular pieces in harmony after the rules of counterpoint were fixed. **1926** C. Gray in Gray & Heseltine *Carlo Gesualdo* i. 10 They would spend whole nights out in the bay [of Naples], singing *villotte*, and madrigals. **1942** *Bull. Amer. Musicological Soc.* Apr. 9 The period of the *villota* is dated back by Torrefranca..to the first half of the 15th century... The theory is founded..chiefly upon a group of eleven *villote*..now preserved in the Marciana in Venice. **1958** [see *canzonetta]. **1980** *Early Music* Jan. 104/2 The *Libro primo* contains a wide variety of text forms including the madrigal, villotta, canzone, and ballata-madrigal.

vim. (Earlier example.)

1843 *Yale Lit. Mag.* VIII. 406 He would have acted out his real nature with all the *vim* and pathos which heroes always manifest in like circumstances.

Vim[2] (vim). Also **vim.** [Cf. Vim in Dict.] The proprietary name of a brand of detergent. Hence **vim** *v. trans.*, to clean with Vim.

1894 *Trade Marks Jrnl.* 6 June 484/1 *Vim*... All goods included in Class 48. Lever Brothers, Limited, Port Sunlight, near Birkenhead; soap manufacturers. **1926** R. Macaulay *Crewe Train* ii. x. 179 Denham looked vaguely at the vim. 'For the sink,' her aunt explained. 'And I've got you a..brush. That's to scrub the vim in with.' 'Thank you very much,' said Denham, doubtfully; 'I don't know how much I shall vim the sink, though.' **1948** 'N. Shute' *No Highway* viii. 205 Shirley collected scrubbers and soap and Vim and a dustpan and brush. **1968** *Listener* 14 Mar. 359/1 The real battle, though, is with the Saturday-night dancers who put down French chalk to make the floor slippery. The badminton players retaliate —of old with Vim or a mixture of tea-leaves, sand and water. **1976** T. Sharpe *Wilt* ii. 11 She got the washing-up done..and the bath Vimmed.

‖ **vimana** (vimā·nă). *India.* [Skr.] **1.** The central tower enclosing the shrine in an Indian temple.

1863 *Chambers's Encycl.* V. 552/1 The temples consist of the temple or vimana, in front of which is the pillared porch or mantopa, [etc.]. **1891** J. Fergusson *Hist. Indian & Eastern Archit.* ii. ii. 221 There is one other peculiarity common to both Jaina and Hindu architecture..It is the form of the towers or spires called Sikras, or Vimanas, which invariably surmount the cells in which the images are placed. **1913** L. D. Barnett *Antiquities of India* x. 233 When for the Buddhist relic is substituted the image or emblem of a god, the apse is changed into a closed cella, with a door, and its place as the holy of holies in the structure is marked by a spire or *śikhara*, forming with the cellā a *vimāna*. **1969** *Hindu Weekly Mag.* (Madras) 3 Aug. p. ii/5 He made a bold departure by making the 'vimana', the tower over the sanctum sanctorum, dominate the entire scheme of temple construction.

2. *Mythol.* A heavenly chariot.

1946 S. Kramrisch *Hindu Temple* II. 344 Their bodies inflated by air race in gusts across the atmosphere, support the Vimānas, the chariots of the gods. **1979** W. H. Canaway *Solid Gold Buddha* xv. 106 I've heard of a *vimana*, a sky chariot. Have you ever thought of using one of those things in the show?

‖ **vin** (væ̃n). [Fr.] Wine: used in various French phrases. **1.** Followed by *de, d'*, etc., designating a wine of a particular classification, district, or method of manufacture, etc., as *vin de Graves, de liqueur*, etc.

1699 M. Lister *Journey to Paris* 161 Those Wines thus in the Must are called..*Vin des Liqueurs*. **1777** P. Thicknesse *Year's Journey* II. xxxvii. 38 Barren as the *Crau* appears to be, it..produces such excellent wine.. that it is called *Vin de Crau*, by way of pre-eminence. **1833** C. Redding *Hist. Mod. Wines* iii. 51 What are called in France *vins du liqueur*, are those in which the saccharine principle has not entirely disappeared during the process of fermentation, and been changed into alcohol. **1842** Browning *Dram. Lyrics* 15/2 For council dinners made rare havock With Claret, Moselle, Vin-de-Grave, Hock. **1911** Vin du Glacier [see *Fendant]. **1939** B. Collier *Catalan France* xii. 273 In the case of wines intended to be drunk as *apéritif* or dessert wines..[the colour] is usually that indicated by the terms '*vin d'or*' or '*doré*'. **1955** *Times* 5 Aug. 7/4 The agreement also foresees an increase in the export of French vins d'appellation to the value of 350M. francs. **1967** A. Lichine *Dict. Wines* 543/2 *Vin de liqueur*. This French term has two meanings. 1. A very sweet wine, such as a rich Sauternes. 2. Wine of approximately 18% of alcohol caused by the addition of brandy and added to Champagne prior to shipment. **1968** R. Amberley *Incitement to Murder* i. 18 The wine..was not an *appellation contrôlée*, or even a *vin de qualité supérieure*; it was..an unpretentious local wine.

2. With (*ppl.*) *adjs.*, describing or purporting to describe wines of a certain quality or prepared in a certain way.

1833 C. Redding *Hist. Mod. Wines* v. 176 At Moulins they make a species called *vin fou*, or mad wine, or rather ..'drunkard's wine'. They fill a small, strong-bound cask, having no bung, with must; this they put into another cask, and plunge it into the vat, from which it is not withdrawn until the fermentation ceases. **1849** Geo. Eliot *Let.* 22 Dec. (1954) I. 324 Dear Sara's letter is..not at all physicky—rather an agreeable draught of *vin sucré*. **1920** G. Saintsbury *Notes on Cellar-Bk.* xi. 162 *Vin brulé*, a very popular beverage in Old France, and a regular Christmas and New Year tradition in the Channel Islands. **1952** V. Wilkins *King Reluctant* i. vi. 94 It is at least drinkable—a *vin bourgeois*. **1958** A. L. Simon *Dict. Wines* 161/2 *Vin chaud*, mulled wine. **1966** P. V. Price *France: Food & Wine Guide* ii. 238 The '*vin fou*' of the Jura is made by bottling either white or rosé wine at the peak of its first fermentation. **1974** *Times* 10 Aug. 11/2 A white Corbières. Vin Vert. Montagne d'Alaric.

3. Special collocations. **vin compris** [lit. 'understood']: (phrase denoting) wine included in the price of a meal or other entertainment; **vin cuit** = *wine cuit* s.v. Cuit, cute; an aperitif wine; **vin de paille** [lit. 'of straw'] = *straw wine* s.v. Straw *sb.*[1] 14 (see quots.); **vin de table** = *table wine* s.v. *Table *sb.* 22; cf. *Tafelwein; **vin d'honneur,** a wine formally offered in honour of a person or persons; the reception at which the wine is offered; **vin doux (naturel)** [lit. 'sweet (unfortified)'], a sweet aperitif wine; **vin du pays** [lit. 'of the country'], a local wine; also *transf.* and *fig.*; **vin gris** [lit. 'grey'], a rosé wine of eastern France; **vin jaune** [lit. 'yellow'], (see quot. 1966); **vin mousseux** [*mousseux], sparkling wine. See also *vin blanc, ordinaire, rosé, rouge** below as main entries.

1889 Kipling *From Sea to Sea* I. xii. 308 The *vin ordinaire* which is *compris*, is good. **1910** G. B. Shaw *Let.* ?16 Apr. (1972) II. 921 Further excitement is provided by the institution of *vin compris* at meals. **1967** *Observer* 14 May 28/3 Vin compris in Camden. 'Free wine tasting,' promised the notice. **1833** C. Redding *Hist. Mod. Wines* vi. 193 Their [*sc.* the Spaniards'] boiled wines, or vins *cuits*, as the French call them, are mingled with other growths. **1962** *Times* 21 Dec. 10/7 The bottle of *vin cuit* or unfermented wine. **1833** C. Redding *Hist. Mod. Wines* iii. 51 The wines called *vins de paille* are so denominated from the grapes being laid for several months upon straw before they are taken to the press. **1964** A. Sichel *Penguin Bk. Wines* iii. 159 *Vins de paille*, the second of the unusual white wines of the Jura, get their name from the straw mats on which the black Poulsard and Trousseau grapes are dried for several weeks before being pressed. **1972** *Country Life* 14 Dec. 1660/1 Nowadays the grapes are normally laid not on mats but on trays... To produce a litre of *vin de paille* takes about 12–15 kilos of grapes. **1948** A. Waugh *Unclouded Summer* vii. 142 They had treated it as a *vin de table*. **1981** *New Yorker* 28 Dec. 49/1, I took a half-litre bottle of Sichel le Cellier, *vin de table, rouge*. **1920** *Punch* 30 June 512/1 The Prime Minister was celebrating the longest—and pretty nearly the hottest—day by a *vin d'honneur* at Boulogne. **1947** E. Waugh *Scott-King's Mod. Europe* 25 Vin d'honneur offered to the delegates by the Municipality of Bellacita. **1978** *Broadcast* 6 Mar. 4/1 A *vin d'honneur* at Television Centre for TFI's news director. **1958** A. L. Simon *Dict. Wines* 162/1 *Vin doux*, grape-juice before it ferments and becomes wine; also used for a 'sweet' wine. **1959** W. James *Word-bk. Wine* 197 *Vins doux naturels*, The Frenchman's logic has strange lacunae;.. the unnatural and fortified sweet wines made in the south-east [*read*: south-west] corner of his own country he blandly calls 'natural sweet wines'. **1981** *Country Life* 23 July 261/3 That delectable Muscat wine, a *vin-doux-naturel*. **1777** P. Thicknesse *Year's Journey* II. xlix. 138, I could not help treating him with a bottle of *vin de païs*. **1822** L. Simond *Switzerland* I. 34 A well-seasoned veal pie, a boeuf-a-la-mode, plenty of the best *vin du pays*, and even a dessert. **1935** *Times Lit. Suppl.* 18 July 459 A renewed zeal for the land of England, and, especially for the savour of its products, of its beef and cheese and fruit and *vin du pays* or cider. **1965** Vin du pays [see *premier cru]. **1833** C. Redding *Hist. Mod. Wines* v. 176 A *vin gris*, a grey or rather brown wine, is made here [*sc.* at Gannat] by leaving the must to ferment for forty-eight hours. **1958** A. L. Simon *Dict. Wines* 162/1 *Vin gris*, light red wine of Lorraine and Alsace, usually made of mixed black and white grapes. **1974** *Times* 18 May (Suppl.) p. i/4 The *vin gris* of the Jura. **1833** C. Redding *Hist. Mod. Wines* iii. 51 The *mousseux* wines of Arbois are called *vins blanc de garde*, and when old, *vin jaune*. **1966** P. V. Price *France: Food & Wine Guide* ii. 239 The two extraordinary wines of the Jura are the *vins jaunes* and *vins de paille*... *Vins jaunes* are made solely from the Sauvignon grape.. taste slightly like sherry, though..they are not fortified in any way. **1981** *Times* 12 Dec. 12/4 The odd Jura wine, Château Chalon..is a *vin jaune*..staying long in cask, wherein a *veil* forms on the surface. **1789** Vin mousseux [see *mousseux]. **1833** C. Redding *Hist. Mod. Wines* ii. 30 Those wines which effervesce (*vins mousseux*) are impregnated deeply with carbonic acid gas. **1972** A. Ross *London Assignment* 47, I..pulled the napkin off the foil-topped bottle. It was a vin mousseux, and not a very good one.

vina. Add: The predominant spelling is now **veena.** Also *attrib.* and *Comb.* (Earlier and further examples.)

1788 W. Jones in *Asiatick Researches* I. 265 His [*sc.* Nared's] invention of the Vìnà, or Indian lute, is thus described. **1848** J. H. Stocqueler *Oriental Interpreter* s.v., *Veena*, an instrument of the guitar kind, with seven metal strings. It is the most ancient musical instrument of the Hindoos, and in good hands is capable of yielding great melody and expression. **1891** [see *tamboura b]. **1896** [see *surbahar]. **1921** H. A. Popley *Music of India* vii. 104 The vīṇā has seven strings, four of which pass over the frets and constitute the main playing strings. **1945** R. K. Narayan *Eng. Teacher* v. 109 He hardly made any special sound or noise, but it was there all the time, a permanent background against which all his speech and gestures occurred, something like the melody of a veena string from which music arises and ends. **1964** S. Marcuse *Mus. Instruments* 564/2 *Vīṇā*.., stringed instr. of ancient and modern India. In modern times the word has become a generic term for chordophones in S. India. **1969** R. Shankar *My Music* i. 34/2 The stringed instrument par excellence..is the *veena* (also spelled *vina*). **1972** P. Holroyde *Indian Mus.* 258 It has been said by Vajnavalkya, the ancient Indian writer and philosopher: 'He who knows the art of veena-playing and sruti shastra can attain God easily.'

vinaigrette. Add: **1. b.** Now *spec.* a dressing of oil and (wine) vinegar, sometimes with herbs (esp. tarragon), used esp. with salads and cold vegetables. In full *vinaigrette sauce* (also F. *sauce vinaigrette*). Also *vinaigrette dressing*.

[**1877** *Cassell's Dict. Cookery* 1091 *Vinaigrette, Sauce à la.* This is a sauce much used in Paris for cold viands.] **1880** M. Parloa *Appledore Cook Bk.* (new ed.) 235 (*recipe*) Vinaigrette sauce. **1906** Mrs. Beeton *Bk. Househ. Managem.* xxxv. 1112 *Vinaigrette sauce*..4 tablespoonfuls of salad-oil, 2 tablespoonfuls of tarragon vinegar, [etc.]. **1936** Lucas & Hume *Au Petit Cordon Bleu* 136 *Sauce vinaigrette*..tarragon vinegar..lemon..shallot..garlic.. mixed herbs..parsley..salad oil [etc.]. **1959** *Listener* 24 Dec. 1135/1 Mushrooms cooked and tossed in vinaigrette dressing. **1974** N. Freeling *Dressing of Diamond* 7 Colette..made vinaigrette for the avocadoes. **1982** T. Fitzgibbon *With Love* ix. 163 Calf's head with vinaigrette sauce, a dish seldom seen today.

vinaya (vinaya·). *Buddhism.* [Skr., lit. 'leading (away)'.] The rules of conduct that regulate Buddhist monastic life. Also *attrib.*

1854 A. Cunningham *Laddk* xiii. 383 Vinaya, or 'Rules of Discipline'. **1881** Rhys Davids & Oldenberg (*title*) Vinaya texts. *Ibid.* p. xiii, Upâli is mentioned as the first among the custodians of the Vinaya. **1924** S. Dutt *Early Buddhist Monachism* i. 15 The Council of Vesali must be dated about the middle of the fourth century B.C., and the compilation of a complex codex of Vinaya rules not much earlier than that date. **1956** R. Pieris *Sinhalese Social Organization* ii. ix. 74 The monks themselves were prohibited by *vinaya* rules from performing these mundane duties. **1979** *Jrnl. Asian Studies* XXXVIII. 671 The *Vinaya*, or Code of Conduct.

‖ **vin blanc** (væ̃n blañ). [Fr.] **1. a.** French white wine.

1792 A. Young *Trav. France* I. 164 He gave me a bottle of excellent *vin blanc mouseux*, made in Auvergne. **1814** M. Birkbeck *Journey through France* 29 Pouilly, [renowned] for its Vin Blanc. **1853** [see *gofer]. **1917** A. G. Empey *Over Top* 313 *Vin blanc*, French white wine made from vinegar. They forgot the red ink. **1958** 'E. Dundy' *Dud Avocado* i. ix. 157 The vin blanc break around eleven o'clock. **1974** *Country Life* 31 Jan. 183/2 An enormous amount..must be sold not as burgundy at all but as plain *vin rouge* or *vin blanc*.

b. *vin blanc cassis* [*cassis], (see quot. 1964).

1964 A. Sichel *Penguin Bk. Wines* iii. 173 A popular liqueur is made from blackcurrants..and is known as Cassis. Mixed with white wine, it is often used as an aperitif drink in France, called 'Vin Blanc Cassis'. **1982** 'R. Grayson' *Montmartre Murders* iv. 31 Gautier was sipping the vin blanc cassis which a waiter had brought him.

2. In various joc. and illiterate spellings. Cf. *plonk *sb.*[2] Chiefly *Mil. slang.*

1919 J. Buchan *Mr. Standfast* viii. 162 A pint and a dram for me. This is better than vongblong and vongrooge ...in those estaminets. **1933** F. Richards *Old Soldiers never Die* vii. 83 An elderly Frenchman from Armentières ..was soon selling beer and ving blong and doing a roaring trade. **1934** Blunden *Choice or Chance* 31 In Pop. we banqueted no doubt On *vin-blong*, malaga-and-stout.

vinblastine (vinblæ·stīn). *Pharm.* [f. mod.L. *Vin-ca*, former generic name (see Periwinkle[1]) + *leuco*)*blast* s.v. Leuco- + -ine[5].] A cytotoxic alkaloid, $C_{46}H_{58}N_4O_9$, obtained from the periwinkle *Catharanthus roseus* and administered intravenously (usu. as the sulphate) in the treatment of lymphomas and other cancers. Cf. *vincristine.

1962 M. R. Karon et al. in *Pediatrics* XXX. 791/1 Several alkaloids were subsequently isolated from the crude extract [of *Vinca rosea*], two of which, leurosine and vinblastine (formerly called vincaleukoblastine), markedly inhibited this tumour. **1974** *Nature* 15 Mar. 239/1 The alkaloids colchicine and vinblastine interfere with the assembly of microtubules, yet they also affect several properties of the surface membrane of animal cells. **1983** *Oxf. Textbk. Med.* II. xix. 96/1 Vincristine..is preferable to vinblastine..in the treatment of leukaemia.

vinca[1] (vi·ŋkă). Also **Vinca.** [mod.L.: see Periwinkle[1].] = Periwinkle[1] 1; **vinca alkaloid,** any of several alkaloids (as vinblastine, vincristine) obtained from a periwinkle.

1868 D. Thomson *Handy Bk. Flower-Garden* ix. 250 For forming permanent edgings to large beds, these Vincas are very effective. **1937** A. T. Johnson *Woodland Garden* vii. 70 The vincas as a tribe are either misused or not used at all. **1961** D. Stuart *Driven* xvii. 170 There had been pride and delight in her eyes when first the oleanders she tended so carefully had bloomed, and all the outside of the camp was a blaze of vincas that same season. **1970** Passmore & Robson *Compan. Med. Stud.* II. xxix. 11/2 Effects peculiar to the vinca alkaloids..include paraesthesia (especially round the mouth), temporary mental depression, loss of deep tendon reflexes and more rarely headaches, psychoses and convulsions. **1978** R. B. Scott *Price's Textbk. Pract. Med.* (ed. 12) iv. 374/1 The vinca alkaloids vinblastine (VLB) and vincristine (VCR).. have a variety of toxic effects other than those on dividing cells.

Vinča[2] (vi·ntʃă). *Archæol.* The name of a village site near Belgrade, used *attrib.* to designate a central Balkan culture of the chalcolithic age. Also *absol.*

1925 V. G. Childe *Dawn European Civilization* 174 The sunken oval huts, the shoe-last celts, the rude clay idols..and the spirals and meanders decorating the vases suffice to attach Vinča I to the contemporary culture of Moravia. **1940** C. F. C. Hawkes *Prehist. Foundations Europe* iv. 93 The early Vinča pottery shows Anatolian tradition. *Ibid.* 94 Some twenty other settlements of the Vinča culture have been identified along the Serbian bank of the Danube. **1974** *Encycl. Brit. Macropædia* II. 613/1 The Vinča sequence is best documented at the eponymous site, situated..east of Belgrade, overlying the Starčevo levels excavated by the Yugoslav archæologist Miloje Vasić, intermittently from 1908 to 1932. Fine Vinča ceramic wares are burnished in orange or black and decorated with a shallow linear channeling. **1977** *Jrnl. R. Soc. Arts* CXXV. 486/1 The mask-like face certainly recalls the conventions of this period in south-east Europe, notably those of the Vinča culture, or Porodin.

Vincennes (væñsen). The name of a château in the town of *Vincennes* (now a suburb of Paris), used *attrib.* and *absol.* to designate porcelain produced there in the eighteenth cent., before the manufactory was transferred to Sèvres (see Sèvres *a.*); also applied to colour characteristic of Vincennes porcelain.

1766 in F. A. Barrett *Worcester Porcelain* (1953) vi. 29 Great improvements in the Worcester manufactory of china... It is hardly surpassed by the Vincennes, and much cheaper. **1874** C. Schreiber *Jrnl.* 13 Aug. (1911) I. 276 A Vincennes cup all but unique. **1900** F. Litchfield *Pottery & Porcelain* vii. 104 The pieces having rich grand colours in claret colour or crimson lake, and in rich deep Vincennes blue. **1925** W. W. Worster tr. *Hannover's Pottery & Porcelain* III. xiii. 279 In the biscuit figures and groups of Vincennes we find..traces of the growing neo-classical movement. **1949** W. B. Honey *European Ceramic Art* i. 40 The incomparable Vincennes gilding, with its soft-looking delicately tooled surface. **1960** H. Hayward *Antique Coll.* 254/1 The earliest Vincennes..includes jardinières, jugs, ice-pails and trays of simple shape. **1978** *Times* 22 July 8/2 A massive white Vincennes pot-pourri vase.

Vincent[2] (vi·nsěnt). *Path.* The name of J. H. *Vincent* (1862–1950), French medical scientist, used in the possessive to designate a painful ulcerative condition of the inside of the mouth or the throat associated with infection with fusiform bacteria and spirochætes (described by him in 1896); also a similar condition of the gums that is accompanied by foul breath and bleeding.

1904 *Lancet* 16 July 135/2 Vincent's angina is a form of pseudo-diphtheria which is associated with, and is probably due to, certain characteristic micro-organisms. **1918** Vincent's disease [see *trench mouth* s.v. *trench *sb.* 9]. **1926** J. L. T. Appleton *Bacterial Infection* xxiv. The condition..goes under a great variety of names, *e.g.*, 'trench mouth', Vincent's gingivitis or stomatitis and ulcero-membranous stomatitis. **1935** *Motion Picture* Nov. 3/2 (Advt.), Keep gum disorders—gingivitis, pyorrhea and Vincent's disease far in the background. **1959** J. Blish *Case of Conscience* i. 11 He saw few of the mixed bacilli and spirochetes which would have indicated a case of ordinary..Vincent's angina—'trench mouth'. **1962** Blake & Trott *Periodontology* xvii. 173 The interdental

cratering which follows longstanding Vincent's infection leads to interdental stagnation. **1983** *Oxf. Textbk. Med.* I. xii. 21/2 Acute ulcerative gingivitis... At times there are shallow necrotic ulcers affecting the oropharyngeal mucosa which shows diffuse erythema; this has been referred to as Vincent's angina.

Vincentian, *a.*[1] Add: (Examples.) Also, of or pertaining to St. Vincent himself.

1896 W. J. Fitzpatrick *Mem. Fr. Healy* i. 10 The Vincentian Seminary, 34, Usher's Quay, Dublin. **1934** G. B. Shaw *Too True to be Good* 23 The latest rediscovery of the Vincentian principle has been made by Mr Ford, who has testified that if you want a staff of helpful persons..you must not give them either title, rank, or uniform. **1982** G. Greene *Monsignor Quixote* i. v. 73 Marquez ..did little entertaining apart from an occasional father from the Vincentian monastery. **1984** *New Yorker* 9 Apr. 52/3 St. John's is a Vincentian school... The Vincentians are trained missionaries.

Vincentian (vinse·nʃăn), *sb.*[2] and *a.*[3] **A.** *sb.* A native or inhabitant of St. Vincent in the West Indies. **B.** *adj.* Of or pertaining to St. Vincent.

1933 *Vincentian* 4 Feb. 2/2 We have much pleasure in publishing in this issue the Report of the Installation of officers of the 'Vincentian Progressive Friendly Society'. *Ibid.*, Vincentians abroad are fondly thinking of home. **1944** *Minutes Legisl. Council St. Vincent 1943* 92 The parent..was to all intents and purposes a Vincentian—who in turn was also from Vincentian parents. **1950** *Caribbean Q.* II. ii. 18 In *Saraband* the light skinned Vincentian wants to marry a dark girl. **1974** *Advocate-News* (Barbados) 21 Feb. 2/1 Vincentians have temporarily thrown off cares of the current fuel crisis. **1980** *Oxf. Diocesan Mag.* Oct. 8/1 At Wycombe some 15% or more of the population is of immigrant origin—Pakistani, Vincentian, Indian, Jamaican, Polish, Italian, Yugoslav.

vinchuca (vintʃu·kă). [Amer. Sp., f. Quechua *wihchuykuk*.] One of several blood-sucking triatomine bugs of Central and S. America, esp. *Triatoma infestans*.

1932 *Listener* 17 Aug. 227/1 The Vinchuca, a horrid looking beetle from half to an inch long..settles on its victims and draws their blood. **1971** P. C. C. Garnham *Progr. Parasitol.* ix. 189 In the opinion of Adler..Darwin contracted Chagas' Disease..when he landed on the coast of Argentina and was bitten by the 'vinchucas' (the triatomid bugs which convey the infection).

vincristine (vinkri·stīn). *Pharm.* [f. as *vinblastine + *leuco*)*cristine* (former name of the drug), perh. f. *leu(kaemia + *-ro-* + *crist(a + -ine[5].] A cytotoxic alkaloid, $C_{46}H_{56}N_4O_{10}$, obtained from the periwinkle *Catharanthus roseus* and administered intravenously (usu. as the sulphate) in the treatment of acute leukæmia and other cancers.

1962 M. R. Karon et al. in *Pediatrics* XXX. 791 (*heading*) A preliminary report on vincristine sulfate—a new active agent for the treatment of acute leukemia. **1977** *Lancet* 19 Feb. 433/1 Complete remission [of acute lymphocytic leukæmia] was achieved on a regimen of vincristine, prednisone, and daunomycin. **1983** [see *vinblastine].

vindaloo (vindălū·). *Cookery.* Also **bindaloo.** [Prob. f. Pg. *vin d'alho* wine and garlic sauce, f. *vinho* *vinho + *alho* garlic.] An Indian curry dish made with meat, fish, or poultry in a sauce of garlic, wine (or vinegar), spices, etc., served with rice.

1888 W. H. Dawe *Wife's Help to Indian Cookery* 65 Vindaloo or Bindaloo—A Portuguese Kárhi... The best Vindaloo is prepared in mustard-oil... Beef and pork, or duck can be made into this excellent curry. **1954** S. Chowdhary *Indian Cooking* 51 Chicken vindaloo is very tasty with peas putao and other vegetable dishes. **1972** R. Hill *Fairly Dangerous Thing* i. iv. 39 A mouthful of Vindaloo prawn. **1982** Barr & York *Official Sloane Ranger Handbk.* 27/1 His macho-masochisms decree that he must order the hottest—Vindaloo, Madras or even Bangalore Phal. **1985** *Listener* 14 Mar. 15/3 Water is suggested as one option for Chinese food (expensive bottled water, of course), Guinness for oysters and lager for chicken vindaloo.

vine, *sb.* Add: **I. 2. d.** A suit of clothes; *pl.*, clothing. *U.S. slang.*

1932 *Evening Sun* (Baltimore) 9 Dec. 31/5 *Vine*, a suit of clothes. **1959** *Esquire* Nov. 70 J *Vines*, clothes. **1964** L. Hairston in J. H. Clarke *Harlem* 285, I..laid out my vine, a clean shirt and things on my bed. **1973** A. Dundes *Mother Wit* 238 I'm going to lay a vine under the Jew's balls for a dime. **1975** *Amer. Speech 1972* XLVII. 152 Without your vines you're nothing but FBI [*sc.* Fat, Black, and Ignorant].

4. c. (Earlier examples.)

1708 E. Cook *Sot-Weed Factor* 19 When sturdy Oaks, and lofty Pines Were level'd with Musmillion Vines. **1831** W. C. Bryant *Marion's Men* 9 We know its walls of thorny vines, its glades of reedy grass.

III. 8. c. *vine-clad* (earlier example).

1824 Lady Blessington *Jrnl.* May in E. Clay *Lady Blessington at Naples* (1979) 102 The vine-clad hills and fertile Campania.

9. vine moth (earlier example).

1840 J. & M. Loudon tr. *Kollar's Treat. Insects* iii. 172 This Vine Moth is not the only species of the family *Tortricidæ* which selects the vine for its food.

vinegar, *sb.* Add: **6. vinegar-fly,** a fruit-fly, *Drosophila melanogaster*; cf. *drosophila; **vinegar stick,** a sword or walking-stick with a vinaigrette (sense 3) fitted into the handle (now *Hist.*); also *transf.*

1902 L. O. Howard *Insect Bk.* 185 They are also called 'vinegar flies', from the fact that their..larvae are frequently found in canned fruits and pickles which have been imperfectly sealed. **1937** *Discovery* Sept. 282/2 There is a type of cockroach found in vinegar breweries, at least two kinds of vinegar fly, and..the vinegar eel. **1979** P. L. G. Bateman *Household Pests* ii. 19 The tiny red-eyed fruit fly *Drosophila melanogaster*, is one of several species associated with fermenting liquids or rotting fruit. They are also known as wine flies, yeast flies or vinegar flies. **1935** A. J. Pollock *Underworld Speaks* 129/2 *Vinegar stick*, a stilletto; a long bladed knife; a sword. **1968** *Canadian Antiques Collector* Nov. 21/1 The forerunner of the vinaigrette was the vinegar stick, a walking stick with a hollow head for a vinegar-soaked sponge. **1979** 'J. Gash' *Grail Tree* vi. 60 You get them in all shapes, even as 'vinegar sticks', where the container is cleverly made into the handle of a sword or walking-stick.

vi·negared, *ppl. a.* [f. Vinegar *v.* + -ed[1].] Treated or flavoured with vinegar.

1861 [see *slaw]. **1892** A. Conan Doyle *Great Shadow* vi. 74 He set a dish of vinegared herrings before him. **1978** R. Condon *Bandicoot* xiii. 54 And a little of the vinegared sparrows of Wakayama.

vinegarish, *a.* Add: Hence **vi·negarishly** *adv.* (In quots. *fig.*)

1890 'Mark Twain' in M. D. L. Landon *Kings of Platform & Pulpit* 354 My companion said vinegarishly, 'Well, well! What do you say now?' **1976** 'J. Ross' *I know what it's like to Die* xviii. 122 'I never ever laid a finger on him. We were very good friends,' he said vinegarishly.

vinega(r)roon (vi:nĭgărū·n). *U.S.* Also **vinagron, vinegarone, vineg(e)rone.** [ad. Amer. Sp. *vinagrón*, f. Sp. *vinagre* Vinegar *sb.* + (depreciatory) augmentative suffix *-ón*.] A large whip scorpion, *Mastigoproctus giganteus*, which is native to southern North America and Mexico, emits a secretion smelling like vinegar when alarmed, and is often wrongly believed to be venomous. Also *transf.*

1853 L. Sitgreaves *Rep. Exped. Zuñi & Colorado Rivers* 34 Frequently did I find in the road that disagreeable-looking object known to the Mexicans as the vinagron. **1889** *Cent. Dict.* 6758/3 Vinegerone..so called on account of the strong vinegar-like odor of an acid secretion noticeable when the creature is alarmed. **1914** *Blackw. Mag.* July 123/1 His late breaking-in, the lengthy vacation,..keep the 'vinegarone' in his [*sc.* a bronco's] composition. **1920** *Glasgow Herald* 8 July 4/2 His animal friends, the chief of whom seem to be vinegrones, some species of bug. **1948** *Pacific Discovery* Mar. 8/2 The vinegaroons, those queer, eight-legged devils of the desert night, raced at top speed in and out of the shadows to gather insect prey. **1978** *Sci. Amer.* Mar. 36/3 Here are.. the acetic-acid-secreting vinegarroons.

viner[3]. For *U.S.* read orig. *U.S.* and add: esp. one used to harvest peas. (Later examples.)

1949 *Nat. Geogr. Mag.* Sept. 331/2 'Viners' that thresh out the peas were working day and night. **1963** *Times* 14 Jan. 13/1 Pea haulm silage, made under cover direct from the viners, plays a major part in the winter feeding. **1981** *Southern Horticulture* (N.Z.) Spring 36/1 In the 1950s the chore of hand-picking peas was removed with the development of a viner, to which the cut plants were cut and carted. **1982** *East Anglian Daily Times* 20 Nov. 5 The largest harvester around—the Mather and Platt SB8000 pea viner, soon to be produced in Suffolk.

Vinerian (vəini°·riăn), *a.* [f. the name of Charles *Viner* (1678–1756), English jurist.] *Vinerian Professor*, the holder of the chair of English Common Law endowed by Viner at the University of Oxford; so *Vinerian Professorship*. Also applied to other posts or studentships (and their holders) similarly endowed by Viner at Oxford, or to the fund administered for this purpose.

1759 (*public notice dated* 18 June), The Vinerian Professor gives this public Notice, that he proposes to begin his..Course of private Lectures. **1765** Blackstone *Commentaries* I. (title-page), Vinerian Professor of Law. *a* **1777** S. Foote *Devil upon Two Sticks* (1778) i. 3 Had you toil'd thro' the laborious page of the Vinerian professor. **1810** *Oxf. Univ. Calendar* 34 (*heading*) Vinerian Professorship of Common Law. **1846** *Ibid.* 58 Vinerian Professorship of Common Law. Charles Viner, Esq by his will..left about 12,000*l.* to the Chancellor, Masters, and Scholars of the University of Oxford, to establish a Professorship,..of the Common Law. **1934** *Law Q. Rev.* L. 196 Sir William Blackstone, the first Vinerian Professor. **1954** *Wadham Coll. Gaz.* Trinity Term 124 Mr. Gotlieb came as a Rhodes Scholar from Manitoba to Christ Church and took a First Class in the B.C.L., also winning the Vinerian Scholarship. **1980** *Statutes, Decrees, & Regulations Univ. Oxf.* 111 Any surplus income of the Vinerian Fund which remains after paying the emoluments of the scholars shall be applied towards defraying the stipend of the Vinerian Professor of English Law.

vineyardist. (Earlier U.S. example.)

1848 *Rep. Comm. Patents 1847* (U.S.) 199 A French wine maker and vineyardist..from Kentucky.

vingerpol (fi·ŋəɹpǫl). *S. Afr.* Also **fingerpoll, -pole.** [Afrikaans, f. Du. *vinger* Finger *sb.* + *pol* tuft tussock.] One of several African species of *Euphorbia* producing succulent branches which resemble fingers.

1875 J. Noble *Descr. Handbk. Cape Colony* 178 The 'vingerpol' is another succulent plant in use: it has great fleshy fingers growing out of a crown, a foot in diameter. **1883** C. Hobson *Farm in Karoo* 196 'Another very curious plant, circle within circle of things like fingers.' 'That is just what is called... "Finger-pole".' **1889** H. A. Bryden *Kloof & Karroo* 258 Spent and foundered oxen,..when fed with fingerpoll, regained vitality. **1932** Watt & Brejer-Brandwijk *Medicinal & Poisonous Plants Southern Afr.* 104 *Euphorbia esculenta* Marl., Vingerpol, a drought-resisting plant of the Karroo; must not be confused with *Euphorbia caput medusae*, also known as Vingerpol. The former is a valuable fodder, the latter poisonous. **1966** E. Palmer *Plains of Camdeboo* xvi. 260, I dreamed I was struggling with a giant Vingerpol that bounced solidly and fleshily upon me.

vingt-et-un, vingt-un. Add: **vingt-un** has become *Obs.* (Earlier example of *β*.)

1772 Duchess of Northumberland *Diary* 7 June (1926) xxxiii. 186 We play'd..at Vingt et un till supper Time.

vingty (væ·ṅti). [-y[6].] Slang abbrev. of prec.

1936 F. J. Whaley *Trouble in College* viii. 129 Perhaps there was a game of poker or vingty on somewhere. **1946** M. Dickens *Happy Prisoner* ii. 34, I asked her to come and play Vingty. **1954** N. Balchin *Last Recoll. my Uncle Charles* i. 16 As you see, a small school of vingty is in progress.

‖ **vinho** (vī·nʸǒ). [Pg., wine: cf. *vino.] Portuguese wine. Used in various collocations, as *vinho branco*, white wine; *vinho corrente* = *vino corriente* s.v. *vino 1 b; *vinho da casa*, house wine; *vinho de consumo*, cheap wine equivalent to *vin ordinaire*; *vinho tinto*, red wine; *vinho verde*, [lit. 'green wine'] young wine that is not allowed to mature.

1835 J. E. Alexander *Sketches in Portugal* iii. 65 Some of the elders..went to a press in the wall and took from it a square bottle of vinho branco, with which they washed the dust from their throats in order that the chaunting might be more effective. *Ibid.* v. 112 In an outhouse were many pipes of 'vinho tinto'. **1955** M. McCarthy in *New Yorker* 5 Feb. 93/1 Hotel and boardinghouse keepers are required to serve a third of a litre of wine (the *vinho da casa*) with every lunch or dinner. **1958** K. Amis *I like it Here* xii. 152 A few glasses of *vinho tinto*. **1962** 'D. Wilson' *Search for Geoffrey Goring* v. 118 Penny copied the men, drinking the white *vinho verde*. **1975** P. V. Price *Taste of Wine* v. 100/1 There are, however, some wines such as the *pétillant* and *vinhos verdes* ('green wines'), which have a natural 'liveliness' or inclination to sparkle. **1980** *Times* 27 Nov. 21/5 Outstanding wines include ..the Beiramar *vinho corrente*. **1980** *Times* 10 June (Portugal Suppl.) p. vi/2 Prices of *vinho de consumo* (vin ordinaire) rosé. *Ibid.*, The northern Minho district which produces vinho verde. **1985** *Times* 26 Mar. 20/2 Organized tours from London will take tourists..to other regions such as the Dão, with its red and white table wines, and the regions of the *vinhos verdes* or green wines.

vining (vəi·niŋ), *vbl. sb.* [f. Vine *v.* + -ing[1].] The separation of certain leguminous crop plants from their vines and pods; *vining pea*, a pea grown for mechanical vining.

1928 Jones & Rosa *Truck Crop Plants* xi. 215 The vines [of peas] are loaded onto racks and hauled to the cannery or a field vining station. **1944** *Sun* (Baltimore) 26 Apr. 5/5 Prisoners..will be employed in vining and husking operations and in handling the cans after processing. **1959** *Times* 7 Sept. 19/2 One large..herd is getting its sustenance mainly from kale and vining pea silage. **1968** *Times* 16 Dec. 7/3 Peas for vining and threshing. **1975** Lockhart & Wiseman *Introd. Crop Husbandry* (ed. 3) iv. 147 Vining and pulling peas are not so dependent on dry weather at harvest.

vink (fiŋk). *S. Afr.* [a. Afrikaans.] Var. *fink *sb.*[1]

1834 [see *fink *sb.*[1]]. **1913** D. Fairbridge *That which hath Been* 30 Yellow vinks shrieked and jabbered on their hanging nests. **1925** *Centenary Bk. S. Afr. Verse* 235 The bird referred to in the text..is evidently a weaver-bird, locally known as Vink. **1958** *Cape Times* 15 July 8/8 Vinks feed on many kinds of tiny parasites and insects.

vino. [Sp. and It., wine; cf. *vin.] **I.** Non-naturalized forms. (vī·nǒ). **1.** With various adjs. designating the quality, origin, etc., of a wine: **a.** *vino blanco* [Sp. = white], *dolce* [It. = sweet], *dulce* [Sp. = sweet], *fino* [Sp. = quality], *locale* [It. = local], *rosso* [It. = red], *secco* [It. = dry], *tinto* [Sp. = red] (cf. Tinto *sb.*[1] in Dict. and Suppl.).

1673 J. Ray *Observations Journey Low-Countries* 487 We left Cordova... This day we first met with red wine again which they call *Vino tinto*. **1846** R. Ford *Gatherings from Spain* xiv. 155 This *amontillado*..is very scarce, since out of a hundred butts of *vino fino*, not more than five will possess its properties. **1901** F. W. Maitland *Let.* 23 Feb. (1965) 224 When Carnival came, our cook took herself off to spend her wages in *vino tinto*. **1902** Belloc

Path to Rome 317, I bought a bottle of a new kind of sweet wine called 'Vino Dolce'. **1911** *Encycl. Brit.* XXVIII. 725/2 Malaga is a sweet wine.. It is generally.. a blend made from *vino dulce* and *vino secco*, together with varying quantities of *vino maestro*, *vino tierno*, *arope* and *color*. **1949** E. POUND *Pisan Cantos* lxxvi. 47 Ex rum-runner (the rum being vino rosso). **1950** E. HEMINGWAY *Across River & into Trees* xxxvii. 226 A bottle of that *vino secco*, from Vesuvius, for the small soles. **1963** *House & Garden* Jan. 56/2 In little inns.. you get a *vino locale* from the carafe. **1977** C. MCCARRY *Secret Lovers* xv. 222 Mushrooms in one place..gazpacho in another..*vino tinto* in all. **1981** A. FRASER *Splash of Red* iv. 44 There's some lovely chilled *vino blanco* in the fridge.

b. Special collocations. **vino corriente** [Sp., lit. 'common, ordinary'] cheap wine equivalent to *vin ordinaire*; **vino cotto** [It., lit. 'cooked'], (see quot. 1965); cf. *vin cuit* s.v. *VIN; **vino crudo** [It., lit. 'raw'], wine in its natural state, not boiled (cf. prec.); **vino de color** [Sp.], a rich sweet wine, used in the blending of sherry and other fortified wines; **vino de pasto** [Sp.], (see quot. 1965); **vino maestro** [Sp., lit.* 'master'], (see quots.); **vino nero** [It., lit. 'black'], dark red wine; **vino santo** [It., lit. 'holy'], a sweet white dessert wine; = *VINSANTO; **vino tierno** [Sp., lit. 'tender'], (see quots.).

1932 E. HEMINGWAY *Death in Afternoon* 338 *Vino corriente* is *vin ordinaire*. **1978** *Times* 21 Nov. 11/1 A humble *vino corriente*. **1673** J. RAY *Observations Journey Low-Countries* 387 The boil'd wine, which they call *Vin Cotto*, seemed to us much stronger than the wine unboil'd, which they call *Vin Crudo*. **1851** C. REDDING *Hist. Mod. Wines* (ed. 3) vii. 209 Their boiled wine, the Italian *Vino Cotto*, or *vin cuit*, as the French call it, is mingled with other growths, as well as by itself, for the sake of deepening colour or improving flavour. **1965** O. A. MENDELSOHN *Dict. Drink & Drinking* 356 *Vino cotto*, Italian name of must..concentrated to a syrup by evaporation (boiling) and used for sweetening Marsala..and similar fortified dessert wines. **1673** *Vino crudo* [see *vino cotto* above]. **1833** C. REDDING *Hist. Mod. Wines* ix. 241 At Bologna they boil most of their wines, which are then called *vino cotto*, the unboiled they call *vino crudo*. **1851** *Ibid.* (ed. 3) 205 All sherry wine is by nature of a pale colour, the darker shades are conferred by age, or by *vino de color*, or boiled wine. **1966** H. W. YOXALL *Fashion of Life* xv. 259 P.X., the *vino de color*..so sweet and heavy as to taste more like a liqueur than a wine. **1870** H. W. Allen *Number 3 St. James's St.* (1950) 184 Sherry..*Vino de pasto*..is a pale dry wine. **1902** A. BENNETT *Grand Babylon Hotel* xxii. 251 The dry sherries of Spain..Manzanilla, and Amontillado, and *Vino de Pasto*. **1965** O. A. MENDELSOHN *Dict. Drink & Drinking* 357 *Vino de pasto*, this Spanish term, literally 'pasture wine', is variable in meaning. It can stand for a 'table wine', one for everyday use.. But it is also used by English wine merchants for a mediocre kind of sherry, pale and not too dry. **1977** J. B. HILTON *Dead-Nettle* x. 85 For strong drink he had laid in..an octavo cask of Vino de Pasto. **1911** *Encycl. Brit.* XXVIII. 725/2 The *vino maestro* consists of a must which has only fermented to a slight degree and which has been 'killed' by the addition of about 17% of alcohol. **1965** O. A. MENDELSOHN *Dict. Drink & Drinking* 357 *Vino maestro*, Spanish name of the extra sweet and alcoholic wines used to tip up the lesser breeds. Especially used in compounding malaga. **1968** *Listener* 29 Feb. 268/1 When the *vino nero* flows in village bars, songs are sung about Messina as they were before the late Sicilian bandit, Salvatore Guiliano). **1686** G. BURNET *Some Lett.* 122 There is near the Lake of Guarda a very extraordinary Wine which they call *Vino Santo*, which drinks like the best sort of Canary, it is not made till Christmas, and from thence it carries the name of Holy Wine. **1833** C. REDDING *Hist. Mod. Wines* ix. 241 On the shores of the Lake of Garda they make a sweet wine..of prime quality, called Vino Santo. **1881** B. HEALEY *Week of Scorpion* v. 81 Their *Vino Santo*, the traditional Tuscan wine of hospitality. **1911** *Encycl. Brit.* XXVIII. 725/2 The *vino tierno* is made by mashing raisins..with water..pressing, and then adding alcohol..to the must. **1965** O. A. MENDELSOHN *Dict. Drink & Drinking* 357 *Vino tierno*, Spanish wine made from partially dried grapes. Used in compounding malaga.

2. Also **bino**. An alcoholic liquor distilled from nipa-palm sap, drunk in the Philippines.

1901 *Army & Navy Jrnl.* 7 Dec. 341/3 The native drinks, ordinarily known as 'vino', manufactured in these islands, contain in a large amount the poisonous principle contained in wood alcohol. **1903** *Census of Philippine Islands* (1905) IV. 119 *Nipa* or *sasâ* (*Nipa fruticans*) is a species of palm..from the sap or *tuba*..a liquor known as *nipa vino*, or *bino*, is extensively distilled. **1964** J. M. GARVAN *Negritos of Philippines* vii. 90 The young fellow hies off and finds a present for her..in the shape of *vino*. A very acceptable form is a large bottle of Filipino *vino*.

II. Naturalized forms. (vī·nō). **3.** Also **veeno**. Wine, esp. of an inferior kind. *slang*.

1919 W. H. DOWNING *Digger Dial.* 56 Vino, wine. **1935** L. DURRELL *Spirit of Place* (1969) 31, I bear up very well under the stacks of local vino I am forced to consume. **1942** BERREY & VAN DEN BARK *Amer. Thes. Slang* §100/14 Wine,..veeno, vino, wino. **1962** R. JEFFRIES *Exhibit No. 13* v. 50 We'd better use up the red vino. **1976** P. CAVE *High Flying Birds* i. 7, I was far too sozzled with vino to take much notice of Lloyd's vino-inspired ramblings.

‖ **vin ordinaire** (væn ǫrdinę̄r). [Fr.; cf. *VIN and *ORDINAIRE *sb.*] Simple French wine for everyday use.

1820 *Edin. Rev.* XXXIII. 344 And why should as large a duty be levied from the *vin ordinaire*, as from those of the first quality? **1852** E. TWISLETON *Let.*

9 Oct. (1928) iv. 56 You would laugh..if you saw me imbibing the Vin Ordinaire here! **1893** SOMERVILLE & 'ROSS' *Vine Country* viii. 159 A long array of dark-haired, white-coiffed women and girls were to be seen..finishing their jugs of *vin ordinaire*. **1926** A. HUXLEY *Essays New & Old* 17 The red wines of Carthage are really delicious, and even the smallest of *vins ordinaires* are very drinkable. **1955** E. POUND *Classic Anthol.* ii. 133 Our wine but vin ordinaire, We share spare Food but with gifts. **1975** *Woman's Jrnl.* Sept. 73/3 We drank a good *vin ordinaire* out of unlabelled bottles.

‖ **vin rosé** (væn roze). [Fr., f. as *VIN + *ROSÉ *sb.*]

1931 E. WAUGH *Remote People* 138 We dined on the roof; a delicious dinner; iced *vin rosé*. **1958** A. L. SIMON *Dict. Wines* 163/1 *Vin rosé* (1) a pink wine made from very ripe black grapes, the skins of which are not allowed to ferment with the wine, so that they only impart to it a pink tinge instead of a dark red colour; (2) A pink wine made from black and white grapes mixed in the pressing; (3) A white wine coloured with cochineal to the degree of pink required. **1981** P. THORNTON *Tendo for Short* xvi. 174 Bread and cheese, a salad, a bowl of fruit, a litre bottle of vin rosé.

‖ **vin rouge** (væn ruʒ). [Fr., f. as *VIN + ROUGE *a.*] **a.** French red wine; = *ROUGE *sb.*[3] 6.

1917 A. G. EMPEY *Over Top* 313 *Vin rouge*, French red wine made from vinegar and red ink. **1922** E. E. CUMMINGS *Enormous Room* vii. 169 A good deal of color in his cheeks and a good deal of vin rouge in his guts. **1931** E. LINKLATER *Juan in America* II. iv. 79 Rumors of a Chianti racket or a corner in *vin rouge* are obviously premature. **1960** *Guardian* 31 Dec. 5/3 The working [French] man drinks..at each meal..a litre of vin rouge ordinaire. **1980** L. ST. CLAIR *Obsessions* iv. 103 A bottle of *vin rouge* and two glasses.

b. In joc. or illit. spelling. Cf. *VIN BLANC 2. (In quot., *Mil. slang.*)

1919 [see *VIN BLANC 2].

‖ **vinsanto** (vin·santo). [It.] = *vino santo* s.v. *VINO 1 b.

1960 A. SICHEL *Penguin Bk. Wines* iii. 203 Although made all over Italy where fine-quality sweet wine can be made from dried grapes, the true Vin Santo Toscano should be made from grapes that have been left on the vine to dry.. Fine Vin Santo is also made from grapes picked in the normal time in October and allowed to dry until January or February and then crushed. **1973** *Country Life* 28 June 1929/3 The sweet dessert Vin Santo, made from dried grapes..is made quite widely in Italy. **1981** M. NASH *Death of Englishman* i. i. 16 The Marshal drank half a litre of red wine..and a drop of *vinsanto* on Sundays.

‖ **vinta** (vi·nta). [ad. Bisaya *binta*.] A kind of canoe used by the Moros in the Philippine Islands.

1900 F. H. SAWYER *Inhabitants of Philippines* xxviii. 362 These vessels rendered good service, and to some extent checked the incursions of the pirates, but they had not the speed to follow up the fast-rowing *vintas* of the Moros, which could always escape from them unless caught in narrow waters. **1906** *Soldier Slang* in C. McGovern *Sarjint Larry an' Friends*, *Paneo*, a native canoe or row boat, usually a single hollowed-out log; propelled by one or two rude oars, with a paddle for a rudder. *Ibid.*, *Vinta*, same as 'Parao'. **1946** G. SYMPSON *Bk. Thousand Things* 47 [Some] vintas, fishing boats hewn from single long logs, were marvels of workmanship and efficiency. **1979** P. DRISCOLL *Pangolin* xvi. 177 The long, slender *vintas* of the Muslim fishermen lay at anchor.

vintage, *sb.* Add: **2*.** *transf.* and *fig.* **a.** The date or period when a person was born or flourished.

1883 *Sunday Mercury* (N.Y.) 23 Sept. 6/4 'I want to wed a man for breach of promise,' said a maiden of the vintage of 1842, coming into a lawyer's office. **1941** F. L. ALLEN *Only Yesterday* vii. 129 Harding had no sooner arrived at the White House than a swarm of practical politicians of the McKinley-Forker vintage reappeared in Washington. **1945** A. L. ROWSE *West-Country Stories* 2 He was..a benevolent physician of such a vintage. **1967** M. ARGYLE *Psychol. Interpersonal Behaviour* iv. 79 In many cases the hero has a well-defined style of social behaviour—compare..the cowboy of 1890 vintage.

b. Hence, the date or period at which a thing was made or produced.

1929 R. S. & H. M. LYND *Middletown* xiv. 199 Examinations questions of the two periods indicate so little change in method and emphasis in teaching that it is almost impossible simply by reading a history examination to tell whether it is of 1890 or 1924 vintage. **1939** E. S. GARDNER *D.A. draws Circle* vii. 105 Mrs Fermal drove up in a rattling car of ancient vintage. **1946** E. O'NEILL *Iceman Cometh* (1947) i. 11 His pointed tan buttoned shoes, faded pink shirt and bright tie belong to the same vintage. **1972** J. ROSSITER *Rope for General Diets* iii. 31 Her accent was a creamy 1960-vintage Roedean.

3. *b. vintage chart*, year (in quots. *fig.*).

1933 T. E. LAWRENCE *Let.* 1 Aug. (1938) 773 Rather a vintage year, for books. **1964** A. LAUNAY *Caviare & After* xv. 104 If you arm yourself with a vintage chart—readily obtained from most wine merchants—you can learn the best years for the best wines. **1973** *Country Life* 15 Nov. 1541/3 Characteristic and impish Lowry of a fine vintage year 1969. **1981** *Times* 7 May 13/3 Britain's vintage chart..showed 1984 to have been a good..vintage year in the 1984.

c. *transf.* Denoting an old style or model of something, esp. a vehicle; **vintage car**, a

motor car made between 1905 (or 1917) and 1930; cf. *veteran car* s.v. *VETERAN *sb.* 3.

1928 M. ARLEN *Lily Christine* ii. 23 You should see mine in London—a vintage Buick. **1933** *Sat. Even. Post* 13 May 4/2 Alfred P. Sloan tries out a vintage model automobile. **1950** T. GUTHRIE in *Plays of Year 1949–50* III. 569 Goes on working at her vintage telephone. **1958** *Listener* 21 Aug. 261/2 The details of a vintage aircraft or an early locomotive. **1965** *Guardian* 6 Nov. 3/8 A 'veteran' is a car made before December 31, 1904. From January 1, 1905 to December 31, 1910, the definition..is 'Edwardian', and from then to 1930, cars in good shape are classified as 'Vintage'. **1979** J. LEASOR *Love & Land Beyond* iii. 32 The familiar vintage car smell of hot cast iron and old metal and carnauba wax polish. **1985** *New Yorker* 27 May 32/1 A debonair gentleman..wearing a vintage tweed jacket.

d. Characteristic of the best period of a person's work, etc.; classic.

1939 *Country Life* 11 Feb. 147/1 With a few minor reservations, this [play] may be recommended as vintage Coward. **1958** E. LANT *World in Ire* 7 My name is Larry Alden and maybe you'll only know it if you have a long memory or a stack of vintage jazz records. **1972** *Guardian* 25 Jan. 9/1 It was pure vintage Deauville to the background of Henry Hall type music. **1977** D. MACKENZIE *Raven & Kamikaze* i. 20 The Pole's..tone was compassionate. 'She is in love with me.' This was vintage Zaleski. 'Of course.' said Raven.

vintage, *v.* **1.** (Later examples.)

1893 SOMERVILLE & 'ROSS' *Vine Country* viii. 153 A little, incredibly bowed woman, who had been vintaging here at Quinault for the last eighty years. **1923** F. STARK *Let.* 16 Sept. (1974) I. 77 The people here are vintaging.. All the country from Cette to this place is vine. **1979** P. V. PRICE *Taste of Wine* v. 94/2 In many vintages, such as 1964, they were making wines quite different from those of the Medoc, because being further south they vintage a little earlier.

vinyl. Add: Also with pronunc. (vai·nil). **1.** Substitute for def.: The organic radical or group $CH_2=CH—$, which is equivalent to a molecule of ethylene with one hydrogen atom removed. Usu. *attrib.*

2. a. = *POLYVINYL b. Freq. *attrib.*

1939 *Nature* 13 May 787/2 Resins more recently developed, such as vinyl and polystyrene, are slowly making headway. **1951** K. MAYER *Artist's Handbk. Materials & Techniques* iii. 152 The most promising synthetic varieties [of resin] at present are the alkyd, acrylic, and vinyl types. **1957** *Times* 18 Nov. 11/2 An unbreakable doll of vinyl, a glamorous young lady with elegant nylon-clad legs, costing guineas. **1962** *B.B.C. Handbk.* 204 L.P. vinyl records giving basic pronunciation rules. **1969** J. H. STICKELMEYER in W. R. R. Park *Plastics Film Technol.* i. 7 Development of shrink wrapping techniques..has brightened the future for vinyls. **1976** P. LIVELY *Stitch in Time* viii. 98 There is paint that is non-drip and paint that is vinyl or emulsion and paint that is specially good for this, that or the other. **1983** *Sci. Amer.* July 54/1 Most synthetic organic materials such as vinyl or polyethylene are electrically insulating.

b. *spec.* a covering material or fabric made of or containing polyvinyl.

[**1902** *Jrnl. Chem. Soc.* LXXXII. i. 256 Sodium vinylacetate is much less soluble in alcohol than sodium crotonate.] **1955** *House & Garden* July 21/2 (Advt.), Dining-chairs upholstered in washable vinyl. **1970** *Interior Design* Dec. 753/1 The walls are papered with a grey and buff vinyl. **1978** S. BRILL *Teamsters* vii. 263 The inside of the tractor..was lined along the doors, the seats, and the dashboard in brown vinyl.

c. As the material of which gramophone records are made (so *piece of vinyl*). Also, a record. *colloq.*

1976 *New Musical Express* 31 July 27/1 A trio of knock out vinyl magnum opi, the like of which are not often to be heard on our sound systems. **1979** *Zigzag* Apr. 26/1 This is as depressing a piece of vinyl as I've listened to in many a full moon. **1977** *Rip Off & Torn* vi. 6/2, I had rushed home, clutching this piece of shiny vinyl to my chest. **1980** *Musicians Only* 26 Apr. 13/1 If they can continuously judge the results on vinyl. **1984** *Sounds* 1 Dec. 24/2 When that song..was released at the beginning of this month as the band's first single, the Immaculate Fools proved they could cut it equally well on vinyl.

3. *attrib.* and *Comb.*, as (sense *2), *vinyl-coated*, *-covered*, *-faced*, *-surfaced* adjs.; (sense 1) **vinyl acetate**, a colourless liquid ester, $CH_3CH·O·CO·CH_3$, used in the production of polyvinyl acetate and other commercially important polymers; **vinyl chloride**, a colourless toxic gas, CH_2CHCl, used in the production of polyvinyl chloride and other commercially important polymers; **vinyl resin**, any resin (*RESIN *sb.* 2*) that is composed of polyvinyl compounds.

1976 *Home & Freezer Digest* Feb. 37/1 The vinyl-coated wallpaper. **1966** *Guardian* 9 Aug. 8/4 A sophisticated traveller, equipped with..vinyl-covered suitcases. **1974** *Encycl. Brit. Micropædia* VIII. 104/2 In addition to its conversion to polyvinyl acetate, vinyl acetate undergoes similar reactions. **1978** *Jrnl. Chem. Soc.* XCV. 891 Vinyl chloride undergoes a change by the action of sunlight, producing an isomeride. **1933** [see polyvinyl chloride s.v. *POLYVINYL a.]. **1975** *Sci. Amer.* Mar. 415/1 Fifty of the train's 116 cars left the tracks.. and a tanker loaded with highly toxic vinyl chloride exploded soon afterward. **1967** *Jane's Surface Skimmer Systems 1967–68* 47/1 The structure is decked in vinyl-coated nylon fabric, which is also used for the skirt. **1970**

A. ROSS *Manchester Thing* 148 Two vinyl-covered kitchen chairs. **1977** 'E. CRISPIN' *Glimpses of Moon* i. 11 The furniture was all modern, from the oak counter..to the green glass-topped tables and the matching vinyl-covered chairs. **1960** *Times* 29 Nov. 2/4 Felt-backed vinyl-faced ..flooring. **1934** W. DAVIS *Adv. of Sci.* xii. 141 The great versatility of the vinyl resins was strikingly demonstrated in a house which the Carbide and Carbon Chemicals Corporation erected. **1942** GETTENS & STOUT *Painting Materials* 3 Chemically, they [sc. polyvinyl resins] are closely related to the vinyl resins. **1943** *Electronic Engin.* XVI. 216/3 The variation of dielectric properties with plasticiser content is examined for the vinyl resins. **1983** *Which?* Oct. 453/1 These are water-thinned paints usually based on vinyl resins. **1978** *Morecambe Guardian* 14 Mar. 6/6 Washable green and cream vinyl-surfaced cupboards are immensely childproof.

vinylidene (vainī·lidīn, vəinai·lidīn). *Chem.* [f. VINYL + *-IDENE.] The bivalent radical $CH_2=C<$; freq. *attrib.* Cf. *POLYVINYLIDENE.

1898 *Jrnl. Chem. Soc.* LXXIV. i. 188 Acetic anhydride converts it first into oxanilide, and subsequently into vinylidene oxanilide. **1930** [see *SARAN]. **1960** *Chem. in Brit.* XVI. 473/1 Vinylidene, iodide or leuco crystal violet are used as the colour formers..., and carbon tetrabromide or iodoform as the activator.

Vinylite (vai·nilait, -ail-). Also **vinylite**. [f. VINYL + *-ITE[1].] A proprietary name for a vinyl resin used esp. in the manufacture of gramophone records.

1929 *Official Gaz.* (U.S. Patent Office) 28 May 852/2 Carbide & Carbon Chemicals Corporation, New York.. Vinylite for artificial resins in powder form or in form of plates or sheets and molding mixtures containing such resins. **1949** *Trade Marks Jrnl.* 25 Sept. 1153/2 Vinylite.. Synthetic resin sold in slabs, sheets, bars and tubes for industrial purposes and moulding materials made of synthetic resins.. Carbide and Carbon Chemicals Corporation ..New York. **1936** [see *PLEXIGLAS]. **1956** A. HUXLEY *Adonis & Alphabet* 123 We must depend on the vinylite record and the slow-speed phonograph for any large-scale revival of spoken literature. **1974** *Encycl. Brit. Macropædia* XVIII. 35/2 In 1948, fine-groove records of clear Vinylite were introduced by Columbia records. **1977** *Gramophone* May 1762/1 The ace up the CBS sleeve, which really made the LP viable, was the changeover from noisy, abrasive, brittle shellac to quiet, soft, unbreakable vinylite for the new disc material.

vinylogous (vaini·lǫgəs), *a. Chem.* [f. VINYL + -LOGUE + -OUS (or directly after *analogous*).] Of, pertaining to, or being compounds that have the same molecular structure except for one or more —CH:CH— groups. Hence **vi·nylogue** (U.S. -log), a vinylogous compound; **viny·logy**, the relationships between vinylogous compounds.

1935 R. C. FUSON in *Chem. Rev.* XVI. 2 It is proposed to term such a group of compounds a vinylogous series. The members of a vinylogous series will then be vinylogs of one another. *Ibid.*, In this series the operation of the principle of vinylogy may be illustrated by a comparison of the esters of saturated fatty acids with their vinylogs, the α, β-unsaturated esters. **1966** *McGraw-Hill Encycl. Sci. & Technol.* XIV. 316/2 The principle of vinylogy may be used to explain why o- and p-nitrochlorobenzene are reactive..in contrast to the corresponding nitroderivatives, whereas the meta isomer is not. **1977** *Nature* 14 Apr. 660/2 The reaction (21)→(22) is not, as stated, a vinylogous Norrish type I reaction. **1978** R. J. BIRCH in Porter & Fitzsimmons *Further Perspectives Organic Chem.* 15 The decarboxylation which only very rarely under mild conditions, a fact which suggests intervention of a β-keto acid or a vinylogue of this compound.

Vinylon (vai·nilǫn). Also **vinylon**. [f. VINYL + -on (perh. after *NYLON).] Any of a class of synthetic fibres made from polyvinyl alcohol treated with formaldehyde which are used esp. in the manufacture of water-resistant fabrics.

1952 *Chem. Abstr.* XLVI. 1260 (*heading*) Pilot-plant manufacture of the synthetic fiber 'Vinylon'. **1955** KIRK & OTHMER *Encycl. Chem. Technol.* XIV. 718 Fibers of polyvinyl alcohol, spun from water solution and treated with formaldehyde, as developed in Japan (Vinylon,..etc.), have good strength, abrasion resistance, and heat resistance. **1965** *Economist* 23 Jan. 362/3 The plant will produce 30 tons of Vinylon fibre a day. **1975** *Textile Res. Jrnl.* XLV. 592/1 Tetoron is a polyester containing a benzene ring; Vinylon has no benzene ring.

Vinyon (vi·nyǫn). [f. VINY(L + *-on*, after *rayon*, *cotton*]. Any of several synthetic fibres which are copolymers of vinyl chloride with other vinyl compounds.

1938 *Newsweek* 14 Nov. 10/2 An important company's chemists are experimenting with another silk-like yarn called 'Vinyon,' for making hosiery, to rival du Pont's new 'nylon'. **1954** KIRK & OTHMER *Encycl. Chem. Technol.* XIII. 831 The strong, heat-bonding properties of Vinyon fibers are utilized for the production of sculptured carpet fabrics, nonwoven fabrics, paper, felts, and other such specialized products... However, inability to withstand boiling water has permanently relegated Vinyon fiber to specialized uses. **1965** A. J. HALL *Standard Handbk. Textiles* (ed. 6) I. 61 The original Vinyon having a low softening point has since been superseded by Vinyon N and Vinyon HH; they are copolymers of vinyl chloride and acrylonitrile. **1978** J. LABARTHE *Elem. Textiles* ix. 177 Vinyon fibers are made by three manufacturers in the United States at the present time.

viol, *sb.*[1] Add: **4.** *viol-string* (earlier example.)

1849 D. G. ROSSETTI *Let.* 8 Oct. (1965) I. 71 The hand trails weak upon the viol-string That sobs.

viola[2]. Add: **2. a.** Also *viola da gambist* = *viol da gambist*.

1977 *Early Music* Apr. 274 For viola da gambists there will be individual tuition.

4. With other distinguishing terms: **viola bastarda** = *lyra viol* s.v. LYRA 5; **viola da braccio** [lit. 'of the arm'], any member of the violin family, as opposed to a *viol da gamba*; *spec.* an alto violin, a viola; **viola pomposa**, an 18th-cent. viola with an additional string.

1724 Viola bastarda [see VIOL *sb.*[1] 2 a]. **1980** *Early Music* Apr. 250 (Advt.), The *viola bastarda* style of playing was a highly developed idiom that involved frequent changes of register. **1864** SANDYS & FORSTER *Hist. Violin* viii. 97 Vincentio Galilei, 'the heavy style of violinist', was..as able writer on music, and in 1582 names the viola da braccio, which he says, was like the lira not many years previously, the viola da gamba, and the violino, but not the violino. **1976** D. MUNROW *Instruments Middle Ages & Renaissance* 90/1 The name *viola da braccio*, which first occurs in 1543, was first used as a generic term like the older viola, but gradually came to refer to members of the new violin family. **1864** SANDYS & FORSTER *Hist. Violin* xii. 164 John Sebastian Bach introduced an instrument he called the viola pomposa, a five-stringed instrument, the viola pomposa or violoncello piccolo. **1954** GROVE *Dict. Mus.* VIII. 810/2 Sachs believes that Bach's 'viola pomposa' was really the violoncello piccolo. **1976** *Gramophone* Apr. 1602/2 The viola pomposa..a cross between cello and a big Tertis-style viola, with a splendidly rounded sound.

violable, *a.* Add: **violabi·lity** [-ITY], the condition of being violable; cf. INVIOLABILITY; *rare*.

1926 *Glasgow Herald* 1 Apr. 8/4 The suggestion..raises the whole question of the violability of contracts.

violarite (vəiǫlā·rait). *Min.* [f. L. *violār-is* of violet + -ITE[1].] A rare isometric sulphide of nickel and iron, Ni_2FeS_4, occurring in massive form with a violet-grey colour and a metallic lustre.

1924 LINDGREN & DAVY in *Econ. Geol.* XIX. 318 It has been shown that the 'polydymite' from Vermilion, Ontario and Key West, Nevada, is not identical with polydymite.. It might conveniently be designated as violarite. **1981** *Proc. Minerology* (*Amer. Inst. Mining Engineers*) 226 The close association of smythite with pentlandite indicates that its nickel content is derived from nickel liberated during the alteration of pentlandite to violarite.

violate, *v.* Add: **8.** To accuse or find (a prisoner on parole) guilty of violating the conditions of parole. *U.S. slang.*

1972 E. E. LANDY *Underground Dict.* 193 *Violate*,.. be returned to prison for a parole violation—eg. *I was violated.* **1973** J. MITFORD *Amer. Prison Business* (1974) xi. 222 If the parolee is 'violated' by his agent (this curious solecism is used, without any sense of irony, by everyone in Corrections, from the parole board to the convict), he is back to Square One. **1974** E. BRAWLEY *Rap* (1975) v. 133 My parole officer violated me on another phony beef and I wound up in the joint again. **1978** H. D. FRANKLIN *Victim as Criminal & Artist* v. 103 Living outside Los Angeles, with life going reasonably well, Braly suddenly found himself with a zealous new parole officer, who threatened to violate him for driving a car, for having a woman spend the night in his apartment, or for writing anything he disapproved of.

violation. Add: **1. b.** Now freq. in the U.S., esp. = an infringement of the law; an infringement of the rules in some sport.

1961 R. WEBSTER. **1973** E. CALDWELL *Annette* (1974) vii. i. 155 Jack immediately began complaining that the officer had no right to follow him and charge him with a traffic violation when he was certain he had not driven through any stop signs or red lights and had been careful to stay under the speed limit. **1976** *Honolulu Star-Bull.* 21 Dec. 18-D/1, I only hope Big Brother can take a lesson from the baseball umpire and use a little restraint and a little in line against those who transgressed out of the VH program because they would be equally as guilty of violations. **1982** L. BLOCK *Eight Million Ways to Die* (1983) iv. 47, I got a violation the last time I'd seen a cop ticket anyone for jaywalking.

violative, *a.* Delete quot. a 1797 (properly dated 1874).

violaxanthin (vəi:ǫlǎzæ·nþin). *Biochem.* [a. G. *viola-xanthin* (Kuhn & Winterstein 1931, in *Ber. d. Deut. Chem. Ges.* LXIV. 327): see VIOLA[1] and XANTHIN.] A xanthophyll, $C_{40}H_{56}O_4$, occurring as a yellow pigment in daffodils and some other plants.

1931 *Chem. Abstr.* XXV. 3351 The petals of the yellow pansy contain..a pigment which can be taken out with petroleum ether and on sapon. yields a well-crystd. hydrocarbon, $C_{40}H_{56}O_4$, violaxanthin (see *LUTEIN). **1977** *Nature* 13 July 160/2 The photosynthetic pigments found in the green alga] *Clodium*] *fragile* (β-carotene, zeaxanthin,..violaxanthin, and..siphonoxanthin..

viol da gamba. Add: Hence **viol da gambist**,

one who plays the *viol da gamba*; a viola da gambist.

1915 E. S. J. VAN DER STRAETEN *Hist. Violoncello* iii. 32 The next viol da gambist whose name has been handed down to posterity is Vincenzo Galilei... He was the father of Galileo Galilei and an excellent lutenist, gambist, composer and a very learned master of musical theory.

violence, *sb.* Add: **1. d.** Now used in political contexts with varying degrees of appropriateness.

1972 *Science* 23 June 1300/3 It is interesting that 38 percent of American men think that burning a draft card is violence, in and of itself; 38 percent think student protest is violence; and 22 percent feel sit-ins are violence. Clearly, many Americans consider acts of dissent, per se, to be violent. **1980** *Ibid.* 1301/1 Only 35 percent of American men define 'police shooting looters' as violence and only 56 percent define 'police beating students' in this manner. **1984** *Times* 2 July 1/8 Mr Scargill said: 'My facts show to me..that the people guilty of intimidation and of violence in this dispute have been: the police.' **1984** *Daily Telegraph* 5 Oct. 20/2 [At the Labour Party Conference] Much violence was done to the word violence, which it appears can be used to describe almost anything you do not care for.

violent, *sb.* Add: Revived in recent use. Also in *Comb.*

1978 *Jrnl. Communication* XXVIII. 180 The number of roles in which characters were the perpetrators of violence (violents), its victims, or both. *Ibid.* 187 The violent-victim ratio ranged from −1.40 in 1973 to −.06 in 1977, suggesting persistently negative but perhaps decreasing risks of general victimization. **1983** J. SCOTT *All Pretty People* ii. 12 The psychiatric wards are filled with violents, and they release them every day.

violer. (Later examples in form *violar*.)

1797 *Encycl. Brit.* XII. 491/2 At this period violars, or performers on the viele or viol,..abounded all over Europe. **1801** BUSBY *Dict. Mus.* s.v., *Violars*, certain practical musicians much encouraged in Provence during the twelfth century; and so named because they performed on the vielle and viol. **1931** M. BARING *In End is my Beginning* iv. 50 Her son was given..a band of four violars.

violet, *sb.*[1] Add: **4. b.** The scent of violets, esp. as used in cosmetics.

1850 in A. DAVIS *Package & Print* (1967) Pl. 8 Essence of Millefleur, Bouquet, Marechalle, Reseda [*sic*], Violet, Tubereuse. **1890–1** *T. Eaton & Co. Catal.* Fall & Winter 42/2 Perfumes..white violet, Italian violet,..satchet powder,..violet. **1898** [see *LILAC 2 c]. **1928–9** *Army & Navy Stores Catal.* 496/3 Natural Flower Perfumes.. Muguet, Rose, Violet. **1982** *Christian Science Monitor* 26 Nov. 8/3 I stocked up on Roger & Gallet hand soaps in violet and sandalwood.

4*. An onion; in *pl.* spring onions, sage and onions. *slang.*

In many dicts. of slang but no contextual examples found.

1890 BARRÈRE & LELAND *Dict. Slang* II. 397/1 *Violets* (common), an euphemism for sage and onions. **1903** FARMER & HENLEY *Slang* VII. 277/1 *Violet* (or *garden-violet*),..An onion: spec. in *pl.* = spring onions used as a salad. Also (a) in pl. = sage-and-onion stuffing. **1929** M. A. GILL *Underworld Slang*, *Violets*, onions. **1946** J. IRVING *Royal Navalese* 182 *Violets*, spring onions.

5. b. *violet-adorned*, *-dewy*, *-rippling*.

1953 C. DAY LEWIS *Italian Visit* 57 Violet-adorned beauties. **1932** BLUNDEN *Halfway House* 73 All fresh and violet-dewy. **1949** — *After Bombing* 29 And some have sung though never seen Melodious, voiceful, violet-rippling, blushful Hippocrene.

6. violet cream, (a) a violet-scented cosmetic cream; (b) a violet-flavoured confection; **violet powder** (earlier example); **violet tea**, an infusion made from dried violet flowers.

1912 J. WEBSTER *Daddy-Long-Legs* 203, I started down town to-day to buy..a jar of violet cream and a cake of Castile soap. **1969** A. CHRISTIE *At Bertram's Hotel* xxi. 294 They were lovely chocolates.. There were some violet creams. That's the nicest of chocolates. The violet creams. **1980** *Times* 19 Apr. 13/3 The most delicious chocolate that has a crystallised violet on top. **1856** DICKENS *Little Dorrit* (1857) II. iii. 358 Her bonnet smelt of violet powder. **1853** MRS. GASKELL *Ruth* III. iii. 115 She..begged her to buy a cup of soothing violet-tea. **1904** *Cassell's New Dict. Cookery* xviii. 1033/1 *Violet tea*, this is a soothing beverage for persons suffering from bronchitis and similar affections.

violet, *a.* Add: **2. b. violet-green swallow**, a dark-coloured swallow with white patches, *Tachycineta thalassina*, found in western North America.

1858 S. BAIRD *Birds Pacific Rail Road* 311 Violet-Green Swallow.. Rocky Mountains to Pacific. **1972** L. HANCOCK *Sleeping Bag* viii. 205 A pair of violet-green swallows built a nest..above our bedroom window.

d. *violet ray*: (*a*) a ray of violet light; (*b*) a ray of ultra-violet light. Also *attrib.* Now *rare*.

1803 *Jrnl. Nat. Philos.* V. 255 He placed muriate of silver without the solar spectrum and next to the violet rays. This oxide became blackened in a very short time, it became still deeper in the violet ray, but fainter in the blue, and so on. **1903** *Sci. Amer.* 20 June 473/1 The so called 'violet ray', which is now so common in literature, should be dropped, and 'blue ray' should be substituted for it, the blue being the predominant visible color in the place. **1925** A. BENNETT *Jrnl.* 7 Feb. (1933) III. 74 He went in to down under the light of his violet-ray lamp (equivalent of sunlight or some such thing).

1929 *Proc. I.R.E.* XVII. 1388 The real difficulty lies in the modernization of household equipment such as electric fans..and violet ray machines. All of these must be designed with elimination of radio interference in mind. **1933** *Amer. Speech* VIII. ii. 55/2 Nature-ray, all right, or O.K., as in the phrase, *I'm feeling nature-ray*. It's obviously is based on violet rays. **1966** J. S. COX *Illustr. Dict. Hairdressing & Wigmaking* 79/2 *High frequency*, an electric current of high voltage and very low amperage, used in scalp treatment. Also called Violet Ray because of the violet light produced in the glass applicator.

4. Special collocation: *violet shift* (Astr.), displacement of spectral lines towards the violet end of the spectrum; decrease in the wavelength of electromagnetic radiation; usu. called a *blue shift*; cf. *RED SHIFT *sb.*

1959 *Listener* 3 Dec. 971/1 The spectroscope will reveal a violet shift for one component, and a red shift for the other, according to the familiar Doppler principle. **1977** J. NARLIKAR *Struct. Universe* iii. 76 Hence negative *z* is interpreted as 'violet shift'.

‖ **violetta** (vi,ole·tă). Pl. **violette**. [It.] 1. = VIOLET *sb.*[3] and VIOLETTE.

1740 J. GRASSINEAU *Mus. Dict.* 327 *Violetta*, or little viol, is in reality, our triple Viol. **1876** STAINER & BARRETT *Dict. Mus. Terms* 447/2 *Violetta*, a little viol. **1979** *Early Music* Jan. 137/3 Kolneder also points out that the *violette*, *viole* and *violette* are regularly written in the bass clef in the lower octave.

2. *violetta marina* (see quots.); cf. *viol d'amore* s.v. VIOL *sb.*[1] 2 b.

1801 BUSBY *Dict. Mus.*, *Violetta Marina*, a stringed instrument not now in use. Supposed to have been similar in shape and tone to the *Viol d'Amour*. It was first invented and first brought out by Signior Castrucci in the year 1732. **1889** GROVE *Dict. Mus.* IV. 267/2 *Violetta marina*, a name found occasionally in the scores of Handel and his contemporaries, probably to designate the Viola d'Amore. **1963** *Listener* 21 Feb. 354/1 The use of the violetta marina in *Orlando*.

‖ **violette de Parme** (vi,olet dǝ parm). [Fr.] = *Parma violet* (b), (c) s.v. *PARMA[1].

1904 E. NESBIT *Phoenix & Carpet* xii. 232 A little scent sachet labelled 'Violettes de Parme'. **1913** 'SAKI' *When William Came* (1914) xiv. 243 All the hangings..bedside *de Parme*, all the furniture, rosewood. **1938** *Fortnum & Mason Price List* 55/1 Violette de Parme Toilet Soap. **1968** M. KAY *Masha* xxv. 252 She..left, the smell of *Violette de Parme* perfume trailing behind her.

violin, *sb.* Add: **1. c.** Similarly, *to play second violin*, to take the subordinate part. *rare*.

1902 G. B. SHAW *Let.* 20 May (1972) II. 271, I don't see Janet playing a silly second moral violin like Judith.

4. *violin concerto*, *sonata*, *-stand*; *violin-player* (earlier examples), *-playing*; **violin spider** [see quot. 1969[2]], a small brown and orange spider, *Loxosceles læta*, whose bite can be fatal to man and which is chiefly found in South America.

1889 GROVE *Dict. Mus.* IV. 293/2 Mozart in his younger years seems to have been great as a violinist than a piano-player, and his Violin Concertos,..are the most valuable compositions in that form. **1934** Violin concerto [see *pianoforte concerto* s.v. *PIANOFORTE b]. **1797** *Encycl. Brit.* XII. 493 The most celebrated violin players of Italy..have been Farina, M. Angelo Rossi,.. **1814** JANE AUSTEN *Mansfield Park* I. xii. 243 The late expensive ..violin-player in the servants' hall. **1813** J. MAYNE *Jrnl.* 23 Jan. (1909) 344 The Romans have no idea of what good violin-playing is. **1929** Y. MENUHIN in D. VILLIERS *Next Year in Jerusalem* 335 The technical points of violin-playing common to the Jew and the gypsy. **1889** GROVE *Dict. Mus.* IV. 288/2 Towards..1630, we find the first compositions containing that which is really the form of the classical Violin Sonata. **1969** *Daily Colonist* (Victoria, B.C.) 8 June 14/1 War on a colony of deadly South American 'violin spiders' in a Sierra Madre Memorial Park was intensified Saturday when the city of Sierra Madre called a commercial exterminator into the battle. **1969** *Pest Control* Oct. 54/2 The violin spider is so named because its cephalothorax has a distinctive shape of a violin on its head. The handle of the violin points toward the abdomen. **1979** *Daily Tel.* 15 Dec. 15/1 Hundreds of thousands of poisonous violin spiders, whose bite can be lethal and for which there is no known antidote, have invaded Johannesburg. **1915** D. H. LAWRENCE *Rainbow* iii. 80 A drawing-room..with a piano and a violin-stand.

Hence **violinic** (*rare*), **violinistic** *adjs.*; **violini·stically** *adv.*

1776 J. HAWKINS *Gen. Hist. Sci. & Pract. Mus.* III. iv. i. 431 The Violini piccoli alla Francese must in strictness signify small violins; and of these there are none now known but that contemptible instrument called the kit. **1889** GROVE *Dict. Mus.* IV. 813 *Violino piccolo*, a violin of small size, but of the ordinary parts and proportions, differing in this respect from the pochette or kit. **1961** *Times* 24 July 7/6 Mr. Carl Pini had taken the trouble to use a *violino piccolo* for the performance. **1979**

Early Music Jan. 3 The instruments..include..violino piccolo (with its characteristically scalloped outline) theorbo-lute, and bass viol.

violist. Add: Also, a player on the viola.

1977 *Gramophone* Nov. 860/2, I prefer some details of phrasing and the tone quality of the old Tchaikovsky Quartet which has Rudolf Barshai as violist. **1978** *Oxford Times* 3 Feb. 14 The splendid viola part reminds one of the fact that the composer was a violist.

viologen (vai,ǫ[u]·lǫdʒen). *Chem.* [f. VIOL(ET *sb.*[1] + -O + -GEN.] Any of several salts of the 1,1'-dialkyl-4,4'-bipyridylium ion, $(-C_5H_4N^+R)_2$, which are used as redox indicators.

1933 MICHAELIS & HILL in *Jrnl. Gen. Physiol.* XVI. 859 The new indicators will be designated as viologens... They differ from other indicators in several respects: in the first place, the color is exhibited by the reduced form, whereas usually the oxidized form is the colored one. **1964** J. W. LINNETT *Electronic Struct. Molecules* viii. 109 The viologens..can be formed by reducing γγ'-dipyridyl, or its derivatives, with sodium. **1980** *Nature* 27 Nov. 321/1 Methyl viologen (paraquat) acts as electron carrier. **1984** *New Scientist* 31 May 23/1 New electrochromic displays (ECDs) based on viologens—close relatives of..paraquat that turn purple when electrochemically reduced—look set for commercialisation.

violoncello. Add: **1. b.** A player on the violoncello. Cf. VIOLIN *sb.* 1.

1863 GEO. ELIOT *Let.* 6 Oct. (1954) III. 456 We have our violoncello, who is full of sensibility.

3. *violoncello piccolo* (see quot.)... **violoncello piccolo**, a small variety of violoncello; cf. *VIOLINO PICCOLO.

1788 J. WOODFORDE *Diary* 19 Sept. (1927) III. 46 There is not perhaps a better [Violoncello player] in the Kingdom. **1889** GROVE *Dict. Mus.* IV. 813/1 *Violoncello piccolo*, a violoncello of the ordinary pattern, but of smaller size and having thinner strings. **1959** *Collins Mus. Encycl.* 706/1 *Violoncello piccolo*, a small-sized cello for which Bach wrote obbligato parts in 9 of his cantatas. **1977** *Gramophone* Oct. 672/1 It falls short in the provision of a cello rather than Bach's own stipulation of a violoncello piccolo.

‖ **violon d'Ingres** (vi,olǫ̃ dæ̃ŋgr). [Fr., lit. 'Ingres' violin'; cf. *INGRES.] An occasional pastime, an activity other than that for which one is well-known or at which one excels.

1931 *Larousse du XXe Siècle* IV. 68/1 *Violon d'Ingres*. Une légende, assez suspecte, prétend que le peintre Ingres était plus fier de son jeu sur le violon, jeu qui était fort ordinaire, que de sa peinture, qui l'avait rendu illustre. **1964** ROBERT *Dict. Langue Française* VI. 1004/1 *Violon d'Ingres*, le talent, ...un artiste, de pratiquer un art qui n'est pas le sien (Ingres peintre aurait excellé au violon). **1952** E. HYAMS *New Statesmanship* 175 President Eisenhower's violon d'Ingres being golf, it left to.. John Foster Dulles to follow to N. Kruschev into the pages of the *New Statesman*. **1968** J. M. WHITE *Nightclimber* v. 31 Fame, if we achieve it, is thrust upon us for a minor years when we have forgotten our violons d'Ingres. **1969** J. POWELL *Barrel of Stones* viii. 73 This weekend, I'm perfectly prepared to play his violin—badly—to visitors..instead of showing them his pictures. **1977** *Listener* 13 Sept. 345/1, I suspect that music was no more than Rousseau's *violon d'Ingres*, and that the composition of *The Village Soothsayer* was a happy fluke.

viomycin (vai,ǫmai·sin). *Pharm.* [f. *vio-*, of unkn. origin + *-MYCIN.] A bacteriostatic antibiotic, $C_{25}H_{43}N_{13}O_{10}$, produced by several species of bacterium which has been given as an alternative drug in the treatment of tuberculosis, usu. by intramuscular injection of the sulphate.

1950 *Sun* (Baltimore) 22 Apr. 15/1 The company [sc. Chas. Pfizer & Co.] just announced a new anti-biotic drug called 'viomycin' [p.] to be effective in the treatment of tuberculosis. **1976** *Sci. Amer.* Dec. 11/1 Kanamycin, capreomycin, and viomycin are aminoglycosides which can be substituted for streptomycin in TB drug therapy in patients hypersensitive to streptomycin or who are infected with bacteria resistant to streptomycin and basic viomycin.

viosterol (vai,ǫsti·rǫl). *Biochem.* [f. *ULTRA)VIO(LET *a.* and *sb.* + *-STEROL.] Calciferol (vitamin D_2) obtained by the irradiation of ergosterol with ultraviolet light.

1929 *Jrnl. Amer. Med. Assoc.* 31 Aug. 693/1 The Council [on Pharmacy and Chemistry] has adopted the term viosterol to designate irradiated ergosterol. **1933** *Discovery* May 160/1 The daily dose of yeast milk required for the prevention of rickets is approximately 24 ounces (120 rat units) of cod liver oil, 1 teaspoonful (200 rat units) of viosterol 10 drops (800 units). **1971** A. A. MICHELE *You don't have to Ache* vii. 162 Today rickets is rare. Vitamin D and viosterol have been given to most youngsters.

V.I.P., **VIP.** **1. a.** An abbrev. f. the initial letters of 'very important person', esp. a high-ranking guest. Freq. *Mil. slang* in early use.

1933 C. MACKENZIE *Water on Brain* viii. 114 'At the moment he has a V.I.P. with him'... Miss Glidden seemed to divine his perplexity, for..she turned round and whispered through a pursed up mouth, 'Very Important Personage'. **1945** *Daily Mirror* 11 Aug. 5/4 Then they

started pouring buckets of water on the crowds below..until a very important person happened along. The VIP got a bucketful all to himself. **1946** *Penguin New Writing* XXVII. 31 He had a V.I.P. temperament. **1946** E. WAUGH *Diary* 31 Mar. (1976) vi. 645, I found I had been categorized VIP—Very Important Person. **1950** Consented to be asked 'Are you a VIP?' **1953** J. PUDNEY *Ring for Luck* 19 A sophisticated traveller, something of a V.I.P., expecting to be met and seen off wherever he went. **1964** MRS. L. B. JOHNSON *White House Diary* 24 Mar. (1970) 98 There were about three hundred people in front of the platform—VIP's from Huntsville. **1980** J. DITTON *Copley's Hawk* iii. 221 For their own safety—as with all VIPs—their going had to be kept a secret.

b. Spelt *vip* (vip), as an acronym. Now *rare*.

1945 *Fortune* Aug. 167 It's been prime and cosmic 'Vips', usually travel in plush C-54's. **1968** *Listener* 22 Aug. 243/1 'Provided you are a Vip,' he added, joining the initials so that they formed a word. I was rather intrigued by this and asked him to explain what he meant by Vip, was he talking about it as a verb or a noun?.. 'A Vip,' he said, 'is what in the days of you British used to be called a burra sahib, a senior official, but now we have been ordered to use this new Hindi word.'

2. *attrib.*, esp. as *V.I.P. lounge*, *treatment*.

1945 *Yank* 24 Aug. 6 The front page story on the nation's biggest VIP program, was given the alpha-plus VIP treatment. **1956** MRS. L. B. JOHNSON *White House Diary* 1 Aug. (1970) 552 We drove up to an entrance and went quickly into a small VIP lounge where they shut the door. **1980** G. MCDONALD *Snakes* i. 3 I explained the situation to the head stewardess in the V.I.P. lounge.

viper. Add: **3*.** One who smokes marijuana or opium, esp. habitually. Also, a heroin addict. Now *rare*.

1938 M. BERGER in *New Yorker* 12 Mar. 36/1 They play special recordings of 'viper' or weed, songs with weird ritualistic themes. *Ibid.*, I was not a viper, which is the Harlem word for a marijuana smoker. **1946** MEZZROW & WOLFE *Really Blues* xii. 252 vipers..the hipper. **1948** *Background of Blues* 21 The addict is a 'viper' who 'likes to smoke' or 'climbs the bush'. **1956** D. WEBB *Line-Up for Crime* iv. 69 Three good shots of heroin and you're a viper for life. A viper is an addict. **1959** MUIRAGH & HARRIS *We Are the Shadow* (1960) i. 19 They said I wasn't just a viper but also a pusher. **1977** *Canadian* 8 Jan. 15/1 He wrote that song in the late Thirties, back when people who smoked marijuana used to be called vipers.

4. a. *viper-green*.

1968 L. DURRELL *Monsalvat* xvi. 300 Among the thickets of reed and sedge, in the ruins of the Lazaretto where the peacock stood clutching its viper-green crest by the occasional dinging trees, you could hear the Fayum crackling of the tree-frogs. **1978** *Daily Star* (Sheffield) 3 Dec. 25/9 (Advt.), 1975 (N) VW *Passat* 1300... Attractive viper-green.

b. *also*, to have an effect like that of a viper's venom. *nonce-use.*

1953 DYLAN THOMAS *Under Milk Wood* (1954) 63 Mr Pugh..plays innocent pieces of viper-music in the music..

viperid (vai·pərid), *sb.* and *a. Zool.* [ad. mod.L. *Viperidæ*, f. L. *vipera* VIPER: see *-ID[3].] **A.** *sb.* A snake of the family Viperidæ, which comprises the true vipers and in most modern classifications the pit vipers, all venomous snakes having hinged maxillary bones, that are toothless except for a fang that is folded back in the mouth when not in use. **B.** *adj.* Of or pertaining to this family.

1909 WEBSTER, *Viperid* a. & n. **1956** L. M. KLAUBER *Rattlesnakes* II. xi. 716 The viperids were evolved directly from primitive aglyphous ancestors, rather than from either opisthoglyphs or proteroglyphs. **1965** R. & D. MORRIS *Men & Snakes* iv. 97 Generally speaking, the viperids are not as dangerous to man as the cobras and mambas. **1969** A. BELLAIRS *Life of Reptiles* I. v. 193 It has been suggested that the viperid and solenoglyphous type of jaw was derived from that of the proteroglyphs. **1977** A. HALLAM *Planet Earth* 272 The viperids..and the viperids..are both known from the Miocene.

vipoma (vaipō·mă). *Path.* [f. VIP s.v. *V 5 b + *-OMA.] A tumour which secretes vasoactive intestinal polypeptide (VIP).

1973 S. R. BLOOM et al. in *Lancet* 7 July 15/2 We suggest that tumours secreting V.I.P. should be termed 'vipomas'. **1980** *Brit. Med. Jrnl.* 31 Nov. 1334/2 Vipoma syndrome. **1983** *Daily Tel.* 10 Dec. 7/5 About that patient's symptoms were unequivocally due to the vipoma of the left kidney.

viræmia (vairī·miă). *Med.* Also (*U.S.*) **viremia**. [f. VIR(US + -*æmia*, after *anæmia*, *leukæmia*, etc.] The condition in which viruses are present in the bloodstream. Hence **viræ·mic** *a.*

1947 DORLAND & MILLER *Med. Dict.* (ed. 21) 1619/1 *Viremia*. **1954** *Jrnl. Pediatrics* XLIV. 20/1 Little is known about the presence of viremia during the course of the disease [sc. poliomyelitis] in man. *Ibid.* 25/2 The infectious process is now almost generally accepted to be initiated by multiplication in the intestinal mucosa, a viremia that may or may not be followed by multiplication of virus in the central nervous system. **1964** M. HYNES *Med. Bacteriol.* (ed. 8) xxv. 383 Infection with poliovirus is almost entirely by way of the intestinal tract. **1979** *Lancet* 29 Apr. 884/1 The possibility that this illness is a consequence of early viræmia should not be overlooked. **1979** *Nature*

17 May 200/2 Chickens, mice or cats neonatally infected with oncornaviruses become viraemic, develop low or no anti-viral immunity and usually succumb to their tumours.

virago. Now usu. with pronunc. (virā·go).

viral (vai·răl), *a.* [f. VIR(US + -AL.] Of the nature of, caused by, or pertaining to a virus or viruses.

1948 *Diagnostic Procedures for Virus & Rickettsial Diseases* (Amer. Public Health Assoc.) 35 Viral agents belonging to the psittacosis group. **1958** *Spectator* [see s.v. *PARROT *sb.* 4]. **1974** R. CROSSMAN *Diaries* (1977) 3 July 1958 She said the viral pneumonia seemed to have cleared up, and the only area of danger was bronchial tubes. **1976** EDINGTON & GILLES *Path. in Tropics* (ed. 2) xi. 577 Although many viruses may cause hepatitis there are two main types—infectious hepatitis and homologous serum jaundice. It is now recommended that these conditions be redesignated viral hepatitis type A and viral hepatitis type B respectively (WHO, 1973). **1983** *New Scientist* 10 Mar. 642/1 If chemicals can be found that inhibit the viral polymerase without doing too much damage to the human polymerase, then the path to anti-viral drugs will have been cleared.

Hence **vi·rally** *adv.*, by a virus or viruses.

1968 *Biochem. & Biophys. Res. Communications* XXXIII. 505 The normal and the virally transformed cells differ in regard to the quantity of glycolipids. **1977** *Proc. R. Soc. Med.* LXX. 532/1 Diabetes is indeed a virally infected cell. **1980** *Nature* 4 Dec. 432/2 Interest in the viral theory was restored as examples of virally-induced tumours were discovered in mammals. **1982** *Proc. Nat. Acad. Sci.* LXXIX. 6822/1 Virally transformed rat embryo cells.

Virchow-Robin space (f-, vi·ǝko ro·bæn). *Anat.* [Named after R. L. K. *Virchow* (1821–1902), German pathologist, and C. P. *Robin* (1821–85), French histologist.] An extension of the subarachnoid space surrounding a blood vessel for a short distance as it enters the brain or the spinal cord.

1890 H. HILL tr. *Obersteiner's Anat. Central Nervous Organs* 205 The intra-adventitial spaces, or the Virchow-Robin spaces. **1936** *Brit. Jrnl. Psychol.* XXVI. 437/2 Between the adventitia of the vessels and the musculatis a considerable space is seen in all stretches and flowing, the adventitial lymph-space (Virchow-Robin space). **1976** *Path. Ann.* XI. 298 (*caption*) Cryptococcosis involving a Virchow-Robin space in the cerebral cortex.

virement (vī^ə·ʒmań, vai^ə·ǝmĕnt). [a. F. *virement*, f. *virer* to turn (cf. VEER *v.*[1]).] A strictly regulated process of transferring items, esp. public funds, from one financial account to another.

[**1873** *Ann. Reg.* 1872 I. 272 A dubious financial transaction, of the nature of what the French call a *virement*, had been the cause.] **1918** *Rep. Select Committee on National Expenditure in Parl. Papers* (H.C. 1867) VII. 15 Such surpluses being regarded in the meantime, at the Treasury by the Appropriation Act, and its exercise is conditioned by Mr Monk's Resolution of 4th and 5th March 1879. **1917** A. J. V. DURELL *Princ. & Pract. System of Control over Parliamentary Grants* i. ii. 39 In the case of the army and navy the powers of virement enable savings on the grants to be realised and taken over to the vote to be applied in making good a deficiency on another vote. **1921** W. S. CHURCHILL in M. Gilbert *Winston S. Churchill* (1977) IV. Compan. iii. 1625 A supplementary estimate will not be necessary if the powers of *virement* are used in the full. **1938** W. I. JENNINGS *Cabinet Government* viii. 133 Virement is thus of two kinds, according as it is between sub-heads or between votes. **1940** *Evening May's Law of Parl.* (ed. 14) xxv. 677 There is no definite statutory authority for this long standing practice. **1958** *Times* 30 May 15/6 Criticism of the practice of virement whereby the Treasury allow Service Departments to use surpluses on Votes to meet deficits on other Votes is made by the Committee of Public Accounts in their second report. **1966** A. H. MARSHALL *Financial Administration in Local Govt.* xxi. 349 Some authorities..permit 'virement'... **1978** *Language for Life* (Dept. Educ. & Sci.) xxi. 349 Heads of departments..trolled under specific 'heads', with no virement of any kind.

viremia, **-ic**, varr. *VIRÆMIA, -IC.

virga (vɜ·ɹgă). Pl. **-ae**. [a. L. *virga* rod.] **1.** *Mus.* A symbol used in plainsong notation; the note designated by this (see quots.).

1908 H. DUNSTAN *Cycl. Dict. Mus.* 439/2 *Virga*, one of the signs in Neum notation. **1925** *Ibid.* (ed. 4) 555/2 *Virga* (*L.*) 'a twig'. A square note with a stem or tail... The plainsong Virga is interpreted as a quaver, 'a tail' for a long note. **1978** *Gramophone* Aug. 365/2 But most of all, these singers had introduced a new lightness of sound..in it. **1947** M. PIERIK *Song of Church* vii. 213 Among the Latin neumes ten have Greek names, ..while eight have Latin names, *clivis*, *pes*, *punctum*, *virga*, *torculus*, [etc.]. **1954** A. HUGHES *Early Medieval Music* xi. 380 The essential difference between the pre-Franconian mensural notation..and the modal notation which preceded it lies in the fact that the single notes..are clearly differentiated..the *virga* for a long note..the virga of plainsong] for a long note. **1978** *Gramophone* Aug. 365/2

2. *Meteorol.* Streaks of precipitation that appear to be attached to the undersurface of a cloud and usually evaporate before reaching the ground. Also *pl.* in same sense.

1947 M. A. GARBELL *Trop. & Equatorial Meteorol.* iii. 48/1 Streaks of falling rain that do not reach the earth, usually called virga (Latin virga, 'rod')—are then observed under that part of the cloud from which precipitation is falling. **1959** R. E. HUSCHKE *Gloss. Meteorol.* 611

Virga is frequently seen trailing from altocumulus and altostratus clouds. **1968** *New Scientist* 4 Jan. 22/1 (*caption*) An untreated cap cloud which that cumulus-has little if any virga or snow plumes. **1979** L. J. BATTAN *Fund. Meteorol.* viii. 152 When the water or ice particles evaporate before reaching the ground, the precipitation is called *virga*.

virger. For *Obs.* read '*Obs.* exc. at certain cathedrals, such as St. Paul's and Winchester' and add later examples.

1975 M. SULLIVAN *Watch How You Go* x. 174 The Canons' Vergers—in his usual spelling 'vergers'. **1978** *Church Times* 1 Sept. 8/2 If the Dean were still preceded by a wanded virger on his way to the divine office. **1983** *Ibid.* 11 Mar. 13/3 To the Dean and Chapter of Winchester invite applications for a virger.

virgin. Add: **2.** Also *b.* [back-formation] *virge* *v. trans.*, to conduct (cathedral dignitaries, etc.) in the manner of a virger.

1975 [see *VIRGER]. **1977** *Church Times* 25 Nov. 8/5 The Deacon is virged from his place to the Altar for the Gospel during the Fanfare.

virgie (vɜ·ɹdʒi), *a.* and *sb.* Colloq. abbrev. of *virgin* *a.* and *sb.* Cf. -IE.

1930 [see *MUCKED *ppl. a.*]. **1965** H. GOLD *Man who was not with It* xxvii. 251 How do you know? 'A virgie is a virgie.'

virgin, *sb.* and *a.* Add: **I. 2. f.** *transf.* A naive, innocent, or inexperienced person. Freq. with adj. indicating sphere of activity. *colloq.*

1953 A. MOOREHEAD *Rum Jungle* iv. 53 A new player [at two-up]..is known as a 'virgin'. **1964** L. DEIGHTON *Funeral in Berlin* xxiii. 173 He had done no previous political work. He described himself as a 'political virgin'. **1970** *Daily Tel.* (Colour Suppl.) 15 May 36/1 They were in competition..called 'Be a Millionaire'... I was an industrial virgin in those days but friends told me to have a go. I entered and won. **1976** E. STEWART *Launch!* 89 'That's a good cause to be a sexual virgin'... 'I'm a computer virgin,' he said. **1981** *Times* 15 Feb. (*Sunday*) 18 The newcomers are aerial virgins going into Ping-pong balls.'

4. c. *Virgin Mary* [after *bloody Mary* s.v. *BLOODY *a.* and *adv.* C. 2], a glass of tomato juice (see quots.). *Chiefly U.S.*

1970 W. GOLDMAN *Magic* ii. 90 Some girl wanted a Virgin Mary. The bartender might. **1977** *Dallas Morning News* 1 Oct. 1/3 In that bar..a Virgin Mary,..A Bloody Mary without the vodka. **1981** T. HEALD *Murder at Moose Jaw* ix. 109 Crombie ordered himself a straight tomato juice with..Worcester. The Colonel did not, Bognor noted with approval, order it to be a 'Virgin Mary'.

5*. A cigarette made of Virginia tobacco. *slang.* Now *Obs.* or *rare*.

1923 J. MANCHON *Le Slang* 329. **1935** C. BROOKS *Frame-Up* iv. 54 You gave me a cigarette... **1940** GRAVES & HODGE *Long Week-End* iii. 43 In the early Twenties..in offering a cigarette-case one would say, 'I hope you don't mind it's only a Virgin.'

II. 17. a. *virgin bush*, bush land not brought under cultivation; *virgin country*, country that has not yet been opened up to the outside world for trade, etc.; *virgin land*, previously uncultivated land, esp. [tr. Russ. *tselina*] in Western Siberia and Kazakhstan, land made the subject of an intensive agricultural programme by the Soviet government since 1954; *virgin soil* (earlier *lit.* and *fig.* examples).

1906 W. B. *Where White Man Treads* 297 A heroic devil who..for the virgin bush. **1966** *Listener* 14 Oct. 581/2 Roustabout..with his father's cattle in the virgin bush. **1982** *Times* 15 Feb. 6/2 The ..gold..celebrated Wuss and it was the virgin forest. **1981** *Daily Express* 7 Nov. 8/4 Great tractors that will take heavy loads over virgin country where there are no roads. **1955** *Britannica Bk. of Year* 460/2 New state farms were to be set up in virgin land..mainly in Kazakhstan and western Siberia, where there were said to be many millions of acres of virgin..land. **1956** E. CRANKSHAW *Listener* 10 Sept. 378/4 Mr. Khrushchev's virgin-lands scheme in Siberia and Kazakhstan. **1967** C. COCKBURN *I, Claud* xxxv. 438 Hardly anyone can be packed off to some social organization or the 'virgin lands' for lounging things up. **1981** O. BERNIER *Pleasure & Privilege* xiii. 222 Every day new wines were laden..virgin soils, and..virgin land. Homesteaders were now at all over the Eastern states. **1828** WEBSTER *Virgin*..fresh; new; unused; as *virgin soil*. **1847** DICKENS *Dombey* (1848) xi. 104 We shall impart a great variety of information to your little friend... Quite a virgin soil, I believe?

b. *virgin wool* (see quot. 1952).

1921 *Daily Colonist* (Victoria, B.C.) 5 Oct. 7/6 (Advt.), Pure virgin wool socks at 45 c a pair. **1952** *Amer. Speech* XXVII. 262 Woolens and worsteds may be described as made from *virgin wool*—any wool that has never before been spun, woven, knitted, felted, or otherwise made into a product. **1977** *New Yorker* 12 Sept. 106/3, 100% virgin Shetland wool from the Shetland Isles. Hand-loomed and fully-fashioned in the U.S.A.

virginalist (vɜ·ɹdʒinălist). [f. VIRGINAL *sb.* + -IST.] A player on the virginals; a composer for the virginals.

1913 J. E. MATTHEW tr. *C. van den Borren's Sources of Keyboard Mus. in Eng.* iv. 66 After Tallis, John Bull is the only virginalist who still used faux-bourdon as a

means of figuration. **1924** M. GLYN *About Eliz. Virginal Mus. & its Composers* ii. 22 Had Orlando Gibbons lived but ten years longer, he might..the great glory of the English Keyboard school must have attained publication... This music contains seeds which yet fructify, unknown even to the later seventeenth century musicians, who were not true descendants of the virginalists. **1942** E. BLOM *Mus. in Eng.* xi. 183 The Elizabethan virginalists, for example, liked to base variations on them. **1984** *Times* 4 Mar. 4/6 The Elizabethan virginalists discovered that possibility and exploited it. **1979** *Early Music* Jan. 60/1 Bull, the greatest of the virginalists, used cross-rhythms which..

virginality. † *Obs.* and add later example.

1965 F. SARGESON *Memoirs of Peon* iv. 67, I was at first, a little awed..by the room's too severe and uncompromising air of virginality.

Virginia. Add: **1. b.** *Virginia bluebell*, cowslip, a perennial herb, *Mertensia virginica*, of the family Boraginaceæ, native to eastern North America and bearing clusters of blue flowers; cf. *Virginian cowslip* s.v. COWSLIP 2 e.

1934 WEBSTER, Virginia bluebell. **1939** *Nat. Geogr. Mag.* Aug. 236/2 The pale-blue flowers and rosy-pink buds of the Virginia bluebells. **1978** *Detroit Free Press* 16 Apr. (Gardening Guide) 12/2 Virginia bluebells..can be planted wherever they are wanted. **1826** W. DARLINGTON *Florula Cestrica* 23 Pulmonaria virginica..Virginia Cowslip. **1901** N. L. BRITTON *Man. Flora Northern States & Canada* 771 Virginia cowslip.. In moist meadows and along streams. **1914** G. L. STONE *Lake Superior* 293 By 1 Sept.. the Virginia cowslips are coming into bloom. **1968** PETERSON & MCKENNY *Field Guide to Wildflowers* 310 Virginia Cowslip... The nodding, trumpetlike flowers are pink in bud.

c. *Virginia rail* (cf. RAIL *sb.*[3]).

1828 C. L. BONAPARTE *Genera N. Amer. Birds* 334 The Virginia Rail..inhabits throughout North America. **1914** *Chambers's Jrnl.* July 439/1 The Virginia rail.. **1964** *Bull. Mass. Audubon Soc.* Mar. 64 At the Sanctuary on this date we might get the Virginia Rail.

d. *Virginia cigarette, ham, tobacco* (later example).

1919 *Honey Pot* I. 1 Advt. facing p. 1, The best Virginia cigarette I have ever smoked. **1934** *Ulysses* iv. 161 [see *MUDLARK *sb.* 1]. **1968** [see *chicken gumbo* s.v. *CHICKEN *sb.*[1] 8]. **1976** *Times* 3 Jan. 11/6 Smells of herbs and Virginia tobacco fill the air. **1980** *Times* 3 Jan. 19/5 Virginia tobacco...: Cheddar cheese and..orange juice, Virginia ham steak from Surrey County.

2. Also, a cigarette made of Virginia tobacco. Cf. *VIRGIN *sb.* 9*.

1964 D. E. MIDDLETON in C. K. Stead *N.Z. Short Stories* (1966) 203 Now we can enjoy another of Blackie's virginias. **1977** R. DENTRY *Encounter at Kharmel* ix. 171 Do you mind if I smoke my Virginia?

Virginian. A. *sb.* **b.** (Earlier examples.)

1654 in *Colonial Rec. N. Carolina* (1886) I. 18 Sir, If you think good to acquaint the States with what is done by two Virginians born, you will honor our country. **1755** W. SMITH *Brief State Province of Pennsylvania* 15 'Tis true our Neighbours, the Virginians, have taken the Alarm, and called on our Assistance.

Vi·rgin I·slander [f. *Virgin Island(s* + -ER[1].] A native or inhabitant of (one of) the Virgin Islands, the westernmost islands of the Lesser Antilles, which are divided between Great Britain and the United States.

1931 *Ann. Rep. Governor Virgin Islands* 1 In replacing Navy personnel of the former administration, more than 40 per cent Virgin Islanders have been appointed to office. **1955** A. A. ALEXANDER in *Ibid.* 4 There are no colleges or schools of business administration here, and many Virgin Islanders who go to the continental United States for higher education do not return. **1975** H. ORR *Rich Girl, Poor Girl* xix. 259 The lilting tones of the native Virgin Islander.

virginity. Add: **3. b.** *transf.* (The appearance of) virtue or integrity; innocence, inexperience. Also with adj. indicating a sphere of activity. Cf. *VIRGIN *sb.* 2 f.

1975 *Listener* 10 Apr. 458/1 She turned down £20,000 for half-a-day's advertising work. 'I'm selling my virginity,' she said. 'At least, that's what they're bidding for.' **1977** *New Yorker* 8 Apr. 17 The President does not want to know... They all want it handled discreetly, for the sake of their political virginity. **1982** *Times* 15 Feb. 4/2 He could claim no political virginity since he had been..a paid-up member of both Conservative and Labour conferences.

virginium (vɜɹdʒi·niǝm). *Obs.* exc. *Hist.* [f. VIRGINIA, name of a state of the U.S. + -IUM.] = *FRANCIUM.

1932 F. ALLISON et al. in *Jrnl. Amer. Chem. Soc.* LIV. 615 We suggest the name virginium and the symbol Va for element 87. **1936** [see *ALABAMINE]. **1938** *Encycl. Brit. Bk. of Year* 145/1 The discovery of element No. 87 has again been reported... The claim of F. Allison..and in 1931 the announced discovery of Virginium. **1975** *Jrnl. Chem. Educ.* LII. 585/2 These two new elements he called virginium, after his native state, and alabamine.

Virgo. Delete ‖, for *Astr.* read *Astr.* and *Astrol.* and add: **2. a.** *attrib.* or as *adj.*, born under or ruled by the sign of Virgo (23 Aug.–22 Sept.).

1894 E. Kirk *Influence of Zodiac* xiv. 121 With proper training these Virgo people may grow into the most powerful spiritual healers. 1928 E. Adams *Astrol.* 74 The mercurial type of Virgo native is a very different object. 1964 L. MacNeice *Astrol.* i. 16 'Are you Virgo?' 'Oh, no, I'm Leo.' 1970 D. Halliday *Dolly & Cookie Bird* ii. 25 ..said to Austin, 'When is your birthday?'..We got it worked out that he was Virgo.

b. A person born under the sign of Virgo.

1917 H. T. Waite *Compend. Natal Astrol.* 45 The good type of Virgo is the most conscientious, methodical.. man. 1924 C. E. O. Carter *Conc. Encycl. Psychol. Astrol.* 114 Virgo, however, is frequently extremely given over to foibles and personal idiosyncrasies which warp the mental outlook. 1968 D. Anrias *Man & Zodiac* vi. 82 Virgos are often witty but seldom positive. 1968 T. Wolfe *Electric Kool-Aid Acid Test* xxi. 294 'When were you born?'.. 'I'm a Virgo'. 1976 *Reader's Digest* June 75, I don't believe in this astrology business. We Virgos aren't easily taken in.

Hence **Vi·rgoan** = *Virgo 2 b.

1946 'Stella Coeli' *Your Fate* xi. 70 The high type of Virgoan makes a good statistician, efficiency expert, [etc.]. 1960 E. Chatelherault *You & Your Stars* ii. 16 Virgoans are among the most fascinating of the Zodiac tribes. 1978 *TV Times* 28 Jan.–3 Feb. 69/3 There may be a special link with a Virgoan.

‖ **virgo intacta** (vɜ̄·rgo intæ·ktă). [L.] A woman of inviolate chastity (in quot. 1922, *transf.* of a man); one who has never had sexual intercourse, a virgin. Freq. in legal contexts.

1726 J. Ayliffe *Parergon Juris Canonici Anglicani* 228 The wife of one Bury was divorc'd from him upon the Score of Frigidity, it appearing that for three years after the Marriage she remain'd *Virgo Intacta* on the Account of the Husband's Impotency. 1839 J. Haggard *Rep. Cases Eccl. Courts* I. 728 If the parties lay together in one bed for so many years, of such age, and the woman is certified to remain *virgo intacta*, there cannot be a stronger presumption that impotency existed, and that it was incurable. 1898 W. S. Churchill *Let.* 31 Mar. in R. S. Churchill *Winston S. Churchill* (1967) I. Compan. ii. 908 She too is to be pitied..as she was not originally *virgo intacta*. 1922 Joyce *Ulysses* 485 I declare him to be *virgo intacta*. 1932 G. B. Shaw *Let.* 26 Apr. in B. Shaw & Mrs. Campbell (1952) 300 Ellen, though she came through with me *virgo intacta*, gave herself away heart and soul without a thought of reserve. 1968 'A. Gilbert' *Night Encounter* iv. 48 It was later shown that she was *virgo intacta*. 1980 J. B. Hilton *Anathema Stone* x. 107 She may have teased his patience... Hence my uncertainty as to whether she was *virgo intacta*.

virgula. **3. a.** For † *Obs. rare* read *rare* and add later example.

1934 Priebsch & Collinson *German Lang.* ii. x. 380 The full stop or, instead, a virgule, i.e. a short slanting stroke (/) is used..to mark the end of a sentence or of a portion of a sentence.

virgule. Add: **1.** Now also in more general use with various functions (see quots.). Cf. *slash sb.[1] 5.

1946 G. Stimpson *Bk. about Thousand Things* 487 The technical name of the short slanting stroke between and and or in the device is *virgule*. 1962 *Gen. Systems* VII. 299/2 Its mate is suffixed with a slant (virgule), thus: 400 How to Silence. 1980 O. M. Riccio *Intimate Art Writing Poetry* v. 138 The vertical lines (virgules) separate the feet that make up the line.

virguncule (vɜ̄ɹgʌ·ŋkiul). *nonce-wd.* [ad. L. *virguncula*, dim. of *virgo* Virgin *sb.*] = Virgin *sb.* 3 a; a young virgin.

1911 Beerbohm *Zuleika D.* vii. 94 There are the virguncules of Somerville and Lady Margaret's Hall; but beauty and the lust for learning have yet to be allied.

virial. Add: **a.** (Further examples.) *virial theorem*, the theorem that for a steady-state system of particles obeying an inverse square law of force, the time-average of the kinetic energy equals the time-average of the virial; or equivalently, that the potential energy is twice the total energy and the kinetic energy is the negative of the total energy.

1904 J. H. Jeans *Dynamical Theory of Gases* vi. 144 The virial depends solely on the forces acting upon the molecules, and not upon the motion of the molecules. 1925 *Phil. Mag.* L. 414 By the extended theorem of the virial, the effect of the pressure of radiation can be ignored, and thus all consequences of the virial theorem which hold in the absence of radiation hold also when radiation is taken into account. 1965 Phillips & Williams *Inorg. Chem.* I. i. 21 A useful theorem known as the Virial Theorem, which holds for all coulombic potential energy systems.., states that the equilibrium binding energy is equal to $\frac{1}{2}\bar{V}$ or $-\bar{T}$, where $\bar{V}$ and $\bar{T}$ are time-average potential and kinetic energies. 1974 *Encycl. Brit. Macropædia* XV. 852/2 Under such conditions the total potential energy of the cluster is exactly twice as great as the combined kinetic energy of all the cluster stars. This relation is known as the virial theorem. 1980 *Nature* 29 May 395/1 The virial theorem is often used to calculate a galaxy's/system's mass from its size and velocity dispersion.

b. *virial coefficient* [tr. G. *virialcoefficient* (H. K. Onnes 1901, in *Arch. néerlandaises des Sci. exactes & nat.* VI. 874)], each of the (temperature-dependent) coefficients of inverse powers of *V* in a polynomial series used to approximate the quantity pV/RT in the equation of state of an ideal gas or similar collection of particles; so *virial equation, expansion*.

1902 *Sci. Abstr.* V. 364 The series development $pv = A + B/v + C/v^2 + D/v^4 + E/v^6 + F/v^8$ is applied.... The coefficients A, B, &c., are termed virial coefficients, and are functions of the temperature. 1953 H. B. G. Casimir in W. Pauli *Niels Bohr* 121 In the theory of an ideal gas interaction between atoms is neglected, but it is possible to calculate successive approximations, the so called virial coefficients. 1967 Condon & Odishaw *Handbk. Physics* (ed. 2) v. iv. 49/1 The virial expansion is one of the cleanest-cut developments in the subject of statistical mechanics. 1967 Margenau & East *Introd. Polymer Chem.* iii. 65 Some idea of the deviations from ideality in these dilute solutions can be obtained by evaluation of the second virial coefficients. 1978 P. W. Atkins *Physical Chem.* i. 39 Conclusions can be drawn from the virial equation of state only by inserting specific values of the coefficients and taking note of their temperature dependence.

viricidal (vairisai·dăl), *a.* [f. Vir(us + -i- + -cide + -al.] Fatal to viruses. So **vi·ricide**[2], a viricidal agent.

1924 *Brit. Jrnl. Exper. Path.* V. 341 Spots in which no naked-eye change whatever has taken place owing to complete neutralization by viricidal serum have been removed for study. 1946 J. R. Porter *Bacterial Chem. & Physiol.* iv. 225 The ideal disinfectant or viricide has not been discovered. 1970 M. Skinner *Old Rectory* 2 Who'd bred a virus which no viricide could cope with. 1981 *Appl. & Environmental Microbiol.* XLII. 469 The results indicate that CAT may be an effective viricide against poliovirus type 2 in an acid medium. *Ibid.*, The viricidal properties of CAT [sc. chloramine-T] and chlorine are compared.

viridescence. Delete *rare* and add later example.

1961 G. Durrell *Whispering Land* v. 123 Here were the vivid greens of the tropics, so many shades and some of such viridescence that they make the green of the English landscape look grey in comparison.

viridescent, *a.* Delete *rare* and add later examples.

1951 S. Beckett *Murphy* viii. 152 The..kites rode steadily..flown by the child. She could just discern them.. For a moment they stood out motionless and black, in a glade of limpid viridescent sky. 1980 P. Hill *Savages* iv. 37 The mainly deciduous trees joined branches overhead so that the two were plunged into a viridescent gloom.

viridin (vi·ridin). *Pharm.* [f. L. *virid-us* Virid *a.*, specific epithet + -in[1].] A crystalline antibiotic with antifungal properties, $C_{19}H_{16}O_6$, derived from the mould *Trichoderma viride*.

1945 Brian & McGowan in *Nature* 4 Aug. 144/2 We have recently found a number of strains which produce another substance, which we propose to name 'viridin', characterized by remarkably high fungistatic activity. 1970 *Nature* 18 July 300/2 *T. viride* may retard the development of other fungi by producing the antibiotics gliotoxin or viridin, although attempts to demonstrate this activity in pot cultures have been unsuccessful. 1978 T. Korzybski et al. *Antibiotics* III. .. Viridin is inactive against staphylococci, colon bacilli, and typhoid bacilli in concentrations up to 100μg/ml.

virile, *a.* Add: **3.** Also *virile part*.

1970 W. Styron *Confessions Nat Turner* iii. 372, I felt my virile part stiffen again beneath my trousers.

virilia (viri·liă). *rare.* [L. *virilia* penis.] The male genitals; the penis.

1962 V. Nabokov *Pale Fire* 123 When stripped and shiny in the mist of the bath house, his bold virilia contrasted harshly with his girlish grace.

virilism. Add: **2.** The state, of a female, of having some male sexual characteristics; also, = *virilization.

1922 Mason & Ayres tr. *Lereboullet's Endocrine Glands* 164 Virilism may appear either in young girls after puberty, or in women after the menopause. 1945 [see *feminism 2]. 1985 *Oxford Textbk. Med.* I. x. 87/1 Congenital adrenal hyperplasia can present in adult life with hirsutism, virilism, and often disturbed menstrual cycles.

virilist (vi·rilist), *sb.* (and *a.*). *nonce-wd.* [f. Virile *a.* + -ist.] A hearty, excessively 'manly' person; one who makes a cult of conventional masculine virtues. Also *attrib.* or as *adj.*

1910 C. E. Montague *Hind let Loose* viii. 145 To give..to the pedant and virilist their several rights of stark lapidary and jolly brutishness. 1922 — *Disenchantment* v. 69 Your virilist chaplain was apt to overdo..his jolly implied disclaimers.

virilization (virilaizē¹·ʃən). *Path.* [f. Viril(e *a.* + -ization.] The pathological development of male sexual characteristics, esp. in a female. Also **vi·rilized** *ppl. a.*, exhibiting virilization; **vi·rilizing** *ppl. a.*, causing virilization.

1938 .. 1977 L. I. Soffer *Dis. Endocrine Glands* xxi. 689 The virilizing syndrome is an affliction essentially of females and preadolescent males. *Ibid.* 699 The Leydig cell tumours produce pseudosexual precocity in the male and virilization in the female. 1964 *Steroids* IV. 140 A virilized woman..with Cushing's syndrome due to an adrenal adenoma, had 390μg/day of testosterone in her urine. 1974 Passmore & Robson *Compan. Med. Stud.* III. xxviii. 50/1 This virilizing ovarian tumour..occurs in young adults. 1976 *Lancet* 13 Nov. 1081/1 This hormone stopped her menses and produced all the features of virilisation. 1983 *Oxf. Textbk. Med.* I. x. 85/2 Excessive hair growth is among the commoner endocrine afflictions of women. The complaint may exist in isolation or it may be one of a constellation of abnormalities found in the virilized state.

virilocal (vi:rilōᵘ·kăl), *a.* *Anthrop.* [f. L. *viri-lis* Virile *a.* + Local *a.* 2.] Pertaining to or designating a woman's residence after marriage in the domicile of her husband. Cf. *patrilocal, *uxorilocal *adjs.*

1948 [see *uxorilocal *a.*]. 1957 V. W. Turner *Schism & Continuity Afr. Soc.* p. xviii, It is possible that hunting, a purely masculine pursuit, and virilocal marriage, which binds together male kin in local descent groups, are parallel expressions of structural opposition between men and women in this matrilineal society. 1969 in Halpert & Story *Christmas Mumming in Newfoundland* 137 A consequence of the rigidly virilocal marriage and settlement pattern.

Hence **viriloca·lity**; **virilo·cally** *adv.*

1957 V. W. Turner *Schism & Continuity Afr. Soc.* p. xviii, Within villages the dominant principles influencing residence were maternal descent and virilocality. *Ibid.* ii. 55 Women are often brought up patrilocally, marry virilocally, and after divorce reside avunculocally until remarriage. 1974 G. H. Gossen in N. Hammond *Mesoamerican Archaeol.* 219 Chamulas live virilocally in dispersed hamlets.

virion (vi·ri,ǫn). *Microbiology.* [a. F. *virion* (A. Lwoff et al. 1959, in *Ann. de l'Inst. Pasteur* XCVII. 286), f. *vir-us* Virus + *-i-* + *-on* *-on[1].] The complete, infective form that a virus has outside a host cell, with a core and a capsid.

1959 *Ann. de l'Inst. Pasteur* XCVII. 288 The viral infective system, the virion, may be considered as a clathrate type of compound in which the genetic component is enclosed in a coat or capsid formed of subunits or capsomeres. 1967 K. M. Smith *Insect Virol.* i. v The virion is, structurally and physiologically, different from any cellular organelle and from any microorganism. 1975 *Sci. Amer.* May 25/2 An electron micrograph of the poliovirus virion (the virus particle) reveals a sphere 17 nanometers (millionths of a millimeter) in diameter... The virion consists only of protein and the nucleic acid RNA.

viro- (vəi·ro), comb. form of Virus (sense *2 b), as in **vi·rogene**, a gene sequence corresponding to the genome of a tumour virus but occurring, normally repressed, in a cell; **viroge·nesis**, the formation or production of viruses; **virogene·tic, -ge·nic** *adjs.*, giving rise to viruses; **virope·xis** [after *colloidopexis*, f. Gr. πῆξις fixing], the process by which a virus particle becomes attached to a cell wall and incorporated into the cell by phagocytosis; **vi·rosome** [*-some[4]], (*a*) a particle of ribonucleoprotein and virus DNA found in the cytoplasm of certain virus-infected cells; (*b*) a liposome into which viral proteins have been introduced.

1969 Virogene [see *oncogene]. 1971 *Nature* 27 Aug. 620/2 [This] suggests that activation of this sarcoma oncogene in the tumour cells is often accompanied by activation of the gs antigen of the virogene. 1976 *Ibid.* 13 May 101/1 African Old World monkeys..and African apes..possess virogenes that are useful in the taxonomic classification of these animals. 1961 Virogenesis [see *virogenetic* adj. below]. 1982 *Virology* CXXI. 296 (*heading*) The nuclear matrix is involved in herpes simplex virogenesis. 1961 *Jrnl. Insect Path.* III. 195 Inasmuch as the virogenetic network (strands) spreads over the entire cell, the virogenesis is neither confined to the cytoplasm nor to the nucleus. 1982 *Jrnl. Urol.* CXXVII. 646/2 The improved graft survival..may be an additional manifestation of unresponsiveness, not only to immunologic but also to virogenetic stimulation. 1956 *Nature* 25 Aug. 412/1 In the midgut cells of Hymenoptera infected with nuclear polyhedroses proteinaceous virogenic stromata form *de novo* in the nuclear sap.... They are networks, and virus rods differentiate within vesicles in their cords. 1982 *Jrnl. Invertebrate Path.* XXXIX. 205/2 The hypertrophy is evident at a very early stage and before any obvious sign of a virus infection, such as the virogenic stroma. 1948 S. F. de St. Groth in *Nature* 21 Aug. 295/1 This phenomenon strongly recalls the uptake by cells of colloidal dyes, and may..be termed 'viropexis'. 1981 *Jrnl. Gen. Virol.* LII. 329 Virosomes were taken up by Vero cells by viropexis with no evidence of fusion. 1970 Dahl & Kates in *Virology* XLII. 453/2 Since viral DNA complexes are quite poorly defined biochemically it seems appropriate to refer to such structures by the general term 'virosomes'.., in analogy with chromosomes, in order to avoid more restricted nomenclature (e.g., viral DNA, DNA 'factories', etc.), which may..create a misleading impression concerning their composition and function. 1975 J. D. Almeida et al. in *Lancet* 8 Nov. 899/2 The surface hæmagglutinin and neuraminidase projections of influenza virus were removed from the viral envelope.. and relocated on the surface of unilamellar liposomes. The resulting structures were..found to resemble the original virus... The name virosome is proposed for these new bodies. 1981 *Jrnl. Gen. Virol.* LII. 329 Virosomes constitute a model system for studies on the organization of virus membrane proteins and lipids and interaction between virus particles and membranes.

viroid (vəi·roid). *Biol.* [f. Vir(us + -oid.] **1.** A virus-like particle. Also *attrib.* or as *adj.* Now *disused* in favour of next sense.

1946 E. Altenburg in *Amer. Naturalist* LXXX. 559 It is conceivable that there exist ultra-microscopic organisms which are akin to viruses but which are useful symbionts, and that these symbionts occur universally within the cells of larger organisms. We might call these supposed symbionts viroids. 1953 S. E. Luria *Gen. Virol.* xviii. 365 Mutations of viroids could also give rise to nontransmissible, abnormal plasmagenes and be responsible..for the tumoral transformation of cells. 1959 *Oxf. Mag.* 26 Feb. 286/2 The relationship between viruses and other 'viroid' particles. 1963 *New Scientist* 20 June 652/1 If blind natural selection could conjure man out of a viroid in a couple of billion years, what could not man's conscious and purposeful efforts achieve?

2. An infectious entity similar to a virus but smaller and consisting of a strand of nucleic acid only, without the protein coat characteristic of a virus.

1971 T. O. Diener in *Virology* XLV. 426/1, I propose the term 'viroid' for such entities. Altenburg (1946) introduced this term to designate hypothetical symbionts, akin to viruses... If, however, the 'viroid' is redefined operationally in modern terms to encompass nucleic acid species with the properties discussed here, the term serves a useful function. To distinguish pathological conditions incited by viroids from those incited by viruses, the term 'viroid disease' is proposed. 1979 *Nature* 4 Jan. 60/2 Viroids are the smallest replicating pathogenic agents known. 1983 *Times* 6 Apr. 16 There is a parallel class of agents which infect plants, the viroids, which consist solely of strands of RNA.

virology (vəirǫ·lŏdʒi). [f. Vir(us + -ology.] The branch of science concerned with the study of viruses.

1935 in Dorland & Miller *Med. Dict.* (ed. 17) 1502/2. 1937 *Ann. Rep.* 1936 59 In virology, two discoveries overshadowed all others. 1957 *New Scientist* 9 May 11/1 This will be the first chair in virology in this country, and its creation marks a stage in the development of virology from the young brother of bacteriology to a fully grown science in its own right. 1978 H. McLeave *Borderline Case* x. (1979) 96 She no longer looked like a virology professor but an attractive, seductive woman.

Hence **virolo·gic** (chiefly *U.S.*), **-lo·gical** *adjs.*; **virolo·gically** *adv.*; **viro·logist**, a specialist in virology.

1946 *Nature* 14 Sept. 363/1 One of the main objects of the symposium, the finding of common grounds of interest between virologists, bacteriologists, mycologists and geneticists, was fully achieved. 1953 S. E. Luria *Gen. Virol.* p. ix, In spite of many important additions to virological literature, no single volume suitable for classroom use has appeared. 1955 *Proc. Soc. Exper. Biol. & Med.* LXXXIX. 438/1 The Detroit-6 strain fulfills certain criteria..for a useful cell strain for virologic research. 1963 *Guardian* 10 Apr. 20/3 Virologists in the research laboratories at Mill Hill were exploring the possibility that some forms of human cancer may be caused by viruses. 1970 *Sci. Jrnl.* Apr. 3/3 The history of virological research is one of the continual discovery of new viruses or of new facts about old viruses. 1972 *Arch. Neurol.* XXVII. 103 (*heading*) Herpes simplex encephalitis. The course in five virologically proven cases. 1977 M. Sokolinsky tr. *Merle's Virility Factor* xii. 248 He was a virologist without great clinical experience. 1980 *Brit. Jrnl. Dermatol.* CIII. Suppl. No. 18. 24 Virological studies were performed on stools and throat swabs in all patients. 1980 *Amer. Jrnl. Epidemiol.* CXII. 487 There were two virologically confirmed cases with complications. 1983 *Amer. Jrnl. Veterinary Res.* XLIV. 64/1 The serologic and virologic responses of all cattle were monitored for 1 year.

virosis (vəiᵊrōᵘ·sis). Pl. **-oses** (-ōᵘ·sīz). [f. Vir(us + -osis.] A virus disease.

1927 *Phytopathology* XVII. 161 Certain aggregates of symptoms in potatoes (*Solanum tuberosum*) are considered to be due to corresponding degeneration diseases or viroses. [*Note*] A name for 'virus diseases' proposed on December 28, 1925, at Lincoln, Nebraska, by Dr. L. R. Jones. 1963 G. Benz in E. A. Steinhaus *Insect Path.* I. x. 325 The most intimate association between the pathogen and its host is found in viroses. 1972 *Biol. Abstr.* LIII. 3729/2 The 'spindle virosis' of *M. melolontha* is characterized by the coexistence of 2 types of cytoplasmic inclusions: spindles and spherules.

virtu, vertu. Add: **4.** The distinctive qualities inherent in a thing or person.

1934 E. Pound *Eleven New Cantos* xxxvi. 27 Or say where it had birth What is its virtu and power. 1949 — *Pisan Cantos* lxxiv. 11 In the light of light is the virtù. 1969 *Listener* 14 Aug. 214/3 Cromwell was shown in the same light—of a *de facto* sovereign come into power thanks to his *virtù*—by Clarendon. 1973 *Times Lit. Suppl.* 1 June 601/3 The pagan *virtù*, the 'civic humanism' of Machiavelli, had become the proud Christian freedom of the Huguenots.

virtual, *a.* (and *sb.*) Add: **4. d.** For 'momentum' read 'moment' and add: *virtual displacement*, any notional, infinitesimal displacement in a mechanical system that is consistent with the constraints of the system; *virtual work*, the work done by a force making a virtual displacement.

1877 G. M. Minchin *Treat. Statics* iv. 61 The virtual work of a force is the product of the force and the projection along its direction of the virtual displacement of its point of application. 1897 A. E. H. Love *Theoret. Mech.* viii. 139 Principle of Virtual Work. The sum of the virtual works of all the forces on a system in equilibrium vanishes in every infinitesimal displacement. 1942 Synge & Griffiths *Princ. Mech.* ii. 60 Although the chief merit of the principle of virtual work lies in the fact that it does not involve the reactions of constraints, nevertheless it can be used to find these reactions should they be required. 1981 R. R. Craig *Structural Dynamics* ii. 28 Use the principle of virtual displacements to derive the equation of motion of the idealized system shown below.

e. *Nucl. Physics* Applied to an excited state of an atomic nucleus which has energy in excess of that needed for the emission of a particle but a lifetime sufficiently long for it to be regarded as a quasi-stationary state.

1931 *Proc. R. Soc.* A. CXXXIII. 228 According to the theory..the emission of α-particles by radio active nuclei is to be explained by the assumption that there exists in the nucleus a 'virtual' level of positive energy, which is occupied by an α-particle. 1955 I. Kaplan *Nucl. Physics* xvi. 368 Each excited state of the compound nucleus, whether bound or virtual, has a certain mean lifetime. 1963 W. E. Burcham *Nucl. Physics* ix. 372 All nuclear levels, except the ground state, can in principle emit radiation, leaving the nucleus in a less highly excited state, and virtual levels can in addition emit particles.

f. *Particle Physics.* Applied to particles and processes that cannot be directly detected and occur over very short intervals of time and space with correspondingly indefinite energy and momenta, which are not necessarily conserved within the time involved.

1949 *Physical Rev.* LXXV. 1305/2 These divergent terms must now be interpreted as renormalization or modification of the electric charge of the proton due to virtual mesons. 1961 W. S. C. Williams *Introd. Elementary Particles* xiii. 341 If the incident photon is 140 Mev and the positron is emitted at 90° with an energy of 100 Mev, then the four-momentum of this virtual electron is about 140 Mev/c. 1971 *Sci. Amer.* June 71/3 Although it may seem that virtual particles violate fundamental conservation laws, the violation is closely delimited to those areas where the uncertainty principle applies. 1973 L. J. Tassie *Physics Elementary Particles* viii. 15 The electron now consists of its 'bare' self together with all its virtual interactions with the electromagnetic field, corresponding to the electron emitting and re-absorbing virtual photons. 1973 *Sci. Amer.* Oct. 110/1 The scattering of the two electrons is described by saying that these particles exchange a virtual photon that transfers momentum from one particle to the other. 1979 D. R. Hofstadter *Gödel, Escher, Bach* (1980) v. 146 To understand how a real, physical electron propagates..the physicist has to be able to take a sort of average of all the infinitely many different possible drawings which involve virtual particles.

g. *Computers.* Not physically existing as such but made by software to appear to do so from the point of view of the program or the user; *spec.* applied to memory that appears to be internal although most of it is external, transfer between the two being made automatically as required.

1959 *Proc. Eastern Joint Computer Conf.* xvi. 82/2 The sole function of the virtual memory is to increase machine speed. 1966 R. Adair et al. *IBM Cambridge Scientific Center Rep.* No. G320-2007 (*title*) A virtual machine system for the 360/40. 1966 *IBM Systems Jrnl.* V. 79 A virtual-storage computer (vsc) can decode addresses that are longer than those of its memory. The longer address is treated..as a virtual address that must be transformed to the actual, shorter memory address... The virtual addressing of the word in external storage triggers a procedure that automatically brings the addressed word into memory. 1972 *Computer Jrnl.* XV. 199/2 Our system runs in a virtual machine, which is implemented by an interpreter. We can therefore easily add new instructions to our virtual hardware, merely by extending the interpreter. 1973 P. B. Hansen *Operating System Princ.* i. 3 An operating system makes a virtual machine available to each user... The simultaneous presence of several users makes the virtual machines much slower than the physical machine. 1981 Pohl & Shaw *Nature of Computation* vi. 198 The Algolic language defines an Algolic virtual machine that may be implemented on a variety of computers. The Algolic machine could be constructed with the following software on a particular machine. 1982 G. Lee *From Hardware to Software* xxvi. 444 In a multiprogramming system, several programs are being executed 'at once'... Thus the operating system has to make available to each user a virtual store, of which he appears to be the sole user. 1983 *80 Microcomputing* Feb. 232/2 Virtual-memory systems have been prevalent in mainframes and large minicomputers for at least a decade. 1985 *Which Computer?* Apr. 54/1 No doubt this is a side effect of using the disc as a virtual memory.

h. Other collocations: *virtual cathode* (Electronics), a part of a space charge or electron beam where the potential is a minimum, so that electrons are repelled and positive ions attracted; *virtual height*, the height of an imaginary reflecting plane surface which in free space would give rise to the same travel time for reflected radio waves as an actual ionospheric layer; *virtual temperature* (Meteorol.) [tr. F. *température virtuelle* (Guldberg & Mohn *Études sur les Mouvements de l'Atmosphère* (1876) I. i. 6)], the temperature that dry air would have to have in order to have the same density as a given body of moist air when at the same pressure.

1937 Virtual cathode [see *suppressor 3]. 1964 *New Scientist* 1 Oct. 29/1 It was found that a virtual cathode could be obtained with a beam current of 3·5 milliamperes or more, and that its relaxation time was in fact inversely proportional to pressure. 1928 *Proc. IRE* XVI. 85 The heights as given in this paper are virtual heights. They are calculated on the assumption that ordinary reflection takes place and that the layer is parallel to the earth's surface. 1967 [see *ionogram 1]. 1975 D. G. Fink *Electronics Engineers' Handbk.* xviii. 107 The reflection process for plane ionosphere is equivalent to mirror-type reflection at a height equal to the virtual height *h'* of reflection of the equivalent vertical frequency. 1910 C. Abbe tr. Guldberg & Mohn in *Smithsonian Misc. Collections* LI. No. 4. 124 We call the quantity *T* the virtual temperature; for dry air the virtual temperature is the same as the absolute temperature. 1957 G. E. Hutchinson *Treat. Limnol.* I. vii. 468 *C* represents the virtual temperature of isothermal circulation prior to the development of stratification in the early summer. 1979 L. J. Battan *Fund. Meteorol.* v. 83 The effects of humidity can be taken into account by employing a quantity called the virtual temperature.

virtue, *sb.* **I. 4. a.** (Later example.)

1980 'J. Melville' *Chrysanthemum Chain* 38 We'll make a virtue of necessity. I'll take charge of the case myself.

virtuefy, *v.* For *rare*⁻¹ read *rare* and add later example.

1884 H. James *Let.* 8 Mar. (1980) III. 36, I am sorry that the divine Daudet is going to virtuefy his *souillon*.

virtuize (vɜ̄·ɹtiuəiz), *v. nonce-wd.* [f. Virtu(e *sb.* + -ize.] *intr.* To behave with conscious propriety; to act in a virtuous manner.

1920 D. H. Lawrence *Touch & Go* ii. 52 If you want me to virtuise and smug with you, you had [better have stayed away].

virtuoso. Add: **3. b.** *transf.*

1921 H. Crane *Let.* 1 Nov. (1965) 69 [Ben] Hecht is a virtuoso and arouses suspicions that one would never feel for Dreiser or Anderson. 1950 E. H. Gombrich *Story of Art* xviii. 268 For him to be an artist was no longer to be a respectable and sedate owner of a workshop: it was to be a 'virtuoso' for whose favour princes and cardinals should compete. 1952 A. Huxley *Let.* c 20 July (1969) 647 On the basis..of what I have seen done..by a man who is probably the greatest living virtuoso in the field of hypnosis..I would advise you very strongly to try hypnosis.

4. Now also passing into adj.

1947 A. Einstein *Mus. Romantic Era* xi. 225 These compositions are intimate confessions, often difficult but never virtuoso. 1952 S. Kauffmann *Tightrope* v. 82 'Look,' he said, staring intently into her eyes, giving a virtuoso performance of sincerity, 'I can't say all this makes me happy.' 1978 J. Updike *Coup* (1979) iv. 129 The virtuoso arabesques of her sullen profile. 1984 *Times* 7 Sept. 20/8 Another virtuoso set of interim figures from Bunzl left the share price up 2p at 305p.

virtuous, *a.* Add: **I. 2. g.** *virtuous circle* [after *vicious circle* s.v. Vicious *a.* 9 in Dict. and Suppl.], a recurring cycle of events, the result of each one being to increase the beneficial effect of the next.

1953 E. Simon *Past Masters* iii. 156 It will be a virtuous circle of publicity attracting helpers and, I trust, supplementary donations, and these begetting more publicity. 1958 *Brit. Jrnl. Sociol.* IX. 163 The child's..range and expression of discriminating verbal responses is fostered by the social structure... A virtuous circle is set up which is continually reinforced. 1982 *Times* 6 May 18/4 The rating reflects the company's virtuous circle—years of store building and modernization leading to productivity gains, which allow it to hold prices lower than its rivals but still make a better margin of 4·5 per cent.

III. 8. *virtuous-seeming* adj.

1959 S. Spender tr. *Schiller's Mary Stuart* iii. iv. 63, I did not hide my sinful deeds behind The false show of a virtuous-seeming face.

‖ **virtute officii** (vɜ̄ɹtiu·tī ǫfi·ʃiī), *adv. phr.* *Law.* [L.] By virtue of (one's) office.

1800 M. Hale *Hist. Pleas of Crown* II. xvii. 147 The court of King's bench may *virtute officii* bail any person brought before them, of what nature soever the crime is, even for treason or murder, as hath been before shewn. 1939 Henty & Redfern *Lewin's Pract. Treat. Law of Trusts* (ed. 14) iv. 298 The powers of trustees may be divided into (1) powers incident to the office of trustee *virtute officii*, including such powers as powers of maintenance and advancement formerly conferred by express proviso but now implied by statute, and (2) such general powers, including powers of appointment, as the settlor may expressly confer upon the trustees. 1982 *Law Q. Rev.* XCVIII. 403 He is a judge of the High Court *virtute officii*.

virucidal (vəirusəi·dăl), *a.* [f. Viru(s + -cide + -al.] = *viricidal *a.* So **vi·rucide** = *viricide[2].

1925 *Jrnl. Exper. Med.* XLII. 533 Animals inoculated with this strain do not become refractory to skin infection with Virus III and their sera do not become virucidal. 1975 *Water Res.* IX. 872/1 The results obtained in this present study did not confirm the greater efficiency of the OCl⁻ ion as a virucide as compared with hypochlorous acid. 1977 *Lancet* 8 Oct. 760/1 Effective treatment will have to include good nursing and the use of steroids.., besides virucidal agents.

virulent, *a.* Add: **4.** *Microbiology.* Of a phage: causing lysis of the host cell immediately after replicating within it, without a period as a prophage; lytic, not lysogenic. [The sense is due to F. Jacob et al. 1953, in *Ann. de l'Inst. Pasteur* LXXXIV. 223, who used F. *virulent*.]

1953, etc. [see *temperate *a.* 8]. 1969 A. M. Campbell *Episomes* i. 2 Phage types which are able to establish lysogenic systems and to reproduce as prophage are called temperate phages, as distinguished from virulent phages which are unable to do so. 1973 R. G. Krueger et al. *Introd. Microbiol.* xviii. 506/1 This type of virus is called a virulent virus, the agent functioning continuously as a lethal intracellular parasite.

viruliferous (viruli·fĕrəs), *a.* [f. L. *vīrul-entia* Virulence + -iferous.] Containing a virus: said esp. of an insect vector.

1933 K. M. Smith *Introd. Study Viruses* 409 Viruliferous. 1937 *Jrnl. Bacteriol.* XXXIV. 132/2 Viruliferous leafhoppers held at..about 32°C. a few days frequently transmit mild strains instead of typical severe yellows. 1967 K. M. Smith *Insect Virol.* xi. 208 Viruliferous aphids were colonized on a virus-immune plant such as Chinese cabbage for 7 days, hemolymph from these aphids was then injected into a series of known virus-free aphids. 1982 *Jrnl. Gen. Virol.* LXI. 187 The planthopper vector.. became viruliferous after injection with nB.

virus. Delete ‖ and add: **2. a.** (Later examples.) Now superseded by the next sense.

1910 Hiss & Zinsser *Text-bk. Bacteriol.* xlvii. 639 Virus dried for eight days was no longer regularly infectious. 1922 *Jrnl. Amer. Med. Assoc.* 11 Feb. 411/1 It was quickly found that the virus floats in a suspending fluid of specific gravity 1·14, while it sinks in a suspending fluid of specific gravity 1·11... To purify it,..it seems best to wash it and centrifugalize it in a suspending fluid just heavier than itself.

b. Pl. **viruses.** An infectious organism that is usu. submicroscopic, can multiply only inside certain living host cells (in many cases causing disease) and is now understood to be a non-cellular structure lacking any intrinsic metabolism and usually comprising a DNA or RNA core inside a protein coat (see also quot. 1977).

Formerly referred to as filterable viruses, their first distinguishing characteristic being the ability to pass through filters that retained bacteria.

[1880 Pasteur in *Compt. Rend.* XCI. 673 Le virus est constitué par un parasite microscopique qu'on multiplie aisément par la culture, en dehors du corps des animaux que le mal peut frapper.] 1881 *Sci. Amer. Suppl.* 4 June 4516/1 M. Pasteur writes: '..The virus is a microscopical parasite, which may be multiplied by cultivation outside of the body of an animal.' 1899 G. Newman *Bacteria* vii. 260 The vaccination in small-pox is an inoculation of the virus of the disease;..the plague and cholera vaccinations are inoculations of pure cultures of living virus from outside the body. 1900 *Jrnl. Compar. Path. & Therapeutics* XIII. 16 The virus of foot-and-mouth disease passes through a Berkefeld filter when it is suspended in a watery liquid. 1906 *Philippine Jrnl. Sci.* I. 583 The length of time during which the virus may remain viable in the soil and in stables is not determined. 1908 *Jrnl. Compar. Path. & Therapeutics* XIII. 59 Filters which are efficient for the arrest of the smallest of the known visible microbes allow the viruses of these diseases to pass through their pores. 1912 *Jrnl. Med. Res.* XXVII. 20 The probable nature of filterable viruses, whether protozoan or bacterial. 1912, etc. [see *filterable *a.*]. 1915 *Lancet* 4 Dec. 1242/2 We do not know for certain the nature of an ultra-microscopic virus. It may be a minute bacterium that will only grow on living material, or it may be a tiny amœba which..thrives on living micro-organisms... It is quite possible that an ultra-microscopic virus belongs somewhere in this vast field of life more lowly organised than the bacterium or amœba. 1929 *Jrnl. Amer. Med. Assoc.* 6 Apr. 1147/1 Throughout this paper the terms filtrable viruses and viruses will be used interchangeably. 1931 *Nature* 10 Oct. 599/2 But a few years ago I think that we should have had no difficulty in accepting three cardinal properties as characterising a virus, namely, invisibility by ordinary microscopic methods, failure to be retained by a filter fine enough to prevent the passage of all visible bacteria, and failure to propagate itself except in the presence of, and perhaps in the interior of, the cells which it infects. 1935 *Science* 28 June 644/1 A crystalline material, which has the properties of tobacco-mosaic virus, has been isolated from the juice of Turkish tobacco plants infected with this virus. *Ibid.* 8 Nov. 443/2 The defining characters of filterable viruses appear to be ultramicroscopic size and obligate parasitism. 1963 J. H. Burn *Drugs, Med. & Man* xix. 188 One view is that these cells contain a virus and the cancer begins when the virus is no longer kept under control. 1972 N. Calder *Restless Earth* iv. 95/2 The Moon was declared free of viruses or spores that might infect the Earth. 1973 [see *mycoplasma]. 1977 S. S. Hughes *Virus* 112 The term 'virus' as used by bacteriologists of the 1880s and 1890s meant simply 'an agent of infectious disease'. This is the usage in Pasteur's dictum: 'Every virus is a microbe.' (1890). 1982 *Sci. Amer.* Apr. 22/3 Viruses..are not included; noncellular, they are mere genetic recipes and are not alive. 1983 W. A. Stevens *Virol. Flowering Plants* i. 3 Bawden (1956) defined a virus as an obligate parasitic pathogen with dimensions less than 200nm. Although possibly adequate in its day, such a definition does not exclude naked nucleic acid pathogens—the viroids, or some mycoplasmas.

c. *colloq.* A virus infection.

1954 C. S. Lewis *Lett. to Amer. Lady* (1969) 24 We mustn't let these modern doctors get us down by calling a cold a *virus* and a sore throat a *streptococcus*.

3. Also in weakened use, an infectious fear, anxiety, etc.

1982 *Economist* 25 Dec. 83/1 The virus quickly spread. First Canada extracted a promise of restraint from Japan. Then West Germany's mercurial economics minister.. hastened to Tokyo.

5. *attrib.* and *Comb.*, as (sense *2 b) *virus disease, infection, particle*; *virus-carried, -containing, -free, -induced, -infected, -like* adjs.; **virus pneumonia,** pneumonia caused by a virus rather than a bacterium.

1958 *Times* 12 June 11/3 The great diversity of virus-carried diseases. Influenza, poliomyelitis, cholera, Australian Q-fever, [etc.]. 1968 *Times* 3 Oct. 13/6 Antiserum ..was found to react positively with a virus-containing extract prepared from dahlias. 1860 Virus disease [in Dict., sense 2]. 1926 *Jrnl. Trop. Med. & Hygiene* XXIX. 19/2 The intricate processes involved in 'virus' diseases of plants and vertebrates. 1978 J. Gardner *Dancing Dodo* xvii. 126 Rift Valley Fever..a virus disease... Usually transmitted to humans by cattle, sheep, other animals: and usually in Asia. 1946 *Nature* 26 Oct. 569/2 The great work of East Malling Research Station in raising and distributing virus-free clonal stocks. *Ibid.* 17 Aug. 217/2 M. B. Shimkin is also critical of the wide extrapolation of observations made on the relatively few virus-induced tumours to the whole range of cancer. *Ibid.* 23 Nov. 735/2 We have grown many thousands of seedlings from seeds which were obtained from virus-infected plants.. and in not a single instance have we found the seedlings diseased. 1924 *Jrnl. Exper. Med.* XL. 773 (*heading*) A filterable virus infection of rabbits. 1965 A. Roudybush *Season for Death* (1966) xxviii. 165 Mrs. Tor was suffering from a virus infection. 1982 R. Rendell *Master of Moor* xv. 165 He was ill, he had a virus infection. 1946 *Nature* 17 Aug. 218/1 They found no evidence of the presence of a rapidly acting virus-like principle associated with the Jensen rat sarcoma. 1972 *Science* 16 June 1225/2 Electron microscopy initially revealed that HBAg consists of virus-like particles approximately 20 nanometers in diameter. 1968 *Brit. Med. Bull.* XXIV. 244/2 They [sc. the results] are consistent with the inactivation of a virus particle by a single interaction with radiation. 1929 *Public Health Rep.* (U.S.) XLIV. 2635 (*heading*) Vaccine virus pneumonia in rabbits. 1977 'J. Bell' *Such Nice Client* xix. 189 She died of a virus pneumonia caught in an epidemic at the hospital. 1971 *New Scientist* 8 Apr. 82/2 The surface antigens also appear to be virus-specific. 1936 *Discovery* Oct. 329/1 Investigation of ten virus strains.

vis, *sb.*[2] Add: **2. f.** *vis comica*, humorous energy; comic force or effect.

1757 S. Foote *Author* i. i. 6 My disposition has, at present, very little of the Vis Comica. 1798 T. Holcroft *Jrnl.* 12 Oct. in *Mem.* (1925) II. v. ii. 184 This character has..not enough of the *vis comica*. 1887 G. M. Hopkins *Lett. to R. Bridges* (1955) 261 In *vis comica*, in fun,..it is not strong: still there is enough to make me laugh aloud sometimes. 1911 Brereton & Rothwell tr. *Bergson's Laughter* ii. 71 In the scene..between Sganarelle and Pancrace, the entire *vis comica* lies in the conflict set up between the idea of Sganarelle..and the obstinacy of the philosopher. 1979 F. Felsenstein in Smollett *Trav.* p. xxv, Smollett's *vis comica* is..sufficiently broad to allow him to laugh at himself.

g. *vis medicatrix* (*naturæ*), the healing power of nature.

1804 *Edin. Rev.* Apr. 186 In this position arose the *vis medicatrix naturæ*, like a fairy queen, to put the wheel in motion. 1904 W. H. Hudson *Green Mansions* xxii. 295 The *vis medicatrix* with which nature helps our weaknesses. 1949 A. Huxley *Let.* 30 July (1969) 601 The news of your mishap was forwarded to us... I do hope that by this time the enforced rest will have given the *vis medicatrix naturae* a chance to get busy.

visa, *sb.* Delete ‖. This form has now replaced Visé *sb.* as the normal term for an entry or note on a passport.

‖ **visagiste** (vīzaʒī·st). Also anglicized as **visagist** (vi·zădʒist). [Fr.] A cosmetic artist.

1958 *Observer* 14 Sept. 11/3 Elizabeth Arden..has brought over to Bond-street from Paris her *visagist*. 1965 *Harper's Bazaar* Aug. 48/2 Guy Nicolet, visagiste supreme from Revlon, designed for her the Leprechaun Look. 1979 *Courier-Mail* (Brisbane) 16 Apr. 16 (Advt.), Our visiting visagists are: [etc.]. 1981 *Times* 10 Nov. 9/1 In our pictures, visagiste Christina Saunders used Estee Lauder's shimmering bronze face powder. 1984 *Listener* 12 Jan. 26/2 The latest..has for heroine a just-post-punk visagiste.

Visākha, var. *Vesak.

‖ **visarga** (visā·ɪgă). Also 9 **viserga.** [Skr., lit. 'emission'.] A sign in the Sanskrit alphabet representing a hard (voiceless) aspiration; also, the sound itself.

1819 H. H. Wilson et al. *Dict. Sanscrit & Eng.* p. xlii, A final *Viserga* or its omission, and a final nasal mark or its omission, are always optional. 1886 *Encycl. Brit.* XXI. 270/1 The Sanskrit alphabet consists of the following sounds... visarga (ḥ) a hard aspirate, standing mostly for original *s* or *r*. 1939 *Year's Work Eng. Stud.* 1937 29 It is a pity that Flasdieck consistently uses the *visarga* (transliterated as *ḥ*) for the final letter of Sanskrit words originally ending in *s*; for only the reader familiar with Sanskrit and its rules of *sandhi* will at once realize that the first person plural of the substantive verb given as Sanskrit *smaḥ*..is to be taken for etymological purposes as equivalent to *smas*. 1953 W. S. Allen *Phonetics in Anc. India* ii. 50 Since these variants [sc. -*ḥ*, -*x*, and -*ɸ*] are not included in the alphabet, special names are devised for them [by the old Indian phoneticians], viz. *visarjanīya* (or later *visarga*) for -*ḥ*, etc... We shall perhaps be giving the most direct and phonetically appropriate translation if we render it by 'off-glide', as referring to the breathy transition from the vowel to silence. 1975 *Language* LI. 120 The first of these rules obligatorily assimilates a dental stop or continuant before any following coronal (dental, palatal, or retroflex) stop... The third converts *s* to visarga before word boundary.

vis-à-vis, *sb.* Add: **2. c.** A counterpart, an opposite number.

1900 J. K. Jerome *Three Men on Bummel* xii. 273 The Vosges peasant has not the unromantic air of contented prosperity that spoils his *vis-a-vis* across the Rhine. 1975 *Publishers Weekly* 21 July 67/1 Middleton's admiration.. [for the U.S. armed services] extends to their vis-à-vis, the Russian military.

visceral, *a.* **1. a.** Delete † *Obs.* and add later examples.

1949 *Scrutiny* XV. 152 A tendency to borrow the mantle of Mr. Wyndham Lewis in attacking the visceral and the formless in art and poetry. 1969 *Listener* 31 July 162/3 Hardly any of them fall into the tragic error of modern straight music, which eschews visceral appeal entirely. 1976 *Publishers Weekly* 1 Mar. 93/1 By accumulating a mass of homely details he gives his story of the death of Sgt. Mullen great visceral power and emotional impact. 1978 J. Irving *World according to Garp* xvi. 321 Hoping that the visceral reality of Garp's language..somehow rescued the book from sheer soap opera.

3. c. *visceral hump*, the dorsal enlargement, containing the viscera, of snails and other gastropod molluscs in which the ventral part is a foot.

1883 *Encycl. Brit.* XVI. 635/2 As the ventral foot is clearly separate from the projecting head, so is this dorsal region, and it is conveniently spoken of as the visceral hump or 'dome' (cupola). 1927 E. Step *Shell Life* (new ed.) i. 23 Within the shell [of the Snail] is..the 'visceral hump' containing most of the internal organs. 1972 M. S. Gardiner *Biol. Invertebrates* xi. 58/2 This and the coiling of the visceral hump has led, in a number of species, to the suppression of organs on one side (usually the right).

5. c. *visceral brain*, those parts of the brain which mediate bodily activity, esp. visceral activity, in response to emotion.

1949 P. D. McLean in *Psychosomatic Med.* XI. 340 (*caption*) The shaded area of cortex represents what was formerly known as the limbic lobe of Broca and subsequently termed the rhinencephalon by Turner. It corresponds to what is arbitrarily referred to in this paper as the visceral brain. 1972 H. J. Eysenck *Psychology is about People* i. 35 Emotionality-stability seems indissolubly linked with the *autonomic* nervous system, which regulates the expression of the emotions, and which in turn is organized and governed by the 'visceral brain'.

viscerally *adv.* (later examples).

1965 H. Kahn *On Escalation* vi. 119 Most individuals.. conclude, at least viscerally, that the dangers are simply too great. 1970 T. Roethke *Sel. Lett.* 173, I teach viscerally: I try to make up for ignorance by energy and enthusiasm. 1982 *Wall St. Jrnl.* 13 Aug. 16/2 Viscerally, your friends recognize that what is involved here is not one issue, but your capacity to govern.

visceralization (vi:sĕrălǝizēⁱ·ʃən). *Path.* [f. Visceral *a.* + -ization.] The spreading of an infection to the viscera; the movement of a pathogen towards the viscera.

1963 *Ann. N.Y. Acad. Sci.* CXIII. 410 Visceralization of the parasites occurs as a further complication. 1969 *Current Topics in Microbiol. & Immunol.* XLVIII. 30 Subinoculation of the organisms responsible for these conditions into laboratory animals is often followed, even in cutaneous types, by extensive visceralization of the infection. 1980 *Jrnl. Exper. Med.* CLII. 605 BALB/c mice have an exceptional susceptibility to *Leishmania tropica* infection such that cutaneous lesions grow without restraint in all cases leading to fatal metastasis and visceralization in normal and x-irradiated..animals.

Also **vi·sceralize** *v. intr.*, to spread to or attack the viscera.

1969 *Current Topics in Microbiol. & Immunol.* XLVIII. 32 Sudanese or some East African infections of *L. donovani* derived from wild rodents may fail to visceralise, while the purely human kala-azar of India is usually fatal. 1974 W. H. R. Lumsden in *Trypanosomiasis & Leishmaniasis* (Ciba Symp.) 8 Organisms tending to 'visceralize' in man (i.e. with a predilection for infecting cells in the viscera, particularly in the spleen). 1981 *Ann. Trop. Med. & Parasitol.* LXXV. 139 The primary lesions of kala-azar may be confused with oriental sore and..some strains with the capacity to visceralize may fail to do so, causing only dermal lesions.

viscero-. Add: **viscerocra·nium** = *splanchnocranium* s.v. *splanchno-; **viscerotro·pic** *a.* *Med.* [*-tropic], tending to attack or affect the viscera; hence **viscerotro·pism.**

1925 J. S. Kingsley *Vertebrate Skeleton* 58 Recently the terms *neurocranium* and *viscerocranium* have been introduced. 1980 *Gray's Anat.* (ed. 36) 141/2 The trabeculae cranii may largely be derived from branchial arch (neural crest) mesoderm, i.e. from the viscerocranium, having been adapted into the cartilaginous or basal part of the neurocranium, or 'brainbox'. 1935 M. Hoskins in *Amer. Jrnl. Trop. Med.* XV. 675 The term 'viscerotropic virus' is used to designate the strain of yellow fever virus which has been carried in *M. rhesus*. 1940 *Nature* 2 Nov. 596/1 Man ordinarily suffers only the viscerotropic attack [in yellow fever]—in his liver, kidneys and heart. 1976 *Ann. Rev. Microbiol.* XXX. 443 In the 17D strain of virus both the viscerotropic and neurotropic properties of the natural virus were markedly reduced. 1940 Viscerotropism [see *neurotropism 2]. 1973 *Acta Virologica* XVII. 241 Street strains displaying a higher viscerotropism can actively multiply in the internal organs.

viscerotome (vi·sĕrotōᵘm). *Med.* [f. Viscero- + *-tome.] An instrument for obtaining post-mortem samples of liver tissue through a puncture in the abdominal wall (avoiding necropsy), used esp. when yellow fever is suspected. Hence **viscero·tomy**, the use of a viscerotome.

1934 *Amer. Jrnl. Hygiene* XIX. 553 The attempt..led one of us (E. R. R[ickard]) to attempt the design of an instrument for the removal of liver tissue without autopsy. This instrument, later christened the 'viscerotome' by Dr. Mario Bião, reached a practicable stage of development within a few weeks. *Ibid.* 555 The opposition of relatives and friends to autopsy is greatly reduced in the case of viscerotomy. 1940 Van Rooyen & Rhodes *Virus Dis. of Man* xxxv. 472 The viscerotome is a metal instrument..resembling a trocar, about 1 cm. square, possessing a sharp, hollow, pointed extremity fitted with a mechanically operated guillotine blade. *Ibid.*, Viscerotomy has revealed the existence of yellow fever in places in which it had hitherto passed unrecognized. 1971 P. C. C. Garnham *Progr. Parasitol.* iii. 24 The disease was scarcely recognized until 1934, when the introduction by Penna.. of a viscerotomy service for the diagnosis of fatal cases of yellow fever revealed its presence in Brazil. *Ibid.*, The brutal instrument, the viscerotome, is plunged through the abdominal wall of the cadaver into the liver.

viscerotonic (vi:sĕrotǫ·nik), *a.* and *sb.* [f. Viscero- + Tonic *a.*] **A.** *adj.* Designating or characteristic of a type of personality which is comfort-loving, sociable, and easy-going, classified by Sheldon as being associated with an endomorphic physique. **B.** *sb.* One having this type of personality. So **viscero·nia** (-tōᵘ·niă), viscerotonic personality or characteristics. Cf. *cerebrotonic, *somatotonic *adjs.* and *sbs.*

1937 [see *cerebrotonic *a.* and *sb.*]. 1938, etc. [see *somatotonic *a.* and *sb.*]. 1956 Goacher & Whigham in E. Pound tr. *Sophocles' Women of Trachis* 60 Stigmatized as a sex crank, as a maniacal visceratonic [*sic*], he [sc. D. H. Lawrence] in fact made an extraordinary effort to re-establish the 'whole' man. 1969 V. de S. Pinto *City that Shone* iii. 68, I suppose that in modern psychological jargon, as a child in those distant Edwardian days, I could be described as an introvert and cerebrotonic living in a world of extroverts, somatotonics and viscerotonics. 1980 F. J. Bruno *Behav. & Life* xi. 379 As might be expected, viscerotonia is said to be correlated with endomorphy. The extreme endomorph with a viscerotonic personality would be a self-indulgent person.

vi:sco,elasti·city. [f. Visco(us *a.*, Visco(sity + Elasticity.] The property of a substance of exhibiting both elastic and viscous behaviour, the application of a constant stress causing an immediate deformation that disappears if the stress is quickly removed but increases for a time and becomes permanent if the stress is maintained.

1944 *Q. Jrnl. Appl. Math.* II. 119 The solution of the first and second boundary value problems of visco-elasticity is reduced to the solution of equivalent boundary value problems of elasticity, and the determination of the response of the visco-elastic material under consideration to a simple shearing stress or a simple shearing strain. 1963 *Times* 27 Mar. 1/6 Viscosity and viscoelasticity of glasses. 1972 *Science* 2 June 1041/3 The viscoelasticity of the solutions would decrease upon continued stirring. 1976 *Nature* 12 Aug. 573/1 The explanation for the observed increase in cavitation erosion rate in polymer solution may still have its origin in polymer viscoelasticity.

So **viscoela·stic** *a.*, exhibiting or pertaining to viscoelasticity.

1944 [see above]. 1949 *Proc. Internat. Congr. Rheology 1948* ii. 24 Flow birefringence in aluminium soap solutions, considered here as a typical visco-elastic system. 1963 *New Scientist* 22 Aug. 382/3 Polymer solutions.. generally show some elasticity, or tendency to preserve their shape, and are therefore called 'viscoelastic'. 1970 P. Sherman *Industr. Rheology* iv. 236 The Instron tester has been used to study the viscoelastic properties of potatoes, peas, and apples. 1973 J. R. Rice in A. C. Palmer *Symposium Role Plasticity in Soil Mech.* 263 Viscoelastic creep of the soil. 1978 *Sci. Amer.* Nov. 145/1 Silicone putty..is such a viscoelastic fluid.

viscometer. Add: (Further examples.)

This, rather than Viscosimeter, is the more usual term.

1886, etc. [see *Saybolt]. 1946 *Nature* 2 Nov. 614/2 Graphs showing the flow characteristics of a grease at medium rates of shear (plunger viscometer) and at high rates of shear (pendulum viscometer) were displayed. 1967 M. Chandler *Ceramics in Mod. World* ii. 61 The commonest instrument for testing the fluid properties of a slip is the torsion viscometer. 1970 *British Printer* June 70/2 Mr Bisset described the functions of viscometers and tack-meters.

Hence **visco·metry**, the measurement of viscosity; the use of a viscometer; **visco·me·tric** *a.*, **-me·trically** *adv.*

1886 B. Redwood in *Jrnl. Soc. Chem. Industry* 29 Mar. 121/2 (*heading*) On viscosimetry, or viscometry. 1931 G. Barr *Monogr. Viscometry* iii. 48 In practical viscometry it is rarely necessary for the investigator to undertake the labour involved in the evaluation of the constants of his instrument ab initio. 1938 *Chem. Abstr.* XXXII. 1091/1 The viscometric results were in excellent agreement with those obtained by using a photoelec. nephelometer. 1946 *Chem. Rev.* XXXIX. 161 The process can be followed viscometrically, polarimetrically, or by chemical determination of the aldose end group produced. 1978 *Nature* 1 June p. xi/2 Further applications are in the field of interface rheology and viscometric determination of the molecular weight of plastic material solutions. 1982 *Lebensmittel-Wissensch. und Technol.* XV. 242 (*heading*) Viscometric study of the dispersion of whole and gelatinised cassava starch. *Ibid.*, The solubilisation of cassava starch in aqueous dimethylsulphoxide has been followed viscometrically. 1982 *Jrnl. Exper. Bot.* XXXIII. 1248 The enzyme procedure and assay, using citrus pectin, described here have enabled us to assay endo-PG by viscometry during a 6 h incubation period.

viscontial (vəikǫ·ntiăl), *a. rare*⁻¹. [f. *viscont* Viscount + -ial; cf. Viscontal, Viscountial *adjs.*] Reminiscent of a viscount.

a 1916 R. Asquith in M. Asquith *Autobiogr.* (1920) I. xiv. 276 You beat your tangled music out Lofty, aloof, viscontial.

viscose, *sb.* For pronunc. read (vi·skōᵘs, -z). Substitute for etym. and def.: [f. Visc(ous *a.* + -ose[2].] **1.** A viscid, orange or brown solution of sodium cellulose xanthate obtained by treating cellulose successively with sodium hydroxide and carbon disulphide, and used to make regenerated cellulose by extruding it into dilute acid and either spinning it into rayon or casting it as film.

1896 C. F. Cross *Brit. Pat.* 4713 Cellulose..requires a very much smaller proportion of alkali to convert it into alkali-cellulose suitable for use in other manufactures such..as that of the substances now known as 'viscose' and 'viscoid'. 1913 Carrier & Martin in G. Martin *Industr. & Manuf. Chem.: Organic* iv. i. 189 Viscose is reconverted into cellulose (1) spontaneously, on long standing; (2) by heating; (3) by treatment with oxidising agents. 1927 T. Woodhouse *Artificial Silk* 30 By far the greater percentage of artificial silk is made from Viscose. 1968 *Kirk-Othmer Encycl. Chem. Technol.* (ed. 2) XVII. 179 After filtration, the viscose is transferred to a ripening tank system.., where it is deaerated and ripened to the desired level of xanthation. 1981 *Chem. Abstr.* 17 Aug. 544/1 Emulsion sausages were made in the usual way. Pork and beef casings and synthetic casings (protein, viscose, and polyamide) were used.

2. Rayon made by the viscose process.

1932 A. Huxley *Brave New World* iii. 58 Her jacket was made of bottle-green acetate cloth with green viscose fur at the cuffs and collar. 1949 W. Garner *Textile Lab. Man.* iii. 77 Acetate is sometimes partly saponified, especially for the printing of mixed fabrics of viscose and acetate. 1960 *Guardian* 28 Sept. 8/6 Lilian, helion, terital and viscose are blended with wool in textiles by the high fashion houses. 1972 *Vogue* June 113/3 Candy pink and white cotton and viscose shirtwaister. 1980 Gohl & Vilensky *Textile Sci.* ii. 32 Viscose and the other two rayon fibres have similar thermal properties to cotton.

3. Special Combs.: **viscose process,** the process for making rayon with viscose as an intermediate product; **viscose rayon,** † **silk** = sense 2 above.

1913 Carrier & Martin in G. Martin *Industr. & Manuf. Chem.: Organic* iv. ix. 214 (*heading*) The viscose process. 1981 *Kirk-Othmer Encycl. Chem. Technol.* (ed. 3) XVI. 108 Considerable research is underway to replace the viscose process and to improve rayon fiber properties. 1930 *Chem. Abstr.* XXIV. 5153 Viscose rayons of 120 and 150 denier were treated with Na_2S solns. of various concns. 1957 *Woman* 16 Nov. 25/3 Viscose rayon is quite easy to do [*sc.* dye] at home. 1974 *Sci. Amer.* Apr. 57/3 The largest single use for dissolving pulp is as the raw material for the viscose-rayon process. 1913 Carrier & Martin in G. Martin *Industr. & Manuf. Chem.: Organic* iv. ix. 214 The viscid solution is then spun into a solution of ammonium chloride which separates out the cellulose again, and gives..'viscose silk'. 1925 *Good Housekeeping* Apr. 142/3 The remaining variety [of artificial silk], Viscose silk, is now being made in enormous quantities.

viscosimeter. Add: (Further examples.) Cf. Viscometer in Dict. and Suppl.

1886, 1925 [see *Saybolt]. 1944 *Acta Med. Scand.* CXVII. 227 The determinations with viscosimeters having varying dimensions of the capillaries gave no real differences even in sera with very high viscosity. 1974 *Nature* 13 Dec. 572/2 This unexpected phenomenon was further investigated in a modified Couette viscosimeter device.

Hence **viscosi·metry** = *viscometry*; **vi:scosime·tric** *a.*, **-me·trically** *adv.*

1886 B. Redwood in *Jrnl. Soc. Chem. Industry* 29 Mar. 121/2 It is my intention..to treat the subject of viscosimetry, or the determination of viscosity, as related to the valuation of oils intended for the lubrication of machinery. 1895 Benedikt & Lewkowitsch *Chem. Anal. Oils, Fats, Waxes* xii. 621 Viscosimetric constants as determined with the new apparatus. 1904 J. Lewkowitsch *Chem. Technol. & Anal. Oils, Fats, & Waxes* (ed. 3) II. xv. 933 If it is desired to examine the unsaponifiable portion of a blended oil viscosimetrically. 1944 *Acta Med. Scand.* CXVII. 227 With the aid of this modified instrument it seems, as if viscosimetry would be a simple and useful clinical method. 1956 *Nature* 17 Mar. 517/1 A viscosimetric investigation into the influence of pH and ionic strength on molecular shape of a structural muscle protein. 1975 *Welding Production* May 11/1 A method is proposed of examining electrode coating mixtures by means of capillary viscosimetry. 1976 *Jrnl. Polymer Sci.* (Polymer Physics Ed.) XIV. 309 (*heading*) Abnormal viscosimetric behavior of homopolymers in mixed solvents and preferential solvation.

viscosity. Add: **1. c.** In scientific use, the tendency of a liquid or gas to resist by internal friction the relative motion of its molecules and hence any change of shape; the magnitude of this, as measured by the force per unit area resisting a flow in which parallel layers unit distance apart have unit speed relative to one another; also called *absolute* or *dynamic viscosity*; *kinematic viscosity*, the dynamic viscosity divided by the density of the fluid.

1866 J. C. Maxwell in *Phil. Trans. R. Soc.* CLVI. 249 The viscosity of a body is the resistance which it offers to a continuous change of form, depending on the rate at which that change is effected. *Ibid.* 254 Suppose that this friction is equal to a tangential force *f* on every square foot, then $f = \mu v/a$, where μ is the coefficient of viscosity, v the velocity of the upper plane, and a the distance between them. 1880 *Proc. London Math. Soc.* XI. 58 If ν be the kinematic viscosity. 1913, etc. [see *poise *sb.*[2]]. 1921 A. W. Judge *Automobile & Aircraft Engines* viii. 309 The kinematical viscosity of air is thirteen times that of water. 1927 Schoder & Dawson *Hydraulics* xvi. 275 The unit of absolute viscosity is called a poise. 1943 R. C. Binder *Fluid Mech.* v. 50 Viscosity = shearing stress/*rate* of shearing strain. Sometimes the foregoing term is called absolute viscosity. Probably a better term would be dynamic viscosity. 1962 J. M. McKelvey *Polymer Processing* ii. 41 In general, liquid viscosities decrease and gas viscosities increase with increasing temperature. 1964 [see *stoke(s *sb.*[4]]. 1979 C. A. Marchaj *Aero-Hydrodynamics of Sailing* ii. 169 The kinematic viscosity of water..needed for Reynolds Number computation at the 'normal' temperature of 15°C is..$1{\cdot}23 \times 10^{-5}$ ft^2/sec.

3. Special Comb.: **viscosity index,** a number expressing the degree to which the viscosity of an oil is unaffected by temperature.

1929 Dean & Davis in *Chem. & Metallurgical Engin.* XXXVI. 618/1 This system of classification permits expressing the viscosity-temperature coefficient of an oil as a simple function of its Saybolt Universal viscosities at 100 and 210 deg. F. This function, hereafter referred to as the 'viscosity index', is independent of the actual viscosity of the oil. 1977 *Lubricants Business* (Shell Internat. Petroleum Co.) 3 These [additives] are chemical compounds which supplement the properties of the mineral base oil; for example, to reduce wear of moving parts..and to improve viscosity index.

viscous, *a.* Add: **2. c.** *viscous flow*, laminar flow (see *laminar *a.* 2 a).

1930 [see *turbulent *a.* 2 c]. 1973 Fox & McDonald *Introd. Fluid Mech.* viii. 311 (*caption*) Details of viscous flow around an airfoil. 1979 [see *stream-line *sb.* 1b].

viscuous, *a.* For † *Obs.* read *rare* and add later examples. Also *transf.*

In quot. 1936 perhaps a misprint.

1932 S. Gibbons *Cold Comfort Farm* iii. 32 Growing with the viscuous light that was invading the sky. 1936 *Jrnl. R. Aeronaut. Soc.* XL. 11 Heat generated in the blades in overcoming viscuous drag also plays a part in determining the distribution of ice.

visé, *sb.* Add: (Earlier example.) Now superseded by Visa *sb.*

1842 E. Lawes *Scamper through Italy & Tyrol* iii. 32 It became necessary to obtain the *visé* of the English Consul, who demands a fee of five francs and six sous.

visé, *v.* (Later *fig.* examples.)

1917 J. Agate *Buzz, Buzz!* 59 Every member of the audience would appear to be in possession of a passport of respectability visé'd by the police. 1927 P. Hammond *But—is it Art* 102 Since these New York premières are so important, let us try to attend one, carefully passported and viséd by a regular New Yorker.

Viséan (vize·ăn), *a.* Also **Visean.** *Geol.* [ad. F. *viséen* (E. F. Dupont 1883, in *Bull. de l'Acad. R. des Sci.*, etc., *de Belgique* V. 223), f. *Visé*, name of a town in Belgium: see -an.] Of, pertaining to, or designating the upper of the two divisions of the Lower Carboniferous (Dinantian) in Europe. Also *absol.*

1905 *Q. Jrnl. Geol. Soc.* LXI. 264 If I am correct in correlating the Lower Viséan of the Belgian Geological Survey with the *Syringothyris*-Zone of the Bristol area, the terms Tournaisian and Viséan, as employed by me in this paper, do not bear their original connotation. *Ibid.* 265 The Tournaisian and Viséan facies are essentially distinct. 1956 W. Edwards in D. L. Linton *Sheffield* 22 The imposing scarp along the north side of the outcrop at Castleton, with its reef-aprons..referred to movements of the Derbyshire 'massif' in relation to the 'basin' north of it, during Visean times. 1969 Bennison & Wright *Geol. Hist. Brit. Isles* ix. 207 Goniatites have shown that only the Lower Limestone Group belongs to the Viséan. 1983 *Rep. Inst. Geol. Sci.* No. 82/12. 1/1 The Carboniferous rocks [exposed at Benburb, County Tyrone] fall within the Viséan Series of the Dinantian.

viseite (vī·ze,əit). *Min.* Also **viséite.** [a. F. *viséite* (J. Mélon (after G. Cesàro) 1942, in *Ann. Soc. géol. Belgique* LXVI. B53), f. as prec.: see -ite[1].] A hydrous basic calcium and aluminium silicophosphate occurring as translucent isometric crystals of a white, bluish, or yellowish colour.

1944 *Chem. Abstr.* XXXVIII. 6244 In the examn. of the minerals which occur with delvauxite at Visé, a new mineral was found to which the name viseite was given. 1975 *Nature* 27 Feb. 722/1 Among the minerals described to have any significant degree of P(V)–Si(IV) partial substitution are viseite, nagatelite and wilkeite.

visgy (vi·zgi). Also **visgie.** *Cornish* and *Devon dial.* = Bisgay.

1777 in *Eng. Dial. Dict.* (1905) VI. 348/2. 1880 Courtney & Couch *Gloss. Cornwall* 62/1 *Visgie*, an agricultural implement, in shape between a mattock and a hammer, for beating down hedges. *Ibid.* 105 Visgy. 1899 'Q.' *Ship of Stars* iv. 35 A knife lay between his wide-planted feet, and a visgy close behind him on a heap of disturbed sand. 1915 *Blackw. Mag.* Apr. 546/1 The day being Sunday, he could not dare to risk outraging public opinion by carrying shovel or visgy through the open streets. 1954 L. Walmsley *Golden Waterwheel* v. 91 For the roots I had a kind of double-bladed mattock, a combination of pick-axe, chopper, adze and hoe, known in Cornwall as a visgy.

‖ **visibilia** (vizībi·liă). [L., neut. pl. of *vīsibilis* Visible *a.* and *sb.*] Things seen; visual images.

1936 C. S. Lewis *Allegory of Love* ii. 45 On the one hand you can start with an immaterial fact, such as the passions which you actually experience, and can then invent *visibilia* to express them. 1982 *Times Lit. Suppl.* 29 Jan. 97/1 Some people see things that they describe as ghosts; we can dispute the name and nature of these visibilia, but not the fact that they are seen. 1982 *PN Rev.* No. 26. 40/1 A theology of the material world..which can offer us a convincingly spiritual account of the *visibilia*.

visibility. Add: **1. d.** *fig.* The degree to which something impinges upon public awareness; prominence.

1958 March & Simon *Organizations* iv. 103 The greater the prestige of the organization..the greater the visibility of the organization. 1975 *New Yorker* 26 May 28/1 These are busy times, and our report doesn't have so very much visibility. 1981 *Nordic Skiing* Jan. 10/3 The purpose of the team is to increase skiing's visibility in this country. 1984 *Observer* 5 Aug. 15/1 From a business standpoint, the visibility Carl receives during the Olympic Games can enhance his value to the companies.

visible, *a.* and *sb.* Add: **A.** *adj.* **1. d.** Also, speech rendered into a visible record by spectrography.

1947 R. K. Potter et al. *Visible Speech* i. 4 A sound spectrographic record for the words 'Visible Speech' is shown... The pattern is a new form of visible speech, a system of natural phonetic symbols translated from speech itself. 1953 [see *sonograph 1].

3. c. *Econ.* Descriptive of or denoting actual goods exported or imported, as opposed to 'invisible' items such as services rendered or received (cf. *invisible *a.* 1 d).

1882 R. Giffen *Use of Import & Export Statistics* vii. 58 As to the increase of our shipping business as a means of accounting for the non-increase of our apparent exports. It is because our invisible exports have been increasing so enormously, that there is less increase of the visible. 1917 J. A. Todd *Mech. Exchange* xiv. 184 Visible and invisible exports alike go to the debit side of the account. 1957 A. C. L. Day *Outl. Monetary Econ.* xxviii. 365 In this way we get a balance of visible trade: quite literally 'visible', because it only includes goods that can actually be seen as they are put on board ship. 1976 *Economist* 16 Oct. 22/2 In August, 1976, Britain ran a visible trade deficit at an annual rate of £3½ billion, compared with a 1970 deficit of £25m.

d. *visible index*: an index so arranged that each item is visible.

1916 E. R. Hudders *Indexing & Filing* (1919) ii. 26 It is not anticipated that the visible index in any of its forms will ever supersede the card index. 1955 V. George in W. Ashworth *Handbk. Special Librarianship* iii. 45 The system most usually employed..is some form of visible index.

6. *fig.* In a position of public prominence; well-known. Cf. *visibility 1 d.

1977 *Chicago Tribune* 2 Oct. ii. 28/1 (Advt.), National leader in health care field has highly visible position available on its corporate headquarters consulting staff. 1978 *Guardian Weekly* 12 Feb. 14/2 The Wilmington Ten affair makes Chavis the most visible of American political prisoners today.

B. *sb.* **1. b.** *pl.* Visible exports or imports.

1962 H. O. Beecheno *Introd. Business Stud.* xv. 143 A country which is running an adverse balance of trade may still have a favourable balance of payments because the gain on 'invisible' items exceeds the loss on 'visibles'. 1968 *Economist* 23 Mar. 64/1 The current account (that is on 'visibles' and 'invisibles' but leaving out movements of investment funds).

3. The visible part of the electromagnetic spectrum.

1962 [see *Nernst a]. 1973 Williams & Fleming *Spectrosc. Methods Org. Chem.* (ed. 2) i. 21 The n→π* transitions of α-diketones in the diketo form give rise to two bands, one in the usual region near 290 nm..and a second..which stretches into the visible in the 340–440 nm region.

visile (vi·zəil), *a.* and *sb.* [f. L. *vīs-us* sight + -ile, after *tactile, audile.*] **A.** *adj.* Responding most readily to visual sensations; thinking predominantly in visual images. **B.** *sb.* A person of this kind; = Visual *sb.* 2, Visualist 1.

1909 [see *audile *a.*]. 1927 J. Adams *Errors in School* 74 This does not imply that the visiles get their knowledge entirely through the eye. 1940 *Brit. Jrnl. Psychol.* XXXI. 56 The theory of the existence of a 'visile' type. *Ibid.* 60 A theory that adolescents who draw men well tend to be of high..intelligence, to be 'visiles' in their thinking. 1954 J. Evans *John Ruskin* 411 He [*sc.* John Ruskin] had..a visile mind. 1960 Menon & Patel *Teaching of Eng. as Foreign Lang.* (ed. 2) iv. 26 Galton classified individuals, with reference to the sense that dominated his memory, into *visiles, audiles* and *kinaesthetic*.

vision, *sb.* Add: **2. b.** Ability to conceive what might be attempted or achieved, esp. in the realm of politics; statesmanlike foresight.

[1904 Chesterton *Napoleon of Notting Hill* ii. iii. 107, I fight for your royal vision, for the great dream you dreamt of the League of the Free Cities.] 1926 Fowler *Mod. Eng. Usage* 695/2 *Vision*, in the sense of statesmanlike foresight or political sagacity, is enjoying a noticeable vogue. 1960 M. Spark *Ballad Peckham Rye* v. 86 'How do you find Weedin?' 'Totally,' Dougal said, 'lacking in vision. It is his fatal flaw. Otherwise quite sane.' 1965 A. J. P. Taylor *Eng. Hist. 1914–45* xvi. 593 Truman, the new president, had none of Roosevelt's vision as international leader. 1973 E. F. Schumacher *Small is Beautiful* i 243 A lack of vision on the part of the socialists themselves. 1982 D. Fraser *Alanbrooke* ix. 217 Churchill had enormous vision. He could and often did impressively surpass his supporters in his imaginative span.

5*. The visual part of a television broadcast, television images collectively; the transmission or reproduction of such images; also, the signal corresponding to them.

1910 H. N. Casson *Hist. Telephone* ix. 287 Some future Carty..may transmit vision as well as speech. 1930 Moseley & Chapple *Television* i. 9 On 9th February, 1928, the public were startled to learn that the Atlantic had been spanned by vision. 1934 J. H. Reyner *Television* x. 109 The radio transmission of sound or vision is usually accomplished by modulating a high-frequency carrier-wave. 1939 *Jrnl. Television Soc.* III. 8/1 Vision is amplified at an intermediate frequency of 13·2 m.c. (vision carrier). 1955 'J. Christopher' *Year of Comet* i. 5 He had followed the usual practice of leaving sound switched on as well as vision. 1959 G. Freeman *Jack would be Gentleman* i. 8 The sound came on a full minute before the vision. 1973 E. G. M. Alkin *Sound with Vision* i. 3 In any entertainment medium in which sound and vision are combined there is a tendency to consider sound as the poor relation.

6. a. *vision-literature, poem, -world*; *vision-seeking* adj.; **vision quest** *N. Amer.*, the attempt to achieve a vision traditionally undertaken by mature men of the Plains Indian tribes, usu. through fasting or self-torture; **vision splendid,** the dream of some glorious imagined time; **vision-telephone** = *videophone* s.v. *video-.

1929 T. S. Eliot *Dante* 67 The *Vita Nuova*,..a sequence of beautiful poems connected by a curious vision-literature. 1961 A. Clarke *Later Poems* 91 The *Aisling*, or Vision poem, in which Ireland was personified, reached its pitch in the eighteenth century. 1922 R. Benedict in *Amer. Anthropol.* XXIV. 3 Three patterns of wide distribution are sometimes taken to characterize the vision quest of the Plains. 1966 D. Aberle *Peyote Relig. among Navaho* xx. 340 Anthropologists have been impressed by the similarities between peyotism and the Plains vision quest. 1971 E. Shorris *Death of Great Spirit* iv. 40 He is a Uwipi, a medicine man and practitioner of the traditional vision quest, and there is not a Sioux Indian within two hundred miles of Pine Ridge who does not regard him with a certain amount of awe. 1922 W. B. Yeats *Trembling of Veil* ii. xiv. 129 Politics, for a vision-seeking man, can be but half achievement. 1807 Wordsworth *Poems* II. 151 The youth, who..still is Nature's priest, And by the vision splendid Is on his way attended. 1895 A. B. Paterson *Man from Snowy River* (1896) 21 And he sees the vision splendid of the sunlit plains extended, And at night the wond'rous glory of the everlasting stars. 1959 X. Herbert *Seven Emus* x. 110 Such was his acting that he took in his audience along with himself, made them share his optimism, his vision splendid. 1972 I. Moffitt *U-Jack Soc.* xii. 199, I sat obediently and listened, and Sir Philip spread his vision splendid of electricity extended—with nuclear power. 1966 *Guardian* 22 Dec. 3/3 The Post Office is exploring the possibility of..a vision-telephone for calls between individuals. 1969 *New Scientist* 16 Oct. 146/3 Big industrial concerns might..find vision telephones helpful for conferences between executives. 1915 D. H. Lawrence *Rainbow* xi. 267 In the vision-world He spoke of Jerusalem, something that did not exist in the everyday world.

b. In sense *5*; *spec.* **vision-mixer,** a person whose job is to switch from one camera to another in television broadcasting or recording; so **vision-mixing** *vbl. sb.*

1935 *Illustr. London News* 23 Feb. 307/1 (*in figure*) Vision control [of a television receiver]. 1937 *Discovery* Nov. 330/1 The incoming vision signal..carries the time-sequence of light-and-shade in the original image. 1938 *Times* 7 Jan. 13/6 Behind him is the key-man, the vision-mixer. 1951 I. Asimov *Stars like Dust* (1958) i. 7 He jabbed at the vision control and the small screen was alive with light. 1953 Amos & Birkinshaw *Television Engin.* I. i. 17 The composite signal obtained by combining a picture with a synchronising signal is known as a vision [1957: video] signal. 1956 *B.B.C. Handbk.* 1957 59 This unit..has its own VHF sound and vision transmitters. 1960 *Daily Tel.* 17 June 13/4 The present [television] centre runs more than 100 vision and 400 sound circuits. 1961 G. Millerson *Technique Television Production* xvi. 296 The television director is his own editor. He may himself carry out the mechanical operation of the video switching console (vision mixing desk), or have..a switcher (vision mixer), follow his instructions. 1972 D. Lees *Zodiac* 30 Zodiac's director, vision mixers, audio-men and camera men were obviously tops. 1979 Zarach & Morris *Television Princ. & Pract.* ii. 8 The signal from the camera, together with the synchronising pulses.., modulate the vision carrier. 1982 A. Road *Dr. Who* 45/2 To the director's..right [in the gallery] are the vision mixer, the producer..and..the technical manager.

visionariness. 1. (Earlier example.)

c 1806 D. Wordsworth *Jrnl.* (1941) I. 286 That visionariness which results from a communion with the unworldliness of nature.

visiophone (vi·ziofōᵘn). [f. L. *vīsio* Vision *sb.* + *-phone.] = *videophone* s.v. *video-.

1971 *New Scientist* 15 July 140/2 It is estimated that 125 Visiophones will be in use in that area [*sc.* Paris] within four years. 1978 *Jrnl. Communication* XXVIII. 151 The audio and video (visiophone) two-way link was reserved..each day for consultations..between staff in Nairobi and Paris.

visit, *sb.* Add: **1. f.** *Dog-Breeding.* A bitch's journey to and her stay with a dog for breeding purposes. Cf. *visit *v.* 8 b.

1867 *Field* 5 Jan. 5/3 The following bitches have been on a visit to Patent. 1887 G. Stabler *Pract. Kennel Guide* xii. 125 Dogs..refuse food during the time of the bitch's visit. 1910 R. Leighton *Dogs* l. 320 A proper mating should be considered at the outset... It is customary for the bitch to be the visitor, and it is well that her visit should extend to two or three days at the least.

3. c. *Billiards* and *Snooker.* A turn of play at the table.

1927 *Times* 22 Feb. 16/3 At his second visit to the table he made a break of 111.

visit, *v.* **8. b.** For † *Obs. rare* read: *rare* exc. in *Dog-Breeding*: To be put to mate with (a dog) or at (a kennel). (Later examples.)

1867 *Field* 5 Jan. 5/3 The following bitches have visited Good Idea at the Blundell Arms Kennels. 1877 G. Stables *Pract. Kennel Guide* xii. 122 In your correspondence with the owner of the bitch that is going to visit your kennel, be sure to let him know that he is to send her..as soon as there are the slightest signs of her coming in season. 1922 R. Leighton *Compl. Bk. Dog* iii. 34, 40 per cent. of prize-bred bitches which visit prize-bred dogs are unproductive.

e. Also const. *at.*

1753 [in Dict.]. 1836 Dickens *Let.* 31 Mar. (1965) I. 144, I cannot..visit at a relation's house from which my father is excluded. 1976 *Stillwater* (Montana) *News* 1 July 2/1 Ruthie Braunstadter of Billings visited at the home of her grandmother. 1978 *Times* 1 Feb. 15/5 If it were only a friendly meal around a common table, I am sure that Catholics would be only too happy to visit at the tables of their friends whenever they were invited.

f. *to visit with*: = sense 8 a in Dict. Now *U.S.*

1850 E. Ruskin *Let.* 18 Jan. in M. Lutyens *Effie in Venice* (1965) ii. 117 They visit with everybody in Venice. 1872 Geo. Eliot *Middlem.* I. i. i. 8 The small group of gentry with whom he visited. 1903 F. Norris *Pit* v. 150 Almost every evening nowadays the Dearborn girls came ..to visit with the Cresslers. 1927 C. A. Lindbergh *We* iv. 57 Perryville, Missouri, where we visited with some of Klink's friends. 1949 M. Lowry *Let.* 1 July (1967) 179 Margerie is flying..to visit—to visit *with*, I believe I should say—her family for a week. 1973 *Black Panther* 16 June 2/1 Seale visited with David Hilliard at Vacaville (Medical Facility) Prison. 1981 C. Potok *Bk. Lights* (1982) vii. 227 He took Karen to a movie, then visited for a while with her family. 1985 *New Yorker* 11 Feb. 70/1 A young man of Ved's age and grade level is visiting with him.

g. *intr.* To talk or chat; to exchange conversation. *U.S.*

1856 M. D. Colt *Diary* 21 Oct. in *Went to Kansas* (1862) xii. 195 Have visited some, but am now in my room again to rest. 1879 A. Tourgée *Fool's Errand* xix. 111 He.. stopped at the Mission-House, visiting with the teachers. 1898 M. Deland *Old Chester Tales* 75 You can eat it while I get out and visit with the minister. 1929 *N.Y. Times Mag.* 20 Oct. 1 Having disposed of a batch of correspondence he was willing to visit in informal..fashion. It was good talk. 1967 'P. Kruger' *Weave Wicked Web* xi. 94, I called to see her around eight... Stella and I visited until close on eleven. 1979 N. Mailer *Executioner's Song* i. xviii. 304 Nicole drove down to the Preliminary Hearing..but they let her visit with Gary for only a moment.

visitable, *a.* **2.** (Further examples of the sense 'worth visiting'.)

1953 *John o'London's Weekly* 12 June 519/1 He was also known as a charming and eminently visitable old gentleman. 1972 *Maclean's Mag.* Mar. 57/3 While it has its share of visitable old buildings Rijeka is more European than most Yugoslav cities. 1983 R. Muir in Muir & Welfare *Nat. Trust Guide Prehistoric & Roman Britain* ii. 27/2 There are few exciting and visitable relics of [Mesolithic] human life apart from caves.

Visitandine. b. (Earlier example.)

1871 A. J. C. Hare *Walks in Rome* I. vi. 277 It has been decided that some remains which exist in the garden of the Villa Mills (now a Convent of Visitandine Nuns) are those of the House of Hortensius.

visitation. Add: **I. 5. d.** A gathering at the home of a deceased person before the funeral takes place; a wake. *U.S.*

1974 *Amer. Speech 1971* XLVI. 70 Another interesting Boston area universal was the use, even among Protestants, of *wake* for what is elsewhere in the state often called a *visitation*. 1976 N. Thornburg *Cutter & Bone* iii. 75 'The goddamn funeral,' he explained. 'Or what do they call it the day before—visitation?' 1983 *Chicago Sun-Times* 27 Sept. 18 An acquaintance in the life insurance business showed up at the visitation and proceeded to solicit business... What kind of a person would latch onto an opportunity to solicit business at a wake?

III. 10. *visitation fee, nun*: earlier examples.

1791 J. Woodforde *Diary* 6 Dec. (1927) III. 316 Recd. of Ditto, my last Visitation Fee, 0.2.6. *a* 1700 [see *Blue Nun* s.v. *blue *a.* 13].

visite. Add: **1.** (Earlier example.)

c 1847 J. S. Coyne *How to settle Accounts with your Laundress* 4 I'll come in my blue *visite* and my native innocence.

‖ **3.** *visite de digestion*, a formal call paid in return for hospitality received.

1908 J. Churchill *Reminisc.* (1973) iii. 44 The writing of ceremonious notes, the leaving of cards, not to speak of *visites de digestion*, which even young men were supposed to pay, took up most afternoons. 1971 L. P. Hartley *Mrs Carteret Receives* 17, I thought it only civil to ask Madame Carteret if we could pay her a farewell visit, not a *visite de digestion*, but just an acknowledgment of her kindness.

visitee. (Earlier example.)

1710 W. King *Let.* 16 Sept. in Swift *Corr.* (1963) I. 176 There are great men here as much out of humour, as you describe your great visitee to have been.

visiting, *vbl. sb.* Add: **4.** *visiting dress* (earlier example); **visiting-book,** (*b*) = *visitors' book* s.v. *visitor 6; **visiting-card**: also in (orig. *Mil.*) *slang* phr. *to leave one's visiting-card*, to leave unpleasant evidence of having been at a place; **visiting hours,** hours when visitors may call, *spec.* to see a person in a hospital or other institution; **visiting rights,** the right to pay or receive visits (to or from a child in the custody of a divorced spouse) or to receive them (while in an institution of any kind).

1848 Thackeray *Van. Fair* lv. 500 The Bishop went and wrote his name in the visiting-book at Gaunt House that very day. 1945 Partridge *Dict. R.A.F. Slang* 60 *Visiting-card*, a bomb. Mostly in *Leave one's visiting card*... As the civilian drops visiting cards into tray or salver, so the airman drops bombs on..enemy-occupied territory. 1953 'E. Crispin' *Fen Country* (1979) 68 All he had to do was to..leave his visiting card [*sc.* an explosive device] and collect his fee. 1972 V. Canning *Rainbird Pattern* vi. 116 He should have brought Albert [*sc.* the dog] in from the car, he at least could have left a visiting card in self-defence. 1825 E. Weeton *Jrnl.* 20 June (1969) II. 397, I am..as neat in my *every day apparel*, as any of my acquaintances; they many of them exceed me in visiting dresses. 1851 *London at Table* i. 36 A cigar is not the indispensable companion of visiting hours. 1897 *Scribner's Mag.* Sept. 384/1 Formal visiting-hours were ignored in the village of Sewanee. 1947 'G. Orwell' *Let.* 31 Dec. in *Coll. Ess.* (1968) IV. 386 I'd love it if you did come & see me... They don't seem very lavish with their visiting hours... I've only been in the hospital abt 10 days. 1979 'D. Kyle' *Green River High* ii. 15, I scribbled a note giving hospital visiting hours. 1971 *Deb. House of Commons* (Canada) 14 Dec. 10 460/1 Has the minister or his department investigated allegations that visiting rights were denied to members of the Black United Front? 1972 H. Kemelman *Monday the Rabbi took Off* xxiv. 159, I was divorced from his mother—he was ten at the time—I had visiting rights, of course. 1982 G. Wagner *Children of Empire* viii. 138 All the homes were surrounded by high walls to keep intruders out and the children in. Visiting rights were restricted.

visiting, *ppl. a.* Add: **1. c.** *visiting fireman* (U.S. slang): a person given especially cordial treatment while visiting an organization or place; a tourist expected to spend freely.

[1855 *Sun* (Baltimore) 25 Oct. 1/6 A company of firemen from Rochester, N.Y.,..continue to receive the attentions of their brother firemen of Baltimore... This evening the visiting firemen will be the guests of the Washington Hose Company.] 1926 S. Lewis *Mantrap* xxi. 265, I.. couldn't keep my hooks off any he-male that blows into town with the visiting firemen! 1936 H. Bernstein *Choose Bright Morning* ii. 20 He never sees people who might have legitimate business with him... But he receives all the visiting firemen. 1945 H. S. Truman in M. Truman *Harry S. Truman* (1972) x. 202 Naturally got pointed out as the visiting fireman and had a kind of reception between acts and afterwards. 1962 A. Davison *In Wake of Gemini* 115 Members of the firm, whose business it was to look after visiting firemen, had the same happy knack. 1964 *Economist* 25 Jan. 313/2 The marked-up tickets are usually sold to 'visiting firemen'. 1972 K. Benton *Spy in Chancery* i. 19 Don't they know the form for visiting firemen?..As the British delegate..I shall make a courtesy visit to the Embassy.

d. *visiting fellow, lecturer, professor*: an academic who accepts an invitation to work at another institution for a fixed term; so *visiting fellowship*, etc.

1950 *Univ. London Gaz.* 28 Oct. 178 (*heading*) Appointment of Visiting Professor of Belgian Studies. *Ibid.*, The Visiting Professorship of Belgian Studies. 1960 A. Huxley *Let.* 12 Nov. (1969) 898, I am here for the moment working as a Visiting Professor at MIT. 1962 *St. Edmund Hall Mag. 1961–2* 3 Mr. Allen was Visiting Lecturer at the University of Iowa. 1963 *Cambridge Univ. Reporter* 24 Apr. 1450 The Board of Managers of the Smuts Memorial Fund invite applications for Smuts Visiting Fellowships in Commonwealth Studies. *Ibid.*, The emoluments of a Visiting Fellow will be a sum not exceeding £1,500. 1973 *Oxf. Mag.* 1 June 2/2 Professor Levitch.. has accepted a Visiting Fellowship at University [College]. 1980 A. Coppel *Hastings Conspiracy* viii. 59 Langton enjoyed a visiting lectureship at the London School of Economics. 1980 M. Drabble *Middle Ground* 156 A visiting professor..on his way to an excavation. 1982 C. Monteith in A. Thwaite *Larkin at Sixty* 41 All Souls—a College where he later stayed in his own right as a Visiting Fellow.

visitor. Add: **4. c.** *Sport.* A member of a visiting team.

1900 W. J. Ford *Cricketer on Cricket* xiv. 162 Setting aside one unpleasant incident of the present trip..our visitors can hardly complain of the treatment they have received at the hands of the players, the public, or the press. 1916, 1930 [see *home *a.* 2 c]. 1976 G. L. Greaves *Over Summers Again* xvi. 105 September 1966 Saw Yorkshire once again..needing an outright win over Kent to give them yet another championship... Trueman and Nicholson..soon had Kent in trouble and the visitors, all out for 119, closed 91 behind.

5. A menstrual discharge; = Visit *sb.* 4. *slang.*

1980 *Quarto* June 3/2 It was shortly before my thirteenth birthday that I first had 'Visitors'. Mother.. had told me about the monthly cycle... My first 'visitor' was a light one. 1983 *Maledicta 1982* VI. 26 Menstruation (females):..visitor. 1984 *New Yorker* 29 Oct. 45/3 Girls used to say they had the curse. Or they had a visitor.

6. *Comb.*, as **visitor centre** orig. *U.S.*, a building in a tourist area in which exhibitions, slide-shows, etc., are displayed as an introduction to the locality; **visitors' book,** a book in which visitors may write their names and addresses, and, sometimes, comments; **visitors' list,** a public list of those making a visit to a place, esp. to a resort; in quot. 1864, a list of official visitors (Visitor 1 b).

1964 P. Jensen *National Parks* i. 18 Park headquarters is also here. A visitor center explains the features of the area. 1979 *Farmington* (New Mexico) *Daily Times* 27 May 3c/6 Visitor centers will tell tourists the speed limits will be enforced. 1979 *United States 1980/81* (Penguin Travel Guides) 239 A 25-minute movie about the Battle of Shiloh is shown in the Visitor Center. 1983 *Leisure, Recreation & Tourism Abstr.* VIII. iv. 205/2 A new accommodation concept, and an existing visitor centre facility. 1846 *Punch* 20 June 278/2 (*heading*) Ibrahim Pacha's visitors' book. 1870 *Ibid.* 1 Oct. 138/2 The Emperor Nicholas.. 'expressed himself much pleased', as the visitors' books have it. 1910 Belloc *Pongo & Bull* vii. 96 The Duke.. had very properly insisted upon the retention of the visitors' book. 1976 L. Deighton *Twinkle, twinkle Little Spy* xvi. 157 There was a visitors' book..a beautiful leather-bound volume..dutifully signed by the Reid-Kennedys' guests. 1984 *Times* 25 May 7/5 (*caption*) All smiles: the Queen leaving the town hall..after signing the visitors' book. 1864 Mrs. Gaskell *Wives & Daughters* (1866) I. i. 8 Though my name has been down on the visitors' list these three years, the countess has never named me in her note... Her ladyship would be as hurt as any one when she did not see Phœbe among the school visitors. 1904 A. E. W. Mason *Truants* II. vi. 105 She.. bought a visitors' list at the kiosk. 1907 E. Glyn *Three Weeks* iv. 53 The simplest thing..seemed to descend into the hall and look at the Visitors' List... There were only a few people in the hotel. 1927 E. Bowen *Hotel* iii. 27 Victor..took up the local paper and began to read the visitors' list.

Visking (vi·skiŋ). Also **visking.** A proprietary term for seamless cellulose tubing used as membranes in dialysis and as edible casings for sausages.

1931 *Trade Marks Jrnl.* 29 July 1050/1 *Visking casing.* ..sausage casings made of cellulose. The Visking Corporation,..Chicago, Illinois, United States of America; manufacturers. 1941 *Official Gaz.* (U.S. Patent Office) 15 Apr. 550/1 The Visking Corporation, Chicago... *Visking* for merchandise protectors—namely seamless cellulose tubes, and artificial sausage casings made of cellulose. 1956 *Nature* 25 Feb. 381/1 These compounds were removed..by dialysis of aqueous solutions of 2–3 mgm. of the venom, in Visking tubing. 1970 *Ibid.* 26 Dec. 1336/2 The oxidized form dialysed through visking tubing. 1977 *Times Educ. Suppl.* 21 Oct. 29/1 This apparatus was designed to simplify the setting up of experiments using visking tubing.

visna (vi·znă). Also **Visna.** *Vet. Sci.* [a. ON. *visna* to wither.] A fatal disease of sheep in which there is progressive demyelination of neurones in the brain and spinal cord. Freq. *attrib.*

1957 *Jrnl. Neuropath. & Exper. Neurol.* XVI. 389 (*heading*) Visna, a demyelinating transmissible disease of sheep. *Ibid.* 393 This experiment obviously does not prove conclusively that Visna-sera will specifically neutralize Visna-virus but it is consistent with such an explanation. 1970 Jubb & Kennedy *Path. Domestic Animals* (ed. 2) I. 269/2 The virus of visna can be transmitted by intra-pulmonary injection. 1982 *Jrnl. Neuroimmunology* III. 140 Icelandic sheep are free of natural infection since visna virus has been eradicated from the whole country.

visor, vizor, *sb.* Add: **1. a.** Also *transf.*

1957 *Time* 2 Sept. 37/2 Simons quickly clamped shut the visor of his space helmet. 1962 W. Schirra in *Into Orbit* 51 Once we are up..we can open up our visor and breathe the cabin air for a bit.

c. A shade for protecting the eyes from unwanted light while not impeding the vision; *spec.* one attached to the top of the windscreen of a motor vehicle or aircraft.

1925 J.-L. Hudon *Lexique Technique* 109 Windshield visor. 1936, etc. [see *sun visor* s.v. *sun *sb.* 13 a]. 1952 *Times* 9 Jan. 3/3 Canberra pilots at Binbrook are testing a new type of visor for protecting crews against the glare at high altitudes. 1957 Rawnsley & Wright *Night Fighter* 57, I peered into the visor [of a radar], trying to accustom my eyes to the dim light. 1973 'E. McBain' *Let's hear It* iii. 41 The visor on the driver's side was down.

visored, *ppl. a.* Add: **2. b.** Of a cap: peaked.

1950 *Manch. Guardian Weekly* 5 Oct. 15/2 The General left the 'plane, dressed in field jacket and visored, brass-encrusted cap. 1961 M. Beadle *These Ruins are Inhabited* (1963) ii. 30 The salesman spread out an impressive..array of..red-bound black blazers, visored caps, red bordered black Rugby socks. 1983 P. Chevalier *Shaft* xvii. 119 He was wearing exquisite yachting whites and a Navy-style visored cap.

vista, *sb.* Add: **5.** Comb.: **vista-dome** *U.S.*, a high glass-sided railway carriage that enables passengers to look at the view from above the normal level of the train. Freq. *attrib.*

1945 *Time* 18 June 78/3 For travelers who like to watch the country go by, Chicago, Burlington & Quincy Railroad announced a new 'Vista Dome' car, to be put in operation next week. 1948 *Sun* (Baltimore) 6 Jan. 6/5 (Advt.), The Vista-Dome affords a wonderful opportunity for camera fans. 1973 [see *Rheingold].

Vistavision (vi·stăviʒən). Also **vista-.** [f. Vista *sb.* + Vision *sb.*] A form of wide-screen cinematography employing standard 35 mm. film in such a way as to give a larger projected image with ordinary methods of projection. Also *fig.*

A proprietary name in the U.S.

1954 *Newsweek* 15 Mar. 104/3 A new filming process in which Paramount Pictures will produce all its films beginning with 'White Christmas'..was demonstrated last week in Hollywood. Called VistaVision, it..is designed primarily to improve wide-screen clarity. 1955 *Official Gaz.* (U.S. Patent Office) 25 Jan. tm 138/2 Paramount Pictures Corporation... *Vistavision* for motion picture films and cameras and parts therefor. 1961 *Guardian* 20 Jan. 7/3 Passion and vistavision ideas are not enough. 1976 *New Musical Express* 12 Feb. 27/5 I'll direct you to a copy of Diana Ross's 'An Evening With' (Motown) a vistavision affair on which Berry Gordy's favourite movie star provides a medley of Motown oldies.

visual, *a.* and *sb.* Add: **A.** *adj.* **2. a.** *visual acuity*, sharpness of vision; *spec.* as measured or expressed in terms of a definite scale (see quot. 1974).

1889 [in Dict.]. 1938 R. L. Rea *Neuro-Ophthalmology* iv. 86 In the early stages [of papillœdema] there may be.. full central visual acuity. 1974 *Encycl. Brit. Macropædia* VII. 104/1 A visual acuity of unity indicates a power of resolving detail subtending one minute of arc at the eye; a visual acuity of two indicates a resolution of one-half minute..of arc.

b. *visual purple* [tr. G. *sehpurpur* (app. first used by W. Kühne 1877, in *Verh. d. Naturhist.-med. Verein zu Heidelberg* I. 484)]: = *rhodopsin.

[1877 *Nature* 1 Feb. 296/1 These first observations of Kühne on the vision-purple (*Sehpurpur*), as he terms it.] 1878 M. Foster *Text Bk. Physiol.* (ed. 2) iii. ii. 415 For the restoration of the visual purple, after it has been destroyed by light, the maintenance of the circulation of the blood through the tissues of the eye is not essential. 1921 *Proc. R. Soc.* B. XCII. 232 A highly dilute visual purple may suffice for the requirements of photopic vision. 1953 *Sci. News* XXX. 116 Although as many as six visual pigments have been recognized in different species, only one, visual purple, has been obtained from the human retina. 1983 *Guardian* 4 Aug. 17/2 Retinaldehyde..is present in the retina of the eye combined with the visual pigment known as visual purple.

5. *visual field, range.*

1880 W. James *Coll. Ess. & Rev.* (1920) 169 He perceives correctly the position of objects in the visual field. 1927 B. Russell *Anal. Matter* xii. 111 The sort of relation that will not do is illustrated if we take *xy*=*zw* to mean that *xy* and *zw* have the same apparent dimensions in the visual field of a certain observer. 1961 G. E. M. Anscombe tr. *Wittgenstein's Notebks. 1914–16* 45 What is a uniformly coloured part of my visual field composed of? 1953 R. Chisholm *Cover of Darkness* iii. 36 Reading the faces of cathode-ray tubes was a small part of the Observer's task. By description and instruction he had to get his Pilot to visual range. 1965 *Observer* 31 Oct. 1/1 'Visual range'—the distance one can see along the runway—is measured and passed to the pilot.

6. d. *visual aid*, illustrative matter designed to supplement written or spoken information; *spec.* in *Educ.* with reference to pictures, models, films, etc., as an aid to learning. orig. *U.S.*

1911 P. Monroe *Cycl. Educ.* V. 734/2 The last century of schoolroom practice has been marked by a great increase in the use of natural objects, models, pictures, maps, charts, and other visual aids. 1938 *Rep. Physical Educ. & Film* (Brit. Film Inst.) 1 The function of the film in education has been defined as that of a visual aid. 1958 *Economist* 29 Nov. 764/1 The visual aids which the party's television programme used to good effect. 1967 Mrs. L. B. Johnson *White House Diary* 14 Mar. (1970) 497 In one room they were using visual-aid machines for faster reading. 1980 E. Blishen *Nest of Teachers* i. ii. 11 That most familiar of visual aids, an extremely tatty blackboard.

e. *visual display* (Computers) = *display *sb.* 1 c; *visual display unit*, a device for displaying on its screen data stored in a computer, and usu. incorporating a keyboard for manipulating the data; abbrev. VDU, vdu s.v. *V 5 b.

1954 *Jrnl. Assoc. Computing Machinery* I. 57/1 Cathode ray tube equipment for providing external visual displays of information stored internally in the computer. 1967 Klerer & Korn *Digital Computer User's Handbk.* i. 77 The importance of the use of machine-produced graphs and other types of visual displays can hardly be over emphasized. 1969 *Computers & Humanities* IV. 83 For those interested in on-line text manipulation, there is special temptation in newer devices like cathode ray tubes (CRTs) or visual displays, or 'scopes', as they are variously called. 1971 J. Anderson in B. de Ferranti *Living with Computer* vii. 59 With..the introduction..of visual display units..there has been a resurgence of interest in applying such techniques to medical recording. 1977 *R.A.F. News* 11–24 May 7/2 When can we have our visual display unit? 1983 *Brit. Med. Jrnl.* 23 July 271 Most of the comments relating to paper records apply to visual displays for data entry. 1984 *Times* 16 Nov. 12/4 The latest scare comes from reports of women who worked on visual display units (VDUs) during pregnancy and went on to deliver handicapped babies or suffer miscarriages.

7. a. (Earlier example.)

1817 Coleridge *Biog. Lit.* I. iv. 76 The change of one visual image for another involves in itself no absurdity.

B. *sb.* **3.** A visual image or display, a picture; *spec.* the visual element of a film or television production. Usu. *pl.*

1951 *Brit. Kinematogr.* XIX. 110/1 A good deal has been done by..'visuals'. 1959 *Times* 4 Mar. 11/7 Exposition is particularly difficult on television—it gets confused by the visuals and the speaker's loose words. 1961 *Listener* 19 Oct. 622/3 As the commentator's voice announced that allocations for shelters by the U.S. Government had been currently increased 700 per cent, we saw a visual of President Kennedy roaring with laughter. 1966 J. Derrick *Teaching Eng. to Immigrants* vii. 229 The Language Master..is a new transistorized machine into which can be fed visuals and reading matter on long cards. 1972 *Observer* 16 Apr. 34/6 Written entries.. should be typed, visuals (a maximum of 3ft square) carefully packed. 1974 'D. Craig' *Dead Liberty* xxi. 125 The Finance lads liked your economics piece... We're getting some graphs and other visuals done. 1984 *Times* 23 Jan. 7/1 There is more porn in the [cinema] subtitles than in the visuals.

visuality. Delete *rare* and 'app. used by Carlyle only' in etym. and add: **2.** (Later example.)

1912 H. Bradley *Let.* in *Corr. Bridges & Bradley* (1940) 96, I do not think I can even say or hear 'bread and butter', or 'dog and cat', without an accompanying underconsciousness of the way in which these words are commonly spelt. (It may be deplorable that any human soul should have got into this degraded state of subjection to black and white visualities.)

3. Vision, sight.

1923 D. H. Lawrence *Birds, Beasts & Flowers* 27 You are everywhere, and I am blind, Sightless among all your visuality, You staring caryatids. 1941 *Illustr. London News* CXCVIII. 37 Immediately in front of the driver is a movable periscope, shown reflecting the other tanks ahead, whereby he can obtain full visuality when the armoured cowl has to be closed when in action.

4. Visual aspect or representation; physical appearance.

1938 R. G. Collingwood *Princ. Art* vii. 144 His [*sc.* Cézanne's] landscapes have lost almost every trace of visuality. 1975 *Physics Bull.* Apr. 165/3 Regretfully, the printed word can hardly do justice to the visuality of the afternoon. 1976 *Times Lit. Suppl.* 5 Mar. 251/1 Notable among the elements of visuality which lead to the effect of amenity is that of scale.

visualizable (vi:ziuăləi·zăb'l), *a.* [f. Visualize *v.* + -able.] **a.** Capable of being visualized. **b.** Capable of being rendered visible.

1956 E. H. Hutten *Lang. Mod. Physics* v. 202 The models may overlap, in part, and they may become more 'abstract', i.e. less visualisable. 1968 *Amer. Speech* XLIII. 5 In the world of reality for which the elements of language are symbols, there are no visualizable objects which correspond to adjective symbols. 1980 *Nature* 8 May 100/1 We think it unlikely that the large visualizable deposits in the older birds are more than epiphenomenal.

visualization. Add: **2.** The action or process of rendering visible.

1926 [see *cholecystography]. 1936 *Amer. Jrnl. Cancer* XXVII. 49 The hexagonal tube..offers distinct advantages with its flat sides permitting good visualization. 1960 *New Scientist* 28 July 305/3 Echo sounding.. is now being applied to the visualization of structures within the body. 1973 *Nature* 17 Aug. 410/1 Direct visualization of biological material at this level would tell us much about the structure and mode of action of macromolecules. 1982 *Listener* 23/30 Dec. 42/3 The cinematic visualisation of the script..belongs entirely to Welles and his technicians.

visualize, *v.* Add: **3.** *trans.* To render visible.

1912 *Moving Picture World* 17 Aug. 646/2 The printed volumes of ancient history have been carefully preserved and why not the film, which is more accurate and which, supplementing the printed story, would visualize the actual occurrences so that all may readily understand? 1925 D. H. Lawrence in *Calendar Mod. Lett.* Dec. 269 And the camera will *visualize* the sunflower far more perfectly than Van Gogh can. 1938 *Q. Jrnl. Med.* XXXI. 462 Inflating the stomach with gas by means of an effervescing drink, a procedure we have found of great help in visualizing the apex [of the heart in radiography]. 1958 *Arch. Neurol. & Psychiatry* LXXIX. 59/1 The dorsal root axons..may then be stained and visualized. 1971 *Daily Tel.* 18 Nov. 5 (Advt.), Research has led to the building of an ultra-sensitive Schlieren apparatus, in which pulsed ultrasound may be visualised both in water and in solids.

visualizer. Add: **2.** *spec.* in *Advertising*, a commercial artist employed to design layouts.

1921 R. S. Durstine *Making Advertisements* ii. 23 Several arrangements suggest themselves immediately if the visualizer has a natural or a trained imagination. 1948 [see *lay-out 1 b]. 1968 M. Butterworth *Walk Softly* ii. 30 She worked as a visualiser in an ad agency. 1981 *West Lancs. Evening Gaz.* 5 June 24 (Advt.), Visualisers/finished artists..required by a rapidly-expanding studio.

visuo-. Add: *visuo-spatial, tactual,* adjs.; *visuo-spatially* adv.; **vi·suomotor** *a.*, pertaining to or involving motor activity as guided by or dependent on sight; **vi:suopsy·chic** *a.*, an epithet of two cortical areas adjacent to the striate cortex, orig. regarded as sites of mental elaboration of visual sense impressions; **vi:suose·nsory** *a.*, pertaining to or involving the visual perception of sensory signals; *spec.* an epithet of the striate cortex (see *striate *a.* 2), as the part of the brain that receives sensory nerve impulses from the eye.

1942 *Anat. Rec.* LXXXIV. 470 (*heading*) Reestablishment of visuomotor coordination by optic nerve regeneration. 1972 *Science* 5 May 536/2 This difference could be observed in splitbrain monkeys executing a visuomotor task with one eye covered. 1900 Visuo-psychic [in Dict.]. 1954 S. Duke-Elder *Parsons' Dis. Eye* (ed. 12) xxix. 500 A lesion of this vessel thus causes a crossed homonymous hemianopia often with disturbances of the visuopsychic areas. 1980 *Gray's Anat.* (ed. 36) vii. 1010/2 Not only the striate cortex (area 17, visuosensory area), but also the para- and peri-striate areas around it (areas 18 and 19—the 'visuopsychic' cortex) receive projection fibres. 1900 Visuo-sensory [in Dict.]. 1907 J. H. Parsons *Dis. Eye* v. 90 This area, which is the primary visual or visuo-sensory area.., is the cortical projection of the corresponding halves of both retinæ. 1980 Visuo-sensory [see *visuopsychic* adj. above]. 1962 *Jrnl. Speech & Hearing Research* Dec. 359/2 Factor 3 is a clear-cut visuospatial factor. Tests..include matching, copying, drawing, and object-assembly tests. *Ibid.* 367/1 Visuo-spatial behavior..involved both visual and sensory processes. 1977 *Lancet* 10 Dec. 1227/2 Intelligence tests indicated severe intellectual deterioration on subtests measuring visuospatial perception. 1939 *Mind* XLVIII. 360 We touch surfaces of things optically known to us, and when we pass from one thing to another we leave them visuo-spatially related behind us. 1932 H. H. Price *Perception* ix. 277 Let us call it a visuo-tactual solid. 1959 J. L. Austin *Sense & Sensibilia* (1962) ii. 8 Visuo-tactual solids.

Vita[1] (vəi·tă). Also **vita.** [L., = 'life'.] A proprietary term for glass which transmits most of the ultraviolet rays of sunlight. Usu. as *Vita-glass, Vitaglass.*

1925 *Trade Marks Jrnl.* 29 July 1649 *Vita*...Glass. Francis Everard Lamplough, 47, Bunbury Road, King's Norton, Birmingham; chemist. 1925 *Spectator* 14 Nov. 869/1 A substitute for quartz... This vitaglass is what I asked for. 1939 O. Lancaster *Homes Sweet Homes* 60 The small tight-shut windows..will be replaced by a wide expanse of hygienic vita glass. 1948 J. Betjeman *Coll. Poems* (1958) 227 And many a cultivated hour they pass In a fine school with walls of vita-glass. 1957 *Oxford Mag.* 17 Oct. 18/2 The well-intended propaganda of Sir Leonard Hill and others about 1930..gave it a new lease of life by preaching that the sun was the best preventative of rickets provided that 'vitaglass' windows were used. 1981 *Times* 21 Feb. 22/3 The old Marine Spa..contained.. a Vita-glass sun lounge.

vita[2] (vī·tă). **1.** ‖ *vita nuova* [It., = new life]. The title of a work by Dante describing his love for Beatrice, used to denote a fresh start or new direction in life, usu. after some powerful emotional experience.

1934 A. J. A. Symons *Quest for Corvo* xii. 164 There seemed no impediment to his *vita nuova*. 1939 L. MacNeice *Autumn Jrnl.* 43 A cultured accent alone will not provide A season ticket to the Vita Nuova. 1975 P. Organ *House on Cheyne Walk* xviii. 163 Not a very good way to begin *la vita nuova*, with more lies.

2. [a. L. *vīta* life.] A biography, the history of a life; *spec.* = *curriculum vitæ* s.v. *curriculum.

1949 Wyndham Lewis *Let.* 3 Sept. (1963) 505 You ought to cut out *cleanly* and implacably the *vita* material from this typescript. 1960 *Encounter* Mar. 82/2 Clifton Fadiman..whose credentials include..a *vita* which concludes as follows:..master of ceremonies on the popular radio programme *Information Please*, and on the television show *This is Show Business*. 1974 H. L. Foster *Ribbin', Jivin', & Playin' Dozens* i. 18 As far as children and their schools are concerned, the reports were simply another exercise for academicians to contribute, primarily, as a source for an additional listing on a vita or a source for citations for future papers, lectures, books, and speeches. 1979 *Amer. Speech* LIV. 257 All these observations..are summarized in the vitas of the informants.

vital, *a.* Add: **I. 4. d.** (Later examples.) Also *transf.*

1949 *Brit. Birds* XLII. 147 Vital statistics from ringed Swallows. 1956 *Newsweek* 23 Jan. 60 New eyes open on a bright wonderful world—and photography makes identification positive, records vital statistics in life's first few minutes. 1958 J. Townsend *Young Devils* ii. 19 A short history of the school plus its vital statistics—i.e. number of boys, teachers, classrooms, subjects and educational standards. 1971 *Brit. Med. Bull.* XXVII. 13/2 The epidemic of iatrogenic deaths in asthmatic children shows the need for continuous monitoring of vital statistics. 1974 *Nature* 22 Mar. 306/1 The vital statistics of this second edition command respect—94 of the 1,562 pages are needed to index its 64 chapters!

e. *vital statistics*, (*a*) (see sense 4 d in Dict.); (*b*) *colloq.*, the measurements of a woman's figure, *spec.* bust, waist, and hips (cf. *statistics 2 a*); similarly *vital measurements.* (Occas. of a man's figure.)

1958 *Observer* 18 May 10/4 To control and vary our vital measurements with changing fashion. 1968 G. Kent *Pictorial Hist. Wrestling* v. 119/1 *Height* 6ft. 1 in. *weight* 15 st. *chest* 48 ins. *biceps* 15 ins. *thigh* 26½ ins. Donald Dinnie and his vital measurements. 1952 C. R. Cooper *Teen-Age Vice* (1959) viii. 125 Regina..wrote haphazardly to men, giving her age and vital statistics. 1957 *Times* 1 Aug. 5/5 Those feminine measurements which have become known, in the entertainment world, as vital statistics. 1966 Wodehouse *Plum Pie* 60 A book like yours always involves a serious risk for the publisher owing to the absence of the Sex Motif, which renders it impossible for him to put a nude female of impressive vital statistics on the jacket. 1975 C. Weston *Susannah Screaming* (1976) xxv. 135 I have her phone number and vital statistics.

f. Of biological stains or their use: used or carried out on living tissue. Cf. *intra vitam* s.v. *intra *prep.* 2, *intravital *a.*

1907 *Chem. Abstr.* I. 734 The character of the vital staining and apparent deposition of the carmine as particles in the body cells of the rabbit was found under physiological conditions to be essentially as earlier described. 1912 *Ibid.* VI. 2453 (*heading*) The resorption of vital coloring matters in the stomach and alimentary canal. 1926 H. M. Carleton *Histol. Technique* xiii. 194 Janus Green.—This dye..may almost be regarded as a specific vital stain for mitochondria. 1946 A. Fischer *Biol. Tissue Cells* iii. 68 A staining of the nucleus by ordinary vital dyes is..always a sure sign of the death of the cell. 1948 *New Biol.* V. 28 Some dyestuffs do not kill the cell, and if they stain specific structures, this process of vital staining can give important information on the living cell. 1956 *Nature* 25 Feb. 387/1 A large body on one side of the nucleus stains directly..with 0·1 per cent aqueous vital red. 1974 *Ibid.* 18 Oct. 572/1 By staining with vital dyes, Bonner..showed that the cells in the anterior third of the grex become stalk cells.

Vitallium (vəitæ·liə̆m). Also **-alium** and with small initial. [f. *vitall-*, of unkn. origin + -ium.] A proprietary term for an alloy of cobalt, chromium, and molybdenum that has a high resistance to abrasion, corrosion, and heat and is used in surgery, dentistry, and engineering.

1935 *Official Gaz.* (U.S. Patent Office) 18 June 520/2 *Vitallium* for cobalt, chromium alloy. Claims use since July 8, 1934. 1947 *Richmond* (Va.) *Times-Dispatch* 13 Sept. 15/1 Short pieces of tubes made of the metal vitallium have been tried [to replace the missing link in arteries]. 1948 *Jrnl. R. Aeronaut. Soc.* LII. 16/2 The ease with which large quantities of these small supercharger blades could be produced to finished size by means of the 'lost wax' casting process, turned attention to 'Vitallium', an alloy which had been used for making special small castings by this method. 1951 *Trade Marks Jrnl.* 1 Aug. 713/2 *Vitallium*... Common metal alloys in the form of nuggets. Austenal Laboratories Incorporated.., New York, United States of America; manufacturers. 1961 *Lancet* 30 Sept. 757/1 At first a vitallium tube was used for the anastomosis. 1979 *Courier-Mail* (Brisbane) 21 July 3/1 Two teeth in the three-year old German Shepherd guard dog's armour now are capped with vitalium one of the toughest of alloys... It was Taz's enthusiasm for his work which took its toll on his teeth.

vitamin (vi·t-, vəi·tămin). orig. **vitamine** (vəi·tămin, vi·t-, -īn). [f. L. *vīt-a* life + Amine, from a mistaken belief about the chemical nature of the compounds (cf. quot. 1920).]

1. a. Any of a diverse group of organic compounds of which small quantities are needed in the diet because they have a distinct biochemical role, often as coenzymes, and cannot be adequately synthesized by the body, so that in most cases a deficiency produces characteristic symptoms or disease.

1912 C. Funk in *Jrnl. State Med.* XX. 342 It is now known that all these diseases, with the exception of pellagra, can be prevented and cured by the addition of certain preventive substances; the deficient substances, which are of the nature of organic bases, we will call 'vitamines'; and we will speak of a beri-beri or scurvy vitamine, which means a substance preventing the special disease. 1915 *Times Lit. Suppl.* 11 Nov. 400/3 The point about vitamines is that without them the animal ceases to grow or becomes diseased on a physiologically pure diet. 1916 McCollum & Kennedy in *Jrnl. Biol. Chem.* XXIV. 493 We would..suggest the desirability of discontinuing the use of the term vitamine, and the substitution of the term fat-soluble A and water-soluble B for the two classes of unknown substances concerned in inducing growth. 1920 J. C. Drummond in *Biochem. Jrnl.* XIV. 660 The criticism usually raised against Funk's word Vitamine is that the termination '-ine' is one strictly employed in chemical nomenclature to denote substances of a basic character, whereas there is no evidence which supports his original idea that these indispensable dietary constituents are amines... The suggestion is now advanced that the final '-e' be dropped, so that the resulting word Vitamin is acceptable under the standard scheme of nomenclature..which permits a neutral substance of undefined composition to bear a name ending in '-in'. If this suggestion is adopted, it is recommended that the somewhat cumbrous nomenclature introduced by McCollum (Fat-soluble A, Water-soluble B), be dropped, and that the substances be spoken of as Vitamin A, B, C, etc. 1932 Metcalf & Flint *Princ. Insect Life* xii. 465 Green plants can..manufacture complicated proteins, carbohydrates, fats, and vitamines from the nitrates, phosphates, sulfates, and water of the soil. 1966 V. B. Wigglesworth *Life of Insects* iv. 75 Cockroaches whose symbionts have been killed out by treatment with antibiotics must have extra vitamins in their diet. 1974 *Daily Tel.* 15 Feb. 17/4 Vitamins have received so much publicity as an essential factor for healthy living that there has been a tendency to forget that in excess some of them at least can be harmful. 1982 F. Ungar in T. M. Devlin *Textbk. Biochem.* xv. 719 Since the rat has this enzyme and ascorbic acid can be synthesized in its tissues, ascorbic acid is not a vitamin for this species.

b. *fig.*

1921 *Spectator* 16 Apr. 492/2 A book..so full of the vitamines of literature. 1971 *Where* Nov. 334/1 A diet only of football annuals would be deficient of almost every known reading vitamin.

2. With following (or occas. preceding) capital letter, denoting a particular vitamin or group of vitamins.

Some designations were abandoned when the substance concerned was shown to be a mixture, was not confirmed as a new vitamin, or became known under a chemical name.

vitamin A, either or both of two closely related fat-soluble vitamins, A_1 and A_2, esp. the former; = *retinol[2]; *vitamin A_1*, an alcohol that is present (as fatty acid esters) in egg-yolk, liver, butter, and milk, is also formed in the body from carotenoids present in green vegetables and stored in the liver, and is a component of the visual pigment rhodopsin; a deficiency of vitamin A_1 leads to night blindness and anæmia and ultimately xerophthalmia and blindness, *vitamin A_2*, the analogous component of the visual pigment porphyropsin in freshwater fish; **vitamin B,** any or all of several chemically unrelated water-soluble vitamins mostly occurring together in liver, cereals, and yeast and discovered by separation from the original 'vitamin B'; so *vitamin B complex* or *group*; **vitaman B_1** = *thiamine 3 a; **vitamin B_2** = *riboflavin; **vitamin B_6,** any or all of the compounds pyridoxine, pyridoxal, and pyridoxamine, esp. the first (the dietary form of the vitamin), deficiency of which is accompanied by symptoms that can include irritability, nervousness, or convulsions; **vitamin B_{12},** cobalamin or any of several derivatives of it, cobalt, containing compounds synthesized by micro-organisms and present in food of animal origin (esp. meat, eggs, and dairy products), a deficiency leading to pernicious anæmia and neuropathy; cf. *extrinsic factor* s.v. *extrinsic *a.* 3 c; **vitamin C,** ascorbic acid; a water-soluble sugar, $C_6H_8O_6$, which is present in citrus fruits, green vegetables, and tomatoes, and in man is required for the synthesis of collagen and the maintenance of connective tissue, its deficiency leading to scurvy; **vitamin D,** each or all of the fat-soluble vitamins that cure or prevent rickets in children and osteomalacia in adults, one or other being required for the correct metabolism of calcium; *spec. vitamin D_2* [named in Ger. by A. Windaus et al. 1931, in *Ann. d. Chem.* CDLXXXIX. 269], a compound, $C_{28}H_{44}O$, made by the ultraviolet irradiation of ergosterol and added to dairy products; also called *calciferol* or *ergocalciferol*; *vitamin D_3* [named in Ger. by A. Windaus et al. 1936, in *Zeitschr. f. physiol. Chem.* CCXLI. 102], a closely related compound, $C_{27}H_{44}O$, formed in the skin by ultraviolet light and present in egg-yolk, liver, and fish-liver oils; also called *cholecalciferol*; **vitamin E** = *tocopherol; **vitamin G** chiefly *U.S.* = *vitamin B_2* above; now *rare*; **vitamin H** [named in Ger. by P. György 1931, in *Zeitschr. f. ärztliche Fortbildung* XXVIII. 379/2, f. *haut* skin] = *biotin; **vitamin K,** either or both of two related fat-soluble derivatives of naphthoquinone, *vitamin K_1* (= *phylloquinone) and *vitamin K_2* (menaquinone, $C_{41}H_{56}O_2$), one or other of which is required for proper clotting of the blood, the former occurring in green vegetables and the latter being synthesized by intestinal bacteria; **vitamin P,** any or all of the flavonoids present in food plants, formerly thought to be necessary in the diet for the integrity of the capillaries.

1920 Vitamin A [see sense 1 a above]. 1937 J. R. Edisbury in *Nature* 7 Aug. 234/1 A substance apparently identical with the 693 mμ chromogen can replace the vitamin A of rhodopsin without loss of physiological function... It..seems desirable provisionally to designate as 'vitamin A_2' the 693 mμ chromogen. 1950 [see *retinene]. 1960, etc. [see *retinol[2]]. 1968 A. White et al. *Princ. Biochem.* (ed. 4) xl. 906 Vitamin A_2 differs from A_1 by having one additional conjugated double bond in the ring. 1976 H. R. Schiffman *Sensation & Perception* xii. 184/2 When the eye is kept in the dark, vitamin A joins with opsin to reconstitute rhodopsin. 1982 S. G. Chaney in T. M. Devlin *Textbk. Biochem.* xxvi. 1202 Vitamin A is also apparently required for mobilization of iron from the liver.

1920 *Brit. Med. Jrnl.* 31 July 151/1 The water-soluble B vitamine in rice polishings is very stable. 1920 [see sense 1 a above]. 1934 *Nature* 31 Mar. 498/2 The view has already been considered..that these two classes of dissimilar skin changes are to be ascribed to a deficiency not only of vitamin B_2 but also of another component of the vitamin B complex. 1953 D. M. Dunlop et al. *Textbk. Med. Treatment* (ed. 6) 398 Three other components of the vitamin B complex have been reported to have therapeutic effects; pantothenic acid.., pyridoxin..and biotin. 1967 *Martindale's Extra Pharmacopoeia* (ed. 25) 122/1 Dried yeast is used for the prevention and treatment of vitamin B deficiency. 1969 R. F. Chapman *Insects* v. 74 The B vitamins thiamine, riboflavin, nicotinic acid, pyridoxine and pantothenic acid are essential to most insects. 1983 A. Tull *Food & Nutrition* i. 22/2 Like the other B vitamins..nicotinic acid is also an important factor in the release of energy from food..by oxidation.

1928 Chick & Roscoe in *Biochem. Jrnl.* XXII. 790 Experiments..confirmed the conclusion of Goldberger and his colleagues..that the water-soluble B vitamin.. had two components. (1) Vitamin B_1, or the antineuritic, less heat-stable vitamin... (2) Vitamin B_2, a more heat-stable vitamin..in the absence of which the animal fails to grow. 1955 *Sci. News Let.* 26 Mar. 194/1 In each group some mothers got vitamin C (ascorbic acid) pills; some got thiamine, or vitamin B-1 pills; some got pills containing thiamine, iron, and riboflavin and niacinamide which are B vitamins; and some got placebos. 1974 Passmore & Robson *Compan. Med. Stud.* III. xxiv. 26/2 Both wet and dry beriberi occur among chronic alcoholics whose diet can be deficient in vitamin B_1.

1928 Vitamin B_2 [see *vitamin B_1* above]. 1933 *Jrnl. Amer. Chem. Soc.* LV. 2927 Several similarities suggest its close relationship [*sc.* that of pantothenic acid] to vitamin G (B_2). 1934 [see *vitamin B* above]. 1967 H. A. Guthrie *Introd. Nutrition* xii. 236/1 Riboflavin, which has also been known as vitamin B_2, vitamin G, and the yellow vitamin, was recognized in 1917 when it became clear that vitamin B retained some growth-promoting properties after its anti-beriberi properties had been destroyed by heat.

1934 P. György in *Nature* 31 Mar. 499/1 We have for the time being named this 'rat pellagra preventive factor' in its narrow sense vitamin B_6. 1955 D. M. Hegsted in F. C. Blanck *Handbk. Food & Agric.* ix. 292 The natural occurring deficiency of vitamin B_6 has probably not been seen in any species other than recent reports of its development in infants fed certain prepared formulas. 1970 [see *pyridoxine]. 1970 [see *pyridoxamine]. 1974 [see *pyridoxal]. 1983 J. Katz in Kaye & Rose *Fund. Internal Med.* cxxxvi. 904/1 Dietary and primary pyridoxine (vitamin B_6) deficiencies are rare.

1948 M. S. Shorb in *Science* 16 Apr. 397/1 A crystalline compound, vitamin B_{12}, has been isolated from liver..and has been shown to be highly active hematopoietically.. upon cases of pernicious anemia. 1950, etc. [see *cobalamin]. 1961 *New Scientist* 23 Feb. 457/2 The cobalt..had a selective action. Experiments showed that it was taken up by the flora of the rumen of sheep and cattle in the synthesis of vitamin B_{12}. 1982 S. G. Chaney in T. M. Devlin *Textbk. Biochem.* xxvi. 1225 The liver stores up to a 6-year supply of vitamin B_{12}. Thus, deficiencies..are extremely rare.

[1919 J. C. Drummond in *Biochem. Jrnl.* XIII. 77 The diet has been seriously, if not totally, deficient in the antiscorbutic factor or 'water-soluble C'.] 1920 [see sense 1 a above]. 1921 *Jrnl. Industr. & Engin. Chem.* Dec. 1115/1 We know that the antiscorbutic vitamine is water-soluble; indeed, it has been called the water-soluble C vitamine. 1942 *Ann. Reg. 1941* 344 Work on prothrombin, vitamins C, K, and P,..greatly advanced knowledge of haemorrhagic diseases. 1966 E. Birney *Selected Poems* ii. 61 A hotelroom all to myself with a fan and a box of Vitamin C. 1983 *Oxf. Textbk. Med.* I. viii. 24/2 There is no convincing evidence for the claims that large doses of vitamin C (4 g or more daily) prevent or decrease severity of the common cold; the evidence for protection against cancer is stronger, but not conclusive.

1921 Funk & Dubin in *Proc. Soc. Exper. Biol. & Med.* XIX. 15 It is possible in most cases to effect an almost quantitative separation of the B-vitamine..from another substance, which we provisionally have called vitamine D. 1928 [see *ergosterol]. 1932 *Chem. Abstr.* XXVI. 1015 The isolation of cryst. vitamin D_1. *Ibid.*, An added note states that vitamin D_2 has been isolated. 1936 *Ibid.* XXX. 6423 The name vitamin D_3 is proposed for this substance. 1953 Fruton & Simmonds *Gen. Biochem.* xxxviii. 903 The term vitamin D_1 has been discarded since the material to which it was first applied has been found to be a mixture of calciferol and several sterols. 1976 H. Campion et al. in B. E. C. Nordin *Calcium, Phosphate & Mineral Metabolism* xii. 445 The compounds known as vitamin D have in common a unique arrangement of three carbon-carbon double bonds, two of which connect a hydroxylated cyclohexane ring to a substituted hydrindane system. 1983 *Oxf. Textbk. Med.* I. x. 44/2 In most respects these vitamin Ds are comparable in their metabolism and their actions.

1925 H. M. Evans in *Proc. Nat. Acad. Sci.* XI. 373 The evidence..is thus conclusively in favor of the existence of a new vitamine or food accessory to which the designation of fat soluble E may be given. [*Note*] We have adopted the letter E as the next serial alphabetic designation, the antirachitic artanine now being known as D. 1948 Martin & Hynes *Clin. Endocrinol.* viii. 157 Vitamin E deficiency..produces loss of sperm-motility in rats and is followed by atrophy of spermatogenic tissue and final loss of the sex instincts. 1968, etc. [see *tocopherol]. 1983 *Oxf. Textbk. Med.* II. xix. 75/2 Vitamin E is necessary to prevent auto-oxidation of the unsaturated fatty acids in the red cell membrane.

1929 Sherman & Sandels in *Proc. Soc. Exper. Biol. & Med.* XXVI. 536 Experiments with reference to the more heat-stable factor of the vitamin B group (factor P-P, vitamin B_2 or G). 1934 *Jrnl. Biol. Chem.* CVI. 433 The vitamin G concentrate..was prepared by extracting hog livers with boiling water, [etc.]. 1949 R. A. & W. A. Gortner *Outl. Biochem.* (ed. 3) xxxvi. 925 Riboflavin (lactoflavin or ovoflavin or vitamin G) is 6,7-dimethyl-isoalloxazine-9-D-riboside. 1967 Vitamin G [see *vitamin B_2* above].

1937 L. E. Booher in *Jrnl. Biol. Chem.* CXIX. 223 The vitamin concerned..is a relatively heat-stable component of the vitamin B complex... It will be referred to here as vitamin H. 1959 Vitamin H [see *egg-white].

1935 H. Dam in *Nature* 27 Apr. 653/1, I therefore suggest the term vitamin K for the antihæmorrhagic factor. 1939 *Jrnl. Amer. Chem. Soc.* LXI. 1295/1 (*heading*) The isolation of vitamins K_1 and K_2. 1939, etc. [see *phylloquinone]. 1947 *Radiology* XLIX. 304/1 Vitamin K and transfusions of whole blood were ineffective in reducing clotting time. 1975 J. Marks *Guide to Vitamins* 71 Domestic animals suffering from warfarin poisoning should be treated with vitamin K_1. 1983 J. Katz in Kaye & Rose *Fund. Internal Med.* cxxxvi. 905/2 A deficiency of vitamin K cannot occur solely from an inadequate diet. Intestinal bacteria synthesize the vitamin.

1936 Rusznyák & Szent-Györgyi in *Nature* 4 July 27/2 We propose to give the name 'vitamin P' to the substance responsible for the action on vascular permeability. 1949 [see *flavonoid]. 1955 [see *citrin]. 1969 *Brit. Med. Jrnl.* 25 Jan. 235/1 No condition representing lack of vitamin P has ever been satisfactorily demonstrated. 1978 F. H. Meyers et al. *Rev. Med. Pharmacol.* (ed. 6) xli. 449/1 The flavonoids or vitamin P are of interest in relation to the question of how drug efficacy is evaluated rather than because of any nutritional effect.

3. *attrib.* and *Comb.*, as *vitamin capsule, cream, deficiency, pill, shot* (*shot *sb.*[1] 7 g (a)), *tablet, therapy*; *vitamin-containing, -enriched, -free, -poor, -rich* adjs.

1958 *Listener* 2 Oct. 523/1 Perhaps all those compact references at the foot of the page..are vitamin capsules of chapter for verse to insatiable scholars. 1965 M. Spark *Mandelbaum Gate* iii. 61 Freddy..carried small red vitamin capsules about with him to swallow after meals taken outside the British Isles. 1921 *Conquest* Sept. 498/2 The only safe rule is this..eat vitamine-containing food on every possible occasion..and avoid, as far as practicable, vitamine-free foods. 1938 *Encycl. Brit. Bk. of Year* 588/1 Preparations that have but recently come to the fore include..the group of hormone and vitamin creams, etc., known collectively as 'biological' preparations. 1979 P. Ferris *Talk to me about England* iii. 133 She took the vitamin cream. 1920 *Brit. Med. Jrnl.* 31 July 147/1 It is ..still a hypothesis that the particular disease depends upon vitamine deficiency. 1946 R. Lehmann *Gypsy's Baby* 149 My *brain*..just doesn't *function* any more. Don't you think it's some vitamin deficiency? 1980 *Times* 15 Dec. 1/3 One of the hunger strikers..is in danger of irreversible loss of sight because of a vitamin deficiency. 1961 *Which?* Oct. 270/1 Wholemeal biscuits with a vitamin-enriched filling. 1921 Vitamin-free [see *vitamin-containing* above]. 1956 *Nature* 11 Feb. 271/1 Aerated cultures of *B. cereus* 569H were grown at 36° on..vitamin-free casein hydrolysate. 1945 N. Mitford *Pursuit of Love* xv. 121 A packing-case full of vitamin pills. 1981 D. Uhnak *False Witness* (1982) xi. 90 The customers..gulped down vitamin pills with swallows of juice. 1973 T. Pynchon *Gravity's Rainbow* (1975) i. 170 Nasty little fangs achop and looking to ulcerate the vitamin-poor tissue they came from. 1923 *Jrnl. Biol. Chem.* LVI. 333 The fat-soluble, vitamin-rich ration. 1944 J. S. Huxley *On Living in Revolution* 31 Sweeping measures of social security and welfare—..subsidized housing and vitamin-rich food for the under-privileged,..and so on. 1971 J. Philips *Escape a Killer* (1972) i. ii. 25, I always feel as if I'd had a vitamin shot when you turn up. 1980 J. Gardner *Garden of Weapons* ii. iii. 139 Mistochenkov looked in startlingly good health. 'It's the vitamin shots they're giving me,' he told Herbie. 1951 M. Kennedy *Lucy Carmichael* ii. i. 76 She..ended by saying I didn't look very well and she would send me some vitamin tablets. 1982 M. McMullen *Better Off Dead* ii. xiv. 171 A little saucer was placed before him with his vitamin tablets, the B-complex and the C. 1969 *Listener* 1 May 627/2 He thinks the cure of Tobit's blindness must have been either a colossal coincidence or a primeval case of vitamin therapy, by the liberal administration of fish-guts. 1972 Vitamin therapy [see *perceptual *a.*].

Hence **vitami·nic** *a.*, pertaining to or containing a vitamin or vitamins; **vi·taminless** *a.* (*rare*); **vita·minous** *a.* (*rare*), vitaminic; also *fig.*

1914 *Nature* 12 Mar. 42/1 Vitaminous foods are fresh milk.., whole grains, potatoes, [etc.]. *Ibid.*, Such vitamineless foods as sterilised milk,..starch, and sugar. **1926** *Chambers's Jrnl.* Apr. 291/2 The milk tends to become less and less valuable from the point of view of vitaminic value. **1931** C. J. Holmes *Gram. Arts* iii. 27 Life..is the essential thing..and we must not starve ourselves of this vitaminous element. **1980** *Acta Vitaminologica et Enzymologica* II. 75 Drug induced avitaminoses are produced more easily and are more severe if the devitaminizing power of the drug and its dosage are high..and the vitaminic status of the patient is not optimal.

vitaminize (vi·tǎminəiz), *v.* [f. prec. + -ize.] *trans.* To add a vitamin or vitamins to (food, esp. food that lacks the vitamin concerned). Also *absol.* and *fig.* Chiefly as **vi·taminized** *ppl. a.* Also **vi:taminiza·tion,** the treatment of food in this way; **vi·taminizing** *ppl. a.*

1930 *Observer* 13 Apr. 7 Adventure, we are told, is the 'vitaminizing element' in history. **1940** *Economist* 27 Apr. 762/2 A diet of vegetables..milk, butter or vitaminised margarine..brown bread and cheese. **1942** *Endeavour* Jan. 30/2 The compulsory 'vitaminization' of margarine that has taken place since the war has merely been an extension to the whole margarine output of a practice already in partial operation. **1944** *Ourselves in Wartime* 153 Margarine was 'vitaminized', so that all categories, even the cheapest varieties, were equal in vitamin value to butter. **1960** 'R. Gordon' *Doctor in Clover* xi. 92 He has taken his vitaminised milk and played *Clair de Lune* twice on the piano. **1968** M. Pyke *Food & Society* iii. 32 Government regulations enforcing the fluoridization of drinking water or the vitaminization of margarine. **1968** M. Woodhouse *Rock Baby* iii. 25 You've a home-grown, vitaminized, tall-walking, all-American boy. **1970** *R. Soc. Health Jrnl.* XC. 24/1 It had been decided to vitaminize to such a level that in respect of A and D the margarine would be equal to summer butter. **1975** *Islander* (Victoria, B.C.) 3 Aug. 8/2 Vitaminized canned juices.

Vitaphone (vəi·tǎfōᵘn). Chiefly *U.S.* Also **vitaphone.** [f. L. *vita* life + Phone *sb.*[1]] A process of sound film recording in which the sound track is recorded on discs and played in synchronization with the projection of the film; also, sound films made by this method. Now *disused.*

1926 *Westm. Gaz.* 20 Sept. 7/2 A method of talking-motion pictures has been developed in America. The invention is called the vitaphone... It is claimed that the synchronisation of picture and voice..is perfect. **1948** M. Quigly *Magic Shadows* xvii. 160 Thirty years later, magic shadow history was made... The event was the premiere of 'The Jazz Singer', starring Al Jolson and presenting the Vitaphone system of talking motion pictures. **1976** *Islander* (Victoria, B.C.) 8 Aug. 13/2 To begin with two types of sound pictures were made—Movietone and Vitaphone. **1977** *Amer. Film* Oct. 27/1 Warner theaters in every major American city were wired for Vitaphone.

vitello-. Add: **vite:lloge·nesis,** the formation of the vitellus; **vite:llogene·tic, -ge·nic** *adjs.* = Vitellogenous *a.*; **vite:lloge·nin** [-in[1]], a blood-borne protein from which the substance of the vitellus is made; **vite·llophag** (-fæg), **-phage** (-fēⁱdʒ) *Ent.* [Gr. -φαγος eating], a nucleus or energid which, during cleavage and the formation of blastoderm, remains in or moves into the vitellus and assimilates it.

1956 *Nature* 11 Feb. 277/1 (*in table*) All [honey-bee ovaries] fully regressed; no commencement of vitellogenesis. **1974** *Ibid.* 4 Jan. 72/2 Some fish are not fully regressed and have ovaries in early stages of vitellogenesis. **1961** *Biol. Abstr.* XXXVI. 1333/1 (*heading*) Vitellogenetic processes of *A. depilans* observed by electron microscopy with further considerations on the Golgi apparatus. **1978** *Nature* 23 Mar. 351/2 The last female revealed no chorionated eggs and only four late vitellogenetic proximal oocytes..in the ovarioles. **1964** *Symp. R. Entomol. Soc.* II. 37 In *Drosophila*, one vitellogenic oocyte is normally found in each ovariole. **1974** *Nature* 4 Jan. 71/2 The induction of sexual behaviour in female goldfish..by injection of ovulated eggs into the ovarian lumen of individuals with vitellogenic ovaries. **1969** *Jrnl. Insect Physiol.* XV. 1279 Two immunochemically discrete protein yolk precursors or vitellogenins appear in the blood of adult *Periplaneta americana* on day 4 or 5 after emergence. **1973** *Nature* 13 July 103/2 Insect yolk proteins, or vitellogenins, are synthesized and secreted by the fat body. **1982** *Sci. Amer.* Nov. 139/3 In the liver [of the garter snake] the phospholipids are incorporated into a lipophosphoprotein known as vitellogenin. **1892** J. P. McMurrich in *Zool. Anzeiger* XV. 274 In the same manner the endoderm cells are excluded from the surface of the egg, but in this case there is an actual immigration, the cells sinking down into the interior of the yolk, and becoming 'vitellophags'. **1904** [in Dict.]. **1935** [see *presumptive *a.* 3 b]. **1978** R. J. Elzinga *Fund. Entomol.* iv. 86 Some of these nuclei remain behind to become vitellophags, cells for metabolizing yolk for embryonic use.

vitex (vəi·teks). [L. name used by Pliny for *Vitex agnus-castus* or a similar shrub, later adopted as a generic name by Linnæus and earlier botanists.] A deciduous shrub or small tree, often aromatic, of the genus of the same name, belonging to the family Verbenaceæ. Cf. Agnus Castus.

1608 Topsell *Serpents* 37 The leaves of *Vitex* being cast on the coales for a fumigation, doe with theyr vapour chase away venomous beastes. **1829** [see *clerodendrum]. **1955** E. Pound *Classic Anthol.* I. 71 Vitex in swamp ground, Branched loveliness..Vitex negundo, casting thy flowers in air. **1976** *Hortus Third* (L. H. Bailey Hortorium) 1161/2 Vitexes do well in any good soil.

vitiating, *ppl. a.* (In Dict. s.v. Vitiate *v.*) (Earlier example.)

1832 J. S. Mill in *Monthly Repos.* VI. 658 After all which has been done to break down these vitiating, soul-debasing prejudices,..where are we now?

vitiligo. Delete 'a species of leprosy' and add: In mod. use, a skin disease whose only manifestation is the post-natal development of sharply defined white patches that tend to grow in size. (Further examples.)

1842 E. Wilson *Pract. & Theoret. Treat. Dis. Skin* xii. 280 Vitiligo is the designation applied to partial leucopathia, or the diminution or absence of pigmentary secretion upon one or more parts of the body. **1855** G. B. Wood *Treat. Pract. Med.* (ed. 4) II. 460 Vitiligo.—This term..has received various applications. Celsus embraced under it different forms of what are now known as lepra and psoriasis, and another affection which is probably identical with the *lupus non exedens* of modern writers. **1874** W. Tay tr. *Hebra & Kaposi's Dis. Skin* III. xli. 174 The want of pigment at one time represents a substantive anomaly of formation, or a morbus sui generis and a wholly insignificant condition (Vitiligo alba levior, Celsus), and, at another time, forms a part of the phenomena of a very intense, constitutional disease, namely, Elephantiasis Græcorum (Vitiligo alba gravior, Celsus). *Ibid.* 176 According to the latter expressions [of Celsus], we might consider vitiligo to mean Psoriasis or Eczema.., or Elephantiasis Græcorum. In fact, these words have led to the great confusion which has existed even up to the present time..as to this matter. *Ibid.* 177 Most writers..have not been able to ignore the original meaning of Vitiligo as indicating a want of pigment, and make use of the expression to designate partial Achroma. **1887** T. M. Anderson *Treat. Dis. Skin* 28 Some cases of true Leprosy may be mistaken for Vitiligo. **1907** *Lancet* 16 Feb. 407/2 Leucoderma or vitiligo is doubtless familiar to you all. **1976** *Ibid.* 4 Dec. 1248/1 Since 1974 she had had psoriasis, vitiligo, and recurrent herpes labialis. **1979** E. L. Rhodes *Dermatology for Physician* ii. 9/1 Although pale patches on the skin may follow eczema, psoriasis, pityriasis rosea and pityriasis versicolor, the patches of vitiligo are absolutely white.

vitrain (vi·trēⁱn). [f. L. *vitr-eus* Vitreous *a.* + *-ain* in *Fusain (sense 2).] A black, highly lustrous, and often brittle type of coal.

1919, etc. [see *clarain]. **1930** [see *durain]. **1975** D. G. Murchison et al. tr. E. Stach's *Textbk. Coal Petrol.* v. 334 The first step..is the crushing of the run-of-mine coal... When a seam contains thick layers of bright coal, the soft and mostly brittle vitrain concentrates in the smalls, below 10 mm.

vitrectomy (vitre·ktŏmi). *Surg.* [f. Vitre(ous *a.* + -ectomy.] The operation of removing the vitreous fluid from the eyeball and replacing it with another fluid.

1968 D. Kasner in *Highlights Ophthalm.* XI. 304 (*heading*) Vitrectomy: a new approach to the management of vitreous. **1975** *Courier-Mail* (Brisbane) 5 Nov. 18/6 She is the first woman in Western Australia to have the operation, known as a vitrectomy, which restored her sight. **1983** *Oxf. Textbk. Med.* I. ix. 32/2 Recently, vitrectomy has developed as a treatment for blindness from unresolved vitreous haemorrhage and as a treatment of retinal detachment subsequent to vitreous haemorrhage.

Vitremanie (vī·trəmani). [f. F. *vitre* window-pane + *manie* fad.] A process of decorating window panes by the application of coloured designs in imitation of stained glass, popular in the Victorian period.

1881 *Sylvia's Bk. Artistic Knicknacks* 324 The art of vitremanie—the decoration of glass windows, &c., by the application of coloured designs so as to resemble stained-glass—is a modern invention. *Ibid.*, Vitremanie may be applied to exclude unpleasing views seen through staircase or other windows. **1961** J. Gloag *Victorian Comfort* ii. 36 A form of window decoration called 'Vitremanie', which was applied to glass 'to exclude unpleasing views'.

Vitreosil (vi·triosil). Also **vitreosil.** [f. Vitreo(us *a.* + Sil(ica.] A proprietary name for vitreous silica.

1909 *Trade Marks Jrnl.* 15 Dec. 2050 *Vitreosil*... Fused silica articles included in Class 15 in the form of tubes, plates, basins..and other shapes for chemical, electrical, ornamental and other similar purposes. The Thermal Syndicate, Limited,..Wallsend-on-Tyne. **1921** *Chambers's Jrnl.* 26 Mar. 272/1 Vitreosil..is composed of pure silica fused in the electric furnace by a special process. **1938** *Jrnl. R. Aeronaut. Soc.* XLII. 822 Hollow tubes of vitreosil were wound internally with spiral heating coils. **1973** *Nature* 27 July 233/1 Each portion was placed on defatted lens paper on a stainless steel mesh grid in a vitreosil dish containing Trowell's T-8 medium.

vitreous, *a.* Add: **1. a.** *vitreous silica*, an amorphous, translucent or transparent form of silica obtained by rapid quenching from the molten state.

1925 J. W. Mellor *Comprehensive Treat. Inorg. & Theoret. Chem.* VI. xl. 288 Various names are applied to the vitrified quartz—thus, quartz-glass, vitreous-silica, fused quartz,..etc. **1971** *Materials & Technol.* II. i. 22 Transparent vitreous silica is used for the production of lenses..and other optical elements which are required to transmit ultraviolet or infra-red rays.

e. *vitreous enamel*: = Enamel *sb.* 1 a, *porcelain enamel* s.v. *Porcelain 5. So *vitreous-enamelled* adj., *vitreous enamelling.*

1916 *Chem. Abstr.* X. 261 (*heading*) Vitreous enamels or glazes for pottery, etc. **1939** Burns & Schuh *Protective Coatings for Metals* xvi. 381 Vitreous or porcelain enamels are essentially fused silicates or glasses holding in suspension a colloidal dispersion of color oxides, opacifiers and gases. **1963** G. S. Brady *Materials Handbk.* (ed. 9) 281 Vitreous enameled metals are used for cooking utensils, signs, chemical tanks and piping, [etc.]. **1977** R. B. Ross *Handbk. Metal Treatments & Testing* 389 Vitreous enamelling was very popular in late nineteenth and early twentieth centuries. With the advent of new materials, for example, aluminium in cooking ware,..the use of vitreous enamel declined and was confined very largely to articles such as baths. **1984** Worthington & Knight *Home Plumbing* 96/1 Modern materials are less resistant to abrasives than the traditional vitreous enamel.

vitric (vi·trik), *a. Geol.* [f. L. *vitr-um* glass + -ic.] Of tuff: composed chiefly of glassy material.

1915 L. V. Pirsson in *Amer. Jrnl. Sci.* CXC. 193 The use of 'vitric' and 'lithic' is suggested instead of the more common 'glassy' and 'stony' in order to avoid the misapprehension that the outward appearance of the material is referred to. **1976** *Nature* 5 Aug. 461/2 The other vertebrate remains are confined to the uppermost 30 m of aeolian tuffs beneath a widespread pale yellow vitric tuff.

vitrine, *sb.* (Earlier example.)

1880 C. Schreiber *Jrnl.* 19 Nov. (1911) II. 328 A small ornament... I believe it to be Spanish, [but]..I only saw it as it lay in the vitrine at the Exposition.

vitrinite (vi·trinəit). [f. *Vitr(ain + *-inite* (f. -in[1] + -ite[1]).] One of the three major kinds of maceral that go to make up humic coal, rich in oxygen and characteristic of vitrain.

1935 M. C. Stopes in *Fuel* XIV. 11 It is now proposed to give to the individual 'macerals' a distinctive set of descriptive names with the termination -inite... The basis of the grouping is very simple and the names logical; and the whole therefore can easily be built up from: Rock types... Vitrain..Fusain..Clarain..Durain. Macerals... Vitrinite..Fusinite..[etc.]. **1955** *Sci. Amer.* July 62/1 The broadest attack on the problem was launched in 1951 by the British Coal Utilization Research Association, which set out to analyze..a series of vitrinites—bright black coal particles—from typical coals. **1964** *Fuel* XLIII. 123 It is..suggested that the reflectance of vitrinite A occurring in the whole coal may provide the best available index of rank. **1978** *Nature* 7 Dec. 598/1 The most common methods of estimating the degree of thermal alteration [of petroleum source rocks] are based on measurements of the residual insoluble organic material in rocks. Some of the more common methods use..electron spin resonance, kerogen coloration and vitrinite reflectance.

Vitrolite (vi·trələit). Also **vitro-.** [f. L. *vitr-um* glass + -o + -lite.] A proprietary name for opal glass.

1937 *Times* 19 Oct. 11/3 A glass floor has been laid, and glass pictures have been hung, and a bathroom lined with vitrolite is being built. **1939** *Trade Marks Jrnl.* 1 Feb. 147/1 *Vitrolite*... Opal and opaline glass. Pilkington Brothers Limited,..Liverpool, 2; glass manufacturers. **1954** *Archit. Rev.* CXVI. 268/1 Such an atmosphere is not created by stove-enamelled hardboard, vitrolite and chromium-plated barrier rails, the language of the cafeteria. **1975** [see *opalite].

Vittel (vite·l). The proprietary name of a type of mineral water obtained from springs in the neighbourhood of the town of Vittel in the Vosges department of France. Also *Vittel water.*

1895 *Army & Navy Co-op. Soc. Price List* 23 Natural mineral waters..Vichy..Vittel..Wiesbaden. **1909** *Trade Marks Jrnl.* 26 May 901 *Vittel*... A natural mineral water. Société Générale des Eaux Minérales de Vittel,..Vittel, Vosges, France. **1974** N. Freeling *Dressing of Diamond* 76 Carafe of Beaujolais and a bottle of Vittel. **1975** —— *What are Bugles blowing For?* i. 4 He..drank a little Vittel water and went to bed. **1979** 'M. Hebden' *Death set to Music* iii. 29, I ought to have had Vittel water, not wine.

vituperous. Delete ? *Obs.* and add: **1, 2.** (Later examples.)

1914 R. M. Jones *Spiritual Reformers 16th & 17th Cent.* v. 69 Schwenckfeld was denounced in the most vituperous language of the period. **1959** *Times Lit. Suppl.* 10 July 409/3 Venetia..watches with relief Karlo's vituperous and vulgar exit.

vivace, *adv.* (and *sb.*). Add: **a.** Also with the adv. used quasi-adjectivally to characterize musical composition.

1922 D. H. Lawrence *Aaron's Rod* xiii. 187 Lady Franks started with a *vivace* Schumann piece. *Ibid.* 188 'I always prefer Schumann in his *vivace* moods,' said Aaron.

b. *as sb.* A passage intended to be performed in this manner (see also quot. 1889 in Dict.).

1683 [see *allegro B, C]. **1922** D. H. Lawrence *Aaron's Rod* xiii. 188 Our Colonel began to..bounce in his chair,..doing a sitting-down jig to the Schumann *vivace.* **1959** *Times* 11 Sept. 10/6 He moved into the ensuing vivace with unhurried moderation. **1976** *Gramophone* Sept. 410/1 The movement's main vivace is a good deal less mettlesome.

vivandier. Add: (Examples of form *vivandière.*)

1848 H. Greville *Diary* 29 June (1883) I. 278 Women went about disguised as vivandières, giving poisoned brandy to the soldiers. **1896** D. Bingham *Recoll. Paris* I. xiii. 221 Madame de Beaulieu..joined the Mobiles of her native province in the capacity of vivandière. **1963** P. Fleming *Kolchak* xiii. 149 They were often companioned by their own or other men's wives or by ladies who played, sometimes usefully, a role roughly corresponding to the vivandière's.

vivarium. Add: **2. b.** Now usu. = Terrarium. (Earlier and further examples.)

1853 Geo. Eliot *Let.* 13 June (1954) II. 103, I was at the Zoological gardens..and saw..the marine queerities in the Vivarium. **1858** H. N. Humphreys *Butterfly Vivarium* i. 2, I am about to describe a novel kind of Vivarium, by means of which another and very distinct class of animal life—that of the 'world of insects'—may be made to exhibit its wonders. **1914** W. P. Westell *Boys' Bk. Pets* xvi. 217 Another type of vivarium may be constructed on similar lines to the rectangular aquarium, only making it a few inches deeper. **1978** *Nature* 19 Oct. 646/2 Lizards were housed in vivaria in a constant temperature room.

vivax (vəi·væks). *Med.* [L., = 'long-lived'.] The specific name of a protozoon of the genus *Plasmodium*, used *absol.* and *attrib.* to denote the organism and the relapsing type of malaria it causes, in which paroxysms occur every third day and which is usually not fatal. Freq. printed in italic.

1930 [see *falciparum]. **1946** *Jrnl. Amer. Med. Assoc.* 20 July 964/1 The patients were military personnel who had acquired vivax infections in the Pacific or Mediterranean theaters of operation. **1955** *Sci. News Let.* 23 July 52/1 The Negro seems to have a general resistance to strains of vivax from all areas. **1958** N. F. Leopold *Life plus 99 Years* xxi. 321 Vivax malaria is also known as tertian malaria because, once the trophozoites have fallen into step and got synchronized, they sporulate every forty-eight hours. **1983** *Oxf. Textbk. Med.* I. v. 394/2 The vivax resistance factor [in many black Africans] is a blood group determinant.

‖ **vive** (vīv), *int.* [Fr., lit. 'may he (she, it) live': cf. Viva *sb.*[1] and *int.*] **1. a.** *vive le roi* (vīv lə rwa) = *long live the king* s.v. Live *v.*[1] 9, used as a general acclamation to or for a sovereign. Similarly *vive l'empereur.*

?**1594** Marlowe *Massacre at Paris* sig. B7, Sound trumpets within, and then all crye Viue le Roy. *a*1700 Evelyn *Diary* an. 1651 (1955) III. 42 The King..saluting the Ladys & Acclamators who had fill'd the Windos with their beauty, & the aire with *Vive Le Roy.* **1768** Sterne *Sentimental Journey* I. 114 Here's a couple of sous for thee—*Vive le Roi!* said the old soldier. **1815** F. Burney *Let.* Mar. in *Jrnls. & Lett.* (1980) VIII. 76 There was no species of enthusiasm, but..moderate cries of Vive le roi. **1848** E. B. Browning *Let.* 4 July (1897) I. v. 375 How did you feel when the cry was raised, 'Vive l'Empereur'? **1861** G. Meredith *Let.* 19 Nov. (1970) I. 115 The Emperor allows her £1000 a year: her mother gets £2000. *Vive l'Empereur!* **1896** C. M. Yonge *Release* ii. iii. 104 Ecstatic cries of 'Vive le Roi' broke from the populace.

b. *vive la bagatelle* (vīv la bagatel), 'success to frivolity or nonsense', an exclamation denoting a carefree attitude to life.

1732 Swift *Let.* 10 July in *Lit. Corr.* (1741) 150 All for want of my Rule, *Vive la bagatelle!* **1760** Sterne *Tristram Shandy* I. xix. 121 The footing of mere whims, and of a *vive la Bagatelle.* **1880** *Girl's Own Paper* 27 Nov. 144/3 *Vive la bagatelle* means 'Success to trifling'. Not a good sentiment. **1915** J. Webster in *Century Mag.* Sept. 492/1, I seem to have covered a lot of paper without telling you much. *Vive la bagatelle!* **1948** M. Allingham *More Work for Undertaker* xxiii. 260 (*heading*) Vive la bagatelle.

c. Used in other phrases denoting extreme approval or enthusiasm or as a cry of acclamation. Also *joc.*

1865 G. Meredith *Let.* 11 Aug. (1970) I. 315 My mind is free, and vive la liberté! **1889** E. Dowson *Let.* 24 Mar. (1967) 55 Have just finished my poulet. Vive la jeunesse: it has actually lasted 3 weeks. **1906** [see *ban *sb.*[4]]. **1914** 'Bartimeus' *Naval Occasions* xi. 81 Well!..Vive le sport! If there were no fools there'd be no fun. **1922** E. E. Cummings *Enormous Room* viii. 176 Vive la bourgeoisie, I said to myself. **1965** V. Canning *Whip Hand* viii. 89 We all want to know where Mrs Vadarci is going, and you have—*vive l'amour*—a special contact there. **1980** C. Ross *Case for Compensation* v. 29 Vive La France. Vive le sport and where's my passport. I'm going.

d. *vive la différence* (vīv la diferɑ̃s), a *joc.* expression denoting approval of the difference between the sexes. Also in extended use. Occas. as *attrib. phr.*

1963 H. Slesar *Bridge of Lions* (1964) xi. 164 It's your stock-in-trade, isn't it, youth and beauty? *Vive la différence*, and all that jazz. **1964** *Guardian* 21 Apr. 18/4 'Vive la différence' Tories are recognizably Tories, and Socialists are demonstrably Socialists. **1969** [see *jockette]. **1970** G. Greer *Female Eunuch* 29 Frenchmen may well cry 'Vive la différence', for it is cultivated unceasingly in all aspects of life. **1978** *Country Life* 3 Aug. 341/3 The control layout extends Citroen's *vive la différence* attitude.

2. *absol.* and as *sb.*

1919 D. H. Lawrence in *Athenæum* 11 Apr. 167/2 There is a new régime, sound of a new *Vive! vive!* **1922** Joyce *Ulysses* 303 The even more excitable foreign delegates cheered vociferously in a medley of cries..hiphip, vive, Allah.

viverrid (vive·rid), *a.* and *sb. Zool.* [ad. mod.L. *Viverridæ*, f. L. *viverra* ferret: see -id[3].] **A.** *adj.* Belonging or pertaining to the family Viverridæ, which comprises civets, genets, and mongooses. **B.** *sb.* A viverrid animal.

1910 H. F. Osborn *Age of Mammals* iii. 197 Still more striking is the presence [in the Oligocene] of the fierce viverrid carnivores (*Amphictis, Herpestes*) of the modern civet and mongoose types. *Ibid.* iv. 259 The mustelids were becoming more numerous..while the Asiatic civets or viverrids (*Viverra*) are becoming somewhat more rare [in the Miocene]. **1964** Parker & Haswell *Text-bk. Zool.* (ed. 7) II. 795 The aberrant *Proteles*, the Aard-wolf or Earth-wolf of Africa is insectivorous... Like the hyænas it is obviously a highly specialised viverrid offshoot. **1976** *Nature* 5 Aug. 464/1 The carnivore fauna [of the Laetolil beds in Tanzania] is characterised by a high percentage of viverrids, constituting 32% of all the carnivore specimens.

‖ **viveur** (vivœ̄r). [Fr., lit. 'a living person'.] One who lives a fashionable and social life; a man of pleasure. Cf. *bon viveur* s.v. *bon *a.*

1845 Thackeray in *Fraser's Mag.* Nov. 591/2 He became a viveur and jolly dog about town. **1901** 'L. Malet' *Hist. Sir Richard Calmady* iii. x. 261 He has the credit of being something of a viveur. He knew not only his Paris, but his Baden-Baden, and his Naples, and various other warm corners where great and good men..congregate. **1921** L. Strachey *Queen Victoria* iii. 70 The middle-aged, hard-faced viveur was addressed by his young hostess. **1943** *Burlington Mag.* Aug. 201/1 Boulestin was secretary to Willy, the well-known Parisian author, journalist and viveur, husband to Colette.

vivianite. In etym. for 'J. G. Vivian' read: John Henry Vivian (1785–1855), of Truro and Swansea.

vivid, *a.* **2. b.** (Later example.)

1981 *Observer* 15 Nov. 2/8 The famous are shown off-guard—for example,..Sue Lawley using vivid language.

vivier (vivie). Delete † *Obs.*[-1] and add recent examples. Also *attrib.*

1964 F. White *West of Rhone* ii. 31 Viviers, an old town,..was once the fish-ponds, the *viviers*, of the bishop. **1973** *Alderney Jrnl.* 8 Oct. 4 Much has been said about the politics of building vivier tanks at the end of the harbour jetty. **1979** *Guardian* 10 Oct. 12/6 The Dutch lorries..that buy shellfish from Harveys, are vivier lorries. They have huge water tanks full of sea water.

vivipary. Add: Also in *Zool.* (Further examples.)

1963 Davis & Golley *Princ. Mammalogy* v. 118 Another approach is..to infer from changes in skull and jaw morphology the presence or absence of homeothermy..and vivipary. **1981** *Entomologica Generalis* VII. 13 Nothing..is known about the time when in the evolution of Aphidina vivipary has developed.

vivotoxin (vəi·vətǫksin). *Biochem.* [f. *vivo*- (cf. Vivi-) + Toxin.] A substance produced in an infected plant and involved in the disease process (see quot. 1953).

1953 Dimond & Waggoner in *Phytopathology* XLIII. 229/2 To clarify this situation a new term is suggested. A vivotoxin is defined as a substance produced in the infected host by the pathogen and/or its host, which functions in the production of disease, but is not itself the initial inciting agent of disease. **1971** *Agric. & Biol. Chem.* XXXV. 618/2 Vivotoxins are considered to be usually nonspecific to plant species. **1981** R. J. Green in M. E. Mace et al. *Fungal Wilt Dis. Plants* i. 11 Acceptance of the concept of the vivotoxin directed investigators to attempt to demonstrate toxins in the vascular tissues of the infected plants.

Viyella (vəi,e·lă). Also **viyella.** The proprietary name of a fabric made from a twilled mixture of cotton and wool; also, a garment made of this material. Freq. *attrib.*

1894 *Trade Marks Jrnl.* 6 June 471 *Viyella*... Yarns of wool, worsted or hair. William Hollins and company (Nottingham), Limited,..Nottingham; spinners. **1912** G. Frankau *One of Us* v. 41 No impious fingers groped in his Viyella. **1921** 'K. Mansfield' *Let.* 29 Aug. (1977) 229 That's the kind of stuff I meant, too. They had both better be lines with silver grey viyella or cashmere. **1932** R. Lehmann *Invitation to Waltz* i. iii. 46 Cream viyella blouses, white piqué tennis skirts. **1942** *R.A.F. Jrnl.* 16 May 33 The pilot was wearing thick vest and pants, a Viyella shirt..and Mae West. **1969** E. McGirr *Entry of Death* iii. 51 She wears thick Viyella pyjamas. **1977** *Lancashire Life* Feb. 23/1 William Hollins' famous collection of mice, dressed in Viyella, that fabulous mixture of wool and cotton that keeps both mice and human mites warm in winter.

Vizeeree, var. *Wazir[2].

vizsla (vi·ʃlă). The name of a town in Hungary used *absol.* to designate a golden-brown pointer with large pendent ears belonging to a breed developed in the region.

1945 C. L. B. Hubbard *Observer's Bk. Dogs* 153 Of the five national breeds of dogs of Hungary, the Vizsla is the sole sporting dog. **1971** F. Hamilton *World Encycl. Dogs* 208 The Vizsla has a lively, elegant and graceful gait. **1976** *Shooting Times & Country Mag.* 9–15 Dec. 52/1 Vizsla pups, excellent pedigree. **1977** J. Wambaugh *Black Marble* (1978) viii. 103 Letter openers bearing the likenesses of Vizslas and Brittany spaniels.

VJ (vī dʒēⁱ). [f. the initial letters of *Victory over Japan.*] Used *attrib.* and *absol.* to denote the victory of the Allied forces over those of Japan during the war of 1939–45; esp. as *VJ-day*, designating either the day upon which Japan ceased fighting (14 August 1945) or the day of Japan's formal surrender (2 September 1945).

1944 [see *VE]. **1945** *Sun* (Baltimore) 15 June 1/8 (*heading*) 'V-J by Christmas' amazes Carlson. **1945** *Yank* 7 Sept. 9 The GIs had managed to keep their VJ spirit bottled up through most of the phony rumors, but when the real thing was popped, the cork popped with a vengeance. **1956** *B.B.C. Handbk. 1957* 169 VJ-day was recalled by the Overseas Services in a series of talks. **1968** *Listener* 19 Dec. 811/3 (*caption*) Revellers on V-J night. **1974** J. Wainwright *Hard Hit* 87 My memories ended with the war... Then—less than a week after V.J.—I married James. **1977** [see *VE]. **1978** *Jrnl. R. Soc. Arts* CXXVI. 252/1 On VJ night I was Station Duty Officer. **1982** T. Fitzgibbon *With Love* xi. 176 A little under a month after the first bomb on Hiroshima being dropped, VJ-Day as it was called was officially celebrated on 2 September 1945.

vladimirite (vlædimī·rəit). *Min.* [ad. Russ. *vladimirit* (f. L. I. Nefedev 1953, in *Zap. Mineral. Obshch.* LXXXII. 317), f. the name *Vladimir*: see -ite[1].] A hydrated arsenate of calcium, $Ca_4H_2(AsO_4)_4.5H_2O$, occurring as colourless acicular monoclinic crystals.

1954 *Mineral. Abstr.* XII. 352 Vladimirite... Found in the zone of oxidation of ore deposits. **1972** [see *sainfeldite]. **1978** *Mineral. Rec.* IX. 73 Vladimirite... Fine, acicular, brilliant, colorless crystals in quartz cavities or with talmessite crusts were found in 1963 at the Irthem mine [in Morocco].

vlakte (fla·ktə). *S. Afr.* Pl. **vlaktes,** †**-en.** [Afrikaans, f. Du.] An extent of flat open country; a plain. Freq. *pl.*

1785 G. Forster tr. *Sparrman's Voy. Cape Good Hope* II. xiv. 222 Animals (probably of the gazel kind) two feet in height, which used..to herd together on the *vlaktens*, or plains. **1852** C. Barter *Dorp & Veld* 82 The plains, or *vlakten*, occupy more than two-thirds of the whole extent of the Sovereignty. **1938** E. Campbell *'Die ou Pad' & Other Rhymes* 5 The vlakte is tamed and strange. **1975** *Sunday Times* (Johannesburg) 21 Sept. 20 There was everyone wondering why the men from the vlaktes, usually so docile, suddenly seemed so fractious.

vlei. Add: Also 8 **valley.** **1.** (Earlier examples.)

1793 C. R. Hopson tr. *Thunberg's Trav. Europe, Afr., & Asia* II. p. xii, A valley is nothing more than a rivulet, which is sometimes over-grown with rushes, and is broad in some places, and narrow in others. **1801** Truter & Somerville in G. M. Theal *Rec. Cape Colony* (1899) IV. 369 We afterwards passed two other vleis or ponds about two hours distant one from another.

3. vlei lourie, loerie, one of several species of coucal of the genus *Centropus*, esp. *C. superciliosus*, found in southern Africa.

1864 T. Baines *Expl. S.-W. Afr.* xv. 391, I shot one of the loosely-feathered birds called in the colony Vlei Lories, or Reed Hawks. **1908** [see *loerie]. **1936** E. L. Gill *First Guide S. Afr. Birds* 108 Burchell's Coucal, Vlei Lourie (Loerie). **1973** *Weekend Post* 28 Apr. 4 Some are just fascinating in their own right, like the vlei louries..and the Diederick cuckoo. **1978** McLachlan & Liversidge *Roberts' Birds S. Afr.* (ed. 4) 253 White-browed Coucal. Vleiloerie. *Centropus superciliosus.*

Vo-Ag (vōᵘ,æg). *U.S. colloq.* Also **Vo. Ag.** Short for *Vocational Agriculture*, agriculture considered as a subject of study for those who intend to make it their profession. Also *attrib.*

1953 *Manch. Guardian Weekly* 5 Feb. 15/2 At Milton, Georgia, the 'Vo. Ag.' instructor took me to see Gill Brown as an example of a leading F.F.A. member... In the cultivation of the 'management' attitude..Vo. Ag. achieves distinctive success. **1976** *Columbus* (Montana) *News* 17 June 1/2 The Montana Vo-Ag teachers have selected Don Owen as the Outstanding Teacher of the 1975–76 year. *Ibid.* 1 July 4/2 Lyle has accepted a position teaching Vo-Ag at a Junior College.

vocab (vōᵘ·kæb). (Examples.)

1900 Farmer *Public School Word-Bk.* 215 *Vocab*,..(Charterhouse)..a vocabulary. **1971** 'M. Innes' *Awkward Lie* v. 94 Beadon should have said 'tailed', not 'trailed'. He hasn't got the vocab right. **1980** K. Follett *Key to Rebecca* xvii. 195 'Have you done your prep?' 'Yes—French vocab.'

vocabulary, *sb.* Add: **4.** *fig.* A set of artistic or stylistic forms, techniques, movements, etc.; the range of such forms, etc., available to a particular person, etc.

1917 G. B. Shaw *How to become Musical Critic* (1960) 291 As far as mere grammar and vocabulary go, there is nothing more in the statue scene from Don Juan, which threw open the whole magic realm of modern orchestration..than in the exquisite little song of Cherubino. **1949** *Ballet Ann.* III. 109/2 Repetitions, lengthy passages, obscure symbols and movements..which ever aim not only at enriching the classic vocabulary..but at uplifting the mind. **1959** *Listener* 10 Dec. 1042/2 The extraordinary vocabulary of the latest works with their untrammelled blossoming in space, their hectic and at times almost hysterical proliferations of bumps, bulges, bags. **1960** *Economist* 22 Oct. 369/2 No country is likely to be able to rely purely on its own 'vocabulary' of styling; but British manufacturers..might be glad to escape the dependence on Italian and Swiss fashion leadership in car bodies. **1967** 'La Meri' *Spanish Dancing* (ed. 2) vii. 86 Important is the fact that with the Pavane a vocabulary of steps appeared for the first time in dance history. **1972** E. Lucie-Smith in Cox & Dyson *20th-Cent. Mind* II. xiv. 474 Cubism is no longer 'analytic' but 'synthetic'. That is, it aims to create, by means of a consistent vocabulary of form, a universe which parallels the real one. **1977** *New Yorker* 25 July 62/2 Together the two enlarged the vocabulary of lawn tennis and laid the foundation for the all-court game. **1977** M. Girouard *Sweetness & Light* vii. 176 During the 1880s the Bedford Park vocabulary was being taken up by speculative builders and put to every kind of use. **1980** *Jrnl. R. Soc. Arts* Feb. 165/1 The vernacular building, employing the simple vocabulary of wall, sloping roof, door and small window, results in a national display of tranquil richness.

vocal, *a.* and *sb.* Add: **A.** *adj.* **2. b.** *vocal line* [*Line *sb.*[2] 7 h].

1934 C. Lambert *Music Ho!* iii. 190 The disparity between the vocal line and its harmonic background [in *Les Noces*]. **1961** [see *line *sb.*[2] 7 h]. **1963** *Listener* 7 Feb. 264/3 The orchestral skill..the command of texture and expressive colour, and the judging of instrumental comment on the vocal line.

6. a. Also *vocal fold, tract.*

1936 *Summ. Doct. Diss. Northwestern Univ.* IV. 183 Many investigations have been carried on to determine how the vocal folds vibrate. **1940** *Bell System Techn. Jrnl.* XIX. 496 Message waves are produced as muscular motions in the vocal tract. **1960** *Jrnl. Speech & Hearing Res.* III. 159/1 An investigator observed the image of the subject's vocal folds in the laryngeal mirror. **1961** *Jrnl. Acoustical Soc. Amer.* XXX. 1725/2 The generally accepted theory of speech production views the speech wave as the result of acoustic excitation of the vocal tract by one or more sources... The characteristics of the glottal source are to a large extent independent of the vocal-tract configuration anterior to the glottis. **1977** D. Fry *Homo Loquens* iii. 30 The process of articulation..depends on continual changes in the shape of the vocal tract, and hence in its acoustic properties. **1981** *Word 1980* XXXI. 152 The disruptions..due to the effects of coupling between the larynx and the supraglottal tract on the rate of vibration of the vocal folds.

b. = Vocalic *a.* 2. (Earlier examples.)

1788 *Asiatick Res.* I. 13 This is the simplest element of articulation, or first vocal sound, concerning which enough has been said: the word *America* begins and ends with it; and its proper symbol therefore is A. **1818** *Trans. Amer. Philos. Soc.* I. 246 The vocal sounds are those which are represented in alphabets by the letters we call vowels.

7. *vocal-auditory* adj.

1958 *Proc. 8th Internat. Congr. Linguists* 754 Despite eloquent pleas,..that writing can and should be considered as basically a visual system independent of the vocal-auditory process, is it likely that any system of writing would be seriously proposed to-day that was not based on some attempt at a systematic correlation with the spoken language? **1981** *Amer. Speech* LVI. 130 Sign language is as adequate for the deaf as any vocal-auditory language is for a hearing person.

B. *sb.* **4. a.** Vocal music; singing.

1928 S. Lewis *Man who knew Coolidge* i. 30 She felt more kind of called to the music line, and she was taking vocal and piano. **1968** *Blues Unlimited* Sept. 26 For strange vocal try to hear Daniel Brown's version.

b. A musical composition written for, or including a part for, the voice; a vocal part.

1934 S. R. Nelson *All about Jazz* iv. 84 A succession of vocals in radio music is sometimes tiresome. **1938** D. Baker *Young Man with Horn* iv. 171 He had them pretty heavily arranged, with a transitional passage before every vocal. **1957** R. Hoggart *Uses of Literacy* viii. 203 Almost all [the records] are 'vocals' and the styles of singing much advanced beyond what is normally heard on the Light Programme of the B.B.C. **1975** A. Ayckbourn *Norman Conquests* 14 He does a dance to the music and sings as the vocal starts. **1983** *Sunday Tel.* 9 Oct. 24/2 The cost of a machine with a complete package of 400 songs with the vocals edited out, is £600.

vocalese (vōᵘkălī·z). *Jazz.* [f. Vocal *a.* and *sb.* + -ese, perh. partly after next.] A style of singing in which singers put words to jazz tunes, esp. to solos previously improvised by jazz musicians. Also = *scat *sb.*[6] a.

1955 L. Feather *Encycl. Jazz* 248 Annie Ross, Eddie Jefferson and others imitated his idea of translating ad lib jazz into 'vocalese'. **1963** *Guardian* 5 July 11/1 Her unusual talent for simulating orchestral sounds (the jazz singing style called 'vocalese'). **1978** [see *scat *sb.*[6] (and *a.*) b]. **1981** *Times* 15 Apr. 16/8 King Pleasure, a pioneer of the style of modern jazz singing known as 'vocalese', has died... Vocalese is the process of putting words to improvized jazz solos.

vocalise (vōᵘkălī·z, vōᵘ·kălīz). *Mus.* Also (in imitation of Vocalize *v.* 5 b) **vocalize.** [a. F. *vocalise*, f. *vocaliser* Vocalize *v.*] **a.** A singing exercise using individual syllables or vowel sounds. **b.** A vocal passage consisting of a melody without words.

1872 *Scribner's Monthly* Feb. 481/2 When the vocalise was ended, I expressed my gratification and my admiration of his method to the maestro. **1949** *Scrutiny* XVI. 77 The Sonata Vocalise is an experiment in the combination of a wordless coloratura part with Medtner's habitual intricate pianism. **1959** *Listener* 2 July 37/3 The second movement..is very much in the spirit of a vocalize. **1972** J. L. Dillard *Black English* vi. 262 If the preacher slows or hesitates, one of the older women may pick up the tempo and propel the performance forward in a *vocalise* which is strikingly like the 'breaks' utilized by New Orleans jazzmen. **1980** *Early Music* Jan. 18/2 For the virtuoso repertory of the 16th and 17th centuries exercises are available. They belong on the music stand for daily vocalises, and not on a bookshelf for theoretical research.

vocalism. 2. b. (Earlier example.)

1854 T. Aufrecht in C. Bunsen *Christianity & Mankind* III. 93 The vowels have..been well preserved... The final consonants in the flexions have remained. The vocalism and consonantism stand in good organic relation to each other.

vocalize, *v.* Add: **5. c.** To utter any vocal sound.

1960 R. D. Laing *Divided Self* i. 20 Studying verbal behaviour in terms of neural processes and the whole apparatus of vocalizing. *a*1961 W. La Barre in *Webster* (1961) s.v., The gorilla is just as likely to thump upon the upper chest..as he is to vocalize. **1972** *Sci. Amer.* Aug. 29/1 The female mallard generally vocalizes at the rate of from zero to four calls per one-minute interval.

vocalize, var. *vocalise.

vocalized, *ppl. a.* (Earlier example.) In *Jazz*, of the tone of an instrument: made to resemble that of the human voice.

1855 tr. *Lepsius' Standard Alphabet* 35 French *j*. This letter is the soft and vocalised sound. **1961** *John o' London's* 7 Dec. 637/1 The great jazz instrumentalists have based their styles on what is known as the 'vocalised' tone. **1970** P. Oliver *Savannah Syncopators* 18 Emphasis is placed on the quality of blues singing and the 'vocalised tone' of jazz instrumentation.

vocation. Add: Hence **voca·tionless** *a.*

1924 *Blackw. Mag.* Apr. 445/2 Apparently we were all vocationless. **1939** A. Clarke *Sister Eucharia* iii. 27, I stood beneath the iron gate, unveiled, Vocationless.

vocational, *a.* (Further examples.)

1910 *Proc. 1st Congr. National Conservation* 164 In conclusion..the remedies:.. Vocational training in high schools. **1927** W. E. Collinson *Contemp. Eng.* 106 The enlistment of the psychologist's help for vocational guidance. **1951**, **1957** [see *guidance 1 c]. **1977** *New Statesman* 2 Sept. 294/3 The IBA is attracted to the..Open College concept—courses in numeracy and literacy, vocational training for school-leavers, and other forms of further education.

Hence also **voca·tionalism,** training for a vocation; educational emphasis on this; **voca·tionalize** *v. trans.*, to direct towards vocational training.

1912 J. H. Moore *Ethics & Education* 33 Schools should not be industrialized merely, nor commercialized, nor professionalized, but vocationalized. **1924** *Glasgow Herald* 1 May 8/6 The primary function of education is wider than mere vocationalism. **1959** *Oxf. Mag.* 11 June 458/2 We are so vocationalised now that we have begun to believe it is our true function to train professional scholars, professional critics and even professional readers. **1971** *Black Scholar* June 65 (Advt.), Vocationalizing higher education. **1979** *Yale Alumni Mag.* Apr. 12/2 Of all the areas in colleges and universities that will feel..the growing vocationalism of the young, the humanities will be hardest hit. **1980** [see *subsidiarity].

‖ **voce** (vō·tʃe). *Mus.* [It. = voice.] Used with qualifying phrases to designate various qualities or registers of the voice, as *voce di gola* (di gō·lă), a throaty or guttural voice; *voce di petto* (di pe·to), the chest register; *voce di testa* (di te·stă), the head register; formerly, the falsetto voice.

1742 tr. *Tosi's Observations on Florid Voice* i. 22 *Voce di Petto* is a full Voice, which comes from the Breast by Strength. *Ibid.*, *Voce di Testa* comes more from the Throat, than from the Breast. **1771** *Voce di petto* [see *portamento]. **1801** Busby *Dict. Mus.*, *Voce di Testa*,..a falsetto, or feigned voice. **1876** Stainer & Barrett *Dict. Mus. Terms* 450/2 *Voce di gola*, a guttural or throaty voice. **1889** G. B. Shaw *London Music in 1888–89* (1937) 229 You are singing with your chest voice, or *voce di petto*, or long reed register, whichever you please to call it. *Ibid.*, You are singing falsetto, or in *voce di gola*, or throat voice, or short reed register. *Ibid.* 230 This high voice is what used to be called *voce di testa* or head voice. **1951** P. A. Duey *Bel Canto in its Golden Age* xi. 124 Tenducci, the Italo-Anglican voice teacher writes briefly:.. Never force the Voice, in order to extend its compass in the *Voce di petto* upwards; but rather to cultivate the *Voce di testa* in what is called Falsetto. **1980** *New Grove Dict. Mus.* XVIII. 690/1 Until the late 18th and even early 19th centuries most tenors in the Italian tradition emphasized the lyrical quality of their top range and, when required, carried their voices with ease into the falsetto register (treatises of the 18th century and earlier call the falsetto range the *voce di testa*). During the first half of the 19th century, however, this traditional Italian manner of singing in the top range fell into disuse, and the expression *voce di testa* came to refer to the normal, much stronger head voice, falsetto being reserved as a special effect only for the very highest notes.

vociferative, *a.* For †*Obs.*[-1] read *rare* and add later example.

1889 G. B. Shaw *London Music in 1888–89* (1937) 167 The riot in the second act would have been better if it had either been sung note for note as written, or, as usual, frankly abandoned as impossible and filled up according to the vociferative fancy of the choristers.

vocoder (vokōᵘ·dəɹ). [f. Vo(ice *sb.* + Code *sb.*[1] + -er[1].] Any of various devices or systems for analysing speech or other sounds to obtain information that may be transmitted in a much reduced frequency band and used to reconstruct the sounds or synthesize new ones.

1939 H. Dudley in *Jrnl. Acoustical Soc. Amer.* XI. 169/1 Speech has been remade..by analyzing a talker's speech for the fundamental speech information and then using this information to remake the speech with a synthesizing device... The apparatus used has been called a 'vocoder'. **1955** *Sci. Amer.* Feb. 95/1 The Vocoder reduces to about one tenth the band width necessary to send speech. **1956** *Jrnl. Acoustical Soc. Amer.* XXVIII. 160/1 In the Scan Vocoder, we scan the envelope of the short time-frequency spectrum, for example, 30 times per second. The scan voltage is then transmitted through only one narrow band frequency channel as an envelope curve. **1970** *Sci. Jrnl.* Mar. 12/4 The main advantage of vocoders is that they require only very low digit rates of the order of 1·2–2·4 kilobits/second. **1976** *New Musical Express* 12 Feb. 24/5 The song's bridge section follows as a motley collection of barks, yelps and whines are fed through the vocoder, a device which synthesises sounds into eerie echo and resonance. **1980** *Musicians Only* 26 Apr. 13/6 The keys line-up includes..a Korg vocoder.

Hence (as a back-formation) **voco·de** *v. trans.*, to transform by means of a vocoder; **voco·ded** *ppl. a.*, produced by a vocoder.

1973 *Lang. & Speech* XVI. 293 The fundamental frequency contour of a 700-msec. vocoded utterance..was systematically varied to produce 72 contours. **1976** *New Musical Express* 12 Feb. 24/5 'Pigs' is the strongest slice of straight ahead rock the Floyd have recorded and the vocoded pig gruntings are fittingly hideous. *Ibid.* 24/6 Waters' voice is again vocoded and emerges sounding unpleasantly cybernetic. **1981** R. Brown *Megalodon* i. 15 An electronic translator that could handle the full range of dolphin whistles and vocode them out as an ersatz humanoid voice.

vocoid (vōᵘ·koid), *a.* and *sb. Linguistics.* [f. Voc(al *a.* and *sb.* + -oid.] **A.** *adj.* Vowel-like; articulated with no obstruction of the air-stream; contrasted with *contoid *a.* **B.** *sb.* A speech sound of this type.

1943 K. L. Pike *Phonetics* v. 78 Vocoid and contoid groups are strictly delineated by the articulatory and acoustic nature of sounds. *Ibid.* vii. 143 The sounds which as a group function most frequently as syllabics are vocoids... Vocoids include practically all sounds which are usually called 'vowels'.., except that 'fricative vowels' are excluded, while 'vowel glides' such as [r], [w], and [y] are included. **1957**, **1958** [see *contoid *a.* and *sb.*]. **1965** *Language* XLI. 476 This results in a variety of unrounded vocoid transitions between labials and following vowels. **1977** *Word 1972* XXVIII. 307 Vocoid symbolism is at most an extremely vague feeling for the appropriateness of certain vocoids to a particular meaning. **1984** *Amer. Speech* LIX. 342 One violation of (most) native English phonotactics was the *r*, which was usually a tap, even word-initially, rather than a vocoid.

vo-de-o-do, var. *vo-do-deo-do.

vodka. Delete ‖, for 'used in Russia' read 'used orig. esp. in Russia', and add: Also, a glass or drink of this.

1969 M. Pugh *Last Place Left* xxix. 209 Pardoe decided to pour himself a vodka. **1981** 'J. Ross' *Dark Blue & Dangerous* xvi. 89 'Another?' he asked her. 'Thank you. A vodka and mixer.'

b. *vodka glass*; **vodka Collins**: see *Collins[2]; **vodka gimlet** [*gimlet *sb.* 1 c], a cocktail made of vodka and lime-juice; **vodka martini,** a martini cocktail in which vodka is substituted for gin; **vodka-tonic,** a drink consisting of vodka and tonic water.

1953 R. & A. London *Cocktails & Snacks* 46 (*heading*) Vodka martini. **1958** 'J. Byrom' *Or be he Dead* viii. 117 The vodka glasses were genuine ones. **1965** 'P. Quentin' *Family Skeletons* i. 4, I ordered a vodka martini. **1969** 'P. Kavanagh' *Such Men are Dangerous* i. 13 The desk men..drink.., in the summer, vodka collinses. **1974** R. B. Parker *God save Child* ix. 61 Can I get you a drink?.. Would you take a vodka gimlet? **1976** J. Hayes *Missing* (1977) iii. 61 She poured another vodka-tonic.

vodkatini (vǫdkătī·ni). Contr. of *vodka martini* s.v. *vodka b.

1955 Williams & Myers *What, When, Where & How to Drink* 146 (*heading*) Vodkatini (Vodka Martini). **1969** 'J. Morris' *Fever Grass* x. 97 You go on and get a table and order a couple of drinks. Mine is a vodkatini, a double. **1979** *Listener* 18 Jan. 80/1 'You're fighting ten pints of beer in every man.'.. The remark applies in essence (be it beer, whisky or vodkatinis) from the roughest club to the plushiest cabaret.

vo-do-deo-do (vōᵘ:dōᵘdi,ōᵘdōᵘ·). Also **vo-do-de-o, vo-de-o-do, vo-do-deo-vo,** etc. A meaningless jazz refrain, used *attrib.* to designate a style of singing or a song characterized by speed, energy, and the repetition of such a refrain or insistent rhythm. Also *fig.*

[**1927** J. Yellen *Crazy Words, Crazy Tune* (song), Sings the same words to ev'ry song. Vo-do-de-o. Vo-do-do-de-o-do.] **1934** C. Lambert *Music Ho!* iii. 209 The most irritating quality about the Vo-dodeo-vo, poop-poop-a-doop school of jazz song is its hysterical emphasis on the fact that the singer is a jazz baby going crazy about jazz rhythms. **1937** N. Coward *Present Indicative* viii. 335 Slick American 'Vo do deo do' musical farces, in which the speed was fast, the action complicated, and the sentimental value negligible. **1946** Mezzrow & Wolfe *Really Blues* viii. 120 All the hi-de-ho, vo-de-o-do, and boop-boop-a-doo howlers that later sprouted up. **1958** V. Bellerby in P. Gammond *Decca Bk. Jazz* xvii. 214 At quick tempos some of her tricks can sound a little dated and imbued with the 'vo-do-de-o-vo' style of the late 'thirties.

vodun (voᵘ·dun). Also **vodu.** [W. Afr. (Dahomey) *vodu* (see quot. 1890).] A fetish, usu. one connected with the snake-worship and other rites practised first in Dahomey, then introduced by slaves esp. to Haiti and Louisiana. Also *attrib.* and *Comb.*, esp. in **vodunhwe, -kwe,** fetish-house. Cf. Voodoo.

1874 J. A. Skertchy *Dahomey as it Is* iii. 54 The name is derived from *Danh*, a snake, and Hweh, a residence. It is sometimes called *Vodun-hweh, i.e.*, the fetiche house. *Ibid.* vii. 154 This shrine is the Vodun-no-Demen, or fetiche-house of Demen. **1890** A. B. Ellis *Ewe-Speaking Peoples* ii. 29 The term *vôdu*..is still used..in the so-called Vaudoo, or Vaudoux worship of the negroes of Hayti..where the old python *culte* of Whydah still survives. *Ibid.*, Vôdu appears to be derived from *vô* (to be afraid), or from *vô* (harmful). **1920** *Encycl. Relig. & Ethics* XI. 400/2 The Voodoo serpent-cult in Haiti and elsewhere reproduces these W. African cults, one of the names of Dañh-sio being Vodunhwe. **1953** *Caribbean Q.* III. i. 39 The compound began with one house..a chapel or vodunkwe (house of the gods), and a tent. **1956** M. Stearns *Story of Jazz* (1957) ii. 20 A photograph by Earl Leaf of a Haitian *vodun* altar. **1963** [see *obeah 2]. **1973** E. Bullins *Theme is Blackness* 9 Black..vodun ritual-ceremony. **1985** *Times Lit. Suppl.* 11 Jan. 28/3 It is fashionable at present to argue that such African legacies as Santeria, Shango, Vodun or Camboulay provide a more promising basis for a Caribbean identity than imported European religions.

voel (voil). *rare.* [W., mutation of *moel* bald.] A bare hill or mountain.

1876 [see *roped *ppl. a.* 2 a].

voertsek, var. *voetsak *int.*

voetganger (fu·tχaŋəɹ). *S. Afr.* Also **foot-ganger.** [Afrikaans, f. Du. 'pedestrian', f. *voet* foot + *ganger* one who goes]. **1.** A locust in its immature wingless stage. Cf. Hopper[1] 2 in Dict. and Suppl.

1824 *S. Afr. Jrnl.* I. 70 The flying locusts are..less dreaded in this colony, than those which have not quite reached that stage of maturity, and are..vulgarly called 'voetgangers'. **1873** F. Boyle *To Cape for Diamonds* 300 We drove through a host of foot-gangers on the trek. These are the larvæ of locust. **1913** J. J. Doke *Secret City* 119 The locusts, the voetgangers, came in their countless millions. **1936** R. Campbell *Mithraic Emblems* 79 See there, and there it gnaws, the Rust—Voet-ganger of the coming swarm. **1966** E. Palmer *Plains of Camdeboo* xv. 244 All the world turns out to devour the voetgangers, but I remember only the white storks which appeared in numbers, and the fowls. When the first voetganger column appeared, our domestic fowls went berserk.

2. A pedestrian; also, (in quot. 1902) an infantryman.

1902 C. R. De Wet *Three Years' War* 410 It was exceedingly difficult for Colonials to rise, for they knew that not only would they have to be *voetgangers*, but also that if they were captured they would be..punished. **1950** L. G. Green *In Land of Afternoon* 144 In some ways the voetganger of a century ago fared better than the modern tramp.

voetsak (fu·tsæk), *int. S. Afr.* Also **footsack, voertsek,** and other varr. [ad. Afrikaans *voertsek, voetsek*, f. Du. *voort zeg ik* be off I say.] A command to leave (addressed esp. to a dog).

1837 J. E. Alexander *Narr. Voy. Observation W. Afr.* I. 351 Dogs attacked us as we approached; but on the cry of '*voortzuk!*' from the master..they left us. **1871** *Cape Monthly Mag.* III. 332 Cries of *foot-sek* with the slashing of a whip and the yelping of a defeated cur. **1890** *Digger's Doggerel* 22 'Bonzela Baas, Inkos!' He tell me 'Hamba, footsack!' **1899** A. Werner *Capt. of Locusts* 194 Be off with you! How dare you annoy a lady like this? *Voetsak!* **1918** C. Garstin *Sunshine Settlers* 50 He learned that when he heard the word 'Mike' he was wanted, and when he heard the word 'footsack' he was not. **1949** O. Walker *Wanton City* 74 What's White civilization in South Africa?.. Social inhibitions imported from Europe? Broken accents from the world's ghettos? Fooie! Voetsak. **1950** *Cape Argus* 3 May 9/5 Betsy's only response was: 'Voetsek, you skollie.' **1963** A. Delius *Day Natal took Off* 9 He heard Bloubakkies yell to Sobisa, 'Hardloop! Run! Go! Voertsek!' **1969** A. Fugard *Boesman & Lena* 42 Few dops and a guitar and its voetsek yesterday and to hell with tomorrow.

Hence **voe·tsak** *v.*, (*a*) *trans.* to chase (a dog) away; (*b*) *intr.* to leave, to go away.

1897 E. Glanville *Tales from Veld* 227, I jes' drop in t'ask you *voetsack* all the dogs outer the place, 'fore I bring him in. **1958** *Cape Times* 4 Dec. 13/8 De La Fontaine said: 'You had better give me my money back.' I told him to *voetsak.* **1971** *The 1820* Apr. 11 But be careful who you tell to 'voetsak'..or be prepared for the consequences!

‖ **vœu** (vø). Pl. **vœux** (vø). [Fr., lit. 'vow, wish': see Vow *sb.*] A recommendation made by an international conference, which is not mandatory.

1917 E. Satow *Guide to Diplomatic Practice* II. xxvi. 142 A recommendation (*vœu*)..respecting the liquor trade was adopted. **1936** R. C. K. Ensor *England, 1870–1914* xii. 420 A preamble like that..enacts nothing. It is only a *vœu.* **1939** H. Nicolson *Diplomacy* 250 *Voeux.* It sometimes happens that a conference wishes to add to its treaty certain 'recommendations' for future good conduct. These are called 'wishes' or 'voeux'. Thus the Hague Peace Conference of 1899 emitted six 'voeux'. These have no binding force upon the signatories. **1961** A. Marder *From Dreadnought to Scapa Flow* I. vi. 133 Sir Edward Fry..presented a pious *voeu*,..declaring it was 'highly desirable that the Governments should resume the serious study of this question'.

vogesite (vōᵘ·dʒezəit). *Petrogr.* [ad. G. *vogesit* (C. H. F. Rosenbusch *Mikrosk. Physiogr.* (1887) II. 315), f. *Vogesen*, G. name of the Vosges Mountains in N.E. France: see -ite[1].] A lamprophyre consisting essentially of phenocrysts of hornblende (or augite) in a groundmass containing potash feldspar.

1891 F. H. Hatch *Introd. Stud. Petrol.: Igneous Rocks* v. 97 A second type of syenitic lamprophyre, containing hornblende instead of mica, is termed vogesite by Rosenbusch. **1937** A. Johannsen *Descr. Petrogr. Igneous Rocks* III. 37 The vogesites are greenish, grayish, or black melanocratic rocks when fresh, and reddish or brownish when weathered. **1976** *Nature* 5 Aug. 462/2 The vogesite lava is composed of approximately 85–90% olivine and augite.

vogue, *sb.* Add: **7.** *attrib.* or as *adj.* Fashionable; currently in vogue; esp. in *vogue word.*

c **1669** Howard & Villiers *Country Gentleman* (1976) i. i. 66 Pox on your Bourdeaux, Burgundie..no more of these vogue names,..get me some ale. **1915** H. L. Wilson *Ruggles of Red Gap* (1917) iii. 49 Decidedly he was not vogue. His hat was remarkable, being of a black felt with high crown and a wide and flopping brim. **1926** Fowler *Mod. Eng. Usage* 697/1 *Vogue-words.* Every now & then a word emerges from obscurity, or even from nothingness or a merely potential & not actual existence, into sudden popularity. **1947** Partridge *Usage & Abusage* 351/1 *Brave new world* is perhaps as much a cliché as it is a vogue-term. **1958** *Listener* 16 Oct. 621/2 Psychosomatic is the vogue-word of today. **1960** Partridge *Charm of Words* i. 47 One of the main differences between vogue-phrase and cliché: the majority of clichés last for generations. **1972** P. D. James *Unsuitable Job for Woman* iii. 106 Typical of the worst kind of academic writing. Contempt for logic; a generous sprinkling of vogue names; spurious profundity. **1978** *Forum on Med.* Apr. 84/1 Clichés and vogue expressions are equally plentiful. **1981** W. Safire in *N.Y. Times Mag.* 22 Feb. 9/1 Vogue words are his specialty [*sc.* Alexander Haig's]. The academic joyword of the 70's was 'exacerbate'. **1982** *Sunday Times* 5 Dec. 55 It was Chinese orders that made the mining machinery companies vogue stocks in the late seventies.

‖ **vogue la galère** (vog la galɛ̄r), *int.* [Fr., lit. 'let the galley be rowed'.] Let's get on with it! Let's give it a go!

[*c* **1525** Wyatt *Coll. Poems* (1969) 59 My fearful trust, 'en vogant la Galère'.] **1744** Lady Hervey *Let.* 20 Oct. (1821) 86 As long as Mrs. Phipps is well, and Mr. Phipps happy, *vogue la galère.* **1822** Scott *Peveril of Peak* IV. xi. 258 '*Vogue la Galere!*' he exclaimed, as the carriage went onward; 'I have sailed through worse perils than this yet.' **1850** Thackeray *Pendennis* II. vi. 63 Eh, *vogue la galère*, I say. It's good sport, Warrington—not winning merely, but playing. **1909** L. Strachey *Let.* 13 Oct. in *V. Woolf & L. Strachey: Letters* (1956) 35 My health seems still to be something of a Mahomet's coffin. However, vogue la galère! *a* **1918** D. H. Lawrence *Phoenix* (1936) v. 594 We'll embark on a new course of education, and *vogue la galère.*

voguey (vōᵘ·gi), *a.* [f. Vogue *sb.* + -y[1].] = next.

1928 *Observer* 22 July 9/2 An achievement..which steers a happy course between the 'arty' and the 'voguey'. **1980** *Listener* 21 Feb. 228/1 'Don't have' has become voguey and displaced 'haven't got'.

voguish (vōᵘ·giʃ), *a.* [f. Vogue *sb.* + -ish[1].] That is in vogue or temporarily fashionable.

1927 *Daily Express* 26 Mar. 8 (Advt.), Hundreds of the voguish Jumper Suits await your selection. **1951** G. Greene *End of Affair* i. ii. 18 A voguish choice of reading matter in the waiting-room..*Harper's Bazaar* and *Life.* **1960** *News Chron.* 11 Oct. 4/3 An obsessive interest in [James] Dean..is voguish. **1980** *Times Lit. Suppl.* 31 Oct. 1220/1 Couples steer impeccably towards 'the smart upper floor' of a voguish restaurant.

Vogul: see Wogul, vogul in Dict. and Suppl.

voice, *sb.* Add: **I. 3.** (Further examples, esp. of the names of radio stations supposedly representing national or local opinion.)

1775 W. H. Drayton in R. W. Gibbes *Doc. Hist. Amer. Revol.* (1855) I. 181 These men..have practised every art, fraud, and misrepresentation, to raise in this Province an opposition to the voice of America. **1925** *Country Gentleman* 22 Aug. 14/1 If you tune in on WSB any evening at 8 or 10:45, you hear 'The voice of the South'—the Atlanta Journal. **1932** G. B. Shaw *Platform & Pulpit* (1962) 246 Mussolini is the most responsible ruler in Europe because he gives his orders with his own voice and not through an imaginary megaphone called 'The Voice of the Italian People'. Mr MacDonald's voice is a National Voice. **1942** *N.Y. Times* 9 Oct. 19/1 With all the short-wave stations under control of two Federal agencies.., it is expected that 'the Voice of America' will tell more nearly the same story. **1956** *B.B.C. Handbk. 1957* 132 The violently anti-British station, 'The Voice of the Arabs'. **1959** *Listener* 24 Sept. 490/2 In war time the Continental European and British colleagues broadcasting from Bush House, London, to Europe, represented not only 'the voice of London' but the true voice of the nations under Hitler's occupation. **1962** E. Snow *Other Side of River* (1963) vi. 51 The main building had a hundred large rooms, each with modern bath and each with an all-wave radio set; on mine I heard the Voice of America from Manila and United States Army broadcasts from Tokyo. **1970** D. Caute *Fanon* iv. 51 In 1956 the Voice of Free Algeria came on the air and within twenty days..the entire stock of radio sets was sold out. **1982** T. Keneally *Schindler's Ark* xxxvii. 397 The radio technicians..listened with an earphone to the 2 p.m. news, from the Voice of London.

II. 8. e. (The utterance of) an invisible guiding or directing spirit (merging with fig. uses as sense 8 c). Chiefly *pl.* *direct voice*: see *direct *a.* 6 i.

1911 G. B. Shaw *Doctor's Dilemma* i. 12 When my patients tell me that theyve made a greater discovery than Harvey, and that they hear voices, I lock them up. **1924** —— *St. Joan* p. xv, Joan must be judged a sane woman in spite of her voices because they never gave her any advice that might not have come to her from her mother wit. **1924** W. Holtby *Crowded Street* xxxv. 260 Things happen against our will. Always being driven and we follow—voices. **1956** 'C. Blackstock' *Dewey Death* vii. 166 Do you think, perhaps, I am mad? Will I hear voices? **1973** A. Morice *Death & Dutiful Daughter* vi. 65, I ought to check up... My voices tell me that all is not..plain sailing there... Still, I may be wrong. **1975** 'J. Lymington' *Spider in Bath* ii. 30 A voice come to me in the night... Me Mum used to 'ave voices. Told 'er all sort er things.

10. In Parliament, etc., *spec.* a vote given with the voice, esp. by like-minded voters in unison; esp. in phr. *to collect the voices*, to take a vote by noting the relative strength of the calls of *ay* and *no* (cf. *voice vote*, sense 14 below).

1844 Erskine May *Law of Parl.* viii. 179 It must be well understood by members that their opinion is to be collected from their voices in the house, and not by a division; and if their voices and their votes should be at variance, the former will be held more binding than the latter. **1898** [see Dict.]. **1924** [see *voice vote*, sense 14 below]. **1976** S. Lloyd *Mr Speaker, Sir* 179 He then 'collects the voices' by saying, 'As many as are of that opinion say aye: of the contrary noe'.

III. 13. a. *voice-producer, -production* (earlier example) sbs.; *voice-activated, -operated* adjs.

1889 G. B. Shaw in *Star* 13 Dec. 2/4 Teacher of voice production. **1931** H. Nicolson *Diary* 6 May (1966) 73 He told me that Joseph Chamberlain was the best voice-producer he had ever heard. **1937** *Bell Telephone Q.* Apr. 113 This voice operated gain adjusting device operates on the speech coming into the control office to iron out the differences between loud voices and weak ones. **1972** *Computers & Humanities* VII. 50 Another hardware development used in teaching languages with computers is..a highly sophisticated (and rather expensive) tape recorder... It features the ability to stop a tape automatically for learner control decisions, voice-activated starting and stopping of recording, and automatic replay after each exercise. **1976** K. Benton *Single Monstrous Act* i. 11, I want Clancy's room bugged. Use a voice-operated mike. **1983** *Listener* 11 Aug. 33/2 It can't be long before voice-activated equipment enters the modern home, making remote controls a thing of the past.

b. *voice-pitch, -quality, -range.*

1932 D. Jones *Outl. Eng. Phonetics* (ed. 3) vi. 23 Some sounds are more sonorous than others, that is to say they carry better or can be heard at a greater distance, when pronounced with the same length, stress, and voice-pitch. *Ibid.* xxxi. 282 When there is more than one stressed syllable, the fall of the last stressed syllable generally begins at a pitch near to that of the initial unstressed syllables, and falls to the lower limit of the voice-range. **1946** K. L. Pike *Intonation Amer. Eng.* iv. 100 These characteristics—relaxed vocal cords, open throat..—are the goals of the trainer of voices... Voice quality..is comprised of such characteristics plus some..other differences. **1964** J. C. Catford in D. Abercrombie et al. *Daniel Jones* 29 Phoneticians should be able to classify 'voice-qualities' and other phonatory activities in as systematic a way as they classify supralaryngeal articulation. **1977** *Early Mus.* Oct. 492/2 It is actually a difficult piece, requiring..voice-ranges quite different from our standard choral division into soprano, alto, tenor, bass. **1977** P. Strevens *New Orientations Teaching of Eng.* xi. 144 Range of voice pitch, voice quality, and several more dimensions.

14. voice-box, (*b*) = *speak-box* s.v. *speak *v.* 36; **voice channel** *Telecommunications*, a channel with a bandwidth sufficiently great to accommodate speech; **voice coil** = *speech coil* s.v. *SPEECH *sb.* 13; also, a similar coil with the converse function in a moving-coil microphone; **voice frequency,** a frequency within the range required for the transmission of speech (commonly taken as 200 or 300 Hz to 3000 Hz or higher); usu. *attrib.*; **voice leading** *U.S.*, = *part-writing* s.v. PART *sb.* (*adv.*) 29; **voice level,** the volume of a voice measured for recording purposes; **voice-part** (earlier example); **voice radio,** a two-way radio; = *TRANSCEIVER; **voice synthesizer,** a synthesizer for producing sounds in imitation of human speech; **voice vote** (orig. *U.S.*), a vote taken by noting the relative strength of the calls of *ay* and *no* (cf. sense 10 above).

1971 'K. ROYCE' *Concrete Boot* i. 18, I rang the bell and a voicebox crackled in my ear. I told him who I was and pushed the door. **1959** *Listener* 18 June 1057/2 Current costs of this means of trans-oceanic telephone communication is about £70 per voice channel per mile. **1971** E. F. SCHOETERS in B. de Ferranti *Living with Computer* viii. 67 To every terminal must be assigned..equipment able to transform computer signals into a form suitable for long-distance transmission over what the Post Office calls 'voice channels'. **1934** *Discovery* Oct. 301/2 Goodmans' new '12 Watt' permanent magnet moving coil speaker.. has a totally enclosed voice coil. **1961** G. A. BRIGGS *A to Z in Audio* 126 The moving coil or dynamic microphone has been widely used... With low impedance voice coil windings..a wide frequency response with good sensitivity can be achieved. **1977** *Rolling Stone* 7 Apr. 87/3 (Advt.), With voice-coils wound with hi-temp wire on aluminum heat-sinking formers to handle the power of modern amps without burnout. **1925** *Telegraph & Telephone Jrnl.* XI. 151/2 Voice frequency telegraph. 200 to 2,000. **1944** *Electronic Engin.* XVI. 360 The transmitters are keyed by means of voice-frequency tone signals. **1975** R. L. FREEMAN *Telecommunication Transmission Handbk.* viii. 389 In practice voice frequency carrier telegraph techniques handle data rates up to 1200 bps. **1934** WEBSTER, Voice leading. **1942** [see *FAUX-BOURDON]. **1980** *Dædalus* Spring 198 The upward gap of a sixth..has here been masked by voice-leading. **1984** *Listener* 30 Aug. 30/4 With the increasingly frequent appearance of tonal triads and ever more connected voice-leading, something very much akin to the use of commonplace words has been creeping into his music. **1962** A. NISBETT *Technique Sound Studio* 258 *Voice level* is the acoustic volume produced by a voice in the studio. **1981** P. NIESEWAND *Word of Gentleman* xxvi. 180 Will you read a couple of sentences so we can get a voice level. **1776** J. HAWKINS *Hist. Music* I. p. lxxix, Instrumental aids to the voice-parts. **1959** C. OGBURN *Marauders* (1960) iv. 132 Tom Senff..had..got hold of a voice radio. **1962** *Sunday Express* 25 Nov. 3/2 Back in the cockpit, Harvey called George Bell on the voice radio. **1974** G. JENKINS *Bridge of Magpies* ii. 39 I've got a special transceiver—voice radio—laid on. RCA Navy job. [**1939, 1958** Speech synthesizer: see *SYNTHESIZER 2.] **1967** *Electronics* 3 Apr. 266/1 A voice synthesizer that Fujitsu Ltd. has developed as an audible output for its computers... When 'good morning' is fed into the computer by typewriter, the printed-out response is 'ohayo gozaimasu' and at the same time the robot-sounding voice of the synthesizer says it in Japanese. **1975** *Minicomputer Forum* 333 A voice synthesizer driven by a minicomputer and coupled to a telephone answering device and an automatic calling unit is an economical way of providing online computer services to the general public. **1979** *Personal Computer World* Nov. 58/3 On show will be two new line printers, a voice synthesizer, a P2 Quick printer and the new TRS-80 Model II. **1985** *Listener* 30 May 23 (Advt.), You won't have to ask the computer for essential information updates. The voice synthesiser will offer them. **1924** E. W. HUGHES *Man. Amer. Parl. Law* x. 241 If the voices and vote on division be at variance, do not agree, the voice vote will bind. **1955** H. W. DONAHUE *How to manage your Meeting* v. 217 In the *voice vote* all those in favor of the motion say *aye* and those opposed say *no*. This is the method usually used in deliberative assemblies. **1976** H. WILSON *Governance of Britain* iii. 55 Perhaps after suggesting a formula which appears to command assent, the prime minister asks 'Cabinet agree?'—technically a voice vote, sometimes just a murmur.

b. voice-over, narration spoken by an unseen narrator on a film or television broadcast; also, the unseen person whose voice is heard; also *attrib.*

1947 H. C. GIPSON *Films in Business & Industry* 284 *Voice-over*, narration-type recording as opposed to live sound. **1964** *Listener* 16 Apr. 629/1 The BBC..trying to find a way of avoiding naturalistic dialogue by the use of images with 'voice over' narration. **1964** *Punch* 21 Oct. 594/1 How would you like to do some voice-overs? **1966** G. N. LEECH *Eng. in Advertising* iv. 37 In the scripts, the speaker of commentaries is variously designated 'commentator', 'announcer' and 'voice over'. **1968** *Listener* 19 Sept. 379/2 In slow motion she loped muzzily through sylvan glades, and it was a matter of judgment whether the voice-over announcer would come on to plug the sexual properties of petrol, hand lotion or tooth-paste. **1974** K. MILLETT *Flying* II. 225 What I'm trying to do is to supply the voiceover for the pictures I make. **1983** *Daily Tel.* 18 Aug. 13/1 The minister gives up the cloth and takes off..to earn modest fame as a voice-over for TV commercials. **1984** *Daily Mail* 17 May 3/1 His [*sc.* Prince Andrew's] recorded voiceover will be heard by a total of around 300,000 spectators and a TV audience of millions.

voice, *v.* Add: **I. 5. d.** = *NARRATE *v.* 1 b.

1961 *Listener* 14 Dec. 1044/2 Scripted by Roger Fulford ..voiced by Andrew Cruickshank in sub-Dimbleby, it recaptured little of the emotion of the Abdication crisis.

1969 *Radio Times* 6 Feb. 31/1 The films we want to see are prized out of Continental cutting rooms by our multi-lingual production team who make translations of the original commentaries. The English versions are then voiced by nationals of each appropriate country.

7. b. Also *transf.*

1933 *Metronome* Mar. 34 Voicing ensembles should be considered entirely differently from voicing separate sax or brass trios.

8. In same sense as part of a phonological change. Also *intr.* to become voiced.

1959 A. CAMPBELL *OE. Gram.* viii. 163 A system by which a voiceless spirant was voiced if the preceding vowel did not bear the main stress. **1964** *Language* XL. 26 The sibilant articulated in close juncture with /l/ but did not voice.

voi·ceprint. [f. VOICE *sb.* + PRINT *sb.*, after *fingerprint*.] A sonogram of a person's voice. Also *attrib.*

1962 *New Scientist* 14 June 598/2 'Voiceprint' identification may eventually supplement that from fingerprints. **1968** M. GUYBON tr. *Solzhenitsyn's First Circle* xxxiii. 198 This was really something quite new—tracing a criminal by his voice-print. **1973** *Daily Tel.* 5 Apr. 15/3 A Californian appeal court has ruled after years of controversy that voiceprints are acceptable as evidence in criminal cases. **1981** I. ST. JAMES *Balfour Conspiracy* ii. 49 The voice print library is still limited.

Hence **voi·ceprinter, voice printer,** an apparatus for producing voiceprints; **voi·ce-printing** *vbl. sb.*, the production or use of voice-prints.

1962 *Nature* 29 Dec. 1253/2 Voiceprinting uses the spectrographic impressions of the utterances of ten words frequently used in telephone conversations. **1966** *Economist* 20 Aug. 734/3 A young Negro..has been brought to trial in Los Angeles..on evidence provided by an electronic device known as a 'voiceprinter'. **1978** S. BRILL *Teamsters* iv. 144 Nine days after the voice printing Faugno disappeared. **1979** J. CROSBY *Party of Year* (1980) i. 13 If you'd just state your name again, for the Voice Printer.

voicespond (voi·spǫnd), *v.* [f. VOICE *sb.* + CORRE)SPOND *v.*] *intr.* To correspond by means of recorded oral messages. So **voi·cespondence, voi·cespondent.**

1954 *Sun* (Baltimore) 1 Feb. 4/1 A new type of friend-to-friend communication—'voicespondence'—is gaining fans in Europe. To 'voicespond' you have to own..a wire or tape recorder. *Ibid.*, Some 'voicespondents' collect recordings of famous voices... These they exchange by re-recording them for others. **1966** *Guardian* 9 Nov. 11/1 The amateur tape recording enthusiast..[may] 'voice-spond' and 'tapespond'..with others at home and abroad. **1973** *Britannica Bk. of Year* (U.S.) 732/3 Voicespondence, communication between persons by an exchange of tape-recorded messages; voicespond, vb.; voicespondent.

voicing, *vbl. sb.* Add: **4. b.** *Jazz.* The tonal quality of a group of musical instruments in an ensemble; a blend of instrumental sound; harmonization.

1946 MEZZROW & WOLFE *Really Blues* ix. 152 We wrote..figuring out the right voicing by playing clarinet duets until it sounded good to us. **1949** L. FEATHER *Inside BeBop* i. 6 He was a third horn, blending the guitar with the tenor and trumpet for three-part voicings that produced a sound new to jazz. **1958** S. RACE in P. Gammond *Decca Bk. Jazz* x. 127 A technique of voicing for one clarinet and four saxes. **1977** *Zigzag* Mar. 5/1 What I do now is go get a piano player who I really like and show him how the song was written, what kind of voicings, etc.

5. In same sense as part of a phonological change.

1959 A. CAMPBELL *OE. Gram.* ix. 180 The voicing of medial spirants was followed by the unvoicing of final spirants.

void, *a.* and *sb.*[1] Add: **A.** *adj.* **10*.** *Cards.* Of a hand: having no cards in a given suit.

1934 F. D. COURTENAY *System Experts Play* (ed. 3) 17 A void suit at a contract declaration is equivalent to an A. **1958** *Listener* 6 Nov. 753/3 One never lives to enjoy the double, for either dummy or partner will be void and will remove the double. **1972** R. MARKUS *Aces & Places* 25 When West showed void he went into a huddle. **1980** [see sense B. 8 below].

B. *sb.*[1] **3. c.** *spec.* (*a*) A defect in a crystal lattice consisting of a space larger than a single vacancy. (*b*) An interatomic space in any crystal lattice.

(*a*) **1947** *Trans. Amer. Inst. Metallurgical Engineers* CLXXI. 136/2 Supposedly these voids do not equal in volume the sum of lattice-volume change plus the direct zinc transfer. **1952** *Proc. Physical Soc.* B. LXV. 522 It remains now to discuss the generation and removal from the lattice of the large number of vacancies which do not form voids. **1965** J.-I. TAKAMURA in R. W. Cahn *Physical Metallurgy* xiv. 722 In these crystals voids are formed at grain boundaries as a result of large amount of strain. **1974** *Physics Bull.* Dec. 582/3 A nice illustration..was the discovery by Evans of the void lattice in molybdenum, voids of a few tens of Å in radius being ordered on a lattice with a spacing of one or two hundred Å.

(*b*) **1964** WERT & THOMSON *Physics of Solids* ii. 32 The fcc structure has small voids (called interstices) between the atoms. **1982** J. V. SMITH *Geom. & Structural Cryst.* v. 145 (*caption*) Tetrahedral and octahedral voids in (*a*) cubic closest-packing and (*b*) hexagonal closest-packing.

8. *Cards.* The absence of any cards in a particular suit in a player's holding.

1933 C. VANDERBILT *Contract Contracted* ii. 20 The Short Suits are the Doubletons, Singletons and Voids. **1944** *Times* 17 May 6/3 Provision was made for revaluation after the first round of bidding to count three points for a void. **1980** M. DUMMETT *Twelve Tarot Games* vii. 125 Singleton Kings are usually even better than voids, unless..there is a danger that one of the opponents is void in [that suit].

9. *Med.* An emptying of the bladder.

1980 *Brit. Med. Jrnl.* 29 Mar. 889/2 Residual urine was estimated..by catheterization after a normal void.

Hence **voi·dward** *adv.* (*rare*).

1920 JOYCE *Nightpiece* in *Poems Penyeach*, As the bleak incense surges, cloud on cloud, voidward from the adoring Waste of souls.

void, *v.* **I. 2. b.** Delete † *Obs.* and add later example.

1876 *Fur, Fin & Feather* Sept. 167 They [*sc.* squirrels] should be voided..and kept several days in cold weather.

7. b. [Examples in *Med.* of sense 'to empty the bladder'.]

1947 STAFFORD & DILLER *Textbk. Surg. for Nurses* xlix. 500 The patient's bladder must be empty; therefore..she should void before being draped for operation. **1966** *Amer. Jrnl. Obstetrics & Gynecol.* XCIV. 796/1 Sixteen patients were investigated who were unable to void following operation. **1977** *Lancet* 21 May 1072/2 The patient could void, but was unable to..

voidage (voi·dėdʒ). [f. VOID *sb.*[1] + -AGE.] Voids collectively; the proportion of a volume occupied by voids.

1946 *Nature* 17 Aug. 236/1 At 1,000°C. and with an inlet total gas rate of 200 litres/hr., sufficient to expand the bed of catalyst particles to a voidage of 0·7 so that the whole bed was agitated, the exit gas analyses..were [etc.]. **1966** R. D. WEST in P. Hepple *Petroleum Supply & Demand* 70 Injection wells scattered across the reservoir to inject the relatively large volumes of water required, not only to replace the oil, but also to fill the voidage left by the natural depletion. **1970** *Nature* 11 Apr. 158/2 The variation in local voidage occurring near the wall of a large cylindrical container filled with a mixture of spheres. **1983** *Chem. & Engin. Sci.* XXXVIII. 339/1 There is little difference between the surface settling rate of a bubbling bed and of a uniformly expanded bed of the same powder.

voider. Add: **6. b.** *Med.* One who passes urine.

1969 *Radiology* XCII. 1178/1 Two of the patients were troubled by fecal retention and soiling and were thus infrequent defecators as well as infrequent voiders. **1974** *Amer. Jrnl. Roentgenol.* CXX. 407/1 ...

voiding, *vbl. sb.* **1. a.** Delete 'Now *rare*' and add: Now only of the bladder or bowel.

In quot. 1976 perh. *concr.* (cf. sense 8 a).

1946 J. S. HUXLEY *Ess. Pop. Sci.* 28 Underlying the voiding of the bladder is a reflex mechanism. **1972** *Jrnl. Urol.* CVIII. 159/1 During voidings, rectal pressure was monitored. **1976** *Lancet* 9 Oct. 773/1 Urine from each voiding was poured into a funnel and allowed to drip. **1980** *Brit. Med. Jrnl.* 29 Mar. 889/2 Radiographic screening of the outflow tract was carried out during voiding.

voidless, *a.* For *rare*[−1] read *rare* and add: **b.** *poet.* Unavoidable.

1908 HARDY *Dynasts* III. III. i. 382 All must prepare to give with gory death In the now voidless battle.

‖ **voilà** (vwala), *imp.* as *prep.* and *int.* [Fr., f. imp. of *voir* to see + *là* there.] There is, are, etc. As *int.*: there it is! there you are! lo! *voilà tout* (tu) [= all], that is all, there is nothing more to do or say.

1739 GRAY *Let.* 12 Apr. (1971) I. 102 The minute we came, voilà Milors Holdernesse, Conway, and his brother. **1778** H. WALPOLE *Let.* 12 Feb. (1955) XXVIII. ... **1801** JANE AUSTEN *Let.* 11 Feb. (1952) 121 Mr Bramston called here... **1884** ... I am painting portraits, voilà tout. **1858** ... **1904** ... Dale asked me if he could call. I said 'Yes.' Perhaps I was wrong. Anyhow, voilà! ... **1973** *Philadelphia Inquirer* (Today Suppl.) 7 Oct. 5/1 Plain junk that becomes—voila!—a memento when a 'Historic Philly' decal is affixed. **1980** *Quilt World* Sept./Oct. 42/3 The batting and backing were then trimmed and the edges of the quilt sewn shut. Voila! a completed quilt. **1981** 'D. JORDAN' *Double Red* xxxiv. 156 Live all your life! That is all I know, voilà tout.

voilette. 1. (Earlier example.)

1842 C. RIDLEY *Let.* 9 Oct. in *Cecilia* (1958) viii. 101 She had on..a blue silk bonnet with a long veil (not a *voilette*).

voir dire. For 'Also 7 **voire**' read 'Also **voire**' and add: Also, an investigation into the truth or admissibility of evidence, held during a trial. Also *attrib.*

1968 *Daily Colonist* (Victoria, B.C.) 13 Sept. 42/3 Police testimony was called and a voir-dire—a trial within a

trial—was held to determine whether a statement from one of the accused was admissible evidence. Immediately after the voir-dire, Magistrate Fraser questioned Dr. Carpenter. **1973** *N.Y. Law Jrnl.* 19 July 12/3 The voire dire examination as to the qualifications of this witness.. revealed that he lacked the background which would qualify him as an expert. **1980** *Times* 18 Jan. 10/4 His Lordship was prepared to assume that the voire dire (that trial within a trial) and the verdict taken together did constitute a final judgement on the same issues. **1981** *Daily Tel.* 26 June 6/3 The purpose of the 'trial within a trial', known to lawyers as a *voir dire*, had been to decide whether Brophy had been induced to make a number of alleged statements [etc.].

‖ **voix** (vwa). [Fr. = voice.] Used in various phrases, as *voix blanche* (blɑ̃ʃ) [lit. 'white voice'], a toneless voice; *voix céleste* (selɛst) [lit. 'heavenly voice'] = *vox angelica* s.v. VOX 2; also *fig.*; *voix d'or* (dɔr) [lit. 'voice of gold'] a rich dramatic voice; *voix grave* (grɑv), a low-pitched or deep voice; *voix moyenne* (mwajɛn) [lit. 'middle voice'] a voice of middle range.

1876 STAINER & BARRETT *Dict. Mus.* 452/1 *Vox angelica*... Called also *Voix céleste*, *unda maris*, &c. **1895** G. B. SHAW *Our Theatres in Nineties* (1932) I. 137 ... **1907** Voix céleste [see *SUABE]. **1920** D. H. LAWRENCE *Lost Girl* x. 232 The muted *voix blanche* came through his lips. **1935** JOYCE *Let.* 19 Mar. (1966) III. 351 ... voix grave ... voix moyenne. **1962** *Listener* 8 Nov. 779/2 Her voice has always lacked intellectual content, and..the later and more ambitious manner comes to seem one-tenth genuine poem-making to nine-tenths *voix d'or* and delphic vapour.

volant, *sb.* **3.** (Earlier example.)

1851 *Harper's Mag.* Jan. 288/1 Five volants are set on full, each being trimmed at a little distance from the edge by a narrow guimpe.

‖ **volapié** (vǫlapi̯e·). [Sp., lit. 'flying foot', f. *volar* to fly + *pie* foot.] In *Bullfighting*, a manner of killing in which the bullfighter runs in to kill a stationary or slowly-moving bull. Also *à volapié* adv. phr., in this manner.

1838 *Q. Rev.* LXI. 418 The 'volapié'..is dangerous but beautiful; the bull is met half way. **1910** *Encycl. Brit.* IV. 790/1 The stroke..is usually given *à volapié* (half running), the *espada* delivering the thrust while stepping forward, the bull usually standing still. **1932** E. HEMINGWAY *Death in Afternoon* xix. 242 ... **1967** MCCORMICK & MASCAREÑAS *Compl. Aficionado* ii. 58 This was a three-quarter sword following a good volapié.

volatile, *a.* Add: **4. c.** Of markets, shares, etc.: showing sharp changes in price or value (merging with uses of sense 5).

1931 *Daily Express* 31 Jan. 2/6 Some volatile issues recorded..advances. **1977** *Time* 19 Dec. ... **1985** *Times* 26 Jan. 23/1 It has been one of the most volatile [Stock Exchange] accounts in recent history.

5. b. *Computers.* Of a memory: retaining data only as long as there is a power supply to it.

1950 W. W. STIFLER et al. *High-Speed Computing Devices* xiv. 305 In a volatile storage medium, like a delay line, retransmission of each signal once during each storage cycle period is required. **1960** O. DOPPING *Computers & Data Processing* x. 136 The flip-flop register is a fast, expensive, volatile memory. **1979** R. MUTCH ... read-only and volatile memories,..and a battery.

volatility. Add: **5.** *Computers.* The property of a memory of not retaining data after the power supply is cut off.

1969 *IEEE Trans. Magnetics* V. 585/1 ... **1981** *Appl. Solid State Sci. Suppl.* ii. A. 122 Many investigations into lessening the impact of the volatility of the semiconductor memory have been made.

volcaniclastic (vǫlkæ:niklæ·stik), *a.* and *sb.* *Geol.* Also **vo:lcano-.** [Blend of VOLCANIC *a.* (and *sb.*) and CLASTIC *a.*] **A.** *adj.* Both volcanic and clastic. **B.** *sb.* A volcaniclastic rock.

1961 R. V. FISHER in *Bull. Geol. Soc. Amer.* LXXII. 1409/1 Only recently..have attempts been made to group all volcanic clastic (termed *volcaniclastic* in this paper) rocks into a single system... Rocks of pyroclastic origin are only one category of volcaniclastic rocks. **1972** F. J. PETTIJOHN et al. *Sand & Sandstone* vii. 261 The erosional volcaniclastics are comparable—in sedimentary structures, geometry, and thickness—to other terrigenous sandstones. **1975** *Nature* 17 Apr. 581/2 The rock types include..a variety of reworked volcanic ashes, volcaniclastic and fluviatile sediments. **1980** *Ibid.* 29 May 289/2 The Savage Mountain Formation consists of up to 5,000 m of..pillow basalt and massive subaerial flows with associated volcaniclastics.

volcanist. Add: **1.** (Further example.) Cf. *VULCANIST 3.

1971 H. KONDO *Moon* (1972) 14 The volcanists have three main arguments for believing that their theory is correct.

volcano, *sb.* Add: **2. b.** In phr. *to sit on a volcano* and varr.

[**1908** L. MITCHELL *New York Idea* i. 32, I feel as if we were all taking tea on the slope of a volcano.] **1909** GALSWORTHY *Silver Box* I. iii. 16 You're sitting upon a volcano, John. **1930** G. B. SHAW *Apple Cart* I. 26 The more I see of the sort of prosperity that comes of leaving our vital industries to big business men as long as they keep your constituents quiet with high wages, the more I feel as if I were sitting on a volcano. **1954** J. WHITING *Marching Song* II. 42 Is it a volcano I'm sitting on and not, as I'd supposed, a dung-hill? **1983** *Listener* 27 Jan. 28/3 ... he exposes to white, middle-class America the nature of the volcano (racism; social injustice; competition; aggression) on which it blithely sits.

3. volcano rabbit, a small, dark brown rabbit, *Romerolagus diazi*, found only in the mountains of central Mexico and very similar to the pika, having short ears and no tail.

1966 J. FISHER et al. *Red Bk.* 54/2 The volcano rabbit has been reduced to... **1974** G. DURRELL *Catch me a Colobus* ix. 173 The Volcano rabbit lives at a very high altitude..in the pine forests.

volcano-. The sb. used as a formative element: **volcanoclastic,** var. *VOLCANICLASTIC *a.* and *sb.*; **vo:lcanoge·nic** *a.* [*-GENIC], of volcanic origin; **volcanotecto·nic,** involving or pertaining to both volcanic and tectonic processes.

1966 E. LEHNER et al. in G. J. Williams *Econ. Geol. N.Z.* xix. 253/2 ... volcanogenic rocks. **1977** A. HALLAM *Planet Earth* 59/2 Glacial or volcanogenic sediments contain coarse particles which can be very variable. **1907** W. H. HOBBS in *Beiträge zur Geophysik* VIII. 224 The seismotectonic lines often intersect lines of volcanoes (volcanotectonic lines) at volcanic vents. **1941** *Univ. Calif. Publ. Geol. Sci.* XXV. 241 Calderas are volcanic and therefore centric forms of destruction, volcano-tectonic engulfments of the roofs of magma reservoirs, regardless of the number of vents involved. **1965** G. J. WILLIAMS *Econ. Geol. N.Z.* xv. 235/2 The relatively high permeability of the pumiceous deposits..in the large volcano-tectonic basins.. **1979** *Nature* 26 July 286/2 Zukwala volcano..lies on an isolated (non-WFB) volcanotectonic line that projects NNE via a Holocene basalt lava chain.

volcanological (vǫlcănǫlǫ·dʒikăl), *a.* [f. VOLCANOLOGY + -ICAL.] = VULCANOLOGICAL *a.* Also **vo:lcanolo·gic** *a.*

1889 *Cent. Dict.*, Volcanological. **1931** G. W. TYRRELL *Volcanoes* i. 12 With the establishment of volcanological observatories on Vesuvius and Etna..the study of volcanoes has entered upon a new phase. **1961** WEBSTER, Volcanologic. ... Numerous stations have been set up [on Vesuvius] at various heights for making volcanological observations. **1982** *Encycl. Brit. Bk. of Year* 307/2 Mt. St. Helens remained active...

volens (vōu·lenz), *ppl. a.* *Law.* [L., f. *velle* to will.] Consenting to a dangerous course of action. Cf. next.

1872 *Law Jrnl.* (Court of Exchequer) XLI. 101/2 Then comes the great difficulty; it is said 'volenti non fit injuria'. It is true that he was 'volens', in the sense that he entered the employment voluntarily. **1951** *Law Rep. House of Lords* (Appeal Cases) 767 He was not volens or careless. **1965** *Mod. Law Rev.* XXVIII. 570 A finding that a workman was *sciens* was sufficient to defeat his claim in tort, even if he was not volens.

‖ **volenti non fit injuria** (vǫle·nti nǫn fit indjuǝ·riă). *Law.* [L.] No injury is done to a willing person: a defence to an action whereby it is claimed that a person who sustained an injury agreed to risk such injury.

1658 E. WINGATE *Maximes of Reason* cxxii. 482 If the Tenant in an Assise of an house deny the Plaintiffe to dine with him in the house, which the Plaintiffe doth accordingly, but doth not clame the house at that time; this is no entry or possession to cause the Assise to abate; because if he had been a stranger, he had been no trespasser for *volenti non fit injuria*. **1743** C. CIBBER *Egotist* 6 *Volenti*, you know, *non fit injuria*. The publick is not obliged to buy. **1872** [see prec.]. **1933** *Law Rep.* (King's Bench Division) II. 299 We are at present disposed to take the view that the respondent is precluded from recovering on the principle volenti non fit injuria. **1980** *Oxf. Compan. Law* 1280/1 *Volenti non fit injuria*.., the principle that an injured person cannot complain of harm done him if he knew of and voluntarily incurred the risk of that harm... The plea is practically excluded in employment cases but is still relevant to cases of participation in dangerous sports.

Volga (vǫ·lgă). The name of a Russian river used *attrib.* in **Volga German,** a member of an ethnic minority living near the Volga from the seventeenth century until the Second World War; also *adj.*, of or pertaining to this people.

1941 *Times* 9 Sept. 3/3 (*heading*) Volga Germans to go. *Ibid.*, The Volga German region is an autonomous republic, with its own supreme council. **1947** W. BACZKOWSKI *Towards Understanding of Russia* v. 199 The following Autonomous Republics were liquidated: the Republic of Volga Germans, the Kalmyk Republic [etc.]. **1973** A. PRICE *October Men* xii. 176 Korbel deserted to the Wehrmacht... Told them he was a Volga German. **1985** *Observer* 28 July 17/1 ... to keep out troublemakers (Volga Germans? Crimean Tartars?).

Volgian (vǫ·lgiăn), *a.* *Geol.* [f. as prec. + -IAN.] Of, pertaining to, or designating a stage of the Upper Jurassic and Lower Cretaceous in Russia. Also *absol.*

1893 A. GEIKIE *Text-bk. Geol.* (ed. 3) 919 Some of the recognized [Jurassic] life-zones of western Europe can be detected in Russia... [These include] Volgian, consisting of green, brown and dark sandstones and sands. **1955** E. NEAVERSON *Stratigraphical Palaeont.* (ed. 2) xii. 475 The lower beds of the Russian Volgian stage, may be correlated with the Upper Kimmeridge Clay of England. **1978** *Nature* 13 July 133/2 The fault-controlled stepwise subemergence of the block-faulted basin culminated in Middle Volgian times with the most important Mesozoic phase of block-faulting in the region.

volitive, *a.* **4.** (Earlier example.)

1846 MONIER WILLIAMS *Elem. Gram. Sanscrit Lang.* vi. 229 Certain roots..take a desiderative form, without exactly yielding a volitive signification.

‖ **Volk, volk** (fǫlk). [G., Du., Afrikaans = nation, people: see FOLK.] **1.** *S. Afr.* **a.** The Afrikaner people.

1880 G. HUDSON *Let.* 1 Dec. in B. Bellairs *Transvaal War* (1885) 424 ... the 'volk' (people) as much as possible. **1928** E. A. WALKER *Hist. S. Afr.* xiii. 515 The Afrikaner Volk proceeded to find itself along cultural and then along political lines. ... **1948** *Press Digest* (S. Afr. Jewish Board of Deputies) No. 2. 16 What the nationally-minded Afrikaner wishes for himself and for his volk, he wishes also for people of other races... **1953** P. ABRAHAMS *Return to Goli* 173 It is not only the Afrikaans-speaking Whites, not only 'Die Volk', who believe in the colour bar. **1979** J. DRUMMOND *Patriots* iii. 24 ... strongholds of the Afrikaner Volk.

b. The Coloured employees of an Afrikaner.

1882 *Cape Q. Rev.* 357 The water to be utilised for medicinal purposes in a bathing-house adjacent, which at present serves as a bathroom for the 'volk' (workmen) of the farm. **1900** B. MITFORD *Aletta* 79 Come this way. My volk will see to your horse. **1939** S. CLOETE *Cape Coloured People* i. 5 In the western Cape a farmer calls his labourers his 'volk'—the same word that Adam Tas used for his slaves, as distinguished from his Hottentot labourers, in the early eighteenth century. **1968** K. MCMAGH *Dinner of Herbs* 30, I get the volk to drag away the carcasses and skin them.

2. The German people, esp. in the ideology of National Socialism.

1933 C. B. HOOVER *Germany enters Third Reich* vi. 175 *Das Volk* is thought of not as a juridical organization to which one might belong by obtaining citizenship rights through naturalization, but as a community held together by the sacred ties of common blood. **1937** *N.Y. Times Mag.* 21 Nov. 2/3 Like most nations of today, what is known as the Deutsche Volk is in itself a hybrid people composed of Germanic, Celtic, Slavic and Lithuanian elements.

3. *Comb.* **a.** In sense 1 (of the Afrikaner people): **vo·lkspele** (-spi°lǝ) *pl.*, Afrikaner folk dances; hence **vo·lkspeler,** a dancer of these; **vo·lkswil** (-vil), the will of the people.

1949 *Cape Times* 24 Sept. 9/2 There would be volkspele demonstrations, recitations and singing. **1972** *Evening Post* (Port Elizabeth) 5 Feb., Dancing that might well be confined to volkspele and other typically South African dances. **1953** *Cape Times* 17 Apr. 9/8 The group of volkspelers..will sail on an overseas tour in to-day's mailship. **1948** *Cape Argus* 2 Dec. ... **1956** *Star* (Johannesburg) 23 Feb. ... the 'volkswil' from members of a minority government.

b. (In English some of these words are used without proper regard for the inflexional endings customary in German itself.) In the possessive in sense 2 (of the people): **Volksdeutsch** (fǫ·lks,doitʃ) *a.*, fem., pl., **-deutsche** (-tʃǝ), belonging to or characteristic of the *Volksdeutsche*, ethnically German; **Volksdeutsche** *sb. pl.*, people of German origin resident outside Germany and Austria; ethnic Germans; **Volksdeutscher,** a member of the *Volksdeutsche*, an ethnic German; also *attrib.*; **Volksgeist** (fǫ·lksgəist), the spirit or genius which marks the thought or feeling of a nation or people; **Volkskammer** (fǫ·lks,-kamǝɹ), the parliament of the German Democratic Republic; **Volksoper** (fǫ·lksōpǝɹ), the light opera house in Vienna; also *transf.*; **Volkspolizei** (fǫ·lks,politsəi), a police force of

the German Democratic Republic; members of this; **Volkspolizist** (fǫ·lks,politsi·st), a member of the *Volkspolizei*; **Volkssturm** (fǫ·lks,ʃturm) a territorial army established in Germany during the last years of the Third Reich; also *attrib.*; **Volkswanderung** (fǫ·lks-vandǝruŋ) = *VÖLKERWANDERUNG.

1952 E. F. DAVIES *Illyrian Venture* x. 177 The house-keeper matron, a fat, grim *Volksdeutsche* woman. **1962** *Listener* 7 June 1004/2 The old Volksdeutsch grannie who has nothing left to do except learn to die. **1981** I. BOLAND tr. *Ginzburg's Within Whirlwind* I. x. 83 She..expostulated in her *Volksdeutsch* dialect that people must be fed before they are turned out to work. **1937** *N.Y. Times Mag.* 21 Nov. 16/3 The Germans of foreign citizenship are called 'Volksdeutsche', or 'racial Germans'. **1944** I. ORIGO *Diary* 7 June in *War in Val d'Orcia* (1947) 193 The Italian patriots..are..to give shelter to *Volksdeutsche* who have deserted from the German Army. **1961** R. KEE *Refugee World* iii. 20 Ethnic Germans (*Volksdeutsche*) living in east and south-east Europe. **1961** WEBSTER, Volksdeutscher. **1978** H. WOUK *War & Remembrance* xlvi. 463 A tall thin horribly pimpled *Volksdeutscher* burglar from Prague. **1981** I. BOLAND tr. *Ginzburg's Within Whirlwind* I. xiii. 108 He was a German. A Crimean *Volksdeutscher*. **1936** WIRTH & SHILS tr. *Mannheim's Ideology & Utopia* ii. 59 During this period the *Volksgeist*, 'folk spirit', comes to represent the historically differentiated elements of consciousness. **1977** D. WATKIN *Morality & Archit.* III. i. 79 Our fundamental concern is..with the impression left by the emphasis on national character or *Volksgeist* as a determinant of artistic style. **1949** *Times* 8 Oct. 4 The German 'People's Council' proclaimed itself in Berlin yesterday to be the provisional Volkskammer, or Lower House, of the Parliament of the new 'German Democratic Republic'. **1977** *Whitaker's Almanack 1978* 865/1 Further amendments came into force on October 7, 1974 after adoption by the *Volkskammer* on September 27, 1974. **1928** J. A. MAHAN *Vienna of Yesterday & Today* ix. 210 Two opera houses, the Staatsoper..and the Volksoper..are in use every day of the season. **1968** *Vogue* 15 Apr. 23/1 Theoretically, the difference between the Wells and the Garden is that between *Volksoper* and *Staatsoper* in Vienna. **1980** *Times* 31 Dec. 3/5 The report rejects the idea that the D'Oyly Carte company should be built up..to become a British 'Volksoper', performing light music and operetta. **1964** L. DEIGHTON *Funeral in Berlin* vi. 38 A volkspolizei troop carrier was parked at the roadside. **1967** R. V. BESTE *Repeat Instructions* (1968) ix. 94 He walked past *Volkspolizei*, who eyed him suspiciously. **1980** A. COPPEL *Hastings Conspiracy* vii. 52 Two uniformed officers of the Volkspolizei..had hurried him to the military airfield at Bützow. **1974** J. D. WHITE *Leipzig Affair* vii. 63 There was a volkspolizist waiting outside my room. **1944** *Nation* (N.Y.) 25 Nov. 653 The *Volkssturm* is necessarily largely composed of men between fifty and sixty, with a sprinkling of sixteen- and seventeen-year-old boys. **1969** A. MARIN *Rise with Wind* ii. 12 He could still remember the..day when a Volkssturm officer dug him out of the rubble. **1855** GEO. ELIOT *Let.* 9 Jan. (1954) II. 190 The subject I now propose is 'Women in Germany'..through the periods of the Volkswanderung and the romantic.. life of the Middle Ages up to our own day.

‖ **Völkerwanderung** (fö·lkǝrva:ndǝruŋ). Pl. **-ungen.** [G., f. *völker* pl. of *VOLK + *wanderung* migration.] A migration of a people or peoples *en masse*, *spec.* that of Germanic and other peoples in Europe during the later Roman Empire and the early Middle Ages. Also *attrib.*

[**1885**: see *Volkswanderung* s.v. *VOLK 3 b.] **1934** A. TOYNBEE *Study of Hist.* II. 97 The last convulsion of the post-Minoan Völkerwanderung along the west coast of Anatolia. **1946** PRIEBSCH & COLLINSON *German Lang.* (ed. 2) II. ii. 24 Subsequent encroachments..by Goths, Franks and others during the Age of Migration (Völkerwanderung) are discussed. **1961** L. F. BROSNAHAN *Sounds of Language* iii. 52 The Germanic expansion of the Völkerwanderung period. **1969** J. MANDER *Static Society* i. 46 The Germanic *Völkerwanderungen* after the fall of the Roman Empire.

‖ **völkisch** (fö·lkiʃ), *a.* [Ger.: see *VOLK, -ISH[1].] Populist, nationalist, racialist.

1939 *Times Lit. Suppl.* 7 Jan. 8/3 Saratoga and Yorktown are more familiar to the British schoolboy than a more *völkisch* school of history would tolerate. **1962** *Listener* 26 Apr. 739/3 As Franco grew strong enough to disembarrass himself of any kind of *völkisch* institutions, he encouraged the Falange to fade away. **1975** *Times Lit. Suppl.* 21 Nov. 1392/2 Whether the *völkisch* ideology and antisemitic sentiment suffered a similar setback or penetrated to other parties... There was much less personal.. continuity of this older generation of *völkisch* leaders.

volkonskoite (vǫlkǫ·nskǫ,əit). *Min.* Also 9 **wolch-, wolk-; volch-.** [ad. G. *wolchonskoit* (A. Kämmerer 1831, in *Jahrb. f. Mineral.*, etc. II. 420), f. the name of Prince 'Wolchonskoy' (? i.e. Volkonsky): see -ITE[1].] A green or bluish-green amorphous clay mineral of the montmorillonite group in which there is some substitution by chromium, now regarded as a variety of nontronite.

1844 J. D. DANA *Syst. Min.* (ed. 2) VI. 533 Wolchonskoite. Amorphous. Dull-shining. **1852** *Amer. Jrnl. Sci.* LXIV. 62 Silicated Chrome (Wolchonskoite?). **1892** E. S. DANA *Dana's Syst. Min.* (ed. 6) 696 Wolchonskoite. ..Volchonskoite. **1953** R. E. GRIM *Clay Mineral.* iv. 58 Substitutions within the octahedral sheet [of montmorillonite] may vary from few to complete... Replacement of aluminum by iron yields nontronite; by chromium, volkonskoite; by zinc, sauconite. **1973** *Norsk Geol. Tidsskr.* LIII. 329 The Cr_2O_3 content of volkonskoite may vary within wide limits, and values up to 15–20% have been reported from Russian occurrences.

Volkslied (fǫ·lks,līt). Also **volkslied.** Pl. **-lieder** (-līdǝɹ). [Ger. and Du., f. *volks* gen. of *VOLK + *LIED, LIED.] **1.** A German folk song; a popular song in a German folk idiom.

1858 GEO. ELIOT *Jrnl.* 23 May in *Life* (1885) II. viii. 41 His wife..sang us some charming Bavarian *Volkslieder*. **1882** F. A. KEMBLE *Rec. Later Life* II. 8 Dr. Charles Follen, known in his own country as Carl Follenius ..wrote some fine spirited Volkslieder. **1934** C. LAMBERT *Music Ho!* iii. 159 The second [*sc.* Glinka's wedding song in five-four time]..freed music from the restricting and lumbering rhythms of the German Volkslied. **1977** *Early Music* Apr. 213/1 Only two songs are true 'volkslieder' with the usual upbeat opening and steady chorale-like melody.

2. *S. Afr.* **a.** *Hist.* The national anthem of the 19th-century Transvaal Republic. **b.** A simple popular folk song.

1874 *Cape Monthly Mag.* 254 We should, in the pages of the *Magazine*, from time to time collect some of the ..'Volks-liedjes' written in the Cape Dutch patois. **1898** C. RAE *Malaboch* 217 *Volkslied*, the Transvaal National Anthem. **1921** *Eastern Province Herald* 18 Jan., The singing of the Volkslied and a vote of thanks to General Hertzog closed the meeting. **1954** H. GIBBS *Background to Bitterness* II. ix. 149 So they had ridden off, farmers, shop assistants, civil servants, lawyers, singing the *Volkslied* as they camped close to the southern frontier with Natal. **1961** *Sunday Times* (Johannesburg) 19 Mar. 10/7 The 16 girls and 14 boys in the choir will give Volksliedjie recitals and volkspele performances.

Volksraad. Add: **1.** Now, the House of Assembly of the Republic of South Africa. (Earlier example.)

1840 in J. C. Chase *Natal* (1843) II. 127 On the 14th [Feb. 1840], the Chief Commandant..caused the following Proclamation to be read..:— I, Andries Wilhelmus Jacobus Pretorius, Chief Commandant..of all the Burghers of the..Volks-Raad of the South African Society of Port Natal..&c &c &c.

2. *Netherlands East Indies.* A council advisory to the Governor-General.

1920 *Manual Netherlands India* (Admiralty) viii. 248 This council, under the name of the *Volksraad*.. is a representative body of 38 members. **1948** A. J. BARNOUW *Making of Mod. Holland* xi. 210 By an act of December 16, 1916, a *Volksraad* or People's Council was set up. **1974** *Encycl. Brit. Macropædia* IX. 488/1 At the end of World War I the Dutch..created the Volksraad (People's Council). Composed of a mixture of appointed and elected representatives of the three racial divisions defined by the government—Dutch, Indonesian, and 'foreign Asiatic'—the Volksraad provided opportunities for debate and criticism but no control over the government of the Indies.

volley, *sb.* Add: **1. c.** (Examples in *Physiol.*)

1928 *Jrnl. Physiol.* LXV 276 The rhythmic discharge is due to a more or less synchronous activity in a large number of the optic nerve fibres. The different ganglion cells have given up their usual independent fire of impulses and have taken to firing volleys. **1968** *Brit. Med. Bull.* XXIV. 253/2 Small localised changes in cortical potential, such as may follow the arrival of a sensory volley from the periphery, are completely lost in the background of much larger voltages produced by the rest of the cortical surface.

6. b. (Earlier example.)

1851 J. PYCROFT *Cricket Field* v. 79 At Woolwich he hit a volley to long field for *nine*.

7. volley-ball, for *U.S.* read 'orig. *U.S.*' and add: also **volleyball**; (examples); also *attrib.*; also, the ball used in this game.

1896 *Physical Education* V. 50/1 Mr. W. G. Morgan of Holyoke, Mass., has developed a game..which is called Volley Ball... The play consists in keeping a ball in motion over a high net,..thus partaking of the character of two games,—tennis and hand ball. **1936** P. FLEMING *News from Tartary* VII. i. 325 In the evening we played tennis with the Russians, or football or volley-ball with the Hunza guard. **1949** *Dziennik Zwiazkowy* (Chicago) 19 Nov. 6/3 Last year two of our volleyball teams finished in a tie for second place. **1976** *Milton Keynes Express* 4 June 39/7 Stoke Mandeville has told the club that any money raised for them will be used to purchase table tennis nets and volleyballs for the National Spinal Injuries Centre.

volley, *v.* Add: **1. c.** Also, *Association Football*, to kick (the ball) before it touches the ground; to score (a goal) in this way.

1909 in WEBSTER. **1972** G. GREEN *Great Moments in Sport: Soccer* viii. 85 He vollied [*sic*] in Bowen's clever lob with half an hour left. **1976** *Scotsman* 25 Nov. 25/1 In 52 minutes Somner volleyed a goal from McQuade's cross. **1976** *Norwich Mercury* 10 Dec., It was he who provided the perfect pass for Mann to volley the ball against the St. Andrews crossbar before the interval.

e. To fire a volley or volleys at.

1908 HARDY *Dynasts* III. VII. ii. 489 Kempt's brigade.. volleys murderously Donzelot's columns..and repulses them.

Volnay (vǫ·lnē[i]). Also 7 **Volne.** [Name of a commune in the department of Côte-d'Or, France.] A red wine of Burgundy, produced near Beaune (formerly also a white wine).

1699 M. LISTER *Journey to Paris* 161 Volne, a pale Champagne, but exceeding brisk upon the Palate...is said to grow upon the very borders of Burgundy, and to participate of the Excellency of both Counties. **1833** [see POMARD]. **1889** [see *CORTON]. **1961** A. WILSON *Old Men at Zoo* i. 48 'They've got two good burgundies here,' he said, 'this and the Volnay '57.' **1979** R. LEWIS *Violent Death* v. 140 A good lunch of roast beef with an excellent Volnay.

‖ **volost** (vǫ·lǫst). [a. Russ. *volost'*.] The smallest rural administrative subdivision in Imperial Russia and the U.S.S.R. (abolished in 1930).

1889 in G. N. CURZON *Russia in Central Asia* v. 114 There are forty *volosts*, or sub-districts, in the Merv circuit. **1920** B. RUSSELL *Pract. & Theory Bolshevism* I. v. 75, I asked in the villages how they were represented on the Volost (the next larger area) or the Gubernia. **1948** J. TOWSTER *Political Power in U.S.S.R.* iv. 66 The Baltic republics..are still divided into *uyezds* and *volosts* as in Imperial Russia. **1959** *Chambers's Encycl.* XIV. 102/2 The terms *okrug* (national enclave), *uyezd* (district) and *volost* (rural area) are obsolescent. **1974** *Encycl. Brit. Macropædia* XVI. 59/1 Kiselev..provided for a measure of self-government under which the mayor of the *volost* (a district grouping several villages or peasant communes) was elected by male householders.

volplane, *v.* (In Dict. s.v. VOLPLANE *sb.*) Add: Also *transf.* (esp. with bird as subj.) and *fig.* Hence **vo·lplaning** *ppl. a.*

1922 'KLAXON' *Heather Mixture* xi. 223 Two cocks [*sc.* pheasants] came over the brow.., and with steady wings volplaned down the line from the right. **1929** E. BOOTH *Stealing through Life* x. 243, I..volplaned down a flight of steps and hit the..sidewalk on my hands and knees. **1936** D. MCCOWAN *Animals Canad. Rockies* xx. 182 You may be fortunate enough to see one of these [*sc.* flying] squirrels as it volplanes gracefully from a high branch to the ground. **1937** *Discovery* Dec. 364/1 Volplaning marsupials. **1970** 'E. LINDALL' *Gathering of Eagles* v. 46 They could all..die together..watched by the eagles, seeing..the huge birds volplaning in.

Volscian, *sb.* **2.** (Earlier example.)

1859 B. W. DWIGHT *Mod. Philol.* I. 187 The Umbro-Samnite Dialects: Umbrian; Samnite or Oscan; Volscian; Marsian.

Volstead (vǫ·lstėd). The name of Andrew J. *Volstead* (1860–1947), American legislator, originator of the legislation to enforce prohibition (sense *4) which was passed in 1919 by the U.S. Congress, used *attrib.* to designate this legislation or the period during which prohibition was in force.

1920 *Current Opinion* Apr. 451/1 The Wet leaders..will wage a campaign for the election..of Congressmen favorable to changes in the Volstead Act. **1935** A. G. MACDONELL *Visit to Amer.* xiii. 229 The Volstead amendment led to some pretty confusion, but the repeal of the Volstead amendment seems to be almost worse. **1949** *Chicago Tribune* 8 Sept. 22/2 In the Volstead era, corrupt politicians talked to exactly the same effect with the leaders of the Anti-Saloon league. **1977** H. FAST *Immigrants* III. 179 The Congress of the United States was overriding President Wilson's veto of the Volstead Act and making Prohibition the law of the land. *Ibid.* 203 This is the finest wine country in the world, and some day the world is going to discover that—if we ever rid ourselves of this lousy Volstead thing.

Hence **Vo·lsteadism,** the policy of the Volstead Act; prohibition.

1920 *Harvey's Weekly* 27 Mar. 11/2 The Republicans would have to stand for Volsteadism or incur defeat. **1947** *Chicago Daily News* 23 Jan. 14/1 Volstead didn't write the law. Nor was he an apostle of the fanaticism called 'Volsteadism'.

volt (vōult), *v.*[2] *literary.* [f. the sb.] **a.** *trans.* To charge (something) as with electricity; to energize; to shock. **b.** *intr.* To travel like an electric current. Hence **vo·lted** *ppl. a.*

1930 R. CAMPBELL *Adamastor* 48 Dainty one, deadly one..Whose coils are volted with electric power. *Ibid.* 61 A starved mongrel... Fierce tremors volted through his bony notches. **1936** —— *Mithraic Emblems* 21 The volted ecstasy outglows A dolphin dying in the noon. **1936** *Times Lit. Suppl.* 28 Mar. 266/4 Give man the grace to find a firm abode,..Not in power's seat that volts the sitter dead. **1942** S. SPENDER *Ruins & Visions* II. 38 Driven by intolerance and volted with lies. **1962** N. COGHILL in Davis & Wrenn *Engl. & Medieval Stud.* 207 All that is finest and most central in this figure is 'made' by a coalescence or fusion of allegory, parable, and symbol, and that is the poetic fact that volts it with imaginative power.

voltage[2]. Add: **a.** (Later *fig.* examples.)

1949 E. POUND *Pisan Cantos* lxxx. 85 Whoi didn't he [*sc.* Padraic Colum] keep on writing poetry at that voltage. **1961** *Times* 24 Mar. 18/4 Walton's first symphony is a work of extraordinarily high voltage. **1978** F. OLBRICH *Desouza pays Price* xx. 125 Beaming the full voltage of her smile at Habib.

b. Also in *fig.* use.

1959 *Listener* 8 Oct. 590/1 Synge's brief but high-voltage career. **1966** N. FREELING *King of Rainy Country* 128 Her high-voltage emotions injected into the situation would doubtless tangle things still further.

c. *Comb.*, as *voltage doubler*, *-regulation*, *regulator*; *voltage-doubling* adj.; **voltage clamp** *Physiol.*, (the application of) a constant voltage maintained across a cell membrane by

artificial means; so **voltage-clamped** *a.*, **-clamping** *vbl. sb.*; **voltage-controlled** *a.*, controllable by varying the applied voltage; **voltage divider** a linear resistor or series of resistors which can be tapped at any intermediate point to obtain a voltage equal to a desired fraction of the voltage applied between its ends.

1952 *Jrnl. Physiol.* CXVII. 504 (*heading*) Mathematical description of membrane current during a voltage clamp. **1964** G. H. HAGGIS et al. *Introd. Molecular Biol.* vi. 159 These ionic permeability changes associated with the action potential have been studied in squid axons and frog muscle cells..with the very elegant 'voltage clamp' technique. **1979** *Acta Protozoologica* XVIII. 183 Instantaneous I/V plots of voltage-clamped deciliated cells lack the negative resistance property characteristic of ciliated cells. **1981** *Nature* 15 Oct. 517 Although the principle is simple enough—to keep the voltage across a cell membrane constant—the technology of voltage-clamping has become a sophisticated exercise in electrical engineering. **1962** SIMPSON & RICHARDS *Physical Princ. Junction Transistors* xiv. 359 External methods of gain control usually employ auxiliary diodes..to act as a voltage-controlled potential divider. **1976** *Electronic Engin.* Nov. 27/1 It [*sc.* a waveform generator circuit] can be used..in many other applications where a voltage controlled waveform is required. **1930** *Sci. Abstr.* B. XXXIII. 846 (Index), Voltage dividers, capacity and resistance types. **1932** *Bureau of Standards Jrnl. Res.* (U.S.) IX. 81 (*heading*) Theory of voltage dividers and their use with cathode ray oscillographs. **1983** *IEEE Trans. Instrumentation & Measurement* XXXII. 33/2 The attenuation reference standard in this system is a high class..inductive voltage divider. **1947** R. LEE *Electronic Transformers & Circuits* iii. 51 (*heading*) Voltage doublers. *Ibid.* (*caption*) Relation of peak sine voltage to d-c voltage in voltage-doubling circuit. **1956** *Nature* 18 Feb. 298/1 Electrical characteristics are next treated, including..descriptions of..voltage-regulation methods and apparatus. **1902** *Encycl. Brit.* XXVIII. 90/2 (*heading*) Voltage regulators. **1962** SIMPSON & RICHARDS *Physical Princ. Junction Transistors* 496 We consider a voltage regulator whose input voltage..and output voltage and current..are connected by the equation [etc.].

Voltaic, *a.* Add: **3. c.** *fig.*

1920 D. H. LAWRENCE *Lost Girl* iii. 45 A superhuman, voltaic force filled her.

Volta·ic, *sb.* and *a.*[2] [f. the name of the West African river *Volta* + -IC.] **A.** *sb.* **a.** A group of Niger-Congo languages of West Africa. **b.** A speaker of one of these languages; a citizen of the Republic of Upper Volta. **B.** *adj.* Of or pertaining to this group of languages, or to this republic or its citizens. Cf. *UPPER VOLTAN *a.* and *sb.*

1939 L. H. GRAY *Foundations of Lang.* xii. 402 The divisions of Sudano-Guinean, according to Delafosse, may now be enumerated:..Voltaic (fifty-three languages) ..apparently possessing inclusive and exclusive pronouns for the first person plural. **1949** L. HOMBURGER *Negro-African Languages* i. 22 The people who speak Voltaic languages number about 3,000,000. **1967** M. SCHLAUCH *Language* ii. 39 Suffixes are also used to indicate nominal classes (e.g. in Voltaic). **1969** [see *SENUFO]. **1970** P. OLIVER *Savannah Syncopators* 66 The use of a drone or continuous humming between sung phrases is also a vocal feature shared by Voltaic peoples and some blues singers. **1976** *New Society* 11 Nov. 293/3 Doing their military service by teaching in Voltaic schools. *Ibid.* 294/3 Educated Voltaics like to discount the CIA theory with one of their own. **1979** *Observer* (Colour Suppl.) 17 June 56/3 France for example in 1976 exported 64 million dollars worth of goods to Upper Volta (43 per cent of Voltaic imports). **1984** *Chicago Sun-Times* 9 Jan. 18 The government is treated with some skepticism by older Voltaics.

Voltairean, Voltairian, *sb.* and *a.* Add:

A. *sb.* β. (Earlier example.)

1842 J. S. MILL *Let.* 10 Jan. in *Wks.* (1963) XIII. 497 We are all either bigots or Voltairians.

B. *adj.* α. (Earlier example.)

1846 BP. OF MEATH in J. O'Connor *Hist. Ireland 1798–1924* (1925) I. viii. 249 Voltairean newspapers.

β. (Earlier example.)

1833 J. S. MILL *Lett.* (1910) I. 75 The little *feuilles* which one buys as one goes into a theatre, are the representatives of the Voltairian philosophy at present.

Voltairianism. Add: (Examples of form *Voltaireanism*.)

1901 *Q. Rev.* CXCIV. 599 Sir Henry Craik pounces with perhaps unnecessary severity upon the flippant Voltaireanism, the dapper self-assurance, and the 'slovenly omniscience' of Jeffrey and his allies. **1942** *Scrutiny* X. 400 Dostoyevsky's denunciation of the deracinated nobility who despised their own language and national character and lived in a froth of Voltaireanism and French *bons mots*.

Voltairish, *a.* For *rare*[−1] read *rare* and add further example.

a **1846** B. R. HAYDON *Autobiogr.* (1927) III. xiii. 252 '*Ah*,' shrugged out monsieur le valet, with a sparkling Voltairish look which every Frenchman has when convicted, as if planning a repartee, '*c'est très vrai*.'

voltammetry (vǫlt-, vōultæ·mėtri). *Physical Chem.* [f. VOLT *sb.* + AM(PÈRE + -METRY.] An electroanalytical technique for establishing the identities and concentrations of various ions in solution (see quot. 1978). Hence **voltamme·tric** *a.*, by means of or employing voltammetry; **voltamme·trically** *adv.*

1940 KOLTHOFF & LAITINEN in *Science* 16 Aug. 152/1 (*heading*) The voltammetric determination of oxygen. *Ibid.*, By the term 'voltammetry' we mean the determination and interpretation of current-voltage curves obtained in electrolysis experiments using a suitable microelectrode as an indicator electrode. *Ibid.* 153/1 Oxygen can also be determined voltammetrically by using a platinum wire microelectrode. **1976** *Nature* 8 July 146/1 An efflux of 5-hydroxyindole acetic acid..into the CSF was observed by the voltammetric monitoring. **1978** *Kirk-Othmer Encycl. Chem. Technol.* (ed. 3) II. 618 Voltammetry is the general procedure of studying the current-voltage behavior of a microelectrode in a system which also contains a large nonpolarizable reference electrode and a large excess of an electrolyte... Polarography is a special case of voltammetry in which the small, polarizable electrode is a capillary tube from which small mercury droplets emerge at regular intervals... The plot of current vs voltage..provides qualitative and quantitative information about the solution composition. **1983** *Jrnl. Chem. Soc. Dalton Trans.* 991 (*heading*) Electron-transfer processes and electrodeposition involving the iron hexacyanoferrates studied voltammetrically.

Volterra (vǫlte·rā). The name of a town in the Tuscany region of Italy, used *attrib.* to designate alabaster quarried there. Hence also **Volte·rran** *a.*

1924 D. H. LAWRENCE in M. Magnus *Mem. Foreign Legion* 20 So at last coming down to the Mercato Nuovo we saw little bowls of Volterra marble, a natural amber colour, for four francs. *a* **1930** —— *Etruscan Places* (1932) vi. 183 Everybody knows Volterra marble—so called—nowadays, because of the translucent bowls of it which hang under the electric lights, as shades, in half the hotels of the world. *Ibid.*, There is no love lost between a Volterran alabaster worker and the lump of pale Volterran earth he turns into pale marketable form.

volume, *sb.* Add: **I. 3. b.** (Earlier example).

1803 M. WILMOT *Let.* 3 May in *Russ. Jrnls.* (1934) I. 13 A sentimental story that speakes Volumes in favour of the Count and his Daughter.

IV. 12. Special Combs.: **volume control,** (*a*) control of the volume of sound, esp. when reproduced or transmitted; (*b*) a knob or other device for achieving this; **volume-density,** the number *of* anything per unit volume; **volume indicator** *Electronics*, a device for measuring the power of a complex electrical signal corresponding to a sound pattern, so as to indicate the volume of the sound that is represented; **volume table** *Forestry*, a set of empirically derived figures relating the volume of timber in a given type of tree or log to measurable parameters such as height and girth, thus enabling such measurements to be used in estimating timber volumes in the field; **volume unit** = VU, vu s.v. *V 5 b.

1927 *Star* 4 June 6/4 Volume Control... Noises, etc. in the loud speaker can be very easily reduced. **1931** T. H. PEAR *Voice & Personality* 78 It [*sc.* a preacher's voice] should be articulate, but with an efficient and graded volume-control. **1933** *Boys' Mag.* July 108/2 One of the all-important components..is the volume control. **1956** *B.B.C. Handbk.* 1957 100 They are also responsible for the volume control and technical quality of programmes leaving studios. **1977** 'E. CRISPIN' *Glimpses of Moon* viii. 152 Titty..was wearing the hearing-aid, so Ling..addressed himself..to her, while she fiddled with volume control. **1956** *Nature* 4 Feb. 226/1 These photographs have been useful for determining..the area-density and volume-density of the flying locusts. **1968** R. A. LYTTELTON *Mysteries Solar Syst.* v. 155 The time that an individual particle would take to cross the cylinder..would be of the order of $2s/V$, and for this time the particle would be contributing to the average volume-density within the cylinder. **1923** *Trans. Amer. Inst. Electr. Engineers* XLII. 77/2 There has been developed a device which is called a 'volume indicator'. This consists of an amplifier detector working into a direct-current meter. **1961** G. MILLERSON *Technique Television Production* i. 16 His [*sc.* the sound mixer's] attention is divided mostly between the flickering needle of his volume indicator and his picture monitor. **1895** W. SCHLICH *Man. Forestry* III. 39 (*heading*) Estimate of volume by means of volume tables. **1902** *Forestry Q.* I. 6 The Bavarian government, in 1846, instituted a very extended study..of the stems of the more important forest trees of that country. The volume tables which resulted..involved a complete analytical measurement of over forty thousand trees. **1981** *Southern Jrnl. Appl. Forestry* V. 186/2 Users of volume tables are cautious about applying them outside the region from which they were derived, because the effects of geographic location are unknown. **1940** *Chambers's Techn. Dict.* 897/1 *VU*, the number of volume-units above or below zero power-level..indicated by the standardised volume-indicator. **1960** *McGraw-Hill Encycl. Sci. & Technol.* XIV. 374/2 Meters which give readings in volume units, called VU meters, are widely used for monitoring radio broadcasts and for sound recording.

volumeless (vǫ·liumlės), *a.* [f. VOLUME *sb.* + -LESS.] Occupying no volume; *spec.* applied to an idealized polymer chain having this property.

1946 *Nature* 26 Oct. 571/1 An ideally flexible and volumeless [polymer] chain. **1967** MARGERISON & EAST *Introd. Polymer Chem.* i. 17 Certain configurations open to the 'volumeless' chain are not available to the real polymer chain. **1974** *Jrnl. Polymer Sci.: Polymer Symposia* No. 46. 97 An expression for Γ, suggested by a well known equation for a regular cubic network of freely jointed volumeless chains, is derived. **1983** [see *VOLUMINAL *a.*].

volumetric, *a.* Add: (Earlier example.) *volumetric efficiency* (Mech.), the ratio of the volume of fluid actually displaced by a piston or plunger to its swept volume.

1857 H. E. ROSCOE tr. *R. Bunsen's Gasometry* 170 The saturated solution..becomes sufficiently diluted..to allow the ammonia to be determined by a volumetric analysis with sulphuric acid. **1912** *Proc. Inst. Automobile Engineers* VI. 78 In the writer's opinion, volumetric efficiency is almost entirely a function of port area. **1930** *Engineering* 11 July 31/2 The standard valve tuning of the E.35 engine is arranged for the best possible volumetric efficiency at 1,500 r.p.m. **1973** A. PARRISH *Mech. Engineer's Ref. Bk.* II. 51 Leakage tends to be greater than in a reciprocating design but as they [*sc.* rotary positive displacement pumps] are often used for viscous fluids such as oil, this effect may be reduced to give a volumetric efficiency of 95%.

volumetrically, *adv.* (Earlier example.)

1863 W. O. MARKHAM tr. *Neubauer & Vogel's Anal. Urine* 305 The determination of the albumen volumetrically by means of ferrocyanide of potassium.

voluminal (vǒliū·minăl), *a.* [f. L. *volūmin-*, *volūmen* (see VOLUME *sb.*) + -AL.] Of, pertaining to, or possessing volume.

1872 *Proc. R. Soc.* XX. 5 The remarkable phenomena of the voluminal conditions at and near the critical point of temperature and pressure. **1925** J. JOLY *Surface-Hist. Earth* vii. 113 A small upward displacement must then be attended with voluminal expansion. **1983** *Nucl. Instruments & Methods* CCV. 211/1 Many works..have been published for volumeless sources such as point isotropic and beam sources. But only a few papers deal with voluminal sources in which self-absorption and self-scattering of gamma rays cannot be disregarded.

voluntarism. Add: **3.** orig. *U.S.* **a.** The principle of relying on voluntary action rather than compulsion; *spec.* with reference to political and trade-union activities. Cf. VOLUNTARYISM 2 in Dict. and Suppl.

1924 S. GOMPERS in *Rep. Proc. 44th Ann. Convention Amer. Federation Labor* 5/2, I want to urge devotion to the fundamentals of human liberty—the principles of voluntarism. **1948** *Sun* (Baltimore) 6 Dec. 12/3 There is reason to think that voluntarism is preferable to compulsion on racial matters as on everything else. **1973** *Times* 15 Jan. 18/6 From an almost exclusive reliance on 'voluntarism', *ie.* the promotion of negotiating procedures drawing at most indirectly upon law, Great Britain has now imposed upon the conduct of unions and employers more formal..regulations. **1983** *Listener* 22 Sept. 5/1 So much for voluntarism, the golden rule of the British TUC.

b. Denoting the involvement of voluntary organizations in social welfare. Also *spec.* in U.K. use = *VOLUNTEERISM 2.

1957 *Observer* 8 Sept. 15/4 This is the essence of the American economic achievement to-day: an emphasis on the importance of the individual... And an increasing knowledge of the techniques by which men can co-operate in 'free collectivism'—voluntary private organisation. Voluntarism in our social structure is tremendously important. Without it the whole system would break down. **1969** *Wall St. Jrnl.* 21 Mar. 1/1 'Voluntarism'—President Nixon's program to enlist the help of private groups in solving social problems. **1979** *Guardian* 10 Aug., Voluntarism is one of the central characteristics of a number of organisations with which the Prince of Wales has specific connections. **1981** *Times Lit. Suppl.* 7 Aug. 911/2 When their representatives attained some municipal power, they clung to a weedy sort of voluntarism and missed their chance of creating active local democracy. **1981** *Times* 7 Sept. 2/4 The Government's emphasis on the role of volunteers in providing social services could be counterproductive... Some Labour councils connected the twin themes of public spending cuts and the strong emphasis on voluntarism. **1982** *Chicago Sun-Times* 31 Aug. 30 Bernardin could be the key to the rebirth of a larger sense of community—at a time when its implicit sense of voluntarism would be in keeping with the Reagan era and with the voluntaristic spirit alive today. **1984** *Listener* 14 June 12/2 We are going back to reliance on 'voluntarism' in this vital field.

voluntaristic, *a.* Add: **2.** Pertaining to voluntary action.

1961 B. R. WILSON *Sects & Society* I. iv. 89 Where voluntaristic activities are concerned the sectarian's feud with society remains latent. **1969** A. STEWART in Ionescu & Gellner *Populism* 194 Populist movements which are basically protest movements are threatened by the anti-institutional, voluntaristic character of grass-roots populism. **1974** B. PEARCE tr. *Amin's Accumulation on World Scale* I. 29 Being voluntaristic in character, development policy draws upon new techniques of economic planning in order to work out the series of choices involved. **1977** *Dædalus* Summer 72 The obligatory component in social relationship yields to the optional or voluntaristic, status gives way to contract. **1982** [see *VOLUNTARISM 3 b].

voluntary, *a.* Add: **1. b.** *Voluntary Service Overseas*, an organization promoting voluntary work by young people (in education, social welfare, etc.) in developing countries; the service so offered or the scheme itself.

1960 *Voluntary Service Overseas* 1 Voluntary Service

Overseas enables as many as possible of these young people to have this opportunity—and, in meeting the needs of others, to deepen their own experience. *Ibid.*, Governments and agencies overseas are asking for volunteers to serve as temporary auxiliaries in many fields—social welfare, schools, youth clubs... It is in response to these requests that Voluntary Service Overseas is sending selected volunteers. 1964 M. Dickson *World Elsewhere* 11 In September 1958 ten young men left Britain for Sarawak. Three flew to Nigeria and two set off for Ghana. All were eighteen years old... They were the spearhead of the scheme which was Voluntary Service Overseas. 1965 *Listener* 7 Jan. 21/2 One finds British young people doing voluntary service overseas in all sorts of out-of-the-way places.

7. d. *voluntary patient*, one who enters a mental hospital without being committed to it.

1930 *Daily Express* 6 Sept. 9/4 Instructions sent by the Board of Control to local authorities regarding the working of the Mental Treatment Act (1930) stipulate that 'mental hospital' is to be substituted for 'asylum', and 'voluntary patient' is to be used instead of 'voluntary boarder'. 1943 G. Greene *Ministry of Fear* iii. i. 167 If only someone would complain—they are all voluntary patients. 1979 J. Thomson *Deadly Relations* xiii. 189, I had a nervous breakdown... I..was sent to a clinic..as a voluntary patient.

9. a. (Later examples.) *spec.* in *Educ.* with reference to schools, etc., maintained by voluntary bodies.

1944 *Act 7 & 8 Geo. VI* c. 31 § 8 Primary and secondary schools maintained by a local education authority, not being nursery schools or special schools, shall, if established by a local education authority..be known as county schools and, if established otherwise than by such an authority, be known as voluntary schools. 1969 L. Tinkham in Cockburn & Blackburn *Student Power* 84 There are now about one hundred Local Education Authority colleges and half as many independent voluntary colleges. 1976 *Star* (Sheffield) 29 Nov. 9/1 Pupils will be transferred to the Perlethorpe Church of England Voluntary Aided Primary School.

voluntaryism. Add: **2.** (Later examples.) Now usu. with reference to voluntary labour. Cf. *volunteerism 2.

1946 *Organisation & Finance Adult Educ.* (Min. Educ.) 36 We are unanimously of the opinion that voluntaryism as exemplified by the Workers' Educational Association is essential if the spirit of adult education is to be preserved. 1967 *Listener* 1 June 709/2 In the light of the present arguments about the prices and incomes policy, Bevin's insistence on what he called 'voluntaryism' is most significant... Bevin resisted strong pressures in parliament to conscript labour on the home front.

volunteer, *sb.* and *a.* **A.** *sb.* **4.** Delete † *Obs.*⁻¹ and add: A self-sown plant. (Later examples.)

1960 *Jrnl. Forestry* LVIII. 402/3 The stand was planted on a 6×6-foot spacing, with some interspersed volunteers. 1978 *New Yorker* 3 July 42/1 Around the buildings ..are some of the tallest volunteers in New York, top-heavy plants.

B. *attrib.* or as *adj.* **1. d.** *Volunteer State*, a nickname for Tennessee (see quot. 1950).

1853 J. C. M. Ramsey *Ann. Tennessee* 116 Thus early did the 'Volunteer State' commence its novitiate in arms. 1950 *Newsweek* 20 Mar. 96/2 A call for 2,800 volunteers [in the Mexican War of 1847] in Tennessee brought out 30,000 men and gave Tennessee its nickname, 'The Volunteer State'. 1973 *Guardian* 14 June 13/2 There was a spectacular..murder deep in the hills of Tennessee..as could only happen in the deepest by-ways of the Volunteer State.

volunteer, *v.* Add: **5.** (Earlier examples.)

1805 Jane Austen *Let.* 27 Aug. (1952) 166 She volunteers, moreover, her love to little Marianne, with the promise of bringing her a doll. 1813 —— *Let.* 14 Oct. (1952) 354, I talk to Cassy about Chawton; she remembers much but does not volunteer on the subject.

6. (Earlier *fig.* example.)

1814 Jane Austen *Mansfield Park* II. ix. 200 Thursday ..opened with more kindness to Fanny than such.. unmanageable days often volunteer.

volunteered, *ppl. a.* (Earlier examples.)

1845 *Times* 1 Nov. 4/4 The members of the press retired..from the hall, into which they had been invited by the volunteered cards of admission from the benchers. 1879 Geo. Eliot *Theophrastus Such* i. 7 The fellow-feeling which should restrain us from turning our volunteered and picked confessions into an act of accusation against others.

volunteerism. Add: **2.** *N. Amer.* The use of volunteer labour, esp. in the social services. Cf. *voluntarism 3 b, *voluntaryism 2.

1977 *New Yorker* 1 Aug. 48/2 Still another productivity proposal is that as the city work force shrinks, volunteerism should be encouraged; citizen volunteers could serve as auxiliary policemen, park attendants, caseworkers, and school aides, for instance. 1979 *Globe & Mail* (Toronto) 28 May S14 A closer examination of voluntary action or 'volunteerism' as it relates to volunteer support staffs in amateur sport. 1983 *United Airlines Mag.* June 48 Minnesota's Corporate Volunteerism Council, a clearinghouse designed to help match community needs with people willing and able to give some time to others.

‖ **volupté** (volüpte). [Fr.: see Volupty.] = Voluptuousness.

1712 M. W. Montagu *Let.* 12 Aug. (1965) I. 135 All things would have contributed to make your Life passe in (the true volupté) a smooth Tranquillity. 1937 L. Bromfield *Rains Came* iii. xxxvi. 435 The lingam, symbol of creation, of *volupté* and strange desires and pleasures. 1962 W. Nowottny *Lang. Poets Use* viii. 216 The elements of *volupté* and masochism in a fervent devotional tradition. 1981 *Times Lit. Suppl.* 13 Feb. 158/3 The narrative's need to shock, its dedicated oscillatings between *volupté* and vomit, are adolescent.

volupty. Delete † *Obs.* and add later example.

1929 S. Leslie *Anglo-Catholic* xv. 216 Edward could see every brick in position with that strange microscopic vision which makes seeing a volupty of the eye.

volutin (vǫ·liŭtin). *Biochem.* Also † **-ine.** [a. G. *volutin* (A. Meyer *Practicum der bot. Bakterienkunde* (1903) xiii. 80), f. L. *volūt-ans* (pres. pple. of *volūtāre* to roll), specific epithet of the bacterium *Spirillum volutans*: see -in[1].] A basophilic compound containing polyphosphate found in metachromatic granules in the cytoplasm and vacuoles of various micro-organisms and fungi.

1908 *Q. Jrnl. Microsc. Sci.* LIII. 298 The granules ('metachromic granules', 'red granules', 'volutine granules', etc.) probably consist of some reserve material. 1947 *Nature* 11 Jan. 63/2 Volutin is a cytological entity which has long been recognised as an important constituent of the yeast cell. 1970 Passmore & Robson *Compan. Med. Stud.* II. xviii. 14/1 Volutin granules.. have a special significance in the genus *Corynebacterium* since they are produced typically by the diphtheria bacillus.

volva (vǫ·lva). [ad. ON. *vǫlva* (also used) (Norw. *volve*) prophetess, sibyl.] In Scandinavian mythology, a prophetess, a soothsayer, a witch.

1889 P. B. Du Chaillu *Viking Age* I. iv. 27 The Völuspa was an inspired poem of a Völva or Sibyl, and embodies the records of the creation of the present world. 1927 E. V. Gordon *Introd. Old Norse* v. 40 It is the most complete description extant of the vǫlva or sibyl of Scandinavian heathen times. 1966 S. Wavell et al. *Trances* xvi. 133 The priestesses of Freyja known as *volva* were the shamans of their time. They wore animal skins, their boots made of calf skin, their gloves lined with the fur of arctic cats, their goddess being drawn in a cat carriage.

volve, *v.* Restrict † *Obs.* to senses 2 and 3 and add: **1.** (Later example.) *rare.*

1910 Galsworthy *Inn Tranq.* in *Nation* 12 Nov. 266/2 If It did not volve and revolve on Itself, It would peter out at one end or the other.

vomit, *sb.* Add: **6.** *Comb.*, as *vomit-green* adj.; **vomit bag** = *sick-bag* s.v. *sick *a.* 12.

1975 D. Lodge *Changing Places* i. 42 The passengers swallow to relieve the pressure on their eardrums, close their eyes, finger their passports and vomit-bags. 1978 D. Murphy *Place Apart* i. 6 The sleazy plastic Lounge Bar had a vomit-green carpet.

vomit, *v.* **4. b.** (Later examples.)

1922 Joyce *Ulysses* 750 If there was a row on you vomit a better face. 1946 E. O'Neill *Iceman Cometh* iii. 182 You'll be saying something soon that will make you vomit your own soul like a drink of nickel rotgut that won't stay down! 1955 E. Bowen *World of Love* x. 185 Bent in two, she vomited laughter; though also, mortified by the exhibition, she let out penitent sobs and moans.

vomiter. 1. (Later example.)

1962 'R. Gordon' *Doctor in Swim* xxiii. 160 'Also,' he added, 'one is being sick into the swimming pool.' 'All right,' I told him wearily. 'I'll do my best.' Perhaps we'd better start with the little vomiter,' I suggested, putting professional things first.

vomitorial, *a.* Restrict *rare*⁻¹ to sense in Dict. and add: **2.** = Vomitory *a.* (in quot. *fig.*). *rare.*

1868 Swinburne *Let.* 1 Jan. (1959) I. 284 What abject and vomitorial rot of Tennyson's is this in Once a Week!

vomitorium. Add: **2.** *erron.* A room in which ancient Romans are alleged to have vomited deliberately during feasts.

1923 A. Huxley *Antic Hay* xviii. 252 There strode in, like a Goth into the elegant marble vomitorium of Petronius Arbiter, a haggard and dishevelled person. 1965 R. Eberhart *Sel. Poems* 40 Good Boy Man! Your innards are put out, From now all space will be your vomitorium.

vo·mitous, *a.* *U.S.* [f. Vomit *sb.* + -ous.] Repugnant, loathsome, nauseating.

1952 S. Kauffmann *Tightrope* xiv. 241 What kind of world is this..having to trudge back from bad restaurants in sickening heat to face mountains of vomitous, meaningless, commercial tripe? 1976 *Publishers Weekly* 24 May 57/2 Those descriptions of methods of murdering human beings as dishes for the table end up being more than a little vomitous.

vomitus (vǫ·mitǝs). [L.] = Vomit *sb.* 2 a. Also *fig.*

1904 J. M. French *Text-bk. Pract. Med.* iii. 724 The presence of HCl, pepsin, and rennet is generally sufficient evidence that the vomitus has come from the stomach. 1948 *Diagnostic Procedures for Virus & Rickettsial Diseases* (Amer. Public Health Assoc.) 11 The material available for examination may be blood, sputum, vomitus, or pleural fluid obtained during illness, or tissues, etc., collected at autopsy. 1973 *Nature* 23 Mar. 265/1 Stools, urine and vomitus were collected for a total of 48 h. 1978 J. Updike *Coup* (1979) v. 211 Fighting down the vomitus of superstitious terror rising in my craw.

vo·mity, *a.* *U.S.* [f. Vomit *sb.* + -y[1].] Redolent of vomit.

1951 J. D. Salinger *Catcher in Rye* xii. 106 The cab I had was a real old one that smelled like someone'd just tossed his cookies in it. I always get those vomity kind of cabs if I go anywhere late at night. 1967 I. A. Baraka in W. King *Black Short Story Anthol.* (1972) 125 Oh, I am drunk and vomity in my room, with only Charley Ventura to understand my grace. 1984 *Listener* 20–27 Dec. 68/3, 80 per cent of the material is American, vomity kind of lampoons of great American institutions such as Advertising and the Presidency.

von Recklinghausen's disease (fǫn re·kliŋ,hauzǝn). *Path.* [Named after Friedrich *von Recklinghausen* (1833–1910), German pathologist.] **1.** A familial disease in which numerous neurofibromas develop on various parts of the body, esp. the skin, the nerve trunks, and the peripheral nerves (described by von Recklinghausen in 1882). Also **Recklinghausen's disease.**

1899, etc. [see *neurofibromatosis]. 1900 H. A. Thomson *Neuroma & Neuro-Fibromatosis* iv. 55 There is not even a good general name for the group which will include all its members, unless we adopt that suggested by certain French authors, viz.—'von Recklinghausen's Disease'. 1935 [see *neurilemoma, neurilemmoma]. 1969 S. Jablonski *Illustr. Dict. Eponymic Syndromes & Dis.* 256/1 (*caption*) Recklinghausen's disease. 1976 *Path. Ann.* XI. 371 A total of 15 schwannomas were studied ultrastructurally. Of these, two were classified as neurofibromas by light microscopy and were associated with von Recklinghausen's disease.

2. A disease in which bones are weakened by diffuse resorption and fibrous replacement of the bone substance as a result of hyperparathyroidism, leading to bowing of long bones and sometimes deformities of the chest and spine (described by von Recklinghausen in 1891); osteitis fibrosa cystica.

1910 *Ann. Surg.* Aug. 163 Crile and Hill interpret..as multiple giant-cell sarcoma, but it seems to me that it belongs to the group of Von Recklinghausen's disease. 1949 *New Gould Med. Dict.* 862/1 Generalized osteitis fibrosa cystica is called von Recklinghausen's disease of the bone. 1966 Wright & Symmers *Systemic Path.* II. xxxii. 1129/1 Some forty years elapsed before it was recognised that von Recklinghausen's disease of the bones was..directly due to an excess of circulating parathyroid hormone. 1984 Tighe & Davies *Pathology* (ed. 4) xxi. 209 These changes produce the 'brown tumours' of osteitis fibrosa cystica (von Recklinghausen's disease of bone) which may be misdiagnosed as giant cell tumours of bone (osteoclastoma).

vonsenite (vǫ·nsĕnait). *Min.* [f. the name of Marcus *Vonsen* (1879–1954), U.S. amateur mineralogist who collected the original material + -ite[1].] A borate of ferrous and ferric iron, $Fe^{II}_2Fe^{III}BO_5$, that occurs as black, lustrous, orthorhombic crystals and is the magnesium-free end-member of a series with ludwigite.

1920 A. S. Eakle in *Amer. Mineralogist* V. 143 Its distinctive difference in composition, and its manifest difference in structural and optical characters from ludwigite, justifies the writer in proposing the new name vonsenite. 1976 *Mineral. Abstr.* XXVII. 279/2 The first Spanish occurrence of vonsenite, end-member of the magnesium and iron borate series, is reported from the 'Monchi' mine [Badajoz].

von Willebrand (von wi·lǝbrænd). *Path.* The name of E. A. von *Willebrand* (1870–1949) Finnish physician, used *attrib.* and in the possessive to designate a hereditary disease described by him that is characterized by prolonged bleeding, chiefly caused by abnormalities of the capillaries and mucous membranes.

1941 *Q. Jrnl. Med.* XXXIV. 18 It [*sc.* athrombocytopenic purpura] has also been called Von Willebrand's disease, Glantzmann's disease, and 'hemogenia'. 1966 Wright & Symmers *Systemic Path.* I. iv. 190/1 In von Willebrand's disease the platelets are morphologically and functionally normal, but the capillaries show both structural and functional changes. 1977 [see *replacement therapy* s.v. *replacement 3].

voodoo, *sb.* Add: Also **vodoo.** **3.** *voodoo adorer, king, queen.*

1868 *De Bow's Rev.* Aug. 724 But may not the agent be met, then, by some such valid objection, on the part of the *Vodoo* adorers? 1872 *New Orleans Times* 28 June, Soon there arrived a skiff containing ten persons, among which was the Voudou queen. 1974 A. Murray *Stomping Blues* ii. 10 Some specific..charm, or talisman, which can be counteracted only with the aid of a voodoo queen or madam (or somewhat less often, a voodoo king, doctor, witchdoctor, or snakedoctor).

voodooist (vū·dūist). [f. Voodoo *sb.* + -ist.] A practitioner of voodoo. Also *fig.*

1929 W. B. Seabrook *Magic Island* 289 The connection which Haitian Voodooists believe exists between Dangbé and Ayida-Ouédo. 1935 *Sun* (Baltimore) 2 Jan. 10/1 Full payment of the bonus will be accomplished..by the addition of the strength of the miracle workers in finance, the monetary voodoists, to the strength of the veterans' bloc. 1958 *Times* 18 Sept. 13/2 On the one side are set a young American officer, his wife..and others of his race, and, on the other, well, the others: the dark, the alien, the voodooists. 1966 G. Greene *Comedians* i. ii. 57 Joseph was a good Catholic as well as a good Voodooist.

voom (vŭm), *int.* *U.S.* [Echoic.] Indicating the sound of an explosion; usu. in *fig.* contexts. Occas. redupl. as *sb.*, the roar of an engine being revved. Cf. *vroom.

1964 in Hamblett & Deverson *Generation X* 115 Ideally, according to them, young people should be seen and not heard. The moment they do something to draw attention to themselves, voom: trouble. 1968 S. J. Perelman in *New Yorker* 23 Nov. 58/1 The inspiration came to me—voom. 1978 *New York* 3 Apr. 56/1 The screeching car brakes, honking horns, the voom-voom of motorcycles, and the thundering of First Avenue buses exhausted me. 1985 *Observer* 12 May 1/1, I thought someone had let off a smoke bomb, then voom it all went off.

‖ **voorbok** (fūᵊ·ɹbǫk). *S. Afr.* Pl. **voorbokke.** [Afrikaans.] A goat which acts as bell-wether to a flock of sheep. Also *fig.* Cf. Bell-wether 2.

1913 C. Pettman *Africanderisms* 540 *Voorbok*... A goat ..is generally used on South African sheep-farms, instead of a bell-wether as in England. 1947 *Cape Argus* 29 Mar. 6 Many English-speaking South Africans regarded him as a man wielding a moderating influence upon racial politics. Why then should he not be useful as a bell-wether ('voorbok') leading United Party sheep into the Herenigde kraal? 1951 *Cape Times* 15 Aug. 2/6 A delivery van ran into a flock of sheep.., killing 25 sheep and the voorbok. 1972 *Daily Dispatch* (East London) 6 May 10 It reminds me of days long ago when farmers trekked their sheep from the Free State to the better winter grazing of the Natal lowveld. Each flock was led by a goat, the voorbok. The sheep..would never cross a river or enter a gate unless led by the more clever voorbok. 1976 *Evening Post* (Port Elizabeth) 20 Nov., The woman..was described in court as one of the 'voorbokke' of the riots who incited children to throw stones at White people's cars.

‖ **voorhuis** (fūᵊ·ɹhōis). *S. Afr.* [Cape Du., a. Du. *voorhuis* fore-part, hall of a house.] In a Cape Dutch house, an entrance hall, or a room into which the front door opens; a front room or sitting-room.

1822 W. J. Burchell *Trav. Interior S.Afr.* i. vi. 118 At about half an hour after nine, all retired to rest; some to a mat on the floor in the *voorhuis* (entrance-room, or hall). 1867 E. L. Layard *Birds S. Afr.* 118 Perhaps one or two would have found their way into the *voorhuis*, or entrance-hall. 1923 [see *reimpje, reimpie]. 1946 *Cape Times Week-end Mag.* 16 Nov. 4 The thatched cottage..had only two rooms; a *voorhuis* and bedroom. 1981 *N. & Q.* June 193/2 The *voorhuis* was the large central space or hall, often running right through the house, off which the other rooms opened.

‖ **voorkamer** (fūᵊ·ɹkāmǝɹ). *S. Afr.* [Afrikaans.] = prec. Occas. partially anglicized as *fore-kamer*.

1775 S. Pigot *Let.* Aug. in 'V. M. Fitzroy' *Dark Bright Land* (1955) i. 16 Papa and Leonora..are at Whisk in mijnheer's fore-kamer as they style it. 1827 G. Thompson *Trav. & Adv. S. Afr.* (ed. 2) I. i. iii. 49, I slept this night in the outer apartment (voorkamer) or sitting-room of the house. 1896 *Cape Argus* 2 Jan. 5 The flash entered at the front doorway, and the shock was felt by all the occupants of the *voorkamer*, fourteen in number. 1940 'B. Knight' *Piping on Wind* iv. xx. 298 The voorkamer had been cleared of its furniture for dancing. 1981 *N. & Q.* June 193/2 The sitting-room or *voorkamer* appears to have been most often one of the smaller rooms, and with a fireplace.

voorlooper. Add: Now usu. **voorloper.** (Earlier and later examples.) Also *fig.*

1837 J. E. Alexander *Western Afr.* I. xiii. 323 Then a long wagon would pass..drawn by a span of ten or fourteen oxen under the guidance of a *voorlooper*, a brown boy, holding occasionally a small rope attached to the horns of the leading bullocks. 1899 G. H. Russel *Under Sjambok* 50 But, have you not got me a driver and voorloper? 1947 *Cape Argus* 22 Feb. 4 We in the pilot train are at first sight everywhere mistaken for the royal train itself. In time..the fact that we are merely the 'voorloper' of royalty will become more generally known. 1974 *Weekend Post* 21 Dec. 2 Clandestine trips were made across the Great Fish River, and it was on these that George Wood became a 'voorloper' of the oxen.

‖ **voorskot** (fūᵊ·ɹskǫt). *S. Afr.* [Afrikaans, f. *voorskiet* to advance (money).] The advance payment for a crop, wool clip, etc., by a farmers' co-operative society or similar body to its members. Also *transf.* Cf. *agterskot.

1948 *Cape Times* 22 Nov. 14 The Land Bank had fixed the *voorskot* price for first grade lucerne at 3s. for 100 lb. 1958 *Ibid.* 6 Nov. 1/1 Importers of raw materials and machinery would..receive 33⅓ per cent. of 1958 issues as *voorskot*. 1961 *Ibid.* 15 Feb. 9/7 A figure of R50m. is being mentioned as a *voorskot* for Bantu homelands. 1974 *Eastern Province Herald* (Port Elizabeth) 9 Sept., The Voorskot for the 1974/75 wool season has been fixed at an average of R1.50 a kilogram for clean wool.

‖ **voorslag** (fūᵊ·ɹslaχ). [Afrikaans.] The lash of a whip.

1833 *Graham's Town Jrnl.* 4 Apr. 3 He then took his whip and in endeavouring to frighten them began using the 'voor slach' of the whip. 1852 C. Barter *Dorp & Veld* v. 43 Putting a new *voorslag* (lash) to the wagon-whip, that its smack might be clear and loud. 1910 J. Buchan *Prester John* iii. 59 He roared with laughter at my way of tying a *voorslag*. 1939 'D. Rame' *Wine of Good Hope* i. v. 71 The great twelve foot driving whip with its thong of eland hide..and its thin, cutting voorslag. 1973 *Farmer's Weekly* (Suppl.) 30 May 37 Voorslags 86c doz.

Vopo (fōᵘ·po). [a. Ger., f. *volkspolizei* people's police.] (A member of) the *Volkspolizei* (see *Volk 3 b).

1954 *Ann. Reg. 1953* 216 Members of the 'people's police', or *Vopos*, threw away their weapons. 1959 *News Chron.* 9 July 4/6 The Vopo—the People's Police. 1966, 1967 [see *Grepo]. 1978 L. Heren *Growing up on The Times* vi. 237 The young Vopo returned with a sergeant.., who returned our passports with an apology.

voraciousness. (Later example.)

1974 R. Helms *Tolkien's World* iv. 80 The Orcs are covered with hair, in part to represent their sexual voraciousness and animality.

vorlage (fōᵊ·ɹlagǝ). [a. Ger. (both senses). Cf. G. *vorlegen* to lean forward.] **1. a.** *Skiing.* (see quot. 1939.) **b.** *pl.* Skiing trousers. Also **vorlagers.**

1939 Webster Add., *Vorlage*... A position in which one leans forward from the ankles without lifting the heels from the skis, keeping the body, as a rule, at least perpendicular to the slope. 1958 *Vogue* Jan. 1 (Advt.), The Winter Sports Shop..has everything for the skier:.. Luis Ebster designs and cuts the vorlages. *Ibid.* 35/1 Tattersall checks of black and caramel match caramel vorlagers. 1959 H. Lantschner *Ski-ing for Beginners* 89 One must..have the necessary Vorlage to take the weight off the skis when necessary. 1961 *Times* 27 Nov. 13/2 'Tailormade' vorlagers in a large range of sizes. 1963 C. Glyn *Don't knock Corners Off* xxii. 184 Our beanies, vorlages, ankle-length tomato coloured pants.

‖ **2.** An original version of a manuscript or a book from which a copy is produced.

1965 K. Malone in Bessinger & Creed *Medieval & Linguistic Stud.* 120, I conceive that our scribe copied as *heol* the *hleo* of his *vorlage*. 1975 *Times Lit. Suppl.* 25 Apr. 462/4 This was first published in a French translation..in 1930 and was not followed by the Amharic original till more than thirty-five years later... More than one Amharic original was in existence, and the published text was not the *Vorlage* used for the 1930 French edition.

vorlaufer (fōᵊ·ɹlaufǝɹ). [ad. G. *vorläufer* precursor, forerunner, f. *vorlaufen* to run on ahead.] A skier who travels a course before a race in order to establish a standard by which the competitors are marked.

1961 *Times* 11 Jan. 16/1 As a result of the *vorlaufers's* times, the announcer..foresaw a winning time in the region of '1·55, 1·60'. 1964 *Times* 4 Feb. 4/1 Miss T. Hecher..had followed down the *vorlaufers*.

vorpal (vǭ·ɹpăl), *a.* A word invented by 'Lewis Carroll' app. with the sense 'keen, deadly'; also in subsequent allusive uses.

1871 'L. Carroll' *Through Looking-Glass* i. 22 The vorpal blade went snicker-snack! 1941 Auden *New Year Let.* ii. 44 Wave at the mechanized barbarian The vorpal sword of an agrarian. 1970 [see *frabjous *a.*].

‖ **vorspiel** (fōᵊ·ɹʃpīl). *Mus.* Pl. **-e.** [Ger., f. *vor* before + *spiel* (see *spiel *sb.*[2]).] A prelude.

1876 Stainer & Barrett *Dict. Mus. Terms* 452/2 *Vorspiel* (*Ger.*), prelude, introductory movement, overture. 1889 G. B. Shaw *London Music 1888–89* (1937) 66 It was not surprising to find a charm in this 'Vorspiel' that is wanting in the empty and violently splendid overture to Rienzi. 1895 G. Kobbé *How to understand Wagner's Ring* (ed. 6) 66 This vorspiel is a masterly representation..of a storm gathering. 1974 *Encycl. Brit. Macropædia* XIII. 680/1 By the time of Bach, the written *Choral-Vorspiel*, or chorale prelude (as opposed to the improvised prelude to the chorale within the Lutheran service), prelude and fugue..were firmly established forms for the organ. 1983 *New Oxf. Compan. Mus.* II. 1949/2 *Vorspiel* (Ger.), 'prelude'. Wagner called the introductions to his operas *Vorspiele*, and described *Das Rheingold* as a *Vorspiel* to the *Ring*.

‖ **vorstellung** (fōᵊ·ɹʃteluŋ). *Psychol.* Pl. **-en.** [Ger.] An image, idea, mental picture, or presentation.

1807–8 Coleridge *Notebks.* (1962) II. i. 3217 Emptiness & absence, silence, darkness as Spinoza observed whose pocket Mr Locke picked of it without after confession, are as positive Vorstellungen as Light, Sound, Image. 1865 J. Grote *Exploratio* i. iv. 60 Knowledge as acquaintance..is the kind of knowledge which we have of a thing by the presentation of it to the senses or the representation of it in a picture or type, a 'vorstellung'. 1890 W. James *Princ. Psychol.* I. ix. 236 A permanently existing 'idea' or 'Vorstellung' which makes its appearance before the footlights of consciousness at periodical intervals, is as mythological an entity as the Jack of Spades. 1959 J. L. Austin *Sense & Sensibilia* (1962) vi. 61 In Berkeley's doctrine there are *only* ideas, in Kant's only *Vorstellungen* (things-in-themselves being not strictly relevant here).

vortex. Add: **1. b.** Also *spec.* (with capital initial) a group of artists practising vorticism (see next).

1913 E. Pound *Let.* 19 Dec. (1950) I. 28 You may get something that you would miss in *The Vortex*. 1914 *Blast* 20 June 8 Do you think Lloyd George has the Vortex in him? 1969 *Listener* 30 Jan. 138/1 The so-called Great English Vortex, i.e. the group of painters, sculptors and writers who..formed and represented the Vorticist movement.

7. vortex shedding, the periodic detachment of vortices from an object in a fluid flow, causing a varying force to be experienced by the object; **vortex sheet,** a region of vortices that is created at the interface of two masses of fluid having different velocities along the interface; **vortex street**: see *street *sb.* 2 g.

1953 A. Roshko *Devel. Turbulent Wakes from Vortex Streets* (U.S. Nat. Advisory Comm. Aeronaut. TN 2913) 45 There is yet no adequate theory of the periodic vortex shedding and it is not clear what is the principal mechanism which determines the frequency. 1975 *Offshore Engineer* Dec. 42/1 Vortex shedding can impose periodic forces on a pipeline. 1982 *New Scientist* 27 May 566/1 Vortex shedding is quite harmless until it begins to interfere with the safety or the function of man-made structures. 1879 H. Lamb *Treat. Math. Theory Motion Fluids* vi. 154 Let us suppose we have a series of vortex-filaments arranged in a thin film over a surface... The infinitely thin film is then called a 'vortex-sheet'. 1926 [see *separation 13**]. 1983 *Jrnl. Fluids Engin.* CV. 53 Nascent vortex strength and position are determined from the Kutta condition so that the nascent vortex has the same strength as a vortex sheet of uniform strength.

vorticism (vǭ·ɹtisiz'm). Also **Vorticism.** [f. L. *vortex*, *vortic-* Vortex + -ism.] A British art-movement of the early twentieth century, characterized by abstractionism and machine-like forms. Also applied to similar tendencies in literature.

1914 E. Pound in *Fortn. Rev.* Sept. 470 The image is not an idea. It is a radiant node or cluster; it is what I can..call a *Vortex*, from which..ideas are constantly rushing... And from this necessity came the name 'vorticism'. 1915 *Drawing* July 56/1 Vorticism..is in reality our old and amusing friend Cubism, but Cubism heavily charged with electricity. 1926 F. V. P. Rutter *Evolution in Mod. Art* v. 105 Impressionism..was succeeded by Fauvism, Cubism, Expressionism, Futurism, Vorticism, and what not. 1962 *Listener* 5 Apr. 599/1 He [*sc.* Ezra Pound] was fighting under the banner of 'vorticism'. 1977 P. L. Fermor *Time of Gifts* iii. 69 Pale woods and plastics were juggled together with stale and pretentious vorticism. 1980 *Daily Tel.* 6 Feb. 18 Nevinson was one of the first artists to use the contemporary techniques of vorticism and futurism for depicting the war [of 1914–18].

vorticist. Add: **2. a.** An exponent of vorticism in art.

1914 *Daily News* 7 July 6 My Scot one morning preached me a fiery sermon on the poetry of lawn tennis, and..I became a Vorticist. 1914 E. Pound *Let.* 10 Nov. (1971) 47 While the vorticists are well-represented, the College does not bind itself to a school. 1915 [see *fauve]. 1930 J. W. Mackail *Largeness in Lit.* 15 Imagists, contortionists, vorticists..have had their little day. 1970 *English Studies* LI. 269 This period [*sc.* 1910–1922] also saw the birth and death of other more obviously revolutionary groups such as the Vorticists, Imagists and Futurists. 1980 *Illustr. London News* Mar. 58/1 Even the brief and feeble spark of abstraction in England, fanned by the Vorticists and almost immediately snuffed out by the First World War, became the subject of intensive investigation.

b. *attrib.* or as *adj.*

1914 *Blast* 20 June 8 We will convert the King if possible. A Vorticist King! Why not? 1914 E. Pound in *Fortn. Rev.* Sept. 461, I shall be..more lucid if I give.. the history of the vorticist art with which I am most intimately connected, that is to say, vorticist poetry. 1923 *Daily Mail* 3 Mar. 12 Mr Wadsworth has all the qualifications of a tempera painter. In his vorticist past he was inclined to transform the world and its people into rigid, machine-like shapes. 1962 *Listener* 5 Apr. 599/1 The imagist and vorticist movements of less than fifty years ago. 1966 A. L. Coburn *Autobiogr.* ix. 102 It was in 1914 that Wyndham Lewis..founded the *avant-garde* movement named by..Ezra Pound 'Vorticist', the ideas of which derived partly from Futurism and partly from Cubism. 1981 V. Glendinning *Edith Sitwell* v. 83 Wyndham Lewis..was five years older than Edith herself; his Vorticist period, the editorship of *Blast*..already lay behind him.

vorticity. Add: (Earlier and later examples.) In mod. use, a vector quantity equal at any given point in a fluid to twice the angular velocity of a small element of fluid about the point.

1888 A. B. Basset *Treat. Hydrodynamics* I. 75 Let P be any point on the axis of one of these [vortex] filaments, dm the mass of the filament which contains P, ω and dS the molecular rotation and cross section of the filament... Then the quantity $\omega dS/dm$ is called the vorticity of the fluid at the point P. 1896 *Proc. London Math. Soc.* XXVII. 14 The component velocities..u and v, are connected with a stream-function Ψ by the relations $u = d\Psi/dy$, $v = -d\Psi/dx$. The vorticity is represented by $\frac{1}{2}\nabla^2\Psi$. 1916 H. Lamb *Hydrodynamics* (ed. 4) iii. 30 The component angular velocities of the rotation being $\frac{1}{2}\xi$, $\frac{1}{2}\eta$, $\frac{1}{2}\zeta$. The vector whose components are ξ, η, ζ may conveniently be called the 'vorticity' of the medium at the point (x, y, z). 1938 L. M. Milne-Thomson *Theoret. Hydrodynamics* xix. 511 Either sense of rotation can be obtained according as the bath is filled with the hot or the cold tap, the fluid from one or the other acquiring opposite vorticities as it moves near the boundary. 1958 *Science* 4 Apr. 731/3 On the surface of a fluid, the vorticity may be observed by following a cork marked with a cross... If the arms of the cross do not rotate, the vorticity is zero; if they do rotate, there is vorticity. 1976 *Nature* 1 Apr. 457/1 A motor vehicle or any other projectile in air produces vortices in its wake, but no net vorticity.

vortograph (vǭ·ɹtǒgrɑf). Also **Vortograph.** [f. Vort(ex + -o + -graph.] An abstract photograph taken with a camera and a vortoscope.

1917 E. Pound *Let.* 24 Jan. (1971) 104 The vortographs are perhaps as interesting as Wadsworth's woodcuts, perhaps not quite as interesting. 1963 *Times* 29 Apr. 7/2 Mr. Coburn..went on to show the parallels between his photography and the work of painters and sculptors at a later period with his astonishing series of 'vortographs'—photographic abstractions very close in feeling to the vorticist paintings of Wyndham Lewis and such early Epstein sculptures as 'The Rock Drill'. 1966 A. L. Coburn *Autobiogr.* ix. 102 Photography depends upon pattern..as well as upon quality of tone and luminosity, and in the Vortograph the design can be adjusted at will. 1982 M. Weaver *Alvin Langdon Coburn* 26 Prismatic, triangular effects appeared in the Vortographs in which abstraction and conception superseded observation and perception.

vortoscope (vǭ·ɹtǒskōᵘp). Also **Vortoscope.** [f. Vort(ex + -o + -scope.] A mirror device used for producing abstract photographs (see quot. 1966).

1917 E. Pound *Let.* 24 Jan. (1950) 104 The vortescope [*sic*] isn't a cinema. It is an attachment to enable a photographer to do sham Picassos. That sarcastic definition probably covers the ground. 1918 —— *Pavannes & Divisions* App. iv. 252 The vortoscope is useless to a man who cannot recognise a beautiful arrangement of forms on a surface. 1966 A. L. Coburn *Autobiogr.* ix. 102, I aspired to make abstract pictures with the camera. For this purpose I devised the Vortoscope late in 1916. This instrument is composed of three mirrors fastened together in the form of a triangle... The mirrors act as a prism splitting the image formed by the lens into segments. 1982 M. Harker in M. Weaver *Alvin Langdon Coburn* 2 He made a Vortoscope in 1916. Based on the principles of the kaleidoscope the instrument was composed of three mirrors fastened together in the form of a triangle.

Vöslauer (fǿslɑu·,əɹ). Also **Voslauer.** [a. Ger., f. *Vöslau* name of a district in Lower Austria: see -er[1].] An Austrian red or white table wine from Vöslau in the Vienna Woods.

1920 [see *Carlowitz]. 1960 *Times* 11 June 11/6 A local red wine, Voslauer, which is among the best in the country.

Vosne Romanée (vōᵘn romane). The name of a commune in the department of Côte-d'Or in France, used *absol.* to designate the wines produced there. Cf. *Romanée.

1930 G. Knowles tr. *P. de Cassagnac's French Wines* vi. 140 The Côte-de-Nuits begins at Dijon. It includes... Flagey, with Grands-Echezeaux; Vosne-Romanée and the estates of Romanée [etc.]. 1952 A. Lichine *Wines of France* 126 Wines from Les Verroilles..are never sold under their own name..while the third *cuvées* are sold simply as Vosne-Romanée. 1963 N. Freeling *Gun before Butter* iii. 164 We'll have some burgundy. I've a Vosne Romanée; just the trick. 1980 E. Leather *Duveen Letter* xiv. 160 Rupert..ordered a bottle of Vosne-Romanée 1961.

vote, *sb.* Add: **8. a.** *vote on account*, a resolution at the close of the financial year to assign a sum of money to a government department as an advance payment before its full annual expenditure is authorized by law.

1859 Erskine May *Law of Parl.* (ed. 4) 531 Votes on account. The entire sums proposed to be granted for particular services, are not always voted at the same time, but a certain sum is occasionally voted on account of such grants. 1910 W. S. Churchill *Let.* 11 Mar. in R. S. Churchill *Winston S. Churchill* (1969) II. Compan. ii. xiii. 992 The Vote on account is the most powerful and the most simple Parliamentary engine by which the House of Commons is assured of its influence upon the Executive Government. 1963 *Economist* 24 Feb. 71/3 Part of the central government's expenditure, as recorded in the 'vote-on-account', is simply of a transfer kind; it includes, for instance, not only such personal payments as family allowances but also grants to local authorities made out of the central tax pool. *a* 1974 R. Crossman *Diaries* (1977) III. 332, I am worried about this year's Vote on Account.

b. *vote of confidence*, a resolution showing majority support for a government, policy, etc. Similarly *vote of no* (or *want of*) *confidence*. Also *fig.*

1846 G. Bentinck in *Hansard Commons* 8 June 182, I should certainly have preferred an Amendment which took the shape of a direct vote of want of confidence in Her Majesty's Ministers. 1870 Ld. Russell *Sel. Speeches* I. 154 Institutions..whose ministers resign on a vote of want of confidence. 1955 *Times* 10 May 14/4 The Government are asking for a vote of confidence. 1962 *Listener* 13 Dec. 1002/1 Why is it that grown men and women, n less than teenagers, are registering this unmistakable vote of no confidence in a society which has in so many ways improved their physical and material conditions of life? 1963 *Ibid.* 14 Feb. 281/2 The government survived..a vote of no confidence in the Parliament. 1976 *Glasgow Herald* 26 Nov. 1/8 Derby County's Scottish manager, Dave Mackay, was dismissed last night after three years at the Baseball ground. He had asked the club's directors for a vote of confidence.

10. *vote-catcher* (examples), *-collector*, *-getter* (examples), *-getting*, *-loser*, *-rigging*, *-splitting*; *vote-orientated*, *-proof* adjs.; *vote-wise* adv.; **vote bank**, in India, a group of people who can be relied upon to vote together in support of the same party.

1963 H. Tinker *Democratic Ideal in Asia* 17 The tribe, caste, or other association represents a 'vote bank'. 1982 *Jrnl. Commonwealth & Comparative Politics* XX. 9 The kinship group became an important vote-bank. 1923 *Weekly Dispatch* 13 May 2/2 The wild men pin their faith to the Capital Levy as a vote-catcher. 1977 *Cork Examiner* 6 June 1/7 Mr Leddin felt that Ald. Lipper was being victimised because he was a 'vote catcher' for the party in Limerick. 1859 Mill *Parl. Reform* 32 Why should the vote-collector make a distinction where the tax-gatherer makes none? 1906 *Springfield* (Mass.) *Weekly Republ.* 1 Nov. 3 He is also a strong campaigner, and has proved himself a vote-getter. 1981 *Times* 13 June 16/7 Mrs Williams..is..the party's outstanding vote-getter. 1984 *Listener* 4 Oct. 5/1 Margaret Thatcher has held doubting Conservatives loyal by a mixture of unrelenting leadership and a formidable reputation as a vote-getter. 1892 *Courier-Jrnl.* (Louisville, Kentucky) 3 Oct. 3/3 Mr. Needy..was made a victim of the vote-getting machine. 1929 *Daily Tel.* 8 Jan. 10/4 Acidulated suggestions of a lack of moral principle one day, want of suppleness the next, and ignorance of the elementary arts of vote-getting the day after do not help the cause. 1963 A. Howard in Sissons & French *Age of Austerity* i. 16 The Conservatives ..seem to have based their whole vote-getting strategy on a fundamental misreading of the nation's mood. 1963 *Guardian* 25 Feb. 8/1 Nationalisation is a fully certified vote-loser when it comes to election campaigning. *a* 1974 R. Crossman *Diaries* (1975) I. 150 Immigration can be the greatest potential vote-loser for the Labour Party. 1971 H. Wilson *Labour Government* xxxvi. 744 Vote-orientated promises to increase most individual spending programmes. 1928 G. B. Shaw *Intelligent Woman's Guide Socialism* lxiii. 291 Those who do understand it will never be unanimous in resisting it; consequently it is voteproof at the parliamentary elections. 1958 *New Statesman* 22 Feb. 230/3 He also ignored the detailed figures I gave of 10 other branches which recorded votes in excess of the number of members entitled to vote in September 1948; nor does he deal..with any other of the numerous illustrations I gave of vote-rigging. 1983 *Listener* 20 Dec. 32/2 A 1981 BBC play about the famous ETU vote-rigging scandal of the Fifties. 1931 *Economist* 17 Oct. 723/1 Vote-splitting will be avoided as far as possible. 1950 W. S. Churchill *In Balance* (1951) 391 Great harm was done to national interests at the General Election by the policy of the Liberal Party in running hopeless or vote-splitting candidatures in hundreds of constituencies. 1959 *Punch* 16 Sept. 158/2 The wing-collared politicians who emancipated women vote-wise probably had minds awhirl. 1969 G. Kirkland in R. Blythe *Akenfield* iv. 86 We are all mainly Socialists vote-wise.

vote, *v.* Add: **3. a.** Also *fig.* Cf. sense 3 c below.

1970 *Globe & Mail* (Toronto) 26 Sept. 23/1 They said its auditorium was poorly designed—dead acoustics, too many seats too distant. But the audiences, voting at the box-office, have continued to come for almost 40 weeks of every season.

c. In *fig.* phr. *to vote with one's feet*, to indicate an opinion by presence in or absence from a country, institution, activity, etc.

1965 *Listener* 16 Dec. 983/1 Politically, emigration provided a useful safety valve, anyone who hated the regime..could just vote with his feet and get out. 1968 *Ibid.* 31 Oct. 567/2 The McCarthy supporters began to realise that if the only choice was Nixon or Humphrey, perhaps the streets were the only true political arenas... On and after 5 November there will be plenty of voting with the feet. 1970 *Guardian* 30 July 9/6 The children.. just vote with their feet and join in. 1972 M. Jones *Life on Dole* vii. 56 Welshmen, in fact, were voting with their feet against the hopelessness that surrounded them. Those who were qualified scattered themselves across England. 1982 *Christian Order* Nov. 546 Uncounted thousands have 'voted with their feet', i.e., have left the Church. 1985 *Times* 7 Jan. 15/1 With another 16,000 to 20,000 miners returning to work an effective majority would have voted with its feet to end the strike.

5. f. With the name of a political party or of a candidate as object.

1926 *Socialist Rev.* Oct. 48 There are still hundreds of thousands of voters who cannot bring themselves to vote Conservative. 1938 *Sun* (Baltimore) 6 Sept. 20 (*caption*) Vote Lewis. 1949 C. P. Snow *Time of Hope* i. iv. 42 He voted radical and she was a vehement tory. 1977 R. Barnard *Blood Brotherhood* i. 11 The parishioners.. voted Conservative, when they did not vote National Front.

g. *to vote stock* or *shares*, to cast votes representing stock held in a company. *U.S.*

1957 R. A. Heinlein in *Mag. of Fantasy & Sci. Fiction* Oct. 30/2, I had assigned the stock to her but she knew.. that I always voted it, that I had had no intention of parting with control of the company. 1978 *N.Y. Times* 29 Mar. d10/1 The state pension agencies can vote the stock but the banks must confer with others first.

9. b. (Later examples.)

c 1850 J. M. Morton *Box & Cox* 23, I vote, Box, that we stick by her. 1880 Trollope *Duke's Children* I. xviii. 222 We've got the trap and the horses..and I vote we make a start. 1927 E. Bowen *Hotel* ii. 15, I vote we take those two on again.

10. b. To present for voting; to record the votes of (electors). *U.S.*

1859 Bartlett *Dict. Amer.* (ed. 2) 98 They..are taken to the polls, and 'voted', as it is called, for the party. 1884 E. W. Nye *Baled Hay* 217, I believe they vote people there who have been dead for centuries. 1904 *N.Y. Even. Post* 8 Nov. 1, 25 men were in line in many places, and they were voted at a rate of nearly one a minute.

voter. Add: **1. c.** *attrib.*, as *voter registrar*, *registration*. *U.S.*

1960 *Nation* (N.Y.) 23 Jan. 72/3 HR7597, introduced by Congressman Powell..provides for the establishment of a Federal Voter Registration Commission. 1964 *Federal Suppl.* CCXXIX. 933/2 The suits filed by the United States against several county voter registrars..were matters of common knowledge throughout the State of Mississippi. 1976 *National Observer* (U.S.) 10 July 5/5 He hopes to raise all he can—$3 million— for the Carter campaign, and raise even more, maybe $8 million, for voter registration and training schools for candidates and things like that.

Votiak, var. *Votyak.

Votic (vōᵘ·tik). [Origin uncertain.] A language of the Finnish group of the Finno-Ugrian family of languages, used by a small number of speakers in the Ingrian region, now in the north-west part of the U.S.S.R. Also **Vo·tian.**

1908 T. G. Tucker *Introd. Nat. Hist. Lang.* 132 Tchudic and the cognate dialects Vepsic and Votic, about Lake Onega. 1933 [see *Ludian]. 1939 L. H. Gray *Foundations of Language* 369 The languages of the Uralic family are as follows:..*Finnish* group: Finnish proper.., Karelian.., Olonetzian, Ingrian, Lüdish, Vepsian, Votian, Esthonian, etc. 1954 Pei & Gaynor *Dict. Linguistics* 229 *Votian*, a member of the Finnish group of the Finno-Ugric (or Uralic) sub-family of the Ural-Altaic family of languages. 1964 *Language* XL. 98 The Veps, Votic, and Estonian cognates..are hardly necessary. 1977 *Ibid.* LIII. 477 The last ten are Yurak Samoyed, Lapp, Judeo-Spanish, Vepsian, Kashubian, Karaim, Ingrian, Livonian, Votic, and Manx Gaelic.

voting, *vbl. sb.* Add: **2.** *voting age*, *urn*.

1937 V. Bartlett *This is my Life* x. 165 That mass of people..saluting with religious intensity the dustbin-like voting urns. 1966 M. Woodhouse *Tree Frog* viii. 60 He must have reached voting age, but you couldn't tell by looking at him. 1978 *Listener* 2 Feb. 145/2 Those recently reaching voting age. 1978 K. J. Dover *Greek Homosexuality* iii. 111 The funnel of the voting-urn used in a lawcourt.

votive, *sb.* Delete *rare*⁻¹ and add later examples.

1975 Y. Yadin *Hazor* iv. 61 We found..an elevated rectangular platform that looked like an open high place, or *bamah*, where the votives..were laid. 1976 *Scotsman* 27 Dec. 7/4 Archaeologists have recently dug up 8395 such terra cotta votives, as they are called.

Votyak (vōᵘ·ti,ak). Also **Votiak.** [Russ.] **1.** A member of a Finno-Ugrian people inhabiting the Udmurt republic in the north-western region of the U.S.S.R. Also *attrib.* or as *adj.*

1841 *Penny Cycl.* XX. 247/1 The Votiakes are settled west of the Permians, on both sides of the upper course of the river Viatka, and in the country about the source of the Kama. 1845 *Encycl. Metrop.* XXV. 866/2 Near Perm..are the Votyáks, who call themselves 'Udhmurd, *i.e.*, Hospitable men. 1938 *N. & Q.* 23 Apr. 291/2 Votyak folk-lore is influenced by the Persians. 1948 D. Diringer *Alphabet* ii. viii. 482 The Votiaks,..living in the Vyatka region. 1968 Bethell & Burg tr. *Solzhenitsyn's Cancer Ward* (1971) i. viii. 110 Now, as he paced up and down the ward, he remembered how the old folk used to die back home on the kama—Russians, Tartars, Votyaks or whatever they were. 1974 *Encycl. Brit. Macropædia* VII. 313/1 In the *lud* sanctuaries of the Votyaks.. worship was performed by members of the family.

2. The language of this people, belonging to the Permian branch of the Finno-Ugrian family.

1878 *Encycl. Brit.* VIII. 700/1 *Finnic* or *Ugrian* represented by..(*b*) Karelian..(*m*) Votiak. 1908, etc. [see *Permian *a.* (*sb.*) 2]. 1932 W. L. Graff *Language & Languages* 406 Votyak (about 400,000) is situated between the Viatka and the Kama. 1951 W. K. Matthews *Languages U.S.S.R.* iii. 24 The Permian branch, which comprises two languages, Zyryan and Votyak (Udmurt) 1977 [see *Vepsian].

voucher, *sb.*[1] Add: **2. d.** A document which can be exchanged for goods or services as token of payment made or promised by the holder or another (see also quot. 1947).

1947 *Sun* (Baltimore) 12 May 2/5 Stefan has gone through a stack of vouchers—expense accounts—from the American Embassy. 1955, etc. [see *Luncheon 3]. 1960 S. Unwin *Truth about a Publisher* ii. xix. 353 The New Zealand Company had not given me an actual ticket..but a voucher instructing their agent to issue me a ticket. 1970 *Phi Delta Kappan* LII. 49 For some time Christopher Jenks has believed that voucher plans offer an exit from the bureaucratic morass in which many major school systems are mired. 1980 *Jrnl. R. Soc. Arts* July 475/1 It could be done through some kind of voucher scheme.

voudoun (vū·dūn), var. of Voodoo *sb.* Also **voudon.**

1939 M. Steedman *Unknown to World: Haiti* xvii. 159 The gods and goddesses of the Voudon pantheon are known as the 'Lois'. 1953 M. Deren *Divine Horsemen* 60 Voudoun (which is the Fons word for god) includes the loa (the Congo word for the spirits) of many nations. *Ibid.* 15 Voudoun is the religion, primarily African in origin, of the vast majority of the Republic of Haiti in the West Indies. 1953 *Times Lit. Suppl.* 2 Aug. 510 A Voudoun, as some purists insist, with little basis, in spelling it. 1958 J. Sykes *Quakers* i. i. 34 Almost any gathering of English citizenry could run to a pitch one associates more with a Haitian voudoun possessed by 'the loa'.

‖ **voulu** (vūlu), *a.* (*sb.*) [Fr., pa. pple. of *vouloir* to wish, want.] Contrived, deliberate, studied. Also as *sb.* (esp. *collect*).

1909 E. Nesbit *Daphne in Fitzroy St.* xiii. 207 Perhaps there's something more delicate, less *voulu*, in our little dinner as it is. The poignant beauty of the incomplete. 1938 E. Bowen *Death of Heart* iii. iv. 389 There is a narrowness about fantasy: it figures only the *voulu* part of the self. 1948 F. R. Leavis *Great Tradition* ii. 81 There is certainly something of that quality [*sc.* insincerity] in *Daniel Deronda*..an element of the tacitly *voulu*. 1957 L. Durrell *Justine* ii. 132 The idea is not spontaneous, but *voulue*. 1959 *Encounter* Dec. 64/1 All de Sade's excesses were '*voulus*'. They were a scientific investigation into sex. 1962 *Listener* 17 May 885/3 Here and there the impression of the *voulu* and the freakish cannot be avoided. 1968 *Punch* 3 Apr. 507/1 They simply give an impression of *voulu* oddity, nearer to a harsher, more explicitly sexual A. E. Coppard than to Carson McCullers. 1974 *Times* 2 Feb. 7/6 But her knowledge of people as such? Sometimes profound, sometimes *voulu*.

Vouvray (vūvrēⁱ·). The name of a commune in the department of Indre-et-Loire in N.W. France, used chiefly *absol.* to designate the wines produced in this and neighbouring communes.

1885 H. James *Little Tour in France* vii. 60 She uncorked for us a bottle of Vouvray mousseux. 1920 G. Saintsbury *Notes on Cellar Bk.* vi. 97, I have said nothing, in the chapter on Champagne, as to Saumur, Vouvray, and the Swiss imitations... Vouvray..seems to me feeble. 1934 E. Waugh *Handful of Dust* v. 246 One of them..got drunk for the first time on sparkling Vouvray. 1955 'W. Mole' *Hammersmith Maggot* vi. 68 Lunch, properly accompanied by a glass of Vouvray. 1981 'A. Cross' *Death in Faculty* xii. 141 I'm going to have a bottle of Vouvray, Clos de Nouys, 1971.

vowel, *sb.* Add: **3.** *vowel-alternation*, *-articulation*, *-consonant* (later examples in sense 'vowel followed by consonant', chiefly *attrib.*), *-length*, *-letter*, *-like* (earlier example), *-loveliness*, *-phoneme*, *-rhyme*, *-sequence*, *-shade*, *sign* (earlier example), *-sound* (earlier example), *-space*, *-symbol*, *-system* (earlier example), *triangle*; *vowel-initial* adj.; **vowel colour**, the precise timbre and quality of a vowel sound; so **vowel colouring**; **vowel diagram**, a diagram showing relative degrees of closeness or openness, front-raising or back-raising of the tongue, in the articulation of individual vowels; **vowel-glide** [cf. Glide *sb.* 4], the gliding movement from one vowel component to another, as in a diphthong; also, = Glide *sb.* 4; **vowel gradation** = Ablaut; **vowel harmony**, a feature of the Finno-Ugric, Turkish, and other languages, whereby successive syllables of words are limited to a particular class of vowel; **vowel height**, the degree to which the tongue is raised or lowered in the pronunciation of a particular vowel; **vowel-laxing**, the enunciation of a vowel with the speech organs relaxed (cf. *Lax *a.* 5 c); **vowel-quality**, the identifying acoustic characteristic of a vowel; **vowel-quantity**, the duration of time needed for the pronunciation of a vowel; **vowel shift**, a phonetic change of vowel or vowels, *spec.* applied to a series of changes between medieval and modern English affecting the long vowels of the standard language; freq. in phr. *great vowel shift*; cf. *Shift *sb.* 14 d.

1951 W. K. Matthews *Languages U.S.S.R.* iv. 57 The already noticed Manchu vowel-alternation to express sex-gender may be paralleled by Evenki examples. 1937 J. R. Firth *Tongues of Men* 42 Instead of referring to the vowel-articulations, we refer to the resulting sounds. 1948 M. Joos *Acoustic Phonetics* ii. 59 It seems that a listener can hear a difference of vowel color [in the [ɛ] of *hotel* spoken by three American speakers] equivalent to a distance of 1 semitone on the formant chart. 1978 *Amer. Speech* LIII. 291 Different arrays of formants constitute different vowel colors. 1939 F. M. Ford *Let.* May (1965) 321 The mere sequence of the vowel coloring of that phrase will give you acute pleasure. 1979 *Archivum Linguisticum 1978* IX. 156 Pajares notes that the loss of laryngeal and the lengthening (and in some cases vowel coloring) of the preceding vowel can also be considered a monophthongization. 1960 *Amer. Speech* XXXV. 227 Vowel-consonant syllables [are] less intelligible than consonant-vowel syllables. 1977 P. Strevens *New Orientations Teaching of Eng.* xii. 151 Permissible consonant-clusters and vowel-consonant sequences are similar. 1932 D. Jones *Outl. Eng. Phonetics* (ed. 3) viii. 36 (*caption*) Fig. 23 Diagram illustrating the Tongue-positions of the eight primary Cardinal Vowels. Fig. 24. A more accurate form of Vowel Diagram. 1976 C. Barber *Early Mod. Eng.* vi. 299 The long and short pure vowels in StE in the middle of our period are shown on the vowel diagrams in Figure 3 and Figure 4. [1856 Vocal glide: see Glide *sb.* 4.] 1878 H. Sweet *Handbk. Phonetics* 63 (*heading*) Initial and Final Vowel-Glides. 1932 D. Jones *Outl. Eng. Phonetics* (ed. 3) 25 Such vowel-glides are often called semi-vowels. 1962 A. C. Gimson *Introd. Pronunc. Eng.* ii. vii. 121 Diphthongal Vowel Glides. The sequences of vocalic elements included under the term 'diphthong' are those which form a glide within one syllable. 1973 J. D. O'Connor *Phonetics* vii. 220 A diphthong..is phonetically a vowel glide or a sequence of two vowel segments which functions as a single phoneme. *Ibid.* 221 Other vowel glides such as /eu, øu, iu, ɔu/ which also occur [in Danish], must be interpreted as vowel + /v/ since they do not occur freely in the same sorts of context as the remaining vowels and diphthongs. 1887 W. W. Skeat *Princ. Eng. Etym.* x. 158 To discover the original Teutonic vowel-gradation ..we must compare with one another the oldest known forms of the verbs in the various Teutonic languages. 1938 *Amer. Speech* XIII. 209 The term 'vowel-change' is evidently a new attempt (like 'vowel gradation'..) to avoid the German word *Ablaut*. 1973 A. H. Sommerstein *Sound Pattern Anc. Greek* ii. 70 As is well known, vowel gradation, or ablaut, was an important feature of proto-Indo-European morphology. [1893 J. Clark *Man. Linguistics* vi. 151 The Ural-Altaic languages..which are dominated by a law of vocalic harmony that, to speak generally, requires that one class of vowels..should obtain in the various syllables of a word.] 1900 H. Sweet *Hist. Lang.* vii. 122 In Finnish, the vowels are divided, from the point of view of vowel-harmony, into the three classes hard, soft and neutral. 1972 *Language* XLVIII. 365 The nature of the rules of vowel harmony is and has been the subject of some discussion... The traditional view is as follows: the vowel harmony rule in Turkish specifies that, for any vowel, the distinctive feature in question is determined by the value of the previous vowel. 1977 *Archivum Linguisticum* VIII. 71 This is obscured by a distinctive feature system which uses a binary classification of [±high] and [±low] to define vowel height. 1949 E. A. Nida *Morphology* (ed. 2) ii. 16 Note that *h-* occurs before the consonant-initial stem *k'ab* and *k* before the vowel-initial stem *akan*. 1978 *Language* LIV. 23 Before apparently vowel-initial nouns like *hèros* and *honte*, we find no elision. 1977 *Stud. in Eng. Lit.: Eng. Number* (Tokyo) 155 Rule (24) succeeds in formally expressing a linguistically significant generalization underlying various vowel-laxing processes. 1980 *English World-Wide* I. 250 The linguistic variables analyzed in the remaining six chapters are: (i) vowel laxing in monosyllabic personal pronouns and a number of non-pronominal forms. 1932 D. Jones *Outl. Eng. Phonetics* (ed. 3) 237 Vowel-length depends to a considerable extent on the rhythm of the sentence. 1977 P. Strevens *New Orientations Teaching of Eng.* xii. 151 The relation between stress and vowel-length is of the same general type. 1846, etc. Vowel-letter [see *Mater lectionis]. 1933 L. Bloomfield *Language* 292 We use the Latin vowel-letters not only in entirely new values..but in inconsistent ways. 1855 tr. *Lepsius' Standard Alphabet* 28 The Indian grammarians, who express the nasalisation..by a vowel-like sign, namely, by placing a dot over the letter. 1910 D. H. Lawrence *Let.* 26 Oct. (1962) I. 67 One can get good Swinburnian consonant music by taking thought, but never Shakespearean vowel-loveliness. 1935 G. K. Zipf *Psycho-Biol. of Lang.* (1936) 316 All the short vowel-phonemes. 1964 C. Barber *Present-Day Eng.* iii. 38 There is no pair of English vowel-phonemes which is distinguished solely by length. 1977 D. Fry *Homo Loquens* ii. 13 In English..about twenty are the vowel phonemes, exemplified by the word group *beat, bit, bet, bat*, etc. 1920 W. Perrett *Peetickay* 23 A notation of vowel-quality, derived from the 'organic' positions of the tongue, lips, etc., already exists in Alexander Melville Bell's *Visible Speech*. 1965 W. S. Allen *Vox Latina* ii. 52 The Latin vowel-quality is vouched for by Italian *borsa*. 1933 L. Bloomfield *Language* 294 For the writers, the *gh* was now a mere silent graph, indicative only of vowel-quantity. 1977 *Archivum Linguisticum* VIII. 93 According to the Introduction, it has two aims: firstly to examine the various factors involved in the vowel-quantity changes that took place between Middle High German and New High German. 1873–4 G. M. Hopkins *Jrnls. & Papers* (1959) 287 Alliteration is initial half-rhyme, 'shothending' is final half-rhyme, assonance is vowel rhyme. 1961 A. Clarke *Later Poems* 89 In simple patterns, the tonic word at the end of the line is supported by a vowel-rhyme in the middle of the next line. 1939 F. M. Ford *Let.* May (1965) 321 The mere vowel sequences of certain passages will be sufficient to call back to you all the associations of your youth. 1965 *Language* XLI. 482 It gives a desirable economy of phonemes, which the vowel-sequence solution does not. 1955 J. R. R. Tolkien *Return of King* 409 The language that they [*sc.* the Ents] had made was..slow, sonorous..; formed of a multiplicity of vowel-shades and distinctions of tone and quantity. 1909 O. Jespersen *Mod. Eng. Gram.* I. viii. 231 The great vowel-shift consists in a general raising of all long vowels. 1933 —— *Essentials Eng. Gram.* iii. 34 The greatest revolution that has taken place in the phonetic system of English is the vowel-shift. 1936 *Essays & Studies* XXI. 10 The great vowel-shift of the late Middle Ages was followed by numerous other changes. 1964 C. Barber *Present-Day Eng.* iii. 51 The great vowel-shift which took place between Middle English and Modern English. 1977 *Canad. Jrnl. Linguistics 1976* XXI. ii. 177 After the application of the degemination rule and the vowel shift rule one obtains [eksīd]. 1849 *Jrnl. Amer. Oriental Soc.* I. 523 We begin with the Vowel-signs, of which there are three in the Persian cuneiform alphabet. 1795 *Monthly Rev.* Aug. 410 All the simple sounds, vowel and consonantal. 1965 W. S. Allen *Vox Latina* ii. 54 There would have been more vowel-space to accommodate the new sound. 1979 *Amer. Speech 1978* LIII. 294 Suppose the available 'vowel space' for a community's language were constrained to the relatively small triangular region bounded by the calculated Neanderthal approximations to the three reference vowels. 1932 D. Jones *Outl. Eng. Phonetics* (ed. 3) 5 The naming of vowel-symbols presents some difficulty. 1957 R. W. Zandvoort *Handbk. Eng. Gram.* 337 Words in silent *e* drop it before endings beginning with a vowel-symbol. 1855 tr. *Lepsius' Standard Alphabet* 11 He contended successfully against the English vowel-system. 1918 D. Jones *Outl. Eng. Phonetics* vi. 16 If we examine the tongue positions of the typical sounds of these five classes [of vowels] we find that the highest points of the tongue lie roughly on the sides of a triangle... This triangle is known as the 'Vowel Triangle'.

vowel, *v.* Add: Hence (nonce-uses) **vo·welled** *ppl. a.*; **vo·welling** *vbl. sb.*

1873–4 G. M. Hopkins *Jrnls. & Papers* (1959) 284 This is given by vowelling, which is either vowelling on (assonance) or vowelling off. *Ibid.* 287 In Norse poetry..the vowels..seen to be sided or intentionally changed, vowelled off. 1934 Dylan Thomas in *Listener* 24 Oct. 691/2 Some let me make you of the vowelled beeches.

vox. Add: **3.** *vox nihili*, a worthless or meaningless word, *spec.* one produced by a scribal or printer's error; used *transf.* to denote abstract concepts whose meaning is deemed to be indistinct when analysed.

1892 in *Stanford Dict. Anglicised Words & Phrases.* 1916 H. W. B. Joseph *Introd. Logic* (ed. 2) xix. 404 Words like *agency* or *power*, on this view, are *voces nihili*; we think we mean something more by them than habitual sequence, but we do not. 1933 *Mind* XLII. 170 That they [actions] cannot be right *merely* as causes of effects..I have already proved, if I have proved that 'rightness' in this sense is a *vox nihili* or a mere synonym for 'causality'. 1960 *Analysis* XXI. 10 'Inductive *inference*' is a *vox nihili*, unless used—as it is sometimes confusingly used—to denote the concrete factual predictions and retrodictions that the mechanic, the seedsman, the palaeontologist and the astronomer make.

vox pop (vǫks pǫp), *sb.* (*a.*) *colloq.* Also **voxpop.** [Abbrev. of *vox populi* s.v. Vox 1.] **a.** Popular opinion as represented by informal comments from members of the general public, esp. when used for broadcasting; statements or interviews of this kind.

1964 Hall & Whannel *Popular Arts* ix. 225 In television..we could include..the use of the brief survey of popular opinion on any topic by means of the posed question (the so-called 'vox pop'). *Ibid.* 257 The short interview and the occasional 'vox pop' of the local news and sports programmes on radio. 1968 *Listener* 8 Feb. 164/2 A BBC camera crew went round Washington collecting vox pops—close-ups of men in the street saying pithily what they think of things. 1972 D. Hurd *Truth Game* 59 Follow him close with the mike and get odds and ends of vox pop. 1975 *Listener* 18 Sept. 377/2 The audience was asked for comments, and very dim the voxpops sounded after the fluency of the two main speakers. 1981 *Times Lit. Suppl.* 25 Dec. 1501/5 There are wonderfully dreadful quotations in these chapters, including paragraphs of vox pop.

b. *attrib.* or as *adj.*

1966 *New Society* 17 Mar. 5/3 A *vox pop* survey I made among 17 housewives of St Ives. 1967 *Punch* 22 Nov. 794/1 Brandish a vox pop microphone in the concrete jungles and the cry..will recur monotonously, interspersed with blank looks. 1973 G. Talbot *Ten Seconds from Now* iii. 35 'Vox pop' interviews on what the king ought to do. 1983 *Listener* 2 June 20/2 His suitability for the box is being questioned by whipper-snappers who would not have rated a vox-pop appearance..in the Fifties. 1985 *Spectator* 5 Jan. 7/3 A recent sermon included a vox pop poll on who in the congregation, average age 70, was familiar with Michael's song 'Thriller'.

‖ **voyagé** (vwayaʒe), *a.* *Ballet.* [Fr., pa. pple. of *voyager* to travel.] Designating a movement in which the pose is held during progression; usu. as *arabesque(s) voyagée(s)*. Also *absol.* as *sb.*

1931 G. W. Beaumont *Dict. Technical Dance Terms* 30 *Voyagé*,..travelled, travelling. E.g., *arabesque voyagée*. 1957 G. B. L. Wilson *Penguin Dict. Ballet* 278 *Voyagé*,.. a movement across the stage during which the dancer holds a particular pose, usually an arabesque, and progresses in a series of hops or small jumps. 1976 *New Yorker* 26 Jan. 94/3 Tharp several times reinforces the point with moments swiped from the classics: the Wilis' arabesques voyagées.

voyageur. (Earlier example.)

1793 J. Macdonell *Diary* 6 June in C. M. Gates *Five Fur-Traders* (1933) 75 Lost the half of this day by rain.. a perfect nuisance to us voyageurs.

‖ **voyant** (vwayañ), *a.* Fem. **voyante** (-añt). [Fr.] Showy, gaudy, flashy; *spec.* of clothes, appearance, etc.

1906 *Punch* 18 Apr. 286/3 'Anno Domini'..whom one would *expect* to be smart and *voyante*, is simply the dowdiest, quietest of mice. 1927 U. M. Lyon *Etiquette* x. 113 Anything *voyant* in ties, deplorable always, is doubly so at his wedding. 1937 G. Frankau *More of Us* xv. 162 Tactful then mentioned he, maybe the hat—The whole costume indeed—was 'rather voyant'. 'Looks the complete tart if it comes to that', Hissed Innocent.

|| **voyant,** *sb.* [Fr., lit 'seer'.] A visionary; one gifted with an especial degree of mental perception.

[**1924** E. Rickword tr. A. Rimbaud in *Rimbaud: Boy & Poet* 203, I say that one must be a *visionary* (voyant), make oneself a *visionary*. The Poet makes himself a *visionary* by a long immense and reasoned *derangement* of *all the senses*.] **1938** *Times Lit. Suppl.* 21 May 353/2 The systematic derangement of all the senses that he [*sc.* Rimbaud] envisaged as one of the duties of the *voyant* or visionary poet has been attempted. **1958** *Spectator* 22 Aug. 257/3 The emblem of this evil society is the head of a dead pig..animated by flies and by the imagination of the *voyant*, Simon. **1972** E. Lucie-Smith in Cox & Dyson *20th-Cent. Mind* II. xiv. 483 A way of linking Dada to an older and more specifically French tradition, that of the *voyant*, which could be traced to nineteenth-century writers such as Rimbaud and Lautréamont.

voyeur (vwayö·ɹ). Fem. **-euse** (öz). [a. Fr., f. *voir* to see.] **1.** A person whose sexual desires are stimulated or satisfied by covert observation of the sex organs or sexual activities of others. Cf. *peeping Tom* s.v. *PEEPING *ppl. a.*[2]; *scopophiliac* s.v. *SCOPOPHILIA.

1900 H. Blanchamp tr. *Féré's Sexual Instinct* vi. 142 The houses of ill-fame have a *clientèle* of 'voyeurs' of both sexes. **1913** E. Jones *Papers on Psycho-Anal.* xv. 341 The patient, who had frequently indulged in pædicatio, was a pronounced *voyeur*. **1937** L. Bromfield *Rains Came* I. xlv. 192 A moment after she had disappeared the Major came out, and like a *voyeur* Ransome regarded him sharply to discover..whether she had made any progress. **1952** S. Kauffmann *Philanderer* (1957) xii. 188 The same Jake Simon who later made him a *voyeur* before puberty. **1956** C. P. Snow *Homecomings* xxxi. 214 Mrs Beauchamp, day-dreaming of a voyeuse's paradise, seeping herself into invisibility. **1972** *Daily Tel.* 7 Jan. 11/4 The hopeful voyeurs were to be disappointed for after this first fine flurry of bosoms and buttocks not a stitch was shed in the rest of the 45 minutes. **1978** C. P. Snow *Realists* viii. 236 He may have been a bit of a voyeur (he took extravagant pleasure in studying a love-object in bed). **1980** *Times Lit. Suppl.* 10 Oct. 1139/2 The film seems to miss a trick.., in that these two never get it together, but most voyeurs should be satisfied with what is otherwise on offer.

2. *transf.* and *fig.*

1956 C. W. Mills *Power Elite* vii. 162 The miser.. is an impotent voyeur of the economic system. **1958** *Times* 2 Oct. 3/2 The play..draws a firm line between those immersed in living and those plaintively watching them at it; and Hauptmann's sympathy goes to the *voyeurs*. **1967** *Spectator* 6 Oct. 404/3 Those films and TV programmes which argue that these are realities only moral cowards would evade seem to me to risk turning us all into *voyeurs* of pain, connoisseurs of private misery. **1978** J. Irving *World according to Garp* iii. 56, I was brought up to be a spectator... I was raised to be a voyeur.

Hence **voyeu·r** *v. intr.*, to obtain gratification in the manner of a voyeur; usu. as *pres. pple.* or *vbl. sb.*

1959 P. Bull *I know Face* xi. 195 She found herself dressing and undressing between two rows of hanging wardrobes, which involved delightful voyeuring for the gents manipulating the lights on the gantry above. **1970** K. Giles *Murder Pluperfect* iii. 72 'My girl,' said the Inspector, shamelessly voyeuring, lusciously and rashly voyeuring. **1980** M. Gilbert *Death of Favourite Girl* xxv. 240 His conscience wasn't all that clear, if he'd been out doing a spot of voyeuring.

voyeurism (vwayö·riz'm). [f. prec. + -ISM.] The state or condition of being a voyeur; scopophilia. Also *transf.* and *fig.*

1924 J. S. van Teslaar tr. *Stekel's Disorders of Instincts & Emotions* II. 341 *Voyeurism*, erotic gratification experienced at looking at another's sexual organs; morbid desire to peep into secrets. **1927** Bryan & Strachey tr. *K. Abraham's Sel. Papers* iii. 83 If sublimation does not take place then a perversion (*voyeurism* and exhibitionism) arises. **1953** E. Podolsky *Encycl. Abberrations* 538/1 Voyeurism takes the place of the normal sexual act. **1958** *Oxf. Mag.* 8 May 409/1 Oxford.. seems to appeal to a disagreeable mixture of envy and voyeurism. **1969** *Daily Tel.* 25 Apr. 19/5 There is a feeling that more and more actresses are being asked to take their clothes off at auditions.. as a form of voyeurism. **1976** *Listener* 25 Nov. 682/3 A beggar's expression captured by a camera costing enough to feed, clothe and house him for..maybe five years. Is this kind of photography good or bad?..Is it just an obscene form of voyeurism, a record of what one privileged class finds quaint or interesting in another? **1980** *Times Lit. Suppl.* 7 Nov. 1258/3 *Rain on the Roof* is not much more than the averagely good television play featuring voyeurism, adultery and sudden bloody death. **1985** *Times* 21 Jan. 3/3 Many senior politicians have been appalled by the fate of the women, whose sexual activities have been exposed in a bout of national voyeurism.

voyeurist (vwayö·rist), *sb.* and *a.* [f. as prec. + -IST.] **A.** *sb.* = *VOYEUR *a.* **B.** *adj.* = *VOYEURISTIC *a.*

1955 J. F. Oliven *Sexual Hygiene & Pathol.* xxi. 424 The severe 'compulsive' voyeurist is..an obsessed individual. **1960** *Guardian* 6 Jan. 1/7 One father..'yielding to his exhibitionist-voyeurist impulses', regularly went off and joined a nudist colony. **1970** G. Greer *Female Eunuch* 166 Girls for their part displayed interest in the detection of love affairs because of the particular voyeurist pleasures they afforded. **1980** *Times Lit. Suppl.* 3 Oct. 1112/3 The cool meticulous detail in which these and other depravities are related..has an obsessive and at times voyeurist tone.

voyeuristic (vwayöri·stik), *a.* [f. as prec. + -ISTIC.] Of or pertaining to voyeurism. Cf. *VOYEURIST *a.*

1929 in Calverton & Schmalhausen *Sex in Civilization* 593 One day when we were stopping at a hotel he made me gratify his voyeuristic tendency. **1960** *Arch. Gen. Psychiatry* III. 317/2 N.T. is interesting because of his unusual voyeuristic habits. **1973** *Daily Tel.* 12 June 14/2 The British censor has made his cutting mark..so that the camera does not linger with quite the same sort of voyeuristic dedication as it did when the original version was shown at the Berlin Film Festival. **1981** *Times* 14 Sept. 18/8 The dangers of 'voyeuristic' leisure centred on the television set. **1984** *Listener* 26 Apr. 38/3 A certain arrogance in their dealings with the Divis Street tenants, a voyeuristic interest in the more sensational aspects of life in Belfast.

Hence **voyeuri·stically** *adv.*

1967 *Spectator* 14 July 48/1 The sexual autobiography ..of a Victorian gentleman who early in life became voyeuristically obsessed. **1982** *New Musical Express* 30 Oct. 28/1 A slick, flashy, voyeuristically nasty horror movie. **1985** *Times* 6 Mar. 10/1 The fact that the majority of AIDS patients are homosexual dramatizes the illness almost voyeuristically for the population at large.

|| **voyou** (vwayū). [Fr.] A street urchin; a lout or hooligan.

1901 *Pall Mall Mag.* Feb. 195 But indeed the Hooligan, under the name of rough,..*voyou*..has been with us long, and not in this country alone. **1913** G. du Maurier *Martian* 61 'La flotte de Passy', as we called the Passy voyous. **1933** *Times Lit. Suppl.* 4 May 308/3 His curious friendship with a *voyou*. **1980** R. Grayson *Monterant Affair* vi. 50 She had fallen in love with one of the 'voyous' of Paris, a brash scoundrel living on his wits.

|| **vozhd** (vǫʒd). [Russ., lit. = 'chief'.] A leader, one who is in supreme authority: applied esp. to the Russian statesman Joseph Stalin (1879–1953).

1940 E. Lyons *Stalin* xxvii. 247 The 'leader of the world proletariat', the greatest representative of creative Marxism', the 'great machinist of locomotive history'... These are typical titles bestowed on the *Vozhd*, the Leader, by officials who owed their jobs and their lease on life to Stalin. **1959** D. W. Treadgold *Twentieth Cent. Russia* xviii. 289 Even the top functionaries were subject to Stalin's supreme power, and the word *Vozhd* (Leader) came to be used openly to acknowledge and proclaim that fact. **1978** *Encounter* Feb. 42/1 The *Vozhd* of Moscow made his exit in triumph.

|| **vrai réseau** (vrę rezo). [Fr., = true net.] The fine net ground used in making Brussels lace; also, any net ground for lace made by needle or bobbins as opposed to machine. Cf. *RÉSEAU 1.

1865 F. B. Palliser *Hist. Lace* vii. 104 It is the fineness of the thread which renders the real Brussels ground (vrai réseau) so costly. **1900** E. Jackson *Hist. Hand-Made Lace* vii. 58 The ground was of the *vrai réseau*, or needle-point mesh, now seldom seen. **1953** M. Powys *Lace & Lace-Making* iv. 29 This specimen has the Fond Rond which was used just before and contemporary with the Vrai Réseau de Valenciennes which is a diamond mesh. **1973** G. Pond *Introd. Lace* iii. 28 Old Brussels lace is of fine texture and rich in design..It was often made..in separate sprays sewn on net, which in the earlier days was hand-made and known as 'Vrai réseau' to distinguish it from the machine-made net.

|| **vraisemblable** (vręsaṅblābl'), *a.* [Fr., f. *vrai* true + *semblable* like.] Believable, likely, plausible. Also *absol.* as *sb.* (esp. *collect.*).

1830 *New Monthly Mag.* XXX. 5/1 How far all this is fact, we have no means of ascertaining, but the vraisemblable is well preserved, and if not true..is bien trouve. **1841** Geo. Eliot *Let.* 13 Nov. (1954) I. 121, I was amused with your sharp remarks on the advertisement... I immediately perceived them to be vraisemblable. **1876** J. D. Hooker *Let.* 17 Aug. in L. Huxley *Life & Lett. J. D. Hooker* (1918) II. xlv. 366 The whole of the vraisemblable of the latter falls before the Darwinian Gospel. **1898** E. Lynn Lynton *Let.* 23 Feb. in G. S. Layard *Mrs. Lynn Linton* (1901) xvii. 228, I have had your sketch... Cleverly done..not very *vraisemblable* in the man's character. **1957** D. Piper *Eng. Face* ii. 37 The portrait that Sittow painted of Henry VII..is completely *vraisemblable*. **1969** *Listener* 24 July 99/1 A nobleman.. reproved an actor for not holding a scythe as a labourer does: he got a lecture on the elements of the *vraisemblable*. Stage-scything is not like scything. **1978** *Times* 22 Aug. 13/4 Sir John Colville's delightful speculations..on what would have happened if George Washington had lost seemed to be remarkably *vraisemblable*.

vraisemblance. 1. (Earlier example.)

1802 Scott *Let.* 8 Sept. (1932) I. 156 To give to fiction itself the charms of truth or at least of *vraisemblance*.

vrbaite (vəbā·ait). *Min.* [ad. Czech *vrbait* (B. Ježek 1912, in *Rozpravy České Akad.* XXI. xxvi. 2), f. the name of K. *Vrba* (1845–1922), Bohemian mineralogist: see -ITE[1].] A sulphide of thallium, mercury, arsenic, and antimony, $Tl_4Hg_3Sb_2As_8S_{20}$, found as dark grey, tabular or prismatic, orthorhombic crystals.

1913 *Mineral. Mag.* XVI. 375 Vrbaite... Found.. embedded in realgar and orpiment from Allchar, Macedonia. **1973** *Mineral. Abstr.* XXIV. 22/2 The structure was determined on vrbaite from Allchar, Macedonia, the type locality... Vrbaite is the first structure with mixed (As, Sb) chains.

|| **vṛddhi** (vri·dhi). Also **vriddhi.** [Skr., lit. 'increase'.] In Sanskrit grammar, the strongest grade of an ablaut-series of vowels; also, the process of phonetic change whereby vowels of the middle grade are strengthened to achieve this grade. Cf. Guna.

1841 [see Guna sb.]. **1916** A. A. Macdonell *Vedic Gram. for Students* 5 Beside the Guṇa syllables appear, but much less frequently, the syllables **ai, au, ār** (**āl** does not occur), which are called Vṛddhi by the same authorities [*sc.* the Indian grammarians] and may be regarded as a lengthened variety of the Guṇa syllables. **1965** G. Y. Shevelov *Prehist. of Slavic* 116 More important features which led to the rise of length in I[ndo-]E[uropean] were the special kind of morphological analogy called *vṛddhi* (the name used by O[ld] I[ndian] grammarians to denote reflexes of long diphthongs in OI) and the loss of laryngeals. **1979** *Trans. Philol. Soc.* 150 It is time to break away from Herzfeld's notion that fortresses are 'pflanzenbewachsener boden' or 'saatland', and that the vriddhi of initial *u* was *ā*.

vriesia (vrī·ziă). [mod.L. (J. Lindley 1843, in *Bot. Reg.* XXIX. 10), f. the name of W. H. de *Vriese* (1806–62) Dutch botanist + -IA[1].] A perennial herbaceous plant of the genus *Vriesia* (family Bromeliaceæ), native to South or Central America and bearing rosettes of linear leaves and spikes of yellow, red, or white flowers.

1843 *Bot. Reg.* XXIX. 10 (*heading*) Parrot-flowered Vriesia. **1902** L. H. Bailey *Cycl. Amer. Hort.* IV. 1957/1 Vriesias are..tropical American stiff-leaved plants. **1979** *Daily Tel.* 17 Nov. 10/1 Another group [of bromeliads], the vriesias, have flowers that at one moment look like tongues of flame, and at others you think they are formed like cuttlefish.

|| **vrille** (vrīl). *Aeronaut.* [Fr., = spin.] A tail-spin; also, a spinning manœuvre engaged in deliberately as part of an aerobatic display.

1918 B. Hall *In the Air* iv. 32 He is trained on the rapid machines and when perfected is sent to the acrobatic school where they are taught all sorts of stunts, such as looping, vrille, tail and wing slips, and all the modern stunts. **1918** W. J. Abbot *Aircraft & Submarines* 115 The modern airplane is naturally so stable that if *not* interfered with it will always attempt to right itself before the dreaded *vrille* occurs. **1919** [see *TAILSPIN *sb.* a]. **1965** *Amer. Speech* XL. 13 /vr-/ can be attested, also in more recent *vrille* and in a proper name.

vroom (vrum). *colloq.* (orig. *U.S.*). Also **varoom.** [Echoic.] The roaring noise of a motor vehicle accelerating or travelling at speed. Also as *v. intr.*, to make such a noise; to travel or accelerate at speed; as *v. trans.*, to rev (an engine) with such a sound. Also reduplicated and as *int.*

1967 M. J. Arlen in *New Yorker* 21 Jan. 76 To go varooming all over the desert in a couple of jeeps. **1968** L. Deighton *Only when I Larf* iv. 48, I..shaved off my moustache. Vroom vroom. **1969** *Time* 2 May 33 The foursome would prefer tough scramblers, 'with big drive sprockets, knobby wheels—and more vroom'. **1970** *Atlantic Monthly* Oct. 75 The trooper lies on his back.. reading *Hot Rod* magazine, hearing the vroom-vroom of engines. **1971** 'H. Calvin' *Poison Chasers* vii. 93 The car stood at the red light with the engine making impatient vroom noises. **1971** *Time* 19 July 59/2 Just give me ten showgirls out here, and varoom, the young guys'll come out of Los Angeles in first gear. **1973** J. Di Mona *Last Man at Arlington* (1974) iv. xv. 195 A bullet varoomed over his head. **1974** *Guardian* 21 Mar. 11/2 A score of big bad BSA's vrooming through a sleepy one-pub town. **1975** D. Lodge *Changing Places* 190 Morris pulled out and varoomed down the wrong side of the road. **1976** B. Jackson *Flameout* iv. 51 A gang of youths sitting astride motorcycles vroomed their engines. **1979** J. Hansen *Skinflick* x. 77 'How about a sports car?.. Let it roll off a cliff and catch fire. Just like TV. Sensational.' 'Varoom!' Spence said. **1984** *New Yorker* 14 May 41/2 It wasn't the *varooms* and the screeching tires that spooked people.

vrouw, vrow. Add: Also **vrou.** *spec.* in S. Afr. (Further examples.)

1798 A. Barnard *Jrnl.* 21 May in A. W. C. Lindsay *Lives of Lindsays* (1849) III. 455 We were received by the vrow, the mistress of the house—O house, unworthy such a mistress! **1827** T. Phillips *Scenes & Occurrences in Albany & Caffer-Land* i. 12 Distance to them is no consideration; the boor puts his *vrouw*..into the waggon,.. and sets off to travel five hundred miles, with as much ease as we should ten in England. **1946** *Cape Times Week-end Mag.* 16 Nov. 27 Yesterday your vrou brought to my store a cupboard that she would sell. **1972** A. A. Telford *Yesterday's Dress* 79 The 'vrouw' is wearing the thick quilted cap.

V-shaped: see V 2 c in Dict. and Suppl.

V-sign. Also **V sign.** [f. *V* = Victory: see *V 5 b.] **1. a.** The letter V used as a written symbol of victory during the war of 1939–45. **b.** The Morse Code representation of this letter or the first four notes of Beethoven's Fifth Symphony, which have the same rhythm. Also *transf.*

1941 W. S. Churchill *Unrelenting Struggle* (1942) 198 The V sign is the symbol of the unconquerable will of the occupied territories, and a portent of the fate awaiting the Nazi tyranny. **1959** *Listener* 24 Sept. 489/2 The start of the 'V-sign' by Victor de Lavelaye of the Belgian Section [of the B.B.C.] in July 1941. **1972** *Times* 26 Sept. 4/8 The old V-sign, first used for the Belgian service in 1941, was a low-frequency signal. **1978** L. Thomas *Ormerod's Landing* v. 95 There came a V-sign knock on the street door.

2. a. = *victory sign* s.v. *VICTORY *sb.* 5. **b.** A similar gesture made with the back of the hand outwards as an obscene gesture of contempt (see also quots. 1948,-1973).

a. 1942 H. Nicolson *Diary* 23 Apr. (1967) 224 Winston ..gives the V-sign to an audience which does not greet him with any tumultuous applause. **1943** A. Moorehead *End in Africa* xiv. 148 They gave the V sign, they shouted and waved flags, the girls kissed the soldiers and the men ran out with bottles of wine and fruit.

b. 1948 Partridge *Dict. Forces' Slang* 202 *V-sign*, the first and middle fingers uplifted in the form of the letter *V*, for victory. When the hand making the sign was jerked upwards, it was a sign of recognition between two acquaintances passing on a road, or of complete disagreement with the statements or actions of another, adequately replacing a vulgar monosyllable of the same significance. **1959** *Spectator* 4 Sept. 297/2 Congreve's *Double Dealer* is a sex-play... To his immediate predecessors the dirty joke was still a political gesture, a deliberately ambiguous V-sign which semaphored both political and marital freedom. **1973** *Daily Tel.* 29 Nov. 3/8 Two 'louts'..taunted him outside his home by shouting obscenities and making V-signs. **1981** B. Hines *Looks & Smiles* 36 They turned round and gave him the V-sign generously with both hands..the gatekeeper hurried inside his office and slammed the door.

vuagnatite (vwā·nyătait). *Min.* [f. the name of M. *Vuagnat* (b. 1922), Swiss petrologist + -ITE[1].] A basic silicate of calcium and aluminium, $CaAl(OH)SiO_4$, found as white, usu. anhedral, orthorhombic crystals.

1976 H. Sarp et al. in *Amer. Mineralogist* LXI. 825 Vuagnatite..occurs with prehnite, hydrogrossular, vesuvianite, and chorite in rodingitic dykes from an ophiolitic zone in the Taurus Mountains, southwest Turkey. **1979** *Mineral. Abstr.* XXX. 413/2 Vuagnatite occurs as veinlets less than 5 mm wide or as small lenses less than 15 mm long in pectolite veins cutting serpentinite in the Toba district, central Japan.

|| **vue d'ensemble** (vü daṅsaṅbl'). [Fr.] A general view; an overall view of matters.

1865 Mill *Auguste Comte* 159 A philosopher's business is with general truths and connected views (vues d'ensemble). **1886** W. James *Let.* 29 Aug. in R. B. Perry *Tht. & Char. W. James* (1935) I. 601, I gained a *vue d'ensemble* of the philosopher. **1975** *Times Lit. Suppl.* 29 Aug. 972/1 The *vue d'ensemble*..shows an enviable mastery of essentials. **1976** *Ibid.* 9 July 843/1 The impetus to crystallize his professional opinions, to create a *vue d'ensemble* of his life's work.

|| **vuelta** (vwe·lta). *Bull-fighting.* [Sp., lit. 'turn, round'.] The triumphal circuit of the ring awarded to the successful matador.

1932 [see *RECORTE]. **1954** H. Casteel *Running of Bulls* vii. 134 If the applause continues he [*sc.* the matador] may make another circuit (dos vueltas). **1967** McCormick & Mascareñas *Compl. Aficionado* ii. 62 This is another sort of beauty, and the matador has his full award, of two ears and three vueltas.

vug. Add: (Further examples.)

1927 in M. Terry *Through Land of Promise* vi. 90 In places there are small 'vugs' containing quartz crystals. **1953** *Amer. Mineralogist* XXXVIII. 7 The vugs are commonly almond-shaped and range in size from 10 cm. down to microscopic dimensions. **1976** *Nature* 29 Apr. 813/1 The eye is captivated by the beautiful scanning electron microscope photographs of iron crystals deposited in vugs.

vuggy *a.* (further examples).

1953 *Amer. Mineralogist* XXXVIII. 7 Some of the replaced rock has a very vuggy texture. **1970** [see *PYROXFERROITE]. **1971** V. A. Firsoff *Gemstones Brit. Isles* iv. 25 Amethyst occurs in druses and 'small vuggy veins' in the former haematite mines of Mwyndy and Garth Ward.

vugular (vɒ·giŭlăɹ), *a.* *Geol.* [f. Vug + -ULAR.] **a.** Containing vugs. **b.** Of the nature of a vug.

1967 *Trans. Soc. Professional Well Log Analysts* P-1 Non-homogeneous carbonates such as vugular limes or dolomites. **1975** G. Anderson *Coring* i. 2 The individual pore may be like a capillary tube or it may be vugular with druzy crystal infilling. **1977** *Géotechnique* XXVII. 435 It is difficult to determine the porosity of a vugular rock sample by conventional techniques because of the problems encountered in measuring the bulk volume of the specimen.

Vuitton (vwītoṅ). In full **Louis Vuitton.** A proprietary name for the products of a French firm making high-quality luggage and other personal items.

1975 *Vogue* Sept. 305/4 (Advt.), A Louis Vuitton in the hand. **1976** *Official Gaz.* (U.S. Patent Office) 18 May TM182/2 *Louis Vuitton.* For luggage and ladies' handbags. **1977** C. Wood *James Bond, Spy who loved Me* vii. 60 His travel-weary Vuitton suitcase. **1978** H. MacInnes *Prelude to Terror* iv. 32 An amazing hodgepodge of luggage..Vuitton combined with cardboard boxes. **1980** *Trade Marks Jrnl.* 6 Feb. 280/2 *Louis Vuitton* ...1,081,942. Handbags, suitcases, trunks, umbrellas, pocket wallets, purses..key cases. **1982** C. Houck *Fashion Encycl.* 220 Vuitton luggage is the most status-ridden luggage in the world, and some of the cachet has seeped over into Vuitton handbags. **1983** *Time* 1 Aug. 40/2 Louis Vuitton handbags.

Vulcanian, *a.* Add: **6.** [f. *Vulcano*, name of one of the Lipari Islands (cf. quot. 1976).] Of, pertaining to, or designating (the stage of) a volcanic eruption characterized by periodic explosive events.

1912 *Amer. Jrnl. Sci.* CLXXXIV. 412 This hypothesis does not..invalidate the division of the eruption into a 'Strombolian' and a 'Vulcanian' phase—on the contrary, it supplies a cause for the great and continued dynamism of the 'Vulcanian' phase... It is interesting to note that the 'Vulcanian' phase, as far as actual, external eruption is concerned, is cooler than the 'Strombolian'. **1944** [see *Peléan *a.*]. **1976** P. Francis *Volcanoes* v. 169 Probably the best known Vulcanian deposit is one produced by an eruption of Vulcano itself in the 1880s. **1985** E. A. K. Middlemost *Magmas & Magmatic Rocks* i. 28/2 Vulcanian eruptions usually begin with a series of phreatic explosions that eject lithic debris from the volcanic conduit.

vulcanist. 3. (Further examples.)

1969 *New Yorker* 19 Apr. 52/2 Baldwin..was the first person to suggest convincingly that meteors might have made many of the moon's craters; previously, almost all selenologists had been vulcanists. **1971** *Nature* 27 Aug. 600/3 Although there is now a vast amount of data on the physics, chemistry and shape of the Moon, the 'vulcanists' and 'impacters' at the symposium on this subject seem to be as widely divided as ever.

vulcanizable, *a.* (Earlier example.)

1860 [see *PARTIAL *a.* 3 g].

vulcanizate (vɒ·lkănəizē[i]t). [f. Vulcanize *v.*, after *filtrate*, *precipitate*, etc.] A material that has been vulcanized.

1942 *Trans. Faraday Soc.* XXXVIII. 345 The vulcanizates. **1959** *Times* 27 Apr. (Rubber Industry Suppl.) p. ii, The general purpose synthetic rubber..does not readily form crystals and so as a gum vulcanizate it is very weak. **1983** *Jrnl. Materials Sci.* XVIII. 515 The abrasion of NR/BR blend vulcanizates has been studied in three different testing machines.

vulgarian, *a.* and *sb.* Add: Hence **vulga·rianism.**

1920 D. H. Lawrence *Lost Girl* x. 243 She saw his modern vulgarianism, and decadence. **1963** *New Society* 21 Nov. 22/1 The rising generations of eggheads is notorious for its vulgarianism, admiring Brigitte, canonizing Marilyn.

|| **vulgarisateur** (vülgarizatör). [Fr.] = *vulgarizer* s.v. Vulgarize *v.*

1933 D. L. Sayers *Murder must Advertise* xiv. 238, I obtained an introduction to her through what..that incomparable *vulgarisateur*, Charles Dickens—abominably calls a mutual friend. **1948** *Mind* LVII. 383 The introductory chapter..consists of a defence of the philosophical *vulgarisateur*. **1955** J. Wain *Interpretations* p. xiii, The making of large generalizations could safely be left to the *vulgarisateurs*. **1971** Ld. Butler *Art of Possible* viii. 157 The less voluble and extrovert Hall, to act as *vulgarisateur* or publicist for his ideas.

|| **vulgarisation** (vülgarizasyoṅ). [Fr.] = Vulgarization 1. Cf. *haute vulgarisation.

1939 *Year's Work Eng. Stud.* 1937 25 J. R. Firth's popular little sketch, *The Tongues of Men*, is an excellent example of that *vulgarisation* by experts which is becoming so fashionable to-day. **1959** *Times Lit. Suppl.* 11 Sept. 520/1 But Porché's later work had to stand up to the essays of..Sartre and Blanchot, and the propitious moment for his *vulgarisation* had come and gone. **1968** J. A. W. Bennett *Chaucer's Bk. of Fame* ii. 75 Witness his book on the Astrolabe, an admirable piece of *vulgarisation* which we ought not to dissociate severely from his poetry and which may reflect the study of astronomy in the Oxford of his day.

vulgarizing, *vbl. sb.* (Example.)

1946 J. S. Huxley *Unesco* ii. 60 They [*sc.* press and radio] have already rendered many disservices—in the vulgarising of taste, in the debasement of intellectual standards.

vulnerability. Add: **c.** *spec.* in *Contract Bridge*, the quality or state of being vulnerable (sense 2 d below).

1927 M. C. Work *Contract Bridge* ii. 10 Trick values, the number of points to a game, game and rubber bonuses..are not affected by vulnerability. **1942** E. Culbertson *Official Bk. Contract Bridge* 239 The consequence of vulnerability is exposure to greater penalties and entitlement to greater premiums. **1975** *Times* 15 Nov. 13/5 In May 1925, Mr Vanderbilt..joined in a rubber of the continental game of plafond. He saw possibilities in the game, added the attraction of vulnerability..and..introduced it to the Whist Club under the name of Contract Bridge.

d. *pl.* The weaknesses in a military defence system.

1972 *Science* 19 May 819/1 The purpose of this conference is to define a comprehensive but balanced assessment of the risks, vulnerabilities, capabilities, and intentions for subversive disruption. **1978** *Time* 3 July 36/3 'There is no doubt in my mind,' he said, 'that the U.S. is the most powerful country in the world.' He admitted 'concern' about 'vulnerabilities' in NATO, but said he felt the U.S. could 'outthink, outdesign and outperform the Soviets with the resources we have and the steady increases we are requesting'.

vulnerable, *a.* Add: **2. d.** *Contract Bridge.* Of or pertaining to the liability of one side to be awarded increased penalties or increased bonuses as result of having won a game.

[**1925** *Work-Whitehead Auction Bridge Bull.* Oct. 5 In circles where high play predominates, there was introduced a variation of the above count, called 'Le Vulnerable'.] **1927** [see *danger-zone* s.v. *DANGER *sb.* c]. **1965** *Listener* 23 Sept. 474/3 Playing the Two Club system, my partner, the dealer, opened a vulnerable No Trump (strong). **1977** M. Kenyon *Rapist* i. 10 Five no trumps! Don't get vulnerable!

vulture, *sb.* Add: **4. d.** *vulture-wise* adj.

1892 Stevenson & Osbourne *Wrecker* xiii. 206 Towards her the taut *Norah Creina*, vulture-wise, wriggled to windward: come from so far to pick her bones. **1906** W. de la Mare *Poems* 82 This beast in one flat hand clasped vulture-wise A glitt'ring image.

vulture, *v.* Delete *rare*[−1] and add later examples. Also *intr.* with *down*, to descend like a vulture.

1922 Joyce *Ulysses* 47 He rooted in the sand..vulturing the dead. **1948** M. Allingham *More Work for Undertaker* xiii. 164 The tax harpies vultured down for death duties. **1955** *Newsweek* 10 Jan. 49/1 Contestants had vultured the library's reference and guidebooks for the names of the mystery towns. **1977** L. Gordon *Eliot's Early Years* 147 From 'Proteus' came the dog vulturing the dead in part I of *The Waste Land*.

Hence **vu·ltured** *ppl. a.* (*poet.*).

1946 Dylan Thomas *Deaths & Entrances* 47 O spiral of ascension From the vultured urn Of the morning Of man.

vulvectomy (vɒlve·ktŏmi). *Surg.* [f. Vulv(a + *-ectomy.] Excision of (part of) the vulva.

1917 *Amer. Jrnl. Obstetr. & Dis. Women* LXXVI. 797 A vulvectomy was done on June 27th and the patient left the hospital August 13th. **1962** *Lancet* 1 Dec. 1171/1 The histological findings of atypical basal-cell hyperplasia in any section can then justify vulvectomy.

vurry (vɒ·ri), *a.* *colloq.* Repr. a supposed Amer. pronunciation of Very *a.* and *adv.*

1889 Kipling *From Sea to Sea* (1899) I. xii. 306 The tourists..call to one another: 'Sa-ay, don't you think it's vurry much the same all along?' **1913** E. Pound *Let.* (1971) 14 Have just discovered another Amur'kn [*sc.* Robert Frost]. *Vurry* Amur'k'n. **1943** C. S. Churchill *Let.* 19 Dec. in M. Soames *Clementine Churchill* (1979) xxii. 344 The food, however, is 'vurry' American. **1972** K. Bonfiglioli *Don't Point that Thing at Me* vii. 65 [He] wished me a vurry, vurry happy visit to the US of A.

vygie (fəi·χi). *S. Afr.* [Afrikaans, f. *vy(g)*, f. Du. *vijg*, *vig* fig + *-ie* diminutive suffix.] Any of several small succulent plants belonging to the genus *Mesembryanthemum* or a closely related genus, or the brightly coloured flower of one of these plants. Also **vy·gebosch** [Afrikaans *vyebossie*].

1795 tr. *C. P. Thunberg's Trav. Europe, Afr., & Asia* II. 35 The field was here of the Carrow kind, and the sheep were said to feed on those succulent plants, the Mesembryanthemums (*vygebosches*). **1892** *Jrnl.* (Grahamstown) 11 July 1/2 (Advt.), The Veld consists of much-sought-after Karoo, Vygebosch,..with Mimosas in the valley. **1931** *Farming in S. Afr.* 393 The mineral and feeding-stuff analyses of a vygie give no clue to its probable palatability. **1950** *Cape Argus* 5 Aug. 7/5 Vygies or sour figs.. blossom in a wide range of colours—scarlet, blue, purple, pink or flaming yellow. **1966** E. Palmer *Plains of Camdeboo* xvii. 271 The little vygies on the flats. **1971** [see *GOUSBLOM].

Vynide (vai·nəid). Also **vynide.** A proprietary name for a plastic used as a substitute for leather in upholstery, clothing, etc.

1943 *Trade Marks Jrnl.* 16 June 257/2 *Vynide*...Leather cloth. Imperial Chemical Industries Limited, Wexham Road, Slough, Buckinghamshire; manufacturers and merchants. **1958** *Archit. Rev.* CXXIV. 395/2 The book-case shelves and backs, and the wall opposite the window are covered in white vynide. **1960** *Housewife* Feb. 58/2 Thankful I brought good country-looking Vynide jacket ..(looks like leather, won't leak, will sponge clean). **1976** *Lancs. Evening Post* 7 Dec. 12/8 (Advt.), Reupholstery. Your three piece suite made as new in moquette or vynide.

vysotskite (vīsǫ·tskəit). *Min.* [ad. Russ. *vysotskit* (Genkin & Zvyagintsev 1962, in *Zap. Vsesoyuz. Min. Obshch.* XCI. 718), f. the name of N. K. *Vysotsky* (d. 1932), Russian mineralogist: see -ITE[1].] A sulphide of palladium and often nickel, (Pd,Ni)S, found as minute silvery tetragonal crystals having a metallic lustre.

1963 *Chem. Abstr.* LVIII. 8777 For the genesis of vysotskite is characteristic the assocn. with Pt ores, pentlandite, millerite, and other sulfides, in effusive andesite-diabase in gabbro-diabase intrusions. **1978** *Amer. Mineralogist* LXIII. 832/1 Vysotskite has since been reported..from the Stillwater Complex of Montana ..and the Lac des Isles deposit, Ontario. *Ibid.* 838/1 We conclude that braggite and vysotskite are simply compositional variants of the same phase, (Pd,Pt,Ni)S... We suggest that the family of minerals be called the braggite series and that all compositions containing less than 10 mole percent PtS be called vysotskite.

W

W. Add: **3.** W, colloq. shortening of W.C.; a lavatory or water-closet; W., women('s size) (WX, extra-large women's size); W.A., Western Australia; WASP (*U.S.*), Women's Airforce Service Pilots; WAT (*Aeronaut.*), Weight and Temperature; WATS (*U.S.*), Wide Area Telephone Service; W.C. (examples of 'West Central'); W.C.C., World Council of Churches; WCT, World Championship Tennis; W.C.T.U. (*N. Amer.*), Women's Christian Temperance Union; W.D., War Department; W.D.C., Woman Detective Constable; W.D.S., Woman Detective Sergeant; W.E.A., Workers' Educational Association; w.e.f., with effect from; W.E.U., Western European Union; wff (*Logic*) = *well-formed formula* s.v. *WELL-FORMED *ppl. a.* c; WFTU, World Federation of Trade Unions; w.h.b., wash-hand basin; W.H.O., World Health Organization; W.I. (also † West India) (examples); W.I., Women's Institute; W.I.Z.O., Wizo (wī·tzo), Women's International Zionist Organization; WKB (*Physics*) [initials of G. Wentzel, H. A. Kramers, and L. Brillouin, who each published papers on the method in 1926], used *attrib.* with reference to a method for obtaining an approximate solution of the Schrödinger equation based on the expansion of the wave function in powers of Planck's constant, *h*; W.L.A., Women's Land Army; W.M.O., World Meteorological Organization; W.O., War Office; W.O., Warrant Officer; W.O.C.(S), waiting on cement (to set); W.O.S.B. (also with pronunc. (wǫ·zbi)), War Office Selection Board; W.O.W., waiting on weather; w.p., weather permitting; WP, word processing; WPA (*U.S.*), Works Progress Administration; W.P.B., w.p.b. (*slang*), waste-paper basket; W.P.C., Woman Police Constable; w.p.m., words per minute; W.R.A.N.S. (rænz) (*Austral.*), Women's Royal Australian Naval Service; hence WRAN, Wran, a member of this; W.R.N.S., Women's Royal Naval Service; cf. *Wren[2]; w.r.t., with respect to; W.R.V.S., Women's Royal Voluntary Service, formerly *W.V.S.; W.S. (earlier example); W.S.P.U., Women's Social and Political Union; W/T, W.T., wireless telegraphy; freq. *attrib.*; w/v (see quot. 1907); W.V.S., Women's Voluntary Service; WW (I. II), World War (One, Two); W.W.W., World Weather Watch. See also *W.A.A.C., *W.A.A.F., *W.A.C., *Wasp, *W.R.A.C., *W.R.A.F., *Wren[2].

1953 Dylan Thomas *Under Milk Wood* (1954) 11 Talking to the lamp-post..Using language..Singing in the W. **1954** A. S. C. Ross in *Neuphilol. Mitteilungen* LV. 41 *W.* either 'the letter W' or 'W.C.' (a frequent non-U expression for 'lavatory'). **1978** E. Malpass *Wind brings up Rain* x. 105 A small garden of weeds, with a cinder path leading to a W. **1926–7, 1974** W [see S.W. s.v. *S 4 a]. **1900** *Morning Post* 12 July 9/4 Town Pro. W.A. **1971** *Sunday Australian* 8 Aug. 5/2 The State shipping service in WA has been made the scapegoat for Government indecision. **1943** *Yank* 24 Sept. 17 WASP, which stands for 'Womens Air Force Service Pilots' is the new official title of women pilots of the AAF. **1957** *Times* 9 Dec. 12/7 There is an additional requirement governing the maximum take-off and landing weights for the altitude and weight prevailing (known as the 'WAT curve'). **1976** B. Lecomber *Dead Weight* xi. 136 Those coins weigh 600 lb..if we pile in three people on top of that that we'll be totally WAT-limited... WAT—Weight and Temperature. **1962** *Fleet Owner* Aug. 73/1 With WATS you pay a monthly fee for a special line over which you can call any telephone number in prescribed States. **1977** *New Yorker* 12 Sept. 68/3 Harry has a WATS line, so Henny called Sadie again, this time for free. **1857** *Punch* 7 Feb. 51/2 Rowland Hill has just divided London's waste of brick by ten... Lawyers, and good *Coram's* Foundlings, All are found in W.C. **1935** E. Farjeon *Nursery in Nineties* i. viii. 63 We found the little oblong envelope..with..the London W.C. postmark. **1948** *Ecumenical Rev.* Autumn 118 The work of two Conferences convened by the Study Department of the W.C.C... is reported in this book. **1971** K. Grubb *Crypts of Power* viii. 168 Many Churches of Asia, Africa and Latin America joined the W.C.C. **1969** *N.Y. Times* 19 Feb. 54/5 Robert A. Briner of W.C.T. said: 'This is a truce.' **1982** *Tennis Today* June 12/3 Dallas had its best field for a few years, which was an effective answer to those who would suggest WCT are on the way out after the split with the Grand Prix. **1888** Lindley & Widney *California of South* 109 The W.C.T.U. was first organized here in 1883. **1956** B. Holiday *Lady sings Blues* (1973) xix. 157 Nobody who has a police record can hold a liquor licence. This was a sop to the WCTU. **1855** *Admiralty Circular* 27 Aug., In future the mark W.D. (War Department), with the Broad Arrow, shall be used for Stores provided by the Ordnance Department. **1920** *Punch* 7 Jan. 9/1 The orders are dead strict against civilians riding in W.D. vehicles. **1942** E. Waugh *Put out More Flags* ii. 149 Patches of rank land marked on the signposts W.D., marked on the maps as numbered training areas. **1976** 'J. Charlton' *Remington Set* xviii. 86, I need something on a three-ton chassis.. ex-WD Bedford would do nicely. **1970** G. F. Newman *Sir, You Bastard* viii. 217 Find two WDC's to accompany the women. **1972** *Police Rev.* 8 Dec. 1597/1 We in Nottinghamshire have not yet decided whether we have D.P.W.s or W.D.C.s or whether they are W.D.S.s or D.P.W. Sergts. **1910** S. A. Barnett *Let.* 10 Mar. in H. Barnett *Canon Barnett* (1918) II. l. 332 The W.E.A. has life. **1936** A. Huxley *Eyeless in Gaza* xxix. 393 W.E.A. lecture-rooms. **1981** D. Rowntree *Dict. Educ.* 350 *Workers' Educational Association* (*WEA*), a UK organisation founded in 1903 to stimulate and satisfy the demand for education among working class people... The WEA now arranges adult education classes and courses throughout the UK (no longer primarily for working class people). **1942** Partridge *Dict. Abbrev.* 101/2 *w.e.f.*, with effect from (a given date). **1954** J. Masters *Bhowani Junction* xxxii. 275 It terminated her duty with my battalion w.e.f. June the twelfth. **1978** T. Allbeury *Lantern Network* viii. 93 The official notification of his promotion to major w.e.f. 1 Jan. 44. **1954** *Times* 11 Nov. 8/7 The first [of the protocols] modifies the older treaty to allow for the adherence of Italy and the German Federal Republic..and emphasizes close cooperation with the North Atlantic Treaty Organization, on which W.E.U. will rely. **a 1974** R. Crossman *Diaries* (1976) II. 401 George [Brown] proposes to go to a meeting of the W.E.U. next Tuesday and wants to make an important declaration there on Europe. **1944** w.f.f. [see *PRENEX *a.*]. **1956** A. Church *Introd. Math. Logic* (rev. ed.) I. i. 70 We shall hereafter use the abbreviations 'wf' for 'well-formed', 'wff' for 'well-formed formula', and 'wffs' for 'well-formed formulas'. **1967** *Encycl. Philos.* V. 2/2 This property will clearly be perceived if a small letter is systematically replaced by any wff. **1971** G. Hunter *Metalogic* i. 4 What things are to be wffs. **1947** *Times* 17 Nov. 5/6 The International Transport Workers' Federation will meet.. to consider further its decision recently taken in Washington not to affiliate to the W.F.T.U. **1978** *Statesman's Year-Bk.* i. 39 The WFTU formally came into existence on 3 Oct. 1945. **1975** *Evening Herald* (Dublin) 8 May 10/3 (Advt.), Cloakroom with w.c. and w.h.b. **1946** *N.Y. Times* 28 July iv. 2/3 Dr Thomas Parran..described the constitution of the WHO as a major contribution to world peace. **1960** *New Statesman* 2 Apr. 478/3 The Americans are..working with the WHO to build hospitals and eliminate malaria. **1977** G. Scott *Hot Pursuit* xi. 98 If this was a WHO team..we could be passing up the one opportunity we had. **1848** *Brit. Army Despatch* 17 Nov. 292/2 Late Captain 2nd W.I. Regt. **1900** *Naval & Mil. Rev.* July 5 Name of Vessel. Pearl. Present Station. N.A. and W.I. **1973** *Advocate-News* (Barbados) 17 Jan. 2/5 (*heading*) Deep concern over W.I. **1928** *Mid-Sussex Times* 18 Dec. 8/1 A tangible result of the W.I. trading scheme. **1940** C. Milburn *Diary* 26 Mar. (1979) 28 A W.I. Produce Meeting..to hear about sugar for jam and the arrangements about getting it for W.I. members. **1965** *New Society* 7 Jan. 5 At Stony Creek, Ontario, in 1897, the first WI meeting was held... It wasn't until the outbreak of World War I that a Canadian widow named Mrs. Watt introduced the WI to Britain... The first meeting was held in September 1915. **1981** J. Scott *Distant View of Death* xii. 158 Jars of home produce to transport to the WI stall set up in the square. **1925** *Zionist Rev.* IX. 50/2 The growing edifice of the W.I.Z.O. **1940** A. Ulitzur *Two Decades of Keren Hayesod* iii. 47 The Women's International Zionist Organisation (*Wizo*). **1978** *Jewish Chron.* 6 Oct. 14/3 Coventry and Leamington Wizo held a three-day nearly new sale and raised £350 for Israel. [**1932** J. L. Dunham in *Physical Rev.* XLI. 713 The Wentzel-Brillouin-Kramers method of handling the wave equation (hereinafter referred to as the W.B.K. method) is very well suited to the calculation of energy levels in heavy systems.] **1935** *Ibid.* XLVII. 748/2 The wave function can be represented by the WKB (Wentzel-Kramers-Brillouin) solution. **1961** Powell & Crasemann *Quantum Mech.* v. 142 The WKB solutions cannot be valid near a classical turning point, where the momentum is zero. **1974** G. Reece tr. *Hund's Hist. Quantum Theory* vii. 97 This was confirmed by later proofs using the Schroedinger equation and the WKB approximation. **1939** *Times* 26 Oct. 11/5 The farmer who is loth to employ one of the W.L.A. may see his more progressive neighbour very well served. **1942** C. Milburn *Diary* 17 Oct. (1979) 155, I did a little W.L.A. work, going down to the hostel with some magazines. **1951** *Bull. Amer. Meteorol. Soc.* XXXII. 239/1 The International Meteorological Organization wound up its affairs in Paris on March 15–17, 1951 and handed over its assets, obligations and numerous resolutions to the new World Meteorological Organisation... Sir Nelson Johnson aptly put it when he said.. 'The IMO is dying, long live the WMO.' **1977** *Whitaker's Almanack 1978* 810/1 World Meteorological Organization (WMO), Geneva.... The present membership is 139 States and 8 Territories. **1860** F. Nightingale *Let.* 8 Dec. in C. Woodham-Smith *Florence Nightingale* (1950) xvi. 358, I do hope you won't have any vain ideas that you can be spared out of the W.O... You are necessary to the reorganising of the W.O. **1914** R. Brooke *Let.* 24 Sept. (1968) 619 He may go out as an interpreter... The W.O. has his name. **1931** *N. & Q.* 5 Dec. 408/2 Who was Capt. Robert Holden, of the 130th Regt.? From Army Lists and W.O. Commission Books, it seems that he was commissioned Lieutenant in the 115th Regt., 14 Nov., 1794. **1887** O. L. Perry *Ranks & Badges H.M. Army & Navy* ix. 102 Such W.O.'s, N.C.O.'s and men as may be specially placed under their orders. **1977** 'O. Jacks' *Autumn Heroes* iv. 65 He's an ex W.O.II in the Paras. **1949** *Amer. Speech* XXIV. 35 The verbal *wocsing*..is derived from the initials W.O.C.S., which stand for *waiting on cement to set* (the written form is often simply *woc* without capitalization or punctuation)... A field worker tells me I must have misspelled the word, as it is always pronounced ['wɒkəsiŋ] in the field. **1974** D. K. Smith in P. L. Moore et al. *Drilling Practices Manual* xvi. 400 Improvements in cements, understanding WOC times, and the use of admixes have reduced WOC time to a few hours under present-day practices. **1945** *Jrnl. R. Army Med. Corps* LXXXIV. 75 W.O.S.B.'s which followed the pattern of the experimental Board. **1978** *Jrnl. R. Soc. Arts* CXXVI. 268/1 The need for skilful selection of officers..led to the WOSB or War Office Selection Board. **1982** S. Raven *Shadows on Grass* (1983) v. 95, I was summoned to a W.O.S.B...which would finally determine whether I was fit to be trained for a Commission. **1967** R. De Sola *Abbrev. Dict.* (rev. ed.) 277/1 *WOW*, waiting on weather. **1975** *Oil & Gas Industry Gloss. Terms* (Bank of Scotland) 5/2 *W.O.W.*, Waiting on Weather; usually applied to mobile offshore drilling platforms but can also refer to other offshore operations. **1889** E. C. Dowson *Let.* 2 Mar. (1967) 43, I shall be at Baker St. as you prefer—w.p. before 3.0, to-morrow. **1931** Joyce *Let.* 17 Apr. (1966) III. 217, I..hope to arrive in London..that evg. (w.p.). **1974** *New Acronyms & Initialisms* (Gale Research Co.), *WP*, word processing. **1980** *City Recorder* 10 Jan. 7/3 In a really small company, the executive who can operate a WP machine can work almost as a one man band. **1936** *N.Y. Herald Tribune* 1 June 8/4 (*heading*) W.P.A. must increase efficiency, says Ridder. **1943** J. S. Huxley *TVA* ix. 69 In the reception room..the mural by a WPA artist helps the guide to explain the project. **1979** *Listener* 16 Aug. 214/1 The Roosevelt WPA project and other attempts at a social art in the Thirties. **1903** *Photogram* X. 320/1 Anonymous letters are strongly objected to, and ..go into the W.P.B. **1934** T. E. Lawrence *Lett.* (1938) 815 Please dump it in the W.P.B. when it begins to bore you. **1939** *War Illustr.* 16 Dec. p. ii/3 Possessing a very large wastepaper basket, I give short shrift to all anonymous correspondents. A glance at any missive signed with a nom-de-plume and into that w.p.b. it goes. **1974** P. Flower *Odd Job* 26 He..let it fall into the w.p.b. **1984** *Times* 20 Jan. 10/1 So presumably it's someone's job to empy wpbs and clean already clean ashtrays for part of the year. **1963** F. B. Fawcett *Cycl. Initials & Abbrev.* 158/2 *WPC*, Woman Police Constable. **1966** L. Southworth *Felon in Disguise* i. 17 'I'll send in the W.P.C. right away.' The Inspector left the room. **1981** 'J. Ashford' *Loss of Culion* xiii. 94 Why hadn't she accepted the offer of having a WPC with her? **1936** G. Dewey *GS Teacher* p. xxiii, Suggested dictation rate: 30 wpm. **1977** *Belfast Tel.* 17 Jan. 17/1 (Advt.), Applicants should have a typing speed of at least 40 w.p.m. and also 80 w.p.m. shorthand. **1945** Baker *Austral. Lang.* viii. 159 *W.R.A.N.S.*, Women's Royal Australian Naval Service (called *Wrans* or *Rons*). **1970** M. Kelly *Spinifex* ii. 30 One of the WRANS hurried over, 'Telephone, sir'. 'Thanks, Betty.' **1919** *Daily Mail Year Bk.* 49/2 The Admiralty must feel justly proud of their W.R.N.S. **1977** *Navy News* Dec. 20/1 Two world wars played a decisive part in the story of the WRNS. The First saw its formation..The Second swung those doors wide to a tremendous range of opportunities. **1956** W. L. Ferrar *Differential Calculus* iii. 33 Let. *z* = *f*(*y*), *y* = *φ*(*x*); let *z* have a finite derivative *f*′(*y*) w.r.t. *y*. **1979** *Nature* 30 Aug. 845/2 A measure *D* of the depth of this minimum is defined as the average VAI near days −2 and +4 (w.r.t. day 0 as the boundary transit time) minus the average VAI near day 1. **1966** *Care & Distribution of WRVS Processed Clothing* (Women's Royal Voluntary Service) p. iii, The reputation built up by WRVS in the war years for skilled care..has been enhanced in post-war years. **1978** *Morecambe Guardian* 14 Mar. 13/1 She holds a long service medal from the WRVS and still retains very strong links with the village of Holme. **1852** Thackeray *Let.* 24 May in J. Brown *Lett.* (1912) 403 Blackwood the W.S. I saw in the Park yesterday. **1907** N. A. Martel in B. Villiers *Case for Women's Suffrage* 145 A young girl.. came from Huddersfield..in the care of one of the leaders of the W.S.P.U. **1967** R. S. Churchill *Winston S. Churchill* II. xi. 402 The WSPU now resorted to stone throwing and later to arson. **1914** W. S. Churchill in M. Gilbert *Winston S. Churchill* (1972) III. Compan. i. 292 A submarine off Sylt report by W/T (or by pigeons) at night to a waiting destroyer that the weather is favourable. **1923** *Man. Seamanship* (Admiralty) II. 23 Visual and wireless messages..sent either to the flag deck or the W.T. office for transmission. **1965** 'J. le Carré' *Looking-Glass War* vii. 84 'He was a trainer in wireless transmission.'..'The WT man?' **1978** R. V. Jones *Most Secret War* xlix. 480 Then I saw its astonishing heading: 'Obscene W/T Traffic.' **1907** *Brit. Pharmaceutical Codex* p. ix, 'w/v' represents 'weight in volume', indicating that a weighed quantity of a solid substance is contained in solution in a measured quantity of liquid. **1973** *Nature* 5 Oct. 267/2 Here I report the use of four inhibitors at concentrations of 0·5% w/v which can each stabilize ethanol in blood. **1939** N. Last *Diary* 4 Sept. in *Nella Last's War* (1983) 11 I've got lots of plans made to spare time so as to work with the W.V.S. **1976** R. Barnard *Little Local Murder* iii. 30 The compère talked..to the head of the town's WVS. **1960** *Acronyms Dict.* (Gale

Research Co.) 210 WWI, World War I. WWII, World War II. **1976** *Greenlist* (J. R. Wrigley Catal.) No. 23. 30 Crascredo—No Joke... (The humour of W.W.1.) **1979** G. F. Newman *List* iii. 25 The bank had been..incorporated in '41, shortly before America entered WWII. **1963** *Times* 29 Apr. 9/1 It was pointed out during the congress that as ancillary measures must take several years the full realization of plans for a world weather watch (W.W.W.) was as yet a distant prospect but the interval would enable scientists to ascertain weather potentialities. **1970** *Sci. Jrnl.* Apr. 48 At the time of its inception in 1967 the WWW plan relied entirely on improving and extending existing systems. **1942** A. P. Jephcott *Girls growing Up* i. 33 My biggest tragedy is I am fat & wear W.X. clothes. **1971** J. Thomson *Not One of Us* (1972) vii. 79 Cardboard boxes were piled everywhere..'Gents' Hose. Assorted Colors.' 'Ladies Knickers. W.X.' **1984** W. Beechey *Rich Mrs Robinson* xii. 103 Customers..[who] thought of themselves as Small Women size when really they were WX.

4. Symbolic uses. **a.** *Genetics.* *W* is used to designate the female-determining sex chromosome in species in which the female rather than the male is the heterogametic sex.

1917 T. H. Morgan in *Amer. Naturalist* LI. 533 In moths in which the female is the heterogametic sex, the Y chromosome (or the W chromosome to use a different nomenclature) is transmitted only by the female line. **1925** *Amer. Naturalist* LIX. 133 The locus of the F genes is in the W-chromosome that descends from mother to daughter. **1964** R. A. Beatty in Armstrong & Marshall *Intersexuality* ii. 35 There is one kind of spermatozoon (Z) in animals with the ZW:ZZ mechanism. **1971** [see *HETEROGAMETIC *a.*]. **1984** *Nature* 23 Feb. 690/3 Natural populations [of the platyfish, *Xiphophorus maculatus*] are polymorphic for three sex factors (W, X and Y)... The XY and YY genotypes usually develop as males and XX, WX and WY become females; WW genotypes do not in fact occur because male gametes can never bear W factors.

b. *Particle Physics.* [Initial letter of *weak*.] *W* is the symbol for a heavy, charged vector boson that is probably the quantum of the weak interaction.

1960 Lee & Yang in *Physical Rev. Lett.* IV. 310 Possible existence of a weakly coupled boson $W^{\pm}$. *Ibid.*, The question of a neutral W^0 will not be examined here. **1971** *New Scientist* 2 Sept. 498/3 The experimental results agree with this theory if the W particle has a mass of about 36 proton masses. **1977** *Dædalus* Fall 32 Exchange of the W produces the familiar weak interactions, like nuclear beta decay. **1982** *Nature* 23 Sept. 295/2 The heavy W and Z bosons..can be produced in p̄p reactions. **1983** *Sci. Amer.* Apr. 62/2 With the discovery of the *W* particle a major goal of experimental elementary-particle physics has been achieved.

Wa (wā), *sb.* (and *a.*) [Native name.] **a.** (Any of) a group of hill-dwelling peoples of eastern Burma and southwestern Yunnan; a member of one of these peoples. **b.** Their (Mon Khmer) language or dialect. Also *attrib.* or as *adj.*

1860 F. Mason *Burmah* (ed. 2) ii. 69 There is a wild tribe called Lawa on the mountains between the Irrawaddy and the Meinan... There is a small settlement of them..in..Amherst; there formerly was one in Tavoy, where the Karens called them *Wa*. **1902** *Encycl. Brit.* XXX. 519/2 It [*sc.* the Burmese state Mang Lön] is the chief state of the Vü or Wa tribes. **1911** *Ibid.* XXVIII. 223/2 They are popularly divided into Wild Was and Tame Was. The Wild Was are remarkable as the best authenticated instance of head hunters in the British Empire. **1942** [see *Palaung *sb.* and *a.*] **1966** D. Wilson *Quarter of Mankind* xvi. 211 They [*sc.* Communist maps] acknowledged the Burmese claim to one piece of territory depicted in Peking as Chinese, but they took an entirely new bite out of the Wa territory shared between the two. **1974** *Encycl. Brit. Micropædia* IV. 272/1 Ethnolinguistic Groups in China..Wa (Kawa).

W.A.A.C., WAAC, Waac (wæk). Also **wack.** Acronym f. the initial letters of *Women's Army Auxiliary Corps* (1917–19); also *U.S.*, the orig. name of the Women's Army Corps (see *W.A.C.), formed in 1942, a member of this. Also *attrib.*

1917 M. Macdonagh *Diary* 29 Mar. in *In London during Great War* (1935) iii. ii. 186 The W.A.A.C.s (Women's Auxiliary Army Corps). *Ibid.*, One of the riddles of the day is: 'Which would you like: a whack on the head or a Waac on your knee?' **1917** W. Owen *Let.* June (1967) 472 The WAAC's now wait on us. It is rather pleasant. **1918** J. M. Grider *Diary* 5 Mar. in *War Birds* (1927) 85 A Waac officer can't walk out with a Tommy any more than an army officer can be seen with a Waac private. **1919** *Times* 31 Dec. 11/4 Queen Mary's Army Auxiliary Corps. To-day the W.A.A.C.'s, to give them the name under which they first earned the gratitude and regard of the Army and the public, cease to have a corporate existence. **1930** A. Bennett *Imperial Palace* lviii. 456 What do you know of the Western Front, my dear? You aren't old enough to have been a Waac. **1935** A. J. Cronin *Stars look Down* ii. xiii. 379 ''S war work,' Stokes suggested with a ribald leer ''S war work wish [*sic*] a wack.' **1942** *Time* 18 May 62/3 WAAC at last... The Women's Army Auxiliary Corps will provide women for the Army's humbler jobs. **1957** H. H. Jenkins *Diction of 'Yank'* (Univ. Florida thesis) vi. 59 Ordnance announced the addition of the WAAC-Cycle, a light-weight, streamlined bike for the ladies. **1976** 'A. Cross' *Question of Max* x. 134 Everyone was just WAACs and Tommies together.

W.A.A.F., WAAF, Waaf (wæf). Acronym f. the initial letters of *Women's Auxiliary Air Force* (1939–48, subsequently reorganized as part of the Women's Royal Air Force); a member of this.

1939 *Daily Tel.* 18 Dec. 12/4 (Advt.), Women between the ages of 18 and 43 are required immediately for enrolment in the W.A.A.F. **1942** J. Pudney *Dispersal Point* 11 Weather at base closes down for the night: And the ash-blonde Waaf is waiting tea. **1948** W. Fortescue *Beauty for Ashes* xxvii. 209 W.A.A.F.s and airmen climbed into a private auto-bus. **1956** R. Macaulay *Towers of Trebizond* xv. 179 Now the women who go with armies are..called Ats and Wrens and Waafs and Wracs. **1969** G. Macbeth *War Quartet* 42, I had chosen a WAAF, Needing her flesh and heat. **1982** J. Aiken *Whisper in Night* 25 They wouldn't allow W.A.A.F. girls to fly. **1984** *Sunday Tel.* 16 Sept. 10/3 The contribution by the WAAF to the anti-aircraft defence of this country lay in manning..many of the balloons of Balloon Command.

waal (wæl, wāl). Also **wal.** Repr. a U.S. colloq. and dial. pronunc. of Well *adv.* 23.

1863 [see *heater-piece* s.v. *HEATER 3]. **1897** Kipling *Captains Courageous* vi. 129 'D'you suppose we can run her blind?' he shouted. 'Wa-al, *I* can,' Disko retorted. **1978** M. Z. Lewin *Silent Salesman* xxxvi. 195 Wal, I guess that'll be all right.

wabe (wē[i]b). A factitious word introduced by 'Lewis Carroll' (see quot. 1855[2]).

1855 'L. Carroll' *Rectory Umbrella & Mischmasch* (1932) 139 The slythy toves Did gyre and gymble in the wabe. *Ibid.* 140 *Wabe*, (derived from the verb to *swab* or *soak*). 'The side of a hill' (from its being *soaked* by the rain). **1871** —— *Through Looking-Glass* 24 The slithy toves Did gyre and gimble in the wabe.

Wabenzi (wabe·nzi). Also **Wabenze, Wa-Benzi.** [Invented to resemble the name of an African people (viz. *Watusi, etc.), with inserted (Mercedes-)*Benz*: see def.] In Africa, 'the Mercedes-Benz tribe': used *joc.* to designate those Black politicians, businessmen, and others whose success is characterized by their ownership or use of a Mercedes-Benz car.

1967 *Economist* 8 Apr. 112/1 Africa is rapidly earning itself the reputation of the world's biggest joke..the 'Wabenze', the new but already well-known African people whose tribal mark is a ministership and a Mercedes. **1972** J. Biggs-Davison *Africa—Hope Deferred* x. 92 The sweeteners of deference and the shining, pennanted cars that have given their name to such new and privileged tribes as the 'Wa-Benzi'. **1978** S. Naipaul *North of South* i. i. 42 'Wabenzi' is the pleasantly jocular term used to describe the nouveau-riche black middle-class... They signal their status by the acquisition..of a Mercedes Benz. Hence, the Wabenzi—the Benz tribe. **1980** *Times* 17 July 11/1 The Wa-Benzi are the new tribe of Mercedes-Benz mounted black bureaucrats and politicos.

|| **wabi** (wā·bi). [Jap.] In Zen Buddhist philosophy, a quality of simple, serene, and solitary beauty of a slightly sombre kind (see also quot. 1962). Cf. *SABI.

1934 D. T. Suzuki *Essays in Zen.* III. vii. 312 This spirit of 'Eternal Loneliness' is something known pre-eminently in Japan. By this spirit..I mean what is popularly known in Japan as 'Sabi' or 'Wabi'. **1962** J. Petrie tr. *Hasumi's Zen in Japanese Art* iv. 51 The essential features of Higashiyama art, extending into all fields, can be summarized in..the idea of 'Wabi', which is supposed to express the highest beauty and can also be carried over into other fields of art. Fundamentally it means poverty, and at the same time simplicity and calm, but it also implies an inexpressible inner joy hidden in deep modesty. Out of 'Wabi' developed harmony, respect, purity, poverty... That is what the special designation of 'Wabi' amounts to: it was the favourite expression of the Haiku masters. **1965** [see *SABI]. **1979** S. Coe in I. Webb *Compl. Guide to Flower & Foliage Arrangement* xvii. 227/3 Then comes *wabi*, a recognition of ordinary things but seen in a very clear, almost transparent light.

waboom, var. Wagenboom in Dict. and Suppl.

WAC (wæk). *U.S.* Also **W.A.C., wac.** Acronym f. the initial letters of *Women's Army Corps*, formed in 1943 (cf. *W.A.A.C.); a member of this.

1943 *Yank* 30 July 18 The fellows sent the girls to the WAC barracks to get dressed while they finished the KP jobs. *Ibid.*, When the first bunch of Wacs arrived here recently a group of GIs made dates with the prettiest one night for the Camp movie. **1956** B. Holiday *Lady sings Blues* (1973) xxiii. 198 There was a war on and no damn way for me to get to England except to swim or join the WACs. **1962** E. Snow *Other Side of River* (1963) xxxix. 290 Women are not recruited for combat training except as parachutists..but a few are used in auxiliary tasks, like WACs. **1976** 'R. B. Dominic' *Murder out of Commission* xiv. 100, I fetched and carried for a WAC from Camp Pendleton.

wack[1] (wæk). *slang* (orig. *U.S.*). Also **whack.** [Prob. back-formation f. *WACKY *a.*] An eccentric or crazy person; a madman, a crackpot.

1938 'E. Queen' *Four of Hearts* (1939) i. 9 All you wacks act this way at first. Them that can take it snaps out of it. **1951** E. Paul *Springtime in Paris* xi. 198 The show place, the rendezvous of eccentrics, Bohemians, playboys, sightseers and international whacks is the St. Germain des Prés quarter. **1959** R. Graves in *Lilliput* Dec. 48/2 'I don't get the joke,' Len grumbled. 'That wack gave me the creeps! One of those 'creative artists' who create chaos'. **1982** G. F. Newman *Men with Guns* xi. 81 The cop shrugged. 'Some wack with a grudge.'

wack[2] (wæk). *dial.* (chiefly *Liverpool*). Also **whack.** [Cf. *WACKER.] A familiar term of address; 'pal', 'mate'.

1963 *Beatles* 18 Said Ringo: 'Do you want an anaesthetic, wack?' **1966** P. Moloney *Plea for Mersey* 50 I'll get you wack! You must come back! **1972** 'K. Royce' *Miniatures Frame* ix. 124 'How's it, whack?' He..eyed me affectionately. **1976** *New Musical Express* 12 Feb. 18/6 Gorra light 'ave yer, wack?

wacker (wæ·kəɹ), *sb.* (and *a.*) *Liverpool dial.* [Origin unknown.] A Liverpudlian; also = *WACK[2]. Also *attrib.* or as *adj.*

1768 T. Boulton *Sailor's Farewell* iii. 32, I was told for certain, that th' king o' th' blacks had as many wenches as would stand i' th' cumpus of seven acres of graund; and, if it be true, he must be a wacker, e'cod! **1966** P. Moloney *Plea for Mersey* 21 The typical wacker is descended from a long line his mother listened to. *Ibid.* 22 We will consider the refinements of Wacker figures of speech later. **1970** A. Ross *Manchester Thing* 122, I can't help you, wacker... Five o'clock shut, on Friday, wack.

wacko (wæ·ko), *a.* and *sb.* *slang* (orig. and chiefly *U.S.*). Also **whacko.** [f. *WACK(Y *a.* + *-O.] **A.** *adj.* Crazy, mad; eccentric.

1977 J. Wambaugh *Black Marble* (1978) ix. 182 What if he doesn't always go whacko when you make him remember the bad old days? **1977** E. Leonard *Unknown Man* v. 46 You never know, the guy's fucking wacko. **1978** C. Black *Asterisk Destiny* iii. 58 Without a sense of balance..and a touch of humor, you could go whacko. **1981** D. Uhnak *False Witness* iv. 39 She's gone slightly wacko politically. **1984** *Miami Herald* 30 Mar. 3E/2 'Anyone in this business relishes the pressure,' Hazzard, 41, said. 'Maybe I'm a little wacko, but I love it.'

B. *sb.* = *WACK[1].

1977 *Telegraph* (Brisbane) 24 Mar. 4/1, I am not a weirdo, a wacko or an eccentric for wanting to do good, honest work on a day to day basis. **1980** *Washington Post* 22 Aug. A1 Billy Carter, the president's brother, testified yesterday that he is not a 'buffoon, a boob or a wacko'. **1982** R. Ludlum *Parsifal Mosaic* xiii. 197 'They catch a whack-o now and then.' 'Whack-o?' 'Someone who's crossed over the mental line, thinks he's someone he's not.'

wacko, var. *WHACKO *int.*

wacky (wæ·ki), *a.* *slang* (orig. *U.S.*). Also **whacky.** [f. Whack *sb.* + -Y[1]: cf. *out of whack* s.v. *WHACK *sb.* 4. For earlier uses ('a fool; left-handed') see *Eng. Dial. Dict.*] Crazy, mad; odd, peculiar. **a.** Of persons.

1935 J. Hargan in *Jrnl. Abnormal Psychol.* XXX. 365 *Wacky*, insane. **1938** J. Digges *Bowleg Bill* 28 They all want to know why he done it, and is he gone clean whacky. **1942** *Sun* (Baltimore) 25 July 8/1 Her grandmother, wackier than she is, haunts the place. **1950** 'S. Ransome' *Deadly Miss Ashley* xv. 172 She might.. leave..her kids to that wacky aunt of theirs. **1964** *Economist* 13 June 1242/2 Departed, prostrate or, in former President Eisenhower's recent phrase, 'a little bit whacky'. **1978** J. Irving *World according to Garp* xix. 420 It did much to drive the Ellen Jamesians even wackier or simply away. **1984** *Observer* (Colour Suppl.) 18 Mar. 7/2 She plays the wacky mother of Debra Winger.

b. Of things or abstract concepts.

1937 *Sun* (Baltimire) 19 Aug. 8/1 This picture is described as 'the wackiest'. **1941** B. Schulberg *What makes Sammy Run?* i. 14 The whole office was afraid of him. I know that sounds wacky. Hardened newspapermen being afraid of a..little office-boy? **1959** S. H. Courtier *Death in Dream Time* x. 141 Your cousin's death was wacky—why go to the trouble of staging an accident? **1969** L. Hellman *Unfinished Woman* iv. 37 The office was a wacky joint in a brownstone house on 48th Street. **1975** D. Lodge *Changing Places* v. 191 A characteristically whacky, yet somehow endearing tenderness for individual liberty. **1984** *Listener* 24 May 39/3 In his fear of death Betjeman's hand shook, and lines were created more from wacky fright than profound or energising contemplation.

Hence **wa·ckiness**, the state or quality of being 'wacky'; craziness, oddness.

1941 *Sun* (Baltimore) 1 Oct. 10/2 Maybe the majority won't think that 'the wonderful bums' [*sc.* the Brooklyn Dodgers] can win, but they will be out by the thousands.. hoping that wackiness will be more than its own reward. **1980** R.-L. Duncan *Brimstone* iv. 80 For all her wackiness, Annie knew how to live.

wad, *sb.*[1] Add: **4. b.** In fig. phr. *to shoot one's wad*, to do all that one can do. Cf. *to have shot one's bolt* s.v. Shoot *v.* 21 b. *colloq.* (chiefly *U.S.*).

1914 *Dialect Notes* IV. 112 *Shoot one's wad*, to do or say what one can. **1970** A. Cameron et al. *Computers & O.E. Concordances* 31 Well, I'm really not an expert on it. I've practically shot my wad. **1971** B. Malamud *Tenants* 8, I want to be thought of as a going concern, not a freak who had published a good first novel and shot his wad.

5. c. A bun, a cake; also, something to eat, a sandwich. *slang* (orig. *Services*).

1919 *War Terms* in *Athenæum* 1 Aug. 695/1 *Wad*, a bun. **1927** T. E. Lawrence *Let.* 8 Feb. (1938) 506 No wads, so I'm able to do without money. **1937** D. Jones *In Parenthesis* 4 We've got too many buns—and all those wads. **1942** G. Kersh *Nine Lives Bill Nelson* i. 3 I'm in a caff, getting a tea 'n' a wad. **1960** 'A. Burgess' *Doctor is Sick* 226 Give us a bob for a cuppa and a wad, guv. **1973** *Guardian* 2 June 13/4 He found himself..in Kashmir sharing a char and wad with Sikh pilots. **1983** *Verbatim* Autumn 8/2 Like a 'pick', a 'wad' is also eaten standing up. A 'wad', however, is a solitary piece of inferior, if not disgusting food. The diner falls upon it with little pleasure, merely to quiet the beast in his belly.

7. wadcutter chiefly *U.S.*, a bullet designed to cut a neat hole in a paper range target.

1957 *Amer. Speech* XXXII. 195 *Wadcutter*, a lead bullet designed to be used on paper targets and having no ogive but abrupt shoulders so that a full caliber hole is punched in a target. **1981** D. Boggis *Time to Betray* xi. 61 A potential opportunity to..loose the five rounds of .32 wadcutter from the Walther GSP precision automatic.

waddy. Add: **1. a.** (Earlier examples.)

[**1788** D. Southwell in *Hist. Rec. New South Wales* (1893) II. 698 *A stick or club*, wad-di or wad-dty.] **1800** J. Hunter in *Hist. Rec. Austral.* (1914) I. ii. 406 They had each a Spear and a Warmaraa and a Waddy. **1807** J. Savage *Some Acct. N.Z.* viii. 52 The men..carry a waddy..in figure somewhat resembling a large battle-dore..usually formed of hard black stone.

b. *transf.* Any club or stick; *spec.* a walking-stick. (Examples.)

1899 'S. Rudd' *On Our Selection* 19 We each carried a kerosine tin, slung like a kettle-drum, and belted it with a waddy. **1936** M. Franklin *All that Swagger* 88 None of that bad language or I'll take a waddy to you. **1972** *Southerly* XXXII. 200 Freeth had offered him the use of his cane—his 'waddy', as he called it—after the first week. **1974** K. Cook *Bloodhouse* 31 She had seen him..smash his ebony waddy hard down on his neck.

waddy[2] (wǫ·di). *U.S. slang.* Also **waddie.** [Origin uncertain.] A cattle rustler; a cowboy, esp. a temporary cowhand.

1897 E. Hough *Story of Cowboy* 279 A genuine rustler was called a 'waddy', a name difficult to trace to its origin. **1927** J. Lomax *Cowboy Songs* 374 He rides a fancy horse, he's a favorite man, Can get more credit than a common waddie can. **1931** W. Rogers in S. K. Gragert *Will Rogers' Weekly Articles* (1982) V. 470 You town waddies know what a Combine is?

wade, *v.* Add: **3. e.** *to wade in*: to make a vigorous or concerted attack on one's opponent; to intervene, esp. vocally; *to wade into* (*colloq.*, orig. *U.S.*): to assail or confront energetically.

1863 B. Harte in *U.S. Sanitary Commission Bull.* (1864) I. vii. 201/1 Phrases such as camps may teach... Such as 'Bully!' 'Them's the peach!' 'Wade in, Sanitary! **1893** H. A. Shands *Some Peculiarities of Speech in Mississippi* 66 *Wade into.* One man is said to *wade into* another when he attacks him very vigorously with either fist or tongue. This phrase is used by all classes. **1904** J. London *Let.* 17 Nov. (1966) 165 The lawyers..waded into me good and hard for the cash. **1905** *N.Y. Even. Post* 2 Sept., When a herd of sheep wades in on a patch of bluebells, they stand still and eat all day. **1928** *Daily Express* 30 July 13/6 Though severely punished by Pattenden's lefts to the face he repeatedly waded in. **1935** D. L. Sayers *Gaudy Night* ii. 33, I don't stop to think... I just wade right in and ask for what I want. **1952** E. F. Davies *Illyrian Venture* vi. 104 Luckily the Germans had not known how easily they could have waded into us. **1967** N. Marsh *Death at Dolphin* v. 125 Don't let it give you a moment's pause... Just you wade in to Conducis. **1976** *Sun* 11 Mar. 11/4 Miss Georgina Burton..waded in with her shopping bag and chased the gang away. **1984** J. Barnes *Flaubert's Parrot* x. 132 The writer must wade into life as into the sea, but only up to the navel.

Wade-Giles (wēⁱd dʒəi·lz). [The names of Sir Thomas Francis *Wade* (1818–95), diplomatist and first Professor of Chinese at the University of Cambridge, and Herbert Allen *Giles* (1845–1935), Wade's successor at Cambridge.] Used *attrib.* and *absol.* to designate a system for the romanization of Chinese script devised by Wade, and subsequently modified by Giles. (Now widely superseded by *Pinyin.) Also *ellipt.* as **Wade.**

[**1871** G. C. Stent *Chinese & Eng. Vocab.* p. vi, The tones are also according to Mr. Wade's system.] **1943** Y. R. Chao in *Mathews's Chinese-Eng. Dict.* (1945) p. ix, Now that the standard has come back to the pronunciation of Peiping, we can almost use the unmodified Wade's *Syllabary* or the more widely used Wade-Giles system. *Ibid.*, In Chauncey Goodrich's *Pocket Dictionary*, the syllables *ko k'o, ho* and *o* of Wade-Giles are given as *kê, k'ê, hê*, and *ê*. **1944** Chang & Maxwell *Conc. Eng.-Chinese Dict.* 10 People have become accustomed to the Wade form of spelling. **1947** C. F. Hockett in *Jrnl. Amer. Oriental Soc.* LXVII. 256/2 The numbering /[1]/ through /[4]/ is as in Wade-Giles Romanization. **1950** J. De Francis *Nationalism & Lang. Reform in China* xii. 200 Those who have learned Wade as their first transcription of Chinese case are annoyed at the French use of another system. **1961** *Amer. Speech* XXXVI. 190 The Wade-Giles system of romanization of the Peking dialect will be followed. **1966** *Guardian* 29 Nov. 8/2 The Chinese system of romanization will certainly displace Wade-Giles eventually. **1971** R. Newnham *About Chinese* 172 Wade-Giles is the oldest and most widely used of all current romanizations, and its chief merit lies in that wide use. **1979** *Time* 2 Apr. 15/2 The changeover was started by Peking (um, er, Beijing) on Jan. 1, when the government of Zhongguo (otherwise known as China) decreed that in all its foreign-language publications Pinyin would replace the traditional Wade-Giles system of romanization.

wadeite (wēⁱ·dəit). *Min.* [f. the name of Arthur *Wade* (1878–1951), British geologist + -ite[1].] A hexagonal silicate of potassium and zirconium, approximately $K_2ZrSi_3O_9$.

1938 Wade & Prider in *Rep. Brit. Assoc. Adv. Sci.* I. 419 Nineteen occurrences of post-Permian volcanic rocks have been found in the West Kimberley area... The rocks are made up..of leucite, phlogopite,..wadeite (a new K-Zr silicate), [etc.]. **1980** *Contrib. Mineral. & Petrol.* LXXII. 191/1 The restricted occurrence of wadeite to rocks of West Kimberley, Australia and Leucite Hills, Wyoming is believed to be due to their high K/Al and Zr contents relative to other high potash rocks.

wadge: entry now merged with Wodge (in Dict and Suppl.).

Wadhamite (wǫ·dæ̆məit). [f. the name of *Wadham* College, founded by Nicholas and Dorothy Wadham in 1612 + -ite[1].] A member of Wadham College, Oxford.

1760 J. Woodforde *Diary* 20 May (1924) I. 14 Hooke, Boteler and myself went to Welch's of Wadham College, where we designed to sup and spend the evening, but our entertainment was thus, one Lobster of a Pound, a halfpennyworth of Bread, and the same of Cheese... N.B. A Wadamite. **1976** A. Powell *Infants of Spring* xi. 191 'But', added the young Wadhamite, 'I've heard he's an absolute fish out of water when he's away from the academic world he's accustomed to.'

wadi, wady. Add: Also *attrib.*

1902 D. G. Hogarth *Nearer East* 139 The palm-lined wadi beds of Jebel Akhdar. *Ibid.* 143 Aromatic scrub and an occasional thorn is all that can be expected in the wadi bottoms. **1980** *Encounter* May 90/2 Off we go, with me stumbling after him over the rough stony ground of the wadi bed.

wading, *vbl. sb.* Add: **b.** *wading pool; wading shoes.*

1866 J. Macgregor *Thousand Miles in Rob Roy Canoe* (ed. 2) i. 10, I took for this tour..canvas wading shoes,..a waterproof overcoat, [etc.]. **1921** Wading pool [see *sand-pile* s.v. *sand *sb.*[1] 9 a]. **1977** *New Yorker* 15 Aug. 42/3 He inflated and filled a plastic wading pool for the children.

Wafd (wǫft). [a. Arab. *wafd* arrival, deputation, in full *al-wafd al-miṣrī* the Egyptian delegation.] The name of an Egyptian nationalist organization formed in 1918 (from 1923, a political party) whose original aims included the establishment of autonomous government in Egypt and the abolition of the monarchy. (The party was dissolved in 1953, but reconstituted as the *New Wafd* in 1978.)

1922 *Times* 24 Jan. 9/1 The Wafd el Masri, or Zaghlul Delegation..has just issued a detailed programme of non-cooperation. *Ibid.* 27 Jan. 11/2 The reconstituted *Wafd*. **1928** E. W. P. Newman *Great Britain in Egypt* x. 239 The Wafd had been constituted in order to lay the case of Egypt before the Peace Conference of Paris. **1932** *Palestine Post* 13 Dec. 7/1 The quarrel between the Wafd and its dissidents continues. **1943** Visct. Wavell *Allenby in Egypt* i. 40 Zaghlul Bey..had not, however, been the originator of the Wafd (or Delegation), as his party came to be known. It was the creation of others, notably men like Mohammed Mahmoud. **1962** *Listener* 15 Nov. 795/1 Saad Zaghloul, the founder and leader of the Wafd party. **1974** *Great Soviet Encycl.* IV. 649/2 Some of the Wafd members unsuccessfully fought the government of Nasser in actions of 1954, 1957, and 1961. **1978** *Facts on File World News Digest* 24 Feb. 131/1 The New Wafd had 24 supporters in the 360-seat People's Assembly.

Wafdist (wǫ·ftist), *a.* and *sb.* [f. prec. + -ist.] **a.** *adj.* Of or pertaining to the Wafd. **b.** *sb.* A member or supporter of the Wafd; an Egyptian Nationalist.

1926 *Glasgow Herald* 3 June 8 Doubts..were entertained of the prospect of the Wafdist leader taking such a moderate course. **1926** *Spectator* 19 June 1032/2 The Wafdists..are likely to be restrained by the knowledge that any too free indulgence of their characteristic tactics ..would quickly make Adly throw up his office in disgust. **1958** J. W. Day *Lady Houston* xv. 235 That is, indeed, a 'Capitulation', and the worst and most fatal, by Great Britain to the Wafdist mob! **1973** *Times* 29 Oct. 17/7 From being a Liberal Constitutionalist in the days of Sarwat, his protector, he became an active Wafdist under Nahhas Pasha.

wafer, *sb.* Add: **1. b.** *ellipt.*, a sandwich of ice-cream between wafers.

1936 N. Coward *Still Life* in *To-night at 8.30* III. i. 48 An old girl..asked if I'd got an ice-cream wafer... What did she think I was, a 'Stop me and buy one?' **1979** *Listener* 6 Sept. 303/2 The vanilla wafer..proved a great healer.

4. (Earlier example.)

1848 C. Brontë in C. Shorter *Charlotte Brontë & her Circle* (1896) vi. 173 She has taken no medicine, but.. Locock's cough wafers, of which she has used about 3 per diem.

5*. *Electronics.* A very thin slice of a semiconductor crystal used as the substrate for solid-state circuitry.

1956 *Bell Syst. Techn. Jrnl.* XXXV. 3 After diffusion the entire surface of the silicon wafer is covered with the diffused n- and p-type layers. **1967** *Electronics* 6 Mar. 25 Litton engineers haven't decided whether to use single or two-layer metalization to interconnect the circuits within the wafers. **1975** D. G. Fink *Electronics Engineers' Handbk.* viii. 84 After the completion of the test sequence, the probe assembly is automatically lifted up and the probes are indexed over to the next chip to be tested on the wafer. **1979** *Maclean's Mag.* 2 Apr. 37/3 'Wafers' containing hundreds of memory chips (each with 64,000 transistors): dispensing liquor, guiding spaceships. **1984** *QL User* Dec. 18 Currently, chips are manufactured in batches on discs of silicon about four inches in diameter called wafers.

6. (sense 1) *wafer-thin* adj.; (sense 3) *wafer box.*

1837 Dickens *Pickw.* liv. 591 Wilkins Flasher, Esquire, was..spearing a wafer-box with a penknife. **1968** *Canad. Antiques Collector* Nov. 26/1 When envelopes were introduced, the wafer box became a stamp box. *a* **1911** D. G. Phillips *Susan Lenox* (1917) I. xxi. 371 Wafer-thin. **1958** *Times* 15 Nov. 11/6 There is dried meat of the Valais cut in wafer-thin slices. **1976** *Times* 1 June 1/1 Pitmen in north Derbyshire are understood to have voted by a wafer-thin majority against continuing wage restraint. **1984** B. Francis *AA Car Duffer's Guide* 63/2 The radiator is made up of a collection of very narrow tubes linked by wafer-thin strips of metal which are exposed to the air-flow.

‖ **Waffen SS** (va·fən es,es). [ad. G. *Waffen-SS*, in full *Waffen-Schutzstaffel* armed defence squadron: see *Schutzstaffel and S.S. s.v. *S 4 a.] In Nazi Germany during the war of 1939–45: the combat units of the S.S.

1943 *Rev. Foreign Press* (F.O. Research Dept.) 18 Oct. 125/1 By a law of the 22nd July, Laval..authorised French citizens to join those units of the Waffen S.S. fighting on the Eastern Front. **1949** 'R. West' *Meaning of Treason* I. vii. 164 He wanted to join the Waffen S.S. and fight the Russians. **1957** *Times Lit. Suppl.* 20 Dec. 771/2 A nucleus of the 'commandoes', but also with parachutists, Waffen S.S. **1966** G. Stein *Waffen SS* p. xxx, At the beginning of World War II the term Waffen SS was unknown. Five years later, prefaced by such adjectives as elite and fanatical, it appeared regularly in Allied war communiqués. By 1940 *Waffen SS* had become the official designation for the combat units of the SS, which had grown from the handful of armed troops maintained by the Reichsführer SS, Heinrich Himmler, for security and ceremonial purposes. *Ibid.* p. xxxi, By 1943, the exigencies of war had forced the Waffen SS to give up some of its exclusiveness. Large numbers of foreigners were recruited or conscripted. **1972** F. Forsyth *Odessa File* ii. 28 He carried the twin-lightning strikes of the Waffen SS on the right collar lapel.

waffle, *sb.*[1] Add. **a.** (Earlier examples.)

1744, 1794 [see sense *b]. **1809** A. Burr *Jrnl.* 26 Aug. (1903) I. 214 Everywhere, too, you get *wafen*; our wafles, and made and eaten in the same way.

b. *waffle frolic*; **waffle-iron** (earlier example); **waffle stomper** *U.S. slang*, a boot or shoe with a heavy, ridged sole.

1744 in *Mag. Amer. Hist.* (1878) II. 442 For my own part I was not a little grieved that so luxurious a feast should come under the name of a wafel frolic. **1794** *S. Carolina State Gaz.* 30 Aug. 1/2 (Advt.), Waffle irons. **1974** *Sunday* (Charleston, S. Carolina) 7 Apr. 10-c/8 'Waffle stompers', should anybody be wondering, are hiking shoes.

c. *Textiles.* Used *attrib.* of a style of fine honeycomb weaving (see *honeycomb *sb.* 5 b). Also applied to the fabric woven and to the effect produced.

[**1948** *Sears, Roebuck Catal.* Fall-Winter 45/2 Cotton Chenille Spread. Sturdy white cotton sheeting tufted with fluffy chenille in regular, even waffle pattern.] **1949** *Good Housek.* Apr. 261/2 Waffle piqué is characterized by a honeycomb weave; it is not a true piqué. **1951** in M. McLuhan *Mech. Bride* (1967) 74 We had it copied faithfully for you in the..white Birdseye waffle pique. **1953** *Sun* (Baltimore) 7 Mar. (B ed.) 3/4 Part of the secret lay in the 'waffle-weave', a..technique which required special machinery... It produces a cloth that looks like a waffle, or a honeycomb. **1961** *Sunday Express* 28 May 14/2 (*caption*) Tops in waffle-knit cotton. **1962** W. Schirra in *Into Orbit* 49 A series of waffle-weave patches on our long john underwear helps to keep the oxygen moving. **1975** G. Howell *In Vogue* 242/2 (*caption*) Elements of the mid-fifties—the 'sloppy Joe',..the waffle cotton shorts. **1979** *Men's Wear* 3 May 24 Texture returns with fine seersuckers and waffle cloth.

waffle, *sb.*[2] Add: Also **woffle. 2.** Now in *gen.* colloq. use, verbose but inconsequential talk or writing; empty verbiage (see also quot. 1937).

[**1866** J. E. Brogden *Provinc. Words Lincs.* 219 *Waffle-bags*, a great talker.] **1937** Partridge *Dict. Slang* 935/1 *Waffle*, nonsense; gossip(ing); incessant or copious talk. **1953** *Times Lit. Suppl.* 13 Mar. 164/1 A little technical detail and a good deal of emotional waffle. **1957** *Economist* 21 Dec. 1041/2 His ability to distinguish the essence and to cut the waffle in any discussion are exceptional. **1961** C. S. Lewis *Let.* 9 May (1966) 298 For a good.. defence of our position against modern woffle... I know nothing better than G. K. Chesterton's *The Everlasting Man.* **1965** *Spectator* 22 Jan. 96/1 There is a special relationship between Britain and the United States, a special relationship more serious than the waffle we get at banquets. **1973** C. Bonington *Next Horizon* i. 20 Cut out the waffle, and let's see your application.

waffle, *v.* Restrict Now *dial.* to sense in Dict.

and add: Also **woffle, wuffle. 2. a.** To waver; to vacillate or equivocate; to 'dither'. orig. *Sc.* and *north. dial.* Now *colloq.* or non-Standard.

1803, etc. [see *waffler 1]. **1868** [see *waffling *ppl. a.*]. **1893–4** R. O. Heslop *Northumberland Words* II. 762 *Waffle*, to waft about, to waver, to walk hesitatingly, to act with indecision. **1898** B. Kirkby *Lakeland Words* 152 Thoo'l waffle aboot an' say owt. **1943** *Horizon* Apr. 234 While I was still waffling I read 'The Mint'. **1961** C. H. D. Todd *Popular Whippet* 51 Have we an ideal or are we just 'waffling around' following this, that and the other. **1976** *Telegraph-Jrnl.* (St. John, New Brunswick) 4 Sept. 17/1 She defends the board against suggestions that it waffled on the issue because of political pressure.

b. To talk (or write) in a verbose but inconsequential manner; to ramble *on.* Also *trans.* with direct speech as obj.

[**1701** J. White *Countryman's Conductor* 128 *Waffling*, all speakers and no hearers.] **1900** Farmer *Public School Word-Bk.* 216 Waffle,..(Durham) to talk nonsense. **1936** R. Lehmann *Weather in Streets* I. v. 83 Nanny would wuffle on, and make me change my stockings. **1937** G. Heyer *They found him Dead* iv. 83 He woffled a whole lot to me about people bothering his life out. **1949** J. B. Priestley *Delight* 15 The wise men woffled on about violence and crime, missing the point by miles. **1957** G. Smith *Friends* 60 'Funny thing js, it seems to improve my game,' waffled Bill amiably. **1960** L. Meynell *Bandaberry* i. 8, I under-estimated his shrewdness... I thought he was waffling. **1982** P. Dickinson *Last House-Party* ii. 22 You can make the correct noises while all the old buffers are woffling on.

waffling *ppl. a.* (later examples); also as *vbl. sb.*

1847 J. Halliday *Rustic Bard* 145 'Tis you I punch at, worthless, wafflin crood. **1868** J. C. Atkinson *Gloss. Cleveland Dial.* 554 A windy, *waffling* soort o' chap, wheea nivver kens his ain mind. **1945** J. Reith *Diary* 11 Apr. (1975) vii. 346 A typical silly, waffling letter which Cranborne had written. **1958** J. Cannan *And be Villain* v. 118 His sharp harsh tones were a joy to hear after the wafflings of the soft voiced Highlander. **1967** *New Yorker* 8 July 67 There will be..a large majority for a waffling resolution..calling for Israel to withdraw from the conquered territories but not passing judgement on the original conquests. **1973** J. Wainwright *Touch of Malice* 231 Waffling, thoughtless stupidity. **1976** *Sunday Mail* (Brisbane) 23 May 44/6 They should not be regarded as the wafflings of people sounding off in the papers.

waffler (wǫ·fləɹ). [f. Waffle *v.* + -er[1].] **1.** An unreliable person; an idler or waverer. Chiefly *Sc.* and *north dial.*

1803 R. Anderson *Ballads in Cumberland Dial.* (1805) 59 Saint Gworge, the girt champion, o' fame and renown, Was nobbet a waffler to Matthew Macree. **1819** in Carlyle *Early Lett.* (1886) I. 215 The waffler did not get his cart home till Monday. **1927** N. Dundas *Castle Adamant* iv. 60 Of course there are wafflers, ye could look or no less in a town of five thousand souls. **1977** B. Langley *Death Stalk* ii. 15 He had an instinctive distrust of all wafflers and ditherers.

2. One who indulges in waffling talk or writing. Cf. *waffle *sb.*[2] 2.

1959 *Viewpoint* July 4 It was a field day for the professional analysts and wafflers. **1968** *Sunday Truth* (Brisbane) 8 Dec. 41 It amazes me that Peter Evans was ever selected to conduct 4QR's Breakfast Show. He must be the original waffler.

waffly (wǫ·fli), *a.* [f. Waffle *sb.*[2] and *v.* + -y[1].] **a.** Wavering, vacillating, imprecise. **b.** Characterized by or indulging in waffling speech or writing.

1890 J. Service *Thir Notandums* xix. 125 Let the waffly body tak ocht I hae written and mak a kirk or a mill o't as pleases himsel'. **1928** A. E. Pease *Dict. Dial. N. Riding* 149 *Waffly*,..wavering, undecided, shaky, not coherent. **1957** *Economist* 23 Nov. 673/2 The Labour party's attitude to the federation has always been waffly. **1964** *Ibid.* 16 May 691/2 Mr. Duncan Sandys could never be called a waffly man. **1978** *Times* 18 Nov. 11/8, I thought the first part about the World Tree and the Green Man a little thin and waffly.

wafty, *a.* Add: **3.** That wafts or moves to and fro in the wind. *nonce-use.*

1922 Joyce *Ulysses* 289 The wafty sycamore, the Lebanonian cedar, the exalted planetree.

wag, *sb.*[2] Add: **3.** Also, *to hop the wag*: see *hop *v.*[1] 6 a.

wag (wæg), *sb.*[3] *Archæol.* [ad. Gael. *uamhag*, dim. of *uamh* cave, hollow: cf. Weem.] In Caithness, an Iron-age galleried structure set partly below ground-level (see quot. 1963).

[**1776** A. Pope in T. Pennant *Tour in Scotl. 1769* (ed. 4) 338 Figures 2 and 3 are what are styled *forest* or *hunting houses*... They consist of a gallery, with a number of small rooms on the sides..made with the vast flags [stones] this country is famous for... Their length is from fifty to sixty feet. These buildings are only in places where the great flags are plentiful. In Glen-Loch are three, and are called by the country people *Uags*.] **1911** A. O. Curle in *Proc. Soc. Antiquaries Scotl.* 11 Dec. 89 To the galleried structure the name 'wag' in former times was evidently applied and still remains in use, though now transferred from the structure to the place or site, *e.g.* 'Wag-more rig', 'Wag-burn' and 'the Wag'. **1921** *Ibid.* 10 Jan. 93 Interesting as the discovery of these post-holes is in the elucidation of the broch construction, the interest it evokes does not stop here. The arrangement at once recalls the plans of the wags or galleried dwellings in Caithness. **1963** *Field Archaeol.* (Ordnance Survey) (ed. 4) 63 In Caithness there is the local variant structure known as a wag... The dwelling part is represented by a strongly-built hut circle of ordinary plan, but to this is added an oval stone-built chamber about twice as big as the hut with its floor excavated somewhat below the general ground level. **1972** E. M. Mackie in *Dark Ages in Highlands* 16 One site at Forse..could well be a pre-broch defensive structure... This is the so-called 'wag' or 'prehistoric cattlefold' excavated by Alexander Curle.

waganga, pl. of *mganga.

wage, *sb.* Add: **4. a.** Simple attrib., as *wage(s) bill, board, -book, claim, clerk, contract, cost, cut, demand, dispute, inflation, labourer, -level, negotiation, packet, payment, policy, push, restraint, -rigidity, -slavery* (earlier example), *-snatch, spiral, structure, system; wage-related, -working* adjs.; **b.** Objective, as *wage-bargainer, -bargaining, control, -fixing*; **c.** Appositive, as *wage-price* attrib., esp. in phr. *wage-price spiral*; **wages council,** any of a number of joint management and employee councils succeeding the trade boards (from 1945), and responsible for determining the conditions of employment in certain trades; **wage differential** = *differential *sb.* 3 b; **wage drift,** the tendency for wages to rise above national rates through local overtime and other agreements; the extent of this increase; **wage freeze,** a temporary fixing of wages at a certain level; cf. *pay freeze* s.v. *pay- 4 and *freeze *sb.*[1] (ii); **wage hike** *N. Amer.*, a wage increase (cf. *hike *sb.* 2); **wage scale,** a graduated scale of wage rates for different levels of work; **wage stop** (also *wages stop*), the limitation of supplementary benefit to the level of the normal wage; hence **wage-stop** *v. trans.*; **wage-stopped** *ppl. a.*; **wage unit** (see quot. 1936).

1968 *Economist* 23 Mar. 16/3 Any wage bargainer worth his salt should be able to dress up a claim to fit the loose criteria 'justifying' a 3½ per cent increase. **1928** *Britain's Industr. Future* (Liberal Industr. Inquiry) iii. xvii. 209 In the boot and shoe trade..it has been thought desirable to keep wage-bargaining apart from the general discussions carried on by the Joint Industrial Council. **1983** *Times* 24 June 7/6 The fear that mass hunger striking will become a common tool of wage bargaining. **1919** M. Beer *Hist. Brit. Socialism* I. ii. v. 169 The total wage bill of the country diminished. **1923** H. W. B. Joseph *Labour Theory of Value in Karl Marx* ii. 49 If all this is true..a capitalist ought not to be indifferent whether he economizes in his wages-bill or in his other expenditure. **1982** T. Keneally *Schindler's Ark* viii. 96 The meeting of his wage bill was the least of Oskar's worries. **1925** *Scribner's Mag.* Oct. 415/2 The regulation of wages has been put in the hands of wage boards. **1930** *Economist* 22 Mar. 651/1 The success of the Wages Boards..should be carefully considered. **1970** *Washington Post* 30 Sept. B 11/5 A 6 per cent raise to 850,000 hard-pressed blue-collar (wage board) people. **1864** Trollope *Can you forgive Her?* (1865) II. vii. 54, I don't suppose I've opened the wages book half a dozen times since last July. **1960** G. E. Evans *Horse in Furrow* ix. 115 These wages-books..contained the lists of jobs allocated to people employed on the farm, and the amount of wages due to them. **1971** H. Wilson *Labour Govt.* xxxviii. 787 One of the main groups of printing unions, SOGAT, put in a large wage-claim. **1921** *Dict. Occup. Terms* (1927) §939 Wages clerk. **1961** *Evening Standard* 18 July 23/5 Wages clerk to head.. wages section. **1921** *Daily Colonist* (Victoria, B.C.) 20 Mar. 1/3 We deny that any wage contract with the miners has been broken. **1977** P. Johnson *Enemies of Society* iii. 39 The medieval manor..yielded to a cash-nexus society based on wage-contracts between freely-negotiating individuals. **1910** *Encycl. Brit.* V. 879/1 This argument, which combined statutory wage control and statutory poor relief, seems to have been firmly embedded in the English legislative mind..till after 1600. **1978** *Economist* 18 Feb. 75/1 The [Argentine] government..has relaxed wage controls on private industry. **1985** *Financial Times* 4 Mar. 6/4 The most apparent difference is in their approach to inflation and wage control. **1958** *Spectator* 4 July 33/3 Wage-cost inflation. **1975** J. De Bres tr. *Mandel's Late Capitalism* vi. 203 Economies in costs will thus always be accompanied in the long-run by a relative decrease in the share of wage costs in the value of the commodity. **1945** *Wages Councils Act* (8 & 9 *Geo. VI c.* 17) 116 An Act to provide for the establishment of wages councils. **1971** *Observer* 7 Nov. 16/6 Many trade unionists argue that the very existence of wages councils discourages poorer workers from joining trade unions. **1984** *Listener* 1 Nov. 11/3 The Wages Councils, providing legal minimum wages and conditions for 2·7 million of Britain's lowest-paid workers..are threatened. **1925** *Daily Herald* 30 June 2/4 (*heading*) Folly of wage-cuts. **1974** J. Aiken *Midnight is Place* iii. 102 Protests about the wage-cut had been nipped in the bud. **1970** *Guardian* 2 Mar. 1/2 The biggest disagreement is about the effect of wage demands. **1950** Wage differential [see *differential *sb.* 3 b]. **1957** *Observer* 6 Oct. 10/2 The pension scheme, by relating superannuation to wages, strengthens the value of wage differentials and hence the incentive to acquire greater skills. **1977** *China Now* Apr.–May 7/1 They were cutting down wage differentials. **1919** in M. Gilbert *Winston S. Churchill* (1977) IV. Compan. I. 498 The moment the revolt advanced over the line of a pure wage dispute, and the strikers were guilty of a serious breach of the law, then was the moment to act. **1955** W. Gaddis *Recognitions* ii. i. 322 Not written to be played by men in worn dinner jackets,..involved in wage disputes. **1963** *Times* 22 Jan. 4/6 This is a comparatively rare phenomenon, 'wage drift' in reverse, and reflects increased short time and lower earnings from bonuses, overtime and so on. *a* **1974** R. Crossman *Diaries* (1976) II. 703 As for the 3½ per cent ceiling, he told me it was quite unrealistic since wage drift by itself probably comes to 3½ per cent. **1928** *Britain's Industr. Future* (Liberal Industr. Inquiry) ii. xvi. 181 The practical considerations which ought to govern the process of wage-fixing. **1942** Wage freeze [see *freeze *sb.*[1] (ii)]. **1967** *Spectator* 1 Sept. 238/2 The wage freeze has been imposed without a murmur of national strike action. **1980** C. Moorehead *Fortune's Hostages* v. 104 The President [of Uruguay]..tried to introduce economic restraint by wage freezes. **1976** *Washington Post* 19 Apr. A 23/4 By last spring, some unions were putting in for wage hikes up to 60 per cent. **1976** *Morecambe Guardian* 7 Dec. 21/7 One woman waving a placard saying 'Wage Inflation' [etc.]. **1957** V. W. Turner *Schism & Continuity in Afr. Society* v. 135 A man can acquire wealth by working in the White economy as a wage labourer. **1928** *Britain's Industr. Future* (Liberal Industr. Inquiry) iii. 139 A wise wage-policy should aim at the highest practicable wage-levels. **1965** J. Meuvret in Glass & Eversley *Population in Hist.* xxi. 517 In normal times wage-levels and price-levels were both very low. **1928** *Britain's Industr. Future* (Liberal Industr. Inquiry) iii. xvi. 188 It is not even enough that the wage-system should be just in itself; it must be visibly and demonstrably just. And this conception ought to inspire the whole system of wage-negotiation. **1974** *Guardian* 23 Jan. 3/1 Today's wage negotiations were conducted with the Frame group. **1951** R. Firth *Elem. Social Organiz.* iv. 140 One point of view is that the size of the wage-packet remains the most important factor still in the incentive to work. *a* **1974** R. Crossman *Diaries* (1976) II. 638 The trade unionists want to see us spending much less on social services so that there'll be more for wage packets. **1923** H. W. B. Joseph *Labour Theory of Value in Karl Marx* vii. 154 In the absence of definite agreements or enactments, we can produce no rule of universal application, to which wage-payments ought to conform. **1928** Wage-policy [see *wage-level* above]. **1965** *New Statesman* 9 Apr. 576/3 The principles (but not the practice) of a wages policy. **1946** *Sun* (Baltimore) 1 Mar. 1/3 The situation, therefore, presents possibilities of a critical test of the administration's new wage price line. **1958** *Engineering* 11 Apr. 460/1 There appeared to be no doubt whatever in Lord Cohen's mind that this painful process of breaking the wage-price spiral had to be adopted. **1977** M. Edelman *Political Lang.* viii. 147 A wage-price freeze from which major industries quickly won exceptions. **1968** *Manch. Guardian Weekly* 14 Mar. 13 It has become fashionable to say that the British..manage to keep a wage-push inflation in times of crippling deflation. **1979** *Dædalus* Spring 53 Other things being equal, the greater the wage push, the tougher and more 'lesson-teaching' a policy is necessary to control inflation. **1963** *Punch* 20 Mar. 398/2 There must..be a wage-related contributory insurance system. **1958** *Listener* 4 Dec. 930/2 Don't think it's a matter of wage restraint only. **1977** M. Edelman *Political Lang.* viii. 153 Employers perceive wage restraint by workers as in the public interest. **1930** W. K. Hancock *Australia* ix. 187 It is apparent that wage-rigidity is less important in its social consequences than in its economic and psychological consequences. **1983** *Economist* 18 June 21/3 Some belief in wage-rigidity is deeply rooted in his work. **1960** F. Lynde *Quickening* 310 His father was deep in the new wage scale submitted by the miners' union. **1979** *Gloss. Terms Work Study* (*B.S.I.*) 27 *Wage scale determination*, the construction of a scale of wages reflecting the relative values of jobs. **1886** D. Donohue *Let.* 21 Oct. in *N.Y. Times* 24 Oct. 1/5 It is therefore but natural that we should vote for a man who proposes to use his best endeavors to bring about legislation by which wage slavery and land monopoly shall be abolished. **1962** *Observer* 25 Feb. 23/3 They were working themselves at the wage-snatch business, with a bit of smash-and-grab on the side. **1964** J. Creasey *Look Three Ways at Murder* xx. 183 There's a hold-up... Wages snatch! **1977** L. Meynell *Hooky gets Wooden Spoon* xii. 149 She wasn't doing anything for the police in the wages snatch. **1948** *Ann. Reg. 1947* 20 A wage spiral was only prevented by distortion which would soon become intolerable. **1940** *Economist* 28 Dec. 799/1 The attempt to carry out a wage stop similar to the price stop had to be abandoned. **1954** E. H. Carr *Interregnum* 73 This quasi-official wages-stop remained in force throughout 1923. **1963** *Guardian* 25 Feb. 11/7 If a man is wage-stopped, he will not get anything at all from the increases. **1971** *Daily Tel.* 16 June 3/3 The wage stop is the system by which men are prevented from drawing more in State benefits than they had earned while in work—even if this means they get less than the commission's official poverty level. **1963** *Times* 18 Feb. 11/6 To suggest..that all the 15,000 wage-stopped families were outsize would, on the board's own figures, be plainly absurd. **1955** *Ann. Reg. 1954* 1 The commission..undertook to make a thorough examination..of the whole wage structure. **1898** *Kansas City* (Missouri) *Star* 18 Dec. 2/1 The ultimate outcome of the labor movement..will be the destruction of the wage system. **1929** D. H. Lawrence *Pansies* 75 Ultimately, we are all busy buying and selling one another. It began with Judas and goes on in the wage-system. **1936** J. M. Keynes *Gen. Theory Employment* iv. 41 The money-wage of a labour-unit we shall call the wage-unit. **1976** F. Zweig *New Acquisitive Society* ii. iv. 106 Economic planning which is based primarily on a 'wage unit' determined by the planning authorities as the pivot of economic planning. **1898** G. B. Shaw *Perfect Wagnerite* 29 The starving wage-working class.

wagenboom. Add: Also 8 **wage-boom; waboom.** (Earlier and later examples.)

1790 E. Helme tr. *Le Vaillant's Trav. Afr.* I. 255 A few paltry woods..had some resemblance to that named *Wage-boom.* **1946** *Cape Times* 5 Feb. 6/8 He is collecting waboom and kreupelbos seeds above the Camps Bay slopes so that the bare areas..can be covered with these handsome proteas. **1972** Palmer & Pitman *Trees S.*

Afr. I. 525 The waboom is one of the tallest of the genus *Protea* and is conspicuous in dry rocky areas of the Cape.

Wagener (vā·gənəɹ). Also *erron.* **Wagner.** The name of Abram *Wagener* (fl. 1796) American farmer, used *absol.* or *attrib.* to designate an apple tree or its fruit belonging to a variety developed on his farm in Penn Yan, New York State (see quot. 1956).

1848 *Trans. N.Y. State Agric. Soc. 1847* VII. 315 The apple for which it is proposed to award the second premium of the society, is called the 'Wagener apple'. **1886** L. H. Bailey *Field Notes on Apple Culture* iv. 21 For winter:..Jonathan,..Northern Spy, Wagener. **1915** A. E. Wilkinson *Apple* ii. 24 The Wagener thus fits in well with the Northern Spy in soil requirements. **1925** E. Hemingway *In our Time* (1930) iv. 45 Nick stopped and picked up a Wagner apple. **1956** *Dict. Gardening* (R. Hort. Soc.) Suppl. 95/2 Wagener..Medium size, round, flattish, irregular, shining yellowish-green, striped and flushed with bright scarlet. **1969** *Oxf. Bk. Food Plants* 48/2 'Wagener'..is an example of many good apples raised in the U.S.A. from European stock.

wagery (wēⁱ·dʒəri). [f. Wage *sb.* + -ery, after *slavery.*] The wage system, wage slavery (esp. as opposed by the guild socialists); wage-earners *collect.*

1917 A. S. Neill *Dominie's Log* xv. 162, I wonder when people will begin to realise what wagery means. When they do begin to realise they will commence the revolution by driving women out of industry. **1917** S. G. Hobson *Guild Principles in War & Peace* ii. 33 Two generations of wagery were to live their squalid life.. before we find singers..grasping the true meaning of industrial oppression. **1972** A. M. Quinton in Cox & Dyson *20th-Cent. Mind* I. iv. 128 Personal fulfilment through work and the production of honest and emotionally satisfying goods could be secured only by the abolition of 'wagery', the wage system in which men sold their labour power unconditionally.

Wagga (wǫ·gă). *Austral. slang.* Also **wagga.** [f. *Wagga Wagga*, the name of a town in New South Wales.] In full, *Wagga blanket, rug.* A blanket or covering made by opening out two sacks, chaff bags, etc., and stitching them together along one edge.

1900 H. Lawson *Darling River* in *Stories* (1964) 1st Ser. 388 The live cinders from the firebox..fell in showers on deck. Every now and again a spark would burn through the Wagga rug of a sleeping shearer. **1938** X. Herbert *Capricornia* (1939) xxviii. 417 The nap..consisted of two greasy bran-sacks, or, as bushmen call them, Wagga Rugs. **1941** Baker *Dict. Austral. Slang* 80 *Wagga blanket*, a rough covering used by tramps. **1944** J. Devanny *By Tropic Sea & Jungle* 156 When you crawl under your wagga you get in one position and aren't game to move. **1969** L. Hadow *Full Cycle* 248 She went to his camp bed. 'Take your wagga, then.' 'No, it's too heavy.' **1978** *Weekend* (Austral.) 25–26 Nov. 11/4 I'm due to slip under my Wagga blankets.

wagger (wæ·gəɹ), *sb.*[2] *slang* (orig. *Oxford University*). More fully, **wagger-pagger (-bagger).** [One of a collection of words jocularly formed by adding *-agger* (see *-er[6]) to the initial consonants of a word or expression, in this case *waste-paper basket.*] A waste-paper basket.

1903 [see *-er[6]]. **1925** O. Jespersen *Mankind, Nation & Individual* viii. 162 There is an interesting class of words with an inserted g:..*wagger pagger bagger* for *waste paper basket.* **1927** W. E. Collinson *Contemp. Eng.* 125 Such playful formations as the Pragger Wagger (the Prince of Wales..) and wagger pagger bagger *waste-paper basket.* **1934** *Neuphilologische Mitteilungen* XXXV. 130 Public-school slang..*wagger* 'waste-paper-basket'. **1961** Partridge *Dict. Slang* Suppl. 1086/2 *Wagger-pagger*, a waste-paper basket. Short for *wagger-pagger-bagger.*

wagging, *vbl. sb.* Add: **2.** Special Comb. **wagging dance** = *waggle dance* s.v. *waggle *sb.* 2. Cf. *wag-tail dance* s.v. *wagtail *sb.* 5.

1950 [see *round *a.* 5 a]. **1967** *Science* 24 Nov. 1072/3 Successful forager bees..inform their hive mates of the location of the feeding place by wagging dances.

waggle, *sb.* Add: **2. waggle dance** [tr. G. *schwänzeltanz* (K. von Frisch 1923, in *Zool. Jahrb., Abt. f. Allgemeine Zool.* XL. 72)], a movement performed by honey-bees at their hive or nest, believed to indicate to other bees the site of a source of food. Cf. *wagging dance* s.v. *wagging *vbl. sb.* 2.

1952 [see *round *a.* 5 a]. **1961** *Guardian* 15 Sept. 6/6 The waggle dance of the hive bee can convey precise indications as to distance and direction of a food source. **1978** *Sci. Amer.* July 80/3 By means of the 'waggle dance' elucidated by von Frisch the bees inform their hive mates of this source of food.

waggle, *v.* Add: **1. e.** *Aeronaut.* To rock (the wings of an aircraft in flight) rapidly from side to side, usu. to convey a signal.

1918 J. M. Grider *Diary* 18 June in *War Birds* (1927) 200 Springs waggled his wings and pointed and we waggled back. **1950** *Sun* (Baltimore) 25 Nov. 2/1 The plane waggled its wings in salute to the troops. **1972** *Daily Express* 29 June 1/1 The Nimrod [aircraft] then waggled its wings and started back.

Wagner (vā·gnəɹ). The name of the German operatic composer Richard *Wagner* (see Wagnerian *a.* and *sb.*) used *attrib.* in **Wagner tuba,** a valved brass instrument resembling a wide-bore horn devised for Wagner (see quots.).

1938 *Oxf. Compan. Mus.* 963/2 What are sometimes heard to-day under the name of 'Wagner Tubas' represent a considerable modification of the [tenor, bass, and double-bass tubas]..., so far as the tenor tuba and bass tuba are concerned, departing more from the French horn model as to bore, and approaching more to that of the saxhorn; they retain..the funnel-shaped mouthpiece of the horns. **1948** *Penguin Music Mag.* June 132 No one explained that the 'Wagner tubas'..were assorted saxhorns, with cup-shaped and not funnel-shaped mouthpieces. **1961** A. Baines *Mus. Instruments* 315 *Wagner tubas*, 'tuba' here is really a misnomer, the proper name for these instruments being *Wagner-Tuben*... These *Tuben*, conceived by Wagner to bridge certain gaps in the brass tone, are made in two sizes, in B♭ and F, at the same pitches respectively as the B♭ and F horns and using the same part of the harmonic series. In bore they are midway between the horn and the saxhorn... They are fitted with four valves, the purpose of the fourth being to improve the intonation of the lower octave. **1965** *Listener* 3 June 837/3 A platoon of Wagner tubas. **1980** *New Grove Dict. Mus.* XX. 152/1 *Wagner tuba*, a special kind of tuba devised by Wagner for the *Ring* with the object of bridging the gap between the horns and trombones... The main body of the instrument is elliptical, like most German tenor horns, with the bell emerging from the top at a slightly oblique angle, while the lower end almost rests on the player's lap.

Wagnerian, *a.* and *sb.* Add: **Wagne·rianly** *adv.*, in a Wagnerian manner; **Wagnerism** (earlier example); **Wagnerite** (earlier example); **Wagneri·tis** *joc.* [*-itis], excessive admiration for Wagner, viewed as a morbid condition; **Wagnerize** (*b*) *intr.* (earlier example, perh. in sense 'to play (in the style of) Wagner).

1852 *Punch* 19 June 252/2 The infection called Wagnerism is spreading most rapidly throughout the musical profession. **1855** Geo. Eliot in *Fraser's Mag.* July 50/2 *Tannhäuser*..is still the music of men and women, as well as of Wagnerites. **1866** Swinburne *Let.* 26 Dec. (1959) I. 215 My art is at a discount here, but yours is idolized; you could Schumannize and Wagnerize among us to some purpose. **1889** G. B. Shaw in *Eng. Illustr. Mag.* Oct. 55 Wagneritis (a disease not uncommon among persons who have discovered the merits of Wagner's music by reading about it, and among those disciples who know no other music than his). **1897** —— in *Sat. Rev.* 13 Nov. 514/1 In 'Hamlet' he is quite enthusiastic about naturalness in the business of the stage, and makes Hamlet hold forth about it quite Wagnerianly. **1977** *N.Y. Rev. Bks.* 26 May 15/3 He recovered from his Wagneritis sooner than most worshippers.

wagon, waggon, *sb.* Add: **9. b.** Short for *patrol-wagon* s.v. Patrol *sb.* 4 in Dict. and Suppl. *slang* (orig. and chiefly *U.S.*).

1890 J. P. Quinn *Fools of Fortune* xii. 404 When a raid is made..enough 'pluggers' are captured to fill one or two wagons and are driven to the nearest police station with much clatter and display. **1926** J. Black *You can't Win* iv. 31 I'll phone for the wagon. We'll have to take them all to the station. **1953** [see *ice *sb.* 2 c].

c. Short for *station-wagon 2. Also more *gen.*, a motor car (*colloq.*).

1955 M. Allingham *Beckoning Lady* vi. 90 'Is this the wagon?' Amanda rose to meet the car. **1968** K. Weatherley *Roo Shooter* 59 When he was ready to go, the wagon wouldn't start. **1975** 'E. Lathen' *By Hook or by Crook* xviii. 173 'Do you want to take my car, Paul?'.. 'The wagon or the sports car?'

10*. Figurative phrases. **a.** *to hitch one's wagon to a star*, to set oneself high aspirations; to aspire to another's admirable example.

1870 Emerson *Society & Solitude* 27 Hitch your wagon to a star. Let us not fag in paltry works... Work..for ..justice, love, freedom, knowledge, utility. **1929** D. H. Lawrence *Paintings of D. H. Lawrence* 22 What you mentally or 'consciously' desire is nine times out of ten impossible: hitch your wagon to a star, and you'll just stay where you are. **1939** R. Boothby *Let.* 27 May in M. Gilbert *Winston S. Churchill* (1976) V. li. 1072 One of the few things in my life of which I am proud is that in all matters of major policy during the past 5 years I have hitched my waggon to your star. **1953** E. Waugh *Love among Ruins* i. 10 'The Minister of Welfare and the Minister of Rest and Culture,' continued the Chief Guide. 'The stars to which we have hitched our waggon.' **1978** J. Hyams *Pool* iv. 44 Overweight and sedentary, he was content to hitch his wagon to David's star.

b. *on the wagon*, abstaining from alcoholic drink, teetotal. See *water-wagon* (*a*) s.v. Water *sb.* 29. orig. *U.S.*

1906 B. J. Taylor *Extra Dry* 14 It is better to have been on and off the Wagon than never to have been on at all. **1917** J. M. Grider *War Birds* (1927) 23 Springs put him on the wagon for a week. **1934** J. T. Farrell *Young Manhood* xiv. 218 'Was he oiled when the accident happened?' 'No, he was on the wagon again.' **1951** L. Hellman *Autumn Garden* iii. 138 A few years ago I'd go on the wagon twice a year. Now..I don't care. **1976** L. Deighton *Twinkle, twinkle, Little Spy* viii. 83 They dug him out of a bar.., stoned out of his mind... He stayed on the wagon for years.

c. *to fix* (someone's) *wagon*, to bring about (a person's) downfall, to spoil (his) chances of success. Cf. Fix *v.* 14 c in Dict. and Suppl. *U.S. slang.*

1951 T. Capote *Grass Harp* i. 13 She said her brother would fix my wagon, which he did; right here at the corner of my mouth I've still got a scar where he hit me. **1959** J. D. Salinger in *New Yorker* 6 June 119/1 What ever became of that stalwart bore Fortinbras? Who eventually fixed *his* wagon? **1978** M. Puzo *Fools Die* xxvii. 322 At least he could fix Merlyn's wagon, Ford was beyond his reach. He tried getting her fixed by organizing a campaign of hate mail from fans.

11. a. *wagon-spoke* (later example), *-tongue* (earlier example).

1940 W. Faulkner *Hamlet* ii. ii. 129 Her companion used the reversed pistol-butt against the wagon-spoke and the brass knuckles of the other two. **1845** J. Palmer *Jrnl. Trav. Rocky Mts.* (1847) 18 Our pilot notified us that this would be our last opportunity to procure timber for axle trees, wagon tongues, etc.

12. wagon-bed (earlier example); **wagon boss** *N. Amer.*, the man in charge of a wagon-train; a wagon-master; **wagon box,** (*a*) *U.S.*, the body of a wagon; a wagon-bed; (*b*) a large storage chest, usu. kept under the front seat of a wagon (also used as an article of domestic furniture); **wagon chest** = *wagon box* (*b*); **wagon-master** (later U.S. examples); **wagon-tent** *S. Afr.*, the tent-like canopy of a covered wagon.

1853 A. S. Knight in *Trans. Oregon Pioneer Assoc.* (1933) 40 There is no ferry here and the men will have to make one out of the tightest wagon-bed. **1873** J. H. Beadle *Undeveloped West* 98 Our 'wagon-boss', absolute monarch of a train while on the road, rejoiced in the name of John Monkins. **1973** R. Symons *Where Wagon Led* i. vi. 92 The wagon boss is an important man. You don't talk to him, but when he talks *to you*, you keep your ears clean. **1810** in *Austin Papers* (1924) I. 168, 13 Wagon Boxes. **1852** M. B. Hudson *S. Afr. Frontier Life* I. 206 We come to a door's been off a long while, Near to which stands a wagon-box sacred to books. **1853** V. Williams in *Trans. Oregon Pioneer Assoc.* (1922) 196 A number of other wagon boxes have been ripped for skiffs and ply singly. **1946** G. Foreman *Last Trek* 250 The camp was devastated by a tornado that carried wagon boxes, camp equipage, and some of the people through the air. **1969** N. W. Parsons *Upon Sagebrush Harp* i. 2 The double wagon box on the stoneboat was our own, and it was full of our household goods. **1827** G. Thompson *Trav. S. Afr.* II. 134 A couple of waggon-chests. **1968** S. Stander *Horse* 75 He sat on the wagon-chest. **1861** Wagon-master [see *forage-master* s.v. *forage *sb.* 4]. **1976** *Billings* (Montana) *Gaz.* 17 June 1-c/5 Actor John MacIntire, known as wagonmaster on the television program, 'Wagon Train'. **1845** *Cape of Good Hope Almanac* (Advt.), Country people can be supplied with..3-inch Canvas for wagontents. **1926** P. Smith *Beadle* 59 The women wore stiffly starched plain white sun-bonnets, like miniature wagon-tents. **1955** W. Robertson *Blue Wagon* viii. 72, I was in the wagon-tent when those kaffirs came.

wagon, waggon, *v.* **1.** (Earlier U.S. example.)

1794 E. Denny *Jrnl.* 18 Oct. (1860) 403 The French had opened the Indian path..and wagoned considerably upon it.

wagoning, waggoning, *vbl. sb.* (Earlier examples.)

1782 in L. P. Summers *Ann. Southwest Virginia* (1929) 770 He ought to be paid..for one day waggoning. **1837** *Southern Lit. Messenger* III. 87 There never was such a time for wagoning before.

‖ **wagon-restaurant** (vagoṅ, rẹstoraṅ). [Fr.: cf. Wagon-lit and Restaurant.] A dining-car on a Continental train.

1910 *Bradshaw's Railway Guide* Jan. 958 Special Wagon Restaurant Corridor Train leaves Paris for Cherbourg on morning of sailing. **1938** H. Nicolson *Let.* 26 Apr. (1966) 337 This man would be delightful if one met him in a *wagon-restaurant* and got into conversation. **1983** W. Blunt *Married to Single Life* vii. 199 Seated in the overheated *wagon-restaurant* of the Continental express.

wag-on-the-wall. *Horology.* Also **wag-at-the-wa',** etc. [Wag *v.*: see *Sc. Nat. Dict.* for other uses of the expression.] A variety of wall-clock with unenclosed pendulum and weights. Also *attrib.*

1825 Jamieson Suppl. II. 637/2 *Wag-at-the-Wa'*,..a name given to a clock, which has no case, frequently used in the country. **1858** G. Roy *Generalship* 55 To take a fancy to a waggitonawa clock. **1904** J. Smith *Handbk. Old Scottish Clockmakers* 45 The poorer members of the community could not afford the price demanded for the long case clocks and would be content with a 'Wag at the Wa'. **1911** N. H. Moore *Old Clock Bk.* 59 Many Dutch works were sent to England without their cases... Such clocks went till the dirt and dust clogged their wheels... Such clocks as these are often called in rustic communities by the quaint name of 'wag-on-the-wall'. **1966** P. Boyle *At Night All Cats are Grey* 221 What was this nagging memory of a wag-o'-the-wall?.. With a bottle of water used as a driving weight.

wagtail, *sb.* Add: **4*.** *Joinery.* A parting-strip (see Parting *vbl. sb.* 7 b) used in the construction of a sash window.

1940 in *Chambers's Techn. Dict.* **1950** M. T. Telling *Carpentry & Joinery* 43 To prevent this happening a long thin sliver of wood called a parting slip or wagtail is placed between them *sc.* the weights] inside the box.

5. wag-tail dance = *waggle dance* s.v. *waggle *sb.* 2; **wagtail kite** *nonce-use*, a toy kite with a wagging or swinging tail.

1949 C. G. Butler *Honeybee* viii. 120 Von Frisch described two dances, a round-dance and a wag-tail dance. **1973** Wag-tail dance [see *round *a.* 5 a]. **1922** Joyce *Ulysses* 495 Ragged barefoot newsboys jogging a wagtail kite.

wahine (wahī·ni). Also 8 **whinie,** 9 **whyenee, wiena, wyeena; wahini.** [Maori, Hawaiian, and other Polynesian languages: cf. *vahine.] **1.** *N.Z.* A Maori woman or wife.

1773 W. Bayly *Jrnl.* in R. McNab *Hist. Rec. N.Z.* (1914) II. 204 Their Whinies (or women) are not regular featured in general as the men. **1807** J. Savage *Some Acct. N.Z.* xi. 74 *Wyeena*, a woman. **1841** *N.Z. Jrnl.* II. No. 32. 92/2 The chiefs and 'Wienas', or wives, partook. **1845** E. J. Wakefield *Adventure N.Z.* I. ii. 29 Having enquired how many [wives] the Kings of England had, he laughed heartily at finding they were not so well provided, and repeatedly counted 'four *wahine*' (women) on his fingers. **1863** V. Lush *Jrnl.* 28 Sept. (1971) 253 By noon the following day all the wahines had gone back to the Pah, cockles and all. **1911** W. H. Koebel *In Maoriland Bush* viii. 128 Two or three Wahines ply the men lustily with..somewhat broad banter. **1944** *Coast to Coast 1943* 105 You meet te nice girl at te dance. Plenty wahine! **1963** *Times* 7 Feb. 12/3 The warriors and their *wahine* turned away 'towards God', as I was told by an old Maori who sat with me on the grass.

2. In Polynesia: = *vahine.

1847 H. Bingham *Residence 21 Yrs. Sandwich Islands* 188 Give our *aloha* to all the new teachers and their *wahines*. **1865** H. W. Baxley *What I saw on West Coast of S. & N. Amer.* 515 The wahine disputes with them the palm of superiority.

3. *Surfing slang.* A girl surfer.

1963 *Surfing Yearbk.* 43/2 *Wahini*, a girl surfer. **1966** *Surfer* VII. 39 There are other things he did on the board, too, especially the full-moon tandem rides with wahines. **1971** *Studies in English* (Univ. Cape Town) II. 26 *Wahine* meaning woman in Hawaiian, a term popular among writers, has never been used in Cape Town.

wahoo. Add: **2.** *U.S.* The winged elm, *Ulmus alata*, or a closely related tree. Also *attrib.*

1770 E. Mease *Jrnl.* 28 Nov. in *Publ. Mississippi Hist. Soc. Centenary Ser.* (1925) V. 62 The Trees which I remark'd to be the largest were Oaks of different kinds, Wahoos, Button Wood, [etc.]. **1832** D. J. Browne *Sylva Amer.* 308 The Wahoo is a stranger to the Middle and Northern States. **1873** *Newton Kansan* 27 Mar. 1/7 One ounce of wahoo (winged-elm) bark, added to a quart of pure whiskey..is very excellent in dyspepsia. **1969** T. H. Everett *Living Trees of World* xiv. 130/2 The wahoo elm..is usually characterized by branchlets having two broad corky wings.

wahoo[2] (wahū·) [Origin unknown.] A large marine fish, *Acanthocybium solandri*, belonging to the family Scombridæ and found in tropical seas.

[**1905** D. S. Jordan *Guide to Study of Fishes* II. xvi. 266 Still larger is the great guaho, or peto, an immense sharp-nosed, swift-swimming mackerel found in the East and West Indies.] **1909** *Cent. Dict.* Suppl. 1429/2 Wahoo.. A common name of *Acanthocybium solandri*, a scombroid fish of tropical seas. **1920** *Outing* July 248/3 He had fished for—and caught—many a most unheard of species, as well as sharks, sword-fish, wahoo, etc. **1940** *Geogr. Jrnl.* XCV. 130 These fish [*sc.* kingfish] which the Americans call wahoo, are beautifully shaped surface feeders. **1979** *United States 1980/81* (Penguin Travel Guides) 517 The white marlin capital of the world sends fishermen out to the Jackspot..for dolphin, bonita, tuna, and wahoo.

wah-wah[1], var. Wow-wow[1].

1883 [see Wow-wow [1]]. **1891** E. Arnold *Seas & Lands* xl. 520 A 'wah-wah' monkey..was a delightful little creature, with very long silky arms. **1939** A. Keith *Land below Wind* xvii. 291 *Wah wah* was the native name for gibbon ape. **1959** 'M. Derby' *Tigress* iii. 145 The birds and the *wah-wahs*..woke. **1964** J. Pope-Hennessy *Verandah* ii. ii. 78 He [*sc.* Hugh Low] once wrote to his daughter Kitty that he loved only two creatures in the world—his wah-wah monkey, Eblis, and herself. **1975** *Blackw. Mag.* June 542/1 The only members of the ape family found in Malaya are the gibbons, the commonest of which is locally known as the *wah-wah*.

wah-wah[2] (wā·wā). Also **wa-wa, wha-wha, wow-wow.** [Echoic.] **1.** *Jazz.* A musical effect achieved on brass instruments by manipulation of a mute and on an electric guitar by means of a pedal controlling output from the amplifier.

1926 H. O. Osgood *So this is Jazz* 98 Then there is the clown of mutes. Leading philologists of the jazz world are at odds over the correct spelling of the name; some favour wow wow, others prefer wha wha, but there is authority for the simplification wa-wa. **1936** *Amer. Mercury* XXXVIII. p. x/2 *Wah-wah*, a brass effect gotten by favoring a bell of a horn with a mute. **1949** R. Blesh *Shining Trumpets* viii. 183 The 'wa-wa', produced by the cupped hand moved in front of the cornet or trumpet bell. **1950** A. Lomax *Mr. Jelly Roll* (1952) 64 There were many other imitations of animal sounds we used—such as the wah-wahs on trumpets and trombones. **1976** *New Musical Express* 12 Feb. 25/2 You can thrill again to..Clapton playing his insidious wah-wah on 'Get Back Home'. **1977** 'E. Crispin' *Glimpses of Moon* iii. 38 One spends his working life in a perpetual seascape, another writing wah-wahs on trumpet parts for people surfacing in mud-baths into which they have comically fallen..and so on.

2. The sound of a baby's crying, or a noise resembling this.

1938 D. Runyon *Furthermore* viii. 156 Something goes wa-wa, like a sheep..because..Isadore is sitting on John Ignatius Junior's doll, which says 'Mamma' when you squeeze it. **1958** Hayward & Harari tr. *Pasternak's Dr. Zhivago* I. vi. 158 'Wa, wa,' yelled the babies all on one note... Only one voice stood out from the others. It was also yelling 'wa, wa',..but it was deeper.

3. *attrib.*, esp. as (sense 1) *wah-wah effect, mute, pedal.*

1925 G. Gershwin *Rhapsody in Blue* (music score) (1939) 1 (*direction*) Wha Wha effect. **1926** *Melody Maker* Feb. 23/2 (Advt.), The wow-wow mute is made of spun aluminium. **1926** Whiteman & McBride *Jazz* ix. 200 The first time I ever heard what I call the wawa mutes, used with the cornet... The players got that effect by inverting glass tumblers over the bells of the instruments. **1930** R. Paget *Human Speech* 240 The so-called 'wow-wow' mutes for cornets and the like. **1935** *Vanity Fair* (N.Y.) Nov. 71/3 The wah-wah notes of intermittently muted brass. **1943** R. Blesh *This is Jazz* 24 This bad tone..like dirty tone, growl or wah-wah trumpet work, is expressive and therefore good from the Jazz point of view. **1954** *Grove's Dict. Mus.* (ed. 5) IV. 601/1 The trumpets and trombones very frequently use peculiar mutes, such as the 'wa wa mute', which can make the instrument weep or laugh rather like a human voice. **1969** *It* 13–28 June 13/2 Frank proved quite conclusively that he's a brilliant guitarist, especially with a 'wah-wah' pedal. **1979** R. L. Simon *Peking Duck* xxiii. 175 It's an old song... Now they use a wah-wah pedal and a Moog Synthesizer.

waiata (wai·ata). *N.Z.* Also 9 **wyata.** [Maori.] A Maori song.

1807 J. Savage *Some Acct. N.Z.* xi. 75 *Wyata*, singing. **1843** E. Dieffenbach *Trav. N.Z.* II. i. iv. 57 E' Waiata is a song of a joyful nature. **1905** W. Baucke *Where White Man Treads* 88 Then Puke..sang a waiata. **1978** P. Grace *Mutuwhenua* xiv. 104 Singing or chanting the waiata at the conclusion of each speech. **1984** *Metro* (Auckland) Jan. 72/3 Indeed the Maori name for the racecourse is Waiatarua (Two Songs) and tradition suggests it was so named because the sound of the flood waters escaping through the underground caverns closely resembled the singing of waiatas.

waif, *sb.*[1] (and *a.*) Add: **A.** *sb.* **3.** *waif-like* adj.

1924 R. Campbell *Flaming Terrapin* ii. 35 Their waif-like corpses on a stormy bed Toss in their deep deliriums. **1962** I. Murdoch *Unofficial Rose* xv. 149 How young she looked, how waif-like.

Hence **wai·fish** *a.*, **wai·fishly** *adv.*

1936 S. Smith *Novel on Yellow Paper* 220 Such.. wispish, waifish progeny. **1977** *Time* 31 Jan. 21/3 The waifish face beneath the jaunty white cap never loses its ethereal Pre-Raphaelite look. **1980** *Times Lit. Suppl.* 20 June 703/1 *Hurricane's* Samoan scenery is enticingly picturesque: doesn't art student Charlotte—waifishly played by Mia Farrow—arrive on her vacation exclaiming 'I can't wait to get out my paintbox'?

waif, *sb.*[2] Add: **2.** (Earlier *Naut.* examples, *spec.* in whaling.) Also *attrib.*, as *waif-pole.*

1839 *Knickerbocker* XIII. 382 Line-tubs, water-kegs, and wafe-poles, were thrown hurriedly into the boats. **1846** T. B. Thorpe *Mysteries of Backwoods* 85 As the waiffe of the whaleman [marks] his victim on the sea. **1851** H. Melville *Moby Dick* III. iii. 27 The allusion to the waifs and waif-poles..necessitates some account of the laws and regulations of the whale fishery.

wail, *v.* Add: **2. c.** Of a jazz musician: to play very well, with great feeling, etc. (*U.S. slang*). Also *U.S. colloq.*, to perform well.

1955 Shapiro & Hentoff *Hear me talkin' to Ya* 231, I revered the amazing Fats Waller, who had lately made a splash wailing on organ at the Lincoln. **1959** *Encounter* June 43/2 The Beat 'cat' approaches the Beat 'chick' with the ritualistic 'Pad me'..the 'chick's', approach to the male is..'I'm frigid,' to which he can either reply 'I'll make you wail' (function) or, otherwise, 'Don't bug.' **1962** [see *blow *v.*[1] 14 e]. **1977** C. McFadden *Serial* (1978) xxxix. 85/2 Despite his back, he was really wailin' when he hung a sharp right into his driveway, pretending Sutton Manor was a picturesque village along the route of the Tour de France.

wailing, *vbl. sb.* **c.** For '**wailing place**.. (also occas. *wailing wall*)' read **Wailing Wall,** the remaining part of the wall of the Second Temple of Jerusalem, destroyed in 70 B.C., revered by Jews as a place of prayer (also † *wailing place*) (earlier and later examples); also *transf.* and *fig.*

1878 J. Fergusson *Temples of Jews* ii. xii. 183 The most interesting particular mentioned by the Pilgrim is the 'Lapis Pertusus', which was then the Wailing Place of the Jews. **1922** Joyce *Ulysses* 532 Darkshawled figures of the circumcised, in sackcloth and ashes, stand by the wailing wall. **1934** C. Lambert *Music Ho!* iii. 212 Tin Pan Alley has become a commercialized Wailing Wall. **1955** W. Gaddis *Recognitions* iii. v. 945 I've just visited the Wailing Wall, and had a good cry. **1963** *Daily Tel.* 23 Dec. 12/2 The dividing wall [in Berlin] becomes a wailing wall at midnight each night. **1980** K. Follett *Key to Rebecca* xxix. 308 Today I went to the Western Wall, which is also called the Wailing Wall.

wainage. 2. (Earlier example.)

1810 G. Chalmers *Caledonia* II. ii. 134 The waynage' or cultivable lands, and meadows of each district or manor, were possessed, and laboured, in separate portions, by the individuals of the manor.

wainscot, *sb.* Add: **6. wainscot chair,** a panel-back chair (see *panel *sb.*[1] 20).

1663 in *Farm & Cottage Inventories of Mid-Essex 1635–1749* (Essex Record Office) (1950) 95 One Wainsscott Chair. **1891** I. W. Lyon *Colonial Furnit. New Eng.* v. 145 The wainscot chairs which figure in the early records were doubtless those made up—back, seats, and all—of wood, the wood being most invariably oak. *Ibid.* 146 Wainscot chairs were quite common in England and Scotland in the seventeenth century. **1925** [see *panel-back* adj. s.v. *panel *sb.*[1] 20]. **1978** P. Van Greenaway *Man called Scavener* i. 11 A long passage lined with Pembrokes, a Wainscot chair, a rare Caquetoire.

‖ **waipiro** (waipiro). *N.Z.* Also **waipera, waipero, waipirau.** [Maori, f. *wai* water + *piro* putrid.] Alcoholic liquor, spirits.

1845 W. Brown *N.Z.* i. iii. 132 Another native.. keeps a grog-shop, and sells his *Waipero*..to *Hourangi* (drunken) pakehas. **1847** S. C. Brees *Pictorial Illustr. N.Z.* 13 The natives..say, it is '*all the same as te waipe-ra*', or spirits. **1863** F. E. Maning *Old N.Z.* xi. 152 He *would* go on shore..to get some water to mix with his *waipiro*, and was not his canoe found next day floating about with his paddle and two empty case-bottles in it? **1881** J. L. Campbell *Poenamo* xi. 341 We had the inevitable percentage of indulgers in 'waipirau'. **1933** *Bulletin* (Sydney) 20 Dec. 20/3 Under the influence of waipiro a Rotorua Maori pinched a sedan car. **1972** P. Newton *Sheep Thief* xx. 169 She had no time for the waipera... 'Charles,' she said... 'You can decide right now. Me or the drink.'

wairakite (wəi·răkəit). *Min.* [f. *Wairakei*, name of a place in New Zealand + -ite[1].] A hydrous silicate of calcium and aluminium, $CaAl_2Si_4O_{12}.2H_2O$, the calcium analogue of analcime, that occurs as colourless to white monoclinic crystals and is common in low-grade metamorphic rocks.

1955 A. Steiner in *Mineral. Mag.* XXX. 691 During the study of hydrothermal alteration of cores recovered from holes drilled at Wairakei..an unusual zeolite mineral was found... The name wairakite is given to the new mineral. **1979** *Amer. Mineralogist* LXIV. 993/1 A marked feature of wairakite crystals is the presence of complex, fine lamellar twinnings.

waist. Add: **3. c.** The middle section of the fuselage of an aeroplane, esp. a bomber. *U.S.*

1942 [see *waist gunner* in sense 7 below]. **1956** *U.S.A.F. Dict.* 560/2 *Waist*, the middle section of an airplane's fuselage. Applied esp. to the middle section of a bomber.

6. Simple attrib., as *waist height, size*; *waist-length* adj.; 'worn from the waist', as *waist petticoat, slip.*

1953 E. Simon *Past Masters* i. i. 21 The tops of the bookshelves which at waist-height ran along the walls. **1944** R. Lehmann *Ballad & Source* i. ii. 14 She wore a waist-length cape called a dolman. **1977** 'L. Egan' *Blind Search* i. 1 Her waist-length brown hair [was] untidily braided. **1939–40** *Army & Navy Stores Catal.* 642/2 *Waist petticoats*..Moiré poplin. **1918** *T. Eaton & Co. Catal.* Spring & Summer 188/1 Women's overalls, made of fine khaki drill... Waist sizes 24 to 38. **1974** *Country Life* 12 Dec. 1903/1 Tartan skirts..in waist sizes 24–30 in. **1955** M. Hall *Let's make some Undies* (Let's Make It Ser.) 44 (*heading*) Straight placket for waist slips and knickers. **1976** T. Stoppard *Dirty Linen* 9 Maddie is.. wearing..a waist-slip which is also pretty, silk and lace, with a slit.

7. waist-gun, a gun set in the waist of an aircraft; also **waist-gunner; waistline,** (*a*) (in Dict., sense 6) (earlier example); (*b*) a person's waist, esp. with reference to its size; (*c*) = sense 1 c in Dict.; (*d*) a notional line running round the body of a motor vehicle at the level of the bottom of the window frames.

1942 *Yank* 7 Oct. 3 The waist gunner sits patiently in the waist of the fuselage. *Ibid.* 23 Dec. 4 From his waist-gun position he interphoned the appearance of two Jap nightfighters off to the right. **1978** J. Irving *World according to Garp* i. 15 Sergeant Garp had experience as.. a waist gunner in the B-17E. *Ibid.* 16 Technical Sergeant Garp, at his waist gun position. **1896** *Woman's Life* 15 Feb. 448/1 Your velvet vest should end in a narrow V at the waist line. **1928** *Daily Mail* 25 July 8/5 Even if you are one of the fortunate few who need not keep a watchful eye on their waistline, you'll eat Vita-Weat for pleasure and health. **1930** R. Campbell *Adamastor* 25 Our Drakensberg's most lofty scalps Would scarcely reach the waist-line of the Alps. **1941** *Ann. Reg. 1940* 254 The Russians..attacked the 'waist-line' [of Finland] from further north. **1959** *Times* 2 Oct. 9/1 These cars..have an entirely new top from the waistline upwards. **1970** J. Philips *Nightmare at Dawn* (1971) i. 41, I don't have to worry about my waistline. **1974** *Country Life* 28 Feb. 446/1 This latest Capri..has a much bigger window area with a lower waistline.

waistcoat. Add: **5.** *waistcoat-piece* (earlier example).

1789 J. Woodforde *Diary* 19 Sept. (1927) III. 193 Gave my Servant Man Ben a Waistcoat Piece.

waistcoateer. 1. (Later *Hist.* examples.)

1916 Joyce *Portrait of Artist* (1969) v. 176 The grave and mocking music of the lutenists or the frank laughter of waistcoateers. **1922** [see *flat-cap 2].

wait, *v.*^{1} Add: **I. 5. d.** Phr. *wait for it*, said (often parenthetically) to create an interval of suspense before imparting something remarkable or amusing, in order to heighten its effect. Also *ironically*. *colloq.* **1930** M. Allingham *Mystery Mile* xviii. 170 'Wait a minute,' said Mr. Knapp. '*Wait* for it... That is just exactly wot I do know.' **1958** S. Hyland *Who goes Hang?* xvi. 74 'You're hiding something...' 'Wait for it my dear.' **1966** 'H. Calvin' *Italian Gadget* ii. 21 We can have a shower and..wait for it, dinner at the Palazzo Capucci. **1979** R. Laidlaw *Lion is Rampant* xviii. 139 The real attack will come from, wait for it, wait for it—anither direction a'thegither.

6. c. *to wait out*: (*a*) *U.S. Baseball*, (of a batter) to force (a pitcher) to throw a maximum number of pitches by refraining from striking at pitches in the hope of getting a base 'on balls', i.e. because they were not pitched over the home-plate; hence (chiefly *U.S. colloq.*), to wait during (a period of time, an event, etc.); to wait for the end of; also, *to wait it out*, to endure a period of waiting; (*b*) = *to wait for* (sense 7). **1909** *Amer. Mag.* Aug. 401 Still Chance commanded: 'Wait—wait him out.' Every batter went to the plate intent on making Donovan pitch as many balls to them as possible. **1936** *Philadelphia Rec.* 31 July 15/1 Alf M. Landon is up there under instructions from the bench to wait 'em out. **1956** H. Kurnitz *Invasion of Privacy* xv. 95 He monopolized the phone in the bedroom, where Shelly and Zorn were waiting it out. **1959** *N.Y. Times Mag.* 11 Oct. 18/1 He is sealed into the container and..lies there in his 'contour couch' and waits out the long countdown. **1966** D. Francis *Flying Finish* x. 131, I retired to the snack bar..to wait out the twenty minutes. **1977** R. Perry *Dead End* i. 9 Unfortunately, I couldn't afford to wait him out. Daley was already overdue.

7. d. *to wait on*: (*d*) *dial.* (esp. *Austral.* and *N.Z.*), to wait for a while, to 'hold on'. Freq. *imp.* **1943** N. Marsh *Colour Scheme* xiii. 228 'Wait on, wait on,' Dikon heard Webley mumble, 'You'll get it back all right.' **1946** K. Tennant *Lost Haven* (1947) x. 138 Wait on, Patsy... Wait on, Patsy. You talk of letting the water out, but what about the fish?—bloody fish'll go too. **1967** 'G. Douglas' *Death went Hunting* xviii. 161 For some reason I can't define, he seems quite willing to accept his present position... As we say in Yorkshire, he appears to be waiting on. **1968** S. L. Elliott *Rusty Bugles* in E. Hanger *Three Austral. Plays* 92 *Andy*: Open the one from Darky. *Gig*: I am blast you... Wait on, can't you?

g. *to wait and see*. (Example with reference to Asquith's use, and earlier and later *attrib.* examples.) Also *wait and see* sb. phr. **1910** *Blackw. Mag.* May 747/1 Mr. Asquith has deemed it not incompatible with the gravity of his office to elude the curiosity of his opponents with the absurd formula, 'Wait and see'. **1969** S. Hyland *Top Bloody Secret* ii. 121 The philosophy of wait and see..was getting him nowhere. **1976** *Scottish Daily Express* 23 Dec. 11/2 The rest of the back row is in a similar state of 'wait and see'. *attrib.* **1870** L. M. Alcott *Old-Fashioned Girl* xiv. 231 She..thought she would 'wait and see.'.. The 'wait-and-see' decision was making her friend unhappy. **1930** [see *flop *adv.* and *int.*]. **1977** A. J. Russell *Pour Hemlock* xii. 147 His was a wait-and-see posture. **1984** *Times* 8 June 20/1 An understandable wait-and-see management caution.

i. (Earlier examples.) **1788** J. Woodforde *Diary* 29 Feb. (1927) III. 8 Mr. Taswell..desired us to send to Mr. Custance that they might not wait dinner for him. **1836** Dickens *Let.* 17 Sept. (1965) I. 174, I hope and trust you did not wait dinner for me.

j. *to be unable* (etc.) *to wait* (*for* or *to do* something), to be very eager or anxious for or to do it. *colloq.* **1938** D. du Maurier *Rebecca* xvi. 232 'Oh, Madam, how exciting,' breathed Clarice. 'I don't know how I am going to wait for the day.' **1958** I. Murdoch *Bell* xix. 234 We're all so excited, we can hardly wait for tomorrow morning. **1971** J. Osborne *West of Suez* ii. i. 63 Can't wait! Let's go home *now*. **1972** *National Observer* (U.S.) 27 May 10/3 Now the young musicians can't wait to get at their instruments.

k. (you) *wait till* (or *until*).., used to imply a threat, warning, etc., or promise of something interesting or exciting, when the specified event has occurred. Also *ellipt.* as *you wait!* **1938** G. Greene *Brighton Rock* iii. i. 99 'I dunno who Brewer is,' Ida said, 'but things seem lively.' 'You wait till the races start,' the man said. 'They'll be lively all right then.' **1953** E. Simon *Past Masters* iii. iii. 154 Parents brazened out their children's tantrums with the key-phase of, 'You wait!'. **1960** S. Barstow *Kind of Loving* ii. ii. 162 Seventy-four guineas, Henry. Seventy-four bloody lovely guineas. Just wait till we tell Mr. Van Huyten about this. **1975** J. Gores *Hammett* xix. 130 She was going over to..tea with the George F. Biltmores! Wait until she wrote her mother about *that!* **1984** *Guardian* 5 Oct. 14/3 Mr Scargill..will soon, very soon—just you wait—emerge triumphant.

9. d. (Earlier example.) **1827** Carlyle tr. L. Tieck in *German Romance* II. 135 Andres waited supper.

II. 14. wait on or **upon** —. **h.** (Later examples.) Cf. Waiter 4 b. **1840** Geo. Eliot *Let.* 20 July (1954) I. 58 That constant waiting on God for instruction and comfort which [the Quietists]..make the sum total of religion. **1931** J. Buchan *Blanket of Dark* xvi. 307 Be still and wait on God. **1979** R. Blythe *View in Winter* ix. 300 One of the best things which all these new changes have brought about is this notion of waiting upon God.

i. Delete † *Obs.* and add later, chiefly *U.S.*, examples. Also, more generally, to wait for. **1817** W. Sewall *Diary* 2 May (1930) 14/1, I proceeded to the Academy and agreed to enter there upon condition that the Professor would wait on me until the next winter. **1865** O. L. Jackson *Colonel's Diary* (1922) 227 We have been waiting on the pay department. **1915** J. Buchan *39 Steps* vi. 135 He..raised his placid eyebrows and waited on me to speak. **1931** *Amer. Speech* VII. 20 *Wait on*, wait for, the Biblical sense. 'When I got there, John was waiting on me.' (Widespread.) **1955** F. O'Connor *Wise Blood* v. 85 He..darted after Hazel Motes. 'Wait on me!' he shouted. **1960** *Observer* 7 Feb. 13/4 The nation waits on the railwaymen, to see if there will be a strike or not. **1984** P. Turnbull *Big Money* ix. 153, 'I was just waiting on you two coming back.' He stood and reached for his coat.

wait-a-while. 1. (In Dict. s.v. Wait-a-bit a.)

2. In Australia, any of several plants with prickles or spiny leaves, esp. *Acacia colletioides*. **1889** J. H. Maiden *Useful Native Plants Austral.* 306 *Acacia colletioides*..'Wait-a-while' (a delicate allusion to the predicament of a traveller desirous of penetrating a belt of it). **1940** F. D. Davison *Woman at Mill* 86 Clumps of prickly wait-a-while gave back the light from millions of small shiny leaves. **1967** [see *Bathurst burr].

waiter. Add: **III. 6. b.** Restrict † *Obs.* to sense in Dict. and add: More recently, (*U.S.*) an attendant upon the bride or groom at a wedding. *Obs.* exc. *dial.* or *Hist.* **1830** in N. E. Eliason *Tarheel Talk* (1956) 303 [She] says she hears you are to be married. She wishes to know..when..as you promised she would be one of the waiters. **1927** *Dialect Notes* V. 470 *Waiter*,..an attendant of the bride or groom at a wedding. **1959** *Daily Progress* (Charlottesville, Va.) 19 Mar. 14/1 Four young men and four young women were appointed to be 'waiters' on the bride and groom.

† 7. d. *Mil.* A soldier, etc., employed as a domestic servant to an officer. *U.S. Obs.* **1828** A. Sherburne *Mem.* i. 20, I was waiter to Mr. Charles Roberts the boatswain. *Ibid.* ii. 26 Part of our officers with five or six waiters..occupied an elegant house. **1861** *Army Regulations* 559 Non-commissioned officers not employed as waiters.

8*. A uniformed attendant on the floor of the Stock Exchange, Lloyd's of London, or other City of London institution. **1887** [see Hammer *v.* 2 d]. **1904** C. Duguid *Stock Exchange* ii. 12 We have also noticed the Waiters' Stands, about twenty in number, placed in various parts of the House, pulpit-like..so that the important announcements which emanate from these stands may be well heard in the House. **1934** F. E. Armstrong *Bk. Stock Exchange* xvii. 356 It is known that dealers used to foregather in the coffee houses of Change Alley, Cornhill. .. Here, probably, it was that 'waiters', plying their trade, and seeking out from the interior fastnesses of these refreshment rooms some particular Stock-jobber wanted by his client, gave the name to the uniformed 'waiter' who calls the Members within the walls of the present Stock Exchange. **1962** A. Sampson *Anat. of Britain* xxiii. 380 Merchant bankers work in a formal atmosphere... Mahogany, black-coated waiters and grandfather clocks set the tone of privacy. *Ibid.* xxv. 400 The Room [at Lloyd's]... When a broker is wanted, one of the 'waiters' (who stand round the room in scarlet uniform) writes his name on a special instrument. **1976** *Times* 22 Apr. (Baltic Exchange Suppl.) p. i/9 As in other City institutions, attendants wear livery and are called waiters.

IV. 12. *Chess.* = *waiting problem* s.v. *waiting *vbl. sb.*^{1} 2 c. **1906** A. C. White *Tours de Force* p. xxxii, If a problem has no threat, it is called a 'waiter', and all the different continuations are variations. **1935** *Encycl. Sports* 150/1 There is no simple waiting move for white. The key is, indeed, a 'waiter', but four of the resulting mates are different from those in the set position. **1963** M. Lipton et al. *Chess Problems* iv. 72 White can neither retain all the set mates as in a Waiter, nor change some of them while still maintaining zugzwang as in a mutate.

waiting, *vbl. sb.*^{1} Add: **2. c. waiting list,** a list of people waiting for appointments, selection for any purpose, or the next chance of obtaining something; **waiting (move) problem** *Chess* (see quots.); **waiting race** (earlier example); **waiting time,** time spent waiting, *spec.* in *Computing* (see quot. 1962) or *Work Study* (see quot. 1979). **1897** *Outing* XXX. 347/2 The Michaux Club is composed of two hundred members, with a large 'waiting list'. **1916** A. Huxley *Let.* 5 Feb. (1969) 90 There is a huge waiting list for the better jobs on all the Govt. Depts. **1921** *Tax Clerks' Jrnl.* IV. 387 The right to a place on the 'waiting list' for a permanent post. **1976** *Times* 23 Feb. 13/1 Local councils were able to do little to reduce over-long waiting lists. **1891** J. Rayner *Chess Problems* 9 Waiting-move problems—*i.e.*, those in which the first move would not lead to a mate if it were not that Black is forced to reply and thereby weaken his position. **1896** *Chess Monthly* May 280 There is another class of waiting problem which is usually found very interesting... A primary position which has every indication that White has but to linger without discouraging the mates as 'set', but..one of the mates has to be given up in exchange for another to be created. **1907** S. S. Blackburne *Chess Problems* 25 *Block problem*—Otherwise called a 'Waiting Problem' is one in which White's first move leads to mate after every reply of Black, in consequence of such reply. **1962** K. S. Howard *One Hundred Years Amer. Two-Move Chess Problem* 5 Another type [of problem]..was the complete block, or waiting-move problem, especially where some of the mating moves apparently set in the initial position were changed by the key, for which Brian Harley..coined the apt name *mutate*. **1868** H. Woodruff *Trotting Horse* xxxvi. 297 Tallman made a waiting-race of it, and pulled Flora back three lengths. **1962** *Gloss Terms Automatic Data Processing* (*B.S.I.*) 85 *Waiting time*, of a store. The time interval between the instant the control unit calls for a transfer of data to or from the store and the instant the transfer commences. **1976** P. R. White *Planning for Public Transport* v. 115 Waiting time is a function of service frequency and thus is not affected directly by spacing. **1979** *Gloss. Terms Work Study* (*B.S.I.*) 16 *Waiting time*, the period of time for which an operator is available for production but is prevented from working.

waiting-man. For † *Obs.* read *Obs.* exc. *U.S.* and add later U.S. examples. **1839** *Southern Lit. Messenger* V. 752/1 The door of the bachelor's hall was assaulted by a repetition of raps, which quickened the steps of Mr. Singlesides' sedate waiting-man. **1884** G. W. Cable *Dr. Sevier* I. x. 73 The speaker ceased as the mulatto waiting-man appeared.

wait-list. orig. and chiefly *U.S.* [Wait *v.*^{1}] = *waiting list* s.v. *waiting *vbl. sb.*^{1} 2 c. **1960** *Amer. Speech* XXXV. 312 There's a wait list on that one too,' one is told somberly by steamship agents. **1977** in H. J. Eysenck *You & Neurosis* vi. 185 On a rating scale of overall improvement, 93 per cent of the behaviour therapy patients contrasted to 77 per cent of the psychotherapy and wait-list patients were considered either improved or recovered. **1979** *United States 1980/81* (Penguin Travel Guides) 16 *How to Use an Airline*; it's full of information on how to make reservations, the difference between waitlist and stand-by, [etc.].

Hence as *v. trans.*, to put (a person) on a waiting list; to draw up a waiting list for places on (an aircraft, ship, etc.); **wait-listing** *vbl. sb.* **1960** *Amer. Speech* XXXV. 312 A boat is said to be *wait listed*... What gloomy fate, then, awaits the prospective passenger who is *wait listed?* **1971** *Financial Mail* (Johannesburg) 26 Feb. 656/2 What about his wait-listing for an earlier flight? **1973** *Daily Tel.* 5 Sept. 6/7 On less frequent services..overbooking is not employed. Instead, 'wait-listing' is effected, and passengers who learn that their flight is full wait at the airport on the chance of a seat becoming available.

waitress. Add: **2. b.** Chiefly *U.S.* A female servant in a private house whose duty is to wait upon those at table (cf. Waiter 7 c); in extended use, a housemaid. **1875** Mrs. Stowe *We & our Neighbors* 323 Maggie was parlor-girl and waitress, and a good one too. **1906** *N.Y. Herald* 5 Mar. 14 A competent girl as waitress and to take charge of parlor floor; private family. **1937** C. Day *Life with Mother* 215 Each brother had his own sacred place where his own toys were kept, except when the waitress cleaned the room and mixed everything up. **1953** A. Christie *Pocket Full of Rye* xi. 32 Gladys Martin is the parlourmaid or waitress, as they like to call themselves nowadays.

3. Comb.: **waitress service,** service by waitresses in a restaurant, opp. *self-service*. **1960** C. Dale *Spring of Love* i. 3 The floor with waitress service was always crowded. **1970** E. McGirr *Death pays Wages* vii. 150 The main room had tables with white cloths and a waitress service.

Hence **wai·tressing,** service as a waitress; hence (as a back-formation) **wai·tress** *v. intr.*, to work as a waitress. **1936** J. B. Priestley *They walk in City* xi. 335 Well, you're a nice-looking girl and you have done some waitressing—so it oughtn't to be hard for you to get a place. **1950** *Landfall* Dec. 308 There were pretty good ones [*sc.* jobs] going now—cooking and waitressing in town. **1974** K. Millett *Flying* (1975) iii. 311, I was a kid from St. Paul waitressing in Glacier Park. **1984** *Times* 15 Oct. 10/4 The waitress..can..make waitressing into a performing art. **1984** *New Yorker* 29 Oct. 43/1 Jenny was waitressing in Denver.

waive, *v.*^{1} Add: **9. b.** *N. Amer.* Of a sports club: to waive its right to buy (a player from another club in the same league). Also *intr.* Cf. *waiver 1 d. **1908** *Evening Star* (Washington, D.C.) 26 Feb. 13/2 The Giants have given Mike Lynch his unconditional release, as all clubs have waived. **1970** *Globe & Mail* (Toronto) 26 Sept. 36/2 John Schneider..finally was cut and..waived through the Canadian Football League. **1979** *Tucson* (Arizona) *Citizen* 20 Sept. 2d/4 Atlanta Hawks—Waived Tim Claxton, guard, and Rickey Brown, forward.

waiver. Add: **1. d.** *spec.* The formal relinquishment by a club in a professional sports (esp. baseball) league of its right to buy the contract of a player from another club in the same league, before he is offered to a club in another league. Freq. *on waivers*. *N. Amer.* **1907** Herrmann & Pulliam in *4th Ann. Rep. Nat. Commission* (1908) 99 The Club never had any intention to let the player go;..they only asked for waivers as a bluff. **1946** *Encycl. Brit.* III. 167/1 There is a fixed price in each league that must be paid by the club obtaining a player by this waiver method. The player is said to be sent by club A to club B on waivers. **1951** *Britannica Bk. of Year* 88/2 The Giants claimed pitcher Jim Hearn from the Cardinals on waivers in mid-season. **1972** J. Mosedale *Football* v. 75 He was put on waivers at the end of the season. **1978** *N.Y. Times* 29 Mar. b4/4 Oakland..put Dick Allen, first baseman, on waivers for purpose of giving him unconditional release.

‖ waka^1 (wa·ka). *N.Z.* Also 9 **walker, wauka.** [Maori.] A Maori canoe (see quots.). **1807** J. Savage *Some Acct. N.Z.* xi. 77 *Wauka*, a canoe. **1834** E. Markham *N.Z. or Recoll. of It* (MS.) 3 Canoes, or in the Native Mourie Tongue, Walker Mouries, or Native boats. **1841** J. C. Bidwill *Rambles in N.Z.* (1952) 61 Those [Canoes] with topsides..are called Wa-kaw, or in common pronunciation 'Walkers'. **1845** R. Taylor *Jrnl.* 19 June (MS.) III. 222 We had a large waka which just held our large party. **1874** W. M. Baines *Narr. E. Crewe* 81 'Whaka' is the native name, or rather the native generic term, for all canoes, of which there are many different kinds, as tete, pekatu, kapapa and others. **1921** H. Guthrie-Smith *Tutira* x. 73 A miniature *waka* or canoe..was moved..from place to place. **1936** *Discovery* Jan. 13/2 A Moriori Waka or fishing boat. **1949** P. Buck *Coming of Maori* (1950) ii. vii. 203 River canoes (*waka tiwai*), also used on inland lakes, consisted merely of the dug-out hull. Seagoing canoes (*waka tete*)..were larger.

‖ waka^2 (wa·ka). [Jap.] **1.** A form of classic Japanese poetry, lyrical in nature and developed from the ancient traditional ballads. [**1880** B. H. Chamberlain *Classical Poetry of Japanese* p. vi, *Kokin Wa-Ka Shifu Uchi-Giki* ('Memoranda Concerning the Collection of Japanese Odes Ancient and Modern'), by Kamo-no-Mabuchi.] **1932** B. L. Suzuki *Nōgaku* 24 The forms sung in the Nō are the *shidai*,..*rongi*, *waka*, and *kiri*. **1948** *Introd. Class Jap. Lit.* p. iv, In the sphere of *waka* poetry also, the *Kokinshū* anthology ..shows the transition from 'sincerity' to 'sentimentality'. **1968** *Encycl. Brit.* XII. 953/2 The poetic form used—the *waka*, deriving from the earlier folk songs—consists of an alternation of five- and seven-syllable lines, without rhyme, stress, metrical pattern or other technical device.

2. A Japanese poem of thirty-one syllables, a *tanka*. **1938** D. T. Suzuki *Zen Buddhism* i. iv. 95 The secret documents also contain a number of *waka*, versified epigrams, in regard to the mastery of swordsmanship. **1956** D. Keene *Anthol. Jap. Lit.* 25 His *waka*—thirty-one syllabled poems—are among the most beautiful and melancholy in the language. **1977** *Times Lit. Suppl.* 15 Apr. 448/4 The most striking qualities of the haiku and the waka.

Wakamba (wakæ·mbă). [Bantu.] = *Kamba. **1888** *Encycl. Brit.* XXIV. 828/1 [Bantu tongues of eastern group] Wa-Kamba, Wa-Nika, Wa-Pokomo. **1935** [see *simba]. **1937** K. Blixen *Out of Afr.* ii. 138 The immigrant Somalis..have a moral code... But the unprejudiced Kikuyu, Wakambas, or Kavirondos..judge you not. **1955** *Times* 13 July 6/5 Many Nyanza and Wakamba tribesmen in the Nairobi area had taken Mau Mau oaths in the last 18 months. **1977** H. Innes *Big Footprints* i. i. 24 'A Masai?' I asked... 'No, no. He's of the Kamba people—the Wakamba.'

wake, *v.* Add: **I. 6. b.** (Further examples, chiefly *Anglo-Irish*.) **1829** M. Edgeworth *Garry Owen* (1832) vii. 46 You were right, dear, from first to last concerning the poor cratur's dead child; she did not want to have it *waked* at all, for she is not that way—not an Irishwoman at all. **1898** F. P. Dunne *Mr. Dooley in Peace & War* 188 They waked th' oldest son in small beer, an' was little thought of. **1959** T. H. White *Godstone & Blackymor* 168 Everybody was trying to amuse Charlie Plunkett. Otherwise, why 'wake' him? **1974** D. Sears *Lark in Clear Air* ix. 117 They waked Holly Dallan in the parlour of the log house where she had been born and reared.

III. 8. c. *to wake snakes*: see also Snake *sb.* 2 d in Dict. and Suppl.

wakeaday, wake-a-day (wē^{i}·kădē^{i}). ?*nonce-wd.* [f. Wake *v.* + Day *sb.*, after Work-aday.] Of life, etc.: such as one wakes up to each day; that is experienced by ordinary people; regular. **1893** G. B. Shaw *Let.* 21 Aug. (1965) I. 401 All these terrible combats..on behalf of Wilde, Pinero, [Frank] Harris &c. &c. &c. belong to Piona and not to the wake-a-day world. **1951** M. McLuhan *Mech. Bride* (1967) 97/2 Synthetic gods and goddesses (stars) appear to assume the roles of our wakeaday existence. **1962** —— *Gutenberg Galaxy* 269 The electric puts the mythic or collective dimension of human experience fully into the conscious wake-a-day world.

wakee-wakee, var. of *wakey-wakey.

waker, *sb.*^{1} Add: **2.** Also *waker-upper* (with qualifying adj.). *colloq.* **1971** 'E. Fenwick' *Impeccable People* vi. 35 Louise..stood quiet, blinking... 'Sorry,..I'm a slow waker-upper. Let's go and have the coffee.'

4. Also *waker-upper*. Cf. sense 8 a of the vb. *colloq.* **1935** *Morgantown* (W.Va.) *Post* 7 Jan. 5/1 Henderson..serves as..waker-upper, and general handy man. **1966** 'D. Shannon' *With a Vengeance* x. 132 He slept so heavily as to need an 'arrangement' with a waker-upper.

5. (Later example.) **1974** D. Sears *Lark in Clear Air* xi. 139 But to hear the wakers at their trade you would think the deceased was Giant MacCaskill.

wa·ke-up, *a.* and *sb.* [f. vbl. phr. *to wake up*: see Wake *v.* 8 a.] **A.** *adj.* † **1.** *wake-up kittle*: prob. the Pacific kittiwake. *Obs.* **1832** B. Morrell *Narr. Four Voyages Chinese Seas* iii. 216 Whale-birds, wake-up-kittles, man-of-war birds, gulls, and tropic-birds.

2. Special collocations: *wake-up call*, a telephone alarm call for awaking a sleeper, usu. in the morning; *wake-up pill* (slang), a pep-pill; *wake-up service*, a telephone service specializing in wake-up calls. **1976** P. Henissart *Winter Quarry* xxii. 219 He obtained his key from the night desk clerk and..left a wake-up call for eight a.m. **1979** *Listener* 30 Aug. 274/1 Her addiction to the sleeping pills and the wake-up pills had, on more than one occasion, led to an overdose. **1946** *Birmingham* (Ala.) *News-Age-Herald* 3 Feb. 9a/3 Because his wakeup service requires that he arise early in the morning, Harvey has acquired the habit of remaining up all night. **1969** *New Yorker* 14 June 31/1 A number of young bachelors use us simply as a wake-up service—we're cheaper than Western Union.

B. *sb.* **1.** = *pigeon-woodpecker* s.v. *pigeon *sb.* 6. **1844** [see *pigeon-woodpecker* s.v. *pigeon *sb.* 6]. **1897** *Scribner's Mag.* June 773/2 The flicker has a long array of names,..like flicker, clape, wake-up,..derived from his notes.

2. [perh. orig. belonging to Awake *a.*] *to be a* (*full*) *wake-up* (also, of several persons, *to be wake-ups*): to be alert or wide awake (*lit.* and *fig.*). Often const. *to*. *Austral.* and *N.Z. slang.* **1930** *Bulletin* (Sydney) 16 Apr. 58/2 'Cripes, you're a full wake-up to that at last, are you?' Snow exclaimed. [**1934** W. S. Howard *You're telling Me!* i. 12 Well, I'm awake-up; they don't get nothing out of *me!*] **1943** F. Sargeson in *Penguin New Writing* XVIII. 68 Now I was a wake-up to what was in Maggie's mind. *Ibid.* 69 The pair of us were wake-ups when we heard somebody coming up the stairs. **1946** M. Trist in Hadgraft & Wilson *Century of Austral. Short Stories* (1963) 214 'Don't you think we ought to have a cow?'..'No, I'm a wake-up to cows', said Dan. 'Saw enough of them when I was a kid.' **1955** D. Niland *Shiralee* 19 A man should have hauled the cap off his head and chucked ninepence into it just to show him who was a wake-up to who. **1960** N. Hilliard *Maori Girl* iii. x. 248, I never knew till now! Well, I'm a wake-up! **1977** C. McCullough *Thorn Birds* ii. 46 When he saw the army lads were a wakeup he was off like a shot.

3. *ellipt.* for *wake-up pill* above. *slang.* **1969** in Howe & Loraine *Environmental Med.* (1973) xvii. 227 Wake-ups and stay-awakes. **1972** *Sunday Sun* (Brisbane) 2 July 14/3 He calls the pep-pills he swallows wake-ups, truck drivers, [etc.].

4. The act of waking a person from sleep, or of being woken from sleep. **1975** *Globe & Mail* (Toronto) 2 June 27/5 Out of that $609 she could easily hire someone to do all the dirty work, such as housekeeping, and leave the easy work such as cooking and wakeups to herself. **1977** M. Herr *Dispatches* (1978) 80 All of my stuff was..ready for the five-o'clock wake-up.

wa·key-wa·key, *sb.*, (*a.*,) and *vbl. phr.* *colloq.* and *slang* (orig. *Services'*). Also **wakee-wakee, waky-waky.** [Reduplicated arbitrary or childish extension of Wake *v.* 7 a, b.] **A.** *sb.* = Reveille. **1941** C. Graves *Thin Blue Line* iii. 46 At the moment all I want to do is to sleep, sleep, sleep. The thought of 'Wakey, wakey' in the morning terrifies me. *a* **1963** J. Lusby in B. James *Austral. Short Stories* (1963) 221 Three to a tent, chums, wakey-wakey 0430 hours.

2. as *adj.* That awakens or induces wakefulness, esp. as *wakey-wakey pill*. **1946** V. Tempest *Near Sun* iii. 29 When the 'wakey-wakey', as the caffeine tablets are called..have been doled out, the crews climb into the W.A.A.F.-driven air-crew buses. **1952** M. Tripp *Faith is Windsock* xi. 172 Bergen swallowed a couple of wakey-wakey tablets, and Flute did likewise. **1977** *Milestones* Spring 15/2 Rally drivers thought they had found the ideal wakey-wakey pill in amphetamines to help them combat fatigue during marathons. **1979** *Guardian* 4 Aug. 22/9 The last ever Butlin's 'wakey-wakey' call goes out to the snoring chalets.

B. *vbl. phr.* used *imp.* Also **wakey, wakey.** 'Wake up', formerly esp. as a command to soldiers at reveille. Often comb. with phr. *rise and shine* (see *rise *v.* 3 c). Now freq. *joc.* **1945** *Gen* 30 June 51/2 The sarnt came round yelling wakey-wakey. **1948** Partridge *Dict. Forces' Slang* 203 *Wakee, wakee!*, orderly Corporal's cry at reveille. 'Wakee, wakee! Rise and shine!' **1953** E. Simon *Past Masters* i. iii. 32 With a start I woke up... 'Waky-waky,' said Gabrielle. **1968** M. Woodhouse *Rock Baby* xxi. 199 'Wakey-wakey,' he said. 'Stand by your beds.' **1973** T. Heald *Unbecoming Habits* v. 149 'Wakey Wakey, rise and shine,' said Brother Barnabas in an unconvincing demonstration of joviality.

‖ wakf, waqf (wǫkf). [Arab. *waqf*.] In Islamic countries, the custom of giving a piece of land, etc., to a religious institution, so that the revenue can be used for pious or charitable purposes; also, the property given in this way. **1836** E. W. Lane *Acct. Manners & Customs of Mod. Egyptians* I. 159 He first imposed a tax (of nearly half the amount of the regular land-tax) upon all land which had become a *wuckf* (or legacy unalienable by law) to any mosque, fountain, public school, &c. **1877** *Encycl. Brit.* VII. 113/2 The regular dervishes live in..khanakahs, or convents, which are endowed with lands or wakf. **1911** *Ibid.* XVII. 413/1 The law and usage of religious foundations in perpetuity (*waqf*, mortmain) became as important in Islam as monastic endowments in mediaeval Europe... It was the only safe way of providing for posterity. **1917** *Chambers's Jrnl.* July 477/1 The partisans are, on the one side, the clergy, who control the *wakfs* (religious endowments). **1976** M. S. Hoque *Hunger* i. iv. 27 Yes, the Karaitali plot I shall give to Judge by wakf. **1976** S. R. Simpson *Land Law & Registration* i. 11 The inalienability of *waqf* land in Islamic countries similarly holds up development.

waking, *vbl. sb.* Add: **5.** Also **waking-time,** the time when one is awake; the moment at which one wakes up. **1959** T. S. Eliot *Elder Statesman* 5 To my wife, To whom I owe the leaping delight That quickens my senses in our wakingtime. **1971** H. Wilson *Labour Govt.* xv. 263, I asked for a sleeping pill, in case I came to at English waking-time, 4.00 a.m. in Washington.

waky-waky, var. *wakey-wakey *sb.*, (*a.*,) and *vbl. phr.*

Walach, Wallach. 2. For *rare*$^{-0}$ read *rare*$^{-1}$ and add example. **1794** J. B. S. Morritt *Let.* 24 June (1914) iii. 51 The languages spoken here are Hungarian, Wallach, Sclavonian and a little German.

Walachian, Wallachian, *sb.* **2.** (Earlier examples.) **1718** M. W. Montagu *Let.* 16 Mar. (1965) I. 390 In Pera they speak Turkish, Greek,..Sclavonian, Walachian, German. **1846** Borrow *Zincali* (ed. 4) iii. 257 The Hungarian Gypsy tongue..contains many words borrowed from the Wallachian.

Walden inversion: see *inversion 5 b.

Waldensian, *sb.* (Earlier examples.) **1604, 1832** [see *Albigensian adj. and sb.]. **1847** J. T. Mullock tr. *St. Alphonsus's Hist. Heresies* I. x. 260 Rainer..for seventeen years was a Waldensian.

Waldenström (vā·ldenstrŏm). *Path.* The name of Jan *Waldenström* (b. 1906), Swedish biochemist, used in the possessive to designate a disease described by him in 1944 (see *macroglobulinæmia). **1961, 1971** [see *macroglobulinæmia]. **1980** *Daily Tel.* 3 Jan. 1/3 The deposed Shah of Iran is suffering from Waldenstrom's Disease, a rare blood disease which killed President Boumedienne of Algeria a year ago.

waldo, Waldo (wǭ·ldo). Pl. **waldos, waldoes.** [f. the name of *Waldo* F. Jones, the inventor of such gadgets in a science-fiction story by Robert Heinlein (see quot. 1942).] A device for handling or manipulating objects by remote control. **1942** 'A. MacDonald' (R. A. Heinlein) in *Astounding Sci. Fiction* 9 Aug. 16 Even the..humanoid gadgets known universally as 'waldoes'..passed through several generations of development..in Waldo's machine shop before he redesigned them for mass production. The first of them..had been designed to enable Waldo to operate a metal lathe. **1966** I. Asimov *Fantastic Voyage* vii. 75 A handling device (a gigantic 'waldo'—so named by the earlier nuclear technicians from a character in a science fiction story of the 1940s). **1978** D. A. Stanwood *Memory of Eva Ryker* iii. 30 The bathyscaphs are both equipped with remote manipulators—the experts call them 'Waldos'—for working under the extreme pressure.

Waldorf salad (wǭ·ldɔːf). [Named after the *Waldorf*-Astoria Hotel in New York, where it was first served.] A salad made from apples, walnuts, lettuce and/or celery, dressed with mayonnaise. **1911** Leiter & Van Bergh *Flower City Cook Bk.* 103 Waldorf salad [recipe follows]. **1930** E. Ferber *Cimarron* 300 She was the first to electrify the ladies of the Twentieth Century Culture Club by serving them Waldorf salad —that abominable mixture of apple cubes, chopped nuts, whipped cream, and mayonnaise. **1979** A. Hailey *Overload* iii. iii. 206 A Waldorf salad, followed by a chicken casserole.

waldrapp (wǭ·ld,ræp). [a. Ger., f. *wald* wood, forest + *rapp* (var. *rappe, rabe*) crow: cf. Raven *sb.*^{1}] The hermit ibis, *Geronticus eremita*, found in parts of North Africa and the Middle East. **1924** W. L. Sclater *Systema Avium Ethiopicarum* 35 Waldrapp..Morocco, Tunis, and Asia Minor; formerly in Switzerland and other parts of southern Europe. **1964** A. L. Thomson *New Dict. Birds* 390/1 The Hermit Ibis (or Waldrapp)..is a bird of dry country, feeding largely on beetles and nesting in colonies on cliffs. **1970** *New Scientist* 3 Sept. 457/1 The waldrapp..has a strong, unpleasant odour. **1976** G. Durrell *Stationary Ark* iv. 81 The Waldrapp..is a medium-sized ibis, with a long curved beak, sombre black plumage.., a bare reddish-coloured face and a strange crest of long feathers on the back of its head.

wale, *sb.*^{1} Add: **5.** Also, = *waling 1. **1886** H. C. Seddon *Builder's Work* i. 7 When the ground is firm enough to admit of it, deep and narrow trenches..may be secured by horizontal planks or wales, placed opposite each other..and kept apart by struts. **1926** A. E. Wynn *Design & Construction of Formwork* viii. 84 Above 6 ft. high there is always danger of the form twisting..from the impact of the concrete if wales are not used. **1964** R. L. Peurifoy *Formwork for Concrete Structures* ix. 158 The bottom wale should be placed not more than 8 in. above the bottom of the form.

Waler^2. Restrict *Anglo-Indian* to sense in Dict. and add: Also **waler. 1.** Also, a type of light Australian horse. **1897** 'R. Boldrewood' *My Run Home* xxxiv. 330, 'I can't imagine any thing but the English thoroughbred worthy to be named in the same day with a high-caste Arab.'.. 'I'll show you a Waler to-morrow that may convert you.' **1900** H. Lawson *Ballad of Cornstalk* in *Coll. Verse* (1967) I. 380 He mounted his waler and rode to the sea (He was sick of the Bush and he longed for a spree). **1945** Baker *Austral. Lang.* iii. 71 *Waler*, originally a horse exported from New South Wales to India (in modern times, a light type of army horse used in Australia). **1963** *Weekly News* (Auckland, N.Z.) 8 May 38/4 The farm hack is a multi-breed horse—a dash of cob, a little thoroughbred, maybe a touch of the Australian 'waler' whose ancestry lies with Dutch and Spanish strains. **1968** J. Kiddell *Euloowirree Walkabout* (1970) xiii. 123 Lord, he exulted, but he's a fast horse. A Waler. One of the descendants of the bush-bred stock, the thoroughbreds of the Outback.

2. (With capital initial.) A native of New South Wales (or of Australia generally). *slang.* **1880** J. Inglis *Our Austral. Cousins* xiii. 159 In the matter of awnings and verandahs the 'Walers' had a grand chance for a bright, cheerful..display. **1924** G. H. A. Willis *Royal Navy as I saw It* 270 We left the *Oceana*, which continued her voyage to Australia with the 'Walers' (as the Australian passengers were called). **1949** *Geogr. Mag.* Feb. 373 *Sydneysider* or *Waler*, a resident of New South Wales.

walia (wā·liă). [a. Amharic.] The Ethiopian ibex, *Capra walie*. Also *walia ibex*. **1932** *Nat. Hist.* XXXII. 69/2 The walia lives in a much worse place. **1965** L. Brown *Ethiopian Episode* ii. 45 The Walia Ibex were..the principal object of my expedition to Semien. **1971** *Observer* (Colour Suppl.) 31 Oct. 19/2 In the rugged Simien Mountains, where the few remaining walia dwell, farmers continue to destroy the animal's habitat. *Ibid.* 20/1 (*caption*) Fewer than 150 walia ibexes remain on earth.

Walian (wē^{i}·liăn), *sb.* and *a.* [f. *Wale-s* + -ian.] *South* (or *North*) *Walian*. **A.** *sb.* A native or inhabitant of South (or North) Wales. **B.** *adj.* (Characteristic) of or pertaining to this region. **1894** *Wales* Aug. 170/1 He was a Methodist preacher, though he was a South Walian. **1924** J. O. Francis *Legend of Welsh* 63 The best story I have heard of Welsh-speaking Englishmen was told me..by a South Walian. **1936** *Antiquity* X. 461 Richards was a South Walian. **1968** D. E. Allen *British Tastes* iv. 100 The stern Protestant ethos, originally adopted by..a people that then resembled the stern North Walian. **1977** D. Jones *My Friend Dylan Thomas* viii. 98, I thought this was a particularly South Walian expression until I read..*The Lore and Language of Schoolchildren*.

waling. Add: **1.** Also, a long horizontal member used to brace the lining of an excavation or the walls of a form. **1891** T. Potter *Concrete* (ed. 2) I. v. 157 If the sides are loose it is better to stand some boards against them, lifting or removing the latter..as the concrete is being filled in. This is assuming they are not planked or waled, and stretchers are employed to keep the waling planks in position. **1893** J. P. Allen *Practical Building Construction* xxii. 349 Where the ground is looser the poling boards must be placed closer together, perpendicularly, with walings, which are placed horizontally inside them. **1932** [see *soldier *sb.* 6*c]. **1966** C. K. Austin *Formwork to Concrete* (ed. 2) 282 The principle on which this system works, is that of a single form consisting of standard steel panels supported by steel channel walings sliding between soldiers against the concrete face. **1972** V. C. Launder *Foundations* x. 132 Vertical planks or baulks are driven into the ground at the sides of the excavation as it proceeds and held vertical by horizontal timbers called walings, strutted across the trench width.

3. *Basket-making.* (The process of weaving) a band of rods which forms a wale (see Wale *sb.*^{1} 6); the wales of a basket. **1912** T. Okey *Art of Basket-Making* vii. 80 The butts of the waling being pieced in at the front. *Ibid.* xi. 135 The waling of the upsett will be begun at each side. **1949** K. S. Woods *Rural Crafts of England* iii. x. 164 The uprights must now be set very firmly in position by several strong rows of waling. **1964** H. Hodges *Artifacts* x. 146 Where three rods were used simultaneously with a crossing..the weave is referred to as waling. **1983** *Daily Tel.* 20 Sept. 15/5 There is a border of 'waling' at the base of the basket.

walk, *sb.*^{1} Add: **I. 1.** *to take a walk*: also, to 'receive one's marching orders', to be dismissed; freq. *imp.* in formulas of impatient dismissal, sometimes in extended form. Also *transf.* Also, to 'walk out' in a labour dispute (*walk *v.*^{1} 12*b). *U.S. slang.* **1871** 'Mark Twain' *Sketches New & Old* (1875) 248 The first time he opened his mouth and was just going to spread himself, his breath took a walk. **1881** —— in *Century Mag.* Nov. 37 They ring out, 'Oh, dry up!' 'Give us a rest!'..'Oh, take a walk!' **1888** in Farmer *Americanisms* (1889) 550/2 The cash returns were so out of proportion to the sales, that Mr. Berry concluded to make a change and Tascott took a walk. **1937** *Amer. Speech* XII. 4 The so-called Jeffersonian Democrats took a walk from their party allegiance. **1946** *Sun* (Baltimore) 19 Nov. 2 (*caption*) Miners 'take a walk'—Black Diamond mine workers at Monongahela, Pa., are shown waiting for a car after quitting their jobs. **1961** *Lebende Sprachen* VI. 99/2 Go take a long walk off a short pier... Go jump in the lake.

e. *Baseball.* = *base on balls* s.v. *base *sb.*^{1} 15 c. Cf. *walk *v.*^{1} 5 q. **1905** *Sporting Life* (Philadelphia) 16 Sept. 2/1 Taking the totals, or hits and walks, and such a famine in tallies would seem impossible, but there are the figures. **1948** *Daily Ardmoreite* (Ardmore, Okla.) 21 Mar. 4/6 Rice, on second from a walk and a sacrifice, crossed the home plate on a fly which Charley Gilbert misjudged. **1967** *Boston Sunday Herald* 7 May (TV Mag.) 14/2 Even the worst of ballplayers can still reach first base on a walk and even get to second should the pitcher throw wild. **1979** *Arizona Daily Star* 5 Aug. c 2/3 Dwight Evans, who had been behind 0–2 in the count, drew a two-out walk.

f. = *sponsored walk* s.v. *sponsored *ppl. a.* 2. **1971** M. Lee *Dying for Fun* xliv. 213 You ought to get one of the newspapers to sponsor it. Walks and demonstrations and things. **1971** *Guardian* 24 May 11/6 From 1965–69, Oxfam depended on walks for up to £175,000 of its year's £2½ millions budget.

4. e. *to win in a walk*: to win easily and without effort. *U.S. colloq.* **1896** Ade *Artie* xii. 106 'Does he stand a good chance of being elected?' 'That's what keeps me guessin'. Two years ago he win in a walk [etc.].' **1903** A. H. Lewis *Boss* 138 He won in a walk. **1936** E. S. Gardner *Case of Sleepwalker's Niece* xiv. 130 The whole thing..gets back to Duncan. If I can break down Duncan's identification I can win the case in a walk.

f. Any dance modelled on or resembling a walk; chiefly as second element of a Comb., as *cake-walk *sb.* 1, *camel-walk* s.v. *camel *sb.* 5, *Lambeth walk* s.v. *Lambeth 3. **1937** [see *Big Apple* s.v. *big *a.* B.2]. **1975** G. Howell *In Vogue* 9/2 In return we get syncopated music, and what to do to it—the Baleta..the Twinkle..the Missouri Walk. **1975** *N.Y. Post* 27 Dec. 23 The new Walk..is an anathema to all this talk of returning romanticism, because you do it alone, without touching.

II. 14. Delete ? *Obs.* and add: Now usu., a postman's round. **1908** *Chambers's Jrnl.* Jan. 102/2 On arriving at the office of delivery letters are at once stamped and sorted to the 'walks' of the postmen. **1977** *Times* 12 July 8/4 A complaint by a postwoman..that she had been prevented from..obtaining a particular postal 'walk'.

IV. 20. (sense 1) *walk shorts*. **1965** *Punch* 17 Nov. 745 Then we equipped ourselves for our new surroundings [*sc.* Australia]. My wife..insisted on polished cotton walk shorts for a reluctant me. **1976** *National Observer* (U.S.) 6 Mar. 11/6 (Advt.), Authentic lederhosen style classic walk shorts for men and women. **1984** *Gainesville* (Florida) *Sun* 30 Mar. 8a (Advt.), Choose from white tennis shorts or solid and patterned walk shorts in polyester and cotton blends.

walk, *v.*^{1} Add: **II. 5. a.** *to walk about*: also *spec.* of an Aboriginal: cf. *walkabout 1. **1908** E. J. Banfield *Confessions of Beachcomber* ii. i. 265 This for Johnny Tritton, before alonga Cooktown; now walk about somewhere down here. Might be catch 'em alonga mainland.

g. *to walk in*: also const. *on* (cf. sense 12* c below); also *spec.* to arrive unexpectedly; to enter premises, etc., with unwonted ease; to succeed against all expectations. **1930** N. Coward *Private Lives* ii. 53 What shall we do if they suddenly walk in on us? **1975** Cowie & Mackin *Oxf. Dict. Current Idiomatic Eng.* I. 356/2 The security is so bad here that anyone could simply walk in and take what he wanted. **1977** P. Hill *Fanatics* 125 If the Christian Democrats put enough candidates up at the next election they'll walk in. **1978** M. Duke *Death of Dandy Dinmont* iv. 39, I couldn't think of anything else to do. I was almost relieved when Hamilton walked in on me.

h. *to walk out*: of a soldier off duty, to go into town on pass. **1911** [implied in *walking-out order* s.v. *walking *vbl. sb.* 1b]. **1955** *Times* 27 July 5/1 In Western Command..young soldiers are now forbidden to 'walk out' when off duty in plain clothes of unorthodox pattern.

j. (Earlier examples.) **1827** A. Moore *Let.* in N. E. Eliason *Tarheel Talk* (1956) 303 [He] has requested to let him have the supreme pleasure of walking out with her. I fear the poor little fellow is pretty far gone, if I may judge from the frequency of his visits. **1849** Dickens *Dav. Copp.* (1850) v. 47 No sweethearts, I b'lieve?..No person walks with her.

n. *to walk off* (an ailment: earlier example); *to walk* (a message or the like) *through*, to take it in person. **1741** Richardson *Pamela* III. xxxvii. 372 'I fear you have sprain'd your Foot—Shall I help you to a Chair?' 'No, no, Sir, I shall walk it off, if I hold by you.' **1981** C. Potok *Bk. of Lights* (1982) v. 144 'How did the major get that memo so quickly, Roger?' 'I walked it through to his desk.'

o. Also, to haul (a balloon) by walking. **1933** *Sun* (Baltimore) 27 Oct. 7/2 The..passengers disembarked..before the huge dirigible was 'walked' into the hangar. **1938** *Times* 7 Sept. 9/1 He watched a crew 'walk' a balloon out of a shed and connect it to a winch for hoisting.

p. *to walk (all) over*—(fig.): to treat (a person) with contempt; also, to defeat (an opponent) decisively. *slang.* **1851** Mrs. Stowe in *National Era* 25 Sept. 1/5 St. Clare wouldn't raise his hand if every one of them walked over him. **1884** 'Mark Twain' *Huck. Finn* xxii. 219 In the North he lets anybody walk over him that wants to, and goes home and prays for a humble spirit to bear it. **1909** R. E. Knowles *Attic Guest* viii. 105 They won't let a pack of negroes walk all over 'em. **1951** N. Mitford *Blessing* i. vi. 65 A woman who lets her husband do exactly as he likes..lets him walk over her, in fact, would never lose him. **1976** E. Dunphy *Only a Game?* i. 34 We played QPR in a public practice game at our place today. And won easily. We walked all over them.

q. *Baseball.* Of a batter: to secure a base on balls. Cf. *walk *sb.*^{1} 1 e. **1867** *Ball Players' Chron.* 14 Nov. 2/4 Goodrich walked to the first on called balls. **1895** *N.Y. Press* 5 July 6/1 The champions harvested a pair of tallies in the second inning. Clarke did not get them over for Kelly, and Joe 'walked'. **1948** *Chicago Tribune* 7 Mar. ii. 1/4 Baker walked, filling the bases. **1979** *Arizona Daily Star* 5 Aug. c-2/5 Alfredo Griffin singled and Bob Bailor walked to start the eighth-inning burst against Rich Wortham, 11–11.

r. *to walk on*: (of a theatrical performer) to go on stage with few if any lines to say. **1893** H. G. McClelland *Jack & Beanstalk* 35 She used to walk on in the comic scenes. **1913** *Confessions of Dancing Girl* vii. 127, I obtained an engagement to 'walk on' in a musical comedy... I had no lines and no part. **1920** [see *super *v.* 1]. **1959** P. Bull *I know Face* i. 17 He had engaged a lot of art school students to what's known as 'walk-on' in the production.

s. *walk good* (imp.): farewell, good luck. *Caribbean.* **1929** M. W. Beckwith *Black Roadways* xiii. 199 'Walk good, me love,' says one to another setting out on a journey. **1953** R. Mais *Hills were Joyful Together* ii. i. 147 You going further, walk good then; walk good, hear? **1979** J. Berry *Fractured Circles* 58 Walk good, Leela, chile.

t. *to walk away from* ——: to leave the scene of (an accident or the like) on one's feet, instead of being carried on a stretcher. Cf. *walking wounded* s.v. *walking *ppl. a.* 3 d. Also *transf.* **1956** W. A. Heflin *U.S. Air Force Dict.* 561/1 *To walk away from an airplane crash or accident*, to survive an accident unhurt or without serious injury. **1966** M. Woodhouse *Tree Frog* vi. 50, I had a..cut..but that was all. Walking away from it, they call it. **1980** J. Wainwright *Eye of Beholder* 130 'Anybody walk away from it?'..'No. One dead, one smashed up.' **1984** *Times* 19 Apr. 19/6 The provisions for bad and doubtful debts is increased..and Mr Pattullo stated confidently 'there is no international or domestic loan that we could not walk away from. It might cause us a red face but it would not harm the bank.'

6. c. (Later example.) **1841** S. S. Arnold *Diary* 5 Jan. in *Proc. Vermont Hist. Soc.* (1940) VIII. 149 It was a friendly interview; but she said that she could not walk with the Ch[urc]h, and wished to be out.

7. a. *to walk through one's part* (later example); also simply *to walk through*, and *fig.* **1857** C. M. Yonge *Dynevor Terrace* I. xii. 195 Her grave, pensive character only attained to walking through her part [in society]. **1889** C. Scott *Drama of Yesterday & To-Day* II. xiv. 442 Often when she is tired to death,..her strength fails her. She walks through the part, as it is called. **1922** Mrs. P. Campbell *Let.* 11 June in *B. Shaw & Mrs. Campbell* (1952) 256, I would like you to come and see *Hedda Gabler*—it would be nice to hear all the abominable things you might say. Some say I 'walk through'.

c. *to walk over* (further example of absol. use); also *to walk over* (an opponent), to be declared the winner of a contest because of the opponent's failure to compete; *to walk away with*, to win (a prize), steal (a show), with ease; *to walk home*, to win a contest with ease. **1903** Wodehouse *Prefect's Uncle* ix. 136 If you'd been there to bowl we should have walked over. **1932** *Sun* (Baltimore) 21 Dec. 12/1 Jack Biener 'Walks' Home... Jack Biener, favorite at $5 for $2, simply spreadeagled the field and won in a common canter by eight lengths. **1951** N. Coward *Star Quality* 139 It had been the..play's provincial try-out..and..Leonora had unquestionably walked away with the show. **1958** *Times* 11 Aug. 2/7 Treorchy—a magnificent choir—walked away with the prize for big choirs. **1973** *Times* 31 July 9/8 Major J. D. E. Edwards beat Driver D. A. Beck, 6–1, 7–5; Major P. K. Sharp walked over Captain J. B. Merritt.

f. Proverbial phrase: *to walk before one can run* (and varr.), to understand elementary points before proceeding to anything more difficult; cf. Creep *v.* 1 b. Also *to run before one can walk.*

1762 [in Dict., sense 7 a]. **1794** G. Washington *Let.* 20 July in *Writings* (1940) XXXIII. 438 We must walk as other countries have done before we can run. **1876** J. Platt *Business* 124 We must learn and be strong enough to walk before we can run. **1927** *Melody Maker* Sept. 923/1 Beginners..commence on their instruments to dabble in so-called 'hot' choruses... They run before they can walk, for, jumping on to what they believe to be 'hot' style, they attempt to cut out all the essential months of study. **1973** *Times* 17 Apr. (Liberia Suppl.) p. ii/5 He does not want Liberia to run before she can walk. **1980** K. Amis *Russian Hide & Seek* iv. 49 At the moment we can't leave it to the English to do anything. We must learn to walk before we can run. **1984** *New Yorker* 11 June 31/1 'We can't turn you into an actor, but don't turn your back on what we have to offer.' What they meant, of course, was that you've got to learn to walk before you try to run.

g. *Jazz.* Of a bass player: to play a walking bass (see *walking *ppl. a.* 9). Also said of the beat. **1951** L. Hughes *Montage of Dream Deferred* 12 Down in the bass That steady beat Walking walking walking Like marching feet. **1952** *Mademoiselle* Dec. 118 And that's the basic jazz beat, that walking beat. Up here in the north all the jazzmen are playing too fast or too slow —nobody walks. **1956** [see *string bass* s.v. *string *sb.* 32]. **1970** *New Yorker* 23 May 88/2 Then Hall soloed, while Gomez 'walked' behind him.

8. a. Also, of a batsman in *Cricket*, to walk towards the pavilion without waiting for the umpire to give him out; also with *out*. **1960** J. Fingleton *Four Chukkas to Australia* v. 48 Three runs later..Graveney should have walked but O'Neill dropped him at third slip. **1964** D. Sheppard *Parson's Pitch* vi. 107, I never saw him not walk out immediately he was caught at the wicket. He never waited for the umpire's decision. **1973** *Advocate-News* (Barbados) 11 Dec. 14/2 Brian Close, captain of the Robins XI, said: 'A batsman who knows he is out should walk. That is the way we play the game.'

b. Restrict † *Obs.* to senses in Dict. In mod. *colloq.* use said of objects presumed borrowed or stolen. **1898** J. D. Brayshaw *Slum Silhouettes* 125 A sack o' taters, or a sieve o' cherries sometimes goes awalkin' if yer don't keep yer eyes skinned. **1978** A. Melville-Ross *Blindfold* xiv. 87 'Get much theft?' 'Lord yes, but only the sort of stuff you'd expect to "walk" anyway.'

c. *to walk off with* (earlier example). **1727** M. W. Montagu *Let.* 23 June (1966) II. 78 All the little money they had..they put into the hands of a roguish Broker who has fairly walk'd off with it.

d. *to walk away from*: to fail to deal with (something); to refuse to become involved with; also *to walk away* simply. **1963** *Life* 8 Feb. 4/1 The Kennedy proposals walk away from most of the tax reform problems. **1973** P. O'Donnell *Silver Mistress* iii. 48 'Why look for trouble?' ..Something comes along, and you just can't turn round and walk away from it.' **1978** S. Sheldon *Bloodline* xx. 237 Roffe and Sons needs an experienced president, Elizabeth... For your own sake, as well as everyone else's, I would like to see you walk away from this. **1981** *Times* 7 Dec. 13/3 Libya and Nigeria started the year trying to maintain prices at a wide premium over the marker, but instead saw buyers walk away. **1983** *Times* 6 May 15/4 No parent which itself took deposits..could expect to walk away from a subsidiary in trouble without risking a loss of confidence on the part of its own depositors.

e. = sense 12* b below. **1976** *National Observer* (U.S.) 14 Aug. 16/3 Our baby sitter founded the Sitters' Union. They get TV, cookies, and root beer, or they walk. *Ibid.* 28 Aug. 16/1 The sentry in the lobby was a representative of Universal Pictures. His mission: to keep an anxious eye out for 'walkers' and try, if he could, to find out why they walked, since there was still time to do some patching on the film. **1978** S. Brill *Teamsters* v. 180 Carey called a strike, and all four thousand of his UPS members walked. **1983** W. F. Nolan *Hammett, Life at Edge* i. 6 He would not lie to keep a job. And he walked.

11. a. Also *to walk guard*. **1930** F. A. Pottle *Stretchers* 51 In France we always wore large automatics when we walked guard.

12. f. To get *into* an awkward situation as a result of one's own unwariness. **1911** G. B. Shaw *Doctor's Dilemma* iii. 60 *Ridgeon*: I don't so much mind your borrowing £10 from one of my guests and £20 from the other—*Walpole*: I walked into it, you know. I offered it. **1942** J. Sweeney in Murdoch & Drake-Brockman *Austral. Short Stories* (1951) 384 No sooner does the gong go for the third than Irish walks into a haymaker. **1978** M. Birmingham *Sleep in Ditch* 188, I had walked into this with my eyes wide open. No one could taunt me with being always right.

12*. walk out. a. To leave a gathering or place without warning, esp. in protest or disapproval; also *fig.* Const. *on*. orig. *Theat.* **1840** W. C. Macready *Diary* 19 Feb. (1912) II. 45 Very much disgusted and irritated by Mr Elton walking out in the last scene. **1897** *Daily Tel.* 24 Feb. 10/3 New York did not take kindly to his new play... I am delighted to find, on the assurance of the author, that though New York 'walked out', Washington 'walked in' and received it warmly. **1936** H. G. Wells *Anat. of Frustration* vi. 55 Suicide..may be represented very attractively as a proud and passionate refusal to drink the cup to the dregs. You 'walk out' as they say in the film world. **1937** M. Levin in A. Cooke *Garbo & Night Watchman* 124, I rarely walk out on a picture, and never want to walk out on a simple programme picture. **1948** W. Charvat in R. Spiller et al. *Lit. Hist. U.S.* I. xxxi. 525 When a 'stout Illinoian' walked out on his lectures, he reflected that 'the people are always right'. **1969** H. Perkin *Key Profession* iii. 103 The A.U.T. delegates to the International Conference walked out the day before Hitler arrived.

b. *spec.* Of an employee: to leave his place of work at short notice, as a form of industrial action. orig. *U.S.* **1894** W. H. Carwardine *Pullman Strike* iv. 37 The men passed the word from one to another to 'walk out', which they did orderly and deliberately. **1937** *Irish Press* 11 Feb. 1/2 (*caption*) Photo shows the nursing and boiler house staffs 'walking out'. **1951** E. Paul *Springtime in Paris* v. 95 Gas workers were about to walk out, and hamstring home cooking to a certain extent. **1979** *Daily Tel.* 27 Nov. 2/2 Last night one of the two lighting and sound engineering crews on BBC-1's 'Nationwide' walked out shortly before the programme was due to go on the air.

c. To desert a partner, esp. a spouse; to withdraw from an agreed arrangement. Const. *on* (cf. sense 5 g above). orig. *U.S.* **1896** *Typographical Jrnl.* (Indianapolis) IX. 232 The Review, Republican daily, 'walked out' on the St. Louis platform. **1921** Wodehouse *Indiscretions of Archie* xii. 127 'Has she walked out on you?' 'Left us flat!' **1937** *Sunday Times* 2 May 7/4 Father Donelly..is a fairy godfather to her after she has walked out on her guardians. **1953** *Sun* (Baltimore) (B ed.) 31 Aug. 14/3 The Southern Conference digs in Tuesday for its first football practice since seven of its greatest powers walked out to form a league of their own. **1962** *New Statesman* 7 Dec. 829/1 What surprises is the famous malleability of the two women: did neither one dream of walking out? **1973** 'E. Peters' *City of Gold & Shadows* i. 6 My father walked out on my mother when I was seven.

III. 20*. To win easily. Cf. sense 7 c in Dict. and Suppl. *slang.* **1937** Partridge *Dict. Slang* 935/2 *Walk*, to win easily: Public Schools' coll.: from ca. 1895. [*Refers to quot. 1903 in sense 7 c above.*] **1976** *Times* 12 Feb. 10/2, I went to the British [championship] thinking I'd walk it... This was a mistake... It was a close shave.

IV. 21. Also *to walk hots* (see quots.). **1958** *Washington Post* 25 Sept. d-4/1 Each got his start walking hots (leading horses to cool them out following a race or a workout). **1976** *New Yorker* 29 Mar. 109/2 He got an after-school job at a ranch, mucking out stalls and 'walking hots', as the chore of cooling out horses who have just worked or raced is called.

22. c. *Baseball.* Of a pitcher: to give (a batter) a base on balls. **1913** *Chicago Record-Herald* 20 Mar. 10/5 Lange walked Kores in the hostile part, then disposed of the next pair on easy infield flies. **1938** *Chicago Tribune* 5 Apr. 19/2 Dobernic walked three batsmen in a row. **1952** B. Malamud *Natural* 78 With two out Schultz weakened, walking one man and handing the next a good enough throw to hit for a sharp single. **1976** *National Observer* (U.S.) 16 Oct., Our pitcher had walked three men in a row and I was invited to take the mound.

24. d. To swing (a gun) so as to describe a straight line on the target with successive hits. **1944** *Sun* (Baltimore) 15 June 2/5, I..aimed for his groin and walked my tommy gun right up his middle and blew him 90 feet away. **1969** I. Kemp *Brit. G.I. in Vietnam* xi. 187 'Charlie' really seemed to be walking that mortar up behind me; he was right on target with his shots.

V. 25. walk-away: also *transf.* and as *adj.*; **walk-back** *U.S. slang*, a rear apartment; **walk-march** *v.*, to march at a walking pace; of cavalry, to proceed at the walk; also as *sb.* **1926** *Amer. Mercury* Dec. 465/2 It [*sc. Variety*] has developed..the following new terms for a [Broadway] success: 'zowied 'em',..'walk-away hit' and 'clicked heavy'. **1958** *Time* (Atlantic ed.) 6 Oct. 16 Turning from a Democratic walkaway into a neck-and-neck sprint. **1978** *Detroit Free Press* 5 Mar. c 4/1 For UCLA, the walkaway winner of the Pac-8 title, the game against U-M will be a final tuneup before going into the NCAA tournament. **1945** L. Shelly *Jive Talk Dict.* 35 *Walk back*, rear apartment. **1973** 'H. Howard' *Highway to Murder* vii. 80 One-o-four Platt Street was a rooming house... The Royales lived in a walk-back at the rear of the lobby. [**1851** *Regulations for Cavalry* 190 Commanding Officer, repeated by Squadron Leaders, '*Walk, Trot*, or *Gallop March*.'] **1904** *Mounted Infantry Training* (Provisional, H.M. Army) iv. 36 The word 'Walk' or 'Trot' will in every case precede the word 'March' when the men are mounted... The word on which they move should be drawn out and rolled thus:—'Walk—M-a-r-r-c-h', and when the pace is to be increased to a trot—'T-r-r-o-t'. **1909** *Mounted Infantry Training* (War Office) iii. 36 On the command, '*Walk—March*'..the whole company moves off..general alignment and cohesion..being secured by the section leaders riding on the same alignment, and at the proper intervals. **1926** T. E. Lawrence *Seven Pillars* (1935) ix. ci. 553 Buxton a moment later called 'Walk-March!' to his men, and the four-hundred camels..started off for Jefer. **1942** C. Barrett *On Wallaby* iv. 88 We flushed many birds as we walk-marched among rocks and stones. *a* **1977** D. Wheatley *Officer & Temp. Gentleman* (1978) xiii. 157 We had to move at a sedate walk-march.

walkable, *a.* **c.** For (*nonce-use*) read *rare* and add later example. **1943** G. G. Coulton *Fourscore Years* ii. 10 For about two years..I was the youngest walkable child.

walkabout (wǭ·kăbaut). [Pidgin Eng., f. Walk *v.*^{1} + About *adv.*] **1.** *Austral.* A periodic migration by a westernized Aboriginal into the bush. Often quasi-*advb.* in phr. *to go walkabout*. Also *transf.* [**1828** *Sydney Gaz.* 2 Jan. 3 When the executioner had adjusted the rope, and was about to pull the cap over his eyes..he said, in a tone of deep feeling,..'Bail more walk about,' meaning that his wanderings were all over.] **1908** Mrs. A. Gunn *We of Never-Never* 218 The day after that was filled in with preparations for a walk-about, and the next again found us camped at Bitter Springs. **1918** C. S. Stow (*title*) The walkabouts of Wur-run-Nah. **1927** M. M. Bennett *Christison* xxv. 227 The manager would give them [*sc.* blackfellows] rations and let them..go off for a month's 'walkabout', picnicking, and fishing at favourite waterholes. **1933** *Bulletin* (Sydney) 9 Aug. 35/3 An old bush abo., then on the 'walk-about'. **1938** X. Herbert *Capricornia* (1939) xxii. 330 Wha' name—you go walkabout? **1940** F. D. Davison in 'B. James' *Austral. Short Stories* (1963) 63 [Heifers] standing motionless..at the end of one of those walk-abouts that cattle running in wild country periodically make. **1950** 'N. Shute' *Town like Alice* iii. 82 These bloody boongs, they're always going walkabout. **1958** L. van der Post *Lost World Kalahari* ix. 203 Even the Bushman born on these few remote European farms..rediscovers the need for a long 'walk-about' in the vast desert around him. **1973** P. O'Donnell *Silver Mistress* i. 15, I like to revert to my childhood ways now and again. Like an aborigine going walk-about. **1984** *Times* 8 Oct. 10/6 First of all the players go walkabout. Then they all return to the stage and sit down. **1985** *Truck & Driver* June 38/2 (*caption*) The crowds get up close to the racing trucks during the Supergrid walkabout.

2. a. A protracted walk or journey, or one that takes in a number of places. **1946** 'M. Innes' *From London Far* i. vii. 58, I proposed a walkabout. We were to stroll through Auld Reekie's dusk together. **1956** *Richmond* (Va.) *Times-Dispatch* 17 May 22 (*heading*) End of a six month walk-about [*sc.* a global tour]. **1974** 'G. Black' *Golden Cockatrice* i. 9, I've been doing a year's exchange teaching in Hong Kong... These academic walkabouts fascinate me.

b. *spec.* An informal stroll through the crowd by a member of the Royal Family or other public figure. **1970** *Daily Tel.* 31 Mar. 15/3 The Queen realised she was on to a winner with her New Zealand 'walkabouts'. **1974** *Listener* 28 Feb. 263/1 Mr Heath's..electioneering by set speech and walkabout tour. **1980** *Church Times* 25 Apr. 2/5 On Saturday the Bishop met local people during a walkabout in the market. **1984** *Listener* 15 Mar. 20/1, I was engaged in a walkabout during last year's Darlington by-election.

walkathon (wǭ·kăþǫn). *colloq.* (orig. and chiefly *U.S.*). [f. Walk(ing *vbl. sb.*^{1} + *-athon.] A long-distance or protracted walk; orig. a competitive one, now esp. one undertaken to raise money for charity. **1932** *Kansas City* (Missouri) *Times* 20 Feb. 26 Sure a hick town is a place where they have enough superhicks to put on a walkathon. **1933** *Sun* (Baltimore) 4 Nov. 6/6 Next Thursday a number of Hagerstowners will enter upon an endurance contest different from any ever tried here before—a walking marathon, or a 'walkathon', as the promoters insist on calling it. The contestants walk forty-five minutes and rest fifteen for as long as they hold out. **1951** M. Kennedy *Lucy Carmichael* v. i. 239 If I had that hall I'd put on a Walkathon... In California..after two or three weeks, when there's only a few couples left in, they chain them, see? So if the girl faints, the feller can't push her under the ropes. **1957** *Daily Mail* 27 Dec. 4/3 If Barber gets to the bottom of the world by plane, why on earth are the other boys—the Hillary-Fuchs expeditions—walking?.. I gather that the walkathon idea is the British contribution to the Geophysical Year. **1963** *Weekly News* (Auckland) 31 July 3/4 (*caption*) Representatives of harriers, scouts, bible classes and old soldiers taking part in the recent 50-mile 'walkathon' in aid of the Murray Halberg Trust for Crippled Children. **1976** *Indian Express* 8 June, Walkathon. Indian athlete Harbans Singh, 28, walked non-stop for 144 hours in the town of Jalalabad. **1979** *Honolulu Advertiser* 8 Jan. b-2/2 Manned booths for various charitable walk-a-thons or walked themselves.

walker, *sb.*^{1} Add: **1. c.** *walker-on*: = *walk-on *sb.* 1 b. **1897** G. B. Shaw *Our Theatres in Nineties* (1932) III. 217 A good deal of the technique acquired by American actors no doubt makes one almost long for the fatuous complacency of the British 'walker-on'. **1936** N. Streatfeild *Ballet Shoes* xiv. 212 Pauline and Petrova were, of course, principals, and as such separated from the ballet and walkers-on. **1948** [see *stand-in 2 a].

d. A person who takes dogs for walks as an occupation; also, one who rears puppies at home for a time before they are returned to kennels. **1930** C. Frederick et al. *Foxhunting* v. 61 The training of..'morale' begins from the day when the puppy returns to kennels from its walker. **1967** C. G. E. Wimhurst *Bk. Working Dogs* vii. 45 The day will come when the walker will have to return his charge to the Kennel. **1977** Hooker & Butterworth *M*A*S*H goes to Moscow* (1979) vii. 80 A dog walker..is a person..who earns his living by collecting a dozen or so dogs and taking them all for a walk at once. **1984** *Leicester Mercury* 28 June 25/2 Until they are 12 months old, the hound puppies are in the charge of the walkers who keep them at their homes.

10. a. Short for *baby-walker* s.v. *baby *sb.* B. 2. **1934** in Webster. **1971** *E. Afr. Standard* (Nairobi) 13 Apr. 14/1 (Advt.), Children's bicycles, walkers, car-seats and play-pens at Cycle Mart and Exchange Ltd.

b. = *walking frame* s.v. ***walking** *vbl. sb.* 4 b.

1941 F. H. Krusen *Physical Medicine* xvi. 648 Various types of walkers..will give the patient firm support through the arms and axillae when he is taking his first hesitant steps. 1971 *Catholic Worker* May 2/3 She is living at Loretto Home for the Aged..and must use a walker since she broke her hip some time ago. 1980 U. Curtiss *Poisoned Orchard* ii. 12 She's in her seventies *and* in a walker.

Walker^2 (wǭ·kəɹ). The name of John W. *Walker* (b. 1802) used *absol.* and *attrib.* to designate an American fox-hound, usually black, white, and tan, belonging to a strain originally developed by the Walker family.

1904 J. A. Graham *Sporting Dogs* ix. 134 The Walkers are chiefly bred by men in Kentucky of that name and have been shipped to nearly every part of America where foxes are found. *Ibid.* 135 The Walker hounds are fast. 1940 W. Faulkner *Hamlet* 58 A man named Houston, heeled by a magnificent grave blue-ticked Walker hound, led a horse up to the blacksmith shop. 1964 J. Griffen *Hunting Dogs Amer.* v. 195 The Walker Hound is the most popular strain of Foxhound in America today. 1980 *Hunting Ann. 1981* 17/1 Caswell has his two best strike hounds in place: Dan is a registered Walker and a fighter; his white-brown-and-black coat is creased here and there with old scars left by the claws and teeth of a bear.

walkie-lookie (wǭ:ki,lu·ki). *colloq.* [f. Look *v.*, after *walkie-talkie*.] = *peepie-creepie.

1946 *Sci. News Let.* 30 Mar. 195 'Walkie-lookie', the picture equivalent of the small remote voice instrument known as 'walkie-talkie', will come from the 'block' system's light-weight, easily portable television camera. 1952, etc. [see *peepie-creepie].

walkies (wǭ·kiz), *sb. pl. colloq.* [f. Walk *sb.*^{1} + -ie.] A childish or jocular form of *walk* used chiefly with reference to dogs. Also as quasi-*advb.* in phr. *to go walkies.*

[1932 *Amer. Speech* VII. 242 [Jazz jargon.] Thus, we have the line 'I Prefer the Walkies', with its last word coined to rhyme with 'talkies'. 1938 A. Woollcott *Let.* 28 Jan. (1946) 163 There are now—God help me!—four black dogs to go walkee with you.] 1939 A. Thirkell *Before Lunch* iv. 93 'Master's stick for walkies,' said Mr. Middleton. 'Fetch stick for walkies.' 1960 J. Stroud *Shorn Lamb* x. 119, I bring Gorm along here sometimes, for his walkies. 1979 T. Barling *Olympic Sleeper* x. 118 That's one stray piece of information... It's gone walkies some-damned-where. 1981 *Sunday Express* (Colour Suppl.) 26 Apr. 13/1 Before long the subject of walkies comes up. People are obsessed, Mrs Woodhouse says, with taking dogs out for walks.

walkie-talkie (wǭ:ki,tǭ·ki). Also **walky-talky.** [f. Walk *v.*^{1} + Talk *v.* + -ie.] **1.** A small radio transmitter and receiver that can be carried on the person to provide two-way communication as one walks.

1939 *Sun* (Balimore) 4 Oct. 24/3 'Walkie-talkie' is the Army Signal Corps' way of speaking of unit S.C.R. 195—a recently developed radio sending and receiving set so small it is carried on the back and one talks while one walks. 1944 *Life* 3 July 12/1 Above the noise of hundreds of bombs..his voice barked through the walkie-talkie. 1945 *China at War* May 38/2 Walkie-talkie sets also are used. 1955 G. Band *Road to Rakaposhi* xvi. 178 If we had taken portable 'walky-talky' sets there would have been no problem. 1970 *Timber Traders Jrnl.* 21 Mar. 53/3 Senior members of the yard staff have walkie-talkie radio sets with which they maintain communication with the sideloader operators. 1973 C. Bonington *Next Horizon* xii. 175 We talked over the problem, on the walky-talky. 1979 *Arizona Daily Star* 5 Aug. a 12/1 Secret Service agents with walkie-talkies paced the auditorium's mezzanine and surrounding grounds.

2. A doll that can be made to walk and talk. Freq. *attrib.* or as *adj.*

1952 *Landfall* VI. 82 Everything he does simply runs true to type like the tricks of a walkie-talkie doll. 1957 J. Frame *Owls do Cry* ii. xxviii. 137, I shall buy her..a sleeping doll, a walkie-talkie that cries and walks. 1958 N. Marsh *Singing in Shrouds* (1959) vi. 107 Las Palmas is known to tourists for its walkie-talkie dolls. 1961 J. Wayne *Day Ceiling fell Down* vii. 72, I thought it was a proper big doll you'd lost... Like one of them walkie-talkies.

walk-in (wǭ·k,in), *a.* and *sb.* [f. Walk *v.*^{1} + In *adv.*] **A.** *adj.* **1.** Pertaining to or designating a person who walks into premises casually or without an appointment; *spec.* applied to (*a*) a thief who walks in rather than breaks in; (*b*) a person who offers his services to a foreign power unsolicited, as by walking into its embassy or consulate.

1928 *Daily Express* 6 Oct. 11/2 A criminal of the type described by the police as the 'walk-in-thief', who, if he encounters anyone, slides gently from the house, with a plausible excuse. 1962 *John o' London's* 25 Jan. 82/3 A housebreaker by day is a walk-in man. 1976 *Billings* (Montana) *Gaz.* 1 July, They will be on duty for walk-in and other services from 10.00 a.m. to midnight daily.. said the executive director of the hospital. 1978 D. Bloodworth *Crosstalk* x. 84 In March we got ourselves a walk-in defector from China. 1979 *United States 1980/81* (Penguin Travel Guides) 45 In larger cities, many hospitals have walk-in clinics designed to serve people who do not really need an emergency service, but who have no place to go for immediate medical attention.

2. a. Of a storage area: large enough to walk into.

1943 *Pioneering in Food Preservation* (New Dominion Ser. No. 37), A small walk-in refrigeration plant for curing meat. 1960 *Farmer & Stockbreeder* 19 Jan. 118/1 (Advt.), Makers of walk-in incubators. 1966 T. Pynchon *Crying of Lot 49* ii. 36 Oedipa skipped into the bathroom, which happened also to have a walk-in closet. 1976 *Evening Post* (Nottingham) 15 Dec. 17/1 (Advt.), Large fitted kitchen, walk-in larder. 1982 A. Mather *Impetuous Masquerade* vii. 105 A matching chest and fitted walk-in closets.

b. Of a cinema, etc.: entered on foot, in contrast to a drive-in. *N. Amer.*

1968 *Amer. Speech* XLIII. 157 He recognized..a desire for something more than the ubiquitous..drive-in and directed me to a walk-in restaurant a few blocks away. 1973 *Daily Colonist* (Victoria, B.C.) 3 Nov. 2/1 We can't even go to a walk-in movie because he falls asleep and snores so loud it is embarrassing. So we usually go to drive-ins where he can sleep peacefully.

c. Of a room: entered directly from a specified area rather than through an intervening passage.

1967 *Coast to Coast 1965–6* 160 Back left of the bar was a row of walk-in walk-out bedrooms surrounded by a veranda. 1971 D. E. Westlake *I gave at Office* (1972) 75 A narrow walk-in kitchen was off the living room through a doorway in the short wall opposite the window.

B. *sb.* **1.** A walk-in closet or cold-storage room (see sense A. 2 a above).

1946 *Daily Progress* (Charlottesville, Va.) 21 Mar. 15/2 Deep Freeze and Frozen Food Storage Walk-In Refrigerators... Down payment required as all walk-ins are custom built to customers' own specifications. 1970 'E. Queen' *Last Woman* iii. 160 His wardrobe closet was a roomy walk-in.

2. A walk-in defector (see sense A. 1 above).

1975 P. Agee *Inside Company* i. 59 The first type is known as the 'walk-in'. The walk-in is a member of the party who..decides to offer his services to the U.S. government. He makes his initial contact..by walking into the U.S. Embassy or Consulate. 1980 E. Behr *Getting Even* ii. 23 The only really satisfactory defectors were walk-ins, as they were known in the trade.

walking, *vbl. sb.*^{1} Add: **1.** *walking on* (or *with*) *two legs*: in modern China, the use of small-scale, local methods in production and education, as well as large-scale or capital-intensive ones; also *attrib.*

1962 E. Snow *Other Side of River* (1963) xxviii. 209 'Walking on two legs' in 1958 meant starting tens of thousands of small brick blast furnaces or 'back-yard' hearths. 1964 Kang Chao in D. J. Dwyer *China Now* (1974) xiii. 252 Another salient feature of the Great Leap movement was a greater emphasis on indigenous methods of production and labour-intensive investment projects. This policy, officially called 'walking with two legs', represented a sharp departure from previous development strategy which had stressed only modern production techniques and large-scale investment projects. 1971 G. P. Jan in S. E. Fraser *Educ. & Communism in China* i. 141 Technical training for adults in the commune schools, a mixture of modern and native methods, was referred to as the policy of 'walking on two legs'. 1977 *China Now* June 4/1 Agricultural machinery for production (walking-on-two-legs tractors, tractors, bulldozers, pumps, harvesters, etc.).

b. *walking in, -on* (in quots. *attrib.*), *-out* (later *attrib.* example), *together.*

1857 C. M. Yonge *Dynevor Terrace* I. xix. 309 Their 'walking together' was recognised. 1911 *Encycl. Brit.* XXVII. 586/2 Undress Uniforms.—In 'walking-out' order most troops wear the tunic. 1931 E. O'Neill *Homecoming* i, in *Mourning becomes Electra* (1932) 40 Hope you don't mind my walking in on you without ceremony. 1948 *Sporting Mirror* 21 May 5/2 At one time he wanted to be an actor and was offered a walking-on part for 'Cavalcade'. 1948 *Jrnl. R. United Service Inst.* XCIII. 470 To encourage recruiting and to make Army service more popular it would certainly seem very necessary to have a smart Walking-out dress. 1982 P. D. James *Skull beneath Skin* vii. 67 She would have had at least a walking-on part in all the plays.

g. Delete †*Obs.* and add later (? *dial.*) examples.

1892 C. M. Yonge *Old Woman's Outlook* 100 The attraction of 'walking' and the gala day were lacking. 1939 F. Thompson *Lark Rise* xv. 269 At the club walkings there were brass bands and processions of all the club members.

4. a. *walking boot* (earlier example), *dress* (earlier and later examples), *shorts, suit, tour* (earlier example); also used to designate farm implements which are operated by someone walking behind or alongside, as *walking cultivator, plough.*

1854 M. S. Cummins *Lamplighter* xxx. 217 To change her slippers for thick walking-boots occupied a few minutes only. 1869 *Rep. Comm. Agric. 1868* (U.S. Dept. Agric.) 417 Field No. 3..[was] cultivated but once, when about a foot high, with a five-toothed walking cultivator. 1753 G. Washington *Diary* 23 Dec. (1925) I. 63, I put myself in an Indian walking dress. *c* 1792 Jane Austen *Catharine* in *Minor Wks.* (1954) 211 She sends me a long account of the new Regency walking dress Lady Susan has given her. 1909 H. G. Wells *Ann Veronica* iii. 72 She was in one of her old walking-dresses. 1868 *Rep. Iowa State Agric. Soc. 1867* 161 [The] ground [is] plowed and harrowed,..and cultivated with riding and walking-plows. 1965 E. L. Myles *Emperor of Peace River* i. xi. 114 Hitched to a small walking-plow, the team plodded back and forth. 1963 Walking shorts [see *easy-care* s.v. *easy *a.* and *adv.* C.d.]. 1880 'Mark Twain' *Tramp Abr.* xi. 102 The knapsacks, the rough walking suits, and the stout walking shoes which we had ordered. 1854 C. M. Yonge *Heartsease* II. ii. xviii. 48 He was going to take a walking tour in Ireland.

b. walking-beam (earlier and later examples); **walking frame,** a free-standing metal frame for use as a walking aid by a person who holds or leans on the top; = ***walker** *sb.*^{1} 10 a, b; **walking-leg,** in certain arthropods, esp. crustaceans, a limb used for walking; **walking-machine,** a mechanical or robotic device attached to a person to enable him to perform duties beyond his normal capacity or strength; **walking-papers:** for *U.S. slang* read *colloq.*; (earlier and later examples); **walking-ticket** (earlier and later examples).

1845 *Knickerbocker* XXV. 63 Some rushed to the upper deck, and climbed up the chain and up the machinery to the walking beam. 1932 *Amer. Speech* VII. 263 The walking-beam, a bar, pivoted in the center, which rocks up and down, actuating the tools in cable-tool drilling or the pumping rods in a well being pumped. 1960 [see *Samson's post 2 b]. 1973 [see *stem *sb.*^{1} 4 j]. 1974 *Petroleum Rev.* XXVIII. 558/1 Each pile handling unit has its own walking beam system for moving from one pile driving position to the next. 1976 M. Machlin *Pipeline* xlii. 461 Lester, he'd been sitting out on the walking beam trying to stab the control head with a joint of pipe screwed in on the casing. 1961 J. O. Wale *Tidy's Massage & Remedial Exercises* (ed. 10) x. 202 The patient may begin with walking frames or rails. 1966 *Physiotherapy* LII. 146/2 Problems to be overcome are: (1) uneven ground; (ii) steep ground; (iii) stairs. Most walking frames can cope with (i) and (ii). 1979 D. Cook *Winter Doves* ii. i. 49 Her child..was bouncing up and down in a walking frame. 1983 S. Radley *Blood on Happy Highway* v. 39 Osteo-arthritis in her hips made it difficult for May to get about now, even with a walking frame. 1909 W. T. Calman in E. R. Lankester *Treat. Zool.* VII. xv. 271 The first pair of legs..are commonly referred to as the chelipeds, and the following four pairs are distinguished as walking-legs. 1932 J. S. Huxley *Probl. Relative Growth* iv. 123 The pereiopod which was enlarged as the male chela being the first of the walking-leg series. 1977 G. F. Warner *Biol. Crabs* v. 75 They burrow by tilting backwards and digging down into the sand with their walking legs. 1971 *Sci. News* 21 Aug. 118/2 Manipulators, walking machines, artificial limbs and man amplifiers—cybernetic machines that perform as appendages of man. 1973 *Britannica Yearbk. Sci. & Future 1972* 102 A four-legged 'walking machine' mimics and amplifies the movements of its human operator. The right front leg of the machine is controlled by the operator's right arm, the left front by his left arm, [etc.]. 1978 D. Murphy *Place Apart* vi. 110 Now he is 'out of work' and has to use a complicated walking-machine provided by the NHS. 1825 S. Woodworth *Forest Rose* i. iv. 18 As for the bumpkin, her lover, he must take his walking papers. 1916 Joyce *Portrait of Artist* 23 You'll get your walking papers [*sc.* discharge from a school infirmary] in the morning when the doctor comes. 1978 L. Pryor *Viper* viii. 141 Hassan gave me my walking papers last night. 1829 in N. E. Eliason *Tarheel Talk* (1956) 124, I heard..she had given Henry a walking ticket. 1888 'R. Boldrewood' *Robbery under Arms* I. x. 124 Every soul about the place..seemed to have got a cheque and a walking-ticket at the same time. 1970 *Islander* (Victoria, B.C.) 24 May 7/2 If a man spat on the floor, he'd get warned. If he did it again he'd get his walking ticket.

walking, *ppl. a.* Add: **1. c.** *walking delegate* (earlier example).

1889 *Lippincott's Monthly Mag.* Feb. 227, I had no experience of strikes, and 'the walking delegate' was not yet stepping westward.

3. *walking doll,* a mechanically operated doll that can be made to move its legs; *walking funeral,* one in which the coffin is wheeled by hand.

1841 Dickens *Barnaby Rudge* ix, in *Master Humphrey's Clock* II. 285, I wish I may only have a walking funeral, and never be buried decent with a mourning-coach and feathers. 1848 Walking funeral [in Dict.]. *a* 1870 in *Country Life* (1973) 29 Nov. 1833/2 The patent Autoperipatetikos or walking-doll. 1886 C. M. Yonge *Chantry House* II. xvi. 158 A Chinese walking doll was sent..for the amusement of Miss Winslow's school children. 1968 *Canad. Antiques Collector* Dec. 14/2 In the 1820's the first walking dolls were introduced. These opened and shut their eyes and said 'mama' and 'papa'. 1983 L. Taylor *Mourning Dress* i. 40 Even amongst the poor there were grades of funerals. The grandest involved the use of a horse-drawn hearse, but much more common was a 'walking funeral' where the coffin was wheeled on its hearse.

d. Of a casualty: able to walk despite his injuries, not in need of a stretcher; chiefly in phr. *walking wounded,* such casualties collectively; also *fig.*

1917 P. Gibbs *Battles of Somme* 265 The long trails of the walking wounded, marvellously brave, wonderfully full of spirits. 1948 G. Greene *Heart of Matter* ii. i. i. 117 We were told to prepare for nine stretcher cases and four walking ones. 1965 *Economist* 4 Dec. 1072/2 Governor Rockefeller..has been one of the political walking-wounded since his unpopular remarriage in 1963. 1972 *Guardian* 23 Dec. 8/1 There are many less conspicuous casualties, the walking wounded of the affluent society.

4. *walking gentleman, lady* (earlier examples).

1794 C. Mathews *Let.* 14 June in A. Mathews *Mem. Charles Mathews* (1838) I. v. 86 And asked me if I would like to play walking gentleman. 1860 G. Vandenhoff *Leaves from Actor's Note-bk.* xii. 196 Susan..was the best walking-lady on the American Stage.

5. *walking encyclopaedia* (examples); also *walking dead.*

1868 L. M. Alcott *Little Women* I. xii. 181 Meg.. considered him a walking encyclopaedia of useful knowledge. 1912 T. Dreiser *Financier* vi. 60 There were many, many wildcat banks, and they were sufficient in number to make the average exchange-counter broker a walking encyclopedia of solvent and insolvent institutions. 1978 *Rugby World* Apr. 28/3 He could..enlist the aid of some other members of the Irish [rugby] squad like Mike Gibson, or Johnny Robbie, or Mick Quinn, who are walking encyclopaedias on English soccer in particular. 1980 J. Gardner *Garden of Weapons* iii. iii. 247 You look terrible. Like the walking dead.

7. b. *walking catfish*: a freshwater catfish, *Clarias batrachus,* native to south-east Asia and Africa, and able to use its fins to crawl over land.

1968 *Orlando* (Florida) *Sentinel* 20 Nov. 8-c/1 The state of Florida surrendered Tuesday in its war with the weird 'walking catfish' and said the ambulating invader..is here to stay. 1984 *USA Today* 6 Apr. 3B/1 Walking catfish were imported from Thailand to Florida in 1960 by a fish farmer who sold them to aquarium owners to keep their tanks clean. They since have multiplied rapidly in the wild. Some people are working on recipes to make walking catfish a more palatable dish.

9. Jazz. *walking bass*: a bass part, often consisting of broken octaves, that goes up and down the scale in 4/4 time in steps or small intervals; so *walking beat.*

1939 L. Hobson *Amer. Jazz Music* 51 String bass, more often plucked or slapped than bowed, usually playing two or four notes per bar or a 'walking' (melodic) bass. 1947 W. Russell in R. de Toledano *Frontiers of Jazz* iv. 61 A rhythmic and a melodic germ motive are developed over a 'walking bass' figure. 1950 [see *stride *sb.* 7 b]. 1952 Walking beat [see *walk *v.*^{1} 7 g]. 1952 B. Ulanov *Hist. Jazz in Amer.* (1958) iv. 29 If you listen carefully to the Ellington recording of 'C Jam Blues', you will hear a definitive example of the walking bass—1234/1234/1234, over and over again. *Ibid.* vi. 67 He was reliable, he put down a good walking beat, he had learned to read music. 1967 *Crescendo* May 18/2 The way he [*sc.* James Moody] wrote led to the invention of what's called the 'walking' bass, which the Americans took up later. 1980 *New Grove Dict. Mus.* III. 39/1 George Thomas, whose *New Orleans Hop Scop Blues* (1911; published 1916) included a walking bass, used the same device.

10. Mining. *walking dragline,* a large dragline supported on movable feet.

1959 *Times Rev. Industry* June 75/1 A..walking dragline..will carry a 10 cu. yd bucket at 184 ft. radius and will strip 15 tons at each bite. 1977 *Bulletin* (Sydney) 22 Jan. 39/1 (Advt.), Queensland is rich in coal, but it has to be wrested from the ground, and it's here where the huge Marion walking dragline is in its element.

walklet (wǭ·klėt). *nonce-wd.* [f. Walk *sb.*^{1} + -let.] A short walk.

1832 J. Romilly *Diary* 21 Feb. (1967) 4 Peter Ouvry & I had a walklet. 1896 A. Beardsley *Let.* c 12 Dec. (1970) 225 Yesterday I ventured out for a walklet in Winter Gardens.

Walkman (wǭ·kmăn). A proprietary name for small battery-operated cassette players and headphones capable of being worn by a person on foot.

1981 *Trade Marks Jrnl.* 25 Mar. 623/2 *Walkman*... Electrical and electronic apparatus and instruments..for transmitting, receiving, tuning, recording and reproducing audio and visual signals..batteries; aerials; loudspeakers; headphones; earphones..cassettes..Sony Kabushiki Kaisha (Sony Corporation)..Tokyo 141, Japan; manufacturers and merchants. 1981 *Japan Times* 31 Dec. 10/1 Sony Walkmans, easy-driving Honda scooters and aluminum household Buddhist altars sold like hotcakes during 1981. 1983 *Chicago Sun-Times* 22 May 9 I'll wear a Walkman in line [for the film] so I won't hear any smart alecks spill the story. 1983 *Sci. Amer.* June 130/1 The smaller zippers would be for products that reflect the current trend toward miniaturization, such as for carrying cases to fit the Walkman type of portable radio. 1983 *Official Gaz.* (U.S. Patent Office) 20 Dec. TM 288/1 Sony Corporation... *Walkman*... For audio tape player, audio tape recorders, radios and headphones. 1984 S. Townsend *Growing Pains A. Mole* 54 They wear red satin side vent running shorts, sleeveless satin vests, white knee socks, Sony Walkman earphones and one gold earring.

wa·lk-off. [f. vbl. phr. *to walk off*: Walk *v.*^{1} 8 c.] An act of walking off: **a.** From a stage.

1936 N. Coward *Red Peppers* in *To-night at 8.30* I. 85 Their exit consists of a neat walk off together, one behind the other, with their telescopes under their arms. 1970 *Time* 5 Jan. 51 The walk-off is the bittersweet image by which..Chaplin wishes to be remembered.

b. From a ship, sports field, etc. Cf. *walk-out *sb.* 1.

1951 *Daily Progress* (Charlottesville, Va.) 13 June 5/1 (*heading*) Walkoff grows at new atomic plant. 1960 *Daily Tel.* 28 July 1/8, 2 liner crews return after walk-off. 1971 *Guardian Weekly* 10 Apr. 6 The 22 Springbok cricketers who held a walk-off at Newlands Ground here yesterday in protest against apartheid in cricket.

wa·lk-on, *sb.* and *a.* [f. vbl. phr. *to walk on*: see Walk *v.*^{1} 5 in Dict. and Suppl.] **A.** *sb.* **1.** *Theat.* **a.** A walk-on part.

1902, 1907 [in Dict. s.v. Walk *v.*^{1} 25]. 1936 R. Lehmann *Weather in Streets* iv. i. 361 Or I might get a walk-on in a film. 1950 *Sun* (Baltimore) 19 May 16/2 It has the all-starriest cast of all time, with all the great ones doing walk-ons and bits. 1971 R. A. Carter *Manhattan Primitive* (1972) xv. 142 She never got another speaking part: a few walk-ons, then she was an extra again.

b. An actor, etc., who has a walk-on part; = *walker-on* s.v. ***walker** *sb.*^{1} 1 c.

1946 *Sun* (Baltimore) 20 May 6/1 It can dispense with the star system..by concentrating on teamwork in which lead actors one night become walk-ons the next. 1964 M. Drabble *Garrick Year* xii. 193 A square, worried-looking girl, who was married to one of the walk-ons. 1980 *Times Lit. Suppl.* 31 Oct. 1228/5 There are several blandly handsome walk-ons, most of them recently out of drama school.

2. *Sport.* A team member without any regular status. *U.S.*

1974 *Plain Dealer* (Cleveland, Ohio) 19 Oct. 2-D/1 East Tech's Mike Lucas, a 6-4 junior college transfer from Arizona, is one of three 'walk-ons' on Ohio State's basketball team. 1980 *New Yorker* 3 Mar. 80 This year..the Highlands team looks a bit too much like the rest of the students for Martinez's taste... One player..did not even come to Highlands on a basketball scholarship—a category of athlete known in the trade as 'walk-ons'. 1981 *Washington Post* 4 Sept. D-1 He was beaten by Marty Davis, 22, who made the University of California tennis team as a walk-on.

B. *adj.* **1.** Pertaining to or designating an airline service for which prior booking is not required. orig. *U.S.*

1961 *Flight* LXXX. 488/2 The airline..earlier this year introduced 'walk-on' services to Chittagong. 1967 J. Gardner *Madrigal* viii. 220 He..asked if there were any direct flights from Manchester to Zurich... At this time of year there should be no difficulty in getting a walk-on booking. 1977 *Times* 29 July 4/3 British Airways intends to compete with Laker Airways' walk on Skytrain air service between London and New York. 1983 *Flight Internat.* 10 Sept. 680/2 Shuttle's walk-on option with guaranteed backup aircraft.

2. Theat. *walk-on part,* etc. = sense A. 1 a above. Also *transf.* and *fig.*

1963 *Times* 25 Feb. 4/1 Before luncheon the New Zealanders were so vigorous, accurate and hostile that Illingworth, Sheppard, and even Dexter could play only walk-on parts. 1973 J. Leasor *Host of Extras* viii. 141 If he heard of any jobs going—walk-on parts, crowd scenes, the odd TV commercial—he passed them on to me. 1976 *Private Eye* 24 Dec. 17/3 The Bank of England is determined it shall be in a supporting, if not walk-on, capacity rather than a starring role. 1977 *Time* 8 Aug. 21/1 Her striking good looks eventually won her some small walk-on parts in German films. 1983 J. Juxon *Lewis & Lewis* xxi. 223 Few..of Parnell's biographers have given Lewis more than a walk-on part in the drama. 1985 *Listener* 24 Jan. 11/3 Salman Rushdie complained that Indians were for the most part given walk-on roles.

wa·lk-out, *sb.* and *a.* [f. vbl. phr. *to walk out*: see Walk *v.*^{1} 5 j, *12*.] **A.** *sb.* **1. a.** A strike, esp. one called at short notice. Also *transf.*

1888 in Farmer *Americanisms* (1889) 550/2 The walk out of brewery employés, decided upon at last night's meeting of the union, was considerable of a fizzle. Less than thirty men left their work. 1894 *Evening Sun* (N.Y.) 10 July 1/6 Some of the big firms have already closed their works in anticipation of a walkout. 1919 [in Dict. s.v. Walk *v.*^{1} 25]. 1957 *Economist* 12 Oct. 129/1 An attempt to organise a walkout by the white pupils failed when less than one in thirty left the school. 1982 *Daily Tel.* 22 July 5/1 Porters, cooks, cleaners and other ancillary staff staged an almost complete walk-out. 1985 *Times* 7 Mar. 11/3 The planners speak of lightning strikes and walkouts.

b. An act of leaving a meeting, etc., as a gesture of protest or disapproval.

1927 *Observer* 1 May 13/4 He has fulfilled the assurance he gave on taking the Chair that he would observe strict impartiality in the conduct of business... It may be that on one or two occasions Mr. Patel's judgement has been at fault, as, for instance, in the dramatic episode of the Swarajist walk-out. 1946 *Richmond* (Va.) *Times-Dispatch* 2 Apr. 1/1 Ala presented Iran's side to the council last week after Russia's dramatic walkout. 1958 *Times* 2 Sept. 7/4 The opening film [*sc.* at a festival]..saw a rather considerable number of walk-outs. 1973 H. Trevelyan *Diplomatic Channels* viii. 125 The pointed absence or walk out on official occasions can be overdone, but is necessary if a speech by the host is directly insulting to the ambassador's country.

2. A love affair.

1934 E. Waugh *Handful of Dust* ii. 46 He's having a terrific walk out with a girl called Sheila Shrub. 1983 P. Devlin *All of us There* v. 57 Ellen has been having a walk-out with his brother Joe for years.

B. *adj.* Of a room or basement: with its own exit door, i.e. not dependent on stairs to another floor. *N. Amer.*

1958 *Washington Post* 16 Aug. B6/5 This home..has.. full walkout basement and a Hotpoint kitchen. 1968 *Globe & Mail* (Toronto) 13 Jan. 42/1 (Advt.), Lower level features striking walk-out family room and sauna. 1978 *Detroit Free Press* 16 Apr. F-10/8 (Advt.), Full finished walkout basement with fireplace.

walk-over. Add: **b.** *transf.* Anything accomplished with great ease.

1902 G. H. Lorimer *Lett. Self-Made Merchant* xv. 216 It wasn't any walk-over to hold the belt in those days. 1931 *Daily Tel.* 21 Jan. 8/4 This makes its acquisition by an American crook a walk-over. 1975 P. Fussell *Gt. War & Mod. Memory* (1977) i. 27 His little sporting contest did have the effect of persuading his men that the attack was going to be..a walkover.

c. *attrib.* or as *adj.*

1936 *Sunday Times* 14 June 4/2 When the law gets them in its clutches, a shady lawyer is allowed to get a walkover verdict of 'Not Guilty'. 1974 *Times* 5 Oct. 14/7 Lauda seemed set for almost a walkover trip to the title (he proved himself the fastest driver..nine times this year).

wa·lk-round, *sb.* and *a.* [f. vbl. phr. *to walk round*: Walk *v.*^{1} 5 a.] **A.** *sb.* = *walk-around* (*b*) s.v. Walk *v.*^{1} 25. *U.S.*

1861 *Temple Bar* May 199 The 'Jim Crow dance'.. soon gave place to better tunes..and 'walk rounds'. 1895 *Century Mag.* Oct. 958/1 His work with the caravan was to sing songs, chiefly darky songs, accompanied by 'hoe-downs' and 'walk rounds'. 1943 *Life* 5 July 80 After the 'walk-round', which served as an entrance for the black-faced showmen, they stood before a semicircle of chairs and waited for the interlocutor's: 'Gentlemen, be seated.'

B. *adj.* Round which one can walk: applied to a self-service shop with goods so arranged.

1974 P. Wright *Lang. Brit. Industry* xiii. 120 Walk-round store (where the goods are so accessible that you may be tempted to buy far more than you need). 1979 *This England* Winter 19/3 Miss Corrie's dim, prim little shop in the market place had survived the competition of walk-round stores and boutiques.

wa·lk-through, *sb.* and *a.* [f. vbl. phr. *to walk through*: Walk *v.*^{1}.] **A.** *sb.* **1.** *Theat.* **a.** A part not requiring the performer to exert himself, one that he may 'walk through' (Walk *v.*^{1} 7 a); *transf.,* an undemanding task.

1944 *Sun* (Baltimore) 15 July 6/1 The principal roles offer no difficulties, permitting Miss Powell to be herself, as always, and giving her leading man a walk-through. 1950 *Ibid.* 13 Oct. 12/2 For Mr. Milland, an actor with 58 photo-plays to his credit, this is a walk-through. 1980 J. Ball *Then came Violence* (1981) iv. 36 We've just had a homicide and this one isn't a domestic walk-through. 1984 *New Yorker* 30 Jan. 82/2 Most skilled politicians take the measure of a crowd and find ways to relate to it specifically. It is as if Glenn were doing a walk-through.

b. A perfunctory or lacklustre performance.

1970 *Globe & Mail* (Toronto) 26 Sept. 23/1 Richard Burton in Hamlet..giving us only an insulting, offhand walk-through.

2. (See quots.)

1959 W. S. Sharps *Dict. Cinematogr.* 91/2 *Dry run,* otherwise walk-through, a full rehearsal for a production, but without cameras. 1974 *Some Technical Terms & Slang* (Granada Television), *Walkthrough,* one stage after the tech run of a drama, but before moving into the studio proper. The cast and certain technicians will meet for a rough rehearsal in the studio set without cameras. 1977 K. T. Orr *Structured Syst. Devel.* i. 5 Today we see discussions of structured walkthroughs, structured design, and structured analysis. 1983 *Dict. Computing* 388/1 *Walk through,* a product review performed by a formal team... There is a clear statement of the contribution that each member of the review team is required to make, and a step-by-step procedure for carrying out the review... The product is..openly debated with a view to uncovering problems or identifying desirable improvements.

B. *adj.* Of a building, etc.: permitting access from either end.

In quot. 1950, for the cows to enter and leave.

1950 *N.Z. Jrnl. Agric.* May 477/3 The walk-through [milking] shed with doors at the rear of the shed in the conventional manner. 1967 *Economist* 8 July p. viii, He [*sc.* the driver] will want to reach back to his load without dismounting and walking round to open the back doors—hence the spate of 'walk-through' vans in Britain. 1981 M. C. Smith *Gorky Park* iii. i. 295 A remarkable walk-through tube that attached the plane directly to the terminal. 1984 *Buses* Nov. 488/3 A sliding door in the fixed section giving a walkthru' layout for the van.

wa·lk-up, *a.* and *sb.* [f. vbl. phr. *to walk up*: Walk *v.*^{1}] **A.** *adj.* **1.** Of an apartment, etc.: that has to be reached by stairs rather than by a lift. Also applied to a building consisting of such apartments. *U.S.*

1919 Mencken *Amer. Lang.* iv. 110 The term *flat* 'is usually in the United States restricted to apartments in houses having no elevator or hall service'. In New York such apartments are commonly called *walk-up apartments.* 1927 E. Glyn *It* iv. 40 Mary had a tiny two-room apartment in a walk-up building in Brooklyn. 1942 *N.Y. Times* 13 Feb. (late City ed.) 13/4 Six five-story walk-up apartment houses. 1946 Mezzrow & Wolfe *Really Blues* (1957) xiii. 235 Some crummy walk-up tenement flat. 1979 H. Kissinger *White House Years* xxi. 906 Huang Hua and I met around six o'clock in the CIA's walk-up apartment in the East Seventies. 1984 *Business Rev. Weekly* (Austral.) 4–10 Feb. 42/3 The Myer development is a three-storey walk-up affair that has appealed to a lower age group than usual for Gold Coast units.

2. That may be approached on the street, without having to go into a building.

1963 C. J. McCall in A. Dundes *Mother Wit* (1973) 421 The colorful window-signs of innumerable 'readers'..cry out from their store-front or walk-up locations. 1972 *Sunday Sun* (Brisbane) 8 Oct. 16/1 The same bandit had approached Miss Avery at her outside walkup window post on Tuesday and demanded $3000.

B. *sb.* **1.** A walk-up apartment or apartment block. *U.S.*

1925 *Scribner's Mag.* Oct. 6/2 Vacation heaves into sight over the horizon..the swirling dust turned into clean sand; the only walk-up a dune; and the total night life two movie theatres. 1942 R. Chandler *High Window* xxv. 149 The kind of dentists who have shabby offices on second-floor walk-ups over stores. 1954 F. P. Keyes *Royal Box* xiii. 165 The friends he had all lived in the identical kind of six-flat walk-up. 1966 R. Stout *Death of Doxy* (1967) i. 6 The person to ask lived on the second floor of a walkup on 52nd Street. 1976 *National Observer* (U.S.) 4 Sept. 1/2 The blue-jeaned couples climbing the stairs to their walk-ups together are most usually the children of affluence. 1980 J. Krantz *Princess Daisy* xxv. 438 Daisy herself lived in a low-rent SoHo walk-up and held down a full-time job.

2. *U.S. Horse-Racing.* The walk of racehorses to a starting line or tape (as opposed to starting gates). Freq. *attrib.*

1938 *Sun* (Baltimore) 1 Nov. 12/1 The only change in the usual order will be a walkup start instead of a start from a gate. 1946 *Richmond* (Va.) *Times-Dispatch* 8 Mar. 23/3 The field might have to be started from a walkup to an old-fashioned web barrier. 1959 *Washington Post* 15 Nov. c 1 The starter who had trouble..at Laurel in the ragged walk-up, says the foreign entries could be taught in five days to break from the stall gates. 1974 *Encycl. Brit. Macropædia* VIII. 1100/2 Some starts are still effected from a barrier that springs upward when actuated by the starter, or in the 'walk-up' fashion, whereby the starter gives a verbal order when the horses are reasonably well aligned.

3. *Shooting.* The act of walking up gamebirds (Walk *v.*^{1} 20); also *transf.* of clay pigeons; a piece of land kept for this purpose.

1972 *Shooting Times & Country Mag.* 27 May 10/3 There are numerous Bowman traps, designed to simulate varying driven game birds and there is a walk-up with 15 traps. 1975 *Times* 1 Sept. 14/6 An increasing number of landowners..let walked-up shoots... On the Speyside walk-up the other day was a Marseille dentist.

wa·lkway. orig. *U.S.* Also **walk-way.** [f. Walk *sb.*^{1} or *v.*^{1} + Way *sb.*^{1}] **1.** = Walk *sb.*^{1} 9 c.

1792 in *Essex Inst. Hist. Coll.* (1865) VII. 37/2 John Sanders..agrees to pave the walk way in front of his Father's Estate. 1816 W. Bentley *Diary* (1914) IV. 405 A walkway for the first time has been raised in the principal streets in the eastern part of the Town. 1904 *N.Y. Evening Post* 14 May 5 A space..sufficient to provide each house with a walkway to the rear. 1976 *Outdoor Living* (N.Z.) I. ii. 42 (*caption*) Quarry tiled walkway and meandering plants leading through a painted brick archway give a pleasant first impression to this home. 1980 *Daily Tel.* 16 June 2/1 Paths affected by housing development should be retained as 'walkways' through the new estate.

2. A pedestrian passageway linking different parts of a building or structure or complex of buildings, esp. one raised above ground-level and separating the users from machinery, traffic, etc. Also, a specially built path taking sightseers through an area of natural beauty or the like.

1928 R. H. Lansburgh *Industr. Management* (ed. 2) xv. 175 Accident prevention by making floors and walkways safe is a big factor in the industrial accident toll. 1933 *Meccano Mag.* Feb. 109/1 There is perfect communication between all members of the crew, a walk-way from end to end of the aeroplane being provided to enable members to change places quickly. 1953 *Archit. Rev.* CXIII. 83 (*caption*) The walkway..passes under the three classroom blocks and links the two main courts. 1959 *Daily Tel.* 13 Mar. 18/4 The lay-out [of London Zoo] will include all-weather covered elevated walkways from the main gate. 1960 *Guardian* 20 May 7/2 First-floor shops, walkways and bridges for pedestrians. 1960 'N. Shute' *Trustee from Toolroom* i. 13 The long wooden walkways above the tidal mud, the yachts moored bows-on in tiers. 1968 *New Scientist* 26 Sept. 640/2 There will be..a moving walkway along the pier to the two terminals [at Heathrow Airport]. 1973 *Times* 18 Oct. (Brazil Suppl.) p. vii/4 Tourists..will have 'walkways' through special jungle reserves which will remain undisturbed. 1979 *United States 1980/81* (Penguin Travel Guide) 250 The newly restored North Beach with landscaped dunes and oceanfront walkway. 1981 *Sci. Amer.* Nov. 32/1 This environment, the home of the most diverse swarm of species,..is currently studied from walkways hung high among the treetops, steel towers rising into the leafy canopy. 1984 *Daily Tel.* 3 July 1/7 Outside there will be Tivoli-style gardens with a walkway along the Thames.

wall, *sb.*^{1} Add: I. **4. g.** *The Wall*: *ellipt.* for *Wailing Wall* s.v. ***wailing** *vbl. sb.* c.

1895 J. Smith *Pilgrimage to Palestine* xvi. 219 The 'Wailing Place of the Jews'..is situated a little to the north... High overhead towered the..stones of the Temple Wall..with the Wall itself..rising to a height of 60 feet... There, with their faces to the wall—kissing the stones,..or joining in a loud chorus of lamentation.. stood a long row of Jews. 1928 *Western or Wailing Wall in Jerusalem* 6 in *Parl. Papers 1928–9* (Cmd. 3229) XV. 105 His Majesty's Government regard it as their duty..to maintain the established Jewish right of access to the pavement in front of the Wall for the purposes of their devotions. 1967 C. Potok *Chosen* xii. 198 He died while praying at the Wall for the Messiah to come and redeem his people. 1973 *Guardian* 21 June 2/7 The present intention is to link the Wall with the historic 'upper city' (now the Jewish Quarter).

h. *The Wall*: *ellipt.* for *Berlin Wall,* the wall surrounding West Berlin and separating it from communist East Berlin and the rest of East Germany (erected in 1961).

1961 *Daily Progress* (Charlottesville, Va.) 20 Oct. 1/1 Here in Berlin communism has created one of the ugliest and most depressing sights on the face of the globe. It is The Wall—the wall of death, the new concrete curtain of communism. 1964 *Ann. Reg. 1963* 225 It was stated that 1,283,918 had crossed the Wall by the time it closed on 5 January 1964. 1977 G. Markstein *Chance Awakening* lxxviii. 243 My father had his legs blown off..when he tried to flee over the Wall.

II. 8. c. In the game of Mah Jong, the arrangement of tiles from which hands are drawn. Cf. *tile *sb.*^{1} 4 b.

1922 *Lit. Digest* 30 Dec. 38 One studies the unfolding of Ma Jung, one detects Eastern cunning to whet the skill, first the building of the 'wall', undoubtedly meaning the great wall of China, one of the seven wonders of the world. 1950 E. Culbertson *Culbertson's Hoyle* 415 *Wall game,* void game by exhaustion of the wall without any declaration of a complete hand. 1974 *Encycl. Brit. Micropædia* VI. 503/3 Thereafter, the other players, in counterclockwise rotation, each draw one tile, which may be the last discarded tile or a loose tile from the 'wall'.

d. *Baseball.* The barrier marking the outer perimeter of the outfield.

1928 G. H. Ruth *Babe Ruth's Own Bk. Baseball* 25 The boys began smacking the fences with long drives, outfielders began playing with their backs to the wall and infielders had to move back on the grass or have their legs torn off with hot drives. *Ibid.* 117 The ball was hit far over his head to the center field wall. 1973 *Internat. Herald Tribune* 15 June 15/3 The closest the Reds had come to a hit was Pete Rose's long drive to left-center in the third that Jim Dwyer caught at the wall.

e. *Football.* A line of defence players who defend their team's goal during a free kick.

1948 Hankinson & Chadder *Soccer for Schools* Plate 29 (*caption*) The usual procedure for the defence to adopt is for a 'wall' of players to block the line of a direct shot for goal. 1965 D. Bacuzzi *How to play Association Football* xv. 56 (*caption*) Notice how the 'wall' of defenders allows the goalkeeper space to see. 1976 E. Dunphy *Only a Game?* v. 149, I scored direct from a free kick. Curled it round the wall into the top corner from just outside the box.

f. *Surfing.* The steep face of a wave before it breaks.

1962 T. Masters *Surfing made Easy* 66 *Wall,* the steep portion of a wave almost ready to break. 1965 J. Pollard *Surfrider* ii. 20 The steep face of that wave is called its 'wall'. 1968 *Surfer Mag.* Jan. 17/1 'Just Ken' probably doesn't know what it is like to ride a well-shaped wall while hanging ten except when he is smoking grass.

9. b. The pastry forming the sides of a pie.

1747 H. Glasse *Art of Cookery* viii. 73 Make a good Standing Crust, let the Wall and Bottom be very thick. 1894 L. Heritage *Cassell's New Universal Cookery Bk.* 785/1 Form the walls of the pie with the left hand. The sides should be smooth and of equal thickness. 1959 *Listener* 22 Jan. 191/2 Lid the flan with pastry, having egged the top of the 'wall'.

III. 14. *to send to the wall* (earlier example).

1881 E. W. Hamilton *Diary* 13 Mar. (1972) I. 115 Lord Bath..is much exercised in his mind as to the Greek question sending to the wall the interests of Servia.

16. *to give a person the wall* (later example).

1869 A. J. Munby *Diary* 24 Dec. in D. Hudson *Munby* (1972) vii. 278 'If a nigger didn't give me the wall, I'd knock him down as soon as look at him!' Here we have the British Philistine.

19*. *to go over the wall* and varr.: (*a*) to go to prison; (*b*) to escape from prison; (*c*) to leave a religious order; (*d*) to defect (to another country). Hence (*e*) *over the wall* adv. phr., escaped from prison; in prison. *slang.*

a. 1917 W. Muir *Observations of Orderly* xiv. 228 He would be observed 'going over the wall' or 'going to stir' (going to detention prison).

b. 1933 *Amer. Speech* VIII. iii. 27/1 *Go over the wall,* escape. 1936 L. Duncan (*title*) Over the wall. *Ibid.* vi. 95 Us guys..pull wires to get jobs as guards, and you convicts go over the wall whenever you can. 1963 *Times* 5 June 16/1 He knew it was an unwritten law that an escape extinguished such a debt, and so he decided to 'go over the wall'. He gave himself up at Clacton-on-Sea. 1974 P. B. Yuill *Hazell plays Solomon* vi. 66 You really think Mancini would've tried to go over the wall?

c. 1949 M. Baldwin (*title*) I leap over the wall. A return to the world after twenty-eight years in a convent. 1970 *Harper's Mag.* Apr. 110 Mr. Vizzard was a Jesuit seminarian who yearned for the world, leapt over the wall, and found what he was looking for in Hollywood. 1979 'E. Anthony' *Grave of Truth* vii. 190 A bride of Christ, eh? What happens if she jumps over the wall..decides she's had enough of convent life..?

d. 1976 M. Butterworth *Remains to be Seen* v. 84 The bloody place [*sc.* the Foreign Office]..has never been the same since Kim [Philby] went over the wall.

e. 1935 A. J. Pollock *Underworld Speaks* 85/2 *Over the wall,* escaped from prison. 1973 G. Beare *Snake on Grave* xxiii. 141 He's out. Over the wall.

19.** *up the wall*: angry, furious; distraught, mad, crazy; esp. in phrs. *to climb, run up the wall*: to become very angry or distraught; *to drive, send* (someone) *up the wall*: to infuriate or put into a frenzy. *colloq.*

1951 S. Kaye-Smith *Mrs. Gailey* 160 Your mother's running up the wall because he came to dinner. 1953 H. Clevely *Public Enemy* xvii. 101 Old Marks 'll climb up the wall if he hears I closed early. 1956 A. Wilson *Anglo-Saxon Att.* ii. ii. 307 You drive me up the wall. What sort of a mess have you got poor Dad into? 1959 *Observer* 21 June 8/8 When they found out he was a Catholic, they were up the wall. 1961 *New Left Rev.* Mar./Apr. 30/1 She was right up the wall, and poor Aunt Ada isn't in any state to help. 1966 'L. Lane' *ABZ of Scouse* 40 Sends me up ther bloody wall. 1970 *New Yorker* 3 Oct. 105/1 Success or failure hardly entered into the picture. It was this kind of argument that drove some.. executives up the wall. 1975 'E. Lathen' *By Hook or by Crook* xiv. 138 The American wife, the sweetie-pie who sends Everett up the wall. 1977 *Chicago Tribune* 2 Oct. vi. 15/2 The prejudice is so acute; that country is up the wall.

19*.** *off the wall* (also with hyphens used *attrib.*): unorthodox, unconventional; instinctive, intuitive, off the cuff. (See also quot. 1966.) Also used advb. *U.S. slang.*

1966 *Current Slang* (Univ. S. Dakota) Summer 3 *Off the wall,* unimpressive... I have a lit. professor who's *off the wall.* 1968–70 *Ibid.* III–IV. 88 *Off the wall,* adj. Unusual; unorthodox; 'crazy'. 1974 *National Rev.* (U.S.) 4 Jan. 47/2 Brian knows how to startle the over-interviewed with off-the-wall questions that get surprising answers: Ever see a ghost? 1975 *San Francisco Chron.* 11 Jan. 12/3 He became suspicious when Dickenson answered extremely complex questions 'off the wall'. 1976 *Time* 5 Apr. 74, 'I just thought it was off-the-wall funny', says Lear. 1977 C. McFadden *Serial* (1978) iv. 14/1 She had decided to play the whole scene off the wall, to just go with the flow... The really authentic thing to do was to act on your impulses. 1977 *Listener* 20 Oct. 498/2 Among the many new sources of cash—it's called 'off the wall' fundraising—I have heard about a tribe of Apaches which ..invested $2 million in the making of..a western. 1982 *Penthouse* Dec. 84 He started talking off the wall about how he should go to El Salvador.

21. a. simple attrib., *wall-decoration, -front*; 'set or fixed against a wall', *wall-cupboard, light, mirror, -panel, -panelling, phone, picture, safe, sconce* (examples), *socket, switch, telephone.*

1961 *Times* 16 Jan. 13/4 It is furnished throughout with tables, chairs, wall-cupboards. 1976 'W. Trevor' *Children of Dynmouth* ii. 51 The commodious wall-cupboards, the scrubbed wooden table. 1867 D. G. Rossetti *Let.* ? 12 Nov. (1965) II. 643 There are sufficient slight representatives of it [*sc.* the severed head] on vases and in wall-decoration of classic times. 1935 *Burlington Mag.* June 272/2 The question of the connexion between the carpet patterns and the wall-decoration remains difficult enough. 1964 *Listener* 3 Dec. 883/1 This is a rare opportunity to see an important High Renaissance wall-decoration. 1923 D. H. Lawrence *Birds, Beasts & Flowers* 115 The black hole, the earth-lipped fissure in the wall-front. 1869 *Bradshaw's Railway Man.* XXI. p. xvi (Advt.), Wall Lights and Mantle Piece Lustres. 1905 E. Wharton *House of Mirth* i. iii. 42 She turned out the wall-lights. 1972 Wall light [see *penlight, pen-light]. 1934 Webster, Wall mirror. 1940 M. Sadleir *Fanny by Gaslight* i. ix. 241 She studied herself in a long wall mirror. 1981 'J. Melville' *Murder has Pretty Face* i. 20 She could see her reflection in a wall mirror. 1880 L. Higgin *Handbk. Embroidery* 62 Design for wall-panel. By Mr. E. Burne-Jones. 1933 *Burlington Mag.* July 22/1 The absence of graining on the wall panels. 1962 A. Nisbett *Technique Sound Studio* 245 Marked acoustic colouration in a studio may be due to the coincidence of dimensional resonances, to wall-panel resonances, or to frequency-selective excessive absorption of sound. 1880 L. Higgin *Handbk. Embroidery* 62 Design for wall-panelling or curtains. 1962 L. Deighton *Ipcress File* xxx. 194 The wall phone rang. 1975 'R. Butler' *Where all Girls are Sweeter* vi. 70, I.. went over to the wall-phone and dialled. 1895 Wall-picture [see *Japanese lantern* s.v. *Japanese *a.* and *sb.* b]. 1966 J. Derrick *Teaching Eng. to Immigrants* v. 195 Wall pictures to match those in the book, with suitable captions printed by the teacher..can be put up in the classroom. 1931 W. Faulkner *Sanctuary* xxiii. 254 The room contained..three over-stuffed chairs, a wall safe. 1978 A. Neave *Nuremburg* iii. 36 Two wall-safes had been found, one in Bertha's bedroom and one in Gustav's dressing-room. 1954 Wall sconce [see *Oscar^2 b]. 1974 J. Aiken *Midnight is Place* i. 11 A few candles..burned flickeringly in the wall sconces. 1890 Slingo & Brooker *Electr. Engin.* xvii. 608 A wall socket..is useful in cases where it is required to place a movable lamp in circuit at one or other of a number of positions. 1977 Wall socket [see *start button* s.v. *start *sb.*^{2} 12]. 1935 D. L. Sayers *Gaudy Night* vi. 114 She found the wall-switch and went down the central corridor of the Annexe. 1981 C. Dexter *Dead of Jericho* xxv. 142 He..turned on the wall switch... But clearly the electricity had been disconnected. 1914 Wall telephone [see *extension 9 h]. 1977 *Rolling Stone* 30 June 73/2 Her kitchen is very white—walls, doors, floors, white appliances... And a white wall telephone.

b. *wall-builder* (earlier example); *wall-girdled, -hung, -mounted* adjs.

1812 E. Weeton *Let.* 15 June in *Jrnl. of Governess* (1969) II. 22, 7d a yard is the price now usually paid to wall-builders. *a* 1930 D. H. Lawrence *Etruscan Places* (1932) 50 Nowhere is far off, in these small wall-girdled cities. 1876 W. Morris *Sigurd* 2 The least of its wall-hung shields. 1970 Wall-hung [see *modular *a.* 1 b]. 1964 R. F. Ficchi *Electr. Interference* v. 67 The installation, however, is quite important:..it must be located within the shielding or 'wall-mounted' through the shield. 1980 *Sunday Times* (Colour Suppl.) 21 Sept. 50 (Advt.), Toilets and bidets can be floor standing or wall mounted.

22. wall bar, one of a set of parallel bars, attached to the wall of a gymnasium, usu. running from floor to ceiling, on which various exercises are performed; **wall bed,** a bed which can be folded up against a wall when not in use; **wallboard,** (a piece of) board, made from wood pulp, fibre, and other materials, used for surfacing walls and ceilings, etc.; **wall bracket,** a bracket (Bracket *sb.* 1 a) which is attached to a wall as a stand or support for a lamp, ornaments, shelves, etc.; **wall-chalker** (later example); hence **wall-chalking; wall chart,** a chart or poster giving information, often in pictures or diagrams, and designed for display on a wall, esp. in a classroom; **wallcovering,** material used to cover and decorate the inside walls of a building (cf. Wall-paper 1 in Dict. and Suppl.); **wall fish** *dial.,* the edible snail, *Helix pomatia*; **wall garden,** a garden surrounded by a wall, or a border planted beside a sheltering wall; **wall-hangings,** also, embroidered, woven or other decorative drapery for display on walls; occas. *sing.*; **wall newspaper,** (*a*) a newspaper produced by an educational institution or place of work, typed or hand-written, and displayed on the wall; (*b*) (esp. in Communist countries) an official newspaper displayed on the wall in public places, esp. in the street; **Wall of Death,** a fairground sideshow in which a motor-cyclist uses gravitational force to ride his motor-cycle around the inside walls of a vertical cylinder; **wall-painting,** a mural, a fresco; **wall pass** *Football* = *one-two* (*c*) s.v. *one 33; **wall plug:** see *plug *sb.* 1 c (cf. *wall socket,* sense 21 above); **wall-pocket,** (*a*) a receptacle for small household items, designed to hang on a wall; (*b*) = *wall vase* below; **wall-poster,** a poster affixed to a public wall; *spec.* = *ta tzu-pao; **wall-side,** (*a*) the side of a wall; (*b*) the side of a pavement, etc., where there is a wall (also *attrib.*); **wall space,** an expanse of unbroken wall surface, esp. one regarded as an area for displaying pictures, etc.; **wall system** *U.S.,* 'a set of shelves often with cabinets or bureaus that can be variously arranged along a wall' (*Webster's 9,000 Words*); **wall unit,** a piece of furniture consisting of various sections and compartments such as shelves and cupboards, and designed to stand against a wall; **wall vase,** a vase with one flat side allowing it to be hung on a wall; **wallwasher,** a type of lighting fixture designed to 'wash' a wall with light (see quot. 1983).

1903 *Handbk. Physical Training* (Admiralty) i. 53 The men are placed with one side towards, and at one pace from the Wall Bars. 1973 M. Russell *Double Hit* ii. 18 I'll be getting back to the wallbars. 1913 *Maclean's Mag.* Oct. 78/1 The Pacific Wall Bed is sanitary in every respect. 1974 Wall bed [see *Murphy^2]. 1925 (*title*) U.S. Government master specification for gypsum wall board. (U.S. Bureau of Standards.) 1933 *Archit. Rev.* LXXIII. p. lviii, The group of materials commonly known as wallboards, but more correctly termed building-boards, may..be classified in five categories:—(1) fibre boards..(2) laminated boards..(3) wood pulp boards.. (4) plaster boards..(5) composite boards. 1942 [see *storyboard* s.v. *story *sb.*^{1} 9]. 1962 A. Lurie *Love & Friendship* iv. 59 Burned, sodden chunks of wallboard lay about. 1978 *Cornish Guardian* 27 Apr. 34/6 (Advt.), Carpet Tiles. Tools. Wallboards. 1909 Webster, Wall bracket. 1926 L. Elmhirst *Notebk.* in M. Young *Elmhirsts of Dartington* (1982) xii. 295 We eliminated the wall brackets and agreed upon..ceiling lights. 1939 [see *standard *sb.* 16 c]. 1951 W. Faulkner *Requiem for Nun* i. ii. 53 Floor-lamp, wall-bracket lamps, a door left enters from the hall. 1976 *Gramophone* Sept. 516/2 The small size can be easily accommodated particularly if the speakers are mounted on wall brackets. ?1932 Dylan Thomas *Sel. Lett.* (1966) 5 We're over-ripe, we night-walkers, cunt-stalkers, wall-chalkers. 1853 Dickens *Bleak Ho.* xli. 409 Lady Dedlock, the wall-chalking and the street-crying would come on directly. 1958 S. Hyland *Who goes Hang?* xl. 189 He was..examining a wall-chart which depicted..the working..of a bicycle. 1980 Wall chart [see *snake *sb.* 4 f]. 1970 *Times* 11 Dec. 16/4 The wallcovering on three walls is glossy yellow p.v.c. 1979 *Tucson Mag.* Apr. 73/2 (Advt.), We offer a multitude of wall coverings and coordinating fabrics. 1950 O. Blakeston *Pink Ribbon* v. 59 They ate snails..in Gloucester, and they called them 'wall fish'. 1980 *Times* 2 Oct. 13/3 The taste of an open mushroom grilled with garlic, parsley and butter is so splendid, and superior to snails given the same treatment, that I would never now dream of bothering to cook that delicacy known in the Mendips as *wallfish.* 1780 J. Woodforde *Diary* 26 Apr. (1924) I. 280 Busy in painting some boarding in my Wall Garden. 1936 *Discovery* Mar. 86/2 Wall Garden, 385 feet long, first planted in 1935 [at the Brooklyn Botanic Garden]. 1967 E. Short *Embroidery & Fabric Collage* i. 9 If the finished piece of embroidery is to be seen from a distance, as for instance a wall hanging. 1979 *Arizona Daily Star* 5 Aug. J 5/7 (Advt.), They've just had a new shipment of Oaxacan Rugs and Wallhangings. 1983 *Listener* 30 June 17/1 By 1972 I needed a small pantechnicon to convey all my books on macrobiotic cookery, my plants, wall-hangings and floor cushions. 1935 N. Mitchison *We have been Warned* iii. 295 She had been shown the wall newspapers of factories and schools. 1937 E. Snow *Red Star over China* viii. v. 293 There was also a wall newspaper in every club, and a committee of soldiers was responsible for keeping it up to date. 1966 J. Derrick *Teaching Eng. to Immigrants* vi. 213 Project work involving the use of aids and apparatus..such as preparing a broadcast or TV programme or involving the class or group in the production of a wall newspaper should be considered. 1978 *China Now* Mar./Apr. 19/3 They keenly contested for the miserable prizes offered in competitions between groups and individuals in sport, literacy, public health, wall newspapers, and 'factory efficiency'. 1979 *N.Y. Rev. Bks.* 25 Oct. 40/3 The hero stands in Yuryatin reading the wall-newspapers. 1946 G. Tyrwhitt-Drake *Eng. Circus & Fair Ground* xviii. 210 Undoubtedly the most thrilling side-show was the 'Wall of Death', first seen here..in 1928. 1959 *Listener* 26 Feb. 371/2 It might..spin quickly round the steep hollow like a rider on the Wall of Death. 1976 'W. Trevor' *Children of Dynmouth* i. 13 The Hall of a Million Mirrors and the Tunnel of Love and Alfonso's and Annabella's Wall of Death were in the process of erection. 1688 Wall painting [see Fresco *sb.* 2]. 1849 *Jrnl. Brit. Archæol. Assoc.* IV. 92 Church decoration of this kind is..not unfrequently brought to light; but specimens of domestic internal wall-painting are of much greater rarity. 1898 A. Beardsley *Let.* 14 Jan. (1970) 424 I'm afraid good books on the wall paintings of Pompeii are costly and beyond my balance. 1933 *Burlington Mag.* Oct. 146/2 These..portraits.. brought Holbein immediate fame and the order to decorate the King's Privy Chamber in Whitehall Palace with wall-paintings. 1964 W. L. Goodman *Hist. Woodworking Tools* 132 A wall-painting from Pompeii. 1958 *Mod. Soccer* (Football Assoc.) v. 41 In Fig. 38 the LB [Left Back], sensing the possibility of a 'wall-pass', quickly backs away. 1973 *Times* 6 Jan. 7/5 The 'wall pass', 'one-two', touch play, 'push and run', call it what you will, we developed it at Tottenham. 1888 Wall plug [see *plug *sb.* 1 c]. 1914 [in Dict., sense 21 a]. 1962 L. Deighton *Ipcress File* xxv. 166 There was a big two kilowatt electric fire plugged into a wall point... It was the work of a minute to switch on the wall plug. 1880 *Scribner's Monthly* Apr. 921/1 The family comb..occupied a convenient wall-pocket at one side of the small kitchen mirror. 1947 E. Bishop in *Nation* (N.Y.) 22 Feb. 215/1 The eighty watt bulb..Lighting as well on heads Of tacks in the wall paper, On a paper wall-pocket, Violet-embossed, glistening With mica flakes. 1957 Mankowitz & Haggar *Conc. Encycl. Eng. Pott. & Porc.* 125/1 Articles for domestic use..included..punch-bowls, wall-pockets and plaques. 1971 L. A. Boger *Dict. World Pott. & Porc.* 365/1 *Wall Pocket,* in ceramics; a decorative object made of faïence, porcelain or pottery, having the shape of a vase, with one side being flat so that it can be hung on a wall. It is also called a wall vase. 1962 E. Cleaver in A. Dundes *Mother Wit* (1973) 14/2 The mass media.. television,..illustrated wall posters [etc.]. 1966 *China Q.* Oct.–Dec. 3 Within a week or two..20,000 'sightseers' were visiting Peking University each day, partly to read the wall-posters, partly to watch and abuse the 'criminals'. 1977 'S. Leys' *Chinese Shadows* (1978) ii. 70 Huge wall posters placed at random throughout the city. ('Increase our Vigilance and Protect our Fatherland!') 1979 'J. le Carré' *Smiley's People* xxvii. 322 Wall posters offering cheap ski holidays. *c* 1400, 1887 Wall side [in Dict., sense 21 a]. 1933 Auden *Poems* (ed. 2) 80 Climbing over to wall-side of bed. 1959 I. & P. Opie *Lore & Lang. Schoolch.* xii. 266 The old custom of making grottoes at the wall-side edge of the pavement. 1889 in *Cent. Dict.,* Wall space. 1898 G. B. Shaw *Arms & Man* iii. 47 The rest of the wall space being occupied by trophies of war and the chase. 1978 *Lancashire Life* Sept. 101/1 True, a new library is being built in Stanley Road, where wall-space for pictures will be made available, but there will be no gallery proper, and no museum whatsoever. 1978 *Detroit Free Press* 2 Apr. 7B/1 (Advt.), Westwood wall system from Southern is a handsome backdrop for any room and provides invaluable storage. 1980 *Christian Sci. Monitor* (Midwestern ed.) 4 Dec. 17/3 Wall systems are the fastest-growing category of furnishings. 1962 *Listener* 11 Jan. 65/1 Making the best use of up-to-date methods, such as the prefabrication of large wall units, which are being successfully used already in half the European countries. 1979 A. B. Emary *Woodworking* xxiv. 104 Many of the units purchased from stores are made from melamine-covered chipboard and since it is easy to obtain the home woodworker will find this material useful when making objects such as wall units and shelving. 1889 in *Cent. Dict.,* Wall vase. 1937 *Burlington Mag.* Dec. p. xx/1, The book contains many valuable hints..such as..the advantages of wall-vases. 1979 I. Webb *Compl. Guide Flower & Foliage Arrangement* vii. 96/2 (*caption*) Two wall vases hold flowers which suit their differing qualities and appearance. 1966 D. Phillips *Lighting* 36 The principal lighting method is by wall washer fittings recessed into the suspended ceiling. 1983 *Homes & Gardens* Nov. 138 Wall-washers have half the aperture closed off and their function is to give an even illumination of one wall surface from skirting to ceiling without lighting the floor.

wallaby. 1. a. Add to the list of genera: also *Wallabia* or *Thylogale.* (Earlier example.)

1826 J. Atkinson *Agric. & Grazing N.S.W.* ii. 24 The wallabee and padgy mellon..inhabit brushes, and afford good sport in the chase.

b. *pl.* (With capital initial.) The name of the Australian international rugby football team.

1908 *Daily Chron.* 28 Sept. 4/6 The 'Wallabies', as the Australian football players..have christened themselves. 1945 Baker *Austral. Lang.* ix. 178 Just as New Zealand football representatives acquired the names *All Blacks, Fernleaves,* and *Kiwis,* so did Australian representatives become known as *Kangaroos, Wallabies* and *Waratahs.* 1975 *Country Life* 4 Dec. 1528/1 The Wallabies are about to play the first international of their tour against Scotland.

2. (Earlier example.)

1865 E. J. Overbury *Wallaby Track* in Stewart & Keesing *Old Bush Songs* (1957) 233 There are others who stick during sheaving, Then shoulder their swags on their back; For the rest of the year they'll be steering On their well-beloved Wallaby Track.

Wallace (wǫ·lăs). The name of Alfred Russel *Wallace* (1823–1913), British naturalist, used *attrib.* or in the possessive to designate concepts originated by him or related to his work, as *Wallace effect,* the evolution of reproductive isolation between sympatric species; *Wallace('s) line,* a hypothetical boundary proposed by Wallace in 1858 as separating the Oriental and the Australasian biogeographical regions.

1966 V. Grant in *Amer. Naturalist* C. 99 It seems fitting and desirable to designate the process of selection for reproductive isolation as the Wallace effect. 1981 *Amer. Jrnl. Bot.* LXVIII. 1247/2 The sympatric origin of isolating mechanisms, the Wallace effect, postulates that those individuals in sympatric populations of the incipient species when crossed with members of the other species waste gametes in so doing, due to sterility or inviability of their progeny. 1868 T. H. Huxley in *Proc. Zool. Soc.* 313 Passing south of India and Indo-Malaisia [*sic*], but north of the Nicobar islands, the boundary in question would coincide with what may be called 'Wallace's line', between the Indian and the Papuan divisions of the Malay archipelago. 1911 *Rep. Brit. Assoc. Adv. Sci.* 435 For fresh-water fishes Wallace's Line is..of fundamental importance. 1957 P. J. Darlington *Zoogeography* vii. 469 A small fraction of the Oriental mammal fauna extends for a considerable distance across Wallace's Line. 1958 A. J. Toynbee *East to West* xv. 44 Nor am I talking here of 'the Wallace line' between the gum trees and marsupial mammals of Australasia and the standard flora and fauna of the rest of the world. 1982 *New Scientist* 3 June 653/1 Examination of the fauna.. on Bali and adjacent islands led him to define the 'Wallace Line'—the division between the Oriental fauna to the west and the Australasian forms to the east and north.

wallah. Add: **1. b.** (Later examples not in Anglo-Indian contexts.)

1940 E. Pound *Cantos* lx. 90 The European church wallahs wonder if this can be reconciled. 1955 *Times* 15 June 3/5 Thousands of the lorries were being worked for 12, 14, 16, and 18 hours a night, with tragic results. The term used on the roads to describe these drivers was 'night-and-day wallahs'. 1977 J. I. M. Stewart *Madonna of Astrolabe* xvii. 248 It's marvellous what these ambulance wallas can do at a pinch. 1982 B. Trapido *Brother of more Famous Jack* xxxvi. 124, I thought briefly of Roger who, being a music wallah, had always made a thing of St. Cecilia's Day.

d. One carrying out a routine administrative job; a civil servant, a bureaucrat. *colloq.*

1965 A. Nicol *Truly Married Woman* 32 There's no end to what you wallahs in the administration would do to show your damned official broadmindedness. 1974 *Courier-Mail* (Brisbane) 7 June 7/3 Some wallahs in Canberra are sitting in air-conditioned offices telling us what has been flooded and what hasn't.

wallah-wallah, walla-walla (wǫ·lă wǫ·lă). [Origin unknown.] In Hong Kong, a small boat used as a ferry for casual traffic.

1957 R. Mason *World of Suzie Wong* i. iii. 33 Ships of every shape and size..tramps, junks, sampans, walla-wallas. 1967 E. Hunt *Danger Game* viii. 152 They rode in a rickshaw, a sampan and a walla walla boat. 1969 J. Bennett *Dragon* i. 5 Back and forth, ply ferry-boats ..and hydrofoils..off to Macao; walla-wallas; scavenger-boats. 1970 D. Dodge *Hatchetman* 8 Sampans, junks, and *wallah-wallahs* caught in the open water were doomed when Tai Fung blew. 1976 W. Marshall *Hatchet Man* i. 2 A wallah-wallah boat transporting a sailor late back from leave worked its way to alongside..a moored destroyer.

wallaroo. (Earlier example.)

1826 J. Atkinson *Agric. & Grazing N.S.W.* ii. 24 There is also found far in the interior another variety [of kangaroo], called wallaroos; they are much larger than any of the others.

wallbanger (wǭ·lbæŋəɹ). *U.S.* Also **W-.** [App. f. Wall *sb.*^{1} + Banger^1.] In full *Harvey Wallbanger.* A cocktail made from vodka or gin and orange juice.

1970 *Gourmet* Nov. 9 (Advt.), Harvey Wallbanger. Fill tall glass with ice cubes. Fill ¾ full with orange juice. Add 1 oz. Vodka. 1972 *New Yorker* 30 Sept. 41/2 A wallbanger is a vodka or a gin or whatever you please with orange juice. 1981 T. Heald *Murder at Moose Jaw* xi. 130 The Mounties..ordered a brace of Harvey Wallbangers. *Ibid.* 131 Smith took a draught of Wallbanger.

walled, *ppl. a.* Add: **2. a.** *walled-in,* enclosed by walls (earlier example).

1777 P. Thicknesse *Year's Journey* II. xlix. 132 *Bonne* is a good town, well walled-in, pleasantly situated.

wallet. Add: **4.** *wallet-carrying* adj.

1929 D. H. Lawrence *Pansies* 76 Men in bowler hats, hurrying And a mingling of wallet-carrying women.

walletful (wǫ·lėtful). Also **wallet-full.** [f. Wallet + -ful.] As much as a wallet (sense 3 a) will hold.

1909 in Webster. 1966 *Guardian* 9 Dec. 8/6 He has a walletful of notes. 1968 L. Durrell *Tunc* ii. 48 'I began to carry around a wallet-full of children... Look.' He tipped out of a wallet a series of grotesque pictures of nude children.

walley, var. *Wally *sb.*^{2} (and *a.*^{3}).

wall-eye. Add: Also **walleye. 3.** For 'the U.S.' substitute 'N. Amer.' and add after 'fishes': esp. the wall-eyed pike, *Stizostedion vitreum.* (Earlier and later examples.)

1876 *Fur, Fin & Feather* Sept. 163/1 All along the Minnesota Division are numerous clear lakes and ponds, teeming with..'wall-eyes' or pike-perch. 1968 [see

*MUSKY sb.[2]]. **1982** *Nature* 16 Sept. 202/2 Walleye fish, *Stizostedion*, eat each other 'tail first', and chains of up to four fish engaged in simultaneous cannibalism have been seen.

wallflower. Add: **1. d.** A perfume derived from the flowers of this plant.

1907 *Yesterday's Shopping* (1969) 521/1 Perfumes... Verbena. Wallflower. White Carnation. *c* **1938** *Fortnum & Mason Price List* 55/2 Wallflower..Wild Rose..Assorted Bath Ovals. **1972** [see *MIGNONETTE 1 d].

wallful (wǭ·lful). [f. WALL sb.[1] + -FUL.] As much as the surface of a wall will hold; the area of an entire side of a wall.

1959 *Times* 3 Jan. 2/6 This very personal little miscellany, with its notable Bacon and wallfuls of Keith Vaughan and Burra. **1973** *Art Internat.* Mar. 52/2 The traces in question consisted of a few wallfuls of photo collages and a couple of films. **1975** 'D. RUTHERFORD' *Mystery Tour* vii. 173 A wallful of luggage lockers.

walling, *vbl. sb.*[2] Add: **3. walling hammer,** a hammer used for dressing stones in a dry wall.

1841 S. BAMFORD *Life of Radical* (ed. 2) I. 84 This person had..threatened to beat in their brains with a walling hammer which he had in his hand. **1881** W. WESTALL *Old Factory* I. x. 136 Them as has walling hammers comes next and next again them wi' stone-breakers' hammers and hand-hammers. **1964** H. HODGES *Artifacts* vii. 109 In dressing, the usual process was to hack away first the worst of the protruding lumps with a walling hammer. **1971** *Country Life* 9 Sept. 618/1 Prizes include.. a walling hammer for the competitor who has travelled the longest distance in order to take part.

wallop, *sb.* **4. a.** Delete 'and *humorous*'. Also *fig.*

1925 T. DREISER *Amer. Trag.* I. II. iii. 171 What a wallop, eh? An' us leavin' him and that girl in the car. **1936** 'J. CURTIS' *Gilt Kid* v. 57 The brandy went down good and packed a real wallop. **1943** E. B. WHITE *Let.* 4 June (1976) 242 K and I got a big wallop out of hearing from you. **1976** *National Observer* (U.S.) 23 Oct. 22/1 Now it assumes a vertical position on the upper half of the page. Better visibility, more wallop. So we think. **1978** S. BRILL *Teamsters* x. 390 Life for Harold Gibbons didn't pack much of a wallop anymore.

c. *colloq.* Alcohol, esp. beer; alcoholic drink.

1933 *Bulletin* (Sydney) 11 Jan. 20/2 This time the wallop was met 40 miles away. **1945** J. B. PRIESTLEY *Three Men in New Suits* viii. 133 It's drink... Booze or wollop... Nine times out of ten..you wake up in the morning..with the usual hangover. **1949** 'G. ORWELL' *Nineteen Eighty-Four* I. viii. 90 When I was a young man, mild beer—wallop we used to call it—was fourpence a pint. **1962** N. MARSH *Hand in Glove* ii. 40 'May Leonard fix mine?.. He knows my kind of wallop.'.. Leonard adroitly mixed two treble Martinis. **1972** L. LAMB *Picture Frame* vi. 56 Mrs Tyler could do nothing to improve the wallop she served at the Hurdlemakers [Inn].

walloper. Add: **4.** *Austral. slang.* A policeman.

1945 BAKER *Austral. Lang.* vii. 137 We also call a policeman a..walloper. **1954** *Coast to Coast 1953–4* 175 A showman he was. Knock-around man. Always two jumps ahead of the wallopers. **1968** D. O'GRADY *Bottle of Sandwiches* 54 Roebourne boasted one pub, one police station with two wallopers in it,..and a hospital. **1981** *Bulletin* (Sydney) 13 Jan. 391, I could quite happily call them my friends. I could never think of them as wallopers.

walloping, *ppl. a.* **2.** (Examples.)

1959 F. ASTAIRE *Steps in Time* (1960) xxviii. 317 *Funny Face* proved a walloping success in the main big-city first runs and particularly at the Music Hall, where it broke several records. **1980** M. RICHLER *Joshua Then & Now* I. ii. 55 Joshua slid behind the bar, which was unattended, and poured himself a walloping cognac.

wallow, *sb.* Add: **1. a.** (Later *fig.* examples.)

1969 J. GROSS *Rise & Fall Man of Lett.* iii. 89 Nor was he at all averse himself, as a reader, to a nice old-fashioned romantic wallow, with wedding-bells ringing out in Chapter the Last. **1975** *Listener* 4 Dec. 746/2 While we are having a wallow, let me suggest that reviewers discussing 'X' films..should be more explicit.

b. Also *transf.*

1963 A. SMITH *Throw out Two Hands* xi. 114 Every so often came a glutinous wallow where some lorry had spent time not only in extricating itself but in deforming the track still further.

c. *fig.* A state of depression or stagnation.

1934 in WEBSTER. **1938** *Times* 19 Jan. 13/5 Cannot some effort be made to rescue from the wallow into which in the provinces it is falling that fine old British institution the Christmas pantomime? **1969** J. GASKELL *Sweet Sweet Summer* 77 And you know how inter-holed and jig-saw-slotted intricate incestuous old showbis is—*one* despairs, they're all down in the black wallow. **1975** *Bookseller* 3 May 2380/2 Lifting the *Sunday Times* novel reviews out of the terrible wallows they have been in for the last few years.

wallow, *v.*[1] **6. d.** (Earlier examples.)

1876 'MARK TWAIN' *Tom Sawyer* x. 97 But if ever I get off this time, I lay I'll just *waller* in Sunday-schools! **1881** —— *Lett. to Publishers* (1967) 136 The Earl's literary excrement charmed me like Fanny Hill. I just wallowed in it.

wall-paper. Add: Also **wallpaper. 1.** (Earlier examples.) Now also made of other materials, such as vinyl. Cf. *wallcovering* s.v. *WALL sb.[1] 22.

1827 DRAKE & MANSFIELD *Cincinnati in 1826* viii. 65 Two Wall Paper Factories, 9 hands. **1847** F. A. KEMBLE *Let.* 12 Dec. in *Rec. Later Life* (1882) III. 300 You know how subject I am even to such an influence as that of a ridiculous wall-paper.

2. *fig.* An unobtrusive background; esp. with reference (chiefly *derog.*) to sound, music, etc. Freq. with preceding *adj.* or in *attrib.* use.

1919 'C. DANE' *Legend* 28 They wearied me... They faded into a mere wall-paper of sound, and I forgot that they were there. **1934** C. LAMBERT *Music Ho!* III. 228 It is a matter of indifference whether jazz is a stimulant..or a piece of mental wallpaper. **1959** A. BAILEY *Making Progress* 157 Cool jazz formed a background, like aural wallpaper. **1966** *Guardian* 4 June 6/4 Of course, we have to have wallpaper talk, in a week's broadcasting, just as some people have to have wallpaper music all the week. **1969** *Listener* 24 Apr. 588/1 The cultural wallpaper of his own childhood was the Australian commercial radio stations. **1977** *Zigzag* June 40/1 The best track, for me, is..'Living in The Band', but most of this is just so much wallpaper. **1980** *Jrnl. R. Soc. Arts* Mar. 223/2 Possibly television is becoming visual wallpaper in much the same way as radio has become audible wallpaper.

Hence **wa·llpaper** *v. trans.* and *intr.*, to cover a wall with wallpaper; also *transf.* and *fig.*; **wa·llpapered** *ppl. a.*

1934 WEBSTER, Wallpaper *v.t.* & *i.* **1956** *Family Handyman Magazine's Painting & Redecorating Bk.* 34/1 If you want to paint formerly wallpapered walls: 1. Use check system just described. **1962** A. NISBETT *Technique Sound Studio* xi. 198 If this technique..does not quite come off it may be possible to wallpaper over the cracks by adding a new overall backing. **1973** *Listener* 29 Nov. 733/1 You could wallpaper St. Paul's with the free brochures. **1975** P. DICKINSON *Lively Dead* xxi. 122 She had planned to wallpaper the back room. **1977** M. FRENCH *Women's Room* (1978) ii. 108 She painted and wallpapered, refinished furniture. **1977** *Listener* 28 Apr. 537/1 Into the Cambridge Union, through the amazingly wallpapered lobby. **1978** R. MILLS *Comprehensive Educ.* 80 We did the paintwork and wallpapered the ceiling. **1979** 'P. O'CONNOR' *Into Strong City* xii. 39 The room is wallpapered with..a section of Soho dressed as artists as distinct from the criminal section who dress in another fashion.

Wallsend. (Earlier example.)

1821 *Times* 23 Jan. 1/4 Coals 46s..warranted Wallsend, 50s. superior.

Wall Street (wǭl strīt). orig. *U.S.* The name of a street in New York City where some of the most important American financial institutions are centred, used: **1.** *absol.* Denoting the American financial world or money-market. Also *transf.*

[**1806** *Balance* (Hudson, N.Y.) V. 228/1 Walking thro' Wall street yesterday morning, I saw a large crowd.] **1841** *Week in Wall St.* p. ix, In the expressive language of Wall-street, he has himself been 'flunked'. **1871** L. M. ALCOTT *Little Men* xiv. 244 The firm [was] broken up... The barn, which was the boys' Wall Street, knew him no more. **1872** B. JERROLD *London* xii. 104 The New York gossip of yesterday, is ours upon our breakfast table. We can almost hear the hum of Wall Street. **1905** G. B. SHAW *Let.* 3 Jan. (1972) II. 497 Finding Capel Court (our Wall St) against his conscience, he became a carpenter. **1949** *Chicago Daily News* 31 May 1/6 Wall Street traced liquidation to the many new uncertainties created by declining production. **1975** *Times* 25 Sept. 8/7 The banking and business quarter—known as Beirut's 'Wall Street'.

2. a. *attrib.*, as *Wall Street broker, method, price*, etc.

1836 *Jamestown* (N.Y.) *Jrnl.* 16 Mar. 1/2 A company—Wall street brokers and speculators—are the applicants for the loan to the New York and Erie Railroad. **1861** in L. C. Baker *Hist. U.S. Secret Service* (1867) v. 100 Such.. is the windy stuff which — uses to draw money out of the Wall Street kings. **1892** A. C. GUNTER *Miss Dividends* 188 All the rest..[had] fallen victims to his imported Wall Street methods. **1935** G. GREENE *England made Me* II. 47 'Put through any long-distance calls.' 'The Wall Street prices?' **1940** W. FAULKNER *Hamlet* III. ii. 240 He figured if we named him Wallstreet Panic it might make him get rich like the folks that run that Wallstreet panic. **1972** R. PERRY *Fall Guy* iv. 78 His immaculate suit would have put many a Wall Street executive to shame. **1981** A. LURIE *Lang. of Costume* iv. 114 In the urban centers of the West..bankers..sometimes adopt an Eastern manner of speech and a Wall Street appearance.

b. Special Comb. **Wall Street crash, crisis,** the collapse of the American stock-market which took place in October 1929.

1929 *N.Y. Times* 26 Oct. 2/8 Commenting on the Wall Street crash of yesterday, the German press unanimously agrees that Germany has no reason to mourn. **1933** R. G. HAWTREY *Trade Depression* iii. 31 The Federal Reserve Banks can hardly be blamed for their policy of credit restriction up to the moment of the Wall Street crisis. **1981** E. LONGFORD *Queen Mother* ii. 38 The slump or depression of the 1930s began with the Wall Street crash of 1929 in America.

Wall Streeter (wǭ·l,strī:təɹ). orig. *U.S.* [f. prec. + -ER[1].] A Wall Street financier, esp. of the New York stock-market.

1885 *Weekly New Mexican Rev.* 15 Jan. 2/2 The Wall streeters and money changers want less money that they may have a better chance to grind the borrowers. **1937** [see *STEM v.[4] 5 b]. **1943** *Sun* (Baltimore) 2 July 17/1 Many Wall Streeters trimmed accounts to be on the safe side. **1967** *Economist* 2 Sept. 788/3 Mr Haack may also have to do something to dampen a speculative attitude which is reminding some Wall Streeters of 1929. **1979** *Listener* 1 Nov. 579/1 Wall Streeters will tell you that, since its [*sc.* the market's] high of 1968, there has been a crash of sorts.

wall-to-wall (stress variable), *a.* (*sb.*, *adv.*) [WALL sb.[1]] **1.** Of carpeting: covering the whole floor of a room; fitted. Also *absol.* as *sb.*

1953 A. UPFIELD *Murder must Wait* ii. 12, I detest wall-to-wall carpets. **1962** A. LURIE *Love & Friendship* vi. 97 Genuine hooked rugs scattered over the wall-to-wall carpet. **1965** G. MCINNES *Road to Gundagai* iii. 39 The room fascinated me because it had two carpets on the floor, a 'wall-to-wall' covered by an Axminster. **1977** C. MCFADDEN *Serial* (1978) xxxvi. 78/2 Harvey hated crawling around on the wall-to-wall looking for contact lenses. **1977** *New Yorker* 15 Aug. 33/1 What impressed me most about that house was the carpeting, which was mercilessly wall-to-wall. **1978** *Meridian Poetry Mag.* Autumn 10 He bestrides the wall-to-wall carpeting Like a colossus.

2. a. Of objects, etc., other than carpeting: extending from one wall to another; providing coverage of an entire space. Also as quasi-*adv.*

1959 *Observer* 12 Apr. 14/5 Underpinned with warmth—a wall-to-wall convection heater. **1977** *Washington Post* 27 Nov. M3/1 Wall-to-wall art—framed prints, posters or one enormous canvas—can cover a whole wall. *Ibid.* 3/2 I designed everything wall-to-wall... The bed,.. being 75 inches long, the distance between two of the walls, is an instant wall-to-wall component.

b. *fig.* Extending from one end or extreme to the other; allowing no unfilled space; ubiquitous.

1967 *New Yorker* 25 Nov. 222 He made a highly successful series of wall-to-wall mood-music recordings. **1973** *Listener* 19 Apr. 522/2 A respite from wall-to-wall Mozart on Radios 3 and 4. **1976** *Patriot-Ledger* (Quincy, Mass.) 10 July 3/1 The state police said it was just wall-to-wall people and wall-to-wall cars. **1977** *Time* 19 Sept. 9/3 When we kick down doors looking for these people at home, we find almost always tons of literature—wall-to-wall Marx and Marcuse. **1982** S. PARETSKY *Indemnity Only* vi. 74 Why would he agree to see me? He'd never heard of me, he has wall-to-wall appointments. **1984** *Listener* 15 Mar. 19/1 Is there any good reason why we should have news bulletins, local and national, every hour on the hour, chat shows..and wall-to-wall discussion programmes? **1984** *New Statesman* 16 Nov. 16/2 Their sponsors include the IBA..and the BBC (in whose Reithian corridors the epithet 'wall-to-wall *Dallas*' was reputedly coined).

wallum (wǫ·lɒm). *Austral.* [Aboriginal.] **a.** A tall evergreen shrub, *Banksia æmula*, common in parts of Australia. **b.** The heathland along the Queensland coast where this is common.

1889 J. H. MAIDEN *Useful Native Plants Austral.* 383 *Banksia marginata, Cav.*,.... 'Honeysuckle'. The 'Wallum' of the aboriginals of Wide Bay (Queensland). **1890** F. M. BAILEY *Catal. Indigenous & Naturalised Plants Queensland* 103/2 Wallum, Banksia æmula. **1965** *Austral. Encycl.* IX. 310/1 In Queensland, the Wallum coastal formation stretching north of Brisbane has a varied heathland type of flora. **1970** W. W. BRYAN in R. M. Moore *Austral. Grasslands* vii. 107 North of Maryborough rainfall in the Wallum is only 40–45 in. **1979** *Sunday Sun* (Brisbane) 24 June 9/6 CSIRO projects on similar tracts of coastal wallum have shown it can be transformed relatively inexpensively by the aerial application of trace elements. **1979** *Sunday Mail Mag.* (Brisbane) 29 July 18/3 The wallum banksia dominates the dune scrub with its magnificent large flowers atop the serrated leaves.

wally (wǫ·li), *sb.*[2] (and *a.*[3]) *slang.* Also **Wally.** [Origin uncertain: perh. the same word as *WOLLY. Said by some to be the dim. of the personal name *Walter*: cf. *CHARLEY, CHARLIE 6. Cf. also WALLYDRAG, WALLYDRAIGLE.] An unfashionable person; one who is foolish, inept, or ineffectual. Also as a mild term of abuse. Also *attrib.* or as *adj.*

1969 *Daily Mirror* 10 Oct. 19/1 *Wally*, out of fashion. **1974** *Times* 8 Aug. 2/4 The successors to the flat-earthers ..are at present encamped on the perimeter of the great concentric stone circles... They choose to be known as the Wallies of Wessex, Wally being a conveniently anonymous umbrella for vulnerable individuals. **1976** *Telegraph* (Brisbane) 8 Oct. 10/4 The Arnolds call anybody who wears conventional clothes, such as jeans or skirts, a Wally. **1979** *Listener* 20 & 27 Dec. 861/1 Cuban heel boots ..are..like a Wally uniform... John Travolta, the Bee-Gees and Boney M are considered Wally acts. **1983** *Evening Star* (Ipswich) 20 June 7/1 He shrugged off Ms. Ford's throw as temperamental but I bet he felt a right wally. **1983** P. INCHBALD *Short Break in Venice* xviii. 177 Keith cracked a joke over the radio and got called a bleeding wally. **1984** *Daily Tel.* 14 June 1/4 'They looked a right load of wallies,' said an eye-witness. **1985** M. STOTT *Before I Go* iv. 77, I shall seem more of a 'wally' to them than ever because I don't know half the leading telly presenters.

walm, *v.* Add: **1. c.** (Later *poet.* example.)

1908 HARDY *Dynasts* III. III. iii. 387 Throats shout 'advance' And forms walm, wallow, and slack suddenly.

walnut.[1] Add: **2. b. black walnut** (earlier examples); **English walnut** (earlier examples); **white walnut** (examples).

1612 R. JOHNSON *New Life Virginea* sig. B3, They cut downe wood for wanscot, blacke walnut tree, Spruce, Cedar & Deale. **1714** J. LAWSON *Hist. Carolina* 99 The Walnut Tree of America is call'd Black Walnut. **1743** J. CLAYTON *Flora Virginica* 190 *Juglans alba*..White Walnuts. **1772** C. CARROLL *Let.* 9 June in *Maryland Hist. Mag.* (1919) XIV. 149 It froze Here last Thursday night ..., it bit the Leaves of the English Walnut tree. **1822** White walnut [see *creek-bottom* s.v. *CREEK sb.[1] 8]. **1876** 'MARK TWAIN' *Tom Sawyer* xvi. 134 Perfectly round white things a trifle smaller than an English walnut. **1912** White walnut [see *OIL-NUT b, c].

4. a. (sense 1) *walnut cake, -oil* (earlier and later examples); (sense 3) *walnut-panelled* adj.

1889 H. A. DE SALIS *Cakes & Confections* 28 Walnut Cake. Rub four ounces of peeled walnuts..with the whites of three eggs, [etc.]. **1936** *New Yorker* 29 Feb. 21/1 His favourite *Linzertorte*, a walnut cake. **1977** F. PARRISH *Fire in Barley* iv. 39 Some ginger biscuits and a slice of walnut cake. **1612** W. STRACHEY *Trav. Virginia* (1953) I. v. 73 A lock of an ell long, which they annoynt often with walnut oyle. **1963** *Times* 9 Feb. 11/3 The vegetables are luscious and the salads delicately dressed with walnut oil. **1984** M. BABSON *Death Swap* xii. 90 A large tin of truffles ..and the inevitable walnut oil. **1934** WEBSTER, Walnut-paneled. **1961** M. BEADLE *These Ruins are Inhabited* (1963) ii. 24 We ducked through a narrow passageway and into the chapel, a walnut-panelled room. **1981** P. NIESEWAND *Word of Gentleman* xxv. 162 Macgregor's office was walnut-panelled.

Walpolian (wǫlpōᵘ·liăn), *a.* and *sb.* Also **-ean.** [f. the name of Robert *Walpole* (1676–1745), politician, and of his son Horace *Walpole* (1717–97), writer and politician + -IAN.] **A.** *adj.* **1.** Of, pertaining to, or characteristic of Horace Walpole or his writings. Also applied to neo-Gothic architecture of a type popularized by Horace Walpole.

1867 H. E. H. JERNINGHAM *Life in French Château* vi. 165 She had seen through the empty-heartedness and true Walpolian sense of gratitude of those she had befriended. **1876** HARDY *Hand of Ethelberta* II. xl. 146 A tendency to talk Walpolean scandal about foreign courts was particularly manifest. **1944** J. LEES-MILNE *Jrnl.* 2 June in *Prophesying Peace* (1977) 71, I was fascinated by the Walpolian Gothic chapel. **1973** *Country Life* 14 June 1726/1 The greatest Walpolian scholar, W. S. Lewis, remarked in 1934 that Strawberry Hill was 'assembled rather than built'.

2. Of, pertaining to, or characteristic of Robert Walpole or his political career.

1909 J. M. RIGG in *Cambr. Mod. Hist.* VI. xiii. 424 He adhered to the Walpolean tradition of an *entente cordiale* with France. **1979** *N. & Q.* Feb. 76/1 His appeal..to 'custom and reason', 'the practice of former times', and 'the original principles of government' could be used equally against American radicals and Walpolian oligarchs. **1983** *Ibid.* Feb. 23/1 More substance still is given to his role as a Walpolean hack.

B. *sb.* An admirer of Horace Walpole.

1927 *Observer* 24 Apr. 6/2 No Walpolian can be satisfied with anything less than the complete Toynbee edition of the Letters.

So **Walpolia·na, -ea·na** (see ANA *suff.*).

1799 (*title*) Walpoliana. **1875** *Encycl. Brit.* I. 785/2 Of the examples England has produced of this species of composition, perhaps the most interesting is the *Walpoliana*, a transcript of the literary conversation of Horace Walpole, Earl of Orford. **1938** *Times Lit. Suppl.* 26 Feb. 136/4 If General Conway's house, grounds, and..views.. are over-praised, they are only the more authentic as Walpoliana. **1959** *Times* 16 May 8/5 There will be many on this side of the Atlantic who will be deeply grieved by the news of the death of Annie Burr, wife of Wilmarth Sheldon ('Lefty') Lewis, the eminent collector and editor of Walpoliana. **1980** *Jrnl. R. Soc. Arts* Feb. 161/2 Dr. Lewis's collector's flair and comfortable means enabled him to build up an unrivalled assemblage of Walpoleana —original Walpole letters and manuscripts, and books from Walpole's own library.

Walpurgis (vælpǒ·ɹgis). The name (St. Walpurga or Walburga) of an 8th-c. Anglo-Saxon saint and missionary in Heidenheim, Germany, used: **a.** in **Walpurgis night** [tr. G. *Walpurgisnacht*], in German folklore and esp. Goethe's *Faust* a feast of the powers of darkness or witches' sabbath (cf. SABBATH 3) celebrated on the Brocken, a peak in the Harz mountains, on 30 April. Also *transf.*, an orgiastic celebration or party.

1823 F. L. GOWER tr. *Goethe's Faust* 233 Walpurgis Night. The Hartz Mountain. Wild and desolate country. **1900** J. FRAZER *Golden Bough* (ed. 2) I. i. 36 The special season for thus promoting the growth of flax..in some places is Candlemas or Walpurgis Night (the eve of May Day). **1933** J. CARY *Amer. Visitor* xvi. 213 A fire which illuminated a kind of Walpurgis night of ragamuffins, soldiers, doxies from the town. **1964** L. DEIGHTON *Funeral in Berlin* xlv. 286, I sank into a vortex of imaginings in which Walpurgis Night and..the smell of petrol.. were indissolubly linked. **1980** *Amer. Scholar* Autumn 456 Walpurgis Night on the eve of Mayday is an excuse for bacchanalia.

b. In other *attrib.* uses with reference to this festival.

1857 C. KINGSLEY *Two Years Ago* III. i. 33 That Walpurgis-dance of the witches and the fiends, which will ..whirl unbidden through a mortal brain. **1899** R. WHITEING *No. 5 John St.* xxix. 300 The Bosses..have come to keep Walpurgis revel. **1979** J. TATE tr. *K. A. Blom's Limits of Pain* xx. 164 The last day of April.. he could mix in the crowds,..warming himself by the Walpurgis bonfires.

‖ **Walpurgisnacht** (valpu·ɹgisnaχt). [Ger.] = prec. a. Also *transf.*

1822 SHELLEY *Let.* 10 Apr. (1964) II. 407, I have only attempted the scenes omitted in this translation, & would send you that of the *Walpurgisnacht*. **1898** W. JAMES *Let.* 9 July (1920) II. 76 The influences of Nature, [etc.]..all fermented within me till it became a regular Walpurgis Nacht. **1909** —— *Pluralistic Universe* i. 21 Nature can have little unity for savages. It is a Walpurgis-nacht procession, a checkered play of light and shadow, a medley of impish and elfish friendly and inimical powers. **1947** *Partisan Rev.* XIV. 102/2 The *Walpurgisnacht* which is Hollywood. **1965** *Guardian* 10 Dec. 6/5, I might still avoid the woods on Walpurgisnacht. **1977** *Time* 3 Jan. 65/2 *Who's Afraid of Virginia Woolf?* An admirable revival..verifies that after 14 years this marital Walpurgisnacht has become part of the permanent canon of U.S. drama.

walpurgite (væ·lpǒɹgəit). *Min.* [ad. G. *walpurgin* (A. Weisbach 1871, in *Neues Jahrb. f. Mineral.*, etc. 870), f. *Walpurgis*, name of the vein in a mine near Schneeberg in Germany where the first specimen was found: see -ITE[1].] A hydrous arsenate of bismuth and uranium found as yellow, translucent or transparent triclinic crystals.

1872 G. J. BRUSH in J. D. Dana *Syst. Min.* (ed. 5) App. I. 16 Walpurgite. **1958** *Bull. U.S. Geol. Survey* No. 1064. 242 Walpurgite is a secondary mineral, found in the oxidized zone of a vein carrying uraninite, cobalt and nickel arsenides, and native bismuth from which the U, Bi, and As content of the mineral has been derived.

Walras (va·lra). The name of Marie Esprit Léon *Walras* (1834–1910), French economist, used in the possessive in *Walras' law* to denote the mathematical theory of general economic equilibrium devised by him. So **Walrasian** (valrē̆ⁱ·siăn) *a.*, of or pertaining to this theory.

1942 O. LANGE in O. Lange et al. *Stud. Math. Econ.* 50, I propose to call this identity *Walras' law* because Walras was the first to recognize its fundamental importance in the formulation of the mathematical theory of prices. **1942** W. JAFFÉ in *Ibid.* 37 The affinities..between these older Walrasian theories and their independently discovered Keynesian counterparts. **1963** *Canad. Jrnl. Economics & Pol. Sci.* XXIX. 535 (*heading*) A note on Marshallian *versus* Walrasian stability conditions. **1964** *Q. Jrnl. Econ.* LXXVIII. 485 Walras' Law..states that the value of all goods demanded, including money, is equal to the value of all goods supplied. **1972** T. SOWELL *Say's Law* i. 34 *Walras' Law* states that the sum of the respective values (quantities times money prices) of goods supplied plus money supplied equals the sum of the respective values (quantities times money prices) of goods demanded plus money demanded. **1972** *Times* 26 Oct. 8/6 Sir John [Hicks]..moved away from the partial equilibrium approach of Alfred Marshall back towards the older Continental Walrasian 'general equilibrium' approach. **1983** *Economics Lett.* XIII. 49 The possibility of transfer paradoxes in Walrasian stable markets.

walrus. Add: **3. walrus moustache,** a large moustache which overhangs the lips (thus resembling the whiskers of a walrus); similarly **walrus whiskers.**

1918 W. OWEN 31 Oct. (1967) 591 An old soldier with a walrus moustache. **1982** T. FITZGIBBON *With Love* I. xii. 88, I remember Conan Doyle as a large man with sad thoughtful eyes and a walrus moustache. **1930** J. DOS PASSOS *42nd Parallel* I. 100 A big man with walrus whiskers.

Walschaerts (væ·l-, wæ·lʃɑ̄ɹts, -ʃāɹts). *Mech.* Also **Walschaert.** The name of Egide *Walschaerts* (1820–1901), Belgian mechanic, used *attrib.* and in the possessive to designate a kind of valve-gear he invented that was used on some steam locomotives (see quot. 1911).

1880 *Proc. Inst. Mech. Engineers* 435 His arrangement ..was simply a modification of the Walschaert gear. **1911** *Encycl. Brit.* XXV. 834/1 A form of radial gear very largely used in locomotives, especially on the continent of Europe, is the Walschaert or Heusinger-Waldegg gear, in which the valve receives its motion in part from the piston cross-head.., and in part from a single eccentric set at right angles to the crank. **1975** B. REED *150 Yrs. Brit. Steam Locomotives* vi. 47/1 Other Group three-cylinder classes..had three separate sets of Walschaerts motion with eccentric drive for the inside set.

Walsingham (wǭ·lsiŋăm). The name of a town in Norfolk, used *attrib.* in *Walsingham Way*, a designation of the Milky Way supposed formerly to have been used as a guide by pilgrims travelling to the shrine of Our Lady of Walsingham.

c **1878** G. M. HOPKINS *Loss of Eurydice* in *Poems* (1967) 75 A starlight-wender of ours would say The marvellous Milk was Walsingham Way. **1922** JOYCE *Ulysses* 685 The waggoner's star: Walsingham way: the chariot of David: the annular cinctures of Saturn. **1973** M. J. PETRY *Herne the Hunter* vi. 71 In certain parts of England, and especially in East Anglia, the Milky Way was also known as the *Walsingham Way*, on account of the hosts of *pilgrims* who passed along the roads..to the shrine of Our Lady at Walsingham in Norfolk.

Walt Disney (wǭlt di·zni). The name of *Walt*er Elias *Disney* (1901–66), American pioneer of cartoon films, used *attrib.* to denote the style of his films or their characters. Cf. *DISNEYESQUE *a.*

1946 R. LEHMANN *Gipsy's Baby* 12 He was a total failure with rabbits, and if he blundered on one in the course of one of his Walt Disney gallops over the fields, he winced. **1962** A. LURIE *Love & Friendship* vi. 107 Bright, cheap, sentimental in a Walt Disney style, it [*sc.* a Christmas card] depicted a group of cherubs carrying packages. **1966** 'K. A. SADDLER' *Gilt Edge* x. 144 The ancient home of the Meldmays..glistened in the moonlight like a Walt Disney castle. **1974** C. HAMPTON *Savages* (1976) ii. 26 Three young men enter. They are all wearing rubber Walt Disney masks.

Walter Mitty: see *MITTY[2].

Walther (wǫ·ltəɹ, ‖va·-). The name of a German firm of firearm manufacturers, used *attrib.* and *absol.* to designate pistols and rifles made by them.

1920 H. B. C. POLLARD *Automatic Pistols* v. 49 The *Walther Selbstlade* pistol is a pocket automatic of 6·355 mm., six shots, German manufacture. **1934** G. BURRARD *Identification of Firearms & Forensic Ballistics* ix. 203, I was informed that the defendant was using a Walther rifle at the time of the accident. **1965** I. FLEMING *Man with Golden Gun* v. 76 The Walther PPK inside the waistband of his trousers. **1968** A. DIMENT *Great Spy Race* vii. 100 He..took a gun, a baby Walther, from its hiding place. **1971** *Guardian* 28 Sept. 24/1 As a point of historical interest, this is the first time that the Walther pistol, a shoulder-holstered gun introduced into the Royal Ulster Constabulary last year..has been fired in action. **1975** [see *LUGER]. **1981** J. B. HILTON *Playground of Death* i. 7 Bielby had been discovered..with a warm-barrelled pistol in his hand..a Walther 7.65 wartime German police model.

Waltonian, *a.* (Earlier example.)

a **1828** T. BEWICK *My Life* (1981) xvii. 165 Some regulations should be laid own as a guide to the fair anglers... I think Waltonian societies would be all-sufficient to do this.

waltz, *sb.* Add: **4.** Something accomplished with ease. *slang.*

1968 *Time* 5 July 38 Though Dancer eased him up at the end, Nevele Pride won in a waltz. **1970** G. F. NEWMAN *Sir, You Bastard* vi. 179 The qualifying examinations conducted through the Civil Service Commission were a waltz.

5. *attrib.* and *Comb.*, as **waltz king** [G. *Walzerkönig*], an epithet applied to the Viennese composer Johann Strauss (1825–99), famous for his waltzes; **waltz-length** *a.*, (of a garment) calf-length.

1908 *Busy Man's Mag.* Feb. 51/2 Johann Strauss, 'the waltz king'. **1938** *Oxf. Compan. Mus.* 1013/1 It seems as likely that such a waltz as the *Blue Danube*, by Johann Strauss the younger, the 'Waltz King', will last for ever as that Beethoven's Fifth Symphony will do so. **1958** *Vogue* Nov. 40 Nightgowns with lace, or ribbons and bows, in waltz length or swirling to the ankles. **1975** *New Yorker* 17 Nov. 134/2 A waltz-length, front-fastened Edwardian robe of snowy cotton, with textured scrolls of guipure on its yokes, is $75.

waltz, *v.* **b.** For 'Chiefly *slang*' read '*colloq.*' and add: Also, to move unconcernedly or boldly, as to waltz *into, off, up* (*to*), etc. (Further examples.)

1887 in *Amer. Speech* (1950) XXV. 39/2 Out on the Mississippi Valley road when his girl waltzed up. **1935** G. HEYER *Death in Stocks* v. 63 Tony seemed to have waltzed off for the night, so I wandered out on my own. **1951** J. FLEMING *Man who looked Back* i. 10 Here she was waltzing off with this nurse and leaving Ray..alone. **1974** F. NOLAN *Oshawa Project* ii. 12 He'll walk any nomination..and waltz into the White House without even having to put up a fight. **1978** N. MARSH *Grave Mistake* viii. 240 P'raps..he'll come waltzing back with a silly grin on his face having been to stay with his auntie. **1979** C. MACLEOD *Luck runs Out* v. 50 If any student or students had..got mixed up in a mess like this, they'd hardly come waltzing up to the President in open assembly and say so. **1980** 'R. B. DOMINIC' *Attending Physician* xxiv. 218 He waltzed off to be the first one in the lounge. **1984** B. FRANCIS *AA Car Duffer's Guide* 6/1 That same afternoon..the vicar's lady waltzes in.

d. *trans.* To transport or convey (something). *U.S. joc.*

1884 'MARK TWAIN' *Huck. Finn* iii. 22 They've got to waltz that palace around over the country wherever you want it. **1901** MERWIN & WEBSTER *Calumet 'K'* xi. 197 He'd call the men off just the same, and leave us to waltz the timbers around all by ourselves.

e. *to waltz Matilda*: see *MATILDA.

waltzer. Add: **c.** A fairground ride (see quots. 1961[1], 1968).

1961 F. C. ROOPE *Come to Fair* iii. 71 *The Waltzer*, an appropriate name for an ingenious machine. The motion of the cars..is both round and round the undulating track, and also a spinning one on their own axis. *Ibid.*, The Waltzer is completely safe for everyone except the most foolhardy riders. **1965** [see *NOAH'S ARK 7]. **1968** D. BRAITHWAITE *Fairground Archit.* 176 *Waltzer*, a member of the switchback family having freely-pivoted tub-shaped cars mounted on the undulating platform. **1975** *Sunday People* 6 July 11/5 As he lurches about on the waltzer or is shoved away in a dodgem car, it's easy to give him a folded bottom note in place of two. **1980** R. HILL *Killing Kindness* vi. 50 He could be on the dodgems, or the waltzer. He helps around when they're busy.

Wampanoag (wāmpănōᵘ·āg), *sb.* and *a.* Also † **Wampanong, -noug.** [Narragansett lit. 'easterners'.] **A.** *sb.* (A member of) an Indian people of south-eastern Massachusetts and the eastern shore of Narragansett Bay. **B.** *adj.* Of, pertaining to, or designating this people.

1676 R. WILLIAMS *Let.* 1 Apr. in B. F. Swan *Copy of Let. of R. Williams* (1971) 11 God has prospered us so that wee had had driven the Wampanoogs with Phillip out of his Countrie. **1782** 'J. H. ST. JOHN DE CRÈVECŒUR' *Lett. from Amer. Farmer* iv. 144 Those powerful tribes..the.. Massachusets, Wamponougs, Nipnets, Tarranteens, &c. **1820** EASTBURN & SANDS *Yamoyden* (1834) I. 16 The Wampanoag from the height Of Haup, who strained his anxious sight [etc.]. **1871** C. M. YONGE *Pioneers & Founders* i. 25 The Sachem nearest to Plymouth had been at the first arrival of the Pilgrim Fathers, Massasoiet, chief of the Wampanongs. **1910** F. W. HODGE *Handbk. Amer. Indians* II. 903/2 Wampanoag... One of the principal tribes of New England. Their proper territory appears to have been the peninsula on the E. shore of Narraganset bay now included in Bristol co., R.I., and the adjacent parts in Bristol co., Mass. **1948** *Sat. Even. Post* 26 June 23/3 The township of Dartmouth had been purchased from Chief Massasoit, of the Wampanoag tribe. **1979** *United States 1980/81* (Penguin Travel Guides) 642 There's also the Wampanoag Indian Museum. **1980** *Smithsonian* Aug. 22 Red Wing's museum has some nice relics from early days when the Wampanoags & Narragansets..& Mohegans & Nipmucs ranged around southern New England. **1983** *Times* 3 Oct. 3/4 (*caption*) Slow Turtle, of the Wampanoag Nation, Massachusetts, at a ceremony on Hounslow Heath yesterday.

wan, *a.* Add: **6.** *wanwood*, faded or decaying woodland (*nonce-wd.*).

c **1880** [see *leafmeal* s.v. *LEAF sb.[1] 17].

wan, *v.* Add: **3.** *trans.* To make pale.

1889 in *Cent. Dict.* **1903** HARDY *Dynasts* I. I. iii. 16 The grey glooms of a ghost-eyed despondency Wanned as with winter the national mind. **1938** W. DE LA MARE *Memory* 96 Miracle..That starry flake Can of its myriads Such wide pastures make, For sun to colour, And for moon to wan.

wan (wǫn, wan), *numeral a.*, *pron.*, etc. [See ONE *numeral a.*, *pron.*, etc.] Repr. dial. pronunc. of *one.*

1651 [see ONE *numeral a.*, *pron.*, etc. A. 7]. **1802** in G. Fraser *Lowland Lore* (1880) 70, I ame to Give to him .. wan half of Corn. **1880** L. PARR *Adam & Eve* II. ix. 191 'Twas past wan. **1907** G. B. SHAW *John Bull's Other Island* IV. 81 Larry cleared six yards backwards at wan jump. **1920** F. P. DUNNE *Mr. Dooley on making a Will* 4 'Twud grieve me if some wan broke into song at th' news. *a* **1966** 'M. NA GOPALEEN' *Best of Myles* (1968) 47 If wan is up, all has to be up. **1973** *Black World* Sept. 64 Di ripes juices fruit is di wan Dat stan longer in di sun. **1973** 'J. PATRICK' *Glasgow Gang Observed* vii. 70 He's the wan they're really efter. **1979** *Bull. Yorks. Dial. Soc.* Summer 14 Wan mornin t schoolmaister wor ill.

‖ **wananchi** (wanæ·ntʃi), *sb. pl.* [Swahili, pl. of *mwananchi* inhabitant, citizen.] The indigenous workers in Kenya and Tanzania; the labouring masses.

1969 *Busara* (Nairobi) II. II. 55 The problem was how to unite and propel a nation *into a prosperous future and not into the African past.* The distinction between these two threads of thought has yet to be clarified to millions of expectant *Wananchi.* **1970** *E. Afr. Standard* 2 Jan. 6/7 Musical concerts..for the entertainment of the *wananchi.* **1974** *Sunday Tel.* 8 Dec. 5/6 For it was only a few weeks ago..that President Nyerere was going about giving the *wananchi* (labouring masses) the good word about how Socialist Tanzania was doing very nicely, than you. **1980** *Standard* (Nairobi) 2 June 7/2 The Machakos District Commissioner..called on *wananchi* and leaders in the district to maintain peace.

‖ **wanax** (wæ·næks). Also **Wanax.** [ad. Gr. ϝάναξ, early and dial. form of ἄναξ lord.] A Mycenæan or Minoan king or ruler.

ϝάναξ is the word for 'king' or 'ruler' in the Linear B tablets; in the Homeric poems ἄναξ and βασιλεύς are both found.

[**1955** L. R. PALMER in *Trans. Philol. Soc. 1954* 37 *ϝάναξ* is used in Greek of the human lord..and of the divine protectors... The Tacitean passage inclines me to the view that the *wa na ka* is a priest-king (*rex*), who stands at the apex of the social hierarchy above the military leader (*dux*).] **1956** VENTRIS & CHADWICK *Documents in Mycenaean Greek* v. 120 A monarchical system of government is proved for both Knossos and Pylos by references to the king (*wanax*); the absence of any further qualification shows that the state knew one king only. **1958** *Listener* 11 Dec. 1004/1 The..more varied documents..have perhaps tempted Professor Webster to assume too easily that the Mycenean *Wanax* was likewise a divine King. **1960** S. DOW *Greeks in Bronze Age* 25 When the destruction of the Twelfth Century took place, the chief of state, the *Wanax*, was evidently completely abolished. **1961** C. G. STARR *Origins Greek Civilization* (1962) ii. 48 The class struggle rose through serfs or slaves..to the retainers and agents of the great king, the *wanax.* **1978** *Antiquaries Jrnl.* LVIII. 22 At the summit [of an Achaean state] was the wanax, the head of the state, chief administrator, high judge, leader of the armed forces, and absolute controller of all economic activity.

wance(t): see *ONCE *adv.* A. δ.

wand, *sb.* Add: **12*.** The straight rigid pipe linking the cleaning head to the hose of a vacuum cleaner.

1940 E. HEMINGWAY *For whom Bell Tolls* xi. 149 The round opening at the end of the wand of a vacuum cleaner. **1967** *Boston Sunday Herald* 9 Apr. (Advt.), Attachments include braided hose, two wands and rug-bare floor tool. **1978** *Choice* (Austral. Consumers' Assoc.) Nov. 378 Most vacuum cleaners these days have moulded plastic bodies. Hoses are plastic or cloth covered, and some are of the 'stretch' variety. Wands and cleaning heads tend to be either plastic, steel or aluminium or a combination of these.

12.** A hand-held electronic device which can be passed over a bar code to read the data it represents and convert them into a computer-compatible form.

1978 *Bookseller* 17 June 3196/1 The light pen, or 'wand', that could read machine-readable codes on books. **1980** *Sci. Amer.* Apr. 111/1 (Advt.), As part of a portable data entry system, the wand can be used to read shelf tags for inventory control and order entry. **1982** *What's New in Computing* Nov. 62/1 Intermec designs and manufactures ..scanning wands for the printing and computer reading of tickets, tags and labels.

wand, *v.* Restrict *Sc.* and *dial.* to senses in Dict. and add: **3.** *trans.* To scan the bar code on (an article) using a wand (*WAND *sb.* 12**).

1979 *Bookseller* 22 & 29 Dec. 2693/2 Books could be 'wanded' at the point of sale and the information stored on cassettes. **1982** *Fortune* 27 Dec. 100/1 Go to Section X, Jack, and wand all the woofers.

Wandale[2]. Add: Also **Wandal.** (Later example.)

1908 W. G. COLLINGWOOD *Scandinavian Brit.* ii. 16 The tribal confederacies of the Baltic shores—Danes, Swedes, Wandals, Burgunds, Bards, Goths.

wander, *sb.* Add: **2.** A gradual change in the orientation of a gyroscope or other spinning body, esp. the earth.

1930 [see *WANDER *v.* 2 g]. **1963** C. A. WILLIAMS *Aircraft Instrument Control Syst.* ii. 23 In moving a gyroscope from the North Pole to the South Pole the apparent wander would change from clockwise to anticlockwise. **1971** *Q. Jrnl. R. Astron. Soc.* XII. 61 (*heading*) Polar wander and/or plate tectonics in the Palaeozoic. **1980** *Nature* 28 Feb. 845/1 The ~1,100 Myr Grenville mobile belt of the Laurentian (North American) Shield yields a record of uplift magnetisations defining a closed apparent polar wander (APW) loop.

wander, *v.* Add: **I. 2. g.** Of a gyroscope or other spinning body: to undergo a gradual change in orientation.

1930 *Engineering* 7 Mar. 323/2 The static gyro-compass did not wander during a week's trials more than about 1 deg..per day. The wandering was always in the same direction,..but the rate of wander varied slightly with alterations of the course. **1958** *Listener* 13 Nov. 779/2 The gyroscope will wander at the slightest hint of imperfection in its manufacture or assembly.

II. 5. Delete 'Now only *poet.*' (Later examples.)

1970 *Globe & Mail* (Toronto) 26 Sept. 29/5 (Advt.), When you are wandering the attractive shops..you can pause for luncheon. **1976** *Times Higher Educ. Suppl.* 12 Nov. 9/2 Born into a London Jewish family in October, 1936, he remembers the excitement of wandering the bomb sites of postwar London. **1981** E. WARD *Baltic Emerald* x. 71 First you wander this place for me, find Suite A.

7. *Comb.*, as **wander-bird** = *WANDERVOGEL; **wander-plug,** a plug which can be fitted into any of a number of sockets in a dry battery; **wander-spirit** = *WANDERLUST; **wander-witted** = *WANDERY (cf. *wandrynge-wytted* s.v. WANDERING *ppl. a.* 2 b).

1924 A. HUXLEY *Little Mexican* 184 Parties of rucksacked Wander-Birds. **1926** —— *Essays New & Old* 157 Of northern Germany it is enough to say that it is the home of the wander-birds. **1923** *Daily Mail* 5 June 13/5 A few high-tension batteries (50 volts with wander-plugs). **1968** Wander plug [see *spade terminal* s.v. *SPADE sb.[1] 5]. **1927** *Observer* 19 June 22/4 There is a drive..which it is the first duty of every motor owner with the wander-spirit properly developed to explore. **1959** *Listener* 29 Jan. 225/2 A wander-witted granddad, a sad bore to his family. **1959** P. FLEMING *Siege at Peking* xiv. 220 The sights he had seen had turned his hair and beard prematurely white and made him wander-witted.

‖ **Wanderjahr** (va·ndəɹyāɹ). Pl. **-jahre.** [Ger.: see WANDER-YEAR.] = WANDER-YEAR.

1893 YONGE & COLERIDGE *Strolling Players* xiv. 117, I had my 'Wanderjahr' once; but I think I want one about every seven years—a metaphorical spree, anyhow. *a* **1910** 'SAKI' *Coll. Short Stories* (1930) 105 The mouse..seemed to be trying to crowd a Wanderjahr into a few strenuous minutes. **1911** J. WARD *Realm of Ends* xviii. 402 Memories that are revived after death or when all the soul's *Wanderjahre* are over. **1922** 'K. MANSFIELD' *Let.* Nov. (1977) 282, I should like very much to know what he intends to do—how he intends to live now his *Wanderjahre* are over. **1952** G. SARTON *Hist. Sci.* I. xvi. 397 When Plato completed his *Wanderjahre*, he felt in him the vocation of teaching. **1982** *Listener* 7 Jan. 9/3 Wilde..in his American *Wanderjahr.*

wanderlust (wǫ·ndəɹlɒst, ‖ va·ndəɹlust). Also **Wanderlust.** [Ger.] An eager desire or fondness for wandering or travelling.

1902 *Athenæum* 4 Jan. 15 If the present reviewer knows anything of the wanderlust and the wallaby [etc.]. **1902** D. G. BRINTON *Basis Social Relations* iv. 113 The goading restlessness which has driven single tribes or groups of tribes into aimless roving. This *Wanderlust* arises as an emotional epidemic. **1928** *Daily Mail* 7 Aug. 12/5 For people who have the wanderlust there are some nice tours around these parts. **1948** *Rep. Native Laws Commission 1946–48* (Dept. Native Affairs, S.Afr.) 43/1 If.. conditions are such that, when their little interval of *wanderlust* is over, they can pass into settled family life, [etc.]. **1969** G. GREENE *Trav. with my Aunt* I. vii. 71 He would get a little restless with the wanderlust. **1981** J. B. HILTON *Surrender Value* xv. 113 The literally thousands of children to whom he had offered their first experience of wanderlust.

Hence **wa·nderluster**; **wa·nderlusting** *a.*

1927 *Sunday Express* 24 Apr. 3/4 The young wanderluster next turned up in Samoa, and went to Bombay to live in a harem. **1936** MENCKEN *Amer. Lang.* (ed. 4) 219 *Wanderlust*..is used much more frequently in the United States along with its derivatives, wanderluster (*Eng. rambler*), *wanderlusting* and *wanderlust-club.* **1952** S. SPENDER *Learning Laughter* 31 The wander-lusting young Israelis. **1985** A. MCCANDLESS *Burke Foundation* ix. 66 Senior NCOs felt a compulsive urge to track down wanderlusting junior officers.

Wanderobo (wǫndərōᵘ·bo). Also **Wandorobo, Wandrobo,** etc. [Native name.] The name of a nomadic hunting tribe of Kenya. Also *attrib.* or as *adj.*

1902 *Encycl. Brit.* XXVII. 611/2 Primitive hunting tribes are the Wandorobo in Masailand and scattered tribes of small stature in various parts. **1935** E. HEMINGWAY *Green Hills Afr.* I. i. 2 We were sitting in the blind that Wanderobo hunters had built. **1937** [see *NANDI *a.* b]. **1967** M. J. COE *Ecol. Alpine Zone Mt. Kenya* 3 The expedition met men of the wandering nomadic Wanderobo tribe at an altitude of 12,000 ft. **1977** H. INNES *Big Footprints* II. ii. 158 He was attacked by the men of the rain forest, the Wandrobo.

‖ **Wandervogel** (va·ndəɹfog'l). Pl. **-vögel** (¶ **-vogel -vogeln**). [G., lit. 'bird of passage'.] A member of the German youth organization founded by H. Hoffmann at the end of the 19th century for the promotion of out-of-door activities, esp. hiking, and folk culture, as a reaction against the materialistic values of middle-class city life. Also *transf.*, a rambler or hiker. Occas. as *v. intr.* So **wa:ndervo·geling** *vbl. sb.*

1924 D. H. LAWRENCE *Let.* 9 Aug. (1932) 606 If it's going to be Youth, then let it be Youth on the warpath, not wandervogeling and piping imitation nature tunes to the taste of milk and chocolate. **1928** —— *Lady Chatterley* i. 3 They sang the wandervogel songs. **1935** J. BUCHAN *House of Four Winds* ii. 49 They were remarkable young men..not in the least like the Wandervögel whom he had met on many German roads, comfortable sunburnt folk out for a holiday. **1944** M. SAMUEL *Harvest in Desert* xviii. 185 In Germany the *Wandervogel*, youthful rebels against the stodginess of middle-class life. **1958** *Listener* 31 July 149/2 The *wandervogel* student of 1912 with his guitar. **1962** P. PURSER *Peregrinnation 22* xlviii. 223 A couple of the young mid-century *wandervogeln* you find in any youth hostel in Europe. **1967** *Listener* 30 Nov. 705/3 Around 1930, alienated and disaffected youth was being manufactured mainly in Germany, where the First War had produced the biggest earthquake. Some of them called themselves the *Wandervögel*, and wandered around Europe with their guitars and their interchangeable girlfriends, living on what they could get wherever sympathisers would accept them. **1978** J. I. M. STEWART *Full Term* xxi. 241 A bunch of juvenile *Wandervögel.*

wandery (wǫ·ndəri), *a.* [f. WANDER *v.* + -Y[1].] Wandering in thought or speech; vague, distant, 'scatty'.

1912 R. BROOKE *Let.* Feb. (1968) 357, I was being frightfully dim and wandery and soliloquitory. **1953** W. DE LA MARE *Private View* p. xi, This vanished creature..was not only more assured, less wandery, more securely based, and more confident. **1976** P. DICKINSON *King & Joker* x. 134 She seemed so tired and wandery.

wander-year. (Earlier example.)

1880 T. W. ALLIES *Life's Decision* i. 6 At Oxford and in my wander-years it was the one subject for which..I not only had no taste, but the most marked repugnance.

wang, var. *WHANG.

wanga (wæ·ŋgă). [ad. Haitian Creole *ouanga* witchcraft, perh. ad. Kimbundu *wanga* witchcraft, or Tshiluba *bwanga* charm, fetish.] Witchcraft, sorcery; a charm or spell; a magical object. Freq. *attrib.*

1851 *Picayune* (New Orleans) 20 July 2/6 The Voudous also threw 'Wanga' or spells into the complainant's yard. **1882** J. W. BUEL *Metropolitan Life Unveiled* 532, I will make a wanga-charm to charm him with; I will make him a phantom, a ghost. **1934** B. A. BOTKIN in W. T. Couch *Culture in South* 585 Fragments of hoodoo and conjuration, whose spells,..grigris, wangas, luck balls, conjure bottles, and conjure, tricken, or goofer bags are the special province of the Negro 'root doctor' or 'hoodoo man'. **1946** R. TALLANT *Voodoo in New Orleans* (1947) 91 Another sort held gunpowder and red pepper; these were *wangas* to be thrown into somebody's path to cause them to get into fights. **1964** 'R. SEVERN' *Blood & Gold* x. 102 It's a *Wanga* dance... Black magic... She must be a *Mambo*.. a *Voodoo* priestess. **1978** W. HJORTSBERG *Falling Angel* (1979) xxxiii. 158 Sounds like some boko's put a powerful wanga on you... A boko is a hungan [*sc.* Obeah priest] who is evil... Wanga's what you'd call an evil curse... a hex, a spell.

wangle, *v.*[2] Add: **1.** Also *refl.* and const. advb. phr.

1922 *Sat. Rev.* 22 Dec. 953 President Wilson had to 'wangle' through Congress an act to restore the British treaty rights. **1942** 'WYNDHAM LEWIS' *Let.* ?26 June (1963) 324 In the last war like yourself I joined the army, instead of wangling myself into some safe job in London. **1961** C. V. WEDGWOOD *Thomas Wentworth, 1st Earl of Strafford* 186 In fact, he would bribe and wangle himself back into the possession of his illicit gains.

2. *intr.* To obtain something or get somewhere by irregular means, scheming, etc.; to use irregular means to accomplish a purpose. *rare.*

1918 *Sat. Even. Post* 19 Oct. 93/1 And wet—always, always wet, unless the weather is clear and the sea is calm—but good sea boats. They [*sc.* submarine chasers] wangle through somehow. They weather it out. **1919** W. DEEPING *Second Youth* xxii. 190 He agitated for a transfer..and to his joy he wangled back to the Cockneys. *Ibid.* xxvi. 222 He also knew that a man who has 'wangled' ..may find himself..thrust rudely into the trenches. **1921** *Glasgow Herald* 16 June 9 We wangled in, an' we'll wangle oot.

3. *trans.* To influence or induce (a person) to do something. *rare.*

1926 G. D. H. & M. COLE *Blatchington Tangle* xxi. 147 He was always on at me just lately to get Lady Blatchington to sell him the rubies. He said I could wangle her. **1928** *Daily Express* 27 Dec. 5/4 Aren't you glad..you wangled the old lady to relinquish the key.

wangler (earlier examples).

1912 E. WALLACE *Private Selby* xxxi. 281 You're a bloomin' wangler, Short. **1915** *John Bull* 27 Mar. 16/2 We regret to see them reduced to the level of vulgar weight wanglers, and so far as the bread business is concerned, we are quite prepared to believe that it was all the fault of a..machine.

wangle (wæ·ŋg'l), *sb. colloq.* [f. WANGLE *v.*[2]] An act of wangling; an irregular or indirect method of working; something dishonestly contrived or manipulated.

1915 *N. & Q.* 23 Jan. 66 The explanation given was: 'It's a wangle between this Office and the Inland Revenue.' **1923** *Westm. Gaz.* 21 Feb. 11/1 Mr. Justice Sargant: You can't expect the Royal Commission to be bound to what might be called a 'wangle' of this sort. **1927** *Observer* 27 Mar. 16/5 Men who sought to achieve nothing by wangle or intrigue. **1943** H. A. SMITH *Life in Putty Knife Factory* xii. 183 They [*sc.* people of New York City] have made a precise science out of the wangle. **1959** E. POUND *Thrones* xcvii. 29 But by that time they found some other wangle. **1977** P. DICKINSON *Walking Dead* II. ii. 114, I worked a wangle. I got a line on the Minister of Tourism.

wanigan (wæ·nigăn). *N. Amer.* Also **wannegan, wannigan. 1.** (See WANGUN a, c.)

1902 S. E. WHITE *Blazed Trail* 323 Outside the cook and cookee were stowing articles in the already loaded wanigan. **1908** [see WANGUN c]. **1911** [see WANGUN a]. **1945** F. J. FUGINA *Lore & Lure of Upper Mississippi* 221 Wanigans were small flatboats, built of natural crook knees and planked lengthwise. **1957** *Bush News* (Port Arthur, Ontario) 1 June 1/1 Three wanegans, a floating mess hall and cook shack, and six drive camps are maintained to keep the wood moving.

2. A cabin, usu. wooden, mounted on runners; = *CABOOSE 3 b.

1949 in *Dict. Canadianisms* (1967) 834/1 The Wannegan has a sled mounted house type body made of light gauge metal, and heavily insulated for use in sub zero temperatures. **1949** *Survey* June 303/2 The industry is based no longer in isolated logging camps, with lonely men living in 'wannigans' dragged through the forests on sled-like runners. **1958** L. CRISLER *Arctic Wild* 133 At Barrow, Cris chained the wolves in front of our wanigan, one of a row of four or five empty wanigans beside the airstrip. They had made up a cat train. A wanigan is a narrow shack on runners: it can be drawn by tractor across snow or tundra. **1966** *Islander* (Victoria, B.C.) 20 Feb. 7/3 While sleeping in a wanigan, a hut on sleds, I was awakened..by native children shouting. **1979** C. KILIAN *Icequake* xii. 211 We'll have to build some wanigans that'll fit on the sledges. Otherwise we'll never be able to shelter everybody.

wank (wæŋk), *sb.* (*a.*) *slang.* Also † **whank.** [Origin unknown.] **1.** Of a male: (an act of) masturbation.

This word and its derivatives are not in polite use.

1948 PARTRIDGE *Dict. Forces' Slang* 203 *Wank-pit*,.. a bed. (Air Force.) **1951** —— *Dict. Slang* (ed. 4) 1220/1 *Whank*, (male) self-abuse: low: from ca. 1870. Perhaps echoic. **1966** P. WILLMOTT *Adolescent Boys* iii. 49 When I was 13, my friends at school asked me if I'd ever had a wank. **1970** T. LEWIS *Jack's Return Home* 18 Valerie Marshbanks showed everybody her knickers and charged a penny a wank, in the bushes, one at a time. **1971** B. W. ALDISS *Soldier Erect* 83 Oh, Christ! And there wasn't even anywhere in this wilderness where you could enjoy a decent sensuous wank—certainly not in the tent or the latrine. Nowhere. **1977** *Sniffin' Glue* July 15 Behind that bog door are you thinkin' readin' or just havin' a wank?

2. An objectionable or contemptible person or thing. Cf. *WANKER 2.

1970 P. LAURIE *Scotland Yard* vi. 158 Fred's counsel is a fat wank. **1973** *Nation Rev.* (Melbourne) 31 Aug. 1434/5 Kenneth S. Jaffrey, that naturopathic scientific wank. **1977** *Time Out* 17–23 June 14/1 Why did Aldrich direct this lengthy wank for Sinatra and Martin just after he'd made a film as interesting as 'Baby Jane'? **1977** *Maledicta* I. 11 He may be called a *jerk*, a *jerkoff* or a *jackoff*; a *wank* (Brit.), or a *whack.*

3. *attrib.* or as *adj.* Likely to induce masturbation.

1976 J. I. M. STEWART *Memorial Service* viii. 122 'He has an enormous great wank picture in his room. Makes the place like a porn shop.' 'I know that picture too... It's wanky all right.' **1981** P. PORTER *Coll. Poems* (1983) 277 But you are talking about love, you'll say. Yes, and I know the difference, Taking down a wank magazine.

wank (wæŋk), *v. slang.* Also † **whank.** [Origin unknown: see prec.] **1.** *intr.* Of a male: to masturbate. Freq. with *off.*

1950 P. TEMPEST *Lag's Lexicon* 229 *Whank, to*, to masturbate. **1951** PARTRIDGE *Dict. Slang* (ed. 4) 1220/1 *Whank*, loosely *wank*, v.i., to masturbate: low: late C. 19–20. Also *whank off.* **1966** P. WILLMOTT *Adolescent Boys* iii. 49 Some boys..think, 'I'm not going to tell anyone in case they think I'm dirty, wanking myself off.' But me and my mates, we tell everybody we wank off. **1969** F. NORMAN *Banana Boy* 108, I am certain that he could have wanked for Britain in the Olympics and won a gold medal with ease. **1977** W. MCILVANNEY *Laidlaw* xli. 192 You've been wankin'... That's no' nice in public places.

2. *trans.* To masturbate (a man). Freq. with *off.*

1966 [see sense 1 above]. **1975** O. SELA *Bengali Inheritance* xi. 94 'You like for me to wank you?'.. 'No... My wife would not like it.' **1980** 'D. KAVANAGH' *Duffy* v. 93 Lots of punters wanted you to wank them off... You'd think that was the one thing they could do for themselves. **1984** J. BARNES *Flaubert's Parrot* iv. 56, I saw a monkey in the street jump on a donkey and try to wank him off.

Hence **wanked-out** *a.*, exhausted (by masturbation).

1973 W. H. CANAWAY *Harry doing Good* ii. 23 A wanked-out shivering wreck. **1976** M. HARTMANN *Leap for Sun* i. 14 I worked on a wanked-out farming magazine.

Wankel (wæ·-, væ·ŋkĕl, va·ŋ-). The name of Felix *Wankel* (b. 1902), German engineer, used *attrib.* and *absol.* to designate a kind of rotary internal-combustion engine that he invented in which an approximately triangular, eccentrically pivoted shaft rotates continuously in a chamber with its corners touching the walls, so forming three combustion spaces that vary in volume as it turns. Also *Comb.*

1961 *Engineering* 19 May 682/1 Many other people have felt that something to replace the conventional engine is needed and a promising design to rear its head has been the Wankel engine. **1966** E. RUDINGER *Consumer's Car Gloss.* 120 The Wankel engine is much smaller than a conventional engine with a similar power output, and runs more smoothly. **1967** *Economist* 2 Sept. 811/1 The motor industry has waited a long time for the first car in production to use the wankel rotary engine. Now the small German manufacturer, NSU, has got there with the Ro 80. *Ibid.* 811/2 Will the wankel be enough to sustain NSU as an independent motor company, or will NSU one day have to close ranks further with Citroen? **1970** *Guardian* 10 Nov. 11/5 Citroën..has built and sold 500 Wankel engined cars. **1972** *Sci. Amer.* Aug. 17/3 Since the Wankel has no valves, it does not need a camshaft, valve lifters and so on, and it requires no more than two spark plugs per rotor. **1979** *Motorcycling Manual* (Motorcycling Monthly) 6/1 Apart from the few Wankel-engined bikes..all current machines have either two-stroke or four-stroke piston engines.

wanker (wæ·ŋkəɹ). *slang.* Also † **whanker.** [f. *WANK *v.* + -ER[1].] **1.** One who masturbates; *wanker's doom*, disability caused by excessive masturbation.

1950 P. TEMPEST *Lag's Lexicon* 229 A man suspected of excessive masturbation is said to be suffering from 'whanker's doom'. **1961** H. S. TURNER *Something Extraordinary* ix. 158 Because Willy always has big black circles round his eyes he is classified as 'pocket billiard player' or 'wanker'. **1971** B. W. ALDISS *Soldier Erect* 86 Failed fucker, failed wanker was an inglorious double billing. **1972** K. BONFIGLIOLI *Don't point that Thing at Me* xiii. 111 Mortdecai Minor, the likeliest candidate for wanker's doom. **1977** *Maledicta* I. 11 He is *whacky*, or *whacked to the gills*; he is *wonked goggle-eyed*, or is suffering from *Wanker's Doom.* The last refer to the old idea or threat that masturbation will make a person go blind or deaf or crazy.

2. An objectionable or contemptible person. Cf. *WANK *sb.* 2.

1972 A. DRAPER *Death Penalty* v. 36 'Get out, you fucker,' screamed a youth... Another said, 'You wanker,' and indulged in a masturbatory gesture. **1976** U. HOLDEN *String Horses* vi. 69 Her kiddies..rarely spoke except to mutter 'Wanker' or something crude. **1978** K. AMIS *Jake's Thing* xii. 123 'Damon, what's a wanker?'..'These days a waster, a shirker, someone who's fixed himself a soft job or an exalted position by means of an undeserved reputation on which he now coasts.' 'Oh. Nothing to do with tossing off then?' 'Well, connected with it, yes, but more metaphorical than literal.' **1981** P. NIESEWAND *Word of Gentleman* xxxii. 222 They're such a bunch of wankers... You can't trust them to do anything properly.

wanking (wæ·ŋkiŋ), *vbl. sb. slang.* Also † **whanking.** [f. *WANK *v.* + -ING[1].] Masturbation. Also with *off.* So *wanking couch, -pit*, a bed.

1951 PARTRIDGE *Dict. Slang* (ed. 4) 1220/1 *Whanking(-)pit*, the Army's form of *whank-pit*: since early 1920's. But used also by R.A.F. **1966** D. HOLBROOK *Flesh Wounds* 39 Wakey, wakey old mate. You want to get smartly outa that wanking pit! **1969** M. PUGH *Last Place Left* xxvi. 192 You rooting goat. Why can't you take them to your own wanking couch? **1971** P. L. CAVE *Chopper* xi. 106 He'd have so many women he could say goodbye to wanking for life. **1977** *Sounds* 9 July 23/3, I don't think the cutting was any more extreme than hanging from the rafters or wanking off on stage or vomiting.

wanky (wæ·ŋki), *a. slang.* [f. *WANK *sb.* or *v.* + -Y[1].] Objectionable, contemptible; masturbatory.

1972 K. BONFIGLIOLI *Don't point that Thing at Me* xix 173, I was now quite calm, the wanky old avenger preparing to kill his man. **1973** M. AMIS *Rachel Papers* 78 No, man, don't get too wanky with her. And cut out all this intellectual shit. **1976** [see *WANK *sb.* (*a.*) 3]. **1977** *Zigzag* Aug. 15/3 We loved that, 'cos it's such a wanky plastic paper and they thought by slagging us early they'd be in first. **1983** W. MCILVANNEY *Papers of Tony Veitch* xxx. 188 I'm not interested in his wanky theories about why he did it—just that he did it, the bastard.

Wan-Li (wæn,lī). Also **Wan Li.** The royal name of Shen Zong, emperor of China 1573–1620, used *attrib.* and *absol.* with reference to pottery and porcelain made during his reign.

1876 HOOPER & PHILLIPS *Man. Marks on Pott. & Porc.* 186 The dark rich blue of the *Kea-tching* and the white granulated surface of the *Wan-leih*, are very choice. **1885** *Jrnl. Peking Oriental Soc.* I. 98 Plain white cups of *Wan li* porcelain were several taels of silver each. **1902** C. MONKHOUSE *Chinese Porc.* I. 36 Some of the blue is fine in colour, as in some pieces which were lent by Mr. William Agnew to the Burlington Fine Arts Club in 1895. One of them has the Wan-li mark. **1915** R. L. HOBSON *Chinese Pott. & Porc.* II. v. 66 Another group of marked Wan Li ware, comprising bowls and dishes with trim neat finish.. has a soft-looking glaze. **1960** H. HAYWARD *Antique Coll.* 313/2 *Wu ts'ai* (*five-colour*) *wares.* This generally implies the characteristic Wan Li period Chinese porcelain wares, painted in underglaze blue and enamel colours (red, green, yellow and purple). **1978** K. BONFIGLIOLI *All Tea in China* i. 12 In London today they are crazy for blue-and-white wares and cannot tell Wan-Li from *De Metalen Pot.*

wanna (wǫ·nă), repr. colloq. pronunc. of *want to* or *want a.* Cf. WANT *v.* 4, 5 and *WANTA.

1896 S. CRANE *Maggie* v. 42, I see he had a still on, an' I didn' wanna give 'im no stuff. **1926–7** *T. Eaton & Co. Catal.* Fall & Winter 305/1 (*song-title*), I 'wanna' go where you go. **1941** B. SCHULBERG *What makes Sammy Run?* vi. 121 Look, do you wanna be smart? **1962** J. D. MACDONALD *Girl* (1974) x. 140, I don't wanna be no nuisance woman. **1971** *Frendz* 21 May 2/1 Right on, Paul! Wanna job? **1980** G. V. HIGGINS *Kennedy for Defence* x. 104 'That's fine by me,' I said, 'if you wanna leave.'

wannegan, wannigan, varr. *WANIGAN.

wanst (wǫnst, wanst), dial. (chiefly *Anglo-Ir.*) var. of ONCE *adv.* (*conj.*). Cf. *ONCE *adv.* A. δ.

1838 W. CARLETON *Fardorougha* (1839) iv. 81 Repeat the words at wanst. **1889** T. E. BROWN *Manx Witch* 80 Jack..bore it wanst, and bore it twicet. **1890** [see *ONCE *adv.* A. δ]. **1936** M. FRANKLIN *All that Swagger* xvi. 153 The young Delacys were all for the police at the end, but Danny said no. 'Wanst you begin law, it never ends.'

want, *sb.*[2] Add: **5. b.** (Further examples.) Also, something that one wishes to have (as opposed to what one needs or requires).

1946 G. M. TREVELYAN *English Social History* (ed. 2) xii. 388 Tea, porcelain and woven cotton goods were now being imported..in such quantities that they came within reach of the mass of the population. They created new wants and..demand was so great that home-manufacturers took to making cotton goods and china ware. **1950** *Sport* 24–30 Mar. 19/4 (Advt.), Clubs..can use our national net-work of experts to provide grounds, fixtures, insurance, fund-raisers and all club wants. **1955** W. J. BATE *Achievement of Samuel Johnson* (1978) ii. 70 General wishes have to localize themselves into definite wants. **1977** *Chicago Tribune* 2 Oct. XII. 1/4 The first-floor laundry room is definitely more a 'need' than a 'want' item nowadays. **1983** J. BARZUN *Stroll with W. James* 280 He did not fall into the trap of supposing that a child's needs are the same as his wants.

9. *want column* (earlier U.S. example); similarly *want ad*(*vertisement*).

1897 *Chicago Record* 1 Mar. 10/4 Record 'want ads' bring results. **1919** MENCKEN *Amer. Lang.* 160 *Ad-writer, want-ad*,..and *ad-man*, are already accepted in technical terminology. **1940** R. CHANDLER *Farewell, my Lovely* vii. 43 The Montgomery killing hadn't even made the want-ad section. **1977** C. MCFADDEN *Serial* (1978) xlv. 96/1 The want ads..no longer solicited secondhand Hide A Beds. **1887** *Courier-Jrnl.* (Louisville, Kentucky) 12 Jan. 5/3 The *World* is treating Mr. Conkling as it treats its circulation and its 'want' advertisements upon occasion. **1919** F. HURST *Humoresque* 116 She took to..scanning want-advertisements as she stood at the news-stand. **1884** E. W. NYE *Baled Hay* 239 The want column of the Chicago *News*..has the following: 'Twelve "frightful examples" wanted.'

want, *v.* Add: **2.** Also, in *Palæography* and *Bibliography*, to lack (a leaf or a page).

1895 M. R. James *Catal. MSS. Sidney Sussex Coll.* 114 Vellum... Cent. xiv, xv... *Collation:* a⁸ (wants 1, 2) b⁸–f⁸. 1976 *Anglo-Saxon England* V. 150, i⁸ wants 1.

5. e. (U.S. examples const. *of*.)

1855 *Knickerbocker* XLV. 136 Salt, Miss? What do you want of salt? 1884 'Mark Twain' *Huck. Finn* xxxv. 357 What do we want of a saw? 1914 G. Atherton *Perch of Devil* II. 246 But what does he want of two cottages?

f. The phr. *want out* is now in general *colloq.* use. (Earlier and later examples.)

1836 J. D. Davidson *Diary* 29 Oct. in *Jrnl. Southern Hist.* (1935) I. 354 He still called out in a plaintive, infant tone, '*I want in*'—'*I want in*'. 1925 *Amer. Speech* I. 149/2, I want out at the bank because this sack of silver is too heavy to pack around with me. 1959 E. Ambler *Passage of Arms* viii. 219 They can keep everything... We just want out. 1973 *Time Out* 2–8 Mar. 14/1 One of the kids who had paid his money..wanted out. 1973 *Nature* 28 Sept. 173/2 Britain may just be weary of industrial growth and may be saying in quite a sophisticated way that it wants out regardless of the cost. 1977 *Zigzag* Mar. 24/1, I had no regrets about leaving... I just wanted out. 1979 A. Hailey *Overload* III. i. 194 Well, I'm not afraid, or proud, or anything any more. I just want out. 1984 *Guardian* 5 Nov. 17/4 In recent weeks the Federal Reserve chairman Mr Paul Volcker has reportedly told friends that he wants out and would be interested in the presidency of the World Bank. 1985 *Times Lit. Suppl.* 25 Jan. 102/2 If you want out, it is just about possible to live, if only internally, a better life.

wa'n't. (Later U.S. examples.)

1893 H. A. Shands *Some Peculiarities of Speech in Mississippi* 67 *Want*, Negro and illiterate white for *was not* or *were not*. 1929 H. W. Odum in A. Dundes *Mother Wit* (1973) 194/1 Wan't no hero, wan't no coward. 1942 W. Faulkner *Go down, Moses* 144 Ah went to yo house last night, but you want dar.

wanta (wǫ·ntă), repr. U.S. colloq. pronunc. of *want to*. See *wanna.

1894 S. Crane in *Arena* Oct. 666 Let up, will yeh? Do yeh wanta kill somebody? 1925 T. Dreiser *Amer. Trag.* (1926) I. i. 7 That oldest boy don't wanta be here. 1946 E. O'Neill *Iceman Cometh* II. 98 Don't let Hickey put no ideas in your nuts if you wanta stay healthy! 1977 *New Yorker* 12 Sept. 54/3 Do fast lines and play lousy—one phrase of 'Smoke Gets in Your Eyes' and then: 'Folks, you wanta drive somebody crazy?'

wa·ntable, *a.* [f. Want *v.* + -able.] Desirable, of a kind likely to become sought after.

1970 *Times* 21 July 7 This collection [of clothes]..is extremely pretty, wantable and smart in a personal private way. 1979 *Daily Tel.* 22 Jan. 11/2 Most wantable garment in the Paris collections, a Fifties' style soft leather dress cinched with an ultra-wide cummerbund in the Claude Montana collection.

wan tan, var. *won ton.

wanted, *ppl. a.* Add: **3.** Special collocations: *wanted list*, a list of persons sought by the police or by a similar agency; also *wanted file*; *wanted poster*, a poster displaying details of a wanted person or persons, usu. under the headline 'Wanted'.

1967 'E. Peters' *Black is Colour* i. 20 Just take a look at 'em!..Every one of 'em straight out of the wanted file. 1964 I. Fleming *You only live Twice* xix. 228 The C.I.A. on whose wanted list I certainly feature. 1968 P. Durst *Badge of Infamy* v. 50, I checked with the International War Crimes commission to see if Von Friede was on the wanted list. 1970 T. Lilley *Projects Section* xiv. 177 There would always be those in the crowd who would recognise the corpse—either from 'Wanted' posters or from personal acquaintance. 1982 A. Price *Old Vengeful* 12 Oh dear! A good likeness... It looks like a prison picture..or maybe a 'Wanted' poster.

wanting, *pres. pple.* Add: **6.** Also *colloq.* (Earlier and later examples.)

1839 Dickens *Nickleby* xxxiv. 331 'He was a little wanting here,' touching his forehead. 1924 J. Buchan *Three Hostages* vi. 85 Had something given in my brain last night..so that now I was what people call 'wanting'? 1976 'J. Bell' *Trouble in Hunter Ward* xvii. 180 Maisie, poor mite, wanting from birth, she was.

wanton, *a.* Add: **9.** *wanton-headed* adj.

1918 D. H. Lawrence *New Poems* 22 By-paths where the wanton-headed flowers doff their hood.

wantum (wǫ·ntǝ̆m). *nonce-wd.* [Blend of Want *sb.*² + Quantum.] Deficiency or desire, considered as something quantifiable.

The quot. alludes to Prov. xxx. 15 The horse-leach hath two daughters, crying, Giue, giue. There are three things that are neuer satisfied, *yea* foure things say not.

1938 S. Beckett *Murphy* 57 The horse leech's daughter is a closed system. Her quantum of wantum cannot vary.

Wanyamwezi (wanyamwē̆·zi). Also **Wanyamwesi.** [Native name, lit. 'people of the moon', hence (prob.) 'people of the West'.] The name of a Bantu people of Tanzania.

1860 R. F. Burton *Lake Regions Central Afr.* II. xii. 4 The correct designation of the inhabitants of Unyamwezi is, therefore, Mnyamwezi in the singular, and Wanyamwezi in the plural: Kinyamwezi is the adjectival form. *Ibid.*, According to the people..in the days of the grandfathers of their grandfathers the last of the Wanyamwezi emperors died. 1911 J. Frazer *Golden Bough: Magic Art* (ed. 3) I. v. 268 The Wanyamwezi, a large tribe of Central Africa, to the south of the Victoria Nyanza. 1947 *E. Afr. Ann.* 1946–7 93/1 The Wanyamwezi have been one of the better known tribes of Tanganyika from the time when Arab slave raiders first set up their centre at Tabora. 1955 Hill & Moffett *Tanganyika* 515 The Wanyamwezi are one of the more virile and hard-working tribes of the Territory, and also one of the largest.

wap, *v.*¹ Add: **1. c.** *intr. slang.* To copulate. *Obs.*

[1567 T. Harman *Caveat* sig. G4ᵛ, He tooke his Iockam in his fambile, and a wapping he went.] 1611 Middleton & Dekker *Roaring Girl* sig. K4ᵛ, Wee'l couch a hogshead vnder the Ruffemans, and there you [sc. Moll Cutpurse] shall wap with me, & Ile niggle with you. 1699 B. E. *New Dict. Canting Crew, Wap*, c. to Lie with a Man. *If she won't wap for a Winne, let her trine for a Make*, c. If she won't Lie with a Man for a Penny, let her Hang for a Half-penny. *Mort wap-apace*, c. a Woman of Experience, or very expert at the Sport. 1725 in Farmer & Henley *Slang* (1903) VII. 293/1 This doxy dell can cut bien whids, And wap well for a win.

wapping *vbl. sb.* (examples in sense *1 c); also as *ppl. a.*

1610 S. Rid *Martin Mark-All* sig. E3, *Nigling*, company keeping with a woman: this word is not vsed now, but *wapping*, and thereof comes the name *wapping morts* Whoores. 1612 Dekker *O Per Se O* sig. O1ᵛ, And wapping Dell, that niggles well, And takes loure for her hire. 1707 J. Shirley *Triumph of Wit* (ed. 5) III. 198 Wapping thou I know do's love.

Wapishana (wapiʃā·na), *sb.* and *a.* Also **Wapishiana, Wapisiana**; 9 **Wapiana, Wapisiano. A.** *sb.* **a.** A member of an Arawakan people of Guyana and Brazil; also this people collectively. **b.** Their language. **B.** *adj.* Of, pertaining to, or designating this people or their language.

1855 H. G. Dalton *Hist. Brit. Guiana* I. ii. 77 The Acosi, Awake, Wapishana..are also the names of several other tribes which have been met with by late travellers. *Ibid.* (caption facing p. 80) The 'Belle' of the Wapisianas. 1868 W. H. Brett *Indian Tribes of Guiana* II. v. 280 Southward of Roraima the great range of the Pacaraima Mountains extends... Beyond it are the savannahs of the Rupununi, inhabited by the *Wabean* or *Wapisiana* nation. 1883 E. F. Im Thurn *Among Indians of Guiana* vi. 165 The languages of the four branches, Warrau, Arawak, Wapiana, and Carib, will be found to be quite distinct from each other. 1924 [see *lap *sb.*¹ 4 c]. 1949 *Caribbean Q.* I. III. 41 Of the two tribes likely to be displaced, one, the Wapisiana, is not likely to be demoralised. 1959 P. Capon *Amongst those Missing* 167 'An Indian, isn't he?' asked Judith. 'I think so. He has the look of a Wapishana.' 1972 S. E. Brock *Jungle Cowboy* xii. 135 *Wai-Wai* is the Wapishana for tapioca.

wappie (wǫ·pi). *West Indies.* Also **wap(p)ee, wappy.** [perh. f. Wap *sb.*¹ + -ie.] A gambling game played with cards.

1943 in Cassidy & Le Page *Dict. Jamaican Eng.* (1967) 462/2 *Wappie*, a game of cards. 1952 S. Selvon *Brighter Sun* ii. 26 The older boys and men play wapee, a gambling game with cards. *Ibid.* vii. 123 Ah play cards—Ah was a rummy test in me days, and don't talk 'bout wappee! Yuh cud play cards? 1952 *Evening News* (Port of Spain) 22 Jan. 12/3 Tom..happens to back the right cards at wappie. 1959 V. S. Naipaul *Miguel St.* i. 12 They played wappee and drank rum. 1968 K. S. La Fortune *Legend of T-Marie* xii. 51 Cab-men sat on benches or on boxes talking politics or playing wappie.

waqf, var. *wakf.

war, *sb.*¹ Add: **I. 3. a.** Freq. used with def. art. to designate a particular war, esp. one in progress or recently ended. Hence *between the wars*, between the war of 1914–18 and that of 1939–45 (cf. *inter-war *a.*). *Sacred War* (earlier and later examples); *War Between the States* (esp. in the use of Southerners), the American Civil War.

1774 Sacred war [see *Phocian *sb.*]. 1814 *Columbian Centinel* 18 June 2/3 The southern war-hirelings say the Administration will continue the War. 1861 *Chicago Tribune* 26 May 1/9, I, Samuel M. Fassett, photographist, ..will continue to take those fine plain photographs for the low sum of one dollar, during the war. 1867 A. H. Stephens (*title*) A constitutional view of the late War between the States. 1911, etc. Sacred war [see *Phocian *a.*]. 1934 *Sun* (Baltimore) 5 June 14/7 There was a time when it was almost worth one's life in the city of Richmond to refer to the Civil War as the Civil War. The Richmonder who held the memories of the sixties close to his heart always called it the War Between the States. 1936 *Punch* 2 Dec. 640/1 Our telephone system is partly British and partly German and Turkish, and all of it served through the War with varying degrees of distinction. 1942 C. S. Lewis *Screwtape Lett.* xv. 76, I had noticed, of course, that the humans were having a lull in their European war—what they naïvely call 'The War'! 1958 'N. Shute' *Rainbow & Rose* i. 3 He lived and worked in England and the Far East all the time between the wars. 1973 R. Thomas *If You can't be Good* vi. 46 The Maurys..had supplied the South with two generals during the War Between the States.

b. Also *war of nerves*; see *nerve *sb.* 8 f; *war of words* (Journalese), a sustained conflict conducted by means of the spoken or printed word; a propaganda war.

[1725 War of words: see sense 1 b in Dict.] 1981 *Times* 10 Oct. 1/7 As the war of words continued in the Tory party Mrs Thatcher arrived back from the Commonwealth Conference. 1984 *Guardian* 8 Mar. 9/1 (*heading*) Vietnamese intensify war of words on Peking.

c. *to carry the war into the enemy's camp* (*into Africa*, etc.): see *carry *v.* 19 b.

d. *war to end war(s)*: a war which is intended to make subsequent wars impossible; usu. *spec.* the war of 1914–18.

[1914 H. G. Wells (*title*) The war that will end war.] 1921 G. B. Shaw *Back to Methuselah* IV. 187 There was a war called the War to End War. In the war which followed it about ten years later, none of the soldiers were killed; but seven of the capital cities of Europe were wiped out of existence. 1932 P. Quennell *Let. to Mrs. V. Woolf* (1933) 17 He can recall barely five or six summers; then the War to End Wars and so good-bye. 1949 E. Benn *Happier Days* vi. 71 If..war debts between nations had been wiped off the slate, and reparations in money never attempted, the 'war to end war' might have achieved its high purpose. 1953 Earl Winterton *Orders of Day* xxiv. 345 The Government of that day and the then leaders of opinion in general had assured us and the nation at large that it was 'a war to end war'. 1967 W. Lippman in W. Safire *New Lang. Politics* (1968) 480/2 Each of the wars to end wars has set the stage for the next war. 1978 E. Malpass *Wind brings up Rain* ix. 99 Now..the War To End War was over.

e. *to have a good war*: to achieve success, satisfaction, or enjoyment during a war. Also with other adjs. Often ironic.

1969 P. Dickinson *Pride of Heroes* I. 49 Harvey Singleton..had a good war. A very good war indeed. After the Raid he was parachuted into France three times. 1970 —— *Seals* ii. 35 He had a very bad war. 1972 P. D. James *Unsuitable Job for Woman* iv. 124 He had what the men call a good war; we'd call it a bad war I dare say, a lot of killing and fighting. 1974 'J. le Carré' *Tinker, Tailor, Soldier, Spy* xviii. 153 He had a dazzling war... The comparison with Lawrence was inevitable.

5. b. For † *Obs.* read *Obs.* exc. as *minister of* (or *for*) *war*, *secretary at war*, *secretary of* (*state for*) *war*.

1693, etc. [see Secretary *sb.*¹ 3 a]. 1802 C. Wilmot *Let.* 15 Nov. in *Irish Peer* (1920) 115 The six Ministers of the Interior, of the Police, of Justice, of Finance, of War, and of foreign affairs. 1867 Crown Princess of Prussia *Let.* 27 Apr. in R. Fulford *Your Dear Letter* (1971) 133 The King wishes for peace..so does the Minister for War. 1877 J. Blackwood *Let.* 21 Dec. in *Geo. Eliot Lett.* (1956) VI. 434, I am happy to say that our Minister of War is I think a man who may be trusted at the helm. 1903 *Ceremonies at Laying Corner Stone of Army War College Building* (U.S. Army Corps of Engineers) 9 The master of ceremonies then introduced the honorable Secretary of War. 1964 *Act Eliz. II* c. 58 § 1 There shall be transferred to a Secretary of State the functions conferred by any enactment on the Minister of Defence, or on the Secretary of State for War or for Air (however styled). 1980 A. Marwick *Illustr. Dict. Brit. Hist.* 64/2 He [sc. Churchill] served the coalition subsequently as secretary of state for war and air (1918–21).

II. 8. a. (In this and the senses that follow the use of the hyphen follows no regular pattern.) *war aim, base, camp, casualty, War Department* (later U.S. examples), *war footing* (earlier examples), *hospital, measure, neurosis* (hence *-neurotic*), *news, period, production, profiteer, propaganda, -psychosis, ration, record, restriction, scare, -service* (modern examples), *situation, strain, surplus* (chiefly attrib.), *victim, -weariness, widow, -word, wound, years, zone* (earlier and later examples).

1918 A. Bennett *Pretty Lady* xxviii. 193 The Germans were discussing their war aims. 1972 *New Yorker* 22 July 66/3 In the My Lai massacre the soldiers abandoned the unrealistic war aims of Dean Rusk and drew their illogical but understandable conclusion..all Vietnamese have to be killed. 1947 *Daily Tel.* 19 Apr. 4/2 Its virtual disappearance yesterday..is probably the only answer to the fear of its use as a war-base once more. 1977 *South China Morning Post* (Hong Kong) 14 Apr. 5/3 Militarist forces clutch at the blocs and war bases they established in Asia. 1832 A. Earle *Narr. Residence N.Z.* (1966) 176 Mr. Hobbs, the Wesleyan missionary,..had visited the war-camp of the assembled chiefs. 1969 G. Macbeth *War Quartet* 25 In our minds A dream of war-camps festered. 1921 G. B. Shaw *Back to Methuselah* II. 88 It was the war casualty lists and the starvation afterwards that finished me up with politics and the Church and everything else except you. 1974 A. Price *Other Paths to Glory* II. ix. 225 The late Turco... Another war casualty? 1866 G. B. McLellan *Let.* 26 Dec. in *McLellan's Own Story* (1887) xii. 221 The entire establishment..was removed to the War Department building, without my knowledge. 1944 *Time* 2 Oct. 19/1 This was strictly a military document drafted by the War Department. 1847 Thackeray *Van. Fair* (1848) xxviii. 242 The armies of the allied powers were all providentially on a war-footing, and ready. 1872 'Mark Twain' *Roughing It* ii. 22 We were reduced to a war-footing. 1860 F. Nightingale *Notes on Nursing* iii. 23, I by no means refer only to..war hospitals, but..to..military hospitals at home, in time of peace. 1982 P. Quennell *Customs & Characters* ii. 33 The French Ambassadress..had promised she would visit a nearby war-hospital. 1808 W. Eaton in R. C. Prentiss *Life W. Eaton* (1813) 414 The Embargo was contemplated as a war measure. 1948 *Rep. Native Laws Comm. 1946–48* (Dept. Native Affairs, S. Afr.) 32/1 A War Measure has been promulgated as a temporary attempt to relieve the situation. 1975 *Toronto Star* 1 Nov. B4/4 Pierre Trudeau invoked the War Measures Act and plunged Canada into a time of arrest without warrant and detention without charge. 1920 *Internat. Jrnl. Psycho-Anal.* I. 283 Freud's introduction gives some of the chief points of view for the psycho-analytical consideration of the war neuroses. 1944 *Yank* 31 Mar. 8 For this reason there is actually no such thing as 'war neurosis', any more than there is 'war malaria' or 'war pneumonia'. 1955 J.

Strachey tr. *Freud's Psycho-Anal. & War Neuroses* in *Compl. Wks.* XVII. 215 With the end of the war the war neurotics, too, disappeared—a final but impressive proof of the psychical causation of their illness. 1857 C. Kingsley *Two Years Ago* II. v. 200, I cannot sit here quietly, listening to the war-news. It makes me mad to be up and doing. 1915 F. H. Burnett *Lost Prince* xiii. 96 [They] sat down to read the morning paper. The war news was bad to read. 1967 C. Potok *Chosen* iii. 59 There was war news all the time, but no one got this excited unless something very special was happening. 1918 H. Crane *Let.* 12 Aug. (1965) 11 All minors,..if drafted at all, will be apprenticed in machine shops, etc., during the war period. 1939 *Ann. Reg.* 1938 260 The declared policy of Senor Negrin to look beyond the war-period to a Spain in which one day the Spaniards on both sides would have to live together. 1965 A. J. P. Taylor *Eng. Hist. 1914–45* xiv. 517 Bomber command claimed the largest share of Great Britain's war-production. 1918, 1975 War-profiteer [see *profiteer *sb.*]. 1918 W. Owen *Let.* 25 Oct. (1967) 588 He had no qualifications for War Propaganda. 1974 *Guardian* 31 Jan. 1/5 War propaganda on both sides was, of course, bad and distorted. 1927 W. E. Collinson *Contemp. Eng.* 103 Symptoms of that war-psychosis, which afflicted us in common with the other belligerent nations. 1953 H. S. Whitman tr. *Janetschek's Emperor Franz Joseph* 302 Don't worry so much, and you will soon be free of your war-psychosis. 1924 D. H. Lawrence in M. Magnus *Mem. Foreign Legion* 16 He yelled for more bread—bread being war-rations and very limited in supply. 1890 E. Custer *Following Guidon* 2 They longed individually and as a regiment for a war 'record'. 1978 F. Maclean *Take Nine Spies* iv. 126 The men were impressed by his war record. 1922 W. J. Locke *Tale of Triona* xxvi. 292 England..awoke to find war restrictions removed,..and petrol to be had. 1938 J. Charlesworth *Law of Negligence* vii. 133 Where a refuge was erected in the middle of the street, and inadequately lighted, so that a taxi-cab collided with it in the dark because..owing to war restrictions, no lights were maintained,..the local authority were held liable. 1894 W. Le Queux *Gt. War in Eng. in 1897* i. 15 War-scares had been plentiful. 1977 *Listener* 10 Feb. 177/2 The war scare in 1938. 1916 J. Bailey *Let.* 8 Oct. (1935) 168 It was a great joy to see you both and King's Weston again, and to admire your wonderful 'war service' and feel that all the beauties of the house and place are being put to such splendid use [as a War Hospital]. 1979 A. Price *Tomorrow's Ghost* iv. 47 They both looked old enough to have seen war service. 1936 C. Day Lewis *We're not going to do Nothing* 29 In an actual war-situation the trade unions are in the key-position. 1914 T. A. Baggs *Back from Front* xx. 94 It is there that human nature, exuberant or impassive under the war-strain, reveals its own true colours once again. 1952 H. Innes *Campbell's Kingdom* I. v. 110 They wore war surplus clothing relieved by bright scarves. 1968 P. Geddes *High Game* viii. 97 When the big round of war surplus prosecutions started, none of the dirt ever stuck to him. 1982 D. Williams *Copper, Gold & Treasure* 16 I'll let you know if he asks me to buy him any war surplus. 1969 *Guardian* 28 Aug. 11/5 The starving war victims. 1917 'Contact' *Airman's Outings* p. xviii, What, then, would be the effect on German war-weariness if giant raids on fortified towns by a hundred or so allied machines were of weekly occurrence? 1976 *Classical Q.* XXVI. 294 Sinon begins the first section of his lying speech with a reference to the death of Palamedes.., the second by describing the war-weariness of the Greeks. 1866 *Ann. Rep. Commissioner Indian Affairs* (U.S.) 164 These last came from Laramie during the winter, and claim to be war-widows. 1922 F. H. Burnett *Robin* v. 44 Slim young war-widows were to be seen in black dresses and veiled small hats with bits of white crape inside their brims. 1978 R. Barnard *Unruly Son* xvii. 186 We moved to London, where she passed as a war widow. 1932 E. Weekley *Words & Names* 21 We have the war-word *Minnie* for the German *minenwerfer*. 1938 E. Ambler *Cause for Alarm* iii. 47 The limp? Probably a war wound. 1981 C. Miller *Childhood in Scotland* 68 He was in continual pain from his war-wound. 1920 W. J. Locke *House of Baltazar* i. 9 The strain of the war years began to tell. 1977 D. Bennett *Jigsaw Man* xiii. 231 They had hidden the microfilm in the same cache they had used during the war years for passing messages. 1914 *Wells Fargo Messenger* Oct. 27/1 A late report from the war zone states that Mr. Gaston has returned to London. 1939 *Daily Tel.* 18 Dec. 6/5 The danger of 'rupture' has been vastly reduced by Congress's prohibition of American ships from entering the war zone. 1971 D. E. Westlake *I gave at Office* 142 The bar..had temporary plywood over its glassless windows, making it look like a correspondents' hangout in a war zone.

b. *war-belt* (earlier and later N. Amer. examples), *-bow, -club* (later examples), *-dress* (earlier example), *material, souvenir*; *war-balloon* (earlier example), *-beacon, -chariot, -pony*; *war-canoe* (earlier example).

1843 *Practical Mechanic* 16 Dec. 114/1 (*heading*) War balloon. 1954 J. R. R. Tolkien *Fellowship of Ring* II. ii. 277 Such light and flame cannot have been seen on Weathertop since the war-beacons of old. 1754 P. Wraxall *Abridgement Indian Affairs* (1915) 242 He calls upon them now to..join us in our Defence & Revenge & presents the Large War Belt to them. 1847 C. Lanman *Summer in Wilderness* 17 Captain James Clarke,..when about to be murdered by a council of Indians.., threw the war-belt in the midst of the savages, with a defying shout. 1965 *Canad. Historical Rev.* June 109 In December [1775] the Iroquois delegation told Philip Schuyler that Johnson after offering them a war belt and hatchet had invited them to 'feast on a Bostonian and drink his blood'. 1934 Webster, War bow. 1958 'W. Henry' *Seven Men at Mimbres Springs* iv. 48 Nothing so guaranteed a safe passage through Apacheland as a coach that would not tip or leave the road when the warbows were bending and the Springfields blasting back at them. 1789 *Loiterer* 18 July 5 A large War Canoe and some small fishing Proas had been forced out to Sea. 1911 Fletcher & Kipling *School Hist. England* i. 15 They [sc. the Celts] rode on war-ponies, and, like the Assyrians in the Bible, they drove war-chariots. 1907 J. W. Schultz *My Life as Indian* xvii. 198 The fleeing men..were overtaken and shot, or brained with war clubs. 1943 R. Peattie *Great Smokies & Blue Ridge* 24 The Cherokee weapons were the ball-headed war club, spears and bows and arrows. 1984 *Listener* 4 Oct. 13/3 They were attacked by Kukukuku..with stone war-clubs. 1724 H. Jones *Present State Virginia* 5 The Seneca Indians in their War Dress may appear as terrible as any of the Sons of Anak. 1881 W. D. Hay *300 Years Hence* iv. 67 The progress of The Final Wars was marked by a whole series of inventions in war material. 1939 *Ann. Reg.* 1938 265 Meanwhile war material from Germany and Italy continued to pour in. 1865 J. Pike *Scout & Ranger* xi. 123 Many had friends..who came after them with wagons; refusing to let them ride their war ponies. 1929 D. H. Lawrence *Pansies* 128 Prancing their knees under their tiny skirts Like war-horses; or war-ponies at least! 1963 L. Deighton *Horse under Water* xliv. 180 An old war-souvenir pistol.

c. *war-tribe*.

1825 P. S. Ogden *Jrnl.* 18 Feb. (1950) 23 The War tribes appear determined that we Shall not want for their Company this year it cannot be otherwise as we are following the main War track. 1909 'Mark Twain' *Is Shakes. Dead?* v. 53 It could have gone soldiering with a war-tribe..and learned soldier-wiles and soldier-ways.

d. (used in war) *war-cheer*.

1970 R. Lowell *Notebk.* 191 Frederick the Great of Prussia's war-cheer, 'Move, you bastards, do you want to live forever?'

e. *war bond, debt, expenditure, gratuity, -insurance* (example), *-loan* (examples), *-price* (earlier example), *relief, savings, -tax* (earlier example).

1918 *Daily Mirror* 12 Nov. 6/4 It bore a poem, titled 'A Message from Mars', eulogising the airmen and urging them to buy War Bonds. 1981 B. Langley *Autumn Tiger* v. 67 A gigantic billboard..urged him to 'Buy War Bonds'. 1865 *Nation* (N.Y.) I. 386 The Reconstructing State Convention of Alabama has pronounced against the repudiation of the war debt of the state. 1924 *Lit. Digest* 9 Feb. 20/2 The whole subject of war debts should undergo a new process of accountancy. 1931 *Keesing's Archives* 76/2 The oppressive problem of war-debts and reparations. 1983 T. Pocock *1945* vii. 241 The British owed their dominions, colonies, and the rest, a war debt of £4,000,000,000. 1931 War expenditure [see *inconsequentialness]. 1945 *Ann. Reg.* 1944 80 All those returning to civil life would receive war gratuities as a reward for their service. 1978 D. Dunlop in D. Abse *My Medical School* 31 Besides ordinary freshmen like myself straight from school, many came up on their war gratuities. 1898 *Amer. Rev. of Reviews* Sept. 322/2 Newspapers were required to bear the..expense of fire, marine, accident, and war insurance. 1848 Mill *Pol. Econ.* II. iii. xxiii. 185 The only instance of the kind in recent history on a scale comparable to that of the war loans, is the absorption of capital in the construction of railways. 1974 *Daily Tel.* 24 June 17/1 'War Loan is a buy when the price equals the yield' was the joke on everyone's lips in Throgmorton Street a couple of years ago. Today it is no longer a joke—almost a reality. On Friday 3½ p.c. War Loan dropped to an all time low of £23½. 1824 *Cobbett's Weekly Reg.* 7 Feb. 354/2 Corn has not reached half the war-price yet. 1940 G. Marx *Let.* 5 Sept. (1967) 25 The proceeds are given to British War Relief and the actors all donate their services. 1919 *Maclean's Mag.* Jan. 55/3 Every man, woman and child in Canada should invest in War-Savings Stamps all the money that he or she can save. 1799 *Times* 1 June 2/3 The Directory have converted his accusation into a War Tax of three per cent. upon all capital.

f. With words that denote works of art, etc., of which the subject is war, as *war-ballad, book, history, -impression, novel, play, poem* (also *poetry*), *propaganda* (see sense 8 a above), *sonnet, story, verse*; *war film, movie, photograph* (also *photography*); also their authors, as *war novelist, photographer, poet*. Cf. also *war artist, picture* in sense 11 below.

1854 'C. Bede' *Further Adventures Verdant Green* ii. 9 What internal evidence does the Odyssey afford, that Homer sold his Trojan war-ballad at three yards an obolus? 1916 W. Owen *Let.* 23 Nov. (1967) 416, I have suddenly seen what I wanted to do with that War Ballad. 1809 M. L. Weems *Life Gen. Francis Marion* (1814) 3, I never dreampt of such a thing as writing a book; and least of all a war book. 1904 J. London *Let.* 4 June (1966) 159 There won't be any war-book so far as I am concerned. 1978 A. Waugh *Best Wine Last* xviii. 235 Starting with *Journey's End*..there had been a spate of war books. 1897 C. M. Hepworth *Animated Photogr.* p. vii (Advt.), War films. 1930 P. Rotha *Film till Now* I. v. 124 Like all war films manufactured in Hollywood, *The Big Parade* carried little of the real spirit of war. 1972 J. Mann *Mrs Knox's Profession* ii. 14 She looked like an amateur vamp in a war film. 1929 E. Linklater *Poet's Pub* ii. 31 He had been offered a knighthood for his official War History of the submarines. 1966 'G. Black' *You want to die, Johnny?* xi. 198 Split-second timing..isn't achieved as often as the writers of popular war histories tend to suggest. 1917 W. Owen *Let.* 11 Mar. (1967) 442 Do you think, now, that I am going to read the *war-impressions* of home-editors? 1914 *N.Y. Times* 14 Jan. 9/6 Real war 'movies' shown... Moving pictures of real warfare were exhibited in the Seventy-first Regiment Armory last night. 1981 J. van de Wetering *Mind-Murders* I. iv. 43 An old war movie that ended well when the bad enemies surrendered. 1923 W. J. Locke *Moordius & Co.* viii. 99 He had not read the marvellous war novel to which he alluded. 1975 D. Lodge *Changing Places* v. 178 Stephen Crane wrote his great war-novel first and experienced war afterwards. 1966 J. Fredericks *Ready to Die* (1968) iv. 20 He made war novelist sound like something not discussed in polite company. 1977 M. Herr *Dispatches* (1978) ii. 18, I can remember..when I was a kid looking at war photographs in *Life*. 1978 R. Gibson *Catal. 20th. Cent. Portraits* (Nat. Portrait Gallery) 36/1 Lee Miller (1907–77), well known as a war photographer. 1980 J. Danziger *Beaton* 46/2 The naïve approach of his war photography. 1896 *Godey's Mag.* Feb. 182/1 The instrument..imitates horses' hoofs with..untiring fidelity in all war-plays. 1915 *Sphere* 26 June 322/2 The production of a war play is a perilous business at the present time. 1972 P. Black *Biggest Aspidistra* I. iv. 41 *Brigade Exchange*, a war play..created by the pre-Nazi German radio. 1857 J. A. Symonds *Let.* May (1967) I. 105 He chiefly talked about ..[Tennyson's] Maud which he considers a true war poem & praises highly. 1917 W. Owen *Let.* 25 Sept. (1967) 496, I send you my two best war Poems. 1978 *Listener* 23 Mar. 378/3 There was a fine slim anthology of war poems read. 1857 C. Kingsley *Two Years Ago* III. vi. 177 The true war poets..have been warriors themselves. Körner and Alcæus fought as well as sang. 1931 E. Blunden in W. Owen *Poems* 39 He [sc. Owen] was, apart from Mr. Sassoon, the greatest of the English war poets. But the term 'war poets' is rather convenient than accurate. 1962 *Listener* 8 Feb. 259/3 By 'war poet' we now automatically assume anti-war poet. This was no tacit assumption in 1916. 1865 *Atlantic Monthly* May 589/1 We have no such war-poetry. 1917 W. Owen *Let.* 27 Nov. (1967) 513, I knew he valued War Poetry before he told me so! 1973 D. Aaron *Unwritten War* IV. x. 152 Their best war poetry tended to be philosophical and personal. 1915 W. S. Churchill in *Times* 26 Apr. 5/5 The very few incomparable war sonnets which he [sc. Rupert Brooke] has left behind. 1864 M. B. Chesnut *Diary* 1 Jan. in C. V. Woodward *M. Chesnut's Civil War* (1981) 524, I mapped out a story of the war. Johnny is the hero... It is to be a war story. 1982 G. Lyall *Conduct of Major Maxim* xv. 145 They weren't interested in her war story, they'd heard a million war stories. 1918 G. Frankau *Judgement of Valhalla* 41 The other Side Being a letter from Major Average..acknowledging a presentation copy of a book of war-verse. 1952 E. Wilson *Shores of Light* 780, I refrained from mentioning her war-verse.

9. *war-winner*; *war-making* (later examples), *preparation, -winning* (also ppl. adj.); *war-bearing* ppl. adj.; *war-hungry* adj.

1936 Dylan Thomas *Twenty-Five Poems* 10 Dumbly and divinely stumbling Over the warbearing line. 1962 E. Snow *Other Side of River* (1963) lx. 456 The Western caricature of the mad-dog war-hungry Chinese. 1833 *Niles' Reg.* XLIV. 148/1 Very few persons questioned the right of congress to lay an embargo, under the war-making power. 1982 Warner & Sandilands *Women beyond Wire* v. 69 The Japanese..could be lethal..their business in the islands was that of professional war-making. 1931 W. S. Churchill *World Crisis* VI. vi. 97 Neglect in the war-preparations. 1947 Crowther & Whiddington *Science at War* I. 49 Manufacturers found it very difficult to give up mass production, in order to make the 200 or so sets 'off', which were often the war-winners. 1978 Ld. Drogheda *Double Harness* xix. 230 He was indeed one of the real war-winners, having done more than anyone to lighten Churchill's load. 1934 V. M. Yeates *Winged Victory* I. xx. 159 Tom was afraid Miller might be feeling his responsibility and want to do an undue amount of war-winning. 1956 *Nature* 11 Feb. 251/1 This was largely due to the efforts of..Sir Henry Tizard, whose foresight resulted in such war-winning devices as radar. 1974 P. Gore-Booth *With Great Truth & Respect* 123 Their object was to go hell-bent for victory with all the stupendous war-winning momentum which the United States developed.

10. *war-battered, -bitten, -bleached, -blinded, -bright, -brutalized, -devastated, -mazed, -orphaned, -ravaged, -scarred, -shattered, -torn, -wounded* (freq. *absol.*), *-wracked* adjs.; *war-laden* adj.

1938 W. B. Yeats *Herne's Egg* ii. 12 A weather-stained, war-battered Old campaigner such as I. 1942 W. S. Churchill *End of Beginning* (1943) 220 We recreated and revivified our war-battered Army. 1931 —— *World Crisis* VI. xxi. 323 A hundred and twenty-five thousand ragged, war-bitten men. 1941 L. B. Lyon *Tomorrow is a Revealing* 28 Eyes, and the ploughshare, baulk at the recovery Of skeletons war-bleached a grave ago. 1960 *Farmer & Stockbreeder* 15 Mar. 99 It is made by Garden Machinery, Ltd., of Slough. Director of the firm,..extreme right, is a war-blinded South African. 1970 R. Lowell *Notebk.* 108 Regret those jousting aristocracies, War-bright. 1939 C. Day Lewis *Child of Misfortune* 144 For one of our war-brutalized soldiery, you have considerable perception. 1938 *Times* 24 Aug. 12/1 The removal of import duties on mining, agricultural, and other machinery, ostensibly designed to facilitate the rehabilitation of war-devastated areas. 1900 W. B. Yeats *Shadowy Waters* 33 War-laden galleys, and armies on white roads. 1940 C. Day Lewis *Poems in Wartime* 10 Along this war-mazed valley. 1954 W. Faulkner *Fable* 46 A wealthy American expatriate ..who was supporting near Paris an asylum for war-orphaned children. 1978 *Poland* May 48/3, In the Fifties, twice as many children were born in Poland than was thought appropriate to the poverty of the war-ravaged country. 1857 J. G. Whittier in *National Era* 11 June 94/5 When each war-scarred Continental,..Waved his rusted sword in welcome. 1980 J. Lees-Milne *Harold Nicolson* vii. 114 Harold Nicolson..considered the choice of war-scarred Paris for the site of a peace-seeking meeting a psychological mistake. 1937 *Daily Tel.* 19 Oct. 15/3 (*heading*) War-shattered shrine restored. 1939 L. Jacobs *Rise of Amer. Film* v. xix. 386 Griffith revealed his superficial understanding of the war by remarking his sets for *Intolerance* had been more impressive than anything he saw in war-torn France and Belgium. 1975 'E. Lathen' *By Hook or by Crook* xxiii. 209 It was imperative to get the children out of their war-torn background. 1950 War-wounded [see *team *v.* 1 b]. 1968 *Guardian* 23 Feb. 10/6 Few, other than surgical cases (including occasional war-wounded) were brought to the hospital. 1909 M. B. Saunders *Litany Lane* iv. 43 Women of prayer who had raised just as many waxen palms to altars, in nunnery and in palace, for many a war-wracked generation. 1976 *Billings* (Montana) *Gaz.* 17 June 1-A/3 The driver took 'the route normally taken' by anyone wishing to cross from the Christian to the Moslem sides of war-wracked Beirut.

11. War Ag., colloq. abbrev. of 'War

Agricultural Committee'; **war artist**, an artist employed to provide paintings of a war; **war baby**, (*a*) one born during a war, esp. an (illegitimate) child of a man on active service; (*b*) *slang*, a young or inexperienced officer; (*c*) *U.S. slang*, a bond or the like which is sold during a war, or which increases in value because of a war; **war bag** *U.S.*, † (*a*) = *war budget (*a*); (*b*) a bag containing money, clothing, or other supplies; **war-bird**, (*b*) *fig.*, a fighting aircraft or airman; **war bonnet**, a head-dress decorated with eagle feathers, worn by American Indians; **War Box** *slang*, the War Office; **war bride**, a woman who marries a man who is on active service or a man (esp. a foreigner) whom she met while he was on active service; **war bridle** *Canad.*, a harsh bridle made by placing a loop of rope round the lower jaw of a horse; **war budget**, (*a*) *U.S.*, a packet carried by American Indians, containing amulets and military trophies; (*b*) (in Dict., sense 8 e); **War Cabinet**, a Cabinet with responsibility for the political decisions of a country during a war; **war cemetery**, a cemetery composed of war graves; **war chest**, (a chest or strong box for) funds used in waging war; freq. used *fig.*, esp. of funds used by a political party to finance an election campaign; **war college** chiefly *U.S.*, a college providing advanced instruction for senior officers of the armed services; **war communism**, an economic policy, based on strict centralized control of the economy, adopted by the Bolsheviks during the Russian Civil War (1917–21); **war-correspondent** (earlier examples); **war crime**, an offence against the rules of war, formerly excluding, but since the 1939–45 war including, any such act performed on the orders of a higher authority; **war criminal**, one who has committed a war crime; **war damage**, damage caused by action taken by or against an enemy during a war; hence **war-damaged** *a.*; **war dead** *pl.*, servicemen who have died on active service; **war diary**, (*a*) a diary recording the experiences of an individual during a war; (*b*) (see quot. 1918); **war dream**, a dream about war; **war economy**, (*a*) a measure taken in order to save money or other resources because of a war; (*b*) an economy, characteristic of wartime, in which a large part of the labour force is engaged in arms production, etc., rather than in the production of goods for export or for civilian use; **war effort**, the effort of a nation to win a war, or of an individual group to contribute to that end; **war fever**, an enthusiasm for war; **war-fighting**, the fighting of wars; also *attrib.*; **warfront**, the foremost part of the field of operations of opposing armies; **war-game** (later examples); also used of any game simulating war, esp. an elaborate game played with model soldiers, or of any exercise by which a military strategy is examined or tested; **war-game** *v. trans.*, to examine or test (a strategy or the like); **war-gamer**, one who plays a war-game; **war-gaming**, the playing of war-games; the use of such games to examine or test strategies; **war gas**, a gas or other chemical agent used in war to produce irritant or poisonous effects; **war generation** a generation which has experienced a war; **war grave**, the grave of a serviceman who died from wounds inflicted, accident occurring, or disease contracted on active service; **war-guilt**, the responsibility for having caused a war; freq. with reference to the claim that Germany had caused the war of 1914–18, which was embodied in an article of the Treaty of Versailles (1919); **war-hatchet** (earlier examples); **war-head** (later examples); also, that of any missile, esp. one deriving its destructive power from the release of nuclear energy; **war hero**, a man who has acted heroically in a war; also **war heroine**; **War House** *slang*, the War Office; **war hysteria**, unhealthy emotion or excitement caused by war; an enthusiasm for war; **war machine**, (*a*) an instrument or weapon of war; (*b*) *transf.*, the military resources of a country organized for waging war; **war marriage**, a marriage taking place in wartime, esp. one in which the bridegroom is on active service; **war medicine** *N. Amer.*, (a form of) magic formerly used by North American Indians to ensure success in war; also *fig.*; **war memorial**, a monument, etc., commemorating those (esp. from a particular locality) killed in a war, and freq. inscribed with their names; **war-mind**, a mind attuned to or desirous of war; hence **war-minded** *a.*, having such a mind; **war-mindedness**; **war museum**, a museum of the history of warfare in general, or of warfare during a particular period; **war orphan**, a child orphaned by war; **war pension**, a pension paid to someone disabled or widowed by war; **war picture**, (*a*) a painting of which the theme is war; (*b*) a photograph of a scene from the theatre of war; also, a documentary film of action from a war, and *transf.* a written account of this; (*c*) a cinematographic film with war as its subject or background (the usual sense); cf. *war film, movie* (sense 8 f above); **warplane**, an aeroplane equipped for fighting, bombing, etc., in war-time; **war refugee**, one who seeks refuge in another country, etc., from the effects of war; a displaced person; **war reporter** = *war-correspondent*; hence **war-reporting**; **war resistance**, opposition to war, pacifism; hence **war resister**, an opponent of war or of a particular war; **war risk** *Insurance* (chiefly *Marine*), a risk of loss, etc., during war-time; freq. in *pl.* and *attrib.*; **war road** *N. Amer.* = War-path a (concr.); **war room**, a room from which a war or part of a war is directed; **war-substantive** *a.* [Substantive *a.* 1 e], confirmed (in a rank) for the duration of a war; **war-talk**, (*b*) talk about war in general; **war toy**, a toy with which a child can play war-games; **war trial**, the trial of a person for a war crime or crimes; cf. *Nuremberg trial(s)* s.v. *Nuremberg 2; **war veteran** orig. *U.S.* = *veteran *sb.* 1 b; **war-weary** *a.*, (*a*) (see sense 10 in Dict.) (earlier example); (*b*) *U.S.*, *spec.* applied to aircraft badly damaged in war-time, and which are withdrawn from service for repair, conversion, or scrapping; also *ellipt.* as *sb.*; **war wedding** = *war marriage; **war work**, special work occasioned by war, and which is intended to advance the war effort; **war-worker**, a person undertaking war work; also *transf.*; **war-worthy** *a.*, suitable for or befitting war; so **war-worthiness**.

1949 E. Coxhead *Wind in West* vi. 152 The farmer I stay with there is a member of the War Ag. 1970 G. E. Evans *Where Beards wag All* ix. 106 When the War Ag. (Agricultural Committee) took over I asked 'em would they send the gyro-tiller. 1890 Kipling *Light that Failed* (1891) xii. 237 Some man unknown who would be employed as war artist by the Central Southern Syndicate. 1981 S. Chitty *Gwen John* ix. 148 Augustus..was in France as a war artist. 1901 E. W. B. Morrison *With Guns in S.A.* xxxiv. 239 (*caption*) Mrs. Jourdain's 'war baby'. 1917 'Contact' *Airman's Outings* 35 Even these war babies (three of them died very gallantly before we re-assembled for breakfast next day) had bottled most of their exuberance. 1917 R. W. Lardner *Gullible's Travels* 83 'You forgot somethin',' she says, 'You forgot them war babies.' Did I tell you about that? Last fall I done a little dabblin' in Crucial Steel. 1935 I. Miller *School Tie* xv. 286 It was possible to join a Junior Training Battalion—commonly known as the War Babies' Brigade—at the age of seventeen and a half. 1948 *Green Bay* (Wisconsin) *Press-Gaz.* 13 July 4/2 The idle rich of Europe..clamored for war while they invested great amounts in American war babies and reaped superlative profits. 1974 G. Butler *Coffin for Canary* ix. 105 Born Belfast, 1944, so she was just a war baby. 1820 *Western Rev.* II. 48 After the action is over, each person returns his war bag to the commander of the party. 1897 A. H. Lewis *Wolfville* 33 S'pose you-alls gropes about in your war-bags an' sees. I'm needin' of a drink mighty bad. 1933 J. V. Allen *Cowboy Lore* I. 6 What's known as the 'war bag' is carried by many of the boys in their beds to protect their wardrobe, tobacco, etc., and may be anything from a flour sack to a rather pretentious container. 1972 F. van W. Mason *Roads to Liberty* 241 Higsby fumbled in his war bag. 1917 G. Frankau *City of Fear* 3 Above, The war-birds beat And whistle. 1936 'R. Hyde' *Passport to Hell* 206 German and British warbirds were mixing it in an aerial free-for-all. 1981 *Pilot* Jan. 23/2 Some of the war-birds flying today are quite bent, cracked, patched up. 1845 J. C. Frémont *Rep. Exploring Exped.* 134 Indians ..with the long red streamers of their war bonnets reaching nearly to the ground. 1928 'Brent of Bin Bin' *Up Country* xxii. 356 Adjusting her widow's cap like a war-bonnet, she arose to her full height of five-feet-one-and-a-half. 1973 A. H. Whiteford *N. Amer. Indian Arts* 151 The flowing war bonnet of the Plains has become the symbol of the American Indian. 1952 M. Allingham *Tiger in Smoke* i. 12 The War Box cited him 'Missing believed killed'. 1969 M. Pugh *Last Place Left* xxix. 213, I flit between Downing Street and the War Box and the Ministry of Defence. 1918 A. Bennett *Pretty Lady* xi. 61 She was becoming hysterical: the special liability of the war-bride. 1939 *Daily Tel.* 18 Dec. 9/5 Silver tea and coffee sets are being bought..as gifts to the many war brides unable to set up homes. 1978 J. Krantz *Scruples* i. 10 She was as alert as a vixen, as humorous as the song by Maurice Chevalier after which her homesick war-bride mother had named her. 1962 J. Onslow *Bowler-hatted Cowboy* xviii. 175 One summer day, they [sc. two horses] came home, gaunt, their heads bloody and scarred from 'war bridles' with which someone had tried to halter-break them. 1813 R. Stuart *Jrnl.* 14 Apr. in *Discovery Oregon Trail* (1935) x. 236 A Pole surpassing in height any put in the roof, is put out at the chimney where are suspended their Medicine Bags and War Budgets carefully concealed in innumerable wrappers. 1916 *Times* 9 Dec. 9/2 It is an immense gain to have the Prime Minister definitely and irrevocably committed to the creation of a small War Cabinet, constantly..devoted to the prosecution of the war. 1940 J. Reith *Diary* 5 Jan. (1975) vi. 237, I asked if the job carried War Cabinet rank and he said no. 1980 P. Fitzgerald *Human Voices* ii. 38, I don't know who authorised him to speak. I understand it was the War Cabinet. 1922 *Encycl. Brit.* XXXII. 953/1 It is possible that the conspicuous success with which Arlington Cemetery has been designed had a share in influencing the Imperial War Graves Commission in the construction of the British war cemeteries on somewhat similar lines. 1982 'J. Gash' *Firefly Gadroon* xv. 145 There's a turning through the woods where the American War Cemetery stands. 1901 War chest [in Dict., sense 8 e]. 1912 W. Deeping *Sincerity* xvi. 124 He had about forty pounds left, no great sum to start a war-chest with. 1932 *Sun* (Baltimore) 30 Aug. 1/6 (*heading*) War chests practically empty, parties curtail on campaign. 1973 R. L. Simon *Big Fix* iv. 34 All the guilt-stricken celebrities contributing to their war chest. 1894 *Abstract of Courses* (Naval War Coll.) 3 The summer course at the Naval War College began on the 13th of June. 1913 R. Meinertzhagen *Diary* 1 Nov. (1960) 56 A joint war college for all branches of Government Services [in India] would be a God-send. 1978 H. Wouk *War & Remembrance* i. 11 Talking in a calm War College vein. 1928 M. Dobb *Russian Econ. Devel. since Revolution* iii. 64 'War communism', accordingly, sprang into life in the 'forcing-house' of a mortal struggle of the new régime—a struggle in which all things were subordinated to military necessity, and the problems of industry were simply regarded as..the problem of military supplies. 1965 B. Pearce tr. *Preobrazhensky's New Economics* 32 The economics of War Communism were those of a state economy of the war-consumption type, when we were not accumulating but were forced to spend our resources. 1844 War-correspondent [see Correspondent *sb.* 4 b]. 1870 A. Maverick *Henry J. Raymond & N.Y. Press* 256 The 'war correspondents' who had been sent out to the battle-fields to represent the newspapers of New York throve and grew famous. 1906 L. Oppenheim *Internat. Law* 264 Violations of rules regarding warfare are war crimes only when committed without an order of the belligerent government concerned. If members of the armed forces commit violations by order of their government, they are not war criminals and may not be punished by the enemy; the latter can, however, resort to reprisals. 1945 *Daily Express* 16 May 1/1 The United Nations War Crimes Commission announced last night: Hermann Goering's name was placed..on the first list of persons charged with war crimes. 1980 *Oxf. Compan. Law* 1288/2 After World War II, three classes of offences against international law came to be regarded as war crimes, crimes against peace, as by planning or waging a war of aggression, conventional war crimes, or violations of the accepted laws or customs of warfare and crimes against humanity, including extermination, enslavement, deportation and other inhumane acts. 1906 War criminal [see *war crime]. 1929 W. S. Churchill *World Crisis* V. viii. 158 An article of the Peace Treaty obliged the Germans to stigmatize all their greatest men and potentates as War Criminals. 1943 *Ann. Reg.* 1942 190 The question of the trial of war criminals ruffled..the ever-growing friendliness between Britain and America. 1981 J. Wainwright *Urge for Justice* 121 An organisation devoted to tracking down war criminals. 1939 *Act 2 & 3 Geo. VI* c. 72 § 4 Where the land comprised in a lease is unfit by reason of war damage, the following provisions ..shall have effect. *Ibid.* § 24 'War damage' means damage caused by, or in repelling, enemy action, or by measures taken to avoid the spreading of the consequences of damage caused by, or in repelling, enemy action. 1950 E. Hyams *From Waste Land* 10 Sharp and hopeful landlords claimed war-damage compensation. 1975 J. Cleary *Safe House* iv. 177 The walls were spattered with bullet and shrapnel marks..all war damage. 1946 *Mind* LV. 380 The Secretary reported appeals from war-damaged libraries in Europe. 1978 *Times* 8 May 9/8 Volunteers.. shivering in war-damaged, makeshift offices. 1969 J. Burmeister *Hot & Copper Sky* i. 17 They're French war dead. From the Indo-China campaign. 1917 W. J. Locke *Red Planet* i. 4 To fill in my time, I first started.. a sort of War Diary. 1918 E. S. Farrow *Dict. Mil. Terms* 657 *War diary*, a record of events kept in campaign by each battalion and higher organization, each ammunition, supply, engineer, and sanitary train. 1937 Kipling *Something of Myself* iii. 49 An accursed Muscovite paper..published the war diaries of Alikhanoff, a Russian General. 1955 E. Waugh *Officers & Gentlemen* 142 Guy chalked the nightly wanderings of the Commandos on.. his map and recorded them next day in the War Diary. 1981 J. Barnett *Firing Squad* ii. 105 The War Diary of Sergeant Michael Lugard. 1918 W. Owen *Let.* 18 Feb. (1967) 534, I confess I bring on what few war dreams I now have, entirely by willingly considering war of an evening. 1919 W. B. Yeats *Cutting of Agate* 16 The Print Room of the British Museum is now closed as a war-economy. 1940 *Economist* 3 Feb. 189/1 The problem of war economy is to man and equip the Forces, to raise output for war and export needs to the utmost and to cut down civilian consumption. 1948 [see *peace economy* s.v. *peace *sb.* 15 d]. 1972 M. J. Bosse *Incident at Naha* 56 Like any girl caught up in a war economy. She had a pimp. 1919 *Maclean's Mag.* Jan. 49 (*heading*) Britain's wonderful war effort. 1934 W. S. Churchill *Marlborough* II. v. 101 Whigs and Tories alike wished the fleet to be used as a part of the main war-effort. 1954 N. Coward *Future Indefinite* iv. vii. 194 A job which..would be of real value to the war effort. 1977 A. Wilson *Strange Ride Rudyard Kipling* vii. 299 He would never have repeated the story lest it weakened our war effort. 1812 J. Steele *Papers* (1924) II. 668 The late report of the Secty. of the

Treasy. will cool the war fever in some. 1908 H. G. Wells *War in Air* vi. 180 To the normal high-strung energy of New York streets was added a touch of war-fever. 1978 N. Gosling *Paris 1900–1914* 187 In 1912 the slowly developing war fever..began to show itself in sinister local symptoms. 1965 H. Kahn *On Escalation* 284 Deterrence-only is the opposite of 'war-fighting'. 1983 *Listener* 10 Feb. 7/1 They are war-fighting weapons with a first-strike capability. 1950 *Sun* (Baltimore) 29 June 1 General MacArthur left for the South Korean warfront today. 1976 *Billings* (Montana) *Gaz.* 11 July 1-A/2 Lebanon's three major warfronts resounded to artillery, rocket and mortar fire. 1910 H. G. Wells *New Machiavelli* (1911) i. iii. 84 The spectacle of volunteer officers fighting the war game in Caxton Hall. 1951 D. Knight *In Deep* (1964) 92 The cadets..carrying out one of the prescribed games under the direction of student squad leaders. 1966 *Punch* 6 July 26/2 Entertaining incidental scenes (the children's war games, the husband's home movies, haggling over the junk) keep interest always alive. 1967 *Guardian* 16 Oct. 8/5 The National Wargame Championships organised by the British Model Soldier Society. 1970 *Time* 5 Oct. 13 At one point Nixon told Kissinger: 'Let's you and me war-game this,' and they worked the plans over to see, as Nixon put it, 'where the weak points might be.' 1975 *Times* 2 June 13/1 Politicians of all parties cooped up in..Westminster have become so absorbed in their own war-games that they have lost touch with the wider world. 1981 *Washington Post* 8 Nov. L1/6 'Well,' Wakko said, 'I've got to go back to work. We're war-gaming an LNW in Monaco.' 1967 *Guardian* 16 Oct. 8/5 One thing only is causing the wargamers concern. There are so many different societies in the field. 1982 M. Leapman *Yankee Doodles* IV. 208 War-gamers are not the only people undertaking such simulations. 1954 McCloskey & Trefethen *Operations Research for Management* I. 15 They used the technique of war-gaming to develop models of possible operations, then 'tested' various tactics and weapons. 1970 *Daily Tel.* (Colour Suppl.) 30 Oct. 43/2 Today war-gaming has reached a point of sophistication where one almost needs a computer to play. 1980 J. McNeil *Spy Game* ix. 96 War gaming is like that, dashed unpredictable. 1934 Webster, War gas. 1939 L. W. Marrison tr. M. Sartori's *War Gases* p. viii, The most efficient war gases are organic compounds, the inorganic compounds which have great toxicity being unsuitable for use owing to their physical and chemical properties. 1974 M. C. Gerald *Pharmacol.* vii. 134 In 1968, 6000 sheep were accidentally killed in Utah, allegedly as a result of exposure to the war gas VX that was being tested by the Army about 17 miles away. 1930 W. S. Maugham *Bread-Winner* i. 18 They were a dreary lot that war generation. 1978 Cadogan & Craig *Women & Children First* ii. 47 The sense of isolation that characterized the war generation. 1917 *Imperial War Conf.* 5 in *Parl. Papers* (Cd. 8566) XXIII. 323 The Conference..humbly prays His Majesty to constitute by Royal Charter an Imperial War Graves Commission. 1945 J. Reith *Let.* 11 June in *Diaries* (1975) vii. 350, I should have thought..you would have welcomed the establishment of an imperial corporation—the first to be achieved (except War Graves). 1981 J. Barnett *Firing Squad* xii. 175 Search all names against..war casualty lists, ditto War Graves Commission. 1922 *Nation* 9 Sept. 758/1 (*heading*) The myth of war-guilt. 1940 W. Temple *Thoughts in Wartime* ii. i. 60 The war-guilt clause, against which many of us have protested. 1971 *Guardian* 5 Aug. 12/3 Concepts of war 'guilt' derived from the Second World War have encouraged some elements of the Left to identify the war in Vietnam with an emerging American fascism. 1981 J. B. Hilton *Surrender Value* xi. 88 And she had been intense: about war-guilt, about Borchert and Kafka. 1760 G. Croghan *Jrnl.* 4 Dec. in R. G. Thwaites *Early Western Trav.* (1904) I. 116 That you [sc. chiefs and warriors].. may..bury the War Hatchet in the Bottomless Pitt. a1818 B. Hawkins *Sk. Creek Country* (1848) 72 He lifts the war hatchet against the nation which has injured them. 1944 *Sun* (Baltimore) 20 June 3/3 Explosive carried in a warhead [of a German robot plane] is equal to a 2,200 pound German bomb. 1955 *Bull. Atomic Sci.* Apr. 168/3 In the not too distant future we can foresee the dominance of intercontinental guided missiles with hydrogen war heads. 1978 R. V. Jones *Most Secret War* xlv. 447, I was now prepared to call everyone else's bluff, and declare for a rocket of 12 tons all-up weight with a 1 ton warhead. 1898 *Kansas City Star* 19 Dec. 1/5 Following are the names of the members of the entertainment committee who received the war heroes. 1953 L. P. Hartley *Go-Between* iv. 59, I already felt violently jealous of Trimingham, and the fact that he was a war-hero did not recommend him to me. 1982 T. Allbeury *Shadow of Shadows* v. 44 Your father was a war hero. He was awarded..the Legion of Honour. 1932 *New Yorker* 9 Jan. 34/1 On the sidewalks are..a few war heroines with nothing to sell. 1979 'D. Kyle' *Green River High* viii. 106 We..read: *War heroine returns to Sarawak*. 1925 Fraser & Gibbons *Soldier & Sailor Words* 300 *War House, the*, General Staff slang for the War Office. 1926 'Sapper' *Final Count* xii. 302 They thought I was mad at the War House. 1978 D. Wheatley *Officer & Temporary Gentleman* iii. 29 A man in control at the War House who had an enormous hold upon the popular imagination. 1940 'G. Orwell' *Inside Whale* 172 The very people who..had sniggered over their own superiority to war hysteria were the ones who rushed..into the mental slum of 1915. 1968 O. Wynd *Sumatra Seven Zero* ii. 20 Birgid is the child of my war hysteria. Her father was a blond Norwegian. 1881 W. D. Hay *300 Years Hence* iv. 70 The last inventions in war-machines. 1914 W. J. Bryan *Mem.* (1925) 390 The allies see peace only in a success so signal as to crush the German war machine. 1979 *Sci. Amer.* Mar. 123/1 With the introduction of catapults, together with other war machines just coming into use in the West, sieges became more effective. 1981 B. Langley *Autumn Tiger* xv. 244 Stalin..was wary of the German war machine. 1921 'C. Dane' *Bill of Divorcement* ii. 54 If it hadn't been for the war—and the war marriages. 1805 W. Clark *Jrnl.* 11 Jan. in *Lewis & Clark Exped.* (1904) I. vi. 247 Some of our Men go to See a War Medeson made at the Village on the opposite Side of the river. 1893 *Chicago Tribune* 28 Apr. 4/1 Gov. Altgeld..proceeded to administer a dose of war medicine he had been making for some time. 1962 E. E. Evans-Pritchard in *Ess. Social Anthropol.* v. 95 He used some of these forms of magic himself, getting old commoners to bring the medicines and perform the rites, except in the case of the war-medicines, which he administered himself, from the large bongo horn in which they were kept. 1912 War memorial [see *late *a.*¹ 3]. 1930 Kipling *Limits & Renewals* (1932) 324 The little cast-iron *poilu*, which seemed to be standard pattern for War memorials in that region. 1980 P. Lively *Judgement Day* v. 55 The starling flew across the nave, crashed into the War Memorial window..and thumped to the ground. 1928 Blunden *Undertones of War* iii. 27 The joyful path away from the line..was full of pictures for my infant war-mind. 1932 H. Crane *Let.* 13 Apr. (1965) 409 Dos Passos has written a very important record of the war and the 'war mind' in 1919. 1936 *Mind* XLV. 289 A society which prefers war to peace and organises itself for success in war, may be rational in the above sense, if (a) the majority of its members are genuinely war-minded, and (b) the small minority of pacifists in its ranks is allowed to express.. its dissenting opinion. 1948 W. Fortescue *Beauty for Ashes* xxii. 172 At intervals it stopped..to allow war-minded little boys to finger the ugly noses of guns. 1936 H. Read *Surrealism* 36 Motives no less irrational than those which promote war-mindedness. 1917 *Times* 20 Feb. 11/3 At No. 6, Avenue de Malakoff..in two spacious first-floor *appartements*..is housed the War Museum which, when complete, will be presented to the French nation. 1967 O. Wynd *Walk Softly, Men Praying* x. 165 An inspirational experience in a war museum. 1979 E. Bercovici *Wolf Trap* 161 The War Museum on the upper fortress fascinated him. 1915 W. Owen *Let.* 29 June (1967) 342 All France is collecting for its War Orphans. 1971 H. McCloy *Question of Time* i. 14 The nuns kept on trying to trace the families of war orphans left in their care. 1930 E. H. Young *Miss Mole* xxv. 224 The little poultry farm which was to supplement..the hero's war pension. 1980 *Daily Tel.* 24 Apr. 14/5 His mother brought him up alone on a war pension plus what she could make by smocking children's clothes. 1883 B. Potter *Jrnl.* 28 Apr. (1966) 39 First we went to the Fine Arts Gallery..to see..the Egyptian war pictures. 1900 (*title*) War pictures. 1914 R. Grau *Theatre of Science* ii. 40 The war pictures released by this company reflected the high aims of a man. 1915 V. Woolf *Diary* 25 Jan. (1977) I. 28 The Picture Palace was a little disappointing—as we never got to the War pictures, after waiting 1 hour & a half. 1946 J. B. Priestley *Bright Day* viii. 246 Honest war pictures, made on the spot here by people who know what it's like. 1978 *Listener* 30 Mar. 410/1 *In Which We Serve* I remember as the best war picture that I have ever seen. 1911 *Flight* 16 Dec. 1078/2 No one has any very definite idea of what the future type of war-plane will be like. 1938 C. Day Lewis *Overtures to Death* 17 Oh, look at the warplanes! Screaming hysteric treble In the long power-dive, like gannets they fall steep. 1967 A. MacLean *Where Eagles Dare* xi. 230 The Mosquito bomber, all engines and plywood, was, he was well aware, the fastest warplane in the world. 1978 *Guardian Weekly* 4 June 16/1 Categories so politically volatile as warplanes. 1942 D. Powell *Time to be Born* vii. 163 Amanda was nobly..adopting a war refugee. 1973 'B. Mather' *Snowline* iii. 33 There are three million war refugees from Bangladesh in West Bengal. 1950 E. H. Gombrich *Story of Art* 118 The final result is possibly more impressive than the accounts of our own war reporters and newsreel men. 1976 S. Hynes *Auden Generation* x. 342 Compared to war-reporting of the Spanish war..*Journey to a War* is superficial and uninformative. 1932 *Week-End Rev.* 19 Nov. 611/2 We appeal to those who wish to take part in a truly practical and effective effort at war resistance to send us a donation. 1935 J. Bell in *We did not Fight* p. xviii, The most active and ardent war resisters..are more likely to take the line of revolutionary action than conscientious objection. 1976 *Pacifist* Jan. 10/1 We remain an organisation of war-resisters. 1920 *Lloyd's List Law Rep.* 22 July 288/2, I am perfectly clear this is a case that is not brought within the War Risks Policy. 1934 Webster, *War risk insurance*, term insurance written by the United States Government for members of the military and naval forces. 1939 *Country Life* 11 Feb. 133/2 An insurance against war risks should be attached to Schedule 'A'. 1974 E. R. H. Ivamy *Marine Insurance* (ed. 2) xv. 219 The term 'war risk' in a marine policy has been held to include a civil war. 1782 in V. W. Howard *Bryan Station Heroes & Heroines* (1932) xii. 144 On the Southward side below where the War road crosses the said fork. 1968 E. Russenholt *Heart of Continent* II. iii. 41 Canadians and Indians follow the old Indian war road. 1914 A. Wilson *Let.* 29 Oct. in M. Gilbert *Winston S. Churchill* (1972) III. Compan. I. 233, I should like to have a room set apart for me near the War room. 1976 J. Lee *Ninth Man* I. 82 The War Room occupied the southwest corner of the main floor in the White House. a1944 K. Douglas *Alamein to Zem Zem* (1946) ii. 13 He had..returned to his war-substantive rank of captain. 1965 *New Statesman* 10 Dec. 919/1 How could a poor war-substantive captain hope to hold his own against someone like Colonel Passy. 1861 M. B. Chesnut *Diary* 23 Apr. in C. V. Woodward *M. Chesnut's Civil War* (1981) 53, Maria—are you crying because all this war talk scares you? 1939 C. Day Lewis *Child of Misfortune* III. i. 264 You're not letting this war-talk throw a scare into you? 1973 M. Kaye *Toy is Born* xv. 144 It is interesting to note that the hue and cry against war toys a few years ago had little effect on Avalon Hill. 1949 R. Chandler *Let.* 25 Feb. (1981) 149 There is an element of hypocrisy in these war trials. 1971 P. D. James *Shroud for Nightingale* vii. 242 He'd seen her before... In Germany. She was in the dock. It was a war trial. 1906 *N.Y. Even. Post* 29 Jan. 1 A guard of honor selected from the ranks of the Spanish war veterans here. 1980 J. McClure *Blood of Englishman* x. 92 War veterans... It'd all gone a bit to their heads. 1895 W. B. Yeats *Poems* 7, I have not yet, war-weary king, Been spoken of with any man. 1945 *Sat. Even. Post* 17 Mar. 20 Thousands of once precious B-17's are now 'war-wearies'. Not worth salvaging, they clutter up foreign and domestic airfields. 1945 *Fortune* Aug. 208 Five war-weary Liberators, described with horrors by their pilots as 'clunkers'. 1915 *Truth* 4 Aug. 181/2 What do we hear from London about war-weddings? 1890 Kipling *Light that Failed* (1891) ii. 24 Do you want me to do war-work? *Ibid.* iv. 64 He has thrown up war work. 1916 A. Huxley *Let.* 2 Mar. (1969) 92 A friend of mine at Magdalen, a Quaker..objected to war-work of any kind, combatant or non-. 1954 W. K. Hancock *Country & Calling* vii. 189 The answer to that difficulty was for my wife to take up paid war work in place of the voluntary work she had been doing in Birmingham. 1977 *Belfast Tel.* 14 Feb. 9/6 Rene, who was in Mackie's on war work, lived with her widowed father and looked after her young brothers. 1915 *Political Q.* May 108 It is not clear whether or no the special.. war-workers..will be permanently shut out of the trades. 1930 Blunden *De Bello Germanico* iv. 41 War-workers varying from whizzbangs to woolly bears. 1978 Cadogan & Craig *Women & Children First* ii. 48 The experiences of the war workers had been thoroughly documented. 1909 *Q. Rev.* Oct. 578 The aim must now be..to seize every opportunity to improve its war-worthiness. 1908 Hardy *Dynasts* III. vii. viii. 510 Ney holds indignantly that such a feint Is not war-worthy.

waragi (wa·rāgi). [ad. Swahili *wargi*.] In Uganda, a potent alcoholic drink made from bananas or cassava.

1916 in C. Griffin *Laws Uganda Protectorate* (1924) II. 1031 'Native liquor' means any liquor prepared or manufactured in accordance with native custom and includes ..Waragi. 1935 Thomas & Scott *Uganda* xix. 295 It is forbidden by the Liquor Ordinance, 1916, that any person shall supply natives with any liquor other than..beer.. and the Rules promulgated under the Ordinance make it an offence..to distil or even to possess the raw cassava spirit known locally as *waragi*. 1973 *Guardian* 13 Jan. 13/3 The ferocious Ugandan spirit Waragi. 1976 *Listener* 19 Aug. 195/3 He had had three tots of Waragi (a highly potent Ugandan gin made from bananas). 1977 *Time* 7 Mar. 18/3 They [sc. Ugandans] would rather have scarce butter or a slab of meat or a bottle of *waragi*, a potent, banana-based liquor.

Warao, var. *Warrau.

warb (wǭɹb). *Austral. slang.* [Perh. f. Warb(le *sb.*²)] A lazy, unkempt, or contemptible person (see also quot. 1959).

1933 L. Robinson in Murdoch & Drake-Brockman *Austral. Short Stories* (1951) 215 We were both of us what, in the back country, are called 'warbs', meaning confirmed and irredeemable loafers. 1959 Baker *Drum* (1960) II. 155 *Warb*, a low-paid manual worker. 2. A dirty or untidy person. 3. A simpleton or fool. 1967 K. Tennant *Tell Morning This* 201 But it's a no-hoper's jail—a lot of old warbs and kids mixed up with coves like Amos the Cannibal and chaps that razors bounce off.

warble, *sb.*¹ Add: **2.** Special combination. **warble tone** *Physics*, a constant amplitude tone whose frequency is cyclically varied between certain limits, used in acoustic measurement to avoid irregularities associated with the use of single frequencies.

1933 *Proc. IRE* XXI. 1183 The introduction of a 'warble tone' into the technique of acoustical measurements..has also supplied a growing interest in this kind of oscillation. 1958 H. J. Gray *Dict. Physics* 527/1 A warble tone can be produced from an oscillator by rotating a small variable condenser in the tuned circuit at a constant speed. 1971 B. J. Smith *Acoustics* v. 96 To attempt to eliminate the variations due to room modes for each frequency a band of noise is used, either in the form of white noise or in warble tones.

warble, *v.*¹ Add: **4. c.** Of telephones (*spec.* Trimphones): to make a distinctive trilling sound.

1965, 1969 [see *Trimphone]. 1973 G. Moffat *Deviant Death* v. 68 The telephone was warbling softly. 1981 T. Barling *Bikini Red North* x. 206 The telephone warbled. ..'You must answer it.'

warbler. Add: **1. c.** *slang.* A female singer.

1946 B. Treadwell *Big Bk. Swing* 125/2 *Warbler*, girl singer. 1961 *Times* 21 Nov. 13/2 Barbara Holt, making her stage debut, displayed the promise of an uncommon warbler. 1981 *TV Picture Life* Mar. 16/1 (*caption*) Regardless of where her love life leads, the warbler is very much in demand for films these days.

d. *colloq.* A telephone which warbles. Cf. *warble *v.*¹ 4 c.

1973 G. Moffat *Deviant Death* vii. 106 'I didn't hear a telephone, did you?' 'It's one of the new warblers.'

Warburg (wǭ·ɹbǝ̆ɹg). *Biochem.* The name of Otto *Warburg* (1883–1970), German biochemist, used *attrib.* and in the possessive to denote apparatus for the study of the metabolism of small pieces of tissue by the manometric measurement of the rate of oxygen consumption and carbon dioxide production, a technique he pioneered.

1930 *Biol. Abstr.* IV. 2745/1 By use of suitable suspensions of *B. coli* in Warburg's apparatus a biological, direct (manometric) method of measuring the hexoses is described. 1946 *Nature* 3 Aug. 155/2 The kinetics of these reductions may be followed manometrically in the Warburg apparatus by carrying out the reactions in an atmosphere of carbon dioxide. 1975 K. Wilson in Williams & Wilson *Biologist's Guide to Princ. & Techniques Pract. Biochem.* viii. 233 (*caption*) Diagrammatic representation of a Warburg manometer.

Warburgian (wǭːɹbɜ̄·ɹgiăn), *a.* [f. the name of Aby *Warburg* (1866–1929), German-Jewish cultural historian + -IAN.] Of, pertaining to, or characteristic of Warburg or his work, or the Warburg Institute, founded (1904) by him in Hamburg as Kulturwissenschaftliche Bibliothek Warburg but subsequently (1933) transferred to London. Hence **Warbu·rgianism.**

1956 A. Wilson *Anglo-Saxon Attitudes* II. ii. 273, I have neither an aesthetic inclination towards it [*sc.* Renaissance paganism] nor a Warburgian interest in the development of myth. **1958** *Times Lit. Suppl.* 23 May 277/1 It is true that when people ask the staff of the Institute what the 'Warburgian Method' really is they sometimes receive the answer that the whole point of 'The Warburg' is that it has no method. **1974** K. Clark *Another Part of Wood* v. 190 The parts of my writing that have given me most satisfaction, for example, the chapter in *The Nude* called 'Pathos', are entirely Warburgian. **1977** *N.Y. Rev. Bks.* 24 Nov. 36/3 In nearly all his studies Gombrich follows the Warburgian practice of studying subject rather than form; but it is a humanized Warburgianism.

warby (wǭ·ɹbi), *a. Austral. slang.* [f. *WARB + -Y¹.] Unprepossessing in appearance or disposition; unkempt; disreputable, contemptible, decrepit.

1941 K. Tennant *Battlers* xviii. 207 'Of all the warby ideas,' he said.., 'the warbiest is you going on your own.' **1949** R. Park *Poor Man's Orange* 181 Yeah, there she was, in a warby kind of blue dress, and low-heeled shoes. **1959** D. Niland *Big Smoke* 183 A warby unshaven young man in working clothes walked through and right up to him at the back. **1965** E. Lambert *Long White Night* 135 That was one of the funniest sights the main street ever saw—my old man's warby old Model A towing Foran's dirty big gleaming new Packard! **1973** J. McNeil *Old Familiar Juice* 74 He's down there whackin' up bumpers with a couple of 'is warby mates.

warcraft. Add: **2.** (Later examples.) Also, a warship.

1918 *Chambers's Jrnl.* May 318/1 With a whisk of her stern, the warcraft stood off. **1927** *Daily Tel.* 1 Mar. 11/3 It is believed here that the British and Japanese acceptances will pave the way for a conference of these Powers with the United States to consider the limitation of warcraft other than capital ships in line with the principles of the Washington Conference. **1930** *Tablet* 16 Aug. 202/2 These scribblers..would have us..cease building war craft.

war-cry. b. (Earlier examples.)

1836 S. Houston *Let.* 25 Apr. in W. B. Dewees *Lett. from Early Settler Texas* (1852) xix. 198 Col. Sherman, with his regiment,..rung the war cry, '*Remember the Alamo*'. **1837** Dickens *Let.* 24 Sept. (1965) I. 312 Of course we refused it—a new agreement and copyright, being the War Cry.

ward, *sb.*² Add: **II. 6. a.** *Ward in Chancery* (earlier example); *ward of Court* (examples).

1790 *Stevens v. Savage* in *Eng. Rep.* (1903) XXX. 277 Stevens having been committed for a contempt, by having married Miss Jeffry, a ward of the Court. **1815** H. Maddock *Treat. Princ. & Pract. High Court of Chancery* I. ii. 264 If a Child, a Ward of Court, would not be safe, the Chancellor would not permit it to go to *Scotland*. **1837** C. Selby (*title*) The Irish Dragoon; or, Wards in Chancery. **1875** A. H. Simpson *Treat. Law & Pract. Infants* viii. 145 The general rule of the Court is that a ward of Court may not be removed out of the jurisdiction. **1928** A. Bicknell *Law & Pract. Infants* iv. 89 To remove a ward of Court from the jurisdiction without proper leave is contempt of Court. **1977** *Jersey Even. Post* 26 July 13/4 If the children were allowed to leave, they could be made wards of Court and could find themselves in a position similar to that of the recent 'tug of love' children.

V. 18. (Later examples.)

1961 R. Shaw *Sun Doctor* I. i. 47 She looked at the two main wards of the hospital—long low rooms divided into cubicles like an old dormitory in an English public school. **1975** I. Illich *Medical Nemesis* vi. 111 Ailments had to be turned into objective diseases. Species had to be clinically defined..so that officials could fit them into wards. **1982** B. Trapido *Brother of More Famous Jack* xlvii. 193, I was required, during the first three months, to spend occasional week-long spells in hospital in a special ward for the observation of problem pregnancies.

19. b. An administrative division of the Mormon Church (the Church of Jesus Christ of Latter-Day Saints).

1859 *Mountaineer* (Salt Lake City) 27 Aug. 2/4 If the water-masters of our district or ward will see that we have a double portion of water during the ensuing week for our garden, we will now agree not to mention them again. **1925** M. R. Werner *Brigham Young* xii. 422 The bishop was in charge of all the families of his ward. **1979** M. P. Leone *Roots Mod. Mormonism* ii. 36 Wards average about seven hundred people.

VII. 25. a. (sense 18) *ward clerk, sister*; (sense 19 a) *ward boss* (earlier example), *committee, committeeman, politician, politics* (examples), *school*; (sense 19 b) *ward teacher*.

1890 T. Roosevelt *Wks.* (1926) XIV. 110 Many forces..combine to produce the ward boss, the district heeler, the boodle alderman. **1964** G. L. Cohen *What's Wrong with Hospitals?* iii. 52 Ward clerks will relieve Sisters of paper work. **1807** *Salmagundi* 2 June 212 The secretaries of the ward committees strut about looking like wooden oracles. **1922** C. E. Merriam *Amer. Party System* 71 Each of the forty-eight Ward Committees [in Philadelphia]..chooses two members of the City Committee. *Ibid.* 70 Each of the thirty-five wards [in Chicago] elects by direct vote of the party in a primary a ward committeeman for a term of four years. **1976** *Honolulu Star-Bull.* 21 Dec. A-11/3 He knew the workings of the ward committeemen, who directed the precinct captains and stood ready to see that the garbage of the faithful voters was picked up and the potholes in their streets were filled. [**1807** *Salmagundi* 24 Feb. 68 He however maintained as mysterious a countenance as a seventh ward politician.] **1860** *Harper's Mag.* June 94/2 'A housebreaker or a ward politician,' thought I. **1976** *National Observer* (U.S.) 1 May B5/1 In the emerging big cities, ward politicians held control of school systems for a time in the late Nineteenth Century. **1883** *Century Mag.* Aug. 581/2 He had been a little alarmed at the sudden irruption of such men as Farnham and his associates into the field of ward politics. **1957** Ward politics [see *poker-game* s.v. *POKER sb.⁴ b]. **1818** *Niles' Reg.* XIV. 174/1 Neither the people, nor their representatives, would agree to the plan of assessment on the wards for the expenses of the ward schools. **1904** G. Stratton-Porter *Freckles* 15 They sent me out to the nearest ward school as long as the law would let them. **1918** H. Matthews in Murdoch & Drake-Brockman *Austral. Short Stories* (1951) 244 'She be a pretty one—the ward sister,' a Tommy patient said to him one evening. **1980** J. O'Faolain *No Country for Young Men* iv. 94 'Good night now, Aunt Judith.' Grainne summoned a ward sister's authoritative manner. **1878** J. H. Beadle *Western Wilds* xxi. 332 The ward teachers had reported every case of real or supposed heresy.

b. **ward aide**, a person employed to do non-medical work in a hospital ward; **ward-book**, delete † and ? and add later examples; **ward-heeling** *a. U.S.*, pertaining to, engaged in, or designating, the activities of a ward-heeler; **ward orderly**, a person employed to assist nurses in a hospital ward; **ward round**, a visit paid by a doctor, or by a group of doctors and medical students, to each of the in-patients in a ward or wards, or under the care of a particular doctor; cf. *ROUND sb.¹ 15.

1965 *Nursing Times* 5 Feb. 172/2 Ten hospitals in the Manchester region have been authorized by the RHB to employ ward aides. **1976** *Billings* (Montana) *Gaz.* 20 June 2-c/1 She is employed as a ward aide at Billings Deaconess Hospital. **1836** J. Paget *Let.* 10 Jan. in *Mem. & Lett. Sir James Paget* (1901) I. iv. 66 The active new apothecary..cures the patients. The ward-books hardly know themselves, the *Mistura Cascarillæ* bottle stopper sticks from disuse, and *Emplastrum Lyttæ* is never mentioned. **1977** *Lancet* 7 May 975/2 The allocation of the patients was done automatically by the ward nurse (by reference to the ward book). **1972** R. Thomas *Porkchoppers* (1974) xxiv. 208 Practical politics, the ward-heeling kind. **1976** *Billings* (Montana) *Gaz.* 1 July, But neither ward-heeling councilmen nor grandstanding mayors can provide the kind of leadership that gets things done. **1980** N. Freeling *Castang's City* xv. 100 Why the hell would there be any integrity in these ward-heeling affairs? **1946** *Nature* 13 July 54/1 The Government and the hospital authorities have agreed upon..the training of more male nurses and the formation of a grade of 'ward orderlies' to assist the nurses. **1971** P. D. James *Shroud for Nightingale* viii. 275 She went..as ward orderly. **1938** *Amer. Speech* XIII. 228/2 Doctors regularly visit the patients under their care and for some reason their visits are called *ward rounds* or *ward walks*. **1963** *Oxford Med. School Gaz.* XV. 81 At some stage during the ward-round he would lead his flock down to the P.M. room. **1977** *Lancet* 5 Feb. 317/1 His ward-rounds were very popular and were attended by a large number of undergraduates and postgraduates.

war-dance. Delete 'by savage tribes' and add earlier examples in sense 'mimetic dance' and *transf.*

1775 J. Trumbull *M'Fingal* I. 1 The true war-dance of Yanky-reels. **1814** Scott *Diary* 7 Aug. in J. G. Lockhart *Life Scott* (1837) III. 163 In a stall pamphlet, called the history of Buckshaven [Fifeshire], it is said those fishers sprung from Danes, and brought with them their *war-dance* or *sword-dance*, and a rude wooden cut of it is given. **1851** *San Francisco Picayune* 24 Oct. 2/2 Having thus satisfied his taste for hat smashing, the tiles were kicked into a heap, and the six joining hands around them went through an extemporary war dance.

wardee (wǭɹdī·). *nonce-wd.* [f. Ward *sb.*² + -EE¹.] An inmate of a hospital ward.

1938 S. Beckett *Murphy* xi. 240 No sound reached him from the adjacent female wards but the infinite variety of those made by the female wardees.

warden, *sb.*¹ Add: **7. c.** An air-raid warden.

1936 [see *AIR-RAID]. **1937** *Lancet* 13 Mar. 669/2 The wardens would probably also be used in connexion with the distribution of civilian respirators. **1951** N. Marsh *Opening Night* vi. 139 'Anyone here done respiration for gassed cases?'..'I can,' said the A.S.M. 'I was a warden.' **1978** E. Malpass *Wind Brings up Rain* xxvi. 232 A tin-hatted Air Raid Warden was hurrying along the street. ..The Warden hurried on.

17. (Earlier examples.)

1855 R. Carboni *Eureka Stockade* 120 A Public Meeting was held... The Resident Warden in the Chair. **1860** S. Davison *Discovery & Geognosy Gold Deposits Austral.* xi. 332 A number of new offices had been created for the gold fields, under the name of 'wardens'.

warden, *v.* Delete *rare*⁻¹ and add: *spec.* to watch over or patrol (a nature reserve, etc.) by or as a warden; also *intr.*, to act as a wild-life warden.

1971 *Times* 7 May 17/7 Having wardened at the Royal Society for the Protection of Birds reserve at Loch Garten, may I enlarge upon the letter from M. S. M. Burns? **1974** *Oxford Times* 4 Jan. 7/6 It would be wardened and visited regularly to prevent vandalism and litter. **1977** *Birds* Spring 40 Philip Coxon has wardened the RSPB Balranald reserve for the last four summers. **1982** *Lakeland Echo* 18 Mar. 5 The eyres are well wardened against egg collectors. **1984** *Natural World* Summer 14/1 Several orchids on a number of sites were wardened, efforts being concentrated on lizard, early spider, military and monkey orchids.

Hence **wa·rdened** *ppl. a.*, **wa·rdening** *vbl. sb.*

1962 *Listener* 1 Mar. 375/2 There is as yet no system of wardening for the valleys... Wardening is limited by statute to places where the Board has access-agreements or owns property. **1971** *Guardian Weekly* 3 July 15 The wardening of Exmoor, for example, with its wide acres of moorland under rising pressure of visitors, is described as rudimentary. **1980** *Birds* Autumn p. v/1 Although the most obviously efficient form of protection is to establish adequately wardened reserves, this is not possible in many cases. **1980** R. Mabey *Common Ground* I. 32 A wardening system was established for the more vulnerable eyries.

wardering (wǭ·ɹdəɹiŋ), *vbl. sb.* [f. Warder *sb.*¹ + -ING¹.] The business of a warder.

1928 *Daily Express* 23 Nov. 10/3 He has also been a prison warder, and told me that wardering is by no means to be despised.

wardite (wǭ·ɹdəit). *Min.* [f. the name of Henry A. *Ward* (1834–1906), U.S. naturalist and dealer + -ITE¹.] A hydrated basic phosphate of sodium and aluminium, $NaAl_3(PO_4)_2(OH)_4.2H_2O$, found as transparent tetragonal crystals.

1896 J. M. Davison in *Amer. Jrnl. Sci.* CLII. 154 A considerable quantity of this mineral has been received by Ward's Natural Science Establishment, of Rochester, N.Y. It is in the form of nodules and shows the mineral in several shades of green, and one specimen of pure milk-white... I would give [it] the name of Wardite, in honor of Prof. Henry A. Ward, of Rochester. **1942** [see *MILLISITE]. **1952** *Amer. Mineralogist* XXXVII. 849 Wardite in well-formed, white to colorless crystals has been found in a pegmatite at Beryl Mountain near West Andover, New Hampshire.

Ward-Leonard (wǭɹd,le·nāɹd). *Electr. Engin.* Also **Ward Leonard.** The name of Harry *Ward Leonard* (1861–1915), U.S. electrical engineer and inventor, used *attrib.* to designate a method of controlling a direct-current motor in which its armature current is supplied by an auxiliary d.c. generator (driven by an a.c. motor), the field current of which is varied to vary the speed of the motor.

1902 H. A. Foster *Electr. Engineer's Pocket-bk.* 737 The motors are controlled by the Ward-Leonard system. **1969** *Jane's Freight Containers 1968–69* 541/1 All three designs have the following features: rack operated traversing gear driven by a motor through Ward Leonard control [etc.]. **1981** Carlson & Gisser *Electr. Engin.* xix. 745 This arrangement, known as the Ward-Leonard system, also provides regenerative braking.

wardless (wǭ·ɹdlės), *a.*² *rare.* [f. Ward *sb.*² 24 a + -LESS.] Of a key: having no wards.

1927 R. A. Freeman *A Certain Dr. Thorndyke* xv. 220 These wardless pin-keys are more subtle than they look.

Wardour-street. Add: **1.** Also in other attrib. phrases.

1896 *Sat. Rev.* 8 Feb. 154/2 Our Wardour Street romancers and whimpering Scotch humourists. **1958** L. Forster in *Aspects of Translation* 20 The peculiar Wardour Street language which some classical scholars used for English a generation or so ago. **1976** *New Yorker* 19 Apr. 118/3 To the difficulty of following Borgese's Wardour Street diction ('Bread and wine needs a man to fight and die'; 'Us enchants he, but eke frightens') Sessions adds that of hearing the words.

2. Used *attrib.* and *absol.* with reference to Wardour Street as a centre of the British film industry.

1920 *Stage Year Bk.* 51 A still more ambitious 'ten million pound' company died even before it became more than a Wardour Street fairy tale. **1927** *Melody Maker* Aug. 818/3 A 'phone message, or note perhaps, from Wardour-street. **1948** *Daily Mail* 7 Feb. 2/5 This has caused a few long faces in Wardour-street. **1958** *Punch* 17 Sept. 360/1 Any Wardour Street film-distributor knows that the public wants a boy-girl story, a happy ending..and sensational spectacle. **1975** *Times* 20 Dec. 9/7 It amazes me how few films we manage to make in a year here: Wardour Street seems to have accepted defeat.

wardrobe. Add: **7.** *wardrobe-maid* (earlier example); **wardrobe trunk**, a travelling trunk which can be used as a wardrobe.

1865 Queen Victoria *Jrnl.* 9 Oct. (1980) 121 The Duchess took me to my room..next to which was one for my wardrobe maid, Mary Andrews. **1890** B. Hall *Turnover Club* 221 In Matt's wardrobe trunks there are very many suits. **1928** S. Lewis *Man who knew Coolidge* I. 39 She pointed out I'd have to get my dress-suit in New York and it wouldn't get wrinkled in a wardrobe trunk. **1979** *Country Life* 16 Aug. 490/3 The murderee.. had in his Louis Vuitton wardrobe trunk, full white-tie evening dress.

ward-room. Add: **1.** (Earlier example.) Since 1948 ward-rooms have been used by warrant officers as well as commissioned officers.

1758 in *Ann. Reg. 1758* (1791) 306/1 On Thursday the 13th inst. at half an hour past one in the afternoon, word was passed into the ward room, by the centry, that the fore part of our ship, the Prince George, was on fire. The lieutenants ran immediately forward. **1948** *Admiralty Order* in *London Gaz.* 25 June 3719/1 We are of the opinion that all officers of the Warrant Officer classes should be regarded as Wardroom Officers for messing purposes and the Warrant Officers Mess as such be abolished.

-wards, *suffix.* Add: **5.** Also f. pers. pron.

1842 Browning *Let.* 13 July in F. G. Kenyon *Robert Browning & Alfred Domett* (1906) 39 Here is a slip just off you-wards and I write at night. **1866** H. H. Furness *Let.* 14 Nov. (1922) I. iii. 156 If you were anyone else than the dear.., kindly fellow that I knew & loved.., I should never dare to write this letter, my manifold sins of omission you-wards would palsy my hand and freeze my ink.

ware, *sb.*³ Add: **3. a.** (Earlier example.)

1741 W. Stephens *Jrnl.* 23 July in *Colonial Rec. Georgia* (1908) IV. Suppl. 199 He had lately drawn his Kiln of Ware, which was baking a second Time.

c. For *Obs.* exc. *dial.* read Now *dial.* and *Trade.* Also in attrib. phr. *ware potato.*

1961 *Ann. Reg. 1960* 509 Heavy imports of new potatoes early in the year depressed the market for old ware potatoes. **1963** *Times* 10 June 7/1 The trade for old ware potatoes in England and Wales is now largely confined to caterers.

wareful, *a.* (Later *poet.* example.)

1937 G. Frankau *More of Us* xv. 160 Slow went his feet, and wareful As federal agent's in some gangster joint.

warehou (wa·rehu). *N.Z.* Also 9 **wareho.** [Maori.] A large marine food-fish, *Seriolella brama*, found near the South Island of New Zealand.

1848 E. J. Wakefield *Handbk. N.Z.* v. 161 The *Wareho* is a fish somewhat resembling the kawai, but of much better flavour. **1886** R. A. Sherrin *Fishes N.Z.* 99 The fish known as trevalli in the Dunedin market is a different fish, allied to the warehou. **1949** P. H. Buck *Coming of Maori* II. viii. 215 *Warehou*..were caught with a rod. **1966** *Encycl. N.Z.* III. 552/1 Warehou..resemble trevally in general shape and colouring (blue-green above, silver below, with a dark blotch just behind the head).

warehouse, *sb.* Add: **1. g.** Phr. *warehouse to warehouse*, used *attrib.* to designate a clause in a cargo insurance policy which provides that the insurance policy applies throughout all of the normal course of transit. Also applied to the policy itself, etc.

1922 *Lloyd's List Law Rep.* 30 Nov. 270/1, I am not sure that in a warehouse to warehouse policy the word 'theft' ought to be limited to theft by violence in the same way as it is in a purely marine policy. **1924** *Ibid.* 24 Apr. 450/2 The policy..incorporates the 'warehouse to warehouse' Clause No. 6 of the Institute Cargo Clauses. **1932** *Law Times Rep.* 27 Aug. 168/2 In my view, the practice has always been that the rule as to ship's papers applies though there is a warehouse to warehouse clause in the policy. **1974** E. R. H. Ivamy *Marine Insurance* (ed. 2) xiii. 121 The general rule is that the risk attaches when the goods are loaded, but may attach beforehand if the policy contains a 'craft' clause or the 'transit' ('warehouse to warehouse') clause. **1982** J. Phillips *Dict. Trading Terms* 65/2 *Transit clause*,..one of the clauses in a marine insurance policy defining the normal course of transit, and including the warehouse to warehouse cover that many traders require.

h. *U.S. colloq.* A large and impersonal institution providing accommodation for mental patients, old people, or poor people.

1970 *Sat. Rev.* (U.S.) 3 Oct. 25/1 We have too many such 'human warehouses', staffed by underpaid and poorly trained personnel. **1972** *Time* 14 Feb. 67 But for most of Willowbrook's residents, the institution is a warehouse, a place capable of providing only shelter and the barest essentials, for those whose families are either unwilling or unable to care for them. **1974** J. Fletcher *Ethics of Genetic Control* 157 We ought to protect our families from the emotional and material burden of such diseased individuals, and from the misery of their simply 'existing' (not *living*) in a nearby 'warehouse' or public institution. **1976** *N.Y. Times* 25 Apr. IV. 1 Subsidized public housing has been anathema to the emigres who now live in the suburbs, where the very mention conjures specters of high-rise 'human warehouses' like those that have been erected in ghetto areas.

warehouse, *v.* Add: **d.** *U.S. colloq.* To place (a person, esp. a mental patient) in a large and impersonal institution.

1972 *Tuscaloosa* (Alabama) *News* 22 Feb. 1 The current federal lawsuit against Partlow State School and Hospital may become a landmark decision in America as the country continues its 'trend away from large custodial institutions where people are warehoused and where they are denied the opportunity to develop their full potential,' Dr. Philip Roos said here Monday. **1979** *Time* 2 Apr. 42/2 Freud's dazzling and complex theory of the mind..came along when American psychiatry was doing little more than warehousing the insane and performing the occasional crude Cuckoo's Nest lobotomy.

e. *Stock Exchange slang.* To buy (shares) as a nominee of another trader, with a view to a take-over. Cf. sense *1 b of the vbl. sb.

1977 *Private Eye* 13 May 17/2 The suggested reward to the Swiss holders for 'warehousing' the ex-Bates shares would be perhaps 10p or more per share profit.

warehousing, *vbl. sb.* For **a, b** read **1, 2** and add: **1. b.** *Stock Exchange slang.* (See quot. 1974.) Cf. *WAREHOUSE *v.* e.

1971 *Daily Tel.* 23 Jan. 14/6 'Warehousing' is an old City practice. **1973** *Times* 8 June 1/3 The memorandum has proposals designed to stop the practice of 'warehousing'. **1974** *Daily Tel.* (Colour Suppl.) 22 Feb. 22/2 'Warehousing'—the technique of building up in collusion a major shareholding in a company behind the cloak of nominee names with a view to a take-over. **1983** *Observer* 27 Mar. 18/9 Not a little 'warehousing' may have proved the prelude to recent attacks from South of the border.

c. *U.S. colloq.* The placing of mental patients or other disadvantaged people in large and impersonal institutions. Cf. *WAREHOUSE *sb.* 1 h.

1973 *National Rev.* (U.S.) 7 Dec. 1259 California's shift from the 'warehousing of the mentally ill' in large state mental institutions has become a model for the nation. **1976** *National Observer* (U.S.) 10 Jan. 2/1 Mental patients have the right to receive community care as an alternative to institutional 'warehousing' or release without care, according to a ruling by a Federal court in Washington, D.C. **1983** *Chicago Sun-Times* 2 Aug. 7 Warehousing became the new 'thing'. Forget about making men better, the theory ran.

wareness. (Later *poet.* example.)

1930 Auden *Poems* 57 In his day-thinking and in his night-thinking Is wareness and is fear of other.

wareshi (ware·ʃi). Also **warishi.** [Origin obscure; variously asserted to be Carib and Arawak.] In Guyana, a type of basket worn on the back and held by a headband round the forehead.

1924 *38th Ann. Rep. U.S. Bureau Amer. Ethnol. 1916–17* xviii. 375 The way soldiers carry their knapsacks ..is the general mode which the Indians adopt..for carrying burdens... The names of these articles: waiyari (Arawak),..and walishi or warishi (Carib.). **1958** J. Carew *Black Midas* vi. 102 Naked Amerindian families in single file, the men in front with wareshis harnessed to their foreheads and shoulders. **1964** C. Henfrey *Gentle People* v. 103 He was reluctant to use a wareshi, an Indian carrier made from vines. **1975** C. F. Gritzner *Guyana in Pictures* v. 59 (*caption*) A prospector ready for the trail carries his gear in a 'warishi' (Amerindian basket) capable of carrying weights of up to 200 pounds.

warfarin (wǭ·ɹfărin). *Pharm.* Also **Warfarin.** [f. *W*isconsin *A*lumni *R*esearch *F*oundation + *-arin*, after Coumarin: see -IN¹.] A water-soluble crystalline anticoagulant used as a selective rodenticide, and as a prophylactic against embolism in the treatment of thrombosis; 3-(3-oxo-1-phenylbutyl)-4-hydroxycoumarin, $C_{19}H_{16}O_4$; (also *warfarin sodium*) the sodium salt of this.

1950 *Chem. Abstr.* XLIV. 7019 *Warfarin*, a coined name for the anticoagulant rodenticidal chemical 3-(α-acetonylbenzyl)-4-hydroxycoumarin. S. A. Rohwer... U.S. Dept. Agr. Interdepartmental Comm. on Pest Control, June 29, 1950, 2 pp. (mimeographed). **1968** *Times* 16 Dec. 7/4 Rats which are immune to warfarin, the most widely used rat poison in the world, are spreading. **1973** *Country Life* 1 Mar. 510/1 The grey squirrel cannot be tolerated.... Shortly it will be permissible to use Warfarin in hoppers. **1976** *Lancet* 25 Dec. 1414/2 Anticoagulant therapy with warfarin sodium was started.

warg (wāɹg). [ad. ON. *vargr* wolf; cf. OE. *wearg* and Wary *sb.* in Dict.] In the tales of J. R. R. Tolkien: a wolf of a particularly evil kind.

1937 J. R. R. Tolkien *Hobbit* vi. 107 Even the wild Wargs (for so the evil wolves over the Edge of the Wild were named) cannot climb trees. **1954** —— *Fellowship of Ring* II. i. 234 There are wargs and werewolves. **1968** *Radio Times* 26 Sept. 48 It [*sc. The Hobbit's* Middle-Earth] is..peopled by dwarves, elves, goblins, giant spiders, trolls, wargs.

war-horse. Add: **b.** (Earlier and later examples.) Also used of veterans of other activities, esp. acting.

1837 R. M. Bird *Nick of Woods* I. 68 Ar'nt thee the Pennsylvanny war-horse, the screamer of the meeting-house? **1867** A. D. Richardson *Beyond Mississippi* 151 That old war horse..threw off the black shaggy bearskin overcoat which he invariably wore. **1910** *National Police Gaz.* (U.S.) 16 July 3/1 She mixed with a lot of old war-horses and really thought she was an actress. **1933** S. Lewis *Ann Vickers* xi. 128 If she had been nervous behind the scenes, this old war-horse, this professional who had played her Lady Macbeth to much worse houses, did not seem nervous now. **1958** *Spectator* 13 June 778/1 Dr. Inge, as an old *Evening Standard* war-horse, has even posthumously some truculent things to say.

c. A tough or determined woman.

1921 R. Fry *Let.* 13 Oct. (1972) II. 514 An old campaigning English lady artist..this excessively repulsive and brick-red old war horse. **1959** A. Christie *Cat among Pigeons* v. 67 A sharp old war-horse, called Miss Chadwick, keeps a wary eye on me.

d. *fig.* Something which is frequently used or very familiar; *spec.* a work of art, esp. music, which is frequently performed.

1947 A. Einstein *Music in Romantic Era* xv. 209 There is a whole series of operatic transcriptions..all pieces that are great technical war-horses. **1952** *Mind* LXI. 495 These verbs..are important because they include such philosophical war-horses as *know, believe* and *deduce*. **1958** J. Flanner in *New Yorker* 27 Sept. 96/2 'Ben Hur'—the old theatrical war horse that has had Ben running..since..1899. **1969** *Listener* 1 May 594/1 Deliberately constructed secondary films fall into various types. There are the old-style 'visual aids', such as the famous war-horse on medieval castles which has for long been cantering round the primary and secondary schools. **1977** *New Yorker* 10 Oct. 37/2, I turned on the radio, and there was Artur Rubinstein, playing my old war-horse, the Saint-Saëns G-Minor Concerto, live from Festival Hall.

wari (wǫrī·, wǫ·ri). Also **awari, warri.** [prob. Ashanti.] A board game, originally West African but now played also in the West Indies and Guyana, a variation of mancala.

1927 G. T. Bennett in R. S. Rattray *Religion & Art in Ashanti* xxxii. 382 Wari..is a game for two players using ..48 pebbles and a board hollowed out into two parallel rows of six cups. **1959** *Chambers's Encycl.* VI. 162/2 *Mancala*, a sedentary game with some likeness to chess or draughts... It was carried by slaves to the West Indies (with its Ashanti name of *wari*) and into Dutch Guinea. **1960** R. C. Bell *Board & Table Games* 120 Awari is a masculine pastime though women occasionally play... Making Awari boards involves spiritual danger, and only old men who have lost a wife are allowed to make them. **1975** *Observer* (Colour Suppl.) 30 Nov. 40/2 Groups of men [in Bridgetown] play 'warri', an old game brought to Barbados from Africa.

Waring¹ (wēᵊ·riŋ). *Math.* The name of Edward *Waring* (1734–98), English mathematician, used in the possessive to designate a conjecture that he published in 1770 and which was proved by Hilbert in 1909, that every integer is equal to the sum of not more than *g* *s*th powers, *g* depending on *s* but not on the integer.

1920 G. H. Hardy *Some Famous Probl. Theory of Numbers* 14, I pass on..to the principal object of my lecture, the very famous problem known..as Waring's Problem. **1920** L. E. Dickson *Hist. Theory of Numbers* II. xxv. 721 E. Maillet proved Waring's theorem for eighth powers. **1940** E. T. Bell *Devel. Math.* xiv. 294 Equally prolific of new analysis and far-reaching theorems in arithmetic was the advance beginning in 1909 with Hilbert's solution of Waring's problem. **1962** C. S. Ogilvy *Tomorrow's Math* v. 92 Nineteen fourth powers are required to make up 79, and it is conjectured that 19 is the answer to Waring's problem for fourth powers.

Waring² (wēᵊ·riŋ). *U.S.* In full: **Waring blender** (also **blendor**). A trade name for a make of food processor, manufactured by Waring Products Corporation, of N.Y. (The device is also used as an agitator in scientific experiments, etc.)

1948 *Amer. Home* June 117/1 Waring blendor. **1950** *House Beautiful* May 188/2 In a Waring Blendor, you mix them in less time... Flips, frappés, fizzes..are easy hot-weather drinks. **1959** [see *HOMOGENIZE *v.* d]. **1960** *Times Lit. Suppl.* 1 July 414/2 'Never nothing like it.' The shops were filled with goods and buyers... In the houses washers, dryers, freezers and refrigerators, air conditioners, vacuum cleaners, Mixmasters, Waring-blenders, television and stereophonic high-fi sets, [etc.]. **1967** M. E. Hale *Biol. Lichens* v. 74 The fragmented mycobiont culture was mixed with the algae in a Waring blender so as to form a suspension. **1977** C. McFadden *Serial* (1978) xxv. 57/1 Last year he gave me this Waring blender.

wa·rlessness. [f. Warless *a.* + -NESS.] Absence of war.

1928 Blunden *Undertones of War* xiv. 158 The sluggish weather and the general silence and warlessness encouraged us to take life easy. *a* **1930** D. H. Lawrence *Last Poems* (1932) 114 Look at the young men under thirty... A certain warlessness even moneylessness, A waiting for the proper touch, not for any word or deed.

war-lord. Add: **b.** [tr. Chinese *jūnfá*.] In China, a military commander who has a regional power base and rules independently of the central government, esp. in the period 1916–28.

1922 *N.Y. Times* 31 Dec. VIII. 12/3 Each provincial Tuchun, or Military Governor, is a little or a big war lord with his own army and his own laws; and his regard for the Peking Government is proportioned inversely to the size of his army and his distance from the capital. **1926** P. Weale *Why China sees Red* ii. 43 The whole war-lord system, which is based on the interception of national revenue at the points where it is levied and on nothing else, was about to be rudely tested. *Ibid.* 44 The Manchurian war-lord was just as sly. **1937** V. Bartlett *This is my Life* xi. 174 Many gentlemen..at the Shanghai Club became indignant over my pronunciation of the names of Chinese war-lords. **1959** *Listener* 5 Feb. 238/2 Many millions more have died since, because of floods, drought, war lords, and civil war. **1962** E. Snow *Other Side of River* (1963) xxxv. 264 The local warlord governor, Yang Hu-ch'eng, was also more interested in keeping the Generalissimo's troops out of Shensi than continuing the civil war. **1978** D. Bloodworth *Crosstalk* vii. 56 Any Chinese general who decided to play the rebel warlord

in those regions..would..prove a dangerous and difficult man to suppress.

Hence **wa·rlordism**, the policies or practices of a war-lord; government by war-lords.

1962 E. Snow *Other Side of River* (1963) xxxix. 285 Throughout his struggles with provincial warlordism, 'communist-bandits', and finally with the Japanese, Chiang remained essentially an old-fashioned militarist. **1966** *New Statesman* 18 Mar. 366/3 Is Indonesia now in for a spell of regional warlordism? **1971** W. F. Dorrill in T. W. Robinson *Cultural Revolution in China* ii. 42 Individual numbers of this journal..charge that P'eng Teh-huai..instituted 'warlordism' (maltreating troops and straining relations between officers and men). **1979** *Guardian* 1 Aug. 5/6 Petty warlordism, protection rackets, the levying of 'taxes'..have alienated the natural allies of the Palestinians.

warm, *a.* (and *sb.*²) Add: **A.** *adj.* **17. a.** *warm-bosomed, -gloved, -seated, -veined*; **b.** **warm boot** *Computers*, a reloading or restart of an operating system, etc., without switching off the computer, esp. when changing programs; also as *v. trans.*, to reload in this way; **warm front** *Meteorol.*, the forward boundary of a mass of advancing warm air.

1980 R. Zaks *CP/M Handbk. with MP/M* i. 22 This combination..produces a 'warm start' (or 'warm boot', or 'system reboot'). A warm start essentially interrupts whatever the computer is doing and starts the operating system over again. *Ibid.* 32 If you are only *reading* from a new diskette..a warm boot is not necessary to introduce the new diskette. **1981** *Your Computer* (Austral.) May–June 100/3 *Warm boot*, to reload the operating system a second or subsequent time. **1983** *Ibid.* May 21/1 After a program has run (on the Commodore 64), CP/M warm-boots—a process that takes about 30 seconds. Once a CP/M program is loaded, it runs just fine. **1983** *Byte* May 28/2 The Model 100..did not need extra time to load the text editor and the document file from the disk and later store the document on disk and reload CP/M (do a warm boot). **1925** Blunden *Eng. Poems* 54 Warm-bosomed hawthorn stands in fruitful rest. **1921** Bjerknes & Solberg in *Geofysiske Publikationer* II. iii. 3 In the first case, the boundary line at the ground will be the front of advancing cold air, or, to introduce a shorter expression, a 'cold front'. In the latter case, the boundary line will be the front of advancing warm air, or simply a 'warm front'. **1969** A. G. Forsdyke *Weather Guide* 49 The rain belt ahead of a warm front is often 200 to 300 miles wide. **1922** Joyce *Ulysses* 536 Hobbledehoy, warm-gloved,..stunned with spent snowballs, struggles to rise. *Ibid.* 265 Sprawled, warmseated, Boylan impatience, ardentbold. **1943** Dylan Thomas *New Poems* 12 A man outside with a billhook,..The warm-veined double of Time.

B. *absol.* and *sb.*² **1. c.** *in(to) the warm*: indoors, out of the cold.

1969 F. Hurt *Death in Mist* vii. 74 I'll just put the spade away and then we can go into the warm. **1974** J. Aiken *Midnight is Place* viii. 231, I don't think Mr Oakapple is ready for more news until we have him sitting in the warm.

2. Also without *British* (*Service*), and (*rarely*) *attrib.*, as *warm-coat*.

1928 Blunden *Undertones of War* iii. 26 My warm-coat was not adequate. **1954** W. Faulkner *Fable* 114 His braces knotted about his waist under his open warm. **1958** *Spectator* 11 July 53/2 A florid gentleman in a military warm.

warm, *v.* **I. 2. b.** For 'Now *rare*' read 'Now usu. with *up*' and add later examples, esp. with sense 'to put (an audience) into a receptive mood'.

1892 'Mark Twain' *Amer. Claimant* xxiv. 236 I'll just go over there and warm up that House of Lords. **1923** *N.Y. Times* 14 Oct. VIII. 4 Warming 'em up—Going on early as a sacrifice for the later acts. **1966** *Guardian* 28 Mar. 3/7 Mrs Bessie Braddock warmed up the meeting for him. **1974** *Times* 21 Jan. 12/8 In the studio..Liew Gardner, the chairman, warmed us up music-hall style and the three panellists appeared. **1982** N. Painting *Reluctant Archer* vii. 113 We might have warmed up his audience for him.

7. Restrict † *Obs.* to senses in Dict. and add: **c.** *to warm the bell* (see quots.). *Naut. slang.*

1924 G. H. A. Willis *Royal Navy as I saw It* 116 'To warm the bell', meaning literally to strike the bell a minute or two before the exact time, was applied to circumstances in which to be in good time was meant. **1925** Fraser & Gibbons *Soldier & Sailor Words* 300 Warming the bell, putting on the clock. Advancing the time illegitimately for some particular reason. **1956** H. W. Edwards *Their Lawful Occasions* xxiv. 130 'Ah! I see. All ready dressed for the shore! I said. 'Warming the bell, eh?' **1976** *Oxf. Compan. Ships & Sea* 924/2 *To warm the bell*..On board warships in the days of sail, time was measured by a half-hour sand-glass. Each time the sand ran through the glass was turned..and the appropriate number of bells struck. It was supposed..that if the glass was warmed the expansion of the neck would allow the sand to run through a little more quickly. Hence..eight bells and the return to one's hammock, would come gratifyingly earlier than it should.

II. 9. c. With *up*. Of a sportsman, etc.: to prepare oneself by light exercise or practice immediately before the start of a contest or other physical exertion.

1883 [see *WARMING *vbl. sb.* 1 d]. **1926** *Amer. Speech* I. 369/2 Pitchers 'warm up' on a plot of grass called 'the bull pen'. **1955** R. Bannister *First Four Minutes* 26, I was warming up on the uneven grass near the track. **1972** J. Mosedale *Football* viii. 118 Ed Neale..used to warm up for the game by breaking beer bottles across his forearm. **1976** J. Snow *Cricket Rebel* 35 Graeme Pollock again. Warming up with his first half century in 70 minutes and dashing to his century [in]..only 35 minutes.

d. With *up*. Of an engine, electrical appliance, etc.: to reach a temperature high enough for efficient working.

1947 A. Ransome *Great Northern?* i. 24 There was a whirr from below as the engine started and a steady throbbing as it was warming up. **1955** A. Budrys in D. Knight *100 Yrs. Sci. Fiction* (1969) 255 Halsey's familiar figure appeared on the screen as the set warmed up. **1958** 'N. Shute' *Rainbow & Rose* i. 18 He went to the transceiver and turned it on to warm up. **1972** *Daily Tel.* 12 Apr. 13/5 The engine warms up rapidly and performs eagerly.

warmable, *a.* For *nonce-wd.* read *rare* and add earlier example.

1839 A. Langton *Jrnl.* in *Gentlewoman Upper Canada* (1950) 98, I must tell you then that the drawing-room is as warmable as ever.

warm-blooded, *a.* Add: **warm-bloo·dedness**, the character or condition of being warm-blooded.

1923 J. S. Huxley in *Cornh. Mag.* Apr. 427 In the birds as in the mammals,..we see the evolution..of physiological characters like warm-bloodedness or efficiency of circulation. **1946** F. E. Zeuner *Dating Past* xii. 84 Warm-bloodedness and many other characters of the mammals are probably the consequence of a single important aromorph. **1982** *N.Y. Times Mag.* 7 Feb. 34/4 High metabolism, a characteristic of the warm-bloodedness of mammals and birds.

warmed, *ppl. a.* Add: Also with *over*, *up*. **a.** (Examples.)

1897 *Private Life of Queen* xvii. 141 The Queen..made a delightful luncheon off warmed-up broth and potatoes she had helped to boil herself. **1916** B. M. Bower *Phantom Herd* 246 A midnight supper of warmed-over coffee and cold bean sandwiches. **1977** H. Greene *FSO-1* vi. 57 Warmed-over moussaka uneaten before him.

b. (Examples.) *like death warmed up*: see *DEATH *sb.* 17 c.

1887 *Nation* (N.Y.) 2 June 465/3 They will be spared the future bitterness of finding..themselves treated to insult and warmed-over excuses. **1950** *New Yorker* 8 Apr. 96/2 *Oggi*..came out with..a warmed-over photograph taken in the Farfa displaced-persons camp. **1965** *Listener* 9 Sept. 391/1 The comfort given by warmed-up studio audiences. **1978** W. Mankowitz *Extraordinary Mr Poe* vi. 186 Verses dedicated *To F*, actually warmed-over lines written for another.

warmer. Add: **3.** *warmer-up*, something that warms oneself or another up; *spec.* (*a*) a preliminary item designed to put an audience in a receptive mood; also, one who presents this; (*b*) a stimulating drink. Also *warmer-upper*, esp. in sense (*b*) above.

1960 *House & Garden* Aug. 70/1 A tipple that has been a warmer-upper for British sailors since before Nelson's day. **1962** *Guardian* 22 Dec. 5/2 As so often it is the warmers-up (in this case Gordon and Bunny Jay) who do the hard work. **1972** *Jazz & Blues* Nov. 28/1 *Shaw nuff* is a very fast warmer-up. **1976** *Ulverston* (Cumbria) *News* 3 Dec. 19/2 As a 'warmer-up' the concert began with a rather pedestrian rendering of the Hebrides overture. **1980** *Times* 9 Dec. 2/1 Stone's Ginger Wine. The original warmer-upper. On its own or as a Stone's Whisky Mac.

warming, *vbl. sb.* Add: **1. d.** With *up*: the action or condition of 'warming up' (see senses 3, 9, and 11 in Dict. and Suppl.). Freq. *attrib.*

1874 J. G. McCoy *Hist. Sk. Cattle Trade* 244 This style, ..of wintering cattle, is called 'Roughing', and the feeding of corn in the spring is termed 'Warming up'. **1883** *Chicago Inter-Ocean* 27 June 5 The players..began their practice play. This is called 'warming up'. **1952** *Times* 22 Feb. 6/4 Jet-engined aircraft have two further advantages—no preliminary 'warming up' is necessary, so they can take off in shorter time. **1960** *Practical Wireless* XXXVI. 429/1 The warming up time for the unit is about 30 seconds.

3. *warming drawer, oven.*

1972 J. Burmeister *Running Scared* vii. 95 Your breakfast will be in the warming drawer if I'm not there. **1950** in M. Cecil *Heroines in Love* (1974) 199 Jocelyn turned out the warming oven. **1973** J. Burrows *Like an Evening Gone* iii. 38 There was a square warming oven... 'For keeping tins and dinner plates hot.'

warmonger. Add (Further examples.) **warmongering** *vbl. sb.* (examples).

1934 *Sun* (Baltimore) 5 Mar. 6/3 Dr. Toynbee differs from Mr. McFadden's war mongers in that he makes a suggestion for avoiding such a catastrophe. **1940** N. Coward *Australia Visited* ii. 9 Many of those in high places..dismissed his [*sc.* Winston Churchill's] eloquent prophetic words as alarmist warmongering. **1940** 'G. Orwell' *Inside Whale* 170 The war-mongering to which the English intelligentsia gave themselves up in the period 1935–9. **1944** Mrs. Belloc Lowndes *Let.* 4 June (1971) 249 Algernon Cecil..regards Churchill as 'a warmonger'. **1955** *Times* 2 Aug. 7/7 They have their war-mongers..of course, but their fervency on Palestine derives as much from a sense of injustice as from wounded pride. **1981** R. Reagan in *N.Y. Rev. Bks.* 25 June 25 It is not warmongering to say that some things are worth dying for.

warm-up. [f. the vbl. phr. *to warm up*: see Warm *v.* 3, 9, and 11 in Dict. and Suppl.] **1.** = Warm *sb.*¹

1878 Mrs. Stowe *Poganuc People* 156 A knot of the talkers were gathered round the stove, having a final talk and warm-up. **1960** H. Pinter *Room* in *Birthday Party* 112 Thank you for the warm-up, Mrs. Hudd. I feel better now. **1974** J. Aiken *Midnight is Place* ix. 257 'Tis a poor lad half drownded—can tha..give him a bit of a roob-down an' a warm-oop?

2. Warmth, the quality of exciting or stimulating. *rare.*

1883 'Mark Twain' *Life on Mississippi* iii. 52 The song didn't seem to have much warm-up to it, somehow.

3. a. The act or process of 'warming up' for a contest, etc., by light exercise or practice. See sense *9 c of the vb.

1915 *Baseball Mag.* Dec. 116/2 Alex, after a short warm-up, vanished from the foreground. **1949** Shurr & Yocom *Mod. Dance* iii. 31 The transition is used only when use of next warm-up is desired. **1951** *Publ. Amer. Dial. Soc.* xvi. 68 Some horses run better after a stiff warm up, others just tire. **1975** *New Yorker* 28 Apr. 32/1 Her second serve was good, was well returned by Jill, and then was driven by Sylvia to Jill's backhand in a way that left her frozen, as it had in the warmup. **1984** *Times* 22 Sept. 3/3 Warm-ups should be taken slowly, as sweating does not mean that muscles are sufficiently stretched for exercises.

b. *transf.* and *fig.*

1943 *Daily Progress* (Charlottesville, Va.) 18 Aug. 1/8 Allied guns boomed in a duel with heavy Axis batteries across the Messina Strait today in a warm-up against the European fortress. **1945** *Sun* (Baltimore) 11 July 9-0/6 Polynesian was running in the Shevlin as a 'warm-up' for Saturday's $50,000 Dwyer. **1958** *Times* 17 Oct. 20/1 It was a slow warm-up that Keenan could scarcely afford. **1976** D. Heffron *Crusty Crossed* xiv. 101 A party in the afternoon, a kind of warm-up to the night's dark devilry.

c. *attrib.*

1943 *Sun* (Baltimore) 22 May 1/7 The bombing of Nauru, Tarawa and other islands in the central Pacific, were warmup attacks. **1945** *Ibid.* 12 Mar. 7-0/2 One of Mexico's leading matadors..sharpens his skill with a 'warm-up' workout..an hour before a..performance. **1958** [see *HOUSE *sb.*¹ 4 h]. **1968** C. Drummond *Death & Leaping Ladies* v. 120 They arrived at Mexico City to play a warm-up match. **1978** L. Pryor *Viper* (1979) viii. 152 The field of cars was allowed one warm-up lap.

4. The act or process of raising the temperature of an engine, electrical appliance, etc., to a level high enough for efficient working. Also *attrib.* See sense *9 d of the vb.

1945 *Sun* (Baltimore) 18 May 1/6 Massed after on the flight deck, engines roaring for the warmup,..were more planes. **1956** H. Kurnitz *Invasion of Privacy* ii. 17 There was an even, monotonous thud emanating from the machine now—the warm-up period. **1958** *Spectator* 1 Aug. 167/2 That infuriating warm-up time necessary for many TV sets. **1966** P. O'Donnell *Sabre-Tooth* xx. 270 The Dove began to taxi forward... He wished there had been a few minutes grace for a warm-up before taking off. **1975** *Physics Bull.* Dec. 550/1 The detector has a warm-up time of 10 s. **1978** *Dumfries Courier* 13 Oct. 11/2 Another refinement the picture tube designers have incorporated is a really quick warm-up which..eliminates that infuriating delay when switching on late for an important programme. **1984** B. Francis *AA Car Duffer's Guide* 63/1 Modern water-cooled systems incorporate a thermostat to give a quick warm-up.

5. a. The 'warming up' of an audience into a receptive mood, esp. before a broadcast programme is recorded or transmitted. See sense *2 b of the vb.

1958 *New Statesman* 15 Mar. 333/1 So that spontaneity shall not degenerate into chaos, the programme is preceded by a half-hour closed-circuit 'warm-up', in which those taking part get to know each others' names and voices. **1970** *Guardian* 14 Feb. 8/4 For this particular show there is an audience..and they arrive at 7.30 p.m. and are given a 'warm up' to get them in the mood for the show. **1983** *Oxford Diocesan Mag.* Aug. 10/2 For pre-service warm-ups, say—a [tape or record of a] full orchestra playing Beethoven's Fifth for Harvest Festival, [etc.].

b. *attrib.*, esp. as *warm-up man.*

1959 R. G. Stern in N. Mailer *Advts. for Myself* (1961) 319 There was no warm-up session except thirty seconds of irrelevant talk which we used for volume control. **1966** *Observer* 30 Oct. 23/4 The warm-up man (an assistant producer) jumps on to the platform. **1974** P. de Vries *Glory of Hummingbird* xii. 159 Falconer regaling..spectators..with some intendedly relaxing 'warmup' chatter. **1979** S. Brett *Comedian Dies* xvi. 149 The audience.. were greeted by..a little-known comedian who had been booked for the occasion as a warm-up man.

6. *U.S.* **a.** A garment designed to keep the wearer warm. **b.** One worn during light exercise or practice; a track suit or track-suit top.

1949 *Sun* (Baltimore) 22 Sept. 7/1 (Advt.), Make a friend of Jack Frost in smart warm-ups! **1969** *Sears Catal.* Spring/Summer 35/1 Sweatshirt warm-ups fit sizes 2 to 6x. **1983** W. Safire in *N.Y. Times Mag.* 12 June 22/2 *Exercise classes* of the 1950's, where one wore *sweat suits*; these are now *workouts* with *warm-ups*.

c. *attrib.* and *Comb.* **warm-up suit**, a track suit. (Cf. also sense 3 c above.)

1945 *Richmond* (Va.) *Times-Dispatch* 9 Jan. 16 When winter comes..it's time for warm-up clothes. **1972** *N.Y. Times* 3 Nov. 44/5 (Advt.), Stretch pants, warm-up pants, skisuits. **1975** *New Yorker* 28 Apr. 31/1 She was wearing a pastel-blue warmup suit over her tennis dress. **1984** *New Yorker* 1 Oct. 30/1 Mr. Sims..wore a Nike cap, a crisp tan warmup suit.

warn, *v.*¹ **6. f.** (Earlier example.)

The expression was finally deleted from the Jockey Club's Rules of Racing in 1969 when the Jockey Club and National Hunt Committee amalgamated, and the course and training grounds at Newmarket were transferred to the Newmarket Estates and Property Co. Ltd.

1845 *Racing Calendar 1844* p. lii, Samuel Rogers and John Braham were warned off the Course and exercising ground at Newmarket.

warning, *vbl. sb.*¹ Add: **4. d.** *the usual warning*: the caution that a police officer making an arrest is bound to give, viz. that anything the suspect says may be taken down and used in evidence against him or her.

[**1919** Gregg & McGrath *Police Constable's Guide* (ed. 3) p. xlix, Persons in custody should not be questioned without the usual caution being first administered.] **1931** 'G. Trevor' *Murder at School* xiii. 255, I gave her the usual warning, of course, but she began to talk, all the same. **1975** 'R. Player' *Let's talk of Graves* v. 181 Holding my warrant in my right hand, I then gave her the usual warning and arrested her.

e. The action or an instance of warning (someone) *off*: see sense 6 of the vb.

1977 K. Benton *Red Hen Conspiracy* ix. 56 He gets an ambiguous warning-off from the Embassy, but that doesn't stop him. **1980** D. Francis *Reflex* xv. 181 They could have half-killed me... All they were truly delivering was a warning off. **1981** *Times Lit. Suppl.* 30 Jan. 108/5 The warning off of the racing correspondent of the *Morning Post*.

5. d. A signal given by means of a siren, etc., to indicate that an air attack is imminent, an air-raid warning. Cf. *ALERT *sb.*¹ 1 b.

1917, 1938 [see *AIR-RAID]. **1940** H. Nicolson *Diary* 7 Sept. (1967) 111 The all-clear sounds at 6, but there is another warning at 8 which actually lasts till 5.30 a.m. **1953** R. Lehmann *Echoing Grove* 284 There are several regulars who bustle along the moment the Warning goes and don't stick their noses out again till the All-clear. **1982** T. Fitzgibbon *With Love* I. viii. 57 The air-raid siren had sounded..but a warning about a week previously had amounted to very little.

12. *warning light*; **warning bell**, (*c*) a bell alerting people to prepare for a meal, etc.; (*d*) *fig.*, an alarm-bell sounded 'in the head', giving a presentiment of danger; **warning gong** *rare* = *dressing-gong* s.v. Dressing *vbl. sb.* 5 a; **warning triangle**, a triangular red frame carried by motorists, and set up on the road as a danger signal to warn approaching drivers of the proximity of a broken-down vehicle or other hazard.

1849 Dickens *Dav. Copp.* (1850) xxi. 212 The warning-bell will ring at nine; the family take breakfast at half-past nine. **1951** E. Coxhead *One Green Bottle* vii. 202 Somewhere, right at the back of her head, there rang a little warning bell. **1951** *Publ. Amer. Dial. Soc.* XVI. 68 *Warning bell*, a signal made to warn that it is time to saddle the horses entered in the next race. **1981** S. Brett *Situation Tragedy* ii. 26 'I'll join you.' Something rang warning bells for Charles. 'Well, no.' **1984** *Times* 5 June 23 (heading) Another warning bell for the secretaries. **1938** D. Smith *Dear Octopus* III. i. 116 Ring the warning gong, will you, dear? **1937** *Motor Catal.* (East London Rubber Co. Ltd.) 131 Ignition replacements..warning lights. **1962** *Which? Car Suppl.* Oct. 130/2 The Fiat 1500 had a particularly comprehensive array of warning lights, including lights to show when choke, side lights and hand brake were in use. **1972** J. Aiken *Butterfly Picnic* i. 6 The encircling mountains outlined by small red warning lights. **1974** 'W. Haggard' *Kinsmen* xiii. 126 The warning lights were out and flashing. Her mother had some absurd new plan. **1971** *Good Motoring* Sept. 23/3 It is essential for motorists to carry warning triangles for use in case of accident or breakdown.

wa·rningfully, *adv. rare.* [f. Warning *vbl. sb.*¹ + -FUL + -LY².] = Warningly *adv.*

1922 Joyce *Ulysses* 197 Quickly, warningfully Buck Mulligan bent down:—The tramper Synge is looking for you, he said, to murder you. **1945** J. Steinbeck *Cannery Row* ix. 53 'I guess everything that comes out of the human mouth is poison,' said Doc warningfully.

warp, *sb.*¹ Add: **V. 8. b.** *Science Fiction.* = *space warp* s.v. *SPACE *sb.*¹ 19.

1936 *Astounding Stories* June 30 AKKA is the symbol for humanity's secret weapon. Its user, with simple instruments, can destroy any object in the universe—by so altering the warp of space that neither matter nor energy can exist. **1954** *Galaxy* Aug. 80/2 Halfway between Earth and Venus there was a sudden shimmer as the Vegan ship slipped out of warp into normal space.

VI. 10. **warp print** = *shadow print* s.v. *SHADOW *sb.* 16; hence **warp-printed** *a.*

1916 *Daily Colonist* (Victoria, B.C.) 23 July 8/7 This offering comprises Fancy Ribbons in warp-print, Dresden, Pompadour and novelty stripe and plain effects. **1968** J. Ironside *Fashion Alphabet* 246 *Shadow* or *Warp print*. The warp yarns are printed with the design before weaving, giving a shadowy print effect. **1957** *Times* 16 Sept. 11/1 The short dress..is in white satin with a small warp-printed design in black.

warp, *v.* Add: **II. 18. c.** *Science Fiction.* To travel through space by way of a space warp.

1946 F. Brown in *Astounding Sci. Fiction* May 129/1 The *Ark*..would warp through space to a point a safe distance outside the Argyle I–II system and come in on rocket power. **1957** T. Sturgeon *Thunder & Roses* 117 Earth was ready for him when he warped in.

warpage. Add: **2.** The extent or result of warping (sense 6).

1950 J. Osborne *Dental Mechanics* (ed. 3) xi. 195 The celluloid bases of 1870..suffered from warpage in the mouth. **1952** *Shell Aviation News* No. 163. 14 (*caption*) Straight-edge laid along surface of wing leading-edge shows that the earlier warpage has been entirely eliminated. **1969** [see *PORTCULLIS]. **1970** R. Lowell *Notebk.* 221 We're warpage in the drift to *fin de siècle*.

war-paint. Add: **2. b.** *colloq.* Cosmetics, make-up.

1869 B. Harte *Luck of Roaring Camp* 84 The stranger ..[brightened] through the color which Red Gulch knew facetiously as her 'war paint'. **1945** L. Shelly *Jive Talk Dict.* 35 *War paint*, cosmetics. **1957** *Landfall* XI. 1. 41 'In a moment,' Sylvia said, clicking open her purse. 'Just a daub of warpaint.' **1970** R. Lowell *Notebk.* 57 If I look for the unbelievably beautiful In a city, it's mostly women in their war-paint. **1977** J. Gardner *Werewolf Trace* x. 91 She had a fresh coat of warpaint and was all breathless.

war-party. Add: **2. a.** (Earlier examples.)

1755 in J. W. Lydekker *Faithful Mohawks* (1938) 84 He had ye care of providing for all ye War-parties of Indians that went thro' that place. *c* **1800** B. Hawkins *Sk. Creek Country* (1848) 70 The war parties all march in Indian file, with the leader in front.

b. *transf.* and *fig.*

1921 C. A. W. Monckton *Some Experiences New Guinea Resident Magistrate* xiv. 293 They had dispatched..a war party..and remorselessly slaughtered the Oobudura. **1946** *Sat. Even. Post* 3 Aug. 81/2 Pass the word back to Mr. Topliff at the gallop! Here's his war party! **1957** V. W. Turner *Schism & Continuity in Afr. Society* ix. 273 He inflicted several defeats on a large war-party of Chokwe. **1969** M. Pugh *Last Place Left* xiv. 99, A solitary raven scout..was summoning the raven war party to the scene. **1976** *Classical Q.* XXVI. 257 The motives of the Spartan war-party are clearly stated.

war-path. b. (Earlier example.)

1880 'Mark Twain' *Tramp Abr.* xxxii. 345 She was on the war path all the evening.

warper. 3. (Earlier example.)

1847 [see *SPEEDER 3].

warping, *vbl. sb.*¹ Add: **10. b.** *warping board.*

1910 L. Hooper *Hand-Loom Weaving* iv. 42 Although the warping board..is very useful for small warps of moderate length,..it would not be convenient for very long warps, or accurate enough for warping several thousands of fine silk threads. **1960** G. Lewis *Handbk. Crafts* 99 The warping board and the mill have the same process in common, the main difference being that when using the board you have to walk backwards and forwards to get your length, whereas when using the mill you revolve it the required number of times.

warrant, *sb.*¹ Add: **II. 15. b.** *warrant of fitness*, a certificate of roadworthiness valid for six months, which must be carried by most classes of motor vehicle in New Zealand. (The equivalent of an M.O.T. test certificate: see *M 5.)

1936 *N.Z. Statutory Regulations 1936–37* (1938) 331 Save as provided in clause (3) hereof, the driver of every motor-vehicle used on a road after the 31st day of March, 1937, shall carry in the vehicle a warrant of fitness... The warrant of fitness shall be issued only by a city authority or a person or firm appointed or approved for the purpose by the Minister. **1948** *N.Z. Law Rep.* 1229, I am convinced that a warrant of fitness does not extend to the drag link or steering assembly of a car. **1953** *Road Code* (N.Z.) 1 Jan. 39 There must be carried on a motor vehicle a Warrant of Fitness issued within the past six months. **1961** B. Crump *Hang on a Minute* 23 Are you aware of the law regarding warrants of fitness for motor vehicles? **1983** *N.Z. Official Yearbk.* XIII. 371 Most lightweight vehicles are required to have a warrant of fitness which can be issued at approved garages, or at testing stations operated by local authorities or the Ministry of Transport.

IV. 18. **warrant card**, a document of authorization and identification carried by a police officer; **warrant chief**: in Nigeria, an African local official, esp. (formerly) one appointed by the colonial power.

1920 H. L. Adam *Police Encycl.* III. iv. 77 All officers in plain clothes are furnished with a 'warrant card', by means of which they can, should they be challenged, at once prove their *bona fides*. **1933** D. L. Sayers *Murder must Advertise* xvii. 296 Beware of..the plain-clothes cop without a warrant-card. **1983** R. Allason *Branch* xii. 169 He stresses to newcomers to the Branch that they hold the same warrant cards as the rest of the Metropolitan Police. **1922** S. M. Grier *Rep. Eastern Provinces by Secretary for Native Affairs* 5 The Native Court Clerk.. conveys to the warrant chiefs instructions sent from the Divisional Officer. **1957** Ld. Hailey *African Survey 1956* viii. 465 The members of these Councils, who came to be known as Warrant Chiefs, were selected from villages within the Council or Court area. **1976** *Daily Times* (Lagos) 12 Oct. 5/2 The role of some warrant chiefs in the selection of the new Owa of Indanre was consistent with the chieftaincy regulations.

warrantee. Add: **2.** Also (*U.S.*), misused for Warranty 1.

1874 J. Wanamaker in *Philadelphia Inquirer* 26 Sept. 8/2 (Advt.), A printed guarantee, bearing the signature of the firm, will accompany each garment as a warrantee. This binds us in every sense, and will be honored as quickly as a good draft of the Government of the United States. **1980** *Verbatim* Autumn 17/1 We have three perfectly good words in English, *guarantee, warranty*, and *guaranty*, all of the same origin and having the same meaning. Now some people are trying to add a fourth, *warrantee*. A TV salesman says, 'We have a great warrantee on this product.'

warrantless, *a.* Restrict *rare* to sense in Dict. and add: **2.** *U.S.* Without judicial authorization; without a search warrant.

1950 F. Frankfurter in *U.S. Reports* CCCXXXIX. 80 To tear 'unreasonable' from the context and history and purpose of the Fourth Amendment..is to make the arrest an incident to an unwarranted search instead of a warrantless search an incident to an arrest. **1968** *U.S. Rep. Cases Supreme Court* 367 The Amendment deserves..a liberal construction in order to protect against warrantless searches of buildings. **1978** *Detroit Free Press* 14 Apr. 1A/4 The activities that are alleged in connection with the Weatherman investigation, I would categorize as warrantless activities.

warrant officer. 1. The rank was abolished in the Royal Navy in 1949.

1949 *Order in Council* (Admiralty) No. 30/C-W. 1 Whereas we are of the opinion that the title 'Warrant Officer' and the term 'Warrant List' in use in the Royal Navy and Royal Marines should no longer be used but be replaced by 'Branch Officer' and 'Branch List' respectively.

Warrau (wārau·). Also 9 **Warow, Worrow; Warraw.** [Native name.] (A member of) an American Indian people inhabiting Guyana, Surinam, and Venezuela; also, the language of this people. Also *attrib.*

The spelling *Warao* has recently emerged as the usual spelling for the language and the people.

1807 H. Bolingbroke *Voy. to Demerary* vii. 151 The Worrows principally inhabit the sea coast lying between the Demerary and Surinam. **1834** W. Hilhouse in *Jrnl. R. Geogr. Soc.* IV. 321 (*heading*) Memoir on the Warow Land of British Guiana. *Ibid.* 324 On these reefs are situated many Indian villages—Warows, Arawacks, [etc.]. **1883** E. F. im Thurn *Among Indians of Guiana* vi. 167 The Warraus are the shortest. **1891** D. G. Brinton *Amer. Race* 354 The Guaraouna or Warrau was, and continues to be, spoken by the tribes of the [Orinoco] delta. **1910** *Encycl. Brit.* XII. 676/2 [British Guiana] The aborigines—Arawaks, Caribs, Wapisianas, Warraus, &c.—..are now estimated at about 6500. **1948** P. Kirchhoff in J. H. Steward *Handbk. S. Amer. Indians* III. 870 The Warrau language constitutes an independent family. *Ibid.* 869 The Warrau live in the intricate Delta of the Orinoco. **1950** J. A. Mason in *Ibid.* VI. 252 The independence of the Warrau linguistic family has been admitted by all authorities. **1965** *Internat. Jrnl. Amer. Linguistics* XXXI. 106 Warao appears to have a large number of compounds. *Ibid.*, The Warao text is divided into numbered sentences without any other punctuation. **1973** *Advocate-News* (Barbados) 18 Dec. 9/5 Several cuckoos have befouled the nests of the Arawaks, Caribs, Wapisianas, Warraws and what-nots. **1978** *Amer. Poetry Rev.* Sept./Oct. 13/3 Roth..gave the title 'The Adventures of Kororomanna' to a long Warao story. **1979** J. Halifax *Shamanic Voices* (1980) i. 28 The Warao shaman also believes the earth to be the world axis.

warren, *sb.* Add: **4.** (Later examples.) Also, any area of living or office space characterized as a mass of passages and (small) rooms. Cf. *rabbit-warren* s.v. *RABBIT *sb.*¹ 3 a.

1922 Joyce *Ulysses* 423 Figures wander, lurk, peer from warrens. **1954** J. R. R. Tolkien *Fellowship of Ring* i. 31 Mr. Frodo left an orphan..brought up anyhow in Brandy Hall. A regular warren, by all accounts. **1977** W. J. Weatherby *Home in Dark* xii. 113 It was a house to hide in: you hardly existed in this anonymous warren. **1980** R. Rendell *Lake of Darkness* iv. 41 A room smaller than this one divided into *three*... It's a real warren.

Warren² (wǫ·rĕn). *Engin.* The name of James *Warren* (fl. 1848), of Middlesex, used *attrib.* and † in the possessive to designate a truss he designed composed of alternately inclined diagonal members joining two horizontal ones, so as to form a series of non-overlapping triangles pointing alternately up and down. [Patented by Warren and Monzani in *Brit. Patent 12,242* (1848).]

[**1852** *Minutes Proc. Inst. Civil Engineers* XI. 12 He had only used the simple triangulation for spans not exceeding 60 feet and generally for shallow girders, but he believed that some girders had been made on Captain Warren's plan for much greater spans.] *Ibid.* 14 Mr. Brunel..said..it was necessary to draw a distinct line of demarcation between the lattice bridge and that kind of construction called Warren's girder. *Ibid.*, The Warren girder was decidedly superior to the lattice bridge. **1866** B. B. Stoney *Theory of Strains in Girders* I. vi. 79 This class of bracing includes girders whose web consists of a simple system of triangles, such as 'Warren's' Girder. **1911** A. Smith *Stresses in Simple Framed Structures* xiv. 83 The Warren truss has a complete system of web members composed..of diagonals. **1952** *Archit. Rev.* CXI. 159/2 Where the roof over the drop room joins that over the mill room there is a reinforced concrete Warren girder spanning 64 ft. and 13 ft. deep. **1967** *Jane's Surface Skimmer Systems 1967–68* 18/2 The lift fan bay forms part of the plenum chamber, the top skin of which is of Warren girder construction.

warrener. Add: **2.** Also *transf.*
1929 R. Bridges *Testament of Beauty* iv. 114 Poor nomads..warreners of the waste.

warrigal, *sb.* (and *a.*) Add: Also 9 **warragle.** **A.** *sb.* **1.** (Earlier examples.)
1838 J. Hawdon *Jrnl. Journey N.S.W. to Adelaide* (1952) 25 We could find no traces of the sheep except in two places, where we could perceive they had been pursued by the Warrigals. 1848 H. W. Haygarth *Recoll. Bush Life Austral.* v. 44 His 'coolie' dogs will awaken him on the approach of a 'warragle', or native dog.
B. *adj.* (Earlier example.)
1855 in Stewart & Keesing *Old Bush Songs* (1957) 164 I'm a warragle fellow that long hath dwelt In the wild interior, nor hath felt, Nor heard, nor seen the pleasures of town.

warring, *ppl. a.* Add: **1. b.** *Warring States*, used to designate the last period (475 B.C. onward) of Chinese history prior to the unification of the country in 221 B.C.
1929 J. Joshua tr. *R. Wilhelm's Short Hist. Chinese Civilization* iii. 133 The period of 'the Warring States'.. is usually regarded as lasting from 403 to 221 B.C. 1935 C. P. Fitzgerald *China* iii. 65 If the story of the period of the Warring States was merely a record of anarchy and violence it would not be worthy of detailed attention in a cultural history. 1937 R. W. Swallow *Anc. Chinese Bronze Mirrors* Pl. 3 (*caption*) Warring States type of mirror. 1963 Kwang-chih Chang *Archaeol. Anc. China* vii. 177 For our purposes, round figures are sufficient: 1100 for the conquest..450 for the beginning of the Warring-States period. 1978 *Nagel's Encycl.-Guide: China* 327 The names of the states of the Warring States period almost all survive as surnames now.

Warrington (wǫ·riŋtən). The name of a town in Cheshire, used chiefly *attrib.* to designate a variety of cross-peen joiner's hammer.
1935 N. R. Rogers *Technol. Woodwork & Metalwork* ix. 96 There are two patterns of cross-pane hammers, Warrington and London Riveting, or Exeter. The former..is made in twelve sizes specified by Nos. 00 to 10 consecutively. 1956 A. P. Morgan *Woodworking Tools* i. 8 The Warrington hammer, and similar types with a tapering 'pene' instead of claws, is more suitable for.. joinery and finer work. 1966 A. W. Lewis *Gloss. Woodworking Terms* 42 The common woodworker's hammer with its cross pane is known as the 'Warrington pattern'. ..A pin hammer is merely a very small Warrington. 1979 A. B. Emary *Woodworking* ii. 14 Although joiners use the Warrington-type hammer, the claw hammer is considered the best for someone who will be doing a wide variety of jobs.

warrior. Add: **6. b.** *warrior-poet.*
1878 O. Wilde *Ravenna* 9 Her warrior-poet, first in song and fight. 1958 O. Caroe *Pathans* xv. 241 The warrior-poet whose words still kindle fire in the hearts of his compatriots.

warry (wǭ·ri), *a.*[2] *poet.*[−1]. [f. War *sb.*[1] + -y[1].] Belligerent, warlike.
1901 W. B. Yeats in Lady Gregory *Ideals in Ireland* 96 She looked 'very strong and warry and fierce, but not wicked'.

Warsaw[2] (wǭ·ɹsǭ). The name of the capital of Poland, used *attrib.* in *Warsaw Pact*, to designate a military alliance of the Soviet Union with certain other European nations (see quot. 1978), formed by the Treaty of Warsaw, signed on 14 May 1955. (Principally established as a Communist counterpart to N.A.T.O.)
1955 *N.Y. Times* 14 May 18/1 It would be a mistake to discount the Warsaw pact. 1962 *Listener* 22 Mar. 500/2 The Warsaw Pact countries. 1976 Ld. Home *Way Wind Blows* xii. 167 There was, however, a running argument among the professionals as to whether the line between the Warsaw Pact and the NATO forces should be thinly held (by a trip-wire) or more strongly manned. 1978 *Internat. Relations Dict.* (U.S. Dept. State Library) 46/2 *The Warsaw Pact*, a multilateral military alliance formed by the Treaty of Warsaw, signed May 14, 1955 by the Soviet Union, Bulgaria, Czechoslovakia, East Germany, Hungary, Poland, Romania, and Albania. Like NATO, the Warsaw Pact has military and civilian institutions.

wart, *sb.* Add: **4. a.** (Later examples, sometimes with implied reference to next sense.)
1934 J. B. Priestley *Eng. Journey* vi. 187 You can meet them, a trifle subdued perhaps but there to the last wart, in the solid downright fiction of my friend, Phyllis Bentley. 1961 *Listener* 2 Nov. 738/2 The Catholic revivalists..the author presents as no doubt they would like to be presented... No warts here, perhaps regrettably. 1982 *Times* 1 Dec. 2/5 It was [the television companies'].. job to hold up mirrors, some of which showed the warts in society.
b. *warts and all*: without concealment of blemishes or unattractive parts (esp. applied to a description or likeness.) Also hyphened as *attrib. phr. colloq.*
[1763 H. Walpole *Anecd. Painting* III. i. 15 Oliver [Cromwell]..said to him, 'Mr Lely, I desire you would use all your skill to paint my picture truly like me, and not flatter me at all; but remark all these roughnesses, pimples, warts and everything as you see me, otherwise I will never pay a farthing for it.'] 1930 W. S. Maugham *Cakes & Ale* xi. 138 Don't you think it would be more interesting if you went the whole hog and drew him warts and all? 1961 *Listener* 21 Sept. 437/1 A convincing warts-and-all likeness of Wingate. 1962 *Sunday Times* 1 Apr. 13/1 The Duke of Edinburgh presents himself warts and all, without blunting the rough edges of efficiency and enthusiasm. 1966 K. Giles *Provenance of Death* iii. 96 In fact you want a run down on Stanisgate, warts and all. Huh? 1974 *Publishers Weekly* 18 Feb. 24 An intimate, in-depth, 'warts-and-all' portrait of our new Vice President. 1976 H. A. Williams *Tensions* vii. 111 God..accepts us, accepts all men, unconditionally, warts and all. 1980 *Times Lit. Suppl.* 12 Sept. 986/3 This book..may disconcert the pious more than it jolts the sceptic, but it has the story, warts, statistics and all.
5. b. An obnoxious or objectionable person. *colloq.*
1896 Ade *Artie* i. 5 There they was, holdin' to this wart. 1925 Wodehouse *Carry on, Jeeves* vii. 167 Sippy had described them as England's premier warts, and it looked to me as if he might be about right. 1948 C. Day Lewis *Otterbury Incident* i. 6 Everyone called him the Wart because he had a huge wart on his left cheek... And because he was a wart. 1977 C. McCullough *Thorn Birds* xvii. 431 Watch your language, you dumb wart! 1984 *N.Y. Times Bk. Rev.* 1 Apr. 33/1 What!..is the old wart going to go on some more about reading?
c. *Naval slang.* A junior midshipman or naval cadet.
1916 [see *crab *sb.*[1] 10**]. 1921 *Blackw. Mag.* July 50/2 They all ignored the six 'warts'. 1962 Granville *Dict. Sailors' Slang* 129/1 *Wart*, naval cadet or junior midshipman, the 'lowest form of Naval life'; an unseemly excrescence.
6. wart disease, a disease of potatoes caused by the fungus *Synchytrium endobioticum* and producing dark pustules on the tubers.
[1903 *Jrnl. R. Hort. Soc.* XXVIII. p. clxxviii, Warty Disease of Potatoes..was introduced from the Continent, and first appeared in Cheshire.] 1915 [see *black scab* s.v. *black *a.* 19]. 1948 W. G. Burton *Potato* v. 103 Potato varieties vary greatly in their susceptibility to attack by wart disease. 1970 H. W. Howard *Genetics of Potato* vii. 46 Breeding for resistance..to wart disease..has been very successful.

war time. Add: **1.** (Later *attrib.* examples.)
1922 C. E. Montague *Disenchantment* viii. 121 Men's friends at home would have the agonies of false alarms added to their normal wartime miseries. 1935 D. L. Sayers *Gaudy Night* ii. 30 The present lot are the real War-time generation. 1945 *ABC of Cookery* (Ministry of Food) 1 It is intended to be used with the Ministry of Food Recipe Leaflets although it is not restricted to war-time cookery. 1965 *New Statesman* 7 May 709/3 Without the exigencies of war, he would need to have war-time powers. But he would lack..the simple war-time criterion of the need 'to maintain production'. 1978 Cadogan & Craig *Women & Children First* viii. 179 Tilli is..suffering wartime shortages bitterly.
2. *U.S. Hist.* Daylight-saving time introduced during World War II (see quots.).
1942 *Time* 9 Feb. 12/3 Franklin Roosevelt decided what he will call wartime daylight saving when it starts next week. Official name: War Time. 1979 *Sci. Amer.* May 39/3 Year-round daylight-saving time, or 'war time', as it was named by President Roosevelt, began on February 9, 1942.

war-whoop. a. (Earlier example.)
1739 W. Stephens *Jrnl.* 21 Dec. in *Colonial Rec. Georgia* (1906) IV. 474 In marching, our Indians set up the war whoop.

wasabi (wa·sābi). [Jap.] A Japanese herb, *Eutrema wasabi*, whose thick root is used in Japanese cooking.
1903 *Bull. Bureau of Plant Industry* (U.S. Dept. Agric.) No. 42. 20 There is a fresh sharpness about Japanese wasabi that not even the finest Austrian sorts of horseradish possess. *Ibid.* 21 For two years the young wasabi plants are cared for in the field. 1972 Y. Lovelock *Vegetable Bk.* 333 The roots of the related toothworts..serve as a horseradish substitute in North America... Wasabi..is a Japanese relation cultivated for the sake of its roots, which are used as a condiment. 1981 J. Melville *Sort of Samurai* xviii. 145 Otani..watched the *sushi* master deftly season the rice-cakes with a dab of tear-jerking *wasabi* paste.

wash, *sb.* Add: **I. 1. b.** *wash and brush-up*, a quick wash together with a tidying of one's hair; also *transf.*, and as *v. trans.* and *intr.*
1899 E. W. Hornung *Amateur Cracksman* 31 We must have a wash-and-brush-up before we go,—for I'm as black as your boot. 1938 G. Greene *Brighton Rock* iii. i. 107, I had to have a wash and brush up. 1941 *Vogue* June 58/2 Wash-and-brush-up your face... Creams are rationed and soap is not. 1956 G. Durrell *Drunken Forest* v. 98 The cockroach ambled about for a bit and then stopped for a quick wash and brush-up. 1959 G. Mitchell *Man who grew Tomatoes* i. 10 I'll wash and brush up, as they say, and be down in ten minutes. 1978 R. H. Lewis *Antiquarian Bks.* viii. 167 Most old books in for rebinding have an automatic 'wash and brush-up'.. before getting decked out in their new finery.
2. d. fig. phr. *to come out in the wash*: (of the truth) to be revealed, become clear; (of a situation, events, etc.) to be resolved or put right eventually. Cf. *washing *vbl. sb.* 8 a.
1903 Kipling *Five Nations* 196 An' it all went into the laundry, But it never came out in the wash. 1917 Wodehouse *Man with Two Left Feet* ii. 29 A sort of fate, what? .. Heredity, and so forth. What's bred in the bone will come out in the wash. 1930 'Brent of Bin Bin' *Ten Creeks Run* xxii. 266 That scandal has been such a long time comin' out in the wash that you must have been mistaken. 1943 N. Coward *This Happy Breed* ii. i. 53 Ethel:..I thought everything was going to be all right... Frank: Don't worry, old girl, it'll all come out in the wash. 1947 'N. Blake' *Minute for Murder* vii. 145 'How on earth could Bill know?'..'Oh, well, it'll all come out in the wash, no doubt.' 1971 J. R. L. Anderson *Reckoning in Ice* vii. 147 When it all comes out in the wash—it does—I suppose the company will meet our fees. 1978 J. Dunn in Hookway & Pettit *Action & Interpretation* 161 All human scientists..practise in the fond hope that the deficiencies of description or the errors and gaps in the intelligibility of record will all come out in the wash.
3. d. A liquid preparation used to protect plants against pests or disease.
1921 *Discovery* May 130/1 The providing of suitable sprays and washes, and other material connected with the checking of plant pests. 1951 *Dict. Gardening* (R. Hort. Soc.) IV. 2081/1 Other washes have had to be derived for use against it [*sc.* the red spider mite]. 1984 *Which?* Apr. 166/3 Dirt or discoloration by lichen and algae on a wall..often isn't harmful... Brush on a sterilising wash.
e. = Sheep-wash *sb.* 1, 2.
1933 *Press* (Christchurch, N.Z.) 25 Nov. 15/7 *Wash*, place and plant for washing sheep. 1965 [see *draining-pen* s.v. *draining *vbl. sb.* 4].
II. 6. b. (*b*) The air current caused by the passage of an aircraft.
1910 R. Ferris *How it Flies* xx. 474 *Wash*, the air-currents flowing out diagonally from the sides of a moving aeroplane. 1931 *Flight* 9 Oct. 1012/1 Certain modifications have been suggested which are intended to reduce the wash from the tailplane on the rudder during a flat spin. 1972 *Daily Tel.* 9 Aug. 4/8 It was not that the hundreds of [helicopter] pilots just overflew the nudist colony, but some flew so low they knocked over tents with the wash from their rotor blades.
d. Also, the removal or displacement of soil by rain and running water (in quot. 1835, a place where this occurs); freq. in *Comb.* with preceding sb., as in *rain-wash* s.v. Rain *sb.*[1] 6 in Dict. and Suppl., *sheet-wash* s.v. *sheet *sb.*[1] 12 b, *soil wash* s.v. *soil *sb.*[1] 10.
1835 J. H. Ingraham *South-West* II. 88 Bermuda grass is used with great success to check the progress of a wash. 1859 *Trans. Illinois Agric. Soc.* III. 412 Land lying in such a position as to protect it from wash..may be kept in constant cultivation. 1913 [see *soil-binding* adj. s.v. *soil *sb.*[1] 9]. 1959 G. H. Dury *Face of Earth* iii. 17 Rivers, surface-wash, and the downhill movement of solid rock combine to remove the substance of the land. 1972 R. J. Small *Study of Landforms* vi. 209 On the upper part of the slope the increased 'erosional' effect of wash away from the crest may tend to produce convexity of profile, and this may be exaggerated..by soil creep.
IV. 12. b. (Earlier example.)
1882 [see *drifting *vbl. sb.* 2].
VI. 15. b. Nonsense, rubbish, 'twaddle'. Cf. senses 11 a *fig.* and 15 a *fig.*, and Hogwash b in Dict. and Suppl.
1913 A. Lunn *Harrovians* xvii. 287 The Housemasters call their Sixth together at intervals and gass 'em... You know the kind of wash. 1933 G. Heyer *Why shoot Butler?* vi. 86 Not strictly the clean potato, is it?..Guest in the man's house, you know. The Public School Spirit, and Playing for the Side, and all that wash.
IX. 21. wash-bag, a small waterproof bag for holding toilet articles; a sponge-bag; **wash coat,** an undercoat, esp. one for improving or preparing the surface rather than giving a colour; **wash-day** (earlier example); **wash-fast** *a.*, that can be washed without losing colour or dye; so **wash-fastness; wash primer,** a wash coat for use on metal; **wash-sale** (earlier example); cf. Washed *ppl. a.* 1 f. See also *wash *v.* 20 b.
1972 E. Thorpe *Night I caught Santa Fé Chief* i. 11, I took the wash-bag out of the grip; the face-cloth was slightly damp. 1980 G. Lord *Fortress* i. 10 She..picked up her wash-bag. She went..to the bathroom. 1951 M. Hess *Paint Film Defects* 385 On wood which has been treated with water-stains the application of wash coats of a 5 to 8 per cent. shellac solution seem [*sic*] to be popular. 1960 *McGraw-Hill Encycl. Sci. & Technol.* X. 593/1 Primers are always pigmented. In clear finishes, the coat which performs this function is described as a sealer, an undercoater, or a wash coat. 1961 J. G. E. Holloway *Mod. Painter & Decorator* (ed. 5) I. 152 Often the application of a full coat of plastic paint will be preceded by a wash coat..to produce a ground of uniform colour. 1846 *Southern Lit. Messenger* XII. 598/1 Thursday is wash-day. 1963 A. J. Hall *Textile Sci.* iv. 184 This method of dyeing can be used to produce wash-fast shades. 1977 *Private Eye* 4 Mar. 21/1 (Advt.), T-shirts and sweater shirts printed to your design in wash-fast dyes, permanent whites, gold, silver or velvet flock finishes. 1962 J. T. Marsh *Self-Smoothing Fabrics* xx. 334 The thermosetting resin..gave moderate recovery but some discolouration and great embrittlement, together with a lack of wash-fastness. 1961 Webster, Wash primer. 1963 *Times* 22 Apr. 6/5 Zinc tetroxychromate is the most commonly used pigment in the so-called 'wash primers' for metals, which concurrently etch and coat the metal surface before the application of other paint coats to ensure good adhesion. 1973 Wash-primer [see *polyvinyl acetal* s.v. *polyvinyl a]. 1848 W. Armstrong *Stocks* 19 These wash sales are of course void between parties. They are of very frequent occurrence and very mischievous.

wash, *v.* Add: **I. 1. f.** (Earlier example of *absol.* use.)
1820 J. Severn *Let.* 17 Dec. in Keats *Lett.* (1958) II. 363, I am obliged to wash up—cook—& read to Keats all day.
(ii) *fig.* To bring to a conclusion; to end or finish (something). *U.S. slang.*
1925 *World* (N.Y.) 25 Oct. ii. 3/1 'That guy might be all right if he washed up [*sc.* washed, cleaned himself],' commented Buck... Just then the stage manager called out: 'What will I do with this act, Mr. Ziegfeld?' 'Wash up him and the bird,' said Flo [Ziegfeld] and that was the last of the Italian and his trained canary... Hype Igoe, the World's sporting writer, heard of the incident..and in commenting..upon Frank Moran, heavy weight pugilist, advised that matchmakers 'wash him up'. The phrase caught the sporting fancy..and has become a colloquial fixture..as a meaty synonym for finals and farewell. 1929 *Sat. Even. Post* 2 Nov. 24/3 'I had an idea,' he explained... 'Just came to me, riding back. I think I know how I can wash it up.'.. He would write it now—tonight! 1940 J. O'Hara *Pal Joey* (1952) 65 They said act of God and fire etc. wash up a contract automatically. 1972 D. Delman *Sudden Death* iv. 98 That man washed himself up with me because he couldn't keep his big, fat, fairy's mouth shut.
h. *fig.* In the game of mah-jong, to shuffle (the tiles).
1926 A. Christie *Murder of Roger Ackroyd* xvi. 196 Caroline got out the Mah Jong box and poured out the tiles upon the table. 'Washing the tiles,' said the colonel. 1929 *Encycl. Brit.* XIV. 677/1 All the tiles are..put face downward on the table and thoroughly shuffled or 'washed'. 1977 'J. le Carré' *Hon. Schoolboy* xiv. 319 Jerry heard a ritual clicking as the habitual mah-jong party washed the pieces before distributing them.
i. *absol.* To wash table utensils as opposed to drying them. Cf. *dry *v.* 1 c, *wipe *v.* 1 c.
1943 L. I. Wilder *Those Happy Golden Years* xxi. 192 Neither of you need worry about the dishes... I'll wash and Grace will wipe. 1958 J. Cannan *And be a Villain* vii. 150 Evadne..insisted on washing while Laura dried. 1962 M. Duffy *That's how it Was* xiii. 115 Billy washes, Arthur wipes, you put away. 1978 *Listener* 13 Apr. 483/1 Let's pack away the tea. I'll wash, you dry.
2. a. *to wash through*, to wash (a garment) by hand, often individually and hastily.
1936 N. Streatfeild *Ballet Shoes* ix. 136 I'll just wash your jersey through. 1968 M. Woodhouse *Rock Baby* xvii. 168 'Isn't that one of my spare shirts?'.. 'Sorry. Yes it is. But I washed it through for you.'
f. (Earlier example.)
1786 J. Woodforde *Diary* 31 May (1926) II. 247, I paid her up to this Day and told her I would get another to wash him.
3. o. *to wash up*: = sense 3 j. *U.S.*
1934 in Webster. 1935 Marsh & Jellett *Nursing-Home Murder* iii. 42 Thoms came into the [operating] theatre. 'We ought to get washed up, sir,' he said. 1947 J. Steinbeck *Wayward Bus* 214 A neatness of a mechanic who has just washed up. 1967 L. Block *Deadly Honeymoon* ii. 26 Wash up and change your clothes. 1979 R. Jaffe *Class Reunion* (1980) i. xi. 145 She was glad when he came out of the bathroom and she could go in and wash up.
p. *to wash one's hands*: euphemistic expression for 'to go to the lavatory'.
1938 I. Goldberg *Wonder of Words* vi. 108 We are invited to wash our hands, or, if we wear dresses, to powder our noses. 1953 R. Warner *Escapade* 119 She pointed to a large oak tree... 'Stupid woman,' said Lady Average. 'If she wants to wash her hands, why doesn't she go to the house?' 1966 G. Greene *Comedians* i. iii. 84 He was out 'washing his hands', as he put it in polite English. 1974 J. Gardner *Return of Moriarty* 291, I wonder if Rosie could, perhaps, take Miss Malloney to, er, to wash her hands.
II. 11. c. *to wash up*: to retrieve (gold) from the riffles, sluices, etc., in which it has collected during washing. Also *absol.*
1869 J. Anderson *Sawney's Lett.* (ed. 2) 27 Now say, what have you 'wash'd up'? Small wages. 1874 A. Bathgate *Colonial Experiences* xi. 142 After some months' hard work, we would wash up, and my mate would go off to sell the gold. 1900 B. Harte *From Sand Hill to Pine* 103 To dig for three or four hours in the morning, smoke their pipes..for an hour at noon, take up their labors again until sunset, when they 'washed up' and gathered sufficient gold to pay for their daily wants, was..the realization of a charming socialist ideal.
IV. 13. d. (Later example.)
1972 J. Wilson *Hide & Seek* vii. 122, I am making a purse..but I pricked my finger and got a bit of blood on it, but..it will wash out.
e. *to wash out* (trans.): to obliterate, cancel, remove.
1580, etc. [in Dict., sense 13 c]. 1932 R. Niebuhr *Moral Man & Immoral Society* (1933) iii. 81 It will prevent the idea of justice, which is a politico-ethical ideal, from becoming a purely political one, with the ethical element washed out. 1983 *Sci. Amer.* Feb. 86/3 Coronal structure hinted at in ordinary photographs is largely washed out by overexposure of the bright inner corona.
(ii) *colloq.* To call off (an event), esp. because of bad weather; to eliminate (a possible course of action). Usu. *pass.*
1917 A. G. Lee *Let.* 25 May in *No Parachute* (1968) 24 Today I have two patrols, one this morning..but after an hour it was washed out through bad weather. 1933 P. MacDonald *Mystery of Dead Police* vii. 49 I'll get that murder charge washed out altogether. 1953 'N. Shute' *In Wet* v. 149 If there's an awful lot of work before we go, I might have to wash it out. 1964 Mrs. L. B. Johnson *White House Diary* 8 Apr. (1970) 104 When the commentator inquired about the possibility of McNamara [being Vice-President] and pointed out that he had been a registered Republican at one time, Lippman rather washed that one out. 1977 *Belfast Tel.* 14 Feb. 22/4 Bangor's first ever mid-week racing fixture..was washed out today when stewards inspected the course and found it waterlogged.
(iii) *Air Force slang.* To kill (an airman) in a crash; to crash (an aircraft); also (const. *of* or *from*), to withdraw (a person) from a course. Usu. *pass.*
1918 J. M. Grider *War Birds* (1926) 87 Wholesale funerals... Six American Naval pilots..thought that Camels were as easy to fly as the Hanriots they had been flying in France and they wouldn't listen to any advice from the instructors here. Three of them were washed out one week. 1927 C. A. Lindbergh *We* v. 73 Coupled with this was the anxiety of waiting for the returns from our examination papers, the failure of any two of which would be sufficient cause for their owner to be washed out from the courses. 1928 *Pop. Sci. Monthly* May 72 That Kiwi washed out the only peppy crate in the outfit. 1942 F. H. Joseph *Lett. home from Brit. at War* 16 Three planes were washed out completely, others damaged. 1943 *Yank* 30 July 18 The air cadet needed only 20 flying hours for his commission when he was suddenly washed out of advanced training and shipped. 1979 M. Hastings *Bomber Command* vi. 145 Owen was washed out of pilot training within a fortnight, and posted to learn to be a navigator.
15. a. Also *fig.*
1929 *Times* 30 Oct. 14/1 Trans-America, which closed last night at 6¼, opened today at 20½..—$840,000,000 ..nominally washed away over-night.
16. (*fig.* example of *to wash up.*)
1947 G. Greene *Nineteen Stories* 74 And so he'd washed up here, under my eyes, sitting all day under the bandstand.
17*. *slang.* To murder. Also with *away.*
1941 in B. A. Botkin *Treas. Amer. Folklore* (1944) i. 124 So Stack, with his gun handle filled with notches, knowed there was a reward out for him for men he had washed away. 1960 'E. McBain' *See them Die* (1963) v. 48 'This Alfredo kid, he not sush a bad guy.' 'He's getting washed and that's it.' 1979 P. Hill *Washermen* xxiv. 54 They had broken the code... The Washermen must be washed.
VI. 19. c. = *launder *v.* 1 b.
1973 *Black Panther* 30 June 2/2 The money had been 'washed' through the Mexican bank passing off as a legal fee to the Mexican lawyer in order to conceal the source of the donation. 1977 B. Freemantle *Charlie Muffin* xii. 127 We *must* wash the money... If that money isn't broken down, Kalenin..just won't cross. 1981 R. Thomas *Mordida Man* xxvii. 237 What was their payoff for washing the money?
20. b. (Not all clearly distinguishable from the Combs. listed in sense 21 of the *sb.* above.) *wash-jug*, *-place* (earlier example), *-rag* (later examples; now *U.S.*); **wash-and-wear** orig. *U.S.*, the property of a garment or fabric of being easily washed, drying readily, and needing no ironing; usu. *attrib.*; **wash-bench** *U.S.*, a bench on which washing is done; **wash-cloth** *U.S.*, a facecloth; **wash-deck** *attrib. Naut.*, used in, or pertaining to, the washing of the deck of a ship; *wash-deck tub* (slang), a small boat, with decks easily washed by the sea; **wash-kettle** *U.S.*, a kettle in which water is heated for washing; **wash-kitchen** (earlier and later examples); **wash-line** chiefly *U.S.* = *washing-line* s.v. *washing *vbl. sb.* 9 a; **wash-pan** *U.S.*, a metal wash bowl; a pan for washing ore; **wash-pen**: also *N.Z.*; **wash sink** *U.S.*, a sink for washing oneself.
1959 *Sears, Roebuck Catal.* Spring/Summer 493/1 Wash and wear suits... Launder by hand or washing machine... Machine-dry or drip-dry. 1966 *Wall St. Jrnl.* 11 Feb. 1/1 Permanent press differs from wash-and-wear... Wash-and-wear doesn't involve baking, but rather depends on the blending of synthetic and natural fibers for wrinkle resistance. Also, wash-and-wear has no permanent crease. 1981 Clark & Swaine *Home Managem.* x. 251 The more sophisticated American machines may have a choice of programmes, e.g...synthetics with pre-wash, wash and wear and cold wash. 1981 M. C. Smith *Gorky Park* i. xii. 164 Parties the Film-Makers Union gave for foreign guests, where the civilized appreciation of a bottle of French perfume or a wash-and-wear skirt was routine. 1843 Wash-bench [see *ironing board* s.v. *ironing *vbl. sb.* 1]. 1884 G. W. Cable *Dr. Sevier* I. xxi. 157 She had reached down and taken from the wash-bench the lump of yellow soap. 1969 N. W. Parsons *Upon Sagebrush Harp* xxiv. 132 The man was wiping his neck on the roller towel hung above the outdoor washbench. 1915 Washcloth [in Dict.]. 1949 M. Mead *Male & Female* xii. 262 A baby's face gently washed with a supersoft wash-cloth. 1978 *Detroit Free Press* 5 Mar. (Parade Suppl.) 14E/1 (Advt.), Bath towel, hand towel & washcloth—all cuddly cotton terry. 1878 E. Wakeman *Log of Ancient Mariner* ii. 41 Him I had lowered from the bow one dark night, into the wash-deck tub, in which he paddled to a little schooner close to. 1884 *Naval Encycl.* 835/1 *Wash-deck Gear*, the brooms, squilgees, holy-stones, buckets, etc., used in washing decks. 1901 F. T. Bullen *Sack of Shakings* 98 One morning, at wash-deck time, when I was prowling around forrard [etc.]. *Ibid.* 115 The wash-deck tub was hauled forrard. 1938 C. S. Forester *Ship of Line* i. 22 His captain's admission that he, too, had baths under the washdeck pump. 1927 W. de la Mare *Stuff & Nonsense* 54 It golloped up bolsters and wash-jugs and chairs. 1946 S. Spender *European Witness* ii. 18 A large enamel wash-jug. 1787 *Kentucky Gaz.* 24 Nov. 2/3 Samuel Blair, Has for sale..a Quantity of excellent.. copper and brass wash kettles. 1882–3 W. Whitman *Specimen Days* 23 They put wash-kettles on the fire, for soup, for coffee. 1973 A. Dundes *Mother Wit* p. xiii, The custom of placing an inverted wash kettle in the center of the floor during a prayer meeting so that the sounds of the singing might go into the pot and thereby not disturb the white folks. 1838 C. Gilman *Recoll. Southern Matron* xxix. 206 Preparations were made for the wedding which she chose to have performed in the wash-kitchen instead of our parlour. 1922 Joyce *Ulysses* 601 It was the daughter of the mother in the washkitchen that was fostersister to the heir of the house. 1890 K. D. Wiggin *Timothy's Quest* 48 There's lots of baby-clothes hanging on the wash-lines. 1952 J. Steinbeck *East of Eden* vii. 56 They stole ..garments from a wash line. 1969 E. H. Pinto *Treen* 157 Wash line winders are usually strictly utilitarian objects, without any collector interest. 1851 Washpan [see *stuff *sb.*[1] 4 d]. 1857 J. D. Borthwick *Three Yrs. in Calif.* 124 A 'prospecter' goes with a pick and shovel, and a wash-pan. 1884 'Mark Twain' *Huck. Finn* xxxvii. 375 We..scratched around and found an old tin washpan. 1946 G. Wilson *Fidelity Folks* 175 The tin washpan or the creek was good enough for that. 1855 H. Phillips *Rockwood Jrnl.* (typescript) 20 John went with him to Wash-pen in afternoon. 1774 J. Woodforde *Diary* 15 Oct. (1924) I. 141, I caught a remarkable large Spider in my Wash Place. 1925 F. Scott Fitzgerald *Great Gatsby* ii. 27 'It's more of an Airedale.' He passed his hand over the brown washrag of a back. 1964 S. Bellow *Herzog* 257 He got into her ears with the washrag as she screamed, cleaned off her face, the nostrils, wiped her mouth. 1978 J. Updike *Coup* (1979) v. 192 Her momma's a washrag and her daddy's a redneck. 1857 *Lawrence* (Kansas) *Republ.* 2 July 4 'Here are all the conveniences for washing,' said the landlord, stepping to a mahogany wash sink and raising the lid. 1873 'Mark Twain' & Warner *Gilded Age* xxix. 270 It was a small room..with a wash-sink in one corner.
c. *wash-gravel* (example).
1860 in *Occas. Papers Univ. Sydney Austral. Lang. Res. Centre* (1966) No. 9. 27 The runs of gold and lower deposits are not traceable to reefs but to made hills, composed principally of water-worn quartz nodules and debris, often cemented together with ferruginous matter, the wash-gravel resting on whitish or yellow felspathic schist.
d. *wash pants.*
1941 *Daily Progress* (Charlottesville, Va.) 3 July 10/7 (Advt.), Mens & boys swim trunks, wash pants, slack suits, sport shirts, etc. 1972 J. Maryland in T. Kochman *Rappin' & Stylin' Out* 210 Red and grown out of wash pants and levis to Oleg Cassini imported mohair suits.

washability. Delete *nonce-wd.* and add later examples.
1958 *Times* 20 Oct. 13/2 This tweed has been woven for the school..and tested for washability. 1970 *Cabinet Maker & Retail Furnisher* 23 Oct. 173/1 Acrylic fibre is nearest to wool in its characteristics with the additional benefit of washability and recovery. 1985 *New Yorker* 20 May 7 (Advt.), The polyester-and-cotton poplin has the convenience of washability.

washable, *a.* (*sb.*) Add: **2. b.** *washable distemper.*
1894 *Country Gentlemen's Catal.* p. v (Advt.), Taylor's Washable Distemper, in all tints. 1926 F. M. Ford *Man could stand Up* i. ii. 27 Washable distemper was not like the poor—always with you. 1939 *Archit. Rev.* LXXXV. 158 (*caption*) Washable distempers.
b. *sb. pl.* Articles of clothing that may be washed without being damaged.
1951 in M. McLuhan *Mech. Bride* (1967) 95 Colors perk up—brighten up—when you suds your washables in Ivory Flakes. 1973 *Philadelphia Inquirer* 7 Oct. 6 (Advt.), Our lovable washables in Arnel triacetate/nylon pretty with satin piping and bow.

washateria, var. *washeteria.

wash-ball. For *rare* read: chiefly *Hist.* (Further examples.)
1936 *Burlington Mag.* Mar. 123/2 A plain globular soap-box (called a washball box in England some fifty years earlier). 1966 T. H. Raddall *Hangman's Beach* ii. x. 150 And now came the grey attendant with wash-ball and towel, and a small wooden tub of steaming water. 1970 *Canad. Antiques Collector* Nov. 16/1 Out of the..cupboard came a silver plated shaving bowl and ewer, while from one small drawer appeared a pair of silver soap or wash ball boxes. 1980 E. Jong *Fanny* ii. xii. 280 The true Royal Chymical Wash-Ball for the beautifying of the Hands and Face.

wash-basin. Delete 'Now chiefly *U.S.*' and add later examples.
1928 Galsworthy *Swan Song* i. v. 35 We must get some more wash-basins in. 1980 *Sunday Times* (Colour Suppl.) 21 Sept. 50 (Advt.), This soft, subtle shade.. reveals the potential..of the range of wash basins. 1984 Worthington & Knight *Home Plumbing* 103/2 The normal height for a wash-basin is 800 mm..from the rim to the floor.

washboard. Add: **I. 3. b.** (Earlier example.)
1845 J. W. Norris *Business Adviser & Gen. Directory Chicago* 95 Manufacturer of the improved zinc wash-boards.
d. A washboard (sense 3 b) used as a percussion instrument; hence, the kind of music produced by bands using this instrument. Freq. *attrib.*
1925 J. Taylor (*title of musical composition*) Wash-board blues. 1933 *Gramophone* June 26/1 Washboard bands are the red-nosed comedians of dance music... They usually play about with delightfully inconsequent

tunes. *Ibid.*, There is a good deal to be said for Yes Suh.. by the Washboard Rhythm Kings... The gent on the washboard is a dexterous performer. 1946, etc. [see *jug *sb.*[1] 1 c]. 1947 R. de Toledano *Frontiers of Jazz* 107 The first washboard was recorded by me. 1961 W. Sansom *Last Hours Sandra Lee* vii. 173 There came a machine slamming of washboard music and wailing voices that sounded as if they were lost in a large, echoing cave. 1973 J. Marks *Mick Jagger* (1974) 56 Washboard was also fair enough, if you wanted that sort of sound... You know.. more or less a primitive sort of Chicago sound. 1973 J. Wainwright *Pride of Pigs* 174 Instead of a drummer there was a young kid with a washboard and thimbles.
4. *fig.* A corrugated surface, esp. of a road.
1934 in Webster. 1950 J. Dempsey *Championship Fighting* 186 Bending exercises are best to develop the stomach muscles into a protective 'washboard' against body blows. 1968 G. Jones *Hist. Vikings* iii. i. 156 Not only a 'washboard' of wide furrows is visible, but even the wheel-tracks of the viking farmer's last carting. 1976 [see *reg[1]]. 1979 *Tucson Mag.* Apr. 51/1 The long gravel drive to its shore..tends to turn into a washboard in the drier summer months.
II. 5. *attrib.* as *adj.* Corrugated, furrowed, esp. as a result of weather and usage.
1913 *Bull. Geol. Soc. Amer.* XXIV. 145 The till sheet over large areas has been rubbed into a fluted or wash-board form on a large scale, but with low relief. 1936 *Trans. R. Soc. Canada* XXX. iv. 10 In the general area where these wash-board moraines were noted there are extensive sand plains. 1949 *Consumer Rep.* Jan. 8/1 Pick out a hubbly surface or a 'washboard' gravel road. 1953 W. Moore *Bring Jubilee* (1955) i. 6 The morasses or wash-board roads which were the only highways. 1968 R. W. Fairbridge *Encycl. Geomorphol.* 989/2 Steep valley slopes [in a rain forest] typically have a washboard topography, a simple parallel consequent drainage system. 1976 *New Motorcycling Monthly* Oct. 6/3 Washboard surfaces would set the rear wheel pattering.

wash-bowl. Add: **2. a.** (Earlier example.)
1816 U. Brown *Jrnl.* in *Maryland Hist. Mag.* (1915) X. 369 His wash-bowl [is] the knot of a tree.
b. *spec.* A vessel in which gold is washed.
1848 in *Essex Inst. Hist. Coll.* (1874) XII. 106, I came from Salem City, With my wash bowl on my knee. 1850 E. Christman *Jrnl.* 3 Apr. in *One Man's Gold* (1930) 131, I am standing in a hole to the depth of my knees, with my pick raised high in the air; my spade and washbowl are lying upon the ground by my side. 1925 E. O'Neill *Desire under Elms* i. iv. 66 Their voices..take up the song of the gold-seekers to the old tune of 'Oh Susannah'.. 'I'm off to Californi-a! With my wash bowl on my knee.'

wash-dirt. (Earlier N.Z. example.)
1862 *Otago: Goldfields & Resources* 20 All the wash-dirt is auriferous.

wash-dish. 2. Delete '?' and add earlier and later examples.
1805 *Austin Papers* (1924) I. 140, 1 Wash dish. 1857 *Quinland* I. ii. i. 275 An iron kettle, which I and all the family used as a common wash-dish before breakfast. 1876 *Rep. Vermont Board Agric.* III. 628 The farmer.. was able to furnish a large maple block to set the wash-dish on. 1891 M. E. Freeman *New England Nun & Other Stories* 271 I've got my tin wash-dish there on the bench.

wa·shdown. [f. vbl. phr. *to wash down*: see Wash *v.* 1 g.] **1.** The, or an, act of washing down; *spec.* an act of washing oneself from top to bottom at a wash-basin as distinct from in a bath or under a shower.
1949 *Amer. Speech* XXIV. 35 Another characteristic of oil-field language is the predominance of compound words. .. The conversion of some verb-adverb combinations into nouns and adjectives are *back-up*, *clean-out*..and *washdown*. 1972 *Bottlers' Year Bk.* 1972–73 40 Taking washdown first—obvious waste can be reduced from running hoses left unattended. 1972 J. Wilson *Hide & Seek* vii. 132, I have a washdown every day and a bath on Sundays. 1979 *Nature* 29 Mar. p. xxx/1 The instrument eliminates bumping and smearing of the sides of the tubes thus eliminating the necessity for washdowns.
2. *attrib.* Of a lavatory: (see quot. 1967). Also *absol.*
1967 *Gloss. Sanitation Terms* (*B.S.I.*) 64 *Washdown W.C. pan*, a W.C. pan in which the excrement falls into the water in the trap and is subsequently removed by the momentum of the flushing water. 1976 *Evening Advertiser* (Swindon) 31 Dec. 4/2 (Advt.), Bathroom suite: consisting of..low level washdown toilet. 1984 Worthington & Knight *Home Plumbing* 96/2 There is a choice of pan styles... The most common is the washdown.

washed, *ppl. a.* Add: **1. f.** (Earlier example.)
1886 *Harper's Mag.* July 205/1 Washed or fictitious sales, or false reports of sales, are also penal offences.
i. Of a carpet: faded, bleached; specially treated so as to soften the colours and impart a sheen.
1911 G. G. Lewis *Pract. Bk. Oriental Rugs* iii. 40 What the trade speaks of as a 'washed' rug is not necessarily a 'doctored' one. There is a legitimate form of washing which is really a finishing process... It merely washes out the surplus color and sets the rest. 1962 C. W. Jacobsen *Oriental Rugs* 121 The danger of your buying both chemically washed and painted rugs today is practically nil. 1969 [see *Peking carpet* s.v. *Pekin 3]. 1970 'D. Halliday' *Dolly & Cookie Bird* ii. 20 My room had..wall-to-wall washed Chinese carpeting in quiet shades of money. 1982 G. F. Newman *Men with Guns* x. 74 Wallechinsky moved agitatedly across the living room.. his expensive, imported shoes making a faint squeak on the washed silk Chinese rug.
2. a. (Earlier example.)
1796 Jane Austen *Let.* 1 Sept. (1952) 9 My new coloured gown is very much washed out, though I charged everybody to take great care of it.
3. (*all*) *washed up*: finished; without prospect of further success or competence; no longer on intimate terms; exhausted, 'washed out'. *slang* (orig. and chiefly *U.S.*).
1923 *N.Y. Times* 9 Sept. vii. 2/1 [Stage slang.] *Washed up*, all through for the night. 1925 *Amer. Speech* I. 36/1 [Stage slang.] How cheery it would be, when family ties begin to irk, to use their honest, 'I'm washed up with you,' to indicate that you hope the breach is permanent. 1933 S. Kingsley *Men in White* (1934) iii. i. 109 I'm washed up with the whole business. 1934 W. Saroyan *Daring Young Man* 38 We're washed up as a race, we're through, it's all over. 1935 J. T. Farrell *Judgment Day* (1945) I. xvi. 312 His lips twisted into a sneer at himself, and he thought that he was just a goddam washed-up has-been. 1942 J. B. Priestley *Black-Out in Gretley* viii. 182 You're too careless, Joe..and now you're all washed up. 1958 C. Williams *Man in Motion* (1959) ii. 23 I'm washed up as a writer. 1980 *Newsweek* 17 Nov. 7/1 Once he was the most underestimated man in American politics—a washed-up movie star, it was said, who was too old, too simple and too far right to be President.
4. *to get washed up*: to get the washing up done.
1950 J. Cannan *Murder Included* i. 12 Have to dine early now... We've only got Mrs Witts in the kitchen, and she likes to get washed up and finished.

washer, *sb.*[1] Add: **5. f.** A machine for washing dishes; a dish-washer.
1958 I. Asimov *Whiff of Death* xvii. 170 Just let me put the dishes in the washer and then we'll go to bed. 1976 H. MacInnes *Agent in Place* xx. 215 She stacked dishes into the washer.
8*. A face-flannel. *Austral.*
1951 D. Cusack *Say No to Death* xxxi. 194 Doreen had given her a washer and a drop of warm water to wash the sleep out of her eyes. 1970 P. White *Vivisector* iv. 236 He was reminded of an old face-washer, often grubby, one of the maids had crocheted for him, in wide mesh.
9. washer-drier, a machine that both washes clothes and dries them.
1968 *Listener* 1 Aug. 130/1 The unappealing image of America as a country..whose highest aspirations were another car in the garage or a bigger washer-dryer. 1971 *Which?* Mar. 72/2 Contracts for automatics cost about £9, for a washer-drier (like a Bendix) about £12. 1983 *The Mag.* Dec. 59/4 Built-in kitchen equipment, including dishwashers and washer-dryers.

washerette (wǫʃəre·t). [f. Washer *sb.*[1] + -ette.] = *launderette.
1968 *Punch* 10 Apr. 521/3 The Mariner he beats his brow; But the Guest is well away. 'She runs a mobile washerette Down here, or so they say.' 1974 *State* (Columbia, S. Carolina) 1 Apr. B-9/6 (Advt.), Philco-Bendix Quality. For the best in washerette profits. 1976 *Honolulu Star-Bull.* 21 Dec. F-3/8 (Advt.), Washerette in prime location. 10 washers & 5 dryers. Good lease.

washer-up (wǫʃərʊ·p). Pl. **washers-up.** [f. vbl. phr. *to wash up* (Wash *v.* 1 f) + -er[1].] One who washes up dishes.
1907 *Daily Chron.* 19 Apr. 12 (Advt.), Woman (young) wanted as washer-up. 1933 D. C. Peel *Life's Enchanted Cup* xiv. 173 An experienced washer-up in canteens. 1960 *House & Garden* Oct. 165/3 Washers-up who seem to spend one third of their life at the kitchen sink. 1978 L. Davidson *Chelsea Murders* viii. 42 They always had the meal together: the patron..the chef, the under-chef, the washer-up.
So **washer-u·pper** *colloq.*
1961 *John o' London's* 21 Sept. 318/3 But Dunham was more than a washer-upper. 1968 C. Drummond *Death & Leaping Ladies* v. 127 The usual two washer-uppers.. who also prepare vegetables. 1977 *New Society* 25 Aug. 381/1 Rubbish-collector, tea boy, washer-upper and so on.

washerwoman. Add: **3. b.** *washerwoman's skin*, skin that is much wrinkled as a result of immersion in water.
1981 C. A. McLaren *Twister over Thames* iv. 65 In the light of my discoveries, notably from the wrinkled nature of the skin—washerwoman's skin, it is called..Master Thacker had been immersed for about twelve hours before ..discovery. 1985 'E. McBain' *Snow White & Rose Red* v. 73 When a body hadn't been submerged too long..the so-called washerwoman's skin wasn't too bad.

washery. Add: Also, a laundry or wash-house. (Earlier example.)
1875 Trollope *Prime Minister* (1876) I. xix. 305 We've got the steam-washery put up.

washeteria (wǫʃətɪᵊ·riă). orig. and chiefly *U.S.* Also **washateria.** [f. Wash *v.* + *-eteria.] **a.** = *launderette.
1959 [see *-teria *suffix*]. 1966 *Sunday Times* 17 July 30 Now that we have grown accustomed to the blandishments of..something called Washeterias, the next step may be drive-in laundries. 1971 L. Gribble *Alias the Victim* i. 17 She wore a skirt and jumper that had the nondescript shapeless look that derives from too many visits to a washeteria. 1976 *San Antonio* (Texas) *Express-News* 27 Nov. 1B/1 T's Washhouse—the only washeteria for 75 miles in any direction. 1982 P. D. James *Skull beneath Skin* xlii. 339 She'll be at the Washateria... She always does her washing on Monday afternoons.
b. *car washeteria*: a self-service car-washing establishment.
1965 *Daily Tel.* 14 May 20/2 (Advt.), The demand from motorists for simple-to-operate inexpensive car washing facilities will go on growing. Here's how the Car Washeteria answers these requirements. 1970 *Times* 24 Dec. 9, I have long grown accustomed to a cafeteria but I now clean my car at a car washeteria.

wash-hand, *a.* **b.** (Earlier examples.)
1789 J. Woodforde *Diary* 17 Nov. (1927) III. 153 Ben returned home about 5 o'clock with..a wash-hand Stand. 1820 G. Colman *XYZ* i. 10 Overset my wash-hand stand;—grop'd to bed in the dark.

wash-house. Add: **2. c.** (Earlier example.)
1856 *Democratic State Jrnl.* (Sacramento, Calif.) 4 Aug. 3/2 There is a washhouse on Font street which seems to imbue its occupants with a desire to quarrel.
d. (Earlier example.)
c 1806 D. Wordsworth *Jrnl.* (1941) I. iv. 236 In the middle of the field is a wash-house, whither the inhabitants of this large town [*sc.* Glasgow], rich and poor, send or carry their linen to be washed.

wa·sh-in. *Aeronaut.* Also **washin.** [After *wash-out 5.] An increase in the angle of incidence of an aeroplane wing towards the tip.
1916 H. Barber *Aeroplane Speaks* 82 The advantages of the wash-in must, of course, be paid for. 1943 *Jrnl. R. Aeronaut. Soc.* XLVII. (Abstr. Section) 70 As the wing tips approach the tunnel wall, both the induced drag corrections and the angles of apparent washin or of apparent twist increase rapidly. 1979 Bertin & Smith *Aerodynamics for Engineers* iii. 83 If the angle of incidence increases toward the tip, the wing has 'washin'.

washiness. 1. (Earlier *fig.* example.)
1763 in *Owen's Weekly Chron.* (1767) 7 May 55 Humanity ..is nothing more than a childish washiness of nature.

washing, *vbl. sb.* Add: **1. h.** (Earlier example of *washing up.*)
1858 Geo. Eliot *Scenes Clerical Life* II. viii. 157 The necessary sum of meals and the consequent 'washing up'.
5. b. *Stockbroking.* (In sense 19 b of the vb.)
1849 *Hunt's Merchant's Mag.* XXI. 118 'Washing' will hardly go down at the board. 1870 J. K. Medbery *Men & Mysteries Wall St.* 138 Washing is where one broker arranges with another to buy a certain stock when he offers it for sale. The bargain is fictitious. 1894 S. Leavitt *Our Money Wars* 287 In 1887..by the process known as 'Washing',—that is, by hiring one set of brokers to buy and another set of brokers to sell,—the price of shares was forced to fifteen times their value.
8. a. Phr. *to come out in the washing*: = *to come out in the wash* s.v. *wash *sb.* 2 d; *to take in one another's washing*: to help one another by buying one another's goods or services, esp. where no new wealth accrues overall; to render mutual services, to be mutually dependent.
1876 Trollope *Prime Minister* IV. xii. 183 The effects which causes will produce,..the manner in which this or that proposition will come out in the washing, do not strike even Cabinet Ministers at a glance. 1889 G. B. Shaw *How to become Mus. Critic* (1960) 147 The inhabitants either live in villas on independent incomes or else by taking in one another's washing and selling confectionery, scrap books, and photographs. 1913 R. Brooke *Let.* 1 Sept. (1968) 501 Most of happiness is because one's friends are happy: so that spiritually—whatever the damned Economists may say—we *do* live by taking in each other's washing. 1937 M. Borden *Black Virgin* iii. 63 Quite half the women she knew were [working] in shops of one sort or another. The only drawback to that being..that they took in each other's washing... Mona and Peg bought Cimmie's clothes. Cimmie bought her hats from Mona, her nighties from Peg. 1959 J. L. Austin *Sense & Sensibilia* (1962) i. 4 These two terms, 'sense-data' and 'material things', live by taking in each other's washing. 1967 G. Sims *Last Best Friend* xviii. 169, I expect you know what dealers are like for selling among themselves, it's rather like taking in each other's washing.
b. *washing-up*, table utensils awaiting washing up.
1972 J. McClure *Caterpillar Cop* xiii. 211 Lisbet had eaten and stacked the washing up ready for the girl. 1977 P. Hill *Fanatics* 6 He made his bed but left the washing-up in the sink.
9. a. *washing-room* (earlier example), *soap*; **washing basket,** a basket for holding articles newly washed or waiting to be washed; **washing bat** *Hist.* (see quot. 1898); **washing-book** (earlier example); **washing-line** = *clothes-line* s.v. Clothes *sb. pl.* 4; **washing-machine** (earlier and later examples); **washing powder,** a cleansing agent in powder form for adding to the water used for washing household linen; **washing-stand** (earlier examples); **washing-stool,** a stool used when washing.
1947 M. Morris in 'B. James' *Austral. Short Stories* (1963) 362 She went down to the lines, walking heavily with her washing-basket full. 1967 *Listener* 17 Aug. 204/3, I..used to fill a big washing-basket with books and bring it downstairs as often as I wanted to. 1898 *Eng. Dial. Dict.* s.v. *Bat*, The washing bat was used to beat the dirty clothes after they had been 'put to soak'. 1969 E. H. Pinto *Treen* 149/2 A woman hitting a man with a washing bat is carved on an oak misericord of 1401, in Carlisle Cathedral... It shows that early washing bats were more shovel-shaped, with wider, shorter blades.

1983 *Daily Tel.* 16 Nov. 15/3 These finely decorated washing bats..were once used to beat the washing. 1868 W. Collins *Moonstone* I. xiii. 216 Before we begin, I should like..to have the washing-book... I want to be able to account next for all the linen in the house, and for all the linen sent to the wash. 1939 L. MacNeice *Autumn Jrnl.* viii. 33 But Life was comfortable, life was fine With two in a bed and patchwork cushions And checks and tassels on the washing-line. 1961 J. Stroud *Touch & Go* iv. 45 Any idea where I can get a washing-line post? 1978 J. Thomson *Question of Identity* xii. 115 Betty Lovell was pegging out sheets on a washing-line. c 1754 in *Hermathena* (1965) ci. 40 Things to be done... Morning Caps made..Curtains Chairs Carpets for Dublin ..Washing Machine. a 1780 *Rake's Progress* (1977) 9 Enter Porter with a Washing Machine, puts it down—Enter Beat'em, pursued by Washerwomen, who beat him & break his washing machine. 1780 *Brit. Patent* 1269 1 Washing machine. Rogerson's specification... My invention of an entire new machine called a laundry, for the purpose of washing and pressing all sorts of household linen. 1897 *Sears, Roebuck Catal.* 139/2 Washing Machines... The Electric Washers are warranted to be well made. 1944 A. Huxley *Let.* 10 Apr. (1969) 503 Fully furnished and equipped down to an electric washing machine. 1975 *Sunday Times* 16 Nov. 44/3, I also kept rushing to the other side of the room trying to empty rubbish into the washing machine. 1869 J. G. Fuller *Uncle John's Flower-Gatherers* 182 The old Prof..calls salt 'chloride of Sodium' and sets me thinking of washing powders. 1895 [see *blueing, bluing *vbl. sb.* 2]. 1969 I. & P. Opie *Children's Games* ii. 104 The mother..asks the shopkeeper for household goods, such as..some washing-powder. 1977 A. Wilson *Strange Ride of R. Kipling* v. 244 Those who find to their surprise that washing powders wash whiter. 1838 F. A. Kemble *Let.* in *Rec. Later Life* (1882) I. 175 One towel was considered all that was requisite not even for each individual, but for each washing-room. 1720 J. Steuart *Letter-Bk.* (1915) 121 Six barells washing soap. 1947 *Daily Gleaner* (Kingston, Jamaica) 5 Nov. 3/3 (*heading*) No washing soap this weekend. 1789 J. Woodforde *Diary* 13 Nov. (1927) III. 151 Bought this day..one new Mohogany Washing-Stand. 1799 *Times* 1 June 4/1 Dressing and washing stands. 1868 Dickens in *All Year Round* 19 Dec. 62/2, I found a man, his wife, and four children, sitting at a washing stool by way of table, at their dinner.
b. The phr. *washing up* (see sense 1 h in Dict. and Suppl.) in *Comb.*, as *washing-up bowl*, *machine*, *water*; **washing-up cloth,** a square of loose-weave fabric for washing dishes, etc.; **washing-up liquid,** liquid detergent for adding to washing-up water.
1938 N. Streatfeild *Circus is Coming* vii. 98 They hurried back to the washing-up bowl. Santa..put a cup in the water. 1983 D. Clark *Vicious Circle* i. 19 Marian had emptied the washing up bowl and mopped down the draining board. 1973 L. Cooper *Tea on Sunday* xvi. 136, I wouldn't myself trust poor Charlot to sell a row of washing-up cloths. 1975 G. Seymour *Harry's Game* iii. 49 The publican pushed the washing-up cloth..across the wooden bar. 1971 C. Bonington *Annapurna South Face* 254 Washing-up liquid. 1980 P. Hill *Savages* vii. 135 Found your true occupation then?..Washin' up... What made you go mad with the washin' up liquid? 1930 *Daily Tel.* 1 Dec. 23/7 (Advt.), Electric washing-up machines..will be sold by auction. 1972 C. Drummond *Death at Bar* i. 7 He has three helpers in the kitchen, one working the washing-up machine. 1932 S. Gibbons *Cold Comfort Farm* x. 137 Niver put my liddle pretty [mop] in that gurt old greasy washin'-up water. 1981 J. Wainwright *All on Summer's Day* 161 It hasn't a head on it [*sc.* beer]... An' it tastes like washing-up water.

Washington (wǫ·ʃiŋtən). *U.S.* [Name of George *Washington* (1732–99), first president of the United States of America.] **1.** *Washington lily*: a tall lily, *Lilium washingtonianum*, that grows in the mountains of the Pacific Coast of N. America and bears white flowers.
[1859 A. Kellogg in *Hesperian* (San Francisco) III. 340 (*caption*) Lady Washington lily. 1863 *Proc. Calif. Acad. Nat. Sci.* II. 13 Dr. Kellogg also exhibited a drawing and growing specimens of a new species of lily from the Sierra Nevada... *L. washingtonianum* (Kellogg) Lady Washington Lily... Stem erect, glabrous, three to four feet high, two or more flowers on peduncles four to five inches long.] 1869 J. Muir *My First Summer in Sierra* (1911) 43 Here and there a Washington lily may be seen nodding above its even surface. 1937 [see *Shasta lily* s.v. *Shasta *a.* 2].
2. *Washington pie*: †(*a*) some kind of pie; (*b*) a light cake made of sponge layers with a jam or jelly († or cream) filling.
1878 N. A. Donnelley *Lakeside Cook Bk.* 33/1 Washington pie. 1904 F. M. Farmer *Boston Cooking-School Cook Bk.* xxxi. 421 Mix and bake Cream Pie. Put raspberry jam or jelly between layers. 1905 J. C. Lincoln *Partners of Tide* i. 10 Won't you have somethin' to eat? One of them turnovers or some Washington pie, or somethin', hey? 1968 E. R. Buckler *Ox Bells & Fireflies* xv. 231 The taste of the Christmas ribbon candy, and the fleeting taste of the good-luck thimble and the birthday wedge of Washington pie.

Washingtonia (wǫʃiŋtōᵘ·niă). [mod. L., f. prec. + -ia[1].] Either of two species of fan palm of the genus of this name, *Washingtonia filifera* and *W. robusta*, found in California and Mexico and elsewhere.
1945 J. L. Marshall *Santa Fe* 188 There were date palms and New Washingtonia palms from the Colorado desert.., but they were tricks of irrigation. 1949 *Los Angeles Times Home Mag.* 14 Aug. 32/4 There is today an impressive row of Washingtonias..standing like sentinels to direct the way to the ocean's cove park. 1976 *Hortus Third* (L. H. Bailey Hortorium) 1168/1 Washingtonias are extensively grown in Calif., often for street planting, and are common along the Gulf Coast, and in Fla., as well as in other parts of the world, particularly those with Mediterranean type of climate.

Washingtonian (wǫʃiŋtōᵘ·niăn), *sb.* and *a.* [f. as prec. + -ian.] **A.** *sb.* **†1.** A supporter or admirer of George Washington and his political standpoint. *Obs.*
1789 *Mass. Centinel* (Boston) 11 Feb. 172/3 The following gentlemen are chosen Electors... All firm federalists —and Washingtonians. 1816 M. B. Smith *Forty Yrs. Washington Society* (1906) 134 If I was a Washingtonian you might say I worshipped the sun.
2. *Hist.* A member of an American temperance society founded in 1840.
1842 *Joliet* (Illinois) *Courier* 2 Feb. 1/4 A meeting of Washingtonians was held at the methodist church on tuesday evening the 18th inst. 1891 *Cycl. Temperance & Prohibition* 203/1 The 'Washingtonians' originated in the conversion into a temperance society in April, 1840, of a Baltimore drinking club..of six men. 1947 F. D. Downey *Lusty Forefathers* 299 The fervor of the Washingtonians, founded by self-redeemed drunkards, felled thousands of apple trees, the fruits of which might otherwise..have become hard cider.
3. An inhabitant of Washington, D.C., or of the state of Washington.
1852 M. Eastman *Aunt Phillis's Cabin* 234 The beautiful prospect, to which Washingtonians are so much accustomed that they are too apt not to notice it. 1892 *Irrigation Age* 1 May 31/3 Washingtonians know a blessing when they see it. 1945 *Maryland Conservationist* Summer-Fall 7/2 The Potomac River..is very heavily fished by Washingtonians. 1972 *Times* 22 Dec. 5/6 Other Washingtonians, who do not have Air Force One to call on to fly them off to a viewing area, have booked motel rooms in York. 1978 G. Vidal *Kalki* x. 227 'It's my fault,' she apologized. 'I'm the Washingtonian. I should have told you.' 1980 *Blair & Ketchum's Country Jrnl.* Oct. 19/3 Recently a Washingtonian informed us there's real danger that the federal government will collapse.
B. *adj.* **1.** Of, pertaining to, or characteristic of George Washington or his politics.
1812 *Salem Gaz.* 22 May 3/1 The political character of Vermont is really Washingtonian. 1858 A. Lincoln *Coll. Wks.* (1953) III. 19 Fighting it in the Jeffersonian, Washingtonian, and Madisonian fashion. 1902 'Mark Twain' in *N. Amer. Rev.* May 614 The Washingtonian character would not have been built. 1919 G. B. Shaw *Peace Conference Hints* ii. 21 The United States were still in the Washingtonian phase of non-intervention.
2. Of or pertaining to the Washingtonian Temperance Society or the practice of temperance that it advocated.
1842 *Knickerbocker* XX. 298 The festival was conducted on the Washingtonian principle. 1880 *Harper's Mag.* Jan. 191/2 Its influence is still visible in the Washingtonian homes which usefully supplement the charities of our large cities. 1911 *Encycl. Brit.* XXVI. 579/2 In the United States a flash of enthusiasm of a similar character, but on a smaller scale, known as the Washingtonian movement, had appeared.
3. Of, pertaining to, or characteristic of an inhabitant of Washington, D.C., or the state of Washington.
1961 in Webster. 1977 L. Meyer *Capitol Crime* iv. 26 There was nothing particularly Washingtonian about Reilly... Washington is full of people..who once had.. ideas about changing things.

Washita (wǫ·ʃitǭ). The name *Washita* (see below) used *attrib.* and *absol.* to designate: **a.** [Fort *Washita*, Oklahoma.] (The rocks of) a subdivision of the Cretaceous in the central-southern U.S.A. **b.** [Ouachita or *Washita* Mountains, Arkansas.] A porous variety of novaculite used for sharpening cutting tools.
1860 *Trans. Acad. Sci. St. Louis* I. 586 Washita limestone.—This important member of our Cretaceous System is made up of a nearly white, yellow, gray and blue limestone. 1885 [see *Arkansas[1]]. 1963 A. Geikie *Text-bk. Geol.* (ed. 4) II. vi. iii. 1212 The Texas Lower Cretaceous deposits..have been divided into three formations... The highest formation, termed the Washita, consists of four groups. 1938 A. Durst *Wood Carving* 18 The largest stone shown is the Washita oilstone; the others are various slips for the smaller tools. 1958 J. R. Biggs *Woodcuts* 32 The knife should be kept sharp on the India or Washita stone. 1974 *Encycl. Brit. Micropædia* X. 562/2 The Washita Group is the uppermost member of the Comanche Series.

Washoe (wǫ·ʃo, -ʃū). Also **Washo.** [Washoe *wá·šiw* Washoe Indian, Washoe Indians.] **1.** (A member of) a North American Indian people inhabiting the area around Lake Tahoe on the border of California and Nevada. Also *attrib.*
1846 J. Clyman *Jrnl.* 4 May in C. L. Camp *J. Clyman, Frontiersman* (1960) 206 The tribe we are now passing through call themselves as well as understood Washee. *Ibid.* 8 May 207 But as the tribe of natives inhabiting this stream and the adjacent country call themselves the Washew tribe and nation I think it would [be] correct to call the stream by the same name viz Waushee River. 1860 H. De Groot *Sk. Washoe Silver Mines* 15/2 To this place, on the approach of cold weather, not only the Washoes, whose territory lies adjacent, but also bands of the Pah-Utahs, from a greater distance repair. 1879 H. R. Mighels *Sage Brush Leaves* 225 We saw..a..young Washoe brave escorting his wife and mother-in-law up towards the campoody. 1903 *Out West* Apr. 439 The colors in Washoe baskets are all natural. 1947 *Desert Mag.* Dec. 32/3 The Washoes, of whom less than 1,000 remain, live in valleys along the Sierra Nevada. 1976 *Amer. Speech* 1974 XLIX. 287 The Indian nomenclature belongs principally to three major Numic-speaking groups (Shoshone, Northern Paiute, and Southern Paiute) and to two non-Numic tribes, the Washo and the Mohave.
2. The region inhabited by the Washoe Indians; hence, a nickname for Nevada. Also *attrib.*
1856 *San Francisco Bull.* 26 May 3/2 The rumored trouble from the Indians of Carson, Wash-hoe and Walker's valley, is entirely without foundation. 1872 'Mark Twain' *Roughing It* xxi. 160 Washoe is a pet nickname for Nevada. 1896 C. H. Shinn *Story of Mine* 110 Not merchants these..but a brave, honest outdoor race whose huge Washoe wagons were the forerunners of the railroads. 1947 *Amer. Weekly* 2 Nov. 21/3 Back in Washoe, later to become the state of Nevada, the Comstock Lode waited—waited to be discovered.
3. The language of the Washoe Indians, one of the Hokan group of languages.
1882 *Mag. Amer. Hist.* Apr. 255 The area of the Washo languages borders to the west on the Maidu. 1921 E. Sapir *Language* iv. 81 Washo..an Indian language of Nevada. 1968 *Language* XLIV. 822, I am thinking of the contrast with the vowel harmony patterns in Washo.
4. Special Combs.: **Washoe canary** *U.S. colloq.*, a burro; **Washoe zephyr,** a strong west wind that blows in Nevada.
1867 *Daily Territorial Enterprise* (Virginia City, Nevada) 8 Mar. 3/1 The discouraged 'Washoe canary' refuses to cheer us with its tuneful warblings. 1877 W. Wright *Hist. Big Bonanza* 114 A queer genius thus described the donkey, called by everybody in that region, 'The Washoe Canary'. 1947 *Sat. Rev.* (U.S.) 10 May 32/2 He is..discoursing on the Washoe zephyr and the Washoe canary. 1865 *Washoe* (Nevada) *Times* 4 Mar. 3/2 We have heard of hail and chain-lightning, etc., but ye gentle Washoe zephyr can discount all and everything in that line. 1872 'Mark Twain' *Roughing It* xxi. 159 According to custom the daily 'Washoe Zephyr' set in; a soaring dust-drift..came with it, and the capital of the Nevada Territory disappeared from view. 1947 Washoe zephyr [see *Washoe canary* above].

wa·sh-off. [f. vbl. phr. *to wash off* (Wash *v.* 15 a).] **a.** Material that is washed off. **b.** The process or fact of being washed off.
1979 *Nature* 17 May 180/2 The pollutants are the wash-off of fertilisers and pesticides from the surrounding banks, the paddy and orchard fields, and the famous Mughal gardens. 1984 *Gardening from 'Which?'* June 181/1 Surface wash-off..can occur with other chemicals.

washomat (wǫ·ʃomæt). *colloq.* Also **wash-o-mat.** [f. Wash *v.* + -o + *-mat; cf. *Laundromat.] = *launderette.
1959 *Times* 30 Dec. 8/7 Twenty-storey skyscrapers were shooting up..among the Chinese laundries, the wash-o-mats. 1983 W. Garner *Think Big, think Dirty* ix. 135 Suttin used a local washomat, launderette, what you will. 1984 T. Hillerman *Ghostway* (1985) vii. 31 Laundries are magnets... The people at the Shiprock Economy Wash-O-Mat would know their customers.

wash-out. Add: **1. b.** *Biol.* and *Med.* The removal of material, esp. from a physiological system, by means of a fluid; the fluid used for, or matter removed by, this.
1955 in *Shorter Oxf. Eng. Dict.* (ed. 3) Add. 1966 *Jrnl. Lab. & Clinical Med.* LXVI. 856 The rapidity of indicator washout..should yield information regarding fluid flow patterns within the cardiovascular system. 1971 *Nature* 23 July 266/2 With a closed system technique in which a nitrogen-free atmosphere was made with helium-oxygen washout, a substantial increase in the concentration of nitrogen was measured. 1977 *Lancet* 8 Oct. 745/1 In a check of the wash-out from swabs and surgical drapes with measured volumes of blood the maximum error recorded was +2.8% by volume. 1980 *Nature* 17 Jan. 265/1 Spontaneous synaptic activity was detected in most myotubes within 1–2 min of curare washout.
c. *Meteorol.* The removal of particles from the air by falling water droplets. Cf. *rain-out 2.
1955 A. C. Chamberlain *Aspects of Trav. & Deposition of Aerosol & Vapour Clouds* (A.E.R.E. Doc. HP/R 1261) 2 Consideration is given to the four mechanisms by which aerosol particles and vapours are removed from the atmosphere, namely: (a) Sedimentation, (b) Impaction, (c) Diffusion.., (d) Wash out by rain. 1974 [see *rain-out 2]. 1980 Iribarne & Cho *Atmospheric Physics* ii. 26 The water cycle is important in the cleansing of the atmosphere by two mechanisms: Rainout... Washout—This is the name given to the elimination of gases by dissolution and of aerosol particles through capture by falling water drops.
3. (Earlier example.)
1873 *Newton Kansan* 29 May 3/2 Owing to a wash out on the Cottonwood last Sunday night, we had no train from the east until Tuesday afternoon.
4. b. *slang.* A useless or unsuccessful person; *spec.* in *Air Force slang*, a person who is eliminated from a course of training.
1918 J. M. Grider *War Birds* (1926) 65 Yesterday was washout day so we all went into town and threw a party at the Court. 1925 [see *napoo *int.*, *a.*, and *v.*]. 1927 C. A. Lindbergh *We* 115 We waited for the almost weekly list of washouts to be published. 1929 D. H. Lawrence *Pansies* 126 Now it's a country of..young wash-outs pretending to be in love with death. 1936 J. B. Priestley

They walk in City v. 115 Gregory Porson was no good, a blighter, . . a worm, a wash-out. **1950** *Chicago Tribune* 27 Apr. IV. 1/5 But there's a redeeming feature about the washouts, etc. The Wrigley field tenants still have their perfect percentage based on three straight victories. **1973** *Times* 23 Apr. 4/7, I think I'm a washout.

c. *Air Force slang.* A wrecked aeroplane.

1928 *Pop. Sci. Monthly* May 72/1 Damage to a plane is spoken of as a 'crash' and if beyond repair as a 'washout'. **1972** *in Amer. Speech* 1972 (1975) XLVII. 114 Wash out—complete wreckage.

5. *Aeronaut.* A decrease in the angle of incidence of an aeroplane wing towards the tip. [Perh. f. *WASH *sb.* 6 b (*b*), but cf. WASH *v.* 2 k, *13 e and quot. 1916¹.]

1913 *Flight* 25 Jan. 87/2 The fact of them [an aeroplane's wing tips] appearing to be negative does not necessarily imply more than a 'wash-out'. **1916** H. BARBER *Aeroplane Speaks* 25 [The Angle of Incidence. . is sometimes decreased or 'washed out' towards the Wing-tips.] *Ibid.* 87 The wash-out also renders the ailerons. . more effective. **1939** *Jrnl. R. Aeronaut. Soc.* XLIII. 791 The designer has available in this connection not only the use of tip slots, but the use of washout and of varying aerofoil section along the span. **1979** BERTIN & SMITH *Aerodynamics for Engineers* iii. 83 The wings of numerous subsonic aircraft have wash out to control the spanwise lift distribution.

wash-pot. Add: **5.** A vessel in which to wash clothes over a fire; a wash-boiler. *U.S.*

1926 B. ISBELL in J. F. Dobie *Rainbow in Morning* (1965) 105 At a later dance a large wash-pot of coffee, surrounded with ample tin cups, was kept boiling under a live oak tree in the yard. **1940** W. FAULKNER *Hamlet* i. 11 The women surrounded by laden clotheslines and tubs and blackened wash pots. **1944** T. D. CLARK *Pills, Petticoats & Plows* xviii. 330 The devil had him imprisoned under an upturned washpot. **1952** F. O'CONNOR *Wise Blood* iii. 61 His mother was standing by the washpot in the yard.

washroom (wǫ·ʃrūm). orig. and chiefly *N. Amer.* [f. WASH *v.* + ROOM *sb.*¹] **a.** A room equipped for washing oneself. **b.** A lavatory or W.C.

1806 in *Mass. Hist. Soc. Coll.* (1809) X. 77 They have usually two good rooms in front, bed-rooms, kitchen, washroom, and other convenient apartments in the rear. **1855** 'P. PAXTON' *Capt. Priest* 160 Finally a long washroom completed the odd assemblage. **1878** R. T. COOKE *Happy Dodd* 193 He. . fairly went down on his knees to her in the wash-room. **1892**, **1908** [in Dict. s.v. WASH *v.* 20 b]. **1941** *Sun* (Baltimore) 9 Aug. 4/4 Sanitation facilities . . include washrooms. **1952** S. SPENDER *Learning Laughter* 140 The indoor separate lavatory and washroom with shower. **1962** A. LURIE *Love & Friendship* iv. 75 The john one makes in washrooms. **1978** *Detroit Free Press* 5 Mar. [Parade Suppl.] 12/3 A queasy stomach sends him dashing for the washroom on bumpy plane flights. **1980** *Ottawa Citizen* 16 June 33/2 Washroom sex can mean a hand reaching for a groin underneath toilet cubicles or it can involve two men in the cubicle.

washstand. **1.** (Earlier examples.)

1789 J. WOODFORDE *Diary* 23 Nov. (1927) III. 151 For the . . Mohogany Wash-stand 1.10.6. **1808** E. WESTON *Let.* 12 Nov. in *Jrnl. of Governess* (1969) I. 126 A box I sent by 9, with a bed, a chest of drawers, two chairs, and a washstand in it.

wash-tub. Add: **2.** *Comb.* **wash-tub (bass)**, a wash-tub converted into a musical instrument like a double-bass by stretching a string across it.

1968 *Blues Unlimited* Nov. 8 We recorded Dewey Corley, who. . plays kazoo and washtub bass. **1970** *Western Folklore* XXIX. 229 Two kinds of plucked one-stringed instruments are known to Negroes in America today. One is the familiar one-stringed bass, sometimes called a 'washtub bass' or 'gutbucket' from the materials of its construction. **1972** *Time* 17 Apr. 39/3 He played a washtub in a group named the Five Hip Cats.

wash-up. Add: **1. c.** A wash; an act of washing oneself. *N. Amer.*

1887 B. HARTE *Millionaire & Devil's Ford* i. 176 You boys can go there for a general wash-up. **1937** C. MATH-EWSON *Second Base Sloan* 64 They. . dropped from the car and went back to the station for a wash-up. **1968** *Globe & Mail* (Toronto) 13 Feb. 8-12/7 The shutdown of the assembly line for 30 minutes each shift—for rest and wash-up.

wash-work. Add: **2.** (Earlier example.)

1893 A. BEARDSLEY *Let.* c15 Feb. (1970) 44 Nobody gave me credit for caricature and wash-work, but I have blossomed out into both styles.

washy, *a.* Add: **6.** Also, exhausted, washed-out.

1922 JOYCE *Ulysses* 734, I looked a bit washy.

wasp, *sb.*¹ Add: **1.** (Examples of the dial. and joc. var. *wops*.)

1893 A. LANG *Prince Ricardo of Pantouflia* vii. 128 'Hang that wops!' said Prince Ricardo. . when it buzzed in his ear. **1908** O. SEAMAN *Wearing of Whisker* in *Salvage* 82 Trained like the ampelopsis, That happy haunt of woolly bears and wopses. **1932** E. STEP *Bees, Wasps, Ants* 81 Wopses, what eat up all our fruit. **1937** D. L. SAYERS *Busman's Honeymoon* xv. 308 Out comes me lord, and they wos all on to 'im like wopses round a jam-pot.

4*. (With capital initial.) A kind of flame-thrower developed by the British army during the war of 1939–45.

1944 *Hutchinson's Pict. Hist. War* 12 Apr.–26 Sept. 467 Like that of the Crocodile, the range of the Wasp is upwards of 150 yards. Fitted to the standard bullet-proof carrier, it is a terrorising weapon. **1965** *Listener* 11 Nov. 763/3 There was this Bren gun carrier with the flame thrower, sir, a Wasp I believe they're called, and I thought it was my duty to see how it worked. **1975** *Incendiary Weapons* (Stockholm Internat. Peace Research Inst.) v. 38 British and Canadian engineers developed a lighter mechanized flamethrower, called the Ronson. . . This was the forerunner of the Wasp (Mark 1) flame gun of which 1000 were ordered and went into production in March 1943.

5. a. *wasp-grub* (earlier and later examples); *wasp-striped* adj.

1678 Wasp-grub [see *wasp-maggot* in Dict.]. **1919** J. MASEFIELD *Reynard* 61 Brocks eat wasp-grubs. **1952** P. ATKEY *Juniper Rock* x. 87 Wasp-striped. . the helicopter reappeared. **1974** E. AMBLER *Dr. Frigo* i. 49 A black butler in a wasp-striped waistcoat.

Wasp (wǫsp), *sb.*² orig. and chiefly *U.S.* Also **WASP.** [Acronym f. the initial letters of *W*hite *A*nglo-*S*axon *P*rotestant.] A member of the American white Protestant middle or upper class descended from early European settlers in the U.S. Freq. *derog.* Also *attrib.* or as *adj.*

1962 E. B. PALMORE in *Amer. Jrnl. Sociol.* LXVII. 442/2 For the sake of brevity we will use the nickname 'Wasp' for this group, from the initial letters of 'White Anglo-Saxon Protestants'. **1963** *Times* 2 May 15/5 There is such a thing as a 'Human Engineering Laboratory'; whether a man is a Wasp (white Anglo-Saxon Protestant) can decide his career. **1963** *New Statesman* 10 May 716/2 This year's executive model will be over six feet tall, clean-shaven, lean, and with large fleshy ears. . . He should try to be or pretend to be a WASP (White Anglo-Saxon Protestant) and ought to have gone to an Ivy League college, preferably Princeton. **1964** E. D. BALTZELL *Protestant Establishment* (1965) i. 9, I should first like to show how the aristocratic process still worked quite well in the case of the family of Abraham Lincoln, and especially how the WASP establishment authoritatively retained the leadership of American society in the generation of Robert Todd Lincoln. **1968** *Times Lit. Suppl.* 4 Apr. 329/1 The Jew can choose to leave his ghetto by 'passing' or by breaking the more and more flimsy barriers put up by Wasp (and non-Wasp) anti-Semitism, but the Negro cannot. **1971** M. MCCARTHY *Birds of Amer.* 71 He was the only older WASP Peter knew. **1977** *Time* 19 Dec. 66/2 United States Secretary of State Felix John Vandenberg—slim, silver-haired, tallish, Wasp—speaks with 'the lingering trace of a British accent, which had been acquired at Eton and Oxford'. **1978** *Jrnl. R. Soc. Arts* CXXVI. 276/1 Can what one calls a WASP properly and without any discrimination select an Asian? **1979** R. JAFFE *Class Reunion* (1980) I. i. 37 Daphne's father was the senior partner of the leading prestigious Wasp law firm in New York.

wasp, *v.* Restrict *nonce-wd.* to sense in Dict. and add: **2.** *intr.* Const. *around, about.* To dart about in the manner of a wasp, in an irritating, noisy, or tenacious fashion.

1967 G. F. FIENNES *I tried to run a Railway* iii. 22 Most nights brought an intruder bumbling overhead with one of our fighters wasping around looking for him. **1981** B. HEALEY *Week of Scorpion* iii. 45 'It must be very unpleasant for her.' 'No doubt. . . But have you considered how you'll make it any less so by having the police wasping about your own parts?' **1981** B. FREEMANTLE *Vietnam Agenda* (1982) xx. 152 Traffic wasped around the plaza.

Wa·spdom. [f. *WASP *sb.*² + -DOM.] The characteristics, beliefs, etc., of American white Protestant 'Wasps'.

1969 *Time* 17 Jan. 21 Thus Roman Catholics like William Buckley, Eugene McCarthy and Ted Kennedy are pushed toward Waspdom by their associations, professions and life styles. **1976** *Time* 27 Sept. 49 *Noblesse oblige* has yielded to *bourgeoisie oblige*—even at the country club, traditionally the most closely guarded bastion of upper-class Southern Waspdom.

waspie (wǫ·spi). [f. WASP *sb.*¹ + -IE.] A ladies' corset designed to make the waist appear very small; a belt of similar design. Cf. *wasp-waist* s.v. WASP *sb.*¹ 5 b.

1957 *Housewife* Sept. 89/2 A 'waspie' to whittle your waist. **1962** P. BRACKEN *I hate to housekeep Bk.* (1963) xi. 124 It it's a snug-fitting dress, and you know you'll be wearing your waspie with it, but don't want to be stuck with it or in it while you shop, put it in your big handbag too, and change in the fitting room. **1970** *Daily Tel.* 20 Apr. 15 The belt of the season is undoubtedly the laced-up waspie in suède or leather. **1976** *Times* 27 May 6/4 There were plenty [of women] still to heave a sigh of relief (if their waspies would let them).

waspily (wǫ·spili), *adv.* [f. WASPY *a.* + -LY².] = WASPISHLY *adv.*

1854 B. P. SHILLABER *Life & Sayings Mrs. Partington* 231 'Because she is a low, vile creature of the town,' said she, waspily. **1928** *Sunday Dispatch* 29 July 2/3 'That's what I'm doing. Turning over a new leaf—and I'm going to do well.' 'Yeah? And what comes after that?' snapped the P.C.M. waspily.

Waspish (wǫ·spiʃ), *a.*² orig. and chiefly *U.S.* Also **Wasp-ish, WASPish.** [f. *WASP *sb.*² + -ISH¹.] Of, pertaining to, or characteristic of American white Protestant 'Wasps'. Cf. *WASPY *a.*²

1968 *Listener* 27 June 843/3 Charles Newman, making a most impressive debut, gives in *New Axis* a picture of a community that is diametrically the opposite of Mr Baldwin's Harlem: that of an upper-middle-class WASPish suburb in an Illinois dormitory town. **1969** *Times Lit. Suppl.* 31 May 591/2 Postwar antisemitism in America. . . Echoes of haughty Waspish or Protestant hostility. . . [this] silly novel about the marriage in 1946 of a sexy, clever Jewish girl from the west side of New York to a handsome booby from an aristocratic Long Island family. **1978** E. TIDYMAN *Table Stakes* ii. 26 His WASPish good looks. **1983** *Times* 8 July 7/5 He is Scorsese's contemporary, but from a different, Wasp-ish social group.

Waspy (wǫ·spi), *a.*² orig. and chiefly *U.S.* Also **WASPy.** [f. *WASP *sb.*² + -Y¹.] = *WASPISH *a.*²

1968 *N.Y. Times* 23 July 41/1 Black-power stalwarts suspiciously eying the Waspy surroundings. **1975** *Publishers Weekly* 9 June 61/3 Max Herschel, a high-powered industrialist in his 60s, coarse but with all the trappings of class, has lost his. . mistress Bones to Steven Routledge, who is the antithesis of Max, WASPy, idealistic, a failed novelist. **1978** *Detroit Free Press* 23 Apr. 383 A slim, Waspy, forty-year-old ex-debutante.

wassa (matter): see *WHASSA.

wassat, repr. *colloq.* pronunc. of *what's that.* Cf. WHAT *pron.* I.

1967 J. WAINWRIGHT *Talent for Murder* 83 'Is that usual? Or, is it unusual?' 'Wassat?' **1976** C. WILCOX *To die Elsewhere* i. 5 'I'll have to do it alone.' 'Wassat?' said Heyward. **1981** 'J. ROSS' *Dark Blue & Dangerous* xxi. 117 'Wassat, sir?' There was deep incomprehension in his expression.

wasser (wǫ·zəɪ). Also **was-ser.** [f. WAS, 1st and 3rd pers. sing. pa. t. of BE *v.* + -s + -ER¹.] = HAS-BEEN *sb.*

1924 KIPLING *Debits & Credits* 313 Bert was for it, with a few remarks from the pals on the pavement on the bench. **1936** F. M. FORD *Let.* 6 Sept. (1965) 259 But I don't want to be harmed by people calling me a lousy old wasser.

Wassermann (‖ va·səɪman, wǫ·səmən). *Med.* Also (*erron.*) **-man.** The name of August Paul *Wassermann* (1866–1925), German bacteriologist, used *attrib.* with reference to a test for syphilis devised by him in 1906, in which antibodies to the causative organism are detected by a complement-fixation test. Also *absol.*, the Wassermann test.

1909 *Jrnl. Exper. Med.* XI. 401 Unless a great simplification of the Wassermann test can be devised, it will never attain the usefulness it deserves. *Ibid.* 401 Cases which were negative to the Wassermann and weakly or quite often strongly positive to the new modification. **1915** J. REMSEN etc. . . but the Wassermann test was met with. **1952** 'N. SHUTE' *Far Country* 214 A Wassermann test would be interesting, and possibly positive. **1970** PASSMORE & ROBSON *Compan. Med. Stud.* II. xxii. 16/2 The test depends upon the reagin. . . antigen mixed with dilutions of the patient's serum in the presence of guinea-pig complement. **1976** H. KEMELMAN *Wednesday Rabbi got Wet* xvii. 71 When did you last had a checkup, she had asked him to do a Wassermann, too, because 'Joe was out of town on business and you never know how it is when men go out of town.'

wastage. **1. c.** (Example.)

1954 M. BERESFORD *Lost Villages* v. 165 Rokeby and Mortham on the Tees had not recovered from their wastage by the Scots in the fourteenth century.

d. (*a*) The loss of students through failure to complete a course of study or training; (*b*) the loss of employees by any means other than dismissal, esp. by retirement or resignation. Freq. as *natural wastage.*

1919 M. GREENWOOD in *Jrnl. R. Stat. Soc.* LXXXII. 187 Our industrial 'death' rate would then merely be the rate at which entrants to a trade pass out of it, or. . with a. . narrower circle, the rates of departure from particular factories, i.e. 'deaths' or wastage rates for different factories will be *prima facie* measures of the efficiencies of the respective factories. **1944** *Min. of Fuel & Power Statistics Digest from 1938* 6 in *Parl. Papers* 1943–44 (Cmd. 6538) VIII. 153 Net. . natural wastage. . [Note] This is the gross natural wastage less the normal juvenile recruitment. **1948** *Ann. Rep. Nat. Coal Board 1947* iv. 45 in *Parl. Papers* 1947–8 X. 387 The manpower target set for the Board. . was. . 730,000 men. . . This meant a net increase. . of 40,000. . and, since wastage was estimated at 60/65,000 men over the year, a recruitment of 100,000 was needed. **1952** *Sec. Educ.* etc. . . wastage. . **1956** *School Sci. Rev.* June 375 The Rector of Imperial College, Dr. R. P. Linstead, in a lecture last October, said (speaking of what he called 'academic wastage'), 'This academic wastage makes itself shown in different universities, but in this College much of the wastage occurs during the first year.' **1958** *Technology* May 66/2 The question of wastage in apprenticeship. **1963** *Higher Educ.: Rep. Comm. under Ld. Robbins 1961–3* 20 in *Parl. Papers* 1962–3 (Cmnd. 2154) XI. 639 We discount all those who begin courses but do not successfully complete them. This is commonly described as 'wastage'—a term that we adopt for reasons of conformity but that we regard as carrying misleading implications. Wastage rates in higher education have not varied much in recent years. **1975** *Times* 25 Nov. 1/3 Nursing

staff. . were liable to 'natural wastage'. **1979** 'J. LE CARRÉ' *Smiley's People* (1980) v. 65 He resigned of his own accord. . , part of the wastage rate that gets everyone so worried. **1983** *Financial Times* 23 Apr. I. 34 The savings which the bank is seeking will involve natural wastage, retraining, redeployment and some measure of redundancy.

waste, *sb.* Add: **III. 11. c.** (Earlier example.)

1864 *Chambers's Jrnl.* 16 July 460/2 Smith the driver. . standing upon the foot-plate of No. 69, leisurely attempting to remove the surplus oil from his black hands with a very suspicious piece of 'waste'.

12. c. Waste water, effluent; *spec.* that which is free of excrement. Cf. *SOIL *sb.*³ 7.

The distinction between waste and soil is commoner in the combs. with -*pipe*.

1886 *Encycl. Brit.* XXI. 715/2 To connect a water-closet soil-pipe with sinks and basins. . is to multiply possibilities for the spread of disease within the house, and it is strongly advisable to convey the 'waste' from them by a separate pipe. **1913** E. H. BLAKE *Drainage & Sanitation* vii. 239 We may next consider the kinds of 'waste-pipe' or. . rubber-type rain-water pipes. . soil pipes taking the wastes from closets and housemaids' sinks, and pipes taking the wastes from baths, lavatories, and sinks. **1959** GOODIN & DOWNING *Domestic Sanitation* v. 127 Sanitary fitments, may be divided broadly into those intended to receive the wastes of the human body, and those designed for dirty, soapy or greasy water. **1973** [see *SOIL *sb.*³ 7]. **1976** *National Observer* (U.S.) 28 Aug. 12/3 One theory not in the report was that forests were being damaged by a mixture of wastes and a contamination of the water system. **1978** T. PETTIT *Home Plumbing* x. 151/2 Waste from WCs is discharged into the soil and vent system of pipework.

IV. 13. (sense 11) *waste disposal.*

1968 R. POWDRILL *Vocab. Land Planning* iv. 64 Industrial land does not include. . land for. . waste disposal (where this occupies a significant area detached from the industrial process). **1977** P. JOHNSON *Enemies of Society* vii. 91 We now have a good idea of the adverse damage done in the areas of Lake Baikal, the Volga, the Caspian and the Aral seas, and other Soviet areas of industrial waste-disposal.

14. waste-basket (earlier example); **waste disposal unit** (see quot. 1967); **waste-disposer** = *waste disposal unit* above; **waste-gate,** (*b*) *Engin.*, a device in a turbocharger which regulates the pressure at which exhaust gases pass to the turbine by opening or closing a vent to the external atmosphere; **waste-heap,** (*a*) a pile of refuse matter; (*b*) *transf.* in *Cards*, a pile of cards formed from the accumulation during the course of a game of those which cannot be played (cf. *rubbish heap* s.v. RUBBISH *sb.* (and *a.*) 5); **waste heat**, heat produced as the by-product of some process; *waste-heat boiler*, a boiler employing this; **waste maker** [f. title of book: see quot. 1961], a manufacturer of consumer goods that are intended not to be durable or to be partially wasted so that the demand for new goods is kept high; **waste mould,** in *Sculpture*, a simple negative mould which has to be broken to release the cast inside it; so **waste moulding**; cf. *piece mould* s.v. PIECE *sb.* 23; **waste-pipe** (further examples); also *spec.* a pipe for the drainage of effluent from sinks, baths, etc., in contrast to a soil-pipe; **waste-plug** = *PLUG *sb.* 2 b, † 2 k; **waste silk**, the inferior silk from the outside of cocoons and from cocoons out of which the moths have been allowed to escape.

1889 'MARK TWAIN' *Lett.* (1917) II. xxix. 514 Send me the pages with your corrections on them, and waste-basket the rest. **1967** *Gloss. Sanitation Terms* (*B.S.I.*) 63 *Waste disposal unit*, an electrically operated mechanical device for reducing kitchen garbage into fragments small enough to be flushed into the drainage system. **1976** W. BESTE *Repeat Instructions* vi. 64 His first conscious memory. . [of] the kitchen stuffing the torn pieces into the waste-disposal unit. **1977** *Evening Post* (Nottingham) 27 Jan. 14/4 (Advt.), Lounge, hall, fitted cupboards, fully fitted kitchen with waste disposal unit. **1982** *Which?* Mar. 82/1 There are obvious advantages. . in being able to get rid of kitchen scraps straight down the drain, with a waste-disposer. **1980** A. N. WILSON *Healing Art* vi. 62 A new sun lounge, . . a waste-disposer. **1948** *Shell Aviation News* No. 115. 19/1 The closed wastegate limitation is the condition where all of the available exhaust gas energy is required to drive the compressor, and all of the exhaust gas is directed through the turbine. **1981** *Pop. Hot Rodding* Feb. 22/1 It will be a relatively simple matter of welding in sections of bent tubing to make the necessary connections (including a waste-gate, should one be required). **1983** *Which?* Dec. 559/1 So a valve is needed—the 'wastegate'—which opens when the pressure is at its maximum safe level to divert some of the exhaust gases away from the turbine. **1873**, etc. Waste heap [in Dict., sense 13]. **1892** 'L. HOFFMANN' *Illustr. Bk. Patience Games* 4 If its [*sc.* the card's] nature does not allow of its being so played, it is laid face upwards in front of the player, the cards so deposited being known as the 'waste-' or 'rubbish-heap'. **1913** —— *Sel. Patience Games* 5 The cards so dealt with being known as the 'waste-heap' or 'rubbish-heap'. **1975** *Way to Play* 145/3 *Waste pile* or *heap*, cards from the stock that cannot immediately be played onto the layout are sometimes placed face up in one or more waste piles, to be brought back into the game as appropriate. **1908**

A. G. KING *Pract. Steam & Hot Water Heating* xxvii. 343 When no waste heat is available, an ordinary type of pipe heater may be used. **1930** *Engineering* 8 Aug. 188/3 The utilization of exhaust gases in waste-heat boilers had improved the efficiency of the large gas engine. **1972** R. G. KAZMANN *Mod. Hydrol.* (ed. 2) iv. 230 The remainder of this energy, 'waste heat', must be disposed of into the immediate environment of the power plant. **1982** W. F. OWEN *Energy in Wastewater Treatment* xii. 281 Three basic types of heat recovery equipment are typically used in wastewater heat recovery systems: shell-and-tube exchangers, waste heat boilers, and heat wheels. **1961** V. PACKARD *Waste Makers* v. 48 In some cases the consumers have no choice but to be waste makers because of the products available and sold to them. Many paste pots come with brushes built into the cover, and the brushes fail. . to reach the bottom. . . This millions of 'empty' paste jars are thrown away with a few spoonfuls of paste still in them. **1970** G. JACKSON *Let.* 17 June in *Soledad Brother* (1971) 282 You dig, we want the waste makers to. . **1929** F. J. GLASS *Modelling & Sculpture* vii. 73 You now proceed to chip away the whole portion of the waste mould. **1971** *Daily Tel.* (Colour Suppl.) 5 Mar. 8/2 From this little clay was impressed—a 'waste-mould'—is taken, from which a plaster cast is made. **1971** A. TOFT *Modelling & Sculpture* xii. 133 The name 'waste moulding' implies that the mould is only made to serve the purpose of taking one cast. **1928** H. H. SHEFFIELD *Sculpture* iii. 73 The method of taking plaster is chipped away so that the mould is destroyed. **1876**, etc. Waste-pipe [see *soil-pipe* s.v. *SOIL *sb.*³ 8]. **1946** E. MOLLOY *Plumbing & Gas-Fitting* x. 122/2 For a waste pipe from a bath, sink, bidet, or lavatory basin discharging into a soil pipe from a water-closet, or a waste pipe from a sink, the Model By-laws specify a suitable trap adequately secured against destruction of the water seal. **1982** A. TOFT *Modelling & Sculpture* viii. 178 The waste pipe in the slab, there is a considerable length of pipe between it and the bottom of the basin. **1882** S. S. HELLYER *Lett.* etc. *Sc. & Art Sanitary Plumbing* v. 193 These waste-plugs. . **1877** G. E. WARING *Sanitary Condition City & Country Houses* 79 If the waste plug is operated by a handle rising the slab, there is a considerable length of pipe between it and the bottom of the basin. **1883** S. S. HELLYER *Lett.* *Sc. & Art Sanitary Plumbing* v. 193 These waste-plugs. . **1965** *Waste plug* to prevent the overflow. **1965** *Waste plug* [see *PLUG *sb.* 2 b]. **1970** *Encycl. Brit.* XVII. 466/2 Before you begin to wind, you must prepare your cocoons. . stripping them of that waste silk that surrounds them, and which served to fasten them to the twigs. **1875** [see SPUN *ppl. a.* 1 a]. **1921** Waste silk [see *SCHAPPE].

waste, *a.* Add: **9. c.** *waste water* (further examples); also *spec.*, water that has been used in some industrial process; also, sewage. (Freq. written as one word.)

1976 *Billings* (Montana) *Gaz.* 27 June 6-c/3 The liquidation process could give up to 4.8 million gallons of wastewater per day and . . there could be another 110 acre feet of wastewater piped through the coal scrubbing process. **1977** *N.Z. Herald* 8 Jan. 2-12/7 (Advt.), City. . N.K. Morrison—to discharge wastewater from State Highway No 1 into Alexandra Creek headwaters in connection with quarrying operations at Cuthill, Albany. **1979** *Arizona Daily Star* 1 Apr. D 3/1 Stephen J. Tencza, former project engineer with the Pima County Department of Wastewater Management, has joined John S. Collins & Associates as a project engineer for the firm's sanitary engineering department.

waste, *v.* Add: **4. c.** To beat up, kill, murder (someone); to devastate a place, to kill its inhabitants. *slang* (orig. and chiefly *U.S.*).

1964 P. MARSHALL *Let.* in J. H. Clarke *Harlem* 317 Stomping and wasting our Little People. *Ibid.* 319 You. . president. . since Duke got wasted! **1966** J. M. BREWER in *A Dundes Mother Wit* (1973) 240/2, I wasted (punched) one of the studs. **1971** *Guardian* 2 Apr. 11/2 'The intention to waste' My Lai. **1973** C. WESTON *Susannah Screaming* xxix. 147 They wasted Barrett because they didn't know he was undercover. **1977** *Courier-Mail* (Brisbane) 2 Sept. 1/4, I think Simone has been wasted (killed) by the southern heavies. **1983** M. C. SMITH *Gorky Park* i. xii. 177 You want to go chasing after the guy who wasted your detective.

14. Delete † *Obs.* (Later examples.) Also *const. away.*

1881 'MARK TWAIN' *Prince & Pauper* xvii. 187 The afternoon wasted away. **1961** M. CALLAGHAN *Thomas Gaunt* 52 He was. . the man you had to go to. **1978** J. A. MICHENER *Chesapeake* 282 The boys were only eight and seven, but already the years were wasting.

wasted, *ppl. a.* Add: **5.** Intoxicated (from drink or drugs). Chiefly used outside the U.K.

1968–70 *Current Slang* (Univ. S. Dakota) III–IV. 135 *Wasted*, drunk. **1974** *Sunday Mail* (Brisbane) 24 Feb. 19/1 The hippies. . are either wasted or sleeping. **1979** J. GUNN in G. W. Turner *Good Austral. Eng.* iii. 56 Being under the influence is. . *turned on*, *wasted*.

wasteland. Add to etym.: This compound is now indistinguishable from collocations of the adjective (sense 1), and all further examples are added here. Add to def.: **1. b.** Land (esp. that which is surrounded by developed land) not used or unfit for cultivation or building and allowed to run wild.

1922 W. J. LOCKE *Tale of Triona* x. 116 They walked. . through the maze of new and distressingly decorous avenues, some finished, others petering out. . into placarded building lots or waste land. **1933** *Archit. Rev.* LXXIV. 166/1 Farm and meadow, hedge and coppice are all part of a system which the urban dweller takes for granted but which agricultural decay can only transform. . into wasteland, flood and marsh. **1969** *Daily Tel.* 5 Sept. 18 Within a stone's throw of the Guildhall are wasteland areas created by the bombing of 25 years ago. **1974** A. J. HUXLEY *Plant & Planet* xxix. 359 Land sterilization by building and industrial use, and the subsequent creation of wastelands. **1980** *Daily Tel.* 23 July 2/2 A con-

ference on wasteland, organised in London by Thames Television.

c. *spec.* a waterless or treeless region, a desert. (Not distinguishable from some examples in Dict., sense 1 a.)

1966 F. HERBERT *Dune* 399 The stranger might think nothing could live or grow in the open here, that this was the true wasteland that had never been fertile and never would be. **1969** *Daily Tel.* (Colour Suppl.) 3 Oct. 20/4 Horses cannot last long in that pitiless wasteland where no rain has fallen for the last five years. **1979** P. THEROUX *Old Patagonian Express* (1980) vii. 131 We were in a waterless desert: no sign of the river in this parched wasteland.

d. *transf.* and *fig.*, sometimes with allusion to T. S. Eliot's poem *The Waste Land* (1922).

1868 TROLLOPE *Phineas Finn* (1869) I. xxxvii. 310 Young members. . who are green from the waste lands and road-sides of private life. **1932** H. NICOLSON *Let.* 15 Apr. (1966) 114 He [*sc.* Sir Oswald Mosley] will be edged gradually into becoming a revolutionary—and to that waste land I cannot follow him. **1934** C. LAMBERT *Music Ho!* v. 281 The composer finds himself in a spiritual waste land. **1964** S. BELLOW *Herzog* 75 The commonplaces of the Wasteland outlook. **1972** *Sat. Rev.* (U.S.) 27 May 7/1 Television's sad wasteland. **1981** *Encycl. Brit. Macrop.* XXII. 35 Above all, there is one fundamental chasm which divides this terminological wasteland. **1981** *N.Y. Rev. Bks.* 5 Nov. 34 The once proud and efficient, public school system of the United States. . has turned into a wasteland.

2. *N.Z.* (See quot. 1875.)

1844 F. MATHEW *Reports* (typescript) II. 18 Any attempt on his part to assert the right of the Crown, over the wastelands, that are known as Waste Lands would have immediately been attended with serious results. **1875** J. VOGEL *Official Handbk. N.Z.* 103 Public—or as they are called, 'waste'—lands are sold on several principles. **1930** L. G. D. ACLAND *Early Canterbury Runs* 1st Ser. viii. 205 Godley. . was very much against letting the 'waste lands' in large areas.

wa-stemaster. Also **Wastemaster.** [f. WASTE *sb.* 11 + MASTER *sb.*¹] The name of a type of waste disposal unit. (In quot. 1963 a similar imaginary or futuristic device.)

A proprietary name in the U.S.

1946 *Official Gaz.* (U.S. Patent Office) 5 Nov. 15/1 Wastemaster; Lockley Machine Co. For electrically driven garbage grinders. Claims use since June 24, 1945. **1949** *House Beautiful* Feb. 85 Choose from these disposers. . . Wastemaster—Lockley Machine Co. **1963** *Punch* 16 Oct. 569/1 The wastemasters were able to come into their own and people began buying—for. . scrap. **1966** *New Statesman* 13 May 700/4 (Advt.), Mais. . . 2 beds, k & b, waste-master. **1973** A. HOLDEN *Girl on Beach* 12, I suppose you've got a washing machine and a clothes-drier and a wastemaster and a pop-up toaster. **1981** I. A. GORDON in *N.Z. Listener* 27 June 86 You know all about dishmasters and maybe even wastemasters.

waste-paper. Add: **1. a.** (Earlier *fig.* example.)

1869 *Bradshaw's Railway Man.* XXI. 361 The conventions of 1858 and 1860. . would thus seem to have become waste paper.

b. waste-paper basket: also *fig.*

1923 J. S. HUXLEY *Ess. Biologist* vi. 217 The waste-paper basket of civilization. **1969** *Listener* etc. **1930** D. H. LAWRENCE *Last Poems* (1932) 120 All that we know is nothing, we are merely crammed waste-paper baskets unless we are in touch with that which laughs at all our knowing. **1979** *London Rev. Bks.* 25 Oct. 23/2 Marriage, the Webbs agreed, was the waste-paper basket of the emotions.

waster, *sb.*¹ Add: **II. 6. a.** (Further examples applied to pottery.)

1878 L. JEWITT *Ceramic Art Gt. Brit.* I. iv. 76 A kiln. . and around the remains of which were many vessels—'wasters' as they would be technically called—of various kinds. **1928** W. B. HONEY *Old Eng. Porcelain* i. 15 Nothing short of an undoubted 'waster' can establish conclusively that a particular type was made on the factory site in question. **1950** D. T. D. CLARKE *Roman Pottery Kiln* 4 Pots of similar grey ware had been found at Market Overton. . which are undoubtedly wasters. **1961** M. KELLY *Spoil Kill* ii. 81 Pots that were twisted, misshapen and collapsed. . . Wasters, they called them. **1974** *Canad. Antiques Collector* Sept.–Oct. 27/2 There are remains of many little potteries scattered all over southern Ontario, and mounds of 'wasters' or broken or discarded ware.

wasting, *ppl. a.* Add: **2. a.** *wasting asset* (see quot. 1974).

1930 *Economist* 26 Apr. 951/2 Dividends to shareholders—representing return of capital, since gold mines are a wasting asset—have shown much less than a corresponding rate of increase. **1953** *Times* 31 Oct. 2/7 The cost will be almost competitive with coal-based electrical power, and it would be contrary to all experience if the cost comparison did not turn steadily in nuclear energy's favour, particularly when coal is a wasting asset and steadily more costly in real terms to extract. **1970** *Guardian* 3 Nov., Dr. Beeching breathed his kiss of death on the Inverness to Kyle [railway line] in 1963; since then it has been regarded as a wasting asset. **1974** *Terminol. Managem. & Financial Accountancy* (Inst. Cost & Managem. Accountants) 65 *Wasting assets*, assets of a fixed nature which are gradually consumed or exhausted in the process of earning income (e.g. mines or quarries).

wastrel, *sb.* and *a.* Add: Hence **wa·streldom, wa·strelism,** extravagance, esp. with reference to government spending.

1906 *Westmorland Gaz.* 14 Mar. 3/2 He was, indeed, heard to say that under our present Fiscal system the wine of our manhood was oozing out while the water of wastrelism was oozing in. Sir Wilfrid Lawson was. . somewhat annoyed to think that the best of all drinks should be confounded with wastrelism. **1906** *Daily Mail* 3 Nov. 4/2 All the old strongholds of Wastreldom have been carried. St. Pancras has thrown out its Wastrels; Islington will have none of them. **1919** *Ibid.* 15 Oct. 6/2 Such arguments in the interest of wastrelism have been abundantly refuted. **1931** J. BUCHAN *Blanket of Dark* 102 It was hard to believe that this was a gathering of the kings of wastreldom.

wat⁴. (Earlier example.)

1844 *Chinese Repository* XIII. 204 The Siamese are in the habit of burning their dead, and the place selected for this purpose is near the wats.

watch, *sb.* **I. 2.** For '† *Obs.* except in *watch-night*' read 'Now *rare arch.* exc. in *watch-night*' and add later example.

1877 *Life St. Willibrord* i. 7 There for some years he served God in fast, watch, and prayer, in a little cell.

IV. 21. e. *silly* (or *crazy*) *as a two-bob watch*: extremely stupid, mad. *Austral. slang.*

1954 P. GLADWIN *Long Beat Home* 72 There now, I clean forgot. I'm getting silly as a two-bob watch. **1963** L. GLASSOP *Rats in New Guinea* 192 Crazy as a two-bob watch. Should have reported it, corporal. **1972** J. DE HOOG *Skid Row Dossier* 75 'Don't buy him a beer, Johnny, he's silly as a two bob watch', someone advised as he tapped me on the shoulder.

VII. 26. (sense 6) *watch temple*; (sense 21) *watchband, -strap.*

1924 *Sears, Roebuck Catal.* 410/1 Watch-band, woven wire. For watches. **1970** N. ARMSTRONG et al. *First on Moon* iii. 64, I found the little card. . and stuck it under his watchband. **1925** *Watchmaker, Jeweler, Silversmith & Optician* LI. 138 (Advt.), The finest range of watch straps. **1977** P. THEROUX *Consul's File* 126 The woman [was] covered with welts the shape of watch-straps. **1921** D. H. LAWRENCE *Sea & Sardinia* ii. 66 Venus of the aborigines, from her watch-temple looking at Africa.

27. watch cap (later examples); **watch-care** (earlier example); **watch-chain**: hence **watch-chained** *a.*; **watch-charm** *U.S.*, a small ornament that may hang from a watch-chain; **watch-keeper**: also in gen. use, one who keeps watch or acts as a look-out; hence **watch-keeping** *sb.* and *a.*; **watch-list**, a list of items or names which require close surveillance, esp. for legal or political reasons; **watch-night,** (*b*) *W. Afr.*, a night-watchman; **watch-pocket** (earlier example); **watch-riband** (earlier example); **watch-ribbon** = *watch-riband*; **watch room,** a room from or within which a watch is kept.

1966 T. PYNCHON *Crying of Lot 49* v. 114 He opened his door and found an aged bum with a knitted watch cap on his head. **1977** J. F. FIXX *Compl. Bk. Running* xii. 139 What I've found best is a simple wool hat, the kind sailors call a watch cap. **1984** *New Yorker* 31 Dec. 64/3 Bearded young fishermen, with ponytails coming out from beneath their knitted watch caps. **1845** *Indiana Mag. Hist.* (1927) XXIII. 152 Very much. . depends on the. . preachers and leaders, who have the after watch-care of the persons that are brought into the church. **1944** BLUNDEN *Cricket Country* i. 12 One or two watch-chained elders. **1976** 'R. GORDON' *Doctor on Job* xi. 93 Frock-coated and gold watch-chained. **1898** H. S. CANFIELD *Maid of Frontier* 15 [How's] the little baby I gave the watch-charm to? **1929** W. FAULKNER *Sound & Fury* 99 Father brought back a watch-charm from the Saint Louis Fair. **1979** *Arizona Daily Star* 1 Apr. c4/2 His uniform number and watch-charm size aren't the only distinctive things. **1981** J. R. L. ANDERSON *Death in High Latitude* ix. 141 We must have someone on watch. . . And the watchkeeper must have a rifle. **1946** C. S. FORESTER *Lord Hornblower* iv. 32 Nine years as a captain had not eradicated the habits acquired during a dozen years as a watchkeeping officer. **1977** *Proc. R. Soc. Med.* LXX. 485/2 A sailor reported recurrent unilateral swelling after watchkeeping. **1974** P. GORE-BOOTH *With Great Truth & Respect* 238 There had been three categories of control, namely embargo, quantitative restriction and a 'watch-list'. **1982** 'J. PENN' *Notice of Death* xix. 178 The Watch List. . a list the officials have at ports of entry, of people who are wanted for one thing or another. **1953** G. M. DURRELL *Overloaded Ark* vi. 113, I engaged what in the Cameroons is known as Watchnight. . . There were two reasons for engaging a night watchman: the first. . to wake me up. The second. . was that he patrolled the edge of the compound. . for driver ant columns. **1975** J. WYLLIE *Butterfly Flood* (1977) xxxi. 148 The watch night. Where is the watch night? . . The nightwatchman was lying in his reclining chair. **1831** M. WHALLEY *Let.* 15 Feb. in J. Constable *Corr.* (1962) I. 260 Thanks for . . a pair of tastefully & beautifully made watch pockets. **1832** *Chambers's Edin. Jrnl.* I. 82/2 Having changed the wear. . of a silk watch-riband for a chain. **1804** BYRON *Let.* 29 Aug. (1973) I. 51 Do not overlook my watch ribbon, and purse as I wish to Carry them with me. **1827** DISRAELI *Vivian Grey* III. v. vii. 131 Here are Eau de Cologne, violet soap, and watch-ribbons. **1850** *Ann. Sci. Discovery* 73 On the top of the structure [*sc.* a lighthouse] is the watch-room, and lantern, or light-room. **1883** 'MARK TWAIN' *Life on Mississippi* 338 One of the two establishments where the Government keeps and watches corpses until the doctors decide that they are permanently dead. . . A wire led to a bell in a watch-room yonder. . where a watchman sits always alert. **1943** *Gloss. Terms Telecomm.* (*B.S.I.*) 85 *Watch room*, the room, at the fire station, which is continuously staffed and contains the alarm-recording apparatus, call-bell keys and telephonic

intercommunicating equipment. **1977** *Stornoway Gaz.* 27 Aug. 1/4 The Commission also want to renovate the existing harbour watchroom to provide welfare facilities for dock workers, a new watchroom, and public toilets. **1979** N. WALLINGTON *Fireman!* iii. 37, I saw. . the watchroom into which came all the emergency calls.

watch, *v.* Add: **I. 4. g.** For (U.S. *colloq.*) read (*colloq.*, orig. *U.S.*). (Earlier and further examples.)

1845 J. J. HOOPER *Some Adventures Simon Suggs* ix. 115 He determined therefore to 'watch out' and keep himself 'whole' in a pecuniary point of view if possible. **1909** J. MASEFIELD *Tragedy of Nan* I. 21 You better watch out she don't tread a thy corns. **1957** R. LAWLER *Summer of 17th Doll* I. i. 19 If you don't watch out, you're gunna start hating the poor bloke before he even gets here.

† h. *to watch in*: to watch a television programme. *Obs.*

1928 *Daily News* 17 Dec. 6/4 (*heading*) Pictures by Wireless. Where to 'watch-in' this week.

II. 12. d. To exercise care, caution, or restraint about (something). *to watch one's step*: see *STEP *sb.* 10 b.

1837 H. MARTINEAU *Society in Amer.* I. II. 187 The valetudinarians of the place. . watch their own and each others' weight. **1958** 'CASTLE' & HAILEY *Flight into Danger* i. 23 You'd better watch that waistline, Pete. **1963** 'W. HAGGARD' *High Wire* iv. 40 Rex said deliberately: 'I have to watch champagne.' 'Really? But this one won't damage you.' **1969** G. CROUDACE *Blackadder* iv. 32 Pauline ate nothing but a small, lightly-grilled steak and a lettuce leaf; it was obvious that she had to watch her weight. **1976** M. MACHLIN *Pipeline* lvii. 573 Just watch your mouth. . . A man died here tonight. **1981** G. MARKSTEIN *Ultimate Issue* 26 It was a comparatively shabby office. . . Euram Marketing gave a distinct impression of watching the pennies.

e. *to watch it*: to be careful. Freq. as imp. (as a threat or warning). *colloq.*

1916 'TAFFRAIL' *Pincher Martin* vi. 100 'Don't yer go lendin' money to any other blokes wot ain't fit to be trusted.' 'I'll watch it.' **1943** K. TENNANT *Ride on Stranger* xviii. 203 You're not getting any younger yourself, kid. You want to watch it. **1954** W. FAULKNER *Fable* 334 'All right,' the corporal said. 'Watch it now.' **1966** 'W. COOPER' *Mem. New Man* II. iv. 144, I replied in the proletarian vernacular of the times. 'Watch it!' **1978** D. BLOODWORTH *Crosstalk* xxii. 170 We really do have to watch it a bit. Thank God we're officially engaged.

watcha (wǫ·tʃă), repr. a colloq. or vulgar pronunciation of *what do* (or *are* or *have*) *you*? Similarly **watcher.** See *WHATCHA.

1926 D. L. SAYERS *Clouds of Witness* vi. 134 Hullo! Watcher mean, nap? I had got hold of a most important train of thought. **1966** R. PETRIE *Dead Loss* xiv. 108 Watcha gonna do, Jeff? You can't leave her here. **1967** E. & M. A. RADFORD *No Reason for Murder* xxiii. 167 Watcher want with Waddy? **1969** in Halpert & Story *Christmas Mumming in Newfoundland* 95 Watcha got there, Uncle?

watcha, *int.*: var. *WOTCHER *int.*

1977 'J. GASH' *Judas Pair* iii. 38 'Watcha, Lovejoy.' 'Come in, Tinker.'

watchable, *a.* Add: *spec.*, that may be watched with pleasure or profit at the theatre or cinema, or on television. (Examples.)

1933 *Punch* 11 Jan. 51/2 There were two other dancing items: Messrs. Lee, Lee, Lee and Lee, who were very watchable and possessed in their pianist an accomplished clown; and Mr. Jerry Cole. **1970** *Daily Tel.* 12 Oct. 11 That rather unfortunate category among television documentaries—worthy but not terribly watchable. **1972** *Ibid.* 22 Nov. 15/2 He was also needed. . to have the stage dressed in something more watchable than brown paper and dustsheets. **1978** *Chicago* June 130/1 Schlafly. . was named one of the most influential and watchable women in America in two national polls taken in 1977. **1983** *Times* 26 Oct. 18/1 A very watchable programme.

Hence **wa:tchabi·lity.**

1966 *Guardian* 29 Mar. 10/3 For sheer watchability, the Liberals have just about won. **1975** *Daily Tel.* 10 Apr. 16/5 If watchability is part of the business of TV documentary, . . then 'The Fight Against Slavery'. . is surely a winner.

watch-dog. Add: **c.** Also *fig.*, esp. applied to a commission or other group appointed as a safeguard against abuses by the authorities (in government, foreign policy, etc.), business interests, etc.

1873 [see *scrub-woman* s.v. *SCRUB *v.*¹ 5]. **1947** *Toronto Daily Tribune* 21 Dec. 12/1 [Vishinsky] spoke. . , raking the United States for. . proposing that the assembly send a semi-permanent UN 'watchdog' commission. **1960** *Daily Tel.* 22 Aug. 1/1 He demanded the dispatch to his country of a United Nations watch-dog force. **1969** *Listener* 3 July 5/1 The watchdog function of the press depends. . on the effects of publicity on a politician's own colleagues and opponents. **1977** *Time* 30 May 51/2 Last March the International Trade Commission, a six-member watchdog group appointed by the President, recommended an increase in the tariffs on Japanese sets from 5% to 25%. **1984** *Daily Tel.* 9 Oct. 8/2 The Post Office was accused yesterday by its consumer watchdog body of putting profit before service to customers. **1985** *Which?* Feb. 58/1 Our regular checks over the years (and those of other 'watchdogs' such as Trading Standards Officers) have consistently revealed a depressingly low standard of

workmanship in garages. **1985** *Times* 31 May 13/7 Although climbing may be the most non-bureaucratic of sports it does need a watchdog guarding its interests.

wa·tchdog, *v.* [f. the sb.] *trans.* To attend, follow, or guard (a person); to maintain surveillance over (an activity, situation, etc.). Hence **wa·tchdogging** *vbl. sb.*

1902 C. J. C. HYNE *Mr. Horrocks, Purser* 70 There seems to be a whole regiment of ragamuffins on board here watchdogging her. **1962** *Punch* 20 June 933/1 The rigours of public watchdogging. **1964** R. F. FICCHI *Electrical Interference* ii. 9 The only way the system problem can be resolved is to have a completely formalized program to 'watch dog' the problem through to its solution. **1968** *Economist* 13 July 17/1 The job of watchdogging the constitution is second nature to Tories. **1977** M. HERR *Dispatches* 191 We [*sc.* war correspondents] were. . there to watchdog the day. **1980** A. SKINNER *Mind's Eye* xiv. 196 Three more. . went down with the strange attacks, and he and Holbrook watchdogged them through.

watcher. Add: **h.** With preceding *sb.* in objective relation, freq. hyphenated: an expert student or one who follows the affairs of (a particular person, country, or institution). *colloq.* (orig. *U.S.*).

1966 *N.Y. Times* 27 Dec. 10/3 That was the major question of the 'China watchers' as Mao observed his birthday without public fanfare yesterday. **1971** *N.Y. Post* 20 Dec. 40 The Nixon-watchers. . would construe Kissinger's return to Cambridge as a potential calamity. **1972** *Jrnl. Linguistics* VIII. I. 125 MIT-watchers have lately been puzzled by a division within the movement. **1974** *Guardian* 26 Mar. 16/1 As one experienced Brussels-watcher put it yesterday, Mr Peart has discovered how flexible the Treaty of Accession can be. **1974** Kremlin-watcher [see *KREMLIN]. **1975** *Times* 19 June 14/8 The meeting. . achieved a level of futility which left even hardened EEC watchers groggy. **1976** *New Yorker* 8 Mar. 27/3 When Confucius first came under attack, professional China-watchers speculated that the criticism had allegorical overtones. **1978** *Guardian Weekly* 23 Apr. 17/3 A veteran Shah-watcher said, 'He realises the extent of present and future problems.'

† i. *watcher-in.* A television viewer. Cf. *WATCH *v.* 4 h. *Obs.*

1928 *Daily News* 17 Dec. 6/4 Watchers-in will be able to judge for themselves the value of picture transmissions during this week.

watcher *int.*: see *WATCHA.

water, *sb.* Add: **I. 1. f.** *water over the dam* or *under the bridge* (*dyke, mill* and varr.): past events which it is unprofitable to revive or discuss; a way of saying 'a long time has passed'.

1913 *Wireless World* I. 34/1 Much water has flowed under London Bridge since those days. **1914** KIPLING *Let.* 15 Sept. in Ld. Birkenhead *Rudyard Kipling* (1978) xviii. 279 Your articles. . are a little too remote. . but of course—much water, or shall we say much blood, has flowed under the bridges since they were written. **1920** GALSWORTHY *In Chancery* I. i. 5 It was a startling remark—showing in a flash what a lot of water had run under bridges since the death of Aunt Ann in '88. **1940** *Nation* (N.Y.) 16 Mar. 364 Last year's results are water under the mill. **1940** *Amer. Guardian* 11 Oct. 1 All that is water over the dam. **1955** C. S. FORESTER *Good Shepherd* 91 He should not have brought the men to battle stations at all. . . But that was water over the dam; no time for regrets at present. **1964** E. F. BECKENBACH *Appl. Combinatorial Math.* x. 292 But these fees are water under the bridge, for we have paid them and they will never be returned to us. **1969** 'W. HAGGARD' *Doubtful Disciple* xi. 116 'Where did he get it?' 'From Seyer, I think—we can't escape that. But that's water over the dam by now.' **1976** *Glasgow Herald* 26 Nov. 5/4 Does he look back in anger as a result of his wartime experiences? 'Not now—perhaps a little to start with. But a lot of water has gone under the dyke since then.' **1977** J. THOMSON *Case Closed* iii. 43 It's been quite a time since we last met. Water under the bridge, as they say. **1981** *Encounter* Oct. 7 You don't want to let any of that business bother you. . . Water under the bridge. Just accept the fact, file it away.

3. b. (Earlier nonce-combination.)

1812 H. C. ROBINSON *Jrnl.* 13 May in E. J. Morley *Blake, Coleridge, Wordsworth, Lamb, Etc.* (1922) 50 Bar. Field called Wilson '*Wordsworth & Water*'.

6. c. *deep waters*: now usu. in phr. *in deep water*(*s*).

1861 T. HUGHES *Tom Brown at Oxford* I. v. 83 Tom felt greatly relieved, as he was beginning to find himself in rather deep water. **1867** TROLLOPE *Last Chron. Barset* I. xxxiv. 294 Once he had been very nearly in deep water because Mrs. Proudie had taken it in dudgeon that a certain young rector, who had been left a widower, had a very pretty governess for his children. **1933** N. COWARD *Design for Living* II. ii. 62 Our lives are diametrically opposed to ordinary social conventions; and it's no use grabbing at those conventions to hold us up when we find we're in deep water. **1950** D. LESSING *Grass is Singing* viii. 157 He stubbornly went his own way, feeling as if she had encouraged him to swim in deep waters beyond his strength, and then left him to his own devices.

III. 17. b. *water on the knee*: an excessive accumulation of fluid in the knee joint.

1890 A. JAMES *Diary* 7 Nov. (1965) 151 They [*sc.* the English] call water on the knee, fluid on the joint. **1902** W. S. CHURCHILL *Let.* 9 Oct. in R. S. Churchill *Winston S. Churchill* (1969) II. Compan. I. 167 He is laid up with Water on the Knee. **1976** *Liverpool Echo* 6 Dec. 1/8 A lone

Canadian sailor spent 17 days at sea clinging to his capsized trimaran before he was rescued suffering only from water on the knee.

IV. 20. b. (Earlier var. example.)

1775 *London Mag.* Nov. 556/2 She has an eye of that quick and brilliant water, that it penetrates and darts through the person it looks on.

VII. 24. a. *water-bowl, -flask, -jug* (earlier examples), *-jugjar, -sack, -skin* (earlier example), *-tin, -trough* (later examples).

1779 in *Dict. Amer. Eng.* (1944) IV. 2452/2,2 water jugs. **1807** R. SOUTHEY *Lett. from England* I. xiv. 161 A compact kind of chest holds the bason, the soap, the tooth brush, and water glass. . . The water-jug and water-bottle stand below. **1821** Water-skin [see *GIRBA]. **1855** *Poultry Chron.* II. 432 The price charged for these pens . . will be 6s. 6d. each, complete, with wooden bottom, water tin, and separate cloth. **1864** J. A. GRANT *Walk across Afr.* 430 Here there is a bend in the Nile, and we were able to fill all our water-sacks afresh. **1920** Water-tin [see *SOUVENIR *v.* 3]. **1922** JOYCE *Ulysses* 491 A man. . passes. . hugging a full waterjugjar. **1926** D. H. LAWRENCE *David* xiv. 106 Here is the pouch and the water-flask. **1935** C. WINCHESTER *Railway Wonders of World* II. 933 There are eleven sets of water-troughs on the West Coast main line between Euston and Glasgow. **1960** J. R. ACKERLEY *We think the World of You* 89 On the floor was Evie's water-bowl and the vegetable remains of her dinner of yesterday. **1970** J. H. B. PEEL *Country Talk* vi. 111 Water-troughs in far fields were tepid. **1979** M. MCMULLEN *But Nellie was so Nice* (1981) I. i. 12 She. . filled Titania's water bowl, gave jealous George an extra stroking.

b. *water-supply* (earlier example), *-well.*

1849 J. SIMON *City Medical Rep.* No. 1, in E. R. Pike *Human Documents of Victorian Golden Age* (1967) 280 It may be doubted, too, whether. . the tenants' water supply can be pronounced good. **1921** *Review of Reviews* Aug. 157/2 Hardships of desert travelling, when great gulfs of distance lie between one water well and another. **1976** *Billings* (Montana) *Gaz.* 30 June 5-D/1 (Advt.), Water well drilling.

c. *water tank, tender, truck.*

1834 Water tank [in Dict., sense 24 a]. **1894** *Outing* June 172 Four or five rough-looking men. . were clustered about the water-tank. **1957** G. V. BLACKSTONE *Hist. Brit. Fire Service* xxv. 443 The evolution of the water tender from the mobile dam unit produced to deal with the water shortages at air-raid fires. **1958** P. SCOTT *Mark of Warrior* II. 122 The water truck rendezvoused. Most of the chaps' bottles were empty. **1967** O. RUHEN in *Coast to Coast 1965–6* 193, I threw down my swag near the water-tank. **1974** D. SEARS *Lark in Clear Air* i. 20 We stopped at every gopher-hole and water-tank and badger-hill. **1976** *Billings* (Montana) *Gaz.* 6 July 8-A/2 The county water truck was dispatched at 4.09 p.m. . . The truck was refilled with water, before the fire was controlled. **1978** *Dumfries Courier* 13 Oct. 17/4 Two water tenders from Dumfries brought the blaze under control.

d. *water-cutting.*

1862 E. HODDER *Memories N.Z. Life* 117 We were glad to. . lie down closely together in one of the dry water-cuttings.

e. *water-feed.*

1914 *Chambers's Jrnl.* May 334/2 The. . lamp. . works automatically, the water-feed to the carbide being drop by drop.

g. *water-seal* (earlier examples).

1877 G. E. WARING *Sanitary Conditions City & Country Houses* 31 These gases. . have. . the power of passing almost unretarded and unchanged through the water-seal traps. *Ibid.* 36 The water-seal is a trap in more senses than one.

i. *water-porridge, -toast.*

1838 A. LANGTON *Jrnl.* in *Gentlewoman Upper Canada* (1950) 80 This morning the same party assembled to tea, coffee, and water porridge—a great favourite with most of the backwoodsmen. **1843** *Ainsworth's Mag.* V. 60 My dinner was pudding or pastry, and if these failed, there was a substitute known by the name of *water-toast*. . . A slice of bread was toasted, and dipped. . in boiling water, and then buttered and sugared. **1848** GEO. ELIOT *Let.* 8 Mar. (1954) I. 255 The sympathy in Ireland seems at present only of the water toast kind. **1947** W. DE LA MARE *Coll. Stories for Children* 61 A bowl of water-porridge, using up for it the last pinch of meal.

j. *water-time.*

1925 R. GRAVES *Welchman's Hose* 9 Our feeding and our water-time, Our breeding and our slaughter-time.

k. *water-land, -point.*

1811 W. H. MARSHALL *Rev. Rep. to Board Agric. from Eastern Dept. Eng.* 10 The term *Water-Lands* may be deemed a solecism. But when it represents lands that have not only been formed by water, but are liable to be annihilated, and their place reoccupied, by the same element—it is surely allowable as a *Technical Term*, to convey a joint idea of 'Fens' and 'Marshes'. **1954** J. R. R. TOLKIEN *Two Towers* iv. 79 The Entwives. . saw the. . green herbs in the waterlands in summer. **1946** U. KRIGE *Way Out* xv. 194 They were taken by the Germans at the water-point. **1964** J. HILLABY *Journey to Jade Sea* 167 The Balessa Kulal, a famous water-point farther down the *lugga*.

l. *water-blowball, -brook* (later examples), *-cloud, -drip, -fount* (example), *-gush, -leak, -mist, -ripple, -song, -sphere, -spray, -surface, -swirl, -talk, -world* (later example).

c **1865** G. M. HOPKINS *Poems* (1967) 122 When lily-yellow is the west. Say, o'er it hangs a water-cloud And ravell'd into strings of rain. **1869** *Bradshaw's Railway Man.* XXI. 242 The dock. . measures 600 feet in length and 300 in width, giving an area of water surface of upwards of four acres. **1876** 'MARK TWAIN' *Tom Sawyer* xxxiii. 260 It was the treasure-box. . occupying a snug little cavern, along with. . some. . rubbish well-soaked with the water-drip. **1888** G. M. HOPKINS *Poems* (1967) 198 All in froth and water-blowballs. **1904** W. DE LA MARE *Henry Brocken* vi. 59 Ears that have heard only. . dismal water-songs, and the yelp and quarrel and night-voice of unseen hosts in the forests. **1916** BLUNDEN *Pastorals* 30 Mocked by the white wings of the water-swirl. **1920** J. MASEFIELD *Enslaved* 34 The melancholy water-drip alone Broke silence near me. **1921** W. DE LA MARE *Veil* 32 Ice on the waterbrooks their clear chimes dumbing. **1922** JOYCE *Ulysses* 698 A rockery with waterspray. **1928** BLUNDEN *Undertones of War* ix. 98 Ahead, the German front line could not be clearly seen, the water-mist and the smoke veiling it. **1928** E. SITWELL *Five Poems* 4 The water-ripples like mosaics gold. **1935** W. STEVENS in *Southern Rev.* I. 80 The statue stood in stars like water-spheres. *a* **1944** K. DOUGLAS *Alamein to Zem Zem* (1946) 52 Few Crusader tanks would run for more than two days in action without developing either an oil-leak or a water-leak. **1952** L. MACNEICE *Ten Burnt Offerings* x. 60 The water-talk ends; the scrawl on the sky Smudges and fades. **1954** J. BETJEMAN *Few Late Chrysanthemums* 66 Back into what a water-world Of waving weed and waiting claws? **1955** E. POUND *Classic Anthol.* II. 113 Naught stands higher than mount, Nor is hollow deeper than water-fount. *a* **1957** R. CAMPBELL *Coll. Poems* (1960) III. 19 Amidst dead calms collapsing water-gushes, And distances cascading to the deeps. **1961** K. REISZ *Technique Film Editing* (ed. 9) ii. 137 Tiny bugs skim over the water-surface. **1967** *Jane's Surface Skimmer Systems 1967–68* 14/1 The vehicle has been used primarily for testing over land, water and ice, to investigate. . seaworthiness, water spray problems, [etc.]. **1977** *Daily Tel.* 9 July 12 Lands where hearts desire the waterbrooks and wells are of crucial importance. **1984** A. C. & A. DUXBURY *Introd. World's Oceans* iv. 120 Water molecules arrange themselves at any water surface to form a weak elastic membrane.

m. *water-stair.*

1920 T. S. ELIOT *Ara Vos Prec* 15 Princess Volupine extends A meagre, blue-nailed, phthisic hand to climb the waterstair.

n. *water-ballet, -music* (later examples), *-pageant* (later examples), *-song, -sonnet, -sports.*

1888 Water-music [see *PADDLING *vbl. sb.*¹]. **1888** L. A. SMITH *Music of Waters* 83 The verses and tune of this water-song. . follow. **1892** Water-pageant [see *BARGEMASTER]. **1920** L. & N. SHEFFIELD *Swimming Simplified* viii. 150 Water sports afford a varied source of amusements. **1940** L. MACNEICE *Last Ditch* 9 And as the twilight filtered on the heather Water-music filled the air. **1944** BLUNDEN *Shells by Stream* 7 Trilling still with finch or lark Or water-sonnet. **1948** T. WILDER *Ides of March* II. 143 When the water-ballet was ended, Caesar's party rose to go in search of the Queen. **1968** Water ballet [see *SYNCHRONIZED *ppl. a.*]. **1976** *National Observer* (U.S.) 30 Oct. 3/3 We enjoy water sports. **1976** *Evening Post* (Nottingham) 15 Dec. 1 Mr Michael Hammond said he had not received a reply from the County Council to an offer of assistance in organising a water pageant on the day of the Queen's visit. **1979** *Tucson Mag.* Apr. 57/1 There are. . dance exercises, water ballet classes, separate massage facilities and other services. **1980** *Early Music* Jan. 50/1 It seems likely that Weiss, Buffardin, and Hebenstreit provided 'water music' for the royal couple on the barge.

p. *water-brother, -wayfarer.*

1965 AUDEN *About House* (1966) 16, I should like To be to my water-brethren as a spell Of fine weather. **1923** D. H. LAWRENCE *Birds, Beasts & Flowers* 99 Fishes. . Outsiders. Water-wayfarers.

q. *water-fay, -ghost, -king, -shape.*

1878 O. WILDE *Ravenna* 5 And down the river, like a flame of blue, Keen as an arrow flies the water-king. **1916** BLUNDEN *Pastorals* 32 That you shall come upon the water-fays. **1925** —— *English Poems* 90 The water-shapes steal towards his groaning drone. **1928** *Oxford Poetry* 10 Many strong men had passed the ford, nor known the presence of that jeering water-ghost Denying their true conquest of the stream.

25. a. *water-divining, -dowsing, -holding, -retaining*; also with *sbs.* and *adjs.*, as *water-retention, -retentive.*

1898 K. L. PARKER in Murdoch & Drake-Brockman *Austral. Short Stories* (1951) 2 We thought you. . would. . know of a goolahgool, or water-holding tree. **1909** O. LODGE *Survival of Man* ix. 128 Just as people occasionally seem able to become cognisant of facts or events by means ordinarily closed to them,—a phenomenon which appears akin to the water-dowsing faculty and to the 'homing' instincts of animals,—so sometimes they can write poetry or solve problems beyond their normal capacity. **1913** W. OWEN *Let.* 13 Nov. (1967) 211 What it costs to make a Scout. For instance:. . Belt. . Staff (Water-divining, extra). **1930** T. S. ELIOT tr. *St-J. Perse's Anabasis* 63 Consecration of stones perfectly round, water-dowsing in dead places. **1936** *Discovery* Jan. 24/1 They [*sc.* the soils] are all highly permeable and with a small water-retaining capacity. **1946** *Nature* 13 July 58/1 The lower part of the profile consists. . either of re-cemented chalk or compact sand, both of which are very water-retentive. **1952** L. MACNEICE *Ten Burnt Offerings* IV. i. 38 A sage whose water-divining mind Will twitch to the smallest drop. **1958** *Yearbk. Agric.* 1957 (U.S. Dept. Agric.) 770/2 *Water retention*, the physical property of soil that is based on surface force action and that makes it necessary to do work in order to remove water from soil pores and from soil surface. **1969** G. BECKER in Krishna & Weesner *Biol. Termites* I. xi. 367 The water content of the soil should be slightly below its water-holding capacity. **1973** D. ROBERTSON *Survive Savage Sea* II. 73 The unpalatable, brackish water could be administered rectally in the form of water retention enemas. **1979** *Arizona Daily Star* 5 Aug. (Advt. Section) 4/1 Applicant should have at least 3 years experience in the design of hydraulic, water-retaining structures.

b. *water-fearer, -lover, -tender.*

1842 T. W. HARRIS *Insects Injurious to Vegetation* 11 The water-lovers (Hydrophilidæ). . act the useful part of scavengers. **1884** *U.S. Navy Exec. Ord.* 31 Dec., New ratings are hereby established as follows: boiler maker, water tenders, oilers. *a* **1885** G. M. HOPKINS *Poems* (1967) 192 Fallers in dreadful frothpits, waterfearers wild. **1922** JOYCE *Ulysses* 655 Bloom, waterlover, drawer of water. **1948** H. C. NICHOLS *Voice at Sea* x. 145 This water-tender, or boss stoker as he was sometimes called, was the devil himself.

26. a. *water-cut, -hidden, -marrowed, -pillared, -saturated, -sealed, -shafted, -smoothed, -sorted, -sprinkled, -wattled, -wound*; also with *adjs.*, as *water-dispersible, -poor, -rich.*

1876 G. M. HOPKINS *Poems* (1967) 65 And mazy sands all water-wattled. **1922** J. A. DUNN *Man Trap* iv. 49 Most of the young men nowadays are water-marrowed pups. **1924** G. A. BURLS *Cost of Power Production by Internal-Combustion Engines* iv. 23 A gas-holder of the familiar inverted water-sealed type. **1925** E. SITWELL *Troy Park* 19 A water-hidden sound. **1927** V. WOOLF in *Forum* May 704 Her cowardice; her mean, water-sprinkled blood. **1928** 'BRENT OF BIN BIN' *Up Country* xix. 325 The water running on to the water-smoothed stones. **1935** L. MACNEICE *Poems* 21 Set these against your water-shafted air of amethyst and moonstone, the horses' feet like bells of hair. **1939** J. STEINBECK *Grapes of Wrath* i. 1 In the water-cut gullies the earth dusted down in dry little streams. **1939** DYLAN THOMAS *Map of Love* 6 Down the stacked sea and water-pillared shade. **1946** *Nature* 6 July 14/1 For an ordered system, the expansion on wetting is lateral for a water-rich system, and perpendicular for a water-poor system. *Ibid.* 9 Nov. 675/1 The extent of development obtained is related to the type and throughput of solvent, which is normally 10 ml. of water-saturated ether per strip. **1946** DYLAN THOMAS *Deaths & Entrances* 63 Turns the moon-chained and water-wound Metropolis of fishes. **1959** *Times* 30 Nov. 18/7 Water-dispersible powders. **1959** A. H. MCLINTOCK *Descr. Atlas N.Z.* 36 Soils derived from fine volcanic ash and watersorted derivatives. **1978** J. M. BROWN in *Further Perspectives Organic Chem.* (CIBA) 149 The selective terminal addition to squalene in water-rich solvents. **1979** *Amer. Jrnl. Trop. Med. & Hygiene* XXVIII. 1014/2 The Department of Health of the Philippine government. . has campaigned since 1950 for the installation of cheap water-sealed toilets.

b. *water-rippling* ppl. adj.; *water-dripping* vbl. sb.

1922 T. S. ELIOT *Waste Land* (1923) 33 The hermit-thrush. . Its 'water-dripping song' is justly celebrated. **1924** E. SITWELL *Sleeping Beauty* xx. 77 A lady sang through water-rippling leaves. **1927** —— *Rustic Elegies* 76 And the trees' vast waterfalls Echoed this water-dripping song like flashing bright bird-calls.

27. *water-dwelling, -haunting, -living, -standing* ppl. adjs. Also *water-gifted* adj.

a **1882** EMERSON *Poet* in *Compl. Wks.* (1883) IX. III. 309 Methought like water-haunting birds Divers or dippers were his words. **1919** H. G. WELLS *Outl. Hist.* 12/1 Water-living creatures which are always under water, wave the freely exposed gills by which they breathe in the water. **1934** WEBSTER, Water-dwelling. **1936** *Discovery* Jan. 7/2 South American water-dwelling frogs. **1937** *Ibid.* Aug. 252/1 The larger water-haunting birds. **1942** W. FAULKNER *Go down, Moses* 340 The planters. . had wrested from the impenetrable water-standing cane and cypress. . cotton patches which. . became fields and then plantations. **1960** T. HUGHES *Lupercal* 46 Four-legged yet water-gifted.

28. *water-chilly, -clear, -cold, -dark, -eager, -flowing, -green* (earlier example), *-precious* adjs.

1884 *Girl's Own Paper* 30 Aug. 762/3 A large range of colour. . primrose, water-green, beige. **1910** W. DE LA MARE *Three Mulla-Mulgars* xxi. 275 An odd little water-clear song. **1923** D. H. LAWRENCE *Birds, Beasts & Flowers* 94 Water-eager eyes. **1923** Water-precious [see *red-gold* s.v. *RED *a.* 13]. **1924** E. SITWELL *Sleeping Beauty* 27 'Midst brightly perfumed water-flowing Eighteenth-century silks. **1925** —— *Troy Park* 59 A flaxen lily Water-chilly. **1928** —— *Five Poems* 3 That cloud of gold, Its kernel, crackling amber water-cold. **1945** —— *Song of Cold* 11 One of the Dead who lay Beneath the earth, like the water-dark, the water-thin Effigy of Osiris. **1961** R. D. BAKER *Essent. Path.* xix. 527 The adenoma has a yellowish or gray, soft cut surface. . . Microscopically it is composed of chief cells, although occasionally of the so-called 'water-clear' cells.

29. water-base *a.*, having water as the main ingredient; **water-based** *a.*, (*a*) operating from ships; (*b*) = prec.; **water body,** a body of water forming a physiographical feature, as a sea, reservoir, etc.; **water-bomber** *Canad.*, an aircraft used for extinguishing forest fires by dropping water on them; hence **water-bomb** *v. trans.*; so **water-bombing** *vbl. sb.*; **water-boy,** (*d*) chiefly *U.S.*, a boy or man who carries or takes round drinking water; **water-bus,** a motor-boat or steamer carrying paying passengers as part of a scheduled service; **water cannon,** a device for shooting a jet of water at high pressure, esp. to disperse crowds, etc.; **water clerk,** an employee of a ship's chandler; **water content,** (amount of) water contained in some material; **water-cool** *v. trans.*, to cool (an engine, etc.) by circulating water; hence **water-cooling** *vbl. sb.* and *ppl. a.*; so **water-cooled** *ppl. a.*; **water-cooler,** a vessel or container in which water is kept cool; *spec. U.S.*, a tank of cooled drinking water in a place of work; **water cushion,** a depth of water that acts to lessen the impact or force of something; **water drive** *Oil Industry*, the use of water to force oil out of a reservoir rock; **water drum**, a drum containing water, or placed in water, and played as a musical instrument; **water fountain**, a drinking fountain; **water-garden**, (*b*) (earlier example); hence **water-gardening**; **water-gilt** *a.* (earlier example); **water injection**, (*a*) *Oil Industry*, the forcing of water into a reservoir formation, esp. as a technique of secondary recovery (cf. *WATER-FLOODING *vbl. sb.*); (*b*) *Aeronaut.*, the injection of water into the cylinders of a piston engine with the fuel, to cool the charge, or into the air intake of a jet engine, to cool the air, so as to increase engine efficiency in either case; **water-insoluble** *a.*, insoluble in water; **water intoxication** *Med.*, a condition resulting from the intake of too much water, leading progressively to drowsiness and unsteadiness, confusion, convulsions, coma, and death; so **water-intoxicated** *a.*; **water-keeper** one who guards a tract of water against poachers; cf. GAMEKEEPER; **water-lain** *a. Geol.* = *water-laid* adj. (*b*) in Dict.; **water mass** *Oceanogr.*, a large body of sea water that remains distinguished by its temperature and salinity from surrounding water; **water park**, a recreational area comprising stretches of fresh water that may be used for boating, etc.; **water-polo**, a game played by teams of swimmers, usu. in a rectangular pool with goal-posts, using a ball similar to a football; **water-power** (earlier example); **water-privilege** (earlier example); **water-repellent** *a.*, not easily penetrated by water though porous; *sb.*, an agent conferring this property; so **water-repellency**; **water-resistant** *a.*, waterproof or water-repellent; so **water-resistance**; **water resources**, natural sources of water available for man's use; freq. *attrib.* (in *sing.* or *pl.*); **water-rights** (also in *sing.*), the right to the use of water in a tract of land (cf. *water-privilege* (*a*) in Dict.); **water-sail**, (*b*) a sail lowered into the water to act as a sea anchor; **water-shear, -shier** (earlier example); **water silk**, watered silk (see WATERED *ppl. a.* 5); a garment of this material; also *fig.*; **water-softener**, an apparatus for making hard water soft by chemical means; also, a chemical used for this purpose; hence **water-softening** *vbl. sb.* and *ppl. a.*; **water-soluble** *a.*, soluble in water; so **water-solubility**; **water-splash** (earlier example); **water-spot** *v. intr.*, of fabric: to be liable to show permanently any mark made on it by a drop of water; hence **water-spotting** *vbl. sb.*, the condition of showing such a mark; **water sprout** = WATER-SHOOT 1 in Dict. and Suppl.; **water-stain**, (*a*) a stain made on a surface by contact with water; (*b*) (see quot. 1940); **water-stop**: restrict † to sense in Dict. and add: (*b*) a place where a traveller or a train may stop for water; (*c*) a sealant to prevent water from leaking through joints (see quot. 1951); **water taxi**, a small boat used for casual passenger traffic on rivers, canals, etc.; **water toothpick** = *water-pick* (see *WATER PIK); **water-torture**, a form of torture in which the victim is made to endure an incessant drip of water on the head (see also quot. 1928); also *fig.*; **water treatment**, (*a*) = WATER-CURE; (*b*) = *water torture* above; **water-tube**, (*b*) each of the tubes carrying water through a water-tube boiler; **water-tube boiler**, delete 'marine'; (further example); **water tunnel** (see quot. 1969); **water-wagon**, (*a*): for *U.S.* read orig. *U.S.* and add later examples of *slang* use (cf. *WAGON, WAGGON *sb.* 10* b).

1949 *Sci. Digest* Dec. 93 Water-base paints have been in use for years, but in the past they could not be washed without coming off. **1975** *McGraw-Hill Yearbk. Sci. & Technol.* 300/2 Because of the cost of oil-base drilling fluids.., research is continuing so that water-base drilling fluids that can provide the properties needed..can be developed. **1955** *Sci. News Let.* 15 Jan. 37/2 The new big swept-wing plane..was described as a 'truly water-based aircraft' capable of operating in high waves and marginal sea conditions. **1981** *Sci. Amer.* Aug. 85/1 Oil-based graphites tend to give off fumes, and so the latest trend in the forging industry is toward the use of water-based graphites. **1897** Water body [see *sheet-flood* s.v. *SHEET *sb.*[1] 12 b]. **1974** R. H. BRITTON in R. Goodier *Natural Environment of Shetland* 123 The most abundant type of standing waterbody on Shetland, (accounting for 27·4% of the total), is a dystrophic loch of area less than 1 hectare. **1975** *Sci. Amer.* June 88/1 (Advt.), There are many excellent shots of the planes in action, 2 of which are shown here water-bombing the blaze. **1961** *Canada Month* 6 Oct. 42/3 The Grove and Tsus fires..could have been kept small if water bombers and helicopters had been available from the beginning. **1975** *Globe & Mail* (Toronto) 20 Aug. 8/6 He said 40 ministry firefighters, another force of 48 junior rangers,..a ministry water bomber and commercial aircraft..are all being used. **1959** *Time* (Canada ed.) 17 Aug. 13/3 Ontario has a new gadget that makes water-bombing more efficient than ever: snorkel-like water intakes on tanks attached to the aircraft floats that enable the planes to load up as they taxi along lake surfaces. **1859** *Harper's Mag.* Apr. 712/1 The 'water-boy' in his first round found me standing by the stove. **1903** *Congress. Directory* 116 He assisted himself in securing an education by working as a 'water boy' on the railroad. **1965** J. A. MICHENER *Source* (1966) 408 Trumpeters, drummers, waterboys and cooks followed in a compact nest, protected by many soldiers, and not until this stupendous preamble had passed did the actual fighting men appear. **1972** J. MOSEDALE *Football* viii. 106 Hubbard was convinced that Dobie was illegally sending in plays by arranging cups on the water tray. He kicked over the tray every time the water boy came on the field. **1979** P. THEROUX *Old Patagonian Express* xii. 185 Praising the quarterback, mocking the water-boy. **1929** A. P. HERBERT in *Times* 16 Nov. 8/1 Let no one tell us that the Londoner 'does not want' the water-bus. There was not much evidence that he wanted the Tubes..before he got them. **1940** *Economist* 21 Sept. 368/2 The waterbus service, which London Transport has been running..between Westminster and Woolwich and Greenwich, has won a popularity... It is an emergency experiment that may survive both air-raids and the war itself. **1981** T. HOLME *Funeral of Gondolas* i. 9 The Rialto Bridge humped regally over the Grand Canal... A water-bus ..churned water in mid-stream. **1964** L. DEIGHTON *Funeral in Berlin* xlii. 251 An American M.P. shouted, 'You want a goddam water-cannon to wash you across the sidewalk, fella?' **1982** *Listener* 23 & 30 Dec. 5/3 People had begun to take to the streets, defying martial law, tear-gas, water-cannon and bullets. **1898** *Barbados Freight Rep.* in F. Holm-Petersen *Fra Sejl til Diesel* (1951) I. 335 Our Water Clerk will meet the Vessels on their entering the bay. **1973** P. THEROUX *Saint Jack* xiii. 156, I knew my job as a water-clerk..and pored over the shipping pages of the *Straits Times*. **1946** *Nature* 21 Dec. 899/2 Vitamin-like substances..in cells with assumed water-contents of 80 per cent. **1957** G. E. HUTCHINSON *Treat. Limnol.* I. iv. 222 The water content of the major part of the lithosphere..is unknown. **1969** Water content [see *water-holding*, sense 25 a above]. **1909** WEBSTER, Water-cool v.t. **1915** W. E. DOMMETT *Submarine Vessels* v. 50 Due to the high temperatures, it is necessary to not only water-cool the cylinder, but also to cool the piston. **1971** *Physics Bull.* July 401/2 The essential difference.. is well illustrated in this example by the need to water-cool the measuring head because of high levels of radiant heat. **1905** Water-cooled [in Dict., sense 26 a]. **1984** B. FRANCIS *AA Car Duffer's Guide* 63/1 Modern water-cooled systems incorporate a thermostat. **1984** *Mining Jrnl.* 4 May 297 China has also expressed interest in Japan's PNC heavy water moderated, light water-cooled advanced thermal reactor. **1846** *Catholic Herald* (Philadelphia) 30 Aug. 272/4 Refrigerators, water coolers, and filterers. **1899** [in Dict., sense 25 c]. **1955** R. P. JHABVALA *To whom she Will* xxiii. 168 An earthenware water-cooler. **1969** *Canad. Antiques Collector* Aug. 6/1 (Advt.), Salt glazed water cooler with incised blue decoration... Each side has riderless horse. **1978** *Amer. Speech* LIII. 5 Slang..is inexact and meaningless, but these are characteristics it shares with much of standard English, as fifteen minutes at a water cooler or a political rally will show. **1910** R. FERRIS *How it Flies* x. 198 To keep the cylinder cool enough to be serviceable, two methods are in use: the air-cooling system and the water-cooling system. **1934** *Physical Rev.* XLV. 608/1 At 3,000,000 volts, ion currents of ¼ microampere are readily obtainable and it is probable that water-cooling of the accelerators and the bombarded targets will result in considerably larger utilizable currents. **1985** *Dirt Bike* Mar. 39/2 They wanted watercooling for reasons of reliability. **1907** H. BROWN *Irrigation* vi. 120 A pressure due to 6½ feet head of water, of which 3 feet is balanced by the water cushion on the floor. **1955** S. LELIAVSKY *Irrigation & Hydraulic Design* I. ii. 270 Finally, we must mention the method of scour prevention which in earlier times was known as the 'stilling pool' or the 'water cushion', and which is now described..as the 'cistern'. **1972** L. M. HARRIS *Introd. Deepwater Floating Drilling Operations* xvi. 169 The larger the water cushion [in a well], the larger the surface pressures required to unload it. **1938** H. S. GIBSON in A. E. Dunstan et al. *Sci. of Petroleum* I. xi. 538/1 The employment of the water drive as a process has received considerable attention. **1973** C. J. MAY in Hobson & Pohl *Mod. Petroleum Technol.* (ed. 4) v. 165 When recovery was mostly by water drive, the percentage extraction varied from 24 to 78, and the recovery per acre-foot of sand from 242 to 1165 brl. **1923** C. M. BARBEAU *Indian Days in Canad. Rockies* 120 A deep sound from the seer's lodge startled the people, a sound like that of a large water drum. **1955** M. BARBEAU *Tree of Dreams* 95 The thumping of water drums at times startled the bleak places, now covered with a mantle of snow. **1970** P. OLIVER *Savannah Syncopators* 106 Outstanding xylophone orchestras, *kora*, water-drum, harp and harp-lute. **1973** A. H. WHITEFORD *North Amer. Indian Arts* 103 The Iroquois also made water drums with plugs to change the level of the water and the tone of the drum. **1946** M. C. SELF *Horseman's Encycl.* 429 Automatic water fountains are most useful for the stable. **1975** 'E. LATHEN' *By Hook or by Crook* xxi. 194 A salt-tablet dispenser beside every water fountain. **1899** S. R. HOLE *Our Gardens* x. 214 Where there are pools of water...you may have the beautiful Water Garden. **1938** F. PERRY *Water Gardening* p. ix, When I first took an interest in water gardening the dearth of literature upon the subject was the cause of much disappointment. **1982** B. CHATTO *Damp Garden* vi. 98 Water gardening on the landscaping scale is quite a different affair from the marginal plantings. *a* **1776** DUCHESS OF NORTHUMBERLAND *Diary* in *Country Life* (1974) 7 Feb. 251/3 The Water Gilt Locks..were made at Birmingham. **1940** *Jrnl. R. Aeronaut. Soc.* XLIV. 590 Will water injection become practical? **1943** *Oil & Gas Jrnl.* 11 Nov. 230/2 Because an adequate low-cost water supply was available at Midway, the economic possibilities of experimental water injection were clear. **1947** *Shell Aviation News* No. 106. 21/1 If required water injection apparatus can be included to give additional power for take-off. **1970** M. SMITH *Aviation Fuels* ix. 62 By means of water injection, heat is taken away from the cylinders, pistons and exhaust valves, and removed out of harm's way by the steam so formed, which discharges from the exhaust system. **1982** Water injection [see *WATER-FLOODING *vbl. sb.*]. **1946** *Nature* 16 Nov. 709/2 The black, water-insoluble pigment. **1937** *Amer. Jrnl. Physiol.* CXIX. 557 The serum sodium and chloride levels are low in the water intoxicated animal. **1922** *Physiol. Rev.* II. 158 (*heading*) The effects of excessive ingestion: water intoxication. **1974** PASSMORE & ROBSON *Compan. Med. Stud.* III. xlix. 8/2 Water intoxication should be considered in any patient with unexplained cerebral dysfunction, particularly if the individual has recently undergone an operation. **1898** J. MACMANUS *Bend of Road* 118 The wather-keeper..was shot be the poochers on the Dhrowes river. **1920** D. H. LAWRENCE *Women in Love* xiv. 202 He called at the water-keeper's cottage and took the key of the sluice. **1959** *New Biol.* XXIX. 14 Any given bed of water-lain rock is usually formed of relatively uniform particles; wind-deposited sands show a similar sorting. **1977** A. HALLAM *Planet Earth* 55/1 The word 'dune' is often thought to refer only to those forms built by wind, but it is now used equally for water-lain structures. **1912** MURRAY & HJORT *Depths of Ocean* v. 261 From a study of the distribution of salinity and temperature the average direction of the drift of the water-masses may be deduced. **1976** *Nature* 19 Feb. 606/2 No influx of the Atlantic watermass into northeast Icelandic waters was observed in June 1975. **1928** *Observer* 15 July 12 In the north-eastern workmen's district of Floridsdorf,..a new 'water-park' has just been opened. It..includes two ponds, connected by canals, so that some kind of 'little Venice' has been created. **1972** *Daily Tel.* 4 Sept. 8 A national water park that would include the Norfolk Broads and the Fens is being planned. **1884** Water polo [see POLO 2]. **1888** [in Dict. s.v. WATER *sb.* 24 n]. **1919** W. T. GRENFELL *Labrador Doctor* iii. 61 Our water polo games were also a great feature here, the water being warm. **1923** T. SACHS *Compl. Swimmer* (ed. 2) 151 Water polo differs from other games, not only in the irrelevancy of its name, but also in being a game that was made to order. **1980** *Guinness Bk. Records* 334/2 Water Polo was developed in England as 'Water Soccer' in 1869. **1817** M. BIRKBECK *Notes Journey Amer.* 119 Water-mills, or, in defect of water-power, steam-mills rise on the nearest navigable stream. **1804** *New Hampshire Probate Rec.* (1916) III. 755 We set off to Deborah Shackford,..the Water Privilege belonging to said Estate. **1942** *Chem. Abstr.* XXXVI. 8428/1 (Index), Water repellency. (*See also* Waterproofness.) **1955** *Industr. & Engin. Chem.* Sept. 1980/1 Some [finishes] add functional qualities, such as water repellency and shrink resistance. **1972** *McGraw-Hill Yearbk. Sci. & Technol.* 217/2 Water birds..were generally regarded as having attained perfection in water repellency. **1922** *Encycl. Brit.* XXX. 59/1 The surface of the dope should be water-repellent. **1952** R. A. PINGREE in H. C. Speel *Textile Chemicals & Auxiliaries* xx. 408 An acetate rayon fabric may tolerate a water repellent which is unacceptable to either viscose rayon or cotton. **1974** P. DE VRIES *Glory of Hummingbird* (1975) xv. 231 Testing a line of water-repellent trenchcoats. **1980** GOHL & VILENSKY *Textile Sci.* vii. 132 Water repellent finishes can be divided into two categories: 1. water-proof finishes... 2. water resistant finishes which delay the absorption and penetration of water and..allow a degree of permeability to air. **1935** C. ELLIS *Chem. Synthetic Resins* I. xxxi. 656 A variation of this method, designed to improve the water-resistance of the resins. **1966** *McGraw-Hill Encycl. Sci. & Technol.* XIII. 541/1 Linseed oil..was used to impart water- and weather-resistance. **1934** WEBSTER, Water-resistant. **1946** *Nature* 19 Oct. 562/2 The water-proofing of soft fibre boards, surface treatment with a paraffin wax emulsion giving a water-resistant surface. **1971** C. BONINGTON *Annapurna South Face* 243 For high-altitude work it is not essential that it [*sc.* a garment] should be waterproof, but it needs to be water-resistant. **1980** Water-resistant [see *water-repellent* adj. and sb. above]. **1913** W. MCCULLOH *Conservation of Water* v. 99 The work done by the State of New York toward the..conservation of her water resources is an exemplification of what may be done by..other states. **1971** P. GRESSWELL *Environment* 218 Twenty-nine River Authorities..are responsible for husbanding the water resources of entire river basins, from springs and streams to final exit into the sea. **1976** *National Observer* (U.S.) 10 Apr. 1/3 A water-resources engineer with the World Bank. **1980** *Jrnl. R. Soc. Arts* Feb. 129/2 Workshops dealt with subjects ranging from water resource management to human resource management. **1793** *Columbian Museum* (Philadelphia) Jan. 16 The purchase of the land, including the farm buildings..and water rights, &c. would probably be at fifteen dollars per acre. **1891** R. WALLACE *Rural Econ. Austral. & N.Z.* xiii. 213 The question of water-right is also one urgently demanding legislation in Victoria. **1920** W. H. MALLOCK *Mem. Life & Lit.* xii. 175 He would..have had to buy from the neighbouring peasants certain way-leaves and water-rights. **1950** *N.Y. Times* 20 Apr. 1/3 He also said that bill threatened their water rights. **1925** A. B. ARMITAGE *Cadet to Commodore* 15 We dropped down on the tide, aided by a large watersail over the stern, to Garden Reach. **1933** P. MITCHELL *Deep Water* 201 [At Calcutta] the mud pilot..brought with him a water sail, like a small royal with the foot weighted and a rope fastened to each corner; this was lowered over the stern into the water. **1769** A. MENZIES *Rep. to Commissioners for Annexed Estates* (1973) 83 There is not only soil but climate to fight against, as it lies within half a mile of the water shire of Scotland. **1852** QUEEN VICTORIA *Jrnl.* 10 Sept. (1980) 74, I wore a white bonnet, a grey water silk, and..my plaid scarf over my shoulder. **1926** *Glasgow Herald* 1 Apr. 8/4 The copies are bound in magnificent leather volumes with sheets of water-silks

1953 C. DAY LEWIS *Italian Visit* iv. 47 Rosetted oxen move..The loose-kneed watersilk gait of Priestesses vowed to Love. **1906** *Engineering* 21 Dec. 834/2 In this combined grease-eliminator and make-up-water-softener there is, we understand, very little matter precipitated. **1930** L. MUNDAY *Mounty's Wife* xvii. 207 This has since been all changed owing to the installation of water softeners. **1974** *Encycl. Brit. Micropædia* X. 575/2 Water softeners usually consist of zeolite or an ion-exchange resin in a tank connected directly into the water system. **1974** *Trade Names Dict.* I. 85/1 Calgonite—Water softener. **1909** *Cent. Dict. Suppl.* 1435/3 Other types of water-softening machines employ the same general principles. **1929** A. R. MARTIN *Water Softening: Base-Exchange or Zeolite Process* 6 The base-exchange process for water softening should be considered in relation to the quality of the water to be softened. **1964** N. G. CLARK *Mod. Org. Chem.* xii. 249 It [*sc.* ethylene diamine] is used as a solvent, and as an intermediate for agricultural chemicals and 'water-softening' agents. **1979** *Jrnl. R. Soc. Arts* Dec. 59/1 Presumably water-solubility is the property which militates against the accumulation of sizeable deposits of potash (and soda) at, or near, the Earth's surface. **1922** Water-soluble [see *fat-soluble* adj. s.v. *FAT *sb.*[1] 6 c]. **1978** *N.Y. Times* 30 Mar. c7/4 Keep feeding lettuce with a weak application of water-soluble plant food. **1835** DICKENS *Let.* 18 Dec. (1965) I. 109 Our driver..ingeniously drove the party into a 'water-splash'. **1950** *N.Z. Jrnl. Agric.* Feb. 193/2 Delustred acetate rayons also water spot very readily; therefore,.. they should never be sprinkled with water and ironed directly, or glazed patches which are difficult to remove will be formed. **1964** *McCall's Sewing* viii. 118/1 Test a scrap of your fabric first to be sure that it does not water-spot. *Ibid.* 118/2 Press [silk] with..a piece of tissue paper next to the fabric to prevent water-spotting. **1944** U. P. HEDRICK *Fruits for Home Garden* vi. 51 Such shoots are called suckers, or water-sprouts, and numbers of them are deleterious. **1976** *Yankee* Apr. 137/2 Thinning out old watersprouts and suckers will promote better fruiting. **1913** E. F. BENSON *Thorley Weir* i. 26 There was something so completely satisfying and suitable in this rough river-dress that he would not have added any embellishment to it, nor have expunged a single water-stain or sun-bleach. **1940** *Chambers's Techn. Dict.* 901/2 *Water stain*.., a stain for wood, consisting of colouring matter dissolved in water. **1966** A. W. LEWIS *Gloss. Woodworking Terms* 114 *Water stain*, colouring matter or dyes dissolved in water and used to stain wood to the required shade. **1972** *Gloss. Terms Timber (B.S.I.)* 16 *Water stain*, discoloration caused by water coming into contact with the surface of the converted timber. **1912** *Chambers's Jrnl.* Apr. 220/2 The towns that cluster about these desert 'water-stops'. **1945** F. H. HUBBARD *Railroad Avenue* ii. 9 Consolidation engines have enormous tenders, but the water stops on Casey's division were far apart, and the water in his tank often would drop rather low if the engine were hauling a heavy train. **1951** *Archit. Rev.* CX. 345 (Advt.), The problem of making watertight expansion joints in flat concrete roofs can be solved by the use of Expandite rubber waterstops. These waterstops were developed in the first place for sealing joints in hydraulic structures and are widely used by hydraulic engineers. **1968** *Punch* 18 Dec. 873/1 My wife keeps urging me to diversify into coolants, sealants, waterstops, mastics and corrosion inhibitors. **1928** *Observer* 8 Apr. 6 A Budapest company has just been granted permission.. to place water-taxis for hire on the Danube. **1974** LD. MANCROFT *Chinaman in Bath* xxxix. 182 The water-taxis which could do so much to ease London's traffic problems. **1978** *New York* 3 Apr. 12/1 (Advt.), Water taxis across our incredibly blue waters to other islands. **1976** *National Observer* (U.S.) 3 July 13/6 (Advt.), Portable water toothpick for travel and the office. **1928** G. B. SHAW *Intelligent Woman's Guide Socialism* lxxxi. 415 The water torture of the Inquisition, in which the fluid was poured down the victims' throats until they were bloated to death. **1946** 'R. WEST' *Train of Powder* (1955) 8 Nuremberg..was also the water-torture, boredom falling drop by drop on the same spot on the soul. **1974** L. DEIGHTON *Spy Story* x. 105 What am I supposed to do, give them the water torture? **1976** *Times* 30 Aug. 8/5 Any individual.. is worn down by the Chinese water torture of daily stress. **1983** *Daily Tel.* 12 May 4/8 A former county sheriff..and three former deputies were charged in Houston with using water torture to extract confessions from prisoners. **1862** M. D. COLT *Went to Kansas* 197 The homeopathic physician was called to see her yesterday, so she is under his treatment, while I have the privilege of giving her all the water treatment I choose. **1966** 'G. BLACK' *You want to die, Johnny?* x. 180 That character who had apologised to himself for using the water treatment on Lee Wat. *a* **1877** KNIGHT *Dict. Mech.* III. 2646/2 When the fire passes *through* the tubes, they are properly flues. The term is, however, applied to pipes, whether water-tubes or fire-tubes, below a certain diameter. **1963** *Times* 24 May (Suppl.) p. iv/7 It had 64 Babcock and Wilcox watertube boilers and eight sets of turbines coupled to three-phase 33⅓-cycle alternators generating at 11,000 volts. **1971** B. SCHARF *Engin. & its Lang.* xiv. 204 In forced circulation boilers there is no need for the boiler drum to be above the water tubes. **1940** H. E. BAUGHMAN *Aviation Dict.* 187/2 *Water Tunnel*, a device similar to a wind tunnel, but using water as the fluid in which models are tested. **1965** *New Scientist* 5 Aug. 333/2 The commissioning..of the..outdoor test tank, and the impending completion of a flume (water tunnel), equips Hovercraft Development Limited with an entire range of skirt-testing apparatus. **1969** *Gloss. Aeronaut. & Astronaut. Terms (B.S.I.)* iv. 18 *Water tunnel*, an apparatus for producing a controlled stream of water or other liquid for fluid dynamic experiments. **1927** *Daily Express* 3 Feb. 5/2 'Have a drink,' said Roger.., but Red stayed his hand at the decanter. 'Water-wagon?' asked Roger, surprised. **1928** G. B. SHAW *Intelligent Woman's Guide Socialism* lxxix. 397 The vast majority of modern drinkers..do not miss the extra efficiency they would enjoy on the water wagon. **1934** *Bulletin* (Sydney) 23 May 41/2 Excuse Harrie drinking soft stuff... He's on the water-waggon at present. **1971** R. DENTRY *Encounter at Kharmel* xii. 216 What, no grog? Are you supposed to be on the water waggon, skip?

30. water-moccasin, for '(see MOCCASIN 3)' read: a venomous aquatic pit viper, *Agkistrodon piscivorus*, found in the southern United States; also, one of several harmless water snakes resembling *A. piscivorus*; = *cottonmouth* s.v. COTTON *sb.*[1] 10 and MOCCASIN 3; **water mongoose**, a dark brown mongoose, *Atilax paludinosus*, found in marshes and near rivers in central and southern Africa; **water monitor** = *Nile monitor* s.v. *NILE; **water-skater** = *water-strider* below; **water-strider** = *pond-skater* s.v. *POND *sb.* 4; **water tortoise**, an aquatic tortoise of the family Pelomedusidæ, native to Africa or South America, esp. *Pelomedusa subrufa*.

1821 T. NUTTALL *Jrnl. Trav. Arkansa* ix. 216 The other frequents water, and is called the water-mockasin. **1842, 1853** [see MOCCASIN 3]. **1884** 'MARK TWAIN' *Huck. Finn* xviii. 169 If you'll come down into de swamp I'll show you a whole stack of water-moccasins. **1931** [see *COPPERHEAD 1b]. **1976** *National Observer* (U.S.) 7 Aug. 8/5 Most of his patients have been bitten by one of the three pit vipers—rattlesnakes, copperhead moccasins, or water moccasins. **1919** F. W. FITZSIMONS *Nat. Hist. S. Afr.: Mammals* II. 26 The Water Mungoose is common throughout South Africa, and extends north as far as the Equator. **1971** *Stand. Encycl. S. Afr.* III. 246/2 The marsh or water mongoose..inhabits thick vegetation along river-banks and marshes over the greater part of Africa south of the Sahara. **1947** J. STEVENSON-HAMILTON *Wild Life S. Afr.* xxxv. 316 *Varanus niloticus*, or the water monitor, is found in reeds and rocks, close to the water's edge. **1974** M. HASTINGS *Dragon Island* v. 47 The water monitor can grow to more than seven feet. **1941** STEINBECK & RICKETTS *Sea of Cortez* xvi. 164 We had sat beside the little pool and watched..the water-skaters. **1977** *Country Life* 26 May 1394/3 All you had to do was brush aside the green weed and the spider-like 'water skaters' and bow down to slake your thirst. **1952** J. CLEGG *Freshwater Life Brit. Isles* 197 The Pond Skaters or Water Striders..are larger creatures. **1973** Water-strider [see **pond-skater*]. **1835** A. SMITH *Diary* 11 Mar. (1939) I. 309 Water has been getting more scarce every year. They eat the water tortoise. **1935** *Discovery* Nov. 330/2 Water tortoises..lay basking on the lily pads. **1974** *Stand. Encycl. S. Afr.* X. 527/1 In Southern Africa there are both land and water tortoises.

31. *water-fruit*; **water-ash**, substitute for def.: any of several North American ash trees, esp. *Fraxinus caroliniana*; also = *box elder* s.v. BOX *sb.*[1] 3 b; (earlier and later examples); **water-blossom** = *WATER-BLOOM 2; **water-elm**, substitute for def.: = *PLANER-TREE; (examples); **water hyacinth**, an aquatic herb, *Eichhornia crassipes*, of the family Pontederiaceæ, native to tropical America and bearing large blue flowers; **water maple**: see also *red maple* s.v. RED *a.* 17 d; (earlier and later examples).

1709 J. LAWSON *New Voy. Carolina* 93 The Water-Ash is brittle. **1819** E. DANA *Geogr. Sk.* 171 The soil is.. thickly covered with timber; such as various species of oak and water ash. **1958** G. A. PETRIDES *Field Guide to Trees & Shrubs* 35 Water Ash... A small tree of southern swamps. **1884** W. PHILLIPS in *Trans. Shropshire Archæol. Soc.* VII. 285 Though the appearance of the 'water-blossom' has often been observed and examined, very little is known of the causes from which it originates. **1906** *Rep. Brit. Assoc. Adv. Sci.* 759 *Microcystis roseo-persicinus*..formed a striking pink 'water blossom'. **1820** J. C. GILLELAND *Ohio & Mississippi Pilot* 257 Water elm in marshes, generally in the rear of rich bottoms. **1903** 'O. HENRY' in *McClure's Mag.* Dec. 144/1 [I] noticed a rabbit-hawk sitting on a dead limb in a water-elm. **1930, 1976** Water-elm [see *PLANER-TREE]. **1930** T. S. ELIOT tr. *St.-J. Perse's Anabasis* 65 Eaters of insects, of water fruits. **1897** H. J. WEBBER in *Bull. U.S. Dept. Agric.: Bot.* XVIII. 13 The water hyacinth is becoming a serious menace to navigation. **1927** E. THOMPSON *Indian Day* xiv. 111 A colony of water hyacinth had rooted itself..where deep water still remained. **1979** *London Rev. Bks.* 25 Oct. 5/3 In this closing scene, the water hyacinths proceed towards the sea. **1803** A. ELLICOTT *Jrnl.* x. 284 Water maple..is met with as high as the Wabash. **1912** I. S. COBB *Back Home* 137 [The] walk..[was] shaded well all the way by water maples.

water, *v.* Add: **I. 3. a.** Also const. *in*.

1885 G. NICHOLSON *Illustr. Dict. Gardening* I. 217/2 The plants [*sc.* Brussels sprouts]..should be watered-in when planted. **1958** *Listener* 21 Aug. 275/2 By taking off well-ripened shoots..and placing three shoots in three-inch pots..watering them in..and keeping them perfectly airtight and shaded, I have got a good percentage to root. **1982** *Times* 22 May 9/4, I always like to water-in my lawn fertilizers.

II. 16. Delete † *Obs.* and add later example.

1966 M. CATTO *Bird on Wing* i. 15 He went into the lav. The Major could hear him watering.

water-bag. Add: **1. b.** *Austral.* A bag, freq. of canvas, used for carrying water, esp. on journeys in dry areas.

1892 [see *SHIRALEE]. **1903** 'T. COLLINS' *Such is Life* (1937) i. 47 Helping myself to a drink from the water-bag under the rear of Thompson's wagon. **1944** *Living off Land* iii. 57 Don't carry a reserve supply of water in a waterbag; evaporation alone can dry it all up in a few hours in hot weather. **1969** P. A. SMITH *Folklore Austral. Railwaymen* 19 The water was no good. We'd bring it out in boiling buckets and then fill the gang's water bags.

water-balance. Add: **4.** (Written *water balance*.) Equilibrium between water intake and water loss.

1922 *Physiol. Rev.* II. 158 In diabetes insipidus, the water balance is set at a level higher than normal. **1957** G. E. HUTCHINSON *Treat. Limnol.* I. iv. 228 (*heading*) Water balance of the hydrosphere and of continental surfaces. **1974** PASSMORE & ROBSON *Compan. Med. Stud.* III. xlix. 9/2 The driving force of thirst normally ensures the minimum daily intake of water needed to maintain water balance.

water-bed. Add: **3. a.** (Earlier examples.)

1844 A. N. BROWN *Let.* 28 Nov. in *Church Missionary Soc. Archives: New Zealand Mission* (MS.), Our kind Bishop had him removed to his House the day he was taken ill and planned a water bed to place him on. *c* **1848** J. R. PLANCHÉ *Extravaganzas* (1879) III. 294 And something like the sort o' bed That Dr Arnolt designates a water bed.

b. A plastic mattress filled with water, designed as an alternative to a conventional bed. Also *fig.*

1970 *Time* 7 Sept. 42 His efforts to improve it led him [*sc.* C. P. Hall] to a much splashier creation... It is the water bed. **1971** *Guardian* 5 Oct. 10/1 The water-bed is simply an ultra-sonically welded vinyl bag filled..with around 150 gallons of water, and contained in a wooden frame which rests on the floor or can be raised to conventional bed height. **1977** *Time Out* 21 Jan. 43 (*caption*) 'The Odyssey', a soft, billowing water-bed of a piece that's performed at the Albert Hall on Tuesday by no less than ten keyboard players plus tympani and vocals. **1978** J. IRVING *World according to Garp* xi. 205 Garp slumps back on the water bed,..the bed rolls like a small sea. **1984** *Listener* 19 July 17/4 (Advt.), For couples who want to get away together..4 poster beds, water beds—with whirlpool bath en suite.

water-bloom (wǭ·təɹblūm). [f. WATER *sb.* + BLOOM *sb.*[1]] **1.** A flower living in water. *poet. rare.*

1820 [in Dict. s.v. WATER *sb.* 31].

2. [tr. G. *wasserblüte*.] (See quot. 1957.) Cf. *RED-WATER 4.

1927, 1948 [see *BREAKING *vbl. sb.* 2 c]. **1957** *New Biol.* XXIII. 86 'Water-bloom'..describes the discoloration of the waters of ponds and lakes, sometimes slow-flowing rivers and occasionally vast areas of the sea by a superabundance of free-floating, microscopic, plant and, in rare cases, animal life. **1963** [see *PERIDINIAN].

waterbok. (Earlier example.)

1835 A. SMITH *Diary* 20 Sept. (1940) II. 240 The flesh coarse-grained and exactly like the flesh of the waterbok.

water bottle. Add: **4.** = *hot-water bottle* s.v. *HOT WATER 1 b.

1840 J. ROMILLY *Diary* 31 Dec. (1967) 207 Lodge burnt his leg with a water bottle & laid up. **1905** S. WEYMAN *Starvecrow Farm* xxv. 230 That was not the day of bedroom fires, or rubber water-bottles. **1929** W. FAULKNER *Sound & Fury* 334 Dilsey reached the top of the stairs and took the water bottle. 'I'll fix hit in a minute... I gwine build de fire myself.'

water-buck. (Earlier example.)

1839 W. C. HARRIS *Wild Sports S. Afr.* 186 A water buck..is about the size of an ass and of somewhat browner colour.

Waterbury (wǭ·təɹbŭri). The name of a town in Connecticut, U.S.A., used *attrib.* or *absol.* to designate a low-priced watch or clock of a type manufactured there.

1884 *Official Gaz.* (U.S. Patent Office) 6 May 526/1 *The Waterbury Watch Company*, Waterbury, Conn. Application filed July 21, 1883. 'The words "*The Waterbury*".' **1887** KIPLING *Plain Tales from Hills* (1888) 73 Platte,.. being poor, had a Waterbury watch and a plain leather guard. **1890** B. HALL *Turnover Club* 16 The Reporter drew from its resting place the Waterbury chronometer. **1893** SOMERVILLE & 'ROSS' *Vine Country* vi. 105 My cousin with some trouble disinterred the Waterbury. **1908** *Sears, Roebuck Catal.* 348/1 Our new eight-day Waterbury Clock. *c* **1909** D. H. LAWRENCE *Collier's Friday Night* (1934) iii. 76 He..takes a Waterbury watch with a brass chain from the wall beside the book-case. **1920** F. E. GREEN *Hist. Eng. Agricultural Labourer 1870–1920* iv. 93 He turned up..on Sunday with a fashionable billycock, a walking stick and a Waterbury. **1939** JOYCE *Finnegans Wake* 290 O Shee who then (4.32 M.P.,..according to all three doctors waterburies) [etc.]. **1963** C. MACKENZIE *My Life & Times* II. 42, I kept surreptitiously looking at my ten-and-sixpenny Waterbury watch.

wa·ter-cart, *v.* [f. the sb.] **a.** *trans.* To provide with water-carts. **b.** *intr.* To weep. *slang.*

1851 DICKENS in *Househ. Words* 2 Aug. 433/1 The great metropolis is..so much more water-carted..than it usually is. **1920** W. DE MORGAN *Old Man's Youth* (1921) xxix. 282 She makes believe she knew Gromp, and I know she didn't. She'd watercart.

water-colour. Add: **5.** *water-colour drawing* (earlier example), *exhibition* (earlier example).

1811 JANE AUSTEN *Let.* 25 Apr. (1952) 275 Henry has been to the Watercolour Exhibition, which open'd on Monday. **1824** J. ARROWSMITH *Let.* 19 June in J. Constable *Corr.* (1966) IV. 184, I by this post write to give directions to be sent to me with my watercolour drawings.

Hence **wa·ter-colour** *v. intr.*, to paint with water-colours; **water-colouring** *ppl. a.*

1855 D. G. ROSSETTI *Let.* 23 Jan. (1965) I. 239, I have been water-colouring again, somewhat against the grain. **1928** R. FRY *Let.* 4 Apr. (1972) II. 621 It was a dismal audience of retired colonel water-colourists and their water-colouring daughters. **1935** A. HUXLEY *Lett.* (1969) 393 Are you oiling or water-colouring or gouaching?

water-coloured, *a.* **2.** (Later example.)

a **1941** V. WOOLF *Haunted House* (1943) 91 Water-coloured rings set in pearls.

water-cracker. Add: **3.** (Examples.) Now *Obs.*

1825 *Missouri Intelligencer* 4 June 3/4 Ward and Parker have just received for sale at their grocery and liquor store molasses, water crackers, [etc.]. **1899** S. T. RORER *Bread & Bread Making* 57 It [*sc.* the dough] may be rolled out into very thin sheets, cut into square crackers, pricked with a fork, baked in a moderate oven, producing what are called Virginia biscuits or water crackers.

water-cress. Add: **3.** *water-cress sandwich, soup.*

1911 W. J. LOCKE *Glory of Clementina Wing* xxi. 310 He picked up from a plate a little three-cornered water-cress sandwich. **1978** P. MOYES *Who is Simon Warwick?* xii. 143 Tea had been laid out..silver teapot and milk jug, wafer-thin watercress sandwiches. **1923** W. G. R. FRANCILLON *Good Cookery* (ed. 2) 102 (*heading*) Watercress soup. **1980** *Redbook* Oct. 234/1, I am intent on the food—water cress soup and a salad of butter lettuce.

water-cure. Add: **2.** In the Philippines, torture by forcing a person to drink large quantities of water in a short time.

1902 'MARK TWAIN' in *N. Amer. Rev.* May 623 The torturing of Filipinos by the awful 'water-cure'..to make them confess. **1976** *New Yorker* 3 May 96/2 The Civil Liberties Union of the Philippines..accused the police and the military of employing 'water cures' and electric shocks as well as beatings with clubs.

water-dog. 3. a. For '† *Obs.* (? exc. dial.)' read '*Obs.* exc. *W. Afr.*' (Later example.)

1960 M. MACDONALD in G. Maxwell *Ring of Bright Water* xii. 164 A Senior African joined the group. 'This be the piccin of water-dog,' he intoned.

watered, *ppl. a.* Add: **4. a.** Also, *watered-down* (now the usual form). Cf. sense 4 b in Dict.

1976 P. CAVE *High Flying Birds* iv. 51 He matched Lorna glass for glass, chucking the stuff down as though it was watered-down Coke. **1978** A. GILCHRIST *Cod Wars* ii. 7 A watered-down beer, unworthy of the name, is legally sold.

b. (Earlier example.)

1889 'MARK TWAIN' *Connecticut Yankee* iv. 55 Everybody took in all this bosh..and never smiled or seemed to notice that there was any discrepancy between these watered statistics and me.

water-engine. 1. (Later example.)

1864 TROLLOPE *Small Ho. at Allington* II. xi. 113 Calling for the police when there is a row in the house is like summoning the water-engines when the soot is on fire in the kitchen chimney.

water-engineer. Restrict † to sense in Dict. and add: **b.** An official charged with the management of the water-supply of a district.

1937 *Discovery* Feb. 60/2 Mr Walters, a leading water engineer.

waterer. 6. Read: A container used for supplying water to animals or plants.

1891 [in Dict.]. **1958** *Times Lit. Suppl.* 21 Nov. p. xii/5 Would you make a plant waterer or a..pea-shooter? **1960** *Farmer & Stockbreeder* 19 Jan. (Suppl.) 31/3 With deep litter, the use of a droppings pit and waterers over it [*sc.* in a poultry house] are a further help. **1970** *Jrnl. Gen. Psychol.* Oct. 240 The food cups and waterers [for the rats] were replaced with clean containers. **1981** *Farmstead Mag.* Winter 85 (Advt.), Includes choosing proper breeds [of rabbits], housing, feeders, waterers, feeding and diets, rabbit management, disease prevention, marketing.

waterfall. Add: **5.** (Earlier and later examples.)

1848 J. H. NEWMAN *Let.* 3 Sept. (1962) XII. 268 He.. looks a striking man in his Jesuit dress, though what his cut may be with a French coat and satin waterfall I can't tell. **1914** *Lit. Digest* 22 Aug. 328/2 A waterfall..is, in addition, a scarf or necktie with long drooping ends, or a chignon with pendent curls.

6. (Earlier U.S. example.)

1859 C. C. RICHARDS *Village Life in Amer.* (1912) 119, I wore my new waterfall for the first time.

7. In a woman's garment, a fall of material or attached decoration (orig. with reference to bustles).

1886 *Girl's Own Paper* 27 Feb. 339/3 Patterns already issued:—polonaise with waterfall back. **1925** *Dialect Notes* V. 346 *Waterfall, n.*, a form of bustle for hanging the dress round the body. **1950** *New Yorker* 25 Mar. 84/2 What she calls 'waterfalls'—cascades of bustle drapery—add interest to the backs of all types of clothes. **1970** *Trafford Spring & Summer Catal.* 4 The dress is fully lined, and a waterfall frill, edged with dainty braid, tops a gently flared skirt. **1981** *Times* 30 July 2/8 The Queen's coat and dress, with its waterfall of pleated crêpe de chine.

8. *Ceramics.* A stream of glaze material used for application to the upper surfaces of tiles (see quot. 1964). Freq. *attrib.*

1961 *Brit. Ceramics Abstracts* 197 Methods of glazing wall-tiles... The second part [deals] with the 'waterfall' process of glazing. **1964** A. E. DODD *Dict. Ceramics* 308 *Waterfall process*, a method for the application of glaze materials to a ceramic body by mechanically conveying the ware through a continuously flowing (recirculated) vertical stream of the glaze suspension. The process is used in the glazing of wall tiles. **1967** M. CHANDLER *Ceramics in Mod. World* iii. 99 (*caption*) 'Waterfalling' of wall tiles. Tiles to be glazed on one side only are conveyed mechanically..through a waterfall of glaze slip that covers only their upper surfaces and edges. **1973** C. W. PARMELEE *Ceramic Glazes* (ed. 3) v. 149 Each color is dried between applications and the final station applies the glaze by waterfall. *Ibid.*, Tiles are placed on a conveyor..and passed under the waterfall machine.

Hence as *v. intr.*, to tumble or cascade in the manner of a waterfall; as *v. trans.*, to cover with a stream of running liquid; so **wa·ter-falled** *a.*, having a waterfall (also *fig.*); **wa·terfalling** *ppl. a.* and *vbl. sb.*

1944 G. BARKER *Eros in Dogma* 34 Who has not seen, over The waterfalling hair at the shoulder of Life, Death from his own face staring out of a glass? **1957** J. FRAME *Owls do Cry* (1958) xxxviii. 171 Gaping idiot mouth waterfalled with slobber. **1967** [see *WATERFALL 8]. **1967** M. CHANDLER *Ceramics in Mod. World* iii. 99 Spray that misses the articles goes onto a wall at the back of the booth that is continuously 'waterfalled'. **1968** B. HINES *Kestrel for Knave* 128 One boy posed Eros-like, and allowed a jet of water to play into his palm and waterfall out on to the tiles of the drying area. **1976** *Shooting Times & Country Mag.* 16–22 Dec. 26/2 The exotic venue was.. the 1,036 acre Wild Life Park at Cricket St Thomas with the waterfalled Bedpool Lakes reposing in the partly wooded valley below. **1980** J. GARDNER *Garden of Weapons* II. iv. 154 The rope ladder..dislodged, waterfalling on to the balcony.

water-flood. Add: **4.** *Oil Industry.* Water-flooding; an instance of this. Freq. *attrib.*

1928 *Trans. AIME Petroleum Div.* LXXVII. 334, I was interested..to see what the pumping problem would be if it was later decided to mine a water-flood property. **1946** *Producers Monthly* Dec. 20/1 Throughout the water flood, measurements were taken of the pressure distribution. **1977** *Offshore Engineer* Apr. 20/1 (Advt.), In waterflood and other really tough uses, the McEvoy Super Mudwonder proves better than valves B or C, or any other competitor.

wa·ter-flooding, *vbl. sb. Oil Industry.* [f. WATER *sb.* + FLOODING *vbl. sb.*] The injection of water into a reservoir rock through wells drilled for the purpose, in order to force oil into neighbouring production wells.

1928 *Trans. AIME Petroleum Div.* LXXVII. 391 Of some value in water-flooding are water analyses and.. data of sand conditions. **1938** H. S. GIBSON in A. E. Dunstan et al. *Sci. of Petroleum* I. xi. 535/1 In the Bradford field of Pennsylvania..water flooding as a production practice was legalized in 1921. **1940** [see *SECONDARY *a.* 5 k]. **1961** *Economist* 2 Dec. 956/1 The..water..will be injected around the flanks of the [oil] field, and not in the centre as in most 'waterflooding' projects. **1982** D. R. SKINNER *Introd. Petroleum Production* II. i. 2 The first methods of secondary recovery utilized gas injection and a primitive form of water injection—saltwater disposal. Later, water injection or secondary waterflooding became the dominant form of secondary recovery.

Waterford (wǭ·təɹfǭɹd). The name of a city in the south-east of Ireland, used *attrib.* and *absol.* to designate glassware first manufactured there in the eighteenth and nineteenth centuries, esp. drinking glasses and chandeliers.

1783 *Dublin Evening Post* 4 Oct. in M. S. D. Westropp *Irish Glass* (1920) iii. 69 Waterford Glass House. George and William Penrose having established an extensive glass manufacture in this city, their friends and the public may be supplied with all kinds of plain and cut flint glass. **1852** J. F. MAGUIRE *Industrial Movement in Ireland* 102 The Waterford glass-house was in active work up to the year 1845, and had earned the highest reputation for its glass throughout the country. **1898** *Sale Catal.* in *Country Life* (1974) 26 Sept. 854/2 Waterford and Cork cut glass. **1917** W. J. LOCKE *Red Planet* xxi. 269 A precious old Waterford claret jug. **1936** M. KENNEDY *Together & Apart* iii. 211 Come and look at this glass... It's Waterford. **1940** J. CARY *Charley is my Darling* lx. 324 A magnificent seven-branch candlestick of Waterford glass. **1964** MRS. L. B. JOHNSON *White House Diary* 25 Apr. (1970) 125, I described the Savonnerie rug, the Waterford chandelier. **1977** *Daily Colonist* (Victoria, B.C.) 26 Oct. 7/3 (Advt.), Waterford crystal, a treasure hunter's delight, now on..sale. **1978** J. CARROLL *Mortal Friends* I. ii. 22 The few pieces of Waterford in the village had been loaned and were out, filled with jams and preserves.

waterfowl. Add: Also **waterfowler** *U.S.*, a hunter of wildfowl; **waterfowling** *U.S.*, the hunting of wildfowl.

1976 *National Observer* (U.S.) 27 Nov. 14/4 The appeal of hunting for even old men was best described by a 68-year-old friend and lifelong waterfowler who shared a blind with me at Izembek several years ago. **1980** *Outdoor Life* (U.S.) (Northeast ed.) Oct. 80/3 Top-quality decoys are more important in puddler hunting than any other kind of waterfowling.

Wa·tergate[3]. The name of a building in Washington D.C., containing the national headquarters of the Democratic Party, which was burgled on 17 June 1972 by persons connected with the Republican administration; used *attrib.* and *absol.* with reference to this event and the circumstances leading to the resignation of President R. M. Nixon in 1974. Also *transf.*, denoting a political or commercial scandal on a large scale.

The element *-gate* has been used in recent times, preceded by the name of a relevant person, etc., to denote an actual or alleged scandal usually of a political or commercial kind. Examples are to be found in *The Second Barnhart Dict. of New English* (1980): *Koreagate* (1976), *Lancegate* (1977), *Muldergate* (1979), etc. The formative element is also discussed in *American Speech* (1978) LIII. 215–17.

1972 *New Republic* 19 Aug. 4 The very name, 'the Watergate caper', tells how funny it is. **1972** *Time* 28 Aug. 20 By coming down hard on Mitchell, the Democrats hope they can make Watergate a devasting—and durable —campaign issue. **1973** *Freedom* 2 June 1/2 Anything the Americans do we can do better. We have produced our own miniature Watergate plus two new magic ingredients —sex and drugs. **1973** 'H. PENTECOST' *Beautiful Dead* (1975) III. i. 141 That's the way it goes in any kind of criminal conspiracy to hide the truth. It's like a Watergate. More and more people become involved, more and more crimes are committed to hide an initial truth. **1974** E. AMBLER *Dr. Frigo* II. 101 A Central American Watergate you want now? **1974** *Times* 27 July 15/7 Your reviewer..sums up the current Chichester production of *Oedipus Tyrannus* as a 'Theban Watergate drama'... Watergate conspiracies were..frequent in the history of classical Greece. **1976** *Washington Post* 26 Jan. A 2/6 'The news has been inundated by a financial Watergate of leaked disclosures of troubled banks and bank holding companies,' said Reuss. **1977** M. EDELMAN *Political Lang.* viii. 150 Watergate, the Pentagon Papers, and revelations of the deception of Congress by officials of the executive branch and by intelligence agencies have made us sensitive to lying in high places. **1980** J. MELVILLE *Chrysanthemum Chain* 175 What about a Watergate style investigation? **1982** M. RUSSELL *Rainblast* xiii. 125 The damage is done. Fleet Street gets the signal and..everyone's sniffing the stench of another Watergate.

Hence as *v. intr.* (and *trans.*), to take part in political conspiracy or in activities directly or indirectly associated with Watergate, as the use of hidden listening devices, concealment of corrupt activities, suppression of evidence, etc. Hence **Wa·tergater**; **Wa·tergating** *vbl. sb.*

1973 *Birmingham* (Alabama) *Post-Herald* 28 Apr. A-4/3 In the political machinery of the future we may hear of a political party 'Watergating' another party. *Ibid.* A-4/4 They will be asking the voters to remember 'Watergate', for the GOP's might be Watergating again. **1973** *Black Panther* 12 May 2/3 The issue is not how high *up* Watergate goes. The issue is how far *down* watergating goes. The bug on our telephones, and yours, is the issue. *Ibid.* 21 July 2/3 (*heading*) David Hilliard, victim of Watergaters. **1974** N. FREELING *Dressing of Diamond* 20 'And if you try to keep something from the press they're apt to be vengeful.' 'You mean Watergating.' 'Right... It's easily concluded that an effort is being made to cover something, for personal or political considerations.' **1975** G. SEYMOUR *Harry's Game* vi. 91 Not much eavesdropping in here. Need to Watergate the place. **1976** *Publishers Weekly* 1 Mar. 9/1 'When I [*sc.* Archibald Cox] was Watergating, for instance' (this appears to be a coinage of his own, but he rolls it off the tongue as if it were common usage). **1976** *Times Lit. Suppl.* 2 July 815/2 His followers perform experiments that are sadistic, pointless, and repetitive, and are given to watergating the evidence ('Not that there was anything *wrong* on those tapes'). **1979** *N.Y. Rev. Bks.* 8 Feb. 10/4 Perhaps what Jerry Ford should have done on the day he pardoned Nixon was to pardon all the Watergaters and all the Vietnam resisters in one controversially magnanimous act.

water-hole. Add: **3.** *Astr.* The part of the radio spectrum between 1420 MHz (at which hydrogen atoms radiate) and about 1660 MHz (at which hydroxyl radicals radiate). [So called because hydrogen and hydroxyl are the constituents of water.]

1976 *Sci. News* 28 Feb. 132 Between the hydrogen and hydroxyl (OH) bands..lies the 'water hole',..which.. offers a frequency less drowned in deep-space static. **1979** *Daily Tel.* 10 Nov. 11/1 Scientists have argued that..the radio water-hole will be the most logical communications medium for intelligences across the universe.

water-horse. Add: **3.** *Newfoundland.* (Earlier and later examples.)

1777 G. CARTWRIGHT *Jrnl.* 5 Aug. (1792) II. 242 Fourteen quintals of fish were washed, the water-horse was carried out, and the green fish were spread. **1966** A. R. SCAMMELL *My Newfoundland* 49 The Blanchard women were spreading out green, waterhorse fish.

Waterhouse–Friderichsen (wǭ·təɹhaus,-frī·dĕriksĕn). *Path.* The names of Rupert *Waterhouse* (1873–1958), English physician, and Carl *Friderichsen* (b. 1886), Danish physician, used *attrib.* to designate a fulminating meningococcal septicæmia with hæmorrhagic destruction of the adrenal cortex that occurs chiefly in children and is fatal within hours if not promptly treated.

1936 *Jrnl. Amer. Med. Assoc.* 16 May 1715/2 The disease now known as the Waterhouse-Friderichsen syndrome was recognised as an entity by Little in 1901. **1974** J. D. MAYNARD in R. M. Kirk et al. *Surgery* xii. 256 Precipitating factors [of adrenal insufficiency] are septicaemia, meningococcal meningitis (Waterhouse Friderichsen syndrome..), severe abdominal trauma, [etc.].

watering, *vbl. sb.* Add: **III. 22. watering hole,** (*a*) = WATER-HOLE 1 a; (*b*) *slang*, a place where refreshment (esp. alcohol) is available, as a bar, hotel, etc. (see also quot. 1972).

1882 Watering hole [in Dict., sense 21]. **1972** B. RODGERS *Queens' Vernacular* 209 *Watering hole* [*spot*] (Brit gay sl), neighborhood cruising locale, usually a park grounds or a bar. **1975** *Telegraph* (Brisbane) 30 Oct. 35/5 The Jindalee Hotel..is a great watering hole. **1981** J. BARNETT *Firing Squad* viii. 77 'I always visit a watering hole on the way home... Cheers.' Palmer took a slow drink. **1984** *Gainesville* (Florida) *Sun* 30 Mar. 1D/5 In a simpler time, players and fans mingled at local watering holes, drinking beers together and becoming friends.

wateringly, *adv.* (Later example.)

1967 P. WHITE in *Coast to Coast 1965–6* 227 She.. gasped, looking wateringly at the bundle.

watering-place. 3. b. (Earlier example.)

1817 JANE AUSTEN *Sanditon* vi, in *Minor Works* (1954) 389 The very quietest part of a Watering-place Day.

waterish, *a.* **10.** Delete † *Obs.* and add later U.S. example.

1832 J. P. KENNEDY *Swallow Barn* I. xxvii. 283 His constitution resembles that waterish gravelly soil you see sometimes around a spring.

waterless, *a.* Add: **c.** Of processes, apparatus, etc.: employing or needing no water.

1930 *Engineering* 6 June 742/3 One advantage of the waterless gasholder was found to be the absence of a succession of sudden increases and decreases in pressure. **1951** *Good Housek. Home Encycl.* 313/2 The waterless cooker consists of a fairly deep aluminium pan, which is divided into compartments. **1971** *British Printer* Aug. 80/2 The need to maintain a critical ink-water balance.. may be overcome by the development of waterless offset-litho.

waterlog, *v.* Add: **1.** Also *intr.* for *pass.*

1780 in Laughton & Heddon *Great Storms* (1927) v. 109 The ship began to waterlog.

2. Also *intr.* for *pass.*

1950 *N.Z. Jrnl. Agric.* Mar. 224/1 Heavy soils which waterlog in wet spells should be avoided, as the plants will not tolerate 'wet feet' for long.

Waterloo. Add: *to meet one's Waterloo* (examples).

1859 W. PHILLIPS *Lesson of Hour* 11 Every man meets his Waterloo at last. **1905** A. CONAN DOYLE *Return of Sherlock Holmes* 356 We have not yet met our Waterloo, Watson, but this is our Marengo. **1961** C. MCCULLERS *Clock without Hands* iii. 67, I felt right then and there I had met my Waterloo. **1982** *Times* 20 Oct. 19/6 The main fount of the economic nightmare now engulfing the world has met with its Waterloo.

c. Waterloo ball, a frivolous entertainment preceding a serious occurrence (with reference to a ball given in Brussels by the Duchess of Richmond on the eve of the Battle of Waterloo); **Waterloo church** (see quots. 1938, 1961); **Waterloo cracker** (earlier example); **Waterloo Cup**, in Coursing, a race held annually at Altcar, near Liverpool.

1954 P. TOYNBEE *Friends Apart* xi. 132, I now see these dances as a succession of Waterloo Balls, a rapturously gay company, rapturously doomed. **1968** *Listener* 11 July 44/3 Is it a Waterloo Ball we are invited to watch, a permanent party on the eve of great events? **1938** B. F. L. CLARKE *Church Builders of 19th C.* ii. 24 In 1828 the Church Building Society was incorporated by Parliament. The same Act abolished Church Briefs... The churches built as the result of the Act were officially known as the Commissioners' Churches; the general public called them Waterloo Churches. **1961** M. H. PORT *Six Hundred New Churches* ii. 28 The popular misnomer, 'Waterloo Churches', became applied to those built with the aid of the first parliamentary grant. [*Note*] The term is sometimes applied only to the four churches built in the old parish of Lambeth (St John, Waterloo Bridge Road; St Luke, Norwood; St Mark, Kennington; St Matthew, Brixton). This may originate from St John's having been commonly termed..'Waterloo Church', which refers, of course, to its site. **1972** N. PEVSNER *Some Archit. Writers 19th c.* v. 29 This was the heyday of the Greek Revival and the so-called Waterloo churches. **1833** CARLYLE *Sart. Res.* in *Fraser's Mag.* Nov. 583/2 What vehicle of that sort have we, except *Fraser's Magazine*? A vehicle all strewed (figuratively speaking) with the maddest Waterloo-Crackers. **1836** *Bell's Life* 20 Mar. 3/5 *Waterloo Cup*, eight dogs...Deciding course.— Melanie beat Unicus, and won the Cup. **1898** [see STAKE *sb.*[2] 3]. **1939** R. GODDEN *Black Narcissus* xx. 182 Call in your mad dog, Clo. Is it the Waterloo Cup you've entered it for? **1983** *Times* 3 Mar. 3/2 More than 300 demonstrators yesterday disrupted the second day of the Waterloo Cup hare coursing meeting in Lydiate Field, Lancashire.

wa·terly, *adv. rare.* [f. WATER *sb.* + -LY[2].] After the manner of water.

1918 D. H. LAWRENCE *New Poems* 62 You undine-clear and pearly, soullessly cool And waterly The pool for my limbs to fathom.

water-mark, watermark, *sb.* Add: **6.** *watermark disease*, a disease of the cricket-bat willow, *Salix alba* var. *cærulea*, caused by the bacterium *Erwinia salicis* and producing dying-back in the crown of the tree and stains in the wood.

1924 W. R. DAY in *Oxf. Forestry Mem.* III (*title*) The watermark disease of the cricket-bat willow. **1950** *Q. Jrnl. Forestry* XLIV. 106 Watermark disease of the cricket bat willow..spreads rapidly both upwards and downwards in the wood vessels. **1976** *Eastern Even. News* (Norwich) 9 Dec. 15/4 This was to cut the cost of services for an inspector on 'watermark disease' of willow trees.

water-miller. Delete ? *Obs.* and add later examples.

1909 E. DANIELS in *Cambr. Mod. Hist.* VI. xx. 714 Arnold, a water-miller in the neighbourhood of Züllichau in the Neumark. **1977** *New Scientist* 14 Apr. 84/1 Water millers in the 1970s can reasonably claim to be the oppressed and persecuted class among a wide range of water users.

Water Pik (wǭ·təɹ pik). Also **Water-Pik, water-pick.** A device for cleaning the teeth by directing a jet of water at them.

Water Pik is a proprietary name in the U.S.

1963 *Official Gaz.* (U.S. Patent Office) 15 Oct. TM 109 Aqua-Tec Corporation, Fort Collins, Colo. Filed Dec. 14, 1962. *Water Pik*. For readily portable oral hygiene appliance utilizing a pulsed jet of water for massaging the gums and cleaning spaces adjacent to the teeth. First use Oct. 14, 1961. **1971** *Consumer Rep.* June 387/2 Many people incorrectly use the term *Water Pik* as a generic name for dental irrigators, but there are a number of other brands on the market. Our project included 11 electric irrigators (of which three were *Water Pik*). **1971** *Better Homes & Gardens* Nov. 136/2 To use the oral irrigating device, place water in the reservoir, hold the water pick in the mouth and turn the pick on. **1978** G. GREENE *Human Factor* I. iii. 44 He turned on his electric water-pick... 'Amusing little gadget, that of yours... I suppose it really is better than an ordinary toothbrush?' **1978** J. UPDIKE *Coup* (1979) ii. 78 Her new possessions—..mechanical beauty aids, a hair-blower and a Water-Pik—were overwhelming the little pisé-walled room.

Hence **wa·ter-pick** *v. trans.*

1976 P. DICKSON *Electronic Battlefield* i. 10 Frank was brushing and water-picking his teeth.

water-pit. Add: **c.** = WATER-HOLE 1 b. *poet.*

1920 BLUNDEN *Waggoner* 19 The great pike lies.. Watching the waterpit sheer-shelving dark.

water-plane. Add: **3.** (Earlier example.) Now *Obs.* Cf. *SEAPLANE.

1912 *Q. Rev.* July 248 With this [float] Curtiss' water-plane was the first to quit the sea under its own power.

waterproofer. Add: **2.** = WATERPROOFING *vbl. sb.* 2 b.

1909 WEBSTER, *Water-proofer*... Also, a waterproofing material, as for roofs. **1923** E. G. BLAKE *Damp Walls* xii. 173 Many of these compounds completely substantiate the claims which are made for them as integral water-proofers by the manufacturers. **1951** *Archit. Rev.* CX. 402/3 To provide a waterproof backing for external tiling Tretol liquid cement waterproofer has been used. **1974** I. H. SEELEY *Building Technol.* iv. 60/1 Colourless waterproofers can make a wall surface water-repellent and less porous, without much change in the appearance.

wa·terproofness. [f. WATERPROOF *a.* + -NESS.] The state or condition of being waterproof.

1934 in WEBSTER. **1943** P. I. SMITH *Synthetic Adhesives* iv. 48 Straight ethyl cellulose cements and ethyl modified adhesives are now being employed... They offer advantages over certain other cellulose thermo-plastic adhesives by reason of their great toughness, waterproofness and heat sealing properties. **1952** *Times* 14 Mar. 6/1 It would then become an offence to describe wrongly such qualities as 'waterproofness' or resistance to fading. **1968** G. MAXWELL *Raven seek thy Brother* xiii.163 The contact of human hand can remove the 'water-proofness' of a seabird's plumage. **1978** *Detroit Free Press* 16 Apr. E 8/3 The old style boot greases were developed during World War II for government use. Waterproofness of combat boots was the main objective.

water-rat. Add: **3.** *Grand Order of Water Rats*, a philanthropic show-business society (see quot. 1951). Also *ellipt.*, as *Water Rats*.

1910 *Era Annual* 47 July..Memorabilia of Theatrical Events... Water Rats Motor Run to Brighton, 1909. **1935** A. HADDON *Story of Music Hall* xix. 90 Elvin.. founded the Grand Order of Water Rats, which led to the institution of nearly every existing artistes' organisation. **1951** *Oxf. Compan. Theatre* 835/1 *Water Rats, the Grand Order of*, a British association of members of the music-hall profession which originated in 1889 from an up-river (Thames) party to celebrate the successes of the Water Rat, a racing pony. **1956** *Golden Jubilee Bk. Show Business* (Variety Artistes Fed.) 17/3 There is the Grand Order of Water Rats..also the 'sister' section of the Water Rats—the Lady Ratlings. **1977** T. HEALD *Just Desserts* iii. 56 He..found a large lavatory decked out in Victorian tiles, original newspaper cartoons ..and a document certifying that Escoffier Savarin Smith was a member of the Grand Order of Water Rats.

wa·ter-rot, *v. U.S.* [See ROT *v.* 4 c; cf. WATER-RET *v.*] = WATER-RET *v.* So **wa·ter-rotted** *ppl. a.*, **wa·ter-rotting** *vbl. sb.*; hence **wa·ter-rot** *sb.*

1705 *Boston News-Let.* 30 July 2/2 For Hemp Water rotted,..per Ton,..Six Pounds. **1759** *Newport* (Rhode Island) *Mercury* 8 May 1/2 Some prefer Water-rotting [for flax]. **1838** H. COLMAN *1st Rep. Agric. Mass.* (Mass. Agric. Survey) 72 The superintendent states that the water-rotted flax of Scotland is much superior to ours, which is dew-rotted. **1843** *Amer. Pioneer* II. 450 The manner of making ropes of linn bark, was to cut the bark in strips..and water-rot it in the same manner as rotting flax or hemp. **1848** *Rep. U.S. Comm. Patents 1847* 246 It is of great importance in raising flax..that the water rot and dew rot must be used together to produce a fine white fibre. **1883** R. L. ALLEN *New Amer. Farm Bk.* (rev. ed.) 252 The best plan for water rotting of hemp is in vats under cover, the water in which is kept at an equable temperature.

water sapphire. [Cf. *SAPHIR D'EAU.] **a.** A colourless variety of native sapphire (sense 1 b). **b.** A type of clear blue quartz (quot. 1829). **c.** = *SAPHIR D'EAU.

1698, etc. [in Dict. s.v. SAPPHIRE 1 b]. **1829, 1850** [in Dict. s.v. WATER *sb.* 29]. **1925, 1936** [see *SAPHIR D'EAU]. **1979** HURLBUT & SWITZER *Gemology* xiii. 182/2 Iolite is the sapphire-blue gem variety of the mineral known by mineralogists as cordierite... 'Water sapphire', or 'saphir d'eau' are misleading names early applied to the gem.

water-scape[2]. Delete *rare* and add further examples. Also, a picture of such scenery.

1886 G. M. HOPKINS *Lett. to R. Bridges* (1935) 240, I may be able some day to let you have a better waterscape from the Dargle or somewhere near Dublin. **1908** CHESTERTON *All Things Considered* 31 The landscape (or waterscape) of my own romantic town. **1955** R. CHURCH *Over Bridge* i. 4 His aim was to establish a *real* aquarium, with fresh water-scape scenery of..mosses, rondures, recesses, mournful weeds. **1979** *Sci. Amer.* Nov. 39/1 Good maps show the virgin waterscape, the snow depth, the detail of the complex arteries that bring the water to San Francisco and to Los Angeles.

watershed[1]. Add: **1.** (Earlier *fig.* example.) Also *attrib.* in *fig.* use.

1878 LONGFELLOW *Kéramos* i. 87 Midnight! the outpost of advancing day!.. The watershed of Time, from which the streams of Yesterday and To-morrow take their way. **1962** *B.B.C. Handbk.* 32 The BBC's Code of Practice on Violence, its new 9.30 p.m. watershed policy, its intention to distinguish those programmes which it thinks unsuitable, and perhaps more important, suitable for children.. show that both sides are aware of the problem. **1973** T. TOBIN *Lett. G. Ade* 2 While a journalist in Chicago, Ade became one of the more astute chroniclers of the daily preoccupations of ordinary people who were living through the 'watershed period'. **1980** *Listener* 29 May 68/3 *On the Town*, which [Gene] Kelly himself describes as a watershed picture, and which opened up the musical to location shooting.

water-shoot. 1. Delete † *Obs.* and add after root:, trunk, or main branch. (Later examples.)

1953 D. M. WRIGHT *Dwarf Fruit Trees* vi. 53 All that is necessary is the annual cutting out of..strongly growing 'water shoots'. **1965** *Sunday Mail Mag.* (Brisbane) 26 Sept. 15 Water shoots are strongly growing stems, usually sappy and developing rapidly.

wa·tersider. *Austral.* and *N.Z.* [f. WATER-SIDE + -ER[1].] A dockside worker.

1914 *Evening Post* (Wellington, N.Z.) 4 Feb. 10 For some time certain sections of watersiders who term themselves loyal men have been making periodic attacks on certain officials of the harbour board. **1925** *Glasgow Herald* 28 Oct. 12 The Entor is berthed at Fremantle under a police guard, with watersiders discharging her. **1949** F. SARGESON *I saw it in my Dream* xiv. 160, I see they're having trouble with those watersiders again. **1969** *Age* (Melbourne) 24 May 3/3 The watersiders' vigilance officer ..went aboard and warned the mate that if the ship sailed she would be declared 'black' in all Australian ports. **1969** *Northern Territory News* 11 July 8/2 Darwin watersiders now officially have a guaranteed minimum wage of $200 a month, but will continue to press for permanent employment with a minimum weekly wage. **1984** *N.Z. Listener* 10–16 Mar. 22/2 There, tattooed watersiders drink away the profits of a crown and anchor session down at the wharves.

wa·ter-ski, *sb.* Also **water ski.** [SKI *sb.*] **1.** One of a pair of skis enabling the wearer to skim the surface of water when towed by a motor-boat. Cf. *SKI *sb.* 1 c.

1931 *N.Y. Times* 3 May xx. 15/4 In 1928 a young Viennese student and amateur skier named Joseph Krupka, observing a long-legged waterfly racing over the surface of a rain-barrel, conceived the idea of skiing over water as he had done over snow. That night he worked out on paper the construction of a water ski. **1931** *Times Educ. Suppl.* 5 Sept. p. i/3 A schoolmaster.. crossed the Channel..on a pair of water skis. **1952** D.

DODGE *To catch Thief* v. 149 A speedboat roared by... A man and a girl on water skis rode the wake. **1973** L. MEYNELL *Thirteen Trumpeters* iv. 54 I'll lay odds *damage to one pair of water skis* turns up on my bill.

2. *attrib.* (repr. *water-skiing* vbl. sb.).

1931 *N.Y. Times* 3 May xx. 15/4 The 1929 water ski champion, Herr Pribitzer of the water-ski rescue section ..has attained speeds of more than twelve miles an hour. **1955** *Sports Illustr.* 1 Aug. 27 California water-ski clothes show a penchant for dressing to match from trunks to skis. **1960** *Housewife* May 54/1, I left England..with four pairs of water-skis..to start a water-ski school. **1981** *Beautiful Brit. Columbia* Spring 44/2 Here you can rent autos, water-ski boats, [etc].

wa·ter-ski, *v.* Also **water ski.** [f. the sb.] *intr.* To skim over the surface of water on water-skis. Cf. *SKI *v.* 1 b.

1953 *Time* 19 Jan. 85/3 Aristotle Socrates Onassis is a Greek-born Argentine who water-skis in the best international circles. **1960** *Sunday Express* 3 July 19/2 What does it cost to water ski? **1971** P. GRESSWELL *Environment* 217 It is not possible to fish and water-ski in the same place. **1979** R. JAFFE *Class Reunion* (1980) II. x. 285 They both liked to water-ski.

wa·ter-skier. Also **water skier.** [f. prec. + -ER[1].] One who water-skis. Cf. *SKIER (*b*).

1931 *N.Y. Times* 3 May xx. 15/4 The water skier can carry about sixty pounds additional weight. **1935** *Lit. Digest* 13 Apr. 46/3 Water-skiers who are of championship caliber can hold one ski above the head and swing the free leg back and forth. **1958** X. FIELDING *Corsair Country* vii. 139 He..would certainly resent its [*sc.* the mole's] present use as a landing-jetty not for galleys but for water-skiers. **1978** P. PORTER *Cost of Seriousness* 34 Up river the water skiers puff and plane.

wa·ter-skiing, *vbl. sb.* Also **water skiing, ski-ing.** [f. as prec. + -ING[1].] The action of the verb, esp. as a sport. Cf. *SKI-ING *vbl. sb.* 2.

1931 *N.Y. Times* 3 May xx. 15/4 Water skiing is beginning to eclipse canoeing. **1935** *Lit. Digest* 13 April 46/3 Georges Ducros, European water-skiing champion,..hopes to popularize his speciality here this summer. **1948** M. MURRAY *King & Corpse* (1949) xix. 214 Water ski-ing at night is a lot more exciting than in the daytime. **1960** *Times* 15 Aug. 12/5 (*heading*) British girl fifth in water-skiing. **1973** 'G. BLACK' *Bitter Tea* ii. 27 Water ski-ing equipment..including a couple of scuba masks.

watertight, *a.* Add: **2.** (Earlier example in *sing.*)

1867 J. T. THOMSON *Rambles with Philosopher* iii. 14 A shoemaker..beating time with his mallet on the hob-nails of an old *water-tight* that he was repairing.

water-washed, *pa. pple.* and *ppl. a.* (Earlier example.)

1826 J. G. WHITTIER *Exile's Departure* in *Poet. Wks.* (1898) 521/1 The forest-crown'd hill and the water-wash'd strand.

water-wave, *sb.* Add: **3.** (Examples.) orig. *U.S.*

1882 *Harper's Mag.* Nov. 877/2 She is pasting down her wetted hair into a semblance of the 'water-waves' of fashionable society. **1923** E. F. WYATT *Invis. Gods* I. i. 7 His grandmother..bending over him her water waves and pearl powder. **1958** J. CANNAN *And be a Villain* i. 5 Passing a clean white hand over the water-waves of his naturally fuzzy hair. **1972** *Vogue* June Special 68/2 Hair in sleek water waves, by Jean Louis David, Paris.

water-waved *ppl. a.*: also of hair: set, with water, into waves; **wa·ter-wave** *v. trans.*, to set (hair), with water, in waves.

1928 A. HUXLEY *Point Counter Point* xxviii. 453 She readjusted a water-waved lock of hair. **1962** E. SNOW *Other Side of River* (1963) lxvii. 513 A little girl with large saucy black eyes and beribboned hair her mother must have water-waved. **1975** J. GORES *Hammet* (1976) xix. 130 Goodie had spent a dollar..to have her blond ringlets water-waved by Georgia.

wa·ter-waving, *vbl. sb.* [f. WATER-WAVE *sb.* + -ING[1].] **1.** The wavy or 'watered' appearance imparted to silk and other fabrics by pressing two pieces together.

1894 J. E. DAVIS *Elem. Mod. Dressmaking* v. 94 Beetled plain linings generally show a marking like a slight water-waving on the surface.

2. A method of waving hair with water.

1925 *Daily Tel.* 13 May 20/5 (Advt.), Wanted, smart man... Must be thoroughly competent in perm. waving, Marcel and water waving. **1927** *Daily Express* 30 Nov. 13/3 She can give lessons in water waving, face massage, and chiropody. **1932** [see *finger-waving* vbl. sb. s.v. *FINGER *sb.* 15].

water-way. 4. (Earlier examples.)

1797 G. IMLAY *Topogr. Descr. W. Territory N. Amer.* (ed. 3) 34 Major Willis..found 1300 yards clear water-way between the lower beaches or counter-shores of the banks on both sides of the river. **1832** J. S. MILL *Let.* 24 May in *Wks.* (1963) XII. 99 Yet if such glimpses are numerous, some general tendency shall predominate even in the few furlongs of water-way which they may chance to disclose.

wa·ter-wings, *sb. pl.* Add: **4.** Inflatable floats which may be fixed to the upper arms of persons learning to swim, in order to give increased buoyancy.

1907 *Yesterday's Shopping* (1969) 323/3 All Water Wings..support a man as easily as a boy..on just the level at which a person can swim or float comfortably. **1922** WODEHOUSE *Girl on Boat* v. 102 A little undersized shrimp of a fellow with a green face and ears like water-wings. **1948** 'P. WOODRUFF' *Whatever Dies* 151 You could take it, if you would be bold and throw away your water-wings. **1976** 'G. BLACK' *Moon for Killers* iii. 37 Gloria's daughter was..dressed for the water..with..a pair of water wings.

wa·ter-witching, *vbl. sb. U.S.* [f. WATER-WITCH + -ING[1].] = WATER-FINDING *vbl. sb.* Hence **wa·ter-witch** *v. intr. rare* (in quot. *transf.*).

1877 H. RUEDE *Sod-House Days* (1937) 196 Talking with Hoot about digging wells, etc., he told me about one new neighbor, Diegel, who has a firm faith in water witching. **1947** R. BEDICHEK *Adv. with Texas Naturalist* x. 119, I followed for a short distance an immense sow..her nose barely skimming the sand, apparently searching for something. It turned out that she was water-witching.. for presently she began rooting. Soon her..snout unearthed a spring of crystal-clear water. **1968** S. E. ROBERTS *Of Us & Oxen* ii. 18 Several homesteaders..had had to dig a number of wells before striking water, even though these wells had been located by 'water-witching'. **1976** *National Observer* (U.S.) 14 Aug. 11/1 One of our neighbors practiced the art of water witching.

waterwork. Add: **3. a.** (Later examples.)

1803 M. WILMOT *Let.* 6 Aug. in *Russ. Jrnls.* (1934) I. 34 The Gardens..so extensive and so beautiful and the Water works are beyond description. **1842** W. F. TOLMIE *Jrnl.* 7 May (1963) 359 As the great display of waterworks was to take place at 5, I finally decided on remaining.

b. For *nonce-use* read *rare* and add later example with reference to rain.

1931 D. L. SAYERS *Five Red Herrings* xvii. 179 'It's not raining... Better than yesterday.'.. 'Tons better. Really,..you'd think they'd turned on the water-works yesterday on purpose to spoil my sketching-party.'

c. *pl.* The urinary system. *euphem.*

1902 F. W. MAITLAND *Let.* 6 July (1965) 249, I gather from Albutt that the immediate cause of death was, as A. put it, 'in the water works'. **1922** JOYCE *Ulysses* 355 Cissy..came back with her tongue out and said uncle said his waterworks were out of order. **1959** E. AMBLER *Passage of Arms* v. 124 Little scotch, lot of soda. Got to keep the old waterworks going in this climate. **1977** W. HILDICK *Loop* xxix. 205 I'd been plagued for a long time..by—well—let's call it water-works trouble.

Watford (wǫ·tfǫ̆ɹd). The name of a town on the N.W. edge of the London conurbation, used with allusion to the view attributed to Londoners that north of the metropolis there is nothing of any significance to English national or cultural life.

1973 G. TALBOT *Ten Seconds from Now* (1974) iii. 37 A man who, until he made the journey from London, thought that woad began at Watford. **1977** *Times* 17 Jan. 13/6 As an Englishman from north of Watford I sometimes suspect that the harmony [in Turkey] is.. greater than..between different regional and social sectors in this country. **1980** D. WILLIAMS *Murder for Treasure* ii. 17 'Provincial visitors'..broadly covered anyone normally domiciled north of Watford. **1983** *Oxford Diocesan Mag.* July 13/2 The urgent need is to unite the country... This does not mean watery consensus politics. It does mean recognising that.. Britain does not end at Watford. **1984** *Sunday Tel.* 21 Oct. 9/6 Yorkshire is becoming the most militant county in England... There's the rub. The line between North and South is not just drawn at Watford. Scargill is of the South.

Watson[1] (wǫ·tsən). The name of the doctor who was the stolid, faithful assistant and foil to the detective Sherlock Holmes in the stories of A. Conan Doyle, used allusively of one who acts similarly as a stooge or audience, esp. for a detective.

1927 R. A. KNOX *Three Taps* vi. 52 Watson-work meant that Angela tried to suggest new ideas to her husband under a mask of carefully assumed stupidity. **1946** D. L. SAYERS *Unpopular Opinions* 188 The story is told by the detective's *fidus Achates* or (to use the modern term) his Watson. **1953** WODEHOUSE *Performing Flea* 18, I wonder what an oesteopath does if a patient suddenly comes apart in his hands. ('Quick, Watson, the seccotine!'). **1958** B. HAMILTON *Too Much of Water* x. 228 I've been..a sort of yes-man or Watson to him. *a* **1976** A. CHRISTIE *Autobiogr.* (1977) v. v. 282, I was..tied to two people: Hercule Poirot and his Watson, Captain Hastings. **1981** CRAIG & CADOGAN *Lady Investigates* ii. 48 Mrs Lucilla Wiggins, the 'Watson' figure in the cases of Mrs Herlock Shomes.

Watson[2] (wǫ·tsən). *Austral.* [Allegedly the name of two Australian brothers who were noted betters about the turn of the century: cf. S. J. Baker *Austral. Lang.* (ed. 2, 1966) xii. 273–4.] In catch-phr. *to bet like the Watsons*: to wager large sums.

1949 L. GLASSOP *Lucky Palmer* 163 Bet well ? You bet like the Watsons. **1954** T. RONAN *Vision Splendid* 76 The survey-party is chequed-up to the skies and while they've got it they'll bet like the Watsons. **1967** F. HARDY *Billy Borker yarns Again* 140 I'd bet like the Watsons meself if I had a million quid in the bank.

Watson–Crick (wǫ·tsən,kri·k). *Biochem.* The names of James D. *Watson* (b. 1928), U.S. biochemist, and Francis H. C. *Crick* (b. 1916), English biochemist, used *attrib.* with reference to the pairing of adenine with thymine (or uracil) and of guanine with cytosine in the two strands of a double helix, described by them in 1953.

1964 G. H. HAGGIS et al. *Introd. Molecular Biol.* ix. 229 The Watson-Crick pairs are purine-pyrimidine pairs. **1966** T. H. JUKES *Molecules & Evolution* i. 5 When cells divide and multiply, this sequence [of bases in DNA] replicates itself enzymatically through the Watson-Crick complementary pairing mechanism. **1976** *Nature* 23 Sept. 289/1 Much of the pairing at the third codon position also involves the normal Watson-Crick base pairs, A–U, and G–C, but for several codon–tRNA interactions, non-Watson–Crick pairs are clearly required.

watsonia (wǫtsōᵘ·niă). *Bot.* Also **Watsonia.** [mod.L. (P. Miller *Gardener's Dict.* (ed. 7, 1759) s.v.), f. the name of Sir William *Watson* (1715–87), Scottish naturalist + -IA[1].] A bulbous plant of the genus of this name (family Iridaceæ), native to South Africa and bearing spikes of white, pink, or red flowers similar to gladioli.

1801 *Curtis's Bot. Mag.* XV. 533 (*heading*) Aletris-like Watsonia. **1843** M. EDGEWORTH *Lett. from Eng.* (1971) 595 The most beautiful flowers..Gladiolis and red and white Watsonia. **1902** L. H. BAILEY *Cycl. Amer. Hort.* IV. 1971/1 Great interest has been aroused in Watsonias recently. **1948** A. PATON *Cry, Beloved Country* I. iii. 20 Here..grow the blue agapanthus, the wild watsonia, and the red-hot poker. **1971** P. M. SYNGE *Collins Guide to Bulbs* (ed. 2) 308 All watsonias are easily raised from seed.

Watsonian (wǫtsōᵘ·niăn), *a.* [f. the surname *Watson* + -IAN.] Of, pertaining to, or characteristic of someone called Watson; *spec.* **a.** J. B. Watson (1878–1958), U.S. behavioural psychologist; **b.** Dr. Watson of the Sherlock Holmes stories (see *WATSON[1]).

a. 1923 K. DUNLAP *Old & New Viewpoints in Psychol.* (1925) ii. 54 'Behaviorism' now means the Watsonian view exclusively. **1927** B. RUSSELL *Outl. Philos.* iii. 41 His [*sc.* the ape's] behaviour when he had once realised that one stick could be made by joining the two was scarcely Watsonian. **1928** R. FROST *Let.* 22 June (1964) 188 The logic of religion by nice gradations outside of Catholicism in Protestantism, outside of Protestantism in agnosticism, and finally outside of agnosticism in Watsonian behaviorism. **1960** [see *REINFORCEMENT 3 c]. **1968** P. MCKELLAR *Experience & Behav.* viii. 218 The stultifying limitation which Watsonian behaviourism imposed upon psychology. **1983** *Brit. Jrnl. Psychol.* LXXIV. 301 Certain influential modern psychologists have singled out attention as an aspect of cognition that fell early to the Watsonian onslaught.

b. 1940 E. BENTLEY *Those Days* ix. 250 Its [*sc.* the Sherlock Holmes Society's] annual dinners..were followed by discussions on points of Watsonian chronology. **1960** *Encounter* Mar. 66/2 Just one more case of Watsonian blackmail. **1968** *Listener* 4 July 22/1 What the elements of magic in the Holmes stories were, and are, I have never seen analysed. The stilted Watsonian style of sentence structure must be one major element. **1981** CRAIG & CADOGAN *Lady Investigates* i. 32 There is no Watsonian colleague to..assist with her cases.

watt. Add: **b. watt-second,** a unit of energy equal to one joule, being the energy consumed at a rate of one watt during one second; hence (abbreviated as) **wattsec.**

1962 *New Scientist* 18 Jan. 157/1 The term 'wattsec' is..a common one among radio engineers while 'joule' appears to be seldom used in practice. **1893** G. KAPP *Dynamos, Alternators, Transformers* ii. 42 This unit [*sc.* the erg] is also inconveniently small, and for practical purposes it is customary to employ a unit 10,000,000 times as great—namely, the 'watt-second' or 'joule'. **1981** *Sci. Amer.* Dec. 37/1 A photon with a frequency in the visible region of the electromagnetic spectrum has little energy: less than 10^{-18} joule, or watt-second.

wattage (wǫ·tėdʒ). [f. WATT + -AGE.] **a.** An amount of electrical power, esp. the operating power of a lamp, appliance, etc., expressed in watts; *colloq.*, electricity, electrical illumination.

1903 *Electr. World & Engineer* 27 June 1095 Dividing the kilowatt-hours mentioned by said number of lamps shows an average per lamp at station of 463·8 watts, deducting from which 9 per cent. for line loss, shows a net wattage per lamp at lamp terminal of 422. **1933** *Times* 5 Dec. (Electricity Supply Suppl.) p. xxiii/1 Employing lamps of comparatively low wattage. **1953** J. CARY *Except the Lord* xviii. 81 The quivering lanterns, the jumping flames..gave an elation to our saturnalia that would not have been kindled by frigid festoons of wattage. **1962** L. DEIGHTON *Ipcress File* xxv. 165 A light of low wattage glowed in the main hall. **1976** A. HOPE *Hi-Fi Handbk.* 29 That's a fairly reasonable minimum wattage rating per channel for a stereo amplifier. **1977** *Time* 21 Nov. 49/1 This Christmas may well burn record wattage. **1984** *Verbatim* Summer 4/2 From the ceiling hang bare light bulbs of such low wattage that the interior is scarcely discernible.

b. *fig.*

1964 *Listener* 10 Dec. 952/1 Peter Cushing's stolid Churchill was thrown in the shade by the high wattage of this acting. **1980** D. FRANCIS *Reflex* xvii. 199 She had a powerful attraction..with the full wattage switched my way.

Watteau. Add: **Watteau pleat** (earlier example).

1873 *Young Englishwoman* Oct. 490/2, I have made a Princesse dress with a Watteau pleat and flounce of eleven yards of print.

Watteauesque (wǫto,e·sk), *a.* [f. WATTEAU + -ESQUE.] Suggestive of or in the style of Watteau.

1925 *Glasgow Herald* 25 Apr. 8/3 Opposite might hang a Watteauesque diversion representing an impossibly light and brilliantly fanciful landscape. **1966** N. MARSH *Black Beech & Honeydew* xi. 256 *Twelfth Night* seemed to work its own miracle... Feste, all frill and Watteau-esque stripes, was enchantingly realised. **1974** *Country Life* 7 Feb. 257/1 Throughout the decade 1720–30 Mercier both painted and engraved Watteauesque subjects.

wattle, *sb.*[1] Add: **II. 4. d.** *wattle extract.*

1955 *Times* 30 June 18/2 The price of South African wattle extract remained the same during 1954 as it was during 1953 and 1952. **1969** T. C. THORSTENSEN *Pract. Leather Technol.* ix. 141 The main source of wattle extract is the *Acacia mollissima*, or Black Wattle.

Watusi (watū·si). Also **Watussi, Watutsi.** [Native name.] **1.** The name of a minority racial group in Rwanda and Burundi, probably of Ethiopic or Nilotic origin, which formerly dominated the majority Hutu people; a member of this group. Also *attrib.* or as *adj.* Also called *TUTSI.

1899 H. A. NESBITT tr. *P. Kollmann's Victoria Nyanza* ii. 13 None of the large cattle adorned with magnificent horns which we find everywhere else among the Wahuma and Watussi are to be noticed in Uganda. **1937** *John o'London's* 5 Feb. 765/1 The Watussi still preserve certain rites and customs analogous to those which are known to have been operative in Ancient Egypt. **1959** A. MOOREHEAD *No Room in Ark* ii. 66 The Watusi are celebrated hunters, very tall and lithe. **1960** *Guardian* 7 June 4/1 Watutsi pastoralists..are encircling the Birunga. **1976** D. TOPOLSKI *Muzungu* xv. 227 Noel was a Watutsi and had been adopted at an early age by a farmer from New Zealand.

2. (Also with small initial.) A popular dance of the 1960s. Also as *v. intr.*

1964 *Time* 20 Mar. 62/3 (*caption*) Watutsiing at Whisky à Gogo. *Ibid.*, A pretty eyeful slaps on new records and dances it all by herself. That way, it's called the Watutsi. **1965** [see *hip-swinging* adj. s.v. *HIP *sb.*[1] 4 b]. **1966** *Punch* 20 July 116/1 They..fed on lotus and daiquiri, they frugged and watutsied. **1966** T. PYNCHON *Crying of Lot 49* v. 105 Nefastis had been watching on his TV set a bunch of kids dancing some kind of a Watusi. **1966** H. NIELSON *After Midnight* (1967) xvi. 213 She could teach you to watusi and swim. **1974** *Encycl. Brit. Micropædia* X. 213/3 Dances evolved from the twist, such as the frug, the jerk, and the watusi, were invariably performed by shaking the pelvis. **1979** R. JAFFE *Class Reunion* (1980) II. vi. 244 They danced the frug, the swim, and the watusi.

Waughian (wǭ·iăn), *a.* [f. the name *Waugh* + -IAN.] Of, pertaining to, or characteristic of the English novelist Evelyn Waugh (1903–66) or his writing.

1960 *Times Lit. Suppl.* 2 Dec. 783/3 Smyth strays into 1960 from some distant Waughian era. **1976** *Ibid.* 1 Oct. 1229/4 Sniping on Firbankian and Waughian lines. **1977** *Time* 19 Dec. 11/1 A country without government? The description immediately calls to mind a Waughian Third World kingdom.

Also **Waughism** (wǭ·iz'm), (*a*) the ideas or style characteristic of Waugh, or those portrayed in his novels; (*b*) a word or expression characteristic of Waugh.

1934 [see *RUSSELLISM]. **1976** A. POWELL *Infants of Spring* x. 167 The sessions he devoted to ragging ('mocking', to use a Waughism..) were likely to take certain routine forms.

wave, *sb.* Add: **I. 2. c.** Also, one of military vehicles or aircraft.

1943 R. V. JONES *Most Secret War* (1978) xli. 382 Longer raids will always be liable to attacks on their last waves whenever fighters can fly. **1951** O. BERTHOUD tr. *Clostermann's Big Show* I. 38 The airfield at Triqueville ..was going to be bombed in force by two waves of 72 Marauders. **1982** *Daily Tel.* 12 Oct. 17/8 The fly past will take place in two waves—a slow one consisting of five formations of helicopters.., then the fixed-wing aircraft, again in five formations.

3. b. Also, a sharp increase in the extent or degree *of* some phenomenon; cf. *crime wave* s.v. *CRIME *sb.* 4.

1910 *Sat. Even. Post* 29 Oct. 46/2 A good many 'waves of crime' occur in the imagination of newspapers. **1920** *Times* 21 Jan. 12/1 The probability of a 'wave' of crime after the war has been foreseen and foretold by students of social problems. **1927** *New Republic* 21 Sept. 109/2 The Metropolitan Life Insurance Company, commenting on the alleged suicide wave among young people, reports ..that the suicide rate for the population as a whole is decreasing. **1958** W. S. CHURCHILL *Hist. Eng.-Speaking Peoples* IV. x. i. 10 Throughout the country a fresh wave of demonstrations followed. **1971** *Daily Tel.* 13 Apr. 6 The pace of dismissals is accelerating as companies strive to restore profits after the massive wave of inflation in costs and wages.

c. *the wave of the future*: the inevitable future fashion or trend; the coming thing.

2. b. *U.S. slang.* A gramophone record or phonograph cylinder.

1936 *Metronome* Feb. 21/4 *Waxing*, phonograph record. **1941** *Jazz Information* Nov. 26/1 I'll stick to the Victor-Bluebird waxings, for they are the best. **1965** [see *FRUG]. **1981** *Listener* 1 Jan. 30/1 Buddy Bolden..is said to have recorded a phonograph cylinder, somewhere around 1902... The rumoured waxing has never been discovered.

wax-light. (Earlier example.)

1600 M. HOBY *Diary* 18 Sept. (1930) 145, I reed, praied, was busie about waxe lightes, and then I dined.

waxwork. Add: **3.** (Earlier examples.)

1763 BOSWELL *Jrnl.* 4 July in *London Jrnl.* (1950) 289, I went and saw Mrs. Salmon's famous wax-work in Fleet Street. **1773** DUCHESS OF NORTHUMBERLAND *Diary* (1926) 207 The Tower, the Venetian Lady, Westminster Abbey, Salmon's Waxworks.

4. (Earlier example.)

1818 W. P. C. BARTON *Compendium Floræ Philadelphicæ* I. 128 Wax Work. A climbing plant frequently reaching the tops of trees.

waxy, *a.*[1] Add: **5.** *Comb.*, as *waxy-faced, -looking, -skinned, -white.*

1846 Waxy-faced [in Dict., sense 3 a]. **1927** *Scots Observer* 1 Jan. 3/1 Barefoot women..some with waxy-faced infants in their arms. **1859** D. BUNCE *Trav. with Dr. Leichhardt* vi. 44 A race of plants having waxy-looking berries. **1871** [in Dict., sense 3 a]. **1964** S. DUKE-ELDER *Parsons' Dis. Eye* (ed. 14) xxii. 328 Oedema is usually not marked, but all over the posterior pole there tend to gather hard, white or yellow, waxy-looking patches of exudates. **1930** J. DOS PASSOS *42nd Parallel* I. 34 He was a sharpnosed waxyskinned young man. **1929** W. FAULKNER *Sanctuary* xxiii. 261 His nostrils were waxy white with rage. **1954** A. G. L. HELLYER *Encycl. Garden Work* 452 *Kindingeri*, waxy-white, reddish at base.

way, *sb.*[1] Add: **I. 1. f.** *Way Out sign.*

1972 L. MEYNELL *Death by Arrangement* xiii. 178 Hooky was forced to follow him along an uncomfortably empty platform towards the 'Way Out' sign. **1982** J. O'FAOLAIN *Obedient Wife* iii. 59 Just follow the Way Out signs.

II. 4. c. *everything coming* (or *going*) *one's way*: everything happening in one's favour; *to know one's way around* (or *about*): (later example); also, to have shrewdness born of experience; *to go separate ways*: to cease to work or operate together and follow different paths.

1903 G. B. SHAW *Maxims for Revolutionists* in *Man & Superman* 242 In moments of progress the noble succeed, because things are going their way. **1903** *Red Bk.* June 167/2 Everything was coming his way. **1921** GALSWORTHY *To Let* I. v. 52 'That's a young woman who knows her way about,' he said. **1932** KIPLING *Limits & Renewals* 385 They must be enjoying themselves now at a theatre. Everything's coming their way. [**1935** N. L. MCCLUNG *Clearing in West* xix. 152 Jack knew his way around, having been here many times the winter before... He drove straight down to Pacific Street to the Farmer's Home.] **1938** *Times* 1 Jan. 10/4 Much was certainly lost when the original directorate decided to go separate ways, as from the separation both parties seem to have suffered. **1943** O. HAMMERSTEIN *Oklahoma!* (1947) 18, I got a beautiful feelin' Ev'rythin's goin' my way. **1957** 'R. WEST' *Fountain Overflows* xi. 241 And whatever you could say about my Ma, you couldn't say she didn't know her way about. **1968** H. R. F. KEATING *Inspector Ghote hunts Peacock* vii. 95 That girl was a pretty slick chick.. She knew her way around. **1973** G. JENKINS *Cleft of Stars* v. 60, I cursed that old rifle for letting me down at the moment when everything was going my way.

7. j. Colloq. phr. *on the* (or *one's*) *way out* (or *down*): going down in status, position, estimation, or favour; similarly with *in* or *up*, expressing the opposite sense.

1937 *Time* 25 Jan. 12/3 Every time one of them has called on the President and emerged smiling, rumor has whispered throughout Washington that the other was 'on his way out'. **1938** *Sat. Rev.* (U.S.) 17 Sept. 17/1 The thrill of being on the way up, of being prominent, being envied. **1938** H. L. ICKES *Diary* 5 Nov. (1955) II. 497 France is but little better than a third-rate power and is on the way down. **1955** A. L. ROWSE *Expansion Eliz. Eng.* i. 27 The Scottish king could well afford to make the concession: she was on her way out, he was on his way in. **1960** *Guardian* 9 Dec. 8/5 Sunrise yellows and pinks are definitely on the way in. **1962** in R. Jarrell *Sad Heart at Supermarket* 92 Poetry is on the way out! **1975** D. BAGLEY *Snow Tiger* xx. 163 This is the last job I'll hold as chief engineer. If I lose it I'll be on the way down—I'll be assistant to some smart young guy who is on his way up. **1980** A. SCHOLEFIELD *Berlin Blind* I. 5 Calland was a good-looking young man on the way up.

8. a. *a little goes a long way* and varr.: see GO *v.* 43 c, *LITTLE *sb.* 4.

c. *by a long way* (earlier example); *to go a long* or *great way*: also, to be in agreement *with* someone; *all the way*: completely; cf. senses 8 e, f below.

1850 LADY LYTTLETON *Let.* 12 June (1912) 401, I cannot *quite* enter into his politics... But a very great way I go along with him. **1859** DARWIN *Let.* Nov. (1887) II. vi. 224 Also from Quatrefages, who is inclined to go a long way with us. **1859** T. HUGHES *Tom Brown at Oxf.* iv, in *Macmillan's Mag.* Dec. 102/1 He is more of a gentleman by a long way than most. **1973** 'N. CARTER' *Spanish Connection* x. 112 I'm saying I can't buy your story all the way, Corelli.

d. *all the way from — to —*: (*a*) throughout the specified interval, at every point in it; (*b*) *U.S.*, (estimated, etc.) at any amount between the specified quantities.

(*a*) **1791** [in Dict., sense 8 a]. **1966** *Listener* 5 May 643/2 The peak age [for crime] is during the last year at school... The rate is fairly high all the way from twelve to twenty.

(*b*) **1878** J. H. BEADLE *Western Wilds* xxxi. 493 The value of the booty taken has been estimated all the way from $150,000 to $300,000. **1931** G. T. CLARK *Leland Stanford* xi. 365 The amount said to have been wagered ..has been variously stated all the way from $5,000 to $50,000.

e. *to go all the way, the whole way*: (*a*) to continue a course of action to its conclusion; *spec.* (*slang*), to engage in sexual intercourse (*with* someone), as opposed just to fondling; (*b*) to agree completely *with* someone.

1915 J. C. POWYS *Visions & Revisions* 12 If you lack the courage, or the variability, to *go all the way* with very different masters, and to let your constructive consistency take care of itself, you may become, perhaps, an admirable moralist; you will never be a clairvoyant critic. **1922** H. J. LASKI in *Holmes-Laski Lett.* (1953) I. 412, I can't go all the way with it, for if it was as a business man that the tyrant found the path to power I should have thought there would have been mention of it in Aristotle. **1924** P. MARKS *Plastic Age* xiv. 151 'Wonder if Janet would have gone the whole way,' flitted across his mind. **1927** H. T. LOWE-PORTER tr. *Mann's Magic Mountain* I. iii. 78 'Am I right?' 'You certainly are, I can go all the way with you there.' **1961** L. P. HARTLEY *Two for River* 49 I'd sooner go the whole way with somebody than natter with them at a tea-table. **1970** W. J. BURLEY *To kill Cat* x. 186 The things we found in her room! I mean it was obvious she was going all the way and her not fifteen! **1975** *Listener* 30 Oct. 574/4, I am not sure that I go all the way with Mr Miller in some of his analysis. **1979** R. JAFFE *Class Reunion* (1980) I. i. 24 She would go to medical school... She didn't know if she would have the guts to go all the way: intern, resident, actually practice medicine. *Ibid.* vi. 86 They would do as much as they could without either removing the rest of her clothes or going all the way.

f. *to come* or *go a long way* (with personal subj.: for impersonal subj. see GO *v.* 43 c, d): to achieve much, to make much progress; *to have a long way to go*, etc., to be far short of some accomplishment; so *a long way from*, far short of, much inferior to.

1917 H. J. LASKI in *Holmes-Laski Lett.* (1953) I. 121 Your bretheren [*sic*].. have still a long way to go before they understand the meaning of a certain dissent in *Adair v. U.S.* **1922** W. S. MAUGHAM in *Pearson's Mag.* Oct. 320/2 He had come a long way since then. **1925** *New Yorker* 5 Sept. 11/3 Which is another way of saying that he will go a very long way. **1933** F. BALDWIN *Innocent Bystander* viii. 150 Sherry had a long way to travel before she would be a Fontanne or a Cornell. **1935** H. L. MENCKEN *Let.* 4 Jan. (1961) 386 You must yet go a long way, of course, before you are eligible to it. **1940** *Chatelaine* July 37/2 Pat and Rosemary have come a long way; they started their careers with Fred Waring's Pennsylvanians and ended up stars in one of the largest motion picture studios in the world. **1957** *Practical Wireless* XXXIII. 684/2 The system is a very long way from Hi-Fi, but is sufficient for the transmission of speech. **1966** *Seventeen* July 140/3 Society seems to have come a long way since the days of the Puritans, and now we're up to topless bathing suits. **1977** P. BAELZ *Ethics & Belief* vii. 79 Man has still a long way to go before he exercises his freedom responsibly and responsively.

9. b. *that way*: spec. (*a*) homosexual; (*b*) (const. *about*) in love or infatuated; also (in general sense) *that way inclined, to get that way.*

1603 [see INCLINED *ppl. a.* 3 a]. **1859** T. HUGHES *Tom Brown at Oxf.* (1861) I. x. 163 Both 'smalls' and 'greats' are sufficiently distant to be altogether ignored, if we are that way inclined. **1916** A. BENNETT *These Twain* xx. 518 'He simply hates doing a thing straight off.' 'Yes, he is rather that way inclined.' **1922** C. SANDBURG *Slabs of Sunburnt West* 6 How do you get that way? **a 1960** E. M. FORSTER *Maurice* (1971) vii. 42 The Greeks, or most of them, were that way inclined, and to omit it is to omit the mainstay of Athenian society. *Ibid.* xii. 62 In his second year he met Risley, himself 'that way'. **1960** WENTWORTH & FLEXNER *Dict. Amer. Slang* 540/2 *That way*, in love. Usu. in 'They are that way about each other.' c. 1940. **1961** V. SACKVILLE-WEST *No Signposts in Sea* 82 If I were that way inclined, which I am *not*, I can imagine falling in love with him myself. **1965** J. P. CARSTAIRS *Concrete Kimono* xxiii. 207 Sharon. Be reasonable. I thought you were 'that way' about Roderick. **1966** 'J. HACKSTON' *Father clears Out* 91 It got that way in the end that I used to look forward to seeing Nolan and his team come lumbering down from the north. **a 1967** J. R. ACKERLEY *My Father & Myself* (1968) xvi. 185, I divined that he was homosexual, or as we put it, 'one of us,' 'that way', 'so', or 'queer'.

d. *the other way about* (earlier examples); also *the other way, the other way around.*

1834 *Edin. Rev.* Oct. 83 The fault, in the present instance, is the other way. **1879** R. L. STEVENSON in *Cornh. Mag.* Oct. 412 He [*sc.* Burns] was 'constantly the victim of some fair enslaver'—at least, when it was not the other way about. **1925** *New Yorker* 28 Mar. 25/2 It is just as good the other way around. **1963** *Christian Century* 9 Jan. 49/1 But the Augustinian, and biblical, position addresses man the other way around: Find God and you will find yourself.

f. *to have it both ways* and varr.: to have advantages from two opposed or contradictory standpoints; to make use of alternatives or contradictions as it suits one.

1914 G. B. SHAW *Fanny's First Play* II. 191 Then I suppose what I did was not evil; or else I was set free for evil as well as good. As father says, you cant have anything both ways at once. **1926** F. M. FORD *Man could stand Up* I. ii. 33 An oafish thing to do! To take a schoolgirl..just past the age of consent, out all night in a dog-cart... You'd think any man who *was* a man would have avoided that! Most men knew enough to know that the Woman Pays.., the school-girl too! But they get it both ways. **1964** C. HODDER-WILLIAMS *Main Experiment* vii. 73 'It was only folklore.'.. 'Yes, but you can't have it both ways. If it frightens you it must mean something.' **1967** *Listener* 13 Apr. 489/2 So our dual terminology helps us to have it both ways.

III. 12. c. *to have one's way*: *spec.* to have sexual intercourse *with* (used chiefly of a man).

1915 CONRAD *Victory* IV. xii. 399 If I had taken you by the throat this morning and had my way with you, I should never have known what you are. **1939** *Ottawa Jrnl.* 23 Aug. 15/2 He might destroy Phil..to have his way with Joan. **1961** W. BROWN *Bedeviled* 113 Although she struggled, she was no match for him and he had 'had his way with her'. **1980** E. JONG *Fanny* III. xiii. 440 Thus could Anne Bonny defend herself when she did not fancy a Man, but when she fancied one, she also had her Way with him.

14. a. (Further examples.)

1860 GEO. ELIOT *Mill on Floss* II. III. vii. 107 I'm not a-defending him, in no way, for being so hot about th' erigation. **1952** M. R. RINEHART *Pool* (1953) xx. 170 It's no way to talk about a sister, but I've had about all I can take.

d. For *no way* see NOWAY *adv.* in Dict. and Suppl.

h. (Earlier example.) Now also in weakened use: a principle or activity that governs all one's actions; a dominating interest or occupation.

1605 SHAKS. *Macbeth* v. iii. 24 My way of life Is falne into the Seare, the yellow Leafe. **1917** H. J. LASKI in *Holmes-Laski Lett.* (1953) I. 104 Education is a way of life and not the collection of information. **1938** *Amer. Jrnl. Sociol.* XLIV. 1 (*heading*) Urbanism as a way of life. **1957** *Times Lit. Suppl.* 1 Nov. 650/5 Democracy has not yet been accepted as a way of life in Germany. **1970** *Daily Tel.* 4 Sept. 5/2 The investigation..is expected to ask searching questions into the safety of supertankers, which have now become a 'way of life'. **1974** *Times* 8 Mar. 23/5 (*heading*) Hong-kong... Where a quick profit is a way of life. **1981** Q. CRISP *How to become a Virgin* 188 Giving talks to American universities could easily be converted into a way of life.

j. *there is no way* (with dependent clause) (colloq.): there is no possibility that; cf. NOWAY *adv.* in Dict. and Suppl.

1975 *New Yorker* 1 Dec. 55/2 There is no way a losing candidate can pick himself up and pretend nothing has happened to him. **1977** *Daily Tel.* 10 Oct. 12/8 We are operating an emergency service and there is no way we would strike and let the old folk down. **1978** S. BRILL *Teamsters* x. 380 There was no way he would support the President's reelection. **1978** G. A. SHEEHAN *Running & Being* xiii. 188 He had recognized the bald head and there was no way I was going to beat him.

l. (*in*) *one way or* (or *and*) *another*: by any of various methods, for any of various reasons, in any of various respects. Cf. sense 9 e in Dict.

1861 T. HUGHES *Tom Brown at Oxf.* I. iii. 40 Being a good whist and billiard player, and not a bad jockey, he managed in one way or another to make his young friends pay well for the honour of his acquaintance. **1923** R. MACAULAY *Told by Idiot* I. v. 23 One way and another, what with papa's friends and mamma's and the children's, a good deal of life flowed into the..house. **1955** L. P. HARTLEY *Perfect Woman* viii. 79 What, after all, had he to tell Alec that mattered so much, one way or another? **1965** M. ALLINGHAM *Mind Readers* xix. 211 We're in for a very busy time, my lad, one way and another. **1973** *Listener* 15 Nov. 661/3, I was quite well educated, one way and another. **1979** A. PRICE *Tomorrow's Ghost* xii. 216 Captain Fitzgibbon wouldn't come back from that last Ulster tour, one way or another.

m. *way of looking at it* or *things*: the (personal) perspective from which one views a situation or event, esp. as regards attitudes brought to it or implications seen in it; a point of view.

[**1845**: see LOOK *v.* 3 a. **1861** T. HUGHES *Tom Brown at Oxf.* III. v. 95 Mary's habits, and thoughts, and ways of looking at and judging of people and things, were much changed.] **1881** H. JAMES *Portrait of Lady* I. xvii. 218, I can't make out that what he tells me about the royal family is much to their credit; but he says that's only my peculiar way of looking at it. **1893** 'L. CARROLL' *Sylvie & Bruno Concluded* ii. 27 It's a new way of looking at it—to me..but it seems a *true* way, also. **1905** E. G. WHITE *Ministry of Healing* 483 We differ so widely in disposition, habits, education, that our ways of looking at things vary. **1911** D. H. LAWRENCE *White Peacock* II. i. 202 It's one way of looking at things. **1963** D. LESSING *Man & Two Women* 141, I mean to say, you've got to take the rough with the smooth, and there's no other way of looking at it.

n. *the way*: so that, with the result that. *Ir.*

1899 SOMERVILLE & 'ROSS' *Some Experiences Irish R.M.* v. 118 A couple o' dhraws o' th' ash plant across the butt o' the tail, the way I wouldn't blind her. **1912** J. STEPHENS *Crock of Gold* xiv. 225 Be sure and hold him tight..the way we can have a good look at him. **1924** R. MACAULAY *Orphan Island* i. 16 We must see about fermenting some of this fruit-juice, the way we'll get something fit to drink.

o. *it's this way*: a colloq. formula introducing an oral explanation.

1905 'O. HENRY' *Strictly Business* (1910) But it's this way: Suppose you're a Fifth Avenue millionaire, soaring high. **1938** T. CALDWELL *Dynasty of Death* (1939) I. 385 Well, it's this way, Paul: you see, a number of us must stay at home to make the guns..for our soldiers to use. **1961** *East Anglian Mag.* July 505 'It's this way,' he expounded to his cronies in the know. **1970** [see *LOSE *v.*[1] 9 b].

15. (*in*) *more ways than one.*

a 1600 HOOKER *Wks.* (1836) III. 796 That justice exacteth punishment for offending, even after their offences be forgiven them, there is, as it seemeth, proof sufficient more ways than one. **1816** *Edin. Rev.* Dec. 464 The foreign Commissioners had not yet reached St Helena, whose presence in the island may justly have alarmed Sir Hudson, in more ways than one, for the safety of his prisoner. **1895** HARDY *Jude* v. viii. 398 Her loss was a loss to me in more ways than one! **1952** M. R. RINEHART *Swimming Pool* xiv. 129 It's a dead end, in more ways than one.

16. a. (*to put*) *in the* (or *a, that*) *way*: (to make) pregnant; cf. *in the family way* s.v. FAMILY 10 b.

1817 JANE AUSTEN *Let.* 23 Mar. (1952) 488 Mrs Clement too is in that way again. I am quite tired of so many Children. **1831** MRS. ARBUTHNOT *Let.* 18 Feb. in C. Arbuthnot *Corr.* (1941) 140 Young Ly. C. is *not in a way*; the old housekeeper wrote to her something about it, & she wrote back word, 'God's will be done.' **1867** QUEEN VICTORIA *Let.* 25 Oct. in R. Fulford *Your Dear Letter* (1971) 155 Dearest child, why did you not tell me, your own mother, when you first began being in that way? **1960** *Pick of Today's Short Stories* XI. 215 They'd both eloped... 'I'll bet he put her in the way'. **1980** J. ROSE *Elizabeth Fry* iii. 44 She suspected herself of being pregnant, 'in the way' as she called it.

19. a. *in a small way* (earlier example).

1833 CARLYLE in *Fraser's Mag.* July 27/1 Next, however, as another more lasting resource, he forges; at first in a small way.

b. *in a big way*: on a large scale, intensively; (*colloq.*) with great enthusiasm or display; very much, very well. orig. *U.S.*

[**1903** *Dialect Notes* II. 306 *Big way* (to get in a), *v. phr.*, to become excited. 'The preacher got in a *big way* and you could hear him a mile.'] **1927** F. HARRIS *My Life & Loves* III. v. 69, I meant to take up the whole problem of journalism in a big way when I came back. **1932** *N.Y. Times Bk. Rev.* 10 Jan. 17/3 A gangster who calls himself Napoleon and who goes in for crime in a big way. **1936** H. L. ICKES *Diary* 30 June (1955) I. 626 The speech went over in a big way. **1943** J. S. HUXLEY *TVA* vii. 51 Over half a million acres..of fishable water..are already being taken advantage of in a big way. **1950** C. MACINNES *To Victors* II. 211, I could go for her in a big way. **1955** A. HUXLEY *Let.* 18 Mar. (1969) 738 Amanita muscaria..he thinks will open the doors of ESP in a big way. **1958** *Times Rev. Industry* Dec. 57/2 Users of traditional materials are looking to the..use of..plastics in a big way. **1980** A. MARS-JONES *Lantern Lect. & Other Stories* (1981) 10 The Trust people played hard-to-get until he started bequeathing the property..to the Welsh nationalists. *Then* they sat up and took notice In A Big Way.

IV. 23. c. For *dial.* read *dial.* and *U.S.* Also *a ways* (unqualified).

1927 W. FAULKNER *Mosquitoes* 202 I'll carry you a ways, until we get somewhere. **1933** BLOOMFIELD *Language* ii. 40 A speaker can be heard only a short ways and only for an instant or two. **1938** T. WILDER *Our Town* I. 52 Can I walk along a ways with you? **1976** *New Yorker* 15 Mar. 67/1 As it passed over the ridge to land on the other side it hit a tree quite a ways up. **1979** N. MAILER *Executioner's Song* (1980) I. xxix. 445 Though I suppose at some point in the future..he may be eligible for parole, that's a long long ways away.

V. 27. pay one's way. b. (Earlier example.)

1803 G. COLMAN *John Bull* II. iii. 22, I earned my fair profits; I paid my fair way.

34. in the way. (See also sense 16 a above.)

35. j. *in way of* (Naut.): = *in wake of* s.v. WAKE *sb.*[2] 4 b (*a*).

1950 L. BAKER *Design Marine Water-Tube Boilers* ix. 144 *Arboring*, a term applied to the removal of tube plate material in way of tube ends to reduce the unsupported length of tube inside the drum. **1957** *Shipping World* 21 Aug. 153/2 Longitudinal centre-line bulkheads are provided in all the main and tweendeck holds,..and in way of the hatch openings there are steel supports for removable wooden longitudinal bulkheads. **1960** *Fishing Gaz.* (N.Y.) 15 Mar. 18/1 The propeller shaft has 5/8" thick centrifugally cast liners or wearing sleeves shrunk on in way of the stern bearing and stuffing box.

36. on the way (this form only): *spec.* (*colloq.*) (*a*) pregnant; (*b*) (of a child) conceived but not yet born.

(*a*) **1588** [in Dict., sense 36 a]. **1865** A. MACDONALD *Let.* Apr. in A. W. Baldwin *Macdonald Sisters* (1960) vi. 100 Poor Mrs Hughes..is 'on the way again, blest if she ain't'. **1941** E. WELTY *Curtain of Green* (1943) 39, I bet you another Jax that lady's three months on the way.

(*b*) [**1858**: cf. sense 38 b in Dict.] **1896** KIPLING *Day's Work* (1898) 271 I'm a married man, an' my fourth's on the ways [*sic*] now, she says. **1919** V. WOOLF *Night & Day* ix. 121 He has two children, and another on the way. **1961** G. GREENE *Burnt-Out Case* vi. i. 183, I think I have a baby on the way... He doesn't want one. **1983** R. RENDELL *Speaker of Mandarin* v. 69 We've..four simply adorable grandchildren with another on the way.

b. imp. (*be*) *on your way*: go away, get going; also (*U.S.*), 'get away' (GET *v.* 54 b in Dict. and Suppl.). *colloq.* (orig. *U.S.*).

1903 'O. HENRY' *Trimmed Lamp* (1907) 236 Be on your way, Freddie. **1929** WODEHOUSE *Gentleman of Leisure* xi. 86 'We're going down with him to the country today, Spike, so be ready.' 'On your way, boss. What's dat?' **1974** 'P. B. YUILL' *Bornless Keeper* xii. 113 Toddle off back to Victoria... On your way, amigo.

c. (*I am*) *on my way*: a formula used to express the speaker's intention of hurrying or of making an immediate departure.

1919 in N. I. White *Amer. Negro Folk-Songs* ii. 124 Lord I'm on my way... Lord I'm on my way. **1948** G. VIDAL *City & Pillar* ii. 35 'I'm on my way,' said Jim. **1971** 'L. EGAN' *Malicious Mischief* (1972) ix. 158 'Call just in..—they had a prowler over on Jackson... He was armed.'.. 'I'm on my way!' snapped Varallo. **1972** J. PHILIPS *Vanishing Senator* I. iv. 37 'Step on it, will you?' 'On my way,' Peter said. **1978** A. PRICE *'44 Vintage* xii. 152 If it's all the same to you, m'sieur, we'll be on our way.

38. under way. Now freq. as one word: see *UNDERWAY *adv.*

VI. 39. a. *way-end* (earlier example); **c.** *way-weary* adj.

1869 W. BARNES *Early England & Saxon-English* 106 When the railway was taken into the hands of more learned men, we had... the *terminus* instead of the rail-end, or way-end, or outending. **1916** BLUNDEN *Harbingers* 60 Wayweary traveller, with your broad bright eyes. **1926** T. E. LAWRENCE *Seven Pillars* (1935) 5 Love, the way-weary, groped to your body.

40 a. way-freight *N. Amer.*, goods that are picked up or set down at intermediate stopping places on a railway or shipping route; also, a train carrying such freight; **way letter** (earlier examples); **way-place** *U.S.*, a stopping place on a road or railway; a wayside hostelry or an intermediate station; **way-point**, substitute for *U.S.* and def.: orig. *U.S.*, a stopping-place on a journey; (earlier and later examples); also, (on an air journey) the computer-checked coordinates of each stage of a long flight; **way-port**, a port which normally serves as a port of call rather than as an ultimate destination; **way-stop**, chiefly *U.S.*, an intermediate stopping place on a journey; also *fig.*; **way train** *U.S.*, a train which stops at intermediate stations on a railway; a stopping train.

1833 *Niles' Reg.* XLIV. 260/2 The hatch..was open to get out a lot of way-freight. **1875** 'MARK TWAIN' in *Atlantic Monthly* Aug. 191 No way-freights and no way-passengers were allowed, for the racers would stop only at the largest towns. **1898** H. E. HAMBLEN *Gen. Manager's Story* 37 The way freight..had crossed over to load some freight. **1977** *Islander* (Victoria, B.C.) 8 May 2/3 The discharging of cargo and the loading of way-freight. **1773** H. FINLAY *Jrnl.* 11 Nov. (1867) 38 Way letters he makes his own perquisite. **1851** E. BOWEN *U.S. Post-Office Guide* 47 On the letters brought by a mail carrier to be mailed, called way-letters, one cent. is to be charged in addition to the usual postage. **1849** H. MELVILLE *Jrnl. Visit to London* (1948) 67 In a fit of the nightmare was going to stop at a way-place, taking it for the place of my destination. **1883** 'MARK TWAIN' *Life on Mississippi* lii. 512 She got out of the cars at a way-place. **1880** *Harper's Mag.* Dec. 53 The Ohio is plied by a line of Cincinnati and Pittsburgh packets, and by smaller craft earning a precarious existence between 'way' points. **1899** J. LONDON *Let.* 12 Sept. (1966) 54 And to-morrow I start out on that postponed trip of mine to Stanford University and Mt. Hamilton, to say nothing of way points. **1971** *Flying* Apr. 29/1 (Advt.), Computer and waypoint selector lets you fly direct to a destination hundreds of miles away.. forget about airways dog-legs. **1983** *Times* 6 Sept. 26/2 They plot course by typing..a series of 'way-points' into the computer. Such way-points occur every four hundred miles, so even if one were wrong, the next should put aircraft back on course. **1984** *Sunday Times* 20 May 34/2 The co-ordinates of the different 'waypoints', or intermediate stages along the flight..were checked and found correct. **1897** 'MARK TWAIN' *Following Equator* xxxii. 303 A good many of us got ashore at the first way-port to seek another ship. **1901** *Daily Colonist* (Victoria, B.C.) 11 Oct. 3/2 The steamer Princess Louise..has been tied up for repairs, and there will be no steamer leaving for the canneries and way ports of the north until Monday at least. **1927** *Blackw. Mag.* Mar. 330/2 'A way-port!' he sighed, after ordering coffee. 'It's turned out to be a terminus for a good many fellows like me.' **1961** WEBSTER, Way-stop. **1969** *Islander* (Victoria, B.C.) 17 Aug. 12/3 She [*sc.* a steamboat] makes various way-stops on each trip and as Skipper McMinn says—'We stop for anyone who jumps up and down on the shore and gives us a holler.' **1981** *Southern Horticulture* (N.Z.) Spring 13 A town that's now making it. Martinborough was once just a way stop on the road to Pirinoa. **1983** C. G. HART *Rich die Young* iv. 45 Pat was using the Academy as a way-stop while he tried to break into the movies. **1873** 'MARK TWAIN' & WARNER *Gilded Age* xxix. 269 Next morning..he descended, sleepy and sore, from a way-train. **1920** S. LEWIS *Main Street* 22 The hordes of the way-trains were not altogether new to Carol.

way, *adv.* Restrict *Obs.* exc. *Sc., north.*, and *U.S.* to senses 1, 2 b, and 3, and add: **2. a.** (Later examples.)

1927 BARONESS ORCZY *Sir Percy hits Back* v. 35 The three men had become mere specks, 'way down the road. **1959** *Times Lit. Suppl.* 16 Oct. 589/3 You are feet deep in snow and the temperature is way below zero. **1972** *Guardian* 17 Nov. 1/6 The census figures confirm..that unemployment is way above the official figure. **1979** R. PERRY *Bishop's Pawn* v. 76 You're way off course... It's back on Unter den Linden.

b. With *down, over.* For *way back, in, off, out, up* see main entries below.

1851 E. S. WORTLEY *Trav. in U.S.* xxiii. 138 The trading and wealthy cities of far off Alabama and Louisiana, 'way down south'. **1854** Way down [in Dict., wrongly placed in sense 2 a]. **1866** *Atlantic Monthly* May 640 Nor these ain't metters thet with pol'tics swings, But goes 'way down amongst the roots o' things. **1850** L. H. GARRARD *Wah-to-Yah* xvii. 222 Calyforny! way over yonder!

c. *fig.* Much, far. *U.S.*

1941 L. I. WILDER *Little Town on Prairie* v. 34 'I wonder how much it costs,' said Ma. ''Way too much for ordinary folks,' said Pa. **1957** *New Yorker* 2 Nov. 105/2 Go by plane, train or ship. Arrive way sooner—relaxed! **1977** *Rolling Stone* 24 Mar., He was a country & western singer and he drank way too much.

wayang (wa·yaŋ). [Javanese *wajang, wayang.*] **1.** In Indonesia and Malaysia, a theatrical performance employing puppets or human dancers; *spec.* a type of Javanese shadow puppet play.

1808 *Asiatick Res.* X. iii. 181 Scenic exhibitions termed 'Wayang-wayang', were till lately, very common in the peninsula of Malayu [*sic*]. **1853** *Jrnl. Indian Archipelago* VII. 271 A very large collection of Wayang figures, cut out of hide. **1881** *Encycl. Brit.* XIII. 604/2 The wayangs or puppet plays, in which grotesque figures of gilded leather are moved by the performer, who recites the appropriate speeches, and as occasion demands plays the part of chorus. **1910** [see *shadow-play* s.v. SHADOW *sb.* 16]. **1934** H. W. PONDER *Java Pageant* iv. 49 The stories have been immortalized..by the 'Wayang', or native theatre, which for countless generations has been so much a part of Javanese life. **1965** L. PALMIER *Indonesia* xi. 171 The wayang performance has always had a magic and religious significance. **1973** D. MAY *Laughter in Djakarta* iv. 71 Tonight he was going..to see a *wayang*, an old Hindu-Javanese story acted and danced by a famous company from central Java. **1978** N. FREELING *Night Lords* xvi. 72 The wayang dollies depended..upon the skill and suppleness of the manipulator's fingers.

2. With qualifying term (see quot. 1969), as **wayang kulit** [Javanese *kulit* skin, leather], the Javanese shadow puppet play.

1893 W. B. WORSFOLD *Visit to Java* x. 178, I had an opportunity of witnessing..the wayang *klitik*, in which the puppets are exhibited themselves to the audience instead of being made to project shadows. **1894** J. D. VAUGHAN in N. B. Dennys *Descr. Dict. Brit. Malaya* 324 In a puppet show the figures are seen and in this their shadows are. The show is called *wayang kulit*, or leather puppets. **1936** G. B. GARDNER *Keris* i. 21 The figures are usually grotesque, of the *wayang kulit* type. **1965** *Economist* 13 Mar. 1149/3 The traditional Indonesian *wayang purwa* puppet theatre show, usually performed at night, in which the vague shadows of leather puppets are fuzzily silhouetted on a screen. **1966** D. FORBES *Heart of Malaya* viii. 100 Figures that moved across the frame of the doorway looked like silhouettes in the *wayang kulit*, the shadow play. **1969** A. R. PHILPOTT *Dict. Puppetry* 277 The precise nature of the performance is distinguished by a qualifying term—e.g. wayang purwa, wayang gedog,.. wayang wong—the last being a performance by human actors... The qualifying term may indicate type of puppet or the type of plays. **1976** *Times* 31 Aug. (Malaysia Suppl.) p. iv/4, Both farmers and fishermen are largely confined to their houses during the monsoon season, when some of them cut silhouette puppets from cowhide for the traditional Wayang Kulit shadow play—a long procession of characters, based on the Indian epic *The Ramayana.*

way-back, *adv.*, *a.*, and *sb. colloq.* (chiefly *U.S.*, *Austral.*, and *N.Z.*). Forms: see below. [f. WAY *adv.* + BACK *adv.*] **A.** *adv.* Usu. written **way back. 1.** Far away; in or *from* a remote rural area.

1855 *Merry's Museum* XXIX. 58, I did not know that there was such a fine Magazine,..for I live 'way back in the woods'. **1884** *Boston Globe* Oct., His unkempt hair, gawky appearance, and homespun suit..all bespoke the citizen from wayback. **1889** KIPLING *From Sea to Sea* (1899) II. xxvi. 28 'You'll see the salmon-wheels 'fore long,' said a man who lived 'way back on the Washoogle'. **1916** G. THORNTON *Wowser* 85 The condition of those living 'wayback'. **1930** M. R. E. BLAIR *By Pacific Waters* ii. 12 I'm sure they are getting a storm way-back. **1936** 'F. GERALD' *Millionaire in Memories* iii. 106 Blackall was much the same as any other township 'way back'. **1976** *National Observer* (U.S.) 6 Nov., 'We needed something to let people know that there was a church there,' says Rector Mart Gayland Pool. 'We sat way back on a slope near a golf course, and our facility looked more like a bowling alley or a Kinney shoe store.'

2. A long time ago; *from way back*, since a long time ago; hence, through and through.

1887 *Lantern* (New Orleans) 30 July 3/3 Nick is a walker from way back. **1889** [see WAY *adv.* 2 b]. **1889** KIPLING *From Sea to Sea* (1899) I. xxii. 432, I am an American by birth—an American from way back. **1892** 'MARK TWAIN' *Amer. Claimant* 167, I tell you, he's an artist from way back! **1907** [see WAY *adv.* 2 b]. **1923** J. H. COOK *50 Yrs. Old Frontier* III. 227 This occurred 'way back', when the Indians had no horses. **1939** *Country Life* 11 Feb. p. xxxvi/1 The 'Rule for Anchoresses', written way back in the thirteenth century. **1948** *Sporting Mirror* 19 Nov. 6/2 Way back in September I said that Norwich City have got what it takes and would be a power this season. **1969** J. D. A. WIDDOWSON in Halpert & Story *Christmas Mumming in Newfoundland* 218 The real mummers we used to have way back. **1970** J. LENNON in J. Wenner *Lennon Remembered* (1972) 144 He not only knew my work, and the lyrics that I had written but he also understood them, and from *way* back. **1978** R. HILL *Pinch of Snuff* ix. 87 'You know Burkill, sir?' he asked. 'From way back.'

B. *adj.* Usu. written **way-back, wayback. 1.** Of long ago.

1885 *Santa Fé Weekly New Mexican* 10 Sept. 1/2 A writer in giving a description of the plains of Kansas in the 'way back' time, speaks of it as a 'wilderness of grasses'. **1962** *Observer* 17 June 3/6 [U.S. writer] He had an ancestor cult. There was a painting of a way-back ancestor wearing a woollen waistcoat.

2. Hailing from, or located in, a remote rural area.

1887, 1918 [see WAY *adv.* 2 b]. **1933** *Bulletin* (Sydney) 24 May 21/2 Mrs. Smallbeer sold her very wayback pub. **1950** K. S. PRICHARD *Winged Seeds* xv. 165 Sally understood that this visit to a way-back station was an experience for them. **1951** *Landfall* V. 278 Miss Dane.. imagined herself a pioneer of progress in a wayback community. **1973** M. CARNEGIE *Friday Mount* 218 No wonder some of the way-back towns had that look.

C. *sb.* Form as for the adj. **1.** A person inhabiting or coming from a remote district.

1890 E. CUSTER *Following Guidon* xvii. 261 We were, in Western terms, 'waybacks from wayback'. **1912** R. S. TAIT *Scotty Mac, Shearer* 125 At a group in front of him a thimble-rigger was expending much eloquence to induce a party of waybacks to relieve him of his surplus cash. **1924** *Contemp. Rev.* Aug. 236 The well-known resourcefulness and sense of location of the present-day way-backs of Australia..are due..to lessons taught the pioneers by the natives.

2. Remote rural areas; *spec.* the Australian outback.

1890 [see prec. sense]. **1901** F. J. GILLEN *Diary* 5 Oct. (1968) 277 The station is built on a stony rise on the western bank of the Newcastle and like most stock stations in the 'wayback' there has been no attempt made to improve the appearance of the surroundings. **1925** H. H. COOK *Far Flung* v. 39 Some 50 miles in the 'way-back'. **1933** *Bulletin* (Sydney) 27 Sept. 20/4 Waltzing Matilda in the waybacks of S.A. I came upon a well-found homestead.

way-bill. 2. (Earlier example.)

1851 W. KELLY *Excursion to California* I. x. 172 According to our way-bill, and comparing it with our computed travel, we could not be far off the next crossing of the Platte.

way-·in, *a. slang.* [f. WAY *adv.* + IN *adv.*, after *WAY OUT *a.*] Conventional; fashionable, sophisticated.

1960 *N.Y. Times Mag.* 12 June 19/1 A famous lady columnist with a way-out taste in millinery but a way-in taste in film fare. *Ibid.* 78/4 Many artists..have sought refuge in..way-in or way-out religious conversions. **1967** *Punch* 29 Nov. 817/2 There's a real way-in guy looking like how a guy on *The Times Saturday Review* ought to look like.

waying (wē[i]·iŋ), *vbl. sb. poet. nonce-wd.* [f. WAY *v.* + -ING[1].] A going away; departure.

1922 HARDY *Late Lyrics & Earlier* 120 So, with this saying, 'Good-bye, good-bye,' We speed their waying.

way-leave, wayleave. Add: More widely, a right of way granted by the owner of land to a particular body and for a particular purpose, often in return for payment; also, a document conferring the right. (Further examples.)

1928 H. G. WELLS *Way World is Going* xi. 127 The exploitation of the air, as a means of..available travel, is hopeless without..secure wayleaves over Europe. **1930** *Times* 22 Mar. 19/2 We have also practically completed the Scottish overhead 'Grid' system for the Central Electricity Board... Owing to the difficulty of securing permission for wayleaves our work was necessarily interfered with. **1960** *Farmer & Stockbreeder* 1 Mar. 91/1 The Central Electricity Generating Board has no legal right of entry on your land without your permission or until such time as you have signed a wayleave. **1963** *Times* 24 May 16/6 A wayleave for a future subway linking the triangle with the Radcliffe Infirmary and passing under the Banbury and the Woodstock roads.

attrib. **1960** *Times Rev. Industry* Feb. 75/1 There is no wayleave payment. **1971** P. GRESSWELL *Environment* 215 Landowners are in a strong position to influence power line proposals through granting or refusing wayleave consent.

way-mark, *v.* [f. the sb.] *trans.* To provide or identify (a path) with waymarks. Hence **way-marked** *ppl. a.*, **way-marking** *vbl. sb.*

1960 *Guardian* 5 Nov. 6/3 The Ramblers' Association.. is cleft..over the question of 'waymarking' paths and routes by means of blobs of coloured paint. *Ibid.* 9 Nov. 8/3 The plan to provide 'waymarking' on some popular British fells. **1973** *Village* Autumn 51 Walkers on the Continent or in America find themselves well supplied with long-distance trails and with waymarked paths in tourist areas. **1982** *Walker's Britain* 328/2 There are still several stretches where the waymarking is incomplete because public rights of way have not been obtained. **1983** *Out of Town* July 72/2 Much effort and imagination has gone into waymarking the walks.

way off, *adv.* and *a.* orig. *U.S.* [f. WAY *adv.* + OFF *adv.*] **A.** *adv.* Far away.

1853 G. C. HILL *Dovecote* 29, I found her 'way off in them woods yonder! **1897** [in Dict. s.v. WAY *adv.* 2 b]. **1897** KIPLING *Captains Courageous* iii. 60 'Way off yander's the *Day's Eye.* **1929** D. H. LAWRENCE *Pansies* 72 Out of the soul's middle to the middle-most sun, way-off, or in every atom. **1952** *Manch. Guardian Weekly* 22 May 7/2 Way off to the right was a beery Irishman.

2. Far from the intended target; greatly mistaken, quite wrong.

1892 *Harper's Mag.* Feb. 438/2 The papers are generally 'way off in some things. **1906** A. H. LEWIS *Confessions of Detective* I. i. 10 'You're dead wrong, Doc!' broke in Mugsey... 'You're 'way off.' **1962** J. GLENN in *Into Orbit* 209 My attitude indications on the instrument panel are way off. **1973** 'H. HOWARD' *Highway to Murder* x. 118 Seemed first impressions could be 'way off. **1977** H. KAPLAN *Damascus Cover* (1978) iv. 33 On the pistol range Ari discovered..that his aim was way off.

B. *adj.* **1.** Usu. **way-·off.** Distant.

1870 A. D. WHITNEY *We Girls* (1871) xi. 229 General and Mrs. Ingleside..had come from their way-off, beautiful Wisconsin home. **1928** D. H. LAWRENCE *Woman rode Away* 93 The way-off things like the sun.

way out, *adv.*, *a.*, and *sb.* orig. *U.S.* [f. WAY *adv.* + OUT *adv.*] **A.** *adv.* Far out, far away. *colloq.*

1868 G. A. CUSTER *Let.* 20 Feb. in E. Custer *Following Guidon* (1890) 53 They had braved the perils..in order to bring us, 'way out here, news from our loved ones. **1882** *Congress. Rec.* 9 Mar. 1758/1 Instead of that they go way out to Peoria, Illinois. **1933** L. I. WILDER *Farmer Boy* xvi. 119 This country..goes 'way out beyond Kansas.. down to the Pacific Ocean. **1944** M. LASKI *Love on Supertax* xi. 102 After a..journey by 'bus, Clarissa alighted way out on the North Circular Road. **1977** *Chicago Tribune* 2 Oct. v. 2/2 There is more clout in the Senate, you see, than way out west away from the mainstream.

B. *adj.* Also **way-ou·t. 1.** Far removed from reality or from convention; extreme; progressive, avant-garde, advanced. *slang.*

1958 G. LEA *Somewhere there's Music* xix. 164, I turn on [*sc.* smoke marijuana] a little and I get way out. **1959** *Encounter* June 42/2 The ego-ideal of the Beatnik is the 'cool hipster'—..the man who is 'way out'. **1959** N. MAILER *Advts. for Myself* (1961) 296 Mine [*sc.* my hypothesis] is interesting, mine is way out (on the avenue of the mystery along the road to 'It') but still I am just one cat in a world of cool cats. **1961** *Life* (Internat. ed.) 8 May 44 The way-out world of micro-electronics. **1964** J. DUNBAR in Hamblett & Deverson *Generation X* 179 One thing I like about Cambridge, people don't try to be too way out. At places like Oxford, or Reading, I've seen blokes going around barefoot and wearing ear-rings. **1967** *Times* 13 Apr. 23 The 'way out' field of nuclear fusion. **1972** J. PHILIPS *Vanishing Senator* III. iii. 162 Vardon thought up a way-out scheme to commit a murder. **1979** *Dædalus* Spring 141 These [*sc.* Radio Città in Bologna and Radio Alice] were 'way-out' examples of local radio. **1982** BARR & YORK *Official Sloane Ranger Handbk.* 38/1 She would not wear Art Nouveau jewels (too way out).

2. Greatly mistaken. *slang.*

1959 M. SUMMERTON *Small Wilderness* viii. 112 You're way out, Puss. If anybody has got a perfect alibi..it's Cliff. **1965** *New Statesman* 7 May 721/3, I would not presume to argue with Edward Hyams on matters botanical, but he is way out on snakes.

C. *sb.* A person who holds extreme or unconventional views.

1968 *Manch. Guardian Weekly* 17 Oct. 19 The drop-outs have copped out, the redskins have bitten the dust, the way-outs have faced the nitty-gritty (truth). **1970** *Encycl. Sci. Suppl.* (Grolier) 158 Two years ago we were considered way-outs. Nobody knew what conservation meant.

Hence **way-ou·tness** *slang*, unconventionality.

1965 *Tablet* 27 Nov. 1327/1 Some way-outness prevents him from acquiescing in the wisdom of acceptance. **1968** R. V. BESTE *Repeat Instructions* i. 10 That's the kind of way-outness that should have made Security suspicious. **1979** S. SMITH *Survivor* xvii. 185 Young people of all degrees of 'way-outness'.

ways and means. Add: **2. a.** Also *attrib.*

1867 *Oregon State Jrnl.* 5 Jan. 2/2 The Ways and Means Committee decided to postpone an action on Mr. Boutwell's bill. **1919** *Lit. Digest* 22 Mar. 21/2 Mr. Fordney, of Michigan,..will probably be..Chairman of the Ways and Means Committee. **1973** B. J. SIMS *Suppl. to Sergeant on Stamp Duties* (ed. 6) A55 This section and the associated Ways and Means Resolution provides powers whereby changes in stamp duty may be given effect to by means of a Budget Resolution. **1977** *Time* 12 Dec. 34/2 Al Ullman, the House Ways and Means chairman, has been pleading with Carter for a 'minimalist' rather than a 'maximalist' tax bill.

wayside. Add: **a.** Phr. *to fall by the wayside* [after Luke viii. 5: see quot. 1526 in Dict.], to fail to stay the course, to drop out.

1965 *New Statesman* 7 May 719/1 Responding to persuasion, young wives go back into teaching... Some toughen and survive, others fall by the wayside. **1973** *Times* 24 May 10/1 She went to San Diego for a world junior tournament last year and fell by the wayside only because of a marker's error. **1977** L. T. MILIC in Bond & McLeod *Newslett. to Newspapers* I. 41 As society changes, so must the tone of a publication or it falls by the wayside.

b. *wayside pulpit*, a board, usu. placed outside a place of worship, displaying a religious text or maxim.

1925 *Advertising World* Dec. 302/2 (*heading*) How the 'wayside pulpit' scheme was organized. **1932** Q. D. LEAVIS *Fiction & Reading Public* II. iv. 193 An inspection of the slogans displayed on Wayside Pulpits..reveals that they are largely devoted to denunciation of an attitude described as pessimistic. **1976** *Church Times* 20 Feb. 11/1 'My greed is another's need' and 'Live simply that others may simply live' have become familiar slogans on wayside pulpits and in parish magazines. **1981** F. INGLIS *Promise of Happiness* iii. 93 We most of us *know* when we are lying. My wayside pulpit-point is that we no longer care very much.

way station. *U.S.* Also **way-station. 1. a.** (In Dict. s.v. WAY *sb.*[1] 40.) Also *transf.*

1850 *Ann. Rep. Railroad Corp. Massachusetts 1849* 21 Way stations for express trains. **1854** *Harper's Mag.* VIII. 566/2 The boats touched at most of the prominent towns on the river, to land such passengers as might desire to disembark at 'way-stations'. **1855** [see *LOCK *v.*[1] 7 d]. **1856**, etc. [in Dict.]. **1912** F. J. HASKIN *Amer. Govt.* 210 Cities which are to-day mere way stations on the international routes of trade will grow into rich world centers. **1914** *Sat. Even. Post* 4 Apr. 52/2 At ten o'clock that night at a way-station the kid was ditched. **1934** A. WOOLLCOTT *While Rome Burns* 258 To fly by way of Ottawa, Point Barrow, and other way-stations. **1944** *Daily Progress* (Charlottesville, Va.) 21 July 1/8 The island [*sc.* Guam].. formerly served as a way-station on the trans-Pacific airroute to the East. **1976** N. THORNBURG *Cutter & Bone* vi. 134 It was..a way station carefully restored and preserved to offer at least a semblance of its original state. **1984** *New Yorker* 9 July 43/2 The United States may..use Moroccan facilities..as way stations for combat troops destined for service in, say, the Middle East.

b. *fig.*, or in fig. context.

1892 *Congress. Rec.* 23 Mar. 2462/2 'Will the gentleman allow me to ask him a question?'..'The gentleman will excuse me. On a fast schedule I can not stop at way stations.' **1926** [see *TECTAL *a.*]. **1948** MENJOU & MUSSELMAN *It took Nine Tailors* 10 She thought a theater was just a way station on the road to perdition. **1960** BRUNER & KLEIN in Kaplan & Wapner *Perspectives in Psychol. Theory* 65 There are corticofugal impulses that go down through the reticular formation to program selectivity of intake by way stations in the sensory system. **1973** *Sci. Amer.* July 52/1 Lymphocytes are found in high concentrations in the lymph nodes, way stations along the lymphatic vessels. **1978** W. GARNER *Möbius Trip* (1979) i. 12 The Belgravia apartment was not a home... It was a way station. **1982** H. KISSINGER *Yrs. of Upheaval* viii. 302 The cease-fire was merely a tactic, a way station toward their objective of taking over the whole of Indochina by force.

way up, *adv.* and *a.* orig. *U.S.* [f. WAY *adv.* + UP *adv.*] **A.** *adv.* Far up.

1851 'E. WETHERELL' *Wide, Wide World* I. xii. 150 Do you live 'way up there? **1862** O. W. NORTON *Army Lett.* 100 A minister of the gospel who was so wonderfully.. war-like way up in Erie. **1901** [in Dict. s.v. WAY *adv.* 2 a]. **1946** K. TENNANT *Lost Haven* (1947) vii. 95 If you owned a bath, it put you 'way, 'way up in the world. **1956** H. KURNITZ *Invasion of Privacy* xiv. 92 She came down the stairs and I thought that was funny because they live way up top. **1972** *National Observer* (N.Y.) 27 May 14/1 Way up in the high-level social circles are those who are so rich they never, ever talk about money.

B. *adj.* Usu. **way-·up.** Excellent, first-class; of high social standing. *slang* (chiefly *U.S.*).

1887 F. FRANCIS *Saddle & Mocassin* 81 A real way-up cook, who could make chile-con-carne, tamales, and all the best Mexican dishes. **1902** KIPLING *Traffics & Discoveries* (1904) 17 He's a way-up barrister when he's at home. **1909** 'O. HENRY' *Roads of Destiny* xviii. 299, I want to be manager of something way up—like a railroad or a diamond trust or an automobile factory.

way-up (wē[i]ʌp), *sb. Geol.* [f. phr. (*which, this, the right*, etc.) *way up* (WAY *sb.*[1] 9 a, UP *adv.*[2] 9 a).] Orientation as regards which part is uppermost or was deposited last.

1958 *Q. Jrnl. Geol. Soc.* CXIII. 364 Way-up has been determined from graded bedding. **1969** BENNISON & WRIGHT *Geol. Hist. Brit. Isles* ii. 34 Certain characters, such as..the way-up of shells, provide evidence of bottom conditions and the strength of currents. **1982** COLLINSON & THOMPSON *Sedimentary Structures* vi. 97/1 Aeolian structures could help to establish way-up in highly dipping sequences.

Wazir[2] (wāzī[ə]·ɪ). Also **Waziri**; 9 **Vazíri, Vizeeree, Wuzeerá**, etc. A member of a Pathan people of north-west Pakistan; also, this people collectively. Also *attrib.* or as *adj.*

1815 M. ELPHINSTONE *Caubul* III. iii. 385 The Vizeerees are said to be tall and muscular. **1838** in *Parl. Papers 1839* XL. 29 The Vezeeree territory. **1842** C. MASSON *Journeys* I. vii. 100 A few Vazíri huts. **1847** H. B. EDWARDES *Diary* 3 Dec. (1911) 156 A very extensive triangular *Thull*, or sand, occupied by the Vizeeree tribes. **1851** R. G. LATHAM *Ethnol. Brit. Colonies* iv. 182 The mountaineers around them—the *Vizeri*—are a pure branch. **1873**, etc. [see *MAHSUD]. **1924** LD. RONALDSHAY *India* vii. 76 Our dealings with the Mahsud Waziris. **1957** B. J. GOULD *Jewel in Lotus* x. 134 The 'girls' were Wazir tribesmen and the welcome they received was a salvo of rapid fire. **1977** J. CLEARY *High Road to China* v. 163 The Mahsuds and Wazirs have..shot down several of our machines. *Ibid.* 175 The four Waziri..remained behind.

wch, wch., abbrev. of WHICH *pron.*

1739 T. CORAM *Let.* 15 Sept. in R. McClure *Coram's Children* (1981) iii. 31 The Attorny & Solicitor Generals Fees..wch they had for Examining the Proposals. **1793** C. BURNEY *Let.* 10 July in F. Burney *Jrnls. & Lett.* (1972) II. 171, I dread the evils into wch their sudden union wd involve them. **1811** SHELLEY *Let.* 6 Jan. (1964) I. 37 Ignorant of the refinements in Love, wch. can only be attained by solitary reflexion.

wd, wd., abbrev. of *would* s.v. WILL *v.*[1]

1793 [see *WCH, WCH.]. **1800** [see *WHOLESALE *v.*]. **1811** SHELLEY *Let.* 12 Jan. (1964) I. 44 Wd. that I cd. believe them to be as [they] are represented. **1888** H. O. D. DAVIDSON *Let.* 12 July in R. S. Churchill *Winston S. Churchill* (1967) I. Compan. I. v. 169, I thought it wd do him good to spend a day with you. **1910** [see *week-endize* vb. s.v. *WEEK-END]. **1930** E. POUND *XXX Cantos* vi. 24 They wd. have given him poison But for the shape of his sword-hilt. **1981** J. STUBBS *Ironmaster* xiv. 188 There wd be an Outcry if the wedding was to fall in the month of May.

we, *pron.* Add: **1. h.** Used in conjunction with *they* to allude to the tension between two mutually exclusive groups or categories of people, or their opposing interests. Cf. *them and us* s.v. *THEM *pers. pron.* 1 e.

[**1884** F. MAURICE *F. D. Maurice* II. xvii. 531 That division of all men into the two classes of 'we' and 'the rest',..which we most of us adopt.] **1926** KIPLING *Debits & Credits* 327 Would you believe it?—They look upon We As only a sort of They! **1942** H. HAYCRAFT *Murder for Pleasure* xv. 316 The..fundamental contradiction between the We and They in government. **1948** R. H. TAWNEY in F. A. Iremonger *William Temple* v. 88 The 'We and They' complex..could not survive continuous co-operation with colleagues..whose experience of life was quite different from his own. **1965** *Observer* (Colour Suppl.) 25 Apr. 13/1 The children have a chance of learning to deal with adults, without the we–they feeling. **1973** *Guardian* 30 Mar. 14/2 Anything which emphasises the 'we' and 'they' of the situation will drive men..towards the extremes of politics.

weak, *a.* Add: **2. b.** *weak sister*, an ineffectual or unreliable person (or either sex); a person of weak character; also *transf. colloq.* (orig. *U.S.*).

1857 *Call* (San Francisco) 3 May 1/1 G. W. Swerzy..is a 'weak sister' and a rather 'bad egg'. **1866** GEO. ELIOT *Felix Holt* I. iv. 95 'I lack grace to deal with these weak sisters,' said the minister. **1949** R. B. WEST *Rocky Mt. Cities* 311 The morning *Rocky Mountain News*..dawdled along as one of the weakest sisters in the Scripps–Howard string [of newspapers]. **1955** E. BERCKMAN *Beckoning Dream* (1956) xix. 135 Luanna was a softie.., a weak sister. She's the one you'd expect to go all to pieces, and she did. **1976** 'J. ROSS' *I know what it's like to Die* xvii. 116 Birdsell was a weak sister... He was..known to be greedy and a physical coward.

12. f. (*b*) *Math.* Of a mathematical entity or concept: implying less than others of its kind; defined by fewer conditions.

1950 W. FELLER *Introd. Probability Theory* I. viii. 157 The strong law of large numbers was first formulated by Cantelli (1917)... Like the weak law, it is only a very special case of a general theorem on random variables. **1964** A. P. & W. ROBERTSON *Topological Vector Spaces* p. vii, It often clarifies results in the theory of normed spaces, especially those concerned with the weak topology, to regard them as particular cases of more general results about topological vector spaces. **1971** G. HIGMAN in Powell & Higman *Finite Simple Groups* vi. 211 We then define E_n to be the weakest equivalence relation on Ω_n satisfying the following three conditions. **1979** *Proc. London Math. Soc.* XXXVIII. 439 Let L be the collection of minimal edge-sets of paths which join *s* to *s'* or *t* to *t'*. Then L has the weak MFMC property (but..in general, not the strong).

k. *Physics.* Applied to one of the four known kinds of force between particles, which is effective only at distances less than about 10^{-15} cm., is very much weaker than the electromagnetic and the strong interactions, and conserves neither strangeness, parity, nor isospin.

[**1953** M. GELL-MANN in *Physical Rev.* XCII. 833/2 Let us suppose that both 'ordinary particles' (nucleons and pions) and 'new unstable particles'..have interactions of three kinds:.. (iii) Other charge-dependent interactions, which we take to be very weak.] **1954** [see *STRONG *a.* 10 e]. **1968** M. S. LIVINGSTON *Particle Physics* vii. 139 Parity conservation is violated in this weak interaction. **1972** G. L. WICK *Elementary Particles* iii. 43 Typical weak interactions are nuclear beta decay and the slow decays of elementary particles. **1976** *Sci. Amer.* Jan. 45/1 The weak force affects every particle but one, the exception being the photon. **1982** *Ann. Reg. 1981* 385 Theorists had already inferred that electromagnetism and the weak force were two extremes of the same thing.

15. d. Of flour: made from soft wheat, so that it contains relatively less gluten and more starch, rises less with yeast, and is less cohesive. Of wheat: soft.

1889 J. BLANDY *Bakers' Guide* (ed. 2) 65 It is very important that young bakers should know how to buy, and blend flour for bread-making; and we..direct them to use a hard dry flour in the sponge, with an eighth part of soft weak flour..to feed the yeast. **1914** *Times* 8 June 16/5 Some flours, among which that from English wheat may be taken as a familiar example, produce small flat close-textured and 'runny' loaves. These are called weak flours. **1924** J. STEWART *Bread & Bread Baking* ii. 15 Weak flours are..important in pastry baking and confectionery. **1951** *Good Housek. Home Encycl.* 466/2 'Soft' or 'weak' wheats contain less gluten and more starch. **1962** *Listener* 22 Mar. 510/1 As a general rule, 'strong' flour contains more nitrogen than 'weak' flour.

e. Of the mixture in an internal-combustion engine: = *LEAN *a.* 4 f.

1918 V. W. PAGÉ *Aviation Engines* iii. 114 A rich mixture ignites much quicker than a weak mixture. **1948** 'N. SHUTE' *No Highway* ix. 217 Mr. Honey's estimate of the time to tailplane failure, under normal weak mixture cruising conditions, was 1,440 hours. **1981** R. BACON *Two-Strike Tuning* ix. 63 All two-stroke engines are sensitive to mixture strength. Too rich and the power and acceleration suffer, while if too weak the result can be a seized or holed piston.

16. b. *weak link*: the weakest or least dependable of a number of interdependent items; also in *Proverb.* Cf. *weak point* (*a*), sense 16 e in Dict.

1868 *Cornh. Mag.* XVII. 295 A chain is no stronger than its weakest link. **1885** [see sense 16 a in Dict.]. **1926** H. CRANE *Let.* 12 Aug. (1965) 272 Mrs. Simpson was enormously pleased at your postcard; and I with your praise of the Dedication. You generally do pick the weakest link; that verse has bothered me. **1942** I. S. SHRIBER *Body for Bill* (1947) xiv. 183 'Stansfield was a pretty weak individual... He was threatening to give the whole thing away.'.. 'A chain is no stronger than its weakest link, remember?' **1965** M. FRAYN *Tin Men* xv. 88 An unreliable man, Haugh, a weak link in the team. **1975** R. LEWIS *Double Take* iv. 138 All right, he lifted the heart drug, but we..picked out the weak link in Ward too. **1984** *Bookseller* 8 Sept. 1049/2, I sometimes seek comfort in the old adage that a chain is only as strong as its weakest link.

e. (*a*) (Earlier example.)

1865 MILL *Auguste Comte* 126 M. Littré..is a disciple only of the Cours de Philosophie Positive, and can see the weak points even in that.

20. Of a variant pronunciation: usual in contexts where the word is unstressed.

1890 H. SWEET *Primer Spoken Eng.* 13 Words that occur very frequently with weak stress often develope a weak form by the side of the original strong one. **1917** D. JONES *Eng. Pronouncing Dict.* p. xix, Circumstances exist in which strong forms occur unstressed, but in no case does a weak form occur stressed. **1982** J. C. WELLS *Accents of Eng.* I. iii. 227 In many accents the pronoun *you* has a weak form /jə/ (conventionally spellable *ya* in the United States, but *yer* in non-rhotic-oriented England).

21. b. (Earlier example.)

1833 *Philol. Museum* II. 385 No weak verb ever in process of time became strong, while strong verbs do become weak.

21*. Similative phrases in which *weak* may have any of various meanings. (See also sense 5 a in Dict.)

1535 BIBLE (Coverdale) *Ezek.* vii. 17 All knees shalbe weake as the water [**1611** (A.V.) weake as water]. **1874** TROLLOPE *Phineas Redux* II. xxx. 244 She would not sin... Having so resolved, she became weak as water. **1926** J. BUCHAN *Dancing Floor* I. ii. 54 We were as weak as kittens, but..extraordinarily happy. **1980** A. PRICE *Hour of Donkey* ix. 123 He must have been as weak as a kitten, with all the blood he'd lost. **1983** J. WAINWRIGHT *Their Evil Ways* v. 154, I think you're mad... Mad and as weak as water.

22. a. *weak-fleshed, -limned, -principled, -skinned, -stressed* adjs.

1967 Weak-fleshed [see *raw-jawed* adj. s.v. *RAW *a.* 9]. **a 1918** W. OWEN *Poems* (1963) 90 The weak-limned hour when sick men's sighs are drained. **1913** D. H. LAWRENCE *Sons & Lovers* ix. 246 It was the nose and eyes of her own mother's people—good-looking, weak-principled folk. **1933** DYLAN THOMAS *Lett.* (1966) 72 And unless you want to regard the man [*sc.* D. H. Lawrence] as a vain, weak-skinned, egocentric, domineering little charlatan, *don't* borrow the book. **1898** H. SWEET *New Eng. Gram.* II. 32 If three strong-stressed words come together—especially in immediate succession, but also with intervening weak-stressed words — the stress of the middle word is often reduced. **1966** *English Studies* XLVII. 83 In languages that use interrogatives as indefinites, such as Dutch, the latter are always weak-stressed.

weakfish. (Earlier example.)

1791 J. BAXTER *Jrnl.* 21 June in *Amer. Speech* (1965) XL. 200 Went a fishing..had 350 weekfish.

weak-kneed, *a.* (Earlier example.)

1863 *Rio Abajo Press* (Albuquerque, N.M.) 24 Feb. 2 But we must forego these comforts and conveniences, because our legislators are too weak-kneed to enact a tax law.

weakly, *adv.* Add: **8.** *Math.* In a way that implies less or that satisfies fewer conditions (cf. *WEAK *a.* 12 f (*b*)).

1955 M. LOÈVE *Probability Theory* ix. 443 If all sequences T^nX are weakly compact, then it is easily proved that sup $\|T^n\| < \infty$. **1972** R. J. WILSON *Introd. Graph Theory* vii. 102 A digraph *D* is said to be connected (or weakly-connected) if it cannot be expressed as the union of two disjoint digraphs.

weak-minded, *a.* **1.** (Earlier example.)

1716 T. CAVE *Let.* 5 Aug. in M. M. Verney *Verney Lett.* (1930) II. xxii. 43 We all continue in perfect sanity of body but very weak-minded this hott weather, as you'l judg by this Scrawll.

weald. Add: **3. wealdsman**, an inhabitant of the Weald.

1928 *Daily Tel.* 29 May 8/5 The flares signalled her triumph to the wealdsmen of Sussex.

Wealden, *a.* Add: **2. b.** Applied to a style of timber house built in the Weald in the late medieval and Tudor periods (see quots. 1961, 1963).

1956 *Archaeol. Jrnl.* CXII. 93 In Kent the aisled hall was replaced by the type of building often called the 'Wealden' house, though it has a wider distribution than that. **1961** M. W. BARLEY *Eng. Farmhouse & Cottage* I. ii. 27 The Wealden house has a central hall open to the roof with a storeyed wing at one end or both. The whole is under a continuous roof, but the storeyed wing is jettied out, to overhang the ground floor by a foot or more. **1963** S. E. RIGOLD in Foster & Alcock *Culture & Environment* xiii. 351 The characteristic hall-house of Kent and Sussex has a unitary hipped roof..covering both the hall and the two-storeyed ends... When the upper stories are jettied laterally, the façade of the hall between them is relatively recessed, and the lateral wall-plate of the jettied ends is carried right across..in front of the wall-plate of the hall... This is the so-called 'Wealden house'. **1974** *Country Life* 14 Feb. 312/4 A yeoman farmer's house of the Wealden type that existed in the mid 15th century.

we-all (wī-ǭl), *pron. U.S. dial.* [f. WE *pron.* + ALL *a.*] Used in place of WE *pron.*

1875 'MARK TWAIN' *Let.* 23 Nov. (1917) I. xv. 268 We-all send love to you-all. **1905** A. V. CULBERTSON *Banjo Talks* 25 Ter do lak we-all in de pas'. **1926** E. M. ROBERTS *Time of Man* vii. 266 His wife was young..and we-all made a jolly set. **1949** *Chicago Tribune* 27 Feb. VII. 6/6 Did we-all see Smokey hold? **1964** *N.Y. Times Mag.* 23 Aug. 62/2 *Soul brother*, Negro; also referred to as..*we-all,..the people.*

So **we-·all's**, our; ours.

1887 *Scribner's Mag.* Oct. 475/1 O Lawd, 'lighten we-all's unnerstandin's. **1893** H. A. SHANDS *Some Peculiarities of Speech in Mississippi* 67 'That house is we all's' means that the house belongs to all of us. **1905** A. V. CULBERTSON *Banjo Talks* 160 Bin settin' yer..lak dis, So I be sut'n dat I doan' miss De train dat teck me back ter we-all's place.

wealth. Add: **6.** *wealth-creating, -holder, -making; wealth-fantasy*; **wealth tax**, a tax levied on the basis of a person's capital or financial assets.

1964 M. MCLUHAN *Understanding Media* (1967) x. 110 In the Roman world the army was the work force of a mechanized wealth-creating process. **1940** 'G. ORWELL' in *Horizon* Mar. 181 This kind of thing is a perfectly deliberate incitement to wealth-fantasy. **1957** A. C. L. DAY *Outl. Monetary Econ.* ii. 19 Similar considerations will influence many other wealth-holders. **1980** *TWA Ambassador* Oct. 14/1 Much of the growth in the number of wealthholders in the United States has been among the affluent, but clearly non-wealthy middle-class and upper-middle-class households. **1964** M. MCLUHAN *Understanding Media* (1967) x. 111 The Roman Army as a mobile, industrial wealth-making force. **1963** *Daily Tel.* 22 Feb. 26/6 (*heading*) Wealth tax statement next week. **1974** *Guardian* 23 Mar. 12/1 The Labour Party..suggested an annual wealth tax starting at 1 per cent on £50,000, and running up to 5 per cent on £400,000 and more. **1976** *Jrnl. R. Soc. Arts* Mar. 200/1 Until the final form of wealth tax has been decided it is not possible to decide whether it will be fiscally neutral as regards private woodlands.

wealthily, *adv.* (Later example.)

c **1885** E. DICKINSON *Poems* (1955) III. 1123 Ill it becometh me to dwell so wealthily When at my very Door are those possessing more, In abject poverty.

wealthy, *a.* Add: **6.** (With capital initial.) Name of a N. American variety of late-ripening, red-skinned cooking or dessert apple.

1869 C. DOWNING *Fruits & Fruit Trees Amer.* (ed. 2) ix. 398 Wealthy. A new variety originated by Peter M. Gideon, near St. Paul, Minn., from seed gathered in Maine about 1860... Fruit medium, oblate or roundish oblate, whitish yellow ground, shaded with deep rich crimson. **1921** *Daily Colonist* (Victoria, B.C.) 7 Oct. 7/4 Seal of Quality Groceries. No. 1 Wealthy Apples, excellent, sound stock, in boxes containing about 40 lbs. **1944** *Chicago Daily News* 25 Sept. 13/3 Right now Wealthies or Maiden's Blush are the choice varieties for cooking or pie. **1975** *New Yorker* 11 Aug. 39/1 The five apples so suddenly swept from the general market were the Baldwin, the Wealthy, the Golden Grimes, the Ben Davis, and the Black Twig.

weaner. 2. For (? *Australian*) read 'orig. *Austral.* and *N.Z.*' and add: Also, a pig or any other animal weaned during the current year. Also *attrib.* (Earlier and further examples.)

1865 R. HENNING *Let.* 19 Aug. (1966) 208 He takes the heaviest flock of all, 2,200 weaners. **1928** *Daily Express* 3 Feb. 5/2 Instruction is given in..the rearing of calves for stock, for market as 'weaners', and for sale as veal. **1950** *N.Z. Jrnl. Agric.* Jan. 63/1 Pigs sold as weaners through saleyards. **1958** *Times* 29 Sept. 2/7 Though weaner prices fell..during 1957/58, high prices for fat pigs are still reflected in the store market. **1969** T. WEST in R. Blythe *Akenfield* xi. 179, I want to produce weaner-pigs... Are you interested? **1981–2** *Deer Farmer* (N.Z.) Summer 10/3 The trial of disbudding of 80 to 100 weaner bucks will start in March next year. *Ibid.* 16/2 A herd of 300 New Zealand wapiti and wapiti-red hybrid females, is run at Criffel, along with 200 bulls, and 30 male and 30 female New Zealand wapiti weaners. **1984** *N.Z. Farmer* 12 Apr. 12/1 The cattle policy is to winter all weaners.

weapon, *sb.* Add: **2. c.** Also *with his own weapon(s).*

1781 JOHNSON *Lives of Poets: Congreve* 17 He is very angry, and, hoping to conquer Collier with his own weapons, allows himself in the use of every term of contumely and contempt. **1850** C. KINGSLEY *Alton Locke* II. xvi. 259 Try no more to meet Mammon with his own weapons, but commit your cause to Him who judges righteously. **1897** B. Stoker *Dracula* xxii. 306 He has chosen this earth because it has been holy. Thus we defeat him with his own weapon, for we make it more holy still.

3. Delete † *Obs.* and add later examples. *coarse slang.*

1762 T. Bridges *Homer Travestie* iv. 189 She guides his weapon where she lists: Nay more, a touch of her soft hand, If fallen down, will make him stand. **1922** Joyce *Ulysses* 529 Well for you, you muff, if you had that weapon with knobs and lumps and warts all over it. **1972** H. & R. Greenwald *Sex-Life Lett.* (1974) 279 This sexual thrill still comes over me whenever I see a horse flashing his weapon, and although I feel guilty and try to look away, I usually look as long as decently possible.

4. simple attrib., as *weapon-hoard, -pit, -power*; *weapon-like* adj.; objective, as *weapon-delivery* (also *attrib.*); *weapon-making* vbl. sb.; **weapon(s)-carrier**, a vehicle or aircraft specially designed for carrying weapons; **weapons-grade** *a.*, applied to fissile material of suitable quality for making nuclear weapons; **weapon(s) system** orig. *U.S.*, a military weapon together with all the equipment required to make use of it, such as detection and control apparatus, a launcher, and a delivery vehicle; **weapon-training** *vbl. sb.*, training in the use of weapons.

1938 *19th Cent.* Feb. 195 The possibilities of the new weapon-carrier in the air. **1947** *Horizon* Sept. 207 A weapons-carrier flew by loaded with G.I.s. **1963** *Daily Tel.* 10 Oct. 15/8 TSR 2, the RAF's tactical and strategical bomber,..has the most secret and sophisticated all-weather weapons delivery system in the world. **1977** *R.A.F. News* 22 June–5 July 2/5 Live ammunition, bombs and missiles were, in general, carried only by aircraft flying sorties which would culminate in weapon delivery at one of the recognised ranges. **1961** *Ann. Reg. 1960* 152 Both countries..should each set aside 30,000 kilograms of weapons-grade uranium 235, as a first step in their transfer to peaceful uses. **1977** N. Freeling *Gadget* I. 11 It won't make a bomb... It's nowhere near weapons grade. **1955** J. R. R. Tolkien *Return of King* 50 There are no great weapon-hoards here, lord. **1922** D. H. Lawrence *England, my England* (1924) 36 Not what we would call love, but a weapon-like kinship. **1936** Auden *Look, Stranger!* 67 Built by the conscious-stricken, the weapon-making, By us. *a* **1944** K. Douglas *Alamein to Zem Zem* (1946) 29 Looking down..at a weapon-pit beside us, I saw a Libyan soldier reclining there. **1958** M. K. Joseph *I'll soldier no More* ix. 161 The neat row of weapon-pits so convenient for sleeping. **1937** L. Hart *Europe in Arms* xvii. 238 At night..an increase of man-power in proportion to weapon-power is desirable. **1956** W. A. Heflin *U.S. Air Force Dict.* 564/2 Weapon system. *Also* weapons system. **1958** *Engineering* 11 Apr. 450/1 The 'weapon system concept'—that is, a unified integration of airframe, weapon-firing control, and enemy identification. **1977** M. Edelman *Polit. Lang.* viii. 147 Americans and Russians are constantly told that the other is ahead in this or that weapons system **1980** R. L. Duncan *Brimstone* x. 244 There were always 10,000 diverse elements which were required to come together at precisely the right instant, the total testing of a weapons system and a defense against that system. **1945** E. Waugh *Brideshead Revisited* 9 Had I put in the names of two corporals for the weapon-training course? **1979** R. Perry *Bishop's Pawn* vi. 106 Their weapon training had been sadly neglected... In killing Moss, they'd used sufficient ammunition to decimate a small army.

weaponeer (wepənɪ̄ə·ɪ). *U.S.* [f. prec. + -eer.] **a.** One who has charge of a weapon of war prior to its deployment.

Orig. used *spec.* of nuclear weapons.

1945 in *Amer. Speech* (1947) XXII. 149/2 Here are the names of the Superfortress crew which carried the atomic bomb to Japan... Naval observer and 'weaponeer', Capt. William S. Parsons. **1952** *Word Study* Feb. 4/1 United States atomic weaponeers probably will set off the world's greatest explosion within the next few days.

b. An expert or specialist in the development of weapons of war.

1979 *New Yorker* 13 Aug. 67/1 Robert Wilson..head of experimental physics at Los Alamos; Philip Morrison, who had gone..to take care of the bombs..; Richard Feynman, who had been in charge of one of the computing sections... I fitted in with this bunch of weaponeers. **1982** *Christian Sci. Monitor* 22 June 22/1 Probably the most interesting to the weaponeers of the great powers was the testing of US vs. Soviet fighter planes.

So **weaponee·ring** *vbl. sb.*, the development and production of weapons of war.

1955 *Bull. Atomic Sci.* Sept. 239/3 Nearly half of our expenditure for research and development each year goes to weaponeering. **1960** *N.Y. Times Mag.* 29 May 20 You must..make those essential advances in the state of the weaponeering art which will most intercept any possible aggression by any potential enemy. **1976** *Aviation Week* 19 Jan. 54/1 It's easier to conceptualize bombs, weaponeering and 'hard kill' missiles than the wizardry of electronic detection and countermeasures equipment.

weaponization (wepənaizē¹·ʃən). *U.S.* [f. as prec. + -ization.] The process of equipping with weapons of war, or adapting something for use as a weapon. So **wea·ponized** *ppl. a.*

1969 *U.S. Dept. Defense Appropriation Bill 1970* (91st Congress 1 Sess., House Rep. 698) 72 Aircraft Weaponization (advanced and exploratory development) was reduced. **1973** *Black Panther* 7 Apr. 10/2 Observers worry that the fledgling crime protection industry may follow a similar line, with huge profits made in this industry supporting a new domestic lobby for more weaponized solutions to the anger of the ghetto, the barrio and the poor white hollow. **1976** *Aviation Week* 1 Nov. 19/2 He said the Navy plans to build a 3,000-ton 'weaponized' surface-effect ship. **1982** *N.Y. Times* 23 Mar. A-14/2 Chemical warfare agents were..weaponized with Soviet assistance in Laos, Vietnam and Kampuchea. **1983** *Aviation Week* 17 Jan. 27/2 He cited an alarming trend toward the weaponization of space as the reason for his new emphasis on the civilian uses of space.

weaponry. Delete *rare* and add: Now esp. weapons of war. (Further examples.)

1956 C. W. Mills *Power Elite* viii. 184 The new weaponry has been developed as a 'first line of defense'. **1959** *Times Lit. Suppl.* 16 Jan. 27/2 War has become inevitable because of the weaponry which both sides are amassing. **1961** *New Left Rev.* Jan.–Feb. 49/1 Modern weaponry advances faster than any other branch of technology. **1972** *Daily Tel.* 27 May 1/1 The Russians have deployed 64 defensive Galesh missiles around Moscow since 1969, but the Americans are more advanced in this weaponry. **1977** *Irish Times* 8 June 5/2 The US should put aside hopes of outstripping the Soviet Union in doomsday weaponry.

fig. **1966** *New Statesman* 23 Dec. 942/1 Its weaponry of surprises..allows Miss Smith free movement in areas where the clichés lie thick.

wear, *sb.* Add: **1. b.** *the worse for wear* (further examples). Also, drunk. See also Worse *a.* 4 c.

1857 C. M. Yonge *Dynevor Terrace* II. xvii. 270 His boots were less polished..and he looked somewhat the worse for wear. **1936** J. Buchan *Island of Sheep* iii. 44 When I came across him in Persia..he was rather the worse for wear. **1966** D. Francis *Flying Finish* x. 127 You'll be kicking your heels about the airport for a few hours. Don't let any of them get...er...the worse for wear. **1977** M. Allen *Spence in Petal Park* xix. 87 The blonde girl left about ten, looking distinctly the worse for wear. **1982** T. Berger *Reinhart's Women* iii. 46 The vehicles in view were routine automobiles, two of them the worse for wear, with dents and rust and jagged antenna-stems.

6. c. *wear-and-tear pigment* (Biochem.) [tr. G. *abnutzungspigment*]: a pigment that accumulates in cells as they age.

1928 *Amer. Jrnl. Path.* IV. 236 This pigment (lipofuscin) was recognized by Lubarsch who classified it along with melanin as 'wear-and-tear' pigment (*Abnutzungspigmente*). **1943** *Jrnl. Compar. Neurol.* LXXVIII. 45 It would be better to abandon the attempt to group all these substances [from different tissues] under the common term 'wear-and-tear' pigment. **1968** [see *lipofuscin* s.v. *lipo-].

IV. 9. a. *attrib.* and *Comb.*, as *wear-resistance, -resisting*; *wear-proof, -resistant, -resisting* adjs.; **wear-dated** *a.* (see quot. 1968).

1897 *Sears, Roebuck Catal.* 183/2 This suit..is regular wear-resisting goods. **1921** *Daily Colonist* (Victoria, B.C.) 9 Apr. 7/1 Wearproof Suits for Boys. Specially selected materials and extra care in the making of these suits enable them to give every wearing satisfaction. **1943** *U.S. Patents Q.* LVI. 494/2 Wear-resisting, anti-fatigue, and shock-absorbing properties are relative and not absolute terms. **1946** *Nature* 5 Oct. 476/2 The advantage of these new processes is that the weight and wear-resistance of the fabrics are increased. **1960** *Farmer & Stockbreeder* 29 Mar. (Suppl.) 4/1 A wear-resistant outer lining. **1968** J. Ironside *Fashion Alphabet* 102 *Wear-dated*, officially recognised guaranteeing the length of normal wear of garments. **1978** *Detroit Free Press* 16 Apr. (Detroit Suppl.) 14 (Advt.), Choice of covers includes lush acrylic velvet by Monsanto with two-year wear-dated guarantee.

wear, *v.*[1] Add: **I. 4. c.** *to wear the trousers*: see *trousers *sb. pl.* 2 c.

8. c. To tolerate, accept, or agree to (a proposal, etc.). Usu. in negative with *it* as obj.

1925 Fraser & Gibbons *Soldier & Sailor Words* 301 *To wear*, to put up with, *e.g.*, 'I won't wear it', I won't stand it. **1945** C. H. Ward-Jackson *It's a Piece of Cake!* (ed. 2) 61 *Wear it. Agree to it, accept it.* Thus, 'I've asked the old man for permission to keep the bar open an hour later, but he won't wear it.' **1950** C. MacInnes *To Victors the Spoils* I. 70 'The new Captain would never wear it.' 'Then it's up to you to put it to him the right way.' **1961** E. Waugh *Unconditional Surrender* III. i. 203 Presently the Lieutenant came across to them bearing a cigar-case. 'I can't wear them myself,' he said. **1970** P. H. Johnson *Honours Board* 61 The mother said this was very kind but that Peter would never—she was given to girlish slang—'wear it'. **1981** S. Jackman *Game of Soldiers* I. 53 No.. Watson wouldn't wear it... Nor will I.

10. c. With *on* or *upon*. Of a circumstance: to affect (a person) adversely; to fatigue or debilitate. Cf. Weigh *v.*[1] 22 a.

1864 O. W. Norton *Army Lett.* (1903) 245, I did not suppose these things were serious enough to wear upon your health. **1901** 'L. Malet' *Hist. Sir R. Calmady* v. ii. 389, I have had a detestable winter... It wore upon me. It demoralised me. **1915** D. H. Lawrence *Let.* ?29 Apr. (1962) 335 You can't imagine how it wears on one, having at every moment to resist this established world.

III. 14. c. *to wear thin* (*fig.* examples).

1931 J. van Druten *London Wall* II. i. 69 You must have known yourself it's [*sc.* a love affair] been wearing hin. **1942** J. S. Huxley in *Polit. Q.* XIII. 384 The inter-war disputation between the 'have' and the 'have-not' powers is wearing a bit thin. **1982** G. M. Fraser *Flashman & Redskins* 52, I knew Susie's first good opinion of Spring had worn thin.

V. 19. a. *to wear late* (earlier example.)

1826 Scott *Woodstock* I. iii. 98 It wears late, and gets dark.

wearabi·lity. [f. Wearable *a.* + -ity.] The capability of being worn or of enduring wear; suitability for wear; durability.

1927 *Daily Express* 5 Sept. 5/2 A series of distinctive models to demonstrate individuality with wearability. **1958** *Vogue* Dec. 108 This is the sweater you must have this autumn for its soft lines, its beautiful wearability. **1966** *Chem. & Engin. News* 19 Dec. 13/2 Corfam sells in the U.K. for about the same as top-quality shoe leather, with which Du Pont compares its product for wearability under flexing and for 'breathability'. **1981** *Observer* (Colour Suppl.) 12 Apr. 75 (Advt.), A 100% cotton shirt with all the comfort and wearability men associate with cotton.

wearing, *ppl. a.* Add: **3.** *wearing course* (Highway Engin.) (see quot. 1954).

1940 *Gloss. Highway Engin. Terms* (*B.S.I.*) 28 *Wearing course*, the layer of material applied to form the carriageway. **1977** *Bitumen* (Shell Internat. Petroleum Co.) 2/4 The traffic load is carried mainly by the base layer, while the wearing course provides a waterproof non-skid cover which resists traffic wear.

wearisomeness. 1. (Later example.)

1882 E. W. Hamilton *Diary* 25 Nov. (1972) I. 364 R.G. evidently wants to get out of harness also, partly from a feeling of wearisomeness and partly from bad health.

wea·r-out. *rare.* Also **wearout.** [f. vbl. phr. *to wear out*: see Wear *v.*[1] 9, 14.] Wearing out; destruction or damage from use.

1897 C. T. Davis *Manuf. Leather* (ed. 2) 270 There is absolutely no wear-out to any of the working parts of the bed. **1979** *Arizona Daily Star* 5 Aug. c 4/1 (Advt.), They are..protected against premature wearout during the warranty period.

weary, *a.* Add: **1. f.** *Weary Willie*: see *Tired Tim* s.v. *tired *ppl. a.*[1] 1 c.

1896 *Illustr. Chips* 16 May 1/3 *Lazy Larry*: 'Watcher doin', Willie?' *Weary Willie*: 'Oh, jest wipin out a little debt I owe.' **1901** *Munsey's Mag.* Sept. 884/2 Dan had not been gone a day when the first Weary Willy appeared and demanded pie, with a horrid leer. **1906** E. Dyson *Fact'ry 'Ands* vii. 76 'Garn,' he said, 'no airs. Yer Weary Willie's brother Sam, *halias* Ther Frequent Sleeper, [etc.].' **1909** *Punch* 20 Jan. 46 (*caption to picture of two tramps*) *Weary Willie*: I'd sooner walk up 'ill than I would down, any day—it do throw yer into yer boots so. **1927**, etc. [see *tired *ppl. a.*[1] 1 c]. **1929** *Amer. Speech* IV. 345 *Weary Willie*, a tramp who usually hikes it and is too tired to work. **1972** [see *tired *ppl. a.*[1] 1 c].

III. 9. *weary-brained, -eyed* adjs.

1898 G. B. Shaw *Let.* 1 May (1972) II. 38, I finish the book at a sitting, as I don't want to be weary-brained when Charlotte comes. **1930** J. Masefield *Wanderer of Liverpool* 24 Weary-eyed men came on deck.

weasel, *sb.* Add: **5.** (Earlier example.)

1845 in C. Cist *Cincinnati Misc.* I. 240 The inhabitants of..S. Carolina [are called] Weasels.

5*. A tracked vehicle capable of travelling over difficult terrain; *spec.* (*a*) a light cargo and personnel carrier (*U.S. Mil.*); (*b*) a snow tractor (see quot. 1958).

1944 *Yank* 4 Aug. 17/2 Cargo carrier M29, nicknamed the Weasel, is now in full production. **1949** [see *snow-buggy* s.v. *snow *sb.*[1] 7 b]. **1958** *Times* 11 Nov. 6/7 The Weasel—one of the snow vehicles used by Sir Vivian Fuchs on his trans-Antarctic journey—was invented by a civilian, the late Mr. Geoffrey Pyke. **1964** 'J. H. Roberts' *Q Document* (1965) ix. 206 The hotels operated what the student referred to as 'wesaru'—which..was the Japanese way of pronouncing 'weasel', a cross between a jeep and an army tank—to carry the guests. **1980** *Globe & Laurel* July/Aug. 227/2 We in fact lost two vehicles doing this, one being my own command vehicle which was a cargo LVT with a weasel in the back.

5.** An equivocal statement or claim, esp. one used in an intentionally misleading advertisement. See *weasel word*, sense 6 b below.

1959 T. Griffith *Waist-High Culture* (1960) 83 The answer may have to be a 'weasel', the phrasing that avoids or begs the question. **1963** D. Ogilvy *Confessions Advert. Man* xi. 155, I plead guilty to one act of *suggestio falsi*—what Madison Avenue calls a 'weasel' **1975** *Idle Moments* (Austral.) Dec. 26/2 The' weasels' are so cleverly written, so subtle, you hardly notice them at all.

6. a. *weasel-mind*; *weasel-eyed* adj.

1922 Weasel-eyed [see *peanut-brained* adj. s.v. *peanut b]. **1985** C. FitzGibbon *Love lies a Loss* v. 59 The weasel-eyed creditors lined the dock. **1923** *Chambers's Jrnl.* Feb. 88/1 Simon would have dallied by the way, his weasel-mind alert to draw news of the hindering from this Heseltine.

b. weasel word orig. *U.S.*, an equivocating or ambiguous word which takes away the force or meaning of the concept being expressed; hence **weasel-worded** *a.*

1900 S. Chaplin in *Century Mag.* June 306/2 'The public should be protected—' 'Duly protected,' said Gamage, 'That's always a good weasel word.' **1916** *N.Y. Times* 1 June 1/2 Colonel Roosevelt began the day's speechmaking by opening his guns upon President Wilson... He accused Mr. Wilson of using 'weaselwords' in advocating universal military training, but 'only the compulsion of the spirit of America'. A weasel, the Colonel explained, would suck all the meat out of an egg

and leave it an empty shell. **1939** *Florida* (Federal Writers' Project) I. 125 There were no 'weasel word' qualifications, such as 'it is alleged'. **1952** G. Sarton *Hist. Sci.* I. xvi. 404 It is perhaps a little ambiguous to call them idealists. [*Note*] The weasel word idealist is sometimes understood as the opposite of realist. **1977** P. Johnson *Enemies of Society* viii. 108 Whereas in the spheres of advertising, education and economics the use of weasel words tends to be towards gross overstatement, in the field of killing and mass destruction, the tendency is to understate and minimize. **1923** Ld. Charnwood *Theodore Roosevelt* x. 215 It is even comically reminiscent of the writer's own criticisms later of Mr. Wilson's 'weasel-worded' phrases. **1981** *N.Y. Times* 29 Mar. 4/1 The facts it contained did not support what one official termed the agency's 'weasel-worded' conclusion.

c. *attrib.* or as *adj.*, after *weasel word*, above. Of a statement, etc.: equivocating, ambiguous, quibbling.

1912 T. Roosevelt in *Outlook* 27 July 662/2 The weasel sentence about States' rights could well have been suggested by the astuteness of Mr. Bryan's fellow-Democrat Mr. Ryan. **1965** M. Naylor *Your Money* x. 59 If..other things remain equal..,the price will rise... It is now time to deal with that weasel qualification, 'other things being equal'. **1974** R. M. Pirsig *Zen & Art of Motorcycle Maintenance* (1976) iv. xxviii. 337 The whole business seemed to many of them merely a new and pretentious jargon of weasel concepts. **1979** *Financial Rev.* (Melbourne) 27 Apr. 2 The probability is that the commission will deliver another of its weasel judgements, recognising merit on all sides.

wea·sel, *v.* *colloq.* (orig. *U.S.*). [f. the sb.] **1. a.** *trans.* To render (a word, phrase, etc.) ambiguous or equivocal; to remove or detract from (its meaning) intentionally.

1900 *Century Mag.* June 305/2 I've seen him take his pen, and go through a proposed plank or resolution, and weasel every flat-footed word in it. **1919** T. Roosevelt in *Maine, my State* (Maine Writers Research Club) 20 'His words weasel the meaning of the words in front of them,' said David, 'just like a weasel when he sucks the meat out of an egg and leaves nothing but the shell'.

b. *intr.* To equivocate or prevaricate, to use weasel words.

1956 [see *dirt *sb.* 6 e]. **1963** D. Ogilvy *Confessions Advert. Man* v. 99 If you tell lies, weasel, you do your client a disservice. **1972** C. Weston *Poor, Poor Ophelia* xxv. 151 He listened to the younger detective weaseling at the other end.

2. a. To extricate oneself from or get *out of* a place in the manner of a weasel. Also with *in* (with movement in the opposite direction).

1925 J. Bone *London Perambulator* 162 How to weasel out of London, north, south, east and west, with the fewest possible obstructions. **1963** T. Pynchon *V.* vii. 171, I was always weaseling in, you know, on some show where you wouldn't expect to find naval personnel. **1968** P. Dickinson *Skin Deep* ix. 176 Pibble weaseled out of the car and ran across the road.

b. To escape from or extricate oneself *out* (*of* a situation, obligation, etc.), esp. dishonourably; to welsh *on*. Also with *one's way*.

1956 *Washington Post* 7 Aug., For this country to weasel on its obligation would be both to fracture the Atlantic alliance and to engage in the most offensive and immoral sort of appeasement. **1962** N. Maxwell *Witch-Doctor's Apprentice* ii. 10, I wanted to commit myself publicly to it so that it would be hard to weasel out after only a day or two. **1973** *New Yorker* 3 Mar. 85/1 Canterbury is one of the 'decadent' communities that gradually weaseled out of the Shaker strictures against ornament and luxury. **1978** M. Puzo *Fools Die* xxi. 239 A real fucking claim agent weaseling out of his obligations. **1980** *Logophile* IV. i. 46/1 It required weaseling his way into the confidence of his bank-manager. **1981** *Spectator* 6 June 16/2 Jilly Cooper was too kind-hearted to name those who weaseled out of the exercise.

3. *trans.* To obtain or extract (something) *out of* another, esp. by cunning.

1975 L. Deighton *Yesterday's Spy* xii. 99 He..'weaseled' luggage for the boat-train passengers and was not above stealing the occasional camera. **1975** *Observer* 30 Nov. 22/4 My sole achievement was weaselling a medical certificate out of my G.P.

Hence **wea·selling** *ppl.a.* and *vbl. sb.*

1956 H. Kurnitz *Invasion of Privacy* xii. 81 Never could get along with lawyers... Bunch of weaseling doubletalkers. **1969** *Listener* 31 July 132/2 'Legitimate puffery' is often plain lying. At best, it encourages 'weaselling'—the use of meaningless and unverifiable formulae like 'Bloggo is better'; at worst, it is demonstrably fraudulent. **1978** *N.Y. Times* 30 Mar. D 18/1 Arum says there has been heavy pressure for a return bout from all over the world. He offers this as justification for his weaseling out of a commitment to match Spinks with Ken Norton.

weaselish (wī·z'liʃ), *a.* *rare.* [f. Weasel *sb.* + -ish[1].] = Weaselly *a.*

1923 D. H. Lawrence *Captain's Doll* xiv, in *Ladybird* 225 The driver, who was thin and weaselish.

weaselly, *a.* Add: Also **weas(e)ly.**

1973 M. Amis *Rachel Papers* 128 He was wearing a fashionable black polo-neck jersey (fashionable, that is, among the weasly middle-aged) whose sleeves he was rolling down. **1982** Barr & York *Official Sloane Ranger Handbk.* 17/2 What a funny little weasely face he has!

weather, *sb.* Add: **I. 1. c.** (Further examples.) Also *spec.* (*Lit.*), applied to an intellectual climate, state of mind, etc.

1909 H. James *Roderick Hudson* (rev. ed.) vii. 147 He supposed that these changes of intellectual weather.. were the lot of every poet. **1922** G. Santayana *Soliloquies in England* 30 What governs the Englishman is his inner atmosphere, the weather in his soul. **1927** T. Wilder *Bridge of San Luis Rey* 17 Such authors live always in the noble weather of their own minds. **1962** K. Allott *Penguin Bk. Contemp. Verse* 18 A short introduction giving explicit attention to the poetic 'weather' of each of the last four decades.

2. f. (Earlier example.)

1827 *Austin Papers* (1924) I. 1622 The fredonians is all here rather under the wether.

h. (Earlier example.)

1829 P. Egan *Boxiana* 2nd Ser. II. 302 Asking of no favours from the clerk of the weather to keep off 'the pitiless pelting storm', as their greasy jackets were proof against all watery attacks.

j. *above* (or *over*) *the weather* (Aeronaut.), above the range of weather conditions acting at ground-level; above the clouds.

1944 *Aviation* Feb. 497/1 The plane climbs..to fly 'over the weather'. **1958** *Listener* 16 Oct. 593/1 It was said that they [*sc.* accidents] had destroyed all prospect of carrying passengers at speeds not far short of the speed of sound, far above the weather, at heights of 35,000 feet.

II. 6. a. *weather bulletin, -cast* (later example), *-journal, -lorist, -map* (earlier example), *prediction, report* (examples), *-saw, -screen*.

1926 R. Macaulay *Crewe Train* II. viii. 157 She asked Arnold..to tell her when the weather bulletin came on; that was normally the only part of the programme to which she cared to listen. **1980** P. Moyes *Angel Death* xv. 198 The weather bulletin..advised guests that Hurricane Beatrice was..moving at a brisk fourteen knots. **1980** *Time* 17 Mar. 37/1 A native American art form, the television weathercast. **1868** G. M. Hopkins *Jrnls. & Papers* (1959) 189 Henceforth I keep no regular weather-journal but only notes. **1905** *Westm. Gaz.* 21 Aug. 10/1 A remarkable dearth of acorns..which, according to the weather lorists, is a favourable augury for the coming weather. **1877** Weather map [see *facsimile 2 a]. **1909** Weather prediction [see *gaffe]. **1951** M. McLuhan *Mech. Bride* (1967) 75/1 The comment is given in the style of stock-market operations or weather predictions. **1863** R. Fitzroy *Weather Bk.* 349 Local changes should be indicated to observers..by due attention to the published Weather Reports. **1939** T. S. Eliot *Family Reunion* II. i. 97 And now it is nearly time for the news We must listen to the weather report. **1980** A. E. Fisher *Midnight Men* vii. 78 He could do without unfavourable weather reports. **1871** G. M. Hopkins 6 Aug. *Jrnls. & Papers* (1959) 213 The common weather-saw about the rainbow. **1914** 'Bartimeus' *Naval Occasions* xx. 181 The men on the bridge ducked their heads as..a shower of spray drifted over the weather-screens. **1977** P. Smalley *Trove* ii. 84 The triple-panel weather screen was fitted with heavy duty wipers.

b. *weather-caster* (so *-casting*), *-forecaster*.

c **1904** *Encycl. Dict.* Suppl., Weather-caster. **1965** *Punch* 5 May 660/2 His great ambition in life is to be a TV weathercaster. **1980** *Time* 17 Mar. 37 TV weathercasters have been much mocked for their polyester jocularity. *Ibid.* (*heading*) The wonderful art of weathercasting. **1900** *Nature* 29 Nov. 110/2 Disappointing..from the viewpoint of the weather forecaster. **1981** *Times* 9 Dec. 1 The weather forecasters were critized..for not giving enough warning..of the snowfall.

c. *weather-roughened*.

1897 W. B. Yeats *Secret Rose* 187 Her dark, weather-roughened skin.

d. *weather-resistant, -resisting*. Also *weather-resistance*.

1894 [see *roofing (*vbl.*) *sb.* 1 b]. **1934** *Archit. Rev.* LXXVI. 16/1 Many years of use have proved the method satisfactory, both as a weather-resistant and as insulation. **1942** *E. African Ann.* 1941–2 98 (Advt.), Anti-rust paint..durable, elastic, weather-resisting. **1967** M. Chandler *Ceramics in Mod. World* iv. 117 Another property that makes both porcelain and glass insulators particularly suitable for high-voltage insulators is their weather-resistance. **1970** *New Yorker* 3 Oct. 27/2 You can bolt on anything from redwood to weather-resistant aluminum.

7. weather balloon, a balloon sent up to provide meteorological information, either by the course it takes or by means of instruments it carries; **weather bureau** *U.S.*, an agency (*spec.* one established by the Government) which observes and reports on weather conditions; **weather centre**, an office which provides weather information and analysis; *spec.* in U.K., part of the Meteorological Office; **weather clerk** = *clerk of the weather* s.v. Clerk *sb.* 6 c; **weathercoat**, a weather-proof coat, a raincoat; **weather cycle**, a recurring pattern of weather or of some tendency in the weather; **weather-fast** *a.*, secure against the weather; **weather-man**, delete † and add: (*a*) (later examples); now also *spec.* one who presents a weather forecast on radio, television, etc.; (*b*) (freq. with capital initial and in *pl.*) (a member of) a violent revolutionary group in the U.S. (see quot. 1970); cf. *Weather Underground* below; **weather modification**, the deliberate alteration of the weather in an area; **weather-monger** (later example); **weather plane**, an aeroplane designed to collect data on weather conditions at high altitudes; † **weather-prophecy** *Obs.*, the foretelling of the weather; **weather radar**, radar used for meteorological investigations (e.g. of rain); **weather satellite**, a satellite especially equipped to observe weather conditions and to provide meteorological information; **weather ship**, a ship serving as a weather station; **weather station**, a meteorological observation post; **weather-strip** orig. *U.S.* (examples); **weather-stripped** *ppl. a.*; **weather-stripping** *vbl. sb.*, material used to weather-strip a door, window, etc.; the process of applying this; **Weather Underground**, the revolutionary organization formed by the Weathermen (see above); **weather window** *Oil Industry*, a brief interval in the year when the weather is calm enough to allow construction, loading, etc., operations to be carried out at sea; **weather woman**, (*a*) (with capital initial) a female member of the revolutionary Weatherman organization; (*b*) a woman who presents a weather forecast on radio or television.

1940 *War Illustr.* 19 Jan. 614/3 (*caption*) Finnish soldiers are investigating weather conditions by sending up a weather balloon. **1979** J. Gribbin *Weather Force* vii. 160 (*caption*) Russian scientists..prepare to launch a flock of weather balloons, which will radio back information about conditions in the atmosphere's lower levels. **1871** *Harper's Mag.* Aug. 401/1 In the year 1857 Lieutenant M. F. Maury..appealed to the public and Congress, through the press, urging the establishment of a storm and weather bureau. **1890** *U.S. Statutes* XXVI. 653 The civilian duties now performed by the Signal Corps of the Army shall hereafter devolve upon a bureau to be known as the Weather Bureau. **1950** *Los Angeles Times* 12 Feb. 1/4 Weather Bureau figures show that ·34 inch fell during the rainstorm. **1978** S. Sheldon *Bloodline* iv. 71 July turned out to be the rainiest month in the history of the French weather bureau. [**1959** *Times* 19 Aug. 8/7 The Air Ministry Meteorological Office is to open a 'weather shop' where the public may call in person at the new home of the London forecasting office at Princes House, Kingsway.]. **1961** *A.A. Handbk.* 17 'Weather Centres' staffed by the Meteorological Office are open in London, in Glasgow, and in Manchester. **1973** C. Bonington *Next Horizon* xiii. 185, I..went through the daily ritual of getting the weather forecast. This entailed 'phoning..the weather centre in London. **1877** 'Mark Twain' *New England Weather* in *Index* (Boston) 11 Jan. 16/2 It must be raw apprentices in the weather-clerk's factory who experiment and learn how in New England.., and then are promoted to make weather for countries that require a good article. **1898** H. S. Canfield *Maid of Frontier* 111, I wouldn't have a weather clerk inside of me for any thing. **1897** J. L. Allen *Choir Invisible* x. 132 He got up at last and wrapped his weather-coat about him. **1930** *Daily Express* 6 Oct. 13/5 (*caption*) Real Harris tweed weathercoat. **1978** *Sunday Times* 21 May 1/6 (Advt.), A pure silk wrap-around weathercoat..to protect you from summer showers..£165. **1930** *Engineering* 31 Jan. 148/2 Based upon a weather cycle or period of almost fourteen years. **1910** J. Farnol *Broad Highway* I. xxiv, It was somewhat roughly put together, but still very strong, and seemed, save for the roof, weather-fast. **1901** Weather man [see *hot wave* s.v. *hot *a.* 12]. **1944** *Sun* (Baltimore) 15 Nov. 11/2 Nobody ever gets anywhere telling the weatherman how to behave. **1952** W. Stevens *Let.* 26 June (1967) 757 It did not go below 85° in N.Y. last night according to the weather man. **1970** *Guardian* 28 Oct. 13/3 The Weathermen have been in existence for just over a year, since the SDS [*sc.* Students for a Democratic Society] split of June, 1969... The Weathermen got their name from a line in a Bob Dylan song: 'You don't need a weatherman to know which way the wind blows.' **1971** *Times* 15 Jan. 12/6 Could this country have acquired an Anglicized offshoot of the American Weatherman—or Weathermen as these violent urban guerillas are less accurately but probably more widely known? **1979** R. Perry *Bishop's Pawn* i. 23 The West was agreed that the IRA, the Weathermen, the Red Army Faction..were composed of criminals, terrorists and murderers. **1983** *Listener* 14 July 17/3 We asked the weatherman, Jack Scott, to demonstrate some of those extraordinary regional variations for us. **1951** *U.S. Congr. Senate Committee Interior Hearings* Apr. 152 Weather modification on a small scale, such as protection against frost..is known to be possible. **1968** *Times* 1 Nov. 6/6 Russian research on methods of reducing damage to crops by hailstorms is being examined seriously in the United States, according to a National Science Foundation report on last year's activities in weather modification. **1977** *Time* 7 Mar. 55/1 The Governors also agreed to create a task force that could channel such requests for aid and coordinate weather-modification (cloud seeding) programs. **1911** J. G. Frazer *Golden Bough: Magic Art* (ed. 3) I. iv. 227 Wizards, doctors, weather-mongers, prophets. **1962** *Listener* 18 Oct. 632/2 The 'Coliseum of cloud' that a weatherplane captured for us. **1976** *Evening Post* (Nottingham) 13 Dec. 7/2 Experts..identified it as a crashed weather plane which sends wind and temperature conditions from a height of 90,000 feet. **1843** Mill *Logic* I. III. iv. 389 The reliance on astrology, or on the weather-prophecies in almanacs. **1946** *1st Technical Rep. Weather Radar Research* (Mass. Inst. Technol. Dept. Meteorol.) (AD 54113) 3 (*heading*) Weather-radar observations at M.I.T.'s Radiation Laboratory. **1979** *Atmosphere-Ocean* XVII. 78 The radar data were obtained from the McGill Weather Radar located just outside Montreal. **1960** *Aeroplane* XCIX. 90/2 After taking 22,952 photographs of the Earth's cloud cover, Tiros I, the World's first weather satellite, has ended its useful life..after the satellite's electronics had suffered a failure. **1976** L. Deighton *Twinkle, twinkle, Little Spy* xi. 115 His factories make complicated junk for communications satellites... And there are weather satellites too. **1946**

Shell Aviation News No. 100. 6/3 A proposal by the Search and Rescue Committee that weather ships should be maintained in the North Atlantic for meteorological observations. **1978** *Nature* 1 June 407/1 Following the withdrawal of US weatherships in 1973, it is the only regularly reporting deep ocean (3,000 m) station in the North Atlantic north of the tropics and south of 50° N. **1895** *Funk's Stand. Dict.*, Weather station. **1953** *Encounter* Nov. 7/1 Japan gets its weather from China, but no weather reports—at least not until the Japanese experts again manage to break the code of the Chinese weather-stations. **1981** 'E. Lathen' *Going for Gold* vii. 87, I was on to the weather station... The forecasters are talking about the blizzard of the century. **1847** *Rep. Comm. Patents 1846* (U.S.) 94 One patent has been granted for improvement in fences, and another for a weather strip for doors. **1921** *Daily Colonist* (Victoria, B.C.) 25 Oct. 6/6 (Advt.), Weather Strip—'Stormproof', 24 feet in box. **1970** K. Ball *Fiat 600, 600D Autobook* xii. 143/2 The front windscreen and rear window are secured in place by a special weatherstrip. **1985** *Times* 19 July 13/4 In windy winter conditions the windloading presses the door up against the weatherstrip. **1908** I. N. Stevens *Liberators* 8 The wind that shook the windows, weather-stripped as they were, crept into the room. **1945** Nelson & Wright *Tomorrow's House* xiii. 147/2 A heavy flush door, weather-stripped,..would..reduce the direct transmission of sound. **1942** *Archit. Rev.* XCI. 99/3 The windows are pine with aluminium weatherstripping. **1959** 'S. Ransome' *I'll die for You* xii. 144 A part of the weather stripping was loose, and in a heavy rain it leaked. **1975** *Globe & Mail* (Toronto) 14 Nov. 2/5 As for weather-stripping, Mrs. Macdonald said their house doesn't need it because of extra insulation and double windows. **1972** *National Observer* (U.S.) 27 May 10/2 The..'Weather Underground', which boasts that it is responsible for so many of these bombings, is down to only 15 or 20 members now, according to sources in the House Internal Security Committee. **1982** H. Kissinger *Years of Upheaval* iv. 89 The terrorism of the Weather Underground. **1974** *Petroleum Rev.* XXVIII. 787/1 The weather-window is normally reckoned to last into September. **1983** *Sunday Times* 6 Mar. 69/4 It's been said that Esso's development of artificial islands has not merely opened the weather window further but ripped it off its hinges. **1971** *Times* 15 Jan. 12/7 Only one unconnected Weatherwoman has since been traced. **1973** *Daily Tel.* 14 Dec. 3/3 BBC Television is to have its first weather woman. She is Miss Barbara Edwards,..who at present reads weather forecasts on radio. **1982** *Times* 28 May 9/3 Diana Arp..was from a very wealthy family and became a Weather woman, making bombs.

8. weather deck, a deck exposed to the weather; the uppermost unprotected deck, other than the forecastle, bridge, and poop; **weather-dodger** *slang*, a screen on the bridge of a ship, affording protection from the weather; **weather-mark** *Sailing*, a mark on a racing course towards which boats sail into the wind.

1909 *Man. Seamanship* (Admiralty) II. xx. 409 Wood planking is now fitted only to the weather decks. **1973** H. Gruppe *Truxton Cipher* (1974) xiii. 135 Tolley..disappeared down the weather-deck ladder. **1924** R. Clements *Gipsy of Horn* v. 84 One was..in comparative comfort under the lee of the weather-dodger. **1894** *Outing* XXIV. 36/2 The 'Una' turned the weather-mark with a lead of nearly half an hour. **1963** *Times* 8 June 5/1 By the weather mark Andromeda was in front.

wea·therable, *a.* [f. Weather *v.* + -able.] Capable of withstanding the effects of the weather. Also **wea:therabi·lity.**

1961 Webster, Weatherability. **1963** H. R. Clauser *Encycl. Engin. Materials & Processes* 662/1 The poor weatherability of some of the man-made fibers can be overcome by the use of special finishing and/or coating treatments. **1972** J. G. Cruickshank *Soil Geogr.* ii. 52 Even where the parent material is rock in place, the weatherable minerals—or what has survived complete chemical weathering—are of continuing importance for soil profile development. **1979** *New Scientist* 17 May 547 Silicones in paint prevent colour fading, and give better long-term weatherability and heat resistance.

weatherboard. Add: **1. c.** A weatherboarded dwelling or other building. *Austral.*

1925 'H. H. Richardson' *Way Home* II. i. 123 Jerry and his bride had made ready their tiny weatherboard. **1935** L. Mann *Human Drift* xxxvi. 238 Magnificently the two stories of Geelong weatherboards, new that year, overlooked on the ridge. **1975** D. Malouf *Johnno* ii. 28 But our one-storeyed weatherboard wasn't the only one to be fortified. The whole city had taken on the aspect of an armed camp.

weathercock, *sb.* Add: **3. c.** *Aeronaut.* Used *attrib.* and as *adj.* with reference to the tendency of an aircraft to turn away from the set compass direction into the relative wind.

1898 *Aeronaut. Jrnl.* Jan. 6/2 The little vertical rudder has a little bit of a weather-cock action. If the wind suddenly shifts, it swings the machine round to head the wind. **1916** G. C. Loening *Military Aeroplanes* xii. 166 Some tendency to head into the relative wind is necessary. This is obtained by having enough rudder or fin surface aft to bring the directional center back of the c.g. and is called 'weathercock' stability. **1928** N. Macmillan *Art of Flying* vii. 74 A seaplane is usually directionally-stable to the wind on the water with engine stopped, a quality which is described as being 'weathercock'... If the seaplane is weathercock, she will ride head to wind and drift backward. **1935** C. G. Burge *Compl. Bk. Aviation* 616 Weathercock instability. **1945** *Jrnl. R. Aeronaut. Soc.* XL. 461/2 Weathercock stability should be reasonably high, to cut down the magnitude of the inadvertent skid. **1970** T. Hacker *Flight Stability & Control* vii. 160 The variation of the vertical tail area will influence the weathercock stability.

weathercock, *v.* Add: **1.** (Later example.)

1824 *Blackw. Edin. Mag.* June 703/2 No change of times can ever change that feeling in me. Let others weather-cock it to and fro as they please.

4. *intr.* **a.** *Naut.* Of a ship: to (tend to) head into the wind.

1952 Smith & Violet *Wind calls Tune* i. 19 She would weathercock head to seas and, under mizzen and foresail, she would sail well without anybody at the tiller. **1974** *Tel.* (Brisbane) 4 June 8/5 In these conditions the boats had a strong tendency to weathercock, making it exceptionally difficult to pick up buoys.

b. *intr.* (See quot. 1956.) Also *trans.*, to cause to weathercock.

1956 W. A. Heflin *U.S. Air Force Dict.* 565/2 *Weathercock.* Of an airplane, rocket, or similar body: to align, or attempt to align, its longitudinal axis with the direction of the wind. **1976** 'M. Albrand' *Taste of Terror* v. 38 That terrible cross-wind caught the plane's high vertical stabilizer and began to weathercock it into the wind.

Hence **wea·thercocking** *vbl. sb.*

1952 Smith & Violet *Wind calls Tune* vi. 59 We had to get the mizzen as flat as possible to make it more efficient as a 'weather cocking' influence. **1985** *Flight* 4 May 23/3 The butterfly tail..actually eliminates weathercocking in crosswinds.

weather-eye. Add: *to keep one's weather-eye open*, etc. (earlier and further examples).

1829 P. Egan *Boxiana* 2nd Ser. II. 505 Lads of the Fancy, who always keep their 'weather eye' up towards business, *i.e.* looking after the blunt. **1846** Dickens *Dombey* x. 89 Joe is awake, Ma'am... Josh has his weather-eye open, Sir. **1851** H. Melville *Moby Dick* I. xxxiv. 253 All whale-ships' standing orders, 'Keep your weather eye open, and sing out every time'. **1916** 'Taffrail' *Pincher Martin* iv. 53 Even the captain and the navigator..did not seem to be taking things very seriously, though in reality they both had their weather-eyes very much lifting. **1959** B. Wootton *Soc. Sci.* 10, I have tried to keep a weather eye open to what is going on elsewhere: references to work in other areas..will be found scattered through these pages. **1981** *Times* 7 Apr. 6/7 The Russians..are keeping a weathereye on other navies.

weathering, *vbl. sb.* Add: **2.** (Later examples.) Also *weathering ground*.

1855 Salvin & Broderick *Falconry in Brit. Isles* iii. 44 In the morning..they may be..placed upon blocks in the open air, without their hoods; this is called 'weathering'. **1960** M. Woodford *Man. Falconry* ii. 9 The weathering ground should be a level, well-drained area of short grass protected from the prevailing winds. **1978** *Country Life* 27 July 215/3 Hawking enthusiasts will be delighted to find a well-stocked weathering ground at Kinmount.

wea·therize, *v. U.S.* [f. Weather *sb.* + -ize.] *trans.* To make weatherproof; *spec.* to render (a building) impervious to the effects of weather, by insulation, double-glazing, etc. Also **wea·therized** *ppl. a.*; **wea·therizing** *vbl. sb.*; **wea:theriza·tion.**

1946 *Woman's Home Compan.* Mar. 117/3 (Advt.), Weather-Bird Shoes for boys and girls are weatherized. **1946** *Collier's* 9 Nov. 92 (Advt.), A brand-new portable [radio] with new weatherized case. **1976** *Laurel* (Montana) *Outlook* 9 June 3/2 A project which could provide insulation and other weatherization materials... A coordinator for the Weatherizing program. **1977** *Time* 4 Apr. 52/2 Plants and office buildings will have to be weatherized to prevent energy loss. **1980** *New Age* (U.S.) Oct. 38/1 In Fitchburg, Massachusetts, people were so scared of freezing last winter that they dreamed up a program of door-to-door citizen action which was supposed to weatherize half their houses in five weeks. **1983** *Hyde Park* (Chicago) *Herald* 6 July 21 The city's weatherization program makes it worthwhile.

Weather-ometer (weðərǫ·mītəɹ). Also **Weatherometer, weatherometer.** [f. Weather *sb.* + -ometer.] A proprietary name for a device which subjects substances to simulated weather conditions in order to determine their weather-resistance.

1929 *Official Gaz.* (U.S. Patent Office) 11 June 308/2 Atlas Electric Devices Company, Chicago, Ill... Weather-ometer. For Apparatus for Testing the Effect of Weather Upon the Surfaces of Objects. Claims use since Dec. 3, 1926. **1937** *Nature* 28 Aug. 353/1 For accelerated weathering tests on various paints the weatherometer..has been used. **1951** [see *fade-ometer]. **1965** *New Scientist* 1 July 25/3 A cyclic weatherometer developed as a screening test for new types of timber preservatives and water repellants.

weatherproof, *a.* (*sb.*). Add: **b.** *sb.* Also, a weatherproof coat; a raincoat.

1925 *Studio* (Art & Publicity Special Autumn No.) (Advt. suppl.) p. ix, Aquascutum pure wool weatherproofs come in one quality only—the best. **1935** *Times* 21 Feb. 14/2 [The Queen] saw the colour first in a crêpe de Chine weatherproof made with a cape. **1962** *N.Y. Times* 16 Nov. 3 A & F presents a selection of Burberry topcoats, overcoats and weatherproofs.

Hence as *v. trans.*, to make weatherproof or impervious to weather; **wea·therproofed** *ppl. a.*; **wea·therproofness.**

1926 T. E. Lawrence *Seven Pillars* (1935) vi. lxxix. 435 The Indians weather-proofed their own north-west rooms. **1929** *Punch* 17 Apr. p. xxii/2 Weatherproofed by Burberrys, their value as overcoats is enhanced in every way. **1933** *Times Lit. Suppl.* 7 Dec. 874/3 There is something animal—..in their strength, endurance, weatherproofness and capacity to live rough. **1963** *N.Y. Times* 1 Dec. 12 The 'Stormster'..all-climate coat... Weather-proofed cotton-poplin..lined. **1969** *Jane's Freight Containers 1968–69* 9/2 It is recommended that the test for weatherproofness (Test No. 9) be made last. **1976** *West Lancs. Evening Gaz.* 8 Dec. 11/9 The revolutionary roof coating system designed to protect, insulate and weatherproof your home.

weather-stained, *ppl. a.* (Earlier example.)

1849 Thoreau *Week Concord Riv.* 57 Town records, old, tattered, time-worn, weather-stained chronicles.

weather-wise, *a.* Add: **a.** Also *absol.* as *sb.*

1922 Joyce *Ulysses* 390 Biggish swollen clouds to be seen as the night increased and the weatherwise poring up at them.

wea·therwise, *adv.* orig. *U.S.* [f. Weather *sb.* + Wise *sb.*[1] II.] As regards the weather.

1963 [see *personnel-wise* advb. s.v. *personnel 3]. **1971** *Country Life* 21 Oct. 1084/1 October is usually a very good month weatherwise. **1975** *Budget* (Sugarcreek, Ohio) 20 Mar. 15/6 Weatherwise the last week has been real nice.

Weatings (wī·tiŋz), *sb. pl.* Also **w(h)eatings.** [f. respelling of Wheat *sb.* + *-ings*, as in *middlings* s.v. Middling *sb.* 3 c.] A proprietary name for the residue of the milling of wheat (the sharps: see Sharp *sb.* 9), used as food for farm animals.

1931 *Trade Marks Jrnl.* 30 Sept. 1329/1 Weatings... Fine wheat offals (for food). The M.M.A., Limited,.. London E.C. 3. **1937** C. Forman *Pig Breeding & Feeding* vii. 79 These wheatings are advertised as not more than 5 per cent fibre. **1937** E. M. Rickard-Bell *Handbk. Mod. Pig Farming* vii. 81 For the weatings we can substitute up to 10 per cent. with coconut cake. **1960** *Farmer & Stockbreeder* 8 Mar. 5/1 Weatings and bran both quoted at £25 per ton ex mill. **1976** G. Johnson *Profitable Pig Farming* (ed. 5) viii. 124 Sharps, weatings or middlings have not been included in any of the above selection of suitable rations.

weave, *v.*[2] Add: **1. b.** Also of a wild animal in captivity.

1934 [implied in *weaving *vbl. sb.*[2] 2]. **1973** G. Durrell *Beasts in my Belfry* vii. 140, I could only presume that he [*sc.* a buffalo doing a sort of waltz] put on this performance for the same reason that a lion paces up and down its cage or a polar bear or an elephant will weave from side to side—a soothing and interesting habit to pass the time till the next meal.

c. *R.A.F.* Of an aircraft or its pilot: to fly a weaving course, usu. in attempting to avoid enemy planes or anti-aircraft fire. Also *transf.* (in this use *slang*).

1941 *Battle of Britain August–October 1940* (Min. of Information) 13 Enemy bomber formations were..protected by a box of fighters, some of which flew slightly above to a flank or in rear,..and..others weaving in and out between the sub-formations of the bombers. **1942** *R.A.F. Jrnl.* 3 Oct. 22, I took my Edgar Wallace from the billet and weaved out to the Nissen hut. **1943** P. Brennan et al. *Spitfires over Malta* 26 The remaining four 110's at once broke, and began weaving, each steering a different course home. **1952** [see *predict *v.* 4]. **1973** N. Monsarrat *Kappillan of Malta* 38 The sky above him seemed full of planes, weaving and circling like the flies in the wine-shop.

d. *to get weaving*, to apply oneself briskly to something; to 'get a move on'. Cf. *to get cracking* s.v. *crack *v.* 22 b. *colloq.* (orig. *R.A.F.*).

1942 *R.A.F. Jrnl.* 30 May 19 We'll knock that..place off the map. Get weaving. **1942** T. Rattigan *Flare Path* III. 83 We'd better get weaving, or we'll find this chemist feller has gone to lunch. **1948** A. Baron *From City, from Plough* v. 36 There won' 'alf be a queue at the NAAFI... Let's get weaving. **1959** M. Pugh *Chancer* 49 When he gets weaving, the dashboard ashtray does three thousand revs. **1964** R. Braddon *Year of Angry Rabbit* i. 8 And this call's costing me a ruddy fortune so how's about you hang up and get weaving, eh? **1971** B. W. Aldiss *Soldier Erect* 185 Pack your night things in a small pack and get weaving, while I lay on transport.

weaver[1]. Add: **2. b.** *Boxing.* A boxer who weaves from side to side as a tactical move. Cf. Weave *v.*[2] 4.

1829 P. Egan *Boxiana* 2nd Ser. II. 165 They again became weavers, till they measured their lengths upon the ground, Warren again undermost. **1950** J. Dempsey *Championship Fighting* xi. 54 'Bobbers and weavers'—chaps who come in bobbing low and weaving from side to side.

c. *R.A.F. slang.* A pilot (or aircraft) pursuing a weaving course. Cf. *weave *v.*[2] 1 c.

1942 in Forbes & Allen *Ten Fighter Boys* 47, I called up the C.O. and said I would like to take up position as a 'weaver'. **1956** J. E. Johnson *Wing Leader* iv. 53 Some of our squadrons provided two weavers in an attempt to guard themselves from the bounce. The weavers flew above the squadron and continually weaved and criss-crossed.

d. A driver of a motor vehicle who moves continuously from lane to lane, esp. in order to pass other vehicles. *colloq.*

1960 *Amer. Speech* XXXV. 312 For a long time we have

heard of the *weaver*, the driver who shifts from lane to lane in driving. **1973** *Telegraph* (Brisbane) 13 Sept. 30/1 Then we have the weaver, who careers from lane to lane, passing other cars right and left.

5*. *Basket-making.* Any cane which is woven between the stakes of a basket.

1897 A. Firth *Cane Basket Work* ii. 17 *Weavers*, the canes which are placed alternately behind and before the spokes. **1960** E. Legg *Country Baskets* 27 If you will just bear in mind the names of the parts—sticks or stakes, and weavers—..you will never forget the different grades required... The stakes must be stouter or stronger than the weavers.

6. a. weaver ant, a tropical ant, esp. one of the genus *Oecophylla*, which builds nests of leaves fastened together by the silk of its own larvæ.

1913 *Ann. Rep. Smithsonian Inst. 1912* 456 The highly interesting weaver ants..use their larvæ as weaver's shuttles. **1977** *Sci. Amer.* Dec. 146/1 Weaver ants are extremely abundant, aggressive and territorial.

weaving, *vbl. sb.*[2] Add: **2.** The side-to-side movement by an animal of its head and neck.

1934 Miller & Robertson *Pract. Animal Husbandry* 59 Weaving is a nervous habit acquired by many wild animals in captivity (especially bears), and occasionally by horses. **1973** G. Durrell *Beasts in my Belfry* iv. 70 Sam had a habit—not uncommon in bears—which is called weaving.

weaving, *ppl. a.* Add: Hence **wea·vingly** *adv.*

1945 L. MacNeice in *Horizon* Nov. 295 Loom of wind Weavingly laughingly leavingly weepingly. **1959** C. Ogburn *Marauders* (1960) vi. 183 Overby ran erect, like a halfback,..carrying his rifle weavingly before him as if it were a football.

web, *sb.* Add: **I. 5. b.** Also, a continuously moving plastic sheet or film.

1958 E. G. Fisher *Extrusion of Plastics* vii. 96 The sheet passes through one or two pairs of nip rolls which draw the web through the take-off. **1974** J. H. Briston *Plastics Films* xv. 191 The web of material is controlled at all stages of the wrapping operation and cut-off only takes place when the product has been fully enclosed.

III. 14. a. Also, a longitudinal vertical member joining the upper and lower components of a wooden rib, spar, or beam in an aircraft.

1909 *Flight* 11 Sept. 553/2 The rib for a double-surfaced deck [*sc.* wing] is more elaborate in construction, and is itself stiffened with 'webs'. **1918** *Ibid.* 25 July 830/2 Each of the spars is built up of spruce flanges, connected on front and rear faces by three-ply webs, the whole forming a box. **1919** Pippard & Pritchard *Aeroplane Struct.* xvii. 199 The load is transmitted by shear across the web portion of the rib, and so this portion must be very strongly attached to the web of the spar. **1962** *Flight Handbk.* (ed. 6) iii. 45 Platz's wing employed box spars with plywood sides (webs) and heavier wooden tops and bottoms (booms).

h. *pl.* Snow-shoes. *N. Amer.*

1923 *Beaver* Jan. 145 It is said they still enjoy an occasional zestful tramp on the 'webs' over leagues of new Manitoba snow. **1939** K. Pinkerton *Wilderness Wife* ix. 103 After breakfast we went out to slip on our webs. **1966** M. E. & O. Murie *Wapiti Wilderness* xviii. 223 Snowshoes, or 'webs' as the Jackson Hole people call them, were the tried and true aids.

IV. 18. Objective, as *web-spinning* adj.; similative, as *websoft* adj.; **web-fed** *a. Printing* = *reel-fed* adj. s.v. *reel *sb.*[1] 6; **web-nest**, a filmy tissue enclosing a group of certain caterpillars or web-spinners (sense *b); cf. *web-worm*; **web offset**, offset lithographic printing on a continuous reel of paper (cf. sense *5 b); freq. *attrib.*; **web-spinner**, (*a*) a web-spinning spider; (*b*) a brownish gregarious insect of the order Embioptera, the females of which are wingless.

1949 Melcher & Larrick *Printing & Promotion Handbk.* 358/1 Newspapers are printed on web-fed rotary presses. **1965** Zigrosser & Gaehde *Guide Coll. Orig. Prints* iv. 73 In the rotogravure process the plate is in the form of a copperplated cylinder, functioning in a web-fed rotary press for long and speedy runs of printing. **1895** W. Schlich *Man. Forestry* IV. 279 The caterpillars, enclosed in the common web-nest, first gnaw the upper side of the leaves. **1903** *Biol. Bull.* IV. 100 They [*sc.* Embiidæ] still spin their web-nests. **1959** *Times Lit. Suppl.* 15 May 296/3 America, the land of newspapers on the giant scale,..is making great strides..with web-offset colour on newsprint. **1967** E. Chambers *Photolitho-Offset* xv. 238 Web offset can be defined as a method of lithographic printing in..one or more colours on one or both sides of a web of paper in a single operation. **1981** *Printing World* 28 Jan. 13/1 It was printed on a new Timson T32 web offset press. *a* **1915** Joyce *Giacomo Joyce* (1968) 7, I hold the websoft edges of her gown. **1923** *Jrnl. & Proc. R. Soc. Western Australia* IX. i. 61 The Order Embioptera, or Web-spinners, is a small but very distinct and isolated group of insects. **1941** J. S. Huxley *Uniqueness of Man* ix. 193 In the spiders, we find a very interesting difference between the hunters and the web-spinners. **1944** *Proc. U.S. Nat. Mus.* XCIV. 401 (*title*) A revision of the Embioptera, or web-spinners, of the New World. **1972** Swan & Papp *Common Insects N. Amer.* vi. 85 Embiids or Webspinners..occur mostly in the tropics and subtropics. **1904** W. H. Hudson *Green Mansions* v. 57 This was no web-spinning, sedentary spider. **1946** *Nature* 2 Nov. 630/2 The web-spinning Tineid caterpillar..is usually an inhabitant of hawthorn bushes.

weber. Add: **a.** (Earlier examples of the sense 'coulomb'.) Also † **veber.**

1872 *Telegr. Jrnl.* I. 11/1 The current through *a b* will then be equal to one British Association unit of current, or one veber per second. **1874** [see *farad 2].

† b. A unit of magnetic pole strength, equal to the pole strength that produces a field of 1 oersted ('gauss') at 1 centimetre. *Obs.*

1882 R. Clausius in *Phil. Mag.* XIII. 394, I would take leave to propose that for it [*sc.* the unit of magnetism] the name 'weber' be introduced. **1892** G. F. Barker *Physics* IV. iii. 645 The unit pole just now defined..is called a weber and the unit field which it produces at unit distance is called a gauss.

c. The M.K.S. unit of magnetic flux (now incorporated in the International System of Units), equal to 100 million maxwells; one volt-second. Symbol Wb.

1891 *Trans. Amer. Inst. Electr. Engineers* VIII. 542 We can start by defining 10^8 lines of force as the unit of induction and call it, say, a weber. **1895** *Rep. Brit. Assoc. Adv. Sci.* 196 The Committee..recommend for tentative adoption the following terminology:—— 1. That as a unit for magnetic field, a hundred million 'c.g.s. lines' be called a weber. **1935** *Proc. Nat. Acad. Sci.* XXI. 579 Various International Electrical Congresses..and since then, the I.E.C., have adopted, by successive steps, the well-known series of nine practical electromagnetic units (ohm, volt, ampere, farad, coulomb, joule, watt, henry and weber). **1963** G. Troup *Masers & Lasers* (ed. 2) v. 67 For magnetic dipole transitions, $|p_{mn}|$ is of the order of a Bohr magneton ($1·2 \times 10^{-29}$ weber metre). **1971** I. G. Gass et al. *Understanding Earth* iv. 71/2 The [magnetic] field varies over the Earth from 7×10^{-5} to $2·5 \times 10^{-5}$ weber metre^{-2} (Wb m^{-2}).

Weber[2] (vē¹·bəɹ). *Physiol.* [The name of Ernst H. *Weber* (1795–1878), German physiologist and anatomist.] **1.** *Weber's law*: the observation made by Weber that the increase in a stimulus that is just noticeable is a constant proportion (the *Weber fraction* or *ratio*) of the initial stimulus, for any one sense.

1872 [see *psycho-physical *a.*]. **1890** W. James *Princ. Psychol.* I. 537 So much for a general account of what Fechner calls Weber's law. **1935** *Brit. Jrnl. Psychol.* XXVI. 226 The obvious fact that Weber's Law is disobeyed both for very high and for very low light intensities leads to the clearly apparent result that..the various greys at either end of Dr. Ostwald's scale are never even approximately equidistant. **1938** Stevens & Davis *Hearing* iv. 149 We are not here concerned with the constancy of the Weber fraction $\Delta I/I$, which was *another* of Fechner's assumptions, but only with..the ability of an added just-noticeable difference to contribute always the same increment to the total subjective effect. **1952** I. J. Hirsh *Measurement of Hearing* vii. 194 We conclude that Weber's Law holds (for white noise) from about 20 to 100 db Sensation Level. **1970** *Jrnl. Gen. Psychol.* LXXXII. 38 The results of the present experiment suggest that the use of contrasting St[andard] and Co[mparison] vibrotactile stimuli will result in relatively small Weber ratios.

2. *Weber-Fechner law* [*Fechner] = *Weber's law* above; also, the related statement that the strength of a sensation is proportional to the logarithm of the stimulus causing it.

1891 A. D. Waller *Introd. Human Physiol.* xv. 536 (*heading*) The Weber-Fechner law. **1968** Passmore & Robson *Compan. Med. Stud.* I. xxiv. 30/2 To increase the loudness by steps that seem equal to the listener the intensity of sound must be multiplied by approximately the same factor for each step (Weber–Fechner relation).

Weberian (vē¹bī°·riăn), *a.*[2] [f. the name of Carl Maria von *Weber* (1786–1826), German composer + -ian.] Of, pertaining to, or characteristic of Weber or his music.

1958 *Listener* 28 Aug. 321/3 Many of the typically Weberian passages in dotted notes. **1978** *Gramophone* June 56/2, I hope I have not led you to undertstimate the quality of so much of the music—the Weberian horns at the start of No. 1.

Weberian (vē¹bī°·riăn), *a.*[3] [f. the name of Max *Weber* (1864–1920), German sociologist and political economist + -ian.] Of, pertaining to, or characteristic of Weber, his philosophy, or his writings.

1959 G. D. Mitchell *Sociol.* 114 The importance of this Weberian analysis is that a system of stratification is seen to depend upon the economic character of society, the way prestige is distributed, and the kind of legal and political institutions it possesses. **1969** P. A. Robinson *Freudian Left* 6, I consider the Weberian ideal of a *wertfreien* science misguided. **1975** *Times Lit. Suppl.* 1 Aug. 880/5 The account of the transformation of Wolof war-leaders into bureaucrats, bringing Weberian theory to bear, is a new departure. **1979** L. Lerner *Love & Marriage* vi. 231, I shall use 'sublimation' essentially in its Weberian meaning.

Webernesque (vē¹bəɹne·sk), *a.* [f. the name of Anton von *Webern* (1883–1945), Austrian composer + -esque.] Characteristic of the music of Webern. So **We·bernish** *a.*, somewhat resembling the music of Webern.

1959 *Listener* 28 May 960/3 The ejaculatory, disjointed, Webernish texture. **1961** *Times* 13 Oct. 18/5 The instrumentation is carried out in characteristically Webernesque fashion. **1976** *Gramophone* Sept. 437/3 The surface is.. rather Webernesque.

Weber number (vei·bəɹ, we·bəɹ). *Physics.* Also **Weber's number.** [tr. G. *webersche zahl* (F. Eisner 1932, in Wien & Harms *Handb. d. Experimentalphysik* IV. IV. 225), f. the name of Moritz *Weber* (1871–1951), German naval engineer, who first formulated it in 1919.] A dimensionless quantity used in the study of surface tension, bubbles, and waves, usu. expressed as $\rho l v^2/\gamma$ or the reciprocal of this, where γ is the surface tension of the fluid, ρ its density, l the characteristic length, and v the velocity of the fluid or of waves in the fluid; also, the square root of either quantity.

1937 O'Brien & Hickox *Appl. Fluid Mech.* v. 149 Generalized Representation of Weir Coefficients... The first three terms [in the coefficient of discharge] merely represent the geometrical similarity including the roughness. The other two are Weber's number and Reynolds' number. **1946** H. Rouse *Elem. Mech. of Fluids* x. 322 The radical form of this ratio $W = V/\sqrt{(\sigma/\rho L)}$ is known as the Weber number. **1977** R. A. Duckworth *Mech. of Fluids* vii. 149 In ship testing, the influence of surface tension forces, for which the Weber number, $(\sigma/\rho V^2 l)$, provides the criterion for dynamic stability, may be ignored because..such forces are negligible compared with the viscous and gravitational forces. **1979** A. L. Lydersen *Fluid Flow & Heat Transfer* ii. 39 Examples of force ratios are..inertia forces/surface forces = .. $\rho L V^2/\gamma$ = We = Weber number. **1979** *Sci. Amer.* Aug. 170/3 When the beaker is made to create water sheets in the middle range of Weber numbers, the ridge sends out curved waves.

web-foot. Add: **3. b.** (Earlier example.)

1845 W. Whitman in *Amer. Speech* (1961) XXXVI. 297 Oregonese [are called] *Webfoots*.

c. An infantryman. *U.S. slang.*

1867 B. W. Duke *Hist. Morgan's Cavalry* xiv. 400 When the 'webfeet' called us 'buttermilk rangers', we did not get angry with them. **1917** J. Morgan *Recoll. Rebel Reefer* 210, I was only a poor 'webfoot'.

web-footed, *a.* Add: Also *fig.* (in quots., *Mil.*).

1945 *Tee Emm* (Air Ministry) V. 49 We feel that the web-footed boys [i.e. Coastal Command] are not the only ones who may fall into the water out East. **1980** *Globe & Laurel* July/Aug. 226/1 The Small Raids Wing of the newly established Amphibious School based at Fort Cumberland and RMB Eastney, which also absorbed former web-footed units like RM Det 385.

Webley (we·bli). The proprietary name of various types of revolver and other small arms, etc., originally made by the firm of P. Webley and Son. Also *absol.*

1889 *Field* 4 May 618/1 After an exhaustive trial..the new Webley revolver..has been adopted by the English Government. **1897** *Sears, Roebuck Catal.* 586/2, 44 caliber, center fire, Webley cartridge. **1911** *Encycl. Brit.* XXI. 656/1 In the Webley the bolt is upon the standing breech and grips the extremity of the hinged barrel. **1920** *Trade Marks Jrnl.* 24 Mar. 567 Webley... Small-arms. Webley & Scott, Limited,..Birmingham; small arms manufacturers. **1923** *Official Gaz.* (U.S. Patent Office) 16 Jan. 502/2 Revolvers, automatic pistols, and other small arms. Webley & Scott Limited, Birmingham, England. **1937** S. Smith *Good Time was had by All* 53 A Webley, service pattern, is a useful weapon. **1979** P. Cosgrave *Three Colonels* vi. 137 Haddad was wearing a heavy Webley ostentatiously on his hip.

Webster[2] (we·bstəɹ). The name of Noah *Webster* (1758–1843), the American lexicographer, used *absol.* to designate his Dictionary (first published in 1828), and any of its later revisions and abridgements (now published by G. and C. Merriam Co., Springfield, Mass.).

1843 *Quincy* (Illinois) *Herald* 17 Mar. 4/1 (*heading*) Definitions not found in Webster. **1928** *Amer. Speech* IV. 90 We use *Webster* (in America) colloquially as synonymous with *dictionary*. **1950** [see *Oxford 2]. **1978** *Amer. Speech* LIII. 70 Many 'Websters' later, came the controversial dictionary edited by Philip B. Gove and published in 1961 by the G. and C. Merriam Company of Springfield, Massachusetts, to whom Webster's heirs had sold the rights to the dictionary in 1843.

Websterian (webstī°·riăn), *a.*[1] [f. the name of Noah *Webster* (see prec.) + -ian.] Of, pertaining to, or characteristic of Webster's Dictionary (see prec.) or any of its later versions or abridgements.

1874 B. F. Taylor *World on Wheels* 28 Websterian 'probabilities' says that is not the derivation of 'scale' at all. **1897** *Bookman* Nov. 201 We are quite sure that if the English were to adopt the Websterian spelling, Professor Matthews would very, very soon experience a conservative reaction. **1935** A. C. Baugh *Hist. Eng. Lang.* xi. 442 James Fenimore Cooper..was pronouncing..*beard* as *berd* or *baird* (another Websterian pronunciation). **1962** *New Yorker* 10 Mar. 156/2 Nearly all the books that come off the press..are edited and printed in accordance with Websterian usage. **1979** *Amer. Speech* LIV. 13 Thus was the battle joined between the traditional grammarians and those who might well be called the Websterian grammarians.

Webste·rian, *a.*[2] [f. the name of the English dramatist John *Webster* (1580–1625) + -ian.]

Of, pertaining to, or characteristic of Webster or his plays.
1928 *Observer* 12 Feb. 4 With the great Websterian passages there is mingled a great deal of melodramatic fustian. **1940** M. Lowry *Let.* 27 July (1967) 32 It is 'original' if you fear for past Websterian, not to say Miltonian, minor lacks of ethics on my part. **1950** *Scrutiny* XVII. 152 The other (*The Insatiate Countess*) a rather Websterian account of a promiscuous noblewoman's destruction. **1971** *Guardian* 19 Jan. 8/5 The true Websterian frisson is achieved on an empty stage.

websterite² (we·bstəraɪt). *Petrogr.* [f. *Webster*, name of a village in N. Carolina + -ITE¹.] An ultramafic intrusive igneous rock composed essentially of orthorhombic and monoclinic pyroxenes.
1890 G. H. Williams in *Amer. Geologist* VI. 44 The specimens obtained by Mr. Merrill from Webster are such admirable representatives of the bronzite-diopside rocks that the name Websterite is suggested..for them; and there seems to be no impropriety in extending this term over all the massive intrusives which are composed entirely of monoclinic and orthorhombic pyroxene. **1978** S. R. Nockolds in S. R. Nockolds et al. *Petrol. for Students* xiii. 148 Pyroxenite with both ortho- and clinopyroxene ('websterite') is the commonest variety, found both in connection with some great layered igneous bodies..and in small intrusions.

Wechsler (we·kslər). *Psychol.* The name of David *Wechsler* (b. 1896), American psychologist, used *attrib.* or as *adj.* in connection with various intelligence tests devised by him (and in use since 1939), esp. the *Wechsler-Bellevue Intelligence Scale*, the *Wechsler Intelligence Scale for Children* (WISC), and the *Wechsler Adult Intelligence Scale* (WAIS). Also *absol.*
1942 *Amer. Jrnl. Psychol.* LV. 608 The Wechsler-Bellevue Intelligence Scale, an individual scale adapted to ages 10–60 and consisting of 5 verbal and 5 performance-tests, with vocabulary-test as alternate. **1954** A. Anastasi *Psychol. Testing* xii. 306 Two scales for measuring 'general intelligence', namely, the Wechsler-Bellevue Intelligence Scale and the Wechsler Intelligence Scale for Children. *Ibid.*, One of the primary objectives of the Wechsler-Bellevue is to provide an intelligence test suitable for adults. **1961** *New Scientist* 16 Mar. 665/3 Special psychological tests, including the well-known Wechsler Memory Scale, were employed to assess memory. **1970** *Jrnl. Gen. Psychol.* July 69 The tasks of the Wechsler..are not composed of stimuli all of the same degree of structuredness either. **1981** D. Statt *Dict. Human Behav.* 129 *Wechsler Adult Intelligence Scale*, one of the most widely used intelligence tests for adults, combining performance and verbal ability testing.

Weddell (we·dĕl). The name of James *Weddell* (1787–1834), Scottish navigator, used *attrib.* or in the possessive as **Weddell('s) seal**, to designate a large brown Antarctic seal, *Leptonychotes weddellii*, first recorded by him and named in his honour in 1826. Cf. *sea-leopard* s.v. SEA *sb.* 23 b. Also *absol.*
1902 G. E. H. Barrett-Hamilton in *Rep. Coll. Nat. Hist. Voy. 'Southern Cross'* i. 19 Weddell's seal is probably of wide distribution. **1923** F. Wild *Shackleton's Last Voy.* vi. 118 Near by a fat Weddell seal lay asleep. **1938** [see *CRAB-EATER 3]. **1971** R. Sale *Man who raised Hell* I. i. 14 He was trying to tag a Weddell mother seal... Weddells are big animals, going to twelve feet. **1979** *Nature* 11 Jan. 87/3 The Weddell seal has adapted to fill a specialised niche—the inshore fast ice zone—where it winters under the ice, keeping breathing holes open with its teeth.

weddellite (we·dĕlaɪt). *Min.* [f. the name of the *Weddell* Sea, Antarctica, where it was first found + -ITE¹.] A hydrated calcium oxalate, $CaC_2O_4.2H_2O$, which occurs as colourless tetragonal crystals and is common in calculi.
1942 Frondel & Prien in *Science* 24 Apr. 431/1 Stones [formed in the body] composed wholly of carbonate-apatite are relatively rare. Whewellite ($CaC_2O_4 \cdot H_2O$), weddellite ($CaC_2O_4 \cdot 2H_2O$) and especially struvite..are ordinarily present. **1961** *Lancet* 26 Aug. 452/1 (*caption*) The large crystals in the centre of the field were of calcium oxalate dihydrate (weddellite). **1976** *Chem. Abstr.* 20 Sept. 190/1 (*heading*) Whewellite ($CaC_2O_4 \cdot H_2O$) and weddellite ($CaC_2O_4 \cdot 2H_2O$) from the Upper Cretaceous limestone near Zminj in Istria (Croatia, Yugoslavia).

wedding, *vbl. sb.* Add: **2. b.** *Golden Wedding* (earlier example).
1850 J. von Tautphœus *Initials* I. v. 121 'Perhaps you have no golden or silver wedding in England?' 'I confess I never heard of any thing of the kind,' said Hamilton... 'To be fifty years married and to have a golden wedding, is a sort of event in a family.'
4. a. *wedding-anniversary* (later example), *-coat* (earlier example), *-dress* (earlier example), *-journey* (earlier example), *photo*, *photograph*, *-present* (earlier example), *-suit* (later examples), *-supper* (later example), *tour*, *trip*, *visit*.
1971 'D. Halliday' *Dolly & Doctor Bird* iii. 33 It's our wedding anniversary. **1838** M. R. Walker *Let.* 27 July in C. M. Drury *Elkanah & Mary Walker* (1940) iii. 92, I was glad indeed to see him safe & sound with his cloths sadly rent having lost his coat his *wedding coat*. **1801** M. Edgeworth *Belinda* II. xxxi. 357 Something must be left to the imagination. Positively I will not describe wedding dresses. **1841** Thackeray *Gt. Hoggarty Diamond* in *Fraser's Mag.* Nov. 598/1 She lent us her chariot for the wedding journey. **1966** P. Willmott *Adolescent Boys* iii. 46 They showed us different wedding photos. **1956** *Focal Encycl. Photogr.* 1273/2 A good set of wedding photographs calls for a surprising amount of planning and organization. **1972** C. Fremlin *Appointment with Yesterday* x. 76 Cora should have received the wedding photograph. **1854** C. Schreiber *Jrnl.* 27 May (1952) 35 My wedding present to Felicia, a white and gold bookcase. **1807** Wedding suit [see *family tree* s.v. *FAMILY *sb.* 11]. **1959** W. Faulkner *Mansion* xiii. 296 Maybe he picked it up along with that-ere white wedding suit. **1961** New Eng. Bible *Rev.* xix. 9 Happy are those who are invited to the wedding-supper of the Lamb! **1847** C. M. Yonge *Scenes & Characters* xxxi. 340 The wedding had been hurried on, and the wedding-tour was shortened. **1955** C. Carrington *Rudyard Kipling* ix. 201 It will not be necessary to retrace Rudyard's steps throughout his wedding tour. **1855** Dickens *Dorrit* (1857) I. ii. 17 A clerical English husband..on a wedding trip with his young wife. **1925** F. Scott Fitzgerald *Great Gatsby* viii. 183 Tom and Daisy were still on their wedding trip. **1974** *News & Reporter* (Chester, S. Carolina) 22 Apr. 4-A/8 After a wedding trip to Charlottesville, Virginia the couple will reside on Morreene Road, Durham. **1794** J. Woodforde *Diary* 7 Apr. (1929) IV. 103 Mr. & Mrs. Carbould are gone out for a few days to make a wedding visit to Mrs. Carbould's Brother at Castor near Yarmouth. **1872** Geo. Eliot *Middlem.* II. iii. xxviii. 89 Through the next weeks there would be wedding visits received and given.
b. wedding band *U.S.* = *wedding-ring*; **wedding-breakfast**, delete 'formerly' and add earlier and later examples; cf. *BREAKFAST *sb.* 2 b; **wedding-bush**, a shrub of the genus *Ricinocarpos*, of the family Euphorbiaceæ, esp. *R. pinifolius*, which is native to eastern Australia and bears clusters of fragrant white flowers; **wedding-cake**: also *fig.*, esp. applied *attrib.* (often somewhat dismissively) to a sumptuously ornate style of architecture, and (also *absol.*) to buildings in this style; (later examples); **wedding canopy** *Judaism* = *CHUPPAH; **wedding-cards** (earlier example); also in *sing.*; **wedding group**, (a photograph of) a wedding party; **wedding list**, a list of acceptable wedding gifts for guests to consult and act upon; **wedding party**, the assemblage of persons at a wedding; **wedding reception**, a party at which the wedding guests are formally greeted and entertained after the marriage ceremony; cf. *RECEPTION 2 d; **wedding-ring**: also, a ring similarly presented by the bride to the bridegroom, and worn afterwards by him; later *fig.* examples, both *attrib.*
1946 R. Lyle *Mademoiselle's Handbk. Bridal Consultants* vi. 79 Wedding bands have in centuries past been made of iron as well as gold. **1977** 'E. McBain' *Long Time no See* i. 16 On the third finger of his left hand, there was a wedding band. **1850** Thackeray *Pendennis* II. xv. 144 There is a wedding breakfast. **1962** *Sunday Times* 11 Nov. 25/4 They married in church, and had a wedding breakfast but no honeymoon. **1923** *Census of Plants of Victoria* (Field Naturalists' Club of Victoria) 41 Wedding Bush. **1961** Wedding-bush [see *MANNA¹ 9]. **1925** F. Scott Fitzgerald *Great Gatsby* i. 8 A breeze..blew curtains,..twisting them up toward the frosted wedding-cake of the ceiling. **1949** H. G. Alsberg *Amer. Guide* 105 Kennebunk ..Among numerous old mansions are..Wedding Cake H. (c. 1825), with most unusual Vict. trimmings. **1968** *N.Y. City* (Michelin Tire Corp.) 16 Others [sc. skyscrapers], sometimes known as *wedding cakes*, were covered with ornate sculpture in the 1900 '*Gigi style*'. **1971** J. Willett in A. Bullock *20th Cent.* 242/1 A massive piece of wedding-cake architecture (..fortunately never built) for the projected Palace of Soviets in Moscow. **1892** I. Zangwill *Childr. Ghetto* I. 37 The hope was expressed that Mr. and Mrs. Belcovitch would like..to see their daughters' daughters under the *Chuppah*, or wedding canopy. **1978** I. B. Singer *Shosha* iv. 74 You should lead your daughter to a black wedding canopy! **1847** C. M. Yonge *Scenes & Characters* ii. 9 She was putting her sister's wedding cards into their shining envelopes. **1953** Dylan Thomas *Let.* 31 Mar. (1966) 401 Had your wedding card. Congratulations. **1861** Geo. Eliot *Silas Marner* 361 The wedding group had passed on..to the humbler part of the village. **1930** R. Lehmann *Note in Music* v. 190 The texts, the wedding group, and the photograph of a grave on the wall. **1979** J. Adam Smith *John Buchan* 51 People in Scotland, she said, might like to have wedding groups, but she doubted if it was ever done in London. **1981** *Times* 7 Mar. (Bride & Home Suppl.) p. vi/1 Today's brides..confine romance to their wedding dress and..Roneo their less personal requirements... The stores still keeping wedding lists do provide a valuable service. **1873** C. M. Yonge *Pillars of House* IV. xlvii. 340 The blow was known to all that sad wedding party. **1877** W. S. Gilbert *Engaged* (?1883) II. 31, I have ordered four flys for the wedding party. **1979** J. Gardner *Nostradamus Traitor* xxxv. 165 Another alert. This time the wedding party had to take shelter. **1871** A. J. Munby *Diary* 4 Nov. in D. Hudson *Munby* (1972) 299 We, Council and Students..offered this evening a wedding reception to Litchfield and his bride. **1965** [see *STUNG *ppl. a.* 2]. **1978** S. Sheldon *Bloodline* xlv. 374 After the ceremony there was a wedding reception at the Baur-au-Lac. **1953** Dylan Thomas *Under Milk Wood* (1954) 74 His mother..with her wedding-ring waist and bust like a black-clothed dining-table suffers in her stays. **1980** *Quilt World* Sept./Oct. 23/1 Sheri's mother feels that it is a single wedding ring quilt because when four blocks are pieced together a ring is formed.

‖ **wedeln** (vē·dəln), *sb. Skiing.* Also **wedel.** [a. G. *wedeln* (in same sense).] A skiing technique using a swaying movement of the hips to make short parallel turns (see *PARALLEL *a.* 1 b). Also *attrib.*
1957 J. Ceely tr. *Joubet & Vuarnet's Ski ABC* ii. 106 The skier effects a real wedeln when he is able to minimize the second step of the above progression. *Ibid.* iii. 130 In linked turns or wedeln, planting the pole adds favorable *rhythm* to the movements. **1958** *Times* 1 Mar. 7/6 The reverse shoulder technique of *wedeln*..or *godille* as the French call it, has become so universal among first-class racers of all nations that no one using the classical rotation can hope to be placed in a serious race. **1963** *New Yorker* 2 Feb. 37 Down he goes, anyway, his wedeln so crisp he never seems to care at all for the configuration of the snow. **1963** *Amer. Speech* XXXVIII. 208 Other attributive uses are *wedel track* and *wedel turn*. **1973** R. Hayes *Hungarian Game* xxxii. 195 We had skied nearly to the base of the mountain and Mary Louise shot over a drop... I did a quick wedeln series and looked for her. **1974** H. Evans et al. *We learned to Ski* 44 Those who can do the long radius parallel turn learn to make shorter linked parallel turns—the wedel. **1975** *Oxf. Compan. Sports & Games* 965/2 Between 1956 and 1958, the Austrian, Sailer, won seven out of eight world titles,.. and he introduced the *wedeln* style to a wide public.

‖ **wedeln** (vē·dəln), *v. Skiing.* Also (app. more commonly) **wedel.** [a. G. *wedeln* (in same sense), lit. to wag (the tail).] *intr.* To use the wedeln technique in skiing. Also *transf.* in *Skateboarding.*
1961 *Times* 7 Jan. 7/7 Sign-posted high roads which shepherd each carload of uninspired humanity down from the heights..wedelning as best they can. **1963** *Amer. Speech* XXXVIII. 208 *Wedel; wedeln, v.* to execute a special kind of short parallel turns close to the fall line in quick succession. **1968** *Time* 15 Nov. 49 They wedeled down the 1,200-ft. slope or slammed through the slalom course. **1973** P. A. Whitney *Snowfire* iii. 36, I don't care..whether my skis are expertly parallel. I've never been able to wedel. **1977** *Skateboard Special* Sept. 3/1 Special slalom decks have a raised centre to make it easier to 'pump' or 'wedel'.
Hence **we·del(l)ing** *vbl. sb.*
1977 *Skateboard Special* Sept. 2/2 *Wedeling*, a sophisticated method of reducing speed while zig-zagging down steep hills. *Ibid.* 5/1 Slalom decks..are often slightly raised in the middle to make pumping and wedeling easier. **1979** N. Slater *Falcon* ix. 161 His *wedelling* turns between each traverse were smothered in a flurry of snow.

wedge, *sb.* Add: **2. b.** *the thin end of the wedge* (earlier examples).
1856 C. Fox *Jrnl.* 8 Nov. in *Memories Old Friends* (1882) xxii. 308 Beware, Englishmen, of the tendencies to hierarchy in your country when the thin end of the wedge is introduced: it will work its way on to all this. **1867** *Hansard Commons* 27 June 615 The thin end of the wedge.
5. i. A v-shaped sign used in various musical and other notations (see quots.).
1893 E. M. Thompson *Handbk. Greek & Latin Palæogr.* v. 68 The paragraph-mark was not..uniformly the horizontal stroke; the wedge > ..and similar forms were employed. **1970** *Language* XLVI. 78 Wedges printed after vowel symbols, e.g. [a^ a> a˅ a<], indicate raising, backing, lowering, and fronting. **1980** *Early Music* July 401/1 The most fascinating [signs] are the wedges indicating *crescendo*, *diminuendo* and *messa da voce* on single long notes: ◂, ▸, ◆, and a passage with second-position fingerings.
j. *Golf.* A golf club with a wedge-shaped head, used for lofting the ball at approach shots, or (= *sand wedge* s.v. *SAND *sb.*² 10 a) out of a bunker, etc. Also, a shot made with a wedge.
[**1924** J. White *Easier Golf* iv. 100 What I attempt to do is to use this heel [of a club]..as a wedge, and by driving this into the sand behind the ball I create sufficient disturbance to force the ball out of any lie.] **1937** [see *sand wedge* s.v. *SAND *sb.*² 10 a]. **1952** *Chambers's Jrnl.* May 300/1 Basil walked moodily off the tee, and after five minutes' search found his ball embedded in a patch of the foulest rough on the course, hacked it out with his wedge, and, playing two odd to the green, lost the hole. **1961** *Times* 1 July 4/1 He..played an overcautious wedge at the Royal. **1975** *Daily Tel.* (Colour Suppl.) 12 Sept. 9/4 Putting is out; most golfers carry just a driver, a four-wood, mid-iron and wedge.
k. A wedge heel; a wedge-soled shoe. See sense 9 b below. *colloq.*
1959 *Chambers's 20th Cent. Dict.* Add. **1965** R. Hardwick *Plotters* (1966) xi. 102 Stretch pants, wedges, and a leghorn hat. **1968** J. Ironside *Fashion Alphabet* 137 *Wedge*, a solid heel joined to the sole in one solid piece. **1976** *Washington Post* 19 Apr. A12/3 (Advt.), Casual style wedges in Oxford and slip-on styles. **1983** *Times* 14 July 11/3 Gladiator straps on stacked wooden wedge..£44.50.
l. A hair style in which the ends of the hair are slightly graduated so that they form a series of wedges. orig. *U.S.*
1976 *Time* 19 Apr. 69 There are many variations on the new wedge. Stylists at the Paul McGregor shops in New York and Los Angeles have shaped the back of the cut into three inverted pyramids. **1977** *Daily News* (Perth, Austral.) 19 Jan. 6/4 After she became a headliner, Dorothy's hairdo, called the wedge, sent girls rushing off to hairdressers to duplicate the look. **1985** *Hair* Summer 78 (*caption*), Short, sculptured sweeping version of the wedge has classy clout in the form of a pink flash.

9. a. *wedge-blade*; *wedge-balancing* adj.
1921 D. H. Lawrence *Tortoises* 19 Four rowing limbs, and one wedge-balancing head. **1917** —— *Look! We have come Through!* 113 The fine, fine wind... Like a fine, an exquisite chisel, a wedge-blade inserted.
b. *attrib.* and *Comb.* Designating a wedge-shaped heel extended under the instep of a woman's shoe (also, the sole which includes this), or a shoe having such a heel. Freq. as *wedge-heel*, *shoe*, *sole*; *wedge-heeled*, *-soled* adjs. Cf. sense 5 k above.
1939 M. B. Picken *Lang. Fashion* 164/3 *Wedge-soled*, having a wedge-shaped piece making a solid sole, flat on the ground from heel to toe. **1940** Graves & Hodge *Long Week-End* xxi. 375 A high-heeled fancy shoe..and a wedge-heeled streamlined type. **1940** *Manch. Guardian Weekly* 11 Oct. 259 Today's displays of courts..and wedge-heel, and all other of the creations of the fashion-designer, give no indication..of what was really a welcome weeding out. **1940** O. Nash in *New Yorker* 23 Nov. 18/2 Let us give thanks that women's wedge shoes weren't invented until they were. **1942** in C. W. Cunnington *Eng. Women's Clothing in Present Cent.* (1952) viii. 271 Practical [shoes], with flatter heels, square toed and wedge-soled. **1951** [see *CREEPER 1 d]. **1957** R. Hoggart *Uses of Literacy* iv. 102 Mail-order firms advertise fancy wedge-shoes. **1975** D. Beaty *Electric Train* 153 Painted faces clumping up..on six-inch wedge shoes. **1983** P. Devlin *All of us Here* x. 112 Her daughter, in a new permanently pleated skirt, wedge-heeled shoes.
10. wedge-tail *Austral.*, the wedge-tailed eagle (see *wedge-tailed* adj.); = *EAGLE-HAWK 2; **wedge tent** = A TENT.
1935 A. C. Chisholm *Bird Wonders Austral.* x. 102 The Wedge-tail is a formidable foe for any native mammal. **1965** [see *EAGLE-HAWK 2]. **1974** D. Stuart *Prince of my Country* ii. 9 Watching the long effortless circling of the wedgetail high in the air. **1977** *Times Lit. Suppl.* 21 Jan. 76/2 Australia is the only place in the whole world where the wedgetail eagle is known. **1862** O. W. Norton *Army Lett.* (1903) 49 We used to sleep on the ground or on pine boughs when we had the small wedge tents. **1940** G. W. Martin *Mod. Camping Guide* v. 86 The wedge tent, known also as the A tent, is a popular model with explorers and other outdoorsmen who want something a little larger than a tiny crawl-in tent. **1980** D. T. Roscoe *Your Bk. Camping* ('Your Bk.' Ser.) ii. 22 Wedge tents.. are designed to save weight and bulk and to withstand wind better when the smaller end is pitched directly into it.

wedge, *v.*¹ Add: **6.** *to wedge out* (Geol.): = *thin out* s.v. THIN *v.*¹ 2 a; = *lens out* s.v. *LENS *v.*
1819 [implied in *wedging out* s.v. *WEDGING *vbl. sb.* 3*]. **1839** R. I. Murchison *Silurian Syst.* 140 Limestone..can be traced tapering away from a central mass to thin extremities, which really wedge out between the coal grits and the older deposits. **1945** *Bull. Amer. Assoc. Petroleum Geologists* XXIX. 1563 The distinction from the Permeability Trap Reservoirs is made by restricting the Pinch-Out Trap Reservoirs to types located in such stratigraphic intervals or zones which actually wedge out. **1966** *Earth-Sci. Rev.* I. 163 Ignimbrites tend to wedge out against or thin over topographic highs. **1979** *Nature* 27 Sept. 267/1 These nappes wedge out and converge to the west and seem to represent a telescoping of Lower Palæozoic Facies.

wedged (wedʒd), *ppl. a.*² [f. WEDGE *v.*² + -ED¹.] Of wet clay: that has been wedged to expel air-bubbles before it is worked.
1903 in J. Burnett *Useful Toil* (1974) III. 298 From balls of wedged or refined clay he made the pancake-like shapes of clay which he had to use in making the next set of plates. **1967** M. Chandler *Ceramics in Mod. World* ii. 63 However well pugged or well *wedged* (hand worked to eliminate air) his clay may be, the ceramic sculptor cannot be certain that it contains no *blebs*.

wedger (we·dʒəɹ). [f. WEDGE *v.*² + -ER¹.] A workman who wedges clay to expel air-bubbles from it.
1881 *Harper's Mag.* Feb. 361/2 The 'wedger' takes a lump of..twenty pounds, lays it on the kneading block before him, [etc.]. **1921** *Dict. Occup. Terms* (1927) §104 *Wedger, clay wedger*, cuts lump of clay into pieces or wedges with wire cutter, and beats one piece of clay vigorously against another to make it homogeneous and free of air bubbles, for use in manufacture of highest class ware.

wedgie (we·dʒi). *colloq.* Also **wedgy.** [f. WEDGE *sb.* + -Y⁶.] A wedge-heeled shoe (see *WEDGE *sb.* 9 b); more recently, also *spec.* one with a built-up or 'stacked' sole. Usu. in *pl.*
1940 *Women's Wear Daily* 15 Mar. 18/1 New streamlined wedgies that make the foot look smaller. **1943** *Consumers' Res. Bull.* (U.S.) Dec. 20/2 Brown leather 'wedgie' with bottom-sole of thin leather. **1950** [see *sling-back* s.v. *SLING *sb.*³ 7]. **1962** A. Huxley *Island* ix. 133 Soft Platform Wedgies in Wide Widths. **1974** E. Brawley *Rap* (1975) I. v. 83 She always wore those white wedgies, old Marie-Claude, claimed they eased her feet on the job.

wedging, *vbl. sb.* Add: **2.** Delete 'jutting *out* or' and transfer quot. 1819 to sense 3* below.
3*. *Geol.* With *out*: the narrowing of a stratum or the like to the point of extinction. Cf. *WEDGE *v.*¹ 6.
1819 [in Dict., sense 2]. **1883** W. S. Gresley *Gloss. Coal-Mining* 283 *Wedging out*, cropping or thinning out.

Wedgwood. Add: **1. a.** Now a proprietary name both in the U.K. (since 1876) and the U.S. (since 1906).
1876 *Trade Marks Jrnl.* 2 Aug. 325 Wedgwood. Godfrey Wedgwood..trading under the Firm of Josiah Wedgwood and Sons..Pottery Manufacturers... Pots. 2823. 28th Feb. 1876. **1906** *Official Gaz.* (U.S. Patent Office) 4 Dec. 1671/1 Wedgwood. Particular description of goods.—Porcelain, stoneware, and earthenware, including Jasper.
b. (Later *absol.* examples, = Wedgwood blue.)
1923 *Daily Mail* 19 Feb. 1 (Advt.), Coloured Jap silks (36 inches wide)... In pink, coral, wedgwood, helio, [etc.]. **1974** P. Dickinson *Poison Oracle* ii. 41 Her angry Wedgwood eyes held his.

wee, *a.* Add: **e.** *Wee Frees*: also *transf.*
1953 Earl Winterton *Orders of Day* viii. 92 In 1919.. both the Labour and Liberal Oppositions were small and ineffective. The latter, facetiously known as the 'Wee Frees'..split into two halves led by Sir Donald Maclean and Mr. George Lambert respectively. **1966** *Punch* 20 July 123/3 His account of the way in which the Wahabis—the Calvinistic 'wee frees' of Islam—are surrendering to the worst of Western culture is a lively moral tale in itself. **1979** H. Wilson *Final Term* i. 10 The smell of power..was in their nostrils, for the first time since the 'Wee Frees', the Samuelite Liberals, had left the 1931 Coalition Government.
f. *the wee (small) hours* = *small hours* s.v. HOUR 3 b. *colloq.*
[**1787** Burns *Death & Dr. Hornbook* in *Poems & Songs* (1968) I. 84 The auld kirk-hammer strak the bell Some wee, short hour ayont the *twal*.] **1849** C. Brontë *Shirley* II. ii. 73 She followed the steps of the night, on its pathway of stars, far into the 'wee sma' hours ayont the twal.' **1859** [see HOUR 3 b]. *a* **1891** H. Melville *To Major John Gention* in *Compl. Wks.* (1924) XIII. 366 In the wee hours..how affluent hast thou been on that theme. **1932** 'L. G. Gibbon' *Sunset Song* 232 They'd another long dram, and they argued far into the wee, small hours. **1949** P. Michaels *This Perverse Generation* v. 44 No one has a persistent inner compulsion to..talk about silly things in crowded, stuffy, little night-club rooms at wee hours of the morning. **1966** 'H. MacDiarmid' *Company I've Kept* viii. 193 We walked back..in the 'wee sma' 'oors' of the following morning. **1979** *United States 1980/81* (Penguin Travel Guides) 278 Several acts keep the place hopping from dinner time until the wee hours.

wee (wī), *sb.*² *colloq.* [See next.] = *WEE-WEE *sb.* 1.
1968 R. Clapperton *No News on Monday* vi. 61 Wanda is downstairs having a wee. **1971** P. Purser *Holy Father's Navy* ii. 12 Hurry up, I want to do a wee. **1973** *Punch* 1 Aug. 139/1 Thought I saw someone comin' in de gate, but it only a dog havin' a wee on de magnolia.

wee (wī), *v. colloq.* [Echoic: see *WEE-WEE *v.*] *intr.* To urinate. Also *refl.* = *WET *v.* 5 d.
1934 Dylan Thomas *Let.* 11 May (1966) 128 Wee on the sun that he bloody well shines not. **1970** *Guardian* 13 July 9/2 Ladies always have to walk a mile and they'll wee themselves if they don't find something soon. **1976** *West Lancs. Evening Gaz.* 15 Dec., She hit her daughter because she kept on 'weeing' all the time. **1983** *Daily Mail* 30 May 17/5 Our headmaster told us that any boy caught short should if absolutely necessary wee into an empty milk bottle.

weed, *sb.*¹ Add: **3. c.** Marijuana; a marijuana cigarette. *slang* (orig. *U.S.*).
1929 *Amer. Speech* IV. 345 *Weed*, marijuana cigarette. (A Mexican drug.) **1932**, **1933** [see *REEFER¹ 3]. **1939** C. R. Cooper *Designs in Scarlet* viii. 145 There are certain centers in which marihuana offers more of a menace.. The peddling of the weed is dependent upon persons.. content with small profits. **1949** R. Chandler *Little Sister* xxiv. 173 They were looking for..a suitcase full of weed. **1955** J. Kerouac in *Paris Rev.* Winter 14 You could smell tea, weed, I mean marijuana, floating in the air. **1965** W. Soyinka *Road* 24 Say Tokyo reaches out a stick of weed to him which he accepts behind his back. **1978** R. Hill *Pinch of Snuff* iii. 28 She *might* be on the game. Or she might have a bit of weed about the place. Or anything.
5. b. Hence, a small, feeble, or contemptible person; a 'weedy' type (see *WEEDY *a.*¹ 4 b). *slang.*
1953 Willans & Searle *Down with Skool!* (1954) 12 There are a grate number of other weeds and wets about the place. **1959** I. & P. Opie *Lore & Lang. Schoolch.* ix. 170 *Little 'uns*..Tiny Tim, Tom Thumb, tot, and, very common, weed. **1960** M. Spark *Bachelors* x. 179 She's mad in love with that little weed Patrick Seton. **1970** *Times* 4 Mar. 13/5 A girl torn between a brainy weed and a moronic body-builder. **1982** Barr & York *Official Sloane Ranger Handbk.* 71/1 Don't try too hard: swots are weeds.
7. a. *weed control*, *-drift*, *-growth*, *-life*.
1923 Kipling *Irish Guards in Great War* I. 222 They broke and disappeared in the rank weed-growth. **1934** A. Huxley *Beyond Mexique Bay* 249 Lawrence wrote eloquently of Oaxaca and Lake Chapala,..of the merits of that rank weed-life of the natural man. **1948** S. B. Whitehead *Reclaiming your Garden* ii. 17 Chemicals.. have a selective action in weed control. **1951** *Sport* 7–13 Jan. 8/1 It had the drawback of increasing weed-growth. ?**1953** Dylan Thomas *Sel. Lett.* (1966) 416 In the slimy squid-rows of the sea there's such a weed-drift and clamour of old plankton drinkers. **1966** *Listener* 1 Sept. 304/1 The improvement of cultivation techniques to make weed control possible.
b. *weed-clogged*, *-covered*, *-laden*, *-mantled*, *-ridden*, *-sodden*, *-woven* adjs.
1898 O. Wilde *Ballad Reading Gaol* 12 Every day Crawled like a weed-clogged wave. **1910** W. de la Mare *Three Mulla-Mulgars* xxiii. 305 Nod's raft swam last across this weed-mantled lagoon. **1922** Joyce *Ulysses* 699 In loose allwool garments..trundling a weedladen wheelbarrow without excessive fatigue at sunset amid the scent of newmown hay. **1938** W. de la Mare *Memory* 29 Roofless and eyeless, weed-sodden, dank, old, cold. **1944** Blunden *Shells by Stream* 15 Weed-woven the shallowing pool. **1953** L. Kuper *Living in Towns* 300 Waste land and weed-covered pit banks. **1968** E. A. McCourt *Saskatchewan* xvii. 195 The inscriptions on the wooden crosses in the weed-ridden settlement graveyard. **1972** R. Adams *Watership Down* xxxii. 245 The rough, weed-covered ground of the combe sloped away below them.
c. *weed-cutter* (earlier example); *weed-waving*, *-winding* adjs.; also *weed-free* adj.
1850 *Rep. Comm. Patents 1849* (U.S.) I. 297, I likewise claim the combination of the adjustive weed cutter and leveler F. **1879** G. M. Hopkins *Poems* (1967) 78 On meadow and river and wind-wandering weed-winding bank. **1940** J. Betjeman *Old Lights for New Chancels* 33 The light skiff is push'd from the weed-waving shore. **1960** *Farmer & Stockbreeder* 22 Mar. 56/3 It is the modern alternative..to secure weed-free cropping. **1981** M. Sellers *From Eternity to Here* i. 7 A well-planned and weed-free vegetable garden.
d. weedhead *slang* (chiefly *U.S.*), one who is addicted to marijuana; a marijuana smoker (see *HEAD *sb.* 7 e); **weed inspector**, an official in charge of controlling the growth of noxious weeds; **weed-killer**, for def. read: something that kills weeds, *spec.* any of various chemical preparations used for killing weeds; liquid, powder, etc., of this kind; (earlier and later examples); also *fig.*
1952 *Amer. Speech* XXVII. 30 *Weedhead*,..marijuana smoker. **1966** C. Himes *Heat is On* xx. 157 Weedheads jabbered and gesticulated. **1973** R. Parkes *Guardians* x. 178 The students that spat and jeered at authority, the weed-heads that threw acid in a vicar's eyes. **1909** *Chambers's Jrnl.* Nov. 702/2 The writer, in his capacity as weed-inspector, has had..to direct the settlers to plough under entire fields of growing grain in order to eradicate such weeds. **1974** D. Sears *Lark in Clear Air* iv. 48 The one in overalls was Brad-Awl Callum, weed inspector, pound-keeper, truant officer and County Constable all rolled into one. **1745** W. Ellis *Agric. Improv'd* in *Mod. Husbandman* July xi. 88 Such Wheat commonly runs up into such rank Stalks, as to become great Weed-killers. **1901** M. Franklin *My Brilliant Career* xxi. 179 Every one would be wanting to engage me as the great noxious weed-killer and poisonous insect exterminator if I made away with you. **1929** Wodehouse *Mr. Mulliner Speaking* i. 23 What this woman needed was a fluid ounce of weed-killer. **1967** B. Patten *Little Johnny's Confession* 46 Who poured weedkiller over your innocence? *a* **1974** R. Crossman *Diaries* (1976) II. 291 She only talked to Anne and myself about a new weedkiller called Paraquat, pointing out of the window to a field where it was being used.

weed, *v.* Add: **4. e.** *trans.* To perform a process of selecting from (a collection of documents, a file, etc.), rejecting those items which are unimportant or not worth retaining; to select (papers, etc.) in this manner. Also, to select (papers, etc.) in order to withhold them from general inspection. Also with *out.*
1874 H. Taylor *Let.* 30 Sept. in J. Brown *Lett.* (1912) 412, I am not sure that when I admired his poetry most, I did not admire his letters more, but like almost every collection of letters they should be weeded. **1946** A. E. Porter *Let.* 26 Aug. in P. Moon *Transfer of Power* (1979) VIII. 309 Dick Tottenham, who compiled most of the files and is now on leave preparatory to retirement, would be singularly well qualified to weed them out and present you with those which are in fact worth keeping. **1977** *Times* 31 Aug. 4/1 The Ministry of Defence has begun a review of the way classified papers are written, filed, preserved (or 'weeded').
f. *to weed down*, to reduce (a list of candidates, etc.) to a short-list.
1942 *R.A.F. Jrnl.* 18 Apr. 6 Those who pass are further weeded down by a ballot. **1962** J. Glenn in *Into Orbit* 17 NASA asked us to take a series of tests which would help weed us down further.
6. *Angling.* Of a trout: to bury itself in weeds when hooked (*refl.* and *intr.*). Also, to entangle (the fish) in weeds (const. in *pass.*).
1889 F. M. Halford *Dry-Fly Fishing* ix. 211 Do not be afraid of weedy places. Trout..seldom weed at night, probably because they do not see the angler. *Ibid.* 224 When a fish is once weeded, the following tactics are what I suggest. **1960** *Times* 2 July 11/2 The risk of allowing a fish to weed itself is much reduced..if the weedbed lies upstream. *Ibid.*, More good trout have been lost in attempts to prevent them reaching weedbeds than have been lost in efforts to get them out once they have become weeded.

weeded (wī·dĕd), *a.*² *arch.* [f. WEED *sb.*² + -ED².] Dressed in widow's weeds.
1895 Hardy *Jude* v. vii. 369 Having ascertained thus much the immensely weeded widow retraced her steps. **1971** 'A. Burgess' *MF* i. 18 An untrustworthy young man in black spoke to the frail weeded widow.

weeder. Add: **3. b.** *spec.* A person employed by a government department to weed documents, letters, etc.
In quots. 1976 and 1984 the process is one of selecting papers in order to withhold or suppress them rather than to discard them.
1976 Summers & Mangold *File on Tsar* v. xx. 254 Before papers get anywhere near the Public Record Office the weeders, for that is what they are called in government circles, have first bite; beavering away in Whitehall basements, they plough through the documents accumulated by all government ministries, deciding what should be suppressed. **1983** W. Garner *Think Big, Think Dirty* vii. 101 The weeders worked in Central Registry, thinning down documentation as it grew bulky, repetitive or outdated. **1984** *Times* 23 May 10/3 Highly secret work, so secret that the weeders will see that it is not disclosed, even after 30 years.

weedery (wī·dəri), *sb.*² *poet. nonce-wd.* [f. WEED *sb.*² + -ERY.] Mourning garments.
1908 Hardy *Dynasts* III. vi. ii. 459 Even as he For whom thou wear'st that filial weedery Was waylaid by my tipstaff nine years since.

weedicide (wī·disaɪd). [f. WEED *sb.*¹ + -I- + -CIDE.] A chemical preparation designed as a weed-killer; weed-killer.
1934 *Jrnl. Dept. Agric. W. Austral.* XI. 521 Sodium arsenite has been employed extensively..as a weedicide, and..proved the most successful weed killer. **1951** *Chambers's Jrnl.* Aug. 457/1 Agricultural scientists..have set themselves the task of finding efficient straight weedicides. **1975** *N.Z. Jrnl. Agric.* Sept. 21/1 Maize farmers.. spend a great deal of time and money on ground preparation, seed, fertilisers and weedicide. **1979** *Indian Express* 10 Apr. 10/4 As preparatory cultivations are carried out to control weeds only, efforts should be made to use weedicides for it whenever necessary.

weeding, *vbl. sb.* Add: **2. b.** (Later examples.) Also with *out.*
1916 *Kinematograph Year Bk.* 36 (*heading*) The weeding-out process. **1932** *New Yorker* 14 May 56/2, I should like to hear some of the manuscripts that the weeding-out committee considered inferior to the final five. **1953** J. S. Huxley *Evolution in Action* ii. 42 This weeding-out process. **1968** *North Amer. Rev.* CVII. 589 In one case, and in one case only, could this weeding [of a library's collection] be properly made. **1977** *Times* 31 Aug. 4/1 Valuable material may have been destroyed during 'weeding'. **1981** D. Francis *Twice Shy* xii. 139 He disagreed strongly with my weeding-out policy, even though I'd..discussed every dud [horse] to be discarded.

weedy, *a.*¹ Add: **3.** (Earlier example.)
1867 *Trans. Illinois Agric. Soc.* VI. 638 *Weedy*, flour made from wheat that has come in contact with a noxious weed imparting an unpleasant smell.
4. b. (Later examples.) Also without reference to physical qualities: feeble, half-hearted, weak; lacking firmness or strength.
1922 E. M. Forster *Life to Come* (1975) 73 The chief had developed into an affable and rather weedy Christian with a good knowledge of English. **1971** *Petticoat* 24 July 4/3 I'm just too weedy, I don't threaten to expose them. **1982** Barr & York *Official Sloane Ranger Handbk.* 122/1 At whatever grade, they are fearless ('Be weedy in front of the wops?').
c. Of things.
1950 *Chambers's Jrnl.* Mar. 131/1 He must have conveyed this to her in one way or another, because soon afterwards she sent back his presents, and the weedy friendship which might have blossomed into a beautiful love was nipped in the bud.
5. *weedy-brown* adj.; *weedy-slow* adv.
1958 J. Betjeman *Coll. Poems* 263 A mounting arch of water weedy-brown. **1923** E. Sitwell *Bucolic Comedies* 38 In the Castles drowned long ago Where the empty years pass weedy-slow.
Hence **wee·diness**, the quality or state of being 'weedy'; lack of physical presence; feebleness.
1924 Wodehouse *Bill the Conqueror* v. 103 He had a certain weediness, a lack of thews and sinews. **1925** W. Deeping *Sorrell & Son* viii. 80 He realized that in spite of the other man's weediness he was a competitor who was to be respected.

week, *sb.* Add: **1. b.** (Later examples.) Also, a week during which some event takes place, either annually or on one occasion only; a week during which attention is focused on a particular topic for promotional, charitable, etc., purposes.
1761 G. Colman *Jealous Wife* II. 21 How unlucky it is, that this damn'd Accident shou'd happen in the New-market Week! *c* **1810** W. Hickey *Mem.* (1960) xx. 326 We had previously enjoyed private lodgings at Epsom.. for the race week. **1890** W. Booth *In Darkest Eng.* II. vi. 231 Self-denial..[which] the Soldiers of the Salvation Army practice every year in Self Denial Week. **1917** *Wells Fargo Messenger* V. 133/2 (*caption*) Native sons of the Golden West begin 'Prune Week' ceremonies. **1965** 'E. McBain' *He who Hesitates* iii. 34 'How come everybody's so eager to join me this morning?' Roger asked. 'Who knows?... Maybe it's national brotherhood week.' **1977** *Lancashire Life* Mar. 99/1 He won a prize for the best money-raising poster during Warships Week.
3. Now also meaning five working days, from Monday to Friday inclusive, as opposed to the weekend; *three-day week*: see *THREE III. 2.
5. d. *week in, week out*: see IN *adv.* 2.
e. *week-to-week* (attrib. phr.), continuing or recurring in successive weeks; continual. Cf. To *prep.* 6 c.
1959 *New Statesman* 24 Jan. 92/2 The lightning flashed above Sinai, and in its glare, the starry-eyed observer, remote from the week-to-week grind of party work, saw a great machine, whirring smoothly to life. **1981** J. Sutherland *Bestsellers* i. 15 The bestseller lists... Their week-to-week attention singles out sensational books of the moment.
6. c. Also † *a week of Saturdays*, an indefinite period, a long period. Cf. *Month of Sundays* s.v. MONTH¹ 3 f.
1831 *Constellation* 8 Jan. 57/3 No, you couldn't gess a week of Saturdays and so I'll tell you—cause it is *unginteel*.

weekday. Add: **3. a.** Now also used to mean a day of the week other than Saturday or Sunday.
1976 P. R. White *Planning for Public Transport* ii. 32 On weekdays (Mondays to Fridays), to which many urban surveys are confined, many trips are made within a confined time budget.

week-end. Add: **1. a.** (Earlier examples). Also, the end of a week. *long week-end*: see *LONG *a.*¹ 18.
1638 in *Victoria County Hist. Yorkshire* (1912) II. 415/2 The greatest weight of the said exaction will fall upon very poor people..who making every week a coarse kersey and being compelled to sell the same at the week end..are nevertheless constrained to yield one half penny apiece. **1793** W. B. Stevens *Jrnl.* 27 Feb. (1965) I. 70 Wrote to Dewe that I would put on my seven league boots next weekend and stretch my course to Appleby. **1870** *Food Jrnl.* 1 Mar. 97 'Week-end', that is from Saturday until Monday,—it may be a later day in the week if the money and credit hold out,—is the season of dissipation.
2. *attrib.* **a.** For use at week-ends, as *week-end bag*, *case*, *cottage*, etc.; occurring at or for the duration of a week-end, as *week-end course*, *leave*, *school*, *war*, etc.
1887, **1896** [sense a in Dict.]. **1911** 'Saki' in *Bystander* 12 Apr. 70/2 I've seen a week-end cottage near Dorking that I should rather like to buy. **1924** Kipling *Debits & Credits* (1926) 316 You live like a home defence Brigadier, besides week-end leaf [*sic*]. **1934** Webster, Weekend case. **1944** J. S. Huxley *On Living in Revolution* 117 Short weekend courses and longer 'study workshops' for teachers. **1946** C. McCullers *Member of Wedding* I. 10 At dark John Henry came to the back door with a little week-end bag. **1952** 'M. Cost' *Hour Awaits* 136 She carried a small, smart week-end case of crocodile. **1958** O. Caroe *Pathans* xxv. 418 Roos-Keppel did indeed start his tour of office with two small operations the scale of which is sufficiently shown by the fact that one of them was called the week-end war. **1962** J. Braine *Life at Top* x. 136, I came to each and every week-end school here. **1967** R. V. Beste *Repeat Instructions* (1968) xiv. 148, I haven't a week-end cottage. **1973** *Guardian* 22 Jan. 1/1 Lieutenant Mark Phillips, on weekend leave from Germany, went hunting on Saturday with Princess Anne. **1982** J. O'Faolain *Obedient Wife* ii. 42 A weekend bag packed with scent, toothbrush and so forth.
b. Carrying out a specified activity or fulfilling a specified role only at week-ends or for pleasure (sometimes with the implication 'casual, amateur'); e.g. *week-end Air Force*, *athlete*, *father*, *gardener*, *motorist*, *sailor*, *soldier*, *writer*. Cf. *SUNDAY *sb.* 3.
1935 *Discovery* Oct. 314/1 The publisher suggests that *English Earth* will interest equally the farmer and the week-end motorist. **1941** *Time* 30 June 31/1 Week-end athletes should be careful. **1943** C. H. Ward-Jackson *Piece of Cake* 62 *Week-end Air Force, The*, the Auxiliary Air Force. It was formed in 1925, its officers and men being citizens who gave week-ends and other part-time to their duties. **1959** *Listener* 12 Mar. 461/1 None of your present week-end, or decade-end writers realize that the present agitators are the sons of a former day, of which these writers presumably know nothing. **1962** *Jrnl. Family Law* Fall 104 The law is, in fact, uncertain as to the rights of the ex-spouse. Consequently, it seems even more uncertain about the privileges which the law should extend to the children with 'week-end fathers'. **1970** *New Yorker* 3 Oct. 40/3 Will you look at them weekend soldiers. **1974** *Harper's & Queen* Sept. 117 Humphrey Brooke was only a weekend gardener until..he decided to retire. **1976** 'D. Halliday' *Dolly & Nanny Bird* ix. 120 The boat was..full of tanned, husky weekend sailors.
week-ending: also as *ppl. a.*, (in the habit of) spending the week-end away from home; **week-endize** *v. intr.*, to spend a week-end away from home (*nonce-wd.*); **week-endy** *a.*, suggestive of the week-end.
1910 R. Bridges *Let.* 21 June in Bridges & Bradley *Corr.* (1940) 77, I was glad to get your note saying that you wd week-endize. **1930** J. B. Priestley *Angel Pavement* xi. 567 She spent the next few minutes getting from the bus to the station, which was very crowded and week-endy. **1934** Webster, Weekending *adj.* **1947** J. Hayward *Prose Lit. since 1939* 36 *The Long Week-end*—an ironical title which only the 'week-ending' Englishman will appreciate. **1973** A. Grey *Some put their Trust in Chariots* xiii. 72 Weekending French families setting out in their saloons for the countryside. **1976** G. Ewart *No Fool* II. 61 Contrariwise, would you admire what's trendy (you were a fashion once yourself) or see virtue in what's sutburban or weekendy?

weekender. Add: **2. a.** A week-end cottage (see *WEEK-END 2 a). *colloq.*
1941 Baker *Dict. Austral. Slang* 81 *Weekender*, a week-end holiday cottage or shack. **1967** E. Hunt *Danger Game* iii. 59 He had a week-ender at Palm Beach and a lovely car. **1976** *Billings* (Montana) *Gaz.* 16 June 10-c/5 (Advt.), Small, clean..home in Red Lodge. Perfect week-ender any season. **1983** *Austral. Women's Weekly* Aug. 21/2 There is, says Pauline, constant confusion over.. shacks and holiday homes. 'A shack in NSW is a tumble-down wreck. In NSW you have a weekender.'
b. A bag large enough to carry everything needed for a week-end away from home; a week-end bag. *U.S. colloq.*
1961 *Harper's Bazaar* June 24/2 A capacious week-ender in pale coffee-bean hide. **1980** *TWA Ambassador* Oct. 99/1 You'll use the Kluge Bag like a week-ender too, because it's just as easy to carry on and a whole lot better.

weekly, *a.* and *sb.* Add: **A.** *adj.* **2. a.** *weekly boat*: a coaster on which the crew is paid by the week. Cf. *monthly boat* s.v. *MONTHLY *a.* 4.
1927 F. H. Shaw *Knocking Around* 4 She was a weekly boat, where the crew provided their own food. **1946** J. Batten *Dirty Little Collier* 12 These weekly boat men are quite used to slipping home for a few hours.
b. *weekly boarder*: a school pupil who boards at the school during the week and returns home at week-ends.
1841 Dickens *Let.* 6 Mar. (1969) II. 225, I will send [him] as a weekly boarder to the best school I can find in the place... I say weekly boarder, because I should wish the boy..to be at home from Saturday night to Monday Morning. **1973** J. Mann *Only Security* i. 9 It had a school where Clovis might be a weekly boarder.
B. *sb.* (Earlier examples.)
1833 [see *MONTHLY *sb.* 2]. **1845** Dickens *Let.* 26 July (1977) IV. 337 A notion..has occurred to me in connexion with our abandoned little weekly.

weeksite (wī·ksaɪt). *Min.* [f. the name of Alice M. D. *Weeks* (b. 1909), U.S. geologist + -ITE¹.] A hydrated silicate and oxide of uranium and potassium, $K_2(UO_2)_2(Si_2O_5)_3 \cdot 4H_2O$, found as soft, yellow orthorhombic crystals.
1960 W. F. Outerbridge et al. in *Amer. Mineralogist* XLV. 43 Weeksite at the Autunite No. 8 claim occurs as small spherulites of yellow radiating crystals. **1978** *Mineral. Abstr.* XXIX. 232/1 Weeksite..is described for the first time from France in the uranium deposit of Les Bois Noirs.

weeny, *a.* Add: **2.** Special collocation: **weeny-bopper** *colloq.* [after *teeny-bopper*], a very young (esp. female) pop fan (sometimes notionally of a younger age group than a teeny-bopper, but the two terms are freq. interchangeable).
1972 *Daily Express* 2 Nov. 16 Britain's 'weeny-boppers' are in danger of dancing their way to incurable deafness. **1975** *Evening News* 5 July 16/5 Being a weeny-bopper can be a problem when it comes to clothes... Our model, Karen, nearly 13, got her mum to take her round the stores. **1976** M. Butterworth *Festival!* viii. 133 A couple of weenie-boppers at a pop concert.

weeny (wī·ni) *sb.*¹ Also **weenie.** [f. the adj.]
1. *colloq.* A very young child.
1844 C. Ridley *Let.* Oct. (1958) xv. 180, I must tell you now about the chicks [i.e. children]. Little weeny is growing visibly. **1973** 'D. Halliday' *Dolly & Starry Bird* vi. 82 Not Mr. Paladrini who was so nice to the weenies? **1977** *Ottawa Citizen* 29 June 7/1 Our five-year-old granddaughter keeps asking when the trip is going to begin. Travelling with weenies is something that Mama and I have done for most of our lives.
2. *U.S. slang.* **a.** A girl; an effeminate man. **b.** An objectionable person.
1929 [see *PHOOEY *int.* (*sb.*)]. **1963** *Amer. Speech* XXXVIII. 171 Some of the less frequent..phrases [for an effeminate young man] are: *dink*..and *weenie*. **1964** *Ibid.* XXXIX. 118 Free variation..is also established in the interchangeability of the names of smaller animals for socially unacceptable persons: *toad*, *squirrel*, and *shrimp* all serve for the zoologically unsound but all-inclusive *weenie*.

weeny (wī·ni), *sb.*² *U.S. slang.* Also **weeney**, **weenie.** Var. *WIENIE. Cf. *WINNY.
1906 *Dialect Notes* III. 163 *Weenie*..Wiener Wurst, sausage. 'Hot Weenies.' **1935** *Amer. Speech* X. 159 'Weinies', 'Wienies', and 'Weenies' are also for sale. **1960** C. Hamblett in *Pick of Today's Short Stories* 136 The simplest basics... Marilyn Monroe, hot dogs, weenies. **1981** P. Theroux *Mosquito Coast* viii. 72 Father said, '..I've got other weenies to roast.' And he went back to his maps.

weep, *sb.* Add: **1. b.** Also *the weeps*: a fit of weeping or melancholy. Also *transf.*
1922 Joyce *Ulysses* 297 And Bob Doran starts doing the weeps about Paddy Dignam. **1937** 'G. Orwell' *Let.* 31 July in *Coll. Ess.* (1968) I. 280, I heard from Murry who seemed in the weeps about something. **1982** *Medico-Legal Jrnl.* L. 10 Leaving his Counsel to do what we call 'The weeps', i.e. to plead in mitigation of sentence.

weep, *v.* Add: **I. 4. a.** (Further examples.)
1972 *Guardian* 9 Feb. 6/8 The freighter Autolycus..was ordered to leave the harbour..after her cargo of nitroglycerine was found to be 'weeping'. **1978** C. Egleton *Mills Bomb* ix. 91 Gelignite... Dangerous? Only if they were handling jelly that was beginning to weep. **1982** *Times* 6 Jan. 7/7 Stirring the loyalists up was like..playing with jelly that had been weeping for two years.

weeper. Add: **6.** = *WEEPIE. *colloq.*
1934 'N. Bell' *Winding Road* xvii. 445 A few hot-eared scribes have written weepers about such things and kidded themselves they'd done the trick. **1949** M. Laski in *Sketch* 21 Dec. 551/2 Every magazine..reckons to print at least one weeper every Christmas. **1977** *New Yorker* 8 Aug. 10/1 Irene Dunn does the suffering in this version of the Fannie Hurst classic weeper about a woman who loves a selfish married man. **1984** *Miami Herald* 30 Mar. 7D/4 We have been hearing that romance is making a comeback in the movies, and here it is: the old-fashioned, meet-ya-when-this-war-is-over weeper.

weepie (wī·pi). *colloq.* Also **weepy.** [f. WEEP *v.* + -IE; cf. *TALKIE, etc.] A sentimental film, story, play, etc.; a 'tear-jerker'.
1928 *Sunday Dispatch* 23 Dec. 12/2 There are undoubtedly times when a film calculated to raise buckets of tears has its appeal. Someone recently christened this type of picture..a 'weepie'. **1948** *Sunday Pictorial* 18 July 11/4 'If Winter Comes' (Empire) is a re-make of the famous weepie novel. **1958** *Spectator* 18 July 85/3 *Chicken Soup with Barley*..is an East End Jewish weepie. **1962** *Times* 14 May 14/1 An all-out weepy is the only possible description for *Girl in a Birdcage*. **1977** D. Francis *Risk* xvi. 207 How much easier if the miscreant would confess..a sentimental solution..which happened only in weepie films. **1984** *Listener* 5 July 35/1 'Weepies' come in all shapes and sizes. Ostensibly they are women's pictures, because only women cry at them, right?

weeping, *ppl. a.* Add: **6.** *weeping cherry*, *myall* (earlier examples), *rose.*
1824 'A. Singleton' *Lett. from South & West* 62 The weeping-cherry..bears blossoms when a part of the fruit is ripe. **1859** D. Bunce *Trav. Dr. Leichhardt* 91 Many species of *Acacia* made their appearance, including the celebrated Weeping Myall. **1869** S. R. Hole *Bk. about Roses* viii. 124 They may soon be trained into Weeping Roses. **1889** Weeping myall [see *BOREE²]. **1951** *Dict. Gardening* (R. Hort. Soc.) IV. 1825/2 Newly planted bush and standard (not weeping) roses should be pruned back. **1969** *Better Homes & Gardens* (U.S.) Apr. 83 *Weeping cherry*, bright pink blossoms are artistically spaced along the gracefully hanging branches.

weeping willow. Add: **2.** Rhyming slang for 'pillow'. Now *rare* or *Obs.*
1880 D. W. Barrett *Life & Work among Navvies* (ed. 2) II. ii. 43, I have been told..that often when a man is ready to retire to rest, he will inform his mates, 'That he's done his lot for the day, and is goin' to lay his pen'oth o' bread (head) on the weeping-willow (pillow).' **1944** N. Streatfeild *Curtain Up* x. 123 Time young Holly was in bed... Hannah wants your head on your weeping willow, pillow to you.

weeshy, *a.* (Earlier example.)
1825 [see *DEESHY *a.*].

Weetabix (wī·tăbiks). The proprietary name of a breakfast cereal in the form of thick crumbly biscuits made from wheat.
1936 *Official Gaz.* (U.S. Patent Office) 3 Mar. 31/1 Olive Cross, Washington, D.C. Filed Jan. 11, 1936. *Weetabix* for crushed wheat breakfast food biscuits. **1938** *Trade Marks Jrnl.* 17 Aug. 1009/2 *Weetabix*... Wholemeal biscuits. Weetabix Limited,..Kettering, Northamptonshire; manufacturers. **1974** R. Rendell *Face of Trespass* i. 14 He..poured milk over a couple of Weetabix and sat down to eat his breakfast. **1981** F. Inglis *Promise of Happiness* ii. 47, I changed loyalties..from cornflakes to Weetabix.

wee-wee, var. WI-WI².

wee-wee (wī·wī), *v. colloq.* [Echoic: freq. as a child's word.] *intr.* To urinate.
1930 C. Beaton *Diary* Dec. in *Wandering Years* (1961) ix. 192 Young men..hurried into the garden to wee-wee. **1944** D. Welch *In Youth is Pleasure* vi. 113 Orvil hated Guy... He was glad that the dog had wee-wee'd on Guy's expensive jacket. **1954** D. Abse *Ash on Young Man's Sleeve* 167, I suddenly rushed into the sea..and wee-weed in the water for a joke. **1960** L. Pincus *Marriage* ii. 110 Mrs. Robinson felt disgusted at the thought of 'the tail' touching her, 'because it wee-wee'd'. **1975** 'J. Lymington' *Spider in Bath* ii. 40 Show us the way to the bar and the lady would probably like to wee-wee.

wee-wee (wī·wī), *sb. colloq.* [f. as prec.: cf. *PEE-PEE².] **1.** Urine; an act of urination; *to do (have, etc.) a wee-wee, to go (make) wee-wee*, to urinate. Cf. WATER *sb.* 18 a.
1937 Partridge *Dict. Slang* 943/1 *Wee-wee*, a urination; esp. *do a wee-wee*..late C. 19–20. **1938** *Life & Health* Sept. 19/2 If he wants to urinate, it's, 'Mommie, wee, wee, duty.' **1948** A. N. Keith *Three came Home* x. 184 Our barrack..smelled of kids, pots, and wee-wee. **1955** J. P. Donleavy *Ginger Man* xvi. 175 If you'll excuse me a moment, I must make wee wee. **1961** J. Stroud *Touch & Go* xiii. 138 He made his voice deliberately offensive: 'He's gone to do a wee-wee.' **1969** M. Braithwaite *Never sleep Three in Bed* xvii. 76 Hub and I had to go wee wee, and when advised by Mother that we were to go upstairs instead of in the back yard we took the stairs two at a time to see if such a thing could be true. **1982** J. Scott *Uprush of Mayhem* vi. 67 When he needed a wee-wee he did it in a corner of the hut.
2. A penis.
1964 W. & J. Breedlove *Swap Clubs* xv. 233 Our grandmothers, wielding butcher knives, threatened to cut off our fathers' wee-wees if they didn't stop playing with them. **1972** *Screw* 12 June 16/3 [The] self-righteous defender of what he thought to be his threatened wee wee, could not contain his machismo.

Wegener¹ (vē̆·gənəɹ). *Geol.* The name of Alfred *Wegener* (1880–1930), German geophysicist, used in the possessive with reference to the theory of continental drift which he first published in 1912.
1922 *Living Age* 10 June 657 Professor F. E. Weiss.. writes in the Manchester Guardian that Professor Wegener's theory 'constitutes a good working hypothesis, and the striking simplicity with which it allows many phenomena to be explained will greatly stimulate further enquiry'. **1926** [see *continental drift* s.v. *CONTINENTAL *a.* 1 d]. **1963** *Sci. Amer.* Apr. 90/1 Between 1920 and 1922 Wegener's hypothesis excited great controversy. **1982** *Nature* 23/30 Dec. 681/2 Jeffreys refers with regret to the defection of Sir Arthur Holmes to 'Wegener's theory'.
Hence **Wegene·rian** *a.*
1960 *Bull. Amer. Assoc. Petroleum Geologists* XLIV. 245 (*caption*) Early tertiary paleography and trans-Atlantic migration route for shelf benthos according to Wegenerian hypothesis of continental drift. **1967** *Oceanogr. & Marine Biol.* V. 340 Some zoologists are inclined to accept the Wegenerian idea that the Atlantic Ocean is indeed young, and of no greater antiquity than the Cretaceous. **1980** *Guardian* 20 Nov. 13/8 Pangea..probably existed for a few hundred million years before it..began to break up to generate Wegenerian drift.

Wegener² (vē̆·gənəɹ). *Path.* [The name of F. *Wegener*, 20th-c. German physician.] *Wegener's granulomatosis*: an often fatal disease characterized by granulomatosis of the respiratory tract and necrotizing blood-vessels.
1948 *Acta Path. & Microbiol. Scandinav.* XXV. 582 Clinically, this case agreed well with the previously described cases of Wegener's granulomatosis. **1957** *Thorax* XII. 57/1 The syndrome has been known as Wegener's granulomatosis since his detailed description of three cases in 1936 and 1939. **1977** *Daily Colonist* (Victoria, B.C.) 4 Dec. 27/6 An autopsy showed that my grandmother died of a very rare disease called Wegener's granulomatosis.

weggebobble (we·dʒɪbɒb'l). *nonce-wd.* Humorous alteration of VEGETABLE *sb.*
1922 Joyce *Ulysses* 163 Coming from the vegetarian. Only weggebobbles and fruit.

‖ **Wehmut** (vē·mūt). [Ger.] Sadness, melancholy, wistfulness, nostalgia.
1907 M. A. von Arnim *Fräulein Schmidt & Mr. Anstruther* xlii. 130 What I feel when I listen to music is chiefly *Wehmut*, and I don't think much of *Wehmut*... It is a forlorn thing, made up mostly of vague ingredients—vague yearnings, vague regrets, vague dissatisfactions. **1920** D. H. Lawrence *Let.* in C. Mackenzie *My Life & Times* (1966) V. 170, I get a sort of Wehmut. Quoi faire! **1933** J. Hilton *Lost Horizon* 20 A sort of universal sadness,..something remote or impersonal, a *Wehmut* or *Weltschmerz*, or whatever the Germans call it.

wehrlite (vēᵊ·ɹlaɪt, wə̄·ɹlaɪt). [f. the name of Adolf *Wehrle* (1795–1835), Austrian Councillor of Mines: see -ITE¹.] **1.** *Petrogr.* [ad. G. *wehrlit* (F. von Kobell *Grundzüge d. Mineral.* (1838) iii. 313).] A peridotite mainly consisting of olivine and monoclinic pyroxene with common accessory opaque oxides.
1861 H. W. Bristow *Gloss. Mineral.* 404/1 *Wehrlite*, the name given by von Kobell to a massive granular mineral, which is probably a variety of Lievrite. **1913** J. P. Iddings *Igneous Rocks* II. i. vi. 316 Wehrlite is closely associated with olivine-gabbro, into which it grades with increasing feldspar. **1979** *Nature* 7 June 489/2 The cumulate peridotites are recognised in thin sections which show serpentinised wehrlite with 50% clinopyroxene.
2. *Min.* [a. F. *wehrlite* (J. J. N. Huot *Nouveau Man. complet de Minéral.* (1841) I. 188).] A native rhombohedral alloy of bismuth and tellurium occurring as tin-white to steel-grey foliated masses.
1874 *Amer. Jrnl. Sci.* CVIII. 259 The following minerals are positive (+), or negative (−), in contact with copper:..Wehrlite $Bi_4(TeS)_4$ +. **1931** *Chem. Abstr.* XXV. 1762 Eleven Te minerals are found in Hungary... Tetradymite,..wehrlite..and stützite are described. **1972** *Doklady Earth Sci.* CC. 167/1 Wehrlite is a sulfur-free bismuth telluride whose composition and position in the classification of minerals are interpreted differently by various authors. Wehrlite was first described at the Deutsch Pilsen deposit, Hungary.

‖ **Wehrmacht** (vē·ɹmaχt). Now *Hist.* [Ger., lit. 'defence force'.] The name used for the German armed forces between 1921 and 1945. Also *attrib.*
1935 in *Documents on German Foreign Policy* (1959) III. 1008 Ranks and badges of rank of the members of the Reich Air Ministry..will assume a military character and resemble those of the Wehrmacht. **1945** *Daily Mirror* 8 May 3/2 The Wehrmacht commander announced that he did not recognise what he described as the 'armistice'. **1945** 'G. Orwell' in *Tribune* 9 Nov. 10/3 He..attempted to pass himself off as an ordinary soldier of the Wehrmacht. **1959** M. Crosland tr. *J. Rovan's Germany* 176 Young Germans..begin travelling around the country armed with a stewpan and an old Wehrmacht kitbag. **1965** *English Studies* XLVI. 226 The Wehrmacht mentality is plain, the terrible joyous uplift in the fact of killing, the terrible satisfaction to be extracted from a sense of the irresistible. **1978** L. Heren *Growing up on The Times* vii. 257 Of all the armies which fought in the second world war, the *Wehrmacht* was undoubtedly the best, perhaps the best in history.

Wei (wēⁱ). The name of a Chinese dynasty, esp. one of the partial dynasties ruling (in the north of China) from the mid-4th to the mid-6th cent. A.D. Also used *attrib.* and *absol.* to designate works of art, esp. sculpture and pottery, produced in the period.
1894 G. N. Curzon *Probl. Far East* viii. 276 The Inner Wall is attributed to the Wei dynasty in A.D. 542. **1913** R. Fry *Let.* 31 May (1972) II. 368 I've just seen a show in Paris full of the most amazing things; among them the finest Wei Dynasty statues from somewhere away in the west of China. **1952** Watson-Gandy & Gordon tr. *Grousset's Rise & Splendour Chinese Empire* xv. 111 Their dynasty was called Wei (often referred to as T'o-pa Wei to distinguish it from the Three Kingdoms Wei). *Ibid.* 114 Wei sculpture has been called the Far Eastern equivalent of the Romanesque and Gothic sculptures which developed six and eight centuries later in Europe. **1960** C. Winick *Dict. Anthropol.* 556/1 *Wei*,..unglazed pottery often decorated with pigment and tomb figures of the Wei dynasty in China, A.D. 368–557. **1982** M. Young *Elmhirsts of Dartington* viii. 196 Ancient Chinese pots of the Han, Wei, Tang and Sung dynasties.

weibullite (vəi·buləit). *Min.* [ad. Sw. *weibullit* (G. Flink 1910, in *Ark. f. Kemi, Mineral. och Geol.* III. xxxv. 4), f. the name of K. O. M. *Weibull* (1856–1923), Swedish mineralogist: see -ITE¹.] A mineral containing lead, bismuth, selenium, and sulphur and occurring as grey crystals at Falun, Sweden; now regarded as an orthorhombic species but for long of uncertain status.
1913 *Mineral. Mag.* XVI. 375 Weibullite... This mineral, from Falun, Sweden, was described by M. Weibull in 1885 as a seleniferous variety of galenobismutite. **1980** *Amer. Mineralogist* LXV. 789 Studies of specimens from Falun, Sweden, reported to contain the two minerals weibullite and wittite have established at least *three* selenium-rich bismuth-lead sulphosalts of similar chemistry. A nomenclature is proposed in which two of the minerals retain the old names weibullite and wittite, while the third is given the new name nordströmite.

‖ **wei ch'i** (wēⁱ tʃi). Also **wei chi.** [Chinese *wéiqí*, f. *wei* to surround + *qi* chess.] A traditional Chinese board game of territorial possession, equivalent to *GO *sb.*²
1871 *Jrnl. North-China Branch R. Asiatic Soc.* VI. 107 The chessman of the *wei ch'i*..are generally *not* made of ivory, but of stone or a glassy substance. **1892** *Ibid.* XXVI. 80 *Wei-ch'i* is considered *par excellence* the game of the literary class, while Chess is the favourite diversion of military men. **1911** H. F. Cheshire *Handbk. Goh or Wei Chi* 148 To the cultured classes..'Wei Chi',..was, and is an almost essential part of their life. **1917** *Encycl. Sinica* 594/1 *Wei ch'i, surrounding chess*, often called chess by foreigners... The square board on which it is played is divided by eighteen lines each way, making 324 squares. The play..is at the points where the lines cut or meet. **1969** R. C. Bell *Board & Table Games* II. iii. 59 The eighteenth-century Chinese painting on glass reproduced as the frontispiece, shows two ladies playing Wei-ch'i on an antique form of board of 13 × 11 squares. **1976** *New Yorker* 26 Jan. 23/2 The pattern recalls the Chinese game of wei-ch'i (known here by its Japanese name, *go*), in which a player captures his opponent's counters by surrounding them with his own.

Weichsel (vəi·ksĕl). *Geol.* The German form of the name of the river Vistula in Poland, used *attrib.* and *absol.* to designate the fourth and final Pleistocene glaciation in northern Europe, corresponding to the Würm glaciation of the Alps. Hence **Weichse·lian** *a.* (also *absol.*).
1934 R. A. Daly *Changing World of Ice Age* i. 29 Recently four Glacial stages have been traced in Germany. Figure 19 shows the nested moraines of three, named in order of decreasing age, Elster, Saale, and Weichsel. **1968** Embleton & King *Glacial & Periglacial Geomorphol.* 10 The links with North America, apart from the correlation of the Wisconsinan with the Würm/Weichselian, are more tenuous. **1969** Bennison & Wright *Geol. Hist. Brit. Isles* xvi. 361 After the early Weichselian glaciation the climate varied considerably. **1974** *Encycl. Brit. Micropædia* X. 598/3 The Weichsel Glacial Stage has been divided into at least two main cold phases,..separated from each other by a period of more moderate climatic conditions, an interstadial. **1979** *Jrnl. Arid Environments* II. 293 Five successive units of time, respectively the Palaeozoic, the Mesozoic, the Tertiary, the Quaternary, and the late Weichselian and Flandrian.

weigh, *sb.*² (Earlier example.)
1777 E. Draper *Let.* 25 Aug. in *N. & Q.* (1944) 15 July 28/1, I can assure you on the authority of Mr. Sullivan, that he saw him underweigh in the Bessborough and for the East Indies several Weeks ago.

weigh, *v.*[1] Add: **II. 7. b*.** *to weigh* (someone) *against gold* (or *silver*): to perform the Indian ceremony in which (a rajah, etc.) is weighed and his weight in gold (or silver) distributed as largesse.

1969 J. Ovington *Voy. Suratt.* 179 The Moguls are sometimes weighed against Silver. **1934** *Times* 25 Aug. 13/2 The Maharajah..will be weighed against gold... The gold-weigh ceremony is usually performed with gold supplied by the person being weighed... This amount will be distributed in charity. **1936** *Times* 14 Jan. 13/6 At this Durbar the Aga Khan will be weighed against gold, and it is expected that 20,000 guests will attend the function.

d. *to weigh off*: to punish; to convict or sentence. *slang* (orig. *Mil.*). Now chiefly *Criminals'*.

1925 Fraser & Gibbons *Soldier & Sailor Words* 301 *Weighed off, to be*, to be brought up before an officer and punished. **1945** *Tee Emm* (Air Ministry) V. 54 P.O. Prune will have to investigate and deal with a charge..and possibly weigh off the first delinquent of his service career. **1958** F. Norman *Bang to Rights* i. 22 You just got weighed off yesterday? **1963** T. & P. Morris *Pentonville* ii. 20 One young man..commented that he had been 'weighed off at X Assizes by some old geezer togged up like Father Christmas'. **1978** B. Norman *To nick Good Body* x. 81 Another was in custody..waiting to be weighed off.

e. *to weigh in*: to weigh (an air passenger's luggage) before departure; to subject (a passenger) to this procedure. See *excess luggage* s.v. Excess 6 b.

1934 Rhys-Williams *Diary* 1 Aug. (MS.), Left Eaton Place at 4.30 p.m. for Victoria, where we were 'weighed-in', and had our luggage weighed and labelled. **1961** L. Deighton *Ipcress File* v. 30 She weighed in my wardrobe case. **1970** *New Yorker* 16 May 41/2 The porter..takes her bag and follows her to the desk to have it weighed in.

f. Angling. *to weigh in*: of an angler, to have (one's catch) officially weighed at the end of a competition. Also *absol.*

[**1928**: see *weigher 3 a]. **1949** *Club Anglers' Jrnl.* Nov. 14/1 The river fished well and the winner weighed-in 6 lb. 4 oz. 12 drm. **1972** *Match Rules* (Nat. Fed. Anglers) in E. Marshall-Hardy *Angling Ways* (1973) xxxix. 306 No competitor may have his catch weighed in who has litter lying on the banks of his swim. **1976** *Wymondham & Attleborough Express* 17 Dec. 22/5 Only 10..competitors weighed in,..but..Frank Kilbourn..float fished to take 9–2 of good roach.

9. a. *to weigh in*: also, of a boxer (turning the scales *at* a particular weight) before a fight. Hence in general colloq. use.

1909 'O. Henry' *Roads of Destiny* xviii. 307 He was six feet four and weighed in at 135. **1931** *Daily Express* 13 Oct. 1/7 Both boxers weighed in this afternoon. **1958** S. Wilcox *3 Days Running* vii. 79 When at last I was able..to 'weigh-in'..I weighed five pounds more than at the beginning of the day. **1966** *Aviation Week & Space Technol.* 5 Dec. 6/1 The complete inertial package weighs in at only 14 pounds. **1979** *SLR Camera* Mar. 35/1 The compact 'Zuiko' 1000mm measures just 26 inches and weighs in at around eight and a half pounds.

c. fig. *to weigh in*: to bring one's weight or influence to bear; to enter a forceful contribution to a discussion, etc. *colloq.*

1909 G. B. Shaw *Let.* 31 July (1972) II. 854, I want you to ask the Chief Rabbi to weigh in. **1919** Beerbohm *Seven Men* 147 A few weeks later the Anglo-Indians weigh in. In due course we have the help of our Australian cousins. **1938** E. Bowen *Death of Heart* iii. iii. 378 The telephone crisis..had been the moment for Lilian to weigh in. **1956** A. L. Rowse *Early Churchills* 221 The Princess Anne, in her constant rôle of fairy godmother to the Marlboroughs, weighed in; nor could it have been done with more tact and good feeling. **1976** *Milton Keynes Express* 4 June 38/4 Sim weighed in with 4–27 off eight overs.

d. To launch *into* and attack (a person, etc.). Also *fig. colloq.*

1941 Baker *Dict. Austral. Slang* 81 *Weigh into someone*, to attack, wade into a person in a fight. **1976** F. Warner *Killing Time* i. 8, I survived the war,..and then, if I was a minute after 9.30 in the evening, my Mother would weigh into me.

weighed, *ppl. a.* **3.** (Later example.)

1935 S. Spender in *London Mercury* May 8 The tall girl with her weighed-down head.

weigher. Add: **3. a.** Also *weigher-in*.

1906 A. H. Lewis *Confessions of Detective* i. i. 7 I've been weigher-in at the boxing tournaments for over ten years. **1928** *Daily Tel.* 25 Sept. 12/5 It is a rule that every fish caught must be kept alive, and after being weighed must be put back into the water by the official weigher-in. **1982** J. A. Sharwood in *Occasional Papers Univ. Sydney Austral. Lang. Res. Centre No. 20.* 20 He may be inclined to complain to the *boundary rider*..about *the bloke with the short arm* being employed as the *weigher-in* there.

wei·gh-in. [f. the vbl. phr. *to weigh in*: see *weigh *v.*[1] 7, 9.] **1.** *Boxing.* The weighing-in of a boxer before a fight. See sense *9 a of the vb.

1939 *Sun* (Baltimore) 28 June 15/1 The general might bring the subject up at the weigh-in and instruction period for the two fighters at noon to-morrow. **1946** *Daily Progress* (Charlottesville, Va.) 18 June 7/4 Louis expected to remain at Pompton Lakes until just before the weigh-in. **1952** L. A. G. Strong *Darling Tom* 138 At the weigh-in, each man had received instructions to scow at the other. **1974** H. L. Foster *Ribbin', Jivin', & Playin' Dozens* vi. 251 Boxers at weigh-ins usually attempt to psych one another.

2. *Angling.* The weighing of the anglers' catch at the end of a competition. See sense *7 f of the vb.

1949 *Club Anglers' Jrnl.* Oct. 14/2 A good weigh-in, topped by a splendid bag of roach. **1971** *Rocquaine Regatta: Programme* (Guernsey) 7 Event 1–8.00 a.m. Sea Angling. Start at Imperial Hotel 8.00 a.m.—weigh-in 10.30 a.m. **1972** *Shooting Times & Country Mag.* 1 July 16/1 The weigh-in caused great excitement as individual catches were very close.

weight, *sb.*[1] Add: **II. 8. a.** *weight for weight*: also (with hyphens) used *attrib.*

1964 W. G. Smith *Allergy & Tissue Metabolism* vi. 71 In the perfused cat hind limb it is as active as acetylcholine on a weight-for-weight basis. **1968** *Times* 3 Dec. 10/8 Female rats were given daily doses reckoned to be about eight times as powerful on a weight-for-weight basis as those taken by the tribeswomen. **1974** *Brit. Med. Jrnl.* 19 Jan. 107/2 Special care needs to be taken over the use of Lanoxin brand [of digitalis], which is now twice as potent on a weight-for-weight basis as formerly.

c. *to lose weight*: to become thinner or less corpulent; *to put on weight*: see Put *v.*[1] 46 f.

1961 M. Spark *Prime of Miss Jean Brodie* iv. 114 She had lost weight through her sad passion for Mr. Lloyd. **1970** M. Patten *Bedsitter Cookery* 89/1 Most sensible people today are anxious to keep a slim figure and a well-planned diet is an essential towards either losing weight or maintaining a good weight. **1982** J. Mann et al. *Diabetics' Diet Bk.* i. 20 To lose weight you should aim to have only 1,300 calories a day.

10. c. *to pull* (*one's*) *weight*: see *pull *v.* 15 b; *to throw* (*chuck*, etc.) *one's weight about* or *around*: to assert oneself or one's authority, esp. in an objectionable way; to act officiously. *colloq.*

1917 A. G. Empey *From Fire Step* 31 Don't chuck your weight about until you've been up the line and learnt something. **1922** C. E. Montague *Disenchantment* viii. 104 Some typically stupid English General..was clearly throwing his weight about, as they say, without any real understanding of anything. **1926** S. Jameson *Three Kingdoms* xii. 348 'Come to that,' he said, 'Isabel has more right than any of you to fling her weight about.' **1941** J. P. Marquand *H. M. Pulham, Esq.* i. 10 Bill King..always used to say that Bo-jo was a bastard, a big bastard. Perhaps he meant that Bo-jo sometimes threw his weight around. **1955** E. Hillary *High Adventure* 163 A big, strong, swanking chap who had thrown his weight about a good deal lower down. **1966** N. Marsh *Death at Dolphin* (1967) vi. 145 Why hadn't he put his foot down?.. He should have thrown his weight about. **1982** 'M. Hebden' *Pel & Staghound* xv. 176 Madame Rensselaer seemed to enjoy throwing her weight about.

d. *atomic weight* (earlier and later examples); = *atomic mass* s.v. *atomic *a.* 1; similarly *molecular weight*, the relative molecular mass of a molecule, equal to the sum of the atomic weights of the constituent atoms.

1820, etc. Atomic weight [in Suppl. s.v. *atomic *a.* 1]. **1872** *Jrnl. Chem. Soc.* XXV. 949 The relative molecular weights of ether, alcohol and water. **1950** *Sci. News* XV. 88 Blue hæmocyanin... This molecule is the largest of any known substance, having a molecular weight of several millions. **1978** P. W. Atkins *Physical Chem.* 11 We can determine how many elementary units we have by measuring the mass of the sample..and knowing the relative molecular mass (R.M.M., the 'molecular weight').

11. b. Phr. *to take the weight off* (*one's feet*): to sit down and rest. Cf. *to take a load off* (*one's feet*) s.v. *load *sb.* 3 h. *colloq.*

1936 'J. Tey' *Shilling for Candles* ix. 100 Waiters like to take the weight off their feet for a little. **1960** L. Davidson *Night of Wenceslas* i. 19 We were at the seat now. 'Like to take the weight off?' I said. **1965** A. Roudybush *Season for Death* (1966) xxxii. 190, I stepped into the library..to take the weight off my feet for a minute. **1973** H. Miller *Open City* xv. 168 Sit down, take the weight off.

III. 15. b. (Earlier and later examples.) More widely, a multiplying factor associated with each of a series of numerical quantities, esp. ones that are added together.

1825 *Phil. Mag.* LXV. 167 The arithmetical mean of a set of observations..is the particular case when the weights a, a', a" etc. are all equal, and the sum of the errors is equal to zero. **1868** J. C. Watson *Theoret. Astron.* vii. 372 The relative accuracy of two or more observed values of a quantity may be expressed by means of what are called their weights. **1935** Pauling & Wilson *Introd. Quantum Mech.* iv. 100 The degree of degeneracy (the number of independent wave functions associated with a given energy level) is often called the quantum weight of the level. **1940** G. Crowther *Outl. Money* iii. 93 For some sorts of index numbers, weighting is essential... Weights that are correct at one time may be incorrect at other times. **1949** *Economist* 8 Oct. 775/1 If the estimate of the change in productivity had been based on calculations using post-war weights they would have indicated a larger increase in productivity in the United States. **1970** O. Dopping *Computers & Data Processing* ii. 32 The number 491 means 4 × 100 + 9 × 10 + 1. The weights are 100, 10, and 1, respectively. **1983** *Personal Computer World* Dec. 142/2 When each digit is multiplied by its weight the sum of the products, including the check digit, whose weight is 1, is exactly divisible by 11 in a valid [Standard Book] number.

19. (Later examples.)

1938 'G. Orwell' *Homage to Catalonia* v. 65 The Communist Party, with Soviet Russia behind it, had thrown its whole weight against revolution. **1951** C. P. Snow *Masters* xxxiii. 268, I can't do as much as I should like, but I shall throw in my weight wherever I can. **1976** *Southern Even. Echo* (Southampton) 11 Nov. 1/1 The floodgates opened on the fluoridation of water supplies.. when the..Regional Health Authority put their weight behind the scheme.

IV. 21. d. A measure of an illegal drug; hence, the drug. Also without article. *slang.*

1971 *Frendz* 21 May 11/2 Avoid carrying weight late at night. **1972** *Listener* 23 Mar. 359/3 Your hash dealer is usually a friendly happy freak who's managed..to buy himself a weight and he deals it out to his friends. **1978** S. Wilson *Dealer's Move* i. 13 Neil was taking colossal risks, there'd be up to thirty weights sitting in the flat at one time.

VI. 25. a. *weight-gain, limit; weight-conscious* adj.; **weight belt**, a belt to which weights are attached, designed to help divers and underwater swimmers stay submerged; **weight cloth** (earlier example); also *fig.*; **weight function** *Physics*, a function that specifies the weight (sense *15 b) of some quantity; **weight training**, a method of physical training involving the use of weights.

[**1943** *Diving Man.* (U.S. Navy Dept.) x. 150 Next, the weighted belt is fastened on.] **1955** R. & B. Carrier *Dive* iv. 111 Weight belts should also provide for interchangeable weights to regulate buoyancy as needed. **1966** 'L. Holton' *Out of Depths* (1967) xii. 115 'Here. I'll show you a diver's gear.'.. He held up a weight belt. **1978** A. P. Balder *Sport Diving* ii. 14 The purpose of the weight belt is to help the diver achieve the weightless state. **1887** Kipling *Plain Tales* (1888) 144 You can arrange the race with regard to 'Shackles' only. So long as you don't bury him under weight-cloths, I don't mind. *Ibid.* 181 Maybe, Fate's weight-cloths are breaking his heart. **1974** *Radio Times* 28 Feb. 280/2 A nervously weight-conscious society. **1930** Weight function [see *orthogonalize *v.*]. **1974** G. Reece tr. *Hund's Hist. Quantum Theory* ii. 33 He made use of a weight function *G(E/ν)* for the enumeration of states. **1956** *Nature* 3 Mar. 423/2 (*caption*) Average weight-gains of animals fed on lime-treated maize. **1981–2** *Deer Farmer* (N.Z.) Summer 3/1 (Wapiti) Crosses with reds; the hybrids also produce good velvet and weight-gains. **1961** *Engineering* 21 July 72/1 The new 1½ litre 'weight limit' formula came into effect..in May. **1955** O. State *Weight Training for Athletics* i. 17 *Weight training*..implies training with light weights..for the purpose of improving one's performance in a particular sport. **1957** Duncan & Bone *Oxf. Pocket Bk. Athletic Training* (ed. 2) iii. 24 Weight training may now be regarded as an *essential* part of athletic training. **1976** E. Dunphy *Only a Game?* v. 147 We have to do weights. I don't believe in weight-training.

b. *weight-bearing* (sb. and adj.), *-lifter, -puller, -putting, -reducing* (sb. and adj.), *-throwing*.

1954 Martin & Hynes *Clin. Endocrinol.* (ed. 2) ii. 50 Osteoarthritis of the hips, knees and spine develops from undue strains of excessive weight-bearing as life advances. **1959** *Manch. Guardian* 9 July 5/7 We cannot even tell whether the heavy walls of the new buildings are the weight-bearing structures they look to be. **1977** P. A. Ring in *Bone & Joint Dis.* (Brit. Med. Assoc.) 83 It may be better to strive for union in the relatively young patient, even at the risk of a period of protected weight-bearing. [**1884** *Nat. Police Gaz.* (U.S.) 12 Jan. 13/2 Alonzo Hiwanda, the..champion heavy-weight lifter.] **1897** *Ibid.* 26 May 3/4 Bothwell, of Glasgow, is well known as a powerful man, besides a weight lifter. **1955** R. Bannister *First Four Minutes* 112 The waddling gait and breathlessness of a muscle-bound weight-lifter. **1980** *Sunday Times* (Colour Suppl.) 14 Sept. 99/2 A weightlifter complained that he did not like showing his tattoos in public. **1868** H. Woodruff *Trotting Horse* xxiii. 200 The weight-pullers..are of medium weight. **1900** A. E. T. Watson *Young Sportsman* 84 *Weight putting*—The weight should weigh 16 lbs., and in England must be of iron. **1948** *Sporting Mirror* 21 May 14/3 Giles had never done any weight putting when he went to Germany with the army. **1922** *Times* 7 Oct. 13/5 Grilling is the great essential of the weight-reducing diet. **1958** F. C. Avis *Boxing Ref. Dict.*, *Weight reducing*, taking off superfluous weight, often by means of vapour baths, sweating exercises, etc. **1978** N. Marsh *Grave Mistake* ii. 42 She tried..to get Verity to fix a day when she would come to a weight-reducing luncheon. **1901** J. P. Paret *Woman's Bk. Sports* 163 Weight-throwing has four or five variations. **1960** *Times* 29 Apr. 16/6 The investigation covered swimming, track running, weight-throwing.

weight, *v.* Add: **2. d.** Also, to treat (the components of any numerical quantity) similarly.

1927 C. Spearman *Abilities of Man* App. p. xviii, We urgently require to know how the single tests should be relatively 'weighted' in their combination. **1971** I. G. Gass et al. *Understanding Earth* v. 82 The individual data were weighted according to quality, so that a poorly determined result makes a smaller contribution to the mean than a precisely determined value. **1976** *Daily Record* (Glasgow) 30 Nov., Replies were weighted by age and General Election voting to make sure they were representative of all Record readers. **1977** *Whitaker's Almanack 1978* 1219 In working out the [cost-of-living] index figure, the price changes are 'weighted'—that is, given different degrees of importance—in accordance with the pattern of consumption of the average family.

4. a. (Earlier example.)

1734 J. Steuart *Letter-Bk.* (1915) 378 Your meall to be weighted with the common standard weights of Mariebrugh.

weightage (wē·itėdʒ). Chiefly *Pol.* or in *Pol.* contexts. [f. Weight *v.* + -age.] The assignment of a weighting factor to compensate for some (numerical) disadvantage, esp. in favour of a sparsely populated area, or to a minority party, interest, etc.; the amount so added. See *weighting *vbl. sb.* 3.

1906 in A. Husain *Fazl-i-Husain* (1946) vi. 96 Weightage, not by numerical strength but by political importance and value of the contribution made to the defence of the Empire. **1937** *Times* 24 Dec. 13/3 The Liberals [in Romania] only managed to secure 38 per cent. of the votes... They do not, therefore, qualify for the 'weightage' provided by the law of 1926. **1949** I. Jennings *Constitution of Ceylon* ii. 189 In agricultural countries like South Africa and Australia..the rural population, on whom the wealth of the country largely depends, must be given some weightage against the more concentrated and more highly organized urban population. **1957** L. F. R. Williams *State of Israel* 159 Does the present plan give these fractional groups a weightage in public affairs.. which their relative unimportance cannot justify? **1971** *Queen's College* (Oxford) *Record* Dec. 22 The geographical distribution of men who have gone down does not reflect quite the same northern weightage as do schools of origin. **1980** *Sunday Mail* (Brisbane) 9 Nov. 25/1 The National Party yesterday reaffirmed its policy of electoral distribution based on the electoral 'weightage' principle.

weighted, *ppl. a.* Add: **2. c.** *weighted* is similarly used of numerical quantities other than averages. (Earlier and later examples.)

1845 *Encycl. Metrop.* II. 443 We may..call the constant *c* the specific weight of the observations to which it applies, and ΣcA ÷ Σc the weighted mean. **1962** A. Nisbett *Technique Sound Studio* 277 Quoted noise levels are sometimes 'weighted' against bass according to standard loudness contours. Weighted and unweighted measurements may differ by 20 dB or more at low frequencies. **1970** G. K. Woodgate *Elem. Atomic Struct.* vii. 137 The identity..simply states that the weighted mean of the energies of the levels belonging to a term coincides with the energy of the unperturbed term. **1972** *Times* 27 Sept. 2/2 (*heading*) 'Weighted' vote at Labour conference suggested.

weighting, *vbl. sb.* Add: **3.** The assignment of weights (Weight *sb.* 15 b in Dict. and Suppl.); the weights so used.

1905 [in Dict., sense 1 a]. **1940** [see *weight *sb.* 15 b]. **1965** Phillips & Williams *Inorg. Chem.* I. iii. 75 One general technique for improving on the first approximation by the V.B. method is to include additional structures..in the complete wave function, with appropriate weighting factors. **1972** *Guardian* 18 Feb. 13/5 Rents and eating out..accounts for 22 per cent of the 'weighting' of the [retail price] index. **1983** *Personal Computer World* Dec. 142/2 Each of the eight digits of the base number is given a weighting.

4. An amount added to a salary for a special reason; esp. *London weighting*, that paid to compensate for the higher cost of living in the London area.

1946 *Scheme of Conditions of Service* (National Joint Council for Local Authorities' Admin., Profess., Techn. & Clerical Services) 19 The salary scales shall be weighted, as follows, in favour of officers employed in the London area:..£20 weighting with proportionate weighting of female scales. **1952** *Times* 25 Jan. 2/7 The wording of the statement relating to the award did not make clear whether the recommended increase was meant to be over and above the weekly 10s. 'weighting allowance' granted to London [fire]men. **1958** *Times* 5 Dec. 3/2 Salary (including London Weighting) according to age and experience. **1976** *Broadcast* Dec. 17/1 Brief consideration was given to a number of items including: weighting for major towns and cities other than London, London weighting, interest on monies owed to staff, [etc.]. **1982** *Daily Tel.* 20 Oct. 1/1 The government is to end the annual publication of the London weighting index which is widely used as the basis for calculating extra payments for employees in London.

weightless, *a.* Add: **a.** Also (of a body having mass), not apparently acted on by gravity, either because the gravitational field is locally weak, or because both the body and its surroundings are freely and equally accelerating under the field (as in an orbiting satellite).

1929 *Science Wonder Q.* Fall 55/2 Do you mean that.. we will be weightless as soon as you..set the lever at zero? **1950** *Jrnl. Aviation Med.* XXI. 396/2 A body is weightless as soon as it is allowed to move freely under the influence of gravity and of its own inertia. **1953** A. C. Clarke *Prelude to Space* v. 28 The perfect [spaceship] pilot..must be capable of operating efficiently..when he was 'weightless'. **1978** *Nature* 20 July 236/1 We report here the results of an experiment in the weightless environment of space. **1983** A. Mason *Illusionist* i. 15 A man who could command his body to float weightless through the air could not command the necessities of life.

weightlessness (further examples).

1929 *Science Wonder Q.* Fall 58/2 If they had not already been accustomed to weightlessness, the first heedless step would have carried them far from the ship. **1932** D. Lasser *Conquest of Space* xiii. 192 The terrors of weightlessness. **1959** *Observer* 31 May 1/4 The two monkeys spent a number of minutes in a condition of weightlessness or zero G. **1974** R. Adams *Shardik* x. 74 Her stance gave a curious impression of weightlessness, as though she might actually be about to float down into the hollow. **1983** *Brit. Med. Jrnl.* 13 Aug. 479/2 The most important vestibular disturbance encountered in weightlessness is motion sickness.

Weight Watcher. orig. *U.S.* Also **weight-watcher.** [Weight *sb.*[1]: see *watch *v.* 12 d.] **1.** *pl.* A proprietary name used esp. for (members of) an organization, Weight Watchers International Inc., formed to promote dietary control as a means of slimming, or any of its associated clubs. Occas. in *sing.*

1961 *Official Gaz.* (U.S. Patent Office) 28 Feb. tm 130/2 The Low Calorie Candy Co., Inc... Filed Mar. 14, 1960. *Weight Watcher.* For dessert and pie mixes sold in combination packages... First use Feb. 1, 1960. **1964** *N.Y. Herald Tribune* 1 Nov. 11 4/3 Weight Watchers is an Alcoholics Anonymous for compulsive eaters. *Ibid.* (*caption*) Novice Weight Watchers line up for the first session of a new group. **1966** J. Nidetch (*title*) The Weight Watchers Cookbook. **1966** *Sunday Tel.* 28 Aug. 2/4 A Long Island housewife has successfully tackled the American problem of compulsive eating with her movement Weight Watchers Incorporated, which now has hundreds of thousands of members in 15 States. **1967** *Business Week* 4 Mar. 106/3 They..transplant the Weight Watcher idea..under a system of franchises. **1967** *Trade Marks Jrnl.* 25 Oct. 1628/2 *Weight Watchers...* Books, printed publications, stationery, diaries, printed matter and pocket portfolios..all containing data pertaining to dieting and weight control. Weight Watchers International Inc. **1967** *Official Gaz.* (U.S. Patent Office) 5 Sept. tm 21/2 I. B. Kleinert Rubber Company, New York... *Weight Watcher.* For girdles. **1968** *Ibid.* 16 Jan. tm 133/1 Weight Watchers International, Inc... Filed Sept. 26, 1967. *Weight Watchers...* For indicating membership in the applicant association. **1968** *Ibid.* 18 June tm 146/1 *Weight Watchers.* For planning, executing and supervising diet programs by means of group meetings, courses [etc.]... First use May 15, 1963. **1972** J. Anderson in *Clin. Endocrinol.* (1973) 80 Group therapy should not be derided and is probably one of the main reasons for the apparent success of 'Weight-Watchers' clubs. **1977** P. Hill *Liars* xi. 144 You should join Weightwatchers... They'd slim you down in no time. **1978** *Cornish Guardian* 27 Apr. 12/5 Mr. Arthur spoke of the health campaign in arranging a 'fun and jog' for all ages, involving the local Weight Watchers, Keep Fit Classes. **1984** S. Moody *Penny Dreadful* xi. 144 Half the world starved while the other half joined Weight Watchers.

2. (With small initials.) A person who tries to lose weight, esp. by dieting; one who is weight-conscious.

1966 *Family Circle* Jan. (recto front cover), Meal plans for weight-watchers. **1968** *Sunday Times* 10 Nov. 35 Italians are not exactly the keenest weight-watchers in the world and hardly let a day go by without forking into the pasta. **1970** *Guardian* 6 June 12/5 What the intelligent weight watcher needs is a diet which keeps her fit. **1971** *New Scientist* 4 Feb. 231/3 The hint to weight-watchers is obvious: keep food out of sight, and it should stay out of mind. **1983** *Daily Tel.* 9 Feb. 17/4 The old message that it was good for weight watchers to eat a slice of cheese on a cream cracker has gone out of the window.

Hence **wei·ght-watching** *ppl. a.* and *vbl. sb.*

1970 *Globe & Mail* (Toronto) 26 Sept. 29/2 (Advt.), Luncheon here is to be recommended..and for the weight watching girls..there are tempting light suggestions. **1978** *Dumfries Courier* 20 Oct. 28/3 Martin..decided that dieting or weight-watching were not for him. **1982** W. J. Burley *Wycliffe's Wild-Goose Chase* iv. 66 'Beef casserole with boiled potatoes and carrots.' Good! But not for weight watching.

Weil (vail). *Path.* [The name of H. A. *Weil* (1848–1916), German physician, who described the disease in 1886 (*Deutsch. Archiv f. Klin. Med.* XXXIX. 210).] *Weil's disease*, a severe, sometimes fatal, form of leptospirosis that is characterized by fever, jaundice, and muscle pains and is acquired by infection from the urine of rats.

1889 *Brit. Med. Jrnl.* 6 July 11/1 (*heading*) Notes on a case of Weil's disease. **1934** [see *leptospirosis]. **1961** R. D. Baker *Essent. Path.* ix. 200 The mortality of Weil's disease (spirochetal jaundice), the common leptospirosis of man, is about 10 per cent. **1977** C. McCullough *Thorn Birds* xii. 276 He looked thin, wrinkled and yellow... 'What is Weil's disease, Luke?'.. 'Oh, it's just some sort of jaundice most cutters get sooner or later. The cane rats carry it.'

Weil–Felix (vail,fī·liks). *Med.* [The names of Edmund *Weil* (1880–1922), Austrian physician, and Arthur *Felix* (1887–1956), Polish bacteriologist, who described the reaction in 1916 (*Wien. klin. Wochenschr.* XXIX. 33).] *Weil–Felix reaction*: an agglutination reaction which takes place when serum from a patient infected with typhus is added to certain strains of bacteria of the genus *Proteus*, used as a diagnostic test for the disease.

1919 *Public Health Rep.* (U.S.) XXXI. 2446 The Weil-Felix reaction..has recently come into use as a means of diagnosing typhus fever. **1956** *Nature* 11 Feb. 257/2 The Weil-Felix reaction..proved of immense value in the differential diagnosis of typhus from typhoid and other fevers of unknown origin, and stimulated a great deal of research to explain why it was possible to obtain a specific agglutination reaction with an organism playing apparently no part in the causation of the disease. **1978** *Jrnl. R. Soc. Med.* LXXI. 509 The Weil–Felix reaction, which is the only generally available diagnostic test, failed to detect over 50% of proven cases in several series.

Weimar (vai·mār). The name of a city in Thuringia, Germany, where the democratic constitution under which Germany was governed from 1919 until the start of the Third Reich in 1933 was drawn up. Used *attrib.* and *absol.* with reference to the political, social, and cultural aspects of Germany during this period, esp. in phr. *Weimar Republic*.

1932 *Internat. Affairs* XI. 770 The return pure and simple to the Weimar system. **1934** H. P. Greenwood *German Revolution* iii. 39 The National Assembly at Weimar epitomised..the whole Weimar Republic. **1958** *Listener* 20 Nov. 828/2 The liberal fancies of the Weimar Republic. **1963** W. H. Chamberlin *German Phoenix* xiii. 244 In distinct contrast to the Weimar period, a political system based on free elections.. has had time to strike deep roots. **1968** P. Gay *Weimar Culture* (1969) p. xiv, The dazzling array of these exiles..tempts us to idealize Weimar as unique, a culture without strains.., a true golden age... But to construct this flawless ideal is to trivialize the achievements of the Weimar Renaissance. **1974** W. Laqueur *Weimar* vii. 224 There was..a light side to Weimar culture: Fritzi Massary and Richard Tauber, Marlene Dietrich and the *White Horse Inn*. *Ibid.* ix. 273 The Weimar revival reached its apogee in the late 1960s with the rise of the New Left. **1982** S. G. Duff *Parting of Ways* vii. 74 The Weimar Government tried to meet..obligations by printing money.

Weimaraner (vaimārā·nəɹ, wai-). [Ger., f. *Weimar* (see prec.) + *-aner* (something) of this place, region, etc.] A (breed of) grey, short-coated, drop-eared pointer, which was originally bred as a hunting dog in the Weimar region.

1943 *Amer. Kennel Gaz.* Jan. 77/1 The admission to registration of the Weimaraner..brings to 109 the breeds now recognized as pure-bred. **1952** L. Ross *Picture* 35 He had..a pen for eight Weimaraner puppies. **1954** *Time* 1 Mar. 19/1 Republican speechwriters came to a point like so many Weimaraners last November. **1968** *Globe & Mail* (Toronto) 17 Feb. 49/5 (Advt.), Weimaraners are medium size sporting dogs. **1979** *Daily Mail* 26 Oct. 25/2 The upper middles have recently taken to foreign breeds—weimaraners and rotweilers.

weiner (wī·nəɹ), var. of *wiener *sb.* a. Cf. *weeny *sb.*[2] *N. Amer.*

1961 in Webster. **1965** P. Tamony *Americanisms* (typescript) No. 10. 8 In Frankfurt small sausages were termed *Wein*; this turns up in American colloquialism as *weiner* and *weenie*. **1973** H. Nielsen *Severed Key* iv. 45 We got a little cold beer and some weiners, and we thought we'd have us a picnic. **1980** J. M. Bickham *Regensburg Legacy* iv. 56 The hotel supper..sauerkraut, weiners, green beans.

‖ **Weinstube** (vai·nʃtūbə). [Ger., f. *wein* Wine + *stube* *Stube, room.] A small German wine-bar or tavern. Cf. *bierstube.

1899 F. Norris *McTeague* 126 Its place was taken by a German saloon, called a 'Wein Stube'. **1936** C. Beaton *Diary* Sept. in *Wandering Years* (1961) xvi. 297 We sit in the Weinstube drinking white wine. **1946** S. Spender *European Witness* 138 The Weinstube was one of those German drinking cellars which resemble a chapel. **1969** K. Benton *Twenty-Fourth Level* vi. 101 He dined in a noisy, cheerful German *Weinstube*. **1981** L. Deighton *XPD* vi. 36 The wreckage of a German Weinstube.

‖ **Wein, Weib, und Gesang** (vain vaip unt gəza·ŋ), *phr.* [Ger.] Wine, women, and song, proverbially considered the essential ingredients for carefree entertainment and pleasure by men.

First popularized as the title of a Strauss waltz (1869). Strauss prob. took it from the anon. couplet found in the Luther room at Wartburg: Wer nicht liebt Wein, Weib und Gesang /Der bleibt ein Narr sein Leben lang (see Wine *sb.*[1] 1 f (b), quot. 1862).

1885 G. B. Shaw in *Dramatic Rev.* 27 June 341/1 The '*Wein, Weib, und Gesang*' waltzes which the Inventions Council offer us as the flower of modern European music. **1924** G. B. Stern *Tents of Israel* vi. 87 Franz..was a typically Viennese Rakonitz, in the famous Wein-Weib-und-Gesang style. **1935** C. Isherwood *Mr. Norris changes Trains* iii. 38, I shall..prepare myself to enjoy an evening of *Wein, Weib, und Gesang*. More particularly *Wein*. **1959** M. Crosland tr. *J. Rovan's Germany* 21 The famous trilogy '*Wein, Weib und Gesang*' (wine, woman and song) which form the subject of numerous drinking songs on both sides of the Rhine.

weir, *sb.* Add: **6.** *weir-stream* (earlier example).

1889 J. K. Jerome *Three Men in Boat* ix. 143 We might have somehow got into the weir stream, and be making for the falls.

weird, *a.* Add: **4. b.** Colloq. phr. *weird and wonderful*, marvellous in a strange or eccentric way; both remarkable and peculiar or unfathomable; exotic, outlandish. Freq. *iron.* or *derog.*

1859 J. H. Stirling in *Meliora* Oct. 231 These [poems] are doubtless meant to be very weird and wonderful, but they are mere breath, and..barren as the wind. **1886** O. Wilde in *Pall Mall Gaz.* 1 Feb. 5/1 There is psychology of a weird and wonderful kind. **1908** T. E. Lawrence *Let.* 9 Aug. (1954) 70 Their food is weird and wonderful. **1946** Visct. Knebworth *Boxing* xiv. 176 The beginner so often gets the idea that he is going to do the most weird and wonderful movements. **1962** *Friend* 3 Aug. 947/1 Nearly all the weird and wonderful decorations were provided by a decorator member of the club. **1978** S. Naipaul *North of South* ii. vi. 227 A weird and wonderful place is Jo'burg.

weirdie (wiᵊ·ɹdi). *slang.* Also **weirdy.** [f. Weird *a.* + -ie.] **1.** An odd or unconventional person; one who is considered 'weird'; *spec.* applied to any young man with long hair and a beard. Freq. in *pl.*

1894 A. S. Robertson *Provost o' Glendookie* 101 'He's awa without his curran' loaf.' 'He's a weerdie.' **1949** W. R. Burnett *Asphalt Jungle* (1950) ii. 19 Cobby.. thought to himself: 'He's a weirdy, all right.' **1954** 'P. Quentin' *Wife of Ronald Sheldon* vii. 57 God, is that one a weirdie!.. There was a cobweb in her hair. **1959** *Listener* 3 Dec. 975/1 The weirdies that Kerouac seems always to meet wandering and muttering in the small hours. **1960, 1961** [see *beardie 2]. **1962** *Punch* 14 Feb. 268/2 One [bedsitter]..advertiser..added 'No Weirdies either'. **1966** *Daily Tel.* 17 Nov. 18/8 There was not an unwashed bearded weirdie in sight! **1974** K. Millett *Flying* (1975) i. 94 I'm not a friend, just the visiting weirdie.

2. Something that is 'weird', fantastic, bizarre, or grotesque. Freq. applied to a film, book, etc.

1948 *Astounding Sci. Fiction* Jan. 15 The *Cosmos* had one of its feature writers compose a weirdie about a world consisting of beings of pure mind. **1962** *Listener* 14 June 1043/3 *The Lake Lovers* is a weirdie. **1968** *Blues Unlimited* Nov. 25 Country Jim is a weirdie. **1969** R. Petrie *Despatch of Dove* i. 26 No mistake, it was a weirdy of a day.

weirdo (wiᵊ·ɹdōᵘ), *sb.* and *a. slang.* [f. Weird *a.* + *-o[2].] **A.** *sb.* = *weirdie 1.

1955 L. Feather *Encycl. Jazz* 347 *Weird-o*, a weird person. **1958** *Observer* 13 Apr. 15/3 He is worried by Press reports which represent him as 'a weirdo—there is another word for it'. **1967** *Courier-Mail* (Brisbane) 13 Apr. 22/3 Another set of weirdos using a slick philosophy of revolt against the established order as camouflage for a lazy or corrupt existence. **1972** J. McClure *Caterpillar Cop* iv. 45 A shock-haired, bearded weirdo in a tartan dressing-gown and wellington boots. **1976** M. Machlin *Pipeline* xli. 448 We are near the village and I go back a lot, but like I said, they all treat me like I was some weirdo. **1981** *London Rev. Bks.* 3–16 Sept. 3 Santa Fe is acknowledged as a milieu of aesthetes and weirdos. **1984** *Melody Maker* 6 Oct. 34/4 This record is for the real weirdos.

B. *adj.* Bizarre, eccentric, odd.

1962 *Sunday Times* 5 Aug. 20/6 Frankly, I'm sick of your whole weirdo line. Leave me alone. **1969** C. Burke *God is Beautiful, Man* (1970) 46 About halfway through the party a real weirdo thing happened. **1974** M. Moore *Silver Birch Country* 43 The lady I'm looking after is a dear old duck, completely weirdo, but she's got a terrible sense of humour, and I like her. **1979** *Tucson* (Arizona) *Citizen* 20 Sept. 2A/6 It..makes us sound like some sort of weirdo fanatics opposed to all medicine.

weisenheimer, var. *wisenheimer.

Weissenberg (vai·sənbɜɹg). [The name of Karl *Weissenberg* (b. 1893), Austrian-born physicist.] **a.** *Cryst.* Used *attrib.* with reference to a technique of single-crystal X-ray diffraction introduced by him, in which a metal shield allows the diffracted X-rays to produce only one set of parallel lines of spots which are recorded over the whole of the photographic film by rotating it synchronously with the crystal, enabling the Miller indices and other crystal parameters to be easily obtained.

1934 W. P. Davey *Study of Crystal Struct. & its Applic.* vii. 205 The Weissenberg camera may be used..in the indexing of diffraction spots which are so thickly clustered on layer lines as to require otherwise a large number of oscillation photographs. **1962** Weissenberg photograph [see *salesite]. **1976** D. Sherwood *Crystals, X-rays & Proteins* xiv. 507 To obtain full three-dimensional information, we may take a series of photographs corresponding to each layer line in which we are interested. This was realised by K. Weissenberg, and in 1924, he published an experimental method which enables this to be done. This is now known as the Weissenberg method.

b. Physics. *Weissenberg effect*: an effect observed when a visco-elastic liquid is stirred, when the liquid rises in the centre and climbs the stirring rod rather than forming a concave surface like normal fluids.

1949 M. Reiner et al. in *Jrnl. Soc. Chem. Industry* LXVIII. 327/2 When a vertical rod is rotated in certain viscous elastic liquids, the liquid climbs up the rod and when a disc is rotated in such a liquid near the bottom of a beaker containing it, the liquid is drawn radially towards the centre... Freeman and Weissenberg claim that the first experimental observations were made by Weissenberg and Russell... In view of this it would seem justifiable to describe such phenomena as 'the Weissenberg Effect'. **1978** *Sci. Amer.* Nov. 148/2 If you would like to produce the Weissenberg effect, you might use a mixing bowl mounted on a turntable.

weissite[2] (vai·sait). *Min.* [See quot. 1927 and -ite[1].] A copper telluride, Cu_5Te_3, occurring (often in association with rickardite) as bluish black or bluish grey pseudocubic crystals with a metallic lustre that darken on exposure to air.

1927 W. P. Crawford in *Amer. Jrnl. Sci.* CCXIII. 346 The mineral is named Weissite after the late Loui Weiss, owner of the Good Hope mine [at Vulcan, Colorado]. **1980** *Mineral. Abstr.* XXXI. 354/1 Rickardite and weissite occur with pyrite, tetrahedrite, chalcopyrite, and altaite in this orebody in andesites and dacite porphyrites.

‖ **Weisswurst** (vai·svuɹst). [Ger., f. *weiss* white + *wurst* sausage.] (A) whitish German sausage made chiefly of veal.

1963 I. Fleming *On H.M. Secret Service* xxvi. 289 At the Franziskaner Keller..they ate mounds of Weisswurst and drank four steins of beer each. **1970** *Sat. Rev.* (U.S.) 12 Sept. 107/3 Germany's never-stale variety of sausages, particularly..the delicate Weisswurst. **1983** *N.Y. Times* 7 Sept. c7 Weisswurst, the great specialty of Munich, is generally made with veal, pork, bread crumbs, nutmeg, salt, pepper and, sometimes, lemon peel or parsley. It is also stuffed into casings before cooking.

welcome, *sb.*[1], *int.* and *a.* Add: **B. 4. d.** *you are* (or *you're*) *welcome*: a polite formula used in response to an expression of thanks.

[**1907** W. W. Jacobs *Short Cruises* ii. 34 'Thank you,' said the girl, with a pleasant smile. 'You're quite welcome,' said the skipper.] **1960** *Times* 14 Sept. 12/7 The coloured lift attendant in South Carolina who had that attractive way of saying, almost singing, 'You're welcome' whenever we thanked her. **1977** P. Dickinson *Walking Dead* i. iv. 55 'Thanks,' said Foxe..'You're welcome,' said Dreiser. **1980** A. E. Fisher *Midnight Men* viii. 93 He dialled Directory Enquiries and asked the girl if she had a number. She gave him one and told him he was welcome.

e. *you are welcome* (*to* something): said ironically of something one is glad to be without.

1937 A. Thirkell *Summer Half* i. 10 'Fine Old English Gentleman,' said the applicant enthusiastically. 'You are welcome to him.' **1969** J. N. Smith *Is he Dead, Miss ffinch?* xviii. 118 My Uncle Len and Aunty Marge live there in a caravan. (They're welcome.)

C. 5. d. *welcome aboard*, said (in allusion to nautical usage) as a joc. greeting to someone joining a particular group, enterprise, etc.

1962 J. D. MacDonald *Key to Suite* ii. 28 He put Hubbard's material in the envelope, hesitated, then scrawled across the front of it, 'Welcome aboard!' **1970** J. Sangster *Touchfeather, Too* iii. 79 We headed across the cool green lawn to the clubhouse... I was introduced to the manager..[who] had once served..in the navy. 'Welcome aboard, Miss Touchfeather,' he said. **1977** 'J. le Carré' *Hon. Schoolboy* v. 111 'Welcome aboard,' said Guillam... They had reached the fifth floor.

welcome, *sb.*[2] Add: **2. a.** Also freq. *attrib.*

1955 B. Pym *Less than Angels* vi. 71 We aren't getting on very fast with your welcome-home party. **1966** B. Kimenye *Kalasanda Revisited* 32 He would take the creature up to the Musaka's and ask Miriamu to prepare it as a special 'welcome home' supper for Yosefu. **1974** M. Birmingham *You can help Me* viii. 178 It gives me the chance to be part of your welcome-home committee.

c. *dial.* A bell tolled on the occasion of a person's death. Also *transf.*

1878 F. Kilvert *Jrnl.* 25 Dec. (1977) 328 The Welcome Home, as it chimed softly and slowly to greet the little pilgrim coming to his rest, sounded bleared and muffled through the thick snowy air. **1948** F. Thompson *Still glides Stream* xii. 226 It was a small, homely procession which..accompanied Reuben on the last of his many journeys. The silvery sweet strain of a robin threaded the silence. 'The welcome home!' said Mrs. Finch.

6. *attrib.* and *Comb.*, as **welcome mat** *U.S. colloq.*, a mat put out to greet welcome visitors; chiefly used in *fig.* phrs. to indicate a friendly welcome (cf. *red carpet* s.v. *red *a.* 19 a); **welcome song**: *spec.* the first ode composed by Purcell (see quots.); **welcome wagon** *N. Amer.*, a car bringing gifts and samples from local merchants to newcomers in a community; also *fig.*

1951 I. Asimov *Foundation* (1953) ii. v. 73, I certainly don't intend to lay down the welcome mat. **1963** M. McCarthy *Group* iii. 50, I can only ask you to come here whenever you're in town. The welcome mat will be out. **1978** R. Nixon *Mem.* 622 The door will not only be open—I've been weaving a welcome mat. **1681** H. Purcell (*song-title*) A Welcome Song in the year 1681 for the King. **1883** Grove *Dict. Mus.* III. 47/1 In 1680.. [Purcell] produced the first of his numerous odes, viz. 'An Ode or Welcome Song for his Royal Highness [the Duke of York] on his return from Scotland'. **1942** Welcome song [see *birthday ode* s.v. *birthday 3]. **1961** Webster, Welcome wagon. **1970** A. Toffler *Future Shock* (1971) vi. 104 We have in many American suburbs a commercial 'Welcome Wagon' service that accelerates the process by introducing newcomers to the chief stores and agencies in the community. **1971** H. T. Walden *Anchorage Northeast* 13 The welcome-wagon type of hospitality is not here. **1976** *Times Lit. Suppl.* 2 Jan. 4/3 He was never serious about moving to America, fearing perhaps that the American welcome wagon, by killing his habit of resentment with kindness and pelf, might stultify his..literary gifts.

welcomeness. (Later example.)

1977 [see *picnic party* s.v. *picnic *sb.* 3].

weld, *sb.*[2] Add: **3. weld decay,** (increased susceptibility to) corrosion in chromium-nickel stainless steel that has been kept at 600° to 900°C for a time (as in welding), owing to the precipitation of chromium carbide and the consequent lowering of the chromium content; **weld pool,** the pool of molten metal formed about a joint in welding.

1932 E. Gregory *Metallurgy* vii. 275 The heating of alloy steels of the 18 per cent chromium, 8 per cent nickel type in the range 650°–900° C. greatly decreases their corrosion resistance... This phenomenon is known as weld-decay. **1973** A. Parrish *Mech. Engineer's Ref. Bk.* v. 74 This local depletion of chromium causes lack of passivity in acid corrodants with consequent attack along grain boundaries (weld decay). **1964** W. Steeds *Engin. Materials, Machine Tools & Processes* (ed. 4) vii. 169 With coated electrodes too high a current..makes control of the weld pool difficult. **1975** Bram & Downs *Manuf. Technol.* ii. 55 The arc and the weld pool are protected from atmospheric contamination.

weld, *v.* Add: **2. b.** (*b*) (Earlier example.)

1802 J. Playfair *Illustr. Huttonian Theory* 283 The *line* of separation..has, on the whole, been marked out with great precision; and, though the stones have been firmly united, or, as one may say, welded one upon another, yet, when a fresh fracture was obtained, the stratified and unstratified parts have rarely failed to be distinguished.

weldability. (Earlier example.)

1869 H. S. Osborn *Metallurgy Iron & Steel* iii. 85 There is a degree of weldability in platinum which causes that metal to be classified with iron as a weldable metal.

weldable, *a.* (Earlier example.)

1855 D. Lardner *Hand-bk. Nat. Philos.: Hydrostatics* iii. v. 304 Weldable metals.—The metals capable of being welded soften before they are fused.

welded, *ppl. a.* Add: **3.** *Geol.* **a.** Applied to pyroclastic rock formed by the union of small, heat-softened particles.

[**1802**: see *weld *v.* 2 b (*b*). **1899** J. P. Iddings in A. Hague et al. *Geol. Yellowstone Nat. Park* ii. x. 406 The mass is compact of glass, but it consists of irregularly shaped streaks and patches of different color. These twist and curve about one another and appear like a perfectly welded mass of strips or ribbons and irregular fragments of variously colored glass.] **1909** —— *Igneous Rocks* I. i. viii. 333 These examples of welded pumice are from rhyolitic lavas in the Yellowstone National Park. **1935** *Trans. Amer. Geophysical Union* 309 Although commonly and perhaps generally associated with deposits of light-colored volcanic ash of rhyolitic composition, the welded tuff is not confined to this association but occurs also on older rocks. **1962, 1970** [see *ignimbrite]. **1977** A. Hallam *Planet Earth* 74/1 There may also be intercalations of submarine pillow lavas or welded tuffs indicative of volcanic islands.

b. Applied to an intimate, close-fitting contact between two bodies of rock that have not been heat-softened or tectonically disrupted.

1939 *Q. Jrnl. Geol. Soc.* XCV. 354 The contact is, as usual, welded, and the base of the overlying sediments consists of current-bedded, brown-weathering, fine sandstone. **1948** R. R. Shrock *Sequence in Layered Rocks* ii. 55 There are examples..where an entire geological system is represented by the hiatus along the welded contact. **1976** *Jrnl. Geol. Soc.* CXXXII. 125 The contact of the slump sheets with the overlying mega-beds is welded, i.e. depositional fit is present.

welding, *vbl. sb.*[1] Add: **3. welding torch,** a blow-pipe used in welding.

1921 *Engineering Index 1920* 417/2 A new carriage for welding torches. **1975** Bram & Downs *Manuf. Technol.* ii. 42 The welding torch used with these systems must be designed to aspirate the acetylene.

weldment (we·ldmĕnt). [f. Weld *v.* + -ment.] A unit consisting of pieces welded together.

1945 in Webster Add. **1950** *Engineering* 10 Feb. 149/1 In fabricated 'weldments'..it might not be necessary to stress-relieve. **1962** *B.S.I. News* Nov. 21/1 Fabricators may be taking a grave risk when they accept orders for weldments..furnished to inadequate specifications. **1979** *Railway Age* 31 Dec. 50/2 New features include..use of alloy steel castings in place of weldments.

Weldmesh (we·ldmeʃ). Also **weldmesh.** [f. Weld *sb.*[2] or *v.* + Mesh *sb.*] The proprietary name of wire mesh formed by welding together two series of parallel wires crossing at right angles. Freq. *attrib.*

1935 *Trade Marks Jrnl.* 24 Apr. 516/1 *Weldmesh*... screens, partitions, guards, frames, sieves and seatings, all made of welded steel wire. The British Reinforced Concrete Engineering Co. Limited,..Stafford; manufacturers and merchants. **1957** *Archit. Rev.* CXXI. 116 The balcony rail is of weld-mesh panels, with teak handrail. **1971** *Country Life* 11 Nov. 1255/1 It [*sc.* a cage] is made of weldmesh, used for tiger cages. **1978** *Cornish Guardian* 27 Apr. 10/3 (Advt.), 8 ft weldmesh pig lamps. **1984** *Times* 28 July 2/3 The committee recommended that the chain-link fence should be supplemented by weldmesh fencing which is more difficult to breach.

welfare, *sb.* Add: **3*. a.** The maintenance of members of a group or community in a state of (esp. physical and economic) well-being, esp. as provided for and organized by legislation or social effort. See also sense 4 in Dict. and Suppl.

1918 [see *rest room* s.v. *rest *sb.*[1] 14 a]. **1965** A. J. P. Taylor *Eng. Hist. 1914–45* iv. 121 Free treatment of venereal disease was the sole innovation in 'welfare' directly attributable to the first World War. **1968** M. Pyke *Food & Society* v. 66 And a Western community converted to the principles of welfare will supply vitamins and much else without requiring profit. **1977** M. French *Women's Room* (1978) ii. 139 Welfare..was starting to be a big thing. A lot of Puerto Ricans coming up to New York to get a free handout.

b. *ellipt.* (Usu. with capital initial.) A welfare centre or office; (the officials of) a welfare department.

1928 [see *pally]. **1960** D. Lessing *In Pursuit of English* iv. 135 Once she asked Welfare if Aurora could go to a council nursery. **1972** J. Mann *Mrs. Knox's Profession* vii. 57 That poor mite... The mother didn't ought to leave it like that... They ought to get the welfare to that woman. **1984** *Observer* (Colour Suppl.) 18 Mar. 6/2 First I rang the Welfare to make sure they would get me a flat.

c. = *welfare benefit*, sense 4 below. Esp. in phr. *on* (*the*) *welfare*. orig. *U.S.*

1946 [see *snag *v.*[2] 4]. **1964** S. M. Miller in I. L. Horowitz *New Sociol.* 295 Women on welfare strongly demonstrated against the cessation of allowances. **1970** *Toronto Daily Star* 24 Sept. 1/1 People receiving welfare in Metro broke all previous records. **1974** K. Millett *Flying* (1975) ii. 141 Half the people I know feed on welfare. **1976** F. Zweig *New Acquisitive Society* ii. v. 111 If a man with four children on supplementary benefits would only be better off working if he could earn about £75 a week, it would..need a very conscientious man, keen on his work, to resist the temptation to stay on welfare.

4. *welfare centre, clinic, committee* (example), *department, office, officer, service, work* (earlier example); also, provided by the State for those in need, as *welfare benefit, cheque, food, milk*; subsisting on benefits provided by the State, as *welfare family, mother*; **welfare capitalism**, a capitalist system seeking to combine a desire for profits with concern for the welfare of its employees; **welfare fund**, a fund or funds from which payments are made in time of sickness, etc.; **welfare hotel** *U.S.*, a hotel in which people on welfare are housed until more permanent quarters can be found for them; **welfare roll** *N. Amer.*, a list of those entitled to welfare benefits from the State.

1977 M. Edelman *Political Lang.* vii. 125 Through disorder the poor have increased welfare benefits in the United States. **1960** *New Left Rev.* Sept.–Oct. 10/2 The very real achievements of 'welfare capitalism'. **1978** P. Bailey *Leisure & Class in Victorian Eng.* ii. 43 Robert Owen's New Lanark mills had included an annexe comprising a school, museum, music hall and ballroom..and there were several other examples of this kind of welfare capitalism. **1917** *New Witness* 28 June 202/1 It is continually stated that Maternity Clinics and Infant Welfare Centres have met with the greatest success in France. **1941** J. S. Huxley *On Living in Revolution* (1944) 21 Communal feeding centres, crèches or welfare centres. **1947** *Sun* (Baltimore) 6 May 22/8 The cashing of four welfare checks sent to one of her male inmates after his death. **1976** *Billings* (Montana) *Gaz.* 17 June 1–D/2 When he got his Social Security and welfare checks. **1937** 'G. Orwell' *Road to Wigan Pier* v. 93 The baby was getting its weekly packets of milk from the Welfare Clinic. **1952** *Oxf. Jun. Encycl.* X. 282/1 The welfare committees also provide welfare services for the blind, deaf, dumb and crippled. **1922** S. Lewis *Babbitt* ii. 17, I wonder if I could get one of the department-stores to let me put in a welfare-department. **1977** M. Edelman *Political Lang.* v. 79 A welfare department or education department bears a name that is even less adequate in defining the priorities to which it must respond. **1977** *New Yorker* 27 June 88/3 The Spencers were far from being a welfare family; the house cost forty thousand dollars, which they could afford. **1948** *Ann. Reg. 1947* 487 There were..subsidies on animal feedingstuffs, welfare foods, milk in schools. **1958** Welfare food [see *family allowance* s.v. *family *sb.* 11]. **1947** *Ann. Reg. 1946* 205 Mr Lewis [*sc.* a trade-union leader]..refused even to discuss his demands until he had been granted a 'welfare fund' financed by a royalty on coal. **1978** S. Sheldon *Bloodline* xxxix. 348 They were listed if they had paid taxes or drawn unemployment insurance or welfare funds. **1971** *Times* 8 Jan. 5/1 The scandal of the 'welfare hotels' [in New York] where the city places homeless families has been simmering for many weeks. **1977** *N.Y. Rev. Bks.* 15 Sept. 3/1 One welfare hotel in New York, the scene of repeated mayhem, is next to the local police station. **1958** Welfare milk [see *family allowance* s.v. *family *sb.* 11]. **a 1974** R. Crossman *Diaries* (1976) II. 560 Increased family allowances with supplementary allowances and children's tax allowances, the price of school meals, the price of welfare milk, [etc.]. **1971** *N.Y. Times* 9 June 43 The needs of..the pressured pensioner, the welfare mother and the harried commuter in our cities involve the whole range of vital urban services. **1978** *Guardian Weekly* 8 Jan. 16/4 Welfare mothers should get off the dole and go to work. **1976** *National Observer* (U.S.) 12 June 4/3 I'd rather see people employed than at the welfare office. **1944** Welfare officer [see *say *v.*[1] 1 c]. **1963** T. Parker *Unknown Citizen* ii. 58, I have been to see the C.A.C.A. welfare officer responsible for Smith's after-care. **1970** *Toronto Daily Star* 24 Sept. 1/1 The number of family units on the welfare rolls has more than doubled in the past year. **1979** *United States 1980/81* (Penguin Travel Guides) 300 A city where the list of places that offer poor boy sandwiches is probably longer than the welfare rolls. **1952** Welfare service see *welfare committee* above]. **1903** *Review of Reviews* July 79/1 The term 'industrial betterment', or 'welfare work', is used in a wider sense to include all of those services which an employer may render to his work people over and above the payment of wages. It has even been used to include the provision of homes for employees, kindergartens, schoolhouses [etc.].

we·lfare state. Also **Welfare State.** [State *sb.* 30.] A country in which the welfare of members of the community is underwritten by means of State-run social services.

The term is sometimes said to have been coined by Sir Alfred Zimmern in the 1930s, but it has not been traced in his published writings.

1941 W. Temple *Citizen & Churchman* ii. 35 We have.. seen that in place of the conception of the Power-State we are led to that of the Welfare-State. **1948** *Economist* 24 Jan. 135/2 The welfare state got its start in the deepest depression this country [*sc.* the U.S.] has ever known. **1950** *Times* 19 May 5/3 This is one of the achievements for which the 'welfare State', with its vast apparatus of taxation, subsidies, family allowances, school meals, and other services, can claim credit. **1959** B. Wootton *Social Sci. & Social Path.* 16 The myth of the 'welfare state' has turned the minds of investigators away from the study of material want. **1967** M. Drabble *Jerusalem the Golden* ix. 214 She wondered whether she should fall on her knees and thank..the Welfare State. **1976** *New Yorker* 22 Mar. 48/3 They came because Sweden, more than any other country in the industrial West, is a workers' state—not a Socialist state..but a stunning experiment in welfare-state capitalism. **1977** M. Walker *National Front* 8 A scholarship kid who went to grammar schools and won a scholarship to Oxford because of the 1944 Education Act and the Welfare State. **1984** *Listener* 22 Mar. 9/3 Repeated assertions that the welfare state is to be dismantled and the counter-assertion that Mrs Thatcher's administration is committed to maintaining welfare expenditure. **1985** *New Statesman* 9 Aug. 3/1 The welfare state will have to retain the power to take away children from danger.

Hence **we:lfare-sta·tism** orig. *U.S.*, the social conditions or organization associated with a welfare state; **we:lfare-sta·tist**, one who advocates such organization. Cf. *welfarism, *welfarist 1.

1949 in *Amer. Speech* (1957) XXXII. 296 Welfare statism advanced slowly and gradually. **1958** *Spectator* 31 Jan. 143/1 The combined effects of high taxation, continuous inflation, welfare statism and the tempting array of consumer goods. *Ibid.* 13 June 761/1 The demands of the welfare-statists in Congress for increased Government spending and tax cuts are now only occasionally heard. **1971** P. Worsthorne *Socialist Myth* v. 76 Labour Governments have to make too many economic concessions to the rich.., which places very strict limits on the scope and scale of welfare-statism. **1980** *London Rev. Bks.* 3–16 July 15/3 'Welfare statism' came into being in the 1940s... It consisted in..the extension to the whole population of adequate health, education and social insurance.

welfarism (we·lfeᵊriz'm). orig. *U.S.* [f. Welfare *sb.* + -ism.] The principles or policies associated with a welfare state; also = *welfare-statism.

1949 *Life* 25 July 17/2 There must be safeguards so that welfarism does not end in economic or political tyranny. **1961** *Engineering* 17 Feb. 249/1 All Germans.. also agree that the term 'welfarism' is likely to have widely different meanings in the U.K. and in Germany. **1962** *Times* 23 Jan. 9/7 Text-books..are slanted towards welfarism, socialism, and world government. **1968** P. B. Austin *On being Swedish* ix. 72 'The Swede,' said old Sundbärg, before welfarism or modern Swedish prosperity were born or thought of, 'would not mind Sweden being prosperous, providing that no one in it were better off than himself.' **1976** R. Delmar in Mitchell & Oakley *Rights & Wrongs of Women* ix. 283 State regulation of the family through welfarism (family allowances etc.).. means that the working class..has the law available to them. **1984** *Times* 21 Apr. 9/2 The proposition that the ills of Britain..are more or less the fault of capitalism is no more self-evident than the proposition that they are more or less the fault of welfarism.

welfarist (we·lfeᵊrist), *sb.* (*a.*) [f. as prec. + -ist.] **1.** One who is concerned with welfare, esp. that of animals. Also *attrib.* or as *adj.*

1941 I. Brown in *Manch. Guardian Weekly* 14 Mar. 214/3 There is in this country an enormous and semi-official bureaucracy which deals in social welfare and all manner of educational and uplifting matters. It is rapidly developing a jargon of its own... The Uplifters and Welfarists are assaulting us with the Scorpions of their Jargantuan. **1979** *Country Life* 15 Feb. 428/1 How does one define 'Animal Welfare'?.. To some the term 'welfarist' is almost one of contempt. **1980** *Times* 13 Feb. 3 The union has failed on our behalf to respond to the lies and unfair allegations of the welfarists. **1980** *Observer* (Colour Suppl.) 16 Mar. 41/3 The animal welfare movement—or the 'welfarist' movement as it is now widely called. **1983** *Sci. Amer.* Aug. 5/2 The pork producer of today generally is an animal protectionist and an animal welfarist in the best practical sense.

2. orig. *U.S.* One who supports the principles or policies associated with a welfare state. Also *attrib.* or as *adj.* Cf. *welfare-statist.

1968 *New Yorker* 12 Oct. 201 The welfare state..is not incompatible with the police state. George Wallace is a welfarist, and so is Mayor Daley. **1977** *Daily Tel.* 30 June 16 Questions arise which, in recent years, have been rendered almost taboo by the near-universal acceptance of facile welfarist orthodoxy. **1985** *Daily Tel.* 6 May 14/1 It has often been said..that he is the prisoner of a Californian perspective which scorns the moribund welfarist states of Western Europe for the growth and dynamism of the Pacific basin area.

well, *sb.*[1] Add: **8. a.** (Earlier example in sense 'the open space in which a lift operates'.)

1890 B. Hall *Turnover Club* viii. 87 But Gean hustled the man out to the elevator shaft and dropped him into the well beneath.

d. = *orchestra pit* s.v. *orchestra 4.

1933 P. Godfrey *Back-Stage* i. 15 The orchestra are in position in the 'well'. **1951** *Oxf. Compan. Theatre* 836/2 The Orchestra Well for the accommodation of the theatre musicians is in front of and below the stage itself.

11. c. Also *spec.* in *Ceramics*, the depressed central portion of a plate, saucer, or dish.

1937 *Crockery & Glass Jrnl.* Nov. 28 The Fleurette shape..with flower grouping in the well and repeated on the shoulder. **1971** *Country Life* 21 Oct. 1055/1 The saucer is decorated with a circular medallion of The Bull and the Mouse, its well with four sprigs of flowers, the bowl with the Bull and the Frogs.

11*. *Physics.* = *potential well* s.v. *potential *sb.* 4 c.

1942 *Rep. Progress Physics* VIII. 302 The..way to estimate the depth of the well is to postulate that the binding energy of the least strongly bound particle shall be equal to the experimental value for this quantity. **1972** *Sci. Amer.* Apr. 27/1 The original aim was to create a well so deep (from 10 to 20 million volts deep) that the ion-ion collisions could be energetic enough for nuclear transmutations to occur.

12. *well-conductor, -pulley, -site*; *well-like* adj. (earlier example).

1974 *BP Shield Internat.* Oct. 2/1 The jacket..wraps round the well-conductors which go down into the ground. **1854** Thoreau *Walden* 195 We have one other pond just like this, White Pond..but I do not know a third of this pure and well-like character. **1940** W. Faulkner *Hamlet* i. i. 18 He had already begun to hear the mournful..plaint of a rusted well-pulley. **1972** L. M. Harris *Introd. Deepwater Floating Drilling Operations* iii. 22 The wellsite geologist should..provide technical assistance to the drilling supervisor. **1979** *Jrnl. R. Soc. Arts* CXXVII. 406/2 This led in due time to the adoption of a system of deep ditches around the various well sites.

13. well-fire = *well-grate* below; **well-grate** (see quot. 1910); **well-kick**, the exerting by an oil-well of pressure in excess of that of the drilling fluid pumped into it, leading to loss of circulation; **well-sweep** (earlier example).

1906 *Studio* XXXVIII. 127/1 Well-fires are used in all the rooms. **1910** *Encycl. Brit.* XII. 378/2 In the closing years of the 19th century a 'well-grate' was invented, in which the fire burns upon the hearth, combustion being aided by an air-chamber below. **1927** W. E. Collinson *Contemp. Eng.* 90 Fires..are more often well-grates (i.e. low) than basket-grates (with hobs for the kettle). **1972** L. M. Harris *Introd. Deepwater Floating Drilling Operations* x. 97 Closing in around the drill pipe and circulating a conventional well kick. **1974** P. L. Moore et al. *Drilling Practices Manual* xi. 277 Failure to recognize a well kick could be disastrous. **1836** C. A. Goodrich *Universal Traveller* (ed. 2) i. i. 27 Here and there, by the side of the older houses, may be seen a well-sweep, a primitive contrivance to draw up water by a pole, which is attached to a beam, moving up and down on an axle.

well, *a.* **5. c.** For 'Now only *U.S.*' read 'Now chiefly *U.S.*' and add: Also *well-baby*, used *attrib.* to designate clinics or health care arrangements for routine checking of healthy children, as a form of preventive medicine; **well woman** (usu., with hyphen, *attrib.*), a woman who has undergone satisfactory gynæcological tests.

1921 *Daily Colonist* (Victoria, B.C.) 5 Oct. 6/4 A well-baby clinic will be held at the Saanich Health Centre.. An invitation is extended to all mothers to bring their infants. **1963** *Jrnl. Amer. Med. Assoc.* 2 Nov. 459/1 She had been advised to bring the child to the well-baby clinic. **1976** G. E. Godber *Brit. National Health Service* i. 5 Many of the elected councils of cities and counties had not used their powers to provide antenatal and well-baby care. **1981** 'G. Gaunt' *Incomer* xxvi. 173 That Thursday's well-baby clinic functioned with the slick accuracy of a Jesuit mass. **1977** *PEN Broadsheet* No. 3. 3/3 A range of leaflets on contraceptive methods, well-women care, sex-related diseases. **1980** *Brit. Med. Jrnl.* 29 Mar. 958/2 The new hospital will have a 40-bed gynaecological unit with a wide range of outpatient services and a specially designed 'well woman clinic' to provide urgently needed health screening. **1984** S. Townsend *Growing Pains A. Mole* 13 She says she needs the money for her 'Well Woman' test. She is having primary and secondary sexual organs checked. **1985** *Observer* (Colour Suppl.) 14 Apr. 23/3 She looked every inch the part of the world's most glamorous well-woman.

well, *adv.* Add: **II. 5. a.** *well taken*: of a point in an argument, aptly or judiciously raised. orig. *U.S.*

To be distinguished from the phr. *to take* (someone's) *point*, where *take* = to understand the significance of.

1863 A. Lincoln *Coll. Works* (1953) VI. 245 The point made in your paper is well taken. **1907** *Nation* (N.Y.) 14 Feb. 146 One of Mr. Hearst's points seems to us well taken. **1936** E. B. White in *New Yorker* 14 Mar. 16/2 The question is well taken. **1943** [see *point *sb.*[1] 28 c].

IV. 18. e. *to be* (or *get*) *well away*: to have (or obtain) a good start over one's pursuers; usu.

fig., to make good progress in an activity (esp. drinking). *colloq.*
1910 *Glasgow Herald* 7 Feb. 13/2 From the drop out Andrew got well away but Henry pulled him up. 1927 W. E. COLLINSON *Contemp. Eng.* 101 Expressions..like..'he's well away' (he has got into his stride or into swing whether in tackling a meal or a flirtation or in drink). 1947 'N. SHUTE' *Chequer Board* iii. 51 Bristow had a bottle of whisky and I had one of gin so we were well away. 1950 J. CANNAN *Murder Included* vii. 141 'What's happened?'.. 'If we knew that, we should be well away.' 1956 C. BLACKSTOCK *Dewey Death* vi. 128 He paused again, but Sergeant Robins saw no point in making any comment; the young gentleman was plainly well away. 1973 J. PORTER *It's Murder with Dover* viii. 72 Many great men..[can] drop off to sleep at any time..and Chief Inspector was no exception. He was well away by the time MacGregor climbed back into the car. 1984 A. CARTER *Nights at Circus* III. v. 233 The Colonel..overcomes his resistance to vodka to such an extent he is soon well away and sings songs of Old Kentucky.

V. 22. c. With intensive (usu. *slang*) adv. or adj. prefixed, as *bloody well*, *damn well*, etc.
1884 [see *BLEEDING *ppl. a.* 6]. 1898 [see JOLLY *adv.* 2]. 1903 [in Dict., sense 22 (*a*)]. 1921 E. O'NEILL *Emperor Jones* i. 160 Ring the bell now an' you'll bloody well see what I means. 1928 E. WAUGH *Decline & Fall* III. iii. 240, I should bleeding well say there was. 1933 [see *RUDDY *adv.*]. 1941 [see *DAMN *a.* and *adv.*]. 1943 D. WELCH *Maiden Voyage* iii. 14 Someone in the next cell was shouting, 'Bloody well let me out, you bastards.' 1962 L. R. BANKS *End to Running* I. vi. 96 Because actually, as a matter of fact, don't y'know, I'm not sodding well coming.

VII. 28. *well-vascularized.*
1959 W. ANDREW *Textbk. Compar. Histol.* iv. 113 (*caption*) The 'hairs', covered with well-vascularized epidermis, may be as much as 20 mm. long.

well-adjusted, *ppl. a.* Add: **b.** With reference to emotional adaptation. Cf. *ADJUSTED *ppl. a.* 4.
1939 L. MACNEICE *Autumn Jrnl.* xii. 49 A civilised, articulate and well-adjusted Community where the mind is given its due But the body is not distrusted. 1940 [see *ADJUSTED *ppl. a.*]. 1952 C. P. BLACKER *Eugenics: Galton & After* 307 Able and well-adjusted children. 1977 E. AMBLER *Send no More Roses* i. 11 The Able Criminal..may be presumed..to be emotionally stable and 'well-adjusted'.

well-aged, *a.* For '† *Obs.*' read 'Now *rare*' and add later example.
1979 [see *road-bed* s.v. *ROAD *sb.* 9 b].

well-aired, *ppl. a.* **2.** (Earlier example.)
1789 H. NEWDIGATE *Let.* 1 July in A. E. Newdigate-Newdegate *Cheverels of Cheverel Manor* (1898) vi. 89, I came home to a well-air'd Comfortable Bed.

well-argued, *ppl. a.* (Later examples.)
1964 K. G. GRUBB *Layman looks at Church* iv. 100 Laymen make contributions in Assembly debates..which are often serious and well-argued. 1975 *Amer. Speech* 1971 XLVI. 261 Vennemann's paper is a well-argued proposal for modifying marking theory in TG grammar.

well-arti·culated, *ppl. a.*
1951 W. K. MATTHEWS *Languages U.S.S.R.* iii. 33 Mordvin..shares with Cheremiss a well-articulated system of word-formation. 1977 *Jrnl. Commonw. & Compar. Politics* XV. 5 Pre-unionists parties in some colonies and republics of South Africa were functionable, rather well-articulated units.

well-atte·nded, *ppl. a.* Of a meeting: attended by a large number of people.
1946 *Nature* 21 Dec. 918/1 Dr. W. H. Taylor presided over a well-attended meeting. 1979 G. POTTINGER *Secretaries of State for Scotland 1926–76* xix. 193 The Scottish Covenant Movement..held two well-attended assemblies in 1947 and 1948.

well-behaved, *ppl. a.* Add: **2.** *Math.* Applied to different entities with varying implications as to their susceptibility to manipulation, as continuity or differentiability (of a function), convergence (of a series).
1939 C. B. BOYER *Concepts of Calculus* vi. 246 Inasmuch as Euler restricted himself to well-behaved functions, he did not become involved in those subtle difficulties connected with the notions of infinity. 1965 PATTERSON & RUTHERFORD *Elem. Abstr. Algebra* iii. 60 Of the two operations in a ring *R*, addition is 'well-behaved' in that it satisfies the commutative and associative laws and there exist an identity element and inverses... Multiplication is not so well-behaved. 1968 FOX & MAYERS *Computing Methods for Scientists & Engineers* i. 13 Since $x = 0$, and $J_0(x)$ is perfectly 'well-behaved', then also $y'(0) = 0$.

3. Of a computer program: communicating with hardware via standard operating system calls rather than directly, and therefore able to be used on different machines.
1984 *Austral. Microcomputer Mag.* Jan. 42/1 The disk drives can read and write IBM-PC format disks, and of several programs tried on the system, those that were 'well behaved' worked and those, such as word processors, that tend to directly address the machine's hardware would not always work. 1984 *Austral. Personal Computer* May 65/3 PC mode handles all well-behaved programs... In the case of direct hardware calls, problems usually arise if a 'not quite IBM-compatible' machine is used.

well-braced, *ppl. a.* (Earlier lit. example.)
1785 T. DWIGHT *Conquest of Canäan* VI. 141 The well-brac'd buckler glittered o'er his breast.

well-bred, *ppl. a.* **2.** (Earlier example.)
1805 *Times* 7 Nov. 1/1 To be sold..four capital well-bred hunters.

well-bree·ched, *ppl. a.* Prosperous, well-to-do. Cf. BREECHED *ppl. a.* 4.
1821 P. EGAN *Life in London* II. ii. 178 Jerry is in Tip Street upon this occasion, and the Mollishers are all nutty upon him, putting it about, one to another, that he is a well breeched Swell. 1860 HOTTEN *Dict. Slang* (ed. 2) 104 *Breeched, or to have the bags off*, to have plenty of money; 'to be well breeched', to be in good circumstances. 1968 P. SCOTT *Day of Scorpion* I. ii. 64 Her first husband..died well-breeched. 1980 *Jrnl. R. Soc. Arts* Apr. 266/1 That they had been able to become rich or well-breeched in a worldly sense is incidental. 1985 *Times* 22 Mar. 23/8 Britoil presents the spectacle of a fairly well-breeched oil group.

well-ca·red-for, *ppl. a.* [See CARE *v.* 3.]
1942 W. FAULKNER *Go down, Moses* 129 He..watched Lucas cross the Square,..erect beneath the old, fine, well-cared-for hat. 1959 J. CARY *Captive & Free* 116 An envied minority of cherished, well-bred, well-educated, well-cared-for families. 1979 V. L. PANDIT *Scope of Happiness* xx. 136 A lovely well-cared-for garden.

well-caulked, *ppl. a.* (Later example.)
1969 *Jane's Freight Containers 1968–69* 464/2 Liberal use of adhesives and sealants to provide a well-caulked box.

well-chosen, *ppl. a.* Add: **b.** Freq. in phr. *a few well-chosen words*, a short and telling speech or piece of writing. Also *ironically*.
1854 *Harper's Mag.* Feb. 423/2 Thomas Carlyle..has excelled all his contemporaries in the graphic pictures which he has painted in a few well-chosen and expressive words. 1912 BEERBOHM *Christmas Garland* 46 You figure him at the gate, shaking hands all round, and speaking perhaps a few well-chosen words about the future. 1957 D. ROBINS *Noble One* xix. 177, I can and shall go down and settle *her* with a few well-chosen words.

well-concea·led, *ppl. a.*
1925 F. SCOTT FITZGERALD *Great Gatsby* vii. 139 He put out his..hand with well-concealed dislike. 'I'm glad to see you, sir.' 1982 *N.Y. Times* 3 Feb. A 8/1 This correspondent visited the wellconcealed underground location of Venceremos radio in the mountains.

well-concei·ved, *ppl. a.*
1836 J. S. MILL in *London Rev.* II. 368 A well-conceived and well-executed work of fiction. 1862 —— *Pol. Econ.* (ed. 5) II. v. ii. 395 Mr. Hubbard..whose well-conceived plan wants little of being..an approximation to a just settlement. 1919 W. S. CHURCHILL *Let.* 20 Jan. in M. Gilbert *Winston S. Churchill* (1977) IV. Compan. I. 472 Executive Heads of well-conceived Branches. 1979 *Jrnl. R. Soc. Arts* Apr. 259/2 The latter Council has gradually built up a series of well-sited, well-conceived and well-run centres.

well-conditioned, *a.* Add: **2. c.** *Surveying* and *Math.* Such that a small error in measurement or change in data gives rise to only a small change in the calculated result.
1882 J. L. ROBINSON *Treat. Marine Surveying* viii. 141 If the equilateral triangles are not obtainable, then they must be as 'well-conditioned' as possible, i.e. the angles must lie between 30° and 75°. 1952 D. R. HARTREE *Numerical Analysis* viii. 155 The normal equations are less well-conditioned than the original equations. 1973 C. W. GEAR *Introd. Computer Sci.* vi. 261 Is the problem of computing the hypotenuse of a right-angled triangle, given the other two sides, well-conditioned?

well-constructed, *ppl. a.* Add: (Later examples.) Hence **well-constru·ctedness.**
1893 W. S. GILBERT *Utopia (Limited)* I. 25 Oh admirable art! Oh neatly-planned intention! Oh happy intervention—Oh well-constructed plot! 1915 A. J. BALFOUR in M. Gilbert *Winston S. Churchill* (1972) III. Compan. II. 1281 Well-constructed trenches. 1975 *Studies in Eng. Lit.: Eng. Number* (Tokyo) 138 The defect of Mr. Halliburton's well-constructed volume, it seems to me, lies in its well-constructedness.

well-co-o·rdinated, *ppl. a.*
1940 W. FAULKNER *Hamlet* II. ii. 127 Precocious, well-co-ordinated and quick to learn..he acquired enough credits in three years to enter college. 1983 *N.Y. Times* 7 May I. 14/5 A buoyant and wellcoordinated performance of the Vivaldi.

well-covered, *ppl. a.* Add: In quot. 1884 in Dict., used in sense 'thickly covered with flesh'. Hence in *colloq.* use of a person: plump, corpulent. (Later examples.) Cf. *WELL-UPHOLSTERED *ppl. a.*
1943 D. WELCH *Maiden Voyage* xvii. 136 A mild, well-covered person, with crinkly hair and rather piggy eyes. 1972 'E. FERRARS' *Breath of Suspicion* I. 7 He was rosy, bland and very well-covered.

well-cra·fted, *ppl. a.* [*CRAFT *v.* 2.]
1976 *Nature* 15 July 169/3 A well-crafted murder mystery. 1983 *Listener* 16 June 29/3 It is gripping, slick, well-crafted and beautifully shot.

well-crea·med, *ppl. a.*
1922 JOYCE *Ulysses* 527 Your wellcreamed braceletted hands.

well-designed, *ppl. a.* (Later examples.)
1934 *Archit. Rev.* LXXV. 39/1 Well-designed manufactures are, after all, not so rare in England. 1975 *Language for Life* (Dept. Educ. & Sci.) xxvi. 540 There is evidence in some schools that well-designed measures can be successful.

well-dete·rmined, *ppl. a.*
1905 W. JAMES *Ess. Radical Empiricism* (1912) iv. 128 Its successors differ from it in another well-determined way. 1968 FOX & MAYERS *Computing Methods for Scientists & Engineers* iii. 50 If the problem is well-conditioned the constants of this combination are well-determined.

well-do·cumented, *ppl. a.* Supported or attested by much documentary evidence.
1937 *Burlington Mag.* Apr. 156/1, 1465, in which year he [*sc.* Bellini] produced his first well-documented painting. 1946 *Nature* 30 Nov. 770/1 Well-documented studies of the Tehuelche and Puelche of Patagonia. 1978 *Jrnl. R. Soc. Med.* LXXI. 697/1 The well-documented increases in prescribing represent therapeutic irresponsibility.

well-doer. (Later example.)
1961 NEW ENG. BIBLE *John* iii. 11 The well-doer is a child of God; the evil-doer has never seen God.

well(-)done, *ppl. a.* **3.** (Earlier example.)
1747 H. GLASSE *Art of Cookery* i. 4 Pork must be well done, or it is apt to Surfeit.

well-drai·ned, *ppl. a.*
1871 W. ROBINSON *Subtropical Garden* II. 198 A well-drained, sandy soil..is the best for this plant. 1912 J. W. WHITE *Flora of Bristol* 251 The well-drained banks of railway lines. 1960 *Farmer & Stockbreeder* 15 Mar. (Suppl.) 10/3 Most herbs will grow in any ordinary well-drained ground.

well-endowed, *ppl. a.* Add: Also *spec.*, with reference to sexual potency or size of sexual organs. *colloq.*
1951 N. MONTSARRAT *Cruel Sea* v. 302 'I'm not rich.'.. 'You are doubtless well-endowed... It's better, really... A lot of women think so.' 1968 in H. & R. Greenwald *Sex Life Lett.* (1974) 79 By the age of ten, my member was already larger than that of a well-endowed adult. 1983 *Maledicta 1982* VI. 157 He says he has been well endowed and you must know this means 'heavily equipped sexually'.

‖ **Wellentheorie** (ve·lĕnte,ori). *Philol.* [G., f. *welle* wave + *theorie* THEORY[1].] The theory that linguistic changes spread like waves over a speech-area and the dialects of adjacent districts resemble each other most; = *wave theory* (*b*) s.v. *WAVE *sb.* 10.
1939 L. H. GRAY *Foundations of Lang.* ii. 42 To account for the spread and relationship of languages, two main hypotheses have been advanced, both, it is true, primarily for the Indo-European group, but, in principle, equally applicable to all others. These are the *pedigree-theory* (*Stammbaumtheorie*) advanced by August Schleicher in 1866 and the *wave-theory* (*Wellentheorie*) proposed by Johannes Schmidt in 1872. 1964 R. H. ROBINS *Gen. Linguistics* viii. 349 The theory of common characteristics resulting from the spreading of linguistic features 'in waves' over adjacent dialects within a family is called the 'Wellentheorie'. 1965 [see *STAMMBAUM]. 1975 *Amer. Speech* 1971 XLVI. 254 Real change..may in some sense be systematic (as in the analysis of style and social variables and the Wellentheorie of linguistic geographers).

well-entre·nched, *ppl. a.*
1929 W. S. CHURCHILL *World Crisis* V. xix. 432 We had a rather restricted but well-entrenched and well-wired position. 1979 P. BUCKLAND *Factory of Grievances* v. 108 A handful of well-entrenched free-traders.

well-equipped, *ppl. a.* (Earlier example.)
a 1854 MILL *Early Draft Autobiogr.* (1961) 120 A well-equipped ship.

Wellerian, *a.* (Earlier example.)
1839 DICKENS *Let.* 25 Jan. (1965) I. 359 Your agreement is—in Wellerian phraseology—gammon.

Wellerism. Add: Usu. *spec.*, a form of comparison in which a familiar saying or proverb is identified, often punningly, with what was said by someone in a specified but humorously inapposite situation. (Earlier and later examples.)
1839 *Boston Morning Post* 9 Jan. 2/2 Wellerisms.—'It does one's heart good to look at you,' as the fox said to the chickens, when he found he couldn't get over the barn-yard wall, to eat them. 1854 *Yankee Notions* III. 142 (*heading*) Phrenological Wellerisms. 1931 A. TAYLOR *Proverb* IV. 219 Wellerisms involving a temporal clause, e.g. '*Much noise and little wool,' said the Devil when he sheared a pig*, are largely used of women. 1959 [see *knock-knock* sb., v., and int. s.v. *KNOCK-.] 1975 *New Society* 25 Dec. 685/1 Sam Weller has joined Dr Spooner and fathered the wellerism: 'Meet you at the corner as one wall said to another,' but the wellerism can also be transformed into a riddle.

well-e·xecuted, *ppl. a.*
1836 [see *WELL-CONCEIVED *ppl. a.*]. 1978 R. LUDLUM *Holcroft Covenant* xxxvii. 425 It was a well-executed trap.

well-expre·ssed, *ppl. a.*
1845 MILL in *Westm. Rev.* XLIII. 331 It is a well-thought and well-expressed explanation. 1943 *Mind* LII. 354 What he [*sc.* Roger Bacon] has to say on *privatio* is sound and well-expressed.

well-farmed, *ppl. a.*
1848 MILL *Pol. Econ.* I. I. xii. 214 The careful cultivation of a well farmed district. 1955 P. C. WINTERTON *Fifty Tumultuous Years* 216 Well-farmed land.

well-fixed, *ppl. a.* Add: **2.** Reasonably affluent, comfortably off. *U.S. colloq.*
1822 A. D. MURPHEY *Let.* 22 Aug. in *Papers* (1914) I. 263 His Brother is well fixed, has one of the best tracts of land in Tennessee, and is growing rich *fast*. 1912 N. M. WOODROW *Sally Salt* 228 I'm well to do, Hilda, well fixed in the world. 1952 J. STEINBECK *East of Eden* ix. 78 He was better than well fixed—he was rich. 1970 J. BLACKBURN *Land of Promise* i. 11 He..had become 'well fixed', as the family colloquialism expressed moderate wealth.

well-formed, *ppl. a.* Add: **b.** Also *spec.*, formed according to stated grammatical rules.
1961 A. G. OETTINGER in *Proc. Symposia Appl. Math.* XII. 104 One important common problem is that of obtaining an algorithm for distinguishing sentences from nonsentences or..well-formed strings from not well-formed strings. 1969 R. A. HALL in *Neuphilologische Mitteilungen* LXX. 204 The oft-repeated claim that all 'well-formed' sentences of a language are derivable from a single syntactic kernel is clearly unfounded. 1980 *Amer. Speech* LV. 90 'Le crabmeat cocktail' is well-formed in that *cocktail* has been borrowed into French as a masculine noun.

c. *Logic.* Applied to any sequence of symbols conforming to the formation rules of a logical system. Esp. as *well-formed formula*.
1936 A. CHURCH in *Amer. Jrnl. Math.* LVIII. 346 We select a particular list of symbols... And we define the word *formula* to mean any finite sequence of symbols out of this list. The terms *well-formed formula*, [etc.],..are then defined by induction. 1954 I. M. COPI *Symbolic Logic* vi. 184 In a logistic system..any formulas which on the intended interpretation do *not* become significant statements are *not* well formed formulas. 1967 *Encycl. Philos.* V. 22/2 A formal language *L* is given by specifying (*a*) a list of *symbols* of *L* and (*b*) a set of *formation rules* for combining these symbols into acceptable, or well-formed, expressions (terms, formulas, sentences) of *L*. 1978 A. G. HAMILTON *Logic for Mathematicians* iii. 53 The use of parentheses in well-formed formulas is precisely given in the definition.

Hence **well-fo·rmedness** (chiefly in *Linguistics*).
1957 *Encycl. Brit.* XIV. 306/1 The condition of well-formedness [in logical formulas]. 1961 N. CHOMSKY in *Word* XVII. 221 Information as to whether the sequence of phones is a properly formed or *grammatical* sentence and if not, in what respect it deviates from well-formedness. 1970 J. P. THORNE in J. Lyons *New Horizons in Linguistics* ix. 186 Between..extremes of well-formedness occur sentences of varying degrees of grammaticalness. 1979 F. KERMODE *Genesis of Secrecy* iii. 64 We depend upon well-formedness..in written language.

well-fo·rmulated, *ppl. a.*
1968 C. G. KUPER *Introd. Theory Superconductivity* i. 2 A well-formulated problem. 1977 P. STREVENS *New Orientations Teaching of Eng.* ii. 23 Well-formulated propositions.

well(-)founded, *ppl. a.* Add: Hence **well-fou·ndedness.**
1920 *Glasgow Herald* 25 Feb. 9/3 There can be no discussion..with regard to..the well-foundedness of the charges. 1970 B. BREWSTER tr. *Althusser & Balibar's Reading Capital* (1975) III. i. 212 Interpretations..whose well-foundedness will, I hope, emerge later in the paper.

well-gowned, *ppl. a.*
1920 E. POUND *Hugh Selwyn Mauberley* 20 Doubtful, somewhat, of the value Of well-gowned approbation Of literary effort. 1975 S. MILLIGAN in C. Allen *Plain Tales from Raj* xv. 158 Always very pale and very beautiful and well-gowned and never moving very fast if they were on horses.

well-head. Add: **3. b.** The structure surmounting an oil- or gas-well. Freq. *attrib.*
1951 *U.S. Rep. 1950* CCCXL. 180 The issue in this case is the power of a state to fix prices at the wellhead on natural gas. 1969 *Times* 16 Dec. (Bahrain Suppl.) p. iv/3 (Advt.), Well-head structures for undersea drilling. 1972 L. M. HARRIS *Introd. Deepwater Floating Drilling Operations* ix. 93 The wellhead is installed on top of, and run with, the conductor pipe. 1983 *Fortune* 13 June 60/2 For each $1 drop in the wellhead price, producers' net income falls only 25 cents.

well-heeled, *ppl. a.*: see *HEELED *ppl. a.* 2 b.

well-hung, *ppl. a.* Add: **1. a.** Chiefly (now always) in *spec.* sense, having large genitals. (Later examples.)
1868 *Index Expurgatorius of Martial* 3 In Rome well-hung youths made a good profit by their amours. 1958 L. DURRELL *Balthazar* v. 103 They love a well-hung diplomat. 1977 D. WILES *Death Flight* xviii. 177 Hey, man... You sure is well hung for a priest!

wellie: see *WELLY.

well-in, *adj. phr.* Add: (Earlier example.)
1845 T. McCOMBIE *Arabin* 241 They had a pretty little farm, and were well in.

Wellington. Add: Also **wellington. 1. a.** *Wellington boot*: in recent use = sense 2 b below.
1971 [see *DELPHINIUM b]. 1980 L. LEWIS *Private Life of Country House* vi. 79 Snowboots..were virtually superseded by rubber wellington boots, which I first saw when I was about eleven [i.e. *c.* 1920].

b. *Wellington hat* (earlier example, applied to a type of lady's hat).
1815 [see *half-dress* s.v. *HALF- II. n].

c. *Wellington chest* (*of drawers*): a tall narrow chest of drawers used for keeping specimens. Occas. ellipt. as *Wellington*.
1953 'N. BLAKE' *Dreadful Hollow* iv. 50 There was the wellington to which Stanford Blick had directed him. Nigel opened one of its drawers. 1960 H. HAYWARD *Antique Coll.* 304/2 *Wellington chest*, a tall narrow chest containing about a dozen drawers which can be locked by a single hinged flap securing all the drawers. 1971 *Country Life* 7 Oct. (Suppl.) 23 (Advt.), A small antique mahogany Wellington chest of drawers measuring only 19½ inches wide, 14 inches deep and 41 inches high. 1982 'J. GASH' *Firefly Gadroon* v. 53 There's a space where I used to have my Wellington chest before I flogged it for bread.

2. b. A waterproof boot usu. reaching the knee, worn in wet or muddy conditions. Usu. *pl.*
1907 *Yesterday's Shopping* (1969) 326/1 Black glazed rubber boots. Ladies' Wellingtons. 1944 D. WELCH *Jrnl.* 25 Jan. (1973) 107 He wore an old thick jersey, and grey flannels tucked into Wellingtons. 1984 *Brian Mills Catal.* Spring & Summer 337/4 Waterproof wellington in PVC.

well-i·ntegrated, *ppl. a.*
1943 *Mind* LII. 127 A well-integrated body of..laws. 1976 B. GIBSON *Birmingham Bombs* ix. 74 They got on with their neighbours and most of them were regarded as well-integrated members of the community.

wellish, *adv.* For *dial.* read *dial.* and *colloq.* and add further examples.
1830 H. COCKBURN *Let.* 30 Dec. (1932) 30 How are you? All wellish here. 1856 D. G. ROSSETTI *Let.* 6 Mar. (1965) I. 293, I fancy it will pay wellish, too. 1934 E. BOWEN *Cat Jumps* 230 '[Do you] know him fearfully well?' 'Wellish,' said Rachel.

well-kempt, *ppl. a.* Delete † *Obs.* and add later examples. Also tidy, well cared for.
1922 JOYCE *Ulysses* 212 A wellkempt head, newbarbered. 1934 P. FLEMING *One's Company* II. xviii. 306 The numerous temples were so well-kempt. 1973 *Observer* (Colour Suppl.) 23 Sept. 43/1 Tall, improbably athletic, handsome, bearded, he looked like a well-kempt prophet.

well-known, *ppl. a.* Add: Hence **well-kno·wnness.**
1961 D. J. BOORSTIN *Image* viii. 162 The star system..puts a premium on well-knownness for its own sake. 1978 J. PEARSON *Façades* xxiii. 401 A celebrity has been defined by Daniel Boorstin as a person who is known for his well-knownness. 1984 *Listener* 15 Mar. 31/2 Telly persons are, as it were, well known for their well-known-ness.

well-languaged, *a.* For '† *Obs.*' read 'Now *arch.*' and add later examples.
1938 *Times Lit. Suppl.* 20 Aug. 548/2 Sir Robert Cordell, whose well-languaged seventeenth-century will..makes interesting reading. 1954 L. HOTSON *First Night of 'Twelfth Night'* i 29 The well-languaged Elizabeth kept helping Lord Grey interpret the gist of it.

well-lit, *ppl. a.*
1866 GEO. ELIOT *Felix Holt* III. xlix. 249 Esther had to seat herself in the..drawing-room, in a well-lit solitude. 1931 W. S. CHURCHILL *World Crisis* VI. xii. 179 Hindenburg, who was waiting in the 'well-lit station hall' at Hanover. 1965 *Motor* 17 July 5/1 Dipped headlights are probably of benefit on less well-lit roads.

well-made, *ppl. a.* Add: **2.** Also *well-made play* [tr. F. *pièce bien faite*], a type of play written according to a prescribed formula and aiming at neatness of plot and dramatic incident rather than profundity of characterization, truth to nature, etc.
[1895 G. B. SHAW in *Sat. Rev.* 2 Nov. 576/1 Then 'The New Magdalen' was a fashionable and well-made piece.] 1910 —— *Brieux* 13 Commercially, the classic play was supplanted by a nuisance which was not a failure: to wit, the 'well made play' of Scribe and his school. The manufacture of well made plays is not an art: it is an industry. 1962 *Listener* 11 Oct. 574/1 The merits and the demerits of a Galsworthian 'well-made' play.

well-meant, *ppl. a.* Add: **c.** Of persons: well-meaning. *rare.*
1849 C. BRONTË *Shirley* I. vii. 153 A well-looked, well-meant, and on the whole, well-dispositioned girl.

well-me·rited, *ppl. a.*
1827 J. S. MILL in *Arch. für Sozialwissensch.* LXII. 456 Reformed patriots whose exertions in the cause of liberty have at length earned the well-merited award of a place. 1899 BELLOC *Moral Alphabet* 48 So he wrote, without stopping, for several days In terms of extreme but well-merited praise. 1946 *Nature* 26 Oct. 576/2 Under his wise guidance a well-merited reputation both in teaching and in research was rapidly built up.

well-mo·dulated, *ppl. a.*
1845 POE in *Broadway Jrnl.* II. 354/2 The rhythm is..well-modulated. 1934 C. LAMBERT *Music Ho!* IV. 235 Mr. Christopher Stone whose well-modulated voice has doubtless given pleasure to millions.

well-mo·tivated, *ppl. a.*
1965 N. CHOMSKY *Aspects of Theory of Syntax* 210 A stronger but rather well-motivated condition is proposed by Postal. 1977 P. STREVENS *New Orientations Teaching of Eng.* v. 59 A theory of language teaching, on the other hand, with well-motivated links to specific areas of linguistics and psychology—*that* might be attractive.

well-natured, *a.* For *Obs.* read *rare* and add: **1. a.** (Later example.)
1921 GALSWORTHY *To Let* II. i. 119 He went as a well-natured dog goes for a walk with its mistress, leaving a choice mutton-bone on the lawn.

well-oiled, *ppl. a.*: see *OILED *ppl. a.* 3.

well-o·rchestrated, *ppl. a.*
1947 A. EINSTEIN *Mus. Romantic Era* xvii. 315 The opinion that has become current with critics and with the public that the *Capriccio* is an unusually well-orchestrated piece is wrong: the Capriccio is a splendid composition for orchestra. 1979 *Wall St. Jrnl.* 20 Dec. 6/3 Why has such a well-orchestrated masking of political repressions been necessary?

well-ordered, *ppl. a.* Add: **2.** *Math.* [tr. G. *wohlgeordnet* (G. Cantor in *Math. Ann.* (1883) XXI. 548, (1898) XLIX. 207).] Of an ordered set: having the property that every non-empty subset of it has a first or least element.
1902 *Amer. Jrnl. Math.* II. 384 The usual notation $\alpha+\beta+\gamma+$..is only suitable for a finite, or at least for a well-ordered set of numbers. 1931 P. DIENES *Taylor Series* iii. 95 All these definitions..extend to every type of transfinite sequence (well-ordered set) of numbers. 1975 R. A. SILVERMAN tr. *Kolmogorov & Fomin's Introd. Real Analysis* i. 23 The set *M* of rational numbers in the interval [0, 1] is ordered but not well-ordered.

well-ordering, *vbl. sb.* Add: In *Math.*, the property of being well-ordered; also, a well-ordered set. Freq. *attrib.*
1941 BIRKHOFF & MACLANE *Survey Mod. Algebra* i. 9 The integers have one further important property, not characteristically algebraic and not shared by other number systems. This is the well-ordering principle. 1963 W. V. QUINE *Set Theory* 145 Well-orderings are notable for their exemplary behavior. 1966 *Math. Rev.* XXXI. 6/1 Only denumerable well-orderings are considered. 1970 A. G. HOWSON *Handbk. Terms Algebra & Analysis* xvi. 80 Zermelo's Well-Ordering Theorem states that if *X* is any set whatsoever, then there exists a well-ordering of *X*.

So **well-o·rder** *v. trans.*, to arrange the elements of (a set) in such an order as to produce a well-ordered set.
1944 *Ann. Math. Stud.* XIII. 117 For a finite set can be well-ordered in obvious fashion and hence must obey the axiom of choice. 1966 *McGraw-Hill Encycl. Sci. & Technol.* XII. 206/1 One considers the collection *Z* of all subsets of *X*, selects a point $x_\alpha = f(z_\alpha)$ from each element z_α of *Z*, and well orders *X* so that [etc.].

well-pa·dded, *ppl. a.* Provided with sufficient padding. Also *transf.* and *fig.*
1933 J. BUCHAN *Prince of Captivity* II. ii. 190 What has become of the nice, easy-going, well-padded people with soft voices and wide smiles. 1945 WYNDHAM LEWIS *Let.* 29 June (1963) 384 However well-padded with dollar-bills you might be. 1976 'R. GORDON' *Doctor on Job* vii. 62 Mr Clapper rocked back in his well-padded chair.

well-plucked, *a.* *colloq.* [PLUCKED *a.*] Plucky, fearless.
1873 M. A. BARKER *Station Amusements in N.Z.* 148 He was a well-plucked one... He told me it took five mortal hours to come the last mile. 1936 M. DE LA ROCHE *Whiteoak Harvest* xviii. 230 'That gray one is a 'andful for you'... 'I'm not afraid.' 'You're wot I call a well-plucked 'un.'

we·ll-point. *Civil Engin.* [WELL *sb.*[1]] One of a system of pipes sunk into the ground around an excavated area in order to lower the water-table. Hence as *v. trans.*, to supply with well-points.
1951 *Engineering* 4 May 536/3 The consulting engineers considered..that lowering of the ground-water level by means of well-points would be the most efficient method of keeping the excavation dry. 1958 J. S. SCOTT *Dict. Civil Engin.* 409 An excavation in sand cannot become quick if effectively wellpointed. 1971 O. H. BULLITT *Search for Sybaris* xvii. 169 The most effective way of dewatering an area..would be by means of a well-point system.

well-pu·blicized, *ppl. a.*
1973 G. TALBOT *Ten Seconds from Now* xii. 165 The Queen's well-publicised love of horses. 1979 *Jrnl. R. Soc. Arts* CXXVII. 645/2 The discussion should be open and well-publicized.

well-reasoned, *ppl. a.* (Earlier example.)
1834 J. S. MILL in *Monthly Repos.* VIII. 593 Lord Brougham..delivered a firm, steady, and well-reasoned opinion.

well-resea·rched, *ppl. a.*
1958 *Times Lit. Suppl.* 15 Aug. 13 (Advt.), This well-researched book has been highly acclaimed. 1966 [see *RESEARCHED *ppl. a.* 2]. 1978 *Jrnl. R. Soc. Med.* LXXI. 697/1 There is a respectable weight of well-researched evidence.

well-re·sted, *ppl. a.*
1890 W. JAMES *Princ. Psychol.* II. xvii. 15 Successive contrast..can be avoided only by carefully fixating with the well-rested eye a point of one's field. 1965 J. A. MICHENER *Source* (1966) 798 At dusk on the afternoon of Tuesday, April 13, the Palmach men roused Ilana and her two well-rested companions.

well-scrubbed, *ppl. a.*
1916 JOYCE *Portrait of Artist* (1969) ii. 64 A well-scrubbed kitchen. 1949 C. FRY *Lady's not for Burning* II. 49 Always fornicate Between clean sheets and spit on a well-scrubbed floor. 1976 'R. GORDON' *Doctor on Job* iii. 18 A small, round, well-scrubbed looking man.

well-shaved, *ppl. a.* = WELL-SHAVEN *ppl. a.*
1940 W. FAULKNER *Hamlet* I. i. 15 He just sat there..well-shaved and clean in his perfectly clean faded shirt.

well-shod, *ppl. a.* (Later example.)
1916 JOYCE *Portrait of Artist* (1969) v. 227 He..passed out, his wellshod feet sounding flatly on the floor.

Wellsian (we·lziăn), *a.* (and *sb.*) Also **Wellsean.** [f. the name of H. G. *Wells* (1866–1946) + -IAN.] Of, pertaining to, or resembling the ideas and writings of H. G. Wells, esp. in his science fiction, social comment, etc. Occas. as *sb.*, a devotee or follower of H. G. Wells.
1912 *Westm. Gaz.* 9 Nov. 4/2 The delightful comments on the Wellsian philosophy. 1916 J. FREEMAN *Moderns* 93 The extension of such a rigid word as morality, until it includes its own contradictions, is typically Wellsean. 1916 G. B. SHAW *Pygmalion* 200 The new-born Wellsian had to find her bearings almost as ridiculously as a baby. 1923 A. HUXLEY *Antic Hay* iv. 55 'Let me put you down for a couple of pairs [of pneumatic trousers].' Mr. Mercaptan shook his head. 'Too Wellsian,' he said. 'Too horribly Utopian.' 1946 R. G. COLLINGWOOD *Idea of Hist.* 252 Some Wellsian machine for looking backwards through time. 1962 E. SNOW *Other Side of River* (1963) lxvi. 503 One Wellsian exhibit showed the long-range plan of a canal and river system which would virtually encircle China. 1977 M. DRABBLE *Ice Age* I. 69 He thought of a Wellsian paradise, a Welwyn Garden City, with neat boxes.

well-si·ted, *ppl. a.*
1925 J. G. MACLEOD in *Oxf. Poetry* 27 Or that old solemn cormorant Who sits like a well-sited statue carved in black. 1979 [see *WELL-CONCEIVED *ppl. a.*].

well-spa·ced, *ppl. a.* Of items that are neither too close nor too far apart from each other.
1939 M. SPRING RICE *Working-Class Wives* iv. 70 She..has four children, well-spaced. 1962 E. SNOW *Other Side of River* (1963) lxxv. 576 Large blocks of well-spaced three-story apartments. 1977 N. SAHGAL *Situation in New Delhi* xiii. 128 Sometimes during a Cabinet meeting he said not more than five well-spaced words of one syllable each.

well(-)spoken, *ppl. a.* **3.** (Later example.)
1963 WODEHOUSE *Stiff Upper Lip, Jeeves* ii. 18 Knowing where I could get a couple of tickets for a well-spoken-of musical.

well-stru·ctured, *ppl. a.*
1974 tr. *Wertheim's Evolution & Revolution* 97 To provide a model of an integrated, well-structured whole in which any conflict could be smoothed down. 1978 *Language* LIV. 205 There is a well-structured account for Korean by Martin 1962.

well(-)sui·ted, *ppl. a.* **1.** [SUIT *v.* 10 b.]
The examples in quots. 1771 and 1837 can equally be analysed as *well* adv. + *suited* rather than as a ppl. adj.
1771, 1837 [see SUIT *v.*[1] 10 b]. 1950 T. S. ELIOT *Cocktail Party* II. 109, I consider That you are exceptionally well-suited to each other.

2. [SUITED *ppl. a.* 2.]
1855 W. WHITMAN *Leaves of Grass* 69 Do you think I could walk pleasantly and well-suited toward annihilation? 1980 D. K. CAMERON *Willie Gavin* xiii. 127 His likeness..shows him as well-suited as any eminent Edwardian, the wide skirt of the jacket falling away from a high button.

well-tanned, *ppl. a.* Add: Now chiefly with sense 'tanned by the sun, sunburnt'.
1867 W. WHITMAN *Leaves of Grass* (ed. 4) XLVII. 89 Preferring..those well-tann'd to those that keep out of the sun. 1892 'MARK TWAIN' *Amer. Claimant* xvi. 164 He had a well-tanned complexion. 1916 JOYCE *Portrait of Artist* (1969) ii. 60 Uncle Charles was a hale old man with a welltanned skin.

well-tempered, *ppl. a.* **4.** (Earlier example.)
1820 tr. *J. N. Forkel's Life John Sebastian Bach* 93 The well-tempered clavichord; or preludes and fugues in all the keys.

well-turned, *ppl. a.* Add: **5.** *well turned-out*: smartly dressed, well-groomed.
1903 [see *CLASSY *a.*]. 1919 E. P. OPPENHEIM *Strange Case Mr. Jocelyn Thew* I. ii. 17 A very distinguished-looking and exceedingly well-turned-out caller. 1976 G. EWART *No Fool* I. 20 Well-turned-out were the waisted women.

well-upho·lstered, *ppl. a.* Having soft and thick upholstery; usu. *transf.* in *colloq.* use, plump, 'well-covered'.
1932 H. R. WAKEFIELD *Ghost Stories* 44 Mrs. C., a handsome well-upholstered matron, had a shrewd Scottish flair for entertainment. 1939 H. HODGE *Cab, Sir?* v. 48 She looks a bit second-hand, though she's well-upholstered. 1964 in Hamblett & Deverson *Generation X* 171, I have a preference for dark girls with warm brown eyes and I like them well-upholstered. 1971 E. LEMARCHAND *Death on Doomsday* ii. 26 Well-upholstered elderly women.

well-weaponed, *ppl. a.* (Later example.)
1917 KIPLING *Diversity of Creatures* 405 To the sound of trumpets shall their seed restore my Cities, Wealthy and well-weaponed.

well-willing, *a.* **a.** (Later example.)
1931 *Times Lit. Suppl.* 20 Aug. 631/1 Not even the most well-willing of critics can escape from its two main flaws.

well-written, *ppl. a.* (Later example.)
1911 G. B. SHAW *Blanco Posnet* Pref. 360 An able and well-written statement of the case.

welly (we·li). Also **wellie.** Abbrev. of *WELLINGTON 2 b. **a.** *colloq.* A wellington boot. Also *Comb.*, as *welly-boot*.
1961 *Guardian* 2 June 9/5 The ground floor we converted back into a hall, for coats and wellies, etc. 1971 J. OSBORNE *West of Suez* I. 42 Huge surgeons who tower over you in green and rubber wellies. 1972 D. HASTON *In High Places* i. 15 We only have welly boots and gym-shoes. 1976 *Printing World* 30 Apr. 8/2 Wellington, who, as we all know, has a boot named after him... The influence of the child is apparent today as I gather it is now widely known as a 'wellie'. 1982 S. RADLEY *Talent for Destruction* iii. 20 Perhaps it wasn't done for a parson to wear welly boots under his cassock.

b. *slang.* A kick, acceleration. Also *fig.*
1977 *Daily Mirror* 10 May 23/1 The girl they call 'Daredevil Divi' gave the car a bit more wellie. In racing language, this meant she was stepping on the accelerator. 1979 *Guardian* 12 Feb. 20/3 The tactic most likely to succeed in the conditions was the long welly upfield. 1983 D. GETHIN *Wyatt & Moresby Legacy* xxv. 174 'When I say go, give it some welly... Go.'.. Explosions sounded.

Hence **welly** *v. trans.*, to kick or trip up with one's foot. *slang.*
1966 F. SHAW et al. *Lern Yerself Scouse* 48 *Eee wuz wellied*, he was kicked. 1980 D. MORRIS *Tribal Words* (typescript), *Wellied*, heavily tackled and brought down. To *welly* is to make a savage tackle, boot-first... 'He must expect to be wellied a few times' is a typical expression.

wels (wels, vels). [a. Ger. *wels.*] = *SHEAT-FISH.
1880 A. C. L. G. GÜNTHER *Introd. Study of Fishes* 565 The species which has given the name to the whole family is the 'Wels' of the Germans, *Silurus glanis*. 1905 D. S. JORDAN *Guide to Study of Fishes* II. ix. 182 The huge sheatfish, or wels,..next to the sturgeon, is the largest river fish in Europe. 1931 J. R. NORMAN *Hist. Fishes* vii. 138 The Wels or Glanis..of Europe normally feeds on fishes, frogs, and crustaceans. 1969 A. WHEELER *Fishes Brit. Isles & N.W. Europe* 221/2 Wels have been introduced in a number of private lakes in southern England.

Welsbach (ve·lzbaχ). The name of Carl Auer Freiherr von *Welsbach* (1858–1929), Austrian chemist and engineer, used *attrib.* to designate the gas mantle (MANTLE *sb.* 5 g), invented by him, and the lamps employing it.
A proprietary name in the U.S.
1887 *Pall Mall Gaz.* 18 Mar. 12/1 The Welsbach burner would produce a purer light..with two cubic feet of gas per hour. 1901 *Daily Colonist* (Victoria, B.C.) 1 Oct. 5/4 The Victoria Gas Co., Ltd., are now installing complete Welsbach Lamps free of cost, charging the nominal sum of 5 cents per lamp per month for mantle renewal. 1912 A. BENNETT *Matador* 308 The Welsbach incandescent mantles on the chandelier saved thirty per cent in gas-bills while increasing the light by fifty per cent. 1964 *Official Gaz.* (U.S. Patent Office) 21 July TM 119/2 The Welsbach Corporation, Philadelphia, Pa... *Welsbach*. For gas lamps..and gas mantles. 1980 *Sci. Amer.* May 123/2 The Welsbach mantle, a type of gas burner widely used for home lighting around the turn of the century, proved to be a rich source of N rays.

Welsh, *a.* and *sb.* Add: *adj.* **2. b. Welsh flannel** (earlier example); **Welsh mutton** (earlier example).
1771 SMOLLETT *Hump. Cl.* I. 126 These sums she has more than doubled, by..dealing in cheese and Welsh flannel, the produce of his flocks and dairy. *Ibid.* II. 228, I dined upon a delicate leg of Velsh mutton and cully-flower.

c. Welsh Black, a black-coated ox or cow of a breed originally developed in north Wales, now usually kept for both meat and milk production; **Welsh cattle** (earlier example); **Welsh hound,** a dog similar to an English foxhound but wire-haired; **Welsh mountain (sheep),** a small, hardy sheep of a breed developed in high regions of Wales; **Welsh terrier,** a stocky, rough-coated, usually black and tan terrier with a square muzzle and drop ears, belonging to a breed originally developed in Wales to hunt vermin.
1919 K. J. J. MACKENZIE *Cattle* xi. 149 Welsh Black Cattle. Some 20 years ago there were two distinct types of black cattle in Wales. 1953 A. FRASER *Beef Cattle Husbandry* ix. 134 The modern Welsh Black is the outcome of two types or breeds. 1977 *S. Wales Guardian* 27 Oct. 2/3 (Advt.), Welsh Black cow, 5th calver, with Cross-Friesian Steer calf at heel. 1747 H. GLASSE *Art of Cookery* xiii. 129 Take the leg of a fat but small Beef, the Fat Scotch or Welch Cattle is best. 1893 R. B. LEE *Mod. Dogs (Sporting Division)* iv. 86 The harrier is oftener coarser in his coat than the foxhound, which may be ascribed to crossing with a rough Welsh hound that I believe is still to be found in some parts of the principality. 1930 J. D. D. EVANS in C. Frederick et al. *Foxhunting* xxxi. 306 The Welsh hound is probably in some degree of later origin. 1973 *Country Life* 27 Dec. 2173/2 The three Welsh hounds he is shown holding have long, hairy coats, whitish-grey in colour. 1899 W. J. MALDEN *Sheep Raising & Shepherding* ii. 5 Such breeds as the Welsh Mountain..have soft short wool. 1960 [see *EXMOOR]. 1979 *Country Life* 7 June 1769/3 A black version of the Welsh Mountain sheep. 1885 *Kennel Chron. & Pedigree Register* VI. 161/1 Welsh terriers.—*Dog*: 2nd, W. C. Whiskin (Welsh Dick). 1894 R. B. LEE *Mod. Dogs (Terriers)* x. 231 The dog of which I write as a Welsh terrier was unknown until some eight years..ago. Then he appeared in some of our shows; he was given a place in the Stud Book; a club was formed in 1886 to look after his welfare. 1950 A. C. SMITH *Dogs since 1900* xi. 193 In 1942 Welsh Terriers came into prominence in the United States. 1980 E. LEATHER *Duveen Letter* i. 13 The Welsh terrier removed himself from the tapestry-covered Gainsborough chair.

d. Welsh cake, a kind of individual spicy cake made in Wales with currants and ginger; **welshcomb** *v. trans.*, to comb one's hair by using one's thumb and fingers instead of a comb; **Welsh dragon,** a heraldic dragon as the emblem of Wales; also *fig.*; **Welsh Nationalist,** someone wanting home rule for Wales; *spec.* a member of the Welsh Nationalist Party; **Welsh Office,** an administrative department of the British Government with responsibility for Welsh affairs; the building where this is housed; **Welsh Wales** *colloq.*, the parts of Wales where Welsh culture is especially strong or which are most distinctively Welsh; **Welsh wizard** or **Wizard,** a nickname for David Lloyd George (1863–1945), Welsh politician and British prime minister.
1932 DYLAN THOMAS *Sel. Lett.* (1966) 6 Mother has made Welshcakes. 1975 B. MEYRICK *Behind Light* xiv. 184 Sandwiches, spice buns, Welsh cakes and bread and margarine. 1980 B. FREEMAN *First catch your Peacock* viii. 163 *Welsh Cakes* (*Pice ar y maen*). There are two ways of making these traditional little spicy cakes. The most usual is on the bakestone or griddle, which produces them in a rather dry, biscuity form... The other method is to make them..in a Dutch oven, and this produces cakes which are firm on the outside, soft and melting within. 1922 JOYCE *Ulysses* 125 He took off his silk hat and..welshcombed his hair with raking fingers. 1971 'A. BURGESS' *MF* x. 111 Dressed and welshcombed, I pocketed my luggage and went downstairs. 1857 C. M. YONGE *Dynevor Terrace* I. xxi. 349 If she had let the Welsh dragon show his teeth in style, he would only have had to make unpleasant apologies. 1909 A. C. FOX-DAVIES *Compl. Guide Heraldry* xiii. 225 His Majesty the King has recently added the Welsh dragon differenced by a label of three points argent as an additional badge to the achievement of His Royal Highness the Prince of Wales. 1980 *Times* 16 Dec. 3/6 They..see the Welsh dragon flying over public buildings in the principality. 1891 *Dod's Parl. Compan. 1890–91* 246 George, David Lloyd... 'A Welsh Nationalist', supporting 'Home Rule', 'Temperance', 'Disestablishment', and other items in the programme of the Advanced Liberal Party. 1925 *North Wales Observer* 13 Aug. 5/3 The proposals of the new Welsh Nationalist Party were calmly received..at the Baptist Church, Pwllheli, on Thursday. 1937 W. H. JONES *What is happening in Wales?* iii. 9 What makes the Welsh Nationalist movement essentially a literary movement? 1966 M. WOODHOUSE *Tree Frog* xviii. 131 'Patriotism dictates,' he said. 'Don't give me that,' I said. 'I'm a Welsh Nationalist.' 1972 *Guardian* 19 June 24/8 Welsh Nationalists intend to set up a mobile pirate radio station. [1964 *Times* 20 Nov. 6/7 Mr. Harold Wilson, Prime Minister.., in a statement defining the responsibilities of the Secretary of State for Wales, said:— .. The interests of Wales are now represented in the Cabinet by the Secretary of State. .. He will have a Welsh office in Cardiff..and a small ministerial office in London.] *Ibid.* 22 Dec. 10/4 Mr. Griffiths, Secretary of State for Wales, yesterday opened the first permanent home of the Welsh Office in London. 1976 *S. Wales Echo* 26 Nov., Demonstrations by angry parents outside the Welsh Office now seem certain after South Glamorgan County Council's controversial decision to turn a Cardiff high school into a Welsh-speaking secondary. 1976 *Flintshire Leader* 10 Dec. 1/2 The call for a crossing did not meet Welsh Office conditions because there were not enough houses or people living in the village. 1977 *Western Mail* (Cardiff)

5 Mar. 3/2 The council has had to go to the Welsh Office for money for industrial development. 1954 G. DANIEL *Welcome Death* x. 125 'The Vale of Glamorgan is legally Wales, isn't it, although no one speaks any Welsh here?' 'Quite right... This is Wales, if not Welsh Wales'. 1971 D. AYERST *Guardian* xxi. 290 In the hills behind the coastal resorts, Welsh Wales with its roots in the chapels had an intense cultural life of its own. 1983 A. BEEVOR *Faustian Pact.* i. 8 The sight of the sheep-cropped hills... There was Welsh Wales. 1917 M. MACDONAGH *Diary* 9 July in *London during Gt. War.* (1935) III. iv. 203 Is not 'L.G.' known as the 'Welsh Wizard'. 1922 *National Rev.* July 652 We are told..that several of His Majesty's Ministers are acutely uncomfortable in the impasse into which they have allowed themselves to be manœuvred by the Welsh Wizard. 1976 W. J. BURLEY *Wycliffe & Schoolgirls* vii. 126 There had never been a real statesman since the little Welsh wizard.

welsh, *v.* Restrict *Racing* to sense in Dict., for † **welch** read **welch** and add: **2.** *intr.* Const. *on*. To fail to carry out one's promise to (a person); to fail to keep (an obligation).
1932 H. CRANE *Let.* ?Jan. (1965) 395, I really can't welsh on Eyler Simpson (who is equally responsible, since he signed the lease with me). 1971 H. WOUK *Winds of War* i. 4 The real shadow on this couple was that Commander Henry thought Rhoda had welshed on their courtship understanding. 1974 *Socialist Worker* 2 Nov. 5/1 When the brothers were captured on a bank raid, the British government welched on them, dropped them like a hot penny. 1978 *Lancashire Life* Apr. 73/4 Very few people welsh on paying their taxi fare. 1982 T. KENEALLY *Schindler's Ark* v. 72 Across his desk..had crossed copies of angry SS memoranda addressed to army officials and complaining that the army was welching on its arrangement.

Welshie, var. *WELSHY *sb.*

Welshwoman. (Later example.)
1964 C. MACKENZIE *My Life & Times* III. 282 Nurse Williams..was a plump little Welshwoman from Rhyl.

Welshy (we·lʃi), *sb.* *slang.* Also **Welshie.** [f. WELSH *a.* and *sb.* + -Y[6].] A Welshman or Welshwoman.
1951 E. COXHEAD *One Green Bottle* ii. 46 You'd think I was a Welshy by my name, Gwen Evans, but I'm a proper Cockney. 1978 E. MALPASS *Wind brings up Rain* ii. 23 Alice was still staring at her unblinking. 'You deep little Welshie', she said.

welt, *sb.*[1] Add: **7. welt pocket,** a slot pocket having a welt on the lower edge that extends upward to cover the slit.
1932 D. C. MINTER *Mod. Needlecraft* 134/1 Welt pocket... Mark opening with tack 4½ inches long. Cut welt 5 inches long and width required. 1978 *Detroit Free Press* 5 Mar. A 16/2 (Advt.), Single-breasted styling with self-belt, epaulets, welt pockets.

welt (welt), *sb.*[3] *dial.* [Origin unknown.] The practice by which some members of a gang of dockers take an unauthorized break while the rest work, and so turn and turn about. Hence as *v. intr.*, to take a break thus; **we·lting** *vbl. sb.*[3]
1964 *Guardian* 7 Dec. 4/1 It is 1 30 p.m...and the afternoon 'welt' is settling in. *Ibid.*, 'Welting'—the practice whereby part of a gang takes an unofficial break while the rest continues working—is firmly entrenched in Liverpool. 1965 *Wall St. Jrnl.* 13 Jan. 11 A visitor dropping into almost any waterfront cafe beside the forest of ship masts rising along the River Mersey in Liverpool will quickly discover one reason. He'll find stevedores 'welting' or enjoying their stout and porter at the bar on company time. Work crews allow members to 'welt' on a share-and-share alike basis. 1967 *Economist* 4 Nov. 490/1 Ending the welt would make possible higher basic rates of pay. 1977 *Guardian* 4 Sept. 2/2 In some docks..there is still tension because of managers' attempts..to crack down on the 'welt'.

‖ **Weltanschauung** (veltanʃau·uŋ). Also with small initial. Pl. **-ungen.** [Ger., f. *welt* WORLD *sb.* + *anschauung* perception.] A particular philosophy or view of life; a concept of the world held by an individual or a group; = *world-view* s.v. WORLD *sb.* 25.
1868 W. JAMES *Let.* in R. B. Perry *Tht. & Char. W. James* (1935) I. viii. 160, I remember your saying..that the characteristic of the Greek '*Weltanschauung*' was its optimism. 1906 *Nature* 10 May 26/2 In 1863 Haeckel entered the lists as the champion of the evolutionist 'Weltanschauung'. 1917 A. S. PRINGLE-PATTISON *Idea of God* iv. 69 The intimate appreciation of living experience forms the basis of the whole *Weltanschauung* which he [*sc.* Bergson] offers us. 1934 M. BODKIN *Archetypal Patterns in Poetry* 326 A man's philosophy..is his *Weltanschauung*—the individual vision, or perspective of reality. 1938 E. QUINN *Mission of Austria* iii. 30 Both Catholicism and National Socialism are *Weltanschauungen*. 1952 G. SARTON *Hist. Sci.* I. iv. 121 The creation of that astrological *Weltanschauung* which dominated late ancient and medieval thought and is not yet extinct today proves the survival through the Dark Interlude of some astronomical ideas of immemorial antiquity. 1958 J. JOCZ *Theol. of Election* i. 11 The impact of the Christian *Weltanschauung* upon the Jewish mind. 1972 *Science* 2 June 988/1 The main reason why evolutionism..made such slow progress is that it was the replacement of one entire *weltanschauung* by a different one. 1978 N. JARDINE in Hookway & Pettit *Action & Interpretation* 124 Speakers of different cultures, having different *Weltanschauungen*, ideologies, interests, paradigms, etc.

‖ **Weltansicht** (ve·ltanziχᵊt). *rare.* [Ger., f. as prec. + *ansicht* view.] A world view.
1892 W. JAMES *Let.* 19 Sept. (1920) I. 324, I realized how exactly a philosophic *Weltansicht* resembles that from the top of a mountain. 1977 *Archivum Linguisticum* VIII. 49 The following statement that every language represents a 'Weltansicht' (world view) of its own also proves the significance of the non-uniformity of languages for von Humboldt.

‖ **Weltbild** (ve·ltbilt). [Ger., f. as prec. + *bild* picture.] A view of life.
1934 L. MUMFORD in W. Frank et al. *Amer. & Alfred Stieglitz* ii. 47 Stieglitz..helped restore those values that had been left out of the narrow *Weltbild* of his contemporaries. 1963 J. LYONS *Structural Semantics* iii. 40 The suggestion that we are influenced in our 'Weltbild' by the language we have been brought up to speak is open to different interpretations. 1973 I. ROBINSON *Survival of Eng.* iii. 82 The automatic wage-increase is so much taken for granted as a necessary part of the world that it is perhaps more *Weltbild* than what is usually thought of as language. 1983 *Bull. Amer. Acad. Arts & Sci.* Oct. 35 Mahler's *Weltbild* bore the marks not only of intellectual populism but of his class and ethnic origins.

welter, *sb.*[3] **1. b.** *Welter Stake* (earlier example).
1820 *Sporting Mag.* VI. 2/1 A capital gentleman jockey for a *Welter* stake.

welter weight. 1. b. (Earlier example.)
1832 *Q. Rev.* XLVII. 240 'Out upon this great carcass of mine,' says one of the best of the welter-weights.

‖ **Weltliteratur** (ve·lt,litĕrătūᵊ:ɹ). Also **-litteratur.** [Ger., f. *welt* WORLD *sb.* + *literatur* LITERATURE.] A literature of all nations and peoples; a universal literature.
[1827 GOETHE *Gespräche mit Eckermann* (1836) I. 325 National-Literatur will jetzt nicht viel sagen, die Epoche der Welt-Literatur ist an der Zeit.] 1913 E. POUND *Let.* 7 Nov. (1971) 24 Until 'we' accept what I've been insisting on for a decade, i.e., a universal standard which pays no attention to time or country—a Weltlitteratur standard—there is no hope. 1962 *Listener* 6 Sept. 358/2 It gives a new meaning to Goethe's dream of *Weltliteratur*. 1974 *Times Lit. Suppl.* 17 May 526/2 A truism, that Pound and Eliot were rebels against American provincialism and proponents of a *Weltliteratur*.

‖ **Weltpolitik** (ve·ltpǫliti:k). [Ger., f. as prec. + *politik* politics.] International politics; world affairs from a political standpoint; a particular country's policy towards the world at large.
1903 J. BAILEY *Diary* 19 Feb. (1935) 90 Lord George Hamilton said..that the one thing needed to make the U.S.A. friends with us was their going into *Welt Politik*. 1905 D. M. WALLACE *Russia* II. xxxviii. 428 Never, perhaps, has the construction of a single line [*sc.* the Trans-Siberian Railway] produced such deep and lasting changes in the sphere of *Weltpolitik*. 1941 *N. & Q.* 26 July 43/2 The ineffectiveness of the Kruger telegram opened German eyes to the need of a fleet if Germany was to play a successful, or even a real, part in *Weltpolitik*. 1979 G. ST. AUBYN *Edward VII* vii. 363 The Triple Entente was a direct reaction to Germany's *Weltpolitik*.

‖ **Weltschmerz** (ve·ltʃmĕᵊɹts). Also **welt-schmerz.** [Ger., f. as prec. + *schmerz* pain.] A weary or pessimistic feeling about life; an apathetic or vaguely yearning attitude.
1875 J. A. SYMONDS *Renaissance in Italy* I. iv 232 The Weltschmerz did not exist for the men of the Renaissance. 1896 W. CALDWELL *Schopenhauer's System* 523 His philosophy is a study of the *Weltschmerz* that we all feel at times. 1923 A. HUXLEY *Let.* 2 Sept. (1969) 218, I have also been having a..jaundice lying on my liver, which reduced me to a fearful state of weltschmerz and incapacity to do anything. 1935 C. ISHERWOOD *Mr. Norris changes Trains* ix. 150 'What's the matter?' I asked. 'Things in general... The state of this wicked world. A touch of *Weltschmerz*, that's all.' 1947 [see *ACCIDIA]. 1960 C. GEERTZ *Religion of Java* vi. 75 He said..the young good ones..die early, as a kind of reward, for it is a good thing to be dead. He spoke happily, not in any *Weltschmerz* mood. 1965 W. GOLDING *Hot Gates* 136 The sadness, the *weltschmerz* resulting from the constant movement of the ship. 1981 J. D. MACDONALD *Free Fall in Crimson* i. 6 'It is like *weltschmerz*.' 'Which, as you have so often told me, is homesickness for a place you have never seen.'

wen[1]. Add: **3.** Comb.: **wen-man** *nonce-wd.*, a city-dweller.
1937 AUDEN *Lett. from Iceland* viii. 102 The mountain-snob is a Wordsworthian fruit... He calls all those who live in cities wen-men.

wen, repr. a pronunc. of WHEN *adv.* (*conj.*, *sb.*) in dialect or in uneducated speech.
1893 H. A. SHANDS *Some Peculiarities of Speech in Mississippi* 67 *Wen*, sometimes used by illiterate whites and negroes for *when*. 1901 M. FRANKLIN *My Brilliant Career* iii. 16 It puts me in mind ev the time wen the black fellers made the gins do all the work. 1952 [see *QUEEN *sb.* 5 e]. 1979 *Amer. Speech* LIV. 67 W'en you see the fire come from the brimstone..this earth ain' gon' be burnin'.

wend, *v.*[1] **15.** (Later example without poss. pron.)
1883 [see FAIRWAY].

Wendic, *a.* and *sb.* (Earlier examples.)
1848 *Rep. Brit. Assoc. Adv. Sci. 1847* 267 The Old Slavonic of the Bible and of Nestor, the Russian, Servian, Croatic, and Wendic. 1856 MAX MÜLLER in *Oxford Ess.* i. 14 We know nothing of the Arian race, before it was broken up into different nationalities, such as Indian, German, Greek, Roman, Windic, Teutonic, and Celtic.

wendigo, var. *WINDIGO.

Wendy house. Also with small initial or hyphen. [Named after the small house built around Wendy in J. M. Barrie's play *Peter Pan* (1904).] A small house-like structure for children to play in.
1949 M. ATKINSON *Junior School Community* 11 Wendy house—made by a joiner: two large pieces of plywood hinged together. 1957 *Listener* 9 May 743 There is a Wendy-house in the corner [of the class room]. 1971 *Where* Dec. 356/1 They have performed wonders in getting the bus and re-equipping it with ladders, a wendy house and even a telephone. 1977 J. McCLURE *Sunday Hangman* viii. 80 The rocking horse was legless.., the pedal car was a write-off, and the Wendy house had been trampled flat.

wenge (we·ŋge). [Local name in Zaïre.] The dark brown timber of *Millettia laurentii*, a tree of the family Leguminosæ found in Central Africa.
1963 *House & Garden* Feb. 60 (*caption*) Seating unit series..teak and wenge frame, latex foam cushions. 1972 *Handbk. Hardwoods* (Building Res. Establishment) (ed. 2) 164 The timber wengé..from Zaire is generally similar in appearance and properties to panga panga.

‖ **wên jên** (wən ʒən, wən rən). Also with hyphen. [Chinese *wénrén* man of letters, f. *wén* writing + *rén* (*jên* in Wade-Giles) man.] Chinese men of letters.
1958 W. WILLETTS *Chinese Art* II. vii. 509 From early Ming times until quite recently, the history of Chinese painting has been written by that highly articulate but alarmingly unanimous body of people, the *wên jên* or *literati*. 1970 *Oxf. Compan. Art* 232/2 Though *wên-jên*, many of them were also professional painters.

‖ **wen li** (wən li). Also **wenli** and with capital initial. [Chinese *wén lǐ* grammar, literary style, f. *wén* writing + *lǐ* texture, reason.] = *WEN-YEN.
The synonymy is based on a misconception of the Chinese meaning, and does not exist in that language.
1887 *Chinese Times* 11 June 502/1 The bishop's style, call it *Wen li* or *Mandarin*, is admirably clear and idiomatic. 1917 S. COULING *Encycl. Sinica* 597/2 The term *Wên li* is now in constant use especially among foreigners, to denote the Chinese literary style, which differs in degrees of conciseness or obscurity, and hence is sometimes divided into 'high' and 'low' *Wên li*. 1972 E. A. NIDA *Bk. of Thousand Tongues* 70/1 Wenli was a written language which could be used throughout the whole of China. 1977 C. F. & F. M. VOEGELIN *Classification & Index World's Lang.* 114 The descendant of an earlier form of Chinese known as Wen Yen or Wen Li continues to be used by all educated Chinese..for special purposes.

Wenlock. Add: In mod. use, the name of the middle of three divisions of the Silurian, lying below the Ludlovian and above the Valentian (Llandoverian); used *attrib.* and *absol.* (Further examples.)
1946 [see *LUDLOVIAN *a.*]. 1969 BENNISON & WRIGHT *Geol. Hist. Brit. Isles* vi. 116 A great deal of relatively recent work has been done on the limestone and shale shelf-sea facies of Wenlock and Ludlow age in the Welsh Borders. 1979 R. ANDERTON et al. *Dynamic Stratigr. Brit. Isles* vii. 96/1 Turbidity currents deposited sands in a separate turbidite zone during the Wenlock and lower Ludlow.

Wenlockian, *a.* Add: Also, of or belonging to the Wenlock series. Freq. *absol.*
1946 [see *LUDLOVIAN *a.*]. 1969 BENNISON & WRIGHT *Geol. Hist. Brit. Isles* vi. 124 The thickness of the Wenlockian strata may amount to about 25,000 feet. 1974 *Encycl. Brit. Macropædia* XVI. 774/2 The Llandoverian..saw the beginning of a sharp distinction between basin graptolitic shale facies and calcareous and..sandy shelf facies. This distinction was well marked in the Wenlockian.

Wensleydale. Add: **b.** Also, a white cheese (see quot. 1963).
1963 A. L. SIMON *Guide Good Food & Wines* 648/2 The best-known variety of Wensleydale cheese, cylindrical in shape, like Stilton, but of smaller dimensions, which grows 'blue' when ripe, like Stilton... The other sort of Wensleydale cheese is a flat-shaped, white cheese which is eaten fresh and does not generally go blue. 1985 D. CLARK *Performance* iv. 105 Brawn and Wensleydale cheese sandwiches.

‖ **wen-yen** (wən yen). Also **wenyan, wenyen,** and with capital initial. [Chinese *wényán*, f. *wén* writing + *yán* speech, words.] The traditional literary language or style of China, superseded in the twentieth century by *PAI-HUA.

1936 N. WALES in E. Snow *Living China* 336 Until 1917 there existed in..stalemate three fairly distinct strata of literature: (1) the ancient cult of the *literati* in the dead *wen-yen* classical written language, ..(2) the healthy parvenu *pai-hua*, 'plain speech', literature of the people in the spoken language, and..(3) the story-tellers' literature in the provincial dialects. **1964** *Anthropol. Linguistics* Mar. 31 Many words which require two characters in Han Chinese can be written with one character in Wenyen. **1968** [see *PAI-HUA]. **1969** *Language* XLV. 690 One is the Classical Chinese, *wényán*, which has been used from antiquity up to recent years. **1980** *Times Lit. Suppl.* 27 June 725/1 He has been engaged..in an immense study of the ancient Chinese classics written in the elegant but archaic *wenyen* Chinese favoured by old-fashioned scholars—a language almost as remote from present-day speech as Latin from the modern European vernaculars.

Werdnig–Hoffmann (vɜ·ɹdnig). *Path.* The names of Guido *Werdnig*, 19th-century Austrian neurologist, and Johann *Hoffmann* (*HOFFMANN 3), who described the disease in 1890 and 1893 respectively, used in the possessive and *attrib.* to designate a fatal familial disease that is present at birth or develops soon afterwards and is characterized by muscular atrophy, paralysis, and loss of sucking ability.

1903 *Trans. Clin. Soc.* XXXVI. 226 (*heading*) Three cases of family progressive spinal muscular atrophy (Werdnig-Hoffman [*sic*] type). **1920** *Brain* XLIII. 170 The case was scarcely one of amyotonia congenita, but rather was related to Werdnig-Hoffmann's progressive muscular atrophy, in spite of there being no obvious element of heredity. **1978** *Arch. Dis. Childhood* LIII. 921/1 Werdnig-Hoffmann disease—the acute severe infantile form of spinal muscular atrophy—often presents in the neonatal period with profound weakness.

were-. Add: **we·re-jaguar,** in Olmec mythology, a creature partly human and partly feline.

1967 L. DEUEL *Conquistadors without Swords* xviii. 235 Today,..more than 400 years after the Spanish Conquest and 2,000..years since its origin, the were-jaguar, the *nawal*, is still invoked to frighten children who will not go to sleep. **1967** E. P. BENSON *Maya World* ii. 24 Olmec art is full of creatures who are part human and part feline... Often they are a combination of human infant and jaguar. They are called 'were-jaguars'. **1979** E. ABRAMS tr. *H. Stierlin's Precolombian Civilizations* 68 This werejaguar figure tenoned into the wall of the pyramid at Chavín.

werewolf. Add: **2*.** A member of a right-wing paramilitary German underground resistance movement.

1945 in *Amer. Speech* (1949) XXIV. 289/2 It boasted that..underground killers—'Werewolves'—had carried out the sentence. **1946** E. LINKLATER *Private Angelo* xxi. 266 A company of Free Austrians who..handed him over to a ridiculous little party of people who called themselves Werewolves. **1950** C. MACINNES *To Victors the Spoils* I. 111 Isn't it going to be dangerous..? What about the Gestapo and the werewolves? **1982** C. THOMAS *Jade Tiger* 48 The subject matter of the interrogation—local conditions, Werewolf units, SS and Gestapo individuals' whereabouts.

Werner (wɜ·ɹnəɹ, ‖ vɛ̄·rnər). *Path.* [The name of Carl W. O. *Werner* (b. 1879), German physician, who described the syndrome in 1904.] *Werner's syndrome*: a rare hereditary syndrome whose symptoms include short stature, endocrine and vascular disorders, and premature ageing and death.

1934 OPPENHEIMER & KUGEL in *Trans. Assoc. Amer. Physicians* XLIX. 359 After careful consideration we have selected the patronymic name, Werner's syndrome, rather than Rothmund's syndrome, for on reading Rothmund's original paper (1868)..we are convinced that he described a quite different condition. **1962** A. SORSBY in A. Pirie *Lens Metabolism Rel. Cataract* 298 The association of cataract with such affections as..Werner's syndrome has been known for many years. **1980** *Practitioner* Nov. 1170/2 Rapid whitening of scalp hair associated with rapid ageing of the face..are features reported in Werner's syndrome, a heredo-familial disease in young adults who age rapidly.

Wernicke (v-, wɜ·ɹnikə). *Path.* The name of Karl *Wernicke* (1848–1905), German neurologist, used in the possessive to designate: **a.** A neurological disorder in which there is an inability to understand speech and, usually, to speak sensibly, caused by a lesion of *Wernicke's area*, an area of the cerebral cortex comprising parts of the temporal and parietal lobes.

1887 VICKERY & KNAPP tr. *Strümpell's Text-bk. Med.* 679 The word, when it is heard, may fail to call up the appropriate mental image. Kussmaul has given this condition the name of word deafness (Wernicke's sensory aphasia). The patient is not really deaf, for he hears everything, but he no longer understands what he hears, and has forgotten what the words signify. **1907** *Practitioner* Oct. 545 In the Aphasia of Broca..the cases..closely resemble those of Wernicke's aphasia, with the difference that, in Broca's aphasia, the patient cannot speak. **1908** A. GORDON *Dis. Nervous Syst.* vii. 118 Pierre Marie..holds that aphasia..is caused by a lesion in the lenticular nucleus and in Wernicke's zone; the latter comprises the following portions: supra-marginal gyrus, angular gyrus, the posterior portions of the first two temporal convolutions. **1965** [see *LOGORRHŒA, LOGORRHEA]. **1976** *New Yorker* 15 Nov. 152/2 There are two areas of the cortex that have been shown to be directly involved in speaking. Those areas—known since the late nineteenth century as Broca's area and Wernicke's area—are on the side of the brain (usually the left) that is dominant for speech. **1979** *Sci. Amer.* Sept. 161/1 In Wernicke's aphasia speech is phonetically and even grammatically normal, but it is semantically deviant.

b. An encephalopathy caused by vitamin B_1 deficiency and characterized by mental confusion and uncontrolled movements, esp. of the eyes. So *Wernicke–Korsakoff* [see *KORSAKOFF], applied to Wernicke's syndrome and Korsakoff's syndrome when both are present in an individual.

1910 E. E. SOUTHARD in Osler & McCrae *Syst. Med.* VII. xiii. 631 (*heading*) Hemorrhagic superior polioencephalitis (Wernicke's disease). *Ibid.*, The non-alcoholic and the alcoholic forms of Wernicke's disease are considered. **1939** *Jrnl. Path. & Bacteriol.* XLVIII. 259 We suggest therefore that, as in chronic alcoholism so in pregnancy, B_1 deficiency may play a part in producing Wernicke's encephalopathy as well as polyneuritis. **1966** *Trans. Amer. Neurol. Assoc.* XCI. 31 The Wernicke-Korsakoff syndrome is..both a clinical and a pathological entity. **1978** *Sci. Amer.* Oct. 76/3 The Wernicke-Korsakoff syndrome is a neurological disorder that begins with an acute phase characterized by palsy and poor muscular coordination; with treatment the acute phase gives way to a chronic phase, Korsakoff's psychosis, characterized by severe amnesia.

werrit, *v.* Add: Also **we·rriting** *ppl. a.*

1808 E. WEETON *Let.* 5 Oct. in *Jrnl. of Governess* (1969) I. 111, I was laughed at, or found I had displeased. I had a most werreting life of it. **1865** [in Dict.].

‖ **wertfrei** (vē°·ɹtfrəi), *a.* [Ger., f. *wert* value, WORTH *sb.*[1] + *frei* FREE *a.*] Free of value-judgements; morally neutral. Hence **we·rtfreiheit** (also with capital initial) [-HOOD], the quality of being *wertfrei*.

1909 W. M. URBAN *Valuation* xiv. 422 The more neutral or '*wertfrei*' judgments of science. **1944** H. A. HODGES *Wilhelm Dilthey* v. 80 It is generally recognized that the natural sciences have no interest in judgments of value. Their *Wertfreiheit* is one of their most treasured attributes. **1964** ROUSSEAS & FARGANIS in I. L. Horowitz *New Sociol.* 289 Max Weber distinguishes between science as being *wertfrei* and *wertlos*. *Wertfrei* is defined as being free from prevailing passion and prejudice. **1975** *Times Lit. Suppl.* 25 July 848/1 What specially distinguishes Ontology, American Style, however, is that it is far more *wertfrei*, uncommitted, cool, detached and technical. c **1978** C. R. TAME *Against New Mercantilism* 8 No one would abandon the heritage of *wertfreiheit* in economics and the social sciences.

Wesak, var. *VESAK.

‖ **Wesen** (vē·zən). *rare.* [Ger.] **a.** A person's nature (as shown in characteristic behaviour).

1854–5 GEO. ELIOT in J. W. Cross *George Eliot's Life* (1885) I. vi. 353 Fräulein Solmar is..probably between fifty and sixty, but of that agreeable *Wesen* which is so free from anything startling in person or manner. **1884** MRS. H. WARD *Miss Bretherton* i. 10 And then her *Wesen* is so attractive; she is such a frank, unspoilt, good-hearted creature.

b. The distinctive nature or essence of anything.

1959 *Listener* 22 Oct. 689/2, I believe myself that it is only in the totality of its historic manifestations that Christianity can be understood, and that so long as it does survive, its *Wesen*, its nature, will continue to reveal new potentialities.

Wessex (we·seks). [OE. *West Seaxe* West Saxons.] **1.** The name of a kingdom in south-west England in Anglo-Saxon times, used by Thomas Hardy as the name of the county in which his stories are set (corresponding approximately to Dorset, Somerset, Hampshire, and Wiltshire) and since used as a name for south-west England or this part of it.

1868 W. BARNES *Poems of Rural Life in Common Eng.* Pref., As I think that some people, beyond the bounds of Wessex, would allow me the pleasure of believing that they have deemed..my homely poems in our Dorset mother-speech to be worthy of their reading, I have written a few of a like kind, in common English. **1874** HARDY in *Cornh. Mag.* Nov. 624 Greenhill was the Nijnii Novgorod of Wessex; and the busiest..day of the whole statute number was the day of the sheep-fair. **1876** *Examiner* 15 July 794/1 The Wessex man knows that these passages have in them the real ring, all equally true to life and scenery. **1938** *Proc. Prehistoric Soc.* IV. 52 The work..was..undertaken with a view to examining the cultures of the geographical area usually comprised in the term 'Wessex' in the period immediately following the Beaker phase. **1979** *N. & Q.* June 193/2 All [volumes] share a chronology of the life and works, Hardy's General Preface to the Wessex Edition, and notes on Wessex and Wessex names.

2. *attrib.* = *SADDLEBACK *sb.* 4 h.

1919, etc. [see *SADDLEBACK *sb.* 4 h]. **1919** [see *KILLER 4 b]. **1978** A. WILLIAMS *Backyard Pig Farming* iv. 27 There used to be a Wessex Saddleback originating in Dorset; it had black back legs.

3. *Archæol.* Of, pertaining to, or designating an Early Bronze Age culture in southern England, *c* 2000–1500 B.C., represented by grave-goods of native and European provenance.

1938 S. PIGGOTT in *Proc. Prehistoric Soc.* IV. 52 Many elements..here described as typical of the Wessex Culture of the Bronze Age are..found in associations which are late and outside the main culture-area. **1954** *Antiquity* Mar. 28 The axes..of this broad-butted type,..characteristic..of Piggott's Wessex Culture. **1963** E. S. WOOD *Collins Field Guide to Archaeol.* I. iv. 64 The Wessex nobility seem to have come from Germany. **1975** *Times Lit. Suppl.* 14 Mar. 282/1 The unlikelihood of Mycenaean influence on Stonehenge and the 'Wessex culture' of southern England. **1983** P. A. CROWL *Prehist. Britain* i. 33 The Wessex urns..lacking the refinement one would have expected of this..culture.

west, *adv.*, *sb.*[1], and *a.* Add: **B. 3.** *Bridge.* (With capital initial.) The player sitting opposite and partnering East, and having South to his right.

1926 [see *EAST *sb.* 4]. **1958** *Listener* 2 Oct. 541/1 West was a good enough player to have a chance of succeeding. **1974** *Country Life* 3 Oct. 975/1 Warned off Hearts and Clubs, West had to lead a Spade or a Diamond.

C. *sb.* **3. b.** (Earlier example.) See also *WILD WEST.

1796 G. WASHINGTON in *Claypoole's Amer. Daily Advertiser* 19 Sept. 2/2 The West derives from the East supplies requisite to its growth and comfort.

e. (With capital initial.) The non-Communist states of Europe and America.

1946 H. NICOLSON *Diary* 22 Aug. (1967) 75 He is convinced that the Russians wish to dominate the world... The only way in which the West can counter this is to pool their philosophy of liberalism, put up a united front. **1951,** etc. [see *EAST *sb.* 2 b]. **1957** *Ann. Reg.* *1956* 228 Some 5,000 citizens a week continued to flee to the West. **1964** M. MCLUHAN *Understanding Media* (1967) ii. 40 Competitive sports between Russia and the West. **1979** T. BENN *Arguments for Socialism* i. 38 It is not only in the West that Marxism is seen as one of the main sources of democratic socialist philosophy.

D. *adj.* **1. c.** *the West Bank*, a region west of the River Jordan and north-west of the Dead Sea which became part of Jordan in 1948 and was occupied by Israel in the Arab–Israeli War of 1967; hence *West Banker*, an inhabitant of the West Bank.

1967 *Times* 3 Aug. 16/7 Making the Israeli pound legal tender side by side with the Jordanian dinar on the Israeli-occupied West bank should go a long way towards increasing imports of goods from Britain. *Ibid.* 10 Aug. 7/1 Even those Israelis who would gladly abandon Sinai, Gaza and the West Bank would prefer to keep the Syrian heights overlooking the Sea of Galilee. **1968** *N.Y. Times* 22 Dec. IV. 4/6 The many interviews given by King Hussein who often refers to granting more self-government to the West Bankers. **1972** *Guardian* 10 Apr. 11/3 Hussein..is seeking to prevent disaffected West Bankers..from reversing their links with the Hashemite throne. **1978** *Internat. Relations Dict.* (U.S. Dept. State Library) 35/2 These groups rejected..the establishment of a Palestinian state on the West Bank. **1983** 'J. LE CARRÉ' *Little Drummer Girl* ii. 32 Miss Bach had been talking wistfully of taking up the wagon-trail life of a West Bank settler.

2. b. *West Briton*: † (*a*) a native of Wales; (*b*) a native of Ireland; in mod. use, (chiefly derogatory) one who favours a close political connection with Great Britain; hence *West Britonism*.

1712 P. LEIGH *Life S. Wenefride* 46 Whatever this incredulous Age may think of..our Saint's Return to Life; it appear'd so evident to the West Britains..that many Pagan People..came..to receive Baptism. **1816** J. GIFFARD *Let. to Sir Robert Peel* 19 Mar. (Brit. Library Add. MSS. 40,253, f. 258), The periphrastic Title of the United Kingdom of Great Britain and Ireland..goes out of its way—to remind people that they were once disunited and to keep them so—had the whole been called by one common name *Britain*—we should have had the Inhabitants proud of the glorious Title *Britons* and we West *Britons* would have been as much conciliated and attached as the North Britons are. **1836** D. O'CONNELL in J. O'Connor *Hist. Ireland 1798–1924* (1925) I. vii. 226 The people of Ireland are ready to become portion of the Empire, provided they be made so in reality and not in name alone; they are ready to become a kind of West Britons, if made so in benefits and justice; but if not, we are Irishmen again. **18..**, **1909** [in Dict.]. **1910** D. HYDE in R. M. Dorson *Peasant Customs* (1968) II. 718 The men who..while protesting..against West Britonism, have helped..to assimilate us to England and the English. **1918** West Britonism [see *SHONEEN]. **1925** in J. O'CONNOR *Hist. Ireland 1798–1924* II. xxiv. 368 The American friends of Irish liberty are both grieved and resentful at some of the recent exhibitions given there of the revival of West Britonism. **1944** JOYCE *Stephen Hero* xvii. 54 No West-Briton could speak worse of his Countrymen. **1960** C. C. O'BRIEN *Shaping of Mod. Ireland* 19 When Moran and his friends talked of West Britons they had in mind, I imagine, some archetype of a dentist's wife who collected crests, ate kedgeree for breakfast and displayed on her mantelpiece a portrait of the Dear Queen. **1962** B. INGLIS *West Briton* viii. 143, I never heard of West Briton being used except pejoratively. **1972** C. C. O'BRIEN *States of Ireland* iv. 77

Protestant loyalists—that is to say, most Protestants—also came inevitably under attack, usually as West Britons.

c. (Further examples.)

1824 COLLIER & MACCARTHY (*title*) West-African sketches. **1863** *Irish People* 5 Dec. 24/3 The West-British press chimed in. **1865** R. F. BURTON *Wit & Wisdom from W. Afr.* iii. 121 The practical selfishness and feelinglessness of the wild West African, who, when, tamed by slavery, becomes one of the most tender of men. **1925** J. O'CONNOR *Hist. Ireland 1798–1924* II. xxiv. 373 People dance the same dances as were the fashion in the old West-British days. **1950** *New Yorker* 16 Sept. 83/1 They [*sc.* the Germans] think that German rearmament is inevitable, and suggest that a sort of Foreign Legion..be activated immediately, in which all West European men willing to go to war against Communism could volunteer. **1958** *Listener* 11 Dec. 977/2 He [*sc.* Herr Brandt] is coming gradually to symbolize for many West Berliners their determination to remain free. **1969** A. MARIN *Rise with Wind* (1970) vi. 75 Clay sank into a chair, his eyes fixed coldly on the West German. **1973** *Times* 27 Nov. 9/1 (*heading*) Fewer West Berliners visit the East. **1976** W. LAQUEUR in D. Villiers *Next Year in Jerusalem* 86 The non-Jewish Jew is a specifically West European phenomenon. **1976** M. BIRMINGHAM *Heat of Sun* iii. 34 To build a house in his home town, to which all West Africans dream of retiring. **1981** J. JOHNSTON *Christmas Tree* 114 It wasn't that I objected to De Valera's neutrality... I had no political feelings of being West British..no Crown fever. **1983** *Spectator* 14 May 8/1 It is unsettling to find a pillar of West German industry collecting Nazi memorabilia.

d. With abstract sbs. derived from the sbs. and adjs. of prec. sense.

1895 *Dundalk Examiner* 24 Aug. 2/6 A slogan cry which would..sound the death-knell of ascendancy and West Britishism in this country? **1971** J. SPENCER *Eng. Lang. W. Afr.* 28 There is certainly a sufficiency of terms and expressions peculiar to the use of English in this region to justify the term West Africanism. **1980** *English World-Wide* I. 1. 76 AVE [*sc.* African Vernacular English] is ..characterized by a vocabulary adapted to its environment, which shows itself in oft-quoted West Africanisms.

west coast. 1. The western coast of a country or region; in some cases with capital initials as a proper name. Also *attrib.*

1377, 1689, 1801 [in Dict. s.v. WEST *a.* 1 a]. **1845** *N.Z. Jrnl.* 13 Sept. 234/2 Of the west coast of the Middle Island, commonly called by the whalers the 'West Side', we heard a good deal. **1850** *Calif. Courier* (San Francisco) 2 Dec. 2/1 Our position here on the West Coast has been and still is a peculiar one. **1862** *Jrnl. R. Geogr. Soc.* XXXII. 294 Arrangements entered into with the Provincial Government of Nelson for the survey of the West Coast district of that province. **1897** M. KINGSLEY *Trav. W. Afr.* 8 Sound knowledge..collected during an acquaintance with the West Coast of over thirty years. **1926** A. HUXLEY *Jesting Pilate* iv. 287 The stranger coming to the West Coast will be astonished by the amount of casual embracement. **1959** A. MCLINTOCK *Descr. Atlas N.Z.* p. xvi, The road..here swings in a northerly direction..towards Arthur's Pass, 3,020ft, and thence to the West Coast. **1971** *Country Life* 9 Dec. 1642/2 Last year..the west-coast herrings proved to be the only plentiful supply in northern Europe. **1977** H. FAST *Immigrants* II. 88 We're the lifeline, the West Coast, San Francisco.

2. (Usu. with capital initials.) Used *attrib.* with reference to a style of modern jazz playing that was centred on Los Angeles in the 1950s, typified by small ensembles, technical sophistication, and elaborate writing. orig. *U.S.*

Cf. *West Coaster* below, quot. 1954.

1954 *Downbeat* 7 Apr. 6/1 The latest example of this thinking-by-pigeonholes is the attempt to convince the populace that there is a growing west coast school of jazz. *Ibid.* 19 May 16/3 Nat Hentoff's comments on 'west coast jazz' aroused considerable comment. **1959** *News Chron.* 12 Aug. 6/5 He is not only benevolent about West Coast jazz but aware of its technical ins and outs. **1961** *Times* 4 Feb. 11/5 Music of considerable variety, ranging from some vigorous Dixieland..to West Coast Jazz (with Palm Court cello). **1962** *Melody Maker* 21 July 7/1 Some of the 1954 tracks have a nostalgic, almost dated, appeal, in the writing, it is so typical of West Coast jazz of the time—neat, smooth and often very clever. **1980** *New Grove Dict. Mus.* XX. 371/2 Miles Davis's 1949–50 recordings were an initial influence as is shown by such archetypal West Coast performances as Rogers's *Didi* (1951).

3. Used *attrib.* to designate a kind of large rear-view mirror (see quot. 1963). orig. *U.S.*

1963 *Amer. Speech* XXXVIII. 46 *West coast mirror*,..a large, square, rear-view mirror attached to the side of a cab. **1968** *Globe & Mail* (Toronto) 17 Feb. 41 (Advt.), Mercury ¾ ton pickup... 4 speed transmission, west coast mirrors. **1980** *Truck & Bus Transportation* (Austral.) Mar. 96/1 All-round vision is generally good, but Cronulla Carrying have gone one step further by replacing the meagre standard mirrors with the efficient west-coast type.

Hence **West Coaster,** (*a*) one who lives on the West Coast; *spec.* (*N.Z.*) = *COASTER 3 c; (*b*) a player or devotee of West Coast jazz.

1896 *N.Z. Alpine Jrnl.* II. 157 He was..not a native born West Coaster. **1936** 'R. HYDE' *Passport to Hell* ii. 54 He washed shirts for the brawny West Coasters. **1941** O. DUFF *N.Z. Now* v. 71 The people are never 'Southlanders'..as the people of the West Coast are 'West Coasters'. **1949** M. STEEN *Twilight on Floods* II. ii. 198 Eighty-five per cent West Coasters die of fever, or return home total wrecks! **1954** *Time* 1 Feb. 38/2 Today, the liveliest center of developing jazz is California... The West Coasters include such names as..Shelly Manne.., Shorty Rogers.., Gerry Mulligan and Stan Getz.., Dave Brubeck. **1958** K. GOODWIN in P. Gammond *Decca Bk. Jazz* xiii. 148 Groups from four to nine pieces have been most popular among the West Coasters. **1974** M. BRAITHWAITE *Ontario* ii. 7 It is nonsense to maintain that there are no special characteristics of Canadians from different regions of the country. West Coasters *are* different from those who live on the East Coast. **1977** *Times* 16 May 8/7 Their effect on humourless West Coasters [*sc.* in California] was..devastating.

west end. Add: **2.** (Earlier example.)

1776 *Gazetteer & New Daily Advertiser* 11 Sept. in Bond & McLeod *Newslett. to Newspapers* (1977) III. 186 A gentleman in a certain coffeehouse at the Westend of town.

b. The theatres of the West End, or their personnel.

1894 *Theatre* Oct. 155 The influence of the west end is felt both in the cheaper London houses and throughout the provinces. **1979** *Listener* 16 Aug. 206/3 No one wanted a National Theatre. The West End didn't want it because they feared a new rival.

4. (Examples relating to West End theatres.) Also passing into *adj.*

1890 G. B. SHAW *London Music 1888–89* (1937) 322 The more commercial atmosphere of the West-end theatre. **1890** O. WILDE *Pict. Dorian Gray* iii, in *Lippincott's Monthly Mag.* July 29, I will take a West-End theatre and bring her out properly. **1928** A. HUXLEY *Point Counter Point* x. 159 So well travelled, so brilliantly cosmopolitan and West-End. **1936** N. COWARD *To-Night at 8.30* I. 103 If you're so bloody West End why the hell did you leave it? **1954** 'M. COST' *Invitation from Minerva* 171, I got my first West-End engagement. Since then, I've never looked back. **1983** S. VIZINCZEY *Innocent Millionaire* iii. 14 Occasionally his London agent got him a part in the West End production of an American play.

west-ender (earlier example); **west-endian, -endy** *adjs.*, characteristic or suggestive of a west end, *spec.* that of London.

1833 *Chambers's Jrnl.* 30 Mar. 66/2 There have been instances of 'west-enders' going on a tour of discovery.. within the precincts of Wapping. **1856** J. M. LUDLOW *Let.* Nov. in C. L. Graves *Life & Lett. A. Macmillan* (1910) ii. 91 [A London shop] more West-endian than Bell's or Nutt's. **1911** J. BONE *Edin. Revisited* i. 12 A minister of the Gospel from the West Coast identified Edinburgh as an 'east-windy, west-endy city'. **1959** *New Chron.* 25 July 4/5 Most of it proved too precious and West Endy for television.

westerliness. (Example.)

1927 [see *EASTERLINESS]

western, *a.* and *sb.*[1] Add: **A.** *adj.* **3. a.** *Western Approaches*, the area of sea immediately to the west of Britain; *Western Islands* = *Western Isle* (*a*), (*b*); *Western Isle*, (*a*) *pl.*, the Hebrides; cf. *west isles* s.v. WEST *a.* 1 c; (*b*) *pl.*, the Azores; † (*c*) Ireland (*rare*$^{-1}$); *Western Ocean*, the Atlantic.

1697 W. DAMPIER *New Voyage round World* v. 107 The most remarkable places that I did ever hear of for their breeding, is at an Island in the West Indies called Caimanes, and the Isle Ascention in the Western Ocean. **1758** J. ARMSTRONG *Let.* 21 Oct. in *N. & Q.* (1979) Feb. 44/2, I hope you have had an agreeable View of the Western Isles. **1760** F. FAUQUIER *Let.* 28 Oct. in G. Reese *Official Papers* (1980) 422 The Vessel is cleared out for Gibraltar, and then under pretence of being drove by stress of Weather into Madeira or some of the western Isles. **1775** JOHNSON (*title*) Journey to the Western Islands of Scotland. **1776** GIBBON *Decl. & Fall* I. i. 5 The western isle might be improved into a valuable possession. **1805** in *Naval Documents U.S. Wars with Barbary Powers* (U.S. Office Naval Rec.) (1944) V. 366 It is my opinion, she is competent to be sent across the Western Ocean, and should it be deemed necessary to send her to the Mediterranean, she could be speedily equipped. **1810** J. E. CALDWELL *Tour through Part of Virginia* (ed. 2) (1951) 47 The Azores, or Western Islands, are nine in number. **1870** Western Islands [in Dict.]. **1920** *Times* 1 Mar. 9/3 The title of Admiral Sir Reginald Tupper..has now been changed to Commander-in-Chief of the Western Approaches. There are three captains, R.N., in charge of Naval Areas under his orders. Captain Denis B. Crampton..commands the Irish Sea Area; Captain William D. Church..the Kingstown Area; and Captain E. G. Lowther-Crofton..the Buncrana Area. **1935** J. MASEFIELD *Victorious Troy* 8 Did you ever see a storm, a real storm, a Western Ocean Hurricane? **1946** W. S. CHURCHILL *Secret Session Speeches* 38 The powerful reinforcement of large-range aircraft..which were sent.. to the Western Approaches are now active. **1961** G. FOULSER *Seaman's Voice* ii. 33 Western Ocean gales are notorious for their ferocity. **1976** *Mariner's Mirror* LXII. 177 Sometimes homeward-bound convoys would be routed away from the Western Approaches, the Bay of Biscay, and the English Channel. **1976** *Scotsman* 15 Dec., The rents of the 2000 local authority houses in the Western Isles are to be increased by £39 a year from April. **1979** *N. & Q.* Feb. 44/2 We can now be fairly certain that John Wilkes..also made a journey to the Western Islands of Scotland.

f. *Western Front*, the front in Belgium and northern France in the wars of 1914–18 and 1939–45.

[**1914** *Parl. Deb. Written Answers* (*Commons*) 12 Nov. 167 The British casualties in the Western area of the war up to 31st October are, approximately, 57,000. **1914** M. HANKEY *Memo.* 28 Dec. in M. Gilbert *Winston S. Churchill* (1972) III. Compan. I. 337 The remarkable deadlock which has occurred in the western theatre of war.] **1914** LLOYD GEORGE *Memo.* 31 Dec. in *Ibid.* 352 These objects cannot be accomplished by attacks on the Western Front. **1915** *Times* 7 Jan. 9/6 Lord Kitchener..explained that the operations on the Western front have for some time resolved themselves into a state of siege warfare. **1915** A. BENNETT (*title*) Over there: war scenes on the Western Front. **1917** *Weekly Dispatch* 3 June 1/3 (*heading*) Mystery of the Western Front. **1939** *War Weekly* 3 Nov. 35/1 The incalculable factor on the Western Front is the mind of Hitler. **1983** P. A. CROWL *Intelligent Traveller's Guide Historic Britain* x. 502 What was lost were thirty-six Czech divisions delivered to Hitler free of charge at Munich plus a greatly improved German position on her western front.

g. *Western American*: = *general American* s.v. *GENERAL *a.* 2 a.

1919 G. P. KRAPP *Pronunc. of Standard Eng. in Amer.* 147 If your own speech is of the Eastern American type, transcribe a passage illustrating it into Western American speech. **1925** —— *Eng. Lang. in Amer.* II. i. 30 The consonant *r* is more distinctly sounded in northern British and Western American. **1936** MENCKEN *Amer. Lang.* (ed. 4) vii. 358 The chief characters of Western, or General American and of New England and Southern American have been indicated. **1959** L. M. MYERS *Guide to Amer. Eng.* (ed. 2) ii. 30 Three major dialect areas have long been recognised in American English—New England, Western or General American, and Southern. Linguistic geographers now prefer the terms Northern, Midland, and Southern.

h. *Western European Union*, an association formed in 1955 from the former Western Union, with the addition of Italy and (West) Germany, in order to coordinate defence and promote cooperation in economic matters and (until 1960) in social and cultural ones; abbrev. W.E.U. s.v. *W 3; *Western Union*, an association of West European nations (Belgium, France, Luxembourg, the Netherlands, and the United Kingdom) which was formed in 1948 for purposes of military and economic cooperation and became the Western European Union in 1955.

1948 E. BEVIN in *Hansard Commons* 22 Jan. 390 The European Recovery Programme brought all this to a head, and made us all face up to the problem of the future organisation. We did not press the Western Union..in the hope that when we got the German and Austrian peace settlements, agreement between the Four Powers would close the breach between East and West. **1950** *Times* 22 Aug. 4/6 About 50 squadrons of fighters and bombers will take part in the first Western Union air defence exercise. **1954** *Times* 11 Nov. 8/7 Some of the preparations for bringing the Western European Union into existence have been made by a committee in London. **1973** B. COCKS *European Parliament* viii. 67 The Assembly of the international organisation known as Western European Union has a close relationship with the Council of Europe since its entire membership is composed of the Representatives from the seven WEU countries to the Consultative Assembly. **1974** P. GORE-BOOTH *With Great Truth & Respect* 351 Mr Brown has recounted the strategem he employed at a Ministerial Western Union meeting to ensure that the application was formally presented in a way that could not run into procedural objections. **1976** J. WHEELER-BENNETT *Friends, Enemies & Sovereigns* iv. 125 He..expressed considerable concern as to what the effect would be of Germany's rearmament in accordance with her membership of the European Defence Community (or, as it turned out to be, the Western European Union).

4. a. In mod. use also *spec.* (*a*) applied to the countries of western Europe that opposed Germany in the wars of 1914–18 and 1939–45; (*b*) of, pertaining to, or designating the non-Communist states of Europe and America.

(*a*) **1914** *Times* 23 Nov. 9/2 The appearance of Turkey as the ally of Germany and Austria against the Western Powers and Russia necessarily put an end to negotiations between Sofia and Constantinople. **1917** I. F. MARCOSSON *Rebirth of Russia* viii. 141 German imperialism, after having defeated our Western Allies, will turn against us the whole power of its arms. **1938** E. AMBLER *Cause for Alarm* viii. 132 The Nazis and the Fascists..agreed to present a united front to the Western powers. **1940** *Economist* 13 Jan. 51/2 The outbreak of open hostilities between the U.S.S.R. and the Western Powers. **1974** *Encycl. Brit. Macropædia* XIX. 958/1 The Russian Revolution of March..1917 dismayed the western Allies and delighted the Central Powers. *Ibid.* 1006/1 The western Allies' 'Operation Overlord'..took place on June 6, 1944.

(*b*) **1918** *Times* 4 June 5/2 The greatest question in the world to-day is whether Russia is to be abandoned, or whether she is to be saved; whether Western ideals are to prevail in the country whose potential power will be the balance in history. **1947** *Ann. Reg.* *1946* 218 A pointed appeal to the Russian people to regard their two Western Allies [*sc.* Great Britain and the United States] as the only blot on the Soviet horizon. *Ibid.* 219 The need [of Russia] for an American loan and the consequent recognition of the desirability of making some concessions to the Western Powers. **1956** *B.B.C. Handbk.* *1957* 60 The jamming..of certain language transmissions of the BBC and other Western Bloc broadcasters. **1959** *Daily Tel.* 18 Dec. 1 Expectations of some progress in Western politics rose in Paris to-night on the eve of the 'Western Summit' meetings which will take place here this week-end. **1982** *Ann. Reg.* *1981* 67 The fourth [proposal] called on the Soviet Union to accept Western plans for reducing the risks of surprise attack.

c. *western man* (also with either one or two initial capitals): man as shaped by the culture and civilization of Western Europe and North America.

1909 CHESTERTON *Orthodoxy* i. 14 An active and

imaginative life,..a life such as western man at any rate always seems to have desired. **1927** WYNDHAM LEWIS (*title*) Time and western man. **1962** E. CLEAVER in A. Dundes *Mother Wit* (1973) 10/2 The traditional judgments which Western Man has made..are now..the cause of very serious maladjustments in our society and.. the world at large. **1970** C. C. O'BRIEN *Camus* i. 27 The role often claimed for Camus, as an expression of the conscience of Western man. **1981** *Times Lit. Suppl.* 20 Mar. 321/2 Edward Bond's cloudy gropings towards a view of Western man.

5. a. (Earlier example.)

1794 T. COOPER *Some Information respecting Amer.* 8 These parts..furnish yearly a very considerable number of emigrants to the middle and western states.

b. (Earlier examples.) *western equine encephalitis* or *encephalomyelitis*, a mosquito-borne viral encephalitis in the U.S., South America, and eastern Europe that affects chiefly horses but also people and is sometimes fatal, esp. to children; *western roll* (Athletics), a method of high-jumping in which the athlete jumps from the inside foot, swings up the other leg, and rolls across the bar on his side; *Western saddle* (see quot. 1946); *western sandwich* (N. Amer.), a sandwich in which the filling is an omelette containing onion and ham.

1703 in *Mass. Hist. Soc. Coll.* (1838) 3rd Ser. VII. 61 Letters from Piscataqua come in the Western mail. **1784** G. WASHINGTON *Diary* 4 Oct. (1925) II. 326 The Western Settlers—from my own observation—stand as it were on a pivot. **1929** G. M. BUTLER *Mod. Athletics* viii. 108 (*caption*) The 'western roll'. **1933** *Proc. Soc. Exper. Biol. & Med.* XXXI. 217 (*heading*) A serological difference between Eastern and Western Equine Encephalomyelitis virus. **1946** M. C. SELF *Horseman's Encycl.* 346 The Western or cowboy saddle which is similar to that used.. in all countries where men spend long hours in the saddle ..is characterized by its deep seat, high cantle and pommel... The stirrups are set about midway and the cowboy rides with an almost straight leg. **1959** M. CALLAGHAN in R. Weaver *Canad. Short Stories* (1968) 2nd Ser. 8 He thought of having a western sandwich in the café across the road from the hotel. **1959** *Jrnl. Infectious Dis.* CV. 295 A similar situation may exist with respect to western equine encephalitis infection in swine. **1961** *Canad. Jrnl. Microbiol.* VII. 295 Western equine encephalitis has been a disease of public health importance in Saskatchewan since it was first recognized in 1935. **1964** Western roll [see *SCISSORS *sb. pl.* 2 b]. **1964** M. MCLUHAN *Understanding Media* (1967) xxxi. 341 The varied and rough textures of Western saddles, clothes, hides. **1973** D. HUGHES *Along Side Road* xx. 155 He stopped to have a western sandwich and a cup of tea. **1976** *Western Mail* (Cardiff) 27 Nov. (Advt.), Just arrived from America: New selection of Western saddles. **1978** G. WRIGHT *Illustr. Handbk. Sporting Terms* 23/3 The western roll is rarely practised today, the favoured methods being the straddle and the flop. **1983** *Amer. Rev. Respiratory Dis.* CXXVII. 132/3 We report here a patient with western equine encephalitis who developed hypoventilation. **1983** *Amer. Jrnl. Trop. Med. & Hygiene* XXXII. 1130 *Culex tarsalis* was a less competent vector of western equine encephalomyelitis (WEE) virus after 2–3 weeks' extrinsic incubation at 32°C than after incubation at 18° or 25°C.

c. *western hemlock*, a conifer, *Tsuga heterophylla*, native to the western coast of North America; also, its light brown timber; *western red cedar*, a large columnar conifer, *Thuja plicata*, native to western North America; also its reddish-brown timber; *western white pine*, a pine with grey-green needles, *Pinus monticola*, native to high ground in western North America; *western yellow pine* = *PONDEROSA.

1869 *Amer. Naturalist* III. 410 Western White Pine... I found scattered trees of this beautiful species on the highest parts of the Rocky Mountains. **1886** J. MACOUN *Catal. Canad. Plants* I. 461 *J*[*uniperus*] *occidentalis*, Hook. Western Red Cedar... I place all our western 'red cedar' under this species. *Ibid.* 471 *T*[*suga*] *Mertensiana*, Carr. Western Hemlock... In the Selkirk Mountains it is a tall, beautiful tree, over 150 feet high. **1901** *World's Work* July 888/2 The wood of the western yellow pine.. is used by them for mine timbers. **1905** *Bull. Bureau of Forestry* (U.S. Dept. Agric.) LXVI. 33 The rock pine, western red cedar,..have, without doubt, come down from the Rocky Mountains. **1908** N. L. BRITTON *N. Amer. Trees* 67 Western Hemlock..grows in rich, moist soil. **1923** Western yellow pine [see *heavy-wooded pine* s.v. *HEAVY *a.*[1] 30]. **1957** *Handbk. of Softwoods* (Forest Products Res. Lab.) 22 Consignments of western hemlock frequently contain a percentage of fir. **1963** Western hemlock [see *KAPUR]. **1969** Western red cedar [see *RED CEDAR a]. **1977** *Weekly Times* (Melbourne) 19 Jan. 4/1 (Advt.), The following is included in your kit: Council plans and specifications, flooring,..glazed Western Red Cedar windows, [etc.]. **1978** W. H. HARLOW et al. *Textbk. Dendrology* (ed. 6) 66 Western white pine was first observed along the banks of the Columbia and Spokane Rivers in 1831.

d. (Also with capital initial.) Applied to the films and novels called 'westerns' (see sense B. 4 below). orig. *U.S.*

[**1909** *Moving Picture World* 6 Nov. 638 The success of their Western series of last year was abundantly satisfying and added greatly to the reputation of the firm. *Ibid.*, Western subjects, in which the wild and woolly plays the leading part, have won immense popularity.] **1910** *Ibid.* 21 May 834/1 It is almost impossible to criticize these Wild Western films, because cowboys are likely to do almost anything. **1913** *Moving Picture Ann.* *1912* 29 Many film makers still turn out great quantitites of so-called Western and Indian pictures. **1931** *Ann. Reg.* *1930* 49 Garry [*sic*] Cooper has revived the popularity of Western pictures. **1959** *News Chron.* 5 Aug. 6/4 Many Western novels are abominably written. **1967** M. ARGYLE *Psychol. Interpersonal Behaviour* i. 28 An indirect form of aggression will occur, which may consist of..mere aggression in fantasy, such as watching western films or wrestling matches. **1974** *Encycl. Brit. Micropædia* X. 624/2 The western film can be dated from *The Great Train Robbery* (1905).

9. *Comb.* (chiefly in sense 4 a), as *western-educated*, *-European*, *-style*, *-trained*, *-type* adjs.

1933 N. WALN *House of Exile* I. vi. 96 A Western educated woman doctor. **1974** M. FIDO *R. Kipling* 50/2 Kipling['s]..generation made 'Bengali' almost a synonym for 'western-educated Indian', and always used the word with a touch of contempt. **1949** M. MEAD *Male & Female* vi. 132 Almost any Balinese male placed in a series of western-European males would look 'feminine'. **1969** 'E. LATHEN' *When in Greece* ii. 17 Greece..was an associate in the Common Market, which would bring every Western European banker into the picture. **1895** *Montgomery Ward Catal.* Spring & Summer 330/1 Three-Horn Western Style Side Saddle. **1953** *Archit. Rev.* CXIV. 255/2 Peking of course offers a complete contrast to such cities as Shanghai where large areas have been covered with western-style multi-storey buildings. **1977** P. JOHNSON *Enemies of Society* xi. 160 The abolition of western-style academic research, and the substitution of acupuncture for standard medical practice. **1962** E. SNOW *Other Side of River* (1963) xiii. 309 Since 1958 all Western-trained doctors have been required to devote at least six months to the study of Chinese medicine. **1958** *Times* 13 Aug. 12/4 In West Africa they hold to the traditional styles, though the more sophisticated often keep western-type frocks in their wardrobe.

B. *sb.* **4.** (Also with capital initial.) A film or novel belonging to a distinct genre in which life in the American West in the nineteenth century is portrayed, usu. through idealized stock situations and characters, esp. cattlemen (cowboys) and gun-fights. Cf. sense A. 5 d above. orig. *U.S.*

1912 *Moving Picture World* 27 July 306 (Advt.), 'The Fight at The Mill'... A powerful Western, distinctly unusual among typical 'Westerns' containing a beautiful story and a dashing Indian battle that will interest and instruct. **1915** [see *SUPE]. **1918** *Wells Fargo Messenger* VI. 178/1 What would the good old 'Western' be without the historic Wells Fargo stage coach and its treasure box? **1923** *Time* 11 June 15/1 Love stories are their first choice, comedies second, society life as known to the De Mille brothers third, and then come the Westerns. **1927** *Sat. Rev. Lit.* 15 Oct. 232 (Advt.), The Gun-Slinger by George M. Johnson. A Western with the kick of a .45. **1930** *Publishers' Weekly* 8 Feb. 689 (Advt.), Five sure-shot Westerns... Salesmen and booksellers everywhere report to us a keen and growing demand for this type of story. **1954** E. E. CUMMINGS *Let.* 9 Mar. (1969) 227 'William S. Hart' was a vastly popular..hero of our early Westerns: i.e. melodramatic movies featuring terrific battles between noble & wicked hardriding sharpshooting super cowboys. **1958** *Listener* 9 Jan. 60/1 Twenty-five per cent of the best television time is still given over to 'Westerns'. **1962** L. DEIGHTON *Ipcress File* xviii. 110 On the army table were a few books; German grammar.. two paperback westerns. **1977** B. PYM *Quartet in Autumn* xviii. 171 Watching a Western on the other channel.

C. *adv.* *Equestrianism.* In the manner of a cowboy; in a relaxed style with a deep-seated saddle and almost straight legs.

1972 *Country Life* 5 Oct. 817/1 In a year Lady Sarah was so used to riding Western that she found it almost strange to revert to English for hunting. **1980** *Times* 28 June 13/7 Riding western is not like riding Badminton style.

westerner. Add: **2.** (Earlier example.)

1880 W. JAMES in *Atlantic Monthly* Oct. 449/2 Not to fall back on the gods, where a proximate principle may be found, has with us Westerners long since become the sign of an efficient..intellect.

4. *Hist.* An advocate of or believer in the concentration of forces on the Western Front during the war of 1914–18.

1928 F. B. MAURICE *Rawlinson of Trent* p. xi, Upon the problems of the Great War, Rawlinson has naturally much light to throw. Who was right, the Easterner or the Westerner? **1931** W. S. CHURCHILL *World Crisis* VI. xix. 282 Falkenhayn was a convinced and inveterate 'Westerner'. **1960** *Times Lit. Suppl.* 20 May 318/3 Captain Falls is a firm 'Westerner' although he believes that the Dardanelles enterprise was 'a well-inspired venture'. **1977** G. H. CASSAR *Kitchener* xiv. 295 Kitchener's difficulties were exacerbated by the two prevailing schools of strategical thought, the Easterners and the Westerners.

5. a. *Hist.* A 19th-century Russian who adopted or advocated Western attitudes and behaviour.

1949 I. DEUTSCHER *Stalin* vi. 207 But Lenin remained a 'Westerner' in several senses. **1950** E. H. CARR *Bolshevik Revolution* I. i. 8 The westerners held that it was the destiny of Russia, as a backward country, to learn from the west.

b. One belonging to the non-Communist West.

1964 M. MCLUHAN *Understanding Media* (1967) xxi. 222 Are we to suppose that this kind of media illiteracy is characteristic only of Westerners, and that Russians know how to correct the bias of the medium? **1975** P. THEROUX *Great Railway Bazaar* xxx. 330, I was now the only Westener on the train.

westernism. 1. (Earlier example.)

1838 *Knickerbocker* XI. 447, I now recollect but few specimens of Jack's *westernisms*, and these I think were not his best.

westernize, *v.* Add: **b.** *intr.* To become western in character. *rare.*

1903 L. F. WARD *Pure Sociol.* 33 Some of the nations of the East, notably Japan, are rapidly westernizing.

Hence **we·sternizer,** one who makes a country or culture more Western.

1935 *Times Lit. Suppl.* 2 May 287/2 French and English incursions..entered [Afghanistan] from the East, and..carried Dravidian ideas with them against that tide of Westernizers of whom Alexander was one of the earliest. **1958** *Listener* 27 Nov. 864/2 Arab Westernizers. **1964** *Economist* 13 June 1251/2 Ch'en Tu-hsiu was a westerniser. **1976** *Times Lit. Suppl.* 23 Apr. 490/3 The dispute between Slavophiles and Westernizers, originally a literary controversy, spawned a vast secondary literature, first in Russia and then in the wider world.

westernly, *adv.* Restrict † *Obs.* to sense in Dict. and add: **2.** In a Western manner.

1588 [see KENTISHLY *adv.*]. **1976** M. H. KINGSTON *Woman Warrior* (1977) 59 The one [Chinese] faculty member in the western suit smiles westernly. **1982** C. THOMAS *Jade Tiger* iii. 59 A stylised, dignified, almost Westernly-handsome Chinese. **1983** *Christian Science Monitor* 15 Feb. 12/1 Dressing more stylishly—and often, more Westernly—is a preoccupation... Young Russians pass quick and critical judgement on the clothes of their peers.

westernness (we·stəɹn,nès). Also with capital initial. [f. WESTERN *a.* + -NESS.] The quality of belonging to a Western country or culture, or having Western attitudes and ideas.

1953 *Essays in Crit.* III. 132 We become aware..in the work of Conrad of the co-existence of two 'moralities': that derived from a simple tradition of 'Westerness'.. and that derived from an awareness of the force..of '*égoïsme*' in a decaying order. **1977** P. LASLETT *Family Life & Illicit Love in Earlier Generations* i. 13 Westernness or any other cultural attribute.

westerveldite (we·stəɹveldəit). *Min.* [f. the name of Jan *Westerveld* (1905–62), Dutch geologist + -ITE[1].] An orthorhombic arsenide of iron or iron and nickel (see quot. 1972).

1972 I. S. OEN et al. in *Amer. Mineralogist* LVII. 354 A Co-bearing nickel-rich iron monoarsenide corresponding in composition and structure to a Ni-rich member of the orthorhombic FeAs-(Fe,Ni)As solid solution series of synthetic alloys occurs in chromite-niccolite ores from La Gallega, Spain. The name westerveldite is proposed for minerals in this solid solution series. **1977** *Neues Jahrb. f. Mineral. Abhandlungen* CXXX. 209 Recently westerveldite with the ideal composition FeAs was found at four localities within the Ilímaussaq alkaline intrusion in South Greenland.

West Highland. Used *attrib.* and *absol.* to designate animals associated with the West Highlands of Scotland, as (*a*) a breed of cattle also called kyloes (see KYLOE); (*b*) a kind of terrier (see *POLTALLOCH).

1875 *Encycl. Brit.* I. 389/1 The Kyloes or West Highland cattle are the most prominent of this group [*sc.* mountain breeds]. **1906** *Our Dogs* 15 Sept. 548/1 White West Highland Terriers.—Here truly may be written there is ample room for improvement. **1910** 'SAKI' *Reginald in Russia* 28 A lady..was expressing to me.. her interest in West Highland terriers. **1950** [see *POLTALLOCH]. **1953** A. FRASER *Beef Cattle Husbandry* ix. 122 This breed, sometimes called the 'West Highland' or 'Kyloe', is descended from the native breed of the Scottish highlands. **1968** P. DICKINSON *Skin Deep* ix. 174 She's a nice lass, Roedean, breeds West Highlands down at Sonning. **1976** *Daily Record* (Glasgow) 4 Dec. 25/3 (Advt.), West Highland pups..suitable for Christmas. **1978** *Country Life* 24 Aug. 489/3 The White West Highland was let out of the front door.

So **West Highlander,** a kyloe.

1832 *Chambers's Edin. Jrnl.* I. 70/2 On many farms I observed a variety..resembling the heavy class of our West Highlanders. **1882** [see KYLOE]. **1979** *Country Life* 22 Nov. 1950/1 No breed of cattle more perfectly graces its native habitat than the West Highlander.

westie (we·sti). Also **westy.** [f. WEST *a.* + -IE, -Y[6].] A West Highland white terrier.

1959 *Observer* 1 Feb. 12 The 'Westie' is one of the few terriers to maintain its position. **1978** *Country Life* 24 Aug. 489/3 The Westy returned and sat down. **1979** *Daily Mail* 26 Oct. 25/4 Mr Definitely-Disgusting.. [is] particularly partial to 'Westies' as he calls West Highlands.

West Indian, *sb.* and *a.* Add: Also (as one word) **Westindian. 1.** *sb.* **c.** A person of West Indian ancestry.

1928 *Times* 25 June 5/1 The out-cricket of the West Indians on Saturday indicates that the compliment, which has been paid to them, has not been prematurely offered. **1957** *Times* 18 Feb. 4/7 Statistics show that more than 26,000 West Indians migrated to Britain during 1956. **1961** *Ann. Reg.* *1960* 122 In his message to West Indians on Christmas Day the Prime Minister of the Federation, Sir Grantly Adams, spoke of West Indian unity. **1971** *Observer* 21 Feb. 5/1 The Scots West Indian... Born in Edinburgh, he has Jamaican blood from his father's side. **1973** *Montserrat Mirror* 23 Mar. 5/2 Trinidad is the home of calypso but this form of music belongs to all

Westindians. **1981** *Westindian World* 31 July 2/3 My findings demolished a myth about Westindians.

Hence **West Indianness,** West Indian quality or character.

1953 *Caribbean Q.* III. 111. 181 We are..still trying to discover what..makes us characteristically West Indian, or if you like, what is the essence of our West Indianness. **1972** RAMCHAND & GRAY *West Indian Poetry* 89 The West Indian poet's confidence about his West Indian-ness.

Westinghouse (we·stiŋhaus). The name of George *Westinghouse* (1846–1914), U.S. inventor and manufacturer, used *attrib.* and *absol.* to designate a kind of air brake he invented in 1868 for use on railway trains, operated by compressed air on a fail-safe principle.

a **1877** KNIGHT *Dict. Mech.* I. 356/1 The Westinghouse Atmospheric Brake..was patented in 1869, and has been adopted on many railway lines in the United States and Europe. **1886** *Encycl. Brit.* XX. 248/2 The Westinghouse brake was greatly in advance of previously existing systems. **1933** *Times Lit. Suppl.* 2 Nov. 738/3 Many will regret the gradual abandonment of the Westinghouse brake on steam-hauled trains. **1949** D. M. DAVIN *Roads from Home* I. v. 75, I jammed on the Westinghouse, saying to myself I'd look a fine bloody fool if I'd stopped the train for nothing. **1967** G. F. FIENNES *I tried to run a Railway* iv. 40 Only the Stratford District had kept the Westinghouse brake.

Hence **Westinghou·sian** *a.* (*fig. nonce-use*).

1948 V. NABOKOV in *New Yorker* 31 July 20/1 The train stopped with a long-drawn Westinghousian sigh.

Westmark (we·st-, ‖ ve·stmɑɹk). Also **west-mark, west mark.** [Ger., f. *west* WEST *sb.*[1] + *mark* MARK *sb.*[2] 2 c.] The currency unit of West (formerly western) Germany, as distinguished from the *OSTMARK of East Germany.

1948 *Times* 2 Sept. 4/6 In view of the report that the east-mark is going to be recognized Berliners have been getting rid of the west-mark. **1959** [see *OSTMARK]. **1964** L. DEIGHTON *Funeral in Berlin* v. 31 'How much money are you carrying?' I spread the few Westmarks and English pounds on the desk. **1980** A. SCHOLEFIELD *Berlin Blind* iii. 129, I have postcards in the bus which you may buy with Westmarks. **1980** *Times Lit. Suppl.* 31 Oct. 1230/3 In Germany..the Bundesrepublik offers 10,000 west-marks as the Thomas Mann prize, and the DDR 18,000 east-marks for the Heinrich Mann.

Westminster. Add: **2.** The Palace of Westminster; hence, Parliament, of which the Palace is the seat. Freq. *attrib.*

The present Palace of Westminster (built 1840–67) is more commonly known as the Houses of Parliament.

1807 *Morning Chron.* 13 Apr. 3/2 The Westminster Company of Independent Performers being lately dissolved. **1869** TROLLOPE *Phineas Finn* II. lxxiv. 306 The girl whom he loved..better even than Westminster and Downing Street. **1918** G. FRANKAU *One of Them* xvii. 127 What art thou, Westminister? A caucused lobby? An oratorial-acrobatic stadium..? Or art indeed the Common Weal's palladium? **1961** S. A. DE SMITH in *Jrnl. Commonwealth Political Stud.* I. 3 In its narrow sense the Westminster Model can be said to mean a constitutional system in which the head of state is not the effective head of government; [etc.]. **1972** *Guardian* 11 July 13/8 It is sometimes suggested that what de Gaulle did for France in Algeria, Westminster should do for Britain in Ulster. **1977** *Time* 27 June 20/1 One of the most frequently heard catch phrases has to do with moving away from the 'Westminster system' of parliamentary representation toward some form of presidential or federal system.

b. *Westminster chimes* or *quarters*: the pattern of chimes struck at successive quarters by Big Ben in the Palace of Westminster, and used for other clocks and (more recently) door chimes; it uses four bells struck in five different four-note sequences, each of which occurs twice in the course of an hour.

1860 E. B. DENISON *Rudimentary Treat. Clocks* (ed. 4) p. vii, Cambridge and Westminster chimes. *Ibid.* 191 A very grand G hour bell to the BAGD bells of the peel, on which the Cambridge and Westminster quarters might then be struck. **1923** W. I. MILHAM *Time & Timekeepers* xvii. 298 To make and place a clock.. striking the hours and Westminster quarters on five bells. **1924** *Eng. Clocks & Watches* (Horol. Jrnl.) 42 The clock can be fitted with Whittington and Westminster Chimes. **1962** V. NABOKOV *Pale Fire* 43 Four hundred thousand times The tall clock with the hoarse Westminster chimes Has marked our common hour. **1967** 'R. SIMONS' *Taxed to Death* iv. 63 When Wace pressed the bell-button they heard Westminster chimes ringing in the hall. **1980** *New Grove Dict. Mus.* IV. 244/2 The best known of all clock chimes, the Westminster Quarters.., was derived from a quatrain in Handel's *Messiah*. In 1794 William Crotch wrote four variations on the fifth and sixth bars of 'I know that my Redeemer liveth'..for the new Cambridge University clock in Great St Mary's Church. They were accepted, and in 1845 were copied on the Royal Exchange clock, London. **1981** *Country Life* 12 Feb. 362 (Advt.), A superb clock..the three chimes, Westminster, Whittington and Winchester, obtainable at will.

West Nile. *Med.* [f. WEST *a.* + name of the river *Nile*.] Used *attrib.* to designate a mosquito-borne virus and the disease it causes, usu. a mild fever but sometimes a fatal encephalitis.

1940 K. C. SMITHBURN et al. in *Amer. Jrnl. Tropical Med.* XX. 471 The purpose of this paper is to report the isolation of one such [infective] agent, which we call the West Nile virus, and to describe some of its properties. **1955** *Sci. Amer.* Mar. 64/3 He concluded that West Nile fever was predominantly a disease of childhood. **1961** M. HYNES *Med. Bacteriol.* (ed. 7) xxv. 392 Antigenically related viruses have a similar ecology..in central Africa (West Nile encephalitis). **1983** *Oxf. Textbk. Med.* I. v. 104/2 Recognizable disease due to West Nile virus infection has been observed in Israel... No vaccine is yet available.

West of England. Also with hyphens. The name of a region of England, used *attrib.* and *absol.* to designate high-quality woollen broadcloth for which it has long been noted.

1843 *Penny Cycl.* XXVII. 555/2 In the West of England..each workman confines himself exclusively to a particular branch of the manufacture; and this has been supposed to have led to the excellence of the West of England cloth. **1882** *Queen* 23 Dec. (Advt.), Homespuns, Tweeds,..West of England cloths. **1936** 'N. BLAKE' *Thou Shell of Death* i. 8 His waistcoat..[of] West-of-England cloth. **1972** E. KERRIDGE in J. G. Jenkins *Wool Textile Industry in Gt. Brit.* 33/1 The same [*sc.* a more intensive use of capital and skilled labour] was true to a lesser extent..of the new superfine West of Englands. **1976** F. GREENLAND *Misericordia Drop* I. vii. 49 He wore ..grey West-of-England flannels.

Weston (we·stən). The name of Edward *Weston* (1850–1936) English-born electrical engineer, used *attrib.* **a.** In *Electr.*, designating a primary cell with electrodes of mercury and of cadmium amalgam and electrolyte of cadmium sulphate, used as a standard voltage source for calibrating electrical instruments.

1901 [see *METASTABILITY]. **1963** G. L. PICKARD *Descriptive Physical Oceanogr.* vi. 94 The Weston cell has a limited sensitivity. **1972** *Physics Bull.* Jan. 40/2 The Bureau is..interested in the Josephson effect for possible use in electrical standards, although for the moment the standard Weston cell is preferred for convenience.

b. In *Photogr.*, designating an obsolescent system of film speeds based on exposure meters made by the Weston Electrical Instrument Company or its successors.

1940 *Chambers's Techn. Dict.* 905/2 Weston film-speed. **1950** W. F. BERG *Exposure* 167 The introduction of Weston speed figures was of considerable importance. **1963** JERRARD & MCNEILL *Dict. Sci. Units* 105 The Weston number, the British Standards Institute speed number and the American Standards Association system also indicate the speed of the emulsion at its maximum sensitivity.

Westphalian, *a.* and *sb.* For **a, b** read **A, B** and add: **A.** *adj.* **2.** *Geol.* [ad. F. *westphalien* (A. de Lapparent *Traité de Géologie* (ed. 3, 1893) 819).] Of or belonging to a stratigraphic division of the Upper Carboniferous in Europe, above the Namurian and below the Stephanian. Also *absol.*

1901 [see *STEPHANIAN *a.*]. **1915** C. SCHUCHERT *Text-bk. Geol.* II. xl. 729 The Coal Measures formation is again divided into two series, the earlier half, or Middle Carboniferous, being widely known as the Westphalian.. when coal bearing. **1969** BENNISON & WRIGHT *Geol. Hist. Brit. Isles* ix. 221 In Britain the greater part of the Coal Measures belongs to the Westphalian. **1976** *Nature* 22 July 277/1 The age of this late retrogression..is synchronous with the intrusion of younger Variscan granites, a major break in sedimentation and the main phase of Hercynian folding (Sudetic phase: Westphalian in age).

west side. Add: **c.** Also **West Side.** That district of New York City which lies on the west side of Manhattan. Also *attrib.* *U.S.*

1858 *Harper's Mag.* July 283/2 As our friend entered the door a well-known 'West side' operator made his bid. **1903** *Ibid.* July 213 The abysmal craving of New Yorkers —West Side or East Side—is for friends. **1958** A. LAURENTS (*title of play*) West Side story. **1976** BOTHAM & DONNELLY *Valentino* iii. 25 He left the West Side and moved into a stable garret adjoining the home of millionaire Cornelius Bliss. **1981** J. VALIN *Dead Letter* xix. 182 That region of worn houses..that is the westside ghetto.

West Sider. *U.S.* [-ER[1].] A resident of Manhattan's West Side.

1903 *N.Y. Even. Post* 14 Nov. 4 The persistence with which the West Siders have followed up this question of the Broadway trees. **1914** G. ATHERTON *Perch of Devil* I. 2 Ida, forced..to accept employment with a fashionable dressmaker and consumed with envy of the 'West Siders' whose measurements she took. **1980** *N.Y. Times* 21 July A-8/2 Because the West Sider usually has so much money and time tied up in his lawn, plants and trees..he finds himself in a constant struggle to keep growing things from turning brown.

west wind, west-wind. Add: **b.** (Usu. with capital initials.) One of the four 'tiles' or discs called winds in the game of mah-jong; the player who takes this tile at the beginning of the game and sits opposite East Wind, or a player who succeeds him in being so designated.

1922 R. E. LINDSELL *Ma-Cheuk or Mah-Jongg*, 13 East discards a West wind. North..exposes a pair of West winds. **1952** M. STEEN *Phoenix Rising* vii. 154 She slid the West Wind on to her ebony rack. **1960** [see *EAST WIND b]. **1976** R. C. BELL *Discovering Mah-jong* 15 The third round is West Wind's round.

westy, var. *WESTIE.

wet, *sb.*[1] Add: **6.** A 'wet' person (see *WET *a.* 15 b); *spec.* a politician with liberal or middle-of-the-road views on controversial issues (often applied to members of the Conservative Party opposed to the monetarist policies of Margaret Thatcher).

1931 F. L. ALLEN *Only Yesterday* x. 254 The Government putting wood alcohol and other poisons into industrial alcohol to prevent its diversion, and the wets thereupon charging the Government with murder. **1933** D. L. MURRAY *Eng. Family Robinson* vii. 159 He's quite right... You *are* a wet! Who does pay regularly? **1939** G. HEYER *No Wind of Blame* xvi. 299 He's a regular wet, that chap: doesn't hold with blood sports. **1948** C. DAY LEWIS *Otterbury Incident* ix. 111 Don't be a wet. We'll get off all right. **1961** C. WILLOCK *Death in Covert* xi. 201 'That wet,' said fford, reverting to a schoolboy expression. 'Wet he may be, but he knows about lighters.' **1974** I. MURDOCH *Sacred & Profane Love Machine* 76 You've made me into a bloody wet. I'm a fighter and you've made me into a weak person. **1976** S. BARSTOW *Right True End* III. xii. 180 She likes to throw out these challenges that put me to the test and make me feel a weak-kneed wet. **1980** B. W. ALDISS *Life in West* ii. 42 He's a bit of a wet, but quite a sound art-historian. **1980** *Sunday Tel.* 6 Apr. 9 At least Sir Ian Gilmour and other political wets do not have their hair pulled. **1980** *Times* 7 Apr. 9/1 Mr James Prior, Secretary of State for Employment, is described in one Sunday paper as 'the champion of the Tory wets'. *Ibid.*, Who..are to be counted among the wets? The answer seems to be anybody who crosses the Prime Minister in fashioning a particular policy. **1980** W. WHITELAW in *Observer* 23 Nov. 11, I don't really know what a wet is. **1983** *Age* (Melbourne) 5 Oct. 13 [Of U.K. politics] In contrast to the expansionist, protectionist and welfare-oriented Wets, the Dries stand for small government, economic rationality and individual responsibility.

7. *U.S. slang.* = *wetback* s.v. *WET *a.* 20.

1973 *Daily Tel.* (Colour Suppl.) 16 Feb. 13/1 In the past, unscrupulous employers would employ a 'wet' for a month, then denounce him to the Immigration authorities before pay day. **1979** *Time* 8 Oct. 33/1 A group of 'wets', or 'undocumented workers', as official jargon calls them. Most of the Mexican aliens are poor, frightened and docile people whose only crime is seeking to find work and a better life in the U.S. **1979** G. SWARTHOUT *Skeletons* 104 Why doesn't this [system] detect every wet who puts a toe across the line?

wet, *sb.*[2] Add: **2.** *slang.* Urination, the act of urinating; urine. *rare.*

1925, 1975 [see *WET *v.* 17].

wet, *a.* Add: **2. f.** (Later examples.) Freq. with def. article and also with capital initial. *colloq.* (chiefly *Austral.*).

1908 MRS. A. GUNN *We of Never-Never* i. 5 He..wired an inane suggestion about waiting till after the Wet. **1934** *Bulletin* (Sydney) 29 Aug. 20/4 In the 'wet' it became a miniature lake at which one cocky's horses were wont to drink. **1941** I. L. IDRIESS *Great Boomerang* vii. 51 An early and heavy wet would set in that would spill water for a thousand miles south-west. **1968** S. L. ELLIOTT *Rusty Bugles* in E. Hanger *Three Austral. Plays* I. ii. 41 That's what everyone tells me. Wait until you've done a Wet. **1981** P. CAREY *Bliss* iii. 135 Each year when the wet ended she found herself looking forward to it again.

4. e. Applied to a removable liner for the cylinder of an internal-combustion engine that has cooling water flowing between it and the cylinder wall.

1935 *Jrnl. R. Aeronaut. Soc.* XXXIX. 470 The four cylinders 63 m/m. bore by 120 m/m. stroke were steel jacketed, wet liners, having four valves per cylinder. **1959** *Motor* 14 Oct. 304/2 Cylinder blocks with individual wet liners of cast iron. **1975** M. J. NUNNEY *Automotive Engine* iii. 94 Positive sealing arrangements must be made with wet cylinder liners to prevent leakage of coolant into the crankcase. **1981** H. E. ELLINGER *Automotive Engines* x. 157/2 Coolant flows around the cylinder sleeve, so this type of sleeve is called a wet sleeve.

5. d. *to get wet*: to lose one's temper, become angry. *Austral. slang* (? *Obs.*).

1898 *Bulletin* (Sydney) 17 Dec. Red Page, To *get narked* is to lose your temper; also expressed by *getting dead wet*. **1916** C. J. DENNIS *Songs of Sentimental Bloke* 42 Romeo gits wet as 'ell. **1945** BAKER *Austral. Lang.* 121 A man in a temper is said..*to get wet*.

e. *to get* (someone) *wet*: to gain the upper hand over; to have at one's mercy. *N.Z. slang.*

c **1926** 'MIXER' *Transport Workers' Song Bk.* 29 He skites about in-fighting. Stick to him, Mick; you've got him wet. **1941** *Coast to Coast 1941* 124 'Got you wet, haven't they?' He flung the remark over his shoulder as he went over to his bed. **1945** F. SARGESON *When Wind Blows* vi. 40 Now we've got 'em wet.

f. Of those activities of intelligence organizations, esp. of the K.G.B., that involve assassination. *slang.*

1972 A. PRICE *Col. Butler's Wolf* vi. 58 The Russian slang for Spetsburo Thirteen was *Mokryye Dela*—'The department of wet affairs'..and to get wet was the feared, inevitable fate of traitors pursued by the special bureau. 1975 J. Grady Shadow of Condor ii. 47 'The courier made other mistakes... It was a wet affair.'.. Ryzhov like to use the old KGB liquid euphemism for executions. 1980 J. Gardner Garden of Weapons II. vii. 191 He had seen men killed: and killed them himself: he had directed 'wet operations', as they used to be called.

15. b. Inept, ineffectual, effete; also as quasi-adv. and in comb. *wet fish*, a wet individual, a 'drip'. Also *spec.* in *Politics* (see quots. 1981 and 1983). Cf. *WET sb.1 6.

1916 'Taffrail' Pincher Martin ii. 27 I'll give yer a clip 'longside the ear'ole if you ain't careful. Don't act so wet. 1924 P. Marks Plastic Age 94 They attended a performance of Shaw's 'Candida' given by the Dramatic Society and voted it a 'wet' show. Ibid. 192 A man is wet if he isn't a 'regular guy'; he is wet if he isn't 'smooth'; he is wet if he has intellectual interests..; and he is wet.. if he is utterly stupid. 1938 E. Bowen Death of Heart II. iv. 239 Cecil is so wet! Coming early like that, then sticking round like that. 1944 A. Christie Towards Zero 86 Audrey marry that wet fish? She's a lot too good for that. 1963 Wet fish [see *MOOSE1 c]. 1969 K. Amis Green Man iv. 180 The Jesus of the Gospels can be a bit of a wet liberal at times. 1973 P. O'Donnell Silver Mistress iv. 74 Don't talk wet, Jan. There's nothing you could do. 1980 Times Lit. Suppl. 28 Nov. 1355/2 The contrast between the splendid façade and the rather wet interior of the man [sc. Havelock Ellis], who was kind and gentle and distinguished, but also distressingly absent, indifferent and faint.

1981 Observer 26 July 12/3 The term 'Wet' was originally used by Mrs Thatcher, who meant it in the old sense of 'soppy', as in 'What do you mean the unions won't like it, Jim? Don't be so wet.' It meant feeble, liable to take the easy option, lacking intellectual and political hardness. Like so many insults, it was gleefully adopted by its victims, and so came by its present meaning of liberal, leftish, anti-ideological. 1982 Listener 23/30 Dec. 6/3 In considering the promotion of wet (or wettish) Ministers, she will tell herself that Pope was right. 1983 Age (Melbourne) 5 Oct. 13 Britain's Tory Prime Minister, Mrs Margaret Thatcher, began this vogue terminology by contemptously dismissing dewy-eyed dissenters from her arid Right-wing policies as 'wet'.

c. *all wet*: mistaken, completely wrong. orig. and chiefly *U.S.*

1923 N.Y. Times 9 Sept. VII. 2/1 All wet, all wrong. 1931 Kansas City Times 29 Aug., Alfalfa Bill Murray may be 'all wet' in his state-line bridge and oil production controversies. 1940 G. Ade Let. 5 June (1973) 221 Regarding the Rotary Clubs, I..am an honorary member. I think the organization is alright and that Sinclair Lewis was all wet when he tried to poke fun at the small town booster. 1941 E. B. White Let. Summer (1976) 216, I haven't had much time to think things over and I am probably all wet on a lot of things in here. 1951 A. Baron Rosie Hogarth 282 You're all wet if you think I'm giving up that easy.

d. *wet behind the ears*: see *EAR sb.1 1 c.

16. c. For *U.S.* read 'orig. and chiefly *U.S.*' (Earlier and later examples.) Hence as quasi-adv. in phr. *to go* or *vote wet*. Cf. *DRY a. 11 a.

1870, etc. [see *DRY a. 11 a]. 1888 North American (Philadelphia) 3 Apr. 1/1 Forty-nine counties have voted 'dry', and thirty-three 'wet'... Thirteen of twenty towns went 'dry', and seven 'wet'. 1954 K. Amis Lucky Jim 109 The still recent tradition of a 'wet' Summer Ball. 1974 Times 7 Oct. 4/1 Flintshire, Radnorshire, Breconshire.. voted to go wet. Ibid. 4/2 That poll ended the curious situation of one inn which straddled on the wet-dry border... The public bar was dry and empty, but the lounge bar was wet and crowded.

d. (Earlier and later examples.)

1888 Battle Creek (Michigan) Weekly Jrnl. 29 Feb., This is the first great victory for the 'wets'. 1896 [see *DRY sb. 5]. 1968 Daily Tel. 8 Nov. 1/4 The 'wets' gained three counties..in the Welsh referendum on Sunday drinking.

17. b. Designating chemical tests and analysis involving the use of solvents or other liquids; = HUMID a. c; so *wet-chemical* adj. Cf. WAY sb.1 14 c.

1800, etc. [see sense 17 in Dict.]. 1858 Phil. Mag. XVI. 331 This method is particularly adapted..when the substances of this group occur in so small quantities that they are no longer recognizable in the wet way. 1932 F. Soddy Interpretation of Atom xv. 253 Almost all the ordinary chemical tests for the common elements, by which they are identified in the ordinary reactions of 'wet' analysis, are not tests for the elements, but for their ions. 1967 Electronics 6 Mar. 29 (Advt.), You can be sure of a complete refinery service... Including, under one roof..laboratory facilities for wet chemical analysis and electrolytic methods of analysis. 1973 Nature 8 June 365/1 Since the Second World War, physical methods of analysis..have increasingly displaced wet chemistry from the industrial routine analytical laboratory. 1977 New Scientist 17 Feb. 384/1 Traditional methods of detecting nitrogen oxides as air pollutants monitor the change in colour of an acid permanganate solution as the oxides are absorbed. These wet-chemical methods.. require relatively large samples of gases.

18*. Of natural gas: containing significant amounts of the vapour of higher hydrocarbons.

1926 Daily Colonist (Victoria, B.C.) 18 July 16/7 Wet gas flow of 3,000,000 feet a day was struck at McLeod No. 2 well in Turner Valley last night. 1948 Petroleum Handbk. (Shell Petroleum Co. Ltd.) (ed. 3) ix. 154 Gases produced in contact with oil can be either 'dry' or 'wet', depending on the nature of the crude oil and the method of separating the gas from the oil. 1982 Shell Briefing Service No. 5. 5/2 LPG is essentially a mixture of propane and butane stored at ambient temperature under moderate pressure. It can be derived from the gas associated with crude oil or from 'wet' natural gas directly at the well.

19. a. *wet-plucked.*

1960 Farmer & Stockbreeder 19 Jan. Suppl. 41/3 At slaughter the birds are all..wet-plucked by machine and then eviscerated. 1969 R. Adlard in R. Blythe Akenfield xiv. 234 The feathers are no use because the chickens [in factory farms] are wet-plucked, so there is only a mess.

b. *wetmouthed.*

1951 Dylan Thomas Sel. Lett. (1966) 352 [Fresh recruits] see before them in the hot moonlight wet-mouthed Persian girls from the bazaar.

20. wetback orig. and chiefly *U.S.*, an illegal immigrant who crosses the Rio Grande from Mexico to the U.S.; also *attrib.* and *transf.*; **wet bar** *N. Amer.*, a bar or counter in a private house from which alcoholic drinks are served; **wet bob**: so **wet bob** *v. intr.*; **wet-bobbing** *vbl. sb.*; **wet diggings** orig. *U.S.*, gold diggings in or near a river or stream; cf. *dry diggings* s.v. *DRY a. C. 3; **wet dream**, an erotic dream which causes a man or boy to have an involuntary sexual orgasm during sleep; also *fig.*; **wet-eared** = *wet behind the ears* s.v. *EAR sb.1 1 c; **wet end**, that end of a paper-making or drying machine into which the wet material is passed; **wet lease** (see quot. 1979); so **wet lease** *v. trans.*, **wet-leased** *ppl. a.*; **wet leg** *slang*, a self-pitying person; **wet look** [LOOK sb. 2 f], an appearance of a wet or shiny surface; usu. *attrib.*, esp. of fabrics (see quot. 1968); **wet pack**, a compact waterproof bag which folds or rolls up and is designed for carrying toilet articles; **wet-point** *a.*, of villages, settlements, etc.: having an available water supply; **wet process**, a manufacturing process involving the use of water or other liquid; freq. *attrib.*; **wet rent**, a levy paid to a brewery by a publican in a tied public house in proportion to the amount of beer sold (see also quot. 1907); **wet shave**, a shave (SHAVE sb.2 2) carried out with the aid of a razor, soap, and water as opp. to a (usu. electric) razor alone; so **wet shaver**, someone who shaves by this method; **wet shaving** *vbl. sb.*; **wet smack** *slang* (chiefly *U.S.*), a spoil-sport; **wet spinning**, (*a*) spinning of natural fibres when they are wet from passage through a water bath; (*b*) spinning of man-made fibres in which the spinneret extrudes the streams of liquid into a coagulating bath; so **wet-spin** *v. trans.*, **wet-spun** *ppl. a.*; **wet strength**, the strength of paper and textiles when wet; **wet suit**, a suit, usu. of rubber, worn by divers, surfers, etc., to protect them from the cold; hence **wet-suited** *a.*; **wet time**, in the building trade, time during which work cannot be carried out owing to bad weather; **wet trade** (see quots.); **wet-weather** *a.*, (*a*) associated with or occurring in rainy weather; (*b*) designed for use in rainy weather; **wet-white**, liquid white theatrical make-up; **wet wing** *Aeronaut.* (see quot. 1969); usu. *attrib.*

1929 Foreign Affairs Oct. 101 The peon walks or swims across..and is welcomed by his countrymen here as a 'wet back'. 1972 Observer (Colour Suppl.) 28 May 28/1 Last year in California alone, border patrols turned back 27,000 wetbacks (the contemptuous name derives from their practice of swimming the Rio Grande to reach the US). 1978 N.Y. Times Mag. 23 July 23/2 Wetbacks (a derogation of Mexicans swimming the Rio Grande to slip into the U.S.) became illegal aliens, and are now referred to as undocumented persons. 1979 Guardian 8 June 5/2 Illegal migrants from South China..are getting into Hong Kong..usually swimming the last part of the trip. The total of Chinese 'wetbacks' intercepted..in the first week of June alone came to 3,722. 1982 T. Beattie Diamonds xii. 100 It might be that wetback job I did... But they can't prove anything. 1968 Globe & Mail (Toronto) 15 Jan. 23/6 (Advt.), Panelled family room, games room, wet bar. Real executive home! 1978 R. Thomas Chinaman's Chance xx. 206 Ploughman turned to find Reginald Simms standing by a small wet bar across the room. 1884 J. Montagu Let. Mar. in Troubridge & Marshall John Ld. Montagu of Beaulieu (1930) 31, I have been out wet-bobbing several times and am getting coached. 1901 G. Frankau Eton Echoes 40 (heading) Wet Bobbing. 1926 Spectator 3 July 11/1 Any alternative summer game or sport..such as is provided by 'wet-bobbing' at a school like Eton. 1849 J. Wyld Geogr. & Mineral. Notes 21 The works are divided into two classes,—Dry Diggings and Wet Diggings. 1862 J. L. C. Richardson Sk. Otago 48 See how the wet diggings will pay in the summer time. 1935 E. B. Buckbee Saga Old Tuolumne 11 He worked ceaselessly throughout the day lifting gold from the 'wet diggings'. 1965 G. J. Williams Econ. Geol. N.Z. vii. 72/1 The conglomerates accumulated on the slopes of the mountains are the proper field for the 'dry diggings', while from the gravel and sand of the beds of rivers and smaller streams the gold is obtained by 'wet diggings'. 1851 W. Acton Pract. Treatise Diseases of Urinary Organs (ed. 2) I. ii. 226 Spermatorrhœa..is known..as nocturnal or diurnal emissions, pollutions, wet-dreams, [etc.]. 1921 H. Crane Let. 11 Feb. (1965) 53 The wet-dream explosions of Virgil Jordan and McAlmon. Their talk is all right—but what is true of it has been said adequately before. 1946 B. Marshall George Brown's Schooldays 170 Well, what are you standing there looking like a wet dream for? 1963 A. Heron Towards Quaker View of Sex ii. 16 It is at this stage that nocturnal emissions or 'wet dreams' as they are often called, are frequently the first clear sign of sexual maturity in the boy. 1971 B. W. Aldiss Soldier Erect 10 Jesus, what a wet dream of a party that was! 1978 A. Neave Nuremburg viii. 86 He was said by the prosecution to have boasted to his chauffeur of nightly wet dreams and exhibited the semen to prove it. 1967 E. McGirr Hearse with Horses iii. 50 If a race was fixed they wouldn't need a wet-eared kid mixed up with it. 1971 F. Forsyth Day of Jackal I. i. 21 Apart from a few wet-eared ninnies who refused to come, Rodin led his entire battalion into the military putsch of April 1961. 1888 Cross & Bevan Text-bk. Paper-Making x. 154 This part of the machine, which is called the 'wet-end', is placed at a slight slope. 1927 T. Woodhouse Artif. Silk iii. 25 The wet pulp is now run on to the feed end, usually termed the 'wet-end', of the drying machine. 1962 Wet end [see dry end s.v. *DRY a. C. 3]. 1962 Aeroplane & Astronautics CII. 88/2 Philippine Air Lines has wet-leased (i.e., aircraft plus flight crew) a Boeing 707 from Pan American. 1977 Indian Express 18 May 1/2 The Airbus will be either wet leased or chartered by Air-India. 1979 Daily Tel. 8 June 36/6 Aircraft can be leased by the hour, day, week, month, quarterly or longer on a 'dry' lease which means that crews are not provided, or on a 'wet' lease which means that the owner of the aircraft also supplies crew and, in some cases, the necessary fuel. 1978 Observer 29 Jan. 1/5 These too will have to be taken out of service for modifications, and their place taken by 'wet leased' foreign aircraft (that is, planes taken complete with their own crews). 1922 D. H. Lawrence Let.? 12 Oct. (1962) II. 726 Being too much of a wet-leg, as they say in England, nakedly to enter into the battle. 1929 —— Pansies 124 It is strange to think of the Annas, the Vronskys, the Pierres, all the Tolstoyan lot Wiped out... And the Tchekov wimbly-wambly wet-legs all wiped out. 1981 Times Lit. Suppl. 3 July 745/1 We know how much Auden hated wet-legs, how constantly he repeated his many litanies of his own good fortune. 1968 J. Ironside Fashion Alphabet 102 The 'Wet Look' is a chemical finish to fabrics to make them appear shiny and wet. 1969 Times 24 Nov. 16/2 Natural coloured python or wet-look patent are the most fashionable finishes for day. 1970 D. Uhnak Ledger (1971) ix. 114 Her lips, shining with a wet-look lipstick, quivered. 1971 Daily Tel. 2 Feb. 11 (caption) The chair and stool covered in white wet-look fabric. 1981 Westindian World 31 July 14/2 (Advt.), Hot & cold straightening, curly perm, wet look. 1928–9 Army & Navy Stores Catal. 419/1 Wet Pack. Fitted with comb and nail file, etc. Size closed, 5½ × 4½ in... Pigskin 12/–. 1974 Harrods Christmas Catal. 18/2 For travelling men... Two wet packs with waterproof linings. 1920 Wet-point [see dry-point village s.v. *DRY a. C.3]. 1969 G. C. Dickinson Maps & Air Photographs xiv. 216 (heading) 'Wet-point' sites—i.e. places with an available water supply. 1909 Webster, Wet... Chem., etc. Employing, or done by means of, or in the presence of, water or other liquid... The wet process or way. 1930 Engineering 3 Jan. 18/3 The Assano Portland Cement Company's works at Nishitama. This is a wet-process plant. 1945 H. D. Smyth Gen. Acct. Devel. Atomic Energy Mil. Purposes vii. 75 Study of product recovery processes as a whole (wet processes, physical methods). 1969 Wet process [see *BY-PRODUCT b]. 1907 F. E. E. Bell At Works v. 122 Some of the yearly benefit clubs of which the head-quarters are at public-houses demand..an extra contribution, from 1d. to 3d., what is called the 'wet rent', which is quite deliberately allowed for drink each meeting-night. 1967 Economist 29 Apr. 480/2 The Jones board has implicitly accused the brewers of subsidising too many low volume country pubs, by charging less than the market rents but rather more for their beer, a practice known in the trade as a 'wet rent'. In actual fact, wet rents are steadily becoming proportionately less important, and the brewer's idea is to protect the publican against the ups and downs of trade by charging him, in effect, a rent that varies slightly with beer sales, thus identifying his interest more closely with that of the brewer. 1978 Times 3 May 19/6 The brewers ..continue phasing out 'wet' rents under which a tenant pays more or less to the brewery according to the amount of beer sold through the pub. 1976 NBR Marketplace (Wellington, N.Z.) III. 2/1 Something over 50 per cent of the estimated 900,000 regular shavers in New Zealand prefer a wet shave start to the day. Ibid. 2/3 About 94 per cent of wet shavers use the safety razor with double-edge blades or the modern single-edge blade systems. 1964 Financial Times 25 Feb. 11/8 The chief obstacle at present is the wet-shaving industry's promotion of the new stainless steel blades. 1980 'D. Kavanagh' Duffy iii. 44 They only took the television set and his electric razor... He went back to wet shaving. 1927 Amer. Speech III. 221 Wet smack,..something unsatisfactory; applies particularly to an individual who spoils a party; a kill-joy. 1929 Wodehouse Mr. Mulliner Speaking i. 33 The man is beyond question a flat tyre and a wet smack. 1977 Maledicta Summer 17 If she is actually frigid, she's a wet smack. 1963 A. J. Hall Textile Sci. ii. 75 Some of these polymers are soluble in organic solvents..and thus allow the preparation of solutions which can be dry spun..or wet spun—that is, extruded into a coagulating bath. 1973 Materials & Technol. VI. iv. 328 Polyacrylonitrile solutions have been wet-spun..into a coagulating bath. 1864 Wet spinning [see dry spinning s.v. *DRY a. C.3]. 1927 T. Woodhouse Artif. Silk 28 The coagulation by means of liquid of any kind has given rise to the term 'wet spinning', whereas the term 'dry-spinning' has been applied in all cases where the solvent is vaporized. 1969 A. J. Hall Stand. Handbk. Textiles (ed. 7) iii. 127 In wet spinning the roving is led through a trough of hot water..so that the fibres are softened. 1973 Materials & Technol. VI. iv. 295 Another method of taking up the wet spun yarn. 1960 R. W. Marks Dymaxion World of B. Fuller 59/1 Even in 1954 Kraft paper having exceptional 'wet tensile strength' had been developed—'wet strength' meaning the ability of the paper to retain its structural quality when saturated. 1962 J. T. Marsh Self-Smoothing Fabrics xiv. 211 These examples of dimensional stability are of some consequence, and indeed of great consequence with fibres of regenerated cellulose whose low wet-strength is a serious defect but one which is remedied by the crease-resisting process. 1973 Nature 27 Apr. 588/1 Cross linking has been used for over thirty years in making 'wet strength' papers. 1955 Wet suit [see dry suit s.v. *DRY a. C.3]. 1964 Skin Diver Oct. 19 An American skin diver aboard an Irish fishing boat..had a difficult time convincing the skipper that his 'wet' suit would save a man's life if he fell into the freezing water. 1970 Daily Tel. (Colour Suppl.) 18 Sept. 12 On deck three of us, clad in rubber wetsuits, prepared to slip over the side. 1972 Islander (Victoria, B.C.) 4 June 16/1 The wet suit, worn to keep the diver warm, is almost a necessity in these northern waters. 1984 S. Townsend Growing Pains A. Mole 78 She looked dead erotic in her black wetsuit and crash helmet. 1972 Nat. Geographic Oct. 584 Wet-suited author examines the giant wraparound grin of a right whale. 1978 D. Williams Treasure up in Smoke xix. 174 The alerted wet-suited figure had..waded to the beach. 1938 Times 5 May 10/4 For nearly 20 years the building trade operatives have..claimed that for uncontrollable irregularities of employment..there should be a scheme of compensation for loss of earnings. The phrase which they used to focus the claim was 'payment for wet time'. 1952 Economist 12 July 118/2 Steel erectors on American building sites do not enjoy either a guaranteed week or payment for 'wet-time'. 1962 Listener 26 July 154/3 The 'do-it-yourself' enthusiast who is preparing to tackle garden operations involving the use of cement, lime, and water—the so-called 'wet trades'. 1973 Times 24 Feb. 13/1 The shortage of skilled workers, particularly in the 'wet trades' of bricklaying and plastering. 1858 T. S. Woodward Let. 20 Dec. in Reminisc. (1939) 157 Fortunately, we found a little wet-weather spring near the top. 1901 [see WET a. 2 e]. 1922 M. A. von Arnim Enchanted April i. 8 Big grey eyes almost disappearing under a smashed-down wet-weather hat. 1934 M. V. Hughes London Child iii. 28 The boys were off on some long wet-weather tramp. 1978 'D. Rutherford' Collision Course 182 I'm gambling on rain... We're giving you wet-weather tyres. 1922 M. Arlen Piracy III. xiv. 256 Just look how depraved they are! They are covered with verdigris, but they call it wet-white! 1976 'D. Fletcher' Don't whistle 'Macbeth' 51 The first time I sang Elvira, I had to cover myself from head to toe with wet-white. [1958 Flying Rev. Oct. 37/1 Scheduled to Supplement earlier Stratofortresses currently serving with the Strategic Air Command, the B-52G employs a 'wet' integral-tank wing which substantially increases the bomber's unrefuelled range.] 1961 Flight LXXIX. 818/2 These new 'wet wing' versions, with greatly increased weight and machined-plank wing skins, have suffered local stresses greater than any experienced with the earlier versions of lower weight and performance. 1969 New Scientist 25 Sept. (Microbes in Industry Suppl.) 23/2 In modern 'wet-wing' aircraft, such as Concorde, the fuel is simply pumped into the wings which are coated internally with sealants. In older aircraft..the fuel is contained in rubber bags in the wings.

wet, *v.* Add: **I. 1. b.** *Sci.* Of a liquid: to cover or penetrate (a substance or object) readily, so that a small quantity spreads uniformly over it rather than lying as droplets upon it.

A common criterion of wetting is the angle that the surface of a droplet makes, at its point of contact, with the surface on which it rests (as measured through the liquid): the liquid is said to wet or not to wet the surface according as the angle is less or greater than 90 degrees.

1855 D. Lardner Hand-bk. Nat. Philos.: Hydrostatics I. v. 69 If a liquid be poured into a vessel whose sides are of such a nature as to be wetted by it, the liquid.. will be curved upwards near the points where it touches the side. 1884 A. Daniell Text Bk. Princ. Physics xi. 246 Objects which are wetted by the liquid in which they float are thus apparently attracted by it; those which are not so are apparently repelled. 1932 Phytopathology XXII. 926 The presence of an appreciable quantity of sodium hydroxide..increases the ease with which the leaves can be 'wetted' in the solution. 1967 M. Chandler Ceramics in Mod. World vi. 171 Silicon nitride..is not wetted by molten metals. 1974 Encycl. Brit. Macropædia XI. 782/1 The adhesion of water to glass at an air-water-glass interface is greater than the cohesion of water, and hence water is said to wet the glass... The cohesion of mercury is greater than its adhesion to glass and it does not wet the glass. 1978 Nature 20 July 237/1 An example of this is a droplet of the liquid resting on a solid surface. When the contact angle is less than 90°, the liquid wets the solid. Molten beryllium does not wet BeO because the contact angle exceeds 90°.

5. c. Also *fig.*, to become excited or upset (as if about to void urine in one's clothes).

1979 'M. Underwood' Smooth Justice i. 35 There are quite a few people who'll wet their pants if I get sent down. 1981 A. Price Soldier no More 184 We did see the Histories season at Stratford, I grant you. But I don't remember any schoolgirls wetting their pants next to me.

d. *refl.* To urinate involuntarily. Also *fig.* (cf. sense 5 c above).

1922 Joyce Ulysses 730 What do I care with it dropping out of me and that black closed breeches he made me buy takes you half an hour to let them down wetting all myself. 1970 G. F. Newman Sir, You Bastard 258 The Sunday editors would wet themselves; they liked nothing better than a sordid purge in an institution. 1976 Times Lit. Suppl. 30 Jan. 100/5 She also sweats, weeps, vomits and wets herself.

7. a. Also, *to wet* (one's) *beak, beard.*

1939 T. S. Eliot Old Possum's Bk. Pract. Cats 16 For to the Bell at Hampton he had gone to wet his beard. 1978 J. Carroll Mortal Friends I. v. 53 Is there a public house here where a fellow could wet his beak?

e. (Earlier example.)

1783 J. Woodforde Diary 9 Oct. (1926) II. 97 With the latter I walked to the Swan and there wetted with him that is, drank a glass of Wine.

8. b. *to wet the baby's head* and varr.: to drink to celebrate the birth of a child. *colloq.*

1885 W. Westall Old Factory xxiv. 161 'We'll wet little Mabel's head with some of it.' 'What mean you?'.. 'Why my wife was brought to bed last night of a little lass as we are going to call Mabel, and I'd like us to drink to her health. That's what we call wetting a child's head in these parts.' 1924 Lawrence & Skinner Boy in Bush xiv. 210 Come along in—all welcome!—an' wet the baby's eye. 1953 E. Simon Past Masters III. v. 173 At the party given to 'wet the baby's head' the McGillivrays' friends and relations produced only large and expensive gifts. 1970 Guardian 2 May 3/7 If he had not been wetting the baby's head, and so been slightly above proof, he might have run for it.

10. b. For '*dial.*' read '*dial.* and *colloq.*' (Earlier and later examples.) Also with tea-leaves as *obj.*

1902 Cornh. Mag. Dec. 776, I ha' wetted th' tea pretty nigh half-an-hour ago. 1939 Joyce Finnegans Wake 585 You never wet the tea! 1944 M. Laski Love on Supertax viii. 77 Make yourself at home, and I'll just wet the tea-leaves. 1978 I. Murdoch Sea 419 'I'll wet the tea,' said Hartley and disappeared into the kitchen.

II. 17. To urinate. Also *fig.*

1925 D. H. Lawrence Novel in Reflections on Death of Porcupine 122 But see old Leo Tolstoi wetting on the flame. As if even his wet were absolute! 1935 V. Woolf Let. 21 June (1979) V. 403 The marmoset is just about to wet on my shoulder. 1954 J. Steinbeck Sweet Thursday xiv. 82 House-broken dogs wet on the parlor rug. 1975 J. Cleary Safe House ii. 71 The children want to wet... Come on, love. Have your wet.

III. 18. The vb. stem in comb., as **wet-bed** = *bed-wetter* s.v. *BED sb. 19.

1934 'J. Spenser' Limey breaks In iv. 61, I lay awake for so long that I heard the night watchman come to call the wet-beds. 1960 J. Stroud Shorn Lamb xviii. 204 Does he enurete?.. I've got four chronic wet-beds already.

weta (we·tā). *N.Z.* [Maori.] Any of several wingless orthopteran insects of the genus *Deinacrida*, *Pachyrhamma*, or *Hemideina*.

1843 E. Dieffenbach Trav. N.Z II. III. 396 Weta—an insect so called. 1857 C. Hursthouse N.Z. I. v. 123 The Weta, a suspicious-looking scorpion-like creature, apparently replete with 'high concocted venom', but perfectly harmless. 1863 S. Butler First Year in Canterbury Settlement ix. 441 One of the ugliest-looking creatures..is called 'weta', and is of tawny scorpion-like colour, with long antennae and great eyes, and nasty squashy-looking body, with (I think) six legs. 1888 Trans. N.Z. Inst. XXI. 41 Not a sound was heard in that lonely forest, except..the sharp noise produced by the weta. 1949 [see *HUHU]. 1961 R. Park Hole in Hill (1962) xiv. 115 A giant glistening black insect..waving its antennae... 'It's only a weta!' 1975 E. Hillary Nothing Venture, Nothing Win ii. 40 When I was making up the bed I found a huge weta..in one of the blankets.

wetland (we·tlænd). [f. WET a. + LAND sb.] An area of land that is usually saturated with water, often a marsh or swamp. Also *attrib.* Also *pl.* (sometimes const. as *sing.*).

1743 M. Catesby Nat. Hist. Carolina II. p. iv, On this wet Land grows a variety of Evergreen Trees and Shrubs. 1847 H. Howe Hist. Coll. Ohio 98 'Wet land'..by judicious cultivation..rapidly improves in fertility. 1955 Sci. News Let. 29 Oct. 281/2 The wetland partridge is about twice the size of the valley quail. 1965 New Scientist 17 June 763/3 Wetlands are defined to include marshes, bogs, swamps and any still water less than six metres deep. 1969 Nature 19 Apr. 239/2 Wetland ecosystems in the limited sense of this work are defined as ecosystems with a watertable, above, at or very near the substrate surface, the substrate remaining saturated throughout the year. 1979 Daily Tel. 25 Oct. 11/4 Plans to protect the Somerset wetlands—an area of rare wildlife, whose future has been in dispute—are to be prepared. 1980 National Trust Spring 16/3 We intend to preserve this swamp area in its natural state as a haven not only for wetland flora but also for birds and animals. 1985 Daily Hampshire Gaz. (Northampton, Mass.) 9 Aug. 17/5 Under state law construction can not take place on wetlands unless there are plans to replace the wetlands.

wetly, *adv.* (Examples in sense *15 b of the adj.)

1975 Times Lit. Suppl. 12 Dec. 1486/3 Raffles's relations with his accomplice, Boswell, and room-mate, the wet and unendearing Bunny. Bunny..went to prison..to shield A.J., who rather let him down; but Bunny wetly forgave him. 1978 Guardian Weekly 12 Mar. 21/3 A peaceable fellow-officer, played slightly wetly by the American Beau Bridges.

Wetmore (we·tmōəɹ). The name of Alexander *Wetmore* (1886–1978), American ornithologist, used *attrib.* in *Wetmore order* to designate the system of bird classification developed by him.

1965 Jrnl. Lancs. Dial. Soc. Jan. 5 They are listed in Wetmore Order, i.e. the system of classification which is generally accepted in modern bird books. 1979 Nature 29 Mar. 490/1 Indeed Wetmore's name has become a household word among ornithologists, for his classification of the birds of the world, forming the basis for the well known 'Wetmore order', has, with modifications necessitated by new knowledge, become the generally accepted arrangement, as adopted in Peters' Checklist of Birds of the World and other authoritative compilations.

wetness. Add: **c.** Feebleness, ineptness. Cf. *WET a. 15 b.

1977 Times 29 Sept. 4/1 It was surely not Liberal 'wetness', as the incoming president, Mr. Gruffydd Evans, termed their traditional virtues of niceness and fair-mindedness. 1981 R. D. Edwards Corridors of Death vii. 33 A profession which regards loyalty as weakness and decency as wetness. 1983 Times 4 Nov. 7/1 The idea that an ally has a right to independent judgment is too easily dismissed as what Mrs Thatcher might describe..as wetness.

wettable, *a.* Add: (Examples corresponding to *WET v. 1 b.)

1903 W. R. Fisher tr. Schimper's Plant-Geogr. III. i. 225 The foliage in a constantly humid climate is as a rule easily wettable. 1955 [see *CAPTAN]. 1976 McGraw-Hill Yearbk. Sci. & Technol. 208/2 Spraying the wettable powder or flowable formulation results in distribution patterns that are particulate in form rather than lamellar.

Hence **wettabi·lity**, the property of being wettable; the degree to which something may be wetted (*WET v. 1 b).

1913 Chem. Abstr. VII. 3441 The wetting of glass by different pairs of liquids was investigated with special reference to the effect of prolonged wetting upon the 'wettability'. 1933 Amer. Jrnl. Sci. XXV. 329 The 'wettability' of the particle in water. 1973 Nature 2 Mar. 14/1 The wettability of leaf surfaces is influenced by the fine structure and chemical composition of the wax. 1977 J. L. Harper Population Biol. Plants xi. 379 Larrea affects the water relations of the soil surrounding the plants, reducing soil wettability.

wetted, *ppl. a.* Add: **2.** *Aeronaut.* Of an aircraft surface: in contact with the moving airflow.

1916 F. W. Lanchester Flying Machine from Engin. Standpoint 110 If the direct resistance is properly assessed on the basis of 'wetted' surface, whether we call it a surface coefficient or a skin-frictional coefficient is merely a question of terminology. 1958 Observer 11 May 13/4 The plan form of the 'nicked delta' or delta with the inner middle rear part cut out, was adopted to reduce to a minimum the 'wetted' area or part over which air flows, without reducing the part that does useful work. 1983 D. Stinton Design of Aeroplane v. 208 When an aeroplane is very clean and highly streamlined, parasite drag may be attributed to skin friction drag... For this we need to work out wetted area.

wetting, *vbl. sb.* Add: **4*.** Urination, usu. resulting from incontinence or stress.

1943 [see *SOILING vbl. sb.1 1 b]. 1960 I. Bennett Delinquent & Neurotic Children viii. 252 Soiling, wetting, and difficult behaviour.

wetting, *ppl. a.* Add: **a.** (Examples corresponding to *WET v. 1 b.)

1948 Nature 28 Feb. 313/2 A non-polar 'wetting' liquid such as carbon tetrachloride is floated on mercury in the reservoir beneath the U-tube. 1980 Brit. Med. Jrnl. 18 Oct. 1047/2 A wetting solution coats a hard lens with a chemical that permits water to spread and form a surface that is less traumatic to eye tissues.

b. *wetting agent*, a chemical that can be added to a liquid to reduce its surface tension and make it more effective at wetting.

1927 Chem. Abstr. XXI. 414 The Na salts of the products may be used as wetting agents. 1950 Engineering 5 May 517/1 The success of Teepol as a wetting agent is due to its effectiveness in reducing the surface tension of water. 1977 J. Hedgecoe Photographer's Handbk. 39 Rinsing the film in 'wetting agent' (weak detergent) at the end of washing helps to prevent the accumulation of drops of water.

wetware (we·twēəɹ). [f. WET a. after *hardware, software*.] Chemical materials organized so as to perform arithmetic or logical operations; brain substance, as having this ability.

1975 Nature 23 Oct. 634/1 An electronic computer is made up of hardware and software; a chemical automaton needs an additional component, a chemical reaction system which might be called 'wetware'. 1977 N.Y. Times 8 May 1. 34/6 Computer scientists have lately begun talking about 'wetware', which is the human brain. 1984 Times Lit. Suppl. 14 Dec. 1442/3 There is no obvious reason why biological 'wetware' should be any better at impuing internal structures with semantic significance than silicon 'hardware'. 1985 Listener 10 Jan. 9/3 The whole claim of strong AI is that the physical and chemical hardware or wetware of the system are quite irrelevant.

we-uns (wī·ɒnz), *pron.* *U.S. dial.* Also **we uns, we'uns.** [f. WE pron. + *uns*, dial. var. *ones* (ONE pron.).] Used in place of WE or US prons.

1864 Harper's Mag. Dec. 16/2 'What for you uns', said they, in their barbaric dialect, 'come down here to fight we uns?' 1865 O. L. Jackson Colonel's Diary (1922) 208 If we 'uns were to go down to Goldbro or Raleigh, do you think we 'uns could get any old creetur?.. A horse or a mule? 1907 H. B. Wright Shepherd of Hills xi. 105 You can get out o' these hills an' be somebody like we'uns. Ibid. xii. 109 He sure talks so we'uns can understand. 1913 H. Kephart Our Southern Highlanders xiii. 286 Let's we-uns all go over to youerunses house. 1938 C. H. Matschat Suwannee River iii. 53 We-uns'll..light for home.

Weymouth. Add: **2.** *Horseriding.* Designating a type of curb bit (see quot. 1963) or a double bridle comprising this bit and a snaffle with two sets of reins.

1792 T. H. Morland Every Man his Own Judge 70 A Weymouth bridle, with bit, and bradoon, is in my opinion preferable to any other sort for the road. 1919 R. S. Timmis Notes on Riding & Driving iii. 39 A few good snaffles, both racing and exercising (with cross-pieces outside the cheek), and a Weymouth or a Pelham bit and bridoon are all that are necessary for practically all horses. 1938 F. C. Hitchcock To Horse! ix. 274 The only objection to a double bridle is that its use entails two separate mouthpieces in the horse's mouth. The usual pattern bit used is called the Weymouth. 1946 M. C. Self Horseman's Encycl. 433 The Weymouth bridle consists of a bit and bridoon. It is the bridle most usually used for the finished saddle horse. 1963 Bloodgood & Santini Horseman's Dict. 21 Weymouth or Ward Union bit, bit consisting of straight, moderately long cheek-pieces, stationary or sliding mouthpiece, either straight or with a slight Muller or Cambridge port. Simplest of all curb bits and most generally used. 1965 C. E. G. Hope Riding v. 59 The curb used with a double bridle, known as the Weymouth or Ward Union, invariably has a plain port mouth.

wh. 1. (Also **wh-, wh'**) Informal written abbrev. of WHICH a. and *pron.*, in relative use.

c 1858 E. Dickinson Poems (1955) I. 16 Sleep is the station grand Down wh', on either hand The hosts of witness stand! 1865 Hardy Let. 20 Oct. (1978) I. 5 You will know wh. part of the Abbey I mean if you think of Salisbury Cathedral & of the row of small arches over the large arches. 1889 W. Whitman Daybks. & Notebks. (1978) II. 528 Paid $3.15 for insurance for $300 on stock at 1213 Filbert st. wh- is continued on to Feb: 8 '90.

2. *Linguistics.* A symbol representing an interrogative or relative pronoun (most of which begin with the digraph *wh*). Freq. *Comb.*, as *wh-clause, -question, -transformation*, etc.

1957 N. Chomsky Syntactic Struct. vii. 69 In the morphophonemics of English we shall have rules: wh+he→/huw/, wh+him→/huwm/, wh+it→/wat/. 1960 Internat. Jrnl. Amer. Linguistics XXVI. III. ii. 36 Again adopting Chomsky's excellent analysis of interrogative sentences, we wish now to produce those structures introduced by words beginning, for the most part, with wh-, i.e. WH-questions and their affirmative counterparts, relative clauses and question-word clauses. 1962 N. Chomsky in 3rd Texas Conf. Probl. Linguistic Anal. in Eng. 147 Application of the wh-transformation is conditional on the interrogative transformation 3. 1964 Language XL. 5 The interrogative specifier Wh can remain unattached or can have attached to it (indicated by +) various elements of the sentence. 1966 G. N. Leech Eng. in Advertising vi. 61 It is the type which consists of an embedded wh- clause. 1975 J. Goulet Oh's Profit v. 31 Trabasso: You mean he demonstrates nonkernel sentences? Liedlich: Passive, interrogative, imperative, and he's beginning to get the hang of WH-subordination, too. 1980 B. Newman in Bible Translator XXXI. 326 Lengthy sentences are found throughout NIV... Romans 1. 1–4 is a single sentence consisting of 72 words, involving at least one case of ellipsis, one dash, one colon, two 'wh'-clauses, and several appositions. 1981 R. Burchfield Spoken Word 30 Wh-questions: Which hotel is he staying at? Who are you voting for?

whack, *sb.* Add: **2. b.** Also more generally, a sharing-up or distribution.

1896 Ade Artie xii. 107 He hadn't been in on the whack-up six weeks till he was wearing one o' them bicycle lamps in his neck-tie. 1912 R. A. Wason Friar Tuck xi. 85 'What ya goin' to kill her with?' he asked, his eyes dancin' like an Injun's at the beef whack-up.

c. *U.S.* A bargain or agreement. Esp. in phr. *it's* (or *that's*) *a whack.*

1860 Johnson's Orig. Comic Songs (ed. 2) 45, I axed her for to marry me, she said it was a whack. 1876 'Mark Twain' Tom Sawyer vi. 70 'I'll stay if you will.' 'Good—that's a whack'. 1884 J. Hay Bread-Winners x. 149 Say the word, and it's a whack. 1903 A. D. McFaul Ike Glidden in Maine xviii. 146 'I'll guarantee to get him to take you to Grand Menan with him.' 'It's a whack,' said Jim. 1911 Dialect Notes III. 540 Whack,..an agreement, a 'go'; e.g., 'That's a whack!'

4. *out of whack*: disordered, malfunctioning; out of order or alignment. Cf. *WACKY a. Chiefly *U.S.* **a.** Of a person or a part of the body.

1885 C. A. Siringo Texas Cowboy v. 33, I was too weak to walk that far on account of my back being out of whack. 1899 Ade Doc' Horne viii. 79 My stomach seems to be out of whack. 1903 A. M. Binstead Pitcher in Paradise vi. 146 At last he utterly gets his thinker out of whack an' goes back to the villa. 1918 H. A. Vachell Some Happenings xii. 205 His liver is out of whack and no mistake. 1969 'V. Packer' Don't rely on Gemini (1970) xviii. 150 Margaret had had symptoms of early menopause last winter: that had thrown her way out of whack, could conceivably explain such erratic and erotic behaviour.

b. Of a mechanism.

1906 McClure's Mag. Feb. 34 Being able to get at any part of the mechanism which may be 'out of whack' is important. 1934 D. Hammett Thin Man xi. 77 The phone in the apartment was out of whack. 1949 Time 30 May 53/2 With normal vibration a lot of them would have gone out of whack. 1975 New Yorker 28 Apr. 40/3 He sends no message on the tape recorder to the little boys, because they have already put the machine out of whack. 1985 Mail on Sunday (Colour Suppl.) 3 Mar. 20/2 The body's like an automobile. You have to rest and repair it, not run with the motor out of whack.

c. *fig.*

1952 C. Armstrong Black-Eyed Stranger xiv. 117 Ambielli's got principles. They are a little off, slightly out of whack. 1973 in G. Gibson Eleven Canad. Novelists 123, I don't know whether it is because my own sense of sexual timing or whatever is out of 'whack' with everybody else's. 1975 M. Amis Dead Babies v. 33 Everything is out of whack at Appleseed Rectory; its rooms are without bearing and without certainty. 1978 S. Brill Teamsters vi. 250 In the next decade..the bad loans and poor investment management would..start to throw the cash-flow projections out of whack.

whack, *v.* Add: **3.** (Examples with *up*.)

1893 H. A. Shands Some Peculiarities of Speech in Mississippi 77 Whack up, an expression employed by all classes, probably as semi-slang, to mean to divide, to share. 1961 Coast to Coast 1959–60 126 I'll whack up the breakfast, then, and see how poor bloody Bill's getting on. 1981 Amer. Speech LVI. 27 The DARE project has turned up whack it up, whack up (two informants for each).

4. *intr.* With *off*: to masturbate. *U.S. slang.*

1969 P. Roth Portnoy's Complaint 78 Did I mention that when I was fifteen I took it out of my pants and whacked off on the 107 bus from New York? 1969 Listener 17 Apr. 538/3 Fellatio with the Monkey does not present the same practical difficulties as whacking off in Momma's bathroom. 1977 Transatlantic Rev. LX. 36 'What-in-hell you do for sex anyway?' he asked the boy one night. 'Whack off into the tin pot where they keep the mashed potatoes?'

whack, var. *WACK.

whacked (hwækd), *ppl. a.* *slang.* [f. WHACK v. + -ED1.] **1.** Tired out, exhausted.

1919 Athenæum 15 Aug. 759/1 'Whacked to the wide' means to be tired out. 1952 J. Cannon Body in Beck v. 82 He had been on the job since dawn, was whacked and must call it a day. 1960 L. Meynell Bandaberry vi. 100 I'm whacked. How far have we done? 1976 J. Snow Cricket Rebel 118 In addition to recovering from the injury to my right hand I was whacked when I arrived back in England from the MCC tour.

2. *whacked out*: mad, crazy; *spec.* intoxicated with drugs. Cf. *WACKY a. *U.S.*

1969 Current Slang (Univ. S. Dakota) Summer 17 Whacked out, unorthodox; inclined toward foolish acts. 1969 'V. Packer' Don't rely on Gemini (1970) i. 8 You Cancers are whacked out because the moon rules you. 1975 High Times Dec. 68/3 Then there's the pilot who was whacked out of his skull and landed a hundred-grand rented Cessna 411 gear up in Las Vegas. 1980 W. Safire in N.Y. Times Mag. 14 Sept. 11/2 In America, the term 'whacked-out' is current, as an intensified form of 'spaced out' or 'zonked out', meaning soft-headed after prolonged and excessive use of drugs.

whacking, *vbl. sb.* Add: **1. b.** *transf.* A beating or defeat in a contest; a 'thrashing'.

1951 Sport 27 Apr.–3 May 3/2 If Wednesday gets a whacking from Spurs this time..they can hardly grumble.

whacking, *ppl. a.* (Earlier examples.)

1806 J. Davis Post Captain iv. 19 She looks..like a whacking frigate. 1819 J. Thomson Poems (ed. 2) 201 A whakin' fee gets tauld them down for sorry haet, I trow.

whacko (stress variable), *int.* Also **wacko.** [f. WHACK sb. + *-O2.] An exclamation of delight or excitement: Splendid! Excellent! Hurrah!

1941 Baker Dict. Austral. Slang 81 Whacko, good! Hurrah! A popular ejaculation. 1944 A. F. Bruno Desert Daze 7 Chips let his [Mills bomb] go..Whacko! 1961 J. Maclaren-Ross Doomsday Book I. vii. 74 'All's well,' Marsh said elatedly... 'Whacko!' Eustace cried. 1967 Southerly XXVII. 75 This is the message. 'Home Friday. Wacko.' 1978 L. Davidson Chelsea Murders xvii. 94 After all it was only two days to—whacko!—Monday.

whacko, whacky, varr. *WACKO, *WACKY.

whadd(a)ya, etc., repr. colloq. pronunc. of 'what do you'.

1927 C. Hopley-Woolrich Children of Ritz iv. 72 Gaffney shook hands. 'Whaddya know, eh?' he greeted him breezily. 1945 A. Kober Parm Me 33 'Waddeya mean,' she says back to me. 1952 E. Wilson Equations of Love 34 'Well, whaddaya know!' exclaimed Mort. 1955 W. Gaddis Recognitions II. ix. 446 Whadda you say? 1967 J. Wainwright Talent for Murder 205 Wadya mean, copper? 1975 New Yorker 20 Oct. 38/2 'Hey, hey, whaddya say!' he shouts. 1981 J. D. MacDonald Free Fall in Crimson xiii. 147 'Peter Kesner in, please?' 'Whaddaya want with him?'

‖ **whakapapa** (fakāpa·pa). [Maori.] (Maori) genealogy; a genealogical table.

1960 N. Hilliard in C. K. Stead N.Z. Short Stories (1966) 249 You don't even know your own whakapapa! 1966 Encycl. N.Z. II. 438/2 Most Maori traditional narrative includes some whakapapa or genealogical record of a connection between the characters in the story. The web of the tale is often so entwined as to require the explanation afforded by the whakapapa. 1974 N.Z. Listener 20 July 10/4, I can..make a brief and somewhat tentative reference to the links in our whakapapa, between our tribes they are a long way back. 1975 D. Bagley Snow Tiger iv. 53 Turi's whakapapa stick, his most prized possession,..which gave his ancestry.

whale, *sb.* Add: **1. b.** (*a*) **black whale** (earlier example).

1831 in R. McNab Old Whaling Days (1913) i. 3 The black whales visit the bays and coasts of New Zealand for the purpose of calving.

(*b*) (Earlier example of *right-whaling*.)

1849 H. Melville Mardi I. i. 5 This horrid and indecent Right Whaling..is as the butchery of white bears upon blank Greenland icebergs.

5. c. *a whale of*: for (U.S.) read (orig. U.S.) and add later examples.

1921 Chambers's Jrnl. May 308/1 He had come here to have one whale of a time. 1938 G. Heyer Blunt Instrument iii. 45 It doesn't look such a whale of a case to me. 1954 J. B. Priestley Magicians i. 15 An equally dashing, whale-of-a-fellow, R.A.F. type. 1963 N. Marsh Dead Water (1964) i. 22 She's having a whale of a time with Mr. Joyce. 1980 B. Castle Castle Diaries 363 They regaled us with drinks and a superb buffet and we had a whale of a time.

6. a. *whale-cry, -ground, -hole, -hunt, -steak* (examples); (as weapons, etc.) *whale-pike, -pole, -rope* (earlier example); *whale-blue, -mouthed* (later example), *-shaped* adjs. **b.** **whale-louse** (later examples).

1946 Dylan Thomas Deaths & Entrances 55 The coast Blackened with birds took a last look At his thrashing hair and whale-blue eye. 1851 H. Melville Moby Dick II. xii. 78 The ancient whale-cry upon first sighting a whale from the mast-head. 1851 Whale-ground [see oil-ship s.v. *OIL sb.1 6 e]. 1897 Kipling Captains Courageous v. 111 'Whale-hole.'..He had led them to the edge of the barren Whale-deep, the blank hole of the Grand Bank. 1851 H. Melville Moby Dick III. xix. 134 The far different nature of the whale-hunt. 1916 R. C. Andrews Whale Hunting xxi. 248 This growth [on the snout of the right whale] is produced by whale lice. 1952 J. Fisher Fulmar xviii. 423 He writes of the fulmars 'searching out' whale-lice. 1972 Nat. Geographic Oct. 579 (caption) Communities of whale lice.. some of them half an inch long, cling to the growths [on the head of each right whale]. 1952 L. MacNeice Ten Burnt Offerings ii. 23 The whale-mouthed arch the bones of the future. 1851 H. Melville Moby Dick II. xii. 94 The valiant Captain danced up and down with a whale-pike. Ibid. xiv. 124 The flag of capture lazily hanging from the whale-pole inserted into his spout-hole. 1849 —— Redburn I. xx. 194 Coiled away in a tub. Like a whale-rope. 1930 Times Educ. Suppl. 25 Oct. p. iv/1 In the whale-shaped head is a window for the driver. 1978 M. Puzo Fools Die ii. 16 White-dotted red square dice were dazzling flying fish over the whale-shaped crap tables. 1851 H. Melville Moby Dick II. xii. 166 Don't I always say that to be good, a whale-steak must be tough? 1969 Listener 14 Aug. 206/3 What did we eat? Well, we ate whale-steak for one thing.

whale, *v.*2 **2.** (Earlier examples.)

a 1852 F. M. Whitcher Widow Bedott Papers (1883) vi. 67 You remember that one that come round a spell ago a whalin' away about human rights. 1886 Harper's Mag. July 322/1 In tones of wrath..he whaled it at his opponent throughout the fifteen minutes alloted to him.

whaleback. Add: **3.** More widely, any land form or land mass likened to the back of a whale; *spec.* (*a*) = *ROCHE MOUTONNÉE; (*b*) an elongated sand dune.

1913 Proc. Geologists' Assoc. XXIV. 247 A characteristic rounded form resembling..the 'whale-back' of glaciated areas. 1918 Geogr. Jrnl. LI. 23 In these whalebacks and crescents the cross-section that has the longest base passes through the summit of the dune. 1928 Chambers's Jrnl. Jan. 1/2 Behind all, a dim whale-back that might be Stroma, or Ultima Thule. 1933 Geogr. Jrnl. LXXXII. 125 A whaleback is a flat-topped ridge of sand anything up to 100 miles in length, of the order of half a mile wide, and up to 100 feet high. 1952 V. Canning House of Seven Flies xi. 155 A long stretch of sand..the long ridges of wave marks from the last tide shadowed across the rising whaleback. 1955 Geogr. Jrnl. CXXI. 476 Some British whalebacks are undoubtedly roches moutonnées, some others, and probably many tropical examples, are genetically related to tors. 1974 M. Gilbert Flash Point x. 82 Behind the whaleback of Kinder Low and Edale Head a storm was brewing up. 1977 A. Hallam Planet Earth 67/2 Where folds can be traced in three dimensions, it is found that the structures die out along their fold axis and, where suitably exposed, form whalebacks.

whalebacked *a.* (earlier example).

1869 'Mark Twain' Innoc. Abroad 441 We can see th long, whale-backed ridge of Mount Hermon projecting above the eastern hills.

whalebone, *sb.* Add: **wha·leboning**, a beating with a piece of whalebone (sense 3).

1851 H. Melville Moby Dick I. xxx. 205 Only a whaleboning that he gave me—not a base kick.

whaler. Add: **4.** Also **waler.** [ellipt. f. *Murrumbidgee w(h)aler* s.v. *MURRUMBIDGEE: (see also quot. 1945).] A tramp or 'sundowner'. *Austral. slang.*

1883 R. E. N. Twopeny Town Life in Australia 244 A 'waler' is a bushman who is 'on the loaf'. He 'humps his drum', or 'swag', and 'starts on the wallaby track'. 1886 F. Cowan Australia 31 The Whaler: of the Murrumbidgee and the Darling; when it suits his pleasure and convenience, a dolce-far-niente outcast in the fertile valleys of the rivers named, beyond the running of a warrant or a writ. 1903 'T. Collins' Such is Life 4 Willoughby, who was travelling loose with Thompson and Cooper, was a whaler. 1945 Baker Austral. Lang. v. 102 According to an old-timer correspondent: 'They were so apt to lie about the size of the 'whales' they caught that a generic name for this class of unemployable traveller came into being.' This explanation is open to some doubt... In our early days New South Welsh horses exported to India for army use were known as walers. The original Murrumbidgee whalers may therefore have been N.S.W. tramps... Blood brethren of the whaler (this spelling is retained because tradition holds mainly to the 'whale' theory)..are the Domain dosser, [etc.]. 1963 A. Marshall In Mine Own Heart (1964) xx. 164 The whaler, a term that had originated from the name given to those swagmen who in the early days spent their time moving up and down the Murrumbidgee River..now applied to those who walked from town to town in preference to jumping trains. 1965 B. Wannan Fair Go, Spinner II. 53 After drinking some Wilcannia beer, a whaler I once saw got up and started to fight with himself.

4. Special Combs. **whalerman** = WHALER 1; **whaler shark**, any of several sharks of the genus *Galeolemma*, found in Australasian waters.

1891 R. L. Stevenson In South Seas (1896) I. xiii. 128 Captain Chase, they called him, an old whaler-man. 1963 Times 18 May 9/7 The first big bang was at night and the Norwegian whalermen heard it six miles away. 1974 G. Jenkins Bridge of Magpies ii. 33 Old whalermen's graves in New England. [1882 J. E. Tenison-Woods Fish New South Wales iv. 92 The following list [of sharks] includes all that are known to occur in our seas:..the Whaler, [etc.].] 1937 Z. Grey Amer. Angler in Austral. vii. 70 Among the trawlers it was not unusual to see a dozen whaler sharks all in a bunch. 1972 Islander (Victoria, B.C.) 9 Apr. 7/1 A whaler shark darting over the reef flat with a sudden burst of speed.

whaling, *vbl. sb.*1 Add: **1. b.** (Earlier examples.)

1722 New-England Courant 18 June 2/2 Huffey of Nantucket..went out from thence on the Whaling Account. 1767 M. Cutler in W. P. & J. P. Cutler Life & Corr. M. Cutler (1888) I. i. 19 Our whaling vessels sailed for the Western Islands. 1782 'J. H. St John de Crèvecœur' Lett. from Amer. Farmer v. 158 They have greatly cheapened the fitting out of their whaling fleets.

2. Comb. **whaling station**, a land base where whales which have been caught are flenched and rendered.

1874 C. M. Scammon Marine Mammals N.W. Coast N. Amer. III. v. 247 At the point where the enormous carcass was stripped of its fat, arose the 'whaling station', whose trypots were set in rude furnaces..and capacious vats were made of planks, to receive the blubber. 1930 L. G. D. Acland Early Canterbury Runs 1st Ser. vi. 116 The country..was accessible on foot from the old shore whaling station. 1963 L. Diack Labrador Nurse v. 25 One night we tied up at a whaling station. 1977 C. McCullough Thorn Birds ii. 20 The eleven men..came out at the whaling station of Hobart.

wham (hwæm), *sb.*2 *colloq.* [Echoic. Cf. next.] **1.** A heavy blow; the sound of a heavy blow (or of an explosion, etc.). Also, a resounding success, a 'knock-out'; an attempt *at* something (cf. WHACK sb. 1 b).

1923 N.Y. Times 9 Sept. VII. 2 Wham, a success, a knock-out. 1924 Dialect Notes V. 257 Onomatopoetic words.. bam, blam, ca-blam, slam, wham, zam..(all = sound of blow). 1949 J. R. Cole It was so Late 90 The occasional echoing wham of a charge of gelly. 1957 'J. Wyndham' Midwich Cuckoos iv. 38 Might be a good idea to have a wham at it. 1973 C. Bonington Next Horizon viii. 121 Have another try... This time the peg held, another half-dozen whams of the hammer, and it was in to the hilt.

2. As *int.* or *adv.*: with a wham.

1924 E. Hemingway In our Time 10 The bull rammed him wham against the wall. 1934 J. M. Cain Postman always rings Twice xi. 126 And then, wham, I pleaded her guilty. 1948 'J. Tey' Franchise Affair viii. 90 They go that short step too far and wham! out comes that business-like paw. 1958 P. Mortimer Daddy's gone a-Hunting xxx. 170 He..walked through the front door and wham. She did it with a clock. 1965 M. Frayn Tin Men xxvi. 144 When the iron was hot, wham!—he would come out like a tiger and knock it for six. 1975 A. Ayckbourn Norman Conquests 5 It was just wham, thump and there we both were on the rug.

wham (hwæm), *v.* *colloq.* [Echoic. Cf. prec.] **1.** *trans.* To strike violently; to propel with great force, by hitting, throwing, kicking, etc. Also *fig.*

1925 Sat. Even. Post 14 Feb. 16 The wow finish, properly, is the legitimate successor to the old apple-sauce flag-waving finish for whamming an audience. 1930 E. Ferber Cimarron xxi. 349 Standing Bear whams it out so straight and so far that he makes the [golf] pro look like a ping-pong player. 1933 J. Thurber My Life & Hard Times iv. 57 She..picked up a shoe, and whammed it through a pane of glass across the narrow space that separated the two houses. 1950 A. Buckeridge Jennings goes to School xii. 239 You must have put all your weight behind it, or you wouldn't have gone down flat like that, after you'd whammed it in. 1951 Sport 7–13 Jan. 15/2 Basically the same team, which had been languishing generally on the wrong side,..whammed in six against Derry. 1962–3 E. Birney Sel. Poems (1966) II. 59 Nine shoeboys wham their boxes Slap at my newshined feet. 1971 C. Bonington Annapurna South Face xiii. 161 He whammed in the ice-hammer, pulled up on it, kicking with the two front points of his crampons into the ice.

2. *intr.* To pound or strike violently; to move with speed, violence, or noise. Also *fig.*

1948 W. C. Williams in Poetry June 147 Each time he'd swing the axe and I heard it wham into the wood, I'd let out a wild cackle of delight. 1948 D. Ballantyne Cunninghams 135 The nausea rushes that made his head wham. 1954 A. C. Clarke Silence Please in Tales from White Hart (1957) 3 Bert's blast whammed overhead. 1962 K. Kesey One flew over Cuckoo's Nest I. 51 The black boy whammed flat against the wall and stuck, then slid down to the floor. 1973 P. White Eye of Storm xii. 586 A partition of the door still in motion whammed against an ear and sent his hat spinning. 1980 Daily Tel. 27 Aug. 2 (Advt.), The incredible Casio FX39..whams through complicated equations and elementary statistical formulae.

wham-bam, -bang, *adv.* (or *int.*), *a.*, (*sb.*). [f. *WHAM sb.2 2 + *BAM int. or BANG v. 8.]

A. *adv.* or *int.* With a wham and a bang: used to denote a sudden or forceful effect (*lit.* or *fig.*); *spec.* with reference to sexual intercourse conducted quickly and without tenderness, esp. in phr. *wham, bam, thank you ma'am.*

1956 B. Holiday Lady sings Blues (1973) ii. 22 With my regular white customers, it was a cinch. They had wives and kids to go home to. When they came to see me it was wham, bang, they gave me the money and were gone. Ibid. 25 'I thought I was giving you a chance,' she spouted at me. 'But you turned out to be a girl of bad character.' Wham, bang, four months she handed me. 1971 S. Firestone Dialectic of Sex vi. 152 Men are interested in nothing but a screw (wham, bam, thank you M'am!). 1977 Ripped & Torn VI. 9/1 We play a set that starts at the beginning and goes Wham Bam Bam to the end. 1977 Playgirl May 13/1 Not all men are 'wham bam thank you ma'am' types.

B. *adj.* Loud, violent, forceful (see also quot. 1960).

1960 Wentworth & Flexner Dict. Amer. Slang 573/1 Wham-bam, quick(ly) and rough(ly); displaying more energy than finesse. 1976 Publishers Weekly 4 Oct. 65/2 Harbinson on Elvis..ticks off the outrageous 'Hound Dog' of pop music in wham-bam style. 1977 Listener 20 Oct. 498/3 Screenwriters..know that it's mainly wham-bang shock effect that sells.

C. *sb.* A sudden, violent effect.

1975 Listener 17 July 68/3 Now it is the big wham-bang of sudden [price] rises.

whammer (hwæ·məɹ). *Mountaineering.* [f. *WHAM v. + -ER1.] A kind of piton hammer.

1971 [see piton hammer s.v. *PITON 3]. 1974 H. MacInnes Climb to Lost World ix. 142 The Whillans 'Whammer'—a multi-purpose piton hammer.

whammo (hwæmōu: stress variable), *int.* Also **whamo.** [f. *WHAM sb.2 + *-O2.] = *WHAM sb.2 2; an exclamation suggesting a sudden violent blow or surprising event, etc.

1932 Fitchburg (Mass.) Sentinel 7 May, But the heavy comes to, and Whamo! The boy goes down bam. 1945 Record (Philadelphia) 4 July 11/1 'Ring out the tidings, Grandpa!' and the old gent spit on his hands, and Whammo! went the Liberty Bell. 1959 N. Mailer Advts. for Myself (1961) 97 They meet again in New York.. and whammo do they get together. I mean drinking and making love, nothing can stop them. 1969 P. Dickinson Pride of Heroes I. 45 Everyone a bit nervy for about a fortnight, and then, whammo, something happens. 1981 Daily Express 24 July 14/4, I put the telephone down and whammo! Another 'little twinge'.

whammy (hwæ·mi). *colloq.* (orig. and chiefly *U.S.*). [f. *WHAM sb.2 + -Y6.] An evil influence or 'hex'. From the 1950s, often with reference to the comic strip Li'l Abner (see quot. 1951), esp. in phr. *a double whammy* and varr. Hence, an intense or powerful look, etc.; something effective, upsetting, problematic, etc.

1940 J. R. Tunis Kid from Tomkinsville x. 151 Interest round the field now centered in the Kid's chances for a no-hit game... On the bench everyone realized it too, but everyone kept discreetly quiet on account of the Whammy. Mustn't put the Whammy on him! 1951 Al Capp's Li'l Abner July, Evil-Eye Fleegle is th' name, an' th' 'whammy' is my game. Mudder Nature endowed me wit' eyes which can putrefy citizens t' th' spot!.. There is th' 'single whammy'! That, friend, is th' full, pure power o' one o' my evil eyes! It's dynamite, friend, an' I do not t'row it around lightly!.. And, lastly—th' 'double whammy'—namely, th' full power o' both eyes—which I hopes I never hafta use. 1952 B. Malamud Natural 75 They were afflicted with more than the usual number of hexes and whammies and practised all sorts of magic to undo them. 1964 J. Masters Trial at Monomoy ii. 66 You heard that our local witch has put her whammy on you now? 1970 Daily Tel. (Colour Suppl.) 30 Oct. 19/2 That smile, a huge, sweet, melting smile, a whammy of an MRA smile, a West-Coast switched-on-sincere smile, which envelops just everybody. 1976 New Yorker 16 Feb. 107/1 In the Germany scenes, Wertmuller achieves the effect of liveliness through one whammy after another. The starving prisoners in the camp are beaten and murdered to the tune of the Ride of the Valkyries. 1979 C. James Pillars of Hercules III. xi. 122 Holmes was a nonconformist in a conformist age, yet still won all the conformist rewards. It was a double whammy.

whang, *sb.*1 Restrict *Sc.* and *dial.* to senses in Dict. and add: Also **wang. 3.** The penis. *slang* (orig. and chiefly *U.S.*).

1935 H. L. Davis Honey in Horn iii. 34 Leave them horses alone or I'll cut your whang off. 1949 H. Miller Sexus viii. 250 You say he's got a terrific wang, Bill. I don't know how he ever gets it in there. 1952 N. Mailer Barbary Shore x. 89 Guinevere..went on at length with one of her inexhaustible stories about a lover and his whang. 1959 M. Richler Apprenticeship D. Kravitz I. x. 60 He's got a whang that could choke a horse. I know, we had a leak together once. 1969 K. Vonnegut Slaughterhouse-Five v. 115 Montana was naked, and so was Billy, of course. He had a tremendous wang. 1981 G. Hammond Revenge Game ix. 102 Maybe you're not as ready with your whang as you were, or maybe you couldn't keep it up——.

whang, *v.*1 Restrict *Sc.* and *dial.* to other senses in Dict. and add: Also **wang. 1. b.** *trans.* and *intr.* (Later examples.) *dial.* and *colloq.*

1905 in Eng. Dial. Dict. VI. 439/2 He wanged a stone at me. 1914 C. Mackenzie Sinister Street II. III. i. 500 The governor wanged them into my lap. 1965 Punch 22 Sept. 420/1 Anybody wanting to wang up a skyscraper or indeed any building of size and importance will have to publish a comprehensible model or drawing. 1980 D. Bogarde Gentle Occupation ix. 249 Suddenly a stone spun out of nowhere and whanged harmlessly against the bonnet of the car. 1984 New Yorker 23 Apr. 80/3 Bad bush pilots..cross the margins of heavy weather and whang into mountains. Ibid. 29 Oct. 140/3 Mondale was ready for him and whanged the line back.

whang, *v.*2 Add: Also **wang. a.** Also, as of the noise of a loudspeaker, the speed of a car, etc.

1952 Observer 2 Nov. 3/5 The words from the loudspeaker wang back from the quiet village houses, but the doors remain closed. 1977 Motor 19 Feb. 24/1 You rush from the pits just as the leading Porsches wang past.

whangdoodle, whang-doodle. *N. Amer.* Also **whangydoodle.** [Fanciful.] **a.** An imaginary creature. **b.** Something unspecified, a 'thingummy'. (See also quot. 1904.)

1858 H. T. Lewis in Salem (Illinois) Advocate 27 Jan. 1/2 They shall..flee unto the mountains of Hepsidam, where the lion roareth and the wang-doodle [Yankee Notions Feb. 52/1 whang doodle] mourneth for his first-born. 1870 Punchinello (N.Y.) 9 Apr. 30/1 In his own State the Ku-Klux raged, together with the fierce whang-doodle. 1890 Boston Jrnl. 24 July 2/5 The rougher element among the boys..formed the 'Whang Doodle' Club for the purpose of settling him. 1904 R. F. Foster Practical Poker 11 The 'whangdoodle', a round of compulsory jack-pots after a big hand has been shown. 1923 Nation (N.Y.) 22 Aug. 179 The downtrodden and oppressed force on the Gazette..has to live with this old whang-doodle and listen to his preterbacious scandulations. 1931 Amer. Speech VI. 259 Whangdoodle, whangydoodle, what have you, whatsis, whatsit, [etc.]. 1979 Globe & Mail (Toronto) 24 Jan. 6/1 In 1976... A new company sprang to the fore in Quebec... PQ Productions claimed to have invented the whangdoodle.

whangee. (Earlier example.)

1790 in W. Roughead Bad Companions (1930) 6 He.. sometimes wears a cocked hat,..and generally carries a Wangee cane in his hand.

whanger (hwæ·ŋəɹ), *sb.*1 *U.S. slang.* Also **wanger.** [f. WHANG v.1 + -ER1.] = *WHANG sb.1 3.

1939 J. Steinbeck Grapes of Wrath ii. 14 An' there we spied a nigger, with a trigger that was bigger than an elephant's proboscis or the whanger of a whale. 1976 M. Machlin Pipeline xiv. 160 She didn't get the idea so fast, so he whipped the old wanger out of his union suit and laid it on the table in front of her.

whank, whanker, occas. varr. *WANK, *WANKER.

whare. Add: Also with pronunc. (fa·re). Also 9 **wurrie. 1.** (Earlier examples.)

1807 J. Savage Some Acct. N.Z. xi. 77 Wurrie, a house, or hut. 1817 J. L. Nicholas Voyage to N.Z. I. xii. 352 A young woman..beckoned to me to accompany her to her warree or hut.

2. Hence *gen.*, a hut or shed; *spec.* on a sheep station, a building where the hands sleep or eat. Also with defining word.

1853 A. S. Atkinson Jrnl. 26 Oct. in Richmond-Atkinson Papers (1960) I. 135 James and I went to the site chosen for our new whare. 1853 J. M. Richmond Let. 11 Nov. in Ibid. 133 Their 'wharre', as it is called, is a most romantic tho' not v. commodious dwelling;..it is in fact a roof on the ground, thatched with nikau, a palm, the only one in N.Z. 1891 R. Wallace Rural Econ. Austral. & N.Z. xv. 225 Pioneering, or cutting a place out of the bush and building a log 'whare', is extremely rough and lonely work. 1904 'G. B. Lancaster' Sons o' Men 4 He scudded across the tussock flat to the eating-wharé; burst open the door, and cast the word loose on the boys. 1926 A. F. Webb in D. M. Davin N.Z. Short Stories (1953) 205 We had dinner at twelve and made a plum duff because there was time to cook it while we were all about the whare. 1939 J. Mulgan Man Alone viii. 95 You'll be sleeping in the whare down there... There's no room in the house. 1963 B. Pearson Coal Flat vii. 141 Eventually Miss Dane said: 'Time I got back to the whare.' 1972 M. Shadbolt

Strangers & Journeys i. 29 They found he had built a whare. A one-room shack of roughly-split timber.

3. Special combinations. **whare puni**, a (Maori) family sleeping-house (see also quot. 1911); **whare runanga** [*RUNANGA], a Maori council chamber.

1911 W. H. Koebel *In Maoriland Bush* xx. 262 It is regrettable that the interpretations of the carvings upon the beams and panels of the old *whare-punis* or meeting-houses have been lost. **1926** H. Guthrie-Smith *Tutira* (ed. 2) 86 A *whare-puni* or sleeping-house. **1950** *N.Z. Jrnl. Agric.* May 502/2 The great michi, the barge boards of a whare-puni (a sleeping house). **1891** R. Wallace *Rural Econ. Austral. & N.Z.* xiv. 218 A special house of assembly, the *whare runanga*..is set apart in which to receive and entertain strangers. **1910** J. Cowan *Maoris of N.Z.* xii. 163 Most Maori villages of any importance contain at least one *whare-whakairo*, a large house..used as the communal assembly hall, council-place (*whare-runanga*),..and guest-house (*whare-manuhiri*). **1955** W. J. Phillips *Maori Carving Illustr.* 40 The Assembly House or whare runanga is often well adorned with carvings.

wharf, *sb.*[1] Add: **3.** *wharf-labourer, -master* (later examples), *-shed*; **wharf crane**, a crane fixed in position on a wharf (see quot. 1968); a wharf-side crane; **wharf-lumper** *Austral.* [Lumper *sb.* 1 a], a wharf-labourer; **wharf-rat** (earlier examples in both senses).

1893 K. P. Dahlstrom tr. *Weisbach & Herrmann's Mech. Hoisting Machinery* vi. 243 The ordinary wharf crane with capacity to lift 100 to 200 cwt. **1903** J. Horner *Elem. Treat. Hoisting Machinery* xvii. 195 There is a class of fixed jib cranes which have no other name than that which designates the nature of their service, fixed wharf cranes... But by the term wharf crane, a broad type only is understood. **1968** *Gloss. Terms Materials Handling* (B.S.I.) iv. 14 *Dockside or wharf crane*, a jib crane designed for loading and unloading ships, consisting of a full or semi-portal, fixed or rail mounted, supporting a revolving superstructure and jib. **1890** *Evening Post* (Wellington, N.Z.) 11 July 2 A wharf-labourer who stands charged with the theft of an oil skin coat... The accused was at work discharging coal on the Mawhera. *a* **1948** L. G. D. Acland *Early Canterbury Runs* (1951) xi. 321 Trouble with wharf labourers.. kept them six weeks in Auckland. **1906** E. Dyson *Fact'ry 'Ands* iii. 39 Three weeks..later, Sarah was married to a wharf-lumper..and Fuzzy's dream of love was over. **1951** V. Palmer in *Landfall* V. 292 In Victoria..it was read by nearly everybody, from wharflumpers to politicians. **1836** J. M. Peck *New Guide for Emigrants to West* xii. 320 The following, from the register of a wharf master, will exhibit the commerce for 1835. **1968** M. M. Sibley *Port of Houston* iii. 59 Wharfmaster Daniel G. Wheeler reported that in that year [*sc.* 1844] 6.892 bales passed over the Houston wharves. **1823** J. F. Cooper *Pilot* ii. i. 13 To burrow like a rabbit, or jump from hole to hole, like a wharf-rat. **1836** *Franklin Repository* (Chambersburg, Pa.) 4 Oct. 1/3 I've an idea, my man, that you are one of the wharf rats; and, if so, the less lip you give me the better. **1952** R. Finlayson *Schooner came to Atia* xi. 61 In the..market place by the wharfshed.

wharfie (hwǭ·ɹfi). *Austral.* and *N.Z. colloq.* [f. Wharf *sb.*[1] + -ie.] A wharf-labourer; a stevedore or docker.

1912 *Lone Hand* 1 May 40 The best testimonial to Hughes' ability is the fact that he has so often swayed the unruly 'wharfies', and controlled their organisation for so long. **1926** J. Devanny *Lenore Divine* vii. 47 Imagine Holly haranguing the wharfies from the soap-box. **1928** *Bulletin* (Sydney) 21 Mar. 12/1 'Twas Bill the wharfie grinned and stuck his hook into his belt. **1938** W. E. Dexter *Rope Yarns* 234 Ships arrived [at Melbourne] with general cargo—oddments from a needle to an anchor—and were looked upon as legitimate prey by the warfies and lumpers. **1949** D. M. Davin *Roads from Home* 226 They..watched the wharfies unloading. **1963** B. Pearson *Coal Flat* xx. 355 Sid Holland would put those bloody wharfies and miners in their place. **1978** B. Mason in *Islands* (N.Z.) Aug. 18 But one of his wharfie mates had given him a ticket for his birthday. **1981** *National Times* (Austral.) 25–31 Jan. 24/2 A lazy wharfie would be known as 'the Judge' because he was always sitting on a case, and another 'the London Fog' because he would never lift.

Wharncliffe (hwǭ·ɹnklif). The name of James Archibald Stuart-Wortley-Mackenzie, 1st Baron *Wharncliffe* (1776–1845), used *attrib.* and in the possessive to designate a standing order in Parliament which requires the directors of a company wishing to promote any private Bill for the extension of the company's powers to secure the consent of its members or shareholders, or a meeting at which this consent is sought.

1846 *Hansard Lords* 23 Apr. 874 Their Lordships had already required further securities, in particular cases, by the Order called Lord Wharncliffe's Order, which required in the case of established companies, if they.. demanded powers beyond their original powers—that..a meeting consisting of three-fifths of the company should have sanctioned the proposed alteration. **1851** Erskine May *Law of Parl.* (ed. 2) xxvii. 560 It is directed by an order commonly known as 'Lord Wharncliffe's order'. **1887** F. Clifford *Hist. Private Bill Legislation* II. xx. 784 In order to prevent directors of companies from promoting Bills without the knowledge or sanction of shareholders, the House of Lords framed, in 1846, a series of Orders, under which 'Wharncliffe meetings', as they were afterwards termed, must be held, to consider each Bill so promoted. **1923** *Daily Mail* 24 Feb. 3/2 Your approval will be asked at a Wharncliffe Meeting which will be held in the near future. **1948** O. C. Williams *Hist. Devel. Private Bill Procedure* I. vi. 166 The Wharncliffe Order, then, was *not* first framed in 1846, as Clifford says, but in 1838; it was *not* one of a series of orders, but developed into a series by processes of division and addition; its object was *not* simply to prevent directors of companies from promoting bills without the knowledge and sanction of shareholders but to prevent the promotion of bills to obtain further powers (especially to construct branch lines—always a speculative project) without such sanction.

whassa (hwǫ·sǎ). Also **wassa.** Repr. colloq. or careless pronunc. of 'what is the (matter, etc.)'. Cf. *WHAT'SA MATTER.

1906 E. Nesbit *Railway Children* ii. 28 Roberta woke Phyllis... 'Wassermarrer?' asked Phyllis. **1942** Berrey & Van den Bark *Amer. Thes. Slang* § 256/14 *What's the matter?*..wha'sa mat?, wha'sa matter?..wazzo maro? **1951** C. M. Kornbluth *Marching Morons* in *Best of C. M. Kornbluth* (1977) 156 'Wassamatter?' snorted her husband. **1967** 'W. Wright' *Shadows don't Bleed* v. 84 Whassa matter? You don' believe me? **1973** J. Drummond *Bang! Bang! You're Dead!* xxv. 86 'Wassermatter?' 'I want to join, is all.' **1978** M. Kenyon *Deep Pocket* ix. 102 'Whassa time?' A quarter to three.

what, *pron.*, *a.*[1], *adv.*, *conj.*, *int.* (*sb.*) Add: **A. I.** *pron.* **5. b.** Also *what say?* (*slang*, orig. *U.S.*), what did (or do) you say? shall we? (cf. *SAY *v.*[1] 2 m); *what's with..?* (*colloq.*, orig. and chiefly *U.S.*), what's the matter with..?, what has happened to? (see also quot. 1962).

1825 J. Neal *Bro. Jonathan* I. 357 'Was he hurt, uncle Harwood?' 'What say?' **1855** W. G. Simms *Forayers* 52 What 'say, boys—won't a back-and-rush of the nags do it? **1934** S. Lewis *Work of Art* 294, I think it would be fun to run up the Hudson to Ye Bunche of Grapes some noon. What say? **1966** *New Yorker* 24 Dec. 25 What say we skip a few 'fa-la-la's'? **1972** 'B. Graeme' *Tomorrow's Yesterday* iii. 32 What say we have coffee at home for once?

1940 J. O'Hara *Pal Joey* 125 Nick what's with the free food? Explain. **1960** 'E. McBain' *Killer's Choice* ix. 97 'What's with this kosher bit?' he asked. 'Get me some butter.' **1962** *Amer. Speech* XXXVII. 203 The elliptical 'What's with..?' (*Vos iz mit..?*) also has occurred, not only in the sense of 'What's new?'..but also as a substitute for 'What's the matter with..?' (a sense common in Yiddish). **1976** *National Observer* (U.S.) 24 Jan. 1/1 But it's not easy, because an interloper keeps asking depressing questions—such as: How do you feel about the state of the country? What's with the economy? **1977** H. Fast *Immigrants* iii. 172 There are ways to find out what's with Jake.

c. *what about it?*: an enquiry as to the course of action to be adopted.

1927 H. A. Vachell *Dew of Sea* 259 Your head keeper says we must have two guns apiece. Now—what about it? **1935** D. L. Sayers *Gaudy Night* viii. 163 'I say,' said Mr. Farringdon, '..you simply *must* come.'.. 'What about it?' said Harriet, deferring to Mr. Pomfret.

8. b. (Later examples of *what all* without the negative verb.)

1901 H. Sutcliffe *Barbara Cunliffe* 9 Some reckon Tib helps him wi' his black magic, an' his turning stones to gold, an' what all. **1942** W. Faulkner *Go down, Moses* 76 My mind gonter change about whatall I seed. **1947** E. Meynell *Sussex* ix. 225 Sheep are most unfortunate creatures in the infections to which they are liable—the foot rot..and the liver-fluke and what all. **1957** J. Kerouac *On Road* (1958) iii. iv. 202 Weariness and.. sorrow and what-all was on his mind. **1962** A. Lurie *Love & Friendship* vii. 123 That old Mr Higginson... Got his house full of bird dirt and what-all.

d. Also *that's what*; *you know what?*

1902 *Dialect Notes* II. 247 *That's what*, exclamation of affirmation or assent. **1908** L. M. Montgomery *Anne of Green Gables* xv. 165 She'll..be ready enough to go back of her own accord, that's what. **1960** H. Pinter *Room* 98 You know what though? It looks a bit better. It's not so windy. **1965** 'Lauchmonen' *Old Thom's Harvest* ii. 29 'You don't bring anybody.'..'Oh—oh. That's what.' **1982** H. Engel *Murder on Location* xviii. 164 'You know what, Chris?' 'What?' 'You know it stinks to heaven as well as I do.'

11. a. (Further examples of *what for* introducing a clause.) (A non-standard use.)

1948 E. Waugh *Loved One* 51 What for you want new ideas? **1984** J. Platt et al. *New Englishes* vii. 127 *What for* you want to do that?

c. Also, *to show* (someone) *what for*: to make him take notice; to show who is in charge.

1960 N. Hilliard *Maori Girl* 142 If a man came here I took a liking to, I'd show him what-for soon enough. **1966** *Listener* 18 Aug. 229/1 The stereotype of 'the wily oriental gentleman..the half-civilized levantine..the type of fellow who must be shown what for'.

C. I. *pron.* **3. b.** (Later examples.)

1966 I. Murdoch *Time of Angels* iii. 32 Pattie resented too, what before she had scarcely noticed, Carel's assumption that Muriel and Elizabeth were socially her superiors. **1970** C. W. K. Mundle *Critique Linguistic Philos.* 16 Their claim is sometimes that so and so is..incorrect English, sometimes (what is very different) that it is absurd or meaningless. **1976** *Times* 7 June 14/6 To this he added, what could hardly be encouraging to other aspirants to riches, 'In building a large fortune it pays to be born at the right time.'

4. e. *what have you*: anything else (similar) that there may be, or that one can think of. orig. *U.S.*

1925 *New Yorker* 10 Oct. 28/2 New Yorker, Newarker, or what have you? **1930** H. Craddock *Savoy Cocktail Bk.* i. 113 Fill the said tumbler with Water, Ginger Ale, or What Have You, until almost to the top. **1944** Auden *For Time Being* (1945) 115 Disguising himself as a swan or a bull or a shower of rain or what-have-you. **1956** A. Wilson *Anglo-Saxon Attitudes* 163 Too busy or going on leave or what-have-you. **1968** K. Weatherly *Roo Shooter* 110, I must have been away about two hours, buying the stores and what-have-you. **1973** C. Bonington *Next Horizon* iv. 66, I was eager to snatch at every opportunity to get myself established as a writer, film-maker, what-have-you, in an effort to find a clearly defined career.

5. b. (Earlier example of *not but what.*)

c **1883** E. Fitzgerald *Let.* in A. C. Benson *E. Fitzgerald* (1905) i. 19 Never having read his father's [poems] ..till drawn to them by me... Not but what he loved and admired his father in every shape but that.

6. Likewise with *as.*

1960 M. Spark *Ballad Peckham Rye* iii. 29 'He's the same as what we are,' Dixie said. **1966** P. Willmott *Adolescent Boys* ii. 26 They're all about the same age as what we are.

II. *adj.* **10.** (Later examples of *what time* used as a deliberate archaism.) Also, *what while* (poet.).

1861 D. G. Rossetti tr. *Dante's Vita Nuova* in *Early Italian Poets* ii. 299 It is your fickleness..makes me tremble thus What while a lady greets me with her eyes. *c* **1882** G. M. Hopkins *Poems* (1967) 93 Walked with the wind what while we slept. **1936** W. B. Pemberton *Carteret* ix. 253 He and his brother were quietly mobilising their forces what time a rollicking..Carteret dictated optimistic despatches. **1945** R. Hargreaves *Enemy at Gate* 24 One side getting ready for the next time what-time the other as carefully and methodically prepared themselves to fight the last.

whata, var. *FUTTAH.

whatcha (wǫ·tʃǎ), repr. a colloq. or vulgar pronunciation of *what do* (or *are*, or *have*) *you*? See *WATCHA.

1934 J. T. Farrell *Calico Shoes* 43 H'lo, baby! Whatcha say, kid! **1966** M. & G. Gordon *Undercover Cat prowls Again* (1967) v. 44 Whatcha getting me today, Tim? **1973** *Black World* June 65 Awwwwh, Baby what'cha done to me-ee. **1978** 'M. Craig' *Were he a Stranger* xvi. 128 'Whatcha want?'..'We're looking for a man,' Ted called.

whatchamacallit (wǫ·tʃămăkǭ:lit), repr. a pronunciation of *what-you-may-call-it* (see What-d'ye-call-'em, etc. γ). Chiefly *U.S.*

[**1928** M. Ostenso *Mad Carews* xii. 160 It's your—whatcha-may-call-it—your dowry!] **1942** Berrey & Van den Bark *Amer. Thes. Slang* § 75/4 *Contrivance*.. *gadget*, whatchamacallit. **1974** R. B. Parker *God save Child* (1975) ii. 13 A pet whatchamacallit... Guinea pig. **1979** *Globe & Mail* (Toronto) 24 Jan. 6/2 Wouldn't everyone feel silly if it turned out..that the whang-doodle was just a whatchamacallit with speed stripes?

whatever, *pron.* and *a.* Add: **3. c.** *adv.* Whatever may be the case, at all events. *dial.* (and *colloq.*).

1870 'R. Piketah' *Forness Folk* 15, I cuddent leave t' pleass whativver wi'out seein' her. **1900** 'A. Raine' *Garthowen* 93 She's got a tidy pair of ankles, whatever. **1933** 'R. Connor' *Girl from Glengarry* 120, I am doing my utmost whatever. **1960** R. Williams *Border Country* i. ii. 58 What do it matter it's down?.. He is Will whatever. **1962** *Amer. N. & Q.* I. 15/1 Whatever, from the early 1700s to the present day..it was the musical that struck root as an indigenous form. **1980** *New Musical Express* 12 Jan. 33/1 Whatever, the myth looks momentous in its sleek new American threads.

4. c. *or whatever*: used after a noun (or nouns) to suggest that some other unspecified term might be employed instead, as being more usual, preferable for any reason, or more applicable; or something similar; or the like. *colloq.*

1905 W. James *Let.* 25 Apr. (1920) II. 225 Poor Professor De Sanctis, the Vice President or Secretary or whatever. **1913** E. Pound *Let.* 7 Nov. (1971) 24 *If* Chicago (or the U.S.A. or whatever) will slough off its provincialism, if it will begin to be aware of Paris (or of any other centre save London),..there is no reason for Chicago or *Poetry* or whatever not being the standard. **1917** H. James *Sense of Past* ii. 83 One of those concentrated terms of pious self-dedication or whatever by which the aspirants of the ages of faith used to earn their knighthood. **1958** P. Scott *Mark of Warrior* ii. 167 I'd get on to battalion or brigade, or whatever, and tell 'em. **1964** [see *Kiwanis]. **1975** I. Murdoch *Word Child* 47 And even if we are all thoughts in the mind of God or whatever why should you be able to become God? **1981** 'M. Innes' *Lord Mullion's Secret* ii. 21 There isn't a handy second title around. Viscount Tom Noddy, or whatever. **1984** J. Barnes *Flaubert's Parrot* x. 129 Bourgeois monarchy, or bureaucratised totalitarianism, or anarchy, or whatever.

d. Similarly replacing other parts of speech.

1947 *Periodical* XXVII. 93 It was in one with a brown (or whatever) cover. **1976** *Church Times* 6 Aug. 9/1 Now that the Archbishop of Canterbury has 'relinquished', 'delegated' or whatever his metropolitical authority to the local Arab Anglicans.., is it not proper [etc.].

what-ho (hwǫt,hǭᵘ·), *a.* [The exclamation *what ho!* s.v. What B. I. 3 a used as an adj.] Superior, smart, stylish; designating the type

of person supposed to use the exclamation, esp. the heartier kind of officer and gentleman.

1937 in *Amer. Speech* (1938) XIII. 239/1 At the time of their installation the elevators at the Ritz Carlton were considered the What-ho-iest in town. **1973** *Times Lit. Suppl.* 21 Sept. 1074/3 Those who survive to be relieved by what-ho young soldiers in scarlet tunics and white bandoliers. **1977** *Time* 22 Aug. 37/1 The Legionnaires are a carefully assorted lot,..a soulful French musician, a what-ho English blueblood, a hulking Russian who once guarded the Czar's family, and so on.

what-if (hwǫ·t,if), *a.* and *sb.* [An extended use of the phr. *what if..?*: see What A. I. 5 a.] (That involves) speculation as to what might have been, had antecedent conditions been different; an instance of this.

1973 *Nation Rev.* (Melbourne) 31 Aug. 1455/1 The whatif game, a futile exercise in hindsight, poses such unanswerable questions as whatif Romulus and Remus had fallen foul of a hostile..wolf? **1974** J. Irving *158-Pound Marriage* (1980) iii. 57 Joseph Stalin..was himself a figure surrounded by a horde of *what if's*. **1977** *New Yorker* 29 Aug. 66/2 'The Eagle Has Landed', one of the current unalarming terror films about what-if, the what-if being, in this case, what if the Germans had tried to capture Churchill? **1982** *Times* 14 Jan. (Information Technology Suppl.) p. v/4 *What-If Games.* Computer software is available which allows users to change one variable in a set of data, and see how this affects all the other variables. **1984** *Computers in Teaching* No. 1. 29 All of the models are theoretically realistic and allow the undergraduate student to pose the 'What if..?' kinds of question, which are so difficult to answer without recourse to simulation.

what-is-it, var. *WHATSIT.

what-like, *interrog. a.* (*sb.*) Add: **a.** Also, of what kind or character. (Earlier and further examples.)

1719 A. Ramsay *Poems* (1945) I. 214 To speer what like a Carlie is he. **1810** R. H. Cromek *Remains of Nithsdale & Galloway Song* 37 What like may your lassie be? **1865** Dickens *Mut. Fr.* ii. iii. 16 She knows Miss Abbey of old, remind her, and she knows what-like the home, and what-like the friend, is likely to turn out. **1905** in *Eng. Dial. Dict.* VI. 443/1 What-like hat had he on? **1953** 'N. Blake' *Dreadful Hollow* xii. 150 Now you tell me—what like are the Blick laddies?

Whatman. Add: A proprietary term. (Later examples.)

[**1876** *Trade Marks Jrnl.* 8 Nov. 748 *J. Whatman*... W. and R. Balston, Springfield, Maidstone, Kent; Paper Manufacturers... Writing paper and drawing paper.] **1916** *Ibid.* 26 July 763 W. E. R. Balston Limited Genuine Whatman folded filter papers. **1976** *Ibid.* 28 Apr. 890/2 *Whatman*..paper and paper articles..not including printed publications. Whatman Limited, Springfield Mill, Maidstone, Kent, Paper manufacturers.

what-not, whatnot. Add: **1. a.** Also, 'whatever you like to call it'.

1876 E. W. Heap *Diary* 11 June in *Publ. Amer. Dial. Soc.* (1969) lii. 56 [We] all started out on our grand excursion Picnic fishing party or what not.

c. Used as a euphemism for something the speaker does not wish to name.

1964 in Hamblett & Deverson *Generation X* 85 By the time I was fourteen I'd been a court witness in an indecent exposure case after an Indian doctor had been caught flashing his whatnot at me in an Adventure Playground. **1977** *Custom Car* Nov. 28/2 What ho, a twace of the fairer whatnot in the old Panther eh? **1977** M. Riley *Ideal Friend* iv. 30 She said..tapping the Cellophane-covered éclairs, 'I don't know about you but these always put me in mind of nignogs' whatnots.'

Hence **wha·tnotism**, any or every kind of 'ism' (as a final term in an enumeration of 'isms').

1915 Galsworthy in *Fortn. Rev.* 1 Nov. 928 What is wanted in a work of art is an..adequate correspondence between fancy and form..so that one shall not be distracted by its naturalism, mysticism, cubism, whatnotism. **1951** Koestler *Age of Longing* i. v. 113 Those convicted of formalism, neo-Kantianism,..and whatnotism were given their deserved punishment.

what'sa matter (wǫ·tsǎ,mæ·təɹ). Also **what-samatter**, etc. Repr. colloq. or careless pronunciation of 'what is the matter?' Cf. *WHASSA.

1935 *New Yorker* 12 Jan. 18/3 What'sa metta with Kitty Shapiro? **1950** *Commentary* Sept. 255/2 'Whatsamatter,' he shouted at Rosa, 'you want her blood?' **1960** 'E. McBain' *Heckler* (1962) xiv. 136 Whattsa matter? Something wrong? **1977** D. E. Westlake *Nobody's Perfect* (1978) ii. i. 102 Whatsa matter don't you wanna go home!!!

what's-her-face (wǫ·tsəɹfẽⁱs), occas. U.S. var. *what's-her-name* s.v. What's-his-name. Cf. *WHATSISFACE.

1980 in S. Terkel *Amer. Dreams* 5 Several times during my year as what's-her-face I had seen the movie *The Sting.*

what's-his-name. Add: Also **whatsisname**; **what's-your-name** (later example).

1942 W. Faulkner *Go down, Moses* 87 Is that so? Look here, Mister What's-your-name—. **1943** K. Tennant *Ride on Stranger* x. 114 And then Bleeby.. accusing me..of turning Whatsisname against him. **1979** S. Wilson *Glad Hand* i. i. 11 *Marilyn.* What is going on? *Brian.* Same old thing: raising the whatsis-name—the Antichrist.

whatsisface (wǫ·tsizfẽⁱs), U.S. var. What's-his-name.

1967 *Current Slang* (Univ. S. Dakota) Spring 5, [1964] *What's his face*, one whose name is forgotten. **1977** J. Wambaugh *Black Marble* (1978) vi. 79 They're having another Save Harry Whatzisface party there today. **1978** *N.Y. Times Mag.* 23 July 23/3 The derivation of some vogue phrases is a mystery:.. What visual need caused the unforgettable 'whatsisname' to become whatsisface?

whatsit (wǫ·tsit). *colloq.* Also **what-is-it, what's-it**, and (*U.S.*) **whassit.** [The phr. 'what is it?' used as a sb.: see What A. I. 1.] = What's-his-name, What's-its-name, etc. (used variously of a person or thing); a 'thingummy'.

a **1882** *Philad. Times* in *Dict. Americanisms* (1951) II. 1855/1 The two negro girls, who figure as 'what-is-its', are paid $200 a week. **1898** J. D. Brayshaw *Slum Silhouettes* 158 'Now,' said Joe, 'who says pudden? Mister What's It—a little piece?' **1922** S. Lewis *Babbitt* vi. 77 He's a what-is-it from Columbia. **1931** *Kansas City Times* 29 Sept., A Whassit. Excitement..Friday afternoon was caused by the appearance of an insect which [etc.]. **1954** P. Frankau *Wreath for Enemy* iii. v. 215, I couldn't even walk along the passage to the whatsit. **1979** P. Alexander *Show me Hero* xvii. 178 Suddenly you're a man. Not just because you happen to have a couple of whatsits, but because you *feel* it. **1984** B. Francis *AA Car Duffer's Guide* 14 Do you think I ought to check the strength of the whatsit—electrolyte—while I'm at it?

wha:t-the-he·ll, *v. slang.* [The phr. 'what the hell?' used as an expression of irritation: see Hell *sb.* 9.] *intr.* To exclaim 'what the hell..?'; to make an angry demand for an explanation.

1924 Wodehouse *Leave it to Psmith* x. 211 While everybody's cutting up and what-the-helling. **1939** H. Hodge *Cab, Sir?* 181 The yawper, of course, is convinced that if he doesn't what-the-hell a little..we shall deliberately go a long way round. **1963** Wodehouse *Stiff Upper Lip, Jeeves* xvii. 132 This telephone call was Aunt Dahlia what-the-helling.

wha:t-the-he·ll, *adj. phr. slang.* [The phr. *what the hell!* s.v. *HELL *sb.* 9.] Casual, insouciant, devil-may-care.

1968 *Listener* 1 Aug. 130/1 Much of this is due to his casual 'What the hell?' attitude to the over-familiar pomposities of public life. **1977** *Time* 19 Dec. 47/2 The only real stumbling block is fear of failure. In cooking you've got to have a what-the-hell attitude.

whau (fau, wau). *N.Z.* Also **wou.** [Maori.] A shrub or small tree, *Entelea arborescens*, of the family Tiliaceæ, native to New Zealand and bearing serrate leaves and clusters of small white flowers; = *CORKWOOD 2. Also, the light wood of this tree.

1840 J. S. Polack *Manners & Customs New Zealanders* 263 The *Pongo* and *Wou*..are varieties of the cork-tree. **1868** *Trans. N.Z. Inst.* I. iii. 35 For floats, the light wood of the small tree Whau..was used. **1889, 1946** [see *corkwood 2]. **1980** J. T. Salmon *Native Trees N.Z.* 120 Whau is an attractive New Zealand tree with an unusual, almost tropical appearance.

wheal, *sb.*[3] (Later example.)

1857 C. M. Yonge *Dynevor Terrace* I. iii. 38 The last unfortunate wheal failed when the rope broke.

wheat, *sb.* Add: **2. b.** The pale gold colour of ripe wheat. Also *wheat-gold.*

1915 Wodehouse *Something Fresh* iii. 83 Joan Valentine was a tall girl, with wheat-gold hair. **1965** [see *French roll* s.v. *French 3 b]. **1970** *New Yorker* 8 Aug. 1 (Advt.), Great embroidered coat of cotton-polyester in wheat with pumpernickel trim. **1977** M. Herr *Dispatches* 175 He was wearing a denim workshirt and wheat jeans. **1983** *Harrods Mag.* Spring & Summer 104/2 Cotton trousers in White, Wheat, Slate Blue or Navy. **1984** H. Hirt *Heat of Winter* i. 2 His face was.. very fair—what the Indian matrimonial advertisements describe as a 'wheat' complexion.

4. a. *wheat-acre, -belt* (Belt *sb.*[1] 5 a in Dict. and Suppl.), *-bran* (later example), *-bread* (later examples), *-breeder, -breeding, -cake, cocky* Austral. (Cocky *sb.*[2] 2), *country, -farm, -farmer, -farming, -feed, futures* (Future *sb.* 6), *-lumper, -lumping, ranch, rancher, -straw, -wine*; *wheat-bellied, -blazing* adjs.; *wheat-midge* (earlier and later examples). **b. wheat berry** (earlier example); **wheat-bird** (later examples); also, in North America, the horned lark, *Eremophila alpestris*; **wheat bulb fly**, the larva of a muscid fly, *Hylemyia coarctata*, which attacks the base of wheat stems; **wheat-duck** (examples); **wheatflakes** *sb. pl.* (orig. *U.S.*), a breakfast cereal made from flaked and flavoured wheat (cf. *cornflakes* sb. pl. s.v. *corn *sb.*[1] 11); **wheat germ**, the embryo of the wheat grain, extracted during milling, and valued as a source of vitamins; **wheat-grass**, (*b*) a creeping perennial grass of the genus *Agropyron*; **wheat roll**, a roll made of wheatmeal bread; **Wheat State**, in the U.S., a popular nickname for Kansas or Minnesota; also used of South Australia.

1876 G. M. Hopkins *Poems* (1967) 177 The blue wheat-acre is underneath. **1922** Joyce *Ulysses* 196 Eve. Naked wheatbellied sin. **1863** *Harper's Mag.* Oct. 718/1 The enterprising town..is the wheat-market for a considerable section of the wheat-belt of the state. **1910** *Chambers's Jrnl.* Mar. 205/2 The laying out of ready-made farms in the wheat-belts of North-West Canada. **1980** *Jrnl. R. Soc. Arts* Mar. 175/2 In the wheat belts in the USA and Australia there were..large areas still undeveloped. **1848** *Rep. Comm. Patents 1847* (U.S.) 373 Taking the outer coating or bran from the wheat berry previous to grinding produces the following important results. **1865** Wheat bird [see *Peabody]. **1917** T. G. Pearson *Birds Amer.* ii. 212 Horned Lark.. [also called] Prairie Bird; Road Trotter; Wheat Bird. **1937** Blunden *Elegy* 60 seek the wide wheat-blazing plain. **1946** *Nature* 31 Aug. 293/1 The fungus was grown in various modifications of Czapek–Dox medium with addition of manganese sulphate, in some cases with.. autoclaved wheat-bran extract. **1862** M. D. Colt *Went to Kansas* 83, I live entirely on food made of corn.. leaving the wheat bread for grand-ma and grand-pa. **1880** [see *southern *a.* 4 c]. **1978** *Listener* 10 Aug. 180/3 Oatcakes, potato cakes and wheat bread were cooked deliciously on a griddle. **1912** *Rep. 13th Meeting Australasian Assoc. Adv. Sci.* 536 (*heading*) The realization of the aims of William J. Farrer, wheat breeder. **1974** *Encycl. Brit. Macropædia* III. 1157/2 Wheat breeders regularly produce new varieties. **1898** W. J. Farrer *Let.* 30 Aug. in R. Archer *William James Farrer* (1949) xiv. 109, I should continue to carry on the wheat-breeding work at Lambrigg. **1965** *Austral. Encycl.* IX. 284b (*caption*) Wheat-breeding plots at the Temora Experiment Farm. **1883** E. A. Ormerod *Rep. Observations Injurious Insects 1882* 20 Wheat-bulb fly..was observable early in March. **1921** *Jrnl. Agric. Sci.* XI. 98 Wheat-bulb fly..does not appear to do much harm in a wet, cold, or damp summer. **1975** *N.Z. Jrnl. Agric.* Sept. 67/1 The topics covered in the first few months of the scheme include cereal mildew, wheat bulb fly, yellow rust. **1772** M. Patten *Diary* (1903) 293 His wife baked a parcel of Wheat Cakes for me when I went up to Cockermouth. **1865** A. D. Whitney *Gayworthys* 218 There are wheat-cakes and maple syrup for your breakfast. **1981** J. Dunning *Deadline* (1982) xix. 191 Trudy fixed him a breakfast of eggs and bacon and wheat cakes. **1933** *Bulletin* (Sydney) 1 Mar. 13 A good, typical S. Australian public man—a wheat cocky. **1941** K. Tennant *Battlers* xxi. 228 Like many another broken 'wheat cocky'..Jim might be packing his kids and wife into his old truck any time now. **1776** *New-York Gaz.* 24 June 3/3 To be Sold..a very good Grist-Mill.. in a very good Wheat Country. **1890** *Stock Grower & Farmer* 29 Mar. 5/3 The panhandle country..is a fine wheat country. **1979** Tanous & Rubinstein *Wheat Killing* (1980) ix. 55 We were in the flat wheat country drained by the Missouri River. **1888** G. Trumbull *Names & Portraits of Birds* 21 He found this species [*sc.* the American widgeon] in enormous flocks on the wheat-fields, and..it was there called the wheat-duck. **1917** T. G. Pearson *Birds Amer.* I. 120 Wheat Duck..is very fond of wild celery. **1980** *Hunting Ann. 1981* 40/3 A wigeon in one region would be called a baldpate in another area or wheatduck in another. **1958** *Publ. Amer. Dial. Soc.* xxx. 6 Farmers who have large wheat-farms. **1870** *Rep. Comm. Agric. 1869* (U.S. Dept. Agric.) 5 The wheat farmer..is not joyous over his market returns. **1959** *Cape Times* 18 July 2/5 Wheat farmers welcomed the good rains. **1965** *Austral. Encycl.* IX. 285/2 The most spectacular change in wheat-farming practice in recent years. **1932** *Daily Tel.* 8 Oct. 4/2 Oats quiet of sale... Millers' wheatfeed quiet. **1960** *Farmer & Stockbreeder* 29 Mar. 4/2 Demand for wheatfeed is steady. **1903** *Bull. Maine Agric. Exper. Station* No. 84, 143 Fruen's Best Wheat Flakes, 'made from the best Pacific Coast White Wheat'. **1939** G. Greene *Lawless Roads* ii. 45 He looked up from his dry wheat flakes. **1970** M. Kelly *Spinifex* ii. 37 A grocer size wheatflakes box. **1908** 'O. Henry' *Strictly Business* 252 After I had taken some $9,000,000 out of the soap business I made the rest in corn and wheat futures. **1979** Tanous & Rubinstein *Wheat Killing* (1980) ii. 13 The rise in the wheat price will mean a fortune to them if they own the wheat futures. **1897** *Sears, Roebuck Catal.* 15/3 Wheat Germ Meal... Cooked in 5 minutes. **1933** *Discovery* May 160/1 The richest source of vitamin E is wheat germ. **1980** *Sunday Times* (Colour Suppl.) 20 Jan. 57/3 Wheatgerm Loaf. A good hearty farmhouse loaf. **1762** W. Hudson *Flora Anglica* 45 Common Wheat-grass, Dog's-grass, Quick-grass or Couch-grass. **1871** *Harper's Mag.* July 187/2 Among the more important of these plants the wheat-grass stands pre-eminent. **1968** F. W. Gould *Grass Systematics* 186 Several species of *Agropyron* are important forage grasses on western rangelands, outstanding among which are.. bluebunch wheatgrass, and..western wheatgrass. **1934** *Bulletin* (Sydney) 3 Jan. 14/3 Harrison, a Wallendbeen (N.S.W.) wheat-lumper, carried 1170 bags of wheat the other day. **1957** Wheat-lumping [see *ring *v.*[1] 11]. **1840** J. & M. Loudon tr. *Kollar's Treat. Insects* ii. 123 The Wheat Midge... The perfect insect has a distant [*sic*] resemblance to the common midge, but is smaller. **1931** K. M. Smith *Textbk. Agric. Entomol.* xi. 169 Wheat midge was especially destructive in 1926 in the eastern counties. **1874** Wheat ranch [see *sheep ranch* s.v. *sheep *sb.* 7 c]. **1947** *Mazama* Dec. 1/1 An overnight trip to the 500-acre wheat and stock ranch..near Wamic. **1947** *Chicago Tribune* 1 Nov. 11/4 A former life term prisoner..admitted the..slaying of a retired Canadian wheat rancher. **1977** J. Gillis *Killers of Starfish* (1979) v. 32 Maybe he was a big wheat rancher. **1962** E. Snow *Other Side of River* (1963) lix. 447 The dining room I saw was serving wheat rolls, turnips, cabbage and spinach which looked adequate and wholesome. **1978** H. McLeave *Borderline Case* (1979)

iv. 49 Shigo brought hot coffee, wheat rolls baked on the spot, butter. **1911** D. Malloch *Resawed Fables* 65 He had a Friend in the Retail Lumber Business..and he sent him enough Money to get Home to the Wheat State. **1945** Baker *Austral. Lang.* x. 187 Popular names for the various Australian states are:..South Australia: the Wheat State. **1950** R. Meyer *Festivals U.S.A.* 225 Kansas is sometimes called the Wheat State, but it is more familiarly known as the Sunflower State. *c* **1903** O. Read in *Library Southern Lit.* (1909) X. 4374 The Squatter, with his wheat-straw beard, his hay hair and his autumn leaf complexion. **1941** L. B. Lyon *Tomorrow is Revealing* 44 A son with a bird's glint, and wheat-straw hair. **1954** E. Pound *Cantos* liii. 281 With gold cup of wheat-wine that he go afield to spring ploughing. **1982** C. Thomas *Jade Tiger* 52 Wheat wine, almost pure alcohol.

wheatear[1]. Add: **2.** A pattern in embroidery, lace, weaving, etc., or an ornament in wood-carving, etc., resembling an ear of wheat.

1882 [in Dict.]. **1911** *Encycl. Brit.* XIII. 306/1 The backs of Hepplewhite chairs were often adorned with galleries and festoons of wheat-ears or pointed fern leaves. **1919** T. Wright *Romance Lace Pillow* ix. 83 The ancient pattern called the *Wheat-ear and Cornflower*.. is still made. **1955** R. W. Millar tr. *Daniel-Rops' Jesus in His Time* ix. 366 Heavy columns of porphyry with rather ungainly capitals carved with grapes and wheat-ears. **1957** Simpson & Weir *Weaver's Craft* xii. 151 The patterns most generally used for tweeds are..Twill.. Goose eye..Wheatear. **1977** *Penguin Dict. Decorative Arts* 374/2 Typical of the [Hepplewhite] style are.. wheat-ears with which the central splat of shield-back and other chairs are decorated.

wheaten, *a.* Delete 'Now *rare.*' and add: **1.** (Later examples.)

1922 Joyce *Ulysses* 503 Wheatenmeal with honey and nutmeg. **1944** L. Mumford *Condition of Man* vi. 202 Wheaten bread.

4. Of a pale honey colour. *Wheaten terrier*, a soft-coated terrier belonging to a breed originally developed in Ireland and distinguished by its pale golden wavy coat. Also *absol.* as *sb.* denoting the dog (also, the colour).

1943 *Our Dogs* 5 Mar. 234/5 The soft-coated Wheaten Terrier has now been recognised by the English Kennel Club. *Ibid.* 19 Mar. 281/2 There were 30 entries of Wheatens at the Irish Kennel Club show. **1945** C. L. B. Hubbard *Observer's Bk. Dogs* 146 Wheaten-coloured Terriers of soft coats have existed in Ireland for a considerable time. **1959** *Times* 14 Aug. 1/7 (Advt.), Soft-coated Wheaten Terrier puppies. **1971** F. Hamilton *World Encycl. Dogs* 480 The mature Wheaten is an attractive, compact, well-built dog, strong and energetic. **1975** *Country Life* 6 Feb. 311 The Border terrier..coat of either wheaten, red, grizzle and tan, or blue and tan.

Wheaties (hwī·tiz), *sb. pl.* Also **wheaties.** [f. Wheat *sb.* + -ie.] The name of a breakfast cereal made from wheat.

A proprietary name in the United States.

1925 *Official Gaz.* (U.S. Patent Office) 24 Mar. 738/2 Washburn Crosby Company, Minneapolis, Minn. *Wheaties* ..Cereal food products. **1935** *Good Housekeeping* (N.Y.) June (Advt., rear cover), Wheaties bring real whole wheat..in a ready-to-serve form that children adore. **1952** *Galaxy* Sept. 50/1 An institution is where they put Aunt Maggy when she started collecting Wheaties in a stamp album. **1968** *Shakes. Q.* Winter 38 For Ryot..to tempt Youthe with archery, would be as if Gluttony tried to corrupt his appetite with a bowl of wheaties. **1981** I. St. James *Balfour Conspiracy* vii. 241 Their morning wheaties came wrapped in a..Press Release.

Wheatstone. Substitute for def. of *Wheatstone('s) bridge*: a simple circuit for measuring a resistance by connecting it so as to form a quadrilateral with three known resistances and applying a voltage between a pair of opposite corners: a galvanometer connected between the other two corners registers no current when the ratios of the two pairs of adjacent resistances are equal. (Further examples.)

1901 *Phil. Trans. R. Soc.* A. CXCVI. 29 The two grids A and A′ formed two arms of a Wheatstone bridge. **1953** A. Smith *Blind White Fish in Persia* i. 25 He had with him..a Wheatstone's bridge—a gadget of complex appearance for measuring the electrical resistance of the soil. **1979** E. N. Lurch *Electric Circuit Fundamentals* ix. 296 The Wheatstone bridge..is the elementary bridge circuit that is the prototype of all the more complex bridges used in electric circuit analysis.

b. Used in the possessive, *attrib.*, and *absol.* to denote forms of electric telegraph invented by Wheatstone.

1858 Faraday in *Notices of Proc. R. Inst.* II. 555 (*heading*) On Wheatstone's electric telegraph in relation to science. **1881** *Ibid.* IX. 302, I found that one of our Wheatstone instruments was actually working at the rate of 180 or 190 words a minute. **1898** [in Dict.]. **1922** Glazebrook *Dict. Appl. Physics* II. 788/2 Wheatstone simplex circuits are used extensively for the transmission of press telegrams to all parts of Great Britain. **1949** *Brit. Jrnl. Psychol.* XL. 37 A Wheatstone transmitter was driven at 6 and 14 words a minute by a variable speed gear.

wheaty, *a.* Restrict † *Obs.* to sense in Dict. and add: **b.** Of or pertaining to wheat.

1933 L. I. Wilder *Farmer Boy* vii. 47 The wheaty smell of new bread.

whee (hwī), *int.* [Echoic.] An exclamation of joy, exhilaration, astonishment, etc. Occas. as *sb.*, a high-pitched sound resembling this.

[**1918** E. A. Mackintosh *War, the Liberator* iv. 145 Whee-ee-ee-errump! The air was full of dust and smoke from a little way up the trench.] **1920** S. Lewis in *Sat. Even. Post* 11 Dec. 10/1 'Whee!'..He sprang up, posed like the Statue of Liberty, hurled a pillow at her. **1944** E. S. Gardner *Case of Black-Eyed Blonde* (1948) xx. 205 The little boy shrieked with delight. 'Whee-eee-ee,' he cried. **1960** M. Macdonald in G. Maxwell *Ring of Bright Water* xii. 168 Her [*sc.* the otter's] basic conversational vocabulary was a high-pitched whistling 'Whee'. With loud and soft, short and long and other variations of 'Whees' she had quite a lot to say. **1978** N. Freeling *Night Lords* v. 24 Whee, thought Castang: he has worked fast. **1981** *Studia Mystica* IV. iv. 34 A cry of 'Whee!'

whee (hwī), *v.* [f. prec.] **a.** *trans.* With *up.* To stimulate, excite. *U.S. colloq.* **b.** *intr.* To utter a high-pitched sound.

1949 *Sat. Even. Post* 3 Dec. 3/3 And did that whee him up to do his Christmas shopping early? **1960** M. Macdonald in G. Maxwell *Ring of Bright Water* xii. 171 She [*sc.* an otter] would lie on the floor..'wheeing' plaintively. **1966** *N.Y. Times* 17 Apr. v. 2, I was all wheed up, feeling great, I knew I would win. **1971** *Atlantic Monthly* Mar. 36, I got so patriotically wheed up that I ended by calling for three cheers for General Douglas MacArthur.

wheel, *sb.* Add: **II. 3. h.** (Later examples.)

1929 J. B. Priestley *Good Compan.* i. ii. 67 For the next hour she sat at the wheel under his tuition. **1972** T. P. McMahon *Issue of Bishop's Blood* (1973) xvi. 229 The long-haul truckers drove themselves right into a ditch after too many hours at the wheel.

7. a. (Earlier examples.)

1880 *Scribner's Monthly* Feb. 483/1 A few possessors of the birotate chariot, numbering some forty odd, enjoyed a 'wheel around the Hub'. **1882** *Wheelman* I. 13 'I love my wheel,' he said, 'as the yachtsman loves his boat.'

b. *to be on someone's wheel*: to be close behind someone, to be on his track; to put pressure on someone (to do something). *slang* (chiefly *Austral.*).

1941 V. Davis *Phenomena in Crime* vi. 78 Don't come here if there's a busy on your wheel! **1954** V. Kelly *Shadow* 89 Down there the cops'll give you a go. Here they're on your wheel all the time. **1959** A. Upfield *Bony & Mouse* 104 I'll be ready for it. I'm going to be right on Tony's wheel when it happens. **1969** O. White *Under Iron Rainbow* 118 The inspector's been on my wheel to trace him.

c. *pl.* A car. *slang* (orig. *U.S.*).

1959 *Esquire* Nov. 70 J, *Wheels*, car. **1970** K. Platt *Pushbutton Butterfly* (1971) v. 51 'Can I drive you to where you're agitating today?' 'Beautiful. I don't have wheels,' he said. **1971** 'H. Carmichael' *Quiet Woman* iv. 33 I'd be out and about if I had wheels. Damn car won't be ready until tomorrow. **1982** G. Lyall *Conduct of Major Maxim* xxv. 222 'Did you find me some wheels?'.. 'Yep: a Renault 16TX.'

7*. *we had one but the wheel came off*, joc. phr. used to indicate that the speaker has not understood the subject of the foregoing conversation.

1937 Partridge *Dict. Slang* 366/1 *Had one and* (or *but*) *the wheel came off* (*,we*), a lower-class and military c.p. directed at an unintelligible speaker or speech. **1974** P. Wright *Lang. Brit. Industry* xiv. 128 If asked for something foolish, you can say,..'I had one but the wheels came off.'

III. 8. b. *U.S. slang.* A dollar; = *CART-WHEEL 2.

1807 H. Tufts in E. Pearson *Autobiogr. of Criminal* (1930) ii. iv. 293 *Wheel*, a dollar. **1825** J. Neal *Brother Jonathan* I. 160, I shows him a double handful o' the royal goold; the ginooine yeller stuff—wheels. **1902** W. N. Harben *Abner Daniel* 143 How will fifteen hundred round wheels strike you? **1907** C. E. Mulford *Bar-20* v. 47, I paid twenty wheels for that eight years ago.

c. A whole cheese, flan, or other food which is made with a circular form but may be cut into sections. Usu. with *of* and defining term.

1977 *New Yorker* 3 Oct. 53/3 A feast of varied delicacies, its principal ornament a small wheel of Camembert. **1978** C. Conran *British Cooking* 233/2 On May Day in the city of Gloucester a huge golden wheel of cheese, festively garlanded, used to be carried in procession round the town. **1978** *Neiman-Marcus Christmas Bk.* 93 A full three pound wheel, covered with protective black wax. **1982** M. Babson *Death warmed Up* viii. 75 The wheels of pizza and quiche lorraine in the makeshift rack. **1985** *Sci. Amer.* May 67/1 The semisoft, blue-mold cheese is made from sheep's milk and formed into wheels weighing about 2·5 kilograms (5·5 pounds) each.

IV. 11. a. Also in phr. *the wheel has come full circle* and varr. (in allusion to Shakes. *King Lear* v. iii. 174), the same situation has come about again, things have returned to their original position.

1944 W. S. Maugham *Razor's Edge* v. 176 The wheel comes full circle... There was a time when the black sheep of the family was sent from my country to America; now apparently he's sent from your country [*sc.* America] to Europe. **1954** J. A. Sheard *Words we Use* iv. 158 Old English had a derivative noun, *godspellere*, but this..was later replaced by a foreign loan-word,..*evangelist*... But in recent years the wheel has come full circle, and by a new process of derivation the Americans have their *hot gospellers*! **1966** W. H. Lewis in *Lett. C. S. Lewis* 24 The wheel had come full circle: once again we were together in the little end room at home. **1977** J. Crosby *Company of Friends* xviii. 117 The wheel was coming full circle. The public was fed to the teeth with disclosure. It yearned for the security of secrecy.

12. b. *on wheels.* (*a*) (later examples); (*c*) used as an intensive: in the extreme.

1914 [see *job *sb.*[2] 4 d]. **1943** S. Lewis *Gideon Planish* 127 Looks just like a sweet little ivory statue, but is she hell on wheels! **1958** M. Dickens *Man Overboard* iv. 59 It was his wife. She's a bitch on wheels, from what he tells me. **1970** 'D. Halliday' *Dolly & Cookie Bird* iv. 42 Look at the time... If you're going to show us your balloons, you'll have to do it on wheels. **1978** S. Brill *Teamsters* vii 275 In the 1930s and '40s and into the '50s, truck driving was sweatshop labor on wheels. **1980** N. Freeling *Castang's City* xxv. 174 Local wine-shipper in quite a high-class way... The business ran on wheels.

d. *silly as a wheel*: extremely silly. *Austral. slang.*

1952 T. A. G. Hungerford *Ridge & River* 57 Oscar was sound, but silly as a wheel. **1966** J. Morrison in *Coast to Coast 1965–66* 157, I warned Rose. She was as silly as a wheel, too, but a man's got to do what he can to protect his daughters.

13. c. *to see* (*what makes*) *the wheels go round* and varr: to see how things work; chiefly *fig.* with reference to the operation of a business, organization, etc. *colloq.*

[**1876** J. Habberton *Helen's Babies* 11 'I want to see the wheels go round,' said Budge.] **1923** R. D. Paine *Comrades of Rolling Ocean* ix. 160, I want to watch a supercargo and see his wheels go round. **1922** *Broadcaster* Oct. 149/1 The natural indifference of the fair sex to any knowledge of what 'makes the wheels go round.' **1979** P. Levi *Head in Soup* iii. 58 How amateur we were. Those who know how the wheels turn are always bored. **1980** N. Freeling *Castang's City* xv. 100 There's any amount of what makes the wheels go round... Feather-bedding and barrel-rolling.

d. = *big shot* s.v. *shot *sb.*[1] 22 c. Cf. *big wheel* (*b*) s.v. *big *a.* B. 2. *slang* (orig. and chiefly *U.S.*).

1933 *Amer. Speech* VIII. ii. 55/2 *Wheels*, substitute for big shots, leaders of a gang. **1956** B. Holiday *Lady sings Blues* (1973) xviii. 149 After I got to be a wheel in the kitchen, I used to take care of Marietta by saving her the best of the food. **1963** J. N. Harris *Weird World Wes Beattie* (1966) iii. 36 Well, in business, Howie is a sort of minor wheel... He owns pieces of things. Radio stations, a commercial film company, a night club. **1975** *Globe & Mail* (Toronto) 12 Sept. 27/8 If politicians and business people and other wheels don't like it, he couldn't care less. **1980** A. Fox *Kingfisher Scream* vi. 94 Some Pentagon wheel's flying in and Don feels he has to travel up there with him.

15. c. In *Rugby Football.* (See quot. 1897.)

1897 [see *back *a.* 1]. **1927** H. Walpole *Jeremy at Crale* xvi. 284 Back they went, down again, the ball flung in. The Callendar forwards had it and manœuvred the finest wheel of the match, swinging round against all opposition.

V. 17. a. Of or pertaining to a wheel, etc., as *wheel alignment, arch, brake, trim*; furnished with a wheel, etc., as *wheel hoe, loader.*

1908 *Motor Man.* (ed. 10) vi. 165 (*heading*) To test wheel alignment. **1971** 'D. Rutherford' *Clear Fast Lane* 70 I'm not going on till we've had the wheel-alignment checked. There's..a service station ten kilometres on. **1935** *Automobile & Carriage Builders' Jrnl.* Mar. 45 (*caption*) Details of the wheelarch, scuttle ventilator and rear locker. **1983** *Buses* Feb. 57/1 Longitudinal seating provided over the rear wheelarches. **1936** *Discovery* July 228/2 The average pilot regards his wheel brakes as an assistance to ground taxying rather than as a means of arresting his run. **1974** *Encycl. Brit. Macropædia* I. 378/1 Wheel brakes are generally hydraulically operated. **1858** C. Flint *Milch Cows* 193 In weeding, a little wheel-hoe is invaluable. **1911** *Daily Colonist* (Victoria, B.C.) 22 Apr. 9/6 (Advt.), We carry a full line of Garden Drills, Double and Single Wheel Hoes. **1971** Wheel loader [see *dozer[4]]. **1976** *Ilkeston Advertiser* 10 Dec. 3/4 (Advt.), Morris 1800 (Princess style) Saloon (1975 'P')... Fitted radio, wheel-trims etc. **1983** *Which?* Sept. 435/2 However, all of the 'MG' wheel trims fell off at some time.

b. Objective, as *wheel-changing*; instrumental, as *wheel-driven, -worn* (later example) adjs.

1974 *Harrod's Christmas Catal.* 69/2 Lotus..scale model: complete with wheel brace for wheel changing. **1972** Wheel-driven [see *salt *a.*[1] 1 c]. **1944** Blunden *Shells by Stream* 13 Its kingdom is the farm, the farmer's lane Its wheelworn churchway from the lonely road.

18. wheel arrangement, the relative positioning of driving wheels and idle wheels on a locomotive; **wheel-back** (earlier example); also a chair with such a back; **wheel balance** *Mech.*, an even distribution of mass about the axis of a wheel so that it rotates without wobbling or vibrating; so **wheel balancing**, the process of achieving this for the wheels of a motor vehicle; **wheel bay** = *wheel well* below; **wheel brace**, (*a*) a tool for screwing and unscrewing nuts on the wheel of a vehicle; (*b*) a kind of hand drill worked by the turning of a wheel; **wheel car**, a simple farm-cart (see quot. 1931); **wheel clamp**, a clamp designed to be locked to one of the wheels of an illegally

parked motor vehicle to immobilize it; hence as *v. trans.*; so **wheel clamping** *vbl. sb.* and *ppl. a.*; **wheel-dog** *Canad.*, the dog harnessed nearest to the sleigh in a dog team; **wheel-dwelling, -hut** *Archæol.* = *WHEEL-HOUSE 3; **wheel landing** *Aeronaut.*, a landing (of an aircraft with a tail-wheel or tail-skid) in which the main wheels touch down first, followed by the tail; **wheel pants** *Aeronaut.* (see quot. 1956); **wheel-pit**, (*a*) (earlier example); **wheel-set**, a pair of wheels attached to an axle; **wheel slip**, the failure of the wheels of a vehicle to grip the surface on which they are travelling, so that they slip instead of rolling; also (*rare*) as *v. intr.*; **wheel spin**, the spinning of the wheels of a vehicle, caused by hard acceleration of the engine combined with the failure of the wheels to take a grip on a slippery surface; **wheel-tax** (earlier example); **wheel well**, the recess, under the wing of a vehicle, into which the wheel fits, or, on an aircraft, into which the landing gear is retracted; **wheel-window** (earlier example); **wheel wobble**, vibration of the wheels of a vehicle in motion, usu. when travelling at some speed; also *fig.*

1912 *Railway Mag.* Mar. 203/1 Of the total number of engines mentioned, 80 were of the 2–2–2–0 wheel arrangement. **1966** K. Möller *Amer. & Brit. Railway English* 43, 0–8–0 is a notation for a wheel arrangement of eight driving wheels and no leading or trailing wheels. A great number of wheel arrangements have special names, originating with the railway on which they were first used. **1902** W. H. Hackett *Decorative Furnit.* xi. 133 A set of six typical Heppelwhite [*sic*] chairs..had 'wheel' backs, on taper legs, with cross stretchers. **1927** W. E. Collinson *Contemp. Eng.* 90 The revival of Welsh dressers, wheel-backs and ladderbacks. **1968** J. Arnold *Shell Bk. Country Crafts* 133 In the hand-made 'wheel-back' there is the vestigial hub in the centre, but for economy this is absent from the factory-made splat. **1946** W. H. Crouse *Automotive Mech.* xxiv. 533 Wheel balance can be checked in several ways. **1962** *Which? Car Suppl.* Oct. 139/2 Severe vibration developed throughout the car at speeds over 65 mph and a further check of wheel balance failed to overcome this. **1951** I. Frazee et al. *Automotive Suspensions* x. 276 (*heading*) Wheel balancing. **1977** Wheel balancing [see *shimmy *sb.*[2] 2]. **1976** 'A. Hall' *Kobra Manifesto* xv. 200 The problem was to keep my body arched against the curved top of the wheelbay, giving me a chance of escaping the wheels when they slammed home and locked. **1920** *Motor Man.* (ed. 23) xiv. 144 The most popular form of fitting is by means of separate bolts carried on the fixed hub, to which the pressed-steel wheel is held by a number of capped nuts which can be detached by a wheel brace. **1964** F. Preston *Man makes Hole* 6/2 The hand drill or wheel brace is of fairly recent origin, although it derives from the so-called bevel drill widely made in Germany. **1974** Wheel brace [see *wheel changing*, sense 17 b above]. **1975** R. A. Salaman *Dict. Tools* 187/2 *Drill, hand* (Wheel Brace..).. The modern form of Hand Drill was an American innovation of about 1870 which reached this country about the turn of the century. **1984** B. Francis *AA Car Duffer's Guide* 38/2 Then, with the wheelbrace, slacken off all the wheel nuts about half a turn. **1931** *Antiquity* June 185 The special features of the wheel-car are..(a) the great length..of the body..; (b) the position of the axle tree *above*..the main beams of the frame; (c) the bumpers..; and (d) the embryo cart structure. **1968** J. Arnold *Shell Bk. Country Crafts* xi. 163 Round about Clun, they knew it [*sc.* a gambo] as a wheel-car. **1980** *Daily Tel.* 2 Jan. 3 (*heading*) Car park offenders face wheel clamps. **1981** *Times* 19 Nov. 3 Illegal parking in London has become so widespread that the Government may bow to police demands to be allowed to use wheel clamps to immobilize offending vehicles. **1983** *Daily Tel.* 14 July 19/1 Cars belonging to diplomats will no longer be wheel-clamped. **1980** *Daily Tel.* 2 Jan. 3/2 More officers will be available to tow away dangerously parked vehicles and the use of the 'Denver Shoe' a wheel clamping device, is being considered. **1983** *Sunday Tel.* 15 May 3/1 The wheel clamping team..will consist of one sergeant, eight police constables, 28 vehicle removal officers and eight traffic wardens. **1922** G. C. F. Pringle *Tillicums* 85, I put a smaller dog..in the lead and hitched Steal up next the sleigh as my 'wheel-dog'. **1965** A. V. Wilson *No Man stands Alone* 29, I firmly believe that the 'wheel-dog', next to the sled, can upset one any time he wishes. **1931** V. G. Childe *Skara Brae* vii. 174 The Jarlshof hut..illustrates..the normal construction of a wheel-dwelling. *Ibid.* 173 The relics from the wheel-huts round the broch of Jarlshof cannot be distinguished from those from the broch itself. **1928** N. Macmillan *Art of Flying* x. 142 With many aeroplanes..too slow speed of approach makes the elevator unable to apply the necessary load quickly enough. This results in a wheel landing with the tail up. **1942** *R.A.F. Jrnl.* 13 June 8 The second pilot..said that the impact was no worse than a bad wheel landing. **1956** *U.S. Air Force Dict.* 567/1 *Wheel pants*, a set of streamlined fairings around each wheel in certain fixed landing gears. **1971** *Flying* Apr. 40/1 The 172 did appear with clean, 175-style wheel-pants. **1850** S. Judd *R. Edney* iii. 43 The subordinate branches were carried on below, under the 'bed' or main floor of the mill, near the wheel-pit. **1969** *Jane's Freight Containers 1968–69* 102/1 Provision has to be made for wheelsets to be available at the destination terminal. **1980** *Sci. Amer.* Aug. 33/3 A revolutionary bogie design which prevents wear on the wheel flange as the train negotiates a bend. No longer is the wheelset banging from side to side, abrading the flanges. **1945** H. J. Massingham *Wisdom of Fields* x. 193 A tractor was ploughing in the stubble... The machine was wheel-slipping. Wheel-slip means [*sc.* causes] soil-panning and winter-souring of it;..it means waste. **1960** *Times* 14 Mar. 21/2 Wheel-slip, with wheeled tractors, can be an intolerable nuisance. **1983** *Austral. Transport* Feb. 21/2 The locomotives are equipped with greatly improved wheel slip controls. **1928** *Daily Tel.* 11 Sept. 15/6 He took a grassy approach too wide, and had a wheel spin, from which he cleverly recovered. **1937** *Times* 13 Apr. p. xvi/4 On a sandy gradient of about 1 in 2, the vehicle was stopped and started, ascending and descending. No wheel spin or baulking was noticed. **1956** *Railway Mag.* Nov. 722/1 Steps already had been taken to check wheelspin. **1979** R. Lewis *Violent Death* i. 7 The van came lurching..from under the trees... Wheel-spin threw up a mist of pine needles. **1780** A. Young *Tour in Ireland* ii. xvii. 75 Taxes are inconsiderable, for there is no land tax, no poor rates..only half a wheel tax. **1959** F. D. Adams *Aeronaut. Dict.* 181/2 *Wheel well*, a recess or hollow in a wing, fuselage, etc. for a retractable landing-gear wheel. **1961** N. D. van Sickle *Mod. Airmanship* (ed. 2) iv. 103 Having the openings closed while the gear is down is advantageous because the tires will not throw foreign matter into the wheel well during ground movement. **1974** *Hot Rod Yearbk.* XIV. 219/1 Epoxy paint..was also sprayed on rear inner wheel wells. **1975** *Times* 31 Mar. 4/1 (*caption*) A Boeing 727 arrives at Saigon airport..with the body of a South Vietnamese soldier hanging from the wheelwell. **1821** M. Browne *Jrnl.* 2 May in *Diary of Girl in France* (1905) 25 There are [in Amiens Cathedral] two pretty painted wheel-windows. **1930** *Engineering* 7 Feb. 163/1 Concerned with problems of suspension and wheel wobble and shimmy. **1961** *Times* 17 Jan. 14/1 Apart from an alarming wheelwobble early in their second innings West Indies have had by far the best of today's play. **1978** A. Waugh *Best Wine Last* vii. 56 It [*sc.* a car] developed a wheel wobble at between forty-eight and fifty miles an hour.

wheel, *v.* Add: **II. 8. c.** *colloq.* To bring (someone) *in*, as for an interview, meeting, performance, etc. Also, with similar meaning, const. *on, out.* Also *fig.*

1970 *New Yorker* 28 Feb. 29/3 The Administration wheels out what are at the moment issues..which everyone can agree on. **1977** M. Allen *Spence in Petal Park* xviii. 78 Wheel Prendergast straight in when he arrives. **1978** *Daily Mirror* 12 Jan. 2/4 The agreed quota for Japanese car imports in 1977 should be wheeled out again for 1978. **1983** *Listener* 20 Oct. 27/1 'Celebrities' were wheeled in before a studio audience. **1984** *Times* 9 Feb. 11/5 This new element is wheeled in when cousins come to stay. *Ibid.* 15 Aug. 11/1 Kenny Everett..was wheeled on with other celebrities to warm up a Conservative rally for the Leader in the course of last year's general election campaign here. *Ibid.* 8 Dec. 6/4 Although his field is limited to southern France from the fourteenth to the eighteenth century, the French media wheels him out to make pronouncements on Giscard's reign, his reservations about Mitterand's regime, or whether Nazi war criminals like Barbie should be executed.

10. b. To drive a car slowly, as when manœuvring into or out of a car park.

1962 R. Unekis *Chase* (1963) vii. 20 Grozzo wheeled the Olds into the big parking area. **1974** N. Freeling *Dressing of Diamond* 182 It was Castang who wheeled the Citroën out of the parking-lot. **1976** *National Observer* (U.S.) 11 Dec. 8/2, I wheeled the bright-blue test car into a parking space.

wheel and dea·l, *v. colloq.* (orig. and chiefly *U.S.*). Pa. t. **wheeled and dealed.** [f. *WHEEL *sb.* 13 d + Deal *v.*] **1.** *intr.* To engage in scheming or shrewd bargaining, esp. of a political or commercial nature. Cf. *WHEELER-DEALER.

1961 Webster, *Wheel and deal*, to take the part of a leader or wheel and to take charge of affairs or arrangements (showed the town how an absolute dictator wheels and deals—*Newsweek*). **1962** 'K. Orvis' *Damned & Destroyed* vii. 53 You don't act like you really know where to wheel-and-deal yet. **1967** *National Observer* (U.S.) 3 July 13 Reagan could break loose votes from the Democrats if he really wanted to. You can always wheel-and-deal and get at least part of what you want. **1967** *Listener* 24 Aug. 250/1 Frost is wheeling and dealing off camera. **1974** *Publishers Weekly* 24 June 58/1 Lads who..wheeled and dealed with megacorporations. **1976** J. I. M. Stewart *Memorial Service* xiv. 234 What's in the wind is a little quiet wheeling and dealing about the black sheep of the family.'..'Why should the Provost wheel and deal about you?'

2. *trans.* To obtain by scheming; to deal or bargain in.

1971 R. Dentry *Encounter at Kharmel* xii. 217 In other words, if we agree to shut up, you'll wheel and deal some pin money for us. **1974** *Scottish Daily Express* 16 Apr. 4/3 Dr. Henry Kissinger, due here shortly to wheel and deal hi way to stilling the guns. **1979** *Tucson Mag.* Mar. 49/1 The Inn was originally built to wheel and deal Arizona land.

Hence **wheeling and dea·ling** *vbl. sb.* Cf. *wheeler-dealing* vbl. sb. s.v. *WHEELER-DEALER.

1969 *Listener* 9 Jan. 60/3 As a proof of his mischievous intentions, he [*sc.* Henry II] invites the devious King of France to this gathering. It would need a Norman Mailer to describe accurately the wheeling and dealing that follows. **1976** [see sense 1 above]. **1981** *Beautiful Brit. Columbia* Summer 9 When silver prices rode the crest and silver-miners and promoters flocked to the Slocan.. the streets of New Denver..were alive with wheeling and dealing.

whee·lbarrowing, *vbl. sb.* [f. Wheelbarrow *sb.* + -ing[1].] **1.** Conveyance in a wheelbarrow.

1893 [in Dict. s.v. Wheelbarrow *sb.*].

2. *Aeronaut.* Landing where only the nose-wheel (of an aircraft with a tricycle under-carriage) is in contact with the ground.

1977 *Flight* 13 Aug. 480/2 We found the nose could be held up. We thought that there could be little excuse for nosewheel landings or wheelbarrowing. **1983** D. Stinton *Design of Aeroplane* x. 370 Sloppier landing habits, failure to get the tail down on landing..cause too many ballooning, wheelbarrowing and mishandling incidents.

wheel-chair. (In Dict. s.v. Wheel *sb.* 18.) Also **wheelchair, wheel chair.** Substitute for def.: **1.** A chair on wheels used by invalids or the disabled; also = *Bath-chair* s.v. Bath *sb.*[2] 2. (Further examples.)

1817 Jane Austen *Let.* ? May (1952) 497, I..am to..be promoted to a wheel-chair as the weather serves. **1841** Thackeray *Second Funeral of Napoleon* iii. 77 A servant passes, pushing..a shabby wheel-chair. **1890** *Harper's Mag.* Mar. 613/1 Mrs. Aydler flitted backward and forward in her wheel chair. **1958** *Times* 15 Sept. 11/1, I was in no condition to register the details of arriving in the hospital ward in a wheelchair. **1977** *Whitaker's Almanack 1978* 584/2 Britain banned a team of 5 White and 5 Black young Rhodesians confined to wheelchairs from coming to the U.K. to participate in games for the disabled at Stoke Mandeville, Bucks. **1981** *Sunday Times* 14 June 86 A lot of people seem slightly on edge at meeting me in a wheelchair but my main aim is to dispel this straight away.

2. *attrib.* and *Comb.*, as *wheelchair patient, wounded*; *wheelchair-bound, -ridden* adjs.; also with reference to sporting events for those confined to wheelchairs, as *wheelchair athlete, games, Olympics, slalom.*

1972 Wheelchair athlete [see *wheelchair games* below]. **1981** *Daily Mail* 17 June 38/5, I am fortunate enough not to be wheelchair-bound. **1972** R. C. Adams et al. *Games, Sports & Exercises* (ed. 2) iii. 10/2, 1960..marked the first time the wheelchair games were held along with the Olympic Games... Wheelchair athletes from all over the globe were greeted by the Pope. **1972** *Even. Telegram* (St. John's, Newfoundland) 5 Aug. 18/1 The Wheelchair Olympics..were started after the Second World War for crippled war veterans. They have since become games for athletes with spinal cord injuries or those who have been paralysed by polio. **1970** *Stoke Mandeville Dict. Managem. Paraplegic Patients* 4 Occasionally indicated on the stiff hip or knee of wheelchair-patients. **1968** T. Stoppard *Real Inspector Hound* 16 Magnus, the wheelchair-ridden half-brother of her ladyship's husband Lord Albert Muldoon. **1964** *Times* 12 Nov. 11/3 Carol Bryant..won a gold medal in the wheelchair slalom. **1982** *Daily Tel.* 5 Oct. 15/1 He would not change his mind about excluding 'wheelchair wounded' from the Falklands victory parade.

Hence **whee·l-chaired** *a.*, in or confined to a wheelchair.

1938 *Amer. Speech* XIII. 196 Wheel-chaired. **1973** *Times* 11 Apr. 8/6 A mini-bus is used in one of our split-site comprehensives to carry wheelchaired pupils from building to building. **1977** *New Scientist* 3 Mar. 499/1 Another council decision..caused the AAAS's wheelchaired delegates to wheel themselves out of the room in disgust.

wheeled, *a.* Add: **1. a.** Also freq. in *wheeled chair*, = *WHEEL-CHAIR.

1847 Dickens *Dombey & Son* (1848) xxx. 300 Withers the page, released..from the propulsion of the wheeled-chair. **1911** F. H. Burnett *Secret Garden* xx. 213 The strongest footman in the house carried Colin down-stairs and put him in his wheeled chair. **1981** J. Mann *Funeral Sites* i. 11 I'm too fat for an operation, they say... It's endurance or a wheeled chair.

whee·l-engraving. [f. Wheel *sb.* + Engraving *vbl. sb.*] The art or craft of engraving patterns, etc., on glass by means of a rotating copper wheel and an abrasive mixture of emery and oil, sand and water, or the like.

1884 Wheel engraving [in Dict. s.v. Wheel *sb.* 17 b]. **1929** W. Buckley *Diamond Engraved Glasses of 16th Cent.* 8 Wheel engraving became common before the end of the 17th century and superseded the use of the diamond as the usual method of engraving glasses. **1957** *Encycl. Brit.* X. 413/1 By the end of the [17th] century this type of diamond point work was superseded in popularity by wheel engraving. **1975** *Oxf. Compan. Decorative Arts* 403/2 English glass..was greatly in demand in the Netherlands for the flourishing school of wheel-engraving in the German manner.

Hence (as a back-formation) **whee·l-engrave** *v. trans.*; also **whee·l-engraved** *ppl. a.*; **whee·l-engraver.**

1926 B. Rackham in W. Buckley *European Glass* p. x, The wonderful skill of technique shown in the finest German wheel-engraved glasses may make up..for the loss of nobility involved. **1937** *Burlington Mag.* Nov. 221/1 The lines are crudely wheel-engraved. **1961** E. M. Elville *Collector's Dict. Glass* 55/2 There are quite a number of diamond point and wheel engravers today who have constant employment in such work. **1972** *Country Life* 30 Nov. 1499/1 A later tumbler in the same sale was wheel-engraved with a bust portrait of Prince Charles Edward. **1979** *Homes & Gardens* June 83/1 Often these declare their purpose at a glance, with wheel-engraved patterns of the ale brewer's hop leaves and barley.

wheeler. Add: **II. 13.** (Earlier and later examples.)

1886 H. Baumann *Londinismen* 232/1 *Wheeler*, velocipedist. **1929** *Newport & Market Drayton Advertiser* 28

June 3/4 Shropshire wheelers. Sunday's run to Bala. **1979** *Guardian* 8 Aug. 7/4 A posh camping shop..no shop for humble wheelers.

wheeler-dea·ler. *colloq.* (orig. and chiefly *U.S.*). [f. *WHEEL AND DEAL *v.*: cf. -ER[1].] A schemer, esp. in business or politics; one who wheels and deals (see quot. 1960).

1960 Wentworth & Flexner *Dict. Amer. Slang* 574/2 *Wheeler-dealer*..one who wheels and deals; an adroit, quick-witted, scheming person; a person with many business or social interests. **1963** *Economist* 24 Aug. 666/1 Two Dallas oil millionaires,..described as 'a pair of old-line Texas wheeler-dealers'. **1968** G. Wyckoff *Image Candidates* iii. 20 Ted had been known as the Wheeler-Dealer because he always had important phone calls to make from his hotel room when we arrived on location. **1973** *Guardian* 23 Mar. 15/3 Old-style American corruption of the wheeler-dealer variety. **1978** L. Heren *Growing up on The Times* ix. 301 He [*sc.* Lyndon Johnson] was a shop-soiled old politico, a wheeler dealer, and past master of consensus politics.

Hence **wheeler-dea·ling**, the activity of a wheeler-dealer; = *wheeling and dealing* vbl. sb. s.v. *WHEEL AND DEAL *v.*

1968 *N.Y. Times* 1 Feb. 42 Wondering what wheeler-dealing was going on over telephone lines among the various managements. **1976** *Listener* 5 Feb. 135/3 To ensure that..councils (or the community) got the upper hand in any wheeler-dealing between councils and developers. **1979** *Daily Tel.* 27 Apr. 36/6 Mr Prior also stressed that voting liberal could mean another 'hung' Parliament which would mean more 'wheeler dealing' at Westminster. **1984** *Listener* 22 Mar. 4/1 It is here in the intricate wheeler-dealing of the Common Market, the Atlantic alliance and international trade and finance that the 'diplomatic' reputations of the future will be made.

wheel-house. Add: Also **wheel house, wheelhouse. 1.** (Earlier and later examples.)

1835 J. H. Ingraham *South-West* I. xxiii. 247 The pilot (as the helms-man is here called) stands in his lonely wheel-house. **1846** A. Hawkins *Let.* 20 Nov. in N. E. Eliason *Tarheel Talk* (1956) 304 The carpenter in coming out of the wheel house whare he had been mending the wheel dropd the candle on some loose cotton. **1906** 'Mark Twain' *Autobiogr.* I. 310 Rush..astern to the solitary lifeboat lashed aft the wheelhouse on the port side. **1976** *Southern Even. Echo* (Southampton) 2 Nov. 20/1 On the roof of the main building is a full-size replica of a ship's wheelhouse which is used for training.

2. b. = Round-house *sb.* 3 a.

1971 [see *round-house *sb.* 3 a].

3. *Archæol.* A circular stone dwelling of the late Iron Age of a type widespread in northern and western Scotland, having partition walls radiating from the centre. Cf. *wheel-dwelling, -hut* s.v. *WHEEL *sb.* 18.

1935 V. G. Childe *Prehist. Scotland* x. 217 A wheel-house is in essence a walled area roughly circular or oval.. divided up into a number of voussoir-shaped rooms or compartments by radial walls arranged like the spokes of a wheel. **1957** T. C. Lethbridge *Gogmagog* viii. 149 A considerable mass of pottery of what must be considered Iron Age culture, though of Roman date, has now been recovered from brochs and wheel-houses in the Hebrides. **1970** Bray & Trump *Dict. Archaeol.* 254/2 Wheelhouses survived well into the Roman period as dwellings and farmhouses.

wheelie (hwī·li). *slang.* Also (*rare*) **wheely.** [f. Wheel *sb.* + -ie.] **1.** orig. *U.S.* **a.** The stunt of raising the front wheel off the ground while riding a bicycle or motor-cycle.

1966 *N.Y. Times* 12 Nov. 45 A popular sport for young bicycle riders is 'doing a wheelie'. This means lifting the front wheel off the ground and balancing on the rear wheel alone. **1969** *Oz* Apr. 31/1 Odd Job tries to do a few wheelies but can't quite get the front wheel off the ground. **1975** *Courier-Mail* (Brisbane) 6 Dec. 2/4 Wheelies are no trouble to the world 125cc motocross champion... But..it's a technique he uses strictly off the road. **1982** *Daily Tel.* 25 June 6/7 Acquitting a motor-cyclist.. of careless driving by performing a 'wheelie'—driving with his front wheel off the road. **1985** *Daily Mail* 6 Apr. 25/1 That's the bike seen on TV with crash-hatted kids doing wheelies.

b. *transf.* In skateboarding, the stunt of riding on only one pair of wheels, with either the nose or tail of the board in the air. (In quot. 1978 further *transf.*)

1976 A. Cassorla *Skateboarder's Bible* 10 Many of the tricks now popular originated that year, including nose and tail wheelies. **1978** *Evening Standard* 10 May 8/1 Wheelies centre rolls to a close. A pioneer skateboard centre is closing through lack of support. **1979** W. Jones in *Voices 20* Autumn 43 Down the subway he doth go.. Out the other end and into a wheely.

2. A sharp U-turn made by a motor vehicle, causing skidding of the wheels. orig. *Austral.*

1973 *Sunday Mail* (Brisbane) 29 July 3/1 'Hoons' felt free to do 'wheelies'—making U-turns at high speed to make their tyres scream. **1977** *Custom Car* Nov. 63/2 One of the most hair-raising wheelies I've ever seen.. necessitating an extensive chassis rebuild. **1982** J. S. Borthwick *Case of Hook-Billed Kites* xxxvi. 122 Tom did a wheelie into Route 77.

3. *Austral.* A person in or confined to a wheelchair.

1977 *Courier-Mail* (Brisbane) 14 Nov. 2/3 Terry Valentine braces the wheelchair as wheelie-shotputter Rene Ahrens..prepares to make his toss. **1978** *Sunday Mail* (Brisbane) 10 Sept. 34/3 So many places and things are inaccessible to the 'wheelie'. **1981** *Telegraph* (Brisbane) 14 Jan. 44/2 The wheelie symbol on selected parking spaces..is often ignored.

wheeling, *vbl. sb.* Add: **e.** (Earlier example in sense 'travelling in a wheeled vehicle'). Also, the condition of a road suitable for the passage of (wheeled) vehicles.

1850 L. Sawyer *Jrnl.* 3 July in *Way Sketches* (1926) v. 81 Our road lay over deep barren sand, which rendered the wheeling very difficult. **1864** *Ret. Agric. Soc. Maine* 52 The January thaw..gives us bare hills and wheeling. **1873** 'G. Hamilton' *Twelve Miles* ii. 25 He told her he would give her a sleigh-ride when it came wheeling.

j. *wheeling and dealing*: see *WHEEL AND DEAL *v.* Also **wheeling-dealing** *vbl. sb* and *ppl. a.*

1973 *Guardian* 21 June 2/3 The cliché of the moment is that Mr Brezhnev is..a sort of wheeling-dealing gladhander. **1977** *Guardian Weekly* 28 Aug. 22/4 The wheeling-dealing that went on within the tribunal when it came to considering the verdicts.

wheel-lock. Add: **3. b.** = Lock *sb.*[2] 15.

1927 *Observer* 20 Feb. 21/3 The majority of our fire-fighting appliances are not provided with adequate wheel-locks.

wheelman. Add: **1.** (Earlier examples.)

1865 *Oregon State Jrnl.* 12 Aug. 2/5 The wheelman says that large fragments of the bottom and a part of the rudder were afterwards seen alongside the wreck. **1866** 'Mark Twain' *Lett. from Hawaii* (1967) 195 Four other gentlemen and the wheelsman were all assembled on the little after portion of the deck.

3. A driver, *spec.* (*Criminals' slang*) the driver of a getaway vehicle. orig. *U.S.*

1935 J. Hargan *Gloss. Prison Lang.* 8 *Wheelman*, driver of a getaway car. **1962** 'K. Orvis' *Damned & Destroyed* xii. 80 Later on,..he began driving a cab. Also being a wheel-man for the mobs. **1967** M. Procter *Exercise Hoodwink* vi. 44 The thieves had locked themselves in, and when they had emptied the safe, they had phoned for their wheel man like calling for a taxi. **1975** *Publishers Weekly* 27 Jan. 283/3 (Advt.), When a young California hoodlum blew the brains out of a helpless store clerk last year Ken Pestana was the unwitting wheelman waiting outside.

wheelwork. Add: **c.** Applied to the works of a watch or other time-piece. Also *fig.*

1843 *Penny Cycl.* XXVII. 107/2 In a repeater there is an additional train of wheels between the frame-plates, called the runners, or little wheel-work. **1868** *Chambers's Encycl.* X. 82/1 The arrangement of the wheel-work in a watch. **1890** W. James *Princ. Psychol.* II. xix. 112 There seem no good grounds for supposing this additional wheelwork in the mind. **1981** *Times* 2 May 15/4 Examination of the instrument reveals nothing inconsistent with it being a bench-timer, possibly contrived from an early frame and wheel work.

wheelwrighting. (Earlier example.)

1883 *Rep. Indian Affairs* (U.S.) 253 Carpentry, harness-making, wheelwrighting.

whelk[1]. Add: **d. whelk-stall** (later examples); freq. in phr. *to be unable to run a whelk stall* and varr., to be incompetent, esp. in business; **whelk-tingle,** substitute for def.: = *TINGLE *sb.*[3]; (later example).

1894 J. Burns in *South-Western Star* 13 Jan. 3/4 From whom am I to take my marching orders? From men who fancy they are Admirable Crichtons,..but who have not got sufficient brains and ability to run a whelk stall? **1928** 'N. Shute' *So Disdained* iv. 159 If you try to run him as a manager as well, then your luck'll be out... He couldn't run a whelk stall to make it pay. **1960** C. Storr *Marianne & Mark* iii. 42 The beaches and the whelk stalls. **1965** O. Manning *Friends & Heroes* xiv. 148, I said we were all disgusted at the way the School had gone down; and I said things would be no better under Callard. I said Callard couldn't run a whelk-stall. **1966** *Hansard Commons* 22 Nov. 1283 The great majority of hon. Gentlemen opposite have not the qualifications to run a whelk stall profitably. **1980** M. Drabble *Middle Ground* 25 Which should she pick?..vain Albert from the whelk stall? **1981** *Financial Times* 1 Apr. 15/2 None of them [*sc.* the 364 economists] has had enough practical experience to run the proverbial whelk stall. **1959** *Times* 25 Aug. 5/6 A survey of the damage done on English oyster beds by the American whelk tingle..has just been concluded.

whelp, *sb.*[2] Erron. for Welt *sb.*[1]

1912 in *Dialect Notes* III. 593 She whipped the horse till she raised great whelps on him. **1952** *Publ. Amer. Dial. Soc.* xvii. 34 Time was in the upcountry when the teacher would, with a hickory, raise whelps on the legs of a recalcitrant pupil. **1962** W. Faulkner *Reivers* viii. 181 How the hell did Sugar Boy ever let him get this far without at least one whelp on him? **1980** *Verbatim* Autumn 17/2 A quite common mispronunciation is 'whelp' for 'welt': 'He has some big whelps on his arm.'

whelphood. (Earlier example.)

1847 E. Brontë *Wuthering Heights* I. xiii. 325 It [*sc.* a dog] had spent its whelphood at the Grange.

whelping, *vbl. sb.* Add: *whelping ice* (see quots.).

c **1900** J. P. Howley in *Regional Lang. Stud.—Newfoundland* (1978) viii. 23 Whelping ice. The part of an ice field where they [*sc.* seals] bring forth their young. **1919** W. T. Grenfell *Labrador Doctor* (1920) ix. 174 The smoother, whiter variety known as 'whelping ice'—that is, the Arctic shore ice..on which the seals give birth to their pups. **1969** H. Horwood *Newfoundland* xii. 83 The drift ice where they [*sc.* seals] give birth to their young is the whelping ice.

when, *adv.* (*conj.*, *sb.*). Add: **I. 2.** Also as a reply to the formula *say when.*

1911 *Maclean's Mag.* Oct. 297/2 'Say when?' I held the glass with a shaking hand: 'When.' **1931** A. Powell *Afternoon Men* I. 13 'Say when, sir,' said the waiter. 'When,' said Pringle. **1948** E. Waugh *Loved One* 3 'When,' he added aside to the young man, who helped him to whisky. 'Right up with soda, please.'

II. 4. a. Also with ellipsis of following clause: in the past, in the old days. *N. Amer. colloq.*

1962 M. Richler in *Kenyon Rev.* Winter 88 Six months from now..I'll be saying I knew you when. **1968** H. Waugh '*30' Manhattan East* (1969) 163 She needn't try those airs with me. I knew her when. **1984** M. Hinxman *Night they murdered Chelsea* viii. 65 The Hearst newspaper group are even flying in Gloria Beesley to cover the case. She knew Charlotte when.

whenabouts. Add: Also, the approximate time at which a thing happened. Also *interrog.*

1952 V. Gollancz *My Dear Timothy* 74 By eighteen hundred and ninety-eight, when or whenabouts I first heard it. **1966** A. E. Lindop *I start Counting* xx. 247 'When? Whenabouts?' '..Lemme see. Be about four o'clock in the afternoon.'

whenceness (hwe·nsnės). *rare.* [f. Whence *adv.*, *conj.* (*sb.*) + -ness.] The place or source from which something comes or arises; place of origin.

1922 Joyce *Ulysses* 388 All is hidden when we would backward see from what region of remoteness the whatness of our whoness hath fetched his whenceness. **1980** *Dædalus* Spring 248 Given the group's oblique disposition, the 'whenceness' of this saved soul seems unmistakable.

whenever, *adv.*, *conj.* Add: **I. 1.** Also as adv. with loss of relative force: at whatever time. *colloq.*

1917 H. James *Ivory Tower* in *Amer. Novels* III. v. 221, I said to myself..three weeks ago, or whenever, that it wasn't for that I was going to come over. **1982** J. D. MacDonald *Cinnamon Skin* xi. 107 Maybe we can leave it that you can come over to Lauderdale whenever.

where, *adv.* and *conj.* Add: **II. 10. c.** In U.S. use freq. equivalent to That *conj.* (see also quot. 1931).

1927 E. O'Neill *Marco Millions* II. ii. 122, I can see where I'll have to be telling her what to do every second. **1931** G. O. Curme *Syntax* 245 This old use of *where* with the force of a noun + *in which* is still heard in colloquial speech: 'This morning I read in the Tribune *where* (in the literary language *an account in which*) *a boy killed his father.*' **1938** D. Runyon *Furthermore* iii. 51, I see by the papers where three Brooklyn citizens are scragged. **1958** T. Capote *Breakfast at Tiffany's* 110 [I] had read where the Trawlers were countersuing for divorce. **1965** *New Yorker* 15 May 45, I see where the St. Regis has changed hands again. **1976** *National Observer* (U.S.) 14 Aug. 2/4, I can see where people might think that Kelley doesn't know what's going on in his own organization.

d. In *colloq.* phr. *where it's* (*he's*, *she's*) *at*: the true or essential nature of a situation (or person); the true state of affairs; a place of central activity. Cf. *at *prep.* 1 d. orig. *U.S.*

1903 [see *at *prep.* 1 d]. **1965** *Daily Mail* 2 Oct. 5/2 What's the phrase you use for being in touch?.. Where it's at. **1967** *Listener* 26 Oct. 522/3 As Dylan says, 'I'll let you be in my dream, if I can be in yours.' I think I know where he's at. **1971** *Melody Maker* 9 Oct. 17/5 The musicians frequently became frustrated..not really believing their own bands were where it was at. **1974** R. M. Pirsig *Zen & Art of Motorcycle Maintenance* x. 117 That, today, is where it is at, and will continue to be at for a long time to come. **1977** W. J. Weatherby *Home in Dark* xiii. 69 She was always a housewife at heart. She just took too long to find out where she was at.

e. U.S. dial. *to where*, to or at a point, position, etc., such that; to such an extent that. Occas. with omission of *to.*

1933 M. K. Rawlings *South Moon Under* xvi. 157 Is your loggin' to where you kin leave it for a whiles? **1938** —— *Yearling* xvi. 181 My grand-pappy got hisself stung oncet to where he was in the bed a fortnight. **1960** H. Lee *To kill Mockingbird* xi. 109 Having developed my talent to where I could throw up a stick and almost catch it coming down. **1969** B. K. Green *Wild Cow Tales* 247, I would pitch a rope over a steer's neck and give it a whip-like motion to where the knot would come back under his neck on the ground back on my side. **1974** N. Guidici in S. Terkel *Working* vi. 316, I want to have enough money where I wouldn't have to be a bum on the street.

11. c. *to the point where*, to a situation, condition, extent, etc., such that.

1938 F. Scott Fitzgerald *Let.* 22 Feb. (1964) 569 If t ever came to the point where you thought you ought to lay up under medical care, his is the sanitarium which I should choose. **1960** *Radio Amateur's Handbk.* (ed. 37) 190/2 Adjust the potentiometer..to the point where the oscillator cannot be heard between dots and dashes at normal keying speed. **1968** Chomsky & Halle *Sound Pattern Eng.* 329 Our investigations of these features have not progressed to a point where a discussion in print would be useful. **1970** P. Whittle *Probability* v. 100 Models which can be simple, without being idealized to the point where they have no practical value.

12. b. (Later example.)

1929 R. A. Cram *Catholic Church & Art* iv. 57 Where the pagan architecture had been an *exterior* art..and where Roman and Byzantine art had striven to achieve space in its simplest form, the North worked for interior space.

IV. 15. b. *whereamong* (later example), *wherewithout* (later example).

1929 R. Bridges *Testament of Beauty* I. 17 Whereamong hath the sceptic honourable place. **1899** Beerbohm *More* 95 Mere masses of colour, crude intensity of conception, wherewithout posters fail, were quite unnecessary.

whereof, *adv.* Add: **II. 5. b.** Phr. *to know whereof one speaks* (or *writes*, etc.): to know what one is talking about, to speak from experience.

1922 H. Van Loon *Story of Mankind* xliii. 256 He [*sc.* Erasmus] had travelled a great deal and knew whereof he wrote. **1967** R. Stein *Great Cars* 165/1 Ettore Bugatti knew whereof he spoke when he advised people griping about hard starting to keep their cars in heated garages. **1975** *Publishers Weekly* 24 Mar. 42/2 Fischer has been a lifelong reporter on public affairs and was on LBJ's Commission on Rural Poverty, so he knows whereof he writes.

whereso, *adv.*, *conj.* **4.** (Later *arch.* example.)

1889 'Mark Twain' *Connecticut Yankee* xi. 130 Whereso if ye be minded..it is in the east.

wheresomever. 3. (Later example.)

1837 R. Bird *Nick of Woods* iii. 43 He haunts about our woods..and kills 'em [*sc.* Indians] wheresomever he catches 'em.

whereto, *adv.* **I. 2.** (Later *arch.* example.)

1900 A. Meynell *John Ruskin* vii. 119 Whereto, then, is the persuasion of this book directed?

wherever, *adv.*, *conj.* **2. b.** For 'Now *rare* or *Obs.*' read: Now usu. preceded by *or*, whatever place. (Later examples.) Cf. *whatever *pron.* and *a.* 4 c; *whenever *adv.*, *conj.* 1. *colloq.*

1917 H. James *Sense of Past* iv. iii. 242 A pot of about the size..of that one..on the cabinet or wherever. **1952** A. Hocking *Best Laid Plans* vi. 90 This committee will now adjourn to The Pig and Whistle or wherever, for liquid refreshment. **1976** *Bookseller* 24 Apr. 2088/1 A jet flies off to London, New York, Paris, or wherever. **1978** *Jrnl. R. Soc. Arts* CXXVI. 220/1 In times of trouble, whether in Cyprus, or Karachi, or wherever, you would see everybody tuning in to the BBC. **1981** W. Bronk *Life Supports* 31 The terrible world where hollow catastrophe Hangs wherever.

whet, *v.* Add: **3.** Also with *up.*

1823 T. Bewick *Memoir* (1975) xii. 105 The extreme interest I had always felt in the hope of administering to the pleasures & amusement of youth..whetted me up & stimulated me to proceed.

absol. **1877** 'Mark Twain' in *Atlantic Monthly* Nov. 586/2 You see 'em begin to whet up whenever they smell argument in the air. **1893** —— in *Century Mag.* Jan. 342/2 The people were still in the drawing-room, whetting up for dinner.

whew, *sb.*[1] Add: **2. b.** *dial.* A factory hooter.

1869 J. Hartley *Halifax Clock Almanack* 48 Yond's th' whew, soa we mun goa an' do another bit for th' maister. **1929** J. B. Priestley *Good Compan.* iv. 118 Bruddersford has an elaborate system of factory buzzers —usually known as whews. **1934** —— *Eng. Journey* vi. 194 Time for them had been marked by the sound of its [*sc.* the mill's] hooter—locally known as a 'whew'.

whew, *sb.*[2] *dial.* [f. Whew *v.*[2]] A hurry; *esp.* in phr. *all of a whew*, in a hurry, impatient or excited.

1905 in *Eng. Dial. Dict.* VI. 453/1 Sec a whew he's in. **1922** A. Brown *Old Crow* xi. 119 He wants me to go down in his river pastur', choppin'. All of a whew to git at it.

whey, *sb.* Add: **3.** *wheyhued, -pale* (examples), *-sour* adjs.; **wheygoose** *nonce-wd.*, used as a term of opprobrium.

1949 C. Fry *Lady's not for Burning* I. 8 What shall I do With this nattering wheygoose, Alizon? Shall I knock him down? *a* **1915** Joyce *Giacomo Joyce* (1968) 2 Smitten by the hot creamy light, grey wheyhued shadows under the jawbones. **1916** —— *Portrait of Artist* v. 193 He saw in a moment the student's wheypale face. **1978** H. Wouk *War & Remembrance* xiii. 129 Ascher's whey-pale face wanly lit up at the comparison. **1922** Joyce *Ulysses* 29 With her weak blood and wheysour milk she had fed him.

which, *a.* and *pron.* **B. I. 2. b.** For 'humorous' read '*dial.* or humorous' and add later examples.

1835 A. Parker *Trip to West & Texas* 88 Ask a question, and if they do not understand you, they reply '*which?*' **1910** P. W. Joyce *Eng. as we speak it in Ireland* 348 When a person does not quite catch what another says, there is generally a query... Our people often express this query by the single word 'which?' **1938** W. Faulkner *Unvanquished* 83 Yankee say, 'Sartoris, John Sartoris,' and Marse John say, 'Which? Say which?' **1950** —— *Coll. Stories* 752 'Here,' Weddel said, extending the tumbler... The Negro stopped... 'Which?' he said. He looked at the glass.

III. 7. c. (Further examples.) Now very common in spoken English.

Quot. 1950 is a mixed construction. Cf. And *conj.* 11 a.

1902 H. James *Wings of Dove* I. iv. 85 He imaged it—which was enough as some proved vanity. **1950** Patterson & Conrad *Scottsboro Boy* II. v. 122 He..said, 'Haven't I told you black sons of bitches about talking after bed hours?' 'I wasn't talking,' I said. And which I wasn't. **1981** *London Rev. Bks.* 19 Feb.–4 Mar. 9/2 To be fair, Frances Partridge is concerned in this book to put the record straight on the central episode of Carrington's suicide: to emphasize Ralph Partridge's fear that this would happen, and his desperate efforts to avert it. Which is reasonable enough.

whi·ch-a-way, *pron.* *U.S. colloq.* and *dial.* [Cf. Which *a.* and *pron.* B. 12; *every which way* s.v. Every *a.* 1 f in Dict. and Suppl.] Which way, in what direction. Cf. *that-a-way *adv.*

1909 *Dialect Notes* III. 381 *Which-a-way*,..which way. **1938** M. K. Rawlings *Yearling* i. 13 Which-a-way will we begin huntin' him? **1968** O. Spann in P. Oliver *Screening Blues* iii. 125 Well, you know I'm so mad this morning, don't know whichaway to go.

whichever, *a.* and *pron.* Add: (Later examples of a person.)

1880 Hardy *Trumpet-Major* I. ix. 178 Whichever of us she likes best, he shall take her home. **1919** G. B. Shaw *Inca of Perusalem* in *Heartbreak House* 205 The Inca is to come and look at me, and pick out whichever of his sons he thinks will suit.

whi·chway(s, *adv.* Chiefly *U.S.* = *every which way* s.v. Every *a.* 1 f in Dict. and Suppl. Often prec. by *all.*

1961 in Webster, Leaving her towel and brush and comb lying whichway. **1968** 'J. Welcome' *Hell is where you find It* xiv. 166 He told me they [*sc.* drugs] took everyone all whichways. If you'd ever had a drink or two before, you want a lot more where you were on the pills kick—sometimes. **1978** *People's Friend* 13 May 19/1 She pictured the scene and winced at the idea of Gregory seeing her without make-up, her hair all-which-way from the steam. **1975** 'Miss Read' *Battles at Thrush Green* i. 16 What chance is there of pushin' a mower up these 'ere paths with the graves all going which-way? *Ibid.* xix. 223 He was on a bike far too big for him—sawing away he was, wobbling all whichways.

whick, *a.* North. var. of Quick *a.* Cf. *wick *a.*[2]

c **1760** W. Hutton *Dialogue in Vulg. Lang. Storth & Arnside* (*c* 1900) 4 Was It whick, says Ta? **1790** [see Quick *a.* 2]. **1879**— [see *Eng. Dial. Dict.*].

whicker, *v.* Add: **2.** (Earlier examples.)

1753 J. Poulter *Discoveries* (ed. 5) 7 The Horse, as soon as the others past began to whicker, so that we were obliged to gag him.

3. To make a sound as of something hurtling through or beating the air.

1926 *Spectator* 28 Aug. 313/2 Bid Jove send down a thunderbolt to whicker through the sky. **1965** G. Maxwell *House of Elrig* xiii. 167 My aunt's black-and-white nun pigeons whickered past my window and drank at the bird-table.

whicker *sb.*: also, the sound of something beating the air; hence also **whi·ckering** *vbl. sb.* and *ppl. a.*

1899 Somerville & 'Ross' *Some Experiences Irish R.M.* xi. 277 A pale, yellow foal sprinted up beside us, with shrill whickerings of joy. **1920** J. Masefield *Right Royal* 73 Far over his head with a whicker of wings Came a wisp of five snipe from a field full of springs. **1937** E. Sitwell *I live under Black Sun* I. iii. 48 The door of her room..opened with a dark strawy noise like the wickering voice of a bear. **1940** H. Spring *Fame is Spur* i. 11 And so great was the silence that the whickering of banners could be heard. **1965** G. Maxwell *House of Elrig* ii. 27 Black rock cliffs with deep mysterious caves full of the whicker of rock-pigeons' wings.

Whieldon (hwī·ldǒn). The name of Thomas *Whieldon* (1719–95), Staffordshire potter, used *attrib.* to designate the kind of coloured earthenware made in his factory (founded 1740). Also *Comb.*, as *Whieldon-type* adj., resembling this ware.

1869 C. Schreiber *Jrnl.* 1 Oct. (1911) I. 42 One Wheildon Ware plate. **1900** F. Litchfield *Pott. & Porc.* vii. 317 Whieldon ware is peculiarly light and the articles well potted. **1929** H. Read *Staffordshire Pottery Figures* Pl. 14 (*caption*) The term 'Whieldon type' [is used] when the figure depends entirely for its decoration on coloured glazes. **1942** *Burlington Mag.* Oct. 260/1 Most dangerous are the increasingly skilful fakes of Astbury and Whieldon figures. **1968** *Canad. Antiques Collector* July 13/1 What is Whieldon Ware? This is a term referring to all types of ware of a mottled, cloudy or splashed character. **1978** *Times* 28 Jan. 12/6 (*caption*) A Whieldon-type teapot and cover, c. 1765. **1983** *Country Life* 2 Dec. (Suppl.) 72/2 A Whieldon pottery horse, decorated in underglaze colours of green, yellow and brown.

whiff, *sb.*[1] Add: **I. 1. d.** *U.S. slang.* A miss, a failure to hit (a ball).

1952 *N.Y. Herald Tribune* 15 May 21/6 On the first tee he took a careful stance and then fanned the air four times. After the fourth whiff he growled, 'This is the hardest course I ever played.'

6. (Earlier example.)

1800 M. Edgeworth *Parent's Assistant* (ed. 3) VI. 158 Lean on my arm, madam, and we'll have you in and at home in a whiff.

whiff, *v.*[1] Add: **6.** *U.S. slang.* **a.** *intr.* Of a batter in Baseball or a golfer: to miss the ball. Cf. *fan *v.* 8 b.

1913 *Wells Fargo Messenger* I. 93/2 When he has to line 'er out he does, but he doesn't whiff at random. **1926** *Amer. Speech* I. 369/2 He [*sc.* a baseball player] 'whiffs' when he fails to hit. **1942** Berrey & Van den Bark *Amer. Thes. Slang* § 677/34 *Miss the ball*,..whiff.

b. = *fan *v.* 8 a.

1914 R. Lardner in *Sat. Even. Post* 7 Mar. 7/2, I whiffed eight men in five innings in Frisco yesterday. **1951** in Wentworth & Flexner *Dict. Amer. Slang* (1960) 575/1 Vic Raschi whiffed twelve batters in gaining his 15th win of the year. **1941** *Nebraska State Jrnl.* 20 June (*heading*), Hurler whiffs 20.

whiffet. Add: **1.** (Earlier example of *whiffet dog.*)

1848 *Ladies' Repository* VIII. 315 The best protection to a house, with a family in it that can be named—that is, a little, barking, noisy, cowardly, whiffet dog.

whiffle, *sb.* Add: **3.** A soft sound as of gently moving air or water.

1972 F. Ford *Atush Inlet* i. 9 Their subdued cries could be heard faintly against the gentle whiffle of falling water. **1976** J. Crosby *Snake* (1977) xxx. 179 She listened to..the soft whiffle of her breathing.

4. *Comb.*, as **w(h)iffle-ball** *U.S.*, a light hollow ball (orig. designed for golf practice in a confined space), used for playing a variety of Baseball; also the game itself; **whiffle-minded** *a.* (*U.K.* and *U.S. dial.*), changeable fickle.

1965 F. Knebel *Night of Camp David* xvii. 273 The boys of Saybrook were playing whiffle ball. **1970** *Time* 25 May 43 [David Eisenhower]..passing the afternoon playing wiffle ball on the south lawn of his father-in-law's White House. **1970** *New Yorker* 11 July 20 Kids playing with whiffleballs and baseballs. **1976** Woodward & Bernstein *Final Days* 242 He would get a whiffle-ball game going on the White House tennis court. **1980** *N.Y. Sunday News Mag.* 2 Mar. 12/2, I would chase the whiffleball across the street. **1902** H. F. Day *Pine Tree Ballads* 47 Hate to act so whiffle-minded, but my father used to say, 'Men would sometimes change opinions; mules would stick the same old way'. **1905** in *Eng. Dial. Dict.* VI. 456/1 'e's so w'iffle-minded—'e dunna know 'is own mind two minutes together.

whiffled (hwi·f'ld), *a.* *slang.* [Origin obscure: cf. Squiffy *a.*] Intoxicated, drunk.

1927 Wodehouse *Meet Mr. Mulliner* vi. 191 Intoxicated? The word did not express it by a mile. He was oiled, boiled, fried..whiffled, sozzled, and blotto. **1930** —— *Very Good, Jeeves!* ii. 46 'Have you forgotten that I did thirty days..for punching a policeman..on Boat-Race night?' 'But you were whiffled at the time.' **1956** J. D. Carr *Patrick Butler for Defence* xiv. 157 Helen.. was much too clear-headed..ever to let herself get whiffled.

whiffletree. (Earlier example.)

1842 W. P. Hawes *Sporting Scenes* II. 69 Our whiffletree became detached from the vehicle, and fell upon the horse's heels.

whiffling, *vbl. sb.*[1] (s.v. Whiffle *v.*[1] in Dict.) Add: **2.** Vacillation, evasion.

1841 J. F. Cooper *Deerslayer* I. i. 23, I would carry the gal off to the Mohawk by force, make her marry me in spite of her whiffling. **1906** *Springfield* (Mass.) *Weekly Republ.* 18 Oct. 3 This outcome of a week of doubt and whiffling will be viewed with mixed emotions. **1984** *Daily Tel.* 13 Feb. 12/5 When first I heard these whifflings, a couple of years ago, I thought they must be satiric.

whifflow (hwi·flō[u]). *Naut. slang.* [Fanciful formation.] (See quot. 1961.)

1961 F. H. Burgess *Dict. Sailing* 222 *Whifflow*, an unnamed gadget; used when a proper name is forgotten. **1971** 'A. Burgess' *MF* v. 64 The cabin was still a mess of smashed and battered whifflows.

whiffy, *a.* Add: (Earlier and later examples.) Also *fig.*

1849 H. Melville *Mardi* II. xxvii. 109 A pithy, whiffy sentence or two. **1934** R. Macaulay *Going Abroad* xi. 77 'A bit whiffy,' Hero said, as they passed among the cottages that encircled the muddy..pool. **1962** Auden *Dyer's Hand* (1963) 520, I have always found the atmosphere of *Twelfth Night* a bit whiffy. **1978** *Birds* Summer 45/2 The area is dusty and whiffy with lorries arriving to tip every four minutes.

Whig, *sb.*[2] and *a.* Add: **C. Whig historian,** a historian who interprets history as the continuing and inevitable victory of progress over reaction; **Whig history,** history written by or from the point of view of a Whig historian.

1924 G. B. Shaw *Saint Joan* Pref. p. x, Her [*sc.* Joan's] ideal biographer..must understand the Middle Ages ..much more intimately than our Whig historians have ever understood them. **1980** H. Trevelyan *Public &*

Private 149 George Macaulay Trevelyan..was essentially a Whig historian, thus continuing the family tradition derived from his father and his kinsman Macaulay. **1931** H. C. Butterfield *Whig Interpretation of Hist.* i. 6 The truth is that there is a tendency for all history to veer over into whig history. **1973** *Listener* 28 June 869/1 Macaulay..wrote consciously Whig history: yet.. enunciated the principles of historical criticism which explains why Whig history is a distortion.

Whiggish, *a.*[2] Add: (Later example with reference to historical interpretation: see *Whig historian, history* s.v. *Whig *sb.*[2] and *a.* C.)

1975 *Times Lit. Suppl.* 28 Nov. 1404/3 The danger, ever-present in women's history (as in labour history) of whiggish perspectives: of self-indulgently allowing enthusiasm for the women's cause today to obstruct sensitive understanding of women's situation yesterday.

Whiggishly *adv.* (later examples in corresponding sense).

1975 *University* (Princeton Univ.) Winter 4/1 This is not to say whiggishly that science at any juncture has been the only description of physical reality that was historically possible. **1980** *Times Lit. Suppl.* 25 July 837/1 Authors tend to win a place in the history of social and political thought by making what is usually, and whiggishly, referred to as a 'theoretical contribution'.

while, *sb.* **II. 6. b.** Restrict †*Obs.* to senses in Dict. and add: Also *U.S.*, a long time.

1836 T. C. Haliburton *Clockmaker* (1837) 1st Ser. xvi. 136 You'll search one while..afore you'll find a man that..is equal to one of your free and enlightened citizens. **1852** Mrs. Stowe *Uncle Tom's Cabin* I. xi. 159 I'd mark him..so that he'd carry it *one* while. **1897** 'Mark Twain' *Following Equator* liii. 511 If India knows about nothing else American, she knows about those, and will keep them in mind one while.

while, *conj.* Add: **4. a.** (Later example.)

1918 W. de la Mare *Motley* 68 How do the days press on, and lay Their fallen locks at evening down, Whileas the stars in darkness play.

C. while-you-wait *adj.* or *adv. phr.* (orig. *U.S.*), designating a service that is performed immediately (as opp. to one for which the customer must leave his property and collect it later); also *fig.*; also *absol.* as *sb.*, an establishment providing such a service; freq. (in advertisements) spelt *while-u-wait.*

1929 *Amer. Speech* V. 24 Those who are selling 'service' are fond of using expressions of this sort:..Shine While U Wait, Hats Cleaned While U Wait, [etc.]. **1936** Mencken *Amer. Lang.* (ed. 4) 209 Q-room.., While-U-wait, and Bar-B-Q.., all of them familiar signs. **1965** H. Gold *Man who was not with It* xxix. 271 We were at a low office block.., shoe repair and while-you-wait. **1972** *Guardian* 11 July 10/6 The..catalogue essay..is a masterpiece of myth-making, art history while-u-wait. **1972** *Times* 9 Aug. 12/7 (*caption*) A while-you-wait parts replacement service. **1977** *Evening Gaz.* (Middlesbrough) 11 Jan. 13/2 (Advt.), M.O.T. test while-u-wait.

whilie (hwəi·li). *Sc. dial.* Also **whiley, whyllie.** [f. While *sb.* + -ie.] A short time.

1819 J. Burness *Plays* 29 Master Clinton is out a whyllie syne. **1908** *Old-Lore Misc.* I. iv. 183 After it was burned a whiley. **1920** J. L. Waugh *Heroes* 18 When ye've been a whilie here. **1951** N. M. Gunn *Well at World's End* xxiv. 214 'Be quiet!' she said... 'I just came for a whilie to see them.' **1981** G. Hammond *Revenge Game* vi. 55 The inspector's house..was let to a retired couple for a whilie, but they moved away up to Inverness.

whillaloo. Add: Also 7 **fuillilaloo.** Also as *int.* and *v. intr.* Cf. *pillaloo *sb.* (*int.*).

1663 R. Head *Hic et Ubique* I. vi. 18 Enter Patrick crying..Fuillilaloo! **1977** *Times Lit. Suppl.* 22 Apr. 480/4 All those uncles trailing their coats and shouting whillaloo and clear the way. **1899** [see *pillaloo *sb.* (*int.*)].

Whillans (hwi·lănz). The name of Don *Whillans* (1933–85), mountaineer, used *attrib.* of objects devised by him for the assistance of climbers, as *Whillans box*, a kind of frame tent (see quot. 1971[1]); *Whillans harness*, a harness designed for use by mountaineers climbing fixed ropes; *Whillans whammer*, a kind of peg hammer or piton.

1971 C. Bonington *Annapurna South Face* v. 66 At this stage, Base Camp was no more than a staging camp, with a two-man tent and two Whillans Boxes... We had been worried that ordinary tents might prove inadequate on the very steep ground... Don had come up with the solution, designing a prefabricated, box-like structure, with a framework of timber and an outer covering of proofed nylon. *Ibid.* xii. 147 This [*sc.* being on an ice cliff] was one place where Dougal found a use for the Whillans Whammer, Don's space-age climbing tool, for its squat, triangular-shaped pick proved ideal for this type of ice. **1972** D. Haston *In High Places* xi. 118 Huddled into the back of the Whillans box—a super-strong frame tent designed by Don—we could only sit and wait. **1974** H. MacInnes *Climb to Lost World* ix. 137 It was ironic that the designer of the Whillans harness was the only member of the expedition who didn't have..one which would fit his ample girth. **1978** P. Gillman *Fitness on Foot* v. 73 The Willans harness is manufactured by the Lancashire company Troll.

whimp, *v.* Add: (Later example.) Hence as *sb.*

1925 *Blackw. Mag.* Aug. 169/2 'Don't whimp,' I said to Irene. 'I am not whimping, daddy.' *Ibid.* 173/1 This was something beyond a whimp.

whimper, *sb.* Add: **2.** *not with a bang but a whimper*: see *bang *sb.*[1] 2 a.

whimsily, *adv.* (In Dict. s.v. Whimsy, whimsey *sb.* (*a.*).) (Later example.)

1980 P. Moyes *Angel Death* xviii. 237 The whimsily-drawn pamphlet which they gave to visitors.

whimsy, whimsey, *sb.* (*a.*) Add: **7. b.** A small object made by a glass-maker or potter for his own amusement.

1938 A. Fleming *Scottish & Jacobite Glass* ix. 109 Dame Fashion..seems to settle upon glass as a favourite and satisfactory medium of decoration. Other 'wimsies' are cheap little fantastic groups of figures, fruit and flowers delicately made from a tube modelled by a tool with infallible dexterity. **1976** *Canadian Collector* (Toronto) Mar.–Apr. 23/1 We were able to locate several more examples of the whimseys produced by the last potter.

whimsy-whamsy. Add: (Earlier and later examples.) Also *attrib.*

1807 Southey *Lett. from England* III. lix. 109 An old Welsh baronet..chose some years ago to set up a heresy of his own... He himself called it Rational Whist; his friends, in a word of contemptuous fabrication, denominated it his *whimsy-whamsy*. **1931** *Time & Tide* 26 Sept. 1118 Have we not whimsy-whamsy authors of our own without importing the too, too, quaint devices of foreign playwrights? **1945** S. Lewis *Cass Timberlane* xl. 302 Sure, the jolly little playboy, and underneath his whimsy-whamsy, he's the coldest-hearted rich-man's lawyer. **1951** McLuhan *Mech. Bride* (1967) 101/1 It is not a laughter or comedy to be compared with the whimsy-whamsy article of James Thurber.

whin[1]. Add: **4. whin-mill,** a mill for crushing whin for horse-feed.

1793 in *Trans. Buchan Field Club* (1935) XIV. 76 Carrying wood for the whine mile. **1893** C. A. Mollyson *Parish of Fordoun* 188 With a plentiful supply of oilcake and other nutritious feeding stuffs there is no place now for the whin-mill. **1957** E. E. Evans *Irish Folk Ways* viii. 110 The knocking stones..where the whins were 'melled' with a wooden maul, are sometimes to be seen in the farmyard, and there were a few water-driven 'whin-mills'.

whine, *sb.* (Examples referring to the sound of machinery.)

1928 E. Wallace *Double* xx. 295 They heard the whine of a car draw up on the ground below. **1942** W. Faulkner *Go down, Moses* 143 The air pulsed with..the whine and clang of the saw. **1962** *Which? Car Suppl.* Oct. 127/2 Other noises of which our drivers complained were rear axle whine in all the cars.

whine, *v.* **1. c.** (Examples referring to machinery.)

1962 *Which? Car Suppl.* Oct. 140/1 Engine always whined when started from cold. **1972** *Daily Tel.* 16 May 9 Two minutes after the jet engines whine to a standstill she walked slowly down the special lateral gangplank. **1974** S. Middleton *Holiday* iv. 42 Lawn-mowers whined.

whinge, *sb.* For '*Sc.* and *dial.*' read 'orig. *Sc.* and *dial.*' (Later examples.) Now *esp.* a peevish complaint.

1938 S. Beckett *Murphy* iii. 37 He threw his voice into an infant's whinge. 'I cudden do anything, Maaaammy.' **1947** I. L. Idriess *Isles of Despair* xxxviii. 254 The bull [whale] complained with a stupid little grumbling whinge and edged a few yards farther away. **1963** [see *peanut 2 b]. **1973** P. White *Eye of Storm* i. 64 'You're so *unfair*!' A whinge developed through a moan into a downright blub. **1981** *Listener* 4 June 749/1 This is not just an envious whinge. **1982** J. Thomson *To make a Killing* xiii. 231, I knew bloody well he'd shop me and make a fuss... He'd already had a whinge about the rubbish I'd left. **1985** *Times* 10 Jan. 10/6 In my one-but-last whinge I was going on about the burdensome duties of The Talk.

whinge, *v.* For '*Sc.* and *north. dial.*' read 'orig. *Sc.* and *north. dial.*' and add: Now also **winge.** *Esp.* to complain peevishly. (Further examples.) **whinging** (also **w(h)ingeing**) *vbl. sb.* and *ppl. a.* (further examples).

1907 J. M. Synge *Let.* 31 Mar. (1971) 121 Forgive this contemptible sort of whinging. I am so lonely and miserable I cant help it. **1922** Joyce *Ulysses* 10 You crossed her last wish in death and yet you sulk with me because I don't whinge like some hired mute from Lalouette's. **1946** K. Tennant *Lost Haven* (1947) xvii. 272 She had lifted up her brief skirt..to exhibit her sand-fly bites. .. 'You don't want to whinge about them... You had a good time, didn't you?' **1955** S. Beckett *Molloy* II. 172, I forgot that my son would be at my side,..whinging for food. **1965** *Listener* 2 Sept. 339/2 There is a stinging phrase in use, 'wingeing Poms' (translate into 'complaining English'). **1969** *Advertiser* (Adelaide) 12 May 5/4 Stop whingeing and give a bloke a go, mates. **1973** B. Bainbridge *Dressmaker* 8 If that girl didn't stop her wingeing, the neighbours would be banging on the wall. **1983** *Times Lit. Suppl.* 11 Mar. 236/1 In 1849, Arnold whinged to Clough that the age was '..unpoetical'. **1983** *Sunday Times* 31 July 33/1 'What sort of people do Australians hate most?' 'The whingeing Pom... Poms that come over and do nothing but whinge.' **1984** *Times* 20 Jan. 10/7 This is not the month for whingeing criticisms. **1984** *Sunday Times* 9 Dec. 7/1 All must drill most Tuesday nights..and not whinge when the trousers of their best suits are crumpled and smutted under the uniform.

whinger, *sb.*[2] For '*Sc.* or *dial.*' read 'orig. *Sc.* or *dial.*' and add later examples.

1934 *Bulletin* (Sydney) 27 June 11/2 Touching the query about 'whinger'.., 'winjer' was accepted slang for 'grumbler' at Q. Uni. a few years ago, and probably still is. I have seldom heard it elsewhere, and no one who uses it seems to know the derivation. **1959** I. & P. Opie *Lore & Lang. Schoolch.* x. 186 Other local terms for crying... In Dublin the usual word is 'whinging', hence 'whinger', a term also still used in Cumberland, and occasionally heard in Liverpool. **1983** *Listener* 14 Apr. 17/2 Certainly, no whinger like me will ever turn Simon into a dissident.

whiny, *a.* Add: *whin(e)y pin(e)y*: so *whiney-pine* vb. intr., to make whining noises.

1920 'K. Mansfield' *Let.* 27 Sept. (1977) 182 Two infant wasps..each caught hold of a side of a *leaf* and began to tug... They became furious. They whimpered, whiney-pined—snatched at each other—wouldn't give way.

whio (fi·o, wī·o). *N.Z.* Also 9 **wihu, wio.** [Maori.] The blue duck, *Hymenolaimus malacorhynchos*, native to New Zealand.

1847 T. Brunner *Jrnl.* 2 Apr. in N. M. Taylor *Early Travellers N.Z.* (1959) 272 Shot a *wihu*, or blue duck. **1855** R. Taylor *Te Ika a Maui* xxv. 407 *Wio*,..the blue duck, is found abundantly in the mountain streams of the south part of the North Island..; it takes its name from its cry. **1880** J. C. Crawford *Trav. N.Z. & Austral.* 122 At Kai-inanga, Deighton shot a pair of *whio*, or blue-ducks. **1966** *Encycl. N.Z.* I. 499/2 The most peculiar is undoubtedly the blue duck, mountain duck, or *whio*.

whip, *sb.* Add: **I. 1. d.** *a fair crack of the whip* (colloq.): a fair chance to participate or act.

1929 K. S. Prichard *Coonardoo* 179 I'll see you get a fair crack of the whip now, Mr. Watt. **1944** L. Glassop *We were Rats* 2, I am sorry to have to tell you that the Lord's had a fair crack of the whip and He's missed the bus. **1957** *Technology* Oct. 271/1 We should give the technical high school a trial..with a fair crack of the whip when the talent is being handed round. **1971** *Radio Times* 19 Aug. 50/1 It is the first time in 4½ years that those opposing the present abortion law have been given a really fair crack of the whip on a B.B.C. panel.

2. b. = *whip aerial* s.v. *whip- 1 c.

1940 *Electronics* July 68/2 The whip is used to increase the capacitance and to carry some current to greater heights. **1960** *Practical Wireless* XXXVI. 342/2 The aerial is an 8 ft. 'whip' which is swung into the vertical on arrival at a stopping place, being attached to the side of the caravan permanently, on an insulator. **1976** *S9* (N.Y.) Feb. 34/1 They are factory pretuned..and will take up to 500 watts of power, radiating from a 46-inch stainless steel whip.

3. c. For *dial.* read *dial.*, *Austral.*, and *N.Z.* (Earlier and later examples.)

1888 G. G. B. Sproat *Rose o' Dalma Linn* 242 He'll hae whups o' tabacca. **1897** I. Scott *How I stole 10,000 Sheep* vii. 29, I was glad to hear Jim come cantering up with 'whips' of bread, cheese, beer and horse-feed. **1928** 'Brent of Bin Bin' *Up Country* xi. 183 Whips of room for us both. **1948** R. Finlayson *Tidal Creek* I. vi. 59 'Didn't think old Podder would ever bother about that bit of land,' says Uncle Ted. 'Got whips of land.' **1961** G. Farwell *Vanishing Australian* 182 Then you want capital—whips of it.

4. b. *Printing.* A compositor who sets type speedily.

1890 Barrère & Leland *Dict. Slang* II. 409/1 *Whip*... (Printers), quick setter of type. *a* **1974** P. Evett in J. Burnett *Useful Toil* (1974) III. 333, I was put into the piece 'ship' on the paper, where I can truly say I held my own, though I was no whip. **1978** *Times Lit. Suppl.* 15 Sept. 1022/4 An average compositor at that time would have set a thousand characters or ens an hour, and a 'whip', or fast setter on piece-work, would have set upwards of fifteen hundred.

6. (Earlier example.)

1850 Thackeray *Pendennis* II. vi. 52 Captain Raff, the honourable member for Epsom,..retired after the last Goodwood races, having accepted, as Mr. Hotspur, the whip of the party, said, a mission to the Levant.

7. a. (Earlier example of *whip-up.*)

1884 E. W. Hamilton *Diary* 2 May (1972) II. 608 It was carried..by a majority of 2 to 1, owing no doubt in great measure to the whip-up which the Prince of Wales had made.

b. Now usu. *whip-round* (not restricted to charitable collections). (Further examples.)

1874 Hotten *Slang Dict.* 339 Whip-round. **1887** *Echo* 23 Nov. 4/4 Neighbours, who knew that she had no money, instituted a 'whip round', and soon raised the necessary amount. **1948** M. Laski *Tory Heaven* i. 12 The whip-round for garments and the ladies' little cries when they were told that clothes were rationed at home. **1977** *Centuryan* (Office Cleaning Services) Christmas 2/3 It appears a whip-round for the drinks was suggested. **1980** A. Morice *Death in Round* xiv. 107 She..handed over the money that had been raised by the whip round. **1985** *Times* 14 June 5 The extra money will have to be found by a nonrepayable whip-round among member states.

d. *the whip*: the discipline that goes with being a member of a party in Parliament; an MP's membership of a party.

1950 Theimer & Campbell *Encycl. World Politics* 458/2 To decline the whip is a method of resignation from the party. **1955** *Times* 24 May 15/1 Some effort had been made to arrive at a non-intervention arrangement, but it broke down when Mr. Walker was asked if he would accept the Conservative whip. **1966** *Listener* 25 Aug. 289/1 If he is a member of the Labour Party, he is bound by the standing orders of the Parliamentary Labour Party... To defy the standing orders may involve the withdrawal of the whip. **1980** B. Castle *Castle Diaries* 12 The bitterness intensified when, in October 1971, sixty-nine Labour MPs, headed by Roy Jenkins, defied the Labour whip and voted for Mr. Heath's motion.

9. b. *Cricket.* A whipping or springy action of the batsman's or bowler's wrist in playing or delivering the ball.

1903 [see *flick *sb.*[1] 1 d]. **1923** *Cricketer Ann. 1922–3* 78 Kilner bowls left hand slow..has a good action with a nice 'whip' in it.

III. 15*. A fairground roundabout in which a continuous revolving chain carries a number of cars or tubs round an oval track, the tubs being pivoted so as to swing freely about their point of attachment to the chain.

A proprietary name in the U.S.

1925 Wodehouse *Carry on, Jeeves!* vi. 152, I could hardly drag him away from the Whip, and as for the Switchback, he looked like spending the rest of his life on it. **1937** Hull & Whitlock *Far-Distant Oxus* xx. 277 Bridget, Anthony, and Peter went off for a ride on the 'Whip'. **1969** L. Moody *Ruthless Ones* ix. 96 They went into the fun fair and tried the big dipper, the wheel, the whip. **1976** *Official Gaz.* (U.S. Patent Office) 8 June TM89 A. G. Mangels Co., Inc., Bay Shore, N.Y... *Whip.* For carnival type amusement ride... First use since at least as early as 1914. **1979** C. Wood *Bond & Moonraker* v. 61 'The Whip' of his childhood days, but revolving at a speed that would have..hurled it half-way across the fairground.

whip, *v.* Add: **I. 2. d.** To pinch or steal, to make off with; †to swindle. *slang* (orig. *Criminals'*).

1859 G. W. Matsell *Vocabulum* 95 *Whipped*, cheated out of a share, or equal part of the plunder. **1904** 'No. 1500' *Life in Sing Sing* xiii. 259 *Holding the mark till the tool whips his stone.* Engaging a person's attention till the thief succeeds in stealing his diamond. **1946** G. Kersh *Clean, Bright & Slightly Oiled* ii. 11 Hi, you, you give me back that dog-end you whipped. **1958** M. K. Joseph *I'll soldier no More* 19 'Where's your hat, Barnett?'.. 'Dunno, Someone musta whipped it.' **1976** A. Miller *Inside Outside* xi. 173 One of them was rightly furious as the escaper had whipped (stolen) his overcoat. **1981** P. O'Donnell *Xanadu Talisman* ix. 182 The Shah must've whipped this... Stashed it away in a Swiss bank.

II. 7. b. *intr.* Of cream: to be capable of being whipped.

1943 *Mod. Lang. Notes* LVIII. 13 Cream *whips* quickly. **1979** A. Parker *County Recipe Notebk.* viii. 108 Single cream..will not whip.

12. Phr. *to whip one's weight in wildcats* and varr.: (to be able) to fight vigorously; to be fit and strong. Chiefly *U.S.* ? *Obs.*

1828 *Spirit of Seventy-Six* (Frankfort, Kentucky) 17 Jan. 3/5, I can ride upon a streak of lightning, whip my weight in wild cats. **1834** [in Dict.]. **1852** H. C. Watson *Nights in Block-House* 20 Not as long as I can whip my weight in catamounts or bar, I'll never give in. **1870** G. H. Lewes *Let.* 17 May in *Geo. Eliot Lett.* (1956) V. 96 We hope to see you both come back ready to 'whip your weight in polecats'. You will not find us in that vigorous condition! **1878** [see *once *adv.* A. y].

14. (Earlier example.)

1742 H. Finch *Let.* 18 Nov. in P. D. G. Thomas *House of Commons in 18th Cent.* (1971) vi. 114 The Whigs for once in their lives have whipped in better than the Tories.

16. a. (*h*) *Austral.* and *N.Z.* To complain or moan.

1892 *Bulletin* (Sydney) 7 May 10/3 Now he only 'whips the cat' at the bottom of the Carlton poll. **1909** H. Thompson *Ballads about Business* 12 You could make tenners den like vinkin', dough Now you are vippin' der cat. **1911** *Triad* 10 June 18 Tell him [*sc.* a misled person] he has leave to go and whip the cat. **1948** V. Palmer *Golconda* xiii. 194 If there's anything wants doing you've only got to ask Macy Donovan... And he makes light of it, too. No whipping the cat: no setting himself up as a little tin god.

whip-. Add: **1. a.** *whip-flick, -stroke, -thong* (earlier example). **b.** *whip-cracking*: also as *adj.*; *whip-minder, -scarred, -smacking* (earlier example). **c. whip aerial, antenna,** an aerial in the form of a flexible wire or rod with a connection at one end; **whip-club** (earlier example); **whip-grass** (earlier example); **whip-scorpion,** substitute for def.: an arachnid of the order Pedipalpida, having a flattened abdomen and long flagella attached to the first pair of legs; (examples); **whip-stall** *Aeronaut.*, a stall in which an aircraft changes suddenly from a nose-up attitude to a nose-down one.

1941 *Electronics* Jan. 60/2 It was necessary to vary the height of the whip aerial which was mounted on the top of the solenoid. **1979** A. Jute *Reverse Negative* (1980) 42 The car had a prominent whip aerial of the kind police mobile patrols use mounted on its rear fender. **1943** F. E. Terman *Radio Engineers' Handbk.* 1019/2 (Index), Whip antennas. **1974** R. B. Parker *Godwulf Manuscript* xviii. 141 An aggressively nondescript car made noticeable by the big whip antenna folded forward over the roof and clipped down. **1808** *Monthly Pantheon* I. 416/1 A new new Whip-club is now about to be established. **1934** Webster, Whip-cracking, *adj.* **1939** R. Campbell *Flowering Rifle* I. 17 For whom I sent the gay whip-cracking words To round them up in flabbergasted herds. **1976** A. Murray *Stomping Blues* ix. 166 He also behaves for all the world like a whip-cracking trail driver. **1960** C. Day Lewis *Buried Day* vi. 126 Our instructor..was a ..man with..a word of command like a whip-flick. **1976** *Sunday Sun* (Brisbane) 23 May 115/1 *The whip flick.* Done in a tight finish. You flick your whip across the other horse's nose, up goes his head and you have the advantage in the photo. **1814** O. Rich *Synopsis Genera Amer. Plants* 106 Scleria... Whip-Grass. **1928** R. Nevill *Romantic London* vii. 143 A quaint old-world calling..was that of 'whip-minder'; a number of people formerly making a living by looking after the whips of drivers of vehicles, while the latter were engaged on pleasure or business. **1961** *Times* 23 Mar. 17/3 When Covent Garden boasted 'whip minders'. **1849** J. R. Lowell *King Retro* in *Nat. Anti-Slavery Standard* 10 May 199/2 From whip-scarred flesh the soul can soar To him who made and sees us. **1966** R. Hayden in S. Henderson *Understanding New Black Poetry* (1973) 158 Harriet Tubman, Woman of earth, whipscarred, A summoning, a shining. **1912** J. H. Comstock *Spider Book* i. 16 The common name whip-scorpions was doubtless suggested by the slender caudal appendages of the Thelyphonidæ. **1981** *Sci. Amer.* Dec. 32/1 Watch out for the fungus-ridden whip scorpion and the vampire bats. **1845** Poe in *Broadway Jrnl.* 2 Aug. 60/2 The 'Katherine and Petruchio' of Niblo's, is absolutely beneath contempt —a mere jumble of unmeaning rant, fuss, whip-smacking, crockery-cracking, and other Tom-Foolery. **1927** C. A. Lindbergh *We* ii. 33 For an instant we hung motionless in the perfect position for a whipstall. **1936** *Aircraft Engin.* Apr. 111/1 The mechanics of the whip-stall or uncontrollable nose dive are simple. **1953** C. A. Lindbergh *Spirit of St. Louis* II. vi. 326 A whipstall at 1500 feet, with nothing but needles by which to orient myself! **1889** 'Mark Twain' *Connecticut Yankee* xxvii. 354 A precaution which had been suggested by the whip-stroke that had fallen to my share. **1958** L. Durrell *Mountolive* xii. 229 Nessun felt the heat of the whip-stroke on his hand though the lash had not touched him. **1827** *Hallowell* (Maine) *Gaz.* 20 June 4/5 They have also received a large supply of..Whips and Whipthongs.

2. b. whip-pan *Cinemat.* and *Television*, a panning movement fast enough to give a blurred picture; also as *v. intr.*

1960 D. Davis *Grammar of Television Production* 33 The 'whip pan', a device..whereby the camera sees one object, then pans very quickly and sees another, is.. legitimate because it does what the eye does and blurs the intervening detail. **1965** P. Jones *Technique of Television Cameraman* x. 136 Some television directors..instruct the cameraman to whip pan across a scene, but cut to the next static shot on another camera before the pan has ceased. **1979** *Observer* 26 Aug. 20/8 The cameras..zooming in and out, whip-panning, busying about the place looking for new angles. **1980** *Times Lit. Suppl.* 3 Oct. 1098/3 Unlike Brian De Palma in his movie version of King's earlier novel, *Carrie*, Kubrick doesn't use whip-pans, sudden zooms on neck-wrenching shocks: the horrors are revealed discreetly, almost lovingly.

whipcord, *sb.* **2. b.** (Earlier example.)

1895 *Montgomery Ward Catal.* Spring & Summer 4/2 All Wool Black Whipcord Suiting..shows fine raised satin finished cords running diagonally through the cloth.

whipcord, *v.* Delete (*nonce-wd.*), substitute quot. 1784 for quot. 1811 in Dict., and add later example.

1784 R. Robinson *Jrnl.* 26 May in *Belfast Monthly Mag.* (1809) June 435/1 Whip-corded the boys' plough whips. **1863** Mrs. Gaskell *Cousin Phillis* i, in *Cornh. Mag.* Nov. 627 He has often to whip-cord the plough-whips.

whip-hand. Add: **1.** (Earlier example.)

1806 *Ann. Reg. 1804* 413/1 For a morning's ride this might be *complimentary*; but it was here depriving me of the *whip hand.*

2. Now usu. without *of.*

1947 *Sun* (Baltimore) 22 Dec. 2/1 Its objective will be to outstrip the Marshall plan and so to gain the political whiphand over Europe. **1951** F. Yerby *Woman called Fancy* (1952) xvi. 302 In that election year of 1894, the white vote was so hopelessly divided that the blacks.. held the whip hand. **1974** *Howard Jrnl.* XIV. 49 The white population who have for so many centuries held, both literally and metaphorically, the whip hand. **1977** M. Thatcher in *Observer* 25 Sept. 10/1 If trade unions hold the whip hand, upon whose back does the lash fall? **1985** *Times* 26 Jan. 21/2 Sir Owen, however, still has the whip hand: he has the money and can bail the banks out of the whole complex exercise.

whip-lash, *sb.* Add: Also **whiplash. 3.** An injury to the head, neck, or spine caused by the head's being dashed to and fro on the less mobile trunk when a seated person is jerked forwards or backwards, as in a car accident. Usu. *attrib.*, esp. in *whiplash injury.*

1955 *Jrnl. Amer. Med. Assoc.* 5 Nov. 983/2 Poor seat-design accounts for thousands of so-called whiplash injuries. **1962** *Times* 23 Jan. 5/5 They discount arguments that 'whiplash' injury is common among safety belt wearers. **1971** H. Pacy *Road Accidents* i. 21 In damage to rear of vehicle think of whiplash. **1975** *Year Bk. Ear, Nose & Throat* 14 This article documents another useful study regarding the effects of cervical spine trauma, or 'whiplash' trauma. **1977** *Woman's Day* (Austral.) 24 Oct. 47/1 My husand had a car accident at the beginning of last year and received a whiplash injury. **1983** *Which?* Sept. 402/3 If no effective head restraint is provided, the head tends to get left behind, causing major bending and straining of the neck—'whiplash' injuries.

whiplash (hwi·plæʃ), *v.* [f. prec. sb.] **1.** *trans.* **a.** To inflict sudden or severe harm on.

1957 A. MacNab *Bulls of Iberia* viii. 83 The bull's trajectory is accordingly also bent in an arc... The bull is not now being violently whip-lashed as in the 'benders', but is being smoothly worn down. **1975** *Business Week* 14 July 50 Whether such a complex plan can be managed effectively, or whether it will be whiplashed by the short-term interest of elected officials and mired in a new superbureaucracy is perhaps the most important unanswered question. **1980** *N.Y. Times* 28 June 9/5 Much of the playing was perfunctory. Mr. Getz had a ghastly time, whiplashed between feedback and reed trouble that led to a classic climactic squeak. **1982** *Christian Sci. Monitor* 5 Oct. B 2/2 Oil field service companies have been 'whiplashed' as profit-starved major oil companies have sharply cut back drilling programs.

b. To jerk in a contrary direction; *spec.* to cause a whiplash injury to.

1971 *Daily Colonist* (Victoria, B.C.) 27 May 55/1 Parents who shake their babies in a fit of temper are threatening their lives, a surgeon has warned. Severe shaking can 'whiplash' the baby's head, causing blood clots on the brain. **1980** *Washington Post Mag.* 30 Nov. 53/1 (*caption*) The final solution to the problem of the hook on your tape measure slipping off the edge of the credenza to which you've attached it, whiplashing the tape into your eye. **1982** J. Gardner *For Special Services* xiii. 133 The force of impact had whiplashed the man's head, breaking his neck.

2. *intr.* To move suddenly and forcefully, like a whip that is cracked. Also *fig.*

1963 *Lebende Sprachen* VIII. 169/3 [Drivers' vocabulary.] To whiplash. 1. his head whiplashed. 2. the trailer whiplashed. **1971** *Daily Tel.* 13 Dec. 3/2 The Environment Department is investigating methods used to fix posts for motorway crash barriers. It fears that if they have not been planted deep enough, a crash might uproot them and allow a stretch of high-tension metal barrier to 'whiplash' across the carriageways. **1972** D. Delman *Sudden Death* (1973) iii. 77 He set us against each other. And he figures..one of us..is going to whiplash with something he can use. **1977** *Washington Post Mag.* 27 Nov. 40/3 Conservatives say they can't do or say anything because it will hurt their careers. It's like the old back-lash has whiplashed. **1983** *Washington Post* 20 Feb. G3/6 The cable that catches the planes when they come in snapped. It whiplashed around the deck and caught the Chief in the spine. **1983** D. Boggis *Women they sent to Fight* xxxviii. 220 Margaret released her... Żelaszny whiplashed round..terrified.

whi·pless, *a.* [-less.] Of a Member of Parliament: having resigned, or having been deprived of, the whip.

1962 *Guardian* 13 Dec. 2/3 Mr Emrys Hughes, the 'whipless' Labour member for South Ayrshire. **1967** R. Butt *Power of Parliament* xii. 317 Opinions varied as to how far the 'Whipless' MPs suffered from social or other pressures from the parties they had deserted during their period in isolation. **1976** *Times* 23 Feb. 13/2 A small and whipless group of independent Labour MPs.

whipped, whipt, *ppl. a.* Add: (The spelling whipt is now *arch.*) **8.** Subject to a Parliamentary whip.

1970 P. G. Richards *Parliament & Conscience* iii. 60 This was duly debated and defeated on a straight party whipped vote. **1976** *Times Lit. Suppl.* 12 Mar. 300/2 In the 1970–74 Parliament, two thirds of the Conservative members voted against their party whip on at least one occasion, and one (Enoch Powell) did so in 113 whipped divisions. **1981** Marsh & Chambers *Abortion Politics* vii. 194 MPs..are subject to constituency and interest group pressure on whipped issues.

whipperginnie. Add: **3.** (Later *arch.* example.)

1923 R. Graves *Whipperginny* 45 The minds of these two princes Were of such subtlety and such nimbleness That Whipperginny on the fall of a card Changed to Bézique or Cribbage or Piquet.

whipper-in. 1. (Ear er *fig.* example.)

c **1771** S. Foote *Maid of Bath* Prol. p. vi, To change the figure—formerly I've been To straggling follies only *whipper-in.*

whi·pper-snap, *v.* [Back-formation f. Whipper-snapper.] *intr.* To behave like a whipper-snapper; to be impertinent. Hence **whi·pper-snapping** (and varr.) *vbl. sb.* and *ppl. a.*

1908 W. De Morgan *Somehow Good* xi. 100 The lines they might elect to whipper-snap on were not to be those of sentimental nonsense. **1913** D. H. Lawrence *Sons & Lovers* xiii. 349 Think I'm goin' ter have *you* whipperty-snappin' round? **1925** —— *Refl. Death Porcupine* 231 Oh, the universe has a terrible hole in the middle of it, an oubliette for all of you, whipper-snappering mongrels. **1973** *Times* 11 Apr. 13/5 Jackie Rea, a former champion, there to match whipper-snapping with an old dog's tricks.

whippet. Add: **4. b.** Usually *attrib.* as *whippet tank.*

1918 E. W. Farrow *Dict. Mil. Terms* 664 *Whippet tank*, an English armored car equipped with caterpillar treads. **1920** J. C. F. Fuller *Tanks in Gt. War* 176

March 26 [1918] is an interesting date in the history of the Tank Corps, for, on the afternoon of this day, the Whippet Tanks made their debut. **1938** G. Greene in *Spectator* 22 July 139/2 Whippet tanks—camouflaged as in war. **1946** *New Yorker* 9 Mar. 83/1 Solid city blocks of whippet and giant tanks.

whippiness (hwi·pinés). [f. Whippy *a.* + -ness.] Pliable quality; flexibility.

1881 *Sportsman's Year-bk.* 70 Some successful anglers use the two extremes of whippiness and stiffness [in fly-rods]. **1913** W. E. Dommett *Motor Car Mech.* 130 This control has the objection..that the gear box has to be long with possible whippiness of the shafting. **1975** *Daily Tel.* 5 Apr. 8/1 All you have to do is to choose a branch of sufficient whippiness to be pulled down to the ground and fixed there.

whipping, *vbl. sb.* Add: **1. a.** (Further U.S. *fig.* example.)

1835 F. A. Chardon *Jrnl.* 10 July (1932) 37 Went to the Medicine dance last—Came back late and got a whipping from my Wife for my bad behaviour.

f. The action of stirring *up* strong feelings or the like (see Whip *v.* 13).

1952 C. Day Lewis *Grand Manner* 12 This whipping-up of words into a frenzy. **1955** *Times* 18 July 4/7 He had emphasized that the 'whipping up' of public opinion against South Africa..would..estrange the great majority of South Africans. **1959** *Daily Tel.* 29 Dec. 6 This prospect suggests a possible explanation for his deliberate whipping-up of patriotic frenzy.

4. a. whipping-block, a block on which offenders are laid to be whipped; **whipping cream,** a grade of cream suitable for whipping; **whipping-house** *U.S.*, a building in which at one time Blacks were whipped.

a **1877** Swinburne *Lesbia Brandon* (1952) 504 He..begged..that he might not be hoist across the whipping-block by a servant. **1953** R. Graves *Poems* 26 And taught St. Dominic's to baulk At gown and hood and whipping-block. **1924** *Techn. Bull. N.Y. State Agric. Exper. Station* No. 113. 3 Good whipping cream gave a reduced volume of whipped cream when compared to poor whipping cream. **1978** *Chicago* June 248/2 Pure whipping cream for coffee and batters. **1852** Mrs. Stowe *Uncle Tom's Cabin* II. xxix. 147 It was the universal custom to send women and young girls to whipping-houses.

d. whipping side *Austral.* (see quot. 1965).

1957 Stewart & Keesing *Old Bush Songs* ix. 259 You see our ringer already turned and he's on the whipping side. **1965** J. S. Gunn *Terminol. Shearing Industry* II. 37 *Whipping side,* the name given to the last side of the sheep to be shorn and the blow here is down from the shoulder.

Whipple (hwi·p'l). *Path.* [The name of George Hoyt *Whipple* (b. 1878), U.S. pathologist, who described the disease in 1907 (*Bull. Johns Hopkins Hosp.* XVIII. 382).] *Whipple's disease*: = *intestinal lipodystrophy* s.v. *LIPODYSTROPHY.

[**1939** *Amer. Jrnl. Path.* XV. 483 (*heading*) Malabsorption of fat (intestinal lipodystrophy of Whipple).] **1945** *Ibid.* XXI. 1079 The pathologic findings in Whipple's disease are characterized by deposits of fat and fatty acids in the small intestine and mesenteric lymph nodes. **1978** *Price's Textbk. Pract. Med.* (ed. 12) VIII. 652/2 Whipple's disease, characterized by fever, joint pains, wasting, and diarrhoea, can initially be suspected [in cases of Crohn's disease].

whip-poor-will. (Earlier and later examples.)

1709 J. Lawson *New Voy. Carolina* 146 Whippoo-Will, so nam'd, because it makes those Words exactly. **1920** W. D. Howells *Vacation of Kelwyns* 42 The whippoorwills..whirred through the cool, damp air. **1938** J. M. Gordon *Canad. Mosaic* (1939) ix. 218 Dreamily lying with prairie for pillow Clear I hear calling the lone whip-poor-will. **1960** R. T. Peterson *Field Guide Birds of Texas* 134 Whip-poor-will... Best known by its vigorous cry repeated in endless succession at night.

whipsaw, *sb.* Add: **2.** *fig.* Something that is disadvantageous in two ways. orig. and chiefly *U.S.*

1873 *Kansas Mag.* Mar. 232/1 There was fifteen hundred on the turn—seven hundred and fifty on each side of it—and the run was tray, ace; a whipsaw. **1929** L. F. Carr *Amer. Challenged* 79 The whip-saw of paying high prices for what they bought and being forced to receive low prices for what they sold. **1967** *Listener* 23 Nov. 656/3 The wage push..and the rising interest rates..have together caught the American economy in a cruel and sharp whipsaw... The worst sort of inflation of costs and the worst sort of deflation of values. **1977** *Time* 25 July 48/3 By the spring of 1974, the whipsaw effect of recession and rising costs—particularly for oil which fuels 80% of Con Ed's generating capacity—left the company strapped.

whip-saw *v.* (earlier and later examples); hence **whip-sawing** *vbl. sb.* (lit. and *fig.*).

1842 *Amer. Pioneer* I. 83 Dwellinghouses, made of wood, whip-sawed into timbers, four inches thick, and of the requisite width and length. **1873** *Kansas Mag.* June 497/1 On the next Budd whipsawed him, and that closed that deal. **1885** *Mag. Amer. Hist.* May 496/1 *Whip-sawing,* the acceptance of fees or bribes from two opposing persons or parties. **1903** *Sun* (N.Y.) 8 Nov. 10 The speculators have subjected themselves to the process known in Wall Street as whipsawing, that is, they have bought when the market was strong and sold when the market was weak, and found each time that they bought at the top and sold at the bottom. **1930** H. A. Innis *Fur Trade in Canada* II. v. 140 Men were engaged in cutting, squaring, whipsawing, and hauling timber for the construction and repair of the forts. **1957** *Listener* 12 Dec. 970/1 Mr [Adlai] Stevenson has been whip-sawed by conflicting advice. **1958** F. G. Slaughter *Daybreak* III. xiii. 176 The tendency to whipsaw all society into robots who work, think and eat alike is hardly an end product of intelligence. **1969** D. Bagley *Spoilers* ii. 58 'Okay, so you've whipsawed me,' said Follet sourly. **1975** *Weekend Mag.* (Montreal) 12/1 Whip-sawing, industrial relations slang for the union practice of wringing a high settlement from a weak company and then using that settlement as a floor for bargaining with a big company, was rampant. **1976** *Billings* (Montana) *Gaz.* 27 June 8-c/3 A major problem occurs when one small union negotiates a salary increase for its workers and all other state employes in the same job classification want the same increase. The effect is to 'whipsaw' the state between their competing demands. **1979** C. E. Schorske *Fin-de-Siècle Vienna* vii. 351 Schoenberg whipsaws us upward out of the crepuscular calm.

whip-snake. Add: Also, in southern Africa, a grass or sand snake of the genus *Psammophis*, esp. *P. notostictus.*

1880 J. Nixon *Among Boers* ii. 45 He had all but put his foot upon a whip snake. These snakes are small and slender, being only as thick as the little finger. **1912** F. W. Fitzsimons *Snakes S. Afr.* (new ed.) 462 The Grass and Sand Snake, when alarmed, glide off over the stunted herbage... Colonists know these as Whip Snakes. **1952** *Cape Times* 24 Apr. 9/2 As schoolboys were walking along the road..they saw a whipsnake.

whip-stitch, *sb.* Add: **2.** (Earlier example.) Also (without *at*), each item without exception; 'every last thing'.

1824 P. Hobby *Life of Marion* p. i, What can one do, when one's friends are..calling out at every whipstitch..'Well, but sir, where's Marion?' **1888** F. R. Stockton *Dusantes* III. 130 Every whip-stitch of his bag and baggage shall be trundled after him.

whipstock. Add: **4*.** *Oil Industry.* A long, tapered steel wedge which can be placed at the bottom of a hole to cause the drilling bit to deviate sideways, e.g. in directional drilling.

1903 *Dialect Notes* II. 345 *Whip-stock, n.*, an implement used in drilling past a set of tools when fast. **1935** *Econ. Geol.* XXX. 740 Controlled deflection of a hole toward a specific objective through the use of whipstocks, knuckle-joints and improved methods of well surveying has given added flexibility to rotary tools. **1973** J. W. Jenner in Hobson & Pohl *Mod. Petroleum Technol.* (ed. 4) iv. 128 The whipstock had to be pulled out of the hole and reset every 25 ft or so..in order to achieve sufficient angle build up.

whirl, *sb.* Add: **11.** *colloq.* (orig. *U.S.*). An attempt, esp. an initial or tentative attempt. Freq. in phr. *give it* (and varr.) *a whirl.* Cf. *BURL *sb.*[2] 2.

1884 C. B. Lewis *Sawed-Off Sk.* 277 After licking the best man in his own camp he came down to give us a whirl. **1889** 'Mark Twain' *Connecticut Yankee* xix. 234 No sound and legitimate business can be established on a basis of speculation. A successful whirl in the knight-errantry line... It's just a corner in pork, that's all, and you can't make anything else out of it. **1904** 'O. Henry' *Sixes & Sevens* (1911) 75 I'd been saving up for a year to give this New York a whirl. **1922** S. Lewis *Babbitt* vi. 90 But—I wish I could've had a whirl at law and politics. **1923** Wodehouse in *Strand Mag.* Apr. 335 Jeeves, if he cared to take a whirl at it, could be Prime Minister or something to-morrow. **1949** A. Miller *Death of Salesman* I. 66 Come on up. Tell that to Dad. Let's give him a whirl. **1965** K. Roberts in J. Carnell *New Writings in S-F* III. 127 I'm going up again next weekend. Give it another whirl. **1979** S. Wilson *Greenish Man* 11 You've nothing to lose. Give it a whirl, try it for a month. **1985** *Times* 28 Feb. 20/2 John Syer came to me and said he could help... So I thought I would give it a whirl.

whirligig, *sb.* **2.** (*b*) (Earlier example.)

1816 E. Weeton *Let.* 22 May in *Jrnl. of Governess* (1969) II. 145 Large caravans enter the town with..wooden horses, whirligigs, gambling tables, barrel organs.

whirling, *vbl. sb.* Add: **4.** *attrib.* and *Comb.*, as *whirling speed*; **whirling disease,** a disease of trout caused by the parasitic sporozoan *Myxosoma cerebralis,* which affects the balance of the fish it attacks.

1961 J. I. Lengy et al. tr. A. V. Uspenskaya in G. P. Petrushevski *Parasites & Dis. Fish* 47 One of the most dangerous of the known parasitic diseases is the so-called 'whirling disease'. **1962** *Spec. Sci. Rep. U.S. Fish & Wildlife Service* No. 427. 2/2 Whirling disease appeared in brook trout at the Benner Spring Fish Research Station..in 1956. **1982** *Times* 12 Feb. 4/5 Whirling disease..is a parasite which gets into the skull of trout fry, causing a fish to lose its balance so that it swims round and round until it eventually dies. **1894** *Phil. Trans. R. Soc.* CLXXXV. 283 The whirling speed was taken to be at the commencement of whirl, that is to say, the lowest speed at which the shaft definitely whirled.

whirlpool[2]. Add: **1. c.** A bath or pool with underwater jets of hot, usu. aerated, water, used for purposes of physiotherapy or relaxation; also, a pumping unit for producing such jets. orig. *U.S.*

1975 *Sports Illustrated* Aug. 41 Sportswriters let you know they are on such intimate terms with an athlete that they can interview him in 'the whirlpool'. *Ibid.* 42/3 Last year Underwriters' Laboratories..dropped its long-standing approval of portable whirlpools whose unit, motor and all, goes into the water. **1976** *Billings* (Montana) *Gaz.* 16 June 8-A/3 (Advt.), Relaxing, soothing whirlpool... Directional nozzle, aerator, timer, handle. **1978** *Detroit Free Press* 5 Mar. D17/2 (Advt.), Entertainment and dancing in the Wharf Lounge. Indoor pool, sauna, whirlpool, charming restaurants. **1980** D. Williams *Murder for Treasure* ix. 86 The heated indoor whirlpool..measures twelve feet across. **1985** *Brit. Med. Jrnl.* 6 Apr. 1024/1 The whirlpool or Jacuzzi is a North American invention which has flourished there since the early 1970s but has only recently been introduced into Britain. Hot water is agitated mechanically through pressurised jets in a large tub and gives the bather a pleasurable sensation. Bathing in the company of others is usual.

2. b. (In sense 1 c above.)

1950 *Life & Health* Oct. 8/2 Whirlpool baths have stimulating effects. **1972** 'E. Lathen' *Murder without Icing* (1973) xii. 113 It's more like a country club..saunas and whirlpool baths, cocktail bars and singles nights. **1975** *Sports Illustrated* Aug. 42/2 Doctors generally agree on the salubrious effects of whirlpool therapy. **1978** *Official Gaz.* (U.S. Patent Office) 13 June TM 88/1 Jacuzzi Bros Inc., Little Rock, Ark... *Jacuzzi*... For hydro-therapy products..therapeutic whirlpool baths and parts thereof. **1984** *Miami Herald* 6 Apr. 15D/5 (Advt.), Bayside pool, sundeck and whirlpool spa. **1984** *Listener* 19 July 17/4 (Advt.), For couples who want to get away together..4 poster beds, water beds—with whirlpool bath en suite.

whirlwind, *sb.* Add: **3.** *spec.* applied to something done in great haste.

1865 [in Dict.]. **1942** [see *Romance *sb.* 5]. **1952** J. L. Waten *Alien Son* 87 Auntie Fanny lived her own life, never commenting on her husband's whirlwind comings and goings. **1969** 'D. Shannon' *Crime on their Hands* vii. 99 We only got engaged last week. It was a whirlwind romance. **1977** D. E. Westlake *Nobody's Perfect* 65 Jet-setter Arnold Chauncey, just back from his whirlwind tour of Brasilia. **1984** *Times* 20 Feb. 10/2 His whirlwind investigation of NHS management.

whirlwind *v.* (earlier example).

1894 'Mark Twain' *Let.* 22 Dec. (1917) II. 617 These salvation-notions that were whirl-winding through my head.

whirly-. Add: **whirly-whirly,** (*a*) a dentist's drill (*nonce-use*); (*b*) *Austral.*, a whirling air current or dust cloud.

1928 A. P. Herbert *Trials of Topsy* xii. 73 He *thrust* the whirly-whirly *inch* by *inch* into the very *dome* of a girl's head. **1930** V. Palmer *Men are Human* xiii. 112 A cool breeze..raised little whirly-whirlies of dust. **1959** *Listener* 15 Jan. 113/1 The dust whirls and capers into fantastic whirlie-whirlies. **1972** *Southerly* XXXII. 4 A small whirly-whirly swept down the verandah, lifting dust and lolly papers in a mini-spiral.

whirlybird (hwɜ·ɹlibɜɹd). *slang.* (orig. *U.S.*). [f. Whirly- + Bird *sb.*] A helicopter.

1951 *Air Facts* 1 July 30/1 The biggest untold story out of Korea is of a few score unarmed American helicopters and a handful of pilots who have flown themselves and their 'whirlybirds' into military history. **1959** *Sunday Times* 26 Apr. 31/4 The noise which piston-engined 'whirly-birds' inevitably make causes no nuisance to workers in London. **1983** *Chicago Sun-Times* 28 July 82/1 The Bellwood-based whirlybird company has asked Civil Aeronautics Board approval to operate regularly scheduled flights.

whirra (hwi·rǎ). [f. Whirr, whir *sb.*] A whirring sound that varies in quality. (The examples are *Austral.*)

1929 K. S. Prichard *Coonardoo* i. 14 The pigeon flew off with a whirra of grey silken wings. **1969** H. Willard in P. A. Smith *Folklore Austral. Railwaymen* 174 Within two minutes whirra whirra whirra, the spears and boomerangs were coming over our heads.

whirry, *v.* **2.** (Later example.)

1920 Blunden *Waggoner* 24, I whirry through the dark.

whirry (hwɜ·ri), *a.* [f. Whirr, whir *sb.* + -y[1].] Characterized by, or of the nature of, a whirr.

1936 E. Dark *Return to Coolami* xiv. 142 There are the locusts beginning... A nice noise, whirry, hot, drowsy. **1982** *Financial Times* 25 Sept. 6/6 Intal suffered from the disadvantage that it could only be taken by the patient by means of a rather complex, whirry machine called the 'Spinhaler'.

whish, *v.*[1] **2.** (Later examples.)

1929 R. Graves *Good-bye to all That* xiii. 153, I heard one shell whish-whishing towards me. **1939** L. MacNeice *Autumn Jrnl.* xiv. 54 The wheels whished in the wet. **1959** R. Bradbury *Day it rained Forever* 214 Wouldn't it be nice to take a Sunday walk the way we used to do, with your silk parasol and your long dress whishing along?

whisk, *v.* **1. b.** (Later examples.)

1916 'Boyd Cable' *Action Front* 12, I heard..something else goin' *whisk* like a cane switched past your ear. **1919** H. Walpole *Secret City* II. v. 353 A beautiful fruit just within his grasp... He's going to taste it, when whisk! it's gone.

whisker, *sb.*[1] Add: **4. d.** Phr. *to have whiskers* and varr.: (of news, a subject, etc.) to be no

longer novel or fresh; similarly *to grow whiskers.* Also concr., of food: to become contaminated with mould.

1935 D. L. Sayers *Gaudy Night* viii. 182 That old story... It's got whiskers on it—it's six years old. **1951** M. Kennedy *Lucy Carmichael* VII. i. 345, I am putting on Capek's *R.U.R.* But it has got whiskers. It was quite a novelty when it was first put on. **1959** *Times* 6 May 4/6 The subject is beginning to grow whiskers. **1977** D. Francis *Risk* vii. 78 The steak in the fridge had grown whiskers. **1977** D. O'Sullivan in D. Marcus *Best Irish Short Stories* II. 90 'Did I ever tell you the one about the Scotsman and the octopus?'..'It has whiskers.'

5. b. *fig.* A very small distance or amount, a fraction: used chiefly in comparisons. *colloq.* (orig. *U.S.*).

1913 *Dialect Notes* IV. 6 *Whisker, n.*, a little; a trifle. 'Move it just a whisker.' **1953** *Wall St. Jrnl.* 11 Aug. 1/5 The London price is still a whisker below the 30 cents a pound charged by major U.S. producers. **1973** P. O'Donnell *Silver Mistress* i. 13 Sooner or later they would go on a job and not come back... Even in the past year they had come within a whisker of it twice. **1980** *Jrnl. R. Soc. Arts* Mar. 236/2 In these storms at sea, sunsets, sunrises, cloud formations and light conditions, Turner was within a whisker of pre-empting the great Monet himself. **1983** *Times* 15 July 18/3 Yesterday the shares rose 2p to 99p—a whisker from the year's high. **1984** *Listener* 14 June 15/3 Someone shoots for goal, and he either misses it by a whisker or by miles.

c. *Electr.* A wire used to form a rectifying contact with the surface of a semiconductor; cf. *cat's whisker* s.v. *Cat *sb.*[1] 18 and 19.

1915, etc. [see *cat(s') whisker* s.v. *Cat *sb.*[1] 18 and 19]. **1949** *Ann. Reg. 1948* 418 By the addition of a second wire whisker touching the germanium within a few thousands of an inch of the first the diode was converted into a triode. **1959** K. Henney *Radio Engin. Handbk.* (ed. 5) ix. 15 These diodes are representative of a family of germanium point-contact diodes using unplated whiskers. **1975** D. G. Fink *Electronics Engineers' Handbk.* ix. 62 Until 1965 point contact diodes were fabricated utilizing moderately low resistivity material with the rectifying contact established by contacting the semiconductor surface with a metal whisker.

6.* A single crystal that has grown in a filamentous form a few microns thick, characterized by a tensile strength much greater than the bulk material and used in quantity as reinforcing agents.

1946 *Monthly Rev. Amer. Electroplaters' Soc.* Jan. 28/1 The growth of needle-crystals on cadmium deposits has caused considerable annoyance in the radio industry. These crystals are known as 'whiskers'. They grow between condenser plates of variable condensers, and, being electrical conductors, actually short-circuit the plates. **1951** *Corrosion* VII. 329/1 An attempt was made to develop whisker growths in the laboratory. **1961** *New Scientist* 28 Dec. 776/3 Whiskers, the hair-like crystals which are far stronger than steel, are now being incorporated in bonding materials: for example, General Electric's silver reinforced with sapphire whiskers. **1973** *Sci. Amer.* July 44/2 Alumina whiskers have a tensile strength of up to three million pounds per square inch and a modulus of 62 million pounds per square inch.

7. whisker pole *Naut.* (see quot. 1976); = sense 6 a in Dict.

1954 *Motor Boating* Dec. 27/1 *Iris* was flying all her kites—main, mizzen, genoa winged out on the whisker pole, and mizzen staysails. **1960** J. J. Rowlands *Spindrift* 204 On the yacht-club float a girl..is rubbing down the last coat of varnish on a whisker pole. **1976** *Oxf. Compan. Ships & Sea* 938/1 *Whisker pole,* a short bearing-out spar used in yachts and sailing dinghies to bear out the clew of the jib on the opposite side of the mainsail when running before the wind, thus obtaining some of the advantage which would be gained in a larger vessel when she sets a spinnaker. **1980** *Yachts & Yachting* 29 Feb. 651/2 'American Express' carried two poles that extended from 11.5 ft to 18 ft plus a standard pole of 7.5 ft. The long ones were used as a spinnaker pole in the collapsed position, as whisker pole in the fully extended position, and as a bowsprit in the 14ft length.

whiskerless *a.* (earlier example).

1843 Dickens *Mart. Chuz.* (1844) ii. 10 His very throat was moral... Serene and whiskerless.

whiskered, *a.* **1. a.** (Earlier example.)

1769 T. Warton *Let.* 10 Oct. in D. Garrick *Private Corr.* (1831) I. 369, I went on board one of the Russian ships, and had the pleasure of being surrounded with a thousand whiskered sailors and soldiers from Archangel.

whiskery, *a.* (In Dict. s.v. Whisker *sb.*[1]) Add: **2.** Suggestive of or resembling whiskers or a whisker; having whiskers.

1927 H. V. Morton *In Search of England* v. 98 The dark room smelt of..that indefinite whiskery smell of old men. **1959** *Times* 4 May 4/5 They fluffed a chip out of whiskery grass. **1984** *Listener* 10 May 3/1 Streets with a few measly roadside stalls peddling second-rate oranges and whiskery root vegetables.

whisky, whiskey, *sb.*[1] Add: In addition to the distinction mentioned in Dict., *whisky* is the usual spelling in Britain and *whiskey* that in the U.S. **a.** *whisky-and-soda* (examples).

1898 G. B. Shaw *Mrs. Warren's Profession* II. 177, I could do with a whisky and soda now very well. **1924** H. Crane *Let.* 30 Nov. (1965) 195 As whiskey and soda was served I quickly revived. **1979** G. St. Aubyn *Edward VII* vii. 316 Offering him a whisky-and-soda and a cigar.

b. *whisky bottle, decanter, glass, -shop* (earlier example); *whisky-drinking* sb. and adj. (earlier examples); *whisky-gold, -soaked, -sodden* (earlier example) adjs.; **whisky-head** *U.S. slang,* one who consumes a great deal of whisky; **whisky-house** *Obs.*, a place where whisky is sold; **whisky mac** (also **Whisky Mac**), whisky and ginger wine mixed in equal proportions; a drink of this; **whisky money** *Hist.*, the proportion of the beer and spirit duty which was allocated to technical education by the Local Taxation (Customs and Excise) Act of 1890; **whisky priest,** an habitually drunken priest; **whisky-skin** *U.S. slang,* a drink containing whisky; **whisky-soda** (not in U.K. use), whisky-and-soda; **whisky sour** orig. *U.S.*, a drink of whisky acidulated with the juice of citrus fruit; **whisky-straight** (earlier example); **whisky voice,** a hoarse or alcoholic voice; **whisky-water** = *whisky-and-water.*

1843 'R. Carlton' *New Purchase* II. lvi. 242 He abstained..from his whiskey bottle. **1981** M. Hatfield *Spy Fever* I. vi. 53 The whisky bottle was still in play, though its contents..had not shrunk catastrophically. **1931** M. Allingham *Police at Funeral* xi. 151 He..shot a hopeless glance at the whisky decanter. **1976** E. Ward *Hanged Man* xxi. 129 Galbraith placed the whisky decanter within reach. **1883** 'Mark Twain' *Life on Mississippi* lviii. 571 Whiskey-drinking, breakdown-dancing rapscallions. **1884** —— *Huck. Finn* xxi. 212 There was considerable whisky drinking going on. **1940** R. Chandler *Farewell my Lovely* xiii. 82 She wore a hat with a crown the size of a whisky glass. **1918** E. Sitwell *Clown's Houses* 15 The sunlight pours all whisky-gold. **1944** S. Bellow *Dangling Man* 179 'Took you in it at last, didn't I!' I exclaimed. 'You damned old whisky-head.' **1968** P. Oliver *Screening Blues* 23 Blues about liquor and the 'whisky-head man', about prostitution, gambling, vagrancy and intended violence, figure in the work of singers of all generations. **1767** *Scots Mag.* Apr. 222 Grant kept a whisky-house. **1835** R. M. Bird *Hawks of Hawk-Hollow* II. 6 You would have some of the wherewithall smuggled up to this identical old woman's whiskey-house! **1960** *Spectator* 14 Oct. 579 It [sc. Stone's Ginger Wine] is a little cloying taken neat, but mixed with an equal quantity of whisky it becomes 'Whisky Mac'. **1961** L. Payne *Nose on my Face* iv. 63, I..said I'd have a whiskey mac. **1976** *Liverpool Echo* 22 Nov. 7/5 A thief stole a £45 cask of whisky mac from an off-licence in Pasture Road, Moreton. **1982** Barr & York *Official Sloane Ranger Handbk.* 92/2 You drink beer, whisky macs, cherry brandy, sloe gin—or neat whisky. **1911** *Encycl. Brit.* XXVI. 495/1 If the 'whisky' money..were found to be well and carefully expended, no future Chancellor would be able to divert it to any other purpose. **1937** G. A. N. Lowndes *Silent Social Revolution* ii. 39 Action taken by the Technical Education Committees of the County Councils..to encourage the formation of classes and guarantee them financial support out of the 'Whiskey Money'. **1973** L. Holcombe *Victorian Ladies at Work* ii. 30 A portion of the 'whisky money', the proceeds from the increased duties on beer and spirits, to be spent on technical education by the county and county borough councils. **1939** G. Greene *Lawless Roads* vi. 161 'He was just what we call a whisky priest.'..He had taken one of his sons to be baptized, but the priest was drunk. **1971** H. C. Rae *Marksman* I. iii. 19 With cheap striped pyjamas buttoned close around his throat Doyle looked like a whisky priest in a penal settlement. **1977** *Times* 4 Aug. 10/5 The communist equivalent of one of those Greeneland fables wherein a whisky priest rallies..to strike a blow for the God he no longer believes in. **1804** Lewis & Clark *Orig. Jrnls. Lewis & Clark Expedition* (1904) I. 10 Such as have made hunting..a pretext to cover their design of visiting a neighbouring whiskey shop. **1856** *Yale Lit. Mag.* XXI. 146 (Th.), Nine whiskey skins, and our spirits rushed together. **1891** *Sunday Times* 22 Feb. 2/3, I heard of the contemplated establishment of a London American club, the scheme of which seemed to comprise unlimited cocktails, whiskey skins, corpse revivers, [etc.]. *a* **1910** 'Mark Twain' *Autobiogr.* (1924) I. 209 Some old whisky-soaked, profane..infidel of a tramp captain. **1978** R. Ludlum *Holcroft Covenant* xiii. 153 Ellis made arrangements for the whiskey-soaked clothes to be picked up by the cleaners and returned by mid afternoon. **1915** H. L. Wilson *Ruggles of Red Gap* 50 Here, Charley, veesky-soda! **1975** O. Sela *Bengali Inheritance* xxv. 220 Shaking heads over their whisky-sodas saying, what could you expect. **1883** 'Mark Twain' *Life on Mississippi* lvi. 548 A harmless whiskey-sodden tramp. **1889** *Cent. Dict.*, Whisky sour. **1904** R. M. Lovett *Richard Gresham* 186 Bring a couple o' whiskey sours there, barkeep. **1975** D. Lodge *Changing Places* iii. 116 The lavish whisky-sours and daiquiris being prepared by the host. **1980** L. Birnbach et al. *Official Preppy Handbk.* 102/2 Tailgate picnics, whiskey sours in the stadium, and the general complexity of the sport guarantee that nobody knows what is going on. **1864** *Congressional Globe* 21 Apr. 1876/2 From the impassioned tone of the gentleman from Illinois..one would suppose that he had been investing in whisky straight. **1964** J. C. Catford in D. Abercombie et al. *Daniel Jones* 32 Simultaneous whisper + voice + creak: one form of 'beery' or 'whisky' voice. **1978** J. Updike *Coup* (1979) vii. 294 The women in the souk, with those long red finger-nails and blue hair in bandanas and those cracked whiskey voices. **1919** 'Etienne' *Strange Tales from Fleet* 5 'Thank you,' said the Captain, 'a whisky water, please.' **1978** T. Willis *Buckingham Palace Connection* i. 7 The ice-machine had broken down and I had to put up with a tepid whisky-water.

whisky, *v.* (In Dict. s.v. Whisky, Whiskey *sb.*[1]) (Earlier *refl.* example.)

1830 G. Colman *Random Rec.* II. 139 Post-boys and waggoners water'd their horses, and *whisky'd* themselves.

whiskyish (hwi·ski,iʃ), *a. rare.* [f. Whisky, whiskey *sb.*[1] + -ish[1].] **a.** Inclined for whisky. **b.** Tainted with whisky.

1929 W. Deeping *Roper's Row* ii. 11 Don't be in a 'urry, my lad. Wait till they're warm. If they're whiskyish, wait till the whisky's got 'em. **1929** E. Bowen *Last September* vi. 73 Some one tried to kiss you with whiskyish breath.

whisper, *sb.* Add: **4.** *whisper-like, -proof* adjs.; **whisper-shot** (further example).

1876 *Gentl. Mag.* Sept. 339 To ascertain whether..our boasted right of asylum was really whisper-proof. **1904** W. H. Hudson *Green Mansions* iii. 45 The mysterious melody began... It was uttered by the same being heard on former occasions..that low, whisper-like talking. **1936** N. Streatfeild *Ballet Shoes* xviii. 278 Petrova looked round to see that Posy was out of whisper-shot. **1964** J. C. Catford in D. Abercombie et al. *Daniel Jones* 37 What *feels* like *breath*..begins to *sound* more whisper-like at rates of flow above about 300 cl/sec.

whisper, *v.* **2.** (Earlier example with the words uttered as obj.)

1836 E. B. Browning *Poet's Vow* v, in *New Monthly Mag.* XLVIII. 217 They whispered oft, 'she sleepeth soft'.

whispering, *vbl. sb.* Add: **4. whispering campaign,** a systematic circulation of rumours, esp. in order to denigrate someone or something (orig. in *U.S. Politics*); **whispering Willie** *slang* (see quots.).

1920 *Nation* (N.Y.) 10 Nov. 517/1 The scandalous underhandedness of the whispering campaign of the Democrats..only prove[s] the spuriousness of all their protestations of belief in equal rights for black and white. **1949** 'R. West' *Meaning of Treason* I. vi. 118 A whispering campaign designed to weaken public confidence. **1962** D. Lessing *Golden Notebk.* I. 139 He was desperately depressed—a whispering campaign around the party and near-party circles, that he was and had been 'A capitalist spy'. **1978** D. Bloodworth *Crosstalk* viii. 69 The Chinese have been mounting a whispering campaign against the Soviet Union, quite distinct from their overt anti-Soviet propaganda. **1918** H. W. McBride *Emma Gees* 135 The..'Whispering Willies' belong to the class of large caliber, long range naval gun shells which pass over the front line so high that only a sort of whistling sound is heard. **1937** Partridge *Dict. Slang* 952/2 *Whispering Willie,* a type of big naval gun used by the Germans: East African campaign of the G.W.

whist, *sb.*[2] **2.** (Earlier example.)

c **1874** D. Boucicault *Shaughraun* (*c* 1884) I. iii. 7/2 Hould your whist now! Wipe your mouth, an' give me a kiss!

whist, *sb.*[3] Add: **a.** *short whist* (earlier example).

1829 E. M. Arnaud *Epitome Whist* 29 The game is won by the party whose score first amounts to ten points in Long Whist, or five in Short Whist.

b. *whist party* (earlier example), *-player* (earlier example); **whist-drive** (further examples); now often as a means of raising funds for charities.

1915 T. Burke *Nights in Town* 179 When I received the invitation to the whist-drive at Surbiton my first thought was, 'Not likely!' **1924** [see *Institute *sb.*[1] 4]. **1959** [see **community centre* s.v. *Community 11]. **1977** *Lancs. Life* Nov. 73/1 They raised the money themselves (with hot pot suppers, whist drives, amateur drama and dances). **1981** G. Markstein *Ultimate Issue* 196 In the lobby of the officers' club the wives were having a whist drive. **1744** S. Fielding *Adv. David Simple* I. II. i. 144 One of the Ladies, who was of the Whist-Party the Night before. *Ibid.* 147 Your Curiosity seems to be fully satisfied with what you have seen of the whist-players.

whistle, *sb.* Add: **1. b.** (*d*) *to blow the whistle on* (a person or thing): to bring an activity to a sharp conclusion, as if by the blast of a whistle; now usu. by informing on (a person) or exposing (an irregularity or crime). Also without *on.*

1934 Wodehouse *Right Ho, Jeeves* xvii. 222 Now that the whistle had been blown on his speech, it seemed to me that there was no longer any need for the strategic retreat which I had been planning. **1953** R. Chandler *Long Good-Bye* vi. 38 Come on, Marlowe. I'm blowing the whistle on you. **1965** *Midnight* 12 July 20/1 More and more frequently though, a whistle is being blown on the more exuberant borrowers. **1978** S. Wilson *Dealer's Move* v. 98 So Arnie and Alfie blew the whistle on you all. What are you going to do about it? **1984** *Gainesville* (Florida) *Sun* 29 Mar. 5A/4 Jim Kirkland, the man who first blew the whistle on Gainesville's deteriorating financial condition, has resigned after less than three months on the job.

d. *whistle and flute*: rhyming slang for 'suit' (Suit *sb.* 19). Chiefly *ellipt.* as *whistle.*

1931 Brophy & Partridge *Songs & Slang of Brit. Soldier, 1914–18* (ed. 3) 375 *Whistle and flute,* a suit (of clothes). **1941** G. Kersh *They die with their Boots Clean* I. 27 He is the one permanent type of Londoner..the..Cockney... To Barker..a suit is a Whistle, or Whistle-an'-Flute. **1960** A. Prior in *Pick of Today's Short*

Stories XI. 180 Half-Nelson lives for clothes... He never keeps a whistle more than a month. **1970** A. Draper *Swansong for Rare Bird* vii. 51 My best whistle was in a big heap on the floor. **1980** 'J. Gash' *Spend Game* ix. 97 'Him with the fancy whistle.' Whistle-and-flute, suit.

4. *whistle-call* (earlier example); **whistle-blower** chiefly *U.S.*, one who 'blows the whistle' on a person or activity (see sense 1 b (d) above), esp. from within an organization; also **whistle-blowing** *vbl. sb.* and *ppl. a.* (lit. and *fig.*); **whistle-language** = *whistle-speech* below; **whistle punk** *N. Amer. Logging,* a workman who sends signals by means of a whistle to those operating a donkey-engine; **whistle-speech,** a system of communication by whistling based on the spoken language, found esp. among peoples of mountainous districts and used to communicate over long distances.

1970 *N.Y. Times* 23 Mar. 40 When they reflect more fully on how well the majority leader handled a whistle-blower and protected their interests. **1983** *New Scientist* 23 June 838/1 A whistleblower who tries to alert his own organisation to a problem and fails will, if he feels strongly enough about the matter, go outside. **1971** *Ibid.* 9 Dec. 69 The Code [of Good Conduct of The British Computer Society] contains secrecy clauses that effectively prohibit Nader style whistle-blowing. **1978** *Monitor* (McAllen, Texas) 21 May 16A/6 He has introduced legislation to protect 'whistleblowing' federal employees from reprisals if they reveal wasteful, illegal or improper government activities. **1980** *Times* 1 Apr. 3/4 The growth in Britain of 'whistle-blowing' journalism (blowing the whistle on the secret parts of the state and its servants by disclosing their activities) would seem to have sealed the fate of the D-notice system. **1983** D. Dunnett *Dolly & Bird of Paradise* vii. 80 Whistle-blowing guys in white helmets. **1746** W. Ellis *Agric. Improv'd* May xvi. 100 In a certain Park, where Pheasants and Partridges come at the Whistle-call. **1956** J. Whatmough *Language* iii. 48 In this book we are not concerned with such departures from true speech as the so-called whistle 'languages' of Mazateco..and of the Canary Islands. **1957** *Amer. Anthropologist* LIX. 487 My direct interest in the subject stems from a brief encounter with a whistle-'language' and a slit-gong xylophone..among the Northern Chins of Burma. **1978** *Maledicta* II. 254 Whistle-Languages: Who knows whether there are insults or other abuses in whistled languages of the Canary Islands, Kuskoy/Turkey, etc.? **1925** *Amer. Speech* I. 136 The 'whistle-punk', who handles the signal wire that runs from the timber to the whistle of the donkey-engine. **1945** B. Macdonald *Egg & I* xiv. 184 Sharp and clear came the whistle punk's signals for a skidder. **1965** M. McIntyre *Place of Quiet Waters* ix. 172 He might get a job as a whistle punk in a logging camp. **1948** *Language* XXIV. 280 (*heading*) Mazateco whistle speech. **1972** Hartmann & Stork *Dict. Lang. & Linguistics* 255/2 Young English children often use whistle speech as a game. **1979** L. Campbell in Campbell & Mithun *Lang. Native Amer.* 958 Whistle speech is shared by Amuzgo, Mazatec,..some Nahua dialects, and Mexican Kickapoo.

whistle, *v.* Add: **II. 9. b.** *to whistle in the dark*: to put on a brave front; to make a pretence of confidence.

1939 [implied in *whistling *vbl. sb.* 1 d]. **1958** *Spectator* 8 Aug. 185/3 At his press conference, Mr. Dulles was whistling bravely in the dark. **1971** 'L. Egan' *Malicious Mischief* (1972) ii. 29 That fellow's whistling in the dark. And I think he knows it. **1983** S. Hill *Woman in Black* 92 'I am finding the whole thing rather a challenge.' 'Mr Kipps..you are whistling in the dark.'

10. (Later example.)

1917 H. A. Vachell *Fishpingle* xii. 236 He hurried on, now doubly assured that Joyce had 'whistled'.

11. To smell unpleasantly or strongly. *slang. rare.*

1935 Auden & Isherwood *Dog beneath Skin* II. v. 113 Wot wouldn't I give fer a bath? Cor! I don't 'alf whistle!

whistleable (hwi·s'l,ăb'l), *a.* [f. Whistle *v.* + -able.] Of a tune, etc.: capable of being whistled; suitable for whistling.

1962 *Guardian* 9 Oct. 7/1 This movement contains no fewer than four very singable and whistleable tunes. **1973** *Daily Tel.* 19 Apr. 10/4, I want to sit light to the glorious fantasy, to let it ride through my mind like a whistleable tune.

whistled, *ppl. a.* Add: **1. a.** (Earlier example.)

1816 J. Heckelwelder *Let.* 24 July in *Trans. Hist. & Lit. Comm. Amer. Philos. Soc.* (1819) I. 396 Where *w* in this language is placed before a vowel, it sounds the same as in English; before a consonant, it represents a whistled sound.

b. *whistled language* or *speech*: = *whistle-language, -speech* s.v. *Whistle *sb.* 4.

1948 *Language* XXIV. 283 Many words and phrases in Mazateco have identical tonal patterns. In the spoken language segmental phonemes usually distinguish tonally identical words and phrases. In the whistled language the absence of the segmental features gives opportunity for ambiguities. **1957** *Archivum Linguisticum* IX. 44 The Silbo Gomero is not the only whistled language in the world..but..is unique in being based not on prosodic but on purely articulatory features. **1978** *Verbatim* Sept. 13/1 Such systems of communication by whistling, based on the language of the user, are conventionally referred to as 'whistled languages' or 'whistled speech'.

3. Drunk, (mildly) intoxicated. *slang* (orig. *Mil.*).

The relationship, if any, to *whistled drunk* in Dict. is obscure.

1938 G. March-Phillipps *Ace High* II. iv. 216 They would be drunk as lords, tight as owls, screwed, canned, whistled. **1942** H. E. Bates *Greatest People in World* 8 He bounced in very late..and then began to eat as if he had returned from a hunting expedition. 'Pretty whistled last night, boys,' he would say. 'Rather off my feed.' **1968** 'O. Mills' *Sundry Fell Designs* viii. 83 He'd taken a skinful aboard, somewhere. He sounded more than a bit whistled. **1979** *Private Eye* 6 July 15/1 We all sidled off to a very nice little snug at the Golden Goose, where..all of us got faintly whistled.

whistler. 2. b. For *Arctomys pruinosus* substitute *Marmota caligata*; = *siffleur. (Earlier and later examples.)

1703 tr. *Lahontan's New Voy. N.-Amer.* I. 110 [We saw] little beasts called Siffleurs or Whistlers. **1866** J. K. Lord *Naturalist in Vancouver Island* II. 195 The Redskin is the whistler's most implacable enemy; he never tires of hunting and trapping the little animal. **1912** *Canad. Alpine Jrnl.* (Special No.) 28 The big hoary marmots are well named 'whistlers' by all mountain climbing people of the Canadian Rockies. **1973** *Islander* (Victoria, B.C.) 4 Feb. 4/3 Here we saw ptarmigan and heard the marmots, or whistlers.

3. b. An atmospheric heard as a whistle that falls in pitch, caused by radio waves generated by lightning and guided by the lines of force of the earth's magnetic field.

1928 *Nature* 17 Nov. 768/1 These observations refer to a peculiar class of atmospheric, which from their musical nature are appropriately termed 'whistlers'. **1963** G. M. B. Dobson *Exploring Atmosphere* viii. 141 It is also possible to get some information about the ionization at very great heights above the earth from the curious phenomenon of 'whistlers' or 'whistling atmospherics'. **1974** [see *magnetospheric *a.*]. **1979** [see *wave guide].

Whistlerian (hwisliə·riăn), *a.* [f. the name *Whistler* + -ian.] Of, pertaining to, or characteristic of the American painter and wit James Abbott McNeill Whistler (1834–1903) or his work; after the style of Whistler.

1891 E. Dowson *Let.* 2 Feb. (1967) 183 Oscar arrived late looking more like his Whistlerian name, in his voluminous dress clothes, than I have ever seen him. **1905** W. J. Locke *Morals of Marcus Ordeyne* ii. 19 A sort of Whistlerian nocturne of golden fog! **1927** Chesterton *Robert Louis Stevenson* iv. 92 We talk of some Whistlerian satire as a squib; but satire can only shine in the dark. **1956** D. Jones *Let.* 24 Aug. in R. Hague *Dai Greatcoat* (1980) III. 170 I'm glad to have found one surviving little oil study from this almost Whistlerian world. **1960** *Times* 15 Nov. 16/5 This is all very well in an aesthetic and Whistlerian sense. **1979** S. Weintraub *London Yankees* vi. 180 His [*sc.* Sargent's] icily elegant and Whistlerian portrait..of Madame Judith Gautreau.

Also **Whistlerism** (hwi·slərız'm), the style or æsthetic theory of Whistler; **Whi·stlerish** *a.*

1912 C. Rowley *50 Yrs. Work without Wages* 147 We discussed the prevailing fashions of Whistlerism, Impressionism, and..post-Impressionism. **1918** G. B. Shaw *Pen Portraits* (1932) 40 They are art for art's sake: the political variety of Whistlerism. **1979** *Times Lit. Suppl.* 23 Nov. 10/4 The subject of the Thames..her Whistlerish visual evocations.

whi·stle-stop, *sb.* orig. *U.S.* Also **whistle stop.** [Whistle *sb.*] **1.** A small station or town at which trains do not stop unless requested by a signal given on a whistle.

1934 M. H. Weseen *Dict. Amer. Slang* 418 *Whistle Stop,* a small town. **1944** *Sat. Rev.* (U.S.) 2 Sept. 2/4 The frank..and..challenging story of the men of the U.S. Foreign Service who represent America in the whistle-stops of the world. **1948** *N.Y. Times* 7 Sept. 18/8 President Truman told a railroad station crowd here tonight that 'before this campaign is over I expect to visit every whistle stop in the United States'. **1949** *Time* 9 May 29/3 To protest making Electra a whistle stop for express trains, he had thousands of plastic whistles molded in the shape of locomotives. **1949** 'H. Robbins' *Dream Merchants* (1950) 290 He thought Rock had been acting strangely yesterday when they had been married at that whistle stop just inside the California border. **1957** B. Hutchison *Canada* 217 The railway traveler sees only the dismal villages of the main line, the whistle stops around a wooden grain elevator..and a garage. **1965** S. G. Lawrence *40 Yrs. on Yukon Telegraph* xvii. 102 The railway company only recognized the town as a whistle stop.

2. One of a series of rapid, superficial visits.

1952 *Manch. Guardian Weekly* 7 Aug. 3 Truman opens his trap at the first whistle-stop. *Ibid.* 9 Oct. 3 As for Mr Truman's contribution by whistle-stop, his speeches have been..violently abusive. **1959** *Observer* 2 Aug. 9/4 We have gone on the marathon round of the dress shows (making whistle stops at breakfast and lunchtime at the smaller houses). **1976** *Courier-Jrnl.* (Louisville, Kentucky) 17 Sept. A2/1 President Ford is making a three-day tour..that will include a series of 'whistlestops' aboard a Mississippi riverboat.

3. Used *attrib.* to designate a journey with a lot of brief halts; *spec.* one by a campaigning politician that takes in many undistinguished places in this way. Also *fig.*

1949 *Time* 6 June 22/1 Louis Johnson..raised enough money..to pay for Harry Truman's whistle-stop campaign. **1952** *Manch. Guardian Weekly* 18 Sept. 3/1 On the whistle-stop tour down California's Central Valley. **1959** *Manch. Guardian* 23 July 1/3 The Queen and the Duke of Edinburgh..continued their 'whistle stop' journey to Moose Jaw. **1972** G. Durrell *Catch me a Colobus* v. 94 Our whistle-stop tours of the villages round about had paid dividends and when we went to visit them again we rarely came away empty handed. **1973** M. Truman *Harry S. Truman* i. 1 We had left Independence, Missouri, earlier in the day, and made a whistle-stop visit to Junction City, Kansas, at 11:05 p.m. **1976** *Times Lit. Suppl.* 23 July 904/4 She goes on to a whistle-stop history of attitudes to female inversion from Ancient Greece to the present day. **1978** *Broadcast* 17 July 15/1 BBC Radio 1's Roadshow set off again this week with a seven-week whistle stop tour of Britain's holiday resorts. **1981** *N. & Q.* Dec. 556/1 The result is an unremitting whistle-stop tour through barren regions.

Hence **whi·stle-stop** *v.*, (*a*) *trans.*, to travel through (a region) on a whistle-stop tour (*rare*); (*b*) *intr.*, to make a whistle-stop tour; also **whi·stle-stopping** *vbl. sb.*

1952 *News* (Birmingham, Ala.) 26 July 1/3 In a sort of swan song to the Democratic Party as its leader, he offered to whistle-stop the country for his successor. *Ibid.* 23 Sept. 14/5 Ike Eisenhower had settled down to whistle-stopping. **1952** *Time* 13 Oct. 23/3 In Michigan last week, nearly 100,000 people turned out to see Eisenhower as he whistlestopped across the state. **1957** *Ann. Reg. 1956* 183 Most of the 'whistle-stopping' was left to the assiduous Mr. Nixon. **1959** *Observer* 12 July 4/7 The Queen and the Duke of Edinburgh have been whistle-stopping their way across British Columbia. **1964** J. Reston in M. McLuhan *Understanding Media* xxxii. 339 Everybody's now whistle-stopping through somebody else's country, usually ours. **1972** *Observer* 23 Apr. 6/4 Italian politicians are whistle-stopping around the country this weekend in..the..election campaign. **1978** *Guardian* 14 Dec. 15/4 Howard Jarvis..the Messiah of taxpayers..has whistle-stopped across the United States with..226 events every 10 days.

whistling, *vbl. sb.* Add: **1. d.** In *fig. phr. whistling in the dark*: see *Whistle *v.* 9 b.

1939 *Time* 18 Dec. 21/3 Since precious little German trade can be sailed, submarined or flown overseas, writing about 'new possibilities'..sounded like official whistling in the dark. **1968** J. M. White *Nightclimber* xix. 132 He, like me, hated and feared being carried in this ship, for all his whistling in the dark. **1977** *Listener* 10 Feb. 169/3 Lenin and his wife..were not above a little whistling in the dark to keep up their spirits.

whistling, *ppl. a.* Add: **1. a.** *whistling kettle,* a kettle fitted with a device that emits a whistle as the water boils.

1928–9 *Army & Navy Stores Catal.* 173/4 *The Whistling kettle.* When the water boils the kettle whistles. **1961** J. Stroud *Touch & Go* iv. 43 The whistling kettle..burst into an unnerving shriek. **1974** R. Ingham *Yoris* xx. 63 She put a small whistling kettle on the gas ring.

c. *Mil.* Designating a missile which makes a whistling sound in flight, or a gun from which such missiles are fired. Freq. in the nicknames of these.

1864 J. Brobst *Let.* 28 May in M. B. Roth *Well, Mary: Civil War Lett. Wisconsin Volunteer* (1960) iv. 67 We dare not show our heads unless we want them to send one of their whistling jimmies at us. **1902** J. Milne *Epistles of Atkins* iv. 67 At Ladysmith 'Sighing Sarah' and 'Whistling Willie' proclaim their own shots from Umbalwana. **1926** T. E. Lawrence *Seven Pillars* xcv. 507 The aeroplanes circled round in their cold-blooded way, to drop whistling bombs into its trenches. **1948** W. White *Man called White* 256 Three heavy German guns which the Americans nicknamed 'The Anzio Express' and 'Whistling Willies'.

2. b. *whistling atmospheric*: = *Whistler 3 b.

1953 *Phil. Trans. R. Soc.* A. CCXLVI. 128 The main facts of observation concerning the whistling atmospherics..are summarized above. **1959** Davies & Palmer *Radio Stud. Universe* ix. 174 Storey at Cambridge in 1952..was investigating a phenomenon known as whistling atmospherics or simply 'whistlers' which are groups of radio waves at audio frequencies (15 kc/s). **1963** [see *whistler 3 b].

4. b. whistling thorn, a small prickly tree, *Acacia drepanolobium* or *A. zanzibarica,* found in East Africa.

1949 R. O. Williams *Useful & Ornamental Plants in Zanzibar* 102 *Acacia zanzibarica*..Coast Whistling Thorn. A thorny tree..bearing balls of bright yellow flowers. **1966** C. A. W. Guggisberg *S.O.S. Rhino* iii. 53 The rapid spread of the whistling thorn over vast areas..is probably a result of the reduction..of this animal! **1976** K. Thackeray *Crownbird* i. 9 The cab was full of whistling thorn, and swarming with red ants.

Whit (hwit). [The first element of Whitsun, Whit Sunday, etc.] = Whitsun 1, Whitsuntide. **Whit walk,** a Whitsuntide event in which church congregations walk in procession through the streets.

For *Whit Monday*, etc., and *Whit-week* see s.v. Whit Sunday, Whitsunday.

1959 I. & P. Opie *Lore & Lang. Schoolch.* xii. 232 Well-dressing,..and the children's great Whit walks in Manchester and elsewhere, are undoubtedly exciting occasions in the lives of local youngsters. **1963** *Times* 5 June 7/1 (*heading*) Whit road toll down by 19. **1976** Nichols & Armstrong *Workers Divided* 108 This morning our foreman told us that we've got to work Whit Bank Holiday. **1978** P. Bailey *Leisure & Class in Victorian Eng.* ii. 46 The popular holiday ritual of the Whit walks—street processions of witness complete with flags and decorations and marching bands. **1979** *Guardian*

30 May 10/8 In Lancashire... Whit Walks in most towns and many villages.

whitbed (hwi·tbed). Also 9 **white bed.** [f. White *a.* + Bed *sb.*] One of the upper beds of Portland Stone, lying next below the roach; stone from this, valued as a clean freestone for building.

1829 T. Webster in *Trans. Geol. Soc.* II. 38, I obtained from the quarrymen the thickness of the several beds, and the names by which they distinguished them from each other: but..these local appellations are not used by London architects and builders, the whole together passing here under the name of Purbeck stone only. The following is a list of the strata of limestone... 13. White bed, excellent. **1860** R. Damon *Handbk. Geol. Weymouth* 78 Whit Bed or Upper Tier.—This bed, the best stone that the island produces in point of quality, is of a whitish brown colour when first raised, but becomes paler on parting with its quarry water. **1911** *Encycl. Brit.* XXII. 122/1 The Portland limestones have been much in demand for building purposes; at Portland the 'Top Roach', the 'Whit Bed' or top freestone, and the 'Best Bed' (or Base Bed) are the best known. **1925** J. Bone *London Perambulator* ii. 32 You can see shell imprints on the freshly cut whitbed stone on the top of the new Bush Building. **1934** *Archit. Rev.* LXXV. 27 (*caption*) A 'close up' of the polished Portland stone wall sheathing—a new compact crystalline limestone, with a lovely fossil formation, discovered under the tiers of whitbed at the Portland quarries. **1936** [see *curf]. **1980** *Univ. Coll. London Bull.* Mar. 2/2 The thickest and most sought after unit in the local sequence is the Whitbed, noted for its homogeneity stemming from its lack of fossils or flint masses which may blemish other horizons above or below in the succession.

whitchet, var. *Wichert.

white, *sb.* Add: **9. b.** (*b*) (Later examples.)

1922 E. Raymond *Tell England* II. iv. 207 All honest boys, we know, fancy themselves in their whites. **1974** K. Millett *Flying* (1975) I. 101 Rich playing championship tennis..in his whites. **1978** G. McDonald *Fletch's Fortune* (1979) xiv. 96 Stop at the pro shop... We'll fix you up with a racket and balls... Have whites?

d. *pl.* White articles of washing.

1962 *Which?* Aug. 231/2 The programme you choose for the washing you want to do ('whites', for example, or 'delicate fabrics' are possible settings on both machines) automatically determines washing and spin drying times. **1979** A. Price *Tomorrow's Ghost* xiii. 229 It used to be right dirty rain... Woman couldn't put her whites out..when it was raining.

10. a. (Later example of *pl.* use.) Also (*sing.*) in general sense, money (*slang*).

1903 A. M. Binstead *Pitcher in Paradise* viii. 204 Again and again the needy one implored his obdurate chum to shake out at least a deuce of whites. **1960** [see *Cabbage *sb.*[1] 1 e].

11. (Later examples.)

1961 [see *Red *sb.*[1] 5 b]. **1972** 'W. Haggard' *Protectors* ix. 111 He..had drunk most of a bottle of wine. He had discovered the local whites with pleasure. **1978** T. L. Smith *Money War* III. 182 He would have the filet of sole amandine... He couldn't quite make up his mind which of the wonderful whites to choose to go with it.

13. *poor whites* (examples); also *sing.* and *fig.*

1819 W. Faux *Jrnl.* 28 July in *Memorable Days in Amer.* (1823) 118 The poor white, or white poor, in Maryland,..scarcely ever work. **1833** in *Maryland Hist. Mag.* (1918) XIII. 338 The poor whites at the South are not as well off in their physical condition as the slaves, and hardly as respectable. **1886** J. A. Froude *Oceana* xviii. 326 When he dies, the Maori and the poor whites in New Zealand will have lost their truest friend. **1896** R. Wallace *Farming Industries of Cape Colony* 406 The so-called 'poor whites' are chiefly the descendants of French protestant refugees, and, in some districts, of early Dutch settlers. **1934** A. N. J. Den Hollander in W. T. Couch *Culture in South* xx. 414 In discriminating southern speech, it was not used to include all white persons who were poor... The 'poor-whites' were those who were both poor and conspicuously lacking in the common social virtues and especially fell short of the standard in certain economic qualities. **1958** L. van der Post *Lost World Kalahari* iii. 56 All who worked for my grandfather no matter whether Griqua, Hottentot,..Cape-coloured or poor white, were ultimately held in equal affection. **1974** 'J. le Carré' *Tinker Tailor* i. 9 Jim Prideaux was a 'poor white of the teaching community.

14. (*b*) Also, the white ball in pool.

1981 P. Quinn *Tackle Pool* ii. 25 If the white is at point A it must be played into the black almost full ball.

15. c. A white diamond.

1878 [see Off colour, off-colour *phr.* and *a.* 2]. **1895** [see *Bywater]. **1928** [see *Bye *sb.* 3]. **1972** V. Canning *Rainbird Pattern* xi. 227 The diamonds were genuine,..blue whites, fine whites and whites. **1973** *Times* 25 Aug. 17/3 The (more or less) accepted English classes run thus in descending order: (1) finest fine white or river *alias* blue-white; (2) fine white; (3) commercial white.

d. A white ostrich-feather.

1881 A. Douglass *Ostrich Farming S. Afr.* xiii. 81 The cocks' quill feathers..he will..sort first... Prime whites, first whites, second whites, tipped whites. **1890** A. Martin *Home Life on Ostrich Farm* vi. 103 A large and magnificent bunch of *wing*-feathers, the finest and longest of 'prime whites'.

e. *slang.* Morphine. Cf. *white stuff* s.v. *White *a.* 11 e.

1914 Jackson & Hellyer *Vocab. Criminal Slang* 87 *White, noun,* current amongst morphine habitues. Morphine. Example: 'How many times a day are you shooting the white?' **1977** N. Adam *Triplehip Cracksman* iii. 32 By 1965 they were growing poppies for half the world's white.

f. White bread; a white loaf. *colloq.*

1960 Wentworth & Flexner *Dict. Amer. Slang* 576/1 *White,* white bread. **1974** 'A. Gilbert' *Nice Little Killing* iv. 55 Last of all came the baker... Leave a small white to be on the safe side. **1977** D. E. Westlake *Nobody's Perfect* 45 A luncheon-loaf sandwich on white with mayo in his left hand. **1978** R. Westall *Devil on Road* vi. 35, I got thick-sliced white and corned-beef.

g. An amphetamine tablet. *slang.*

1967 [see *Pill *sb.*[1] 1 d]. **1969** *Observer* 21 Dec. 1/1 The street pusher with his 'wanna score some whites (Benzedrine)? Dollar a roll.' **1972** H. C. Rae *Shooting Gallery* I. 19 He had anticipated a rash of arrests for possession of brown drugs and amphetamines—but not this, not a straight leap into the lethal whites.

17. e. Phr. *white-on-white,* used *attrib.* to designate articles made of white cloth with a white woven-in design; also *fig.*

1955 W. Gaddis *Recognitions* II. vii. 572 A bow tie of propeller proportions stood out over extra-length collar bills on a white-on-white shirt. **1958** J. Blish *Case of Conscience* xi. 113 'Why don't you give *me* a chance?' Michelis said raggedly. Then he turned white-on-white. **1976** A. Goldman in D. Villiers *Next Year in Jerusalem* 221 Perhaps it was radio..that forced American humor in the thirties to enter a phase of white-on-white neutrality. **1978** *Detroit Free Press* 5 Mar. D9/2 The Smithsonian Institution has several white-on-white quilts done in this manner.

19. a. For 'an Italian Ghibelline' read 'a member of one of the two factions into which the Guelphs split (see *Black *sb.* 8 a)'. Now, a member of any of various counter-revolutionary or strongly conservative parties.

1802, etc. [see *Black *sb.* 8]. **1942** 'A. Bridge' *Frontier Passage* i. 6 There were a few Whites in Madrid..and..they had a pretty thin time of it. **1954** B. & R. North tr. *M. Duverger's Pol. Parties* II. i. 216 In small French villages public opinion spontaneously distinguishes between 'Whites' and 'Reds', 'clerical' and 'anti-clerical'. **1965** M. Michael tr. *Myrdal's Rep. from Chinese Village* (1967) IV. 186, I joined the Young Pioneers. There we had classes about which districts were red and liberated and which were held by the Whites or the Japanese.

b. *spec.* An opponent of the Bolsheviks during the Russian Civil War (1918–21).

1921 F. McCullagh *Prisoner of Reds* iii. 26 A few miles off, on the west, was a large force of whites, which intended to advance on Krasnoyarsk that night. **1924** E. G. Jellicoe *Playing the Game* xiii. 224 Expeditionary Armies of Britain and the United States, invaded Northern Russia..in order to link up with Russian Whites against Russian Bolsheviks. **1944** M. Laski *Love on Supertax* ix. 86 She *is* Russian... Her parents were Whites who fled to Paris just before the October Revolution. **1950** E. H. Carr *Bolshevik Revolution* I. 325 In all these regions the ultimate effect of the civil war waged by the 'whites' with foreign backing had been to consolidate the prestige..of the Russian Soviet Government. **1964** L. Deighton *Funeral in Berlin* 318 Chekist operators... Originally these were an anti-sabotage, anti-revolutionary force..during the civil war..empowered to..execute Whites, or Reds who were getting a little bleached. **1976** [see *Red *sb.*[1] 6 b].

22. *in the white* (further examples).

1957 *N.Z. Timber Jrnl.* Aug. 59/2 *In the white,* applied to finished furniture ready for polishing or other treatment. **1965** *Wireless World* July 9 (Advt.), This range includes..ready-assembled cabinets in the white for finish to own requirements. **1968** J. Arnold *Shell Bk. Country Crafts* 130 Factory-made chairs are often dispatched for later finishing, 'in the white' they call it. **1971** *Country Life* 10 June 1416/2 James Giles..bought consignments of Worcester porcelain in the white for decorating to commission. **1981** *Sci. Amer.* Oct. 134/3 Violinmakers often say that a violin sounds better in the white than it does after it is varnished.

23. white(s)-only *a.*, reserved for white people.

1968 *Listener* 18 July 86/3 In 1958, the Court of Appeal supported the Musicians' Union in their boycott of a whites-only dance-hall in dear old Wolverhampton. **1971** *Sunday Times* (Johannesburg) 28 Mar. 1/3 It was a Whites-only compartment. **1971** *Guardian* 29 Sept. 19/2 In Salisbury [Rhodesia], there are perhaps half a dozen 'Whites Only' signs—mainly on public lavatories. **1980** *English World-Wide* I. 1. 55 In the 1950's Nassau's whites-only schools, cinemas, and restaurants were desegregated.

White (hwɔit), *sb.*[2] The name of Gilbert *White* (1720–93), English naturalist, used in the possessive to designate *Zoothera dauma,* a yellowish-brown and white thrush with black markings native to Asia, eastern Europe, and Australia, and orig. named *T. whitei* in his honour by T. Eyton (1836).

1836 T. Eyton *Hist. Rarer Brit. Birds* 93 The general colour of White's Thrush, on the upper surface, is ochraceous yellow. **1893** *Ibis* 371 (*heading*) On the occurrence of White's thrush in European Russia. **1954** D. A. Bannerman *Birds Brit. Isles* III. 165 In its plumage White's thrush is characterized by the very prominent black crescentic markings.

white, *a.* Add: **1. e.** *whiter than white*: extremely white; freq. *fig.*

In mod. use popularized as an advertising slogan for Persil soap-powder.

[**1592**: see White *sb.* 17 a.] **1924** *O.E.D.* s.v. *White* sb. 23, Exceeding or surpassing white, 'whiter than white'. **1949** D. Smith *I capture Castle* vii. 95 The strangeness of her face: that look she has of belonging to a whiter-than-white race. **1962** *Daily Tel.* 28 June 1/3 He is said to have said that the report made out the BBC to be 'whiter than white'. **1974** 'A. Garve' *File on Lester* vii. 31 Where their leaders are concerned, the masses are puritan—they expect standards of personal behaviour whiter than white. **1979** K. Bonfiglioli *After you with Pistol* xxii. 180 My knuckles were now Whiter-Than-White.

f. *Sci.* and *techn.* Applied to (non-optical) radiation, esp. sound and X-rays, having approximately equal intensities at all the frequencies of its range; esp. *white noise* (also *fig.*).

This use arises by analogy with the spectral composition of white light.

1922 *Nature* 1 Apr. 414/2 Just as the spectrum of a hot body normally consists of a continuous spectrum of white light, together with certain spectrum lines the wavelengths of which are characteristic of the radiating material, so an element emitting X-rays not only gives out 'white' radiation, but superposes its characteristic lines on the general spectrum. **1943** *Jrnl. Aeronaut. Sci.* X. 129/1 Inside the plane it is different; there all frequencies added together at once are heard, producing a noise which is to sound what white light is to light... That white noise is annoying needs little argument. **1948** *Bell Syst. Technical Jrnl.* XXVII. 642 If the noise is itself white..the result reduces to the formula proved previously. **1959** *Lancet* 12 Sept. 342/2 'White-sound' generators, which blind out extraneous noises, are unsatisfactory [for use in perceptual isolation experiments]. **1976** *Jrnl. R. Soc. Arts* CXXIV. 588/2 The proportion of power converted into the more penetrating 'bremsstrahlung'—or 'white' radiation—is approximately proportional to the atomic number of the target material. **1977** P. B. & J. S. Medawar *Life Science* i. 14 When the noise signals are so subdued, random and heterogeneous that their pretensions to conveying information are negligible, we may speak of 'white noise', e.g. the sound—as of innumerable mice eating Rice Crispies—that sometimes accompanies long-distance telephone calls. **1980** P. Way *Icarus* ix. 57 Maybe they *could* listen in, even through the white noise of the running water. **1984** *Mail on Sunday* (Colour Suppl.) 2 Dec. 6/2 (Advt.), At standard or even very low listening levels, you will never be harassed by hum or white noise.

2. e. [tr. It. *voce bianca* white voice.] Of a singing voice or its sound: lacking any emotional coloration (such as may be imparted by vibrato). Also *transf.* Cf. *voix blanche* s.v. *voix.

1884 F. Niecks *Dict. Mus. Terms* 257 *Voce bianca* (It.), lit. 'white voice'. The female and children's voices, and also some bright-sounding instruments, are thus called. **1904** S. Joyce *Dublin Diary* (1962) 39, I called McCormack's voice 'a white voice'—it is a male contralto. **1921** L. Tetrazzini *My Life of Song* xix. 316 Be careful not to simulate too broad a smile. Too wide a smile often accompanies what is called 'the white voice'. This is a voice production where a head resonance alone is employed, without sufficient of the appoggio or enough of the mouth resonance to give the tone a vital quality. This 'white voice' should be thoroughly understood, and is one of the many shades of tone a singer can use at times..to produce certain atmospheric effects. For instance, in the mad scene in *Lucia,* the use of the 'white voice' suggests the babbling of the mad woman, as the same voice..in the last act of *La Boheme* suggests utter physical exhaustion, and the approach of death. An entire voice production on this colourless line, however, would always lack the brilliancy and the vitality which inspires enthusiasm. **1951** W. Morum *Gabriel* I. iv. 56 That vibrato..[is] no use for symphony work. In the big orchestras the trumpeter employs what we call a *white* tone. A pure tone. **1957** V. Nabokov *Pnin* 182 'I want a last piece of advice from you,' said Liza in what the French call a 'white' voice. **1961** *Times* 28 Sept. 16/1 An attractive, brightly ringing voice, rather white at the top but pleasantly dark below. **1975** *Gramophone* Dec. 1075/1 The soprano, Emma Kirkby, produces a 'white tone' which is scarcely distinguishable from that of a choir boy in some items, and this makes for a commendable purity of intonation. **1976** *Times* 8 Nov. 8/6 Where another team might produce a remote, 'white' sound, without vibrato..the Amadeus [Quartet] permitted a more human, warm tone. **1981** Ld. Harewood *Tongs & Bones* xiii. 209 He contented himself for the first act with accurate, small-scale singing in a rather small, white voice.

f. Of a drink of coffee: with milk or cream added.

[**1900** G. Bell *Let.* 25 May (1927) I. 113 Besides the bitter black coffee, we were handed cups of what they [*sc.* Hasineh Arabs] called 'white coffee'—hot water, much sweetened and flavoured with almonds.] **1925** X. M. Boulestin *Conduct of Kitchen* 10 It is somewhat distressing..to have to stop at the coffee-stall on the way home for an honest sandwich and a cup of 'white' coffee. **1940** *Punch* 6 May (Summer No., unpaginated) (*caption*), Please don't hesitate to say if you prefer your coffee white. **1982** H. Shaw *Death of Don* i. 3 'Black or white, Master?' 'White, please.'..They took their coffee and brandy and sat down.

4. a. *poor white folks* (also *folk*), *trash* (examples); so *poor white, poor-white* as compound adj. (not always contemptuous, and in wider use, esp. in *S. Afr.*); also *fig.* Hence *poor-white-folksy, -trashy* adjs. Cf. Trash *sb.*[1] 4 in Dict. and Suppl., *White *sb.* 13.

1821 *Austin Papers* (1924) I. 446 My friend could probably take with him about twenty negroes and perhaps a poor white family consisting of a man and his wife. **1833** [see *Trash *sb.*[1] 4]. **1836** J. K. Paulding *Slavery in U.S.* 205 The slave of a gentleman universally considers himself a superior being to 'poor white folks'. **1864** *Harper's Mag.* Aug. 412/2, I wouldn't do my hair in a three strand braid on no account; it is too poor-white-folksy for me. **1911** *Chambers's Jrnl.* Jan. 6/1 An effort has also been made to enrol men of the 'poor white' class in the police force, for which they appear well adapted. **1949** *Race Relations in S. Afr.* 413 It was not until 1898 that the first organized effort at their rehabilitation was made. In that year the Dutch Reformed Church in the Cape Colony established the Kakamas Labour Settlement for 'Poor White' families. **1951** H. Giles *Harbin's Ridge* 63 He never had been much account. Always content just to make out, which we considered poor-white-trashy in our parts. **1958** *New Statesman* 1 Feb. 143/1 In *The Hamlet* Faulkner describes the infiltration out of nowhere into..that sequestered poor-white corner of Yoknapatawpha County, Mississippi, of the Snopes family. **1958** *Times Lit. Suppl.* 13 June 328/5 Mr. Chase's thesis allows us to see them [*sc.* many popular American novels] as, so to say, poor-white relations of incomparably more distinguished works, relations that all the same show, in however degenerate a way, similar fundamental responses to the nature of American experience. **1958** A. Jackson *Trader on Veld* 43 As a matter of course, every property was divided equally among the owner's sons upon his death... Few things contributed more effectively to the creation of a Poor White class than did this usage. **1979** J. Drummond *Patriots* xv. 77 He'd been poor, the son of a poor-white farmer.

c. Of or pertaining to white people.

1852, etc. [in Dict., sense 4 a]. **1868** *N.Y. Herald* 4 July 5/2 The registered white vote has been very greatly increased. **1933**, etc. [see *white jazz*, sense 11 e below]. **1937** L. & E. Dowling tr. *H. Panassié's Hot Jazz* ii. 28 White musicians were playing..a so-called 'white' hot style intended to compete with the other style. **1944** *Living off Land* iv. 64 Natives are always hunting the coast for food. Also, there will be cattle stations, or white camps along the coast. **1959** 'F. Newton' *Jazz Scene* iv. 70 The most characteristically 'white' style in the history of jazz. **1965** F. Symington *Tuktu* 59 Most missionaries tried to teach their charges how to cope with the 'white' culture and economy. **1977** *Times of Swaziland* 25 Feb. 12 (Advt.), Farming estate... Strategically situated in centre of largest white area of popular Natal Midlands. **1984** J. McClure *Artful Egg* xi. 156 A couple..who affected sophisticated white manners and even spoke English with an almost white accent.

5. b. *to bleed white*: (*b*) (examples).

1935 *Sabbath School Worker* Nov. 6/1 'There are too many appeals for money', the people are 'bled white', and 'we can't give another penny'. **1945** R. Chandler in *R. Chandler Speaking* (1966) 113 It is the writers' own weakness as craftsmen that permits the superior egos to bleed them white of initiative, imagination, and integrity. **1982** 'W. Haggard' *Mischief-Makers* i. 16 Her husband had been a wealthy man, the lady's solicitors sharp and ruthless, and her husband had been bled white to get rid of her.

6. a. *white nun*, a Cistercian nun (cf. White monk).

1877 J. Penderel-Brodhurst *Guide to Boscobel* v. 20 Whiteladies... The name is derived from the circumstance that the house was once a Priory of Cistercian or White Nuns. **1954** A. Seton *Katherine* xxxii. 536 Katherine..surveyed the two nuns... White nuns, Cistercians, shrouded in snowy wimples and habits.

b. (Further examples.) In recent use applied to the Kuomintang in China and to the Christian Democrats in Italy.

1937 E. Snow *Red Star over China* i. i. 21 To get in touch with Communists in the 'White' areas [of China] was extremely difficult. **1952** [see *Kuomintang]. **1965** C. D. Eby *Siege of Alcázar* (1966) iii. 63 In less than forty-eight hours the Alcázar had become a solitary White island in the middle of a raging Red sea. **1965** M. Michael tr. *Myrdal's Rep. from Chinese Village* (1967) iii. 131 My father was taken by the white bandits and beheaded. **1967** C. Seton-Watson *Italy from Liberalism to Fascism* xii. 514 A left wing, led by Miglioli, the pacifist and 'white' trade unionist, called for a Christian proletarian party that would make capitalism its main enemy. **1973** P. A. Allum *Politics & Society in Post-War Naples* 326 The DC and PCI are heirs to particular Italian subcultures, the Catholic and the marxist. Both.. ensure..the electoral strengths of both parties in North and Centre, and above all in those regions (e.g. the 'white' provinces of the NE and 'red' provinces of the Centre, etc.) where they organise specific populations.

7. c. Of propaganda: truthful.

1965 B. Sweet-Escott *Baker Street Irregular* i. 29 The Ministry of Information..confined itself to straight or 'white' propaganda in neutral and friendly countries. **1976** [see *propaganda 3].

11. a. In names of animals. **white † admirable, admiral** [Admiral *sb.* 6], a dark-coloured butterfly, *Limenitis camilla*, with white markings; **white-bird,** (*c*) (without hyphen) in Irish folklore, a bird of fairyland; **white egret** = *white heron* (*a*) below; **white-fly,** a small bug of the family Aleyrodidæ, usually covered with pale, powdery wax, esp. *Trialeurodes vaporariorum*, which is a pest of greenhouse plants; **white fox,** a small fox, *Alopex lagopus*, native to northern Canada, Greenland, and Iceland, which has white fur in winter; also, the fur of this animal; **white goat** = *Rocky Mountain goat* s.v. *Rocky *a.*[1] 1 b; **white heron** (usu. qualified by *great*), (*a*) the common egret, *Egretta alba*, a large white bird with a yellow bill and dark legs found in parts of Europe, Asia, North Africa, the Americas, and Australasia; (*b*) a white subspecies of the great blue heron, *Ardea herodias*, found in Florida; **white owl:** see Owl *sb.* 1 b; **white perch** *U.S.*: see Perch *sb.*[1] 2; (examples); **white pointer:** see *pointer 12; **white rhino(ceros),** a large, wide-mouthed rhinoceros, *Ceratotherium simum*, native to parts of Sudan, Uganda, and South Africa; **white steenbras,** a large marine food fish, *Lithognathus lithognathus*, found in coastal regions of South Africa; **white whale** = Beluga 2.

b. In names of plants. *white grape* (example); **white ash,** (*a*) add: *esp.* a North American ash, *Fraxinus americana*; (earlier and later examples); *white-ash breeze* (earlier example); **white-bark pine,** a pine with pale, flaky bark, *Pinus albicaulis*, native to north-western North America; **white birch:** see Birch *sb.* 1 b; (earlier and later examples); **white box,** either of two Australian trees, the evergreen *Bursaria spinosa*, which bears clusters of fragrant white flowers, or a box eucalypt, *Eucalyptus albens*, which has pale leaves; **white campion:** see Campion[2]; **white cedar,** (*a*) any of several North American conifers, esp. one of the genus *Chamæcyparis*; (*b*) *Austral.*, a name used for species of *Melia*, deciduous trees native to the East Indies and Australia; **white clover:** see Clover 1 b; (later examples); **white elm,** the American elm, *Ulmus americana*; also, the European elm, *Ulmus lævis*, which resembles it closely; **white fir,** any of several North American firs, esp. *Abies concolor*, native to the south-western United States; **white mangrove:** see Mangrove[1] 2; **white maple,** any of several maples with pale bark, esp. the silver maple, *Acer saccharinum*, or the mountain maple, *A. spicatum*; **white mulberry,** a round-topped mulberry, *Morus alba*, or its white or pink fruit; **white oak,** any of several species of North American oak, esp. *Quercus alba*, which is native to the eastern part of the continent; also, the wood of this tree; **white poplar,** (*a*) (see Poplar 1 b); (*b*) *N. Amer.*, the aspen, *Populus tremuloides*; (*c*) *N. Amer.*, the tulip-tree, *Liriodendron tulipifera*; **white-rot:** see Rot *sb.*[1] 2 c and sense 11 e below; **white spruce,** a spruce with bluish foliage, *Picea glauca*, native to North America; **white walnut** = Butter-nut 1.

c. In names of minerals, chemical products, etc. *white clay*; **white brick,** (*b*) a hard, durable variety of brick made from gault; **white cast iron** = White iron b; cf. *grey* (*cast*) *iron* s.v. *Grey, gray *a.* 8 c; **white earth,** earth material (as clay) that is light-coloured; *spec.* in *Painting*, a white earth-colour; **white oil,** (*a*) crude oil that is pale in colour; (*b*) a colourless petroleum distillate; *spec.* a highly refined heavy distillate used medicinally and in the food and plastic industries; **white phosphorus,** (*a*) the white opaque incrustation that forms on phosphorus when it is kept under water (? *obs.*); (*b*) the ordinary allotrope of phosphorus, a translucent waxy whitish or yellowish solid which unlike red phosphorus is poisonous and very reactive; **white precipitate,** either of two mercuric amidochlorides obtained by treating mercuric chloride with ammonia: *fusible white precipitate*, $HgCl_2(NH_3)_2$, and *infusible white precipitate*, $HgClNH_2$ (ammoniated mercury), obtained when there is excess ammonia and used in ointments against worm infection; **white rust,** a white coating that forms on zinc in air, consisting of some or all of the oxide, hydroxide, and carbonate (see also sense 11 e below); **white sapphire,** a variety of corundum that is colourless owing to the absence of the impurities responsible for the blue colour of ordinary sapphire; **white spirit,** a volatile colourless liquid distillate of petroleum that boils between about 150°C and 200°C and is widely used as a paint thinner and solvent; **white tin,** (*a*) refined metallic tin, in contrast to black tin; (*b*) the ordinary allotrope of tin, in contrast to grey tin.

d. white cell = *white corpuscle* in Dict.; **white finger(s)** = *Raynaud's phenomenon* s.v. *Raynaud; also *attrib.*; **white scour,** a disease of calves, freq. due to infection with *E. coli*, causing severe diarrhœa, dehydration, and often death.

1798 E. Donovan *Nat. Hist. Brit. Insects* VII. 75 The White Admirable Butterfly feeds upon the common honey suckle or woodbine. **1906** R. South *Butterflies Brit. Isles* ii. 59 The White Admiral (*Limenitis sibylla*). The 'White Admirable Butterfly', as it was called by some of the older English entomologists, needs only to be seen to be at once recognized. **1717** J. Petiver *Papilionum Britanniæ Icones* 1/2 in *Opera* (1764) II. vii, White Admiral. Found about Dullidge and Wickham near Croyden, as also at Henly upon Thames. **1826** J. Curtis *Brit. Entomol.* III. 124 (*heading*) Limenitis camilla. The White Admiral. **1857** H. T. Stainton *Man. Brit. Butterflies & Moths* I. 33 White Admiral..Blackish brown, with a broad white band crossing the centre of the wings. **1922** V. Woolf *Jacob's Room* ii. 36 He had seen a white admiral circling higher and higher round an oak tree, but he had never caught it. **1968** *Oxf. Bk. Insects* 46/2 White Admiral..belongs to the same family as the Fritillaries. **1683** *Coll. New Hampshire Hist. Soc.* (1866) VIII. 146 [They] did feloniously..use about one cord of white ash. **1784** [see *red ash* s.v. *Red *a.* 17 d]. **1851** H. Melville *Moby Dick* II. xxxix. 262 There she slides, now! Hurrah for the white-ash breeze! **1950** *Chicago Tribune* 27 Apr. iii. 10/1 Under each cutting spindle is placed a block of white ash. **1908** N. L. Britton *N. Amer. Trees* 12 White Bark Pine..a rather small tree of alpine habitat. **1949** *Sierra Club Bull.* Dec. 24, I knew I could not get any sleep just by crawling under the low branches of a white-bark pine. **1974** *Blackw. Mag.* Oct. 307/1 Stands of whitebark pine mingled with spruce..provide welcome shade. **1789** J. Morse *Amer. Geogr.* 197 On the high lands are..beech and white birch. **1961** H. MacLennan *Rivers of Canada* 48 Otherwise nothing but the immense low forest of spruce with the occasional splash of white birch. **1980** *Family Handyman* Sept. 63/2 Because hardwoods are more dense, there is more energy in a cord of oak, say, than a cord of white birch or white pine. **1892** W. B. Yeats *Countess Kathleen* 106 (*title*) The white birds. **1894** —— *Land of Heart's Desire* 41 The Child (*from the door*): White bird, white bird, come with me, little bird! Maire Bruin: She calls my soul! **1940** E. Pound *Cantos* lvi. 60 May the white birds remember this warrior. **1909** A. E. Mack *Bush Calendar* 67 Flowers blooming [in January]. *Bursaria spinosa*. White box or black thorn. **1923** *Census of Plants of Victoria* (Field Naturalists Club of Victoria) 46 *Eucalyptus albens* Miquel White Box. **1936** F. Clune *Roaming round Darling* xvii. 161 White box is a good burning wood, sheds a brown bark in springtime, then has a white surface. **1946** K. Tennant *Lost Haven* (1947) xvii. 289 'What are those trees down there?'..'Blueberry ash, white box, whipwood.' **1965** *Austral. Encycl.* III. 406/2 White box..having pallid glaucescent foliage. **1845** J. H. Parker *Gloss. Terms Archit.* (ed. 4) I. 72 In colour they are paler than ordinary red bricks, but are redder than the common white brick of Suffolk. **1969** R. Blythe *Akenfield* ix. 140 The school at Akenfield..is a stark, knife-edged building constructed of Suffolk white-brick, [etc.]. **1979** *Guardian* 10 July 19/7 The famous Suffolk whitebrick which Georgian architects favoured for the region's grander houses. **1795** White cast iron [see Iron *sb.*[1] 2]. **1967** A. H. Cottrell *Introd. Metall.* xxv. 518 In white cast irons, which are usually made by limiting the content of graphite-forming elements such as silicon to low levels.., all the carbon exists as cementite and the name white refers to the bright fracture produced by this brittle constituent. **1674** J. Josselyn *Acct. Two Voy. New England* 67 The white Cedar is a stately Tree. **1709** [see *pitch pine]. **1856** [see Cedar 3]. **1884** A. Nilson *Timber Trees New South Wales* 97 White Cedar.—An elegant tree. **1908** E. J. Banfield *Confessions of Beachcomber* i. i. 20 The white cedar..is a welcome and not unworthy substitute in appearance and perfume for English lilac. **1941** *Sun* (Baltimore) 12 Aug. 15/2 The hulls will be of two thicknesses, mahogany over white cedar. **1980** P. Moyes *Angel of Death* xi. 148 A fallen tree—a biggish white cedar with a trunk about a foot in diameter. **1861** *Q. Jrnl. Microsc. Sci.* I. 167 Colourless corpuscles, or white cells, exist in the blood in a comparatively small number. **1885, 1968** White cell [see *red cell 1]. **1480** White clay [see Clay *sb.* 1 a]. **1783** J. Wedgwood *Let.* 13 Oct. (1965) 272 Having seen a specimen of fine white clay..and being told it came from the Apalachian mountains..I was so delighted with the appearance of this beautiful raw material..that..I determined upon sending an agent to the spot. **1852** in *Proc. Amer. Antiquarian Soc.* (1933) XLIII. 373 Mr Nichols..has in contemplation the purchase of a tract of land containing a mine of white or China Clay. **1790** S. Deane *New-England Farmer* 58/2 Red and white clover are the only sorts known and esteemed in this country. **1884** F. J. Lloyd *Science of Agriculture* xv. 268 White or Dutch clover..is a well-known variety of good feeding quality. **1977** J. L. Harper *Population Biol. Plants* i. 25 The useful measure is the number of leaves, for example in..white clover. **1832** W. A. Ferris *Diary* 8 May in *Life in Rocky Mts.* (1940) xxv. 143 It is sometimes found in various parts of the country, and is sometimes called 'white earth'. **1910** A. P. Laurie *Materials of Painter's Craft* iv. 42 Then among the whites we have a large number of white earths, of which chalk is of course the most important. **1969** R. Mayer *Dict. Art Terms & Techniques* 430/1 Because of its clarity and high absorbency, white earth is well suited for, and in limited use as, a base for certain lakes. **1835** J. J. Audubon *Ornith. Biogr.* III. 137 [The Louisiana Heron] is at all seasons a social bird, moving about in company with the Blue Heron or the White Egret. **1872** E. Coues *Key to N. Amer. Birds* 267 Genus *Ardea* Linnæus... Great White Egret, White Heron... *Egretta*...Little White Egret, Snowy Heron... *Candidissima*. **1939** *Florida* (Federal Writers' Project) i. 26 The handsome white egret, once nearly extinct, is now protected. **1957** D. A. Bannerman *Birds Brit. Isles* VI. 68 The great white egret, though included in the genus *Egretta*, is in habit more like the grey heron and the purple heron and less like the egrets. **1770** White elm [in Dict.]. **1860** *Trans. Illinois Agric. Soc.* IV. 451 The White Elm..is not good timber—is hard to split. **1948** *N.W. Ohio Q.* Winter 10 It was made of a strip of white elm bark about one foot wide. **1981** *Sci. Amer.* Aug. 40/3 Most European elms, including the European white elm (*U. laevis*), the English elm (*U. procera*) and the various cultivars of the species *U. carpinifolia* are also susceptible.

1939 Stedman *Med. Dict.* (ed. 14) 1231/2 *White fingers*, an occupational disease occurring in operators of pneumatic hammers who are exposed to cold, affecting usually the fingers of the left hand. **1947** [see *pneumatic *a.* (*sb.*) 1 a]. **1971** *New Scientist* 15 Apr. 154/3 Researchers have found bone softening in chain saw operators and there is also the 'white fingers' complaint, with fingers going cold and numb. **1973** [see *Raynaud]. **1978** *Kingston* (Ontario) *Whig-Standard* 18 July 15/2 Regular users of chain saws, grinders, and pneumatic hammers, drills and chisels often develop a condition called 'white finger' disease. **1850** A. J. Allen *Ten Years in Oregon* v. 52 They found the red and white fir, spoken of by Clark and Lewis. **1897, 1913** [see *grand fir* s.v. *grand *a.* 12]. **1948** *Pacific Discovery* Mar. 7/2 Sequoias become established most easily..on cool north and east slopes with sugar pine and white fir. **1900** *Technical Ser. Div. Entomol., U.S. Dept. Agric.* No. 8 10 The minute 'white-flies'..may be flying around. **1925** A. D. Imms *Gen. Textbk. Entomol.* 360 The 'white flies' are a much neglected group. **1946** *Nature* 14 Dec. 852/1 A few [gall midges] are carnivorous, preying upon..white-flies, other gall midges and the like. **1981** *Farmstead Mag.* Winter 35/3 Mites, aphids and whiteflies sometimes bother greenhouse cukes, but don't let this panic you into using a chemical spray. **1696** in H. Kelsey *Papers* (1929) 54 They [*sc.* Indians] brought nothing but 2 white fox skins. **1774** [in Dict.]. **1862** *Canad. Naturalist* May 138 White foxes have been killed on the south shore of Great Slave Lake. **1926** *Daily Colonist* (Victoria, B.C.) 22 July 19/6 The average price for white fox was $34.85. **1930** R. W. Service *Coll. Verse* 269 Fur had they, white fox, marten, mink, to trade. **1969** *Beaver* (Winnipeg) Summer 10/2 The white fox is the principal source of fur income at Rankin [Inlet] as elsewhere. **1877** C. Hallock *Sportman's Gazetteer* 40 The White Goat is confined to the loftiest peaks of the Rocky Mountains. **1936** D. McCowan *Animals Canad. Rockies* xiv. 122 A full grown male White goat has a body length of about five feet, stands approximately forty inches high at the shoulder and is from two hundred to two hundred and fifty pounds in weight. **1846–7** Thoreau *Walden* (1957) 114 The red pine and the black ash, the white grape and the yellow violet. **1624**, etc. White heron [see Heron, hern 1 b]. **1813** A. Wilson *Amer. Ornithol.* VII. 106 The opportunities which I have ..had, of observing them with the train..from its first appearance to its full growth, satisfies me that the Great White Heron with, and that without the long plumes are one and the same species, in different periods of age. **1846**, etc. [see *kotuku]. **1917** *Auk* XXXIV. 86 The Great White Heron is of more social habits than the Blue Heron. **1939** *Florida* (Federal Writers' Project) i. 26 The great white heron, a Florida native, nests on the keys. **1957** D. A. Bannerman *Birds Brit. Isles* VI. 71 The great white heron is..very rare in North Africa in winter west of Egyptian territory. **1964** A. L. Thomson *New Dict. Birds* 367/2 The Great White Heron *A*[*rdea*] '*occidentalis*' of Florida..may be no more than a local population of a colour phase. **1966** *Encycl. N.Z.* I. 209/1 Swamp and lake-edge birds include the rare white heron.. (*Egretta alba*). **1966** P. Sherlock *West Indian Folk-Tales* 57 A flock of white herons flew across the river. **1774** in *Rep. Bd. Trustees Publ. Archives Nova Scotia* (1945) 34 This town..affords a great store of fine timber.. white and black ash; white mapple; rock mapple. **1832** D. J. Browne *Sylva Amer.* 101 The white maple puts forth green and yellow flowers early in the spring. **1916** E. T. Seton *Woodcraft Man.* 291 Silver Maple, White or Soft Maple..usually a little smaller than the Sugar Maple. **1981** *Publ. Amer. Dial. Soc.* lxvii. 37 Sugar maple... All of the Iron Range respondents [in Minnesota] use the generic *maple*... On the Mesabi single instances of *soft maple, sugar maple*, and *white maple* are recorded. **1610** *True Decl. Estate of Virginia* 55 There are innumerable White Mulberry trees. **1737** J. Wesley *Jrnl.* 2 Dec. (1910) I. 402 The white mulberry is not good to eat. **1850** [see Mulberry 1]. **1957** M. Hadfield *Brit. Trees* 248 The white mulberry..(so called from its white or pinkish, insipid fruit), is cultivated as the principal food plant of the caterpillar of the silkworm moth. **1975** E. Wigginton *Foxfire 3* 276 Dried white mulberries were used as a substitute for raisins or figs. **1634** White oak [see *red oak* s.v. *red *a.* 17 d]. **1770** [in Dict.]. **1873** 'Mark Twain' & Warner *Gilded Age* xvii. 163 You kin git all the rails you want outen my white-oak timber over thar. **1883** J. Macaulay *Grey Hawk* iii. 44 The banks on both sides are covered with poplar and white oak and other trees, which grow to a considerable size. **1930** W. Faulkner *As I lay Dying* 132 Tull take and cut them two big whiteoaks. **1941** *Sun* (Baltimore) 12 Aug. 17/2 They had in storage enough Dorchester county white oak to construct keels and frames. **1975** White oak [see *pin oak* s.v. *pin *sb.*[1] 18]. **1913** V. B. Lewes *Oil Fuel* 38 In some parts of the world small deposits of what are called 'white oil' are..found. **1919** *Electric Jrnl.* XVI. 336/2 Lectroseal transformer oil and white oil of paraffin have marked absorption. **1925** A. B. Thompson *Oil-Field Explor. & Development* I. xi. 504 The so-called 'white' oils occasionally encountered in oil-fields are usually transparent and amber or sherry tinted, and are evidently filtration products of darker varieties commonly found in the neighbourhood. **1938** F. M. Archibald in A. E. Dunstan et al. *Sci. of Petroleum* IV. 2838/1 Petrolatum liquidum is the highest grade of white oil. **1977** *Lubricants Business* (Shell Internat. Petroleum Co.) 4 Technical and medicinal white oils are also important. **1775** A. Burnaby *Trav. N. Amer.* 15 These waters are stored with incredible quantities of fish, such as sheeps-heads, rock-fish, drums, white perch. **1851** T. A. Burke *Polly Peablossom's Wedding* 129 The trout and white perch bit beautifully. **1949** *Sat. Even. Post* 12 Mar. 46/4 About the best fun was going out to the pond after white perch. **1849** H. Watts tr. *Gmelin's Hand-bk. Chem.* II. v. 107 Phosphorus, kept under water ... gradually becomes covered with an opaque crust which..afterwards turns white... This white phosphorus retains its original appearance when dried over oil of vitriol. **1865** *Chem. News* 24 Nov. 251/1 He establishes that white phosphorus is neither a hydrate nor an allotropic state of ordinary phosphorus,..but that it is, in fact, merely ordinary phosphorus irregularly corroded on the surface by the action of air dissolved in the water.

1884 Frankland & Japp *Inorg. Chem.* xxx. 371 Amorphous phosphorus, prepared by any of the above methods, invariably contains a small quantity of white phosphorus, the presence of which renders the product dangerously inflammable. **1976** *New Yorker* 15 Mar. 80/3 Two white-phosphorus rounds were exploded over the landing zone to indicate the 'all clear'. **1774** in L. Peyton *Adventures my Grandfather* (1867) 137 The forest of Kentucky consists of yellow and white poplar, black walnut, [etc.]. **1785** H. Marshall *Arbustrum Amer.* 78 *Liriodendron*..generally known by the name of Tulip-tree, or White and Yellow Poplar. **1908** C. Mair *Through Mackenzie Basin* 81 It was well timbered, with white birch and spruce, and white poplars between the belts of jackpine. **1825** *Phil. Mag.* LXV. 127 With common salt I obtained the same results, mercury remaining, and white precipitate being formed. **1880** *Encycl. Brit.* XI. 318 White precipitate..is used medicinally. **1923** J. W. Mellor *Compr. Treat. Inorg. & Theoret. Chem.* IV. xxxi. 786 Conversely, infusible white precipitate is converted back to fusible white precipitate by the action of a solution of ammonium chloride in liquid ammonia. **1956** J. Anderson tr. *Remy's Treat. Inorg. Chem.* II. ix. 474 The most important example of an ammonia addition compound is the 'fusible white precipitate', and important examples of mercury-substituted ammonia or ammonium derivatives are the 'infusible white precipitate' and 'Millon's base'. **1838** W. Cornwallis Harris *Wild Sports S. Afr.* xix. 184 A pair of white Rhinoceroses opposed our descent. **1941** J. S. Huxley *Uniqueness of Man* viii. 184 The numerous creatures which would have become extinct but for vigorous protection, such as the white rhinoceros. **1972** *Islander* (Victoria, B.C.) 9 July 2/3 In South Africa the white rhino has increased from 20 to 2,000. **1981** P. Turnbull *Deep & Crisp & Even* vii. 125 He stuck out like a white rhino at a tea party. **1932** *Proc. Roy. Soc.* A. CXXXVII. 132 Reports of the occurrence of zinc-coated products have been called by the industry 'White Rust'. **1976** A. R. L. Chivers in L. L. Shreir *Corrosion* (ed. 2) I. 4.63 'White rust'..is a voluminous white deposit. **1896** White sapphire: see Sapphire 1 b]. **1884** E. W. Streeter *Precious Stones & Gems* (ed. 4) iii. 160 The varieties of Precious Corundum are termed by jewellers the Oriental Sapphire, the Oriental Ruby, the Opalescent Ruby, the Star Ruby, the Green, the Yellow, and the White Sapphires, all of which are 'precious'. **1903** J. Spencer tr. *Bauer's Precious Stones* III. 266 Zircon and corundum ('white sapphire'). **1941** B. W. Anderson *Gem Testing for Jewellers* ix. 88 Perfectly colourless corundum, usually referred to as 'White Sapphire', is not common in nature... Synthetic corundum..is, however, manufactured on a large scale. **1744** W. Ellis *Mod. Husbandman* Jan. xii. 79 He lost several of his Flock by the Gripes and the White-scour. **1887** W. Housman *Cattle* vii. 231 Inflammation of the stomach and bowels [of calves]..is also commonly spoken of as 'white scour'. **1963** *Times* 17 May 3/7 [Advt.] Today we can protect your calves from pneumonia and white scour. **1980** *Chem. Abstr.* XCV. 3961 Eight categories of light, medium and heavy types of petroleum distillate (white spirit) were prepared and compared with turpentine as to boiling range. **1977** *Reader's Digest Bk. Do-It-Yourself Skills & Techniques* 11/2 For cleaning out oil-based paints, wash the brush in white spirit. **1791** C. Cartwright *Jrnl.* Aug. (1792) I. 30 About four miles above, are several small white whale were sporting in the river. **1834** [see Whale sb. 1 b]. **1923** *Beaver* Mar. 290/1 The white whales were as plentiful as ever. **1957** ... **1963** ... White whales were ... (1792) [etc.] ... a rock ... that's ... spruces. **1823** [see Spruce *sb.* 4 b]. **1963** K. Kroetsch *Alberta* vi. 164 Heavy stands of white spruce grow on the islands in Astotin Lake—the kind of spruce that covered much of the vicinity perhaps five hundred years ago. **1977** *New Yorker* 2 May 35/1 He thought that the white spruce and other species could live farther north. **1801** White steenbras [see Steenbras]. **1949** [see *mussel sb. 4]. **1959** [see *fool-hooked* s.v. *fool *adv.*] ... **1974** *Sunday Times* (Johannesburg) 6 Apr. 26/3 White steenbras ... Prevalent ... One of the best-known angling-fishes in Southern Africa. **1802** White tin [see Tin *sb.* 1 b]. **1962** A. Findlay tr. *Ostwald's Princ. Inorg. Chem.* xii. 720 Common or white tin, [etc.]. White tin ... its allotropic ... **1950** *Glasgow Herald* ... of the white tin ... **1950** ... 1960 ... White tin is ... **1704** *Dana's Syst. Min.* (ed. 7) I. 127 Solid white tin (β-tin) by contact with grey tin (α-tin), alters to a grey powder ('tin plague'). **1950, 1965** [see *grey tin* s.v. *grey, gray *a.* 8 c]. **1973** J. J. Lagowski *Mod. Inorg. Chem.* xi. 334 (*caption*) Each atom in white tin is surrounded by six other atoms arranged in a distorted octahedral structure. **1743** J. Clayton *Flora Virginica* 190 *Juglans alba*... White Walnuts. **1822** J. Woods *Two Years' Residence Eng. Prairie* 228 White-walnut, or butter-nut, and black-walnut, are not so good as the English walnut. **1916** E. T. Seton *Woodcraft Man.* 275 White Walnut, Oil Nut, or Butternut..rarely 100 feet high. **1958** G. A. Petrides *Field Guide Trees & Shrubs* 86 Butternut.. also known as White Walnut, wood lighter in colour than that of its more valuable relative. **1969** *Beaver* ... 6–7 July in *Papers* (1929) 88, 2 hands..brought news of a white whale drove ashore. **1834** [see Whale *sb.* 1 b]. **1923** *Beaver* June 340/2 Indian reports were received that porpoises, or white whales, were..making excursions..up a certain creek. **1978** *Weekend Mag.* (Toronto) 22 July 16/1 White whales or belugas..become progressively lighter with age. **1985** *Times* 8 Aug. 15/8 Soviet seamen trying to save a large pod of white whales..trapped in the ice.

e. **White Africa,** the white inhabitants of Africa; the parts of Africa ruled by white people; **White Army,** any of the armies which opposed the Bolsheviks during the Russian Civil War (1918–21); also, a group which opposed the Red Guards in Finland in 1918; **White Australia,** used *attrib.* and *absol.* to designate a policy of restricting immigration into Australia to white people; **white backlash,** resentment felt by white people against demands made by, or concessions made to, black people; hence **white backlasher; white book:** *spec.* (with capital initials) the book first published in 1882 as *Annual Chancery Practice* and now entitled *Supreme Court Practice*; (further examples); **white cane** = *white stick 3; **white chauvinism** orig. *U.S.*, a white person's excessively high regard for his own race; so **white chauvinist; white Christmas,** a snowy Christmas; **white coal:** more commonly, flowing water as a source of energy; also, electricity; cf. *white fuel* (*a*) below; (earlier and later examples); **white dominion,** a dominion (sense *2 b) in which the majority of the inhabitants are white; **white dwarf** *Astr.*, a small, faint, very dense star (usu. but not necessarily white in colour) lying below the main sequence, and representing the stable phase assumed by stars having less than 1·4 solar masses when their nuclear reactions cease; (not regarded as a type of dwarf: see quot. 1978 and cf. *dwarf *sb.* 2 b); **white elephant:** see Elephant 2 in Dict. and Suppl.; **white embroidery,** white-thread embroidery on a white ground; = *white work* below; **White English,** term occas. used in contrast to Black English, in the sense 'the English of white speakers'; **White Father,** (*a*) a white man regarded as protecting or controlling people of another race; (*b*) [tr. F. *Père Blanc*], a member of the Society of Missionaries of Africa, a Roman Catholic order founded in Algiers in 1868; **whiteflow** (earlier examples); **white flag,** (*c*) used *attrib.* (with capital initials) to designate a communist group active in Burma since 1946; **white flight** chiefly *U.S.*, the migration of white people from inner-city areas (esp. those with a large black population) to the suburbs; **white folk(s), white-folk(s),** applied by U.S. Blacks to white people; **white frost,** hoar-frost; **white fuel,** (*a*) flowing water as a source of power; cf. *white coal* above; (*b*) lead-free petrol; **white goods,** (*a*) domestic linens, as sheets, towels, etc. (now not necessarily white); (*b*) electrical goods that are conventionally white, such as washing machines and refrigerators; **White Guard, Guardist,** (*a*) a member of a force which fought for the Finnish government against left-wing insurgents in the civil war of 1918; (*b*) a member of a counter-revolutionary force in Russia during the civil war of 1918–21; also *transf.*; **White Highlands,** an area in Western Kenya formerly (1904–59) reserved for Europeans; hence **White Highlander; white hole** *Astr.* [opp. *black hole*], a celestial object which expands outwards from a space–time singularity emitting energy, in the manner of a time-reversed black hole; **white hope,** orig., a white boxer who might beat Jack Johnson, the first Black to be world heavyweight champion (1908–15); hence, a person who, or a thing which, it is hoped will achieve much or on whom or which hopes are centred; **White Hun,** a member of a nomadic people of uncertain origin, also called Ephthalites or Hephthalites, who lived in Bactria in the fifth and sixth centuries A.D.; **white hunter,** a white man who hunts big game professionally; **white jazz,** jazz as played by white musicians; **white knight,** (*a*) (with allusion to a character in *Through the Looking-Glass*), an enthusiastic but ineffectual person; (*b*) a hero or champion; *spec.* (*Stock Exchange slang*) a company that comes to the aid of one facing an unwelcome take-over bid; **White Lady,** (*a*) a cocktail made of two parts of dry gin, one of orange liqueur, and one of lemon juice; (*b*) *Austral. slang*, a drink of methylated spirits, sometimes mixed with another ingredient; **white land** *slang*, open land that is not designated for development or change of use, or on which development is not allowed (so called from its being uncoloured on planning maps); **white level** *Television*, the signal level corresponding to the maximum brightness in transmitted pictures; **white lightning** *slang* (orig. *U.S.*), (*a*) inferior or illicitly distilled whisky; (*b*) a kind of LSD; **white list** *colloq.* [after Black list in Dict. and Suppl.], a list of people or things considered acceptable; **white market** [after

*black market], authorized dealing in things that are rationed or of which the supply is otherwise restricted; **white meter,** a meter that registers off-peak consumption of electricity; **white mule** *U.S. slang*, a potent colourless alcoholic drink; *spec.* illicitly distilled whisky; **white Negro,** (*a*) a Negro, or a person with Negro ancestry, who has a pale or albino complexion; (*b*) a white person who defends the rights or interests of Negroes, or identifies with them; (*c*) *nonce-use* (see quot. 1949); **white nigger** *slang* (chiefly *U.S.*), (*a*) a derogatory term for a white person who does menial labour; (*b*) a Negro who is regarded as deferring to white people or accepting a role prescribed by them; (*c*) (see quot. 1970); **white night,** (*b*) a night when it is never properly dark, as in high latitudes in summer; **white note,** (*b*) a note corresponding to a white key on a keyboard; = Natural *sb.* 7 a; **white paper,** (*c*) *spec.* (with capital initials) before 1940, an Order Paper of the House of Commons which was a corrected and revised version of one issued earlier the same day (a Blue Paper); (*d*) (with capital initials) a government publication presented to Parliament and having white covers rather than blue ones (usu. less bulky than those with blue covers); *esp.* one outlining proposed legislation or stating policy; also *transf.*; *White Paper candidate* Naval slang (see quot. 1962); **white plague,** tuberculosis; cf. **white death,** *gen.* in Dict.; **white port,** port wine made from white grapes; **whitepox** [see quot. 1972], epithet of a pox-virus isolated from monkeys that is very similar to the smallpox virus; **white-print,** a document printed in white on a dark ground; **white rabbit, White Rabbit,** used with allusion to the White Rabbit in Lewis Carroll's *Alice's Adventures in Wonderland* (1865), who was running because he was in danger of being late; also as *adj.* and *adv. phr.*; **white racism,** belief in the superiority of the white race, leading to antagonism towards people of other races; hence **white racist** *a.* and *sb.*; **White Rajah,** any of the three Rajahs belonging to the English family of Brooke who ruled Sarawak between 1841 and 1941; also *transf.*; **white ribboner,** one who wears a white ribbon as a badge of temperance; a teetotaller; **white room,** a clean and dust-free room used for the assembly, repair, or storage of spacecraft or delicate mechanisms; **white rot,** any of several fungal diseases of wood or living plants indicated by white patches of decay or mould (see also Rot *sb.*[1] 2 c); **white rum,** a colourless variety of rum; **white rust,** a fungus disease of certain plants indicated by white blisters on leaves or stems, esp. one caused by *Albugo candida* affecting cruciferous plants or one caused by *Puccinia horiana* affecting chrysanthemums (see also sense 11 c above); **white sale,** a shop sale of white goods and household linen; **white settler,** (*a*) a white inhabitant of a non-white territory; (*b*) *transf.* (see quot. 1976); **White Sister,** a nun wearing a white habit; *spec.* a member of the Congregation of the Missionary Sisters of Our Lady of Africa, founded in 1869 to assist the White Fathers, or of the Congregation of the Daughters of the Holy Ghost, founded in 1706 in Brittany; **white slave** (earlier and later examples); *spec.* a prostitute, esp. one trapped into prostitution by others; also (with hyphen) as *v. trans.*, to sell or trap (a girl) into enforced prostitution, esp. abroad; **white-slaving** *vbl. sb.*; **white soup,** soup made with white stock; **white stock,** stock made with chicken, veal, or pork; **white stone, whitestone,** (*b*) a colourless gemstone; (*c*) a form of rendering; **whitestone** *v. trans.*, to whiten with stone (cf. Hearthstone *sb.* and *v.*); **white stuff** *slang* (chiefly *U.S.*), morphine, heroin, or cocaine (cf. *white *sb.* 15 e); **white supremacism,** a doctrine or the practice of white supremacy; hence **white supremacist** *sb.* and *a.*; **white supremacy,** domination by white races over non-white, esp. black, races; **white supremacist** *rare* = *white supremacist sb.* above; **White Terror** (see Terror *sb.* 5; further examples); also *spec.* a similar period in Hungary in 1919–20 and in China in the years following 1927; **white tie,** *spec.* a man's white bow-tie worn with a black tailcoat; also *ellipt.*, a man's formal evening dress including a white tie; *freq. attrib.*; **white trash:** see Trash *sb.*[1] 4 in Dict. and Suppl.; **white war,** war without bloodshed; economic warfare; **white ware** (earlier example of sense 'white earthenware'); **white way** *U.S.* (usu. with capital initials), a brilliantly lit city street; *spec.* (usu. *Great White Way*), the part of Broadway either side of Times Square, the heart of New York's theatre district, or a similar street in any other town; **whitewear** = *white goods* (*a*) above; **white wedding,** a wedding at which the bride wears a traditional white dress; **white whisky,** colourless whisky; *spec.* (*N. Amer.*) home-made or illicit whisky; **white work,** embroidery worked in white thread on a white ground.

1920 J. Buchan *Prester John* xxii. 353 The amnesty came..and white Africa drew breath again. **1974** A. Williams *Gentleman Traitor* i. 16 The armies and police forces of White Africa. **1918** White Army [in Dict., sense 6 b]. **1960** C. Mackenzie *Great Fortune* ii. 109 I make you look like a White Army officer. **1977** J. Cleary *High Road* iii. 95 He had come out of Russia, a cavalry commander in one of the White Armies. **1921** *White Australia* (in Dict., sense 4 a). **1930** W. K. Hancock *Australia* iv. 77 The policy of White Australia is the indispensable condition of every other Australian policy. **1979** *Guardian* 3 Jan. 7/2 Mr Gough Whitlam's Labour Government abolished the 'white Australia' policy five years ago. **1964** *Courier-Mail* (Brisbane) 29 July 2 Gold water is no racist, but there's little doubt that his support prefers to win votes from the 'white backlash', the so-far unmeasured resentment among white voters to some of the negro demonstrations and riots. **1974** *Spartanburg* (S. Carolina) *Herald* 25 Apr. 12/1 He said many white backlash had developed against aboriginal advancement programs. **1966** *Economist* 17 Sept. 1130/2 The result leaves 'white backlashers' little choice in November.. Mr Peabody, a staunch liberal..will oppose a Negro, Mr Edward Brooke, who has won the Republican senatorial nomination. **1968** *Listener* 7 Nov. 623/1 The spies converge on Shaefer, and the homely white-backlashers already lay their fan. **1965** J. Dennis *Young Man's Guide to Law* xiii. 150 Great industry has to be employed with the practice books—'the White Book' (the High Court practice) and the 'Green Book' (the County Court practice). **1981** I. H. Jacob *Supreme Court Pract.* i. p. vii, It may fairly be claimed that the year 1982 is the hundredth anniversary of the White Book. **1927** *Times* 8 June 7/5 (Advt.), Nowadays it takes more than a white cane to help blind people. **1980** D. Mackenzie *Raven & Paperhanger* vi. 85 There's a special place for blind men. And you get a white cane. **1940** *Political Affairs* XXV. 935/1 The corrupting influence of white chauvinism has operated to maintain the most harmful division in the ranks of American labor. **1951** W. Z. Foster *Outl. Polit. Hist. U.S.* xxxiv. 563 White chauvinism—race hatred—has been, and still is, a question of hard cash to the big capitalists and landowners of the United States. *Ibid.*, Much of the race prejudice that does exist among the Latin American peoples is due to the corrupting attitudes of white chauvinists (diplomats, tourists, and businessmen) from the United States. **1984** *Washington Post* 26 Feb. 10/1 White chauvinism in 1982 writing has in large part replaced the tentative thrust toward 'ethnomusical' and sociology made during the 1960s to 1970s. **1857** C. Kingsley *Two Years Ago* III. ii. 305 We shall have a White Christmas, I expect. Snow's coming. **1913** *Collier's* 13 Dec. 8 (*heading*) A White Christmas. **1942** I. Berlin (*song-title*) White Christmas. **1976** *Weekend Echo* (Liverpool) 24 Dec. 1/2 The weather men say the big shiver could bring our first white Christmas for ten years. **1885** *Nepean* (Manitoba) *Star* 11 Aug. 2/1 ... we should those interested with the people's money ... embezzle... the least portion of that money, under colour of black coal or white coal, ditch contracts, or any other pretext. **1963** *Daily Tel.* 18 Sept. 18 All may not think electricity the best heating or cooking or even lighting agent. But it is the cleanest and simplest and deserves its nick-name of 'white coal'. **1971** *Nat. Geographic* July 35/2 Many former waterfalls now slip submissively through penstocks and turbines, and this abundant 'white coal' has drastically altered the country's age-old fish-forest-and-farm economy. **1966** *Guardian* 6 Sept. 8/4 Assuming that the crumbling process would continue, Britain would be left with the 'white dominions'. **1973** C. Carrington in Kipling *Compl. Barrack-Room Ballads* 48 After..1871, there were no British regular troops in the new 'White Dominions'. **1917** H. Wilson *Strange Ride R. Kipling* v. 233 Canada was the white dominion that Kipling had known longest. **1924** *Monthly Notices R. Astron. Soc.* LXXXIV. 320 The white dwarfs Sirius (comes) and O₂ Eridani. **1925** *Nature* 5 Dec. 834/1 Invoked to decide the truth of a suspicion of transcendently high density in the 'white dwarf' stars, it (sc. Einstein's theory of gravitation) has decided that in the companion of Sirius matter is compressed to the almost incredible density of a ton to the cubic inch. **1935** B. Russell *Relig. & Sci.* viii. 217 If some of these things happen first, we shall in any case be all destroyed when the sun explodes and becomes a cold white dwarf. **1969** *Listener* 2 Jan. 20/3 White dwarfs are stars like the Sun which have collapsed into a sphere the size of a planet. **1978** Pasachoff & Kutner *University Astron.* x. 283 Do not confuse the term 'white dwarf' with the term 'dwarf'. The former refers to the dead state, while the latter refers to normal stars on the main sequence. **1876** Geo. Eliot *Dan. Der.* II. iv. xxxi. 223 Gwendolen..had a piece of white embroidery which on examination would have shown many false stitches. **1931** A. N. Arthur *Embroidery Bk.* viii. 83 Bullion knots are frequently used in white embroidery. **1971** S.

Levey *Discovering Embroidery of Nineteenth Cent.* 9 White embroidery flourished during the early part of the century. **1974** *Florida FL Reporter* XII. 3/3 Black English, a language on its own right..is as entirely, if not entirely, rooted in *White English* dialect usage. **1974** *Newsl. Amer. Dial. Soc.* Nov. 44 Intonation patterns of Black English were studied and compared with those occurring in White English and formal Black English. **1835** C. F. Hoffman *Winter in West* I. 251 The unfortunate agent..was sent off at the approach of the Indians as their friend and 'father',—the reply being,—'that the White Father would ..' **1886** B. C. Cockes *Cardinal Lavigerie* i. 5 The White Fathers—a name given to the Algerian missionaries on account of their wearing the long white robe of the Arab. **1894** *Harper's Mag.* Sept. 516/2 The White Father has sent me. **1976** *Telegraph* (Brisbane) 18 Sept. 2/2 The people are decried as are the 'White Fathers'—those who control our destiny. **1977** R. Lucas *tr. De Foucauld's Lett. from Desert* vii. 130 The Apostolic Prefect will probably tell him to spend a few days at Maison-Carrée, near Algiers, the mother-house of the White Fathers. **1832** Whiteflow [see *buddere *a.*] **1853** C. B. Hall in T. P. Hughes *Left Wall, Pioneers* (1895) 218 My back boy..showed me three or four bodies, partially concealed by logs. There were numerous tracks of horses round about. He explained the occurrence in his way—'I believe blackfellow bimbulee about that.' Then whitefellow gilbert and put 'em along o' fire.' **1949** *New Statesman* 12 Feb. 147/1 The story begins last March. Then the Communists (White Flag and Red Flag), who had already gone underground, began guerilla warfare on Government treasuries. **1959** *Listener* 18 June 1051/1 The so-called White Flag Communists who followed the Stalinist line. **1969** *Far Eastern Econ. Rev.* 10 Oct. 108/3 [see *Hin *a.* 3a]. **1969** *New Republic* 22 July 19/2 School quality is a far more important factor than racial feeling in this white flight from desegregated schools. **1975** *Political Sci. Q.* 675/1 White flight from cities has been a much studied phenomenon in the last decade. **1978** *Sci. News* 23 Sept. 216 Previous studies of this so-called 'white flight' phenomenon have been criticized for not taking into account the type of desegregation involved and for ignoring other factors that might have induced white families to leave the central city anyway. **1929** W. Faulkner *Sound & Fury* 107 If it hadn't been for [Negroes] we'd have to work like whitefolks. **1932** E. Caldwell *Tobacco Road* xv. 179 What's the matter with your automobile, white-folks? **1973** *Black World* May 20 [in his essays on whitefolk, Du Bois shows] two specific historical occurrences which reflect the paradoxes, lies and hypocrisy of white civilization. **1981** A. Mackay *Death on River* (1982) 120 She dressed conservatively. White folks' clothes, she thought wryly. **1582, 1655** [in Dict.]. **1740** (*see* Hoar-frost]. **1975** W. Fleming *Snal* 14 Mar. In N. D. Merenus *Trav. Amer. Colonies* (1916) 634 Monday night there was a small white frost. **1835** J. Martin *Comprehensive Descr. Virginia* 66 Our white frosts are gentle and sharp, in being a simple dew slightly congealed. **1913** F. Soddy *Matter & Energy* v. 135 The 'white fuel' of the Norwegian hill-sides. **1983** *Daily Tel.* 27 Nov. 20/7 Italy has..greater advantages for the development of 'white fuel' by more than any single river. **1968** *New Scientist* 6 Feb. 19 When the catalyst is used with 'white' or unleaded fuel (as on motorised trucks for indoor use in factories) this difficulty does not arise. **1920** Y. Laton & G. *Caks* White Goods & Muslinunder Sale 12 These prices for Shirt Waists and Wrappers are special for the White Goods Sale only. **1948** L. J. Wilder *These Happy Golden Years* xxxi. 276 Mary brought nothing the white goods, Ma and Laura discussed Laura's dresses. **1966** *Economist* 8 Oct. 158/1 Refrigerators, deep freezers, washing machines, clothes dryers, and other so-called 'white goods'. **1976** *Which?* Mar. 61/1 Electrical equipment..includes things like washing machines and fridges (what the trade calls white goods) as well as TVs and audio shops (the trade calls brown goods). Specialist TV and audio shops don't normally sell white goods. **1981** *Times* 9 Mar. 19/1 An abiding problem for the white goods manufacturers is the high level of imports. **1922** White Guard [see *Red Guard 1 b]. **1931** G. Hicks *Sm. Marx Gor* v. 67 ... I. L. Horowitz *Masses in Lat. Amer.* xiii. 454 ... labour leaders on the spot and were ... in charge of ... the 'white guards'. **1958** ... *Ibid.* 497 'White guards'..were groups of armed men hired by the landowners ... **1962** Trevelyan *Worlds Apart* xxiii. 267 The house was lucky to escape destruction during the Revolution, when it was said to have been for a time in the front line as a White Guard post opposite the Kremlin. **1974** J. White tr. *Poulantzas's Fascism & Dictatorship* iv. iii. 210 The fascist parties were ... closely identified with the Russian White Guards, ... **1931** in J. Degras *Soviet Documents on Foreign Policy* (1952) II. ... A White Terror by the White Guardists... **1963** ... 'but what is he? A reactionary? A White Guardist?' **1971** S. Talbott tr. *Khrushchev Remembers* 1. 15 Our army won many important victories over the White-Guardist ... when the White Guards returned from Mt Kenya. ... **1976** P. Driscoll *Barboza Credentials* i. 1. 24 I had seen Kenya..proper in spite of the White Highlands. **1955** E. Huxley *White Man's Country* I. ix. 208 In East Africa the settlers' principal anxiety was that Indians would permeate the relatively small area of land suitable for colonisation—'the white highlands'. **1957** W. M. Hailey *Afr. Survey* (rev. ed.) xi. 729 The reservation of the White Highlands for Europeans prevented the presence of Africans on farms in areas which the Europeans would normally have found relief from congestion. **1978** S. Naipaul *North of South* i. iv. 114 Soil erosion had been one of the great settler obsessions. The battle against it had become part and parcel of the battle for civilization, providing a powerful argument for the preservation of the status quo in the 'White' Highlands. **1971** *Nature Physical Sci.* 3 May 20/1 Black holes..are related in a genitive manner to 'white holes', believed to be singularities from which matter and energy emerge. **1977** *N.Y. Rev. Bks.* 29 Sept. 22/2 There

is speculation..that every black hole is joined to a 'white hole'—a hole that gushes energy instead of absorbing it. **1911** *Daily Colonist* (Victoria, B.C.) 28 Apr. 12/4 A New York promoter has succeeded in arranging for a contest between Albert Palmer, New York's most prominent white hope, and Carl Morris, the giant locomotive engineer. **1912** I. S. Cobb *Back Home* 233 Judge Priest was a celebrity, holding the limelight to the virtual exclusion of grand opera stars, favourite sons, white hopes, [and] defaulters. **1929** *Observer* 10 Nov. 13/6 In the south, bounded on the Black Sea and liberally furnished with British material, Denikin and his Cossacks were the 'white hope' of the anti-Bolshevists. **1941** L. Berkins *Up from Mauding War* iii. 20 We was a hell of a lot better when they never came up, Joe Louis.. was a champion of white hope (thus a critic had described him) of English music. **1948** *Times* 13 July 40/2 Ibid of the Negro race, and to explore the white hope that the old cry for a 'white hope' never came up, Joe Louis.. was a champion white hope. **1952** D. McCardle, *Allingham Tiger in Smoke* iv. 81 Detective Coleman had been one of Luke's white hopes. **1960** *Daily Tel.* 6 Oct. 12 In the immediate post-war years cheap and almost limitless atomic power was the white hope of a small island sadly short of raw materials. **1979** *Nature* 23 Aug. 658/1 Interferon is the great white hope of cancer therapy. **1787** Gibbon *Decl. & F.* III. xxvi. 584 The White Huns, a name which they derived from the change of their complexions. **1966** H. Yule *Cathay & Way Thither* i. p. lv, The Yeuchi..who became known to the West as the Indoscythians, and at a later date as the White Huns. **1965** G. Wheeler *Soviet Central Asia* i. 5 In the fifth century Southern Turkestan was conquered by the Ephthalites or White Huns. **1945** N. Mitford *Pursuit of Love* xi. 90 She's happier now, isn't she, with her white hunter? **1964** D. Varaday *Gara-Yaka* xiv. 214 Two white hunters lay in wait, and each shot one of the pride. **1980** G. M. Fraser *Mr American* xxvi. 546 He's an elephant hunter for trade..a white hunter, a sport with. **1933** (*word-title*) White jazz. **1946** R. Blesh *Shining Trumpets* 1. 23 No better-disguised white jazz except a chaotic pastiche of Chicago-style jazz. **1950** [see *schmaltz 2]. **1976** J. Berendt *Jazz Lab* xi. There seem to have been white musicians almost from the start. 'Papa' Jack Laine led bands in New Orleans from 1891. He is known as the 'father of white jazz'. **1895** M. Kingsley *Jrnl.* 23 May in *Tran.* (1897) vi. 110 The chief..bows with a jerk that causes the populace to shout, and the fine old man from Gaboon was as much to me as ever had been the White Knight in 'Alice in Wonderland'. **1956** N. Marsh *Off with his Head* (1957) ii. 41 'I believe I have made a really significant discovery.' Oh, Lord! thought the White Knight. **1968** *Kip* in A. Pixie in *Mer. Capital* 8 Oct. 1 ... The White Knight of the group ... **1976** J. Phillips *Backlash* (1977) iii. ii. 230 Woody would like it to be thought that he was the white knight to my damsel in distress. **1979** *N.Y. Times Mag.* 30 Sept. 24/4 The Kangers' problems stemmed from the habit that.. the team's general manager..had of hiring intellectual cronies to call the plays, and replacing them with himself when they failed—a kind of 'white knight' complex. **1982** *Guardian* 30 Oct. 15/1 Thomas Tilling emerged yesterday as the White Knight appointed by Sirec to save the Ever Ready battery maker from the unwelcome attentions of Hanson Trust. **1930** H. Craddock *Savoy Cocktail Bk.* 175 White Lady Cocktail. ¼ Lemon Juice, ¼ Cointreau, ½ Dry Gin. **1939** Stevens. N. Tennant *Tiburon* 19 I've cooked over and fought men in the corner lying stupefied over a mixture of 'white lady'—boiled methylated spirit with a dash of boot polish and iodine. **1953** D. Hamilton *So Sad, so Fresh* xviii. 117 He indicated a cocktail cabinet..and produced the two 'White Ladies'. **1964** *Telegraph* (Brisbane) 24 Sept. 5/2 Aborigines..used to swill cheap 'white lady'..and other concoctions in 'White Lady'—a fiendish brew of methylated spirit and powdered milk. **1970** *Bulletin* (Sydney) 7 Sept. 20 Sunday night is the White Lady at the Binh, mixin' it with Coke.' ... I. Jarvis ... Fairer than ... **1966** *Guardian* 14 July 8/5 How much 'white land' the planning authorities have left between the limits of development shown on the town map and the beginning of green belt. **1971** P. Gresswell *Environment* 229 Open country with land uses ..not dealt with in the written statement or shown on the map is designated as white land. **1976** *Times* 23 Feb. 24/8 White land is land that is, and will remain, in established agricultural use. **1949** *Electronic Engin.* XXI. 50 The white level ... 1962 ... Maximum white level shall be 15 per cent or less of the peak carrier amplitude. **1953** Amos & Birkinshaw *Television Engin.* I. i. 17 White level may be positive or negative with respect to black level. **1963** J. Gomersall *Fund. Television Servicing* i. 5 Television standard standard expresses the percentage of video signal in terms of percentage of white level. **1921** *Double Dealer* July 20/2 The men lean or sit on the counter and talk politics, fish and wages, or more enthusiastically, the devastating and soul-searing qualities of the current 'white lightning', 'white mule', or just plain 'corn', as the local moonshine whiskey is called. **1940** C. McCullers *Heart is Lonely Hunter* ii. iv. 119 He had a pint of bootleg white lightning. **1969** *Times* 9 Dec. (Taiwan Suppl.) p. ii/3 The distillery's main product is kaoliang, a potent liquor made out of Quemoy-grown sorghum and known as 'White Lightning'. **1972** *Village Voice* (N.Y.) 1 June 77/3 Ellen..unfolded some foil and said she contained three tabs of Owsley's original 'white lightning', the Moulton-Rothschild of LSD. **1975** B. Garfield *Hopscotch* xii. 128 It was white lightning country and the backhill bootleggers were numerous. **1979** R. L. Simon *Peking Duck* ii. 57 ... 'white list' is the final stroke—the white flag held up to Liberalism at the moment when its army is on the verge of victory over it. **1939** *Country Life* 11 Feb. p. xxii/1 Orb..[ ] Furs.—Avoid those tortured to death. Buy only those named on the Fur Crusade White List. **1977** *Lancet*

30 Apr. 963/1 One idea is a 'white list' of preferred drugs or a list of excluded drugs for which the N.H.S. would not expect to pay. **1980** *Listener* 23 Oct. 553/1 She was now bopping legally and mournfully on the white market. **1968** P. Oliver *Screening Blues* vi. 181 There appears to have been no discrimination, and no indication in..the catalogues, nor any apparent attempt to secure a white market for Negro records of this [sc. pornographic] character. **1973** *Times* 28 Dec. 1/4 A feature of the system would be a white meter. **1978** ... 1963 *Times* 1 Oct. 9/3 It is connected to a separate wiring circuit and an offpeak or white meter. **1974** *Ecologist* Oct. 299/1 Measures such as low-tariff electricity ('White Meter') in off-peak periods only, have already made consumers prepared to group their demand for an intermittent supply. **1889** H. H. McConnell *Five Years a Cavalryman* 60 About this time I first became acquainted with 'white mule', ... 'white mule whiskey'. **1928** *Collier's* 29 Dec. 8/1 What do you think about a bunch of boys and girls..stealin' a keg of white mule from a dealer? **1942** W. Faulkner *Go down, Moses* 106 Gets himself a whole gallon of good corn whiskey, ... white-mule whisky. **1973** *Globe & Mail* (Toronto) 23 Feb. 5/8 At other times..the stuff would..be white mule... There is..a little of the old wine-maker like the whiteskin mule..once so respected in the Ozarks. **1765** *Phil. Trans.* ... *R. Soc.* LV. 45 (*heading*) An Account of a White Negro shewn before the Royal Society. **1790** W. Wilberforce *Jrnl.* 6 Apr. in R. I. & S. Wilberforce *Life W. Wilberforce* (1838) I. vi. 264 Hard at work on Slave Trade evidence all day with ... Dickson. **1824** J. Goodridge *Notes Settlement & Indian Wars W. Parts Virginia & Pennsylvania* 33 ... 1838 R. I. & S. Wilberforce *Life W. Wilberforce* I. vi. 95 Messrs. Clarkson, Dickson, &c. [are] ... **1958** ... 'white negroes'. **1928** 'M. Tensas' *Odd Leaves Life Louisiana 'Swamp Doctor'* ... **1949** Koestler *Promise & Fulfilment* i. vii. 69 The Jewish Defense organization (like) ... a white Negro, ... **1957** N. Mailer in *Dissent* IV. 279 The hipster had absorbed the existentialist synapses of the Negro, and for practical purposes could be considered a white Negro. **1960** G. E. Wilson *John Clarkson* ix. 51 Both Clarksons were counted among the activists whom Pitt in a rare jest called the 'white Negroes'. **1839** R. M. Bird *Nick of Woods* I. 170 ... 'white nigger'... **1950** ... **1863** ... F. Douglass *Reconstruction* ... **1964** ... 'The white nigger' ... **1972** ... white niggers ... **1977** ... the 'white nigger' ... **1970** ... **1911** ... white nights ... **1955** ... White notes ... **1924** ... White Paper ... **1936** ... Hansard ... White Paper ... **1956** ... **1962** ... Naval slang ... White Paper candidate, candidate for a temporary (wartime) commission in the RNVR. The White Paper was passed in Parliament for a scheme of promotion for suitable ratings who had served on the lower deck for at least three months in a ship at sea in time of war. **1967** *Listener* 8 June 739/2, I first hastened to read President Nyerere's 'White Paper' on education. **1971** H. Wilson *Labour Govt. 1964–70* xiii. 238 We published the White Paper. **1950** Kerr & James *Wavy Navy* 235 These ratings had been recommended for commissions as 'White Paper' candidates by their Commanding Officers and recommended to the Admiralty. **1955** *Times* 16 June 8/5 The text of the two agreements would be published in the White Paper and laid before Parliament as White Papers. **1962** Granville *Dict. Sailors' Slang* 132/2 *White Paper candidate*, candidate for a temporary (wartime) commission in the RNVR. ... **1909** ... *Daily Colonist* ... white plague ... **1966** ... *Dædalus* ... **1976** ... white plague ... **1909** ... **1913** ... the white port ... **1967** ... **1972** ... whitepox ... ... **1934** ... Margaret ... *Spell of London* 88 ... **1961** L. Mumford *City in Hist.* xvii. 547 As early as the eighteenth century..London had moved..into ... horrible white plague. **1723** J. Nott *Cook's & Confectioner's Dict.* sig. Pp. 7v, Two Gallons of White-port Wine. **1842** Mew & Ashton

Drinks of World 100 of White Ports the best are *Muscatel de Jesus*.. and the *Lachryma Christi*. **1920** G. Saintsbury *Notes on Cellar-Bk.* iii. 31 White Port.. I think nearly deserving of the curse above pronounced on sparkling claret. **1978** M. Walker *Infiltrator* xii. 144, I was given a glass of white port. **1972** Gispen & Brand-Saathof in *Bull. World Health Org.* XLVI. 591/1 The parental Copenhagen virus..came..to be called the white-pox pock forming virus particles by mutation. *Ibid.*, The occurrence of white pox viruses in healthy monkeys cannot be explained by the instability noted above. **1977** *Brit. Med. Jrnl.* 26 Feb. 530/1 Viruses are yet indistinguishable from variola virus were isolated from the kidneys of six healthy monkeys and two rodents; these have been termed 'whitepox' viruses. **1979** *Nature* 24 May 295/2 Vaccinia, cowpox and camelpox viruses lack continuous transmissibility in man, but the situation with monkeypox virus and whitepox viruses requires further comment. **1959** H. Levinson *White Cipher* 84 He memorized the details like a draughtsman reading a white-print. **1967** *White-print* [see *Ozalid]. **1930** R. Lehmann *Note in Music* ix. 154 'I must hurry, I must hurry,' she said..like the White Rabbit,..and without another thought. **1979** S. Brett *Comedian Dies* v. 52 Her pretty little face looked anxious... 'Oh, um. If you'll excuse me, I've got to..' 'You said, white-rabbiting away. J. Elliott *Cowards of her Dreams* iii. 144 Off he went, scuttling..., White Rabbit late for a date. **1970** *Rep. 20th Ann. Round Table Meeting Linguistics & Lang. Stud.* 223 Because most middle-class whites feel that Negro dialect is inherently 'bad' and that its speakers are somehow beneath them, they react negatively to it.. White racism and racist-exploitation of Negroes. **1973** *Black Panther* 17 Mar. 6/3 No charges were pressed against any of the White racists..pointed ... ... cops who were clearly racist ... because the Navy felt that they 'they' (Blacks) had been in the wrong! **1977** M. Walker *National Front* ii. 14 White racism... the ... 1981 ... A history of Sarawak under its two White Rajahs. **1962** *New Statesman* 1 July 2/1 ... reminds us that the Malay ... **1974** *Radio Times* 19 Mar. 37/3 The story of the last White Rajah of Sarawak. **1887** *Fig. Ma Ind.* ... **1929** 'O. Henry' *Trimmed Lamp* ... 'The demon rum'.. is the white ribboners miscall whiskey. **1947** *Daily Tel.* 3 July 11/7 The National British Women's Total Abstinence Union, which has 6,000 members, still wears a white ribbon bow in the form of a badge to its followers, who are known as the 'white ribboners'. **1961** *Aeroplane & Astronautics* CI. 684/1 The new factory incorporates the latest production methods and in view of the development of special-purpose connectors—particularly in the micro-miniature field—a 'white room' is being fitted out so as to give the cleanest manufacturing conditions for this type of component. **1965** *Life* 5 Nov. 111/4 The capsule itself will be stored in the pristine solitude of a 'white room' near Cape Kennedy until Schirra and Stafford are ready to go again. **1970** N. Armstrong et al. *First on Moon* iii. 66 Two of these five members of the close-out crew were in the 'white room' on swing arm No. 9. **1908** M. C. Cooke *Fungoid Pests of Cultivated Plants* 135 White rot of Grapes...occurs on the fruit, leaves, and young shoots. **1946** Cartwright & Findlay *Decay of Timber* iv. 48 Decomposition of wood by fungi is of two main types, which have been described as brown rots and white rots respectively... In a white rot all the components of the wood, including the lignin, are decomposed. **1951** *Dict. Gardening* (R. Hort. Soc.) III. 1446/1 White rot is of the utmost concern to the fungus *Sclerotium cepivorum*; onions... affected plants are seen to have rotten roots... the bulb is covered with a very white, fluffy mycelium. **1969** G. Becker in Krishna & Weesner *Biol. Termites* I. xii. 356 A large number of mold fung, white-rot fungi, and bacteria can produce toxic substances. **1963** E. Wyndham *Hills of Hebron* xvi. 198 He set a half-bottle.. ... for a few bottles of white rum. **1972** *Times* 29 Aug. 10/1 White rum promises to be the spirit of the 1970s. **1981** M. J. Berkeley in *Jrnl. R. Hort. Soc.* III. 266 White rust, so common on cruciferous plants. **1937** F. D. Heald *Introd. Plant Path.* vi. 77 The presence or development of white rust on the leaves ... **1981** *Daily Tel.* 16 May 19/2 Several of the most persistent problems in Britain today ... chrysanthemum white rust. **1914** *Photo-Era* XXXIII. 168 (*caption*) A spring white-sale. **1919** *New Yorker* 10 Oct. 158/2 The great annual white sale... **1927** K. Blixen *Out of Afr.* iv. 298 Kikosch was a young Native in the service of a young white settler of Molo. **1969** J. Mander *Static Society* vi. 134 A White Settler minority, as in Rhodesia, ... keep power if it is sufficiently organized to ... **1974** ... white settler in the Highlands ... **1976** *Listener* 3 June 716/2 'White settlers' is a phrase commonly used ... to describe the incoming community in the Highlands ... **1980** J. ... sb. 4 a]. **1974** *Daily News* (Tanzania) 23 July 1/2 The three-day occupation of the main radio station was led by criminals (white settlers) protesting at the independence settlement, and the subsequent fighting and looting in the city's African suburbs, have shattered Portuguese hopes of a peaceful and amicable transfer of power in the colony. **1976** *Listener* 3 June 716/2 'White settlers' is a phrase commonly used in the Highlands to describe the incoming community in the Highlands to describe refugees from that well-known rat-race who buy any old wreck of a cottage, spend a lot of money to tart it up, [etc.]. **1899** White Sister [in Dict., sense 6 a]. **1912** W. H. Barker *Wayfaring in France* vi. 305 It was a White Sister kneeling and praying. **1908** *Catholic Times* 6 Mar. 11/2 We have the White in the Katanga many missions..and everywhere are White Fathers, religious women (White Sisters). **1957** C. D. Kittler *White Fathers* vi. 81 The answer, Lavigerie realized, would be a society of women to be the White Sisters. **1789** *Deb. Congress U.S.* 13 May (1834) 350 He hoped it would comprehend the white slaves as well as black, who were imported from all the jails of Europe. **1807** Southey *Lett. from England* III. xxxviii. 150 Let us leave to England..the distinction.. of being the white slaves of the rest of the world, and doing for it all its dirty work. **1913** C. Pankhurst

Great Scourge p. viii, Regulation of vice and enforced medical inspection of the White Slaves. **1917** A. Huxley *Let.* May (1969) 125, I am safe from these body-snatchers, kidnappers, baby killers and white slave traffickers, the Recruiters. **1970** 'J. Quartermain' *Diamond Hook* xvi. 99 If you stop me.., I'll white-slave Jessie to South America. **1977** D. Wheatley *Young Man Said* x. 147 Was she white-slaved—a fate which befell more than a few girls of her type and class in those days? **1960** D. Lessing *In Pursuit of English* i. 16 A father-figure.. with a [*sic*] strong white-slaving propensities. **1723** J. Nott *Cook's & Confectioner's Dict.* sig. L8v, To make White Cullis... Use this with White Soops and Ragoos. **1813** Jane Austen *Pride & Prejudice* I. xi. 123 As for the ball..as soon as Nicholls has made white soup enough I shall send round my cards. **1977** J. Aiken *Five-Minute Marriage* vi. 95 Next week Mrs. Andrews really must start making white soup; and I must write.. to Gunter's about the ices. **1853** R. Riddell *Indian Domestic Econ.* 63 Take three quarts of good white stock. **1905** *Tasty Dishes* (new ed.) 10, 3 pints of white stock. **1960** *Good Housek. Cookery Bk.* (rev. ed.) 196/1 Vegetable water or stock made from bones should be used for gravies and brown sauces; milk, or milk and white stock for white sauces. **1861** H. W. Bristow *Gloss. Mineral.* 320/2 When cut for jewelry, it [*sc.* rock crystal] is called by lapidaries, 'white stone'. **1937** *Burlington Mag.* Nov. p. xix/2 A gold ring with a telling portrait carved in whitestone. **1941** F. Thompson *Over to Candleford* vi. 98 She kept the whole of the fair-sized house cleaned and polished and whitestoned. **1963** *Times* 11 June 15/4 The whitestone and glass frontage gives the impression of verticality. **1978** R. Doliner *On the Edge* (1979) iv. 62 A whitestone Italian Renaissance mansion on Sixty-third Street. **1908** J. M. Sullivan *Criminal Slang* 27 *White stuff*, morphine. **1915** G. Bronson-Howard *God's Man* I. iv. 39 There's quite a trade in laudanum... The 'White Stuff's' on the up-and-up too. **1953** W. Burroughs *Junkie* (1972) xiii. 129, I had never been able to drink before when I was on the junk, or junk-sick. But eating hop is different from shooting the white stuff. You can mix hop and lush. **1967** N. Lucas *C.I.D.* x. 135 Luckier still not to have graduated from pep pills to..'The White Stuff'..heroin. **1958** *Listener* 12 June 967/1 The steady propulsion towards white supremacism. **1979** *Daily Tel.* 5 Sept. 6 He [*sc.* Ian Smith] is well aware that the noisy and active minority who regard him as the totem of white supremacism will call for his blood. **1959** *New Statesman* 30 May 751, I have wondered..whether there was any link between the demagogues of Notting Hill and white supremacists elsewhere. **1961** *Spectator* 20 Jan. 65 A way of life..that is white-supremacist. **1964** L. Nkosi *Rhythm of Violence* II. i. 26 The White Supremacists will not get away so easily! **1977** *Times* 30 Aug. 10/3 Mr John Tyndall, the National Front's founder and chairman..describes himself as 'an unashamed white supremacist' and regards whites as intellectually.. superior to blacks. **1981** *Times* 17 Mar. 12/3 The real contest in next month's general election will be between the ruling National Party and the white supremacist parties to the right of it. **1902** A. Tourgée *Let.* 15 May in T. L. Gross *Albion W. Tourgée* (1963) viii. 143 It is the very highest form of blasphemy to claim that the idea of 'white supremacy' and the later barbarism which demands race-subjection or extermination is pleasing to God or conformable to the religion of the Man of Nazareth. **1931** W. S. Churchill in J. C. Squire *If it had happened Otherwise* 179 Upon the rebound from this there must inevitably have been a strong reassertion of local white supremacy. **1981** *Times* 18 Mar. 8/6 The Rustenburg constituency..represents some of its most far-right votes for white supremacy. **1958** *Times Lit. Suppl.* 11 July 386/3 They imply that, had the Supreme Court said 'desegregate by the so-and-so of this year', and had President Eisenhower backed the Courts to the limit, the Southern white suprematists would not have 'fought back', or would only have done so unsuccessfully. **1971** J. Bishop *Days of Martin Luther King, Jr.* iv. 332 A few White Suprematists said that there was no doubt that the bombing was the work of..a militant black who wanted to incite his people to riot. **1920** *Glasgow Herald* 7 May 9 A report on the alleged existence of a 'White Terror' in Hungary has been issued in the form of a White Paper. **1965** J. Ch'ên *Mao & Chinese Revolution* (1967) I. v. 125 The less hesitant Wuhan declared the CCP [*sc.* Chinese Communist Party] outlawed on that fateful day of 13 July 1927. The so-called White Terror thus began and chaos ensued. **1970** G. Huizer in I. L. Horowitz *Masses in Lat. Amer.* xiii. 480 The secretary-general..noted the continuous struggle of the CNC [*sc.* National Confederation of Peasants] against the 'white terror' of landowners and caciques. **1977** *Time* 21 Mar. 26/2 In the 1930s leftists lived in constant fear of the so-called White Terror imposed by the [Chinese] Nationalist secret police. **1980** *Times Lit. Suppl.* 25 Apr. 471/2 After the collapse of [Béla] Kun's regime, the White Terror raged, but Korda somehow survived unharmed. **1853** 'C. Bede' *Adventures Verdant Green* vii. 65 You are going to wine with Smalls this evening... I suppose you would go properly dressed,—white tie, kids, and that sort of thing, eh? **1930** M. Kennedy *Fool of Family* xx. 208 'Is it a grand party?' asked Caryl nervously... 'I mean is it white tie?' explained Caryl. 'Oh yes, of course it's white tie.' 'Then it is grand.' **1936** [see *black tie* s.v. *BLACK *a.* 19]. **1942** D. Powell *Time to be Born* iv. 83, I will give a white-tie dinner for eighteen. **1981** Ld. Harewood *Tongs & Bones* i. 29 The glamour of the occasion impressed me greatly—I was probably the only person in the boxes not in a white tie. **1932** H. G. Wells *Work, Wealth & Happiness* xii. 607 Tariff obstruction at this higher level is, for all practical ends, *war at the frontier*, White War, the chronic as distinguished from that acute form in which invasion, bomb, bayonet and poison gas play leading parts, which more emphatic sort of warfare we may call Red War. **1939** *New Statesman* 3 June 878/2 Armament firms will boom more conspicuously, but the promised Government limitation of earnings or special taxations of 'whitewar' profits must deprive the armament or semi-armament equities of their usual attraction. **1776** J. Wedgwood *Let.* 14 Jan. (1965) 189 But for *Usefull China*, or such a *white-ware* as you mention, I must beg a longer time. **1909** *Sat. Even. Post* 20 Feb. 8/1 Start at Fifty-ninth Street and walk down what the Manhattanese call the 'Great White Way'. **1920** S. Lewis *Main Street* 416 Then, glory of glories, the town put in a White Way. **1933** E. Caldwell *God's Little Acre* xi. 170 Out of the grey darkness of the building the girl suddenly appeared in the glow of the whiteway lights. **1939** *Florida* (Federal Writers' Project) II. 259 Central Avenue [in St. Petersburg], the city's 'White Way', extends rulerlike for 7 miles across the peninsula. **1977** *Washington Post* 30 Jan. E-1/1 When dancer-choreographer Merce Cunningham..appeared.. at New York's Minskoff Theater recently..one might have supposed..that the Great White Way had suddenly gone avant-garde. **1980** *N.Y. Times* 10 Dec. A-14/2 'Welcome to Boston's Great White Way,' the sign on a theater marquee pridefully proclaimed. **1905** H. G. Wells *Kipps* I. ii. 40 Cretonnes, chintzes, and the like; serviettes, and all the bright hard whitewear of a well-ordered house. **1949** N. Mitford *Love in Cold Climate* I. xvi. 170 She was awfully old for a white wedding, thirty or something terrible. **1962** *Daily Herald* 8 Jan. 6/8, I had a lovely white wedding... Given my time again I would cheerfully splash everything on one. **1976** *Listener* 29 July 105/1 Young black girls [in Soweto] now demand white weddings with lots of bridesmaids and floating veils. **1901** G. Parker *Right of Way* 23 Rouge Gosselin flung off his glass of white whisky, and threw after it another glass of cold water. **1957** W. Faulkner *Town* xxiv. 357 Ratliff..took a pint bottle of white whiskey from inside his shirt. **1968** 'N. Blake' *Private Wound* i. 17 Padraig, another Jamieson for Mr. Eyre. The white whiskey, mind, this time. Did y'ever try Jamieson's white? **1863** Mrs. Gaskell *Sylvia's Lovers* III. i. 2 Sitting in the dark parlour..and doing 'white work', was..wearying to her. **1936**, etc. [see *Mountmellick]. **1967, 1975** White work [see *Richelieu].

12. a. *white-fiery, -yellow.*

1876 G. M. Hopkins *Wreck of Deutschland* xiii, in *Poems* (1967) 55 Wiry and white-fiery and whirlwind-swivellèd snow. **1922** Joyce *Ulysses* 179 He..felt a slack fold of his belly. But I know it's whiteyellow.

b. *white-enamelled, -flattened, -heaped, -pointed, -quartered, -spread; white-shining, -steaming* adjs.; **white-burning** *a.*, applied to clay that gives a white product when fired; **white-dominated** *a.*, dominated by white people.

1965 G. J. Williams *Econ. Geol. N.Z.* xx. 359/2 The clays so formed are plastic, refractory and white-burning. **1967** M. Chandler *Ceramics in Mod. World* ii. 49 A small proportion of more plastic white-burning clay is sometimes included. **1960** White-dominated [see *question mark 2]. **1981** *Listener* 31 Dec. 810/1 Blacks tend to regard journalists as part of the white-dominated, Establishment-prone media. **1915** 'Bartimeus' *Tall Ship* iii. 51 Forward, the white-enamelled bulkhead was pierced by two entrances. **1918** D. H. Lawrence *New Poems* 47 Oh, masquerader, With a hard face white-enamelled. **1922** Joyce *Ulysses* 86 Nose whiteflattened against the pane. *Ibid.* 39 Belly without blemish, bulging big, a buckler of taut vellum, no, whiteheaped corn. **1948** D. Ballantyne *Cunninghams* 165 The dark blue sea, white-pointed by the wave tops. **1962** White-quartered [see *pink-scrolled* s.v. *PINK *a.*[1] C. b]. **1851** J. G. Whittier *Benedicite* in *Nat. Era* 16 Oct. 166/5 God's love—unchanging, pure, and true—The Paraclete white-shining through His peace. *a* **1973** J. R. R. Tolkien *Silmarillion* (1977) 262 A city white-shining on a distant shore. **1918** D. H. Lawrence *New Poems* 26 Daisies that waken all mistaken white-spread in expectancy. **1921** R. Graves *Pier-Glass* 26 And a white-steaming mist Obscures desire.

c. *white-aproned, -barked, -bloomed, -blossomed, -bodied, -coned, -curtained, -fanged, -fronted* (later examples), *-gaitered, -glanced, -hooded, -jacketed, -maned* (later examples), *-naped, -necked, -polled, -smocked, -souled, -spatted, -stockinged, -tied, -tiled, -tilted* [Tilt *sb.*[1]], *-walled* (later examples); **white-arsed** *slang*, a term of abuse; **white-backed vulture**, an African vulture of the genus *Pseudogyps*; **white-breasted** (later examples); *white-breasted nuthatch*, a North American nuthatch, *Sitta carolinensis*; **white-crowned**, having a white crown; *white-crowned sparrow*, a North American sparrow, *Zonotrichia leucophrys*; **white-elephantine**, of the nature of a white elephant; uselessly splendid; **white-floured**, with the face whitened by flour; **white-throated sparrow** = *Peabody.

1868 J. G. Whittier in *Atlantic Monthly* 3 Jan. 1 Bare-armed..she came, White-aproned, from her dairy. **1977** J. Gillis *Killers of Starfish* x. 76 A white-aproned waiter appeared..bearing little plates of cheese squares. **1922** Joyce *Ulysses* 587 He's a whitearsed bugger. **1975** *Daily Colonist* (Victoria, B.C.) 18 May 1/1 Delegates.. sat in shocked silence when an Indian leader accused them of being 'white-arsed Liberals'. **1884** R. B. Sharpe *Layard's Birds S. Afr.* (ed. 2) 794 African White-backed Vulture... General colour deep brown. **1964** D. Varaday *Gara-Yaka* ix. 78 The sitters were white-backed vultures, the most common in this area [by the Limpopo]. They were so dark brown in parts that they looked dirty, but their lighter parts appeared immaculate in contrast. **1779** *U.S. Mag.* (Philadelphia) Feb. 85 The lowly mangrove fond of wat'ry soil; The white barked gregory rising high in air. **1948** White-barked [see *Engelmann]. **1922** Blunden *Shepherd* 43 From white-bloomed plum. **1911** J. Masefield *Everlasting Mercy* 79 That white-blossomed pond. **1904** W. B. Yeats *King's Threshold* 55 It was praise of that great race That would be haughty, mirthful, and white-bodied. **1808** A. Wilson *Amer. Ornithol.* I. 41 The White-breasted Nuthatch is common almost everywhere in the woods of North America. **1946** G. Stimpson *Bk. about Thousand Things* 491 The white-breasted cormorant is largely responsible for the production of the vast guano deposits on the islands off the coast of Peru. **1972** L. Hancock *There's a Seal in my Sleeping Bag* i. 14 Searching for the white-breasted sea eagle. **1980** *Northeast Woods & Waters* Dec. 23/2 Hairy and downy woodpeckers, white-breasted nuthatches..attack the suet on the old pear tree. **1920** Blunden *Waggoner* 40 Smoke's light blue pennants coil From white-coned oasts. **1836** R. King *Narrative of Journey* II. 196 The *fringilla leucophrys*, or white-crowned finch..perched on the topmost branch. **1839** W. B. O. Peabody *Rep. Ornithol. Mass.* 32 The White-crowned sparrow..is one of the finest of this family of birds. **1894** B. Torrey *Florida Sketch-bk.* 235, I discovered..perched at the top of the oak, tossing back his head and warbling—a white-crowned sparrow. **1975** *Nature* 18 Sept. 182/1 The Californian scrub habitat is occasionally devastated by fire, so that the white-crowned sparrow population is reduced to a few birds living in isolated patches of surviving scrub. **1977** *New Yorker* 19 Sept. 123/1 The twisted fig tree, the almond, not yet white-crowned, the slow tendrils of grape reaching into the sky are companions for a time. **1914** D. H. Lawrence *Widowing of Mrs. Holroyd* I. i. 3 At the back is a white-curtained window. **1959** *Economist* 28 Mar. 1152/1 The white elephantine palace by the lake at Geneva may be good enough for the foreign ministers. **1971** A. Sampson *New Anat. of Britain* xvii. 335 Sir John Hill..had applied quite drastic economies to its white-elephantine operations [*sc.* those of the Atomic Energy Authority]. **1952** C. Day Lewis tr. *Virgil's Aeneid* xi. 254 His head was helmeted in a wolf's mask Whose gaping mouth with its white-fanged jaws served for a visor. **1925** E. Sitwell *Troy Park* 21, I saw the white-floured zanies go. **1908** E. J. Banfield *Confessions of Beachcomber* I. iii. 98 White-fronted Heron, *Notophoyx novæ-hollandiæ*. **1909** A. E. Mack *Bush Calendar* 23 Birds breeding in September... *Ephthianura albifrons*. White-fronted chat. **1955** E. Pound *Classic Anthol.* I. 60 Chariots, rank on rank With white-fronted horses. **1971** *Country Life* 27 May 1292/3 The famous Wexford Slobs, main winter headquarters of the Greenland race of white-fronted geese. **1922** Joyce *Ulysses* 558 His nag, stumbling on whitegaitered feet, jogs along the rocky road. **1930** Blunden *Poems* 290 Those white-glanced pools. **1900** W. S. Churchill in *Morning Post* 17 Feb. 8/1 White-hooded, red-crossed ambulance waggons. **1927** A. Clarke *Son of Learning* II. 38 The Abbot said There is a barrel of white-hooded ale Here. **1910** W. J. Locke *Simon the Jester* xxiii. 323 White-jacketed waiters darting to and fro. **1980** H. R. F. Keating *Murder of Maharajah* xiii. 156 White-jacketed Goan bearers. **1883** W. Whitman *Daybks. & Notebks.* (1978) II. 319 The sea-beach and surf—its myriad ranks like furious white-maned racers, urged by demoniac emulation to the goal, the shore. **1955** E. Pound *Classic Anthol.* IV. 212 White-maned black stallions Pull with due order. **1932** *Discovery* July 232/2 The white-naped ravens and the mountain buzzards swing overhead. **1975** *New Yorker* 24 Mar. 34/2 The white-naped crane, fifteen hundred left, fifty in zoos. **1912** J. Stevenson-Hamilton *Animal Life Afr.* xvii. 299 The ravens are represented by the white-necked raven (*Corvultur albicollis*) in the south.. of the Ethiopian region. **1965** G. B. Schaller *Year of Gorilla* vii. 161 The most regular visitors to our meadow were a pair of white-necked ravens, lovely birds with iridescent black plumage and a striking white collar around the neck. **1968** *Sunday Mail Mag.* (Brisbane) 8 Sept. 6/1 Only two species of seals now live on the southern Australian coast-line, namely the white-necked hair-seal [etc.]. **1922** Joyce *Ulysses* 537 Staggering Bob, a whitepolled calf, thrusts a ruminating head.. through the foliage. *Ibid.* 102 The whitesmocked priest came after him tidying his stole with one hand. **1973** M. Amis *Rachel Papers* 186 There—round-eyed, white-smocked and spotless—was Rachel. **1874** J. G. Whittier *Sumner* in *Memorial to Charles Sumner* 100 He never brought His conscience to the public mart; But lived himself the truth he taught, White-souled, clean-handed, pure of heart. **1902** G. W. E. Russell *Londoner's Log-Book* iii. 40 Sir William Harcourt as the white-souled champion of spiritual religion. **1922** White-spatted [see *slew-foot]. **1934** Dylan Thomas *Let.* 14 Jan. (1966) 93 The white-spatted representatives of a social system tha has, for too many years, used its bowler hat for the one purpose of keeping its ears apart. **1916** E. Pound *Lustra* 48 Her white-stockinged feet. **1957** J. Agee *Death in Family* III. xvii. 284 Catherine stood..looking at the skirt and at her white-stockinged feet. **1811** A. Wilson *Amer. Ornithol.* III. 51 White-Throated Sparrow.. [winters] in most of the states south of New England. **1865**, etc. White-throated sparrow [see *Peabody]. **1977** *New Yorker* 5 Sept. 23/3 Dozens of white-throated sparrows..have appeared among the cattails. **1848** A. H. Clough *Bothie of Tober-na-Vuolich* i. 5 The Tutor.. White-tied, clerical. **1972** A. Roudybush *Sybaritic Death* (1974) ii. 5 Tail-coated, white-tied and silk-hatted men. **1924** G. B. Stern *Tents of Israel* xiii. 182 I've wanted things, too... Hundreds of baths; baths in white-tiled rooms, and not skimping the hot water. **1978** T. Gifford *Glendower Legacy* (1979) 53 An ancient wino was mopping one corner of the long, narrow, white-tiled floor. **1939** F. Thompson *Lark Rise* i. 2 The baker's little old white-tilted van. **1958** *Punch* 21 May 670/3 Dunlop white-walled tyres, white pedals, and white pump. **1985** A. McCandless *Burke Foundation* i. 4 White-walled houses with red-tiled roofs.

d. (*a*) *white-duck*; (*c*) **white-shoe** *slang* (chiefly *U.S.*), effeminate, immature; **white telephone**, (of a film) telling an unrealistic story set in elegant surroundings; **white-wall**, (of a tyre) having white sidewalls.

1849 White duck [see Duck *sb.*[2] 3]. **1925** H. Crane *Let.* 19 Aug. (1965) 214 White undershirt and loose white duck pants. **1966** in *Islands* (N.Z.) (1978) Aug. 93 White-duck curtains..Hang at the windows. **1957** J. D. Salinger *Zooey* in *New Yorker* 4 May 62/2 Phooey, I say, on all white-shoe college boys who edit their campus literary magazines. Give me an honest con man any day. **1974** G. Jenkins *Bridge of Magpies* vi. 85 What sort of white-shoe captain are you? **1975** *N.Y. Times* 22 Sept. 33/1 Covert operations can be stripped from the CIA... So can such monkey business as dropping simulated poison cannisters in the New York subways—the games of white-shoe boys who never grew up. **1958** *Oxf. Mag.* 22 May 462/2 Then from Italy, which had hitherto only produced 'white telephone' films, came this simple, humble and extremely moving story. **1975** *New Yorker* 5 May 24/3 This is an icy high-minded white-telephone movie. **1953** L. Z. Hobson *Celebrity* viii. 116 A Buick Roadmaster... Fully equipped, radio, heater, white wall tyres. **1965** *Punch* 20 Oct. 567/2 Then, I shall buy this year's model, too, my beloved,..with whitewall tyres and a cigar-lighter. **1978** *Listener* 2 Feb. 158/2 When film makers go 'period', as they did for *Chinatown*, the bulky Buicks and Oldsmobiles have to be lovingly rebuilt, white-wall tyres, teeth-like radiator grills and bonnet 'ventiports', almost from scratch.

e. †**white-choker** *slang*, a clergyman (cf. Choker 2); so **whitechokerism**; **white-hat**, (*b*) *U.S. Naval slang*, an enlisted man; (*c*) *slang* (orig. *U.S.*), a good man; a hero; **white-leg** (examples); **whitewall**, a white-wall tyre (see sense 12 d above).

1903 A. H. Lewis *Boss* xxi. 292 It's that same Reverend Bronson who gives Melting Moses th' office to dog me. I'll put Mr. Whitechoker onto my opinion of th' racket. **1912** A. Bennett *Matador of Five Towns* 100 You belong to that Methody lot... I seed you talking to them white-chokers. **1866** J. R. Lowell *Let.* 10 Apr. (1894) I. 361, I don't understand your English taste for what you call 'respectability' (I should call it 'whitechokerism'), thinking, as I do, that the one thing worth striving for in this world is a state founded on pure manhood. **1956** E. N. Rogers *Queenie's Brood* 241 There's a white hat out here who has gone crazy. **1975** *Courier-Mail* (Brisbane) 28 Feb. 5/2 Laver's the last of the white hats (the good guys who wear the white hats in cowboy movies). **1975** W. Safire *Before the Fall* III. vii. 191 Nixon and Haldeman clung to the original game plan..against the urging of..Garment, and other 'white hats'. **1978** *Guardian Weekly* 15 Jan. 18/2 His judgments of the men he dealt with... The white hats are Truman [etc.]. A prime villain is Britain's postwar foreign secretary. **1811** R. Hooper *Lexicon-Medicum* (new ed.) 615/2 *Phlegmasia dolens*... By the Germans it is called Œdema lacteum, and by the English the white leg. **1899** [see *milk leg* s.v. Milk *sb.* 10 a]. **1939** M. Spring-Rice *Working-Class Wives* v. 122 She is very anæmic, has 'whiteleg', constipation and piles. **1976** *Lancet* 27 Nov. 1197/2 Iliac-vein thrombosis or 'white leg' affects the left side more commonly than the right. **1982** P. Barker *Union Street* 250 After our May was born she never walked properly again. She had what they called the white leg. **1958** *Autocar* 31 Oct. 675 (*caption*) Bentley Flying Spur, sans fins, sans whitewalls, sans tinsel. **1968** *Globe & Mail* (Toronto) 5 Feb. 26/5 (Advt.), Hardtop, big 6 with automatic, radio and whitewalls. **1978** *Detroit Free Press* 5 Mar. C7/2 (Advt.), A built-in Scuff Bar that helps keep whitewalls white.

f. *white-breast.*

1933 D. H. Lawrence *Mod. Lover* (1934) 11 The fallow flickered over with pink gleams of birds white-breasting the sunset.

white, *v.*[1] Add: **1. b.** Const. *out*. Of vision: to become impaired by exposure to a sudden bright light (see also quot. 1981). Also *trans.*, to 'blind' (an audience in a theatre) by such means.

1978 'A. Stuart' *Vicious Circles* 22 At once my eyes whited out—as disoriented by the brilliant evening sun as a bat caught by a searchlight. **1981** *Times Lit. Suppl.* 30 Jan. 112/1 As the women lie down to sleep in the hot summer morning, the stage lights white out to mime the atomic fireball. **1983** *Listener* 3 Feb. 32/3 In Bristol the Little Theatre performs the stage play, using lasers and whiting-out audiences.

2. c. Delete † and add later example.

1972 E. Wigginton *Foxfire Bk.* 181 And it was the sulfur that whited the apples, and they had a little sulfur flavor.

f. *to white out*: to obscure or cover with something white, esp. a white fluid used by typists. Also *fig.*

1975 J. Butcher *Copy-Editing* iii. 25 If you want to cancel an underlining for italic, white it out, or put two or three short lines through it, not a wavy line. **1978** M. Duffy *Housespy* vi. 141 Its long shop window was whited out. **1982** R. Leigh *Girl with Bright Head* xi. 74 There's also a couple of places where she has had to white out mistakes and type over them. **1983** 'J. le Carré' *Little Drummer Girl* xiii. 224 She drove with her mind whited out and her thoughts deliberately foreshortened. **1984** *Times Lit. Suppl.* 13 July 771/3 The embarrassed printer explained that he'd whited the little dot out, thinking that it was a dust spot.

g. To make *up* (an actor) to look white.

1977 R. Barnard *Death on High C's* xv. 148 He was already 'whited up' for the part of the Duke of Mantua... He must look odd, with his deadly white colouring and negroid lips.

white ant, *sb.* [f. White *a.* + Ant.] **1.** (See Ant 3.)

1684 Locke *Jrnl.* 17 Nov. in K. Dewhurst *Locke* (1963) 265 Told me of a sort of white ants that there mightily infests them. **1699**, etc. [see White *a.* 11 a]. **1729**, etc. [see Ant 3]. **1908** E. J. Banfield *Confessions of Beachcomber* I. vii. 227 The 'white ant' (which is not an ant).. would literally eat us out of house and home. **1928** R. Campbell *Wayzgoose* i. 20 White-ants and borers, turning boards to dust. **1938** X. Herbert *Capricornia* (1939) viii. 102 The white-ants have eaten the wheels of my buckboard. **1974** D. Stuart *Prince of my Country* v. 40 The wind and the rain and the white ants will level the camps.

2. With allusion to the supposed destruction of the brain by white ants, implying loss of sanity, sense, or intelligence. *Austral. slang.*

1908 H. Fletcher *Dads & Dan: between Smokes* 64 It wants a fool or a very sane cove indeed ter live in ther lonely bush an' keep ther white ants out o' his napper. **1926** L. G. E. Gee *Bushtracks & Goldfields* 65 And so he rambles on..and in the unsteady glance of his honest, old eyes and his disconnected speech, I read the mark of the Australian solitudes—'white ants' they call it up north. **1938** H. Drake-Brockman *Men without Wives* 27 '"Get the white ants?" What do you mean?' 'Go ratty. Mad.' **1948** V. Palmer *Golconda* vii. 49 They had a definite respect for Christy. He might have a few kinks ..but there was something dinkum about him, and if there were white ants behind his forehead they had a lot of work ahead of them. *a* **1951** E. Hill in Murdoch & Drake-Brockman *Austral. Short Stories* (1951) 292 My brownie days are over... I reckon I've got white ants.

white-a·nt, *v.* Chiefly *Austral.* [f. prec.] *trans.* To destroy in the manner of termites or white ants; to undermine, eat away, or sabotage.

1925 *Glasgow Herald* 14 Nov. 9/6 The extremists..have deliberately 'white-anted' the Labour movement..and squandered the funds of the wealthy unions. **1952** L. Overacker *Austral. Party System* vi. 182 The Communists have 'white anted' the unions, elected their members to offices in the Miners' Federation, the waterside workers' and ironworkers' unions, and developed 'shop committees' as basic units in the factories. **1962** R. Wallis *Point of Origin* 96 After hearing..about me.. he decided he'd have to do his duty as a gentleman and tell Rockdale he was being white-anted. **1968** D. Ireland *Chantic Bird* xi. 102 Television had white-anted their audiences, and they had to use the place for other things besides films.

Hence **white-a·nted** *ppl. a.*, **white-a·nting** *vbl. sb.*

1936 F. Clune *Roaming round Darling* xx. 205 The piece of the boat is five feet long and is made of soft wood, badly white-anted. **1945** Baker *Austral. Lang.* xiv. 245 White-anting. **1950** D. Cusack *Comets Soon Pass* in *Three Austral. Three-Act Plays* II. ii. 55 *Dr. John.* Each man must find his own pole to swing to. I have found mine. *Mrs. Ellington-Brown.* I think that's too wonderful, so mystic. *Talbot.* White-anting society! *Jack Smith.* Too mystic for my taste, Doc. I think you've got to get out and fight for things. **1973** *Sydney Morning Herald* 30 Aug. 6/4 We are promised largesse in the form of harbourside parks in the same breath as the white-anting of a remote scenic gorge is sanctioned.

white bear. Chiefly *N. Amer.* **a.** = *polar bear* s.v. *Polar *a.* b.

1600 Hakluyt *Princ. Navigations* III. 6 The soile is barren in some places,..but it is full of white beares. **1823** *Canad. Mag.* I. 394 The great white bear takes refuge in the most icy climates. **1860** P. H. Gosse *Romance Nat. Hist.* 62 The white bear, seated on a solitary iceberg in the Polar Sea. **1953** W. B. Mowery *Tales of Mounted Police* 149 [He had] several livid weals across his left cheek where a white bear had once clawed him.

b. A grizzly bear (*Ursus horribilis*) in a light-coloured phase.

1791 J. Long *Voy. Indian Interpreter* 95 The large white bear, commonly called the grisly bear, is a very dangerous animal. **1852** J. Reynolds *Hist. Illinois* 172 He was destroyed there [in the Rocky Mountains] by a white bear. **1952** J. Jennings *Strange Brigade* (1954) 105 There were also red deer or *biche*, and white bears and white partridges.

white-bea·rded, *a.* [White *a.* 12 c.] Having a white beard. **a.** Of a man.

1596 [in Dict. s.v. White *a.* 12 c]. **1914** D. H. Lawrence *Widowing of Mrs. Holroyd* III. 81 A little stout, white-bearded man.

b. Of wheat.

1788 G. Washington *Diary* 8 Sept. (1925) III. 417 Also sowing..one bushel of the White bearded Wheat sent me by Beale Boardly. **1850** *Rep. U.S. Comm. Patents 1849: Agric.* 132 The white-bearded wheat, a valuable kind less liable to total failure than any other; not very popular with millers.

c. *fig.*

1920 E. Sitwell *Wooden Pegasus* 100 And, mourners too, white-bearded seas Walk slowly by them as they come. **1960** *Farmer & Stockbreeder* 19 Jan. (Suppl.) 1/1 Waves came solid green and white-bearded, like frost giants racing.

whiteboard (hwəi·tbōəɹd). [f. White *a.* after *blackboard*.] A white surface for use like a blackboard but accepting felt-tipped pens and wax crayons.

1966 'W. Cooper' *Mem. New Man* II. v. 160 He.. went to the blackboard. (Actually it was an up-to-date plastic white-board, on which one wrote with a coloured wax crayon.) **1977** *Times Educ. Suppl.* 21 Oct. 28/1 (Advt.), They are whiteboards that stay white, year after year. **1978** J. McNeil *Consultant* ix. 106 They came to a meeting room... The walls were bare except for a white-board. **1985** *Times Educ. Suppl.* 19 July 20/5 We should also bear in mind that partially-sighted pupils often fare better if a white-board is used rather than a blackboard.

white boy, whiteboy. **1.** Delete † *Obs.* and add later example.

1919 T. S. Eliot *Let.* 9 July in *Waste Land Drafts* (1971) p. xvii, The small public which *I* could bring to it [*sc.* the *Egoist*] now reads the *Athenaeum* every week. There I am a sort of white boy; I have a longish critical review about three weeks out of four.

whi·tecap, *v.* *U.S.* [f. the sb., sense 4.] *trans.* To commit an outrage upon (a person) in the style of the whitecaps. Chiefly as **whi·tecapping** *vbl. sb.* Also **whi·tecapper.**

1895 T. Roosevelt in *Century Mag.* Nov. 72/2 The lawbreaker, whether he be lyncher or whitecapper, or merely the liquor-seller who desires to drive an illegal business. **1900** M. Nicholson *Hoosiers* 45 The milder form of outlawry, known as 'white-capping', has also been practised in Indiana occasionally. **1904** *N.Y. Even. Post* 28 Jan. 9 The Mississippi has voted Gov. Vardaman a special appropriation to enable him to suppress the 'white cappers'. **1908** D. G. Phillips *Old Wives for New* iv. 68 If he wasn't such a wonderful doctor he'd have been white-capped long ago—tarred and feathered and railed out of town. **1943** A. G. Powell *I can go Home Again* 167 During the short time I served as county judge, a series of 'whitecappings', directed against Negroes, occurred in the lower part of the county. **1970** [see *Ku-Klux 1 a].

whi·tecapping, *a.* *rare*$^{-1}$. [f. as prec. + -ing[2].] Covering with or as with a white cap.

1912 J. London *Son of Sun* v. ii. 175 Their long slopes.. were broken by systems of smaller whitecapping waves.

white cliffs, *sb. pl.* [f. White *a.* + Cliff.] **1.** Chalk cliffs; *spec.* those of Dover, regarded as a symbol of Great Britain.

1879 [see Cliff 1 b]. **1902** Kipling *Just-So-Stories* 7 Take me to my natal-shore and the white-cliffs-of-Albion. **1940** N. Burton (*song*) There'll be blue-birds over the white cliffs of Dover. **1940** R. S. Lambert *Ariel & All his Quality* iii. 84 Full of a mystic vision of Empire..inspired by the sight of the white cliffs of Dover. **1978** M. Kenyon *Deep Pocket* xiv. 181 You'll be deported, you'll never see the White Cliffs again.

2. (With capital initials.) The name of a town in New South Wales, used *attrib.* to designate opals mined there.

1911 C. E. W. Bean *'Dreadnought' of Darling* xxv. 222 The Wilcannia banks live on the White Cliffs opal. **1936** H. P. Whitlock *Story of Gems* x. 127 The White Cliffs opals are not unlike those from Hungary, but they show broader flashes of colour. **1975** R. Webster *Gems in Jewellery* xi. 57 The White Cliffs opal is cream in colour and found in seams in sandstone.

whitecoat. Add: **3.** A doctor or hospital attendant who wears a white coat. (In this sense the form *white coat* is not *Obs.*)

1911 [see *schmerz]. **1932** 'Jock' *Dartmoor from Within* vi. 134 He makes straight for the tub, and 'White Coat' alters his course to cut him off. **1980** *Brit. Med. Jrnl.* 29 Mar. 934/2 We roar into the hospital. White coats run out.

white-collar, *sb.* and *a.* orig. *U.S.* **A.** *sb.* **a.** (As two words.) A white collar regarded as characteristic of a man engaged in non-manual work.

1919 U. Sinclair *Brass Check* xiii. 78 It is a fact with which every union workingman is familiar, that his most bitter despisers are the petty underlings of the business world, the poor office-clerks..who, because they are allowed to wear a white collar.., regard themselves as members of the capitalist class. **1976** M. Hinxman *End of Good Woman* i. 9 Tom emigrated to Canada. Dick put on a white collar and became a bank clerk.

b. A person engaged in non-manual work.

1930 A. P. Herbert *Water Gipsies* iv. 39 That family over there..come here every Thursday of their lives for a little family reunion, and white collars, too, all of them. **1938** W. Smitter *F.O.B. Detroit* 32 It wasn't long before the white-collars up front began taking notice of what was going on on the floor. **1954** E. Pangborn *Mirror for Observers* (1955) I. i. 19 A residential backwater for factory workers, white-collars, transients. **1962** 'K. Orvis' *Damned & Destroyed* i. 12 A pair of white-collars from a near-by St. James Street brokerage office pounded the bar for fresh drinks. **1971** W. J. Burley *Guilt Edged* i. 5 [The] passenger ferry..had made only two return trips, one for the workers at seven-thirty and one for the white-collars at eight-thirty.

B. *adj.* **a.** Of a person: engaged in non-manual, esp. clerical, work.

1921 *Ladies' Home Jrnl.* May 98/4 Urban chain restaurants have accustomed white-collar boys and girls to tasty viands, albeit in limited amounts. **1924** W. McDougall *Ethics & Some Mod. World Probl.* iv. 125 The strata of brain-workers made up the white-collar class or middle classes. **1937** *Atlantic Monthly* Dec. 750/1 Proletarian literature..has been accompanied by books on the white-collar worker, the storekeeper..the scientist, and the millionaire in situations equally disastrous or degrading. **1948** *Chicago Tribune* 3 Apr. II. 1/4 The modern white collar girl wants a job which not only offers opportunities but advances as well. **1959** [see *blue-collar* s.v. *Blue *a.* 13]. **1969** *Times* 30 Apr. 26/6 The first strike action by manual workers against the British Steel Corporation's new policy of white collar union recognition broke out yesterday. **1982** D. Gorham *Victorian Girl* ii. 29 Teachers and nurses..were of less importance numerically than [female] 'white collar' workers.

b. Of work or an occupation: not manual or industrial; *spec.* clerical.

1926 *Amer. Speech* II. 96/2 The uneducated and uneducable found a new field opening to them, and rushed in, to take advantage of the 'white-collar' work. **1937** 'G. Orwell' *Road to Wigan Pier* xi. 205 The typical Socialist..is either a youthful snob-Bolshevik..or, still more typically, a prim little man with a white-collar job. **1962** Auden *Dyer's Hand* (1963) 123 He has a dingy white-collar job. **1979** T. Benn *Arguments for Socialism* i. 41 The definition of a worker is extended to include all wage and salary earners and paves the way for the extension of trade unionism into the realms of clerical white collar, scientific and technical and managerial work.

c. (See quot. 1937.) *U.S.*

1932 [see *dirt farmer* s.v. *Dirt *sb.* 7 d]. **1937** *Amer. Speech* XII. 105 The adjectives *suitcase* and *bonanza* and *whitecollar* are applied in recently developed wheat-farming areas to large owner-farmers who live outside the community and appear during the sowing and harvesting seasons.

d. Applied to a person who takes advantage of the special knowledge or responsibility of his position to commit non-violent, often financial, crimes; also to the crime itself.

[**1932** E. H. Sutherland in *Publ. Amer. Sociol. Soc.* Aug. 60 The financial crimes of the white-collar classes.] **1934** —— *Princ. Criminol.* ii. 32 These white-collar criminaloids..are by far the most dangerous to society.. from the point of view of effects on private property and social institutions. **1964** M. Argyle *Psychol. & Social Probl.* v. 65 Older middle-class people are tempted to commit offences other than theft or violence, and the various kinds of 'white collar crime' are hard to detect—income-tax avoidance, bogus expense claims and complex business illegalities. **1977** *Wandsworth Borough News* 7 Oct. 5/3 Dangerous drivers and white-collar criminals are far more likely to receive lenient treatment than the petty habitual thief. **1984** *Daily Tel.* 12 Nov. 20/2 White-collar crime like fraud is..on the increase.., and the computer has opened enormous vistas of extra opportunity.

white-co·llared, *a.* [-ed[2].] **1.** Wearing a white collar; also *fig.*

1932 H. G. Wells *Work, Wealth & Happiness of Mankind* vii. 237 The black-coated, white-collared clerk. **1947** J. Mulgan *Report on Experience* 18 Ten millions of the rest, bowler-hatted, white-collared, moved in monotonous rhythm. **1951** D. Glover *Sings Harry* 41 It's plain hard hazardous work To work with the white-collared wave.

2. = *White collar *a.* a.

1933 *Sun* (Baltimore) 14 Apr. 4/6 Hands blistered and backs sore from hard physical labor, so-called 'white-collared men' of West Virginia are calling for more and yet more work. **1947** *Hist. 'The Times'* III. v. 117 That public was the great and growing, vigorous 'white-collared' lower-middle class. **1959, 1967** [see *blue-collared* adj. s.v. *Blue *a.* 13]. **1977** M. Green *Children of Sun* (rev. ed.) x. 460 *Lucky Jim*..described a new class on the British scene, the white-collared proletariat, trained technicians but not educated gentlemen.

whited, *ppl. a.* Add: **1.** *whited sepulchre* (earlier example).

1835 J. E. Alexander *Sketches in Portugal* i. 13 What a whited sepulchre we found the city to be!

white-eared, *a.* Add: *white-eared flycatcher*, a monarch flycatcher, *Monarcha leucotis*, found in Australia; *white-eared pheasant*, an eared pheasant, *Crossoptilon crossoptilon*, found in forest regions of eastern Tibet and neighbouring China.

1869 J. Gould *Birds Austral. Suppl.* 12 (*heading*) White-eared Flycatcher. [**1918** W. Beebe *Monogr. Pheasants* I. 187 Once only was a glimpse permitted to us of the wonderful White Eared-pheasants.] **1976** G. Durrell *Stationary Ark* iv. 78 Our chances of establishing the White-eared pheasant in captivity seemed..slim. **1980** G. Pizzey *Field Guide Birds Austral.* 258 White-eared Monarch Flycatcher..has been likened to a miniature Magpie Lark.

whi·te-face, *sb.* and *a.* Also **whiteface**, † **white face. A.** *sb.* † **1.** The widgeon. *Obs. rare.*

1709 [see *bald-face 1].

2. One of a Hereford herd of cattle. Now chiefly *N. Amer.*

1860 [in Dict. s.v. White *a.* 12 e]. **1965** E. McCourt *Road across Canada* 152 Herds of white-faces dot the slopes.., grazing knee-deep in the lush grass of the wide valley-bottoms. **1970** M. G. Eberhart *El Rancho Rio* (1971) viii. 85 I've been breeding white-faces—pure-bred Herefords.

3. White or light-coloured make-up, esp. as worn by a clown, or by a black actor playing a white character. orig. *U.S.*

1895 *N.Y. Dramatic News* 9 Nov. 14/4 Lew Dockstader, in his new white-face act,..will be seen at Keith's, November 18. **1947** *Partisan Rev.* Jan.–Feb. 65 The selection of Canada Lee, a negro in white-face, to play Bosola. **1948** M. Winter in P. Magriel *Chron. Amer. Dance* 53/1 English clowns..returned to whiteface, but kept certain characteristics of blackface performers. **1981** *Times* 11 Apr. 7/1 We find him grovelling on the floor like a mock Othello in whiteface.

B. *adj.* Of an animal: having a white face.

1785 T. Jefferson *Notes Virginia* vi. 126 White face teal. **1961** R. P. Hobson *Rancher takes Wife* xvii. 217 There's a herd of top whiteface cows. **1978** *Detroit Free Press* 5 Mar. C20/5 (Advt.), Whiteface Capuchin male, 2 years old. **1984** *Properties Open in 1984* (National Trust) 40 A Country Park surrounds the property and contains a flock of Whiteface Woodland sheep.

white-footed, *a.* Add: *white-footed mouse*, any of several species of North American mice of the genus *Peromyscus*, esp. *P. maniculatus* or *P. leucopus*.

1857 *Rep. Comm. Patents 1856: Agric.* (U.S.) 86 The food of [the Northern shrike]..consists almost wholly of arvicolae and a few white-footed prairie mice. **1869** *Amer. Naturalist* III. 120 When the axe-man struck the tree, a Whitefooted Mouse..rushed from the nest. **1936** D. McCowan *Animals Canad. Rockies* viii. 68 The White-footed mouse is of medium size and has a silky coat that is dark brown above and light on the underparts. **1977** J. L. Harper *Population Biol. Plants* xv. 465 The seed was collected and buried mainly by white-footed mice.

white gold. 1. † **a.** Platinum. *Obs.*

1764 *Gentl. Mag.* XXXIV. 128/1 (*heading*) A farther Account of a Metal, called Platina, or White Gold. **1798** [in Dict. s.v. White *a.* 11 c].

b. A name applied to various silvery-coloured alloys of gold with nickel, palladium, platinum, or silver.

1893 *Funk's Stand. Dict.* s.v. *Gold*, *White gold*, an alloy of about five parts of silver to one of gold. **1921** *Daily Colonist* (Victoria, B.C.) 12 Mar. 2/1 Modern Wedding Rings... White Gold Rings, $10. **1940** *Chambers's Techn. Dict.* 382/1 *White gold* is usually an alloy with nickel, but as used in dentistry this alloy contains platinum or palladium. **1946** G. Stimpson *Bk. about Thousand Things* 242 White gold jewelry nowadays differs from regulation gold, not in the quality or the quantity of gold used, but in the kind of alloy. **1956** J. N. Anderson *Appl. Dental Materials* vii. 78 Casting alloys [of gold] containing a large amount of palladium together with silver are called white golds. **1971** *Nature* 18 June 443/1 Two basically gold-silver alloys were known to the Greeks and Romans: (1) 'white gold'—a haphazard mixture derived from grains of weathered auriferous ore; and (2) 'electrum'—an alloy of controlled quality. **1974** *Country Life* 26 Dec. 2002/3 Platinum has largely been superseded by white gold as a jewellery metal. **1980** 'E. McBain' *Ghosts* ii. 28 One rope choker of eighteen-karat yellow and white gold.

2. Any white substance regarded as valuable.

1966 *Times* 28 Feb. (Canada Suppl.) p. xi/1 Most of the subterranean 'white gold' [*sc.* potash] lies beneath Saskatchewan. **1974** G. Jenkins *Bridge of Magpies* ii. 27 My job was to police the Sperregebiet from the sea... There is..on this God-forsaken shore: a string of rocky little inshore islands coated in bird guano—white gold, they call it.

white-haired, *a.* Add: **2.** *white-haired boy*, a favourite. *colloq.* Cf. White-headed *a.* 2 b.

1910 *Nat. Police Gaz.* (U.S.) 29 Jan. 3 (*heading*) The white haired boy. *Ibid.* 3/4 He, this white-haired Willie-boy, really wanted her to become his wife. **1923** H. C. Bailey *Mr. Fortune's Practice* ii. 38 His mother's white-haired boy, he is. Not 'alf. **1936** J. Dos Passos *Big Money* 496 You're the whitehaired boy around here. **1977** I. Shaw *Beggarman, Thief* II. iii. 147 Rudy..was the white-haired boy of the family.

Whitehall[1] (hwəi·thǭl). **1.** The name of a street in London, used to designate the government offices situated there, or the civil service in general.

1827 *Morning Post* 24 Mar. 3/2 In consequence of that accommodation, an equal amount of Exchequer Bills before lodged at Whitehall are afterwards to be locked up in Threadneedle-street. **1850** *Daily News* 13 Mar. 5/2 The infection of..'Christian Socialism' is spreading to Whitehall. **1910** *Times Educ. Suppl.* 6 Sept. 13/2 The introductory remarks..breathe a very different spirit from that of earlier official utterances of Whitehall. **1946** C. S. Forester *Lord Hornblower* ix. 79 Heaven only knew what Whitehall and Downing Street would say. **1958** *Radio Times* 23 Feb. 6/1 This is a tense story of sea warfare... The mess-room talk is most authentic..and Mr. White is obviously familiar with his 'Whitehall types'. **1977** *Listener* 7 Apr. 442/1 British Leyland..is almost entirely a Whitehall creation.

2. Special Combs.: **Whitehall farce**, any of a series of bedroom farces produced at the Whitehall Theatre, London, esp. those presented between 1950 and 1967 by Brian Rix; **Whitehall Warrior** *slang*, a civil servant; an officer in the armed forces employed in administration rather than on active service.

1966 N. Marsh *Black Beech & Honeydew* x. 233 My uncle..was like a Professor in a Whitehall farce. **1966** *Guardian* 20 Aug. 4/4 Blackpool..fulfils a social need. Like a Whitehall farce, it dictates its own terms and makes general criticism futile. **1976** M. Gilbert *Night of Twelfth* v. 42 The play [*sc.* *Twelfth Night*]..[is] a love story mixed up with a Whitehall farce. **1973** K. Giles *File on Death* vii. 174 I'm Quarles, a battered old Whitehall Warrior. **1976** W. White *Long Silence* vii. 57, I didn't want anybody to think I was a chairbound officer, a Whitehall Warrior. **1978** P. O'Donnell *Dragon's Claw* v. 81 Roger was a Whitehall Warrior until he retired.

Hence **Whitehalle·se**, jargon regarded as typical of the civil service; **Whiteha·llism**, attitudes or personnel regarded as typical of the civil service.

1915 Ld. Esher *Let.* 21 Oct. in M. Gilbert *Winston S. Churchill* (1972) III. Compan. II. 1232 In the Navy.. there seems to be a trifle too much of 'Whitehallism'. **1940** *Manch. Guardian Weekly* 15 Mar. 216 Shortage of paper may now prompt economy of speech, and if it sloughs away some of the pomposities of business English and Whitehallese, which is slightly more correct in its heavy Latinity but just as lacking in sense and suppleness, the war may be said to have done us a little good. **1958** *Times* 15 Nov. 8/3 At the time he said the Government were snubbing Wales and there was no prospect of 'Whitehallism' ever understanding Welsh aspirations. **1975** *Economist* 15 Feb. 115/2 In spite of long exposure to Whitehallese, she writes in English. He lapses far too often into gobbledygook. **1984** *Guardian* 1 Jan. 4 Phrases that take two words to say what one used to—'check out', 'meet with', 'consult with' (although Mrs Thatcher may shortly ban the last one from Whitehallese).

† **Whiteha·ll**[2]. *U.S. Obs.* The name of a district of New York, used *attrib.* to designate a type of rowing-boat. So † **Whiteha·ller**, one who uses a Whitehall boat.

1828 J. F. Cooper *Notions of Americans* I. 40 The latter [*sc.* New York boatmen], it appears, are of a class of watermen, that are renowned in this country, under the name of Whitehallers. **1835** C. J. Latrobe *Rambler in N. Amer.* i. 25 The light skilfully managed wherry of the Whitehaller. **1849** H. Melville *Redburn* II. xxix. 289 The Whitehall boats were around us. **1890** N. P. Langford *Vigilante Days* II. 129 To attempt the passage..in a whitehall boat would be madness.

whitehead[1], *sb.* Add: **4.** Also **white head. a.** A disorder in which the scalp is covered with white spots or crusts. **b.** [After *black-head 4.] A white or white-topped pustule.

1911 *Trans. S. Afr. Med. Congr.: 12th Meeting* 165 The so-called 'white head', so often seen in Bechuanaland..is a pustular syphilide affecting the scalp. The pustules..tend to coalesce, forming thick whitish crusts. ..In some cases the head becomes covered, giving the appearance of a solid white cap. **1922** *Brit. Jrnl. Dermatol. & Syphilis* XXXIV. 267 The scalp condition we are about to discuss is known to the natives..by the name of wit kop, dikwakwadi, or white head. **1940** Becker & Obermayer *Mod. Dermatol. & Syphilol.* xxix. 521 Milium or 'white head' is the name for a tiny, pearly-white globular lesion..with a shiny, translucent surface. **1978** Parsons & Sommers *Gynecology* (ed. 2) xx. 308/1 Closed comedones ('whiteheads') are the precursors of the inflammatory papules and pustules that commonly occur. **1982** P. M. Margolin *Last Innocent Man* I. vii. 70 The boy's right hand raised slowly and began to pick at a whitehead on his cheek.

Whitehead[2]. Add: (Earlier example.) Also *attrib.*

1877 *Sci. Amer.* 2 June 337/2 The Whitehead torpedo can be made to go at the rate of 20 knots for 1,000 yards, and at any depth that is desired from 1 foot to 30 feet.

white-headed, *a.* Add: **2. b.** (Later examples.) Not now restricted to Ireland.

1933 A. Christie *Lord Edgware Dies* xxii. 186 You're positively convinced now that Ronald Marsh is a white-headed boy who can do no wrong. **1954** T. S. Eliot *Confidential Clerk* II. 60 Perhaps you think it would be bad for your prospects Now that you're Claude's white-headed boy.

Whiteheadian (hwəithe·diăn), *a.* [f. the name *Whitehead* + -ian.] Of, pertaining to, or characteristic of the English mathematician and philosopher A. N. Whitehead (1861–1947) or his ideas.

1943 *Mind* LII. 68 The Whiteheadian attempt to reconcile permanence and flux, time and eternity. **1977** *Church Times* 25 Feb. 6/5 Process theology has taken over the Whiteheadian scheme and used it as a means of expounding the Christian faith. **1978** *Christian* IV. IV. 328 The last point..follows if it is true—as the Whiteheadian conceptuality to which I happen to subscribe would say—that 'a thing is what a thing *does*'.

white-heart, *sb.* and *a.* For **1, 2** read **A, B** and add: **A.** *sb.* **1.** (Earlier example.)

1707 [see *black heart* s.v. *Black *a.* 19].

2. *Metallurgy.* Malleable cast iron made by keeping white iron at a high temperature for several days in an oxidizing environment, so as to remove the carbon from the surface layers and increase the ductility and strength. Freq. *attrib.*

1925 *Jrnl. Iron & Steel Inst.* CXII. 433 The material examined consisted of metal from the open-hearth furnace only, corresponding to that used for whiteheart castings and blackheart castings in Europe and America. **1949** [see Réaumur b]. **1960** [see *graphitize *v.* 1 b]. **1968** A. H. Cottrell *Introd. Metallurgy* xxv. 519 Whiteheart malleable iron is made by heating the casting in an oxidizing environment.

B. *adj.* (Earlier example.)

1747 H. Glasse *Art of Cookery* iv. 57 Take a fine White-heart Cabbage.

white horse. 5. (Examples.)

1846 J. R. Browne *Etchings of Whaling Cruise* 130 The white, hard blocks, containing but little oil, and which are found near the small, and at the flukes, are called 'white horse'. **1851** H. Melville *Moby Dick* II. xcix. 173 White-horse,..obtained from the tapering part of the fish, and also from the thicker portions of his flukes.

white house. [House *sb.*[1]] **1.** (With capital initials.) **a.** The popular name for the official residence of the President of the United States at Washington; hence, the President or his office.

1811 F. J. Jackson *Let.* 24 Apr. in H. Adams *Documents New-Eng. Federalism* (1877) 385 [Foster] goes..to act as a sort of political conductor to attract the lightning that may issue from the clouds round the Capitol and the White House at Washington. **1812** A. Bigelow *Let.* 18 Mar. in *Proc. Amer. Antiquarian Soc.* (1930) XL. 331 There is much trouble at the white house, as we call it, I mean the President's. **1833** [in Dict. s.v. White *a.* 11 e]. **1884** *Century Mag.* Apr. 803/1 There is no building quite as satisfying to my eye as the White House. **1927** S. Bent *Ballyhoo* iii. 80 Conversationally they referred to the 'White House Spokesman', when he existed, as the Executive Larynx or the Presidential Ghost. **1950** *Daily Ardmoreite* (Ardmore, Okla.) 14 Feb. 1/7 The White House said no further action on its part is contemplated at this time. **1958** *New Statesman* 11 Jan. 30/1 He has no influence in the White House, and in recent months the requests for his advice have been little more than perfunctory. **1977** M. Edelman *Political Lang.* vi. 111 The White House tapes exemplify this common form of public language.

b. *transf.* Applied to other buildings serving as official residences.

1860 *Southern Enterprise* 3 Oct. 2/5 He announces himself, in the event of Lincoln's election, as candidate for '*the White House*' of the independent State of Georgia! **1878** *Trans. Illinois Dept. Agric.* XIV. 146 Tecumseh had his thousands of braves encamped above and below Vincennes, Indiana, where Gen. Harrison occupied the 'White House' of this great Northwest. **1947** F. D. Downey *Our Lusty Forefathers* 101 George Washington had been elected President, inaugurated in New York, and had established his 'White House' at No. 3 Cherry Street. **1974** *Encycl. Brit. Micropædia* VIII. 844/1 It [*sc.* San Clemente] gained national prominence in 1969 when Pres. Richard M. Nixon purchased property there for use as a summer White House. **1975** *Caribbean Contact* Feb. 16/1 Speaking with Mr. Ebenezer Joshua at his 'white house' home overlooking the prison compound—as I did during my visit to St. Vincent to cover the recent election—one immediately appreciates the feelings of St. Vincent's radical and disenchanted youth.

2. *Sc.* In north-western Scotland and the Hebrides, a house built of mortared stone; *spec.* one having single-thickness walls cemented with lime mortar. Cf. *black house 2. *Obs.* exc. *Hist.*

1824 J. Macculloch *Highlands & Western Isles* I. 112 The true white house consists of masonry and slate..but the heteroclite, 'kind of white house', is covered with thatch, and, what is much more essential, possesses a chimney. **1870** *Proc. Soc. Antiq. Scotland* VII. 154 The distinctive terms for a house built with lime-mortar, or without it, remain the same... In the northern islands it is still a White-house, and in the Western Highlands it is Tigh-gal. **1955** A. Geddes *Isle of Lewis & Harris* i. 27 Here and there new houses stand out, the 'white houses' (*tighean geala*)..are usually grey... These dwellings have generally been stone-built and were often slate-roofed. **1974** *Northern Stud.* IV. 22 The 1924 [Crofters] Act afforded an opportunity to improve housing and in the next decade the 'white house' began to replace the traditional 'black house' or 'taigh dubh'.

white iron. b. For 'a large proportion of carbon' read 'most or all of its carbon'.

white line, white-line, *sb.* Add: **2. a.** (Later examples.)

1863 [see *em]. **1960** G. A. Glaister *Gloss. Bk* 28/2 *Blank line*, a line which is filled with quads, leads, or blank slugs; a white line in which no letters or other type characters appear.

b. *Engraving.* An engraved line which prints white; the art or technique of using such lines.

1884 H. A. Dobson *Thos. Bewick & his Pupils* 145 The other difference, of which Bewick is said to be the inventor, consisted in the employment of what is known technically as 'whiteline'. **1906** A. Hayden *Chats on Old Prints* iii. 86 He [*sc.* Bewick] was not the inventor of the white line, but he used it freely and adapted his designs accordingly. **1924** H. Furst *Modern Woodcut* i. 10 The black line method keeps the woodprint..in a servile reproductive state, the onus of design falling..on the original designer... The problem of the designer in white line is an entirely different one, requiring..more forethought. **1938** F. Weitenkampf *Illustrated Bk.* ii. 52 For the woodcutter the white line was assuredly an easier method of producing tonal effect than elaborate cross-hatching in black. **1973** *Times* 31 July 10/6 His first two prints were from wood blocks, the wood-engraver's 'white line' being used with decision in the silhouetted *Reclining Nude* of 1931.

3*. Alcohol as a drink; also, one who drinks alcohol. *U.S. slang.*

1908 J. M. Sullivan *Criminal Slang* 27 *White-line*, an alcohol drinker. **1914** Jackson & Hellyer *Vocab. Criminal Slang* 88 *White line, white lime.* Current amongst yeggs and hoboes. Alcohol. Example: 'You'll have to go to the croker and get a stiff for the white line.' **1926** J. Black *You can't Win* vi. 66 'A four-bit micky, a fifty-cent bottle of alcohol—Dr. Hall, white line,' he translated in disgust. **1926** *Flynn's* 16 Jan. 640/1 All we could glom was a shot of white line.

3.** A narrow white strip painted on the road surface to guide or direct motorists; *esp.* one that separates adjacent traffic lanes.

1924 *Oxford Times* 29 Aug. 9/4 The experiment of the white line, which has proved so successful in encouraging the careful driving of motors round corners in Worcestershire, might with advantage be tried in this district. **1930** *Motor* 10 June 892/2 We do think that observations might be directed at white line offences where they occur in really dangerous places. **1971** *Daily Tel.* (Colour Suppl.) 22 Oct. 25/3 White lines broken but close together can mean a corner or hillcrest is coming. **1976** *Evening Chron.* (Newcastle) 26 Nov., Mr. Cook ruled that the council was guilty of maladministration because it could have speeded up the painting of white line markings.

4. white-line dart (moth), *Euxoa tritici*; (earlier and later examples).

1840 J. & M. Loudon tr. *Kollar's Treat. Insects* II. 102 The White-line Dart Moth... A moth injurious to buckwheat and autumn-sown grain. **1948** W. J. Stokoe *Caterpillars Brit. Moths* I. 178 The White-line Dart..is widely distributed.

white man. Add: **2. a.** (Later examples of spelling *whiteman*.) *the white man's burden*: see *burden *sb.* 2 a; *the white man's grave*, equatorial West Africa considered particularly unhealthy for white people.

1836 F. H. Rankin *White Man's Grave* I. p. viii, [Sierra Leone] bears the terrific and poetic title of the 'White Man's Grave'. **1897** M. H. Kingsley *Trav. W. Afr.* 2 My friends..said, 'Oh, you can't possibly go there; that's where Sierra Leone is, the white man's grave, you know.' **1924** Maurice & Arthur *Life Ld. Wolseley* iv. 65 The Gold Coast had well earned the name of 'The White Man's Grave'. **1938** X. Herbert *Capricornia* (1939) iii. 24 The whitemen left the hunting to the [Australian] natives. **1944** F. Clune *Red Heart* 19, I dug up his body, souvenired his false teeth and diaries, and reburied him in whiteman fashion. **1952** P. Atkey *Juniper Rock* xiv. 127, I was a bride at eighteen... I went out to the white man's grave. **1956** A. Sampson *Drum* xi. 156 As whites regard Africans as natives or boys, not people or men so Africans never describe whitemen (which they spell, significantly, in one word), as *abantu*, or people. **1970** G. F. Newman *Sir, You Bastard* ii. 67 The street in Hammersmith where Whitmarsh lodged was so overrun with immigrants that an English-speaking whiteman was a latterday Livingstone.

white meat, whitemeat. Restrict *Obs.* exc. *dial.* to senses a, c, and d and add: **b.** Also *sing.*

1973 *Guardian* 19 Mar. 7/5 People in Britain ate less red meat last year, but more 'white' meat such as pork and poultry. **1975** P. V. Price *Taste of Wine* vii. 134/1 White meat is generally taken to include pork, veal, chicken and turkey, while the 'reds' are beef, lamb, duck and goose. **1985** *Which?* Feb. 54/2 In general, nutritionists recommend that the average British diet should contain less fat and more fibre... They say a good way to achieve this change is to eat more white meat (especially poultry) and fish in place of red meat.

2. *slang* (chiefly *U.S.*). White women considered as sexual partners or conquests. Cf. *meat *sb.* 3 e.

[**1937** *Printers' Ink Monthly* May 45/3 *White meat*, an actress.] **1940** 'J. Crad' *Traders in Women* v. 134 The.. liner took me to Shanghai, and here once again I met the European procurer and selling of 'white meat'. **1972** [see *home *sb.*[1] 13*]. **1976** M. Maguire *Scratchproof* x. 152 I'm off white meat. I have a good thing going with a negro film editor. **1982** J. Philips *Target for Tragedy* (1983) I. iii. 52 Some stranger who sees a piece of white meat he thinks might come his way.

whitener. 2. (Later example.)

1971 D. Potter *Brit. Eliz. Stamps* ii. 22 Optical whitener is used in the making of the paper.

white-out (hwəi·taut). Also **whiteout.** [f. phr. *to white out* (cf. *white *v.*[1]), by analogy with *black-out.] **1. a.** *N. Amer.* A heavy snow-storm, a blizzard.

1942 *Sun* (Baltimore) 30 Mar. 8 (*caption*) Whiteout. **1980** *Sat. Rev.* (U.S.) May 66 Blizzards—white-outs they call them here [*sc.* in Labrador]—bring snow that whirls and thrashes and blinds, stinging noses, and cabins disappear in the whiteness.

b. A condition in which neither shadows nor the horizon can be seen and physical features are lost in the background, caused by an evenness of lighting such as sometimes occurs in cloud or in snow- or ice-covered regions.

1946 *Sun* (Baltimore) 20 Apr. 7/2 Hedine, of the United States Weather Bureau at Winnemucca, Nevada, described the 'Arctic whiteout' today, defining it as a condition of the snow country wherein all land features are camouflaged, 'blending earth and sky so that the horizon and all landmarks are indistinguishable'. **1955** *Sci. Amer.* Apr. 54/3 Lieutenant John P. Moore, a Navy pilot, was killed when his helicopter crashed during a 'white-out'. This condition, one of the chief hazards of Antarctic travel, occurs when sunlight diffuses through a solid overcast. **1959** V. Fuchs *Antarctic Adv.* vii. 96 A whiteout is something like a blackout in reverse... No surface irregularities in the snow are visible in the diffused, opaque light, but a dark object like a man or a vehicle may be clearly seen. **1966** F. Hoyle *October 1st* viii. 95 It was impossible to know whether you were looking ten yards..or even a hundred miles. The effect..was far more weird than the kind of white-out you sometimes get on a snowfield in the mountains. **1976** M. Machlin *Pipeline* i. 7 White-out had set in just after Takolik had seen The-Man-Who-Hides. **1980** *Daily Tel.* 30 July 16 Bad weather, including white-outs caused by low cloud, hampered the early stages and Bonington is reported as saying he can well understand why the peak has never been climbed before. **1984** *Times* 5 Jan. 9/4 When whiteout exists, by the interaction of sunlight, snow, cloud and reflection, it induces the belief in a pilot that he is flying over flat terrain with unlimited forward visibility.

2. A white liquid that can be brushed on to paper to obliterate marks and provide a white surface on which to type or write afresh.

1977 L. O'Donnell *Aftershock* xiii. 180 You changed the date... Did you cover the original entries with a strip of paper, or did you use white-out? **1984** *New Yorker* 23 Jan. 44/2 A Chinese version of typists' white-out.

white pine. [White *a.* 11 b.] **1.** *N. Amer.* Any of several North American pines, esp. *Pinus strobus*, which is native to eastern and central parts of the continent; also, the pale soft wood of such a tree. Cf. Weymouth.

1682 *Early Rec. Providence, Rhode Island* (1899) XIV. 113 From ye said heape of stones to range north..to a great white pine. **1767** *Quebec Gaz.* 8 Dec. 3/1 They are hereby forbid to cut down..White Pine..on the lands above described. **1785**, etc. [see Pine *sb.*[2] 2]. **1948** *Reader's Digest* Jan. 68/2 Of all American woods none has been more significant than white pine. **1961** H. MacLennan *Rivers of Canada* 97 When Wright surveyed the Ottawa forests he found an abundance of white pine standing two hundred feet tall. **1973** A. H. Whiteford *North Amer. Indian Arts* 107 White pine bark canoes were made by the Kutenai. **1974** M. Braithwaite *Ontario* ix. 134 All about him..were great stands of white pines, tall trees stretching as straight as a ruler.

2. *Austral.* A tree belonging to any of several species of *Podocarpus* or *Callitris*. Cf. Kahikatea, *podocarp.

1855, etc. [see Pine *sb.*[2] 2]. **1975** D. Bagley *Snow Tiger* ii. 33 Gone were the stands of tall white pine and cedar, of kahikatea and kohekohe.

3. Special Combs.: **white pine blister (rust)**, a rust disease of certain pines, caused by the fungus *Cronartium ribicola*, which spends part of its life cycle on gooseberry or currant bushes; **white pine weevil**, the larva of a brown beetle, *Pissodes strobi*, which tunnels in new shoots of certain pines.

1911 *Bull. U.S. Bureau Plant Industry* No. 206. 9 The white-pine blister rust now imported into this country from Germany is caused by a heterœcious fungus. **1974** M. Hoyt *Thirty Miles* vi. 66 We had currant bushes..before anybody knew they were an intermediate host to..white-pine blister. **1905** *Bull. Forestry Bureau* (U.S. Dept. Agric.) No. 63. 14 The white pine weevil..is a reddish-brown snout beetle. **1976** *Columbus* (Montana) *News* 27 May (Joliet Suppl.) 4/5 Whitepine weevil..can kill twigs and branches of some evergreens.

White Russian, *sb.* and *a.* [f. White *a.* + Russian *sb.* and *a.*; cf. *Belorussian *a.* and *sb.*] **A.** *sb.* **1. a.** The Russian dialect or language spoken in Belorussia, a district in the western part of Russia which is now one of the constituent republics of the Soviet Union.

1850 'Talvi' *Hist. View Lang. & Lit. Slavic Nations* II. i. 51 The White-Russian is the dialect spoken in Lithuania and a portion of White Russia, especially Volhynia. **1932** C. A. Phillips tr. *H. von Eckardt's Russia* vi. ii. 475 Up to the sixteenth century the Lithuanian Grand Princes and boyars regarded White Russian as their language. **1949** Entwistle & Morison *Russian & Slavonic Lang.* i. 30 White Russian became a chancery language, not a literary tongue. **1960** W. K. Matthews *Russian Hist. Gram.* I. ii. 34 There are seven groups of Slavonic languages, viz. the East Slavonic (Russian, White Russian, and Ukrainian) [etc.].

b. A native or inhabitant of Belorussia.

1886 [see Russian *sb.* 1 a]. **1912** [in Dict. s.v. White *a.* 11 e]. **1918** R. Wilton *Russia's Agony* i. 9 The White Russians, a comparatively small section of the Northern Slav people, inhabiting Smolensk and the upper reaches of the Dnieper. **1960** W. K. Matthews *Russian Hist. Gram.* III. ii. 309 The grammatical treatises of the time.. are mainly the work of non-Russian scholars—White Russians and Ukrainians.

2. = *white *sb.* 19 b.

1927 *Daily Tel.* 29 Mar. 11/6 The White Russians in the Northern Army..were purely soldiers, while the Reds were carrying on propaganda. **1930** *Times* 17 Mar. 12/6 Yesterday afternoon 'White' Russians, most of them women, made a demonstration at the offices of Amtorg Trading Corporation, the Soviet's American commercial agency. **1943** tr. *N. Basseches's Unknown Army* iv. 59 If they were in the territory of the Soviets, they either broke through in time to join the White Russians or they were caught by the Red mobilization. **1973** 'D. Halliday' *Dolly & Starry Bird* xii. 184 Innes was sitting looking at the Director like a White Russian receiving word of Biological Ajax. **1976** *New Yorker* 15 Nov. 39/1, I was surrounded by..some of those privileged White Russians who abandoned their first-adopted countries of Europe to come to the States at the onset of the Second World War.

B. *adj.* **1.** Of or pertaining to Belorussia or its people.

1886 *Encycl. Brit.* XXI. 71/1 In 1879 in European Russia,—exclusive of six Lithuanian and White Russian governments,—42,530 persons were tried before the courts. **1918** Trofimov & Scott *Handbk. Russian* I. 1. 4 The White Russian dialect covers the smallest area of all the Russian dialects. **1926** L. H. Guest *New Russia* i. 21 The following are Independent Republics: The Ukrainian Socialist Soviet Republic..The White Russian Socialist Soviet Republic. **1944** [see *Belorussian *a.* and *sb.*]. **1960** W. K. Matthews *Russian Hist. Gram.* II. xii. 276 In the latter part of the seventeenth century the influence of the White Russian and Ukrainian scholars and writers began to be felt.

2. Of or pertaining to the Whites in the Russian Civil War.

1929 W. S. Churchill *World Crisis* V. xii. 247 We have seen them [sc. the Czechs] already in October 1918.. exasperated by White Russian mismanagement. 1957 P. Kemp *Mine were of Trouble* iii. 39 The Requetés were raising two squadrons in Seville, under a White Russian colonel named Alkon. 1964 R. Perry *World of Tiger* ii. 24 The White Russian hunter Yankovsky. 1974 *Encycl. Brit. Macropædia* XVI. 70/2 The Red Army..drove him [sc. Wrangel] and his army into exile. There remained only the Japanese and White Russian forces in eastern Siberia.

whitesmithery. (Earlier example.)
1812 *Niles' Weekly Reg.* 25 Jan. 390/2 Emery..is an article of the first consequence in the cotton and woolen manufactures, and in white smithery.

whitesmithing (hwəi·t,smiþiŋ), *vbl. sb.* [f. Whitesmith¹ + -ing¹.] = Whitesmithery.
1835 *Lexington* (Kentucky) *Observer* 10 June, White-smithing. Frederick Klaiber lately from Germany..has just commenced the above business. 1900 *Daily Chron.* 2 Jan. 3/1 Part of the bench at which the missionary-explorer learnt whitesmithing is exhibited.

white stick. Add: **3.** A white walking-stick carried by a blind person both as a distinguishing feature and to locate obstacles. Cf. *white cane* s.v. *white *a.* 11 e.
1961 *A.A. Handbk.* 20 Responsible blind welfare organizations strongly recommend all blind persons to carry a white stick. 1967 S. Beckett *Stories & Texts for Nothing* viii. 110 But what is this I see, and how, a white stick and an ear-trumpet. 1974 *Times* 21 Feb. 10 His first perilous adventures with the white stick. 1978 'H. Carmichael' *Life Cycle* xiv. 150 The man who doesn't admire you shouldn't be allowed out in the street without a white stick.

white-tail. **2.** For *Cariacus* read *Odocoileus.* (Earlier and later examples.)
1872 R. G. McClellan *Golden State* 241 There are several varieties: the mule-deer, black-tail, antelope, and white-tail. 1936 D. McCowan *Animals Canad. Rockies* vii. 59 The hoofs of the wapiti and whitetail deer are too small to propel these animals through the water with any great speed. 1968 R. Kroetsch *Alberta* i. 12 White-tails and mule deer and mallards and grouse tumble before the unerring aim. 1980 *Hunting Ann. 1981* 29/1 My mountain-hunting buddy..came down out of his renowned mule deer country..to join me in a search of a big Colorado whitetail.

white-tailed (hwəi·t,tēⁱld), *a.* [White *a.* 12 c.] Having a white tail; *white-tailed deer* = White-tail 2 in Dict. and Suppl.; *white-tailed eagle*, the European sea eagle (see Eagle *sb.* 1 a); *white-tailed gnu*, the common (as distinct from the brindled) gnu, *Connochætes gnou*; *white-tailed ptarmigan*, a ptarmigan, *Lagopus leucurus*, found in western North America.
1642 [in Dict. s.v. White *a.* 12 c]. 1678 J. Ray tr. *Willughby's Ornithol.* ii. 61 Of the Pygarg or white-tail'd Eagle. 1832 W. A. Ferris *Jrnl.* 27 Aug. in *Life in Rocky Mts.* (1940) 131 In the afternoon..we killed a white-tailed fawn. 1887 White-tailed deer [in Dict. s.v. White *a.* 12 c]. 1889 *Cent. Dict.*, (caption s.v. *Gnu*) Common or White-tailed Gnu. *Ibid.*, White-tailed ptarmigan. 1912 J. Stevenson-Hamilton *Animal Life Afr.* vii. 106 There is no more remarkable beast, either in appearance or manners,..than the white-tailed gnu. 1926 F. C. R. Jourdain in J. J. Walker *Nat. Hist. Oxford Distr.* 146 White-tailed Eagle..has occurred on Wantage Downs. 1941 J. S. Huxley *Uniqueness of Man* viii. 184 Other species now exist only in captivity. Such are the beautiful and fantastic white-tailed gnus. 1948 A. L. Rand *Mammals Eastern Rockies* 208 When the alarmed white-tailed deer goes bounding away, its tail usually stands straight up, and it is a great snowy banner that leaves no doubt of identity. 1968 R. Kroetsch *Alberta* ii. 59 Soon after, we saw two white-tailed deer, just on the timberline. 1973 *Islander* (Victoria, B.C.) 1 Apr. 2/2, I have found myself staring at a whitetailed ptarmigan in the high mountains believing that he was snow. 1981 *Birds* Autumn 55/3 (caption) White-tailed eagle—last bred in Britain in 1916... A scheme to reintroduce these birds to Scotland is now underway.

whitethroat. **2.** (Earlier and later examples.)
a 1862 Thoreau *Maine Woods* (1864) 198 We heard the white throats along the shore. 1916 D. C. Scott *Poems* (1927) 47 A rocky islet followed With one lone poplar and a single nest Of white-throat-sparrows that took no rest. 1939 [see *Peabody]. 1978 A. Lampman *Lyrics of Earth* 32 The white-throat's distant descant with slow stress Note after note upon the noonday falls.

whitewash, *sb.* Add: **2.** (Earlier *attrib.* examples.)
1814 *Austin Papers* (1924) I. 240, 1 White Wash Brush. 1848 D. G. Rossetti *Let.* 20 Jan. (1965) I. 34 All my traps have been moved up into an attic, to make room for ladders, whitewash-pails, and such-like gear.
4. (Further examples.) For *U.S. colloq.* read *colloq.* (orig. *U.S.*). Also, a victory in a series of games of which the opponents fail to win any.
1867 *N.Y. Clipper* 31 Aug. 164/2 The first 'whitewash' of the [baseball] game was drawn by the Mutuals. 1874 *State Jrnl.* (Lincoln, Nebraska) 26 June 4/1 The second match game of croquet took place yesterday morning, and resulted in a second whitewash for the latter named gentleman. 1961 *Times* 4 May 4/6 Miss Truman who yesterday allowed Mrs. Cawthorn but 23 points in what the players of darts would term a 'whitewash'. 1962 *Times* 26 May 3/5 England nearly scored a white-wash over France..only the victory of G. Mourgue d'Algue standing between them and a 12–0 lead on the first day. 1977 *Evening Gaz.* (Middlesbrough) 11 Jan. 14/1 Only one whitewash this week in the Friendly League. 1978 *Rugby World* Apr. 4/1 Scotland must be bitterly disappointed that they have suffered their first whitewash for ten years.
†**5.** *slang.* (See quot. 1864.) Cf. Whitewasher 3. *Obs.*
1864 Hotten *Slang Dict.* 270 *Whitewash*, a glass of sherry as a finale, after drinking port and claret. 1879 Trollope *John Caldigate* III. x. 142 'Take another glass of port, old boy.' Bagwax did take another glass, finishing the bottle... 'Take a drop of whitewash to wind up, and then we'll join the ladies.'
6. *Comb.*: **whitewash gum**, either of two eucalypts with powdery white bark, *Eucalyptus apodophylla* and *E. terminalis*, found in northern and central Australia.
[1926 J. M. Black *Flora S. Austral.* iii. 420 *E[ucalyptus] terminalis*... Whitewashed gum; bloodwood.] 1934 *Bulletin* (Sydney) 2 May 21/2 The whitewash gum..forms a striking feature of the landscape about Alice Springs. 1965 *Austral. Encycl.* III. 406/2 Whitebark or 'whitewash gum'..of Arnhem Land has perfectly smooth trunks covered with a white mealy 'bloom' that rubs off when touched.

whitewash, *v.* **3.** For *U.S. colloq.* read *colloq.* (orig. *U.S.*). Also *loosely*, to beat by a large margin. (Earlier and later examples.)
1867 *Chicago Republican* 6 July 2/6 The Unions were whitewashed 3 times, and the Forest Citys 5 times. 1972 *Korea Times* 19 Nov. 1/5 Husky south Korean girls white-washed Thailand 106–17..in the second game. 1981 R. Lewis *Seek for Justice* vi. 193 He took the first game [of darts]... He all but whitewashed Freddy in the second.

whitewashed, *ppl. a.* Add: **3.** *whitewashed American, Yank,* or *Yankee*, a person who affects American manners, or who has spent a short time in America; also *transf.*
1855 in *Occas. Papers Univ. Sydney Austral. Lang. Res. Centre* (1966) No. 10. 26 'I have heard people say they would like to see us clear altogether of British rule.'.. 'Have you heard that said here?'—'Yes, by a few of those disaffected persons; very few; they are generally what are termed "white-washed Yankees".' 1898 A. J. Boyd *Shellback* 73 He was not one of the low, bullying, half-Irish, half-American sort of men who are called 'whitewashed Yankees'. 1926 W. S. Dill *Long Day* 147 This particular story concerns a 'white-washed American', *i.e.* a native of Canada who had been naturalized in the United States and then secured repatriation in his own country. 1938 F. A. Worsley *First Voy. in Square-Rigged Ship* 82 Whitewashed Yanks (Europeans who had served a voyage in American ships or spent a short period in the States) were numerous. 1970 J. F. Leavitt *Wake of Coasters* 62/2 Some of the schooners in later years were 'white-washed yankees': American built vessels kept under U.S. registry but with the controlling interest actually owned across the border in New Brunswick or Nova Scotia.

whitewasher. Add: **1.** (Earlier example.)
1733 *S. Carolina Gaz.* 24 Feb., He's a Bricklayer, Plaisterer and White-washer.
2. (Earlier example.)
1820 M. Wilmot *Let.* 27 Sept. (1935) 84 On recollection his cause is too *good* to be successful in such clever hands as her [*sc.* Queen Caroline's] whitewashers.

whiteweed. (Earlier example.)
1803 in *Mass. Hist. Soc. Coll.* (1804) IX. 200 On the upland and meadows grow burdens grass, ribwort, white weed, [etc.].

Whitgiftian (hwitgi·ftiăn), *sb.* and *a.* [f. *Whitgift* + -ian.] **A.** *sb.* A pupil or former pupil of Whitgift School, Croydon. **B.** *adj.* Of, pertaining to, or characteristic of John Whitgift (*c* 1530–1604), Archbishop of Canterbury and founder of Whitgift School.
1880 *Whitgift Mag.* Jan. 13/2 We were glad to notice among the Chorus several 'Old Whitgiftians'. 1905 (*title*) The Whitgiftian. [*Previously* The Whitgift Magazine.] 1962 *Hist. Mag. Protestant Episcopal Church U.S.* XXXI. 128 The picturesquely rhetorical phrase of F. W. Maitland has been considered the most decent dismissal of the whole Whitgiftian flavour: 'a remorseless predestinarian'. 1967 P. Collison *Elizabethan Puritan Movem.* v. i. 245 Of this generation of clergy, few with minds of their own would subscribe to the Whitgiftian formula without a qualm. 1977 P. Clark *Eng. Provincial Soc. from Reformation to Revolution* v. 184 In the county [of Kent]..the Whitgiftian reaction caused a marked polarisation between moderate Puritans and conformist Presbyterians on the one hand, and less respectable radicals and separatists on the other.

whither, *adv.* Add: **2. b.** Followed by a single word or short phrase.
1982 *English Studies* LXV. 90 The recently recycled interrogative adverb *whither* (as in *Whither Democracy?*).

whiting, *sb.* Add: **1. b.** (*d*) *blue whiting*, an oceanic fish of the cod family, *Micromesistius poutassou*, found in north-western Europe and the Mediterranean; = Poutassou.
1959 A. C. Hardy *Open Sea* II. xi. 229 The blue whiting..lives over the deep water off the edge of the continental shelf. 1974 *Guardian* 20 Mar. 11/1 The blue whiting..cod-like in taste and texture, slender in shape, about a foot long. 1977 *Grimsby Even. Tel.* 5 May 8/4 Certainly its size makes it an easier fish to process than the more publicised blue whiting.

whitishness. (Later example.)
1929 S. Leslie *Anglo-Catholic* xii. 158 In the lamplight he noticed her deathliness of hue, the whitishness of lead-poisoning.

whitleather. Add: **1. a.** α. (Later example.)
1960 G. E. Evans *Horse in Furrow* xvii. 213 Sidney Austin, the harness-maker, still uses strips of whitleather to repair..the collars of farm-horses.
b. (Later example.)
1913 D. H. Lawrence *Love Poems & Others* 44 A widow o forty-five As has sludged like a horse all her life, Till 'er's tough as whit-leather.

Whitley (hwi·tli). The name of J. H. *Whitley* (1866–1935), chairman of a committee set up in 1916 to consider relations between employers and employees, used *attrib.* with reference to the recommendations of this committee concerning good industrial relations, etc.
1917 in *State Service* (1969) Sept. 226/3 The application of the Whitley report should be extended to occupations of a purely commercial or clerical character. 1919 *Manch. Guardian* 11 Feb. 7/3 (*heading*) The Whitley councils. 1923 *Daily Mail* 29 Jan. 7 Mistress and maid should be their own Whitley Council. 1924 *Glasgow Herald* 20 Sept. 11 During the war and after the war Whitley bodies were set up in industries which up till then had nothing of the kind. *Ibid.*, The Whitley machinery could be used to discuss reduction in wages. 1928 *Britain's Industr. Future* (Liberal Industr. Inquiry) iii. v. 174 It is important to understand the causes of the limited degree of success which has attended the Whitley scheme. 1976 *Star* (Sheffield) 3 Dec. 5/2 He recommends that a national forum be set up where Ministers can discuss policies with staff representatives, along with new regional Whitley Councils, and local committees.
Hence **Whi·tleyism**, the use of Whitley Councils or similar methods for dealing with relations between employers and employees.
1919 *Manch. Guardian* 28 Feb. 14/4 Judge Parry..criticised the bureaucracy for its failure to apply the principles of 'Whitleyism' to departments of the Government service. 1928 *Daily Tel.* 14 Aug. 10/6 There has now followed a striking development, completing the destruction of Whitleyism in the Post Office. 1969 *State Service* Sept. 226/1 (*heading*) Whitleyism in the Civil Service.

whitlockite (hwi·tlǫkəit). *Min.* [f. the name of Herbert P. *Whitlock* (1868–1948), U.S. mineralogist + -ite¹.] A calcium hydrogen phosphate containing ferrous iron and magnesium, $Ca_9(Mg,Fe)H(PO_4)_7$, found as transparent or translucent rhombohedral crystals of various colours and often occurring in dental calculi.
1940 C. Frondel in *Program & Abstr. 21st Ann. Meeting Mineral. Soc. Amer.* 7 Whitlockite is anhydrous calcium triphosphate..with Ca substituted for by Mg.. and Fe... The mineral is named after Herbert P. Whitlock,..at present Curator of Minerals and Gems in the American Museum of Natural History. 1971 *Nature* 3 Dec. 264/1 The rock contains relatively small amounts of the phases that we have found in other Apollo basalts (..whitlockite, baddeleyite). 1979 Williams & Elliott *Dental Biochem.* xii. 226 Whitlockite is more common in subgingival compared with supragingival calculus.

Whitmanesque (hwitmăne·sk), *a.* [f. the name *Whitman* + -esque.] Characteristic or suggestive of Walt Whitman (1819–92), U.S. poet, or of his poetry.
1882 *Good Lit.* Sept. 2 Clever persons can manufacture Whitmanesque verse quite equal to the average of the original. 1901 E. Crosby *Edward Carpenter* 6 The long series of poems in Towards Democracy is with few exceptions written in the Whitmanesque meter, or lack of meter. 1913 W. De La Mare in *Edin. Rev.* Jan. 193 Eloquence and facility are the danger of Whitmanesque verse of this nature. 1934 C. Lambert *Music Ho!* v. 281 There is very little Whitmanesque acceptance of life about the artist of today. 1957 P. Wildeblood *Main Chance* 152 They have a Whitmanesque simplicity that we've quite lost. 1977 *Time* 1 Aug. 50/2 Such a collage has an effect of Whitmanesque tenderness.
So (mostly somewhat *nonce*) **Whitmane·se**, the characteristic style or diction of Whitman; **Whitma·nia**, (*a*) [-mania], (a punning word for) exaggerated admiration for Whitman; (*b*) [-ia¹], writings pertaining to Whitman; **Whitma·niac**, a devotee of Whitman; **Whitma·nian** *a.* = *Whitmanesque *a.*; **Whitma·nian** *sb.*, an admirer or imitator of Whitman; **Whi·tmanish** *a.* = *Whitman-

esque *a.*; **Whi·tmanism**, Whitman's metrical or poetical style; a feature of this; **Whi·tmanist**, **Whi·tmanite**, a Whitmanian; **Whi·tmanize** *v. intr.*, to write in the manner of Whitman; **Whitma·nnic** *a.* = *Whitmanesque *a.*
1887 Whitmania [see *Brontëan *a.*]. 1887 M. Berenson *Let.* 6 Jan. in Strachey & Samuels *M. Berenson* (1983) ii. 36, I was a Whitmanite at Smith College. 1889 *Pall Mall Gaz.* 25 Jan. 3/2 Having thus to a certain degree settled upon what one might call the *technique* of Whitmanism, he began to brood upon the nature of that spirit that was to give life to the strange form. 1893 R. Le Gallienne *Retrosp. Rev.* (1896) I. 213 'I see twenty-two young men from Foster's watching me, and the trousers of the twenty-two young men' is irresistible Whitmanese. 1894 *Nation* 7 June 433/1 One of the worst of Whitmanisms, the interlarding of foreign words. 1902 *Academy* 16 Aug. 173/1 Mr. Moody does not Whitmanise on the one hand, or follow the outworn Tennysonian convention on the other. 1906 *Dial* (Chicago) 1 Mar. 144/2 Much of the conversation reported is trivial to all but ardent Whitmanites. 1918 *Cambr. Hist. Amer. Lit.* II. iii. i. 267 Whitmanism..has already had the ironical fate of developing something not unlike a cult. 1934 *Times Lit. Suppl.* 30 Aug. 586/3 Before Rossetti established himself publicly as the principal English Whitmanist, 'Leaves of Grass' had been the subject of several reviews. a 1930 D. H. Lawrence *Phoenix* (1936) 269 Whitmanish 'adhesiveness' of the social creature. 1948 L. Spitzer *Linguistics & Lit. Hist.* 218 The first [*sc.* the old alexandrine], Claudel replaced by the Biblical and Whitmanian verset. 1953 A. Alpers *Katherine Mansfield* 124 Thus reminded that she had a country of her own, Katherine addressed to Wyspiański another of her Whitmanish declamations. 1959 *Times Lit. Suppl.* 16 Oct. 594/4 A foreword by Mr. Charles E. Feinberg, the noted Whitmaniac of Detroit. a 1960 E. M. Forster *Maurice* (1971) 217 Edward Carpenter..was..a Whitmannic poet whose nobility exceeded his strength. 1964 *New Statesman* 13 Mar. 414/3 The presses groan with Whitmania. 1977 *Listener* 30 June 866/3 The Fabian Society..sprang from an idealistic society called the Fellowship of the New Life, much influenced by the Whitmanian, Edward Carpenter.

whitmoreite (hwi·tmǫrəit). *Min.* [f. the name of Robert W. *Whitmore* (b. 1936), U.S. mineral collector + -ite¹.] A secondary hydrated basic phosphate of ferric and ferrous iron, $Fe^{2+}Fe^{3+}_2(PO_4)_2(OH)_2.4H_2O$, found as twinned monoclinic crystals of a brownish colour.
1974 P. B. Moore et al. in *Amer. Mineralogist* LIX. 900/2 Whitmoreite occurs as thin acicular crystals five to ten times as long as they are thick, which range from 0·1 to 2 mm in length. 1979 *Mineral. Abstr.* XXX. 450/1 The occurrence and parageneses of the following newly recognized secondary phosphates in the pegmatite of Hagendorf, West Germany, are recorded: whitmoreite, schoonerite, [etc.].

Whitstable (hwi·tstăb'l). The name of a coastal town in Kent, used *attrib.* and *absol.* to designate oysters bred there.
1883 *Queen* 20 Oct. (Advt.), Any others that are advertised at a low price..*cannot possibly* be the *genuine* Medina or Whitstable Oysters. 1940 A. L. Simon *Conc. Encycl. Gastron.* II. 69/2 Most Whitstable oysters to-day are..relaid Brittanys or Belons. The oysters known as *Royal Whitstables* are, however, genuine natives, taken from a breeding ground the boundaries of which were settled by law about 1900. 1960 *Times* 2 Nov. 13/6 They may not be Whitstables, but they are oysters of a kind. 1971 *Vogue* 15 Sept. 43/1 One of London's best fishmongers. Ask for fresh sardines, Whitstable oysters, game. 1973 'J. Sturrock' *Wicked Way to Die* x. 142 They've got as fine a barrel of Whitstables here as ever I've seen.

whitter, *v.* Restrict *Sc.* to senses in Dict. and add: **3.** Now usu. in form **witter**. To chatter or mutter; to grumble; to speak with annoying lengthiness on trivial matters. Occas. *trans.* Freq. const. *on*. Hence **w(h)i·ttering** *ppl. a. colloq.* (orig. *Sc.* and *dial.*).
1808 A. Scott *Poems* 82 The winking swankies whitter, An' fondly ee some female band. 1854 A. E. Baker *Northamptonshire Gloss.*, *Whitter*, to murmur, to grumble, to complain... 'Don't *whitter* so.' 1886 R. E. G. Cole *Gloss. Words S.W. Lincs.* 168, I witter my-sen at times, and my husband tells me I'm a regular wittering old woman. 1925 E. C. Smith *Mang Howes* 21 A clecken o guidweives at a gairdeen-yett whuttert ti other whan they eyed iz. 1959 [see *suicide blonde* s.v. *suicide *sb.*² d]. 1966 'O. Mills' *Enemies of Bride* ii. 16 You might..try making the tea, instead of wittering on about Cordon Bleu methods. 1973 *Where* Jan. 13/2 Don't whitter away at every item [on the agenda], giving up at the first unsatisfactory explanation. Make your choice of issue, then take your time. 1981 R. D. Edwards *Corridors of Death* i. 4 The questions which those who had spotted him as the man-in-the-know were wittering at him. *Ibid.* xxxvi. 164 It wasn't like Robert to witter on like this. 1982 *Observer* 3 Oct. 9/2 If I wasn't going to hear the Tories wittering on in Brighton this week, I'd be in Frankfurt listening to publishers wittering on at the annual Book Fair. 1983 *Listener* 20 Jan. 5/2 A really wittering, patronising speech programme is a worse insult to the intelligence than the most fatuous disc jockey.

whittle, *v.*² Add: **I. 1. a.** Also with *down* (cf. sense 2).
1972 D. Bloodworth *Any Number can Play* xii. 103 A young orang..tried poking it [*sc.* a hole in a log] with a twig that was too thick, then whittled down the twig. 1979 J. Harvey *Plate Shop* xv. 72 Ted put his feet up on the tin waste-paper box..and started absorbedly hewing and whittling a pencil down to the stub.
II. 4. *intr.* To worry or fret. Occas. *trans. dial.*
1880 *N. & Q.* 6 Mar. 205/2 When I was a boy my mother daily used this word to express fidgetiness or uneasiness. 'What are you *whittling* about?' seems to ring in my ears at this moment. 1913 D. H. Lawrence *Sons & Lovers* viii. 202 'How do you think I'm going to manage?' 'Well, it won't make it any better to whittle about it.' 1984 *Daily Tel.* 23 Oct. 10/3 'I'm whittled to death about the future of the mining industry.' These, or words like these, are attributed to Mr. Michael Eaton, the new character in the long-running serial story of the mining dispute.

whittled *ppl. a.*: also *whittled-down*.
1961 A. Brownjohn in E. Lucie-Smith *Brit. Poetry since 1945* (1970) 266 Farmers call hillocks And ponds.. By the first words to hand; a heavy, whittled-down Simplicity meets the need. 1962 E. Snow *Other Side of River* (1963) xxiv. 183 If any of these somewhat blind guesses are right, the whittled-down results still remain impressive. 1980 M. Booth *Bad Track* ii. 34 Long streets with a whittled-down green, a church..a pub.

Whitworth. Add: **b.** Used *attrib.* and † in the possessive to designate a series of screw threads proposed by Whitworth in 1841 (and later additions to it), fasteners having one of these threads, and tools for use with the fasteners.
[1841 *Proc. Inst. Civil Engineers* I. 157 (*heading*) 'On an uniform system of screw threads.' By Joseph Whitworth.] 1877 *Calvert's Mechanics' Almanack* 4 The terms, 'Whitworth's Threads', 'Whitworth Taps'..have sprung from the lips of all concerned with the iron trade. 1916 *Proc. Inst. Automobile Engineers* XI. 176 We all imagined that the Whitworth system for the larger sizes of screws was so perfect that there would be no difficulty in making sure that Whitworth nuts would fit Whitworth bolts sufficiently well for all practical purposes. 1968 J. Arnold *Shell Bk. Country Crafts* 160 Until the advent of the Whitworth thread, which standardized threads all over the country, it was the practice for smiths to tap their own threads. 1970 *Kay & Co.* (Worcester) *Catal. 1970–71* Autumn/Winter 770/2 Ring Spanners... Available in Whitworth, AF or Metric sizes. 1972 *Practical Motorist* Oct. 209/1 Whitworth fasteners are no longer in general use, although you will encounter them on older cars.

whity, *sb.* Add: Also **whitey** (the usual form), **whitie**. **1.** (Also with capital initial.) Also, a white woman; white people collectively. Freq. derog. *slang* (chiefly *Blacks'*).
1942 Berrey & Van den Bark *Amer. Thes. Slang* § 385/2 *White person*,..whitie. 1952 S. Selvon *Brighter Sun* iv. 61 A white-skinned girl..was called 'Whitey cockroach!' 1964 *Time* 31 July 12/3 Harlem..is where the white man is no longer the 'ofay' but 'Mr. Charlie' or 'the man', and mostly 'whitey', derived from the Black Nationalist talk of 'the blue-eyed white devil'. 1967 C. Drummond *Death at Furlong Post* xi. 138 Get to hell away from me! You Whities stink! 1968 *Times Lit. Suppl.* 4 Apr. 329/2 The world of 'Whitey' into which these Negroes no longer want to be integrated. 1971 A. King *One Love* 19 There's a Whitey in every Black man that has to come out, or die, before he's ever himself. 1972 R. K. Smith *Ransom* i. 24 We're gonna hit Whitey and hit him again. 1976 *Listener* 15 Apr. 462/1 There is a pub in south London where black intellectuals meet, and if you happen to be a white man, the landlord—who is of West Indian origin—delights in calling you 'whitie'. As far as he is concerned, it's all in fun. 1977 *New Yorker* 26 Sept. 131/1 He's no more than a trivial whitey to be squished. 1980 *Amer. Speech* LV. 211 It encompassed a protest of whitey's 'theft' of yet another style of jazz—swing.

whizgig. (Earlier example.)
1821 M. Edgeworth *Let.* 22 Nov. (1971) 279 A Whizgig for Pakenham in my next.

whizz, whiz, *sb.*¹ Add: **1. b.** The practice of picking pockets (chiefly in phr. *on the whizz*); a pickpocket. *slang* (orig. and chiefly *U.S.*).
1925 E. Jervis *25 Yrs. in Six Prisons* i. 17 Some of the boys are 'on the whiz' (pickpockets). 1931 *Amer. Speech* VII. 117 *Whiz*, n. A pick-pocket. 1936 'J. Curtis' *Gilt Kid* 245 They might pinch him for being on the whizz. 1963 T. Tullett *Inside Interpol* xii. 162 The pickpocket, known in the underworld as the 'whiz'..is always a specialist.
3. An act of urination. *slang.*
1971 D. Clark *Sick to Death* i. 21 She could have left him alone..while she went for a whizz or changed her clothes.
4. *attrib.* and *Comb.*, as **whizz-boy, -man** *slang*, a pickpocket; **whizz-mob** *slang*, a gang of pickpockets.
1931 M. Allingham *Police at Funeral* vii. 95 How many murders do we get in this class... It's navvies, whizz-boys, car thieves..who run off the rails and commit murder. 1938 F. Chester *Shot Full* xxv. 285, I used to frequent a number of public-houses, used by 'the boys', as criminals are known among the English. There were screwsmen,..'whizz-men',..and 'drag-men'. 1959 *Listener* 12 Mar. 485/1 The quick-fingered craft of those whom the Elizabethans called nips and we call whizz boys. 1932 'S. Wood' *Shades Prison House* xix. 278 There one may rub shoulders with..thieves of every type, whiz-men, burglars, car-bandits. 1929 G. Dilnot *Triumphs of Detection* iv. 47 A 'wizz mob' which operated, say, at Hammersmith Broadway, would immediately suspend business..if they saw a local detective in the vicinity. 1941 J. Phelan *Murder by Numbers* v. 53 'Putting a smother they call it... Crowd cover up something...' 'I see. Like a whizz mob—pickpockets, I mean.' 1955 D. Webb *Deadline for Crime* iii. 52 Provincial police forces looked to him for help when they wanted their towns cleared of the 'whiz mob', as English pickpockets are known in the underworld.

whizz, whiz (hwiz), *sb.*² Also **wiz** (wiz). *slang* (orig. *U.S.*). [Perh. identical with Whizz *sb.*¹, but in sense 1 b also regarded as f. Wiz(ard *sb.*] **1. a.** Something very remarkable.
1908 G. H. Lorimer *Jack Spurlock* vii. 157 It is not only a whiz, but a hummah! You are in on the ground flo' of King Solomon's Mines, Limited. 1920 F. Scott Fitzgerald *This Side of Paradise* i. ii. 45 'Wonderful night.' 'It's a whiz.' 1959 *Times* 7 Dec. 13/3 Here are some of the gifts I have given to children in recent years: a massive iron key that could surely unlock the deepest dungeon in Nottingham Castle and makes a whizz of a paper-weight.
b. A person who is wonderfully skilful or talented in some respect.
1914 'High Jinks, Jr.' *Choice Slang* 20 A person is designated as a 'Whizz' when he has exceptional ability along one or more lines. 1921 H. Crane *Let.* 1 Oct. (1965) 66, I..have a strong notion that as a copy writer I will eventually make a 'whiz'. 1924 W. M. Raine *Troubled Waters* xiii. 142 Millie done fixed my game laig up with that ointment good as new. I want to tell you-all that girl is a wiz. 1928 S. Lewis *Man who knew Coolidge* i. 36 He thinks he's such a wiz at cars, but..he couldn't locate that squeak. 1948 A. Huxley *Ape & Essence* (1949) 69 He's an absolute whizz at Malicious Animal Magnetism. 1962 E. B. White *Let.* 13 July (1976) 493 You chose a real whiz..when you picked me for your grammarian. 1978 S. Brill *Teamsters* vi. 211 Malnik was well known..as an associate of long-time mob financial wiz Meyer Lansky. 1982 *Financial Times* 22 June 9/2 He has since become a whizz at ping pong. 1984 *New Yorker* 9 July 35/3 Little Nick Silver, a math whiz from Toughkenamon..was the youngest kid at camp. 1984 *Times* 18 Oct. 14/6, I have a whizz of an accountant who will probably arrange things.
2. Comb.: **whiz(z)-kid**, an exceptionally successful or brilliant young person, esp. in politics or business; hence **whiz(z)-kiddery**, the phenomenon of whizz-kids; the style or mode of work of a whizz-kid.
1960 *Time* 21 Nov. 100/1 The 'Whiz Kids'—as the team soon was known. 1962 *Economist* 22 Dec. 1202/1 Critics..regard President Kennedy as a quiz-kid surrounded by whiz-kids. 1966 Ogilvy & Anderson *Excursions in Number Theory* ix. 103 Zerah Colburn was an early nineteenth-century mathematical whiz-kid. 1967 *Economist* 24 June 1353/3 The whole programme has been a curious hotch-potch of whiz-kidery, preconceived theory and painstaking trial-and-error pragmatism. 1976 *Observer* 26 Sept. 8/1 Many in the institutions—banks insurance companies, investment funds, Stock Exchange and so on—who swallowed whizz-kiddery in its myriad forms. 1976 *Time* 27 Dec. 13/1 The sin of Whiz Kid Shelepin was that he tried to build a political base from which to promote his own post-Brezhnev candidacy for the top post. 1977 M. Drabble *Ice Age* i. 27 Anthony was watching unedited film of an interview with Len the property whizz kid. 1977 *Times* 4 July 22/3 Whizz-kiddery is out; gravitas is in. 1980 *Times Lit. Suppl.* 3 Oct. 1079/5 Editors are a humble and obscure race; lacking the glamour of the whiz-kids and wheeler-dealers.., they are rarely seen in polite society, their names unknown to the columns of *The Bookseller*. 1981 *Sunday Express* 25 Jan. 17/1 Prime Minister Margaret Thatcher will meet Britain's latest whizz-kid inventor when she hosts a unique gathering of inventors and financiers at Downing Street tomorrow. 1985 *Times* 22 June 9/5 We have often been tempted to listen to the siren voices of those who would advise..the latest radical church services, guitars and steel bands and other forms of whizkiddery.

whizz, whiz, *v.* Add: **4.** *intr.* To urinate. *slang.*
1929 D. H. Lawrence *Pansies* 24, I wish I was a gentleman As full of wet as a watering-can To whizz in the eye of a police-man. 1976 R. B. Parker *Promised Land* vii. 37, I wondered if anyone had ever whizzed on Allan Pinkerton's shoe.

whizz, whiz, *int.* Add: Cf. *gee whiz(z *int.*

whizz-bang (hwi·zbæŋ), *int.*, *sb.*, and *a.* *slang.* Also **whiz-bang**, without hyphen, and as two words. [f. Whizz, whiz *v.* or *int.* + Bang *sb.*¹] **A.** *int.* Expressing a whizzing sound that ends with a thud or explosion, such as may be heard as a bullet or shell strikes a target.
1836 Dickens *Pickw.* (1837) ii. 9 Fired a musket.. rushed into wine shop..back again—whiz, bang. c 1838 C. Mathews in M. R. Booth *Eng. Plays of 19th Cent.* (1973) IV. 133 She called in a farmer..Who loaded his blunderbuss..Whizz, bang! Lord, I thought I was murdered outright. 1920 [in Dict. s.v. Whizz, whiz *int.* and *adv.* b].
B. *sb.* **1.** (In Dict. s.v. Whizz, whiz *int.* and *adv.* b.)
1915 [in Dict.]. 1918 W. Owen *Poems* (1920) 16 What murk of air remained stank old, and sour With fumes of whizz-bangs. 1923 Kipling *Irish Guards in Gt. War* I. 143 Three men killed in the line by a single whizz-bang.

1968 J. R. Ackerley *My Father & Myself* vi. 51 In 1918, just before the Armistice, he was killed by a whizz-bang. 1979 S. Wilson *Vampire* ii. 56 Those guns. Those ever present guns. Eighty-eights. Whizz-bangs. None of us need to be reminded of the names.
2. A resounding success; a marvel.
1916 in *Amer. Speech 1972* (1975) XLVII. 116 Masson is a whizbang at getting up the kind of food that makes the troops want to fight. 1944 T. H. Wisdom *Triumph over Tunisia* 182 The raid was a whizz-bang, the R.A.F. expression denoting something highly successful. 1978 M. Puzo *Fools Die* xvi. 169 These were the sharpest kids in America, the future business giants, judges, show business whizbangs. 1983 *Listener* 14 July 37/1 George Stevens..knew how to make box-office whizz-bangs but not very interesting movies.
3. A firework that jumps around making a whizzing noise and periodic bangs.
1960 J. Lodwick *Asparagus Trench* 53, I carried.. whizz-bang fireworks, harmless but disconcerting pyrotechnical trivia these, by reason of their strange gyrations. 1983 D. Lambert *Judas Code* iii. 55 He lit three more firecrackers—Whizz Bangs they were called.
C. *adj.* **a.** Excellent. **b.** Fast-paced, very lively; spectacular.
1959 I. & P. Opie *Lore & Lang. Schoolch.* ix. 161 Other superlatives currently in favour are:..swell, whizzing, whiz-bang, whizzo. 1963 *Economist* 5 Jan. 28/1 Americans are often the first to admit that sometimes a whiz-bang quality about their methods tends to upset their friends. 1965 *Listener* 16 Sept. 431/1 I'm not suggesting that programmes on the arts should be as whizz-bang as *The Dick Van Dyke Show*, but I do suspect that Drama and Light Entertainment could teach them a lot. 1967 *Spectator* 8 Dec. 725/2 A sculptor whose inventions..are made for prolonged contemplation when much work is made for whizzbang impact. 1972 *National Observer* (U.S.) 27 May 20/5 Bernstein inclines to brisk tempos; it would be interesting to see a regiment actually try marching to his whiz-bang 'Stars and Stripes Forever'. 1984 *Listener* 5 Jan. 8/3 As for home-grown, whizzbang, laugh-a-line comedy—Channel 4, where are you?
Hence as *v. trans.*, to shoot whizz-bangs at; **whizz-banged** *ppl. a.*
1918 G. Frankau *One of Them* ix. 66 How oft, in some wild Western whizz-banged dug-out..Has my soul flown from Staff-emitted paper To the glad days, when from my purse I'd lug out That last fat stake. 1919 *King's Royal Rifle Corps Chron.* 1916 139 This line was whizz-banged heartily. 1928 Blunden *Undertones of War* iv. 35 Some of us were just in time, when next the enemy gunners whizzbanged here, to jump down from the fire-step into a dugout stairway.

whizzer. Add: **2.** Something or someone extraordinary or wonderful; a 'stunner'. *slang.*
1888 E. L. Dorsey *Midshipman Bob* i. x. 93 'Foretop-gallant studdingsail-boom-tricing-line-block strap-thimble.' Ain't that a whizzer? 1947 'N. Blake' *Minute for Murder* v. 98, I must say she was a whizzer in those days. 1976 *Zigzag* Apr. 28/1 'She's long' features Bill's best guitar solo (despite many other whizzers). 1977 'J. Gash' *Judas Pair* viii. 95 It's a whizzer... I've found a cased set.
3. A pickpocket. *slang.*
1925 N. Lucas *Autobiogr. Crook* vii. 108 The stalls of theatres at matinees are sometimes patronized by 'whizzers'. 1941 V. Davis *Phenomena in Crime* xiv. 195 There are a score of girl 'whizzers' in London who can get a man's pocket wallet..with conjuring skill. 1974 R. Edwards *Dixon of Dock Green* 17 It was also a right place for 'whizzers'—pick-pockets.
4. *on a whizzer*: on a drinking spree. *N. Amer. slang.*
1910 B. Edwards *Best of Bob Edwards* (1975) v. 104 He was only off on a little bit of a whizzer. 1936 *Univ. Texas Stud. in Eng.* XVI. 51 A number of phrases with go refer to the act of 'getting drunk': one may *go on..a whizzer.*

whizzing, *vbl. sb.* Add: **2.** Pick-pocketing. *slang.*
1925 N. Lucas *Autobiogr. Crook* vii. 98 My pals went in for every known form of getting other people's property. .. 'Drumming', 'parlor jumping', 'whizzing'. 1941 V. Davis *Phenomena in Crime* xv. 209 Nearly all classes of 'whizzing' take place on the 'shove-up' principle.

whizzing, *ppl. a.* Add: **2.** Excellent, 'smashing'. *slang.*
1953 [see *knock-out *sb.* 4]. 1959 [see *whizz-bang *a.*].

whizzo, wizzo (hwi·zo, wi·zo), *int.* and *a. slang.* [f. Whizz, whiz *sb.*¹ + *-o².] **A.** *int.* An exclamation expressing delight.
1905 in *Engl. Dial. Dict.* 1943 *Penguin New Writing* XVI. 28 Wizzoh! No night fighters! 1954 D. Ames *Crime, Gentlemen, Please* xxi. 123 'It's really a little surprise for the kiddies.' 'Whizzo!' cried Anna, grabbing it. 1959 J. Verney *Friday's Tunnel* xxviii. 269 Friday.. yelled, 'Oh, whizzo!'
B. *adj.* Excellent, wonderful.
1948 *R.A.F. Rev.* Jan. 20/2 It's whizzo when you get a fried egg sunny-side-up for tea. 1948 I. Brown *No Idle Words* 97 A father who told his son that he had..arranged for the boy to visit Norway received the following answer: 'Absolutely wizard, flash, whizz-o, grand, lovely to beetle up to Norway.' 1955 M. Allingham *Beckoning Lady* xiii. 185, I wanted to look at some wizzo lettering on..the Tomb. 1968 *Listener* 19 Dec. 810/3 The Squadron-Leader and I decided to give a party—what the Squadron-Leader called a proper whizzo party with marks on the ceiling.

whizzo (hwi·zo), *sb. slang.* [f. *whizz, whiz *sb.*² + *-o².] = *whizz, whiz *sb.*² 1.
1977 *Daily Express* 29 Mar. 20/3 Keyboard whizzo Keith Emerson uses his [side of an album] for a neo-classical piano concerto, accompanied by the London Philharmonic Orchestra. 1981 *Sydney Mirror* 2 July 8/4 Electronics whizzo Dick Smith..aims to become the taxman's friend in another way.

who, *pron.* (*sb.*) Add: **I. 4. a.** *Who's Who*: (earlier and later *transf.* examples); also *fig.* Also, *you and who else?*: a contemptuous expression of incredulity, conveying scepticism about a person's ability to do some past or threatened deed, esp. of violence.
1917 *Wells Fargo Messenger* V. 183/2 The Messenger is no 'Who's who'. 1917 *National Police Gaz.* (U.S.) 18 Aug. 2/4 We don't believe that Ed W. Dunn's latest effusion would win a place for him in the poet's 'Who's who!' corner. 1929 'E. Queen' *Roman Hat Mystery* xviii. 260 'Forget, and I'll dip you into the East River.' 'You and who else?' breathed Djuna. 1929 *Times Lit. Suppl.* 18 Apr. 308/2 First he [*sc.* the biographer] gets out of the way the 'Who's Who' of Wallace Williamson's career in a terse opening chapter. 1951 P. Branch *Lion in Cellar* iii. 38 ''Oo creased 'im?' he asked... 'I did,' he said firmly... 'You an' 'oo else?' he jeered. 1962 W. Nowottny *Lang. Poets Use* ii. 34 Whilst using obituary or *Who's Who* language, it [*sc.* the diction] subtly detaches itself from the social attitudes such language is normally associated with. 1971 A. Morice *Murder in Married Life* xiii. 124 Julian: 'Then I'll throw you out.' Murderer: 'You and who else, ha ha.' 1974 *Advocate-News* (Barbados) 19 Feb. 12/1 The list of batsmen to come is straight out of the 'who's who' of attacking cricketers. 1981 *Country Life* 16 July 205/4 *Women in History*..is a sort of *Who's Who* and *Who Was Who* of women who..should be known.
b. *who does what(?)*: which person will do which task; *esp.* (in a demarcation dispute) members of which trade union will do a certain job.
1922 H. S. Walpole *Cathedral* ii. iii. 194 But who's going to decide who does what?.. We're not much in the sewing line. 1960 *Guardian* 13 Sept. 3/2 A who-does-what dispute between the Amalgamated Engineering Union and the Electrical Trades Union. 1962 *Economist* 13 Oct. 118/1 The squabble over who-does-what. 1962 *Daily Tel.* 28 Nov. 1/1 The Trades Union Congress will seek to settle future 'Who does what?' demarcation disputes with quick and decisive action. 1979 *Now!* 21–27 Sept. 60/1 The £100 million complex has stood idle, paralysed by an inter-union 'who does what' row over 42 jobs.
¶ **5.** (Later examples.)
1941 V. Woolf *Between Acts* 101 Who was she looking for? 1958 *Observer* 6 Apr. 3 (*heading*) Who do you want to save? 1966 I. Murdoch *Time of Angels* x. 106 Who, after all, could I possibly be in love with? 1969 *Listener* 13 Nov. 664/1 One of the policemen..went up to him and almost shouted: 'Who do you think you're talking to?' 1980 J. Gerson *Assassination Run* ii. 35 The days of Philby and Blake when no one knew who to trust.
II. ¶ **13.** For the note read simply: Still common colloquially. (Examples of the use described in Dict. as *rare* or *obs.*)
1979 *Globe & Mail* (Toronto) 27 Aug. 14/5 They come to see Bowser, who they equate not only with The Fonz, but the Cookie Monster and Mork from Ork, too. 1984 *Times* 6 Feb. 12/3 Just over half..of our sample who we assessed as working class concurred.
III. 14. c. (Later example.)
1955 *Bull. Atomic Sci.* June 228/3 The 'who' and 'why' of ethical judgments may lie in the realm of metaphysics; but the 'how' are phenomena in the natural world.

whoa, *int.* **2.** (Earlier example.)
1843 [see *haw *int.*² and *sb.*⁴].

who-all (hū·ǫl), *pron. U.S. dial.* Also **who all**. [f. Who *pron.* + All *a.*] Used for Who *pron.* in interrogative and relative functions (with sing. as well as pl. sense).
1899 B. W. Green *Word-bk. Virginia Folk-Speech* 424 Who-all *interrog.* Meaning all who: as '*Who all* were there.' 1905 A. V. Culbertson *Banjo Talks* 15, I ain' care who-all come dis way! 1916 R. Frost *Let.* 21 Mar. (1964) 27, I wish I could remember..who-all I've baptized into my heresies. 1938 M. K. Rawlings *Yearling* vii. 67 Jody asked brashly, 'Who-all's your sweetheart?' 1938 J. Stuart *Beyond Dark Hills* vii. 184 Will you get up and tell the student group just why you were out there..and who all were with you? 1944 in *Amer. Speech* (1946) XXI. 52 We always said, as the town [*sc.* Hawley, Minnesota] still does, 'Who-all was there?' and 'What all did you do?' Many of the Irish also use 'who-all' and 'what-all'.

whodunit (hūdʌ·nit). *colloq.* Also **whodunnit** and (*rare*) other varr. [repr. *who done* (= illiterate for *did*) *it?*] A story or other work of fiction about the solving of a mystery, esp. a murder; a detective or murder story. Occas. used for 'who did it' in other contexts.
1930 D. Gordon in *News of Bks.* (U.S.) July 10 *Half-Mast Murder*, by Milward Kennedy—A satisfactory *whodunit*. 1942 G. Mitchell *Laurels are Poison* vi. 61 That was another case of Oo-dun-it. Or was it? 1943 *Britannia & Eve* Feb. 16/1 Clifton Fadiman..moved in with an intellectual slap-stick show, which could be appreciated equally by professors and the public for 'who-dun-it?' books. 1951 M. McLuhan *Mech. Bride* (1967) 104/2 Would the thriller fan be abashed to learn that the whodunit anticipated the techniques of modern science and art? 1959 'A. Gilbert' *Death takes Wife* xiii. 173 The whodunnit writers have got us all educated. 1961 *Times* 26 July 15/5 A new 'whodunit'..is to be produced at St. Martin's theatre. 1971 Wodehouse *Much Obliged, Jeeves* vii. 69, I..go in mostly for who-dun-its and novels of suspense. For the who-dun-it Agatha Christie is always a safe bet. 1971 E. Lemarchand *Death on Doomsday* xi. 169, I think sleeping dogs will be let lie, provided we can establish whodunit. 1975 *New Yorker* 21 Apr. 2/1 (Advt.), *Equus*—A brilliant psychological whodunit by Peter Shaffer. 1980 *Times Lit. Suppl.* 30 May 615/5 In the whodunnit, we are conditioned to look for not the most obvious but the *least likely* suspect.
Hence **whodu·n(n)itry**, material or writing such as occurs in a 'whodunit'.
1961 *Daily Tel.* 18 Dec. 10/4 'The Judge and his Hangman' on BBC television last night. This is whodunitry with undertones. 1966 *Punch* 8 June 859/2 His *The Weekend Girls*..settles for whodunitry rather than sociology. 1972 *Daily Tel.* 4 Apr. 9/8 There is no sexual element whatever, and..it doesn't dabble in whodunnitry.

whole, *a.*, *sb.*, *adv.*, (*int.*). Add: **D. 1. whole caboodle**: see *caboodle; **wholefood**, unrefined food containing no artificial additives; an article or kind of such food; **whole hog**: also (usu. with hyphen) *attrib.* as *adj.*, thorough-going, out-and-out; *whole-hogger* (further examples); *whole-hogging* adj. = *whole-hog* adj.; *whole-hoggism* (earlier examples); **whole kit and boiling**, etc.: see *kit *sb.*¹ 3; **whole meal**: also (*colloq.*), a wholemeal loaf; **whole milk**, milk from which no constituents have been removed; also *attrib.*; **whole nine yards** *U.S. colloq.*, everything, the whole lot; also as *adv.*, all the way; **whole tone** *Mus.* = *whole note* (*a*) in Dict.; *whole-tone scale* (see quot. 1928); freq. with reference to compositions based on this scale, particularly those of Debussy; **whole wheat**, wheat which has not been deprived of some constituents by sifting; usu. *attrib.* (with hyphen or as one word), designating flour or foodstuffs made from this.
1960 *Mother Earth* Oct. 341 We should like to hear from further growers who may have available supplies of wholefood, especially winter salads, parsnips [etc.]. 1971 *It* 2–16 June 23/3 (Advt.), The Country Bizarre is a little seasonal magazine on traditions, crafts..whole food culture, poetry, drawings. 1978 *Peace News* 25 Aug. 19/3 (Advt.), If you are interested in wholefoods, running a shop collectively and a political awareness of food please contact us. 1980 *Times* 21 Feb. 12/3 The longest lunch queues in London now are for wholefood... Vegetarian restaurants and health food shops are not new. What is changing is their style. 1829 *Virginia Herald* (Fredericksburg) 28 Mar. 2/3 Of late he has shown a disposition to become 'a whole hog man'. 1855 I. C. Pray *Mem. J. G. Bennett* 141 James Gordon Bennett..is a thoroughgoing, 'whole-hog' Jackson man. 1935 *Planning* 23 Apr. 8 Once you start planning you cannot stop half-way, and whole-hog planning means tyranny. 1956 N. Pevsner *Englishness of Eng. Art* iii. 61 In the architecture of about 1900 there is in England the fresh yet friendly and human style of Voysey, not the whole-hog throwing overboard of all traditions as in Frank Lloyd Wright in America. 1977 *Rolling Stone* 30 June 69/2 My guess is that few white Rhodesian soldiers out there in the bush are wholehog white supremacists anymore. 1904 *Daily Chron.* 28 July 5/6 The country is sick of the whole-hoggers, the half-hoggers,..and the whole lot of them. 1907 E. Nesbit *Enchanted Castle* xi. 333 Your ancestors were whole-hoggers. They have done the thing as it should be done—every detail attended to. 1920 D. H. Lawrence *Women in Love* xxix. 438 He is such a whole-hogger. 1923 R. Macaulay *Told by Idiot* i. xvii. 60 Stanley was like that—enthusiastic, headlong, a deep plunger, a whole-hogger. 1966 *Listener* 26 May 749/1 In the matter of theatre censorship, I am a whole-hogger. 1934 C. Lambert *Music Ho!* v. 301 He [*sc.* Berg] cannot be described as a wholehogging atonalist. 1943 Wyndham Lewis *Let.* 24 Nov. (1963) 370 He is a whole-hogging Thomist. 1960 *Guardian* 27 June 7/2 Whole-hogging festival visitors. 1838 *Carlisle Patriot* 18 Aug. 2/5 The quaint version which the *Times* gave the other day of 'whole hoggism'. 1848 *Blackw. Mag.* July 54 Purge the land of moderatism and anti-whole-hog-ism. 1967 Wholemeal [see *Hovis]. 1983 A. T. Ellis *Other Side of Fire* xvi. 102 Small white, small wholemeal and a couple of croissants. 1970 *Kenya Farmer* Feb. 9/2 We send 110 gallons whole milk per day to Eldoret and separate all the rest for rearing stock. 1977 *Lancet* 19 Feb. 388/1 Sensitivity to cow's whole milk was investigated in six patients. 1982 P. Rance *Great Brit. Cheese Bk.* i. v. 97 These wholemilk cheeses, traditional in this area, vary considerably. 1970 *Word Watching* Apr. 7/2 *Whole nine yards*, the entire thing. 1981 *Washington Post* 16 Jan. (Weekend sect.) 20/3 A Japanese disaster film, *Virus*, goes the whole nine yards, showing the city as a deserted freeway underpass. 1983 *Aviation Week* 7 Mar. 46/2 The Army came out and gave us the whole nine yards on how they use space systems. 1897 J. S. Shedlock tr. *Riemann's Dict. Mus.* 863/1 *Whole-tone*, the larger of the two progressions by tone within the fundamental scale. 1928 *Melody Maker* Feb. 209/3 The Whole Tone Scale.. is composed entirely of intervals of a Tone, thus having only *seven degrees* between its Tonic and its Octave. It has only come into use quite recently and is employed by the school devoting itself to..'futuristic' harmony. 1934 [see *eleventh *sb.* 2]. 1935 G. Abraham *Stud. Russ. Mus.* iv. 77 Dargomïzhsky's fondness for the sharpened fifth of the scale, for the augmented triad which is, so to speak, the 'common chord' of the whole-tone scale.

1952 B. Ulanov *Hist. Jazz in Amer.* (1958) 284 The augmented chords and whole-tone melodies reveal their Debussyan source more clearly. 1977 *Time* 21 Mar. 62/3 His inclusion of Russian fold music, Turkish airs, even the whole-tone scale from the Orient (more than half a century before Debussy) suggests that he was exceptionally curious and openminded. 1903 Wholewheat bread [see *peanut butter* s.v. *peanut b]. 1946 *Sun* (Baltimore) 14 Feb. 14/1 As everybody knows, whole-wheat bread is more nutritious than white bread. 1971 *Times* 11 Sept. 10/4 The distinction between *galettes* (made from buckwheat or wholewheat) and crêpes. 1980 *Sunday Times* (Colour Suppl.) 20 Jan. 57/1 The most basic, natural loaves of all, contain 100 per cent whole wheat flour.
2. a. *whole-body, -grain* (examples), *-house, -word.*
1947 *Radiology* XLIX. 283/1 To determine whether a daily dose of whole-body irradiation when given over a period of several hours produced the same injury as when given within minutes. 1961 *Lancet* 7 Oct. 784/2 Modifications..would require interference with the normal whole-body response to injury. 1977 *Times* 10 Sept. 2/1 Patients from several London hospitals are being sent to BUPA's medical centre to be X-rayed by their EMI whole-body scanner. 1983 P. Niesewand *Scimitar* xx. 566 Lyle and Ross were..subjected to everything from lumbar punctures and sperm tests to whole body scans. 1960 *Farmer & Stockbreeder* 15 Mar. (Suppl.) 10/1 Second-class protein..is found in whole-grain cereals, nuts, lentils and soya beans. 1976 *Woman's Day* (U.S.) Nov. 158/2 Unleavened whole-grain bread should be served generously to assure that your family fills up on fat- and cholesterol-free foods. 1985 *N.Y. Times Mag.* 6 Jan. 6/4 Popular among runners of marathons who stuff themselves with whole-grain pasta before trotting off to the day's race. 1952 *Archit. Rev.* CXI. 212/2 The Radiation 'whole-house' warming system. 1976 *National Observer* (U.S.) 19 June 8 (Advt.), Your Trane Comfort Corps consultant is a full-time specialist in whole-house air conditioning. 1964 P. A. D. MacCarthy in D. Abercrombie et al. *Daniel Jones* 157 This in turn facilitates the recognition of whole-word patterns. 1975 *Language for Life* (Dept. Educ. & Sci.) xxvi. 521 Word recognition is not merely a matter of learning unique whole-word forms. 1980 *Redbook* Oct. 220/1 Most important, the teaching of beginning reading was dominated by the 'whole-word' or 'look-say' method, in which children learned to recognize entire words, rather than by the method of 'phonics' in which they learned to sound out letters and groups of letters.
d. whole-earther *colloq.*, somebody who is actively concerned about the protection and wise use of natural resources and wildlife; **whole-life** *a.*, pertaining to or designating an insurance policy for which the premiums are payable until the death of the insured person; **whole-number rule** *Physics*, the empirical law that the atomic weights of the elements are mostly close to being whole numbers; **whole rock** *a. Geol.*, designating the use of a complete rock sample in an analytical procedure, as distinct from the individual minerals composing it.
1975 *Times* 5 Aug. 12/7 The 'amenity lobby'..includes a new wave of 'whole earthers': notably the Conservation Society founded in 1966..and Friends of the Earth. 1980 *Blair & Ketchum's Country Jrnl.* Oct. 67/1 It includes..neo-Jeffersonians, back-to-the-landers, whole-earthers, communists, and neopioneers seeking to revive old country ways. 1845 *Williams's Directory of Leeds* 46 (Advt.), One-third of the 'Whole Life' Premium may remain unpaid..as a Debt upon the Policy. 1881 *Harper's Mag.* Jan. 79/1 Never take a whole-life policy to embarrass the declining and unproductive years of life. 1977 *National Observer* (U.S.) 15 Jan. 9/2 Whole life—also called cash-value, straight, permanent, ordinary and endowment life—combines insurance protection with a savings or endowment plan. [1919 F. W. Aston in *Nature* 18 Dec. 393/2 Of more than forty different values of atomic and molecular mass so far measured all, without a single exception, fall on whole numbers.] 1923 E. N. da C. Andrade *Structure of Atom* vii. 111 The whole number rule allows us to suppose that all nuclei are built up of the same mass elements, *i.e.* protons. 1967 Oldenberg & Holladay *Introd. Atomic & Nuclear Physics* (ed. 4) xvi. 238 The great simplification was finally introduced through the whole-number rule, which indicates a few fundamental particles as building blocks of all matter. [1955 *Bull. Geol. Soc. Amer.* LXVI. 1711 Approximate minimum ages have been determined for the Cranberry gneiss..and Henderson gneiss by measuring A^{40}/K^{40} ratios on samples of the whole rock.] 1964 *Geochem. Internat.* I. 739/2 It was decided to determine the age of the granites by the Rb-Sr method on whole rock samples. 1979 A. W. Hofmann in Jäger & Hunziker *Isotope Geol.* 215 The evidence for a Caledonian age of the pre-Hercynian gneisses rests in part on two whole-rock Rb-Sr isochrons.

wholesale, *sb.*, *a.*, *adv.* Add: **4. c.** As *sb.*, a wholesale dealer or organization.
1851 [see *retail *sb.*¹ (and *a.*) 3]. 1884 [see Retail *sb.*¹ (and *a.*) 3]. 1928 *Daily Express* 29 May 7/4 The ability of the wholesales to adopt methods of mass production..must be lessened.
wholesale *v.* (earlier and later examples); also *absol.*; hence **who·lesaling** *vbl. sb.* and *ppl. a.*
1800 M. L. Weems *Let.* 17 Dec. in *M. L. Weems: Wks. & Ways* (1929) II. 152 But for this I wd instantly wholesale my books & quit the business forever. 1837 Dickens in *Bentley's Misc.* Oct. 413 We have been prevailed upon to allow this number of our Miscellany to be retailed to the public, or wholesaled to the trade, without any advance upon our usual price. 1881 *Oregon State Jrnl.* 1 Jan. 7 We are prepared to Wholesale and Retail Cheaper than any place in this city. 1906 S. E. Sparling *Introd. Business Organization* xi. 254 In the trade jobbing is virtually synonymous with wholesaling. 1926 N. S. B. Gras in Crump & Jacob *Legacy of Middle Ages* 440 Although many merchants might prefer the wholesale trade, they were not allowed to be exclusively wholesaling merchants. 1962 R. B. Fuller *Epic Poem on Industrialization* 134 'Science News Service' An industrial syndicate Wholesaling to publishers Reported thirty thousand technical innovations. 1972 *Vogue* Jan. 12/2 They wholesale to many shops. 1975 'E. Lathen' *By Hook or by Crook* xiv. 137 Gregory takes care of the wholesaling in this country. Paul runs the retail stores. 1982 *Electr. Wholesaler* Sept. 40/1 He started a general electrical wholesaling firm. 1984 *Listener* 23 Feb. 9/2 There is the jobber, wholesaling shares and making money out of the margin.

wholescale (hōᵘ·lskēⁱl), *a.* [f. Whole *a.* + Scale *sb.*³, influenced by Wholesale *sb.*, *a.*, *adv.*] = Wholesale *a.* 5. Cf. *full-scale.
1960 B. Bergonzi in F. Kermode *Living Milton* x. 168 Leavis's case..is not a mere critical reappraisal of Milton, but a whole-scale demolition. 1983 M. Edwardes *Back from Brink* v. 76 If we were going to run into this sort of problem over £22 million of investment in one factory, how could we contemplate a wholescale modernisation and new product programme across BL, running into hundreds of millions of pounds in dozens of locations? 1984 *Amer. Banker* 5 June 3/1 For middle-level executives, there will be some 'shifting, but not on a wholescale scale', he said.

wholewise, *adv.* Add: Also as quasi-*adj.*
1937 *Mind* XLVI. 252 The whole-wise working of the organism is further illustrated by the 'privileged postures' which we take up as a convenient background to various performances.

wholism (hōᵘ·liz'm). [Alteration of *holism, after Whole *sb.*] The doctrine or belief that wholes must be studied as such, and that the parts can only be understood in relation to the wholes to which they belong; the doctrine that evolutionary forces tend towards the forming of new and more complex wholes; = *holism.
1939 J. E. Boodin *Social Mind.* p. vii, Two conceptions ..have recently been emphasized in philosophy and social theory, namely creative synthesis or emergence and wholism or gestaltism... Wholism means that..events can be understood only as figuring in a whole or gestalt. 1941 *Mind* L. 394 Boodin is fully justified in claiming both that he thought and wrote in the spirit of 'creative synthesis' and 'wholism', before these terms had been invented or had, at any rate, become popular. 1962 R. & H. Hauser *Fraternal Soc.* 9 The keynote of their work is 'Wholism'. 1981 *Amer. Jrnl. Clin. Biofeedback* IV. 33 The biofeedback experience also highlights the concept of wholism.
Hence **who·list** *a.* and *sb.*, **wholi·stic** *a.*, **wholi·stically** *adv.*
1941 *Mind* L. 397 As everyone knows who has studied the use of the concept of 'creative synthesis', and, in general, all 'wholistic' types of philosophy, thinkers of this school are not content to describe the Universe merely as making and unmaking wholes of various sorts. 1956 J. S. Bruner et al. in J. S. Bruner *Beyond Information Given* (1974) ix. 163 We shall refer to the ideal strategy just described as the wholist strategy. 1962 R. & H. Hauser *Fraternal Soc.* ii. ii. 121 As wholists we ask, is not all this..activity..useless. *Ibid.* iv. 181 Our approach to the problems of violence is wholistic. 1964 F. H. Blum in I. L. Horowitz *New Sociol.* 166 Being concerned with the totality of the human situation, he [*sc.* Mills] dealt with them [*sc.* key problems] wholistically. 1972 L. S. Hearnsaw in Cox & Dyson *20th-Cent. Mind* I. vii. 232 Between the wars a new brand of psychology was born, the psychology of personality, wholistic in its presuppositions. 1974 H. J. Klausmeier et al. *Conceptual Learning & Devel.* iii. 67 We rarely receive information in a nice sequence of positive instances so that we may adopt a wholist strategy. 1980 R. Herink *Psychotherapy Handbk.* 698 Wholistic therapy.

wholly, *adv.* Add: **1.** (Later example of *wholely.*)
1915 D. H. Lawrence *Rainbow* xii. 327 Then, and then only..could he act wholely, without cynicism and unreality.
3. Comb.: **wholly-owned** *a.*, applied to a company all of whose shares are owned by another company.
1964 *Financial Times* 11 Feb. 12/1 The directors..have decided to give the holders of Ordinary shares the opportunity of acquiring an interest in the wholly-owned subsidary. 1972 *Accountant* 21 Sept. 360/1 The UK company is a subsidiary—although not wholly-owned. 1976 *Scotsman* 20 Nov. 3/2 The plan is recommended by the boards of all the companies, who will become wholly-owned subsidiaries of the new Malaysian group.

whomp (hwǫmp), *sb. colloq.* (orig. and chiefly *U.S.*). [Echoic.] **a.** A heavy, low sound. **b.** A heavy blow; also *fig.*
1926 *Blackw. Mag.* May 595/2 Ever think of Piccadilly in the evening, and the 'whomp' of an orchestra starting up in some theatre? 1970 J. H. Gray *Boy from Winnipeg* 145 We got some special whomps just in case we had sneaked anything. 1977 R. L. Duncan *Temple Dogs* i. iii. 104 Corbett realized that he had heard a sound, a kind of muted whomp and the Colonel had been shot. 1979 *Washington Post* 4 Oct. A15/2 Liberal and conservative journals are good at least once a year for a whomp at the fat, spoiled, arrogant and pricey world they believe the average bureaucrat to live in. 1983 *Ibid.* 16 Oct. G4/4 He recruited bassist Tony Butler and drummer Mark Brzezicki. The massive and dramatic rhythmic whomp they provide reflects their studio work.

whomp (hwǫmp), *v. colloq.* (orig. and chiefly *U.S.*). [f. the sb.] **1.** *trans.* **a.** To defeat decisively. **b.** To strike (a person) hard, to hit, thump.
1952 *Britannica Bk. of Year* 667/1 *Whomp*, to defeat decisively. 1973 'D. Shannon' *Spring of Violence* xi. 194 If you did something wrong at school you got whomped. 1979 D. Anthony *Long Hard Cure* ix. 79 He had a history of whomping women. 1984 *New Yorker* 1 Oct. 113/1 Tuggle keeps whomping us on the skull.
2. *trans.* With *up*. **a.** To produce quickly, with little preparation or planning.
1955 T. Taylor *Grand Inquest* ix. 241 This procedural paraphernalia was, to borrow Al Capp's apt expression, stricly 'whomped up'. 1957 *New Yorker* 23 Nov. 67/1, I remember the agreement very well. The two of you whomped it up the day after Bob got his overseas orders. 1961 J. Steinbeck *Winter of our Discontent* 190 Wives whomping up a last-ditch dinner. 1980 *Christian Sci. Monitor* 22 May B-16/3 When people ask questions about things I really don't know the answer to..the temptation is to put on my sage mantle and whomp up something.
b. To arouse or stir up (feeling, a disturbance, etc.).
1961 in Webster. 1970 *Daily Colonist* (Victoria, B.C.) 5 May 1/3 Antiwar groups held rallies at dozens of colleges and universities..to whomp up student interest in a national student strike during the closing weeks of the academic year. 1975 M. Amis *Dead Babies* xv. 74 To his hopelessness and grief, Philboyd could not act immediately; time was—when there'd have been enough tubby little rednecks like himself still living in Tara—they could have pitched right in there and whomped up a storm.
3. *intr.* To fall with a 'whomp'.
1960 *New Scientist* 14 Apr. 933/1 The Sunday edition of the *New York Times*..whomped to the floor outside my apartment door.

whoness (hū·nės). *rare.* [f. Who *pron.* (*sb.*) + -ness.] **a.** That which makes a person who he is. **b.** The state of being an isolated individual.
1922 [see *whenceness]. 1931 *Times Lit. Suppl.* 28 May 422/4 A crisis of spiritual rebirth in which the personal will submit only after long struggle to an ineluctable impersonal destiny..thus escaping from the anguish of 'whoness'..into the peace of 'wholeness'.

whoof, *int.* (*sb.*). Add: **2.** Also **woof** (wŭf, wuf). (Expressing) a sound like that of a sudden expulsion of air (less sibilant than 'whoosh').
1921 [in Dict.]. 1921 'K. Mansfield' *Scrapbk.* (1939) 182 The heavy baize door swung to with a 'woof'. 1936 Wodehouse *Laughing Gas* vii. 88 He came over to the arm-chair and sank into it with a luxurious whoof. 1945 *Penguin New Writing* XXIII. 10 Her great guns swing up... Then woof! with a sheet of flame that hides the ship she's hurled a packet of one-ton bricks at something out of sight. 1966 R. H. Rimmer *Harrad Experiment* (1967) 34 Woof! I'm pooped.

whoof, *v.* (in Dict. with Whoof *int.* (*sb.*)). Add: **2.** To make a sound as of air being expelled. Also *fig.*
1966 'L. Lane' *ABZ of Scouse* 117 *Whoof*, to pass wind. 1978 J. Updike *Coup* (1979) vi. 248 He took up a hand mike.., *whoofed* into it experimentally. 1979 *Homes & Gardens* June 126/1 'I am getting seriously worried about prices. They are going to whoof, like this.' And his arm rose at a steep angle from his desk and pointed somewhere in the direction of the stratosphere.

whoofle (hwū·f'l, hwu·f'l), *v.* [Echoic; cf. Whoof *int.* (*sb.*, *v.*) and Whuffle *v.*] *intr.* To make a snorting, gurgling, or snuffling sound; (in quot. 1902 *trans.*, to take *up* with such a sound). Hence **whoo·fling** *vbl. sb.*
1902 H. F. Day *Pine Tree Ballads* 225 To have him fill his saucer and go whoofling up his tea. 1934 L. A. G. Strong *Don Juan & Wheelbarrow* 156 A whoofling and puffing behind him announced Joey. 1944 'Brahms' & 'Simon' *No Nightingales* vi. 27 Mr. Blount sighed his content and moved the ledger to one side. He moved it over the chessboard and upset all the pieces. General Burlap whoofled.

whoom (hwum), *v.* [Echoic.] *intr.* To make a resonant booming or rushing sound. Hence as *sb.*
1936 L. Durrell *Spirit of Place* (1969) 41 Wild pigeon whoomed over. 1942 D. M. Crook *Spitfire Pilot* 90 The deep 'whoom' of a bursting bomb could be heard. 1956 C. D. Simak *Strangers in Universe* (1958) 21 He heard another jet whoom upward from the field. 1956 B. Holiday *Lady sings Blues* (1973) xi. 105 There was a whoom and this big tree crashed over with a wham and a bang.

whoompf (hwum(p)f), *int.* (*sb.*) Also **whoomph**, etc. [Echoic.] (Expressing) a sudden, violent rushing sound, as when a quantity of flammable material bursts into

flame. Cf. the synonymous *woomph *int.* (*sb.*, *adv.*).
1958 'W. Henry' *Seven Men at Mimbres Springs* xv. 170 Then whoomff! land on it with all fours. 1962 *John o' London's* 6 Dec. 527/2 The whole place goes up in flames.., whoomph. 1973 D. Lees *Rape of Quiet Town* vii. 122 A rending crash of metal and a whoompf of flame. 1983 J. Mann *No Man's Island* xi. 145 Check for gas leaks. Light a match and—whoomph.

whoop, *sb.* Add: **1. c.** Slang phrases (orig. and chiefly *U.S.*): *a whoop and a holler* (and varr.): a short distance; *not to care a whoop* (and varr.): not to care one bit; to be indifferent.
[1753 C. Gist *Jrnl.* 27 Dec. (1893) 85 We grew uneasy, and then he said two whoops might be heard to his cabin.] 1815 [see sense 1 a in Dict.]. 1904 *Baltimore American* 30 Aug. 6 The voting public as a whole doesn't care a whoop about the question. 1908 J. London *Let.* 27 Oct. (1966) 268, I don't care a whoop in high water whether you get married..or not. 1920 E. H. Jones *Road to En-Dor* (ed. 2) xxvii. 313, I don't believe Enver Pasha cares two whoops whether I've had syphilis or not. 1924 Wodehouse *Bill the Conqueror* vii. 141 'It isn't as if she cared a hang about him.' 'Doesn't she?' 'Not a whoop.' 1936 E. B. White *Let.* 24 Dec. (1976) 145, I don't give a whoop about dignity. 1951 L. Craig *Singing Hills* 155 They lived in a cabin which Miriam said was three whoops and two hollers away. 1957 J. Agee *Death in Family* ii. x. 157, I wouldn't give a whoop if you got blind drunk, best thing you could do. 1974 D. Sears *Lark in Clear Air* i. 14 A string of hounds..were only a whoop and a bellow behind father.

whoop, *v.* Add: Also with pronuncs. (hwŭp, wŭp). **1. e.** *whoop it up*: for 'U.S. slang' read colloq. (orig. U.S.)'; also, to stir up political enthusiasm; also *whoop things up*. (Further examples.)
1888 *Century Mag.* May 156 His rival is a prominent politician, with an abundance of party workers to 'whoop it up' for him. 1891 B. Harte *First Family Tasajara* i. 8 What did we whoop things up here last spring to elect Kennedy to the legislation [sic] for? 1935 Wodehouse *Luck of Bodkins* iii. 37 You didn't by any chance..whoop it up with those mysterious foreign adventuresses who haunt those parts? 1951 E. Paul *Springtime in Paris* ii. 19, I supposed that elsewhere in France there might be as many young enthusiasts whooping it up for De Gaulle. 1954 B. Hecht *Child of Century* iv. 230 Sherwood [Anderson] would be able to whoop it up for me in a half-dozen periodicals which had come to consider his word as artistic law. 1956 'J. Wyndham' *Seeds of Time* 136 Thousands of trippers whooping it up with pandemonium for most of the night. 1959 'N. Blake' *Widow's Cruise* 93 Some premonition seemed to cast its shadow over the revellers, in spite of Mr. Bentinck-Jones's efforts to whoop things up. 1983 *Listener* 8 Sept. 24/2 The broadcasting moguls and their groupies whooped it up in Edinburgh and other select watering holes.
f. *whoop up* (trans.): to arouse enthusiasm for; to promote or praise with vigour; also, to give a boost to.
1885 *South Florida Sentinel* (Orlando) 5 Aug. 3/3 Whoop up Florida to those Yankees. 1893 [see *stand-off *sb.* 3]. 1904 *Sun* (N.Y.) 8 Sept. 10 The bail was reduced to $10,000, but was whooped up to $15,000 when Larry was re-arrested. 1950 *Sun* (Baltimore) 6 Nov. 3/2 Spokesmen for each party whooped up interest in the outcome. 1970 *Globe & Mail* (Toronto) 26 Sept. 6/5 All human progress, even in morals, has been the work of men who have doubted the current moral values, not of men who have whooped them up and tried to enforce them. 1976 *Listener* 23 Sept. 375/1 If there was any temptation to whoop the original up into contemporary shape, he resisted it. 1983 *Listener* 14 July 19/2 It somehow won that year's Prix Italia,..which so immensely whooped me up that I galloped down to Venice to collect.

whoop-de-do (hū:pdidū·, hw-, w-). *U.S. colloq.* Also **whoop-de-doo**, etc. [A fanciful extension of Whoop *v.* or *whoops *int.*] A fuss, bustle, or commotion; a 'to-do'; *spec.* in *Motor-cycling*, a very bumpy stretch of road.
[1895 S. Crane *Red Badge of Courage* xvi. 160 'Whoop-a-dadee,' said a man, 'here we are! Everybody fightin'.'] 1929 W. Faulkner *Sound & Fury* 321 But I cant have all this whoop-de-do and sulking at mealtimes. 1949 S. Lewis *God Seeker* vi. 34 But what's the use of a loud-mouthed evangelical like your Reverend Chippler,..with his..general circus whoop-tee-do? 1962 J. Steinbeck *Trav. with Charlie* 186 This is not patriotic whoop-de-do; it is a carefully observed fact. 1976 B. Kaysing *Fell's Beginner's Guide to Motorcycling* 256 *Whoop-de-doo*, a road that goes up and down like a roller coaster track. 1980 *Dirt Bike* Oct. 15/1 Very soon we were all lying beside the road, for even though the road looked good at first, it was plagued with whoopdiedoos, and we came into them a little hot. 1981 *Verbatim* Spring 24/1 There was many an angry powwow and much whoop-de-do, but in the end, of course, the bigwigs won. 1985 *Dirt Bike* Mar. 27/1 Through whoopdedos it takes a full stroke without bottoming harshly and keeps giving you maximum ground contact.

whoopee (see below), *int.* and *sb.* [f. Whoop *int.* + -ee².] **A.** *int.* (hw-, wupī·) An exclamation of exuberant joy. Cf. *hoop-ee *int.*
1862 *Harper's Mag.* July 282/1 He yelled at the top of his voice, 'Whoopee! Whiskey only twenty-five cents a gallon!' 1890 Kipling *Barrack-Room Ballads* (1892) 32 Whoopee! Tear 'im, puppy! 1895 *Outing* XXVI. 428/2 John's 'whoopee' had caused a little ebon..to set open the gates. 1932 B. C. Plowright *For Groupers Only* iii. 23 Whoopee!! this is great news! 1974 *Listener* 19 Sept. 355/3 You take your second MB..and once you've passed this—whoopee! You're virtually guaranteed to qualify.
B. *sb.* (hw-, wu·pī; hw-, wū·pī) Exuberant or boisterous merry-making; revelry; † a lively or rowdy party; phr. *to make whoopee*, to indulge in such behaviour; (in quot. 1928, to behave amorously). Cf. *whoop-up. *colloq.*
1928 G. Kahn *Makin' Whoopee* (song), Another bride, another June, Another sunny honeymoon, Another season, another reason for making whoopee! 1929 *Punch* 24 July 86/2 A London hostess, writing to a gossip page, said—'I am giving a Whoopee. Do come to it.' 1930 *Sat. Even. Post* 13 Dec. 25/1 Novelists portray him as the gin-drinking patron saint of whoopee. 1930 E. Waugh *Vile Bodies* iv. 51 Noel and Audrey are having a little whoopee on Saturday evening. 1933 Dylan Thomas *Poems* (1971) 84 Even heaven has a smell Of putrefying angels who Make deadly whoopee in the blue. 1938 F. D. Sharpe *Sharpe of Flying Squad* ii. 27 Boys and girls at the end of an evening's 'whoopee', would come out of a night club and take the first car they saw for a joy-ride. 1945 M. Soames *Let.* 24 July in *Clementine Churchill* (1979) xxiv. 385 The evening broke up about midnight, in a general atmosphere of whoopee and goodwill. 1949 F. Swinnerton *Doctor's Wife comes to Stay* 109 'I thought you and Mother would make whoopee here—' 'Whoopee!' muttered the Doctor. 'Disgusting word for a disgusting occupation!' 'Oh, just noisy hopelessness,' explained Rex. 'Despair set to rhythm.' 1972 D. Francis *Smokescreen* ii. 26 We had left the bright lights, the adulation, and the whoopee, and gone to live in the country. 1976 *Times Lit. Suppl.* 13 Aug. 1009/5 Frustrated laughers, dancers and makers of whoopee. 1984 Q. Crisp *Manners from Heaven* vii. 74 'It often happens that when we think we're making whoopee we're only making a whoops! instead,' I replied.
2. *Comb.*: **whoopee cushion**, a cushion which when sat upon emits a sound like that of the breaking of wind.
1960 *Spectator* 3 June 804 The comically battered face of a whoopee cushion. 1975 P. Theroux *Great Railway Bazaar* viii. 98 These people..were as hard to silence as whoopee cushions. 1977 *Sunday Times* (Colour Suppl.) 6 June 42/3 Andrew..has..a taste for practical jokes..slipping whoopee-cushions where his father or mother was likely to sit.

whooper. Add: **a.** Also **whooper-up.**
1904 *N.Y. Times* 4 July 1 The only candidate who has back of him a boom which is not characterized by 'whooper-up' methods. 1909 J. R. Ware *Passing Eng.* 266/1 *Whooperups*.., inferior, noisy singers. 1932 H. Crane *Let.* 12 Apr. (1965) 408 They're generally preferable to all the trained and professional strummers and whoopers-up I've ever heard.
b. Also, = *whooping crane* s.v. Whooping *ppl. a.* in Dict. and Suppl.
1860 *Southern Cultivator* XVIII. 324 Here [in Florida] is found every grade, kind, size, and color..from the beautiful little morning Dove..to the tall Whooper, of 5 or 6 feet high. 1838 C. H. Matschat *Suwannee River* 286 It is the favorite haunt of the gray whoopers. 1979 *Time* 2 Apr. 23/3 Whatever he felt about the whooper, Carter appreciated the award, which recognized his support for environmental protection and recreation.

whooping, *ppl. a.* Add: **a.** **whooping crane** (earlier and later examples).
1731, 1837 Whooping crane [see Hooping *ppl. a.**]. 1879 N. H. Bishop *Four Months in Sneak-Box* 108 Whooping-cranes..in little flocks, dotted the grassy prairies. 1938 M. K. Rawlings *Yearling* x. 94 He pointed. 'The whoopin' cranes is dancin'.' 1976 *Daily Colonist* (Victoria, B.C.) 7 May 10/6 A scraggly-looking whooping crane chick, hatched this week at the government wildlife centre here.
c. *fig.* Unusually large; whopping; also, very noisy, wild, uproarious. Also as quasi-*adv.*, hugely, immensely. *slang* (chiefly *U.S.*).
1866 'Mark Twain' *Let.* 30 July (1917) I. v. 115 The first few days we came at a whooping gait. 1906 E. Dyson *Fact'ry 'Ands* vii. 88 Odgson..was then lyin' in ther City cells, whoopin' delirious. 1939 G. Ade *Let.* 7 June (1973) 211 Let's make each one of these parties a whooping success. 1969 Fabian & Byrne *Groupie* (1970) xiii. 94 They unstrap me and shoot two whooping great penicillin injections into my backside.

whoops (hwŭps, hwups), *int.* [Var. of *oops.] An exclamation of dismay or surprise, usu. upon stumbling, or realizing an obvious mistake. Also **whoo·psie(-daisy)** *int.* = *upsidaisy.
1925 *New Yorker* 26 Sept. 8/2 (caption) Whoopsie Daisy! 1937 E. Pound *Let.* Jan. (1971) 287 Whoops! And do I envy you. I do. 1957 J. Kerouac *On Road* (1958) ii. viii. 159 Whoops, I thought I was on the wrong side of the road. 1969 C. Armstrong *Seven Seats to Moon* xiii. 126 The woman said, 'Whoopsie', and her strong hand came under his armpit. 1973 G. Talbot *Ten Seconds from Now* xii. 161, I was appalled at the tape playback to hear that I had punctuated my commentary by a 'whoops!' every minute or so. 1980 G. M. Fraser *Mr American* xviii. 328 'Whoops!' said Pip... 'Claridge's, eh? That's what I like to hear!'

whoop-up (hū·pʊp, hw-, w-). Chiefly *N. Amer.* f. vbl. phr. *to whoop it up*: see Whoop *v.* 1 e.] An instance of 'whooping it up'; a noisy celebration or party; revelry.
1913 Cowie *Company of Adventurers* 319 As soon as the general 'whoop-up' began, all the traders..packed up their outfits snugly and retired. 1927 *Daily Express* 5 Oct. 3/3 The Ward Room is—apart from the 'whoops up' natural to lonely men—noted for its air of sober responsibility. 1953 D. Cushman *Stay away, Joe* 22 Ain't you gonna have no dance, no rodeo, whoop-up? 1968 E. S. Russenholt *Heart of Continent* iii. ix. 153 For 'whoop-up juice' they [sc. whisky traders] reclaim the rifles Indian hunters have just bought with a year's hunting. 1976 D. Heffron *Crusty Crossed* xiv. 101, I thought it quite..sensible of Big Point to have one great annual public whoop-up in which to give a little exercise to the witch and devil of one's soul.

whoosh, *v.* Add: Also **woosh. 1.** (Later examples.) Also, to move rapidly with a rushing sound.
1909 H. G. Wells *Tono-Bungay* ii. ii. 163 Make it all slick, and then make it woosh. 1922 D. H. Lawrence *Aaron's Rod* xxi. 306 You want to whoosh off in a nice little love-whoosh and lose yourself. 1966 I. Jefferies *House-Surgeon* viii. 156 The blood was wooshing in and Bernard nodded.
2. *trans.* To cause to move rapidly with a rushing sound. Also *fig.* Const. *up*, to enliven.
1909 H. G. Wells *Tono-Bungay* ii. ii. 162 A Real Live Thing! Wooshing it up! Making it buzz and spin! 1920 D. H. Lawrence *Touch & Go* 7 A system of vacuum tubes for whooshing Bradburys about from one to the other. 1956 W. Sansom *Loving Eye* 102 Cars wooshed water-spray on the wet macadam. 1968 B. Hines *Kestrel for Knave* 27 He whooshed the curtains open and switched the light off. 1971 *Sunday Express* (Johannesburg) 28 Mar. (Home Jrnl.) 2/2 (Advt.), Removable Fibre-fill padlets whoosh you into high young curves, naturally. 1982 *Nature* 13 May 91/1 Chrétien will be the first Western astronaut to be whooshed into space by a Soviet rocket.

whoosh, *sb.* Add: Also **woosh.** (Earlier and later examples.) Also, an exclamation 'whoosh!'; a movement accompanied by a rushing sound; a gushing or 'whooshing' style.
1880 'Mark Twain' *Tramp Abroad* xx. 194 He fetched a prodigious 'Whoosh!' to relieve his lungs. 1909 H. G. Wells *Tono-Bungay* iii. iv. 391 Once or twice before you've stepped in—with that sort of Woosh of yours. 1934 —— *Exper. in Autobiogr.* i. i. 37 Just because of that constitutional apathy it will be characteristically free from individual Woosh. 1962 'R. Gordon' *Doctor in Swim* i. 9 We sat for a moment listening to the woosh of the jets. 1976 *Globe & Mail* (Toronto) 8 Nov. 16/6 When I develop a mental picture of the person I'm affecting, my objectivity goes out the window in a woosh of sympathy. 1984 *Listener* 14 June 32/3 We may be used to the idea of pressing buttons on commercial synthesisers and summoning whooshes of space-age sound.

whoosh, *int.* Also **woosh.** [f. the vb.] An exclamation evocative of or accompanying a sudden explosive rushing sound or movement.
1899 S. R. Crockett *Kit Kennedy* xxxvii. 261 The cravin' wad juist bank up like a water ahint a dam—and then—whoosh, awa' she gaed. 1909 H. G. Wells *Tono-Bungay* ii. ii. 162 That's you, steady and long and piling-up,—then, wo-oo-oo-oo-osh. 1927 *Blackw. Mag.* Apr. 488/1 John said, 'Woosh! some armful. Look out for the eggs.' 1936 'R. Hyde' *Check to your King* 69 The Princess..shouts 'Whoosh!' 1949 Dylan Thomas *Let.* 13 Oct. (1966) 328 Bills and demand notes, at me like badgers, whoosh! 1965 *Family Circle* Oct. 13/1 Plain lonesome? Whoosh, it's a friend. 1977 *Sounds* 9 July 19/2 Onstage we just go like, woosh!

whoosy, whoozy, varr. Woozy *a.* in Dict. and Suppl.

whop, *sb.* Add: Also **wop.**
1895 Kipling in *Youth's Compan.* 19 Sept. 442/4 Then he..drew up with a doleful wop! wop! wop! by the side of the great forty-five-ton, six-wheel coupled,..Number Twenty-five. 1899 W. S. Churchill *River War* I. xiii. 423 The wop! of the distant explosion came back, like the echo of the report.

whore, *sb.* Add: **1. c.** A male prostitute; any promiscuous or unprincipled person. (Esp. as a term of abuse.)
1633 [see Worm *sb.* 10 b]. 1906 J. Joyce *Let.* 19 Aug. (1966) II. 152 He began to shout..when the lazy whores of priests began to chant. 1957 P. Kemp *Mine were of Trouble* vi. 108 Lyall would interrupt with..'But surely you can't expect the Irish to be any use in Spain? There aren't any hedges here for them to shoot from behind.'.. Lawler would storm out, shouting: 'Ye great buckin' whore!' 1968 E. Gaines in A. Chapman *New Black Voices* (1972) 103 'You hear me whore?' 'I might be a whore, but I'm not a merciless killer,' he said. 1976 *New Yorker* 12 Jan. 73/2 Gig Young can play the top whore in 'The Killer Elite' because his sad eyes suggest that he has no expectations and no illusions left about anything.
4. *whore-like* adj. (later example); **whore-hunt** *v.* (later example); **whoremistress**, a brothel-keeper; **whore's egg** *N. Amer.* (chiefly *Newfoundland*) = Sea-urchin 1; **whore-shop** *slang*, a brothel.
1931 R. Campbell *Georgiad* i. 15 Lovelorn poets..troop whore-hunting down the country lanes. 1974 H. J. Parker *View from Boys* 213 'A right scrubber' is a girl who's rough-looking, whorelike. 1922 Joyce *Ulysses*

515 Bella Cohen, a massive whoremistress enters. 1969 A. Marin *Rise with Wind* xii. 154 Consejo..works for a whoremistress we call Tía Concha. 1829 T. C. Haliburton *Hist. & Statist. Acct. Nova Scotia* II. ix. 405 Shell fish. Whore's egg. 1930 *Amer. Speech* V. 393 *Whore's egg*,..a small spring crustacean esteemed by the Italians as a delicacy. 1948 Z. N. Hurston *Seraph on Suwanee* 296 That damn whore's egg! Ruin you if only one spine gets into your hand. 1972 E. Staebler *Cape Breton Harbour* ix. 85 You be careful when you's swimming that you don't step on a whore's egg, they sea urchins is full o' prickles will give you a fester. 1938 V. S. Pritchett *You make your own Life* 79 What a town like this wants is a couple of good whore shops and a factory. 1972 A. MacVicar *Golden Venus Affair* vi. 67, I hate The Golden Venus... It's just a whoreshop.

whore, *v.* Add: **1. c.** *intr.* fig. To pursue or seek *after* (something false or unworthy). In allusion to Exod. xxxiv. 15. Cf. Whoring *vbl. sb.* (quot. 1535).
1913 E. Pound *Let.* 13 Aug. (1971) 21 The unspeakable vulgo will I suppose hear of him [sc. F. M. Hueffer] after our deaths. In the meantime they whore after their Bennetts and their Galsworthys and their unspeakable canaille. 1937 J. M. Murry *Necessity of Pacifism* 24 The intelligence of Socialism went a-whoring after the strange gods of Russia. 1970 R. Long in A. Chapman *New Black Voices* (1972) 421 The University was whoring after strange gods, they all seemed to say: technology, athletics, materialism. 1972 *Language* XLVIII. 425, I do not accept Chomsky's conception of social scientists as universally whoring after the surface features of other sciences, neglecting all fundamental problems, and taking refuge in spurious precision and trivialities.
2. For ? *Obs.* read Now *rare* and add later example.
1969 A. Hunter *Gently Coloured* iii. 33 Some friend squeezing you dry, whoring your sister.

whore-house. 1. For † *Obs.* read: Revived in recent (chiefly *U.S.*) use. (Later examples.)
1909 in J. A. & A. Lomax *Amer. Ballads & Folk Songs* (1960) 104 Frankie went down to de whore-house, Rang de whore-house bell, Says 'Tell me, is my lovin' Albert here? Caze Frankie's gwine to raise some hell—Oh, he's my man, but he's a-doin' me wrong.' 1935 J. Steinbeck *Tortilla Flat* i. 25 'Pilon!..I am an heir! I own two houses.' 'Whore houses?' Pilon asked hopefully. 1951 J. Masters *Nightrunners Bengal* i. vi. 82 Every one knew her as the madam of a high-grade whore-house. 1978 G. Greene *Human Factor* iii. iii. 123 'If you want to fuck a black whore,' Captain Van Donck interrupted with impatience, 'why don't you go to a whore-house in Lesotho or Swaziland?' 1982 *Times* 22 May 8/1 Prospectors came by the thousand, saloons and whore-houses were erected.
2. *attrib.* and *Comb.* **a.** Simple attrib., as *whore-house bell, owner, perfume, scum.* **b.** Designating or pertaining to a style of music, esp. jazz, played in brothels, as *whore-house music, piano.* **c.** *Comb.*, as *whore-house madam* = *madam 3 c (*d*).
1909 Whore-house bell [see sense 1 above]. 1938 D. Baker *Young Man with Horn* iii. i. 141 You certainly play whorehouse piano, fella, and nigger whorehouse at that. 1946 R. Blesh *Shining Trumpets* xiii. 295 This rich and earthy piano playing, called by extreme jazz purists with an ear for the picturesquely accurate, 'whore-house piano'. 1949 R. Chandler *Little Sister* xvii. 110 A very cheap grade of whore-house perfume. 1954 W. Faulkner *Fable* 379 Shoot now, you whorehouse scum. 1956 B. Holiday *Lady sings Blues* (1973) i. 8, I guess I'm not the only one who heard their first good jazz in a whorehouse... A lot of white people first heard jazz in places like Alice Dean's, and they helped label jazz 'whorehouse music'. 1975 G. V. Higgins *City on Hill* vi. 150 When you try to talk about something else, it's like trying to discuss cryogenics with a whorehouse madam.

whoremaster. Delete bracketed etymological note and *Obs.* or *arch.*, and add: **2.** *spec.* A procurer or pimp.
1864 in Webster. 1922 E. E. Cummings *Enormous Room* vii. 163 Now I must tell you what happened to the poor Spanish Whoremaster. 1964 in Hamblett & Deverson *Generation X* 94 Johnny knew just when to corrupt and when to give the old ego a boost. He's one of the great whoremasters of all time, working on the principle of the carrot and the stick. 1977 M. T. Bloom *13th Man* (1978) vii. 133 The newcomers had little money and..they got tempted by the whoremasters.

whoreson. For *Obs.* or *arch.* read Now *arch.* and add: **a.** (Later examples.)
1926 [see *ingle *sb.*²]. 1975 *Weekend Mag.* (Montreal) 1 Nov. 21/1 If the whoreson who dropped his socks into the chamber pot and sold the results to a lantern jaws like you is not at a rope's end since this fortnight, there is no justice left on earth!
b. (Later example.)
1909 E. Pound *Exultations* 14 You whoreson dog, Papiols, come!

Whorfian (hwǭ·ɹfiăn), *a.* [f. the name of the American linguist Benjamin Lee *Whorf* (1897–1941) + -ian.] Designating the views and theories of B. L. Whorf, esp. in *Whorfian hypothesis*, the theory that one's perception of the world is influenced or determined by the structure of one's native language (also *Whorf hypothesis*). Cf. *Sapir-Whorf hypothesis.
1957 R. K. Merton *Social Theory* (rev. ed.) ii. 92 It is the extreme Whorfian position which Joshua Whatmough attacks. 1963 J. Lyons *Structural Semantics* iii. 40 The view expressed in this quotation from Sapir has been championed more recently by Whorf, and has come to be known within linguistics as the 'Whorfian hypothesis'. 1964 R. H. Robins *Gen. Linguistics* ii. 80 This is part of what has come to be known as the 'Whorf hypothesis'. 1968 M. Black *Labyrinth of Lang.* iv. 75 Some interesting attempts have been made to determine the validity of Whorfian ideas. 1978 *Language* LIV. 167 His chapter on personal context contains discussions of the Whorfian hypothesis.
Hence **Who·rfianism**, a Whorfian conception; Whorf's theories regarded *collect.*
1963 J. Lyons *Structural Semantics* iii. 40 For a strong and convincing attack on the more extreme aspects of 'Whorfianism', cf. M. Black, 'Linguistic relativity'. 1978 *Language* LIV. 267 Parry's notion that formulaic language imposes formulaic thought is a kind of Whorfianism run wild.

whorish, *a.* For *rare* or *Obs.* read *arch.* and add: **1. a.** (Later examples.)
1948 D. Welch *Brave & Cruel* 245 Mary..had nothing to take her mind from the hideous picture of a breast pump, a whorish wife and an idiot baby. 1981 V. Canning *Boy on Platform One* iv. 60 Whorish..the word swam gently into his mind. Whore, too, she was.
b. (Later examples.)
1942 D. Welch *Jrnl.* 30 Aug. (1952) 7 When we had..pushed back the whorish, dirty red satin curtain. 1967 A. Laski *Seven Other Years* iv. 56 It was a charming dress..virginal in colour, whorish in cut. 1980 A. E. Fisher *Midnight Men* iv. 45 Bathrooms should reflect..the woman of the house... I'd like a sort of whorish pink.
whorishly *adv.* (later example).
1977 *Listener* 25 Aug. 246/3 The gratuitous violence, slotted whoreishly into the sequences.

whorl, *sb.* Add: **3. b.** A configuration in finger-prints.
1880, etc. [see *loop *sb.*¹ 4 h]. 1954 F. Cherrill *Cherrill of Yard* vi. 62, I noticed particularly the patterns on the ends of the fingers, for they were of the whorl type. 1977 *Sci. Amer.* Dec. 141/1 The resulting patterns are known to the dermatologist respectively as loops, triradii and whorls.

whory, *a.* For *rare* read Formerly *rare* and add later examples.
1955 J. Kerouac in *Paris Rev.* Winter 11 The whorey smell of a big city. 1967 K. Giles *Death & Mr. Prettyman* i. 18 London was as whorey then as now. 1976 *New Yorker* 15 Nov. 180/3 At twenty she was taking care not only of herself and her child but also of a tubercular half sister, also cast off by their whorey mother. 1980 N. Freeling *Castang's City* xxii. 148 It's not a whory setup; three of these dames live there and they're secretarial types.

whose, *pron.* **3.** (Later examples.)
1906 Conrad *Mirror of Sea* vii. 33 A newspaper of sound principles, but whose staff will persist in 'casting' anchors. 1927 E. Bowen *Hotel* vi. 57 She looked down..and saw a little house, with a blue door whose colour delighted her. 1958 I. Murdoch *Bell* iv. 47 Toby..marvelled at this light which is no light..and whose strength is seen only in the sharpness of cast shadows. 1968 J. Lyons *Introd. Theoretical Linguistics* 55 Whether there are, or could be, two languages whose vocabularies are to no degree whatsoever isomorphic with one another is a question with which we need not be concerned. 1981 I. McEwan *Comfort of Strangers* ix. 122 There were pictures whose context she understood immediately.

whosis (hū·zis). Also **whoosis.** [Colloq. contraction of 'who is this?' (Who *pron.*); in quot. 1923, perh. repr. 'whose is this?'] 'What's-his-name', 'so-and-so'. Often following a title, as *Mr. Whosis.*
1923 J. E. Baxter *Locker Room Ballads* 8 That number one's a Big League Green As slick as Whoosis Vaseline. 1939 R. Stout *Some buried Caesar* vi. 72 He..introduced himself as Mr. Whosis, Assistant District Attorney. 1953 G. W. Brace *Spire* xxiii. 229, I suppose..you mean he should go and see your precious Dr. Whoosis? 1962 J. D. Salinger *Franny & Zooey* 130 How was the script? Did it come? You said Whosis—Mr. LeSage or whatever his name is—was going to drop it off with the doorman. 1965 I. Fleming *Man with Golden Gun* vi. 89 Don't forget one thing, Mister Whoosis. I rile mighty easy.

whosit (hū·zit). Also **whoosit, whoozit, whozit.** [Colloq. contraction of 'who is it' (Who *pron.*).] = prec.
1948 'P. Quentin' *Run to Death* xxi. 156 Ye Old Antique Shoppe with little leaded glass panes. Mother Whosit's Chicken Kitchen. 1951 *Blue Bk. Mag.* Jan. 24/3 Mr. Whoozit—please come quickly. 1951 'J. Tey' *Daughter of Time* viii. 112 Someone, say, insists that Lady Whoosit never had a child. 1967 O. Norton *Now lying Dead* i. 9 That's what I've got to work out... Like Angela Whoosit was telling us. 1977 J. Fleming *Every Inch a Lady* i. i. 7 Arrival..of Mrs Whozit, the lady help.

whump (hwʌmp), *v.* Also **wump.** [f. as next.] **1.** *intr.* To make a dull thudding sound; to move with a 'whump'; to bang or thump; to strike (with a thud).
1897 E. Terry *Let.* 5 Feb. in *Ellen Terry & Shaw* (1931) 126 Not a single speech do I know yet, and my head is thumping and wumping. 1928 *Blackw. Mag.* Jan. 5/1 The look-out sentry..whumped twice, briskly, on his hand-gong. 1939 *Life* 11 Dec. 26 Taft of Ohio sturdily whumped at the New Deal's 'insane deficit policy'. 1981 B. Granger *Schism* xi. 89 The windshield wipers whumped, whumped slowly across the streaky glass.
2. *trans.* To strike heavily or with a 'whump'.
1974 D. E. Westlake *Help* (1975) iii. 20, I would then adjust the rubber stamp.., wump it onto the stamp pad, wump it onto the envelope. 1976 *National Observer* (U.S.) 24 Jan. 19/3 What had been lost at Waterloo and Sedan could be won back by whumping mud forts in the Sahel.
Hence **whu·mping** *vbl. sb.* and *ppl. a.*
1928 *Blackw. Mag.* Jan. 2/2 The occasional whumping and booming of war-gongs. 1977 P. Dickinson *Walking Dead* ii. viii. 206 There was a slow, wumping explosion.

whump (hwʌmp), *sb.* (and *int.*) Also **wump.** [Echoic: cf. *bump, thump*, etc.] A dull thudding sound, as of a body landing heavily. Also *int.* Cf. *whomp.
1915 D. O. Barnett *Let.* 6 May (1915) 130 Then there was a wump over beyond, and a young howitzer shell went zip over my trench. 1922 *Chambers's Jrnl.* 7 Oct. 707/1 The globe suddenly swung in a long arc across some hidden gully in the bottom and fetched up with a stunning 'wump' on the slope of the other side. 1926 Galsworthy *Escape* i. ii. 32 Still—up on the ladder and down with a whump—it hits 'em [sc. gentlemen] harder than it does the others. 1930 C. R. Samson *Fights & Flights* ii. iv. 181 'Wump' fell a second bomb. 1967 *Boston Herald* 1 Apr. 20/2 (caption) Whump! 1976 *New Yorker* 8 Mar. 106/2, I heard this funny sound: a kind of *whummpp.*

whunk (hwʌnk), *sb.* (and *v.*) *rare.* Also **whonk.** [Echoic.] A dull hollow sound, as of a bullet striking something. Also as *v. intr.*, to strike with a 'whunk'.
App. only in the work of Hemingway.
1935 E. Hemingway *Green Hills Afr.* ii. iii. 53 We had both heard the whunk of the bullet. *Ibid.* ii. iv. 76, I heard the whonk of the bullet. 1936 —— in *Hearst's Internat.* Sept. 168/1 He heard a *whunk* that meant that the bullet was home. *Ibid.* 170/3 Hearing the bullets whunk into him.

whup (hwʌp), Sc. var. of Whip *v.* (q.v.). Also *U.S. colloq.* and *dial.* (Examples, esp. of senses 6, 12 of the vb.) Also **whu·pping** *vbl. sb.*
1893 H. A. Shands *Some Peculiarities of Speech in Mississippi* 68 *Whup*,..Negro for *whip.* 1906 *Dialect Notes* III. 164 *Whup*,..to vanquish, to punish, to tire. 'That whups me.' 1929 W. Faulkner *Sanctuary* (1981) xi. 132 You done whupped him. 1939 J. Steinbeck *Grapes of Wrath* xxviii. 504 Whyn't ya whup her, Ma?.. Go on, give her a whup. 1948 A. Lomax in A. Dundes *Mother Wit* (1973) 481/1 Give him a whuppin'. 1950 Patterson & Conrad *Scottsboro Boy* iii. ii. 193, I told the warden I was not guilty of the charge and didn't want to be whupped. 1968 *Punch* 25 Sept. 451/2 The Matt Dillon urge to 'whup' the Commies. 1972 J. Gores *Dead Skip* v. 31 He might have come after Bart..because he wanted to whup a nigger? 1974 W. Garner *Big enough Wreath* vii. 94 You swore there never was a whupping could make you holler.

whuss (hwʌs). *U.S.* (chiefly Black English) *colloq.* abbrev. of 'what is..?'
1935 Z. N. Hurston *Mules & Men* (1970) i. iii. 74 Whuss de matter, Jack? 1938 C. Himes *Pork Chop Paradise* in *Black on Black* (1973) 174 Whuss yo' name? 1977 *Rolling Stone* 16 June 11/2 Whuss happnin'?

whut (hwʌt), *U.S. dial.* and Black English var. of What *pron.*
1909 *Dialect Notes* III. 387 *Whut*, what. A common pronunciation. 1929 W. Faulkner *Sanctuary* (1981) viii. 94, I couldn't tell and wouldn't even keer whut I was eatin. 1936 M. Mitchell *Gone with Wind* v. lix. 996 Miss Melly, you know whut he done? 1961 J. Jahn in A. Dundes *Mother Wit* (1973) 101/1 Whut makes yore head so red. 1973 *Black World* July 62/2 Well, so whut.

why, *adv.* (*sb.*, *int.*) Add: **I. 1. d.** With the negative form of the simple present tense in formulating a positive suggestion, as 'why don't I (we, etc.)..?'
1949 D. Smith *I capture Castle* xii. 212 Why don't I drive you over to hear it now? 1974 G. Mitchell *Winking at Brim* vi. 54 Mummy brought a couple of thermos flasks... Why don't I go and collect one? 1982 R. Doyle *Havana Special* vii. 182 Why don't I stop by her compartment..and see how she is?
3. b. Duplicated in phr. *why, oh why..?*, as an emphatic interrogative, expressing dismay, disapproval, or complete lack of comprehension of another's actions; 'why on earth..?'
1865 M. Arnold *Let.* 23 July (1895) I. 294 Why, oh, why do not you and Edward come to the Black Forest and join us? *a* 1884 T. H. Huxley *Let.* in *Henry Bristow Ltd. Catal.* (1981) No. 269.14 My students..cannot get copies of the second edition of the Biology book. Why oh why was it not ready by October. 1934 N. Marsh *Man lay Dead* xi. 194 Why, oh why, did the murderer sound the gong? 1961 'E. Lathen' *Banking on Death* (1962) viii. 70 Why, oh why, had she been so bitchy to his wife? 1975 *Times* 8 Mar. 13/7 Why, Oh why does Rolls-Royce..name its latest product after an area of French marshland?
VI. 9. *Comb.* **why-question**, a question inquiring after the reason for something; one

which is introduced by the word 'why'. Cf. *wh 2.
1973 A. Dundes *Mother Wit* 568 'Why' questions are always difficult to answer. 1978 *Language* LIV. 71 A *that*-clause can be the basis of a why-question when it is assigned the semantic status of the volunteered stance of the subject of the verb whose complement it is.
Hence as *v. intr.*, to ask the question 'why?' (chiefly as *pres. ppl.*); also **why-ing** *vbl. sb.*
1926 H. Pearson *Whispering Gallery* ii. 19, I made the mistake of doing or dying, but at the same time why-ing. 1928 D. H. Lawrence *Phoenix II* (1968) 520 Why indeed? But once you start whying, there's no end to it. 1932 E. M. Brent-Dyer *Chalet Girls in Camp* i. 24 'Why?'.. 'For goodness' sake don't start why-ing, Rix!' 1959 *Times Lit. Suppl.* 2 Jan. 7/1 (heading) Howing and whying.

whydunit (hwəidʌ·nit). *slang.* Also **why-dunnit.** [f. Why *adv.* + *(who)dunit.] A story, play, or film in which the main interest lies in the detection of the motive for some crime or other action.
1968 *Guardian* 1 May 7/2 Patricia Highsmith..writes why-dunnits rather than who-dunnits, psychological thrillers. 1970 *Homes & Gardens* Feb. 122/2 This novel isn't so much a whodunit as a whydunit, with a revolver doing the deed. 1984 *Listener* 17 May 35/3 It is a terrific whodunnit, constantly hinting at whydunnit, at the ethical squalor of all the participants.

whyness (hwəi·nès). [f. Why *adv.* + -ness.] That which causes a thing to be as it is; the essential reason for something. Cf. Whatness.
1896 R. Fry *Lett.* (1972) I. 116 You who..care nothing about the whyness of the what. 1932 *Times Lit. Suppl.* 20 Oct. 765/3 But it is the whatness not the whyness of things that matters. 1950 *Mind* LIX. 405 Logical empiricism indicates the 'howness' of the world, but not the 'whyness'. 1962 *Time* 11 May 70 Teacher Foote reports that -ness added to nouns, pronouns, verbs, and phrases—a custom thought until now to be mostly whimsical, as in *whyness*, and *everydayness*—has become popular among distinctly unjocose people.

wibbly-wobbly, *a.* Add: Also as *sb.*, in phr. *all of a wibbly-wobbly. nonce-use.*
1922 Joyce *Ulysses* 399 Bless me,..I'm all of a wibbly-wobbly.

wichert (wi·tʃəɹt). *dial.* Also **whitchet, witchert,** etc. [Orig. uncertain; perh. repr. a local pronunc. of 'white earth'.] A variety of chalk marl subsoil found near the Chilterns in Buckinghamshire, which is mixed with chopped straw and used locally for walling.
1912 *R. Comm. Hist. Monuments: Buckinghamshire, South* 342 Wichert or Whitchet (white earth).—A local term for a kind of white marl or mud found at Haddenham, Dinton, and in the district, and used unburnt mixed with chopped straw for walling. 1916 C. F. Innocent *Devel. Eng. Building Construction* x. 136 In Buckinghamshire, where the walls were built of a kind of white clay called 'witchit', found about eighteen inches below the surface of the ground. 1929 H. Harman *Bucks. Dial.* 165 The wichert (or whitchet), which is a kind of white marl found locally, is laid in heaps beside the line of the intended wall and well soaked with water. When the stonework is in position, the wichert is turned and short straw is trodden into it; the purpose of this is merely to keep it fairly compact whilst it is wet. 1942 W. Rose *Good Neighbours* iv. 42 He also made the curious three-pronged forks, with flat tines, with which the masoners built the wichert walls of the village. 1951 P. Oyler *Feeding Ourselves* iii. 32 Hand-made bricks and tiles, stone and thatch, cob or wichert cannot be out of place in the scenery from which they come. 1958 *Records of Buckinghamshire* XVI. iii. 136 It is considered most probable that the main walls were built of the local chalk mud charged with chopped straw, known in these parts as *witchert.* 1977 *Oxford Times* (S.E. ed.) 4 Mar. 1 The cottage was built of Witchert, a sophisticated sort of mud, of which there are many examples in Haddenham. The material crumbles when demolished.

wichetty, var. Witchetty in Dict. and Suppl.

Wichita (wi·tʃətǭ). Also 9 **Wichataw.** [f. the name of *Wichita*, in Kansas.] **a.** (A member of) a Caddoan Indian people of southern central N. America (now Oklahoma, formerly also Kansas and Texas). **b.** Their language. Also *attrib.*
1841 H. S. Foote *Texas & Texans* I. xiv. 299 There are several other remnants of tribes in Texas..the Wichataws, who live far North, on the Brassos. 1883 W. F. Cody in B. A. Botkin *Treas. Amer. Folklore* (1944) Music. Enter a group of Wichita Indians. 1960 R. W. Marks *Dymaxion World of B. Fuller* 37/1 When the first Wichita house finally was opened to the public, many were struck by its spaciousness and air of luxury. 1965 *Language* XLI. 84 The consonantal opposition nasal/oral occurs in all the languages of the world except Wichita. 1978 *Ibid.* LIV. 503 Of the three American Indian linguistic groupings discussed here, those of the Caddoan family, spoken in the Southern Plains, are perhaps the least described; of the surviving languages—Caddo, Wichita, and Pawnee—none is being learned by children.

wichuraiana (witʃu:rəi,ā·nă). [a. specific epithet of *Rosa wichuraiana* (F. Crépin 1886, in *Bull. Soc. Bot. Belgique* XXV. 189), f. the name of Max Ernst *Wichura* (1817–66), German botanist + *-iana.] A climbing, almost evergreen, rose belonging to the species *Rosa wichuraiana*, which is native to eastern Asia and bears white flowers, or one of many cultivars developed from it, usually distinguished by small glossy leaves and flowers in clusters. Freq. *attrib.*
1907 [see *Penzance]. 1913 [see *Dorothy Perkins]. 1923 *Daily Mail* 10 Mar. 15 The top growth of roses, excepting..wichuraianas, should be severely pruned in the first season. 1945 G. M. Taylor *Roses* xv. 76 Some of the Wichuraianas..will cover a wall very quickly. 1960 *News Chron.* 6 Aug. 6/1 The wichuraiana ramblers are a case in point. 1962 R. Page *Education of Gardener* vi. 199 Among rambler roses I like to use the wichuraiana varieties.

wick, *sb.*¹ For sense numbers read **1. a, b, c,** and **3,** and add: **1. d.** In fig. phr. *to turn the wick up* (or *down*), to open (or close) the throttle of an engine; to accelerate (or decelerate). *colloq.*
1948 [see *throttle *sb.* 4 b]. 1965 Priestley & Wisdom *Good Driving* iii. 28 The gas pedal can be likened to the wick of an oil lamp. Turn it up and you get more light... Indeed it is a simile much used by motor cyclists who talk of 'turning the wick up' as a more graphic and descriptive way of saying 'I accelerated'.
2. a. *to get on* (*one's*) *wick*, to irritate or annoy (a person); to exasperate; to get on one's nerves (*nerve *sb.* 8 e). *colloq.*
It is sometimes suggested that both this and the next sense derive from (Hampton) *Wick*, rhyming slang for Prick *sb.* 17. See Partridge and Wick *sb.*² 2.
1945 *Penguin New Writing* XXVI. 56 Parades and bullshit get on his wick. 1958 K. Amis *I like it Here* 32 But I wish he wouldn't think he'd got the right to knock the English. That's what really gets on my wick. 1961 'B. Wells' *Day Earth caught Fire* iv. 54 'Strewth, these licensing laws get on your wick, don't they,' they grumbled. 1977 K. Benton *Red Hen Conspiracy* iii. 22 The way you talk about Pat gets on my wick. 1984 B. Francis *AA Car Duffer's Guide* 6/2 Gets on my wick, she do.
b. *to dip* (*one's*) *wick*: of a man, to engage in sexual intercourse. *slang.*
1958 J. Carew *Black Midas* vi. 96 'Come on!' Santos bellowed. 'If every time you dip your wick you going to fall in love, then God help you!' Belle jumped out of bed and pulled on her dress. 1969 D. Niland *Dead Men Running* iv. 159 When you're starved for a woman dip your wick, and the starvation's gone. 1971 B. W. Aldiss *Soldier Erect* 111 Di asked, 'You don't feel like a bit of a bunk-up this evening, Stubby, by any chance?' 'A bit of what?' 'Dipping your wick, man!' 1981 R. Barnard *Sheer Torture* xiii. 137 None of your barmaids or local peasant wenches for Pete. He's very calculating where he dips his wick.

wick, *a.*² *North.* var. of Quick *a.* Cf. *whick *a.*
1848 Mrs. Gaskell *Mary Barton* I. viii. 127 In th' Infirmary..there be good chaps there to a man, while he's wick, whate'er they may be about cutting him up at after. 1911 F. H. Burnett *Secret Garden* xi. 105 'It's as wick as you or me,' he said;..Martha had told her that 'wick' meant 'alive' or 'lively'. 1970 'J. Herriot' *If only they could Talk* ix. 69 This 'oss is as wick as an eel. 1972 *Observer* 23 Apr. 23/4 Knott is, to use a Yorkshire expression, 'wick', but wick cricketers are rare these days. 1978 *Lancashire Life* Oct. 99/1 Granny Martha Mosscropp, approaching her century and as wick as a flea, had known in girlhood the enclosed life of Victorian Ramstwistle.

wicked, *a.*¹ (*sb.*, *adv.*). Add: **I.** *adj.* **1. c.** Designating a stock evil character in a fairy-tale, as *Wicked Fairy, Stepmother, Uncle*, etc. Freq. *transf.*
1897 Kipling *Stalky & Co.* (1899) 39 He owned a soft, slow smile which well suited the part of the Wicked Uncle. 1906 *Sleeping Beauty* ('Tales for Little People' ed.) 8/2 'That looks like the wicked fairy, I'm sure,' said his majesty to himself. 1946 A. Huxley *Let.* 26 May (1969) 544 That blessing and curse of cleverness, with which the Fairy Godmother, who is also the Wicked Fairy, endowed me. 1978 M. Babson *Tightrope for Three* xv. 78 He could not see Lillian in the classic 'wicked stepmother' situation. 1982 'J. Melville' *Painted Castle* i. 21 If you left Tad out of consideration, uncomfortable things were apt to happen. He had a touch of the Wicked Fairy about him.
3. b. Excellent, splendid; remarkable. *slang* (orig. *U.S.*).
1920 F. Scott Fitzgerald *This Side of Paradise* i. iii. 119 'Tell 'em to play "Admiration"!' shouted Sloane... 'Phoebe and I are going to shake a wicked calf.' 1977 *Western Mail* (Cardiff) 5 Mar. 8/2 He could, as I say, sidestep off either foot, but what sped him on was a wicked acceleration over 20 yards.
II. *absol.* or as *sb.* **4. a.** Also in phrs. *no peace for the wicked*: see *peace *sb.* 14*; *no rest for the wicked.*
1935 Marsh & Jellett *Nursing-Home Murder* iv. 57 The throat specialist..remarked: 'No rest for the wicked, nurse.' 1958 A. Sillitoe *Sat. Night & Sunday Morning* i. 20 'No rest for the wicked,' she laughed. 1965 T. Capote *In Cold Blood* iv. 321, I wish you'd send me earplugs. Only they wouldn't allow me to have them. No rest for the wicked, I guess. 1979 M. Babson *So soon done For* vii. 54 'I wish I could take some time and get away. But there's no rest for the weary.' 'Or the wicked.'

wicket. Add: **3. a.** *double wicket* (earlier examples); *to keep wicket* (earlier example).
1773 J. Burnby *Kentish Cricketers* 14 Davis, who loves a Game of Cricket, And shines whene'er he keeps the Wicket. 1778 *Coventry Mercury* 6 July 3/4 On Tuesday last..a Cricket Match, (full set at double wicket) was played between the Wappenbury and Coventry players. 1801 J. Strutt *Sports & Pastimes Eng.* II. iii. 83 Cricket...This game which is played with the bat and ball, consists of single and double wicket.
d. Fig. phrs.: *to be on a good wicket*, to be in an advantageous or favourable position; *to bat* (or *be*) *on a sticky wicket*: see *sticky *a.*² 1 c.
1941 *Punch* 24 Dec. 551/1, I wondered why I was so anxious to conceal my age; for the old are on a good wicket. 1961 *Listener* 2 Nov. 737/2 Perhaps the most satisfactory contributions are those of Lord Birkett, who is on a good wicket in describing the change in legal attitudes to obscenity, and Dr. Robert Gosling. 1977 *Verbatim* Dec. 3/2 To be on a good wicket is, like being on a good pitch, to 'be in a good spot'. To be on a good wicket with someone is to 'be in favor' with him.
6. (sense 3) **wicket-bag, -taker.**
1916 Joyce *Portrait of Artist* (1969) 90 A team of cricketers passed,..one of them carrying the long green wicketbag. 1962 *Times* 20 June 4/1 In the second Test match..Coldwell wins his [cap] as the season's premier wicket-taker. 1976 J. Snow *Cricket Rebel* 76 Barry Knight had been the main wicket-taker in the West Indies first innings with four.

wicket-keep. (Earlier example.)
1867 *J. Lillywhite's Cricketer's Compan.* 107 [He] promises very well as a wicket-keep.

wicket-keeping. Add: (Earlier examples.)
1826 F. Reynolds *Life & Times* II. xiv. 170 No man could..surpass..Hammond in wicket keeping. 1833 *Sporting Mag.* LXXXII. 353/2 The wicket-keeping of Wenman.
Hence (as a back-formation) **wi·cket-keep** *v. intr.*, to keep wicket; **wi·cket-keeping** *ppl. a.*
1891 W. G. Grace *Cricket* v. 138 Lillywhite was bowling and I was wicket-keeping. 1955 I. Peebles *Ashes* i. 13 Two wicket-keeping batsmen. 1976 *Milton Keynes Express* 25 June 49/5 In his nine games for the club so far this season the wicket keeping batsman has claimed 19 victims.

wicking, *sb.* (Earlier example.)
1847 *Rep. Comm. Patents 1846* (U.S.) 220 This is combined with a small tube within it, through which the wicking is introduced, to cause the tallow to unite around the wicking.

wickiup, var. Wickyup in Dict. and Suppl.

wi·ckless, *a.* [f. Wick *sb.*¹ + -less.] That burns without a wick; not fitted with a wick.
1899 T. Eaton & Co. *Catal.* 'Summer Needs' 13 (heading) Wickless blue flame oil stoves. 1924 *Chambers's Jrnl.* Sept. 638/1 In all wickless stoves and lamps particles of carbon are deposited in the vaporiser. 1950 [see *gun-flash* s.v. *gun *sb.* 14 a].

wickmanite (wi·kmănəit). *Min.* [f. the name of F.-E. *Wickman* (b. 1915), Swedish mineralogist + -ite¹.] A cubic hydroxide of manganese and tin, $MnSn(OH)_6$, found as small yellowish or colourless octahedral crystals.
1967 Moore & Smith in *Arkiv för Mineral. och Geol.* IV. 398 Wickmanite..is of interest since it is the first tin mineral reported from Långban. 1977 *Canad. Mineralogist* XV. 437/1, Wickmanite, $MnSn(OH)_6$, schoenfliesite, $MgSn(OH)_6$, and manganoan schoenfliesite occur in separate low-temperature parageneses in hydrothermally mineralized skarns at Pitkäranta, Karelia.

wicky. b. (Examples.)
1804 A. F. M. Willich *Domestic Encycl.* I. 53/1 The plant [sc. Andromeda] is there [sc. in the southern states] called 'wickie'. 1901 C. T. Mohr *Plant Life Alabama* 654 Wicky..Louisianian area..Low sandy pine barrens.

wickyup. The standard form is now *wickiup.* (Further examples.)
1876 *Sun* (N.Y.) 10 May 2/6 Come up and see me at my wickiup in Montana. 1930 E. Ferber *Cimarron* i. 11 He was raised in a tepee; a wickiup had been his bedroom, a blanket his robe. 1959 E. Tunis *Indians* 110/1 In winter the Diggers put their wickiups over pits for additional warmth, just as the Basketmakers did. 1973 'P. Buchanan' *Requiem of Sharks* xi. 111 In the slang of the Pascagoula Indians, she was built like a brick wickiup.

wid (wid), repr. colloq. and dial. pronunc. of With *prep.*
[See With *prep.* for pre-16th cent. examples.] 1869 S. H. Bradford *Scenes in Life Harriet Tubman* 26 Jesus will go wid you. 1884 D. Boucicault *Shaughraun* i. i. 3 Never fear, I'll be even wid your honour yet. 1895 Baines & Smiley in A. Dundes *Mother Wit* (1973) 256/2 You an' I was sittin' at de table wid but one dish ob soup. 1897 Kipling *Capt. Cour.* iii. 77 We do be condescending to honour the second half wid our presence. 1935 [see *Jack *sb.*¹ 2 e]. 1953 K. Tennant *Joyful Condemned*

xii. 106 What's up wid yuh now? 1978 J. Irving *World according to Garp* xix. 432 She didda lot for people wid complicated lives.

Widal (vi-, widā·l). *Path.* The name of G. F. *Widal* (1862–1929), French physician, used *attrib.* and in the possessive to designate an agglutination test for typhoid and other *Salmonella* infections described by him in 1896.
1899 H. M. Biggs in *Typhoid Fever* (N.Y. State Med. Assoc.) 272 A negative result from the Widal test cannot be regarded as having much significance. *Ibid.*, While the Widal reaction may be very late in its appearance,..when present it is of the greatest possible value in the diagnosis. 1908 *Practitioner* Sept. 423 The absence of spots, Widal's reaction,..and the presence of a marked leucocytosis and of localised pain, will settle any doubt. 1974 R. M. Kirk et al. *Surgery* ii. 35/2 This is the basis of Widal's reaction..for the diagnosis of typhoid, and it can also be used to detect dysenteric infection. 1976 *Lancet* 20 Nov. 1143/2 Stool microscopy, stool culture (including viral culture), and a Widal test failed to establish a diagnosis.

widdle (wi·d'l), *sb.*² *colloq.* [Echoic: cf. *piddle *sb.*, *wee *sb.*², and next.] An act of urination.
1954 J. Pudney *Smallest Room* 36 The *wee-wee*, the *widdle*, the *pee-pee*, and the *piddle.* 1969 D. Clarke *Nobody's Perfect* iii. 77, I hardly ever saw him unless I..wanted a widdle. 1977 A. Coren *Lady from Stalingrad Mansions* 63 Love is..mekkin' sure yer betrothed 'as a pensionable position wi' luncheon vouchers an 'gets out of 'is bath when he wants a widdle.

widdle (wi·d'l), *v.*³ *colloq.* [f. as prec.: cf. Piddle *v.* 2.] *intr.* To make water, to urinate. Hence **wi·ddling** *ppl. a.*
[1956 G. Durrell *My Family & Other Animals* xi. 143 Larry's suggestion that [the puppies]..be called Widdle and the Puke was greeted with disgust by Mother.] 1968 *Listener* 13 June 785/2 *Work is a Four-Letter Word*, with its short-term expedients (including some inferior pop music and a widdling dog), is grimly unrisible. 1970 'R. Gordon' *Doctor on Boil* ii. 15 From some of the receptacles you physicians produce, you seem to imagine a camel could widdle through the eye of a needle. 1974 'D. Meiring' *President Plan* xii. 95 Martinez was practically widdling with excitement. 1983 W. Harriss *Bay Psalm Bk. Murder* ix. 82 He headed straight for me... I damn near widdled.

widdy-widdy-way (wi:di,wi:di,wē¹·). *dial.* Also **widdy(-way),** etc. [A rhyme used during the game: see also in *Eng. Dial. Dict.* and quot. 1969 source below.] A children's game of tag.
1846 J. R. Planché *Invisible Prince* i. ii. 9 And hail the scenes where I was wont to play At marbles, hopscotch, hoop, and widdy way. 1859 *Games & Sports for Young Boys* 1 *Widdy.* This is a very spirited game, and is peculiarly adapted for wintry weather. 1893–4 R. O. Heslop *Northumb. Words* II. 788 *Widdy-widdy-way*,..a boys' game. Two boys start hand in hand from a 'bay', and endeavour to touch their opponents. Anyone touched must return with them to the bay and join hands with the first to make a fresh sally... If the chain of hands be broken, the sally has proved a failure, and each outsider endeavours to capture and ride in triumph on the back of one of his quondam pursuers. 1897 H. G. Wells *Plattner Story* 250 Figures kept moving from one line to another, like children playing at Widdy, Widdy Way. 1969 I. & P. Opie *Children's Games in Street & Playground* ii. 94 In Peckham Rye the game is known as 'Chain Widdy'.

wide, *sb.* Add: **3.** (Earlier example.)
1846 W. Denison *Cricket* 5 The parties deliver beyond their natural powers; control of the ball is thus lost, and a 'wide' is the consequence.
4. *to the wide*: to the extreme; entirely, utterly. Used in various slang phrs., as *blind* (*broke, dead, out*, etc.) *to the wide*; *done to the wide*: see *do *v.* 11 e.
1915 G. Frankau *Tid'apa* iii. 19 'Blind, blind to the wide.' It was shaky, his hand on the dipper-bar, As the water slopped over, gurgling, from its Ali-baba jar. 1920 Wodehouse *Jill the Reckless* xiv. 208 Here was a girl who seemed to like him although under the impression that he was broke to the wide. *a* 1936 Kipling *Something of Myself* (1937) vi. 155, I have seen a Horse Battery 'dead to the wide' come in at midnight in raging rain. 1946 *Coast to Coast 1945* 29 Now yer broke to the wide—I'd rather yer died. 1958 F. C. Avis *Boxing Dict.* 96 *Out to the wide*, completely unconscious. 1959 L. Lee *Cider with Rosie* 90 Wake up, lamb... He's wacked to the wide. Let's try and carry him up. 1963 M. Duggan in C. K. Stead *N.Z. Short Stories* (1966) 97 Honest, simple and broke to the wide.

wide, *a.* Add: **I. 1. b.** Also (Austral.) *the wide brown land*, Australia; *wide open spaces*: see *open *a.* 8 a.
1914 D. Mackellar *Witch Maid* 29 Her beauty and her terror—The wide brown land for me. 1934 J. & G. Mackaness (*title*) The wide brown land. 1973 *Australian* 4 May 11 Migrants are staying away in droves from the widest and brownest part of this wide, brown land.
II. 5. c. As the final element in comb. with sbs. which denote regions, organizations, etc., as World-wide, and *country-wide, nation-wide, state-wide* (see at first element in Suppl.), in the sense 'as wide as the——' or 'extending throughout the whole——'.
III. 11. c. Quot. 1887 may be redated 1879 (*Macmillan's Mag.* Oct. 502/1). Also, shrewd, sharp-witted; (dishonestly) cunning or knowledgeable; skilled in sharp practice; engaging in shady dealings. See also *wide boy*, sense 12 c below. (Later examples.)
1928 E. Wallace *Gunner* xxviii. 226 You can handle these swells, Danty, and you're wide enough to keep yourself out of trouble. 1938 F. D. Sharpe *Sharpe of Flying Squad* i. 13 Underworld men and women..refer to themselves as 'wide people' or 'one of us'. They're a colourful, rascally lot these 'wide 'uns'. 1956 T. Huddleston *Naught for your Comfort* ii. 28 He must become a 'tsotsi', a cosh-boy, a wide-guy—because at least there's excitement that way, while it lasts. 1981 *Event* 16 Oct. 101/3 They've never struck me as a bunch of wide-persons.
IV. 12. a. *wide-armed* (earlier example), *-beaked, -bellied, -lapped, -legged, -margined, -shouldered, -sleeved* (later examples), *-waked, -windowed.* **c.** **wide-angle** *a.* (examples of *wide-angle lens*); also in extended use and as *sb.*; **wide-aperture** *a.*, applied to (an instrument having) an objective lens of large diameter; **wide-band** *a.*, capable of transmitting or handling signals in a wide frequency band; **wide-bodied** *a.*, of a large jet aeroplane: having a wide fuselage (cf. *jumbo 1 b); also **wide-body** (usu. *attrib.*); **wide boy** *slang*, one who lives by his wits, often dishonestly; one who engages in petty-criminal activities, a 'spiv'; cf. sense *11 c; **wide-cut** *a. Oil Industry*, involving or produced by fractional distillation over a wide temperature range, or the fraction so obtained (see quots. 1958, 1966); **wide-eyed** *a.*: also *fig.*; **wide gauge** *Railways* = Broad gauge; **wide-leafed** *a.* (earlier example, *transf.* of a table); **wide-meshed** *a.*, of a net: having wide meshes or interstices; (in quots., *fig.* of a survey); **wide receiver** *Amer. Football*, a pass receiver who stands several yards to the side of an offensive formation; cf. *flanker *sb.*¹ 3 d, *receiver¹ 1 c (*a*); **wide-scale** *a.*, that occurs on a wide scale; extensive; cf. *large-scale* adj. s.v. *large *a.* 15 c; **wide screen,** a cinema screen which presents a wide field of vision in proportion to its height (see quot. 1957); freq. *attrib.*; **wide-spectrum** *a.*, (*a*) *fig.*, effective against a wide range of organisms; = *broad-spectrum* s.v. *broad *a.* D. 2; (*b*) *lit.*, characterized by light of a wide range of wavelengths; **wide-wale** a., of fabrics, esp. corduroy: broad-ribbed.
1897 C. M. Hepworth *Animated Photogr.* xiii. 97 The use of a wide-angle lens..is..abominable in connection with the production of a living photograph. 1947 H. Lewis *Photogr. Today* 53 On analysing my shots I usually find that 70 per cent. have been taken with a 5 cm. lens, 2 per cent. with a long-focus lens, and the rest with a 3.5 cm. wide-angle lens. 1955 *Mademoiselle* Mar. 113 *Oklahoma!* is made in 'fabulous new Todd-AO wide-angle, large-screen process'. 1965 C. Forsyte *Double Death* iii. 22 He kept most of his attention on the special wide-angle driving mirror that raked the traffic on his tail. 1974 J. Irving *158-Pound Marriage* i. 11 Forget the wide-angle. (I see Edith and Severin Winter only in close-ups.) 1983 *Which?* Sept. 388/3 A zoom lens lets you move in from a wide-angle view to a closer shot. 1958 *Amateur Photographer* 31 Dec. 914/2 For colour work a wide-aperture lens is invaluable. 1966 D. G. Brandon *Mod. Techniques Metallogr.* i. 57 With a wide-aperture telescope..there is no loss of brightness on magnification. 1869 J. R. Lowell *Poet. Works* (1912) 415 The friend of all the winds, wide-armed he towers. 1935 *Wireless Engineer* XII. 251/1 A means of examining the behaviour of wide-band amplifiers when supplied with transient input waves. 1967 E. Chambers *Photolitho-Offset* iv. 42 Although this ideal is not fully realised the fact remains that very acceptable results can be obtained using either wide-band (trichromatic) or narrow-cut filters. 1982 *Economist* 6 Mar. 25/2 The government wants Britain's cities to be cabled quickly with wideband cable. 1807 J. Barlow *Columbiad* iii. 131 The wide-beak'd hawk, that now beholds me die, soon with his cowering train my flesh shall tear. 1921 D. H. Lawrence *Birds, Beasts & Flowers* 30 An enormously wide-beaked mouth. 1921 W. de la Mare *Veil* 6 Dipped the wide-bellied boat. 1980 *Jrnl. R. Soc. Med.* LXXIII. 7 A wide-bellied, ungainly but functional ambulance. 1970 *Times* 4 Sept. (Aviation Suppl.) p. i/4 About £200m. is being requested to get the proposed BAC 3-11 wide-bodied, 250-seater subsonic airliner off the ground. 1983 *Times* 12 Feb. 20/8 Western airlines..were not allowed to fly wide-bodied jets such as the Airbus into Moscow until the Russians had developed their own Il 86. 1968 *Flight Internat.* 14 Nov. 777/1 BAC foresees a demand for standards matching the high-capacity wide-body aircraft of the long-haul routes on short/medium-haul routes. 1979 T. Gifford *Hollywood Gothic* (1980) xxx. 308 The wide-bodies slid down..into the bustle of Los Angeles International Airport. 1983 *Listener* 9 June 6/2 Only two companies are now producing wide-body airliners. 1937 R. Westerby *Wide Boys never Work* 232 Jim was turning, or had already turned, into a Smart Aleck, a Wide Boy, a despiser of the Mugs who worked. 1947 *People* 22 June 5/3 It seems the wide boys are trying to muscle in and buy these dogs to put against one another in private fights. 1952 'J. Tey' *Singing Sands* iv. 57 He was a wide boy. Wide boys don't want trouble. 1960 V. Gielgud *To Bed at Noon* iii. i. 159 Blackmailed—for the murder? Not even the widest of the local wide-boys could have got on to it. 1976 J. O'Connor *Eleventh Commandment* iii. 38 All the wide boys thought I had gone mad when they saw me in khaki. 1958 *Chambers's Techn. Dict.* 1027/2 *Wide-cut fuel*.., low octane petrol (gasoline) obtained from wide-cut distillation used in turbojets in order to conserve kerosene. 1966 *McGraw-Hill Encycl. Sci. & Technol.* X. 54/2 Petroleum is separated by distillation into fractions designated as (1) straight-run gasoline..; (2) middle distillate..; (3) wide-cut gas oil, which boils at about 345–540°C., and (4) residual oil. 1982 *Fuelling Aviation* (Shell Internat. Petroleum Co. Ltd.) 5/2 The military wide-cut fuel is called JP-4 and this is the major fuel for the airforces of the world. 1923 D. H. Lawrence *Birds, Beasts & Flowers* 109 The human soul is fated to wide-eyed responsibility In life. 1983 L. Deighton *Berlin Game* ix. 95 You ask him all those wide-eyed innocent questions about making profits from cheap labour. 1841 Wide gauge [see Wide *a.* 5 a]. 1982 S. G. Duff *Parting of Ways* iv. 43 We all boarded the train for Moscow, changing onto the wide-gauge railway at the Soviet frontier. 1856 J. G. Whittier *Poet. Works* (1898) 353/1 Pacific rolls his waves a-land, From many a wide-lapped port and land-locked bay. 1928 Blunden *Japanese Garland* 19 Fine fields, wide-lapped, whose loveliest-born Day's first bright cohort finds. 1779 P. Freneau *House of Night* in *U.S. Mag.* Aug. 356 A wide-leaf'd table stood on either side. 1938 R. Graves *Coll. Poems* 28 The wide-legged robin with his breast aglow. 1889 O. Wilde in *Fortn. Rev.* Jan. 43 Book-bindings, and early editions, and wide-margined proofs. 1938 *Dialect Notes* VI. 626 Professor A. H. Marckwardt..has begun a wide-meshed survey of the Great Lakes region and the Ohio River valley. 1980 *English World-Wide* I. i. 28 Unfortunately..his survey is even more wide-meshed than Orton's. 1968 *Redskins* 17 Nov. 77/3 Depth at wide receiver is strong, too, in rookie Dennis Homan. 1981 *Washington Post* 8 Apr. d1 We will have to take the best athlete available... That could be an offensive lineman, a running back or a wide receiver. 1958 G. Lienhardt in Middleton & Tait *Tribes without Rulers* 108 There was little wide-scale co-operation against the common enemies. 1980 *Daily Tel.* 26 May 6/7 By confining the emergency arrangements as far as possible to the Bristol line, BR has avoided widescale timetable changes. 1931 *Ann. Reg. 1930* ii. 48 The Wide Screen is still only a matter for experiment, as standardisation has not yet been achieved. 1932 *Ibid. 1931* 47 The 'Wide Screen' invention, though perfected, was not offered to the public by the big producing concerns, seeing that it would involve the studios in huge expenditure. 1953 *Manch. Guardian* 13 Aug. 4/7 Hollywood..has decided to coast for the present on a compromise between 3-D and Cinemascope—namely on the less spectacular development known as Wide Screen. 1957 *Encycl. Brit.* XV. 862/1 Basically 'wide screen' means any departure from the screen proportions fixed by Edison and his contemporaries at 4 to 3 (or 1·33 to 1); *i.e.*, three units high for every four wide... If the aspect ratio were to be changed there was only one practical way—screens would have to be wider. 1967 H. Harrison *Technicolor Time Machine* (1968) iii. 27 An accurate, full-length, wide-screen, realistic, low-budget, high-quality historical. 1976 *National Observer* (U.S.) 16 Oct., Imagine, right before your eyes on the wide screen, the stern of the Titanic..comes shooting out of the water as if the projector had been reversed. 1935 Kipling *Two Forewords* 19 But *thou*, O Nakhoda, art young and wide-shouldered. 1973 T. Pynchon *Gravity's Rainbow* i. 127 A few women in clinking boots and wide-shouldered swagger coats, but no children. 1926 D. H. Lawrence *David* viii. 63 Takes off striped coat, or wide-sleeved tunic. 1980 *Catal. Fine Chinese Ceramics* (Sotheby, Hong Kong) 214 A Jade Carving of a lady wearing a wide-sleeved robe. 1959 S. Duke-Elder *Parsons' Dis. Eye* (ed. 13) xv. 175 One of the wide-spectrum antibiotic drugs such as the tetracyclines. 1972 *Country Life* 25 May 1351/1 The farmer uses..a wide-spectrum weedkiller, which is a mixture of chemicals designed to control a whole range of weeds. 1977 J. L. Harper *Population Biol. Plants* x. 321 They inserted wide-spectrum fluorescent tubes between the rows of a close canopied crop of soyabeans. 1982 *Sci. Amer.* Mar. 98 Snakes of two families can detect and localize sources of infrared radiation. Infrared and visible-light information are integrated in the brain to yield a unique wide-spectrum picture of the world. 1856 J. G. Whittier *Poet. Works* (1898) 52/2 With steeds of fire and steam, Wide waked Today leaves Yesterday behind him like a dream. 1957 M. R. Picken *Fashion Dict.* 374/2 *Wide-wale serge*, serge with broad diagonal weave. 1980 L. Birnbach et al. *Official Preppy Handbk.* 98 Wide wale corduroy pants. 1869 J. R. Lowell *Poet. Works* (1917) A life wide-windowed, shining all abroad, Or curtains drawn to shield from sight profane. 1970 *Daily Tel.* 30 Apr. 17 A wide-windowed bar parlour.

wide, *adv.* Add: **3*.** Transferred senses of *wide open.* **a.** *Boxing*, etc. Fully exposed to assault; unprotected, off one's guard. Freq. *fig.*, esp. in phr. *to leave* (*lay*, etc.) (oneself) *wide open.*
1915 E. Corri *30 Yrs. Boxing Referee* 150 Johnny Summers..in an unguarded moment, left himself wide open and encountered one of the most decisive knock-out punches I ever saw. 1941 B. Schulberg *What makes Sammy Run?* i. 14 You never find me going in for favors... It leaves you wide open. 1948 'N. Shute' *No Highway* vi. 148 Honey lays himself wide open to that sort of thing. 1966 'A. Hall' *9th Directive* iv. 42 One fine day he would catch me wide open and slam me down.
b. Of an issue, case: not circumscribed or prejudiced by conditions; unrestricted (in its implications, effects, etc.); not resolved or decided; *spec.* of a police investigation.

1963 'J. Melville' *Burning is Substitute* iii. 51 Charmian suddenly had the feeling that this affair..was wide open, could reach anywhere. 1970 *Daily Tel.* 10 July 19 The fate of Penguin Publishing Company is still wide open. 1973 J. Thomson *Death Cap* iii. 41 They're the only people who so far have entered the case... As far as I'm concerned, it's still wide open. 1982 C. Aird *Last Respects* xiii. 137 It's [sc. a murder enquiry] what you might call wide open still... You'll have to look on it as a challenge.

6. **wide-ranging** (later examples), **-rolling** (earlier example), **-sweeping**; **wide-apart**; **wide-open** *a.*, (*b*) *U.S.*, of a town: not oppressed by laws or law enforcement.

1941 E. Bowen *Look at all those Roses* 39 The wide-apart birch-trees. 1983 T. Hughes in *Listener* 21 Apr. 27/1 They have a chirruppy, chicken-sweet expression With goo-goo starlet wide-apart eyes. 1892 *Harper's Mag.* June 103/1 It is what they call in Montana 'a wide-open town'. 1975 J. Gores *Hammett* xi. 79 He has been elected three times because the citizens want a wide-open town. 1958 *Times Lit. Suppl.* 7 Feb. 76/4 A representative anthology, which is so wide-ranging in its material..that its final effect is rather of confusion than of enlightenment. 1980 B. Hill in *Beautiful Brit. Columbia* Summer 39 The wide-ranging sheep that are one of the island's main farm products provide the source of wool for local weavers. 1785 T. Dwight *Conquest of Canäan* xi. 295 Wide-rolling dust the neighbouring concave fills. 1924 *Motor* 14 Oct. 491 (*caption*) One of the two wide-sweeping bankings on the new speedway at Montlhery, near Paris. 1979 *Jrnl. R. Soc. Arts* CXXVII. 409/2 Wessex..will therefore not be subjected to wide-sweeping environmental problems.

wide-awake, *sb.* Add: **2.** (Earlier example.)

1877 R. L. Price *Two Americas* iv. 57 Sea-gulls and wide-awakes hovered in hundreds over the water.

3. (Earlier example.)

1865 Dickens *Mut. Fr.* II. iii. xii. 111 You have been told that he might pull through it.., Wide-Awake; have you?

wide-mouth, *a.* Add: **B.** *sb.* One who speaks loudly, boastfully, or without restraint. Cf. *big mouth* s.v. *Big *a.* B. 2.

1959 I. & P. Opie *Lore & Lang. Schoolch.* x. 189 The tell tale is christened..a tout, traitor, quisling, or wide-mouth. 1978 *Times Lit. Suppl.* 28 Apr. 462/2 You feel, frequently, like booing him for a bighead and a wide-mouth.

widger (wi·dʒəɹ). Also *erron.* **wigger.** [See quot. 1956.] A gardening tool consisting of a small strip of metal with a shallow furrow down the centre, used as a miniature trowel to move seedlings, cultivate pot plants, etc.

1956 *Dict. Gardening* (R. Hort. Soc.) Suppl. 333/1 The original widger..is the shape of a small spatula... The name 'Widger' was transferred to this horticultural tool by Mr. Clarence Elliott; it comes from one of a series of nonsense definitions used to test the memorizing ability of British Naval Cadets. 1962 *Listener* 22 Nov. 887/3 A stainless steel 'wigger' is a very useful little gadget for pricking out seedlings. 1973 *Country Life* 5 Apr. 940/1 Seedlings..I move..with a widger—an invaluable narrow-bladed, long trowel-like hand tool.

widget (wi·dʒėt). orig. *U.S.* [Perh. alteration of *Gadget.] An indefinite name for a gadget or mechanical contrivance, esp. a small manufactured item.

1931 *Amer. Speech* VI. 259 Widget. 1937 E. Lyons *Assignment in Utopia* (1938) iii. iii. 299 Every time the percentage of widgets turned out by her factory rose her features shone. 1961 *N.Y. Times Mag.* 26 Nov. 109 Widgets by wire... Suppose something goes wrong with your Westinghouse appliance. What then? Your Westinghouse dealer can get you any new part faster than anyone around. He orders it by special telegraph line. 1966 'E. Lathen' *Murder makes Wheels go Round* i. 5 The corporation would..go about its business of producing bigger and better widgets. 1973 S. Alsop *Stay of Execution* (1974) ii. 193, I asked Joe if the widget could be protected by patent. Joe said..it couldn't... A few weeks ago,..IBM began to make its own widget, cheaper and better serviced than the Intercomp widget. 1982 Barr & York *Official Sloane Ranger Handbk.* 11/1 You never have to see the industry ('widget factories') and commerce ('selling brushes') that make the money.

widgie (wi·dʒi). *Austral.* and *N.Z.* Also **weegie.** [Origin uncertain.] An Australasian teddy-girl, the female counterpart of a *bodgie.

1950 *Sun* (Austral.) 5 July 19 There'll be..prizes for the most colorfully dressed 'bodgy' and 'weegie'. 1956 S. Hope *Diggers' Paradise* 86 A popular district with bodgies and widgies is 'the Cross'. 1965 E. Brown *Big Man* xix. 168 A mob of bodgies and widgies on a camping holiday. 1977 *Times* 13 May 14/1 Gang delinquency..has made its mark around the world..in Australia the bodgies and widgies.

Widmanstätten (vi·t-, wi·dmänʃtetən, -st-). Also **-staetten, -statten.** The name of A. J. Widmanstätten (1754–1849), Austrian scientist, used *attrib.* with reference to an orderly pattern of intersecting bands seen in some meteorites and steels when a polished section is etched, attributed to the crystallization or precipitation of a new solid phase along the crystal planes of a parent solid phase.

1861 *Q. Jrnl. Geol. Soc.* XVII. ii. 9 The plane of the cutting shows strings of whitish colour indicative of crystalline structure ('Widmannstetten's figures'). 1881 L. Fletcher *Guide Meteorites Brit. Museum* 16 The want of homogeneity in meteoric iron is beautifully shown by the 'Widmanstätten' figures. 1927 *Jrnl. Iron & Steel Inst.* CXVI. 584 He points out the difference between the needle-like structure of the ferrite and the Widmanstätten structure. 1971 I. G. Gass et al. *Understanding Earth* viii. 116/2 Those irons..usually show the Widmanstaetten structure.

Hence † **Widmanstä·ttian** *a.*

1842 *Amer. Jrnl. Sci.* XLIII. 359 The powder arranged itself in directions coinciding with the Widmanstättian figures. 1883 *Phil. Trans. R. Soc.* CLXXIII. 889 It is the constituent of nickel-iron which forms the fine lines constituting the Wiedmannstättian figures, and not schreibersite, as usually stated in writings on the etched figures of meteoric iron. 1886 [see *Neumann].

widow, *sb.*¹ Add: **1. d.** (Later allusive examples.)

1952 W. M. Miller in *Galaxy* Nov. 153/1 It was different if the business-widow called on a couple. Then the lone male could retire. 1965 *Guardian* 30 July 10/5 One Scottish TA unit, aware of the dangers of creating 'TA widows' opens its bar on drill nights to wives and girl-friends. 1973 R. Busby *Pattern of Violence* ii. 32 You tell her to come and see me. We police widows have got to stick together. 1980 *Financial Rev.* (Austral.) 14 Jan. 8/2 Dick Smith's resident computer expert..said the key-boards, screens, printers and central processors are giving birth to a new social problem, 'computer widows'.

h. *The Widow* (*of* or *at Windsor*): a familiar epithet for Queen Victoria, whose husband predeceased her by forty years. orig. chiefly *Services'.*

1888 Kipling *Private Learoyd's Story* in *Soldiers Three* 14 They tell me t' Widdy herself is fond of a good dog. 1890 —— *The Widow at Windsor* in *Barrack-Room Ballads* (1892) 39 Then 'ere's to the Widow at Windsor, An' 'ere's to the stores an' the guns, The men an' the 'orses what makes up the forces O' Missis Victorier's sons. 1900 *Captain* III. 235/1 The design..shows the Queen as a widow—the 'Widow of Windsor'. 1932 *Times* 12 Feb. 14/2 'The Widow' (as we subalterns had irreverently nick-named the Empress of India). 1964 E. Longford *Victoria R.I.* xxxvi. 562 She died just after half-past six... The famous 'hush' which had always surrounded 'The Widow at Windsor' was shattered. 1980 R. Hall *Lovers on Nile* xiv. 216 The 'Widow of Windsor' would feel herself justified in having ostracized Sam and Florence.

3. c. (Earlier example.)

1781 Boswell *Jrnl.* 28 Apr. (1977) 333 He [sc. Lord Townshend] had called Sir Joshua, 'Will you give us one cool bottle of claret?' They were taking away the former. 'No,' said Lord Townshend, 'Let us first take the widow.'

d. *Typogr.* A short line at the end of a paragraph, esp. one which is set at the top of a page or column, or which contains only (part of) one word, and is therefore considered unsightly.

[1904 *Man. Rules Compositors S. S. McClure Co.* 25 All running heads are to be set one nonpareil from the body, unless otherwise instructed. Care must be taken to overcome 'rivers', and to this end indiscriminate division of words is allowed. Care should also be exercised to overcome 'widdies' at the top of pages.] 1925 [see *Slug *sb.*¹ 4 f]. 1932 P. Van D. Stern *Introd. Typogr.* ii. 15 When a single word runs over, it is often desirable to alter the copy..so that the words can be run back. Single words standing in a line are called 'widows'. 1948 *Bull. N.Y. Public Library* Jan. 3 Early in 1936, H. M. Lydenberg..began a quiet, and not quite humorless, investigation into the origin and identity of the typographical 'widow', that awful slattern of the printed page. 1954 M. Laski in *Author* Winter 30/2 It is a common experience, when working for *Vogue*, to be asked to add a few words to a paragraph so as to avoid unsightly 'widows' or single-word lines. 1963 D. Ogilvy *Confessions Advertising Man* vii. 124 It has been discovered that 'widows' increase readership, except at the bottom of a column. 1980 B. Crutchley *To be Printer* 55 Our best customers were those who looked to us..to..print well, which meant avoiding 'widows'.

e. *five-fingered* (also *dry-mouthed*) *widow*, in phrs. alluding to the act of masturbation. *slang* (chiefly *Services'*).

1971 B. W. Aldiss *Soldier Erect* 44 In there [sc. the 'shithouse'], behind the stable-like door of one compartment or another, I went to a regular evening rendezvous with my dry-mouthed widow. 1975 C. Allen *Plain Tales from Raj* xv. 159 Many turned, as a last resort, to the 'five-fingered widow'.

4. a. *widow mother* (later example). **c.** *widow-making* (later example). **d.** **widow-maker,** (*b*) *N. Amer. slang*, a dead branch caught high in a tree which may fall on someone below; (*c*) *slang* [tr. G. *witwemacher*], a nickname for the Lockheed F-104 Starfighter strike and reconnaissance aircraft (see quot. 1975); also (*U.S. Mil.*), a grenade launcher; **widow-man** *dial.* = Widower¹ 1 a.

1945 M. H. Allee *Smoke Jumper* iv. 47 He remembered the Kid's caution about widow-makers, limbs falling from high overhead. 1965 M. McIntyre *Place of Quiet Waters* ix. 163 Now's the time to look out for widow-makers... Don't you go walking about in the woods when she's blowing like this. 1975 *Times* 26 Sept. 7/2 The loss of 178 aircraft in Germany earned it [sc. the Starfighter] the title of 'Witwemacher', or widow-maker. 1976 *Courier-Mail* (Brisbane) 13 Feb. 4/4 They opened up with automatic rifle fire, a Browning machine-gun, and 66 anti-tank rockets, and a 'widow-maker'—a grenade launcher. 1876 G. M. Hopkins *Wreck of Deutschland* xiii, in *Poems* (1967) 55 The widow-making unchilding unfathering deeps. 1887 T. E. Brown *Doctor & Other Poems* 35 Sir John, it appears, Was a widda man. 1946 C. McCullers *Member of Wedding* ii. 43 He was a widowman, for her mother had died the very day that she was born—and, as a widow-man, set in his ways. 1937 C. Day Lewis *Starting Point* 36 Theo's getting as fussy as a widow-mother.

5. **widow's cruse** *fig.* (with allusion to I Kings xvii. 12–16) [see Cruse], a supply which, though apparently meagre, is, or seems to be, inexhaustible; **widow's walk** chiefly *N. Amer.*, a rectangular balustraded platform (characteristic of New England architectural styles in the 18th and 19th cent.) built on top of the roof of a house, esp. for providing an unimpeded view of the sea (see quot. 1978).

1816 Scott *Old Mortality* in *Tales my Landlord* 1st Ser. IV. xii. 268 'Can you lodge a stranger for a night?' 'I can, sir, if he will be pleased with the widow's cake and the widow's cruize.' 1915 D. H. Lawrence *Phoenix II* (1968) 382 Wherein..is the immortality, in the constant occupation of the nest, the widow's cruse, or in the surpassing of the phoenix? 1977 *Jrnl. R. Soc. Arts* CXXV. 463/1 Information is infinitely reproducible without diminishing it: it is a veritable widow's cruse. 1939 S. Chamberlain *Nantucket* 25 Variously termed a 'Captain's Walk', or the 'Widow's Walk', it is just 'The Walk' in Nantucket. 1961 J. Steinbeck *Winter of our Discontent* i. 14 The fine old house,..his great-grandfather's,..with..Adam decorations and a widow's walk on the roof. 1978 J. A. Michener *Chesapeake* 463 The name *widow's walk* derived from romantic tales of those loyal women who continued to keep watch for a ship that had long since gone to the bottom of some coral sea.

widower¹. Add: **1. c.** The counterpart of Widow *sb.*¹ 1 d in allusive use, as *football widower*, etc. *colloq.*

1969 *Listener* 17 Apr. 534/3 He's a football widower because I'm the one who's always trooping away to football matches. 1971 A. Nixon *Attack on Vienna* xi. 109 Mr Fletcher had had a quiet drink with another bridge widower. 1973 *Guardian* 25 May 11/2 The age of golf widowers is developing.

width. Add: **4. a.** (Earlier example.)

1872 D. G. Rossetti *Let.* 26 Sept. (1967) III. 1076 It would be quite enough to make four curtains..6 ft ½ wide (or under would do if more convenient with the widths of the velvet).

b. The width of a swimming-bath taken as a measure of the distance swum. Cf. *Length *sb.* 4 d.

1930 *Swimming Instruction* (Amateur Swimming Assoc.) 55 From this stage the class should proceed to swim..several widths, legs only, using supports or 'Dog Paddle'. 1971 *Daily Tel.* 17 Nov. 3/6 Mrs Annie Oakley, 86,..has been presented with a certificate for swimming two widths at Soundwell Swimming Club. 1981 H. Engel *Ransom Game* (1982) xxx. 198 She went off the board again... She did two lengths to each of my widths.

Wiedemann–Franz (vī·dəman frɑnts). *Physics.* [The names of G. H. *Wiedemann* (1826–99) and R. *Franz* (1827–1902), German physicists, who published the law in 1853 (*Ann. d. Physik* LXXXIX. 497).] *Wiedemann–Franz law*: the law that at any given temperature the ratio of the thermal to the electrical conductivity has approximately the same value for all metallic elements; *Wiedemann–Franz ratio*: this ratio or the Lorenz ratio (see *Lorenz).

1924 J. R. Partington in H. S. Taylor *Treat. Physical Chem.* I. xi. 490 The Wiedemann-Franz law is only approximate. 1966 Phillips & Williams *Inorg. Chem.* II. xix. 23 The transition metals are also good conductors of heat. On the free-electron theory it is predicted that there should be a direct relation between the thermal, k, and electrical, σ, conductivities, the Wiedemann–Franz ratio. 1975 D. G. Fink *Electronics Engineers' Handbk.* vi. 10 The Wiedemann-Frantz [*sic*] ratio L is defined as $L = k/\sigma T$. 1975 *Jrnl. Low Temperature Physics* XX. 691 The Lorenz numbers..are very close to the theoretical value..predicted by the Wiedemann–Franz law for pure electronic heat conduction limited by impurities.

wiederkom (vī·dəɹkǫm). Also **wiederkomm** and with capital initial. [ad. F. *vidrecome* goblet (Robert, 1752), ult. ad. G. *wiederkommen* to return, come again (see quots.).] A tall, cylindrical, German drinking-vessel, made of (usu. coloured or painted) glass.

The Eng. form reflects the original Ger. derivation. F. *vidrecome* is sometimes understood as a fanciful corruption of G. *willkomm* loving-cup, but this suggestion appears to be unsubstantiated.

1878 A. Nesbitt *Descr. Catal. Glass Vessels S. Kensington Museum* p. cxxiii, The cylindrical drinking vessels, generally called wiederkoms..are sometimes very large, some being as much as 20 in. in height. 1881 C. C. Harrison *Woman's Handiwork* iii. 229 For side-board decoration, the Wiederkom or 'come again' drinking-cups in emerald-hued [Bohemian] glass, have always been popular. 1897 A. Hartshorne *Old Eng. Glass* xv. 82 A

glass called a 'Wiederkom' was one which was filled, passed round the table..and 'came again' empty. 1907 E. Dillon *Glass* xvi. 266 The term *wiederkomm* given by so many English..writers to the large broad forms [of drinking-glasses], is unknown in Germany, so that I think the expression may be definitely abandoned and replaced by the word *humpen* or *willkomm humpen.* 1926 N. H. Moore *Old Glass* i. 81 The huge glasses known as 'Will-kommen'..were originally used by a host to welcome his newly arrived guest... The term 'Wiederkom' by which these glasses are known in England is a misnomer. 1946 W. B. Honey *Glass* vi. 75 The *Stangenglas*..became popular in a modified form as the *Willkomm*, or 'greeting-glass' (sometimes mistakenly called a *Wiederkom*), in thin metal. 1977 H. Newman *Illustr. Dict. Glass* 342 *Wiederkomm (humpen)*,..literally, come again beaker. A term used by A. Nesbit [*sic*]..in referring to *Willkomm (humpen)* and many other types of enamelled *Humpen*... It is a term not used by German museums.

Wien¹ (vīn). *Physics.* The name of Wilhelm Wien (1864–1928), German physicist, used *attrib.* and in the possessive to denote (*a*) an approximation to Planck's law that holds at short wavelengths, according to which the flux of radiant energy of wavelength λ emitted by a black body at temperature T is proportional to $1/\lambda^5 \exp(hc/\lambda kT)$; (*b*) the displacement law (sense (i) s.v. *Displacement 2 e).

1899 *Astrophysical Jrnl.* X. 40 My observations..make it seem possible that the law derived by W. Wien represents the emission of 'the absolutely black body'. In Wien's formula, J = [etc.]. 1900 *Sci. Abstr.* III. 383 Wien's laws, according to which the wave-length of the maximum radiation is inversely proportional to the absolute temperature, and the corresponding maximum energy proportional to the fifth power of the absolute temperature, were confirmed. 1904 Wien's displacement law [see *Displacement 2 e]. 1948 [see *Rayleigh–Jeans]. 1963 G. L. Pickard *Descriptive Physical Oceanogr.* v. 52 According to Wien's Law this energy is concentrated round a wavelength of 0·5 μ. 1978 Pasachoff & Kutner *University Astron.* ii. 24 From Wien's displacement law, we can see that the colors of stars in the sky are telling us something about their temperatures.

Wien² (vīn). [The name of M. C. Wien (1866–1938), German physicist.] **a.** *Electr.* *Wien bridge*: an alternating-current bridge circuit devised by Wien which is used to measure capacitance (or frequency) in terms of resistance and frequency (or capacitance); *Wien bridge oscillator*: an oscillator based on this circuit.

1922 Glazebrook *Dict. Appl. Physics* II. 1029/1 *Wien bridge*, for the measurement of the capacity and power factor of a condenser. 1957 *Practical Wireless* XXXIII. 709/2 In Fig. 2 a simple bridge for capacitor testing is given. This is known as a Wien bridge. 1967 *Electronics* 6 Mar. 63/3 (Advt.), Here, the RA-240 is used in the design of a highly stable, uncompensated Wien bridge oscillator. 1979 R. Hamilton *Electronics for Technicians* vii. 157 The other type of R–C oscillator to be considered is known as a Wien bridge oscillator. (The name derives from a type of bridge circuit used in measurements, the feedback network of the oscillator being part of a balanced Wien bridge.)

b. *Physical Chem.* *Wien effect*: the increase in the electrical conductivity of an electrolytic solution as the field strength is increased.

1934 R. P. Bell *Electrolytes* v. 90 The increase in conductivity [with frequency] is..of the same order of magnitude as the Wien effect. 1978 P. W. Atkins *Physical Chem.* xxv. 832 The Wien effect is the observation of higher mobilities [of ions] at higher electric fields. (There are two Wien effects. The first Wien effect is the one just described; the second Wien effect is the enhancement of the degree of ionization of an ionogen, or weak electrolyte, by the applied field.)

wiener (vī·nəɹ), *a.* and *sb.* Also *erron.* **weiner.** [a. Ger., of Vienna.] **A.** *adj.* **1.** *wiener schnitzel*: see *Schnitzel.

2. *wienerwurst* [G. *wurst* sausage]: = *Vienna sausage* s.v. *Vienna 1 a. *U.S.*

1889 *Gallup* (New Mexico) *Gleaner* 27 Mar. 3/3 We..are willing to bet our unpaid debts, against a weiner-wurst [*sic*] that the modest local of the Democrat blushed more than the bride when he saw her in the diaphanous costume he describes. 1899 F. Norris *McTeague* v. 75 The lunch baskets were emptied... There were wienerwurst and frankfurter sausages. 1949 *Los Angeles Times* 15 May ii. 5/2 I've never lamped a crooked dime, not e'en a wienerwurst.

‖**3.** *Wiener Kreis* [G. *kreis* circle]: = *Vienna Circle* s.v. *Vienna 1 a.

[1929 (*title*) Wissenschaftliche Weltauffassung. Der Wiener Kreis (E. Mach Verein).] 1932 *Jrnl. Philos.* XXIX. 122 The philosophy of the 'Wiener Kreis' is unique in that it was formulated for the first time in a concise joint manifesto. 1950 B. Russell *Logic & Knowl.* (1956) 370 Wittgenstein's *Tractatus Logico-Philosophicus*..provided a stimulus which helped in the formation of the 'Wiener Kreis', where logical positivism first took the form of a definite school. 1964 *New Statesman* 10 Apr. 574/2 The early chapters are full of perception theory, Chicago School aesthetics, *Wienerkreis* linguistics, ..Platonic misconceptions and all stations to the Hochschule at Ulm.

B. *sb.* **a.** = *wiener wurst* above. Cf. *Weiner. *N. Amer.*

1904 H. R. Martin *Tillie* iii. 34 I'm havin' fried smashed potatoes and wieners. 1935 *Motion Picture* Nov. 79/1 Hot dogs are just wiener sausages! 1970 S. J. Perelman *Baby, it's Cold Inside* 81 Platters of smoking hot wieners flanked by creamed spinach.

b. *Comb.* **wiener roast** *N. Amer.*, a barbecue at which wieners are cooked and served.

1920 *Outing* July–Aug. 245/1 All over France they introduced the women war workers to American hikes, wiener roasts, camping in the open, and games of all sorts. 1970 J. H. Gray *Boy from Winnipeg* 53 Snowshoeing clubs..given to walking for miles after every snow, usually stopping for a wiener roast somewhere along the way.

wienie (wī·ni). *N. Amer. slang.* Also *erron.* **weinie.** [f. *Wiener *a.* and *sb.*: see -ie.] = *Wiener *sb.* a. Cf. *Weeny *sb.*², *Winny.

1911 *Daily Colonist* (Victoria, B.C.) 26 Apr. 7/3 Weinies, pretzels and coffee were served and German relishes. 1919 U. Sinclair *Jimmie Higgins* xx. 195 Mocking soldier-boys, who made merry..over sauerkraut and..'wienies', otherwise known as 'hot dogs'. 1940 R. Chandler *Farewell, my Lovely* xxxiv. 161, I spotted him [sc. the hot dog man] in a white barbecue stand tickling wienies with a long fork. 1959 E. Ambler *Passage of Arms* iv. 103 The barman opened up a can of weinies. 1977 *Time* 4 July 37/1 One man's meat is another man's corndog (a wienie impaled on a stick and dunked in a bubbling cornbread batter).

wife, *sb.* Add: **1. c.** *Wife of Bath*, one of the pilgrims in Chaucer's *Canterbury Tales*; used allusively (usu. *attrib.*), chiefly with reference to sexual appetite and outspokenness.

1926 A. Huxley *Essays New & Old* 178 Her comments on the connubial state were so very Juliet's Nurse, so positively Wife-of-Bath, that we were made to feel quite early Victorian. 1946 'J. Tey' *Miss Pym Disposes* xviii. 185 The wide flat hat planked slightly to the back of her head on top of her wimple—Wife of Bath fashion—gave her an air of innocent astonishment. 1974 K. Millett *Flying* (1975) ii. 183 Alison sings, a great lusty Wife of Bath woman. 1978 R. Rendell *Sleeping Life* iii. 23 Horrifyingly, she added, with a Wife of Bath look, remembering the old dance, 'Wouldn't be for sex, not so likely.'

2. b. (*d*) *wife and mother*, a conventional epithet describing a woman who shows a zealous devotion to her family (now also somewhat *joc.*).

1798 Mrs. Inchbald *Lovers' Vows* (ed. 3) ii. iii. 31 Go to Amelia—explain to her the duties of a wife and of a mother. 1850 Mrs. Gaskell *Let.* Apr. (1966) 108 One of my mes is a true Christian..another of my mes is a wife and mother, and highly delighted at the delight of everyone else in the house. 1911 G. B. Shaw *Getting Married* 196 She's a born wife and mother, maam. Thats why my children all ran away from home. 1930 A. Christie *Murder at Vicarage* xxxii. 252 I'm going to be a real 'wife and mother' (as they say in books). 1974 M. Cecil *Heroines in Love* v. 128 They could remain devoted wives and mothers and do their bit for the Cause.

(*e*) preceded by an adj. or sb. denoting the husband's occupation (freq. of a Mil. character, as *navy, service wife*), and esp. connoting a wife who fulfils official expectations of this role.

1951 'J. Tey' *Daughter of Time* i. 9 The present Valerie or Angela or Cecile..must be a naval wife. 1975 'J. Bell' *Victim* i. 17 All the vulgar arrogance of an overseas army wife between the wars. 1981 P. McCutchan *Shard calls Tune* ii. 18 Beth had been a police wife..and a Foreign Office Security wife... She knew she mustn't ask where Simon was going.

c. (*b*) *a wife in every port*, a licence or indulgence (jocularly) said to be enjoyed by sailors.

1761 I. Bickerstaffe *Thomas & Sally* i. iii. 5 'Tis pretty sport, for one that gets a wife at ev'ry port. 1907 *Punch* 22 May 365/2 (*caption*) *Admiral.* And what made you wish to become a sailor, my boy? *Navy Candidate* (*in perfect good faith*). Because he's got a wife in every port, sir. 1933 Somerville & 'Ross' *Smile & Tear* xi. 132 'The wife in every port', supposed to be the perquisite of sailors, is no more than the constant aspiration of every self-respecting dog.

g. The passive member of a homosexual partnership. *slang.*

1883 W. A. Hammond *Sexual Impotence in Male* i. 57 The one who was in this disgusting arrangement to act the part of 'husband' came to his 'wife's' bed and remained there during the night. 1957 Danforth & Horan *D.A.'s Man* (1958) i. 3 He's got a new girl. His 'wife' went home last week. 1978 J. Hyams *Pool* xiii. 199 The group's leader [a homosexual]..made his 'wife' head of production.

5. b. (*a*) **wife-basher**; (*b*) **wife-bashing, -battering, -beating** (earlier example). **d.** **wife-bound** *a.* (later example); **wife-swapping,** the interchange of marital partners for sexual purposes within a social group; hence **wife-swap** *sb.* (occas. as *v. intr.*); **wife-swapper.**

1909 *Practitioner* Dec. 828 Poisoning conducted on these lines..resembles the action of the wife-basher, who attacks his victim with a poker... The wife-basher, however, is aware of the obviousness of his crime. 1979 J. Wainwright *Tension* 98 She walked to the home of the 'wife-basher'..and..went into action. 1978 —— *Thief of Time* 221 'Why should some wandering female run to the nearest doctor?'..'Wife-bashing. Unwanted pregnancy. A score of reasons.' 1978 *Times* 16 Feb. 4/7 Wife-battering is most likely to occur among couples with a family history of violence. 1856 Geo. Eliot *Let.* 18 Jan. (1954) II. 225 A Petition..that married women may have a legal right to their own earnings, as a counteractive to wife-beating and other evils. 1820 Keats *Let.* 28 Jan. (1958) II. 247 Henry is wife-bound in Cambden Town there is no getting him out. 1976 *Private Eye* 24 Dec. 8/2 M. Phillipe Dannat..told the magistrates of Nice that he had assaulted M. Georges David at a wife-swap rendezvous in the hills above St. Tropez. 1978 F. Weldon *Praxis* xx. 184 They played strip poker; they wife-swapped. 1969 C. Himes *Blind Man with Pistol* x. 111 Wife swappers, gang fuckers, seekers of depravity. 1959 M. Pugh *Chancer* xiv. 170 He began to discuss the wife-swapping parties held locally... 'But how do they get away with it here?... The town's so small. I would have thought that after six months..they'd have to convene a mass divorce trial.' 1967 W. & J. Breedlove *Swinging Set* x. 119 They brought up the subject of 'wife-swapping' with four other couples. 1976 'W. Trevor' *Children of Dynmouth* iii. 77 There was wife-swapping every Saturday night at parties on the new estate.

wifie. (Earlier examples.)

1786 Burns *Poems, chiefly in Scottish Dial.* 126 His clean hearth-stone, his thrifty Wifie's smile. 1819 Keats *Let.* 18 Sept. (1931) II. 439, I intend to write a letter to you[r] Wifie.

wifing (wəi·fiŋ). *rare.* [f. Wife *sb.* + -ing¹.] The activity or condition of being a wife or housewife.

1905 G. B. Shaw *Let.* 3 Jan. (1972) II. 499 As to ordinary domestic mothering and wifing she [sc. Shaw's mother] was utterly unfitted for the sentiment of it. 1952 S. Kauffmann *Philanderer* (1953) ii. 28 If there was one kind of wife he didn't want..it was one who made a career of 'wifing'.

wig, *sb.*³ Add: **1. e.** *Austral. Sheepshearing.* The wool of a sheep growing around the eyes and on top of the head, removed during shearing. Cf. *Topknot 1 b.

a 1964 H. P. Tritton in R. Ward *Penguin Bk. Austral. Ballads* (1964) 228 Two blows to chip away the wig. 1972 J. S. Gunn in G. W. Turner *Good Austral. Eng.* iii. 61 One thing I did notice about shearing was..two terms for the one idea..for example *rouseabout/shedhand*..*topknot/wig*.

4. (Earlier example.)

1789 J. Woodforde *Diary* 1 Feb. (1927) III. 81 Thomas Carr dined with our Folks in Kitchen. Gave him a tolerable good Wigg.

5. **wig-picker** *U.S. slang*, a psychiatrist; **wig-stand,** a support, usu. of wood or porcelain, comprising a base and rounded stem upon which a wig may rest when not in use (cf. *wig-block*).

1961 *Amer. Speech* XXXVI. 147 *Wig picker*,..a psychiatrist. 1971 M. McCarthy *Birds of Amer.* 153 Was I afraid of what a wig-picker might say? 1883 R. W. Procter *Barber's Shop* (rev. ed.) xix. 189 Here is the lost one's original epitaph (with the wig-stand and block to match)..The Barber's Epitaph. 1911 O. Onions *Widdershins* i. 18 A couple of mushroom-shaped old wooden wig-stands. 1970 *Country Life* 17–24 Dec. 1245/2 Hand-painted wig stands from Dodo Designs.

wiglet (later examples).

1964 *Sun-Herald* (Brisbane) 21 June 56/3 Wiglets, or half wigs start from 11 gns and full wigs are from 32 gns. 1979 L. Kallen *Introducing C. B. Greenfield* xi. 131 A stand bearing wigs and wiglets.

wig, *v.*² Add: **3.** *intr.* With *out.* To be overcome by extreme emotion; to be stimulated to the point of imbalance; to go mad, 'freak out'. *U.S. slang.*

1955 *Amer. Speech* XXX. 305 He wigged out at the prof's gag. 1968 P. Welles *Babyhip* xx. 139 'The Boss Pornographers,' he said, 'it's LSD Music, to wig-out by.' 1975 *Time* 27 Oct. 70/3 Some in the startled crowd recall him saying, 'The company is now in God's hands.' One executive wondered if Goshorn had 'wigged out'. 1978 J. Gores *Gone, no Forwarding* (1979) xi. 69 Kearney was going to wig out when the expense voucher for $100 worth of cocaine came in.

Hence **wigged-out** *ppl. a.*

1977 *New Yorker* 24 Oct. 152/2 The lunacies..just function as part of a normally wigged-out mode of existence. 1980 *San Francisco Bay Guardian* 16–23 Oct. 21/2 It's a barbed, wigged-out satire on hypocrisy and authoritarian therapy via the problem of alcoholism.

wigger, var. *Widger.

wiggle, *v.* **1.** (Later *fig.* example.)

1927 H. A. Vachell *Dew of Sea* 260, I must wiggle out of the mess.

wiggle, *sb.* (In Dict. s.v. Wiggle *v.*) Add: **1.** (Earlier examples.) *to get a wiggle on* (earlier example).

1816 J. K. Paulding *Lett. from South* I. 235 They suffered their hair to grow into a mighty bunch behind, and walked with the genuine Rutland wiggle; that is to say, on tiptoe, and with a most portentous extension of the hinder-parts. 1869 L. M. Alcott *Little Women* II. xxiv. 355 Rob's footstool had a wiggle in its uneven legs. 1896 *Inlander* Jan. 147 *Get a wiggle on you*, hurry up; bestir yourself.

2. = Wiggler in Dict. and Suppl.

1831 T. Buttrick *Voy., Trav., & Discoveries* 78 The

water was very bad... After straining it would still exhibit live insects, which they call wiggles.

3. A wavy line drawn by a pen, pencil, etc.

1942 *Punch* 12 Aug. 127/1 An old envelope bearing the regimental Paymaster's stamp, partly obliterated by adhesive tape, and the word 'Confidential' crossed out with a wiggle in pencil. 1967 R. D. Mattuck *Guide to Feynman Diagrams in Many-Body Problem* iv. 63 The majority of writers draw the above interaction with a dashed line... However, we shall always use the wiggle.

wiggler. (In Dict. s.v. Wiggle *v.*) Add: **1.** Applied esp. to the larva of a mosquito. (Earlier and later examples.)

1859 J. R. Bartlett *Dict. Americanisms* (ed. 2) 492 *Waggletail*, the larva of the mosquito, etc.; also called a *wiggler*. 1938 J. Steinbeck *Long Valley* 186 The mosquito wigglers tumbling up and down, end over end, in the water. 1969 K. M. Wells *Owl Pen Reader* iv. 365 In a month or two this bog hole will be full of wigglers.

2. *Physics.* A magnet designed to make a beam of particles in an accelerator describe a sinusoidal path in order to increase the amount of radiation they produce. Freq. *attrib.*

1974 *IEEE Trans. Nucl. Sci.* XX. 984/1 This proposal projects the installation of many additional beam runs and 'wiggler' magnets. 1981 *Science* 16 Oct. 316/2 A wiggler is a special magnet that fits into one of the straight sections of a storage ring between the bending magnets. The wiggler bends the electrons into a sine wave-shaped path whose local radius of curvature is smaller than that of the smooth circular arc of the bending magnets.

wiggletail. (In Dict. s.v. Wiggle *v.*) (Earlier example.)

1855 *Chicago Times* 9 Aug. 4/6 The mosquito proceeds from the animalcule commonly termed the wiggle-tail.

wiggle-waggle, *sb.* Add: **2.** = *Cake-walk *sb.* 2. Also *wiggle-woggle.*

1910 *Penny Guide Japan-British Exhib.* 25 Fun on the Wiggle Waggle. 1923 R. Macaulay *Told by Idiot* iii. xxi. 256 The establishment of the White City at Shepherd's Bush, with the Franco-British Exhibition..and flip-flaps, switchbacks, wiggle-woggles, and scenic railways. 1938 'G. Orwell' *Homage to Catalonia* xii. 254 A dreadful thing called the Wiggle-Woggle at the White City Exhibition.

wiggy, *a.* Add: **2.** [f. *Wig *v.*² 3.] Mad, crazy, 'freaky'. *U.S. slang.*

1963 L. Deighton *Horse under Water* xxii. 96, I just got some new jazz records from the States, Ace. Pretty wiggy. 1972 *Last Whole Earth Catalog* (Portola Inst.) 31/1 Traditionally considerations such as his—economics, organizations, the future—turn a prophet's soul terrible and dark or at least partially wiggy. 1978 *Amer. Poetry Rev.* Nov./Dec. 26/2 'Poor devil,' she added, 'he blew the star's fuse when we went wiggy for the Thin Man on a cross.'

wigwam. Add: **d.** A pyramidal frame-work of bamboo and similar poles used to support beans, sweet peas, etc.

1971 H. Evans *How to cheat at Gardening* viii. 120 If you must have sticks, tie them in threes, wigwam fashion. *Ibid.* 121 (*caption*) Easiest way of arranging bean-poles—the wigwam. 1978 A. Huxley *Illustr. Hist. Gardening* v. 150/2 We continue to grow beans and the like..on wigwams formed of bamboo canes.

† **wilch** (wiltʃ, wilʃ). *Hist. Suffolk dial.* Also **wilsh.** [Origin unknown.] A bottle-shaped wicker strainer formerly used in brewing to strain the liquid from grains of steeped malt. Cf. Thead.

1823 E. Moor *Suffolk Words & Phrases* 484 *Wilch*, the sediment or lees of beer, home-made wine,..also a brewing utensil. 1830 R. Forby *Vocab. East Anglia* II. 375 *Wilch*,..the wicker strainer set upright in the mash-tub, to prevent the grains from running off with the wirt. 1956 G. E. Evans *Ask Fellows who cut Hay* v. 61 The utensils used in the brewing were..a *wilch* (or *wilsh*), a bottle-shaped appliance made of wicker. (The wilsh was a filter used when straining off the liquid or wort from the mash of steeped malt.) 1962 A. Jobson *Window in Suffolk* i. 28 The brewing tackle would be housed in a large shed or outhouse set apart for that purpose, and would include the tubs or keelers, the wilches, mash sticks,..mallets, spigots and taps.

wilco (wi·lkō), *int.* orig. *Mil. slang.* Also **willco.** Abbrev. of '*will co*mply', used to express acceptance of instructions, esp. those received by radio or telephone. Cf. *will do* s.v. *Will *v.*¹ 11 c.

1946 F. Hamann *Air Words* 56 *Willco*, will comply; will do. 1948 A. M. Taylor *Lang. World War II* (ed. 2) 221 *Wilco*, radio term for 'will comply'. Used throughout the services and also taken up by civilians. 1961 H. Waugh *Road Block* i. 12 'Roger, wilco, and out,' the staticky voice sang. 1972 D. Hart-Davis *Spider in Morning* ii. 28 'If it happens again, hold your breath.' 'Wilco.' 1977 D. Beaty *Excellency* xvii. 190 'Please clear the runway quickly for the President's Starjet!'..'Wilco,' he said.

wild, *a.* and *sb.* Add: **A.** *adj.* **I. 3. a.** *wild silk*, silk produced by wild silkworms or an imitation of this made from short silk fibres. (Earlier and later examples.)

1876 [see Tusser 1 a]. 1883 [in Dict.]. 1896 [see Tusser 1 b]. 1911 *Daily Colonist* (Victoria, B.C.) 28 Apr. 14/2 The cargo was made up as follows: Raw silk, 960 bales; wild silk, 49 bales. 1963 R. Himmel *It's Murder, Maguire* vii. 46, I always suspected him of wearing wild silk underwear. 1972 J. Aiken *Butterfly Picnic* ix. 162 Her white wild-silk bikini.

II. 6. d. *to run wild*, (*c*) of an oil-well, to release uncontrollable quantities of fluid or gas. Also *to blow wild.*

1925 [see *relief well* s.v. *Relief² 9 b]. 1931 *Times* 18 Feb. 15/6 When the wells 'blow wild' the city is enveloped in a dark spray of oil. 1975 L. Crook *Oil Terms* 35 *Blow out*, a situation where a well becomes out of control due to the fluids from the formation 'blowing wild' at the surface.

11. b. Also const. *for.*

a 1817 Jane Austen *Persuasion* (1818) III. vi. 107 The girls were wild for dancing. 1937 J. T. Farrell *Fellow Countrymen* 184 He imagined that she was his woman... She was saying she was crazy about him... She was wild for him.

c. Also † const. *after.* (Earlier and further examples.)

a 1817 Jane Austen *Persuasion* (1818) IV. vii. 134 The men are all wild after Miss Elliot. 1865 R. Henning *Let.* 21 Oct. (1966) 214 The whole family are wild after music.

d. *like wild*: with passionate eagerness, with great excitement. Cf. *like mad* s.v. Mad *a.* 1 c.

1674 C. Stewkeley *Let.* 4 May in M. M. Verney *Mem.* (1899) IV. vii. 225 Ursula..hath bin at all the Salsbury rasis, dancing like wild with Mr Clarks. 1962 *Radio Times* 17 May 43 Should he [sc. a jazz musician] 'blow' with feeling, or great excitement ('like wild') [etc.].

13. a. Also in phr. *in* or *beyond one's wildest dreams*, in or beyond one's most fantastic or unrestrained imaginings or expectations.

1961 C. McCullers *Clock without Hands* x. 203 In his wildest dreams he could not associate Johnny with danger. 1969 *Listener* 24 July 123/3 The programme has succeeded beyond its instigators' wildest dreams. 1984 *Tampa* (Florida) *Tribune* 5 Apr. 6c/2 You know, it's hard to believe I'm really here. It's beyond my wildest dreams that I'd be managing a team that I once played for.

14. c. *U.S. slang.* Remarkable, unusual, exciting. Used as a general term of approbation.

1955 L. Feather *Encycl. Jazz* x. 347/2 *Wild*, adj., remarkable, exciting. 1960 [see *Lay *v.*¹ 55 l]. 1968 *Listener* 22 Aug. 236/3 Los Angeles is so wild they should just let it swing and see what happens. 1978 *Hot Car* June 103/5 Naugahyde..has long been the favourite amongst Stateside rodders because of its stretchy qualities, amazing range of colours (including some wild marble-like effects).

15. b. Of a playing card: having any rank chosen by the player holding it. Also *fig.* See also *wild card*, sense 16 below.

1927 *Auction Bridge Mag.* May 26/1 These are played with all the twos as jokers and usually known as 'Deuces Wild'. 1940 O. Jacoby *On Poker* x. 139 Any card or cards may be counted as wild, in which case they have the same rights as jokers. 1963 E. Linington *Death of Busybody* vi. 72 Don't tell me, a tie-up. Look, Luis, let's not call every card in the deck wild, for God's sake. 1973 M. Catto *Sam Casanova* vi. 109 Think of the amazing variations of the game [sc. poker]! Five-Card stud. Seven-Card Draw with Joker wild.

III. 16. **wild card,** (*a*) (see sense 15 b above); also *fig.*; (*b*) *Sport* (orig. *U.S.*), a player or team chosen for a tournament at the discretion of the organizers after the regular places have been taken up; freq. *attrib.*; (*c*) *Computers*, a character that will match any character or combination of characters in a file name, etc.; **wild cherry**: see Cherry *sb.* 3 a; (examples): **wild duck,** a duck belonging to any of numerous undomesticated species; **wild garden,** a group of hardy plants, exotic or native, in an informal setting, designed to look as natural as possible; hence **wild gardener, gardening; wild geranium** *S. Afr.* = Geranium 2; **wild ginger,** in North America, any of several plants of the genus *Asarum*, esp. *A. canadense*, or, in India, a wild plant of the genus *Zingiber*; **wild grape,** a wild species of *Vitis* or its fruit; **wild lime**: see Lime *sb.*² b; also, in Australia, = *Kumquat 2; (examples); **wild orange**: see Orange *sb.*¹ 3; also, in Australia, any of several species of *Capparis* or *Canthium*; in South Africa = *Kaffir orange* s.v. *Kaffir 4; (examples); **wild parsnip**: see Parsnip 2; also, = Cow-parsnip; also, a poisonous plant of the family Umbelliferæ, esp., in North America, the water hemlock, *Cicuta maculata*, or, in Australia, *Trachymene glaucifolia*; (later examples); **wild party,** a boisterous, unchecked, or dissolute party; **wild pig** = *Captain Cooker; **wild pitch** *Baseball*, a pitch which is not hit by the batter and cannot be stopped by the catcher, enabling a base-runner to advance; hence as *v. trans.*, to enable (a runner) to advance in this way; **wild plum**: see Plum *sb.* 3; (earlier and later examples); **wild rice,** an aquatic grass, *Zizania aquatica*, native to North America, having seeds resembling rice and used as food; (earlier and later examples); **wild rye**: see Rye *sb.*¹ 2 c; a North American grass of the genus *Elymus*; (earlier and later examples); **wild talent,** any of various psychic powers such as extrasensory perception, telepathy, telekinesis, etc.; **wild track** *Cinematogr.* (see quot. 1940); **wild well,** an oil well which is out of control and blowing oil or gas from the borehole (cf. sense 6 d (*c*) above).

1940 O. Jacoby *On Poker* x. 138 The Bug, three sixes and a ten merely count as three sixes since the Bug is not strictly a wild card. 1970 *New Yorker* 3 Oct. 34/3 The other thirteen games..will be 'wild-card' encounters, to be played on alternate Monday nights. 1971 *Guardian* 17 June 12/6 Kennedy is the wild card in the 1972 Deck, as the Nixon men see it. 1976 M. Nelson *Crusoe Test* iii. 35 The joker. The wild card. The card the holder can use as he pleases. 1976 *Sunday Mail* (Brisbane) 15 Aug. 3/11 Renee was not ranked high enough to be accepted on her standard of play, but she could be nominated as the 'wild card'—a crowd pleaser. 1977 *Hongkong Standard* 14 Apr. 11/2 Fifteen-year-old Betty Newfield of the US reached the second round by defeating Marlie Buehler of Australia 4–6, 6–0, 7–5 after getting into the draw as a wild card. 1981 *Washington Post* 18 Mar. d3 The conference championship games are now played on the home field of the competitor that has the best season record, unless it's a wild-card team. 1984 *Times* 21 Sept. 19/6 The wild card in the BPCC pack is Mr Maxwell's dual role as head of both BPCC and Mirror Group Newspapers. 1984 K. Buckner et al. *Using UCSD p-System* vi. 56 The wildcard '?' should be used to remove several files from a disk. *Ibid.* xv. 156 the WILD unit makes available wild card pattern matching on string variables. 1985 *Personal Computer World* Feb. 244/1 (Advt.), Powerful wild cards permit editing of categories of file name in one instruction. 1666 *Brief Descr. Province Carolina* 4 There are many sorts of fruit Trees, as Vines, Medlars, Peach, Wild Cherries. 1784 W. Walton *Narr. Captivity B. Gilbert* 81 They were under the Necessity of eating wild Cherries. 1899 S. O. Jewett *Queen's Twin* 81 She had a sprig of wild-cherry blossom in her dress. 1972 G. Chadbund *Flowering Cherries* 11 Wild cherries occur naturally on chalky soil. 1538, 1676 Wild duck [in Dict., sense 1 a]. 1723 J. Nott *Cook's & Confectioner's Dict.* sig. M6, Draw and truss your Wild Ducks, parboil them, and half roast them. 1881 O. Wilde *Poems* 115 The water-rat.. Made for the wild-duck's nest. *a* 1916 'Saki' *Toys of Peace* (1919) 82 By the time they had arrived at the wild duck course it was beginning to be a rather expensive lunch. 1852 C. M. Yonge *Two Guardians* iii. 29 Strangers would..think her wild garden a collection of weeds. 1925 J. Buchan *John Macnab* xiii. 268 An expert from Kew..had made a wonderful wild garden. 1980 A. Wilson *Setting World on Fire* ii. vi. 170 It's your garden parties that are ridiculous... And Rosemary's famous wild garden. 1966 'J. Berrisford' *Wild Garden* x. 117 The wild gardener who is also a plantsman..may grow the meconopses. 1870 W. Robinson *Wild Garden* i. 19 It [sc. Caucasian comfrey] will soon run about, exterminate the weeds, and prove quite a lesson in wild and natural gardening. 1911 *Daily Colonist* (Victoria, B.C.) 30 Apr. (Mag. Section) 3/4 The cult of wild gardening is apt to run into the same kind of excesses as the pursuit of the simple life. 1978 A. J. Huxley *Illustr. Hist. Gardening* ix. 309 William Robinson and Gertrude Jekyll..preached a return to more naturalistic and even 'wild' gardening. 1840 Wild geranium [see *Ivy-berry b]. 1966 E. Palmer *Plains of Camdeboo* xvii. 281 Here and there are Pelargoniums—wild geraniums to us. 1804 M. Lewis *Jrnl.* 1 June in *Orig. Jrnls. Lewis & Clark Exped.* (1905) VI. iv. 154 Wild ginger grows in rich bottom land. 1866 [see Ginger *sb.* 2 b]. 1964 R. Perry *World of Tiger* xi. 160 The Great Indian rhino..feeding on the succulent shoots of marsh reeds and especially the wild ginger. 1973 M. Crowell *Greener Pastures* 187 We recognize the wild ginger. 1763 G. Milligen-Johnston *Short Descr. Prov. S. Carolina* (1770) 9 Wild Grapes grow on this Land. 1843 [see *Guarri]. 1929 M. de la Roche *Whiteoaks* xvi. 202 The jewelled leaves of the wild grape..scarcely dried before another dew. 1958 G. A. Petrides *Field Guide to Trees & Shrubs* 114 A number of cultivated varieties have been developed from wild grapes. 1767 P. Collinson *Let.* 31 July in W. Darlington *Mem. J. Bartram & H. Marshall* (1849) 292 The Wild Lime..is a singular plant. 1832 D. J. Browne *Sylva Amer.* 221 In Georgia this tree is known by the name of Sour Tupelo and Wild Lime. 1863 R. Henning *Let.* 26 Nov. (1966) 147 We went out to pick some wild limes for preserving. They are a little fruit about the size of a large gooseberry, but in colour, taste, smell and shape exactly like a small lemon. 1965 Wild lime [see *Kumquat 2]. 1969 T. H. Everett *Living Trees of World* xxi. 209/1 The wild-lime (*Zanthoxylum fagara*) of Florida, Mexico, the West Indies..is an evergreen species. 1802 J. Drayton *View S. Carolina* 8 Small rising grounds sometimes present themselves, on which grow..wild orange. 1858 J. A. Warder *Hedges & Evergreens* 44 Our beautiful Wild Orange..is much planted about Southern residences, for hedges. 1932 [see *Klapper]. 1936 F. Clune *Roaming round Darling* xvii. 165 The wild orange, ten feet high, dark green brittle leaves, large yellow-stemmed flowers, and bearing fruit as big as tennis-balls, with pomegranate seeds inside. 1969 T. H. Everett *Living Trees of World* xx. 172/1 Known as wild-orange and mock-orange, it [sc. *Prunus caroliniana*] has creamy white flowers and glossy black fruits. 1790 *Trans. Amer. Philos. Soc.* III. 234, I have heard this poisonous herb, called by the names of Wild-Carrot, Wild-Parsnep,..and Mock-Eel-Root. 1807 [see

musquash-root s.v. *Musquash 3]. 1889 J. H. Maiden *Useful Native Plants Austral.* 142 The sudden death of numbers of cattle in the vicinity of Dandenong..was attributed to their having eaten a plant known as the wild parsnip. 1932 J. W. Winson *Weather & Wings* 51 The poison is described further as being 'wild-parsnip', 'cowbane', [etc.]. 1955 *Arctic Terms* 88/1 *Wild parsnip*. The cow parsnip. 1965 *Austral. Encycl.* VIII. 546/2 The wild parsnip of inland plains, does seem to be responsible for stock losses. 1925 F. Scott Fitzgerald *Lett.* (1964) 295 It is true I saved McAlmon from a beating he probably deserved and that we went on some wild parties in London with a certain Marchioness of Milford Haven. 1970 'D. Halliday' *Dolly & Cookie Bird* iii. 35 He was probably just afraid of the talk. It was rather a wild party. 1840 W. Deans *Let.* 30 Oct. in J. Deans *Pioneers of Canterbury* (1937) i. 29, I will visit it [sc. Palliser Bay] in company with 50 or 60 natives who are going to hunt wild pigs. 1930 L. G. D. Acland *Early Canterbury Runs* 1st Ser. x. 237 Stonyhurst has always been a great place for wild pigs. 1977 C. McCullough *Thorn Birds* iv. 75 Wild pigs frightened of nothing, savage and flesh-eating, black hairy things the size of fully grown cows. 1867 *Ball Players' Chron.* 4 July 1/2 Zeller,..getting round on a passed ball and wild pitch, came home on another passed ball. 1970 *Washington Post* 30 Sept. d1/8 In the first game, young Bob Grich led off the home 10th with a single and Coleman wild-pitched him to second base. 1979 *Arizona Daily Star* 1 Apr. c6/4 Greg Laing walked in the bottom of the eighth and scored on a wild pitch. 1709 J. Lawson *New Voy. Carolina* 105 The wild Plums of America are of several sorts. 1838 E. Flagg *Far West* II. 177 Endless thickets of the wild plum..were to be seen. 1863 R. Henning *Let.* 26 Nov. (1966) 146 We sat down under the shade of a wild-plum tree. *Ibid.*, They are not bad, those wild plums; they are about the size of a medlar, quite black in colour, and when ripe they taste very like sloes. 1925 Z. A. Tilghman *Dugout* 56 Fan being gone after some wild plums down the creek. 1951 W. Faulkner *Requiem for Nun* iii. 213 A mere dusty widening of the trace, trail, pathway in a forest of oak and ash and..wild plum. 1748 H. Ellis *Voy. Hudson's-Bay* 170 By the Sides of Lakes and Rivers there is abundance of wild Rice. 1778 J. Carver *Trav. N.-Amer.* 522 Wild Rice..grows in the greatest plenty throughout the interior parts of North America. 1911 G. S. Porter *Harvester* vi. 94 Wild rice..he had planted for the birds. 1934 H. Miller *Tropic of Cancer* 47 They were eating too. A young chicken with wild rice. 1980 *Times Lit. Suppl.* 26 Sept. 1064/5 The paper..was full of reports of discontent around Ompah at overcropping of wildrice. 1984 *Times* 13 June 9/4 Wild rice is not really rice at all but the seeds of a grass that grows wild along the waters-edge of lakes in Minnesota, Wisconsin and southern Canada. 1751 C. Gist *Jrnl.* 27 Jan. (1893) 43 The wild Rye appeared very green and flourishing. 1968 F. W. Gould *Grass Systematics* 181 Widespread and variable in the United States are *Elymus canadensis* L., Canada wildrye, and *E. virginicus* L., Virginia wildrye. 1944 A. Huxley *Let.* 28 July (1969) 510 The fact of what Charles Fort calls 'wild talents' is admitted by all open-minded people. 1960 K. Amis *New Maps of Hell* (1961) iv. 98 A new type of human being, sometimes outré in appearance, more often gifted with the 'wild talent' which has become a science fiction catch-phrase and convention. 1940 *Chambers's Techn. Dict.* 908/2 *Wild track*, a sound-track which is recorded independently of any photographic track or mute, but is destined to be used in editing a sound-film. 1964 Hall & Whannell *Popular Arts* ix. 258 The..combined use of wild-track voices with counter-pointing visual images. 1980 'P. Loraine' *Lions' Ransom* i. iii. 51 Fox was..making a 'wild-track' of Busai's morning birdsong. 1915 Redwood & Eastlake *Petroleum Technologist's Pocket-bk.* iv. 244 '*Wild' well.* This term is used to denote a well which produces such quantities of oil or gas, or both, under such high pressure that it is either impossible to bring it under control or it is only controlled when a very considerable time has elapsed after the oil or gas has been met with. 1977 *Sunday Times* 24 Apr. 1/2 If the wild well..is not brought under control within the next 24 hours, the fight could last for weeks, months even.

17. a. *wild-caught* adj.; **b.** *wild-brained* (example), *-coloured, -haired* (earlier example), *-hearted, -winged* (later examples) adjs.; **d.** in nonce *poet.* uses, as *wild-worst, -worth.*

1894 'Mark Twain' in *Harper's Mag.* Oct. (1914) 675/2 Wild-brained martyrdom was succeeded by uprising and organization. 1949 *Amer. Speech* XXIV. 98 American mink may be either wild-caught or ranch-raised. 1970 Saunders & Phelps in H. W. Mulligan *Afr. Trypanosomiases* xiv. 329 The ovaries of wild-caught females..can be used. 1954 M. K. Wilson tr. *K. Z. Lorenz's Man meets Dog* (1964) xix. 176 The striped markings in the face of the 'wild-coloured' cat enhance the least movements of the facial skin. 1872 J. G. Whittier in *Atlantic Monthly* Apr. 474 The wild-haired Bacchant's yell. 1904 W. De La Mare *Henry Brocken* viii. 83 Beasts of a long-sharpened sagacity, wild-hearted, rebellious. 1916 Joyce *Portrait of Artist* (1969) iv. 171 He was alone and young and wilful and wild-hearted. 1906 Hardy *Dynasts* II. i. v. 161 A straggler merely he... But they decide, At last, to post his news, wild-winged or no. 1936 L. B. Lyon *Bright Feather Fading* 45 The wild-winged bliss. 1876 G. M. Hopkins *Poems* (1967) 59 The cross to her she calls Christ to her, christens her wild-worst Best. *c* 1878 *Ibid.* 75 Only the breathing temple and fleet Life, this wildworth blown so sweet.

B. *sb.* **4.** Phr. *to play the wild*: to behave in a careless or reckless manner; to play havoc with. *U.S.*

1849 J. B. Jones *Wild Western Scenes* i. 10 But love can play the 'wild' with any young man. 1911 R. D. Saunders *Col. Todhunter* ix. 143 I'm shorely glad to get home. I been playin' the wild in St. Louis.

wild cat. Add: **3. b.** (Earlier example.)

1861 'Mark Twain' *Lett.* (1917) I. iii. 54 'Wild cat' isn't worth ten cents.

c. An exploratory oil-well, drilled where there is only a possibility of success. Cf. *Wild *a.* 6 d, *wild well* s.v. *Wild *a.* 16.

1877 *Sci. Amer.* 22 Dec. 387/3 A large number of 'wildcats', or test wells, have gone down off the eastern edge of the defined line, but with very few exceptions they have proved to be dusters. 1943 *Jrnl. Sedimentary Petrol.* XIII. 111/2 Both deep, off-structure wildcats and and field wells are important. 1977 *Offshore Engineer* May 39/1 Esso is drilling in the deepest water off Egypt's Mediterranean coast with a second wildcat in 470m of water 100km off Alexandria.

d. Illicitly distilled whisky. Cf. *wild-cat whisky*, sense 4 b in Dict.

1887 A. A. Brown *Lumbering on Cumberland* vii. 80 Mr. Kearney alighted and tendered us a drink from his bottle of 'wild cat'. 1945 M. Lyon *Fresh from Hills* iv. 47 You can keep on a-makin' wildcat till hell freezes over.

e. *ellipt.* for *wildcat strike*, sense 4 c below.

1959 *Daily Mail* 28 Oct. 1 (*heading*) War on the wildcats. 1969 *Guardian* 22 Aug. 9/1 The TUC made their 'solemn and binding declaration' to the Prime Minister about dealing with wildcats. 1978 J. Wainwright *Ripple of Murders* 43 They'd thought he was bluffing... So there's been wildcats and pickets, and lock-ins.

4. b. Also with reference to wildcat strikes (see sense 4 c below).

1959 *Daily Tel.* 31 Dec. 11/2 'Wildcat' risk in bank staffs. 1973 *Black Panther* 29 Sept. 3/3 A majority of the Black workers..voted to reject the union proposal, upholding the original wildcat demand. 1976 M. Machlin *Pipeline* xix. 241 Some people think it was some wildcat members of 798 that set them after the company laid off about a hundred of them.

c. Special Comb.: **wildcat drilling,** the drilling of a wildcat well; **wildcat strike,** a sudden and unofficial strike; hence **wildcat striker**; similarly *wildcat stoppage, walkout*; **wildcat train** *U.S.*, an extra train running in addition to those on the timetable (see quot. 1885); occas. *ellipt.*; similarly *wildcat engine*; **wildcat well** = sense 3 c above.

1937 *Bull. Amer. Assoc. Petroleum Geologists* XXI. 1079 A study of wildcat drilling in the Gulf Coast Plains during 1935 and 1936 indicates that between 7 and 11 per cent of all such holes opened new oil or gas pools, the remaining 93–89 per cent having been dry. These figures speak eloquently of the risk involved in wildcat drilling. 1976 *Offshore Platforms & Pipelining* 60/1 The time span from hard freeze in late autumn to the melt in the late spring leaves an opportunity for no more than about 6,000 ft of wildcat drilling. 1888 *Missouri Republican* 23 Feb., The Montreal night express was thrown from the track..by a wild-cat engine that had been turned loose..by an evil-disposed person. 1891 E. S. Ellis *Check No. 2134* xiii. 88 There was just one chance in a hundred of a wild-cat engine approaching. 1974 *Telegraph* (Brisbane) 5 Feb. 16/1 Freelance truckers entered the fifth day of their wildcat stoppage. 1937 *Sun* (Baltimore) 16 Nov. 3/1 A clause..conceding to the corporation the right to discipline persons responsible for 'wildcat' strikes. 1954 *Encounter* June 7/2 [The workers'] behaviour itself becomes a judgement... It takes the form of slow-downs, a silent war against production standards, and most spectacularly in the violent eruptions of wildcat strikes against 'speed-ups' or changes in the timing of jobs. 1978 S. Brill *Teamsters* v. 179 Carey led a militant wildcat strike over a symbolic issue. 1945 *Chicago Daily News* 10 Dec. 1/9 (*caption*) Would fire or fine wildcat strikers. 1981 M. Nabb *Death of Englishman* ii. 89 He wasn't going to stand by and see his country insulted, disrupted..by wildcat strikers. 1870 *Daily Territorial Enterprise* (Virginia City, Nevada) 22 Oct. 3/1 In company with four or five others, he had gone out on the road upon a hand car, when a 'wild cat train' (an extra train running on no regular time) overtook them. 1885 *Good Words* July 452/1 Every now and then the newspapers allude to 'wild-cat' trains... The 'wild-cat' is the slowest of all trains. It is only used for freight, and reaches its destination when it can, running whenever the line is clear, and shunting when a passenger train is due on the same track. 1942 *Sun* (Baltimore) 26 Sept. 9/6 Our estimated 400..were out in what both union and management termed a 'wildcat' walkout. 1977 *Time* 28 Mar. 46/2 The month-long wildcat walkout by 3,000 precision toolmakers at British Leyland. 1883 Wildcat well [in Dict., sense 4 b]. 1907 *Bull. U.S. Geol. Survey* No. 318. 25 In making maps of subsurface strata in areas that have not been productive, most of the records used for making a convergence sheet must be taken from 'wild-cat wells'. 1975 W. G. Roberts *Quest for Oil* (rev. ed.) iii. 35 It is nowadays extremely rare to hear of anyone sinking a true 'wildcat' well—that is, one drilled simply because someone has a hunch that his patch of ground has oil beneath.

wild-catter. Add: *spec.* (*a*) a prospector who sinks wildcat wells; (later examples); (*b*) a wildcat striker.

1925 A. B. Thompson *Oil-Fields Explor. & Devel.* I. vii. 314 The speculative spirit aroused by the gusher has had much to do with..inspiring the wild-catter to pursue his quest for extended areas of development. 1947 R. Bedichek *Adv. with Texas Naturalist* x. 116, I left the highway, following an old road which led me to the site of some wildcatter's dream and disillusionment. 1966 *Punch* 23 Mar. 404/3 That union refusing membership to a one-armed labourer may have had a point. Just as you don't hit a boy with glasses, even the wildest wildcatter might jib at intimidating this man. 1973 [see *Shoot-'em-up]. 1980 A. Coppel *Hastings Conspiracy* iv. 31 The militant wildcatters or the British air traffic controllers' union. 1981 *Sci. Digest* Aug. 118/3 Like wildcatters bringing in a gusher, a few of the students develop products with gilt-edged possibilities.

wild-catting, *vbl. sb.* Add: *spec.* (*a*) the drilling of a wildcat well; (later examples); (*b*) participation in a wildcat strike; occas. also as *ppl. a.*

1909 W. S. Tower *Story of Oil* v. 66 Many of the most valuable oil deposits..have been revealed by the more or less random process of 'wild-catting'. 1967 *Economist* 18 Mar. 1014/2 Just like wild-catting shop stewards, they [sc. demonstrators] brought into disrepute the cause that they affected to support. 1969 *Guardian* 22 Sept. 15/1 The West Berlin workers decided that anything their Ruhr brothers can do in the way of wildcatting the Berliners can do better. 1972 L. M. Harris *Introd. Deepwater Floating Drilling Operations* xv. 159 Nearly always, a floating drilling operation is a wildcatting venture in an area previously undrilled. 1976 M. Machlin *Pipeline* xxxii. 370 The way things are going these days, I think I might just go back to the wildcatting in East Texas.

Wildean (wəi·ldiǎn), *a.* [f. the name of Oscar *Wilde* (see below) + -an.] Of, pertaining to, or characteristic of the Irish writer Oscar Fingal O'Flahertie Wills Wilde (1854–1900), or his works.

1924 *Nation* 26 Mar. 352/1 Epigrams are his undoing. The Wildean nineties are in his blood. 1937 *Scrutiny* V. 386 Ravel is Wildean, 'witty' in the nineteenth century salon. 1958 R. Williams *Culture & Society* ii. ii. 171 A good example of the Wildean paradox. 1967 *Listener* 6 July 15/1 Social morality is turned on its head with a Wildean comment on one of the film's less violent fatalities: 'Marie's tragic death restored my faith in suicide.' 1977 *Time* 21 Feb. 28/3 They are cold, loveless creatures, incapable of responding to one another except by lobbing epigrams, Wildean in rhythm but not in wit, back and forth.

wildebeest. Add: Also **wildebees.** Generic name *Connochætes taurinus* or *C. gnou.* (Earlier and later examples.)

[1801 J. Barrow *Acct. Trav. S. Afr.* I. iv. 259 The *gnoo* or *wild beast*, as it is called by the Dutch.] 1824 W. J. Burchell *Trav. S. Afr.* II. 109 Wild animals; among which were..many *wilde-beests* or *gnues*. 1929 D. Reitz *Commando* 129 Great herds of zebra, wildebeest, and sable, stood fearlessly gazing at us. 1958 *Cape Times* 13 Aug. 3/4 For the rest it was impala, wildebeest and koodoo. 1970 *Life* Jan. 50 Their faces covered with grotesque tufts of hair, the wildebeest..are the oddest and fiercest-looking antelopes.

wilderness. Add: **3. c.** *in the wilderness* (in allusion to *Numbers* xiv. 33), (*a*) of a politician, political party, etc.: out of office; (*b*) *gen.*, unrecognized, out of favour.

1930 *Economist* 2 Aug. 220/1 For Charles X represented a Restoration of the *Ancien Régime*..which had 'learnt nothing and forgotten nothing' during a quarter of a century in the wilderness. 1958 *Spectator* 6 June 719/3 Parties should liquidate their failures and frustrations in the wilderness, not in power. 1966 *Listener* 5 May 661/2 Richard Baker asked Bernard Keeffe why Mahler, so long in the wilderness as far as England was concerned, is now a box-office success. 1969 *Ibid.* 3 July 12/3 Carmichael has now accepted a junior post in the Panther hierarchy and Rap Brown and Jim Foreman have been driven into the wilderness. 1976 *Southern Even. Echo* (Southampton) 17 Nov. 22/3 If he fails to gain the title he lost to Cain on a cut eye decision, it could mean months in the wilderness and set him back even further. 1984 *Times* 1 Aug. 17/2 After months in the wilderness, which has seen the price slip from a high of 95½p to a low of 65½p shares of Marley..is [*sic*] back in favour with the institutions.

wild-fire, wildfire. Add: **4. b.** A leaf-spot disease of tobacco, caused by the bacterium *Pseudomonas tabaci*. Also *attrib.*

1918 *Jrnl. Agric. Res.* XII. 451 The disease appeared so quietly, spread so rapidly, and affected the leaves so seriously that it was commonly given the appropriate designation 'wildfire'. 1955 *Sci. News Let.* 29 Jan. 73/2 Immunity to wildfire..was first transferred..from a wild tobacco species. 1971 *Nature* 15 Jan. 174/1 Wildfire disease of tobacco is perhaps the most thoroughly studied of all toxin-mediated plant diseases.

‖ **Wildflysch** (vi·ltfliʃ). *Geol.* Also **wildflysch.** [Ger. (F. J. Kaufmann 1871, in B. Studer *Index der Petrogr.* (1872) 258): see Flysch.] Flysch containing large, irregularly distributed blocks and occupying beds that are distorted.

1929 P. G. H. Boswell tr. *Heritsch's Nappe Theory Alps* iv. 47 A great part of the sheared Flysh has the character of the Wildflysch: dark, puckered and highly micaceous marls with interbedded seams of sandstone, quartzite, limestones, breccias, conglomerates and exotic blocks. 1960 *Bull. Geol. Soc. Amer.* LXXI. 878/2 The general aspect of Wildflysch forcibly suggests submarine slumping and sliding on a large scale. 1963 [see *Famennian *a.*]. 1981 A. Hallam *Facies Interpretation* iv. 82 A thick series of shallow-water carbonate platform deposits..are overlain by a wildflysch unit with a chaotic jumble of limestone blocks in a shaly matrix.

wild life. Also **wildlife, wild-life.** [f. Wild *a.* 1 and 2 + Life *sb.* 1.] **1.** The native fauna and flora of a particular region.

1879 R. Jefferies (*title*) Wild life in a southern county. 1912 A. R. Dugmore (*title*) Wild life and the camera. 1958 *Times Lit. Suppl.* 21 Nov. p. xxii/2 The chance to live among country things and indulge a native passion for wild life (for if you scratch an Englishman you are likely to find a naturalist *manqué*). **1982** *Times* 18 Oct. 20/8 Ancient woods..are especially important for wildlife.

2. a. *attrib.*

1936 *Discovery* June 190 His description of the patient progress of the wild-life photographer has authority behind it. *Ibid.* 191/2 (*heading*) The wild-life film. **1943** J. S. Huxley *TVA* 12 The total range of activities covered by the TVA..includes..wild life conservation. *Ibid.* 54, I spoke earlier of the wild life survey of the region. **1958** *Times Lit. Suppl.* 10 Jan. 24/1 A well-illustrated anthology based on the B.B.C.'s wild-life and naturalist programmes. **1982** G. Hammond *Fair Game* v. 47 The shooting man..needs a rich wildlife scene. **1984** *Guardian* 22 Oct. 3/2 The Liberals estimate that annual compensation to farmers who are preserving scenic land or important wildlife habitats ..is running at the rate of £1 million.

b. *Comb.*, as **wildlife park,** a collection of wild animals kept in conditions as close as possible to their natural ones; **wildlife sanctuary,** an area of land in which hunting, collecting, or any other disturbance of the native fauna and flora is forbidden.

1965 P. Wayre *Wind in Reeds* xvi. 234 We became officially known as the Norfolk Wildlife Park. Britain's first wildlife park was away to a flying start. **1976** P. R. White *Planning for Public Transport* viii. 157 Coach operators' inclusive prices for visits to stately homes and wild-life parks. **1936** D. McCowan *Animals Canad. Rockies* i. 12 A warning to poachers in a wild life sanctuary. **1973** V. Canning *Flight of Wild Goose* iv. 57 The whole of that area was kept as a wild life sanctuary.

Hence **wi·ldlifer,** a person interested in the study and conservation of wild plants and animals.

1963 *Spectator* 8 Feb. 177/3 Children either drop the whole thing or become wildlifers of the intrepid, modern, TV-inspired kind. **1982** G. Hammond *Fair Game* v. 49, I can just picture a bunch of slightly hostile wild-lifers.. running the estate.

wildling. 2. (Earlier example.)

1841 S. Bamford *Passages in Life of Radical* I. xi. 72 All said he was killed... The doctor..approached along an avenue made through those wildlings.

wild man. 1. c. (Earlier examples, in *sing.* and *pl.*)

1905 D. G. Phillips *Plum Tree* 266 And I wished for a 'wild man' as the candidate for governor. **1910** Belloc *Pongo & Bull* xix. 287 The Wild Men on the Opposition side might cheer.

Wild Turkey. The proprietary name of a brand of whisky; a drink or glassful of this.

1949 *Official Gaz.* (U.S. Patent Office) 26 Apr. 1007 Austin Nichols & Co., Incorporated, Brooklyn, N.Y... Wild Turkey... For Whiskey. Claims use since May 29, 1942. **1968** *Trade Marks Jrnl.* 20 Mar. 448/1 Wild Turkey 917,193. Wines, spirits (beverages) and liqueurs. Five Mills Limited, 37 Grafton Way..London W.1; Merchants. **1975** *New Yorker* 20 Jan. 30/1, I know—it took me seven weeks to do a page and a half the last time, but I was into the Wild Turkey then, and you'd be amazed how fast I can write when my pencil can actually form legible letters. **1979** J. Crosby *Party of Year* (1980) viii. 48 He poured Cassidy another Wild Turkey. **1980** J. Krantz *Princess Daisy* xvi. 258 Two bottles of Soave Bolla and one of Wild Turkey bourbon.

wild type. *Genetics.* The type of strain, gene, or characteristic that prevails among individuals in natural conditions, as opposed to an atypical mutant type. Freq. *attrib.* or as *adj.*

[**1913** C. B. Bridges in *Jrnl. Exper. Zool.* XV. 587 When a female [Drosophila] with white eyes is mated to a wild male with red eyes, the daughters have red, and the sons, white eyes.] **1914** —— in *Science* 17 July 107/2 Half of the wild-type daughters..when out-crossed to barred males..gave exceptions as follows:..Exceptions. 5% of both sexes. Wild type ♀; barred ♂. **1932** [see *suppressor 2]. **1946** *Nature* 19 Oct. 558/1 These include wild-type strains [of *E. coli*] with no growth-factor deficiencies, and single mutant types requiring only thiamin or phenylalanine. **1970** *Sci. Amer.* Mar. 103/1 Most mutant genes are nonfunctional or do something very different from wild-type genes, so that they can be easily distinguished. **1970** *Nature* 22 Aug. 806/1 Albinism is monofactorial and is recessive to wild type pigmentation. **1976** *Ann. Rev. Microbiol.* XXX. 90 A number of compounds produce phenocopies of morphological mutants when added to the growth medium of wild type.

Wild West. Also **wild west, wild West. 1.** The western part of the U.S. during its lawless frontier period.

1849 C. Brontë *Shirley* III. xiii. 272 What suggested the wild West to your mind? **1851** [see Wild *a.* 4]. **1898** H. James in *Literature* 30 Apr. 512/1 Has he [*sc.* Bret Harte] continued to distil and dilute the Wild West because the public would only take him as wild and Western? **1903** Chesterton *Robert Browning* v. 111 A gambling hell in the Wild West. **1937** Phillips & Niven *Colour in Canad. Rockies* ix. 61 On my first visit there were many marked qualities of the 'wild west' there. **1977** *Times* 20 Sept. 12/1 The Rio Grande..has been oversold in the legends and songs of the old Wild West.

2. *transf.* and *fig.*

1889 G. B. Shaw *London Music in 1888–89* (1937) 170 Somewhere in the wild west of the Old Brompton Road. **1944** F. Clune *Red Heart* 69 Australia's Wild West, as picturesque as Texas, was buzzing with rumours of raids, hold-ups. **1975** J. O'Faolain *Women in Wall* 11 My setting is the Wild West of an age often called 'Dark'.

3. a. *attrib.*

1922 E. E. Cummings *Let.* 26 Feb. (1969) 82 Attacks by Bedoins, wild-west style, shooting at Dos with rifles. **1922** E. M. Forster *Life to Come* (1972) 100 They passed through the village, on their way back past a cinema, which was giving a Wild West stunt. **1940** 'G. Orwell' in *Horizon* Mar. 193 The Wild West story..with its cattle-rustlers. **1965** A. Nicol *Truly Married Woman* 5 She removed the Wild West novels and romance magazines. **1971** *Advocate-News* (Barbados) 17 Sept. (Guyana Suppl.) p. iv/2 There it will link up at the 'wild west' border town of Lethem with a similar road the Brazilian army engineers are building to connect with Manaus and the Pan-American Highway.

b. Special Comb.: **Wild West show,** a circus or fairground entertainment depicting cowboys and Indians with exhibitions of riding, shooting, etc.; also *fig.*; similarly *Wild West exhibition.*

1885 in B. A. Botkin *Treas. Amer. Folklore* (1944) I. 150 Buffalo Bill's 'Wild West' Prairie exhibition and Rocky Mountain show. **1895** 'Mark Twain' in *N. Amer. Rev.* July 8 A man who could hunt flies with a rifle and command a ducal salary in a Wild West show. **1914** A. Bennett *Price of Love* vii. 133 Skating-rinks, Wild West exhibitons, Dutch auctions. **1937** N. Marsh *Vintage Murder* xxiv. 268 'Shut up. This isn't a Wild West show.' 'You give me the lie!' 'Oh, for God's sake don't go native.' **1976** *Billings* (Montana) *Gaz.* 20 June 8-c/2 Later, the way it worked in the 'wild west' shows of the day, the U.S. cavalry came along, rescued the passengers and drove off the Indians. **1979** J. Wainwright *Duty Elsewhere* vii. 29 'Y'mean—illegal methods?' Something of a wild west show.' 'That's one way of putting it.'

Hence **wild we·stern** (also with initial capitals) *a.*, characteristic of or resembling the Wild West; as *sb.*, a film about the Wild West; = *western B. 4; **Wild We·sterner.**

1864 M. B. Chesnut *Diary* 2 Dec. in C. W. Woodward *M. Chesnut's Civil War* (1981) 682 He had come to take Serena—alone. That is his wild western fashion. **1934** *Cinema Q.* III. iv. 198 'Wild Western' was, almost from the inception of the film, one of its most popular subjects. **1963** I. Fleming *On Her Majesty's Secret Service* xvii. 192 A group of harlequins, Wild Westerners and pirates. **1967** D. Francis *Blood Sport* viii. 95 Jackson preserved its own wild western flavour to the extent of a small authentic stage coach waiting in front of the drug store. **1981** A. Lurie *Lang. of Clothes* iv. 112 At any national convention the Wild Westerners will be the easiest to identify. **1982** W. Mankowitz *Mazeppa* vii. 118 The Menken enjoyed the Washoe wild western atmosphere.

Wilfridian (wilfri·diăn). *Eccl. Hist.* [f. St. *Wilfrid* (634–709) + -ian.] A member of a religious fraternity founded by Father F. W. Faber (1814–63) for his fellow-converts to Roman Catholicism; later united with the oratory of St. Philip Neri, Birmingham.

1847 F. W. Faber *Let.* 23 Sept. in J. W. Bowden *Life & Lett. Frederick William Faber* (1869) viii. 329 The Wilfridians are allowed to work their double work, against ignorance and brutal sin. **1848** J. H. Newman *Let.* 2 Jan. (1962) XII. 144 What seems to me best is..for..St John and you to go to Cotton to take charge of the Wilfridians. **1869** J. W. Bowden *Life & Lett. Frederick William Faber* viii. 295 From the name of the latter Saint they were commonly called Wilfridians. **1928** *St. Wilfrid* xvii. 231 Father Faber did much to make him known and loved... In his first year as a Catholic priest, he and his forty Wilfridians converted a whole parish in Staffordshire. **1981** S. Chitty *Gwen John* ix. 134 Frederick Faber.. became a Catholic and founded the order of Wilfridians at Elton.

wilful, *a.*[1] Add: **1. c.** In nonce Comb. with *wavy.*

1877 [see *meal-drift* s.v. *meal *sb.*[1] 3 a].

Wilhelmine (vi·lhelməin), *a.* [f. the Ger. name *Wilhelm* William (see below) + -ine[1].] Of or pertaining to (the reign of) William II, emperor of Germany 1888–1918. Also **Wilhe·lmian, Wilhelmi·nian** [cf. G. *wilhelminisch*] *adjs.*

1931 C. R. Turner tr. *G. Schultze-Pfaelzer's Hindenburg* xii. 279 A swaggering general of the later Wilhelminian period. **1948** A. Huxley *Ape & Essence* (1949) 21 Floating products of Wilhelmine wealth and culture. **1956** S. Bedford *Legacy* iv. v. 282 The events..shed a queer light on the Wilhelminian era. **1957** *Cassell's German & Eng. Dict.* I. 614/2 *Wilhelminisch, adj.*, of William II, Wilhelmian (*of Germany*). **1962** *Times* 14 Nov. 16/2 The Wilhelmian empire. **1973** *Times Lit. Suppl.* 5 Oct. 1183/1 Like so many intellectuals in Wilhelminian Germany. **1975** *Historical Jrnl.* XVIII. 821 It was in this situation that the role of the Centre party in Wilhelmine politics became decisive. **1979** *Observer* 18 Nov. 35 The mood of fatalism that overtook the Wilhelmine Empire of Germany before the advent of the First World War.

Wilhelmstrasse (vi·lhelm,ʃtrā:sə). [Ger.] The name of a street in Berlin, the site of the German foreign office until 1945; hence used for the pre-war German foreign office and its policies.

1914 in *Conc. Oxf. Dict.* Addenda. **1919** Ld. F. Hamilton *Vanished Pomps of Yesterday* i. 28 The Ambassador took the hint, and that was the last note in Russian that reached the Wilhelmstrasse. **1923** G. Buchanan *My Mission to Russia* I. iv. 45 The reception, however, accorded to this tentative proposal by the Wilhelmstrasse was not encouraging. **1938** H. Nicolson *Diary* 21 Sept. (1966) I. 363 What remains of Czechoslovakia..must subordinate her foreign policy to that of the Wilhelmstrasse. **1956** S. Bedford *Legacy* iii. iv. 143, I can't think what they made of him at the Wilhelmstrasse. **1979** G. St. Aubyn *Edward VII* vii. 316 Chamberlain was resolved to seek partnership with France..should discussions with the Wilhelmstrasse break down.

‖**wili, willi** (vi·li). *Slavonic Mythol.* [Ger. or Fr. *wili, willi*, ad. Serbo-Croat *vila* nymph, fay. Cf. *vila.] (See quot. 1949.) Chiefly used in connection with the ballet *Giselle.*

[**1841** Vernoy de Saint-Georges et al. *Giselle ou Les Wilis* 14 C'est l'heure lugubre où, selon la chronique du pays, les Wilis se rendent à leur salle de bal.] **1949** A. Chujoy *Dance Encycl.* 511/1 *Wilis* (or *Willis*), in Western-Slavic and Eastern-German legends, the spirits of betrothed girls who have died as a result of being jilted by faithless lovers. They came out to dance at night and led the faithless ones to their death by making them dance until they fell dead of exhaustion. **1961** *Times* 2 Oct. 16/6 Miss Jill Bathurst..danced Odette and the wili Giselle. **1963** P. Hansford Johnson *Night & Silence* xx. 138 The cat continued to dance, star of the snowy ballet, with a million Wilis whirling behind him. **1977** *N.Y. Rev. Bks.* 13 Oct. 44/2 She gazes out of the fascinating portrait that Henri Lehmann painted of her in 1843 like some supernatural being, a willi, a peri, or a refined succubus.

wiliwili (wī·liwīli). [Hawaiian.] A coral tree, *Erythrina sandwicensis*, of the family Leguminosæ, native to Hawaii and Tahiti and bearing clusters of orange flowers.

1888 W. F. Hillebrand *Flora Hawaiian Islands* 100 'Wiliwili'..loses its leaves in late summer. **1913** J. F. Rock *Indigenous Trees Hawaiian Islands* 191 The very soft, white wood of the Wiliwili..is still used by the natives for outriggers on their fishing canoes. **1917** *Nature* 20 Sept. 57/2 In the arid regions is found the wiliwili.., a deciduous tree with gnarly growth. **1965** *N.Z. Listener* 17 Dec. 4/2 The Hawaiian chiefs..riding the lighter, balsa-like wili-wili boards.

wilkeite (wi·lki,əit). *Min.* [f. the name of R. M. *Wilke*, 20th-c. U.S. mineral collector + -ite[1].] A silicate and sulphate mineral of the apatite group occurring as translucent pink or yellow hexagonal crystals (see quot. 1982).

1914 Eakle & Rogers in *Amer. Jrnl. Sci.* CLXXXVII. 263 The writers take pleasure in naming it wilkeite in honor of R. M. Wilke, who as a mineral collector and dealer has done much to advance the science of mineralogy. **1937** *Amer. Mineralogist* XXII. 977 It has become necessary to investigate the substitutions of the sort found in wilkeite. **1975** *Nature* 27 Feb. 722/1 Among the minerals described to have any significant degree of P(V)–Si(IV) partial substitution are viseite, nagatelite and wilkeite, an apatite with partial $SiO_4 + SO_4$ substitution for PO_4. **1982** *Amer. Mineralogist* LXVII. 90 Wilkeite is not a valid mineral species, since it is only one of many solid solutions involving the six end-members fluorapatite, hydroxyapatite, chlorapatite, fluorellestadite, hydroxyellestadite, and chlorellestadite.

Wilkism (wi·lkiz'm). [Irreg. f. the name of John *Wilkes* (1727–97), English radical politician + -ism.] The principles or policies associated with John Wilkes.

1769 Mrs. Harris *Let.* 24 Mar. in Earl of Malmesbury *Lett.* (1870) I. 177 The Wilkism, and obscenity of the woman proved the greatest attraction. **1778** J. Witherspoon *Address to Natives Scotl. residing in Amer.* 5 What effect this *Wilkism* (If I may so speak) of many Americans may be supposed to have had upon the minds of gentlemen from Scotland, it is not difficult to explain. **1930** R. Postgate '*That Devil Wilkes*' xiv. 248 (*heading*) The end of Wilkism.

Also **Wi·lkite,** a follower of John Wilkes or his ideas.

a **1797** J. Wilkes in Lincoln & McEwen *Lord Eldon's Anecdote Bk.* (1960) 15, I have nothing to do with such a Man. He was a Wilkite, which I never was. **1917** H. Bleackley *Life Wilkes* xiv. 251 The zeal of hundreds of sturdy Wilkites had oozed away. **1930** R. Postgate '*That Devil Wilkes*' xiv. 258 They knew only that he had shot down Wilkites.

will, *sb.*[1] Add: **II. 5. c.** Chiefly used in the names (often hyphenated) of supposed natural instincts or drives, as *will to art* [tr. G. *wille zur kunst*]; *will to be, believe, live* (earlier example; also *transf.*); *will to* (or † *unto*) *power* [tr. G. *wille zur macht*], in Nietzsche's philosophy and, later, in analytic psychology (esp. A. Adler's individual psychology): the driving force behind all human behaviour which should lead to self-mastery but when frustrated can become the will to dominate others; cf. *power-drive *sb.* 2; (later examples).

1889 G. B. Shaw *How to become a Musical Critic* (1960) 147 Vegetarianism, the higher Buddhism..negation of the Will-to-Live..all these are but samples of what Wagner-

ism involves nowadays. **1891** —— *Let.* 29 July (1965) I. 301 John Robertson seeks for facts that support his will-to-believe that Materialist-Rationalists are the only honest Secularists. **1923** J. van Teslaar tr. *Stekel's Psychoanal.* ii. 61 That 'will to power' means, 'Above all, I want to be loved.'.. Will to power is will to be loved. **1926** Galsworthy *Silver Spoon* i. xiv. 110 Humanity has got to save itself! To save itself—what was that, after all, but expression of 'the will to live'? **1929** H. Read *Staffordshire Pottery Figures* 21 The 'folk' spirit which makes the early salt-glaze..figures so precious as evidences of an innate 'will-to-art'. **1930** D. H. Lawrence *Virgin & Gipsy* ii. 35 Yvette suddenly saw the stony, implacable will-to-power in the old..Granny. **1931** J. S. Huxley *What dare I Think?* iv. 143 Only by banishing the driving force of emotion and the false certitude of the will-to-believe..does she [*sc.* Science] arrive at greater power. **1945** W. de la Mare *Burning-Glass* 12 And naught but his marooned precarious self For questing consciousness and will-to-be. **1948** R. Stagner *Psychol. of Personality* (ed. 2) xv. 288 Adler believed..that the will to power was a fundamental drive, and that it was thwarted by some inferiority. **1963** N. Frye *Romanticism Reconsidered* 14 The tremendous will-to-power finales of Beethoven. **1972** D. V. Tansley *Radionics* iv. 33 The ancient seers of India.. observed that the base chakra was responsible for.. providing a channel for the will-to-be to express itself. **1976** J. Goode in Mitchell & Oakley *Rights & Wrongs of Women* vii. 232 Partly this is based on a will to power, the demand for a totally submissive love. **1977** *Times* 2 Dec. 21/3 The [Crown] agents' 'remarkable will to live' pushed them into critical change in their financial operation in 1966. **1979** E. H. Gombrich *Sense of Order* vii. 193 That 'will to art', which Riegl had conceived as an alternative to the mechanistic explanations of individual motifs, developed into a vitalistic principle underlying the whole history of art. **1985** E. Gellner *Psychoanalytic Movement* i. 27 The Will to Power is a far, far more disturbing, more corrosive idea for human optimism than is the domination of the human psyche by sexuality.

V. 24. a. *will-web.* **c. will-form,** a form on which a will may be made out.

1924 D. H. Lawrence *England, my England* 150 Say I want to see Mr. Whittle as soon as he can, and will he bring a will-form. **1948** 'J. Tey' *Franchise Affair* vi. 56 An old woman..wanted to alter her will... So Robert had taken some new will-forms. **1904** Hardy *Dynasts* I. i. 6 As key-scene to the whole, I first lay bare The Will-webs of thy fearful questioning.

will, *v.*[1] Add: **B. I. 11. c.** *will do* (with omission of *I*): an expression of willingness to carry out a request. Cf. *wilco. *colloq.*

1955 W. Tucker *Wild Talent* xvi. 217 'Paul! Bring my gate pass.'..'Will do.' **1967** L. White *Crimshaw Memorandum* v. 91 'And find out where the bastard was.'..'Will do,' Jim said. **1971** J. Wainwright *Last Buccaneer* ii. 220 'Make sure he comes.' 'Will do,' said the D.D.I. **1981** A. M. Stein *Body for Buddy* ix. 176 'Let me know.' 'Will do,' I said.

II. 29. (Further examples.) Also, could naturally or inevitably be expected to, esp. in the light of one's known character or tendencies. *colloq.*

1919 'C. Dane' *Legend* 43 One never knew what Madala would do next, and yet when she'd done it, one said—'Of course! Just what Madala *would* do!' **1926** C. Mackenzie *Rogues & Vagabonds* 268 'He always pushes me out.' 'He would.' **1930** E. Waugh *Vile Bodies* ix. 150 There's our Lily now. You know how she would go in for being a manicurist. **1932** M. H. Rinehart *Miss Pinkerton* xvi. 164 'We're interested in Monday night, and that's all.'..'You would be!' **1946** H. J. Massingham *Where Man Belongs* iii. 96 He, Ireson told me, 'is the most promising boy of the lot.' He would be. **1963** *Times* 1 July 6/6 Mr. Burge asked: Do you know Lord Astor has made a statement to the police saying that these allegations of yours are absolutely untrue? Miss Rice-Davies: He would, wouldn't he? **1980** 'T. Hinde' *Daymare* i. 8 'Well, he would, wouldn't he,' she says. 'It's what you'd expect of a born capitalist.'

III. 42. d. *I wouldn't know*: see *know *v.* 11 g.

47. b. *wouldn't it?* (ellipt. for *wouldn't it rock you?, wouldn't it root you?*, and similar catch-phrases): an exclamation of annoyance and disgust or (less usually) amusement. *Austral.* and *N.Z. slang.*

1940 *Telegraph* (Sydney) 13 Jan. 4/7 Favorite expression with the troops is, of course, '*Wouldn't it?*'—Short for 'Wouldn't it make you sick?' **1941** *2nd N.Z.E.F. Times* 3 Nov. 6 Well, *wouldn't* it? **1951** Cusack & James *Come in Spinner* 382 Guinea kicked a hassock across the room. 'Wouldn't it!' she muttered furiously, 'wouldn't it!' **1954** J. Cleary *Climate of Courage* xii. 185 'Asking your wife if you can write to her. Wouldn't it?'

Willesden (wi·lzdĕn). The name of a suburb of north-west London, used *attrib.* to denote forms of paper or canvas that have been toughened and waterproofed by being treated with cuprammonium solution.

1895 C. F. Cross et al. *Cellulose* i. 13 Vegetable textile fabrics passed through a bath of the cuprammonium hydroxide are 'surfaced' by the film of gelatinized cellulose... These fabrics are sold under the style or description of 'Willesden' goods; the manufacture being in the hands of a company whose works are situated at Willesden. The company's processes are based on the patents of Drs. J. Scoffern and C. R. A. Wright. **1907** *Yesterday's Shopping* (1969) 283/1 Patent Sleeping Valise... No. 3. Willesden Kharki cotton canvas throughout. No. 4. In Willesden flax canvas. **1911** *Encycl. Brit.* XXIII. 705/1 'Willesden paper'..is cardboard chemically treated to render it tough, waterproof and fire-resisting. **1912** R. A. Freeman *Singing Bone* i. ii. 30 Boscovitch continued to stare up at the little square case covered with Willesden canvas... Thorndyke good-naturedly lifted it down and unlocked it. As a matter of fact he was rather proud of his 'portable laboratory'. **1926–7** *Army & Navy Stores Catal.* 219/3 Willesden Canvas. Suitable for.. awnings, shelters, etc. **1964** J. S. Scott *Dict. Building* 358 *Willesden paper*, a building paper made of cardboard treated with cuprammonium hydroxide to rotproof it. **1982** J. Sherwood *Shot in Arm* xiv. 141 A second-hand Willesden canvas cabin trunk with leather corners.

willi, var. *wili.

William. Add: **2.** An obsolete Dutch coin (see quot. 1893).

1844 T. B. Macaulay *Let.* 9 Oct. (1977) IV. 218 While he was changing me a gold William I got away from the old villain. **1893** R. Bithell *Counting-Ho. Dict.* (rev. ed.) 317 *William*, a gold coin formerly used in Holland, and valued at 10 guilders. Its metallic value was about 16s. 2d. sterling.

3. *slang.* [With a pun on Bill *sb.*[3]] **a.** An account for payment, a bill.

1859 H. J. Byron *Maid & Magpie* ii. 18 When de farmers around are behind in their rent I does little Williams, at sixty per shent. **1903** Farmer & Henley *Slang* VII. 353/2 *To meet sweet William*, to meet a bill on presentation.

b. A dollar note. (See also quot. 1869.) *U.S.*

Sometimes without a capital initial.

1865 *Republican Banner* (Nashville, Tenn.) 5 Oct. 3/1 Will. had to remember the Workhouse in his will to the tune of a 'ten dollar William'. **1869** *Overland Monthly* III. 128 $100 bills were there [*sc.* in Texas] called 'Williams', and $50 bills 'Blue Williams'. **1887** in Wentworth & Flexner *Dict. Amer. Slang* (1960) 580/1 [He] lost his five dollar William. **1927** C. A. Siringo *Riata & Spurs* i. 10 Mr. Myers wrote me..to buy a suit of clothes with the twenty-dollar 'william'.

4. Used *attrib.* to designate the style of architecture, furniture, etc., associated with the reign of monarchs of this name; esp. **William and Mary** (freq. hyphenated), with reference to William III and Mary, joint King and Queen of Great Britain, 1689–94; **William IV,** with reference to William IV, King of Great Britain, 1830–7.

1905 Fenn & Wyllie *Old Eng. Furnit.* vii. 74 The low-backed armchair..was..subsequently displaced by the more dignified and far more comfortable high-backed kind known to us as the 'Stuart' and the 'William and Mary' chair. **1927** *Daily Tel.* 29 Nov. 7/1 Jacobean and William and Mary chests. **1948** D. Welch *Jrnl.* 31 Aug. (1952) 266 Our chairs were William and Mary with high caned backs. **1955** 'W. Mole' *Hammersmith Maggot* iii. 41 A fine set of William IV chairs. **1977** *New Scientist* 3 Mar. 512/1 A William-and-Mary country house in the depths of Somerset. **1982** 'J. Gash' *Firefly Gadroon* i. 13 A blazing row over a William IV davenport desk.

Williamite. Add: **2. b.** Of glass: bearing portraits or emblems of William III, as an indication of anti-Jacobite feelings.

1905 P. Bate *Eng. Table Glass* xii. 105 No. 213 is a Williamite glass bearing the inscription—'*The immortal memory*'; others read, '*To the glorious memory of King William*'. **1936** *Burlington Mag.* Oct. p. xxiii/1 Many specimens of engraved glasses including Jacobite specimens..and Williamite, Volunteer and other inscribed glasses. **1973** *Country Life* 22 Mar. Suppl. 72/2 A rare Williamite glass.

William Morris. = *Morris.

1944 D. Welch *Jrnl.* 26 Oct. (1952) 135 We had tea at Pitt's Cottage, on a William Morris, mortifying sofa. **1962** I. Murdoch *Unofficial Rose* xiv. 128 The bright blue bird-woven William Morris tiles. **1969** S. Sitwell *Gothic Europe* xiii. 155 They are Brussels tapestries..too 'flowered' in the foreground and therefore too 'William Morris' in style. **1981** M. E. Atkins *Palimpsest* ii. 18 You'll splash the wallpaper, it's William Morris.

Hence **Wi·lliam Mo·rrisy** *a.*, resembling, or in the style of, William Morris.

1960 *Times* 29 July 13/5 A William Morrisy life of craftsmanship close to the soil. **1968** 'O. Mills' *Sundry Fell Designs* viii. 88 She's a William Morrisy person; and all she was after was the simple, creative life. **1977** R. Barnard *Blood Brotherhood* x. 113 A William Morrisy stained-glass window.

william-nilliam (wi·lyăm ni·lyăm), *adv.* Humorously extended form of Willy-nilly *adv.*

1907 G. S. Gordon *Let.* 9 Sept. (1943) 23, I have called you sweet girl. But I will not..retract; and so sweet girl you must remain william nilliam. **1917** A. Huxley *Let.* 8 Apr. (1969) 123, I..found myself pushed—almost william-william [*sic*]—into a very nasty and ill-paid job. **1959** P. Bull *I know Face, But...* x. 188 A splendid change from ordinary digs where the plate is plonked in front of you william-nilliam.

Williams[2] (wi·lyămz). *Computers.* [The name of F. C. *Williams* (1911–77), English electrical engineer, who with T. Kilburn described such a tube in 1948.] *Williams tube*: a cathode-ray tube used in some early computers to store and display an array of spots representing bits; so *Williams memory.*

1950 W. W. Stifler et al. *High-Speed Computing Devices* x. 202 The proposed machine will have an electrostatic storage system consisting of a bank of Williams tubes. **1970** O. Dopping *Computers & Data Processing* x. 150 The Williams memory had many weaknesses but until the middle fifties it was the only available memory in the microsecond class apart from the expensive flip-flop registers. **1982** D. P. Siewiorek et al. *Computer Structures* vii. 107/2 The Williams Tube which implemented the control register was also used to hold the present instruction..itself subsequent to its being read out of main store.

willies (wi·liz), *sb. pl.* *slang* (orig. *U.S.*). [Etym. unknown.] *the willies*: a fit of nervous apprehension. Chiefly in phrs. *to give* (someone) *the willies, to get the willies.*

1896 *Dialect Notes* I. 427 *To have the willies*, to be nervous. **1900** G. Bonner *Hard Pan* 99 It just gives me the willies to think of your being down on your luck. **1913** J. London *Valley of Moon* 105 Bert gives me the willies the way he's always lookin' for trouble. **1927** H. A. Vachell *Dew of Sea* 261, I sure got the willies at the thought of meeting you. **1942** G. Kersh *Nine Lives Bill Nelson* ix. 57 It *can* give you the willies when, in broad daylight, you hear a rifle go off. **1953** F. Swinnerton *Month in Gordon Square* 202 Gosh! She was getting the willies. It was awful. **1962** J. Heller *Catch-22* xii. 127 Chief White Halfoat shuddered. 'That guy gives me the willies,' he confessed. **1975** Felton & Fowler *Best, Worst* 277 You can now visit Winchester House. But we wouldn't advise it if you suffer from the willies. **1984** A Carter *Nights at Circus* iii. i. 199 Not that the 'wagon salon' isn't very pleasant, if it don't give you the willies.

will-o'-the-wisp, *v.* Delete † and add later examples.

1926 J. B. Priestley *G. Meredith* v. 129 Woman..is far less likely than man to be Will-o'-the-Wisped away by sheer unreason masquerading as reason. **1954** L. MacNeice *Autumn Sequel* xxvi. 159 Words may will-o'-the-wisp him.

willow, *sb.* Add: **II. 5.** (Earlier example.) Cf. *King Willow* s.v. *king *sb.* 6 b.

1846 J. Martin in *Frederick Lillywhite's Cricket Scores & Biographies* (1863) III. 442 And now the 'willow' see them wield.

III. 6. a. *willow bottom* (later example), *walk* (earlier example), *wand* (later example). **b.** *willow-lined* adj. **c. willow gentian,** a herbaceous perennial, *Gentiana asclepiadea*, native to Europe and bearing deep blue or white flowers in axils along its curving stems; **willow grouse,** (*b*) the ruffed grouse, *Bonasa umbellus*, called thus chiefly in British Columbia; **willow leaf,** also as *adj.* = *willow-leaved* adj., sense 6 b; **willow-leaved pear(-tree),** a tree, *Pyrus salicifolia*, that is related to the pear and is native to S.E. Europe and Asia Minor, bearing long narrow leaves and small fruit and often thorny; **willow oak** (later examples); also, the laurel oak, *Q. laurifolia*; **willow tit (mouse),** a black-headed, buff-coloured European tit, *Parus montanus* (formerly *P. atricapillus*); **willow-ware** (*b*) articles woven from osiers; **willow-wielder** (earlier example).

1962 W. Stegner *Wolf Willow* i. i. 12, I see a black iron bridge, new, that evidently leads some new road off into the willow bottoms. **1883** W. Robinson *Eng. Flower Garden* 136/1 Willow Gentian prefers a sheltered position. **1935** C. Elliott *Rock Garden Plants* 123 The Willow Gentian of sub-alpine woods, grows two feet high, with arched wiry stems, strung along their upper half with fine blue trumpet-flowers. **1962** R. Page *Education of Gardener* xiii. 357, I may choose *Gentiana asclepiadea*, the willow gentian. **1907** J. G. Millais *Newfoundland* 274 The Newfoundland willow grouse..fly in large bodies from one district to another. **1960** *Gulf Islander* (Galiano, B.C.) 23 July 1/1 We listened to the resonant call of the willow grouse. **1961** W. P. Keller *Canada's Wild Glory* ii. 93 The grouse came back, too. These were the willow grouse or as the same bird is known in the east, 'the drummer' or ruffed grouse. **1818** A. Eaton *Man. Bot.* (ed. 2) 447 Willow-leaf golden-rod. **1975** *Country Life* 20 Mar. 699/3 Willow-leaf pears..form charming leafy tapestries. [**1789** W. Aiton *Hortus Kewensis* II. 176 Willow-leav'd Crab Tree. Nat[ive] of the Levant.] **1820** *Bot. Reg.* VI. 514 (*heading*) Willow-leaved pear-tree. **1914** W. J. Bean *Trees & Shrubs Hardy in Brit. Isles* II. 292 Willow-leaved Pear... Branchlets covered with down which is quite white when young. **1980** V. Canning *Fall from Grace* ix. 155 A carpet of silvery *Cineraria maritima* spread under a group of willow-leaved pears. **1946** J. W. Day *Harvest Adventure* x. 159 Half-way between Yarmouth and Acle on that dead straight, willow-lined road. **1813** Willow oak [see *grey oak* s.v. *grey *a.* 8]. **1897** [see *peach oak* s.v. *peach *sb.*[1] 6]. **1949** *Amer. Forests* Sept. 18/3 A tall willow oak drips slender verdant fingers. **1975** *Country Life* 2 Jan. 38/3 More native trees were also planted, notably..the willow oak. **1907** *Brit. Birds* I. 44 The Willow Tit varies a good deal geographically. **1979** C. M. Perrins *Brit. Tits* vii. 60 The Willow Tit was the last British species to be recognized. **1958** *Spectator* 22 Aug. 244/1, I discovered a rare bird, then almost unknown,..at Beckenham—a willow titmouse. **1803** J. Palmer *World as it Goes* II. 14 The carriage entered a willow-walk, terminated by a small antique building. **1954** J. R. R. Tolkien *Fellowship of Ring* iii.

80, I shall be as thin as a willow-wand. **1851** C. Cist *Cincinnati* 172 Baskets, cradles, wagons and other willow-ware. **1880** *Harper's Mag.* June 30/1 We find women employed in making..willow-ware and cane chairs. **1870** *John Wisden's Cricketers' Almanack* 91 With willow wielders like these, it is no wonder Notts holds the high position it does as a batting shire.

f. Short for *willow pattern* (Willow *sb.* 6 c), as *willow cup, plate, pottery*. See also *willow ware* in Dict.

1926 R. Macaulay *Crewe Train* ii. x. 179 It would look jolly with blue willow cups and plates on it. **1928** T. S. Eliot in E. Pound *Sel. Poems* p. xvii, People who like Willow pottery and Chinesische-Turms in Munich and Kew. **1961** M. Beadle *These Ruins are Inhabited* (1963) xi. 142 In contrast, the laburnums..curved earthward with willow-plate grace.

willowy, *a.* Add: Hence **wi·llowily** *adv.*; **wi·llowiness.**

1932 A. Huxley *Brave New World* xi. 192 He put his arm round the Head Mistress's waist. It yielded, willowily. **1972** *Daily Tel.* 13 Mar. 11 Virile shoulders, tapering downward to a more traditional willowiness, are the hallmark of the 1972 Cardin man.

willy, willie, *sb.*[2] Add: **2.** *slang.* An infantile name for the penis. Also *Comb.*, as *willy-warmer.*

1905 *Eng. Dial. Dict.* Suppl. 178/2 *Willy*, the male organ; a slang name for a child's penis. Cum., Wm. **1972** *Listener* 22 June 841/3 The gallant soldier-boys are afflicted with 'syph, darling' ('their willies rot away'). **1975** *Observer* 7 Dec. 27/3 Joky gifts are speechlessly embarrassing; this season's dud is a woolly willy-warmer. **1977** J. Wilson *Making Hate* ix. 113 A younger male [baboon]..fingered its crimson penis... 'It's playing with its *willie*!' Nicky squealed. **1985** P. Angadi *Governess* x. 93 We used to hold each other's willies... We didn't know about sex then.

willy (wi·li), *sb.*[3] [Prob. related to Williwaw: cf. Willy-willy.] In the South Atlantic (Tristan da Cunha): (see quots.).

1832 A. Earle *Narr. Residence N.Z.* (1966) 204 These sudden squalls are called '*Willies*', at least, such is the name given them by the sailors who frequent the island [*sc.* Tristan da Cunha]. **1941** A. B. Crawford *I went to Tristan* xi. 158 A shower is a 'light squall' and 'willies' are eddies of spray above the surface of the sea caused by small whirlwinds. *Ibid.* 268 *Willie*, whirlwind of spray over the sea.

willya (wi·lyă). Repr. colloq. pronunc. of 'will you..?', esp. as a tag after an imperative.

1941 B. Schulberg *What makes Sammy Run?* ix. 241 Willya find out who rang for a messenger boy? **1956** 'E. McBain' *Cop Hater* (1958) iv. 38 Hey, shut up, will-ya? **1968** C. Burke *Elephant across Border* i. 28 Now go away, willya? **1981** G. McDonald *Fletch & Widow Bradley* viii. 28 Write a new story... Only get the competition to print it this time, willya, Fletch?

willyamite (wili,ā·məit). *Min.* [See quot. 1893 and -ite[1].] A sulphide and antimonide of cobalt and nickel, (Co,Ni)SbS, in which Co exceeds Ni, found as white or grey pseudo-cubic crystals having a metallic lustre.

1893 E. F. Pittman in *Jrnl. & Proc. R. Soc. New S. Wales* XXVII. 366, I propose to name the mineral Willyamite (pronounced Willy-ah'-mite) after Willyama the official name of the Broken Hill township, and the aboriginal word meaning a hill with a broken contour. **1976** *Norsk. Geol. Tidsskr.* LVI. 449 Ullmannite, cobaltian ullmannite and willyamite occur as blebs and laths in galena from Espeland mine, Aust-Agder, Norway.

willy wet-leg: see *wet-leg* s.v. *wet *a.* 20.

Wilms (vilmz). *Path.* [The name of M. *Wilms* (1867–1918), German pathologist.] *Wilms('s)* or (erron.) *Wilm's tumour*: a malignant tumour of the kidney that occurs in infants and children.

1910 E. L. Keyes *Dis. Genito-Urinary Organs* 974/2 (Index), Wilms tumor of kidney. **1928** Eisendrath & Rolnick *Text-bk. Urol.* xlviii. 753 These tumors in their pure state have been termed mixed cell or Wilms tumors and are composed of..muscle fibres, blood vessels, cartilage, [etc.]. **1948** R. A. Willis *Path. of Tumours* lx. 925 Many names have been applied to the embryonic renal tumours—'adenosarcoma',..'Wilms's tumour'.. and 'nephroblastoma'... It is now clear that there is but one entity, embryonic renal tumour. **1961** R. D. Baker *Essent. Path.* xiii. 307 Embryonal and mixed tumors are most frequently mixed salivary gland tumors, Wilm's tumor of the kidney, and testicular and teratomatous tumors. **1971** *Brit. Med. Bull.* XXVII. 68/1 Retinoblastomata and Wilms' tumours were found to have shorter latent periods. **1980** *Jrnl. R. Soc. Arts* Jan. 99/1 In rhabdomyosarcoma, Ewing's sarcoma and Wilm's tumour considerably increased survival has been obtained.

wilsh, var. *wilch.

Wilson[1] (wi·lsən). *Path.* [The name of S. A. Kinnier *Wilson* (1878–1937), English neurologist.] *Wilson's disease*: = *hepato-lenticular degeneration* s.v. *hepato-.

1915 Stedman *Med. Dict.* (ed. 3) 1032/2 *Wilson's disease*, progressive degeneration of the lenticular nucleus, occurring as a familial disease associated with cirrhosis of the liver. **1919** *Arch. Internal Med.* XXIV. 497 (*heading*) Progressive lenticular degeneration associated with cirrhosis of the liver (Wilson's disease). **1978** *Brit. Med. Jrnl.* 18 Nov. 1384/2 Though originally described as a neurological disorder with associated cirrhosis, we now recognise hepatolenticular degeneration (Wilson's disease) as a copper storage disorder in which other tissues become affected as the excess copper is released from the liver.

Wilson[2] (wi·lsən). *Physics.* The name of C. T. R. *Wilson* (1869–1959), Scottish physicist, used *attrib.* and in the possessive to designate the cloud chamber (see *cloud *sb.* 12) invented by him.

1917 *Sci. Abstr.* A. XX. 337 Wilson's condensation chamber..was still smaller than the author's small sphere. **1931** *Ann. Reg. 1930* 59 Harkins and Smith took 39,000 photographs of the tracks of 390,000 α-particles in a Wilson cloud chamber containing nitrogen. **1961** *New Scientist* 23 Feb. 474/2 The Wilson cloud chamber had done heroic work for many years in the examination of the tracks of particles, but its limitations in speed and scope were becoming increasingly apparent.

Wilsonian (wilsǒᵘ·niăn), *a.* (and *sb.*) [See -ian.] Pertaining to or characteristic of Woodrow *Wilson* (1856–1924), president of the United States 1913–21, noted for his uncompromising idealism. Also as *sb.*, a follower of Woodrow Wilson.

1921 *Labour Monthly* Sept. 285 In whatever shades or purgatory await our public men after the completion of their labours, a special circle should be reserved for the Old Wilsonians. **1924** *Amer. Mercury* Jan. 53/1 It was at this precise moment in his career that the Wilsonian storming of Valhalla began. **1934** H. G. Wells *Exper. Autobiogr.* ii. ix. 694 The Wilsonian notion of a League. **1962** *Listener* 22 Mar. 524/1 Mr. Tillman remains a starry-eyed Wilsonian. **1980** J. Lees-Milne *Harold Nicolson* I. vii. 116 The French bitterly opposed the very idea of the covenant as Wilsonian idealistic nonsense.

Also **Wi·lsonism,** the policies of Woodrow Wilson.

1920 *Harvey's Weekly* 16 Oct. 13/1 No more time need be lost in following the slush-fund herring trail away from the vital issue of Wilsonism. **1945** Koestler *Yogi & Commissar* iii. i. 125 Movement followed Movement and withered away. Jacobinism,..Wilsonism, the League of Nations..they were all branches of the same tree rooted in the Age of Enlightenment. **1977** *N.Y. Rev. Bks.* 12 May 16/4 But they were fighting to save Wilsonism, if need be from Wilson himself.

wilt, *sb.*[2] (Later examples of *Bot.* sense.)

1918 [see *psylla]. **1946** *Nature* 13 July 56/1, I came across what is apparently a hitherto undescribed wilt disease of the oil palm. **1961** A. Schoenfeld tr. *Stapp's Bact. Plant Pathogens* i. 103 This wilt disease [of beans] can be said with certainty not to occur in Germany. **1981** Buczacki & Harris *Collins Guide to Pests of Garden Plants* 306 Most wilts are caused by Deuteromycete fungi.

wilt, *v.* Add: **2. b.** *Agric.* To leave (mown grass, etc.) to dry partially in the open before putting it in a silo.

1971 *Power Farming* Mar. 9/1 'But,' said Mr. Whitton, 'the loader must be used as part of a system, and it is most essential that the silage be wilted and chopped.' **1974** *BSI News* May 8/3 The process of wilting the crop [of grass] from 75%–85% moisture content down to 55%–65%. **1980** *Daily Tel.* 28 Jan. 10/4 New techniques such as wilting the crop in the field before it is ensiled.

wilting, *vbl. sb.* Add: **wilting coefficient,** the moisture content of the soil (expressed as a percentage of its dry weight) when a plant begins to wilt.

1912 Briggs & Shantz in *Bull. U.S. Bureau Plant Industry* No. 230. 9 It appears advisable to use a more specific term for the moisture content of the soil corresponding to the wilting point of a plant, and we have employed the term 'wilting coefficient' in this sense in the present paper. **1980** *Communications Soil Sci. & Plant Analysis* XI. 843 Studies were performed to determine the wilting coefficient of various selected light tropical soils collected at different locations in..Venezuela.

Wilton[2] (wi·ltən). The name of a farm near Grahamstown, Cape Province, South Africa; used *attrib.* to denote a later Stone Age culture of southern Africa.

1928 A. J. H. Goodwin in *Ann. S. Afr. Mus.* (1929) XXVII. x. 251 Our first knowledge of the Wilton Industry comes from the Cape Peninsula, various crescents, thumbnail scrapers, and the like appearing from a number of kitchen middens and sand-dune sites in this district. **1936** L. S. B. Leakey *Stone Age Afr.* v. 96 In the Wilton culture the most typical tools are,..crescents and other small geometric microliths, together with small double-end and thumb-nail scrapers. **1959** J. D. Clark *Prehist. S. Afr.* ii. 41 The Wilton [culture] is named from the rock-shelter on the farm of that name west of Grahamstown. Its distribution is very wide. **1980** *Cambr. Encycl. Archaeol.* 174/1 The microlithic industries (known as Wilton in eastern and southern Africa) of the early to mid-Holocene.

Wiltshire. Add: (*a*) *Wiltshire Horn(ed)*, (a sheep of) a recently revived breed, distinguished by its very light short wool; (*c*) (earlier examples; also *North Wiltshire*).

1816 Jane Austen *Emma* i. x. 188 She was come in herself for the Stilton cheese, the north Wiltshire, the butter, the cellery, [etc.]. **1863** H. Jones *Jrnl.* 20 Aug. in F. W. Lindsay *Cariboo Dream* (1971) 38, 1 case Gloucester Cheese $75.72 Wiltshire do. $97.35. **1945** J. F. H. Thomas *Sheep* ii. 34 The Wiltshire Horned..is no longer to be found in Wiltshire..; the main area of distribution is now Northamptonshire and Buckinghamshire. **1970** *Observer* 26 Apr. (Colour Suppl.) 36/1 Wiltshire Horn..has no wool but, instead, a thick matted coat. **1977** *Jrnl. R. Soc. Arts* CXXV. 708/1 Supposing, for example, that wool became a nuisance and that it was far better to keep sheep that were wool-less and just had hair. It so happens that we have got such a breed, the Wiltshire Horn, that has a rather hairy coat, and so we can use that.

wimble, *sb.* Add: **3*.** Also **wimbel.** An implement for twisting together strands (esp. of straw) to make rope for tying up hay-trusses, fleeces, etc. Cf. *wimbrel[2].

1863 J. R. Wise *New Forest* 288/1 *Wimble*, an instrument with which to take up faggots or trusses of hay. **1874** Hardy *Far fr. Madding Crowd* I. xxii. 243 Gathering up the fleeces and twisting ropes of wool with a wimble for tying them round. **1886** —— *Mayor Casterbr.* I. i. 2 A rush basket, from which protruded at one end the crutch of a hay-knife, a wimble for hay-bonds being also visible. **1969** E. H. Pinto *Treen* 97 Wimbels are essentially cranked devices for twisting ropes out of straw, formerly required for binding corn stooks.

wimble, *v.*[1] Add: **3.** *trans.* To make (a rope) using a wimble (sense 3*).

1874 Hardy *Far fr. Madding Crowd* I. x. 131 'What have you been doing?' 'Tending thrashing-machine, and wimbling haybonds.'

wimbling *vbl. sb.* (later examples in sense *3 of vb.); **wi·mbler,** one who makes ropes with a wimble.

1964 *Courier-Mail* (Brisbane) 21 Dec., The policeman asked Godfrey Booth: 'Your occupation, sir?' Mr. Booth ..replied 'Cag handed straw wimbler.' Mr. Booth lives in Bobbington, Staffordshire. *Ibid.*, Mr. Booth said: 'When I left school I took up farming, and wimbling took second place.'

Wimbledon (wi·mb'ldən). The name of a district of South London, used *colloq.* to designate the Lawn Tennis Championships on Grass played annually at the All-England Lawn Tennis and Croquet Club there. Also *transf.* and *attrib.*

1907 F. W. Payn *Tennis Topics & Tactics* xvi. 157 The presence of nearly 30,000 spectators in all at the Wimbledon Championships. **1919** *Country Life* 5 July 29/1 Mr. Gore, the former champion, whose twenty-eighth Wimbledon this is. *Ibid.* 12 July 54/2 Whether the 'Victory' Championships will go down to history as 'The Wet Wimbledon' or 'The Wonderful Wimbledon' time alone can tell. **1930** A. P. Herbert *Water Gipsies* xxv. 381 Fay's *livid*—Says she wouldn't marry him now if he gave her free seats for Wimbledon. **1935** *Encycl. Sports* 383/2 No record of the Wimbledon championships would be complete without mention of a few of the winners of the men's and women's doubles championships. **1965** V. Canning *Whip Hand* xii. 134 He's.. Wimbledon standard tennis, Olympic standard swimming. **1971** *Guardian* 17 Nov. 10/3 Alun Owen is an old hand at the ding-dong, the ping-pong, the Wimbledon of sex. **1979** D. Anthony *Long Hard Cure* v. 42 Was that the year you won Wimbledon?

wimble-wamble (wi·mb'l,wǫ·mb'l), *adv.* and *sb. dial.* or *arch.* [Redupl. f. Wamble *sb.* and *v.*] **a.** *adv.* (See quot.) **b.** *sb.* ? The 'general run', the ordinary crowd.

1890 J. D. Robertson *Gloss. Dial. & Archaic Wds. Glos.* 179 *Wimble-wamble, to go sort of, vb.* to roll about in walking. **1937** H. G. Wells *Star Begotten* viii. 142 They will observe how they resemble each other and how they differ from the wimble-wamble of the common world.

wimbly-wambly (wi·mbli,wǫ·mbli), *a. dial.* Also **wimley-wamley.** [Redupl. f. Wambly *a.*] Shaky, unsteady; feeble, effeminate.

1881 *Leeds Loiners' Comic Olmenac* 24, I went all wimley-wamley e me head. **1882** F. W. P. Jago *Ancient Lang. & Dial. Cornwall* 312 I'm all wimbly-wambly. **1929** D. H. Lawrence *Pansies* 113 Flat-chested, crop-headed, chemicalised women, of indeterminate sex, And wimbly-wambly young men, of sex still more indeterminate.

wimbrel[1], var. Whimbrel.

1688 [in Dict., s.v. Whimbrel]. **1898** J. A. Gibbs *Cotswold Village* 102 There are wimbrels and curlews that have been shot here..stuffed and hung up in glass cases.

wimbrel[2], dial. var. *wimble *sb.* 3*.

1939 D. Hartley *Made in England* ii. 76 There are other types [of implement], such as the wimbrel, rather like the spindle of a spinning wheel in principle. **1969** E. H. Pinto *Treen* 97 Other country names for them [*sc.* wimbles] include wimbrels, straw twisters, [etc.].

Wimmera (wi·mərā). The name of a river and the region surrounding it in north-western Victoria, Australia, used *attrib.* in **Wimmera rye-grass** to designate a grass belonging to a drought-resistant variety of *Lolium rigidum* first identified in the area about 1900.

1920 *Proc. R. Soc. Victoria* XXXII. 199 Wimmera Rye

Grass. **1928** R. G. Stapledon *Tour Austral. & N.Z.* ix. 74 The so-called Wimmera rye-grass..first appeared in Victoria, presumably as a stowaway, about thirty years ago. **1934** *Bulletin* (Sydney) 4 Apr. 28/1 Next sow (in N.S.W. anyway) 5 lb. Wimmera rye grass. **1973** Tothill & Hacker *Grasses Southeast Queensland* 197 Wimmera ryegrass..is grown widely in southern Australia. **1977** *Weekly Times* (Melbourne) 19 Jan. 10/5 Wimmera rye grass is regarded as a curse.

wimmin (wi·min). A semi-phonetic spelling of 'women', recently adopted by some feminists as a form not containing the ending *-men*. Also, at an earlier date, occasionally used ironically in other contexts.

1910 H. G. Wells *Hist. Mr Polly* vi. 201 'Wimmin's a toss up,' said Uncle Penstemon. 'Prize packets they are, and you can't tell what's in 'em till you took 'em 'ome and undone 'em. Never was a bachelor married yet that didn't buy a pig in a poke.' **1938** *Snow White & Seven Dwarfs* 31 'Didn't I tell you?' sniffed Grumpy. 'She's crazy. Wimmin! Pah!' **1983** *Observer* 13 Mar. 16/4 Another woman was writing the words of a song... 'We coil and spring we grow and sing we dance with the tree of life we are the serpents of healing and rebirth wimmin have reclaimed the earth'... 'Why 'wimmin'?' I asked... 'We want to spell women in a way that does not spell men.' **1983** *Sunday Times* 10 Apr. 36/3 Return to Greenham Common, view the wool webs, the papier mâché masks, the eccentric re-spelling of words like 'wimmin', the improbable cosiness of the little tents in a landscape of wire fencing and policemen. **1983** *Listener* 14 Apr. 4/1 Meanwhile, what of the Peace Women ('wimmin' in feminist placards) camped outside Greenham Common? **1983** *Private Eye* 22 Apr. 5/2 (*heading*) Wimmin. **1985** *Sunday Tel.* 11 Aug. 13/8 The Greenham women—God bless 'em! (Sorry—I should write 'wimmin', since the word 'women' contains the horrid inclusion of 'men'. Their little eccentricity!)

wimp[1] (wimp). *slang.* [Origin uncertain; perh. an abbreviated corruption of *women*.] A woman or girl.

'*Wimp* was also used as a verb at Oxford *c.* 1917, e.g. *to go wimping*' (M. Marples, *University Slang* (1950), p. 98).

1923 J. Manchon *Le Slang* 338 *Wimp*, femme, fille, donzelle. **1937** Partridge *Dict. Slang* 959/1 *Wimp*, a (young) woman, a girl: from ca. 1920. **1940** [see *Skinny Liz* s.v. *skinny *a.* 6].

wimp[2] (wimp). *slang.* (orig. *U.S.*). [Origin uncertain; perh. f. Whimper (cf. Eng. dial. *wimp* (of a dog) to whine.] A feeble or ineffectual person; one who is spineless or 'wet'. (Used only as a term of abuse or contempt.)

1920 Ade *Hand-made Fables* 97 Next day he sought out the dejected Wimp. **1964** *Amer. Speech* XXXIX. 119 A *baff* is 'a person who does silly things deliberately'; but *wimp* is still mysterious and undefined in my notes. **1966** *Current Slang* Winter 8 *Wimp*, a backward person... He's a real *wimp* on a date. **1970** *New York* 16 Nov. 10/2 That Goodell, he's nothing but a wimp. And this Ottinger, it got so I couldn't stand the sight of him. **1976** *New Mus. Express* 31 July 8/2 Although he's best known here as a fairly muscular MOR wimp,..he has a big reputation as a prodigiously talented multi-media whizz in the States. **1979** T. Gifford *Hollywood Gothic* (1980) xxii. 220 Solly Roth and his wimp of a son.. what a wet bunch that family was. **1981** P. Theroux *Mosquito Coast* vi. 48, I can afford to be robbed... But what about the poor wimps who can't afford it? **1984** *Sunday Tel.* 30 Dec. 15/6 In daily life Ronnie Lee is a wimp. Put him in a balaclava and he thinks he's a he-man. **1985** *She* July 140/2 Masseur! Huh! He sounds a right little wimp.

Wimpey, var. *Wimpy *sb.* 2.

wimpish (wi·mpiʃ), *a.* *slang.* (orig. *U.S.*). [f. *wimp[2] + -ish[1].] Characteristic of a 'wimp'; feeble, ineffectual; snivelling. Freq. of persons.

1925 S. Lewis *Arrowsmith* xxvi. 288 They looked like lunching grocers: brisk featureless young men;..wimpish little men with spectacles, men whose collars did not meet. **1977** *Sounds* 9 July 30/6 The ever so slightly wimpish 'Give Me Some Time'. **1978** J. Irving *World according to Garp* xiii. 255 You call that wimpish asshole and say good-bye. **1982** *Mail on Sunday* 2 May 13/3 The wimpish young schoolmaster. **1983** *Times* 28 May Suppl. 1/5 The only motive for reading it may be to swank about it at literary cocktail parties, which is a wet and wimpish reason. **1985** *Times* 11 Feb. 14/1 The Duke of Edinburgh had adopted a new 'limp' handshake... Expecting something flabby and wimpish, the men got royal bonecrushers.

Hence **wi·mpishness,** ineffectual character or behaviour; feebleness.

1978 *Oxford Times* 6 Jan. 13/4 *Renaissance: Novella* (Warner Bros. K 56422)—It's fashionable to sneer at the 'wimpishness' of delicate music like this but..I like it. **1983** *Times* 9 Mar. 10/1 If Michael Straight is a wimp (weakling or 'wet') as some people allege, the wimpishness is not immediately apparent. **1984** *Daily Express* 17 July 8/1 Now he has raised a coast-to-coast horselaugh by his best ever display of wimpishness. Having sacked the party's incompetent national chairman, he promptly reinstated him in the face of protests.

wimpling, *ppl. a.* Add: **2.** Also *transf.* in *poet.* use.

1877 G. M. Hopkins *Poems* (1967) 69 How he rung upon the rein of a wimpling wing In his ecstasy.

Wimpy (wi·mpi), *sb.* [The name of the cartoon character J. Wellington *Wimpy* in the 'Popeye' cartoon strip, who was often portrayed eating a hamburger.] **1. a.** A proprietary name for a variety of hamburger. Also (*rarely*) in slang use.

1935 *Official Gaz.* (U.S. Patent Office) 21 May 557 Wimpy Grills, Inc., Chicago, Ill... *Wimpy*..For sandwiches, roasted and toasted meats... Claims use since Sept. 12, 1934. **1935** J. Hargan *Gloss. Prison Lang.* 8 *Wimpy*, hamburger. **1943** *American Mercury* Nov. 553/2 Other chow terms in popular use..*wimpies* for hamburgers. **1954** *Trade Marks Jrnl.* 20 Jan. 63/2 Wimpy... B720, 112. Bread rolls containing cooked foods. Edward Vale Gold, 140, North Dearborn, Chicago, Illinois, United States..Manufacturer and Merchant. **1959** *Observer* 8 Nov. 3/1 The bright glossy bars where Wimpies are served are the most striking example of Britain's changing eating habits. **1967** T. Harknett *Two-Way Frame* x. 77, I had a sterile Wimpy and cup of insipid black coffee. **1981** C. Storr *Vicky* vii. 56, I can always go out and get a Wimpy or something.

b. Shortened form of *Wimpy bar*, sense 3 below.

1966 M. Waddell *Otley* xix. 171 We wound up in the Wimpy by the Broadway. **1968** *Listener* 13 June 763/1 Two years of success by any standard and then a break-up. Mecca had opened a dance hall... A Wimpy came and a bowling alley.

2. Usu. in form *Wimpey.* A Wellington bomber aeroplane. *slang.*

1942 *Tee Emm* (Air Ministry) II. 81 You have the mad sort of chauffeur who tries to roll a Wimpey. **1944** 'N. Shute' *Pastoral* i. 1 There was a Wimpey running up one engine, somewhere away out in the middle distance of the aerodrome. **1954** [see *map reference* s.v. *map *sb.*[1] 4].

3. *attrib.* and *Comb.*, as *Wimpyburger, culture, -eating*; **Wimpy Bar,** an establishment where Wimpy hamburgers are sold.

1959 *Observer* 8 Nov. 3/1 At least once a week a 'Wimpy Bar' is being opened somewhere in England. **1966** 'K. A. Saddler' *Gilt Edge* xiii. 180 That night in the Wimpy Bar..Len had summed him up at a glance. He had only agreed to take part in the job to get out of the Wimpy Bar. **1982** M. Gilbert *Final Throw* v. 35 'Let's find somewhere to eat.'.. They found a Wimpy Bar that was open. **1939** *Amer. Speech* XIV. 154/2 *Wimpyburger*, a specially large hamburger sandwich. **1971** Laver & Collins *Education of Tennis Player* xvi. 209 A tea garden..is located outside of Centre Court [at Wimbledon] where the famous strawberries and cream are sold.., along with a concession to the present day: Wimpy-burgers. **1971** 'J. Quartermain' *Man who walked on Diamonds* i. 7 A London throbbing with traffic and Wimpy culture. **1959** *Observer* 8 Nov. 3/1 Behind this wave of Wimpy-eating lies a simple commercial formula.

wimpy (wi·mpi), *a.* *slang* (orig. *U.S.*). [f. *wimp[2] + -y[1].] = *wimpish *a.*

1967 *Current Slang* Spring 5, [1964] *Wimpy*, stupid or sluggish. **1969** *Publ. Amer. Dial. Soc.* li. 16 *Wimpy*, spineless. **1977** D. Linzee *Discretion* (1981) i. 16 'It is exquisite.' 'I think it's kind of a wimpy little picture, personally.' **1977** *Sounds* 9 July 31/1, I vaguely anticipated something wimpy and limpy prattling about San Francisco and love and peace man. **1980** *High Fidelity* June 106/2 The Seventies witnessed macho rock & rollers sneering at singer/songwriters as 'wimpy'. **1984** *Nutshell* (Gainesville ed.) Spring 52/2, I *was* this little wimpy kid in elementary school and high school. **1984** *Melody Maker* 6 Oct. 15/4 Dennis had a brilliant artist last summer, really strong drawings, but now it's gone back to being a bit wimpy.

Hence **wi·mpiness** = *wimpishness.

1982 *Chicago Sun-Times* 1 Sept. 54 (*caption*) Wanna be a political activist? Are you held back by wimpiness?

Wimshurst (wi·mzhəɹst). *Physics.* [The name of James *Wimshurst* (1832–1903), English engineer.] *Wimshurst machine*: an electrostatic generator consisting of two or more counterrotating discs of insulating material mounted close together on a common axle and having a ring of metal sectors around their periphery; each disc has a stationary pair of brushes at diametrically opposite positions which are electrically connected, so that successive sectors become inductively charged as they come into contact with a brush and give up their charge later in the turn as they pass a fixed set of needle points connected to one or other electrode.

1886 R. Wormell *A. von Urbanitzky's Electricity in Service of Man* i. 63 The Wimshurst machine and that of Carré..are the least subject to these defects of any that we are acquainted with. **1978** *Sci. Amer.* Apr. 159/1, I placed my candle between the oppositely charged poles of a Wimshurst machine, the ancient hand-crank generator of high-voltage static electricity.

win, *v.*[1] Add: **3. a.** Also *transf.* in catch-phr. *to win the peace*, to bring about the successful reconstruction of a country defeated in or severely damaged by a war. Hence *win-the-peace* attrib.

1942 H. A. Wallace *Century of Common Man* (1944) 10 As part of the effort to win the peace, I am hoping that what might be called the 'ever normal granary principle' can be established for a number of commodities on a world-wide scale. **1945** *Daily Herald* 31 Aug. 2/1 The nation, girding itself for a supreme win-the-peace endeavour, will derive high encouragement from this enterprise by the mining community. **1950** A. Huxley *Themes & Variations* 243 That the Russians have been 'winning the peace' is due..to the fact that they profess and teach, as absolutely true, a clear-cut philosophy of man and nature. **1962** *Listener* 8 Mar. 402/2 They have also tried to agree that nobody was going to win the peace, but nobody was going to lose it either.

b. Phrs. *you can't win them all; you win some, you lose some*, etc.

1954 R. Chandler *Long Goodbye* xxiv. 122 Take it easy, Doc. You can't win them all. **1966** P. O'Connell *Sabre-Tooth* xiv. 189 You win a few, you lose a few, and it's no good getting sore. **1976** *Times* 23 Nov. 14/1 You look like being saddled with the uninspiring Willy... On the other hand, you seem to have got your way over Mrs. Thatcher's nominee... You win some, you lose some. **1979** K. M. Peyton *Marion's Angels* ix. 151 'It'll be all right,' she said. 'I daresay. You can't win them all.' **1984** *Listener* 1 Nov. 24/3 Academic friends..have found just one definite factual error... Ah, well; win some, lose some.

4. b. Phrs. *you can't win*, said (often in exasperation) to emphasize that whatever one does, it will be judged wrong or insufficient; *you win*, used to concede defeat in argument, etc.

1926 J. Black (*title*) You can't win. **1943** N. Marsh *Colour Scheme* vi. 99 All right... You win. I apologize. **1962** *Redbook* Mar. 44/2 She says I should always be dignified in front of him. Next she hands me the garbage pail and says, 'Take this out.' You can't win, no matter which way you turn! **1976** P. Lively *Stitch in Time* v. 55 'You can't win,' said Martin with sudden gloom, 'when you're the eldest. Whatever you do, you shouldn't have because you're old enough to know better.' **1982** 'S. Woods' *Enter a Gentlewoman* ii. iii. 113 'It's hardly fair to judge other people by one's own principles.' 'All right, you win.'

12. d. *to win out* (as opp. to *out of*): orig. *U.S.*; cf. *to lose out* s.v. *lose *v.*[1] 4 d; (further examples); *to win through* (adv. and prep.) (later examples).

1868 J. C. Atkinson *Gloss. Cleveland Dial.* 575 He's sair an' badly. But t'doctor thinks he'll *win thruff*. **1896** *Voice* 9 Apr. 4/5 McKinley will lead on the first ballot, but 'who will win out' is a different question. **1902** Kipling *Traffics & Discoveries* (1904) 7 But on delusions—as to their winning out next Tuesday week at 9 a.m.—they are—if I may say so—quite British. **1924** Galsworthy *Forest* iii. 78 You, Lockyer—a soldier! One spurt and we'll win out. Come! **1927** *Daily Express* 14 Dec. 13/5 It is good to know that Wodehouse's clever humour has won through. **1931** A. L. Rowse *Politics & Younger Generation* i. 262 If the League can manage to win through the divisions of the post-war world,..it will have established itself. **1947** 'G. Orwell' *Eng. People* 38 The American tendency is to burden every verb with a preposition that adds nothing to its meaning (*win out, lose out, face up to*, etc.). **1959** *Listener* 28 May 958/1 He made many enemies, but finally won through opposition to become one of the most controversial commanders of the last war. **1969** A. J. Meadows *High Firmament* vii. 160 Eventually, the uniformitarian concept won out in both astronomy and geology. **1974** 'M. Innes' *Appleby's Other Story* i. 7 Victorian bankers who won out when all the little local concerns began to be bundled up together. **1977** 'E. Crispin' *Glimpses of Moon* ii. 28, I won through, though... I survived. **1984** *Times Educ. Suppl.* 30 Nov. 28/1 The book has a brisk story and impeccable moral attitudes: gypsies, orphans, teachers and policemen are all good, ordinary people who win out in the end.

winability: see *winnability.

winceyette (winsi,e·t). [f. Wincey + -ette; cf. Flannelette.] A lightweight napped cotton fabric used for nightclothes, etc. (see quot. 1955).

1922 *Daily Mail* 14 Nov. 1 (Advt.), Ponting's offer of shirtings and winceyette. **1955** *Textile Terms & Definitions* in *Jrnl. Textile Inst.* (*Standardisation*) XLVI. 544 *Winceyette*, a light-weight fabric, originally and usually of cotton, raised on both sides, the weave usually being plain or twill. **1962** *Economist* 20 Jan. 253/2 The Nelson [cotton] factory was to begin production of denim, winceyette, cotton wool and surgical dressings in May. **1979** D. Cook *Winter Doves* ii. v. 87 She applied the paste with two fingers, wiping them clean on her Winceyette nightdress.

winch, *sb.*[1] Add: **6.** *winch-machine*; **winch-man:** also, a man lowered by a winch from a helicopter, esp. to rescue people from shipwrecks, etc.

1824 R. Stevenson *Bell Rock Lighthouse* vi. 329 A winch-machine, with wheel, pinion and barrel, round which last the chain was wound. **1946** A. J. Hall *Stand. Handbk. Textiles* iv. 169 The winch machine is essentially a vat..above which is mounted a horizontal winch. **1958** *Times* 23 June 6/1 The girls..were brought up into the helicopter by winchman Sergeant Jim Gilpin. **1977** *R.A.F. News* 27 Apr.–10 May 1/4 Along with winchman FS Roger Lynn he airlifted a two-ton cabin into the garden of a Durham man so that a kidney machine could be installed. **1979** *Globe & Mail* (Toronto) 15 Aug. 1/2 (*caption*) Winchman on a rescue helicopter hangs over a crew member from yacht Ariadne.

Winchester. Add: **I. 1. a.** (*c*) In mod. use (see quots. 1959, 1972). Also *Winchester bottle*.

1862 *Chemist & Druggist* 15 Feb. (Advt. sect.) 37/1

Druggists' Bottles... Winchester, 100 oz. **1959** *Gloss. Packaging Terms (B.S.I.)* 28 *Winchester*, a term applied to round, narrow or wide-mouth bottles usually used for the distribution of chemicals or pharmaceutical products. **1963** *Pharm. Jrnl.* CXCI. 59 The author suggests that the Winchester bottle was thus named by the druggists who utilised it for supplying the [Winchester] hospital's drug orders. **1972** *Bottlers' Year Bk. 1972–73* 423 *Winchester*, a large bottle of variable capacity used for soluble essences, etc., usually containing from about 6 to 10 lb. of the product.

b. *Winchester quart*: (*a*) a quart (2 pints) in Winchester measure; (*b*) *Pharm.*, 4 Imperial pints, i.e. 80 fluid ounces (in quot. 1870, 100 fl. oz.); also, a bottle holding 4 pints.

See *Pharm. Jrnl.* (1963) CXCI. 59 for an argument that in sense (*b*) it is properly 85 fl. oz., a quarter of the new barn gallon of 2½ Imperial gallons.

1742 W. Ellis *Mod. Husbandman* July x. 61 At our Country Towns, they sell a Winchester Quart of Milk.. for a Penny. **1758** *Rep. Comm. House of Commons Weights & Measures* 39 Standard weights and measures in the possession of the Hall-keeper of the Guild-Hall... 1 corn half peck marked 1601. 1 Winchester quart ditto. 1 ditto pint ditto. **1816** P. Kelly *Metrology* 89 The Coal Bushel holds one Winchester quart more than the Winchester bushel [*sc.* 2150.42 cubic inches]; it therefore contains 2217.62 cubic inches. **1870** *Pharm. Jrnl.* XI. 650 Omagh is said to take about 400 Winchester quarts (equal to 250 gallons) [of methylated ether] yearly. **1874** *Ibid.* 2nd Ser. V. 442/1 A Winchester quart (four pints) is first half filled with infusion. **1880** [in Dict., sense 1(*b*)]. **1897** *Chemist & Druggist* 5 June 891/1 The questions on which we should like information are—What is a Winchester quart the fourth of, or how it came to designate a half-gallon? and whether it and the Winchester pint were ever recognised measures? **1963** *Pharm. Jrnl.* CXCI. 60/1 The Winchester quart's success was due, one suspects, to the fact that it is the largest bottle which can conveniently be held in one hand.

2*. *Winchester school*, a southern English style of manuscript illumination of the 10th and 11th cent., originating at Winchester. Also *Winchester manner, style*.

1892 J. H. Middleton *Illuminated MSS. Classical & Mediaeval Times* vii. 101 Another very fine example of the Winchester school of illumination is the manuscript Charter which King Edgar granted to the new minster at Winchester in 966. *Ibid.*, In artistic power this tenth century Winchester school of illuminators appears, for a while at least, to have been foremost in the world. **1910** G. F. Warner in Warner & Wilson *Benedictional of Saint Æthelwold* p. xl, It is an example of Canterbury modification of Winchester style. **1928** E. G. Millar *Eng. Illuminated MSS XIVth & XVth Cent.* ii. 14 The Anglo-Saxon outline draughtsmen of the Winchester and related schools. **1954** M. Rickert *Painting in Britain: Middle Ages* ii. 42 But it is not until the second half of the tenth century that the full force of Carolingian art under Æthelwold's sponsorship of the production of manuscripts resulted in the development at Winchester of the famous Winchester style. **1970** *Oxf. Compan. Art* 559/1 *Winchester School*... Though some splendid manuscripts came from Winchester, books decorated in the 'Winchester' manner were certainly made in other southern English monasteries.

II. 3. b. *Computers.* Used *attrib.* and *absol.* with reference to a hermetically sealed storage device incorporating one or more high-capacity hard discs with heads and sometimes also a drive unit. [So called because the original device was intended to contain two 30 megabyte discs and its IBM number would have been 3030, the same as that of a famous Winchester rifle (which used a 0·30 calibre cartridge containing 0·30 grains of powder).]

1973 *Modern Data* July 60/1 The 'Winchester' Disk... The product of the so-called 'Winchester' project, the eventual nature of the 3340 has been the subject of rumors reported in the trade press. **1976** *Computer Weekly* 26 Aug. 16/6 There are also special cabinets for the Winchester type of disc module—a recording medium that is expensive in itself irrespective of the data stored on it, and that requires extremely careful handling. **1978** *IEEE Trans. Magnetics* XIV. 201/1 An example of the current state of the art in fixed head designs utilizing Winchester technology are the fixed heads used in IBM's 3340 and 3350 disc drives. **1980** *Sci. Amer.* Aug. 117/2 It is now known generically as Winchester technology, that being the code name under which the device was developed at IBM. A Winchester disk memory has one or more rigid disks, either eight or 14 inches in diameter. **1985** *Which Computer?* Apr. 61/2 One machine has twin floppies, the other has a 10MB Winchester.

winchite (wi·ntʃəit). *Min.* [f. the name of H. J. *Winch* + -ite¹.] A blue or violet monoclinic mineral of the amphibole group, approximately $NaCa(Mg,Fe^{2+})_4(Al,Fe^{3+})Si_8O_{22}(OH)_2$.

1906 L. L. Fermor in *Trans. Mining & Geol. Inst. India* I. 79 *Winchite*..is the name which has been bestowed ..upon the blue amphibole... An analysis of this mineral shows it to be closely allied to tremolite in chemical composition. **1980** *Canad. Mineralogist* XVIII. 101/1 In composition, this asbestos probably ranges from a potassian richterite to a potassian winchite.

Winco: see *Wingco.

wind, *sb.*¹ Add: **I. 2. b.** *Mah Jong.* Any of the four compass-positions about the wall of tiles taken up by a player; the player who occupies this place. Also, any of sixteen tiles (four of each sort) representing one of the four winds used in the game.

1922 M. S. Rosenblatt *Majong* 2 There are 4 'Winds'.. and there are 4 pieces of each 'Wind'. **1925** [see *pung *v.*², *sb.*², and *int.*]. **1938** V. L. Cecil *Maajh* 2 Each player took the position of one of the four Winds. **1960** R. C. Bell *Board & Table Games* vi. 152 The tiles are grouped into: Cardinal tiles... Winds... Honour tiles... Minor tiles. *Ibid.* 156 Each wind in turn becomes the wind of the round. The first round is East Wind's. **1979** M. Hammer *Learn to play Mah Jongg* ii. 35 The next step is to evaluate which tiles are more prevalent—odds, evens, winds, singles, pairs.

6. Also *to spit against* (or *into*) *the wind*.

1578 H. Wotton tr. *Yver's Courtlie Controversie* ii. 109 Thou shalte be like him that spitteth againste the winde, whose slaver fleeth in his owne face. **1612** Webster *White Divel* sig. E4, For your names, of Whoore and Murdresse they proceed from you, As if a man should spit against the wind, The filth returne's in's face. **1968** *Guardian* 1 Oct. 8/5 The decision to withdraw our forces.. was inevitable, and Mr Heath is spitting into the wind when he tells Australian audiences that a Conservative Government would go back. **1975** *Times* 10 Nov. 12/4 To adopt a vivid barrack-room expression, it is no good spitting against the wind or shouting against thunder.

II. 8. b. (*a*) (Later *fig.* examples.)

1823 J. Bric *Let.* 22 Feb. in *Corresp. D. O'Connell* (1972) II. 447 You have hit the thing between wind and water and whilst you have justly elevated your own name you have done much for your country. **1967** M. Gilbert *Dust & Heat* iii. 239 Mallinson *must* have guessed what was coming. Nevertheless, it hit him between wind and water.

10. b. *to put the wind up* (a person) (earlier example).

1918 W. Owen *Let.* 11 Oct. (1967) 584 Shells so close that they thoroughly put the wind up a Life Guardsman in the trench with me.

11. c. (Examples of *to slip one's wind*.)

1883 *Gringo & Greaser* 1 Sept. 2/2 He had entirely slipped his wind—for want of which he was buried the 11th ult. **1896** H. Lawson *While Billy Boils* 233 He laid the longest strip [of bark] by the side of the corpse... 'Come on, Brummy,..yer ain't as bad as yer might be, considerin' as it must be three good months since yer slipped yer wind. I spect it was the rum as preserved yer.'

d. *second wind* (earlier example); also *transf.* and *fig.*

1824 *Sporting Mag.* XIV. 166/2 Langan shewed a faint glimpse of second wind, and came up boldly. **1907** W. James *Mem. & Stud.* (1911) x. 229 Everybody knows what it is to 'warm up' to his job. The process of warming up gets particularly striking in the phenomenon known as 'second wind'. **1948** 'J. Tey' *Franchise Affair* i. 15 Perhaps it was the presence of an ally that had heartened her; or perhaps she had just got her second wind. **1963** Mrs. L. B. Johnson *White House Diary* 21 Dec. (1970) 18, I believe I am about to catch my second wind.

12. b. Also *pl.*, wind instruments.

1976 *Early Music* July 293/1 The author seems not to differentiate sufficiently between 'folk' and 'art' instruments of the Middle Ages, and especially when he deals with winds. **1978** P. Griffiths *Conc. Hist. Mod. Music* vii. 102 His [*sc.* Berg's] atonal chamber concerto for piano, violin and thirteen winds..is full of triple formations.

13*. The solar wind (see *solar *a.* 7), or a similar stream of particles emanating uniformly from any other star.

1966 *McGraw-Hill Encycl. Sci. & Technol.* III. 500/2 Presumably the wind is stronger when solar activity is high, but direct observations cannot be made before the next maximum. **1968** *Times* 5 Dec. 8/7 The fascinating region of space where the earth's magnetic field interacts with the 'wind' of atomic particles streaming out from the sun. **1982** *Sci. Amer.* July 83/1 Most stars, including the sun, are known to be losing mass in the form of a stellar wind.

III. 15. Also freq. in formula *wind(s) of*... **a.** (*b*) (later examples); *wind of doctrine* (later examples in allusion to *Eph.* iv. 14); also *fig.*

1907 W. Raleigh *Shakespeare* iv. 108 If once we are foolishly persuaded to go behind the authority of Heminge and Condell..we ..are afloat upon a wild and violent sea, subject to every wind of doctrine. **1913** G. Santayana *Winds of Doctrine* ii. 25 Prevalent winds of doctrine must needs penetrate at last into the cloister. **1926** R. H. Tawney *Relig. & Rise of Capitalism* iii. 179 With such a wind of doctrine in their sails men were not far from the days of complete freedom of contract. **1953** H. Weisinger *Tragedy & Paradox of Fortunate Fall* vi. 267 The winds of new doctrine swept through the streets of Athens and London and left the old and conventional modes of religious thought bare. **1953** E. Coxhead *Midlanders* vii. 158 The winds of want still blew about the world. **1962** *Listener* 26 Apr. 717/1 Ideas..become ossified if they are not exposed to the wind of criticism. **1968** *Globe & Mail* (Toronto) 3 Feb. 10/5 To protect their own lives and those of their children, they will bend with the winds of war.

b. (Further examples.)

1914 T. Dreiser *Titan* xiii. 103, I know all about this. I've seen which way the wind is blowing. **1929** 'E. Queen' *Roman Hat Mystery* xxii. 301 Ellery got his first indication of which way the wind blew during the meeting at the Ives-Pope house. **1957** N. Mitford *Voltaire in Love* x. 115 Thieriot..seeing..that the wind was now blowing in Voltaire's direction, consented..to give the required evidence. **1976** Ld. Home (*title*) The way the wind blows.

c. *spec.* in phr. *wind* (also *winds*) *of change*.

Harold Macmillan (Lord Stockton) delivered his celebrated 'wind of change' address to the South African parliament in Cape Town on 3 Feb. 1960 (see quot.). Our records show a marked increase in the frequency of the phrase after this date.

1905 S. Naidu *Golden Threshold* 97 The wind of change for ever blows Across the tumult of our way. **1927** D. H. Lawrence *Mornings in Mexico* 154 The place of after-life and before-life, where house the winds of change. **1932** J. Clapham *Econ. Hist. Mod. Britain* II. iii. 107 The [gas] companies or municipal works with their comfortable monopoly areas..began to find a little wind of change blowing among their retorts and coke heaps. **1954** J. Masters *Bhowani Junction* xxxix. 345 Then the great changes swept across India and the world, and she had searched, not by deliberate plan but because the wind of change blew through her too, for ways of escape. **1960** H. Macmillan in *Times* 4 Feb. 15/3 The wind of change is blowing through the continent. **1960** *Economist* 15 Oct. 275/2 This is but one way in which the mining complex of De Beers, Anglo American and Rhodesian Anglo American is adapting itself to the winds of change in Africa. **1965** D. Francis *Odds Against* vi. 86 'Is this your own show..or whose?' 'I suppose—mine.' 'Uh-huh... The wind of change, if I read it right?' **1971** *Nature* 26 Nov. 179/1 The universities are also likely to feel some eddies from the winds of change that are swirling around the White House. **1976** 'J. Charlton' *Remington Set* xiv. 69 The winds of change are beginning to blow..and your purpose in life isn't quite as defensible..as it used to be.

16. c. *to sniff the wind*: to try the atmosphere; to examine the prevailing state of affairs before taking action (cf. sense 4).

1972 'R. Crawford' *Whip Hand* i. v. 22 Schuyler sniffed the wind and took his time about it. **1974** 'D. Kyle' *Raft of Swords* viii. 78 'I have no *reason*... I just know.' 'You sniff the wind. Very sensible. What do you smell?' **1977** *Time* 22 Aug. 5/2 Certainly the Labor government and the nation's judiciary system are sniffing the wind.

IV. 20. in the wind. a. (Earlier example.)

1818 'A. Burton' *Johnny Newcome* iii. 175, I did not think..I was so much in drink! But now by th'holy smut I find That cursedly I'm in the wind.

29. with the wind. Now esp. in *fig.* phr. *gone with the wind*: gone completely (as if blown away by the wind), disappeared without trace.

1896 E. Dowson *Verses* 17, I have forgot much, Cynara! gone with the wind. **1918** Galsworthy *First & Last* ix, in *Five Tales* 61 A man, when he drowns, remembers his past. Like the lost poet he had 'gone with the wind'. Now it was for him to be true in his fashion. **1936** M. Mitchell (*title*) Gone with the wind. **1948** W. S. Churchill *Gathering Storm* xix. 271 The services of thirty-five Czech divisions..[were] cast away,..all gone with the wind.

V. 30. a. *wind-dispersal*, *effect*, *-flaw* (Flaw *sb.*²), *-force*, *-puff* (later example), *resistance*, *-rush*, *-shift*, *-song*, *-speed*, *-streak*, *-torrent*, *-walk*, *-wave*, *-well*; *wind-shelter*.

1911 J. A. Thomson *Biology of Seasons* iii. 277 Any structural peculiarity that increases area without increasing weight will aid in wind-dispersal. **1937** Wind effect [see *air position* s.v. *air *sb.*¹ III. 1]. **1941** B. Hellström in *Ingeniörsvetenskapsakad. Handl.* No. 158. 8 A denivellation of the water surface takes place, by which the level of the lake is lowered at the windward and raised at the leeward shore. This denivellation is called the Wind Effect. **1913** J. Masefield *Daffodil Fields* 110 Flicking windflaws fill the air with brine. **1931** E. Linklater *Juan in Amer.* i. 15 A frown on that bland forehead was like the wind-flaw on a saucer of milk that some petulant child has blown across. **1935** *Geogr. Jrnl.* LXXXVI. 533 The most remarkable feature was the great variation in wind-force and direction. **1976** *Islander* (Victoria, B.C.) 14 Nov. 7/2 The seas began to look greyer—but we hadn't had anything more than windforce seven—so far. **1881** G. M. Hopkins *Poems* (1967) 89 A windpuff-bonnet of fawn-froth Turns and twindles. **1934** *Discovery* Dec. 344/2 At a high speed, wind resistance becomes an important factor. *a* **1945** E. R. Eddison *Mezentian Gate* (1958) xxxix. 218 Their pure eyes..turned..to that thunder-laced windrush of darkness which is the heat and unpicturable secret centre of light's and beauty's self. **1976** 'A. Hall' *Kobra Manifesto* xv. 201 The faint scream of the windrush [under an aeroplane at take-off] in the roaring background. **1930** E. Pound *XXX Cantos* viii. 30 With the road leading under the cliff, in the wind-shelter into Tuscany. **1968** G. Maxwell *Raven seek thy Brother* ix. 127 Windshelters..of stone or turf and furnished with artificial nesting sites, are usually colonized immediately [by eider ducks]. **1914** J. Masefield *Philip the King* 53 A sudden windshift snatched us from our graves And drove us north. **1963** *Times* 30 May 14/7 A windshift..brought the nauseating smell of the penguin rookery straight over the camp. **1946** J. W. Day *Harvest Adventure* vi. 83 Rigging drummed and whistled a raw wind-song. **1934** *Discovery* June 150/2 High wind-speeds in relation to aircraft. **1977** J. L. Harper *Population Biol. Plants* x. 323 Turbulence falls off rapidly down through a canopy but is a function of wind speed, even deep in a corn crop. **1930** E. Pound *XXX Cantos* xxvii. 127 Twig where but wind-streak had been. **1973** C. Sagan *Cosmic Connection* (1975) viii. 62 The Mariner 9 photography of the Martian volcanoes, windstreaks, moons, and polar icecaps. **1929** Blunden *Near & Far* 57 Dim stars like snowflakes are fluttering in heaven, Down the cloud-mountains by wind-torrents riven. **1877** G. M. Hopkins *Poems* (1967) 70 Summer ends now; now, barbarous in beauty, the stooks rise Around; up above, what wind-walks! **1900** G. K. Chesterton *Wild Knight* 7 Meadows where the wind-waves pass. **1946** L. D. Stamp *Britain's Struct. & Scenery* vi. 51 The waves of the sea are primarily wind-waves. **1984** A. C. & A. Duxbury *Introd. World's Oceans* viii. 249 Most waves observed at sea are progressive wind waves..generated by the wind. **1936** Dylan Thomas *25 Poems* 23 Why east wind chills and south wind cools Shall not be known till windwell dries.

b. *wind-cheating*, *-screening* adjs.

1963 Bird & Hutton-Stott *Veteran Motor Car* 246 Both had their engines placed..very low down so as to allow the use of flat wind-cheating bodies. **1977** *Lancashire Life* Jan. 81/1 Because of their wind cheating shape and fairly high overall gearing, the Citroen CXs are very economical on long motorway journeys. **1923** Kipling *Land & Sea Tales* 214 She hovers On the summits of wind-screening seas.

c. (Further examples, mainly poetical compounds.) *wind-aided*, *-beat*, *-beaten* (later examples), *-bit*, *-bitten*, *-borne*, *-broken*, *-buffeted*, *-chilled*, *-curled*, *-dappled*, *-driven*, *-flawed*, *-flown*, *-flushed*, *fluted*, *-formed*, *-hardened*, *-heeled*, *-laced*, *-laden*, *-laid*, *-lifted*, *-loved*, *-mastered*, *-milled*, *-perplexed*, *-pollinated*, *-powered*, *-rinsed*, *-ripped*, *-scarred*, *-scoured*, *-scourged*, *-shorn*, *-snatched*, *-sown*, *-spun*, *-stirred*, *-stormed*, *-sucked*, *-thrashed*, *-torn*, *-tossed* (earlier example), *-transported*, *-turned*, *-washed*, *-waved* (later example), *-wrinkled*, *-writhen*.

1959 *Times* 12 Mar. 3/3 Langton kicked another long wind-aided penalty goal. **1978** *Detroit Free Press* 16 Apr. E 3/2 He won the 100-yard dash with a wind-aided performance of 9.5 seconds. **1877** G. M. Hopkins *Poems* (1967) 66 Wind-beat whitebeam! airy abeles set on a flare! **1900** W. B. Yeats *Shadowy Waters* 45 These waste waters and wind-beaten sails. **1973** *Canadian Antiques Collector* Jan.-Feb. 59/1 Inland, behind wind-beaten villages and red capes. **1892** Kipling *Other Verses* 161 In the heel of the wind-bit pier. **1919** J. Masefield *Reynard the Fox* ii. 61 Blown Hilcote Copse, Wind-bitten beech. **1965** F. Sargeson *Memoirs of Peon* ix. 270 The trees..had redeemed a windbitten waste from its native barbarism. **1842** Emerson *Saadi* in *Poems* (1914) 133 To northern lakes fly wind-borne ducks. **1969** Bennison & Wright *Geol. Hist. Brit. Isles* xvi. 368 The brickearths may, however, have been not solely laid down in expanses of water but be in part wind-borne. **1914** J. Masefield *Philip the King* 44 They have died, Far from wind-broken Biscay, far from home. **1901** 'L. Malet' *Hist. R. Calmady* v. i. 383 Heavily-cloaked figures tacking, wind-buffeted, across the grey-black street. **1921** D. H. Lawrence *Tortoises* 25 The autumn, wind-chilled sunshine. **1952** L. MacNeice *Ten Burnt Offerings* 51 Wind-curled fountain, tigerish weir, garrulous rain. **1883** R. Bridges *Prometheus the Firegiver* in *Poet. Works* (1912) 25 Piloting over the wind-dappled blue Of the summer-soothed Aegean. **1920** J. Masefield *Enslaved* 109 The grey sea..cloud-coloured, flat, Wind-dappled from the glen. **1882** W. D. Hay *Brighter Britain!* I. iii. 89 The sun shining on the wind-driven sand that covers them [*sc.* hilltops]. **1967** *Oceanogr. & Marine Biol.* V. 102 The Strait of Dover may accept a wind-driven residual current averaging 3½ miles and occasionally reaching 20 miles per lunar day. **1971** G. M. Brown *Fishermen with Ploughs* 95 A huge wind-flawed mirror. **1938** C. Day Lewis *Overtures to Death* 30 The wind-flown tower. *Ibid.* 55 To reproach you we rise Wind-flushed and early. **1943** —— *Word over All* 15 Wherein the shores Foam-fringed, wind-fluted of the strange earth dwell. **1911** F. O. Bower *Plant-Life* 124 The wind-formed dune takes a very definite crescentic shape styled a Barchan. **1926** D. H. Lawrence *Sun* iv. 17 He was powerless against her rosy, wind-hardened nakedness. **1939** Dylan Thomas *Map of Love* 20 Wind-heeled foot in the hole of a fireball. **1887** G. M. Hopkins *Poems* (1967) 104 Curls Wag or crossbridle, in a wind lifted, windlaced—See his wind lilylocks -laced. **1928** C. Day Lewis *Country Comets* 9 The unconscious dignity Of hills and wind-laden grass. **1965** G. J. Williams *Econ. Geol. N.Z.* ix. 132/2 Both water- and wind-laid blacksand sediments formed. **1924** 'L. Malet' *Dogs of Want* iv. 112 The soft green blur and flickering flames resolved themselves into gently wind-lifted leaves and distant sparkling water. **1936** Auden *Look, Stranger!* 11 Upon wind-loved Rowley. **1945** P. A. Larkin *North Ship* 27 Two tall ships, wind-mastered, wet with light. **1947** Dylan Thomas *In Country Sleep* in *Horizon* Dec. 303 The dew falls on the wind-Milled dust of the apple tree. **1864** G. M. Hopkins *Poems* (1967) 128 His body sway'd upon tiptoes Like a wind-perplexed rose. **1911** F. O. Bower *Plant-Life* 96 As for instance in the Rue (Thalictrum), which has become wind-pollinated. **1968** F. W. Gould *Grass Systematics* i. 7 Grasses..are wind-pollinated. **1976** *Jrnl. R. Soc. Arts* CXXIV. 732/1 It is very logical to feed wind-powered energy in the form of either electricity or direct heat directly into a buffer system and thence to direct use. **1948** L. MacNeice *Holes in Sky* 20 Wind-rinsed plumage of oat-field. **1960** S. Plath *Colossus* (1967) 33 The spindrift Ravelled wind-ripped from the crest of the wave. **1939** S. Spender *Still Centre* 41 Beyond the wind-scarred hill. **1896** Kipling *Seven Seas* 73 Bone-bleached my decks, wind-scoured to the graining. **1980** D. K. Cameron *Willie Gavin* vi. 54 There was hardly a year when the winter ploughs did not turn up an old hunter of that wind-scoured plain. **1898** J. G. Whittier *M. Martin* in *Poet. Works* 67/2 You wind-scourged sand-dunes, cold and bleak. **1924** 'L. Malet' *Dogs of Want* ii. 29 Bare, wind-scourged, rock-strewn slopes. **1867** J. G. Whittier *Poet. Works* (1898) 280/2 Lonely and wind-shorn, wood-forsaken..Lieth the island of Manisees. **1933** W. de la Mare *Lord Fish* 61 Gnarled, wind-shorn trees. **1980** R. Mabey *Common Ground* ii. i. 70 At no more than 590 feet ..above sea level some of its windshorn oaks are reduced to a metre or so in height. **1925** C. Day Lewis *Beechen Vigil* 32 The wind-snatched rumour. **1902** W. Stevens *Jrnl.* 18 Aug. in *Lett.* (1967) 59, I lay under a group of dark cedars near that strange wind-sown cactus with its red blossom. **1922** Blunden *Shepherd* (ed. 2) 74 Windspun leaves burn silver-grey. **1843** J. G. Whittier *Poet. Works* (1898) 388/1 And down again through wind-stirred trees He saw the quivering sunlight play. **1946** R. Macaulay in E. Brontë *Wuthering Heights* p. vi, The lonely, wind-stormed old farmhouse that stood on the heights above Haworth's grey streets. **1946** R. S. Thomas *Stones of Field* 26 The wind-sucked bone shows blue. **1933** Somerville & 'Ross' *Smile & Tear* ix. 98 A few miserable wind-thrashed ash-trees. **1910** Kipling *Rewards & Fairies* 244 The wind-torn breaker-tops. **1957** T. Gunn *Sense of Movement* 58 Not like the fighting boys and wind-torn rooks. **1838** J. R. Lowell *Class Poem* 20 Flapping his raven pinions in the west, The thunder brooding o'er his wind-tost crest. **1946** F. E. Zeuner *Dating Past* iii. 56 Minute grains of wind-transported pollen caught on the wet surface of the bog. **1935** Dylan Thomas in *Life & Lett. To-day* Dec. 75 Doom on deniers at the wind-turned statement. **1971** *Country Life* 8 July 84/1 The raw elements of Millet's compositions, granite walls, dirty-legged cattle,..wind-turned trees. **1912** C. Mackenzie *Carnival* xvi. 186 At such an hour..even Piccadilly Circus stands..wind-washed and noble. **1919** J. Masefield *Reynard the Fox* 92 The wind-washed steeple stood serene. **1928** Blunden *Retreat* 18 The wind-waved bough betrayed the wild sylph glancing. **1925** V. Woolf *Mrs Dalloway* 242 Suddenly she shoots to the surface and sports on the wind-wrinkled waves. **1921** F. B. Young *Black Diamond* ix. 116 They crossed a zone of huge, wind-writhen hawthorns. **1954** J. R. R. Tolkien *Fellowship of Ring* 401 High ridges crowned with wind-writhen firs.

d. *wind-grey*, *-hard*, *-long*, *-raw*, *-smooth*, *-wild*.

c **1944** A. Power *From Old Waterford House* xi. 95, I had seen it under so many moods, from wind-grey to sun-yellow. **1954** W. Faulkner *Fable* 184 Like the wind-hard banner of the old Norman earl. *a* **1890** G. M. Hopkins *Poems* (1967) 180 Or wind-long fleeces on the flock A day off shearing day. **1922** Joyce *Ulysses* 48 About her windraw face her hair trailed. **1929** E. Sitwell *Gold Coast Customs* 38 Wind-smooth fruits. **1936** C. Day Lewis *Noah & Waters* 50 Under the wind-wild sky.

31. wind axis *Aeronaut.*, each of a set of rectangular coordinate axes having their origin in the aircraft and the *x*-axis in the opposite direction to the relative wind; usu. *pl.*; **wind-balanced** *a.*, applied to rotary gun mountings on aircraft having a device which automatically compensates for the turning moment caused by air pressure on the guns; also **wind-balancing** *vbl. sb.*; **wind-bells** *sb. pl.*, slips of glass or porcelain suspended from a frame so as to tinkle against one another in the wind; **wind-blow,** (*a*) a stretch of land eroded by wind; (*b*) (see quot. 1955); (*c*) = *windthrow* below; **windblown bob** [*bob *sb.*¹ 5 b], a bobbed hairstyle popular among women in the 1930s (see quot. 1975); **wind-bracing,** substitute for def.: connecting members designed to stiffen a building or other structure against the wind; the provision of such members; (further examples); **windbreaker,** (*b*) *U.S.* = Wind-break 1; (*c*) *U.S.* (with capital initial) the proprietary name of a kind of shirt or leather blouse; *gen.* (chiefly *N. Amer.*) = *windcheater* (*b*) below; **windburn** [after *sunburn*, etc.], (usu. superficial) inflammation or discoloration of the skin caused by exposure to wind; hence **wind-burned, -burnt** *a.*; **windcap** *Mus.* [tr. G. *windkapsel*] = *reed cap* s.v. *reed *sb.*¹ 13 a; freq. *attrib.*; **wind-channel** = *wind tunnel* below; **windcharger,** a small windmill which generates electricity for a farm, dwelling, etc.; **windcheater,** (*a*) *Golf*, a ball driven low into the wind, *spec.* one played with strong back-spin (see quot. 1909); (*b*) a kind of wind-resistant jacket or blouson; **wind chill,** the cooling effect of moving air on a body; also, = **wind-chill factor*; **wind-chill factor, index,** a measure or scale of the combined effect of low temperature and wind-speed on body temperature (see quot. 1939); **wind chimes** *sb. pl.* = *wind-bells* above; **wind cone** *Aeronaut.* = *wind sock* below; **windcrust** *Mountaineering*, a crust formed on the surface of soft snow by the wind (see quot. 1936); **wind energy,** energy obtained from harnessing the wind; cf. *solar *a.* 4 a; **wind farm,** a group of energy-producing windmills or wind turbines; **wind-firm** *a.*, of a tree: firmly rooted so as to be able to withstand strong winds; hence **wind-firmness**; **wind-flag,** a flag on a shooting-range designed to indicate the direction and force of the wind; **wind-gap:** *spec.* (see quot. 1939); (examples); **wind-god** (later example); **wind-jacket** = *windcheater* (*b*) above; **windjammer:** substitute for def.: *slang*, (*a*) *U.S.* a bugler, bandsman; (*b*) a sailing-vessel (*obs.* exc. *Hist.*); (*c*) *U.S.* a rumour-monger, a loquacious person; (*d*) = *windcheater* (*b*) above; (further examples); hence **windjamming,** (*a*) sailing a windjammer; (*b*) talking, gossiping; (*c*) playing a wind instrument; **wind-lane,** a current on the surface of a body of water, caused by the wind; **wind load** *Engin.*, the force on a structure arising from the impact of wind on it; also **wind loading**; **wind-lop** *Canad.* [Lop *sb.*⁶], a choppy surface on the sea, caused by wind; **wind machine,** (*a*) *spec.* one that blows out relatively warm air for protecting crops against frost (see quot. 1976²); (*b*) in theatrical and other productions, a machine for simulating the sound or other effects of wind; also *fig.*; (further examples); **wind noise,** the sound of the wind against a motor vehicle moving at speed, as heard within the vehicle; **windproof** *a.*, impervious or resistant to wind; used esp. of outer garments; hence *ellipt.* as *sb.*, a windproof garment; **wind-reef** *U.S.*, the semblance of a reef on the surface of a river, caused by the wind; **windrock,** damage to the roots of young plants, caused by the movement of the stem in the wind; also as *v. trans.*; so **wind-rocking**; **wind-scorpion** = Solpugid in Dict. and Suppl.; **wind-screen** (earlier U.S. examples in sense 'windbreak'); **windscreen washer** = *screen-washer* s.v. *screen *sb.*¹ 7 a; **windscreen wiper,** a device (usu. one of a pair) on a motor vehicle for automatic wiping of the outside of the windscreen during rain, snow, etc., usu. consisting of a mechanically or electrically operated moving rubber blade; also one on an aircraft; **wind shadow,** (*a*) *nonce-use*, a ripple caused by the wind on water and having the appearance of a shadow running over it; (*b*) an area behind a moving object where the air is disturbed and its pressure reduced; **wind shear,** a variation in wind velocity along a direction (usu. vertical or horizontal) at right angles to the wind's direction; **windship,** a wind-powered ship; a sailing-ship; **wind-sight,** a special arrangement of the back-sight of a rifle capable of adjustment to compensate for the effect of wind on the bullet; **wind-slab** *Mountaineering*, a thick wind-crust, of a kind liable to slip and create an avalanche; cf. *slab avalanche* s.v. *slab *sb.*¹ 6; **wind-slash,** slash resulting from windthrow; **wind sleeve** *Aeronaut.* = *wind sock* below; **wind sock,** a cloth cone flown from a mast, esp. on an airfield, to indicate the direction of the wind; = *drogue 3 (*c*); **wind-spider** = *wind-scorpion* above; **wind-splitter** *colloq.* (chiefly *U.S.*), something so sharply drawn or so swift as to suggest the notion of splitting the wind; cf. Wind-cutter; so **wind-splitting** *a.*; **wind sprint** *Athletics* (see quot. 1948); **wind-stocking** = *wind sock* above; **wind-stream,** an air-stream, esp. the disturbed air in the wake of an aircraft; **wind stress,** stress or force due to wind; **wind-swept** *a.*, (*a*) (see sense 30 c in Dict.); (*b*) *spec.* of a hair-style, designed to give the appearance of having been blown by the wind (cf. *windblown bob* above); **windthrow,** the uprooting and blowing down of trees by the wind; also (usu. *attrib.*) of timber so uprooted; **wind tunnel,** a tunnel-like apparatus for producing an air-stream of known velocity past models of aircraft, buildings, etc., in order to investigate flow or the effect of wind on the full-size object; also *attrib.*, *transf.*, and *fig.*; **wind turbine,** a turbine driven by wind; an apparatus designed to generate electricity when a large vaned wheel is rotated by the wind; **windway,** (*b*) also in a woodwind instrument; **wind wing** *U.S.*, † an adjustable glass ventilation panel attached to the side of the windscreen of a motor vehicle (*obs.*); a small ventilation window or quarterlight on a motor vehicle.

1932 *Jrnl. R. Aeronaut. Soc.* Mar. 194 Calculations..of a complete model rotated about the wind axis..give a fair approximation to the spinning characteristics of the aeroplane. **1984** F. J. Hale *Introd. Aircraft Performance* i. 4 The wind axes are not body axes; that is, they are not fixed to the aircraft other than at the cg. A change in the direction of flight can change *x* without changing the attitude of the aircraft. **1928** *Daily Tel.* 6 Mar. 6/3 Royalties not exceeding £7,500 to Messrs. Vickers, Ltd., for wind-balanced ring mountings. **1928** G. F. S. Gamble *Story N. Sea Air Station* xiii. 219 A wind-balancing gear was provided which relieved the observer of much fatigue at high altitudes. **1901** 'L. Malet' *Hist. R. Calmady* ii. ii. 105 They pressed him back and back against the base of a seven-storied pagoda, the wind-bells of which jangled far above him from the angles of its tiers of fluted roofs. **1983** *Daily Tel.* 21 Oct. 16/1 Windbells tinkled from the eaves of temples, spreading the Holy Word of Buddha, keeping demons away. **1921** H. Guthrie-Smith *Tutira* xx. 180 The sheep..are returfing the naked windblows. **1944** W. Stevens in *Q. Rev. Lit.* Spring 157 The drivers in the wind-blows cracking whips. **1955** *Britannica Bk. of Year* 489/2 *Wind-blow*, a destructive gale of wind. **1961** *New Scientist* 16 Mar. 662/2 Comparisons have been made of trees on sites where wind-blow has occurred and those where similar trees are stable. **1979** *National Trust* Spring 18/3 Marram grass.. holds the sand together and reduces the effects of wind blow. **1933** N. Waln *House of Exile* iii. i. 187 She had her hair cut in a new fashion which, she told me, was called a windblown bob. **1975** *Fairchild's Dict. Fashion* 262/2 *Wind-blown bob*, popular 1930's woman's hairstyle, cut short and shingled,..so that hair fell softly about the face as if blown by the wind. **1911** Husband & Harby *Structural Engin.* ix. 278 Wind bracing in roofs is employed to counteract the overturning moment of the wind acting on the ends. **1961** *Listener* 28 Sept. 464/1 An enormous funnel of unoccupied space goes from top to bottom of it [*sc.* a skyscraper] in order to provide mere wind bracing for the rest of it. **1974** *Sci. Amer.* Feb. 98/2 The statue [of Liberty] posed a special problem in wind bracing. **1873** J. H. Beadle *Undevel. West* xxxiv. 730 If there is any wind-breaker northwest, between there and Alaska, I had no evidence of it. **1918** *Official Gaz.* (U.S. Patent Office) 5 Nov. 214/2 The Hilker-Wiechers Manufacturing Co., Racine, Wis. *Windbreaker*... Men's shirts for outer wear. **1925** *Ibid.* 13 Jan. 256/2 Gufterman Bros., Inc., St. Paul, Minn... *Windbreaker*... Leather blouses, shirts, [etc.]. **1934** *Beaver* (Winnipeg) June 6/2 The wind-breakers and coats are shown by a series of photographs. **1964** 'R. Macdonald' in H. Q. Masur *Murder most Foul* (1973) 109 A man with a bulky shoulder harness under his brown suede windbreaker. **1985** *Times* 9 Feb. 36/4 The terminal was full of muscular young men in windbreakers and running shoes. **1939** C. Morley *Kitty Foyle* xxx. 313 It was comical to see the dames.. worrying about windburn and sunsquint and brittle nails. **1977** *Birds* Spring 40 All night my skin is hot with windburn, and between my teeth..the salt-sharp flavour of the rain. **1942** J. Steinbeck *Moon is Down* vi. 137 They were windburned and strong..Will Anders and Tom Anders, the fishermen. **1954** 'Bryher' *Fourteenth of October* ii. 17 His eyes were the same blue in his windburnt face. **1981** 'E. Lathen' *Going for Gold* iv. 40 Practicing for the Swiss women's slalom team had left her with windburned cheeks. **1940** C. Sachs *Hist. Musical Instr.* (1942) xv. 320 Wind-cap instruments were first introduced to art music in the fifteenth century. The cromorne was the oldest European instrument with a wind cap. **1970** W. Apel *Harvard Dict. Music* 588/2 More important are the *crumhorns*... Their tube was nearly cylindrical,..and a pierced cap (wind cap) covered the reed so that the player could not touch it. **1980** *Early Music Gaz.* Apr. 13/3 There will be a weekend for players of the recorder, gemshorn, windcap instruments, cornetti and for renaissance dancers in Hutton Hall, near Carlisle. **1918** Cowley & Levy *Aeronautics* iv. 98 A series of experiments are conducted in the wind channel to test the lift and drag for different forms of sections. **1972** *Nature* 18 Aug. 375/1 The secrecy..was lifted in 1919 to reveal .. developments in techniques for scale model testing in wind 'channels' (tunnels). **1946** E. W. Manning *Igloo for Night* 156 We could hear the wind tearing past, and the high screaming whine as it met the wires of the radio masts and the wind-charger. **1949** *Farmer's Weekly* (S. Afr.) 13 July 69/5, I have an old car generator. Can this be converted into a windcharger? **1976** *Sci. Amer.* June 94/3 The introduction of the windcharger in the 1930's brought to remote farms and ranches enough electricity to power radios and a few light bulbs and appliances. **1909** P. A. Vaile *Mod. Golf* xii. 180 The wind-cheater, the ball that skims away over the daisies and then rises gracefully at the end of its flight, to fall sometimes almost dead. **1940** [see *crew neck* s.v. *crew *sb.*¹ 7]. **1956** L. McIntosh *Oxford Folly* 53 Incongruous in his neat suit and tidy hair among the tousled undergraduates in windcheaters or polo-necked sweaters. **1977** G. Peper *Scrambling Golf* ix. 162 One of those low, delayed-rising 'wind-cheaters'. **1982** C. Thomas *Jade Tiger* 195 The first chill of the night, seeping through his thin windcheater, alerted and refreshed him. **1939** P. A. Siple *Adaptations of Explorer to Climate of Antarctica* (Ph.D. diss., Clark University) 166, I therefore propose in this discussion to multiply temperatures in degrees Centigrade below freezing by wind velocity in meters per second, the product of which I shall call the *wind-chill index*. *Ibid.* 177 July exhibited a mean wind-chill of 462·8. **1949** *Jrnl. R. Aeronaut. Soc.* LIII. 1/2 'Wind Chill'... This is the worst form of cold weather to encounter, as not only does it cause exhaustion, low morale, pain and frustration, but in some cases the lack of the will to live when the wind chill factor is unusually high. **1959** R. E. Huschke *Gloss. Meteorol.* 629 *Wind-chill index*—(Also called *wind-chill factor*), the cooling effect of any combination of temperature and wind, expressed as the loss of body heat in kilogram calories per hour per square meter of skin surface. **1963** *New Scientist* 7 Feb. 276/1 Wind chill, which is actually another name for the dry convective cooling power of the atmosphere, is a term descriptive of the cooling effect of air movement and low temperature. **1977** J. F. Fixx *Compl. Bk. Running* xiii. 151 Because of the wind-chill factor, a given temperature feels colder than in still weather. **1985** *Times* 8 Jan. 26/4 Francis Wilson, the BBC weatherman, yesterday introduced *Breakfast Time* viewers to a new and chilly forecasting feature: the 'wind chill factor'. **1958** T. Williams *Orpheus Descending* iii. iii. 85 Someone has entered the confectionery door, out of sight, and the draught of air has set the wind-chimes tinkling wildly. **1976** M. Millar *Ask for Me Tomorrow* (1977) iii. 18 Go..to the glass door and shake the wind chimes good and hard. She's in Marco's room. **1918** *Flight* 2 May 496/1 A wind cone set up in..fields .. near aerodromes, would enable cross-country fliers to know..where to land and in what direction. **1936** E. A. M. Wedderburn *Alpine Climbing* iv. 50 Wind both causes the snow to drift and forms a crust on the powder snow; this is the chief kind of crust found in winter. It is most important to distinguish between sun crust and wind crust. **1955** E. Hillary *High Adventure* 69 The surface here was most unpleasant—a thick wind-crust over deep unstable snow. **1976** *Jrnl. R. Soc. Arts* CXXIV. 731/2 At present the main bar to developing and using wind energy in this country is very high capital costs of equipment. **1980** *Sunday Times* 24 Aug. 4/4 The plan is to set up one (windmill) of medium size as soon as possible to gain experience, and then to establish a 'wind farm', of about ten windmills, each capable of generating a megawatt of electricity. **1982** *Energy Spectrum* (Shell Internat. Petroleum Co. Ltd.) 8/1 Larger systems of more than 1 MW are also envisaged either singly or in 'wind farms' for integration into utility grids. **1895** W. R. Fisher *Schlich's Man. Forestry* IV. iv. iii. 469 The westerly border-trees..have now become so wind-firm that the severance-felling might be widened. **1927** *Forestry* I. 21 To keep plantations wind-firm..initial spacings of the order of six feet..are necessary. **1981** *Southern Horticulture* (N.Z.) Spring 31/2 Containerisation of such material without this evening-out treatment produces trees that are neither wind-firm in the container, nor on the planting site. **1962** *Times* 1 Jan. 6/4 It [*sc.* a tree] was there to increase the wind-firmness of the woodland. **1923** Kipling *Land & Sea Tales* 181 He pointed towards the stiff-tailed wind-flags that stuck out at all sorts of angles as the eddy round the shoulder of the Down caught them. **1889** Wind-gap [see Gap *sb.*¹ 5 b]. **1895** *Geogr. Jrnl.* V. 144 If the land should be raised a few hundred feet, these head-waters would soon be gained by the Trent; and the divide between the successful and defeated systems would be pushed to the notch in the hard Oölite, which would then be a 'wind-gap', instead of a 'water-gap', as the Pennsylvanians say. **1939** *Bull. Geol. Soc. Amer.* L. 1343 The term 'wind gap' is now more commonly restricted to abandoned water gaps, while those gaps not believed to have been former water gaps are designated as 'cols'. **1977** Wind gap [see *river capture* s.v. *river *sb.*¹ 4 d]. **1930** Blunden *Summer's Fancy* 31 They stole away, and heard the windgod trill Winging the corn that to the bright west rolled. **1940** F. Smythe *Adventures of Mountaineer* xi. 201 It was all we could do.. to pull off our wind jackets. **1955** G. Band *Road to Rakaposhi* vii. 87 David and I, who were wearing bright red windjackets,..slipped carefully past. **1880** *United Service* Oct. 458 [The adjutant] watched the roll-call of his 'wind-jammers'. **1892** *Rudder* Sept. 217/1 The deck-hands on the liners contemptuously refer to [sailing vessels] as 'wind-jammers'. **1917** S. Lewis *Job* 209 We do our work and don't howl about like all these socialists and radicals and other wind-jammers. **1930** *Even. Standard* 20 Aug. 2 (Advt.), Thousands of golfers wear the Barker 'Windjammer' and report it to be a splendid garment. **1931** *Amer. Mercury* XXIV. 354/2 [Circus Words.] *Windjammer*, a band musician. **1932** Auden in *Rev. Eng. Stud.* (1978) Aug. 282 My hand was wrung By one bareheaded in a windjammer jacket. **1942** M. Hargrove *See here, Private Hargrove* xlii. 119 *Windjammer*, the bugler. **1976** *Milton Keynes Express* 25 June 4/4 The very popular zip-fronted cotton velour wind-jammer..is also great weekend gear with jeans. **1886** D. Kemp *Man. Yacht & Boat Sailing* (ed. 5) 658/1 Wind jamming. A new-fashioned slang term for sailing by the wind. **1893** *Columbus* (Ohio) *Dispatch* Oct. 5 Could this power of wind-jamming have been saved there would have been some good accruing from the extra session. **1894** *Nautical Mag.* Feb. 102 People would begin to understand the meaning of seamanship as apart from the so-called and much-despised 'wind-jamming'. **1919** S. Lewis *Free Air* 182 You're the worst wind-jamming liar I ever met. **1946** *Seafarers' Log* 18 Jan. 4/1 He really fooled the entire crew, and the Chief Mate was so impressed with his windjamming that he wanted to make him Bosun. **1943** T. Dudley-Gordon *Coastal Command at War* 22 He knows its direction by the 'wind-lanes' on the sea. **1979** *Fisherman's Weekly* 21 June 6/1 Thousands of tiny shucks from the freshly opened beech leaves, blown onto the surface, had collected in floating rafts, and were marking the wind lanes. **1911** Husband & Harby *Structural Engin* ii. 36 (*heading*) Wind load. **1961** *B.S.I. News* Dec. 16/2 Stability requirements for cranes (including consideration of wind loads). **1970** *New Scientist* 17 Sept. 584/2 The BRS project will improve wind-tunnel techniques, as well as increasing knowledge of windloads —which means better, safer, and perhaps more economic building. **1924** P. A. Franklin in Hool & Kinne *Movable & Long-Span Steel Bridges* i. 47 Design machinery for wind loadings as set forth in chapter of design of operating machinery. **1985** *Times* 19 July 13/4 In windy winter conditions the windloading presses the door up against the weatherstrip. **1908** N. Duncan *Every Man for Himself* i. 18 An' the sea was runnin' high—a fussy wind-lop over a swell that broke in big whitecaps. **1974** F. Mowat *Boat who wouldn't Float* xix. 234 The combination of wind-lop and heavy swell produced a motion that was indescribable. **1906** R. A. Streatfeild *Mod. Music & Musicians* xix. 338 The fantastic pieces of musical extravagance that are a special feature of 'Don Quixote', such as the wind machine and the bleating sheep, are thoroughly in keeping with..Strauss's real methods. **1928** D. H. Lawrence *Lady Chatterley's Lover* xvi. 282 So many people, like your famous wind-machine, have only got minds tacked on to their physical corpses. **1928** A. Rose *Stage Effects* 9 Fig. 4 shows a wind machine, as used in many theatres. It is built up in the form of a paddle-wheel. **1962** A. Nisbett *Technique Sound Studio* x. 181 A wind machine consists of a weighted piece of heavy canvas hung over a rotating, slatted drum... A wind machine produces just one sound: wind. And the same goes for thunder sheets. **1976** *Gramophone* Apr. 1611/2 Calling as it does for no fewer than twenty horns.., quadruple woodwind, six trumpets and trombones.., not to mention the windmachine, thunder-machine and numerous other percussion instruments, it is impractical to mount nowadays. **1976** *Upper Valley Progress* (Mission, Texas) 6 Oct. 10/1 (Advt.), With thermal inversion, created by our Tropic Breeze wind machine, crop level temperatures are raised as much as 10 degrees. **1936** Wind-noise [see *fairing *vbl. sb.*³]. **1984** *Buses* Aug. 346/2 Only the wind noise through the roof light..gave any indication of our speed. **1616, 1856** Wind-proof [in Dict., sense 30 b]. **1923** F. Wild *Shackleton's Last Voyage* v. 76 Each man was provided with a fur-lined leather cap, heavy pea-jacket, light windproof jacket, a stout pair of trousers. **1937** F. Smythe *Camp Six* xiv. 150 Too tired..to remove our ice-caked windproofs. **1975** E. Hillary *Nothing venture, Nothing Win* viii. 120 We..crawled out of our tents, dressed in all our warm clothing and windproofs. **1977** *Navy News* July 16/3 Availability of the windproof jacket will lead to the progressive phasing out of personal greatcoats and overcoats. **1875** 'Mark Twain' in *Atlantic Monthly* Mar. 288/1 It wasn't a bluff reef... It wasn't anything but a wind reef. The wind does that. **1969** *Gloss. for Landscape Work (B.S.I.)* v. 19 Wind rock. The loosening of the root ball of a tree or plant through the oscillation of the stem by wind. **1972** S. Emberton *Year in Shrub Garden* iii. 151 Any plants which have..been wind-rocked..must be staked upright. *Ibid.* 181 Roses, bush types—shorten to prevent wind-rocking. **1981** Buczacki & Harris *Collins Guide to Pests of Garden Plants* 486 Windrock very commonly occurs on young trees, shrubs and herbaceous plants with a large top in relation to their root system. **1912** J. H. Comstock *Spider Bk.* 35 The solpugids are exceedingly agile; on this account they have been called wind-scorpions. **1959** *Southwest Rev.* Spring 137/1 An arachnid frequently, and naturally, confused with the true vinegarone is the solpugid—or wind-scorpion, wind-spider, or sun-spider. **1858** J. A. Warder *Hedges & Evergreens* 240 The common Cedar is..much used.. where a quick, permanent, and effective wind-screen is wanted. **1887** *Cent. Mag.* Mar. 740/2 That department.. was nearly surrounded by a wind-screen of hemlock boughs and odd pieces of canvas. **1948** *Autocar* 5 Nov. 1093/1 Trico-Folberth's windscreen washer drives home the lesson ..'None so blind as those who can't see.' **1973** *Country Life* 22 Feb. 468/2 Windscreen washers and wipers are operated by a right-hand steering column stalk. **1984** B. Francis *AA Car Duffer's Guide* 18 I've been fiddling about for ages trying to get the windscreen washers to work properly. **1922** *Motor* 21 Nov. 831/3 (*heading*) An automatic windscreen wiper. **1975** *Daily Tel.* (Colour Suppl.) 4 Apr. 18/2 Peter Wallace, the Flight Engineer, began his safety check, examining everything from the windscreen wipers to the radar. **1985** *Computing* 15 Aug. 25/2 We still have to make do with dreadful windscreen wipers in cars. **1909** D. H. Lawrence in *English Rev.* No. 565, I wait for the baby to wander hither to me, Like a wind-shadow wandering over the water. **1931** *Flight* 25 Dec. 1269/2 He had found that behind the wings of an aeroplane 'wind shadows' existed covering a region of reduced pressure. **1977** J. F. Fixx *Compl. Bk. Running* xvii. 202 It also makes sense to vary your speed in order to take advantage of an opponent's wind shadow. **1951** *Gloss. Aeronaut. Terms (B.S.I.)* iii. 36 Wind shear. **1976** *Sci. Amer.* Nov. 32 For a typical wind shear of one mile per hour per mile of height and an average wind speed of 20 miles per hour, the pattern of fallout 100 miles downwind from ground zero would be about 25 miles wide. **1977** *Time* 18 Apr. 37/3 'Wind shear', created by colliding air masses, was listed as the probable cause of an Eastern 727's crash while landing. **1934** A. J. Villiers (*title*) Last of the wind ships. **1980** *Times* 7 Nov. 21/4 Will the rising price of oil bring back the sailing ship—or windship as it is now called—to the trade routes of the world from which it was largely banished a century ago? **1985** *Tel. Sunday Mag.* 18 Aug. 9/1 At 75 he [*sc.* Jacques Cousteau] is as lean and as trim as his revolutionary new 'wind ship', Alcyone, which he has just sailed successfully—and using less fuel —across the Atlantic. **1923** Kipling *Land & Sea Tales* 182 Give your wind-sight another three degrees, Walters. **1920** Wind slab [see *slab *sb.*¹ 6]. **1936** E. A. M. Wedderburn *Alpine Climbing* iv. 51 A form of wind crust is wind slab. As this causes the worst kind of avalanche it is important to detect it. Its surface is smooth and unfortunately often little wind marked and its colour is matt white or yellowish. Wind slab is often found alternating with patches of softer wind marked snow. **1975** E. Hillary *Nothing venture, Nothing Win* xviii. 286, I had..noticed the debris of two large windslab avalanches nearby. **1978** Y. Chouinard *Climbing Ice* ii. 40 The wind will also scour ridges and deposit some snow on the lee side; this then becomes an unstable mass called wind slab. **1866** *N.Y. Times* 13 Apr., All persons having occasion to..start a fire in any old chopping, wind-slash..[etc.] shall give five days' notice. **1905** *Forestry Bureau Bull.* No. 61. 53 An area upon which the trees have been thrown by the wind... blow down, wind slash. **1971** F. C. Ford-Robertson *Terminol. Forest Science* 244/1 All such material [*sc.* slash] blown down by wind is termed wind slash. **1920** *Flight* 29 Apr. 470/1 Three wind sleeves have been installed at Lyons (Bron) aerodrome... Two of these 'sleeves' are red, and are situated on the western side of the landing-ground. **1939** *Air Ann. Brit. Empire* 371 Pilots of those days mistrusted a wind sleeve, which was difficult to see. **1929** E. W. Dickman *This Aviation Business* 139 It requires more work than to stake out a cow pasture, put up a hangar and wind sock, and announce the opening. **1958** *Woman* 9 Aug. 31/4 Briony walked..along the sands as far as the wind-sock up on the golf-course. **1979** J. Leasor *Love & Land Beyond* vi. 88 A wind sock hung limply on a mast. **1959** Wind-spider [see *wind-scorpion* above]. **1966** C. Sweeney *Scurrying Bush* vi. 88 A large solipugid, a very hairy, fast running arachnid that in Africa is often called a 'hunting spider' or sometimes a 'wind spider'. **1893** M. A. Owen *Voodoo Tales* 28, I seed dem ole win'-splittehs [*sc.* long lean hogs]. **1900** *Daily Express* 13 July 6/6 The wind-splitter ..keeps up a wonderful pace. **1941** I. L. Idriess *Great Boomerang* vii. 51 No 'wind-splitters' nearly as wide across the hips as the forehead. **1890** *Harper's Mag.* Dec. 58/2 A tall thinnish man, with..a white wind-splitting face. **1900** *Daily Express* 13 July 6/6 The 'wind-splitting train' was tested over the line between Baltimore and Washington recently. **1948** Duncan & Bone *Oxf. Pocket Bk. Athletic Training* 35 'Wind-sprints'..consist of covering one or two laps of the track, and in so doing moving up very gradually from walking or slow running into faster running and then reversing the process, which will be repeated several times. **1981** *North-east Woods & Waters* Jan. 19/1 My last ¼ of a mile was done doing wind sprints to help my lungs and heart to weather the beating of what was to come. **1932** D. Garnett *Rabbit in Air* ii. 60 It seemed to me several points different if judged by the factory smoke than if judged by the wind stocking. **1983** P. Devlin *All of Us There* vii. 78 The wind-stocking fluttering to show pilots which way the wind is blowing. **1929** *Oxford Poetry* 1 Let's pick the petals of all joy apart, And launch them uncontrolled on the wind-stream. **1934** *Discovery* June 155/1 The wind-stream is so powerful that a man could not possibly stand against it. **1954** Fisher & Lockley *Sea-Birds* v. 127 The oceanic travellers..spend their time making ground by..excursions (by gravitational falls) into the sheltered trough between the crests of the waves, out of the main wind-stream. **1976** A. White *Long Silence* vii. 59 You don't go out at right angles to the plane or the windstream can spin you. **1884** *Engineering* 5 Sept. 225/1 The position and character of the floor between the girders also materially affect the wind stresses. **1953** *Jrnl. Marine Res.* XII. 249 (*heading*) Wind stress on an artificial pond. **1984** A. C. & A. Duxbury *Introd. World's Oceans* viii. 272 The sea surface slopes, as happens..under wind stress. **1932** *Daily Tel.* 2 Mar. 9/5 Curls have ousted points of straight hair, and the old windswept hair is dead. **1940** Graves & Hodge *Long Week-End* xvi. 280 The 'windswept' coiffure came over from Paris in 1931... The hair was cut short, brushed forward with a swirling movement. **1963** Wodehouse *Stiff Upper Lip, Jeeves* iii. 29 She is..as loony a young shrimp as ever wore a wind-swept hair-do. **1985** *Hair Summer* 64 (*caption*) Windswept layered hair requires mousse. **1939** H. J. Lutz in *Amer. Jrnl. Science* CCXXXVII. 392 This investigation was devoted particularly to the influence of tree windthrow on soil morphology. **1953** *Brit. Commonw. Forest Terminol.* i. 147 *Windthrow*, uprooted by wind; a tree or trees so uprooted. Syn. *Windblow*. **1966** *Brit. Columbia Logging* 3/1 They also can sell in their local areas wind-throw timber and stands threatened with destruction by disease or insects. **1981** *N.Z. Jrnl. Forestry* XXVI. 96 Line transects recording soil depth and percentage windthrow were made through single-aged stands of trees where definite patterns of windthrow occurred. **1911** *Aeronaut. Jrnl.* Oct. 53 The planes were tested in a 'wind tunnel'. *Ibid.* 62 Wind tunnel experiments. **1933** *Jrnl. R. Aeronaut. Soc.* XXXVII. 36 The aerodynamic characteristics were observed on a model of the airship in a wind tunnel. **1961** L. Mumford *City in History* (1966) x. 355 Not by accident did the medieval townsman, seeking protection against winter wind, avoid creating such cruel wind-tunnels as the broad, straight street. **1970** *New Scientist* 23 July 194/2 Wind-tunnel tests..establish airflow patterns over ships. **1974** *Times Lit. Suppl.* 13 Dec. 1410/5 The Weimar Republic was above all a testing time, part of that great German wind tunnel in which . ideas and principles, standards and personalities were subjected to the gale of history. **1983** *Aviation News* 8 Sept. 339/1 To compare inflight data with wind tunnel data for the same aircraft. **1985** *Times* 1 Mar. 3/3 The ultimate aim was C15, in every way a family car, using a very streamlined body already being tested in a wind tunnel. **1909** *Chambers's Jrnl.* Mar. 203/1 A small petrol or oil engine as a standby to be used when there is insufficient wind to drive the wind-turbine. **1946** A. Huxley *Let.* 5 Nov. (1969) 557, I gather that the experimental wind turbine which has been producing fifteen hundred kilowatts in Maine has proved entirely satisfactory. **1982** *Daily Tel.* 17 Nov. 1/6 A £650,000 wind turbine machine..was switched on yesterday by Sir Walter Marshall, chairman of the Central Electricity Generating Board. **1959** Windway [see *flûte-à-bec]. **1979** *Early Music* July 365/1 It is also possible to wash the windway with water and washing-up liquid... This is recommended for very dirty or mouldy windways. [**1933** *Automobile Trade Jrnl.* Nov. 52/1 Such items as windshield wings, either as a part of or separate from front door windows [etc.].] **1934** Wind wing [see *sticker¹ 5 a]. **1951** R. Chandler in Gardiner & Walker *R. Chandler Speaking* (1962) 110 It sounded like old Simpson's Chevvy... He could tell by the broken windwing.

wind, *v.*¹ Add: **II. 20*. wind down. a.** *intr.* To draw gradually to a close.

1952 Dylan Thomas *Coll. Poems* p. ix, This day winding down now At God speeded summer's end. **1977** *Time* 19 Sept. 22/1 Instead of winding down, investigations were being stepped up. **1985** R. Barnard *Disposal of Living* vi. 75 The fête was beginning to wind down then. I *think* Mary was still around.

b. *intr.* for *refl.* Of a person who has been 'screwed up' to a certain pitch or is in a state of tension: to relax, to unwind.

1958 *Observer* 7 Sept. 3/5 He is slowly 'winding down' after his exhausting television shows. **1970** *New Yorker* 24 Oct. 50/1 Even the West Indian was winding down. **1979** *Homes & Gardens* June 77/2 It takes him about two days to wind down. When your husband runs his own firm his stress is very great. **1985** R. Hunter *Fourth Angel* viii. 137 An evening at the theatre and a chance to wind down and relax.

c. *trans.* To open (the window of a vehicle) downwards by rotating a handle. Cf. *wind up*, sense 22 e (*c*) below.

1961 I. Murdoch *Severed Head* viii. 71 The windscreen was becoming opaque... I wound down the window on my side and the cold choking air came in. **1975** D. Lodge *Changing Places* v. 165 Philip stopped at a red light and wound down his window.

d. *fig.* To reduce in scale gradually; to bring (an activity) to an end.

1969 *Washington Post* 16 Apr. A22/2 Very little else is possible before the war is wound down. **1969** *Guardian* 5 Aug. 2/7 The enemy might prefer gradually to 'wind down' the level of combat step by step. **1977** *Rolling Stone* 16 June 56/3 Natalie is pregnant and will wind down her work schedule in anticipation of a fall delivery. **1981** *Daily Tel.* 26 Nov. 21 He might be able simply to wind the business down to a size which becomes manageable again.

21*. wind on. *Photogr.* To turn (the film in a camera) to the next position in readiness for taking another photograph. Also *absol.*

1947 A. Ransome *Great Northern?* xxiii. 289 Dick wound on the film, closed the camera and put it in its case. **1964** 'F. Clifford' *Hunting-Ground* vi. 67 Thirty-six on the film and I'm supposed not to have wound on once. **1982** C. Thomas *Jade Tiger* iii. 66 He adjusted the focus... Click, wind on, click again.

22. wind up. d. (*e*) *intr.* Of a person, etc.: to end up, to finish up (in a certain place or condition); to find oneself eventually. *colloq.*

1918 V. Woolf *Diary* (1979) I. 115, I went to have my tooth finished, winding up for tea at the Club. **1921** E. O'Neill *Emperor Jones* i. 155 When I gits a chance to use it I winds up Emperor in two years. **1942** W. Stevens *Let.* 2 Oct. (1966) 421 The same reasons would prevent her from marrying as long as the war goes on, and..she may wind up as an old maid. **1952** Wodehouse *Barmy in Wonderland* iii. 29 Men who own hotels always wind up in the breadline with holes in their socks. **1968** *Globe & Mail* (Toronto) 17 Feb. 3/1 Canada has made no written request that military equipment sent to the United States should not wind up in Vietnam. **1976** *National Observer* (U.S.) 13 Mar. 9/2 Somebody who wants to get away from it all is likely to wind up in a chalet in a Heidilike village on a mountain. **1980** L. Birnbach et al. *Official Preppy Handbk.* 111/1 Many of these forays.. wind up involving mayhem or destruction of property.

e. (*c*) In reference to a motor vehicle: to close (the window) by rotating a handle. Cf. *wind down*, sense 20* c above.

1970 H. R. F. Keating *Inspector Ghote breaks Egg* ii. 16 He slowly wound up the window of his big car. **1971** P. D. James *Shroud for a Nightingale* i. 16 She wound up the car window and stepped on the accelerator.

wind, *v.*² Add: **II. 6.** To cause (a baby) to bring up wind after feeding; to 'burp'.

1958 *Observer* 19 Oct. 10/6 My five-month-old son, though well fed, thoroughly winded and much loved, delights in yelling loud and long. **1961** *Guardian* 28 June 6/3 Two babies..to feed and wind and change. **1978** D. Murphy *Place Apart* x. 211 Paddy's wife handed him their six-months-old daughter, to be 'winded' while she was undressing their two-year-old son... The baby burped dutifully.

windage. 3. Delete 'also,...' and transfer quot. 1903 to sense *4.

4. The (actual or potential) air resistance of a moving object, esp. a vessel or a rotating machine part; also, the force of the wind on a stationary object.

1897, 1898 [in Dict., sense 2]. **1903** [in Dict., sense 3]. **1909** A. Williams *Engin. Wonders of World* III. 42/2 The designer has to consider how to curve the [propeller] blades so as to give a maximum thrust for a minimum windage. **1948** *Times* 24 Nov. 2/2 A new type of anchor designed to prevent warships, particularly aircraft-carriers with their large windage area, from dragging their anchors. **1953** C. S. Forester *Hornblower & 'Atropos'* 60 Their twelve oars hardly sufficed to control their more than forty feet of length, and the windage of the huge cabin aft was enormous. **1958** *Engineering* 31 Jan. 157/3 The radar aerial..is of parabolic section and slatted to reduce windage. **1961** E. Lightfoot *Moment Distribution* v. 123 Design against windage is important in skyscraper buildings. **1971** *Sci. Amer.* Dec. 7/1 If the time cycle between storage and retrieval is long, most of the stored energy [of the flywheel] is lost in windage and friction. **1977** *Mod. Boating* (Austral.) Jan. 98/1 There is a tremendous amount o windage in that topsides and cabin.

5. Special Comb.: **windage loss,** loss of power through the air resistance of rotating parts.

1922 *Encycl. Brit.* XXX. 35/2 In determining the useful H.P. of rotary engines, 'windage loss'..had first to be determined. **1966** *McGraw-Hill Encycl. Sci. & Technol.* IV. 449/1 Windage loss is relatively large in air-cooled high-speed machines.

wind-bag, windbag. Add: **3.** *Naut. slang.* A sailing ship or 'windjammer'.

1924 R. Clements *Gipsy of Horn* 11 A sailing ship—an old wind-bag, as the young, up-to-date watchkeeper would call it. **1930** J. Masefield *Wanderer of Liverpool* 47 A crowd of windbags moored fore and aft, to buoyed anchors. **1946** W. McFee *In First Watch* i. 15 He had been cook in a windbag and a sailor before the mast.

wind-break, *sb.* **1.** For Chiefly *U.S.* read orig. *U.S.* and add earlier and later examples. Now freq. without hyphen as one word.

1861 *Trans. Ill. Agric. Soc.* IV. 479 These trees, which are valuable as shade and wind-breaks, should be planted. **1910** W. Schlich *Man. Forestry* (ed. 4) II. 123 Species with a thin crown are indifferently adapted for wind breaks. **1934** *Times Educ. Suppl.* 10 Feb. p. iv/3 Whether as ornament or to secure privacy, as a windbreak or to indicate a boundary, a hedge has its place in most gardens. **1950** *N.Z. Jrnl. Agric.* July 5/3 Such a windbreak is very valuable for sheltering dipped sheep. **1962** *Coast to Coast 1961–62* 138 An old limestone place, with a slate roof, and a windbreak of pepper-trees at one side. **1968** *Southerly* XXVIII. 172 The pine trees acted as a wind-break for the solitary house set a hundred yards lower down. **1970** J. H. B. Peel *Country Folk* ii. 38 A hedge serves as a windbreak for crops and as a nesting-place for birds. **1975** *Toronto Star* 27 Dec. D4/1 So except in balmy weather, be sure to tote a plastic or canvas windbreak. **1981** *Farmstead Mag.* Winter 49/1 Some people prefer deciduous trees in all or part of the windbreak.

wi·nd-down. *colloq.* [f. vbl. phr. *to wind down*: see *wind *v.*¹ 20*.] The process of bringing or coming to an end; a gradual reduction in scale.

1969 *Time* 21 Feb. 29 The campaign heralds the official wind-down of the Cultural Revolution, a finale that is to climax in 'all-round victory'. **1971** *New Scientist* 8 July 95/3 The reduction in the number of students.., the cutback in..spending, the cancellation of the SST, the wind-down of space exploration, [etc.]. **1971** *Guardian Weekly* 25 Dec. 3 Even now, with the virtual completion of the wind-down begun in 1970, the military break with Indochina is not complete.

wind-drift. Add: **b.** The action of wind currents, esp. on water.

1898 *Geogr. Jrnl.* June 662 The sand so produced is rounded by wind-drift in an unmistakable manner, the grains being entirely different from those of sea-sand. **1964** *Oceanogr. & Marine Biol.* II. 257 This patch [of effluent] is subject to wind-drift and can be carried directly on to beaches. **1967** *Ibid.* V. 103 Since the cross sections of these two straits are much less than that of the throat of the Celtic Sea much surface wind drift water must return to the Atlantic some other way.

winded, *ppl. a.*¹ **3.** (Earlier example.)

1883 'Mark Twain' *Life on Miss.* iii. 49 They couldn't keep that up very long without getting winded.

windego, obs. var. *windigo.

winder, *sb.*² Add: **2. b.** *fig.* † *spec.* a sentence of transportation for life (*obs. slang*).

1812 J. H. Vaux *Vocab. Flash Lang.* in *Mem.* (1964) 279 A man transported for his natural life, is said..to have *knap'd a winder*. **1836** J. F. O'Connell *Residence in New Holland* 37 Previous convictions and character must have affected his sentence, as it was, in flash phraseology, a winder. **1913** D. H. Lawrence *Sons & Lovers* ix. 243 It's a winder when you have to pour your own tea out—an' nobody to grouse if you team it in your saucer and sup it up.

winder (wi·ndəɹ), *sb.*⁵ Repr. dial. or slovenly pronunc. of Window *sb.*

1684 G. Meriton *Yorks. Dialogue in Praise of Yorks. Ale* 49 Ah Nan steek th' winderboard, & mack it dark. **1838** Dickens *Nickleby* (1839) viii. 69 We go upon the practical mode of teaching, Nickleby;..W-i-n, win, d-e-r, der, winder, a casement. **1877** [see *window bottom* s.v. *window *sb.* 5 d]. **1901** M. Franklin *My Brilliant Career* xxxii. 272 Lizer, shut the winder quick. **1976** *Trans. Yorks. Dial. Soc.* XIV. 37 Ah've just been cleeanin' t'winders.

windfall. Add: **2.** *fig.* Applied (*poet.*⁻¹) to a flood of unexpected light.

1945 Dylan Thomas *Fern Hill* in *Horizon* Oct. 221 And once below a time I lordly had the trees and leaves Trail with daisies and barley Down the rivers of the windfall light.

3. *Econ.* **windfall profit,** unexpectedly large or unforeseen profit; similarly *windfall gain*, *loss*, etc.

1936 J. M. Keynes *Gen. Theory Employment* ii. vi. 57 The change in the value of the equipment, due to unforeseen changes in market values, exceptional obsolescence or destruction..may be called the *windfall loss*. *Ibid.* v. xx. 288 The windfall gain will wholly accrue to those entrepreneurs who happen to possess products at a relatively advanced stage of production. **1951** Sloan & Zurcher *Dict. Economics* 266 *Windfall profit*, a profit in excess of that which can be considered normal. **1973** *Times* 21 Dec. 6/7 A proposal for Congress to impose 'an emergency windfall profits tax' on the oil company. Although President Nixon himself told the American consumer, 'there will be no windfall profits at their expense.' **1977** *N.Y. Rev. Bks.* 26 May 31/4 The shift to free market pricing would give the oil companies windfall profits.

wind-gall². (Earlier example.)

1823 J. F. Cooper *Pilot* I. ii. 19 There be streaked wind-galls in the offing, that speak..plainly..to shorten sail.

Windies (wi·ndiz), *sb. pl.* *colloq.* (orig. *Austral.*). [Contraction of *West Indies*.] West Indians, *spec.* the West Indian cricket team; also, immigrants from the West Indies.

1965 W. Grout *My Country's Keeper* 69 The Australian public..took the Windies to their hearts from that moment. **1971** J. Brunner *Honky in Woodpile* ii. 15 Skinheads and others were out bashing Pakis and Windies. **1976** *Sunday Tel.* 4 Jan. 32 Windies roll with brutal pace beating. **1980** *Economist* 21 June 51/1 Would the West Indies beat England in the first Test match?.. In the cricket match, the 'Windies' scraped home in a nail-biting finish.

windigo (wi·ndigo). Also 9 **weendego(ag, wendigo; wihtigo, witiko,** etc.; and with capital initial. [Ojiwa *wintiko*, pl. *wintikok*; some spellings reflect the Cree cognate *wihtikow*.] In the folklore of the northern Algonquian Indians: a cannibalistic giant, the transformation of a person who has eaten human flesh.

1714 J. Knight *Jrnl.* 7 Oct. in W. Cowan *Papers of Seventh Algonquian Conference 1975* (1976) 21 Some Indians came from Fort Nelson who says they saw a Whitego wch is an Apparition. **1830** E. James *Narr. John Tanner* 316 The Muskegoes, who inhabit the low and cheerless swamps on the borders of Hudson's Bay, and are themselves reproached by the other tribes as cannibals, are said to live in constant fear of the Weendegoag. **1847** J. B. Nevins *Two Voyages* 115 When Windego saw him, he was very angry, and said, 'What do you mean, boy, by coming out and making that noise? I am going to eat you.' **1859** P. Kane *Wanderings of Artist among Indians*

N. Amer. 60 The Weendigoes are looked upon with superstitious dread and horror by all Indians. **1924** Chambers's Jrnl. Mar. 170/1 At midnight they were wakened by what Jacques took to be a wendigo in the woods behind. **1933** J. M. Cooper in *Primitive Man* Jan. 20 The Cree Witiko Psychosis... This peculiar form of mental disturbance is characterized by (1) a craving for human flesh, and (2) a delusion of transformation into a Witiko who has a heart of ice or who vomits ice. **1934** *Jrnl. Abnormal & Social Psychol.* XXIX. 7 The repugnance to food is construed as positive evidence that the person is becoming a 'wihtigo', *i.e.*, a cannibal. **1960** T. Stacey *Brothers M.* II. xxxii. 361 Daudi..was still utterly subjected, as if by some unseen windigo that was withdrawing him to its own element. **1961** O. Nash *Coll. Verse* 425 The Wendigo, The Wendigo! Its eyes are ice and indigo! **1971** *Brit. Med. Bull.* XXVII. 78/1 States of excitement or panic may be so influenced by local conditions as to give the appearance of specific psychoses. . The Windigo psychosis of the Chippewa, Ojibwa and Cree Indians illustrates the way in which such states can develop.

windiness. 1. (Later examples.)
1922 A. Machen *Far off Things* i. 10 Holborn has a certain vastness and windiness about it as the sky grows from black to grey. **1957** G. E. Hutchinson *Treat. Limnol.* I. vii. 446 (*heading*) Windiness and area. **1971** *Nature* 10 Dec. 345/1 The classic loess deposits in China.. can probably only be explained in terms of greater windiness in the China/Gobi Desert area.

winding, *vbl. sb.*[1] Add: **I. 2. b.** (Later example.)
1917 T. S. Eliot *Prufrock & Other Observations* 18 Among the windings of the violins And the ariettes Of cracked cornets.

II. 8. b. *Electr.* An electric conductor that is wound round a magnetic material, esp. (*a*) a coil encircling part of the stator or rotor of an electric motor or generator, or an assembly of such coils connected to form one circuit; (*b*) one forming part of a transformer.
1888 S. P. Thompson *Dynamo-Electric Machinery* (ed. 2) xii. 259 If the successive sections are to be connected up consecutively, then they must be wound.. alternately with right-handed and left-handed windings. **1947** R. Lee *Electronic Transformers & Circuits* v. 141 In step-down transformers the capacitance may be regarded as existing mainly across the primary winding; in step-up transformers, across the secondary winding. **1962** *Newnes Conc. Encycl. Electr. Engin.* 894/1 The simplest type of winding is a field coil around a salient pole.., the coil comprising a number of turns (between one and several thousand) of wire or strip. **1979** Nasar & Unnewehr *Electromechanics & Electric Machines* iii. 67 Transformer windings are constructed of solid or stranded copper or aluminum conductors.

winding, *vbl. sb.*[2] (Later example.)
1940 W. de la Mare *Pleasures & Speculations* 48 The first windings of the Last Trump.

windmill, *sb.* Add: **3. c.** *Cricket.* A style of bowling with a high overarm delivery. ? *Obs.*
1867 *Australasian* 19 Jan. 76/3 A change in bowling was tried, Wardill going on with his 'windmills', *vice* Conway. **1900** W. A. Bettesworth *Walkers of Southgate* 124 Taking his run up to the wicket, swinging his arm in what has been described as 'a windmill action'. **1920** in P. F. Warner *Cricket* 86 Spofforth's windmill deliveries.

4. b. (Examples of phr. *to tilt at windmills.*)
1898 [see Tilter *sb.*[1] 1]. **1937** A. Christie *Death on Nile* xxiv. 238 Rather eccentric..inclined to tilt at windmills. **1978** P. Bryers *Cat Trapper* viii. 57 Mike was into the sort of thing I'd like to be doing... Tilting at windmills?

5*. An airscrew, esp. one of the kind designed by Cierva for the autogiro. Now *Hist.*
1931 Cierva y Cadorniu *Wings of Tomorrow* 88, I designed the blades of the windmill. **1935** *Sun* (Baltimore) 31 Jan. 3/3 This new fast auto gyro will have no propeller. It will tilt its present windmill, gear it to the engine and so get its lift and forward drive. **1949** *Gloss. Aeronaut. Terms* (*B.S.I.*) II. 20 *Windmill*, an airscrew designed to produce power by axial transmission relative to the air.

6. a. (sense 5* above) *windmill aeroplane, (air)plane, rotor, wing*; **b. windmill brake state** *Aeronaut.* (see quot. 1969).
1931 *Statesman* (Calcutta) 5 Dec., The Autogiro or 'windmill' aeroplane has just been put on the public market in this country for the first time. **1928** *Daily Express* 10 Aug. 11/4 A 'windmill' airplane is to fly the channel. **1948** *Jrnl. R. Aeronaut. Soc.* 269/1 In the windmill brake state, the rotor is again working in a regular slipstream. **1969** *Gloss. Aeronaut. & Astronaut. Terms* (*B.S.I.*) v. 19 *Windmill-brake state*, the operating condition of a rotor when the rotor thrust and the axial flow through and outside the rotor disc area are all in the same direction. **1927** *Times* 27 Apr. 16/3 It was agreed to use the word ..aerodyne to designate all heavier-than-air craft, in which category the class name for the windmill plane appears as gyro plane. **1944** H. F. Gregory *Anything Horse can Do* iv. 48 The stick [of the Autogiro] was connected to the hub of the windmill rotor by push-pull tubes and cables. **1931** Cierva y Cadorniu *Wings of Tomorrow* 82 A flying machine with a windmill wing.

windmill, *v.* Add: **b.** *trans.* and *intr.* To move (one's arms or legs) in a manner suggestive of a windmill.
1927 W. E. Collinson *Contemp. Eng.* 18 At this school we had our first taste of fighting or rather windmilling with the arms. **1928** *Daily Express* 6 Nov. 8 They..set about their efforts again, windmilling his arms and legs until he gasped with unconscious exhaustion. **1959** R. Collier *City that wouldn't Die* v. 67 Windmilling your arms to keep the blood coursing. **1979** S. Brett *Comedian Dies* v. 55 Lennie Barber..seemed to lose his balance and sank back, arms windmilling, on to the side of his chair. **1982** W. Boyd *Ice-Cream War* 4 The colonel windmilled his arms and cracked his knuckles.

c. *intr.* Aeronaut. Of the propeller or rotor of an aircraft: to spin unpowered; *to windmill down*, to descend with the rotor spinning; also *fig.*
1934 *Jrnl. R. Aeronaut. Soc.* XXXVIII. 18 Captain Barnwell said..that it could be assumed that when an engine was fully throttled the airscrew was windmilling freely. **1942** *Flight* 26 Mar. 296/2 When an airscrew is wind-milling..the effect is exactly opposite to its normal one. **1958** *Listener* 2 Jan. 10/1 The rotor windmilling freely as on an autogiro. **1963** Sobey & Suggs *Control of Aircraft & Missile Powerplants* vii. 181 The forward flight of the airframe will cause the engine to windmill and create sufficient airflow through the engine to minimize the probability of a hot start. **1976** *Shooting Times & Country Mag.* 16–22 Dec. 29/2, I shot at another goose which peeled off, flew fluttering for a hundred yards, and then windmilled down stone dead. **1978** M. Babson *Tightrope for Three* xxvi. 152 Autorotation was a standard and perfectly safe manoeuvre, the helicopter windmilling down with the pilot still in complete control.

Hence **wi·ndmilling** *ppl. a.* and *vbl. sb.*
1945 *Jrnl. R. Aeronaut. Soc.* XLIX. 716 With the windmilling propeller the changed air-flow over the aircraft does not impair the handling characteristics. **1959** C. A. Meyer in O. E. Lancaster *Jet Propulsion Engines* 149 A typical curve showing the drag of a turbojet engine during windmilling is shown in Fig. C, 11f. **1973** J. Wainwright *Touch of Malice* 98 The young man was obviously a nutter... His slobbing mouth. His windmilling arms. **1978** M. Farren *Feelies* 33 The other girl was spun, flat on her back with windmilling arms and legs.

Windmill Hill. The name of the site of a causeway camp near Avebury, Wilts., type site of the neolithic age in Britain, used to designate the type of culture, pottery, etc., characteristic of that period.
1930 E. C. Curwen in *Antiquity* IV. 26 The varieties of pottery..have been collectively described as the Windmill Hill type. **1947** J. & C. Hawkes *Prehist. Britain* ii. 40 The mausoleum most fashionable among the Windmill Hill people was the long barrow. **1954** S. Piggott *Neolithic Cultures* ii. 18 The Windmill Hill culture was basically that of cattle-breeders. **1963** *Field Archaeol.* (Ordnance Survey) (ed. 4) 33 While 'Windmill Hill' pottery predominates, most other forms of Neolithic pottery are represented. **1971** *World Archaeol.* III. 239 The Q1 skeleton strongly suggested that a member of the Windmill Hill Culture was cut to pieces by this type of large metal weapon. **1983** P. A. Crowl *Intell. Traveller's Guide to Historic Britain* i. 15 Those large mounds of earth thrown up by people of the Windmill Hill culture to cover their dead.

windolite (wi·ndǝlait). Also **Windolite, windowlite.** [f. Window *sb.* + *-lite* (alteration of Light *sb.*).] The name of a transparent material serving as a substitute for glass.
A proprietary name in Australia.
1927 *Glasgow Herald* 1 Feb. 6/4 If the pullets are to be confined..wide shelter boards and glass, or 'windolite', shutters must be fixed. **1935** *Times Educ. Suppl.* 16 Mar. p. iv/1 A sheet of glass or windolite or a small *cloche* will afford them [*sc.* plants] shelter. **1951** 'N. Shute' *Round Bend* 30 An upstairs window was broken and shut up with windowlite tacked over the frame.

wi·nd-on, *a.* *Photogr.* [f. vbl. phr. *to wind on*: see *wind *v.*[1] 21*.] Designating or pertaining to (part of) the mechanism for advancing a film to the next position.
1963 *Listener* 31 Jan. 198/2 As for the camera..the wind-on mechanism jammed. **1976** J. McClure *Rogue Eagle* vi. 101 Oeloefse removed a sliver of film from behind the wind-on spool.

window, *sb.* Add: **3. d.** *Geol.* = *fenster.
1908 H. B. C. Sollas tr. *Suess's Face of Earth* III. viii. 350 This term 'window' has been brought into use by our fellow geologists in Switzerland, and we shall adopt it in this work for those cases in which a subjacent tectonic element is brought to light by erosion. **1927** L. W. Collet *Structure of Alps* II. i. 26 Three windows occur in the Eastern Alps. **1939, 1954** [see *fenster]. **1980** *Sci. Amer.* Oct. 131/2 The presence of sedimentary rocks in the windows of the Blue Ridge indicates that the crystalline rocks there overlie sedimentary material.

e. A transparent panel in a package, through which the contents can be seen; *spec.* on an envelope (see *window-envelope*, sense 5 d).
1914 [see *window-envelope*, sense 5 d below]. **1938** D. E. A. Charlton *Art of Packaging* 94 The latest.. improvement..[paper] napkins visible through a cellophane window. **1952** E. J. Labarre *Dict. Paper & Paper-making* (ed. 2) 90 The use of...envelopes...with cellophane windows is prohibited by most continental postal administrations. **1977** C. McCarry *Secret Lovers* iii. 40 Wilson..flipped the plastic windows to make certain that all the papers were still in the wallet.

f. Freq. with capital initial. Mil. code name: = *chaff *sb.*[1] 6 b.
1942 Ld. Cherwell in *Oxf. Mag.* (1963) 9 May 283/1 If you go into the meeting and try to get 'Window' used, you'll find me and Tizard united against you. **1946** J. P. Baxter *Scientists against Time* vi. 93 The British used Window for the first time over Hamburg on the night of July 24–25, 1943. **1947** [see *chaff *sb.*[1] 6 b]. **1962** A. P. Rowe *Let.* in R. V. Jones *Most Secret War* (1978) iv. 41 What I want to emphasize is that from no one at no time did I hear a breath of anything like window. **1963** D. Irving *Destruction of Dresden* III. iii. 135 The crews of these new bomber formations had been cascading *Window* into the air in copious amounts. **1980** M. Middlebrook *Battle of Hamburg* viii. 125 It is not known which aircraft dropped the first bundle of Window.

g. *Computers.* (i) The screen of a VDU regarded as a means of displaying part of a drawing stored in a computer; the part of a drawing, program, etc., chosen for display.
1966 *Computer Jrnl.* IX. 21/1 The 10-inch square display screen..is treated as a 'window' on to a very large drawing board. *Ibid.* 22/1 The display 'window' can be moved one grid space in any of four directions. **1968** *IBM Systems Jrnl.* VII. 163 A subsequent computation can determine the point at which any line crosses the edge of the window. **1982** J. E. Scott *Introd. Interactive Computer Graphics* vii. 124 The size and location of the window are expressed in user coordinates because the window is specified in relation to the drawing... The dimensions and center point of the viewport are expressed in normalized screen coordinates.
(ii) = *viewport 2.
1974 *AFIPS Conf. Proc.* XLIII. 251/1 The display screen is divisible into rectangular, possibly overlapping 'windows'. **1980** W. Newman in C. E. Vandoni *Eurographics 80*, 4 NAN..uses overlapping windows, in this case in colour but with contents restricted to text. **1983** *MicroComputer Printout* Sept. 57/2 A similar, but more flexible system allows you to split the screen into two windows for viewing different sections of the model at once. **1985** *Acorn User* Feb. 37/2 Windows can be created which can then be rearranged to provide any print format required.

4. b. *to go* (*be thrown*, etc.) *out of the window* (U.S. without *of*), to be abandoned, discarded, or made worthless; also (U.S.) *to be out the window.*
1939 H. L. Ickes *Secret Diary* (1954) III. 3 Steve Early ..said that the 'brain trust was out of the window'. **1945** *Sun* (Baltimore) 1 Oct. 4-0/3 Production of specialty goods—such as birthday and wedding cakes—was 'out the window'. **1946** *Ibid.* 6 July 4/1 As a guide, past experience went out the window early this year when the number of retirements suddenly increased. **1964** S. M. Miller in I. L. Horowitz *New Sociol.* 300 The concept of 'unemployables' was largely thrown out the window. **1968** F. Lundberg *Rich & Super-Rich* iv. 173 As FDR himself said, 'the New Deal is out the window.' **1969** G. Donaldson *Fifteen Men* xi. 184 'The Uncle Louis kissing babies went out of the window this afternoon', said Green. **1977** *Chicago Tribune* 2 Oct. XIII. 24/3 The old rule-of-thumb of putting insulation with a resistance rating of 19 in your attic (R-19) is 'out the window'.

c. A continuous range of electromagnetic wavelengths for which the atmosphere (or some other medium) is relatively transparent.
1949 *Bull. Amer. Meteorol. Soc.* XXX. 233/1 Dr. Buettner indicated as one problem the measurement of solar radiation near 0·21μ, where absorption due to ozone decreases and that due to oxygen increases, forming a 'window' in the solar spectrum. **1969** *Guardian* 6 Feb. 9 The earth's infra-red 'windows' are at wavelengths of 1 to 2 and 8 to 14 microns. **1970** *Nature* 10 Oct. 158/1 This particular solvent is transparent in the region of the neodymium laser wavelength (1·06 μm) and..the absorption spectrum of the resulting solution exhibits a 'window' in this region. **1974** *Sci. Amer.* Apr. 71/3 The most recently opened window on the galactic center is at X-ray wavelengths.

d. = *launch window* s.v. *launch *sb.*[1] 7; *weather window*: see *weather *sb.* 7. Chiefly *U.S.*
1965 [see *launch window* s.v. *launch *sb.*[1] 7]. **1967** *N.Y. Times* 18 Oct. 30/1 The Soviet and American vehicles flew to Venus close together because both were fired during one of the periodic 'windows' for such shots. **1968** *Sci. Jrnl.* Dec. 17/2 Between February and April next year the 'window' will be open for launchings to Mars. **1973** *Times* 15 May 1/5 There will be tomorrow only a 10-minute 'window'—the period in which the rocket must be launched to reach the appropriate orbit. **1977** A. Peccei *Human Quality* ix. 190 This is therefore the time to act. The seventies offer what in space exploration is called a 'window', an opportunity, and probably one of the last ones, for us to launch such an undertaking.

e. Hence used more widely in sense 'a period of time', in phrases *window of opportunity* (or *vulnerability*), esp. with reference to the arms race. orig. *U.S.*
1979 *Hearings U.S. Congr. Sen. Comm. Armed Services* I. 168 We are facing a window of ICBM vulnerability during the period of 1982 to 1986. **1980** *N.Y. Times* 21 Sept. A27/2 To intimidate the Americans with a Soviet 'window of opportunity' to knock out Minuteman missiles. **1981** *Ibid.* 3 Oct. 13/1 Mr. Reagan..enlarged upon the meaning of his oft-repeated theme about the 'window of vulnerability'... The term is generally used to mean the time period in which American land-based missiles are believed to be vulnerable to a surprise Soviet attack. Today, Mr. Reagan said it also applied to Soviet superiority at sea and in Europe. **1982** *Nature* 4 Mar. 5/1 Environmentalists and labour union groups are seizing this 'window of opportunity' between the failure of the last industry challenge and the eventual tightening up of administrative requirements to get as much information on existing pesticides..as they can. **1985** *Sunday Times* 16 June 60/8 Regional bank bosses know that..they must rush to acquire their neighbours, to make the most of their window of opportunity.

5. a. *window-bay* (cf. Bay-window), *-circle*, *-hole* (earlier example), *-hook*, *-pole*, *-shade* (later N. Amer. examples), *-shaft*, *-slit*, *-square*, *-sticker* (*sticker 5 a), *-unit*. **b.** *window-breaker*. **c*.** Appositive, 'that is a window', consisting chiefly of glass, as *window door*, *wall*.
1861 Rossetti *Let.* June (1965) II. 406, I offered to paint figures of some kind on the blank spaces of one of the gallery window bays. **1920** D. H. Lawrence *Women in Love* i. 3 Ursula and Gudrun Brangwen sat..in the window-bay of their father's house..working and talking. **1903** 'O. Henry' in *McClure's Mag.* July 333/1 We'll get that cannon..and fire some window-breakers with it. **1944** Blunden *Cricket Country* i. 14 Marbles, tops of various shape and various function—the window-breaker never was so bad as his name, the peg-top always looked more sinister. **1865** G. M. Hopkins *Poems* (1967) 34 The towers musical, the quiet-walled grove, The window-circles. **1926** D. H. Lawrence *Plumed Serpent* xi. 181 Ramón..closed the window-doors. **1884** 'Mark Twain' *Huck. Finn* xxxiv. 351 When we got to the cabin, we took a look..and on..the north side we found a square window-hole. **1659** [Harvard] *College Book* 1, in *Publ. Colonial Soc. Mass.* (1925) XV. 10 It[em] for window-hookes ..—04[d.]. **1932** Blunden *Face of England* 130 One bird came to the window-hook. **1922** W. B. Yeats *Trembling of Veil* I. xix. 64 A fellow-theosophist once found him hanging from the window-pole. **1984** *New Yorker* 24 Dec. 44/2 Dusk was gathering in the tall windows that needed a window pole to close. **1921** *Daily Colonist* (Victoria, B.C.) 20 Oct. 7/4 If you want window shades for your home, we will be pleased to send our men and give you an estimate. **1978** S. Brill *Teamsters* iv. 127 Though darkened by the drawn windowshade it was a comfortable room. **1918** D. H. Lawrence *New Poems* 54 Petals heaped between the window-shafts In a drift die there. **1880** 'Mark Twain' *Tramp Abr.* xlii. 490 It [*sc.* the Castle of Chillon] has romantic window-slits that let in generous bars of light. **1955** J. R. R. Tolkien *Return of King* vi. i. 184 A door on his left faced a window-slit looking out westward. **1699** J. Wallis *Let.* 10 Oct. in *Private Corresp. Samuel Pepys* (1926) I. 189 The sun-shine does appear with the distinct figure of the window-squares upon the ground within doors. **1956** D. Gascoyne *Night Thoughts* 24 Behind the rows of window-squares. **1963** *Daily Tel.* 3 Dec. 15/4 A new window-sticker and poster campaign. **1962** *Listener* 11 Jan. 63/2 The endlessly repeated small window-units of multi-storey buildings tend to be both boring and overpowering. **1970** *Globe & Mail* (Toronto) 25 Sept. 34/7 (Advt.), Recreation room ..with 'window wall' walkout to patio and garden. **1977** *Chicago Tribune* 2 Oct. XII. 10/2 Two large terraces which can be entered through window walls provide a breathtaking lake view.

d. window bill, a poster or advertisement for display in a window; **window bottom** *dial.* = Window-sill; **window-box** (earlier example); **window-card,** a card to be displayed in a window; **window display,** a display of goods in a shop-window; **window-dress** [back-formation f. *window-dresser*, etc.], (*a*) *intr.* to arrange and display goods to the best advantage in a shop-window; (*b*) *trans.* (in quots. *fig.*: see *window-dressing* (*c*)); **window-envelope** (earlier example); **window garden,** (a display of plants in) flower pots or boxes on a window-ledge or sill; **window-gazing,** staring at the displays in shop-windows, window shopping (see *window-gazer*, sense 5 c); hence (as back-formation) **window-gaze** *v. intr.*; **window guidance,** a form of credit rationing practised by Japanese banks; **window operation** = prec.; (see quot. 1965); **window-pane,** (*c*) *slang*, a monocle; (*d*) in full, *window-pane check*: a kind of large check pattern on clothes; a single square of this; also **window-pane checked** *a.*; **window plant,** (*a*) a plant grown indoors in the light of a window; (*b*) any of several succulent plants of the genera *Mesembryanthemum*, *Lithops*, or closely related genera, which grow almost buried in the ground, with only a transparent section of a leaf visible above it; **window-screen,** (*b*) *U.S.*, a screen of mesh designed to be put across a window-opening to admit air whilst excluding insects, etc.; **window-seat,** also a seat by the window in a train, bus, aeroplane, etc.; **window-shop** *v. intr.*, to go from shop to shop to look at the goods displayed in shop-windows without buying; freq. as *pres. ppl.*; also in extended use and *fig.*; hence **window-shopper; window-shopping** *ppl. a.* and *vbl. sb.*; **window table,** a table (pleasantly situated) by the window in a restaurant, etc.; **window-trimmer** *U.S.* = *window-dresser*; **window-trimming** *U.S.* = *window-dressing*; also *fig.*; **window-washer** (chiefly U.S.), (*a*) = *screen-washer* s.v. *screen *sb.*[1] 7 a; (*b*) one whose job is to wash windows, a window-cleaner; also **window-washing** *vbl. sb.*; **window winder**: in a motor vehicle, a mechanism for opening and shutting the (side-)windows.
1868 *Era Almanack* p. xi, Theatrical posters, window bills, show cards, portraits, &c. **1965** *Spectator* 29 Jan. 124/1 Window-bills went up in streets where they had never formerly been seen. **1877** J. Hartley *Halifax Clock Almanack* 43 Sam made a grab at it, an it flew to th' winder-bottom. **1914** D. H. Lawrence *Prussian Officer* 162 The daffodils in the white window-bottoms shone across the room. **1960** *Times* 24 Oct. 12/6 Altar and every window-bottom would be bright with rosy apples. **1895** 'Mark Twain' in *Harper's Mag.* Dec. 144/1 A watering-pot in her hand and window-boxes of red flowers under its spout. **1905** Window-card [see *cut-out *sb.* 2 c]. **1965** F. Sargeson *Memoirs of Peon* v. 115 There was a window-card that advertised board and lodging. **1897** *Sears, Roebuck Catal.* 689/1 Store Lamp... Just the thing to throw light on a window display. **1930** *Daily Express* 6 Oct. 9/2 A blaze of warm, glowing colours, elaborate window displays ..usher in..the autumn shopping season. **1962** E. Snow *Other Side of River* (1963) lxx. 538 One corner store nearby offered a neat window display of ready-made, well-tailored children's garments. **1913** J. M. Keynes *Indian Currency & Finance* vii. 205 It is scarcely possible..that they should 'window-dress' their balance sheets. **1928** *Britain's Industr. Future* (Liberal Industr. Inquiry) 417 The common practice of 'window-dressing' the published statements by making them refer to the figures of specially selected days instead of the daily averages should be made illegal. **1957** A. C. L. Day *Outl. Monetary Economics* xiii. 177 Each of the four of the Big Five banks which window-dressed its balance sheet made it up on a different day of the week from the others. **1971** D. Clark *Sick to Death* iii. 56 Nobody will let us near a shop to window dress on Saturdays... But on Sundays we get a free run because the shops are shut. **1980** *Daily Tel.* 24 Sept. 17/8 The cheque was part of an elaborate fraud designed to 'window-dress' the balance sheet of a troubled banking company. **1914** *Maclean's Mag.* Dec. 124/1 Use B-E window envelopes. **1884** G. W. Cable *Dr. Sevier* xii. 81 The asylumed window of 'St. Anna's' could glance down into it over their poor little window-gardens. **1980** *News & Observer* (Raleigh, N. Carolina) 28 Oct. WA-2/7 Additions, solariums, greenhouses, window gardens, decks. **1959** *Spectator* 21 Aug. 218/3 As you walk the busy streets and window-gaze. **1968** *Daily Mirror* 20 Aug. 9/4 Take a look at the men's wear section in the chain stores; window-gaze in any man's shop. **1949** M. Steen *Twilight on Floods* IV. vi. 614 Up the Haymarket to Regent Street for an orgy of window-gazing. **1961** R. Graves *More Poems* 42 Window-gazing, at one time or another In the course of travel. **1964** *Econ. Picture of Japan* (Keidanren) IV. 53 For several years after the War, the financial policy of the Bank of Japan was characterized more or less by a qualitative control policy or a selective loans system or a so-called 'window guidance'. **1977** *Ann. Rep. Bank Internat. Settlements* 60 In Japan the authorities kept 'window guidance' ceilings on bank credit expansion in force as a precaution. **1961** *Monthly Econ. Rev.* (Bank of Japan) Sept. 9/1 The practice of the Bank of Japan in giving guidance to the client commercial banks regarding their fund position and operation..has come to be..known as 'window operation'. **1965** H. T. Patrick in W. W. Lockwood *State & Econ. Enterprise in Japan* xii. 609 In 1954..and especially in 1957 and 1961–1962, the Bank of Japan had to resort to direct credit rationing... The term for this is *madoguchi shidō*. The Bank of Japan does not like to have this technique called credit rationing, referring to it instead as 'window operation', a more literal translation. **1923** J. Manchon *Le Slang* 338 *Window-pane*,..un monocle. **1935** D. L. Sayers *Gaudy Night* xvii. 372 Winderpane, we called 'im, along of the eyeglass, but meanin' no disrespect. **1966** Wodehouse *Plum Pie* ix. 249 Freddie no longer wore the monocle. . His father-in-law had happened to ask him one day would he please remove that damned window-pane from his eye. **1966** *Guardian* 28 Sept. 3/3 Trends towards large windowpane checks. **1969** 'O. Bleeck' *Brass Go-Between* (1970) v. 60, I had the chance to admire his fawn trousers with their burnt orange windowpanes. **1978** L. Block *Burglar in Closet* i. 4 My suit was a tropical worsted, a windowpane check in light and dark gray. **1973** *Country Life* 10 May 1330/1 Window pane checked Voile shirt £10.50. **1863** Mrs. Gaskell *Let.* 5 Dec. (1966) 720, I have been waiting..for my cousin Mr Holland to bring me in his list of subscriptions to Mr Parkes' 'booklet' on window-plants. **1895** C. Collins (*title*) Greenhouse and window plants. [**1951** *Dict. Gardening* (R. Hort. Soc.) III. 1290/2 Some species..normally grow buried in the soil with only the upper surface of the leaves exposed; this upper surface is translucent..: such plants are known as 'windowed plants'.] **1971** *Stand. Encycl. S. Afr.* III. 652/2 The amazingly adapted 'window-plant',..almost entirely embedded in the ground, only the transparent apical part of the corpusculum being exposed to the air, allowing light to enter the body of the leaf. **1892** *Vermont Agric. Rep.* XII. 135 Mills manufacturing..furniture and window screens. **1907** *St. Nicholas* May 614/1 We tried to buy wire netting—the sort we use for window screens at home. **1942** W. Faulkner *Go down, Moses* 158 Walks out of the cell toting the door over his head like it was a gauze window-screen. **1926** Kipling *Debits & Credits* 410 They entered the little train... 'Isn't it lucky we've got window-seats?' **1967** O. Hesky *Time for Treason* x. 77 He took a window-seat in the special bus. **1967** E. Hunt *Danger Game* viii. 142 In the plane Elaine was annoyed to find Mrs. Delf had the window seat allotted her. **1981** G. Markstein *Ultimate Issue* 289 The train came into the station, and Verago took a window seat. **1922** S. Lewis *Babbitt* ix. 122 They ate chocolates, went to the motion-pictures, went window-shopping. **1936** B. & S. Spewack *Boy meets Girl* II. iii. 70 Is it true, Mrs. Seabrook, that you and Larry have been window shopping? **1945** G. Endore *Methinks the Lady* ii. 27 Sometimes I went window-shopping with that apartment in mind. **1951** *Landfall* V. 167 'Maybe we could window-shop then?' Wally said. 'Care for a diamond necklace like that one?' **1957** *Times* 12 Nov. (Canada Suppl.) p. xv/3 At weekends carloads of three-generation family groups visit suburban areas to 'window-shop' for somewhere to live. **1973** R. Busby *Pattern of Violence* vi. 96 The office girls came out..and.. joined the phalanxes on the pavements to window-shop outside the big stores. **1934** Webster, Window-shopper. **1951** H. MacInnes *Neither Five nor Three* xxv. 341 A pavement filled with window-shoppers. **1972** P. Marks *Collector's Choice* I. 23 Behind a window..stood a Boudin drawing... The window-shopper smiled. **1955** D. Davie *Brides of Reason* 32 And at our back His eye augments our window-shopping greed. **1956** D. M. Davin *Sullen Bell* xi. 72 The old, innocent pleasure of window-shopping in Regent Street. **1978** *Lancashire Life* Nov. 140/2 (Advt.), Window-shopping may be fun. But instead of admiring from the outside we'd like to welcome you inside. **1936** Kipling *Something of Myself* v. 143, I..was elected to the Athenaeum... I managed to be taken to a delightful window-table [for lunch]. **1957** M. Kennedy *Heroes of Clone* I. v. 46 She and Roy shared a window table. Mundy sat..at the other end of the dining-room. **1979** *Tucson Mag.* Apr. 78/2 Ask for a window table or one on the patio. **1910** *Chambers's Jrnl.* Aug. 512/1 Mr. W. W. Sawyer..was originally a window-trimmer in the cites of Chicago, Milwaukee, and Portland. **1980** *Washington Post* 1 Feb. B4/3 Mr. Van Der Linden began working for Woodward & Lothrop in 1926 as a window trimmer. **1926** *Publishers' Weekly* 22 May 1676/1 Window-trimming. **1984** *N.Y. Times* 21 Mar. D21/3 They even have an Association of Legal Administrators, which is not just window trimming but evidence of the increasing importance of business managers. **1968** *Globe & Mail* (Toronto) 17 Feb. 49/9 (Advt.). 65 Austin.. new tires, window washers. **1970** *Wall St. Jrnl.* 15 June 7/1 Mr. Welk arrived at his office building early and encountered a window-washer. **1977** *New Yorker* 27 June 84/2 One of the cops..had been sent out on a window-washer's platform to talk him into coming down. **1910** W. James *Mem. & Stud.* (1911) xi. 291 To coal and iron mines,..to dishwashing, clothes-washing, and window-washing..would our gilded youths be drafted off. **1950** S. F. Page *Body Engin.* iii. 59 Window pillars and window winders should not be permitted to obstruct the view. **1971** *Sunday Times* (Johannesburg) (Business Sect.) 28 Mar. 4/6 The faults usually consisted of..faulty window winders, loose door handles and sticky locks. **1976** *Derbyshire Times* (Peak ed.) 3 Sept. 20/6 They'll be even more irritated by low-geared window winders.

windowing. Restrict † *Obs.* to sense in Dict. add: **2.** *Computers.* The process of selecting part of a stored image for display or enlargement.
1969 S. Bird in Parslow & Prowse *Computer Graphics* I. 20 The display file is produced from the data structure by the display program package and the transformations such as expansion..and 'windowing' are carried out in the process. **1973** Newman & Sproull *Princ. Interactive Computer Graphics* viii. 154 In the case of the windowing routine, there are various ways in which the page-to-screen transformations may be defined: one obvious way is to specify window and viewport boundaries. **1981** *Internat. Jrnl. Numerical Methods Engin.* XVII. 1110 Only a segment of the first-level window may be delegated to second-level windowing; the rest should remain permanently in operational memory. **1985** *Personal Computer World* Feb. 165/1 It includes functions to deal with the more esoteric of the QL's facilities, such as windowing and general screen-handling.

windowy, *a.* Restrict † *Obs. rare* to sense in Dict. and add: **2.** Having many or large windows.
1863 'G. Hamilton' *Gala-Days* 353 The homes of the students, which seem to have been built..solely to furnish shelter,—angular, formal, stiff, windowy. **1888** *Harper's Mag.* June 130/2 Several large, ugly, windowy wooden bulks grew up for shoe shops.

windrow, *sb.* Add: **c.** (Further examples.) Also *fig.* Used of similar rows of various things not exposed to or caused by the wind.
1948 *Times* 13 Feb. 5/6 Bulldozers then level off the soil and uprooted bush, packing it aside to form banks known as 'windrows' between each contour. **1957** L. Eiseley *Immense Journey* 49 The slowly contracting circle of the water left little windrows of minnows. **1974** *Sci. Amer.* Aug. 21/1 The water soon turned cold again and the fish departed, leaving windrows of dead *Pleuron-codes* along the beaches. **1980** *Ibid.* Oct. 156/2 The soft rock is gathered into long windrows and transferred mechanically to conveyor belts that carry it away to the processing plant.

windrow, *v.* Add: **windrowed** *ppl. a.*: for 1893 citation read '**1851** H. Melville *Moby Dick* I. xli. 311' and add later examples; **wi·ndrower,** a machine for cutting and raking crops into windrows; **wi·ndrowing** *vbl. sb.*
1946 R. Campbell *Talking Bronco* 24 All round the snarled and windrowed sands Expressed the scandal of the waves. **1948** Turner & Johnson *Machines for Farm, Ranch & Plantation* x. 316 Select side-delivery windrowers when cutting grass-seed crops such as alfalfa. **1955** 'P. Janvier' in *Astounding Sci. Fiction* Nov. 68/1 Straggled clumps and windrowed hay..were all that remained of the shrubbery and the lawn. **1970** K. C. Willett in H. W. Mulligan *African Trypanosomiases* xxx. 583 If 'windrowing' (clearing of the felled vegetation into wind-rows) is necessary the cost is greatly increased. **1976** *Columbus* (Montana) *News* 17 June 5 (Advt.), Hay Equipment... Windrowers... Balers.

wind-shake, *sb.* Add: **b.** A shaking (of something) in or by the wind. *poet. nonce-use.*
1939 Dylan Thomas *Map of Love* 12 After the funeral, mule praises, brays, Windshake of sailshaped ears.

windshield (wi·ndʃīld). [f. Wind *sb.*[1] + Shield *sb.*] **a.** Any of various devices for shielding a person or thing from wind; *spec.*

(chiefly *U.S.*) on a motor vehicle = *windscreen* s.v. Wind *sb.*[1] 31.
1902, [in Dict. s.v. Wind *sb.*[1] 30 a]. **1907** *Yesterday's Shopping* (1969) 320D/1 [Coat] Fitted with wind shield and storm cuffs, for driving or motoring. **1911** *N.Y. Times* 16 Oct. 12/7 (Advt.), Speedwell 1911 four-passenger, semi-racer..extraordinary equipment includes top, windshield, shock absorbers, [etc.]. **1924** P. C. Macfarlane *Tongues of Flame* ii. 12 She steadied herself with one hand upon the wind-shield while the other waved to the enthusiastic group of welcomers. **1941** F. H. Joseph *Lett. Home from Britain at War* (1942) 6 We circled the airport three times to allow the captain to clear his windshield of ice by hand. **1946** B. Macdonald *Egg & I* 110 A blast went off almost under the truck and the rocks broke my windshield. **1962** A. Nisbett *Technique Sound Studio* 277 *Windshield*.., shield which fits over microphone and protects diaphragm from 'rattling' by wind, and also contours the microphone for smoother airflow round it. **1978** W. F. Buckley *Stained Glass* 230 Fifty cars, with special passes on their windshields, squatted around the tall, leafless elm trees.

b. *attrib.* and *Comb.* *windshield cleaner*, *scraper*, *squirter*; **windshield wiper** = *windscreen wiper* s.v. *wind *sb.*[1] 31.
1921 *Daily Colonist* (Victoria, B.C.) 13 Mar. 10/6 The Folberth automatic windshield cleaner. **1927** *Sat. Even. Post* 24 Dec. 56/2 Each has extra wide windshield..and windshield wiper. **1955** W. Tucker *Wild Talent* v. 62 Paul could see the rain falling, could see the madly swinging windshield wipers on the waiting cars. **1975** B. Garfield *Hopscotch* xxvi. 275 A combination windshield-scraper and brush. **1976** M. Machlin *Pipeline* ii. 31 The misty, sentimental look on Steele's face disappeared as though a giant windshield wiper had cleared its teary ambience. **1978** *Time* 10 Apr. 22/1 Johnson.. once had to halt his automobile to solve the problem of turning on the windshield squirter.

Windsor. Add: **1. Windsor blue** = *phthalocyanine blue* s.v. *phthalocyanine b; **Windsor knot,** a large, loose knot in a (neck)-tie; so **Windsor-knotted** *a.*; **Windsor pear** (see quots.); **Windsor Red,** the name of a recently introduced type of English cheese containing red wine; **Windsor soap** (earlier example); **Windsor tie** *U.S.*, a broad bias-cut necktie or scarf; **Windsor uniform** (earlier example).
1912 R. Ridgway *Color Standards & Color Nomenclature* 40/1 Windsor blue. **1938** H. Nicolson *Let.* 17 Apr. (1966) 336 His Windsor blue eyes were wistful. **1970** Windsor blue [see *Monastral]. **1953** *Man about Town* Spring 117 (*caption*) How to tie the 'Windsor' knot. **1959** T. Williams *Sweet Bird of Youth* III. 111 He nods slightly, loosening the Windsor-knot of his knitted black silk tie. **1976** J. H. Spencer *Surgenor Campaign* i. 18 The tie was a crisp silver, the sort normally worn only with morning dress and tied in a Windsor knot. **1953** K. Amis *Lucky Jim* ix. 98 His Windsor-knotted silk tie. **1664** J. Evelyn *Kalendarium Hortense* 72 August..Pears. Windsor, Soveraign, Orange, [etc.]. **1860** R. Hogg *Fruit Man.* 221 Windsor... A fine old pear for orchard culture. Ripe in August. It should be gathered before it becomes yellow. **1940** J. Betjeman *Old Lights for New Chancels* 17 Remaining orchards ripening Windsor pears. [**1969** *Vogue* 15 Mar. 65/2 *Red Windsor*, a new British cheese.. basically an English cheddar, gets its pink tinge from an English wine.] **1969–70** *Wine & Food* Dec.–Jan. 11/2 More ideas for cheese gifts... Windsor Red in plain jar, each 10/6. **1982** P. Rance *Gt. British Cheese Bk.* I. ii. 50 These cheeses are made by breaking up Cheddar or Double Gloucester..and..in the case of Windsor Red, pouring wine over the re-milled curd. **1822** B. Haydon *Jrnl.* 16–17 Sept. in *Mem.* (1926) I. 321 A barber who shaved me..so praised his Windsor soap, that I..took six cakes. **1895** *Montgomery Ward Catal.* 95/2 Windsor Ties... Japanese Silk Windsors... Size 4½ × 34 inches. **1912** J. London *Smoke Bellew* 147 He went on dressing,.. tying a Windsor tie in a bow-knot at the throat of his soft cotton shirt. **1968** J. Ironside *Fashion Alphabet* 114 A bias-cut wide tie, usually black, tied in a loose bow in front of the neck—known in America as a Windsor tie. **1781** *Gentl. Mag.* LI. 391/2 The birth-day of the Prince of Wales..was celebrated with extraordinary magnificence... The King, the Prince, the Duke of Cumberland, the great officers of state, and nobility, appeared in the Windsor uniform on this occasion—blue and scarlet.

2. (Further examples.)
1836 T. Power *Impressions of America* I. 440 A bit of old brown Windsor to shave withal. **1840** Thackeray in *Comic Almanack* Nov. 45, I never..knew Naples from brown Windsor. **1895** [see *Windsor tie*, sense 1 above]. **1901** [see *comb-back* s.v. *comb *sb.*[1] 9]. **1939** F. Thompson *Lark Rise* vi. 102 If the father had a special chair..it would be but a rather larger replica of the hard windsors with wooden arms added. **1969** 'J. Morris' *Fever Grass* ii. 21 A small electric fan..and two more Windsors, were the room's only furnishings. **1976** J. Philips *Backlash* (1977) I. ii. 27 Two armchairs, Windsors, for visitors.

windsurf (wi·ndsɒɹf), *v.* orig. *U.S.* [Back-formation f. *Windsurfer: see next.] *intr.* To ride a sailboard; to sailboard. Also **wi·ndsurfing** *vbl. sb.*
1969 *Chr. Sci. Monitor* 17 Nov. 17/1 Depending on the wind and water conditions, older as well as young people can windsurf. *Ibid.*, Windsurfing is new, so new that it's been on the market only within the past month. **1972** *Islander* (Victoria, B.C.) 16 Jan. 3/1 Spreading up and down the west coast is a brand new water sport—windsurfing. **1976** *Southern Even. Echo* (Southampton) 11 Nov. 23/1 Windsurfing, a cross between sailing, surf-riding and high-wire walking, has one big attraction: your boat can be small enough to carry under your arm. **1977** *Austral. Sailing* Jan. 69/2 Young or old, guy or girl, thick or thin, we'll teach you to windsurf in a few short hours. **1980** C. Matthews *Loosely Engaged* 9 Swam, sunbathed and wind-surfed the whole day. **1984** *U.S.A. Today* 6 Apr. 2C/1 But windsurfing—on the Windsurfer—is merely a demonstration sport in the 1984 Olympic Games. **1984** *Times* 25 Aug. 11/3 Earlier this year..an Oxford graduate, aged 25, spent 10 weeks windsurfing clockwise around the coast of Britain.

Windsurfer (wi·ndsɒɹfəɹ). orig. *U.S.* Also **windsurfer.** [f. Wind *sb.*[1] + *surfer.] **1.** The proprietary name in the U.S. of a kind of sailboard.
1969 *Chr. Sci. Monitor* 17 Nov. 17/2 The board segment of the Windsurfer is shaped, with 'slight changes', like a surfboard, though it is heavier, at 37 pounds and longer, at 12 feet. **1974** *Official Gaz.* (U.S. Patent Office) 20 Aug. TM 166/1 Windsurfer, Windsurfing International Inc., Santa Monica, Calif... For sailboats comprising a surf board type hull and a sail. **1981** *Daily Mail* 9 Apr. 39/3 He [*sc.* Hoyle Schweitzer] kept production of his Windsurfer down in order to monitor quality. **1983** *Reader's Dig.* Apr. 132 More Windsurfers have been sold than any class of sailing boat ever. **1984** *Sunday Times* (Colour Suppl.) 28 Oct. 25/2, I wanted to learn more and get myself back on a windsurfer as soon as I could.

2. One who engages in the sport of windsurfing, a sailboarder.
1969 *Chr. Sci. Monitor* 17 Nov. 17/3 The lone windsurfer (one per board) stands near the middle, left foot just ahead of the mast and hands holding tightly to the 'wishbone'. **1977** *Austral. Sailing* Jan. 27/3 Clive Colonso is one of Britain's few expert windsurfers. **1982** *Times* 3 May 5/1 Twenty windsurfers were rescued from the North Sea yesterday. **1984** *Times* 25 Aug. 11/7 Windsurfers tend to be individualists, happy to sail alone.

wind-up, *sb.*[1] and *a.* Add: **A.** *sb.*[1] **2.** *Baseball.* The motions of a pitcher preparing to pitch the ball. Also *fig.* and in other sports.
1931 D. Runyon in *Collier's* 25 Apr. 38/2, I take a good wind-up..but..the ball does not break as I expect. **1936** *Philadelphia Rec.* 30 July 19/1 Blanton is the sort of orator who cannot shorten his pitching motion... He is unable to make a simple motion without taking a full windup. **1951** [see *stretch *sb.* 1 i]. **1974** Mills & Butler *Tackle Badminton* v. 45 The great temptations to be avoided with drop shots are..making an exaggerated wind-up with over-emphasized power, [etc.]. **1976** *Webster's Sports Dict.* 483/2 The windup, which is usually accompanied by a rocking of the body, sets a rhythm which the pitcher follows until the ball is released.

3. a. Material that has become wound round something. **b.** The action of winding or coiling something round something else. **c.** The action of becoming twisted or stressed by the application of torque.
1964 *Gloss. Letterpress Rotary Printing Terms* (B.S.I.) 21 *Wind up*, paper accidentally wrapped round the impression cylinder, plate cylinder, or inking rollers. **1966** J. S. Cox *Illustr. Dict. Hairdressing & Wigmaking* 165/2 *Wind-up*,..the winding of the hair on curlers. **1969** W. R. R. Park *Plastics Film Technol.* ii. 15 This technique..generates a greater percentage of scrap or recycle material than the use of a stationary windup. **1972** *Sci. Amer.* Dec. 51/1 The carriage was pushed back and forth by the spinner, one way during the drawing-twisting operation and the other way during windup. **1975** *Drilling Technol. & Collet Chuck* (Bristol Erikson Ltd.) 4 Since the forces created in any cutting action are never constant, it follows that the amount of torsional 'wind-up' will be continually varying. **1976** G. Robson *Land-Rover* vii. 117 To take care of transmission wind-up..the new car was to have a third, central differential, with a limited-slip mechanism inside it. **1978** *Hot Car* July 89/4 Traction bars..are..bolted by way of U-bolts and brackets to the rear leaf springs of a car such that they prevent wind-up of the rear axle on full-power starts.

B. *adj.* **1.** (Later examples, now used esp. of toys and gramophones.) Also of a window: made to be moved up (to shut) and down (to open) by means of a handle wound with a rotary motion.
1951 *Festival of Britain Catal.* 149/1 Wind-up plate glass window, weatherproof and draughtproof. **1962** E. O'Brien *Lonely Girl* v. 64 The last record lay on the green baize of the wind-up gramophone. **1968** 'E. McBain' *Fuzz* ix. 155 The police in this city are like wind-up toys with keys sticking out of their backs. **1970** *Motoring Which?* July 98/1 A few of these modifications—wind-up windows..also appeared on the ordinary Mini. **1982** N. Painting *Reluctant Archer* vii. 105 There were other gramophones, too. Wind-up ones.

wind-up (wi·nd,ɒp), *sb.*[2] *colloq.* [f. phr. *to get the wind up* s.v. Wind *sb.*[1] 10 b.] A state of nervous anxiety or fear; an occurrence of this.
1917 G. S. Gordon *Let.* 13 Feb. (1943) 69 By that time my runner was showing signs of 'wind-up'... He thought I was very unfeeling, not to go down to a cellar till the shower [of shelling] was over. **1922** *Encycl. Brit.* XXX. 64/1 Many other pilots..have been through the same stages of 'wind-up'. **1931** J. Hilton *Murder at School* x. 204 We were having a smoke... We got an awful wind-up, thinking somebody..might have smelt something. **1952** *Chambers's Jrnl.* Feb. 82/2 Putting on a bold face, but with a fair amount of wind-up, I walked..in the direction the hand pointed to. **1980** A. Price *Hour of Donkey* xiv. 220 Bit of nerves..the old wind-up.

windy, *a.* Add: **II. 8. a.** (Later examples.)
1918 C. J. Biddle *Fighting Airman* (1968) 147 He thought what made the men more 'windy' than anything else..was the thought of..having to lie there all day before being able to get to a doctor. *a* **1948** D. Welch *Voice through Cloud* (1950) iv. 39 He [*sc.* a patient facing an operation] laughed so much that the man with the bandaged ear became exasperated and said, 'Why do you make so much noise? That shows you're windy. If you didn't care, you wouldn't say anything.' **1960** J. R. Ackerley *We think World of You* 123 'E was windy, but I swore it was safe and that nothing could 'appen. **1985** D. Clark *Performance* ii. 40 'Are you feeling windy?' 'Do I look as if I am?'

b. Applied to a frightening or nerve-wracking place or situation. *Services*'.
1919 *Narrative Battery A, 101st Field Artillery* (U.S. Artillery) 118 It was a 'windy' place to be.., as the enemy raked it with machine gun and trench mortar fire all day and night. **1925** Fraser & Gibbons *Soldier & Sailor Words* 305 *Windy Corner*, any place specially dangerous or trying to the nerves on account of enemy fire. **1927** A. M. Sullivan *Old Ireland* xi. 226 All the 'windy corners' of his front. **1928** T. E. Lawrence *Let.* 1 May (1938) 599 Such performances require a manner to carry them off... A windy business.

9. *Windy City* (U.S.), a nickname for Chicago.
1887 *Courier-Jrnl.* (Louisville, Kentucky) 31 Jan. 5/1 An alleged anarchist dynamite plot from the Windy City. **1908** K. McGaffey *Show Girl* 58 Chicago is surely rightly named when they call it the Windy City. **1948** *News-Dispatch* (Michigan) 3 Apr. 9/3 The handsome Windy City youngster has an enormous following. **1979** K. Bonfiglioli *After you with Pistol* xvi. 120 The scent of the Chicago River as it slides greasily under the nine bridges in the centre of the Windy City.

windy (wi·ndi), *sb.*[1] Repr. colloq. and dial. pronunc. of Window *sb.* Cf. *winder *sb.*[5]
1830 W. Carleton *Traits & Stories Irish Peasantry* I. 193 Will you hand me over that other clew out of the windy-stool [= window-sill] there? *c* **1883** D. Boucicault *Shaughraun* II. i. 11 He got sight of my face agin the windy. **1921** V. Jacob *Bonnie Joann* 37 Lowse ye the windy-sneck a wheen. **1977** *Hot Car* Oct. 15/2 Another problem with fitting 'lectric windys to English cars is their narrow door design.

windy (wi·ndi), *sb.*[2] *N. Amer. local slang.* [f. Windy *a.* 6 a.] A tall story; a piece of boasting or exaggeration.
1933 *Amer. Speech* VIII. I. 53/2 *Windy*,..a tall tale, a wildly unreasonable story. **1933** J. V. Allen *Cowboy Lore* III. 60/2 *Telling a windy*, telling a boastful story. **1935** H. L. Davis *Honey in Horn* iii. 24 He could invent windies about his stand-in with the girls.

wine, *sb.*[1] Add: **1. g.** *wine and cheese* (*party*, etc.).
1961 *Daily Tel.* 5 Dec. 9/2 To my mind, the ideal wine and cheese party is given around midday. **1969** *Times* 25 Sept. 27/2 All 550 members of the staff have been invited to a wine and cheese party on that day. **1976** M. Duke *Death at Wedding* xiii. 148 He's gone to the local Labour Party wine-and-cheese do. **1977** B. Pym *Quartet in Autumn* xvii. 155 She did not feel capable of guessing what kind of an evening party, for she could only think of 'wine and cheese' which seemed altogether unworthy of Mr Strong.

2. (Earlier *lit.* examples of *the wine of the country.*)
1817 H. Matthews *Diary of an Invalid* (1820) ii. 39 As much of the wine of the country as you like. **1865** A. Trollope *Can you forgive Her?* II. xxxvi. 287 He had ordered a bottle of Sauterne; but the landlord had thought..that a bottle of ordinary wine of the country would do as well.

4. (Earlier example.)
1857 'C. Bede' *Mr Verdant Green Married* xii. 101 Mr Bouncer..gave his last wine (wherein he produced some 'very old port').

6. (Further example.) Also in *sing.*
1935 W. A. Thorpe *Eng. Glass* iv. 129 Mansell had three grades of 'wines' which in 1639 he described as follows.. 'Ordinary Drinking-Glasses—for Wine.' **1947** *Glass Notes* Dec. 16 *Problem for 1948*, to discover the following:..a facet stem wine with a domed foot. **1974** *Habitat Catal.* 72/2 *Bistro.* Really good value for drinking anything from sherry to sweet stout. 3½ oz sherry 14p, 5 oz wine 15p 8 oz goblet 16p.

6*. Passing into *adj.* A dark red colour.
1895 *Montgomery Ward Catal.* 3/1 Royal Serge, 22 inches wide, in plain, solid colors..Colors: Cardinal, wine, brown. **1923** [see *laurel *sb.*[1] 2 e]. **1950** B. Pym *Some Tame Gazelle* xv. 166 She had visions of herself..in her brown velvet or wine crêpe de Chine. **1981** *Country Life* 22 Jan. 226/3 Feather-stitch grey and wine pullover.

7. a. (*a*) *wine-breath*, *-sap*.
1922 W. B. Yeats *Seven Poems* 1 Being sharpened by his death To drink from the wine-breath While our gross palates drink from the whole wine. **1917** D. H. Lawrence *Look! We have come Through!* 158, I want the fine, kindling wine-sap of spring.
(*b*) *wine basket*, *box*, *cistern*, *-cup* (*lit.* examples), *-district*, *funnel*, *-gourd*, *industry*, *jug*, *-kitchen*, *-land* (later examples), *table*, *-trade* (later examples).
1835 Dickens *Sk. Boz.* (1836) 1st Ser. I. 291 Waiters with wine-baskets in their hands are placing decanters of Sherry down the tables. **1974** *Habitat Catal.* 81/2 Wicker wine basket. For serving fine delicate wines, without disturbing the sediment. **1982** *Daily Tel.* 8 Dec. 17/1 Wine boxes have made buying easier... Wine boxes are generally about £7.50 for three litres. **1984** *Which?* May 195/1 *Which? Wine Monthly* has been testing wine boxes again. A few this time were rather nice..but many were still disappointing, particularly when they'd been opened for a week or so. **1881** W. J. Cripps *College & Corporation*

Plate v. 132 (*caption*) *Wine Cistern*, circa 1701. **1971** *Country Life* 1 Apr. 766/1 His [*sc.* Thomas Heming's] earlier shallow sauce tureens..were echoed in his own 1,457-ounce massive wine cistern for Belton House. **1819** W. Scott *Ivanhoe* I. xiv. 294 He raised ..the wine-cup to his lips. **1910** S. W. Bushell *Chinese Art* II. 17 The poets of the time liken their wine cups to 'disks of thinnest ice'. **1980** *Catal. Fine Chinese Ceramics* (Sotheby, Hong Kong) 75 An incised white dragon Winecup of thinly potted bell shape. **1835** J. E. Alexander *Sk. in Portugal* xi. 260 A considerable reach of the river was also seen to the east and west, and the wine-district in the far distance of Alto Douro. **1976** *National Observer* (U.S.) 4 Dec. 8/3 A bureau spokesman says it could be the first step toward establishment of a national wine-district system similar to that of France. **1838** J. G. Flügel *Compl. Dict. Ger. & Eng. Languages* II. 833/2 *Wein*,..*-trichter*, *m.* wine-funnel. **1981** *Times* 17 Oct. 12/7 A Dundee wine funnel of about 1820 by William Law sold for £680. **1952** L. MacNeice *Ten Burnt Offerings* 56 Did not these whitewashed rooms among wine-gourds, goat-skins, ikons, Include a letter or two with a foreign postmark. **1966** P. V. Price *France* 132 More than three million Frenchmen are engaged in the wine industry and there are about a million and a half wine growers. **1976** R. M. Stern *Will* iv. 24 Prohibition stifled the California wine industry. **1922** Joyce *Ulysses* 142 'Tis the hour, methinks, when the winejug, metaphorically speaking, is most grateful in Ye ancient hostelry. **1981** R. Manheim tr. *G. Grass's Meeting at Telgte* xv. 88 His busy treasure hunting seemed to leave him no free hand for the wine jug. **1924** D. H. Lawrence in M. Magnus *Mem. Foreign Legion* 45 So we went into a little cave of a wine-kitchen to drink a glass of wine. **1963** *Punch* 21 Aug. 280/2 The most northerly German winelands. **1971** *Sunday Times* (Johannesburg) 28 Mar. 25/1, (Advt.), On a southern mountain slope, in the heart of the Stellenbosch winelands, the skills of man and the secrets of nature combine to create five distinctive wines. **1839** Poe in *Burton's Gentleman's Mag.* Oct. 212, I had indulged more freely than usual in the excesses of the wine-table. **1935** *Burlington Mag.* May p. xli/2 A superb wine-table, also tripod, a pair of torchères, *circa* 1760. **1976** *Derbyshire Times* (Peak ed.) 3 Sept. 15/5 (Advt.), Two Walnut pie crust wine tables. **1966** P. V. Price *France* 133 The wine trade in Great Britain consider that the British wine drinker is protected ..by the laws of the country. **1977** *Times* 14 May 13/3 Accommodation will be heavily booked at vintage time by the wine trade.

b. *wine-drinker*, *-importer*, *-shipper*; *wine-loving*, *-making*, *-producing* (earlier example), vbl. sbs. and ppl. adjs.
1935 A. G. Macdonell *Visit to Amer.* x. 183 California could produce a vin ordinaire to sell at thirty or forty [cents]. If she did, she would gradually build up a great community of wine drinkers. **1983** *Listener* 14 July 18/3 Burgundy and Bordeaux are still the British wine-drinker's dream lands. **1959** E. H. Clements *High Tension* ii. 21 My family were wine-importers with offices in London, Edinburgh and Bordeaux. **1981** W. J. Burley *House of Care* i. 6 His job with a firm of London wine importers. **1921** 'L. H. Davison' *Movements in European Hist.* iv. 43 The Romans of Latium were short, dark men of the wine-loving lands. **1980** *Times* 27 Nov. 21/1 A wine-loving traveller. **1814** P. P. Carnell (*title*) A treatise on family winemaking. **1883** V. Lush *Jrnl.* 27 Aug. (1975) 245 If life and health be spared to us, wine-making will become henceforth like jam making and fruit tinning, one of the fixed employments of the Autumn. **1979** A. Maling *Koberg Link* (1980) xxv. 135 'We have vineyards.'..'Tough business, winemaking.' **1846** R. Ford *Gatherings from Spain* xiv. 150 The wine-producing districts. **1949** C. Graves *Ireland Revisited* x. 151 The traditional story about James Lynch Fitzstephen is that he was a leading wine-shipper in Galway. **1972** *Times* (Wines & Spirits Suppl.) 27 Nov. p. viii/5 With their large number of outlets they could go direct to wine growers on the Continent, by-passing the wine shippers.

c. *winebig*, *-drenched*, *-ensanguined*, *-fizzling*, *-flushed*, *-heavy*, *-stained*, *-warm* adjs.
1922 Joyce *Ulysses* 249 John Henry Menton..stared from winebig oyster eyes. **1914** W. B. Yeats *Responsibilities* 2 Those wine-drenched eyes. **1925** H. Acton in *Oxf. Poetry* 4 And we had thought to fashion of our joy Round crackling pearls to pelt our wine-drenched loves. **1928** W. B. Yeats tr. *Sophocles' King Oedipus* 6 And Bacchus' wine-ensanguined face that all the Maenads sing. **1922** Joyce *Ulysses* 420 Come on, you winefizzling..existences! **1912** E. Pound *Ripostes* 26 Wealthy and wine-flushed. **1964** J. Michie tr. *Horace's Odes* I. vii. 33 Set on his wine-flushed brow ..brave garlands of poplar. **1897** W. B. Yeats *Secret Rose* 2 The old and foolish king..snored fitfully in a wine-heavy sleep. **1899** —— *Wind among Reeds* 49 Dwelt among wine-stained wanderers in deep woods. **1983** J. Masters *Man of War* xxiii. 299 A Michelin map spread out on the wine-stained..table. **1953** C. Day Lewis *Italian Visit* vi. 65 When cypresses jetted like fountains of wine-warm air.

8. wine bar, (*a*) a bar or counter in a club, shop, etc., where wine is kept or sold; (*b*) a licensed establishment specializing in the serving of wine (and food); **wine book,** (*a*) a book for keeping records of wines bought and consumed; (*b*) a book about wines; **wine-buff,** a wine enthusiast; **wine butler,** a servant who has charge of the wine-cellar and serves the wine (cf. Butler *sb.* 1 a); **wine coaster** = Coaster 6; **wine-cooler** (later examples); **wine-dance,** a dance performed in celebration of wine; **wine-dark**: occas. (*poet.*) as *sb.*; **wine-dot** [joc. f. Wyandotte] *Austral. slang*, an addict of cheap wine; **wine farm** *S. Afr.*, a farm on which grapes are grown for wine-making and on which wine is frequently made; so **wine farmer; wine fountain,** a large vessel for holding and dispensing wine; **wine-grower** (earlier example); **wine gum** [*gum *sb.*[2] 1 g], a fruit-flavoured sweetmeat made with gelatine; **wine label,** (*b*) the paper label affixed to a bottle of wine, stating its name and provenance; **wine lake,** a stockpile or surplus of wine; **wine list** (now the usual term) = *wine-card*; **wine lodge,** (*a*) = Lodge *sb.* 12 c; (*b*) a licensed establishment selling wine, beer, and soft drinks; **winemanship,** the display of real or pretended knowledge about wine; **wine rack,** a frame with compartments for holding bottles of wine; **wine room,** a bar-room where wine is served; **wine-skin:** also *transf.*; **wine snob** (see quot. 1951); hence **wine-snobbery; wine steward,** a servant responsible for serving wine; **wine-tasting,** testing the quality of wine by tasting; an occasion when this is done; **wine-vault,** (*a*) (earlier example); (*b*) (earlier example); **wine waiter,** a waiter responsible for serving wine; similarly **wine waitress; wine writer,** a person who writes about wine for publication.
1938 R. Graves *Count Belisarius* iii. 65, I was busy at some task behind the wine-bar. **1940** M. Sadleir *Fanny by Gaslight* I. 270 He offered her a job as barmaid... Her new place of business was a girlery as well as a wine-bar. **1976** *Amer. Speech* 1974 XLIX. 117 *Wine bar*, counter in a liquor store, stocked with wines. **1981** B. Knox *Killing in Antiques* iv. 87 Dunbar stopped the car in a side street..just a stone's throw from the wine bar. **1983** *Which?* Dec. (Publications Suppl.), For an accurate description of over 200 wine bars across the country, this section of the book is unbeatable, with critical comments on the range of wines, and an assessment of the food and perceptive summing-up of the atmosphere. **1947** L. G. Green *Tavern of Seas* vii. 59 Documents, wine-books, casks and iron chests..all make a picture of careful work and gay entertainment. **1975/76** *Listener* 25 Dec. & 1 Jan. 891/1 It is easy to mock the pretensions of wine writers... Writing a new wine book is as difficult as building a better mouse-trap. **1976** *Ibid.* 5 Aug. 158/2 One area which beer connoisseurs will have to cultivate in order to approach the influence of wine-buffs—the language of appreciation. **1880** E. W. Hamilton *Diary* 9 May (1972) I. 10, I have been offered and have undertaken the post of Chief Wine Butler to Mr. G., which I hope will secure something rather less nasty in his cellars. **1973** *Times* 25 Aug. 12/6 Wine service is of such skill that it should make the average English wine butler blush. **1956** G. Taylor *Silver* ix. 201 *Wine Coasters*, circular wooden base on baize, with silver sides. **1971** *Country Life* 15 July 183/1 The platform [of a cruet] was encircled with a deep gallery of wood..in the manner of a giant wine coaster. **1848** Wine-cooler [see *en permanence* s.v. *en *prep.*]. **1977** W. M. Spackman *Armful of Warm Girl* 29 The waiter had swooped in rolling a second wine-cooler to set beside the first. **1920** D. H. Lawrence *Touch & Go* I. ii. 29 They begin to sing, dancing meanwhile, in a free little ballet-manner, a wine-dance, dancing separate and then together. **1934** W. B. Yeats tr. *Sophocles' Oedipus at Colonus* in *Coll. Plays* 543 Come praise The wine-dark of the wood's intricacies. **1953** T. A. G. Hungerford *Riverslake* 35 'Is he a wine-dot?' 'Is he hell!.. He's never off it.' **1976** D. Hewett *This Old Man comes rolling Home* 11 Gawd, you smell like an old wine-dot, Laurie. **1923** O. Schreiner in *Cape Times* 18 Aug. 3/1 The sinking valley with its sprinkling of wine-farms. **1970** *Cape Times* 28 Oct. 21/1 (Advt.), Choice Wine Farm in extent 40 morgen. **1984** *Times* 1 Nov. 27/7 He had invited Miss Budd to stay on his wine farm. **1915** G. McC. Theal *Hist. S. Afr. 1795–1872* ii. 36 The British government held out great inducements to South African winefarmers to increase the quantity of their produce. **1889** *Cent. Dict.*, Wine-fountain. **1931** E. Wenham *Domestic Silver* ii. 17 Wine-fountains nearly 4 feet long and 3 feet wide. **1969** E. H. Pinto *Treen* 53/1 The lignum vitae wine fountain..is part of the Burrell Bequest to Glasgow Museum and Art Gallery. **1844** Mill *Ess. Pol. Econ.* i. 45 The wine-growers of France..imagine that free trade would relieve their distress by raising the price of their wine. **1953** Winegum [see *hundred *sb.* and *a.* 7]. **1981** *Times* 1 May 19/3 Energy is stored in plastic pellets, like wine gums. **1954** 'M. Cost' *Invitation from Minerva* 209 On the wine label, above the sycamore..is a coat of arms. **1980** N. Freeling *Castang's City* xxi. 139 People belonged to a multiplicity of little gatherings..Wine-label collectors, neighbourhood betterment leagues. **1974** Wine lake [see *lake *sb.*[4] 1 b]. **1979** *Guardian* 14 Mar. 14/1 Wine-lakes and butter-mountains may be jokes, but they are sick ones. **1984** *Times* 4 Oct. 1/4 The table wines.. have..fared reasonably well, to the distress no doubt of the European community, whose wine lake is already overflowing. **1898** G. B. Shaw *You never can Tell* II. 251 Crampton snatches the wine list rudely from him and irresolutely pretends to read it. **1935** A. G. Macdonell *Visit to America* x. 182, I waved the wine list..and shouted for the wine waiter. **1972** P. V. Price *Eating & Drinking in France* 248 The Nicolas establishments are reliable and the wine lists especially attractive. **1880** Wine lodge [see Lodge *sb.* 12 c]. **1922** Joyce *Ulysses* 392 *Mort aux vaches*, says Frank then in the French language that had been indentured to a brandy shipper that has a winelodge in Bordeaux. **1962** *Guardian* 24 Dec. 4/3 There's the Wine Lodge. You can get a glass of small white Australian for ninepence. **1977** *Punch* 31 Aug.–6 Sept. 345/1 Huge and dingy, the saloon bar looked like a cross between the main hall in the old Euston railway station and one of Yates's less-glamorous Wine Lodges. **1958** *Observer* 11 May 16/4 (*heading*) Winemanship. **1977** T. Heald *Just Desserts* v. 87 A passable imitation of genuine winemanship. **1974** *Habitat Catal.* 82/1 Wine rack. Wood and metal frame. **1981** 'J. Sturrock' *Suicide Most Foul* vii. 129 Wine racks, but..not many bottles. **1865** *Leaves from Diary Celebrated Burglar & Pickpocket* xxxv. 116/1 Several ladies..made their way to where we were sitting, and in the usual wine-room style flung themselves into our lap! **1898** A. Bennett *Man from North* xxvi. 224 Seated in a wine-room or lager-beer hall. **1965** O. Arundell *Sadler's Wells* viii. 102 [In 1825] they made Rosoman's old private house at the New River Head end of the theatre into box-offices, wine-rooms and saloon. **1923** D. H. Lawrence *Birds, Beasts & Flowers* 15 What is it, in the grape turning raisin, In the medlar, in the sorb-apple, Wineskins of brown morbidity. **1928** E. Waugh *Decline & Fall* x. 113 Hullo, Prendy, old wine-skin! How are things with you? **1951** R. Postgate *Plain Man's Guide to Wine* i. 17 A Wine Snob is a man..who uses a knowledge of wine, often imperfect, to impress others with a sense of his superiority. **1977** Wine snob [see *snob *sb.*[1] 3 d]. **1982** 'W. Haggard' *Mischief-Makers* i. 17 He bought it [*sc.* wine] at a multiple grocer but his excellency was not a wine snob. **1966** H. W. Yoxall *Fashion of Life* xxv. 241 There's been much talk recently about wine snobbery, most of it rather stupid. **1898** A. M. Binstead *Pink 'Un & Pelican* iii. 65 'Aha!' cried Swears..'here's a bit o' luck—the wine-steward! Half a dollar is never thrown away on a wine-steward.' **1978** *Chicago* June 237/1 Freddy's the least intimidating and probably most knowledgeable wine steward in town. **1936** 'R. West' *Thinking Reed* vii. 216 He paused..to say in his thick, wine-tasting voice: 'Your wife's looking very pretty!' **1945** E. Waugh *Brideshead Revisited* I. iv. 75 Sebastian had found a book on wine-tasting, and we followed its instructions. **1958** [see *put *v.*[1] 38 f (*a*)]. **1980** *Sunday Times Mag.* 14 Sept. 96 He or she might take a dozen trips abroad each year, attend five or six wine tastings in a week, sample 30 or so bottles of wine a day. **1791** J. Woodforde *Diary* 27 Sept. (1927) III. 301 Mr. J[?] Priest having the keys of his Father's Wine Vaults, I went and tested some Port Wine. **1837** Dickens *Sk. Boz.* 2nd Ser. 73 The old tottering public-house is converted into a spacious and lofty 'wine-vaults'. **1927** C. Connolly *Let.* 11 Feb. in *Romantic Friendship* (1975) 251, I met the Spanish wine waiter. **1969** I. Drummond *Man with Tiny Head* i. 24 The wine-waitress brought the wine-list. **1974** *Times* 9 Oct. 18/7 A race by wine waiters and waitresses..each carrying a tray bearing four glasses and an open bottle of wine. **1975** Wine writer [see *wine book* above].

wine, *v.* Add: **2.** Freq. in collocation with *dine*: cf. sense 3 in Dict. (Later examples, not in a university context.)
1937 L. Hart in R. Rogers *Rodgers & Hart Songbk.* (1951) III. 166 I've wined and dined on mulligan stew, and never wished for turkey. **1961** *Guardian* 10 Nov. 7/1 Mr Delmer dined and wined with the enemies of democracy in Germany. **1981** *N.Z. Listener* 4 July 80/1 Impressive consultants (with many of whom I have wined and dined).

wineberry. Add: **2.** *N.Z.* = *makomako[2].
1889 T. Kirk *Forest Flora N.Z.* 223 The makomako or 'wine-berry' of the settlers was discovered by Banks and Solander. **1910** L. Cockayne *N.Z. Plants & their Story* iii. 37 The wineberry..has distinctly pleasing rosy-coloured flowers. **1966** [see *makomako[2]]. **1971** *N.Z. Listener* 6 Sept. 17/1 There were wineberry trees in the bit of bush.

Winebrennarian (wəinbrĕnē·riăn), *sb.* (and *a.*) *U.S.* Also **Winebrennerian.** [f. the name of John *Winebrenner* (1797–1860), founder member of the sect + *-arian.] A name given to a member of the Church of God, an evangelical sect founded in 1830 in Pennsylvania. Occas. also as *adj.*
1867 W. H. Dixon *New Amer.* II. xxix. 309 No sect escaped this rage for separation..[neither] River Brethren, nor Winebrennarians. **1889** *Cent. Dict.*, Winebrennerian.. *a.* and *n.* **1903** *Christian Advocate* 8 Jan. 7 Christian Scientists..Church of God (Winebrennarian). **1925** T. Dreiser *Amer. Tragedy* (1926) II. II. xlvii. 65 It was the summer seat and gathering place of some small religious organization or group—the Winebrennarians of Pennsylvania. **1974** R. Kern *John Winebrenner* p. vii, It would be regrettable if only 'Winebrennerians' of all sorts read the book.

wine-cellar. Add: **b.** The wine stored in a wine-cellar, esp. with reference to its quality.
1861 Mrs Beeton *Bk. Househ. Managem.* 963 Nothing spreads more rapidly in society than the reputation of a good wine-cellar. **1976** J. Archer *Not a Penny More* viii. 87 James arrived carrying a bottle of Beaune Montée Rouge 1971—even his wine cellar was fast disappearing. **1976** P. G. Winslow *Witch Hill Murder* (1977) ii. 30 Her cook, her wine cellar and the service at her table were not to be matched.

wined, *ppl. a.* Add: **2.** Also const. *up*.
1973 C. Bonington *Next Horizon* ii. 41 'We'll get them well wined up tonight, and persuade them that there's a good route round the back of the Tower,' said Barrie. **1982** *Newsweek* 11 Jan. 26/2, I think they were wined up and looking for a joy ride.

wine-glass. Add: **a.** (Earlier examples = *wineglassful*.)
1786 J. Woodforde *Diary* 18 Aug. (1926) II. 264 She is to take a Wine Glass of the Mixture..every six Hours. **1846** *Jewish Man., or Pract. Information Jewish & Mod. Cookery* i. 2 Pour in a wine-glass of port-wine.

wine-grape. 2. For *U.S.* read orig. *U.S.* and add later examples.
1922 Joyce *Ulysses* 468 Mammoth roses murmur of scarlet winegrapes. **1981** *Times* 7 Feb. 13/4 They do not want any modification of traditional Dao style by the introduction of other European wine grapes.

winey, var. Winy *a.*

wing, *sb.* Add: **II. 5. d.** (*a*) (Earlier example of the plane of an aeroplane); (*b*) (earlier and later examples of a pilot's badge: also *transf.* and *fig.*); (*c*) *slang*, an arm (examples); also *transf.*

1883 W. Aitken *Lays of Line* 65 Cam' an auld sodger yince wha was short o' a wing. 1904 O. & W. Wright *Brit. Pat.* 6732 1 The superposed horizontal surfaces.. formed by stretching cloth upon frames of wood and wire, constitute the 'wings', or supporting part of the apparatus. 1917 'Contact' *Airman's Outings* i. 5 The pilots have passed their tests and been decorated with wings. 1947 *Sun* (Baltimore) 3 Apr. 20/1 He came up with a bad arm during the season, and had been troubled before with it. If the big man's wing behaves this year he should be of considerable value. 1964 J. Cheever *Wapshot Scandal* ii. xxvii. 259 He..began to pitch the eggs... He had a good wing and by heaving the eggs far away.. he was able to divert the..crowd. 1967 *Boston Globe* 22 Mar. 11/1 Wins wings as stewardess for American Airlines. 1976 *Publishers Weekly* 19 Apr. 78/3 Mike Hagen earns his wings as a crop duster in rural Florida.

6. b. Also *spec.* of jumps for horse-riding: see quot. 1953[1].

1953 G. Brooke *Introd. Riding & Stablecraft* 12 Wings to a fence, something in the nature of hurdles placed on either side and at an angle to a fence to prevent a horse from running out to either hand. *Ibid.* iv. 39 It is advisable to start over a small fence with wings. 1960 *Times* 23 July 9/4 The moment to hit the pony is when it is well into the wings and about half a stride from the jump. 1977 J. Kidd *Horse & Pony Man.* iv. 56 When the fence is introduced always place wings or sloping poles on either side to discourage the horse from running out.

e. (Examples of a motor vehicle.)

1928 *Daily Mail* 25 July 9/3 The force of the impact threw the car temporarily out of control, but with its front wings crumpled it continued its dash towards London. 1955 *Times* 10 May 7/7 The visibility forward would be better if it took in the near side front wing, but the rearward view through the 3ft. 9in. wide window is excellent.

9. a. Also in extended use, any more or less separate section of a building, esp. of a hospital or prison.

1908 J. M. Sullivan *Criminal Slang* 27 Wing, a section of a prison. 1959 L. Lee *Cider with Rosie* 132 Hannah Brown was put to bed in the Woman's Wing, and Joseph lay in the Men's. 1967 *Listener* 1 June 718/3 Three weeks later he was back in C wing. 1981 C. Priest *Affirmation* iii. 19, I found a letter from the Governor of Durham Prison, saying that Uncle William had been admitted to the hospital wing.

c. Also in *fig.* phr. *waiting in the wings* and varr., ready to act or make an appearance; (for the moment) taking no part in the action.

1876 H. James in *Atlantic Monthly* Dec. 691/1 The author has given him a mother who..has been kept waiting in the wing, as it were, for many acts. 1946 P. Bottome *Lifeline* iii. 39 We've Churchill waiting in the wings, to take the helm when the storm breaks. 1963 V. Nabokov *Gift* iv. 237 Already famous, he remained as it were in the wings of his busy, talkative thought. 1977 *Sat. Rev.* 3 Sept. 44/1 Despite vast expenditures on research and development..the videodisc is still hovering diffidently in the wings. 1985 *Times* 19 Jan. 21/1 Yesterday's huge jump in the share price suggests there is a buyer in the wings.

11*. *Physics.* A part of a spectral line where the intensity tails off to nothing at either side of it.

1959 *Canad. Jrnl. Physics* XXXVII. 1252 (*caption*) Graph illustrating the dispersion line form for the high-frequency wing of the S(1) line of normal hydrogen at 85° K. 1982 *Sci. Amer.* July 77/3 At positions in the cloud other than the position of the infrared source the broad velocity wings disappeared and the lines had the narrow widths we had originally expected.

III. 18*. *to spread* (*stretch, try*) *one's wings*: to test or develop one's powers; to lead a life of wider scope than hitherto.

1864 G. Meredith *Let.* 1 June (1970) I. 260 One thought my Marie merely trying her wings. 1872 Geo. Eliot *Middlem.* II. iv. xxxiv. 192 He is trying his wings. He is just the sort of young fellow to rise. 1876 Trollope *Prime Minister* III. xx. 332 When I found myself the son-in-law of a very rich man I thought I might spread my wings a bit. 1926 R. H. Tawney *Relig. & Rise of Capitalism* ii. 67 It was in an age of political anarchy that the forces destined to dominate the future tried their wings. 1953 'W. Cooper' *Ever-Interesting Topic* v. ii. 252 He decided to compose music as well as to play it: he began to try his wings as a creative artist, and he found they held him up. 1973 'P. Malloch' *Kickback* xi. 69 'Hagan's stretching his wings a bit.' 'Beginning to feel his weight, is he?' 1978 S. Radley *Death & Maiden* xv. 145 She wanted to spread her wings a bit, meet new people.

19. *wing-and-wing* (earlier examples).

1781 J. Greenwood in *Maryland Hist. Mag.* (1910) V. 129 We were now wing and wing, that is right before the wind. 1828 J. F. Cooper *Red Rover* I. iii. 84 That.. schooner would make more way going wing-and-wing than jammed up on a wind.

19*. *a wing and a prayer*, a joc. form of reference (after quot. 1943) to an emergency landing by an aircraft. Also *fig.* and as *attrib. phr.* in allusion to reliance on hope in desperate situations.

1943 H. Adamson *Comin' in on a Wing & a Prayer* (song), Tho' there's one motor gone, we can still carry on, Comin' In On A Wing And A Pray'r. 1967 *Economist* 3 June 998/2 The ITA's problem is to decide which applicants give most promise of maintaining an improvement over six years... This is largely a wing and a prayer decision. 1971 P. O'Donnell *Impossible Virgin* xii. 250, I reckoned it was better to get kitted up for a proper job rather than come charging down 'ere on a wing and a prayer. 1977 W. Marshall *Thin Air* xii. 150 The co-pilot brought it in... Wing and a prayer! 1980 T. Barling *Goodbye Piccadilly* xvi. 334 The pilot spoke to him... 'This is real wing and a prayer weather.'

IV. 20. a. (In reference to parts, structure, or function) (*a*) *wing-beat*, *-bone*, *-length*, *-shoulder* (later example), *-span*, *-spread*; (*b*) (of aeroplanes) *wing-length*, *-skid*, *-span*, *-spread*, *-stay*; (*d*) (in sense 7) *wing-back*, *commander* (examples); (in sense 7 b) *wing-man* (examples); (*e*) *wing-nut* (later examples).

(*a*) 1826 Wing-bone [in Dict.]. 1897 'N. Blanchan' *Bird Neighbors* 143 Bank Swallow... About an inch shorter than the English sparrow, but apparently much larger because of its wide wing-spread. 1909 *Westm. Gaz.* 2 Nov. 2/3 For wing-beats of great angels we would hear the herdsman's call. 1922 Joyce *Ulysses* 505 Head askew, arches his back and hunched wingshoulders. 1927 *Daily Express* 31 Aug. 8/3 It..is shaped like the wing-bone of a chicken. 1943 A. Clarke *Coll. Plays* (1963) 173 This big wind that filled My wingbones blew me into the trees. 1946 *Nature* 21 Dec. 904/1 The accompanying table shows..the weight in kgm. and wing-length in cm. of the female. 1949 *Brit. Birds* XLII. 187 The wing-span was found to measure nearly four feet, and the length was 21 inches. 1957 *New Yorker* 13 July 22/2 We got over six hundred bats, from insectivorous ones with an eight-inch wingspread to fruit eaters with a five-foot wingspread. 1971 *Sci. Amer.* Dec. 79/3 For aerodynamic reasons large birds have a slow wingbeat. 1977 P. Way *Super-Celeste* 123 The skull and upper bones of the [eagle's] wingspan had..driven like a cannon ball into the pilot's belly.

(*b*) 1908 H. G. Wells *War in Air* x. 317 It had taken only an hour or so to substitute wing stays from the second flying-machine and to replace the nuts he had himself removed. 1910 R. Ferris *How it Flies* xx. 474 Wing skid, a small skid, or runner, placed under the tip of the wings of an aeroplane. 1912 *Q. Rev.* July 231 If the 1000 lb. aeroplane is to travel slower, it must have a larger wing-spread. 1918 Pagé & Montariol *Gloss. Aviation Terms* 33/1 Wing span. 1920 *Flight* XII. 864/1 The Loughead S1 model, as it is called, is a single-seater biplane with a wing span of 28ft. 1975 *Farnborough 76* (Soc. Brit. Aerospace Companies) 30/2 The world's smallest jet aircraft, the Bede BD-5J..with a wing span of only 17ft. 1978 R. Jansson *News Caper* 9 There was the fighter again, flying parallel half a winglength away. 1978 *Sci. Amer.* Nov. 135/1 In 1899 the Wrights built a biplane kite with a five-foot wingspread that embodied their wing-twisting roll control.

(*d*) 1914 *Times* 22 Dec. 4/3 Royal Flying Corps..Wing Commander.—Brev. Maj. H. R. M. Brooke-Popham, Oxf. and Bucks. L.I. 1933 *Time* 13 Nov. 57/1 A wing-back is..a halfback who takes position about a yard and a half behind the line of scrimmage and about the same distance outside his own end. 1942 *Sun* (Baltimore) 26 Jan. 4/1 Baltimore scored first on a pass from Charley Ernst, center forward, to Harry McAdams, newly acquired wingman. 1943 J. B. Priestley *Daylight on Saturday* xxviii. 217 And a real wing-commander came in yesterday and talked to me. 1974 Wingback [see *rush *v.*[2] 6 g]. 1976 *Derbyshire Times* (Peak ed.) 3 Sept. 26/1 Matlock, in contrast, always looked dangerous with Peter Scott, the Fenoughty brothers, Mick and Nick, and wing-man Colin Oxley constantly troubling the Runcorn defence with their speedy breaks.

(*e*) 1910 *Chambers's Jrnl.* May 349/1 The wing-nut on its shaft is released, the detachable rim-wheel placed on the shaft, and the nut replaced. 1971 *Flying* Apr. 26/2 The control and gust locks..are adjustable to fit virtually any light aircraft by means of easy-to-operate wing nuts.

b. *wing-borne*, *-flapping*, *-like* (earlier example), *-shadowed*, *-shattered*, *-stiff*, *-weary* adjs.

1934 Webster, Wing-borne. 1942 S. Smith *Mother, what is Man?* 67 Than earth-born engine-borne, heaven-born wing-borne is better? 1977 *Guardian Weekly* 5 June 3/2 About half the crashes happened when the aircraft was hovering, or in transition from normal wingborne flight. 1915 E. Pound *Cathay* 10 He goes out to Hori, to look at the wing-flapping storks. 1953 N. Tinbergen *Herring Gull's World* xxi. 183 A screaming, wing-flapping tangle. 1795–1804 W. Blake *Vala* vi, in *Compl. Writings* (1972) 318 And the wing-like tent of the Universe, beautiful, surrounding all. 1938 D. Gascoyne *Hölderlin's Madness* 28 The bewildered words which try to tell The tale of his bright night And his wing-shadowed day. 1928 Blunden *Retreat* 60 But now the grey age passes by my faint senses And charm lies wing-shattered or dead. 1945 P. A. Larkin *North Ship* 33 It was your severed image that grew sweeter, That floated, wing-stiff, focussed in the sun. 1868 J. G. Whittier in *Atlantic Monthly* Jan. 1 The sky is hot and hazy, and the wind, Wing-weary with its long flight from the south. 1946 J. W. Day *Harvest Adventure* x. 154 The woodcock come in wing-weary from their North Sea voyagings.

21. wing-back chair = *wing chair*, sense 20 a (*e*) in Dict; also *ellipt.*; **wing-bud**, in insect larvæ, a histoblast from which the wings develop; **wing-clapping**, the production of a noise by a bird slapping its wings against its body; hence **wing-clap** *sb.* and *v. intr.*; **wing collar**, a high stiff shirt collar with the upper corners turned down; **wing-dam** *sb.* (earlier examples); **wing-flap**: see *flap *sb.* 5 e; **wing flutter** *Aeronaut.*, flutter (*flutter *sb.* 1 d) of an aircraft wing; **wing-footed** *a.*, also *fig.*; **wing formula** (see quot. 1964); **wing loading** *Aeronaut.*, the gross weight of an aircraft divided by the total wing area; (in quot. 1912 perh. used differently); cf. *power loading *vbl. sb.* 1; **wing-man**, the pilot of an aircraft which is positioned behind and to one side of the leading aircraft, as in formation for combat; the aircraft itself; **wing mirror**, (*a*) a side mirror (freq. adjustable) on a dressing table; (*b*) a rear-view mirror projecting from the side of a motor vehicle; **wing-over**, of an aircraft or hang-glider (see quot. 1959); **wing-poke** (collar) = *wing collar* above; **wing root** *Aeronaut.*, the part of a wing where it is attached to the fuselage; **wing-shot** *sb.* and *a.* (earlier examples); **wing-tag** *v. trans.*, to attach a distinguishing marker to the wing of a bird; **wing-tip**, (*c*) chiefly *U.S.*, applied *attrib.* to shoes with a toecap having a backward extending point and curving sides, suggestive of the shape of a wing; also *absol.*; **wing-walking**, acrobatic stunts performed on the wings of an aircraft which is airborne, as a public entertainment; **wing-warping**, in early powered flight, the bending or twisting of a wing by means of an attached wire as a method of stabilizing the aeroplane or turning it.

1933 J. Steinbeck *To God Unknown* i. 1 The wing-back chair by the fireplace. 1977 *Chicago Tribune Mag.* 2 Oct. 9/1 (Advt.), The chair that stands still in time—the Classic Wingback with Chippendale legs. 1973 'D. Jordan' *Nile Green* xlvi. 247 She sat in her wingback chair flicking through one of the coffee table books. 1917 R. J. Tillyard *Biol. of Dragonflies* iii. 47 The wing-bud is simply an ectodermal evagination, in the form of a small bag lined internally with hypoderm cells, and externally with the cuticle. 1969 R. F. Chapman *Insects* xxi. 407 A progressive development of the wing buds occurs at each moult. 1941 H. F. Witherby et al. *Handbk. Brit. Birds* IV. 142 Performance [of display-flight by turtle-dove] may be accompanied by wing-clapping. 1964 A. L. Thomson *New Dict. Birds* 631/2 More rattling or clattering wing-claps may be made by pigeons suddenly taking wing when alarmed. 1976 *Country Life* 18 Mar. 672/2 The mechanical production of snaps from the beak may be compared with wing-clapping by birds... The long-eared owl will wing-clap during its spring nuptial flight. 1915 H. L. Wilson *Ruggles of Red Gap* (1917) ii. 33, I chose a shirt of white piqué, a wing collar with small, square-cornered tabs, and a pearl ascot. 1975 *Times* 19 May 12/7 Saturday's guide was Charles E. Lee, a transport historian whose wing collar..enhanced the building's period atmosphere. 1809 T. G. Fessenden *Pills Poetical* 36 All his rhetorick was directed toward election districts, and wingdam bills, and seconding motions. 1863 Wing dam [see *paddock *v.* 2]. 1927 *Daily Tel.* 21 Jan. 10/7 The new theory..suggests that wing-flutter may be more common than has been supposed. 1982 C. L. Ruhlin et al. *Transonic Flutter Study of Wind-Tunnel Model* (NASA Rep. 82-23239) viii. 5/2 Most of the winglet effect on the wing flutter speed was due to the winglet mass, not aerodynamics. 1977 *Time* 22 Aug. 13/1 Wing-footed United Nations Ambassador Andrew Young has been exploring the politically and economically troubled waters of the Caribbean, and soon will attend an anti-apartheid conference in Lagos, Nigeria. 1936 *Brit. Birds* XXX. 226 This specimen..has..a wing formula as follows. 1964 A. L. Thomson *New Dict. Birds* 892/2 Wing formula: a statement of, mainly, the relative lengths of the primary feathers. 1912 *Q. Rev.* July 246 A range of this amount is obtained entirely by proportioning the position of masses, the wing-curve and the wing-loading. 1916 A. W. Judge *Design of Aeroplanes* iii. 29 In current practice the wing loading expressed in pounds per square foot for biplanes is about $0.005V^2$, ..where V is the maximum designed speed in feet per second. 1972 *Times* 19 May 17/4 It cannot be a glider, as it has far too high a wing-loading. 1946 *Sat. Even. Post* 6 Nov. 86/2, I looked to both sides of us. Our two wing men were gone. 1981 S. Dunmore *Ace* i. i. 15 We will fly together... You will be my wingman..to protect my rear end. 1982 *Daily Tel.* 25 May 1/4 He hit two Mirages with Sidewinder missiles while his wingman hit the third in the formation. 1925–6 T. Eaton & Co. *Catal. Fall & Winter* 311/2 Dressing table..triple mirrors, centre one beveled.. two plain wing mirrors. 1948 *Motor* 3 Nov. 396/3 An assortment of wing mirrors. 1959 C. Williams *Man in Motion* vi. 62 The dressing-table with its wing mirrors. 1959 *Motor Manual* (ed. 36) viii. 217 Additional wing mirrors are..very useful, particularly on the off-side as traffic on the point of overtaking is then clearly visible. 1981 M. Nabb *Death of Englishman* iii. i. 143 He..had banged his head on a Carabiniere car wing-mirror. 1928 *Morning Post* 20 Oct. 9/3 One of the passengers..got panicky when the pilot executed a 'wing-over'. 1959 F. D. Adams *Aeronaut. Dict.* 183/2 Wing-over, noun, an airplane maneuver in which the airplane makes a steep zooming climb then banks and turns in the vertical plane into a dive or glide from which the recovery is made at approximately the original altitude and in a direction opposite to the original direction. 1978 A. Welch *Bk. Airsports* i. 9/2 They indulge in 'show-off' flying—fast dives and steep wing-overs—that the simple hang glider was never designed to take. 1905 H. G. Wells *Kipps* iii. i. 351 Kipps wears a grey suit, with a wing poke collar. 1910 —— *Hist. Mr. Polly* i. 13 His collar was chosen from stock, and with projecting corners, technically a 'wing-poke'. 1906 A. Samuelson *Flight-Velocity* i. 12 Near the wing root an outrigger or boom..is fastened. 1966 M. Woodhouse *Tree Frog* xxvi. 195, I jumped down off the wing root..and started to think about search parties. 1875 *Fur, Fin & Feather* 118 Bogardus, champion wing-shot of America, uses Orange Lightning [powder] for trap-shooting. 1878 C. Hallock *Hallock's Amer. Club List & Sportsman's Gloss.* p. xii, *Wing-shot*, a., hit in the wing. *Wing-shot*, n., a shot at birds on the wing;

one who shoots at birds while flying. 1953 Scott & Fisher *Thousand Geese* 215 Five of the young were wing-tagged. 1981 *Animal Behaviour* XXIX. 302/1 Three females and one male were wing-tagged. 1928 *World* (N.Y.) 23 May 4/6 (Advt.), Wing tip oxfords by Horsheim have unusually good style. 1971 *Weekend World* (Johannesburg) 9 May 14/5 (Advt.), Walk tall in the elegant clean lines of a Bostonian wing-tip or genuine handsewn moccasin. 1976 'B. Shelby' *Great Pebble Affair* 45 Get a pair of black wingtip shoes. 1980 M. Gordon *Company of Women* i. ii. 38 The hard, expensive shoes of John F. Kennedy, the shoe with pinholes in the leather, wing tips they were called. 1927 C. A. Lindbergh *We* i. 11 Exhibitions..in which I usually made a jump and did a little wing-walking. 1979 *Sunset* Apr. 3/3 Also awesome is a wing-walking act in which specially trained gymnasts do headstands and other maneuvers on the wings of a W.W.II Stearman biplane as it loops, rolls, and lands. 1910 R. M. Neilson *Aeroplane Patents* 27, 6732 of March 19, 1904.—O. and W. Wright. This is the famous wing-warping patent. 1969 K. Munson *Pioneer Aircraft 1903–14* 7 Wing-warping was not, in itself, an invention of the Wrights; what was significant was their improvement of linking the warp-control cables with a single, hinged rudder.

wing, *v.* Add: **I. 2.** Also (chiefly *U.S.*) with an aircraft as subject, or *transf.* of a passenger, to travel by aircraft. No longer restricted to *poet.* or *rhetorical* use.

1938 *Sun* (Baltimore) 21 July 1/8 (*heading*) English plane wings swiftly over Atlantic. 1973 C. Sagan *Cosmic Connection* (1974) xxviii. 197 A single bit of radio information, sent winging across space to the Earth, would cost far less than a penny. 1977 *Time* 30 May 25/2 As Air Force One winged toward Washington, one Californian was clearly relieved that Carter's visit had been so brief. 1983 *Fortune* 18 Apr. 137/1 Winging into New Hampshire from Los Angeles headquarters aboard an Arco jet one Sunday, Cooper began the next three days at 7:30. 1984 *Times* 4 Aug. 32 The Prince of Wales flies back from Monaco, only to wing off within hours for Papua New Guinea.

5. (Later example with aircraft as subj.)

1976 C. Egleton *State Visit* xiv. 123 The VC 10 winged him back to Heathrow.

II. 10. b. *intr.* *to wing out*: to set a sail on a boom projecting sideways. Hence *winged out* or *wung out*, = *wing-and-wing* s.v. Wing *sb.* 19.

1867 G. E. Clark *Seven Yrs. Sailor's Life* i. 14 Here was I, deep-loaded, winged out, and oft-times flying before the winter blast. 1890 Webster, wung out. 1907 *Rudder* Nov. 827/2 On rounding, the schooners winged out; but..the wind came out of East of South, and they jibed their foresails and trimmed sheets a little. 1956 A. F. Loomis *Hotspur Story* 109 Thither we sailed, mainsail to starboard and staysail wung out. *Ibid.* 214 The wung-out schooner which we had noticed earlier in the afternoon lost the race. 1969 H. Horwood *Newfoundland* x. 71 Tearing down the outside passages with sails 'wung out' before a roaring nor'-wester.

11. (Earlier *intr.* example.) Hence in phr. *to wing it*; now usu. in *slang* use (orig. and chiefly *U.S.*), to improvise; to speak or act without preparation, to make statements on unstudied matters (see also quot. 1950).

1885 *Stage* 21 Aug. 12/2 'To wing'..indicates the capacity to play a rôle without knowing the text, and the word itself came into use from the fact that the artiste frequently received the assistance of a special prompter, who..stood..screened..by a piece of the scenery or a wing. 1933 P. Godfrey *Back-Stage* iii. 39 He must give a performance by 'winging it'—that is, by refreshing his memory for each scene in the wings before he goes on to play it. 1950 *Amer. Speech* XXV. 238/1 *Wing it*, vb., to lay off an approximate 90° angle by eye. 1959 *Esquire* Nov. 70 *Wing*, to do something without preparation. 1970 *Time* 26 Jan. 12 Cox: The resistance put up against us dictates [our] strategy. *Bernstein*: ..You mean you've got to wing it. 1971 *Publishers' Weekly* 6 Dec. 20/2 They can talk about the book, kind of winging it based on the ads, just like other people do with reviews. 1979 *Globe & Mail* (Toronto) 22 Jan. 8/2 Mr. Trudeau came without notes, choosing to wing it, and struggled..unsuccessfully to establish Mr. Leger's resemblance to an owl.

Wingco (wi·ŋko). *R.A.F. slang.* Also **Winco**, **Winko**, and with small initial. Abbrev. of *Wing Commander*: see Wing *sb.* 20 a (*d*).

1941 Michie & Graebner *Lights of Freedom* iii. 45 A cockney member of the ground crew piped: 'Sir, I think the 'Winko' [wing commander] is after the Hun.' 1942 *R.A.F. Jrnl.* 3 Oct. 12 One of them was a Winco and the other two were Army officers. 1943 'T. Dudley-Gordon' *Coastal Command at War* ix. 88 On another raid... The wingco, was leading. 1944 'N. Shute' *Pastoral* iii. 54 Don't let Winco hear him, or he'll get us into trouble. 1957 J. Braine *Room at Top* xii. 120 I've sung that... with Wingcos and Group Captains joining in. 1974 T. Allbeury *Snowball* xxii. 135 Wing Commander Pallin from the Ministry of Defence... I'd like to ask the Wingco to keep himself free to check..on the state of the game in Moscow. 1982 F. Parrish *Snare in Dark* ii. 25 There was a pub..taken over by a retired Wing Commander... The Winco, as he liked to be called, was a ready market.

wing-ding, **wingding** (wi·ŋ,diŋ). Also **whingding**. [Redupl. of Wing *sb.*] **1.** *U.S. slang.* A fit or spasm, esp. as simulated by a drug addict; freq. in phr. *to throw a wing-ding*. Also in weakened sense, a furious outburst.

1927 *Amer. Speech* II. 281/1 *Wingding*, a false illness or fit. 1933 *Ibid.* VIII. ii. 28/1 When an addict who.. cannot obtain dope..becomes desperate, he may throw a wing-ding (feign a highly realistic fit in public) in the hope that the doctor..will administer narcotics to quiet him; professional wing-dingers are addicts who make a practice of obtaining their narcotics in this manner. 1939 R. Chandler *Big Sleep* xxxii. 292 She threw a wingding. Looked like a mild epileptic fit. 1944 *Amer. Speech* XIX. 107 A wing-ding is a particularly explosive fit of rage or frustration (I'm telling you the mate will throw a wing-ding!). 1946 'J. Evans' *Halo in Blood* xiv. 166, I.. watched her take deep unsteady breaths... Her hands were locked together in her lap but that didn't keep them from trembling. 'About a minute,' I said mildly, 'You're going to throw a wing-ding they'll hear in Detroit. You're wound up tighter than a dollar watch.' 1957 V. Packard *Hidden Persuaders* ix. 102 This venture back to the womb touched off a little wingding in advertising circles. 1965 P. Tamony *Americanisms* (typescript) No. 11. 3 It assigned..Winifred Sweet..to throw a wing-ding..in Market Street.

2. *slang* (orig. and chiefly *U.S.*). A wild party; a celebration or social gathering.

1949 *Sat. Even. Post* 5 Mar. 10/3 We are not sure just what the Festival is to be, but some sort of native whingding no doubt. 1955 R. Bradbury *October Country* 18 We would have to arrive when the local Rotary's having its whingding. 1964 *Punch* 15 July 79/3 My invitation to a White House schnapps wingding. 1972 *Sunday Sun* (Brisbane) 6 Aug. 3/4 Last Tuesday was Pat's birthday, so there was a big wing-ding at Maroochydore's posh Surfair pub. 1975 *Listener* 18 Dec. 832/3 The funeral bak'd meats will serve the triple economy of a divorce wing-ding as well. 1979 A. Hailey *Overload* iii. xi. 243 How are you, Nim? Don't see you often at these Jewish wingdings.

Hence **wing-di·nger**, (*a*) (see quot. 1933); (*b*) a pretended fit; a wild outburst.

1933 [see sense 1 above]. 1949 V. J. Monteleone *Criminal Slang* (new ed.) 253 *Wing-dinger* (n.), a pretended fit or spasm; a forced faint. 1976 *Telegraph* (Brisbane) 5 Aug. 39/3 This leads to a wing-dinger of a brawl, when Bobbie's brother..sights the louts who have busted up his father and their truck on the bridge.

winge, var. Whinge *v.* in Dict. and Suppl.

winged, *a.* Add: **3. c. winged bean**, a tropical legume, *Psophocarpus tetragonolobus*, native to south-eastern Asia and cultivated for its edible leaves, winged pods, and tubers; cf. *Goa bean* s.v. Goa[1]; **winged thistle** *N.Z.*, either of two thistles of the genus *Carduus*, *C. tenuiflorus* or *C. pycnocephalus*, which have winged stems.

1910 H. F. Macmillan *Handbk. Trop. Gardening & Planting* 189 *Psophocarpus tetragonolobus.* Winged bean; Goa bean; Manilla bean. 1915 *N.Z. Jrnl. Agric.* 21 June 550 Winged thistle [seed]. About the same size as spear-thistle seed. 1966 *Encycl. N.Z.* III. 599/1 Noxious weeds are here listed... Winged thistle. 1975 *Times* 30 Aug. 12/7 An international panel..[is] recommending a major development effort to turn..the winged bean into a main crop.

winged, *ppl. a.* **1.** (Earlier example.)

1789 *Ess. on Shooting* xiv. 223 He [sc. the dog] should be held in a string, ready to be slipped in case of need, after a winged partridge, or a wounded hare.

wingedness. (In Dict. s.v. Winged *a.*) (Later example.)

1909 W. Bateson *Mendel's Princ. Heredity* i. x. 172 Here we see that the one 'dose' of wingedness—as we may call it—sufficed only to bring the wings to half the full size, and two 'doses' are needed to develop them properly.

wingeing, var. Whinging *vbl. sb.* and *ppl. a.* in Dict. and Suppl.

winger. Add: **2.** Also in *Hockey* and *Lacrosse*, a wing player.

1922 *Daily Mail* 15 Dec. 13 No right winger has more visibly impressed me than Sutcliffe this season. 1969 *West Australian* 5 July 32/3 Allowing winger Kaye Olsen to gain position and put Wembley into attack.

3. *Naut. slang.* **a.** A steward.

1929 F. C. Bowen *Sea Slang* 152 *Winger*, a steward waiting at table, with the class prefixed. 1962 *Times* 26 Apr. 15/2 A winger is a steward on a passenger liner. 1962 [see *blood *sb.* 15 d].

b. A comrade or friend (see also quot. 1977).

1943 *Penguin New Writing* XVII. 46 He had seen his 'winger', his best friend, decapitated. 1948 Partridge *Dict. Forces' Slang* 208 *Winger*, an assistant or 'stooge'. The term has displaced 'raggie' as a name for a pal. (Navy.) 1957 R. Watson-Watt *Three Steps to Victory* xl. 233 Bickell,..Max's *fidus Achates* and 'winger' in M.A.P. and in the wide circles which rippled..out from that most explosive of Ministries. 1977 G. Melly *Rum, Bum & Concertina* v. 57 The expression 'winger' means, at its most innocent, a young seaman who is taken under the wing of a rating or Petty Officer older and more experienced than himself to be shown the ropes.

4. *left-winger*, *right-winger*: see *left wing, *right wing.

wingy (wi·ŋi), *sb. colloq.* [f. Wing *sb.* + -y[6].] A one-armed man; also (with capital initial) used as a nickname. Cf. *wing *sb.* 5 d (*c*).

1880 D. W. Barrett *Navvies* (ed. 2) ii. iii. 49 If a poor fellow..is short of a leg or an arm, 'Peggy' or 'Wingy' is at once affixed to him. 1910 H. Lawson *Stories* (1964) 2nd Ser. 296 Wingy..is a ratty little one-armed man' whose case is usually described in the head-line as 'A 'Armless Case' by one of our great dailies. 1931 'D. Stiff' *Milk & Honey Route* v. 58 Missions are very anxious to recruit the 'wingies' and 'armies', or the one-armed hobos. 1964 T. Ronan *Packhorse & Pearling Boat* 129 As Dad later referred to him as 'Wingy' Collins I presume that he had one arm amputated, or some similar disability.

wink, *sb.*[1] Add: **3. c.** In *Work Study*, a unit of time equivalent to one two-thousandth of a minute. Also *Comb.*, as *wink-counter*. orig. *U.S.*

1937 R. M. Barnes *Motion & Time Study* ix. 72 There are 100 equal divisions on the dial of the clock; therefore, time is indicated directly in 1/2000 of a minute by the large hand. This time interval of 1/2000 of a minute was called a 'wink' by Gilbreth. 1946 R. L. Morrow *Time Study & Motion Economy* ix. 90 The wink-counter..is a small motor driven device, originated by Professor David B. Porter..to be used for both motion and time studies. In appearance it resembles a 'speedometer'. 1961 *Engineering* 15 Sept. 352/1 A very early type of micromotion filming was used by the Gilbreths in the early days of motion study, and the unit of time which they employed, a two thousandth of a minute or a 'wink', is still often used for detailed motion analysis.

wink, *sb.*[4] Shortening of *tiddlywink 2 c. orig. *U.S.*

1890 *Game of Tiddledy Winks* (McLoughlin Bros., New York) 1 Its great interest and success lies in the novel feature of jumping the Winks into the Wink-pot. 1957 *Times* 17 Dec. 9/4 Tiddlywinks does not yet qualify for a 'blue', or even half a one, but it is nice to know that the club has a tie, dark blue with a blue cup and a wink rampant. 1979 F. R. Shapiro *Encycl. Tiddlywinks* 8 The Silver Wink, donated by Prince Philip, is awarded to the winner of an annual elimination tournament for universities.

wink, *v.*[1] Add: **2. c.** Also with advbs.: to go *out* or *off* suddenly; to come *on* suddenly.

1848 Wink out [in Dict.]. 1930 W. Faulkner *As I lay Dying* 244 He locks the door. Dewey Dell is inside. Then the light winks out. 1972 *Sci. Amer.* Jan. 108/3 An observer who is a few miles away but within the shadow sees the star wink off and five minutes later reappear slightly west of the moon. 1979 *Tucson* (Arizona) *Citizen* 20 Sept. 7B/6 (*heading*) 'Buck Rogers' no supernova, but it won't wink out, either. 1982 *Washington Post* 21 Mar. 3/2 Bleuzinski perched on the pool table, leaned forward, and looked directly into the camera. The red light winked on.

wink, *v.*[3] orig. *U.S.* [f. *wink *sb.*[4]] *intr.* To play tiddlywinks. Freq. as *vbl. sb.* Occas. *trans.* (in quot. *fig.*).

1955 V. Nabokov *Lolita* i. v. 26 This is all very interesting, and I daresay you see me already frothing at the mouth in a fit; but no, I am not; I am just winking happy thoughts into a little tiddle cup. 1958 *Sunday Times* 2 Mar. 16/3 While practising secretly, I pulled an important muscle in the second or tiddly joint of my winking finger. 1962 *Boston Globe* 14 Oct. 81 The Crimson tiddlers winked their way to a 23 to 12 victory over a green Purple team. 1979 *Harvard Mag.* May-June 38 They went to many carpet stores to find the perfect surface for winking.

winked, *ppl. a.* Add: *winked-at* (later examples).

1971 J. Brunner *Honky in Woodpile* iii. 24 Some winked-at gambling and smuggling. 1979 *Dædalus* Summer 107 Such genre paintings represent not a random clutter of whimsically winked-at transgressions but an array of symbols encoding quite specific moral instructions.

winkel. (Earlier example.)

1827 G. Thompson *Trav. & Adv. S. Afr.* i. iii. 35 The village contains a couple of small retail shops, or winkels, as they are called.

winker[1]. Add: **4.** A direction indicator on a motor vehicle in the form of a flashing light; = *indicator 3 g.

1951 *Autocar* 2 Nov. 1411/1 Another advantage of the 'winkers' is the fact that no mechanical fault can develop. 1960 *News Chron.* 21 July 6/4 On the M1..there are no curves to cancel the winkers. 1967 *Autocar* 28 Dec. 2/2 The main-beam and winker lights have little, pull-down 'eyelids'. 1970 A. Sillitoe *Start in Life* v. 255, I put on the winkers, swung out, and swept forward.

Hence (in sense 3) **wi·nkered** *a.*

1804 M. Wilmot *Let.* 5 July in *Russ. Jrnls.* (1934) i. 110 Women..dress'd in a sort of winker'd cap of pearls which showes the face very becomingly. 1907 J. M. Synge *Playboy of Western World* iii. 61 That's the playboy on the winkered mule.

winker[2]. Short for *tiddlywinker* (b) s.v. *tiddlywink. orig. *U.S.*

1958 *N.Y. Times* 9 May 28 Tomorrow..the Cantab Winkers play the Oxonian Tiddlers in Oxford. 1965 *Times* 5 Jan. 11c The nation's 'winkers' have two ambitions for 1965. 1979 *Harvard Mag.* May-June 39 Winkers lost interest in recruiting new winkers.

winkle, *sb.* Add: **2.** *slang* (chiefly *juveniles'*). The penis (of a young boy).

1951 Partridge *Dict. Slang* Add. 1223/2 *Winkle, n.*, penis: children's, (young) schoolboys': late C. 19–20. 1970 *Guardian* 3 Feb. 8, I was mildly troubled by the insistence, especially of one headmaster, on the 'proper names for things'. Penis is right and winkle is wrong. 1970 T. Hughes *Crow* 63 O do not chop his winkle off His Mammy cried. 1973 M. Amis *Rachel Papers* 78 'Thanks,' he said to his (new) witch-like girlfriend as she handed him a joint so ill-made that it resembled a baby's winkle.

3. *Comb.*, as **winkle-picker** *slang*, a shoe with a long pointed toe; **winkle-pin** *Mil. slang* = Bayonet 2.

1960 *Spectator* 15 Apr. 553 The incredibly pointed custom-built shoes in which teenagers keep other teenagers at arm's length... The shoes, called winklepickers, look like something out of Grimm's fairy tales. 1960 *News Chron.* 13 Sept. 5/3 The 'winkle picker' high heels and the high spirits have gone. 1978 C. Sykes in R. Buckle *U & Non-U Revisited* 57 The mass-produced variety, popularly known as 'winkle-pickers', were very ugly. 1980 *Bulletin* (Sydney) 6 May 5/3, I had the hairdo, the lairy shirt, the winkle-picker shoes. 1924 Kipling *Debits & Credits* (1926) 314 As his sergeant I had to check him for misusin' his winkle-pin on dirt. 1950 Partridge *Here, There & Everywhere* 62 The bayonet..has many names..[e.g.] *winkle-pin*.

winkle, *v.*[2] *colloq.* (orig. *Mil. slang*). [f. Winkle *sb.*] *trans.* *to winkle out*: to extract or eject (as a winkle from its shell with a pin); to draw forth, find out or elicit.

[1925 Fraser & Gibbons *Soldier & Sailor Words* 306 *Winkle, to*, to capture individual prisoners by stealth... Also,..to steal.] 1942 'M. Home' *House of Shade* ii. 20 What's winkled you out at this goddam hour? 1943 *People* 31 Oct. 1/7 Methodically winkling the Germans out of their strongpoints. 1951 'M. Innes' *Operation Pax* v. vi. 220 Until we winkle out this young man..you and I make not a bad team. 1958 J. Press *Chequer'd Shade* 4 It is illegitimate to compare the far-fetched conjectures of Eliot's commentators with the inside information which we might have winkled out of Donne. 1966 *Listener* 20 Jan. 111/3 It's the cunning of the interviewer that counts most of all, and certainly Mr Muggeridge manages to winkle out some interesting bits and pieces. 1970 *Sunday Mail Mag.* (Brisbane) 17 May 14/1 He could winkle out sin where no other man dreamed it existed. 1976 A. Eden *Another World* vii. 79 Hart's skills were in a sense wasted upon us, and he was at length winkled out to the advantage of a base hospital.

winkler. (In Dict. s.v. Winkle *sb.*) Add: **2.** *slang.* One who assists in the eviction of tenants (see quot. 1970). Cf. *winkle *v.*[2]

1970 *Sunday Times* 15 Nov. 3/5 Plausible, highly-paid 'winklers' who are hired by property companies and landlords to persuade families to leave their rent-controlled tenancies so the homes can be sold at high prices. 1977 *Whig-Standard* (Kingston, Ontario) 29 Sept. 26/3 The tenants said the agents aided by middlemen called 'winklers', had bribed and harassed them to get them to move.

winkling (wi·ŋkliŋ), *vbl. sb.* [f. *winkle *v.*[2] + -ing[1].] The action of the vb., esp. with reference to the removal of tenants from rented accommodation. Also *winkling-out*. Cf. *winkler 2.

1970 *Guardian* 20 Nov. 7/1 'Winkling'—persuading private tenants to quit. 1973 *Daily Tel.* 24 May 11/2 He described the 'winkling' processes followed by certain developers to rid themselves of unwanted but protected tenants. 1974 E. Ambler *Dr Frigo* iii. 195 An army assault team had been called in. In their winkling-out of the defenders they had..made rather a mess of the Palace. 1975 *Listener* 18 Nov. 826/1 Mr Skeaping had been accused of 'winkling'; bribing his tenants to leave so that he could sell the property with vacant possession.

winks (wiŋks). Shortening of *tiddlywinks*: see Tiddlywink 2 b. *U.S.*

1942 R. & L. Freeman *Cavalcade of Toys* xvi. 366 Back in 1903 'Battle Winks' was a popular game. 1962 *Harvard Crimson* 6 Nov. 3/2 (*heading*) Crimson winks squad downs two opponents. 1979 *Technology Rev.* Mar./Apr. B23 Unlike chess, which has limited predictable moves, winks is a game of chance as well as skill, says Mr. Lockwood.

winky. Add: Also **winkie.** (Further examples.)

1846 W. Cross *Disruption* vii. 61 Jimes and mee are going On with owre studdys already like winkie. 1901 M. Franklin *My Brilliant Career* xix. 163 Every one has to obey him like winkie or they can take their beds up and trot off quick and lively. 1923 Kipling *Land & Sea Tales* 115 This Baxter-man..SOS'ed like winkie,

winless (wi·nlės), *a.* *N. Amer.* [f. Win *sb.*[1] + -less.] Characterized by an absence of victories in a series of sporting contests; also, designating a period of time during which no victory was won.

1966 *Daily Progress* (Charlottesville, Va.) 8 June 30/1 People keep telling VMI's Gary McPherson he has the very best winless basketball team in the country. 1970 *Globe & Mail* (Toronto) 25 Sept. 32/3 Winless in 14 previous starts this season, Miss Ella Cinders had little trouble with Sandy Hawley up last Saturday as she galloped to a 12-length win. 1972 J. Mosedale *Football* iii. 36 They..went through another winless season. 1977 *Arab Times* 3 Dec. 9/3 Bechtel now go into their final game against AG & P with both teams winless in what should be quite a battle to see which team makes it out of the cellar.

winnability (wi:năbi·liti). Also **winability.** [f. Winnable *a.*: see -ity.] Capacity for winning or being won.

1972 *New Society* 16 Nov. 400/2 The penal cases committee seems to be heavily influenced by two considerations... The other is the 'winability' of the case; the penal cases committee refers chiefly open-and-shut cases to the disciplinary committee. 1975 W. Safire *Before the Fall* i. iv. 43 'My biggest problem,' Nixon concluded, 'is "Nixon can't win".'.. We discussed ways to build 'winability'. 1979 *Guardian* 2 Nov. 13/6 Kennedy..has that overriding quality—winnability. 1983 *Times* 16 Apr. 2/5 The most detailed work on the winnability of the new seats has been done by a few academics.

Winnebago (winəbēɪ·go), *sb.* (*a.*) [ad. Fox *wi·nepye·ko·ha*, lit. 'person of dirty water', an allusion to the muddy waters of the Fox River below Lake Winnebago, which became clogged with dead fish in the heat of the summer.] **1. a.** (A member of) a Siouan people of eastern Wisconsin. **b.** The language of this people. Also *attrib.* or as *adj.*

1766 J. Carver *Jrnl.* 25 Sept. in J. Parker *Jrnls. J. Carver & Related Documents, 1766–70* (1976) I. 78 Arrivd at the great town of the Winebaygoes. *Ibid.* 79 The town of the Winebagoes is situate on the south east end of an island at the east end of the Winebago Lake. 1827 *Spirit of Seventy-Six* (Frankfort, Kentucky) 2 Aug. 2/1 An express reached here this moment from Galena.. with information of hostilities having been commenced by the Winnebago Indians, on the settlers. 1835 C. F. Hoffman *Winter in West* I. 257 The Winnebago chief.. [had] just left the establishment. 1839 H. R. Schoolcraft *Algic Researches* I. 13 The Winnebagoes are clearly of the Abanic stock. 1860 *Harper's Mag.* Sept. 568/2 As he could not speak Winnebago, the first thing to be done was to find an interpreter. 1881 *Encycl. Brit.* XII. 832/1 The Winnebagoes are a branch of the Dakota family. 1907 L. H. Morgan *Anc. Society* iii. iii. 440 In Winnebago and Achaotina she is 'my sister'. 1910 F. W. Hodge *Handbk. Amer. Indians* II. 958/1 The Winnebago have been known to the whites since 1634, when the Frenchman Nicollet found them in Wisconsin, on Green Bay. 1933 [see *Mandan *a.* and *sb.*]. 1966 A. C. Hardy *Divine Flame* iii. 65 Those tribes belonging to the great Sioux family such as the Omaha, Ponka, Kansas, Dakota, Iowa, Winnebago, etc. 1973 A. H. Whiteford *N. Amer. Indian Arts* 81 Winnebago women's mocassins have a flap over the toe. 1975 *Language* LI. 317 Ferguson..suggests that Hockett's analysis of Winnebago is unusual.

2. Special Comb.: **Winnebago camper**, a motor vehicle with insulated panels used as living accommodation by campers (a proprietary term in the U.S.); also *ellipt.*

1966 *Mobile Home Jrnl.* Oct. (Advt., rear cover), Your most enjoyable travel companion is a Winnebago Pickup Camper Coach. 1970 *Official Gaz.* (U.S. Patent Office) 17 Nov. TM 132 Winnebago Industries, Inc., Forest City, Iowa. Filed Aug. 6, 1969. Winnebago... For Vehicles and Components...—Namely Motor Homes, Travel Trailers, House Trailers, Camper Coaches, [etc.]..First use April 1959. 1975 I. K. Martin *Regan & Manhattan File* 104 Regan..noted..a Winnibago camper parked to the rear. ..The back door of the Winnibago opened.

winner. Add: **2.** (Later examples of 'a thing that scores a success': now *colloq.*) Also, a potentially successful project, enterprise, etc.

1934 *Punch* 14 Nov. 552/1 The growing function of the outlying theatres is to spot winners for the West-End. 1948 M. Laski *Tory Heaven* v. 66 'I'd like to be a land-agent... I do really think I could have made a success of it.' 'I'm sure you would... It sounds like a winner to me.' 1958 *Times* 12 Sept. 13/1 The last crop of new ballets commissioned for the Edinburgh International Ballet company includes one winner, a near miss, and a very honourable mention. 1972 *Sunday Express* 9 Jan. 5/3 The warmth of wool plus good, classic styling, makes this coat a winner this winter. 1976 *Southern Even. Echo* (Southampton) 18 Nov. 4/3 Cyril Berry..must be on a winner with his latest book. 1985 *Woman's Own* 22 June 36/2 The actor believes that combining the strong with the sensitive is exactly what makes a man a winner—on screen and off.

3. *winner-take(s)-all*, attrib. phr. used to denote contests or conflicts in which victory is outright or the successful competitor alone is rewarded; occas. (without hyphens) in non-attrib. use as an idiomatic sentence.

1969 *Listener* 10 Apr. 496/1 When you say war, I think that's what you mean: nations and empires clashing, and there will be one winner and one loser on clear-cut lines. I won, you lost. But here there's not supposed to be, the way I understand it, a winner-take-all-type thing. 1972 *National Observer* (U.S.) 27 May 5/1 Should McGovern win the June 6 California primary with its winner-take-all bag of 271 votes,..he then would be within easy range of a first-ballot nomination at the convention opening on July 10. 1972 *Guardian* 8 June 12/1 In California winner takes all. It will be almost impossible now to deny McGovern the nomination. 1973 *Times* 16 Nov. 1/1 The Government has accepted that there is no way out of a grim, winner-takes-all clash with the National Union of Mineworkers. 1976 'H. Carmichael' *False Evidence* iv. 63 There must've been a worthwhile rakeoff... The outcome was that winner took all. 1978 A. Price *'44 Vintage* xxiii. 264 A winner-takes-all lottery.

winning, *vbl. sb.*[1] Add: **9. winning-chair** (earlier example); **winning-post** (earlier example); also *fig.*; **winning streak** orig. *N. Amer.*, a sequence of successes, esp. in sporting contests.

1835 W. Dyott *Diary* Oct. (1907) II. 212 A platform was erected in front of the winning chair [on Lichfield Race Course] to accommodate the ladies who were to deliver the standards [to a regiment]. 1759 A. Murphy *Let.* 22 July in D. Garrick *Private Corr.* (1831) I. 101 You must judge whether they [sc. horses] are *marketable*, or likely to tire before they come to the winning-post. 1790 T. Wilkinson *Mem.* II. 194 Miss Notable and Miss Prue from the archness and excellent acting of Mrs Abington, seemed to have the decision at the winning post for fame. 1951 *Times* (Weekly Ed.) 30 May 3 Meals are landmarks, milestones which must be passed before the winning-post of bed-time is finally and thankfully reached. 1968 Winning streak [see *goal-tending* vbl. sb. s.v. *goal *sb.* 6]. 1976 *New Yorker* 15 Nov. 162/1 'We're on a winning streak. We're on a hot roll,' one city official said happily last summer, during the triumphal series of events that began with Operation Sail and extended through other local Bicentennial celebrations. 1976 *Cumberland News* 3 Dec., Gilsland's Station Hotel team, playing in the Irthing Valley Sunday League, are still in a winning streak.

winning, *ppl. a.* Add: **2.** In U.S. colloq. use also in superlative.

1974 *State* (Columbia, S. Carolina) 5 Mar. 6-A/7 John Bates, coach of Maryland-Eastern Shore, at 26–1 the winningest college basketball team in the nation. 1979 *Tucson* (Arizona) *Citizen* 20 Sept. 5D/1 Slota defeated Sarah Cap, the winningest active greyhound with 113 career victories. 1985 *Dirt Bike* Mar. 23/2 (Advt.), That's the moment you know what the winningest racers and most satisfied riders know.

winningly, *adv.* (Later examples.)

1934 G. B. Shaw *On Rocks* i. 208 *Sir Arthur* [*winningly*] And do you, Miss Brollikins, feel that you have got nothing? 1980 *Times Lit. Suppl.* 20 June 702/2 Clothes and hair styles [of petty criminals in the 1870s] repay scrutiny: no one ever dressed up or posed winningly for these pictures.

Winnipeg (wi·nɪpeg). The name of the capital of Manitoba, Canada, used *attrib.* in *Winnipeg couch*, a couch convertible into a double bed.

1954 S. M. Russell *Living Earth* 233 He sat on the Winnipeg couch that stood at one end of the room. 1962 J. Onslow *Bowler-Hatted Cowboy* viii. 74 Beneath an old army blanket I drowsed to sleep on my Winnipeg couch. 1973 B. Broadfoot *Ten Lost Years* xiii. 153 A couch, one of those Winnipeg couch things in the living room.

Hence **Wi·nnipegger**, a native or inhabitant of Winnipeg.

1882 G. M. Grant *Picturesque Canada* I. 288 Winnipeggers..never make comparisons with any city smaller than Chicago. 1936 Mencken *Amer. Lang.* (ed. 4) x. 549 Richmonder, Winnipegger, Montrealer, Lynner. 1971 J. Gray *Red Lights* ii. 27 They watched Winnipeggers frantically planting trees all over the place.

winny (wi·ni). *U.S. slang.* Var. *wienie. Also *Comb.*, as *winny-wurst*. Cf. *weeny *sb.*[2]

1867 J. Christison *Crime & Criminals* 37 For a week longer he served at his usual business, which was that of peddling 'winnies', mostly among the saloons. 1914 B. Tarkington *Penrod* xix. 199 Winnies! *Here's* your hot winnies! Hot winny-*wurst*! 1929 T. Wolfe *Look Homeward, Angel* xx. 272 Fortune out of winnies. They're hot, they're hot.

wino (wəi·no). *slang* (orig. *U.S.*). [f. Win(e *sb.*[1] + *-o[2].] An habitual drinker of cheap wine; an alcoholic or drunkard, esp. one who is destitute.

1915 *World* (N.Y.) *Mag.* 9 May 14/3 *Wineoe*, a wine bum; known on the Pacific Coast, especially in California. 1926 J. Black *You can't Win* xii. 153 The wine dumps, where wine bums or 'winos' hung out. 1946 [see *juiced *a.* 2]. 1957 J. Kerouac *On Road* i. i. 9 Working..without pause eight hours a night..in greasy wino pants with a frayed fur-lined jacket and beat shoes that flap. 1958 *Times* 24 Nov. p. viii/5 In fact, Canadians have reserved the term 'wino' for the most reprehensible of their drinkers. 1961 *Guardian* 28 Feb. 8/7 A conglomeration of hop-heads, winos, overworked policemen. 1967 *Sunday Truth* (Austral.) 16 July 28/4 To save gas she washes in public toilets, and if she feels like a drink she has a swig from a wino's bottle at South Brisbane. 1973 'J. Marks' *Mick Jagger* (1974) 106 That sonuvabitch Dean Martin..that lousy wino wop! 1979 *Evening Standard* 2 Mar. 19/4, I am in sympathy with the plea by Mrs A. L. Hughes for the survival of buskers, but feel her attack on 'winos' is both misdirected and lacking in human understanding. 1981 M. Leitch *Silver's City* xii. 103 He saw the winos watching him out of bleary eyes as they huddled on their benches passing their brown bottles to and fro.

Winstonian (winstōᵘ·niăn), *a.* [f. the name of Sir *Winston* Leonard Spencer Churchill, British prime minister 1940–5 and 1951–5: see -ian.] Of, pertaining to, or characteristic of Sir Winston Churchill. Cf. *Churchillian *a.*

1905 W. S. Churchill *Let.* 9 May in R. S. Churchill *Winston S. Churchill* (1969) II. Compan. i. 391 It is vy kind of you to write me such a long letter. It will be carefully preserved among the Winstonian archives. 1945 S. Sassoon *Siegfried's Journey* viii. 79 The Winstonian exposition continued until Eddie reappeared with an apologetic intimation that Lord Fisher was growing restive. 1967 *Guardian* 16 May 8/6 Winstonian echoes to match the countless plaster busts in shop windows.

wint, var. Went.

winter, *sb.*[1] Add: **3. a.** (*b*) Of clothing (further examples).

1628 F. Drake *World Encompassed* 64 Notwithstanding

it was in the height of Summer..we could..haue beene contented to haue kept about vs still our Winter clothes. 1759 G. Clough *Let.* 30 Sept. in *Essex Inst. Hist. Coll.* (1861) III. 104/1 Cold weather..will make us..put on our Winter Clothing. 1838 *Workwoman's Guide* in Walkley & Foster *Crinolines & Crimping Irons* (1978) xi. 165 Care should be taken to separate..winter clothing from that worn in summer. 1870 E. G. E. Ward *Jrnl.* 9 Nov. in D. P. Carew *Many Years, Many Girls* (1967) i. 33, I have been able to-day to send my children in England some winter clothes. 1874 Geo. Eliot *Let.* 16 June (1956) VI. 57 The cold winds..have forced us to put on winter clothing. 1876 C. M. Yonge *Three Brides* I. iii. 35 Her hair and pretty Parisian winter dress arranged to perfection. 1892 *Daily News* 12 Dec. 1/2 (Advt.), Gentlemen's undervests. Winter weight, 32 in. to 48 in. chests. 1934 G. B. Shaw *On Rocks* ii. 226 There is a generous fire in the grate; and the visitors wear winter clothes. 1940 L. I. Wilder *Long Winter* ix. 73 They dressed carefully in their woolen winter dresses. 1979 T. Barling *Olympic Sleeper* xi. 138 He was warm in his winterweight pin-striped suit. 1984 W. Beechey *Rich Mrs. Robinson* xii. 89 He needs some winter vests badly.

(*c*) *winter-while* (poet.).

a 1889 G. M. Hopkins in *Dublin Rev.* (1920) July-Sept. 46 They came from the south, Where winter-while is all forgot.

(*d*) *winter cruise*.

1934 Winter cruise [see *front-pager* s.v. *front *sb.* (and *a.*) 14]. 1976 *Liverpool Echo* 6 Dec. 7/1 Aznar Line are having a record breaking season with their winter cruises out of Liverpool.

c. (Further example.)

1939 *WPA Guide to Florida* (Federal Writers' Project) i. 7 The traveler..may detour inland to discover the hidden winter-vegetable kingdom on the muck lands.

4. (instrumental) *winter-heavy*, *-left*, *-locked*, *-wearied*, *-weary*, *-weighed*; (similative) *winter-blue*, *-cold*, *-white*; ('in or during winter') *winter-felled* (earlier example), *-flowering* (earlier example), *-hardy*, *-sown* (later example); appositive, as *winter-spring*.

1936 R. Frost *Let.* 6 Feb. (1964) 270 And Sirius is a winterbluegreen star. 1958 J. W. Day *Lady Houston* xv. 225 Never had..the winter-blue woods of Kimbolton or the generous warmth of Brampton Park..beckoned more seductively. 1944 E. Sitwell *Green Song* 7 Henry thought me winter-cold When to keep his love I turned from him as the world Turns from the sun. 1742 Ellis *Timber-Tree* II. 13 From whence they infer, that the worm can't breed so soon in a Summer-fell'd Tree, as in a Winter-fell'd one. 1794 Winter-flowering [see Aconite 3]. 1960 *Farmer & Stockbreeder* 5 Jan. 43/1 We may..be given a more winter-hardy, leafier kale. 1975 *Daily Colonist* (Victoria, B.C.) 20 July 22/7 As it is not winter-hardy here, it is commonly grown as an annual. 1920 D. H. Lawrence *Women in Love* xxix. 435 This was an old world she was still journeying through, winter-heavy and dreary. 1955 S. Spender *Coll. Poems 1928–53* ix. 173 Its vermilion seems A Red Admiral's wing, with veins Of lichen and rust, an underwing Of winter-left leaves. 1926 S. Leslie *Cantab* xv. 183 He sobbed like a winter-locked river hastening over the weir at the first warmth of spring. 1946 Dylan Thomas *Deaths & Entrances* 34 Two proud, blacked brothers cry, Winter-locked side by side. 1960 *Farmer & Stockbreeder* 15 Mar. 141/3 Tri-Farmon 41 effectively controls the widest possible range of weeds in winter and spring-sown wheat. 1888 W. D. Hay *Blood* vii. 29 Although it was so late in the winter-spring season, the weather was wild and wintry. 1967 *Oceanogr. & Marine Biol.* XVI. 409 The 'Atlanto-Scandian' winter-spring spawning stocks. 1892 J. G. Whittier *At Sundown* 58 This stormy interlude Gives to our winter-wearied hearts a reason For trustful gratitude. 1917 D. H. Lawrence *Look! we have come Through!* 160 We who are winter-weary in the winter of the world. 1866 J. G. Whittier *Snow-Bound* 46 And woodland paths that wound between Snow drooping pine-boughs winter-weighed. 1915 E. Sitwell *Mother* 16 Her ice-cold breast was winter-white.

5. a. winter annual (see quot. 1900); **winter carnival** *Canad.*, 'an organized winter social activity featuring winter sports, beauty contests, ice-sculpture, etc.' (*Dict. Canadianisms*); **winter coat**, (*a*) the coat of an animal in winter, where this differs from that in summer; (*b*) a (woman's) coat suitable for winter weather; **winter country** *N.Z.*, land where livestock can be wintered; **winter garden**, (*b*) (earlier and later examples); (*c*) a building used for concerts, plays, dances, etc., at a seaside resort; **winter-killed** *pa. pple.* and *a.*, **-killing** (earlier examples); also **winter-kill** *v. intr.*, to become winter-killed; *trans.*, to make winter-killed; also *absol.*; **winter-kill** *sb.*; **winter oil**, edible oil that remains clea. at low temperatures owing to the removal of constituents that would have caused congelation or precipitation; **Winter Olympic Games, Winter Olympics**, international competitive winter sports held under the auspices of the International Olympic Committee, usually every four years; **winter packet** *Canad.* (*Obs.* exc. *Hist.*), a boat or land party carrying mail in winter-time between trading posts; the mail itself; **winter road** *Canad.*, a road or a route used in winter when the ground is frozen or there is snow; **Winter War** (also with small initials), the war between the U.S.S.R. and Finland in 1939–40; **winter woollies**: see *woolly *sb.* 1.

1900 B. D. Jackson *Gloss. Bot. Terms* 290/2 *Winter-annual*, a plant which germinates in autumn, and living through the winter, fruits and dies. 1977 J. L. Harper *Population Biol. Plants* xviii. 547 A single population [of *Papaver dubium*] includes winter annuals and spring annuals. 1884 *Outing* (U.S.) Feb. 400/2 The winter carnival at Montreal, which was so successfully inaugurated last year, will open on February 4. 1973 *Globe & Mail* (Toronto) 13 Jan. 33/1 Most winter carnivals in Ontario rely less heavily on the snowmobile for their fun weekends. 1894 Winter coat [in Dict., sense 3 a (*b*)]. 1920 [see *bear fur* s.v. *bear *sb.*[1] 9]. 1956 R. Macaulay *Towers of Trebizond* viii. 78 The camel..was very smooth, having just shed its winter coat. 1982 C. Fremlin *Parasite Person* xix. 128 Helen felt the warmth of the sun..through her thick winter coat. 1898 Morris *Austral English* 513/1 *Winter country*, in New Zealand (South Island), land so far unaffected by snow that stock is wintered on it. 1912 A. Wall *Century N.Z.'s Praise* 80 Good winter-country, where sweet grasses grow. 1949 P. Newton *High Country Days* 197 Safe country which is saved for the winter is 'winter country'. 1783 T. Blaikie *Diary of Scotch Gardener* (1931) 179 The winter Garden adjoining to the Hott houses, was more Beautifull than Elegant. 1951 *Dict. Gardening* (R. Hort. Soc.) IV. 2282/2 The Winter Garden is usually of sufficient size to allow the central part of the interior being laid out in walks and large beds. 1977 *Lancs. Life* Nov. 81/1 The Palace incorporated a winter garden from which the stage could be seen without spectators needing to go through into the auditorium. 1817 S. Brown *Western Gaz.* 49 That wheat ..never gets winter-killed or smutty. 1845 *Farmers' Cabinet* 15 Jan. 195/1 This blight is not to be confounded with winter-killing. *Ibid.* 15 Feb. 202/2 It is not so hardy as some varieties: it is more subject to winter-kill. 1846 E. Emmons *Agric. N.Y.* I. 281 The grain very rarely winter-kills. 1849 *Ex. Doc. 31st U.S. Congress 1 Sess. House* No. 5. ii. 653 The..snow which lies upon the ground nearly six months in the year would be likely to 'winter-kill' it. 1918 S. S. Visher *Geogr. S. Dakota* 56 Red clover is not a success..largely because it winter-kills. 1945 *Ecol. Monogr.* XV. 343 (*heading*) Limnological conditions in ice-covered lakes, especially as related to winter-kill of fish. 1977 *Chicago Tribune* 2 Oct. xi. 1/4 High nitrogen fertilizer..would only promote late growth that would winterkill. 1980 *Northeast Woods & Waters* Dec. 18/1 Last year's rate of winterkill was lower than usual because of the relatively mild weather conditions. 1894 C. R. A. Wright *Animal & Veg. Oils* xi. 257 Oils that have been thus treated are sometimes termed 'winter oils'. 1920, 1939 Winter oil [see *demargarinated *a.*]. 1970 T. J. Weiss *Food Oils* iii. 59 The solid portion of the oil which had set up in storage tanks in the winter at 40°–42° F..was settled out and removed, leaving an oil which would remain clear when chilled. Cottonseed oils were thus divided into *summer* and *winter* oils. [1928 *Times* 17 Feb. 6/4 The usual clean crisp snow has given place to an earthy slush, and as a result the second celebration of a Winter Olympiad has come to an abrupt standstill.] 1932 *Times* 1 Feb. 7/4 A thaw which has set in threatens to destroy..the third winter Olympic Games. 1936 *Times* 27 Jan. 5/1 The Lake Placid bob-run..provided some of the most exciting spectacles at the 1932 Winter Olympics. 1956 *Times* 6 Jan. 9/5 In 1948..I won a bronze medal in the Winter Olympic Games. 1981 'E. Lathen' *Going for Gold* i. 11 It took the Winter Olympics to keep him in the continental United States in February. 1831 E. Smith *Let.* 25 Nov. in *Champlain Soc. Publ.* (1938) XXIV. 79 Our Winter Packet being now preparing to Travell on to your Quarter, I will not let it go without acknowledging the receipt of your friendly epistle. 1971 J. McDougall *Parsons on Plains* xi. 92 We saw the flicker of a campfire. We found that it was the one winter packet from the east on its way to Edmonton. 1801 A. Mackenzie *Voy. from Montreal* vi. 84 One of the natives who followed us, called it the Winter Road River. 1808 H. Gray *Lett. from Canada* (1809) 254 The country people who first form the winter roads on the snow, direct their *Carioles* by the nearest course where the snow is most level; and they go in as straight a line as possible, to the place where they are destined. 1916 *Yukon Territory* 194 In the summer of 1902 the government built a winter road between Dawson and Whitehorse, a distance of approximately 333 miles. 1971 *Country Life* 24 June 1572/1 We had been told to follow a path until it joined a 'winter road' which would in turn lead us to the marsh. 1973 *Kingston* (Ontario) *Whig-Standard* 26 Jan. 7/2 The winter road over the ice is about four lanes wide and is 'brushed' with evergreens at the sides as a guide during swirling snow storms. 1979 A. M. Tizzard *On Sloping Ground* ix. 130 The bay would be frozen over and there was always a good winter road across Twillingate Island. 1942 F. Owen in W. P. & Z. Coates *Soviet–Finnish Campaign* p. i, What about the Finns? In the Winter War they gained a deserved fame for valour and military skill. 1957 *Times Lit. Suppl.* 11 Oct. 603/2 The so-called 'Winter War', resulting from the Soviet attack on Finland in November, 1939, and ending with the Finnish surrender of March, 1940. 1971 W. H. McNeill in A. Bullock *20th Century* 49/2 The Finns' success in holding the Russians at bay for the long weeks of the so-called 'winter war' (1939–40). 1973 J. Fleming *You won't let me Finish* x. 82 The cook came here during the Winter War.

b. winter-crack (see quot. 1898); **winter duck**, (*a*) (later examples); **winter flounder** (earlier example); **winter gnat** = *winter midge* in Dict.; **winter grape** (earlier and later examples); **winter peach**, the fruit of a peach-tree cultivated in a greenhouse and fruiting during autumn or winter; **winter sleeper**, an animal that hibernates; **winter squash** (earlier and later examples); **winter-sweet** (*b*) a shrub, *Chimonanthus præcox*, of the family Calycanthaceæ, native to China and bearing pale yellow fragrant flowers in winter before the leaves appear.

1877 E. Peacock *Gloss. Words Manley & Corringham, Lincs.* 276/2 *Wintercrack*, a small green plum, the fruit of which ripens very late. 1898 *N. & Q.* 13 Aug. 235/2 A fair-sized round, yellowish plum, only fully ripe in November, is known in Derbyshire as the 'winter-crack'. They are called 'cracks' because with the first frosts the fruit cracks on one side, being then fully ripe. 1914 D. H. Lawrence *Prussian Officer* 282 There were some.. winter-crack trees. 1885, 1917 Winter duck [see *lady-bird 3]. 1814 S. L. Mitchill *Fishes N.Y.* 387 New-York Flatfish..is called the winter flounder. 1899 D. Sharp in Harmer & Shipley *Camb. Nat. Hist.* VI. vii. 473 The winter gnats of the genus *Trichocera* are a fair sample of this sub-family. 1926 A. H. Hamm in J. J. Walker *Nat. Hist. Oxf. District* 257 Four species of 'Winter-gnats' are always common from autumn to spring. 1968 *Oxf. Bk. Insects* 122/1 There are ten British Winter Gnats, which belong to the family Trichoceridae and look like small crane-flies. They get their English name from their way of 'dancing' in large swarms on winter afternoons. 1771 G. Washington *Diary* 20 Nov. (1925) II. 43 Began to Plant Cuttings of the Winter Grape. 1789 [see *frost-grape* s.v. *frost *sb.* 7 c]. 1949 *Amer. Photogr.* Apr. 244/2 Winter grape is one of our commonest species from northern New York to Michigan. 1787 J. Woodforde *Diary* 31 Oct. (1926) II. 354 Mr. Custances Garden brought us this Morning a Basket of Winter Peaches. 1960 I. Wallach *Absence of Cello* (1961) 199 Marian wanted a winter peach. 1709 Winter-sleeper [in Dict., sense 3 a (*c*)]. 1911 J. A. Thomson *Biol. Seasons* iv. 333 A survey of the Winter-sleepers seems to show that the life-saving reaction must have arisen by..natural selection. 1775 *Boston Transcript* 26 Apr. iii. 12/7, I have a fine prospect of a Crop of..winter Squashes this fall. 1969 *Oxf. Bk. Food Plants* 122/2 Winter squashes are cut in the autumn and can be kept for 3 or 4 months or longer. 1893 W. Robinson *Eng. Flower Garden* (ed. 3) 325/2 Winter Sweet is a lovely shrub which in our country requires a wall. It flowers in December and January; beautiful, and of delicious fragrance. 1934 Ld. Berners *First Childhood* ii. 20 Just outside the windows there grew a shrub of the early-blossoming chimonanthus. (Winter-sweet it was called in the days before gardeners grew so refined.) 1955 [see *chimonanthus]. 1980 *Gardener's Dozen* 12 My winter-sweet.. sometimes gets knocked about by the frost and snow.

winter, *v.* Add: **1.** Also (*Canad.*) with *out*.

1870 in C. Wilson *Campbell of Yukon* (1970) 165, I..had long consultations with most of the Freemen, wintering out in this quarter. 1968 E. Russenholt *Heart of Continent* viii. 132 When November [comes]..the population of Assiniboia is 'at home'—excepting only, those hundreds who elect to 'winter out' on the plains and along the waterways. 1970 R. Symons *Broken Snare* xvi. 112 He [*sc.* a steer] had found a bunch of wild horses... So he had wintered out quite happily with his kind hosts.

2. c. With *over*. = *over-winter *v.* 4. Also *intr.*

1979 C. Kilian *Icequake* iv. 42 How are we supposed to winter over on a goddam iceberg? 1982 'E. Lathen' *Green grow Dollars* i. 12 A tomato that could be planted, wintered over, then harvested.

winterer. Add: **1. c.** A hibernating animal.

1930 *Observer* 6 Apr. 24/2 Sudden warmth..may awake a winterer too precociously.

Winterhalter (vi·ntəɹhaltəɹ). The name of Franz Xavier *Winterhalter* (1806–73), German portrait painter of royalty, used *attrib.* to designate things characteristic of his pictures, esp. court settings and a style of women's formal dress.

1923 *Daily Mail* 11 Sept. 11/2 The 'period' dress, Winterhalter or Velasquez, is almost entirely restricted to dinner, or formal afternoon, wear. 1937 H. Nicolson *Helen's Tower* vii. 146 How did she cope with the Winterhalter atmosphere of that decaying court? 1944 'Brahms' & 'Simon' *No Nightingales* xxvii. 155 An..old lady tottering but aristocratic, at grips with a son-in-law too mean to buy a Winterhalter ball-gown for dining *en famille*. 1957 M. B. Picken *Fashion Dict.* 376/2 *Winterhalter*, name applied to costumes characterized by off-shoulder necklines, corseleted waistlines, crinoline skirts with flounces. 1970 R. T. Wilcox *Dict. Costume* 398/2 *Winter-halter*, another term for the crinoline period.

winterim (wi·ntərim), *a.* and *sb.* *U.S.* [Blend of Winter *sb.*[1] and Interim *adv.*, *sb.*, and *adj.*] (Of or pertaining to) a short winter term in some private schools in the U.S., part of which is spent by some pupils on projects away from the school.

1972 *Handbk. Private Schools* (ed. 53) 1079 A four-week 'Winterim' in which juniors and seniors may go off campus and a special program is run for freshmen and sophomores. 1976 *National Observer* (U.S.) 22 May 15/1 (Advt.), Winterim on-and-off campus work/study program—Thorough college preparation. 1979 *N.Y. Times Mag.* 30 Sept. 91/2 (Advt.), Interscholastic and recreational sports—On and off campus Winterim. 1980 L. Birnbach et al. *Official Preppy Handbk.* 52/2 Winterim session spent skiing.

wintering, *vbl. sb.* Add: **II. 5*.** Land where livestock may be wintered.

1937 A. Fraser *Sheep Farming* iv. 46 The ewes can be helped if the hill is cleared as much and as early as possible..by getting the ewe hoggs to their wintering at the earliest possible date. 1978 *Dumfries & Galloway Standard* 21 Oct. 21/3 (Advt.), Wintering wanted for 45 Blackface Ewe Lambs from November till 1st March 1979.

III. 6. *wintering ground* (examples).
1805 Z. M. Pike *Jrnl.* 15 Oct. in *Acct. Expeditions Sources Mississippi* (1810) I. 33 This day's march made me think seriously of our wintering ground. *c* 1890 R. Campbell in C. Wilson *Campbell of Yukon* (1970) xi. 104 Do away with a wintering ground which, from the starvation so frequently experienced there, our men hold in actual dread and abhorrence. 1977 *Monitor* (McAllen, Texas) 9 Jan. B 8/4 The request for Texas biologists to investigate the whitewing wintering grounds came from the International Whitewinged Dove Council.

wi·nterishly, *adv.* [f. Winterish *a.* + -ly[2].] In a manner suggestive of winter.
1905 *Smart Set* Oct. 26/1 Though by good rights the day might have been winterishly cold it was of a balmy mildness.

winterize (wi·ntəraiz), *v.* orig. and chiefly *U.S.* [f. Winter *sb.*[1] + -ize.] *trans.* To adapt or prepare (something) for operation or use in cold weather.
1938 *Amer. Speech* XIII. 160/1 A radio announcer... urged his listeners to have their cars winterized. 1949 A. Hayes *Girl on Flaminia* iv. 90 She sat on the hard canvas cushion of the jeep, and she must have felt exposed. The jeep had not been winterized. 1950 *Jane's Fighting Ships 1950–1* 454 *Arneb* has been refitted for Arctic Service... Other vessels are also to be 'winterized'. 1954 E. W. Eckey *Veg. Fats & Oils* iii. 124 The temperature below which no portion of the oil should be chilled is a little above the cloud point of the oil to be winterized. 1964 S. Bellow *Herzog* 96 This is a fine house you have... Summer only, isn't it? You could winterize it easily. 1973 B. Wright *Four Seasons North* 42 Sam winterized the rifles, wiping away all the lubrication, which would freeze if left in the guns. 1980 L. Auchincloss *House of Prophet* vii. 101 We agreed to take over my parents' summer house in Seal Cove and winterize it.
Hence **wi·nterized** *ppl. a.*, **wi·nterizing** *vbl. sb.*; **winteriza·tion.**
1926 *Jrnl. Oil & Fat Industries* III. 421/1 Some of the cottonseed stearine made by the winterizing process and the hydrogenated oil are used in margarine. 1927 *Oil & Fat Industries* IV. 301/2 Depending upon the length of time winterized oil will stand clear, brilliant and limpid when exposed to the so-called Winter Oil Test, an oil is more or less suitable for..salad oil and mayonnaise manufacturers. 1940 *PM* 29 Nov. 14/3 Instead of having an entire company of trainees, even one 'winterized tent' of trainees, all together,..there will be throughout the 44th Division scads of veterans surrounding each new man. 1940 *Capital* (Topeka) 8 Dec. 2A (Advt.), Buy any car... Plus complete winterization. 1943 *Oil & Soap* July 131/2 The present slow and inefficient winterization process for cottonseed oil. 1955 Smalley & Klohr in F. S. Mallette *Probl. & Control of Air Pollution* xvii. 199 (*caption*) Winterized sulfur recovery plant. 1957 *Encycl. Brit.* VI. 584/2 Winterization is a process by which a portion of the refined and bleached oil is solidified by chilling and filtered off. 1966 *Economist* 22 Jan. 341/3 Builders..are showing considerable interest in 'winterisation'... Over a third now have some degree of lighting on site. 1970 *Toronto Daily Star* 24 Sept. 39/7 (Advt.), Winterized jackets have the accent on fashion with bright colours. 1971 C. Bonington *Annapurna South Face* 286 All the cameras functioned well, and although only the Nikons were winterized, there was no trouble from freezing up. 1978 *N.Y. Times* 30 Mar. B15/2 (Advt.), Two 5 rm winterized cottages. 1979 *Farmington* (New Mexico) *Daily Times* 27 May 6c/3 (Advt.), Experiments on winterizing and solar energy retrofits.

winter quarters. Add: **1.** Also, such a place occupied by any travelling company or by private individuals. (Further examples.)
1714 W. Vickers *Let.* 21 Sept. in M. M. Verney *Verney Lett.* (1930) II. xxi. 16 Miss Pen comes to our winter quarters in the middle of next week. 1897 A. Beardsley *Let.* 29 July (1970) 353 At the end of this week I shall move to Paris, and then into winter quarters as soon as possible. 1939 *Florida* (Federal Writers' Project) III. 395 Bailey circus winter quarters.

winters (wi·ntəɹz), *adv. U.S.* [Pl. of Winter *sb.*[1]] During the winter.
1907 'Mark Twain' *Christian Sci.* II. 235 It can appoint its own furnace-stoker, winters. 1978 *Chicago* June 158/1 Winters, William is the tennis director at the Genesee Valley Tennis Club in Flint.

wi·ntersome, *a. rare*[−1]. [f. Winter *sb.*[1] + -some[1].] = Wintry *a.*
1864 Trollope *Small Ho. Allington* II. xv. 148 The fourteenth of February in London was quite as black, and cold, and as wintersome as it was at Allington.

winter sport. **a.** A sport enjoyed in the winter; *spec.* an outdoor sport on snow or ice, such as skiing or skating. Usu. *pl.*
1828 *Ladies' Mag.* Mar. 141 (*heading*) Children at their winter sports. 1847 C. M. Yonge *Scenes & Characters* xiii. 163 Intent upon the various winter sports in which William and Lord Rotherwood allowed him to share [at an English country house]. 1879 H. Chadwick (*title*) Handbook of winter sports. Embracing: skating (on ice and on rollers,) rink-ball, curling, ice-boating, and American football. 1906 *Dress* Dec. 18/1 Tobogganing, another favorite winter sport introduced from Canada, has many devotees. 1934 F. Scott Fitzgerald *Tender is Night* I. iii. 17 Vivid advertising cards of the railroad companies..winter sports at Chamonix. 1956 A. H. Compton *Atomic Quest* 213 Winter sports, especially skiing, were popular. 1975 *Oxf. Compan. Sports* 950/2 Skibobbing is a new winter sport, combining the virtues of the ski, the bobsleigh, and the velocipede in a downhill run.

b. *attrib.* (in *sing.* and *pl.*).
1908 E. & M. Syers *Bk. Winter Sports* 323 Some notes on winter-sport resorts. 1922 *Cook's Continental Timetable* Jan. 377 (Advt.), Central position. Close to all Winter Sports places. 1950 E. Hemingway *Across River* xxxiii. 207 It would be just like any winter-sports hotel. 1966 N. Freeling *King of Rainy Country* 38 She had gone.. on a wintersport holiday. 1975 *Country Life* 30 Oct. 1192/2 A really beautiful collection of winter sports clothes.

Hence **winter-sport** *v. intr.*, to engage in winter sports; **winter-sporting** *vbl. sb.*
1948 M. Laski *Tory Heaven* ix. 128 I'd like to go winter-sporting in January. 1955 T. H. Pear *Eng. Social Differences* xi. 265 Few who 'winter-sport'..feel that their social rating is thereby raised. 1961 *Times* 13 May 11/2 The famous winter-sporting centres. 1974 *Guardian* 23 Mar. 14/2 If you've fished in Scotland, if you've winter sported in Switzerland, you can do it all in Sweden.

wipe, *sb.* Add: **1. c.** *Cinemat.* and *Television.* An effect in which an existing picture seems to be wiped out by a new one as the boundary between them moves across the screen (the pictures themselves remaining stationary). orig. *wipe-dissolve.*
1933 *Cinema Q.* II. I. 43, I..deplored the constant use of wipe-dissolves to cover the weak continuity. 1934 C. Lambert *Music Ho!* iv. 263 There is no real equivalent in music even of the 'wipe-dissolve' which leads the eye gently but quickly from one scene to another. 1936 A. Brunel *Film Production* 43 It may be argued that wipes are not easily achieved. 1960 *Guardian* 8 June 7/3 A special effects generator..enables 20,000 different shades of 'wipe' to be deployed... It makes a fascinating variety of shapes and devices upon the screen. 1979 *Broadcast* 1 Oct. 54/2 Within the SqueeZoomed sequence of archive TV shots..Tony Rayner inserted two wipes to blue..which allow live 'headline' shots from that day's programme to be chromakeyed in.
2. a. *a wipe in the eye*: a disappointment or rebuff; = *smack in the eye* s.v. *SMACK *sb.*[2] 3 a. Cf. Wipe *v.* 9 d in Dict. and Suppl.
[1644: in Dict.] 1926 T. E. Lawrence *Let.* 6 Apr. (1938) 495 Your statement that the hospital passage would be a wipe in the eye for 19 readers out of 20 puts it out of court. 1949 D. M. Davin *Roads from Home* III. iii. 236 It was a wipe in the eye for John the way he was getting out.
4. a. (Earlier example.)
1708 *Memoirs of John Hall* 23 *Wipe*, a Handkerchief.
b. A disposable piece of soft absorbent cloth or tissue, sometimes impregnated with a cleansing agent, for wiping clean one's hands or anything small.
1971 *Textile Industries* Dec. 50/1 Towels, Covers, Pads, and Wipes. 1974 Hawkey & Bingham *Wild Card* ix. 87 Half-empty boxes of medical wipes lying on the ultra-centrifuge. 1978 'M. Yorke' *Point of Murder* ii. 19 Kate's hands had got oily..but she kept some tissue wipes in the car. 1980 *Chem. in Brit.* XVI. 449/4 For situations where protective gloves are inconvenient, Chicopee has brought out Dermawipe impregnated hand wipes.

wipe, *v.* Add: **1. c.** *absol.* = *DRY *v.* 1 c. Also with *up.* Cf. *WASH *v.* 1 i.
1943, 1962 [see *WASH *v.* 1 i]. 1968 R. V. Beste *Repeat the Instructions* ii. 19 He wiped while Huskion..scrubbed away in the sink. 1974 M. Birmingham *You can help Me* vii. 169, I was helping Mrs Hope wipe up in the kitchen. 1981 A. Wilson in T. Thompson *Edwardian Childhoods* iii. 78 One'd wipe and one'd wash—we didn't make hard work of it.
d. To demagnetize (a ship) by passing a horizontal current-carrying cable up and down the hull; also, to remove a recording from (magnetic tape).
1946 'L. Luard' *Changing Horizons* 145 'No complaints, except she's steel.' 'And not wiped or degaussed,' the Skipper commented. 1947 Crowther & Whiddington *Science at War* 171 He arranged that ships should be 'wiped' with temporary horizontal coils. 1962 R. W. Clark *Rise of Boffins* iv. 95 Demagnetizing the ships..by 'wiping' the sides..with a horizontal cable carrying a strong current. 1962 E. Salter *Voice of Peacock* xx. 203 In the case of auditions, the tape was sometimes wiped so's it could be used again. 1965 D. Francis *Odds Against* x. 137, I wiped the tape clean. 1980 *Listener* 8 May 594/2, I presume the BBC wiped, as they say, the original tape.
2. d. To erase (a magnetic recording, or data stored on a magnetic medium). Freq. with *off, out.*
1900 *Engin. Mag.* XIX. 758/1 When it is desired to wipe out a record, the electromagnet..is attached to a constant battery and run over the wire, thus magnetising it uniformly once more and preparing it to receive a new message. 1934 *Wireless World* 5 Jan. 8/3 When a record is no longer required, the programme recorded on the strip can be 'wiped out'. 1976 *Broadcast* 23 Aug. 10/3 He was staggered at the quantity of programmes in which James MacTaggart had been involved. Most of it has been wiped. 1981 *Times* 4 July 10/3 This is a three-hour reusable tape with an hour's quite sophisticated cabaret already recorded, which you can keep or wipe off. 1984 *Computerworld* 26 Mar. 14/2 If one formats an IBM Personal Computer XT and does not indicate which drive to format, the machine formats the hard disk and wipes out all data on it.
e. (Without *prep.*) To dismiss, reject, repudiate (esp. a person). *Austral.* and *N.Z. slang.*
1941 K. Tennant *Battlers* 196 Giving her money..in the casual manner that wiped her from all consideration as a human being. 1946 *Coast to Coast 1945* 123 Listen pal—your girl wiped you, didn't she? 1948 *Landfall* June 111 Hands in pockets, shoulders hunched, he strode bitterly up the street from the pub. He'd wipe them, have nothing to do with the morons. 1954 T. A. G. Hungerford *Sowers of Wind* 162 She dumped me, wiped me like a dirty nose. 1967 F. Sargeson *Hangover* xiv. 124 If it came to that one of his reasons for wiping university was a senior lecturer who had failed to avoid the same gross error. 1975 R. Beilby *Brown Land Crying* 295 You can wipe that idea, if that's what you're thinking.
6. e. Also, to kill (a person); also with *out. slang.*
1968 J. Philips *Hot Summer Killing* III. i. 129 Is he the one who was wiped earlier tonight in the Molyneaux Hotel? 1969 C. Burke *God is Beautiful, Man* (1970) 47 They decided to find a way to get rid of him, to wipe him out. 1977 *Time* 12 Sept. 40/2 You could be wiped out if you moved a single inch. 1980 J. McClure *Blood of Englishman* ii. 18 Someone tried to wipe Bradshaw... The shot caught him here in the collar-bone.
h. *to wipe* (an expression, esp. a smile) *off a person's* or *one's face*: (to cause him) to cease showing it. *slang.*
[1567: cf. sense 6 d in Dict.] 1895 Conrad *Almayer's Folly* xii. 256 A face from which all feelings and all expression are suddenly wiped off by the hand of unexpected death. [1898: cf. sense 6 c in Dict.] 1935 *Time* 24 June 28/1 Wipe dat smile offen his face! 1936 D. Carnegie *How to win Friends & influence People* II. ii. 99 Bill, you are going to wipe the scowl off that sour puss of yours today. 1972 D. S. Viscott *Making of Psychiatrist* ii. 37 Terry O'Conner seemed to think it was funny but wiped the smile off her face every time her eye caught Larry's. 1977 *Observer* 14 Aug. 3/7 Only one sentence would have wiped the smile off Mason's face. 1978 G. Greene *Human Factor* VI. i. 305 She realised she was smiling at the telephone—thank God, they hadn't yet invented a visual telephone, but all the same she wiped the smile off her face.
i. *pass.* or *intr. Surfing.* To be knocked from one's surfboard. With *out. slang.*
1962 T. Masters *Surfing made Easy* 66 *Wiped out*, getting knocked off of a surfboard, usually by a wave. 1965 [see *LOCKED *ppl. a.* g]. 1966 *Weekly News* (N.Z.) 19 Jan. 6/3 When 'wiping-out' a surfer should try to hold his board. 1968 *Surfer Mag.* Jan. 48/2 Frye misjudged one of his turns high in the curl and wiped-out in the white water.
8*. *Cinemat.* and *Television.* To pass from or *from* one scene to another by means of a wipe; to employ a wipe.
1951 Halas & Privett *How to cartoon for Amateur Films* 118 We now wish to wipe from one scene to another. 1952 *Cinema* 7 Jan. 97/1 [The Director] can cut, fade, wipe or mix at will.
9. d. (Later examples.)
1928 D. L. Sayers *Unpleasantness at Bellona Club* xiv. 168 'I'm glad somebody appreciates me. Anyhow,' he added viciously, 'I bet that's wiped old Pritchard's eye.' 1929 F. M. Ford *Let.* 11 Sept. (1965) 187 He had only got me away from Duckworth in order to wipe Gerald's eye. 1949 N. Mitford *Love in Cold Climate* I. vi. 60 At teatime the village policeman reappeared.., having wiped the eye of all the grand detectives who had come from London in their shiny cars. He produced a perfect jumble-sale heap of objects which had been discarded by the burglars. 1956 'A. Gilbert' *And Death came Too* xiv. 146 Eventually he agreed to take the case (his heel of Achilles being an inability to pass up a chance of wiping the official eye).
e. *to wipe the floor with*: (earlier examples); also *to wipe up the floor* or *ground with.*
1887 *Courier-Jrnl.* (Louisville, Kentucky) 4 Jan. 2/6 Two brothers wipe up the floor with a Missouri newspaper man. 1888 in Farmer & Henley *Slang* (1903) VII. 359 The Scroggin boy was as tough as a dog-wood knot. He'd wipe up the ground with him; he'd walk all over him. 1896 *Dialect Notes* I. 427 *Wipe the floor with*, to defeat. 1897 *Nat. Police Gaz.* 26 May 7/4 Green fairly wiped the floor with Roberts in the first two rounds. 1908 *Magnet* I. I. 3/1 I've wiped up the ground with bigger fellows than you, for far less cheek than you've given me.
10. The vb.-stem in combination, as **wipe-clean** *attrib.* or as *adj.*, designating fabrics or furnishings that may be cleaned simply by wiping.
1962 *N.Y. Times Mag.* 9 Sept. 102 In new wipe-clean Boltaflex Vinyl Suede. 1965 *Economist* 13 Feb. 700/2 The [synthetic] shoes, although theoretically 'wipe-clean', tend to look grubby when unpolished. 1970 *Vogue* Jan. 25/1 Lovely cookers with plain glass tops—put the pan down..and you are on the hob, yet this is a wipe-clean glass surface. 1977 *Austral. House & Garden* Jan. 114 (Advt.), Interiors are wipe clean, white melamine laminate, edged in white P.V.C. 1983 *Which?* Dec. (Publications Suppl.), Both binders are hardwearing and have wipe-clean covers.

wi·peable, *a.* Also **wipable.** [f. Wipe *v.* + -able.] Capable of being wiped.
1926 *Pocket Oxf. Dict.* s.v. *Wipe*, Wipable. 1979 *Personal Computer World* July 23/2 The single unit..has a wipeable light brown plastic housing with a robust typewriter style keyboard and separate numerical keypad. 1981 *Washington Post* 26 Aug. B5/2 For the latest in lunch carriers this fall, go with this wipeable tote decorated with elephants, sailboats or pigs. 1982 *Times* 20 Feb. 11/5 The tape..will be wipeable, leaving a tape..pristine for one's own recording.

wipe-out (wəi·paut). [f. vbl. phr. *to wipe out*: see Wipe *v.* 6.] **1.** *Radio.* The condition in which a strong received signal renders impossible the reception of other signals (either wanted ones or interference).

1921 *Wireless World* IX. 13/1 With radiotelephony the case is worse, as the wipe out is continuous if it occurs at all. 1929 *Encycl. Brit.* IV. 218/2 Within a 'wipe-out' area uninterrupted service can be guaranteed, unless the interference is produced by listeners themselves. 1940 *Amateur Radio Handbk.* (ed. 2) x. 160/1 There are three types of interference that may be caused in neighbouring receivers working on broadcast waves, by an amateur telegraphy transmitter. First, the 'wipe-out' effect, where the signal from the transmitter 'blocks' the receiver due to either the excessive field strength of the transmitter, or the inselectivity of the receiver, or both.
2. *Surfing.* A fall from one's surfboard as a result of a collision with another surfer or a wave. Cf. *WIPE *v.* 6 i. *slang.*
1962 *Austral. Women's Weekly* 24 Oct. Suppl. 3/4 *Wipeout*, a dramatic fall off a board when a rider is trying to catch a wave. 1963 [see *SURFIE]. 1969 *Observer* 3 Aug. 33/1 The biggest danger always lies in a 'wipe-out', with a loose board which may hit the rider or other surfers. 1970 *People* (Austral.) 26 Aug. 20/1 One bad wipeout—at Sunset Beach, Hawaii—earned him broken ribs.
3. Destruction, annihilation; a killing; a crushing defeat; an overwhelming experience. *slang* (orig. *U.S.*).
1968 *Sun* (Baltimore) 7 July 5/2 Charlie is 1810. We had a wipeout... *Translation.* Girl: Charlie is old news. We broke up. 1971 J. Henderson *Copperhead* vi. 71 Less than thirty-six hours to incapacitate 85 percent of the population. The remaining 15 percent would take a little more than a week. Strategically it would be a wipe-out. 1972 *Jazz & Blues* Sept. 8/1 When I heard Art it was a wipeout. He just wiped me out man. 1977 *Daily Mirror* 12 Apr. 27/7 A record 140,000 [motor-cycling] fans have watched the embarrassing wipe-out by 410 points to 379. 1979 L. Meyer *False Front* iii. 24 This is something like the wipeout of a personal fortune. 1984 'M. Hebden' *Pel & Pirates* xviii. 143 Think it was a gang wipe-out, Patron?

wiper. Add: **2. b.** (Earlier examples.)
1826 *Price List* in *Austin Papers* (1924) 1369 To wiper claw for rifle ..50. 1827 J. Kerr *Let.* 27 Feb. in *Ibid.* 1607 Thimble rod and socket end of wipers lost... 50.
c. = *windscreen wiper* s.v. *WIND *sb.* 31. Also *attrib.*, as *wiper blade, switch.*
1929 *Times* 2 Nov. 4/7 The driver has an all-enclosed cab, with..a sloped adjustable screen, with a wiper, in front. 1942 W. Faulkner *Go down, Moses* 337 It was the youngest face of them all,..staring sombrely through the streaming windshield across which the twin wipers flicked and flicked. 1953 L. Durrell *Balthazar* iv. 67 The windscreen became gradually snowed-up and he switched on the wipers to keep it clear. 1959 *Times* 25 Sept. 8/2 Wiper blades dry the screen and then park automatically. 1970 *Motoring Which?* July 93/1 On the 1800s and 1800Ss the lights or wiper switches collapsed in about one in four cars. 1976 H. Kemelman *Wednesday the Rabbi got Wet* xiii. 80 It was coming down so fast that my wipers couldn't handle it.
6. A pivoted arm that automatically rotates through an arc to make electrical contact with any of a curved row of terminals in a telephone exchange; also, the rotary or sliding contact of a potentiometer.
1906 J. Poole *Pract. Telephone Handbk.* (ed. 3) xxx. 483 Opposite the lower part of each 'bank' a short arm is fitted, on the ends of which are 2 springs, which, when the rod is rotated, sweep over and under the strips of contacts, and are, therefore, called 'wipers'... The circular ratchet teeth..enable the vertical rod with the wipers to be raised. 1926 [see *BANK *sb.*[3] 10 b]. 1969 [see *slide-wire* s.v. *SLIDE- a]. 1975 C. D. Todd *Potentiometer Handbk.* vii. 166/2 Many different variations of the mechanical means which moves the wiper across the resistive element are possible. 1976 T. H. Flowers *Introd. Exchange Systems* iii. 82 In the L. M. Ericsson five-hundred-line switch, a stick carrying a set of wipers is rotated..to point in one of twenty-five angular directions,..then the stick is slid linearly outwards for the wipers to engage with one of twenty sets of fixed contacts.
7. *Comb.*: **wiper arm** = sense 6 above.
1933 K. B. Miller *Telephone Theory & Practice* I. 2 The subscriber,..by sending the proper number of impulses over one of his line wires, could cause the wiper arm of his switch to step *up* to the row containing the contact of the line desired and then..to step *around* to engage the particular one. 1967 D. Eadie *Introd. Basic Computer* xv. 348 A full-fledged analog multiplier..can be constructed if we take the pot just described and drive the wiper arm with a servomotor.

wiping, *vbl. sb.* Add: **3. wiping head,** a head (*HEAD *sb.* 11 g) for removing any recording from a magnetic tape or wire; an erase head.
1938 *Jrnl. Inst. Electr. Engineers* LXXXII. 266/2 The wiping head carries a direct current adequate to saturate the tape completely, so that it leaves the head fully magnetized. 1950 [see *record button* s.v. *RECORD *v.* 12 a].

wirble, *sb.* (Later example.)
1932 J. Joyce *Let.* 1 Aug. (1966) III. 251 But what about me in my present wirbel of worries.

wire, *sb.* Add: **II. 2. d.** *U.S.* A wire stretched across and above the track at the start and finish of a racecourse. Freq. in phrases: *down to the wire*, up to, or all the way to, the finishing-line; freq. *transf.* and *fig.*; (*from*) *wire to wire*, from start to finish of a race; also *transf.* and *attrib.*; *under the wire*, at the finishing-line; *fig.*, (to fall) within the limits or scope of something.
1887 *Courier-Journal* (Louisville, Ky.) 5 May 1/1 Eva K., Little Munch..were first under the wire. 1901 'H. McHugh' *Down the Line* 93 Swift often told himself that he could give Marshall P. Wilder six sure-fires and beat him down to the wire. 1920 C. Sandburg *Smoke & Steel* 138 He flashed his heels to other ponies..and hardly ever came under the wire behind the other runners. 1929 M. C. Work *Compl. Contract Bridge* v. 75 There are some hands which may just 'get under the wire' of the above definitions. 1950 *Keowee Courier* (Walhalla, S.C.) 31 Aug. 2/2 Baseball season is coming down to the wire, and the leading teams are about as close as two Scotchmen on bargain day. 1974 *State* (Columbia, S. C.) 15 Feb. 4-B/5 Nicklaus..led from wire to wire in the Hawaiian Open [Golf Tournament]. 1975 *New Yorker* 10 Nov. 137/2 Bertram Firestone's Honest Pleasure wound up his racing for the year with a wire-to-wire victory in the Laurel Futurity last weekend. 1982 'E. Lathen' *Green grow Dollars* xviii. 151 We're going to force Vandam's into court as fast as we can. I think we'll just make it under the wire. 1984 *Miami Herald* 6 Apr. 22A/1 Odds remain good that the Democrats' race will go down to the wire.
†**e.** *the straight wire*: the honest truth; also used without article as a phrase emphasizing the truth of an assertion. *Austral. slang. Obs.*
1892 'J. Miller' *Workingman's Paradise* 203 When it's all over you'll remember what I say and know it's the straight wire. 1909 A. Wright *Rogue's Luck* 70 'Now, no kid, Harry,' said Ned anxiously. 'Straight wire, did you beat him?' 1936 M. Franklin *All that Swagger* xlii. 394 'Will you?' said Humphrey... 'Straight wire, I will.'
f. *to pull one's wire*: see *PULL *v.* 19 j.
3. a. *live wire* (fig.): see *LIVE *a.* 8.
b. (Earlier example.) *crossed wires*: see *CROSS *v.* 5 c.
1846 *Punch* 5 Dec. 238/2 If this plan of Electric Telegraphs for the million should be carried out, the Post Office..might be turned into a central terminus for all the wires.
d. The telephone system; an individual telephone connection. Freq. in phrases *over the wire* (or † *wires*), *on the wire.* Now somewhat old-fashioned.
1902 *Chambers's Jrnl.* Feb. 128/2 A Parisian dentist had discovered a process of 'seeing by wire', which..means that he can while speaking through the telephone *see* his correspondent at the other end of the line of communication. 1925 H. Crane *Let.* 4 June (1965) 207, I did enjoy that talk with you over the wires to Cleveland! Your voice is so much better than ink and paper. 1925 F. Scott Fitzgerald *Great Gatsby* viii. 186, I tried [Gatsby's house] four times; finally an exasperated central told me the wire was being kept open for long distance from Detroit. 1929 'E. Queen' *Roman Hat Mystery* xiv. 211 I'll get the newspaper boys on the wire and ask them to ballyhoo the opening. 1932 [see *NUMBER *sb.* 4 f]. 1935 W. Cather *Lucy Gayheart* I. vii. 51 Every day his concert agent ..called him up as soon as his wire was open. 1947 S. Bellow *Victim* xxiii. 280 He ought to have spoken to Nunez about the broken chain while he was on the wire. 1974 Wodehouse *Aunts aren't Gentlemen* iii. 24 And now for heaven's sake get off the wire, I'm busy.
e. A private warning or message. Chiefly in phr. *to give the wire. slang.*
1925 Fraser & Gibbons *Soldier & Sailor Words* 307 *Wire, to give the*, to give a secret warning. 1930 E. H. Lavine *Third Degree* xvi. 210 The real thieves get 'a wire', and play poker. 1936 J. Curtis *Gilt Kid* vi. 60 He'd been straight with her and had given her the wire right in the beginning. 1972 R. Busby *Reasonable Man* xviii. 161 He gave me the wire that there was a big one coming off.
8. (Earlier example of *to pull the wires.*)
1813 *Deb. & Proc. Congr. U.S.* 5 Jan. (1853) 12th Congr. 2nd Sess. 562 When those who pulled the wires saw fit, they passed away.
IV. 14. b. Also, a wire-haired dachshund.
1938 *Times* 1 Jan. 1/6 (Advt.), Beautiful Corgis,.. Wires, Dachshunds, [etc.]. 1975 *Country Life* 6 Feb. (Advt. Suppl.) 27/1 Puppies for sale..Long Haired Wires.
V. 15. a. *wire basket, mesh* (also *attrib.*), *netting* (earlier example), *trolley.* **b.** *wire-framed, -rimmed.* **c.** *wire-clipper, -milker* (see Milk *v.* 4 d), *-nippers*; *wire-like* (later example).
1845 E. Acton *Mod. Cookery* vii. 196 A wire basket..is convenient for frying parsley and other herbs. 1961 J. Stroud *Touch & Go* vii. 69 The people..carry the same wire baskets round the same Supermarket. 1964 D. Francis *Nerve* ii. 16 There were a few letters..in the wire basket on the inner side of the door. 1977 C. McCullough *Thorn Birds* ii. 31 A piece of boned fish.. fried in the smoking well of liquid fat along with the chips, only in a separate wire basket. 1916 H. L. Wilson *Somewhere in Red Gap* iii. 109 That fresh bunch of campers had a pair of wire clippers in the whip socket. 1971 S. Hill *Strange Meeting* i. 28 The sound of the man sleeping above him in the wire-framed bunk. 1979 B. Malamud *Dubin's Lives* i. 6 She wore wire-framed blue-tinted glasses. 1952 E. Pound *Personae* 180 The wire-like bands of colour involute mount from my fingers. 1932 J. Dos Passos *1919* 271 There were knots of police in blue standing about..outside and inside the wiremesh gates huskylooking young men in khaki. 1944 *Living off Land* v. 109 Tanks..are best dealt with by screening all openings with a protective wire mesh. 1974 T. Hughes in *Listener* 4 Apr. 438/2 The sheep..in her wire-mesh compound. 1899 C. Hyne *Further Adv. Capt. Kettle* vi. 131 The wire-milkers. 1801 in *Deb. & Proc. Congr. U.S.* 21 Dec. (1851) 7th Congr. 2nd Sess. 1292 The books shall be..set up in portable cases ..with wire-netting doors and locks. 1914 B. M. Bower *Flying U Ranch* 168 Want me to go back and get the wire nippers? 1974 A. Lurie *War between Tates* (1977) ix. 185 Slouched down on the sofa.. is a skinny adolescent boy with thick wire-rimmed spectacles. 1976 *Southern Even. Echo* (Southampton) 17 Nov. 13/3 She had separated the meat and cold goods into a bag to stop them contaminating sugar in the wire trolley.
16. wire act, an acrobatic act performed on a tightrope; **wire bed,** (*a*) a bed fitted with a wire spring base or mattress; (*b*) in paper-making, a moving bed of wire over which the pulp is passed, its fibres at this stage beginning to form a web; **wire birch** *Canad.*, a small birch, *Betula populifolia*, which has light-coloured bark and is found in eastern North America; **wire brush,** (*a*) *Jazz* = *BRUSH *sb.*[2] 1 b; (*b*) a brush with stiff wire bristles used in cleaning, esp. for removing rust; hence **wire-brush** *v. trans.*, to clean with a wire brush; so **wire-brushed** *ppl. a.*, **-brushing** *vbl. sb.*; **wire-cartridge** (earlier example); **wire-cutter,** (*a*) (earlier example of term applied to a person); **wire-dancing** (earlier example); **wire edge** (earlier example); **wire-frame** *a.*, (*a*) applied to a picture (usu. computer-generated) in which every edge of an object is depicted, regardless of its visibility on the object itself, and nothing else; also *ellipt.*; (*b*) (of spectacles) having a frame made of wire; **wire ground** (earlier example); **wire-guided** *ppl. a.*, directed (in quot. 1922, carried out) by means of electric signals transmitted along a wire; *spec.* applied to a missile connected to a control point by a wire; **wire house** *U.S.*, a brokerage firm having branch offices connected to its main office by private telephone and telegraph wires; **wire recorder,** an apparatus for magnetically recording sounds, etc., on wire and afterwards reproducing them; so **wire recording,** a recording so made, or the process of making one; **wire-rim** *a.* = *wire-frame* adj. (*b*) above; also *ellipt.*; **wire service** *U.S.*, a news agency that supplies syndicated news by wire to its subscribers; **wire-stitcher,** an automatic stapling machine which takes continuous wire and forms the staples as an integral part of the stapling operation; so **wire-stitching** (also *attrib.*); hence (as back-formations) **wire-stitch** *v. trans.*, **wire-stitched** *ppl. a.*; **wire story** *Journalism*, a story distributed by a wire service; **wire-strainer** *Austral.* and *N.Z.* = *wire-stretcher* below; **wire-stretcher** chiefly *N. Amer.*, a tool for making taut the wire of a fence or the like; **wire-walking** (examples); **wireway,** a channel or duct for enclosing lengths of wiring, esp. one made of sheet metal; ducting of this nature; **wire wheel,** a car wheel having wire spokes (used esp. on sports models); **wire wool,** matted thin wire, used esp. for scouring kitchen utensils; **wire-wound** *a.* (further examples in *Electr.*).
1906 *Variety* 13 Jan. 7/2 The Roses..have a wire act with some good tricks. 1912 C. Mackenzie *Carnival* xi. 136 They did not object to interminable wire-acts, and put up with divination feats of the most exhausting dullness. 1976 *National Observer* (U.S.) 24 Jan. 18/5 You always knew a wire act would open the show. 1882 W. Whitman *Daybks. & Notebks.* (1978) II. 296 Wire beds—829 no. 10th st. 1918 W. Owen *Let.* 31 Oct. (1967) 591 Other officers repose on wire beds behind me. 1962 F. T. Day *Introd. Paper* iv. 38 The wire bed is kept perfectly level while it oscillates to bring about the interlacing of fibres in the pulp. 1917 B. R. Morton *Native Trees of Canada* 68 *Betula Populifolia*..White birch, grey birch, wire birch. 1956 [see *Indian pear* s.v. *INDIAN *a.* 4 b]. 1974 J. Dowell *Look-Off Bear* p. viii, Beyond our grove of pines there were mixed growths of wire-birch, swamp willow, [etc.]. 1927 *Melody Maker* Apr. 389/1 In quiet passages, the wire brush on a Chinese cymbal gives a very pleasing effect. *a* 1935 T. E. Lawrence *Mint* (1955) I. xxii. 76 Our job was to wire-brush and repaint a lot of salvaged sheeting... A good job, it looked. Six of us and six wire brushes. 1957 Manvell & Huntley *Film Music* iii. 98 Wire-brush percussion. 1974 A. Ross *Bradford Business* 76 Even the short heavy bolts had been rubbed up with a wire brush. 1978 J. Wainwright *Thief of Time* 210, I should ..check the trays of seed potatoes... Wire-brush the seed-trays. 1955 *Archit. Rev.* CXVII. 68 Construction: reinforced concrete frame with mainly brick walls, but certain panels of wire-brushed concrete. 1978 *Country Life* 28 Dec. 2212/1 The furniture in wire-brushed carved oak. 1980 *Yachts & Yachting* 29 Feb. 672/2 Lead is perhaps the easiest needing only one coat of undercoat and one of Metallic Primocon between wirebrushing (or rubbing down) and the new antifoul. 1978 E. Gundrey *Simple Plumbing* 22 Rust removal involves wire-brushing. 1839 W. Watt (*title*) Remarks on shooting, in verse, comprehending..the recent and admirable invention of the patent wire cartridge. 1888 in W. P. Webb *Great Plains* (1931) 314 While a man was putting up his fence one day in a hollow a crowd of wire-cutters was cutting it behind him in another hollow. 1785 *Daily Universal Reg.* 1 Jan. 4/1 Must he [*sc.* an editor]...have writers of

tumbling—wire dancing—and hurly burly description? 1807 H. H. Brackenridge *Mod. Chivalry* II. iv. 21 In the course of mixing with good company, the wire edge of art would wear off, and an ease of demeanor be attained. 1963 *AFIPS Conf. Proc.* XXIII. 348 A prototype graphical communications system capable of manipulating straight line, 'wire frame', figures in three-dimensional space is now in operation. 1977 E. Leonard *Unknown Man No. 89* ii. 13 His tinted wire-frame glasses glistened. 1982 Ballard & Brown *Computer Vision* ix. 292 A set of vertices or edges can define many different solids. (It is possible, however, to determine algorithmically all possible polyhedral boundaries described by a three-dimensional wireframe.) 1982 J. E. Scott *Introd. Interactive Computer Graphics* viii. 135 A wire-frame drawing is less pleasing visually, but it is considerably faster for the computer to produce. 1983 *New Scientist* 24 Mar. 819/1 The more detail in a wire frame, the harder it is to understand. 1983 L. Deighton *Berlin Game* viii. 83 Bret put his wire-frame glasses into their case. 1865 F. B. Palliser *Hist. Lace* iii. 27 The honeycomb network or ground..is of various kinds; wire ground, Brussels ground, [etc.]. 1922 *Encycl. Brit.* XXXII. 1022/1 Wire-guided high frequency telegraphy and telephony. 1958 C. C. Adams *Space Flight* 52 The latter two [rockets].. are wire-guided. 1972 [see TOW s.v. *T 6 a]. 1982 *Sci. Amer.* May 103 (Advt.), The wire enables the Hughes TOW to have one of the highest velocities and longest ranges (2·3 miles) of any wire-guided missile in the world. 1982 *Daily Tel.* 10 May 4/3 Two wire-guided torpedoes of the Tigerfish type. 1904 *N.Y. Evening Post* 18 June (Financial Sect.) 1/7 The so-called 'wire house'..is a product of the boom times. 1966 *Economist* 25 June 1436/1 It [*sc.* the New York Stock Exchange] has been firing salvos..about possibly setting up an auxiliary trading floor somewhere in New Jersey... Several larger nation-wide 'wire' houses have already said that they are considering some such plans to relocate. 1982 *Times* 27 July 15/4 United States banks..in the past have left financial futures very much to the brokers and major 'wire houses' such as Bache or Hutton. 1942 *Frontier* Sept. 3/1 (*caption*) The wire sound recorder developed by Armour Research Foundation differs from previous types in the type of recording head. 1957 *Encycl. Brit.* XI. 29/2 Three types of recording systems are in common use: (1) mechanical, as in the disk phonograph; (2) optical, as in the sound film; and (3) magnetic, as in the tape or wire recorder. 1978 'D. Kyle' *Black Camelot* ix. 130 'I think we'd better record this as we go along.'.. The wire recorder had been produced and checked. 1933 *Amer. Speech* VIII. II. 77/1 Similar information should be given for film, strip, and wire recordings. 1943 *Electronics* Oct. 236 (*caption*) A wire recording of the Army Hour is taken off the air by..engineers. 1966 *McGraw-Hill Encycl. Sci. & Technol.* XIV. 518/2 Except for minor details, the techniques and systems used for magnetic wire recording are similar to those for magnetic tape. 1982 *Sunday Tel.* (Colour Suppl.) 21 Nov. 8/2, I just read the words, they were recorded straight on to disc, then transferred on to wire recording tape and that was the end of it. 1977 *Sat. Rev.* 23 July 10/2 Slender, with a trim beard and wire-rim glasses. 1982 J. Valin *Day of Wrath* (1983) xvii. 131 He's got long hair, wears wire rims, muttonchops. 1950 *Mag. of Fantasy & S.F.* Fall 7, I monitored a couple of newscasts; the second one carried a story by another wire service on the domes. 1962 E. Lacy *Freeloaders* viii. 181, I didn't think the *Herald Tribune* used wire services for Europe. 1976 *National Observer* (U.S.) 6 Nov. 24/1 A morning newspaper in the East, using a wire service's totals, had Rockefeller ahead. 1985 *Times* 11 May 21/1 Reuters is keeping a close watch on its troubled American wire service rival, United Press International. 1902 *Census Bull.* (U.S.) No. 216. 65 A..combination folding and wire-stitching machine, which by a continuous and automatic operation takes the sheets from the feeders, and folds, gathers, collates, covers and wire-stitches copies of magazines and pamphlets. 1921 T. J. Wise *Bibliogr. Writings Joseph Conrad* (rev. ed.) I. 27 There are no signatures, the pamphlet being composed of a single halfsheet..issued wire-stitched. 1887 *Courier-Journal* (Louisville, Ky.) 20 Feb. 3/2 Printing-Office... Card Cutter, Wire Stitcher [etc.]. 1967 Wire stitcher [see *SHORT RUN *sb.* 4]. 1881 *Even. News* 26 July 4 (Advt.), Over one hundred machines in motion [in a printing and paper-making exhibition]... Wire stitching, paging, gumming, etc. 1957 *Encycl. Brit.* III. 859/2 The automatic assembling, wire-stitching and covering machine units complete the operation of pamphlet binding. 1943 Wire story [see *HOT-SHOT 1 b]. 1979 J. Crosby *Party of Year* (1980) iii. 25 The foreign desk was behind a glass screen... Feinberg was editing a wire story. 1882 Armstrong & Campbell *Austral. Sheep Husbandry* xviii. 204 Novel Wire Strainer... This instrument..should be made of light iron... Three short spikes, or legs, should be fixed behind, so as to give the instrument a grip of the post as soon as the wire is tightened. 1959 A. Upfield *Bony & Black Virgin* xxiii. 215 The big man studied the method of joining the cut wires... 'Chain wire-strainer was used.' 1975 *N.Z. Jrnl. Agric.* Sept. 57/1 The installation of permanent wire strainers on each strand of a new fence..would overcome this difficulty. *a* 1877 Knight *Dict. Mech.* III. 2797/2 *Wire-stretcher*, a tool for straining lightly telegraph or fence wires. 1954 W. Faulkner *Fable* 187 Cowboy..exterminated from the earth by a tide of men with wire-stretchers and pockets full of staples. 1958 J. G. MacGregor *North-West of 16* ix. 132 Then it [*sc.* barbed wire] had to be tightened with our wire-stretchers (a simple block-and-tackle arrangement) until when you plucked it, it sang like a fiddle string. 1981 *Farmstead Mag.* Winter 43/4, I assemble my supplies for next year's battle: my fence staples, spare wire, my fencing pliers and wire stretcher. 1898 *Pearson's Mag.* Sept. 332 Wire-walking..must always retain a greater amount of fascination. 1920 *Variety* 31 Dec. 124 She learned acrobatics..wire walking and aerial work. 1932 A. L. Abbott *National Electr. Code Handbk.* viii. 97 (*caption*) A length of wireway with hinged cover. *Ibid.* 99 Wireways may extend transversely through dry walls. 1953 H. A. Chinn *Television Broadcasting* xv. 606 In order to provide protection for audio, video, communications, control, and a-c cables..it is customary to instal such cables in conduit, raceways, pipe shafts,..and similar wireways. 1964 R. F. Ficchi *Electrical Interference* x. 192 A bare ground wire rubbing against a chassis or wireway can cause a considerable amount of noise in a system. 1912 *Motor Manual* (ed. 14) (Advt., rear cover), Rudge-Whitworth detachable wire wheels lengthen the life of tyres 70%. 1926 *Daily Colonist* (Victoria, B.C.) 4 July 26/4 (Advt.), Sport Roadster $675 Delivered. Racy streamlines, wire wheels, [etc.]. 1963 [see *GAS *sb.*[1] 5 d]. 1976 N. Thornburg *Cutter & Bone* i. 8 A classic 1948 MG-TC with running board and wire wheels. 1958 J. Cannan *And be Villain* iii. 71 A cupboard where detergents, a reserve of dishcloths and the rolls of wire wool were kept. 1977 'J. le Carré' *Hon. Schoolboy* xii. 264 His hair was like wire wool crimped into small trenches. 1931 *Boys' Mag.* XLV. 125/2 The potentiometer should be a 'Colvern', wire wound. 1946 *Nature* 30 Nov. 799/2 The development of a vitreous enamel coating for fixed wire-wound resistors. 1975 D. G. Fink *Electronics Engineers' Handbk.* vii. 7 Rheostat (Power). These are variable wire-wound resistors used as speed controls.

wire, *v.* Add: **2. a.** (Example of *to wire a cork.*)
1796 M. Edgeworth *Parent's Assistant* (ed. 2) I. 74 Did not I order you..to carry these bottles to the cellar; and did not I charge you to wire the corks?
c. Also, to cover *over* with wire.
1774 J. Wesley *Let.* 26 July (1931) VI. 104 You must..wire over the cupola.
e. Also, to make electrical connections to; to connect electrically *to*; to provide *with* by means of connecting wires; *spec.* to fit with a concealed listening device. Also with *up.* (Earlier and later examples.)
1891 E. I. Bax *Pop. Electr. Lighting* iv. 27 To admit of this the expense of wiring the room will have to be increased. 1923 *Wireless World* 19 May 205/2 It is preferable to wire the valve panel before fixing it to the baseboard. 1960 *Practical Wireless* XXXVI. 393/1 The heater circuit is best wired first, leads being run close against the chassis. 1970 J. Earl *Tuners & Amplifiers* vi. 140 It is not usually difficult to wire the stereo loudspeaker pair and programme sources for the correct left and right channels. 1978 S. Brill *Teamsters* ix. 144 The prosecutor wired Henderson's phone so that there would be tapes of Faugno and Andretta threatening him. 1978 *Australian* 21 Aug. 9/2 One in every five homes with television are wired to a cable system. 1982 *Sci. Amer.* Sept. 68/2 The explosive charge is wired with electric blasting caps and detonated from a safe distance. 1983 J. Fuller *Convergence* xxx. 303 Just tell the truth... We have to wire you up.
f. To incorporate (a facility, etc.) *into* a device by electric wiring. Cf. *WIRED-IN *a.* 2.
1962 *Communications Assoc. Computing Machinery* V. 159/1 A scheme for wiring binary-to-decimal conversion into a machine at a small cost.
8. [Cf. Wire *sb.* 13.] *intr.* To practise pick-pocketing. Also *trans.*, to pick the pocket of. *slang.*
1853 M. Carpenter *Juvenile Delinquents* i. 40 There are..at least ten times as many boys 'wiring' (picking pockets) as when I was young. *Ibid.* iv. 145 If he was bigger he could wire a man of his poke. 1891 'F. W. Carew' *No. 747* xxxv. 414, I used to go wirin' in the main-thoroughfares.

wired, *ppl. a.* Add: **1.** *spec.* of glass.
1908 [in Dict.]. 1930 *Engineering* 12 Dec. 755/1 Plate glass, rolled figured glass, corrugated glass, and wired glass, all produced by rolling. 1979 P. Way *Sunrise* xi. 116 He had pushed through the wired-glass door.
3. Also with *up*, and *fig.*
1946 *Coast to Coast 1945* 216 The gate was thoroughly wired up—three Queensland hitches of No. 8 wire. 1975 *Washington Post* 29 Sept. A-20/4 Let us..concede that point for a moment—although it shouldn't be conceded until it is properly wired up with all sort of qualifications.
6. a. Employing wires or similar physical connections to convey electric signals, *spec.* for television or radio.
1924 *Telegraph & Telephone Jrnl.* XI. 6/1 Here are some extremely interesting particulars regarding 'Wired Radio' Broadcasting. 1930 E. E. Hunt *Audit Amer.* 20 In 1913 there were only 48 wired homes per 1,000 of the non-farm population. 1937 *Wireless World* 2 Dec. 565/2 (*heading*) Wired television. 1958 *Oxford Mail* 26 Aug. 3/6 In a discussion on wired television..Coun. W. G. White asked if the present or any future council was going to bar the television aerial from its housing estates. 1960 Gregory & Van Horn *Automatic Data-Processing Systems* ii. 61 Computers with externally stored programs..get their operating instructions from wired plugboards. 1960 *Electronics Weekly* 30 Nov. 2/4 A wired sound and television service may soon be provided..for Leicester. 1969 *Electr. Communications* XLIV. I. 14/1 The best compromise was sought between wired-logic control, which is very efficient but inflexible, and stored-program control. 1971 *New Scientist* 1 July 19/2 It is often argued that the provision of more radio and TV channels, particularly on the greatly expanded scale of the 'wired city', will lower standards. 1972 *Listener* 6 July 3/2 As America becomes increasingly a wired nation, with cables reaching out already into millions of homes, the channel limitations of over-the-air television are being superseded. 1976 Brzozowski & Yoeli *Digital Networks* ii. 30 Wired logic refers to the capability of tying together the outputs of gates to realize either the AND..or the OR..function without additional hardware.
b. Fitted with, or wearing, an electronic listening device; more fully *wired for sound.* Also *fig. colloq.*
1957 J. D. MacDonald *Man of Affairs* ix. 141 The joint is wired, he says... The next step is cameras and infra red and tape recorders, I guess. 1967 *Boston Sunday Globe* 23 Apr. 18/3 Several agreed with the words of one who said he knew enough 'to stay away from Karafin... He's wired for sound 24 hours a day... He can keep stories out of the paper or get them in.' 1982 G. Lyall *Conduct of Major Maxim* xiv. 129 The very idea of being 'off the record' was nonsense, since the room was almost certainly wired.
c. With *up* in either of prec senses.
1971 *New Scientist* 16 Sept. 614 (*heading*) The visual systems in their brain are wired up to match the visual world that is important to them. 1972 D. Bloodworth *Any Number can Play* xix. 197 That one-sided chat you had with the wired-up Goddess of Mercy. 1982 P. D. James *Skull beneath Skin* xxv. 210 She was glad that she wasn't wired up to a lie machine. 1984 *Listener* 26 July 20/1 Dr Glover came to the orchestra on the strength of being well wired-up with broadcasting contacts.

wired-in (stress variable), *a.* [f. Wired *ppl. a.* + In *adv.*] **1.** Bounded by wire, in the form of netting or fencing. Cf. Wire *v.* 2 c.
1855 [see Wired *ppl. a.* 2]. 1973 J. Thomson *Death Cap* ix. 128 She was feeding the chickens..in the wired-in run. 1975 J. McClure *Snake* ix. 118 Wessels..hid behind the wired-in back of a parked truck.
2. Incorporated in or connected to a device or system by means of wiring. Also *fig.*
1957 C. E. Osgood et al. *Measurement of Meaning* i. 5 Certain stimulus patterns have a 'wired-in' connection with certain behaviour patterns (unconditional reflexes) and additional stimuli have acquired this capacity (conditional reflexes). 1962 *Commun. Assoc. Computing Machinery* V. 159/1 (*heading*) On a wired-in binary-to-decimal conversion scheme. 1975 P. Elbow *Oppositions in Chaucer* v. 120 His predilection for language and thought frees him..from a single, rigid, programmed, or wired-in response to a fox. 1975 *Language for Life* (Dept. Educ. & Sci.) xv. 234 This should be fitted with sound-proof projector booth, wired-in good quality speakers, and a large permanent screen. 1982 *Data Communications* Oct. 119/2 These terminals are dedicated to specific tasks through specific wired-in instructions and have no built-in intelligence. 1984 *Science* 22 June 1304/3 The wired-in semantics of these connections substitutes for the time-consuming interpretation process needed in systems that pass symbolic information.

wire-grass. **2.** (Earlier example.)
1790 W. Bligh *Narr. Mutiny on Board H.M.S. Bounty* 48 In the hollow of the land there grew some wire grass.

wireless, *a.* (*sb.*) For **a, b** read **A, B** and add: **A.** *adj.* (Further examples.) *wireless telegraphy*: also, in British law, used to include wireless telephony; *wireless telephony*: the transmission of speech and other uncoded signals by means of radio waves; = *RADIO-TELEPHONY. Now chiefly *hist.*, having been superseded by *radio*(-).
1898 *Jrnl. Inst. Electr. Engineers* XXVII. 799 The general principles of electric space telegraphy—or wireless telegraphy, as it seems to wish to be called. 1903 C. H. Sewall *Wireless Telegr.* I. 88 Wireless telephony. Telephoning without wires has not gained by the great developments in its sister-art. 1904 *Act 4 Edw. VII* c. 24. §78 The expression 'wireless telegraphy' means any system of communication by telegraph..without the aid of any wire connecting the points from and at which the messages or other communications are sent and received. 1906 S. R. Bottone tr. *Mazzotto's Wireless Telegr. & Teleph.* xi. 390 After the discovery of wireless telegraphy by means of electric waves, many attempts were made to apply the same principle to telephony. 1913 A. H. Verrill *Harper's Wireless Bk.* xiii. 113 There is no reason to suppose that wireless telephony will not soon have the range of wireless telegraphy. 1922 Joyce *Ulysses* 702 A private wireless telegraph which would transmit by dot and dash system the result of a national equine handicap. 1923 A. Huxley *On Margin* 49 If they are a little more up-to-date they adjust their wireless telephone to the right wave-length and listen-in to the fruity contralto at Marconi House, singing 'The Gleaner's Slumber Song'. 1936 G. B. Shaw *Simpleton Unexpected Isles* II. 51 (*stage-direction*) A writing table littered with papers and furnished with a wireless telephone. 1949 *Act 12 & 13 Geo. VI* c. 54. §68 The expression 'wireless telegraphy' means the emitting or receiving, over paths which are not provided by any material substance..of electromagnetic energy of a frequency not exceeding three million megacycles a second, being energy which either—*a*) serves for the conveying of messages, sound or visual images..or *b*) is used in connection with the determination of position, bearing or distance, or for the gaining of information as to the presence..of any object. 1952 [see *sound radio* s.v. *SOUND *sb.*[2] 7 b]. 1960 *Practical Wireless* XXXVI. 403/1 This station, which was built by Guglielmo Marconi, was the first to span the Atlantic with wireless telegraphy. 1981 *Daily Tel.* 12 Feb. 9/1 HMS Inskip, the Navy's wireless-telegraphy station, near Preston, Lancs.
B. *sb.* Substitute for def.: **1.** Wireless telegraphy or (esp.) telephony; sound broadcasting; = *RADIO *sb.* 2 a, b; also, a particular radio station (= *RADIO *sb.* 2 c).
Largely superseded by *radio* exc. in *hist.* contexts.
1903 *N.Y. Commercial Advertiser* 31 Jan. S 2/2 First in this great field of making the 'wireless' a handmaid of commerce is the de Forrest system, which has won the approval also of the United States government. 1904 [in Dict., sense b]. 1915 Grahame-White & Harper *Aircraft in Gt. War* IV. xi. 150 The application of wireless to aeroplanes has, in the British Flying Corps, been studied very carefully. 1922 [see *BROADCASTER 1]. 1924 [see *SUPRANATIONAL *a.*]. 1927 C. Connolly *Let.* 27 Jan. in *Romantic Friendship* (1975) 231 Chesterton is trying to be funny over the wireless. 1932 R. A. Knox *Broadcast

Minds* i. 13 We say 'the' wireless... For the wireless, in England, is a unique force; there is no question of two wirelesses differing, as two newspapers may differ in their outlook. 1939 *Daily Tel.* 18 Dec. 1/5 Moscow wireless claims advances in the Murmansk district. 1951 *Sport* 27 Apr.–3 May 2/1 Photos in the papers and interviews on the wireless. 1952 *Times Rev. of Year* 1 Jan. p. v/2 Sound radio (wireless declined farther towards archaism) has done much during the year. 1971 *Daily Tel.* 20 Jan. 10/3 [The Misses Waters] belong to the days when radio was 'the wireless'. 1978 E. Blishen *Sorry, Dad* II. i. 40 Wireless was still quite raw and improbable. The Dockrees next door had one... It was a crystal set. 1980 *Bookseller* 5 Jan. 23 The table shows the books recorded in December... Wireless and Television..42.
2. A radio-telegram.
1904 *Everybody's Mag.* Aug. 161/2 This is how it is taken down in those unerring short-hand notes of the recording angel and sent by special wireless to the typewriter for His Majesty of the Sulphur Trust. 1911 G. Stratton-Porter *Harvester* xvi. 351 Is Ajax [*sc.* a peacock] now sending a wireless to Ceylon asking for a mate? 1926 Galsworthy in *Scribner's Mag.* Aug. 192/1 Going home to have a look at Kit and send Fleur a wireless, he passed four musicians. 1940 N. Marsh *Surfeit of Lampreys* (1941) ii. 25 The steward gave her two [letters] and a wireless message. She opened the wireless first.
3. Short for *wireless set* (see sense 4 below); a radio.
1927 T. E. Lawrence *Let.* 4 Oct. (1938) 543 We have no wireless, and I don't look at papers. 1933 A. Thirkell *High Rising* i. 16 He could..repair the headmaster's wireless and drive his car. *a* 1944 K. Douglas *Alamein to Zem Zem* (1946) 49 The wirelesses in the new tanks had to be checked. 1954 W. Faulkner *Fable* 166 It's too bad every house he passes don't have a wireless, like ships do. 1971 *Daily Tel.* 13 May 7/1 As a child I used to stand alone in front of that big brown box that used to be called a 'wireless' and conduct symphony concerts. 1972 M. Gilbert *Petrella at Q* (1977) 37 P.C. Owers..summoned assistance on his pocket wireless. 1973 [see *radio play* s.v. *RADIO *sb.* 5 b].
4. *attrib.* and *Comb.*, in senses 1 and 3 above (in U.S., and increasingly in British use, replaced by the corresponding Combs. with *radio*), as *wireless aerial, announcer, battery, broadcast, broadcasting, mast, operator, set, station, transmitter, valve*; in the sense 'transmitted or broadcast by wireless', as *wireless concert, news, play, programme, talk*; **wireless cabinet,** a cabinet incorporating a radio; **wireless licence,** an official permit needed for the possession of a radio; **wireless shack** = *radio shack* s.v. *RADIO *sb.* 7; **wireless silence** = *radio silence* s.v. *RADIO *sb.* 7; **wireless wave** = *radio wave* s.v. *RADIO *sb.* 7.
1924 *Radio Times* 12 Dec. 527 (*caption*) Them's 'is wireless aerials. 1937 *Discovery* Feb. 37/1 The wireless aerial and its stays would become coated with an almost uniform layer of this frost. 1983 C. Dexter *Riddle of Third Mile* i. 12 The voice..of a pre-war wireless announcer... What they called an 'Oxford' accent. 1933 *Radio Times* 14 Apr. 83 (Advt.), Ever Ready wireless batteries. 1930 G. B. Shaw *Apple Cart* p. xi, Last October (1929) I was asked to address the enormous audience created by the new invention of Wireless Broadcast. 1980 J. Lees-Milne *Harold Nicolson* I. xi. 205 On the 2nd May [1923].. he..heard a wireless broadcast over the air for the first time. 1923 *Radio Times* 28 Sept. 6 Wireless broadcasting service. 1930 Kipling *Limits & Renewals* (1932) 208 [He] came back with a couple of cigarettes from the store behind the wireless cabinet. 1923 *Radio Times* 28 Sept. 18/1 Much has..been written on the subject of wireless concerts. 1920 *Radiograph* May 147 (Advt.), Applicants for wireless licence who specify them will not be required to furnish diagrams. 1928 *Melody Maker* Feb. 187 (*heading*) The over-taxed Frenchman! Wireless licences two-pence per annum. 1958 *Whitaker's Almanack 1959* 1098/1 If application is made at any other office which conducts wireless licence business, arrangements can be made for a licence to be issued at a Head Post Office. 1943 F. Thompson *Candleford Green* x. 167 Now..a wireless mast in every back garden. 1948 Wireless mast [see *rev-counter* s.v. *REV *sb.* 2]. *c* 1919 H. C. Witwer *Smile a Minute* IV. iii. 158 Joe, I have just seen the wireless news [received on board ship]. 1926 in *Listener* (1974) 25 Apr. 519 Wireless news will be broadcasted by the B.B.C. 1942 E. Waugh *Put out More Flags* ii. 139 She sat..listening to wireless news from Germany. 1978 P. G. Winslow *Coppergold* 127 Joss's death..had been on the wireless news. 1910 Wireless operator [see sense b in Dict.]. 1929 *Daily Express* 7 Nov. 1/1 Crew. Flight-Captain Rod Schinka..(first pilot)... Herr Niklas..(wireless operator). 1978 F. Maclean *Take Nine Spies* iv. 138 Their mission had been joined by a second wireless-operator. 1929 *Radio Times* 8 Nov. 406/1 *Journey's End* was written for stage representation and is, therefore, not in the more restricted sense a 'wireless play'. 1948 D. Welch *Jrnl.* 7 July (1952) 259 Last week there was a wireless programme on Marie Bashkirtseff. 1923 J. Reith *Diary* 19 Mar. (1975) ii. 131, I was standing talking to them with the wireless set at my back and I pushed the switch. *a* 1944 K. Douglas *Alamein to Zem Zem* (1946) 25 About dusk the wireless sets in all tanks were switched on. 1978 *Dumfries Courier* 13 Oct. 10/4 In the 20 years up to 1950 the radio set—or wireless set as it was affectionately called—was the focal point of home entertainment. 1937 G. H. Grant *Heels of Gale* vi. 59 The wireless shack..had been lifted on board by a crane on the day before sailing and bolted to the wooden planks. 1961 J. Bisset *Commodore* xxvii. 281 On this deck also was the Wireless Shack. 1915 Ld. Fisher 12 Apr. in M. Gilbert *Winston S. Churchill* (1972) III. Compan. I. 793 It seems to me that the positions are all well chosen for all our different activities, and Jellicoe fully alive to wireless silence. *a* 1944 K. Douglas *Alamein to Zem Zem* (1946) 49 Wireless silence was in force. 1909 *Chambers's Jrnl.* July 428/2 At Aldershot..there is a powerful wireless station. 1926 T. E. Lawrence *Seven Pillars* (1935) III. xxix. 170 My life was spent in moving back and forth..to the town, the port, the wireless station. 1978 D. A. Stanwood *Memory of Eva Ryker* xxvi. 251 The Cape Race Wireless Station has a record of the message... It was transmitted from the *Titanic* five minutes before she struck ice. 1930 J. S. Huxley *Bird-Watching & Bird Behaviour* p. vii, Six wireless talks. 1923 *Radio Times* 28 Sept. 26/3 A small wireless transmitter will be installed at the 'Old Vic'. 1978 F. Maclean *Take Nine Spies* iv. 144 Bernhardt had established two wireless transmitters. 1923 *Radio Times* 28 Sept. 33 (Advt.), A new wireless valve. 1915 H. H. Turner *Voyage in Space* vi. 250 Suppose you pretend that a second is itself like a year; divide it into..30 million parts; one of these tiny parts will be about the time that what we may call a 'wireless' wave of electricity takes to vibrate. 1936 *Discovery* Sept. 285/1 The reason why wireless waves travel round the earth's surface instead of disappearing into space has been explained by investigations of the properties of the upper atmosphere. 1960 *Practical Wireless* XXXVI. 403/1 Marconi announced that signals from Poldhu had bridged the Atlantic. .. That news..confounded the many learned critics who had said that wireless waves would never reach beyond the horizon.

wireline (wəi^ə·ɹlain). Also **wire line.** [f. Wire *sb.* + Line *sb.*[2]] **1.** (In Dict. s.v. Wire *sb.* 16.)
2. *Oil Industry.* **a.** A cable for lowering and raising tools and the like in a well shaft. Freq. *attrib.*
1916 A. B. Thompson *Oil-Field Devel. & Petroleum Mining* x. 468 At each pulley a short length of chain is inserted to accommodate the change of direction, and at each well a wire line can be led over the top derrick pulley and direct on to the pump. 1948 *Petroleum Handbk.* (Shell Internat. Petroleum Co.) (ed. 3) v. 92 Where a core head can drill more than 20 feet without getting dull, a retractable or 'wire line' inner-core barrel is used. After a core has been cut, the inner barrel containing the core can be brought to the surface with a wire line lowered inside the drill pipe. 1977 *Offshore Engineer* May 11/1 To prepare the well, the 'Christmas tree'. is.. replaced with a temporary single-pipe ram preventer which allows wireline tools to be introduced.
b. An electric cable used to connect measuring devices in a well with indicating or recording instruments at the surface.
1972 L. M. Harris *Introd. Deepwater Floating Drilling Operations* ii. 6 Logging or other wireline operations. 1974 P. L. Moore et al. *Drilling Practices Manual* xi. 279 The use of wireline logs for determining pore pressures is well documented. 1977 *Offshore Engineer* May 60/3 Various methods for monitoring grouting were used, including a radio-active isotope scheme.., and more conventional wireline temperature surveys.
3. A telegraph or telephone line of wire.
1934 in Webster. 1947 *Trans. Amer. Inst. Electr. Engineers* LXVI. 492/3 The necessary frequency space for wide-band operation usually is not justified economically for wire line operation. 1983 *Mini-Micro Systems* July 240/2 The FCC is currently examining applications from wire-line and non-wire-line carriers for licenses to provide cellular mobile phone service in 30 large U.S. cities.
4. A fishing line of metal wire.
1974 *Encycl. Brit. Macropædia* VII. 374/1 Wire lines created from extruded Monel Metal or stainless steel assist in the sinking of a moving lure... But fish caught on metal or metal core lines are not eligible for International Game Fish Association records. 1984 *Miami Herald* 6 Apr. 9F/6 Boats are limited to four fishing lines... 'We allow use of the kite rig but do not permit use of wireline.'

wireman. Add: **2. b.** A wire-tapper. *colloq.*
1973 *Telegraph* (Brisbane) 24 May 12/4 Watergate conspirator James McCord was one of the best 'wiremen' in the eavesdropping game. 1977 *Time* 21 Feb. 19 He had been one of the most sought-after 'wiremen', or electronic eavesdroppers, in the East, supplying bugging and recording devices to clients on both sides of the law.
3. A journalist working for a telegraphic news agency.
1973 D. May *Laughter in Djakarta* iv. 61 An American wire-man..who lived in Djakarta. 1977 'J. le Carré' *Hon. Schoolboy* xv. 338 Keller was..a wireman..and Jerry knew him from other wars.

wirephoto (wəi^ə·ɹfōʊto). orig. *U.S.* Also with hyphen and as two words. [f. Wire *sb.* + Photo.] A facsimile process for transmitting pictures over telephone lines; also (*colloq.*), a photograph transmitted by this means.
1939 Webster *Add.*, Wirephoto. 1940 *Chicago Daily Tribune* 11 May 1/3 All these pictures were sent..by radio and wirephoto. 1964 M. McLuhan *Understanding Media* II. xx. 203 The newspaper mesh of dots that is called 'wirephoto'. 1972 T. Ardies *This Suitcase* xvii. 189 A picture of the teenaged Helmut Stern..came in last night via wirephoto. 1973 C. Sagan *Cosmic Connection* (1974) xv. 109 The television pictures from *Mariner 9* were radioed from Mars to Earth in much the same way that a newsprint wire-photo is transmitted on Earth. 1981 'D. Shannon' *Murder Most Strange* vii. 147 The wire photo came in..and Higgins took it to show the Ortiz girl and she identified it.
So **wire-photograph.**
1962 *Listener* 12 July 57/1 Cartoonists in the United States rely for day-to-day material on newspaper morgues and wire-photographs. 1968 J. Sangster *Touchfeather* ix. 92 'Come and look at this.' The 'this' was a wire photograph... The transmission hadn't been up to much.

wire-puller. (Earlier example.)
1833 in J. R. Commons et al. *Documentary Hist. Amer. Industr. Soc.* (1910) VIII. 340 Wire-pullers..for the furtherance of..party interest.

wirescape (wəi^ə·ɹskēⁱp). [f. Wire *sb.* + Scape *sb.*[3], after *landscape.*] Scenery, or a scene, dominated by overhead wires and their supports.
1951 *Archit. Rev.* CX. 377 Wire, lots of wire, lining streets, crossing fields, acting as totems in villages and skeleton umbrellas on towns, by reducing the endless variety of the human and natural scene to the common denominator—wirescape—has made a dreadful uniformity out of the world it seeks to unite. 1959 *Times Lit. Suppl.* 9 Jan. 16/1 Each year the demands of the new industrial revolution gnaw away more insistently at the countryside... How many of us realize the Wirescape that impends? 1965 *New Statesman* 5 Nov. 713/1 The visual squalor, of which a notable feature is the appalling wirescape, of New York's periphery. 1969 E. Sandon *View into Village* x. 86 In the street is also to be seen that typically modern feature—an appalling wirescape. 1978 *Gold Coast Bulletin* (Austral.) 29 Sept. 7/1, I think we should be removing these unsightly wirescapes from the central precincts of the city.

wire-tap (wəi^ə·ɹtæp), *sb.* [Back-formation from next.] An act of tapping a telephone line, esp. as a form of surveillance; also, the device by which this is done.
1955 H. Roth *Sleeper* xiv. 113 He had..refrained from any discussion of wire taps or followers. 1963 L. Deighton *Horse under Water* xvii. 72 He has R.N. Signals Gibraltar doing a wire-tap job on me. 1976 *Billings* (Montana) *Gaz.* 11 July 7-A/1 Agents risked doing such things as roughing up antiwar radicals or placing illegal wiretaps. 1978 S. Brill *Teamsters* iii. 103 The wiretaps were to end on March 6. 1982 H. Kissinger *Years of Upheaval* iv. 103 The next morning it became apparent that Nixon had been talking about the wiretap records.

wi·re-tapper, *vbl. sb.* [f. Wire *sb.* + Tapper[1]. Cf. Tap *v.*[1] 2 c in Dict. and Suppl.] One who makes a (usually secret) connection to a telephone or telegraph circuit in order to intercept messages or eavesdrop.
1893 *Blue & Gray* Apr. 313/2 In that band of wire-tappers I had the honor to serve for four years. In 1863 I was appointed or employed as a telegraph operator in the field. 1894 [in Dict. s.v. Wire *sb.* 15 c]. 1906 *N.Y. Tribune* 1 Feb. 8/2 On the day he bet his money, the wiretappers made it appear to Felix..that such sportsmen as James R. Keene and John W. Gates were betting thousands on Old Stone, through them, and advised him to 'get in on the good thing', too. 1910 'O. Henry' *Strictly Business* 36 Who wears the diamonds in this town? Why, Winnie, the Wiretapper's wife. 1929 *U.S. Rep.* CCLXXVII. 453 The wire tapper destroys this privacy... Does not wire tapping involve an 'unreasonable search' of the 'house' and of the 'person'? 1953 'S. Ransome' *Drag Dark* (1954) xv. 146 A call..couched in cryptic terms calculated to fool a wire tapper. 1969 *N.Y. Rev. Books* 2 Jan. 41/2 In the process, Macdonald proves himself more reprehensible than a wiretapper, for he uses intimacy, rather than electronics, to do his dirty work. 1977 'J. D. White' *Salzburg Affair* vi. 52 He could only drive on, find another telephone, in case the M.F.S. wire-tappers had traced the first call.
So **wire-tapping** *vbl. sb.*, the practice or activity of a wire-tapper; (as a back-formation) **wi·re-tap** *v. trans.*, to tap the telephone line of; to monitor (a call) by means of a wire-tap.
1904 *Outing* Dec. 334/1 Despite the habitual exposure in American newspapers of the..'wire-tapping' swindle, the victim continues to be parted from his thousands with painful frequency. 1929 *U.S. Rep.* CCLXXVII. 474 The progress of science in furnishing the Government with means of espionage is not likely to stop with wire-tapping. 1952 W. R. Burnett *Vanity Fair* vi. 56 Chad Bayliss did not want to discuss anything as important..over his apartment house phone... There had been a rash of wire-tapping. 1959 A. Harrington *Life in 'Crystal Palace'* (1960) iv. 75 The private citizen is..being wire-tapped while he is looking for a job. 1973 *Times* 23 May 8/4 His telephone calls had not been wiretapped. 1973 *Black Panther* 21 July 2/3 The FBI, through their wire-tapping,..was practicing electronic surveillance. 1976 *National Observer* (U.S.) 27 Mar. 2/4 The U.S. Army may not wire tap American civilians in foreign countries unless it first gets a warrant from an American judge. 1978 R. Nixon *Memoirs* 388, I authorized Hoover to take the necessary steps—including wiretapping—to investigate the leaks. 1985 *Sunday Times* 20 Jan. 9/4 The U.S. customs officials say that his allegation of illegal wiretapping has been totally disproved.

wirework. Add: **3.** Wire-walking.
1906 *Variety* 3 Mar. 21/2 For sensationalism, the Meers in their wire work make the heart beat quicker. 1928 *Daily Express* 13 June 13/3, I learned acrobatics, wire work, dancing, and juggling.

wire-worker. Add: **2. b.** (Later example.)
1883 C. F. Wilder *Sister Ridnour's Sacrifice* 130 The politician grasps the hand of his wire-worker and tool.
3. = *wire-walker* s.v. Wire *sb.* 16.
1970 M. Kelly *Spinifex* v. 91 She was one of the Flying Volantes, a bloody good wire worker.
wire-working *vbl. sb.* (*b*) (earlier example); also as *ppl. a.*; hence (as a back-formation) **wire-work** *v. trans.*, to influence by pulling wires.
1831 *American* (Harrodsburg, Ky.) 28 Jan. 3/2 One of

the *wire-working* writers in the Union, seems disposed to consider it a little less than treason. **1843** J. Q. Adams *Diary* 23 Mar. (1876) XI. xxii. 343 James Monroe was recalled by President Washington through Thomas Pickering, wireworked by Alexander Hamilton. **1857** B. Hayes *Diary* 11 Sept. (1929) v. 167, I have kept aloof from the wire-working as well as from the more stormy scenes of politics.

wiring, *vbl. sb.* Add: **2. a.** Esp. the electric wires in an apparatus or building. (Further examples.)

1887 *Jrnl. Soc. Telegr.-Engineers* XVI. 182 This would be supplied from central stations (but without any outlay for insulated conductors beyond the 'wiring' of the actual domiciles to the extent rendered necessary by the number and position of the lamps required). **1923** *Wireless World* 5 May 135 (*caption*) The underside of the panel, showing the arrangement of the components and the wiring. **1958** C. Fremlin *Hours before Dawn* iii. 29 A rather cultured way of making me put a two-bar fire in her room... It's a matter of the wiring on the top floor. **1979** V. Capel *Burglar Alarm Systems* x. 107 This is about the worst place, as an intruder could soon silence the alarm by..cutting the bell wiring.

3. wiring diagram, a diagram of the wiring of an electrical installation or device, showing the electrical relationship of connections and components and usu. also their physical disposition; also *fig.*

1946 G. M. Chute *Electronics in Industry* viii. 63 Another kind of diagram... is called the connection or wiring diagram, because it shows the wire connections between the various parts. **1967** 'A. Cordell' *Bright Cantonese* xvii. 203 'There must be a switch assembly somewhere.'.. 'She gave me a wiring diagram.' **1969** *Times* 13 Feb. 10/3 He believes that until more is known about how the wiring diagram is modified, there can be little evidence in support of theories assuming that changes in nerve connexions are the basis of learning and memory. **1979** B. Scaddan *Mod. Electr. Installation* II. iv. 63 A circuit diagram shows how the system *functions*... A wiring diagram shows how the system is to be *wired*, and all components of the circuit should be shown in their correct places.

wirra, *int.* Add: (Earlier example.) **wirrasthru** (earlier example).

1829 G. Griffin *Collegians* I. vii. 153 O, wirra, Eily! this is the black day to your ould father. *c* **1874** D. Boucicault in M. R. Booth *Eng. Plays of 19th Cent.* (1969) II. 190 Ses he, 'You won't see home for two months.' Then I set up a wierasthru. *Ibid.* 214 Oh, weir asthru! What'll I do!

wirra (wi·rā), *sb. Austral.* [Aboriginal.] **1.** A species of acacia, *Acacia salicina*, burnt by Aborigines for its ash; = *Couba.

1906 J. H. Maiden *Wattles & Wattle-Barter* (ed. 3) 90 *A. salicina*, Lindl... Following are some additional aboriginal names..'Wirrha', Cooper's Creek, near Lake Eyre. **1941** I. L. Idriess *Great Boomerang* xiv. 102 Burned leaves of the wirra (a species of acacia, the leaves of which when burned yield a powder of potash).

2. A shallow wooden scoop used by Aborigines.

1935 H. H. Finlayson *Red Centre* vii. 74 A hunting party may decide suddenly to move on to another ground. Without more ado, the men reach for their spears and walk away, and their women follow, carrying no more than a yamstick, a *wirra*, and their youngest child. **1956** *Landfall* June 99 She carried two large pitchi, her four-foot-long mulga digging-stick, which she had freshly sharpened..by charring..; and her wirra, a shallow wood scoop.

wirrah (wi·rā). *Austral.* [Aboriginal.] An Australian saltwater fish, *Acanthistius serratus* (family Serranidæ) that is greenish brown with blue spots.

1882 J. E. Tenison-Woods *Fish & Fisheries N.S.W.* 34 'Wirrah' or Plectropoma is a genus similar to that of Serranus, but armed with a row of spinous teeth on the lower jaw..besides the pair of canines above. **1933** *Bulletin* (Sydney) 11 Jan. 21 We had caught a number of more genteel fish, so we heaved the common wirrahs and leather-jackets overboard. **1978** J. M. Thomson *Field Guide Common Sea & Estuary Fishes Non-Tropical Austral.* 100 Except for the Wirrah most of the rock cods are regarded as good eating.

‖ **Wirt** (virt). In 9 **Wirth.** [Ger.] The landlord of a German inn.

1858 Geo. Eliot *Jrnl.* July in J. W. Cross *Life* (1885) II. viii. 48 The stout, red-faced *Wirth*. **1970** *Guardian* 28 Feb. 12/4 The Swabian *Weinstube* is a cosy place... The *Wirt* will make his round to bid every one of his guests a personal good evening.

‖ **Wirtschaft** (vi·rtʃaft). *rare.* In 9 **Wirth-.** [Ger.] **1.** Domestic economy, housekeeping.

1850 C. M. Yonge *Henrietta's Wish* ii. 21 The house..was very soon pretty and cheerful, and the *wirthschaft*..well ordered and economical. **1889** —— *Reputed Changeling* I. viii. 138 The Doctor..[asked] whether the ladies abroad were given to housewifery. 'The German dames make a great ado about their *Wirthschaft*, as they call it..but as to the result! pah!'

2. [Ger., short for *gastwirtschaft*, f. *gast* Guest *sb.*] = *Wirtshaus.

1903 G. W. Hartley *Wild Sport* iii. 60 They had some beer at a snug little *wirtschaft*, and then they all fished. **1950** E. Hemingway *Across River* xxxviii. 269 What is there to eat in this *Wirtschaft*?

Hence **Wi·rtschaftswunder**, (erron.) **-schaftwunder** (-vu:ndəɹ), the 'economic miracle' of West Germany, i.e. the substantial and lasting recovery in its economic state and standard of living following the war of 1939–45; also *transf.*

1959 *Times* 13 Mar. 16/3 Some of the pot-holes in the present age of prosperity, the *Wirtschaftswunder*, are laid bare in a manner that leaves little to the imagination. **1961** *Times* 10 Apr. 11/6 A *Wirtschaftswunder* incomparably greater than the Federal Republic's. **1965** *Punch* 22 Sept. 411/1 The peoples of both India and Pakistan are shockingly poor... There is no possibility here of miraculous recovery, the *Wirtschaftwunder*. **1980** 'D. Grant' *Emerald Decision* i. 14 My daughters..are products of the Socialist *wirtschaftwunder*. **1983** *London Mag.* July 62/1 The Germany of the mind's eye; not..the concrete cities of the Wirtschaftswunder.

‖ **Wirtshaus** (vi·rtshaus). In 9 **Wirths-.** Pl. **Wirtshäuser** (-hoizəɹ). [Ger.] In German-speaking countries: a hostelry, inn.

1829 C. Wilmot *Jrnl.* 6 Sept. (1935) 330 They..have a Wirthshaus and tables and chairs to eat and drink and be merry. **1967** *Sat. Rev.* 22 Apr. 37/2 They returned home with precious addresses of tiny bistros and *brasseries* and *osterias* and *Wirtshäuser*. **1982** G. Lyall *Conduct of Major Maxim* xx. 179, I have an unfinished beer at the *Wirtshaus*. Would you like to join me?

Wisconsin (wiskɒ·nsin). *Geol.* The name of a state of the north central U.S.A., used *attrib.* and *absol.* to designate (the time of) the fourth and final Pleistocene glaciation of North America, corresponding to the Würm glaciation of the Alps.

[**1894** T. C. Chamberlin in J. Geikie *Gt. Ice Age* (ed. 3) xliii. 763 All this complex is grouped under a single term—the East-Wisconsin formation—because the grounds for a formal subdivision are not yet sufficiently clear.] **1895** *Amer. Naturalist* XXIX. 240 The second, third, and fourth glacial stages of the European Ice age..were probably also time equivalents, respectively, with the Kansan, Iowa, and Wisconsin stages in the United States and Canada. **1896**, etc. [see *Illinoian *a.*]. **1967** E. B. Leopold in Martin & Wright *Pleistocene Extinctions* 235 Taylor considers that the Pliocene and Quaternary climates before the late Wisconsin were much less continental than now. **1981** J. E. Sanders *Princ. Physical Geol.* xiii. 332 The last ice mass to cover the Great Lakes basin arrived during the late Wisconsin Stage, starting 20,000 years ago.

Hence **Wisco·nsinan** *a.* (also *absol.*).

1968 [see *Weichsel]. **1978** *Nature* 8 June 456/2 These volcanics are thought to have been formed under ice of Wisconsinan age. **1981** F. W. Shotton in Neale & Flenley *Quaternary in Brit.* xiii. 142 This is the way the Wisconsinan ice invading the U.S.A. from the Canadian Shield is interpreted.

wisdom. Add: **1. f.** Phr. *in his* (or *its*, etc.) *wisdom*: now usually ironic.

1852 Queen Victoria in *Hansard* CXXIII. 20 To enable the Industry of the Country to meet successfully that unrestricted Competition to which Parliament, in its Wisdom, has decided that it should be subjected. **1863** N. Hawthorne *Our Old Home* 397 Possibly his Lordship thought, in his wisdom, that the good feeling which was sure to be expressed by a company of well-bred Englishmen, at his august and far-famed dinner table, might have an appreciable influence on the grand result. **1930** W. Faulkner *As I lay Dying* 68 If you have no son, it's because the Lord has decreed otherwise in His wisdom. **1974** K. Clark *Another Part of Wood* vi. 232 In the 1930's, when the country was at least ten times as rich as it is today, the Treasury 'in its wisdom' twice found it necessary to cut off our annual purchase grant altogether.

wise, *sb.*¹ Add: **II. 3. b.** (Further examples.) No longer *arch.*

The meaning is 'in the manner of', 'in the..manner'.

1851 H. Melville *Moby Dick* II. xxviii. 192 Ahab..took Stubb's long spade..and striking it into the lower part of the half-suspended mass, placed its other end crutch-wise under one arm. **1854** H. D. Thoreau *Walden* 21 Waiting at evening on the hill-tops for the sky to fall, that I might catch something, though I never caught much, and that, manna-wise, would dissolve again in the sun. **1885** *Cornhill Mag.* Mar. 283 Priests sitting with their legs tucked up tailor-wise, in the attitude of Buddha. **1919** R. Firbank *Valmouth* iv. 52 Flecked with wood shavings, Saint Joseph-wise, it [*sc.* a gown] brought with it suggestions of Eastern men. **1921** Kastner & Charlton *Poetical Wks. of Sir Wm. Alexander* I. p. lvii, The style throughout, Seneca-wise, ought to be magnificent and grave. **1923** R. Macaulay *Told by an Idiot* I. ii. 11 Her mass of chestnut hair parted Rosetti-wise in the middle. **1940** 'Gun Buster' *Return via Dunkirk* II. iv. 117 In a few minutes our vehicles were coiled serpent-wise round the château.

(ii) Used in the same way but with the sense: as regards, in respect of. *colloq.* (orig. *U.S.*).

1942 E. R. Allen in J. J. Mattiello *Protective & Decorative Coatings* II. viii. 252 It should be noted that there are two types of hydrogen atoms positionwise. **1948** *Sat. Rev.* 6 Mar. 16/3 Plotwise, it offers little more or little less of what-happens-next interest than may be found [etc.]. **1958** *Spectator* 10 Jan. 37/2 John Robert Russell, 13th Duke of Bedford..in twelve TV performances, was the greatest, successwise, among the aristocrats. **1958** *Times* 5 Sept. 11/5 An ill-disciplined, over-paid, frustrated youth, whose life chances have been vastly improved moneywise without commensurate social adjustment. **1961** *Far East Film News* (Tokyo) Apr. 5/1, 1961 so far has been UA [*sc.* United Artists] all the way prize-wise with this company taking an even dozen Oscars. **1976** J. I. M. Stewart *Memorial Service* xii. 184 These were a gentle race..desperately worried over the grim state of the market job-wise. **1981** *Gossip* (Holiday Special) 24/1 Acting-wise, I like Katharine Hepburn, Joanne Woodward, Judy Garland and, of course, Marilyn.

wise, *a.* (*sb.*³, *adv.*) Add: **1. a.** *wise old man*; spec. = *Wise man 4.

1940 [see *Shadow *sb.* 1 d]. **1956** R. F. C. Hull tr. *Jung's Symbols of Transformation* in *Coll. Wks.* V. ii. vii. 332 The archetype of the wise old man first appears in the father, being a personification of meaning and spirit in its procreative sense. **1961** G. Adler *Living Symbol* xvii. 397 The more remote and more powerful figure of the 'wise old man' represents a further step..to a higher and more comprehensive wisdom. **1968** 'A. Whitney' *Every Man has his Price* viii. 61 Now he was a wise old man, greatly feared, much respected. **1975** D. Daniell *Interpreter's House* iii. 60 There is a Wise Old Man..blind and of immense strength who..blesses John Burnet. **1977** M. Green *Children of Sun* i. 36 A whole movement focuses passionate values..on them [*sc.* young men]—as opposed to focusing them on the wise old man.

3. b. (*b*) For U.S. colloq. read colloq. (orig. *U.S.*). (Earlier and further examples.)

1896 Ade *Artie* ii. 14, I told him that when he wanted to get wise to what was in my hand all he had to do was to dig up his bit and come in. *Ibid.* xvii. 155 There was somethin' ailed me, but I was n't wise to it. **1913** A. Bennett *Regent* x. 296 'Tell me,..she hasn't got herself arrested yet, has she?' 'No. And she won't!' 'Why not?' 'The police have been put wise.' **1937** G. Heyer *They found Him Dead* ii. 41 Say, sister, get wise to this! You can't put nothin' across on me! **1950** G. Greene *Third Man* ii. 21, I met him my first term at school... He was a year older and knew the ropes. He put me wise to a lot of things. **1955** M. Gilbert *Sky High* xv. 210, I suppose Bill had just about got wise to you. **1977** F. Parrish *Fire in Barley* v. 49 Dan wondered if the arty woman was wise to him.

c. *wise guy* (colloq., orig. *U.S.*): an experienced or knowledgeable man; usu. ironic or derog., a know-all, a wiseacre; someone who makes sarcastic or annoying remarks; also (with reversal of meaning), someone easily duped; also *attrib.*

1896 Ade *Artie* xvi. 150 He was the wise guy and I was the soft mark. **1903** H. Hapgood *Autobiogr. of Thief* iv. 82 When these Rufus's up the State get a Yorker or a wise guy, they'll strip him down to his socks. **1910** W. M. Raine *Bucky O'Connor* ii. 28 You're wise guys, gents, both of yez. **1920** B. Tarkington in *On Plays, Playwrights & Playgoers* (1959) 42 However, they'd made the crowd aware of wise guy superiority. **1922** Wodehouse *Adventures of Sally* xiii. 219 Obviously one of the Wise Guys of whom her friend the sporting office-boy had spoken, he was frankly dissatisfied with the exhibition. **1929** W. T. Scanlon *God have Mercy on Us!* lvi. 331 We had positive orders not to pick up any form of documents and to leave them for the Intelligence Section—the 'Wise Guy Section', as we called it. **1932** [see *Con *v.*]. **1935** [see *Easy *a.* 13 b]. **1941** B. Schulberg *What makes Sammy Run?* (1943) i. 7 Listen, wise guy,..if you found something wrong..why didn't you come and tell me? **1959** C. Williams *Man in Motion* xi. 150 'What're you, a wise guy?' he snarled. **1972** *Village Voice* (N.Y.) 1 June 50/3 The cop..told Rob he didn't think it was funny, portfolio or not, declared that he was a clear-cut wise guy and placed him under arrest. **1976** *National Observer* (U.S.) 7 Aug. 17/1 Kramer and Roberts seem unable to shake off the brittle, knowing, wise-guy tone of voice.

8. b. *wise-ass* sb. and adj. (cf. *Smart-arse, -ass). **c.** *wise-assed.*

1971 *Current Slang* V. iv. 21 *Wise ass*, n., a wise guy. **1972** J. Poyer *Chinese Agenda* iii. 17 Listen to what I have to say, then you can make all the wise-ass remarks you want. **1978** J. Irving *World according to Garp* iv. 66 Benny Potter from New York—a *born* wise-ass. *Ibid.* 67 It was unfortunate that wise-ass Benny Potter was the first to tell Garp the news. **1967** P. Tamony *Americanisms* (typescript) No. 18. 2 A fantastic display of brash male and female wise-assed mediocrity. **1976** 'Trevanian' *Main* xii. 260 Some wiseassed note about the bad luck of getting a parking ticket the same night you get killed.

wise, *v.*² Add: **1.** Freq. const. *on* or *to*. Also *refl.* (Earlier and later examples.)

1905 R. Beach *Pardners* iv. 113, I cast the bad eye on the boys to wise 'em up. **1925** F. Scott Fitzgerald *Great Gatsby* vii. 124, I just got wised up to something funny the last few days. **1929** *Princeton Alumni Weekly* 24 May 982/2 To stick out one's neck is to commit an unpardonable error, to lay oneself open to criticism... A persistent offender should wise up on himself. **1955** W. Gaddis *Recognitions* I. iv. 158 Yeah, you got to wise up to yourself, see? **1960** C. MacInnes *Mr Love & Justice* 26 That's..what I'm wising myself up on. **1971** *Wall St. Jrnl.* (Eastern ed.) 10 Mar. 1/4 Antique dealers are wising up to the growing demand for old radios. **1984** *Listener* 7 June 36/3 'Write a poem about it,' he suggests. 'Wise up, sir,' the new generation tells him.

2. *to wise off* (U.S. slang): to make wisecracks *at* someone.

1943 *Yank* 2 July 10 I'd love to have one of those acting noncoms wise off at me. **1981** 'P. Mallory' *Killing Matter* xiii. 136 He's a real meanie. I wouldn't be wising off at him if I were you.

Hence **wised-up** *ppl. a.*

1926 J. Black *You can't Win* xx. 301, I could make a living without taking tough chances against wised-up city police. **1952** M. McCarthy *Groves of Academe* (1953) x. 205 His wised-up air was as irritating..as Donna's exaggerations. **1973** R. Parkes *Guardians* ix. 172 It's nasty. Very nasty. But at least I'm wised up now.

wisecrack (wəi·zkræk). *colloq.* (orig. *U.S.*). Also **wise crack, wise-crack.** [*Crack *sb.* 5.] A clever, pithy witticism or remark. Also as quasi-*adj.*

1924 G. Ade *Let.* 20 Dec. (1973) 101 When Geraghty came yesterday, both of us had thought of putting in another character, a young wise-crack small town loafer,..who thinks he is very sly. **1925** *Sat. Even. Post* 14 Feb. 44 The Palace, Chicago, will howl at a wise crack, a nifty, that Duluth audiences won't even flag as it flies over their heads. **1950** G. B. Shaw *Buoyant Billions* 98 The satirical humor of Aristophanes, the wisecracks of Confucius, the precepts of the Buddha. **1959** I. & P. Opie *Lore & Lang. Schoolch.* ix. 174 He might..have seen sense in the wisecracks which..scientifically minded boys indulge in..: 'What is the matter?' 'That which occupies space.' **1977** *Rep. Comm. Future of Broadcasting* (Cmnd. 6753) ii. 15 Lord Hill..saw no reason why the BBC should have been expected to apologise for a wisecrack in a satire programme..that if you can see the Prime Minister's lips moving you know he is lying. **1979** R. Jaffe *Class Reunion* (1980) I. viii. 109 Say only nice, polite, ladylike things, no nasty wisecracks.

So **wi·secrack** *v. intr.*, to make wisecracks; also *trans.*, with quoted words as obj.; **wi·secracking** *ppl. a.* and *vbl. sb.*; also **wi·secracker**, one given to making wisecracks.

1915 *Call* (San Francisco) 30 Apr. 17 Wisecrackin' city fellers ain't got nuthin' on you. **1923** *N.Y. Times* 9 Sept. VII. 2 *Wise-cracker*, a city fellow who makes wise remarks. **1924** P. Marks *Plastic Age* 28 The lights flashed on and the crowd filed out, 'wise-cracking' about the picture. *Ibid.* 113 Carl the flippant, the voluble, the 'wise-cracker', lost his tongue. **1927** *New Republic* 12 Oct. 218/2 He has the knack of wise-cracking, and his dialogue is of that slick and well oiled kind that you may meet in good vaudeville. **1939** Joyce *Finnegans Wake* 33 It has been blurtingly bruited by certain wisecrackers..that he suffered from a vile disease. **1940** Graves & Hodge *Long Week-End* xx. 345 Everyday life could be made interesting on the screen without fictitious drama or wise-cracking comment. **1946** *Sat. Rev. Lit.* 2 Nov. 41/1 Both authors wisecrack their way through adverse circumstances. **1949** H. Robbins *Dream Merchants* (1950) 144 Jane saw him come into the office. 'If it ain't the vice-president himself!' she wisecracked. 'How's the picture business?' **1949** G. B. Shaw *Sixteen Self Sketches* xiv. 82 Without him I might have been a mere literary wise-cracker, like Carlyle and Ruskin. **1977** *Rolling Stone* 21 Apr. 58/1 Jim Rockford is a worldly wisecracker. **1979** *Fortune* 21 May 71/1 Fraser wisecracks that the Carter Administration's voluntary wage guidelines have 'self-destructed'. **1982** *Daily Tel.* 11 May 14 He has defended rioters..with such vigour as to alarm his fellow lawyers, one of whom wisecracked to me: 'Who will rid us of this turbulent pest?' **1982** E. North *Ancient Enemies* viii. 105 Liz reminds me..of Tammy, who was tough and wise-cracking.

wise man. **2. c.** Delete '† *Obs.* as a specific sense'. *colloq.* in mod. use.

1959 J. Balogh in H. Thomas *Establishment* 98 In the negotiations on the finance of NATO rearmament in Lisbon the British representative 'wise man' accepted a contribution wholly out of line with the relative capacities to bear the burden. **1969** D. Acheson *Present at Creation* xxxi. 277 Lester Pearson has continually urged the council to set up committees of 'wise men' to find a use for it [*sc.* Article 2 of the North Atlantic Treaty]. **1973** *Times* 5 May 4/4 The appointment of two independent 'wise men' by the United States and the European Community to prepare the ground for the forthcoming round of international trade talks, was suggested. **1983** *Times* 24 Feb. 6/8 A socialist leader..has been nominated to the elite body of nine 'wise men' who form France's Constitutional Council. **1984** *Times* 29 Oct. 1/3 The Gaddafi affair..is unlikely to go away as quickly as the TUC's 'four wise men' monitoring the dispute would wish.

4. (*old*) *wise man*: an archetypal figure appearing in myths, folklore, etc., representing wisdom or meaning, esp., in the theory of C. G. Jung, one of the archetypes of the collective unconscious; cf. *wise old man* s.v. *Wise *a.* (*sb.*³, *adv.*) 1 a.

1692 W. Salmon tr. 'Hermes Trismegistus' in *Practical Physick* v. 203/1 But if thou shalt say, that Wisdom or the Wise Man does Rule or Command among all Mankind. **1940** S. Dell tr. *Jung's Integration of Personality* iii. 87 The magician is the archetype of the old wise man. *Ibid.* 88, I have been content to call it the archetype of the old wise man or of meaning. **1973** J. Singer *Boundaries of Soul* x. 262 The archetype of the mana-personality, an Old Wise Man whose power is born of understanding the timeless life processes.

wisenheimer (wəi·zənhəimər). *U.S. slang.* Also **weisen-, wise-.** [f. Wise *a.* + *-enheimer*, as in German names such as *Oppenheimer*.] A wiseacre, a 'clever dick'. Also *attrib.* or as *adj.*

1904 R. L. McCardell *Show Girl & her Friends* 51 He wants to know some good way to reduce his weight... You don't know any such a way? No? Why, I thought you was a wiseheimer. **1919** Mencken *Amer. Lang.* v. 151 Several years ago *-heimer* had a great vogue in slang, and was rapidly done to death. But *wiseheimer* remains in colloquial use as a facetious synonym for *smart-aleck*, and after awhile it may gradually acquire dignity. **1919** *National Police Gaz.* (U.S.) 4 Jan. 3/1 Cawkins..like a true Wisenheimer, considered the gentle, goosey kind of a beautiful girl the most appetizing. **1922** S. Lewis *Babbitt* xxiv. 287 The wisenheimers grab a look at a fellow's nails when they want to tell if he's a tinhorn or a real gent! **1937** *Amer. Speech* XII. 9/2 Some wiseheimer American newspaper man has picked this up and tagged it onto President Hoover. **1957** J. D. Salinger *Franny & Zooey* (1964) 65 We were nervous..at the statistics on child pedants and academic weisenheimers who grow up into faculty-recreation-room savants. **1959** *Washington Post* 26 Dec. A19/2 Then some wisenheimer from the agency decided we needed a trailer. **1975** *Times Lit. Suppl.* 21 Feb. 185/1 The wisenheimer who gets his ornate come-uppance. **1977** M. Babson *Murder, Murder, Little Star* vii. 50, I shoulda listened to her. But..I was too wisenheimer.

wish, *sb.*¹ Add: **4. wish book** *N. Amer. slang*, a mail-order catalogue; **wish card** *rare*, in fortune-telling, a card which predicts the attainment of a desired end; **wish-dream** [cf. G. *Wunschtraum*], a dream or fantasy that reflects some hidden wish; also *attrib.*; **wish list**, a list of desired objects or occurrences; **wish-thinking** = *wishful thinking* s.v. *Wishful *a.* 2 a.

1933 *Amer. Speech* VIII. 32/1 *Wishbook*, a mail-order catalogue. **1971** *Alberta Hist. Rev.* Summer 25/1 That was when we would look at our 'Wish Book', the mail order catalogue, until it was out of date, and then tear the pages out in the 'House of Parliament'. **1922** Joyce *Ulysses* 761 I'll throw them the 1st thing in the morning till I see if the wishcard comes out. **1934** R. Campbell *Broken Record* i. 9 Wish-dreams might account for this desire to 'headlong-hall' me into the next world. **1945** Koestler *Yogi & Commissar* I. iii. 31 She is not necessarily the wishdream-girl of suburban circulating libraries. **1953** *Encounter* Nov. 25/2 The wish-dream world of the Stockholm peace campaigner. **1966** *New Scientist* 28 July 222/1 Black gold in the back lot must be the standard rags-to-riches wishdream in the United States. **1972** *Times* 30 May 19/2 It had a presently confidential 'wish list' of programmes it would like to see abandoned. **1976** *National Observer* (U.S.) 10 July 9/3 Wholesale replacement of the nation's taxi fleet is hardly at the top of operators' wish lists. **1930** J. Jastrow *Piloting your Life* 170 It would take not a chapter but a volume to describe all the varieties of impediments of thought. If reduced to a schedule they might read like this:..*Wish Thinking*, believing what you hope or want to be true; [etc.]. **1945** R. Knox *God & Atom* viii. 115 Most people who are capable of thinking, and are not deceived by wish-thinking, agree that the world is in ferment. **1958** J. Lodwick *Bid Soldiers Shoot* viii. 268 In matters of wish-thinking..the Victor of the Pacific, MacArthur, had met his match.

wish, *v.* Add: **1. a.** (*c*) *to wish to God*: to wish intensely.

c **1385**, *a* **1562** [in Dict.]. **1932** 'N. Shute' *Lonely Road* vii. 178, I wish to God we'd gone back to the boat. **1941** L. A. G. Strong *Bay* 7, I wish to God I knew how to begin. **1976** *Daily Mirror* 11 Mar. 7/2 Christopher..has been sent to Borstal... His mother said: 'I wish to God we had checked up on him.'

8. To foist or impose (something or someone) *on* (*to*) someone; to endow with at another's wish.

1915 N. L. McClung *In Times like These* x. 164 Women have never chosen the liquor business... It has been wished on them. **1926** *Publishers' Weekly* 22 May 1725/1 Mr. Remington..has not been able to be with us... That is why we are able to wish that good job on him. **1934** E. Waugh *Handful of Dust* i. 20 Who was the old girl you wished on me at that party last night? **1954** 'N. Shute' *Slide Rule* 1 An unwanted kitten that they had wished on to my children. **1962** *Listener* 5 July 14/2 The plan was to build a much larger school than was needed, at the same time 'wishing' additional housing on the village to justify it. **1971** *Guardian* 9 Jan. 13/2 We owed money everywhere... I wouldn't wish that on any of today's young housewives. **1983** M. Babson *Fool for Murder* xviii. 166 It really was most unfair of Uncle Wilmer to wish it on Wanda-Lu... He wasn't the one who'd have to keep the place tidy.

wisha (wi·ʃə), *int.* *Anglo-Ir.* *colloq.* [ad. Ir. *mhuise* indeed (the unlenited form *muise* gives anglicized *musha* Musha).] An exclamation indicating dismay, emphasis, or surprise.

1826 M. Wilmot *Let.* 29 Feb. (1935) 234 O 'wisha' 'wisha', shall I ever arrive at the ball I promised you! **1842** S. Lover *Handy Andy* vi. 70 I'm afeard o' my life to go to bed!..Wisha! but I'd give the world it was mornin'. **1898** J. D. Brayshaw *Slum Silhouettes* 49 Oh, wisha! didn't he break the leg of me wid his stick? **1914** Joyce *Dubliners* 157 'Wisha! wisha,' says I. 'A pound of chops..coming into the Mansion House.' **1936** 'F. O'Connor' *Bones of Contention* 8 Wisha, for goodness' sake will you come down and leave the girl sleep? **1978** 'M. M. Kaye' *Far Pavilions* lviii. 812 Wisha, but it's a gloomy devil you are an' all. **1965** *N. Munster Antiquarian Jrnl.* IX. iv. 186 Wisha, I don't know what to say.

wish-bone. (In Dict. s.v. Wish *sb.*¹) Add: Also **wishbone.** **2.** *Naut.* A boom composed of two halves that curve outward from the mast, on either side of the sail, and in again, the clew of the sail that lies between them being attached to the point where they meet aft. Freq. *attrib.*, designating a sail or a boat with such a boom.

1934 U. Fox *Sailing, Seamanship & Yacht Construction* I. 54 (*caption*) Wishbone gaff. **1935** *Yachting Monthly* Feb. 306/1 The working sail area of the 'wishbone' ketch rig shown totals 2,794 sq. ft. **1954** D. H. C. Phillips-Birt *Rigs & Rigging of Yachts* ii. 60 The wishbone ketch is the result of combining an unusual form of staysail ketch rig devised by Mr. F. Fenger..with the wishbone spar invented by Nathaniel Herreshoff. **1958** *Times* 27 Oct. 10/6 The fishermen..were already launching their dug-out canoes, some under the traditional wish-bone sprit-sail. **1981** B. Webb *Schult's Sailing Dict.* 327/1 Sailboards have wishbone booms. **1984** *Times* 25 Aug. 11 The next stop was to pick a point to steer for, ease the rig (the mast, sail and wishbone) towards me until I was looking through the transparent panel in the sail.

3. A wishbone-shaped element in the independent suspension of a motor vehicle, the two arms of which are hinged to the chassis and their join hinged to the wheel; freq. *attrib.*

1934 *Automobile Engineer* XXIV. 289/3 The American types..do not use the longitudinal radius links, and thus the wheel position depends entirely on the 'wishbone' link bearings. **1959** *Times* 27 Apr. (Rubber Industry Suppl.) p. vi/1 Such bushes are now almost universally used for springs, torque arms, and in some forms of wishbone suspension. **1983** 'D. Rutherford' *Stop at Nothing* ix. 165 It took me an hour and a half to put the Saab to rights. The front wishbone had been seriously distorted.

4. *U.S. Football.* Used *attrib.* and *absol.* to designate an offensive formation in which the full back lines up ahead of the half-backs in an alignment that resembles the shape of a wishbone.

1972 *N.Y. Times* 3 Nov. 48/2 Dartmouth's best hope lies in shutting off Yale's wishbone offense as engineered by Dick Jauron. **1974** *Spartanburg* (S. Carolina) *Herald-Jrnl.* 21 Apr. 134/3 The white team ran the wishbone that UCLA used last year. **1979** *Tucson* (Arizona) *Citizen* 20 Sept. 10D/3 The fullback in the Santa Rita Wishbone has averaged nearly 30 carries per game in the first two contests.

wishful, *a.* **2. a.** Delete *Obs.* or *dial.* In mod. use in weaker sense: expressing or indicative of a wish; chiefly in *wishful thinking*, thinking, esp. belief or expectation, that is influenced by one's wishes to the extent that relevant (consciously) known facts are (subconsciously) ignored or distorted; also as *adj.*; so *wishful thinker*.

1932 *Sat. Rev. Lit.* 2 July 817/4 At two vitally important points Glenn Frank's incisive analysis fades away in a vague realm of hope or even of wishful thinking. **1940** *Illustr. London News* CXCVI. 498/2 The possibility of any relief in that direction can only exist in the minds of wishful-thinkers. **1940** L. D. Weatherhead *This is Victory* ii. 58, I do not mean that that which is believed has no other support than man's wishful thinking. **1941** Auden *New Year Letter* i. 17 Twelve months ago in Brussels I heard the same wishful-thinking sigh. **1942** C. S. Lewis *Screwtape Lett.* ix. 50 It all depends on whether your man is..of the wishful-thinking type who can be assured that all is well. **1951** 'A. Garve' *Murder in Moscow* i. 20 He was a woolly wishful-thinker who happened to be inordinately vain as well. **1958** *Spectator* 6 June 724/1 He [*sc.* a prisoner] hoarded these glimpses of past happiness, rationing his wishful reminiscing to half an hour a day. **1958** *Listener* 25 Sept. 478/1 There are some embarrassingly wishful statements... 'The day when Joyce embarked in Dublin..they were burning his first book... He could see the smoke of the bonfire.'.. *Dubliners* was not burned: it was pulped. **1970** *Guardian* 10 Dec. 4/4 It is hard to reconcile this sort of picture with the one presented by the wishful-thinkers in Saigon. **1974** E. Ambler *Dr Frigo* II. 117 I'm not a wishful-thinking idiot. **1980** *Sunday Times* (Colour Suppl.) 30 Mar. 55/2 An evocation of youth's transitoriness and innocent wishful-thinking.

wi·sh-fulfilment. [tr. G. *wunscherfüllung* (S. Freud *Die Traumdeutung* (1900) i. 64).] The imaginary fulfilment of acknowledged or unconscious wishes in dreams and fantasies; a dream or other event or object in which the fulfilment of a wish is given (usu. imaginary or symbolic) expression.

[**1901** H. Ellis in *Jrnl. Mental Sci.* XLVII. 370 The author [*sc.* Freud] points out that..we may with much more reason regard them [*sc.* dreams] as the protectors of sleep, willing us to repose with an imagined fulfilment of our wishes.] **1908** *Jrnl. Abnormal Psychol.* III. 237 This is a hyperbolic realization of his reveries, corresponding to the wish-fulfilment of the normal dream. **1916** A. A. Brill tr. *Freud's Wit & its Relation to Unconscious* vi. 250 Wider reading circles have contented themselves to reduce the contents [of *The Interpretation of Dreams*]..to a catchword, 'Wish fulfillment'—a term easily remembered and easily abused. **1928** C. H. Dodd *Authority of Bible* iii. 66 Behind the song lies..a longing for the downfall of an implacable enemy, which finds in the picture of disaster a 'wish-fulfilment'. **1939** D. Cecil *Young Melbourne* iv. 106 Living wholly in a wish-fulfilment world of her own creation, she insisted it was the real one. **1953** A. Huxley *Let.* 31 Oct. (1969) 687 The jewelled palaces are partly, no doubt, wish fulfilments—the opposite of everyday experience. **1956** E. L. Mascall *Christian Theol. & Natural Science* vi. 217 More careful examination would enable us to sift out those beliefs which had a rational justification for those which were mere wish-fulfilments. **1958** E. A. Armstrong *Folklore of Birds* iii. 55 It is a man's wish-fulfilment story, as Cinderella is a woman's. **1958** *Punch* 25 June 852/2 She is gentle, charming, childlike, submissive, a young man's wish-fulfilment. **1978** M. Levey *Case of Walter Pater* xiv. 198 All Pater's dreams, nightmares as well as wish-fulfilments, find expression here.

So **wi·sh-fulfilling** *a.*

1922 R. S. Woodworth *Psychol.* xix. 501 They [*sc.* dreams] are 'wish-fulfilling'. **1945** D. L. Moore *Vulgar Heart* i. 14 The wish-fulfilling assumption that such honours will be deserved. **1976** *Listener* 8 Apr. 447/3 A serenity that accepts as self-evident truth what we reject as wish-fulfilling illusion.

wishing, *vbl. sb.* **d.** *wishing-stone* (earlier example).

1859 G. W. Dasent *Pop. Tales from Norse* (ed. 2) p. xcii, Thus, we have *oska-steinn*, wishing-stone, *i.e.* a stone which plays the part of a divining rod.

Wishram (wi·ʃræm). Also 9 **Wish-ham.** [ad. Sahaptin *Wišxam*.] **a.** (A member of) an American Indian people living in the southern part of the state of Washington. **b.** The language spoken by this people, a dialect of Upper Chinook.

[**1836** W. Irving *Astoria* I. 109 We would make special mention of the village of Wish-ram.] **1855** *Rep. Commissioner Indian Affairs 1857* 351 They are divided into three principal bands, namely: the Wish-hams, Click-a-hut, and Skien bands. **1907** *Amer. Anthropologist* IX. 533 The Indians formerly living on the northern shore of Columbia river..are known by their Yakima and Klikitat neighbours..as *Wúcxam*, which, in its anglicized form of Wishram, or Wishham, is their common appellation today. *Ibid.* 535 The Wishram is prevailingly sonant in its use of stops. **1930** Spier & Sapir *Wishram Ethnography* 153 The Wishram were one of the earliest groups known to explorers of the Columbia River Basin... Only a few Wishram still remain. **1962** *Anthropol. & Human Behavior* (Anthropological Soc. Washington) 25 'Cussing out', a Wishram Chinook's English label for a class of aboriginal speech events. **1972** *Language* XLVIII. 378, I have drawn extensively upon my field research on the Wishram–Wasco dialect of Chinook.

Wiskinkie (wiski·ŋki). *U.S.* Also **Wiskinky, Wiskinski.** [Etym. unknown.] The official of the Tammany Society of New York charged with the office of door-keeper.

1800 *Commercial Advertiser* (N.Y.) 3 Jan. 2/3 Tammany Society, in the following order: 1st, The Wiskinkie, supporting the Cap of Liberty veiled in crape. **1843** *New Mirror* 15 Apr. 18/2 They were placed in charge of the Wiskinski of the wigwam. **1864** C. G. Halpine *Life & Adventures of Private Miles O'Reilly* 191 He had been a brave when the present Grand Sachems, Wiskinkies and Sagamores were no more than little papooses. **1905** W. L. Riordon *Plunkitt* 99 Dan Donegan, who used to be the Wiskinskie of the Tammany Society, and received contributions from grateful office holders. **1938** J. W. Norwood *Tammany Legend* xi. 145 The chief of a 'Tribe' was its Sachem; the master of ceremonies, the Sagamore; the Sergeant at Arms, the Wiskinkie. **1967** Connable & Silberfarb *Tigers of Tammany* i. 28 The Wiskinkie (sergeant-at-arms) swore an oath to preserve the Society from 'intruders and eavesdroppers'.

wisp, *v.* Add: **5.** *intr.* Of hair, etc.: to hang or twine in wisps. Hence **wisped** (wispt) *ppl. a.*

1913 W. de la Mare *Peacock Pie* 31 Topknot to love-curl The hair wisps down. **1922** A. S. M. Hutchinson *This Freedom* II. ix. 164 Her face flushed; her hat awry; her hair escaped and wisped about her eyes. **1976** 'A. York' *Dark Passage* iv. 56 Long, straight hair which wisped on her shoulders.

wispy, *a.* Add: Hence **wi·spily** *adv.*

1923 A. Huxley *Antic Hay* 30 His long grey hair floated wispily about his head. **1985** M. Wesley *Harnessing Peacocks* ix. 76 His hair hung wispily round his collar.

‖ **Wissenschaft** (vi·sənʃaft). [Ger.] (The systematic pursuit of) knowledge; learning, scholarship; science. Hence ‖ **Wi·ssenschaftslehre** (-lērə), a theory or philosophy of knowledge (used with reference to the work of J. G. Fichte, author of *Grundlage der gesammten Wissenschaftslehre* (1794)).

1834 F. D. Maurice *Let.* 24 July in F. Maurice *Life* (1884) I. xi. 168 My friend adds, '..They all seem to think *Wissenschaft*..more important than soundness of creed.' **1846** J. D. Moreau *Hist. & Crit. View Speculative Philos.* II. v. 72 Fichte's object was..to erect a system..of rigid scientific knowledge... Hence it was that, in place of 'Philosophy' he assumed the term 'Wissenschaftslehre', as most designative of his great purpose. **1896** W. Caldwell *Schopenhauer's System* ix. 503 That most vicious aspect of German philosophy, so prominent in the Hegelian dialectic and Fichte's *Wissenschaftslehre*, whereby it always seems to be telling us what a fact *must be* before we know what it *is*. **1961** D. G. James *Matt. Arnold* iv. 93 By 'science' he does not, of course, mean 'science' as we now ordinarily understand it; he means *Wissenschaft*, that is, as he says elsewhere, 'knowledge systematically pursued and prized in and for itself'. **1976** B. Williams *Making of Manchester Jewry* iv. 108 The emergence of Theodores as..a major English representative of the German *Wissenschaft*. **1982** C. Clemeau *Ariadne Clue* (1983) xi. 120 These fragments are painstakingly collected..published..and roundly applauded as the choicest fruits of philological *Wissenschaft*.

Wistar (wi·stāɹ, -āɹ). *Med.* and *Biol.* The name of the *Wistar* Institute of Anatomy and Biology, Philadelphia (founded by I. J. Wistar (1827–1905), grandnephew of Caspar Wistar (see Wistaria)), used *attrib.* to designate rats bred from a strain developed at the Institute for laboratory purposes.

1938 *Amer. Jrnl. Cancer* XXXIV. 353 The Wistar rats used in this series were from the Experimental Colony strain of the Wistar Institute. **1956** *Nature* 10 Mar. 453/2 One series of irradiated mice was injected with a suspension of bone-marrow cells of the Wistar rat. **1970** *Jrnl. Gen. Psychol.* LXXXII. 28 This kind of preference is the same as has been found by Wagner..and Rowntree with Long-Evans and Wistar albino rats.

wistaria. Add: **b.** A light blue-purple shade, the colour of wistaria blossom.

1911 *Daily Colonist* (Victoria, B.C.) 5 Apr. 24/1 (Advt.), Important Silk Purchase..in colors of rose, Persian blue.., wisteria. **1927** *Observer* 3 Apr. 25/5 (Advt.), Sunrise, geranium, wisteria, russet. **1974** *Country Life* 25 Apr. 1025/2 Her best colours..are wistaria, ice blue..and lemon yellow.

wit, *sb.* Add: **IV. 14. b.** *wit-writing* (later example); **c.** **wit-wanton** *v.* (later *arch.* examples); **wit-worm**: for † read 'now *rare*'; (later example).

1922 Joyce *Ulysses* 388 And Master Lynch bade him have a care to flout and witwanton. **1922** E. R. Eddison *Worm* xv. 209, I will not suffer mine indignation so to witwanton with fair justice as persuade me to put the wite on Witchland. **1932** F. Scott Fitzgerald *Let.* 2 Aug. (1964) 498, I did not destinate to signify that you were a wiseacre..but..that you were..a longhead,..as are so many epigrammatists, wit-worms, [etc.]. **1947** C. Day Lewis *Poetic Image* ii. 50 The conceits of the Metaphysicals are in a way wit-writing too.

‖ **witblits** (vi·tblits). *S. Afr.* Also **witblitz** and as two words. [Afrikaans, irreg. f. Du. *wit* White + G. *blitz* lightning.] Home-brewed brandy, a strong and colourless raw spirit.

1934 *Sunday Times* (Johannesburg) 8 Apr., 'Wit blits' (white lightning) was the name given to peach brandy in the Johannesburg Magistrate's Court yesterday. **1948** *Cape Times* 21 July 16/3 In cases of snakebite people on the platteland have always run for the witblitz. **1955** L. G. Green *Karoo* ix. 105 Farmers are allowed to distill small quantities of witblits for their own use. *Ibid.*, *Witblits*, of course, is home-distilled dop brandy with a high alcoholic content. **1966** *Economist* 12 Mar. 1044/3 Coloured people are drinking less [in South Africa], particularly the favoured types of plonk known affectionately as *witblits*..and *skokiaan*.

witch, *sb.*² Add: **5. a.** *witch-act* (earlier examples); *witch-burner* (earlier example), *-master*; *witch-burning*, *witch-roasting*. **b.** **witch-ball**, (*b*), a hollow ball of (usu. coloured or silvered) glass, formerly displayed in a house as a charm against witchcraft and now for decorative purposes; **witch bottle**, a stone or glass bottle, filled with urine, nails, hair, etc., which was either burned or heated for the purpose of repelling or breaking a witch's power over her victim; **witch-bowl**, a decorative circular glass bowl; **witch dance**, a ritual dance performed by witches; **witch-hopple** *U.S.* = Hobble-bush; **witch-post**, in Yorkshire, a wooden post, usually of mountain ash, marked with a cross and built into a house as a protection against witches; **witch-smelling**, the smelling out of witches; also *fig.*, witch-hunting; **witch-stone** (earlier example).

1758 M. W. Montague *Let.* 14 Nov. (1967) III. 188, I..am convinced of the necessity of the repeal of the Witch-act (as it is commonly called). **1916** J. H. Yoxall *Collecting Old Glass* v. 38 Witch-balls seem to have been made at Bristol,..at Nailsea..and at Wrockwardine... These balls, it is said, were hung at each door and window, 'to keep the witches out'. **1927** *Daily Express* 22 June 9/4 There is a fashion just now to collect the deep blue or silver glass balls which our forefathers hung about the house to keep witches away. 'Witch balls' they were called. **1952** L. MacNeice *Autumn Leaves* 19 The witch-ball on the stairs. **1978** E. Ellenbogen tr. *Simenon's Maigret & Toy Village* ii. 34 Its houses..its tiny carefully-tended gardens, its clay animals and glass witchballs. **1893** *Jrnl. Brit. Archaeol. Assoc.* XLIX. 267 Nails were formerly placed in an earthen vessel, and buried beneath the floor, near the hearth, to keep away the witches, and to afford protection from the 'evil eye'; hence such vessels were called 'witch-jugs' or 'witch-bottles'. **1908** E. Smith in A. C. Kelway *Memorials Old Essex* 252 An old witch-bottle..found 'below the floor and very near the fireplace'..contained some water, about fourteen horse-nails, and twenty thorns! **1966** G. E. Evans *Pattern under Plough* vi. 74 Under the hearthstone was the spot most frequently chosen to bury the witch-bottle. **1980** *Rescue News* Sept. 2/3 Both pots must have been buried on purpose, perhaps as charms but the bellarmine is thought to be rather too early to be a witch-bottle. **1955** *Times* 13 May 12/5 Along with salt-cellars, ash-trays, witch-bowls, and the curious jam dishes. **1964** G. Sims *Terrible Door* xiv. 75 What looked like a fish-bowl..was a 'witch bowl' with glass fishes suspended from floating glass bubbles in stagnant water. **1892** *Review of Reviews* Feb. 170/1 (*heading*) A plea for the witch-burners. **1909** *Strand Mag.* XXXVIII. 692/1 They had taken to witch-burning. **1928** G. Ade *Let.* 10 July (1973) 135 Sooner or later we should elect a Catholic to the Presidency just to prove that we are living in the 20th century instead of the 18th and that witch-burning and religious persecutions are no longer the pastimes of a free and intelligent people. **1921** M. Murray *Witch-Cult in Western Europe* v. 132 The round dance was..essentially a witch dance. **1971** *Country Life* 9 Sept. 633/3 The round reel survives from the prehistoric witch-dance. **1840** C. F. Hoffman *Greyslaer* II. 44 Tangled thickets of moss wood and wytch-hopple gave now the springy footing the tired hunter loves. **1943** R. Peattie *Great Smokies* 283 This is the hobblebush or witch hobble [*sic*], an abundant high-mountain shrub whose large rounding leaves reach their color peak in September. **1910** Kipling *Rewards & Fairies* 96 'What's a Witch-master?'..'A master of witches, of course.' **1931** V. Randolph in B. A. Botkin *Folk-Say* 86 My pappy follered gunsmithin' mostly, but he was a witch-master too. *a* **1944** J. Ford *Some Reminiscences of Danby Parish* (1953) 96 It was the custom of the Priest to cut the Roman X on the upright oak post which went up to the low ceiling... They came to be called 'Witch Posts'. **1957** E. E. Evans *Irish Folk Ways* v. 64 In north Yorkshire the jamb post, called the witch post, is occasionally found to be covered with designs intended to protect the hearth from evil spirits. **1971** K. Thomas *Religion & Decline Magic* xvii. 543 Other preservatives included 'witch-posts' built into the structure of the house. **1922** Joyce *Ulysses* 202 A Scotch philosophaster with a turn for witchroasting. **1937** H. G. Wells *Star Begotten* vii. 130 Some sort of world-wide witch-smelling for Martians everywhere... You could tell them because *instinctively* you dislike them. **1940** 'G. Orwell' *Inside Whale* 157 Frenzied witch-smellings after 'Bolshevism'. **1953** J. S. Huxley *Evolution in Action* vi. 141 The witch-smelling ordeals of Africa. **1855** G. Borrow *Jrnl.* 23 Aug. in *Exped. to Isle of Man* (1915) 8 What could those witch-stones be?

c. **witch's tit**: in fanciful proverbial phr. (*as*) *cold as a witch's tit*, extremely cold.

1932 Van Wyck Mason *Spider House* xviii. 210 It's cold as a witch's tit outside. **1974** *Times* 17 Aug. 7/3 It was cold as a witch's tit, so I sat there and shivered. **1980** R. L. Duncan *Brimstone* viii. 200 Just listening to a weather report...Albuquerque's clear but cold as a witch's tit.

witch (witʃ), *v.*² *U.S.* [f. Witch *sb.*²] *intr.* and *trans.* To dowse for water with a divining rod. Hence **wi·tcher**, a dowser.

1963 G. Thomson *Crocus Country* xi. 74 The term to 'witch for water' is said to come from the fact that it was usually done with a witch-hazel wand. *Ibid.*, The witcher would walk up and down in the general area where a well was needed, with the ends of a forked hazel twig held firmly in his hands. **1970** J. Blackburn *Land of Promise* ii. 32 The witcher came to a place where the stem of the willow could no longer be held upright. **1978** *Country Life* 7 Dec. 1953/3 He got a well-digger to survey the site... The first driller and others consulted all 'witched' the situation.

witch-doctor. Add: **2.** *Mil. slang.* A psychiatrist.

1966 *Listener* 29 Dec. 960/3, I did not again rub shoulders with..the fraternity until I entered the army.., where they were known..as 'trick-cyclists', 'head-shrinkers', or 'witch-doctors'. **1979** D. Anthony *Long Hard Cure* vi. 58 That sounds like one of your witch doctors at the Retreat.

Hence **witch-doctoring** *vbl. sb.*, **-doctory** (also *fig.*).

1924 Kipling *Debits & Credits* (1926) 182 All the cars I met were 'protected' [with a label] as mine was—till I reached..the limit of the witch-doctoring. **1927** G. B. Shaw in *Sunday Express* 7 Aug. 7/7 The dismal survivals of augury and witch-doctoring. **1944** J. S. Huxley *On Living in Revolution* iv. 45 As irreconcilable as is..witch-doctoring with preventive medicine, or number-mysticism with higher mathematics. **1962** *Observer* 13 May 6/5 'Organisation and management' was for a long time regarded—as one Treasury man put it—as 'third-rate witchdoctory'. **1972** H. A. Williams *True Resurrection* ii. 22 And this was called scientific medicine as opposed to all forms of witch-doctory. **1977** P. Johnson *Enemies of Society* xv. 197 Those who practise psychiatric medicine are in the position of early-nineteenth-century doctors, trying to get round as yet unsolved difficulties by witch-doctoring.

witchert, var. *Wichert.

witchetty. Add: Also (*rare*) **wichetty, widgety.** Substitute for def.: In full *witchetty grub*. A large white grub (the larva of certain moths and other insects) which infests the roots and stem of the *witchetty bush* (= Mulga 1 a in Dict. and Suppl.), from which it is extracted for use as food by Aboriginals and as bait by fisherman. (Further examples.)

1899 *Contemp. Rev.* Mar. 407 In the witchetty grub totem this sacred painting tallies with..a stone kist at Tillicoutry. **1935** H. H. Finlayson *Red Centre* iii. 30 The broad-leafed mulga or witchetty bush, the roots of which harbour a grub beloved by the blacks. **1944** F. Clune *Red Heart* 37 The sun gleamed on a motor-bike beneath a clump of witchetty bushes. **1954** B. Miles *Stars my Blanket* viii. 50 The widgetty grub tree at the roots of which the natives dig for the grubs which are like fat white slugs. **1960** *Times* 5 July 11/7 'Witchetty' bushes, a kind of wattle with bright yellow blossom. **1961** P. White *Riders in Chariot* xi. 373 You look to me..like you was made out of old witchetty grubs. **1962** *Oxf. Univ. Gaz.* 19 Mar. 849/1 It is a pointed, oval, red-ochre-coloured, wooden bullroarer with designs of witchetty-grubs. **1968** M. Pyke *Food & Society* iv. 42 Witchetty grubs are famous as an article of diet eaten by aborigines. **1977** C. McCullough *Thorn Birds* viii. 178 There were witchetty grubs, fat and white and loathsome.

wi·tch-hunt, *sb.* Also **witch hunt, witchhunt.** [Witch *sb.*²] **1.** A search for witches, or for someone suspected or accused of witchcraft.

1885 R. Haggard *K. Solomon's Mines* x. 151 To-night ye will see. It is the great witch-hunt, and many will be

smelt out as wizards and slain. **1927** J. Buchan *Witch Wood* xvi. 272 David had..seen a witch hunt..as a boy—and then there had been a furious and noisy crowd. **1960** D. Hudson *Forgotten King* 17 Elizabethan inns and beards and witch hunts give place to the coffee house and the Restoration theatre. **1975** A. Fraser *Whistler's Lane* 9 Her mind had been..on the witch hunts of the early seventeenth century.

fig. **1915** 'I. Hay' *First Hundred Thousand* xiii. 178 Platoon commanders were bidden to hold a witch hunt, and smell out a chiropodist.

2. a. A single-minded and uncompromising campaign against a group of people with unacceptable views or behaviour, *spec.* communists; *esp.* one regarded as unfair or malicious persecution.

1938 'G. Orwell' *Homage to Catalonia* xi. 241 Rank-and-file Communists everywhere are led away on a senseless witch-hunt after 'Trotskyists'. **1947** *Partisan Rev.* XIV. 344, I don't like Stalin's methods, but I shall never, never join in that witch-hunt. **1950** *Here & Now* (N.Z.) Dec. 8/1 Inside the Labour Party there was a witch-hunt of unbelievable viciousness against the Government's critics. **1958** *Times Lit. Suppl.* 21 Nov. 669/4 The story of a security officer in America in the days when McCarthy witch-hunts were frequent and when communists lurked..under every bed. **1972** *Guardian* 31 Aug. 6/7 Delegates to the annual Conference at the TUC at Brighton next week are urged..not to indulge in a witch hunt..when discussing the 34 affiliated unions which have remained on the register of trade unions. **1976** *Survey* Spring 179 Literary zealots..then took part in the anti-zionist and anti-revisionist witch-hunt. **1977** *Gay News* 24 Mar. 3/1 During the operation—labelled a 'witch-hunt' by the local gay community—28 men were arrested. **1977** *Times* 28 Apr. 2/1 Mr Orme, Minister for Social Security,..said he was not prepared to countenance a witch-hunt against claimants. **1979** A. Price *Tomorrow's Ghost* vii. 120 We must be absolutely fair... This isn't a witch-hunt. **1983** P. Ustinov *My Russia* i. 8 It is fashionable today to conduct a moderate witchhunt for that pro-Soviet bunch of Cambridge undergraduates..who spied for Russia.

b. A campaign *against* an individual.

1960 *Daily Tel.* 29 Jan. 1/2 The Opposition Front Bench do not intend to conduct a 'witch-hunt' against Mr Marples over his business connections. **1973** C. Bonington *Next Horizon* i. 20 The argument had developed into a witch-hunt against Barrie with, I suspect, very little justice. **1977** *Daily Mirror* 30 Mar. 31/1 After the Germans had strolled home 5-1, the controversial Neale accused non-playing captain Peter Simpson of leading a 'witch-hunt' against him.

wi·tch-hu:nter. **1.** = Witch-finder a.

1867 Harland & Wilkinson *Lancs. Folk-Lore* i. 184 Dr. John Webster (who detected Robinson, the Lancashire witch-hunter).

2. One who takes part in or publicly advocates a witch-hunt (sense *2).

1935 *New Republic* 19 June 158 (*heading*) Witch-hunters at work. **1940** H. L. Ickes *Diary* 22 Feb. (1954) III. 139 Probably the witch-hunters are largely responsible for this. No one likes to be called a Communist and yet that is what every liberal has to submit to. **1960** *Encounter* Mar. 78/2 It is..necessary to that public to learn that its witch-hunters are corrupt. **1980** J. O'Faolain *No Country for Young Men* xvi. 347 Obsessed as a pair of witch-hunters, minds zipping along their single track, they challenged him.

wi·tch-hu:nting, *vbl. sb.* **1.** The activity of seeking out witches and obtaining evidence against them.

1640 B. Jonson *Sad Shepherd* ii. vii, in *Workes* II. 149 You speake, Alken, as if you knew the sport of Witch-hunting, Or starting of a Hag. **1885** Rider Haggard *King Solomon's Mines* xv. 249 Ignosi..reaffirmed the promises..that witch-hunting should cease. **1935** B. Russell *Relig. & Sci.* iv. 99 In New England, a fierce outbreak of witch-hunting occurred at the end of the seventeenth century. **1950** Auden *Enchafèd Flood* (1951) ii. 51 The actual horrors of persecution, witch-hunting, and provincial superstition from which they were trying to deliver mankind. **1981** M. Warner *Joan of Arc* v. 114 Double-think is..endemic to the business of witch-hunting, for..the witch-hunter is the alleged witch's most committed believer.

2. Participation in or advocacy of a witch-hunt (sense *2).

1932 J. F. Carter *What we are about to Receive* xviii. 204 Once the election is over..we shall quietly lay aside our witch hunting. **1943** G. Greene *Ministry of Fear* II. i. 125 You can't avoid witch-hunting in war-time. **1968** *Daily Tel.* 16 Nov. 16/3 We have been treated to a plethora of half-truths, innuendo, witch-hunting and ignorance. **1977** *Socialist Press* 2 Mar. 5/5 Despite this combination of administrative hysteria, right wing witch hunting and Stalinist betrayal, the fees issue is still alive at Essex.

Also as *ppl. a.* Hence (as a back-formation) **wi·tch-hunt** *v.*, (*a*) *trans.*, to subject to a witch-hunt (sense *2); (*b*) *intr.*, to take part in a witch-hunt.

1889 W. H. D. Adams *Witch, Warlock & Magician* II. v. 402 Our witch-hunting King offers an explanation of a peculiarity which..our readers have already noted. **1946** *Sun* (Baltimore) 19 July 20/1 The War Department hasn't gone off half-cocked to 'witch hunt, red bait or to bust' unions. **1948** 'J. Tey' *Franchise Affair* v. 49 Give those Midland morons a good excuse and they'll witch-hunt with the best. **1960** *Twentieth Century* Apr. 380 Assorted political personages (including a witch-hunting Senator). **1960** *News Chron.* 29 June 6/7 That uncouth, witch-hunting and paederastic gowk [*sc.* James I]. **1975** *Listener* 4 Dec. 754/2 David Niven..has severe words for it [*sc.* Hollywood]..for letting itself be..libelled and witch-hunted by..gossip columnists. **1980** *Jrnl. R. Soc. Arts* Mar. 180/1 There is..a tendency to witch-hunt when any disaster happens. **1983** W. McIlvanney *Papers of Tony Veitch* x. 59 'When did you join the vigilantes, Jack?' 'Never. I'm not witch-hunting whoever did it.'

witchit, var. *Wichert.

witch knot. Add: **1. b.** A knot tied for the purpose of making or averting a spell.

1884 A. Lang in M. Hunt *Grimm's Household Tales* I. p. xlvi, All over the world savages..tie 'witch-knots'. **1947** A. Runeberg tr. E. H. Meyer in *Witches & Demons in West-European Folk Belief* vii. 95 Witches..twist twigs into witchknots, and leave the fairy rings in the grass after their dances. **1957** E. E. Evans *Irish Folk Ways* xxi. 304 The cow-doctor uses sympathetic magic..by drawing apart over the animal's back the loose ends of a string tied in a complicated witch-knot or 'bat',..which comes undone when pulled.

witchy, *a.* Delete *rare* and add later examples. Also, characteristic or suggestive of a witch.

1968 S. Plath in *Atlantic Monthly* Sept. 54/2 Over the trees at the far side of the Common the..torch flare flattens and recovers under some witchy invisible push. **1975** M. Drabble *Realms of Gold* III. 241 It [*sc.* a figurine] had a witchy, androgynous, yet friendly look. **1976** *Listener* 4 Nov. 590/2 The witchy black of the Grimm forests.

witful. Restrict *rare* to sense 1 and add: **2.** (Later examples.)

1924 *Countries of World* I. 639/2 The women deck these stands with witful skill. **1935** W. de la Mare *Early One Morning* 321 A willing and witful child. **1980** *N.Y. Times* 26 Aug. c-8/5 Ravel's Sonato for Violin and Cello... For pinpoint gestural precision and witful repartee, this sophisticated instrumental dialogue is an unending source of fascination.

witgat (vi·tɣat). *S. Afr.* [Afrikaans, f. *wit* white + *gat* hole.] Any of several trees of the genus *Boscia*, which have pale bark and are found in dry areas of southern Africa, esp. the evergreen *B. albitrunca*. Also *witgatboom* [Afrikaans *boom* tree].

1824 W. J. Burchell *Trav. Int. S. Afr.* II. 18 Their trunks..appeared at a little distance as if they had been whitewashed. From this singular character, they have gained the name of *Wit-gat boom*. **1860** Harvey & Sonder *Flora Capensis* I. 362 A tall tree called *Witgat* by the colonists. **1932** C. Fuller *Louis Trigardt's Trek* 47 We were standing under a *witgatboom*. **1966** E. Palmer *Plains of Camdeboo* iv. 54 Close to this grew a witgat tree not fifteen feet tall but with a thick, seamed, milk-white trunk and a dense grey-green crown of tough little leaves. *Ibid.* xvii. 271 They..made coffee from the roots of the witgat. **1972** Palmer & Pitman *Trees S. Afr.* I. 620 Witgat wood is heavy and tough. **1973** Y. Burgess *Life to Live* 25 Coffee made from the roots of the witgat tree.

with, *prep.*, (*adv.*, *conj.*). Add: **II. 9. e.** *what's with——?* what about, what are the circumstances of?; how are things with, what's the matter with? *colloq.* (orig. *U.S.*).

1940 J. O'Hara *Pal Joey* 125 What's with the free food? Explain. **1962** E. Linington *Extra Kill* viii. 122 He says..'What's with Whalen?' When he hears Whalen's out, he gets mad. **1969** 'V. Packer' *Don't rely on Gemini* (1970) viii. 62 'What's with you and these long baths?' Archie asked. **1978** K. Amis *Jake's Thing* xv. 158 What's with Jake is that he can't get it up any more, and what's with Brenda is she thinks it's her fault for having gotten middle-aged and fat.

22. a. Also in phr. *with us, them,* alive, still living.

1611 Bible *Acts* ix. 39 All the widowes stood by him weeping, and shewing the coats and garments which Dorcas made, while shee was with them. **1961** E. Williams *George* xxiii. 391 That Mr Bellis, he's gone dead, not that he had that much life in him when he was with us, poor fellow. **1966** *Listener* 3 Feb. 166/2 The English church at Shiraz..was built entirely thanks to the enthusiasm of the then incumbent, who was a very learned man (is indeed still with us).

e. *to be with* (a person), to follow his line of reasoning, to keep up with and understand his explanation, instructions, etc. *colloq.*

1900 F. P. Dunne *Mr. Dooley's Philos.* 248 We keep our thoughts fixed upon th' inanity iv th' finite in comparison with th' onthinkable truth with th' ondivided an' onimaginable reality. Boys ar-re ye with me? **1933** *Punch* 8 Feb. 150/3 'Does it look to you as if the Boss keeps his skeletons in any ice-boxes around here?' 'Huh?' she said, not quite with me yet. **1955** N. Fitzgerald *House is Falling* xi. 190 'I'm with you so far,' said Hugh who for the first time was really giving his mind to the problem. **1977** T. Heald *Just Desserts* viii. 188 'She will have to be taken in hand. Which..has been the point of the exercise all along.' 'I'm not with you.'

f. *to be with it*, to be within a particular fashionable or exclusive group or set, to be up-to-date or *au fait* with the latest news, ideas, etc.; to be mentally alert. Also *to get with it*, to become informed or up-to-date, etc. *slang* (orig. *U.S.*). Cf. *With-it *a.*

1931 *Amer. Mercury* Nov. 353/2 *Not with it*, said of an outsider. 'He's not with it.' **1959** R. Condon *Manchurian Cand.* (1960) vii. 108 They are with it, Raymond. Believe me, they are even away ahead of me. **1960** *Guardian* 9 Dec. 13 The new *Time and Tide*, to borrow the language of the teen-ager, is 'with it'. **1961** J. O'Hara in *Assembly* 159 Bud come to see you, especially when you had a chance of winning? Get *with* it, boy. **1971** *Daily Mail* 6 May 24/4 Horne made a strong attempt to get with it. Result: the stronger emphasis on fashionwear. **1976** *Jrnl. R. Soc. Arts* CXXV. 17/1 The need to be in fashion—in the swim—up to date—'with it'—might not have been the least of our driving forces for general progress. **1981** M. Doody in Martin & Mullen *No Alternative* iv. 37 What is 'with it' for one or two generations can seem palpable folly a hundred years later. **1985** W. J. Burley *Wycliffe & Four Jacks* vii. 149 There's an old man, living in a home... He's quite with it—I mean he's mentally alert.

24. d. *with-profit*(*s*) adj., of a life assurance policy: allowing the insured to receive a share of the profits of the insurance company, usu. in the form of a bonus. Also applied to holders of such policies, the associated payments, etc. Cf. *without-profit*(*s*) adj. s.v. *Without *prep.* 7 c.

1924 Tayler & Tyler *Life Assurance* ii. 19 Nearly all the offices transacting life assurance business issue two great classes of policy—(*a*) Those which share in the profits, known as 'with-profit' or 'participating' policies; and (*b*) Those which do not share in the profits, known as 'without-profit' or 'non-participating' policies. **1944** S. d'E. Colam *Life Assurance for Agents* 16 Premiums for with profit policies are larger than for without profit policies. **1950** *Economist* 18 Nov. 840/2 In trying to assess the relative merits of with-profit contracts, the only firm basis of comparison is the actual amount paid. **1961** *Observer* 10 Dec. 4/3 A modest writing-up of book values would add to the good times which with-profit-holders can expect in the future in the form of bumper bonuses. **1965** *Economist* 24 July p. xxii/2 These profits go to the with-profits policyholders, as well as the profits on the with-profits business itself. **1979** *Financial Times* 20 Jan. 7/6 If you are shopping around for a with-profits policy, take a look at how insurance brokers work out projections of maturity value. **1982** *Equity & Law Life Assurance Co. Ann. Rep. 1981* 16 The rate of terminal bonus depends on the year of entry as a with-profit benefit and is applied to the with-profit sum (or annuity) and attaching bonus.

withdrawal. Add: **4. b.** *Psychol.* The state or process of psychic retreat from objective reality or social involvement; also *transf.*

1916 C. E. Long tr. *Jung's Coll. Papers on Anal. Psychol.* vi. 203 Autistic withdrawal into one's own phantasies is what I formerly designated as the obvious overgrowth of the phantasies of the complex. **1937** K. Horney *Neurotic Personality of our Time* v. 98 The fourth means of protection [*sc.* against the basic anxiety] is withdrawal. **1957** P. Lafitte *Person in Psychol.* xii. 181 Withdrawal and regression..are general classes of behaviour as compared to the social specificity of striving for advancement or mendacity. **1970** Touraine & Pécaut in I. L. Horowitz *Masses in Lat. Amer.* iii. 90 At the lowest level, we find *withdrawal*, characterized by a rejection of the industrial world. **1973** *Jrnl. Genetic Psychol.* June 315 One type of children's fantasy is 'withdrawal'.

5. Cessation of use or provision of a drug; *spec.* the interruption of doses of an addictive drug, with resulting craving and physical reactions.

The 1897 example is an isolated one.

1897 [see Injector 2]. **1929** D. Hammett *Dain Curse* xxi. 237 Tears were one of the symptoms of morphine withdrawal. **1965** Wilner & Kassebaum *Narcotics* vi. 96 Withdrawal of morphine by substitution and subsequent withdrawal of methadon. **1972** *Nature* 22 Dec. 443/1 Dr R. Ericsson..suggested that the criteria for a successful antispermatogenic drug were..return of fertility on withdrawal, and normal libido. **1977** *Lancet* 29 Jan. 255/1 Any doctor prescribing or patient receiving this potent drug should consider carefully the effect of withdrawal which has not, to my knowledge, been researched.

6. = *coitus interruptus* s.v. *Coitus.

1889 W. T. Stead *Diary* 20 Jan. in J. W. R. Scott *Life & Death of Newspaper* (1952) xix. 244, I have from the birth of Willie practised simple syringing with water. Of late always withdrawal. **1923** M. C. Stopes *Contraception* iv. 48 Vaginal stimulation consummating the ejaculation after withdrawal, commonly called '*coitus interruptus*'. **1963** M. McCarthy *Group* iii. 66 What method of contraception had been used..? 'Withdrawal,' murmured the doctor. **1978** G. Cunningham *New Woman & Victorian Novel* 6 Such methods as the safe period, the sheath, the sponge..and withdrawal.

7. *attrib.* and *Comb.*, as (sense *5) *withdrawal pain, period, syndrome*, etc.; **withdrawal slip**, a form which must be filled in when withdrawing money from a bank or other place of deposit; **withdrawal symptom**, an unpleasant physiological reaction resulting from the process of ceasing to taken an addictive drug; usu. *pl.*; also *fig.*

1924 *Brit. Jrnl. Inebriety* XXI. 88 The *withdrawal symptoms* of addiction disease. **1929** Light & Torrance in *Arch. Internal Med.* XLIV. 11 The general behavior and symptomatology of these addicts were uniform during the forty-eight hour withdrawal period. *Ibid.* 14 Addicts will admit that when they are unable to obtain drugs and when withdrawal symptoms..become severe, the assurance of an available supply at a considerable distance will cause them to travel..with remarkable speed and efficiency. **1961** *Lancet* 23 Sept. 677/1 A steroid withdrawal syndrome occurs in patients who have stopped corticosteroid therapy. **1962** *Ibid.* 6 Jan. 54/2 Lobeline sulphate closely resembles nicotine in many of its pharmacological actions. Using it as a substitute for tobacco during the withdrawal phase, Dorsey found the results 'encouraging'. **1962** 'K. Orvis' *Damned & Destroyed* xvi. 113, I had seen that stare and twitching frequently enough now..to be able to tag it as an indication of the degree of withdrawal sickness. 1965 *New Statesman* 7 May 716/1 Often these women directly sabotaged the programme. One flushed her son's withdrawal medication down the lavatory. *Ibid.* 3 Dec. 866/2, I asked him how long it was since the withdrawal pains had stopped. 1966 C. E. Israel *Hostages* 94 She wouldn't have had the dreg end of her withdrawal agonies yet. 1967 *Guardian* 7 Dec. 1/6 Methedrine has been used in heroin withdrawal treatment. 1970 G. F. Newman *Sir, You Bastard* v. 140 Morgan was entering the withdrawal stage and would soon be requiring another intravenous dose. 1970 G. Greer *Female Eunuch* 276 Mrs J.S. used up two supplies of pills in all innocence, and then discovered that she had withdrawal symptoms. 1973 'E. McBain' *Let's hear It* xv. 226 On the withdrawal slip before him, he wrote the date, and the number of his account, and then he filled in the amount. 1976 *Times* 18 Oct. 3/7 Sir Harold Wilson..the former Prime Minister..says he has suffered no 'withdrawal symptoms' since resigning. 1979 F. Olbrich *Sweet & Deadly* ix. 110 The bank manager..showed Ramesh the withdrawal slip for four thousand rupees.

withdrawn, *ppl. a.* Add: Also *spec.* in *Psychol.*, characterized by isolation and loss of contact with objective reality. Cf. *WITHDRAWAL 4 b.
1932 *Smith. Coll. Stud. Social Work* III. 145 Only four of the patients..were of the..type often thought to be common among the pre-schizophrenic... Perhaps children of the withdrawn, timid type are only rarely referred to a child guidance clinic. 1950 *Times* 12 May 7/7 The emotional re-education of the severely withdrawn type was at best a tricky business, for which frequent consultation between the school staff and the psychologist was advisable. 1971 H. Kohut *Analysis of Self* ix. 243 He not only would tend to become generally withdrawn..and diffusely depressed, but..also manifested a striking change in his dream pattern.

withdraw·nness. [f. prec. + -NESS.] Withdrawn or retired character.
1927 *Public Opinion* Feb. 102/1 He has the curiosity and interest of a young man and has none of the settled habits and introspection and withdrawnness of an old man. 1976 *Classical Q.* XXVI. 161 A more general characteristic of the Homeric style, the restrained objectivity and aristocratic withdrawnness.

withe, with, *sb.* Add: **5.** *withe axe*; **withe-rod,** a deciduous shrub, *Virburnum nudum*, native to North America and bearing clusters of small white flowers; also, a thin flexible twig from this or a similar shrub; (earlier and later examples).
1819 Keats *Let.* 5 Sept. (1958) II. 156 At the days end his thoughts will run upon a withe axe if he ever had handled one. 1776 G. Cartwright *Jrnl.* 19 Oct. (1792) II. 215 The people came down from the lodge, and brought..a bundle of white-rods [*sic*]. 1846 G. B. Emerson *Rep. Trees & Shrubs growing in Forests Mass.* 364 The Naked Viburnum. Withe Rod... A slender, erect shrub. 1943 R. Peattie *Great Smokies* 265 We recognize the..withe rod..and wintergreen.

wither, *v.*[2] Add: **3. b.** *spec.* in phr. *to wither away*, used with reference to the belief held in Marxist philosophy that when the dictatorship of the proletariat has effected the necessary changes in society, the state will eventually cease to be necessary and will therefore disappear; also used allusively or generally. So *withering away*.
1919 tr. *Lenin's State & Revol.* i. 21 Engels speaks here of the *destruction* of the capitalist State by the proletarian revolution, while the words about its withering away refer to the remains of a *proletarian* State *after* the Socialist revolution. *Ibid.* 22 Only the proletarian State or semi-State withers away after the revolution. 1935 E. Burns tr. *Engels' Anti-Dühring* III. ii. 315 The government of persons is replaced by the administration of things and the direction of the process of production. The state is not 'abolished', it withers away. 1937 *Times* 7 July 17/6 The Marxist theory of the 'withering away' of the State. 1948 M. Laski *Tory Heaven* v. 81 Reynolds is an M.I.5 nark... Eventually, they say, all that sort of thing will just wither away. 1971 *Guardian* 9 Sept. 13/1 Stormont was designed to wither away. It was invented in the hope that the two parts of Ireland would become united within the British Empire. 1980 D. Fernbach tr. *Buci-Glucksmann's Gramsci & State* xii. 285 The transition from an inevitable 'productivist' phase to an integral state thus takes place by way of hegemony and the distant tendential perspective of a withering away of the state. *Ibid.* 289 A state that withers away to the extent that its function withers away.

witherling[1]. For † *Obs.* read *Obs.* exc. *arch.* and add later example.
1922 W. Stevens *Let.* 21 Dec. (1967) 232, I have omitted many things, exercising the most fastidious choice, so far as that was possible among my witherlings.

withers, *sb.* Add: **c.** *wither pad.*
1963 E. H. Edwards *Saddlery* xv. 112 Numnahs and wither pads are used in conjunction with saddles. 1976 *Horse & Hound* 3 Dec. 52 (Advt.), The John Ayres New Zealand Rug... Featuring a sheepskin wither pad.

withershin(s), widdershin(s), *a.* [f. the adv.] Moving in an anticlockwise direction, contrary to the apparent course of the sun (considered as unlucky or sinister); unlucky, ill-fated, relating to the occult.
1926 D. H. Lawrence *Plumed Serpent* vi. 112 She made up her mind, to be alone, and to cut herself off from all the mechanical widdershin contacts. *Ibid.*, He, too, was widdershins, unwinding the sensations of disintegration and anti-life. 1936 Dylan Thomas *Twenty-Five Poems* 16 Shall I still be love's house on the widdershin earth, Woe to the windy mansions at my shelter? 1973 G. M. Brown *Magnus* vi. 112 There is a black joy abroad, a dance of the deadly sins, a withershin rout. 1976 *Early Music* Oct. 399/1 The sentiments and rituals of the court can be grotesquely guyed by the spirits (widdershins dances, sick-caricature mimes to accompany the Sorceress's prophecies and provoke those ho-ho outbursts, etc.).

withholding, *vbl. sb.* Add: **2.** Special Comb.: **withholding rate** *U.S.*, the rate for a withholding tax; **withholding table** *U.S.*, a table showing amounts of tax to be deducted from a dividend payment, salary, etc.; **withholding tax** orig. *U.S.*, a tax deducted at source, *spec.* one levied by some countries on interest or dividends paid to a person resident outside that country.
1972 *Time* 17 Apr. 43/3 Spending has been held back in part because of a colossal blooper by the House Ways and Means Committee in setting the new withholding rates. 1976 *Billings* (Montana) *Gaz.* 30 June 6-A/1 The House unanimously passed and sent to President Ford Tuesday a two-month extension of current lower income tax withholding rates. 1947 *Sun* (Baltimore) 15 May 2/8 The Finance Committee halved the House bill for the current year and made the new withholding tables effective as of July 1. 1940 *U.S. Federal Rep.* 2nd Ser. CXII. 1000/2 Intra-company payments designated as 'interest' would not be so regarded..for the purpose of the withholding tax. *Ibid.*, The principal amount..was due complainant on account of withholding taxes. 1950 *Tax Cases* XXXIII. 346 The Appellant received her arrears of interest as follows: In June, 1943... $18,000 Less: U.S. withholding tax at 30 per cent... 5,400. 1960 I. Wallach *Absence of Cello* (1961) 7 Will you tell me why the hell you never paid the withholding taxes for your employees? 1971 *Financial Mail* (Johannesburg) 26 Feb. 717/1 Interest accruing to non-residents of the Republic is subject to deduction of a withholding tax at the rate of 10 per cent, exemption from the tax having been granted in respect of accruals of interest amounting to R20 or less in any one year. 1979 *Daily Tel.* 27 Oct. 27/1 Many foreign countries have tax laws, which, in principle, require the foreign payer of the dividends or interest to deduct a withholding tax when making the payment to a non-resident. 1984 *Times* 5 Oct. 25/1 The German withholding tax was introduced in 1965 to prevent an overvaluation of the Deutschemark injurious to exports.

within, *prep.* Add: **6. d.** *a story within a story* and varr., a story, performance, etc., complete in itself but occurring within another. Cf. *play within a play* s.v. *PLAY *sb.* 14 a.
1961 Webster s.v. *Within*, A musical within a musical. 1971 J. Gores in 'E. Queen' *Magicians of Mystery* (1976) 162 A new kind of procedural detective story..it uses the dream 'story-within-a-story' which antedates even.. The Vision of Pierce Plowman. 1976 C. Bermant *Coming Home* II. ii. 125 A plump, bespectacled woman..grasped him in a tearful embrace. Was this a drama within a drama, a man who had thought he'd lost his wife and would rather that she had stayed lost? 1978 *Listener* 19 Jan. 86/3 Fitzgerald was featured creating one of his.. short stories... This device allowed for a film within a film. 1984 B. Paul *Renewable Virgin* ii. 38 There was some sort of a crime-within-a-crime just waiting to be discovered.

within (wiðı·n), *sb.* [f. the adv.] That which is within or inside (esp. *fig.*).
1912 J. Stephens *Crock of Gold* xiii. 166 It [*sc.* anger] is not the beneficent blindness which prevents one from seeing without, but it is that desperate darkness which cloaks the within, and hides the heart and the brain from each other's husbandly and wifely recognition. 1938 L. MacNeice *Mod. Poetry* 28 Wyndham Lewis maintains that it is the artist's or writer's business to depict the Without of people and not their Within. 1973 *Times* 26 Nov. 15/8 Having every intention of looking again and again before the exhibition finally departs for its permanent home in the 'Great Within', or wherever—I feel I must compliment the compilers of the excellent catalogue.

within-door, *adv. phr.* (*a.*) For † *Obs.* read Now *rare* and add later examples.
1884 Tennyson *Becket* I. i. 35 They [*sc.* moths] burn themselves *within*-door. 1954 M. Sharp *Gipsy in Parlour* xxi. 200 All female within-door work had been properly done.

with-it (wi·ðit), *a. slang.* [f. vbl. phr. *to be with it*: see *WITH *prep.* 22 f.] Fashionable, up-to-date.
1962 *Listener* 29 Nov. 909/2 Curtain designs for the really with-it 'contemporary home'. 1963 [see *FAR-OUT *a.* b]. 1970 J. G. Vermandel *Dine with Devil* ix. 52 The with-it Mr. Angel enjoyed a more subtle turn of mind. 1977 J. I. M. Stewart *Madonna of Astrolabe* v. 94 The silly woman just thought it a with-it thing to say to a celebrated dramatist.
Hence **wi·th-it-ness.**
1963 *Punch* 22 May 752/1 The headlong rush of..social notabilities to win themselves the TW badge of with-it-ness. a 1974 R. Crossman *Diaries* (1976) II. 445 This has all paid off in terms of the audience ratings where the B.B.C. has been doing well in the last six months, winning the battle for the audience by its with-itness.

withness. (Earlier and later examples.)
1904 W. James *Ess. Rad. Empiricism* (1912) ii. 47 This imperfect intimacy, this bare relation of *withness* between some parts of the sum total of experience and other parts. 1929 A. N. Whitehead *Process & Reality* II. ii. 88 The account..traces back these secondary qualities to their root in physical prehensions expressed by the 'withness of the body'. 1946 *Sci. & Society* X. 244 Conversely union and withness, the *ta-tong* of Chinese thought, has been the aim of democracies. 1962 *Times* 11 Dec. 11/4 The 'withness' is all.

without, *adv., prep., conj.* Add: **B.** *prep.* **III. 7. c.** *without profit(s)* adj., of a life assurance policy: providing normal cover but not allowing the insured to receive a share of the profits of the insurance company. Also applied to the associated funds, business, etc. Cf. *with profit(s)* adj. s.v. *WITH *prep.* 24 d.
1924, 1944 [see *WITH *prep.* (*adv., conj.*) 24 d]. 1960 *Times* 24 Oct. (Financial Rev.) p. xiii/4 For without-profits contracts are tending to come down. 1965 *Economist* 24 July p. xxii, When interest rates are high and there is significant inflation, profits on the without-profits businesses are high, since the premiums were originally fixed on the basis of lower money returns than are now being earned. 1982 *London Life Association Ann. Rep.*, Total without profit funds.
C. *conj.* **2.** Also, chiefly in U.S. dial. use: unless, without its being the case that.
1867 J. R. Lowell *Biglow Papers* 2nd Ser. p. lvii, I don't git much done 'thout I bogue right in along 'th my men. 1903 'T. Collins' *Such is Life* (1937) i. 51 A man shouldn't make a dog of his self without he's well paid for it. That's my religion. 1955 F. O'Connor *Wise Blood* iii. 52 Everything she looked at was that child... She couldn't lie with that man without she saw it. 1962 E. Albee *Who's Afraid of V. Woolf?* (1964) i. 51 Man can put up with only so much without he descends a rung or two on the old evolutionary ladder. 1984 A. Carter *Nights at Circus* ii. 46 No two deaths without a third follows.
D. *sb.* For *nonce-use* read *rare.* (Further example.)
1938 [see *WITHIN *sb.*].

witless, *a.* Add: **7.** Alluding to a state of extreme fear. Esp. in *colloq.* phr. *to be scared witless.*
1975 D. Bagley *Snow Tiger* 19 It's the last job he'll ever have and he's scared witless that he'll lose it. 1982 S. Brett *Murder Unprompted* ii. 19 'How are you feeling?'.. 'Scared witless, darling.'

witness, *sb.* Add: **I. 8. b.** = *Jehovah's Witness* s.v. *JEHOVAH 2. orig. *U.S.*
1931 *Watchtower* 15 Oct. 316/2 If any one does become fearful and ceases to be a witness, he ceases to be of the remnant and of God's anointed or Christ. 1935 *Time* 18 Nov. 59/1 By last week 28 Witnesses of Jehovah had popped up in the U.S. public schools. Cora Foster..faced dismissal after confessing that she, too, was a Witness. 1974 *Watchtower* 15 Jan. 56/1 Suddenly, under religious animosity, the young man whipped out a knife and stabbed the Witness to death. 1980 R. Hill *Spy's Wife* ii. 8 Charity collectors went away happy, and..even Mormons and Witnesses had got enough courtesy to bring them back.
III. 15. witness-stand (earlier example).
1853 Thoreau *Let.* 10 Apr. (1958) 304 Expect no trivial truth from me, unless I am in the witness-stand.

witogie (vitŭ·χi). *S. Afr.* Also **witoogie, witteoogje.** [Afrikaans, f. Du. *wit* white + *oog* eye + *-ie* diminutive suffix.] Any of several birds of the genus *Zosterops* found in southern Africa, esp. *Z. pallidus* (formerly *Z. capensis*).
1867 E. L. Layard *Birds S. Afr.* 116 Zosterops Capensis..Witteoogje, lit. white eye. 1936 E. L. Gill *First Guide S. Afr. Birds* 37 Witogie... The Cape White-eye sings all through the summer. 1949 *Cape Argus* 15 Oct. (Mag. Sect.) 2/7 Those pretty little birds known as white-eyes or witogies..are well-known in most parts of the country as small green or yellowish birds with a characteristic circle of white feathers round each eye. 1957 *Cape Times* 11 Dec. 11/2 This burly bird has had a couple of twittering witoogies in close attendance. 1963 M. Kavanagh *We Merry Peasants* x. 110 The tiny witogies have for their own use a fruit-laden pomegranate tree.

witteboom. (Earlier and later examples.)
1799 A. Barnard *Let.* 4 Apr. in *Lett. Lady Anne Barnard to Henry Dundas* (1973) 185 Her Ladyship..is soon to present the Regiment with their colors [*sic*], in which the Whittebomb [*sic*]..is Happily blended and united with the Royal oak. 1926 C. G. Botha *Our S. Afr.* (1938) 85 Witteboomen..is the name of the well known silver trees found in the Cape Peninsula. 1972 Palmer & Pitman *Trees S. Afr.* I. 493 Witteboom..is believed to grow naturally only in the Cape Peninsula.

witter, *v.*: see *WHITTER *v.* 3.

Wittgensteinian (vi·tgənstəi·niăn), *a.* and *sb.* [f. the name of the Austrian-born philosopher Ludwig *Wittgenstein* (1889–1951) +

-IAN.] **A.** *adj.* Of, pertaining to, or characteristic of Wittgenstein, or his theories or methods. **B.** *sb.* An adherent of Wittgenstein's ideas.
1946 *Mind* LV. 25 Unfortunately, for the outsider there exists no official and adequate statement of the Wittgensteinian technique. *Ibid.*, 'W—ns' will be used for 'Wittgensteinians'. 1954 [see *RUSSELLIAN *a.* and *sb.*]. 1966 D. Jenkins *Educated Soc.* iii. 140 The Wittgensteinian line of concluding that when one cannot profitably speak one must perforce be silent. 1969 T. F. Torrance *Theol. Sci.* i. 19 To use Wittgensteinian language, are these 'images' 'pictures' or 'tools'? 1973 *Listener* 4 Jan. 21/2 A hard-line defence, by a leading Wittgensteinian, of the Pope's pronouncement on contraception. 1980 *Times Lit. Suppl.* 20 June 714/2 Professor Wright..picks up some Wittgensteinian themes and explores how they might be developed. *Ibid.* 714/4 The question of whether ..a Wittgensteinian can defend the distinction between necessary and contingent propositions.

Wittig (vi·tiχ, -ig). *Chem.* The name of Georg Friedrich Karl *Wittig* (b. 1897), German chemist, used *attrib.* to designate various synthetic techniques introduced by him, as **Wittig reaction,** a method for the preparation of substituted alkenes utilizing the action of an alkyl phosphorus ylide on a carbonyl compound (aldehyde or ketone); **Wittig rearrangement,** the conversion of benzyl or allyl ethers in the presence of a strong base to the corresponding secondary or tertiary alcohol.
1951 *Jrnl Amer. Chem. Soc.* LXXIII. 1437 The Wittig rearrangement of benzyl ethers by lithium phenyl. 1956 *Chem. Abstr.* L. 6443 The previous attempt to prep. a model vitamin D triene by the Wittig reaction..has been continued. 1974 Gill & Willis *Pericyclic Reactions* vi. 195 This elegant experiment proved that the migration involves a supra-supra interaction in Wittig rearrangements. 1979 *Sci. Amer.* Dec. 74/1 Vitamin A is synthesized industrially using the Wittig reaction. 1980 *Chem. in Brit.* XVI. 466/3 Vitamin A acetate is produced industrially via the Wittig synthesis.

wittite (vi·təit). *Min.* [ad. Sw. *wittit* (K. Johansson 1924, in *Ark. f. Kemi, Mineral. och Geol.* IX. IX. 2), f. the name of Th. *Witt*, Swedish mining engineer: see -ITE[1].] A mineral containing lead, bismuth, selenium, and sulphur and occurring as grey monoclinic crystals.
1924 *Mineral. Abstr.* II. 340 Wittite resembles molybdenite in appearance. 1980 [see *WEIBULLITE].

wiv (wiv). Representation of a vulg. pronunc. (esp. Cockney) of WITH *prep.* (*adv., conj.*).
1898 J. D. Brayshaw *Slum Silhouettes* 1 Tall an' thin, yer say? Wot, wiv long white 'ands, an' black 'air—Yus! 1933 D. L. Sayers *Murder must Advertise* xix. 332 You'll 'ave 'im steppin' aht ter meet me wiv' a crimson carpet and a bokay. 1981 'J. Gash' *Vatican Rip* iv. 44 Want me to come wiv yer, Lovejoy?

wiwi[1]. (Earlier and later examples.)
1840 J. S. Polack *Manners & Customs New Zealanders* II. 285 Wi-wi, kind of wiry grass that is pulled up in tufts, it also is the produce of the marsh. 1970 Moore & Edgar *Flora N.Z.* II. 59 The Maori name for rushes and rush-like plants is Wiwi.

wi-wi[2]. For *Austral.* read *Austral.* & *N.Z.* and add earlier example.
1841 E. J. Wakefield in *N.Z. Jrnl.* II. XLV. 243/1 Should the Wiwis, or French, kill any of our Chiefs.

Wiyot (wī·yǫt). [Wiyot *wiyat* (people of) the Eel River delta.] An American Indian people formerly living on the coast of northern California; the Macro-Algonquian language of this people. Also *attrib.* or as *adj.*
1851 G. Gibbs *Jrnl.* 9 Sept. in H. R. Schoolcraft *Information respecting Indian Tribes* (1853) III. iv(3). 127 The name given to this people by their neighbors is Wee-yot. 1911 A. L. Kroeber *Lang. Coast Calif.* 384 The Wiyot occupied the Coast from the Bear River mountains north as far as to Little river. *Ibid.*, Wiyot is spoken indistinctly and lacks..phonetic clarity. 1918 *Univ. Calif. Publ. in Amer. Archaeol. & Ethnol.* XIV. III. 232 Cedar though present on Wiyot territory, is not abundant enough for the purposes for which a soft wood is needed. 1925 *Ibid.* XXII. I. 5 There are also some Wiyot living among the Athapascans at Blue Lake. *Ibid.* 6 The individual differences of Wiyot speech will be discussed. 1946 L. Bloomfield in C. Osgood *Linguistic Structures Native Amer.* 201 Two languages of California, Wiyot and Yurok, have been suspected of kinship with Algonquian. 1961 H. Driver *Indians N. Amer.* xiv. 251 From the Wiyot to the Bella Coola, such usufruct was patricentered. 1964 R. H. Robins *Gen. Linguistics* 308 Isolated languages in the western states, Blackfoot..and Wiyot and Yurok (California). 1974 *Encycl. Brit. Micropædia* X. 720/1 Wiyot settlements were located on streams of bays, rather than on the ocean itself. 1977 *Language* LIII. 501/1 Pentland expands on Karl Teeter's suggestion ..that the rule affricating dental stops which is operative in diminutives in Wiyot..as well as in Algonquian, should be ascribed to Proto-Algic.

wiz, var. *WHIZZ, WHIZ *sb.*[2]

wizard, *a.* and *sb.* Add: **A.** *sb.* **2. b.** *Wizard of the North* (examples). Also freq. as *financial wizard*, a person skilled in making money, or in organizing financial affairs.
1869 R. Walton *Random Recoll. Midl. Circuit* 134 Fortunately the 'Wizard of the North' came upon the spot [*sc.* Kenilworth], and 'Henceforth' (as a modern historian has it) 'the ruined place was to be sanctified [etc.]'. 1893 *Ladies' Home Jrnl.* May 27/2 Sir Walter Scott was called 'The Wizard of the North'. 1952 G. Sarton *Hist. Sci.* I. xix. 471 The eunuch, Hermeias, who began his career as a money-changer, was a kind of financial wizard and became very wealthy and powerful. 1967 G. F. Fiennes *I tried to run a Railway* v. 58, I had energy..to be the financial wizard on the parochial church council. 1975 *Times* 24 May 4/7 Judge Kennet..noted that Mr Tzour had been noted as a financial wizard.
† **d.** A professional conjuror. *U.S. Obs.*
1859 L. Wraxall tr. *J. E. Robert-Houdin's Mem.* II. iv. 108 On my arrival in England, a conjuror of the name of Anderson, who assumed the title of Great Wizard of The North, had been performing for a long period at the little Strand Theatre. 1895 *N.Y. Dramatic News* 14 Dec. 6/1 The wonderful record established at the California theatre by Hermann the Great..has finally been broken.. [by] the wonderful wizard [himself].
B. *adj.* **2. b.** *slang.* Excellent, marvellous, very good.
1922 S. Lewis *Babbitt* xvii. 216 The Rev. Dr. John Jennison Drew is a wizard soul-winner. 1932 E. Waugh *Black Mischief* vii. 277 They..righted themselves and stopped dead within a few feet of danger. 'Wizard show that,' remarked the pilot. 1943 J. B. Priestley *Daylight on Saturday* i. 1 The roofs are nicely camouflaged, and the stiff coloured netting..is a wizard show. 1954 [see *SUPER *a.* 3]. 1958 'R. Crompton' *William's Television Show* vii. 189 Gosh, that party of Ginger's last Christmas was wizard. a 1966 'M. na Gopaleen' *Best of Myles* (1968) 25 How awfully wizard being at the theatre with you! 1974 *Times* 17 Aug. 7/3 'How wizard!' they said... 'How absolutely super!'

wizardry. Add: **2.** Also more *loosely*, skill, expertise, or the result of this.
1951 *Sport* 27 Jan.–2 Feb. 3/3 Rounding off the wing wizardry of Finney and Morrison are inside men Horton, Wayman and Bobby Beattie. 1974 W. J. Burley *Death in Stanley Street* viii. 142 Bits of electrical wizardry which must have come from a record player or a television set. 1979 *Arizona Daily Star* 5 Aug. (Comic Suppl.), Peter Parker uses his scientific wizardry.

wizen, *a.* (Earlier example of *Comb.*)
1819 M. Edgeworth *Let.* 17 Apr. (1971) 201 An old thin stupid wizzen looking Mr. Evelyn received us.

wobbegong (wǫ·bıgǫŋ). Also **wobbygong, wobegong.** [Aboriginal name.] A brown carpet shark with buff markings, *Orectolobus maculatus*, found off the coast of Australia.
1852 G. C. Mundy *Our Antipodes* I. xii. 392 The most hideous to behold of the shark tribe is the wobegong, or woe-begone, as the fishermen call it... His broad back is spotted over with leopard-like marks. 1882 J. E. Tenison-Woods *Fish & Fisheries N.S.W.* iv. 94 The wobbegong..is chiefly nocturnal. 1917 *Chambers's Jrnl.* Sept. 588/1 There is also the well-known wobbygong, a creature of extraordinary and beautiful colouring. 1937 Z. Grey *Amer. Angler in Austral.* vi. 53 The most remarkable feature of the wobbegong is his teeth. 1956 S. Hope *Diggers' Paradise* xx. 183 The worst types are the white pointer,..hammerhead and carpet shark, also called the wobbegong. 1981 B. Stonehouse *Sharks* iv. 46 Woggegongs..add to their camouflage by growing fronds that look like seaweed on their faces.

wobble, wabble, *sb.* Add: (The spelling *wabble* is now obsolete.) **2.** *Biochem.* The variable pairing that is possible between a base in a transfer RNA anticodon and the corresponding base in a messenger RNA codon. Freq. *attrib.*
1966 F. H. C. Crick in *Jrnl. Molecular Biol.* XIX. 548 (*heading*) Codon—anticodon pairing: the wobble hypothesis. *Ibid.* 551, I now postulate that in the base-pairing of the third base of the codon there is a certain amount of play, or wobble, such that more than one position of pairing is possible. 1974 *Nature* 22 Feb. 517/2 tRNA$_f^{met}$ of *E. coli*, yeast and mouse ascites tumour cells has the ability to recognise both the codons ApUpG and GpUpG and to thus exhibit code degeneracy or 'wobble' at the third base (3' end) of the anticodon. 1982 K. H. Muench in T. M. Devlin *Textbk. Biochem.* xix. 921 According to the wobble rules 31 different tRNAs would suffice to read the 64 codons.

wobble, wabble, *v.* Add: (The spelling *wabble* is now obsolete.) **5. wobble plate** = *swash-plate* s.v. *SWASH *sb.*[1] 9; freq. *attrib.*
1929 V. W. Pagé *Mod. Aviation Engines* II. xlvi. 1897 A peculiar 'wobble' plate mechanism replaces the usual crankshaft arrangement. *Ibid.* (*caption*) Wobble plate. *Ibid.* (*caption*) A typical example of a 'wobble' plate or barrel type engine. 1943 Wobble-platemeter [see *NUTATE *v.*].

wobbler. Add: **2.** *Mech.* **a.** A projection on a roll in a rolling-mill, by means of which it may be turned.
1904 J. W. Hall in F. W. Harbord *Metallurgy of Steel* xvi. 294 At the outer end of each neck forming part of the casting is a 'wobbler', provided with either three or four prongs or corners, by means of which the roll is driven. 1919 *Jrnl. Inst. Metals* XXII. 383 Rolls.—These are usually of chilled cast iron or hardened steel. For wobbler ends, the four-horn design is the most general in use... A rule is suggested regarding length of spindles and clearance between wobbler and box. 1930 *Engineering* 25 Apr. 539/1 A driving bar socket for the wobbler drive is bolted to the face-plate. 1978 W. L. Roberts *Cold Rolling of Steel* iii. 64 Wobblers are shown in Figure 3-3, and flat roll ends in Figure 3-4.
b. = *wobble plate* s.v. *WOBBLE, WABBLE *v.* 5.
1950 W. E. Wilson *Positive-Displacement Pumps & Fluid Motors* iii. 42 Oil pressure forces the pistons against the nonrotating wobbler. The resultant force is transmitted through ball and roller bearings to the wobbler plate on the shaft and imparts a rotating action to it.
3. *Angling.* A lure that wobbles and does not spin.
1928 E. F. Spence *Pike Fisher* v. 55 The 'shining streak of silver' does not resemble any inhabitant of river or lake, but the 'wobbler' does look something like an injured fish. 1945 [see *LEERVIS]. 1960 M. Sharcott *Place of Many Winds* vii. 120 'I bought a dozen new wobblers,' he says as he lifts the lid to reveal the shiny brass cohoe-spoons. 1977 *Best of Austral. Angler* 49/1 The wobbler and spoon type lure, however, whilst also being highly attractive to trout, are more suited to the physical requirements.

wo·bbly, *sb.* orig. *U.S.* [Origin uncertain.] A member of the Industrial Workers of the World (see *INDUSTRIAL *a.* e).
[1913 *Miners Mag.* 24 Apr. 5 Joe Elton, Sabotist, Syndicalist and fearless I.W.W. with a red, flowing tie, with fire in his eye and fight in his backbone, the I Wobbily-Wobbily organizer..is traveling the country delivering his message. 1914 *Rep. Calif. District Courts of Appeal* (1915) 402 He telegraphed... Send all speakers and wobbles [*sic*] possible... It appeared at the trial that the term 'wobbles' meant members of the I.W.W.] 1914 *Voice of People* (Portland, Oregon) 1 Oct. 2/4 The workers are..asking why the wobblies are not holding meetings. 1921 *Outing* (U.S.) Nov. 94/3, I saw an angel and the devil standing side by side. The devil wore a 'Wobbly' (I.W.W.) button. 1923 *Nation* 5 Sept. 242/2 In Vancouver, in 1911, we had a number of Chinese members and one restaurant keeper would trust any member for meals. He could not pronounce the letter w, but called it wobble, and would ask: 'You I. Wobble Wobble?' and when the card was shown, credit was unlimited. Thereafter the laughing term among us was I. Wobbly Wobbly. 1932 E. Wilson *Devil take Hindmost* xxi. 218 The Wobbly leaders..called the men out of the tunnels. 1948 V. Palmer *Golconda* xxx. 251 And sometimes he [*sc.* the Labour Party candidate] was bothered by young fellows, usually advocates of One Big Union, who tried to lure him into deep water... He came to recognize them in the end and to stop them with light thrusts before they had lured him too far. 'What're you fellows? Wobblies, aren't you? I.W.W.—I Won't Work, but listen to me talk, eh?' 1957 [see I.W.W. s.v. I III]. 1967 A. L. Lloyd *Folk Song in Eng.* v. 387 'The celebrated working man', a song of American origin.. was brought to Durham by a Wobbly collier from Kentucky. 1980 *Times* 21 June 6/1 A poor white American, probably a Wobbly—a member of the Industrial Workers of the World, a group of radical labour unions largely made up of itinerant workers.

wobbulator (wǫ·biŭlatəɹ). *Electronics.* Also **wobulator.** [f. WOBB(LE *sb.* or *v.* + MOD)ULATOR.] A device for producing a signal whose frequency varies rapidly and repeatedly between two limits.
1945 Cooke & Markus *Electronics Dict.* 428/1 Wobbulator. 1958 *Electronic Engin.* XXX. 541 A wobulator for amplitude testing often gives errors because of unwanted signal amplitude variations. 1977 S. W. Amos *Radio, T.V. & Audio Technical Reference Bk.* xxxii. 9 Alignment should not be attempted on any u.h.f. tuner without the use of a u.h.f. wobbulator.
So **wo·bbulated** *ppl. a.*, varied or produced by means of a wobbulator; **wobbula·tion,** repeated variation of a frequency by a wobbulator.
1944 *Electronic Engin.* XVI. 327/3 A powerful high-frequency sound with a wobbulated effect might give the birds a disagreeable sensation. 1957 *Practical Wireless* XXXIII. 569/1 Apply a 10·7 Mc/s signal, wobbulated 300 kc/s, to a test point. 1965 *New Scientist* 15 Apr. 156/1 'Wobbulation' or the sweeping of the modulation frequency through 15 per cent either way of its mean value. 1982 *IEEE Jrnl. Solid-State Circuits* XVII. 671/1 The initial and final frequencies, the wobbulation rate, the ramp amplitude and frequency, and the wobbulation mode are all controlled from the input data.

Wodehousian (wudhau·siăn), *sb.* and *a.* [f. the name of Sir Pelham Greville *Wodehouse* (1881–1975), British author + -IAN.] **A.** *sb.* **a.** A typical character in one of the comic novels of P. G. Wodehouse. **b.** An admirer or an habitual reader of Wodehouse's novels. **B.** *adj.* Pertaining to or characteristic of Wodehouse or of his works.
1931 *Times Lit. Suppl.* 21 May 409/4 Berry Conway and Lord Biskerton, the lads who are after Big Money, are true Wodehousians, stamped with the authentic stamp. 1938 *Ibid.* 12 Feb. 107/3 One may wince at Wodehousian baronets taking lodgers. 1943 *Scrutiny* XI. 288 About every seven pages some Wodehousian character receives a severe and almost mortal shock. 1958 *Times Lit. Suppl.* 14 Feb. 85/4 An up-to-date version of gaily irresponsible Wodehousian farce. 1973 M. Muggeridge *Infernal Grove* iv. 229 The broadcasts, in point of fact, are neither anti- nor pro-German, but just Wodehousian. 1979 *Daily Tel.* 2 Aug. 11/3 With some minor reservations.., this addition to Wodehousian lore can safely be admitted. 1980 *Times* 2 Feb. 7/2 The experienced Wodehousian's heart leaps.

wodge. Substitute for entry:
wodge (wǫdʒ). *colloq.* (orig. *dial.*). Also **wadge.** [Perh. phonæsthetic alteration of *wedge*: cf. WEDGE *sb.* 4 and *Eng. Dial. Dict.*] A bulky mass; a chunk or lump; a wad (of paper).
1860, 1862 [see WADGE]. 1913 E. Pound *Let.* 7 Nov. (1971) 25, I don't want a great wadge of prose, but about double what we have at present. 1922 [in Dict.]. 1949 D. Smith *I capture Castle* ii. viii. 112 You must take only one kind of food on the fork at a time; never a nice comfortable wodge of meat and vegetables together. 1958 Hayward & Harari tr. *Pasternak's Dr. Zhivago* i. vii. 195 He held out a wadge of papers across the hand-rail. 1963 A. Smith *Throw out Two Hands* iii. 39 We strode out into the rain with a wodge of well-stamped supplications. 1977 *Private Eye* 4 Mar. 7/3 True, there's a wadge of self-opinionated dolts who drive around in head scarves and Range Rovers. 1981 *Brit. Med. Jrnl.* 21 Mar. 968/1 A posterior pack is made from a wadge of gauze as large as the end of the patient's thumb, which is rammed tightly into the posterior choana. 1984 *Listener* 6 Dec. 35/1 These tomes are usually given a lively, busy design, with screaming wodges of colour. *Ibid.* 20–27 Dec. 7/2 Cross-headings, the lay reader should know, are those devices used to break a grey wodge of type and encourage you to keep reading.

wodgy, *a.* Add: (Later examples.) Also *fig.*
1928 *Daily Express* 8 June 5/5 Wedding cakes..are fattening and indigestible; they are 'wodgy' to the palate. 1978 *Daily Tel.* 30 Aug. 13 (*caption*) Swirl a piece of totally straight hair high, pin a little wodgy bun of bright crepe paper on over the kerby grips: that's Patrick Ales [*sic*] way. 1979 *Hi-Fi News* Dec. 169/2, I only wish I could be as totally enthusiastic about the recording. At average levels it is fine but sudden fortes come with a wodgy quality that is not at all pleasing; there are too many individual resonances for the ear to cope.

wodginite (wǫ·dʒinəit). *Min.* [f. *Wodgina*, name of a locality in Western Australia + -ITE[1].] A rare oxide of tantalum, niobium, and manganese (usu. also containing tin and iron) which occurs as brown or black monoclinic crystals or grains.
1963 E. H. Nickel et al. in *Canad. Mineralogist* VII. 390 The name wodginite is proposed for a mineral found at two widely separated localities—Wodgina, Australia and Bernic Lake, Manitoba. 1978 *Mineral. Rec.* IX. 18/2 (*caption*) Unusual twin crystal of wodginite/cassiterite. The size is about 6 × 4·5 × 3·5 cm.

woe, *sb.* Add: **1. a.** Freq. in phr. *tale of woe*, a narrative of (one's) misfortunes. Now usu. *joc.*
1790 [in Dict.]. 1882 E. Field *Compl. Tribune Primer* 111 (*heading*) A tale of woe. 1951 *Sport* 16–22 Mar. 4/2 Listen to the tale of woe from Swindon Town, who.. suffered their 16th away league defeat... Swindon..lost goalkeeper Norman Uprichard at a goalless stage of the game. 1967 Baker & Jones *Coffee, Tea or Me?* xvi. 199 Kelman savored a crisp piece of sausage pizza as he thought about our tale of woe. 1973 *Times* 11 June 18/8 Yet it is not all a tale of woe. An entirely new management structure has been brought into force, and Liverpool was the first port in the world to link a computer service to its cargo handling.

woffle, var. *WAFFLE *v.*, *sb.*[2]

wog[2] (wǫg). *slang.* [Origin uncertain: often said to be an acronym, but none of the many suggested etymologies is satisfactorily supported by the evidence.] **1.** A vulgarly offensive name for a foreigner, esp. one of Arab extraction.
1929 F. Bowen *Sea Slang* 153 *Wogs*, lower class Babu shipping clerks on the Indian coast. 1932 R. J. P. Hewison *Essay on Oxford* 5 And here the Ethiop ranks, the wogs, we spy. 1937 F. Stark *Baghdad Sketches* 90 When I return, Nasir fixed me with real malignity in his little placid eyes. 'I knew she wanted me to go,' he said. 'I could see what she was thinking. They call us wogs.' 1942 C. Hollingworth *German Just behind Me* xiii. 258 King Zog Was always considered a bit of a Wog, Until Mussolini quite recently Behaved so indecently. 1944 [see *COME *v.* 38 e]. 1955 E. Waugh *Officers & Gentlemen* ii. 323 He turned up in western Abyssinia leading a group of wogs. 1958 *Times Lit. Suppl.* 11 Apr. p. vi/3 We have travelled some distance from the days when Wogs began at Calais. 1965 [see *COMMIE]. 1982 J. Savarin *Water Hole* i. iv. 42 He hated Arabs... They were all wogs to him.
2. The Arabic language.
1977 P. Raymond *Matter of Assassination* vi. 63, I can't speak Wog and don't seems to be getting anywhere. 1982 'W. Haggard' *Mischief-Makers* xiv. 157 'I've picked up a few words of wog, sir.'.. The driver spoke terrible barrack-room Arabic.
3. a. *attrib.* passing into *adj.*
a 1963 J. Lusby in B. James *Austral. Short Stories* (1963) 236 Wog chappie scuttling around seeking safe side of the beast. 1970 G. F. Newman *Sir, You Bastard* viii. 234 We were hawking, and getting treated like bleeding wog brush salesmen. 1973 *Daily Tel.* 31 May 3/2 Judge Sheldon heard that trouble started..when white girlfriends of coloured soldiers..were taunted by members of the Royal Scots as 'wog lovers'. 1977 *Drive* Sept.–Oct. 112/2 Any foreign car, even a Ferrari or a Mercedes, is a *wog motor*, unless it's a *Yank*.
b. *Comb.* **wogland** *derog.*, a foreign country.
1961 [see *NIG, NIG *sb.*[2]]. 1967 'J. Munro' *Money that Money can't Buy* ii. 24, I don't live in Wogland [*sc.* Spain] because I like it.
Hence also **wo·gger; wo·ggy** *a.*
1922 Joyce *Ulysses* 740 She called him wogger. *Ibid.* 741 She may have noticed her wogger people were always going away. 1973 M. Catto *Sam Casanova* iv. 75, I met some kid in a night-club here, does some sort of Woggy belly-dance. 1979 Reese & Flint *Trick 13* 100 That woggy fellow..was cleaning up.

wog[3]. *Austral. slang.* [Origin uncertain.] A germ or parasite; an insect; an illness or disease. Cf. *BUG *sb.*[2] 4 d.
1934 *Bulletin* (Sydney) 31 Oct. 20/4 Buckley's fluke.. is a wog that enters the nostrils of these snakes during hibernation. 1941 C. Barrett *Coast of Adventure* iii. 51 Jolly little people..popping into old jam tins a miscellany of wogs—from bull-ants to scorpions and centipedes. 1953 A. Upfield *Murder must Wait* xxi. 191 The wogs flying about the light. 1964 R. Braddon *Year Angry Rabbit* i. 9 But find the wog, find the super-myxomatosis, the whatever-it-may-be that kills today's rabbits. 1976 D. Francis *In Frame* viii. 126 A beastly stomach wog, so he couldn't come.

Wogdon (wǫ·gdən). Also erron. **Wogden.** The name of Robert *Wogdon* (fl. 1776–1800), a noted gunsmith, used *absol.* to designate a duelling pistol made by him.
c 1810 W. Hickey *Mem.* (1923) III. 150 By God, Bill, you shall shoot the dirty little rascal through the head. I have a delicate pair of Wogdens that will do his business effectually. 1969 G. Lyall *Venus with Pistol* xxv. 161 Give it a little time for the word to go round that Bert Kemp had matched a pair of fancy Wogdons. 1981 'J. Sturrock' *Suicide Most Foul* vi. 127 Had I been armed with only one of my Wogdons the end would have been different, but a gentleman does not take pistols to a ball.

wo·ggle, *sb.* [Origin unknown: cf. TOGGLE *sb.*] A loop or ring of leather, cord, etc., through which the ends of a Scout's neckerchief are threaded.
1930 *Daily News* 10 May 4/4 Woggles have now become an established part of Scout uniform, and I have seen some very good examples made by Scouts. 1977 *Grimsby Even. Tel.* 27 May 2/7 The woggle—the ring holding the neckerchief in place on the Scout uniform. 1983 J. Deft *Beaver Leader's Handbk.* 32 You must decide yourself whether each new Beaver should be asked to pay for his scarf and woggle, or whether these should be provided by the Colony.

Wogul, vogul. Add: The usual form is now **Vogul. a.** (Earlier and later examples of this form.)
1880 A. H. Sayce *Introd. Sci. Lang.* II. x. 325 The Hungarians were once the neighbours of the savage Voguls of the Ural. 1948 D. Diringer *Alphabet* 483 The Voguls in the Ural mountains. 1975 [see *OSTYAK].
b. The language of this people, belonging to the *OB-UGRIAN group.
1908 T. G. Tucker *Introd. Nat. Hist. Lang.* 133 Ugric, which comprises..Vogul and Ostiak, dialects of a few thousands scattered over a wide region eastward from the Northern Ural and about the Obi River. 1933, etc. [see *OB-UGRIAN]. 1951 W. K. Matthews *Languages U.S.S.R.* iii. 21 The 'primitive' or East Ugrian languages, Ostyak and Vogul. 1980 *Amer. N. & Q.* Oct. 29/1 Marianne Sz. Bakró-Nagy..has pulled together some 500 terms..on the bear as a tabu animal in the Urals among the Ostyak (Khanty) and Vogul (Mansi) speaking peoples.
Wogulian: delete † and add later example.
1925 P. Radin tr. *Vendryès's Language* ii. iii. 118 In Wogulian *mini* 'he goes'..[is] formed like *puri* 'taking'.

Wöhler (vö·ləɹ). *Mech.* [The name of August *Wöhler* (1819–1914), German railway engineer.] *Wöhler test*: a fatigue test in which a horizontal bar is rotated axially while supported at one end and loaded at the other.
[1888 *Rep. Brit. Assoc. Adv. Sci. 1887* 434 (*caption*) Limits of stress from Wöhler's endurance tests.] 1911 *Jrnl. Iron & Steel Inst.* LXXXIV. 655 The testing machines include..a rotary fatigue (Wöhler test) machine. 1948 P. F. Foster *Mech. Testing of Metals & Alloys* x. 203 One objection urged against the Wöhler test is that it is merely a skin test, since the major part of the section is but comparatively lightly stressed. 1980 *Proc. European Offshore Steels Res. Seminar 1978* (Welding Inst.) p. III/P7–2 The test programme included Wöhler tests (fatigue tests) with an alternating load.

wok (wǫk). Also **wock.** [a. Chinese (Cantonese).] A bowl-shaped pan used in Chinese cookery.
1952 D. Y. H. Feng *Joy of Chinese Cooking* i. 37 A well-stocked Chinese kitchen usually has..several convex-bottomed circular pans hammered out of thin iron or copper called *wock*. 1962 E.-M. Wong *Chinese Cookery* i. 4 For versatility and easy handling the wock is indispensable. 1969 *Britannica Bk. of Year* (U.S.) 801/1 *Wok*, a bowl-shaped cooking utensil used especially in the preparation of Chinese food. 1972 *Maclean's Mag.* Mar. 46/3 The Chinese wok technique is my mainstay of cooking, although I do it in a plain frypan. 1973 J. Gores *Final Notice* (1974) x. 61 The pixie-like waitress came in with a boardful of fresh vegetables for the *wok*. 1977 *Sunday Times* (Colour Suppl.) 4 Dec. 20/2 Wok cooking is about to sweep the Western world. 1983 *Listener* 30 June 17/1 By 1972 I needed a small pantechnicon to convey all my books on macrobiotic cookery, my plants, wall-hangings and floor cushions, my astrological tables, women's-lib posters and my wok.

wolchonskoite, wolk-, obs. varr. *VOLKONSKOITE.

wolf, *sb.* Add: **1. c.** The skin or fur of the animal. (Chiefly *attrib.*: see sense 10 a below.)
1805 Lewis & Clark *Orig. Jrnls. Lewis & Clark Expedition* (1904) II. 377, I have also observed some robes among them of beaver, moonox, and small wolves. 1876 *Smithsonian Misc. Coll.* XIII. vi. 69 Furs... Wolf, (*Canis lupus*)—linings, rugs, and robes. 1940 *Chambers's Techn. Dict.* 911/1 *Wolf*, the dressed skin of one of the varieties of wolf. 1974, etc. [see *wolf hat*, etc., sense 10 a below].
3. a. (*b*) Freq. as *Tasmanian wolf*; = THYLACINE.
1941 E. Troughton *Furred Animals Austral.* 50 (*heading*) Tasmanian wolf or tiger. 1966 G. Durrell *Two in Bush* vi. 178 The predators are represented by such things as the Tasmanian Wolf—not a true wolf, of course, but a marsupial, looking remarkably like its counterpart.
3*. = *wolf tree*, sense 10 e below.
1949 *Q. Jrnl. Forestry* XLIII. 127 Most props containing large knots have been prepared from quick-grown heavily branched trees such as wolves. 1966 *Times* 21 Apr. 16/7 Douglas fir plantations nearly always have some undesirable wolves which have to be cut out.
4. c. *slang.* (*a*) A sexually aggressive male; a would-be seducer of women; (*b*) orig. *U.S.*, a male homosexual seducer or one who adopts an active role with a partner.
Occas. applied to a woman: see quot. 1968 s.v. *WOLFESS 2.
(*a*) 1847 Thackeray *Van. F.* (1848) xxxvii. 335 'Rawdon,' said Becky, .'I must have a sheep-dog... I mean a moral shepherd's dog..to keep the wolves off me.' 1862 Mrs. H. Wood *Mrs. Haliburton's Troubles* II. ii. 23, I vowed I'd tell Mark what I had seen and heard, and what sort of a wolf she allowed to make her presents of fine clothes. 1945 S. Lewis *Cass Timberlane* xix. 113 She was innocent, but this Roskinen was a wolf. 1968 [see *KARATE *v.*]. 1973 'E. Peters' *City of Gold & Shadows* ii. 25 He did not look like a wolf, but he did look like a young man with an eye for a girl.
(*b*) 1917 *New Republic* 13 Jan. 293/2 The sodomist, the degenerate, the homosexual wolf. 1931, etc. [see *JOCKER]. 1950 Patterson & Conrad *Scottsboro Boy* ii. 91, I learned men were having men. Old guys, they called them wolves, they saw me looking at this stuff and thought I might be a gal-boy. 1978 K. J. Dover *Greek Homosexuality* ii. 87 In prisons the 'wolf' is the active homosexual, and does not reverse roles with his partners.
9. k. *to throw to the wolves*: to sacrifice (a subordinate, friend, ally, etc.) to one's enemies in order to save oneself.
1927 F. Harris *My Life & Loves* III. x. 146 But if Gladstone had had his letter back, I think the G.O.M. would have thrown Dilke to the wolves. 1958 *Listener* 6 Nov. 743/2 This able and agreeable doctor [*sc.* Lord Addison] was thrown to the wolves by a Prime Minister who had good reason to know that his own position was desperate. 1980 P. Kinsley *Vatchman Switch* xli. 236 If anyone..showed disloyalty he would throw him to the wolves.
l. *lone wolf*: see *LONE *a.* 3 c.
10. a. *wolf chase* (earlier example), *eye, pelt, snow*; also with reference to the skin or fur of the animal, as *wolf coat, collar, hat, jacket*; **b.** *wolf-breeding* adj.; **d.** *wolf-coloured* (later example).
1889 W. B. Yeats *Wanderings of Oisin* 77 Wolf-breeding mountains. 1824 in *Coll. Missouri Hist. Soc.* (1928) VI. 75 Had a wolfe chase. 1977 P. Way *Super-Celeste* ii. 117 She put on her Siberian wolf coat. 1974 *Selfridge Christmas Catal.* 14 Leather coat with wolf collar and hamster lining. 1926 Wolf-coloured [see *KEESHOND]. 1922 Joyce *Ulysses* 434 Her wolfeyes shining. 1974 *Country Life* 3 Oct. 980/2 Natural wolf three-quarter length jacket..worn with a wolf hat. 1976 *Jrnl.* (Newcastle) 26 Nov. (Advt.), Mink coat..also modern wolf jacket with matching fox hat, both coats new. 1923 D. H. Lawrence *Birds, Beasts & Flowers* 200 On to the fur of the wolf-pelt that strews the plain. c 1878 G. M. Hopkins *Poems* (1967) 73 There did storms not mingle?.. wolfsnow, worlds of it, wind there?
e. wolf call *colloq.* (orig. *U.S.*) = *WOLF-WHISTLE; **wolf-cry** [f. vbl. phr. *to cry 'wolf'*: see WOLF *sb.* 9 a] = *false alarm* s.v. *FALSE *a.* 14 c; **wolf cub,** (*a*) (see sense 10 a in Dict.); (*b*) = *CUB *sb.*[1] 2 c; also *fig.*; **wolf pack,** a number of wolves naturally associating as a group, esp. for hunting; also *fig.*, esp. denoting an attacking group of German submarines in the war of 1939–45; **wolf pen** *U.S.*, a strong box made of logs used for trapping wolves; **wolf tree,** a tree that is occupying more space than has been allowed for it, so restricting the growth of its neighbours (cf. sense 3* above); **wolf-willow** *Canada*, any of several shrubs, esp. *Elæagnus commutata*, which has silver-grey foliage.
1948 *Time* 27 Sept. 12/1 Grins, whistles, wolf-calls.. followed her in this exclusively male territory. 1958 *Spectator* 6 June 726/3 The streets are lined by groups of

lounging youths watching the girls go by (but no whistles or wolf-calls). 1915 W. J. Locke *Jaffery* xxii. 315, I have a habit of losing things and setting the household in frantic search,..only to discover that I have had the wretched object in my pocket all the time. So accustomed is Barbara to this wolf-cry that if I came up to her without my head and informed her that I had lost it, she would be profoundly sceptical. 1980 *Listener* 9 Oct. 462/3 The news that the Met season might have to be cancelled..is an annual threat, a wolf-cry. 1916 R. Baden-Powell in *Wolf Cub* Dec. 2/1 Hullo, Wolf Cubs! What swells you are to have a newspaper all to yourselves! 1963 H. Wilson in *Times* 8 May 6/3 If we had to face a really dedicated and trained spy, not an overgrown wolf cub who had gone wrong, then the system would have been wide open in respect of security. 1981 E. Longford *Queen Mother* ii. 35 (*caption*) Wellington: the Duchess of York inspects a pack of wolf cubs. 1895 Wolf pack [in Dict.]. 1941 *Hutchinson's Pict. Hist. War* 9 July–30 Sept. 270/1 The U-boat is now being used as a unit in a flotilla... We had a hint of it a year ago when the Berlin bulletins talked about 'wolf pack' attacks on convoys. 1951 W. Stevens *Let.* 2 Oct. (1967) 731 There is probably a sort of wolf-pack that follows him [*sc.* Hermann Hesse] round. His idea of throwing out a poem or two to slow them up and invite them to devour each other sounds almost like folklore. 1977 *Time* 26 Sept. 9/2 What Andreas Baader and Ulrike Meinhof spawned as a small wolf pack of urban guerrillas has now become a scattered army of vicious malcontents, bent on destroying the society around them. 1980 'D. Grant' *Emerald Decision* vi. 129 They were headed for the perilous North Channel..if they survived the wolfpacks. 1647 in *Watertown* (Mass.) *Rec.* (1894) I. i. 12 The Towne gaue: to John Witherll: there Right in the palisado that inclosed the woulfe pen. 1876 J. S. Ingram *Centennial Expedition* 106 The places of interest are..the Aviary, the Fox Pens, the Wolf-Pens. 1928 R. S. Troup *Silvicultural Systems* xix. 187 The stands..were kept fairly dense in order to promote clean stems, congested thickets being thinned and wolf-trees removed. 1966 D. Waters *Forestry* xviii. 94 Wolf trees are large mis-shapen trees which do not provide good timber. 1889 J. G. Donkin *Trooper & Redskin* 86 The luscious perfume of wolf-willow and wild rose..come scampering on the western breeze. 1948 A. L. Rand *Mammals Eastern Rockies* 90 Wolf-willow clumps, gopher holes, odd stones, aspen bluffs. 1974 M. Laurence *Diviners* 357 There were these thin prairie maples and the wolf willow.
Hence **wo·lflessness** *nonce-wd.* [cf. WOLF *sb.* 5 a], the state of 'not having the wolf at the door', i.e. being free of poverty.
1928 D. H. Lawrence *Rawdon's Roof* 26 The perfect wolflessness of Rawdon's door, the perfect windlessness of Rawdon's roof.

wolf, *v.* Add: **2.** Also without *const.*: cf. *WOLF *sb.* 4 c. Occas. *trans.*
1929 *World's Work* Nov. 40 The college boy (in 1929) knows a smoothie who wolfed on a friend and creamed his lady. 1934 G. & S. Lorimer *Stag Line* vii. 232 No matter how I feel, I wouldn't wolf a brother's girl. 1940 J. O'Hara *Pal Joey* 186, I give with the vocals and wolf around in a nite club.
4. *trans. U.S. Blacks.* (See quots. and cf. *WOOFING *vbl. sb.*) Occas. *intr.* with *at*.
1966 *Urban Education* II. ii. 108 *Wolf*, to make fun of someone. 1969 *Sports Illustr.* 3 Nov. 36/2, I turned round and started wolfing at the guy, and he just strolled off. 1971 E. E. Landy *Underground Dict.* 199 *Wolf v.*, criticize; chop down. 1974 H. L. Foster *Ribbin', Jivin', & Playin' Dozens* iv. 172 *Wolf, wolf'n, woof, woofin, wolf ticket*, can mean anything from making fun of someone to challenging someone to a fight, a powerful person. 1978 *Detroit Free Press* 2 Apr. (Detroit Suppl.) 8/3 'C'mon, man,' they tell Balls, backing down, 'we was just wolfin' ya. We gotta be careful who we sell to.'

wolf-dog. 2. Delete second part of def. and add further (*N. Amer.*) examples.
1896 M. McNaughton *Overland to Cariboo* 47 A large number of wolf-dogs were prowling about. 1953 B. J. Banfill *Labrador Nurse* 19 The mossy grass knolls were dotted with tethered wolf dogs.

wolfeite (wu·lfəit). *Min.* [f. the name of Caleb W. *Wolfe* (1908–80), U.S. crystallographer + -ITE[1].] A basic phosphate of ferrous iron and bivalent manganese, $(Fe^{2+},Mn^{2+})_2(PO_4)(OH)$, that occurs as transparent or translucent monoclinic crystals and forms a series with triploidite.
1949 C. Frondel in *Amer. Mineralogist* XXXIV. 694 The name wolfeite is proposed for the mineral and is particularly appropriate in view of Professor Wolfe's studies of iron and managanese phosphates from Palermo and other localities. 1951 C. Palache et al. *Dana's Syst. Min.* (ed. 7) II. 853 The names triploidite and wolfeite are applied to those parts of the series with Mn > Fe and Fe > Mn, respectively. 1979 *Mineral. Mag.* XLIII. 507/1 Unlike wolfeite from the Palermo pegmatite, the wolfeite at Thackaringa [Australia] does not appear to be an alteration phase of a pre-existing phosphate and is a late stage phase in the core of the pegmatite.

wolfer. Restrict *rare* to sense 2 and add earlier and later examples of sense 1.
1872 *Rep. Indian Affairs 1871* 410 A regular stampede took place out of that section of the country of 'Wolfers' and whiskey traders. 1930 C. M. MacInnes *In Shadow of Rockies* 66 Even more disreputable than the whisky traders were the wolfers. 1976 *Times* 8 July 16/4 The ghoulish wolfers poisoned the plains with strychnine. 1973 R. Wiebe *Temptations of Big Bear* II. iii. 101 Settlers ripping up land and knocking down trees and wolfers dashing about scattering poison and killing wolves and buffalo.

wolfess. Add: **2.** A woman who is sexually aggressive; a woman who seeks to seduce men.
1945 *Bulletin* (Philadelphia) 27 Nov. 42/1 A nice girl hasn't got a chance with a wolfess around. 1968 *Word Study* Dec. 4/2 *Wolf* suggests the sexually aggressive female as well as the sexually agressive male, though it is often expanded to *wolfess* when applied to a woman.

Wolffian, *a.* Add: *Wolffian ridge*, each of two longitudinal ridges on either side of the embryo on which the limb buds arise.
1874 Foster & Balfour *Elements Embryol.* I. vi. 143 The somatopleure..is raised up..into a low rounded ridge which runs along nearly the whole length of the embryo from the neck to the tail... This ridge..is known as the Wolffian ridge. 1931 A. Robinson *Cunningham's Text-bk. Anat.* (ed. 6) 76 By the end of the first month.. the Wolffian ridges have appeared. 1976 N. J. Berrill *Development* xiv. 309 In amniotes the combined epidermal-mesenchymal thickening extends as a horizontal ridge along each side of the body (the Wolffian ridges). The intermediate part of the ridge later disappears, leaving anterior and posterior regions as the definitive limb areas.

Wolfian, *a.*[2] (Earlier example.)
1824 De Quincey in *London Mag.* Jan. 5/1 Was the Iliad the work of one mind, or (on the Wolfian hypothesis) of many?

wolfish, *a.* Add: **2. b.** In sense 8 b of WOLF *sb. rare.*
1889 *Grove's Dict. Mus.* IV. 89/1 Bad Tenors [*sc.* tenor violins] are worse than bad violins; they are unequal and 'wolfish'.
4. *wolfish-looking* (earlier example).
1820 Scott *Ivanhoe* I. i. 14 A rugged wolfish-looking dog..half mastiff, half greyhound.

Wolf–Rayet (vǫ·lf,rę·yę, wulf-). *Astr.* The names of C. J. É. *Wolf* (1827–1918) and G. A. P. *Rayet* (1839–1906), French astronomers, used *attrib.* to denote any of a class of hot white-to-blue stars (first described by them in 1867) which are characterized by bright, broad spectral lines due to hydrogen, helium, carbon, or nitrogen and are believed to be short-lived and unstable.
1890 A. M. Clerke *Syst. of Stars* v. 71 Accurate measurements of the three original Wolf-Rayet stars.. were made. 1930 R. H. Baker *Astron.* ix. 356 The Wolf-Rayet stars are distinguished from the other Class O stars by the great width of the bright lines in their spectra. 1978 Pasachoff & Kutner *University Astron.* ii. 40 Astronomers think that the emission in Wolf-Rayet stars comes from shells of material that the star has ejected into the space surrounding it.

wo·lf-whistle. *colloq.* Also **wolf whistle.** [f. *WOLF *sb.* 4 c + WHISTLE *sb.* 3.] A distinctive whistle from a man expressing sexual admiration for a woman; also *transf.*
1952 *Time* 21 Jan. 29/3 No one took exception to U.S.N. wolf-whistles at the señoritas. 1953 N. Balchin *Sundry Creditors* 46 Some vulgar female person let out a low wolf-whistle as she passed him. 1958 *Daily Express* 13 Mar. 8/5 She heard one kid give a wolf whistle, and his chum exclaim: 'Coo, what a smashing car!' 1960 A. Kimmins *Lugs O'Leary* i. 11 They passed the pretty probationer. Lugs gave her a low wolf whistle. 1971 *New Scientist* 29 Apr. 246/1 A young housewife..recently asked for a reduction in the rates of her residence because of nuisance from wolf-whistles. 1980 'T. Hinde' *Daymare* vi. 61 Bob Smiles whistles at him..a hideous wolf-whistle.
Hence as *v.*, to utter a wolf-whistle (at); **wo·lf-whistling** *vbl. sb.* and *ppl. a.*
1955 *Sun* (Baltimore) 2 Sept. 1/5 The Governor of Mississippi today called for a complete investigation of the kidnap-killing of a Negro youth who allegedly wolf-whistled at a white woman. 1958 L. Little *Dear Boys* 222 They had their heads and shoulders hanging dangerously out of the windows [of a coach], wolf-whistling the odd bints on the pavements. 1958 *Times* 2 Sept. 11/7 Surrounded as I am by thousands of barking dogs, wailing cats, and wolf-whistling budgerigars. 1961 Wodehouse *Ice in Bedroom* 41 Dolly Molloy unquestionably took the eye... Wolf-whistling is of course prohibited in the lobby of Barribault's Hotel so none of those present attempted this form of homage. 1976 J. Grenfell *Joyce Grenfell requests the Pleasure* i. 17 An American sailor wolf-whistled at her. 1981 G. Petrie *Tondeau of Chartres* i. 19 Julie and Elaine took their bows to a cacophony of wolf-whistling.

wolfy (wu·lfi), *a. U.S.* [f. WOLF *sb.* + -Y[1].] Wolf-like; characterized by, or suggestive of the presence of, wolves; ferocious, uncivilized.
1828 *Western Souvenir 1829* 314 'Couldn't you take a pack or two of wolves along?' said Pete, sneeringly. 'We can spare you a small gang. It's mighty wolfy about here.' 1831 J. K. Paulding *Lion of West* (1954) ii. 54 Well, I hadn't had a fight for as much as ten days—felt as though I must kiver myself up in a salt bin to keep—'so wolfy' about the head and shoulders. 1838 B. Drake *Tales & Sk., from Queen City* 36, I say, Mr. Jack-of-knaves, it looks rather wolfy in these parts. 1927 C. M. Russell *Trails plowed Under* 114 This talk makes the whole bunch wolfy.

Wollaston (wu·lăstən). *Physics.* [The name of W. H. *Wollaston* (1766–1828), English physicist.] *Wollaston('s) prism*: a prism made by cementing together two prisms of calcite or quartz with their optic axes perpendicular to each other and to the incident light, which is thereby separated into two diverging beams of polarized light.
1890 T. Preston *Theory of Light* xi. 258 Wollaston's Prism.—This prism differs from that of Rochon only in that the optic axis of the first prism ABD is parallel to the face AB, so that it is merely Rochon's prism turned through a right angle. 1970 *Nature* 18 July 264/2 The Wollaston prism, which splits the incident starlight into two beams, polarized in mutually perpendicular planes, was placed in an adjustable mounting above the spectrograph slit.

wolly (wǫ·li). *slang.* Also **wally.** [Origin unknown: cf. *WALLY *sb.*[2]] A uniformed policeman, esp. a constable. Cf. *WOOLLY *sb.* 3.
1970 G. F. Newman *Sir, you Bastard* 8 The wollies were out in their cars, patrolling for drunks and discontents. 1977 'D. Cory' *Bennett* ii. 69 The doorman.. mentioned it to one of our wollies on the beat. 1983 J. B. Hilton *Asking Price* v. 33 These traffic Wollies make sure it all goes down, once they've licked their pencils.

Wolof (wǒu·lǫf), *sb.* and *a.* Also **Jolof** (yǒu·lǫf), **Woloff,** etc. [Native name.] **A.** *sb.* **a.** (A member of) an African people of Senegal and the Gambia. **b.** The language of this people, belonging to the Niger-Congo family.
1745 F. Moore in *New Gen. Coll. Voy.* II. 227/2 The Natives, who were Jolloifs, had taken his chief Mate and Surgeon Prisoners. 1823 Mrs. H. Kilham (*title*) African lessons. Wolof and English. 1848 *Rep. Brit. Assoc. Adv. Sci.* 162 Here the Woloff has not only no particular affinities, but fewer miscellaneous ones than any other language. 1883, etc. [see *PEULH *sb.* and *a.*]. 1908 T. G. Tucker *Introd. Nat. Hist. Lang.* 147 This process is common in Wolof. 1930 C. G. Seligman *Races of Afr.* iii. 58 The lower and middle portions of the Senegal River form the ethnic divide between Hamites and Negroes. Immediately south of the river the latter are represented by the Wolof (or Jolof). 1961 *Guardian* 25 Nov. 7 Senegal combines at least six different tribes... The largest is the Wolof numbering about 800,000. 1972 J. L. Dillard *Black Eng.* iii. 74 The widespread use of Wolof, which seems to have a special *lingua franca* status among West African languages, in the thirteen colonies. 1976 *Times* 20 Nov. 11/4 The third largest group are the Wollofs, whose women..are..among the most beautiful in Africa.
B. *adj.* **a.** *gen.*
1828 Mrs. H. Kilham *Specimens African Languages Sierra Leone* p. vii, In the Jolof Language there are two sounds of the guttural kind. 1865 R. F. Burton *Wit & Wisdom in W. Afr.* i. 2 (*heading*) Proverbs in the Wolof tongue. 1879 J. A. Farrer *Primitive Manners & Customs* iii. 92 The Wolof proverb, that 'lies, though many, will be caught by Truth as soon as she rises up'. 1918 *Harvard Afr. Studies* II. 98 De Rochebrune distinguishes the gluteal accumulation of fat commonly found in Wolof women and girls from true steatopygia. 1961 F. G. Cassidy *Jamaica Talk* vii. 146 The word *juke*, as in the American *juke box*..has been traced by Turner to Wolof *jug*, to misconduct oneself. 1977 J. Wyllie *To catch a Viper* (1979) vii. 45 The original African ethnic groupings such as those represented by the Yoruba, the old nations of Ashanti, Dahomey and Goshi and the Mossi, Mende, Mandinka and Wolof peoples.
b. Special collocation: **jollof rice** (see quot. 1982).
1959 [see *FUFU]. 1966 C. Achebe *Man of People* ii. 25 Whenever you allowed him a say in this matter he invariably came up with Jollof rice—his favourite dish. 1982 G. M. Dalgish *Dict. Africanisms* 76/1 *Jollof rice*, a West African dish, a stew of fish, chicken or beef, tomatoes, onions, rice, and chili peppers; said to be ultimately of Sierra Leone origin.

wölsendorfite (vö·lzĕndǫɹfəit). *Min.* [a. F. *wölsendorfite* (J. Protas 1957, in *Compt. Rend.* CCXLIV. 2942), f. *Wölsendorf*, name of a locality in Bavaria: see -ITE[1].] A hydrated oxide of lead, calcium, and uranium, $(Pb,Ca)U_2O_7.2H_2O$, found as orange or red orthorhombic crystals.
1957 *Chem. Abstr.* LI. 13659 (*heading*) Wölsendorfite, a new uranium mineral. 1975 *Mineral Abstr.* XXVI. 321/1 Red and orange aggregates of wölsendorfite are described from the oxidation zone of a U-Mo ore deposit [in the U.S.S.R.].

Wolstonian (wulstǒu·niăn), *a. Geol.* [f. *Wolston*, name of the village in Warwickshire where the type site is situated: see -IAN.] Epithet of the penultimate Pleistocene glaciation in Britain (identified with the Saale of continental Europe), and of a stratigraphic stage of the Pleistocene lying above the Hoxnian and below the Ipswichian; of or belonging to this stage or glacial. Also *absol.*
1969 *Proc. Geol. Soc.* Aug. 152 It is recommended that for the Pleistocene and Holocene of the British Isles the following ages/stages be adopted as a regional scale... Pleistocene: Devensian, Ipswichian, Wolstonian [etc.]. 1975 *Nature* 9 Oct. 478/2 The Sugworth deposit lies topographically well above the Hanborough Terrace

which has been ascribed both to the Hoxnian Interglacial and, more recently, to an early part of the Wolstonian. 1981 F. W. Shotton in Neale & Flenley Quaternary in Brit. xiii. 143 This makes the gravels post-Hoxnian or at least late-Hoxnian, and the overlying Wetton Till even later. So if the latter is pre-Devensian, it can only be Wolstonian.

Wolves (wulvz). [pl. of Wolf sb.] Colloq. name for Wolverhampton Wanderers Football Club.

1908 O. Seaman Salvage 140, I hardly care at all Whether the Wolves break up the Throstles' wings. 1923 Racing Record 10 Feb. 3/2 Bradford City I take to defeat the Wolves. 1960 [see *slate sb.¹ 2 b]. 1978 P. Bailey Leisure & Class in Victorian Eng. vi. 139 A Church of England school team in Wolverhampton, later the Wolves.

wolvish, a. Delete † Obs. and add: **1.** (Later examples.)

1911 D. H. Lawrence White Peacock I. vi. 97 There was a report of two grey wolvish dogs. a 1945 E. R. Eddison Mezentian Gate (1958) i. 16 'Nay, read it if you please: I had it but five minutes since.' And with a wolvish look he tossed the letter upon the table. 1978 J. Updike Coup (1979) vii. 267 She tugged at his hand with that fretful, proprietorial impatience of her wolvish race.

2. Also Comb., as wolvish-looking adj.

1954 J. R. R. Tolkien Fellowship of Ring iv. 101 Two wolvish-looking dogs sniffed at him suspiciously, and snarled.

wolvishness. Delete † and add later example.

a 1945 E. R. Eddison Mezentian Gate (1958) xxxvii. 191 God shield us from women on our councils of war... Besides, I mistrust Parry wolvishness. And bitch-wolf was ever more fell than dog-wolf.

woman, sb. Add: **I. 1. i.** new woman (earlier example).

1893 G. Gissing Odd Women I. viii. 235 A strong character, of course. More decidedly one of the new women than you yourself—isn't she?

j. woman about town: see *town sb. 8 b; woman in the street: see *street sb. 3 g; woman of letters (earlier example); woman of the people. Also woman-to-woman: cf. Man sb.¹ 4 h in Dict. and Suppl.

1818 'T. Brown' Brighton; or The Steyne I. i. 20 She passes for a woman of letters. [1859 Lytton What will he do with It? III. vi. ix. 302 Talking thus, Arabella forgot the relationship of pupil and teacher; it was a woman to woman—girl to girl—friend to friend.] 1907 G. B. Shaw Major Barbara iii. 274, I thought she was a woman of the people, and that a marriage with a professor of Greek would be far beyond the wildest social ambitions of her rank. 1929 J. B. Priestley Good Companions iii. v. 586 'I should think you have news,' said Susie, smiling and being tremendously woman-to-woman. 1933 Kipling Souvenirs of France ii. 44 A woman of the people led her away. 1935 E. Bowen House in Paris ii. iii. 114 Her round slate-blue eyes rolled in a woman-to-woman way. 1962 'L. Grex' Terror wears Smile iv. 58 'There's not going to be any nonsense. You know what I mean.' Those words had been uttered in a quiet woman-to-woman chat some weeks before. 1982 H. Innes Black Tide v. iii. 272 If I told her, woman-to-woman, the sort of person Karen was... Perhaps she'd understand then.

l. Also hell hath no fury like a woman scorned and var. (see Scorn v. 3 a, quot. 1697). Hence woman scorned, used allusively; also as attrib. phr.

1868 J. G. Saxe Poems 291 In classic authors we are often warned, There's naught so savage as a 'woman scorned'. [1886 M. H. E. Bates Chamber over Gate xxvi. 363 You know 'Hell hath no fury', etc. If your wife should ever wake up to the true state of the case..I'm afraid she'd be an ugly customer.] 1927 W. Johnston Affair in Duplex 9B 60 The old idea of 'the woman scorned' on which Chilton had been relying this time failed utterly. 1932 W. McFee Harbourmaster xxi. 333 They say hell hath no fury like a woman scorned. 1940 G. H. Coxe Glass Triangle x. 126 If you really want to know who could have wanted to kill him, you might as well start with me... You've heard that one about hell having no fury like a woman scorned? Well, that was me. 1942 N. Marsh Death & Dancing Footman iii. 49 He broke out into..merriment calculated..to arouse in Chloris the pangs proper to a woman scorned. 1967 —— Death at Dolphin viii. 222 She really does bear out the Woman Scorned crack. She is..not all that charitably disposed at any time. 1971 'J. Ripley' Davis doesn't live here any More 124 The discarded mistress—the 'woman scorned'—motive. 1973 I. Murdoch Black Prince 330 'Hell hath no fury like a woman scorned.' In a way I might have been flattered.

II. 6. a. woman-city, -eye, -flesh, -godhead, -haunt, -luck.

1880 G. Meredith Trag. Comedians II. v. 83 Exactly what his appreciation, in womanflesh, would lead him to fix on. 1915 D. H. Lawrence Rainbow iii. 81 He must get out of this oppressive, shut-down, woman-haunt. 1922 Joyce Ulysses 468 Under it lies the womancity. Ibid. 155 Mrs Breen's womaneyes said melancholily. 1923 D. H. Lawrence Ladybird 53 She could not finally believe in her own woman-godhead. 1946 Dylan Thomas Deaths & Entrances 40 The next-door sea dispelled Frogs and satans and woman-luck. 1971 V. Canning Firecrest vi. 83 He put his arm round her shoulder..and felt through silk the warmth and firmness of woman flesh.

b. (a) woman doctor, driver, -help, journalist, officer, p.c., police officer, -savage, teacher.

1853 Dickens Child's Hist. Eng. xxix, in Househ. Words 12 Mar. 48/2 Edward was now sinking in a rapid decline... They handed him over to a woman-doctor who pretended to be able to cure it. 1899 W. James Talks to Teachers on Psychol. ii. i. 227 What our girl-students and woman-teachers most need..is..the toning-down of their moral tensions. 1902 A. Bennett in Academy 21 June 635/1 The average woman-journalist is the most loyal, earnest and teachable person under the sun. 1911 D. H. Lawrence White Peacock i. ix. 158 At any rate the extra woman-help came. 1921 R. Macaulay Dangerous Ages iii. 65 Pamela bright and cool and firm, like a woman doctor. 1930 D. H. Lawrence A Propos of Lady Chatterley's Lover 25 We read of the woman-savage who wore three overcoats on top of one another to excite her man. 1968 R. L. Fish Bridge that went Nowhere iv. 44, I might have known it would be a woman driver! 1972 L. Lamb Picture Frame xviii. 154 A woman p.c. was clearing an outside drain. 1973 'B. Mather' Snowline x. 121 I'll send a couple of woman officers along. 1976 R. Lewis Witness my Death i. 36 You've shown all the worst traits that can be expected in a woman doctor. 1976 Southern Even. Echo (Southampton) 11 Nov. 32/5 A chase through rush-hour crowds ended with a suspected shoplifter escaping into the darkness..as he was pursued by a woman police officer. 1982 D. Mackenzie Raven's Revenge xi. 104 A small car stopped... The woman driver was already crossing the pavement, house-keys in hand. 1982 A. Brookner Providence ix. 108, I wonder why they didn't send a woman teacher.

c. objective, woman-scorner (example); woman-worshipping adj.; woman-hating adj. and sb.; similative, woman-breasted adj.

1928 W. B. Yeats tr. Sophocles' King Oedipus 42 Oedipus overcame the woman-breasted Fate. 1946 Dylan Thomas Deaths & Entrances 30 The woman breasted and the heaven headed Bird. 1939 W. Fortescue There's Rosemary xii. 87, I even got personal praise from a woman-hating Don who acted as dramatic critic for the 'Varsity magazine. 1973 E. Taylor Serpent under It xi. 172 He was really an old dear—all that woman-hating stuff was just a pose. 1935 Auden & Isherwood Dog beneath Skin i. v. 57 But perhaps you're a woman-scorner. 1921 D. H. Lawrence Sea & Sardinia iii. 114 Woman-worshipping Don Juans.

7. woman-hour, an hour's work done by a woman; **woman-movement** (earlier example); **woman-power,** (a) the exercise of authority by women; (b) the number of women available for work; the power of women in work; **woman question,** a controversy over the rights of women, esp. that in the nineteenth century; **woman trouble** colloq., (a) U.S., gynæcological problems (cf. Trouble sb. 4); (b) difficulties caused to a man by a relationship with a woman or women; **woman-year,** a year of a woman's life; esp. used as a cumulative measure in medical tests carried out on a number of different women.

1961 Guardian 23 Mar. 24/6, I shudder to think how many man-hours and woman-hours are spent..typing, 'Dear So-and-So'. 1979 M. McCarthy Cannibals & Missionaries i. 4 I'm doing a serious study of the woman-hours expended in this family. 1883 Harper's Mag. Aug. 468/2 Whether the great progress of the 'woman movement'..is due to the agitation of 'woman's rights', or proceeds in spite of it. 1927 Amer. Jrnl. Philol. XLVIII. 201 In the kingdoms established by the Successors in Egypt and in Syria women appear as co-rulers with their husbands and as regents during the minority of a son or the exile of a husband. This woman-power is sometimes regarded as a Macedonian tradition. 1938 Lancet 5 Nov. 1071/2 If the aim is to use the woman-power of the countryside to the best advantage it will often be best for the women themselves to organise their work collectively. 1941 New Yorker 29 Mar. 46/3 The long-threatened drive to bring womanpower into the war effort. 1968 Ramparts May 8 If you had a cover on Black Power like your cover on Woman Power, it would be a picture of a sharecropper with a harmonica in one hand and a piece of watermelon in the other. 1973 Black World June 32 Woman power Is Black power Is Human power Is Always feeling. 1976 H. Wilson Governance of Britain 7 Ernest Bevin..was in charge of the industrial and military mobilization of Britain's manpower and womanpower. 1977 Spare Rib July 23 We're desperately short of woman-power on all areas of the magazine. 1984 Sunday Times (Colour Suppl.) 28 Oct. 80/2 The more progressive industries are beginning to realise that they can't afford to waste Britain's womanpower. 1857 Geo. Eliot Let. 21 Sept. (1954) II. 383 Quite delivered from any necessity of giving a judgment on the Woman Question or of reading newspapers about the Indian Mutinies. 1884 T. Stanton (title) The woman question in Europe. 1930 J. Collier His Monkey Wife ix. 119 A belated essayist on the Woman Question. 1981 R. Tremain Cupboard ii. 31 The attitude of men to the Woman Question. 1959 N. Mailer Advts. for Myself (1961) 95 She comes to see him about something or other, woman trouble maybe, and he seduces her in his medical chambers. 1967 J. Irwin Murderous Welcome vi. 51 She repudiated strongly any suggestion of discord between husband and wife and poured scorn on the mere idea of woman-trouble. 1977 L. Meynell Hooky gets Wooden Spoon xi. 127 'What's happening to Len Carron these days?'..'Woman trouble.' 1959 Science 10 July 81/2 Sixteen certain and one probable pregnancy occurred in women taking the medication; this represented a rate of 2·7 pregnancies per 100 woman-years.

8. woman's magazine, a magazine designed primarily for women; also (colloq.) woman's mag; freq. attrib., esp. alluding to superficiality or stereotypical attitudes regarded as associated with such productions; **woman's movement** = woman-movement, sense 7 above; **woman's page,** a page of a newspaper devoted to topics intended to be of special interest to women; **woman's woman,** a woman whose qualities are appreciated by other women, a woman who is popular with other women (cf. man's man s.v. *man sb.¹ 21); **woman's work,** work traditionally undertaken by women.

1912 Magazine Maker Sept. 7 (title) Making a woman's magazine. 1944 U. Orange Company in Evening ii. 34 The woman's magazine short story market may be a footling one. 1958 Observer 23 Feb. 14/2 The play..fools about at woman's-mag-whimsy level of moral convention. Ibid. 25 May 16/7 A Miss Lonelyhearts on a foundering woman's mag. 1958 Spectator 13 June 768/2 Ophelia's infatuated woman's magazine royalism rings with a specially hollow tinkle in Glen Byam Shaw's production of Hamlet. 1959 J. Braine Vodi xxi. 231 Honestly, that woman's magazine stuff, just the sort of advice these damned aunties give. 1974 Times Lit. Suppl. 3 May 483/2 Without these pages of imaginative grace, the novel would be dangerously close to the woman's magazine level of romantic fiction with which it persistently flirts. [1881 E. C. Stanton et al. Hist. Woman Suffrage I. xiv. 577 The 'Woman's Rights' Movement is a practical one.] 1894 J. E. Schmahl in Englishwoman's Rev. 16 Apr. 90 For many years Maria Deraismes was looked upon as the sole and undisputed head of the woman's movement in France. 1906 C. P. Gilman Women & Economics (ed. 5) iii. 49 So utterly has the status of woman been accepted as a sexual one that it has remained for the woman's movement of the nineteenth century to devote much contention to the claim that women are persons! 1933 E. Waugh Scoop i. i. 15 Those carefree days when he had edited the Woman's Page. 1952 M. Steen Phoenix Rising iv. 87 An editor or a publisher or a woman's page-writer. 1971 Woman's page [see *sexism]. 1886 Kipling Plain Tales (1888) 47 Mrs. Hauksbee was honest..and, but for her love of mischief, would have been a woman's woman. 1923 G. Atherton Black Oxen xl. 246 She had never been a 'woman's woman', and it was patent that, as ever, she was far more animated in the company of men. 1976 'M. Nelson' Crusoe Test i. 12 Elegant was the word for Carla Bayer... She was not a woman's woman. [1670 Woman's work: see Woman sb. 1 l.] 1890 A. J. Armstrong Ingleside Musings 139 Thae bairns are just a woman's wark To keep them clean an' tidy. 1971 K. Millett Sexual Politics (1972) i. ii. 39 The 'woman's work' in which some two thirds of female population..are engaged is work that is not paid for.

9. a. women doctors (later example), drivers, friends, journalists, ministers, priests, professors, students.

1878 Harper's Mag. Mar. 602/2 The established physicians shook their heads. They never believed in 'women doctors'. 1896 C. L. Dodgson (title) Resident women-students. 1930 A. Bennett Imperial Palace xi. 63 A strong sex-bias which had persuaded him that women-drivers were capable of any enormity. 1935 D. L. Sayers Gaudy Night vii. 147 There are much better ways of enjoying Oxford than fooling round..with the women students. 1941 J. D. Carr Case of Constant Suicides xi. 142 What kind of a professor are you, anyway? Running around with women professors from other colleges. 1956 A. S. C. Ross in M. Black Importance of Lang. (1962) 97 The custom is now obsolescent, save perhaps between close women-friends. 1967 L. Meynell Mauve Front Door xi. 138, I switched off the engine and climbed out of the cab; bloody women drivers, I thought. 1971 Guardian 15 Apr. 11/1 The diocese of Hong Kong, the only diocese out of 300 to have stated openly its support for the ordination of women priests. Ibid. 11/2 In the Congregationalist church..it can no longer be considered particularly unusual to have women ministers. 1975 Ibid. 21 Jan. 6/2 Creches should be provided at places where women journalists work. 1981 'J. Ross' Dark Blue & Dangerous iv. 22 Did he have friends? Women friends? 1981 'A. Cross' Death in Faculty ix. 106 Most women students..don't really believe women professors actually exist.

c. women liberators, women's liberationists (see sense 10 below); **womenpower** = woman-power, sense 7 above; **women-wise** adv., in the fashion or way of women; **women-years,** pl. of woman-year, sense 7 above.

1969 Time 21 Nov. 53 Women Liberators at Atlantic City. 1968 Ramparts Feb. 28 Her concept of women power, then, is assimilation to achieve a grey-flannel equality for the purpose of bettering women's estate in society by having them beat the hell out of..the men at the Establishment game. 1978 Daily Tel. 3 May 17/2 Consultants to advise businesses on..how to..improve their use of manpower (or womenpower for that matter). 1930 Women-wise [see man-wise s.v. *man sb. 20 a]. 1977 Lancet 29 Oct. 922/1 The latest report is based on 206 689 women-years of observation.

10. women's college, a university college that admits only women as students; **women's group,** a group formed for the discussion or furtherance of the interests of women; **Women's Institute,** an organization of women in rural areas who meet regularly and engage in various social and cultural activities; **women's liberation,** the liberation of women from subservient social status and all forms of sexism; also (usu. with cap. initial) a militant movement with these aims; also abbrev. as **women's lib** (cf. *lib); hence **women's libber** [*libber], liberationist; **women's magazine** = woman's magazine, sense 8 above; similarly women's mag; **women's movement,** (a) = woman-movement s.v. Woman sb. 7; (b) the women's liberation movement (see above);

women's page = woman's page, sense 8 above; **women's room** U.S. = ladies s.v. *lady sb. 4 f; **Women's Rural Institute:** see *rural a. 7; **women's studies** orig. U.S., academic studies concerning women, their role in society, etc.; **women's wear,** clothing for women; **women's work** = woman's work, sense 8 above.

1867 Geo. Eliot Let. 22 Nov. (1956) IV. 401 There is a scheme on foot for a women's college, or rather university..to be in connection with the Cambridge university. 1920 A. N. Whitehead Concept Nature i. 7 Suppose that the expositor is in London, say in Regent's Park and in Bedford College, the great women's college which is situated in that park. 1948 M. Laski Tory Heaven vii. 92 The women's colleges at Oxford..became reserved for B's. 1984 'A. Cross' Sweet Death xi. 129 Can you offer me one reason for women's colleges these days? 1968 Ramparts May 8 Your attitude was condescending throughout, and your analysis of radical women's groups..amounted to a movement fashion report. 1977 Evening Post (Nottingham) 27 Jan. 4/4 He is a solicitor..and while in Paris he was invited to talk to an American women's group. 1906 Rep. Women's Institutes Province of Ontario 69 After placing before them the aims and objects of the Women's Institute..the warden kindly assured us that he was..in sympathy with our work. 1909 Jrnl. Home Econ. I. 161 The usual practice has been to hold the Women's Institute on the same day..as the institute for men. 1912 Rep. Farm & Agricultural Schools & Colleges in France, Germany, & Belgium (Board of Educ.) 20 The success which has followed the formation of Women's Institutes in Poland, the United States, and Canada incited some social and agricultural reformers in Belgium to organise similar means of education. Ibid. 20 The Women's Institute is an association of farmers' wives, daughters, and sisters who meet periodically..to hear lectures, read papers, and study books on professional subjects. 1921 Daily Colonist (Victoria, B.C.) 20 Mar. 32/2 The first general meeting of the Victoria Women's Institute..was held Friday afternoon. 1924 [see *scout sb.⁴ 2 c (a)]. 1935 N. & Q. 2 Feb. 87/2 The Women's Institute of the old village of Cambo. 1982 Daily Tel. 25 Nov. 14/6 With reference to the letter from a Women's Institute member in Yorkshire regarding the splitting up of the Yorkshire federation [etc.]. 1969 Time 21 Nov. 15 'My twelve-year-old son has been hearing a lot about Women's Lib lately,' says Ruth. 1971 Times 17 July 5/8 Ecology..one Texas paper said recently, has replaced women's lib as the people's favourite cause. 1976 P. & W. Proctor Women in Pulpit i. 10 The earring criticism soon subsided as they got used to her and saw that she wasn't interested in staging women's lib demonstrations. 1971 Women's libber [see *libber]. 1973 Lancet 24 Feb. 419/1 We have heard a lot lately from unhappy ones of one kind and another—the women's libbers, the occasional bleat from the male side, [etc.]. 1978 J. Galway Autobiogr. ii. 16 It sounds now like a good case for the women's libbers but my mother was essentially a happy woman. 1985 Observer (Colour Suppl.) 14 Apr. 23/2 What the women's libbers don't realise is that what they denigrate as domestic drudgery is exactly the right sort of thing to be doing when you are mucking around with the under-twos. 1966 New Left Rev. Nov./Dec. 12 Fourier was the most ardent and voluminous advocate of women's liberation and of sexual freedom among the early socialists. 1967 New Left Notes 10 July 4/1 The SDS National Convention adopts the following statement and program as written by the Women's Liberation Workshop. 1974 L. Deighton Spy Story v. 55 Women's Liberation..planned to march to Westminster. 1978 S. Sheldon Bloodline iv. 64 There were rumors that Hélène Roffe was an advocate of the women's liberation movement. 1969 Leviathan (Berkeley, Calif.) June 43/1 The radical women's liberationists believe that the true extent of women's oppression can be revealed and fought only if the women's liberation movement is dominated by working class women. 1979 R. Rendell Means of Evil 147 Sheila singing lustily, Sylvia, the Women's Liberationist, with less assurance as if she doubted the ethics of lending her support to so..sexist a ceremony. 1959 Observer 18 Jan. 19/2 The daydream world of women's mags. 1942 D. Powell Time to be Born ii. 42 All Vicky could do was to read the women's magazines and discover how other heroines had solved this problem. 1960 K. Amis New Maps of Hell iii. 81 Art is mentioned..with a frequency not even paralleled in women's-magazine stories. 1981 J. B. Hilton Playground of Death vi. 72 There were shopping notes, memos..a pile of women's magazines. 1902 H. Blackburn Women's Suffrage vi. 107 The Married Women's Property Bill occupied the main attention of those engaged in the women's movement. 1944 G. Myrdal American Dilemma App. 5 The women's movement got much of its public support by reason of its affiliation with the Abolitionist movement. 1968 Ramparts Feb. 31 The most active of the new radical women's movements is in Berkeley—which should surprise no one. 1985 Observer (Colour Suppl.) 14 Apr. 18/3 She champions women who simply want to stay at home with their babies without a lot of pressure from the women's movement. 1929 E. Linklater Poet's Pub xiii. 154 She had had two or three articles accepted for the women's page of the Daily Day. 1980 M. Babson Dangerous to Know i. 8 There weren't all that many openings for Women's Page Editors around the Street these days. 1961 Webster, Women's room. 1977 M. French (title) The women's room. 1981 'A. Cross' Death in Tenured Position i. 5 The women's room on the ground floor. 1972 Newsweek 10 Dec. 124/3 In the classroom, many women think less of competing with men than of learning about themselves. 'Women's studies' was nearly unknown before 1970; now 78 institutions have complete women's studies programs. 1976 Spare Rib Nov. 14/2 Last year there was one women's studies course here. This year four are planned. 1919 Mencken Amer. Lang. iv. 121 Women's wear, in English shops, is always ladies' wear. 1980 Times 22 Jan. 9/6 Austin Reed is planning to open a womenswear area in the Regent Street branch. 1974 R. Adams Shardik xxv. 209, I can't help wondering why he trims lamps at noon. Or why he trims lamps at all, if it comes to that, seeing it's women's work and he has that girl to help him. 1977 Undercurrents June–July 41/3 Spinning, weaving, knitting, crocheting, sewing are women's work and are called crafts.

womanfully, adv. (Earlier example.)

1821 M. Edgeworth Let. 29 Jan. (1971) 236, I will trample..upon all the prickles of the impossibilities and flatten them womanfully.

woman-hater. (Further examples.)

1847 [see woman-spiter s.v. Woman sb. 6 c]. 1951 M. McLuhan Mech. Bride (1967) 99/1 The glamour business..is crammed with both women-haters and men-haters of dubious sex polarity. 1982 J. Fox White Mischief xxi. 247 Colvile was a great woman hater.

womanly, a. Add: **1.** Freq. in phr. womanly woman.

1387–8 [in Dict.]. 1872 (title) Woman's rights and the wife at home. By a womanly woman. 1891 G. B. Shaw Quintessence of Ibsenism iii. 34 In real life a self-sacrificing woman, or, as Mr. Stead would put it, a womanly woman, is not only taken advantage of, but disliked as well for her pains. 1926 Wodehouse Heart of Goof iii. 96 The least you can do, as a good womanly woman, is to have a capable lawyer watching your interests. 1958 M. Kennedy Outlaws on Parnassus xii. 196 Telemachus..upbraids his mother for not behaving like a womanly woman. 1978 J. Porter Dead Easy for Dover vii. 78 The Brigadier..only exists to make Madame more credible as a womanly woman.

womanness. Add: Also, the quality of being a woman. (Later examples.)

1926 D. H. Lawrence Plumed Serpent ii. 29 They hated her mechanically for the very fact that she was a woman. They hated her womanness. 1971 Guardian 2 Dec. 11/4 There may be something about 'woman-ness' which has to be defined in its own terms, not by reference to maleness.

woman's rights, women's rights. Add: (Further examples.)

1632 (title) The lawes resolutions of womens rights, or the lawes provision for woemen. 1842 E. Pease Let. 29 Mar. [MS. in Library of Society of Friends, London], I believe that the Chartists generally hold the doctrine of the equality of woman's rights. 1859 A. J. Munby Diary 28 Jan. (1972) 18 Those who prate of women's rights, if they knew their own meaning, would honour such mighty daughters of the plough. 1876 H. James Roderick Hudson xi. 378 Five unmarried sisters, one of whom gave lyceum-lectures on woman's rights. 1897 E. Bellamy Equality xx. 119 There was a great stir about women's rights, but the programme then announced was by no means revolutionary. 1902 G. B. Shaw Let. 20 June (1972) II. 276 The Woman's Rights young lady. 1963 B. Friedan Feminine Mystique i. 28 The fact is that no one today is muttering angrily about 'women's rights', even though more and more women have gone to college. 1977 Socialist Press 2 Mar. 4/1 The latest in a series of non-political jamborees organised around the question of women's rights took place at Alexandra Palace last Saturday. 1982 N.Y. Times Mag. 1 Aug. 6 Campaigning by the women's-rights movement to purge the English language of sexism.

woman's righter (earlier U.S. example).

1858 J. J. Boyer Let. 9 July in Lawrence (Kansas) Republican 15 July 1/4 Capt. Holmes and lady joined us. She is a regular woman's righter, wears the Bloomer, and was quite indignant when informed that she was not allowed to stand on guard.

womanthrope (wu·mănþrōᵘp). [Joc. formation f. Woman sb. + -thrope as in Misanthrope.] A hater of women.

1891 O. Wilde Intentions 126 They would become confirmed misanthropes, or if I may borrow a phrase from one of the pretty Newnham graduates, confirmed womanthropes for the rest of their lives. 1902 'Coldstreamer' Ballads of Boer War iv. 35 I'm quite willing for to be Wot scholards calls a 'womanthrope'.

womb, sb. Add: **2.** Also in phr. womb-to-tomb, esp. used attrib. to denote procedures, etc., which span a lifetime. Cf. cradle-to-grave s.v. *cradle sb. 2.

1964 A. Wykes Gambling i. 8 During our womb-to-tomb progress we never stop gambling, for we cannot know the outcome of each of the many decisions we have to make every day. 1967 McLuhan & Fiore Medium is Massage 12 Electrical information devices for universal, tyrannical womb-to-tomb surveillance are causing a very serious dilemma between our claim to privacy and the community's need to know. 1968 G. Jackson Let. 29 June in Soledad Brother (1971) 163 From the womb to the tomb this plays in our minds. We are not worth more than the amount of capital we can raise. 1979 Bookseller 23 June 2830/3 Kane and Abel..is a womb-to-tomb tale.

5. (Mostly poet.) womb-element, -fruit, -land, -life; womb-fibrilled, -like adjs.; womb-ward adv.

1923 D. H. Lawrence Birds, Beasts & Flowers 94 Who lies with the waters of his silent passion, womb-element?—Fish in the waters. 1923 Womb-fibrilled [see *inturned ppl. a.]. 1922 Wombfruit [see *quickening vbl. sb.]. 1930 A. Huxley Vulgarity in Literature iv. 16 Those yearning popular songs which are the national anthems of Wombland. 1876 G. M. Hopkins Wr. Deutschland vii, in Poems (1967) 53 Warm-laid grave of a womb-life grey. a 1930 D. H. Lawrence Last Poems (1932) 308 The shell-like, womb-like, convoluted shadow. 1981 J. Wainwright All on a Summer's Day 24 An Interview Room..is womb-like in its complete isolation. 1923 D. H. Lawrence Birds, Beasts & Flowers 19 There was a flower that flowered inward, womb-ward.

Womble (wǫ·mb'l). [Shortening of Wombledon, fanciful alteration of *Wimbledon.] An imaginary animal depicted as inhabiting Wimbledon Common (see quot. 1968). Also, a soft toy representing this creature. Also transf.

1968 E. Beresford Wombles (1974) (dust-jacket), The Wombles are a bit like teddy bears to look at but they have real claws and live underneath Wimbledon Common and devote their lives to 'tidying up' all the things those untidy Human Beings leave behind. 1975 Sunday Express 15 June 6/3 In addition to the Wombles pop group and the TV series, there are Womble jigsaws, Womble dolls, Womble T-shirts, Womble pillow cases... Now..there are Womble-approved crisps. 1977 Lancashire Life Dec. 77/2 Now she had locked herself in the ladies' with five rubber frogs and a selection of plastic Wombles and was refusing to come out. 1978 Times 26 Aug. 3 (caption) Members of the Outset youth service group felling and clearing dead trees on Wimbledon Common. These 'Wombles' have also assisted in pond clearing. 1982 Buses Sept. 393/2 The customers being predominantly senior citizens or 'wombles' in the London Transport vernacular.

womby, a. Delete rare and add later examples. Also fig.

1934 Dylan Thomas Let. 2 May (1966) 117 I've a good mind to ferret an Old Boys' Society, & read them the waxiest and wombiest efforts that I've got. 1951 W. Sansom Face of Innocence xi. 150 The sense of round shapes about—the rounder mouldings of stone, the curves of boats..roundness, if you like, that is womby. 1977 Time 5 Sept. 44/1 It [sc. a van] is self-contained and self-containing, and its womby little room is packed with the motherly comforts of home.

womenfolk. (Earlier U.S. example of womenfolks.)

1851 J. J. Hooper Widow Rugby's Husband 50 Such wimmen folks.

wo·menish, a. [f. women, pl. of Woman sb. + -ish¹.] Of, pertaining to, or characteristic of, women. Also in phr. wine and womenish: cf. Wine sb.¹ f (b).

1892 S. Hale Let. 28 Apr. (1919) 273 The day was so hot that it smelled perspiration of emigrant women-ish. 1920 D. H. Lawrence Let. 4 Jan. (1962) I. 606 At midnight the Monty crowd ordered champagne and tried to look wine and womenish.

womens (wi·minz). U.S. dial. Non-standard pl. of Woman sb.

1928 [see *spade sb.³ 3 a]. 1945 L. Saxon et al. Gumbo Ya-Ya i. 8 'It's damn funny' Fisher sniffed, 'how womens is.' 1967 Boston Sunday Herald 26 Apr. (Comic Section), Sometime I think wimmens is the strongest gender. 1970 R. D. Abrahams Positively Black iv. 84, I got so many womens I cannot call they name.

womerah, womerar, varr. Woomera.

womoonless (wumū·nlės), a. nonce-word. [See def.] Joc. combination of Womanless a. and Moonless a.

1922 Joyce Ulysses 278 Croak of vast manless moonless womoonless marsh.

won (wǫn, wǭn), sb. [ad. Korean wŏn in same sense.] The basic monetary unit of (North and South) Korea.

1950 Times 16 Nov. 7/7 Between June 25 and October 1 expenditure totalled 28,000m. won,..while revenue amounted to about 1,000m. won. 1952 R. Cutforth Korean Reporter x. 85, I gave him 20,000 won, patted him on the back and said goodbye. 1981 'A. Hall' Pekin Target vii. 62, I declared 100,000 won and asked where I could change pounds sterling. 1984 Times 25 Jan. 6/8 Twelve South Korean herb medicine dealers were arrested for selling 1bn won (£850,000) worth of false cures.

wonder, sb. Add: **I. 1. a.** seven wonders of the world: so eighth wonder of the world (used hyperbolically of any impressive object, etc.).

1831 M. Edgeworth Let. 20 Jan. (1971) 473 A..spoiled child of 30 whose mother and father having not been able to conceal from him that they think him the 8th wonder of the world have at last brought him to acquiesce in their opinion. 1930 Amer. Speech VI. Eighth wonder of the world... Ford runabout. 1977 H. Fast Immigrants iv. 267, I rode the first cable car on California Street... The Eighth Wonder of the World.

d. (Later examples.) boneless wonder, a gymnast; fig., someone or something lacking 'backbone'; chinless wonder: see *chinless a. b.

1931 W. S. Churchill in Hansard Commons 28 Jan. 1022, I remember, when I was a child, being taken to the celebrated Barnum's Circus... The exhibit on the programme which I most desired to see was the one described as 'The Boneless Wonder'. My parents judged that that spectacle would be too revolting and demoralising for my youthful eyes, and I have waited 50 years to see the boneless wonder sitting on the Treasury Bench. 1946 Boy wonder [see *Chattertonian a. and sb.]. 1951 'J. Tey' Daughter of Time xiv. 186 The spectacle of Dr. Gairdner

trying to make his facts fit his theory was the most entertaining thing in gymnastics that Grant had witnessed... As a contortionist Dr. Gairdner was the original boneless wonder. 1963 Guardian 15 Feb. 20/6 One of those boneless wonders that go by the name of 'Observer' editorials. 1967 M. Shulman Kill 3 iv. ii. 168 Reconciling more contradictory positions than could be broken up by a boneless wonder on a trapeze.

6. g. (Earlier examples.)

1782 Boswell Jrnl. 16 July in Boswell, Laird of Auchinleck (1977) 456 While she was out, my father and Lady Auchinleck called, for a wonder. 1811 Princess Charlotte Let. 13 Nov. (1949) 12 Soon for a wonder I plucked up courage & went in.

i. wonders will never cease: that is indeed surprising; now freq. ironic.

1828 T. Creevey Let. 11 Feb. in Creevey's Life & Times (1934) xii. 258 Off he went with, 'Well, Creevey, wonders will never cease!' I met Lord Bathurst at the Duke of Buccleuch's [etc.]. 1837 Dickens Pickwick Papers xlv. 489 Vonders vill never cease... I'm wery much mistaken if that 'ere Jingle worn't a doin' somethin' in the vater-cart vay! 1902 Conrad Typhoon xxiv. 191 'Solomon says wonders will never cease,' cried Mrs. Rout joyously. 1962 M. Summerton Nightingale at Noon (1963) viii. 105, I offered: I'll help you...' She..gave me a cheeky grin. 'Hear that! Wonders will never cease!' 1974 A. Price Other Paths to Glory i. vii. 88 Wonders will never cease... Early Tudor—practically untouched.

III. 9. a. Further examples: wonder-beauty, boy, -child, drug, -gleam, goal, -horse, -look, -woman.

1921 D. H. Lawrence Sea & Sardinia v. 210 Real fresh wonder-beauty all around. 1922 —— Aaron's Rod xviii. 269 The glimmer of the open flower, the wonder-look, still lasted. 1927 E. O'Neill Marco Millions iii. i. 167 Worth while your waiting, eh?.. Yes, my wonder boy! 1927 A. Conan Doyle Case-Bk. Sherlock Holmes 15 A wonder-woman in every way. 1929 R. Bridges Test. Beauty iv. 188 The shifting hues that sanctify the silent dawn with wonder-gleams. 1938 Encycl. Brit. Bk. of Year 38/2 The one signed work in the series..was the wonder-child of the project. 1939 Time 14 Aug. 50/2 Sulfanilamide, the 'wonder drug', introduced into the U.S. in 1936, is credited with remarkable cures. 1939 Wonder horse [see *second-guesser]. 1939 Joyce Finnegans Wake 395 You know her, our angel being, one of romance's fadeless wonderwomen. 1948 Wonder drug [see *subtilin]. 1958 P. Scott Mark of Warrior i. 82 Old Ramsay's something of a wonder boy. He'll be top cadet of the course. 1975 Daily Tel. 18 June 2/8 Experts..began work on the vaccines following the failure of the post-war 'wonder drugs' such as sulphonamides and penicillin to wipe out these two diseases. 1976 West Lancs. Evening Gaz. 15 Dec. 1. 4/7 He scored a superb hat trick with a wonder goal to round it off. 1976 Liverpool Echo 23 Nov. 7/1 Southport's golden sands, world famous as the training track of wonderhorse Red Rum. 1980 'R. B. Dominic' Attending Physician xiv. 117 Senator Gerald Ewell was a Democrat... 'What's Wonder Boy done this time?' demanded Tony. 1980 I. Hunter Malcom Muggeridge iv. 59 Various bizarre proposals to sort out and rearrange our genes so that everyone will become superman and wonderwoman. 1985 Times 2 Jan. 15/2 The word from the market is that a replacement 'wonder drug' is now in clinical trials.

b. Further examples: wonder-life, song.

1895 Kipling Seven Seas (1896) 84 The everlasting Wonder Song of Youth! 1929 R. Bridges Test. Beauty iii. 105 With what other numberless wonder-lives of the Saints they wrote.

10. a. (instrumental) wonder-ridden, -stricken (earlier example), -wide adjs.

1916 D. H. Lawrence Amores 76, I see each shadow start with recognition, and I Am wonder-ridden. 1818 Shelley Laon & Cythna v. xliii. 114 The morning's golden mist, Which now the wonder-stricken breezes kist With their cold lips, fled. 1922 Joyce Ulysses 530 Milly Bloom..calls, her young eyes wonderwide.

b. wonder-fine.

1929 R. Bridges Test. Beauty i. 29 Not to these look we with grateful pleasur or satisfaction of soul, wonder-fine tho' they be.

11. wonder-man: also in weakened sense, a man whose achievements are admired; **wonder rabbi,** in the Chasidic movement, a *tsaddik; **Wonder State** U.S., a nickname for the state of Arkansas.

1933 Amer. Speech VIII. iii. 39/2 Wonderman. Foreign fighters are often thus described [by sports writers]. 1935 Wodehouse Luck of Bodkins xv. 173 They get the idea that they are sort of wonder-men who can just look around and find talent where nobody else would suspect it. 1961 Catholic Herald 23 June 3/1 (heading) De Gaulle, hero and wonderman. 1962 A. Sampson Anat. of Britain xxvii. 450 In Whitehall he had the reputation of a wonderman, and had even been tipped by some as an eventual head of the Treasury. 1907 I. Zangwill Ghetto Comedies 409 We Chassidim have no fear. Our wonder-rabbi has power over all the spheres. 1970 C. Kersh Aggravations of Minnie Ashe i. 11 [Her] father had been a wonder rabbi in some obscure Jewish village in Galicia—a worker of miracles. 1923 Gen. Acts Arkansas 804 Be It Resolved by the Senate of the State of Arkansas... That hereafter Arkansas shall be known and styled 'The Wonder State'.

wo·nderlandish, a. [f. Wonderland + -ish¹.] Seemingly enchanted.

1929 J. B. Priestley Good Companions iii. v. 590 He was beginning to feel wonderlandish again, what with Mr. Memsworth and the champagne.

wonderling. For † Obs. read rare and add later example.

1913 A. O'Connor Poems 8 Sweet wonderlings Of passing fancy, slight, too slight for birth, Yet dazzlingly alive with sudden mirth.

wonderstone (wʋ·ndəɪstōᵘn). [f. Wonder sb. + Stone sb.] **1.** (See quots.)

1824 Trans. Geol. Soc. I. 295 In the neighbourhood of Wells, and at Bleydon near the Bristol Channel, it [sc. the conglomerate] forms a beautiful breccia, called wonder-stone, consisting of yellow transparent crystals of carbonate of lime, disseminated equably through a dark-red earthy dolomite. 1887 H. B. Woodward Geol. England & Wales (ed. 2) ii. 232 The road to Wookey Hole..and that leading to Dulcot..show in places in the Red Marl a bed termed the 'Wonder Stone'.

2. A soft bluish-grey rock of volcanic origin in South Africa that takes a high polish.

1936 Mineral Resources Union S. Afr. (Geol. Survey, Union S. Afr.) (ed. 2) 299 Wonderstone appears to be one of the most indestructible of building stones. 1952 Archit. Rev. CXI. 329/1 Some of the work in 1936 by Henry Moore in African wonderstone and by Barbara Hepworth was very derivative, in the best sense, of the shapes assumed by..the stone and pebble plants of the African Karoo. 1952 L. MacNeice Ten Burnt Offerings 84 He was..Firm as a Rameses in African wonderstone. 1975 Stand. Encycl. S. Afr. XI. 490/1 Large quantities of high quality wonderstone are available in the Dominion Reef System, 10 km north of Ottosdal in the Western Transvaal.

wonder-working, ppl. a. Add: Also as vbl. sb.

1900 F. T. Elworthy Horns of Honour iii. 180 There is some confusion about the several stories told of its [sc. the hand of glory's] wonder-working.

wongai (wǫ·ŋgai). Austral. [Aboriginal name.] A name used in the islands off the north coast of Australia for the jujube, Zizyphus jujuba. Also wongai tree.

1947 I. L. Idriess Isles of Despair xxii. 146 All the yams and berries and wongais will soon ripen. 1959 K. Tennant All Proud Tribesmen vii. 84 Wongai trees..have big dark berries like dates. 1968 Courier-Mail (Brisbane) 10 July 2/8 Thursday Island..has an informal charm and friendliness which easily tempts one to bite into the sticky fruit of the wongai tree. If you eat this fruit, legend says, you are bound to return.

wonga-wonga. Add: **1.** For picata substitute melanoleuca. (Earlier example.)

1821 L. Macquarie Jrnl. of Tours (1956) 223 Major Morisett has most kindly sent his young friend Lachlan the following very handsome presents of pets; vizt. four black swans..and one wanga-wanga pigeon.

2. In full wonga(-wonga) vine. An evergreen climber, Pandorea pandorana, of the family Bignoniaceæ, native to Australia and bearing panicles of pale yellow or pink flowers.

1895 J. H. Maiden Flowering Plants & Ferns N.S.W. 33 The Wonga Wonga Vine... A tall, woody, glabrous climber with more or less twining branches. 1936 F. Clune Roaming round Darling xvii. 162 Another shrub was the wonga-wonga vine. It has white flowers, and the blacks used to hollow the stems and make whistles from them. 1946 K. Tennant Lost Haven (1947) xvii. 278 The bells of the Wonga vine, milky with plum velvet in their throats.

wongi (wǫ·ŋgi). Austral. local. [Aboriginal.] A talk or chat; a speech.

1929 K. S. Prichard Coonardoo xxv. 243 He..had seen smoke..and come in for a bit of sugar and a wongie. 1939 X. Herbert Capricornia xxiv. 354 'Give's a drink Joe —— I've rasped me old throat raw.'..'By cripes, Andy, that was a great wongi.' 1969 L. Hadow Full Cycle 178 If he asks you to have one..well, he's out for a wongi.

wonk¹ (wǫŋk). Also **wunk.** [Said to repr. pronunc. of Chin. huáng gǒu yellow dog.] In China, a dog. Also wonk dog.

1900 H. A. Giles Gloss. Subjects Far East (ed. 3) 318 Wunk,..yellow dog. A term commonly applied by foreigners to the ordinary Chinese dog. From the Ningpo pronunciation wou kyi, of the above two characters. 1909 J. O. P. Bland Houseboat Days in China vii. 78 Particularly around the great cities you find him of the modern type, sporting a muzzle-loader..and a half-trained wonk. 1939 'A. Bridge' Four-Part Setting i. 4 Away in the Chinese village a wonk dog bayed at the moon. Ibid. 5 There are all these Chinks and wonks about—you oughtn't to go alone. 1967 'A. Cordell' Bright Cantonese vi. 67 Starving wonk dogs, the scavengers of China.

wonk² (wǫŋk). slang. [In sense 1 related to *wonky a. The other senses may represent different words.] **†1.** In phr. all of a wonk, nervous, upset. Obs. rare.

1918 [see *doodah 1].

2. Naut. (See quots.)

1929 F. C. Bowen Sea Slang 153 Wonk, a useless hand, or a young naval cadet who has not yet learnt the elements of his job. 1962 W. Granville Dict. Sailors' Slang 134/1 Wonk, midshipman.

3. Austral. **a.** A white person.

1938 X. Herbert Capricornia 252 He went to the Dagoes and Roughs of second-class and won their friendship by..telling them how he had been cast out by the Wonks of the saloon. 1959 Baker Drum 157 Wonk, a white man or white woman. Aborigines (esp. half-castes) use this pejorative much as whites use the word boong to denote an aboriginal.

b. An effeminate or homosexual man.

1945 Baker Austral. Lang. vi. 123 An effeminate male is a..gussie, spurge and wonk. 1970 P. White Vivisector 213 I'd have to have a chauffeur to drive me about—with a good body—just for show, though. I wouldn't mind if the chauffeur was a wonk.

4. U.S. A disparaging term for a studious or hard-working person.

1962 Sports Illustrated 17 Dec. 21 A wonk, sometimes called a 'turkey' or a 'lunch', roughly corresponds to the 'meatball' of a decade ago. 1970 E. Segal Love Story 32 Who could Jenny be talking to that was worth appropriating moments set aside for a date with me? Some musical wonk? 1980 N.Y. Times Mag. 20 July 8 At Harvard the excessively studious student is derided as a 'wonk', which Amy Berman, Harvard '79, fancifully suggests may be 'know' spelled backward. (In British slang, 'wonky' means 'unsteady'.)

wonky (wǫ·ŋki), a. slang. [Obscure: the G. element wankel- has similar force.] Of a person: shaky, groggy; unstable. Of a thing: faulty, unsound; unreliable.

1919 Ld. Northcliffe Let. in Hist. The Times (1952) IV. i. xi. 507 Am weak, and wonky, as the telephone girls say, after a bad morning with the subscribers. 1923 H. C. Bailey Mr Fortune's Practice iii. 81 'Who runs the "Daily Watchman?"'.. It's the wonkiest print on the market.'.. 'You said "on the market"... Corrupt?' 'Well, naturally.' 1925 E. Wallace Strange Countess ix. 83 Financial adviser to some heads of departments, whose accounts went a little wonky. 1929 P. Gibbs Hidden City xvi. 79 It had made his heart jump in a wonky sort of way. 1932 Kipling Limits & Renewals 127 Haman's headlight's wonky. Something must have happened. 1957 Listener 11 July 67 Despite the perfection of isolated lines and phrases..most of the poems seem slightly out of shape, wonky, as if the kiln had not been hot enough. 1958 Observer 23 Feb. 15/5 Would they really have sent her on a dangerous mission with an ankle still wonky from an old parachute fall? 1981 Times Lit. Suppl. 5 June 633/3 The vast majority of murderers are ipso facto acutely wonky, and most frequently wonky in dispiriting and unimaginative ways. 1983 D. Boggis Woman they sent to Fight ii. 17 The window fitted badly, and her chair was..wonky with one short leg. 1984 A. Carter Nights at Circus ii. vii. 156 'How's the wonky arm?' she enquired. He showed his sling.

Hence **wo·nkiness.**

1982 Times 29 Apr. 10/8 Do not be disconcerted by its [sc. a book's] wonkiness of style.

wonst: see *once adv. A. δ.

wontedly, adv. Delete 'Now rare or Obs.' and add later examples.

1913 W. Owen Let. 4 Jan. (1967) 175 The Vicar's presence (taciturn instead of wontedly gay)..sat heavy on my soul the night. 1980 N.Y. Times 12 Dec. c-30/3 On Sunday at 7.30 p.m., South Street's 19th-century surroundings will be wontedly hushed, except at Bowne. 1981 Guardian 4 Oct. 21 Dexter was wontedly tough on some of the actors.

won ton (wǫn tǫn). Also **wan tan, wun tun,** and as one word. [Chinese (Cantonese), Pinyin húntun.] A small round roll or pocket of dough containing a savoury filling, eaten alone (after being deep-fried) or boiled in soup (won ton soup).

1948 R. W. Dana Where to eat in New York 66 The theater and night-club performers drop in late for chicken egg foo yong, a won ton soup, squab, or Chinese steak. 1952 W. Y. Hong Chinese Cook Bk. 37 Pick up about ¼ teaspoon of meat (or fish) mixture and place on one corner of the won-ton skin. Roll almost to the opposite corner... The usual serving is 12 to 15 won-tons per person. Ibid. 38 Chicken Won-Ton Soup... 8 cups super soup stock... 2 cups chicken meat... 90 pieces of won-ton. 1956 'E. McBain' Cop Hater (1958) xx. 172 The wonton soup was crisp with Chinese vegetables... The wontons were brown and crisp. 1972 K. Lo Chinese Food i. 54 Well made wuntuns floating in clear soups resemble clouds. 1976 Times 20 Aug. 12/8 Two spring rolls for 25p, skins for Wan Tan dumpling at 60p a lb. 1976 Time 27 Sept. 63/2 The Chinese deep-fry everything from shrimp toast and wontons to beef and chicken.

woo (wū), v.² [Origin unknown.] = *mah jong v.

1922 H. Sterling Standard Rules & Instr. Chinese Game Mah Chang (ed. 4) 6 Experienced players invariably prefer to 'Woo' quickly with a small score rather than fail to 'Woo' at all. Ibid. 12 The Woo hand adds 2 points if he Woo with only one possible piece..or if he hold one of his last pair and Woo with the other. 1943 K. S. Whitehead Mah Jong Chinese Way § 59 A player whose hand contains a false set has a 'foul hand', and cannot woo. 1973 J. Scarne Scarne's Encycl. Games xxiii. 451 When he completes his hand, four sets and a pair, a player may woo or mah-jongg by showing his whole hand. He wins the deal, ending play.

woo (wū), sb. [f. Woo v.¹] (A spell of) caressing or love-making; esp. in phr. to pitch a woo: see *pitch v.¹ 17 d.

1937, etc. [see *pitch v.¹ 17 d]. 1938 N. Marsh Artists in Crime ix. 120 Hello, you two, what are you up to? Having a woo or something? 1959 —— Singing in Shrouds vi. 111 A pair of tango dancers..strutted and stalked..and frowned ineffably at each other. 'What an angry woo,' Tim said. 1968 Guardian 27 Nov. 9/6 Couples making woo in motor-cars should be careful not to rock them too much.

wood, sb.¹ Add: **I. 5. i.** to have the wood on (a person) and varr.: to have the upper hand, to have a hold on. Austral. and N.Z. colloq. Cf. to have the goods on s.v. *goods sb. pl. 2 b.

c 1926 'Mixer' Transport Workers' Song Bk. 7, I hold the 'wood' on those who work. 1944 J. H. Fullarton Troop Target vi. xxii. 168 Then we've taken another hiding. And I thought we had the wood on Jerry today. 1954 T. A. G. Hungerford Sowers of Wind xxi. 264 Can't you realize I've got the wood on you? You've got two minutes. 1965 L. Haylen Big Red i. 55 It was another of her occasions of fear: she liked having the wood on you. 1974 D. Stuart Prince of My Country ix. 66 Father stands up. 'Look, Marney... Get down and be civil or shut up and get to hell out of it'! Mr Marney dismounts. Mr Molloy pours tea and makes room on the bench. It looks as if Father has the wood on this sour old man right from the start.

II. 7. c. (Earlier example.)

1822 Sunday Times 20 Oct. 1/2 (Advt.), The long established system of serving wine from the wood, in full measures.

g. A golf club with a wooden head; a shot made with such a club (more commonly wood shot).

1915 A. W. Tillinghast Cobble Valley Golf Yarns 75 Hodge couldn't quite get there with two from his wood. 1927 Jones & Keeler Down Fairway xv. 203 For the drive with the wood, and for all normal wood shots, I play the ball opposite the arch of the left foot. 1928 Evening News 5 May 8/3, I do not think another professional golfer in America is hitting such terrific tee shots and full woods off the fairway as Gene. 1952 W. J. Cox Play Better Golf xi. 54 The normal flight of the ball from a No. 4 wood is high. 1971 'D. Halliday' Dolly & Doctor Bird viii. 104 Lady Edgecombe..hit her first ball..a good third of the distance, nicely placed for a wood shot fairly close to the green. 1977 Times 17 June 28/1 (Advt.), Uxbridge Golf Centre... 4 woods, Nos 1, 3, 4, 5 and Irons 3–9.

h. The wooden frame or handle of a racquet, with reference to a shot in which these parts are accidentally used instead of the strings.

1955 Times 30 June 4/1 Could Nielsen save the set? He did after a lucky one off the wood had been a help. 1961 [see double-fault vb. s.v. *double a. C. 3]. 1974 Mills & Butler Tackle Badminton ii. 27 A fault can occur even when the shuttle is struck by the wood.

8. f. dead wood: see Dead wood in Dict. and Suppl.

g. to touch wood: see Touch v. 29.

III. 9. b. (a) wood-end (later example), -ride (later examples), -riding, shadow, -song (later example), -wonder; wood-child, -rhapsodist, -woman; (b) wood smoke (earlier example); wood box (earlier example). **c.** (b) wood-carrier (later example), -chopper (earlier example), -sculptor; wood-chopping (earlier and later examples; cf. wood-chop, sense 10 a below); wood sculpture. **d.** (a) wood-retreat, -well; (b) in sense 7 c, as wood port. **e.** (a) wood-grown, -lost adjs.; (b) wood-cased (earlier example), -feeding, -fired, -tongued adjs. **h.** similative, as wood-green, -wild adjs.

1850 S. Judd R. Edney ix. 135 The Old Man romanced with the fire, making it seem how he could graduate it exactly to the necessities of the room, and the state of the wood-box. 1921 Daily Colonist (Victoria, B.C.) 8 Oct. 9/1 (Advt.), Before you put on your slippers fill up one of our strong, attractive, useful, tidy Wood Carriers. It holds about six pieces of stove wood. 1892 W. B. Yeats Countess Kathleen 71 Between the pepper-pot And wood-cased hour glass. 1925 Blunden Eng. Poems 86 The wood-child with man's torture racked Dares seek him out, if he'll retract. 1779 Mass. Hist. Soc. Coll. (1814) II. 458 The Century discov[er]ed a man creeping towards the wood choppers. 1845 Thoreau Jrnl. 14 July in Writings (1906) VII. 367 He was going to do his woodchopping. 1933 Bulletin (Sydney) 23 Aug. 35/3 Woodchopping..is a fine, healthy and manly sport. 1919 J. Masefield Reynard 69 The wood-end rang with the clear voice crying. 1946 Nature 9 Nov. 644/2 Protozoa and bacteria are essential for digestion in the wood-feeding termites. 1974 W. Trager in K. Elliott et al. Trypanosomiasis & Leishmaniasis 247 Hypermastigote flagellates of the wood-feeding roach Cryptocercus..have a whole variety of sexual phenomena. 1956 Railway Mag. Mar. 163/1 The wood-fired locomotives were never very efficient. 1978 M. Duffy Housespy vi. 157 I've built a wood-fired kiln. 1807 J. Barlow Columbiad v. 169 The sandy stream-bank and the woodgreen plain Raise into sight the new made seats of man. 1925 E. Sitwell et al. Poor Young People 10 His wood-green laughter. 1922 W. B. Yeats Trembling of Veil 135 Little wood-grown islands. 1956 R. Macaulay Towers of Trebizond xiii. 142 The white-walled, red-roofed town and the wood-grown height beyond it. 1916 Blunden Pastorals 15 Voices of wood-lost winds. 1972 House & Garden Feb. 109/4 Each shipment of wood ports will have a continuity of quality... Ruby, tawny and white ports are all matured in wood. 1909 T. S. Eliot in Harvard Advocate 26 Jan. 135 As if one should meet A pensive lamia in some wood-retreat. 1885 W. B. Yeats in Dublin Univ. Rev. May 82/1 The birds that nestle in the leaves are sad, Poor sad wood-rhapsodists. 1928 Blunden Retreat 36 And wood-rides never reach the glittering gate. 1972 R. Adams Watership Down vii. 24 The head moved slowly, taking in the dusky lengths of the wood-ride in both directions. 1934 Blunden Mind's Eye 154 An abundant round of skilful practical doings, from the wagon-shed to the wood-riding. 1943 N. & Q. 9 Oct. 234 Wood-riding, green way across a wood. Northants. 1968 Canad. Antiques Collector Aug. 13/3 Quevillon, one of the leading wood-sculptors of the early 19th century, worked at Longueil from 1818 to 1821. 1977 Belfast Tel. 27 Jan. 10/7 It's a new oak prie-dieu..and it has taken wood sculptor Billy Graham and joiner Tommy Simons 120 man-hours to turn it out. 1974 Saturday (Charleston, S. Carolina) 20 Apr. 5-A/2 (Advt.), Children up to 15 are encouraged to come and participate free in learning to paint, make jewelery, wood sculpture and other crafts with all materials free. 1922 Joyce Ulysses 11 Woodshadows floated silently by through the morning peace. 1747 H. Glasse Art of Cookery ii. 42 Hang it up in a Chimney where Wood-Smoke is. 1930 T. S. Eliot Marina, Those who suffer the ecstasy of the animals, meaning Death Are become unsubstantial, reduced by a wind, A breath of pine, and the woodsong fog By this grace dissolved in place. 1938 Dylan Thomas Map of Love (1939) 13 But I, Ann's bard on a raised hearth, call all The seas to service that her wood-tongued virtue Babble like a bellbuoy over the hymning heads. 1920 E. Sitwell Wooden Pegasus 106 Dark wood-wells. 1953 —— Gardeners & Astronomers 37 And is blown by the bright air Upon your wood-wild April-soft long hair. 1903 W. B. Yeats In Seven Woods 21 And the wood-woman whose lover was changed to a blue-eyed hawk. 1925 Blunden Eng. Poems 92 Oh could it but be held by these wood-wonders.

10. a. wood-and-water joey Austral. slang, an odd job man; **wood-block,** (b) Mus., a hollow wooden block used as a percussion instrument; cf. Chinese block s.v. *Chinese a. 2 and temple block s.v. *temple sb.¹ 6 c; **wood-burner,** (a) a locomotive that is fuelled with wood; (b) a wood-burning stove or fire; **wood-burning** a., using wood as fuel; **wood-butcher** (earlier example); **woodchip,** a chip of wood; also (in full woodchip paper), wallpaper with woodchips, etc., embedded in it to give an uneven appearance; woodchip board = chipboard s.v. *chip sb.¹ 9; **wood-chop** Austral. and N.Z., a wood-chopping contest; † **wood-corder** U.S., a town official responsible for stacking cut wood for sale into standard 'cords' or piles; **wood-flat** U.S., a raft or flat-bottomed boat used for transporting wood by water; **wood-free** a. Paper-making, made free from mechanical wood, though not necessarily from chemical wood; also as sb., a wood-free paper; **wood-heap** Austral. = wood-pile in Dict.; **wood-lot:** for U.S. read orig. U.S. and add earlier and later examples; **wood-mote** (later examples); **wood-pile,** (b) phr. a nigger in the woodpile: see *nigger sb. 1 d; also in allusive and euphemistic varr.; (c) Mus. slang, a xylophone; **wood-ranger:** for U.S. read orig. and chiefly U.S.; (earlier and later examples); **wood ray** Bot. (see quot. 1933); **wood-road,** a track or rough road through woods; **wood rot,** a fungal disease that causes wood to rot; so **wood-rotting** a.; **wood-saw** (examples); **wood-sawyer,** (a) (earlier and later examples); **wood-skin** (later examples); in full woodskin canoe; **wood-wind:** also, an individual instrument of this kind (now often made of some other material); **wood-wing** Theatr., a wing which is shaped and decorated so as to represent a tree or trees.

1887 All Year Round 30 July 67/2 A 'wood-and-water Joey' is a hanger about hotels, and a doer of odd jobs. 1930 V. Palmer Passage i. v. 42, I wanted you to be something different from a wood-and-water joey, earning a few pounds here and there. 1966 Woman's Day (Sydney) 31 Oct., He is a 'wood and water joey'—the lad who does the odd jobs around the homestead. 1930 Étude Sept. 620 (caption) The drummer in a modern theater orchestra uses the assortment of instruments here shown. There are .Trap Console, Italian Tam Tam, and Wood Block. 1969 Listener 23 Jan. 121/2 The viola player also plays woodblock, and the viola and cello bow a suspended cymbal. 1972 Jazz & Blues Oct. 28/2 The drummer accompanies on the drums, with woodblocks used to give tonal contrast. 1901 World's Work (N.Y.) Dec. 1518/2, I began when there was nothing but wood-burners, big flaming smokestacks, and all that. 1965 G. McInnes Road to Gundagai v. 81 A gas stove and an old fashioned woodburner. 1980 Sunday Times (Colour Suppl.) 30 Mar. 69/3 Finland's last wood-burner steams through an Arctic Circle blizzard. 1951 W. Faulkner Requiem for Nun iii. 225 The light-wheeled bulb-stacked wood-burning engines shrieking among the swamps. Ibid. 251 The intractable and obsolescent of the town who still insisted on wood-burning ranges. 1960 Times 20 Oct. 15/2 A wood-burning river steamer. 1980 A. E. Fisher Midnight Men xv. 187 Sarah's studio..was warm..with a big wood-burning stove. 1883 Sporting Life 27 May 4/3 What has he done to the New York Clipper's wood butcher that he should be thus caricatured? 1958 Times Rev. Industry Dec. 61/3 Information service, covering all aspects of the production of wood-chip board. 1973 Nation Rev. (Melbourne) 31 Aug. (Suppl.) 1/1 The impending threat to Australia's native forests from intensive forestry, and particularly from woodchip projects. 1976 Dumfries & Galloway Standard 25 Dec. 4 (Advt.), Top quality woodchip reduced from 49p to only 39p roll. 1976 Milton Keynes Express 18 June 14/1 (Advt.), Woodchip paper—ideal for overpainting. Sale price 37p. 1977 Abingdon Herald 2 June 9/2 (Advt.), Fine quality woodchip (ideal for overpainting) only 39p per roll. 1918 Bulletin (Sydney) 16 May 48/2 Bill Lucas will chop against a local champion... After the wood-chop five rounds between. 1934 T. Wood Cobbers xvi. 191, I saw a good wood-chop and some tumultuous steer-riding. 1964 Courier-Mail (Brisbane) 27 July 8/5 It will be dearer at this year's Show if you want to just drop in to see one or two woodchops. 1681 Rep. Record Commissioners City of Boston (1881) VII. 143 Chosen Overseers of Wood Corders. 1781 First Records Baltimore Town (1905) 43 The Commissers had it [sc. an oath] administred to him and afterwards appointed him Wood-corder. 1850 Knickerbocker XXXVI. 105 When he has a long wand, he is a wood-corder. 1785 in Maryland Hist. Mag. (1925) XX. 42 He hath gone up and down frequently in battans, scows and wood-flats. 1838 Jrnl. & Register (Columbus, Ohio) 27 Apr. 2/5 There were no boats at hand except a few large and unmanageable wood flats which were carried to the relief of the sufferers..by the few persons on the shore. 1883 'Mark Twain' Life on Mississippi 237 The Pennsylvania was creeping along,..towing a wood-flat which was fast being emptied. 1904 Jrnl. Soc. Chem. Industry 15 Jan. 34/2 (heading) Manufacture of wood-free cardboard for printing. 1966 Economist 24 Sept. 1269/1 The mill will make..good quality 'wood-frees'. 1979 Morning News (Karachi) 24 May 5/2 This variation is applied for .woodfree and mechanical pulp. 1943 K. Tennant Ride on Stranger (1968) iii. 21 Get back to the wood-heap. 1966 'J. Hackston' Father clears Out 77 Father was out at the woodheap chopping Mother's wood for her. 1658 Suffolk (Mass.) Deeds (1885) III. No. 174, I heeretofore purchased..all the rights to any wood Lott. 1706 Town Records (Manchester, Mass.) (1889) I. 115 It is Voted and agreed to lay out 50 or 60 Acors of land at the west end of our common for a wood lot. 1975 N.Z. Jrnl. Agric. Sept. 25/2 Burning requires fuel, but..piles of branches from the woodlots .are soon used up. 1976 Shooting Times & Country Mag. 18–24 Nov. 28/2 Not that Jim wouldn't shoot a woodcock that got up in front of him, or a pheasant from the plough between a couple of woodlots. 1900 J. Nisbet Our Forests & Woodlands i. 29 In the Charter of 1217 provision was made for a Court of Attachment or 'Woodmote' being held every forty days... Like the Woodmote, the Swainmote was originally held at irregular times. 1978 Lancashire Life Apr. 27/2 One named Ughtred Hodgkinson attended a woodmote at Whitewell in Bowland in 1570. 1936 Metronome Feb. 61/2 Wood pile, xylophone. 1936 W. Stevens Let. 13 May (1967) 311, I agree that there is something wrong in the woodpile. 1951 Time 22 Oct. 69 Red Norvo kept salting his half-hour stands with such tunes as .he used to rap out on his 'woodpile' (xylophone) with Paul Whiteman's band 20 years ago. 1977 'J. D. White' Salzburg Affair xvi. 139 He was the odd man out, the African in the woodpile. 1734 in Acct. Progress Colony Georgia (1741) App. v. 51 [The French] have Five hundred Men in Pay, constantly employed as Wood-Rangers, to keep their neighbouring Indians in Subjection. 1915 W. B. Yeats Reveries (1916) 137, I could not sleep from my fear of the wood-ranger. 1933 Trop. Woods XXXVI. 3 Wood ray or xylem ray, the part of a ray internal to the cambium. 1975 Sci. Amer. July 102/2 Among the components of the cambium are what are called ray initials; the continuation of a ray initial down into the sapwood of a stem, a branch or a trunk is known as a wood ray. 1821 J. F. Cooper Spy (1831) vii. 81 The English captain took the advice of this mysterious being and finding a wood-road..turned down its direction. 1891 Century Mag. Apr. 921, I moved camp, following the wood-road to the summit. 1954 C. Bruce Channel Shore 89 In early winters he and James had cut firewood there and hauled it out over the wood road he had swamped, and up the main road, home. 1926 Rev. Appl. Mycol. V. 521 The winter draws attention to the misleading impression created by the use of the term 'branch canker' for two totally distinct types of injury: one caused by the attacks of such organisms as Macrophoma theicola, and the other resulting from a wood rot. 1931 E. E. Hubert Outline of Forest Path. xi. 449 The classification of wood rots is largely based upon the colour changes produced in wood by fungi. The discolorations produced by wood-rot and sap-stain fungi..are responsible for a large part of the loss due to degrade in lumber. 1973 C. Bonington Next Horizon viii. 128 The garden bounded by a high hedge with an old wooden seat, softened with age and wood-rot. 1918 Wood-rotting [see sap-rot s.v. *sap sb.¹ 7 a]. 1971 P. H. B. Talbot Princ. Fungal Taxonomy i. 17 One can only conjecture how different the course of history might have been if the British fleet had not been laid low at times by the action of wood-rotting fungi. 1816 Austin Papers (1924) I. 264, 1 Wood Saw. 1884 'Mark Twain' Huck. Finn vi. 39, I found an old rusty wood-saw without any handle. 1815 North Amer. Rev. II. 143 Deaths by Violence... In New York Mr. John Wood, killed in the street by Patrick Hart, a wood-sawyer, with a stick of wood. 1891 M. E. Wilkins New Eng. Nun 43 Matilda's antecedents had come of wood-sawyers and garden-laborers. 1904 W. H. Hudson Green Mansions xxi. 289 Some compassionate voyager would let me share his wood-skin. 1934 E. Waugh Handful of Dust v. 287 The canoes were made of woodskin... They worked patiently but clumsily; one woodskin was split in getting it off the trunk. 1958 J. Carew Wild Coast iii. 44 He had to fetch his woodskins from Honey Reef. 1966 P. Sherlock West Indian Folk-Tales 37 Each morning the men of the tribe went out in their woodskin canoes. 1922 Joyce Ulysses 280 Double-basses, helpless, gashes in their sides. Woodwinds mooing cows. 1926 Whiteman & McBride Jazz ix. 195 Musicians recognize four general classes of instruments in speaking of the orchestra—strings, wood winds, brasses, and the battery of traps. 1967 T. Stoppard Rosencrantz & Guildenstern are Dead iii. 83 One of the sailors has pursed his lips against a woodwind. 1978 Early Music July 333/2 Vivaldi had to rely on Austrian and German makers for the newer woodwinds. 1933 P. Godfrey Back-Stage i. 19 Wood-wings are lugged into position. 1974 D. Smith Look back with Love xvi. 164 One of these quick-changes occurred during my first scene, and to cover it, I had..a short soliloquy, halfway through which a glance into the wood-wings showed me that our leading man was still three-quarters Lesurques when he should have been seven-eighths Dubosc.

b. wood bee (later examples), moth (later example); **wood buffalo** (later examples);

wood-cat, (*b*) (earlier example); **wood-duck** (earlier and later examples); **wood-grouse,** (*b*) (examples); **wood grub,** the larva of any of several wood-boring insects; † **wood hog** *U.S.*, a variety of pig which feeds in woods; **wood hoopoe,** any of several birds of the genus *Phœniculus* (or the family Phœniculidæ), native to Africa and distinguished by blue and green plumage and a long tail; **wood-leopard (moth):** = *leopard-moth* s.v. LEOPARD 6 b; (earlier example); **wood(s)-pussy** *N. Amer. colloq.*, a skunk.

1836 *Southern Lit. Messenger* II. 96 The wood-bee revels on their sweets. **1953** A. CLARKE *Coll. Plays* (1963) 344 The wood-bees court tangles of dew. **1897** E. COUES *New Light on Early Hist. Greater Northwest* II. xviii. 622 They are the wood buffalo, more shy and wild than those on the plains. **1961** W. P. KELLER *Canada's Wild Glory* v. 274 One small pocket of pure wood buffalo persist in a remote corner of the area, and plans are afoot to establish new sanctuaries for these. **1972** Wood buffalo [see *plain(s) buffalo* s.v. *PLAIN *sb.*[1] 10]. **1791** J. LONG *Voyages* 41 The country every where abounds with wild animals, particularly..otters, martins, minx, wood cats, racoons, [etc.]. **1777** Wood duck [see *NARRAGANSETT 2]. **1911** C. E. W. BEAN *'Dreadnought' of Darling* vi. 57 Wood duck ..are really not duck at all, but Queensland goose. **1980** *Outdoor Life* (U.S.) (Northeast ed.) Oct. 80/1 Grain-fed mallards or pintails are superb table fare, as are wood ducks fattened on acorns. **1838** T. NEED *Six Years in Bush* iv. 30 And the woods with partridges, wood-grouse, black squirrels and occasionally a turkey. *a* **1861** T. WINTHROP *John Brent* (1862) xxii. 245 The brace of wood grouse he had shot that morning. **1917** T. G. PEARSON *Birds Amer.* II. 14 Hudsonian Spruce Partridge. *Canachites canadensis canadensis*... Wood Grouse; Wood Partridge. **1956** *Numbers* (Wellington, N.Z.) May 8 The rotten wood..split lengthwise and fell apart, baring the wet sawdust tunnels of woodgrubs. **1964** R. BRADDON *Year Angry Rabbit* (1967) xx. 158 Her husband fed their child with a wriggling wood grub. **1805** R. PARKINSON *Tour in Amer.* 290 The real American hog is what is termed a wood-hog: they are long in the leg, narrow on the back, [etc.]. **1840** *Cultivator* VII. 81 The next fall, *mast* was plenty, and 'wood hogs' were fat. **1908** HAAGNER & IVY *Sk. S. Afr. Bird-Life* 26 The Wood Hoopoes..are represented in South Africa by two well-marked species. **1953** R. CAMPBELL *Mamba's Precipice* xi. 115 A whole flock of wood-hoopoes with scarlet beaks and silk-shot, glossy, green and purple feathers were raising the most amazing din in the tree. **1964** A. L. THOMSON *New Dict. Birds* 894/2 The wood-hoopoes..are very unlike the true hoopoes in general appearance. **1819** Wood leopard-moth [see LEOPARD 6 b]. **1916** A. HUXLEY *Burning Wheel* 24 Mottled and grey and brown they pass, The wood-moths, wheeling, fluttering. **1899** F. D. BERGEN *Anim. & Plant Lore* 61 Wood pussy, skunk. **1950** *Chicago Daily News* 16 Feb. 5/1 Miss Bennett paid $35 for the deodorized house-broken wood pussy. **1972** *Islander* (Victoria, B.C.) 18 June 9/2 You would never have known that said woods pussy had met its doom and left so many 'scents' behind in its will.

c. wood betony, (*b*) *N. Amer.*, a kind of lousewort, *Pedicularis canadensis*; **wood sanicle:** see SANICLE 1.

1886 *Harper's Mag.* Dec. 99/1 The wood-betony, it is called—to select its worthier title—a common early flower of our woods. **1976** *Hortus Third* (L. H. Bailey Hortorium) 832/1 Wood betony. Pubescent per., to 1½ ft.. Spring. Que. to Fla., W. to Tex. and n. Mex. **1793** J. SOWERBY *Eng. Bot.* II. 98 (*table*) Sanicula europæa Wood Sanicle... Common enough in woods, growing among dead leaves of trees. **1857** A. PRATT *Flowering Plants & Ferns* III. 12 S[*anicula*] *Europæa* (Wood Sanicle). **1961** R. W. BUTCHER *New Illustr. Brit. Flora* I. 816 The Wood Sanicle is a perennial plant with erect, ribbed stems.

d. *pl.* used *attrib.* in senses 2 or 3, as **woods boss** *N. Amer. Lumbering*, a foreman in charge of lumberjacks; **woods colt** *U.S. colloq.*, a horse of unknown paternity; also, a foundling; an illegitimate child.

1928 C. PERRY *Two Reds of Travoy* 44 'He's a scrapper from way back. Sort of a bully in the village, I guess.' 'Derosier's woods boss,' breathed Gwen. **1946** K. TENNANT *Lost Haven* (1947) xiv. 231 Alec strolled ashore to talk with the 'woods boss'. **1970** *Islander* (Victoria B.C.) 17 May 6/3 Pete Haramboure became manager and his son, John, woods boss. **1895** *Dial. Notes* I. 395 *Woods colt*, foundling. Winchester, Ky. **1903** *Ibid.* II. 337 *Woods colt*, a horse of unknown paternity. Also applied to a person of illegitimate birth. **1913** [see *OUTSIDER 1 c]. **1959** W. FAULKNER *Mansion* i. 4 Will Varner was going to have to marry her off..quick, if he didn't want a woods colt in his back yard next grass.

Wood (wud), *sb.*[3] The name of B. *Wood* (see quot. 1860[1]) used in the possessive to designate an easily melted alloy consisting of bismuth, lead, tin, and cadmium in decreasing proportions and used esp. for soldering.

Patented by Wood in *U.S. Patent* 27,590 (1860).

1860 *Amer. Jrnl. Sci.* LXXX. 271 [New 'fusible metal'. —Dr. B. Wood of Nashville, Tenn., has secured a patent (Weekly Scientific Artizan, Cincinnati, May 5th, 1860,) for an alloy composed of cadmium, tin, lead and bismuth, which fuses at a temperature between 150° and 160° F.] *Ibid.* 272 We have had time only to repeat a few of Dr. Wood's interesting experiments.. The alloy made by fusing together two parts of cadmium, two parts tin, four parts lead and eight parts bismuth melts at a temperature varying not far from 70° C. (158° F.) It may appropriately be called 'Wood's fusible metal'.—Eds. **1876** *Jrnl. Chem. Soc.* XXX. 592 The author then describes the method adopted by himself to measure the volumes of the four following fusible alloys at temperatures between 0° and 120°:—... III. Wood's alloy, the composition of which is represented by the formula $Bi_4Pb_2Cd_2Sn_1$. **1947** J. C. RICH *Materials & Methods of Sculpture* vi. 192 Wood's metal is rarely employed sculpturally although the material could be used as a casting medium because of its low melting point. **1974** *Nature* 11 Oct. 506/2 One eye was centred on a projection perimeter..and the visuotectal representation for that eye on the right tectum mapped with a Woodsmetal microelectrode.

Wood (wud), *sb.*[4] *Med.* The name of Robert W. *Wood* (1868–1955), U.S. physicist, used in the possessive to designate (*a*) a special glass that is opaque to visible light but transmits ultraviolet, and (*b*) ultraviolet light obtained by using this glass as a filter to remove visible components.

1925 *Index Medicus* X. 988/1 Experimental tumours studied by Wood's light. **1927** *Brit. Jrnl. Actinotherapy* Jan. 24/2 The healthy scalp under Wood's light gives only a feeble fluorescence of a dark violet colour. **1927** *Brit. Jrnl. Dermatol. & Syphilis* XXXIX. 352 Wood's glass costs about 1*s.* 6*d.* per square inch, but only a small piece is required. **1951** WHITBY & HYNES *Med. Bacteriol.* (ed. 5) xiv. 261 The microscope is illuminated by a mercury-vapour lamp with a Wood's glass filter which transmits ultraviolet but not visible light. **1958** *New Biol.* XXVII. 56 In 1925 two French workers discovered that *Microsporum*-infected hairs showed a very characteristic greenish fluorescence in ultra-violet light which had been filtered through glass containing nickel oxide, the so-called Wood's Light. **1961** R. D. BAKER *Essent. Path.* ix. 223 (*caption*) A Negro child developed papular white scaly oval lesions.. The involved regions fluoresced with Wood's light. **1983** *Oxf. Textbk. Med.* I. v. 371/1 When large numbers of children are involved, screening of scalp infections with a filtered ultra-violet (Wood's light) lamp is useful.

woodbine. Add: **2*. a.** (Normally **Woodbine.**) A proprietary name for a brand of cheap cigarettes; a cigarette of this brand.

[**1886** *Trade Marks Jrnl.* 6 Jan. 8 *Wild Woodbine Cigarettes*. W. D. & H. O. Wills, Bristol & London.] **1907** *Ibid.* 11 Sept. 1602 *Woodbine*... Tobacco whether manufactured or unmanufactured. The Imperial Tobacco Company.., Bedminster, Bristol. **1910** *Sessions Paper of Central Criminal Court* 16 Nov. 18 Prisoner asked for a packet of Woodbine cigarettes (1 d). **1914** *Autocar* 21 Nov. 736/1 'Woodbine! This is a bit of luck,' he exclaimed, taking a cigarette. **1924** H. DE SÉLINCOURT *Cricket Match* ii. 21 He picked a woodbine out of its paper on the mantelpiece. **1939** JOYCE *Finnegans Wake* 587 First a couple of Mountjoys and nutty woodbines..in the snug at the Cambridge Arms. **1970** B. CARTLAND *We Danced All Night* vii. 198 It was only during the war [of 1914–18] that Tommies had got used to the cheaper type, especially Woodbines. **1979** 'P. O'CONNOR' *Into Strong City* x. 29 A Woodbine cigarette found in my pocket. **1983** J. CROALL *Neill* viii. 148 As long as..he had enough for a packet of Woodbines, he was fine.

b. An Englishman, esp. a soldier, considered as a habitual smoker of Woodbine cigarettes. *Austral. slang.*

1919 W. H. DOWNING *Digger Dial.* 54 *Woodbine*, an English soldier, so called from the name of a cheap brand of cigarette favored by Englishmen. **1937** E. HILL *Water into Gold* 192 Bagtown became 'Woodbine Ave'..so-called for the number of English settlers in residence. [**1978** R. BEILBY *Gunner* 43 'Inglesi,' he grinned. 'Pommies. Chooms. 'Bines. That's what we call them.']

woodchuck. Add: **a.** For *Arctomys* substitute *Marmota*. (Earlier example.)

1674 *Cal. State Papers, Amer. & W. Indies* (1889) VII. 581 The natural inhabitants of the woods, hills, and swamps, are..rabbits, hares, and woodchucks.

b. *woodchuck hole.*

1853 H. D. THOREAU *Jrnl.* 29 Mar. (1949) V. 62 Looking at the mouth of a woodchuck-hole..[I see] that those places are sprinkled with..salt-shaped masses of frost. **1974** P. GZOWSKI *Bk. about this Country* 43/2 The hay wagon had dropped into a woodchuck hole.

woodcock, *sb.* **3. d.** (Earlier example.)

1861 MRS. BEETON *Bk. Househ. Managem.* xxxiii. 822 (*heading*) Scotch woodcock.

Wood Cree. Also **Woods Cree.** [f. WOOD *sb.*[1] + *CREE *sb.* and *a.*: a shortening of earlier *Strong* (also *Thick*) *Woods Cree*, tr. a Wood Cree name.] **1. a.** One of the major divisions of the Cree Indians, inhabiting woodland areas of Saskatchewan and Manitoba in Canada. **b.** A member of this people. Cf. *PLAINS CREE.

1885 *Boston Jrnl.* 23 June 1/8 The Wood Crees have gone back to get a cache of provisions. **1910** F. W. HODGE *Handbk. Amer. Indians* II. 414/1 *Sakawithiniwuk* ('people of the woods'). The Wood Cree, one of the several divisions of the Cree. **1947** *Beaver* June 15/1 The Wood Crees, who were in the minority, were, in general, less troublesome. **1972** [see *PLAINS CREE].

2. The language spoken by the Wood Cree.

1958 R. A. LOGAN *Cree Lang.* 4 One dialect (Moose Cree) uses the sound of L where another dialect (Northern or Woods Cree) uses the sound of TH.. Only Northern Cree uses TH. **1978** D. H. PENTLAND in Cook & Kaye *Ling. Stud. Native Canada* 190 Woods Cree is now spoken by about five per cent of the total number of Cree Speakers.

wooden, *a.* Add: **I. 3. b.** *U.S.* = WOODED *ppl. a.* ? *Obs.*

1816 U. BROWN *Jrnl.* 15 Aug. in *Maryland Hist. Mag.* (1910) X. 358 To Smith field a Wooden Town in a Wooden Country & a wooden bred set of Tavern-keepers. **1843** [see sense 3 in Dict.]. **1891** M. E. RYAN *Pagan of Alleghanies* i. 12 And then there are others more seldom seen, the women from the 'wooden' country of the interior.

II. 7. wooden spoon. Hence **wooden-spooner, -spoonist,** a competitor who is awarded the 'wooden spoon'; a loser.

1927 *Daily Express* 23 Mar. 13/3 Champions and wooden spoonists of the Isthmian League last season were opposed on the Civil Service ground at Chiswick. **1954** J. FINGLETON *Ashes crown Year* 275 Somerset were wooden-spooners last summer and will be so again. **1973** *Nation Rev.* (Melbourne) 31 Aug. 1442/3 4BH slips to fourth place in the five station market, with perennial wooden spooners, 4BK, only 2000 listeners behind. **1975** *Globe & Mail* (Toronto) 26 May S5/1 England won the British soccer championship..with Wales, once again the wooden spoonists. **1981** *Daily Mail* 25 Nov. 30 (*heading*) A flat rate from the wooden spoonists.

9. wooden cross *Mil. slang*, a wooden cross on a serviceman's grave; hence, death in action regarded ironically as an award of merit; **wooden kimono** *U.S. slang*, a coffin; **wooden nickel** (or **money**) *U.S. slang*, a worthless or counterfeit coin; chiefly in *fig.* phr. *to take a wooden nickel* and varr., to be swindled or fooled; **wooden nutmeg:** see NUTMEG 1 b; **wooden overcoat:** see *OVERCOAT; **wooden pear** (earlier example); **wooden suit** *slang*, a coffin; **wooden wedding:** for *U.S.* read orig. *U.S.* and add earlier and later examples.

1917 A. G. EMPEY *Over Top* 314 *Wooden Cross*, two pieces of wood in the form of a cross placed at the head of a Tommy's grave. **1919** in *Amer. Speech 1972* (1975) XLVII. 117 Seven of the 'Blue Tails' went down to get their Wooden Crosses. **1949** A. MURPHY *To Hell & Back* xvi. 195 There is no other branch of the army that offers so many chances for the Purple Heart, the Distinguished Wooden Cross, the Royal Order of the Mattress Covers. **1926** MAINES & GRANT *Wise-Crack Dict.* 15 *Wooden kimona*, case for cold storage. **1946** MEZZROW & WOLFE *Really Blues* ii. 19, I expected the man to turn up..with his tape measure to outfit me with a wooden kimono. **1915** C. MATHEWSON *Catcher Craig* ii. 25 He was instructed.. not to take any wooden money. **1922** S. LEWIS *Babbitt* v. 67 S'long! Don't take any wooden money. **1971** F. P. GROVE *Tales from Margin* 27 'Well,' said Walt, 'be good, fellah! Don't take any wooden money!' Even this cheap vulgarity irritated her now. **1927** *Amer. Speech* III. 132 [College slang.] Not to 'take any wooden nickels', in other words, to be alert. **1937** L. HELLMAN *Diary* 23 Oct. in *Unfinished Woman* (1969) viii. 100 Luis and I got to Madrid. He said I was not to take any wooden nickels. **1964** in Hamblett & Deverson *Generation X* 90 Then one night I met Johnny, one of the biggest sharks in the Mayfair aquarium, and that was the end of that. Nobody ever sold Johnny a wooden nickel. **1971** M. TORRIE *Bismarck Herrings* ii. 29 Having advised her..not to accept any wooden nickels, [he] drove back. **1971** R. DENTRY *Encounter at Kharmel* iii. 58 There hadn't been a tribal rising worth a wooden nickel since the Partition troubles died down. **1860** G. BENNETT *Gatherings of Naturalist* 322 The Wooden Pear-tree of the colonists..is peculiar to Australia. **1968** W. GARNER *Deep, Deep Freeze* xx. 188 Any mistake on his part could win him the prize of the wooden suit. **1972** J. S. HALL *Sayings from Old Smoky* 42 When a guy comes and steals my stuff, he better be ready for a wooden suit or Boot Hill. **1870** D. MACRAE *Americans at Home* II. 293 The fifth anniversary is called the wooden wedding.. The presents suitable to this anniversary are of wood. **1918** H. BARNETT *Canon Barnett* I. xiv. 162 In 1893 we decided to commemorate our wooden wedding by a congregational party.

III. 10. woodentop *slang*, (*a*) a uniformed policeman; (*b*) a dim-wit.

1981 J. WAINWRIGHT *All on Summer's Day* 96 I'm a copper. An ordinary flatfoot... A real old woodentop. That's me. **1983** A. BEEVOR *Faustian Pact* v. 33 They've even got the bleeding Army out... Bunch of woodentops from Chelsea barracks. **1984** *Listener* 16 Feb. 24/3 A policeman who is called a 'butter boy' or a 'wolly' must be something like a 'woodentop'.

wooden *v.*, (*b*) *Austral.* and *N.Z. slang*, to render insensible; to knock unconscious; also const. *out*; **woodenness** (earlier examples).

1904 'G. B. LANCASTER' *Sons o' Men* 252 He'll wooden more of you out if you scare him. *c* **1926** 'MIXER' *Transport Workers' Song Bk.* 126 It [*sc.* a block of ice] 'woodened' him out, and he lay there quite flat. **1952** M. ALLINGHAM *Tiger in Smoke* xi. 184 If you 'ad only woodened 'er, we'd have 'ad all the time in the world. **1974** *Southerly* XXXIV. 145 If you can't wooden 'em [*sc.* kangaroos] at a 'undred yards with one I.C.I. bullet, you're not tryin'! **1854** H. D. THOREAU *Walden* 356 Many concentric layers of woodenness in the dead dry leaf of society. **1860** F. W. FABER *Precious Blood* ii. 66 Considerable dryness, stiffness, woodenness,..would characterize this philanthropic city.

woodhenge (wudhe·ndʒ). *Archæol.* [f. WOOD *sb.*[1], after STONEHENGE.] A henge (a prehistoric circular bank enclosing a circular ditch) believed to have contained a circular timber structure, as represented by a ring of post holes; *spec.* (with capital initial) and orig., the proper name of the first example of this kind to be discovered, near Stonehenge.

1927 M. E. CUNNINGTON in *Antiquity* Mar. (*caption to plate between pp.* 92 *and* 93) 'Woodhenge': oblique view from the south. **1927** *Times* 28 Nov. 17/5 Woodhenge is

assigned to the Early Iron Age, say, 500 B.C., with a claim that the close correspondence in lay-out proves it to be a prototype in wood of Stonehenge. **1933** W. A. DUTT *Norfolk* (ed. 8) 60 One of the remarkable prehistoric circles known as 'Woodhenges'. **1935** *Nature* 7 Sept. 365/1 The generic term 'Woodhenge' was first used by Mrs. M. E. Cunnington to describe the circle near Amesbury with wooden uprights in place of stone, which she excavated in 1926 and 1928. The Norfolk Woodhenge, which was discovered from the air in 1929, was known from air photographs to be a striking example of the type. **1939** JOYCE *Finnegans Wake* 596 The Diggins, Woodhenge, as to hang out at. **1951** [see *HENGE[2]]. **1970** *Sci. Amer.* May 58 The four largest henge monuments in England, each surrounded by earthworks measuring more than 1,000 feet in diameter, are Avebury and three woodhenges. **1977** *Griffith Observer* (Griffith Observatory, Los Angeles) May 14/2 The Cahotian circles bore a superficial resemblance to neolithic timber structures like Woodhenge, near Stonehenge... Wittry therefore dubbed Circle 2 as an 'American Woodhenge'.

woodhouseite (wu·dhausəit). *Min.* [f. the name of C. D. *Woodhouse*, 20th-c. U.S. mineral collector + -ITE[1].] A hydrated sulphate and phosphate of calcium and aluminium, $CaAl_3(PO_4)(SO_4)(OH)_6$, found as colourless rhombohedral crystals and belonging to the beudantite group.

1937 D. M. LEMMON in *Amer. Mineralogist* XXII. 943 Woodhouseite is a late hydrothermal mineral lining vugs in quartz veins that cross the andalusite zones [in a Californian deposit]. **1980** *Mineral. Mag.* XXXI. 318/2 Heating such minerals as apatite..and woodhouseite to 600°–800° followed by quenching destabilizes these phosphates and allows more complete extraction of their uranium content.

woodie (wu·di) [-IE], *colloq.* abbrev. of WOOD-PIGEON.

1947 *Contemp. Rev.* June 368 Their habits were not so regular as the woodies. **1960** *Farmer & Stockbreeder* 23 Feb. 57/3 These birds eat wood pigeons' eggs... Some keen observers estimate that something like 80 per cent of the 'woodies'' eggs that are laid are so destroyed. **1972** *Shooting Times & Country Mag.* 4 Mar. 17/3 There shouldn't be many woodies remaining in the area. **1980** G. HAMMOND *Reward Game* ix. 129 Do you want sixty-odd woodies for the freezer?

woodie, var. *WOODY *sb.*

woodland. Add: **1. b. woodland caribou,** a northern caribou, *Rangifer tarandus*, found in forested areas of Canada.

1854 MAYNE REID *Young Voyageurs* 154 He had killed three caribou, of the large variety known as 'woodland caribou'. **1879** [in Dict.]. **1921** *Daily Colonist* (Victoria, B.C.) 30 Oct. 21/1 The only caribou I've ever hunted were in the Kootenays, woodland caribou. **1965** F. SYMINGTON *Tuktu* 44 The woodland caribou eats about the same forage as the barren-ground caribou.

2. *Archæol.* (With capital initial.) The name of a culture that existed in eastern North America between approximately 1000 B.C. and A.D. 1000, characterized by agriculture, hunting, burial mounds, and a distinctive style of pottery.

1917 C. WISSLER *Amer. Indian* xiv. 219 We now come to the so-called Eastern Woodland area, the characterization of which is difficult. **1946** *Nature* 2 Nov. 615/2 A single mound-group belongs to a later phase, the Middle Mississippi, and the village site and one mound are ascribed to the Woodland-culture pattern, probably still later. **1967** *Listener* 2 Mar. 290/2 Most of the characteristic traits of the late, *i.e.*, the Woodland, period are found in incipient form in the late Archaic, and it is a period about which it is difficult to generalize. **1977** G. CLARK *World Prehistory* (ed. 3) ix. 408 Hunting and fishing continued to play significant roles..even during the terminal phase of the Woodland culture (A.D. 900–1300).

Hence **woo·dlanded** *ppl. a.*, covered with woodland; **woodlander:** also, a plant whose natural habitat is in woodland.

1945 J. BETJEMAN *New Bats in Old Belfries* 6 By roads 'not adopted', by woodlanded ways She drove to the club in the late summer haze. **1948** W. ARNOLD-FORSTER *Shrubs for Milder Counties* iv. 113 D[*aphne*] *Blagayana*. A dwarf woodlander, evergreen. **1974** *Country Life* 12 Dec. 1896/1 American woodlanders, such as shortias..and erythroniums revel in it [*sc.* beech leaf-mould]. **1982** *Garden* CVII. 487/2 All [clintonias] are woodlanders or shade plants.

woodpecker. Add: **3.** *U.S.* and *Austral. Mil. slang.* A machine-gun.

1898 J. H. PARKER *Hist. Gatling Gun Detachment* vii. 127 Goin' to let the woodpeckers go off? **1932** J. DOS PASSOS *1919* 410 The shrill bullets combing the air and the sorehead woodpeckers the machineguns mud cooties gasmasks and the itch. **1945** BAKER *Austral. Lang.* viii. 157 *Woodpecker*, a Japanese .77 machine-gun. **1945** *Yank* 27 July 7 The Japs opened up with what sounded like dual-purpose 75s, 20-mm pompoms and woodpeckers. **1976** G. MARKHAM *Japanese Infantry Weapons World War Two* 41 The popular Taishō 3rd year type machine-gun.. was introduced in 1914. Its peculiar stuttering fire earned the gun the Australian nickname of 'woodpecker' (or 'woodchopper').

Woodruff[2] (wu·drɒf). *Mech.* [See quot. 1892.] *Woodruff key:* a key whose cross-section is part circular (to fit into a curved keyway in a shaft) and part rectangular.

1892 P. BENJAMIN *Mod. Mechanism* 455 [The Woodruff System of Keying.—The Woodruff Manufacturing Co., of Hartford, Conn., has brought out a novel system of keying.] *Ibid.* 924 (Index), Woodruff keys. **1923** C. D. ALBERT *Machine Design Drawing Room Probl.* i. 63 Woodruff keys are quite extensively used in machine tools and in machine construction generally. **1976** *New Motorcycling Monthly* Oct. 34/1 Remove Woodruff key from its slot in magneto shaft.

woodruffite (wu·drɒfəit). *Min.* [See quot. 1953 and -ITE[1].] A hydrated oxide of zinc and manganese, $(Zn,Mn^{2+})Mn^{4+}O_7.1–2H_2O$, found as black or grey monoclinic crystals.

1953 C. FRONDEL in *Amer. Mineralogist* XXXVIII. 769 The name woodruffite is proposed for this species after Samuel Woodruff (deceased), for many years employed as a miner by the New Jersey Zinc Company. **1979** *Ibid.* LXIV. 1214/1 The spectrum of woodruffite shows it to be a structural analog of todorokite, as has been assumed from the similarity of their X-ray powder patterns.

woodshed (wu·dʃed), *sb.* Also **wood shed, wood-shed.** [f. WOOD *sb.*[1] + SHED *sb.*[2]] **1.** A shed for storing wood, esp. for fuel. Also *euphem.*, a lavatory.

1844 [in Dict. s.v. WOOD *sb.*[1] 9 b]. **1854** H. D. THOREAU *Walden* 54, I have also a small wood-shed adjoining, made chiefly of the stuff which was left after building the house. **1868** N. HAWTHORNE *Passages from Amer. Notebks.* II. 9 We have been employed partly in an augean labor of clearing out a wood-shed. **1921** W. DE LA MARE *Mem. Midget* ii. 10 Pollie had gone to the woodshed to fetch kindling. **1940** W. FAULKNER *Hamlet* II. ii. 129 Serve you right for keeping a mare like that in a woodshed. **1974** M. HOYT *Thirty Miles* i. 1 The plumbing wasn't. Its place was taken by a small building known by the somewhat less-than-frank title of 'woodshed'.

2. *fig.* **a.** Phr. *to take into the woodshed* and varr.: to reprimand or punish. *N. Amer. colloq.*

From the old tradition of giving a child a spanking in the woodshed, i.e. not in the presence of others.

1907 *St. Nicholas* July 826/2 He could save himself and most of his companions from unpleasant reckonings in various and sundry woodsheds. **1949** *Time* 18 Apr. 22/2 If you don't do what we tell you to do we are going to take you out into the woodshed. **1966** *Toronto Daily Star* 21 Dec. 14 (*heading*) Taking the Senator to the woodshed. **1983** *Chicago Sun-Times* 16 July 34 Assuming the Fed is traditionally pliant, why does not Reagan simply take Volcker to the woodshed and tell him to ease up?

b. Phr. *something nasty in the woodshed:* see *NASTY *a.* 7. Also in allusive varr.

1940 AUDEN *Another Time* 111 What was it, Ernst, that your shadow unwittingly said? O did the child see something horrid in the woodshed Long ago? **1958** *Times Lit. Suppl.* 17 Jan. 30/1 Mr Amis does not, however, present Garnet Bowen as a case-history, whose dislike of foreign parts could be explained on a woodshed basis. **1959** *Listener* 8 Jan. 78/3 As the leading Torquemada, Miss Margaret Lane clearly felt some obligation to strive to uncover something—well—interesting in the woodshed.

c. *Mus. slang.* As a place where a musician may, or should, practise in private (see also quot. 1937).

1937 *Printers' Ink Monthly* May 45/3 *Wood shed*, a severe rehearsal. **1946** *Hollywood Note* June 4 T.D. [*sc.* Tommy Dorsey] goes back to the woodshed. **1977** *Rolling Stone* 16 June 66/2 Leavell's playing won't scare many jazz pianists into the woodshed.

woodshed (wu·dʃed), *v. Mus. slang.* [f. the sb.] *trans.* and *intr.* To practise or rehearse, esp. privately (see also quot. 1978).

1936 L. ARMSTRONG *Swing that Music* 71 We used to practice together, 'wood-shed' as we say (from the old-time way of going out into the wood-shed to practice a new song). **1946** MEZZROW & WOLFE *Really the Blues* viii. 108 I'll have to woodshed this thing awhile so I can get straight with you all. **1950** BLESH & JANIS *They all played Ragtime* (1958) x. 203, I would hear the tunes and, to make sure, go home and 'woodshed' them in every key, put them in major and minor and all the ninth chords. **1968** A. YOUNG in A. Chapman *New Black Voices* (1972) Drew's got an alto [horn].. Drew dont hardly touch it, he too busy woodsheddin his drums. **1978** *Amer. Speech* 1975 L. 302 [Jargon of barber-shop singing.] *Woodshed*, work out the harmony parts (to a known melody) by ear; sing as a group for the first time..; improvise (an interpretation).

Hence **woo·dshedding** *vbl. sb.*, (*a*) the dispensing of punishment; (*b*) the practice or rehearsal of music; (*c*) spontaneous or improvised barber-shop singing.

1940 *Amer. Speech* XV. 205 *Woodshedding*, disciplinary action. **1946** MEZZROW & WOLFE *Really the Blues* ix. 151 Instead of woodshedding, he went out after the big money with the primitive equipment he had when he started. **1955** SHAPIRO & HENTOFF *Hear me talkin' to Ya* xi. 190 It was here that the term 'woodshedding' originated. When one of the gang wanted to rehearse his part, he would go off into the woods and practice. **1956** S. LONGSTREET *Real Jazz* xiii. 101 Bix [Beiderbecke] did plenty of woodshedding, playing alone, to some recording on the family Victrola. **1973** T. PYNCHON *Gravity's Rainbow* I. 129 No head falsetto here but complete, out of the honest breast, a baritone voice brought over years of woodshedding up to this range. **1974** *Harmonizer* Jan.-Feb. 18/2 Woodshedding is not a 'spectator sport'—only participants can fully enjoy it. **1976** *Times* 27 Sept. 12/4 Spontaneous barbershopping is known as woodshedding, because a woodshed is as good a place as any to burst into sudden song.

Woodsia (wu·dziă). *Bot.* Also **woodsia.** [mod.L., f. the name of Joseph *Woods* (1776–1864), architect and botanist + -IA[1].] A fern of the genus of this name (family Polypodiaceæ), comprising small, rock-loving, tufted plants found in mountainous parts of Britain and other temperate regions and in the Arctic.

1815 R. BROWN in *Trans. Linnean Soc.* XI. 171 This genus I have named in honour of my friend Mr. Joseph Woods... The character distinguishing *Woodsia*..consists in its involucrum being inserted under the group of capsules..the *sorus*, which it completely surrounds at the base; while it is in every stage open at top. **1848** T. MOORE *Handbk. Brit. Ferns* iii. 37 The Woodsias have no especial claim to be esteemed for their elegance. *a* **1894** W. FALCONER *Let.* in W. Robinson *Wild Garden* (ed. 4) viii. 83 Woodsias, tiny Aspleniums, and other Ferns. **1908** E. STEP *Wayside & Woodland Ferns* 55 Oblong Woodsia (*Woodsia ilvensis*). This species differs but slightly from the Alpine Woodsia, and some botanists deny its distinctness. **1961** R. W. BUTCHER *New Illustrated Brit. Flora* I. 176 *Woodsia alpina*... This Northern Woodsia occurs very rarely on mountain cliffs and rocks in a few places in Scotland and N. Wales.

woodsy, *a.* For *U.S.* read orig. and chiefly *U.S.* and add later examples.

1973 *Times Lit. Suppl.* 27 Apr. 472/5 The merely woodsy setting of the keeper's activities and of his meetings with Connie. **1977** *Daily Tel.* 9 Apr. 7/3 Scatter them on woodsy bits of the garden. **1981** 'D. JORDAN' *Double Red* xxi. 92 The same perfume, a cool tinge of something woodsy. **1985** *Dirt Bike* Mar. 44/2 Duane Summers powers the XC up a woodsy trail.

Woodward-Hoffmann (wu·dwǫɹd hǫ·fmæ̃n). *Chem.* The names of Robert Burns *Woodward* (1917–79), U.S. chemist, and Roald *Hoffmann* (b. 1937), Polish-born U.S. chemist, used *attrib.* with reference to a series of generalized symmetry selection rules first proposed by them in 1965 which predict whether a particular pericyclic reaction will be allowed under the given conditions.

1968 *Jrnl. Amer. Chem. Soc.* XC. 1920/2 A considerable amount of research has been directed toward exploring the validity and extent of applicability of the 'Woodward-Hoffmann rules'. **1974** GILL & WILLIS *Pericyclic Reactions* iv. 100 (*heading*) The general Woodward-Hoffmann rule for pericyclic reactions. **1980** M. ORCHIN et al. *Vocab. Org. Chem.* x. 330 Sigmatropic shifts. Here the 'migrating bond' is treated as though it were heterolytically cleaved, then the Woodward-Hoffmann rules are applied.

woodwaxen. (Later example of spelling *woadwaxen*.)

1946 G. STIMPSON *Thousand Things* 50 It is supposed that the original green cloth made at Kendal by the Flemish weavers was colored with a dye obtained from the plant known as woadwaxen.

woodwork, wood-work. Add: **1. c.** *Assoc. Football slang.* The frame of the goalposts.

1960 *Times* 21 Nov. 4/2 Three more times they hit Bonetti's woodwork. **1977** *Grimsby Even. Tel.* 5 May 18/6 Twice in the first half, Scunthorpe hit the Bradford woodwork.

d. Phr. *to come* or *crawl out of the woodwork* and varr., to come out of hiding; to emerge from obscurity. So *to crawl (back) into the woodwork* and varr., to disappear into obscurity.

1964 'E. LATHEN' *Accounting for Murder* (1965) vii. 59 These nutboys start crawling out of the woodwork. **1973** *Times Lit. Suppl.* 9 Feb. 154/4 The Nazi elite faded into the woodwork without waiting to be removed, making it tempting to say that denazification should have been left to the Germans. **1973** *Current Affairs Bull.* (Sydney) Aug. 31/1 They are the new Australian playwrights and they are coming out of the woodwork everywhere. **1974** 'M. INNES' *Appleby's Other Story* iv. 30 At least we can tell this bloody wog to crawl back into the woodwork. **1977** C. MCCULLOUGH *Thorn Birds* xii. 289 Funny how the men in my life all scuttle off into the woodwork, isn't it? **1979** 'J. LE CARRÉ' *Smiley's People* (1980) iii. 39 George Smiley, sometime Chief of the Secret Service..had one night come out of the woodwork to peer at some dead foreigner. **1984** *Broadcast* 7 Dec. 27/1 The imminence of a BBC licence increase application brings the advertising agencies out of the woodwork.

3. c. Forestry, work done in woods.

1738 W. ELLIS *Timber-Tree Improved* I. 24 There is in my Neighbourhood a Man that is..often imployed in Wood-Work. **1904** G. A. B. DEWAR *Glamour of Earth* x. 243 Making a good and sure living..and filling an honourable post in wood work, to our surprise he took one day a strange step: flung up his work and migrated.. to the town.

4. *attrib.*

1959 I. & P. OPIE *Lore & Lang. Schoolch.* xvii. 362 The gardening master is commonly 'Spuds', the woodwork teacher is 'Chips'. **1980** E. BLISHEN *Nest of Teachers* I. iv. 22 The woodwork master..insisted that I come with him to his woodwork centre.

woodworking: also, forestry.

1950 *New Yorker* 26 Aug. 71/1 Woodworking firms are making a candid twelve-per-cent profit. **1951** R. FIRTH *Elem. Social Organization* ii. 51 Their introduced steel tools must have materially lightened the labour of woodworking and clearing of brush-wood in agriculture.

woody (wu·di), *sb. slang* (orig. *Surfing*). Chiefly *U.S.* Also **woodie.** [f. WOOD *sb.*[1] + -Y[6].] An estate car with timber-framed sides.

1961 *Surfer Q.* Winter 34 (*caption*) A 'woodie' piled high. Photographer Larry Stephens..challenges anyone to produce a picture of a surf car with more boards than his. **1969** *Surf Internat.* (Austral.) I. xi. 13 Nat and Paul push the woodie, it's stoked too, an' finally blows its gasket. **1973** J. MARKS *Mick Jagger* 87 Ramada Inn, the Yucca Hotel, Holiday Inn and the Seven Seas Hyatt Lodge—all in a row—saluting Dominique and the horde of military foundlings who plunge along Hotel Circle in borrowed *woodies*—those immaculate 1954 jobs by Ford: sturdy station wagons, armoured in wood. **1980** L. BIRNBACH et al. *Official Preppy Handbk.* 18/1 The other children were quite happy with their little red wagons; she would accept nothing but a woody.

woof, *int.* and *sb.*[2] Add: **2.** Var. WHOOF *int.* (*sb.*) in Dict. and Suppl.

3. Low-frequency sound of poor quality from a loudspeaker.

1961 in WEBSTER. **1962** *Listener* 22 Nov. 882/1 It isn't only technicians who can justifiably complain about too much tweet and woof. **1978** *Gramophone* Jan. 1298/3 They..are every bit the equal of the LPs, a beautifully warm and detailed orchestral tapestry, with..a richly resonant bass (without too much 'woof').

woof, *v.*[2] (In Dict. with WOOF *int.* and *sb.*[2]) **1.** (Later examples.)

1932 E. M. BRENT-DYER *Chalet Girls in Camp* vi. 97 Rufus..crossed the meadow at his best pace, woofing indignantly at intervals. **1955** V. NABOKOV *Lolita* II. xxviii. 171 A nondescript cur came out from behind the house, stopped in surprise, and started good-naturedly woof-woofing at me, his eyes slit, his shaggy belly all muddy, and then walked about a little and woofed once more. **1974** *Publishers Weekly* 5 Aug. 53/3 His attempt suggests a puppy woofing at a caterpillar—but keeping a safe distance.

2. *U.S. Blacks' slang.* **a.** *intr.* To talk (or, *trans.*, to say) in an ostentatious or aggressive manner.

1934 *Amer. Speech* IX. 290/1 [Negro slang.] *Woof*, to talk much and loudly and yet say little of consequence. **1935** Z. N. HURSTON *Mules & Men* I. iv. 86 The men would crowd in and buy soft drinks and woof at me, the stranger, but I knew I wasn't getting on. **1941** *Life* 27 Jan. 78 To reinforce a statement, a sub-deb says, *I ain't woofin'*..which means 'I'm not fooling'. **1941** *Direction* Summer 15/2 Stack got all big at the nose and woofed: 'All right, boss, you either fixes me up with that gin, or I pulls down this bar!'. **1972** J. WAMBAUGH *Blue Knight* vi. 86 He was woofing me, because he winked at the blond kid. **1974** H. L. FOSTER *Ribbin', Jivin', & Playin' Dozens* iv. 140 A student might say, 'Mr. Foster, he's woofin' on me.' This may have meant anything from he is challenging me to a fight, to he is making fun of my clothing or my mother.

Hence **woo·fing** *vbl. sb.* and *ppl. a.*

1942 *Amer. Mercury* July 96/2 *Woofing*, aimless talk, as a dog barks on a moonlight night. **1969** H. R. BROWN in H. L. Foster *Ribbin', Jivin', & Playin' Dozens* (1974) v. 179 Those young brothers came out of this woofing, diddy-bopping and raising hell period. **1973** B. G. COOKE in T. Kochman *Rappin' & Stylin' Out* 45 'Woofing' is a style of bragging and boasting about how 'bad' one is and is sometimes used by males and females when rapping to each other. **1975** *Today's Education* Sept./Oct. 54 Some of the woofin' has been precipitated by Whites trying to hustle Blacks out of goods and materials which have been promised or which are rightfully theirs. **1977** *Time* 14 Nov. 90/3 Cosby, who has one of the great faces of the Western world, is the best thing in this woofin', shuckin' film.

woofer (wū·fəɹ, wu·fəɹ). [f. prec. + -ER[1].] **1.** *U.S. Blacks' slang.* (See quot. 1934.)

1934 *Amer. Speech* IX. 289/1 *Woofer*, applied to one who talks constantly, loudly, and in a convincing manner, but who says very little. **1935** Z. N. HURSTON *Mules & Men* I. iv. 88, I want outside to join the woofers, since I seemed to have no standing among the dancers. **1974** H. L. FOSTER *Ribbin', Jivin', & Playin' Dozens* v. 202 The woofer may also move his body in a menacing way to make his woof more threatening and intimidating.

2. A loudspeaker designed to reproduce accurately low-frequency sounds whilst being relatively unresponsive to those of higher frequency. Cf. *SQUAWKER 3, *TWEETER.

1935 K. HENNEY *Radio Engin. Handbk.* (ed. 2) xxiv. 830 Wide range of frequency response is sometimes secured by using as many as three groups of speakers: low ('woofers'), medium, and high ('tweeters'). **1959** *Consumer Rep.* Sept. 453/1 The Best Buy..is much the better value of the two tweeters when coupled to a check-rated woofer. **1964** M. MCLUHAN *Understanding Media* (1967) xxxi. 348 It is like a badly wired woofer in a hi-fi circuit that produces a tremendous flutter in the bottom. **1979** *Arizona Daily Star* 5 Aug. I. 10/1 (Advt.), Big savings on a great sounding speaker. 8″ woofer for deep bass and 2½″ tweeter for clear highs.

woofits (wū·fits). *slang.* [Origin unknown.] An unwell feeling, esp. in the head; moody depression.

1918 J. M. GRIDER *War Birds* (1927) 96 Curtis says he is suffering from the Woofits, that dread disease that comes from overeating and underdrinking. *Ibid.* 207, I drank too much coffee before getting up and I'm as nervous as a kitten now. Must be getting the Woofits. **1932** *Amer. Speech* VII. 338 [Johns Hopkins jargon.] *Woofits*, ailment that comes with 'the morning after the night before'. **1958** 'N. SHUTE' *Rainbow & Rose* 100 Getting the woofits now, because I don't sleep so good.

woofter (wu·ftəɹ, wū·ftəɹ). *slang.* Also **wooftah.** [Fanciful alteration of *POOFTER.] = *POOFTER.

1977 *Private Eye* 8 July 5/1 The headshrinker had been reduced to a nervous wreck, and was prepared to dismiss the rabidly heterosexual Tynan as a wooftah. **1980** A. N. WILSON *Healing Art* iv. 47 The two young woofters in the pub.

woofy (wu·fi), *a.*[2] [f. WOOF *int.*, *sb.*[2], and *v.*[2] + -Y[1].] Of reproduced sound: having too much bass, or bass that is indistinct.

1932 J. H. REYNER *Mod. Radio Communication* (ed. 4) xx. 204 We shall experience a loss of the upper frequencies, the reproduction lacking brilliance and sounding 'woofy'. **1975** *Gramophone* Nov. 819/2, I prefer the sound of the horns..on the Decca Ace of Diamonds record, a much cleaner sound than the rather 'woofy' quality on the new record.

woofy, *a.*[3] [Relationship to WOOF *sb.*[1] and *sb.*[2] not clear.] Gruff; densely-textured.

1960 C. P. SNOW *Affair* xl. 371 The hairline which, when he was drunk, separated the diffuse and woofy benevolence from a suspicion of all mankind. **1976–7** *Art N.Z.* Dec./Jan. 15/1 She would have none of the delirious woofy mango-swamp muck of the then Auckland School. **1983** R. SUTCLIFF *Blue Remembered Hills* xvi. 124 A moustache..not of the woofy RAF variety but more akin to the kind worn by sergeant-majors.

wool, *sb.* Add: **1. g.** (*b*) (Earlier examples.) Also † *to spread* (etc.). For *U.S.* read orig. *U.S.* (*g*) *all wool and a yard wide* and varr., of excellent quality; thoroughly sound or honourable. (*h*) *wool away!* (Austral. and N.Z.) (see quot. 1965). (*i*) *to lose one's wool* (slang), to lose one's temper; similarly *to keep one's wool.*

(*b*) **1839** *Jamestown* (N.Y.) *Jrnl.* 24 Apr. 1/6 That lawyer has been trying to spread the wool over your eyes. **1842** *Spirit of Times* (Phila.) 29 Sept. (Th.), Look sharp, or they'll pull wool over your eyes.

(*g*) **1882** G. W. PECK *Peck's Sunshine* 85 You want to pick out (as the 'boss combination girl' of Rock Co.) a thoroughbred, that is, all wool, a yard wide. **1909** [see *LALLAPALOOSA]. **1913** J. LONDON *Valley of Moon* 60 You're a live one, all wool, a yard long and a yard wide. **1963** L. MEYNELL *Virgin Luck* v. 114 It didn't seem to matter so much with people as decent as that one. She was all wool and a yard wide, that one. **1974** 'A. GILBERT' *Nice Little Killing* iii. 40 No one will ever catch her..with an alibi all wool and a yard wide.

(*h*) *c* **1897** D. MCK. WRIGHT in A. E. Woodhouse *N.Z. Farm & Station Verse* (1950) 33 Wool away! Wool away is the cry And the merry game of busting is begun. **1949** P. NEWTON *High Country Days* v. 53 The call of 'wool away' had lagging fleecies dashing to rescue fleeces before the shearer would be out with his next sheep. **1965** J. S. GUNN *Terminol. Shearing Industry* II. 38 *Wool away*, the call of a shearer who wants the picker-up to carry away a fleece. This has to be done after each sheep, and fleeces are not left lying around on the floor while another sheep is being shorn.

(*i*) **1926** 'A. BERKELEY' *Wychford Poisoning Case* v. 48 'All right,' Alec said soothingly. 'Keep your wool on.' **1944** D. WELCH *In Youth is Pleasure* v. 87 Dennis said a lot more, growing increasingly vicious with each new sentence... 'My dear, don't lose your wool,' she said, mimicking old-fashioned schoolboy slang. **1959** [see *RAG *sb.*[1] 3 c]. **1967** O. NORTON *Now lying Dead* vi. 108, I lost my wool then.

3. a. Also, a woollen garment.

1933 H. ALLEN *Anthony Adverse* II. IV. xxv. 354 'I *am* a little cold after all,' said Father Xavier, looking at the fire regretfully. 'A second till I change into my wool.' **1952** M. LASKI *Village* v. 94 The beige silk frock could at last be discarded for a really not-too-bad navy blue wool. **1975** BYFIELD & TEDESCHI *Solemn High Murder* (1976) i. 10 Mueller had taken away his rumpled suit, leaving his heavier wools hanging in the open closet. **1978** S. BRILL *Teamsters* ix. 340 The custom-tailored wools that might.. have made him look like a well-heeled Wall Street lawyer.

5. a. *wool basket, bin, -hat* (see sense 5 d below), *shop.* **b.** *wool-classer, -grower* (later examples). **c.** *wool-lined* (earlier example), *-woven* adjs.

c **1878** J. ALBERY *Dram. Wks.* (1939) II. 300 Fawley places a note in Haidee's wool-basket. **1965** J. S. GUNN *Terminol. Shearing Industry* II. 38 *Wool basket*. There are several of these containers into which various locks and bellies are thrown to be baled up separately. **1933** *Press* (Christchurch, N.Z.) 30 Dec. 13/7 *Wool bins*, open compartments like stalls in a stable, where wool is stacked by classes until it is pressed. **1974** D. STUART *Prince of My Country* i. 3 The woolbins loom broad and tall, the press towers above them, there are bales in squat heaps. **1892** W. E. SWANTON *Notes on N.Z.* ii. 96 There is the wool classer with his assistant rollers. **1911** W. H. KOEBEL *In Maoriland Bush* viii. 122 The wool-classer takes his stand before the sorting table. **1968** *Guardian* 29 Feb. 14/3 Ian Redpath (Victoria), 26. Wool classer. Opening bat. **1921** *Daily Colonist* (Victoria, B.C.) 11 Oct. 6/3 Mr. Vernon, a wool-grower of Albert Head. **1962** *Economist* 31 Mar. 1275/1 Australian woolgrowers stand to earn £24 million more this season. **1971** *Sunday Australian* 8 Aug. 1/5 Half of Australia's 93,000 woolgrowers will get less than $600 each from the Federal Government's new wool subsidy. **1824** E. WEETON *Let.* 22 May (1969) II. 270 My wool-lined beaver gloves. **1923** *Harmsworth's 'Best Way' Series No. 95* 15/1 Ask at any wool shop for 'Beehive' Recipe Card No. 50 (price 2d.). **1943** A. CHRISTIE *Moving Finger* xiii. 147 She was knitting—ever so vexed she'd run out of wool... So I ran her in, dropped her at the wool shop. **1983** C. BOWDER *Birth Rites* I. 38 The colour's a bit unusual. It was a discontinued line in my local wool-shop. **1888** G. M. HOPKINS *Poems* (1967) 198 No more: off with—down he dings His bleached both and wool-woven wear.

d. wool alien, a plant introduced into a country by means of imported wool containing its seed; **wool-blind** *Austral.* and *N.Z.*, (of a sheep) having its sight obscured by its growth of wool; also *ellipt.*; hence **wool-blindness; wool church,** one of the English churches built or modified out of the wealth produced by the Tudor wool trade; **wool clip** = CLIP *sb.*[2] 2 b; **wool-clipper,** a clipper for carrying wool; **wool-dyed** *a.* (earlier and *fig.* examples); **wool-fat** (earlier example); **wool-grease** (earlier example); **wool hat,** (*a*) a hat made of coarse wool; † (*b*) *U.S.*, a supporter of the Democratic Party (*obs.*); (*c*) *U.S.*, a small farmer, or an unsophisticated or conservative countryman, from the South; also (senses (*b*) and (*c*)) *wool hat boy*; **wool king** *Austral.* and *N.Z. colloq.*, a wealthy or large-scale sheep farmer; **Woolmark,** an international quality symbol for wool instituted by the International Wool Secretariat; also *transf.*; **wool-press** (earlier example); **wool presser,** one who operates a wool press; **wool-pulling** *vbl. sb.*, (*a*) the removal of wool from a sheepskin; (*b*) the act of pulling the wool over a person's eyes; deception; **wool-shed:** also *N.Z.* (earlier and later examples); **wool table** *Austral.* and *N.Z.* (see quot. 1965); **wool team** *Austral.* and *N.Z.*, a team of draught animals for transporting wool; **wool-track** *Austral.*, a track along which wool was conveyed to a port; **wool-wax,** (*a*) = SUINT; (*b*) = LANOLIN.

1919 HAYWARD & DRUCE *Adventive Flora Tweedside* p. xxi, It must not..be assumed that all the wool aliens will disappear. **1961** *Proc. Bot. Soc. Brit. Isles* IV. 221 The party visited a railway siding in the same county, and further wool aliens were found... On enquiry he found that wool waste ('shoddy') was unloaded at the sidings and delivered to local farmers for use as a manure, and when this was followed up foreign weeds were found to be plentiful in their fields. **1976** *B.S.B.I. News* Sept. 22 J. R. Palmer searched hop fields near Wateringbury (Kent) and also found wool aliens present although no 'shoddy' has been used here for at least four years. **1933, 1953** Wool-blind [see *eye-clip* vb. s.v. *EYE *sb.*[1] 28]. **1955** J. MORRISON in B. James *Austral. Short Stories* (1963) 158 Worse than pushing a mob of wool-blinds up the ramp of a shearing shed. **1965** J. S. GUNN *Terminol. Shearing Industry* II. 37 The wig is removed with all the wool during shearing..because, if it is not done, the sheep may become 'wool-blind' before shearing time. **1950** *N.Z. Jrnl. Agric.* Oct. 349/3 Through eye clipping, wool blindness is avoided. **1936** M. ALLIS *Eng. Prelude* xxxiii. 252 Long Melford..with the stately 'wool' church, a miracle of lace in stone and flint. **1950** H. J. MASSINGHAM *Curious Traveller* ix. 175 Wild nature is the architect in Pembrokeshire and the massive castles..bear the same architectural relation to cliff and mountain as..the wool-churches of the Cotswolds. **1976** *Cambridge Independent Press* 16 Dec. 1. 3/5 An interesting talk, illustrated with coloured slides, on the wool churches of East Anglia was given. **1862** *Rep. Comm. Patents 1861: Agric.* 131 The wool-clip of New England commands a ready cash market in Boston. **1893** [in Dict., sense 5 a]. **1977** *Weekly Times* (Melbourne) 19 Jan. 3/4 The Corporation had put proposals to the Minister for Primary Industry..to acquire the export portion of the Australian wool clip. **1984** *N.Z. Farmer* 12 Apr. 12/1 The wool clip never strayed far from about 5kg per sheep wintered. **1984** *Oxf. Illustr. Hist. Brit.* iii. 160 Their reserves of liquid capital enabled Italian companies to offer attractive terms. They could not only buy an abbey's entire wool clip for the current year; they could also buy it for years in advance. **1903** C. PROTHEROE *Life in Mercantile Marine* 4 The *Chatto* was a full-rigged ship of a thousand odd tons, in reality a wool-clipper, but being winter time, she was now loaded with tallow and grain. **1924** J. MASEFIELD *Sard Harker* 37 The wool-clippers and big four-masters were being squeezed out. **1832** *Niles' Weekly Reg.* XLIII. 65/2 Messrs. Randolf and Ritchie who are chiefs of the 'wool-dyed democrats' of the present day. **1904** *Charlotte* (N. Carolina) *Observer* 19 June 2 Higginson is one of the old abolition gang, is wool-dyed and blind. **1875** *Chem. News* 15 Jan. 26/2 The question as to the composition of the wool-fat could not be fully solved. *Ibid.*, We have..examined two fresh kinds of wool-grease. **1794** T. COXE *View U.S.* 314 Wool hats, of Winchester make, are in much repute. **1828** *Western Intelligencer* (Hamilton, Ohio) 3 Oct. 3/1 Thus has Mr. Woods endeavored to gain the votes of the wool hats as he terms his Jackson friends in Washington. **1836** *Western Hemisphere* (Columbus, Ohio) 3 Aug. 1/7 The very men whom a few years ago they called the 'ragged wool hat boys' and 'Tories', they are now seeking to attach to their [Whig] party!! **1856** [in Dict., sense 5 a]. **1880** *Harper's Mag.* Dec. 159 An old 'wool-hat' came along with a cart drawn by a single ox. **1898** B. H. YOUNG *Hist. Jessamine County, Kentucky* 163 They made wool hats. **1927** K. EUBANK *Horse & Buggy Days* 170, I was a smart boy from town, and this particular guy thought I was a wool-hat boy. **1942** J. A. RICE *I came out of Eighteenth Cent.* ii. 95 South Carolinians liked their hatred to be personal, and the 'Wool Hat Boys' whooped with delight when Tillman ripped the hide off the 'Columbia Ring' and the Charlestonian gentlemen. **1942** *Time* 21 Sept. 19/3 Georgia's 'wool-hat' boys (small farmers). **1960** *Spectator* 2 Sept. 332 The 'wool-hats' (i.e. dyed-in-the-wool segregationists) of the rural Tobacco Roads who fear negro competition. **1980** *Washington Star* 29 Sept. A13/5 Carter knows that when..Eugene Talmadge

shouted about 'state's rights', the 'woolhats' of Georgia knew what he was saying. **1889** G. R. HART *Stray Leaves from Early Hist. Canterbury* iii. 19 Founders of the present race of wool kings in many parts of Canterbury. *a* **1922** H. LAWSON in *Penguin Bk. Austral. Ballads* (1964) 156 These are men who died to make the Wool-Kings rich. **1964** *Wool Future* Sept. 1/1 Woolmark, the international quality symbol for pure new wool, will be seen in British and Irish shops for the first time this month. **1980** *Times* 8 July 10/5 To get the Woolmark seal of approval you can only have a minute percentage of gorse..or whatever still stuck on the yarn. **1983** D. DUNNETT *Dolly & Bird of Paradise* xi. 141 One or two sheep..with no barbed wire in sight to ruffle their gorgeous Woolmark. **1846** C. J. PHARAZYN *Jrnl.* 21 Dec. (MS., Turnbull Libr., Wellington, N.Z.) 67 Employed all day at Watarangi assisting in packing fleeces. George making wool press. **1892** W. E. SWANTON *Notes on N.Z.* ii. 96 There is..the wool presser and his mate to bale up the wool. **1847** J. S. ROBB *Squatter Life* 16 In short I'm up to the whole 'wool pulling' system. **1885** Wool-pulling [in Dict., sense 5 b]. **1971** D. BAGLEY *Freedom Trap* iii. 59, I was given permission to start correspondence courses.. It was all a bit of wool-pulling to make them think Rearden was reconciled to his fate. **1846** C. J. PHARAZYN *Jrnl.* 11 Dec. (MS., Turnbull Libr., Wellington, N.Z.) 67 Counted rams after breakfast. George finished washing penn at river with Robin and Teddy and self to wool shed at Watarangi and finished the same. **1977** *N.Z. Herald* 8 Jan. 4-7/7 (Advt.), Four-brm home, 3-stand woolshed, barn, yards, airstrip. **1865** M. A. BARKER *Let.* 1 Dec. in *Station Life N.Z.* (1870) 32 We next inspected the wool tables, to which two boys were incessantly bringing armfuls of rolled-up fleeces. **1950** *N.Z. Jrnl. Agric.* Oct. 310/2 The scrubbing of the shearing board and the wool table is an essential practice. **1965** J. S. GUNN *Terminol. Shearing Industry* II. 40 *Wool table*, a table of spaced ridged lateral slats on which the fleece is rolled and skirted and the pieces picked. Any loose locks fall through to be picked up. **1865** R. HENNING *Let.* 18 Feb. (1952) 82/3 Biddulph..has sent both the bullock-drivers to the Port with the wool-teams. **1959** H. P. TRITTON *Time means Tucker* v. 41/2 Yarragrin..was also famous as a camp for the wool-teams, coming in from the north-west. **1903** 'T. COLLINS' *Such is Life* (1937) vi. 317 These wool-tracks, that knew him so well, will know him no more again for ever. **1959** J. WRIGHT *Generations of Men* (1960) xvii. 217 They followed a line through the trees that led southward across the road, once an important wool-track to the coastal ports. **1911** *Encycl. Brit.* XX. 51/2 An exceptional position [among animal waxes] is occupied by wool wax, the main constituent of the natural wool fat which covers the hair of sheep.. Wool fat is now being purified on a large scale and brought into commerce, under the name of lanolin, as an ointment. **1943** *Thorpe's Dict. Appl. Chem.* (ed. 4) VI. 135/2 Wool grease (wool fat..) is the crude mixture of wool wax and fatty acids recovered from the soapy liquor used for the scouring of raw wool. *Ibid.*, Crude wool grease is used as a lubricant..; some is refined for use as 'lanolin' (pure wool wax) in..cosmetics,..rust preventives, etc. **1954** [see *DEGRAS, DÉGRAS b]. **1956** *Nature* 10 Mar. 470/1 Further work was carried out..on the formation of polyacrylonitrile in wool and on suint and wool wax. **1964** N. G. CLARK *Mod. Organic Chem.* xvii. 340 Wool wax occurs to the extent of 20 to 30 per cent in raw sheep's wool. **1966** GETTENS & STOUT *Painting Materials* 81 Wool Wax..is the natural grease from the fleece of sheep.

wool, *v.* **2.** (Earlier example.)

c **1831** A. LINCOLN in H. Binns *Life Lincoln* (1927) 34, I never use tussle and scuffle. I don't like this wooling and pulling.

wool-carding, *vbl. sb.* and *ppl. a.* (Earlier example.)

1806 *Balance* (Hudson, N.Y.) V. 288/2, I was, lately, much pleased with seeing a wool-carding machine in operation.

Woolfian (wu·lfiăn), *a.* and *sb.* [f. the name *Woolf* + -IAN.] **A.** *adj.* Of, pertaining to, or characteristic of Virginia Woolf (1882–1941), English writer, or her work. **B.** *sb.* An admirer or devotee of Virginia Woolf. *rare.*

1936 *Scrutiny* V. 183 The more discerning might have noticed that it [*sc. A Note in Music*] was drawn not from life but from *Jacob's Room* and *Mrs. Dalloway*, a solemn exercise in Woolfian style and structure. **1944** E. H. W. MEYERSTEIN *Let.* 24 Nov. (1959) 296, I read the proofs of *Mrs. Dalloway* years ago which a Woolfian gave a friend of mine. **1977** W. HILDICK *Loop* viii. 40, I don't think I'd ever used the word 'lark' like that..before. It was..something I'd picked up in my Woolfian researches.

woo·lgathersome, *a. nonce-wd.* [-SOME[1].] Suggestive of wool-gathering.

1922 C. E. MONTAGUE *Disenchantment* vii. 91 The average German soldier, the docile blond with yellow hair, long skull, and blue, woolgathersome eyes.

woollenize (wu·lĕnəiz), *v.* [f. WOOLLEN *a.* and *sb.* + -IZE.] *trans.* To impart to (vegetable fibres) the appearance and texture of wool. Hence **woo·llenizing** *vbl. sb.*

1890 *Times* 19 Aug. 10/4 The various processes to be carried out at the model fibre factory..comprise.. cottonizing and woollenizing fibres to imitate fine cotton or wool. **1927** *Daily Tel.* 21 June 8 (Advt.), Successful fancy cloth effects have..been attained by the application of mercerising and woollenising processes to these yarns.

woolleny (wu·lĕni), *a. rare.* [f. WOOLLEN *a.* and *sb.* + -Y[1].] Made of or resembling woollen cloth.

1863 'G. HAMILTON' *Gala-Days* 41 I have a veil—none of your woolleny gruff fabrics.

Woollies (wu·liz). Also **Wooleys, Woolies, Woollys.** *colloq.* name for a shop bearing the name of F. W. *Woolworth* PLC (cf. *WOOLWORTH); occas. used for the company itself.

1939 *Airman's Gaz.* Dec., At Woollies store they congregate For powders, creams and lotions. **1957** R. HOGGART *Uses of Literacy* v. 120 Popular shops (with 'Wooley's'—Woolworth's—a clear favourite with working-class people). **1962** *Guardian* 24 Dec. 4/2 Some of them are shoplifting: not Woollies combs, but watches and rings. **1971** *Daily Tel.* 16 July 17 'Woolies' says that profits for the first half included a surplus of £261,000 on property sales. **1980** J. DITTON *Copley's Hunch* I. ii. 33 It's a good one. Not one of your Woollie's specials at a tanner a throw.

woollily (wu·lili), *adv.* Also **woolily.** [f. WOOLLY *a.* (*sb.*) + -LY[2].] In a way lacking in clarity or incisiveness.

1937 *Daily Express* 5 Feb. 13/2 Since nothing reads more woollily than descriptions of colours..I'll skip a list of new shades. **1979** *Guardian* 26 Oct. 2/8 Mr Atkins [was] saying a thing woollily not once..but twice.

woolly, *a.* (*sb.*) Add: **3. c. woolly bear,** (*a*) (further examples); also *spec.* the larva of the carpet beetle; freq. *attrib.*; (*b*) *Mil. slang* (see quots.); **woolly mammoth** = MAMMOTH *sb.* 1; **woolly worm** *U.S.*, a hairy caterpillar.

1909 WEBSTER, *Wooly worm*, the larva of any sawfly that covers itself with a white woolly secretion. **1911** E. FERBER *Dawn O'Hara* ii. 19 I'd eat wooly worms if I thought they might benefit me. **1915** *War Illustr.* 31 July 546/2 The German high-explosive shell, known to our men by the nickname of the 'Woolly Bear',..detonates with a cloud of thick white smoke. **1918** H. W. MCBRIDE *Emma Gees* 135 'Woolly Bear' is the name given to a large, high explosive shell, with a time fuse, which bursts overhead, giving out a dense black smoke. **1923** KIPLING *Irish Guards in Gt. War* II. 82 They were drenched with a five hours' bombardment of 4.2's and 'woolly bears'. **1933** A. S. ROMER *Vertebr. Paleontol.* xix. 376 The woolly mammoth was a form adapted to cold climates. **1940** R. G. RUSSELL *101st Field Artillery 1917–19* 94 A German 150-millimetre battery fired 'woolly bears', time-fuse shells, which burst too high to do any harm. **1950** *N.Z. Jrnl. Agric.* Nov. 478/3 The most common enemy of cinerarias in New Zealand is the 'woolly-bear' caterpillar, the larva of the magpie moth. **1951** *Good Housek. Home Encycl.* 324/2 *Woolly bear*. This is the grub of a small beetle which..congregates in hot airing cupboards. **1961** Woolly bear [see *carpet beetle* s.v. *CARPET *sb.* 5]. **1969** BENNISON & WRIGHT *Geol. Hist. Brit. Isles* xvi. 359 The presence of either the woolly mammoth or the reindeer does not necessarily indicate an arctic climate. **1972** E. WIGGINTON *Foxfire Bk.* 209 The woolly worm tells of a bad winter if: there are a lot of them crawling about. **1974** A. DILLARD *Pilgrim at Tinker Creek* xiv. 247 Woolly bears, those orange-and-black-banded furry caterpillars of the Isabella moth, were on the move. **1976** *Islander* (Victoria, B.C.) 16 May 6/1 The woolly mammoth..roamed the tundra areas. **1980** *Blair & Ketchum's Country Jrnl.* Oct. 28/2 October is the month when the woolly bear caterpillar, sometimes called a fuzzy-wuzzy or woolly worm, can be seen crossing country roads. **1983** *Listener* 27 Oct. 16/3 Our wall-to wall carpets attract the 'woolly bear' grubs of the carpet beetle.

d. (Earlier example.) Also *transf.*, and as *woolly* simply.

1884 A. J. SOWELL *Rangers & Pioneers of Texas* xi. 330 Occasionally, in some Western village, you will hear a voice ring out on the night air..'Wild and woolly',.. and then you may expect a few shots from a revolver. It is a cowboy out on a little spree. **1891** M. E. RYAN *Told in Hills* III. iv. 191 Let us 'move our freight', 'hit the breeze', or any other term of the woolly West that means action. **1907** S. E. WHITE *Arizona Nights* viii. 130 'Who's your woolly friend', the shiny Jew asks of the girls. **1940** R. S. LAMBERT *Ariel & All his Quality* viii. 197 [They] looked with scepticism upon a plan which they regarded as wild and woolly.

7. *woolly-witted* adj.

1927 *Observer* 6 Nov. 15/1 The managerial attitude towards producers is at present woolly-witted. **1949** ST. J. ERVINE *Craigavon* II. lvii. 273 That woolly-witted insurrectionist.

B. *sb.* **1.** *winter woollies*, warm underwear (not necessarily of wool); freq. *joc.* Also *fig.*

1926 WODEHOUSE *Heart of Goof* vi. 194 His mother had bought him a new set of winter woollies which felt like horsehair. **1933** DYLAN THOMAS *Sel. Lett.* (1966) 24 Catch him [*sc.* Wordsworth]..walking the hills with a daffodil pressed to his lips, and his winter woollies tickling his chest. **1964** *Observer* 13 Sept. 11/3 If we wear the winter woollies of traditional trade unionism against the hot sun of automation, we may sweat it out instead of thinking it out. **1974** *Nature* 18 Oct. 569/1 The dinosaurs' unsatisfied need was not so much for laxatives as for winter woollies!

2. A sheep; esp. (*Austral.* and *N.Z.*) one before shearing. *U.S.*, *Austral.*, and *N.Z. colloq.*

1910 J. G. NEIHARDT *River & I* iii. 92 In Scotland when a feller sees a sheepman coming down the road with his sheep, he says: 'Behold the gentle shepherd with his fleecy flock!'.. In Montana, that same feller says..'Look at that crazy blankety-blank with his woolies!' **1930** *Bulletin* (Sydney) 2 Apr. 23/1 We curse the stubborn woollies..as the sweating shearers tussle. **1935** H. DAVIS *Honey in Horn* xi. 162 She had a little short-bodied guitar of the kind that Mexican sheep herders used to carry around behind their saddles to entertain the woolies with. **1949** F. SARGESON *I saw It in my Dream* II. xiii. 111 White dots that you could tell were both sheep and lambs; and they were so white it was easy to tell that they weren't woollies any more. **1972** P. NEWTON *Sheep Thief* vi. 48 The biggest proportion proved to be Totara sheep of mixed ages, three of them woollies.

3. [Cf. *WOLLY.] A uniformed policeman. *slang.*

1965 R. E. RIDGWAY in B. Wannan *Fair Go, Spinner* II. 66 Later on, as the station expanded and more 'woollies' were added, the shed grew accordingly. **1975** *Listener* 6 Feb. 163/2 Sir Robert Mark...saw its [*sc.* the CID's] members behaving as if they could walk on water, and looking down on the 'woollies' who had to plod the beat in uniform. **1978** 'B. GRAEME' *Double Trouble* xv. 191 One of the woollies blew his whistle. **1984** *Private Eye* 20 Apr. 6/2 A small army of 'Woollies'—CID slang for uniformed officers—were summoned.

Woolpit (wu·lpit). The name of a village in Suffolk, used *attrib.* in *Woolpit brick*, a pale-coloured brick made from earth there.

1887 J. E. TAYLOR *Tourist's Guide Suffolk* 103 One of its chief industries is brickmaking, for the bed of brick-earth here makes a beautiful stone known everywhere as 'Woolpit Brick'. **1966** G. E. EVANS *Pattern under Plough* ii. 36 The Tudor front of the house was given a severe façade of Woolpit brick.

Woolton (wu·ltən). The title of F. J. Marquis (1883–1964), 1st Earl of *Woolton*, used in (*Lord*) *Woolton pie*, a vegetable pie publicized when he was wartime Minister of Food.

1941 *Food Facts for Kitchen Front* 44 *Lord Woolton Pie*. The ingredients of this pie can be varied according to the vegetables in season. Potato, swede, cauliflower and carrot make a good mixture. **1955** E. WAUGH *Officers & Gentlemen* II. vi. 291 The London crowd shuffled past, surfeited with tea and Woolton pies. *a* **1969** O. SITWELL *Queen Mary* (1974) 47 At luncheon..we found a new and rather horrible war-time dish had been prepared for us, called Lord Woolton Pie. **1981** *Times* 16 Mar. 12/7 A wartime diet of Woolton Pie and whalemeat.

wool-winder. Add: **2.** A frame on which wool is wound.

1969 *Canad. Antiques Collector* Aug. 20/2 An old wool-winder, still in its original state, stands to the left. **1976** *Evening Post* (Nottingham) 15 Dec. 20/6 (Advt.), Knitmaster 321 Punchcard Knitting machine, with worktable, woolwinder, punch etc.

Woolworth (wu·lwɜ̄ɹþ). The name of the retailing company (orig. sixpenny store) F. W. *Woolworth* PLC, used *attrib.* to designate low-priced goods regarded as typical of its merchandise.

1931 *Times Lit. Suppl.* 5 Nov. 862/3 Miss Helen Simpson is refreshingly modern with her 'Woolworth craze' at Oxford as the real villain. **1932** AUDEN in *Rev. Eng. Stud.* (1978) XXIX. 292 Moving woodenly like a woolworth doll, A lady came in clothes so ugly..that the eye was cruel. **1939** T. S. ELIOT *Old Possum's Bk. Pract. Cats* 22 One of the girls Suddenly missed her Woolworth pearls. **1948** 'J. TEY' *Franchise Affair* vii. 69 Some Woolworth plants in the gardens. **1974** *Guardian* 21 Mar. 10, I am never likely to be involved in a study of..Lehar and can only look on all of it as real Woolworth stuff (in contrast to, say, Mozart, who is vintage Fortnum and Mason). **1980** I. ST. JAMES *Money Stones* III. iii. 106 You couldn't swing a Woolworth watch on diamonds found here.

Hence **Woolwo·rthian** *a.*

1933 R. W. CHAMBERS *Whatever Love Is* xx. 286 A strange, stark, snowless, dingy Christmastide with a Woolworthian cheapness about it. **1937** P. THORNTON *Dead Puppets Dance* II. vii. 144 There were hundreds of people at Kučevište, and many of them wore..high-heeled shoes and Woolworthian stockings if their husbands could afford..them. **1978** *Washington Post* 27 Aug. C-38/2 Nowadays the plates are chipped and the tables lit by Woolworthian gilded pierced tin lamps flickering with candles.

woomph (wŭmf, wumf), *int.* (*sb.*, *adv.*) *slang.* Also **woomf.** [Imitative.] (Expressing) a sound similar to a 'whoof' (*WHOOF *int.* (*sb.*) 2) but with a deeper or more resonant component. Cf. the synonymous *WHOOMPF *int.* (*sb.*).

1955 LD. WINTERTON *Fifty Tumultuous Years* 82 Two old gentlemen were dozing in their chairs when the 'Woomph', 'Boomph' of a bomb simultaneously woke them up. **1979** R. FIENNES *Hell on Ice* ix. 144 There was a sudden *woomf* as the fumes and the liquid ignited. **1982** S. BRETT *Murder Unprompted* xiv. 133 He threw a cushion, which went woomph into the side of Charles's head.

woon. Add: Also **wun.**

1898 H. F. HALL *Soul of People* vii. 92 Outside Mandalay the country was governed by *wuns* or governors. **1972** A. T. Q. STEWART *Pagoda War* ix. 112 (*caption*) Thibaw's envoy, the *wun* with his golden umbrella, stands beside him.

woop woop (wu·p,wup). *Austral.* and *N.Z.* Also **woop-woop(s, wop-wop.** [Sham Aboriginal (but see below).] **a.** A jocular name for a remote rural town or district; also (without *the* and with capital initials) as the name of an imaginary place in a remote area.

One suggestion is that the term is derived from the 'geelorious town o' Whoop-Up' in E. L. Wheeler's *Deadwood Dick on Deck* (1878), where 'Whoop-Up' is the name of a back-country American goldmining town.

1926 'J. Doone' *Timely Tips for New Australians* 23 *Woop Woop*, a humorous method of alluding to the country districts used most frequently in New South Wales. 1928 A. Wright *Good Recovery* 34 They're chasin' Murraba out along the Woop Woop Road, or somewhere. 1930 *Bulletin* (Sydney) 1 Jan. 28/2 'Who on earth is she?' gasped the visitor from Woop-Woop. 1958 J. Lindsay *Life Rarely Tells* 213 Next morning he'd rush away. 'Off to the Woop-woop!' Somewhere in the backblocks that meant. 1960 N. Hilliard *Maori Girl* iii. i. 174 'Where do you come from?' 'Up in the wilds—the woop-woops, Taranaki.' 1963 *Truth* (Wellington, N.Z.) 8 Oct., A job was found right out in the wop-wops. 1970 *N.Z. Listener* 21 Dec. 51/2 While you're out in the woop-woops next time, spare a thought for the local farmer. 1975 *Courier-Mail* (Brisbane) 26 Mar. 12/6 Police feared they would be transferred to 'Woop Woop'.

b. An inhabitant of such a place; a country bumpkin. *rare.*

1936 M. Franklin *All that Swagger* 472 Adrienne was no blob or woop-woop. 1950 *Coast to Coast 1949–50* 201 I'll make a fair dinkum woop-woop out of you in no time.

woosh, var. Whoosh *v.*, *sb.* in Dict. and Suppl.

Wooster (wū·stəɹ). The name of Bertie *Wooster*, an amiable, vacuous, young man about town in the novels of P. G. Wodehouse, used allusively. Also *attrib.*

1939 Auden & Isherwood *Journey to a War* i. 44 He was so much more subtle, more intelligent than his cultivated Bertie Wooster drawl. 1960 *New Statesman* 24 Sept. 424/2 Very young men about town..quite amiable in a Bertie Wooster sort of way. 1963 R. H. Morrieson *Scarecrow* (1964) xii. 135 Her husband, the aged Wooster type whom Angela and I had encountered.

Also as *v. intr.*, to behave in the manner of Bertie Wooster; **Woo·sterish** *a.*; **Woo·sterism,** a remark or action characteristic of Wooster.

1959 *Observer* 26 Apr. 23/5 Harold..lives in a pretty Woosterish way. 1964 *Punch* 17 June 906/3 Frolicsome nitwits woostering in well-heeled suburbia. 1969 *Times* 5 May 23/3 Dapper, cheerful young men without the disdain of the real Jeeves when provoked by inane Woosterisms. 1978 *Country Life* 14 Dec. 2103/1 A ridiculous pink velvet dog with inane Woosterish eyes. 1979 K. Bonfiglioli *After you with Pistol* xvii. 130 We Woostered away for a while, giggling slightly. *Ibid.*, While we idly bandied these Woosterisms..he slid a scribbling-pad across the desk. 1983 *Times* 17 Oct. 15/4 This show..reducing Olivia to a charm-school hostess, Andrew to a Woosterian silly ass and..Toby to a bar-fly.

woozy, *a.* Add: Also **whoosy, whoozy, woozey.** For *U.S. slang* read *colloq.* (orig. *U.S.*). **1.** Further examples.)

1915 Wodehouse *Psmith, Journalist* xvi. 114 'He's still woozy,' said the Kid. 'Still—what exactly, Comrade Brady?' 'In the air,' explained the Kid. 'Bats in the belfry. Dizzy.' 1929 Kipling *Limits & Renewals* (1932) 356 He had kept himself going on rum sometimes, and was woozy when the pinch came. 1937 *Black Mask* Jan. 24/2, I got hit. It made me woozey for a minute. 1952 B. Malamud *Natural* 17 He got up whoozy and walked, finding it hard to believe his eyes. 1961 J. B. Priestley *Saturn over Water* iii. 29 The woozy state I was in. 1977 M. Hinxman *One-Way Cemetery* xix. 139 He'd have phrased it more delicately if he hadn't felt quite so whoosy. 1978 *Daily Tel.* 17 Jan. 17/2 Liquid lunches can leave a man weak and woozy late in the afternoon, drinkers were told.

2. Representing or marked by muddled thinking or unclear expression; lacking rigour or discipline; sloppy.

1941 Auden *New Year Let.* ii. 37 All vague idealistic art..Is up his alley, and his pigeon The woozier species of religion. 1961 *Catholic Gaz.* May 129/2 To Dickens, Christmas meant a debauch of vague and woozy sentiment. 1970 Auden in *New Yorker* 21 Feb. 118/1 Like Ruskin, he can at times write sentences which I would call 'woozy'; that is to say, too dependent upon some private symbolism of his own to be altogether comprehensible to others. 1971 *Daily Tel.* 15 Mar. 13/4 One wonders if it is not simply the drink that has made so many Irish writers bury their poetic insights beneath so much that is garrulous, maudlin and whoozy. 1975 *New Yorker* 3 Feb. 84/2 There are gaps in the plot and woozy lapses in time. 1977 *Rolling Stone* 24 Mar. 41/2 She supports the old male stereotype of woman as overwhelmingly physical, instinctual, and her writing is too woozy for me. 1977 *N.Y. Rev. Bks.* 9 June 16/3 The other poem of 1939 [by Auden], with its 'affirming flame', is woozy too. 1980 *Times Lit. Suppl.* 25 Apr. 470/5 A level of woozy tautology.

Hence **woo·zily** *adv.*, **woo·ziness.**

a1911 D. G. Phillips *Susan Lenox* (1917) I. xxi. 395 'Shut up!' cried the drunken man... He caught them each by an arm, stared woozily at Etta. 1924 *Black Mask* Nov. 48/2 This thing had fallen on me while my nerves were ragged from three days of boozing... [Now] my wooziness had passed. 1937 *Ibid.* Jan. 23/2, I shook my head woozily. 1937 Auden *Let.* in Auden & MacNeice *Lett. from Iceland* 221 Landscape's so dull if you haven't Lawrence's wonderful wooziness. 1967 *Listener* 9 Feb. 193/2 Staring woozily at a wine flask. 1977 C. Isherwood *Christopher & his Kind* xii. 181 Much of what Christopher called Wystan's wooziness was essentially religious in context. 1984 *Observer* 19 Feb. 25/5 Later in life she more stubbornly shut herself off from the world's demands behind the defences of deafness, bad English and Benedictine-fuelled wooziness.

wop (wǫp), *sb.*¹ and *a.* *slang* (orig. *U.S.*). Also **Wop.** [Origin uncertain; perh. ad. It. dial. *guappo* bold, showy, ruffian, f. Sp. *guapo* bold, dandy, f. L. *vappa* sour wine, worthless fellow.] **A.** *sb.* **a.** An Italian or other southern European, esp. as an immigrant or foreign visitor (see also quot. 1914). Now considered *offensive.*

[1912 A. Train *Courts, Criminals & Camorra* ix. 232 There is a society of criminal young men in New York City... They are known by the euphonious name of 'Waps' or 'Jacks'. These are young Italian-Americans who allow themselves to be supported by one or two women... They form one variety of the many gangs that infest the city.] 1914 Jackson & Hellyer *Vocab. Criminal Slang* 88 *Wop*, noun. Used principally in the east. An ignorant person; a foreigner; an impossible character. .. Example: 'You couldn't find a jitney with a search warrant in this bunch of wops.' 1915 Wodehouse *Psmith, Journalist* xix. 138 He's a wop, kid... A wop. A dago... An Italian. 1924 E. Hemingway *In our Time* 17 Wops, said Boyle, I can tell wops a mile off. 1930 G. B. Shaw *Apple Cart* ii. 78 *Lysistrata.* What they call an American is only a wop pretending to be a Pilgrim Father. He is no more Uncle Jonathan than you are John Bull. *Magnus.* Yes: we live in a world of wops, all melting into one another. 1940 N. Mitford *Pigeon Pie* i. 7 Luke's Italian was far more affected than that of any native wop. 1942 *R.A.F. Jrnl.* 13 June 26 The pilots.. suggested that the 'Wops were yellow' or that they could not 'take it through cloud'. 1952 E. F. Davies *Illyrian Venture* ii. 26 We had breakfast in the mess tent, waited on by a cheerful wop. 1973 'I. Drummond' *Jaws of Watchdog* ii. 26 Sandro dived into the pool... 'You great fat clumsy Wop,' said Jenny, 'you've put my cigarette out.'

b. The Italian language.

1937 [see *for all I know* s.v. *know *v.* 11 g]. 1938 E. Pound *Let.* 6 May (1971) 313 'Sardinia is Barbagia' don't seem either English or Wop. 1982 A. Melville-Ross *Trigger* xx. 225 There's a lot of chat in Wop which I doesn't understand.

B. *adj.* Italian.

1938 E. Pound *Let.* 8 Jan. (1971) 303 'Praedis': I don't care how you spell your wop painters. 1940 [see *kike]. 1941 C. E. Milburn *Diary* 15 Feb. (1979) 83 We have dropped parachutists in Italy... A very nice surprise for our 'Wop' enemy! 1955 E. Waugh *Officers & Gentlemen* 326 You'll find her full of wop prisoners. 1983 S. F. X. Dean *It can't be my Grave* iv. 64 Are you telling me, you Wop son of a bitch, that I can't get into my father's lift?

wop (wǫp), *sb.*² *R.A.F. slang.* [Acronym from *w(ireless op(erator* (cf. *op³ 2 b).] A radio operator.

1939 *Airman's Gaz.* Dec., You have a choice of three.. trades—..W.O.M.,..W.E.M.,..and the W/Op. which rude people twist into Wop (most unfairly). 1957 R. Barker *Ship-Busters* iv. 70 Wireless operator/air gunners ..most of the wop/A.G.s..came straight from gunnery school.

wops, dial. and joc. var. Wasp *sb.* in Dict. and Suppl.

Worcester. Add: **1.** Also *ellipt.* for *Worcester sauce.*

1889 J. K. Jerome *Three Men in Boat* ii. 22 If Harris's eyes fill with tears,..it is because Harris has been eating raw onions, or has put too much Worcester over his chop. 1981 T. Heald *Murder at Moose Jaw* ix. 103 A straight tomato juice with a liberal splashing of Worcester.

2. Used *attrib.* (with *Pearmain* or *apple*) and *absol.* to designate an early, slightly conical red-skinned eating apple belonging to a variety introduced to cultivation about 1875 by Richard Smith, a Worcester nurseryman.

1877 *Garden* 13 Oct. 344/1 Worcester Pearmain Apple. —Of this beautifully coloured, fully flavoured new Apple, Mr. Richard Smith, of Worcester, has sent us samples. 1929 E. A. Bunyard *Anat. Dessert* 6 The really ripe Worcester has character..and the Raspberry flavour distinctive. 1958 *Listener* 27 Nov. 903/1 Four medium-sized Worcester Pearmain apples. 1936 H. V. Taylor *Apples of England* iii. 43 Worcester Pearmain..became the most important of the commercial early dessert apples. 1982 R. Holles *Sun Blight* i. 6 She walked along.. inspecting the piles of Worcester apples. *Ibid.* ix. 98 In the Portobello Road..mounds of apples grinned at him, Worcesters, Granny Smiths, Golden Delicious.

Worcesterberry (wu·stəɹberi). Also **worcester-.** [f. prec. + Berry *sb.*¹] A small black gooseberry of the North American species *Ribes divaricatum*, once believed to be a hybrid of the blackcurrant and the gooseberry and sold as such by a Worcester nurseryman.

1923 *Amat. Gardening* 27 Oct. p. ix/3 (Advt.), Worcesterberry. Cross between gooseberry and black currant. Fruiting trees. 3/- each. Richard Smith & Co... Worcester. 1926 *Observer* 5 Sept. 9/2 New fruits, such as the logan and most excellent worcesterberry, multiply almost yearly. 1969 *Oxf. Bk. Food Plants* 80/2 The so-called 'Worcesterberry'..seems first to have been sold by a nurseryman in Worcester who thought it was a black currant gooseberry hybrid. 1980 *Amat. Gardening* 25 Oct. 20/2 If spraying fails, or if you do not like doing it, why not try growing the Worcesterberry instead?

Worcestershire. (Earlier example of *Worcestershire sauce.*)

1843 *Naval & Military Gaz.* 1 Apr. 208/2 (Advt.), Lea and Perrin's 'Worcestershire Sauce', prepared from a recipe of a nobleman in the county.

word, *sb.* Add: **I. 1. d.** *too —— for words:* —— to an extent that cannot adequately be described. *colloq.*

1913 *Vanity Fair* Nov. 65 New York is beginning to look too smart and clean for words. 1928 E. O'Neill *Strange Interlude* viii. 289 But for Gordon to..propose marriage — it's too idiotic for words! 1937 J. Mercer *Too Marvellous for Words* (song), You're just too marvellous, too marvellous for words.

2. c. Also *a word in your ear* (colloq.): a brief message for you in confidence.

1599 Shakes. *Much Ado* iv. ii. 27 Come you hither, sirrah; a word in your ear, sir. 1838 Dickens *Let.* 25 Jan. (1965) I. 360 A word in your ear. Macready objected to Talfourd's play. 1980 *Daily Tel.* 9 May 18 Salome and Kumba [*sc.* two gorillas] would like a word in your ear.

5. *to have words* (examples); also *to have a word.*

1777 W. Mawhood *Diary* 24 Aug. in *Publ. Cath. Rec. Soc.* (1956) L. 117 Came to Town to breakfast had words with Mrs. Mawhood. 1839 Dickens *Nicholas Nickleby* xlviii. 480 'We were a very happy little company, Johnson,' said poor Crummles. 'You and I never had a word.' 1901 'Zack' *White Cottage* 37 Have you and Mark had wuds? 1910 King George V in H. Nicolson *George V* (1952) vii. 105, I have lost my best friend & the best of fathers. I never had a word with him in his life. 1935 Z. N. Hurston *Mules & Men* ii. vi. 287 Celestine is not mad any more about the word we had last week.

6. a. (Further examples in *sing.* without article.)

1948 'H. Green' *Concluding* 205 Word had gone round that at last they were engaged. 1958 M. L. King *Stride toward Freedom* iii. 45 The arrest..was becoming public knowledge. Word of it spread around the community like uncontrolled fire. 1983 *Times* 16 Sept. 16/2 Word is the Government is offering a fixed price for small investors. 1984 *Times* 14 June 22/2 Word in the market suggests Mr Holmes a'Court may be prepared to sell on his stake.

b. Delete Now *rare* or *Obs.* and substitute: Now usu. in phr. *the word is that* (..). Chiefly *U.S.*

1963 R. Jessup *Cincinnati Kid* iv. 55 Money is beginning to show for you against The Man, Kid... The word is..that you're good enough to take Lancey, if anybody can. 1965 P. O'Donnell *Modesty Blaise* vii. 82, I know of him. The word is that he's good. 1982 P. Lovesey *False Inspector Dew* iv. 153 The word is that the captain will be speaking to us.

II. 12. a. Also *four-letter word*: see *four-letter* adj. s.v. *Four C. 2.

f. *Telegr.* Any of the sequences of a prescribed fixed number of characters (including a space) in a telegraphic message that has been coded or redivided for transmission.

1897 J. Nicolson *Telegraphic Signals* ii. 20 Artificial letter-grouping, mathematically called 'words', or permutations,..is referred to in a pamphlet by the French cryptographist, M. le Marquis de Viaris..as a substitute for telegraphic codes composed of dictionary words. 1911 *Encycl. Brit.* XXVI. 521/2 An experimental printer constructed about 1908 by the British Post Office, operated ..at the rate of 210 words (1260 letters) per minute. 1976 R. N. Renton *Telegraphy* i. 14/2 The 'telegraph word' is taken as an arbitrary 5-letter word with one letter-space, making six characters in all.

g. *Math.* An ordered sequence of generators of a group.

1952 S. C. Kleene *Introd. Metamath.* xiii. 382 A finite sequence of zero or more (occurrences of) the letters, we call a word. 1971 G. Higman in Powell & Higman *Finite Simple Groups* vi. 212 Any word in the n_i and their inverses determines a partial map of the set of equivalence classes into itself. 1972 M. Kline *Math. Thought* xlix. 1141 There may be relations among the generators, and these would be of the form $F_j(A_i) = 1$; that is, a word or combination of words equals the identity element of the group. 1981 *Sci. Amer.* Mar. 26/1 A lovely 'pretty pattern' called the 6-*U* state..can be reached from the start position by way of the word *L'R²F'L'B'UBLFRU'RLR₂F₂U₂R₂*.

h. *Computers.* A consecutive string of bits that can be transferred and stored as a unit (see quot. 1969); *machine word*, a word of the length appropriate for a particular fixed word-length computer.

1946 [see *write *v.* 3 h]. 1948 *Proc. R. Soc.* A. CXCV. 272 Certain of these numbers or 'words' are read, one after another, as orders. 1954 *Computers & Automation* Dec. 16/1 *Machine word*, a unit of information of a standard number of characters, which a machine regularly handles in each register. 1964 F. L. Westwater *Electronic Computers* ix. 140 The basic unit of internal storage is called a 'word', which may contain either instructions or data. 1969 P. B. Jordain *Condensed Computer Encycl.* 566 Computers with words less than 9 bits long call the words bytes, characters, or digits (decimal). 1970 A. Cameron et al. *Computers & O.E. Concordances* 58 It is heavily dependent upon fitting *x* number of characters into each machine word, a problem we cannot get around easily. 1980 C. S. French *Computer Sci.* vi. 24 The number of bits in each location (word), known as the word length will depend on the make and model of computer.

III. 15. b. (Earlier example.)

1841 Mrs. Gaskell *Lett.* (1966) 44 My word! authorship brings them in a pretty penny.

19. Also *word of mouth* sb. phr., oral communication, oral publicity.

1934 in Webster. 1951 B. Schulberg *Disenchanted* vi. 67 He tells everyone..that you're one of his favorite American authors and..that kind of word of mouth

ain't bad. 1967 B. Whitaker *Of Mice & Murder* xiii. 147, 'I wonder how he heard it was for sale.' 'Word of mouth, I suppose.' 1980 'D. Kavanagh' *Duffy* iii. 43 The only way to get successful..was to work at being really efficient and then hope for word-of-mouth to back you up. 1984 A. Brookner *Hotel du Lac* i. 14 The only publicity from which the hotel could not distance itself was the word of mouth recommendations of patrons of long standing.

20. b. *word by word*: also *spec.*, in alphabetization; opp. *letter by letter* (see *Letter *sb.* 1 c).

1938 L. M. Harrod *Librarians' Gloss.* 12 There are two methods [of alphabetization] in use: 1, 'letter by letter'; 2, 'word by word', or 'nothing before something'. In the former method 'Newton' *precedes*, in the latter it *follows*, 'New York'. 1951 *British Standard Alphabetical Arrangement* (B.S.I.) 6 Items having the same first word shall be arranged in the alphabetical order of the second word, those with no second word standing first. Similarly those having two words in common are arranged in the alphabetical order of their third word and so on. The whole group thus arranged shall precede any word alphabetically qualified to follow the first word of the group. (This is known as the 'word-by-word' or 'nothing-before-something' principle.) 1979 *Amer. Speech 1976* LI. 149 This dictionary uses word-by-word rather than letter-by-letter alphabetizing.

IV. 29. a. Simple attrib., as *word-boundary, -break, -combination, -division, -element, -end, -ending, -family, -form, -function, -game* (also *fig.*), *-idea, -memory* (earlier example), *-music* (examples), *-order* (earlier and later examples), *-pattern, -patterning, -position, -sound, -status, -stem, -store, -stress, -structure, -study, -taboo, -tone, -usage, -value*; (with agent-n. or the like), as *word-artist, -merchant, -musician, -smith* (later examples); *word-smithing*; also *word-based, -like* adjs. **b.** Instrumental, as *word-drunk* adj. **c.** Objective, as *word-coiner*; *word-finding, -hunting* (later example), *-making* (earlier examples; see also sense d below), *-setting* (Set *v.* 73), *-twisting* sbs.; *word-choice, -creation, -formation* (earlier example); also *word-formational, -formative* adjs. **d. word association** *Psychol.*, a psycho-diagnostic technique based on analysis of a person's reactions to the presentation of stimulus words, esp. with regard to the (subconscious) contents and type of the immediate associations formed, reaction time, etc.; more generally, the associations connected with certain words; freq. *attrib.*; **word-base** *Philol.*, the simple word from which its derivatives and inflected forms arise; a root morpheme, etc.; **word-blind** *a.* (earlier example); **word-category** *Linguistics* = *word-class; **word-class** *Linguistics* [cf. G. *wortklasse*], a category of words of similar form or function; esp. applied to parts of speech; **word-count,** a statistical study of *word frequency* (see below); **word-field** *Linguistics*, a group of lexical items seen as associated in meaning because occurring in similar contexts; **word-final** *a.*, occurring at the end of a word; also as *sb.*, a letter or sound occurring in this position; hence **word-finally** *adv.*; cf. *word-initial* adj. below; **word frequency,** the relative frequency with which a word is used in a given text or corpus; **word geography,** the study of the regional distribution of words and phrases, or a book treating of this; hence **word-geographical** *a.*; **word-hoard** (earlier example as a conscious Anglo-Saxonism); recently in general use, the words used by a person or group of people, vocabulary; also, a source or store of words; **word-index,** a list of the words used by a given author or in a given work (or corpus) with reference to the passages in which they occur, but without quotations (cf. Concordance *sb.* 6 b); **word-initial** *a.*, occurring at the beginning of a word; also as *sb.*, a letter or sound occurring in this position; hence **word-initially** *adv.*; **word-internally** *adv.*, = *word-medially* below; **word-ladder,** a puzzle in which a word has to be converted into another of equal length by being taken through a series of word-changes, each word differing by one letter from the last; also called *doublets*; **word length** *Computers*, the number of bits, digits, etc., in a word (sense 12 g above); **word-magic** *Anthropol.*, magic thought to be exerted by the knowledge or use of the proper name or term for something, or the supposed magical property residing in such a name; also *transf.*; **word-making and word-taking,** a game played with lettered cards, app. a forerunner of the modern Lexicon or Scrabble; **word mark,** (*a*) a real or invented word used as a trade mark; (*b*) *Computers*, a bit that takes a different value according as the character containing it does or does not begin (or end) a word; a character containing such a bit; **word-medial** *a.*, occurring in the middle of a word; hence **word-medially** *adv.*; **word method** *Educ.*, a method of teaching pupils to read in which they are taught to recognize words as complete units before learning the letters or syllables which compose them; the 'look-and-say' method (see *Look *v.* 47); **word-painted** *a.*, (*a*) decorated or adorned with words; (*b*) 'painted' or described vividly in words; **word-pair,** a pair of words resembling each other in sound or form; **word-palatogram** (see quot. 1948); **word-picture** (earlier example); **word-play** (earlier and later examples); **word problem** *Math.*, the problem of determining whether two different products are equal, or two sequences of operations are equivalent; **word processing** [cf. G. *textverarbeitung* text processing], the storing and organizing of texts by electronic means, *spec.* by a word processor; hence (as a back-formation) **word-process** *v. trans.*, to edit, produce, etc., using a word processor; **word-processed** *ppl. a.*; **word processor,** a keyboard device incorporating a computer programmed to store, amend, and format text that is keyed in, a printer to print it automatically, and often also a screen to display it; **word recognition** *Educ.*, the process or faculty of perceiving words in reading and identifying them with the ideas they represent; **word-salad,** a type of speech indicative of advanced schizophrenia in which random words and phrases are mixed together unintelligibly; also *fig.*; **word-sign,** something used to represent a word; *spec.* a graphic character representing a complete word; esp., in Egyptian hieroglyphics, etc. = *Logogram 2 b; **word-stock,** (in Dict., sense 29 a) the sum of words available to a language, dialect, etc.; vocabulary; also *fig.*; (later examples); **word-symbol,** a word used as a sign or symbol; *spec.* = *Logogram 2 b; **word time** *Computers*, the time between the reading of the first bits of successive words; **word-type,** (*a*) a word used to symbolize or represent an idea; (*b*) *Philos.* (see quot. 1936); (*c*) a word forming a distinct item in a vocabulary; **word-watch** *v. intr.*, to observe linguistic usage, esp. with regard to changes and innovations; also **word-watcher**; **word-watching** *vbl. sb.*; **word-wrap** [cf. *Wrap-around *sb.* 3] (see quot. 1982); so **word-wrapping** *vbl. sb.*; **word-writing,** Bloomfield's term for ideographic writing.

1933 Dylan Thomas *Let.* Sept. (1966) 20 Mr. Neuburg has payed you a..compliment. 'One of the most exquisite word-artists of our day.' 1945 C. Bax *Vintage Verse* iii. 95 This faultless word-artist [*sc.* Milton]..was buried in St. Giles's, Cripplegate. 1910 *Rev. Neurol. & Psychiatry* VIII. 641 (*title*) The practical value of the word-association method in the treatment of the psycho-neuroses. 1918 M. D. Eder *Jung's Stud. in Word-Association* p. v, We owe to Dr. Jung..the application of the association method to *unconscious* mental processes... These studies in word association have now acquired a permanent place in the historical development of this [*sc.* psychoanalytical] theory. 1946 A. Christie *Hollow* xxvi. 221 'What is there about that interests you so, M. Poirot?..' 'Association—a point of the psychology.' 'Word association?' Horse and cart. Rocking horse?' 1952 C. P. Blacker *Eugenics: Galton & After* 52 Galton also tried out on himself an elaborate system of word-association tests. His method was different..from that later popularized by C. G. Jung, but the underlying idea was the same. 1971 J. Elsom *Theatre outside London* vii. 126 These discussions were linked with word-association games. 1977 *Canad. Jrnl. Linguistics 1976* XXI. ii. 200 'Navaho word associations', which examines the role of grammatical form-classes in word association tests, has no subject under the age of seventeen. 1931 C. L'E. Ewen *Hist. Surnames* xiv. 360 The root of Skr. and Pers. *yuvan* 'young' may well be one of the word-bases of Ewan, Owen [etc.]. 1956 *Essays & Studies* IX. 98 The appearance of the same word-base in various forms (polyptoton): (al) ar knitt & onyd in this onyng, & made holy in this holyhede. 1964 J. Vachek in D. Abercrombie et al. *Daniel Jones* 195 Some ModE affixes (especially the word-formative ones) are..more easily separable from their word-bases than others. 1963 J. Lyons *Structural Semantics* ii. 11 A word-based grammar seems to be more satisfactory than a morpheme-based grammar for the description of languages of the 'inflecting' type. 1890 W. James *Princ. Psychol.* I. ii. 55 If this order of association be ingrained and habitual in that individual, injury to his *visual* centres will make him not only word-blind, but aphasic as well. 1933 L. Bloomfield *Language* xxiii. 419 At the time of the loss of *-n*, the language did not distinguish word-boundaries in the manner of present-day English. 1978 *Language* LIV. 21 The well-known word-boundary phenomena of French, such as liaison, elision, and *h*-aspiré. 1968 J. R. Biggs *Basic Typogr.* 41/1 Spacing is consistent but the right-hand margin is irregular. This does not always reduce word breaks very much. 1980 B. Crutchley *To be a Printer* 55 Bad word-breaks at the end of lines and similar shoddiness. 1938 B. L. Whorf in *Language* XIV. 275 *Word Category*; a category (overt or covert or mixed) which delimits one of a primary hierarchy of word classes each of limited membership (not coterminous with entire vocabulary), e.g. the familiar 'parts of speech' of Indo-European and many other languages, vs. *Modulus Category*; one which modifies, either any word of the vocabulary, or any word already allocated to a delimited class, e.g. voices, aspects, cases. 1964 J. Vachek in D. Abercrombie et al. *Daniel Jones* 195 The comparative suffix *-er* does not imply the change of the word-category of the basic word, while the agentive *-er* necessarily does so. 1941 E. Blunden *Thomas Hardy* xi. 236 An innovation of word-choice. 1914 L. Bloomfield *Introd. Study Lang.* iv. 108 (*title*) Word-classes. *Ibid.* iv. 109 Other word classes which are not expressed by formational similarity. 1924 O. Jespersen *Philos. Gram.* iv. 61 We have a great many words which can belong to one word-class only: *admiration, society, life* can only be substantives [etc.]. 1953 C. E. Bazell *Linguistic Form* vi. 76 The so-called parts of speech (still more inappropriately word-classes) are classes of stem-morpheme. 1973 *Computers & Humanities* VII. 159 Lyne's resolution of word class problems (is *y* an adverb or a pronoun?) uses a rather complex algorithm. 1935 *Vanity Fair* Nov. 38/1 There appears to be little leakage of their vernacular into even so ambitious a word-coiner as *Variety*. 1981 *Verbatim* Spring 14/2 Since I started publishing *Verbatim*, these and other word-coiners have been sending their creations to me, pressing for recognition. 1864 W. D. Whitney in *Ann. Rep. Bd. Regents Smithsonian Inst. 1863* 108 The conditions of that ancient period, and the degree in which they could quicken the now sluggish processes of word-combination and formation are beyond our ken. 1932 A. Gardiner *Theory of Speech & Lang.* iii. 158 Syntax..may be defined as the study of the forms both of the sentence itself and of all free word-combinations which enter into it. 1930 *Proc. Brit. Acad.* XVI. 147 There are prejudices, preferences, analyses, comparisons, statistics, verse-tests, word-counts, sense of style, poetical feelings, intuitions—but we must not call all this evidence. 1937 Palmer & Hornby *Thousand-Word Eng.* 11 The compiler has recourse to statistics of word-frequency; he organizes a 'word-count'. 1957 *Eng. Lang. Teaching* (British Council) XII. 1.10 There are valuable word-counts which give a clear picture of the relative importance of specific words in our total lexicon. 1980 *Amer. Speech 1977* LII. 7 A computerized word count. 1884 *Amer. Jrnl. Philol.* July 187 That species of word-creation commonly designated as parasynthetic. 1952 W. D. Jacobs *William Barnes, Linguist* 7 Are there means of word-creation and actual words themselves in the writings of Barnes by which English might well profit? 1914 L. Bloomfield in *Trans. Amer. Philol. Assoc.* XLV. 66 We have many instances of the writing of uneducated people..in which the word-division is entirely wrong. 1929 K. Sisam in *S.P.E. Tract* XXXIII. 441 A respectably printed American book..is not noticeably different in [typographical] word-division from a contemporary English book. 1976 *Classical Q.* XXVI. 95 His word-divisions are probably his own and without any authority. 1912 Kipling *Diversity of Creatures* (1917) 23 Word-drunk people. 1964 *Punch* 15 Apr. 575/1 The word-drunk Don Adriano. 1928 O. Jespersen *Internat. Lang.* ii. 121 In Novial the elements are separate words, in Esp[eranto]—Ido inseparable word-elements. 1964 C. Barber *Ling. Change Present-Day Eng.* iv. 78 These [new learned words] are usually formed from Latin or Greek word-elements. 1965 W. S. Allen *Vox Latina* i. 35 It [s] is not voiced between vowels or at word-end as in English *roses*. 1878 W. Barnes *Outl. Eng. Speechcraft* 83 (*heading*) The power of the word-endings [*sc.* suffixes]. 1966 J. Derrick *Teaching English to Immigrants* vi. 210 Meaning is conveyed... with a reduced form of grammar—word-endings are left off, structural words omitted, etc. 1926 Fowler *Mod. Eng. Usage* 553/2 A phonetically consistent method is in English peculiarly hard to reconcile with the keeping together of word families. 1978 *Language* LIV. 237/1 Finding translation equivalents and association of morphological word-families. 1952 H. Basilius in *Word* VIII. 103 Jost Trier's study of the German word-fields relating to the concept reason, its powers and qualities. 1965 *Amer. Speech* XL. 62 *Job* is not identical with *Arbeit*: it stands at the lowest level of this word-field. 1918 A. W. Aron in C. Hockett *Leonard Bloomfield Anthol.* (1970) 58 These variations in word-initial [in Irish] do depend on the phonetic character of the original preceding word-final. 1949 E. A. Nida *Morphology* (ed. 2) ii. 24 Only the word-final tones are indicated. 1977 G. P. Delahunty in D. Ó Muirithe *Eng. Lang. in Ireland* 132 Devoicing of word-final voiced consonants. 1983 *Word* XXXIV. 149 We find that in modern Swabian they all occur in word-final position. 1965 *Canad. Jrnl. Linguistics* Fall 64 Those [vowel segments] which occur both before consonants and word-finally. 1978 *Language* LIV. 443 High vowels are dropped after heavy stems word-finally. 1955 R. Jakobson in Saporta & Bastian *Psycholinguistics* (1961) 423/2 The more difficulties he has with word-finding in the proper sense of this neuropsychiatric term; that is, difficulties with spontaneous selection of words. 1874 H. Bendall tr. *A. Schleicher's Compar. Gram.* 3 The Semitic, which is not akin to the Indo-European, has more word-forms. 1952 *Mind* LXI. 239 As society discovers.. that judgments imputing responsibility..are never justified, the word-form 'responsibility' comes to change its meaning. 1967 D. G. Hays *Introd. Computational Linguistics* ii. 21 The plan is to record, before each word form, the number of cells it occupies. 1856 W. D. Whitney in *Jrnl. Amer. Oriental Soc.* V. 197 The work..exhibits.. the phenomena of the agreement and disagreement of the Greek and Sanskrit accentuation, throughout the departments of declension, conjugation and word-formation. 1948 L. Spitzer *Linguistics & Lit. Hist.* ii. 81 Thus Cervantes has expressed his perspectivistic vision

in a word-formational pattern of the Renaissance reserved for hybrids. 1979 *Dictionaries* I. 18 Lexicographers very often assume a kind of word-formational capacity or knowledge when they list derivatives as run-on entries. 1964 Word-formative [see *word-base* above]. 1928 B. Q. Morgan *German Frequency Word Bk.* p. ix, It was our original intention to publish the figures for word frequency, group frequency, and basic frequency. 1951 *Archivum Linguisticum* III. ii. 123 (*heading*) Word-frequency in Norwegian. 1974 Bedford & Dilligan *Concordance Poems Alexander Pope* II. 669/1 A six-page analytic table showing word-frequency distribution and the ratio between each word and the number of its occurrences. 1912 L. Bloomfield in C. Hockett *Leonard Bloomfield Anthol.* (1970) 35 The relation of word-form to word-function. 1910 R. B. Stern *Neighborhood Entertainments* vi. 263 (*heading*) Some word games [describes anagrams, logomachy, etc.]. 1922 S. Lewis *Babbitt* xviii. 227 Word-games in which you were an Adjective or a Quality. 1929 A. C. S. Ashmore (*title*) Word games and word puzzles. 1934 *Mind* XLIII. 117 Those who are reluctant to regard philosophy as mere mystery-mongering or as an academic word-game. 1953 E. Coxhead *Midlanders* vi. 143, I never could do word-games or crosswords. 1975 R. Lewis *Double Take* iv. 124 We aren't here to play word games... If this is the way you want to conduct the discussion we might as well call it off. 1974 Passmore & Robson *Compan. Med. Stud.* III. xxxiv. 8/1 Wittgenstein described speech as a 'word game', implying that language follows defined rules similar to those which govern sports. 1962 A. McIntosh in Davis & Wrenn *Eng. & Medieval Stud.* 240 Word-geographical criteria..may well therefore turn out to provide the only practicable line of attack. 1921 E. C. Roedder in *Jrnl. Eng. & Ger. Philol.* XX. 183 The finding and fixing of the isolectic lines is a task of word geography. 1949 H. Kurath (*title*) A word geography of the Eastern United States. 1980 *Amer. Speech* LV. 195 Its lexicology is sketched here as a development of conventional word geography. 1869 W. Barnes *Early England & Saxon-English* 130 A Hoard, as herd, is a kind of gathering of any kind of things, as..Word-hoard—Vocabulary. 1961 Webster *Pref.* 6a/1 Books consulted in the Springfield City Library whose librarians have..given the editorial staff..access to its large and valuable word-hoard. 1966 *Listener* 24 Nov. 779/2 Thomas was immensely proud of his bulging word-hoard. 1975 P. Fussell *Great War & Mod. Memory* (1977) ii. 49 *Lousy with*, meaning *full of*,..entered the colloquial word-hoard around 1915. 1935 A. Huxley *Let.* 13 Jan. (1969) 389 Nothing is more inclined to keep me awake than word-hunting. 1902 E. W. Scripture *Elem. Exper. Phonetics* x. 150 A word-idea should be learned as parts of various courses of thought in order to form the necessary language associations. 1922 D. H. Lawrence *Aaron's Rod* xiii. 175 Even his deepest ideas were not word-ideas, his very thoughts were not composed of words and ideal concepts. 1937 M. L. Hanley (*title*) Word index to James Joyce's *Ulysses*. *Ibid.* p. iii, In 1931..I had made a word index and partial concordance to the B-Text of *Piers Plowman*. 1960 *Amer. Speech* XXXV. 215 A lamentable deficiency is the lack of a full word index. 1918 Word-initial [see *word-final* above]. 1926 L. Bloomfield in Saporta & Bastian *Psycholinguistics* (1961) 29/2 English word-initial [st-]. 1949 E. A. Nida *Morphology* (ed. 2) ii. 16 Word-initial prevowel glottal stops. 1981 *N. & Q.* Oct. 398/1 A large number of unexplained intersubstitutions of *c* and *g* in word-initial position. 1973 A. H. Sommerstein *Sound Pattern Anc. Gr.* iii. 103 Not all rules with left-hand environments can apply word-initially. 1964 D. Ward in D. Abercrombie et al. *Daniel Jones* 385 The substitution of corresponding voiceless phonemes for all voiced phonemes except sonants in word-final position and word-internally before voiceless consonants. 1945 *N. & Q.* 6 Oct. 151/1 In the amusement sections of newspapers it [*sc.* the game or puzzle called 'Doublets'] is usually referred to as 'Word-Ladder'. 1958 *Birmingham Mail* 27 Jan. 6/7 Today's puzzle is for your shortest Word Ladder from *Head* to *Body*. It can be done in five easy steps without using any unusual words. 1982 D. Parlett *Penguin Bk. Word Games* 98 Word Ladders. 1951 *Proc. IRE* XXXIX. 277/2 Digital computers commonly use a fixed word length (that is, a fixed number of characters) which is a characteristic of each computer. 1970 O. Dopping *Computers & Data Processing* vi. 98 The word length is usually chosen in such a way that a numeric operand or result in general can be stored in one memory cell with 'normal' precision. 1960 J. B. Carroll in Saporta & Bastian *Psycholinguistics* (1961) 338/2 A single word or word-like utterance. 1923 Ogden & Richards *Meaning of Meaning* ii. 42 The earlier writers are full of the relics of primitive word-magic. To classify things is to name them..to know their names is to have power over their souls. 1938 S. Chase *Tyranny of Words* iv. 37 Here, to follow Malinowski, we note the seeds of word magic, in which *the name gives power over the person or thing it signifies*. 1960 H. Read *Forms of Things Unknown* vii. 121 The name he chooses is magically apt, and in word-magic we must acknowledge the primordial intensive aspect of poetry. a1856 J. Stoddart *Glossology* (1858) x. 231 *Onomatopoeia*. The literal signification of the term..is nothing more than 'word-making'. 1867 W. D. Whitney *Lang. & Study of Lang.* 116 All word-making by combination..is closely analogous with phrase-making. 1935 *Encycl. Sports, Games & Pastimes* 396/1 The games played with these [cardboard] letters are very numerous. Word making and word taking may be mentioned first. 1952 G. Raverat *Period Piece* xii. 243 Our chief intellectual exercise [*c.* 1900] was the Letter Game; Word-making and word-taking. 1980 L. Lewis *Private Life of Country House* ii. 20 My mother.. scattered on the floor the cardboard letter squares from a game called 'Wordmaking and Wordtaking'. 1902 *Encycl. Brit.* XXXIII. 387/1 The registration of 'word marks' was first provided for by the Trade Marks Act, 1883. In that statute, however, clause (d) read 'a fancy word or words not in common use'. 1964 T. W. McRae *Impact of Computers on Accounting* i. 11 The programmer then looks at the data..and divides up the storage locations as required by setting a 'wordmark' in the far right-hand position of each variable-length storage word. 1969 P. B. Jordain *Condensed Computer Encycl.* 567 Each instruction must begin with a word mark [in a variable word-length computer]. 1976 *Century of Trade Marks* (Patent Office) i. 1/2 These were early instances of word marks, though usually in conventional form—for example, CATIM (Cati manu: from the hand of Cato), and OFALBIN (officina Albini: the workshop of Albinus) —whereas earlier marks had been almost invariably devices or ideographic symbols. 1980 J. Frates *Introd. Computer* vi. 165 The storage location of the last piece of data will store a word mark in addition to the character to indicate to the computer that it has reached the end of that particular item of data. 1949 E. A. Nida *Morphology* (ed. 2) 293 (*heading*) Reduction of word-medial consonant clusters. 1963 J. Lyons *Structural Semantics* iv. 68 There is an opposition.., in word-initial and word-medial position, between the voiced and the voiceless plosives. 1968 *Language* XLIV. 532 A variety of further clusters occurred word-medially. 1890 W. James *Princ. Psychol.* I. xvi. 684 'Ataxic' and 'amnesic' aphasia, 'word-deafness', and 'associative aphasia' are all practical losses of word-memory. 1920 *Punch* 7 Jan. 9/2 The word-merchant [*sc.* a journalist] was laughing at us all the time. 1977 *Grimsby Even. Tel.* 14 May 7/4 He [*sc.* Malcolm Muggeridge] is the best word merchant of our time. [1879 *First Infant Reader* in *Mod. School Reader Series* p. 2 The experience of many years has convinced us that a judicious combination of the Word and Phonic methods of teaching Children to read is the best.] 1932 in E. Blyton *Mod. Teaching in Infant School* v. 62 The Word Method, or the Look-and-say Method. 1981 D. Rowntree *Dict. Educ.* 350 *Word method*, teaching a pupil to read by getting him to recognize whole words right from the start. 1855 Geo. Eliot in *Westm. Rev.* Oct. 596 As long as the English language is spoken, the word-music of Tennyson must charm the ear. 1895 G. B. Shaw *Our Theatres in Nineties* (1932) I. 77 M. Maeterlinck's fragile word-music. 1962 *Observer* 22 Apr. 23/4 The ailing cause of Shakespeare-designed-to-be-read-as-word-music. 1895 'Mark Twain' in *N. Amer. Rev.* July 11 This is [James Fenimore] Cooper. He was not a word-musician. His ear was satisfied with the *approximate* word. 1892 H. Sweet *New Eng. Gram.* I. §113. 42 We find the same grammatical relation expressed..sometimes by word-order. 1958 *Aspects of Translation* 35 Inflexions and grammar impose a more rigorous word-order on the French language than on English. 1965 O. Funke in *English Studies* Feb. 58 The transition from paradigmatic to syntagmatic (prepositional or word-order) expression. 1973 *Archivum Linguisticum* IV. 28 The comparative fixity of word-order relative to grammatical function is the conditional par excellence for the supposed superfluity of gender distinctions. 1870 J. G. Whittier in *Atlantic* Apr. 467 Not by the page word-painted Let life be banned or sainted. 1937 'C. Caudwell' *Illusion & Reality* v. 100 The word-painted lands of the nightingale, of the Grecian urn, of Baiae's isle. 1936 G. K. Zipf *Psycho-Biol. of Lang.* iv. 134 Word-pairs like *submit* and *remit*, or *accuse* and *excuse*. 1964 J. Vachek in D. Abercrombie et al. *Daniel Jones* 194 The ModE word-pair *longer* [lɔŋgə] (comparative of *long*): *longer* [lɔŋə] (the noun of agent derived from the verb *to long*). 1948 J. R. Firth *Papers in Linguistics 1934–51* (1957) xi. 150 Palatograms here presented..are *word-palatograms*. That is to say, they are used for the abstraction of articulatory contact and possibly also of movement from suitably selected words taken as whole utterances. 1957 Word-palatogram [see *kymography]. 1912 M. Beerbohm *Christmas Garland* 66 Intensive vision has this Mr. Hardy, With a dark skill in weaving word-patterns. 1938 L. MacNeice *Mod. Poetry* ii. 40 The normal business of poetry is the conveying of information through certain kinds of word-patterns. 1957 C. E. Bazell in *Miscelanea Homenaje a André Martinet* I. 27 Some languages confer word-status by integrating a unit into the particular word-pattern. 1951 R. Firth *Elem. Social Organiz.* vii. 221 It does this..by response to the aesthetic qualities of the word-patterning and imagery used. 1851 J. Brown *Let.* 23 June (1912) 119, I wish you would paint some word pictures, some studies from Nature as you take your drives. 1855 H. Martineau *Autobiogr.* (1877) I. 397 An opportunity..for paradox, and word-play. 1962 W. Nowottny *Lang. Poets Use* v. 99 The next [chapter] will discuss some aspects of conspicuous word-play—or, to use a more convenient term —of 'verbal schemes'. 1967 *Sci. Amer.* Sept 268/1 The double acrostic..was..the most popular form of word-play in English-speaking countries throughout the last quarter of the 19th century and until the end of World War I. 1967 C. L. Wrenn *Word & Symbol* 13 The medieval love of riddling and word-play was occasionally displayed by Anglo-Saxon versifiers. 1982 I. Hamilton *Robert Lowell* (1983) ii. 18 Argument pursued for the sake of wit and wordplay rather than for any just or true solution. 1894 O. Jespersen *Progress in Lang.* iv. 93 Is it beneficial to a language to have a free word-position? 1961 *Brno Studies* III. 46 The number of word-positions in which the correlation could be utilized. 1947 *Jrnl. Symbolic Logic* XII. 90 The word problem for semigroups. 1972 M. Kline *Math. Thought* xlix. 1143 For one defining relation Wilhelm Magnus (1907–) showed that the word problem is solvable. But the general problem is not. 1984 *Which Micro* Dec. 73 (Advt.), For word processing letters in professional type. 1985 *Daily Tel.* 10 June 11/1 To a newspaper reporter, the ability to wordprocess stories on aircraft and in hotel bedrooms must truly be a boon. 1984 *N. & Q.* Dec. 552/1 This text (excepting the chapter-notes and bibliographies) reproduces a word-processed typescript. 1970 *Administrative Management* Nov. 36/3 'Word processing', a concept that combines the dictating and typing functions into a centralized system. 1977 *Times* 12 Sept. 5 Word processing can already be seen to be at the forefront of the next revolution in the office... The keyboard of the word-processing typewriter..is standard but typing on it produces not only a paper copy but also a magnetic recording which can be automatically searched and edited. 1970 *Administrative Management* Nov. 37/1 In 1970..ITEL..introduced its 'Word Processor'. 1974 *Ibid.* Mar. 48/2 The multi-functional word processor can handle a whole range of text manipulations. 1977 *N.Y. Times* 1 Jan. 22 Word processors call up documents, page by page and line by line, on cathode ray screens for editing. They print out finished versions automatically or send them via telephone lines to distant points. 1979 *Daily Tel.* 24 Dec. 16/3 Word-processors show their greatest strengths when they are used to produce long reports which need to be constantly altered and with any typing which must be word perfect. 1981 *Times Lit. Suppl.* 22 May 588 (Advt.), Manuscripts typed, edited, corrected and indexed by word processor. 1984 D. Lodge *Small World* ii. ii. 121 A roomful of secretaries..would wait patiently beside their word-processors, ready to type..his latest reflections. 1928 *Funk's New Stand. Dict.*, Word recognition. 1956 T. W. Clymer in R. H. Beck *Three R's Plus* 139 Word-recognition skills have been mentioned... Context clues are the quickest and easiest of the word-recognition techniques. 1978 J. Baron in W. K. Estes *Handbk. Learning & Cognitive Processes* VI. iv. 159 Let us assume that one mechanism of word recognition in reading involves activation of a semantic code directly from a letter or spelling-pattern code. 1915 *Stedman's Med. Dict.* (ed. 3) 1034/2 *Word-salad*, a term applied by Forel to the jumble of meaningless words uttered by a patient suffering from catatonia. 1930 L. E. Hinsie *Schizophrenia* ii. 28 The symptomatology is ordinarily not at all bizarre; there is not the scattering of thought, nor the 'word-salad'. 1960 R. D. Laing *Divided Self* xi. 215 Her 'word-salad' seemed to be the result of a number of quasi-autonomous partial systems striving to give expression to themselves out of the same mouth at the same time. 1976 N. Postman *Crazy Talk* 228 The exorbitant fee one must pay..is made to seem plausible by a word salad of imposing proportions. 1960 *Times* 18 Jan. 3/1 And so many composers have turned gratefully to word-setting. 1985 *Times* 13 June 13/8 Why is word-setting generally so difficult for English composers? 1900 *Jrnl. Anthrop. Inst.* XXX. 136 As regards word-signs, in general the connection of the meaning of the word with the picture is obvious enough once it is pointed out. *Ibid.*, The connecting link between the picture and its word-sign value. 1908 G. K. Chesterton *Man who was Thursday* ix. 165 It did not take him long to learn how he might convey simple messages by what would seem to be idle taps upon a table or knee... 'We must have several word-signs,..words that we are likely to want.' 1941 *Language* XVII. 149 In my opinion, *ideogram* and *word-sign* are not interchangeable terms, either in Egyptian or English. 1964 P. A. D. MacCarthy in D. Abercrombie et al. *Daniel Jones* 162 Four uniliteral 'word-signs' for *the, of, and, to*, are standard. 1961 *Observer* 23 July 19/3 Gunther has lost none of his old skill as a wordsmith: the adjectives come pouring out in a torrent of enthusiasm, bristling with brackets, dashes, afterthoughts, and statistics. 1968 *Daily Tel.* 7 Nov. 23/1 We already know John Updike as a resourceful wordsmith, a fine writer. 1976 *Bookseller* 4 Sept. 1645/1 That ubiquitous wordsmith A. J. P. Taylor. 1958 Wordsmithing [see *copy-writing* vbl. sb. s.v. *copy *sb.* C]. 1981 *Maledicta* V. 346 Enjoys word-smithing, learning languages, and the study of names. 1925 I. A. Richards *Princ. Lit. Criticism* xvi. 119 Many people are able to imagine word-sounds with greater delicacy..than they can utter them. 1951 N. M. Gunn *Well at World's End* xxx. 295 She was making word-sounds, lapping him about. 1937 A. H. Gardiner in *Mélanges Ling. et Phil. offerts à J. van Ginneken* 310 It seems necessary, as between the different classes [of proper names], to assign independent word-status further only to classes II and V. 1982 *Papers Dict. Soc. N. Amer. 1977* 67 In such cases the problem of word-status is involved. 1948 L. Spitzer *Linguistics & Lit. Hist.* i. 7 This French word-family..was a blend of at least two word-stems. 1962 W. Nowottny *Lang. Poets Use* v. 100 Syntactical patterns and individual words (or word-stems). 1911 L. Bloomfield in C. Hockett *Bloomfield Anthol.* (1970) 29 The farther back we look into the history of any IE. language, the more diversified and concrete a word stock do we find. 1940 J. H. Jagger *Eng. in Future* i. 25 About two per cent. of the word-stock consists of words of this sort [*sc.* slang words that have entered Standard English] made since the Norman conquest. 1973 *Computers & Humanities* VII. 195 Normal discourse draws upon a word-stock which in any theorizing must be treated as infinite. 1863 W. Barnes *Gram. & Gloss. Dorset Dial.* 9 In searching the word-stores of the provincial speech-forms of English, we cannot but behold what a wealth of stems we have overlooked at home. 1971 J. Z. Young *Introd. Study Man* xxxv. 489 [The region behind the superior temporal gyrus] has also been called 'the area of ideational speech' or indeed 'word store'. 1924 H. E. Palmer *Gram. Spoken Eng.* i. 6 *Word-stress.* (In the opinion of the author the term *syllable-stress* would be more appropriate.) This term is used with reference to a syllable (in a word of more than one syllable) which is susceptible of receiving one of the four nucleus-tones. 1953 C. E. Bazell *Linguistic Form* viii. 100 Word-stress is therefore highly marginal in language. 1966 J. Derrick *Teaching Eng. to Immigrants* iii. 111 This distribution of stress in the individual word, 'word stress' as it is called, is a basic difficulty for the foreign learner. 1951 *Essays & Studies* IV. 123 Alliteration, assonance.. can be considered as markers or signals of word-structure or of the word-process in the sentence. 1975 *Language for Life* (Dept. Educ. & Sci.) xi. 183 Their attention should constantly be drawn to details of word structure. 1940 *Amer. Speech* XV. 183 Mrs. Ernst presents for her pupils ..the essentials of word-study. 1979 *N. & Q.* June 245/2 The author of a word-study. 1904 Word-symbol [see *ideogenetic* a. s.v. *ideo-]. 1933 L. Bloomfield *Language* xvii. 287 In the writings of other languages, where words are of various lengths, we find word-symbols used for phonetically similar parts of longer words. 1955 G. A. Kelly *Psychol. Personal Constructs* 459 A preverbal construct is one which continues to be used even though it has no consistent word symbol. 1923 Ogden & Richards *Meaning of Meaning* ii. 37 In Fraser's *Golden Bough* numerous examples of word taboos are collected. 1978 *Language* LIV. 217 The importance of word taboo on the basis of more recent linguistic-anthropological work. 1954 *Computers & Automation* May 22/1 Word time. 1969 P. B. Jordain *Condensed Computer Encycl.* 567 All activities must be calculated in multiples or submultiples of word time or cycle time. 1894 O. Jespersen *Progress in Lang.* ix. 340 So much for

word-tones; now for the sentence melody. 1928 *Proc. Brit. Acad.* XIV. 354 The four word-tones used in the Mandarin language of Peking to keep otherwise identical words apart. 1964 *Archivum Linguisticum* XVI. 81 He gives some interesting examples of how word-tones and sentence intonation may combine. 1920 D. H. Lawrence *Women in Love* xxiii. 339, I know your dodges. I am not taken in by your word-twisting. 1959 I. & P. Opie *Lore & Lang. Schoolch.* xiv. 320 By using slang, local dialect,..word-twistings, codes and sign language, children communicate with each other in ways which outsiders are unable to understand. 1911 S. S. Colvin *Learning Process* (1931) vii. 108 The word-types of images, as can be readily seen, are symbolic; they stand for concrete realities, which, however, generally are not revived in connection with the symbol. 1936 *Jrnl. Philos.* 17 Dec. 702 Let us call a 'word-type' a class or kind of defining character of a class of tokens which are similar to one another in certain essential aspects. 1961 *Brno Studies* III. 33 Mathesius laid special stress on the part played in English complex condensation cases by three types of nominal forms derived from verbal bases... The word-types will be referred to as..condensers. 1976 *Biometrika* LXIII. 435 How many word types did Shakespeare actually know? 1924 R. M. Ogden tr. *K. Koffka's Growth of Mind* v. 270 A difference in the serial order of the correct word-usage must then depend..upon a difference in the colour-phenomenon itself. 1971 *Jrnl. Gen. Psychol.* Apr. 188 Creativity and originality were not measured by any of the three types of word-usage. 1932 Faucett & Maki (*title*) Study of English word-values. *Ibid.* 8 This book will..be useful to those interested..in fixing a graded vocabulary scale for supplementary readers,..in helping teachers and students to develop a sense of word-values [etc.]. 1938 I. Goldberg *Wonder of Words* xx. 438 For & has the phonetic value of *et*, but it has the word-value of *and*. 1968 *Listener* 25 Apr. 525/1 What happens if in turn we word-watch on Mr. Davie? Could it be that to use the word 'histrionic' ten times in one short article is itself somewhat histrionic? 1973 *N.Y. Times* 7 May 39/1 Word watchers introduce a note of good sense..to the action and passion of their times. 1980 *Amer. Speech* LV. 77, *-gate* has undergone some developments that should interest word-watchers. 1981 *Verbatim* VII. iii. 20/2 Collectors are needed to find quotations for a new dictionary of American slang... The chief motivation for volunteering should be interest in specialized word-watching. 1977 *IEEE Trans. Professional Communication* June 14/2 One very useful feature..is called word wrap. 1982 A. J. Meadows et al *Dict. New Information Technol.* 193/1 *Word-wrap*, a word processing term. It refers to the way in which a partially typed word is moved to a new line if its length proves too much to fit into the existing line. 1983 *Austral. Personal Computer* Sept. 124/2 Automatic wordwrap operates at the end of a screen line (40 chars). 1984 *Computing Today* May 93 (*heading*) Word-wrapping. 1985 *Listener* 25 Apr. 38/1 Word-wrapping, that puts in the ends of lines automatically on reaching the right-hand side of the screen, was another counter-creative feature. 1933 L. Bloomfield *Language* xvii. 285 A better name [for ideographic writing]..would be word-writing or logographic writing. 1942 —— in C. Hockett *Bloomfield Anthol.* (1970) 385 In word writing each word is represented by a conventional sign... Chinese writing is the most perfect system of this kind.

word, *v.* Add: **4. d.** *intr.* for *pass.* To admit translation into words. *poet.*⁻¹ (after Wear *v.*¹ 15).

1935 L. MacNeice *Poems* 26 My dream will word well — But will not wear well.

5. To speak to, accost; to tell, pass word to. Also, to rebuke or tell off. *Austral. slang.*

1906 E. Dyson *Fact'ry 'Ands* i. 2 I'll word 'em [girls] when they pass again. 1916 C. J. Dennis *Songs Sentimental Bloke* 30 I met 'im on the quite, An' worded 'im about a small affair. 1936 N. Marsh *Death in Ecstasy* vi. 79 He looks more like a regular dick. An' yet if I worded him maybe he'd talk back like a bud's guide to society stuff. 1945 Baker *Austral. Lang.* vi. 121 He..*words* him, rebukes him. 1967 K. S. Pritchard *Subtle Flame* 234 Ted worded a mate of his on the *Western Star*.

wordage. Delete *rare* and add later examples. In recent use also, an amount of words written or spoken; the number of words in a document.

1926 *Glasgow Herald* 19 May 7/2 Managers of the great news agencies..have never placed upon the cables so large a wordage concerning any British domestic event. 1958 *Times* 26 Aug. 5/7 Telegrams will carry a fixed charge irrespective of wordage. 1966 *Punch* 21 Sept. 455/2 Obviously he took his correspondence as seriously as he took his Journals and his daily wordage of publishable prose. 1975 G. Howell *In Vogue* 64/1 [Tallulah Bankhead] spoke seventy thousand words a day—the wordage of *War and Peace* over a weekend. 1985 *Univ. Cape Town Studies in English* Feb. 61 Scientists and scholars who have published in abundance are actively solicited by editors, bookmen, and publishers for still more wordage to be put into print.

wordlore. Add: **b.** (Earlier example.) Hence **wo·rdlorist.**

1861 *Trans. Philol. Soc. 1860–1* 154 A perfect Dictionary must not only be a complete Repertory, but also an available Directory within the whole province of word-lore (*wort-lehre* as distinguished from *satz-lehre*). 1929 *N. & Q.* 15 June 419/2 No word-lorist who studies place-names.

wordly, *a.* (Later example.)

1927 M. Sadleir *Trollope* 370 This fact indicates.. two of his personal qualities..his wordly proficiency and his good manners.

wordmonger. In recent use also without contemptuous overtones.

1916 *Daily News* 8 Nov. in E. Weekley *Etymol. Dict. Mod. Eng.* (1921) 944 Professor Weekley is well known to our readers as the most entertaining of living word-mongers. 1981 V. Glendinning *Edith Sitwell* 4 She is a poet of dream and vision, a musical wordmonger.

wordmongery (earlier example).

1881 Max Müller tr. *Kant's Critique Pure Reason* II. ii. iii. 223 There remains nothing but mere wordmongery.

wordsman (wɔ̄·ɹdzmæn). = Wordman. So **wo·rdsmanship.**

1959 I. & P. Opie *Lore & Lang. Schoolch.* iii. 50 Wordsmanship. It is common practice to snub a companion who makes irritating use of words such as 'Well!' 'What?' and 'Eh?' 1962 *Canadian Intelligence Service* XII. ii. 3/1 The U.S. Senate Security Subcommittee recently issued a report entitled *Wordsmanship; Semantics as a Communist Weapon*. 1981 W. Safire in *N.Y. Times Mag.* 27 Apr. 18/3 Robert Burchfield, chief editor of the Oxford Dictionaries, known to wordsmen as 'Superlex'. 1984 *N.Y. Times Mag.* 30 Dec. 6/2 The occasion was a gathering at the library of renowned scholars, including a babble of wordsmen, to mark the 200th anniversary of the death of Samuel Johnson.

wordster (wɔ̄·ɹdstəɹ). *nonce-wd.* [f. Word *sb.* + -ster.] One who deals in or handles words: (*a*) one who indulges in talk rather than action; (*b*) a skilful user of words; (*c*) a student of words and their meanings.

1917 H. A. Jones *Pacifists* [Dedication], *Dedicated* To the tribe of Wordsters, Pedants, Fanatics, and Impossibilists, who so rabidly pursued an ignoble peace, that they helped to provoke a disastrous war. 1965 *English Studies* XLVI. 465 [The suffix *-ster*] may serve the function of condensing long words such as 'philologist' and 'lexicographer' into short *wordster*. 1971 'J. Quartermain' *Man who walked on Diamonds* ii. 14 The..brilliant wordster, always good for the *bon mot*. 1976 *Verbatim* Dec. 8/2 As an amateur wordster, my personal lexicon contains lengthy lists of various types of words.

Wordsworthian, *sb.* and *a.* Add: **b.** *adj.* (Earlier examples.)

1817 W. Whewell *Let.* in M. Moorman *William Wordsworth* (1965) II. ix. 325 His [*sc.* Coleridge's] critique on the Daffodils might serve as a model for similar strictures on all Wordsworth's Wordsworthian poems. 1845 A. de Vere *Let.* 28 Sept. in *Recoll.* (1897) x. 204 You are a greater admirer of the special Wordsworthian genius.

Wo:rdsworthia·na [*-iana suffix], things connected with Wordsworth, writings about Wordsworth; **Wordsworthy** *a. colloq.*, typical or suggestive of Wordsworth.

1889 W. Knight (*title*) Wordsworthiana: papers read to Wordsworth Society. 1938 S. Beckett *Murphy* v. 106 They [*sc.* sheep] seemed in rather better form, less Wordsworthy. 1983 *London Rev. Bks.* 7–20 July 18/4 Recent items of Wordsworthiana include *The Visionary Company*.

work, *sb.* Add: **I. 1. b.** (Later examples.)

1739 J. Wesley *Doctrine of Salvation* 5 Because all Men are Sinners against God, and Breakers of his Law, therefore can no Man by his Works be justified, and made righteous before God. 1883 W. C. Dowding *Luther & his Work* 6 We are accounted righteous before God only for the merits of our Lord Jesus Christ, by faith; and not from our own works or deservings. 1906 W. Walker *John Calvin* xv. 415 Calvin..leaves room for a conception of 'works' as strenuous..as any claimed by the Roman communion. 1963 E. P. Thompson *Making of Eng. Working Class* xi. 364 How, then, to keep grace? Not by good works, since Wesley had elevated faith above works:.. Works were the snares of pride and the best works were mingled with the dross of sin; although.. works might be a *sign* of grace. 1972 Q. Bell *Virginia Woolf* I. i. 4 The Clapham Sect was concerned with works rather than with faith.

d. (Further examples.)

1749 J. Cleland *Mem. Woman of Pleasure* II. 120 All this was not the work of the fourth part of a minute. 1871 Hardy *Desperate Remedies* II. ii. 74 To bring him out and lay him on a bank was the work of an instant. 1927 C. Asquith *Black Cap* 73 To light his candle and put on his dressing-gown and slippers was the work of a moment.

5. c. *slang.* A criminal act or activity. Cf. Job *sb.*² 1 b.

1812 J. H. Vaux *Vocab. Flash. Lang.* in *Mem.* (1964) 279 An offender having been detected in the very fact..is ..said to have been *grab'd at work*. 1865 in *Comments on Etym.* (1983) XIII. iii.–iv. 17 We..surrounded her from observation while at 'work'. 1926 J. Black *You can't Win* xxi. 338 Coppers located 'work' for burglars and stalled for them while they worked. *Ibid.* xxiv. 379 That kind of 'work' is unprofessional, unnatural, and disgusting. 1963 T. Tullett *Inside Interpol* xiv. 192 If he netted only about 200 guilders he would start 'work' again in a week.

8. (Earlier example.)

1832 W. Whewell *First Princ. Mech.* iv. 52 The *work done* does not depend on the pressure alone. *Ibid.* 53 The work done by a machine may be represented as certain pressures exerted through certain spaces.

II. 18. b. Phr. *in the works* = *in the pipeline* s.v. *Pipe-line *sb.* b. *N. Amer.*

1973 *Globe & Mail* (Toronto) 12 July 2/3 In his statement, Mr. Cote said he had been informed during the election campaign that a 'telegraph organization was in the works in certain ridings of the South Shore'. 1976 *National Observer* (U.S.) 16 Oct. 10/3 As might be expected, a movie deal is in the works. 1979 *Tucson Mag.* Jan. 10/3 Actually there is a sequel in the works and the project was begun as a two film package. 1984 *National Times* (Austral.) 2 Nov. 41/2 There are, of course, follow-up books in the works.

20. a. (Later *colloq.* examples, also applied to persons.)

1906 E. Dyson *Fact'ry 'Ands* xv. 197 'Ceptin fer er hunexpected wail he jerked out iv 'is works now 'n 'again, that cat was just er livin' silence. 1884 'Mark Twain' *Huck. Finn* xxxii. 333 Here we're a running on this way, and you hain't told me a word about Sis, nor any of them. Now I'll rest my works a little, and you start up yourn. 1885 —— in *Century Mag.* Dec. 196/1 Then it would bray—..spreading its jaws till you could see down to its works. It was a disagreeable animal.

b. *slang* (orig. *U.S.*). *the* (*whole*) *works*, the whole lot, everything; *esp.* in phrases, *to give* (*or tell*) *the works*: to tell the whole story; *to shoot the works*: see *Shoot *v.* 23 j; *to give* (*someone*) *the works*: to give (him) a rough time, *spec.* to murder; also, to give (someone) the full treatment (not necessarily unpleasant); *to get the works*: to receive severe punishment, reprimand, adverse criticism, etc.

1899 J. London *Let.* 18 May (1966) 38, I..quite enjoyed the thought of saying good-bye to the whole works. 1920 *Collier's* 5 June 36/3 'I ain't trying to jimmy into your most intimate affairs, but is they—is they a girl?'.. He..sat down..and gimme the works. 1927 *Vanity Fair* XXIX. 134/2 'Giving a guy the works' is handing someone a raw deal. 1928 *Amer. Mercury* Apr. 429/2 One-Lung here squealed, an' I got the works for two years —poundin' rocks wit' a sledge. 1929 C. F. Coe *Hooch* vii. 156 This man never was bumped here at all. They gave him the works some place a long way off. 1930 *Daily Express* 23 May 11/3 Threatening that unless the money was produced somebody would 'get the works'. 1934 Wodehouse *Right Ho, Jeeves* ix. 111 Heave a couple of sighs. Grab her hand. And give her the works. Right. 1936 J. Steinbeck *In Dubious Battle* iii. 53 Tell him the works. 1969 E. Bagnold *Autobiogr.* xii. 236 The Chinese Prime Minister is a better play than *The Chalk Garden* but it didn't get the works. 1979 L. Kallen *Introducing C. B. Greenfield* xiv. 193, I have uncovered a sensational story that is crying to be written... Best-seller list, movie, the works.

c. *pl.* A drug addict's equipment for taking drugs. *U.S. slang.*

1934 L. Berg *Revelations of Prison Doctor* iv. 42 All became adept in the use of 'the works'; this was a syringe and needle. 1951 *N.Y. Times* 15 June 14/3 Do they ask you if you want 'the works' when you're buying needles? 1953 W. Burroughs *Junkie* xiv. 140, I went into the bathroom to get my works. Needle, dropper, and a piece of cotton.

III. 27. out of work: (earlier modern example and earlier and later *attrib.* and *sb.* examples); hence *out-of-worker*; *out-of-workness.*

1864 J. O'Neil *Diary* 10 Apr. in J. Burnett *Useful Toil* (1974) 1. 85 One half of the time I was out of work and the other I had to work as hard as ever I wrought in my life. 1885 *Marine Engineer* 1 Sept. 137/2 'Out-of-work benefit' came to £57,000. 1888 [see Out of III.]. 1894 A. Morrison *Tales of Mean Streets* 48 The advent of a flush sailor..disposed to treat out-of-workers. 1903 A. McNeill *Egregious Engl.* v. 49 Out-of-workness is.. the most fearful thing in life that can happen to an Englishman. 1913 A. Spender in H. Barnett *Canon S. A. Barnett* (1918) II. xlvi. 273 How to tide over the winter for the out-of-work docker. 1939 'G. Orwell' *Coming Up for Air* II. ix. 153 We'd suddenly changed from gentlemen..into miserable out-of-works whom nobody wanted. 1955 M. Gilbert *Sky High* viii. 105 He was an out-of-work actor. 1974 R. Butler *Buffalo Hook* v. 45 I'm just an out-of-work who wants to stay that way.

30. Also with *for* in colloq. phr. and var. (Further examples.)

1843 Dickens *Christmas Carol* ii. 61 Old Fezziwig stood out to dance with Mrs. Fezziwig. Top couple too; with a good stiff piece of work cut out for them. 1862 Trollope *Orley Farm* II. xxxi. 247 Then Mr. Chaffanbrass rose ..and every one knew that his work was cut out for him. 1874 Hardy *Far from Madding Crowd* II. xxii. 276 What with one thing and another, I see that my work is well cut out for me. 1893 R. L. Stevenson *Catriona* vii. 71 'Ye'll find your work cut out for ye to establish that,' quoth she. 1899 E. W. Hornung *Amateur Cracksman* 43 'We shall have our work cut out,' was all I said. 1927 R. Austin Freeman *Magic Casket* vii. 212 'You will have your work cut out,' I remarked, 'to trace that man. The potter's description was pretty vague.' 1951 *Sport* 27 Jan.–2 Feb. 9/3 The Quakers will have their work cut out to keep the bigger clubs away.

32. b. In trivial sense, esp. in phrases *carry on with, get on with, keep up*, etc. *the good work.*

1920 'Sapper' *Bull-Dog Drummond* xii. 309 Vallance Nestor carried the good work on. 1938 G. Greene *Brighton Rock* vii. iv. 347 Drink up. We better get on with the good work. a1953 E. O'Neill *Long Day's Journey* (1956) 1. 17 So keep up the good work, Mary.

IV. 34. a. Simple *attrib.*, as *work-boat, -chant, -life, light, -load* (*Load *sb.* 4 c), *-norm* (*Norm 1 c), *-plan, -site, -song, -tool* (later example), *-week*; (of persons), *work-gang, -lass, -person*; (of animals), *work-stock* (earlier examples); (of clothes) worn for work, as *work boot, -clothes, pants, -shirt, -shoe, -wear*. **c.** *work-gnarled, -hard, -soiled, -thickened, -weary* adjs. **d. work-and-back** *Printing* = *sheet-work* s.v. Sheet *sb.*¹ 12 b; **work and tumble, work and turn,** and **work and twist,** methods of printing the second side of a sheet of paper from the same forme as the first (see quots.); **work book,** (*a*) in a business firm etc., a book containing a record of daily duties, work (to be) done, etc.; (*b*) (chiefly *U.S.*), a book in which are set out problems to be worked out, questions to be answered, etc.; **work-box** (earlier examples); **work camp** orig. *U.S.*, (*a*) a camp organized for a work project, esp. by volunteers serving the community; (*b*) = *labour camp* s.v. *LABOUR, LABOR *sb.* 8; **work card,** (*a*) a card issued by one's employer and serving as a kind of identity document; (*b*) a pupil's card on which are set out questions to be answered, problems to be worked, etc. (cf. *work-book* (*b*) above); **work ethic**, work seen as virtuous in itself, a term usu. connected with Protestant attitudes and deriving from Max Weber's thesis on the origins of modern capitalism (cf. *Protestant ethic* s.v. *PROTESTANT *a.* 1 b); **work experience**, work projects arranged for the purpose of providing experience of employment, esp. for school-leavers; **work-fellow** (later examples); **work flow**, in an office or industrial organization, the sequence of processes through which a piece of work passes from initiation to completion; **work-force**, the workers or employees collectively, usu. of a particular firm or industry; **work furlough** *U.S.*, leave of absence from prison by day in order to continue in one's daily work; **work group**, (*a*) a group of people in a factory or the like who customarily work together; (*b*) = *work party* (*a*) below; **work-hand** (earlier example); **work-harden** *v. Metall.*, to toughen (a metal) by cold-working; also *intr.*, to become tough as a result of cold-working; so **work-hardened** *ppl. a.*, **work-hardening** *vbl. sb.*; **work head**, (*a*) = HEADSTOCK 1 b; (*b*) an interchangeable working attachment for a powered implement or tool; **work-horse**, (*a*) a horse used for work on a farm (later example); (*b*) *fig.*, a machine, person, etc., that dependably performs arduous labour; **work-in-progress**, work undertaken but not completed, esp. (*a*) in commerce (see quot. 1978); (*b*) in the arts; **work measurement** (see quot. 1979); **work-minded** *a.*, eager to work hard; eager to go out to work; hence **work-mindedness**; **work name**, an alias used by someone engaged in secret intelligence work; **work party**, (*a*) a group of people who come together to carry out a piece of work of mutual or social benefit; (*b*) = *working party* s.v. *WORKING *ppl. a.* 2 c (*d*); **work permit**, a document representing official permission to take a job in a foreign country; **workpiece**, the object which is worked on with a machine or tool; **work point**, in the People's Republic of China, a unit used in calculating wages due, based on the quality and quantity of work done; **work rate** *Football*, the extent to which a player contributes towards the fatiguing running and chasing in a game; **work release** *U.S.* = *work furlough* above; **work-rule** *U.S.*, one of a set of regulations governing working procedures, conditions, etc., in a business or industry; **work-sharing**, short-time working by all employees within an industry intended to prevent redundancies when there is an excess of available man-power; **work-sheet,** (*a*) = *work-book* (*a*) above; also *fig.*; (*b*) *U.S.*, a questionnaire; (*c*) a paper on which are recorded notes, calculations, etc., relating to work in progress; (*d*) a list of exercises, problems, etc., to be worked by a student (cf. *work-book* (*b*) above); **work-shy** *a.*; also *absol.* as *sb.*; **work-space,** (*a*) *Computers* = *working storage* s.v. *WORKING *vbl. sb.* 16 b; (*b*) space (for people) to work in; **work study,** (*a*) investigation of the methods of working in a business, etc., with the aim of increasing output and efficiency; (*b*) used *attrib.* with reference to schemes of combining work and study established in Communist China; **work surface** = *work-top* below; cf. *working surface* s.v. *WORKING *vbl. sb.* 16 b; **work-table** (earlier example); **work-team,** (*a*) a team of draught-horses, oxen, etc.; (*b*) a team of people who work together, a work group, *spec.* in Communist China any of the working units making up a commune; **work-top**, a table or other flat surface suitable for working on, *esp.* in a kitchen.

1959 L. M. Harrod *Librarians' Gloss.* (ed. 2) 249 *Sheet work*, printing one side of a sheet of paper from an 'inner forme' and the other from an 'outer forme'. Also called 'work and back'. 1967 E. Chambers *Photolitho-Offset* ii. 18 Sheet work is the term used to indicate that two formes are used to print the sheet, sometimes called 'work and back'. 1931 H. Jahn *Hand Composition* xvi. [illegible] The work-and-turn method [illegible]. 1959 L. M. Harrod *Librarians' Gloss.* (ed. 2) 296 *Work and tumble*, the method of printing the second side of a sheet of paper by turning it over in its narrow direction and feeding it into a printing machine to print the reverse side. 1888 Work and turn [see *sheet-work* s.v. SHEET *sb.* 12 b]. [illegible] The sheet may be turned over after the first side has been printed, and the same edge of the sheet fed to the grippers for a second impression. This is termed 'work-and-turn'. 1931 H. Jahn *Hand Composition* xvi. 263 The Dexter standard jobbing folder..makes thirteen different folds adapted to work-and-turn and sheetwise forms. 1964 Work-and-turn [see *half-sheet* (*b*) s.v. *HALF- II. 2]. 1930 *20th Cent. Encycl. Printing* viii. 253 The work and twist form is used for the printing of ruled work where the vertical and horizontal rules of a sheet are printed with one impression. 1968 *Gloss. Terms Offset Lithogr. Printing* (B.S.I.) 25 *Work and twist*, printing one side of the sheet, then reversing the sideway edges and front and back edges of the sheet, and printing the same side again with the same printing plate. 1941 H. I. Chapelle *Boatbuilding* [illegible] 1977 *Washington Post* [illegible] A fleet of workboats is continually dredging a passage channel. [illegible] The work book is derived directly from the final storyboard, and is an analysis of each shot and sequence in terms of frame-by-frame basis. [illegible] *Work book*, a book containing the exercises and problems [illegible]. 1975 *Publishers Weekly* 17 Nov. 98/1 With its workbook approach of charts, graphs, questionnaires and the like, this book will put some women off. 1976 *National Observer* (U.S.) 1 May 16/1 At first wearing a suit, then gradually assuming work boots and old clothes. 1605 P. Erondell *French Garden* sig. E3 verso, I have but my silver Bodkin, my little work-box. [illegible] 1940 [illegible] 26 Everything was spread about..in a tiny apartment..boxes of work-boxes, netting-cases, etc. 1933 Nichols & Glasser *Work Camps for America* 13 The types of participants in work camps vary according to the purpose of the camps and the organizations which control them. 1949 F. L. Wright *Autobiogr.* (rev. ed.) IV. 309 Now by way of an architect's work-camp comes fresh adventures in the desert. 1964 M. Banton *Policeman in Community* iii. 56 They stopped to talk with a youth they knew who was serving a sentence in a work camp. 1970 *Honey* June 106/3 International Work Camps are held in most countries. Some camps offer paid work in forestry or farming. [illegible] than specific projects that are usually related to social service. 1981 'W. Haggard' *Money Men* i. 18 The Gestapo..had sent him back to Germany to a work camp where he'd been starved for a month. 1982 *Listener* 11 Oct. 26/1 An elaborate process of deception was instituted, by which the evacuation of Jews to the death-camps was disguised as a 'resettlement' into 'work-camps'. 1950 M. Levin *Eva* [illegible] I managed to get a blank German work card. 1967 J. Derrick *Teaching Eng. to Immigrants* 239 The work cards, picture-cards, wall pictures and flashcards which accompany the course, are also recommended. 1975 A. Watson *Living in China* iv. 99 The work card issued by the place of employment is an important means of social identification. 1980 *Daily Tel.* 12 Feb. 6/8 Instead of text books the scheme uses a series of work-cards... These..are distributed to the pupils according to their ability. 1946 R. Blesh *Shining Trumpets* i. iii. 57 Perhaps the most familiar form of work-chant is the vendor's street cry. 1967 A. L. Lloyd *Folk Song in Eng.* i. 54 The work-chants of Portland quarrymen. 1901 'Mark Twain' in *Century Mag.* Nov. 26/2 Tommy was..in his dreadful work-clothes. 1978 F. Weldon *Praxis* xx. 170 She changed out of her work clothes. 1959 *Past & Present* xv. 44 Weber also asserted that Calvinistic Protestantism was an indispensable precondition of the development of a capitalistic work ethic. 1973 P. A. Whitney *Snowfire* vi. 112 The work ethic, you mean?..my little Puritan. 1980 *Jrnl. R. Soc. Arts* July 468/1 *They* are showing the way which we should follow—if only we were not 'locked on to' some puritanical work ethic. 1975 *Whitaker's Almanack 1976* 1041/2 In recent years there has been a marked growth in the provision of 'work-experience' schemes which involve the participation of pupils in the work of industrial, commercial and other firms. 1983 *Fortune* 16 May 112/3 CETA's primary approach, providing 'work experience' in temporary public service jobs, does no good at all, presumably because the jobs don't lead anywhere. Work experience has big payoffs, however, when it comes in the form of on-the-job training for a permanent position in the private sector. *a* 1890 J. H. Newman *Meditations* (1893) II. 289 We thus pray.. for our associates and work-fellows. 1903 C. Coleridge *Charlotte M. Yonge* x. 276 She was the most delightful comrade, workfellow, or playfellow. 1950 I. A. Herrman *Office Methods, Systems, & Procedures* vii. 131 Work flow diagrams are effective in solving various kinds of problems. 1976 *National Observer* (U.S.) 19 June 2/4 Byrd is a master of legislative detail with a reputation as a fair-minded manager who accelerates the work flow. 1961 *Times* 30 May 13/7 Books come off no assembly line. The raw material is provided by a notoriously undisciplined work-force. 1982 *Daily Tel.* 18 Nov. 2/1 They were non-unionised because 'that is the wish of the majority of the work force'. 1957 *Statutes of California 1956 & 1957* II. MDLXXX. 2933 If the court so directs that the prisoner be permitted to continue in his regular employment, the work furlough administrator shall arrange for a continuation of such employment so far as possible without interruption. *Ibid.* 2934 This section shall be known as the 'Work Furlough Rehabilitation Law.' 1970 *Criminology* May 63 Work furlough has been used..for felons as well as misdemeanants. 1980 *New Age* (U.S.) Oct. 15/2 Both men received suspended jail sentences and three years probation; each will serve about a month in a work furlough program or community service and must undergo psychiatric counseling. 1948 *Common Ground* Summer 41/2 Their plans sound the work-gang chorus. [illegible] A work gang..cleaning up the litter along the road. 1913 D. H. Lawrence *Sons & Lovers* x. 256 She began to spare her hands. They, too, were work-gnarled now. 1957 J. Kerouac *On Road* [illegible] A wiry..man with work-gnarled hands. 1964 J. A. C. Brown *Social Psychol. of Industry* iv. 114 A factory or a society is not ordinarily a mass of isolated individuals..; it is an integrated pattern of primary work-groups. 1980 *Ann. Reg. 1959* 381 All of them [*sc.* sects] laid emphasis on activity: dancing, propagation of the faith, and free labour with work-groups building and clearing temples. 1972 M. Argyle *Social Psychol. of Work* ix. 213 Many studies have shown that job satisfaction is affected by relationships in the work group. 1834 W. Sewall *Diary* (1930) 154 He was an excellent work hand. 1930 W. Faulkner *Light in August* i. 14 The gray woman not plump and not thin, manhard, workhard, in a serviceable gray garment. 1924 *Engineer* 7 Mar. 249/1 Metal rolled cold or drawn through dies may be work-hardened to an extent rendering it unfit for further working. *Ibid.* 251/2 It may be that its [*sc.* 'browning's] real function is to reduce the tendency of the steel to work-harden. 1966 *New Scientist* 16 Mar. 671/1 The ability of the material to work-harden [illegible]. 1972 *Mineral. Mag.* XXXIII. [illegible] Naturally deformed galena has been 'work-hardened' by tectonic movement. 1964 R. F. De Garmo et al. *Materials & Processes in Manuf.* (ed. 6) ii. 41 When most materials are plastically deformed, they work-harden; that is, they become harder and the yield-point stress is raised. 1924 *Jrnl. Iron & Steel Inst.* CX. 431 The abnormally low value of the yield-point stress in..both quench-hardened and work-hardened steels. 1973 J. G. Tweeddale *Materials Technol.* II. iv. [illegible] Cold drawing can impart a good surface finish and accurate size to a product and leaves the material in a work-hardened condition which is often desirable. 1924 *Engineer* 7 Mar. 248/2 (*heading*) Work-hardening of metals and the Herbert tester. 1973 J. G. Tweeddale *Materials Technol.* II. iv. [illegible] The work-hardening capacity [illegible]. 1930 *Engineering* 25 Apr. 538/3 A pair of flat ways for the work-head, tailstock and journal supports. 1960 *Farmer & Stockbreeder* 12 Jan. 121/1 (Advt.), Other workheads, quickly interchangeable with standard tools, include..hedge cutters, pruning saw, etc. 1964 S. Crawford *Basic Engin. Processes* vii. 194 The work-head..is a self-contained unit carried at the opposite end of the table to the tailstock. 1949 *Sun* (Baltimore) 3 Oct. 2/6 This caliber howitzer has been classed as..the 'work horse' of the Army. 1966 *Electronics* 3 Oct. 34 The satellites will be launched on improved versions of the workhorse Delta vehicle. 1973 *Listener* 20 Dec. 841/2 Gerald Ford..has been known..as the most dependable of Republican work-horses. 1981 E. Eberl *Ransom Game* (1982) xxii. 133 The big barn doors. The entrance to the lower floor, where the cows and work horses used to be. 1983 *Habitat Catal. 1982/83* [illegible] A real workhorse of a desk. [illegible] 1920 *Times* 25 Mar. 24/5 'Work' in progress, less instalments thereon, is £141,069, against £47,351 in the previous year. 1952 [illegible] (1967) 450 It is clear that the place of the finished book will be important in your long work-in-progress. 1976 P. Israel *French Kiss* (1977) ii. 24 The work-in-progress on one of the easels. 1978 J. Kellock *Elements of Accounting* x. 176 *Work-in-progress*, is the value of incomplete work in the factory and is usually computed on the following basis: the cost of materials and production labour plus the proportion of indirect expenses chargeable to the work up to its present stage of manufacture. 1920 D. H. Lawrence *Lost Girl* vi. 98 Yet it was always packed with colliers and work-lasses. 1946 J. W. Day *Harvest Adventure* vii. 112 The pond..is tenanted by tame decoy ducks, pinioned, whose work-life is to lure the wild birds down. 1977 *National Observer* (U.S.) 1 Jan. 1/2 Men..committed to serious and demanding work lives. 1947 J. Steinbeck *Wayward Bus* 13 Get the work-light on the long cord connected. 1977 *Chicago Tribune* 2 Oct. I. 53 (Advt.), Self-cleaning oven.. Fluorescent worklight. 1946 Work-load [see *LOAD *sb.* 4 c]. 1962 *Listener* 4 Jan. 4/1 The application of time study to speed and tighten up the work load. 1962 J. Glenn in *Into Orbit* 195, I pushed and pulled thirty times at the bungee cord which permitted me to exercise with a known workload. 1978 G. A. Sheehan *Running & Being* xv. 210, I shift to shorter steps..to maintain the same workload. 1985 *Times* 18 Jan. 5/1 He is one of more than 1,000 teachers whose workload is being analysed by the National Union of Teachers in a survey to be published next week. 1948 (*title*) Manual of procedures: work measurement in public works offices (U.S. Bureau of Yards & Docks). 1969 J. Argenti *Managem. Techniques* 271 The procedure used in Work Measurement to determine the amount of labour required to do a job is to time how long it takes the average man to perform each element of the job. 1979 *Gloss. Terms Work Study* (B.S.I.) 2 *Work measurement*, the application of techniques designed to establish the time for a qualified worker to carry out a task at a defined rate of working. 1954 *Encounter* Sept. 33/1 The Army requires..work-minded people who try to do a good job of whatever they're told to do. 1968 *Economist* 11 May 46/1 The more 'work-minded' a mum in her later years, the less available she is to look after her daughter's kids. 1960 *Encounter* Nov.

27/1 William H. Whyte, Jr.,..points out that some large corporations, worried about the decline in 'work-mindedness', are seeking to substitute an ideology of corporate loyalty. 1977 'J. le Carré' *Hon. Schoolboy* iii. 54 Karla ..was the workname of the Soviet case officer who had recruited Bill Haydon..and had the running of him. 1980 *Encounter* Feb. 14/2 At the local level, there must always be potential disputes between workers and management over redundancy, work-norms, wage-differentials..and so on. 1980 *Times* 24 May 14/7 You can poke fun at life under Communism—how to fiddle your work-norms. 1927 *Amer. Speech* II. 366/2 The man had on his work pants this morning. [illegible] Denim workpants slung low on his hips. 1957 R. Frankenberg *Village on Border* xii. [illegible] They [*sc.* women] also work together in sewing groups and work-parties preparing material for sale-of-work. [illegible] V. W. Turner *Schism & Continuity in an African Society* [illegible] The cutting and clearing of bush..may involve a cooperative work-party..of kin and neighbours. 1981 [illegible] Many considered the shock of being drafted to a work party every bit as bad as being arrested. 1969 *Globe & Mail* (Toronto) [illegible] work permits [illegible]. 1971 *Times* 25 Feb. 4/1 A Commonwealth citizen wishing to work here in future will need a work permit for employment in a specific place for a fixed initial period. 1983 *Daily Tel.* 12 May 4/8 Work permit clamp. The Singapore Government has tightened up on employment permits for foreigners. 1807 *Monthly Mag.* 1 Feb. 67/1 [illegible] 1934 Webster, Workpiece. 1949 *Tool Engineers Handbk.* (Amer. Soc. Tool Engineers) [illegible] Improper workpiece locating can readily result in excessive troubles and spoilage. 1952 *Economist* 6 Dec. [illegible] The guiding wheel is charged negatively and the workpiece positively. 1958 *Sci. Amer.* Nov. 110/2 The surface of the workpiece undergoes much more heating in abrasive machining than in conventional machining. [illegible] 1976 [illegible] Work point [illegible]. [illegible] For most co-operatives the distribution of income [illegible]. [illegible] 1980 *Times* 26 Feb. [illegible] work rate [illegible]. [illegible] 1983 *Observer* [illegible] The work-release prisoner [illegible]. [illegible] work-rules [illegible]. [illegible] work-sharing [illegible]. 1925 S. Lewis *Arrowsmith* [illegible] created by Gunter's public statements that in principle the Government is opposed to work-sharing in the motor-car industry. 1925 S. Lewis *Arrowsmith* xxxix. 427 'I'll find out from my wife what dates we have already and telephone you tomorrow evening'. 'So you let the Old Woman keep the work-sheet for you, huh?' 1930 *Dialect Notes* VI. 73 Professor Jud made a number of specific suggestions regarding mechanical features of the work-sheets. 1958 *Listener* 31 July 155/1 Sorted away in the stacks are some 5,000 sets of poets' worksheets, the notes, drafts, revisions. 1966 J. Derrick *Teaching Eng. to Immigrants* 238 A new type of course, designed for eight-year-old foreign learners, consisting of gramophone records, teacher's notes, and pupils' 'working scripts' which are work-sheets and meant to be expendable. 1967 R. Bregzis in Cox & Grose *Organization & Handling Bibl. Rec. by Computer* v. 118 NUC catalogues and other files or catalogues are checked as necessary, and all information [is] recorded on a catalogue worksheet. 1975 *Language for Life* (Dept. Educ. & Sci.) x. 145 It is even less likely to happen where children work individually through assignment cards or work sheets. 1976 P. Alexander *Death of Thin-Skinned Animal* xx. 206 Look, if you haven't done any maintenance there, boyo, how come I've got a work-sheet for renewing thirty foot of ogee guttering and repointing the bloody gable end? 1981 *Amer. Speech 1977* LII. 167 He employed a variable questionnaire based upon the New England short worksheets. 1923 *Dialect Notes* V. 235 A work shirt made of crossbarred cotton cloth. 1980 *Daily Tel.* 19 Nov. 15/8 (Advt.), Fisherman's smock. Original workshirt of local fishermen. 1965 H. I. Ansoff *Corporate Strategy* (1968) vi. 96 Royal Little has built the successful Textron Corp. composed of consumer electronics, textiles, helicopters, work shoes, and satellite motors. 1980 D. E. Westlake *Castle in Air* vi. 63 Manuel was dressed in rough corduroy trousers,.. heavy workshoes, and a coarse cotton shirt. 1928 *Daily Express* 2 Apr. 7/4 To make the lot of the work-shy as favourable as that of the worker. 1983 *Times* 12 Oct. 14/5 The Gravediggers' Union, understandably affronted by having their members portrayed as drunken workshies. 1975 *BP Shield Internat.* May 5/4 Up to three teams of divers may be maintained under these conditions to permit a 24-hour per day diving operation at the work-site. 1980 *Daily Tel.* 20 Mar. 28 (Advt.), He must be an engineer, between 30 and 50, perfectly fluent in French and English, with overseas experience in work-site construction. 1932 W. Faulkner *Light in August* ii. 28 The men in faded and work-soiled overalls. 1911 *Jrnl. Amer. Folk-lore* XXIV. 379 Like the other songs, the work-songs give a keen insight into the negro's real self. [illegible] 1977 *Listener* 25 Aug. 244/3 The persistent play with these four notes suggests the call and response of the work song. [illegible] Such a code would mean building an automatic translating system into existing designs of computers, thereby reducing their 'work space'. 1979 Kraft & Toy *Mini/Microcomputer Hardware Design* viii. [illegible] Its general register set is placed in main storage and realized as a 16-word area of memory that is considered a workspace. 1979 *Tucson* (Arizona) *Citizen* 20 Sept. [illegible] Several high-rise government office buildings with a combined work space of perhaps 100,000 to 150,000 square feet. 1985 *Which Computer?* Apr. 53/1 Even on a fully configured IBM PC..you can find yourself running out of workspace. 1877 *Rep. Indian Affairs* [illegible] had..broken down our light Indian work-stock. 1883 'Mark Twain' *Life on Mississippi* [illegible] stock, ..horses and mules were housed in a place of safety. 1951 (*title*) The implications of work study (Imperial Chemical Industries Ltd.). 1962 E. Snow *Other Side of River* (1963) xxx. 235 Part-time and work-study middle schools are discussed in the next chapter. 1965 J. Ch'en *Mao & Chinese Revolution* [illegible] In August 1920 Mao and others founded a Russian Affairs study group as well as sponsoring a Work-study Scheme for students to go to Russia. *Ibid.* [illegible] Work-study students who had just returned from France or Russia. 1978 *Cornish Guardian* 27 Apr. 27 (Advt.), Applicants would be expected to have at least two years' practical experience in the Work Study and O & M field. 1971 *House & Garden* Dec. [illegible] Spotlights illuminate work surfaces and dining-area. [illegible] They were inside a kitchen [illegible] work surfaces, refrigerator, an electric cooker gleamed. [illegible] 1979 *Guardian Weekly* [illegible] work-team [illegible]. [illegible] 1967 *Good Housekeeping* [illegible] upstairs showroom shown in three displays of door and worktop colour—the kitchen of rounded edges—doors and worktops in a host of colours and textures. 1984 *Which?* Oct. 458/3 If you are going to use tiles on a worktop, check that they have good scratch resistance. 1967 *St. Andrews Citizen* 25 Feb. 5/4 Men's workwear. Full range mens overalls..trousers, jackets, coats and boilersuits. 1981 *Daily Tel.* 22 Sept. 9 (Advt.), The workwear rental company. 1853 C. Brontë *Villette* I. v. 82 A brief holiday, permitted for once to work-weary faculties. 1935 *Economist* 26 Oct. 802/1 Hourly wage rates are a little higher;..the work-week is a little longer. 1980 *News & Observer* (Raleigh, N. Carolina) 28 Oct. 21/3 Layoffs in a local furniture plant and shortened workweeks at county textile mills.

35. *attrib.* and *Comb.* with *works* (sense 18), as *works bus, canteen, club, kitchen, manager, outing*; **works committee, council,** a committee of workers or their representatives, formed for joint discussions with employers.

1969 R. Blythe *Akenfield* iv. 80 Works-bus waiting to carry him from door to site. 1980 A. Townsin *Blue Triangle* iii. 48/1 A second works bus. 1963 A. Howard in Sissons & French *Age of Austerity* i. 17 The Naafi and the works-canteen. 1978 J. B. Hilton *Some run Crooked* iii. 19 He ate his midday meal in a works canteen. 1908 *Mod. Business* Aug. 69/1 Any surplus is devoted to some charity or to some of the works clubs. 1917 *Interim Rep. on Joint Standing Industr. Councils* 4 in *Parl. Papers 1917–18* (Cd. 8606) XVIII. 415 We are of opinion that.. Works Committees, representative of the management and of the workers employed, should be instituted..to act in close co-operation with the district and national machinery. 1966 T. Lupton *Managem. & Social Sci.* iii. 63 In a small firm, Joint Consultation might take place in a Works Committee. 1925 *Glasgow Herald* 31 July 5 The most important is the Works Council Law of 1920, which requires a works council to be set up in each establishment employing 20 persons or more. 1977 *Times* 22 Sept. 2/8 The need to develop industrial democracy on the shop floor through works councils. 1908 *Mod. Business* Aug. 69/1 Another valuable outlet for its energies is the management of a Works Kitchen. 1918 A. Bennett *Pretty Lady* xxvii. 177 I used to take their part against the works-manager. 1976 *Derbyshire Times* (Peak ed.) 17 Sept. 3/7 Mr. Marshall (26), works manager, ..escaped unhurt. 1943 J. B. Priestley *Daylight on Saturday* viii. 47 His bus ride to the factory..took on the air of a works outing. 1974 *Times* 23 May 664/2 A works outing to Blackpool.

work, *v.* **B. I. 1. c.** For † *Obs.* read *Obs.* exc. in *Freemasonry*, and add later examples.

1884 W. J. Hughan *Origin Eng. Rite Freemasonry* i. 5 It seems difficult to understand how any one conversant with their noble Histories can cherish the fancy that the Craft..and other degrees were worked by our ancient brethren during the seventeenth century. 1903 J. T. Lawrence *Masonic Jurisprudence & Symbolism* viii. 74 What generally takes place in a lodge of instruction is that the lectures, or sections of them, are worked, officers to conduct the same being appointed at a previous meeting. 1954 W. Hannah *Christian by Degrees* iv. 63 The 26th degree known as Prince of Mercy (not worked in England) also regards Hiram as a type of Christ in His death and resurrection. 1978 *Lochaber News* 31 Mar. 2/7 An EA Degree was worked and was well received by the Brethren present.

10. Esp. in phr. *to work havoc*, where the pa. t. *wrought* is common (though it is often interpreted as the pa. t. of *wreak*: cf. *WREAK *v.* 8 b).

1900, 1908 [see *HAVOC *sb.* 2]. 1978 *Washington Post* 30 Nov. A-24/2 Settlers who are prone to California dreaming,..and on whom..the anything goes atmosphere and the wide-open spaces work havoc. 1983 *National Law Jrnl.* (U.S.) 4 July 14/2 With hard disk technology..power failures can often work havoc. 1984 *Financial Times* 4 June 11/2 [illegible] a decade of inflation had wrought havoc with its portfolio of fixed interest mortgages.

11. b. *colloq.* To arrange, engineer, or bring about. Usu. const. *it*.

1889 E. Dowson *Let.* 1 Mar. (1967) 42 If you can possibly work it meet me somewhere to-morrow. 1911 G. B. Shaw *Doctor's Dilemma* iii. 57 The way to work it is this: I'll postdate the cheque next October. 1953 K. Tennant *Joyful Condemned* xxii. [illegible] I guess I can work it. 1962 Wodehouse *Service with Smile* xi. 177 Uncle Fred, did you work this? 1975 D. Lodge *Changing Places* i. 17 Masters (who was Chairman) was prepared to work it for Philip if he was interested.

12. h. (Earlier examples.) Also *fig.*

1727 'E. Dorrington' *Hermit* 11. 121 He sees..Hay-makers, going to work..and reapers to make their harvest. 1743 [see PASSAGE *sb.* 4 b]. 1803 D. Wordsworth *Jrnl.* 25 Aug. (1941) I. 257 He was just come from work. I do not think that he [illegible] with him, for he had worked his passage over. 1934 G. B. Shaw *Village Wooing* 113, I have no time for talk. I have to work my passage. 1958 *Oxf. Mag.* 15 May 448/2 Italy, liberated piecemeal and 'working her passage' to the improved status of the Hyde Park Declaration and the New Deal for Italy. 1973 *Times* 20 Mar. 13/2 One of the greatest boons in packing is choosing which shoes to take... They are heavy..and do not really work their passage.

[illegible] (of a thief, esp. a pickpocket.

1865 *Leaves from Diary Celebrated Burglar & Pickpocket* xvi. [illegible] They agreed, upon their discharge, to 'work' together. *Ibid.* [illegible] Joe edged himself into the Scotch Boy's 'mob'..and 'worked' with them. 1882 J. D. McCabe *New York* 501 Even vessels lying at anchor in the harbor, are [illegible] by thieves. 1905 E. Wallace *Four Just Men* viii. 153 The night being comparatively young, Billy decided to work the trams. 1930 —— *Lady of Ascot* i. 19 It's the same crowd that has been working country houses for weeks. 1938 F. D. Sharpe *Sharpe of Flying Squad* xvi. 181 They [*sc.* pickpockets] used to go off in busloads..to 'work' various districts of London. 1951 W. C. Williams *Autobiogr.* xlv. 299 He had been a fur thief working the big department stores. 1963 T. Tullett *Inside Interpol* xii. 171 A Pole.. last caught in August, 1957, working a crowd in Geneva.

14. d. (Earlier example.)

1884 'Mark Twain' *Huck. Finn* xix. 183 Preachin's my line, too; and workin' camp-meetin's.

16. a. Also of a locomotive engine, to pull (a train).

1912 [in Dict.]. 1982 *Railway Mag.* Nov. 508/1 A replacement..powered the train as far as Carnforth where another '47' was later provided to work it forward.

b. To herd (sheep, cattle, etc.). Also *intr.* for *pass.* Chiefly *Austral.* and *N.Z.*

1930 L. G. D. Acland *Early Canterbury Runs* 1st Ser. i. 5 The practice was for a shepherd to go round the boundary once or twice a day, and at night work the sheep below one of the river terraces to camp. 1946 F. Davison *Dusty* (Foreword), Sheep dogs..working lost flocks in the mountain gullies. 1950 *N.Z. Jrnl. Agric.* July 5/2 Sheep work and draft best on a slight up-grade. 1961 B. Crump *Hang on a Minute* 87 With Jack working along the top of the ridge and Sam half-way down the side they worked all the sheep off that side of the valley. 1976 *Evening Post* (Bristol) 23 Apr. 24/9 (Advt.), Border collie bitch starting to work cattle.

19. a. *spec.* in *N.Z.*, to use (a dog) for the purpose of herding sheep or cattle.

1878 E. S. Elwell *Boy Colonists* 48 Fricker..[was] delighted to shew the 'new chum' how to work a cattle dog. 1928 P. T. Kenway *Pioneering in Poverty Bay* viii. 56 It was said of the Highland shepherd in New Zealand, that he would..work his dogs, getting in stray sheep, every day for a month.

II. 23. d. *to work like a charm*: see *CHARM *sb.*¹ 1 c.

24. a. *to work like a beaver, horse, nigger*: see these words. Similarly *to work like a dog, to work one's tail off*.

1926 [see *PERISH *v.* 1 e]. 1969, etc. [see *TAIL *sb.*¹ 5 a]. 1976–7 *Sea Spray* (N.Z.) Dec./Jan. 95/2 These lads have worked like dogs all winter.

26. a. Also *out of* (a place), to use it as a base, office, etc., for work; *to* (a person), to be responsible to as one's immediate superior or supervisor.

1941 B. Schulberg *What makes Sammy Run?* xii. 300 She's turned pro... She's working out of Gladys'. 1961 B. Fergusson *Watery Maze* xiv. 360 The Forward Officer (Bombardment) working to H.M.S. *Roberts* was killed with his signaller. 1972 *Where* Sept. 263/1 Registration officers work to the Registrar General. 1975 I. Murdoch *Word Child* 6, I worked to a man called Duncan, now briefly seconded to the Home Office. 1976 M. Delving *China Expert* i. 12 He had no shop but worked out of the small, comfortable house he had bought. 1979 P. Cosgrave *Three Colonels* viii. 174 They had all worked either to Davies..or Morgan... None had come in contact with the head of the department.

27. a. For '*Obs.* exc. as in b' read '*Obs.* exc. as in b, c, and d'.

c. *slang.* (See quot. 1839.) Cf. sense 12 i (*e*) above.

1839 H. Brandon in W. A. Miles *Poverty, Mendicity & Crime* 166/1 *Work*, to rob, or act in any way according to the divers occupations of thieves, &c. 1882 *Sydney Slang Dict.* 10/2 We went to the gaff that night and tried to work. 1955 *Publ. Amer. Dial. Soc.* xxiv. 70 Some Americans [*sc.* pickpockets]..are front workers..; that is, they can and do work facing the victim. 1963 T. Tullett *Inside Interpol* x. 150 Huffman 'worked' for a short time in Rome, where he defrauded several shopkeepers.

d. *to work to rule*: to follow the rules of one's occupational duties punctiliously in order to reduce efficiency, usu. as a form of protest in an industrial dispute. So *work-to-rule* attrib. phr.; also as *sb.* Similarly, in the professions, *work-to-contract*.

[1940 *Ann. Reg. 1939* 310 A 'ca' canny' movement—called 'work to rules'—among the [railway] employees.] 1950 *Ann. Reg. 1949* 40 The delegates replied by ordering a general work-to-rule 44-hour week..unless claims were settled. 1952 *News Chron.* 13 Mar. 5/7 That conductor was working to rule... All passengers must be seated before moving off; no overtaking of other buses; and no efforts to make up lost time. 1958 *Times* 4 Aug. 6/4 A report that prison officers..were working to rule in protest against the report that prisoners there had been assaulted. 1959 *Daily Tel.* 21 Nov. 1/5 The work-to-rule and shut-down were expected to be carried out in Manchester, Birmingham, Liverpool and other provincial cities. 1960 *Guardian* 13 June 1/6 A 'work-to-rule' plan instituted by members of the Amalgamated Engineering Union after pay negotiations..had broken down. 1962 *Spectator* 26 Jan. 96 What about lesser sanctions—go-slows, work-to-rules and overtime bans? 1967 R. Whitehead in Wills & Yearsley *Handbk. Managem. Technol.* 69 The system would fail even more often if the staff stuck rigidly to the rules. We see the results when they 'work to rule', as it is. 1969 *Daily Tel.* 19 Apr. 23/3 Members of the London Schoolmasters' Association will 'work to contract' next term because of the two weeks' suspension without pay earlier in the year of 22 teachers. 1972 'M. Sinclair' *Norslag* x. 82 A work-to-rule among ground staff had led to some flights being delayed. 1975 *Times* 13 Jan. 15/1 Instead of wholesale industrial action by most of the [medical] profession, we are left with the consultants and their 'work-to-contract'.

III. 36. work in. c. To co-operate or get along *with*.

1915 E. Fenwick *Diary* 14 Oct. in *Elsie Fenwick in Flanders* (1981) 89, I had tried so hard to work in with her. 1960 M. Spark *Ballad Peckham Rye* viii. 181 If Mr. Druce thought I was working in with you, he'd kill me. 1974 O. Manning *Rain Forest* i. ix. 101, I am a very fast learner, and I work in well with Mr. Axelrod.

37. work off. c. Also, to get rid of, palm off, pass off. Occas. *refl.*

1813 M. L. Weems *Wks. & Ways* (1929) III. 92 The Maps..may be work[ed] off and in time to give you bank interest. 1884 Kipling *Let.* 21 Nov. in C. Carrington *Rudyard Kipling* (1955) iv. 58 I've been writing a story... I'm trying to work it off on some alien paper to get myself pice thereby. 1897 'O. Thanet' *Missionary Sheriff* 7 The lightning-rods ain't in it with this last scheme—working his self off as a Methodist parson. 1900 'Mark Twain' *Speeches* (1910) 164 He has not written as many plays as I have, but he has had that God-given talent, which I lack, of working them off on the manager. 1948 V. Palmer *Golconda* viii. 58 Corney had been skiting about his claim for months, and everyone knew it was a duffer, but he hung on in the hope of working it off on someone.

38. work out. a. Also *refl.*

1906 *Jrnl. Abnormal Psychol.* I. 37 We might properly say that the 'uncompleted emotion'..could be given an opportunity to work itself out.

l. Also *gen.* to practise, take exercise, rehearse.

1929 *Cosmopolitan* Aug. 72/2 Feet's feet take up so much room when he is on the floor that only two other dancers can work out at the same time. 1948 G. Vidal *City & Pillar* II. ix. 264 Jim worked out in the YMCA. 1965 C. Brown *Manchild in Promised Land* viii. 221 I'd go up to the gym and work out for a little while, and I wasn't tired any more. 1973 R. L. Simon *Big Fix* (1974) xv. 110, I sat..watching the members of the *Teatro Comunal* work out. 1980 J. Ball *Then came Violence* xiv. 117 He belonged to a health club where he worked out regularly. 1984 *Daily Tel.* 30 Apr. 15/7 He does not look his 59 years. Perhaps it helps that he had his face lifted twice, works out with weights and had synthetic implants in his jaw.

38*. work over. *slang.* To beat up, thrash (a person).

1927 *Dialect Notes* V. 467 *Work one over*, to resort to violence in the third degree inquisition of the police. 1934 D. Hammett *Thin Man* viii. 37 Morelli's face was a mess: the coppers had worked him over a little just for the fun of it. 1947 *Partisan Rev.* XIV. 329 The crooked cop can't look at Marlowe without a self-revealing yen to 'work him over'. 1970 *Daily Tel.* 11 Dec. 1/1 An engineer was followed into a sub-station by two men who threatened to 'work him over'. 1978 R. Perry *Dutch Courage* ii. 23 Alan held me and Bernard worked me over.

39. work up. n. *U.S. Med.* (See quot.) Cf. *WORK-UP 2.

1961 *Amer. Speech* XXXVI. 145 *Work up*, to perform a series of diagnostic procedures (X-rays, laboratory blood tests, electro-cardiograms, and so forth).

workably (wɜ̄·ɹkăbli), *adv.* [f. WORK *v.* + -ABLY.] In a workable way; so as to be workable.

1943 Nissen & Bergmann *Cineplastic Operations on Stumps of Upper Extremity* iii. 36 Three sutures are placed, one in each of the free corners..and one in the centre edge of the pediculated flap for the purpose of holding it workably tense. 1971 *Nature* 12 Mar. 69/2 Should there not..be some kind of rebate for bulk delivery or, more workably, a charge for single delivery which is mitigated by bulk delivery?

workaholic (wɜ̄·ɹkăhǫlik). *colloq.* (orig. *U.S.*). [f. WORK *sb.*, after *alcoholic*.] One who is addicted to work, or who voluntarily works excessively hard and unusually long hours. Also *attrib.* or as *adj.*

1968 W. E. Oates in *Pastoral Psychology* Oct. 16 (*heading*) On being a 'workaholic'. 1971 —— (*title*) Confessions of a workaholic. 1973 *Bulletin* (Sydney) 25 Aug. 45/2 The workaholic, as an addict is called, neglects his family, withdraws from social life, and loses interest in sex. 1974 *Daily Colonist* (Victoria, B.C.) 17 July 18/8 Often the workaholic boss threatens the health and welfare of those unfortunate enough to work for him. 1976 *S. Wales Echo* 27 Nov. 6/9 At all costs you should avoid becoming a 'Workaholic'... You should leave your work behind with the office. 1981 *Time* 13 May 67/3 Unlike their workaholic American cousins Europeans tend to see lengthy vacations as somehow part of the natural order of things. 1984 *Guardian* 22 Oct. 11/4 They're concerned about the pressures of their jobs, which demand that they become workaholics.

Hence **wo·rkaholism**, the condition of being a workaholic.

1968 W. E. Oates in *Pastoral Psychol.* Oct. 16/2, I have dubbed this addiction of myself and my fellow ministers as 'workaholism'. 1971 —— *Confessions of a Workaholic* i. 1 Workaholism is a word which I have invented... It means addiction to work, the compulsion or the uncontrollable need to work incessantly. 1981 *Farmstead Mag.* Winter 23/2 For them it requires no effort of will to go off energy-saving appliances,..leave off gluttony on home-grown foods or security schemes, give up workaholism. 1983 *Sunday Tel.* (Colour Suppl.) 20 Feb. 14/4 We talked about..workaholism, autobiography and Isaac Asimov.

workalike (wɜ̄·ɹkăləik), *a.* and *sb.* Also **work-alike.** [f. WORK *v.* + ALIKE *adv.*] **A.** *adj.* Of a computer: able to use the software of another machine and behaving in the same way when the software is used. **B.** *sb.* A workalike computer.

1981 *Infoworld* 13 Apr. 54 (*heading*) PMC-80: TRS-80 'workalike' computer. *Ibid.* 54/1 Personal Micro Computers has renamed the Video Genie PMC-80, and uses the phrase 'TRS-80 workalike' in some of its advertising. 1983 *Austral. Personal Computer* Aug. 90/2 Most software writers recognise the existence of IBM workalike machines and therefore attempt to avoid nucleus calls to direct device I/O. 1983 *Popular Computing* Dec. 83/2 The Ace is an Apple II workalike that accepts Apple II software, disk drives, and..add-on cards. 1985 *Daily Tel.* 9 Sept. 2 (Advt.), A true, 16-bit, μPD8086 chip. (Not the humble 8088 of so many IBM work-alikes.)

workaway (wɜ̄·ɹkăwēɪ). *U.S.* [f. WORK *v.* + *-a-* + WAY *sb.*¹] One who works his passage on a ship.

1906 *Federal Reporter* CXXXIX. 92 He authorized the mate to take four men as workaways to earn their passage from Nome to Tocoma. 1933 M. Pell *S.S. Utah* 58 The workaway, a quiet young Swede, also went. 1945 *Seafarers' Log* 20 July 3/3 From there [*sc.* from Honolulu] they were sent as workaways back to San Francisco. 1973 *Art Internat.* Mar. 100/2 If one didn't have the price, one could present one's self to a ship's purser and ask for a job as a workaway.

worked, *ppl. a.* **5.** (Earlier examples of *worked-out, -up*.)

1831 P. Egan *Show Folks* 41 Like a well-worked up scene on the stage. 1864 'Mark Twain' in *Californian* Nov. 5 We admire his mature judgment in selling out of a worked-out mine.

worker. Add: **2. e.** (Earlier example.)

1873 'Mark Twain' *Gilded Age* xliv. 399 In Washington he was..clerk of two house committees, a 'worker' in politics.

5. worker-director, a worker who is also on the board of directors of a firm; **worker participation** [*PARTICIPATION 2 b], participation of workers in the management of the firms or industries for which they work; **worker-peasant,** used *attrib.* with reference to co-operation between urban and rural communities in Communist China; similarly **worker-peasant-soldier**; **worker-priest,** orig. a Roman Catholic priest in post-war France who earned his living as a factory-worker or the like; now more widely, a priest who engages in secular work for part of his time.

1968 *Economist* 3 Aug. 53 The proposal—that worker-directors should be put on the boards of a number of nationalised industries..—is a waste of time. 1980 *Whitaker's Almanack 1981* 583/2 Sir Keith Joseph announced the ending of the Post Office worker-director experiment. 1973 *Guardian* 19 June 17/4 Mr Heath's.. flourish of the worker participation banner. 1978 *Jrnl. R. Soc. Arts* CXXVI. 326/2 'Industrial democracy', probably better called 'worker participation'. [1937 E. Snow *Red Star over China* iv. vi. 173 The Chinese Workers' and Peasants' Revolutionary Committee was organized about this time.] 1962 —— *Other Side of River* (1963) xviii. 135 This he managed to do by means of postulating the existence of a 'rural proletariat' and a worker-peasant army under the leadership of the Communist party itself acting as the vanguard of the true (urban) proletariat. 1968 in Gray & Cavendish *Chinese Communism in Crisis* 210 All enterprises with suitable conditions should introduce in a big way the worker-peasant labour system. 1976 tr. Tuan Jui-hsia in *Yenan Seeds & Other Stories* 19 Spring was very much in the air in the Worker-Peasant-Soldier Theatre. 1949 *Commonweal* 29 July 385/2 We must bow our heads in deep humility before the heroism of the worker priest. 1959 *Manch. Guardian* 4 Aug. 3/3 Five worker-priests.. believed to be the only ones in the Church of England.. do manual work, live on their factory earnings and receive no ecclesiastical stipends. 1970 *Daily Tel.* 29 Dec. 11/3 The Anglican church makes a distinction between worker-priests, who exercise a priestly function at their workplace, and priest-workers, who do an ordinary job, then return to parish work, usually as curate, in their 'spare time'. 1984 *Times* 28 May 10/8, I could not help wondering how it would be received by the congregations of, say, a worker priest in Nicaragua.

6. *Comb.* with *workers'*, as (sense 2 c) *workers' committee, control, flat*; **Workers' Educational Association,** the name of an organization founded in 1903 to provide evening classes and tutorials in economic, political, and liberal studies, originally for working people.

1965 J. Kolaja *Workers' Councils* II. 28 The highly skilled workers' committee, and the apartments committee were considered temporary. 1972 M. Argyle *Social Psychol. of Work* viii. 217 The workers' committees have worked well, and members have acted responsibly on them. 1928 *Britain's Industrial Future* (Liberal Party) III. xviii. 228 Consultation with a body of workers will improve and strengthen it [*sc.* a business]; anything that can accurately be described as 'workers' control' will destroy it. 1974 *Times* 5 Apr. 16/5 As a trades unionist, I am in favour of workers' control. 1903 Workers' Educational Association [in Dict., sense 2 c]. 1936 *N. & Q.* 11 July 19/2 What is gained from the University Extension, the Workers' Educational Association, or a year of study at Ruskin College? 1980 J. L. Thompson *Adult Education for Change* 22 The Workers' Educational Association provision has in many respects become barely distinguishable from that promoted by the universities, despite its roots in workers' education and political and economic studies. 1932 S. Jameson *Single Heart* v. 127 She was able..to buy a slum estate in Evan's constituency and build on it blocks of workers' flats. 1982 S. Grant Duff *Parting of Ways* vi. 54 The shelling of the workers' flats in Vienna in February 1934.

workerist (wɜ̄·ɹkərist), *a.* and *sb.* [f. WORKER + -IST.] **A.** *adj.* Of, pertaining to, or characteristic of a worker-oriented view of society; (too) sympathetic to the role of labour in the class struggle. **B.** *sb.* One who adopts workerist values; *spec.* applied (somewhat *derog.*) to a member of the middle or upper classes who espouses the cause of the working class.

1959 W. Birmingham tr. J. Daniélou in *Cross Currents* Fall 381/2 The workerist conception locates poverty on the level of the standard of living. The 'collectivist' locates poverty on the level of private property. 1981 *Filmnews* (Austral.) May 6/2 The practice of Cinema Action was criticised as workerist for assuming that there was a unified working class. 1984 *Sunday Tel.* 2 Dec. 21/2 Oxford, long the home of what is now known as the 'workerist' (public school student turning very Left-wing). 1985 *Daily Tel.* 11 Feb. 12/2 The genuine proletarians of the hard Left who regard him [*sc.* Mr. Benn] as 'workerist'.

workfare (wɜ̄·ɹkfeəɹ). orig. and chiefly *U.S.* [f. WORK *sb.*, after *welfare*.] A policy of requiring recipients of welfare money to do some work in exchange for this benefit.

1968 *Harper's* July 71 One of Evers' programs is what he calls workfare; he has said that everybody ought to work for what he gets. 1969 R. Nixon in *Washington Post* 9 Aug. 1/2 What America needs now is not more welfare but more 'workfare'. 1978 *Globe & Mail* (Toronto) 28 Nov. 5/3 Mr. Walker, one of the most conservative politicians in the Legislature..thinks they would support a pilot project in an interested municipality, although he is not claiming Government support for workfare. 1981 *Daily Tel.* 26 May 5 Two California towns..are at the forefront of a movement to implement 'workfare'—projects aimed at forcing welfare recipients to do some labour in exchange for taxpayers' money. 1985 *Times* 12 Feb. 14/5 There must be a real inducement to work. In the US a number of states have introduced 'Workfare' to complement welfare.

work function. *Physics.* [f. WORK *sb.* + FUNCTION *sb.*] **1.** The minimum quantity of energy, characteristic of the material concerned, which is required to remove an electron to infinity from the surface of a solid (usu. a metal). Symbol: ϕ.

1923 *Proc. R. Soc.* A. CIV. 637 The estimations of the photo-electric work function..are in all cases a good deal greater than the corresponding values of the thermionic work function. 1950 A. Kolin *Physics* xxxv. 715 The work function can be determined by finding the threshold frequency ν_0 below which no photoelectrons are emitted. 1972 *Sci. Amer.* Mar. 53/1 When two metals are placed in contact, electrons pass from one to the other because of the difference in the metals' quantum-mechanical work functions. This process continues until an equilibrium is reached.

2. A thermodynamic property of a system: its internal energy minus the product of its temperature and entropy. Symbol: A.

1929 R. H. Fowler *Statistical Mech.* iv. 96, $k \log K(T)$ for the crystal is the thermodynamic function known as Planck's characteristic function, and $-kT \log K(T)$ is the more usual work function..A. 1937 P. S. Epstein *Textbk. Thermodynamics* v. 92 When the temperature and the volume of a system are kept constant, its work function has a tendency to decrease. 1978 P. W. Atkins *Physical Chem.* v. 139 If we know the value of ΔA for a change we can also state the maximum amount of work the system can do. This relation is the reason why *A* is sometimes called the maximum work function, or the work function.

workhouse. Add: **3.** (Examples.)

1653 *Boston Rec.* (1886) X. 26 The setting up of a Bridewell or Workehouse for Prisoners Malefactors &.. poore people. 1772 A. G. Winslow *Diary* 25 Feb. (1895) 36 She..soon got into the workhouse for new misdemeanours. 1870 'Mark Twain' *Curious Dream* (1872) 83 Eggs..so unwholesome that the city physician seldom or never orders them for the workhouse. 1964 *Federal Probation* Dec. 8/2 The Workhouse receives and releases the work-release prisoner any time during the day or night, depending on his working hours.

4. b. workhouse sheeting (examples).

1875 L. Troubridge *Life amongst Troubridges* (1966) 116 A Workhouse sheeting jacket, body and *tablier*.. to wear with dark blue frilled petticoat and sleeves. 1880 [see *BOLTON].

work-in (wɜ̄·ɹkin). [f. WORK *v.* + *-IN³.] A form of protest, usu. against threatened closure of a factory, etc., in which workers occupy the work-place and continue working. Also *transf.*

1968 *Punch* 6 Mar. 327/1 Student protest reached an all-time high with Leicester University's plan to stage a '24-hour work-in'. 1973 *Times Lit. Suppl.* 30 Nov. 1469/1 The series of work-ins or sit-ins in the past two or three years in which workers have occupied factories in pursuit of wage claims or as a refusal to accept redundancy notices. 1976 [see *SIT-IN *sb.* 1]. 1983 *Daily Tel.* 23 June 19/3 A judge strongly attacked the police for staging a dawn raid to break up the work-in.

working, *vbl. sb.* Add: **I. 1. b.** *working to rule*, the action of strictly observing the limits of one's occupational duties; also = *work-to-rule* sb. (see *WORK *v.* 27 d).

1927 W. E. Collinson *Contemp. Eng.* 84 The inconveniences of lightning-strikes, ca' canny policy (deliberate restriction of output) and working-to-rule. 1951 *Engineering* 2 Nov. 568/3 The overtime ban and working to rule have remained in force. *Ibid.*, Similar working-to-rule methods..were put into operation by lightermen at the Port of London. 1958 *Times* 19 Aug. 8/3 To what extent the 'working to rule' will apply will depend on the attitude of individual busmen. 1964 M. Argyle *Psychol. & Social Probl.* xiv. 172 Anti-organization practices such as restriction of output, unofficial strikes and working to rule. *a* 1974 R. Crossman *Diaries* (1976) II. 686, I tried to make them realize..what I meant by quietism. I suggested that it meant a 'non-enthusiastic execution'—working to rule, shall we say?

2. b. *spec.* in *Freemasonry*, (the performance of) a rite, a system of ritual.

1884 W. J. Hughan *Orig. Eng. Rite Freemasonry* p. iii, Although under various Grand Lodges the details of the working differ, the landmarks remain practically identical. 1903 J. T. Lawrence *Masonic Jurisprudence & Symbolism* viii. 75 The more important one [*sc.* duty] is to see that ceremonies are conducted in accordance with working sanctioned by the Grand Lodge of England. 1932 S. M. Hills *Freemason's Craft* viii. 64 The Articles of the Union..stipulated that there should henceforth be perfect unity of working, and the Lodge of Reconciliation was formed..to agree upon a working.

5. f. Of a bus, train, etc.

1978 M. Keeley et al. *Birmingham City Transport* 181 City—Bull Ring—Coventry Road—Lyndon End. Short working of 94. 1982 *Railway Mag.* Nov. 508/1 A reader who visited Scarborough..noted a wide variety of locomotive classes in use on summer-holiday workings.

III. 15. *working-out*: spec. in *Mus.* = DEVELOPMENT 10; *working-over* (*slang*) = *GOING OVER, GOING-OVER 2 b and c.

1889 Grove *Dict. Mus.* IV. 486/2 *Working-out* (also called Free Fantasia; and Development; Durchführung), the central division of a movement in Binary form, such as commonly occupies the first place in a modern sonata or symphony. 1936 *Discovery* Apr. 124/1 The music of Bach, with its perfect counterpoint and logical working-out. 1948 *Penguin Music Mag.* VII. 43 The..sleight-of-hand that Grieg saw fit to employ in a 'working-out'. 1960 C. Hamblett in J. Pudney *Pick of Today's Short Stories* XI. 143 The cops frisked him..hoping he would put up a fight, so they could give him a working-over first. 1964 L. Deighton *Funeral in Berlin* viii. 55 A girl with too much make-up..gave her eyebrows a working over.

IV. 16. a. *working arrangement, bee, model.* **b. working-box** (earlier example); **working card** *U.S. obs.* = *union-card* s.v. *UNION *sb.*¹ 11 c; **working copy,** a copy of a book or other document used or annotated by someone working on its contents; **working dinner, lunch,** a dinner or lunch at which those present discuss business; **working order** (earlier example); **working plan,** a plan serving as the basis for the construction of a building, management of a project, etc.; *spec.* in *Forestry* (see quots. 1895, 1926); **working space** = *working storage*; cf. *work-space* (*b*) s.v. *WORK *sb.* 34 d; **working storage,** part of a computer's memory that is used by a program for the storage of intermediate results or other temporary items; **working surface** = *work-top* s.v. *WORK *sb.* 34 d; **working title,** a provisional title given to a book, film, or other work before the final title is settled; **working top** = *work top* s.v. *WORK *sb.* 34 d.

1854 *Household Narrative* Apr. 80/2 He stated..that the more complete fusion of capital into one company ought not to be sanctioned, but that sort of combination known as working arrangements should be encouraged. 1904 *Windsor Mag.* June 16/1 A simple working arrangement is usual based on a percentage division of the gross receipts between the two. 1970 *New Yorker* 29 Aug. 45/1 Jews and pagans would never get to Heaven, with the exception of..Moses, who had a close working arrangement with Allah. 1883 'A Lady' *Facts: or, Exper. Recent Colonist in N.Z.* viii. 68 The ladies of the community.. meet for a common cause..Working bees are then got up. 1956 W. R. Bird *Off-Trail in Nova Scotia* i. 23 We were told much of working bees, barn raisings, the making of maple sugar. 1778 J. Woodforde *Diary* 9 Sept. (1924) I. 235 It..looks when covered like a working Box for Ladies. 1872 *Pacific States Enterprise* (San Francisco) 16 Mar. 3/1 They have adopted the 'working card' system. 1874 *Internat. Typogr. Union Proc.* 34 Subordinate Unions are recommended to..enforce the 'working-card' system. 1896 *Ibid.* 35/2 It was agreed to issue him a working card. 1923 *Proc. 43rd Convention Amer. Fed. of Labor* 324/2 This resolution..affects such other organizations as they seek to affect with an exchange of working cards and other courtesies. 1897 W. C. Hazlitt *Confessions of Collector* vi. 100, I would gladly pay him a guinea for it, and find him a working copy into the bargain. 1967 E. R. Lannon in Cox & Grose *Organization & Handling Bibl. Rec. by Computer* iv. 95 We can.. print the dictionary in two forms; one form is referred to as the 'Working Copy' edition, intended for the use of our own editorial staff. 1970 *Daily Tel.* 22 Sept. 1/8 Union chiefs and chairmen of five nationalised industries had a 'working dinner'..last night. 1964 *Guardian* 27 Oct. 18/6 After these meetings there was a 'working' lunch at the British Embassy. 1966 H. Moore *On Sculpture* 247 The first maquette for the wood Interior Exterior Forms was produced in 1951, later the same year I made the working model (24½ in. high), which was cast into bronze. 1982 *Sunday Tel.* (Colour Suppl.) 14 Nov. 33/1 Carvings were plastercast, moulds were taken and durable metal-alloy working models were made. 1845 *Knickerbocker* XXVI. 410 The use of steam-pumps is requisite night and day, to keep them [*sc.* mines] in working order. 1880 'Mark Twain' *Tramp Abroad* xxxiv. 370 The ghastly desolation of the place was as tremendously complete as if Doré had furnished the working-plans for it. 1895 W. Schlich *Man. Forestry* III. III. 173 Forest working plans regulate, according to time and locality, the management of forests in such a manner, that the objects of the industry are as fully as possible realized. 1926 Tansley & Chipp *Study of Vegetation* xi. 255 The Working Plan forms..a scheme for exploiting the forest whereby regeneration will keep pace with exploitation. 1983 *National Trust* Spring 10/1 Constructive, planned woodland management only became possible after the war, and our oldest working plans are now barely thirty years old. 1954 Working space [see *surface sterilization* s.v. *SURFACE *sb.* 6 c]. 1973 M. Woodhouse *Blue Bone* vii. 63 The converted hold..had contained a full-sized billiards table. There was, at least, plenty of working space. 1954 *Computers & Automation* Dec. 23/1 Working storage... Like a worksheet in pencil and paper calculation. 1971 N. Chapin *Computers* xv. 445 In the operand data structures, programers commonly distinguish between constants that are not part of the program, working storage (for intermediate results and status or progress indicators), and input-output buffer areas. 1983 D. H. Sandars *Computers Today* v. 113 This total earnings figure is copied (instruction, o8) in address 15, which is the working storage area. 1962 A. Wise *Death's-Head* iii. 23 The electric percolator standing on the formica working-surface. 1970 *Which?* Sept. 279/1 Three of the small freezers..had laminated tops you could use as a working surface. 1940 R. Chandler *Let.* 27 June in Gardiner & Walker *Raymond Chandler Speaking* (1962) 211 The title of my book is not *The Second Murderer*. I used that for a while as a working title, but I didn't like it. 1977 G. Fisher *Villain of Piece* iii. 32, I was now busy turning the whole caboodle into a series of four articles... I gave it a working title: 'My Life with Britain's Top Villain'. 1959 *Housewife* June 70/2 Table-top refrigerators are popular because they give an additional working top. 1980 D. Clark *Golden Rain* v. 115 'Which cupboard please?' 'The last one under the working-top on the left.'

working, *ppl. a.* Add: **2. a.** Also *spec.* of a girl or woman: that goes out to earn a living rather than remain at home; *working girl*: also *euphem.*, a prostitute (*U.S. slang*).

1865 Dickens *Mut. Fr.* II. IV. vi. 209, I am removed from you and your family by being a working girl. 1889 [see *SISTER *sb.* 2]. 1913 J. Vaizey *College Girl* ix. 119 'I shall have to earn money myself, so I want to pass all the exams. I can.' The Percivals stared... They had never met a prospective *working* girl before! 1933 D. C. Peel *Life's Enchanted Cup* xv. 183 It was, perhaps, because she had known what it was to study in the intervals of tending children..that she could sympathise with working mothers. 1963 *Times* 2 Jan. 10/3 Many Australians had to take two jobs to make ends meet, and 'there were any number of working wives'. 1970 O. Norton *Dead on Prediction* i. 7 I'm going to be a working girl again now. Doing some articles for *Mercia*, for a start. 1978 F. Weldon *Praxis* xx. 174 Praxis gave up her job: Ivor did not want a working wife. 1979 *Arizona Daily Star* 5 Aug. J 3/2 There were studies showing juvenile delinquents springing from single-parent or working-mother homes. 1968 *Current Slang* (Univ. S. Dakota) Fall 52 *Working girl*, n., a prostitute. 1971 *N.Y. Times* 9 Aug. 33/5 They call themselves 'working girls'... Their work is a 'business', or even..a 'social service'... By the prostitute's code, prostitution is moral. 1984 *Chicago Sun-Times* 26 Mar. 12 U.S. Prostitutes has estimated that thousands of 'working girls' will travel to San Francisco for business generated by the convention.

c. Restrict *Mil.* to sense in Dict. and add: (*b*) a group of women, meeting to do work, esp. sewing, for a good cause; (*c*) a committee appointed to examine and report on a particular subject and to make recommendations based on its findings; (*d*) a group of prisoners engaged on outdoor work, freq. outside the perimeter of the prison.

1876 C. M. Yonge *Three Brides* I. viii. 127 Cecil had offered to take Anne to see the working party, and let her assist thereat. 1900 —— *Modern Broods* vii. 72 The parish room, where the ladies were to hold a working party for the missions. 1946 *Times* 10 Jan. 2/3 The 'working party' is a device for securing the best possible guidance on the policies that should be adopted to bring an industry to the highest pitch of efficiency under private enterprise. 1948 *Hansard Commons: Written Answers* 8 Mar. 112 The Working Party on the Turn-Round of Shipping was set up to examine the causes of delay... Teams from the Working Party have visited the major ports. 1963 T. Parker *Unknown Citizen* v. 136 Charlie was out of the prison that day, on a working party at a farm some miles away. 1976 L. Kennedy *Presumption of Innocence* i. 46 One day at Parkhurst he walked away from an outside working-party... Although he was wearing prison overalls, no one paid any attention. 1981 E. Longford *Queen Mother* v. 82 The Queen and her working party met twice a week to make surgical dressings and comforters for the troops. 1982 *Church Times* 12 Nov. 1/3 The General Synod's Board of Education has set up a working party on independent schools which will look at the Independent sector in education 'from a Christian perspective'.

d. Also of dogs used for hunting, herding, guard duties, etc. Also *fig.*

1897 *Blackw. Mag.* June 744/1 Notwithstanding the care most people take to buy pups of 'good working parents', it is the blood that tells. 1936 *Times Lit. Suppl.* 25 Jan. 73/2 The American husband (in fiction) is losing his working-dog quality, his ambition to toil. 1947 C. L. B. Hubbard (*title*) Working dogs of the world. 1982 G. Hammond *Fair Game* xi. 99 [The dog] went for the pigeon..and fetched it back... Miss Wyper was overwhelmed. This was her first introduction to the truly fulfilled dog, the working dog doing the job for which it was bred.

4. (Later example.)

1934 Dylan Thomas in *New Verse* No. 9. 12 The dry Sargossa of the tomb Gives up its dust to such a working sea.

7. (Earlier examples of *working model*.)

1770 J. Ferguson *Introd. Electricity* 134 A working model of the great crane at *Bristol*. 1822 C. F. Partington *Hist. & Descr. Account Steam Engine* i. 13 In the following year a working model of the above engine was submitted to the Royal Society.

working class. Add: **a.** (Earlier examples.)

1789 J. Gray in G. Dempster *Discourse containing a Summary of Proceedings of Soc. for extending Fisheries & improving Sea Coasts of Gt. Britain* 50 More spacious plots of ground..may be allowed to the clergyman and schoolmaster, and to other persons superior to the working class. 1795 J. Aikin *Descr. Country 30–40 Miles round Manchester* 262 Houses for the working class are not procured without difficulty.

b. (Earlier examples.)

1839 J. S. Mill in *Westm. Rev.* Apr. 497 The Working Men's Association..who represent the best and most enlightened aspect of working-class Radicalism. 1849 F. D. Maurice *Let.* 3 Mar. in J. F. Maurice *Life F. D. Maurice* (1884) I. xxv. 513 Thank you very much for entering so heartily into my working class meetings.

working-day. 2. (Earlier example.)

1853 *Hogg's Instructor* X. 282/2 To grant the Saturday afternoon holiday, and to limit the duration of every other working day within a certain definite period of time, not exceeding twelve hours, including the proper interval for meals.

working-man. Add: **b.** *Comb.* in the possessive, denoting institutions established for working men, as *working man's* (or *men's*) *association, club, college, institute.*
1839 Working men's association [see *WORKING CLASS b]. **1844** *Lexington* (Ky.) *Observer* 2 Oct. 3/1 The Working Men's Clay Club of this city will hold an adjourned meeting. **1861** MRS. GASKELL *Let.* 16 Apr. (1966) 650 He has..established a small working man's Club, with the help of a *low* Church curate. **1961** *Economist* 30 Dec. 1270/1 The Working Men's Club and Institute Union achieves its centenary next year. **1976** *National Observer* (U.S.) 18 Dec. 18/3 It opens in a Manchester secondary-school classroom, where a sextet of aspiring stand-up comics has assembled for their Big Chance, an audition before a real..agent at a local working-man's club. **1856** C. FOX *Let.* 27 June in *Jrnls.* (1972) 223 Oxford... I was delighted to hear of their successful experiment to unite Town and Gown by a Working Man's College. **1921** G. B. SHAW *Back to Methuselah* II. 61, I was asked to deliver an address to the students at the Working Men's College. **1971** G. STEINER *Bluebeard's Castle* iii. 61 The categories of schooling and public enlightenment—the lyceum, the public library, the working men's college. **1882** F. A. KEMBLE *Records of Later Life* III. 293 A reading that I gave for the Working Men's Institute in Brighton. **1980** E. BLISHEN *Nest of Teachers* I. i. 4 A Working Men's Institute..established in the last century.

workman. Add: **6.** *Comb.* in the possessive, denoting things (esp. transport) provided for workmen (sense 1), as *workman's* (or *workmen's*) *bus, club, compensation, train, tram.*
1965 A. PRIOR *Interrogators* xi. 198 An early morning workman's bus roared and rumbled. **1980** 'D. GRANT' *Emerald Decision* i. 20 Workmen's bus, heading for the harbour. **1911** G. B. SHAW *Doctor's Dilemma* I. 22 Except for the workmen's clubs, my patients are all clerks and shopmen. **1921** *Daily Colonist* (Victoria, B.C.) 12 Mar. 6/4 Compensation was paid to him by the Workmen's Compensation Board and he was later discharged as having recovered. **1940** *Economist* 3 Feb. 198/2 An all-round increase in the rates payable for workmen's compensation. **1872** JERROLD *London* p. ix, The workman's train and the crowds pressing over London Bridge. **1975** P. MCCUTCHAN *Very Big Bang* xi. 105 'They'll call back when the current's off.' 'No more workmen's trains?'..'Not on this section.' **1906** *Jackson's Oxford Jrnl.* 8 Sept. 6/6 At 5.10 the workmen's tram joined in the procession. **1970** S. ALEXANDER *St. Giles's Fair* 19 The workman's tram passed through St. Giles at five ten a.m.

work-out. Add: **a.** Also more widely, an exercise session, practice, or test. (Earlier and later examples.)
1909 R. A. WASON *Happy Hawkins* 161, I expect to give it a fair good work-out before I'm through with it. **1923** H. C. WITWER *Fighting Blood* iii. 96 I ain't going to get no gym workout. **1938** M. K. RAWLINGS *Yearling* ix. 98 'Will we take both dogs?' 'Nobody but old Julia. She ain't had a work-out since she was hurt. A slow hunt'll do her good.' **1952** *Sun* (Baltimore) 3 Mar. 28/3 The United States Air Force Filter Center here had its first surprise workout yesterday. **1960** *Sunday Express* 27 Nov. 14/3 Work-out gymnasia with..general slimming equipment. **1963** A. ROSS *Australia 63* vii. 128 Both teams had work-outs at the Oval, over-watered pitches making net practice impracticable. **1972** *Daily Tel.* 6 Sept. 6/8, I am not suggesting that old people should do strenuous physical work-outs, but stiff joints and muscles may be helped by properly supervised exercises. **1979** *Tucson Mag.* Apr. 56/1 (Advt.), A multitude of weight training systems designed to accommodate your kind of workout room. **1981** J. FONDA *Jane Fonda's Workout Bk.* (1982) 22 She took me to an exercise class that put me through the most vigorous and thorough workout I had ever had.
b. *fig.*
1934 J. O'HARA *Appointment in Samarra* i. 17 Four of the young men had had work-outs with her off the dance floor, and as a result Constance was not a virgin. **1941** *Punch* 10 Sept. 222 This passage of rich prose is designed as a sort of test or work-out for the new alphabet. **1958** B. HAMILTON *Too Much of Water* viii. 161 A public work-out would test audience reaction. **1967** *Melody Maker* 29 Apr. 10 (Advt.), The totally new snare drum. This one you must see! Get round to your dealer and give it a workout. **1977** *New Yorker* 24 Oct. 173/1 The Villa-Lobos provides a thorough workout for the bassoon.

wo·rkover. [f. WORK *v.* + OVER *adv.*] The repair or maintenance of an oil well.
1976 *Offshore Platforms & Pipelining* 232/3 A conventional wellhead is installed, and the wells are completed by a workover rig. **1977** *Financial Times* 1 Apr. 11/5 Some have suggested a well work-over every three years; others say once every 15 years will be sufficient. **1977** *Offshore Engineer* Aug. 12/3 The workover programme on B-14 well is now complete. **1980** *Daily Tel.* 11 Sept. 29 (Advt.), Jack Up Rig in the Southern North Sea on Viking workovers and new developments. **1985** *New Yorker* 22 Apr. 51/1 The auction business thrived: deep rigs, workover rigs.

workshop. Add: **2. a.** A meeting for discussion, study, experiment, etc., orig. in education or the arts, but now in any field; an organization or group established for this purpose.
1937 *N.Y. Times* 1 Aug. VI. 5/3 The major requirement for admission to this Summer workshop is an approved project for which the applicant seeks aid and advice. **1938** L. MACNEICE *Mod. Poetry* xi. 200 The communist poet, Maiakovski, established a 'word workshop'..to supply all revolutionaries with 'any quantity of poetry desired'. **1952** L. ROSS *Picture* (1953) 21 The elder Reinhardt..came to Hollywood in 1934... For the next five years, he ran a Hollywood school known as Max Reinhardt's Workshop. **1959** *Ottawa Citizen* 14 Sept. 6/1 At a conference or 'workshop' on road safety sponsored by the Ontario Department of Transport recently, there was general agreement that much more must be done to improve driving standards. **1961** in *B.B.C. Handbk.* (1962) 36, I want to see a Television Workshop—a regular period in which everyone feels he can have a go without having to mind too much whether he is successful straight off. **1967** P. MCGIRR *Murder is Absurd* ii. 33 In college Kenny joined the..drama workshop and began work on a play. **1972** *Computers & Humanities* VII. 96 The participants then divided into four workshops and, after five intensive meetings, reconvened to present their findings at the fourth and final plenary session. **1984** *Times* 17 Mar. 15/8 Priority bookings for their tastings, wine workshops and special dinners.
b. *attrib.*
1937 *N.Y. Times* 1 Aug. VI. 5/4 The importance of the workshop idea to American education. **1968** *Globe & Mail* (Toronto) 3 Feb. B 2/3 Local residents considered..17 consumer protection items suggested by workshop groups conducted on Thursday. **1976** S. BRETT *So much Blood* ii. 25 The Masonic Hall was not free for Charles to rehearse in... Michael Vanderzee had just started a workshop session... Charles..had no objection to..workshop techniques. They were useful exercises for actors. **1983** *National Trust* Spring 24/1 In the morning, group discussions were led by the Company's seven actor/teachers in a 'workshop' atmosphere concentrating on the social history of the early eighteenth century.

work station. Also **wo·rkstation.** [f. WORK *sb.* + STATION *sb.*] **1.** A location at which one stage in the manufacture or assembly of a product is carried out before it is moved on for the next stage.
1950 T. M. LANDY *Production Planning & Control* vii. 161 The assigning of work stations by the planners is largely influenced by these considerations: costs, centralization of work within a section..and existing labor load. **1980** M. P. GROOVER *Automation, Production Systems & Computer-Aided Manufacturing* iv. 66 An automated flow line consists of several machines or workstations which are linked together by work handling devices that transfer parts between the stations.
2. A desk with a computer terminal and keyboard; the terminal itself.
1977 *Which Computer?* Sept. 25 Another cavil on the workstation is the absence of a 'home' key to return the cursor to the head of a new page. **1981** *Office* June 23/1 Wordcom 70 can be used as a shared facility system with up to eight work-stations using the cartridge disk. In this way a number of departments can have access to a single database. **1983** *What's New in Computing* Jan. 53 (Advt.), Featured above in teak: printer stand, linking quadrant and 3' by 2' workstation with monitor shelf. **1985** *Which Computer?* Apr. 122/3 It can then be viewed by users at any workstation attached to the system.

work-to-contract, work-to-rule: see *WORK *v.* 27 d.

wo·rk-up. [f. vbl. phr. to *work up*: see WORK *v.* 39.] **1.** *Printing.* A piece of spacing material that works loose in the forme and prints a smudge, or the mark so printed. Also, an instance of this.
1948 R. R. KARCH *Graphic Arts Procedures* x. 270 The pressfeeder watches the sheets carefully so that if *work-ups* appear he will not spoil the run. A work-up is spacing material that has risen in the form so that it prints on the sheet. **1950** D. G. HYMES *Production in Advertising* vii. 279 Alert pressmen can..stop the press, and hammer down the workups after only a few sheets have been spoiled. **1967** KARCH & BUBER *Offset Processes* ii. 24 The letterpress printer may be plagued by 'work-up'.
2. *Med.* A diagnostic examination of a patient. orig. *U.S.*
1961 in WEBSTER. **1966** *Current Diagnosis* 206/2 (*heading*) Diagnostic work-up for patients with diastolic hypertension. **1972** *Sci. Amer.* Aug. 71/3 A mother who was told that her child would be 'admitted for a work-up' did not realise that he was to be hospitalized. **1977** *New Yorker* 12 Sept. 103/1 We gave him the usual workup, including neurological and ear-nose-and-throat evaluations. **1978** PARSONS & SOMMERS *Gynecol.* (ed. 2) xx. 313/2 A few cases where the borderline between virilism and hirsutism is hazy will benefit from an extensive work-up.
3. *Chem.* The experimental procedures followed to separate and purify substances for analysis or the products of a chemical reaction.
1967 *Chem. Abstr.* LXVII. 10187/2 (*heading*) Workup of chlorohydrin process waste products. **1971** *Canad. Jrnl. Chem.* XLIX. 2467/1 The resulting mixture was then heated to reflux for 1 h to complete the acetylation and gave after usual work-up a quantitative yield of [compound] 12. **1978** *Jrnl. Amer. Chem. Soc.* C. 3548 Water added during work-up serving as the proton source at C-1.
4. The process of bringing a ship into seaworthy condition.
1971 *Daily Colonist* (Victoria, B.C.) 27 May 35/4 Canadian forces ships work on a 20-month cycle basis containing four parts: refit, trials, workups, and operational. **1978** *Navy News* Oct. 2/6 After trials and work-up the Achilles will return to Chatham for Christmas.

workwise (wɜ̄·ɪkwaiz), *adv.* [f. WORK *sb.* + -WISE.] As far as work is concerned.
1962 *Punch* 6 June 864/1 Workwise, your future is clear. **1979** *Yale Alumni Mag.* Apr. (Suppl.) cn32/1 Although Carol has kept herself busy (workwise) she managed to spend Thanksgiving in Portugal. **1981** S. JACKMAN *Game of Soldiers* I. ii. 33 The unit ticks over well enough work-wise... Trouble is, there's not enough work.

world, *sb.* Add: **II. 7. g.** *broke to the world*: see *BROKE *ppl. a.* 3; (*it's a*) *small world*: see *SMALL *a.* 3 b; (*on*) *top of the world*: see *TOP *sb.*[1] 16.
17. b. *to get up in the world, to go down in the world*: (earlier examples).
1791 J. WOODFORDE *Diary* 20 Mar. (1927) III. 257 John Greaves, my Carpenter..married about 2 Years or more ago, to a Servant Maid of Mrs. Lombe's..and lived very happy together and daily getting up in the World. **1837** J. S. MILL *Let.* 6 Aug. in *Works* (1963) XII. 346 To alter their style of living and go (as the vulgar phrase is) down in the world.
IV. 19. a world. a. *a world of good* (earlier example).
1791 F. BURNEY *Jrnl.* Sept. (1972) I. 57 The Water has done me a World of good—I drink it at morning & Noon regularly.
20. the world. e. *to think the world of* (earlier U.S. example).
[**1852** H. B. STOWE *Uncle Tom's Cabin* II. xxxiv. 206 He had a cousin come to New Orleans, who was his particular friend,—he thought all the world of him.] **1892** 'MARK TWAIN' *Amer. Claimant* iii. 24 They..think the world of Mulberry.
f. *woman of the world* (earlier examples).
1780 F. BURNEY *Diary* Apr. (1904) I. VII. 328 She is an easy, chatty, sensible woman of the world. **1822** M. EDGEWORTH *Let.* 10 Apr. (1971) 393 Lady Clare is a painted—made up—vulgar thorough going woman of the world.
22*. this world. a. *out of this world*: (i) superlatively good, fine beyond description; beautiful, delightful, wonderful. Also as adv. and attrib. phrases. *colloq.* and *slang* (orig. *U.S. Jazz*).
1928 R. FISHER *Walls of Jericho* 303 *Out (of) this world*, beyond mortal experience or belief. **1931** *Inter-State Tattler* 17 Dec. 12 Alberta Hunter..warbles out of this world. **1935** *Swing Music* July 114/2 Benny's clarinet playing here is out-of-this-world for beauty of tone. **1946** *Sat. Rev. Lit.* (U.S.) 19 Oct. 25/3 Petarded on his own cliché And violently hurled, Should be the Joe whose one bon mot Is 'It's out of this world!' **1952** G. WILSON *Julien Ware* 36 A slender, graceful, out-of-this-world bridge Claud...had been. **1957** J. BRAINE *Room at Top* vi. 51 You've got a lovely part. Out of this world. **1972** J. ROSSITER *Rope for General Dietz* v. 61 She gave me the skinned fruit... With Cointreau poured on, mine tasted out of this world.
(ii) In neutral or derogatory contexts: unworldly; quite remarkable; also, incredibly bad or repulsive.
1941 B. SCHULBERG *What makes Sammy Run?* vii. 149 The gallery was in a funny little bungalow with an easy-going, out-of-this-world atmosphere. **1951** 'A. GARVE' *Murder in Moscow* ii. 32 They hate our guts, and the way they behave is out of this world. **1958** *Oxford Mail* 27 Aug. 6/1 The worst part of a woman's magazine..is the fiction. Stories about quite impossible people in out-of-this-world situations. **1963** P. WILLMOTT *Evolution of Community* viii. 92 The L.C.C.'s wallpapers..are very antiquated, out of this world.
b. *the* (personal or other proper name, pl.) *of this world*: people (countries, etc.) considered to represent the type specified; people, etc., like (sb. sing.). *colloq.* Freq. somewhat *derog.*
1960 J. STROUD *Shorn Lamb* iv. 44 He's settling... We're quite used to the Egberts of this world. **1969** M. PUGH *Last Place Left* xiv. 106 The Pardoes of this world would always brownnose to the landed gentry. **1972** *Observer* 20 Feb. 11/3 There is a limit on how far the Libyas of this world can bid up the price of oil.
23. a. *world-construction, cruise, -end* (attrib.), *events, -formula, government, -image, -model, -outlook, première, principle* (earlier example), *record, sorrow* (earlier example), *-structure, -system, -theory, tour, -will.* (Not clearly distinguishable from some of the examples in sense 24 b in Dict. and Suppl.)
1906 W. R. INGE *Truth & Falsehood in Relig.* 115 Science has no commission to produce an ideal world-construction on a materialistic basis. **1933** World cruise [see *CRUISE *sb.* 1 a]. **1977** A. C. H. SMITH *Jericho Gun* iv. 54 Let's take a world cruise. **1896** KIPLING *Seven Seas* p. vii, I was born in her gate... Where the world-end steamers wait. **1940** J. PEDERSEN *Israel* II. IV. 559 Like Isaiah he [*sc.* Jeremiah]..undertakes to interpret world-events. **1888** J. ROYCE *Let.* 21 May in R. B. Perry *Thought & Char. W. James* (1935) I. 800, I have largely straightened out the big metaphysical tangle about continuity, freedom, and the world-formula. **1907** W. JAMES *Pragmatism* ii. 50 The whole function of philosophy ought to be to find out what definite difference it will make to you and me, at definite instants of our life, if this world-formula or that world-formula be the true one. **1915** N. L. MCCLUNG *In Times like These* ix. 153 The problems of discovery have been solved; the problems of colonization are being solved, and when the war is over the problem of world government will be solved. **1958** B. W. ALDISS *Non-Stop* IV. v. 241 The ship

is in an orbit round Earth and there it must stay. That was the edict of the World Government. **1981** *Washington Post* 18 Mar. 1 He would never be part of an organization that advocated world government. **1936** *Discovery* May 162/1 The Determinists have created for themselves an intellectual structure which represents a world-image or rather a physical world-image. **1949** G. J. WHITROW *Structure of Universe* v. 75 In order to obtain some picture of the universe as a whole, we must construct a world-model which will reproduce satisfactorily the properties of this observable (limited region of space and time). **1915** (*serial title*) World outlook. **1929** *New Statesman* 31 Aug. 628/1 All poetic genius has always fumbled instinctively for a world-outlook in which everything has significance at all times. **1976** tr. Shih Min in *Yenan Seeds & Other Stories* 75 Remould your world-outlook and steel yourself into a self-aware revolutionary. **1934** WEBSTER, World première. **1948** *Daily Ardmoreite* (Ardmore, Okla.) 7 July 1/5 'Return of the Bad Man' will open a three day engagement in Ardmore..just one day after its world premier. **1981** LD. HAREWOOD *Tongs & Bones* ix. 150 He..put on several important world premières of British operas. **1854** GEO. ELIOT tr. *L. Feuerbach's Essence of Christianity* x. 101 Individual subjectivity..is regarded as the highest essence—the omnipotent world-principle. **1909** G. B. SHAW *Pen Portraits & Reviews* (1931) 236 In his stories of mystery and imagination Poe created a world-record for the English language. **1976** *Daily Tel.* 20 July 1/5 Cornelia Ender won 100m women's freestyle gold model in world record 55.65 secs. **1868** GEO. ELIOT *Spanish Gypsy* II. 173 Silva had thought To melt hard bitter grief by fellowship With the world-sorrow trembling in his ear In Pablo's voice. **1920** A. S. EDDINGTON *Space, Time & Gravit.* ix. 150 The world-structure is not of a kind which can be traced in an exact way by mesh-systems, and in any large region the mesh-system drawn must be considered arbitrary. **1874** G. H. LEWES *Problems* I. 85 Our parochial system will sometimes be favourably contrasted with the results of their world-system. **1977** P. JOHNSON *Enemies of Society* ii. 12 We have characterized its [*sc.* Freedom's] development into the Roman world-system as essentially a liberal economic process, presided over by a night-watchman state. **1834** J. S. MILL in *Monthly Repos.* VIII. 657 They are probably as sincere..as they are capable of being, in any creed, or world-theory, or abstract principle. **1960** W. V. O. QUINE *Word & Object* i. 24 The saving consideration is that we continue to take seriously our own..aggregate science, our own particular world-theory or loose total fabric of quasi-theories, whatever it may be. **1958** J. POPE-HENNESSY in P. Quennell *Lonely Business* (1981) III. 210 Maps of his world-tours on the walls. **1971** 'G. BLACK' *Tome for Pirates* v. 84 A very slow world tour from which he returned with reluctance. **1891** G. B. SHAW *Quintessence of Ibsenism* iv. 70 The world-will shall answer for Julian's soul. **1892** J. ROYCE *Spirit Mod. Philos.* 239 We ourselves are embodiments of the world-will.
b. Objective, as *world-changer, -girdler, -saver, -wielder*; *world-building, -making* (earlier example); *world-beating, -creating, -destroying, -devouring* (later example), *-embracing* (earlier U.S. example), *-enfolding, -girdling, -lifting* (example), *-mothering, -renouncing* (examples), *-shattering, -surrounding* (earlier example), *-transforming, -troubling, -wielding* adjs. **c.** Instrumental, as *world-besotted, -forgotten, -read* adjs. **d.** In other adverbial uses: (*a*) *world-lost, -minded* (so *-mindedness*) adjs.; (*b*) *world-famed* (earlier example), *-renowned* (earlier example) adjs.
1928 *Sunday Express* 24 June 20/4 The way he flashed the passing shot wide of Higgs..was world-beating stuff. **1977** *Daily Mail* 24 Sept. 15/1 The BBC..never became the really 'world-beating station that I would like it to have been'. **1932** W. B. YEATS *Words for Music* 15 Imitate him if you dare, World-besotted traveller; he Served human liberty. **1920** A. S. EDDINGTON *Space, Time & Gravit.* x. 160 It might seem that this kind of fantastic world-building can have little to do with practical problems. **1891** W. JAMES *Let.* 30 Jan. (1920) I. 305 Verily you are the stuff of which world-changers are made! **1854** GEO. ELIOT tr. *Feuerbach's Essence Chr.* xxii. 218 The world-creating activity in itself negatives every determinate activity. **1909** G. K. CHESTERTON *Orthodoxy* (ed. 2) iv. 92 We count on the ordinary course of things... We risk the remote possibility of a miracle as we do that of a poisoned pancake or a world-destroying comet. **1938** DYLAN THOMAS *Let.* 1 June (1966) 199 A world-devouring ghost creature bit out the horror of tomorrow from a gentleman's loins. **1807** J. BARLOW *Columbiad* IV. 155 The world-embracing scope That prompts his genius and expands his hope. **1928** W. B. YEATS tr. *Sophocles' King Oedipus* 6 For all is world-enfolding sea. **1858** World-famed [see *seven-cubit* s.v. *SEVEN *a.* and *sb.* C. 2]. **1861** *Westm. Rev.* LXXVI. 281 Such a world-forgotten village as Raveloe. **1941** I. L. IDRIESS *Great Boomerang* iv. 29 This man's dream was to become a peaceful and world-forgotten patriarch. **1892** *Outing* (U.S.) Mar. 447/1 They probably learned enough about it to make them treat the next world-girdler with high respect. **1934** A. WOOLLCOTT *While Rome Burns* 93 Twenty such world-girdling tales. **1978** H. WOUK *War & Remembrance* xxi. 209 All the members of a world-girdling alliance were attacking us. **1894** KIPLING *Seven Seas* (1896) 45 O' that warld-liftin' joy no after-fall could vex. **1854** J. G. WHITTIER in *Nat. Era* 17 Aug. 130/4 New-born, the world-lost anchorite A man became! **1941** T. WOLFE *Hills Beyond* iii. 235 He abandoned finally the world-lost fastnesses of Zebulon for the more urban settlement of Libya Hill. *a* **1776** HUME *Dialogues conc. Nat. Relig.* (1779) v. 61 A slow, but continued improvement carried on during infinite ages in the art of world-making. **1945** G. MURPHY *Human Nature & Enduring Peace* xvi. 241 What we mean by world-minded education. We mean education for intelligent world citizenship. **1979** *Amer. Speech* 1976 LI. 79 Germans have traditionally been world-minded, receptive to foreign influences. **1926** *Religious Education* Apr. 190 Character is not a cause of world-mindedness, it is a result of world-mindedness and many other attributes. **1960** A. BJERSTEDT in *Jrnl. Conflict Resolution* IV. 185 (*title*) Ego-involved world-mindedness. **1883** G. M. HOPKINS *Poems* (1967) 93 Wild air, world-mothering air, Nestling me everywhere. **1912** HARDY *Jude the Obscure* p. x, An influential article..printed in a world-read journal. **1910** W. MONTGOMERY tr. *A. Schweitzer's Quest Hist. Jesus* xvi. 247 Inexhaustible reserves of world-renouncing, world-contemning sayings. **1964** C. S. LEWIS *Discarded Image* iii. 47 A world-renouncing, ascetic, and mystical character then marked the most eminent Pagans. **1831** CARLYLE in *Foreign Q. Rev.* Oct. 372 The wild, deep, and now world-renowned, *Legend of Faust*, belongs to a somewhat later date. **1952** B. WOLFE *Limbo* IV. 214 A 'messianic complex', an urge to be a 'world-saver'. **1974** M. TIPPETT *Moving into Aquarius* II. 155 During those years there have been huge and world-shattering events in which I have been inevitably caught up. **1817** SHELLEY *To Constantia Singing* in *Posthumous Poems* (1824) 144 Whilst, like the world-surrounding air, thy song Flows on. **1935** W. B. YEATS *Full Moon in March* 68 What sacred drama through her body heaved When world-transforming Charlemagne was conceived? **1895** —— *Poems* 259 And shook at Invar Amargin The hearts of the world-troubling seamen. **1881** G. M. HOPKINS in *Note-Bks. & Papers* (1937) 346 Satan..is the κοσμοκράτωρ, the worldwielder. **1887** —— *Poems* (1967) 70 And the azurous hung hills are his world-wielding shoulder.
24. Passing into adj. **b.** (Further examples.) (Not clearly distinguishable from some of the examples in sense 23 a in Dict. and Suppl.)
1833 J. S. MILL in *Monthly Repos.* VII. 510 The most stirring scenes of that mighty world-drama, under his pen turn flat, cold, and spiritless. **1879** G. H. LEWES *Stud. Psychol.* ix. 162 The World-process has been assigned to a Soul of the World. **1890** W. MORRIS *News from Nowhere* xv. 129 They had gradually created..a most elaborate system of buying and selling, which has been called The world-market. **1898** G. B. SHAW *Perfect Wagnerite* 14 He is trusting to another great world-force, the Lie. **1904** W. JAMES *Ess. Radical Empiricism* (1912) i. 8 Experience, at this rate, would be much like a paint of which the world pictures were made. **1910** A. G. SPINK *National Game* 309 (*heading*) World champions. **1914** G. FRANKAU *Poet. Wks.* (1923) I. 185 Battlers for world-peace, slaves of Honour's lamp. **1920** B. RUSSELL *Pract. & Theory Bolshevism* IX. 109 The following passages [from article by Lenin] seemed to me..illuminating:—The present world-situation in politics places on the order of the day the dictatorship of the proletariat. **1921** J. C. MAXWELL GARNETT (*title*) Education and world citizenship. **1921** D. H. LAWRENCE *Sea & Sardinia* v. 163 Will the last waves of enlightenment and world-unity break over them [*sc.* the Sardinians] and wash away the stocking-caps? **1923** —— *Fantasia of Unconscious* 15 No more little Excelsiors crying world-brotherhood. **1927** A. CECIL *Brit. Foreign Secretaries* vii. 353 Aberdeen..would have felt all the talk about German world-domination too journalistically sensational to be politically probable. **1929** J. BUCHAN *Courts of Morning* 14 America... could not take a big hand in world affairs... She had too much to do at home. **1930** J. H. RANDALL (*title*) A world community. **1932** A. G. HERBERT tr. *Nygren's Agape & Eros* I. vi. 146 For Plotinus the whole world process is summed up in the double conception of the out-going of all things from the One..and the return of all things to the One. **1936** *Mind* XLV. 460 It was left to the Stoics to elaborate the conception of the world-state and of the world-citizen. **1936** World-culture [see *CULTURE *sb.* 5 d]. **1937** 'G. ORWELL' *Road to Wigan Pier* xii. 247 It is quite easy to imagine a world-society, economically collectivist. **1940** World opinion [see *APPEASE *v.* 2 c]. **1943** E. M. W. TILLYARD (*title*) The Elizabethan world picture. **1945** AUDEN *For Time Being* 89 Instead of Country Fair, there is World Market. **1946** J. S. HUXLEY *Unesco* i. 17 The task of unifying the world mind. **1948** L. SPITZER *Linguistics & Literary History* 220 This is also the main idea of Claudel's Spanish Catholic world-drama *Le soulier de satin*. **1954** 'M. COST' *Invitation from Minerva* 218 The world-press..was hourly dominated by bulletins of their plight. **1959** *New Yorker* 24 Oct. 185/1 Joyce..gave her..world rights to publish and sell 'Ulysses'. **1962** *Listener* 22 Mar. 498/1 Goethe, it will be remembered, spoke of *Weltkultur*... The great Russian component of world culture is as individual as our own or the French. **1966** S. BEER *Decision & Control* xv. 391 As usual, the study begins with a..world situation. **1967** P. D. JAMES *Unnatural Causes* I. xv. 116 He was completely unconcerned with world affairs. **1974** I. WALLERSTEIN (*title*) The modern world system..and the origins of the European world-economy in the 16th century. **1977** P. JOHNSON *Enemies of Society* v. 61 Britain expanded this initial overseas foothold by the Navigation Acts..which..constituted the beginning of an English-controlled world market. **1981** *Listener* 2 July 3/1 American columnists..are querying whether Reagan is not hazarding world peace. **1981** N. TUCKER *Child & Book* iv. 108 In Enid Blyton's work, this excessively simple world picture is carried to extremes. **1983** *Times* 14 May 8/3 The power of world opinion is a vital adjunct to non-violence.
25. world-all [tr. G. *weltall*], the world considered as a unit; the universe; **world-auxiliary**, a language (esp. an invented one) which may be used as a standard means of communication between speakers from different language communities throughout the world; cf. *auxiliary language* s.v. *AUXILIARY *a.* 2 a; **World Bank**, an international banking organization established to control the distribution of economic aid between member nations, *spec.*: † (*a*) the Bank for International Settlements, established through the League of Nations at Basle in 1930 (*obs.*); (*b*) the International Bank for Reconstruction and Development, affiliated to the United Nations and operational since 1946; **world-class** *a.*, applied to persons or things regarded as outstanding throughout the world; **World Court**, the International Court of Justice (formerly, the Permanent Court of International Justice, 1921–45), established in 1946 as the principal judicial arm of the United Nations; **World Cup**, in Assoc. Football, a quadrennial competition amongst national teams for the Jules Rimet trophy, first contested in 1930; also *transf.* in other sports; **world English** = *Standard English* s.v. *STANDARD *a.* 3 e; **world fair** = *world's fair*, sense 26 a below; **world ground**, the reality, or principle, that underlies the world; **World Health Organization**, an international body established in 1948 to promote co-operation between nations to improve health conditions (abbrev. *W.H.O.*: see *W 3); **world-historical** *a.* (earlier example); **world-language,** (*b*) (earlier example); **world-line** *Physics* and *Philos.* [tr. G. *weltlinie* (H. Minkowski, as for *world-point*)], the succession of points in space-time that are occupied by a particle; **world literature** [cf. G. *weltliteratur*], (*a*) a body of work drawn from many nations and recognized as literature throughout the world; (*b*) (the sum of) the literature of the world; **world-point** *Physics* and *Philos.* [tr. G. *weltpunkt* (H. Minkowski in H. A. Lorentz et al. *Das Relativitätsprinzip* (1913) 57)], a point in space-time; a particular point in space at a particular instant of time; **world-policy, -politics** [cf. G. *weltpolitik*]: also **world-political** *a.*; **world-ranking** *a.*, that ranks among the best in the world; **World Series**, a series of games contested annually as a play-off between the champions of the two major baseball leagues in the U.S.; also *transf.*; **World Service**, a B.B.C. radio service with a strong content of news and current affairs, broadcast principally for English-speaking listeners overseas (formerly called the *Overseas Service*); **world-spirit**, (*b*) (earlier example).
1847 J. D. MORELL *Hist. Philos.* (ed. 2) I. II. 369 Fichte founded a subjective idealism in which the me was the *world-all*. **1925** R. M. OGDEN tr. *Koffka's Growth of Mind* 347 For a child there is as yet no single world-all. **1927** E. S. PANKHURST *Delphos* v. 49 The world-auxiliary, used by everyone as a second language, will obviate the need for any other language save the native one. **1930** *Business Week* 28 May 9/2 French shares for the World Bank were offered publicly this week. **1943** *N.Y. Times* 5 Apr. 6/3 Senator Elmer Thomas said..that the establishment of a world bank founded on a standard international coin was inevitable. **1944** *St. Louis* (Missouri) *Post-Dispatch* 23 July 6A/3 (*heading*) Russia agrees to boost quota in World Bank. **1973** 'D. JORDAN' *Nile Green* xi. 47 The statistical boys..had worked out all the flaws in the World Bank report. **1950** *Sport* 22–28 Sept. 14/2 Such is the magnetism of world class heavyweight boxers! **1973** *Daily Express* 11 May 22/1 Keegan, looking every inch a world-class player..scored with a spectacular header. **1973** *Guardian* 22 Oct. 21/4 The timescale of astronomers is human and to those in world class research..too short to waste much time on bumbledom. **1979** *Beautiful British Columbia* Winter 40 A total of more than $100 million is expected to be spent at Whistler..to make the area competitive with world-class resorts in the United States and Europe. **1927** *New Republic* 21 Sept. 110/2 Our reservations to the resolution adhering to the World Court were received with an apathy which was next door to hostility. **1946** *N.Y. Times* 7 Feb. 8/2 (*heading*) 15 Judges elected for World Court. **1984** *Times* 10 Apr. 1/6 The Reagan Administration said yesterday it believed the World Court in The Hague did not have jurisdiction. **1950** B. WRIGHT *Captain of England* xvii. 154 For the first time in my experience the words 'World Cup' began to come into the discussions footballers have when they meet, and as the four British home associations had re-entered F.I.F.A., it was only to be expected that they would..enter this competition. **1954** *Times* 8 Nov. 10/1 (*heading*) Rugby League results..World Cup. **1967** *Times* 9 Nov. 16/4 (*heading*) World Cup golf. **1978** R. WESTALL *Devil on Road* xx. 185 An action-replay of a World Cup goal on telly. **1927** K. MALONE in *Amer. Speech* II. 323/2 He..warns against a slavish conformity to the dictionary, i.e., to the prescriptions of standard English, or world-English, as some people call it. **1980** *English World-Wide* I. I. 80 The categories or types of AVE..can be seen as existing across a scale having 'World English' ('book English', 'standard English', 'teachers' preferred English', &c.), at one end, and a national variety most distinct from it at the other. **1899** C. STUMPF *Let.* 8 Sept. in R. B. Perry *Thought & Char. W. James* (1935) II. 193 The tumult of a world fair—even the thought of it makes me nervous! **1978** P. BOARDMAN *Worlds of Patrick Geddes* vi. 179 A brief..account of the city [*sc.* Paris]..from pre-Roman times up to the greatest of world fairs. **1898** W. JAMES in *Psychol. Rev.* July 424 The world is evidently more complex than we are accustomed to think it, the 'absolute world-ground', in particular, being farther off (as Mr. F. C. S. Schiller has well pointed out) than it is the wont either of the usual empiricisms or of the usual idealisms to think it. **1948** *Scot. Jrnl. Theol.* I. 121 The most that science, working with its concepts of causation on a different level, can offer is a world-ground, or mind-energy at work in the world. **1946** *N.Y. Times* 28 June 9/1 The vanquished nations..with their large health problems, have acute need of the World Health Organization that the United Nations is creating. **1977** *New Scientist* 7 Apr. 3/2 All the more horrendous, then, are the statistics which the World Health Organisation has published to publicise World Health Day, which falls today (7 April). **1854** C. C. J. BUNSEN *Outl. Philos. Universal Hist.* I. 64 Both these researches, the philosophical and the world-historical, will be reserved for the second volume of our sketch. **1867** W. D. WHITNEY *Lang. & Study Lang.* xii. 469 If we expect..that our tongue become one day a world-language, understood and employed on every continent.., then it is our bounden duty [etc.]. **1916** *Monthly Notices R. Astron. Soc.* LXXVI. 700 The points of space occupied by a given material point at successive times form in the four-dimensional time-space a continuity of one dimension, which is called the world-line of the point. **1946** *Mind* LV. 146 The intersection of the world-lines AM and TM. **1962** *Listener* 27 Dec. 1095/3 According to Einstein's General Theory of Relativity it seems that objects which appear to be responding to the pull of gravity..are simply following the shortest available world-line through the space-time continuum. **1976** *Nature* 1–8 Jan. 30/2 The collapsing star is represented by some of its (time-like) worldlines including the worldline of its centre. **1831** CARLYLE in *Edin. Rev.* CV. 179 Instead of isolated, mutually repulsive National Literatures, a World Literature may one day be looked for? **1908** P. E. MORE *Shelburne Essays* V. 140 Longfellow brought from Germany the ideal of a world literature which should absorb the best of all lands. **1949** WELLEK & WARREN *Theory of Lit.* v. 41 The term 'world literature', a translation of Goethe's Weltliteratur, is perhaps needlessly grandiose. **1963** *English Studies* XLIV. 148 He has noted the widespread occurrence of the bond-story of *The Merchant of Venice* in world-literature. **1923** PERRETT & JEFFERY tr. H. Minkowski in *Lorentz's Princ. Relativity* v. 76 A point of space at a point of time, that is, a system of values *x, y, z, t*, I will call a world-point. The multiplicity of all thinkable *x, y, z, t* systems of values we will christen the world. **1930** L. SILBERSTEIN *Size of Universe* I. 1 The event thus localized in space..and in time..is called a worldpoint. **1967** R. A. GEORGE tr. *Carnap's Logical Structure of World* iv. 194 The points of n-dimensional, real-number space, we call *world points*. **1975** R. ADLER et al. *Introd. Gen. Relativity* (ed. 2) iv. 122 An event is a point in four-space: a world-point. **1936** World-political [see *anti-Comintern* s.v. *ANTI-[1] 3 a]. **1958** S. SPENDER *Engaged in Writing* vii. 133 He was the first world-political, international, intellectual man. **1970** *Daily Tel.* 19 Aug. 10/4 Here in Britain we have two world-ranking centres of radio astronomy. **1913** *Collier's* 4 Oct. 5/1 In this next impending world-series carnival between Giants and Athletics we have had the hunch [etc.]. **1951** *Time* 12 Mar. 59 For Norwegians..the Holmenkollen is the World Series, and stars such as Hoel and Björnstad are Norway's Di Maggios and Musials. **1973** M. WOODHOUSE *Blue Bone* iii. 20 'We could have played half the World Series by now.'..'Yes, we take three days to play a game of cricket.' **1966** *B.B.C. Handbk.* 83 The *World Service* addresses itself to those who understand English, wherever they happen to be—listeners throughout the Commonwealth and English-speaking people in other countries. **1981** *Times* 22 Jan. 8/7 The embassy press officer..was waiting for news from the World Service of the BBC. **1846** G. H. LEWES *Biogr. Hist. Philos.* IV. 212 The World-Spirit (*Weltgeist*) has at last succeeded in freeing himself from all incumbrances.
26. In the possessive. **a.** In senses corresponding to those at 24 and 25, as *world's championship, record, Series*; **world's fair** orig. *U.S.*, an international exposition of arts, science, industry, and agriculture.
1888 *Spaulding's Base-Ball Guide* 47 In 1887 the world's championship series had become an established supplementary series of contests. **1910** A. G. SPINK *National Game* 312 The world's championship. **1850** *New-England Farmer* II. 413 The State Board of Agriculture are making up a collection of samples of Indian corn for the World's Fair. **1908** E. TERRY *Story of my Life* xii. 280, I had loved the Chicago of the Lake with the white buildings of the World's Fair shining on it. **1982** J. S. BORTHWICK *Case of Hook-Billed Kites* iv. 193 Like those rides at world's fairs..where you sit in a little car that draws you through different habitats. **1893** *Outing* (U.S.) XXII. 154/2 He has..held the world's record in the pole vault for distance. **1905** *Sporting Life* 7 Oct. 3/1 Jack Sheridan and Hank O'Day have been appointed to umpire the world's series. **1925** F. SCOTT FITZGERALD *Great Gatsby* iv. 88 He's the man who fixed the World's Series back in 1919. **1965** F. O. DU PRE *U.S. Air Force Biogr. Dict.* 58/1 He won the Schneider Cup Race—the World's Series of seaplane racing—in 1925, with an average speed of 232 mph.
b. In hyperbolical phr. *the world's worst* (...), the very worst or most incompetent. *colloq.*
1921 T. WOLFE *Let.* 13 Nov. (1956) 22 'The Woman of Bronze', the world's worst play. **1929** J. B. PRIESTLEY *Good Companions* II. i. 248 She was easily the world's worst as a pianist. **1933** L. EINSTEIN in O. W. Holmes *Holmes-Einstein Lett.* (1964) 352, I hasten to add that they are the world's most famous bridge players and she the world's worst! **1954** R. BISSELL *High Water* i. 11 He shaved every other day and of all the Second Mates in the company they could have dumped on me he was the world's worst. **1962** C. DRAPER *Mad Major* iv. 88, I am probably the world's worst dancer.
c. In colloq. phr. *one of the world's workers*, an industrious person. Freq. in neg. contexts.
1933 [see *PULL *v.* 24 f]. **1964** D. GRAY *Devil wore Scarlet* x. 91 'Mr. Weston isn't one of the world's workers, exactly,' said Mary. **1976** G. MOFFAT *Short Time to Live* ii. 20 Jackson..is *not* one of the world's workers, as you must have noticed.

wo·rlded, (*ppl.*) *a.* *rare* (chiefly *poet.*). [f. WORLD *sb.* or *v.* + -ED.] Containing worlds. Also with qualifying word.
1885 TENNYSON *Tiresias* 167 The fires that arch this dusky dot—Yon myriad-worlded way. **1907** 'MARK TWAIN' in *Harper's Mag.* Dec. 44/2, I think there is such a planet..in one of the thinly worlded corners of the universe. **1934** DYLAN THOMAS *18 Poems* 31 How light the sleeping on this soily star, How deep the waking in the worlded clouds.

world-power. Add: **2.** (Earlier example.) Also *transf.* of a person.
1860 [see WORLD *sb.* 23 a]. **1900** *Congress Rec.* 29 Jan. 1259/1 We have become a 'world power'.

World War, world war. [f. WORLD *sb.* + WAR *sb.*[1]; cf. G. *weltkrieg*.] **1.** A war involving many important nations; *spec.* that of 1914–18 or of 1939–45.
1909 [see WORLD *sb.* 24 b]. **1914** B. VAUGHAN *What of To-Day?* xii. 103 What the South African War failed to teach I really believe this world-war will bring home to us. **1921** A. HUXLEY *Crome Yellow* ix. 82 Armageddon, that world war with which the Second Coming is to be so closely associated. **1949** G. A. BIRMINGHAM *Laura's Bishop* 171 To call this, when it comes, a world war is to minimize its importance. It will be worse than a world war. **1978** I. B. SINGER *Shosha* v. 90 Let's snatch a little peace before another world war breaks out.
2. In the designation of a particular (real or hypothetical) war, as *First World War, World War (No.) I* (or *One*): see *FIRST *a.* C. 2 a; *Second World War, World War (No.) II* (or *Two*): a subsequent world war, *spec.* that of 1939–45: see *SECOND *a.* 7 a; *Third World War, World War (No.) III* (or *Three*): see *THIRD *a.* 5. Also *transf.*
1919 *Manch. Guardian* 18 Feb. 10/2 (*heading*) World War No 2. **1939** *Time* 11 Sept. 38/1 Some of the diplomatic juggling which last week ended in World War II was old-fashioned international jockeying for power. *Ibid.* 18 Sept. 10/2 Exports of arms, munitions and related materials in World War I amounted..to only 25% of total exports to the Allies. **1945** DUKE OF BEDFORD *Let.* 16 Apr. in B. Russell *Autobiogr.* (1969) III. i. 44 You will have to postpone your visit until the brief interlude between this war & world-war no 3. **1947** *Time & Tide* 29 Nov. 1269/2 The despair and cynicism that followed what it has now become fashionable to call World War One. **1948** N. WIENER *Cybernetics* 7 When I came to the Institute after World War No. 1 [etc.]. **1959** *N.Z. Listener* 17 Apr. 6/1 Clearly the meaning of the treaties in case of wars which can be limited is somewhat different from the meaning they have in the event of World War III. **1963** D. BROUN *Egypt's Choice* i. 12 'During World War Deuce, sir,' the Colonel said. **1968** K. BIRD *Smash Glass Image* viii. 102 Rattling their rifles as if they were fighting World War Three. **1968** *Listener* 12 Dec. 787/3 When World War Two broke out on 3 September 1939, Monnet remembered his World War One experiences. **1976** P. R. WHITE *Planning for Public Transport* iv. 72 As early as World War One some minor stations and routes were closed.

world-wide, *a.* Add: Now freq. **worldwide**. **b.** as *adv.* (Later examples.)
1953 *Reader's Digest* July 27 World-wide, three million dogs have already safely got this..vaccine. **1972** *Nature* 24 Mar. 184/3 Workers in the field number no more than about fifty, worldwide. **1980** *Bookseller* 14 June 2528/1 (Advt.), Subscriptions manager required for expanding publishing business trading worldwide.

worm, *sb.* Add: **I. 3. e.** transf. and fig. phr. *worm's-eye view* [after *bird's-eye view* (BIRD'S-EYE *a.* 3); see also *EYE VIEW], a view taken as from the standpoint of a worm, i.e. from ground-level; a revealing or detailed perspective of a subject. Also *worm's-eye map* (Geol.) (see quot. 1972).
1908 [see *EYE VIEW, EYE-VIEW]. **1933** *Archit. Rev.* LXXIII. 67/2 The illustration is a worm's-eye view of a corner of the building. **1945** A. HUXLEY *Time must have Stop* xiv. 145 He..looked..up at the statue above him. What a curious worm's-eye view of a goddess! **1951** KRUMBEIN & SLOSS *Stratigr. & Sedimentation* xiii. 421 Such paleogeologic maps, in which the observer looks upward at the base of a higher unit, have been called worm's eye maps. **1960** *John o' London's* 14 Apr. 428/3 His 'worm's eye view' of Dublin was beginning to give way to the great vision of a major artist. **1964** *Bull. Amer. Assoc. Petroleum Geologists* XLVIII. 1187/2 A lap-out map, commonly known as a 'worm's eye' map, is a special method of paleogeologic expression where post-unconformity geologic relations are portrayed. **1972** *Gloss. Geol.* (Amer. Geol. Inst.) 797/1 *Worm's-eye map*, (*a*) a term applied..in reference to the pattern of formations that would be visible to an observer looking upward at the bottom of the rocks overlying a given surface. (*b*) A map showing overlap of sediments. **1982** A. PRICE *Old Vengeful* ix. 147 This is the worm's-eye view of what you seek. If you wish for the eagle's-eye view, you must go to Paris.
8. c. In colloq. phr. (*to open*) *a can of worms*, (to address) a complex and largely unexamined problem or state of affairs the investigation of which is likely to cause much trouble or scandal.
1962 *Times* 21 Feb. 12/4 He..knew that he had opened the bidding on what is sometimes called 'a can of worms'. **1969** *N. Dakota Law Rev.* XLV. 215 Counsel can..better comprehend..the domestic can-of-worms that appears in so many delinquency and neglect cases. **1973** *Times* 22 May 16/5 Mr Berger has opened, in the old American phrase, a fine can of worms. He is suggesting that an impeached President, should he be found guilty, could appeal to the Supreme Court. **1976** L. BERNSTEIN *Unanswered Question* vi. 418 There are so many of those 'underlying strings'..waiting to be tied up; so many cans of worms have been opened, and a lot of those slippery little beasts are still wriggling around. **1984** A. PRICE *Sion Crossing* vii. 137 Oliver isn't up to this sort of thing. And this is my can of worms.
IV. 17. a. *wormfinger*; instrumental, as *worm-chewed, -laid* adjs.; parasynthetic, as *worm-faced* adj.
1927 D. H. LAWRENCE *Mornings in Mexico* 28 Rattling the worm-chewed window-frames. **1934** DYLAN THOMAS *Let.* 12 Apr. (1966) 105 Avaunt, you worm-faced fellows of the night. **1922** JOYCE *Ulysses* 550 Jogging, mocks them with thumb and wriggling wormfingers. **1933** C. S. LEWIS *Pilgrim's Regress* 248 Once the worm-laid egg broke the wood.
b. *worm-tackle* (earlier example).
1835 *Chambers's Edin. Jrnl.* Jan. 390/3 First of all, the worm-tackle. For this, sizeable hooks..are generally preferred.
c. *worm-syrup* (later examples).
1897 *Sears, Roebuck Catal.* 27/2 *Worm syrup*..for expelling worms from children. **1972** E. WIGGINTON *Foxfire Bk.* 247 Take 'worm syrup' which is made by boiling Jerusalem oak and pine root together.
d. *worm-geared* adj.
1936 *Discovery* Aug. 238/2 It [*sc.* the camera] is loaded into position on the plane with a worm-geared winch and pulley system. **1973** *Gloss. Terms Materials Handling* (*B.S.I.*) VI. 16 *Worm geared chain pulley block*,..mechanical advantage is obtained chiefly by..use of a worm wheel and worm.
f. *worm-killer*, a preparation for destroying garden worms.
1915 H. H. THOMAS *Gardening for Amateurs* I. 22/1 Proprietary worm-killers can also be obtained, and these must always be employed as directed. **1959** *Times* 7 Mar. 9/1 There are always the lead arsenate wormkillers.

worm, *v.* Add: **II. 4. b.** To treat (an animal) with a preparation designed to free it of parasitic worms.
1932 N. MITFORD *Christmas Pudding* xi. 179 Lady Bobbin spoke to those about her of horses, hounds, and such obscure eventualities as going to ground..and being thoroughly well wormed. **1940** W. FAULKNER *Hamlet* IV. i. 276 He drenched and wormed and..drew the teeth of horses and mules. **1961** C. H. D. TODD *Pop. Whippet* 69 Having decided upon your puppy..ask if it has been wormed. **1978** *Detroit Free Press* 5 Mar. C 20/3 (Advt.), Collie Pups..wormed, pet or show.
IV. 13. b. To wind packing strips between (the cores of a multicore electric cable) so as to give a more nearly circular cross-section; also, to wind (conductors) together to form such a cable.
1909 COYLE & HOWE *Electric Cables* ii. 112 Prior to impregnating, the paper-insulated cores are laid up together and wormed with jute. **1953** C. C. BARNES *Power Cables* i. 6 The laid-up cores are wormed into circular formation and are armoured overall. **1982** KING & HALFTER *Underground Power Cables* ii. 31 These solid-type multicore cables are of belted construction, in which the conductors are separately paper-insulated, 'wormed' together and the interstices filled with a packing or filling of fibrous material in order to obtain a circular section.

wormer. Add: **4.** A preparation used to rid animals of worm infestations.
1934 in WEBSTER. **1971** *Farmer & Stockbreeder* 23 Feb. 45 (Advt.), An ideal wormer. It is highly effective against all roundworms in the gut. **1980** *Kenya Veterinarian* June p.v (Advt.), There's only one total spectrum wormer for sheep and cattle.

wormery (wɜ̄·ɪməri). [f. WORM *sb.* + *-ERY 2 b.] A place or container in which worms are kept.
1952 *Britannica Bk. Year* 667/1 *Wormery*, a place for breeding worms. **1972** *Daily Tel.* 10 July 14 We kept a wormery and during our observations we found the worms had pulled leaves down under the soil. **1980** M. DRABBLE *Middle Ground* 129 She complained about the smell of the hamster and rabbits, and thought the wormery disgusting.

worm-hole. Add: Also **wormhole**. **2.** *Physics.* A hypothetical interconnection between widely separated regions of space-time.
1957 MISNER & WHEELER in *Ann. Physics* II. 532 This analysis forces one to consider situations..where there is a net flux of lines of force through what topologists would call a handle of the multiply-connected space and what physicists might perhaps be excused for more vividly terming a 'wormhole'. **1978** PASACHOFF & KUTNER *University Astron.* xii. 326 Thus, in principle, mass that disappears in a black hole may emerge somewhere else. If the somewhere else is a distinct region in our universe, the connection is called a wormhole. **1981** P. DAVIES *Edge of Infinity* ix. 179 The quantum disturbance will be so severe that even the topology of spacetime will alter. Instead of a 'bumpy sheet', it will display a foam-like structure, full of worm-holes and bridges.

wormhood. (Later *fig.* example.)
1917 LD. BRAYE *Lines in Verse & Fable* 63 Extinction of all influence and fame, And abject knowledge of my wormhood.

worming, *vbl. sb.* Add: **2. b.** Treatment administered to rid an animal of parasitic worms.
1936 J. Z. RINE *Dog Owner's Man.* vi. 99 All worming may prove more effective if preceded by a twenty-four-hour diet of buttermilk. **1947** *New Biol.* III. 69 The cost and labour of rounding them [*sc.* sheep] up for this periodical 'worming' may be..great. **1981** *Times* 22 May 3/2 Worker cats needed neutering, vaccinating, worming, regular feeding.
6. b. The action of worming electric cables; also *concr.*, material used for this.
1909 COYLE & HOWE *Electric Cables* ii. 112 The specific gravity of the worming jute. **1949** *Proc. Inst. Electr. Engineers* XCVI. II. 633/1 Would the author indicate..in what sizes additional worming or padding is required to allow the inclusion of the 0.0225-in² conductor and yet produce a good cable design? **1962** P. DUNSHEATH *Hist. Elect. Engin.* xvi. 259 Much attention was given to such refinements as..construction of wormings.

wormish, *a.* For † *Obs.* read *rare* and add: (Later *fig.* example.) Also **wo·rmishness**, wormish or craven behaviour.
1923 V. WOOLF *Let.* 1 Apr. (1977) III. 26 Murry wrote me a wormish letter, by the way, about the differences between us, and our memories and so on. **1925** H. NICOLSON *Let.* 23 July in J. Lees-Milne *Harold Nicolson* (1980) I. xi. 239 My wormishness to Elizabeth.

worn, *ppl. a.* Add: **2. a.** Also *Comb.*, as *worn-looking* adj.
1918 MRS. B. LOWNDES *Out of War?* 48 Stern worn-looking man. **1978** D. MURPHY *Place Apart* vi. 120 They seem so watchful and worn-looking.

worried, *ppl. a.* Add: Also *Comb.*, as *worried-looking* adj.
1942 'N. SHUTE' *Pied Piper* 26 Howard saw him the first Saturday that he was there, a sandy-haired, worried-looking man of forty-five or so. **1982** T. HOLME *Devil & Dolce Vita* vii. 49 She's been a bit worried-looking... Distracted.
Hence **wo·rriedly** *adv.*, in a worried or distressed manner, concernedly.
1924 'L. MALET' *Dogs of Want* v. 125 She worriedly wondered whether green isn't a more trying colour than blue when you get hot. **1952** S. KAUFFMANN *Philanderer* (1953) xii. 195 'That's wonderful.' He looked at her worriedly. 'Only it's got to be your decision. Your responsibility.' **1976** 'R. GORDON' *Doctor on Job* iii. 18 'It won't take long, will it?' he asked worriedly.

worriment. (Earlier example.)
1833 S. SMITH *Life & Writings Major J. Downing* 161 I've had a good many head-flaws and worriments in my life time.

worrisome, *a.* Add: (Earlier example.)
1845 W. G. SIMMS *Wigwam & Cabin* 1st Ser. viii. 107, I..followed the old man into the house, with my feelings getting more and more strange and worrisome at every moment.
worrisomely *adv.* (later examples).
1973 *Newsweek* 23 Apr. 22/1 Three simultaneous crises..that seemed worrisomely different from those of the past. **1981** G. MCDONALD *Fletch & Widow Bradley* xiii. 51 Charley is a worrisomely tight man... Anything out of the ordinary rattles him.

worry, *sb.* Add: **4.** Special Combs. **worry beads**, a string of beads manipulated by the fingers as a means of occupying one's hands and calming the nerves; **worry lines**, lines or wrinkles on the forehead supposedly formed by a habitual expression of worry.
1964 in M. McLuhan *Understanding Media* viii. 78 You will notice that many Greek men..spend a lot of time counting the beads of what appear to be amber rosaries... They are *komboloia* or 'worry beads'. **1978** G. GREENE *Human Factor* v. i. 233 The man had a rosary in his lap and seemed to be using it like a chain of worry beads. **1985** *Observer* 3 Feb. 19/3 Sheikh Yamani, worry beads to hand, sums up the general feeling of unease as OPEC last week managed to preserve its fragile unity. **1972** 'J. QUARTERMAIN' *Rock of Diamond* xvi. 99 Worry lines creased his forehead. **1982** L. CODY *Bad Company* xiv. 102 There was grey in her hair and worry lines between her brows.

worry, *v.* Add: **5. c.** (Earlier example with *down.*) Also without *adv.* (*phr.*), to worry about (a problem, etc.) (*U.S. colloq.*).
1811 JANE AUSTEN *Sense & S.* II. x. 186 She was sometimes worried down by officious condolence to rate good-breeding as more indispensable to comfort than good-nature. **1959** N. MAILER *Advts. for Myself* (1961) 119 He had always asked too many questions, he had worried the task too severely. **1963** *N. & Q.* Dec. 443/1, I shall not worry the distinction between *alba* and *aube*. **1978** T. L. SMITH *Money War* I. 17 He had worried the chance meeting on the flight home.
7. c. Also in colloq. phrases, as *I should worry*: see *SHALL *v.* 18 d; *not to worry*: see *NOT *adv.* 4.
8. b. *to worry along* (earlier example).
1871 'MARK TWAIN' *Screamers* xxix. 146 My friend, you seem to know pretty much all the tunes there are, and you worry along first rate.
9. worryguts *dial.* and *colloq.* = next; freq. as a term of address; **worry wart** *colloq.* (chiefly *U.S.*), an inveterate worrier, one who frets unnecessarily.
1932 *Somerset Year Bk.* 83 The missis, who be a prapper worryguts. **1966** O. NORTON *School of Liars* iv. 72 He laughed. 'Worryguts!' 'I wasn't worried. I was just trying to be efficient.' **1982** D. PHILLIPS *Coconut Kiss* ix. 94 It's all right..isn't it?' I asked. ''Course it is, Worryguts,' said Vera. **1956** I. BELKNAP *Human Problems of State Mental Hospitals* x. 177 The persevering, nagging delusional group—who were termed 'worry warts', 'nuisances', 'bird dogs', in the attendants' slang. **1974** J. HELLER *Something Happened* 445 'Don't be such a worry wart.' 'Don't use that phrase. It makes my skin prickle.'

worse, *sb.* **3. c.** (Later examples.)
1739–40 RICHARDSON *Pamela* (1740) I. xxvii. 85 How easy it is to go from bad to worse, when once People give way to Vice. **1894** SOMERVILLE & 'ROSS' *Real Charlotte* I. vii. 87 The land went from bad to worse. [**1930** F. A. POTTLE *Stretchers* 64 The weather was warm, and if worse came to worst, we could encamp in our pup tents where we were.] **1961** NEW ENG. BIBLE *2 Tim.* iii. 13 Wicked men and charlatans will make progress from bad to worse.

worship, *sb.* Add: **III. 10.** *worship service.*
1954 *Grove's Dict. Mus.* (ed. 5) VIII. 10/2 In the worship services of those groups among which the urban urge has been less evident, they [*sc.* gospel songs] have been immensely useful now for nearly a century. **1978** R. M. NIXON *Mem.* 538 On our first Sunday in the White House we held the first White House worship service in the East Room.

worship, *v.* Add: **1. b.** (Earlier examples of phr. *to worship the ground* (one) *walks* or *treads on.*)
1848 A. BRONTË *Tenant of Wildfell Hall* II. viii. 147 As to looking askance to another woman—he's safe enough for that..he worships the very ground I tread on. **1854** DICKENS *Hard Times* III. iii. 287 There are ladies—born ladies..who next to worship the ground I walk on.

wo·rsification. *rare.* [Humorous corruption of *versification*, as if f. WORSE *a.* and -FICATION.] The composition of bad verses; poor versification.
1849 J. R. LOWELL in *Mass. Q. Rev.* Dec. 51 Since we have found fault with some of what we may be allowed to call the worsification, we should say that the prose work is done conscientiously and neatly. **1908** *Let. to F. J. Furnivall* 27 Nov., The worsification of the poetry written in younger days is far more complete and thorough in the Italian and French poets [*sc.* Tasso and Ronsard] than in the English one [*sc.* Langland].

worst, *a.* Add: **2. c.** (Later example.) Also, *the worst way.*
1904 *N.Y. Tribune* 26 June (Illustr. Suppl.) 4/4 'So you want to go to Cuba, do you?' asked Colonel Roosevelt. 'I do, worst kind,' replied McShane. **1914** G. ATHERTON *Perch of Devil* I. 55, I need new duds the worst way.
d. *worst-case* adj. phr.: that is or pertains to the worst of a number of possibilities.
1964 R. F. FICCHI *Electrical Interference* ii. 18 It is first assumed, using a worst-case analysis technique, that the mean beam of the receiver and transmitter antenna are in direct line of sight. **1979** R. LITTELL *Debriefing* v. 88 Worst-case contingency planning is still the basis of scenario construction. **1980** *Times* 18 Jan. 14/1 Analysts believe that the Kremlin drew up a 'worst-case' scenario which took into account both an embargo on American grain and a threat to the Moscow Olympics. **1985** *Harper's Mag.* Jan. 68/2 Pickens could spin off a royalty trust, perhaps sell the downstream operations. . Such a move had been possible all along, but it was obviously the worst-case method of going about the task.

worst, *adv.* Add: **c. worst-seller**, a book distinguished commercially by its low sales (opp. *best seller* s.v. *BEST *adv.* 3 b); the writer of such a book; also **worst-selling** *a.*
1924 O. SITWELL *Triple Fugue* 73 Could a written testimonial be obtained from the shades of..Dryden,..Gray, Keats..and from their heirs, the worst-sellers of to-day, it is probable that the purport..would be found..to be remarkably alike in every case. **1925** V. WOOLF *Common Reader* 262 There is..the best-seller public and the worst-seller public. **1933** T. E. LAWRENCE *Let.* 17 Dec. (1938) 783, I confess to a lively apprehension of that potential worst-seller of yours. **1956** A. HUXLEY *Adonis & Alphabet* 120 If there were no 'angels', there would be no worst-selling literature to leaven the enormous lump of intellectual and artistic conformity. **1980** 'J. GASH' *Spend Game* vi. 65 A tatty copy of the world's worst-seller like Dr Chase's book.

worsted, *sb.* Add: **2. c.** *ellipt.* for a garment made of worsted cloth; a worsted jacket or suit.
1962 L. DEIGHTON *Ipcress File* i. 9, I struggled into the dark worsted and my only establishment tie. **1972** K. BONFIGLIOLI *Don't point that Thing at Me* iii. 20, I put on a dashing little tropical-weight worsted, curly-brimmed coker and a pair of buckskins. **1975** *Times* 8 Jan. 12/7 Behaving in a manner more suited to the canvas jacket than the charcoal-grey worsted.
5. worsted work, embroidery done with worsted yarn on canvas; an example of this; hence **worsted-worked** *a.*
c **1702** C. FIENNES *Journeys* (1947) 277 One with a half bedstead as the new mode, dimity with fine shades of worsted works well made up. **1826** M. WILMOT *Let.* 25 Sept. (1935) 250, I do *worsted work*..but..my eyes are too weak to count the threads of any but coarse canvas... What do you make of yours? Foot stools! cushions! bell pulls! **1888** MRS. H. WARD *Robert Elsmere* I. I. x. 280 His wife, whose head was bent close over her worsted work. **1853** MRS. GASKELL *Cranford* viii. 116 Carlo lay on the worsted-worked rug.

wort, *sb.*[1] Add: **4.** *wort-blue* adj.
1933 AUDEN in *Rev. Eng. Stud.* (1978) Aug. 304 Wound round neck the wort-blue tie.

‖ **Wörter und Sachen** (vö·ɹtəɹ unt za·χĕn). [Ger., words and things (R. Meringer *Indogermanische Forschung* (1904) XVI. 101).] (See quot. 1964.) Freq. *attrib.*
1937 J. ORR tr. *Iordan's Introd. Romance Linguistics* i. 73 Their researches imply the investigation of a variety of cultural influences and exchanges which entitles them to be associated..with the Wörter und Sachen movement. **1957** *Language* XXXIII. 54 The twin methods of 'Wörter-und-Sachen' and 'Sachen-und-Wörter', devised half a century ago by..Meringer and Schuchardt..require no formal introduction at this late date. **1964** R. H. ROBINS *Gen. Linguistics* 79 A special aspect of dialect study is known as *Wörter und Sachen*... This involves the detailed study in different dialects of the forms of words relating to material objects and processes. **1976** *Amer. Speech* 1973 XLVIII. 164 The Wörter-und-Sachen technique has been used for five decades by Swiss and German dialectologists to probe the relationship between words and the objects and processes they describe.

worth, *a.* Add: **I. 3. c.** (Earlier example.)
1875 'MARK TWAIN' *Sk. New & Old* 310 We shall fly our comet for all it is worth.
3. d. *for what it is worth*, a dismissive phr. intimating that something (esp. an accompanying statement) is of uncertain or little value. Often *parenthetically*.
1888 J. ROSS *Three Generations* 228 There is my opinion; I give it for what it is worth. **1922** F. HARRIS *My Life & Loves* I. xv. 327 However, the fact is so peculiar that I insert it here for what it may be worth. **1952** M. NORTON *Borrowers* xx. 157 'Well,' she conceded at last. 'I'll tell you. For what it's worth.' **1962** A. HUXLEY *Let.* 19 Jan. (1969) 928 Laura brought up the idea suddenly and it fired my imagination. So here, for what it is worth, it is. **1979** J. JOHNSTON *Old Jest* 152 You're going to have to decide which side you're on. Nancy, for what it's worth, seems to have made her decision.

worthy, *sb.* **1. c.** (Later *transf.* example.)
1906 G. B. SHAW *Let.* 29 Sept. (1972) II. 657 The points we cannot accept. These are .2. The triumvirates, on the ground that the nine worthies cannot be found to take the responsibility.

wot (wǫt), non-standard written form of WHAT *pron.*, *a.*[1], etc.
1829 [see SLAP-UP *a.* b]. **1865** [see WATER *sb.* 6 f]. **1898** [see *CHIVVY *sb.*]. **1925** [see *GARDEN *sb.* 1 f]. **1949** E. POUND *Pisan Cantos* lxxvii. 50 I'll tell you wot izza comin'. **1972** 'H. CARMICHAEL' *Naked to Grave* v. 60 He's going to have a tough job convincing the police he wasn't the one wot done it.
b. In phr. *wot, no* —— *?*: orig. (in the war of 1939–45) a catchphrase protesting against shortages, written as the caption accompanying a Chad (see *CHAD); now also in extended humorous use.
1945 *Sunday Express* 2 Dec. 2/3 Chad is the Watcher... He peers over walls and asks, 'Wot, no...?' **1946, 1950** [see *CHAD, CHAD]. **1958** J. TOWNSEND *Young Devils* ii. 16 A rusty drawing-pin supported an old Teachers' Union notice. It had scribbled across it 'Wot, no money?' **1979** K. CONLON *Move in Game* I. v. 64 Joanna sent a postcard which said, 'Wot no tulle and confetti?'

wotcher (wǫ·tʃəɹ), *int.* Colloq. corruption of 'what cheer?' (CHEER *sb.* 3 b), a familiar greeting.
1894 A. CHEVALIER *Humorous Songs* 4 'Wot cher!' all the neighbours cried, 'Who're yer goin' to meet, Bill?' **1899** *North-China Herald* 13 Nov. 962/3 (Advt.), 'Wot Cher, Mate?' may be a rough form of salutation. **1928** *Granta* 2 Nov. 71 (*caption*) Wotcher! **1954** J. MASTERS *Bhowani Junction* xxxii. 279 Howland waved violently to Victoria..and shouted, 'Wotcher, Vicky!' **1980** 'J. GASH' *Spend Game* xvi. 162 'Hello, Lovejoy.' 'Wotcher, love.'

would, pa. t. of WILL *v.*[1]

woulda (wu·dă), repr. U.S. dial. pronunc. of 'would have'.
1913 *Dialect Notes* IV. 6 *Would a went*, would have gone. **1925** T. DREISER *Amer. Trag.* (1926) I. I. xii. 83, I coulda chucked my job, and I woulda. **1952** B. MALAMUD *Natural* 103 If it was something serious you woulda been caught long ago. **1978** G. VAUGHAN *Belgrade Drop* x. 63 Security woulda got her. She'd never have the chance to tell DI.

would-be, *a.* Add: **c.** (Earlier example.)
1813 JANE AUSTEN *Let.* 11 Oct. (1952) 343 A large, ungenteel Woman, with self-satisfied & would-be elegant manners.
e. Used *predicatively*: mannered, pretentious. (App. restricted to the works of D. H. Lawrence.)
1922 D. H. LAWRENCE *Lett.* (1932) 556 These drawings are so completely without irony, so crass, so strained and so would-be. **1928** *Ibid.* 751 James Joyce bores me stiff—too terribly would-be and done-on-purpose, utterly without spontaneity or real life. **1932** A. HUXLEY in *Lett. D. H. Lawrence* p. xvii, The symphony oppressed

him; it was too big, too elaborate, too carefully and consciously worked out, too 'would-be'—to use a characteristic Lawrencian expression. He was quite determined that none of his writings should be 'would-be'. He allowed them to flower as they liked from the depths of his being.

wound, *sb.* Add: **8. a.** *wound-fever* (earlier example); *wound-dressing* (concr.: cf. 8 b in Dict.). **b.** *wound-healing.*

1887 T. Longmore in J. B. Hamilton *Trans. Internat. Med. Congress, 9th Session* III. 117 Primary wound dressings shall be available at all times and in all places. 1959 *First-Aid Boxes in Factories Order* 21 May in *Stat. Instruments 1959* (1960) I. 1266 A sufficient number (not less than twelve) of adhesive wound dressings of an approved type and of assorted sizes. 1976 D. Francis *In Frame* xi. 159 The outer bandages proved to be large strong pieces of linen..just below my shoulder blade, a large padded wound dressing. 1863 L. M. Alcott *Hospital Sk.* iv. 51, I..recognized a certain Pennsylvania gentleman, whose wound-fever had taken a turn for the worse. 1949 M. Mead *Male & Female* x. 216 The resistance against certain diseases, the wound-healing capacities of a whole people, may depend upon the meticulousness with which they use learned, not specific inherent, capacities. 1964 *Oceanogr. & Marine Biol.* II. 409 Under conditions of wound-healing and repair the normal inhibitor of melanogenesis present in the rest of the animal is overcome or absent.

9. **wound-herb** = Woundwort; **wound hormone** [tr. G. *wundhormon* (G. Haberlandt 1921, in *Sitzungsber. d. Preuss. Akad. d. Wissensch.* 222)], a substance that is produced in a plant in response to a wound and stimulates healing; cf. *traumatic acid* s.v. *traumatic *a.* 3; **wound stump** = Cicatrix 2; **wound-tumour disease,** a plant disease marked by tumours on roots, stems, or leaves and enlargement of veins and caused by the *wound-tumour virus, Aureogenus magnivena,* which is transmitted by leaf-hoppers.

1597, etc. Wound-herb [in Dict., sense 8 a]. 1955 A. L. Rowse *Expansion of Elizabethan England* i. 6 Scottish practitioners flocked..to gather simples and wound-herbs. 1977 *Irish Press* 29 Sept. 10/1 The Yarrow was principally used by herbalists as a wound-herb. 1921 *Chem. Abstr.* XV. 2914 Exptl. evidence exists that the action of a wound as a stimulus in exciting cell division is due to decompn. products of the mechanically injured or dead cells. These products function as wound hormones. 1966 R. M. Devlin *Plant Physiol.* xvii. 427 Most plant tissues do not respond to traumatic acid, suggesting that it may be a specific wound hormone for bean-pod tissue. 1923 D. H. Lawrence *Birds, Beasts & Flowers* 52 Yet see him fling himself abroad in fresh abandon From the small wound-stump. 1945 L. M. Black in *Amer. Jrnl. Bot.* XXXII. 408/1 It now seems that the terms 'wound-tumor virus' and 'wound-tumor disease' may be more appropriate and distinctive. 1967 K. M. Smith *Insect Virol.* xi. 219 A quick method of detecting the wound-tumor virus in the leafhopper is by staining the hemolymph smears of the insect with the D (dialysis) conjugates.

woundable. (Later examples.)

1975 *Times Lit. Suppl.* 21 Mar. 293/1 When trauma appears on the scene.. Scully shows that he is as woundable as the next boy. 1976 *Sunday Times* (Colour Suppl.) 22 Feb. 39/1 He is also shy, loyal and surprisingly woundable.

wou·nd-down, *a.* [f. *wound,* pa. pple. of Wind *v.*¹ + Down *adv.*] That has undergone winding down (see *wind *v.*¹ 20*); that has been lowered by winding.

1939 Dylan Thomas *Map of Love* 20 The wound-down cough of the blood-counting clock. 1974 P. McCutchan *Call for Simon Shard* xii. 110 He sniffed through the wound-down window. 1984 W. Garner *Rats' Alley* xi. 218 He could hear its [*sc.* the helicopter's] racket through the wound-down window of the car.

wound-up, *a.* Add: (Earlier example.) Also in sense 22 a of Wind *v.*¹

1788 J. Woodforde *Diary* 27 Nov. (1927) III. 68 Ben returned about 4 o'clock this Aft. completely wound up, eat no Dinner but went directly to sleep in a Chair. 1973 A. Grey *Some put their Trust in Chariots* v. 22 The wound-up windows of the car.

woundward (wū·ndwǫ̈ɹd), *a. nonce-wd.* [f. Wound *sb.*: see -ward.] Towards wounds or wounding.

1946 Dylan Thomas *Deaths & Entrances* 48 The woundward flight of the ancient Young from the canyons of oblivion!

wourali, var. Woorali.

wove, *ppl. a.* and *sb.* **1. b.** (Earlier example.)

1806 H. Fourdrinier *Brit. Patent 2951* 4 A number of moulds of the description called laid or wove, any number of which..are capable of forming one long mould.

woven, *ppl. a.* **2.** (Later example of Comb.)

1904 Hardy *Dynasts* I. 5 O woven-winged squadrons of Toulon..draw westward Ere Nelson be near!

wow, *sb.*¹ Add: **2.** Fluctuations in pitch in reproduced sound that are sufficiently slow to be heard as such in long notes; a property in a reproducer that gives rise to this, esp. uneven speed.

1932 *Wireless World* 16 Mar. 277/2 Wobble or 'wow'—to use the expressive American term..is not so troublesome nowadays, most modern sound cameras having anti-wow mechanism. 1942 *Electronic Engin.* XIV. 640/1 The principal snag of sub-standard projectors, that of speed variations, is well cared for in so far as relates to low-frequency variations, which are known as 'wow' and not, as stated, warble. 1960 K. Amis *Take Girl like You* xxiii. 271 Every couple of weeks Graham found out some new way of reducing distortion or filtering off surface noise or eliminating wow. 1968 *New Scientist* 20 June 615/1 The Lick Observatory conclusion was entirely spurious, the result of undetected 'wow' in the tape recorder used. 1971 *Wireless World* Oct. 478/1 Wow can be caused by a badly eccentric or warped record. 1982 *Listener* 16 Dec. 34/2 Insist on listening to some music, preferably piano music that shows up wow and flutter especially well.

wow (wau), *sb.*³ and *a.* *slang* (orig. *U.S.*). [f. Wow *int.*] **A.** *sb.* A sensational success. Freq. const. *of.*

1920 *Collier's* 11 Dec. 21/1 In Round Five they stalled some more... The sixth innin' was a wow! 1926 [see *stop *v.* 21 c]. 1927 Wodehouse *Small Bachelor* vi. 94 'A friend of mine tipped me off that this company was a wow'. 'A what?' 'A winner. He said it was going to be big and advised me to come in on the ground floor.' 1944 S. Bellow *Dangling Man* 54 What a wow of a finish. 1954 C. Churchill *Let.* 1 Sept. in M. Soames *Clementine Churchill* (1979) xxvi. 445 Mr Graham Sutherland is a 'Wow'. He really is a most attractive man. 1962 V. Connaught *Secret Heart of Princess Alexandra* vii. 73 From that moment forward, she was a wow with every Australian in the land. 1983 D. Francis *Danger* xvii. 236 Chattering guests all having a wow of a time.

B. *adj.* Exciting or expressing admiration and delight.

1921 *Variety* 9 Dec. 31 The *wow* comedy song... '*Say It With Liquor*'. 1962 *John o' London's* 1 Mar. 211/2 A chorus of wow reviews from international critics. 1972 *Daily Colonist* (Victoria, B.C.) 13 Feb. 27/4 Two-foot-high letters inviting you to buy Vitamin E capsules, often at wow potencies, plaster the fronts of drug stores.

wow (wau), *v.*² *slang* (orig. *U.S.*). [f. Wow *int.*, *sb.³] *trans.* To make enthusiastic, to impress or excite greatly (esp. an audience).

1924 *Variety* 24 Dec. 14/5 He doesn't wow 'em at any time and seems misplaced in the show. 1938 E. B. White *Let.* 20 Dec. (1976) 191 Your Hollywood visit piece (which I had never read) wowed me. 1949 *Time* 19 Sept. 45/3 She wowed them with a dramatic reading of the death scene from *Romeo and Juliet.* 1950 Blesh & Janis *They all played Ragtime* ii. 44 The ragtime pianists were already 'wowing' their audiences with syncopated renderings of the classics. 1961 *Sunday Express* 12 Mar. 14/5 Yet *another* new look..wowed London last week. 1980 *Times* 17 June 13/3 They are unlikely to wow anybody who does not already respond to Burne-Jones's rather wan charms. 1984 *Daily Tel.* 25 Sept. 11/4 Mr Macdonald, who supplied the off-screen commentary for this year's Channel 4 coverage of the SDP conference, had the bright notion of training up a novice speaker who would wow them at Buxton.

wow, *int.* Restrict 'Chiefly *Sc.*' to sense 1 and add: **2.** (Later examples.) Now chiefly expressing astonishment or admiration.

1896 Ade *Artie* 8 'The girls—wow!' 'Beauties, eh?' 'Lollypaloozers!' 1931 R. Campbell *Georgiad* i. 11 Bang on your nose my spectacles appear And (Wow!) an ear-ring slits my tender ear. 1941 J. D. Carr *Case of Constant Suicides* i. 17 A brown-haired girl..straightened up to stare at him. 'Wow!' said Alan inaudibly. 1962 E. Cleaver in A. Dundes *Mother Wit* (1973) 20/1 Wow, what a sight that would be! 1980 'R. B. Dominic' *Attending Physician* xxiv. 217 'Wow!' Mike Isham whistled reverently. 'No wonder she was willing to murder.'

Also **wowee·** († **wowey**) *int.*, in the same sense.

1921 S. Ford *Inez & Trilby May* xvi. 279 Think of the row that will start when it comes out that this is an inside job, with a princess playing the star part. Wowey! 1963 *Mad Mag.* July 23/2 Boy! Wow-wee! That's quite an exciting evening line-up! 1975 R. H. Rimmer *Premar Experiments* (1976) iii. 216 Bren was jubilant. 'My big sister is pregnant. Wow-ee! Unbelievable! Fantastic!' 1981 R. Barnard *Sheer Torture* iii. 29 He had served on the Arts Council Music Panel, 1958–60. Wowee!

wowser. For *Australia* read 'orig. *Austral.* (now chiefly *Austral.* and *N.Z.*)' and add: Also **wowzer.** Perhaps most commonly in sense 'a fanatical or determined opponent of intoxicating drink'. (Earlier and later examples.)

1899 *Truth* 8 Oct. 5 Willoughby 'Wowsers' Worried. The 'Talent' get a 'Turn'... Ten young men were fined.. for having behaved in a riotous manner on the Military Rd. 1918 *Chrons. N.Z.E.F.* 79/2 The Wowzers look with disdain upon the 'Come and have a spot, old boy' kind of welcome. 1937 R. A. Knox *Double Cross Purposes* ii. 40, I hope to God the old wowzer didn't see we'd been at it. He's as sharp as a weasel. 1939 R. Campbell *Flowering Rifle* i. 18 Where wowsers may discharge their wondrous lore Who'll 'fight for peace', and yet disarm for war. 1939 X. Herbert *Capricornia* xviii. 233 Only ones't got a victory outer that flamin' war was the blasted wowsers! 1941 C. Barrett *Coast of Adventure* iii. 56 Men without vision or liberal views. We might call them 'wowsers' today. 1949 D. M. Davin *Roads from Home* i. v. 72 The good old days were gone. And it wasn't only the wowsers that had spoilt everything. 1961 *N.Y. Times* 12 Feb. 36/1 A relentless wowser (prohibitionist)..banned saloons. 1963 *Economist* 3 Aug. 422/2 Some alien wowser such as Senator Goldwater. 1970 *N.Z. Listener* 12 Oct. 13/2 A bit late to find that out, you snobbish ratbag wowser. 1975 D. Stuart *Walk, Trot, Canter & Die* xxiii. 137 But now in me old age I'm a wowser an' I don't have to worry about anything. I'd have a bottle of whisky in the packs... but most trips it wouldn't even get opened, so you can't say I was a drunkard. 1977 *Times* 18 Mar. 18/7 This country's pattern of..licensing hours..is the work of wowzers of every description. 1981 M. Gee *Meg* xiv. 142 You won't drink with me, will you? Didn't think so. I'm stuck in a family of wowsers. 1982 *Times* 7 Dec. 12/6 These authoritarian wowsers would like to see a law forbidding anybody to watch any programme they disapprove of.

b. *attrib.* and *Comb.*

1934 *Bulletin* (Sydney) 12 Sept. 11/4 That morbid sex curiosity which is the curse of wowser-ridden communities. 1936 R. Campbell *Mithraic Emblems* 125 Let Spender over wowser-problems fret. 1966 G. W. Turner *Eng. Lang. Austral. & N.Z.* i. 22 The wowser tradition is stronger in New Zealand, where there is less of the bush tradition; it usually takes the form of a cautious anxiety about life rather than open Puritanism. 1969 *Age* (Melbourne) 24 May 2/5 Sir Henry, while making his attack on NSW and its poker machines, maintained that Victoria was not a 'wowser' State. 1978 P. H. Johnson *Good Husband* xxii. 190 There were 'no smoking' notices, at which Maisie demurred... 'I find this wowser activity uncomfortable.'

Hence **wow·serish** *a.*, of the nature of a wowser; puritanical; **wow·serism,** the practice or beliefs of a wowser or wowsers.

1909 Wowserism [in Dict.]. 1933 F. Clune *Try Anything Once* 122 They had lost their dash and grown wowserish. 1936 M. Franklin *All that Swagger* l. 377 Novels which.. provoked attention by their unreticent details of bodily functionings and provided capital sport when flung in the face of *fin de siècle* wowserism. 1966 *New Statesman* 4 Mar. 289/2 After a time, I warmed even to Stennis and Morse; though the first was narrow and wowserish. 1966 G. W. Turner *Eng. Lang. Austral. & N.Z.* i. 21 Wowserism is prohibition elevated into a philosophy. 1971 *Observer* 14 Mar. 15/2 Australia's peculiar 'wowserism', the often unpredictable censorship of films, books and shows by the authorities, is largely a function of older Australia. 1983 *Age* (Melbourne) 3 Dec. 11/2 Coming hard on the heels of the casino inquiry, which also recommended in the negative, the Government's decision on poker machines may give it a puritanical or wowserish image. 1984 *Daily Tel.* 24 Aug. 10/4 When Pierre Trudeau visited Australia in 1970 he told the natives: 'You have wowserism; we have Toronto.'

wow-wow, var. *wah-wah².

W particle: see *W 4 b.

W.R.A.C. Also **WRAC** and (*colloq.*) **Wrac** (ræk). [f. initial letters of its name.] The Women's Royal Army Corps, formed as the women's corps of the British Army in 1949 to replace the A.T.S.; also, a member of this corps.

1949 *Times* 1 Feb. 2/3 (*heading*) Inauguration of W.R.A.C. and W.R.A.F. *Ibid.*, The King has given orders for the following appointments in the W.R.A.C.:—The Queen to be Commandant-in-Chief the Women's Royal Army Corps. 1950 *R.A.F. Rev.* Sept. 7/2 They have defeated the W.R.A.C. in every championship shoot since the women's inter-Services championships were started in 1947. 1956 R. Macaulay *Towers of Trebizond* xv. 179 Now the women who go with armies are not encouraged to be so useful to them, they are called Ats and Wrens and Waafs and Wracs and are kept behind the battle lines and are only a small consolation to the troops. 1982 *Whitaker's Almanack 1983* 482 Retirement Benefits... The annual rates for W.R.A.C. were given; these apply to equivalent ranks in all Services. 1984 *Daily Tel.* 7 Apr. 8/1 'You are following in the footsteps of great men,' said the Duchess, Controller-Commandant of the WRAC.

wracked, erron. var. Racked *ppl. a.*³ (Later examples.)

1974 *Times Lit. Suppl.* 19 Apr. 417/4 Lowry could be sodden, sullen, wracked with shame and remorse: a figure of total anguish. 1974 A. Davis *Autobiography* iv. 279 During the few months of our friendship, I don't think I realized how wracked he must have been by that decade of accumulated frustrations ,by that terrible sense of impotence.

W.R.A.F. Also **WRAF** and (*colloq.*) **Wraf** (ræf). [f. initial letters of its name.] The Women's Royal Air Force (1918–20, reformed 1949), the women's corps of the Royal Air Force; also, a member of this.

1918 *Times* 3 Apr. 4/4 No women for the 'mobile' branch will be enrolled at present. Candidates for posts as officers should apply to the W.R.A.F. Inquiry Office,.. Strand. 1921 *Spectator* 4 June 719/2 Sketches with pen and pencil of the duties of the 'Wrafs'. 1955 *Times* 23 May 6/6 Five women holding W.R.A.F. commissions, all with university degrees, are serving as technical officers in the R.A.F. 1977 *R.A.F. News* 11–24 May 2/5 A charity disco at Newton enabled the WRAF girls there to raise £58. 1983 *Daily Tel.* 22 Aug. 8/5 Dame Felicity Peake was director of the WRAF from its inception in 1949.

wraggle-taggle, var. *raggle-taggle *a.* and *sb.*

wraith, *sb.* Add: **4.** *wraith-ship.*

1924 V. F. Boyson *Falkland Islands* viii. 181 Dimly as she came, so she passed away, as though in very truth the wraithship said to appear at every British naval fight.

wraithly (rēⁱ·þli), *a. rare.* [f. Wraith *sb.* + -ly¹.] Resembling a wraith, wraith-like.

1909 M. B. Saunders *Litany Lane* i. iv. 41 The tinkle-tinkle of a wraithly Tom Moore singing flowery love-songs at fluted-silked pianos.

wrangle, *sb.* Add: **1.** Also *fig.*

1931 Blunden *To Themis* 22 Rumour multiplies the wrangle of wheels and clash of hoofs abroad.

wrangle, *v.* **4.** Delete † *Obs. rare* and add: To obtain by wrangling. (Later example.)

1934 in Webster. 1976 *National Observer* (U.S.) 31 Jan. 1/3 The pall of snowdrifts and ice would have impeded reinforcements' marching even if Arnold had been able to wrangle help from American Brig. Gen. David Wooster, a procrastinator who then occupied Montreal.

7. (Earlier and later examples.)

1899 F. Remington *Sundown Leflare* 11 De herd, which was more horses..dan ten men kin wrangle. 1952 H. Innes *Campbell's Kingdom* i. ii. 38 He wrangles a bunch of horses and acts as packer for the visitors in the summer

wrangled, *ppl. a.*¹ Add: Cf. *rangled.

wrangler. Add: **3.** Also **Wrangler.** A proprietary name for jeans. Freq. *pl.* orig. *U.S.*

1947 *Official Gaz.* (U.S. Patent Office) 16 Dec. 395/2 Blue Bell, Inc., Greensboro, N.C... *Wrangler.* For Western style dungarees and pants. Claims use since Jan. 19, 1929. 1963 *Trade Marks Jrnl.* 16 Oct. 1478/1 *Wrangler*... Articles of protective clothing..and articles of sports clothing... Kilgour & Walker Limited.., Aberdeen; manufacturers. 1966 *Ibid.* 7 Sept. 1313/2 *Wrangler*... Articles of protective clothing..; trousers, jeans [etc.]. Blue Bell, Inc.., Greensboro, State of North Carolina. 1972 *New Society* 13 Apr. 68/2, I suggested.. that the girls ..might like to travel in jeans and change into their best gear when we stopped for lunch. My idea.. produced..a unanimous decision..that they were going in their wranglers anyway. 1978 D. Bloodworth *Crosstalk* xxvii. 211 A tall blond youth in wranglers and ringlets. 1981 C. Watson *Bishop in Back Seat* xxi. 131 He was in his regulation costume: Wrangler jeans, soiled Stetson.

wrap, *sb.* Add: **1. c.** Material used for wrapping, esp. very thin plastic film.

1930 *Food Industries* Jan. 13/1 [Cellophane] was first introduced into this country from France as a wrap for candy. *Ibid.*, The transparent type of wrap proved decidedly popular. 1958 *Chain Store Age* Apr. 168/3 The Aluminum Company of America is now packaging Alcoa wrap in a 'flat pak'. 1976 'O. Jacks' *Assassination Day* v. 81 He bundled the notes up into foil wrap, put two in..his refrigerator. 1977 *Time* 14 Mar. 39/3 The thin sail (ordinary plastic kitchen wrap is five times thicker) would be coated with an aluminum reflecting layer on the side that will face the sun. 1979 *Sci. Amer.* Jan. 131/1 You could also build up layers of stretched plastic food wrap. 1980 *Outdoor Life* (U.S.) (Northeast ed.) Oct. 126/2 Cover an exposed drain valve with plastic wrap to prevent road slush from accumulating.

4. *pl.* In *fig.* phrases referring to concealment or disuse, as *under* or *in wraps,* concealed; in abeyance; *to take* or *pull the wraps off,* to disclose; to bring back into use.

1939 *Sun* (Baltimore) 18 Dec. 3/6 The fact that the belligerents have kept their air power under wraps almost from the beginning reveals more than meets the eye. 1950 'S. Ransome' *Deadly Miss Ashley* xv. 178 You grabbed his notebooks and tried..to keep them under wraps. 1956 A. H. Compton *Atomic Quest* 49 Some members of the committee were insistent that the entire uranium project should be put in wraps for the duration. 1964 *New Society* 15 Feb. 21/2 The Government took the wraps off its plan for regenerating the Northeast. 1965 Mrs. L. B. Johnson *White House Diary* 20 Jan. (1970) 226 Next Hubert [Humphrey] stepped forward—for once his exuberance was under wraps. 1973 *Times* 4 Oct. 4/3 Only now was the truth about battered wives being revealed 'because the wraps have just been pulled off a taboo subject'. 1978 *Dumfries Courier* 20 Oct. 11/1 Show visitors will see numerous others which are still under wraps until nearer the Show. 1984 *Times* 14 May 7/2 This week will see the wraps coming off another popular project, where Britain wants to be seen leading the way.

5. *Cinemat.* and *Television.* The end of a session of filming or recording.

1974 M. Ayrton *Midas Consequence* i. 63 Other cars are heard starting up out of shot and the lights on the pergola go off so I assume it's a wrap and the crew is listening to the director saying something consequential and busy about tomorrow's call. 1980 J. Krantz *Princess Daisy* xii. 191 'Right, it's a wrap.'.. The large lights, cameras, sound equipment and other tools of the trade were quickly stowed away. 1983 *Listener* 23 June 18/2 The director says: 'Cut! Thank you, Ben, that's a wrap —there is no more filming.'

6. Special Combs. (see also Wrap- in Dict. and Suppl.): **wrap party** *Cinemat.*, a party held to celebrate the completion of filming; **wrap reel, wheel,** a large revolving framework on which yarn can be wound and measured.

1978 J. Krantz *Scruples* xiii. 388 Work on *Mirrors* finished on schedule, on Friday, August 23rd, and the wrap party was scheduled for the next night. 1978 *Morning Bulletin* (Rockhampton, Queensland) 3 Apr. 6/4 At the wrap party..Syl played host in a three-piece white suit. 1889 G. E. Davis *Sizing & Mildew in Cotton Goods* ii. 20 In order to test the fineness of yarn a wrap reel is used, measuring 54 inches in circumference, which is so arranged that by a single turn of a handle two complete revolutions are given to the reel itself. 1928 W. L. Balls *Studies of Quality in Cotton* xii. 235 The lea was wound upon wrap-reels of varying diameter. 1956 S. E. Ellacott *Spinning & Weaving* 36/2 Arkwright's wrap reel for measuring hanks of yarn (840 yards) was a six-armed revolving star with a winch handle. 1953 Wrap wheel [see *rice *sb.*¹ 4]. 1969 E. H. Pinto *Treen* 318 After yarn had been spun on a spindle or spinning wheel, it was transferred on to a wrap or clock wheel, a rotary instrument which skeined and measured it. The original wrap wheel is always said to have been invented by Richard Arkwright.

wrap, *v.* Add: **I. 6*.** *to wrap up* (fig.) **a.** *trans.* To put an end to, bring to completion; also, to defeat; *to wrap it up,* to stop doing something. *slang.*

1926 T. E. Lawrence *Seven Pillars* (1935) iii. xxxvi. 213 The British were wrapping up the Arabs on all sides—at Aden, at Gaza, at Bagdad. 1937 *Amer. Legion Monthly* May 9/1 Only one shot to finish before midnight and we'd wrap it up in thirteen days. 1949 A. Miller *Death of Salesman* ii. 128 To hell with whose fault it is... Let's just wrap it up, heh? 1957 J. Osborne *Look back in Anger* i. 25 Wrap it up, will you? Stop ringing those bells! 1957 P. Frank *Seven Days to Never* ii. vii. 80, I guess that wraps it up for tonight... I don't know of anything else we can do. 1960 G. Sanders *Mem. Professional Cad* ii. iii. 127 'Wrap it up,' he would shout. 1976 *Billings* (Montana) *Gaz.* 2 July 1-c/1 Nastase wrapped up Ramirez, 6-2, 9-7, 6-3. 1984 *Times* 14 Mar. 2/1 (*heading*) Labour MPs advised to wrap up their muck raking.

b. *intr.* To stop talking. Freq. as imp. *slang.*

1943 Hunt & Pringle *Service Slang* 70 *Wrap up,* stop talking. Or, get ready to go home. 1945 C. H. Ward-Jackson *Piece of Cake* (ed. 2) 63 *Wrap up!,* Be quiet! Pipe down! 1958 F. Norman *Bang to Rights* i. 49 Why dont you rap up. 1959 'O. Mills' *Stairway to Murder* iv. 44 'Geoff, wrap up about the jigsaws,' Charles entreated him.

6.** *intr. Cinemat.* and *Television.* To finish filming or recording.

1976 in B. Armstrong *Gloss. TV Terms* 94. 1983 *London Mag.* Aug./Sept. 30 We wrapped on schedule, three days later... The movie got terrible reviews.

II. 10. *to wrap* oneself *(a)round* (an item of food or drink): to eat or drink it. Occas. with non-*refl.* direct obj., to make (another) eat or drink. *colloq.*

1880 J. C. Harris *Uncle Remus* xv. 219 She cut me off er slishe..an' I sot down on de steps an' wrop myse'f roun' de whole blessid chunk. 1927 D. L. Sayers *Unnatural Death* xii. 136 Lord Peter, having wrapped himself affectionately round an abnormal quantity of bacon and eggs, strolled out. 1946 K. Tennant *Lost Haven* (1947) xviii. 305, I bet they had to wrap Alec round a few beers before they got him up to the mark. 1959 G. Endore *Detour through Devon* 3 Wrap yourself around a cup of coffee. 1962 A. Lejeune *Duel in Shadows* ii. 25, I shall be glad to get indoors and wrap myself round a large drink.

11. To crash (a vehicle) into a stationary object. Const. *around, round. slang.*

1950 J. D. MacDonald *Brass Cupcake* (1955) x. 105, I took a car off the street and wrapped it around an oak tree. 1958 'J. Brogan' *Cummings Report* xvii. 183 Steady, or you'll have us both wrapped round a telegraph pole if you're not careful. 1969 L. G. Arthur in A. E. Wilkerson *Rights of Children* (1973) 132 If a child wraps a stolen car around a telephone pole, is $2,000.00 restitution..an excessive fine? 1984 *Times* 19 May 8/1 The men towing the boat from one training venue to another wrapped it round a traffic light.

wrap-. Add: **2.** *spec.* Designating a garment to be wrapped about the body for warmth, or a wraparound garment (see *wraparound *a.* 1 a).

1845 *Ainsworth's Mag.* VII. 499 A wrap-cloak, or sheet, being thrown penance-ways over the head and shoulders. 1887 Wrap shawl [in Dict. s.v. Wrap *sb.* 2 a]. 1928 *Daily Mail* 9 Aug. 3/4 A.. bathing dress worn under a well cut wrap-coat of crêpe de Chine. 1976 *New Yorker* 8 Mar. 1 (Advt.), Reversible wrap skirt lined in red and white. 1980 L. Birnbach et al. *Official Preppy Handbk.* 131/1 Diane von Furstenberg wrap dress. 1982 Barr & York *Official Sloane Ranger Handbk.* 30/1 A belted wool wrap coat: this can be tweedy, checked or camel. 1984 *Chicago Sun-Times* 25 Jan. 33/1 By 1976 she had designed the wrap dress, which became an overnight success.

wraparound (ræ·pāraund), *sb.* and *a.* Also **wrap-around, wrapround, wrap-round.** [f. Wrap *v.* + Around *adv.* and *prep.*] **A.** *sb.*

1. A garment that is thrown or wrapped round the body; a wraparound garment (see sense B. 1 a below).

1877 [see Wrap-]. 1959 *Vogue Pattern Bk.* June–July 25 A wrap-around that buttons into place... The sort of comfort-with-elegance dress. 1973 *Harrods Christmas Catal.* 27 Casual wrap-around in figured polyester.

2. A fastening or label that wraps round a bottle.

1953 *Federal Suppl.* CXVIII. 182/2 The Guardian Seal, made of aluminum, attached to the bottle and requiring no secondary closure;..wrap arounds, a laminated foil attached to a paper with an adhesive on the inside giving the same decorative features as a cellulose band; [etc.]. 1966 J. Aiken *Trouble with Product X* iii. 43, I switched over to copy for a counter-card, window bill and wrap-around for Bom, the Meat'n Milk Drink. 1970 K. Platt *Pushbutton Butterfly* (1971) xiii. 149 He..came back with a bottle. He unfurled the plastic wraparound and broke it open.

3. *Computers.* The procedure or facility by which a linear sequence of memory locations or positions on a screen is treated cyclically, so that when the last has been counted or occupied the first is returned to automatically (on the line below in the case of screen displays). Also *transf.*

1965 E. A. Weiss *Programming the IBM 1620* ii. 21 This wrap-around feature does not apply to addresses used in commands. 1970 O. Dopping *Computers & Data Processing* vi. 101 Even when the index register has no room for sign, subtraction can be done by utilizing the cyclic character of the addressing system ('wraparound'). 1979 J. E. Rowley *Mechanised In-House Information Syst.* i. 76 Display can be improved by..contextual wrap-around, or bringing the end of the title to the left hand side of a keyword and then marking the end of the title with a display device.

B. *adj.* **1.** *Fashion.* **a.** Of a garment: that is open all the way down, wraps around the body, and is fastened usu. by tying. Also, of a belt.

1937 M. Levin *Old Bunch* 13 She dressed spiffy with wrap-around sport skirts. 1938 'E. Queen' *Four of Hearts* (1939) x. 143 A silver lamé hostess-gown with a trailing wrap-around skirt over Turkish trousers. 1945 [see *pop-over 2]. 1951 *Rep. Patent, Design & Trade Mark Cases* (U.S.) LXVIII. 256 Mr. Percival contended that the Applicants' skirt was a novel 'wrap round' skirt. 1957 *U.S. Supreme Court Rep.* 2nd Ser. I. 880/1, 1 Purple wool imitation lambskin wrap around short coat. 1972 J. Godey *Three Worlds* ii. 19 The man was wearing a wraparound coat of some synthetic fur. 1976 *Scotsman* 20 Nov. (Weekend Suppl.) 4/3 (Advt.), Right: The unlined, hooded style of winter coat with wrap-around belt. 1976 T. Stoppard *Dirty Linen* 9 Maddie is..wearing..a wrap-round skirt, quite short. 1982 Barr & York *Official Sloane Ranger Handbk.* 42/1 There are a few basic lines that continue practically for ever, like the pre-war wraparound double-breasted and the basic City three-piece.

b. Of sunglasses, goggles, etc.: that have lenses which extend around the side of the head.

1966 T. Pynchon *Crying of Lot 49* iii. 57 There stood Di Presso, in a skin-diving suit and wraparound shades. 1968 A. Williams *Brotherhood* i. iv. 50 He was no longer wearing his wrap-around dark glasses. 1976 *National Observer* (U.S.) 27 Mar. 15 (Advt.), Space-age wraparound sunglasses make all others obsolete!

2. a. That extends round a corner, esp. of parts of a building or parts of a motor vehicle.

1954 *Archit. Rev.* CXVI. 92/1 The wrap-round cornices are used to tie it back to the side façade. 1957 H. Roth *Shadow of Lady* xiv. 103 A sturdy little car..even if without American flourishes, like 'wrap-around' windscreens. 1959 *Motor* 21 Jan. 951/1 Protective wrap-around bumpers. 1966 *Daily Tel.* 9 Nov. 12/5 Recent years' flirtation with vestigial fins, wrap-round rear windows and other gimmicks. 1972 [see *wet-suited* adj. s.v. *wet *a.* 20]. 1972 *Village Voice* (N.Y.) 1 June 87/4 (Advt.), Huge liv rm & wraparound sundeck. 1976 *Glasgow Herald* 26 Nov. 19/5 Visibility from the reclining driver's seat is outstanding, with a very wide and deep 'wrap around' rear window. 1980 J. Strouse *Alice James* (1981) xiii. 221 The buildings..are..ornamented with scrollwork, wraparound porches, archways.

b. Of a cinema screen: having a greater sideways extent than normal; subtending a large angle at the audience. Similarly of a view. Also *fig.*

1950 *Pop. Sci. Monthly* Aug. 75 You're not just looking *at* this 'wrap-around' movie show—you're in it! 1968 *Globe & Mail* (Toronto) 17 Feb. 25 (Advt.), Look down on the world from Stop. 33. The room with the wrap-around view. 1968 *Tel.* (Brisbane) 18 Sept. 34/5 His book is entitled *The Invasion*... Hay contents himself with a brief, wraparound picture of the real invasion. 1972 J. McClure *Caterpillar Cop* i. 7 The Big Romance soon to be filmed in fabulous Technicolor on a wrap-round screen.

3. *gen.* That surrounds or encompasses.

1957 *Times Survey Brit. Aviation* Sept. 2/4 The missile has a two-stage propulsion system, consisting of wrap-around boosts and a sustainer rocket motor. 1964 M. McLuhan *Understanding Media* (1967) xxviii. 300 Stereo sound..is 'all-around' or 'wrap-around' sound. 1967 *Maclean's Mag.* Sept. 14 The Volkswagen is the German's ideal image of space: it's a wraparound, secure little thing. 1970 *Gloss. Aeronaut. & Astronaut. Terms* (*B.S.I.*) vi. 3 *Wrap-round boost,* a number of boost rocket motor assemblies located externally along the sides of the missile body. 1978 *Detroit Free Press* 5 Mar. A10 (Advt.), Wrap-around heat encircles the sides of the crockware. 1979 *Jrnl. R. Soc. Arts* CXXVII. 655/2 The 21 new rooms and the lower gallery..provide a more than adequate wrap-around environment for the works on display. 1981 *Times* 27 Jan. 1/8 His wrap-around bodyguards leapt from the flower-beds..., shouldering reporters and Saudi policemen from his path so that..it was simply not possible to see more than his eyebrows and hair.

4. *Printing.* Designating a flexible relief printing plate which is wrapped round the cylinder of a rotary press, and machines or methods which employ one.

1959 *Brit. Printer* Dec. 116/3 (Index), Wrap-around, Harris letterpress rotary. 1962 [see *rotary *a.* 2 b]. 1963 *Publishers' Weekly* 5 Aug. 87/1 Wrap-around relief printing—direct and indirect—is..being used increasingly... Most of the printing is for labels and packaging materials.

1972 A. Tyrrell *Basics of Reprography* xiv. 216 In this way curved or wrap-around printing formes can be prepared by reprographic methods, from relief surfaces in the flat. 1983 A. Campbell *Designer's Handbk.* viii. 128 A modern development of letterpress printing is the wrap-around rotary press, which prints from a one-piece shallow relief plate fastened around a press cylinder.

5. *Publishing.* Designating (*a*) a book cover made from a single sheet of material; (*b*) a jacket whose design extends from front to back without being divided by the spine.

1968 G. A. Stevenson *Graphic Arts Encycl.* 418 *Wrap-around cover,* soft cover used to bind or hold a booklet, brochure, etc. It consists of one sheet of stock that forms both front and back covers. Any type of mechanical binding may be used. 1972 *N.Y. Law Jrnl.* 24 Oct. 4/2 It must have a wrap-around cover page containing the usual information. 1979 *Bookseller* 23 June 2829/2 Mr. Paton thought that designers did not give nearly enough consideration to the potential of the spine. In this respect he was suspicious of the wrapround jacket.

6. *U.S. Finance.* Used with reference to (*a*) a mortgage which continues when the mortgaged property is sold, the repayments to the original lender being made by a new lender who also provides the additional funds needed for the purchase; (*b*) a tax-deferral scheme in which the interest on certain investments goes into paying the premiums for an annuity.

1968 *Federal Suppl.* CCXCII. 594 Midwestern would exchange a $2·2 million wrap around note, secured by certain hotel properties. 1971 *Legal Bull.* (U.S.) Sept. 185 Wrap-around mortgage financing. 1977 *National Observer* (U.S.) 22 Jan. 8/2 The days may be numbered for a popular tax shelter known as the investment, or 'wrap-around', annuity. 1979 *Arizona Daily Star* 5 Aug. (Advt. Section) 17/8 Priced below appraisal. Will consider a wrap-around deed or trust. 1981 *U.S. Federal Reg.* 7 Apr. 20875/1 Official Staff Interpretation FC-0146.. treats 'wrap-around' loans as the equivalent of refinancings.

wrap-over (ræ·pōᵘvəɹ), *sb.* and *a.* Also **wrap over, wrapover.** [f. Wrap *v.* + Over *adv.*] **A.** *sb.* Part of something, usu. a garment, that overlaps another part of itself.

1935 *Times* 21 Oct. 11/3 There is a good wrapover on the skirt of the coat. 1960 *Vogue Pattern Bk.* Early Autumn 37 Divided skirt..concealed by a wrap-over at front and back.

B. *adj.* **a.** Of a garment: having a wrap-over. **b.** Overlapping.

1960 *Guardian* 19 Feb. 8/7 A wrap-over petal skirt. 1973 *Harrods Christmas Catal.* 24/1 Wrap-over coat from Italy. 1979 *Nature* 19 Apr. p. xvii/3 The simple laboratory stirrer..has a wrapover top to provide spillage protection. 1979 *Homes & Gardens* June 103/1 The wrapover skirt, very fashionable this summer, is made of four flared panels, lightly gathered at the waist.

wrapped, *ppl. a.* Add: **I. 2. b.** (Earlier example.)

1793 F. Burney *Let.* 24 Feb. (1972) II. 26, I live a Wrapt up Invalide, close to the Fire side.

3*. Enclosed in a wrapping; *spec.* pre-packaged.

1957 M. Summerton *Sunset Hour* xiii. 186 The loaf of wrapped bread. 1963 L. Deighton *Horse under Water* vi. 31 Plastic spoons and large wrapped sugar segments. 1976 *Times* 13 Aug. 2/6 The familiar wrapped and sliced white loaf still accounts for more than half of bread sales. 1984 C. Curzon *Masks & Faces* viii. 90 He fetched the wrapped loaf and filled the toaster.

wrapper, *sb.* Add: **I. 1. d.** (Earlier example.)

1853 *Heal & Son Catal.: Illustr. Catal. Bedsteads* 5 The plain Quilts..are applicable wherever extra warmth is required, either as a wrapper in the carriage, or as an extra covering on the bed.

4. a. Also *wrapper leaf.*

1944 [see *run *sb.*¹ 20 e]. 1978 D. Williams *Treasure up in Smoke* v. 50 All hand-made cigars consist of a thick core of compressed tobacco leaf, a binder.., and finally a wrapper leaf.

wrapper, *v.* Add: **2.** Also *absol.* or *intr.*

1934 H. G. Wells *Exper. Autobiogr.* I. iv. 151 Half an hour before closing time we began to put away for the last time and 'wrapper up'.

wrapping, *vbl. sb.* Add: **3. wrapping-paper** (earlier examples).

?1715 Pope *Let.* in *Corr.* (1956) I. 317 If the Fruit is not so good as I wish, let the Gallantry of this Wrapping-paper make up for it. 1768 J. Lyndon *Let.* 17 June in *Rec. Colony of Rhode Island* (1861) VI. 548 One paper mill, at which is manufactured wrapping, package and other coarse paper. 1789 *Deb. Congr. U.S.* (1834) 1st Congress 1 Sess. App. 2130 The several duties shall be aid on the following goods... On all writing, printing and wrapping paper.

wrap-round: see *wraparound *sb.* and *a.*

wrap-up (ræ·pɒp), *sb.* and *a.* [f. vbl. phr. *to wrap up*: see *wrap *v.* 6* a.] **A.** *sb.* **1. a.** An easily satisfied customer; an easy sale. **b.** Any easy task.

1938 *Amer. Speech* XIII. 150/2 *Wrap-up,* an easy sale. Also a customer easily satisfied. 1940 'E. Queen' *New Adventures* 284 Not too tough. A wrap-up. 1952 *N.Y. Times Mag.* 21 Sept. 58/3 The ideal customer is known as a 'wrap-up', which is self-explanatory.

2. A summary or résumé, esp. of news; a conclusion.

1960 Wentworth & Flexner *Dict. Amer. Slang* 588/2 *Wrap-up,*..a conclusion, an ending; a summary and conclusion. 1961 *Times Lit. Suppl.* 13 Oct. 677/4 Finally, Mr. Kalb gives us what he calls the wrap-up [of a book]. 1969 *New Yorker* 30 Aug. 20 (*caption*) Suddenly..a wrap-up of the highlights of my life flashed before my eyes. 1973 H. Gruppe *Truxton Cipher* xv. 155 'I have no further questions for you.'.. This was the wrap-up. Harry knew ..he would be handed over for court-martial. 1975 *New Yorker* 5 May 128/2 NBC presented a thirty-minute 'special report' on Cambodia, which consisted mainly of a wrapup of NBC's regular news footage of the previous week. 1980 U. Curtiss *Poisoned Orchard* ix. 97 The wrap-up of a job I've been working on will have to be done tonight. 1981 *Daily Mail* 18 May 19/4 Last night's wrap-up saw the old soldier with the black eye-patch welcomed by desert Bedouins. 1985 *Village Voice* (N.Y.) 8 Jan. 39/2 Only in his final wrap-up does he concede that power can also take the form of creative 'attention, or love'.

B. *adj.* That concludes or sums up.

1968 Mrs. L. B. Johnson *White House Diary* 9 Apr. (1970) 658 Here on the courthouse square at Gonzales was the wrap-up scene, an official good-by for our five-day adventure. 1976 *Publishers Weekly* 27 Sept. 82/2 A long wrap-up section amplifies this practical aspect of their book. 1977 *Church Times* 7 Apr. 2/4 A two-day centennial programme... The Archbishop will..address the wrap-up banquet. 1980 *Jewish Chron.* 21 Mar. 23/3 This is what is described as a 'wrap-up' volume, summarising, in non-technical language, what has been learned. 1980 *Quilt World* Sept./Oct. 16/3 The group session wound down with a light monologue..on 'Fifty Ways to Lose Business', and then small group discussions were held, followed by a wrap-up session.

wrassle, U.S. dial. var. Wrestle *v.* in Dict. and Suppl.

wrath, *sb.* Add: Now usu. with pronunc. (rɒþ). **1. d.** Also *fig.* in phr. *like the wrath of God,* dreadful, terrible; dreadfully, terribly.

1936 J. Buchan *Island of Sheep* xii. 224 The winds..in the Norlands can blow like the wrath of God. 1955 M. Allingham *Beckoning Lady* xiii. 178 Fancy coming home like the wrath of God and starting a fight. 1967 'R. Foley' *Fear of Stranger* (1968) v. 56 You look like the wrath of God, Kay... No flesh on you to speak of. 1982 'W. R. Duncan' *Queen's Messenger* ii. 19 Are you ill? You look like the wrath of God.

6. *wrath-bearing.*

1920 T. S. Eliot *Ara Vos Prec* 12 These tears are shaken from the wrath-bearing tree.

wreak, *v.* Add: **III. 8. b.** Esp. in phr. *to wreak havoc.* (For *wrought havoc* see *work *v.* 10.)

1926 A. Christie *Murder of R. Ackroyd* xx. 239 Annie is not allowed to wreak havoc with a dustpan and brush. 1976 B. Fell *America B.C.* viii. 101 The storm waves could surely wreak more havoc upon the timbered hulls of Phoenician galleys than on the steel plates of modern ships. 1978 C. Rayner *Long Acre* vii. 70 Fenton, well aware of the havoc he was wreaking in poor Miss Emma's heart, wickedly fed her passion for him. 1983 *Times* 21 Nov. 7/7 Moko, the banana disease, has already wreaked havoc on the trade. 1984 *Daily Tel.* 5 Nov. 20/2 The feared shake-out in microcomputer manufacturing..will wreak havoc in the industry.

wreck, *sb.*¹ Add: **II. 9. c.** *N. Amer.* A road or railway accident.

1912 J. Sandilands *Western Canad. Dict. & Phrase-Bk.*, *Wreck,* the word to apply to a railway accident; or, more correctly, train wreck. 1974 *Evening Herald* (Rock Hill, S. Carolina) 19 Apr. 11/4 In spite of the reduction in accidents, the sergeant said, 'We still have too many wrecks.' 1979 N. Mailer *Executioner's Song* (1980) i. xviii. 304 On the drive back to Springville, she was dreaming away and got in a wreck. Nobody was hurt but the car.

d. The death of a large number of pelagic birds, usually as the result of a storm.

1936 *Brit. Birds* XXIX. 327 In January, 1915, there was a great oil 'wreck' of Scoters. 1971 *New Scientist* 8 Apr. 69/1 There have been similar wrecks in the past, and the report mentions 11 for guillemots and the auks in the last century.

wrecked, *ppl. a.* Add: **2. c.** Intoxicated; under the influence of drugs. *U.S. slang.*

1968–70 *Current Slang* (Univ. S. Dakota) III–IV. 139 *Wrecked,* intoxicated. 1973 D. Lang *Freaks* 63, I could not get it on, could not get it on, not unless I was, one: totally wrecked; and, two: had to have a gun in my hand.

wrecker¹. Add: **3.** A demolition worker.

1958 J. Thurber in *Atlantic Monthly* Feb. 52/1 Jacob Volk, a building wrecker..who tore down two hundred and fifty big structures in Manhattan. 1968 *Globe & Mail* (Toronto) 17 Feb. 6/3, I was saddened to discover wreckers' hoardings surrounding the Royal Bank building at 10-12 King Street East. 1977 H. Fast *Immigrants* vi. 352 He stood on the corner of California Street on Nob Hill watching the wreckers take the Seldon mansion apart, stone by stone.

wrecker². Add: **2. a.** (Earlier example.)

1789 O. Equiano *Life* II. viii. 57 They met with this little sloop, called a wrecker; their employment in those seas being to look after wrecks

b. A railway vehicle with a crane or hoist for removing crashed trains or similar obstructions; also, a breakdown truck. Also *attrib.*

1904 *Booklovers' Mag.* May 663 This special train has been dubbed the 'Wrecker'. Really it is a relief train, ready to respond to any call for aid in case of accident. 1955 V. Nabokov *Lolita* II. xxx. 188 Around midnight, a wrecker dragged my car out. 1970 *Globe & Mail* (Toronto) 28 Sept. 31/6 (*caption*) Ward has picked Maggie up..in a wrecker! 1973 *Amer. Speech 1969* XLIV. 257 The wrecker train..has a flatcar with a crane. 1978 *Detroit Free Press* 16 Apr. 18/10 (Advt.), Business offered... Car wash, wrecker service and..service station. 1980 R. L. Duncan *Brimstone* vi. 126 A couple of police cars and a wrecker pulling apart three cars that had tailgated.

wrecking, *vbl. sb.*¹ Add: **1.** (Further examples.)

1940 *Construction Methods* Apr. 110/2 Wrecking is, in reality, construction in reverse gear... At the Louisville, Ky. East End slum clearance project..there were approximately 480 buildings to wreck... The Cleveland Wrecking Co. has had many large contracts of this type. 1972 *Times* 10 Mar. (Suppl.) p. ii/1 (Advt.), The most experienced firm in the U.K. in the wrecking of blast furnaces.

3. *attrib.*, as *wrecking company*; **wrecking ball,** a large, heavy metal ball which, hung from a crane, may be swung into a building to demolish it; **wrecking bar,** an iron bar with one end chisel-shaped for prising and the other bent and split to form a claw.

1952 *Business Week* 19 July 33/2 Instead of using a one-ton wrecking ball at the end of a 60-ft. beam, the building must be knocked down..with a 16-lb. sledge hammer. 1977 *Rolling Stone* 21 Apr. 34/4 It's the laugh of a man who just watched a wrecking ball smash his house to splinters so a new freeway could go through. 1984 *New Yorker* 20 Feb. 50/2 The wrecking ball bursts through the wall with the bookshelves, scattering the works of famous authors. 1924 *Sears, Roebuck Catal.* No. 148. 866/3 Wrecking Bars. Forged steel 24-inch, 30c. 1947 *Construction Methods* Mar. 88/2 To minimize damage to material during removal, the contractor developed his own tools as supplements to the standard wrecking bar and claw hammer. 1940 Wrecking company [see sense 1 above]. 1976 *National Observer* (U.S.) 3 Apr. 7/1 A wrecking company recently signed a contract to level all 30 of the remaining 11-story buildings.

wrecking, *vbl. sb.*² Add: **1.** (Further example.)

1969 *Sydney Morning Herald* 24 May 63/2 (Advt.), Jaguars, wrecking now. Continually dismantling 2·4, 3·4 and 3·8.

2. b. (Examples relating to motor vehicles.)

1939 [see *out-city *a.*]. 1968 *Globe & Mail* (Toronto) 13 Feb. 32/2 (Advt.), Well established wrecking business with living quarters on large lot fronting on Napanee River.

wrecking, *ppl. a.* Add: **1.** *wrecking amendment* (Pol.), one designed to defeat the purpose of the bill concerned.

1967 M. Pinto-Duschinsky *Polit. Thought of Lord Salisbury* vii. 145 Its very moderation led directly to the passing of a wrecking amendment by Lord John Russell, who favoured a different and much more far-reaching measure. 1979 H. Wilson *Final Term* ix. 189 Again the Conservatives, with considerable Labour support, moved 'wrecking' or near-wrecking amendments.

wrecky (re·ki), *a.* [f. Wreck *sb.*¹ + -y¹.] Broken-down; debilitated.

1925 *Brit. Weekly* 17 Sept. 541/1 After which you are left a wreck and probably remain wrecky next day. 1973 M. Amis *Rachel Papers* 15, I had a well-earned-half at the pub and chatted with the landlord and his wrecky wife.

wren. Add: **2*.** A woman, esp. a young woman. *U.S. slang.*

1920 S. Lewis *Main Street* 388 Some tank, that wren! Ha, ha, ha! 1927 *Amer. Speech* III. 167/1 *Dame, frail, skirt, Jane, wren, broad,* girl. 1929 A. Conan Doyle *Maracot Deep* 198 Scanlan has..married his wren in Philadelphia. 1946 B. Treadwell *Big. Bk. Swing* 125/2 *Wren,* small, fickle young girl. 1982 M. McMullen *Until Death do us Part* (1983) 9 Midge was, in her quiet unobtrusive way, a perfect marvel of efficiency, 'My dear wren,' Jane sometimes called her.

3. *wren-king*; *wren-nested* adj.; **wren-warbler,** any of several warblers of the genus *Primia,* found in tropical Africa or Asia; also, a brightly coloured wren of the subfamily Malurinæ, found in Australasia.

1965 Auden *About House* (1966) 13 From gallery-grave and the hunt of a wren-king to Low Mass and trailer camp is hardly a tick by the carbon clock. 1925 Blunden *Eng. Poems* 104 Wren-nested hedges. 1924 E. C. S. Baker *Fauna Brit. India: Birds* (ed. 2) II. 530 The Ashy Wren-Warbler breeds from March to September. 1931 *Discovery* May 141/2 The tiny new wren warbler..a wee mite of a bird with a tail almost as long as its body. 1955 Mackworth-Praed & Grant *Birds E. & N.E. Afr.* 392 Wren-warblers..occur in both woodland and thorn-scrub. 1974 I. Rowley *Bird Life* vi. 68 Most *Malurus* have distinctive and attractive songs so that the name 'wren-warbler' is an apt one.

Wren² (ren). Also **wren.** [f. three of the initial letters of the name of the Service,

made into a singular noun.] A member of the Women's Royal Naval Service, formed in 1917; also (*pl.*), the Service itself.

1918 [see *penguin 2 b]. 1927 *Glasgow Herald* 15 Apr. 7 The war years with all their Waacs and Wrens and Wrafs, seem now to be immeasurably far off. 1940 *War Illustr.* 5 Jan. 558 At all Naval depots 'Wrens' are now doing work as clerks, cooks and in many other capacities, thus relieving men for more active work. 1946 'Tackline' (*title*) You met such nice girls in the Wrens. 1956 [see *W.R.A.C.]. 1979 D. Gurr *Troika* xi. 75, I..reported to the Admiralty... 'Captain Jackson's office, please?' 'Second floor, sir.' A good bust and a smile on the duty wren cheered me up.

Hence **Wre·nnery** *joc.*, a building used to accommodate Wrens.

1943 Hunt & Pringle *Service Slang* 70 *Wrennery,* billets of the 'Jenny Wrens'. 1945 'N. Shute' *Most Secret* 124, I shall be living in the Wrennery and coming out to Dittisham every day. 1959 P. McCutchan *Storm South* ii. 41 Where did you pick up all this insight into human nature—in the Navy? Serve in a Wrennery or something, did you? 1964 *Navy News* July 5/4 The work included.. the building of a Wrennery to accommodate 200 Wrens.

Wrenaissance (renēⁱ·säns, -aṅs). *Archit.* [f. the name of Sir Christopher *Wren* (see Wrenean *a.*) after Renaissance.] An architectural style modelled on or influenced by that of Wren, esp. as represented by some of the work of Sir Edwin Lutyens.

1942 R. Lutyens *Sir Edwin Lutyens* iii. 40 Gothic and Renaissance ('Wrenaissance'! as father has punned..) are both architectures of meaning. 1944 *Archit. Rev.* XCV. p. xlvi/1 We cannot..allow ourselves, out of affection for a great man [*sc.* Lutyens] and out of admiration for his highly personal style, to be saddled with a Wrenaissance London as a monument to a period which the first world war brought to a murderous close. 1967 *Time* (Atlantic ed.) 26 May 45/1 Frederick Gibberd... extended a piazza to roof over an English Wrenaissance crypt built in the 1930s. 1980 M. Lutyens *Edwin Lutyens* iv. 62 In 1906, with the building of Heathcote at Ilkley, Yorkshire, for Mr Ernest Hemingway..he reached what he called his 'Wrenaissance'; Wren thereafter became his lodestar. 1981 *Times* 12 Feb. 17/4 That heavy, florid..brick and stone style christened 'Wrenaissance'.

wrench, *sb.*² Add: **6. wrench fault** *Geol.* = *strike-slip fault* s.v. *strike *sb.* 20.

1951 E. M. Anderson *Dynamics of Faulting* (ed. 2) i. 2 The term 'Blatt' will be translated as wrench fault. 1977 A. Hallam *Planet Earth* 61/1 Major wrench faults (e.g., the Great Glen Fault of Scotland) exhibit displacements of 100km..or more.

wrenching, *vb. sb.* Add: **1. c.** *N.Z.* = *root-pruning* s.v. Root *sb.*¹ 22.

1950 *N.Z. Jrnl. Agric.* July 55/1 Toward the end of August root crops..tend to run to seed. This growth can be retarded considerably by wrenching..pushing a fork or spade into the soil..and..levering the roots up slightly. This breaks the extreme end of the taproot.

Wrenean, *a.* Add: Also **Wrenian.**

1944 *Burlington Mag.* Oct. 260/1 The complete reinstatement of a destroyed Wrenian interior. 1973 *Country Life* 13 Dec. 2017/3 The style is about half correct Classical or Wrenian and half Victorian Italianate.

wrenlet (re·nlėt). [f. Wren + -let.] A young wren.

1858 *Chambers's Jrnl.* Aug. 82/2 It affords a cradle to near a score of wrenlets. 1927 *Observer* 22 May 23/2 In a knot in the rope was a wren's nest, with two or three wrenlets visible inside.

wrestle, *v.* Add: Also, in *U.S. dial.*, with pronunc. (ræ·s'l). For 8 **rassle** read 8– (latterly *U.S. dial.*) **rassle**; also β (chiefly *U.S. dial.*) **rastle, wrassle.** **I. 1. a.** (Further example.)

1974 *Black World* Jan. 56/2 He might be stronger'n me and he might wrassle the best, but I got his waters on, all right.

b. (Further examples.)

1940 L. I. Wilder *Long Winter* xvi. 150 I've spent this whole morning rasseling with that dumb horse. 1941 *Harper's* Feb. 329/2 Leaving..us to rassle with the bear. 1962 W. Faulkner *Reivers* v. 101 'What you been doing? wrassling with hogs?' 'We got in a mudhole.'

c. (Further example.)

1936 in P. Oliver *Screening Blues* (1968) vi. 189 If you keep on rasslin' you gonna make me break my needle off.

II. 6. a. (Further examples.)

1940 *Sat. Even. Post* 22 June 39/2 He could rassle any three men. 1968 *Listener* 30 May 702/3 In this picture he rides horses, climbs mountains and wrassles Indian chiefs. 1976 *Ibid.* 24 June 817/1 Thrown to the ground and wrastled by the brutally handsome Provo.

b. (Further example.)

1893 N. K. Griggs *Lyrics of Lariat* 46 A Maverick daisy he saw—..And so He rastled it low And gave it a touch of his brand.

7. Delete *rare*⁻¹ and add: More widely, to move (something inanimate or inert) with physical force. Const. various preps. and advbs.

1970 J. Dickey *Deliverance* 186, I..dropped down on one knee and wrestled him across my shoulders in the fireman's carry from boy scout days. 1973 M. Amis *Rachel Papers* 38 Eventually he wrestled all the string and paper into an armful-sized bundle and forced it down the Aga. 1973 M. Woodhouse *Blue Bone* vii. 63 We wrestled the crates down through the forward hatch. 1975 *Globe & Mail* (Toronto) 11 Sept. 2/7 Agents wrestled a ·45 calibre automatic pistol out of her hands. 1976 A. Price *War Game* i. viii. 142 The same hand, strong and supple.., had once wrestled a bomb-laden Lancaster into the air. 1976 Scott & Koski *Walk-In* x. 53 He wrestled one of the line of overhead doors up. 1981 J. D. MacDonald *Free Fall in Crimson* xiv. 159 A truck pulled up... Two men hopped out and started to wrestle the wicker basket out of the back.

wrestler. Add: Also *U.S.* **rassler, wrassler.**

1. (Further examples.)

1900 F. P. Dunne *Mr. Dooley's Philos.* 207 He was a gr-reat rassler an' whin he had a full Nelson on th' foolish man that wint again him, he used to say, 'Dear me, am I breakin' ye'er neck, I hope so.' 1941 J. Thurber in *Sat. Even. Post* 5 Apr. 10/3 'Wrasslers,' says Magrew, cold-like, 'that's what I've got for a ball club, Mr. Du Monville, wrasslers—and not very good wrasslers at that.'

2. (Further example.)

1964 N. Mailer in *Esquire* Nov. 170/4 He was just another hog-wrassler of rhetoric.

wrestling, *vbl. sb.* (Examples of *U.S. dial.* forms: cf. *wrestle *v.*)

1901 W. Churchill *Crisis* ii. vi. 162 He an' de Colonel done commence wrastlin' 'bout a man name o' Linkum [*sc.* Lincoln]. 1930 *Amer. Speech* V. 494 Did any boy of the latter part of the last century ever *wrestle*? Ozarkers are not peculiar in their *rasseling.* 1974 *Plain Dealer* (Cleveland, Ohio) 27 Oct. 2-c/1 Even the second Ali-Frazier fight became a financial success for the promoters after Ali put on his rassling match with Frazier during that TV interview. 1975 *New Yorker* 1 Sept. 21/3 Mr. Hayes betrayed this when he mispronounced 'wrestling'; he called it 'wrestling'. Most self-respecting promoters, like Sid Morse, of Saginaw, Michigan, know that the correct pronunciation is 'rasslin'. 1976 L. Deighton *Twinkle, twinkle, Little Spy* xiii. 133 Crude Yankee wrassling, was it? Not the kind of cricket you play at Lord's?

wretched, *a.* Add: **1. b.** Phr. *wretched of the earth* [tr. F. *damnés de la terre* (F. Fanon 1961, as book title)].

1965 C. Farrington tr. F. Fanon (*title*) The wretched of the earth. 1970 *Guardian* 21 Aug. 11/3 Are the refugees, the most wretched of the earth, fertile ground for revolutionary activity on a mass scale? 1979 *Country Life* 11 Oct. 1236/3 Our own native 'wretched of the earth', the alcoholics, the inadequates, the very poor. 1983 C. Driver *British at Table* viii. 140 Protein alone cannot rescue the wretched of the earth.

wriggle, *v.* Add: **1. c.** Also with quasi-obj. *to wriggle it,* to move with a wriggling motion.

1922 Joyce *Ulysses* 477 Come on, boys! Wriggle it, girls!

wriggle-. Add: **wriggle-work** (see quot. 1960); cf. *wriggled work* (next entry).

1960 *Connoisseur's Handbk. Antique Collecting* 312/2 *Wrigglework,* a form of engraved decoration on pewter and silver, employing a zig-zag line cut by rocking a gouge from side to side in its progress. Used in conjunction with line engraving at certain periods, principally late in 17th cent. 1975 *Oxf. Compan. Decorative Arts* 616/2 The flat-lid tankard shows the pewterer's craft at its best. The plain drum ..was sometimes decorated with 'wriggle-work' designs of symbolic or commemorative significance. 1982 'J. Gash' *Firefly Gadroon* v. 62 The plate..was wriggle-work... This was a William III plate, with..a rim decorated by engraved wriggles.

wriggled, *ppl. a.* (In Dict. s.v. Wriggle v.) Add: *wriggled work* = *wriggle-work* s.v. *wriggle-.

1906 N. H. Moore *Old Pewter* i. 22 The tool which makes the wriggled work is of the nature of a chisel. 1955 R. F. Michaelis *Antique Pewter* ix. 86 English pewter from the best period, i.e. the 17th century,..displays remarkably fine applied decoration..in the form of 'wriggled-work' engraving.

wring, *sb.*² Add: **4.** Comb.: *wring-world.*

1885 G. M. Hopkins *Poems* (1967) 99 But ah, but O thou terrible, why wouldst thou rude on me Thy wring-world right foot rock?

wring, *v.* Add: **IV. 22.** Comb.: † **wring-jaw** *U.S. slang,* rough cider.

a 1775 [see *'simmon *sb.*²]. 1845 J. F. Cooper *Chainbearer* I. iii. 46 'To get a sup of cider for old Jaap.'.. His weakness in favour of wring-jaw being a well-established failing.

wringer. Add: **6. b.** Fig. phr. *to put through the wringer* and varr.: to try or test (a person or, *rarely,* a thing); *esp.* to subject to severe questioning. *slang* (orig. *U.S.*).

1942 *Sun* (Baltimore) 20 June 15/1 With its capitalization put through the wringer through reorganization..the Erie board voted a payment of 550 cents a share. 1950 T. Sterling *House without Door* (1951) xviii. 196 Every one of them was being blackmailed..except one woman, and she was put through the wringer another way. 1965 J. Philips *Twisted People* iv. 56 We felt..that there had been sabotage... Everybody..was put through a wringer by the CIA. 1972 L. Lamb *Picture Frame* xv. 133 Do you think we had better pick him up and put him through the wringer? 1977 D. Francis *Risk* xiv. 184 If I hadn't recently been through so many wringers..I wouldn't have given it another thought. 1984 *Times* 3 July 12/6 Not since the controversial Bishop of Durham..has an episcopal appointee been put through the wringer in this fashion.

wrinkle, *sb.*¹ Add: **I. 3. b.** *spec.* A minor difficulty or irregularity; a snag; freq. in phr. *to iron out the wrinkles.*

1966 D. F. Jones *Colossus* i. 15 As a project it's practically finished, we can't find any more wrinkles to iron out; we've checked and checked again. 1975 *Economist* 22 Feb. 92 The way for the east Europeans to reach western markets without accumulating further huge trade deficits is to import skills which can be exported in hardware. This also enables them to iron out wrinkles in their own system without having to embark on risky economic reforms. 1979 *Guardian* 30 Aug. 3/6 The BBC wanted to make certain advances in technical practices... Wrinkles still remained. 1984 *New Yorker* 14 May 43 Willa had sold her story to Universal Pictures and was in California ironing out some wrinkles in the deal.

II. 8. a. Esp. (*U.S.*) in phr. *a new wrinkle.*

1941 W. C. Handy *Father of Blues* iv. 35 In addition to twirling their batons, they added the new wrinkle of tossing them back and forth to each other as they marched. 1969 *Wall St. Jrnl.* 12 Aug. 3/3 The idea for the briefings, a new wrinkle in selling Presidential policy, came from White House communications director Herbert Klein. 1978 *New York* 3 Apr. 37/3 In his budget proposals, the president came up with an investment tax credit with a new wrinkle: If passed by Congress, it will apply to structures as well as to capital equipment. 1984 *Gainesville* (Florida) *Sun* 29 Mar. 4A (*cartoon*) This guy Chernenko should put a new wrinkle in Russian politics.

III. 10. *winkle-free, -proof, -resistant* adjs.; **wrinkle ridge** *Astr.*, one of the long, irregular ridges that can be seen on the maria of the moon and Mars.

1963 *New Yorker* 8 June 74 (Advt.), Stay neat and wrinkle-free all day. 1978 *Detroit Free Press* 16 Apr. 9A (Advt.), Sheets in wrinkle-free and easy-care cotton/polyester percale. 1957 *Economist* 31 Aug. 685/2 The steady introduction of new fibres..new chemicals..to render cloth..wrinkle-proof. 1957 M. B. Picken *Fashion Dict.* 382/2 Wrinkle-resistant. 1969 *Sears Catal.* Spring/Summer 20 Perma Prest for great no-iron, wrinkle-resistant performance. 1944 J. E. Spurr *Geol. applied to Selenology* I. viii. 60 The wrinkle-ridges are distinct from faults in appearance... The ridges on the surface of the mare are not straight; they are curving, branching, imbricating, plaited. 1971 I. G. Gass et al. *Understanding Earth* vii. 106 (*caption*) The wrinkle ridges on the surface of Mare Tranquilitatis. 1978 *Sci. Amer.* Mar. 81/1 The Viking orbiter photographs show that much of the surface of Mars retains crisp topographic detail: lava flows, wrinkle ridges and crater ejecta stand out in sharp relief.

wrinkly, *a.* Add: **B.** *sb.* Also **wrinklie.** An old or middle-aged person. *slang.*

1972 A. Birchall *Living in Landscape* 51 (*heading*) What do we do with the Wrinklies? 1976 *Times* 31 Aug. 10/8 A Henley reader ..says that her teenage daughter reserves 'wrinklie' for the 60-year-old generation. The reader and her husband..are known as 'oldies'. 1980 *Times* 28 Oct. 12/6 It's pointless to go on a CND march. They're all wrinklies. 1982 Barr & York *Official Sloane Ranger Handbk.* 159/3 *Wrinkly* n., middle-aged Sloane—between 40 and 50. 1983 *Church Times* 11 Mar. 13/2, I am a wrinkly whose monthly cheque from the Church Commissioners is labelled 'Diocesan Dignitary'.

wrist. Add: **1. c.** (Earlier example.)

1803 D. Wordsworth *Jrnl.* 10 Jan. (1941) I. 188 Worked all day—petticoats—Mrs. C.'s wrists.

5. a. *wrist-bangle*; in the names of devices worn on the wrist, as *wrist compass, radio,* etc.

1922 Joyce *Ulysses* 432 Fiercely she slaps his haunch, her goldcurb wristbangles angriling. 1983 D. Hart-Davis *Fire Falcon* xxiii. 272 His only means of steering was his wrist compass. 1972 *Times* 3 Nov. 33/3 It is a world first, it enables the memorable 'wrist radio' label of the Dick Tracy strip cartoons to become reality. 1984 *Listener* 17 May 36/3 You have the Snoop-Mobile, a wrist-radio, a list of suspects and information about each of them. 1984 *Tampa* (Florida) *Tribune* 5 Apr. 6B/4 Cellular mobile radio telephone service..could be the forerunner of Dick Tracy-like wrist telephones. 1972 D. Bloodworth *Any Number can Play* x. 81 I'm going to grow up into a millionaire cowboy with a two-way wrist-television and a formula car.

c. *wrist-play* (earlier example), *-stroke* (earlier example).

1851 J. Pycroft *Cricket Field* vii. 141 All that is required is, straight play and a free wrist... Without wrist play there can be no good style of batting. 1888 R. H. Lyttelton in Steel & Lyttelton *Cricket* ii. 61 The cut..requires a very strong use of the wrist, and, like all wrist strokes, charms the spectator by accomplishing great results at the expense of apparently little effort.

d. wrist-length *a.*, (*a*) (of a glove) reaching as far as the wrist; (*b*) (see quot. 1957); **wrist-slap** *slang,* a mild rebuke; so **wrist-slapping; wrist-spin** *Cricket,* spin imparted to a ball by the wrist; cf. *finger-spin* s.v. *finger *sb.* 15; so **wrist-spinner, wrist-spinning** *vbl. sb.*; **wrist-wrestling,** a contest of strength between two people, each trying to force the arm of the other person backwards (strictly by interlocking thumbs instead of gripping hands); arm-wrestling; so **wrist-wrestler.**

1935 E. Farjeon *Nursery in Nineties* iv. iii. 172 Long evening gloves and wrist-length, kid and suède gloves! 1957 M. B. Picken *Fashion Dict.* 382/2 *Wrist length,* length of coat or other garment, taken with arms hanging at sides, which reaches to wrist. **1963** *Guardian* 1 Feb. 9/7 Jackets are either straight and short to the hips or straight to wrist-length. **1977** M. Edelman *Political Lang.* viii. 148 Antitrust laws similarly sanction mergers and pricing agreements, with occasional token wrist slaps to keep the symbolism pure. **1979** *Time* 13 Aug. 36/3 Critical as the investigators may have been of the utility, the NRC itself got a wrist slap from Congress. **1958** *Times* 24 May 4/2 This unusual example of mass wrist-slapping has been going on for a week. **1979** N. Slater *Falcon* viii. 141 There was no sherry decanter in evidence, no coffee... This was turning out to be a right old wrist-slapping session. **1960** E. W. Swanton *West Indies Revisited* iii. 49 He is reputedly unhappy against wrist-spin. **1977** *New Society* 3 Feb. 246/2 Raffles was a leg-break bowler: can wrist-spin ever be *really* kosher? **1957** T. Bailey *Cricket Bk.* vi. 66/1 Bruce Dooland..clearly showed what destruction a top-class wrist-spinner can achieve in Championship cricket. **1977** *Listener* 5 May 588/1 Garfield Sobers—finger-spinner, wrist-spinner, seam-bowler. **1963** T. E. Bailey *Improve your Cricket* i. 31 Slow bowlers [from overseas] are more frequently of the wrist-spinning variety. **1978** *Detroit Free Press* 16 Apr. 1A/1 My uncle Gerald is a pretty good wrist wrestler. **1973** *N.Y. Times* 29 July x. 4/5 Wrist wrestling, also known as arm wrestling, has its real roots in Petaluma, Calif., where the world championship matches have been televised on ABC's Wide World of Sports for the past four years. **1978** *Maclean's Mag.* 12 June 62 It seems perfectly natural for a wristwrestling championship to be held in Timmins.

wristband. Add: **4.** In sport, a strip of material worn round the wrist to absorb perspiration.

1969 *New Yorker* 14 June 68/3 Ashe wipes his forehead with his wristband. **1984** *Oxford Times* 29 Feb. 3/7 (Advt.), Headband and wristband pack—£1.79.

wristy, *a.* Delete *Cricket* and add earlier and later examples.

1867 *Australasian* 9 Mar. 300/2 Fowler, pretty wristy style, but not a very safe one. **1936** Wodehouse *Laughing Gas* xxii. 242, I remember..wondering how the dickens a female of her slight build and apparently fragile physique could possibly get that wristy follow-through into her shots. **1955** *Times* 18 July 12/3 With the deftest turn to leg and wristy punches through the covers he seemed set for an imposing score. **1959** *Times* 26 June 4/2 Krishnan's forte lay in his lovely wristy stop volleys. **1977** *World of Cricket Monthly* June 41/2 There was no prodding or pushing but a free swing on wristy cut. **1980** R. Hill *Spy's Wife* xix. 148 He downed his pale spirit in one quick wristy movement.

Hence **wri·stily** *adv.*

1963 A. Ross *Australia 63* iii. 82 Barrington cut wristily to beat third man. **1972** P. Brent *Godmen of India* ii. 33 Women polish the earthen floors, bending straight-legged as they work wristily away with a flat stone.

writable, *a.* Add: Also **writeable. 1.** (Later examples.)

1913 G. B. Shaw *Let.* 26 Mar. (1952) 104, I have written everything that is writeable: The rest must be viva voce. **1926** B. Karlgren *Philol. & Anc. China* vii. 159 If the literary language is based on the modern colloquial language, it should..also be writable phonetically. **1970** *IEEE Trans. Computers* XIX. 710/2 The control memory of LX-1 is read-only with respect to the microprogram, but is externally writeable. **1983** *Sci. Amer.* Mar. 43/2 Endowing a computer with a writable control store is a way of removing the hardware barrier from user microprogramming.

write, *v.* Add: **I. 1. d.** Freq. in phr. *to be* (or *have*) *written all over* a person.

a **1899** [in Dict.]. **1914** 'I. Hay' *Knight on Wheels* (ed. 2) xxix. 292 It must be written all over me if you can spot it... Yes, you are right... I'm in love. **1967** G. F. Fiennes *I tried to run Railway* iii. 28 He had horse written all over him. **1979** J. Gardner *Nostradamus Traitor* vi. 20 One was with her... Had DDR written all over him.

3. d. Also in analogous *fig.* phrases, as *writ double, small*, etc.

1951 E. Barker *Princ. Social & Political Theory* 1. 39 Corporativism may be defined as syndicalism writ double. **1959** *Times* 25 Feb. 11/2 This year's Defence White Paper..is last year's writ quietly. **1961** *Observer* 23 Apr. 5/2 In a curious way he's [*sc.* Sir Isaac Hayward's] an amalgam, writ small, of Attlee, Morrison and Bevin. **1967** *Listener* 8 June 762/1 J. P. Donleavy's *The Saddest Summer of Samuel S.* is just *The Ginger Man* writ smaller and smaller.

h. *Computers.* To enter (an item of data) *in, into, on*, or *to* a storage medium (esp. a disc or tape) or a location in store; to enter data in or on (a storage medium). Also *absol.* Cf. *to read in* s.v. *READ *v.* 6 f.

1946 Goldstine & von Neumann in J. von Neumann *Coll. Wks.* (1963) V. 28 In 'writing' a word into the memory, it is similarly not only the time effectively consumed in 'writing' which matters, but also the time needed to 'find' the specified location in the memory. *Ibid.*, A number that is to be written, i.e. stored, has to be placed at a definite, possibly inconvenient place in the memory. **1948** *Math. Tables & Other Aids to Computation* III. 123 The machines will be able to read from, or write on, the tapes. **1953** B. V. Bowden *Faster than Thought* iv. 95 He proposed to make it impossible to write into a store unless it contained zero. **1966** *McGraw-Hill Encycl. Sci. & Technol.* IV. 188/1 The store instruction selects an address through the selection circuit for writing the contents of the accumulator in the memory location specified. **1970** [see *READ *v.* 5 h]. **1973** C. W. Gear *Introd. Computer Sci.* iv. 161 A typical large computer system has many readers and printers... Usually several different jobs are being read and several different outputs are being written at the same time. **1980** *Sci. Amer.* Aug. 114/1 The head that writes the data can also be used to read it. **1980** S. Hockey *Guide Computer Applications in Humanities* ii. 28 Information can only be written to the tape when this ring is in place.

i. Of a recording device: to produce (a graphical record).

1949 [see *MAREOGRAM]. **1975** *Nature* 6 Feb. 423/1 Our predicted signals do not resemble those of typical creep events as written by creepmeters.

j. To sit or take (a written examination). Chiefly *S. Afr.*

1958 *Cape Argus* 7 Nov. 3/3 Several women attended the course but Miss — was the only one to write the course examinations. **1971** *Sunday Express* (Johannesburg) 28 Mar. (Home Jrnl.) 14/2 My daughter is writing Matric this year. **1974** *Advocate-News* (Barbados) 19 Feb. 1/1 Students from Government primary schools will now write the Common Entrance Examination at their respective schools.

7. a. Also, to insert (provisions, etc.) *into* a law, agreement, etc.

1962 *Listener* 25 Jan. 155/1 All sorts of safeguards have been written into the agreements. **1962** *Rep. Comm. Broadcasting 1960* 138 in *Parl. Papers 1961–2* (Cmnd. 1753) IX. 259 A suitable form of words to this effect should be written into the new Charter. **1967** *N.Y. Herald Tribune* (International ed.) 11–12 Feb. 3 The 25th Amendment to the Constitution, spelling out procedure for the vice-president to serve as acting president when the president is disabled, was written into law to-day.

12*. = Underwrite *v.*[1] 2 b.

1882 'F. Anstey' *Vice Versâ* xvi. 298 They talked of 'risks', of someone who had only been 'writing' a year and was doing seven thousand a week,..and of the uselessness of 'writing five hundred on everything'. **1931** *Times* 14 Mar. 12/6 Not all insurance companies have felt justified in writing the risks. **1967** *Listener* 6 July 14/3 The company was still writing insurance in eleven American states. **1976** *Daily Tel.* 1 Nov. 16/2 Settlement of any claims will cost up to 50 p.c. more than had been expected when the risk was written.

13. write down. f. To write (a literary work) in a style adapted to the level of readers of supposedly inferior intelligence or taste. Cf. sense 21 c below.

1876 C. M. Yonge *Womankind* xxviii. 243 Books..which do not dwarf the mind as a series of books written *down* are apt to do.

14. write in. b. To send (suggestions, etc.) in written form to an organization. Cf. sense 22 c below.

1928 *Publishers' Weekly* 14 July 183 The customers..were not slow about writing in their suggestions. *a* **1961** J. Britton in Webster, Teachers are encouraged to write in their requests.

c. To insert (the name of an unlisted person) on a ballot-paper or the like, as the candidate of one's choice. *U.S.*

1932 *Sun* (Baltimore) 23 Aug. 2/2 He knew nothing of the circulation of cards in the Middle West urging voters to write in Smith's name on the Presidential ballot. **1944** *Greeley* (Colorado) *Daily Tribune* 16 Sept. 2/2, I greatly appreciate the good will expressed and effort expended by the friends who wrote in my name as candidate for County Judge on the Republican primary ballot. **1957** *Ann. Reg. 1956* 174 Democratic voters 'wrote in' their preference for Mr. Stevenson over Senator Kefauver in the proportions of 8 to 5. **1968** *New Yorker* 9 Mar. 32 You know who I wrote in? You, Earl.

15. a. write off. Now freq. *fig.*, to dismiss from consideration as insignificant or irrelevant.

1957 P. Lafitte *Person in Psychol.* 44 The psychologist..can write off the difficulty as not falling within the scope of scientific method. **1963** L. MacNeice *Var Parable* (1965) i. 21 A suspension of antipathy towards its author's attitude will give weight to what otherwise might be written off as whimsical. **1973** *Times* 20 Oct. 18/6 He is part of me and I of him. I find that painful. Perhaps he does too. But we cannot write each other off. **1984** A. Smith *Mind* v. xv. 297 To cover all possibilities so that the seemingly dead are not written off medically and therefore legally, before their time. **1985** *Times* 11 Jan. 12/6 All this is part of an exercise..to help girls be more assertive and self-confident about their educational potential: to stop writing themselves off as mere future wives and mothers.

c. *slang* (orig. *Air Force*). To damage beyond repair, wreck (an aeroplane, motor vehicle, etc.).

[**1922** *Flight* 27 July 423/1 In another way, it may be stated that, should the work of the Committee lead to a reduction by one of the aeroplanes written off per year as a result of crashes, [etc.].] **1931** *Ibid.* 23 Jan. 80/1 The D.H.37..got down without much damage. The D.H.9.C..was less fortunate, and was written off in a forced landing. **1942** N. Balchin *Darkness falls from Air* ii. 44 They seemed to be dropping a hell of a lot of stuff... I saw next morning that they'd written off a pub in Notting Hill. **1973** C. Bonington *Next Horizon* xi. 166 She had crashed the car twice, writing it off completely on the second occasion. **1982** *Daily Tel.* 27 Oct. 3/6 He..wrecked his lorry and two cars; pulled out in front of a van and wrote that off too.

16. b. write out. (Earlier example.)

1817 *Blackw. Mag.* I. 519/2 We have heard fears expressed, that Miss Edgeworth might have written herself out.

c. To eliminate or contrive the temporary absence of (a character, etc., in a long-running radio or television serial), with the story-line written so as to account for it.

1967 *Listener* 13 Apr. 503/2 That [*sc.* the *Forsyte*] *Saga* is now more than half way through (Saturdays, BBC-2)... Some of the old characters have been written out. **1969** *Photoplay* Jan. 64/2 Being 'written out' of 'Peyton Place' is no disgrace. It has happened to other fine players. **1971** O. Norton *Corpse-Bird Cries* i. 2 You got them to write you out for a bit. **1982** A. Road *Dr. Who: Making of TV Series* 16/1 Eric Saward was asked..in the course of his story..to 'write out' the Doctor's sonic screwdriver. **1984** 'M. Innes' *Carson's Conspiracy* xiv. 149 Appleby took a searching look at her and—as it might be expressed—wrote her out of the story.

18. e. write up. (Earlier example.)

1726 [see sense 13 c in Dict.].

III. 21. c. Esp. as *to write down*, to adapt one's literary style to the level of readers of supposedly inferior intelligence or taste; freq. const. *to*. Cf. sense 13 f above.

1809 [in Dict.]. **1851** *Househ. Words* 11 Jan. 372/2 Mr. Blackbrook and his disciples are hapless materialists, verse-makers without a sense of the beautiful. They are patronised by those to whom they write down. **1861** [in Dict.]. **1903** A. Bennett *Truth about Author* xii. 150, I had entered into a compact with myself that I would never 'write down' to the public in a long fiction. **1921** *Sci. Amer.* Nov. 20/1 The Editor both 'writes down' and 'writes up'... He may translate the Einstein theories into the nontechnical phrases of everyday life. **1944** L. MacNeice *Christopher Columbus* 9 The inference that to hold the attention..a writer has got to 'write down'. (By writing down I mean pandering—writing by standards which the writer considers low.) **1960** *Guardian* 25 Feb. 6/4 English writers seem to write down to their readers and American writers write as if addressing their equals.

22. b. (*b*) (Further examples.)

The use is more widespread than is indicated in Dict.

1864 Newman *Apologia* vi. 346 When friends wrote me on the subject, I either did not deny or I confessed it. **1892** G. & W. Grossmith *Diary of Nobody* iii. 41, I wrote Merton to that effect. **1900** [see Phone *sb.*[2] and *v.*]. **1905** Havelock Ellis *Stud. Psychol. Sex* IV. 239 She wrote me saying that she could not see me any more. **1922** C. Mackenzie *Altar Steps* xxiii. 263, I will write you again when I have seen Father Burrowes. **1924** —— *Old Men of Sea* xi. 175, I shall write Mr. Hibben about that little joke. **1928** D. L. Sayers *Ld. Peter views Body* iv. 74 He wrote me yesterday and said he'd accidentally left a bag in the cloakroom. **1953** Woodhouse *Performing Flea* 69 She is going to find out about quarantine and then write me. **1955** J. P. Donleavy *Ginger Man* xi. 104, I haven't. You can't blame me. I'm sorry I wrote your father. I'm sorry for it. **1968** *Globe & Mail* (Toronto) 17 Feb. B3 (Advt.), For free literature describing the..accommodation..write [*address given*]. **1973** *Black Panther* 17 Nov. 10/2 It is circulating an impeachment petition nationwide while encouraging all citizens to write their congressmen. **1974** I. Murdoch *Sacred & Profane Love Machine* 35, I wrote you all about California—quite long letters—about the animals and so on. **1977** I. Shaw *Beggarman, Thief* I. i. 2 He lives in Chicago now and writes me often.

c. write in (earlier *Theatr.* example). Also in gen. use, to send a written comment, request, etc., to an organization. Cf. sense 14 b above.

1849 *Theatrical Programme & Entr'Acte* 23 July 59/2 The time that elapsed between his last application to Drury-lane and his appearance was many months, for he 'wrote in', as it is termed from Exeter about the early part of the summer of 1813, acted first in London, January 26th 1814. **1931** *Publishers' Weekly* 5 Dec. 2471/2 The 'Brooklyn Eagle', however, complains bitterly about distribution: 'About seven hundred readers have written in—ever since that squib of ours appeared..asking where they can put their hands on one.' **1949** N. Marsh *Swing, Brother, Swing* ix. 209 It's a mystery, that paper... The types that write in are amazing. **1957** M. McCarthy in *New Yorker* 23 Mar. 76/2, I wrote in for a Vogue pattern to make a tennis dress. **1972** *Listener* 28 Dec. 904/3 If anyone else doesn't know... write in and I'll explain. **1977** *Broadcast* 13 June 10/1 The chap who writes in about a programme.

d. *to write home about*: see *HOME *adv.* 7 d.

V. 27. *Computers.* The infin. used *attrib.* and in *Comb.* with the sense 'writing': **write-permit ring** (pɜ·ɹmit), a ring which has to be inserted in the hub of a tape reel before the tape can be written to or erased; **write-protect** *v. trans.*, to protect (a disc) from accidental writing or erasure, as by removing the cover from a notch in its envelope; also as *sb. attrib.*, designating such a notch, etc.

1951 *Proc. Inst. Electr. Engineers* XCVIII. ii. 15/2 A number can be..written in via the 'write' terminal. **1958** *Communications Assoc. Computing Machinery* Feb. 30 In each track, and separated by ·005" from the associated read head, is the 'write head'. **1961** L. W. Hein *Electronic Data Processing for Business* v. 77 The reflective spot activates the switch only if the tape unit is in write status, that is, information is being written on the tape. **1964** F. L. Westwater *Electronic Computers* iv. 78 In the early applications valves were used to supply the relatively large read and write currents. **1965**, etc. [see *READ *v.* 22]. **1970** A. Chandor *Dict. Computers* 402 Write permit ring. **1980** C. S. French *Computer Sci.* xii. 62 Each record is written onto tape in response to a 'write instruction'. **1980** S. Hockey *Guide Computer Applications in Humanities* ii. 27 One way of ensuring that this does not happen is to use..a write permit ring, a plastic ring which can be inserted in the back of a tape reel. **1981** *Your Computer* May–June 100/3 *Write protect*, to remove the cover from the notch in a floppy disk so that it cannot be written on. **1983** *Austral. Personal Computer* Aug. 104/2 The only other hole in the disk envelope of importance to the user is the write protect notch... This notch must be covered up by an opaque material to write protect the disk.

writeable, var. Writable *a.* in Dict. and Suppl.

wri·te-back. [f. the vbl. phr. *to write back*.] The process of restoring to profit a provision for bad or doubtful debts previously made against profits and no longer required.

1979 *Financial Times* 1 Jan. 21/5 In future the clearers will only make provision for the taxes actually expected to become payable in the foreseeable future. This will result in write-backs of substantial amounts. **1980** *Times* 5 Aug. 17/7 Only then will it emerge whether there are any writebacks to profits arising from the results of the first six months of this year. **1985** *Times* 14 Mar. 14/6 Large write-backs of provisions made in earlier years and no longer needed kept the total..to only £4m in 1981.

wri·te-down. [f. the vbl. phr. *to write down*: see Write *v.* 13 e.] A reduction in the estimated or book value of assets.

1932 *Daily Express* 28 Jan. 10/7 This has involved a write-down of securities. **1955** *Times* 10 May 19/2 Consequent write-downs which, with the higher prices now ruling for footwear, are necessarily more severe than in the past. **1972** *Accountant* 23 Mar. 383/2 A large part of AEI's profit shortfall was attributable to stock and work-in-progress write-downs. **1978** S. Brill *Teamsters* vi. 253 Executive Director Shannon, while only conceding 'possible' write-downs.. **1979** *N.Y. Times* 13 Sept. D1/2 The Polaroid Corporation..would take a $68 million write-down in the third quarter.

wri·te-in. [f. the vbl. phr. *to write in*: see Write *v.* 14 c, 22 c. For sense 2 (first part of def.), cf. also *-IN[3].] **1.** The name of an unlisted candidate inserted by a voter on a ballot-paper, etc., as the candidate of his choice; a vote cast for such a candidate, or the act of voting in this way. Freq. *attrib.* orig. and chiefly *U.S.*

1932 *Sun* (Baltimore) 23 Aug. 2/2 (*heading*) Smith's office denies all knowledge of write-in cards being circulated. **1934** *Ibid.* 4 May 2/1 In that election McKee, although not even a candidate, received a 'write-in' vote of nearly a quarter of a million. **1937** *Ibid.* 18 Sept. 10/3 The really significant item in the returns from New York..is the extraordinary number of 'write-ins' for Mayor La Guardia on Democrat ballots. **1962** *Chicago Tribune* 2 Apr. 40/1 Such an attempted write-in for any candidate would result in many spoiled ballots. **1970** *Times* 30 Dec. 10223/1 Stevenson got a tremendous write-in vote.. Johnson *White House Diary* 10 Mar. (1970) 84 There was a creditable number of write-in votes for Attorney General Kennedy for Vice President. **1977** *Daily Colonist* (Victoria, B.C.) 1 Nov. 1/8 Another man who was out of town when filing closed says he'll gladly be a write-in candidate for councilman. **1982** *Daily Tel.* 13 Aug. 4/1 Three other spaces will be left for 'write-ins'.

2. A protest in the form of mass letters of complaint; also, an invitation from a radio broadcast to its listeners to write in and express their views. Cf. *PHONE-IN.

1972 *Listener* 15 June 780/3 I'm proposing a mass write-in.. **1981** Mason & Chambers *Abortion Politics* i. 27 Both organizations lobbied in Parliament and both organized meetings and write-in campaigns in the constituencies. **1981** *Church Times* 7 Aug. 16/3 In 1976, when the BBC's *Sunday* programme ran a write-in on the subject of a Graham mission, 15,000 listeners replied. **1984** *Times* 10 Feb. 14/8 There has been a 'strong response' to a direct-mail campaign asking pitmen to complete a write-in slip.

wri·te-off. [f. the vbl. phr. *to write off*: see Write *v.* 15 a in Dict. and Suppl., *WRITE *v.* 15 c.] **1.** (In Dict. s.v. Write *v.* 15.)

2. (In Dict. s.v. Write *sb.*[3] 1 b.) Substitute for def.: The cancellation from an account of a bad debt, worthless asset, etc.; an asset so treated; an amount cancelled or lost. (Later examples.)

1957 *Times Lit. Suppl.* 20 Dec. 766/5 It is the point made earlier that the need for resources wherewith to replace the automation machines, the need for more rapid write-offs of those machines once installed.. **1976** *Daily Tel.* 26 Apr. 13/4 Above all, can share-holders be given an analysis of write-offs and a pre-tax comparison for the two halves? **1982** S. Bellow *Dean's December* iv. 96 'Have you ever gotten a penny out of it?' 'I got tax write-offs.' **1984** *Times* 23 May 20/8 In competitor countries the comparison varies between one sort of asset and another but, in general, our write-off periods will be comparable with those overseas.

3. a. An aeroplane, motor vehicle, etc., so badly damaged as to be not worth repair; a wreck; an act of irreparably damaging. orig. *Air Force.*

1918 J. M. Grider *War Birds* (1927) 89 He wasn't hurt but the Spad [*sc.* an aeroplane] was a write-off. **1927** E. W. Springs *Nocturne Militaire* 250 Both machines were complete write-offs and the only things they salvaged were the magnetos. **1944** [see *OVERSHOOT *sb.*]. **1953** J. Trench *Docken Dead* xiii. 199 'What have you done to my car?' 'I'm afraid it's rather a write-off, sir, as far as immediate use goes, anyway.' **1963** N. Marsh *Dead Water* (1964) iv. 149 The hotel launch was still jammed... A complete write-off, it was thought. **1971** *E. Afr. Standard* (Nairobi) 10 Apr. 8/5 The car I tried was well run in and had been rebuilt from a write-off condition. **1977** *Offshore Engineer* June 13/1 The Ekofisk incident has served to sharpen the risk perception of something more serious—like a major explosion, causing a platform write-off, and pollution seepage.

b. Something or someone dismissed as worthless or ineffectual; a failure.

1960 *Guardian* 16 Sept. 7/5 On English television this [film] would have been a write-off. **1966** *Listener* 17 Feb. 257/3, I find the new record of Rossini's *Stabat Mater*..an almost complete write-off.. the pregnant woman's a write-off from the sex appeal point of view after the fourth month. **1982** M. Russell *All Part of Service* ii. 15 If it involves such an effort, I think I might just as well regard this evening as a total write-off. **1984** A. Brookner *Hotel du Lac* v. 65 The day would be a write-off.

writer. Add: **3. c.** *writer's block* [*BLOCK *sb.* 19 e], a periodic lack of inspiration afflicting creative writers; *writer's writer*: a writer whose appeal is primarily to his fellow writers (cf. *poet's poet* s.v. *POET 1 c).

1950 E. Bergler *Writer & Psychoanal.* vi. 113 Writer's block sets in the moment the inner conscience rejects the alibi and substitute alibi. **1966** C. Baxt *Queer Kind of Death* xi. 150 Seth has had a serious writer's block for almost two years now. **1975** M. Bradbury *History Man* ii. 69 This book..has decidedly not gone well. I've had what they call writer's block. The words won't come. **1983** *Listener* 13 Jan. 12/3 Graham Greene relies heavily on the unconscious to get round 'writer's block' as he revealed in an interview with Nigel Lewis.

1941 'G. Orwell' in *Listener* 12 June 841/1 Hopkins is what people call a 'writer's writer'. He appeals to people who are professionally interested in problems of technique. **1958** *Sunday Times* 13 Apr. 9/5 She [*sc.* Ivy Compton-Burnett] is in the first place 'a writers' writer', because she is fascinated by words and phrases as such. **1980** *Times Lit. Suppl.* 12 Sept. 991/2 Nigel Williams is a writer's writer.

d. *writer-in-residence*: a writer given a residential post in a university, etc., in order to share his professional insights. Cf. *poet-in-residence* s.v. *POET 1 c, *RESIDENCE *sb.* 2 b.

1957 J. D. Salinger *Zooey* in *New Yorker* 4 May 33/3 The second-eldest child, Buddy, was what is known in campus-catalog parlance as 'writer-in-residence' at a girls' junior college in upper New York State. **1972** [see *RESIDENCE 2 b]. **1980** *Times Lit. Suppl.* 2 May 496/3 The initiative to hold a poetry festival to celebrate his [*sc.* Basil Bunting's] birthday came from Tom Pickard, writer-in-residence for this year at the University of Warwick.

writerly (rai·təli), *a.* [f. Writer + -ly[1], after *painterly*.] Appropriate to, characteristic or worthy of a professional writer or literary man; consciously literary.

1957 *Times Lit. Suppl.* 16 Aug. p. xxxvi/2 Serious Canadian writers at present are not inclined to concentrate upon the writerly virtues. **1958** *Spectator* 24 Jan. 114/1 A clever and writerly book. **1980** *Sensible Words*..Dryden..sees his writerly obligations in a new light. **1985** *Listener* 23 Jan. 367/3 James Saunders dissecting writerly old age and the onset of what appears to be creative impotence.

wri·te-up. (In Dict. s.v. Write *sb.*[3] 2.) For Orig. (and chiefly) *U.S.* read orig. *U.S.* and add: Now more loosely, any journalistic account or review, whether favourable or not. (Earlier and later examples.)

1885 *Weekly New Mexican Rev.* 19 Feb. 4/1, I have prepared quite an extensive 'write-up' of the resources of this country. **1920** *Chambers's Jrnl.* July 431/1 The 'write-up' and the interviews are prominent features. **1939** Wodehouse *Uncle Fred in Springtime*.. *The Morning Leader* gave it a fine write-up. **1948** *Penguin Music Mag.* Oct. 29 He (or rather she) studies music with an eye sooner or later to engagement, a photograph in the 'press', and a 'write-up', as they call all criticism appearing in print. **1965** *New Statesman* 14 May 753/2 The *Stern* reporter gave [Prince] Philip one of the warmest write-ups he ever got. **1985** *Contact* (Pre-School Playgroups Assoc.) Feb. 14/2 They prepared a paper summarising their findings... This was circulated to the press, who gave a good write-up.

writhen, *ppl. a.* Add: **1. c.** *spec.* Of antique glass or silver: having spirally twisted ornamentation.

1919 M. Percival *Glass Collector* viii. 115 The..writhen glasses of funnel shapes are survivors of the old Venetian tradition. **1935** *Burlington Mag.* Oct. 150/1 Light vertical flutings, matching the bowl and wrythen in delivery. **1960** *Times* 9 Feb. 20/7 A Henry VIII spoon with writhen finial. **1970** *Canad. Antiques Collector* Oct. 17 The jugs and bowls are decorated with writhen ornament. **1981** *Times* 31 Oct. 8/6 An Edward IV wrythen-knop spoon made in London about 1463.

Hence **wry·thening** *vbl. sb.* (see quot. 1960.)

1960 H. Hayward *Antique Coll.* 313/2 *Wrythening*, diagonally twisted or swirled ribbing or fluting on the bowl or stem of a glass vessel. **1967** S. Crompton *Eng. Glass* ii. 209 (*caption*) Jug in bottle-green glass with white wrythening.

writhing, *vbl. sb.* Add: **1.** Also *spec.*, in old glass: cf. *WRITHEN *ppl. a.*

1926 G. R. Francis *Old Eng. Drinking Glasses* p. xxxi, Wrythen, external decoration of the bowl by twisting or wrything while still hot. **1929** W. A. Thorpe *Hist. Eng. & Irish Glass* v. 166 The wrything was done by twisting the paraison while it was being blown.

writing, *vbl. sb.* Add: **I. 1. d.** *Computers.* The process of causing an item of data to be entered into a store or recorded in or on a storage medium.

1946 [see *WRITE *v.* 3 h]. **1970** O. Dopping *Computers & Data Processing* xv. 244 Most writing errors on tape, like most reading errors, are caused by small particles of dust between tape and magnetic head, and in most cases the dust particle is removed..if the writing is repeated. **1973** C. W. Gear *Introd. Computer Sci.* iv. 164 A computer input/output controller..sequences the reading and writing of characters.

3. d. (Later examples with *down*: cf. *WRITE *v.* 21 c.) Also with *up*.

1951 E. E. Evans-Pritchard *Social Anthropol.* v. 88 In this writing-up side of his work the social anthropologist faces a serious difficulty. **1960** *Guardian* 24 June 10/7 Simplification without 'writing down' can serve a good purpose. **1972** *Listener* 9 Nov. 644 The slick writing-down that many professional children's writers indulge in.

II. 6. Phr. *the writing on the wall* (with allusion to *Daniel* v. 5 and 25–28): warning signs of impending disaster, misfortune, etc.

? **1720** Swift *Poet. Wks.* (1736) 93 A baited Banker thus desponds, From his own Hand foresees his Fall; They have his Soul who have his Bonds; 'Tis like the Writing on the Wall. [**1837**: in Dict.]. **1884** [in Dict.]. **1906** (*title*) The writing on the wall. **1949** E. Coxhead *Wind in West* viii. 211 Just try to see the thing with..your famous detachment, and you'll soon recognise the writing on the wall. You've had your fling. **1965** *Listener* 2 Dec. 925/3 The 'eighties and 'nineties were the Golden Age [of music hall]; and in 1905 the writing was on the wall... Musical comedy, the cinema, television all hastened the decline. **1978** *Lancashire Life* Mar. 50/1, I was a fool not to see the writing on the wall when textile machinery manufacturers were rushing all over the world erecting spinning and weaving machinery.

7. b. Also *spec.* = Hagiographa *sb. pl.*

1909 [Dict., in note s.v. Prophet *sb.* 3]. **1941** R. H. Pfeiffer *Introd. to O.T.* (ed. 5) iv. 61 The third division of the Hebrew Bible, following the Law and the Prophets, is simply called the 'Writings' (Hebrew, *Ketûbîm*) or 'Hagiographa' (sacred writings), because it consists of a miscellany of independent books. **1976** *Church Times* 30 Jan. 6/1 He begins with Ecclesiastes and some of the other books from the Writings—namely, Proverbs, Job and the Psalms.

III. 12. a. *writing-chair* (later examples), *-speed, -system.*

1906 P. Macquoid *Hist. Eng. Furnit.* III. iv. 139 (*caption*) Mahogany writing-chair. **1946** H. P. Maynard in W. S. Knickerbocker *20th-Cent. Eng.* 188 One of the most important skills is that of writing-speed. **1953** C. F. Hockett in Saporta & Bastian *Psycholinguistics* (1961) 58/1 In devising a writing-system one can..eliminate a symbol needed earlier. **1979** *Country Life* 14 June 1910/3 Corner chair appears to be the usual description nowadays: they used to be known as writing chairs. **1980** *English World-Wide* I. i. 20 Almost all Native American writing systems developed..for bilingual education purposes utilize writing systems similar to English.

b. *writing-brush, implement* (earlier example).

1854 Dickens *Hard Times* ii. i. 134 Mrs. Sparsit was conscious that by coming in the evening-tide among the desks and writing implements, she shed a feminine..grace. **1921** H. E. Palmer *Princ. Language-Study* vii. 85 Suppose we wish to make Chinese characters with a native writing-brush. **1978** *China Now* Mar./Apr. 34/3 The following suggested items can be obtained from shops which cater for local Chinese communities: wok (a deep frying pan), abacus, Chinese writing brush.

f. *writing clerk* (earlier example).

1772 J. Wedgwood *Let.* 28 Sept. (1965) 136 What were all the rest of the Writing Clerks doing not to observe this.

13. writing bed, a board or level surface for writing on; **writing block,** (*a*) [*BLOCK *sb.* 10 c] a pad of writing-paper; (*b*) [*BLOCK *sb.* 19 e] = *writer's block* s.v. *WRITER 3; **writing-case** (earlier example); **writing-pad,** (*a*) a blotter serving as a surface for writing on, sometimes (quot. 1895) furnished with writing materials, etc.; (*b*) a pad (Pad *sb.*[3] 4) of note-paper; **writing slider** (see quot. 1969); **writing speed,** (*a*) *Electronics*, the maximum speed at which the electron beam can scan the screen of a cathode-ray tube and its path still be recorded by the excitation of phosphors or on photographic film; (*b*) the effective speed of videotape past a head when the rotation of the head is taken into account.

1911 *Daily Colonist* (Victoria, B.C.) 14 Apr. 3/5 (Advt.), Writing Desks. Just the thing for the home, nicely finished in Imperial Golden Oak, large writing bed with enclosed pigeon holes for papers. **1971** *New Yorker* 8 May 3 Open, it's a sewing table. Closed, it's a writing bed..a fine mahogany Sheraton. **1913** 'S. Rohmer' *Mystery of Dr Fu-Manchu* xxix. 294 For this dreary vigil I had come prepared with a bunch of rough notes, a writing-block, and a fountain pen. **1950** E. Bergler *Writer & Psychoanal.* p. ix, They came with only one purpose in mind—to be cured of their 'writing block' (a euphemism for sterility of productivity). **1977** *N.Y. Rev. Bks.* 15 Sept. 36/3 This connection would also help to explain Darwin's long delay in publishing his theory (certainly he had no 'writing block'). **1983** T. Allbeury *Pay Any Price* xvii. 179 As Randall sat down the consultant took out a pen and reached for a writing block. **1813** M. Edgeworth *Let.* 19 Apr. (1971) 25 Mrs. Sneyd and Emma have given me a most convenient red morocco writing case. **1865** Writing-pad [in Dict. s.v. Pad *sb.*[3] 4]. **1895** *Army & Navy Co-op. Soc. Price List* 598 The 'York' Knee Writing pad. Contains Safety Ink, Scissors, Paper Knife,..pockets filled with Note Paper and Envelopes [etc.]. **1906** E. Johnston *Writing & Illuminating* ii. 50 Under the writing-paper there should be a '*writing-pad*', consisting of one or two sheets of blotting-paper. **1917** *Harrods General Catal.* 297/2 Harrods Writing Pads. 100 Sheets. No. 1. Large 8vo., Thick Cream Wove, Plain, 8 × 5. O/6. **1972** 'W. Haggard' *Protectors* i. 3 Scobell had risen behind his desk. There was a writing-pad on it, a pen—nothing more. **1803** T. Sheraton *Cabinet Dict.* 261 The top drawer [of a lobby chest] is usually divided into two; and sometimes there is a writing slider which draws out under the top. **1969** J. Gloag *Short Dict. Furnit.* (rev. ed.) 731 *Writing slider*, a sliding shelf, made to draw out beneath the top of a chest of drawers. **1933** R. A. W. Watt et al. *Applic. Cathode Ray Oscillograph* ii. 37 This statement should not..be read as indicating the limiting 'writing speed' of the oscillograph used. **1954** Lewis & Wells *Millimicrosecond Pulse Techniques* vi. 196 A small spot size is required in order that the deflection sensitivity [of the cathode ray tube] may be high, but if the size is reduced too much there will not be sufficient brightness to give the necessary writing speed. **1981** I. Hickman *Oscilloscopes* vii. 95 Writing speed is defined as the maximum speed at which a spot, passing once across the tube face, can be photographed under specified conditions. **1983** E. Trundle *Beginner's Guide Videocassette Recorders* i. 3 This was simply the idea of moving the record or replay heads rapidly over the surface of a slowly-moving tape to achieve the necessary high 'writing' speed. **1984** *What Video?* Aug. 24/1 The soundtrack was being recorded at a writing speed of 580 cm per second—that's over 15 times the speed of professional studio recordings.

writing-box. Add: Also, a small portable writing-desk. Cf. Writing-desk 2.

1960 H. Hayward *Antique Coll.* 312/2 *Writing-box stands*: the post-Restoration writing-box might be placed on a stand instead of a table. This had gate-legs and the fall-down flap of the box opened on to them for writing. **1971** *Country Life* 1 July 22/1 Portable table-desks or writing-boxes were recorded in the 15th century.

writing-desk. 1. (Earlier *attrib.* example.)

1807 Jane Austen *Let.* 8 Feb. (1952) 178 She is now talking away..& examining the Treasures of my Writing-desk drawer.

writing tablet. Also, a pad (Pad *sb.*[3] 4) of paper for making notes, etc.; = *TABLET *sb.* 1 e.

1895 *Army & Navy Co-op. Soc. Price List* 619 Writing Tablets. 'The Remember.' Bound in long-grained, polished French Morocco. The Refills are made to slip in and out of case. **1917** *Harrods General Catal.* 296/2 Writing tablets. For Scribbling Memoranda and for School Purposes.

written, *ppl. a.* Add: **1. c.** Expressed in due literary form.

1909 J. R. Ware *Passing Eng.* 183/2 *Not enough written* (Authors', 1870), not sufficiently corrected for style. **1922** F. M. Ford *Let.* 14 Aug. (1965) 141 *Felicity Chimney* is a much more ambitious matter. The only thing that is wrong with it is that it is too written. **1963** *Times Lit. Suppl.* 1 Mar. 154/1 The writing is slipshod and frequently repetitive; in fact, as Henry James would say, it is not 'written' at all.

2. b. Also with *on*.

1948 A. N. Keith *Three came Home* vii. 124 They were constantly looking for my papers, written-on or otherwise. **1955** E. Bowen *World of Love* iv. 77 The written-on blue envelope.

5. b. (Earlier and later examples with *up*.) Also with *off*.

1890 G. B. Shaw in *Star* 7 Mar. 2/3 The imagination of the public has undoubtedly been strongly seized by the spectacle of the much-written-up Tosca at the height of its prosperity. **1961** *Sunday Express* 12 Feb. 9/2 He accuses them of..selling 'written off' car wrecks. **1964** *Times Rev. Industry & Technol.* Feb. 9/1 The old British Lion Corporation..was..wound up in 1954 with a written-off loss of £2,969,000. **1972** *Listener* 6 July 3/2 *Catch 44*..is becoming one of the most written-up television projects in America.

c. *written out*: *spec.* of a writer, that has exhausted his creative capacity. Cf. Write *v.* 16 b.

a **1911** D. G. Phillips *Susan Lenox* (1917) II. xii. 355 He's had several failures... They say he's written out. **1959** C. Williams *Man in Motion* ii. 21 Suzy Patton, the has-been. The written-out writer. **1978** A. Powell *Messengers of Day* vii. 108 After a lifetime of work a novelist can possibly be 'written out'.

wrong, *a.* and *adv.* Add: **A.** *adj.* **II. 4. c.** *Criminals' slang.* Untrustworthy, unreliable; not sympathetic to or co-operative with criminals. Cf. *RIGHT *a.* 8 e.

1908 J. M. Sullivan *Criminal Slang* 27 *Wrong*, man too familiar with police; not to be trusted. **1928** E. Booth in *Amer. Mercury* May 81/2 Aw, don't rap [i.e. speak indiscreetly] to that guy; he's wrong. **1953** W. Burroughs *Junkie* (1972) vi. 58 By and large, the reason a man can't score is because he is known to be 'wrong'. **1955** D. W. Maurer *Whiz Mob* ix. 130 He [*sc.* a pick-pocket] tries to avoid those cities or those districts which are known to be *wrong*, or where the police will not have any part of protecting him. *Ibid.* 140 He was what thieves call a *wrong copper*; that is, he did not take the *fix*.

5. d. Of a painting: having an erroneous attribution.

1969 C. Irving *Fake!* (1970) xiv. 173 It's an ugly thing ..when you have to tell a client he's bought a fake. Of all things in this business..the thing I dislike most is being called in to tell if a painting is right or wrong... Fernand..brazenly offered Juviler a genuine Rouault in exchange for the 'wrong' Dufy. **1979** *Daily Tel.* 28 Feb. 10/2 There are huge numbers of 'wrong' paintings and other works on the market, not strictly fakes, although they often become fakes when resold with the intention to deceive. 'Forty per cent. of the pictures we see are wrong,' said Mr Peter Nahum, the Victorian paintings expert at Sotheby's Belgravia. 'They are wrongly attributed, have a false signature or are genuine contemporary copies.' **1983** *Sunday Times* 10 July 2/3 They invited Ronald Alley, deputy director of the Tate to inspect the pictures. He pronounced them 'wrong'—in art world parlance, fakes.

7. b. *to catch* (a person) *on the wrong foot, to get off*, etc., *on the wrong foot*: see *FOOT *sb.* 29.

f. Mus. *wrong note*: a note such as one would not expect in a given key, a discordant note. Freq. *attrib.*

1934 C. Lambert *Music Ho!* ii. 127 The spicing up of a simple harmonic basis by the addition of what are popularly—and rightly—known as 'wrong notes', such as we find in Auric. **1946** C. Mason in A. L. Bacharach *Brit. Music* x. 139 The 'wrong-note lyricism' of Prokofiev's Third Piano Concerto is as vulgar as the street tunes it distorts. **1958** *Listener* 16 Oct. 623/3 The fierce new musical idioms that had been developed by those whom he [*sc.* Vaughan Williams] called 'wrong-note' composers. **1979** *Oxf. Jun. Compan. Mus.* (ed. 2) 269/2 His [*sc.* Poulenc's] style was neo-classical, full of unexpected twists and delightful 'wrong-note' harmonies.

10. a. *spec.* Of the side (of a highway) reserved for oncoming traffic (in Great Britain the right-hand side, in most other countries the left). Also *joc.* with reference to roads in countries having the opposite system to one's own.

1838 [in Dict.]. **1914** M. Beerbohm in *Eng. Rev.* Dec. 19 Our car..*was*, for an instant, full on the wrong side of the road. **1933** A. G. Macdonell *England, their England* xv. 264 A motor-bicycle..had been taking the natural advantage of its speed..to pass the limousine at fifty-five miles an hour on the wrong side at a blind corner. **1965** L. Sands *Something to Hide* ii. 31 'We..usually winter abroad.' 'Very nice too! If you can get used to driving on the wrong side.' **1972** *Guardian* 27 Nov. 12/7, 750,000 British drivers took their cars abroad last year... Many..accidents..[were] possibly caused by confusion over driving on the 'wrong' side of the road.

B. *adv.* **4. b.** *to get* (someone) *wrong*: to misunderstand a person's meaning or intentions, to misinterpret someone. *slang* (orig. *U.S.*).

1927 Dunning & Abbott *Broadway* (1928) xix. 196 'Ever been accused of murder?'..'Don't get me wrong—that stuff ain't in my line.' **1934** T. Wilder *Heaven's my Destination* ii. 39 Don't get her wrong. **1942** Wodehouse *Money in Bank* (1946) xii. 91 We got Soapy all wrong, Chimp. He's explained everything. **1966** *Listener* 20 Oct. 561/1 Old L.B.J. is riddled with anxiety over the thought that we shall go to our graves having got him all wrong. **1968** *Ibid.* 5 Sept. 308/2 Stuart Hood's review of Harold Nicolson's last volume of *Diaries*..ends sympathetically, but begins with a devastating attack on my father's 'snobbishness'... I think, not merely as his son and editor, that they have got him wrong, and are a little uncertain what snobbishness really means. **1974** N. Freeling *Dressing of Diamond* 200 Don't get me wrong; there's no offence meant.

6. c. wrong-reading *a.*, such as can only be read after being first reversed by a mirror.

1955, 1967 [see *right-reading* s.v. *RIGHT *adv.* 16 b].

wrong-foot, *v. trans.* (Stress variable.) [f. Wrong *a.* + Foot *sb.*] **1.** In tennis, football, etc.: (by deceptive play) to cause (an opponent) to have his balance on the wrong foot.

1928 [implied at the vbl. sb. below]. **1959** *Times* 7 Sept. 15/5 Viney and Hasty caught the defence wrong-footed. **1960** E. S. & W. J. Higham *High Speed Rugby* xi. 136 You could pick up the ball as though to go one side, and then, having picked up the ball, swing to the other side... It will wrong-foot the attackers, thereby giving you more time for your kick. **1960** *Times* 29 Nov. 17/4 Truman found himself being wrong-footed by masked drop-shots [in squash rackets]. **1967** J. Potter *Foul Play* (1968) ii. 28 The younger Fitch was holding forth about his patent method of wrong-footing full backs. **1976** Dexter & Makins *Testkill* 21 Abbott, playing back instinctively, was wrong-footed, bat adrift in his hands.

2. *fig.* To disconcert by an unexpected move; to catch unprepared.

1957 F. Hoyle *Black Cloud* iv. 79 'Let me tell you..that the Government has made enquiries and we are not at all satisfied with the accuracy of your report.' Kingsley was wrong-footed. **1963** 'W. Haggard' *High Wire* xii. 130 The tall man's technique was precisely calculated to put him at a disadvantage... Somehow they could always wrong-foot you. **1971** A. Hunter *Gently at Gallop* xii. 128 They sensed he was close, and they were trying to wrong-foot him. **1983** *Listener* 6 Jan. 5/1 What happens in Washington, Moscow and Geneva will leave British political leaders moving quickly in order not to be wrong-footed. **1984** *Daily Tel.* 5 July 1/2 A walk-out would wrong-foot the union in its endeavour to appear ready for negotiations at all times.

Hence **wrong-footing** *vbl. sb.* and *ppl. a.*

1928 *Daily Tel.* 7 Aug. 12/3 His ground strokes had not the same speed and polish as Austin's, nor could he steer all his volleys into the same wrong-footing area. **1971** Laver & Collins *Educ. of Tennis Player* xvi. 216 Wrong-footing is hitting to the place your opponent has just vacated. **1980** *Sunday Times* 17 Apr. 42/5 They emerge as wry, reflective, deliberately wrong-footing to outsiders.

wrongo (rǫ·ŋo, rǫ·ŋgo). *slang* (chiefly *U.S.*). Also **wronggo.** [f. Wrong *a.* + *-O[2].] A bad, dishonest, or untrustworthy person; a 'wrong 'un'. Also, a counterfeit coin.

1937 'J. Curtis' *You're in Racket Too* xxvi. 264 'Sure it ain't duff?' 'Never brought you nothing that was a wrongo yet, did I?' **1938** J. H. O'Hara *Hope of Heaven* x. 139 If I ever saw a wronggo, that Henderson is it. **1953** T. Runyon *In for Life* v. 87 Tailoring skill was never called on more than in the case of wrongos. A well-known and disliked phony was likely to get a specially made suit not designed to please. **1968** L. W. Robinson *Assassin* vii. 78 Phyllis Carr was a wrongo from the beginning. **1985** 'D. Raymond' *Devil's Home* xxxiv. 163 I've had my eye on both of you..and you look like a couple of wrongos to me.

wrongousness. For †*Obs. rare* read '*Obs.* exc. as *nonce-wd.* (in form *wrongeousness*, after *righteousness*)' and add later example.

1923 D. H. Lawrence *Kangaroo* xvi. 333 The heroic effort to carry out the old righteousness becomes at last sheer wrongeousness.

wrong-slot (stress variable), *v.* [Wrong *adv.* + Slot *v.*[2]] In rally driving: to take the wrong road.

1961 *Motoring News* 23 Nov. 10/3 This was a special stage and..Walter had, in effect, only wrong-slotted. **1963** P. Drackett *Motor Rallying* iii. 36 We wrong-slotted just before the Gavia and dropped three. **1968** [see *ROLL *v.*[2] 18 a].

Hence **wrong-slotting** *vbl. sb.*

1963 P. Drackett *Motor Rallying* iii. 45 Amazing the number of competitors who lose marks because they fail to take into account the extra miles they accrued after wrong-slotting.

wrong 'un (rǫ·ŋ'n). *slang.* Also occas. in standard form **wrong one.** [f. Wrong *a.* + Un, 'un[2].] **1.** *Horseracing.* (In Dict. s.v. Wrong *a.* 6 b, c.) Also *fig.*

1889, 1895 [see Wrong *a.* 6 b, c]. **1935** H. Spring *Rachel Rosing* xxv. 301 Hansford had never been known to tip a wrong 'un.

2. A bad, dishonest, or untrustworthy person, a rogue or crook; one who has gone wrong (see Wrong *adv.* 2 b).

1892 I. Zangwill *Childr. Ghetto* I. xi. 243 'What! aren't you *froom*?' she said... 'No, I'm a regular wrong 'un,' he replied. 'As for phylacteries, I almost forget how to lay them.' **1896** [in Dict. s.v. Wrong *a.* 4 a]. **1902** *Daily Tel.* 11 Feb. 10/7 A welsher can be had up for fraud, and anyone who is known as a wrong one is excluded from the racecourse. **1908** A. Bennett *Old Wives' Tale* iii. ii. 295 She was a tremendous—er—wrong 'un here in the forties. Made a lot of money. **1920** D. H. Lawrence *Lost Girl* vi. 107 The policeman was now convinced the man was a wrong-'un. **1925** E. Wallace *King by Night* xiii. 63, I don't suppose there's a 'wrong one' in London that you don't know. **1951** L. P. Hartley *My Fellow Devils* 194 She could not..expect him to confess in so many words that he was a wrong 'un. **1978** J. B. Hilton *Some run Crooked* ii. 15 It seemed quite a hobby with her—Teds, and drop-outs and wrong 'uns.

3. *Cricket.* **a.** A ball that calls for defensive play on the part of the batsman. **b.** *spec.* = *GOOGLY *sb.*

1897 K. S. Ranjitsinhji *Jubilee Bk. Cricket* iii. 118 Stockwell steadies himself after this, and will not pick another 'wrong 'un'. **1911** E. W. Ballantine in *Even. News* 18 Dec. 1/3, I see Hordern got Woolley with a 'wrong 'un'. For the benefit of those who may not grasp what a 'wrong 'un' is, it may here be stated that the 'wrong 'un' is the ball which breaks the opposite way to that indicated by the bowler's action. **1931** *Daily Tel.* 31 July 17/2 Weir deserved better of fate than to be bowled by Peebles' wrong 'un in the last over of the day. **1956** N. Cardus *Close of Play* 142 These devices were of no use against the mysterious and seemingly illogical spin of the 'wrong 'un'. **1977** *World of Cricket Monthly* June 42/2 Neil showed who was the boss by repeatedly stepping out and driving Gupte's wrong 'uns to the fence.

4. A counterfeit coin.

1899 C. Rook *Hooligan Nights* iv. 58 Billy the Snide produced a wrong 'un, and bade young Alf plant it at a big house near the Walk.

wrongways (rǫ·ŋwēiz), *adv. nonce-wd.* [f. Wrong *sb.*[2] + -ways.] In the direction of wrong-doing.

1922 Joyce *Ulysses* 389 She beguiled him wrongways from the true path by her flatteries.

wrong-wise, *adv.* (Later example.)

1903 A. H. Lewis *Boss* i. 6, I found such stimulus [*sc.* beatings with hickory] to go much against the grain and to grievously rub wrong-wise the fur of my fancy.

wrought, *ppl. a.* **II. 9.** (Later example.)

1962 Auden *Dyer's Hand* (1963) 508 Whereupon we are shown Antony talking to his friends in a wrought-up state of self-dramatization and self-pity.

wrung, *ppl. a.* Add: **1.** Also with *out*.

1976 *Times* 26 Jan. 12/4 Feeling like 60 wrung-out dishrags we stumbled out. **1979** P. Wallage *Restoration Post-War Cars* ii. 26/2 Wiping it off with a wrung-out cloth.

3. c. *wrung out*: completely exhausted.

1962 A. Lurie *Love & Friendship* iv. 68 Say, you do look kinda peaked... Not real bad, just kinda wrung out. **1975** *New Yorker* 29 Dec. 15/2 Although she caromed around her office there at enormous speed, she claimed to be 'wrung out'.

wry, *a.* and *adv.* Add: **C.** *wry-angled, -formed.*

1906 Hardy *Dynasts* II. v. vi. 781 What lewdness lip those wry-formed phantoms there? **1937** Wry-angled [see *mountain-roofed* a. s.v. *MOUNTAIN 7 e].

wrying, *vbl. sb.*[2] **3.** (Later example.)

1879 G. M. Hopkins *Let.* 9 Apr. (1935) 78 It seems to me to hit the mark it aims at without any wrying.

Wu (wu). [Chinese *wú*.] Used *attrib.* of a group of Chinese dialects spoken in Shanghai, the south of Jiangsu province, and most parts of Zhejiang province, China. Also *absol.*

1908 M. Kennelly tr. *Richard's Comprehensive Geogr. Chinese Empire* v. 348 The Ngeu..or Wu Dialects, comprising: 1—The Wênchow dialect. 2—The Ningpo dialect. 3—The Sungkiang or Shanghai dialect. **1943** *China Handbk.* i. 30 The Wu group is spoken south of the Yangtze in Kiangsu... It is characterized by the preservation of the ancient voiced stops as aspirated voiced consonants. **1948** R. A. D. Forrest *Chinese Lang.* xi. 224 The Wu language has in modern times a greater than average hostility to final consonants... Where palatalisation occurs in Wu before a..we regularly find the occlusive beside the fricative. **1964** [see *MIN *a.* and *sb.*[6]]. **1977** 'S. Leys' *Chinese Shadows* (1978) ii. 95 The gracious Wu dialect used in south Kiansu and north Chekiang. **1978** *Whitaker's Almanack 1979* 838/1 The Chinese language has many dialects, Cantonese, Hakka,.. Wu (Shanghai) and the northern dialect.

‖ **Wufan** (wu·fan). Also **wu-fan.** [Chinese *wŭfăn*, f. *wŭ* five + *făn* anti-, against.] Used *attrib.* to designate an official campaign launched in China in 1952 against bribery, tax evasion, theft of state property, skimping on work and cheating on materials, and theft of state economic information.

1956 *Contemporary China 1955* I. 63 The *wu-fan* movement against the 'five vices' in the private sector: bribing, tax evasion, theft of government property, cheating on government contracts, and stealing economic information from government sources for private speculation. **1966** D. Wilson *Quarter of Mankind* iii. 35 Within a few months the second campaign followed, this time the *wufan* or 'five-antis' movement to root out bribery, tax evasion, theft of state property, theft of state economic secrets and embezzlement in carrying out state contracts. **1966, 1971** [see *SANFAN].

wuff, *v.* (Later examples.)

1928 D. H. Lawrence *Lady Chatterley's Lover* xii. 197 The dog wuffed softly, slowly wagging her tail. **1932** E. M. Brent-Dyer *Chalet Girls in Camp* vi. 84 'That dog just worships you, Jo.'.. Rufus 'wuffed' joyfully.

wuffer (wʊ·fəɹ). *rare*[−1]. [f. Wuff *v.* + -er[1].] A dog with a loud, deep bark.

1923 D. H. Lawrence *Ladybird* 245 The white cool monster was a Siberian steppe-dog. Alexander wondered what the steppes made of such a wuffer.

Wulfilian (wulfi·liăn), *a.* [f. Gothic *Wulfila* Ulfilas (see below) + -ian.] Of or pertaining to Ulfilas (311–382), missionary, translator of the Bible into Gothic, and inventor of the Gothic alphabet.

1926 G. W. S. Friedrichsen *Gothic Version Gospels* 144 It is quite clear, however, that this is no Wulfilian text. **1968** *Language* XLIV. 731 Thus there was no confusion between the two graphemes in Wulfilian orthography.

Wulfrunian (wulfru·niăn). [f. the name of *Wulfrun*, the 10th-century lady of the manor from whose name *Wolverhampton* is derived + -ian.] An inhabitant of Wolverhampton (see also quot. 1959).

1959 J. W. Godsell *I was no Lady* iii. 50 My English husband had attended the ancient and exclusive Gothic-arched Wolverhampton Grammar School, thereby becoming a Wulfrunian. **1974** *Times* 12 Nov. 15/7 There are 'self-made' Wulfrunians of many centuries on our roll. **1979** *Times* 26 Nov. 4/6 'Are you proud to live in Wolverhampton? Or do you, when asked where you come from, ..try to change the subject?' Those are the searching questions that Wulfrunians are being asked by their borough council.

wump (wʊmp). *slang* (somewhat *rare*). [Origin unknown.] A foolish or feeble person.

1908 [see *motor-bicyclist* s.v. *MOTOR A. *sb.* 6]. **1934** R. Nichols *Fisbo* 31 Hail to thee, thou much sniffed at by superior Persons and all wowsers, wumps and knock-knees.

wump, var. *WHUMP *v.*, *WHUMP *sb.* (and *int.*).

wumph (wʊmf). [Echoic; cf. *WHUMP *sb.* (and *int.*).] A sudden deep sound, as of the impact of a soft, heavy object.

1913 *Daily News & Leader* 15 Aug. 5, I was ashamed of the heavy 'wumph' with which I landed on the other side amid the nettles. **1924** *Glasgow Herald* 20 Dec. 4/2 The female [bittern]..sometimes answers back with a subdued but exciting 'wumph'. **1967** *Punch* 3 May 640/1 The whiplash crack from the shock wave of the small fighters we were flying is no guide to the wumph of an airliner weighing up to 300 tons. **1971** 'A. Diment' *Think Inc.* viii. 141 A deep wumph as the fuel oil..caught fire.

wun, var. Woon in Dict. and Suppl.

‖ **Wunderkind** (vu·ndəɹkint). Also **wunderkind.** Pl. **Wunderkinder, wunderkinds.** [Ger., lit. = wonder child.] **a.** A highly talented child, a child prodigy, esp. in music.

1891 G. B. Shaw in *World* 23 Dec. 15/2 Every generation produces its infant Raphaels and infant Rosciuses, and *Wunderkinder* who can perform all the childish feats of Mozart. **1913** W. J. Locke *Stella Maris* iii. 28 You call her Ariel, or Syrinx, or a Sprite of the Sea, or a Wunderkind whose original trail of glory-cloud has not faded into the light of common day. **1923** D. H. Lawrence *Stud. Classic Amer. Lit.* (1924) 102 The absolute duplicity of that blue-eyed *Wunderkind* of a Nathaniel. **1931** *N. & Q.* 3 Jan. 16/1 A great many instances of *Wunderkinder* were brought together by the late Dr. Leonard George Guthrie, in his Fitzpatrick Lectures to the Royal College of Physicians (1907), entitled 'Contributions to the study of Precocity in Children', privately printed, 1921. **1947** A. Einstein *Music in Romantic Era* xv. 213 Chopin was a *wunderkind*, both as virtuoso and composer. **1973** L. Heren *Growing up Poor in London* iii. 65 Again I suppose that the [school] orchestra was better than most of its kind because of the Jews. Some took private lessons, and were regarded as *Wunderkinder* by their parents. **1984** P. Rose *Parallel Lives* (1985) 81 His career at the Royal Academy school was impressive; indeed he was something of a *wunderkind* in the art world.

b. A talented or successful young man, a 'whizz-kid'. Also *transf.*

1930 E. Culbertson *Contract Bridge Blue Bk.* xvii. 227 He [*sc.* a bridge player] may belong to a proud class of *wunderkinder* who 'never need a book' or who 'have no system'. **1940** H. G. Wells *Babes in Darkling Wood* i. i. 31 He was in the habit of calling his host and hostess 'The ultimate generation, the last and so far the best'. They were, he said, his '*Wunderkinds*'. **1972** [see *KAPUT *a.*]. **1975** *New Yorker* 25 Aug. 50/3 Zen, the colt by Damascus that horsemen say will be the *Wunderkind* of the season, ran a temperature before the Sanford Stakes and was scratched. **1982** R. Ludlum *Parsifal Mosaic* xx. 320 He's received a fair amount of media exposure—the thirty-year old *wunderkind*.

Wundtian (vu·ntiăn), *a.* and *sb. Psychol.* [f. the name of the German psychologist Wilhelm *Wundt* (1832–1920), + -ian.] **A.** *adj.* Of or pertaining to the school of experimental and physiological psychology founded in Leipzig by Wundt or to his ideas or methods. **B.** *sb.* A follower of Wundt, one who adopts his ideas or methods.

1890 W. James *Princ. Psychol.* I. iii. 93 The facts, however, do not seem to me to warrant even this amount of fidelity to the original Wundtian position. **1932**, etc. [see *HERBARTIAN *a.* and *sb.*]. **1945** *Mind* LIV. 215 The later Wundtians, particularly and most ingeniously E. B. Titchener, tried to save sensationism. **1972** H. J. Eysenck *Encycl. Psychol.* II. 61/1 Külpe..and his students ..used a more molar type of introspection, believing that the Wundtian approach was too atomistic.

wunnerful (wʊ·nəɹful). Also **wonnerful, wunnaful.** Repr. *dial.* or *U.S.* pronunc. of Wonderful *a.*, (*sb.*), and *adv.*

1924 H. de Selincourt *Cricket Match* vi. 190 Wonnerful wholesome stuff celery, they say. **1930** M. Allingham *Mystery Mile* xiii. 122, I be a wunnerful smart old man. **1945** A. Kober *Parm Me* 16 Certainly sounds like a wunnaful pickcha, Jen. **1977** *Sounds* 9 July 8/1 Mink's main claim for the credibility stakes is that he's been produced by Jack Nitzche, who scores pretty high in the living legend section of this wunnerful business.

wunst: see *ONCE *adv.* A. δ.

wun tun, var. *WON TON.

wurley. Add: Also **whirlie**; in *pl.* **wurlies.**

1934 A. Russell *Tramp Royal in Wild Austral.* x. 78 The camp was made up of a cluster of spinifex-covered wurlies. **1936** I. L. Idriess *Cattle King* vi. 51 A hundred warriors were lazing about their wurlies, sleeping the midday peace away. **1954** B. Miles *Stars my Blanket* xi. 76 A huddle of wurlies and a yapping throng of lean kangaroo dogs..told us the blacks were camped. **1959** A. Upfield *Bony & Black Virgin* x. 80 Several whirlies of bark and odd sheets of corrugated iron and hessian bags, inhabited by aborigines. **1961** *Times* 19 July 12/6 We found them [*sc.* aborigines]..sitting outside their whirlies.

Wurlitzer (wɜ̄·ɹlitsəɹ). The proprietary name of various musical instruments made by the Rudolf Wurlitzer Company, *spec.* a type of large electric organ, or a player-piano. Freq. *attrib.*

1925 T. Dreiser *Amer. Trag.* I. i. xvii. 124 A Victrola and Wurlitzer player-piano furnished the necessary music. **1926** *Official Gaz.* (U.S. Patent Office) 23 Mar. 840/2 *Wurlitzer.* Particular description of goods.—pianos, player pianos,..organs,..banjos,..bassoons..and parts of such musical instruments. Claims use since Jan. 1, 1857. **1926** *Trade Marks Jrnl.* 12 May 1108 *Wurlitzer*... Musical instruments. The Rudolf Wurlitzer Company.., Cincinnati, State of Ohio, United States of America; manufacturers. **1930** C. Beaton *Diary* Dec. in *Wandering Years* (1961) ix. 200 Religious jazz played through a Wurlitzer. **1956** A. Huxley *Adonis & Alphabet* 231 The tail-coated organist at the console of his Wurlitzer. **1975** *Guardian* 20 Jan. 9/6 They play Wurlitzer music not inspiring to skate to. **1980** *Times* 1 Oct. 12/6 The daily organ recital on what must be the world's best known Mighty Wurlitzer.

Würm (vüɹm). *Geol.* The former name of a lake (the Starnberger See) in Bavaria, adopted by A. Penck (in Penck & Brückner *Die Alpen im Eiszeitalter* (1909) I. i. 110) and used *attrib.* to designate the fourth and final Pleistocene glaciation in the Alps; also *absol.* Cf. *RISS.

1910, etc. [see *RISS]. **1968** [see *WEICHSEL]. **1972** *Sci. Amer.* Mar. 60/2 The very numerous remains found in the Dragon's Lair were evidently deposited there during the final Pleistocene ice advance, the 60,000-year Würm glaciation that ended some 12,000 years ago. **1977** G. Clark *World Prehistory* (ed. 3) xi. 455 Access was easier during the last glaciation, but only noticeably so during its colder phases, that is during the Early and Main Würm, to use designations taken from the Swiss Alpine sequence.

Hence **Wü·rmian** *a.* (also *absol.*).

1927 Peake & Fleure *Hunters & Artists* iv. 40 It is..possible that the Capsian industry had passed from Tunis through Sicily to Italy during, or even before, the Würmian glaciation. **1967** *Oceanogr. & Marine Biol.* V. 453 One may conclude that the cooling of the water during the Würmian was very slight in the Eastern Basin [of the Mediterranean]. *Ibid.*, The shallow-water Würmian beds.

wurra (wʊ·rə), *int. Anglo-Ir.* [Ir. (*a*) *Mhuire* (O) Mary.] An exclamation of grief or despair.

1898 J. D. Brayshaw *Slum Silhouettes* 21 Oh! wurra, wurra, that I should live to see him stiff and cowld. **1936** M. Mitchell *Gone with Wind* x. 206 He groaned. 'Wurra the day!' **1952** E. O'Neill *Moon for Misbegotten* ii. 91 I'm not like you, owning up I'm beaten and crying wurra wurra like a coward.

wurst (wɜ̄ɹst, v-). Also **worsht, wourst.** [a. Ger.] Sausage, esp. of the German type; a German sausage. Also *transf.*

1855 [see *KRAUT 1]. **1868** *Amer. Odd Fellow* VII. 403/1 Sausages, or 'wourst', as they call them, are made in links. **1892** I. Zangwill *Childr. Ghetto* I. xvii. 59 Mrs. Hyams fried a piece of *Worsht* for Miriam's supper and put it into the oven to keep hot. **1939** C. Isherwood *Goodbye to Berlin* 221 There were plates of ham and cold cut wurst. **1955** T. H. Pear *Eng. Social Differences* 181 Liver-sausage and similar 'wursts'. **1966** L. Davidson *Long Way to Shiloh* x. 142 We finished the *wurst* sandwiches. **1967** *New Scientist* 10 Aug. 281/2 The name given to the dam was apparently chosen because of the obliging way in which this huge water-filled wurst..returns to its original unostentatious shape when deflated. **1977** *Drive* May–June 124/2 In the South, one shop had invented its own delicacy: Satan's Sizzlers, 'the great curry wurst'.

Württemberger (vü·ɹtĕmbɜ̄ɹgəɹ). Also **Wurtemberger, -burger, Wur-.** [a. Ger., f. *Württemberg*, the name of a former state in S.W. Germany (now part of the *Land* of Baden-Württemberg) + -er[1].] A native or inhabitant of Württemberg.

1896 G. A. Henty *Through Russian Snows* x. 190 The seven battalions of Spaniards, Wurtembergers, and men from the Duchy of Baden. **1926** F. M. Ford *Man could stand Up* ii. i. 78 Those blessed Wurtembergers would never that day get out of their trenches. **1934** W. S. Churchill *Marlborough* II. xxiv. 539 The Württembergers and Westphalians were only now approaching. **1938** C. V. Wedgwood *Thirty Years War* viii. 384 Of two thousand Württembergers who joined Horn in 1632 at least half deserted in less than a month. *a* **1974** J. Pope-Hennessy in *Lonely Business* (1981) iii. 267 To meet young Württembergers. **1977** N. Freeling *Gadget* iv. 182 'You're a Wurtemburger aren't you?' 'Badener, from Karlsruhe.' **1978** L. Deighton *SS-GB* xxxviii. 333 We should have polished off all you bloody Württembergers in 1918.

wurtzilite (vɜ̄·ɹtsiləit). *Min.* [f. the name of Henry *Wurtz* (1828–1910), U.S. mineralogist + -i- + -LITE.] A black, massive, sectile, asphaltic pyrobitumen produced by the metamorphosis of petroleum.

1889 W. P. BLAKE in *Engin. & Mining Jrnl.* XLVIII. 542/2 In now proposing for it the name *Wurtzilite* I desire to compliment my friend, Dr. Henry Wurtz, of New York who in 1865 described the mineral to which he gave the name grahamite. **1918** [see *pyrobituminous* adj. s.v. *PYRO- 2]. **1965** [see *IMPSONITE]. **1979** J. M. HUNT *Petroleum Geochem. & Geol.* viii. 403 Wurtzilite appears to be a more indurated polymer with an origin similar to elaterite.

‖ **wushu** (wū·ʃū·). Also **wu shu** and with capital initial. [Chinese *wǔshù*, f. *wǔ* military + *shù* technique, art.] The Chinese martial arts.

1973 P. J. SEYBOLT *Revolutionary Educ. in China* xxii. 252 In the gymnasium,..others, wielding swords and spears, were practicing the traditional Chinese *wushu*. **1975** *Times* 27 June 5/5 Some energetic youngsters are practising *wushu* (military art). **1977** O. SCHELL *China* (1978) III. 241 He has just come from *wu shu* practice (martial arts). **1978** CHOW & SPANGLER *Kung Fu* p. xii, Today Wu Shu remains the official term for martial arts in the People's Republic of China, although the emphasis is on its use as a national sport 'to serve the people' in the promotion of health. **1979** *Tel.* (Brisbane) 31 May 2/3 A few lessons in Wushu.

wuss (wʊs). Repr. *colloq.* or *dial.* pronunc. of WORSE *a.* and *sb.*, or *adv.*

1862 A. J. MUNBY *Diary* 22 Mar. in D. Hudson *Munby* (1972) 117 That's wuss than a day's work, that is. **1869** J. GREENWOOD *Seven Curses of London* vi. 91 She'll tell you that, wuss luck, I've got in co. with some bad uns. **1894** [see *STRAIGHT *adv.* 6]. **1896** A. MORRISON *Child of Jago* vi. 61 Nobody's none the wuss for me knowin' about em. **1936** M. MITCHELL *Gone with Wind* lix. 994 It been awful! An' it's gwine be wuss. **1945** J. RHYS-WILLIAMS *Stern Daughter* xv. 97 Lucky if it aint no wuss, Sister.

wüstite (vü·stɔit). *Chem.* Also **wustite.** [ad. G. *wüstit* (R. Schenck et al. 1927, in *Zeitschr. f. anorg. Chemie* CLXVI. 141), f. the name of F. *Wüst*, German metallurgist: see -ITE¹.] An isometric solid solution of magnetite (Fe_3O_4) in iron oxide (FeO).

1928 *Chem. Abstr.* XXII. 566 The existence of 2 solid solns. is shown: (1) a soln. of small quantities of Fe_3O_4 in FeO for which the name '*Wüstite*' is coined, [etc.]. **1957** *Jrnl. Iron & Steel Inst.* CLXXXVII. 78/1 The gradient of the iron content of these layers was determined, and the wüstite composition at the iron–wüstite and wüstite–magnetite phase boundaries was used to define the wüstite area. **1977** *Nature* 6 Oct. 500/1 There are few reports on the direct reduction of wustite by carbon according to an autocatalytic mechanism. **1980** *Ibid.* 30 Oct. 778/1 The extraterrestrial nature of a specific spherule can be confirmed if it contains wüstite—a metastable iron oxide formed at high temperatures and low oxygen partial pressures. Wüstite slowly decomposes into α-iron and magnetite..and is thus almost unknown in nature.

wuther, *sb.* and *v.* Add: Hence **wu·thering** *vbl. sb. rare.*

1879 G. M. HOPKINS *Poems* (1967) 80 If a wuthering of his palmy snow-pinions scatter a colossal smile Off him, but meaning motion fans fresh our wits with wonder. **1951** J. STRACHEY *Man on Pier* 20 The routine hours that are without inspiration in a day—those spent in buying stamps for letters, in filing receipts, in the dreary wuthering of machineries, in the changings from place to place.

‖ **wu ts'ai** (wū· tsai·). Also **wucai, Wu ts'ai.** [Chinese *wǔcǎi*, f. *wǔ* five + *cǎi* colour.] Polychrome; polychrome decoration in enamels applied to porcelain; porcelain with polychrome decoration esp. of the Ming and Qing dynasties.

1904 E. DILLON *Porcelain* vii. 101 We come again to a pentad of colour—not, however, quite the same as the *wu-tsai* of Wan-li times. **1906** S. W. BUSHELL *Chinese Art* II. viii. 32 The ordinary class of polychrome (*wu ts'ai*) decoration of the Ming period. **1915** R. L. HOBSON *Chinese Pott. & Porc.* II. ii. 8 There are the beautiful barrel-shaped seats, some with openwork ground, the designs filled in with colours (*wu ts'ai*). **1964** M. MEDLEY *Handbk. Chinese Art* 88/1 *Wu-ts'ai*.., a term applied to porcelains of the Ming and Ch'ing Dynasties decorated in overglaze enamel colours, and often with coarsely-handled under-glaze blue. **1971** L. A. BOGER *Dict. World Pott. & Porc.* 115/1 Wu ts'ai, which is practically a Chinese way of saying polychrome, is most commonly applied to a decoration comprising designs painted in enamel colors. **1980** *Catal. Fine Chinese Ceramics* (Sotheby, Hong Kong) 62 A fine pair of wucai (wu ts'ai) square Dishes of shallow flared form with brown-edged rims.

‖ **wu-wei** (wu·wē¹·). Also **Woo-wei, wu wei.** [Chinese *wúwéi*, f. *wú* no, without + *wéi* doing, action.] **a.** The Taoist doctrine of letting things follow their own course. **b.** *Hist.* In China, the name of a minor sect.

1859 J. EDKINS *Relig. Condition of Chinese* xiv. 260 One of the most interesting among the minor sects in China is that called the Woo-wei-keaon. It is an off-shoot from Buddhism. The words Woo-wei, mean non-action. These words are, in China, a favourite philosophical phrase, used by all schools of a contemplative or mystic tendency. The Taouists, who spoke of the Eternal Reason which underlies all existences, held that it could be understood, and the perfection of our nature reached only by rest, by stillness physical and mental, by abstaining from external methods of improvement, and by disbelief in their efficacy. This they called Woo-wei, 'to do nothing'. **1917** *Encycl. Sinica* 545/1 Confucius believed in the power of human nature to remain upright *if properly taught*; Lao Tzû believed it would keep straight *if left to itself*. This is his famous doctrine of *Wu-wei*, (Inaction or Non-assertion). *Ibid.* 609/2 *Wu Wei* or Non-Action Society. A secret sect, variously stated as having been founded by disciples of Lao Tzû towards the end of the Chou dynasty, by Lo Huai, the originator of the Lung Hua and Hsien T'ien sects, who lived in the 15th and 16th centuries, and to have been begun three hundred years ago... Its members are described by Edkins as 'a kind of reformed Buddhists'. **1934** A. D. WALEY *Way & its Power* iii. 145 He slips in..*wu-wei*, 'non-activity', *i.e.* rule through *tê* ('virtue', 'power') acquired in trance. **1965** C. & W. CHAI *Humanist Way in Ancient China* 56 To govern by *Wu-wei* (inaction or noninterference), Shun was the one! **1970** H. G. CREEL *What is Taoism?* i. 9 The mere idea of all this toiling for immortality is repugnant to that of *wu wei*, not striving. **1975** C.-Y. CHANG *Tao* lxxiii. 194 *Wu-wei* does not mean that one does not act. It means that one acts but is free from ulterior motives.

wuz (wʊz). Also **wus.** Repr. *colloq.*, *dial.*, or *vulg.* pronunc. of *was.*

1886 F. H. BURNETT *Little Lord Fauntleroy* xi. 222 The rooms wus locked up 'n' empty. **1901** M. FRANKLIN *My Brilliant Career* iii. 16 Some of us wuz always good fer a toon on the concertina. **1945** J. RHYS-WILLIAMS *Stern Daughter* xv. 97 Sister wuz wunnerful good to me back at Wipers in 1917. **1966** *Listener* 15 Sept. 397/2 We were promised a discussion about the programme on BBC-2's *Late Night Line-Up*; we wuz robbed, the discussion hardly mentioned the programme at all. **1973** C. HIMES *Black on Black* 168 Mah belly feels lak mah throat wus cut. **1976** *Observer* 22 Aug. 5 (Advt.), Wor lad wuz in a reet steeut.

Wuzeerá, var. *WAZIR².

wuzzy (wʊ·zi), *a. colloq.* Confused, fuddled, vague. Cf. WOOZY, MUZZY, *adjs.*

1896 *Dialect Notes* I. 427 *Wuzzy*,..confused. **1921** E. A. J. B. LYTTON *Let.* 10 Mar. in Ld. Lytton *Antony* (1935) iii. 74, I am very nearly mad, I am quite slowly turning wuzzy. **1937** J. B. PRIESTLEY *Two Time Plays* 79, I can't remember... I'm—a bit—wuzzy.

Wyandotte. Substitute for entry.
Wyandot (wɔi·ăndǫt). Also † **Wayandott, Wyandot(t)e.** [ad. F. *Ouendat*, ad. Huron *Wendat*.] **1.** (A member of) a North American Indian people belonging to the Huron nation and originally living in Ontario; the language of this people. Also *attrib.* or as *adj.*

1749 J. HAMILTON *Let.* 2 Oct. in *Documents Colonial Hist. New-York* (1855) VI. 531 The Twitchwees & Wayandotts..for two or three years past have dealt largely with our Traders. **1785** T. JEFFERSON *Notes Virginia* xi. 187 Tribes..Wyandots..Near Fort St. Joseph's and Detroit. **1786** [see *HURON]. **1789** [see *SAUK]. **1804** *Maryland Hist. Mag.* IV. 6 The Indian chief, Tarhie, a Wyandote..[was] hunting bears. **1826** J. F. COOPER *Last of Mohicans* III. ii. 46 What will our fathers think the tribes of the Wyandots have become? **1837** [see *SHAWNEE *sb.* 1]. **1913** A. S. PALMER *Samson-Saga* xiv. 167 The Wyandot Indians have a like myth. **1965** *Canad. Jrnl. Linguistics* Spring 105 He [*sc.* Sapir] knew something about Wyandotte. **1979** B. A. LEITCH *Conc. Dict. Indian Tribes N. Amer.* 189 By the 1970s Wyandot were among the leading citizens of Ottawa County, Oklahoma, numbering about 1,000.

2. (Usu. with spelling **Wyandotte.**) One of a breed of medium-sized domestic fowls, of American origin.

1884, etc. [in Dict.].

Wyatt (wɔi·ăt). The name of the architect and designer James *Wyatt* (1746–1813), used *attrib.* to designate buildings or architectural features designed by him or characteristic of his Gothic Revival style.

1819 M. EDGEWORTH *Let.* 4 May (1971) 206 A most comfortable sitting room (scarlet cloth and black furniture—Large Wyatt window plate glass—tables most comfortable). **1936** A. DALE *James Wyatt* vi. 32 The Wyatt window, a tripartite aperture similar to the Venetian window. **1962** *House & Garden* Dec. 63/2 Sienna marble columns (..from a Wyatt house in Somerset). **1973** *Country Life* 18 Jan. 152/3 It is unusual for a Philadelphia house... There is a vaguely Wyatt feeling about the whole concept.

Hence **Wyatte·sque, Wy·attish** adjs.

1942 J. LEES-MILNE *Jrnl.* 18 Jan. in *Ancestral Voices* (1975) 10 A terrible house.., with only a vestige of the eighteenth century in the central stairwell, where there is a trace of Wyattesque or Adamesque treatment, a frieze with ram's skulls. **1946** —— *Diary* 23 Nov. (1983) 108, I found the stairwell actually more Wyattish than Adamatic. **1973** *Country Life* 20 Sept. 776/3 Wyattesque decoration.

wye². Add: *spec.* (*a*) *Plumbing.* A short pipe with a branch joining it at an acute angle. (*b*) *Electr. Engin.* = *STAR *sb.*¹ 12 i.

a **1877** KNIGHT *Dict. Mech.* III. 2823 *Wye*... A name applied to a stem or pipe with branches. **1916** C. E. MAGNUSSON *Alternating Currents* ix. 97 If the three circuits be connected as shown..it is called a Star or Wye connection. **1964** R. F. FICCHI *Electr. Interference* x. 200 With a source that is wye-connected, the system neutral is readily available. **1978** K. W. SESSIONS *Homeowner's Handbk. Plumbing & Repair* iv. 145 (*caption*) Some cast-iron soil-pipe 90° wye branches. **1980** SLEMON & STRAUGHEN *Electric Machines* ii. 143 Three similar single-phase transformers may be connected to give 3-phase transformation, and since the primary and secondary windings may be connected either in delta or in wye, there are four possible combinations of connections.

Wykhamite (wi·kăməit). [f. the name William of *Wykeham* (see WYKEHAMIST *sb.* and *a.*] = WYKEHAMIST *sb.*

1828 M. R. MITFORD *Our Village* III. 203 Two or three more of our young Etonians and Wykhamites. **1972** *Daily Tel.* 1 June 21/3 He is a Wykhamite and Balliol man who was originally destined for the Civil Service.

wyn, wynn² (win). Now the usual forms of WEN².

[**1892** S. A. BROOKE *Hist. Early Eng. Lit.* II. xxiii. 201 W. was sometimes taken to mean *Wyn, joy*, and sometimes *Wen, hope.* **1910** F. TUPPER *Riddles of Exeter Bk.* 234 W always demands the interpretation *Wyn*, a rendering of the rune sustained by the Anglo Saxon alphabet in the Salzburg MS.] **1912** A. J. WYATT *Old Eng. Riddles* p. xxxix, The commoner Anglian runes.. ƿ w wynn (joy). **1955** *Jrnl. Eng. & Gmc. Philol.* LIV. 6 In later Old English *fuþorcs*, *wyn* and *wen* are generally confused, owing to some extent..to the semantic link existing between the two words, although the name of the W-rune was unquestionably *wyn.* **1965** C. BARBER *Flux of Lang.* vii. 129 The runic symbol 'wynn' was used for the Old English *w* sound. **1978** *Norfolk Archaeol.* XXXVII. 56, G offers one or two forms hardly explicable except as corruptions of original spellings with *wynn* for *w*.

wynd. Add: **1. d.** *transf.*

1952 DYLAN THOMAS *Coll. Poems* 170 Small fishes glide Through wynds and shells of drowned Ship towns to pastures of otters.

wysiwyg (wi·ziwig). Also **WYSIWYG.** [Acronym (see quots. 1984).] (See quots.)

1982 *Byte* Apr. 264/2 'What you see is what you get' (or WYSIWYG) refers to the situation in which the display screen portrays an accurate rendition of the printed page. **1982** *Economist* 1 May 8 If he wishes to converse with computer buffs, he will have to cope with neologisms such as 'wysiwyg' (what you see is what you get), pronounced 'whizziwig'. **1984** *Sci. Amer.* Sept. 54/3 Perhaps the most important principle is WYSIWYG ('What you see is what you get'): the image on the screen is always a faithful representation of the user's illusion. *Ibid.* 135/1 The resulting interface between the computer and the user would then fall into the class of interfaces known as WYSIWYG, which stands for 'What you see is what you get'.

X

X. Add: **I. 1. a.** *x-* (rarely *X-*)*height* (Typogr.), the height of a printed lower-case *x*, esp. as representative of the size of the fount to which it belongs.

1945 J. C. TARR *Printing To-day* 177 *x-height*, the height of lower-case letters (excluding descenders and ascenders), i.e. the height of a lower-case x. **1959** T. HARROD *Librar. Gloss.* (ed. 2) 99 *Descender*,..that part of a lower-case letter which extends below the 'x' height. **1964** P. A. D. MACCARTHY in D. Abercrombie et al. *Daniel Jones* 160 Any attempt to preserve traditional usage by having only x-height letters for vowels. **1978** J. LEWIS *Typography* ii. 79 Typefaces with a large x-height are more suitable for an age accustomed to reading sans serif signs... VIP Palatino is another large x-height typeface.

b. *X chair*, a chair in which the underframe resembles the letter X in shape; so *X-frame* (usu. *attrib.*).

1904 P. MACQUOID *Hist. Eng. Furnit.* iii. 52 It is a very common error to assign all these 'X' chairs to foreign importation. **1928** [see *SAVONAROLA 2]. **1945** *Burlington Mag.* May 110/2 These X chairs throughout the sixteenth century, when made for the homes of the wealthy, were covered with rich cloths of gold, velvets, and silks. **1955** R. FASTNEDGE *Eng. Furnit. Styles* 33 (*caption*) Arm chair of X-frame construction... Early seventeenth century. **1961** [illegible] *Antique Eng. Furnit.* 20 Another form of chair, of different origin from the boxchair, was the X chair... Both chairs are assigned to about the same period—that is the middle of the sixteenth century. **1976** X-frame [see *SADDLE SEAT 3 b].

d. Used to mark a location on a map or the like; esp. in phr. *X marks the spot* and varr.

1813 M. EDGEWORTH *Let.* 16 May (1971) 59 The three crosses X mark the three places where we were let in. **1918** J. M. BARRIE *Echoes of War* 5 In the rough sketch drawn for to-morrow's press, 'Street in which the criminal resided'..you will find Mrs. Dowey's home therein marked with a X. **1928** R. KNOX *Footsteps at Lock* iv. 36, I wish I could be there, to see you diving in the mud on the spot marked with an X. **1968** B. NORMAN *Hounds of Sparta* ix. 64 A message from our alcoholic friend. X seems to mark the spot where he lives.

II. 3. a. *axis of x* (example); now always *x-axis*; also *transf.*; *X-cut* adj. (Electronics), of, pertaining to, or designating a quartz crystal cut in a plane normal to its X-axis; *X-plate* (Electronics), each of a pair of electrodes in an oscilloscope that control the horizontal movement of the spot across the screen.

1885 J. CASEY *Treat. Analytical Geom.* ii. 22 If the equation of the line contains no *x*, it is parallel to the axis of *x*; and if it contains no *y*, it is parallel to the axis of *y*. **1886** W. B. SMITH *Elem. Co-Ordinate Geom.* I. i. 10 *OX*, *OY*, are called Co-ordinate Axes, or axes of *X* and *Y*, or *X*- and *Y*-axes. **1929** *Internat. Critical Tables* (U.S. Nat. Res. Council) VI. 211/1 The *x*-axis coincides with the crystallographic *c*-axis of 3-fold symmetry, the *y*-axis is ⊥ to a face of the hexagonal first order prism, and, in dextro crystals, the + direction of the *x*-axis is outward through one of the faces..of the trigonal pyramid. **1930** W. G. CADY in *Proc. IRE* XVIII. 2139 We consider first the manner of indicating the orientation of the more common 'cuts' [in quartz crystals]... In the first case, we have the cut variously referred to in the literature as 'Curie cut'..or 'normal cut'... However, a still more concise term would be the 'X-cut', denoting a plate the normal to whose face, and hence for which the applied electric field, is parallel to an X-axis. Similarly, the term 'Y-cut' would apply to the second type of quartz plate, which has hitherto been referred to as the '30-deg. cut' or 'parallel cut'. **1933** J. H. MORECROFT *Electron Tubes* xii. 337 The velocity of [compression] wave travel is different in the Y axis direction from that along the X axis. *Ibid.* 338 An X-cut plate has a negative temperature coefficient, i.e. the frequency of oscillation decreases as the temperature rises... The Y-cut plates have a positive coefficient. **1934** J. H. REYNER *Television* vii. 71 We then apply a suitable periodic voltage across the X plates which spreads the trace-out at right angles and produces a pattern on the screen. **1945** *Electronic Engin.* XVII. 723 These two equations define the components of the velocity of the spot along the X and Y axes. **1946** *Ibid.* XVIII. 23/1 A D.C. connexion must be made between the output of the time base and the X plates of the tube. **1969** *Funk & Wagnalls Dict. Electronics* 170/1 *X-axis*, in a quartz crystal, a reference axis chosen so as to connect two opposite vertices of its hexagonal cross section; one of the axes showing the greatest electrical activity. **1973** S. K. STEIN *Calculus & Analytic Geom.* ii. 26 Far to the right and to the left the graph gets closer and closer to the *x* axis without ever touching it. **1978** D. T. REES *Cathode Ray Oscilloscope* 9 A voltage applied to the X-plates will deflect the beam sideways. **1982** *IEEE Trans. Industr. Electronics* XXIX. 158/1 The rotated *X*-cut orientation has been found to be optimum from the viewpoint of its frequency versus temperature and pressure characteristics. **1983** V. M. RISTIC *Princ. Acoustic Devices* vi. 180 The relationship between the natural axes *a*, *b*, *c* and the crystallographic axes *X*, *Y*, *Z* must be known in order to use the proper constants. These relationships, for each crystal system, have been adopted by convention. Various piezoelectric, elastic, and other constants of a particular crystal specimen are evaluated in terms of *X*, *Y*, *Z* axes.

c. *Genetics.* (Now always as a capital.) [First used in German by H. Henking 1891, in *Zeitschr. f. wissensch. Zool.* LI. 706.] The symbol of the *X CHROMOSOME. So **X-linked** (stress variable) *a.*, being or determined by a gene that is carried on the X chromosome.

1902 T. H. MONTGOMERY in *Trans. Amer. Philos. Soc.* XX. 177 One of these [three chromosomes of *Protenor belfragei*], that designated *x* in Figs. 119–123, imposes by its relatively very large volume... We shall call this the 'chromosome *x*'. **1902** *Biol. Bull.* Dec. 29 (*caption*) All the chromosomes, including the accessory (*x*), show indications of a longitudinal split. **1909** E. B. WILSON in *Science* 8 Jan. 57/1 In all the species half the spermatozoa are characterized by the presence of a special nuclear element which I shall call the 'X-element', while the other half fail to receive this element. **1910** [see *METEROGAMETIC *a.*]. **1921** *Biol. Bull.* Jan. 118 The case of the aphids and phylloxerans has been the strongest argument for the hypothesis that two Z chromosomes (in homogametic sex) [illegible] **1968** M. W. STRICKBERGER *Genetics* xii. 216 In some instances, both compound X's and compound Y's may be found together in the same species. An extreme example of compound sex chromosomes occurs in the beetle *Blaps polychresta*, where the male has 12 X's and 6 Y's in addition to 18 autosomes. **1977** N. V. ROTHWELL *Human Genetics* iv. 83 One important point to note is that a male never passes an X-linked gene to his sons. **1983** *Oxf. Textbk. Med.* I. iv. 16/2 The triple X female with 47 chromosomes shows very little physical abnormality... It is possible that only one X is working in any cell at a given time. **1983** [see *Y CHROMOSOME].

f. *x-chaser*, etc.: a naval officer proficient in examinations or good at his work (see also quots. 1946, 1962). *slang.*

1904 'VANDERDECKEN' *Mod. Officer of Watch* vi. 64 To get on at sea it is not necessary to be an X hunter, a man may be a smart officer without ever having been near enough to an X to drop salt on its tail. **1912** 'AURORA' *Jock Scott, Midshipman* i. 4 He was what we called an *x* catcher; in fact, he passed out of the *Britannia* a midshipman and was wearing his patches the day he left. **1916** 'TAFFRAIL' *Pincher Martin* v. 71 He was an *x*-chaser, in that he had done remarkably well in all his different examinations. **1946** J. IRVING *Royal Navalese* 190 *X chaser*, a mathematically minded man: a theoretician. Also, a navigating officer who has qualified..as the navigator of a First Class ship. **1962** A. G. COURSE *Dict. Naut. Terms* 215 *X chaser*, a meticulous navigator in the Royal Navy.

g. In the analysis of games of Bridge x represents a card between 2 and 9, inclusive.

1920 A. G. L. OWEN *Mod. Bridge* II. 56 A similar position is this:—Z xxx A King xx BJ 10 xx YAQ. If A leads his King, Y makes Ace and Queen. **1933** C. VANDYCK *Contract Contracted* i. 10 x = any small card... An easy way of remembering the Kx and Qx in different suits is to think of it as the *Grand Marriage.* **1959** REESE & DORMER *Bridge Player's Dict.* 14 East holds..Kxxxx. **1972** *Country Life* 4 May 1119/3 The trump finesse could not gain, even if East held Q xx.

h. *x-question* (Linguistics) (see quots.)

1924 JESPERSEN *Philos. Gram.* xxii. 303 In the other kind of questions we have an unknown 'quantity'... We may therefore use the well-known symbol x for the unknown and the term *x-question* for a question aiming at finding out what x stands for. **1957** S. POTTER *Mod. Linguistics* iii. 71 Tune 1 falls after the turn. It is used in completed statements, in direct commands, and in special or x-questions which cannot be answered by 'yes' or 'no' and which are generally introduced by an interrogative pronoun or adverb. **1964** M. CHAPALLAZ in D. Abercrombie et al. *Daniel Jones* 306 X-questions, that is, questions beginning with a specific interrogative word.

i. *Genetics.* (Now written as lower case.) A symbol representing the lowest number of chromosomes which make up a genome; freq. with preceding number, designating the number of sets of these in a cell, or in each cell of an organism.

1924 *Hereditas* V. 144 Summarizing our results on the chromosome set in *C[arex] pilulifera*, we may now state that this species has 9 chromosomes (X) of which there are 3 long, 4 medium and 2 short ones. *Ibid.* 161 In *Triticum* Sakamura..and Sax..found one species with 14 chromosomes (2X), four species with 28 and three with 42 chromosomes. **1932** C. D. DARLINGTON *Recent Adv. Cytol.* iii. 61 Since a zygote usually receives two similar sets of chromosomes from its two parental gametes, their number is conventionally referred to as 2*n*; where the chromosomes pair regularly at meiosis they therefore form n pairs. Now in a particular individual these 2*n* chromosomes may consist of three sets or four sets of chromosomes relative to its own parents or ancestors. In the present work, therefore, the 'basic number' of this ancestral set is distinguished by the sign *x*. Thus in *Triticum vulgaer* $2n = 42$ and $x = 7$, the somatic chromosome number is therefore hexaploid (6*x*). **1979** A. F. DYER *Investigating Chromosomes* ii. 47/2 *Rosa canina* ($2n = 5x = 35$ = AABCD). **1980** J. SCHULZ-SCHAEFFER *Cytogenetics* vii. 122 Very often, ploidy levels are erroneously reported for n-numbers. But the number reserved for ploidy levels is the x-number or basic genome number (2*x*, 4*x*, 6*x*, etc.).

j. *X factor* (Mil. colloq.), the aspects of a serviceman's life that have no civilian equivalent; pay made in recognition of these.

1969 *Second Rep. Pay Armed Forces* (Nat. Board for Prices & Incomes) vi. 21 in *Parl. Papers* (Cmnd. 4079) 517 There are special conditions of employment...common to all servicemen and which..make it more uncertain and on occasions more hazardous than the normal..employment in civilian life... The elements..constitute, what we have termed the X factor. **1979** *Navy News* May 48/3 The Ministry of Defence have proposed a substantial increase in the X factor across the board on the grounds that the elements that make up the justification for it have shifted to the disadvantage of the Services.

5*. Used to represent a kiss, esp. in the subscription to a letter.

1763 G. WHITE *Lett.* (1901) I. vii. 132, I am with many a xxxxxxx and many a Pater noster and Ave Maria, Gil. White. **1894** W. S. CHURCHILL *Let.* 24 Mar. in R. S. Churchill *Winston S. Churchill* I. Compan. i. (1967) vii. 456 Please excuse bad writing as I am in an awful hurry. (Many kisses.) xxx WSC. [illegible] **1953** DYLAN THOMAS *Under Milk Wood* (1954) 41 Yours for ever. Then twenty-one X's. **1982** C. FREMLIN *Parasite Person* vi. 40 A row of 'X's', hurried kisses, all he had time to scribble.

5.** *X-band*: the range of microwave frequencies around 10,000 megahertz, used in radar transmission.

1946 *Radar: Summary Rep. & Harp Project* (U.S. Nat. Defense Res. Comm., Div. 14) 144/2 *X-band*, refers to wavelengths around 3 cm. **1952** [see *S-band* s.v. *S. 12]. **1976** *Sci. Amer.* June 72/1 Most spacecraft now transmit to the earth a second radio signal at an X-band frequency (8·5 gigahertz).

5*.** *Cinemat.* X is used to denote films classified as suitable for adults only, or to which only those older than a certain age are to be admitted; so *X-rated* adj. (hence *X-rate* vb. trans.), *X-rating* vbl. sb. Also *fig.*

In Britain replaced by *15* and *18* in 1983.

1950 *Rep. Departm. Comm. on Children & Cinema* 64 in *Parl. Papers* (Cmd. 7945) VII. 238 We recommend that a new category of films be established (which might be called 'X') from which children under 16 should be entirely excluded. **1950** *Times* 14 July 8/4 The X certificates..will cover films other than those of a 'horrific' character, which are 'wholly adult in conception and treatment'. **1956** 'M. INNES' *Appleby plays Chicken* I. xvii. 139 'I'm going up.' 'You're doing nothing of the sort. It's X Certificate stuff, my boy, and not for general exhibition. There's a high-up copper who says so.' **1958** *Times* 9 July 6/3 Mr. Davie..has his 'X' certificate pictures..in which his obsessional imagery has taken on an existence, outside the vague allusiveness of the paint, which is too specific for comfort. **1970** *N.Y. Times Index* 1248/2 Panel of 3 Fed judges rules Penna's new law forbidding showing of previews of x-rated movies. **1972** *Daily Colonist* (Victoria, B.C.) 6 Feb. 2/3 There was only one explicit scene—the incest sequence—which caused the film to get an X (no one under 17 admitted). **1973** M. AMIS *Rachel Papers* 136 Sebastian had gone into Oxford to see an X film ('any X film' he said) and to moon around looking for girls with his spotty mates. **1974** *Florida FL Reporter* XIII. 35/3 'Community standards' should determine whether X-rated movies should be allowed to be shown or not. **1974** *Newsweek* 20 May 23 His communicators..kept insisting that the transcripts actually clear the President of any crime more grievous than using X-rated language and thinking unsavory thoughts. **1976** *Publishers Weekly* 24 May 54/3 Most readers will surely X-rate the author's dicta; only the far-out minority will accept them. **1981** *TV Picture Life* Mar. 6/1 For it was daytime TV shows, or 'soaps' as they are affectionately called, that first explored the 'X'-rated areas of life. **1983** *Guardian* 15 Oct. 10/7 In America..X-rating is used only for out-and-out porn.

5**.** *X-C* (or *XC*) *skiing* (N. Amer.) with pronunc. (krǫs)), cross-country skiing.

1972 [see *ski-touring* s.v. *SKI *sb.* 2 b]. **1976** *National Observer* (U.S.) 13 Mar. 11/1 Alpine and XC skiing. **1977** *N.Y. Rev. Books* 14 Apr. 42/4 (Advt.), Midwest Photographer, 33, likes bike rides, hikes, x-c skiing, concerts, theater,..seeks woman friend.

III. 6. (Later examples of *Xtian, -ity.*)

1811 Xtianity [see *INERASABLE *a.*]. **1845** M. ARNOLD *Let.* Mar. (1932) 55 When Tait had well observed that strict Calvinism devoted 1000s of mankind to be eternally,—and paused—I, with, I trust the true Xtian Simplicity suggested '—'. **1915** A. HUXLEY *Let.* Oct. (1969) 79 The ethics are identical with Xtian ethics. **1940** E. POUND *Cantos* lviii. 74 They drove the Xtians out of Japan. **1966** D. JONES *Let.* 8–16 June in R. Hague *Dai Greatcoat* (1980) IV. 223 All chaps should be awfully good ..is..more or less what the present notion of Xtianity boils down to.

7. b. In commercial (esp. *U.S.*) use put for the final *-cks* (or *-cs*) of (esp. monosyllabic) words, as *CLOX, *PIX², *SNAX, *SOX.

x (eks), *v.* Pa. t. **x-ed, x'd.** **1.** (In Dict. s.v. X.) *rare*⁻¹.

2. *trans.* To obliterate (a typewritten character) by typing 'x' over it; to cross *out* in this way; = *EX *v.* Also *fig.*

1942 W. STEVENS *Let.* 28 Jan. (1967) 400, I felt that..you had x-ed me out. *c* **1945** U. TROUBRIDGE *Life & Death Radclyffe Hall* (1961) 71 As she dictated she continued to polish and the typist had always to be prepared to 'X' out at demand any word or sentence. **1958** C. BAKER *Friend in Power* vi. 163 He set the capital key and X'd the sentence through. **1969** J. N. CHANCE *Abel Coincidence* iii. 54 You should x it off your card. **1977** J. AIKEN *Last Movement* ii. 39 She crossed out that line, x-ing it vigorously to ensure its illegibility. **1978** H. KEMELMAN *Thursday the Rabbi walked Out* xxi. 89 You want me to make the correction on my typewriter? I can x it out.

Hence **x-ed** (*out*) *ppl. a.*, **x-·ing** (*out*) *vbl. sb.*

1966 *Punch* 31 Aug. 310/1 There shall be no 'X-ing out' of rival goods with black crosses. **1969** M. LAND *Quicksand* 59 He knew the uneven lines of his portable and the X'd-out words would annoy Dave Winters. **1982** M. MCMULLEN *Until Death do us Part* (1983) v. 29 A sheet of manila paper... A good deal of X-ing out to be seen.

Xanadu (zæ·nădū). [Poetic ad. *Xandu*, i.e. Shang-tu, the Mongol city founded by Kublai Khan.] A place suggestive of the Xanadu portrayed in Coleridge's poem *Kubla Khan*, with its dream-like magnificence and luxury.

[**1625** *Purchas his Pilgrimes* III. I. iv. 80 Xandu, which the great Chan Cublay..built; erecting..a maruellous..palace of marble. **1816** S. T. COLERIDGE *Kubla Khan* 55 In Xanadu did Kubla Khan A stately pleasure-dome decree.] **1948** 'J. TEY' *Franchise Affair* i. 7 To that douce country lawyer..Scotland Yard was as exotic as Xanadu, Hollywood, or parachuting. **1958** M. KENNEDY *Outlaws on Parnassus* xi. 165 Desirable readers..do not expect Xanadu to put them in mind of Yarmouth. **1962** *Holiday* Aug. 70/1 It was only about half an hour's drive to the Xanadu of *le facteur* Cheval. **1969** *Guardian* 12 Nov. 5/7 Bob's double-tiered hideaway..overlooking the fairy-lit battlements of his Xanadu in Mayfair. **1972** K. BONFIGLIOLI *Don't point that Thing at Me* viii. 76 The Ambassador was at some Xanadu-like golf-links far away. **1977** *Time* 25 July 2/1 We have lived in Southern California for twelve years and watched nearly everything encapsulate itself within a plastic bubble; not only giant 'pop Xanadus' like Sea World and Universal Studios, but also miniature golf courses, shopping centers and finally the American home.

Xanga, var. *SHANG.

xanthan (zæ·nþæn). *Chem.* Also **xantham.** [f. XANTH(O- + -AN; *-am* is unexplained.] A powdery polysaccharide composed of glucose, mannose, and glucuronic acid, produced by the bacterium *Xanthomonas campestris* and used in drilling muds and the food industry. Usu. as *xanthan gum.*

1964 *Australasian Jrnl. Pharmacy* XLV. Suppl. No. 19. S80/1 The material is known variously as Polysaccharide B-1459, Xantham Gum, Corn Sugar Gum [etc.]. **1970** *Biotechnol. & Bioengin.* XII. 75 Previous publications from this Laboratory have reported batch fermentations to produce a biopolymer, xanthan, using *Xanthomonas campestris.* **1972** *Materials & Technol.* V. 44 Xanthan gum is used in oil well drilling muds, because of its heat stability and tolerance of salts. **1982** *S. Afr. Food Rev.* IX. 515/1 Since Xanthan gum solutions have a high viscosity at rest..xanthan gum is widely used in food systems as a stabiliser for emulsions.

xanthate. Add: **a.** More widely, a salt or ester of any acid of the form RO·CS·SH, where R is an alkyl or similar radical. (Further examples.)

1895 CROSS & BEVAN *Cellulose* III. 248 The precipitated xanthate may be treated with solutions of suitable salts of the heavy metals, and the cellulose xanthate of the metals prepared. **1945** *Jrnl. Chem. Soc.* 666 Debus..showed that *O*-ethyl *S*-ethyl xanthate, EtO·CS·SEt, could be readily prepared by the interaction of potassium xanthate and ethyl iodide. **1951** C. R. NOLLER *Chem. Org. Compounds* xvi. 311 The sodium alkyl xanthates are used as collecting agents in the flotation process for the concentration of ores. **1974** *Encycl. Brit. Micropædia* X. 777/1 The most important group of xanthates are the sodium salts produced from cellulose.

xanthation (zænþē¹·ʃən). *Chem.* [f. XANTHATE + -TION.] A stage in the viscose process for making rayon, in which alkali-cellulose is treated with carbon disulphide to form cellulose xanthate.

1927 T. WOODHOUSE *Artificial Silk* iv. 30 The larger quantity of Viscose silk is made from wood-pulp... This process is termed Xanthation. **1962** J. T. MARSH *Self-Smoothing Fabrics* iv. 30 When xanthation is complete, the solid is dissolved in dilute caustic soda. **1978** *Nature* 12 Oct. 530/1 Mechanically pulped wood or papers are not generally useful, as the structure resists solution after xanthation.

So **xa·nthate** *v. trans.*, to cause to undergo xanthation; also *absol.*; **xantha·ted** *ppl. a.*, **xantha·ting** *vbl. sb.*

1938 *Thorpe's Dict. Appl. Chem.* (ed. 4) II. 465/2 Surplus carbon disulphide is removed..and the xanthated crumbs are then dispersed by churning with NaOH solution. **1952** *U.S. Patent* 2,592,355 1 In manufacturing viscose-rayon fiber, a low cellulose, high alkali and low viscosity is prepared..by xanthating and dissolving same in usual way. *Ibid.* 4 Without subjecting to aging, it is xanthated in a xanthating apparatus. **1962** J. T. MARSH *Self-Smoothing Fabrics* iv. 30 The crumbs of alkali-cellulose are then xanthated by treatment with carbon disulphide in a rotating churn for about 3 to 4 hr. **1964** V. E. YARSLEY et al. *Cellulosic Plastics* x. 149 The xanthated cellulose. *Ibid.*, The aged alkali-cellulose is charged to the xanthating churns, which rotate slowly. **1978** *Jrnl. Appl. Polymer Sci.* XXII. 897 Starch polyampholytes (xanthated starch amines)..were prepared, characterized, and evaluated as wet- and dry-strength agents in paper handsheets. **1983** *Jrnl. Macromolecular Sci.: Chem.* A. XX. 218 Partially xanthated cellulose.

xanthene (zæ·nþīn). *Chem.* Also **-en.** [f. XANTHO- + -ENE.] A tricyclic crystalline compound, $O(C_6H_4)_2CH_2$, derivatives of which are used as brilliant, often fluorescent, dyes. Usu. *attrib.*

1898 *Jrnl. Chem. Soc.* LXXIV. I. 643 Xanthen gives a *dichloro*-derivative, $C_{13}H_8Cl_2O$. **1902** *Chem. News* 17 Jan. 36/1 The mono-halogenated derivatives of the xanthene series possess basic properties. **1947** L. S. PRATT *Chem. & Physics Org. Pigments* viii. 189 The rhodamines are representative of the basic-type xanthene colors, and the fluoresceins, eosins, phloxins, erythrosins, and rose bengals are representative of the acid type. **1959** [see *EUXANTHIC *a.*]. **1971** R. L. M. ALLEN *Colour Chem.* viii. 120 Xanthene dyes containing salicylic acid or other metallisable structures are manufactured as mordant dyes.

xanthic, *a.* Add: **1. b.** More widely, any acid of the general formula RO·CS·SH or RO·CS·SR'. (Further examples.)

1945 *Jrnl. Chem. Soc.* 666 To interpret results obtained in other aspects of the chemistry of xanthic acid derivatives.., we found it necessary..to systematise current knowledge concerning the isomeric compounds RO·CS·SR', R'O·CS·SR, RS·CO·SR', the first two being esters of xanthic acids and the last an ester of *sym.* dithiocarbonic acid. **1956** KIRK & OTHMER *Encycl. Chem. Technol.* XV. 150 The relatively unimportant xanthic acids are unstable, colorless or yellow oils, and have been known, on occasion, to decompose with explosive violence.

xanthine. Add: **1. b.** Any of several substituted derivatives of xanthine.

1956 I. FINAR *Org. Chem.* II. xvi. 613 Three important methylated xanthines that occur naturally are caffeine, theobromine and theophylline. **1974** M. C. GERALD *Pharmacol.* xv. 280 The xanthines..have proved to be valuable drugs for the treatment of such respiratory diseases as asthma, bronchitis, and emphysema.

2. *xanthine oxidase*, an enzyme catalysing the oxidation of hypoxanthine to xanthine and of xanthine to uric acid.

1905 *Jrnl. Chem. Soc.* LXXXVIII. II. 271 The presence of oxygen is necessary to obtain in liver extract uric acid from the purine bases it contains, or that are added to it. The uric acid found comes almost exclusively from xanthine, and is due to a ferment, xanthine-oxydase. **1983** *Oxf. Textbk. Med.* I. IX. 79/1 Not all patients with xanthine stones have xanthine oxidase deficiency.

xantho-. Add: **2. xanthochro·mia** *Med.* [Gr. χρῶμα colour], (*a*) (see quot. 1894); = XANTHOCHROIA; (*b*) a yellowish discoloration of the cerebrospinal fluid as a result of haemorrhage in the spinal cord or brain; hence **xanthochroma·tic, -chro·mic** *adjs.*; **xa·ntho-derm** (also **Xa·ntho-**) [Gr. δέρμ-α skin], a person of a yellow-skinned (mongoloid) race; **xa·nthophore** *Zool.* [a. G. *xanthophor* (R. Keller 1895, in *Arch. f. Physiol.* LXI. 148): see -PHORE], a cell (as in an animal's skin) containing a yellow pigment; **xantho·pterin** *Chem.* [a. G. *xanthopterin* (Wieland & Schopf 1925, in *Ber. d. Deut. Chem. Ges.* LVIII. 2179): see *PTERIN], a yellow pterin present in the wings of some butterflies and moths and in the urine of mammals and forming leucopterin upon oxidation; 2-amino-4,6-dihydroxypterine, $H_2NC_6HN_4(OH)_2$.

1922 *Arch. Neurol. & Psychiatry* VIII. 24 Elsberg and Rochfort in a study of ninety-two cases of chronic diseases of the spinal cord found xanthochromatic cerebrospinal fluid in fourteen instances. **1969** EDINGTON & GILLES *Path. in Tropics* ii. 79 [In congenital toxoplasmosis] the protein in the cerebrospinal fluid is increased and may be xanthochromatic. **1894** G. M. GOULD *Dict. Med.* 1622/2 *Xanthochromia*, a persistent condition of yellow skin, resembling but not identical with jaundice. **1905** —— *Dict. of New Med. Terms* 568/1 *Xanthochromia*, Tuffier and Miliau's [*read* Milian's] name (1902) for the yellow hemorrhagic discoloration of the cephalorachidian fluid, diagnostic of hemorrhage of the neuraxis. **1912** *Lancet* 7 Sept. 685/2 On the value of a quantitative albumin estimation of the cerebro-spinal fluid (with special reference to the syndrome of massive coagulation and xanthochromia). **1977** *Ibid.* 24–31 Dec. 1352/1 There were no cells in the C.S.F. and no xanthochromia. **1952** F. A. ELLIOTT et al. *Clin. Neurol.* ix. 184 Xanthochromic fluid bleaches on exposure to daylight. **1979** *Jrnl. Neurosurg.* LI. 352/1 The presence of subarachnoid hemorrhage (SAH) is diagnostically confirmed by the detection of bloody and/or xanthochromic cerebrospinal fluid. **1924**, **1935** Xanthoderm [see *melanoderm* sb. and adj. s.v. *MELANO-]. **1935** [see *leucoderm* s.v. *LEUCO-]. **1977** *Scripta Medica* L. 35 By and large, Melanoderms and Xanthoderms have either black or brown hair and there is not enough variation to be of practical interest. **1903** *Proc. Amer. Acad. Arts & Sci.* XXXIX. 261 The two remaining types of pigment bodies in the chameleon, erythrophores and xanthophores, were not identified in Anolis. **1948** [see *NEUROHUMOUR]. **1965** LEE & KNOWLES *Animal Hormones* x. 127 The hormone MSH [*sc.* melanocyte stimulating hormone] not only acts on the melanophores, but also on the xanthophores and erythrophores. **1974** D. & M. WEBSTER *Compar. Vertebr. Morphol.* viii. 173 Other chromatophores, called xanthophores, contain carotenoid and pteridine pigments and cause much of the yellow-to-red coloration. **1926** *Chem. Abstr.* XX. 902 The residue was rubbed up 4 times with H_2O and centrifuged and the crude dirty yellow pasty pigment (xanthopterin (I)) extd. with 20% HCl and pptd. with NaOAc. **1974** *Encycl. Brit. Macropædia* IV. 922/2 Xanthopterin occurs in human urine.

xanthoma. Add: Pl. **xanthomas, xanthomata.** Also, esp. in mod. use, such a patch or tubercle. Freq. with mod.L. adjs. (Earlier and further examples.)

1869 *Jrnl. Cutaneous Med.* III. 241 (*heading*) On xanthoma, or vitiligoidea. *Ibid.* 317 After entry into hospital, patches of xanthoma developed in both eyelids. **1874** W. TAY tr. *Hebra & Kaposi's Dis. Skin* III. 345 There are two forms of the disease—1st, it occurs in the form of yellow patches—Xanthoma planum..; 2nd, in the form of tubercles—Xanthoma tuberosum. **1896** N. WALKER tr. *Unna's Histopath. Dis. Skin* 945 The xanthoma of the eyelid may gradually develop protuberances, without giving up its own peculiar histological character. **1949** R. L. & R. L. SUTTON *Handbk. Dis. Skin* 385 Tuberous xanthomas are occasionally solitary. **1961** *Lancet* 12 Aug. 341/2 All her plasma-lipids tended to return to normal values..and her xanthomata completely disappeared. **1968** A. ROOK et al. *Textbk. Dermatol.* II. xxxviii. 1229/1 Xanthoma disseminatum is a rare histiocytic proliferative disorder characterized by widespread cutaneous xanthomata..but usually without evidence of systemic disturbance of lipid metabolism. **1974** S. L. ROBBINS *Pathologic Basis Dis.* xxx. 1410/1 These xanthomas may be widespread and occur in varied forms, such as xanthoma tuberosum.

Hence **xanthomato·sis**, a metabolic disorder marked by the accumulation of excess lipid and by the presence of multiple and widespread xanthoma.

1900 DORLAND *Med. Dict.* 764/1 Xanthomatosis. **1923** *Brit. Jrnl. Dermatol.* XXXV. 90 Xanthomatosis is probably a process of infiltration dependent primarily upon hypercholesterolæmia. **1961** [see *hypercholesterolæmic* adj. s.v. *HYPER- IV]. **1983** *Oxf. Textbk. Med.* I. IX. 113/1 The relationship between cerebrotendinous xanthomatosis and spinal cholesterosis..is uncertain.

xanthophyll. Add: **a.** Now recognized as an oxygenated carotenoid identical with LUTEIN. (Further examples.)

1934 *Science* 25 May 488/2 Xanthophyll (lutein) appears to be structurally related to α-carotene and zeaxanthin to β-carotene. **1945** *Biol. Rev.* XX. 115/1 Following a convention which is being more widely adopted the specific pigment which has been termed 'xanthophyll' is here called 'lutein'. **1964** E. J. H. CORNER *Life of Plants* i. 5 The orange-yellow carotin and the yellow xanthophyll..cause the yellow colour of those parts of variegated leaves unable through some deficiency to make chlorophyll.

b. [After R. Kuhn et al. 1931, in *Zeitschr. f. physiol. Chem.* CXCVII. 141.] Any of a group of yellow pigments (as lutein and violaxanthin) that are oxygenated carotenoids.

1931 *Chem. Abstr.* XXV. 3659 The term xanthophyll should be used to designate the entire group of OH-contg. carotenoids with 40 C atoms. The individuals thus far known are: lutein and zeaxanthin.., violaxanthin.., and fucoxanthin. **1952** *Chem. & Engin. News* 7 Jan. 104/2 Two of the rules on carotenoids adopted at the London Conference of 1947 were revised to give the following text:.. The name 'xanthophyll' is a group name..for carotene derivatives of natural origin which are soluble in alcohol and are not saponifiable. **1955** G. M. SMITH *Cryptogamic Bot.* (ed. 2) I. ii. 12 There are several xanthophylls [in the Chlorophyta] not found in other algae, and of these lutein is the most abundant. **1976** *Monitor* (McAllen, Texas) 27 Sept. 4A/2 Brown tannin pigments blend with xanthophylls to produce yellow-gold and gold-brown leaves.

Hence **xanthophy·llic** *a.*, of or containing xanthophyll.

1941 *Biol. Bull.* LXXX. 451 In the herbivores, the echinoids too contained some xanthophyllic pigments without exception. **1982** *Monitor* (McAllen, Texas) 2 Apr. 6-C/2 Miss Arden claims she now has a substance, extracted from a xanthophyllic-carotin mixture, which will drain the last vestige of visibility from succeeding generations of frogs but she feels it would serve no useful purpose.

xanthoxenite (zænþozī·nəit). *Min.* [ad. G. *xanthoxen* (Laubmann & Steinmetz 1920, in *Zeitschr. f. Krist.* LV. 580), f. Gr. ξανθο- XANTHO-, after G. *cacoxen* CACOXENITE: see -ITE¹.] A hydrated basic phosphate of calcium and ferric iron, $Ca_4Fe^{3+}{}_2(PO_4)_4(OH)_2 \cdot 3H_2O$, occurring as yellow translucent triclinic crystals that are soft and have a waxy appearance.

1920 *Jrnl. Chem. Soc.* CXVIII. II. 698 Xanthoxenite, a new species from Rabenstein, occurring as small, wax-yellow, monoclinic crystals..intimately associated with dufrenite and cacoxenite. **1949** *Amer. Mineralogist* XXXIV. 698 Xanthoxenite occurs abundantly although inconspicuously at the Palermo mine as one of the last formed of the hydrothermal reworked products of triphylite. **1978** *Mineral. Mag.* XLII. 309/1 Xanthoxenite of Laubmann and Steinmetz (1920) is probably stewartite (in part)... The xanthoxenite of Frondel (1949) is proposed as the species type. It is triclinic.

xanthoxin (zænþǫ·ksin). *Biochem.* [f. *VIOLA)XANTH(IN + OX- + -IN¹.] A photo-oxidation product, $C_{15}H_{22}O_3$, of violaxanthin that occurs in certain plant tissues as a growth inhibitor.

1970 TAYLOR & BURDEN in *Nature* 18 July 302/2 We have extracted a neutral growth inhibitor from the seedlings of dwarf bean..and wheat... The name xanthoxin is now proposed for the inhibitor, the activity of which in the *cis,trans* configuration is comparable with the known naturally occurring inhibitor abscisic acid (ABA). **1980** *Physiologia Plantarum* XLIX. 309/1 The inhibitory effect of both ABA and xanthoxin on total lateral root length was mainly due to their suppression of primordia emergence and their strong inhibition of elongation.

Xavante (ʃava·nti). Also **Chavante, Shavante.** [a. Port., of uncertain origin.] **a.** (A member of) any of several groups of semi-nomadic Indians of the interior savanna of central and east-central Brazil, esp. the Akwẽ-Xavante. **b.** The Ge language of the Akwẽ-Xavante, or the language of any group called Xavante. Also *attrib.* as *adj.*

1904 H. VON IHERING *Anthropol. State S. Paulo, Brazil* 10 Of all the Indians of S. Paulo, the Chavantes are the darkest, and the most backward. **1927** K. G. GRUBB *Lowland Indians Amazonia* vii. 121 The Chavante, irreconcilably hostile, occupy the River Manso or das Mortes. **1950** J. B. D'AVILA in J. H. Steward *Handbk. S. Amer. Indians* VI. 76 To the same [*sc.* Ge] group belong the Shavante, between the Araguaya and Tocantins Rivers;..the Sherente and Craho. **1950** J. A. MASON in *Ibid.* 299 Four groups of Southern Brazil of very different linguistic affinities are known to the Brazilian natives by the name *Chavanté*... Three of them..form small independent (provisionally) families; the fourth is a *Ge* language. **1971** J. S. WEINER *Man's Natural Hist.* v. 221 'Micro-evolution' comparable to that found in the Xavante villages obtains in other South American aboriginal tribes. Ibid. 222 The extraordinary contrast between the high standard of physical fitness and stamina of the young Xavante and his later health and life expectation. **1978** *Sunday Times* (Colour Suppl.) 18 June 33 (*caption*) Xavante Indians in their village close to the River of the Dead..held the white off until recently. **1983** *Word* XXXIV. 61 A number of Brazilian Amazon languages (e.g., Apuriña, Urubû, Xavante, and Nadeb) also rely on both pragmatic and syntactic considerations in determining linear order.

Xaverian (zē¹vi³·riăn), *a.* and *sb.* [f. the name *Xav(i)er* + -IAN.] **A.** *adj.* **a.** Of, pertaining to, or designating a teaching order of Roman Catholic monks founded in 1839 and named in honour of St. Francis Xavier. **b.** Of or pertaining to St. Francis Xavier (1506–56), Spanish missionary. **B.** *sb.* A Brother of the Xaverian order.

1882 PABISCH & BYRNE tr. *Alzog's Man. Univ. Church Hist.* IV. 333 The Xaverian Brothers, founded at Bruges.. in 1839, and introduced into the United States..in 1854, have under their charge, Mt. St. Joseph's College, Carrollton, Md. **1912** *Catholic Encycl.* XIII. 284/2 After holding the office of Superior General of the Xaverians for twenty-seven years. **1915** C. C. MARTINDALE *In God's Army* I. 118 The whole Xaverian history had been one of deliberate ambition. **1931** M. YEO *St. Francis Xavier* vi. 69 One sentence seems to have the true Xaverian ring. **1967** *New Catholic Encycl.* XIV. 1058/1 The Xaverian Brothers played an important role in the development of Catholic education. *Ibid.* 1058/2 In the U.S. the Xaverians staffed most of the parish schools of Louisville.

X chromosome. *Genetics.* Also † **x chromosome.** [*X 3 c.] A chromosome with different morphology and properties from others in the complement, now recognized as a sex chromosome occurring in both sexes of a species, man and other mammals having one in the somatic cells of the male and two in those of the female.

[**1902**], **1911** [see *X 3 c]. **1933** R. H. WOLCOTT *Animal Biol.* lxxiii. 537 In fowls and in moths females have either one *x*-chromosome or both an *x*-chromosome and a *y*-chromosome, while the males have the two *x*-chromosomes. **1961** P. GRAY *Encycl. Biol. Sci.* 232/1 In most species of Spiders there are two different kinds of X-chromosomes but no Y, so that the males have X_1X_2 and the female $X_1X_1X_2X_2$. **1966** *Lancet* 24 Dec. 1397/1 The small size of the Y chromosome relative to the X chromosome has been attributed to the gradual loss of genetic material not concerned with sex determination. **1983** M. B. ZALESKI et al. *Immunogenetics* ii. 36 The *X* chromosome carries a set of genes that determines a wide variety of traits that do not necessarily affect the sex of the organism.

X disease. [X 3.] **1.** *Path.* (chiefly *Austral.*) A disease now identified as Murray Valley encephalitis (see *MURRAY VALLEY).

1918 *Med. Jrnl. Australia* 6 Apr. 278/2 On 25th August, 1917, I published in the *Journal* an account of an epidemic at Broken Hill of what is now called the 'X' disease. **1951** *Ibid.* 2 June 800/1 Our case differs from the classical picture of X disease..in the lack of cellular infiltration in the brain. **1964** [see *Q FEVER]. **1983** *Oxf. Textbk. Med.* I. v. 101/1 Murray Valley (formerly) Australia encephalitis. This disease was originally called Australian X disease.

2. *Vet. Sci.* = *blue comb* (*disease*) s.v. *BLUE *a.* 13. (Now known to be caused by a mycotoxin.)

1950 [see *blue comb (disease)* s.v. *BLUE *a.* 13]. **1961** *New Scientist* 17 Aug. 403/3 A consignment of groundnut meal from Brazil contained an agent toxic to young turkeys, which proved responsible for the death of some 100,000 birds from the hitherto mysterious X disease... The exact nature of the toxic principle has still not been found.

xeno-. Add: **xenoa·ntibody** *Immunol.*, an antibody produced in response to a xeno-antigen; **xenoa·ntigen** *Immunol.*, a xenogeneic antigen; so **xe:noantige·nic** *a.*; **xenoa·nti-serum** *Immunol.*, an antiserum rich in xeno-antibodies; **xenobio·tic** *sb.* and *a.* [BIOTIC *a.*], (designating) a substance foreign to the body; **xe·noblast** *Geol.* [a. G. *xenoblast* (F. Becke 1903, in *Compt. Rend. IX. Congr. Géol. Internat.* (1904) II. 564): see -BLAST] (see quot. 1920); hence **xenobla·stic** *a.*; **xeno·cracy** [-CRACY], a ruling body of foreigners; **xe·no-cryst** *Geol.*, a crystal not derived from the magma that gave rise to the igneous rock containing it; hence **xenocry·stal, -cry·stic** *adjs.*; **xe:nodiagno·sis** *Med.* [ad. F. *xéno-diagnostic* (E. Brumpt 1914, in *Bull. de la Soc. de Path. Exotique* VII. 706)], a diagnostic procedure in which clean, laboratory-bred vectors of a disease are allowed to feed on the individual or material that may be infected and are then examined for the pathogen; hence **xe:nodiagno·stic** *a.*; **xenoglo·ssia, xe·no-glossy** (Gr. γλῶσσα tongue], the practice or faculty of using intelligibly a language one has not learnt; **xe·nograft** *Med.*, a graft of tissue between individuals of different species; = *HETEROGRAFT; **xenola·lia** [Gr. -λαλία speaking, after GLOSSOLALIA] = *xenoglossia* above; **xe·nolith** *Geol.* [-LITH], a piece of rock in an igneous mass which differs from its surroundings and is considered to have been picked up by and incorporated into the mass when the latter was in the form of magma; hence **xenoli·thic** *a.*, containing xenoliths; also, occurring as a xenolith; **xe·nophil(e** *a.* [-PHIL, -PHILE], fond of or attracted by foreign things or people; also as *sb.*, such a person; hence **xenophi·lia**, the state of being xenophile; **xenophi·liac** *a.* [-AC] = *xenophil(e* adj. above; **xenophi·lic, xeno·philous** *adjs.* = *xenophil(e* adj.; **xenothe·rmal** *a. Petrol.*, applied to mineral deposits formed by hydrothermal action at high temperatures but at a shallow depth; **xenotro·pic** *a. Microbiology* [*-TROPIC], (of a virus) present in a host species in an inactive form and only able to infect and replicate in organisms of other species; hence **xenotro·pism.**

1974 *Brit. Jrnl. Cancer* XXX. 304/1 Gel filtration was used to show that the tumour specific xenoantibody responsible for protection was not IgM but was in the IgG fraction. **1984** *Human Immunol.* X. 57 Xenoantibodies to idiotypes of the anti-HLA-A2, A28 MoAb CR11-351 were isolated from an antiserum raised in rabbit #81. **1975** *Nature* 24 Apr. 716/2 An important characteristic of the immune system is the ability to discriminate between antigens expressed on normal tissues within the individual and the many foreign antigens expressed on normal tissues of other species (xenoantigens) and even on normal tissues of members of the same species (alloantigens). **1984** *Jrnl. Immunol.* CXXXII. 2522/1 An increase in specific antigenic activity for the *Rana*-specific xenoantigen. **1973** *Tissue Antigens* III. 5/1 Some of the eluted fractions possessed both alloantigenic and xenoantigenic activity. *Ibid.* 18/2 Strain-discriminating effects of xenoantisera were reported..over thirty years ago. **1978** *Nature* 26 Oct. 711/1 Experiments with xenoantiserum to murine tissues provided the first, fortuitous indication that Thy-1 included an hitherto unrecognised specificity. **1965** H. S. MASON et al. in *Federation Proc.* XXIV. 1172 (*heading*) Microsomal mixed-function oxidations: the metabolism of xenobiotics. *Ibid.*, We would like to call the components of this chemical environment which are foreign to the metabolic network of an organism 'xenobiotic' compounds. **1975** WILLIAMS & WILSON *Biologist's Guide to Princ. & Techniques Pract. Biochem.* i. 13 In order to study the metabolism of a xenobiotic, it is advantageous to administer it in an isotopically-labelled form. **1981** *Internat. Jrnl. Environ. Stud.* XVII. 11/2 Many xenobiotic substances reduce blood concentrations of one or more vitamins. **1920** A. HOLMES *Nomencl. Petrol.* 241 *Xenoblast*, a term applied to crystals which have grown during metamorphism without the development of their characteristic faces. **1962** Xenoblast [see *IDIOBLAST 3]. **1931** A. JOHANNSEN *Descr. Petrogr. Igneous Rocks* I. 232 *Xenoblastic*, a texture in metamorphic rocks corresponding to the xenomorphic in igneous rocks. The crystals lack proper crystal faces. **1980** *Mineral. Mag.* XLIII. 781/1 In thin-section all the felsic grains are seen to be xenoblastic. **1965** E. WEBER in Rogger & Weber *European Right* 507 Everywhere Eminescu looked..he saw foreigners and cryptoforeigners; the intelligentsia, recruited from men who had inherited their character and ideas from Greek or Bulgarian forebears; the ruling Liberals, who drew their manners and policies from the Seine, the Spree, and the Bosphorus... Altogether one vast xenocracy. **1975** H. LUKE in K. M. Setton *Hist. Crusades* III. xi. 394 While it is unlikely that the Cypriote peasantry under the Lusignan kingdom were politically worse off..than the peasantry of other Near Eastern countries.., it is not surprising that by the end of the Venetian occupation they had come to conceive..a profound hatred of the Latin xenocracy. **1894** W. J. SOLLAS in *Trans. R. Irish Acad.* XXX. 493 As a distinctive appelation appropriate to the crystals, both of pyroxene and of plagioclase felspar, which have found their way from the gabbro into the granophyre, the term 'xenocrysts' may be employed. Correspondingly included fragments of the whole rock may be called 'xenoliths'. **1964** G. A. JOPLIN *Petrogr. Austral. Igneous Rocks* ii. 26 Xenocrysts commonly show some resorption or corona indicating that they have reacted with the magma. **1983** *Jrnl. Geol.* XCI. 277 Xenocrysts and xenoliths from three Ithaca kimberlite localities are consistent with derivation from mantle depths of less than 150 km. **1963** *Amer. Mineralogist* XLVIII. 172 Classification of Kerguelen rocks is to a large extent dependent on the amount of xenocrystal material present. **1981** *Jrnl. Geophysical Res.* LXXXVI. 10515 This particular granite cannot be a primary magma... It is a possible product of partial fusion of pelitic rocks between about 20 km and 40 km depth..and xenocrystal muscovite or sillimanite from the source rocks. **1978** *Nature* 19 Oct. 640/1 The phlogopite is derived from the parental magma or its derivatives and is not a xenocrystic phase. **1947** *Anales del Instituto de Medicina Regional* (Tucuman) II. 60 The authors describe..artificial xenodiagnosis for cases in which it is not possible to perform it directly upon patients. **1976** *Nature* 15 July 215/2 Using only male bugs of a susceptible stock for xenodiagnosis should enhance the sensitivity of this diagnostic test. **1955** *O Hospital* (Rio de Janeiro) XLVII. 187 The authors make a comparison between the positivity of xeno-diagnostic tests performed in two ways. **1974** R. ZELEDÓN in K. Elliott et al. *Trypanosomiasis & Leishmaniasis* 58 New xenodiagnostic tests in an endemic area of Chagas' disease in Costa Rica. **1978** *Amer. Speech* LIII. 67 Samarin would call the event of persons speaking in a language unknown to them xenoglossia, something different from glossolalia. **1981** *Times Lit. Suppl.* 3 July 765/3 The traditional view was that, while at Corinth glossolalia had occurred, meaning that there were lexically non-communicative utterances, at Pentecost what occurred was xenoglossia, utterance in an actual foreign language. **1914** A. TEIXEIRA DE MATTOS tr. *Maeterlinck's Unknown Guest* III. 101 Xenoglossy is well known not to be unusual in automatic writing; sometimes even the 'automatist' speaks or writes languages of which he is completely ignorant. **1932** I. EMERSON tr. E. Bozzano (*title*) Polyglot mediumship (xenoglossy). **1980** *Brit. Med. Jrnl.* 9 Aug. 432/2 The investigators are reported to regard her xeno-glossy (ability to speak a foreign language without having learnt it) as a truly paranormal experience. **1961** *Nature* 25 Mar. 1024/2 Grafts between species..of less general interest..have been called xenografts or heterografts. **1974** R. M. KIRK et al. *Surgery* ii. 35 A xenograft..is poorly tolerated by the recipient. **1977** *Proc. R. Soc. Med.* LXX. 480/2 A xenograft in one patient was unsuccessful. **1984** *Times* 21 Aug. 4/1 Surgeons have a choice of artificial valves, or those made from human or animal tissue. The latter, xenografts, are silent and rendered rejection-proof. **1978** D. CHRISTIE-MURRAY *Voices from Gods* xii. 167 There appears to be no evidence of genuine, responsive xenolalia (that is, intelligent conversation carried on in a recognized language completely unknown to the speaker) in any native culture studied by anthropologists. **1981** *Times* 8 Oct. 15/5 There is a distinction between glossolalia (paranormal speaking in tongues) and xenolalia (paranormal speaking in allegedly foreign languages). **1894** Xenolith [see *xenocryst* above]. **1942** [see *METASOMATIZE *v.*]. **1956** 'H. MACDIARMID' *Stony Limits & Scots Unbound* 41 Ultra-basic xenoliths that make men look midges. **1975** *Nature* 10 Apr. 489/1 Xenoliths thought to represent material from the deeper parts of the upper mantle are brought to the surface in kimberlite magmas. **1900** *Q. Jrnl. Geol. Soc.* LVI. 665 The whole of the cliffs around Annestown Bay are composed of the widespread pink and greenish xenolithic felsites. **1930** PEACH & HORNE *Geol. Scotl.* ii. 67 Grey igneous gneiss is interposed..and is xenolithic. One of the pale xenoliths was found to consist of malacolite..and green hornblende. **1980** *Sci. Amer.* May 97/1 With the exception of xenolithic..fragments of mantle that are occasionally brought to the surface by kimberlite pipes and some basalt formations, direct sampling of the upper mantle is impossible. **1934** WEBSTER, Xenophile, -phil *adjs.* **1945** W. PLOMER *Dorking Thigh* 10 And in fancy dress she lingers With a locket in her fingers Containing a curl from That xenophil Greek. **1948** *Penguin New Writing* XXXIV. 128 New York and Los Angeles seem to be replacing Paris as the goal of Colombian xenophiles. **1968** P. B. AUSTIN *On being Swedish* xx. 149 Almost in spite of herself, she becomes a xenophile. **1959** *Times* 3 Dec. 15/7 Among the subjects covered by books in our stock are X-rays, xenophilia, [etc.]. **1964** *New Statesman* 1 May 694/1 The uniters are more prone to xenophilia. Literature being the least international of the arts, those who wish to infuse it with music and colour tend to be internationalists. **1982** *Times* 7 Aug. 8/5 When will Americans realise how xenophiliac their short order cuisine is—hamburgers ..which Hamburg would not understand, French fries incomprehensible to the French. **1974** *Encycl. Brit. Macropædia* X. 309/2 Tension between the xenophobic (fear of strangers) and xenophilic (love of strangers) in postexilic Judaism was finally resolved some two centuries later. **1984** *New Yorker* 16 Jan. 32/2 Even the most xenophilous among us may feel a twinge of alarm. **1935** A. F. BUDDINGTON in *Econ. Geol.* XXX. 209 The writer therefore proposes the term *xenothermal* for these deposits, ..suggestive of the peculiar textures for the normal high-temperature mineral assemblages involved, of the abnormal association of high temperature with shallow depth, and of the 'telescoped' character of many of the deposits. **1976** *Nature* 10 June 482/2 The tin mineralisation of Missouri bears a resemblance to the subvolcanic (xenothermal) deposits of Japan. **1973** J. A. Levy in *Science* 14 Dec. 1151/2 The results show that this NZB type virus is endogenous in other strains of mice and is xenotropic; that is, it grows only in cells foreign to the host. **1978** *Nature* 30 Mar. 456/2 Defective type C RNA tumour viruses which are genetic recombinants between ecotropic and xenotropic viruses have been described and suggested to be the real transforming agents during the course of viral-induced lymphatic leukaemia. **1974** *Ibid.* 22 Mar. 279/3 Another example of xenotropism is the endogenous feline virus which when activated replicates in human cells. **1982** *Jrnl. Virology* XLIII. 472 (*heading*) Monoclonal antibody to spleen focus-forming virus-encoded gp52 provides a probe for the amino-terminal region of retroviral envelope proteins that confers dual tropism and xenotropism.

xenogeneic (zenodʒɛni·ik, -ē¹·ik), *a. Immunol.* [f. Xeno- + Gr. γενε-ά race, stock + -ic.] Derived from an individual of a different species.

1961 P. A. Gorer et al. in *Nature* 25 Mar. 1025/1 'Hetero-specific' has been used before and cannot be said to be illogical; but it is a Greco-Latin hybrid and we feel that 'xenogeneic' goes well with the other two terms we have suggested and is perhaps preferable. **1969** *Ibid.* 27 Sept. 1376/1 Until recently all procedures that were effective in prolonging the lives of allografts..were usually ineffective for sustaining xenogeneic grafts. **1977** *Lancet* 21 May 1105/2 Patients with widespread metastatic melanoma were treated with..xenogeneic anti-melanoma immunoglobulin. **1981** *Jrnl. Immunol.* CXXVI. 2397/1 Xenogeneic anti-Id antibodies.

xenon. Add: (Further examples.) Also *attrib.* and *Comb.*

1938 *Ann. Reg. 1937* 358 Laporte..showed that white light could be obtained from a xenon tube through which brief but very intense currents were passed. **1957** T. L. J. Bentley *Man. Miniat. Camera* (ed. 5) v. 74 The practical uses of electronic flash sets are governed by the characteristics of the modern xenon-filled flash tube. **1959** *New Scientist* 1 Jan. 12/1 The main lighting was provided by a pulsed xenon compact source lamp, and its intensity was of the order of 50,000 foot candles. **1962** *Newnes Conc. Encycl. Nucl. Energy* 877/1 Xenon is of importance in nuclear technology because several radioactive isotopes of the element are found among the fission products. One of these, Xe^{135},..gives rise to fission-product poisoning of the reactor. **1963** [see *react v.¹ 1 c]. **1971** Xenon arc [see *spectrally *adv.* 2]. **1971** *Sci. Amer.* Oct. 92/3 The compounds they studied were two chlorides of xenon, XeCl$_2$ and XeCl$_4$, which they produced indirectly by synthesizing analogous compounds where radioactive iodine 129 took the place of xenon... It will be recalled that xenon used to be called an inert, or 'noble', gas because it was thought to be chemically completely unreactive. **1976** *Daily Times* (Lagos) 8 July 17/1 Most film theatres in this country operating in 35mm have such poor and antiquated projection equipment that a good 16mm heavy duty projector equipped with a xenon lamp will certainly achieve better picture and sound quality.

xenophobia (zenŏfōᵘ·biă). Also **xe·nophoby, zenophobia** (both *rare*). [f. Xeno- + Gr. φόβ-ος fear + -ia¹, -y³.] A deep antipathy to foreigners.

1909 Xenophoby [in Dict. s.v. Xeno-]. **1919** *Nation* 20 Dec. 800/1 We are often told in criticism of the Nationalist movements in Egypt, Turkey, Persia, and China that legitimate agitation for self-government and democratic institutions is marred by xenophobia. **1934** R. Macaulay *Going Abroad* xxix. 249 Violent and inhospitable outbursts of xenophoby have..characterised them [*sc.* the Basques] from their first appearance in history. **1936** E. Waugh *Waugh in Abyssinia* i. 34 The zenophobia of the people was an insuperable barrier to all free co-operation. **1940** E. Pound *Cantos* lvi. 67 Showed no zenophobia. **1963** *Economist* 1 June 908/1 The mild xenophobia..which informed such *Punch* lines as ''e's a stranger: 'eave 'arf a brick at 'im'. **1971** H. Macmillan *Riding Storm* ii. 49 This kind of isolationism or economic nationalism, amounting to xenophobia, seized all nations, great and small, from time to time. **1976** N. Roberts *Face of France* iv. 49 Eight per cent of France's total working population is immigrant... Here were all the conditions needed for the release of latent xenophobia.

Hence **xenopho·bic** *a.*, pertaining to or exhibiting xenophobia; **xenopho·bically** *adv.*; also **xe·nophobe**, a xenophobic person; also as *adj.*

1912 Xenophobic [in Dict. s.v. Xeno-]. **1922** *Mail* 24 May 327/1 The Afghans are said to be suspicious of foreigners, even to be xenophobes. **1937** D. B. Wyndham-Lewis in L. Russell *Press Gang!* 245 Grey, scrawny, xenophobe, oinophil NY chilled-steel tycoon. **1951** H. Arendt *Burden of our Time* 1. i. 3 The identification of antisemitism with rampant nationalism and its xenophobic outbursts. **1956** P. Jennings *Model Oddlies* 34 The kind of London pub which..has a more closely-knit, xenophobe clientèle than the remotest village hostelry. **1977** T. Heald *Just Desserts* vii. 146 It wasn't that he was ..a xenophobe..but the foreignness was obtrusive. **1978** *Listener* 8 June 724/2 Xenophobically named after the old Roman province, the Dacia is, in fact, a licence-built French Renault. **1980** *Times Lit. Suppl.* 22 Aug. 937/5 At that moment [*sc.* the start of a major war], for discreditable xenophobic reasons, Haldane's reputation as a War Minister sank to its nadir, but from 1918 onwards it has always been high. **1983** P. Lively *Perfect Happiness* vi. 72 A stubborn and unfashionably xenophobic refusal to attempt foreign languages. **1983** N. Freeling *Back of North Wind* 77 'Another bloody foreigner! I hate a lot of foreigners,' said Castang xenophobically.

Xenopus (ze·nŏpŏs). [mod.L. (coined in Ger. by J. G. Wagler 1827, in *Isis von Oken* XX. 726/2), f. Xeno- + Gr. πους, ποδ- foot.) A toad of the African genus *Xenopus*, which has claws on its digits and which was formerly used in pregnancy testing as it produces eggs when injected with the urine of a pregnant woman; a clawed toad.

1890 *Proc. Sci. Meetings Zool. Soc.* 70 *Xenopus* is a most admirable swimmer, and remarkable for the manner in which it remains poised for a long time immediately under the surface of the water. **1955** [see *œstradiol]. **1974** *Encycl. Brit. Macropædia* XI. 803/2 In the toad *Xenopus*, each group of hair cells in a neuromast connects to its own nerve fibre.

xerarch (ze·rāɹk), *a. Ecol.* [f. Xer(o- + Gr. ἀρχ-ή beginning.] Of a plant succession: having its origin in a dry habitat.

1913, 1960 [see *hydrarch *a.*]. **1973** P. A. Colinvaux *Introd. Ecol.* vi. 75 Later American work..sought to explain the existence of beech-maple forests on old sand dunes near the southern shore of Lake Michigan as resulting from a particularly dramatic xerarch succession.

xeric (zi³·rik), *a. Ecol.* [f. Xer(o- + -ic.] Having or characterized by a scanty amount of moisture.

1926, etc. [see *hydric *a.*²]. **1932** Fuller & Conard tr. *Braun-Blanquet's Plant Sociol.* v. 115 In spite of this temporary excess of precipitation, the vegetation of the southern Cévennes has many xeric features. **1967** M. E. Hale *Biol. Lichens* vii. 94 Xeric savannas and mesic maple woods. **1979** *Jrnl. Arid Environments* II. 255 This soil represents an edaphically xeric condition.

xero-. Add: **xeroderma pigmentosum** *Path.* [L. *pigmentōsus* pigmented], a rare, hereditary disorder in which skin exposed to the ultraviolet light of the sun becomes discoloured and swollen, chronic injury leading in childhood to cancer and often death; **xe·romorphy** *Bot.* [Gr. μορφή shape, form], the possession by a plant of features characteristic of a xerophilous plant; hence **xeromo·rphic** *a.*; also **xe·romorph**, a xeromorphic plant; **xerophil(e** *sb.*: also as *adj.*; [ad. F. *xérophile sb.* (J. Thurmann in *Essai de phytostatique* (1849) I. xiii. 268)]; (earlier and later examples); **xerophi·lic** *a.* = *xerophilous* adj. in Dict. and Suppl.; **xerophilous** *a.*: also *Zool.*; **xe·rosere** *Ecol.* [*sere *sb.*²], a plant succession having its origin in a dry habitat.

1884 *Medico-Chir. Trans.* LXVII. 169 (*heading*) Three cases of xeroderma pigmentosum, Kaposi or atrophoderma pigmentosum. **1952** C. P. Blacker *Eugenics* 248 Among these genes are those believed to determine retinitis pigmentosa, a severe disease of the eye leading to blindness; epidermolysis bullosa and xeroderma pigmentosum, both diseases of the skin. **1975** *Sci. Amer.* Nov. 68/2 An inherited defect in the enzymes that repair DNA damaged by ultraviolet light, called *xeroderma pigmentosum*, leads to multiple skin cancers. **1934** Webster, Xeromorph. **1953** *Sci. News* XXVII. 10 The term 'xerophyte' is now limited to those plants able to endure conditions of drought, while salt-marsh plants are known as 'xeromorphs'. **1981** *Austral. Jrnl. Bot.* XXIX. 518 *G*[*revillea*] *annulifera* has the nutritional characteristics common to xeromorphs plus strategies to channel nutrients very efficiently to its seeds. **1909** E. Warming *Œcol. of Plants* xvi. 445 Xeromorphic structures such as thick cuticle, soil and the xeromorphic structure which has been described. **1938** Weaver & Clements *Plant Ecol.* (ed. 2) xvi. 445 Xeromorphic structures such as thick cuticle, waxy covering, or abundant development of hairs have little value in directly reducing the rate of transpiration of xerophytes so long as the stomata are open. **1974** *Nature* 26 Apr. 807/2 The *Hybanthus* shrubs conspicuous by their apparent lack of xeromorphic adaptation to a climate which becomes increasingly arid towards the continental interior. **1909** E. Warming *Œcol. of Plants* xlvi. 194 This xeromorphy of plants growing on wet moor-soil occurs all the world over. **1963** *Nature* 30 Nov. 909/2 It is well known that xeromorphy is of physiological importance to the aquatic plants which exhibit it and certain tissues become altered in relation to environment. **1980** *Bot. Jrnl. Linnean Soc.* LXXX. 319 Most of the variable characters are related to xeromorphy and are taxonomically useful within the framework of the present classification. **1878** Xerophile [see *hygrophilous *a.*]. **1921** H. Printz *Vegetation of Siberian-Mongolian Frontiers* 14 At a short distance from the river..the xerophile typical steppe vegetation predominates. **1936** *Hereditas* XXI. 290 *Viola crassa* is xerophile and alpine. **1985** *Times* 4 Jan. 12/2 The site was then abandoned, a conclusion that Dr Evans bases on the high diversity of snail species and the general paucity of xerophile species. **1961** Webster, Xerophilic. **1965** B. E. Freeman tr. *Vandel's Biospeleology* xiii. 213 The Tenebrionidae are mainly xerophilic insects. **1972** *Science* 19 May 788/1 He is equally good when describing various kinds of vegetation, from the aquatic to the xerophilic. **1968** *Jrnl. Zool.* CLV. 365 Many xerophilous species occupying desert or semi-arid areas have adapted their breeding physiology to take advantage of the unpredictable and sporadic rainfall that may occur at any time. **1926** Xerosere [see *hydrosere* s.v. *hydro-]. **1952** P. W. Richards *Tropical Rain Forest* xii. 287 Opportunities for observing xeroseres in the tropics are..not infrequent, particularly in regions of volcanic activity.

xerocopy (zi³·ro-, ze·rokopi). [f. next + Copy *sb.* (*a.*).] A xerographic copy; a photocopy. Hence **xe·rocopying** *vbl. sb.*

1963 *Fortune* Sept. 225/2 (Advt.), Which is the 5 c xerocopy? **1964** *Economist* 22 Aug. 695/1 (Advt.), Xerocopying is an essential part of modern business communication. **1966** *English Studies* XLVII. 152 Most of the items are also available as paperbound xerocopies. **1971** *Fremdsprachen* XV. 278 Dry permanent xerocopies at a ratio of 14·5 × are automatically reproduced on single cut sheets of ordinary unsensitised paper.

xerography (zi³r-, zěro·grăfi). [f. Xero- + -graphy, after *photography*.] A dry copying process in which an electrically charged surface retains both the charge and a pigmented powder on areas not illuminated by light from bright parts of the document, so that a permanent copy may be immediately obtained by placing paper on the surface and applying heat to fuse the powder to it; photocopying.

1948 *N.Y. Times* 23 Oct. 17/8 A revolutionary process of inkless printing..was announced yesterday... Invented by Chester F. Carlson, a New York lawyer, and known as 'Xerography', this basic addition to the graphic arts reproduces pictures and text at a speed of 1,200 a minute. **1957** *Technology* July 164/2 A new copying process, Xerography, attracted constant attention. **1962** *Daily Tel.* 30 Oct. 20 (Advt.), Xerography is capturing a growing share of a market in office copying. **1967** McLuhan & Fiore *Medium is Massage* 123 Xerography ..heralds the times of instant publishing. **1970** A. Toffler *Future Shock* (1971) xii. 280 Advances in offset printing and xerography have radically lowered the costs of short-run publishing. **1976** *Globe & Mail* (Toronto) 1 Nov. 17/4 Asked whether xerography might become the next big movement in art, Bidner insisted on caution.

Hence **xerogra·phic** *a.*, **xerogra·phically** *adv.*

1948 *N.Y. Times* 23 Oct. 17/8 Even an unskilled person can make good Xerographic prints easily. **1958** *Times Lit. Suppl.* 5 Dec. 712/3 The enormous possibilities of applying the xerographic process to microfilms of books. **1968** *U.S. Patent 3,413,716* 1 A xerographically formed pattern of chemical resist is placed on the conductive material. **1976** *Nature* 22 Jan. 204/2 We sampled the air in an unventilated room..housing a xerographic machine. **1979** *Lore & Lang.* Jan. 14 This recitation has been printed many times and is common in the xerographically transmitted broadsheets. **1982** *Trans. Yorks. Dialect Soc.* LXXXII. 49 (Advt.), Available in microform and xerographic form.

xeroradiography (zi³:ro-, ze:rorē¹diǫ·grăfi). [f. prec. + *radiography.] A xerographic process for obtaining an X-ray picture, the X-rays impinging on an electrically charged surface like light in conventional xerography. Hence **xerora·diograph**; **xe:roradiogra·phic** *a.*

1950 *Non-Destructive Testing* IX. 1. 11/1 Xeroradiography is a rapid, low-cost, all-electric method of obtaining permanent X-Ray images. *Ibid.*, The xeroradiographic plate..may be made flexible, and placed in intimate contact with the curved surface of the test object. **1955** *Amer. Jrnl. Roentgenol.* LXXIII. 7/1 The xeroradiographic plate consists of a sheet of metal..on which has been deposited ..a thin layer of selenium. *Ibid.* 8/1 (*caption*) Aluminum step-wedge xeroradiographs. **1960** *Aeroplane* XCIX. 722/2 The inspection of light-alloy castings, hidden structures and assemblies, is stated to be among the applications of xeroradiographic equipment. **1972** *Lancet* 2 Dec. 1186/1 A feature of the xeroradiograph is the remarkably sharp, clearcut 'edge effect' of the powder pattern. **1975** *Nature* 25 Sept. 276/2 The discharging of a selenium plate by X rays was embodied in the original Xerox patents but, despite some investigation in the 1950s, 'xeroradiography' has not been widely used.

Xerox (zi³·rǫks, ze·rǫks). Also **xerox.** [Invented word f. *xero(graphy.] A proprietary name for photocopiers (see quots. 1952, 1953); also used *loosely* (*attrib.* and *absol.*) to denote any photocopy.

1952 *Trade Marks Jrnl.* 19 Aug. 748/2 *Xerox*... Electrophotographic copying machines and apparatus for fusing powder images onto paper in connection with electrophotographic copying machines. The Haloid Company. **1953** *Official Gaz.* (U.S. Patent Office) 12 May 327/2 The Haloid Company, Rochester, N.Y.... *Xerox* for electrophotographic copying machines, cameras, plates... Claims, use since June 22, 1949. **1966** *Economist* 15 Oct. 299/3 In most American offices executives instruct subordinates to 'make me a Xerox of this report' rather than 'make me a copy of it'. **1972** M. Williams *Inside Number 10* xi. 289 The Rank Organization in Brighton installed a xerox copying machine in the office and we also had an electric duplicating machine. **1975** D. Lodge *Changing Places* iii. 128, I enclose a Xerox of the anonymous letter. **1976** M. Machlin *Pipeline* xxvii. 317 It had appeared in Xerox form on bulletin boards in most of the administrative offices of Denali. **1977** M. French *Women's Room* (1978) iii. 272 They had a terrible fight one evening in the Xerox room of the library. **1979** *Author* XC. 157 Reprography..—xerox photocopying. **1980** *London Rev. Bks.* 15 May 8/3 How will the industry cope with new technologies, like tele-ordering, new EEC copyright complications and piracy both in the Middle East and (as we all guiltily know) in every xerox room in the British Isles? **1981** P. Roth *Zuckerman Unbound* 52 Virtually all they had left in common was the rented Xerox machine.

fig. **1979** *Nature* 15 Mar. 209/3 The set is often referred to as 'the first generation of elementary particles'... Nature seems to have made xerox copies: a second generation (μ^- ν' and cs) and possibly a third.

xerox (zi³·rǫks, ze·rǫks), *v.* Also **Xerox.** [f.

prec.] *trans.* To reproduce by xerography; to photocopy.

1966 'E. V. Cunningham' *Helen* iv. 40 Anything you want copies of, why we'll Xerox it out. **1967** McLuhan & Fiore *Medium is Massage* 123 Custom-make your own book by simply Xeroxing a chapter from this one, a chapter from that one—instant steal! **1978** *Globe & Mail* (Toronto) *Weekend Mag.* 21 Oct. 14/3 He xeroxes the menu himself and stamps out to post it on the notice-board. **1979** N. Mailer *Executioner's Song* II. iv. 555 Tamera had gone to work at 5 A.M. and spent six hours Xeroxing Gary's letters. **1982** P. M. Margolin *Last Innocent Man* II. iii. 115 He had Xeroxed the clippings for David.

Hence **xe·roxed** *ppl. a.*, **xe·roxing** *vbl. sb.* (both also with capital initial).

1965 *New Society* 14 Oct. 28/2 Inadequate library provision..has forced university teachers to..prepare.. roneoed or xeroxed material. **1973** *Jrnl. Social Psychol.* LX. 32 The Xeroxing, which made each sheet look less as though it had been individually prepared, was done to prevent distortion of the ratings. **1976** S. Lloyd *Mr Speaker, Sir* vii. 154 During July 1974, we frequently had to rely on Xeroxed copies of the order paper, amendments, and the Official Report. **1978** K. Amis *Jake's Thing* xiv. 139 He..reached for the xeroxed sheet on the bedside table. **1978** *Early Music* Oct. 605/3, I wonder how publishers who want to remain in business will react to Mr Rooley's complaint about xeroxing rules. **1980** E. Behr *Getting Even* ii. 21 The xeroxing propensities of.. one of the top confidential secretaries. **1982** *Amer. Speech* LVII. 268 The dry duplicating process known as Xeroxing has gained great popularity. **1984** *Monitor* (McAllen, Texas) 15 Jan. 16A/6 Conference participants are also urged to bring xeroxed copies of their utility bills. **1985** A. S. Byatt *Still Life* 4 Schoolgirls were dutifully filling in xeroxed, hand-written, one-word-answerable questionnaires.

Xhosa (kōᵘ·ză, kǭ·ză, -să), *sb.* and *a.* Also 9 Koossa, etc.: 9- **Xosa.** [Their own name for themselves.] **A.** *sb.* **a.** A member of any of several related tribes in Cape Province, South Africa, that form part of the Nguni branch of the Bantu; such people collectively.

1801 J. Barrow *Trav. Interior S. Afr.* I. iii. 219 The Kaffers call themselves *Koussie*, which word is pronounced by the Hottentots with a strong palatal stroke of the tongue on the first syllable. **1812** A. Plumptre tr. *Lichtenstein's Trav. Southern Afr.* I. xviii. 250 The tribe of which I mean more particularly to speak call themselves Koossas, or Kaussas. **1827** G. Thompson *Trav. Southern Afr.* App. No. 1. 439 The national appellation of the Southern Caffers is *'Amakosa*, the singular of which is *Kosa*. **1881** *Encycl. Brit.* XIII. 818/2 The Ama-Fengus are regarded both by the Ama-Zulus and Ama-Xosas as slaves or out-castes, without any right to the freedom and privilege of true-born Kaffres. **1948** B. G. M. Sundkler *Bantu Prophets in S. Afr.* iv. 96 The Zionist prophets, operating in Zululand, are in many cases not Zulus, but Sotho, Xhosa, or sometimes men from..Nyasaland. **1981** A. Paton *Towards Mountain* xxii. 189 The suggestion was made at a white staff meeting that we should try allotting separate dormitories to Xhosas, Zulu, Basotho, and so on.

b. The Nguni language of the Xhosas, a tonal language of the Bantu family very similar to Zulu.

1872 W. J. Davis *Dict. Kaffir Lang.* p. vi, The author of this, the first Dictionary of Xosa Kaffir. **1928** *Africa* I. 479 Xosa is established as the literary language of Kaffraria and the Transkei. **1970** *Cape Times* 28 Oct. 2/6 Four hundred people of all races attended prayers for racial harmony, held at St. George's Cathedral, Cape Town last night. The service—conducted alternately in English, Afrikaans and Xhosa—formed part of 24 hours of continuous prayer. **1977** *N.Y. Rev. Bks.* 4 Aug. 41/2 After testifying for four days in his native Xhosa, he asked if he could address the court in English. **1983** M. Kenyon *Free-Range Wife* i. 17 His English was so weird..he might well, as compensation, have flawless French, German, Finnish, Xhosa too.

B. *adj.* Pertaining to or designating the Xhosas or their language.

1812 A. Plumptre tr. *Lichtenstein's Trav. Southern Afr.* I. App. sig. a2ʳ In the Koossa dialect *u* is the vowel that occurs the most frequently. **1872** W. J. Davis *Dict. Kaffir Lang.* p. v, When a word is both Xosa and Zulu in form and signification, both these letters are prefixed to its meaning. **1935** *Critic* (Cape Town) Oct. 1 From the Xhosa verb *bonga* is derived the noun *isibongo*. **1948** B. G. Sundkler *Bantu Prophets S. Afr.* ii. 42 The Order of Ethiopia can hardly be said to have succeeded in attracting the broad masses of Ethiopians... It has remained exclusively Xhosa. **1963** *Times Lit. Suppl.* 17 May 354 A successful Xhosa cattle breeder. **1975** *Cricketer* May 15/2 A running commentary in the Xhosa language kept Bantu spectators in the picture.

xi (sai, zai, ksai, gzai). *Particle Physics.* [Gr. ξεῖ, ξῖ, name of the fourteenth letter (Ξ, ξ) of the Greek alphabet.] Either of a pair of hyperons (and their antiparticles) having a mass of approximately 1320 MeV, spin of ½, hypercharge −1, isospin ½, and even parity, which on decaying usually produce a lambda particle and a pion. Freq. represented by Ξ. Usu. *attrib.*

1954 Gell-Mann & Pais in *Proc. Glasgow Conf. Nucl. & Meson Physics* (1955) 344 It is interesting to consider the 'cascade particle', which we shall call Ξ^- and which has the decay scheme $\Xi^- \to \Lambda^0 + \pi^- + (\sim 65$ MeV). **1964** *New Scientist* 20 Feb. 460/3 According to the conservation of baryons and hypercharge, the omega-minus should be produced in collisions between K-minus mesons and protons and should decay (weakly, with a change of hypercharge) to a xi-particle and pion, or to a lambda-particle and K-minus. **1974** *Nature* 13 Dec. 524/1 The elementary particles nearly obey a number of simple and elegant symmetry rules. Many useful approximate calculations can be made assuming that the symmetries are perfect—for instance..by assuming that the proton and neutron have the same interactions as the lambda, sigma and xi hyperons. **1980** J. S. Trefil *From Atoms to Quarks* vii. 109 There are two xi particles, one neutral and one with a negative charge.

X-irradia·tion. [f. X (rays + Irradiation.] **a.** Irradiation with X-rays. **b.** X-rays, X-radiation.

1956 *Nature* 11 Feb. 287/1 A visible coloration is produced upon X-irradiation of the above silver-free base glass. **1962** O. Hockwin in A. Pirie *Lens Metabolism Rel. Cataract* 422 The application of a high dose of X-irradiation causes changes of lens metabolism immediately after the end of irradiation. **1964** G. H. Haggis et al. *Introd. Molecular Biol.* vii. 196 Chromosome breaks induced during leptotene by X-irradiation should..show up later as injuries to both daughter chromatids. **1976** *Internat. Jrnl. Radiation Biol.* XXIX. 367 X-irradiation..was delivered on four abdominal fields over 15 days.

Hence **X-irra·diate** *v. trans.*, to irradiate with X-rays; **X-irra·diated** *ppl. a.*

1958 *Jrnl. Compar. & Physiol. Psychol.* LI. 178 (*heading*) Maze learning in pre- and neonatally X-irradiated rats. **1971** *Nature* 9 Apr. 367/1 The solution was to X-irradiate the haploid sperms of one species. **1978** *Jrnl. Exper. Med.* CXLVIII. 15 Labeling studies in animals X-irradiated with hind-limb shielding gave a Kupffer cell labeling index of 5–10% of the normal values. **1980** *Photochem. & Photobiol.* XXXII. 183 With both treatment regimens, tumor susceptibility could be transferred to X-irradiated recipients with lymphoid cells.

xography (zǫ·grăfi). [f. X (of unknown significance) + -ography.] A photographic process producing images with a three-dimensional effect (see quots.). So **xo·graph**, an image of this kind.

1965 R. R. Karch *Graphic Arts Procedures* (ed. 3) xiii. 336 Xography provides three-dimensional printing which can be seen without benefit of special reading glasses. **1974** *Encycl. Brit. Macropædia* XIV. 1059/1 In the 1960s a three-dimensional print was developed, essentially an illustration bearing two views, superimposed, of the same image taken from slightly different angles, on a transparent mount striped with a multitude of imperceptible parallel strips (Xograph process). **1977** J. Hedgecoe *Photographer's Handbk.* 301 The basis of the xography process is a cylindrical 'lenticular' screen..which is built into the back of the camera just in front of the film plane.

X organ. *Zool.* Also **x organ** and with hyphen. [After G. *organ X*, *X-organ* (B. Hanström 1931, in *Zeitschr. f. Morphol. u. Ökol. d. Tiere* XXIII. 200, 202), so called because indicated by the letter X in a diagram published by G. Bellonci 1882, in *Mem. dell' Accad. Sci. dell' Ist. di Bologna* III. 419 ff.] A group of neurosecretory cells in the eye-stalk of some crustaceans, one of the secretions of which inhibits the production of moulting hormone by the Y organ.

1938 [see *sinus gland* s.v. *sinus 6]. **1959** W. Andrew *Textbk. Compar. Histol.* xiii. 513 There is a definite pathway..by which the axons of the nerve cells of the x-organ travel to the sinus gland. **1975** *Sci. Amer.* Feb. 74/1 The eyes of crabs are mounted on movable stalks. The stalks also house a neuroendocrine unit called the X-organ sinus-gland complex, which secretes a hormone that causes the pigments to disperse within the chromatophores.

Xosa, var. *Xhosa *sb.* and *a.*

X-radiation. (In Dict. s.v. X rays.) (Earlier and later examples.)

1896 *Strand* July 108/1 If a solid object is placed in the path of this stream..it may become the seat of the production of that which is..variously known as Röntgen radiation or X-radiation. **1974** *Physics Bull.* Dec. 581/2 The developing use of synchrotrons as an intense source of x radiation.

X-ra·diograph. [After X rays.] A radiograph made using X-rays. So **X-radio·graphy.**

1899 G. B. Shaw *Let.* 12 Apr. in *B. Shaw & Mrs. Campbell* (1952) 11, I have just had an Xradiograph [*sic*] taken; and lo! perfectly mended solid bone. **1948** *Endeavour* VII. 110/2 An X-radiograph [of the picture] made after cleaning revealed the density in the region near the proper right hand. **1961** M. Levy *Studio Dict. Art Terms* 121 X-radiography. **1963** B. Fozard *Instrumentation Nucl. Reactors* i. 7 The effect of radiation upon the crystals of silver bromide in a photographic emulsion is well known and the process is used in X- and gamma-radiography. **1975** *Nature* 26 June 697/1 Skeletal X radiography is a powerful tool for studying both modern and fossil corals. **1983** *National Gallery News* Oct. 4/1 This is a long and delicate process which involves.. taking X-radiographs of the painting to ascertain its original state.

X rays, *sb. pl.* Add: Also **x-rays. I. 1. a.** Now known to be a form of electromagnetic radiation of wavelength less than that of short-wave ultraviolet light (i.e. less than about 4 to 40 nm). (Later examples.)

X-rays are often defined as being produced by deceleration of charged particles (esp. electrons) or by electron transitions in atoms, in contrast to the otherwise similar *gamma rays* which arise from radioactive decay of nuclei. Since gamma rays tend to be of shorter wavelength than most X-rays, they have been classified variously either as very short X-rays, or as constituting a separate class of very short wavelengths beyond X-rays. Cf. *gamma rays* s.v. *gamma 1 c (ii).

1930 [see *gamma rays* s.v. *gamma 1 c (ii)]. **1948** *Sci. News* VII. 35 Procedure for the generation of x-rays..is to accelerate a beam of electrons to the required energy, and then stop it by allowing the electrons to hit a metal target. **1958** Condon & Odishaw *Handbk. Physics* vii. viii. 118/2 Although there is no precise definition of the high-energy limit of the energy of quanta called X rays, this term is usually restricted to radiations of fewer than several million electron volts of energy, above which the radiation is referred to as γ radiation. **1971** D. W. Sciama *Mod. Cosmol.* ii. 31 The distinction between X- and γ-rays is a somewhat arbitrary one, but.. we may take the dividing line to be an energy of 100 keV. **1983** *Oxf. Textbk. Med.* I. vi. 86/1 Gamma-rays. These are identical in properties to X-rays but are produced by the spontaneous disintegration of radioactive atoms.

b. (Earlier example.)

1896 *Boston Med. & Surg. Jrnl.* CXXXV. 610/2, I am.. nursing an X-ray finger.

c. (*X-ray.*) Used *attrib.* with reference to primitive pictures in which some representation is given of the insides of people and animals.

1940 L. Adam *Primitive Art* xiv. 119 Characteristic of the northern territory are the so-called 'X-ray' drawings —a special variety of naturalistic art..whereby the artist, when drawing human beings and animals, represents the inner parts of the body. **1956** C. P. Mountford *Rec. Amer.-Austral. Scientific Exped. Arnhem Land* I. iii. 112 Nor do the subjects of the X-ray artists show any movement. It is essentially a static art. **1959** E. A. Fisher *Anglo-Saxon Archit. & Sculpture* 88 Still earlier in really primitive art, and among some primitive races today, the same idea of showing the inside, the backbone, ribs and internal organs..is found, e.g. in the so-called X-ray drawings of some of the natives of the Melanesian area of the Pacific. **1977** G. Clark *World Prehist.* (ed. 3) xi. 479 The X-ray figures of Arnhem Land..may well reflect exotic influences.

II. The sing. form, *X-ray.*

2. (An) examination of a person in which an X-ray picture is taken.

1933 V. Brittain *Testament of Youth* viii. 409 Ten patients..were for immediate operation; a dozen more were for X-ray; several were likely to hæmorrhage at any moment. **1960** 'R. Gordon' *Doctor in Clover* vi. 47, I popped her in a taxi and drove her round to the casualty entrance at St Swithin's, where Miles organised X-rays. **1983** *Daily Tel.* 13 Oct. 18, I injured my back..and after a great deal of pain and misery I had an X-ray.

3. = *X-radiograph.

1934 in Webster. **1942** M. Dickens *One Pair of Feet* viii. 164 Siddons was subjected to all the indignities and discomforts of gastric investigation... X-ray after X-ray was taken and various different diets tried. **1957** 'R. Gordon' *Doctor in Love* xii. 107, I had a difficult x-ray which I thought she could help me interpret. **1969** *Ithaca Jrnl.* 27 Nov. 30 Some doctors try to head off possible litigation by ordering unnecessary and expensive x-rays. **1977** *Daily Tel.* 2 Mar. 19/5 It was not known when Miss Doris Hunt..swallowed the spoon, first seen on an X-ray last year.

4. (Without article.) An X-ray department in a hospital.

1955 'R. Gordon' *Doctor at Large* xiv. 143 Apart from the nurses, there were the buxom dieticians, the cheerful girls in X-ray, the neat secretaries, [etc.]. **1974** 'H. Carmichael' *Motive* x. 116 Dr Egan had gone to X-ray but would be back soon. **1978** J. Irving *World according to Garp* ii. 28 Her books..outgrew the shelf space and slid into the main infirmary..and into X-ray.

III. 5. a. Applied to instruments and techniques using or producing X-rays, as *X-ray analysis, microscope, microscopy, spectrograph, spectrometer, spectroscopy* (hence *X-ray spectroscopic*), *telescope*; **X-ray astronomy,** the branch of astronomy concerned with the X-ray emissions of heavenly bodies; so **X-ray astronomer.**

1924 *Econ. Geol.* XIX. 1 Physicists have developed X-ray analysis in recent years to such a point that the internal structure of any opaque crystalline substance can be determined. **1980** P. Luger *Mod. X-Ray Analysis* vi. 288 The method which has been used from the earliest days of X-ray analysis is the drawing of contours obtained from the electron density map. **1969** *Times* 13 May 14/1 X-ray astronomers have to use rockets or balloons to carry their instruments above as much of the atmosphere as possible. **1982** *Jrnl. Brit. Interplanetary Soc.* XXXV. 291 Coded aperture techniques have been employed to great advantage by X-ray astronomers. **1963** *Daily Tel.* 6 June 21 The first news was given today by the United States at the International Space Research meeting in Warsaw of a new kind of astonomy, X-ray astronomy. **1964** *Space Res.* IV. 966 (*heading*) X-ray astronomy. **1970** *Sci. Jrnl.* May 17/4 It is some years since the accidental discovery of the X-ray source Sco X-1 ushered in the new field of X-ray astronomy. **1979** *Jrnl. R. Soc. Arts* CXXVII. 580/2 X-ray astronomy took its first real spurt forward in 1970 with the launch of NASA's small 'Uhuru' X-ray satellite. **1948** *Jrnl. Optical Soc. Amer.* XXXVIII. 774/1 It is clear that an x-ray microscope is now a definite possibility. *Ibid.* 766/1 A satisfactory x-ray microscopy would open up fields of investigation closed to the optical microscope because of its limited resolution. **1966** *McGraw-Hill Encycl. Sci. & Technol.* VIII. 368b/1 The x-ray microscopes used in microradiography utilize x-

radiation to form images of resolution in the 0·2–2·0 micron..range. *Ibid.*, With this method, x-ray microscopy has not only become competitive with direct light microscopy in resolving power but has gained important advantages. **1981** *Rev. Sci. Instruments* LII. 211/2 The construction of a photoelectron x-ray microscope appears to be feasible and, moreover, this instrument should fairly fulfill the requirements demanded for the extension of x-ray microscopy. **1925** G. A. Lindsay tr. *M. Siegbahn's Spectrosc. X-Rays* iv. 92 The evidence essential for a better answer to these questions was afforded by X-ray spectrographs and Röntgen tubes with hot cathodes. **1983** *IEEE Trans. Nucl. Sci.* XXX. 491 (*heading*) High throughput non-dispersive hard X-ray spectrograph. [*Note*] 'Spectrograph' and 'spectrometer' are used interchangeably in this paper. **1915** W. H. & W. L. Bragg *X Rays & Crystal Structure* iii. 22 The X-ray spectrometer has already determined both the absolute wave lengths of various types of X-radiation and the arrangement of the atoms in several crystals. **1955** *Sci. News Let.* 23 July 1 The electronic assayer is known as an X-ray spectrometer, and the technique, which can be applied to most minerals, is called fluorescent X-ray analysis. **1977** A. Hallam *Planet Earth* 18/1 The X-ray spectrometer mapped aluminium:silicon ratios and delineated the extent of the aluminous highlands. **1925** G. A. Lindsay tr. *M. Siegbahn's Spectrosc. X-Rays* iii. 33 In order to meet such demands of X-ray spectroscopic work, the firm of Emil Gundelach..has constructed tubes embodying a slight modification. *Ibid.* 52 This rotating crystal method was first employed in X-ray spectroscopy by de Broglie. **1966** J. G. Brown *X-Rays & their Applic.* vii. 124 The methods of X-ray spectroscopy which depend on crystal diffraction are essentially methods of measuring the Bragg angle. **1983** *Jrnl. Physics B* XVI. L77 (*heading*) The determination of parameters of recombining laser-producing plasmas by means of x-ray spectroscopy. *Ibid.*, Among the various methods of plasma diagnostics the most preferable in this case are the x-ray spectroscopic methods. **1963** *Daily Tel.* 6 June 21/4 The X-ray telescope has to be raised above the earth's atmosphere and a technique found for focusing X-rays. **1978** Pasachoff & Kutner *University Astron.* xii. 333 It still looks as though Cygnus X-1 represents the first observational detection of a black hole. The new generation of x-ray telescopes should tell us more.

b. Applied to astronomical bodies that emit X-rays in detectable or significant quantities, as *X-ray nova, pulsar, star*; **X-ray burster,** a cosmic source of intermittent, short-lived, powerful bursts of X-rays, typically lasting about a second.

1976 *Nature* 17 June 542/1 This object..is perhaps the most enigmatic 'X-ray burster' found to date. **1978** Pasachoff & Kutner *University Astron.* xxv. 632 (*caption*) Half a dozen of the strong x-ray sources in the galactic bulge region have been possibly identified with x-ray bursters. **1970** *Sci. Jrnl.* Apr. 64 The second example of an 'X-ray nova' was reported during the summer of 1969, based on observations from two US Vela satellites built to monitor man-made nuclear explosions in space. **1977** *Dædalus* Fall 53 One particular kind of variable X-ray stars are the *transient sources*, sometimes called X-ray novae. **1969** G. Fritz et al. in *Science* 9 May 709/1 We wish to report the discovery of an x-ray pulsar in the general direction of the Crab Nebula. The data were obtained during an Aerobee rocket flight on 13 March 1969. **1978** Pasachoff & Kutner *University Astron.* xi. 315 (*caption*) The x-ray pulsar SMC X-1 in the Small Magellanic Cloud has a period of 0·716 sec. **1964** *Daily Tel.* 31 Mar. 17/1 Evidence of the existence of a previously unknown kind of star has been collected by American astronomers. Using an Aerobee rocket..they found two distinct 'X-ray stars'. **1977** *Sci. Amer.* Oct. 42/2 One can say with considerable certainty that X-ray stars are dense remnants of stars that have exhausted their supply of nuclear energy and have collapsed under the attractive force of their own gravity.

6. Special Combs. (of sing. form *X-ray*): **X-ray crystallography,** the study of crystals and their structure by means of X-ray diffraction techniques; hence **X-ray crystallographer; X-ray dermatitis,** dermatitis caused by X-rays; **X-ray diffraction,** the diffraction of X-rays by the regularly spaced atoms of a crystalline material, esp. as a technique of X-ray crystallography; **X-ray eyes,** the apparent ability to see beyond an outward form or through opaque material; very perceptive discernment; also, *X-ray scrutiny*; **X-ray spectrum,** a graph of intensity against wavelength or frequency for the X-rays absorbed or emitted by a material; **X-ray tube,** an electron tube for generating X-rays by accelerating the electrons to high energies and causing them to strike a target that is also the anode, producing the X-rays.

1962 *Listener* 10 May 809/1 Botanists or X-ray crystallographers. **1971** J. W. Jeffery *Methods X-Ray Crystallogr.* p. v, When one is generation 3..of X-ray crystallographers, one cannot expect to do much direct quarrying in that famous paper whose product, the Bernal chart, is known to every crystallographer. **1930** R. W. James *X-Ray Crystallogr.* i. 2 X-ray crystallography is descended from, and constantly uses, the results of the older crystallography. **1968** M. Pyke *Food & Society* vii. 101 A few milligrams of vitamin B$_{12}$ had been isolated and the nature of its complex molecule established by a combination of advanced organic chemistry and X-ray crystallography. **1977** R. S. Drago *Physical Methods in Chem.* xvii. 589 Using X-ray crystallography, one can generally determine the precise composition and atomic arrangement of almost any molecule. **1900** *Philadelphia Med. Jrnl.* V. 187 The believers in the electrical cause of x-ray dermatitis are found to be exclusively physicians. **1908** [in Dict., sense b]. **1959** *Med. Jrnl. Australia* I. 290/1 In November, 1897, Sylvanus Thompson described X-ray dermatitis, and in the following year the Röntgen Society set up a committee to investigate possible hazards. **1977** *Acta Dermato-Venereologica* LVII. 487/1 The fluid from spontaneous blisters in 15 patients with various bullous dermatoses, such as..X-ray dermatitis, all contained measurable amounts of activity. **1924** *Econ. Geol.* XIX. 15 A complete description of X-ray diffraction has been given by the Braggs. **1950** *Sci. News* XV. 139 The techniques of X-ray diffraction, which have contributed so much to the understanding of the inner structure of metals and alloys. **1969** *Times* 28 Jan. 6/6 X-ray diffraction techniques..show that the orientation of the graphite crystals in the fibre depends on the degree to which it is stretched during the heat treatment. **1982** R. M. Schultz in T. M. Devlin *Textbk. Biochem.* ii. 73 The most important of the techniques for the study of a protein's secondary, tertiary, and quaternary structure is x-ray diffraction. **1939** M. Allingham *Mr. Campion & Others* ii. 36 The conviction that he had actually encountered a man with X-ray eyes at last. **1971** 'R. Macdonald' *Underground Man* xxvi. 186 I've been feeling watched, tonight. Drawing the curtains doesn't really help. Whatever it is out there has X-ray eyes. Call it God, or call it the Devil, it hardly matters. **1976** J. Wainwright *Who goes Next?* 205 He didn't have X-ray eyes. He couldn't see through the sides of a parked van. **1982** M. Millar *Mermaid* vi. 68, I got X-ray eyes when it comes to people's weaknesses. **1899** G. Allen *Miss Cayley's Adventures* ii. 45 She looked me through and through again with her X-ray scrutiny. **1925** G. A. Lindsay tr. *M. Siegbahn's Spectrosc. X-Rays* vi. 150 The X-ray spectrum is an atomic property... The frequencies involved in X-ray spectra are very great. **1983** *Jrnl. Physics B* XVI. L79 The large dimension of the spectrograph slit..allowed us to observe the x-ray spectra of H- and He-like F VIII and F IX ions. **1896** *Nature* 12 Nov. 31/2 Unexplained variations in the behaviour of the X-ray tubes. **1974** *Encycl. Brit. Macropædia* XIV. 345/1 Small gamma-ray sources are placed in areas inaccessible to X-ray tubes such as inside pipelines.

X-ray *v.* (earlier example); also *fig.*; hence **X-rayed** (e·ks-) *ppl. a.*, **X-raying** (e·ks-) *vbl. sb.*

1899 *Bristol Med.-Chir. Jrnl.* XVII. 234 Cases that have to be X-rayed in their bedrooms. **1920** *Glasgow Herald* 5 Oct. 7/2 Advantage is claimed for 'X-raying' over sterilization. **1941** *Cold Spring Harbor Symp. Quantitative Biol.* IX. 156/1 X-rayed Bar-M2 males (4000r) were crossed to females with the genes scute vermilion forked and carnation. **1973** *Sci. Amer.* Apr. 30/3 The only way to 'X ray' the earth for such fine structure is to use short-period seismic waves that interact with the boundary of the core. **1977** W. Marshall *Thin Air* i. 6 It took two hours for the poor old Japs to even clear body search... Let alone the X-raying of freight and luggage. **1977** *Time* 22 Aug. 30/3 The treaty will have to be X-rayed by the university. **1977** *New Yorker* 26 Sept. 43/3 She..could feel the plaster armoring his X-rayed ribs. **1980** J. O'Faolain *No Country for Young Men* v. 102 The men were terrifying. She remembered their eyes X-raying her clothes.

Xtal, xtal (kri·stăl). [Cf. X 6.] Abbrev. of Crystal *sb.* and *a.*

1957 *Practical Wireless* XXXIII. 454/1 Remove Xtal unit by cutting off all leads at the source. **1971** *Gramophone* July 257/2 The crystal pickup button is labelled 'xtal' which is engineer's shorthand.

xylary (zai·lări), *a. Bot.* [f. Xyl(em + -ary².] Of, pertaining to, or constituting xylem.

1953 K. Esau *Plant Anat.* xv. 367 The terms phloic procambium and xylary procambium..may be used to stress the early differentiation of the meristem into the two parts. **1973** *Nature* 13 Apr. 479/2 Between 2000 h and 2400 h the developing xylary tissues will rapidly withdraw water from the immediately adjacent developing phloem tissues.

xylitol (zai·litǫl). *Chem.* [ad. G. *xylit* (Fischer & Stahel 1891, in *Ber. d. Deut. Chem. Ges.* XXIV. 538), f. *xyl-ose* Xylose + *-it*, -ite¹: see -ol.] A sweet, crystalline, pentahydric alcohol, $CH_2OH(CHOH)_3CH_2OH$, derived from xylose and present in some plant tissues.

1891 *Jrnl. Chem. Soc.* LX. 1. 668 Xylose yields a new compound, xylitol, which forms a non-crystallisable syrup. **1962** R. van Heyningen in A. Pirie *Lens Metabolism Rel. Cataract* 402 The lens is the only mammalian tissue in which xylitol has been found. **1977** *Daily Colonist* (Victoria, B.C.) 1 Dec. 11/1 A storm of adverse publicity over the natural sweetener xylitol isn't going to deter major chewing gum manufacturers from marketing products containing the substance.

xylo (zai·lo). Colloq. abbrev. of Xylonite.

1926 *Ironmonger* 16 Jan. (Suppl.) 50 (Advt.), This knife is made of stainless steel in four qualities with wood, xylo and ebonite handles. **1962** L. S. Sasieni *Optical Dispensing* ix. 218 Run the thumb-nail round between the xylo and the metal.

xylo-. Add: **xy·lulose** *Chem.* [*-ulose²], a keto pentose that corresponds to the aldo pentose xylose and occurs in the urine of pentosurics.

1936 Levene & Tipson in *Jrnl. Biol. Chem.* CXV. 731 For simplicity it would seem desirable to term the ketose '*d*-xylulose', in conformity with the nomenclature accepted for other keto sugars. **1964, 1968** [see *pentosuric *a.* and *sb.*]. **1974** B. S. Hartley in Carlile & Skehel *Evolution in Microbial World* 170 A mutant of this organism constitutive for ribitol dehydrogenase grows on xylitol by utilising a side specificity of this enzyme to produce xylulose.

xylographica (zailogræ·fikă), *sb. pl.* [mod.L., f. Xylographic *a.* after *typographica *sb. pl.*] Block-books, woodcuts, and the like; xylographic matter.

1931 M. B. Stillwell *Incunabula & Americana* iii. i. 177 Blockbooks, although produced during the same period [as incunabula], are those printed *a full page at a time*, each page being printed from a wooden block... Such books are sometimes called 'xylographic books', or 'xylographica', meaning that they are wood-engraved throughout. **1952** J. Carter *ABC for Bk.-Collectors* 37 Block books, or xylographica, as produced in Europe are presumed..to have preceded the invention of printing from movable metal types. **1982** *Times Lit. Suppl.* 6 Aug. 866/1 Volume I will carry descriptions of block-books and other xylographica.

xylocaine (zai·lokē¹n). *Pharm.* [f. Xylo- + *-caine*, after Cocaine.] = *lignocaine.

1946 Löfgren & Lundqvist in *Svensk Kemisk. Tidskrift* LVIII. 208 The compound ω-diethylamino-2,6-dimethylacetanilide is called xylocaine (LL 30) and is an ideal local anaesthetic, which seems to be superior to procaine in every respect. **1954** [see *lignocaine]. **1961** *Times* 4 Feb. 12/2, I injected a total of 30 c.c. of xylocaine. **1978** *Jrnl. R. Soc. Med.* LXXI. 320/1 Shortening the duration of the convulsive activity in the brain by administering xylocaine before the inducing shock does reduce the therapeutic effects.

xylophonist (zailǫ·fŏnist). [f. Xylophone + -ist.] One who plays a xylophone.

1927 *Daily Tel.* 10 May 12/1 Teddy Brown the xylophonist. **1952** B. Ulanov *Hist. Jazz in Amer.* (1958) xvii. 207 He became a xylophonist. **1976** *Gramophone* Aug. 345/1 There is a really brilliant xylophonist in 'Fossils'.

xylorimba (zai·lŏrimbă). [f. Xylo(phone + *ma)rimba.] (See quots. 1938, 1980.)

1938 *Oxf. Compan. Mus.* 1026/2 *Xylorimba*, an American form of lightweight marimba. **1961** *Times* 10 Nov. 18/7 The icy, scientific precision of xylorimba. **1978** P. Griffiths *Conc. Hist. Mod. Music* ix. 137 Xylorimba and percussion suggest the influence of black African music. **1980** *New Grove Dict. Music* XX. 564/2 *Xylorimba*... An instrument of the xylophone family with a compass sufficiently large to embrace the low-sounding bars of the Marimba and the highest-sounding bars of the xylophone.

Y

Y. Add: **3. a. Y cross,** (*a*) (example); hence **Y-crossed** *ppl. a.*; **Y-front,** a proprietary term for men's underwear, used esp. to denote close-fitting briefs with Y-shaped seaming at the front; freq. as *sb. pl.*, briefs of this kind; **Y gun** *U.S.*, an anti-submarine gun with two firing arms for discharging depth charges; **Y junction,** a junction at which a road forks into two branches, or one road joins another at an angle different from 90 degrees.

1881 G. G. Scott *Ess. Hist. Eng. Church Archit.* 114 St. Regnobert's chasuble at Bayeux, and St. Thomas's at Sens, are examples of the use of the Y cross in France. *Ibid.*, The Y-crossed vestment of Ruben's picture. **1953** *Trade Marks Jrnl.* 17 June 526/1 *Y-front*... Pants and vests, all for men. Lyle and Scott Limited,..Hawick, Scotland; manufacturers. **1959** H. Hobson *Mission House Murder* xxix. 188 Here I was, in my athlete's vest and Y-front briefs. **1961** *Harper's Bazaar* May 103/1 The demand for a T-shirt and a Y-front in Act II. **1976** T. Stoppard *Dirty Linen* 23 He produces..a large pair of Y-front pants. **1978** M. Page *Pilate Plot* (1979) xii. 183 He stripped to his Y-fronts and plunged into the pool. **1918** *Ann. Rep. Secretary U.S. Navy Dept.* 56 A new gun known as the 'Y' gun has been designed and built especially for firing depth charges. **1937** *Jane's Fighting Ships* 471 Y-gun or Depth Charge Projector. **1961** *Guardian* 18 Sept. 3/4 Local Y junctions where drivers.. expect others prophetically to divine the route they are about to take. **1982** M. Duke *Flashpoint* xxv. 182 When he came to a Y-junction he made a sharp turn right.

4. a. *axis of y* (example); now always *y-axis*; also *transf.*; *Y-cut* adj. (Electronics), of, pertaining to, or designating a quartz crystal cut in a plane normal to its Y-axis; *Y-plate* (Electronics), each of a pair of electrodes in an oscilloscope that control the vertical movement of the spot on the screen.

1885, etc. [see *X 3 a]. **1930**, etc. Y-cut [see *X-cut* adj. s.v. *X 3 a]. **1934** J. H. Reyner *Television* vii. 71 In the ordinary applications of the tube we apply the voltage to be examined across one pair of plates (usually termed the Y plates) which causes the spot to be elongated into a line. **1945** *Electronic Engin.* XVII. 723 These two equations define the components of the velocity of the spot along the X and Y axes. **1946** *Ibid.* XVIII. 23/2 The signal [may be] fed to one Y plate and a pulse derived from the anode of V$_4$ fed to the other Y plate. **1965** J. R. Frederick *Ultrasonic Engin.* iv. 65 If the second digit is 4, 5, or 6 this refers to a shear strain around the *x*, *y*, or *z* axes, respectively. **1969** Maddox & Davies *Elem. Functions* i. 12 The graph is a straight line parallel to the *y*-axis and situated 2 units to the right of the *y*-axis. **1976** *Appl. Physics Lett.* XXIX. 76/1 Our measurements were made on a polished Y-cut single-crystal quartz substrate with a pair of aluminum thin-film interdigital transducers ..with orientation for wave propagation along the *X* axis. **1978** D. T. Rees *Cathode Ray Oscilloscope* 9 A voltage applied to the Y-plates will move the beam and the spot in a vertical direction. **1979** Faux & Pratt *Computational Geom.* i. 18 The most familiar equation of a straight line is $y = mx + c$, in which *m* is the slope, and *c* the intercept on the *y*-axis.

b. *Genetics.* (Now always as a capital.) [After *X 3 e.] The symbol of the *Y chromosome. So **Y-linked** (stress variable) *a.*, being or determined by a gene that is carried on the Y chromosome.

1909 E. B. Wilson in *Science* 8 Jan. 57/2 The X-element..appears as a 'large idiochromosome' which has a synaptic mate... The latter chromosome, or its homologue, I shall designate as the 'Y-element'. **1910** *Amer. Naturalist* XLIV. 491 We should..imagine that when a sperm bearing a Y enters an egg a male results. **1911**, etc. [see *X 3 e]. **1917** *Amer. Naturalist* LI. 534 Y or W linked or plastid inheritance. **1949** Darlington & Mather *Elem. Genetics* ii. 51 In *Drosophila* and in man, there are completely Y-linked genes without any allelomorph in the X. **1981** *Heredity* XLVII. 238 A majority of male secondary sexual colour patterns are Y-linked [in the guppy]. **1983** M. B. Zaleski et al. *Immunogenetics* ii. 36 For essentially all mammalian species discussed in this book, females are *X/X* and males are *X/Y*... Other types of sex determination are also known to exist in various species and are called *Protenor* (females are *X/X* and males are *X/O*), *Abraxas* (females are *X/Y* and males are *X/X*) and haploidy-diploidy (males are haploid and sterile, whereas females are diploid and fertile).

7. Y (*colloq.*, chiefly *U.S.*), short for YMCA or YWCA; Y, yuan; YA (*U.S.*), young adult; YAG: see *yag; Y.E., Your Excellency; Y.F.C., Young Farmers Club (formerly Clubs); YHA, Youth Hostels Association; YIG, yttrium iron garnet; Y.M. (*colloq.*), short for YMCA; also, a YMCA hostel; Y.M.C.A.: also, a hostel run by the YMCA; YOP, Youth Opportunities Programme; also, a young person taking part in this scheme; Y.P., young prisoner; yr, year; yr., yr, your; yrs, yours; YTS, Youth Training Scheme; Y.W. (*colloq.*), short for YWCA; also, a YWCA hostel; Y.W.C.A.: also, a hostel run by the YWCA.

1915 *Dialect Notes* IV. 236 [College slang.] *Y*, abbreviation for the college Y.M.C.A. **1945** N. L. McClung *Stream runs Fast* xxvii. 259 Mary would have to have her bath at the 'Y'. **1956** H. Kurnitz *Invasion of Privacy* ii. 22 She's married to this English G.I. she met at the Y, where they have the service dances. **1977** *New Yorker* 27 June 35/3 Rose..did not yet have a place to live; she was staying at the Y. **1962** in E. Snow *Other Side of River* (1963) lxv. 495 The State invested Y. 37,000,000 in the livelihood needs and productive capital construction. **1973** *Times* 21 Mar. (China Trade Suppl.) p. iii/6 The basic unit of renminbi..is the yuan, represented by the symbol Y. **1974** *Publishers Weekly* 7 Oct. 63/1 A powerful and tragic book, 'Betrayed' is as much for adults as YAs. **1870** *Weekly Standard* (Buenos Aires) 9 Mar. 14/2, I beg to communicate to Y.E. the following despatch of Gen. Camara. **1945** G. Cunningham *Let.* 27 Nov. in N. Mitchell *Sir George Cunningham* (1968) v. 117, I feel terribly for Y.E. and for Claude A. in this. **1931** *Young Farmer* Mar. 77/1 The speakers..had motored many miles to come to this..third Y.F.C. meeting. **1960** *Farmer & Stockbreeder* 8 Mar. 65/1 This competition..is also of special interest to Y.F.C.s. **1982** *Financial Times* 30 Apr. 25/6 The YFC movement. **1931** *Ruc-Sac* July 18/1 The Y.H.A. (the Youth Hostels Association of Great Britain) have sent us an advance copy of their handbook. **1982** R. Hill *Who guards Prince* iv. vi. 228 The best British equivalent [hotel] in terms of remoteness and height would be a YHA hut. **1959** *Physical Rev. Lett.* II. 499/1 (*caption*) The specific heat per unit volume of polycrystalline YIG analyzed into its two components. **1975** D. G. Fink *Electronics Engineers' Handbk.* xiii. 113 In addition to mechanical tuning, both YIG and varactor tuning techniques are applicable. **1913** Y.M. [see *lowbrow, low-brow *a.*]. **1916** W. Owen *Let.* 1 Feb. (1967) 377 We are refused admission to the Y.M. or Canteen. **1931** R. Campbell *Georgiad* i. 25 Androgyno..Well on the road... Half way to Georgiana's Y.M. hostel. **1920** S. Lewis *Main St.* xxi. 257, I wish there were a Y.M.C.A. here, so I could take up regular exercise. **1931** R. Campbell *Georgiad* i. 17 Like some Y.M. or W.C.A. It welcomes waifs whom love has cast away. **1956** R. Macaulay *Towers of Trebizond* xiv. 157 There would be a Y.M.C.A and a Y.W.C.A., where billiards and boxing would be played. **1978** *Times Higher Educ. Suppl.* 5 May 28/5 Young people..eligible for the YOP in Cardiff..will be asked for their views. **1983** *Financial Times* 12 Apr. 19/3 One can employ a school leaver on the Youth Opportunities Scheme (YOP), the cost of whom is reimbursed by the MSC. All I had to do was take on a YOP as a personal assistant. **1952** *Chambers's Jrnl.* 1 June 356 They go all sentimental over the Y.P.s. **1976** H. Ferguson *Confessions of Long Distance Acid Head* 56 Ashford..is also used as a place where young offenders who have done Borstal, and a Borstall [*sic*] re-call, serve their sentence. These are known as Y.P.'s (young prisoners). **1880** W. Whitman *Daybks. & Notebks.* (1978) I. 172 Robert Norris..28 yrs old. **1942** W. Faulkner *Go down, Moses* 264 Percavil Brownly 26 yr Old. cleark. **1968** E. Knight in S. Henderson *Understanding New Black Poetry* (1973) III. 326 Last yr Like a salmon quitting The cold ocean..I hitchhiked. **1772** J. Knyveton *Jrnl.* 12 June in E. Gray *Man Midwife* (1946) i. 59 The two rooms and the closet will furnish yr. obdt. with lecture rooms and office. **1811** Shelley *Let.* 3 Jan. (1964) I. 35 Not that I like yr. heroine. **1876** Ld. Beaconsfield *Let.* 13 Sept. in R. S. Churchill *Winston S. Churchill* (1967) I. Compan. 1. 54, I earnestly hope that these arrangements may be consistent with Yr Grace's decision to accept the high office of the Queen's Representative in Ireland. **1973** *Black World* Sept. 84 Ever get tired of people playing with yr life? **1811** Shelley *Let.* 11 Jan. (1964) I. 38 Yr's with affection. PBS. **1922** Joyce *Ulysses* 740 Yrs affly xxxxx. **1932** W. Faulkner *Light in August* xviii. 412 Given it toe barer yrs truly. **1984** *Times* 17 Nov. 2/7 The YTS is not available for many 17-year-olds. **1985** *Times Educ. Suppl.* 9 Aug. 4/1 Thus, all one's instincts and reflexes impel one to support YTS. **1937** Partridge *Dict. Slang* 968/1 *Y.W.*...The Young Women's Christian Association. **1979** M. Soames *Clementine Churchill* xxvi. 424 Clementine never severed her links with the 'Y.W.'. **1931, 1956** Y.W.C.A. [see Y.M.C.A. above]. **1961** *Times* 10 Oct. 16/1 A jolie-laide innocent looking for the Y.W.C.A.

8. In *Particle Physics*, Y denotes the hypercharge quantum number of sub-atomic particles.

1956 [see *hypercharge]. **1974** S. Gasiorowicz *Quantum Physics* xxvi. 443 The missing $I = 0$, $Y = 0$ pseudoscalar meson was found in the examination of $\pi^+\pi^-\pi^0$ masses in bubble-chamber pictures.

Y' (y). **a.** Abbrev. of Ye *pers. pron.*, q.v., sense A. b.

b. Repr. a spoken abbrev. of You *pers. pron.* as subject, esp. in phrases of the type *you know, you see.*

1859 Geo. Eliot *Adam Bede* I. 1. i. 7 Ye might get religion, and that 'ud be the best day's earnings y'ever made. **1889** Kipling *From Sea to Sea* (1899) I. 341 As a bell, y' know, it's rather a failure... They don't ring it properly. **1932** J. B. Priestley *Faraway* i. 49 Y'see, you know about the island. **1953** E. Coxhead *Midlanders* vii. 168 Y'know..I just wanted a spot of land to farm. **1981** J. Wainwright *All on Summer's Day* 14 But y'know ... Sopworth money also meant Sopworth bobbying.

-y⁶, -ie. Add: Also appended to surnames to form a familiar name.

1941 J. Hilton *Random Harvest* v. 352 She ran into his arms calling out: 'Oh, Smithy—Smithy—it may not be too late.' **1958** A. Hackney *Private Life* ix. 84 'Who's Old Kitey?' 'A Mr. Kite. He's our shop steward.' *Ibid.* xvii. 171 'Mr. Cox arranged it.'.. 'You've got to be careful with old Coxy.'

ya² (yā, yă) repr. *U.S.* and *dial.* pronunc. of (*a*) You *pers. pron.*; (*b*) Your *poss. adj.*

(*a*) **1941** B. Schulberg *What makes Sammy Run?* xi. 193 Sunset Club... They've got a new dinge band there that'll kill ya. **1959** N. Mailer *Advts. for Myself* (1961) 46 Stay off the railroads, they bleed ya dry. **1973** E. Bullins *Theme is Blackness* 64 Won't ya please move out of the way, honey? **1980** *Dirt Bike* Oct. 8/1 All of you mini riders can start drooling now, because as soon as we can get our gloves on one, we'll test it for ya.

(*b*) **1946** K. Tennant *Lost Haven* (1947) iii. 50 What ya done with ya collar? **1970** K. Platt *Pushbutton Butterfly* xi. 138 He waved the Luger. 'On ya feet.' I got up. **1973** *Black World* Oct. 56/2 Put ya hands on 'a table.

yaa-boo, var. *yah boo *int.* (and *sb.*).

yaas (yæs, yās), repr. a drawled pronunc. of Yes *adv.*, esp. in U.S. speech.

1893 H. A. Shands *Some Peculiarities of Speech in Mississippi* 68 Yaas, the almost universal pronunciation of *yes*. **1895** A. W. Pinero *Second Mrs. Tanqueray* III. 104 *Paula.* Your wife? *Sir George.* Yaas—Birdie. **1913** Kipling *Diversity of Creatures* (1917) 285 I've become an Episcopalian since I married. Ya-as. **1931** W. Faulkner in *Harper's Mag.* Sept. 401/2 'Yaas,' the man in overalls said in a dry, drawling tone. 'Yaas. It got caught.'

ya bass (yæ bæs), repr. Sc. rendering of 'You bastard.'

1968 *Daily Tel.* (Colour Suppl.) 4 Oct. 21/2 We saw a film, *The Terror of the Tongs*... When we came out, we just started shouting: 'Tongs! Tongs, ya bass!' **1973** Boyd & Parkes *Dark Number* vii. 72 There was a crate of dead wine bottles... *Vino Fino* from Argentina, ya bass. **1974** *Punch* 6 Mar. p. v/1 That archetypal graffito, the Glaswegian *Ya bass.*

yabba (yæ·bă). *Jamaica.* Also **yabah.** [Jamaican.] A large wooden or earthenware vessel used for cookery or storage.

1889 *Victoria Q.* (Kingston, Jamaica) May 50 The familiar 'Yabba' or earthen vessel. **1929** M. W. Beckwith *Black Roadways* 27 Earthen bowls, hand turned and covered with a rude glaze, are always to be had in the Kingston market, but they are more rare in the hills where the old-time 'yabba' is being supplanted by tinware. **1953** *Caribbean Anthol. Short Stories* 103 She took a handful of cornmeal out of the wooden yabba. **1953** R. Mais *Hills were Joyful Together* I. i. 11 Ras..spat in an earthenware yabba set beside him for the purpose. **1959** A. Salkey *Quality of Violence* iv. 56 He picked up an empty *yabah* which was resting on a Bible... He..let it slip from his hands. It..splintered in jagged pieces.

yabber, *sb.* (Earlier example.)

1855 R. Carboni *Eureka Stockade* iv. 5 There was further a great waste of yabber-yabber about the diggers not being represented in the Legislative Council.

yabby (yæ·bi). *Austral.* Also **yabbie, yappy.** [Aboriginal.] **a.** A small, edible freshwater crayfish found in the eastern part of Australia, esp. one of the genus *Charax*. **b.** A burrowing, prawn-like, littoral crustacean of the order Thalassinidea.

1886 F. A. Hagenauer in E. M. Curr *Austral. Race* III. 554/1 Crayfish—yappy. **1894** *Argus* (Melbourne) 6 Oct. 11/2 Small crayfish, called 'yabbies'..may be found all over Australia, both in large and small lagoons. **1897** *Australasian* 30 Jan. 224/4 The bait used is 'yabby', a small crayfish found in the sand on the beach at low tide. **1930** V. Palmer *Men are Human* vi. 49 The yabbies they had cooked in a treacle-tin down by the lagoon. **1944** *Living off Land* iii. 49 Digging out these burrows will secure yabbies... The yabbie can be toasted. **1963** A. Upfield *Madman's Bend* xvi. 130 Been in the water days, by the look of him. Yabbies been at him too. No blood, but hole's here. **1966** *Courier-Mail* (Brisbane) 1 Oct. 2/6 This tasty fresh water crustacean is more frequently called the Yabby, which immediately brings about a confusion with the much-used marine Yabby. **1973** P. White *Eye of Storm* x. 490 Used to come here after yabbies. I'd like to poke around a bit. **1977** *Caravan World* (Austral.) Jan. 105/1 A restaurant serving fresh, succulent yabbies has opened near the Murray Mouth.

ya·bby, *v. Austral.* [f. prec. sb.] *intr.* To hunt for yabbies. So **ya·bbying** *vbl. sb.*

1934 *Bulletin* (Sydney) 24 Oct. 21/2 Here's a sport for those who..forget their bait when yabbying. **1941** K. Tennant *Battlers* v. 53 He..asked about the boys. 'They've gone yabbying.' **1964** *Sunday Mail* (Brisbane) 5 Apr., He participated personally in yabbying forays.

ya boo, var. *yah boo *int.* (and *sb.*).

yacca, var. *yakka.

yacht, *sb.* Add: **b.** *yacht-club* (earlier and later examples), *marina*; **yacht basin,** a dock constructed for the mooring of yachts; a marina; **yacht broker,** a dealer in yachts; so **yacht brokerage; yacht-yard,** a yard where yachts are built or repaired.

1929 *Motorboat* 10 Mar. 20/1 For many years there has been much talk of public yacht basins. **1952** P. Atkey *Juniper Rock* i. 2 An engine breakdown..had compelled Roy to take the *Marsouin* limping into the yacht basin at Marseilles. **1981** L. Deighton *XPD* xii. 102 One of the parking places near the yacht basin. **1882** *Yachting Q.* July (Advt.), Cox and King, Yacht brokers & Yachting auctioneers. **1982** N. J. Crisp *Brink* ix. 187 The yacht brokers were still in business... The pubs were still full of yachting types. **1974** J. Dimona *Last Man at Arlington* i. 51 In Nassau..he had managed to set up a profitable yacht brokerage. **1834** G. Crabbe Jr. in *Poet. Wks. G. Crabbe* I. 13 A party of amateur sailors was formed—the yacht club of Aldborough. **1981** L. Deighton *XPD* xii. 101 The Marina del Rey..has the swanky yacht club as a centre-piece. **1973** 'A. York' *Captivator* ii. 32 The ab, sloop put into Cuxhaven... It entered the yacht marina there, secured a berth. **1983** P. Ferris *Distant Country* ii. 15 The yacht marina..would reopen with brand-new quays and pontoons. **1933** 'L. Luard' *All Hands* 236 The proprietor of a yacht-yard. **1980** P. Moyes *Angel Death* xx. 248 We have to get down there..to the yacht yard. There's something wrong.

yachtie (yǫ·ti), *sb. colloq.* (chiefly *Austral.* and *N.Z.*). Also **yachty.** [f. Yacht *sb.* + -ie.] A yachtsman.

1943 *Amer. Speech* XVIII. 88 [New Zealand English.] Yachty. **1965** G. McInnes *Road to Gundagai* x. 184 One should never..leave them [*sc.* bottles] bobbing about the bay the way the careless 'yachties' did. **1972** *Sat. Rev.* 21 Oct. 62 Dunnie's a yachtie, he lives up on the cliff, and he's a rich guy. **1976–7** *Sea Spray* (N.Z.) Dec./Jan. 35/2 (Advt.), Laid polyester rope. The yachties' workhorse. **1977** *Herald* (Melbourne) 17 Jan. 19/2 They are really to cater for the spillover from Port Phillip Bay as more and more yachties realise that they can be much nearer to the blue water both inside and outside. **1977** *Pacific Islands Monthly* Feb. 25 (*heading*) Slugged yachtie.

yachting, *vbl. sb.* Add: In *attrib.* use applied *esp.* to garments designed for use on yachts.

1873 *Young Englishwoman* June 280/2, I always wear yachting shoes without heels, made of the white canvas, and with leather toes and straps, as now worn by boating and yachting gentlemen. **1894** *Country Gentlemen's Catal.* 155/1 Blue cloth yachting caps, from 8/6. **1931** E. F. Benson *Mapp & Lucia* vii. 200 Contempt for Georgie..had been intensified by the sight of his yachting cap. **1976** 'A. Garve' *Home to Roost* iii. 37 He was wearing..a battered old yachting cap. **1983** G. Thompson *Nobody cared for Kate* iv. 26 Maggie..had dressed for the barge trip... She wore a yachting suit.

yachty (yǫ·ti), *a. colloq.* [f. Yacht *sb.* + -y[1].] Of or pertaining to yachts.

1950 E. Allcard *Single-Handed Passage* 177 Don't use 'yachty' fittings. They are only good for the Solent. **1983** *Times* 19 Aug. 24/1 The yachts displace a lot of money which slops into the town to support..galleries selling yachty prints.

yack, *sb.* Add: **2.** *slang.* Also **yak. a.** Incessant talk of a trivial or boring nature. Freq. reduplicated and as *int.*

1958 J. Cannan *And be Villain* i. 5 That blasted Primrose will have arrived by now—yak yak yak, talking Eve into God knows what. **1965** W. King in *Black Short Story Anthol.* (1972) 306 Then, through the yak-yak, Mac began to tell us the 'unusual something' that he had just recently done. **1972** C. Weston *Poor, Poor Ophelia* (1973) xxxii. 205 A lot of yak in the news about the missing boot. **1983** N. Freeling *Back of North Wind* 99 The sudden head-down butt jabbed into someone's face, is a highly effective way of putting a stop to his yack.

b. An accent or tone of voice.

1957 M. Spark *Comforters* vi. 138 He fiddled with the tape machine... The voice came with an exaggerated soppy yak: 'Caroline, darling.' **1975** *Camping & Trailering Guide* Oct. 33/1 You will hear French Canadian patois mixed with Ontario English, down-Maine tones and New York yak.

yack (yæk), *v. slang.* Also **yak.** [Echoic, or f. prec. sb.] *intr.* To engage in trivial or unduly persistent conversation; to chatter.

1950 'P. Quentin' *Follower* v. 38 Yakked a lot. Know how she is. Talk your ear off... She yakked on about you being in South America. **1955** 'H. Robbins' *Stone for Danny Fisher* i. ix. 65, I was busy yakking with a broad. **1956** W. H. Whyte *Organization Man* (1957) xxvi. 359 Four or five of the girls and their kids will be yakking away. **1958** T. Roethke *Let.* 16 Aug. (1970) 220 I've spent nearly the whole of three sessions with my doctor yacking about you. **1958** L. Whishaw *As far as You'll take Me* iv. 47 Didn't have any breakfast, too busy yacking about women. **1977** J. Porter *Who the Heck is Sylvia?* vi. 57 She should have acted first and yacked about it afterwards. **1981** J. Trenhaile *Kyril* xix. 142 Those two will yak all day.

Hence **ya·cking, yakking** *vbl. sb.* and *ppl. a.*

1959 'S. Ransome' *I'll die for You* ii. 29 She could sit at any bar.., with an ear out for all the yakking that goes on, and pick up any amount of gossip. **1971** *Time* 7 June 31/1 [The ability] to switch away from yakking actresses and the necessity of having yakking starlets for the ratings. **1976** N. Freeling *Lake Isle* xxvii. 192 Shopkeepers..yelling in the street at yacking housewives.

1977 *N.Y. Rev. Bks.* 12 May 6/4 The sound which emerges from the pages is that of the yakking of debutantes.

yacker (yæ·kəɹ), *sb. slang.* Also **yakker.** [Echoic.] **1.** Talk, conversation, chatter. *Austral.*

1882 *Sydney Slang Dict.* 9/1 *Yacker*, talk. **1973** P. White *Eye of Storm* vii. 306 She wished it had been a hospital, when she could have produced a chart, handed over.., and swept off without further yakker. *Ibid.* ix. 441 Couldn't get on with me work—not with all the yakker that was goin' on in 'ere.

2. [f. *yack *v.* + -er[1].] **a.** A chatterbox or gossip. **b.** = *yack *sb.* 2 b.

1959 G. Mitchell *Man who grew Tomatoes* vi. 87 Now that will be enough ammunition for the yakkers, if there are any present. **1960** 'R. Simons' *Frame for Murder* viii. 99 'What sort of a bloke?' 'Tall. Flashy dressed. Got a slight American yakker.' **1973** *Tel.* (Brisbane) 13 Sept. 30/1 Last, but hardly least, is the yakker, who talks continually to his passengers. **1984** *N.Y. Times* 28 Aug. B-2/5 She just brought the parrot along for the ride... He was quite a yakker.

yacker (yæ·kəɹ), *v. slang.* [f. *yack *v.* + -er[5].] = *yack *v.*

1961 S. Price *Just for Record* x. 109 They're yackering away in the kitchen. **1982** *Financial Times* 20 Aug. 11/4 'Yellow Polka-Dot Bikini'—one of the scratchy 78s..—yackers melodiously while the characters gallivant through daytime Calcutta.

yacket (yæ·kėt), *v. slang.* [f. *yacket(y *int.*] *intr.* = *yack *v.* So **ya·cketing** *ppl. a.*

[**1953**: see *yackety-yack(et)ing *vbl. sb.* and *ppl. a.*] **1958** P. de Vries *Mackerel Plaza* 153 All this yacketing oratory. **1969** *New Yorker* 29 Nov. 51/3 We warn them, we yacket away night and day..but they never learn.

yackety (yæ·kėti), *int. slang.* Also **yackity, yaketty, yakkety, yakkity.** [Echoic.] Expressing the sound of incessant chatter. Usu. reduplicated or with *ya(c)k.*

1953 Berrey & Van den Bark *Amer. Thes. Slang* (1954) § 189/2 Idle talk; chatter... Yackety-yack,.. yackety-yackety. **1955** M. Millar *Beast in View* vi. 93 She..parks herself in the phone booth, and there she sits, yackity, yackity, yackity. **1959** S. Gibbons *Pink Front Door* iii. 36 She went yaketty-yak to Katy. **1959** L. Smith *One Hour* (1960) xviii. 231 The TV going yakkety yak. **1959** 'J. Welcome' *Stop at Nothing* vii. 113 Mildred.. has been going yackety yack-yack about you to some purpose. **1976** *Billings* (Montana) *Gaz.* 26 June 5-B/5 I'm talking away, yakkity yak, and he started gasping for breath. **1982** D. Bagley *Windfall* xxix. 288 The Sergeant ..only talks when he has something to say. Everybody else goes yacketty-yack all the time.

Also **ya:ckety-ya·ck(ety** *sb.* = *yack *sb.* 2 a; *v. intr.* = *yack *v.*; **ya:ckety-ya·ck(et)ing** *vbl. sb.* and *ppl. a.*

1953 M. Dickens *No More Meadows* iv. 185 Our laundry's full of yackety-yacketing women this morning. **1958** *Observer* 15 June 15/7 A muddle-headed momma.. who knows no better than to drive away her husband..by constant yackety-yack and pleas to stay at home. **1959** *Woman* 6 June 10/1 For once the place will be free of giggles and girlish yakitty-yak. **1960** *N.Y. Times Bk. Rev.* 3 July 1 (*heading*) How to dig the hips' yackety-yak. **1976** *Billings* (Montana) *Gaz.* 26 June B-14/8 A man working alone could never have enjoyed 'chewing the fat and the yakety yaking and drinking coffee'.

yaffle, *sb.*[2] Add: **2.** The call of the green woodpecker.

1955 D. A. Bannerman *Birds Brit. Isles* IV. 77 It [*sc.* the green woodpecker] is a bird which attracts attention by its loud cry or 'yaffle'. **1976** *Southern Even. Echo* (Southampton) 12 Nov. 18/4 The yaffle of a green woodpecker and needle-like sounds from the tits and goldcrests.

yaffle (yæ·f'l, yǫ·f'l), *v.*[2] *dial.* Also **yoffle.** [Echoic.] *intr.* To eat or drink, esp. noisily or greedily.

1788 Grose *Dict. Vulgar T.*, *Yaffling*, eating. *a* **1821** J. W. Masters *Dick & Sal at Canterbury Fair* in Parish & Shaw *Dict. Kentish Dialect.* (1887) p. xx, Sa when we lickt de platters out, An yoffled down de beer. *a* **1935** T. E. Lawrence *Mint* (1955) viii. 30 You bloody swaddies can't half yaffle.

yag (yæg). Also **YAG.** [f. the initial letters of *yttrium aluminium garnet.*] A synthetic crystal of yttrium aluminium garnet, used in certain lasers and as a simulated diamond in jewellery.

1964 *Q. Rep. Bell Telephone Lab.* No. 2 (AD 439–628). 3 The splitting of the ⁴F₃..manifolds of the trivalent Europium and Terbium ions..has been used to calculate crystal field parameters for Eu and Tb in YAG. **1964** *Appl. Physics Lett.* V. 201/2 Predictions of the laser behavior were tested on singly and doubly doped YAG crystals. **1971** *Daily Tel.* 22 Nov. 13 Have you got a yag yet? They are the latest form of simulated diamonds. Almost as hard as a sapphire, which is almost as hard as a diamond, they are 'grown' in America, brought over here to be faceted, set in 18-carat gold in Hatton Garden and sold as Diamonairs. **1977** *Jrnl. R. Soc. Arts* CXXV. 779/2 The YAG laser..is used routinely..for trimming resistors. **1979** [see *simulant *sb.*]. **1984** *National Geographic* Mar. 341/2 A flash lamp can excite atoms of neodymium, a rare earth, in a rod of YAG.

‖ **yagé** (yā·ʒe, yāhē·). Also **yage, yajé.** [Amer. Sp.] **a.** A South American liana of the genus *Banisteriopsis* used by the Indians to make a hallucinogenic infusion. **b.** The drink made from this.

1924 C. W. Domville-Fife *Among Wild Tribes of Amazons* xvi. 229 A curious potion is made from a plant called *yagé*. **1931** *Jrnl. Washington Acad. Sci.* XXI. 487 One of the most interesting plants found in the region of the upper courses of the Putumayo and Caquetá Rivers [in Colombia] is the *yagé*. The Indians make a beverage from either the wild or cultivated *yagé*, boiling it in a large earthenware vessel an entire day... They add to the *yagé* the leaves and the young shoots of the *oco yagé* or *chagro panga*.., and it is the addition of this plant which produces the 'bluish aureole' of their visions. **1945** F. R. Fosberg in F. Verdoorn *Plants & Plant Sci. Lat. Amer.* 28/1 Two other plants..are used by the Amazonian Indians for their narcotic effect. *Maikoa* (*Brugmansia arborea*) and *yagé* or *caapi* (*Banisteriopsis caapi*) both produce hallucinations. **1953** W. Burroughs *Junkie* xv. 149 Maybe I will find in yage what I was looking for in junk and weed and coke. **1960** *Spectator* 29 July 176 Yage, a vine-bark stew..supposedly invests the user with telepathic powers. **1969** *Science* 17 Jan. 253/2 The Siona of today frequently mix *Datura* leaves with *Banisteriopsis* in preparing yajé. **1975** *High Times* Dec. 80/1 You trek 900 miles overland into the Amazon jungle to sample yagé in its natural habitat. **1977** Lewis & Elvin-Lewis *Med. Bot.* xviii. 413/1 No more interesting or complex narcotic drink can be found than ayahuasca, caapi, or yaje.

yager. b. Substitute for def.: An obsolete kind of rifle. Also *yager rifle.* (Earlier and later examples.)

1817 E. P. Fordham *Narr. Trav.* (1906) 141 Sent the two P—s..for the yager rifle, and the Wallet. **1826** T. Flint *Francis Berrian* I. 60 Their trade with the Americans supplied them with rifles and yagers. *a* **1918** G. Stuart *40 Yrs. on Frontier* (1925) I. 187 Nine Pipes.. came to get a nipple to put on his Yager rifle.

Yaghan, var. *Yahgan.

Yaghnobi (yɑgnōᵘ·bi). Also **Yagnobi.** A modern Iranian language spoken by the Yaghnobs in parts of Tadzhikistan.

1932 W. L. Graff *Lang. & Languages* x. 372 Many.. Iranian dialects..are spoken over large areas of Asia. The chief among them are the Caspian, the Kurdish,..the Yagnobi, and the Ossetic dialect. **1960** *Language* XLIV. 281 The alternative is to assume the more traditional view that Sogdian is a descendant of Avestan, and Yaghnobi of Sogdian. **1974** *Encycl. Brit. Macropædia* IX. 451/2 Yaghnobi is still spoken by a small number of people southeast of Samarkand. It has two main dialects.

Yagi (yā·gi). *Broadcasting.* The name of Hidetsugu *Yagi* (b. 1886), Japanese scholar and electrical engineer, used *attrib.* and *absol.* to designate a highly directional aerial that he invented (*Proc. IRE* (1928) XVI. 715) for receiving or transmitting VHF or UHF waves within a narrow frequency band, consisting of a number of short rods mounted transversely on an insulating support that points towards the signal source.

1943 *Gloss. Terms Telecomm.* (*B.S.I.*) 66 The term *Yagi aerial*, which relates to a particular form of end-fire array, should not be used as a generic term for all end-fire arrays. **1950** *Austral. Jrnl. Sci. Res.* A. III. 20 The array of nine Yagis in three groups of three, one wavelength apart, is fixed on an equatorial mounting. **1951** A. C. Clarke *Sands of Mars* iv. 42 He produced a rough sketch of a simple Yagi aerial. **1960** *Practical Wireless* XXXVI. 362/2 (Advt.), 2 Metre beam 5 element W.S. Yagi. **1975** L. Deighton *Yesterday's Spy* xi. 88 Did he think we needed the eight Yagi aerials for TV?

Yagnobi, var. *Yaghnobi.

yagona: see *yanggona.

yah, *adv.* (Earlier example, representing Lancashire speech.)

1863 A. J. Munby *Diary* 20 Aug. in D. Hudson *Munby* (1972) 170 'Dus Jaan Brahn work here?' 'Yah!' said some of the maidens: 'Aye!' said others.

yah boo (yā bū), *int.* (and *sb.*). *slang* (orig. *children's*). Also **ya(a) boo** and with hyphen. [f. Yah *int.*[1] + Boo *int.*] An exclamation of scorn or derision. Also *attrib.* and *transf.*

1921 H. Williamson *Beautiful Years* 83 Willie and Jack, scorning to reply to the yaa-boos of the retreating urchins, were licking their hurts. **1926** 'R. Crompton' *William the Conqueror* i. 16 'Yah—boo, softie!' he called over the wall. **1961** *Times* 7 Apr. 20/7 It [*sc.* a boo] is far from a pretty sound—rather moronic, in fact—and smacks too much of the 'yah boo' which used to be the height of brilliant preparatory school repartee. **1968** [see *suck *sb.*[1] 11]. **1973** *Times* 28 Dec. 8/6 People..who are tired of the 'Yah-boo' school of debate. *a* **1976** A. Christie *Autobiogr.* (1977) ii. ii. 76 Two small boys arrived..preparing as usual to say, 'Yah. Boo. Shan't go.' **1981** *London Rev. Bks.* 19 Nov.–2 Dec. 12/2 The impatient rejection of 'Ya! Boo!' politics.

Hence **yah boo sucks** [*suck *sb.*[1] 11], used similarly.

1980 'A. Skinner' *Mind's Eye* xiii. 181 Ya boo sucks to anyone who was interested. **1983** *Listener* 22 Dec. 62/2 This is neither a tranche of free advertising, nor a yah-boo-sucks to temperance advocates.

Yahgan (yā·gän). Also **Yaghan, -ane.** [App. a native name.] (A member of) one of the three indigenous peoples of Tierra del Fuego, found in the most southerly part of the islands. Also *attrib.* or as *adj.*

1884 *Proc. R. Geogr. Soc.* VI. 348 The natives, who inhabit Tierra del Fuego and this archipelago of Cape Horn, belong to three principal tribes:..(iii.) the Yaghanes, inhabiting both banks of the Beagle Passage and all the islands in the south of the archipelago. *Ibid.* 349 The Yaghane..passes his time squatting in his hut. **1961** G. Clark *World Prehistory* ix. 238 (*heading*) Yahgan, Ona and Alacu peoples of Tierra del Fuego. *Ibid.*, The Yahgan and their neighbours the Ona and Alacu maintained down to modern times the most southerly settlements of mankind. **1972** *Bk. of Thousand Tongues* (rev. ed.) 459/1 The Yahgan language, which once comprised five mutually intelligible dialects, is related to no other known tongue. **1974** *Encycl. Brit. Macropædia* VIII. 1159/2 The Yahgan are canoe-using fishermen and shellfish gatherers. **1983** *Times* 9 July (Saturday Suppl.) 2/8 Of the poor, ill-clad and wretched Yahganes there is only one left.

yahoo, *sb.* Add: **1.** (Further examples.) In mod. use, a person lacking cultivation or sensibility, a philistine; a lout, a hooligan.

1912 J. Sandilands *Western Canad. Dict. & Phrase-Bk.*, *Yahoo*, a lout from the back-country, an ignoramus, a know-nothing. **1914** 'I. Hay' *Lighter Side School Life* iii. 83 You must not behave like a yahoo in my mathematical set. **1943** J. Lees-Milne *Ancestral Voices* (1975) 200, I took the young yahoo..to the station. **1968** *Courier-Mail* (Brisbane) 21 May 6/4 A Brisbane boat owner has complained that 'young yahoos' are stripping cars left at bayside boat ramps. **1977** *New Yorker* 20 June 56/2 Sarge Waller—who is, among other things, a professional riverman—later commented on Cook and Ulvi's journey and described them as 'yahoos'.

2. = Wild man 2.

c **1810–20** *Handbill*, During the Fair... Two surprising large Yohoes; or, Wild Men of the Woods, being the most Wonderful of the kind ever Exhibited. **1814** *Lincoln, Rutland & Stamford Mercury* 22 Apr. 3/5 Just arrived, and to be seen in a commodious booth, in the Crown and Anchor Yard, Lincoln... The Great Yahoo, or Wild Man of the Woods.

3. *Austral.* [Perh. a different word.] A probably mythical creature resembling a big hairy man, said to haunt eastern Australia. Cf. *yowie[2].

1842 in G. C. Joyner *Hairy Man South Eastern Austral.* (1977) 5 A contested point has long existed among Australian naturalists whether or not such an animal as the Yahoo existed. **1844** L. A. Meredith *Notes & Sketches N.S.W.* x. 95 They have an *evil* spirit, which causes them great terror, whom they call 'Yahoo', or 'Devil-devil.' **1876** *Austral. Town & Country Jrnl.* 4 Nov. 729 For many years past it had been believed by the settlers of that wild part of the country, that the Walla Walla scrub was inhabited by a monster called 'the hairy man of the wood', or what all the blacks stand so much in dread of—the Yahoo. **1937** *Mankind* II. iv. 91 [J.N., a Kumtangerai, told me that big hairy men lived in the scrub at Nana Glen (North Coast, N.S.W.), and were called by the native Jarrā-wahu.] *Ibid.*, In the Mudgee district..a scrubby place was reputed to be the abode of a 'Yahu', and a resident in the Maitland district told me a 'Yahu' was reputed to live in thick scrub there. Each said he was a big hairy man.

yahoodom: also, behaviour characteristic of a yahoo; (earlier example).

1890 Kipling *Let.* in C. Carrington *Rudyard Kipling* (1955) vii. 162 The grotesque Yahoodom of nipping pieces off a half-presented foetus and slamming it into the market.

yahrzeit (yā·ɹtsait). Also **jahr-; yore-, yort-** (yǫ·ɹ-); and with capital initial. [Yiddish, f. MHG *jârzît* anniversary, f. OHG. *jâr* Year + *zît* time.] Among Jews, the anniversary of the death of someone, esp. a parent.

1852 *Asmonean* 10 Dec. 91/1 Men, who..when they have Yahrzeit, will go to congregational meetings, and oppose most violently every reform measure. **1876** Geo. Eliot *Dan. Deronda* II. iv. xxxiv. 359 'Your mother has been a widow a long while, perhaps.'..'Ay, ay, it's a good many *yore-zeit* since I had to manage for her and myself,' said Cohen. **1881** *Fraser's Mag.* Apr. 496 On the *Jahrzeit* Maier takes the boy to the synagogue. **1917** E. Ferber *Fanny Herself* 200 There would be no *Yahrzeit* light burning for twenty-four hours. **1964** W. Markfield *To an Early Grave* (1965) x. 177 They buried my brother-in-law... In two and a half weeks, God spare us, he'll have his *yahrzeit*. **1971** I. B. Singer *Isaac Bashevis Singer Reader* 300 Their 'questions' inevitably concerned the observance of a *yortzeit*.

Yahtzee (yā·tsi). Chiefly *U.S.* [f. Yacht *sb.*] The proprietary name of a game (orig. 'the yacht game') played with dice and a score sheet.

1957 *Official Gaz.* (U.S. Patent Office) 1 Jan. TM 12/2 E. S. Lowe Company, Inc., New York... *Yahtzee.* For poker dice games. **1970** *Trade Marks Jrnl.* 8 Apr. 563/1 *Yahtzee*... Boxed games; dice, boards, counters and cards (other than ordinary playing cards), all for games. E. S. Lowe Company, Inc... New York. **1973** M. Kaye *Toy is Born* 56 Initially popular as a men's game—probably because of the gambling mystique of dice—Yahtzee went on to fascinate women's clubs. **1974** *N.Y. Times* 10 Nov. ii. 37 We sat with a few friends and played Yahtzee all night long. **1977** *Monitor* (McAllen, Texas) 19 June 5c/1 Following the dinner, members met..and spent the evening listening to music and playing Yahtzee.

Yahudi (yāhū·di). *slang* (orig. and chiefly *U.S.*). Also 9 **Yahooda, -ee, Yehoodi; Yehudi.** [Arab. *yahūdī*, Heb. *yehudi* Jew sb.] **a.** A Jew; Jews. **b.** *attrib.* as *adj.* Jewish.

1823 C. M. Doughty *Trav. Arabia Deserta* II. 382 When I was trafficking in Irâk, I had dealings with a certain Yahûdi. **1858** *Asmonean* 19 Mar. 180/3 We are credibly informed that not less than eight hundred families of Yehoodim..are utterly destitute. **1862** J. A. Grant *Jrnl.* 14 Aug. in *Walk across Africa* (1864) xi. 264 Frij and all Seedees believe that the Jews, or Yahoodee, living in Calcutta, seize people, and tie them up by the heels till blood falls from them into a dish. **1900** G. Ade *More Fables* 117 The flip Yahooda, with the City Education and Thirty Centuries of Commercial Training.., saw that here was a Chance to work off some Old Stock. **1930** R. L. Strauss *Amer. Remnant* 117 Dot *mensch* certainly knew his bissness..and look at all the *Yehudim* here too! **1932** [see *Hubshee *sb.* and *a.*]. **1959** I. Jefferies *Thirteen Days* i. 16 As far as the Yehudis were concerned I knew the dirt that was being done. *Ibid.* vi. 83 We ate well, and drank a good Yehudi wine. **1977** *Washington Post* 17 June C-5/3, I see the hate in your eyes, you Yahudi (Jewish) whore, and when we go to work on you, you'll be sorry.

Yahweh. Add: Now the usual form of the word among scholars.

1885 *Studia Biblica* I. 3 Delitzsch..propounds the following theory. The forms *Yahu, Yah*,..are of foreign origin. The form *Yahweh*, on the other hand, is distinctively Hebrew. **1913** H. W. Robinson *Relig. Ideas of Old Test.* iii. 53 No certain evidence for the pre-Mosaic use of the form Yahweh..seems yet to have been brought forward. **1936** W. L. Wardle *Hist. & Relig. Israel* viii. 147 God reveals to Moses as something previously unknown that his name is Yahweh. **1958** S. Godman tr. *Noth's Hist. Israel* i. ii. 99 In Jos. xxiv, there is no mention at all of sacrifices, but rather of a profession of faith in Yahweh. **1973** D. J. Wiseman *Peoples Old Testament Times* p. xxi, The Sinai covenant was the seal of Yahweh's choice of the people. **1984** *Church Times* 23 Nov. 7/2 He is a New Testament scholar concerned to expound..the growth of the biblical idea of Yahweh as the God who cares for the poor.

yair (yēᵊɹ), *Austral.* var. of *yeah *adv.*

1953 A. Upfield *Venom House* i. 3 'You manage all right without brakes?' 'Yair. Nothing wrong with the ruddy engine to ease her up.' **1959** S. H. Courtier *Death in Dream Time* xii. 164 'My coat still there?' 'Yair.' **1964** R. Braddon *Year Angry Rabbit* (1967) i. 8 'You can virtually rig the ballot in two of our most important rural divisions?' 'Yair,' said Alfill. **1977** C. McCullough *Thorn Birds* xv. 344 Yair, but in the glasshouse youse don't keep getting shot at. **1980** *Herald* (Melbourne) (City ed.) 14 Apr. 2/2 Yair, but it was only in Melbourne.

yajé, var. *yagé.

yak. a. For *Poephagus* read *Bos* and add earlier example.

1795 *Asiatick Res.* IV. 351 The Yak of Tartary..is about the heighth of an English bull.

b. *yak-herd*; **yak butter,** butter made from the milk of the yak; **yak lace** (earlier example).

1962 L. Davidson *Rose of Tibet* v. 87 He had bought tea bricks..and a large cake of yak butter. **1980** *Times* 12 Aug. 10/1 Crowds file through the little shrine rooms lit by flickering lamps of yak butter. **1958** *Illustr. London News* 13 Dec. 1041/1 Greatest of all the village-festivals was the *Dumje*, celebrated in early July, before the villagers dispersed with their yak-herds to the high pastures. **1872** *Young Englishwoman* Nov. 606/1 The trimming consists of a narrow passementrie border and black yak lace.

yak, var. *yack *sb.*, *v.*

‖ **yakdan** (yæ·kdān). Also 9 **yakhdan.** [Pers. *yakhdān* ice-house, (also) portmanteau, f. *yakh* ice + *dān*, (affix denoting) what holds or contains anything.] In Iran, a trunk or portmanteau.

1824 J. J. Morier *Adv. Hajji Baba* II. vii. 112, I was in want of a pair of *yakhdans*, or trunks. **1922** *Blackw. Mag.* June 761/1 The bachelor's rule should be never to possess anything which he cannot squeeze into a yakdan. **1954** J. Masters *Bhowani Junction* xxxvii. 320 Then he delved in a yakdan and brought out some food I'd had cooked. **1978** 'M. M. Kaye' *Far Pavilions* i. 13 Four locked *yakdans* containing botanical specimens

yaketty, var. *yackety *int.* **yakhdan,** var. *yakdan.

Yakima (yæ·kimă), *sb.* and *a.* Also 9 **Yacama; Iakima.** [Native name.] **A.** *sb.* **a.** (A member of) a group of American Indians who lived in the area of the Columbia and Yakima Rivers in south central Washington before their confinement, with other tribes, on the Yakima Reservation in 1858; since then, (a member of) the Indians of this reservation. **b.** The language of the Yakima, a member of the Sahaptin group. **B.** *adj.* Pertaining to or designating the Yakima.

1852 S. Eastman in H. R. Schoolcraft *Indian Tribes U.S.* (1853) III. *facing* p. 96 Yackimas. **1855** *N.Y. Herald* 15 Nov. 2/3 All the Yacama Indians are in the field, and the war has fairly begun. **1857** *Spirit of Times* 4 July 277/3, I have..become a sub in one of Uncle Sam's regular regiments engaged in the late campaigns against the Yakima. **1940** M. W. Smith *Puyallup-Nisqually* 20 The two Sahaptin dialects, Klickitat and Yakima. **1946** J. T. Adams *Album Amer. Hist.* III. 84 In 1855 The Yakimas attacked parties of prospectors. **1973** A. H. Whiteford *N. Amer. Indian Arts* 59 The Yakima..make baskets with thick, hard coils of cedar root strips. **1976** *Billings* (Montana) *Gaz.* 1 July 3-B/2 Indians..led by Sid Mills, a Yakima Indian from Washington. **1977** [see *rankle *v.* 5 b].

‖ **yakitori** (yækitō·ri). Also **yaki-tori.** [Jap., f. *yaki* toasting, grilling + *tori* bird.] A Japanese dish consisting of pieces of chicken grilled on a skewer.

1962 M. Doi *Art of Jap. Cookery* 69 Yaki-tori... Chicken Meat..cut into mouthfuls..and soak ten minutes in tare... Place skewered chicken directly over fire and broil. **1970** J. Kirkup *Japan behind Fan* 14 Yakitori stands selling bamboo skewers of roasted bits of chicken and liver. **1983** *Daily Tel.* 10 Nov. 17/2 We have planned a modest meal..making the main course one of two Japanese skewered specialities: one called Yakitori.

yakka (yæ·kă). *Austral. slang.* Also **yacca, yacka, yacker, yakker.** [Aboriginal.] Work, toil; esp. in phr. *hard yakka.*

1888 *Boomerang* 14 Jan. 13 The Brisbane wharf labourers..are so accustomed to hard yacca that they will not be happy for a single day without it. **1898** *Bulletin* (Sydney) 8 Oct. 31/2 Some [wagons] ain't for 'yacker', some's lookin' for 'graft', and some's 'after a job'. **1906** [see *after *colloq.* abbrev.]. **1909** H. Thompson *Ballads about Business* 91 You'll be sure to get some yacker and more country you will see. **1930** W. Hersey in L. W. Fry *Tales by Australians* 133 It'll be a circular tour of everyone..did real hard honest yakker. **1944** *Coast to Coast 1943* 121 They'd been shoved in the background, told they weren't wanted, all the hard yacca put upon them in the home. **1946** K. Tennant *Lost Haven* (1947) 139, I vote we leave the Methodist part alone, and go and clear up where somebody's pitched a tent on some bought out of our yacka. **1948** V. Palmer *Golconda* iv. 28 If there's a cove on this field making money at anything but hard yakker it isn't Macy Donovan. **1968** *Courier-Mail* (Brisbane) 13 Nov. 6/7 Australian scholarships have always been hard yakka. **1982** *Nation Review* (Austral.) 2 Aug. 29/1 He imposes some yakka.

yakker, var. *yacker *sb.*, *yakka. **yakkety,** var. *yackety *int.* **yakking** var. *yacking *vbl. sb.* **yakkity,** var. *yackety *int.*

‖ **yaksha** (ya·kʃă). *Indian Mythol.* Also **yaksa,** and with capital initial. Fem. **yakshi, yakshi·ṇi.** [Skr. *yakṣa*, fem. *yakṣī, yakṣiṇī.*] **a.** Any of a class of demi-gods or nature spirits, often inoffensive tutelary guardians of a place; *esp.* one attendant upon Kubera, the god of wealth. **b.** A statue or carving representing one of these.

1785 C. Wilkins tr. *Bhăgvăt-Gēētā* xi. 92 The *Gāndhārvs* and the *Yākshās*, with holy tribes of *Soors*, all stand gazing on thee, and all alike amazed! **1810** E. Moor *Hindu Pantheon* 276 His servants and companions are the *Yakshas* and *Guhyakas*, into whose forms transmigrate the souls of those men who in this life are addicted to sordid and base passions, or absorbed in worldly prosperity. **1882** A. Barth *Religions of India* v. 169 Civa..sits enthroned on Kailāsa, the fabulous mountain of the North..surrounded and waited on by the Yakshas. **1928** A. K. Coomaraswamy *Yakșas* I. 17 The essential element of a Yakṣa holystead is a stone table or altar, placed beneath the tree sacred to the Yakṣa. **1931** *Times Lit. Suppl.* 24 Dec. 1042/1 The important part played by yakshas in Buddhist religion and art is incontestable. **1963** *Times* 12 Feb. 11/1 The highest price of the afternoon was..paid..for a fifth century..carving of a Yakshi or tree nymph, probably part of a pillar. **1971** *Illustr. Weekly India* 11 Apr. 32/1 Each Tirthankara is identified by the tree under which the vows of asceticism were taken and the attendant Yaksha and Yakshini. **1977** *Jrnl. R. Soc. Arts* CXXV. 570/2 This seated *yakṣa*, perhaps Kubera, from Nagpur District..has only recently come to light.

Yakut (yæku·t), *sb.* and *a.* Also 8 **Yakouti, Yakuty,** 9 **Yakute.** [Russ.] **A.** *sb.* **a.** (A member of) a Mongoloid people of north-eastern Siberia who now form most of the population of the Yakutsk Republic of the Soviet Union.

1763 J. Bell *Travels* I. 240 The Yakuty differ little from the Tongusians. **1797** *Encycl. Brit.* XVI. 570/2 Besides these, there are in the Russian dominions the Nagay Tartars;..the Yakouti. **1890** J. G. Frazer *Golden Bough* I. i. 26 When the day is hot and a Yakut has a long way to go. **1974** T. P. Whitney tr. *Solzhenitsyn's Gulag Archipelago* I. ii. 51 The Yakuts were imprisoned. **1981** M. C. Smith *Gorky Park* i. vii. 107 Some twenty-odd Russians and Yakuts surrounding a small group of Westerners and Japanese.

b. The language of the Yakuts, an Altaic one usually placed in the Turkic group.

1908 T. G. Tucker *Introd. Nat. Hist. of Lang.* viii. 134 The linguistic connection within this group is very close, the languages of the extremes, Turkish and Yakut, for instance, being at least as distinctly related as English and German. **1951** W. K. Matthews *Languages U.S.S.R.* ii. 8 Before the Revolution Yukagir was proscribed in favour of Yakut and Russian. **1976** 'S. Harvester'

Siberian Road xiv. 165 The middle-aged woman translated what he said into a language he took to be Yakut.

B. *adj.* Pertaining to or designating the Yakuts.

1854 Max Müller in C. Bunsen *Outl. Philos. Universal Hist.* I. 279 The Yakute dialect became separated at a very early time from the still undivided Turko-Tataric speech. **1887** *Encycl. Brit.* XXII. 9/1 The Tunguses.. occupy as their hunting-grounds an immense region on the high plateau and its slopes to the Amur, but their limits are yearly becoming more and more circumscribed both by Russian gold-diggers and by Yakut settlers. **1963** V. Nabokov *Gift* iv. 267 Making clumsy paper boats for Yakut children. **1981** I. Boland tr. *Ginzburg's Within Whirlwind* II. iv. 218 He was a Yakut boy—or at least his mother was Yakut.

‖ **yakuza** (yākū·zā). Also **yakusa.** [Jap., f. *ya* eight + *ku* nine + *za*, *sa* three (see below).] A Japanese gangster or racketeer; usu. in *pl.* sense, such people collectively.

8–9–3 is the worst set of cards in a player's hand at a gambling game: hence, the worst sort (K. Koike).

1964 *Newsweek* 14 Sept. 42/2 The youngsters had to listen to boss Sakamoto expound on the noble traditions of the *yakuza*, as gangsters are called in Japan. **1971** *Ibid.* 22 Mar. 42/1 The yakuza, or gangster, is an enduring feature of Japanese life. **1975** *New Yorker* 24 Mar. 98/2 The yakuza are the Japanese gangsters who in recent years have moved from gambling, drugs and prostitution into shakedown rackets. **1977** J. van de Wetering *Japanese Corpse* v. 53 Amsterdam is full of Japanese... Even their gangsters seem to be here, the yakusa. **1979** *Honolulu Advertiser* 8 Jan. A-6/1 Yakuza—the Japanese Mafia—are thriving in Hawaii as members of a crime syndicate.

Yale[2] (yēⁱl). [f. the name of the company founded by Yale (see def.).] A proprietary name for locks and keys, used esp. to denote a lock with a cylindrical barrel that can be turned only when a key with a specially serrated edge is inserted so as to displace a number of pins by the correct distances (invented by Linus Yale, Jr. (1821–68), U.S. locksmith).

1869 *Price List of Yale Lock Manfg. Co.* 3 The manufacturers of the Yale Locks desire to say a few words in relation to the recent changes in their organization, location and prices. **1875** *Iron Age* 1 July 10/1 (Advt.), Yale locks for all uses. **1885** *Trade Marks Jrnl.* 15 Apr. 346 *Yale.* The Yale and Towne Manufacturing Company (Incorporated), Stamford, Connecticut... Locks. Bronze hardware. **1895** *Montgomery Ward Catal.* Spring & Summer 375/2 Yale pattern night lock. *Ibid.* 382/1 Yale padlocks. **1907** *Official Gaz.* (U.S. Patent Office) 7 May 437/1 The Yale & Towne Mfg. Co., Stamford, Conn... *Yale*... Locks and keys. **1920** W. J. Locke *House of Baltazar* xxv. 306 The little brass Yale latchkey. **1930** W. de la Mare *On the Edge* 11 The Yale gear-key which usually lay in the little recess to the left of the dash-board was missing. **1949** M. Mead *Male & Female* v. 114 Each small family is so isolated from others that no one knows how peculiar or how usual are..the behaviours that are shut behind each Yale lock. **1974** *Encycl. Brit. Macropædia* XI. 12/1 Magnetic forces can be used in locks working on the Yale principle. **1974** S. B. Hough *Fear Fortune, Father* vi. 45 It was a Yale key and the back door had a Yale lock. **1976** C. G. Smith *Let.* in *Daily Mail* 28 May 35/2 Yale is a registered trade-mark and in no way refers to one particular lock. Since 1949 we have specialised in marketing automatic, semi-automatic and double-locking night-latches, engineered to foil the glass-breaking and bolt-forcing intruder.

b. *ellipt.*

1918 G. Frankau *One of Them* xxx. 235 Where each man's latchkey fits his neighbours' Yales. **1954** R. Macaulay *Let.* 14 Mar. in *Last Lett. to Friend* (1962) 148, I have changed my lock, but they can pick 'Yales'. **1983** *Oxford Consumer* Autumn 11/2 Your cylinder-lock (what people are apt to call a 'Yale', whether it is actually made by the Yale Co. Ltd. or not).

‖ **yali** (yālí·). [ad. Turk. *yalı* shore, waterside residence, f. Gk αἰγιαλός sea-shore.] A house of a type found on the shore of the Bosporus.

1962 J. Fleming *When I grow Rich* i. 9 Most large wooden houses on the Bosphorus are called either *yalis* or palaces. **1976** *Times* 28 Feb. (Turkish Suppl.) p. i/3 Ahead..flows the deep blue Bosporus. From its either shore rise green hills..dotted with..those lovely old wooden houses called *yalis*. **1978** S. Sheldon *Bloodline* i. 15 Not the tourist Istanbul..but the out-of-the-way places.., the *yalis*, and the small markets beyond the *souks*.

Yalie (yēⁱ·li). *U.S. colloq.* [f. *Yale* + -ie.] A student or graduate of Yale University.

1969 *Newsweek* 12 May 71/1 One new Yalie..attended Harvard summer school and found time to visit New Haven to look Yale over. **1970** E. Segal *Love Story* xviii. 114 He's a Yalie, Ol... A total Yalie. College *and* Med School. **1972** *Nature* 4 Feb. 290/2 The availability of feminine companionship during the Monday-to-Friday period during which Yalies have traditionally gone without. **1983** *Washington Post* 21 Nov. B15/3 FDR, as editor of the Crimson, is said to have once gotten out a special post-victory edition, sticking it to the Yalies just as the fans were leaving the stadium.

y'all (yǭl). *U.S. dial.* Also **yall.** Abbrev. *you-all *pers. pron.*

1909 *Dialect Notes* III. 390 Where are yall goin'? **1928** *Amer. Speech* IV. 103, I heard a young lady, in greeting a group of her friends, say, 'How're y'all this morning?' **1935** *Scribner's Mag.* Feb. 120/2 Ah ain' gwi' be wid yall long. **1944** C. Himes *Black on Black* (1973) 199 'Bout how much ken y'all pick, shawty? **1968** E. J. Gaines *Bloodline* 134 Unc' Toby won't feel right if y'all don't eat his lovely food. **1971** *Black World* Apr. 55 Yall woulda held a stop watch on God-all-mighty. **1982** J. S. Borthwick *Case of Hook-Billed Kites* xxxiv. 114 Yes, Doctor. You'll be in the breakfast room. Y'all have a nice day.

yaller, var. Yellow *a.* and *sb.* in Dict. and Suppl.

yam[1]. Add: **3. yam house,** a building in which to store yams.

1910 C. G. Seligmann *Melanesians Brit. New Guinea* xlix. 672 The number of yam houses makes each hamlet look larger than it really is. **1949** M. Mead *Male & Female* ix. 190 Among the Trobriand Islanders, each man fills the yam-house of his sister, not that of his wife.

yam (yæm), *v. dial.* Also **nyam.** [Derived through W. Indian from W. African words such as Hausa *nama* flesh, meat, Swahili *nyama* meat, Fulah *nyama* to eat; ult. the same word as Yam[1].] *trans.* To eat, esp. with relish.

1725 *New Canting Dict.*, *Yam*, to eat heartily, to stuff lustily. **1801** T. Dancer *Medical Assistant* 174 [Dirt-eaters] display as much curiosity and nicety in their choice of the earth they yam, as snuff-takers or smokers in the kind of tobacco they make use of. **1816** M. G. Lewis *Jrnl.* (1834) 256 There's rice in the pot, take it, and yam-yamme. **1841** Jamieson *Scottish Dict.*, *Nyam*, to chew. **1846** *Swell's Night Guide* 136/1 *Yam*, to eat hearty. **1862** W. G. Hamley *Captain Clutterbuck's Champagne* iv. 68 They purchased the congenial [sugar-cane] plant, and *nyaming* greedily its fibre, were entranced. **1864** Hotten *Slang Dict.* 273 *Yam*, to eat. This word is used by the lowest class all over the world; by the Wapping sailor, West India negro, or Chinese coolie. **1905** *Eng. Dialect Dict.* VI. 563/2 *Yam*, to eat greedily and with noise; to chew. **1970** C. Major *Dict. Afro-Amer. Slang* 125 *Yam*, to eat.

Also as *sb.*[3], food.

1788 P. Marsden *Acct. Island Jamaica* 49 The negroes say, the black parroquets are good for *yam*, i.e. good to eat. **1828** *Marly; or Life of Planter in Jamaica* (ed. 2) 13 Eh! Mosquitoes hab grandy nyamn on dat new buckra! **1835** R. R. Madden *Twelvemonths Residence W. Indies* I. 188 Him want no nyam, no clothes, no sleep. **1903** Farmer & Henley *Slang* VII. 368/2 *Yam* (nautical), food. **1953** *Caribbean Q.* III. iii. 176 That was a wicked Jamaican lizard 'mash up him common-law wife for mout'ful of nyam'.

‖ **Yamato** (yamā·to). [Jap., = 'Japan'.] **1.** The style or school of art in Japan which culminated in the 12th and 13th centuries and dealt with Japanese subjects in a distinctively Japanese (rather than Chinese) way. Usu. as **Yamato-e** († -we)[*e* picture]; also **-ryū** [*-ryū* style, orig. stream, school].

1879 *Trans. Asiatic Soc. Japan* VII. 345 Motomitsu is spoken of as the originator of the *Yamato-we*. *Ibid.* 346 Takanobu was a pupil of the Yamato riu. **1880** T. W. Cutler *Gram. Jap. Ornament* 5 In the thirteenth [century] was founded the Yamato, or Japanese school. **1911** *Encycl. Brit.* XV. 174/1 It did not take shape as a school until the beginning of the 11th century..; it then became known as *Yamato-ryū*, a title which two centuries later was changed to that of *Tosa*, on the occasion of one of its masters..assuming that appellation as a family name. **1935** K. Toda *Jap. Scroll Painting* ii. 19 Another important example of early *Yamato-e* is a series of wall paintings at Hōryūji. **1970** *Oxf. Compan. Art* 607/2 The demand for a more refined art from the aristocratic society of Kyoto, combined with the decline of the T'ang dynasty in China, encouraged the Japanese to incorporate more native elements in their art, especially in the picture scrolls..of the *yamato-e* style. **1980** R. Illing *Art Jap. Prints* vi. 87 In the 1720s Okumura Masanobu and Shigenaga produced prints of landscape views, rather in the style of the older *yamato-e* album paintings, showing the *Omi hakkei*.

2. *Yamato-damashii*: the Japanese spirit.

1942 *R.A.F. Jrnl.* 13 June 6/2 He will be filled with what is called *yamato damashi* [sic] or the pure spirit of Japan. **1957** *Encycl. Brit.* XII. 954O/1 The Japanese.. have been profoundly influenced by a specific type of ideology—sometimes known as *Yamato Damashii* (the soul of Japan), which was partly embodied in..the code of the warrior knight. **1974** in A. Murakami *Romanized Japanese* (1979) 23 And there was no doubt then that the warrior code..*Bushido*..and the Japanese spirit.. *Yamato-damashii*..steeled the Japanese soldier.

yammer, *sb.* Restrict *Sc.* and *dial.* to 'a cry [etc].' and add examples of sense 'a loud noise, a din'.

1932 'L. G. Gibbon' *Sunset Song* 16 There the din of the gulls is a yammer night and day. **1978** *Poetry* Mar. 328 They huddle, and their tabled ground rejoices To the flat yammer of their American voices. **1984** *Washington Post* 11 June B4/3 Diamanda Gala's score, consisting of shrieks, yammers, gasps and vocal but incoherent hysterics.

yammer, *v.* Restrict *Obs.* exc. *Sc.* and *dial.* to senses 1 and 3 and add: **2.** (Further examples.)

1932 'L. G. Gibbon' *Sunset Song* 38 Her five bairns were all yammering blue murder at the same minute. **1952** W. R. Burnett *Vanity Row* viii. 73 Joe Sert yammering and getting purple in the face. **1958** 'W. Henry' *Seven Men at Mimbres Springs* xi. 125 Somewhere off in the eastern hills a coyote yammered with the crazed wildness which never fails to startle the oldest listener. **1959** W. H. Canaway *Seal* i. 16 His guts clanked and yammered like air-locked water-pipes. **1970** C. Sandburg *Compl. Poems* 372 They banged their spoons and bowls on the table And went on yammering for more to eat. **1980** F. Weldon *Puffball* 202 In the kitchen..Mabs' children yammered and cowered and snivelled and were slapped and shouted at. **1984** *Times* 5 Nov. 13/6 Just when women are yammering to be the hand that holds the briefcase..here's this little upstart letting the side down.

yammering *vbl. sb.* and *ppl. a.* (further examples in sense 2 of the vb.).

1937 [see *skriking *vbl. sb.*]. **1940** L. MacNeice *Poems 1925–40* 249 The city's Yammering fire alarms. **1969** M. Braithwaite *Never sleep Three in Bed* ix. 105 There was always so much yammering from us kids that no adult had a chance to say anything. **1977** *Time* 31 Jan. 43/3 Visitors..shepherded round the Acropolis by yammering guides.

yamun, yamen. (Earlier example.)

1747 *Astley's New Gen. Coll. Voy.* IV. i. vi. 275 Each Magistrate, great or small, has his Tribunal, or *Ya-men.*

Yana (yā·nā). [a. Central and Northern Yana (men's speech) *ya·na* person, people.] **a.** The language of the Yana Indians, a member of the Hokan group.

[**1888** *6th Ann. Rep. Bur. Amer. Ethnol.* p. xxxvii, Work was begun on the Nosa language (Yanan family) at Redding, Cal.] **1891** J. W. Powell in *7th Ann. Rep. Bureau Amer. Ethnol.* 135 Yanan Family. Derivation: Yana means 'people' in the Yanan language. **1903** *Amer. Anthropologist* V. 18 Yana shows so few similarities to other languages that it cannot be included in any group. **1913** [see *Hokan]. **1933** Bloomfield *Language* iii. 46 The differences between the two sets of Yana forms can be stated by means of a fairly complex set of rules. **1956** J. Lotz in Saporta & Bastian *Psycholinguistics* (1961) 12/1 In Yana, an Indian language of California, men and women use an entirely different vocabulary. **1971** *Language* XLVII. 831 Although most shifts move only one degree up the scale, Yana and Luiseño advance two degrees in shifting *l > n* and *r > ð* respectively.

b. (A member of) an American Indian people formerly living in northern California. Also *attrib.*

1910 *Univ. Calif. Publ. Amer. Archeol. & Ethnol.* IX. 3 These boundaries are somewhat uncertain, it remaining doubtful whether the Yanas reached the Sacramento. **1933** Bloomfield *Language* iii. 46 The classical instance is that of the Carib Indians; a recently authenticated one is the language of the Yana Indians in northern California. **1962** *Guardian* 23 Feb. 7/3 In 1911..in California..a man..was identified as a survivor of a subtribe of Yana Indians thought to be extinct. **1974** *Encycl. Brit. Micropædia* X. 796/1 The last known Yahi survivor..died in 1916. Other Yana, if they survive, are intermixed with other northern Californian Indians.

yandy (yæ·ndi), *v. Austral.* [Aboriginal.] *trans.* To separate (grass seed) *from* refuse by shaking the mixture in a special way; to separate (ore) similarly or by winnowing. Hence **ya·ndying** *vbl. sb.*

1933 C. Fenner *Bunyips & Billabongs* vi. 158 When a gin has collected a coolamon..full of seed she has also a good deal of sand, dust, grass and leaves. But by shaking and twisting the coolamon in a particularly skilful way an almost perfect separation is made. This art of separation is called 'yandying'. **1937** E. Hill *Great Austral. Loneliness* vi. 50 The black woman... can yandy infinitesimal grass-seeds from their husks for the camp breakfast. **1944** M. J. O'Reilly *Bowyangs & Boomerangs* 48 Yandying, in blackfellow language, means shake-about. It is the natives' method of separating the grass seeds from the husks. **1962** D. Stuart *Yaralie* i. 9 While her mother had sat resting from her work of yandying and specking..she had wandered about. **1975** *National Geographic* Feb. 166 [Native Australian] women often earn money by 'yandying'—winnowing by tossing panfuls of ore into the wind to separate dirt from tin or gold. **1978** O. White *Silent Reach* xi. 113 The only tin that comes out of that country is what the gins yandi—dry blow by hand out of a coolamon.

yandy (yæ·ndi), *sb. Austral.* [from prec.] (See quot. 1959.)

1959 D. Stuart *Yandy* 158 *Yandy, tjardoo*: long shallow oval dish, of wood sometimes, but now almost always of sheet-iron, in which mineral is separated from the alluvial rubbish by means of a complicated racking action. **1962** —— *Yaralie* i. 12 Her mother and father had worked mightily, with the pick and shovel, and the yandy, and the loaming dish for days.

‖ **yang** (yæŋ). Also **Yang.** [Chinese *yáng* yang, sun, positive, male genitals.] In Chinese philosophy, the masculine or positive principle (characterized by light, warmth, dryness, activity, etc.) of the two opposing cosmic forces into which creative energy divides and whose fusion in physical matter brings the phenomenal world into being. Also *attrib.* or as *adj.* Cf. *yin.

1671 J. Ogilby tr. *Montanus' Atlas Chinensis* II. 549 The Chineses by these Strokes..declare..how much each Form or Sign receives from the two fore-mention'd Beginnings of Yn or Yang. **1736** R. Brookes tr. *Du

Halde's Gen. Hist. China* III. 357 The Chinese lay down two natural Principles of Life, vital Heat and radical Moisture..: They give the Name of Yang to the vital Heat, and that of Yn to the radical moisture. **1836** J. F. Davis *Chinese* II. xii. 65 The *Tae-keih* is said to have produced the *Yâng* and *Yin*, the active and passive, or male and female principle, and these last to have produced all things. **1845**, etc. [see *T'ai Chi 1]. **1871** A. B. Mitford *Tales of Old Japan* I. 150 The Chinese doctrine of the Yang and Yin, the male and female influences pervading all creation. **1934** R. Fry *Let.* 12 Apr. (1972) II. 690 Later on they discovered that the god of the furnace being male..to throw the wife in..alone sufficed to provide the Yang and the Yin. **1958** W. Willetts *Chinese Art* I. iv. 271 Light is the essence of *yang*. **1963** 'R. Erskine' *Passion Flowers in Italy* ix. 125 Giorgio looked even taller than usual..and (to all seeming) quite adequately Yang, whatever Consolata might say. **1969** *New Scientist* 10 July 53/1 For balance, the rational needs the irrational, the intellect must mesh with the emotions, the *yang* needs the *yin*. **1971** F. Mann *Acupuncture* (ed. 2) v. 63 In the treatment of disease, if Yang is hot and over-abundant, thus injuring the Yin fluid..the surplus Yang can be decreased by a method called 'cooling what is hot'. **1971** *Guardian* 18 Dec. 9/2 A macrobiotic diet..was the way Zen monks cooked... Foods were divided into Yin things and Yang things... Yang is meat for instance. **1980** *Holistic Health News* (Berkeley, Calif., Holistic Health Center) Sept./Oct. 1/3 In the past 300 years with the rise of empirical science, modern technology, property orientation, and the decline of the sacred, we have seen the creation of an extremely yang, overmasculine world-view.

b. Comb.: **yang-yin** = *yin-yang* s.v. *yin b.

1959 R. F. C. Hull tr. *Jung's Aion* in *Coll. Wks.* IX. ii. v. 58 This vision..might easily be a description of a genuine yang-yin relationship. **1968** E. B. Irving *Reading of Beowulf* iv. 179 The poem does seem to have something of a Yang-Yin structure to it..: as the kind of heroic achievement that Beowulf represents nears its end..self-destructiveness..—the negative side of the heroic ideal—comes into clearer and clearer focus. **1975** *New Yorker* 26 May 32/3 When I was a kid, the Technocrats used to drive these gray cars with the yang/yin symbol on the door.

‖ **yangban** (yæ·ŋbæn). Also **yang-ban, yang ban; yangpan;** and with capital initial. [ad. Korean ¹*yāngpan*, f. ¹*yāng* both, a pair + *pan* social class.] **a.** The former ruling class in Korea. **b.** A member of this; an aristocrat or gentleman; (see also quot. 1972).

1898 I. L. Bird *Korea & her Neighbours* I. iv. 60 The youths who swing and lounge on sunny afternoons along the broad streets, aping the gait of *yang-bans*, are aspirants for official position. **1904** W. E. Griffis *Corea* (ed. 7) xlix. 443 In the Land of Morning Radiance there is a governing minority consisting of about one-tenth of the whole population. These, the Yangban..living on ancient privilege and prerogative and virtually paying no taxes or tolls, prey upon the common people. **1906** H. B. Hulbert *Passing of Korea* ii. 47 The common people constantly went down in the scale and the so-called *yangban* went up, until a condition of things was reached which formed the limit of the people's endurance. **1908** G. T. Ladd *In Korea with Marquis Ito* xii. 292 As for the Yangban, on no account will he do manual work. **1952** C. Osgood *Koreans* viii. 147 The taking of concubines by rich husbands was a commonplace and, in the event that a son was not born to the legal wife..it became an almost inevitable procedure on the part of a Yangpan of distinguished family. **1972** P. M. Bartz *South Korea* iv. 46/2 The civil service was known as *Tongban* (eastern class) and the military as *Soban* (western class), and the two together as *Yangban*, a word later used generally to refer to the nobility, and today used by women of the middle class in polite reference to their husbands or other men. **1977** *Korea Jrnl.* Dec. 21/2 All the citizens had to read the intentions of the ruling *yangban* class and Confucian scholars.

‖ **yang ch'in** (yæŋ tʃin). Also — **ching, jin, kin,** and as one word. [Chinese *yángqín*, f. *yáng* high-sounding or *yáng* foreign + *qín* musical instrument, zither.] A Chinese musical instrument similar to the dulcimer.

1876 Stainer & Barrett *Dict. Mus. Terms* 455/1 *Yang Kin*, a Chinese instrument furnished with brass strings, which are struck with two small hammers, like a dulcimer. **1934** *Jrnl. R. Asiatic Soc.* Apr. 334 This would appear to have been similar to the dulcimer which is known to-day in China as the *yang-ch'in*. **1962** E. Snow *Other Side of River* (1963) lxxiii. 566 They may specialize in piano, violin, cello, flute, or one of the standard Chinese strings: *p'i-p'a, yang-ch'in, yueh-ch'in*, and others. **1970** R. D. Taring *Daughter of Tibet* i. 2 His favourite instrument was the Chinese *yangjin*, which has strings like a harp and is beaten with two bamboo sticks. **1974** *Early Music* Oct. 250/2 One recent Chinese recording has the Yang Ching as an obbligato instrument backed by the classical orchestra. **1980** *New Grove Dict. Music* IV. 275/2 The *yang-ch'in* is used in many types of popular music, including various styles of regional opera, sung narratives and solos.

‖ **Yang Dipertuan** (yæŋ dipɜːtū·ān). Also 9 **Iang de Pertuan; Yang di-Pertuan,** etc. [Malay, lit. 'he who is tuan': see *tuan.] In Malaysia, a king, an acknowledged ruler. Also with adjs., as *Yang Dipertuan Agung* [Malay *agung* principal], *besar* [Malay *besar* important].

1834 P. J. Begbie *Malayan Peninsula* iv. 140 Rajah Alli..could not view the appointment of Rajah Laboo to the office of the Iang de Pertuan *Besar*, with any other feelings than those of great distrust. **1907** F. Swettenham *Brit. Malaya* vi. 131 These places..were placed under the general control of a Raja from Mênangkâbau, in Sumatra, with the title Yang di Pertuan. **1947** R. Winstedt *Malays* 51 In Negri Sembilan, custom prescribed that only the Yang di-pertuan (or Ruler) could have four wives. **1972** *Straits Times* 23 Nov. 13/3 The hotel doormen should stop using the doormen's costume. ..The head-dress..is like the one worn by the Yang Dipertuan Agung.

‖ **yanggona** (yæŋgōᵘ·nā). Also (in Fiji) **yaqona** (with the same pronunc.). (Other spellings recorded below are 'South Sea solecisms' (G. B. Milner).) [Fijian.] The Fijian name for Kava.

1858 T. Williams *Fiji* I. ii. 24 The leading men drink *yaqona* with the king elect. **1879** *Encycl. Brit.* IX. 156/2 The use of the kava root, here called yanggona.., was introduced, it is said, from Tonga. **1913** R. Brooke *Let.* 15 Dec. (1968) 545 *Yagona* (pron. Yangona) is the drink: same as Samoan Kava. It is made by pounding up a root, and is non-intoxicant, though *slightly* narcotic. **1922** A. B. Brewster *Hill Tribes of Fiji* xviii. 179 Christening now takes the place of the old custom of sprinkling the children with water from the *yangona* bowls. **1953** G. K. Roth *Fijian Way of Life* iii. 114 The vessel from which *yanggona* is regularly drunk nowadays is a cup made from the distal half of a coconut shell cut laterally. **1977** *Times* 17 Feb. 16/5 The Queen was..presented with..the potent locally-brewed drink called Yaqona. **1983** *Guardian Weekly* 25 Sept. 12/4 The great council of chiefs—the supreme body for Fijian affairs—met over a bowl of *yakona*.

yang jin, yang kin, varr. *yang ch'in.

‖ **yang-ko** (yæŋko·, ‖ yaŋgə). Also **yangko** and as two words. [Chinese *yānggē*, f. *yāng* seedling, sprout + *gē* song.] A type of folk-dance popular in northern China.

1954 *Folk Arts of New China* 30 Well-known forms of folk art like the *yangko* dances of the Shansi peasants or the Yangtse River boatmen's songs. **1967** J. R. Levenson in A. Feuerwerker et al. *Approaches to Mod. Chinese Hist.* 278 Communists might trip the Shensi light fantastic, the *yang-ko*, partly to get themselves into Shensi—and partly to get Shensi into China. **1973** R. F. S. Yang in Yuan-li Wu *China* 750 Since its Yenan days, the Communist party has utilized the *yang ko* (songs in sprouting time), an improvised version of folksinging and folkdancing, as a very useful propaganda weapon. **1975** C. P. Mackerras *Chinese Theatre in Mod. Times* x. 165 The first [phase]..was characterized by an emphasis on the local peasant drama of the Communist base area in northern Shensi (in particular, the small-scale song-and-dance form called *yang-ko*).

Yang–Mills (yæŋ,milz). *Physics.* The names of C. N. *Yang* (b. 1922), Chinese-born physicist, and R. L. *Mills* (b. 1927), U.S. physicist, used *attrib.* with reference to a class of gauge theories with non-Abelian gauge invariance.

1961 *Bull. Amer. Physical Soc.* VI. 59/1 The Yang–Mills field has been quantized in the gauge $b_0 = 0$ (a similar analysis also holding in the gauges $b_1 = 0$ and $b_3 = 0$). **1977** *Nature* 21 July 207/2 There is wide agreement as to the ingredients in a prospective theory of hadrons: quarks, colour, gluons and the Yang–Mills interaction, to name the most obvious. **1979** *Sci. Amer.* Feb. 88/3 A Yang–Mills field is the essential element in a theory that seems to unify two of the four fundamental forces of nature, the weak force and the electromagnetic one. **1981** M. Gell-Mann in J. H. Mulvey *Nature of Matter* viii. 180 The theory of QCD and the theory of QFD..belong to the same class of theories. They are called Yang–Mills theories.

yangona: see *yanggona. **yangpan,** var. *yangban.

Yang-shao (yæŋ ʃau). *Archæol.* Also **Yang Shao, Yang shao.** The name of a village in the Henan province of China, used *attrib.* and *absol.* to designate a Neolithic Chinese culture (*c.* 5000–3000 B.C.), and its artefacts, evidence of which was first discovered there in 1921.

1923 J. G. Andersson *Early Chinese Culture* 31, I propose that we coin a local term and name it from the type locality *The Yang Shao culture*. **1948** A. L. Kroeber *Anthropol.* (rev. ed.) xvii. 735 Various early polychrome wares of the West have been spoken of as 'similar' to Yang-shao. **1965** T. R. Tregear *Geogr. of China* ii. 47 Yang Shao pottery has several unique shapes. *Ibid.*, Yang Shao Man lived on the loess plateau at a time when the water-table must have been much higher than it is today. **1973** *Genius of China* 48/2 The pigment used in decorating Yang-shao bowls..was generally applied directly on the burnished clay surface. **1978** *Nagel's Encycl.-Guide: China* 106 Proof of cultivation of cereals.. exists from the Yang shao period onwards.

yank, *sb.*[1] Add: **1. b.** For *dial.* and *U.S.* read *colloq.* (orig. *dial.* and *U.S.*). Also *fig.*

1906 *N.Y. Globe* 20 Aug. 6 Here is a fantastic proposition from Germany, which takes one back with an unpleasant yank into the middle ages.

Yank, *sb.*[2] Add: **2.** An American car.

1959 *Listener* 4 June 982/1 The young labourer..will invest his cash in buying a car 'on the 'ire'—not a modest second-hand British product but a 'big Yank'. **1977** *Hot Car* Oct. 11/3 It's not raunchy like a yank but it sure is clean and ripe for customising.

yank, *v.* For *dial.* and *U.S.* read orig. *dial.* and *U.S.* and add: **1. a.** (Further examples.)

1950 R. Macaulay *World My Wilderness* xii. 102 His companion, a younger man with less of the Gael in his aspect and speech, jumped down into the copse,..and yanked her to her feet. **1964** F. Chichester *Lonely Sea & Sky* xii. 129, I kept the seaplane on the surface, planing until I thought it was going as fast as it could, when I yanked the stick back hard, to pull her off suddenly. **1966** *Listener* 14 Apr. 534/1 Any incident, from three youths yanking a cigarette machine off a wall to the mods' and rockers' riots, qualifies as 'gang delinquency'. **1968** B. Hines *Kestrel for Knave* 57 Crossley grabbed a boy by the arm and began to yank him into the open. **1977** C. McCullough *Thorn Birds* i. ii. 35 Fee's muscular arm yanked the brush ruthlessly through knots and tangles until Meggie's eyes watered. **1983** *Austral. Personal Computer* Aug. 62/1 If you want the disks back..you cannot just yank them out.

b. (Further examples.)

1896 G. B. Shaw *Let.* 7 Dec. in *Ellen Terry & Shaw* (1931) 139 Hearing that Janet..had no refuge but the Solferino, she promptly went to that haunt, yanked Janet ..out of it,.. and delivered her punctually..for the performance. **1922** Joyce *Ulysses* 421 Alexander J. Christ Dowie, that's yanked to glory most half this planet from 'Frisco Beach to Vladivostok. **1948** *Sunday Pictorial* 18 July 16/6 In the end attendants had to dive in and yank them out. **1977** J. I. M. Stewart *Madonna of Astrolabe* xiv. 197, I had to yank him out of Oxford—a shocking place, if Cambridge is anything to go by.

c. To withdraw (a theatrical show, an advertisement, etc.); to cancel. *U.S. colloq.*

1940 *Amer. Speech* XV. 205/1 *Yank*, to withdraw, usually because of poor attendance. **1976** *Time* 27 Sept. 65/1 The paper..ticked off 24 local real estate advertisers with a dispiriting account of development along a local lake; they have since yanked their ads. **1978** *Chicago* June 12/1 The *Tribune* flung up more flak for Greene in ads on TV and at the top of page one (it yanked an Arts & Fun ad that repeated the 'prostitute' column).

2. (Further examples.) Usu. const. *at.*

1906 'O. Henry' in *Munsey's Mag.* Aug. 556/2 (1961) ix. 131 The drawer stuck, and he yanked at it savagely. **1957** J. Kerouac *On Road* i. ix. 55, I yanked at the window; it was nailed. **1977** C. McCullough *Thorn Birds* i. 8 She..bagan to comb Agnes's hair... She was yanking inexpertly at a large knot. **1981** *Sunday Express* (Colour Suppl.) 19 July/2 Suddenly Sally/Julie yanks at the neckline of her dress.

Yankee, *sb.* and *a.* Add: **A.** *sb.* **6.** = *Yankee jib* in sense C. b. below.

1912 Heckstall-Smith & Du Boulay *Compl. Yachtsman* vi. 152 The 'Yankee' is a strong pulling sail. **1953** *Yachting* June 48 We handed the yankee in favor of the working jib and forestops'l. **1967** J. Anderson *Vinland Voyage* 211 Peter decided to use the No. 2 yankee, leaving the big No. 1 to its proper job of pulling forward. **1974** *Islander* (Victoria, B.C.) 11 Aug. 11/1 We were lost without the mizzen. With motor and yankee we inched our way..forward.

7. *Horse-racing.* A composite bet on four or more horses, composed of doubles, trebles, and one or more accumulators.

1967 C. Cockburn *I, Claud* xxxiii. 404, I stepped into the betting-shop and placed the type of bet known as a 'Yankee' on four of the races... I was able to collect.. over £72 for the twenty-two shillings I had bet. **1970** *Guardian* 17 Apr. 12/3, I have..won in 4-, 5- and 6-horse yankees sums of up to £200. **1981** B. Hines *Looks & Smiles* 184, I won it on the horses. Me and Phil had a Yankee up.

C. b. Yankee bet *Horse-racing* = sense A. 7 above; **Yankee jib (topsail),** a large jib topsail used in light winds, set on the topmast stay.

1964 A. Wykes *Gambling* viii. 194 (*caption*) The 'Yankee bet' (a permutation bet covering four horses) that can be made with off-course bookmakers in Britain. **1976** *Daily Record* (Glasgow) 29 Nov. 23/5 Yankee bet: Six doubles, four trebles and an accumulator—Pikey (12.0 Windsor), Escapologist (1.45 Wolverhampton), Corriegholl (2.15 Wolverhampton), Heidelberg (3.0 Windsor). [**1904** B. Heckstall-Smith *Dixon Kemp's Man. of Yacht & Boat Sailing* (ed. 10) v. 94 The sheeting of a modern large jackyard topsail requires a master hand's attention, especially when it is fitted 'Yankee fashion', having three sheets, as very many now are—namely, the main topsail sheet, the outer and inner sheets on the ends of the jack-yard.] **1912** Heckstall-Smith & Du Boulay *Compl. Yachtsman* vi. 152 A useful sail is the Yankee jib-topsail. This is the largest or balloon jib-topsail, and the modern and most efficient form of balloon jib-topsail is cut, like all modern head-sails should be, very high in the clew. **1928** *Daily Mail* 9 Aug. 19/6 There is a Yankee jib which, as one sail, covers more than the combined area of jib and foresail. **1939** U. Fox *Crest of Wave* 145 We had settled down with the large Yankee jib topsail set in the place of the double clewed jib. **1976** *Yachts & Yachting* 20 Aug. 339/3 At 30 knots across the deck she dropped her yankee jib and kept going under staysail and heavily reefed main.

Hence **Yankeeish** *a.* (examples); **Ya·nkeeness,** Yankee character.

1818 H. C. Robinson *Diary* 30 Apr. (1967) 58 Allston has a mild manner, a soft voice, and a sentimental air with him, not at all Yankyish. **1830** *Collegian* (Cambridge, Mass.) Apr. 117 Comparisons are generally 'odorous', particularly Yankeeish, and decidedly condemned by Captain Basil Hall. **1909** 'O. Henry' *Roads of Destiny* xxi. 352 Any Yankeeness I may have is geographical.

Yankee Doodle. Add: **2.** (Earlier example.)

1787 J. F. Bryant *Verses* 15 And we'll give the Yankee-doodles a dowse in the jaws.

Hence **Yankeedoodledo-dom** *nonce-wd.* = Yankeedom; **Yankeedoodledom** (earlier example).
1843 T. Carlyle *Let.* 3 July in *Lett. Charles Dickens* (1974) III. 542/1 The last *Chuzzlewit* on Yankeedoodle-dodom is capital. We read it with loud assent. **1845** P. Hone *Diary* 20 May (1889) II. 248 The ladies of this family (natives though they be of Yankee-doodle-dom) seem to possess, in a high degree, the power of capturing the aristocracy of England.

Yanqui (yæ·ŋki), *a.* and *sb.* [a. Sp. *Yanqui* Yankee *sb.* and *a.*] = Yankee *sb.* and *a.*: used esp. in Latin American contexts.
1929 [see *parrot-house]. **1937** [see *Panaman *sb.* and *a.*]. **1952** *Caribbean Q.* II. iv. 9 Latin America replied with denunciations of what is called 'Yanqui Imperialism'. **1969** *Guardian* 6 Oct. 11/2 The Mexicans..naturally were loath to prosecute their own kind..on evidence collected by the Yanquis. **1975** *New Yorker* 30 June 23/3 Pelé, the King of Soccer, the Black Pearl, lured from Brazil by four million seventeen hundred thousand Yanqui dollars and his own sense of duty, was here to play his first game for Warner Communications' New York Cosmos. **1976** *Times Lit. Suppl.* 6 Aug. 989/4 Roosevelt's seizure of Panama led him to publish his archetypal anti-Yanqui diatribe. **1982** J. D. MacDonald *Cinnamon Skin* xxiv. 257 It was a childish game with them, to effortlessly outdistance the heaving sweating Yanquis.

yantra (ya·ntrā). [a. Skr. *yantra* device or mechanism for holding or fastening, f. *yam* to hold, support.] A geometrical diagram used as an aid to meditation in tantric worship; any object used similarly.
1877 M. Williams *Hinduism* ix. 129 As to the *Yantras* these are mystical diagrams—generally combinations of triangular figures. **1928** E. B. Havell *Indian Sculpture & Painting* (ed. 2) ii. 16 To this day *yantras*, or geometric symbols, are used in higher Brahmanical ritual in preference to images of the Hindu pantheon. **1946** H. Zimmer in J. Campbell *Myths & Symbols in Indian Art* iv. 141 In Hindu devotional tradition, 'yantra' is the general term for instruments of worship, namely, idols, pictures, or geometrical diagrams. **1980** *Dædalus* Spring 123 The mouse child uses the image as a yantra to meditate on nothing and infinity. **1982** *N.Y. Times* 7 Nov. xxi. 35/2 An ecstatic state..would be achieved..by contemplation of the mandalas and yantras, abstract diagrams of cosmic forces.

Yao (yau), *a.*[1] and *sb.*[1] Also 9 **Yaou.** [Native name.] **A.** *adj.* Of, pertaining to, or designating a mountain-dwelling people of the Guangxi, Hunan, Yunnan, Guangdong, and Guizhou provinces of China and northern parts of Vietnam.
1834 C. Gutzlaff *Sketch of Chinese Hist.* I. i. 30 In the mountains of Kwang-tung and Kwang-se live great numbers of the Meaou and Yaou tribes, who appear to be the aborigines of the country. **1897** *Fortn. Rev.* July 104, I will now close the subject by saying a few words touching the language of the Kakhyens. Like the Chinese, Annamese, Shan, Miao-tsze, Yao, and Burmese languages, it is monosyllabic and tonal. **1976** R. Condon *Whisper of Axe* I. xv. 90 The KMT troops..took Shan or Yao or Lahu wives. **1977** Yin Ming *United & Equal* 3 In the 1930s the reactionary warlords perpetrated massacres of the Hui people in Kansu and the Miao and Yao peoples in Kwangsi.
B. *sb.* **a.** The Yao people. **b.** The language of the Yao.
1883 *Encycl. Brit.* XVI. 224/1 The Yaou-jin, or Goblin clan, are said to have books, which, though they are now unable to read, they still regard with reverent awe. **1901** E. H. Parker *China* i. 7 In the southern portion of the eastern half there are still a few independent..tribes, known as Yao or Miao. **1939** [see *Miao *sb.* and *a.*]. **1948** R. A. D. Forrest *Chinese Lang.* v. 89 In Indo-China the Yao (akin to the Miao) are also known as Man. *Ibid.* 91 Miao and Yao were assigned by..Schmidt, along with Khamti, Shan, and Ahom, to the northern group of T'ai languages. **1982** B. Hook *Cambridge Encycl. China* 102/1 Because of culture-contact they were..familiar with the notion of writing, and consequently came to employ a neighbouring language, using its script as their own written language. This happened with the Yao, who wrote their poems and hymns in a..modified form of Han Chinese.

Yao (yau), *sb.*[2] and *a.*[2] [Native name.] **A.** *sb.* **a.** (A member of) a Bantu people found east and south of Lake Nyasa in East Africa. **b.** The language of the Yao.
1894 *Rep. Admin. Brit. Central Africa* 23 in *Parl. Papers* (C. 7504) LVII. 771 As the coast people and Arabs began to penetrate East Central Africa they came in contact with the Yao, who, from his predatory nature, took to the idea of slave-raiding with real appreciation. **1916** *Blackw. Mag.* Apr. 551/1 The Africans were Yaos, little men, affectionately termed 'Golliwogs' by their British officers. **1924** [see *Nyanja *sb.* and *a.*]. **1957** Ld. Hailey *African Survey 1956* iii. 97 Nyanja is the lingua franca throughout Nyasaland..though Tumbuka has also been recently recognized as the medium in the Northern Province, and Yao in the Yao-speaking areas. **1974** *Encycl. Brit. Micropædia* X. 799/3 Through Arab contact most Yao are Muslims.
B. *adj.* Pertaining to or designating a Yao or the Yao.
1910 *Encycl. Brit.* III. 360/1 The extensive *Yao* genus of languages stretches from just behind the coast of the Lindi settlements..to the north-east shores of Lake Nyasa. **1955** M. Gluckman *Custom & Conflict in Afr.* iv. 96 When a Yao headman in Nyasaland is installed his taste for human meat is tested in the installation ceremony: because Yao witches eat the corpses of those they kill. **1974** *Encycl. Brit. Micropædia* X. 799/3 Yao social life features annual initiation ceremonies involving circumcisions for boys.

yap, *sb.*[1] Add: **1. b.** A fool, someone easily taken in; also, an uncultured or unsophisticated person. *dial.* and *U.S. slang.*
c **1894** C. H. Hoyt *Texas Steer* (1899) III. 6 Instead of his being the only 'yap', as he calls it, in Congress there were about two hundred other members. **1895** W. Stevens *Let.* 4 Aug. (1967) 6 Paul and Several other Yaps are up with them. **1898** B. Kirkby *Lakeland Words* 157 *Yap*, a chap 'at's a bit ov o gomeril. **1901** 'Flynt' & 'Walton' *Powers that Prey* I. iii. 21 This yap from the country. *Ibid.* III. i. 60 I've seen those yaps come to town an' throw up their hands at sights that a Bowery kid wouldn't drop a cigarette snipe to see. **1915** W. Churchill *Far Country* xxiv. 452 The yaps that listen to him don't understand him, but somehow he gets under their skins. **1926** J. Black *You can't Win* iv. 36 You are just the kind of a yap that gets up in the middle of the night and hides his money so carefully that he has to have a policeman find it for him in the morning. **1977** *New Musical Express* 12 Feb. 12/1 Then this yap starts yowling about anarchy,..and eventually the record seems to end in the middle.
2. b. The mouth. *U.S. slang.*
1900 *Dialect Notes* II. 70 *Yap*, the mouth. **1937** J. Weidman *I can get it for you Wholesale* i. 8 Every time you open your yap to say something. **1959** N. Mailer *Advts. for Myself* (1961) 43 There was a guy screaming his yap off next to him,..holding his face. **1977** H. Fast *Immigrants* IV. 243 They know that if they open their yaps, we'll close them down.
c. Idle or loquacious talk; chatter; = Yawp, yaup *sb.* b. *slang.*
1907 *Dialect Notes* III. 204 *Yap*, offensive or superfluous talk. 'Shut up your yap.' **1926** Kipling *Debits & Credits* 314 He'd preserved it in his head through all those weeks..o' Bert's yap. **1928** A. P. Herbert *Trials of Topsy* 133 All this *pragmatical* yap about *tea* being a necessity and *beer* being a vice. **1945** *Coast to Coast 1944* 1 It wasn't much fun listening to all that yap when it really didn't mean a thing. **1968** K. Weatherly *Roo Shooter* 21 Never mind that yap. Where's the tucker?
d. A chat. *slang.*
1930 Sayers & 'Eustace' *Documents in Case* 145 I'd like to have a yap with somebody who talks my language. **1957** R. Lawler *Summer of Seventeenth Doll* II. i. 66 Real ear-basher he is, always on for a yap.

Yap (yæp), *sb.*[2] (See quots.)
1984 *Chicago Sun-Times* 25 Mar. (Views) 7/1 Yumpies, young upwardly mobile professionals, a.k.a. Yaps (young aspiring professionals). **1984** *Sunday Times* (Colour Suppl.) 28 Oct. 12/3 Phillips' Yaps believe in vigorous self-advancement, jogging and BMWs. **1985** *Times* 9 Feb. 11/1 (*heading*) Yaps, or Young Aspiring Professionals are brash, bright and bound for the top.

yap, *v.* Add: **2. b.** To talk idly or loquaciously; to chatter. Also *trans.*, with quoted words as obj. *slang* (orig. *dial.*).
1886 F. T. Elworthy *W. Somerset Word-Bk.* 844 *Yappy*,..to chatter. The use of the word is distinctly depreciatory. Mind yer work, and neet bide there yappin. **1893** Dartnell & Goddard *Gloss. Words Wilts.* 185 *Yap, Yop*, to talk noisily. 'What be a yopping there for?' **1898** R. Blakeborough *Wit, Character, Folklore, & Customs N. Riding Yorks.* 473 *Yap*, to talk foolishly. **1899** S. Crane *Monster* xviii. 76, I told him to keep his trap shut. But then you know how he'll go all over town yapping about the thing. **1922** S. Lewis *Babbitt* x. 141 He hands me the cold-boiled stare and yaps, 'I dunno, friend, I'll see.' **1937** J. Weidman *I can get it for you Wholesale* iii. 28 You've been yapping away. **1946** K. Tennant *Lost Haven* (1947) xix. 315 Len wished Alec wouldn't yap so much. **1963** *Australasian Post* 14 Mar. 51/2 If you want to yap on like a drongo in the DTs..go ahead: be a gig! **1975** *Daily Tel.* 30 June 13/7 A lot of women who are happy to yap away normally, became tongue-tied when they had to talk and drive. **1985** A. T. Ellis *Unexplained Laughter* 49 They end up writing books about it and yapping away on the television.
Hence **ya·ppingly** *adv.*
1924 *Chambers's Jrnl.* Feb. 128/1 Bob danced yappingly around him.

yappy (yæ·pi), *a.* [f. Yap *sb.* or *v.* + -y[1].] **a.** Given to yapping. **b.** Suggestive of a dog's yap.
1909 in *Cent. Dict.* Suppl. **1937** H. T. Miller *Let me die Tuesday* i. 12 You're a yappy kid with no bringing up whatever. **1977** P. Carter *Under Goliath* vi. 31 He had a wee tight face like one of those little dogs that snap at your heels, and he had a voice like one, thin and yappy. **1977** R. Perry *Dead End* ii. 21 One yappy white poodle.
Hence **ya·ppiness.**
1928 *Daily Express* 28 Aug. 4/4 Pekingese..are not addicted to 'yappiness', and thus differ from many toys.

yaqona, var. *yanggona.

Yaqui (yæ·ki), *sb.* and *a.* [a. Sp., earlier *Hiaquis* pl., ad. Yaqui *hiaki.*]
A. *sb.* **a.** (A member of) an Indian people of north-western Mexico.
1861 *Hist. Mag.* V. 164/2 The Indian population is large, and, properly regulated, would be exceedingly useful... The most numerous tribe is that of the Yaquis. **1875** H. H. Bancroft *Native Races of Pacific States* I. v. 575 The Chinipas, Yaquis, Ópatas and Conchos build.. more substantial dwellings of timber and adobes. **1946** *Nature* 13 July 69/2 The Cáhita Indians of western Mexico consist of two surviving groups, the Yaqui and the Mayo. **1953** E. Haugen *Norwegian Lang. in Amer.* II. xiv. 367 An interesting description is available of conditions among the Yaqui, a tribe in Arizona, whose villages were hispanicized by missionaries and political functionaries by the end of the 17th century. **1977** C. Schaeffer tr. *Simenon's Bottom of Bottle* vi. 95 He was a Yaqui, bigger and stronger than I am.
b. The Uto-Aztecan language of the Yaqui.
1911 *Bull. U.S. Bureau Amer. Ethnol.* No. 44. 12 Three dialects—Yaqui, Mayo, and Tehueco—are usually mentioned. **1943** *Amer. Anthropologist* XLV. 428 Yaqui and Mayo stand in the relation of mutually intelligible dialects. **1957** *Publ. Amer. Dial. Soc. 1956* XXVI. 55 In Yaqui the Spanish word *dios* 'God' appears both as *dios* and *lios*.
B. *adj.* Of, pertaining to, or designating the Yaqui.
1861 *Hist. Mag.* V. 165/1 From that day [*sc.* 1609] to this, the Yaqui Indians have retained their pueblos, or towns, along their river, governed by chiefs of their own tribe, appointed by the Spanish and Mexican governments. **1884** H. H. Bancroft *Hist. N. Amer. States* ix. 216 A party of Tehuecos were..sent with two converted Yaqui women. **1943** *Amer. Anthropologist* XLV. 428 A fuller consideration..of Yaqui history. **1964** F. O'Rourke *Mule for Marquesa* (1967) ii. 24 From Nogales he peddled arms and ammunition.., his most valued customers the Yaqui Indians. **1978** *Tucson Mag.* Dec. 104/3 They perform a Yaqui Deer Dance in addition to Aztec dances and a variety of regional folk dances.

yar(r, *v.* (Later examples.)
1866 J. E. Brogden *Provincial Words & Expressions Lincs.* 227 *Yar*, to snarl. **1953** G. M. Durrell *Overloaded Ark* x. 181 A young and foolish bitch..had yapped and yarred herself into a fit of hysterical bravery.

yard, *sb.*[1] Add: **1. f.** *U.S.* A college campus or the area enclosed by its main buildings; *spec.* at Harvard: *the Yard*, the quadrangle formed by the original college buildings.
1637–9 *Harvard Coll. Rec.* in *Publ. Colonial Soc. Mass.* (1925) I. 172 Mr Nathaniel Eatons Account... The frame in the Colledge Yard & digging the cellar. **1841** *Harvard Faculty Orders & Regul.* 6 Collecting in groups round the doors of the College buildings or in the yard [shall be considered a violation of decorum]. **1871** L. H. Bagg *Four Years at Yale* 27 Besides the fourteen buildings already described, the only others within the yard..were the two wooden dwelling-houses. **1902** *Boston Even. Record* 18 Mar. 8/4 (*heading*) Out of the 'Yard'—how the Harvard students have gone to the 'Gold Coast'. **1942** Berrey & Van den Bark *Amer. Thes. Slang* §829.12 *Campus*, camp, orchard,..yard. **1947** *Harvard Alumni Bull.* 12 Apr. 586/2 Few people have likely ever thought of the Yard as a bird sanctuary... What of the Yard? There must be bird records. **1970** 'E. Queen' *Last Woman in his Life* III. 163, I found out the truth about myself in my freshman year at Harvard... There was an episode in a bar, well away from the Yard. **1981** 'D. Jordan' *Double Red* xv. 71 Stumbling across the Yard..after too much Harvard Provision Co. gin.
3. For Now *dial.* read Now chiefly *N. Amer.* and *dial.* and add further examples.
1835 J. H. Ingraham *South-West* II. 88 Striped grass, cultivated in yards at the north. **1877** H. G. Murray *Tom Kittle's Wake* 21 My daughter, Molly tief pass, maam, den go da him yard. **1907** W. Jekyll *Jamaican Song & Story* 163 The immediate surroundings of the house are called the yard. They seldom speak of going to a friend's house. They say they are going to his yard. **1932** 'L. G. Gibbon' *Sunset Song* 97 The berries hung ripe in the yard of the gardener Galt. **1947** J. A. Lomax *Adventures Ballad Hunter* vii. 185 She says, 'Can you cut yards?' an' I says, 'Yes ma'am.' She says, 'Go roun,..to de back.., you'll find a lawn-mower there, and then begin cuttin'.' **1956** G. E. Evans *Ask Fellows who cut Hay* iv. 55 The village was almost entirely self-supporting, most families living on what they grew or reared on their *yards* or allotments. **1980** W. Maxwell *So Long, see you Tomorrow* (1981) ii. 22 The rented house had no yard to speak of.
4. c. *The Yards*, the stockyards where cattle are collected for slaughter, esp. in Chicago. *U.S.*
1865 *Atlantic Monthly* Jan. 83/2 The average weekly expenditure by butchers at the New York yards during the year 1863 was $328,865. **1906** U. Sinclair *Jungle* xv. 170 Already the yards were full of activity. **1935** A. G. Macdonell *Visit to America* vii. 114 As in Chicago, the pride of Omaha is the Stock-yards... I was looking straight down into the Yards. **1974** 'M. Allen' *Super Tour* ii. 57 I've been called all kinds of things ever since I was a kid back of the Yards.
6. (*a*) *yard-broom*; (*b*) *yard-dung* (earlier example); (*c*) **yard-boy** a general labourer; a gardener or gardener's boy (*obs.* exc. *Caribbean*); **yard sale** *U.S.*, a sale of miscellaneous second-hand items held in the garden of a private house; (*d*) *yard-master* (earlier example).
(*a*) **1921** *Blackw. Mag.* Feb. 195/1 Dip an old yard-broom in a bucket of water. **1982** J. Scott *Local Lads* iii. 32 Billy took up an aged, patchily moulted yardbroom. (*b*) **1744** W. Ellis *Mod. Husb.* Jan. xi. 73 He may now carry out his Stable or Yard Dung. (*c*) **1788** J. Woodforde *Diary* 7 Jan. (1927) III. 2 To my Yard Boy, Charles Crossley, for 3 Quarters of a Years Wages pd 0. 15. 9. **1831** C. Farquharson *Jrnl.* 2 Dec. in *Relic of Slavery* (1957) 47 Employed all hands weeding.. along with the yard boys. **1958** S. Selvon *Turn again Tiger* viii. 185, I take the worst job that was going—as a kind of yard-boy by the white people house. **1975** *New Rev.* May 10/2 In and around Port of Spain cooks, ironers and yardboys in attendance. **1976** *Flint* (Michigan) *Jrnl.* 12 July c-5 *Yard sale*—1508 Webber canning jars, screen tent, patterns, books, [etc.]. **1982** M. McMullen *Until Death do us Part* (1983) vii. 46 There was a yard sale down our street.
(*d*) **1864** *Rep. Children's Employment Comm.* 139/1 in *Parl. Papers* XXII. 487/1 Mr Thomas Wheat, yard-master... My duty is to give orders..and manage the work.

yard, *sb.*[2] Add: **9. b.** *by the yard* (earlier example); also, of books or paintings: bought by quantity or size rather than for quality.
1845 J. W. Turner *Razor Strop Man* 3 He was spinning poetical rhyme by the yard; Had Shakespear been living 'twould astonish'd the bard. **1933** J. Betjeman *Ghastly Good Taste* i. 12 The old books..can be sold..by the yard to America as wall decoration. **1976** 'O. Bleeck' *No Questions Asked* ii. 29 He bought fine paintings by the yard and rare books by the case.
11*. *U.S. slang.* One hundred dollars; one thousand dollars; a bill for this amount.
1926 *Amer. Mercury* Dec. 465/2 One hundred dollars is a *century* or a *yard*. **1929** C. F. Coe *Hooch* vi. 130 He slips him $300 an' promises him $700 more if they'll spring him... Baldy..promises to come right to me for the seven yards that make the grand. **1932** *Amer. Speech* VII. 118 *Yard*,..a thousand-dollar bill. **1942** Berrey & Van den Bark *Amer. Thes. Slang* §18.5 (*One*) *G*, *-gee* or *grand*, *thou*, (*one*) *yard*, one thousand. *Ibid.* §467.2 *One C*, *yard*, a hundred dollars. **1979** V. Patrick *Pope of Greenwich Village* vii. 70 You throw a hundred to the guy who makes the loan... He writes the loan for thirteen hundred, you take twelve, and a yard goes south to him.
12. yard goods, fabric sold by the yard; **yard-stick** (earlier example); also *fig.*, a standard of comparison (later examples).
1941 L. I. Wilder *Little Town on Prairie* v. 33 He'll get most of the trade in yard goods, with somebody there in the store making them up into shirts. **1964** M. Laurence *Stone Angel* iv. 113 At the back was the section where yard-goods were sold, and ladies' and children's ready-to-wear garments hanging dejectedly on racks. **1982** S. T. Haymon *Ritual Murder* xix. 134 Patter of the travelling men who sold crockery and yard goods. **1822** in W. R. Alger *Life Edwin Forrest* (1877) I. 100 Furnish me with every particular, especially how our Tid is, and whether she reads with the yard-stick. **1929** *Morning Post* 4 June 15/6 This is considered more effective than a rough comparison by means of tonnage or range... It is hoped that this new American 'yardstick' will be ready for General Dawes when he leaves for London. **1949** *Here & Now* (N.Z.) Oct. 33/2 What yardstick should we use in assessing success or failure in farming? **1960** A. S. Neill *Summerhill* (1962) vi. 334 We all have our standards of values and we measure others by our personal yardstick. **1984** A. Smith *Mind* iv. xiv. 262 Whatever yardstick is used, it is probable that at least a million people on this planet kill themselves each year.

yard, *v.*[1] **1. a.** (Earlier examples.)
1758 in *Essex Inst. Hist. Coll.* (1874) XII. 140 The Dutch here have a nasty practice of yarding their cows in ye Street before their doors. **1826** J. Atkinson *Agric. & Grazing N.S.W.* 66 When they seem pretty well reconciled to the place, they are *bedded* out one night, and *yarded* the next.

yardage[1]. (Earlier example.)
1867 *Trans. Illinois Agric. Soc.* VI. 322 Net cash receipt for yardage, and profit on feed.

yardang (yā·dæŋ). *Physical Geogr.* Also **jardang.** [a. Turk., abl. of *yar* steep bank, precipice.] A sharp, irregular ridge of sand or the like, lying in the direction of the prevailing wind in exposed desert regions and formed by erosion by the wind of adjacent less resistant material.
1904 S. Hedin *Sci. Results Journey in Central Asia* I. xxvii. 439 At intervals furrows or trenches in the clay sub-soil, called *jardangs*, traced between long elevations or ridges, crop up amongst the dunes. **1934** *Bull. Geol. Soc. Amer.* XLV. (*caption facing* p. 160) Looking down one of the narrower wind-scoured troughs, with a sharp yardang (seven feet high) on the left and a higher one on the right. **1970** R. J. Small *Study of Landforms* ix. 301 Probably the only landforms of deserts that can be confidently ascribed to wind abrasion alone are the comparatively unimportant 'yardangs' and allied 'ridge-and-furrow' features. **1979** *Nature* 5 Apr. 535/1 In other regions [of Mars], the surface has been stripped, yardangs have formed, and in general the topography seems to have been largely configured by aeolian activity.

yard-arm, *sb.* Add: **e.** Phr. *when the sun is over the yard-arm* and varr., the time of day when it is permissible to begin drinking.
1899 Kipling *From Sea to Sea* I. xxiv. 454 The American does not drink at meals as a sensible man should... Also he has no decent notions about the sun being over the yard-arm or below the horizon. **1945** J. C. Colcord *Sea Lang. comes Ashore* 211 An officers' quip..is 'When the sun is over the yardarm..it's time to take a drink'. **1964** *Amer. N. & Q.* III. 23/2 Frequent reference is made to the undesirability of drinking before 'the sun has crossed the yardarm'. **1968** 'J. le Carré' *Small Town in Germany* ii. 20 Just one hour till the sun was over the yardarm... He'd have a beer first. **1979** A. Morice *Murder in Outline* iii. 26, I had promised to take a jugful of dry martini with him and Vera..as soon as the sun went over the yardarm.
yard-arm *v.*, (*a*) also *to yard-arm it: transf.* of persons, to fight at close quarters.
1829 P. Egan *Boxiana* 2nd Ser. II. 358 'Long bowls,' said Curtis to Savage, 'will not answer: you must *yard-arm* it with your adversary.'

ya·rdbird. *U.S. slang.* Also **yard bird.** [f. Yard *sb.*[1] + Bird *sb.* (see sense *1 e), perh. after *jail-bird.*] **a.** *Mil.* A recruit, a newly-enlisted serviceman; also, a serviceman under discipline for a misdemeanour; one assigned to menial tasks. Also *transf.*
1941 *Amer. Speech* XVI. 169/2 *Yard bird*, a raw recruit. **1942** [see *red-line *v.*]. **1943** J. Goodell *They sent me to Iceland* 102 With this wealth of jargon we were able to produce a quiz on army slang..for the benefit of the newly arrived men—better known as 'yardbirds'. **1943** *American Mercury* Nov. 552/1 If he's in the Army he's referred to as a *yardbird*..an old Army term for camp-confined newcomers. **1947** *Amer. Speech* XXII. 111 A soldier, sailor, or marine who frequently receives punishments for offenses against the regulations is classified as a *Y.B.* or 'yard bird'. **1965** C. Brown *Manchild in Promised Land* iii. 80 For the next two weeks, K. B. was Claiborne's yardbird. He had to go everywhere Claiborne went from morning till night. He even had to ask Claiborne when he wanted to go to the bathroom.
b. A convict.
1956 S. Longstreet *Real Jazz, Old & New* 148 A yardbird is a low mug. **1980** A. Pearl *Dict. Popular Slang* 189/1 *Yardbird*, a convict..an ex-convict.
c. A worker in a yard (Yard *sb.*[1] 4) (see quots.). *U.S.*
1963 T. Pynchon *V.* xvi. 427 'Yardbirds are the same all over,' Pappy said... The dock workers fled by, jostling them. **1968** *Amer. Speech* XLIII. 290 *Yard bird*, a disabled engineer, fireman, or switchman who may work only within the yard limits. **1971** M. Tak *Truck Talk* 190 *Yard bird*, a driver who spots trailers and moves vehicles around a terminal yard.

yarder (yā·ɹdəɹ). *N. Amer.* [f. Yard *v.*[1] + -er.] A kind of donkey-engine used in logging.
1911 *Pacific Monthly* Apr. 376/2 The hook-tender gives the signal to the engineer of the 'yarder' as the donkey-engine is termed. **1919** *Camp Worker* 2 June 3/3 [There were] two Ledgerwood skidders, one yarder, one swing and one roader. **1942** Berrey & Van den Bark *Amer. Thes. Slang* §512.10 *Yarder*, a donkey engine which hauls logs from where they are felled to the landing or skid road. **1955** *Bush News* (Port Arthur, Ont.) Feb. 7/1 About 1,100 men are involved on the haul plus 450 horses, 125 trucks, ..and 4 yarders. **1979** *Beautiful Brit. Columbia* Fall 37 In some logging areas, mechanical grapple yarders, machines that resemble construction cranes, settle into the forest floor.

ya·rdful. (f. Yard *sb.*[1] + -ful 2.] As much or as many as a yard will hold. Also *fig.*
1860 in Webster. **1960** *Farmer & Stockbreeder* 15 Mar. 95 A yardful of well-finished Hereford-cross beef. **1978** J. L. Hensley *Killing in Gold* xii. 165 I'm surprised you haven't a yardful of law around here.

yardland. Add: **3.** *Comb.*, as *yardland-holder.*
1890 E. W. Watson *Ashmore* 31 The two plough-oxen, the universal outfit of the English yardland-holder.
Hence **ya·rdlander**, a yardland-holder.
1891 *Athenæum* 16 May 632/3 The notes he gives as to the families of the yardlanders are most interesting. **1906** N. J. Hone *Manor* I. 11 In 1279 a yard-lander at Newington, Oxon, was bound to plough an acre of winter tillage. **1964** H. P. R. Finberg *Lucerna* ii. 32 It would obviously make for convenience to group the strips of each yardlander.

yark: see also *york *sb.*[2]

Yarkand (yā·ɹkænd, yāɹkæ·nd). [The name of a river, district, and city in Sinkiang Uighur (formerly Chinese Turkestan), an autonomous region of western China.] **1.** A language or dialect of the central Turkic or Turco-Tatar group of Altaic languages, spoken in the district of Yarkand. Also *attrib.*
1875 [see *Kashgar 1]. **1954** M. A. Pei *Dict. Linguistics* 236 *Yarkand*, an Asiatic language, member of the Central Turkish group of the Altaic sub-family of the Ural-Altaic family of languages.
2. Used *attrib.* and *absol.* to designate a type of Turkoman carpet.
1880 G. C. M. Birdwood *Industrial Arts of India* II. 168 The tree of life represented on modern Yarkand rugs is always a pomegranate tree. **1913** W. A. Hawley *Oriental Rugs* xii. 251 In Plate L..are two of the most typical and interesting stripes of Samarkands and Yarkands... A stripe with simple archaic pattern peculiar to Yarkands is seen in Plate L. **1931** [see *Kashgar 2]. **1967** U. Schürmann *Oriental Carpets* 72 The eastern Turkestan rugs commonly known..as 'Samarkand' come..from the three oasis-cities of Kashgar, Yarkand and Khotan... A small number of silk Yarkands also exist. **1970** J. Franses *European & Oriental Rugs* 146 (*caption*) Eastern Turkestan Yarkand runner in the form of a saph. *Circa* 1800.
3. *Zool.* Used *attrib.* to designate a heavily built kind of red deer with short antlers that is found in Sinkiang Uighur.
1892 *Proc. Zool. Soc.* 116 Mr. W. T. Blanford exhibited two heads..and a skin of the Yarkand Stag. *Ibid.* 117 The name *C*[*ervus*] *yarkandensis* may be applied to the Yarkand and Tarim Deer. **1918** R. Lydekker *Wild Life of World* II. 231 Very distinct from the wapiti type is the Yarkand deer..of the forests of the Tarim Valley. **1982** G. K. Whitehead *Hunting & stalking Deer* iv. 111 Among the most endangered are the Yarkand deer..of Chinese Turkestan.

Yarkandi (yāɹkæ·ndi), *sb.* and *a.* Also 9- **Yarkundi.** [f. prec.] **A.** *sb.* A native or inhabitant of the city or district of Yarkand.
1841 H. H. Wilson *Trav. Moorcroft & Trebeck* I. II. iii. 351 A Yarkandi asserted that an infusion of poppy-heads was employed to render the leaves of the tea adhesive. **1875** in T. D. Forsyth *Report Mission to Yarkund* iii. (*facing page*) 118 (*caption*) Yarkundis. **1901** P. W. Church *Chinese Turkestan* iii. 33 The guileless Yarkandis. **1926** C. P. Skrine *Chinese Central Asia* viii. 106 A charming..and well educated Yarkandi called Murad Qari.. partook of tea. **1981** A. Ali in *Himalayan Jrnl.* XXXVII. 115 The Yarkandis used this route [*sc.* Saser La in the Karakorams] for trade and for going on haj to Mecca.
B. *adj.* Of or pertaining to Yarkand or its people.
1854 A. Cunningham *Ladák* iii. 49 My informants, who were also Yarkandi merchants, stated exactly the reverse. **1893** H. Lansdell *Chinese Central Asia* II. xxxiv. 108 Turdi Akhoon, a Yarkandi merchant..had arrived from India. **1928** 'Ganpat' *Magic Ladakh* xiv. 255 It is a hard life on the Central Asian trade route if you happen to be a hired ponyman travelling with a callous Yarkandi merchant. **1973** M. Bence-Jones *Palaces of Raj* viii. 144 A portly General endeavoured to stop his Yarkundi pony from jumping over the railings.

yarling (yā·ɹliŋ), *ppl. a.* *Midl.* (and *north.*) *dial.* [f. dial. *yarl* vb. to utter a loud discordant sound: see *Eng. Dial. Dict.*] Howling, wailing.
1911 D. H. Lawrence *White Peacock* I. vii. 119, I heard more plainly..the peevish, wailing, yarling cry of some beast in the wood. **1972** Ld. Robens *Ten Year Stint* ii. 32 They were a 'yarling mob'—crude, vulgar and unfit to lead the decent men I know in the pits.

yarmulke (yā·ɹmŭlkə). Also **yarmulka**; (more rarely) **jarmulka, yarmolka**, etc. [ad. Yiddish *yarmolke*, ad. Polish *jarmułka* cap.] A skull-cap worn by male Orthodox Jews at all times, and by other male Jews on religious occasions; = *koppel.
1903 *Jewish Encycl.* IV. 301/2 The so-called Jewish garb of Poland, including even the 'jarmulka' (under-cap), is simply the old Polish costume which the Jews retained. **1929** *Menorah Jrnl.* XVI. 37 Jacob..saw and did not revere the squat figure clumsy in its..*yarmelke*. **1930** M. Gold *Jews without Money* 95 My father took me to the tailor and had made a handsome velvet *yamalka*. **1941** B. Schulberg *What makes Sammy Run?* ix. 232 There in the synagogue,..with his impressive shawl, his *yarmolka* and his great beard, there life was rich. **1957** L. Stern *Midas Touch* I. v. 45 He wore his *yamulka* (skull cap) when it was ritually required. **1962** 'E. McBain' *Empty Hours* 111, I was collecting the prayers shawls and the *yarmelkas*. **1963** T. Pynchon *V.* iv. 97 He went out of his way to cultivate the Tagliacozzi look:..wearing a bushy mustache, pointed beard, sometimes even a skull-cap, his old schoolboy yarmulke. **1966** L. Davidson *Long Way to Shiloh* xv. 221 You have the yarmulkah? Remember to put it on. **1966** H. Kemelman *Saturday Rabbi went Hungry* (1969) i. 14 He wondered if the cantor had put on his robes and tall white yarmulka. **1971** B. Malamud *Tenants* 90 Sam Clemence, a Mephistophelean type in yarmulke and yellow dashiki. **1975** *Church Times* 7 Nov. 5/1 A lively man in a business suit and embroidered *yarmulka*—the little skull-cap worn by orthodox Jews. **1979** 'A. Hailey' *Overload* III. xi. 244 Hardly any of us took a yarmulke. I didn't. Had to borrow one when I went to the Wall in Jerusalem. **1984** *Times* 24 Sept. 4/6 The captain, who wore Israeli army uniform with a red yarmulka fringed with gold on his head.

yarn, *sb.* Add: **2. b.** A chat, a talk. *colloq.* (chiefly *Austral.* and *N.Z.*).
1857 H. W. Harper *Lett. from N.Z.* (1914) iii. 49 This has been a long yarn. **1883** Stevenson *Treasure I.* x. 80 'Come away, Hawkins,' he would say; 'come and have a yarn with John.' **1888** 'R. Boldrewood' *Robbery under Arms* I. xii. 156 After tea father and I and Jim had a long yarn. **1929** K. S. Prichard *Coonardoo* xv. 147 Meenie and Bandogera had taken advantage of her absence to have a smoke and a yarn together at the wood-heap. **1937** D. Cowie *N.Z. from Within* vii. 109 The word is used in its oldest sense. The New Zealander's 'yarn' is the Scotsman's 'news'. **1966** G. W. Turner *Eng. Lang. Austral. & N.Z.* vi. 124 Other counts have been based on written material and the Australian one on spoken. This accounts for the inclusion of *kid* ('child').., *yarn* ('talk').., [etc.]. **1979** B. Moore *Mangan Inheritance* II. 295 We can take it easy for a while and have a real yarn together. **1984** *Times* 11 Sept. 32/8, I still see some of the Roman Catholics in the street..and we have a yarn.
3. *yarn-carrier*; **yarn count** = *count *sb.*[1] 2 b.
1927 T. Woodhouse *Artificial Silk* 100 The yarns now pass to their respective yarn carriers, the function of which is to place the yarn in the path of the single set of sinkers. **1957** *Textile Terms & Definitions* (Textile Inst.) (ed. 3) 109 *Yarn carrier*,..the final element which guides the yarn to the knitting instruments. **1923** Yarn count [see *number *sb.* 6 g]. **1963** Jerrard & McNeill *Dict. Sci. Units* 154 In the textile industry the yarn count or yarn number gives either the mass per unit length or the length per unit mass of a yarn fibre.

yarn, *v.* Add: **1.** Also, to chat or talk. (Further examples.)
1857 *St Leonard's Station Diary* 14 May, in L. R. C MacFarlane *Amuri* (1946) III. 125 Hanging round the station, yarning and sleeping. **1901** M. Franklin *My Brilliant Career* (1966) iii. 3 Too friendly to pay a short

call, they came and sat for hours yarning about nothing in particular. **1939** A. Powell *What's become of Waring* vii. 206 If I..start yarning with him..we shall be late for dinner. **1941** I. L. Idriess *Great Boomerang* xvii. 122 In the whitewashed Birdsville Hotel, low-roofed but with dim, cool rooms, the blokes yarn the time away. **1944** *R.A.F. Jrnl.* Aug. 256 There is practically nothing to do but..yarn with your friends. **1958** L. Durrell *Balthazar* ii. 37 We were sitting at a café yarning. *a* **1966** 'M. Na Gopaleen' *Best of Myles* (1977) 55 He does be yarnin with the brother above in the digs of a Sunday. **1972** M. Shadbolt *Strangers & Journeys* xi. 195 In the town, where men gathered to yarn on street corners. **1977** C. McCullough *Thorn Birds* vi. 120 Their parents yarned over cups of tea, swapped tall stories and books.
† **b.** *trans.* To recount or narrate. *Obs. rare.*
1840 A. Russell *Tour Austral. Colonies* 40 One who can yarn the dangers of the deep so well.

yarooh (yarū·), *int.* Also **yaroo.** A humorous stylized representation of a cry of pain. (One of Billy Bunter's characteristic exclamations: see quots.)
1909 *Magnet* 20 Nov. 4/2 'Oh!' roared Bunter, as Bulstrode's heavy boot biffed on him. 'Ow! Yah! Yarooh!' **1918** *Ibid.* 8 June 12/2 'Don't keep me waiting, or I shall help you on—like that!' 'Yaroooh!' **1940** 'G. Orwell' in *Horizon* Mar. 178 'Oooogh!', 'Grooo!' and 'Yaroo!' (stylized cries of pain). **1953** *Manch. Guardian Weekly* 23 July 11/2 With a 'Yarooh!' on nearly every page William G. Bunter is on the war-path again. **1972** *Guardian* 21 Jan. 1/5 'Yarooh!' they yelled... 'Get out, you cad!' **1977** M. Amis in A. Thwaite *My Oxford* 205 A sign reading 'Yaroo—College Squit!' suspended from my neck.

† **yarraman** (yæ·rămăn). *Obs.* Pl. **yarramen, -mans.** An Australian Aboriginal word for a horse.
1848 H. W. Haygarth *Bush Life in Austral.* x. 108 A stockman..meets some of the blacks, to whom his first question is 'You make a light yarraman belonging to me' (*i.e.* Have you seen my horses?). **1882** A. J. Boyd *Old Colonials* 69 There's seventeen yarramen—call 'em thirty pounds a head. **1905–6** 'T. Collins' *Rigby's Romance* (1971) v. 21 He needn't be frightened o' these yarramans. I got them like lambs. **1930** A. W. Groom *Merry Christmas* xx. 156 'We tie yarraman here,' he suggested. They fastened their horses to low snags on the dead tree. **1959** Baker *Drum* 158 *Yarraman*, an outlaw horse or wildly behaved station hack. **1964** W. S. Ramson in *Southerly* i. 58 Other aboriginal words, *bora*, *coolamon*, *goondie*, *humpy*, and *yarraman*, came from tribes in the Sydney and Moreton Bay districts.

yary (yē·ri), *a.* *dial.* (chiefly *Newfoundland*). Also **yarry.** [var. of Yare *a.*] Quick, sharp; alert, energetic; wary, wide awake; rising early.
1855 *Trans. Philol. Soc.* 38 [Norfolk] *Yary*, brisk. **1863** J. Moreton *Life & Work in Newfoundland* iii. 35 *Yary*, wary. **1868** in *Dict. Newfoundland Eng.* (1982) 622/2 Here we saw a great number of wild geese in the lagoon.. but it was impossible to get within shooting distance of them, these birds are so wild and extremely yarry. **1881** *Even. Telegram* (St. John's, Newfoundland) 20 Sept. 1 We don't find the cruising war-ships of our yarry neighbours the French and the Americans, lying in port for weeks at a time. **1906** N. Duncan *Adventures Billy Topsail* 256 'Hi, b'y! Get yarry (wide awake)!' cried the captain in the morning. **1925** *Dialect Notes* V. 346 *Yary*,..1. energetic; smart. 2. early. 3. wary. **1966** A. R. Scammell *My Newfoundland* 90 That would be Skipper John Elliott, yary as ever, hi-tailing it for Jacob's ground before the Eastern Tickle crowd got the choice berths for the day.

yas (yæs), repr. colloq. and U.S. Blacks' pronunc. of Yes *adv.* See *yassuh *int.*
1887 H. Baumann *Londinismen* 238 *Yas*,..yes. **1909** L. M. Montgomery *Anne of Avonlea* xxiv. 279 'Was Ginger hurt?'..'Yas'm. He was hurt pretty bad. He was killed.' **1927** *N.Y. Times Mag.* 24 Apr. 4/2 Yas, sir. Dat right, sir. **1936** M. Mitchell *Gone with Wind* xxxii. 546 'I suppose you heard Jonas Wilkerson and that Emmie—' 'Yas'm,' said Mammy. **1966** *Keystone Folklore Q.* XI. 85 Caddy always taught the children not to say 'Yas suh, No suh and Yas 'um'..to white folks.

‖ **yashiki** (ya·ʃiki). Also 9 **yaski.** Pl. **yashiki**, (anglicized) **-s.** [Jap., f. *ya* house + *shiki* a space, site.] The residence of a Japanese feudal nobleman, including the palace or mansion and grounds, and the quarters for his retainers.
1727 J. G. Scheuchzer tr. *Kæmpfer's Hist. Japan* II. ix. 486 *Sokkokf Dai Mio Jassiki*, that is, Palaces and houses of the princes and Lords of the Empire.] **1863** R. Alcock *Capital of Tycoon* II. xiii. 280 The Daimios' *Yashkis* are merely a low line of barracks of the same construction, rather higher in the roofs. **1871** A. B. Mitford *Tales of Old Japan* II. 206 The principal yashikis (palaces) of the nobles are for the most part immediately round the Shogun's castle. **1906** R. A. Cram *Impressions Jap. Archit.* iii. 57 The arrangement of these 'yashiki' varied but little: a hollow square..was formed by the barracks for the daimyo's retainers; these barracks were usually two stories in height. **1959** R. Kirkbride *Tamiko* ix. 65 It was at once obvious to him that it was part of the ruins of a magnificent yashiki, destroyed by fire during the war. **1970** J. W. Hall *Japan* x. 170 All daimyo were obliged to build residences (*yashiki*) in Edo where they kept their wives and children.

yassuh (yæ·sʊ), *int.* Chiefly *U.S.* Repr. Black colloq. pronunc. of 'yes, sir' (often somewhat obsequious). Cf. *yas *adv.* + *suh.
1936 M. Mitchell *Gone with Wind* xxiii. 391 'She's not dead? Is she breathing?' 'Yassuh, she breathin.' **1944** C. Himes *Black on Black* (1973) 199 'Take good care of me, Chops,' I said... 'Yassuh.' **1963** Prange & Vitols in A. Dundes *Mother Wit* (1973) 631/2 Yassuh, I sees all dat. **1973** J. Pattinson *Search Warrant* ii. 33 'You live alone here?' 'Yassuh. Jus' me an' my mem'ries.'

Yates (yēits). *Statistics.* [The name of Frank Yates (b. 1902), English statistician, who published the correction in 1934 (*Suppl. Jrnl. R. Statistical Soc.* I. 217).] *Yates'(s) correction*: a correction for the discreteness of the data that is made in the chi-square test when the number of cases in any class is small and there is one degree of freedom, consisting in the subtraction of ½ from each difference when evaluating chi square.
1934 R. A. Fisher *Statistical Methods for Research Workers* (ed. 5) iv. 96 (*heading*) Yates' correction for continuity. **1968** P. A. P. Moran *Introd. Probability Theory* ii. 76 The use of the correction ½ results in a closer numerical approximation and is known as Yates's correction. **1972** *Jrnl. Social Psychol.* LXXXVII. 53 As the expected frequency in some cells was less than 10, Yate's [*sic*] correction for continuity was applied. **1977** R. Holland *Self & Social Context* vi. 211 Cooper..shows with the help of chi-squared and Yates' correction the effectiveness of conjoint family and milieu therapy.

yatter (yæ·təɹ), *v.* *colloq.* (orig. *Sc. dial.*). [Imitative, perh. after Yammer *v.* + Chatter *v.*; cf. also Natter *v.*] *intr.* To talk idly and incessantly; to chatter, or gossip; to gabble; to complain peevishly. Freq. const. (*on*) (*about* something or *at* someone). Occas. *trans.*
1825 Jamieson (Suppl.) II. 703/2 She's ay yatter-yatterin, and never devaulds. **1831** *Gasometer* 457 She yattered about an ugly man that cam' in a fiddle case. **1896** P. A. Graham *Red Scaur* viii. 121 Grace likes to yatter about the days when she bondaged for him. **1919** J. Buchan *Mr. Standfast* i. vi. 122 No company but a wheen ignorant Hielanders that yatter Gawlic. **1942** N. Streatfeild *I ordered Table for Six* 203, I don't like to yatter about flying much. **1950** 'P. Woodruff' *Island of Chamba* 124 As long as the British yatter on about going and don't go, things are bound to get worse in India. **1963** J. N. Harris *Weird World of Wes Beattie* (1964) v. 62 This dear old Betty was yattering at me on Sunday morning when I was hung over to the eyeballs. **1977** J. I. M. Stewart *Madonna of Astrolabe* ii. 48 The confounded thing might tumble around our ears while we yattered.
Hence **ya·ttering** *vbl. sb.* and *ppl. a.*
1859 J. Watson *Living Bards of Border* 193, I winna get up, sae yer yatterin's vain. **1878** R. Ford *Hamespun Days* 105 A thrawart, yatterin', blatterin' mither. **1935** D. Rorie *Lum Hat* 58 Ta'en in By a yatterin' lump o' original sin. **1972** *Hawick News* 7 Jan. 7/4 Their yatterin', like the stream, goes on for ever.

ya·tter, *sb.* *colloq.* (orig. *Sc. dial.*). [See prec.] Idle talk; incessant chatter or gossip.
1827 J. Watt *Poems* 72 Gin ane hae walth to keep him lievin', Nae cravin' body's yater deevin'. **1898** J. Buchan *John Burnet of Barns* II. ix. 188 The shrill yatter of the fishwives. **1935** F. Niven *Flying Years* i. 10 Any yatter of human follies and failings. **1955** E. Pound *Section: Rock-Drill* xciv. 95 To the Odes to escape abstract yatter. **1978** *Sunday Mail* (Brisbane) 28 May 3/4 No one in the Brisbane Valley any longer believes the tourist yatter given out by Government..circles.

yautia (yauti·ă). [Amer. Sp.] In the West Indies, any of various herbaceous perennials of the genus *Xanthosoma*, esp. *X. sagittifolia*, which belong to the arum family and are widely cultivated for their edible tubers; = Tania, tanier, tannier.
1899 W. Dinwiddie *Puerto Rico* xii. 141 The other root,..known commonly as 'yautia', is much cultivated by the peasantry and held in high esteem, being always on sale in the markets. *Ibid.*, From the 'yautia' roots considerable starch is made..and is sold principally for laundry purposes. **1917** L. H. Bailey *Stand. Cycl. Hort.* VI. 3523/1 The corms and cormels (offsets) of some taros, and the cormels of some varieties of yautia, are free from acridity even in the raw state as cultivated in southern United States. **1975** E. L. Ortiz *Best of Caribbean Cooking* 135 Add the yautia, yams, pumpkin, cassava, plantains, salt and Tabasco to taste. **1981** P. Theroux *Mosquito Coast* xvii. 223 Seeing me with some yautia plants..I told them they were yautias and that their roots were as tasty as carrots.

yava. (Earlier example.)
1774 W. Wales *Jrnl.* 26 June in J. Cook *Jrnls.* (1961) II. 846 They [*sc.* the Tongans] brought off with them the Yauva, or pepper-Root.

yaw, *sb.*[1] Add: Angular motion or displacement about a yawing axis. Also *Aeronaut.* and *Astronaut.*
1916 G. C. Loening *Military Aeroplanes* xii. 166 Struts of large fineness ratio..present considerable side surface and affect the directional center, at different angles of yaw. **1935** [see *pitch *sb.*[2] 2 b]. **1950** *Engineering* 3 Mar. 255/2 The Desynn type of transmitter and indicator..is used to transmit to the recording apparatus such variables as control forces, angle of yaw, pressures, etc. [in a prototype aircraft]. **1974** *Physics Bull.* Jan. 11/1 The six component wind tunnel balance..will be able to measure three forces (lift, drag and side force) and three moments (pitch, yaw and roll) on any aircraft model it supports. **1977** *Offshore Engineer* May 44/3 During these tests, the data acquisition system recorded..pitch, roll, heave, surge, sway and yaw of the lay barge, pull and length of mooring cables, and anchor positions. **1978** R. Jansson *News Caper* 7 The Captain manoeuvred the big jet back to stability, damping out yaw and roll.
c. yaw axis = *yawing axis* s.v. *yawing *vbl. sb.*
1959 F. D. Adams *Aeronaut. Dict.* 184/1 Yaw axis. **1962** F. I. Ordway et al. *Basic Astronaut.* ix. 368 Any vehicle motion will take place about three axes... These axes are the yaw axis, the pitch axis, and the roll axis. **1978** *Sci. Amer.* Nov. 137/1 For the first time the machine included a pair of fixed vertical surfaces behind the wings to stabilize motion about the yaw axis.

yaw, *v.*[1] Add: **1. b.** *Aeronaut.* and *Astronaut.* Of an aircraft or spacecraft: to rotate about a vertical axis, to undergo yawing.
1912 *Q. Rev.* July 243 This disposition tends to offer an ever-increasing amount of surface sideways to the air when a turn is begun, thus accentuating the turn initiated by the rudder and causing the craft to yaw. **1935** C. G. Burge *Compl. Bk. Aviation* 108 The forces on the two wing tips are neither steady nor equal, so that the aeroplane tends to roll and yaw. **1964** [see *pitch *v.*[1] 19 f]. **1979** *Daily Tel.* 7 Apr. 3/2 It then yawed to the right, did a barrel roll like a light aircraft starting at an aerial show, and went into a nose-dive.
3. (Examples in *Aeronaut.*)
1920 *Engineering* 8 Oct. 462/2 It was found that the control was not reversed at large angles of incidence up to 20 deg. unless the model was yawed. **1960** Welch & Denes *Go Gliding* i. 20 Moving the left foot forward yaws the glider's nose to the left. **1975** L. J. Clancy *Aerodynamics* xvi. 525 The aircraft is yawed to starboard.

yawing, *vbl. sb.* Add: (Examples in *Aeronaut.* and *Astronaut.*) **yawing axis**, a vertical axis through a ship or aircraft; an axis through a spacecraft normal to both the longitudinal and lateral axes.
1915 A. Fage *Aeroplane* vi. 86 An indifference [on the part of pilots] to yawing, and possibly to rolling, is regarded favourably in many aeronautical circles. **1935** C. G. Burge *Compl. Bk. Aviation* 238/1 This causes a 'yawing' effect in the opposite direction to the turning effect of the rudder [of the plane]. **1953** Yawing axis [see *rolling axis* s.v. *rolling *vbl. sb.*[1] 9 a]. **1975** L. J. Clancy *Aerodynamics* xvi. 525 If the aircraft has a yawing velocity, *r*, this affects the fin incidence in the same way that pitching velocity affects the tail incidence. **1978** *Jrnl. Fluid Mech.* LXXXVII. 533 These passive yawing motions are studied to find their amplitude, the yawing axis and any associated energy dissipation.

yawl (yǭl, yāl), repr. (Southern) U.S. pronunc. of *y'all *pers. pron.*
1919 *Dialect Notes* V. 40 *Yawl*,..you-all. **1938** C. Himes in *Black on Black* (1973) 167 Why doesn't yuh git happy an' praise de Lawd? Doesn' yawl know who Ah is? **1978** J. R. Gaskin in *Sewanee Rev.* LXXXVI. 426 Dillard accounts for *y'awl*, or *you all*, not as the simple concatenation of two English forms.

yawmeter (yǭ·mitəɹ). [f. Yaw *sb.*[1] + -meter.] An instrument used to detect changes in the direction of flow round an aircraft or other body.
1921 *Flight* 28 July 511/1 A new direction and velocity meter (yawmeter) has recently been constructed. **1947** *Jrnl. R. Aeronaut. Soc.* LI. 15/2 We imagine aerofoil and observer to be stationary and the aerofoil to be immersed in a stream of air of speed *V* normal to the span..the direction of the stream being made known to the observer by, for example, a yawmeter. **1969** *Jrnl. Physics E* II. 989/1 Calibration of the instrument as a yawmeter. Most of the tests on the instrument have been made near the outlet end of an open water channel. **1983** *Ibid.* XVI. 231/1 (*heading*) A yawmeter for steady and low-frequency unsteady flows.

yawn, *sb.* Add: **2. c.** *transf.* and in *transf.* contexts, denoting something that induces boredom; a tedious activity. *colloq.*
1889 E. C. Dowson *Let.* 3 Feb. (1967) 32 My dear Moore. Here goes for my accustomed Sunday yawn to you! Thanks for your note. **1974** D. Gray *Dead Give Away* ii. 24 To you it may be one big yawn, or the laugh of a life-time... But to me it's important. **1978** G. A. Sheehan *Running & Being* viii. 102 For them the Super Bowl is three hours of yawns. **1979** *Broadcast* 4 June 8/3, 7 June will be a major event for psephologists..if..a yawn a minute for British voters. **1984** *Times* 3 Oct. 13/1 So much prescription may sound like a recipe for a great gastronomic yawn.
yaw·nfully *adv.*
1914 W. De Morgan *When Ghost meets Ghost* i. xviii. 691 'On my way to Poynders,' said the Countess yawnfully.

yawner. Add: **1. b.** *transf.* Something dreary or boring. *colloq.* (orig. *U.S.*).
1942 Berrey & Van den Bark *Amer. Thes. Slang* §276/2 *Something uninteresting*,..washout, yawner. **1969** A. Glyn *Dragon Variation* ii. 42 The game between him and Wheaton, still to be played, should be a real yawner. **1980** *Globe & Mail* (Toronto) 5 Nov. 17 (*heading*) The Awakening is a real yawner. **1983** *Chicago Sun-Times* 6 Aug. 76 (*heading*) Opener likely to be yawner.

yawp, yaup, *sb.* Add: **b.** Chiefly *U.S.*, sometimes in allusion to Whitman's use. (Earlier and further examples.)
1835 J. H. Ingraham *South-West* I. 29 'Hold your yaup, you youngster you,' roared the old man in reply. **1855** W. Whitman *Leaves of Grass* 55, I sound my barbaric yawp over the roofs of the world. **1870** 'Mark Twain' in *Galaxy* Oct. 571/1 He..ordered me to 'hold my yop'. **1904** *Buffalo* (N.Y.) *Commercial* 25 Aug. 6 When this contest is ended, the insincere and ridiculous yawp about the fierce belligerency of Theodore Roosevelt will be laid away with the other feeble fakes. **1973** *Publishers' Weekly* 26 Mar. 61/3 American readers may miss the experimentation and 'barbaric yawp' of avant-garde American poetry.

yawp, yaup, *v.* Add: **1. b.** To speak foolishly or noisily. *U.S. colloq.*
1872 S. Hale *Let.* 28 Oct. (1919) 90 Perhaps it is just as well, however, not to yawp much about our going *alone*, as it may be considered loose in America. **1926** T. Beer *Mauve Decade* vi. 233 Where the boys who badgered Richard Harding Davis for autographs in 1890 will be yawping over 'Billy Baxter's Letters' in 1900.

yay (yēi), *adv.* Also **yea.** *U.S. slang.* [Prob. f. Yea *adv.*] In phrases *yay big* (or *high*), 'this big', 'this high': freq. accompanied by a gesture indicating the size intended.
1960 Wentworth & Flexner *Dict. Amer. Slang* 591 *Yea big, yea high*, 1. This big, or this high, accompanied with the spreading of the hands to indicate the size; very large, or high, overwhelmingly large or tall. 2. Not very big or high. **1972** T. Kochman *Rappin' & Stylin' Out* 242 Jeff fired on him. He came back and all this was swelled up bout yay big, you know. **1978** P. Theroux *Picture Palace* 259 Why does a daughter of mine, whom I've loved and respected ever since she was yay high, go out of her way to made a jackass of me?

‖ **yayla** (yēi·lā). Also 9 **yaila**; **yaylak.** [Turk.] A summer camping-ground in the mountains or Turkestan used by Kurdish and other semi-nomadic peoples; the encampment pitched there.
1864 A. Vambery *Trav. Central Asia* xvi. 308 A Yaylak (summer abode), that is, near to the hill on the sea-shore. **1896** D. G. Hogarth *Wandering Scholar in Levant* iii. 53 Others come and go, and the place of the summer *yaila* is fixed hard by the village itself. **1953** O. Caroe *Soviet Empire* xi. 181 In the same country in winter the *yaylaks* will be deep under snow. **1975** J. Rathbone *Kill Cure* III. ii. 85 At last they had come to the yayla or summer pasture.

Yayoi (yā·yoi). The name of a quarter in Tokyo, used *attrib.* and *absol.* to designate a type of early Japanese (wheel-thrown) pottery first discovered at this site in 1884, and hence applied to the mainly neolithic culture characterized by this ware. Cf. *Jomon.
1906 N. G. Munro in *Trans. Asiatic Soc. Japan* XXXIV. 24 The pottery..is called *Yayoishiki*, 'Yayoi sort', because it was first encountered in breaking ground at Yoyoi [*sic*] Street in Tokyo. **1931** G. B. Sansom *Japan* I. i. 3 This latter type is known as the Yayoi type, because of certain characteristic earthenware first found in a neolithic site at a place of that name. **1955** *Far Eastern Q.* XIV. 329 It is not entirely clear why so few workers paid any serious attention until extremely recent years to the archaeology of the Yayoi period. **1960** B. Leach *Potter in Japan* vi. 136 Jomon pottery 3,000 B.C. on to Yayoi from A.D.0. **1968** *Encycl. Brit.* XVIII. 523/2 The prehistoric period in Japan is characterized by two principal cultures known as Jōmon and Yayoi... Yayoi pottery seems to have its beginnings in the 3rd century B.C. and is mostly wheel-thrown. **1970** J. W. Hall *Japan* iii. 19 The Yayoi people brought with them the horse and the cow, though not in abundance.

Y chromosome. *Genetics.* Also † **y chromosome.** [*Y 4 b.] A sex chromosome which occurs in only one of the sexes (in man and other mammals, the male) or in some species is absent altogether, its presence or absence in the zygote determining in man and many other species the sex of the organism.
1911 *Biol. Bull.* Jan. 119 We have associated the *X* and *Y* chromosomes of the male with sex-determination, but possibly they have some other meaning. **1933**, etc. [see *X chromosome]. **1965** R. P. Morehead *Human Path.* vi. 170/1 Proved cases of mutation involving the Y chromosome are extremely rare... As Y chromosomes occur only in males, the characteristic pedigree of such mutations should contain only affected males, never affected females, and all affected males should transmit the defect to all their sons. **1974** P. Cave *Dirtiest Picture Postcard* x. 57 You've buttonholed me to give me long and boring lectures upon Germaine Greer, the faulty Y chromosome and the drudgeries of housework and child-bearing. **1982** *Nature* 2 Dec. 404/2 When a Y chromosome is present the foetal gonad, which has the potential to become either ovary or testis, differentiates into a testis and, when absent, into an ovary... Recent genetic analysis shows that this first step is not as simple as first thought and appears to be controlled by an interaction between autosomal genes and a gene(s) on the Y chromosome, for normal male differentiation. **1983** J. R. S. Fincham *Genetics* ii. 72 In grasshoppers and other insects of the order Orthoptera..there is usually no Y-chromosome. The females are XX and the males just X.

yea, *adv.* (*sb.*) Add: **C. yea-and-nay** *a.*, (*a*) (later examples referring to the phr. *Richard Yea-and-Nay*, a nickname for King Richard I); **yea-say** *v.*: hence **yea-saying** *vbl. sb.* (later examples) and *ppl. a.*; **yea-sayer**, one who says 'yea' or who agrees; a person inclined by nature to assent, or to act in a positive manner.
1911 Fletcher & Kipling *School Hist. Eng.* 70 'Richard Yea and Nay', so called because he spoke the truth. **1957** A. Duggan *Devil's Brood* xii. 165 Because he [*sc.* Bertrand de Born] could not persuade Richard to make war at his bidding he gave him the opprobrious nickname Yea-and-Nay. **1934** Webster, Yea-sayer. **1940** 'G. Orwell' *Inside Whale* 176 There are the 'progressives', the yea-sayers, the Shaw-Wells type, always leaping forward to embrace the ego-projections which they mistake for the future. **1972** A. Friedman in Cox & Dyson *20th-Cent. Mind* I. xii. 434 The Wilcoxes..are businessmen, robust, conservative, organized, practical yea-sayers who lead lives of 'telegrams and anger'. **1960** *Partisan Rev.* Fall 609 In literary criticism..artless enthusiasm..has modulated into..more restrained yea-saying. **1960** *Times* 14 Oct. 18/3 Matthew Smith's art, so much..in tune with the traditional, yea-saying materialism of French painting. **1972** *Jrnl. Social Psychol.* LXXXVI. 220 Subjects who obtained scores of 0, 1, and 11, 12 were dropped from the analysis as representing extremes of yeasaying or naysaying.

yeah (yēə), *adv.* *colloq.* (orig. *U.S.*). Repr. a casual pronunc. of Yes *adv.* Cf. *oh yeah.
1905 *Dialect Notes* III. 67 *Yeah, yep*,..variants of *yes*. **1925** F. S. Fitzgerald *Great Gatsby* iv. 87 'That's a very interesting idea.' 'Yeah.' He flipped his sleeves up under his coat. **1936** M. Kennedy *Together & Apart* iv. 293 'You were in Sweden with him last year, weren't you?' 'Yeah.' **1940** *Music Makers* May 37/3 *Yeah, man*, exclamation of assent. **1949** E. Birney *Turvey* 153 'Yeah, yeah, Hayes was pretty hot but the ref—'. **1950** 'D. Divine' *King of Fassarai* xx. 166 'I take it the natives are friendly?'..'Yeah... We had us a party last night.' **1961** J. Heller *Catch-22* (1962) xix. 194 'Will that be all, sir?' asked the chaplain. 'Yeah,' said Colonel Cathcart. 'Unless you've got something else to suggest.' **1977** B. Langley *Death Stalk* ix. 104 'The shooting. That was Tony.' 'Tony?' 'Yeah, he done that.'

year[1]. Add: **3. a.** Also with cardinal number following, denoting a period of a political regime as a means of calendar reckoning.
1933 E. Waugh *Scoop* II. iv. 226 New Calendar. Year One of the Soviet State of Ishmaelia. **1971** *Times Lit. Suppl.* 27 Aug. 1015/2 When he refers to 'the language of the Year III', he means that of the Year II [of the 1870 Commune]. **1972** R. Cobb *Reactions to French Revolution* i. 38 The *coup d'état* of Fructidor year V (Sept. 1797). **1980** P. Van Greenaway *Dissident* vii. 147 The Seventeen Revolution is Year One to our country.
b. (*a*) Freq. with qualifying word, as *financial, fiscal, sabbatical, school, tax year*: see under the first elements. *academic year*: in a school, college, etc., in the Northern hemisphere usu. reckoned from the beginning of the autumn term until the end of the summer term.
1932 *Handbk. Univ. Oxford* 103 An overseas application made..a few weeks before the beginning of the academic year has little or no prospect of success. **1957** *Encycl. Brit.* XXII. 876/2 The master's degree is usually obtained for one academic year of graduate work. **1971** *Morning Star* 28 Dec. 4 Every September boys and girls..return to school..to begin a new academic year. **1983** *Oxf. Univ. Gaz.* 10 Nov. 218/2 The college proposes to elect a distinguished visitor to a Visiting Senior Research Fellowship during the academic year 1984–5.
(*b*) Used *attrib.* or *absol.* with preceding ordinal numeral to denote a student at a particular stage of education. Also *collect.*
1851 B. H. Hall *Coll. College Words & Customs* 266 In the University of Cambridge, Eng., the title of *Second-Year Men*..is given to students during the second year of their residence at the University. **1894** A. Morrison *Tales of Mean Streets* iii. 50 A fourth-year London Hospital student. **1913** J. Vaizey *College Girl* II. xix. 268 One word in your ear! Don't ask a third-year girl to dance with you. **1927** R. Lehmann *Dusty Answer* III. i. 124 I've done six hours every day this vac... Sibyl Jones has done ten hours every day... Third years ought to be more sensible. **1935** D. L. Sayers *Gaudy Night* vii. 139 There are some oddities in the First Year... I expect the Third Year said the same about us..but..I should call the whole of our year pretty sound. **1966** E. H. Jones *Margery Fry* v. 44 Margery..was the obvious choice from the First Year when a committee was formed to arrange a garden-party in May 1895. **1979** D. Brierley *Cold War* iii. 26 Sociology second years from Nanterre. **1982** D. Clark *Doone Walk* viii. 179 He's a Bristol University third-year bloke.
d. Such a period officially designated for special celebration or to focus public attention on a particular object of concern; esp. *Holy Year*, a year so designated by the Pope, now usu. once every 25 years, during which special Indulgences are granted and ceremonies held.
1699 J. Jackson *Let.* 25 Dec. in *Lett. & Second Diary of Samuel Pepys* (1932) 291 Wee made our entry here on Tuesday last, about 23 a clock, and were soon after deafned with the jangling of all the bells of the town, which for severall days, morning and evening, had proclaimed the approach of the Holy Year. **1776** Pius VI (*title*) Instructions & Directions for Gaining the Grand Jubilee of the Holy Year, celebrated at Rome anno 1775, and extended to the universal Church anno 1776, by his Holiness Pius VI. **1858** H. E. Wiseman *Recoll. Last Four Popes* II. iv. 270 The practice has been, that on Ascension Day of the preceding year, the Pope promulgates the Holy Year, or Jubilee. **1900** H. Thurston *Holy Year of Jubilee* ix. 358 During the Holy Year, and also during the time of the extension of the Jubilee to the rest of Christendom, the Holy Father grants extraordinary powers to confessors. **1957** J. S. Huxley *Relig. without Revelation* (rev. ed.) ix. 205 Mass celebrations, like those of the Holy Year or the rallies and parades of Nazism and Communism. **1960** *Stamp Mag.* May 454/1 Commemorative. For World Refugee Year (Overprint on the rest of the 1958 World Exhibition stamps, with surcharge in aid of World Refugee Year 1960). **1965** *Ibid.* Apr. 244/1 The Australian Post Office will issue a stamp this year to commemorate International Co-operation Year. **1971** M. Lee *Dying for Fun* xliv. 212 He..decided to organize and launch Compassion Year. **1974** *Times* 7 Feb. 15/8 In the year 2073..many of the trees planted in Tree Planting Year 1973 will still be with us. **1983** *Out of Town* Dec. 52/4 Those of us who go to church already know that 1984 is Christian Heritage Year.
4. a. (Later example.) Also *spec.* with reference to the vintage of wine.
1864 'J. Ward' *Diary* 22 May in J. Burnett *Useful Toil* (1974) I. 85 Everything looks well in fields and gardens, with every prospect of a good fruit year. **1941** B. Schulberg *What makes Sammy Run?* xi. 206 Laurette..told the waiter to send it back. 'If you haven't *1927*, don't bother. That's the only good year left.' **1967** 'L. Black' *Two Ladies in Verona* x. 161 A bottle of Mumm Cordon Rouge. I leave the year to you, but it'd better be good. **1984** *Sunday Tel.* 20 May 12/8, I bought the wine. 1964 was quite a good year.
7. a. of the year: denoting things or persons considered to be outstanding examples of their kind in a particular year; **the year dot**: see *dot *sb.*[1] 4 c; **the year one**: see One *numeral* 4; **year in (and) year out** (earlier example); **year-on-year** *adj. phr.*: in *Economics*, used with reference to a comparison of figures with corresponding ones for a date twelve months earlier; **year-to-year** *adj. phr* (later examples).
1883 H. James in *Atlantic Monthly* Sept. 316/1 Wherever the traveler goes, in France, he is reminded of this very honorable practice—the purchase by the government of a certain number of 'pictures of the year'. **1936** L. P. Smith *S.P.E. Tract* xlvi. 220 The market-place where the books of the year are sold in large editions. **1968** 'E. Lathen' *Stitch in Time* vi. 46, I hope they haven't confused Wendell Martin with the GP of the year. **1983** *Daily Tel.* 18 Aug. 8/4 A 35-year-old mother..beat 523 competitors to win the London Chamber of Commerce and Industry's award as top secretary of the year. **1830** *Massachusetts Spy* 28 July 4/1 I've been to.. school *year in and year out*. **1976** *Daily Tel.* 20 July 1/5 It is hoped this will show a year on year rise in average earnings of between 14 and 15 per cent. **1982** *Listener* 16 Dec. 27/3 Over a ten-week period from September to November, the year-on-year decline recorded is equivalent to 12 per cent of individuals, or 7 per cent of households. **1962** *Lebende Sprachen* VII. 113/3 Year-to-year growth ratio. **1977** J. L. Harper *Population Biol. Plants* 203 The relative constancy of mean seed weight over a density range in this experiment is particularly interesting because the year to year variation in seed weight is quite large.
c. *to see the New* (*Old*) *Year in* (*out*) and varr.: to stay up until after midnight on 31 December, to celebrate the start of a new year.
1840 Dickens *Let.* ? 18 Dec. (1969) II. 169 Will you dine with us on the last day of the old year—just to see it jollily out. **1875** L. Troubridge *Life Amongst Troubridges* (1966) 134 It's eleven o' clock now, and shall I tell you what we three are doing? Watching the Old Year out and the New Year in. **1916** M. Diver *Desmond's Daughter* III. x. 227 Accepting an invitation to..'see the New Year in' with Thea. **1921** W. de la Mare *Mem. of Midget* xv. 99, I had written..an invitation to herself and Fanny to sit with me and 'see in' the New Year. **1939** H. Nicolson *Diary* 31 Dec. (1967) II. 52, I do not stay to watch the New Year in or the Old Year out. I write this diary at 11.45 and shall not wait.
8. *year-hedged* adj.; **year class**, the individuals of a particular kind of animal (usu. a fish) that were born in any one year.
1936 Dylan Thomas *25 Poems* 41 The year-hedged row is lame with flint, Blunt scythe and water blade. **1910** J. Hjort in *Publications de Circonstance* No. 53. 18 Very characteristic in this respect are the analyses of samples of the typical Norse spring-herring, where the year-class which formed its first winter-ring in 1904 preponderates largely over all the other year-classes. **1958** *Jrnl. Marine Res.* XVII. 505 The population [of sea-urchins] probably consists of four year-classes. **1967** [see *recruit *v.* 3 e]. **1981** *Trans. Amer. Fisheries Soc.* CX. 185/1 By optimizing the yield from dominant year classes, greater yields from the fishery can be realized for all groups involved.

year[2] (yiəɹ, yɜ̄ɹ). Repr. dial. (chiefly U.S.) pronunc. of Ear *sb.*[1]
1863 *Southern Confederacy* (Atlanta, Georgia) 9 May 1/2 You should git the strait of it from one who seed it with his eyes, and hearn it with his years. **1886** *West Somerset Word-Bk.* 845 *Year*.., the ear. **1891** *Dial. Hartland, Devonshire* 122 *Year* (yur), the ear. **1929** W. Faulkner *Sound & Fury* 72, I wish I was young like I use to be, I'd tear them years right off your head. **1935** Z. N. Hurston *Mules & Men* (1970) I. viii. 173 He took and galloped out in de middle of de road right in front of John's horse and laid his years back.

year-book. Add: Now freq. as one word. **2. b.** *U.S.* An album published annually by the graduating class of a high school or college.
 1926 B. Manbert (*title*) Inter-Scholastic Year Book Manual. **1928** A. H. Anderson *School-Built Annual* 21 Don't think you can get out an annual in a week. A yearbook is a year's job. **1939** D. E. Mitchell *Journalism & Life* xxx. 413 The yearbook is a more formal and permanent production than a newspaper. **1972** M. Mead *Blackberry Winter* vii. 83, I did not see Luther again, but he sent me his yearbook.. Luther was four years older than I and a senior in college. **1978** S. Brill *Teamsters* vi. 233 The mob looks at high-school yearbooks and picks a hundred guys who look smart and clean.

yearling, *sb.* and *a.* Add: **A.** *sb.* **3.** *U.S. colloq.* A student in his first year or beginning his second year at college.
 1900 *Dialect Notes* II. 70 *Yearling*, a second year man. **1940** Berrey & Van den Bark *Amer. Thes. Slang* §825/6 *Freshman*...yearling. **1944** *Collier's* 23 Sept. 69/1 His femme fell for a [West Point] yearling.
 4. *Econ.* A yearling bond (see sense 3 of the adj. below).
 1966 *This is Bill-Broking* (Allen, Harvey & Ross Ltd.) 38/1 *Yearlings*, stocks issued by local authorities for a period of a year and quoted either on the stock exchange or in the discount market. **1970** *Daily Tel.* 29 Sept. 17/5 (*heading*) Local authority yearlings at 8p.c. **1977** *Guardian* 19 Apr. 17/2 At the moment the yearlings give a return of 10 per cent which may be lower than what is available on the ordinary bonds, but are flexible. **1981** *Observer* 18 Oct. 20/1 An interesting alternative [to Government stocks] is the local authority negotiable bond—or the 'yearling', so-called because of its one-year term.
 B. *adj.* **3.** *Econ.* Applied to bonds issued by a local authority usu. for one year.
 1964 *Times* 2 Apr. 18/1 Under present conditions a quotation for a yearling bond would mean additional expense. **1969** *Daily Tel.* 12 Apr. 5/7 Most yearling bonds mature in..a year, sometimes two to five years. **1975** *Economist* 19 July 95 The explosion in the yearling bond market. **1977** *Guardian* 19 Apr. 17/2 Yearling bonds.. which come in units of £1,000, are much more flexible—and like the local authority bonds disgorge interest twice a year. They last for 12 months only and then investors have to start again.

yearly, *a.* (*sb.*) Add: **2. d.** *Yearly Meeting*, in the Society of Friends (Quakers), a national assembly held annually to deal with legislation and questions of policy (see *esp.* quot. 1869). Cf. *quarterly-meeting* (*a*) s.v. *quarterly *a.* 3.
 1688 *Testimony for the Lord, & His Truth* (Women Friends, York) 1 Given forth by the Women Friends at their Yearly Meeting at York, being a Tender Salutation of Love to their Friends and Sisters in their several Monthly Meetings. **1714** in *Jrnl. Friends Hist. Soc.* (1918) 28 Thence into Maryland to friends yearly-meeting at Tradaven-Creek..wherein Truth was plentifully afforded to ye bowing of many souls here. **1831** in S. B. Weeks *Southern Quakers & Slavery* (1896) xi. 300 There is not a school in the limits of the [North Carolina] Yearly Meeting that is under the care of a committee of either monthly or preparative meeting. **1869** Beck & Ball *London Friends' Meetings* v. 53 The Yearly Meeting was from its commencement..first and chiefly, a gathering of public Friends (i.e. ministers), to confer together on matters of faith and doctrine... The ministers alone formed the Annual Assembly in London; but in 1677 the invitation for deputies from the Quarterly Meetings was renewed... The representative element has been formally recognised, and thereby the Yearly Meeting has come to its position of legislative importance in the Church. **1923** E. B. Emmott *Short Hist. Quakerism* xi. 171 The proposal for a General Meeting for the whole country (which we now call the Yearly Meeting) came in the first instance from Durham Friends..in 1659. **1949** *Friend* 17 June 497/1 When the Clerk, in Yearly Meeting, announces the Report of the Committee on Accounts, it is quite astonishing to note how many Friends..get up and walk out. **1974** G. Hubbard *Quaker by Convincement* 1. iii. 41 Through the queries answered in writing four times a year by Monthly Meetings, the Yearly Meeting kept a watchful eye on departures from the norm.

yearning, *ppl. a.* Add: Hence (*nonce-wd.*) **yea·rningness.**
 a **1916** H. James *Sense of Past* (1917) 304 The ideal thing for dramatic interest ..would be that there is just one matter in which..he betrays himself, gives himself away..it should..affect her..with but a finer yearningness of interest.

year-round, *a.* and *adv.* [f. the phr. *all the year round* s.v. Round *adv.* 1 e in Dict. and Suppl.] **A.** *adj.* That exists, occurs, is used, etc., all the year round. Also of persons: residing in a place for the whole year.
 1939 *Florida* (Federal Writers' Project) ii. 177 The western district..contains..many homes of year-round residents. **1939** R. Campbell *Flowering Rifle* ii. 38 And thaws numbed strikers from their year-round frost. **1945** Nelson & Wright *Tomorrow's House* x. 111/1 One advantage of the inside bathroom..is that it has year-round ventilation. **1960** *House & Garden* Aug. 40/1 A skilfully-planned year-round family house. **1961** *Guardian* 8 May 8/6 A moment, known increasingly to year-round Londoners. **1976** *Sci. Amer.* July 118/1 Where the diversion of local streams would not provide a year-round supply of water it was necessary to build a reservoir. **1980** P. Moyes *Angel Death* i 7 Year-round sunshine.
 B. *adv.* = *all the year round.*
 1968 *Globe & Mail* (Toronto) 17 Feb. 47/8 Enjoy swimming year-round in the outdoor and indoor pools. **1979** *Wall Street Jrnl.* 20 Dec. 18/1 The yuletide bulge chiefly reflects the greater number of people in the stores at that time. However, the crime is rising year-round. **1981** C. Miller *Childhood in Scotland* 58 Rabbits..were shot year-round.

yeast, *sb.* Add: **1. b.** (Earlier example of *German yeast.*)
 1845 E. Acton *Mod. Cookery* xxviii. 650 German yeast, imported in a solid state, is now much sold in London.
 d. Delete *Path.* and substitute for def.: A fungus that exists predominantly as single cells rather than a mycelium and in which vegetative reproduction takes place by budding or fission. (Further examples.)
 Now not usu. regarded as constituting any particular taxon.
 1906 G. Massee *Text-bk. Fungi* iii. 275 Symbiotic relationship between yeasts and bacteria is not uncommon. **1922** H. Gwynne-Vaughan *Fungi* i. 7 Yeasts and filamentous fungi are abundant in woodland soils. **1930** H. M. Fitzpatrick *Lower Fungi* i. 16 In the lower Ascomycetes the asci are formed without order throughout a mould-like mycelium, or exist as isolated cells as in the yeasts. **1977** R. C. Cooke *Fungi, Man & his Environment* i. 14 Yeasts appear in the Ascomycetes, Basidiomycetes, and Fungi Imperfecti. This is because the term 'yeast' refers to a special mode of growth and does not describe a particular, special assemblage of fungi. **1983** *Oxf. Textbk. Med.* I. v. 372/2 *Candida albicans*... It is a saprophytic yeast often found as a commensal in the mouth and gastrointestinal tract and commonly present in the vagina.
 4. yeast bread, bread made with yeast (i.e. ordinary bread); **yeast-cake,** (*a*) (later examples); (*b*) a cake made light with yeast; **yeast-powder** (later examples).
 1853 *Southern Ladies Bk.* (New Orleans) I. 130 The chicks in the free states live on yeast bread. **1945** *ABC of Cookery* (Ministry of Food) xviii. 67 Nowadays yeast bread is seldom made in the home. **1855** E. Acton *Mod. Cookery* (rev. ed.) xxxi. 604 To test bread that has been cut (or yeast-cakes), press down the crumb..with the thumb. **1897** R. M. Stuart *Simpkinsville* 136 Here, too, had passed pantalet patterns, bits of yeast-cake and preserving-kettles. **1908** *McClure's Mag.* Feb. 421/2 We are to be the yeast-cake for democracy's dough. **1973** *Listener* 20 Sept. 377/2 Tea was served by Auntie Golda.. thick slices of cinnamon-veined yeast-cake. **1857** W. Chandless *Visit Salt Lake* i. vi. 95 Three boxes of yeast-powder (at thirty cents each) to improve our bread. **1876** *Amer. Cycl.* XVI. 777 Yeast powders, or baking powders, substitutes for yeast, used in making bread.

yeast, *v.* Add: Also (*fig.*) with *up*.
 1921 A. Huxley *Crome Yellow* ix. 88 It must inevitably take a long time for Armageddon to ripen, to yeast itself up.

Yeatsian (yēⁱ·tsiăn), *a.* (*sb.*) Also **Yeatsean.** [f. the name of the Irish poet and playwright William Butler *Yeats* (1865–1939) + -ian.] Of, pertaining to, or characteristic of Yeats or his writing. Also as *sb.*, an admirer of Yeats.
 1928 S. O'Casey *Let.* 5 June (1975) I. 261 Since Mr. Yeats has..shouted a lot of things in at O'Casey's window, he shouldn't be surprised,..when he finds O'Casey hammering at the Yeatsian door. **1941** *Scrutiny* IX. 381 Mr. MacNeice does attempt to define the essence of the Yeatsian idiom. **1954** *N. & Q.* CXCIX. 535/2 Yeatsians are doubtless all familiar with Dr. Jeffares' story of the composition of 'The Wheel' at Euston on 17 September 1921. **1959** *Encounter* Nov. 78/2 The misguided Neo-Platonism of some Yeatsians. **1969** *Listener* 13 Mar. 361/2 Though Reisz can give us a Yeatsian despair ('Ah dancer, ah, sweet dancer'), he fails to evoke a Yeatsian gaiety. **1978** *Studies in Eng. Lit.: Eng. Number* (Tokyo) 157 Through his long poetic career, he changed from his early Yeatsean stance to his later self-assurance. **1982** J. Gross in A. Thwaite *Larkin at Sixty* 86 Larkinesque irony..and high Yeatsian romance.

yech (yek, yək), *int. U.S. slang.* Also **yecch, yeck;** (*rare*) **yee(c)ch.** [Imitative. Cf. *yuck *int.*, *sb.*[3], and *a.*] = *yuck *int.*
 1969 C. Burke *God is Beautiful, Man* (1970) 76 Well, Jesus did the craziest thing you ever heard. He took some clay from the road and he spit on it...yeck! **1972** *Even. Telegram* (St. John's, Newfoundland) 23 June 3/1 Yeecchhhh! How messy. A St. John's pigeon just passed his opinion of a Bayman. **1973** *Daily Tel.* (Sydney) 17 May 36/1 She (yecch) pins a plastic gardenia into her hair. **1975** *Hi-Fi Answers* Feb. 41/1 Colouration introduced by the cartridge will be carried right through the system and emerge at the other end added to the inevitable individual speaker character, result...yeech! **1979** 'A. Hailey' *Overload* iv. xviii. 389 As for the food there—yech! **1984** *N.Y. Times* 30 Mar. a. 6/2 We come up with new information and we tell them and they go 'Yecch—what a mess.'

yechy (yə·ki), *a. U.S. slang.* Also **yecchy.** [f. prec. + -y[1]. Cf. *yucky *a.*] = *yucky *a.*
 1969 *Current Slang* (Univ. S. Dakota) Summer 17 *Yeccky*, adj. Extremely unpalatable (of food or drink). **1975** *New Yorker* 16 June 25/1 They thought the green peppers and onions were 'yechy'. **1983** *Washington Post* 23 Dec. 15/1 She goes on and on about him in yecchy voice-overs: He was like 'a licorice billygoat sniffing the wind for sexual sweat'.

yedda (ye·dă). Also **yeddo.** [Origin unknown.] A type of grass used for making straw hats (see quot. 1925). Freq. *attrib.* as *yedda braid, plait, straw*.
 1918 C. R. Aiken *Millinery Dept.* 28 Yedda braid was first produced in Italy, but the Japanese have made an imitation... Yedda is a tall grass grown in Italy, Japan, and the Philippines. **1922** F. Anslow *Pract. Millinery* viii. 93 The Yedda plaits, made of Japanese and Chinese grass, are soft and pliable, and can be dyed in most beautiful shades, of which, perhaps, the mole, grey and blue are the most attractive. **1923** *Sketch* 9 May 300 Dove-grey yedda straw makes this attractive shady hat. **1925** N. Kneeland *Millinery* ii. 17 Yeddo is a tall grass grown in Italy, Japan, and the Philippines. The hats made from this fiber are loosely woven, light, and delicate. **1927** *Daily News* 20 June 2/4 A smart little hat of varnished black yedda straw. **1929** *Millinery Trade Rev.* May 46 A diversity of summer fabrics is being featured, as well as the ever popular straw items, Bakou, balibuntal, yedda, [etc.]. **1957** E. R. E. Langridge *Textbk. Mod. Millinery* vi. 62 Yeddo is a knotted straw made entirely by hand and worked up by the Swiss peasants in their houses.

Yeddo (ye·do). The former name of Tokyo, used *attrib.* to designate materials, etc., originating there (before 1868, in which year the name was changed), as † *Yeddo crepe, poplin;* **Yeddo spruce,** the Japanese spruce, *Picea jezoensis*.
 1866 in A. Adburgham *Shops & Shopping* (1964) xii. 126 Costumes at reduced prices in Yeddo Poplin. [**1906** Elwes & Henry *Trees Gr. Brit. & Ireland* 87 Mayr informed me last year that the Yezo spruce was not introduced into Europe until 1891.] **1932** W. Dallimore in F. J. Chittenden *Conifers in Cultivation* 26 Picea... jezoensis Carr... Yeddo Spruce—N.E. Asia to N. Japan. **1952** A. G. L. Hellyer *Sanders' Encycl. Gardening* (ed. 22) 378 *Picea..jezoensis*, 'Yeddo Spruce', not so hardy as var. *hondoensis*, 80 to 100 ft., Japan. **1960** C. W. Cunnington et al. *Dict. Eng. Costume* 280/2 *Yeddo crepe*, 1880, a cotton fabric thick as linen but soft... *Yeddo poplin*, 1865, of pure llama wool, resembling French merino. **1981** F B. Hora in *Oxf. Encycl. Trees of World* 70/2 *P. jezoensis* Yeddo Spruce. NE Asia, Japan.

yeep (yīp), *v. rare.* [Imitative.] *intr.* To cheep. Hence **yee·ping** *ppl. a.* and *vbl. sb.*
 1834 T. Keightley *Tales & Popular Fictions* i. 11 His feathered charge, who go along *yeeping* and leisurely picking their steps. **1945** B. Macdonald *Egg & I* (1946) iii. 44 Seven hundred and fifty yeeping chicks. *Ibid.* vii. 91 The yelping of a puppy and the stronger, louder yeeping of the chicks.

yeh (ye), *colloq.* or *dial.* var. of Yes or Yea. Cf. *yeah *adv.*
 1920 Galsworthy *Skin Game* i. 13 *Hillcrist.* Did you meet the Jackmans? *Dawker.* Yeh. **1934** J. Franklyn *This Gutter Life* xx. 160 'I hope your son will be happy'. 'Yeh!—I 'opes so lidy.' **1955** W. Gaddis *Recognitions* i. vi. 206 Yeh, I could write a book. **1962** G. E. Evans *Ask the Fellows who cut Hay* (ed. 2) xxv. 235 The affirmative in the East Anglian dialect is 'yeh', which is undoubtedly not a corruption of yes, but a true survival of the early 'ye' ('Let youre ye be ye, and youre naye naye' of Tindal's Bible). **1963** Lennon & McCartney in *Golden Beatles Bk.* (1966) 37 She loves you yeh yeh yeh. **1973** *Black World* Nov. 91 Can you imagine the R&B band funky as they want to be, using a bass figure that kind of remind you of sweetback, yeh sweetback.

Yehudi, var. *Yahudi.

yell, *sb.* Add: **e.** *slang.* Something or someone extremely amusing; a joke, a 'scream'.
 1926 E. Hemingway *Sun also Rises* xvi. 179 'Bill's a yell of laughter,' Mike said. **1938** N. Marsh *Artists in Crime* xv. 238 'Well, of course!' exclaimed Miss O'Dawne, greatly diverted. 'Aren't you a yell!' **1949** E. Coxhead *Wind in West* ii. 32 All these doctors and their ecologists —what a yell. **1970** [see *look *v.* 10 a].

yellow, *a.* and *sb.* Add: **A.** *adj.* **1. d.** (Earlier and further examples.) In U.S. use freq. as *yaller* when applied to light-skinned Blacks.
 1787 *Asiatick Researches* (1790) II. 2 That the Turks have any just reason for holding the coast of Yemen to be a part of India, and calling its inhabitants Yellow Indians. **1834** *Sun* (N.Y.) 20 Mar. 2/2 A huge looking 'yaller gall' was hammering away at the eyes of a small white man.., because he called her a *snow* ball. **1888** L. A. Smith *Music of Waters* 37 Oh, sigh her up, my yaller gals. **1913** *Punch* 19 Feb. 138/3 Believers in the Yellow Peril who wish everyone else to realise the importance of that menace are proposing to bring it home by means of All Yellow Suppers. **1927** [see *monkey-man* s.v. *monkey *sb.* 17 a]. **1937** C. Himes *Black on Black* (1973) 141 The nervous profile of the driver bent low over the wheel. A yellow nigger. **1942** Z. N. Hurston in A. Dundes *Mother Wit* (1973) 28/2, I done slept with yaller women. **1956** J. Barth *Floating Opera* xxviii. 268 Ah fails to unnerstan' How a wuthless, shif'less dahkie such as you, sah, Kin conglomerate de money fo' a Caddylac sedan, Jest to keep yo' yaller gal fren' sweet and true, sah. **1966** *Listener* 17 Mar. 401/1 The setting is..England now, with a cold war and a yellow peril. **1977** C. McCullough *Thorn Birds* xv. 348 But Japan was Asia, part of the Yellow Peril poised like a descending pendulum above Australia's rich, empty, underpopulated pit.
 2. b. Craven, cowardly. *colloq.* (orig. *U.S.*).
 1856 in P. T. Barnum *Struggles & Triumphs* (1869) 400 We never thought your heart was yellow. **1918** J. M. Grider *War Birds* (1927) 264 One of our noblest he-men, a regular fire-eater to hear him tell it, has turned yellow at the front. **1932** E. Wallace *When Gangs came to London* xv. 121 The yellow jury..acquitted 'em on a murder charge. **1950** J. Agee in *Botteghe Oscure* VI. 392 Then something happened that made me know I was scared of them and I admitted to myself: I'm yellow. **1974** *Guardian* 30 Jan. 24/3 It frightens me when moderate voices are taken to be from weak and yellow men. **1977** 'O. Jacks' *Autumn Heroes* xiv. 203 You're yellow scum. You'll fight when the odds are with you.

 c. Of or pertaining to an organization, a policy, or to persons opposed to militant action by a trade union or trade unions. See also *yellow union*, sense C. 1 e below.
 1913 J. A. Estey *Revolutionary Syndicalism* ii. 47 The so-called *syndicats jaunes*, or yellow syndicats, formed in the interest of employers for the purpose of strike-breaking. **1920** *Glasgow Herald* 12 Sept. 7 The railwaymen.. will be content to follow the lead of the General Confederation of labour and stick to the Amsterdam International, which the dictators of Russia have labelled 'yellow'—that is to say 'blackleg'. *Ibid.* 24 Nov. 8/3 The Russian Soviet Republic has insolently rejected it as 'a Congress of yellow leaders who continually betray the fundamental interests of the Labour movement'. **1922** B. G. de Montgomery *Brit. & Continental Labour Policy* vi. 58 The bus-traffic and road-transport were organized by the members of the 'yellow' or anti-strike syndicates, and by the *bourgeois* class. **1939** A. Philip in H. A. Marquand *Organized Labour in Four Continents* 51 The Confederation of Professional Unions, a 'yellow' organization benefiting from employer support. **1972** G. L. Mosse in *Jrnl. Contemp. Hist.* VII. 206 France was regarded as the classical land of yellow trade unionism.
 4. Of or pertaining to a political party whose colour is yellow. Cf. sense 4 of the *sb.*
 1834 F. Witts *Diary* 12 Aug. (1978) 97 The respective parties mustered when the poll was over at their headquarters, the Bell Hotel being the Blue house and the King's Head the yellow. **1874** Trollope *Phineas Redux* I. ii. 14 He remained there for three or four days..staying at the 'Yellow' inn.
 B. *sb.* **1. d.** Cowardice. Cf. sense 2 b of the adj. above.
 1896 G. Ade *Artie* vi. 57 This is how I found that streak of yellow in him. **1914** B. M. Bower *Flying U Ranch* 146, I was just b'ginnin' to think this bunch was gitting all streaked up with yeller.
 2. a. Also *U.S.*, gold. *Obs.*
 1858 *Pike's Peak Guide Bk.* 329 We commenced sending prospecting parties into the mountains, but they returned every night with 'nary yellow'. **1901** M. E. Ryan *That Girl Montana* xviii. 227 She would watch some strange miner dig and wash the soil in his search for the precious yellow.
 c. A yellow ball used in the game of snooker.
 1910 *Encycl. Brit.* III. 938/2 If it is pocketed, the player scores one and is at liberty to play on any of the coloured balls; though in some clubs he is compelled to play on the yellow. **1950** L. H. Dawson *Hoyle's Games Modernized* iii. 346 At the beginning of the game [of snooker] Yellow is placed on the right-hand corner of the D. **1977** *Cleethorpes News* 6 May 29/4 After potting the yellow he more or less forced Barnes to take green, brown and blue.
 d. A golden Labrador.
 1945 C. L. B. Hubbard *Observer's Bk. Dogs* 97 As long as we have bred Labradors we have had yellows. **1973** *Country Life* 8 Feb. (Suppl.) 325/1 Some of the yellows were a light creamy colour.
 3. See also *high yellow* s.v. *high *a.* 21.
 6. *ellipt.* for *yellow alert*, sense C. 1 e below.
 1940 *Mass-Observation Archive* 1 Aug. in Calder & Sheridan *Speak for Yourself* (1984) iii. 78 Soon after eleven we were remarking that it was time we got the yellow, when the telephone went. **1943** [see *red *sb.*[1] 8]. **1949** [see *alert *sb.* 1 b]. **1978** 'G. Vaughan' *Belgrade Drop* xiii. 84 President Turner had been in touch with the other Nato head of state and their forces had gone on yellow.
 C. 1. a. *yellow Labrador;* **yellow-bob,** a shrike-robin, *Eopsaltria australis*, found in forested areas of south-eastern Australia; **yellow snake,** one of several yellowish snakes, esp. a boa, *Epicrates subflavus*, found in the West Indies; **yellow warbler,** one of several North American warblers of the genus *Dendroica*. **b. yellow bean,** the yellow seeds of one of several varieties of soya bean; **yellow birch,** a North American tree, *Betula lutea*, which has yellow or grey bark; **yellow box,** an Australian gum-tree, *Eucalyptus melliodora*, which has yellowish inner bark; **yellow cedar** = *Nootka cypress* s.v. *Nootka *a.* 2; **yellow poplar,** one of several North American softwood trees or their wood, esp. the tulip-tree, *Liriodendron tulipifera*; **yellow yam,** one of several species of *Dioscorea* producing yellow-fleshed tubers; also, the tubers themselves. **c. yellow ground,** kimberlite that is exposed at the surface and has become yellow as a result of atmospheric oxidation; **yellow jack** = *yellow jacket* (*b*), sense C. 2 c in Dict.; **yellow phosphorus** = *white phosphorus* s.v. *white *a.* 11 c; **yellow ware** (earlier and later examples). **e. yellow alert,** an instruction to be prepared for or an initial state of readiness to cope with an emergency (cf. *red alert* s.v. *red *a.* 19 a); **yellow badge,** a badge of identification that Jews have sometimes been required to wear, esp. by the Nazis in Germany (cf. *yellow star* below); **yellow band,** a mark on a lamp-post to indicate that motor vehicles are not permitted to wait in the vicinity; freq. *attrib.*; also = *yellow line* below; **yellow belt,** the belt worn by one who has attained a certain standard of proficiency in judo (see quots.); **yellow book,** (*a*) an official report of government affairs in various European countries; (*b*) a report issued by the Liberal Party in 1928 on the industrial future of Britain; **yellow card,** in Association Football, a card shown by the referee to a player when he is cautioned; † **yellow jacket,** a military decoration in imperial China (*obs.* exc. *Hist.*); **yellow jersey** [tr. F. *maillot jaune*], a jersey awarded to the winner of (a stage of) a cycle race, *esp.* the *Tour de France;* **yellow leaf,** used (in allusion to quot. 1605) to refer to the process of ageing; **yellow light** *U.S.* a yellow-coloured cautionary light in traffic signals (cf. *amber *a.* b); also *fig.*; **yellow line,** a yellow road-marking, usu. parallel to the kerb, indicating that parking of motor vehicles is restricted (though local regulations vary); also **double yellow line** indicating that parking is forbidden; **yellow pages** *sb. pl.* orig. *U.S.*, an index printed on yellow paper; *spec.* the classified section of or supplement to a telephone directory, listing firms, products, and services; **Yellow Pressman,** a journalist or reporter working for the yellow press (cf. sense 3 of the adj.); **yellow rain,** a yellow powder reported as falling through the air in S.E. Asia and causing severe blistering and sometimes death; **yellow rust,** a disease of wheat caused by the fungus *Puccinia glumarum;* **yellow spot** (earlier example); = *macula lutea* s.v. *macula 2; **yellow star,** a piece of yellow cloth bearing the Magen David, which the Nazis required Jews to wear; **yellow streak,** a trait of cowardice; **yellow union** [tr. F. *syndicat jaune*: cf. sense 2 c of the *adj.* above] a union of workers favouring free enterprise and usually opposed to strike action; **yellow warning** = *yellow alert* above.
 1968 *Punch* 28 Aug. 279/1 NATO forces had quickly been placed on 'Yellow Alert'. **1969** *Times* 17 Sept. 1/8 A yellow alert..on hospital beds..means that all cases not in need of immediate attention will not be admitted to hospitals. **1978** 'G. Vaughan' *Belgrade Drop* iii. 23 The United States president..put his missile submarines throughout the world on yellow alert. **1876** Geo. Eliot *Daniel Deronda* III. v. xxxvii. 127 To Deronda just now the name Cohen was equivalent to the ugliest of yellow badges. **1892** I. Zangwill *Childr. Ghetto* I. 2 People who have been living in a ghetto for a couple of centuries are not able..to efface the brands on their souls by putting off the yellow badges. **1942** I. Cohen *Jews in War* iii. 31 *The Yellow Badge* The crowning device for humiliating the Jews was the revival of a mediæval practice. In October 1941, a decree was issued requiring them to wear a yellow armlet marked with the 'Shield of David', which the Jews of Poland had been wearing for the past two years. **1962** Bridger & Wolk *New Jewish Encycl.* 38/1 (*caption*) The yellow badge the Nazis required Jews to wear in Germany and Nazi-occupied countries. **1948** *Times* 13 May 2/3 When the no-waiting order was first introduced in the West End it was announced that permanent signs would be erected as soon as the materials were available, the yellow bands being temporary. **1959** *Times* 8 Dec. 5/6 Vast numbers of cars..are left at the kerbside all day in all parts of central London except the yellow-band streets. **1962** R. Jeffries *Exhibit No. Thirteen* iv. 36 Parked my car in a yellow-band area. **1967** R. Rendell *Wolf to Slaughter* ii. 17 The car drew up... 'Not on the yellow band, Drayton,' Burden said sharply. **1965** H. Burke *Chinese Cooking for Pleasure* 150 In small cans are bamboo shoots,..red bean curd, black bean, yellow soy bean, [etc.]. **1972** Claiborne & Lee *Chinese Cookbk.* (1973) xi. 422 *Bean Sauce*... Also called 'Whole Bean Sauce' or 'Yellow Bean Sauce', this thick sauce is made from yellow beans, flour, and water and is sold in.. tins. **1983** *Observer* 16 Jan. (Colour Suppl.: Living Extra) 7/4 Black beans..and yellow beans are both products of the versatile soya bean. **1941** M. Feldenkrais *Judo* 166 A white belt is worn by beginners, corresponding to the sixth Kyu. The next grade, the fifth, is indicated by a yellow belt. **1979** *Observer Mag.* 17 June 39/1 For several years he went to judo classes, reached yellow-belt standard (three below black belt). **1787** W. Sargent in *Mem. Amer. Acad.* IX. 158 Black and Yellow Birch... The bark of the latter is used by the Indians for making canoes. **1851** J. S. Springer *Forest Life* 23 The general outlines of the Yellow Birch often resemble the Elm. **1943** R. Peattie *Great Smokies* 156 A yellow birch on Whitetop Mountain was found to be seven feet three inches thick. **1974** M. Braithwaite *Ontario* xi. 169 Hemlock, oak, maple, and yellow birch, they were all there, just waiting to be cut down and sawed up into lumber. **1909** A. E. Mack *Bush Calendar* 68 Then a yellow-bob came to visit us. **1965** *Austral. Encycl.* VII. 470/1 Another common and familiar bird is the yellow robin ('yellowbob') of eastern Australia, a species with a breast of bright yellow. **1883** *Pall Mall Gaz.* 5 Dec. 8/1 Paris, Dec. 5.—The first volume of the new Yellow-book on Tonkin affairs. **1897** *Times* 23 Nov. 5/1 The *Berliner Neueste Nachrichten* reminds the French Republic that..in former Yellow-books there is plenty of evidence to show how generous was the help afforded by Germany to France. **1929** D. Lloyd George *We can conquer Unemployment* 3 In the 'Yellow Book', published a year ago, the Liberal Industrial Enquiry presented exhaustive proposals for dealing with the grave unemployment situation with which Britain was, and is still, faced. **1949** *Time* (Atlantic ed.) 14 Feb. 18/1 The Communists issued a 'Yellow Book' containing what they called Mindszenty's written confession. **1983** *Daily Tel.* 24 Nov. 18/4 Had the Liberal Yellow Book been published in 1920 our history might have been different. **1877** F. von Mueller *Introd. to Bot. Teaching at Schools of Victoria* 15 This tree passes by the very unapt vernacular name Yellow Box-tree. **1934** *Bulletin* (Sydney) 31 Jan. 21/2 We lop mainly yellow box. **1977** *Meanjin* (Austral.) XXXVI. 1. 71 I'll..split off kindling wood From the yellow-box log. **1976** *Times* 11 Nov. 12/4 Two Villa men were shown the yellow card for fairly innocuous offences. **1884** *N.Y. Times* 5 Oct. 5/2 Red and yellow cedar..are the other trees most frequently met with. **1910** [see *Alaska]. **1957** *Handbk. of Softwoods* (Forest Prod. Res. Lab.) 61 'Yellow cedar'..is confined to the Pacific Coast area from Alaska south to southern Oregon. **1886** Yellow ground [see *blue ground* s.v. *blue *a.* 13]. **1947** *E. African Ann.* 1946–7 122/1 The portion of a pipe at the surface which has been altered or weathered and is usually of a yellowish colour is known as 'yellow ground', in contrast to the blue-green colour of the unaltered kimberlite or blue ground. **1978** *Sci. Amer.* Apr. 120/3 Most kimberlite exposed at the surface, called 'yellow ground' by miners and prospectors, is severely weathered. **1927** M. M. Bennett *Christison of Lammermoor* iv. 49 These trees called yellowjacks are soft wood, so white ants enclose them with earth walls and eat the wood out. **1943** A. Marshall in *Coast to Coast 1942* 14, I tied the horse to a yellowjack and crept towards the river. **1864** *North-China Herald* 18 June 99/2 [They] being each and all pre-eminent for bravery, contempt for death, and a generous emulation, are invested with the yellow jacket as a reward for their merits. **1878** H. A. Giles *Gloss. Ref. Far East* 84 A yellow *ma-kwa* is a distinction conferred by the Emperor on high officials; sometimes called the *Yellow Jacket*. **1918** H. B. Morse *International Relations Chinese Empire* II. v. 104 On Li Hung-chung was conferred the military distinction of the Yellow Jacket and the civil distinction of Junior Guardian to the heir apparent. **1964** *Guardian* 16 June 6/6 Metcalfe (England) won his seventh yellow jersey with another aggressively defensive ride. **1983** *Times* 1 July 12/5 It's hard enough even to get a ride in the Tour [de France]... To be the raceleader, to wear the yellow jersey, that's almost worth dying for. **1948** C. L. B. Hubbard *Dogs in Britain* xix. 232 The Yellow Labrador sometimes called the Golden Labrador..differs in several respects from the black Labrador. **1974** *Times* 4 May 23/8 (Advt.), Country home urgently wanted for two purebred Yellow Labrador bitches. [**1605**: see Sere, sear *a.*[1] 1 b]. **1913** L. Strachey in *Edin. Rev.* Jan. 68 The radiant creatures of Sceaux had fallen into the yellow leaf. **1935** C. Isherwood *Mr. Norris changes Trains* vii. 107 Yes, I shall be fifty-three... I find it difficult to become accustomed to the thought that the yellow leaf is upon me. **1974** A. A. Thompson *Swiss Legacy* xvi. 157 He guided the Mercedes through the traffic..taking chances.. He ran a yellow light and then a red one. **1977** *N.Y. Rev. Bks.* 27 Oct. 16/4 They only ask a 'yellow' light—the right to proceed with caution. **1965** *Autocar* 24 Sept. 609/1 The leaflet recently published by RoSPA in conjunction with the Ministry of Transport..states that a yellow line by the kerb means no waiting *except* for loading and unloading. **1968** J. Fleming *Kill or Cure* iv. 56 The local police.. allow me to park on the double yellow line with impunity, when absolutely necessary. **1975** J. Symons *Three Pipe Problem* xviii. 180 Traffic wardens can start booking cars on yellow lines after eight o'clock. **1983** *Church Times* 23 Dec. 11/1 Christmas shoppers who had taken the risk of parking on the single yellow line. **1908** *Sears, Roebuck Catal.* (*verso rear cover*), See the yellow pages in back of this book. **1956** R. A. Heinlein in *Mag. of Fantasy & Sci. Fiction* Oct. 51/2 Get me the yellow-pages phone book... I want to check the exact phrasing of a firm name. **1966** D. G. Hays in *Automatic Transl. of Lang.* (NATO Summer School, Venice, 1962) 152 In a telephone book.. in the..'Yellow Pages', the major variable is name of product or service. **1969** *Times* 5 May 26/2 Yellow Pages are the classified guide that will be part of everyone's GPO telephone directory soon. **1982** S. Brett *Murder Unprompted* i. 10 The random selection method of sticking a pin in the 'Theatrical and Variety Agents' section of the Yellow Pages. **1985** *Punch* 23 Jan. 24/2 I started by ringing a few cowboys through the *Yellow Pages*, just to check on prices. **1866** H. E. Roscoe *Less. Elem. Chem.* xv. 133 The weight of red substance produced is exactly equal to that of yellow phosphorus used. **1944** J. A. Timm *Gen. Chem.* xli. 443 Yellow phosphorus is formed when the liquid solidifies. **1774** J. R. Peyton *Let.* 21 July in J. L. Peyton *Adv. My Grandfather* (1867) 127 The forest of Kentucky consists of yellow and white poplar, walnut, red bud. **1876** W. Whitman *Specimen Days* (1882–3) 89 Here is one of my favorites now before me, a fine yellow poplar, quite straight, perhaps 90 feet high. **1955** *Sci. News Let.* 7 May 302/2 The tulip tree is also variously known as tulip poplar, yellow poplar, white-wood and fiddle-tree. **1909** G. K. Chesterton *Tremendous Trifles* 131 The Yellow Pressman seems to have no power of catching the first fresh fact about a man. **1918** S. Sassoon *Counter-Attack* 29 The boys came back... And Yellow-Pressmen thronged the sunlit street To cheer the soldiers who'd refrained from dying. **1979** W. Safire in *N.Y. Times* 13 Dec. A31/5 The Laotians call it 'the yellow rain'. **1981** *N.Y. Times* 24 Nov. C-1/5 The United States has been trying since 1976 to verify reports that chemical weapons, known popularly as 'the yellow rain', are being used against remote villages in Laos, Cambodia and, more recently, Afghanistan. **1982** *Sci. News* 20 Feb. 122/1 Blood samples were drawn from nine individuals supposedly exposed to a 'yellow rain' gas attack in the fall of 1981... Mirocha 'was able to tentatively identify'..a metabolite of the trichothecene mycotoxin T_2, in samples from only two. **1907** *Jrnl. Agri. Sci.* II. 129 He [*sc.* Mr. Biffen] has discovered and grown several wheats which show to a greater or lesser degree immunity to the attacks of *Puccinia glumarum*, Yellow Rust. **1973** *Scotsman* 7 Aug. 4/6 Mr Blakebell was speaking of yellow rust in wheats. **1725** H. Sloane *Voy. Jamaica* II. 325 Serpens major subflavus..The yellow Snake. **1851** P. H. Gosse *Naturalist's Sojourn in Jamaica* 314 A serpent of the Boa kind..is distinguished by the appellation of Yellow Snake. **1860** Mayne Reid *Odd People* 22 The 'Yellow Snake', or South African Cobra. **1868** J. G. Wood *Homes without Hands* iii. 85 A Yellow Snake..is very plentiful in Jamaica and is perfectly harmless to man. **1819** *Phil. Trans. R. Soc.* CIX. 302 The yellow spot of Soemmerring .. is never seen to advantage until this membrane be removed. **1967** *Guardian* 21 Oct. 8/3 'Private Eye' recently labelled me 'D. A. N. Jew'. Now it happens that I haven't the right to claim the yellow star. **1981** *Times Lit. Suppl.* 6 Nov. 1296/5 Germans like Captain Ernst Janger..who declared himself 'ashamed'..when he saw Jews in Paris wearing their yellow stars. **1911** H. S. Harrison *Queed* v. 55 'A yellow streak in him, and we didn't know it!' bellowed the Major. **1977** 'D. MacNeil' *Wolf in Fold* xi. 116 I'm not showing a yellow streak! But we're going to have casualties. **1947** H. W. Ehrmann *French Labor* vii. 126 The CGT could not properly identify the Catholic trade unions with the various movements, in French usage commonly referred to as 'yellow' unions, which were organized under the auspices of the employers. **1957** M. P. Fogarty *Christian Democracy* xv. 192 Widespread support was given to the yellow unions, notably by the clergy. **1970** R. A. H. Robinson *Origins of Franco's Spain* 331 Socialists were also determined that no 'yellow' unions should flourish, eg a strike-threat by the UGT procured the dismissal of 20 members of the *Federación Española de Trabajadores*. **1783** J. Latham *Gen. Synopsis Birds* II. 482 Spotted Yellow Warbler. **1845** S. Judd *Margaret* i. 160 The leafless Butternut, whereon..the yellow warbler made its nest, sprawls its naked arms. **1938** M. Thompson *High Trails* 153 The yellow warbler..and many other birds fill the air with their songs. **1971** *Islander* (Victoria, B.C.) 13 June 13/2 Overhead a pair of yellow warblers trilled out their song. **1785** J. Woodforde *Diary* 7 Nov. (1926) II. 213 To Nancy for 2 new yellow Ware Chamber Pots 1. 0. **1887** *Harper's Mag.* Dec. 31/1 Sometimes a cherry would fall upon her dark braids, and drop thence in among the verdant contents of the yellow-ware bowl. **1967** *Canadian Antiques Collector* Apr. 9/2 During the next ten years the Bells extended their pottery production to include Rockingham and yellow wares. **1963** *Times* 22 Jan. 10/3 The service have issued a 'yellow warning'. This is intended to warn hospitals to cut down on routine admissions so as to make room for emergencies. **1973** *Times* 13 Nov. 1/2 In electricity supply terms a national 'yellow warning'..means possible voltage reductions. **1913** W. Harris *Notes Fruit & Veg. Jamaica* 42 Yellow yam and its varieties belong to *Dioscorea cayennensis*. **1971** [see *negro yam* s.v. *Negro 7]. **1973** N. Farki *Countryman Karl Black* iv. 38 Rice and two pieces of yellow yam in one plate.

 2. a. *yellow-blue, -golden;* also *yellow-fluffy, -gleamy, -pale.*
 1940 W. Faulkner *Hamlet* iv. ii. 328 There were three buzzards soaring against the high yellow-blue. **1916** D. H. Lawrence *Amores* 50 Flutter for a moment, oh the beast is quick and keen,—Extinct one yellow-fluffy spark. *a* **1930** —— *Phoenix* (1936) i. 3 In the yellow-gleamy sunset, wild birds began to whistle faintly. **1946** S. Spender *European Witness* i. 15 In the foreground yellow-golden fields, with above a flat wall of greyish sky. *a* **1963** L. MacNeice *Astrol.* (1964) iii. 96 Leo the yellow-golden fire of organized mentality. *a* **1930** D. H. Lawrence *Last Poems* (1932) 315 Black lamps ..Giving off darkness, blue darkness, upon Demeter's yellow-pale day.
 b. *yellow-beaked* (later example), *-belled, -browed* (later examples), *-covered* (examples), *-crowned* (later examples), *-eyed, -fanged, -finned* (later examples), *-flecked, -headed* (later examples), *-livered, -mottled, -necked, -robed* (earlier example), *-slashed, -slobbered*, adjs.
 1966 E. Palmer *Plains of Camdeboo* xvi. 262 By far the showiest is the yellow-beaked Stapelia, *Stapelia flavirostris*, with dark flowers marked with yellow and ornamented with silver hairs. **1881** O. Wilde *Poems* 122 On this side and on that a rocky cave, Hung with the yellow-belled laburnum stands. **1913** H. K. Swann *Dict. Names Brit. Birds* 264 Yellow-browed Warbler... A Siberian species of Willow Warbler. **1971** *Country Life* 25 Mar. 705/3 Almost any rarity can turn up, such as..yellow-browed warbler. **1849** *Merchants' Mag.* XX. 118 The yellow-covered literature of the day—translations from the French. **1915** H. Young *Hard Knocks* 23 The little yellow covered novels were the cause of it. **1925** J. Ferguson in *Oxf. Poetry* 18 Like stately flowers, yellow-crowned. **1950** *Caribbean Q.* II. iii. 41 Nor were the larger and stronger Yellow crowned Night Herons, to be outdone. **1881** O. Wilde *Poems* 180 The hot jungle where the yellow-eyed huge lions sleep. **1957** T. Hughes *Hawk in Rain* 39 A square-pupilled yellow-eyed look. **1954** J. R. R. Tolkien *Two Towers* iii. iii. 50 It was the yellow-fanged guard. **1908** C. F. Holder *Big Game at Sea* xxiii. 342 The boatmen..called it the 'yellow-finned tuna'... This was in 1904, and ever since the new tuna, with its vivid lemon finlets, has appeared every August or September. **1936** *Zoologica* XXI. 190 The various nominal forms of the yellow-finned tuna belong to the same species. **1920** D. H. Lawrence *Lost Girl* iv. 55 The seam of yellow-flecked coal. **1846** *Ex. Doc. 30th U.S. Congress 1 Sess. House* No. 41. 436 [We saw] large flocks of the yellow headed black bird. **1972** R. & R. Wright *Cariboo Mileposts* 50 Stands of tules or reeds will hold the woven nests of red-wing and yellow-headed blackbirds. **1935** S. Lewis *It can't happen Here* 156 The meanest, lowest, cowardliest gang of yellow-livered, back-slapping, hypocritical guntoters. **1979** *PN Rev.* No. 9. 27/1 O green, green eating out my eyes, A yellow-livered green in a wet light. **1916** Blunden *Harbingers* 64 Toadstools..Yellow, and yellow-mottled red, and black. **1889** *Cent. Dict.* 7016/3 The yellow-necked caterpillar..feeds in communities on the foliage of apple, hickory, and walnut in the United States. **1908** E. J. Banfield *Confessions of Beachcomber* i. iii. 98 Yellow-necked Mangrove Bittern. **1921** G. E. H. Barrett-Hamilton *Hist. Brit. Mammals* II. 547 The Yellow-necked Field Mouse is distinguished from *A. sylvaticus* by its larger size. **1979** *Essex Countryside* XXVII. 72/2 The hoarding habits of yellow-necked mice are well known. **1819** *Methodist Mag.* Oct. 723 We took leave of our yellow-robed acquaintances. **1928** V. Woolf *Orlando* v. 225 The yellow-slashed sky of dawn. **1922** Joyce *Ulysses* 222 Two barefoot urchins, sucking long liquorice laces, halted near him, gaping..with their yellowslobbered mouths.
 c. yellow-back, (*b*) more widely, any cheaply issued or reprinted novel; (earlier and later examples); (*c*) a U.S. currency note having the back coloured yellow; **yellow-fin,** esp. **yellow-fin tuna,** one of several species of *Thunnus*, esp. *T. albacares*, a large fish found in warmer parts of both the Atlantic and Pacific oceans; **Yellow Hat,** *colloq.*, used *attrib.* or *absol.* in *pl.* to denote a Tibetan Buddhist sect (Gelugpa) founded in the fourteenth century by Tsong-kha-pa; **yellow-head,** (*d*) a warbler, *Mohoua ochrocephala*, found in the South Island of New Zealand; **yellow jacket,** (*c*) *slang* (orig. *U.S.*), a pentobarbitone capsule; **yellow-leg,** (*b*) *N. Amer. colloq.* [from the yellow stripe down the side of the breeches], a U.S. cavalryman or a member of the Royal Canadian Mounted Police; **yellow-throat** (later examples); **yellow underwing,** one of several noctuid moths.
 1877 *Living Age* 14 Apr. 128/1 Four days ago Ley and I started down the river on an exploring expedition, and he took it into his head to rope ('lasso', as the yellow-backs have it) a buffalo. **1902** H. L. Wilson *Spenders* xiv. 150 She was dead in love with the nice long yellow-backs that I've piled up. **1928** M. Sadleir *Trollope: a Bibliogr.* 68 In 1868 *The Belton Estate* was issued at two shillings as a 'yellow back'. **1943** *Copper Camp* (Writers' Program, Montana) 37 They occasionally found yellowbacks tucked in their shoes. **1976** T. Eagleton *Crit. & Ideology* ii. 47 The 'yellowback' railway novel is available to a mass public. **1922** *Pacific Fisherman* Feb. 12/1 Each of these new species—bluefin tuna, yellowfin tuna and striped tuna—proved itself well adapted to the same canning method as the albacore. **1975** *Islander* (Victoria, B.C.) 7 Sept. 4/2 They [*sc.* whales] travel in schools in the eastern Pacific, followed by yellow-fin tuna which feed on their leavings. **1747** *Astley's New Gen. Coll. Voy.* IV. ii. iv. 450/2 They being of the yellow Hat, or Chinese party. **1931** C. Bell *Relig. Tibet* viii. 129 With the enthronement of the fifth Dalai Lama as sovereign over the whole country, the power of the Yellow Hats was greatly increased. **1962** H. E. Richardson *Tibet & its Hist.* ii. 40 The Dalai Lamas..owed their appearance to the great religious teacher known as Tsong Khapa (1357–1417), the founder of a new sect, the Gelugpa, popularly called the Yellow Hats. **1978** C. Humphreys *Both Sides Circle* xx. 212 The famous monastery at Ghoom..belongs to the Gelug-pa or Yellow Hat sect of Tibetan Buddhism. **1873** *Trans. N.Z. Inst.* VI. 144 Yellow-head. Average weight of specimens, 1¼ ounce. **1966** *Encycl. N.Z.* I. 206/2 Still widely distributed in the deeper forests are..whitehead ..and yellowhead. **1953, 1969** Yellow jacket [see *nembie]. **1974** M. C. Gerald *Pharmacol.* xi. 205 Short-acting barbiturates such as pentobarbital ('yellow jackets'). **1894** J. A. Frye *Fables of Field & Staff* 109 The 'Yellow-Legs' are always great on dismounted duty. **1943** W. Chase *Sourdough Pot* xix. 120 Numbers of these prisoners were marched down the main street in charge of a Mountie, or 'Yellow-leg', as they were called on account of the yellow strip running down the outside of their trouser leg. **1957** G. Shirreffs *Rio Bravo* (1972) i. 6 He glanced back at the rough country through which they had come, almost as though looking for the beefy figure of Francis Xavier Feeley, one of the best yellowlegs who had ever forked a McClellan. **1974** W. Hunt *North of 53* xix. 139 The 'yellow legs'—the Mounties—were not permissive in law enforcement. **1865** *Atlantic Monthly* XV. 521, I miss in the woods..the Yellow Throat. **1949** V. S. Reid *New Day* i. xxviii. 143 A..yellowthroat warbler whistles back. **1977** *Blair & Ketchum's Country Jrnl.* May 43/1 The yellow-throats will reemerge on other days. **1749,** etc. Yellow underwing [see Underwing 2]. *a* **1941** V. Woolf *Death of Moth* (1942) 9 The commonest yellow-underwing asleep in the shadow of the curtain. **1968** *Oxf. Bk. Insects* 72/1 Like the other Yellow Underwings..this species baffles its enemies by the way it shows its colours.

ye·llow-be:llied, *a.* **1.** Applied to birds or animals having yellow underparts. Also in extended use (in quot. 1909 of an airship).
 1709, 1783 [in Dict. s.v. Yellow *a.* and *sb.* C. 2 c]. **1827** J. L. Williams *View W. Florida* 30 Black-head fly catcher... Yellow-bellied do. **1869** [see *scrunching *vbl. sb.*]. **1908** E. J. Banfield *Confessions of Beachcomber* I. iii. 95 Yellow-bellied Fig-bird. **1909** Kipling *Actions & Reactions* 140 Yellow-bellied ore-flats..punted down leisurely out of the north. **1936** *Discovery* Oct. 307/2 There are many varieties of flying phalanger, the yellow-bellied 25 inches in length with a bushy tail an inch longer than the body. **1942** R. Peattie *Friendly Mts.* 210 A sweet or black birch tree..had been tapped by a yellow-bellied sapsucker. **1971** *Islander* (Victoria, B.C.) 25 July 16/2 Most members of his species (the yellow-bellied marmot) weigh about 13 pounds at maturity.
 2. *fig.* Cowardly, craven. Cf. *yellow *a.* 2 b, *yellow-belly 5. *slang* (orig. *U.S.*).
 1924 P. Marks *Plastic Age* ix. 75 Yellow-bellied quitters. **1930** [see *potato-mouthed a.* s.v. *potato *sb.* 7]. **1943** J. Mitchell *McSorley's Wonderful Saloon* (1946) i. 24 You yellow-bellied jerk. **1965** J. Porter *Dover Two* xiv. 185 Anything to save his own skin, the yellow-bellied rat! **1971** [see *malt *sb.*[2]]. **1979** 'M. Hebden' *Pel & Faceless Corpse* vi. 58 I'm..a yellow-bellied, lily-livered coward.

yellow-belly. Add: **1. b.** (Earlier and later examples.)
 1787 Grose *Provincial Gloss.* s.v. *Lincolnshire*, Yellow bellies. This is an appellation given to persons born in the Fens, who, it is jocularly said, have yellow bellies, like their eels. **1982** *Times* 5 Oct. 4/5 The Lincolnshire 'yellowbellies' of south Humberside..for generations have entertained a healthy disregard for Yorkshire 'tykes' on the north bank of the Humber.
 3. (Earlier and later examples.) *derog.*
 1842 *New Orleans Crescent* 16 Mar. (Extra) 1 God send that they bayonet every 'yellow belly' in the Mexican army. **1845** [see *copperhead 3]. **1934** 'G. Orwell' *Burmese Days* x. 155 They're Eurasians—sons of white fathers and native mothers. Yellow-bellies is our friendly nickname for them. **1966** [see *slant-eye(s)* s.v. *slant *a.* 3].
 5. A coward. Cf. *yellow *a.* 2 b, *yellow-bellied *a.* 2. *slang* (orig. *U.S.*).
 1930 J. Lait *Put on Spot* 215 Yellow-belly. Coward. **1942** Berrey & Van den Bark *Amer. Thes. Slang* §404/3 *Coward*..yellow-back, -belly or guts. **1952** H. Innes *Campbell's Kingdom* iii. ii. 271 What are you?.. A bunch of yellow-bellies to be fooled into hiding away. **1952** J. Steinbeck *East of Eden* 517 I'm a cowardly yellow-belly. **1965** *Austral. Women's Weekly* 20 Jan. 50/5 'Yellowbelly baby... Spoiled-cat crybaby,' Steve yelled at him. **1969** K. M. Wells *Owl Pen Reader* ii. 209 Grandad's knees shook, and he wasn't no yellow-belly either. **1972** 'H. Howard' *Nice Day for Funeral* ix. 128 She'd call me every kind of yellow belly if I suggested throwing in my hand.

yellow-boy. For ? *Obs.* read *Obs.* exc. *Hist.* and add later examples.
 1883 'Mark Twain' *Life on Miss.* xxxvi. 389 A round ten thousand dollars in yellow-boys. **1898** A. M. Binstead *Pink 'Un & Pelican* x. 219 Gazing contemptuously at the yellow-boy, and then at its donor, she cried: 'An' since when has a "new hat" ceased to be a guinea?' **1957** A. Bryant in J. B. Booth *Palmy Days* p. xi, He reconstructs a vanished world: the last age of horses and 'yellow boys'

ye·llowcake. Also **yellow cake.** [f. Yellow *a.* + Cake *sb.*] An oxide of uranium (and other elements) obtained as a yellow precipitate in the processing of uranium ores.
 1950 *Mining Congr. Jrnl.* Oct. 30/2 Through pH adjustment by regulated additions of sulphuric acid virtually all of the uranium is precipitated as 'yellowcake', an artificial carnotite. **1955** Kirk & Othmer *Encycl. Chem. Technol.* XIV. 439 If the uranium and vanadium are present in the carbonate leach liquors in the proper stoichiometric ratio, then neutralization by acid leads to the very complete precipitation of sodium uranyl vanadate, the 'yellow cake' of Colorado Plateau extractive metallurgy. **1971** *Daily Colonist* (Victoria, B.C.) 11 June 7/7 A West German..group has undertaken to find a market for 4 million pounds of uranium oxide (yellow cake) annually. **1977** *Telegraph* (Brisbane) 5 Sept. 5/1 Police and 200 anti-uranium demonstrators traded kicks and punches, when a fresh consignment of yellowcake reached the White Bay container terminal today. **1981** T. Barling *Bikini Red North* i. 34 Will anything positive be done to stop the French selling yellowcake or hardware on the open market?

yellow dog. *U.S.* **1.** A mongrel dog of a yellowish colour.
 c **1770,** etc. [in Dict. s.v. Yellow C. 1 a]. **1873** M. Holley *My Opinions* 237 If I was a yeller dog, she couldn't seem to look down on me any more, and treat me any worse.
 2. *fig.* **a.** A person or thing of no account or of a low type.
 1881 E. W. Nye *Bill Nye & Boomerang* 166 The presiding officer had lost control, and a surging crowd of yellow dogs had the floor. **1903** *Everybody's Mag.* Oct. 562 In a cut-rate combination you are lucky if you get what you pay for. If there are five magazines in the combination, two of them are good. The rest are 'yellow-dogs'. **1924** A. J. Small *Frozen Gold* i. 44 Understand, I won't allow no yellow dog of a Siwash to step over me. **1975** *New Yorker* 8 Dec. 126/2 Calling someone a yellow dog would not imply that the person so called was actually yellow and wagged his tail. It is just the sort of cultural misunderstanding that student-exchange programs were once expected to clear up.
 b. *attrib.*; applied *spec.* to organizations, etc., opposed to trade unionism.
 1894 Kipling *Day's Work* (1898) 71 America's paved with the kind er horse you are—jist plain yaller-dog horse—waitin' ter be whipped inter shape. **1902** —— *Just So Stories* 92 Old Man Kangaroo is being rude to Yellow-Dog Dingo. Yellow-Dog Dingo has been trying to catch Kangaroo all across Australia. **1902** *Mine Workers' Jrnl.* July 1 A yellow dog lease. **1903** *Outlook* 15 Aug. 931/2 In preference to a Tammany 'yellow dog' ticket his organization would support the Fusion candidate. **1920** *Motorman & Conductor* Oct. 34 A yellow dog contract. **1930** *Sun* (Baltimore) 6 May 12/2 The 'yellow dog' contract, requiring men to barter away their right to organize as the price of a job. **1956** *Mag. of Fantasy & Sci. Fiction* Oct. 32/2 It was the latest form of the yellow-dog clause, one in which the employee agrees to refrain from engaging in a competing occupation for five years by letting his former employers pay him cash to option his services on a first-refusal basis. **1976** *Amer. N. & Q.* XIV. 136/1 The committee scrutinized the records and books of the four major New York companies uncovering the existence of the so-called 'yellow dog funds'.

yellow-fish. Add: **b.** In South Africa, one of several freshwater fishes of the genus *Barbus*.
 1834 A. Smith *Diary* 9 Dec. (1939) I. 168 Fish in the pools of this river of two kinds, the flat head and the bearded yellow fish. **1896** H. A. Bryden *Tales S. Afr.* i. 18 The pouch contained..a 'yellow fish', a barbel-like fish of a pound and a half. **1912** J. Stevenson-Hamilton *Animal Life Afr.* xix. 334 The yellow fish..is the commonest fish of South Africa. **1952** [see *kurper]. **1975** *Stand. Encycl. S. Afr.* XI. 563/1 The yellow-fishes..are popular angling fishes, ranging from one kg to over 14 kg in weight.

yellow-haired, *a.* Add: Also *fig.*
 1917 D. H. Lawrence *Look! We have come Through!* 155 Scyllas and yellow-haired hellebore, jonquils, dim anemones.

yellowing, *ppl. a.* Add: Turning yellow, becoming yellow (quots. in Dict.). Also *fig.*
 1961 E. Williams *George* xxi. 339 Divvers—..a routine side-exam based on the yellowing idea that an Oxford graduate must have a solid religious training. **1977** *Guardian Weekly* 10 July 17/4 All that is left is a batch of yellowing declarations of good intentions.

yellowishness. (Later example.)
 1941 E. R. Eddison *Fish Dinner* xi. 184 The cloud-bank was indigo against that yellowishness of the sky.

yellowly, *adv.* Delete *rare* and add later examples.
 1932 W. Faulkner *Light in August* xiv. 314 He saw before daylight a lamp come yellowly alive in the kitchen. **1958** I. Fleming *Dr. No* vi. 78 The centipede was whipping from side to side in its agony... Bond hit it again. It burst open, yellowly. **1968** P. Dickinson *Skin Deep* ix. 165 The home-made candles burnt yellowly. **1978** H. Wouk *War & Remembrance* x. 100 The lights flickered yellowly on.

yellows. Add: **I. 3. b.** (Earlier examples.) **c.** A similar virus or deficiency disease in other plants.
 1808 [see *peach yellows* s.v *peach *sb.*[1] 6]. **1822** S. Deane *New-England Farmer* (ed. 3) 318/2 Peach trees are subject to a disease called the 'Yellows', of which we have seen no particular description. **1926** *Amer. Jrnl. Bot.* XIII. 647 Asters affected with yellows never show mottling. **1933** *Times Lit. Suppl.* 16 Mar. 187/2 A disease of the tea-bush known as 'yellows' is due to a deficiency of sulphur in the soil. **1957** *New Scientist* 8 Aug. 31/3 At least 2,800 tons of beet were lost from yellows infection.. in the Shotley peninsula.

yellow-wood. Add: (Earlier S. Afr. examples.) Also *attrib.*
 1767 in *Country Life* (1973) 7 June 1607/1 A mahogany Commode, with yellow wood ornaments, and Drawers for medals. **1790, 1801** [see *geelhout]. **1871** H. H. Dugmore *Reminisc. Albany Settler* 27 Plain dinners and hearty suppers,..served up in tin dishes on yellow wood benches. **1915** Rider Haggard *Holy Flower* iv. 63 It.. rolled under a great yellow-wood chest. **1952** G. M. Mills *First Ladies of Cape* 42 Cupboards..were usually framed by stinkwood and yellow-wood doors. **1981** *N. & Q.* June 193/2 The most important furnishings of the *groote kamer* were a yellow-wood cupboard and a stinkwood four-poster bed. **1985** *New Yorker* 18 Mar. 65/2 Old yellowwood and stinkwood chests.

Yemeni (ye·mĕni); *sb.* and *a.* [ad. Arab. *yamanī*, f. *Yemen* name of two States in the south-west of the Arabian peninsula.] **A.** *sb.* A native or inhabitant of North Yemen or South Yemen. **B.** *adj.* Of or pertaining to North Yemen, South Yemen, or the inhabitants of either.
 [**1888** C. M. Doughty *Trav. Arabia Deserta* II. 688 *Yémeny*, a man of el-Yémen.] **1916** *Handbk. Arabia* I. vi. 151 The Yemeni is not regarded as particularly fanatical. **1955** *Times* 1 July 11/6 Britain was obliged to deliver a strong Note protesting about border raiding. This pointed out..that in June there had been a serious attack on Mukeiras by a mixed force of Yemeni troops and tribesmen. **1959** W. Thesiger *Arabian Sands* xiii. 247, I had ridden..with two Arab companions and three.. Yemeni pilgrims. *Ibid.*, One of the Yemenis fetched us food from the market. **1959** *Listener* 27 Aug. 308/2 The southern Yemeni provinces. **1968** *Ibid.* 4 Jan. 7/1 A guerrilla attack on the outskirts of Sanaa for a few Yemini riyals. **1973** 'D. Jordan' *Nile Green* xxv. 105 Mara was wearing..a silver medallion... I decided it must be Yemini. **1982** P. Way *Belshazzar's Feast* xix. 208 Would these people have been Arabs, Yemeni Arabs? *Ibid.* xx. 223 The Soviets..have used their East German puppets, the Yemenis, to create trouble. **1982** M. A. Zabarah *Yemen* 53 Through the Yemeni emigrant and the Yemeni commercial entrepreneur, modern ideas were reaching not only city dwellers but also the tribes.

Yemenite (ye·mĕnəit), *sb.* and *a.* [Senses b, c. f. prec. + -ite[1].] **A.** *sb.* † **a.** [f. an earlier form of the personal name.] A member of a family belonging to the tribe of Benjamin. *Obs.* **b.** = *Yemeni *sb.* **c.** A Jew who was, or whose ancestors were, formerly resident in the Yemen.
 1566 Bible i. *Sam.* ix. 21 Am not I the sonne of a Jaminite of the smallest tribe of Israel: and my kyured is the leest of all the kyuredes of the trybe of Ben Jamin. **1568** *Ibid.*, Am not I the sonne of a Jeminite. **1864** J. T. Thomson *Some Glimpses Life Far East* lviii. 323 By race he was a Yemenite. **1902** *Encycl. Brit.* XXV. 518/2 The Yemenites rashly invited Turkish intervention. **1926** tr. *Granovsky's Land Prob. Palestine* i. 18 The history of Jewish colonization in Palestine shows many attempts to create a class of Jewish farm laborers..by establishing small settlements for Yemenites (Jewish immigrants from Southern Arabia). **1935** A. Revusky *Jews in Palestine* xii. 206 The Sephardim and Yemenites, the two main groups of Oriental Jews. **1965** M. Spark *Mandelbaum Gate* iv. 91 Those Arab girls, those Yemenites, Syrians, those Israelites, Samaritans. **1976** C. Bermant *Coming Home* ii. iii. 154 One of those private schools..basically.. for the sons of Arab oil sheiks, but it also housed..tall, lean Persians, tiny, dusky Yemenites, a few bombastic Greeks.
 B. *adj.* Of, pertaining to, or designating a Yemeni Arab or a Yemeni Jew.
 1876 R. D. Osborn *Islam under Arabs* iii. i. 296 The Yemenite Arabs of Syria were known as the Kelbites. **1902** *Encycl. Brit.* XXXI. 329/2 The present Sultan, a descendant of those Yemenite Imams who consolidated Arab power in Zanzibar. **1926** *Sunday at Home* July 636/1 Another industry which has been revived in Jerusalem is the silver filigree work of the Yemenite Jews. **1949** [see *gun moll* s.v. *gun *sb.* 15]. **1955** S. N. Eisenstadt in *Public Opinion Q.* XIX. 156 (*heading*) Communication in a traditional Yemenite community. **1978** *Church Times* 16 June 11/2 In the Flea Market [of Tel Aviv]..the Guide did not want us to miss the sight of Arabs buying from Yemenite salesmen. **1982** D. Wiltse *Wedding Guest* ix. 113 He was an Arab, but he could have been anything from Moroccan to Yemenite.

yen² (yen). *slang* (orig. *U.S.*). Also **yin, ying.** [Prob. of Chinese origin. The most likely etymon is Chinese (Cantonese) *yăn* craving; the forms *yin* and *ying* may reflect the Mandarin pronunciation *yĭn* of the same character. Reinforcement from *yen³ is possible. See also *yen-yen.
 See E. C. Knowlton in *Amer. Speech* (1961) XXXVI. 175–80 for further discussion and documentation of this word, *yen³, and *yen-yen.]
 1. The craving of a drug-addict for his drug (orig. for opium). (See also quots. 1929, 1937, 1974.)
 1876 H. A. Giles *Chinese Sketches* 115 Chinamen ask if an opium-smoker has the *yin* or not; meaning thereby, has he gradually increased his doses of opium until he has established a craving for the drug. **1891** A. W. Douthwaite *Opium Habit* 5 The frequent and regular repetition of this process of stimulation and depression induces the '*ying*', or carving, which is simply a demand by the nervous system for its accustomed stimulant, without which it is unable to properly perform its functions. **1912** D. Lowrie *My Life in Prison* vii. 79, I even saw two or three guys eat chloride o' lime to stop their yen. **1922** E. Murphy *Black Candle* ii. i. 113 When 'the black candle' is ready for lighting and the smoker has the *ying* upon him—that is to say the mad longing for indulgence —the procedure is like this [etc.]. **1929** Light & Torrance in *Arch. Internal Med.* XLIII. 210 If he falls asleep, which is often the case, he falls into a deep slumber well known as the 'yen'. **1933** [see *muggle[2]]. **1937** A. R. Lindesmith *Nature of Opiate Addiction* iv. 107 The drug user does not ordinarily find that his efforts to explain what he means by 'yen' (which signifies both withdrawal symptoms and desire for opiates) are very successful. *Ibid.*, The drug user says, 'he wakes up some morning with a yen.' **1948** F. Brown *Murder can be Fun* v. 78 He hadn't thought Wilkins would know a biological urge from an opium yen. **1974** M. C. Gerald *Pharmacol.* xiii. 251 Symptoms begin within 8 to 12 hours after the last dose... The addict experiences tearing, a running nose, sweating, yawning, and difficulty in sleeping. This restless sleep is commonly referred to as the 'yen'.
 2. *gen.* A craving, a yearning or longing.
 1906 H. Green *Actors' Boarding House* 248 He had a yen to gamble and bet high. **1928** J. O'Connor *Broadway Racketeers* ix. 107 The kid..had a burning yen for champagne and poker. **1932** S. Gibbons *Cold Comfort Farm* xx. 267 Ezra, who had a secret yen for horticulture. **1952** *Here & Now* (N.Z.) Jan. 19 This yen for a dog that will do everything has had a lot to do with the waning of pointers, setters and retrievers. **1961** *Time* 6 Jan. 4/2 The yen of Christian churchmen for achieving church unity is more pathetic then peculiar to behold. **1967** A. Christie *Endless Night* ii. 21 He'd got such a yen for a picture that he managed to get the money together. **1983** *Listener* 7 July 17/3 You write your music because you have a real yen to write it.
 Hence **yen** *v. intr.*, to crave for a drug; to yearn, desire strongly; **ye·nny** *a.*, affected by a craving for drugs.
 1919 Mencken *Amer. Lang.* iii. 93 A great many of them [*sc.* Chinese words] have remained California localisms, among them such verbs as *to yen* (to desire strongly, as a Chinaman desires opium). **1935** N. Ersine *Underworld & Prison Slang* 80 He's yenning for morph. **1936** F. M. Ford *Let.* 6 Sept. (1965) 261 Not that I particularly yen to mention the Deity, but that I believe that publishers should be as sadistically punished as possible. **1936** E. Pound *Let.* Sept. (1971) 282 Am afraid I got 'em stuck with some bad grub, but it was the only place I cd. count on being open... The violin player yenned toward another place, where I thought they wd. git stuck a price. **1953** W. Burroughs *Junkie* (1972) vi. 63 Nick is followed all the time now. You know yourself when a guy is yenning, he doesn't look behind him. He's running. **1975** H. White *Raincoast Chron.* (1976) 147/1 'We brought a bit of shit in with us to taper off on... Too bad she's all gone.' 'Yeah,' said Pat longingly, 'I sure get yenny sometimes.' **1977** *Times* 11 Feb. 12/2 The need for new educational certainties..cannot be met by yenning for the relative simplicities of the old 'elementary' education.

yen³ (yen). *U.S. slang* and *techn.* [Prob. a. Chinese (Cantonese) *yīn* opium, or (Mandarin) *yān* opium: cf. *yen² and *yen-yen.] **1.** Opium.
 1926 J. Colton *Shanghai Gesture* iii. 188 [Servant enters with..opium...] Here's the yen! **1935** A. J. Pollock *Underworld Speaks* 135/1 *Yen in the cheek*, gum opium or yen shee placed and sucked in back of lower teeth which produces comfort to the addict (this is frequently used when traveling in public conveyances). **1942** Berrey & Van den Bark *Amer. Thes. Slang* §509/2 Opium..yen.
 2. *attrib.* and *Comb.*, as (sense 1, with varying degrees of naturalization) **yen hock, hok** and varr., a needle used in the preparation of opium in the form of pills; **yen hop,** an opium pipe; **yen pock, pox,** and varr. (see quots. 1935 and 1959); **yen she(e)** and varr., the deposit of opium ashes formed in the bowl of an opium pipe; also *loosely*, opium; **yen siang, tsiang,** an opium pipe.
 1882 H. H. Kane *Opium-Smoking in Amer. & China* iii. 35 The other articles..for a smoker's outfit are..a needle (*yen hauck*) on the end of which the opium is taken up, 'cooked', and placed over the small opening in the upper surface of the bowl. **1886** T. Byrnes *Professional Criminals Amer.* 385 Among the frequenters of his place could be seen..such noted characters as..'Yen Hock' Harry, who earned his title by stabbing a man with a 'yen hock'. **1909** I. L. Nascher *Wretches of Povertyville* ii. v. 176 The needle or yen hok is merely a short knitting needle, sometimes with a handle. **1926** *Variety* 29 Dec. 7/4 The dopes and hop heads, with their 'stem', 'yen hok', [etc.]. **1955** *U.S. Senate Hearings* (1956) VIII. 4162 *Yen hock*, a long needle-shaped instrument, flat on one end and used to roll the 'pill' and hold same for cooking. **1968–70** *Current Slang* (Univ. S. Dakota) III–IV. 140 *Yen hok, n.*, a slender needle used in preparing opium for smoking. (Drug users' jargon). **1901** C. R. Wooldridge *Hands Up!* 215 It consists of the 'yen hop', or pipe, usually made of a section and a half of heavy bamboo. **1918** F. Hunt *Blown in by Draft* iii. 60 In the rare old fiction days 'corking a pill' had to do with yen hop, today it tells of naught but rolling a cigarette. **1934** *Detective Fiction Weekly* 21 Apr. 114/1 Yen pok, pill of opium after being prepared for smoking. **1935** A. J. Pollock *Underworld Speaks* 135/1 *Yen pock*, a cooked opium pill often eaten by addicts to produce normalcy and temporary relief. **1946** Mezzrow & Wolfe *Really Blues* xiv. 249 We'd..pack along some yen pox (opium pills that you eat). **1955** [see *mud *sb.*[1] 2 d]. **1959** W. Burroughs *Naked Lunch* 12 Yen pox is the ash of smoked opium. **1882** H. H. Kane *Opium-Smoking in Amer. & China* iii. 35 A straight and curved knife for cleaning the bowl of the ash (*yen tshi*) that rapidly collects and renders the pipe foul. **1892** H. Campbell *Darkness & Daylight* xxviii. 565 And a little box of tin held the *yen she* or bits of refuse opium. **1901** C. R. Wooldridge *Hands Up!* 215 The 'yen she gow', or small chisel, for cleaning out the bowl of the pipe. **1912** A. H. Lewis *Apaches of New York* xi. 229 Number-one hop is $87.50 a can, an' yen-chee..not less'n $32. **1918** *Policeman's Monthly* Oct. 16/3 In answer to this, it was learned that fifty-eight began by smoking opium..eight ate morphine, three ate 'yen shee', the ashes of opium, and the remaining cases started by using cocaine and laudanum, or eating opium. **1947** A. Meyers in J. H. Jackson *San Francisco Murders* 291 Liu uttered..a pathetic plea that he be allowed his daily pipe of 'yen-shee' or opium. **1952** J. Steinbeck *East of Eden* xix. 219 Odors from Chinatown, roasting pork and punk and black tobacco and yen shi. **1882** H. H. Kane *Opium-Smoking in Amer. & China* iii. 35 The whole pipe is called the *Yen Tsiang*, or opium pistol. **1909** J. S. Thompson *Chinese* viii. 336 At last its consistency suits. He places the gummy head on the large flute-like pipe, or *yen siang* (smoking pistol).

Yenan (yenæ·n). [Chinese (Pinyin) *Yan'an*.] The name of a town in northern Shaanxi province, China, which was the headquarters of the Chinese Communist Party in the years 1936–49, used *attrib.* to designate this period in the history of the Party, or to describe the principles and policies evolved by it at that time.
 1949 F. C. Jones *Manchuria since 1931* xii. 231 The Yenan régime had no hold upon the country in general before August 1945. **1957** P. S. H. Tang *Communist China Today: Domestic & Foreign Policies* v. 197 Thoroughness of rural regimentation characterized.. administration of the Shensi–Kansu–Ninghsia border region during the Yenan period. **1966** F. Schurmann *Ideology & Organization in Communist China* i. 59 The scattered guerrilla forces of the Yenan period had to report back to headquarters in systematic ways. **1970** E. Snow *Red China Today* (1971) xxxii. 257 After 1966 foreign ballets were seen no more in China; traditional Chinese opera was also overhauled to make it conform to Yenan principles. Mao Tse-tung's dicta, *Talks at the Yenan Forum on Art and Literature*, became the guidelines for stage and screen performances.. **1975** I. C. Y. Hsü *Rise of Mod. China* (ed. 2) xxiv. 714 The heart of the Yenan Way was the perfection of the mass line and the sharpening of revolutionary nationalism in the countryside, which became the twin pillars of Maoism. **1979** *Encounter* Feb. 74/1 The 'Great Leap Forward' led to a re-evaluation of the significance of the Yenan period of the late 1930s and early 1940s.

Yenisei (yeniséⁱ, yenīsēⁱ·). [a. the name of the river *Yenisei* in Siberia.] One of a group of Palæo-Siberian languages belonging to the Finno-Ugric group. Usu. in *Comb.*, esp. as

Yenisei-Ostiak, the designation of this linguistic group.
1888 *Encycl. Brit.* XXIV. 1/1 *Samoyedic. Yurak* and *Yenisei*, White-Sea to the Yenisei. 1908 T. G. TUCKER *Introd. Nat. Hist. Lang.* viii. 149 The Hyberborean speeches of Asia, some of which may or may not form a family, include..Yenissei-Ostiak (a tongue to be distinguished from the Ural-Altaic Ostiak, with which it agrees neither in its roots nor in the principle of vowel-harmony). 1932 W. L. GRAFF *Lang. & Languages* 406 The Yenisei-Ostiak variety is believed to be related to Tibeto-Chinese. 1939 L. H. GRAY *Foundations of Lang.* 369 The languages of the Uralic family are as follows:.. Samoyede group: Yurak, Yenisei-Samoyede, [etc.]. 1948 R. A. D. FORREST *Chinese Lang.* 22 A remarkable outlier of the Sinitic family, and more specifically of the Tibeto-Burman group, is..a group of dialects known as Yenisei-Ostiak and Kottish. They are now spoken by a few villagers far in the north of Siberia, on the river Yenisei, northwards of Yeniseisk. 1951 W. K. MATTHEWS *Languages U.S.S.R.* iii. 17 Yurak (Nenets)..is spoken from the Kanin peninsula to the estuary of the Yenisei river, Yenisei (Enets) along its lower course. 1958 A. S. C. ROSS *Etymology* i. 27 In the language called Yenisei-Ostyak..a variation of a kind very similar to M[oder]n E[nglish]..is found. 1967 [see *NENETS].

yenny, *a.*: see *YEN².

yenta (ye·ntā). *U.S.* Also **yente**, (*rare*) **yenteh.** [Yiddish, orig. a personal name.] A gossip or busybody; a noisy, vulgar person; a scolding woman or shrew.
1923 A. YEZIERSKA *Salome of Tenements* 12 The slattern *yentehs* lounging on the stoops..were transfigured. 1931 B. HECHT *Jew in Love* 122 Jesus God, you talk like a typical *yenta*. 1948 *Commentary* V. 500/1 *Yente* has become synonymous with noisiness and vulgarity, plus implications of rough good-heartedness. 1968 *Encounter* Sept. 27/1 *Yenta*, I am told, was a perfectly acceptably name for a lady, derived from the Italian *gentile*—until some ungracious *yenta* gave it a bad name. 1970 S. ELLIN *Bind* xxiii. 114 A couple of *yentas* got nothing better to do, they'll take a sunbath right by my window. 1975 *New Yorker* 24 Nov. 167/3 It is to the director's credit that she manages to hold down Doris Roberts' performance as the *yente*. 1978 I. B. SINGER *Shosha* ii. 38 You were always ready to trade me for the first available yenta.

yentz (yents), *v.* *U.S. slang.* [Yiddish, f. *yentzen* to copulate.] *trans.* To cheat, to swindle (see also quot. 1939). Also *fig.* Cf. *SCREW *v.* 6 d, 12*.
1930 *Amer. Mercury* Dec. 458/2 *Yentz*, to cheat. 'They try to yentz me out of me end.' 1939 *Amer. Speech* XIV. 240/2 *To yentz*, to cheat; to fornicate. 1969 S. J. PERELMAN in *Holiday* Mar. 104/4 The faintness one characteristically experiences on discovering that he has been yentzed. 1978 J. KRANTZ *Scruples* x. 276 'I don't *yentz* them,' Maggie explained, Coca-Cola-colored eyes all innocence, 'they just yentz themselves and I try not to run out of tape.'

yen-yen (ye·nyen). *U.S. slang.* Also † *inyun* [Prob. ad. Chinese (Cantonese) *yinyǎn* craving for opium, f. *yīn* opium + *yǎn* craving: cf. *YEN² and *YEN³.] A craving for opium, the 'opium-habit'.
1886 T. BYRNES *Professional Criminals Amer.* 385 A fiend suffering with the *inyun* is a man to be avoided. *Ibid.* 384, I was a victim to the opium habit, or, as the Chinese have it, *inyun fun*. 1892 H. CAMPBELL *Darkness & Daylight* xxviii. 569 'I've got the *yen-yen* (opium habit) the worst way', said one woman, 'and must have my pipe every night.' 1904 H. HAPGOOD *Autobiogr. of Thief* x. 207 Perhaps it was the sight or smell of the hop, but anyway I got the yen-yen and shook as in the ague. 1926 J. BLACK *You can't Win* xvii. 238 He [*sc.* the old Chinaman] was shaking with the 'yen yen', the hop habit. 1961 *Amer. Speech* XXXVI. 178 If Cantonese *yen yen* be regarded as the probable source of English *yen-yen* we may assume that the syllables represent the individual etymons for *yen* 'opium' and *yen* 'craving'.

ye olde (yī ōᵘld, ōᵘ·ldi), *a.* [f. *ye* graphic var. of THE *dem. adj.* (see Y (3)) + *OLDE *a.*] Employed to suggest (spurious) antiquity in collocations the other words of which are often also archaistically spelt. Also *absol.* as *sb.*, a building characterized by (spurious) antique furnishings.
1896 W. WROTH *London Pleasure Gardens* i. 56 A modern public-house. 'Ye olde Bagnigge Wells.' 1900 *Confectioners' Union Hand-bk.* 167 Ye olde English toffee. 1919 WODEHOUSE *Damsel in Distress* xxvi. 298 In London, when a gentlewoman becomes distressed..she collects about her two or three other distressed gentlewomen..and starts a tea-shop in the West-End, which she calls Ye Oak Leaf, Ye Olde Willow-Pattern, Ye Linden-Tree, or Ye Snug Harbour, according to personal taste. 1933 [see *OLDE *a.*]. 1951 'M. INNES' *Operation Pax* v. ii. 197 Not a tourist centre. Nothing ye olde. 1972 P. CLEIFE *Slick & Dead* vi. 36 The Inn was the complete trendy-contemporary Ye Olde—all ship's lanterns, copper pans, chintz and candelabra. 1977 *New Yorker* 16 May 107/1 Quincy Market..basically a suburban shopping mall done up in the instant charm of ye olde exposed brick.

yeoman. Add: **2. c.** *yeoman of signals* (earlier example). Also *ellipt.*
1898 KIPLING *Fleet in Being* 82 The Yeoman of Signals came to the captain's cabin at the regulation pace... 'Signal from the flagship, sir.' 1918 T. S. ELIOT *Let.* 13 Nov. in *Waste Land Drafts* (1971) p. xv, I was *sent for* by the Navy Intelligence, who said..that..they would make me a Chief Yeoman and raise me to a commission in a few months. 1978 H. WOUK *War & Remembrance* i. 6 My chief yeoman's got the logs and other records all lined up.

3. *yeoman warder* (later examples).
1947 *Tower of London* (Min. of Works) 13/2 The interior is shown to the public..on application to the Yeoman-Warder on duty. 1979 J. GARDNER *Nostradamus Traitor* i. 2 'You are a Beefeater, yes?' 'Yeoman Warder, Ma'am. Beefeater's a kind of nickname.'

yep, repr. a dial. (esp. U.S.) or vulgar pronunc. of YES. (Earlier and later examples.)
1891 *Harper's Mag.* Nov. 970 He gently and peacefully murmured, 'Yep'. 1907 J. W. SCHULTZ *My Life as an Indian* xxv. 284 'You must cut your hair.' 'Yep.'.. 'An' quit gamblin.' 'Yep.' 1926 J. GALSWORTHY in *Scribner's Mag.* Dec. 581/1 Their 'Yeahs!' and their 'Yeps!' Americans no longer said 'Yes' it seemed! 1962 E. BIRNEY *Sel. Poems* (1966) IV. 99 Yep ain't nothin we kin do. 1967 *Listener* 19 Jan. 96/3 'Would you like a cigarette?' I got a typically aggressive 'Yep'. 1977 *Time* 16 May 39/2 Yep, I let the American people down. 1979 R. JAFFE *Class Reunion* (1980) I. vi. 74 'Oh? You got a new car?'.. 'Yes... Look out the window.'

yer¹ (yɜ̄ɹ), repr. a dial. or vulgar pronunc. of YOU. (Examples.)
1848 [see YOU *pers. pron.* 4 a]. 1856 *Punch* 2 Feb. 41/1 (*caption*) That's how it was, yer see. 1867 [see *YERE]. 1880 A. E. HOUSMAN *Let.* 10 May (1971) 20 Yah! yer aint got no votes! 1916 G. B. SHAW *Pygmalion* I. 106 Theres manners f' yer! Te-oo banches o voylets trod into the mad. 1946 K. TENNANT *Lost Haven* (1947) iii. 50 Hey! Wait a minute... I want to see yer. 1978 *Hot Car* June 93/1 Brian *would* like to get in a plug for Micky Mees who helped with the heavy stuff. Good on yer, Mick.

yer² (yɜ̄ɹ), repr. a dial. or vulgar pronunc. of YOUR. (Examples.)
1814 [see YOUR *poss. pron.* and *a.* 2 a]. 1894 *Jrnl. Amer. Folk-Lore* VII. 148 She is gwine ter keep de house straight and yer britches mended. 1922 JOYCE *Ulysses* 419 Aweel, ye maun e'en gang yer gates. 1954 W. FAULKNER *Fable* 85 'Use yer boot,' the sergeant muttered. 1973 J. SPEIGHT *Thoughts of Chairman Alf* 26 Yer Queen should have a veto to..overrule Parliament... 'Cos she's born to rule. Not like yer Labour rubbish. 1980 *Herald* (Melbourne) 3 Apr. (City ed.) 2 Wouldn't it rot yer footy socks! Someone's about to contest Cazaly's musical mark.

yerba. (Earlier example of *yerba-maté*.)
1839 [see *PARAGUAY 1].

yerba buena (yɜ̄·ɹbā bu,e·nā). [Sp., lit. 'good herb': cf. YERBA.] A trailing perennial herb, *Satureja douglasii* (formerly *Micromeria chamissonis*) of the family Labiatæ, native to western North America and bearing aromatic leaves and white or purplish flowers.
1847 *Calif. Star* (San Francisco) 30 Jan. 2/3 The town [*sc.* San Francisco] takes its name from an herb to be found all around it which is said to make good tea; and possessing medicinal qualities; it is called good herb or Yerba Buena. 1882 B. HARTE *Flip, & Found at Blazing Star* 15 He seized a few of the young tender green leaves of the yerba buena vine..and ate them. 1915 ARMSTRONG & THORNBER *Western Wild Flowers* 436 Yerba Buena, Tea-vine..was used medicinally by California Indians, so it was called 'good herb' by the Mission Fathers, and is still used as a tea. 1935 J. STEINBECK *Tortilla Flat* xv. 261 Tea made from yerba buena will be good. 1975 *Islander* (Victoria, B.C.) 3 Aug. 3/3 Beside the pond speedwell..yerba buena [etc.].

yere, repr. a dial. (esp. U.S.) or vulgar pronunc. of HERE.
1867 *Harper's Mag.* Feb. 274/2 This yere is Colonel N— who wants ter know yer. 1907 J. W. SCHULTZ *My Life as an Indian* xxv. 284 You must..help me run this yere boardin' house. 1929 W. FAULKNER *Sartoris* IV. i. 270 Yere dey is, Cunnel. 1956 [see *JONG¹]. 1973 J. PATTINSON *Search Warrant* ii. 34 Like young folks is these yere days—long hair, beards.

yeri (ye·ri). [Russ.] The name of the Russian vowel ы, the twenty-eighth letter of the Russian alphabet.
1921 E. SAPIR *Language* ix. 212 Both nasalized vowels and the Slavic 'yeri' are demonstrably of secondary origin in Indo-European. 1977 *Word* 1972 XXVIII. 249 The /ɨ/ is a back unrounded vowel, similar to the Russian yeri.

yerk, yark, *sb.* Add: See also *YORK *sb.*²

Yerkish (yɜ̄·ɹkiʃ), *sb.* (and *a.*). [f. the name of R. M. *Yerkes* (1876–1956), U.S. primatologist + -ISH¹.] A sign language for chimpanzees based on geometric symbols, chiefly devised by E. C. von Glasersfeld for experimental purposes and first published in 1973. Also *attrib.* or as *adj.*
1973 D. M. RUMBAUGH et al. in *Behavior Research Methods & Instrumentation* Sept. 385/2 The study program..included the design of the language system (Yerkish). *Ibid.* 387/2 Each correlator links two items that are expressed in the Yerkish phrase or sentence. 1973 *Science* 16 Nov. 731 Each Yerkish word, or 'lexigram', is a distinctive geometric white symbol on a colored background. 1974 E. C. VON GLASERSFELD in *Amer. Jrnl. Computational Linguistics* (microfiche 12), The Yerkish language for nonhuman primates. 1974 *N.Y. Times* 29 May 52/3 Lana's [*sc.* a chimpanzee's] computerized language, called *Yerkish* in honor of the primate center's founder, Dr. Robert M. Yerkes. 1977 E. C. VON GLASERSFELD in D. M. Rumbaugh *Language Learning by Chimpanzee* v. 114 The grammar of Yerkish had to be kept as simple as possible for several reasons. 1980 *Times* 3 July 16/3 'Yerkish', an artificial language especially developed by Professor D. M. Rumbaugh at the Yerkes Primate Research Centre, Emory University, near Atlanta, United States, for a chimpanzee called Lana.

yerra (ye·rā), *int. Anglo-Ir.* Also **yerrah.** [Ir.] An asseverative oath or exclamation.
1892 E. LAWLESS *Grania* II. iii. 32 Yerra! give him his bit and his sup and his bed,..and 'tis all he wants. 1898 M. MACDONAGH *Irish Life & Character* ix. 156 'What's the matter with the old gentleman?' he was asked. 'What's the matter with him! Yerrah, look at the walk of him... Begor, he only touches the ground in an odd place.' 1914 JOYCE *Dubliners* 155 Yerra, sure the little hop-o'-my-thumb has forgotten all about it. 1939 —— *Finnegans Wake* 95 Yerra, why would he heed that old gasometer. 1958 B. BEHAN *Borstal Boy* III. 218 'Yerra, 'tis nothing,' said I, jovial and Irish, but making every move to work on a bit more. 1965 *N. Munster Antiquarian Jrnl.* IX. iv. 186 Yerra there's no good in talking to you. 1977 J. HODGINS *Invention of World* iii. 83 This is no ordinary boy... Yerra, this is a boy apart.

yerse (yɜ̄ɪs), repr. a non-Standard pronunc. of YES. Also **yers.**
1937 PARTRIDGE *Dict. Slang* 970/2 *Yers; yerse*, yes: sol.: C. 19–20. 1965 J. PORTER *Dover Two* vi. 69 Yerse, we was. Only they nabbed us first. 1969 J. GARDNER *Founder Member* ii. 16 'All three of us going to New York..' parried Boysie. 'Yerse,' said Griffin. 1980 P. G. WINSLOW *Counsellor Heart* v. 80 'A plant?'.. 'Yerse.' She described the plant. 1985 M. GILBERT *Long Journey Home* xii. 125 'Was it as obvious as that?' 'Yers. Well, you can always tell.'

yersinia (yɜ̄ɪsi·niā). *Bacteriol.* Also **Yersinia.** Pl. **-ae.** [mod.L. (J. J. Van Loghem 1944, in *Antonie van Leeuwenhoek* X. 15), f. the name of A. E. J. *Yersin* (1863–1943), Swiss-born French bacteriologist + -IA¹.] A bacterium of the genus *Yersinia* (formerly included in *Pasteurella*), which includes Gram-negative rods that are facultative anærobes present in many animals and causing plague and yersiniosis in man.
1967 *Acta Path. & Microbiol. Scand.* LXXI. 384 Morphologically, they were fairly easy to distinguish from Yersinia. 1982 E. A. GORZYNSKI in Milgrom & Flanagan *Med. Microbiol.* xxi. 316/1 Yersiniae are facultative intracellular parasites. 1983 *McGraw-Hill Yearbk. Sci. & Technol.* 482/1 Once isolation of a suspect yersinia has been accomplished, the biochemical identification ensues more logically.

Hence **yersinio·sis** [-OSIS], infection with or a disease caused by yersiniæ (other than *Y. pestis*, the cause of plague), which in man is self-limiting and usu. marked by lymphadenitis of the mesentery and ileitis or by enteritis and occurs chiefly in children and young adults.
1971 *Country Life* 2 Dec. 1530/1 In older leverets, parasitism may be the main problem, and in adults yersiniosis, formerly called pseudotuberculosis, a bacterial disease occurring in either the acute or chronic form, appears to be a common cause of death. 1983 *Brit. Med. Jrnl.* 27 Aug. 593/1 A large outbreak of yersiniosis in 1980 in a boys' school in Dorset was attributed to contact with a pig kept on the school farm.

yes, *adv.* (*sb.*²) Add: **1. c.** *yes and no*, in answer to a question to which it is difficult to reply: partly, perhaps, to a certain degree.
1873 C. M. YONGE *Pillars of House* I. vii. 154 'Do you come from his father?' 'Well—yes and no. His father is still in Oregon; but he and I have always been one.' 1896 'M. RUTHERFORD' *Clara Hopgood* xxii. 212, I said 'yes and no' and there is another side. 1933 W. S. MAUGHAM *Sheppey* III. 75 *Bessie:*..Expecting somebody? *Florrie:* Yes and no. 1964 R. PETRIE *Murder by Precedent* v. 78 'That's why you gave him a home?' he asked. 'Well yes and no.' 1981 B. MURPHY *Enigma Variations* xiii. 137 'Do you believe that if you continue seeing me you'll be damned?' 'Yes and no.'

d. *yes or no*: used *attrib.* (freq. hyphenated) to denote a question, etc., answerable by, or definable in terms of, *yes* or *no*; *spec.* in *Linguistics*. Also *ellipt.* as *yes-no.*
1924 [see *PRONOMINAL *a.* 2]. 1935 [see *OKEY-DOKE *a.*]. 1952 *Mind* LXI. 52 The yes-or-no question is the one we like to ask. 1957 D. L. BOLINGER in *Publ. Amer. Dialect Soc.* XXVIII. 24 Yes-no Qs are essentially true-false Qs. 1961 F. W. HOUSEHOLDER in Saporta & Bastian *Psycholinguistics* 17/1 Beside this graded kind of 'grammaticalness'..there is also an absolute yes-or-no type. 1963, 1964 [see *MORE C. *adv.* 2]. 1966 A. BATTERSBY *Math. in Management* vii. 180 This type of work [*sc.* simulation] lends itself very readily to electronic computation, because it comprises a number of repetitive 'loops' of instructions, controlled by simple 'yes-no' decisions. 1976 H. KEMELMAN *Wednesday Rabbi got Wet* xlvi. 257 The law is not a yes-or-no thing. 1977 *Times* 8 Feb. 17/1 The Government is proposing to put a single yes-no question to the voters of Scotland and Wales.

1979 *Economist* 16 June 98/2 Whitehall's traditional passion for compromise only makes for trouble in a crisis which needs yes-or-no decisions, fast. 1984 *Word* XXXV. 188 Sentences other than declaratives are broken up into a speech act operator (a wh-question operator, a yes-no question operator, a command operator, etc.) and a propositional kernel.

yes, *v.* Add: **2.** *trans.* To say 'yes' to or agree with (someone); to flatter by habitual assent. *U.S.*
1921 R. W. LARDNER in Mencken *Amer. Lang.* (rev. ed.) 393 He..crossed me up. I ast him for a hook and he yessed me and then throwed a fast one. 1928 J. P. MCEVOY *Show Girl* ix. 134 They yes you to death. 1933 *Times Lit. Suppl.* 23 Nov. 844/2 For Mr. Leonard..there is no fun in life as lived in the United States. Too much work..yessing the boss and dud liquor. 1935 [see *RAVE *sb.*³ 2 b]. 1945 S. LEWIS *Cass Timberlane* xvi. 89 A lot of bums are always yessing you..but me and Boone are good-enough friends to tell you the truth. 1983 *N.Y. Times* 23 Oct. 19/6 Mr. Reagan..is unable to get his proposal off the ground..; his aids yes him to death with plans.

yeshiva (yəʃī·vā). Also **yeshiba(h, yeshivah,** and with capital initial. Pl. **yeshivoth, yeshiva(h)s;** also **yeshibot.** [a. Heb. *yĕšībāh*, f. *yashav* to sit.] An Orthodox Jewish college or seminary; a Talmudic academy.
1851 *Living Age* XXIX. 154/2 The hope of seeing him one day decorated with the dignity of rabbi..will impel them cheerfully to make all the sacrifices which his outfit and partial support at the yeshibah (*academy*) entail. 1881 *Encycl. Brit.* XIII. 681/2 The rabbis received their education at the Yeshiboth ('sessions' of academies devoted to the Talmud, the *Shulchan Aruch*, and their commentators). 1904 *Sat. Rev.* 24 Sept. 404/1 The Jewish orphan..is brought up..at a yeshiva, or seminary, where the Talmud is almost the sole object of study. 1916 H. SACHER *Zionism & Jewish Future* 38 They had passed by so swift and drastic a revolution from the *Cheder* and the *Yeshiba* to the Gymnasium and the University, that the only culture they were interested in was German culture. 1926 S. ASH *Kiddush Ha-Shem* x. 88 Famed far and wide were the yeshivahs of Poland. 1949 KOESTLER *Promise & Fulfilment* III. i. 295 Israel's first Prime Minister..and many of the other political leaders, started their education in the *Yeshivot*, the religious schools of Russian Jewry. 1957 *Encycl. Brit.* XIII. 63D/1 Orthodox Judaism created the Rabbi Isaac Elchanan Theological seminary in New York (1896), which developed into the Yeshiva university, a liberal arts college. 1960 L. P. GARTNER *Jewish Immigrant in England, 1870–1914* ii. 39 Yekuthiel Sussmann Schlosser.. arrived in 1852 as an itinerant solicitor for a projected yeshibah in Kalish. 1965 J. A. MICHENER *Source* 461 Sometimes the yeshiva students contrived ingenious answers. *Ibid.* 705 He..tries always to attend one of the yeshivot. 1976 C. BERMANT *Coming Home* I. i. 17 Yeshivah students..had been equipped for a life of prayer, contemplation and study. 1981 *Amer. Speech* LVI. 3 Orthodox Jews are typically strict Sabbath observers who maintain their own religious day schools and yeshivas. 1985 *Listener* 3 Jan. 27/3 He overcomes this obstacle by making a donation to a yeshiva, or religious seminar.

ye·s-man. *colloq.* (orig. *U.S.*). Also **yes man.** [f. YES *adv.* + MAN *sb.*¹] A man who agrees from self-interest or fear with everything put to him by a superior; an obsequious subordinate.
1912 *Century Mag.* July 339/2 We're both yes-men, Edward. We've got to take orders now. 1924 H. C. WITWER in *Cosmopolitan* Apr. 69/2, I thoroughly enjoy.. the yes-men who hang about the executives and hold their jobs by simply being constantly affirmative. 1928 *Sunday Express* 15 July 3/6 Heeney is paying the strictest attention to his instructors, and in this respect is very different from Tunney, who directs his own training, and whose camp associates are all 'Yes' men. 1933 C. DAY LEWIS *Magnetic Mountain* 51 What do they believe in, these yellow yes-men. 1949 [see *closed shop* s.v. *CLOSED *ppl.a.* 3]. 1954 D. UNWIN *Governor's Wife* v. 119 Your administration..favours the boot-lickers, the sycophants, the yes-men who do as they're told and don't make trouble. 1959 [see *BAND-WAGON]. 1973 *Times* 31 Jan. 14/3 This is not a demand for 'yes men' but for common standards. 1979 J. WAINWRIGHT *Duty Elsewhere* i. 8 The heavies and the molls—the pimps and the yes-men.

Hence **ye·s-girl, -woman**, an obsequiously subordinate woman.
1930 K. BRUSH *Young Man of Manhattan* v. 56 Her attitude.., so respectful, so impressed, hardened his heart. 'Just a yes-girl!' he thought bitterly. 1933 S. LEWIS *Ann Vickers* xxxviii. 454 They want to boss a gang of meek yes-women or they want to be received socially, like princesses. 1937 H. G. WELLS *Star Begotten* vii. 128 All his most trusted henchmen, tools, stooges, subordinates, intimates, Watsons, yes-girls. 1942 E. WAUGH *Put out More Flags* i. 17 'That's right,' said the yes-woman. 1950 D. CUSACK *Morning Sacrifice* in 3 *Austral. Three-Act Plays* i. 188 It..Prepares them to accept all, question nothing, and grow into nice well-behaved yes-girls. 1979 M. SOAMES *Clementine Churchill* xvi. 235 Devoted and fiercely loyal, she never became a 'yes-woman'.

yessir (ye·səɹ, yes,sɜ̄·ɹ). *colloq.* [Repr. an informal pronunc. of *yes, sir*: see *SIR *sb.* 8 c.] **a.** = *yes, sir* s.v. *SIR *sb.* 8 c. **b.** A formula of assent to a superior. Occas. as *sb.*, a quick utterance of 'yes, sir'. Cf. *NOSSIR.
There is much variation in the placing of the main stress in both the ordinary use and in the verbal forms.
1913 R. BROOKE *Let.* 17 Sept. (1968) 511 That notebook... I lost it in British Columbia—yessir, isn't it *too* bloody. 1930 [see *NOSSIR]. 1931 M. ALLINGHAM *Look to Lady* xxi. 215 The girl vanished with a startled 'yessir'. 1933 D. L. SAYERS *Murder must Advertise* vi. 102 'Yessir.' Ginger grinned confidentially. 1936 WODEHOUSE *Laughing Gas* xxiv. 262 'And,' said George, 'it's yessir sure enough the picture of the dead wife he loved.' 1962 L. DEIGHTON *Ipcress File* xviii. 118 A couple of 'yessirs' when you know that 'not on your life' is the thing to say. 1968 *Globe & Mail* (Toronto) (Mag.) 17 Feb. 1 A perfect specimen, yessir, the Canadian Hercules. 1973 A. BROINOWSKI *Take One Ambassador* ix. 125 The commanding officer went past, giving you marks for..the crispness of your Yessir. 1982 M. NABB *Death of Dutchman* vii. 143 'Do you understand me?' 'Yessir!'

Hence **yes(-)si·r, yessi·r** *v.*, (*a*) *trans.*, to defer to (someone) as a superior; (*b*) *intr.*, to say 'yes, sir', esp. obsequiously; **yes-si·rring** *vbl. sb.*
1966 *Punch* 21 Sept. 454/2 Yessiring in the office is insufficient, he offers his clammy devotion to the whole family as well. 1968 L. DEIGHTON *Only when I Larf* i. 16 Imagine..yes sirring the boss until superannuation... Not me, man. I'm for the open road. 1977 'J. D. WHITE' *Salzburg Affair* xii. 104 He came in bowing and yes-sirring, although no one..was there to overhear. 1980 H. R. F. KEATING *Murder of Maharajah* xv. 182 You're used to people yessirring you left and right.

yes, siree (yes sɜ̄rī·). *U.S. colloq.* Also **yes sir-ee, yes sirree.** [f. YES *adv.* + SIRREE.] Yes indeed; certainly. Cf. *NO SIREE.
1846 *Dollar Newspaper* (Philadelphia) 1 July 3/4 'Will you take this man to be your lawful husband?' said the Justice; to which she responded with breathless haste, '*Yes, sir-ee*'. 1898 J. C. HARRIS *Tales of Home Folks* 225 Cassy Tatum! Yes, siree! The very gal! 1900 [see SIRREE]. 1924 H. J. LASKI *Let.* 29 July (1953) I. 638 For local colour he added that when you emphatically approve of an opinion you write 'yes, sirree' on it. 1927 J. N. MCILWRAITH *Kinsmen at War* xxvii. 277 Yes, siree, our army's been going ashore plundering and destroying helpless villages all along the sea-coast of the United States. 1956 B. HOLIDAY *Lady sings Blues* (1973) xxiv. 203 Yes siree bob, life is just a bowl of cherries.

yessum (yes'm). *U.S. dial.* contraction of *yes, ma'am* [cf. MA'AM], a polite form of assent addressed to a woman.
1913 *Dialect Notes* IV. 1 *Yessum*, adv., yes, ma'am. 1929 W. FAULKNER *Sound & Fury* 6 'Take his overcoat and overshoes off.'.. 'Yessum.' Versh said. 1938 M. K. RAWLINGS *Yearling* x. 88 'You feel all right?' she asked. 'Yessum. Sort o' weakified.' 1942 W. FAULKNER *Go down, Moses* 11 Miss Sophonsiba said..neighbors just a half day's ride apart ought not to go so long as Uncle Buck, and Uncle Buck said Yessum.

yest, also, an epistolary abbrev. of YESTERDAY.
1894 DUCHESS OF MARLBOROUGH *Let.* 21 May in R. S. Churchill *Winston S. Churchill* (1967) I. Compan. i. vii. 486 Your father returned yest morn and went to Harrow.

yester-. Add: *yester-tempest.*
1888 G. M. HOPKINS *Poems* (1967) 105 Delightfully the bright wind boisterous ropes, wrestles, beats earth bare Of yestertempest's creases.

yesterday, *adv.*, *sb.*, and *a.* Add: **A.** *adv.* **2.** *I was not born yesterday* (earlier examples).
1757 R. DEMERE *Let.* 10 Aug. in W. L. McDowell *Docs. relating to Indian Affairs* (1970) II. 398, I was not born Yesterday. 1837 MARRYAT *Snarleyyow* xii. 62, I was not born yesterday, as the saying is.

3. *fig.*, *colloq.* Used to suggest extreme urgency or impatience, esp. in phr. *to want* (or *need*) *something yesterday.*
1974 *Times* 7 Feb. 14/7 Attacks by Miss Brigid Brophy and her group, whose case was, roughly, that they wanted everything, and wanted it yesterday. 1978 D. A. STANWOOD *Memory of Eva Ryker* xxi. 196, I need the information yesterday. 1980 T. BARLING *Goodbye Piccadilly* xv. 309 Don't ask me... Just get us down there yesterday.

B. *sb.* **3. b.** In the possessive, as *yesterday's man*, a man, esp. a politician, whose career is finished or past its peak.
1966 'G. BLACK' *You want to die, Johnny?* ii. 27 John saw himself as one of yesterday's men, a survivor. 1972 *Guardian* 14 Jan. 13/8 Support for Nkrumuh still remains limited to his fellow tribesmen in the remote South-west and to those who fell off the high-living Fascist bandwagon when he was overthrown. These people are 'yesterday's men' in the eyes of most Ghanaians. 1979 *Jrnl. R. Soc. Arts* CXXVII. 349/2 The politicians..do not know if they will walk out as ministers..or as yesterday's men.

yet, *adv.* Add: **1. f.** Used as an ironic intensive at the end of a sentence, clause, etc. (imitating the use of Yiddish *noch*). *colloq.* (orig. *U.S.*).
1936 *Sat. Even. Post* 19 Dec. 11/3 'The only kinda men I want are ones who wouldn't be afraid to try out for Whiteman.' 'Whiteman yet!' scoffed one. 1943 M. SHULMAN *Barefoot Boy with Cheek* vii. 68 'Not just a little story, but a big story, and on the front page and with pictures.' There were admiring whistles and cries of 'Pictures, yet!' from the audience. 1957 *N.Y. Times* *Times Book Rev.* 17 Mar. 8 The counter-claim was dismissed (with costs yet). 1962 T. MEEHAN in *Sunday Times* 5 Aug. 20/3 And that spooky organ music they got piped in all over the place—E. Power Biggs instead of Muzak, yet. 1972 D. S. VISCOTT *Making of a Psychiatrist* ii. 32 You can bet your Phi Beta Kappa pin, junior year yet, that D. J. Marley knows exactly what to put down. 1975 *Times Lit. Suppl.* 7 Mar. 250/2 If you own a Beaumont and Fletcher folio, don't lend it. At least not to someone who will return it with extensive annotations (in ink yet!). 1980 *Oxford Times* 22 Aug. 13/3 The tracks include..'To Know Him is to Love Him' (with David Bowie on saxophone, yet!).

yeti (ye·ti). Also **Yeti.** [ad. Tibetan *yeh-teh* little manlike animal.] Native (Sherpa) name for a hypothetical ape-like animal whose tracks have supposedly been found in snow on the Himalayan mountains; = *Abominable Snowman* s.v. *ABOMINABLE *a.* 1 c.
1937 *Times* 13 Nov. 13/5 The Sherpas had no hesitation in pronouncing them [*sc.* tracks] to be those of a Snowman or 'Yeti'. 1951 *Times* 6 Dec. 5/7 Sen Tensing immediately pronounced them to be tracks of yetis or Abominable Snowmen... He describes it as half man half beast, about five feet six inches tall, covered with reddish-brown hair but with a hairless face. 1955 [see *Abominable Snowman* s.v. *ABOMINABLE *a.* 1 c]. 1956 C. EVANS *On Climbing* xiii. 185, I had heard that there was a *yeti* scalp there, and I wanted to see it. 1972 J. NAPIER (*title*) Bigfoot: the yeti and sasquatch in myth and reality. 1975 E. HILLARY *Nothing venture Nothing Win* xv. 238 Even in the Thyangboche Monastery—traditionally the source of much Yeti lore and Yeti sightings—we were unable to find anyone who had seen a Yeti.

yew, *sb.* Add: **3.** *yew-frond.*
1922 JOYCE *Ulysses* 491 The walls are tapestried with a paper of yewfronds.

yew (yū), repr. a vulgar pronunc. of YOU. Also **yewall** = *YOU-ALL *pers. pron.*
1890 KIPLING *Abaft Funnel* (1909) 272 'Do *yew* know,' as the Private Secretary said at Simla,..'it's remarkably hard for an Anglo-Indian to get along in England.' 1921 H. WILLIAMSON *Beautiful Years* 204 'Yew wait... Common as dirt, are we?.. Yew wait, young cocky-boy.' 'Yes, yew wait, yew slug-face, bag o' bones.' 1968 A. DIMENT *Great Spy Race* i. 8 How can I help yew? 1977 *Custom Car* Nov. 14/1 Thank yewall. 1981 P. MACDONALD *One Way Street* xix. 189 Yew just scoot, yew an' young John.

yé-yé (yeye), *a.* (*sb.*) Also **Ye-Ye.** [a. Fr., tr. of *yeah-yeah*, redupl. of *YEAH *adv.*] Designating or pertaining to the modern style of music, dress, etc., associated with France in the 1960s. Also applied to persons. Occas. as *sb.*
The redupl. form *yeah-yeah* freq. occurred in Eng. popular songs of the 1960s.
1960 *Daily Tel.* 24 Oct. 9/3, I found one dramatically *yé-yé* shop in old Lyons..which sells only British goods of a somewhat bizarre kind. 1966 *Guardian* 1 Apr. 10/7 In Paris, clothes are still a lot more class-divisive with Ye-Ye girls and debutantes thoroughly opposed. *Ibid.* 7 Apr. 8 Paris once had twenty music halls; now it has two—the Olympia, which caters more for *yé-yé* singers, and Bobino. *Ibid.*, The Bobino. is no place for *yé-yés*. 1967 *Sat. Rev.* 4 Mar. 49 Amplified like yé-yé music. 1968 *N.Y. Times* 3 July 30 The orchestra Chez Régine will play anything 'from yé-yé to regular music'. *Ibid.* 23 July 42 The name of a high-priced haute couture boutique here, run by Arlette Nastat, a forever yé-yé designer. 1972 M. GOLDBERG *Karamanov Equations* xvii. 161 He sipped a coffee and a Cinzano..watching the mini-skirted yé-yé girls and their hairy escorts parading by. 1979 J. WYLLIE *To catch Viper* viii. 49 Coca-Cola signs and bars, fashionable young women in miniskirts and teen-age *yé-yé* boys.

yez. For *Anglo-Irish* read *dial.* (esp. *Anglo-Ir.*) and add: Also **yeez.** (Later examples.)
1842 S. LOVER *Handy Andy* xxxiv. 280 Who are yiz at all, gintlemin? 1884 D. BOUCICAULT *Shaughraun* I. iv. 10/2 Away with yeez—hide! 1901 M. FRANKLIN *My Brilliant Career* xvii. 147, I have the table laid out for both of yez. 1908 J. JOYCE *Let.* 8 Dec. (1966) II. 226, I will send him very gladly if that will make yiz all happy and loving. 1939 —— *Finnegans Wake* 8 Now yiz are in the Willingdone Museyroom. 1962 D. PHILLIPS *Lichty Nichts* 30 Yez ur gitn a rare day. 1966 [see *JACK *sb.*¹ 1 d]. 1969 in Halpert & Story *Christmas Mumming in Newfoundland* 211 Some people will say, 'How many of yez?' and the janneys will shout back, 'Two or three.' 1977 *Transatlantic Rev.* LX. 147 'Aye, OK,' I says. 'How but? Did yiz arrange that afore?' 1977 *Sounds* 9 July 8/5 It's not going to be on general release and yez can't buy it at general outlets.

Yi (ī). [Chinese.] The name of a minority nationality in China, distributed over Yunnan, Sichuan, and Guizhou; = *LOLO. Also, the language of this people. Also *attrib.* or as *adj.*
1960 CHANG-TU HU et al. *China* v. 66 The Yi (Lolo) are located principally in the Liang Shan area on the borders of Szechwan and Yunnan. 1962 E. SNOW *Other Side of River* (1963) lxxviii. 596 Some of the Yi people in Yunnan planted grain as the American Indians did. 1968 [see *LOLO]. 1974 *Encycl. Brit. Micropædia* X. 818/3 The traditional Yi culture includes a primitive hoe-using agriculture, livestock herding, and hunting. 1978 *Nagel's Encycl.-Guide: China* 65 The Tibeto-Burmese group includes Tibetan, Yi (or Lolo), Hani and Tujia. 1979 *China Now* Mar./Apr. 16/1 The Yi people, like many peasant societies, still want lots of children. 1984 *National Geographic* Mar. 290/2 These people number about 750,000 and belong to the larger group of five million Yi scattered over a wider area.

‖**yichus** (yi·kəs, yi·χəs). Also **yiches.** [Yiddish f. Heb. *yiḥūs* pedigree.] Honour, prestige, status.
1907 tr. *Frank's Simon Eichelkatz* 431/2 *Yichus*, aristocracy; good family connection. 1927 *Amer. Mercury* X. 172/2 There was a steady demand from wealthy prospective fathers-in-law for professional men whose titles would add *yiches* (prestige) to their wealth. 1946 *Commentary* May 63/1 On the other hand our butcher's daughters..had even less *yichus*. 1964 S. BELLOW *Herzog* 86, I know you Herzogs and your *Yiches*. Don't give me that hoity-toity. *Ibid.* 141 All branches of the family had the caste madness of *yichus*. No life so barren ..that it didn't have imaginary dignities, honors to come. 1976 C. BERMANT *Coming Home* I. vii. 99 If I could become a doctor he might recover something of his former grace. A doctor meant *yichus*, social status, prestige.

yicker, var. *YIKKER *v.*

Yid. Add: Also † **Yit(t).** Substitute for def.: A (usu. offensive) name for a Jew. (Later examples.)
1874 HOTTEN *Slang Dict.* 344 *Yid*, or *Yit*, a Jew. *Yidden*, the Jewish people. The Jews use these terms very frequently. 1898 [see *SCHLEMIEL]. 1912 G. FRANKAU *One of Us* vi. 53 As the Yid Knows well the slump-signs ere the slump convulses. 1935 [see *CRAP *sb.*¹ 7 b]. 1940 E. POUND *Cantos* lii. 11 Sin drawing vengeance, poor yitts paying for—. 1946 KOESTLER *Thieves in Night* 279, I became a socialist because I hated the poor; and I became a Hebrew because I hated the Yid. 1963 V. NABOKOV *Gift* iii. 180 Then she went and married a yid. 1971 B. MALAMUD *Tenants* 203 Then they go to a synagogue late at night,..and make Yid noises, praying.

Yiddish, *sb.* (*a.*) (Earlier and later examples.)
1875 *New Era* 5 May 285 In fact, this corrupt German is known as Yiddish or Jewish German. 1917 *Edin. Rev.* Apr. 310 In 1903..Yiddish was solemnly proclaimed the Jewish national language. 1938 *Better English* Feb. 50 As a rule, Yiddish-speaking people do not move in higher society. 1939 [see *JUDÆO-, JUDEO-]. 1970 *Language* XLVI. 939 Standard Yiddish is the only variety taught in the schools. 1977 *Rolling Stone* 16 June 43/2 'Never point your gun at someone,' Prince clucked in a Yiddish accent. 1979 *Guardian* 28 May 11/8 The Jewish community of eastern Europe, the Yiddish-speaking, Ashkenazi quasi-state. 1981 G. CLARE *Last Waltz in Vienna* (1982) I. 11 He must also have spoken Yiddish, the language of the ghetto.

Yiddisher. Substitute for entry:
Yi·ddisher, *sb.* and *a.* Also **yiddisher.** [ad. G. *Jüdischer* Jew.] † **Yidisher. A.** *sb.* A Jew. Also *transf.* (cf. JEW *sb.* 2).
1859 MATSELL *Vocabulum* 97 *Yidisher*, a Jew. 1890 [in Dict.]. 1896 E. TURNER *Little Larrikin* xxiv. 292 But why..that agent..refused to take the premium.. beats me..for he's more than a bit of a yiddisher. 1931 R. CAMPBELL *Georgiad* iii. 62 Doctors much to praise in it can see And with the ancient Yiddishers agree. 1933 L. GOLDING *Magnolia Street* I. iii. 59 Can't he mind his own business, now..he's got hold of a stinking Yiddisher? 1976 *Publishers Weekly* 19 Apr. 81/3 With yiddishers, reasonableness and gentle satire, Vorspan.. sets out to convince you that a bit of suffering is good for you.

B. *adj.* Also **Yiddische**, etc. [ad. G. *jüdische* (inflectional form of *jüdisch*).]
1892 I. ZANGWILL *Childr. of Ghetto* I. iii. 94 At least, she would have starved in a Yiddishe country, not in a land of heathens. 1898 A. M. BINSTEAD *Pink 'Un & Pelican* xii. 276 One very enquiring Yiddisher youth stood munching a shoot of celery. 1925 J. YELLEN (*song-title*) My Yiddishe Momma. 1961 P. DE VRIES *Through Fields of Clover* i. 27 Jokes about hot pastrami... To say nothing of Yiddisher Mamas. 1965 D. S. DAVIS *Pale Betrayer* iv. 52 I'm a real *Yiddishe* mama. 1971 C. FICK *Danziger Transcript* 177, I laughed like a Yiddische baby. 1973 *Jewish Chron.* 18 May 15/3 A Clever Yiddisher boy working next door, heard strange noises. 1976 R. SANDERS in D. Villiers *Next Year in Jerusalem* 198 The young Irving Berlin composed both Italian and Yiddisha pastiche. 1979 *Guardian* 22 Mar. 9/1 In Israel..plangent Yiddischer mammas are passé.

Yi·ddishism. orig. *U.S.* [f. YIDDISH *sb.* (*a.*) + -ISM.] **a.** A linguistic feature influenced by or derived from Yiddish. **b.** Advocacy of Yiddish culture and language.
1926 *Amer. Mercury* VII. 207/1 Most Yidgin writers qualify their Yiddishisms with parenthetical English explanations. 1933 in A. A. Roback *Curiosities of Jewish Lit.* viii. 124 Doeblin sees his model in something on the lines of Yiddishism on a world scale. 1938 *Better Eng.* Feb. 50 No one has yet made an attempt to collect all these Yiddishisms into a single collection. 1962 *Amer. Speech* XXXVII. 202 The use of *better* with *should* here is another Yiddishism..repeated in the announcer's next sentence: 'Better we should stop the clock.' 1966 *New Society* 12 May 9/2 The idiom of the New Yorker—Gentile or Jew—is..full of translated Yiddishisms ('I should live so long', 'Who needs it?' 'You should pardon the expression' and 'Now he tells me'.) 1978 *Soviet Jewish Affairs* VIII. 73 Since Tsinberg's claim, there has been debate as to whether the sixteenth, seventeenth and eighteenth century 'Yiddishists' can properly be viewed as forerunners of modern Yiddishism. 1981 *Amer. Speech* LVI. 17 The noun *glitsh* is a Yiddishism..from the verb *glitshn* 'to slide'.

Yi·ddishist, *sb.* (*a.*) [f. YIDDISH *sb.* (*a.*) + -IST.] An adherent or supporter of Yiddishism (sense a); *spec.*, an advocate of the exclusive use of Yiddish by Jews. Also, a student of Yiddish language or literature. Also *attrib.* or as *adj.*
1917 *Edin. Rev.* Apr. 310 Russian Jewry rang for a time with the bitter controversy of the so-called Hebraists and Yiddishists. The Yiddishists..won... In 1903.. Yiddish was solemnly proclaimed the Jewish national language. 1933 *B'nai B'rith Mag.* XLVIII. 32/3 What will the Yiddishists..say to this? 1964 S. BELLOW *Herzog* 108 Her father had been..a member of the Arbeiter-Ring, a Yiddishist. 1970 *Language* XLVI. 938 The title of this book [*sc. The Field of Yiddish*] suggests that it is primarily of interest to Yiddishists. 1971 *Encycl. Judaica* X. 62/2 The Yiddish movement which came at the beginning of the 20th century, laid the accent on Yiddish as the spoken and living language of the vast majority of world Jewry. 1972 H. KEMELMAN *Monday Rabbi took Off* xlvii. 276 'My husband was a Yiddishist,' said Gittel stiffly, 'He did not speak the language out of principle.' 1976 I. B. SINGER in D. Villiers *Next Year in Jerusalem* 62 The leftist Yiddishists tried to identify Yiddish with the social revolution... Russia had promised its Jews a Yiddishist cultural autonomy. 1982 *Lang. Problems & Lang. Planning* VI. II. 208 Yiddishists apply the term [Yiddish] retroactively to all earlier stages of the language, until its very beginnings about a thousand years ago.

‖**Yiddishkeit** (yi·diʃkəit). Also **yiddishkeit.** [ad. Yiddish *yidishkeyt.*] = JEWISHNESS, JUDAISM 1; Yiddishness.
1892 I. ZANGWILL *Childr. of Ghetto* I. iii. 87 All they teach them in the school is English nonsense... The good Yiddishkeit goes to the wall. 1901 M. WOLFENSTEIN *Idylls of Gass* 25 They do say she reads German books with not a word of *Yiddishkeit* (Judaism) in them. 1956 'H. MACDIARMID' *Stony Limits & Scots Unbound* 28 A *Yiddishkeit* crutch. 1966 *New Statesman* 6 May 648/2 The most sophisticated Israelis are those who are often most attached to *yiddishkeit* and many accept the rabbinical regime as part of Jewish culture without which Israel would have no identity. 1976 B. WILLIAMS *Making of Manchester Jewry* xi. 271 A..strictly observant Jewish society based upon the exclusive *Yiddishkeit* of the Eastern European ghetto. 1984 *Listener* 20 Sept. 14/2 With them the immigrants brought their richest possessions: Yiddish and Yiddishkeit—their language and their way of life.

yield, *sb.* Add: **6.** Esp. under a stress greater than the yield stress; also, the stage in the progressive stressing and deformation of a body when the yield stress is reached.
1913 *Proc. R. Soc.* A. LXXXVIII. 464 Yield occurred.. while there was still a large margin of elasticity left in the side bars. 1925 J. CASE *Strength of Materials* xxxiv. 538 The drop of stress which occurs at yield with materials like wrought-iron and mild steel. 1967 J. G. RAMSAY *Folding & Fracturing of Rocks* vi. 314 The stress conditions which initiate plastic yield. 1981 C. HALL *Polymer Materials* iii. 73 The search for improved impact performance has more recently stimulated a similar systematic study of yield and fracture processes in polymers.

7. Special Comb.: **yield table** *Forestry*, a table giving (usu. with other information) the average value or volume of a species of timber that can be expected from unit area of woodland each successive year.
1888 W. WEISE (*title*) Yield tables for the Scotch pine. 1953 H. L. EDLIN *Forester's Handbk.* xiv. 222 Yield tables..show the likely rate of growth and timber yield of tree crops of a certain kind of tree, grown in a certain country. 1980 *Forestry* LIII. 23 These regressions were solved for the appropriate values of volume or volume increment from published yield tables for this species.

yield, *v.* Add: **III. 20.** *spec.* To deform inelastically; to undergo a large increase in strain without a corresponding increase in stress.
1900 *Phil. Mag.* L. 77 The assumption..that the material yields when one of the principal stresses reaches a certain amount. 1927 F. V. WARNOCK *Strength of Materials* iii. 46 At the point *C* the material has yielded a large amount, and the corresponding stress is known as the 'Yield Stress'. 1968 A. H. COTTRELL *Introd. Metallurgy* xxi. 395 Suppose that the central grain..has in fact so yielded (e.g. because of the stress-concentrating effect of a foreign inclusion in it), but its neighbours are still elastic.

IV. 22. **yield gap**, the excess rate of return of long-dated or undated Government stocks over that of ordinary shares; **yield-point**, substitute for def.: (the stress corresponding to) the point on a stress–strain diagram at which the strain begins to increase substantially without a corresponding increase in stress: in some metals differentiated as *upper yield point*, a point at which the stress ceases to increase as the strain increases, prior to a fall to the *lower yield point*, from which the strain increases while the stress remains almost constant at the lower value; also, esp. in *Geol.*, the elastic limit or the yield strength; **yield sign** *U.S.* = *GIVE-WAY SIGN; **yield strength**, in materials that do not exhibit a well-defined yield point, the stress at which (in addition to the elastic deformation) a definite amount of plastic deformation is produced (usu. taken as 0·2 per cent of the unstressed length); **yield stress**, the value of stress at a yield point or at the yield strength.
1959 *Economist* 25 Apr. 353/1 Investors today keep a careful eye on the 'yield gap'—the margin between dividend yields on ordinary shares and long term rates of interest set by the yield on irredeemable Consols. 1981 *Observer* 4 Oct. 19/2 It would also draw attention to the widening yield gap between gilts and shares. [1870 G. BERKELEY in *Exper. Mech. & Other Properties Steel* 4 Within the 'yielding point' of Steel the amount of lengthening from tension, or shortening from compression, produced by equal forces per unit of area is nearly the same.] 1886 K. PEARSON in I. Todhunter *Hist. Theory Elasticity & Strength of Materials* I. 887 When a bar is subjected to increasing traction, a certain stretch is reached after which there is a sudden and rapid increase of stretch... The point at which this change takes place is very marked, and various names have been suggested for it, as the limit of fatigue, the limit of stability, and the break-down point. The latter name brings out the character of the phenomenon, but at the same time suggests a point related to absolute strength or cohesion; I have therefore spoken of this point in the present work as the yield-point. 1919 FULLER & JOHNSTON *Appl. Mech.* II. x. 378 If the material is very ductile a yield point in torsion will appear at a torque somewhat higher than the elastic limit, similar to the yield point in tension. 1967 J. G. RAMSAY *Folding & Fracturing of Rocks* vi. 258 The specimen has..been permanently strained because the elastic limit has been exceeded. The point where this limit is first exceeded is known as the yield point. 1968 A. H. COTTRELL *Introd. Metallurgy* xxi. 390 In some materials..general yielding can begin in a very striking manner with a yield drop in which the applied stress falls, during yielding, from an upper yield point to a lower yield point. 1971 B. SCHARF *Engin. & its Lang.* iv. 23 Many metals such as aluminium, copper and brass have high ductility but no definite yield point (yield stress). 1981 *Pop. Hot Rodding* Feb. 66/1 To delve into this whole subject more deeply, we discussed it with SPS engineer Jack Schmidt, who spoke to us of tensile strength, yield points, and clamping loads. 1951 C. E. RIGGS in *Amer. City* June 133/1 On one street of each open intersection the sign reading *slow yield right of way* is erected... The new 'yield' signs are..of distinctive shape. 1977 J. CHEEVER *Falconer* 48 Putting up traffic signs, speeding signs, yield signs, stop signs. 1935 *Proc. Amer. Soc. Testing Materials* XXXVIII. 1315 *Yield strength*, the stress at which a material exhibits a specified limiting permanent set. 1967 *Times Rev. Industry* Feb. 45/2 Some British orders went overseas because of the inability..to cope with the more difficult combinations of pipe diameter, wall thickness and yield strength. 1982 *Materials Sci. & Engin.* LVI. 10/1 The yield strength of tempered lath martensitic 0·4% C steels is generally independent of packet size. 1913 *Proc. R. Soc.* A. LXXXVIII. 465 The observed stress at yield might..be below the true yield stress. 1954 C. W. MACGREGOR in W. R. Osgood *Residual Stresses in Metals* 110 Local yielding occurred with an applied uniform tensile stress considerably less than the yield stress. 1971 [see *yield-point* above]. 1973 C. R. BARRETT et al. *Princ. Engin. Materials* vi. 208 The applied tensile stress required to induce plastic behavior is known as the elastic limit or yield stress. 1973 J. G. TWEEDDALE *Materials Technol.* I. iv. 81 The yield stress is slightly above the elastic limit since it clearly represents the incidence of gross plastic strain.

yielding, *vbl. sb.* Add: **4.** (Later example.)
1922 W. SCHLICH *Man. Forestry* (ed. 4) I. 97 Timber fit for sawing would begin to be cut about 10 to 15 years later, and by the eightieth year the forests should be in full yielding.

6. (Further examples. Cf. *YIELD *v.* 20.)
1899 J. A. EWING *Strength of Materials* iii. 31 There is.. a well-marked yield point..at which extension goes on for a time through a considerable distance without increase of load. After this the extension becomes less rapid until the final yielding occurs just before rupture. 1961 LUBAHN & FELGAR *Plasticity & Creep in Metals* v. 104 When a piece of metal is loaded in such a way that the elastic stress is non-uniform,..yielding begins at the most highly stressed point. 1973 C. R. BARRETT et al. *Princ. Engin. Materials* viii. 257 The yielding usually starts at a small notch or irregularity in the sample and subsequently propagates throughout the sample.

yielding, *ppl. a.* **2.** (Later example.)
1849 C. LANMAN *Lett. Alleghany Mts.* xx. 159 The yielding wells are somewhat over two hundred feet deep.

yieldless, *a.* Delete † *Obs.* and add later example of sense 'not productive'.
1974 R. ADAMS *Shardik* xxxiv. 285 There was..something sinister about this place, unhusbanded and yieldless in the midst of the abundant land all about.

yieldy, *a.* **a.** Delete *Obs.* and add later U.S. local example.
1933 *Amer. Speech* VIII. I. 53/2 That 'ar west bottom never was much yieldy, nohow.

‖**Yigdal** (yi·gdāl). *Judaism.* Also 9 **Yigdol.** [Heb., = 'may he be magnified', the opening word of the hymn.] A Hebrew hymn, thought to have been composed by Daniel ben Judah (fl. *c* 1300), embodying the thirteen articles of the Jewish faith, and recited at morning prayer and on Sabbath and festival eves.
1845 *Jewish Chron.* 19 Sept. 244/1 The children sang in a beautiful manner the hymn Yigdol (Sabbath Hymn). 1892 I. ZANGWILL *Childr. Ghetto* I. xii. 269 You confound the air of the Passover *Yigdal* with the New Year ditto. 1907 J. JULIAN *Dict. Hymnol.* (ed. 2) II. 1149/2 The hymn [*sc. The God of Abraham Praise*] is a free rendering, with, as Olivers puts it, as decided 'a Christian character' as he could give to it, of the Hebrew *Yigdal* or Doxology, which rehearses in metrical form the thirteen articles of the Hebrew Creed.

Yi Hsing (ī ʃiŋ). Also **I-hsing, Yi-hsing.** [f. the name *Yi Xing* of a town in Jiangsu prov., China.] In full *Yi Hsing yao, Yi Hsing ware.* A type of unglazed stoneware (esp. for teapots) first produced at Yi Xing in the Song dynasty and reaching its height in the later part of the Ming dynasty.
1904 E. DILLON *Porcelain* x. 165 The Yi-hsing yao, made at a place of that name..includes the red unglazed ware. 1910 *Encycl. Brit.* V. 744/2 The manufacture of red teapots, mugs, bowls, cups, &c., in imitation of the Yi-Hsing-Yao was widespread during the late 17th and early 18th centuries under the name of red porcelain. 1915 R. L. HOBSON *Chinese Pott. & Porc.* xv. 178 The Yi-hsing wares in the celebrated Chinese ceramic collection formed by Augustus the Strong at Dresden supplied designs for the fine red stoneware made in the first years of the eighteenth century by Böttger. 1945 (see *BOCCARO]. 1970 *Ashmolean Mus.: Rep. Visitors 1969* 47 Tea-pot, I-hsing brown stoneware, Chinese, 18th century. 1971 L. A. BOGER *Dict. World Pott. & Porc.* 378/2 Yi Hsing Yao..an unglazed stoneware produced at Yi-hsing-hsien in Kiang-su province... Its greatest productive period was during the latter part of the Ming dynasty and during the Ch'ing period.

yike (yəik), *sb.*² *Austral. slang.* [Origin unknown.] An argument, a dispute; a fight, a brawl. Occas. as *v. intr.*
1940 *Mod. Standard Eng. Dict.* (rev. ed.) 697/2 *Yike*, v. to fight. 1945 R. RENE *Mo's Memoirs* 186 There's that tram connie having a yike with a drunk. 1951 D. STIVENS *Jimmy Brockett* 86 It was a pretty good yike while it lasted. 1952 T. A. G. HUNGERFORD *Ridge & River* 213 Don't let's yike about it. 1964 G. JOHNSTON *My Brother Jack* 244 Sorry your party ended up in a yike. 1976 *Sunday Sun* (Brisbane) 11 Apr. 6/2 ALP circles have scoffed at suggestions of a political 'yike' between State Opposition Leader Tom Burns and TLC chief Egerton. 1984 *Business Rev. Weekly* (Australia) 7–13 Jan. 18/1 We have had a couple of small yikes, mainly on things like contract prices.

yikes (yəiks), *int.* *colloq.* [Origin unknown, but cf. YOICKS *int.*] An exclamation of astonishment.
1971 *TV Comic* 5 June 8 Yikes! He's blown out the candles all right..blown them out of the cake! 1973 G. SIMS *Hunters Point* xiii. 115 Holding her nose..and exclaiming: 'Yikes! It seems that a cat has been shut up there.' 1978 *Detroit Free Press* 5 Mar. B 5/1 Yikes! Even Paul Newman loses the woman in this new breed of movies.

yikker (yi·kəɹ), *v.* Also **yicker.** [Echoic, f. *yik* + -ER⁵.] *intr.* Of a bird or other animal: to make repeated short, sharp cries.
1951 *Chambers's Jrnl.* Sept. 528/2 As we pushed through the bilberry bushes, rowan, and stunted Scots fir, a lemming yikkered angrily at the disturbance. 1959 W. K. RICHMOND *Brit. Birds of Prey* ix. 113 Sometimes he yikkers to himself as he goes,..a low, emphatic chatter. 1960 'L. LAMPLUGH' *Sixpenny Runner* xiii. 137 Good track dogs wait..yickering eagerly and ready to fly out. 1964 T. H. WHITE *Goshawk* III. 154 A pair of young badgers..greedily fought for warm milk and sugar out of a champagne bottle, and nipped my ankles yikkering when they were not nipping the rubber teat.

‖**yimkin** (yi·mkin), *adv.* *slang.* Also **yimpkin.** [a. Iraqi Arabic *yimkin.*] Perhaps (see also quot. 1966).
1925 FRASER & GIBBONS *Soldier & Sailor Words* 311 *Yimkin*, perhaps. An Arabic word used colloquially among troops on the Eastern Fronts. 1966 'L. LANE' *ABZ of Scouse* 120 *Yimkin*, nonsense; I don't believe it. 1967 *Sunday Times* (Colour Suppl.) 10 Sept. 46/4 Yimpkin, perhaps. Expressive of extreme scepticism. 'When Tunis falls we're all going home, yimpkin!' (Ar.)

‖**yin** (yin). Also **Yin, Yn.** [Chinese *yin* shade, feminine; the moon.] **a.** In Chinese philosophy, the feminine or negative principle (characterized by dark, wetness, cold, passivity, disintegration, etc.) of the two opposing cosmic forces into which creative energy divides and whose fusion in physical matter brings the phenomenal world into being. Also *attrib.* or as *adj.*, and *transf.* Cf. *YANG.
1671, etc. [see *YANG]. 1845, etc. [see *T'AI CHI 1]. 1850 [see *QI]. 1911 *Encycl. Brit.* XXIII. 68/1 The altar to the Earth is dark and square, on the north side of the city, the region of *yin*, the principle of cold and gloom. 1931 A. U. DILLEY *Oriental Rugs & Carpets* ix. 210 Other primitive motives are..male and female forms called Yin and Yang. 1963 'R. ERSKINE' *Passion Flowers in Italy* xi. 144 The things that woman was doing to us.. More truly Yin than you could believe. 1976 H. FERGUSON *Confessions of Long Distance Acid Head* 17 Lucy was so yin, receptive and feminine, that the passions, slumbering in my bosom, were consciously aroused.

b. *Comb.*, as **yin-yang**, the combination or fusion of the two cosmic forces; freq. *attrib.*, esp. as *yin-yang symbol*, a circle divided by an S-shaped line into a dark and a light segment, representing respectively *yin* and *yang*, each containing a 'seed' of the other.
1850 *Chinese Repository* XIX. 375 The Great Extreme ..is not exterior to or separate from the Yin-yáng. 1934 A. D. WALEY *Way & its Power* App. II. 112 The aim of the *yin-yang* philosophers was not the triumph of Light, but the attainment in human life of perfect balance between the two principles. 1958 W. WILLETTS *Chinese Art* I. iv. 273 The observed behaviour of this stellar couple accorded perfectly with the *yin yang* theory. 1972 MANAKA & URQUHART *Layman's Guide Acupuncture* (1977) I. 32 Their relativity and inseparability are symbolized by the inclusion, in the Chinese yin-yang symbol, of a small portion of each within the other. 1976 C. SOO *Chinese Art of T'ai Chi Ch'uan* iii. 19 When these [outlines] are put together, the Yin-Yang symbol is obtained. 1977 MILLER & SWIFT *Words & Women* iv. 69 The ancient belief that contrasting male and female forces are at work in everything—the yin-yang of dark and light..too easily becomes an adversary concept.

yin, ying, varr. *YEN².

‖**ying ch'ing** (yiŋ tʃiŋ). Also **Ying Ch'ing, Ying ch'ing**, etc. [Chinese, lit. 'shadowy blue'.] A type of glazed porcelain produced in Jiangxi and other provinces, chiefly during the Song dynasty. Freq. *attrib.*
1922 A. L. HETHERINGTON *Early Ceramic Wares China* xix. 139 The ware..with a very translucent, white sugary body and a bluish-white glaze tending to a more pronounced blue..is known as *ying ch'ing yao* .: *ying ching*.. may be translated 'shadowy blue'. 1934 *Burlington Mag.* May 214/1 The *ying ch'ing* species of white porcelain. 1936 *Ibid.* Jan. 10/1 With pale bluish glaze currently known as Ying Ch'ing (misty blue). 1943 [see *SUNG *sb.* b]. 1949 [see *JU]. 1954 H. GARNER *Oriental Blue & White* ii. 9 *Ying ch'ing* (shadow blue) is a thin translucent white porcelain covered with a clear glaze of bluish tint. The term is a modern one, invented by Chinese dealers, which has unfortunately become established in the West. 1977 O. IMPEY *Chinoiserie* II. vii. 89/2 The famous 'Font-hill vase', a Yüan dynasty *ying ch'ing* bottle fitted with a gothic silver-gilt and enamel mount.

Yinglish (yi·ngliʃ), *sb.* (*a.*) orig. *U.S.* [f. YI(DDISH *sb.* (*a.*) + E)NGLISH *sb.*] A jocular name for a blend of English and Yiddish spoken in the United States; a form of English containing many Yiddishisms. Also *attrib.* or as *adj.*
1951 W. & S. SCHACK in *Commentary* Dec. 586/2 A Jewish American theater in which..the material is of mixed nature, and the language neither the King's English nor the *rebbetzin's* Yiddish but a crossbreed that we might call 'Yinglish'. 1953 H. J. GANS in *Amer. Q.* V. 213 (*title*) The 'Yinglish' music of Mickey Katz. *Ibid.* 215 Katz's life is as Yinglish as the concept of a Bar Mitzvah ranch. 1967 *N.Y. Times* 6 Apr. 44 This show.. is a mixture of Yinglish (English with Yiddish) and Yidlish (Yiddish with English). 1968 [see *O.K. *adj.* a]. 1970 L. M. FEINSILVER *Taste of Yiddish* iii. 372 *Yinglish*. This coined term describes English that contains Yiddish idiom, pronunciation and/or intonation. 1974 *Observer* 31 Mar. 39/4 The text, written in Yinglish and American, abounds in euphemisms. 1983 *Listener* 7 July 20/3 One of the joys of the *Oxford American Dictionary* is searching out the progress of Yinglish.

yip (yip), *sb.* orig. *U.S.* [Echoic.] A short high-pitched cry, as from a dog; a shout, an exclamation; a complaint, an expostulation.
1911 H. QUICK *Yellowstone Nights* xii. 303 They chase 'em, with wild whoops an' yips over the undulatin' reservation. 1928 WODEHOUSE *Money for Nothing* ix. 208 If I'd been a life-insurance company I'd have paid up on him without a yip. 1945 B. MACDONALD *Egg & I* xii. 179 The dog began to bark and, guided by his excited yips, I was able to follow the progress of the hunt around the ranch. 1946 E. HODGINS *Mr. Blandings* iii. 48 This is the first faint yip of pain he's drawn. 1962 J. STEINBECK *Travels with Charley* 221 He [*sc.* a dog] ran..and laughed and gave little yips of pure joy. 1971 *Shankar's Weekly* (Delhi) 4 Apr. 5/1 In the old days, there would have been a spate of..discussions... But today? Not a yip.

yip, *v.* Restrict. *Obs.* or *dial.* to sense in Dict. and add: **2.** *intr.* To utter a sharp cry or yelp (with a dog or human being as subj.); to shout; to complain. orig. *U.S.*
1907 K. D. WIGGIN *New Chron. Rebecca* vii. 177 He would walk right up close and cuff 'em if they dared to yip. 1922 S. LEWIS *Babbitt* vii. 99 There's a swell bunch of Lizzie boys and lemon-suckers..that love to fire off their filthy mouths and yip that Mike Monday is vulgar and full of mush. 1927 P. MARKS *Lord of Himself* 15, I love to hear you yip at him. 1945 B. MACDONALD *Egg & I* ii. 176 Sport and the puppy..yapping and yipping at each other. 1963 *New Statesman* 11 Jan. 38/3 Yet how does the victim react? He yips with delight. 1978 S. RADLEY *Death & Maiden* i. 6 [The] Jack Russell terrier had yipped itself to the edge of hysteria.

3. *trans.* To cry or exclaim (with the words spoken as direct obj.). *U.S.*
1927 *Sat. Even. Post* 24 Dec. 84/2 'Hey!' Jim yipped,.. 'Get away from there!' 1974 WODEHOUSE *Aunts aren't Gentlemen* x. 81 'Has he brought it yet?' she yipped.

Hence **yi·pping** *vbl. sb.* and *ppl. a.*; also **yip-yipping.**
1910 *N.Y. Even. Post* 14 Oct. 4 The applause was really deafening..; not yip-yipping..but steady volume of vocal uproar. 1951 J. MASTERS *Nightrunners of Bengal* xv. 195 Women's voices rang clear..on the verandah, and a child's excited yipping. 1956 H. GOLD *Man who was not with It* (1965) xi. 85 The inner yipping of a man who had assaulted the Pittsburgh of his babyhood may have given them obscure desires to kick me. 1960 I. CROSS *Backward Sex* ii. 60, I could not help making a slight yipping noise as I tried to clear my throat. 1977 *New Yorker* 26 Sept. 42/1, I have spoken to those people about that yipping dog. 1980 A. DESAI *Clear Light of Day* i. 3 He [*sc.* a dog] has such a beautiful voice, it's a pleasure to hear him. Not like the yipping and yapping of other people's little lap dogs.

yip (yip), colloq. pronunc. of YES. Cf. YEP.
1934 'R. CROMPTON' *William—the Gangster* iii. 49 'Is that a fair down there?' 'Yip,' answered William. 1954 —— *William & Moon Rocket* i. 13 'That's fixed it,' said the youth. 'Yip,' said the man.

yippee (yipī, stress variable), *sb.* and *int.* orig. *U.S.* Also **yip-ee.** [Perh. connected with HIP *int.* (*sb.*⁴).] An exclamation of delight or excitement.
1920 S. LEWIS *Main St.* 86 She galloped down a block, and as she jumped from a curb across a welter of slush, she gave a student 'Yippee!' 1939 R. CHANDLER *Big Sleep* xii. 80, I was being brought into camp. I was going to yell 'Yippee!' 1947 N. MARSH *Final Curtain* xvi. 246 She said 'Yip-ee' like a cow-girl. 1951 J. FLEMING *Man who looked Back* xvi. 212 He permitted himself a loud 'Yippee!' 1961 *Guardian* 19 Apr. 5/1 Yippee. I've been blooded. It's lovely. 1976 BOTHAM & DONNELLY *Valentino* vii. 51 Rodolpho let rip a great cowboy yippee. 1980 A. CORNELISEN *Strangers & Pilgrims* viii. 162 It's a boy! *A boy!* Yippee!

Hence as *v. intr.*, to make this exclamation; **yippee-ing** *vbl. sb.*
1938 M. K. RAWLINGS *Yearling* xxvi. 351 They capered together and shouted and yippeed until their throats were hoarse. 1963 A. LUBBOCK *Austral. Roundabout* 182 There was bush ballads, and a whistling and yippeeing! 1977 'E. CRISPIN' *Glimpses of Moon* v. 69 Clarence Tully hilloed. His sons yippeed.

yippie, Yippie (yi·pi). orig. *U.S.* Also **yippy.** [f. the initials of *Y*outh *I*nternational *P*arty + -IE, influenced by *HIPPIE, HIPPY *sb.* and *a.*] A member of a group of politically active hippies, orig. in the United States.
1968 *Time* 5 Apr. 55/1 The Yippies—1968's version of the hippies... The term Yippie comes from Youth International Party. 1968 *Listener* 3 Oct. 428/2 One student outlines his own theories to me. 'This whole scene began with Dylan, the Beatles, and of course pot.' Another complains that the militants need a sense of humour and hopes the Yippies move in with their 'politics of ecstasy'. 1968 *Time* 11 Oct. 28 Pierson had infiltrated a yippie group known as the Headhunters, and soon rose to the dizzying position of personal bodyguard to the yippie leader. 1971 *Bulletin* (Sydney) 19 June 15/3 The news that Sydney's Yippies ('Yippie is a fun Revolution') were preparing to play an energetic part in the strenuously humorless Vietnam Moratorium came as a surprise. 1976 *Times* 18 Aug. 4/7 If we're going to save democracy, we've got to put an end to all this yippy filth, these abortions, [etc.]. 1981 J. DUNNING *Deadline* (1982) xiv. 138 Bill Neal was one of those yippie types... One of those bearded nonconformists.

yips (yips), *sb. pl.* *colloq.* [Origin obscure.] In *Golf*, a state of nervousness which causes a player to miss an easy putt in a competition. Usu. with the def. article.
1963 *Times* 10 June 4/2 His left-below-right putting stroke designed to prevent the 'yips', is most effective once it begins to flow. 1972 *Tel.* (Brisbane) 1 Jan. 5/7 Nevertheless, Jones got a dose of what golfers call 'the yips'. 1984 *Times* 21 Sept. 9/4 Golfers suffer from the 'yips', which means that their muscles seize up and freeze when they are faced with a short putt and they cannot play the stroke.

Yishuv (yiʃū·v, yi·ʃuv). Also **Yishub.** [ad. Heb. *yiššūḇ* settlement.] The Jewish community or settlement in Palestine during the nineteenth century and until the formation of the State of Israel in 1948.
1918 *Round Table* VIII. 321 The Jewish *Yishub*, or settlement in Palestine. 1922 *Encycl. Brit.* XXXII. 1130/1 This new Palestinian *Yishub* (settlement), strengthened in the early years of the present century by a number of young men and women who went to Palestine with the ideal of working as labourers on its soil, became the basis of the political success which Zionism achieved during the World War. 1940 *Contemp. Jewish Rec.* III. 599 The Yishub has made available over £P200,000 for emergency purposes. 1940 A. ULITZUR *Two Decades of Keren Hayesod* ii. 40 From the beginning of the Third *Aliyah* until the end of September, 1940..the Yishuv grew by about 432,000 persons. 1949 KOESTLER *Promise & Fulfilment* I. xii. 130 Let the Yishuv, the Diaspora and the whole world know what Bevin, Attlee and their henchmen are preparing for us. 1962 *Observer* 20 May 25/5 The British Zionists with whom I worked—infected, no doubt, by the free life of the Yishuv in Palestine—did not give me an inkling of what goes on in the mind of an orthodox Jew. 1970 I. SIEFF *Memoirs* vi. 107 These Jews of the old *Yishuv* (meaning settlement) were usually advanced in years before they left their country of origin. 1980 *Times Lit. Suppl.* 14 Nov. 1288/5 The extent and scale of the massacres tended to be underestimated for a while by Jews in the West.., by the *Yishuv* in Palestine, and even by Jews in Europe themselves.

yit, obs. form of YET. Add: Still current in *dial.* use. (Examples.)
c 1460 [see YET *adv.* 1]. 1857 H. S. RIDDELL *Book of Psalms in Lowland Scotch* lxviii. 61 Thouch ye hae læyne

amang the pats, yit sall ye be as the wings o' ane dow. **1880** [see *shoot *v.* 23 g]. **1908** Kipling *Lett. of Travel* (1920) 146 There's them that can't see *yit.* **1939** Joyce *Finnegans Wake* 138 An yit he wanna git all his flesch nuemaid. **1977** *Transatlantic Rev.* LX. 147 'Naw', he says. 'Least, no yit.'

Yit(t), varr. Yid in Dict. and Suppl. **yiz,** var. Yez in Dict. and Suppl.

Yizkor (yi·zkǫɹ). Also **Yiskor, Yizcor.** [Heb., lit. 'may he [*sc.* God] remember'.] A memorial service, formally known as *Hazkarat Neshamot,* held by Jews on certain holy days for deceased relatives, martyrs, etc.

1934 *Encycl. Jewish Knowl.* 644/1 *Yizkor,* popular name for Memorial Service, the full title being Hazcarath Neshoma. **1946** *Commentary* Aug. 173/1 They may still be on time to say *Yizcor.* **1956** S. Bellow *Seize Day* v. 86 He asked him whether he had reserved his seat in the synagogue for Yom Kippur... 'Well, you better hurry up if you expect to say Yiskor for your parents.' **1966** H. Kemelman *Saturday Rabbi went Hungry* viii. 46 The Memorial Service for the Dead..the Yizkor service. **1976** C. Bermant *Coming Home* II. v. 189 Four times a year special remembrance prayers, known as *yizkor,* are read in synagogue... The word means memorial.

-yl. Add: Also with pronunc. (ail). For 'F. *-yle* (also *-ule*)' read 'G. *-yl*'; for '(*Ann. de Chimie,* 1832, LI. 286)' read '(*Ann. der Pharm.* (1832) III. 262)'; and for *benzoyle* read *benzoyl.* Now in more formal use in *Organic Chem.* (see quots.).

1952 *Jrnl. Chem. Soc.* 5075 Rule 58.5. Radicals derived from amino-acids which have trivial names in *ine* by removal of OH from all —CH(NH$_2$)·CO$_2$H and related groups will be named by replacing the ending *ine* with *yl.* **1965** *Recommended Names for Chemicals used in Industry* (*B.S.I.*) 11 Univalent radicals derived from cycloalkanes with no side chain are named by replacing the ending '-ane' of the hydrocarbon name by '-yl'. **1966** [see *furyl]. **1971** *Nomencl. Org. Chem.* (I.U.P.A.C.) (ed. 3) A. 5 Univalent radicals derived from saturated unbranched acyclic hydrocarbons by removal of hydrogen from a terminal carbon atom are named by replacing the ending '-ane' of the name of the hydrocarbon by '-yl'. *Ibid.* B. 70 Univalent heterocyclic radicals whose names end in '-yl'. *Ibid.* C. 128 Radicals derived from unsubstituted ring assemblies are named by adding '-yl',..*etc.,* to the name of the assembly.

ylem (ī·lem). *Astr.* [f. med.L. *hȳlem,* acc. of *hȳlē* Hyle.] In the big-bang theory, the primordial matter of the universe, orig. conceived as composed of neutrons at high temperature and density.

[**1390**: see Hyle.]. **1948** R. A. Alpher in *Physical Rev.* LXXIV. 1581/1 Very shortly after the beginning of the universal expansion, the ylem was a gas of neutrons only. [*Note*] According to Webster's New International Dictionary..the word 'ylem' is an obsolete noun meaning 'The primordial substance from which the elements were formed.' It seems highly desirable that a word of so appropriate a meaning be resurrected. **1954** *Sci. Amer.* Mar. 61/2 As the Universe went on expanding and the temperature of ylem dropped, protons and neutrons began to stick together, forming deuterons..and heavier elements. **1959** J. Blish *Clash of Cymbals* vii. 171 The ylem was the primordial flux of neutrons out of which all else emerged. **1974** Frauenfelder & Henley *Subatomic Physics* xviii. 475 It is interpreted as the radiation that is left over from the primordial fireball and thus provides some information about the conditions in the ylem.

ylid (i·lid). *Chem.* Also **ylide** (-aid). [a. G. *ylid* (Wittig & Felletschin 1944, in *Ann. d. Chem.* DLV. 133): see -yl, -id⁴.] Any neutral compound containing a negatively charged carbon atom directly bonded to a positively charged atom of another element (commonly sulphur, phosphorus, or nitrogen). Hence **yli·dic** *a.*

1951 *Chem. Abstr.* XLV. 6166 The reaction of PhLi upon suitable quaternary ammonium salts results in compds. with a semipolar bond between N and adjacent C. These compds. are called ylides. **1970** *Nature* 25 July 335/2 Silylated ylides of phosphorus, arsenic and sulphur, in which the silyl group is both a stabilizing and an efficient leaving group, transferring the ylidic moiety in very mild conditions. **1972** S. J. Weininger *Contemporary Org. Chem.* xviii. 491 Ylids are extremely useful reactive intermediates for syntheses and are the subject of a good deal of research effort at the present time. **1979** *Nature* 15 Nov. 231/3 The zwitterionic ylides were obtained from the reaction of phenyl-lithium with quaternary ammonium halides. **1982** *Heterocycles* XIX. 1849 Cycloimmonium ylides..possessing two electron-withdrawing groups covalently bonded to the ylidic carbon..can be isolated.

-ylidene (-i·l-, -ai·lidīn), *suffix. Chem.* [f. as *-idene.] Used in place of *-idine when the name of the parent compound does not end in *-yl.*

1971 *Nomencl. Org. Chem.* (I.U.P.A.C.) (ed. 3) A. 17 Names of bivalent radicals derived from saturated or unsaturated monocyclic hydrocarbons by removal of two atoms of hydrogen from the same carbon atom of the ring are obtained by replacing the endings '-ane', '-ene', '-yne', by '-ylidene', '-enylidene' and '-ynylidene', respectively.

Yn: see *yin.

-yne (-ain), orig. var. -ine⁵, now used in its own right to denote a triple bond between carbon atoms, as in *butyne, *propyne.

1931 *Jrnl. Chem. Soc.* 1610 Definitive Report of the Committee [of the International Union of Chemistry] for the Reform of Nomenclature in Organic Chemistry... The name of hydrocarbons containing the triple linkage will end in *yne, diyne,* etc.

yo, obs. form of You: In mod. use, repr. dial. pronunc. of *you, your,* esp. in Black English.

1848 Mrs. Gaskell *Mary Barton* I. vi. 90 Yo stop here, and I'll be back in half-an-hour. **1897** [see *shoot *v.* 2 d]. **1931** W. Faulkner *Sanctuary* xxiv. 271 Minnie tapped at the door. 'Here yo dinner.' **1937** C. Himes *Black on Black* (1973) 139 Niggah, ef'n yo is talkin' tuh me, Ah ain' liss'nin'. **1969** R. Fair in A. Chapman *New Black Voices* (1972) 114 Oh, shut yo mouf up man. **1973** *Black World* June 61 Saturday nite take yo shoes off at the door.

yo (yōᵘ), also **yoe,** repr. dial. (esp. U.S.) pronunc. of Ewe *sb.*¹ Cf. Yow(e in Dict. and Suppl.

1891 *Dialect Notes* I. 71 *Yo,* ewe. **1899** B. W. Green *Word-bk. Virginia Folk-Speech* 434 *Yoe, n.*: yow, yeo; eow; yowe; a female sheep. 'Breeding yoes & a Ramm.' **1922** Blunden *Shepherd* 12 While each one came from the poor frightened yoes. **1946** *Amer. Speech* XXI. 98 S. Illinois] *Yo,* ewe.

yob (yǫb). *slang.* [Backslang for Boy *sb.*¹] Orig., a boy, a youth; in mod. use, a lout, a hooligan; (see also quot. 1918).

1859 Hotten *Dict. Slang* 131 *Yob,* a boy. **1886–96** in Farmer & Henley *Slang* (1903) VII. 375/1 And you bet that each gal, not to mention each yob, Didn't care how much ooftish it cost 'em per nob. **1908** A. N. Lyons *Arthur's* II. i. 108 It'd take more'n a yob in a squash 'at to call *me* a blighted sooper. **1918** Farrow *Dict. Milit. Terms* 673 *Yob,* a slang term used by soldiers meaning an officer or one who is easily fooled. **1927** J. C. Goodwin *Crook Pie* iii. 71 A yob shouted: 'Now for yer belts, boys!' and my friend was thrashed. **1930** P. MacDonald *Link* 130 Well, sir, I sez to meself, what does A do? Then I thinks, keep the mob off. So I jest backs into the door of the public, and doesn't let any of the yobs get out. **1957** J. Osborne *Look Back in Anger* I. 15 'Let's go to the pictures.'.. 'And have my enjoyment ruined by the Sunday night yobs in the front row?' **1962** J. Wain *Strike Father Dead* vii. 303 If you're expecting a description of what those yobs did to us, please forget it. **1977** *Western Morning News* 30 Aug. 1/3 One police officer sheltering from a bombardment of missiles behind a plastic shield said: 'It's just a bunch of yobs.' **1984** *Times* 16 Feb. 3/1, I would not want anybody looking at me to think this man is a thick, stupid, illiterate yob.

Hence **yo·bbery,** hooliganism; **yo·bbish** *a.* characteristic of a yob; **yo·bby** *a.,* loutish.

1955 E. Blishen *Roaring Boys* iv. 210 Though I don't hold with Grimes' yobby way of dressing, I think the Edwardian style's a good thing in itself. **1972** *Guardian* 16 Feb. 12/1 The 16-year-old boy..was.. adopting a yobbish air which drove his mother into nerve-twanging tantrums. **1974** *Times Lit. Suppl.* 1 Mar. 215/2 A comparative study of urban adolescent vandalism and upper-class yobbery. **1980** *Observer* 9 Nov. 12/4 It is one thing to deplore the collective yobbery of Mr. Benn's supporters. **1982** E. North *Ancient Enemies* ii. 19 The boys, although none of them as individuals are that dim or yobby, jeered. **1984** *Listener* 27 Sept. 22/1 A club with one of the worst records for drunken yobbery, Tottenham Hotspur,..brazenly makes a deal with the makers of Holstein lager. **1984** *Sunday Tel.* 2 Dec. 20/6 The loony Left should not be confused with that other Left which has been described as the Left of the yobbish tendency.

yobbo (yǫ·bo). *slang.* Also **yobo.** Pl. **yobbos,** occas. **yobboes.** [An extended form of prec.] A lout, a hooligan.

1922 *Contemp. Rev.* CXXII. 368 To him the boys are always the 'yobos'. **1938** *Evening News* 7 Mar. 11/5 A few inverted words have found common acceptance; slop (policeman), yob or yobbo (street rough, an inverted form of boy). **1940** R. Postgate *Verdict of Twelve* I. v. 74 Hardly any one about except a few yobos who had got nothing to do, and hung around in irritated idleness, spitting manfully in the gutter and telling dirty stories. **1955** E. Blishen *Roaring Boys* III. 147 'Yoboes!' said my colleague. 'My God, I wouldn't teach again in a senior school... Their insolence and rowdiness.' **1959** J. Braine *Vodi* xii. 165 Some yobbos in 1916 broke Nisbauer's shop window. **1960** *News Chron.* 9 Feb. 6/6 The local Teddies and yobbos swing their dubious weight behind the strike. **1964** in Hamblett & Deverson *Generation X* 56 Ever since that Profumo lark I've come to the conclusion that we working class yobos, as they like to call us, have less to be ashamed of than those establishment geezers. **1972** T. Stoppard *Jumpers* II. 80 That astronaut yobbo is good for twenty years hard. **1978** [see *street-corner* s.v. *street *sb.* 4 b]. **1982** *Age* (Melbourne) 4 Feb. 9/4 We get the odd guarded comment from the yobbo on the street about our dress.

yock (yǫk), *sb.* (and *v.*) *Theatr. slang* (chiefly *U.S.*). Also **yok.** (Cf. Eng. dial. *yocha* to laugh.] A laugh. Also as *v. intr.* (and quasi-*trans.*).

1938 H. M. Alexander *Strip Tease* 83 'Listen to 'em yock out there,' says the comic...'Yock' is a belly laugh. **1949** *N.Y. Times* 4 Sept. 7/6 It makes me furious when I have a corny line and it gets a yock. **1951** *New Yorker* 12 May 32/3 There'd be Don, yockin' it up like crazy,..he's so hysterical with loyalty laughter. **1957** S. J. Perelman *Road to Miltown* 73 Brother, I've heard some dillies in my day, but that's the payoff... What a yock this'll give the mob at Sardi's! **1957** Wodehouse *Over Seventy* xiii. 134 A few gay observations on the weather and he is ready for the big yoks. **1961** *Daily Mail* 4 Mar. 8 'I'm right in saying that you wouldn't do this to me, madam?' asked the colonel... The producer said, 'You'd have got a helpful yok from the audience there, colonel.' **1965** *New Yorker* 31 July 56/3 A chuckle or even a short, muted yock is acceptable from time to time. **1973** *Publishers Weekly* 26 Feb. 50/1 (Advt.), An hilarious collection of jokes, rhymes, riddles, tongue twisters, teasers and other assorted high-flying nonsense. The riddles and illustrations sprinkled throughout this page give just a brief taste of this yok-filled book. **1975** *New Yorker* 2 June 36/3 'Maybe you were a king in Babylon and I was a Christian slave,' I chaffed him. 'But enough with the yocks.'

yod. Add: Also **yodh. 1.** (Further examples.)

1861 *Grammatography* 10 Yodh. **1958** D. Diringer *Story of Aleph Beth* I. iii. 36 Several letters in the Aḥiram sarcophagus (*aleph, waw,..yodh,..*and *resh*) indicate a cursive rather than a monumental development. **1982** *Canad. Jrnl. Linguistics* XXVII. 76 Part II also has several appendices, dealing with Middle High German and Old Yiddish transcription, with the source of double waw and double yodh.

2. *Phonetics.* Also **jod.** The semi-vowel (y).

1934 M. K. Pope *From Lat. to Mod. Fr.* II. i. 55 The voiced fricative sound *j* (*jod*), heard in *bien* bjẽ. **1954** Pei *Dict. Linguistics* 237 *Yodization,* the changing of a pure-vowel (usually *e* or *i*) in hiatus into the semivowel which in English orthography is usually written *y,* and called *yod* after a letter of the Hebrew alphabet. **1971** *Canad. Jrnl. Linguistics* XVII. 19 Discussion is limited to two aspects of palatization in French: velars followed by front vowels, and clusters of single consonant plus yod. **1982** J. C. Wells *Accents of English* I. ii. 163 The *cure* vowel is frequently preceded by a yod, /j/... GenAm reflects more widespread Yod Dropping than RP and most other British accents. *Ibid.* II. v. 435 Yod Coalescence in stressed syllables is common in Dublin..; *dew* = *due* = *Jew.* **1984** *Word* XXXV. 70 The consequence of this for yod assibilation in Irish English is that it is not found in instances where it might be expected.

yodel, *v.* Add: **yodel(l)ing** *vbl. sb.,* **yodel(l)er** (earlier example of each).

1827 M. Wilmot *Jrnl.* 18 Aug. in *More Lett.* (1935) 288 The wild yodling of two young girls, who suddenly begun [*sic*] to sing their mountain melody. **1880** 'Mark Twain' *Tramp Abroad* xxviii. 289 Now the jodeler appeared.. and..we gave him a franc to jodel some more.

yodization (yǫdəizēɪ·ʃən). *Phonetics.* [f. *yod 2 + -ization.] = *yotization.

1954 [see *yod 2]. **1966** [see *prosody 2*]. **1982** *Times Lit. Suppl.* 3 Sept. 953/5 Such phonological phenomena as yodization and consonantal dissimilation.

yoe: see *yo, repr. Ewe *sb.*¹

yoga. Delete ‖ and add: Now a widespread cult in many countries outside India. (Further examples.)

1934 A. Huxley *Let.* 22 July (1969) 382 Some modification of this yoga technique may provide what's needed.. since it is as..independent of religion..as Freudism—many Indian yogis being in fact atheists. **1937** 'G. Orwell' *Road to Wigan Pier* xiii. 254 If only..every.. creeping Jesus [could be] sent home..to do his yoga exercises quietly! **1967** *Daily Tel.* 1 Feb. 13/1 All kinds of yoga (*the word means 'union' or 'joining'*) are practised, including yoga of action, wisdom, knowledge, devotion, sounds and higher faculties, but far the most common in Britain is hatha yoga, the philosophy of physical well-being. **1977** 'M. Yorke' *Cost of Silence* iv. 31 She had managed..to enrol..in a weekly yoga class. **1977** J. A. Kotarba in Douglas & Johnson *Existential Sociol.* ix. 266 Osteopaths, naturopaths, yoga instructors, acupuncturists, and anyone else who offers hope. **1979** R. Jaffe *Class Reunion* (1980) III. i. 308 She took all sorts of lessons: gourmet cooking,..macramé, origami, yoga. **1982** 'A. J. Quinnell' *Snap Shot* i. 27 She had immersed herself in the expatriate social routine... Ikebana classes on Tuesdays and Yoga on Thursdays. **1984** *Times* 17 Mar. 15/3 Yoga postures..are demonstrated by a small group.

Yogacara (yōᵘgātʃā·rā). [ad. Skr. *yogācāra* (also used), f. *yoga* Yoga + *ā-cāra* conduct, practice.] **a.** A school of Mahayana Buddhism which teaches that only consciousness is real.

1889 M. Monier-Williams *Buddhism* x. 225 Our present concern is..with the growth and development of mystical Buddhism in India.., through its connexion with the system of philosophy called Yoga and Yogācāra. **1932** M. Hiriyanna *Outl. Indian Philos.* ix. 219 Objects according to the Yogācāra are not..encountered by the mind, but are created by it. **1951** E. Conze *Buddhism* vii. 165 The tradition of Yogacara logic is still active in Tibet. *Ibid.,* Together with Buddhism, the Yogacara school disappeared from India about 1,100 A.D. **1960** J. Hewitt *Teach Yourself Yoga* 11 Buddhist meditation utilizes Yoga, especially the Yogacara school which lays emphasis on the trance. **1978** *Pacific Affairs* LI. 513 The author's equation of Yogacara philosophy with 'mind-only doctrine'..is, on the basis of Tibetan writings, open to doubt.

b. An adherent of the Yogacara school.

1915 R. W. Frazer *Indian Thought* ix. 181 This sect was known as Yogâcāras; the only thing they would admit the reality of was a series of thoughts or consciousnesses... As Yogâcāras, teachers of Yoga, they adhered to a Yoga system of philosophy. **1922** S. Dasgupta *Hist. Indian Philos.* I. x. 411 The Yogācāras or idealistic Buddhists..say that since we can come into touch with

knowledge and knowledge alone, what is the use of admitting an external world of objects as the data of sensation determining our knowledge? **1933** E. J. Thomas *Hist. Buddhist Thought* xvii. 246 It was among the Yogāchāras that Tantrism developed.

Hence **Yogaca·rin** = sense b above.

1951 E. Conze *Buddhism* vii. 161 It was the function and purpose of the Yogacarins to give due emphasis to the outlook on the world revealed by withdrawal into trance. **1974** *Encycl. Brit. Micropædia* X. 822/3 Yogācārins were not unique among Buddhists in practicing meditation.

yoghurt, var. Yogurt in Dict. and Suppl.

yogi. Add: **2.** = Yoga.

1925 A. Huxley *Let.* 25 Feb. (1969) 242 His little book..is a sort of explanation of the ethics of Christianity ...Lose one's life in order to gain it. It is the same idea as lies at the bottom of the Yogi system. **1943** D. Powell *Time to be Born* i. 22 He had no pot at all due to his Yogi exercises. **1952** 'R. Gordon' *Doctor in House* iv. 43 There was another medical student there, a man from St. Mary's who kept tropical fish in a tank in his bedroom and practised Yogi. **1960** R. Croft-Cooke *Thief* i. 9 One day he was in a shop with her when she put her bag down on a chair while she looked at something. I don't know how he got out with it, but then I'm not a thief. It seemed like Yogi to me, or at least conjuring.

Hence **yo·gified** *a. nonce-wd.,* treated in a yogic manner; **yogi-man** *poet.,* a devotee of yoga.

1938 L. MacNeice *Mod. Poetry* v. 78 In T. S. Eliot Hyde is the yogi-man. **1938** —— *Earth Compels* 58 It's no go the Yogi-Man, it's no go Blavatsky. *a* **1960** E. M. Forster *Maurice* (1971) 235 Carpenter's yogified mysticism.

Yogi Bear (yōᵘ·gi bēᵊɹ). The name of a popular U.S. cartoon character, used *attrib.* of toys, garments, etc., featuring the bear.

1960 *Newsweek* 18 July 84/2 Yogi Bear, who lives with a small bear-buddy, Boo-Boo, in a national preserve called Jellystone Park, spends a good deal of his time trying to cadge food. **1962** *Punch* 19 Dec. p. xviii, Marshall & Snelgrove's Toy Shop features..Yogi bear hobby sticks. **1963** *Sunday Express* 3 Mar. 15/4 Another baby bit a chunk out of a yogi bear hat. **1974** P. Dickinson *Poison Oracle* 188 The Shaikhah had easily found jeans and a Yogi-Bear tee-shirt to fit her.

yogibo·geybox. *nonce-wd.* [f. Yogi + Bogy, bogey + Box *sb.*²] The paraphernalia of a spiritualist.

1922 Joyce *Ulysses* 189 Yogibogeybox in Dawson Chambers. *Isis Unveiled.* Their Pali-book we tried to pawn... He thrones an Aztec logos, functioning on astral levels, their oversoul mahamahatma. **1965** *Spectator* 15 Jan. 73/1 Yeats, like AE..stood for the whirlpool, Madame Blavatsky and the yogibogeybox.

yogic (yōᵘ·gik), *a.* Also **Yogic.** [f. Yog(a + -ic.] Of or pertaining to yoga.

1921 Streeter & Appasamy *Sadhu* v. 136 The object of the Yogic trance is not the heart but the head. **1946** A. Huxley *Let.* 26 Oct. (1969) 551 In yogic practices there is a form of intense concentration which induces 'false samadhi', or self-hypnosis. **1960** J. Hewitt *Teach Yourself Yoga* 17 The Yogic internal cleansing method of swallowing a long strip of cloth, retaining it for a while in the stomach to absorb bile, etc. **1960** R. P. Jhabvala *Householder* i. 58 Yogic exercises, you know. He's getting quite good at them. **1967** *Listener* 7 Sept. 298/1 A few serious young people squatting in Yogic postures, spines straight, hands receptively cupped. **1980** *Dædalus* Spring 103 Repressed Hindus rejoice in myths of extreme forms of ascetic yogic mortification.

yogin. (In Dict. s.v. Yogi.) (Further examples.)

1959 *Times Lit. Suppl.* 27 Mar. 181/5 It is not to divide Time and Eternity that the yogin undertakes his arduous task of re-creation, but to restore to them their true relation. **1962** A. Huxley *Island* v. 39 The Yogin and the Stoic—two righteous egos who achieve their very considerable results by pretending, systematically, to be somebody else. **1965** P. Wylie *They both were Naked* I. iii. 131 Sitting opposite me on my divan cross-legged as a yogin etc. **1973** J. Blofeld *Secret & Sublime* i. 19 To yogins steeped in the arts of rejuvenation, prolonging life and achieving one of several kinds of immortality, it meant both of these. **1979** R. Cassilis *Arrow of God* IV. vii. 121 A little rent in the seamless garment of self-mastery. On, yogin, heal thyself!

yogini (yōᵘginī·). [a. Skr. *yoginī* (also used), fem. of *yogī,* f. *yoga* Yoga.] **a.** In India, a female demon or sorceress, esp. one of a group attendant on Durga or Siva. **b.** A female yogi.

1883 M. Williams *Relig. Thought & Life in India* vii. 188 Another class of manifestation is that of the Yoginis. These are sometimes represented as eight fairies or sorceresses..sometimes as mere forms of that goddess [*sc.* Durgā], sixty or sixty-five in number. **1910** *Encycl. Brit.* XIII. 512/1 The different classes of sorceresses and ogresses, called *Yoginis, Dakinis* and *Sakinis.* **1928** A. K. Coomaraswamy *Yakṣas* I. 9 The Seven Mothers.., the Sixty-four Joginīs,..and some forms of Devī..must have been Yakṣinīs. **1969** 'R. Farre' *Beckoning Land* xvi. 196 She wore an unbleached off-white sari and her black hair hung loose denoting that she had freed herself from worldly ties and that she was a yogini (a female yogi). **1972** B. N. Sharma *Social & Cultural Hist. N. India* iv. 76 There was a *Yoginī Saṁpradāya* among the *Sāktas* and many persons received the highest knowledge from the female ascetics. **1979** *Telegraph* (Brisbane) 15 Aug. 7/4 Meditation can cure many illnesses caused by mental and physical tension, according to a yogini (female yogi) visiting Brisbane.

yogurt. Delete ‖ and add: Now in the U.K. usu. with pronunc. (yǫ·gət). Also **yoghurt.** Now common in many English-speaking countries as a commercial semi-solid, often flavoured, foodstuff. (Further examples.)

1925 C. H. Browning *Bacterial.* vii. 154 'Yoghurt', which contains very little alcohol, is prepared by the Bulgarians, Greeks and Turks from cow's milk. **1934** E. Waugh *Handful of Dust* i. 13 Mrs. Beaver stood with her back to the fire, eating her morning yoghourt. She held the carton close under her chin and gobbled with a spoon. **1935** G. Freeman *Liberty Man* ii. 32 Miss Parrot, who ate only yoghourt for lunch, would do a quick change into the library. **1970** R. C. Lowell *Notebook* 188 Open books, yogurt traps in the unmade bed. **1982** *Sunday Times* (Colour Suppl.) 14 Sept. 85/4 Stokowski...was concerned with retaining his youth—and Garbo, always a food faddist, was into the 'yoga and yoghurt' experience.

Hence **yo·g(h)urty** *a.* (and varr.), fed on or smeared with yoghurt; containing or being yoghurt.

1981 *Times* 20 June 12/3 The dull and dispirited expressions that lie on their yoghurty faces. **1983** *N.Y. Mag.* 18 July 15 Not everything yoghurty performs magically—certainly, frozen yogurt hasn't in the last few years.

yohimbine. (Later examples.)

1977 E. J. Trimmer et al. *Visual Dict. Sex* (1978) vi. 65 Yohimbine is incorporated in a preparation that many doctors prescribe for patients with loss of sexual desire and ability. **1979** *Nature* 29 Feb. 600/1 (*caption*) Inhibition by yohimbine and idazoxan of the response of human platelets to adrenaline. **1983** *Guardian* 11 Aug. 18/4 Curare blocks acetylcholine receptors, another alkaloid, yohimbine, does the same for noradrenaline. **1984** *Observer* 9 Dec. 14/4 The detailed promotional material claims that the drug—containing yohimbine and strychnine—will improve both erections and sperm.

yok (yǫk). *slang.* [Yiddish, *goy reversed with unvoicing of final consonant.] A pejorative Jewish term for a non-Jew, a Gentile.

1923 ... *Children of Loneliness* ... there like a yok with her eyes in the air! **1960** *Times* 17 May 17/4 Mr. Faulk... said that on February 10, 1958, Mr. Daniels had said to Mr. Lincoln: 'Unless you join me and Mr. Jackson against that bloody Yok I will crush you, smash you and drag you into the gutter.' His Lordship asked the meaning of 'Yok' and was told that it was a Yiddish word meaning a Gentile, a rude way of saying a 'Goy'. A woman member of the jury.— It is not rude. **1962** R. Essex *Hot Potato* 34 My God, this could all be a Nasser plot. And you let this yok into our Intelligence camp! **1979** *Guardian* 11 July 8 ...the art are pretty smashing but then some of the yoks are fabulous. **1981** R. Samuel *East End Underworld* vii. 76 There were five Jewish boys in the gang—I was the only 'Yok'.

yok, var. *yock *sb.* (and *v.*).

yoke, *sb.*¹ Add: **I. 3. b.** (Earlier example.)

1880 *Girl's Own Paper* 20 Mar. 391/3 Make a new yoke and sleeves, and add a scarf to tie round the neck.

4. c. *Electr. Engin.* The part of a magnet or electromagnet that joins the poles or pole-pieces.

1884 S. P. Thompson *Dynamo-Electric Machinery* vii. 145 One such iron mass... is attached solidly to each pole-piece, and the two are united at the top by a still heavier yoke of iron. **1924** A. L. Cook *Elements Electr. Engin.* xi. 88 On the ring-shaped portion or yoke are inwardly projecting cores, which carry the exciting windings and have pole pieces, curved to fit the armature. **1951** R. B. Dome *Television Princ.* ix. 230 Another problem, that of high-voltage surges across the yoke tube during flyback, must be taken into account in designing the components. **1973** G. J. Fink *Electronics Engineers' Handbk.* xi. 22 For DPI deflection one common arrangement is to have the single-axis yoke rotated physically by an external motor..driven by the radar antenna.

d. *Aeronaut.* = *control column* s.v. *control *sb.* 5.

1934 in Webster. **1956** W. A. Heflin *U.S. Air Force Dict.* 576/2 *Yoke,*..a control column, esp. a dual control column. **1973** ... *Encounter at Kharmel* ix. 151 Ed eased back on the yoke, and the nose of the plane came up. **1977** *Miami Herald* 27 Mar. 20/5 A co-pilot immediately pulled back on the yoke.

9. (sense 1) *yoke-weary* adj.; (sense 3 b) *yoke-back;* **yoke-skey:** see also *jukskei; (earlier and later examples).

1895 *Montgomery Ward Catal.* Spring & Summer 278/3 Men's Overshirts. Gathered yoke back. **1981** *Country Life* 22 Jan. 226/2 Mr Tommy Nutter wears..a jacket of honey-toned Donegal tweed, yoke back, single-breasted. **1817** G. Barker *Diary* 29 May (MS.), Made yoke-skeys. **1835** A. Smith *Diary* 31 Mar. (1939) I. 338 The woman looked at her for a moment and then took up a yokeskey. **1934** B. Buchanan *Pioneer Days in Natal* 31 The straight yoke was placed across their necks and secured by the throat strap attached to the wooden yokeskeys. **1948** W. S. Chadwick *Mother Africa hits Back* i. 21 When transport riding in Barotseland I had two good but rather cheeky Zulu drivers. One night after an argument they threatened me with yoke-skeys and I got in the first blow. **1910** Kipling *Rewards & Fairies* 219 And a wet yoke-weary bullock Pushed in through the open door.

yoke (yōᵘk), *sb.*² *Anglo-Irish.* [Origin unknown.] A thing; a thingummy.

1910 P. W. Joyce *English as we speak it in Ireland* xiii. 352 Yoke; any article, contrivance, or apparatus, for use in some work. 'That's a *quare* yoke' (said by a countryman when he first saw a motor car). **1958** N. Fitzgerald *Student Body* v. 69 Dots stooped to pick up the gun... 'Where's the safety-catch on these yokes?' **1979** K. Dowling *Interface: Ireland* i. iii. 16 This fecking yoke is magnet up to here!

yoke, *v.* **8. b.** (Later examples.)

1765 Boswell *Let.* 11 May in *Corr. Boswell & Johnson* (1966) 167 At any trade I shall be as to you make as my Father calls it. **1920** R. Macaulay *Potterism* ii. i. 61 She is yoking together with an unbeliever.

yokel. Add: (Earlier *attrib.* example.) Also quasi-*adv.,* as *yokel-stubborn.*

1829 P. Egan *Boxiana* 2nd Ser. ii. 665 If he stood still, in his yokel attitude, he was laughed at by the spectators. **1935** L. MacNeice *Poems* 15 The moon's glare, Goggling yokel-stubborn.

yokeless (yōᵘ·k'les), *sb. nonce-wd.* [f. Yokel + -ess.] A female yokel.

1925 D. H. Lawrence *Reflect. on Death of Porcupine* 182 As for the yokel, his little stream may have flowed out of commonplace little hills, and been ready to mingle with the streams of any easy, puddly little yokeless.

yoke-mate. Delete Now *rare* and add later, chiefly *fig.,* examples. So **yoke-mated** *a.* (*rare*).

1882 G. M. Hopkins *Let.* 1 Jan. (1956) 162 One of our Fathers, who was..my yokemate on that laborious mission, died there yesterday night. **1911** Fletcher & Kipling *School Hist. England* 223 The very narrow of Youth's dream, and still, yoke-mate of wisest Age that worked her will! **1914** *Nineteenth Cent.* July 111 ... **1917** *Q. Rev.* Apr. 276 The Austrians... to the disgust of their German allies, did most of the fighting against their unequal yoke-mates, the Italians. **1930** R. Bridges *Test. Beauty* iv. 145 Where lay the harness'd bones of the yoke-mated oxen. **1973** L. Russell *Colonial Canada* iv. 46 If one of a pair were lost or killed, the other was usually unfit to work, because it would not function in unison with a new yoke-mate.

yoker. (Later example.)

1913 R. Harris *Boanerges* xxxvii. 341 Since Jason is alone, another pair of twin yokers will come to his assistance.

‖**yokozuna** (yōᵘkǫzū·nă). [Jap., f. *yoko* across + *tsuna,* f. *tsuna* rope, festoon, orig. a sacred straw festoon presented to a champion wrestler.] A grand champion sumo wrestler.

1966 *Manch. Guardian Weekly* 9 June 7 The last six bouts were the most important with two yokozuna (grand champions) and several ozeki (champions) fighting. **1972** *Times* 11 July 6/7 The only Yokozuna or grand champion at present active, Kitanofuji, was unable to take part in the Nagoya tournament because of injury. **1974** *Daily Tel.* (Colour Suppl.) 22 Feb. 40/3 The yokozuna is the embodiment of all the highest virtues of Sumo and no slightly less than brilliant manners will ever be allowed to degrade the status of the Grand Champion.

Yokuts (yōᵘ·kəts), *sb.* and *a.* [Yawelmani Yokuts *yokhoᵒc* of person, people.]

A. *sb.* **a.** (A member of) any of about 40 closely related Indian peoples of central California; these peoples collectively.

1877 *Contrib. N. Amer. Ethnol.* III. 370 At the time of the American advent.. the Yokuts occupied the south bank of the Fresno. **1929** A. H. Gayton *Yokuts & Western Mono Pottery-Making* 249 These sherds are related stylistically to southern California ware rather than to that of the Yokuts. **1973** A. H. Whiteford *N. American Indian Arts* 39 In southern California the baskets of the Yokuts and Mono have thin grass coils.

b. The language family of the Yokuts, or the language of any Yokuts group.

1912 [see *Penutian *sb.* and *a.*]. **1921** E. Sapir *Lang.* iv. 77 In another Indian language, Yokuts, vocalic modifications affect both noun and verb forms. **1965** *Canad. Jrnl. Linguistics* X. 139 Yokuts is a language family that is centered in the San Joaquin Valley and is flanked by languages of the Uto-Aztecan family.

B. *adj.* Of, pertaining to, or designating the Yokuts.

1877 *Contrib. N. Amer. Ethnol.* III. 370 In the Yokuts nation there appears to be more political solidarity.. than is common in the State. **1944** S. S. Newman (*title*) Yokuts language of California. **1974** *Encycl. Brit. Micropædia* X. 824/2 Yokuts ceremonies included puberty rite for boys, involving use of the hallucinogen tolguache, made from jimsonweed.

yolk, *sb.*¹ Add: **5.** Also **yolk-coloured** adj.; **yolk plug,** a mass of yolk cells partly filling the blastopore in the development of certain fish, amphibians, and insects.

1869 G. M. Hopkins *Jrnls. & Papers* (1959) 189 The other evening..there was a slash of glowing yolk-coloured sunset. [**1881** F. M. Balfour *Treat. Compar. Embryol.* II. vii. 102 Shortly after the stage represented in fig. 71 B, the plug of yolk, which fills up the opening of the blastopore, disappears, and the mesenteron communicates freely with the exterior by a small circular blastopore.] **1892** E. L. Mark tr. *O. Hertwig's Text-bk.*

Embryol. of Man & Mammals vi. 117 The inner lamella [of the Triton embryo]..is connected with the mass of yolk-cells..which lies like a wall in front of the blastopore and even projects into it as the Rusconian yolk-plug. **1916** [see *induction 9 b]. **1959** Southwood & Leston *Land & Water Bugs* 229 Fertile eggs [of the bug *Capsus ater*] commence to develop at once and the grey band of the yolk plug forms just below the operculum about 2 weeks after laying. **1980** *Jrnl. Exper. Zool.* CCXIV. 323 YSL [*sc.* yolk syncytial layer] cytoplasm which reaches the yolk plug during epiboly is not a part of the syncytium when gastrulation begins.

yolky, *a.*¹ (Later example of the form *yelky.*)

1918 W. J. Locke *Rough Road* ix. 104 With a hazardous plunge of his spoon he had made a yellow yelky horror of the egg-shell.

Yom Kippur (yǫm ki·pʊə). Also 9 **Jaumkipur.** [a. Heb. *Yōm Kippūr,* f. *yōm* day + *kippūr* atonement.] The Day of Atonement, the annual Jewish fast day of repentance and expiation that ends the Ten Days of Penitence; *Yom Kippur War,* an Arab-Israeli war that began on Yom Kippur on 6 October 1973 and ended in the same month.

1854 *Asmonean* 6 Oct. 198/3, I will tell thee now why the Jews in Gueseno do not wear the hillel at *Jaumkipur.* **1878** *Harper's* Oct. 728/2 This..is a *Shophar,* such as is blown in the synagogues on the Jewish New Year..and the Day of Atonement (Yom Kippur). **1907** I. Zangwill *Ghetto Comedies* 20 With a woman Yom Kippur is a wonder-working day. **1922** Joyce *Ulysses* 149 Their butteries and larders. I'd like to see them do the black fast Yom Kippur. **1942** *Contemp. Jewish Record* IV. 429/2 Last Yom Kippur..Nazi elite guards..evicted Jews from a synagogue in Szczucin. **1973** *Guardian* 29 Oct. 3/5 The Yom Kippur war had given them [*sc.* the superpowers] a greater hold over their respective clients. **1974** *Ann. Reg.* 1973 200 Yom Kippur (the Day of Atonement), when there is no public transport [in Israel] or private or public business and even non-observant Jews stay at home, fell on 6 October. **1980** 'E. Anthony' *Defector* v. 101 We [Russians] made very good use of the opportunities given to us by the Yom Kippur war. **1981** C. MacLeod *Palace Guard* xii. 89 'Have you ever in your life seen anybody eat the way he does?' 'Yes, my Uncle Hymie on the night after Yom Kippur.'

yomp (yǫmp), *v.* orig. *Mil.* [Origin unknown.

The word came into prominence when used by the Royal Marines during the Falklands conflict of 1982. It was subsequently identified with *yump *v.* by correspondents familiar with the terminology of rally driving, but whether correctly or not has not been confirmed.]

a. *intr.* To march with heavy equipment over difficult terrain. Also *fig.* **b.** *trans.* To cover (a certain distance) in this way.

1982 *Daily Tel.* 3 June 36/6 And always in the cold light of the Falklands dawn, the..Marines..have been ready to 'yomp on' for the next stage of the journey. **1982** *Observer* 20 June 3/2 Yomping round the sodden and trackless wastes of the countryside, I found the voters in less martial mood. **1983** *Listener* 20 Jan. 31/2 Mrs. Thatcher may begin yomping..around the hustings considerably sooner. **1983** *Guardian* 15 Apr. 10/1 Our men..who yomped all those miles in the Falklands. **1984** *Sunday Times* 14 Oct. 9/3 So the sweaty soldier yomping is battle ends up with blisters and a pool of water inside the boot.

So **yo·mping** *vbl. sb.* and *ppl. a.*; also **yo·mper.**

1982 *Daily Tel.* 3 June 1/1 (*heading*) Yompers surprise the enemy. *Ibid.,* Yomping they call it in the Royal Marines Commandos. It means marching, humping up to 120 lb. of equipment and all the arms needed for attack at the far end of the trek. **1982** *Standard* 6 Aug. 8/1 (Advt.), Yompers wanted..in teams of 3 for 40 mile mountain 24-hour walk. **1983** *Financial Times* 28 Jan. 16/6 Will robots replace the 'yomping' soldier in the army of the future?

yom tov (yǫm tǫv). Also **yomtov,** 9 **yontef,** and with capital initial(s). [Yiddish, f. Heb. *yōm* day + *ṭōḇ* good.] A Jewish holiday or holy day.

1854 *Asmonean* 7 July 96/2 Now, Shlome's grandfather was a kind of careless man who preferred making 'yontef' throughout the year. **1892** I. Zangwill *Childr. Ghetto* II. 166 He had been so proud of having earned enough money to make a good *Yomtov.* **1933** A. van Son tr. G. van Bruggen in J. Leftwich *Yisröel* 1012 A Jewish woman cannot be too ill to hear the Seder... That's right, it's Yomtov now. **1962** B. Abrahams tr. *Life Glückel of Hameln* iii. 51 Before we eat I must first take my child..some food. It is Yom Tov. **1976** M. Horovitz in D. Villiers *Next Year in Jerusalem* 113 Quite often, in the midst of supra-national poetry events, I find myself involuntarily imagining *yom tov.*

Yomud (yǫ·mʊd, yǫmū·d). Also 9 **Yimoot; Yomut.** [Native name.] **a.** (A member of) an ethnic group of Turkmen people (once a nomadic tribe) inhabiting the Turkmen Soviet Socialist Republic of the U.S.S.R. Also *attrib.*

1834 A. Conolly *Journey India Overland* I. iv. 35 The large tribe of Yimoot occupy the banks of the Goorgaun river. **1864** A. Vámbéry *Trav. Central Asia* xvi. 307 The Yomuts inhabit the East shore of the Caspian Sea and some of its islands. **1889** G. N. Curzon *Russia in Central Asia* viii. 275 The Turkomans under Khivan rule are Yomuds, Chadars, Emrali, Ata and Alili. **1938** E. O. Lorimer tr. *Krist's Alone through Forbidden Land* ii. 38 We had reached Qala-Qaya, the standing camp or village of the Yomut. **1963** L. Krader *Peoples Central Asia* iii. 58 Yomud had an alternative division, based on subsistence type rather than genealogy. **1974** *Even. Stand.* 12 Feb. 48/5 (Advt.), Superb oriental carpets & rugs. Including:.. A special collection of unique Persian Nomadic rugs from the Yomut..and Quashgai tribes.

b. *attrib.* Designating rugs made by this people, often distinguished by a diamond-shaped motif.

1900 J. K. Mumford *Oriental Rugs* xii. 234 One might reasonably say, looking at some of the Yomud rugs, that they had come from the Shirvan or Dagestan looms. **1940** A. B. Thacher *Turkoman Rugs* 24 This type of juval is the aristocrat of Yomud rugs in respect to density of knotting. **1974** *Encycl. Brit. Micropædia* X. 825/2 The large Yomud carpets are entirely of wool or of goat hair.

yon, *dem. adv.* Restrict *Obs.* exc. *dial.* to sense a and add: **b.** *hither and yon*: (non-*dial.* examples).

1903 H. James *Ambassadors* II. v. 53 What carried him hither and yon was an admirable theory that nothing he could do would not be in some manner related to what he fundamentally had on hand. **1939** W. S. Maugham *Christmas Holiday* iv. 94 The wan characters of Chekov's stories drifted hither and yon at the breath of circumstance like dead leaves before the wind. **1963** Bird & Hutton-Stott *Veteran Motor-Car* 52 Untidy chain or belt drives running hither and yon. **1978** *Nature* 27 Apr. 768/2 It is inaccurate, full of fanciful and unilluminating analogies, infuriatingly unsystematic, and skims hither and yon over the surface of the subject.

yonder, *sb.* Restrict (*nonce-use*) to sense in Dict. and add: **2.** [After quot. 1939.] The far and trackless distance; usu. with preceding adj.

1939 R. Crawford *Army Air Corps* (song), Off we go in to the wild blue yonder, Climbing high into the sun. **1948** *N.Y. World Telegram* 30 Dec. 11/6 A pilot..took wing into that wonderful yonder on a training flight. **1967** C. Cockburn *I, Claud* xxxiii. 410 The ex-editor of *The Week* had suddenly appeared out of the deep green yonder of Ireland. **1974** *Times* 26 Feb. 12/3 Mr. Wilson's ..policy for controlling inflation, which consists of holding down prices by law while letting wages go up, up into the wide blue yonder. **1979** 'D. Kyle' *Green River High* viii. 103 My father had vanished into the great green yonder of a million square miles of jungle. **1985** W. Golding *Egyptian Journal* iv. 57 Minya is a centre for scarpering, for fading away, for disappearing into the blue yonder.

yonks (yǫŋks). *slang.* [Origin unknown.] A long time, 'ages'; chiefly in phr. *for yonks.*

1968 *Daily Mirror* 27 Aug. 7/1, I rang singer Julie Driscoll... She said: 'I haven't heard from you for yonks.' **1977** 'J. Gash' *Judas Pair* iv. 54 Any man that says he can remain celibate for yonks on end is not quite telling the truth. **1980** *Oxford Times* 20 June 18 Even though Gabriel left Genesis yonks ago, his music and particularly his vocals remind one of Genesis. **1984** *Listener* 10 May 32/2 The English have been writing poetry for yonks, and have become damned good at it too. **1985** A. Blond *Book Book* ix. 142 Nicholas Bagnall and David Holloway have run the *Telegraph*'s book pages for yonks.

yonnie (yǫ·ni). *Austral. slang.* [Origin unknown.] A small stone; a pebble.

1941 Baker *Dict. Austral. Slang* 84 *Yonnie,* a small stone, a pebble. **1979** *Sun-Herald* (Sydney) 18 Mar. 79 There were two lamp posts..each equipped with one electric bulb... Young couples courting would smash each of them with a well-aimed 'yonnie' on pay night.

yoo-hoo (yū·hū), *int.* (*sb.*) [Cf. Yoho *int.*] A call made to attract attention, esp. to one's arrival or presence; also as *sb.*

1924 *Dialect Notes* V. 280 *Yoo-hoo* (call). **1926** *New Yorker* 2 Jan. 18/3 Yoo-hoo! When did your school let out? **1937** M. Allingham *Case of Late Pig* vii. 49 He opened the breakfast-room door. 'Yoo-hoo!' said someone inside. **1946** A. Marshall in Murdoch & Drake-Brockman *Austral. Short Stories* (1951) 316 There was a faint 'yoo-hoo!' from behind us. We all turned. **1959** L. Lee *Cider with Rosie* 150 I'm coming—yoo-hoo! Just mislaid my gloves. **1959** A. Wesker *Roots* I. 16 Girl's Voice (off): Yoo-hoo! Any one home? **1970** J. Updike *Beck* 182 Mildred..waved an alabaster, muscular arm: 'Yoo-hoo, Henry, over here.' **1973** E. Bullins *Theme is Blackness* 62 Yoo Hoo up there! Someone will be up to give you a hand with the rest of those things. **1983** 'J. Gash' *Sleepers of Erin* viii. 61 Patrick..trilled a roguish yoo-hoo.

Hence **yoo-hoo, yoo--hoo** *v. intr.* and *trans.,* to call 'yoo-hoo!' (to); **yoohoo-ing** *vbl. sb.*

1948 D. Ballantyne *Cunninghams* 14 He..yoohooed for a chair. **1954** *Partisan Rev.* Nov.–Dec. 599 Life, despite their frantic yoohooing, had passed them by. **1957** J. Kerouac *On Road* I. xiii. 88 Then they yoohooed us. **1969** 'E. Lathen' *When in Greece* xxii. 231 Leaving the ladies to their yoohooing, the three men followed the officer's directions. **1978** 'J. Gash' *Gold from Gemini* iii. 24 Patrick yoo-hooed me over to his place. **1982** —— *Firefly Gadroon* xiii. 122 Margaret blew a kiss... Patrick yoohooed.

yop, var. Yawp, yaup *sb.*

yore, *adv.* Add: **5.** *Yore-flood* (nonce-use), the biblical Flood.

1876 G. M. Hopkins *Wreck of Deutschland* xxxii, in *Poems* (1967) 62, I admire thee, master of the tides, Of the Yore-flood.

yorezeit, var. *yahrzeit.

‖**yorgan** (yǫɹgā·n). Also **yorghan.** [Turk.] A quilt.

1914 *Blackw. Mag.* Dec. 759/2 He..brought a heavy quilted *yorghan,* a warm covering for the coldest night. **1962** *Times* 6 June 16/7 Blankets and silk *yorgans*—Turkish eiderdowns.

Y organ. *Zool.* Also **y organ** and with hyphen. [tr. F. *organe Y* (M. Gabe 1953, in *Compt. Rend.* CCXXXVII. 111), after *X organ.] An endocrine gland in certain crustaceans which secretes a hormone that causes moulting.

1959 E. Scharrer in E. Gorbman *Compar. Endocrinol.* 239 The final common path for the resulting decision to molt is represented by the neurosecretory cell which presumably activates the Y-organ. **1965** Lee & Knowles *Animal Hormones* xi. 137 The Y organ is located in the antennary segment of those species which have a maxillary excretory organ, and in the second maxillary segment of those which have an antennary excretory organ. **1973** *Nature* 9 Mar. 133/2 An insect does not enter premoult if its thoracic glands have been removed, nor a crustacean if its y-organs have been removed.

York, *sb.*¹ Add: **1. a.** *York ham* (later examples); **York-Antwerp rules** [adopted at *York* in 1864 and modified at *Antwerp* in 1877], an international set of rules governing the application of general average in marine insurance.

[**1877** Twiss & Jencken in H. D. Jencken *York & Antwerp Rules* 20 The Rules which your committee now bring before you as the basis for a uniform system of General Average for all maritime countries, and to which the title might be given of the 'York and Antwerp Rules', are appended.] **1881** R. Lowndes *Pract. Treat. Law Marine Insurance* vii. 203 The York-Antwerp Rules constitute a sort of international code of general average, not as yet obligatory. **1970** York ham [see *Cambridge sausage* s.v. *Cambridge]. **1974** E. R. H. Ivamy *Marine Insurance* (ed. 2) xiv. 191 The policy almost invariably provides that either a foreign law or the York-Antwerp Rules 1974 shall apply. **1983** *Harrods Mag.* Xmas. 44/1 At Christmas the whole York ham, costing £35. **1984** J. Grigson *Brit. Cookery* 122 Today, York ham has become a generalised term, meaning no more than a mild cured ham.

5. The name of *York Factory,* a trading settlement in northeastern Manitoba, used *attrib.* in **York boat,** a type of inland cargo boat used in Canada between *c* 1790 and 1930.

1864 *Nor' Wester* (Red River Settlement) 26 Apr. 2/5 Gentlemen of practical experience gave their opinion that the present York boats (bateaux) could be used for the transportation of goods from Lake of the Woods to Lake Superior with as little difficulty as is encountered between this place and York Factory. **1909** G. Bryce *Romantic Settlement Lord Selkirk's Colonists* 71 The birch-bark canoe is a mere trifle on the portage, but the heavy York boat capable of carrying three or four tons is a clumsy lugger. **1971** R. Russell *Carlton Trail* 10 The company hired Scotsmen, Metis, and Indians to man brigades of York Boats.

york (yǭɹk), *sb.*² *dial.* Also **yark, yerk.** [Perh. the same word as Yerk, yark *sb.*] Something used to tie a trouser leg beneath the knee. Usu. *pl.*

1905 *Eng. Dialect Dict.* V, *Yark,* a strap or piece of string to fasten the trousers to keep them free from mud. [*Cites a quot. with 'yerks'.*] **1958** K. Etheridge *Welsh Costume* 66 When kneeling at the coal-face, dirt and small coal are apt to get inside the trouser-leg... Tying of the trouser leg just below the knee prevents this. A piece of string, or a leather strap..may be used for this. The strap..is called a 'york'. **1967** *Listener* 19 Oct. 504/3 The 'tyings' or straps worn below the knees..are, or were, used by the South Wales colliers, whose term for them is 'yorks'. **1977** Scollins & Titford *Ey up, mi Duck!* III. 12 Mr. Flint, who played the Fool, wore moleskin trousers tied with 'yorks'. **1984** *Guardian Weekly* 22 Jan. 4/2 They wore corduroy trousers fastened below the knee with leather straps ('yarks' is the technical name).

york, *v.*¹ (Earlier example.)

1882 *Australians in Engl.* 42 Butler was 'yorked' the second ball he received.

york (yǭɹk), *v.*² [f. *york *sb.*² Cf. Yerk, yark *v.*] *trans.* To keep *up* trouser legs by tying them with 'yorks'.

1960 R. Williams *Border Country* 258 The thongs which yorked the trouser legs just below the knees. **1969** M. Harris *Kind of Magic* 178 He always wore thick brown cord trousers yorked up below the knee with leather straps, and his face was the colour of a bit of old leather.

yorker². (Earlier example.)

1861 *Bell's Life* 25 Aug. (Suppl.) 2/1 Buchanan stopped sometime, and bothered the bowlers much, as he would not hit even a 'Yorker'.

yorker³ (yǭ·ɹkəɹ). [f. *york *sb.*² + -er¹.] = *york *sb.*²

1940 H. Spring *Fame is Spur* xiii. 362 Checked mufflers..were at the throats of most [miners], and their trousers were hitched up with yorkers below the knee. **1972** [see *nicky tam].

York gum. [Named after *York,* a town in Western Australia east of Perth.] A gum-tree, *Eucalyptus loxophleba,* of Western Australia, or its timber.

1846 [see *gum *sb.*² 5]. **1889** J. H. Maiden *Useful Native Plants Austral.* ix. 449 Samples of this timber were sent to the Colonial and Indian Exhibition under the name 'York gum' (*E. loxophleba*). **1934** T. Wood *Cobbers* viii. 101 In addition to seeds for me—yates gum and york gum and mallee wattle—a cold turkey sat in the back seat. **1944** *Coast to Coast 1943* 60 The slope..rose towards his own home, a thin tracery of york-gums screening the house. **1965** *Austral. Encycl.* III. 407/1 York gum..and wandoo..are pale-coloured Western Australian timbers.

Yorkie (yǭ·ɹki). Also 9 **Yorky.** [f. York(shire + -ie, -y⁶.] **1.** A Yorkshireman; † a nickname for a Yorkshireman. *colloq.*

1818 P. Egan *Boxiana* (ed. 2) II. 301 Yorky did not appear wholly without judgment. **1938** 'J. Curtis' *They drive by Night* xiii. 144 You're in Yorkshire now... Them Yorkies never was any good. **1950** W. Bird *This is Nova Scotia* 14 The gallant young Yorkie hushed up her every attempt to talk. **1980** 'J. Gash' *Spend Game* x. 107 Joseph Bramah was a Yorkie, and a genius... His legendary lock patent is dated 1784.

2. A Yorkshire terrier. *colloq.*

1950 A. C. Smith *Dogs since 1900* xiii. 321 A bigger Yorkie..is as satisfactory a companion as could be desired. **1967** 'A. Gilbert' *Visitor* vi. 95 'Mrs. Warren is telling me poodles are out!.. 'And Yorkies are in?' I suggested. **1977** *S. Wales Echo* 18 Jan. 12/3 (Advt.), Wanted, Toy Poodles..and Yorkies. **1984** *Hampstead & Highgate Express* 14 Dec. 27/5 Yorkie and Westie pups, inoculated, guaranteed.

Yorkshire. Add: **1. b. Yorkshire chair,** a type of 17th-century upright chair, usu. distinguished by an open backrest and arched cross-rails; **Yorkshire fog,** a perennial greyish-green grass, *Holcus lanatus,* bearing pale green or purplish panicles; **Yorkshire pudding:** now usu. cooked and served as a separate item to accompany roast beef; hence **Yorkshire pud** *colloq.*; **Yorkshire Relish,** the proprietary name of a kind of savoury sauce; **Yorkshire teacake,** a kind of baked yeast teacake occas. made with currants or sultanas; **Yorkshire terrier,** a small, long-coated, tan and blue-grey terrier belonging to a breed developed in the West Riding of Yorkshire about 1870.

[**1900** E. Singleton *Furnit. of Our Forefathers* I. I. 46 There were not so many kinds of single chairs in the seventeenth century... There were two very favourite patterns, the Derbyshire and the Yorkshire... The Yorkshire model .is rather more ornamental.] **1906** W. E. Mallett *Introd. Old Eng. Furnit.* 20 A solid oak chair of the Stuart period. These are often called Yorkshire chairs. **1976** *Southern Even. Echo* (Southampton) 12 Nov. (Advt. Suppl.) 19/3, 4 Yorkshire chairs and 1 carver, £300. **1874** C. C. Babington *Man. Brit. Bot.* (ed. 7) 419 *H. lanatus*... Meadows and pastures... Yorkshire Fog. **1954** C. E. Hubbard *Grasses* 237 'Yorkshire Fog' is generally regarded as a weed, but when young it has some value for grazing. **1977** *New Yorker* 8 Aug. 58/3 The fairways were edged by the wispiest kind of rough—a thin, random collection of fescue, buttercups, plantain shoreweed, Yorkshire fog, dandelions, and assorted other weeds, grasses and wild flowers. **1975** *New Society* 21 Aug. 411/2 The roast beef and Yorkshire pud dinners provided on Sundays. **1976** C. Bermant *Coming Home* II. iv. 169, I was given roast beef and Yorkshire pud. **1877** *Trade Marks Jrnl.* 13 Jan. 78 *Yorkshire Relish*... William Powell, of..the firm of Goodall Backhouse and Co., Leeds..; drysalters and general merchants. **1881** *Cassell's Family Mag.* Dec. 46/1 And two table-spoonfuls of Yorkshire Relish. **1926** *Daily Colonist* (Victoria, B.C.) 17 Jan. 7/7 (Advt.), Genuine Yorkshire Relish. **1877** *Cassell's Dict. Cookery* 1157/2 *Yorkshire tea-cakes*... Two pounds of good flour;..an ounce of German yeast..two well-beaten eggs and six ounces of butter... Leave it in a warm place to rise... Divide it into ten portions. **1945** 'R. Crompton' *William & Brains Trust* v. 84 One of his happiest memories was the Yorkshire tea cakes that his mother used to make. **1977** E. David *Eng. Bread & Yeast Cookery* 488 The following nineteenth-century recipe for plain Yorkshire tea cakes comes from..Marcus Woodward. [**1871** R. Pearson in *Field* 13 May 386/2 Some friends of mine in Yorkshire wish me to put a matter right in connection with dog shows, namely, the classification of a breed of dogs they claim as of Yorkshire origin—those beautiful blue and tan long-haired terriers, now sometimes entered in the Scotch and broken-haired, also rough-haired terrier classes. They with myself think it would much simplify the matter by calling them the 'Yorkshire blue and tan long-haired terrier'.] **1872** 'Stonehenge' *Dogs of Brit. Islands* (ed. 2) v. 108 The silver-grey Yorkshire terrier is not a distinct breed, being merely a paler variety of the blue-tan. **1922** D. Matheson *Terriers* I. 188 Yorkshire Terriers are hardy, game little dogs. **1959** *Observer* 1 Feb. 12 Yorkshire Terrier. Coat should hang straight and evenly down each side. **1971** *Country Life* 6 May 1098/3 Mrs. Huxham makes it clear..how much pleasure and entertainment she gets from her gay little Yorkshire terriers.

c. The name of a breed of white pig, now widely bred for bacon.

1770 Yorkshire hog [in Dict., sense 1 b]. **1845** *Encycl. Metrop.* XXV. 310/1 The Old Yorkshire Pig is by some considered as the very worst of the large varieties, very long legged, weak loined, not of strong constitution, nor good stye pigs, but yet quicker feeders. **1856** J. C. Morton *Cycl. Agric.* II. 942/1 The Old Lincolnshire, or Yorkshire Pig, was one of largest breeds in the kingdom, and probably the worst... The large breed, or *improved* wold pig, has probably undergone as great a change as any, and has become..the most profitable kind we have. **1914** 'Saki' *Beasts & Super-beasts* 24 Tarquin, the huge white Yorkshire boar-pig, had exchanged the narrow limits of his stye for the wider range of the grass paddock. **1974** *Encycl. Brit. Macropædia* X. 1281/2 The Yorkshire pig, which originated early in the 19th century in England, where it was considered a bacon type, is long, lean, and trim with white hair and skin. Found in most countries, this breed is probably the most widely-distributed in the world.

3. (Earlier and later examples of sense 'Yorkshire dialect'.) Also, Yorkshire pudding; a Yorkshire pig; a Yorkshire terrier.

1717 M. W. Montagu *Let.* 1 Apr. (1965) I. 333 'Tis as ridiculous to make use of the expressions commonly us'd, in speaking to a Great Man or a Lady, as it would be to talk broad Yorkshire or Somersetshire in the drawing room. **1898** J. D. Brayshaw *Slum Silhouettes* 235 Now's yer time, gents, fer a nice 'ot dinner, cut off the jint..., an' two wedgetables; plum-tart or Yorkshire—a shilling. **1902** *Encycl. Brit.* XXV. 194/2 The latter [*sc.* recognized breeds] include the Large White, Middle White, and Small White, which were all formerly embraced under the general term of Yorkshires, and are still so called in other countries. **1922** D. Matheson *Terriers* I. 181 Originally designed as a sporting dog, the Yorkshire has gradually passed into a lap-dog. **1956** A. Wilson *Anglo-Saxon Attitudes* I. ii. 41 Theo..said, 'Shame!' in his broadest Yorkshire. **1967** M. Kenyon *Whole Hog* viii. 93 They're Large Whites. You [*sc.* Americans] call them Yorkshire. **1971** R. J. White *Second-Hand Tomb* ix. 95 Rammel. That's Yorkshire for rubbish. **1977** J. Wambaugh *Black Marble* (1978) viii. 105 The handler two stations down had a Yorkshire with a touch-up. **1977** *S. Wales Guardian* 27 Oct. 5/3 Oh the joy of digging into tender roast beef and Yorkshire!

Yorkshireism: also **Yorkshirism.**

1962 P. Bentley *'O Dreams, O Destinations'* iv. 56 Her voice was pleasant, quite devoid of Yorkshirisms. **1971** J. Wainwright *Dig Grave & let him Lie* 128 'Aye.' The open-vowelled Yorkshirism was heavy with self-disgust.

yortzeit, var. *yahrzeit.

Yoruba (yǫ·rubă), *sb.* and *a.* Also 9 **Yarriba.** [Native name.] **A.** *sb.* **a.** The language of the Yorubas, a tonal language of the Kwa group.

1841 *Outl. Vocab. Lang Western & Central Africa* 2 (*heading*) Ako, Eyo, Yabū, or Yarriba. **1843** S. Crowther (*title*) Vocabulary of the Yoruba language. Part I.—English and Yoruba, Part II.—Yoruba and English. **1888** *Encycl. Brit.* XXIV. 755/1 The Bible and several religious treatises have been translated into Yoruba. **1927** E. S. Pankhurst *Delphos* v. 51 Yoruba, one of the African languages, conjugates its verbs as in English, though its vocabulary is entirely different. **1964** *New Statesman* 1 May 681/1 In Accra recently a Nigerian company, under the direction of the brilliant young artist Demas Nwoko, presented a dramatised version, in Yoruba, of Amos Tutuola's *The Palm-Wine Drinkard.* **1972** B. Emecheta *In Ditch* i. 6 The landlady started scolding her husband in Yoruba.

b. (A member of) a Black people of western Nigeria and neighbouring parts.

1843 S. Crowther *Vocab. Yoruba Lang.* p. iii, The Yorubas, like other nations, have always considered themselves the first people in the world. **1897** M. H. Kingsley *Trav. W. Afr.* xiii. 326, I had a set of porters composed of four Bassa boys, two Wei Weis, one Dualla, and two Yorubas. **1937** *Discovery* July 225/1 The Yoruba have only the vaguest tradition of their own past. **1960** *Guardian* 15 July 14/3 The Yorubas live in the largest urban agglomerations to be found in traditional Africa. **1970** P. Oliver *Savannah Syncopators* 32 Between the lands of the Ibo and those of the Ashanti (Akan) of the old Gold Coast (Ghana) lie the domains of the Yoruba of Nigeria and Dahomey.

B. *adj.* Of, pertaining to, or characteristic of this people.

1843 [see sense A. b above]. **1883** *Encycl. Brit.* XVI. 517/1 In the Yoruba lands the Church Missionary Society has 11 stations. **1938** J. Cary *Castle Corner* 357 Four big Yoruba soldiers. **1957** M. Banton *W. Afr. City* viii. 153 A Yoruba secret society called *Engugun.* **1978** *Jrnl. R. Soc. Arts* CXXVI. 366/2 In Nigeria, Yoruba woodcarvers have adapted their techniques to work in concrete.

Hence **Yo·ruban** *sb.* and *a.*

1853 S. Tucker *Abbeokuta* (ed. 2) ii. 15 There were few or no Yorubans brought to Sierra Leone till the year 1822. *Ibid.* viii. 100 It afforded Mr. Townsend the opportunity of becoming in some degree acquainted with the Yoruban language. **1879** J. A. Farrer *Primitive Manners* iii. 89 Captain Burton justly calls attention to the possibility of many Yoruban proverbs being relics of the Moslems. **1936** V. A. Demant *Christian Polity* xi. 193 The thunder-god of the Yorubans is decidedly an earthly king who became a god. **1957** M. Stearns *Story of Jazz* iii. 27 The musical instruments for such occasions consist of Yoruban drums, shaped like hour glasses, and the drumming and singing are in the Yoruban style.

Yoshiwara (yǫʃiwā·rā). Also **yoshiwara.** In Japan, an area (esp. one in Tokyo) where brothels were officially recognized. Freq. *attrib.*

Official recognition was withdrawn in May 1958.

1870 *Fortn. Rev.* Aug. 154 At Yokohama, indeed, and at the other open ports, the women of the Yoshiwara are loud in their invitations to visitors. **1877** A. C. Maclay *Let.* 25 Nov. in *Budget of Lett. from Japan* (1886) xix. 342 That strange institution of the feudal Government of Japan known as the *Yoshiwara* system... This was a system of legalized prostitution..under government patronage. **1896** Kipling *Seven Seas* 70 And tell the Yoshiwara girls to burn a stick for him. **1911** *Daily Colonist* (Victoria, B.C.) 27 Apr. 3/4 The fire..destroyed the new yoshiwara at Tokyo, sweeping away 6555 houses. **1944** H. G. Wells *'42 to '44* 102 The child may be apprenticed to Fagin's academy, it may be sold into the Yoshiwara. **1966** *New Statesman* 2 Dec. 840/1 A short look at *A Harlot's Progress,* and then—if not at a yoshiwara house by a Japanese master of the Floating World—at Botticelli's *Venus.* **1978** *Country Life* 24 Aug. 500/2 The brothels in the red light district of Edo, the Yoshiwara.

yotization (yotəizēɪ·ʃən). *Phonetics.* [f. *yot-,* repr. Iota + -ization.] The prefixing of the semivowel (y) to another sound or syllable; the change of a sound into (y). So **yo·tized** *ppl. a.*

1936 *Bull. School Oriental Stud.* VIII. 525 Peanius.. presents various types of syllable such as those 'quibus copulatur *ja, seu jota*', which we may describe as 'yotized'. *Ibid.* 532 Basic type of articulation..yotization..labio-velarization..nasalization. **1951** [see *prosody 2*].

you, *pers. pron.,* etc. Add: **II. 5. c.** Phr. *you and your—*: a contemptuous, impatient, or good-natured dismissal of the thing or person mentioned. *colloq.*

1607 Shaks. *Coriolanus* IV. vi. 97 You haue made good worke, You and your Apron men. **1837** H. Martineau *Society in America* III. III. i. 80 An old acquintance of Noah's..said.., 'Go, get along, you and your old ark! I don't believe we are going to have much of a shower.' **1899** Kipling *Stalky & Co.* 177 'I was born there... It was called after my uncle.' 'Shut up—you and your uncle!' **1943** J. B. Priestley *Daylight on Saturday* xxii. 172 I've told 'im... 'You an' your Teds!' I told 'im. **1955** E. Blishen *Roaring Boys* I. 27 'Progressing!' He relished it. 'You and your long words!' **1980** P. G. Winslow *Counsellor Heart* xiv. 171 Ah, you and your Colonel. Worms' meat, he is now.

d. *you and who else?*: see *who *pron.* 4 a; *you and yours*: see Yours *poss. pron.* 2 b.

III. 10. b. (Further examples.)

1964 'E. Lathen' *Accounting for Murder* xv. 142 We run tests... Then, once we had really found the real you, we..would try to find a place that provided a challenge to your best creative talents. **1974** *Spartanburg* (S. Carolina) *Herald* 25 Apr. A-5 (Advt.), Vicalten can be your ticket to a newer, slimmer you. **1978** J. Grenfell *Stately as Galleon* 38 Learn..to dance the natural nature way. Let the music through, find the inner you. **1981** *Sci. Amer* July 14/3 For every quantum-mechanical branch point in your life..you have split into two or more you's riding along parallel but disconnected branches of one gigantic universal wave function.

10*. As *adj.*: expressive of or suited to your taste, personality, etc.

1918 R. Fry *Let.* 12 Mar. (1972) II. 425 I've read your *Lucretius*... I feel sure it's both immensely him and also very much you. **1936** U. Orange *Begin Again* xi. 247 'I think it's lovely,' said Jane unkindly, 'So you, somehow.' **1960** N. Marsh *False Scent* viii. 232 The boudoir.. had been created by Bertie... 'Almost indecently *you,* darling!' Bertie had told Miss Bellamy. **1981** M. Spark *Loitering with Intent* ii. 44, I thought your piece was very much you.

11. you-know-what (later examples); **you-know-who, -whom,** a deliberately unnamed person whose identity is apparent to the hearer.

1766 O. Goldsmith *Vicar of Wakefield* II. ix. 143, I danced last night with Lady G—, and could I forget you know whom, I might be perhaps successful. **1796** M. Edgeworth *Parent's Assistant* (ed. 2) I. 174 Do nothing in this till we have consulted *you know who* about whether it's right or wrong. **1912** C. Mackenzie *Carnival* xiii. 167, I don't think I'm jealous of you know who. **1936** O. Nash *Primrose Path* 198 Be my gazelle, my wishing well..But never my you-know-what. **1937** M. Allingham *Dancers in Mourning* xiii. 179 Not a word to Mrs. You Know Who. **1956** L. McIntosh *Oxford Folly* 37 She's madly gay, but hard as nails when it comes to you know what. **1975** *Verbatim* Dec. 1/2 John O'Donnell expressed his delight that 'we're going to arm German panzer divisions for you-know-what'. **1976** *New Yorker* 26 Apr. 99/1 She gives me a pain in my you-know-what. **1978** J. Irving *World according to Garp* xviii. 390 Old You-Know-Who—the Under Toad, that's who, Helen thought. **1981** Q. Crisp *How to become Virgin* vi. 81 Since neither I nor Mr. Hurt .flashed you-know-what before the cameras..we might both by modern standards be considered old-fashioned.

you-all (yū·ǫl, yū,ǭ·l), *pers. pron.* *U.S. dial.* Also **you all.** [f. You *pers. pron.* + All *a.*] Used in place of You *pers. pron.*

Used, with no clear pattern, both as *sing.* and as *pl.*

1824 'A. Singleton' *Lett. South & West* 82 Children learn from the slaves some odd phrases;..as ..will you *all* do this? for, will *one* of you do this? **1871** R. M. Johnston *Dukesborough Tales* vii. 95 You all little fellows was.. skeered. **1875** [see *we-all *pron.*]. **1901** A. H. Rice *Mrs. Wiggs of Cabbage Patch* vi. 81 Some of you all shake down the stove an' pull the door to fer me. **1919** R. Frost *Let.* 24 Mar. (1972) 56 The second thing is to ask you what you-all are thinking of to want me to judge in a lyric contest. **1924** W. M. Raine *Troubled Water* xix. 201 You-all are losing a better man than Missie ever had. **1926** E. Ferber *Show Boat* 299 You-all one of them Suhveys? **1926** *Amer. Speech* II. 476/1, I was born in South Carolina..and have worked on a New Orleans paper for two years, and I have not once in my life heard the expression 'you-all' used except in plural address. **1927** *Ibid.* III. 5 *You-all* certainly is used as singular in the Ozarks—I have heard it daily for weeks at a time. *You-uns*, however,..is nearly always plural in the Ozark country. **1928** *Ibid.* IV. 54 Here in Missouri..I have again and again heard 'you all' used in speaking to one person. **1942** M. K. Rawlings *Cross Creek* xxii. 344, I ain't never been as far as you-all aim to go. **1944** *Amer. Speech* XIX. 147/2 In almost a score of years of residence in North Carolina I have never heard anyone say 'you all', unless the plural was definitely and distinctly intended. **1954** G. Durrell *Three Singles to Adventure* 15 Is youall to catch the Parika train? **1981** *TV Picture Life* Mar. 46/3 Then I walked into this audition, put my feet up on the desk, finished my beer, and drawled, 'What kind of character you-all looking for?'

So **you-all's** and varr., your.

1869 *Overland Monthly* Aug. 131 During the war we all heard enough of 'we-uns' and 'you-uns', but 'you-alls' was to me something fresh. **1887** *Scribner's Mag.* Oct. 478/1 How are you all's little trick? **1929** W. Faulkner *Sound & Fury* 347, I try to obey his wishes for you alls' sakes. **1934** C. Carmer *Stars fell on Alabama* 190 We are honored to have you-all's company.

young, *a.* (*sb.*[1]) Add: **A.** *adj.* **1. b.** *young master*: see Master *sb.*[1] 22 in Dict. and Suppl.; *young 'un* (earlier and later examples); also *youngun*.

c **1810** W. Hickey *Mem.* (1960) iv. 64 So the young'un there wanted to be off, but I said as I how I knew a trick worth two of that. **1922** Joyce *Ulysses* 418 Collar the leather, Youngun. **1940** C. McCullers *Heart is Lonely Hunter* I. iv. 56 A youngun was sitting on the banisters... He had seen her somewhere before. **1981** J. D. MacDonald *Free Fall in Crimson* xix. 219 Here and there are little groups of younguns who know what an original idea tastes like.

(*b*) *young-girlish* adj., *-girlishly* adv.

1928 A. Huxley *Point Counter Point* xi. 174 She flushed with a young-girlishly timid pleasure. **1975** *New Yorker* 28 Apr. 66/3 Her young-girlish way of lowering her eyes with an air of anguish when I asked her what had happened to her affair with the man she said she had loved.

C. 1. a. Also with names of political parties and movements, denoting a young member or (*pl.*) a section organized by and for young members, as *Young Communist, Conservative, Farmers*, etc.; **young** (or **Young**) **America,** (*a*) a slogan used in connection with an expansionist movement within the Democratic Party in the 1840s and 1850s (*obs.* exc. *Hist.*); (*b*) American youth collectively; **Young Turk**: see also Turk[1] 2 e in Dict. and Suppl.

1844 *St. Louis Reveille* 30 Nov. 2/2 No mammoth bank..can form any part of the creed of the *Young America*! **1852** *U.S. Mag. & Democratic Rev.* Feb. 185/2 We are not for all the young men before the country, but only for the bold, active honor and talent of Young America. **1880** 'Mark Twain' *Tramp Abr.* xxxviii. 444 He and the innocent chatterbox whom I met on the Swiss lake are the most unique and interesting specimens of Young America I came across during my foreign tramping. **1924** *Outlook* 10 Sept. 45/1 Young America could with profit leave such affairs alone. **1962** E. Wilson *Patriotic Gore* p. xxii, Douglas..had been the leader of the 'Young America' movement in the Democratic Party, which had favored..the annexation of Mexico, Cuba and..Central America. **1936** J. Bell *Let.* 4 Jan. in *Ess., Poems & Lett.* (1938) II. 294 While I have been writing these last pages, I have been acutely aware of 'the adversary'. He takes the form of an enthusiastic member of the Young Communist League and he bellows incessantly. **1966** 'H. MacDiarmid' *Company I've Kept* viii. 188 Some of the Edinburgh University students, members of the Young Communist League,..came to the rescue. **1982** *Manch. Guardian Weekly* 21 Nov. 10/3 He had a good secondary education and joined the Young Communists in 1923 while he was at a metallurgical school. **1924** *Times* 17 Mar. 13/5 (*heading*) The 'Young Conservatives Union'. *Ibid.*, This union has been formed by Young Conservatives who desire to prove by attaching themselves to constituencies for social service that their Conservative ideal is one which they are prepared to maintain by action. **1938** B. R. Braine in *Torchbearer* Apr. 41/3 The prefix 'Junior' we dislike, and the word 'Imps' infers an extreme and irresponsible youthfulness that is certainly not in accordance with reality..I suggest..that we take steps to become the 'Young Conservative League'. **1944** *Times* 11 July 2/2 The Conservative Party has decided to establish a new Young Conservative movement which will take the place of the Junior Imperial League. **1959** E. H. Clements *High Tension* x. 165 Fiona isn't a Young Conservative! She's a Communist. **1977** J. Wainwright *Do Nothin' till You hear from Me* iv. 52 It is a Young Conservative hop, in a neighbouring town. **1968** P. Jennings *Living Village* 71 The flourishing Young Farmers Club movement. **1981** J. Wainwright *Urge for Justice* I. v. 35 The lads from the local Young Farmers branch put on a New Year's Ball. **1983** *Economist* 21 May 37/1 As usual, the Young Liberals attacked it from a neo-Trotskyite stand. **1980** *Christian Science Monitor* 28 Jan. 12/3, 80,000 pounds—none of which, say the editors, came from abroad. Much of it probably came through the Labour youth movement, Young Socialists, where the tendency reportedly has strong support.

b. young grammarians *sb. pl. Philol.* [tr. Ger.] = *Junggrammatiker *sb. pl.*; cf. *neo-grammarian; so **young-grammarian** *a.*; **young lion,** a young and vigorous man.

1922 Young-grammarian [see *Junggrammatiker *sb. pl.*]. **1947** *Essays & Studies* XXXII. 89 This was bound to shake the young-grammarian theory of the inviolability of the sound laws. **1856** C. M. Yonge *Daisy Chain* ix. 92 Take care of my arm!.. I was..a little in dread of such a young lion! **1917** H. James *Middle Years* iii. 36 Frederic Harrison..one of his [*sc.* Matthew Arnold's] too confidently roaring 'young lions' of the periodical press. **1937** K. Blixen *Out of Africa* v. v. 407 It was curious to hear the young Kikuyu lions speak with reverence and awe of..the old dancers. **1977** *Listener* 17 Feb. 216/2 An orchestra comprising most of the 'young lions' from the home front.

2. a. *young-born* adj.: also *absol.*; *young-seeming* adj. **b.** *young-minded* adj. **young-blood**: delete † and add: revived in *U.S.* as a hyphenated or one-word form of *young blood* (see Blood *sb.* 15).

1946 Mezzrow & Wolfe *Really the Blues* viii. 106 He was a tall blond good-looking youngblood. **1979** *N.Y. Times Mag.* 30 Sept. 28/4 The fault always lay with..veterans rather than the youngbloods that Willis himself had drafted. **1915** D. H. Lawrence *Rainbow* i. 2 Every year throws forward the seed to begetting, and..leaves the young-born on the earth. **1930** *Daily Express* 23 May 10/4 The Italians and the Russians are young-minded. **1951** S. Spender *World within World* 113 An old man... With young-seeming nervous fingers he touched the rim of his glass.

Young (yɒŋ), *sb.*[2] *Physics* and *Mech.* [The name of Thomas *Young* (1773–1829), English physician and physicist.] *Young's modulus*: = *modulus of elasticity* s.v. Modulus 3.

1865 *Proc. R. Soc.* XIV. 293 Young's 'modulus', which has generally been called simply *the* modulus of elasticity of a solid, is the longitudinal traction of a stretched rod or wire of the substance, divided by the extension produced by it. *Ibid.*, Several accurate determinations of Young's modulus have been made upon wires of different substances hung in the College Tower of the University of Glasgow. **1930** *Engineering* 11 Apr. 465/1 The modern theory of the elasticity of isotropic materials makes use of a number of physical constants, all of which are definitely related to Young's Modulus E. **1967** M. Chandler *Ceramics in Mod. World* iv. 118 The harder it is to stretch a material, the higher is its Young's modulus. **1978** *Jrnl. R. Soc. Arts* CXXVI. 683/1 The stresses built up in a structure due to these temperature differences are proportional to E (Young's Modulus), α (the linear coefficient of expansion) and ΔT (the temperature difference).

young (yɒŋ), *v. Geol.* [f. Young *a.*] *intr.* Of a structure or formation: to present the apparently younger side (in a specified direction). Hence **you·nging** *vbl. sb.*

1934 E. B. Bailey in *Q. Jrnl. Geol. Soc.* XC. 469, I have..been forced to coin the barbaric verb 'to young', in the sense 'to present the younger aspect'. **1969** Bennison & Wright *Geol. Hist. Brit. Isles* viii. 164 It seems probable that beds 'young' both southwards and northwards from the St. Austell Granite. **1972** *Nature* 28 Apr. 431/2 The eastward younging of plutons was taken to indicate an eastward migration of the downwelling plate margin. **1975** Tindall & Thornhill *Blandford Rock & Mineral Guide* I. 30 The direction of younging in a single layer of rock can sometimes be established if there is clear evidence of erosion. **1982** Collinson & Thompson *Sedimentary Structures* ii. 9/2 A sequence [of beds] could therefore be reported as 'younging to the east' for example.

youngberry (yɒ·ŋbĕri). Also **Young-.** [f. the name of B. M. *Young* (fl. 1905), U.S. horticulturist, who first produced it + Berry *sb.*[1]] A dewberry derived from a cross made in Louisiana in 1905 between a blackberry and a dewberry; also, a fruit of this plant, similar to a loganberry.

1927 *Calif. Cultivator* 30 July 104 (*heading*) The Youngberry. *Ibid.* 20 Aug. 178/4 As to the Youngberry plant it does not produce a dewberry fruit and if the name dewberry cheapens a most excellent fruit and is misleading as well we can see no reason why, in California, dewberry should not be dropped and Youngberry..adopted. **1935** *Ann. Rep. Oregon State Hort. Soc.* XXVII. 74 Our oldest Youngberries were set in the spring of 1931. **1971** *Post* (Cape ed.) 9 May (Suppl.) 10/4 Ingredients:..Apple Jelly,..2 cups boiling water, 2 cups milk, 1 tin youngberries or loganberries. **1980** *Times* 9 June 6/4 A few untended youngberry bushes are all that remain of the thriving farming community that once lived here [in Zimbabwe].

younger, *a.* (*sb.*) Add: **1. a.** *younger generation*, the next or rising generation, seen in contrast to the current one or one's own.

1896 G. B. Shaw *Our Theatres in Nineties* (1932) II. 289 A fine young woman in rational dress who..treads the boards with no little authority and assurance as one of the younger generation knocking vigorously at the door. **1914** L. Woolf *Wise Virgins* iv. 93 'Most lakes..are repulsive,' ejaculated Harry. 'Ah,' said Mr. Macausland tolerantly, 'there spoke the younger generation.' **1931** R. Campbell *Georgiad* i. 25 Writers of the younger generation. **1939** T. S. Eliot *Family Reunion* I. i. 14 The younger generation Are undoubtedly decadent. **1976** 'J. Fraser' *Who steals my Name?* ii. 22 Mr. Cedric was a member of the younger generation.

b. (Earlier and later examples of its use in the titles of Scottish heirs.)

1567 in R. Pitcairn *Criminal Trials in Scotland* (1833) I. ii. 496 (*heading*) Deposition by John Hay, 3ounger of Tallo. **1982** *Who's Who* 1425/2 Hugh Magnus Macleod, younger of Macleod.

young fogey. Also **young fogy** and with capital initials. [f. Young *a.* + Fogy, Fogey 2.] A young person of noticeably conservative tastes or outlook. Cf. *old fogey* s.v. Fogy, fogey 2.

Though occasionally used at an earlier date in contrast with *old fogey*, the expression did not become common until the 1980s.

c **1909** C. S. Peirce *Coll. Papers* (1935) VI. I. xii. 218, I expect the day will come when another generation of old and young fogies will be equally indisposed to admit that there is any corner of the whole field that I have not turned up. **1929** 'D. Yates' *Maiden Stakes* 11 Fashions, outlook, the spirit and manners of the age—I found the lot beyond me... I was a young fogey. **1980** *Business Week* 6 Oct. 95 Noting a split in the profession between 'old Turks and young fogies', Aaron says: 'The older generation of economists was stimulated by external problems.' **1981** *N.Y. Times* 18 Mar. A27/1 At their worst, conservatives were old fogeys. There were some young conservatives, of course, but they were unbrilliant young fogeys. **1983** *Listener* 27 Jan. 21/3 He implies that this is a consequence of the decline in educational standards of the past decades. Mr. Wilson, though he has many admirable qualities, is a bit of a professional young fogey. **1985** *Times* 16 July 30 Mr. Gorbachov is something of a young fogy, though with an inquiring mind. **1985** S. Lowry *Young Fogey Handbk.* i. 8 The present resurgence of the Young Fogey ties up neatly with the reinvention of the class system that has been going on at least in the South of England ever since Tina Brown revamped *Tatler*.

Young-Helmholtz (yɒŋ he·lmhǫlts). The names of Thomas *Young and von *Helmholtz used *attrib.* to designate the theory that in the eye there are receptors sensitive to one or other of three colours (red, green, violet) and every colour sensation is due to the stimulation of these in different proportions.

1889 in *Cent. Dict.* **1896** [see Trichromatic *a.*]. **1935** *Discovery* July 187/2 In 1801, Dr. Thomas Young (1773–1829),..propounded the theory which now, as the result of the latter work of Helmholtz, commonly bears the name of the Young-Helmholtz theory of colour vision. **1974** *Encycl. Brit. Macropædia* VII. 108/2 All the evidence points to the correctness of the Young-Helmholtz hypothesis with respect to the three colour basis.

young lady. Add: **young ladyship** (earlier example).

1871 [see *grown-up-dom].

youngly, *adv.* **2.** (Later example.)

1922 Joyce *Ulysses* 192 Yes, Mr. Best said youngly, I feel Hamlet quite young.

young man, † youngman. Delete † and add later examples of *youngman*.

Not in standard use.—Ed.

1961 *Evergreen Rev.* July–Aug. 15 There was a youngman I had seen often around Times Square. Like me, he was..hustling. **1963** J. Rechy *City of Night* I. 43 Part of Pete's technique as a hustler was to tell the men he'd been with that he knew other youngmen like himself. **1967** R. McGough in A. Henri et al. *Mersey Sound* 91 Let me die a youngman's death not a clean & inbetween the sheets holywater death. **1977** *Sunday Times* (Lagos) 6 Feb. 3/3 A police dog also bit a youngman who was rushed to the hospital for treatment.

young-mannish *a.* (earlier example).

1854 C. M. Yonge *Heartsease* I. II. xii. 310 He looked more young-mannish and sentimental than he does now.

youngstock (yɒ·ŋstǫk). [f. Young *a.* + Stock *sb.*[1]] Young (domestic) animals.

[**1888** W. Day *Horse* xviii. 250 To put them [*sc.* yearlings] into small paddocks totally unfitted for rearing young stock in is a sad blunder.] **1963** *Times*, 9 May 17/6 Some excellent hunter youngstock classes came before Major Stoddart at the Oxfordshire Show today. **1971** *Pony* Oct. 6/3 There were 325 entries in the classes for pony youngstock. **1979** *Proc. 11th Ann. Convention Amer. Assoc. Bovine Practitioners* 72/2 Cold housing is generally preferred for youngstock. **1982** *Proc. Internat. Conf. Goat Production & Disease* III. 499/1 Perinatal and youngstock mortality data of a 5 year period..of 8 goat breeding farms.

your, *poss. pron.* and *a.* Add: The first pronunc. given in Dict. has been largely superseded by (yǭɹ).

Similarly with Yours *poss. pron.* and related words.

youraballi, var. *juriballi.

yours, *poss. pron.* Add: **1. c.** *yours truly* (earlier *joc.* example).

1833 Dickens *Let.* Oct. (1965) I. 31 Pray give my love to Letitia; 'accept the same from yours truly' as schoolboys say.

2. e. Phr. *what's yours?* (colloq.): what would you like to drink?

1930 Auden *Poems* 12 There's time for a quick one before changing. What's yours? **1982** P. Lovesey *False Inspector Dead* v. 177 What's yours, Inspector?

yourself, *pron.* Add: **II. 3. b.** Added as a retort after repeating something just said to oneself. *colloq.*

1897 H. G. Wells *Plattner Story, & Others* 214 William came up beside her and said, 'Hello!' 'Hello yourself!' she said. **1944** S. Bellow *Dangling Man* 39 'Minna,' I

said, 'Minna yourself!' *a* **1945** C. Williams *Seed of Adam* (1948) 77 *Hell* (*to Grace*) Stop that noise! *Grace* Noise yourself!

4. b. Now esp. in Anglo-Irish.

1805 G. Colman *John Bull* IV. i. 65 Och! and is it yourself I see, at last? **1896** C. M. Yonge *Release* II. ix. 160 And is it yourself, Miss Caroline, as would honour me by sailing under my flag.' **1907** J. M. Synge *Tinker's Wedding* II. 32 And it's yourself is wedding her, Michael Byrne? **1970** N. Marsh *When in Rome* iv. 92 The monk..spoke with a superb brogue... 'Ah, it's yourself again,' he said.

yourt. Add: (Examples of *yurt, yurta*: now the usual forms.) Also, a circular skin- or felt-covered tent, with collapsible frame, used by the nomadic peoples of Siberia and Central Asia. Also *transf.*

1883 S. W. Williams *Middle Kingdom* (rev. ed.) I. iv. 206 A first-class *yurt* is by no means an uncomfortable dwelling, with its furniture, lining, shrine, and hot kettle in the centre. **1904** H. G. C. Swayne *Through Highlands Siberia* vi. 98 Occasional groups of true Kalmak huts, extinguisher-shaped,..felt 'yurtas', which are used by the richer Kalmuks, through I believe they are properly the Kirghiz type of dwelling. **1911** P. T. Etherton *Across Roof of World* iv. 60 The habitations of the Kirghiz are the 'yurts', or 'khirgas', constructed of felt on a wooden framework, with an opening at the top to let out smoke from the fire lighted in the centre. **1924** *Blackw. Mag.* Aug. 255/2 We had seen little camps of their Yurtas, large black tents. **1940** *Antiquity* XIV. 410 Dwelling of the semi-dugout type. This was probably a conical structure similar to a yurt. **1953** J. Masters *Lotus & Wind* xx. 247 Groups of the round black felt tents called yurts dotted the grass. **1974** *Publishers Weekly* 4 Mar. 5/2 (Advt.), Build a yurt. The round house you can make... It's a portable dwelling for people and pets... Inspired by the ancient Mongolian felt-and-hide structures, the yurt is the newest thing in housing. **1978** A. Grey *Chinese Assassin* ii. 38 The yurt camps..were now quickly dispersing..across the steppes. **1979** J. Halifax *Shamanic Voices* (1980) ii. 49 Kyzlasov lived a distance from his village..in an isolated and impoverished yurta constructed of logs and surrounded by birch trees. **1981** *Nordic Skiing* Jan. 51/2 Skiing the system of five yurts set five miles apart in the Sawtooth Mountains is what Leonard Expeditions is all about.

yous (yūz), also **youse,** dial. varr. You *pers. pron.* (with pl. inflection, though used in sing. sense also). Cf. Yez.

1893 S. Crane *Maggie* i. 8 Youse kids makes me tired. *Ibid.* ii. 13 Ah, Jimmie, youse bin fightin' agin. **1901** M. Franklin *My Brilliant Career* xxx. 256 Ye and Lizer can have a little fly round. It'll do yous good. **1907** J. M. Synge *Playboy of Western World* III. 63 Is it mad yous are? **1929** Wodehouse *Gentleman of Leisure* xiv. 110 Say, youse won't want me any more, boss. **1930** J. Dos Passos *42nd Parallel* I. 102 Say, yous guys, this is fellowworker McCreary. **1939** X. Herbert *Capricornia* xiii. 183, I want yous two back here at Black Adder. **1968** S. L. Elliott in E. Hanger *Three Austral. Plays* II. i. 74 Good luck to youse. **1975** [see *store *sb.* 12 a]. **1981** S. Rena *Painless Death* xx. 132 We've a telephone installed in the lounge since yesterday and none of yous even noticed it!

youth. Add: **7.** *youth cult, culture, -day, -group, movement, organization*; *youth-bereft, -charmed, -oriented* ppl. adjs; **Youth Aliyah** [Heb. *ăliyah* ascent], a movement, begun in 1933 for the emigration of young Jews to Palestine; **youth and old age** = Zinnia; **youth camp,** one of the camps of various kinds that were established for young people in Germany under the Nazis; **youth centre,** a building providing social and recreational facilities for young people; **youth club,** a social club provided for the spare-time activities of young people; the premises of such a club; **Youth Employment Service,** an advisory service for school leavers set up in 1948 and superseded by the Careers Advisory Service; also *ellipt.*; so *Youth Employment office, officer*; **youth hostel** [tr. G. *jugendherberge*], a hostel providing cheap overnight accommodation for young travellers and holiday-makers; hence **youth-hostel** *v. intr.*, **-hostelling** *vbl. sb.*; **youth hosteller; youth leader,** a person having charge of young people in a youth club or other youth organization; **Youth Opportunities Programme,** a Government-sponsored service introduced in 1978 to provide temporary work experience for unemployed young people and replaced in 1983 by the *Youth Training Scheme* (see below); **youth orchestra,** an orchestra open only to young musicians; **youth service,** a service, esp. of local government, providing social and recreational facilities for young people; **Youth Training Scheme,** a Government-sponsored scheme introduced in 1983 to replace the *Youth Opportunities Programme* (see above) and offering job experience and training for unemployed school leavers; **youth work,** social work among young people; hence **youth worker.**

1936 H. Szold *Let.* 4 Sept. in M. Lowenthal *Henrietta Szold: Life & Lett.* (1942) xvi. 312 We build homes, we work our heads off over Youth Aliyah and social service. **1968** P. Durst *Badge of Infamy* i. 3 A Youth Aliyah village in a kibbutz south of Tel Aviv. **1975** E. Avriel *Open the Gates!* vi. 62 We had a number of special immigration certificates for youngsters from the separately functioning body of Youth Aliyah. **1889** G. Nicholson *Illustr. Dict. Gardening* IV. 241/2 Zinnia... Youth and Old Age. **1971** *Farmer & Stockbreeder* 16 Feb. 80/1 Zinnias—common name 'youth-and-old-age'—have normally bloomed in late summer. **1911** M. Beerbohm *Zuleika Dobson* xix. 292 As on the towing-path, so on the youth-bereft rafts of the barges, yonder, stood many stupefied elders, staring at the river. [**1936** *Liverpool Echo* 5 Sept. 4/1 (*heading*) A visit to a Hitler youth camp.] **1942** E. Waugh *Put out More Flags* iii. 217 Those absurd instructors who harangued the youth camps. **1975** H. W. Koch *Hitler Youth* ix. 196 Officials of the party and the Hitler Youth participated. Final selection of candidates was made early each year in what was described as 'Youth Camp'. **1942** H. C. Warner *Christian Youth Leadership* ii. 26 We have seen the sudden outcrop of Youth Centres, Youth Service Corps, Juvenile Civil Defence Units, etc. **1958** I. Murdoch *Bell* xx. 251 The opening of a new youth centre. **1976** 'W. Trevor' *Children of Dynmouth* i. 21 The Youth Centre curtains are apparently unavailable for the Easter Fête, dear. **1943** C. Day Lewis *Word over All* 16 Oh youth-charmed hours. **1940** *Times* 19 Sept. 7/3 Youth clubs may be found in all districts of the city. **1955** E. Blishen *Roaring Boys* iv. 249 Some of the blocks..had community centres and youth clubs. **1957** J. Osborne *Entertainer* 28, I was teaching Art to a bunch of Youth Club kids. **1980** P. Lively *Judgement Day* i. 7 He was no good at Youth Clubs and disturbed black teenagers. **1968** *Harper's* Oct. 8/3 That temptation to jump on the youthcult bandwagon is hard to resist. **1976** B. Bova *Multiple Man* vii. 73 Aspen was once a center of the youth cult... Kids from all over the country flocked there. **1980** J. O'Faolain *No Country for Young Men* x. 209 Typical Ireland! They got the youth cult ten years late. **1958** *Listener* 28 Aug. 308/2 We know little or nothing about the motivation of the new youth culture, whose emergence is one of the key phenomena of the 'fifties. **1977** M. Dickstein *Gates of Eden* 289 The best ongoing rock criticism, some of it quite sharp, appeared in periodicals associated with the youth culture of the sixties, such as *Rolling Stone, Creem, Crawdaddy* and the *Village Voice*. **1985** *Isis* 3 May 4/2 Youth culture implies the new, the non-conformist, the intractable, yet these still form themselves into cults with rules to keep you right. **1953** C. Day Lewis *Italian Visit* v. 55 I too gave tongue in my piping youth-days. **1948** *Youth Employment Service* (Min. Labour) 1 Local Education Authorities in England and Wales..may be authorised by means of a scheme approved by the Minister of Labour and National Service to operate a Youth Employment Service. **1966** P. Willmott *Adolescent Boys* vi. 105 In theory, the youth employment service is available to help school leavers find suitable work. In fact, although more of the boys in the sample had got their first job through the Youth Employment Officer than any other single source, they were not a majority. **1976** L. Thomas *Dangerous Davies* iv. 30 She had gone to the youth employment office..to inquire about..becoming a nurse. **1946** Koestler *Thieves in Night* 155 The youth-group for a while looked on critically at us rapturous elders. **1929** *Liverpool Post & Mercury* 13 Dec. 7/5 (*heading*) Youth hostels. *Ibid.*, A meeting..held last evening in Liverpool, passed a resolution in favour of proceeding with the formation of a local Youth Hostel Association for the purpose of providing hostels in North Wales for holiday sojourns on the lines of those already existing on the Continent, in Scotland, and Northumberland. **1948** 'N. Shute' *No Highway* ix. 185 'Are you fond of hiking?' 'We used to do a lot,' he said. 'Staying in Youth Hostels?' she inquired. **1972** D. Devine *Three Green Bottles* I. iv. 39, I was youth-hostelling on a hired bicycle. **1977** *Daily Colonist* (Victoria, B.C.) 5 June 24/4 When the original—and still operating—youth hostel opened its doors in the West German town of Altena 68 years ago, its purpose was to provide inexpensive accommodations for young students on walking trips. **1977** C. McCullough *Thorn Birds* xvii. 441 The typical fate of Australians in England, youth-hosteling on a shoestring, working for a pittance. **1933** *Y.H.A. Rucksack* Summer 42/1 This is written as a challenge to Youth Hostellers everywhere to send us word whether they want such a hostel. **1977** M. Drabble *Ice Age* II. 152 Whenever he went to any of the three pubs..he had to spend his time listening to complaints about the behaviour of delinquent youth hostellers. **1947** (*title*) Youth hostelling abroad. **1959** *Woman* 2 May 3/4 Youth Hostelling *is* great fun, for those who enjoy an outdoor holiday, with a sufficiency of male escorts. **1936** *Liverpool Echo* 5 Sept. 4/1 In the streets [of Germany] we met policemen, storm-troopers, youth-leaders, soldiers,..all in their various uniforms. **1958** *Listener* 21 Aug. 256/1 Youth leaders trained in Spain by Franco's Falange. **1973** 'B. Mather' *Snowline* i. 10 A youth leader at a church club in London. **1921** *Survey* 31 Dec. 487 (*heading*) Youth movement of Germany. **1941** 'G. Orwell' *Lion & Unicorn* I. 16 No party rallies, no Youth Movements. **1982** *Listener* 11 Feb. 3/1 The youth movements and their clashes with the police in the 1960s. **1977** *Dept. of Employment Press Notice* 29 June 1 Up to 230,000 unemployed youngsters each year will have a chance of work experience or training under a new £160 million Youth Opportunities Programme announced today by Mr. Albert Booth, Secretary of State for Employment. **1983** *Sunday Tel.* 16 Oct. 11/4 A Midlands businessman who has successfully employed Jamaican youngsters under the Youth Opportunities Programme. **1948** *Times* 22 Apr. 7/4 The Bath Assembly..opened this afternoon in the Pavilion with a concert given by the National Youth Orchestra. **1972** *Daily Tel.* 18 Jan. 9/5 Leicestershire Schools Symphony Orchestra, a 100-strong unit which has a considerable reputation as a youth orchestra. **1959** *Listener* 2 July 17/2 Local institutions: for example, water boards, youth organizations and universities. **1977** *Gay News* 7–20 Apr. 8/3 It's always been known as a youth-oriented action spot. **1982** S. Brett *Murder Unprompted* v. 51 The new youth-oriented culture. **1943** *Ann. Reg. 1942* 68 The Council..advocated..adequate youth services. **1962** *Guardian* 25 Sept. 6/4 A great impact would be felt if this country had a fully developed..youth service staffed by trained social workers and youth leaders. **1975** *Times* 2 Jan. 3/1 Libraries and the youth services will be among the main victims of cuts..in 1975. **1981** *Hansard Commons* 15 Dec. 153 We are able to ask the Manpower Services Commission to ensure that this new youth training scheme is in full operation by the autumn of 1983. **1983** *Times* 18 Jan. 1/4 The Government is putting a £100-a-head value on the work and training opportunities created by new jobs 'brokers' under the £1000m Youth Training Scheme, which starts in Sept. **1944** *Ann. Reg. 1943* 306 Trust funds..for the development of youth work. **1964** 'J. H. Roberts' *'Q' Document* (1965) iv. 84 She..had pangs of conscience and decided to go into youth work. **1976** *Equals* Dec. 6/1 A youth worker is to be appointed to escort young people to interviews,..and generally support those making an uncertain start in work or training.

youthful, *a.* Add: **4.** Comb. *youthful-looking* adj.

1846 Poe in *Godey's Mag.* July 15/1 The likeness conveys a good general idea of the man, but it is far too stout and youthful-looking for his appearance at present. **1954** W. Faulkner *Fable* 8 A man not so young actually, but rather simply youthful-looking. **1977** *New Yorker* 19 Sept. 66/3 Mrs. McCabe is a youthful-looking woman.

youthify (yū·þifəi), *v.* [f. Youth + -ify.] *trans.* To make (a person) appear more youthful. Hence **you·thifying** *ppl. a.*

1945 H. L. Mencken *Amer. Lang.* Suppl. I. 573 *Beauty-parlor*..was displaced by *beauty-shop*... The girls have produced a considerable vocabulary of elegant terms to designate their operations, *e.g. to youthify*. **1960** *Housewife* May 24/3 She..has both to 'youthify' and age actors. **1976** U. Holden *String Horses* iii. 41 He wasn't such a bad looker.., his summer tan was youthifying.

youthly, *a.* Add: **1.** (Later examples.)

1922 E. R. Eddison *Worm Ouroboros* xxvi. 323 Yet is Corinius..a valiant and puissant soldier,..and this one in his youthly age. **1923** H. J. Laski *Let.* 23 Oct. in *Holmes-Laski Lett.* (1953) I. 553, I have read a new *Early Life of Burke* which is full of good things, especially in its recovery of some youthly essays of his which have not appeared before.

2. (Later example.)

1925 T. Dreiser *Amer. Tragedy* (1926) II. II. xxviii. 347 All seeking a glimpse of the astonishingly youthly slayer.

youthquake (yū·þkwēⁱk). *colloq.* [f. Youth after Earthquake.] The series of radical political and cultural upheavals occurring among students and young people in the 1960s.

1967 *Punch* 8 Nov. 708/2 Mary Quant opened her first Bazaar shop..simultaneously with the first tremors of the youthquake. **1970** R. Neville *Play Power* 18 A unique feature of today's Youthquake—as *Vogue* once dubbed it—is its intense, spontaneous internationalism. **1976** *Sunday Mail* (Brisbane) 5 Sept. 19/3 He's built an empire based on the youthquake.

you-uns (yū·ɒnz), *pron. U.S. dial.* Also 9 **youns**; 20– **you uns.** [f. You *pers. pron.* + *uns*, dial. var. *ones* (One *pron.*).] Used in place of You *pers. pron.*

1810 M. V. H. Dwight *Jrnl.* 10 Nov. in *Journey to Ohio* (1912) 37 Youns is a word I have heard used several times, but what it means I don't know. **1869** [see *you-all's* s.v. *you-all *pers. pron.*]. **1885** 'C. E. Craddock' *Prophet Great Smokey Mountains* 7, I hev no call ter spen' words 'bout sech ez that, with a free-spoken man like you-uns. **1927** *Amer. Speech* II. 345 The paterfamilias questioned solicitously: Did you uns sleep good last night? **1934** W. Faulkner *Dr. Martino* 341 Why did you uns have to stop here? **1941** *Amer. Mercury* June 660/2 'Proud to know ye!' Sam will beam. 'Why, you-uns be a-comin' in ter th' fire an' set a spell.'

yow, *int.* Add: **a.** In mod. Austral. and N.Z. use = Wow *int.* 2 in Dict. and Suppl. (Later examples.)

a **1943** L. Esson in *Penguin Bk. Austral. Ballads* (1964) 232 Not er shutter lifted Since they jugged 'im. Yow! **1978** P. Grace *Mutuwhenua* vi. 35, I know. Matter of fact, some of those girls you went round with. Yow! **1983** 'F. Parrish' *Bait on Hook* viii. 111 The rain came... 'Yow,' said Cedric, and shot back into the pub.

yow (yau), *sb. Austral. slang.* [Origin unknown.] In phr. *to keep yow*, to keep a look-out, esp. in order to protect some criminal activity.

1942 E. Langley *Pea Pickers* xix. 283 You keep yow,..and whistle..if anyone comes along. **1965** G. McInnes *Road to Gundagai* xii. 206 Molly kept a look out ('kept yow', as we used to say).

yow(e, obs. ff. Ewe *sb.*[1]: for 'obs.' read 'obs. (exc. dial)' and add examples.

1903 'T. Collins' *Such is Life* (1937) v. 249 He went out back, Cooper's Creek way, with three thousand Gunbar yowes. **1925** 'H. McDiarmid' *Sangschaw* 2 Ae weet forenicht i' the yow-trummle I saw yon antrin thing. **1978** *Jrnl. Lakeland Dial. Soc.* XL. 14 I' a field wid sum yows in there was a crab apple tree. **1979** L. Derwent *Border Bairn* i. 15 Her brother, the shepherd..accepted me more or less as one of his flock. A yowe or a gimmer, a stirk or a stot.

yowie[2] (yau·i). *Austral.* [Origin unknown.] A large, hairy, man-like creature supposedly inhabiting south-eastern Australia.

1976 *Australasian Express* 17 Sept. 10/4 Since 1795 there have been over 3,000 reported sightings of the Snowy Mountains version of the abominable Snowman nick-named the Yowie. **1980** *Courier-Mail* (Brisbane) 4 Jan. 1/8 The 'yowie', a large hairy animal similar to the Himalayan yeti and American Big Foot, has existed in Aboriginal folklore for thousands of years. **1984** *Truckin' Life* (Austral.) Feb. 80/1 'In search of a Yowie.' That was how Theiss Toyota promoted its media function for the release of its new HiLux [*sc.* type of van]..last year.

yowl, *v.* Hence **yow·ler,** one who or that which yowls (in quot. 1935 applied to a crooner).

1935 Wodehouse *Blandings Castle* v. 120 He's a yowler, and girls always fall for yowlers. They have a glamour. **1966** 'L. Lane' *ABZ of Scouse* 120 *Yowler*, a cat. **1979** *Tucson* (Arizona) *Citizen* 20 Sept. 10A/3 In every airport I stand, sip, sleep, weep, wail and yowl in, I find an equal number of other standers, sippers, sleepers, weepers, wailers and yowlers.

yo-yo (yǝᵘ·yǝᵘ), *sb.* [Origin uncertain, but prob. from one of the Philippines languages.] **1.** Also **Yo-Yo.** A proprietary name for a toy in the form of two conjoined cones or discs with a deep groove between them in which a string is attached and wound, its free end being held so that the toy can be made to fall under its own weight and rise again by its momentum.

1915 *Philippine Craftsman* Dec. 363 Sumpit (blowgun), pana (arrow), and yo-yo, however, are names very generally used throughout the islands. *Ibid.* 364 There is evidently some commercial possibility in Filipino toys, for a patent was recently secured upon the yo-yo by a firm in the United States. **1932** *Trade Marks Jrnl.* 2 Mar. 279/2 Yo Yo... All good included in Class 49 [*i.e.* toys and games equipment]. Henry Clement Conlin,..Vancouver, British Columbia, Canada; merchant. **1932** *Evening Standard* 1 June 17/4 He asked me to hold his hand until he became proficient, and I experienced a queer thrill as I brought his hand slowly up and down to make the Yo-Yo respond to the twitch of the string. **1932** Auden in *Rev. Eng. Stud.* (1978) Aug. 281 In the year of my youth when yoyos came in The carriage was sunny and the Clyde was bright. **1933** D. L. Sayers *Murder must Advertise* vi. 99 Ginger brought a Yo-Yo to the office with him and broke the window in the boys' room practising 'Round the World' in his lunch-hour. **1958** *Observer* 12 Oct. 15/7 The yo-yo craze of the thirties. **1972** *Daily Tel.* 20 Dec. 14 Overall trading was at a low ebb and bored dealers spent a very quiet afternoon in yo-yo competitions and other pre-Christmas pastimes. **1984** *New Yorker* 23 July 76/2 They've got the right string but the wrong yo-yo.

b. The pastime of playing with a Yo-Yo.

1932 *Daily Express* 2 July 3/3 Some boys playing yo-yo attracted the Queen's attention. **1932** *Morning Post* 15 July 10/5 Games similar to Yo-Yo have been played in almost every age.

2. *fig.* or in fig. context. **a.** Something or someone going continually up and down, or to and fro; also, such a motion or fluctuation. Freq. used in comparative phrases referring to emotions or spirits rising and falling like a Yo-Yo.

1958 *Listener* 16 Oct. 623/2 What is it like to be a human yo-yo, driving all day on a ten-minute bus route. **1963** L. Deighton *Horse under Water* xi. 47 Singleton was jumping in and out of the water like a yo-yo. **1973** C. Bonington *Next Horizon* xii. 168 With a bit of luck the constant yo-yo between Scheidegg and the snow cave would be over. **1975** *Times Lit. Suppl.* 12 Dec. 1496/4 Confronted by these dramatic developments, transport policy ceased to be the political yo-yo it had previously been. **1976** J. Grenfell *Joyce Grenfell requests Pleasure* xvi. 230 Our spirits went up and down like yo-yos. **1980** *Times* 13 Sept. 10/5 Alarming yo-yos in the quality of food and service. **1981** *Daily Tel.* 19 Oct. 25/4 (*heading*) Interest rates all a big yo-yo. **1984** S. Townsend *Growing Pains A. Mole* 23 Your emotions are up and down like a yo-yo.

b. A stupid person, a fool. *U.S. slang.*

1970 *New Yorker* 28 Nov. 40 He would leer, and categorize them in a loud, mocking voice. ('Weirdo' was one of his favourite appellations; also 'Freak', 'Yo-Yo', and creep.) **1975** *Ibid.* 20 Jan. 29/1 Some yo-yo of a technician there pulls the control rods out of the core to polish them with Rally wax. **1978** V. Bugliosi *Till Death us do Part* xi. 325 I've got enough problems without some punk yo-yo threatening me.

3. *attrib.* Marked by a continual up-and-down or to-and-fro motion; continually passing from and into a condition. Also *fig.*

1932 *Amer. Speech* VII. 272 *Yo-yo driller*, a cable-tool driller. **1960** *Spectator* 30 Sept. 501 It isn't the industry's fault that we have a yo-yo economy. **1963** T. Pynchon *V.* i. 29 Though they only thought about one another at random, though her yo-yo hand was usually busy at other things, now and then would come the invisible, umbilical tug. **1977** *Lancet* 15 Oct. 792/1 There was no improvement in patients with severe on-off disabilities with freezing and rapid oscillations ('yo-yo' effect). **1979** *Globe & Mail* (Toronto) 22 Jan. 4/3, I want this job because all the jobs I've had have been yo-yo jobs, I've been laid off my job four times in the last six years.

yo-yo, *v.* [f. prec. sb.] **1.** *intr.* To play with a Yo-Yo.

1932 *Daily Express* 30 June 12/6 (*heading*) Do you yo-yo? **1973** *N.Y. Times* 14 Apr. 18/2 The idea is to go where the kids are and to teach them how to yo-yo.

2. *fig.* To move up and down, or between one point and another; to fluctuate.

1967 *Punch* 12 Apr. 514 The hard facts underlying our economic health—as opposed to the headlines and Treasury press releases—just don't yo-yo about like this. **1973** C. Bonington *Next Horizon* xi. 166 In those early stages of the climb we yo-yo'd back and forth between Kleine Scheidegg and the Rock Band. **1976** *Time* 20 Dec. 23/2 He has yo-yoed between 210 and 296 lbs., now carries a bulky 263. **1978** *Sunday Times* 21 May 53/1 City rates of interest have yo-yoed.

Hence **yo-·yoing** *vbl. sb.* and *ppl. a.*; also **yo-·yoer, yo-·yoist,** one who plays or performs with Yo-Yos.

1933 *Spectator* 6 Jan. 23/2 Surely the unkind girl of figure fifty-eight is the precursor of the modern yoyoist. **1947** *Sat. Even. Post* 10 May 12 His Caspar Milquetoast model for timid yo-yoists was lighter, and, he added, prettier. **1963** T. Pynchon *V.* i 30 As it turned out, the New Year's party was to end all yo-yoing. **1967** F. Conroy *Stop-Time* viii. 114 The greatest pleasure in yo-yoing was an abstract pleasure—watching the dramatization of simple physical laws. **1973** *N.Y. Times* 14 Apr. 18/2 When you're a professional yo-yoer, how can you take anything very seriously? **1976** *Alyn & Deeside Observer* 10 Dec. 34/5 It is more volatile than the yo-yoing pound. **1980** *Illustr. London News* Mar. 56/1 The uncomfortable and costly yo-yoing of temperature experienced in light-weight buildings with large areas of window.

yperite (ī·pərəit). [ad. F. *ypérite*, f. *Ypres*, name of the town in Belgium where the gas was first used, in 1917: see -ite[1].] = *mustard gas* s.v. *Mustard *sb.* 3 c.

1919 *Chem. Abstracts* XIII. 3063 (*heading*) Yperite and poisonous gases. **1940** *Times* 17 Aug. 4/4 Dr. Gerard C. Savoy, of Lausanne, has produced a preparation which is the most efficient known antidoe to yperite (mustard gas). *Ibid.*, Fifty per cent. of the yperite present in the wound is destroyed after one hour. **1979** *Microbiology* XLVIII. 246 Nitrous yperite and other alkylating substances with a radiomimetic effect can inhibit cell division in *Escherichia coli* B.

Ypresian (īpre·siăn), *a. Geol.* [ad. F. *ypresien* (A. H. Dumont 1850, in *Bull. de l'Acad. des Sci.*, etc., *de Bruxelles* XVI. II. 368), f. as prec.: see -ian.] Of, pertaining to, or designating the lowest stage of the Eocene in western Europe, lying above the Landenian. Also *absol.*

[**1852** *Q. Jrnl. Geol. Soc.* VIII. 323 The group of tertiary strata which we meet with next in the descending order in Belgium (comprising the 'Systèmes Laekenien, Bruxellien et Ypresien' of M. Dumont) corresponds most nearly..in age with the Barton, Bagshot, and Bracklesham beds.] **1880** J. D. Dana *Man. Geol.* (ed. 3) III. 512 London Clay; Lower Ypresian of Belgium. **1969** Bennison & Wright *Geol. Hist. Brit. Isles* xv. 339 The Ypresian cycle commenced with the basement bed of the London clay. **1977** A. Hallam *Planet Earth* 227 The most commonly accepted stages [of the Eocene] include (from oldest to youngest) the Ypresian, Lutetian and Bartonian.

Yquem (īke·m). The name of Château d'*Yquem*, a vineyard in the Gironde, France, used *absol.* and (*rarely*) *attrib.* to denote a variety of fine, rich Sauternes wine produced and bottled there.

1869 A. Trollope *Phineas Finn* I. viii. 29 They give you a capital little dinner at Moroni's, and they've the best Chateau Yquem in London. **1902** Belloc *Path to Rome* 396 He had..Yquem with his fish, the best Chambertin during the dinner. **1927** C. Connolly *Let.* 11 Feb. in *Romantic Friendship* (1975) 252, I drank..Yquem at 300 francs the bottle. **1959** *Sunday Times* 1 Nov. 32/5, I find Yquem too sweet. **1982** D. Peppercorn *Bordeaux* xi. 315 A good vintage of Yquem is the quintessence of Sauternes.

yrast (i·ræst), *a. Nucl. Physics.* [See quot. 1967.] Pertaining to or designating any nuclear energy level that is the lowest for some value of the spin; *yrast line*, a line on a graph of spin against nuclear rotational energy (or some function of each of these), connecting points representing the various yrast states of a nuclide.

1967 J. B. Grover in *Physical Rev.* CLVII. 832 It has been proposed that the lowest-energy excited state at a given angular momentum be called 'yrast' level for that angular momentum. *Ibid.* 832/1 The yrast levels play a crucial role in deciding the course and outcome of many nuclear reactions. [*Note*] By 'yrast' level of a given nucleus, at a given angular momentum, is meant the level with least energy at that angular momentum. The English language seems not to have a graceful superlative form for adjectives expressing rotation. Professor F. Ruplin (of the Germanic Languages Department of the State University of New York, Stony Brook) suggested the use of the Swedish adjective *yr* for designating these special levels. This word derives from the same Old Norse verb *hvirfla* (to whirl) as the English verb *whirl*, and forms the natural superlative, *yrast*. It can thus be understood to mean 'whirlingest', although literally translated from Swedish it means 'dizziest' or 'most bewildered'. **1971** *Physics Lett.* XXXIVB. 575 Non-axially symmetric deformation of a nucleus is introduced to account for the admixture of the collective rotational bands and the distribution of the yrast levels. **1974** Frauenfelder & Henley *Subatomic Physics* xvi. 435 The yrast line of a nucleus gives *E* as a function of *J*. **1983** *Nucl. Physics* A. CDIII. 421 Precession of the yrast 2†, and 4†, and the second 2‡ states in ²⁴Mg have been measured.

yttro-. Add: **yttrotu·ngstite,** a basic oxide of yttrium and tungsten, $YW_2O_6(OH)_3$, occurring as yellow monoclinic crystals.

1950 E. H. Beard in *Colonial Geol. & Mineral Resources* I. 51 The results show conclusively that the term 'thorotungstite' is inapplicable as a true description of the mineral... It is suggested that the mineral in future be called yttrotungstite. **1971** *Mineral. Mag.* XXXVIII. 262 All the specimens studied consisted mainly of a fine-grained aggregate of pale-yellow earthy yttrotungstite, with rare druses lined by yttrotungstite crystals.

yu (yü). *Archæol.* [Chinese.] An ancient Chinese wine vessel in the form of a small metal pail with a swing handle and a decorative cover, popular in the Shang and Early Zhou periods.

1904 S. W. Bushell *Chinese Art* I. iv. 90 The sacrificial wine vessels illustrated..have been selected as the most ancient pieces in the collection... The first is an ovoid jar (*yu*), of the shape used by the old kings for presents of wine to deserving subjects, with a cover surmounted by a knob, and a loop handle ending in dragons' heads. **1945** P. Ackerman *Ritual Bronzes of Anc. China* iv. 99/1 *Yu* range from six to nine inches high... They often come in pairs. **1973** W. Watson in *Genius of China* (R. Acad.) II. 74/2 The *yu* with its high handle is an invention of late Shang.

Yuan[1] (yu,ā·n, ‖ yue·n). Also 7 **Ivena,** 8 **Ywen,** 8–9 **Yuen; Yüan.** [a. Chinese *yuán*, lit. 'first'.] **a.** The name of the Mongol dynasty established as rulers of all China by Kublai Khan in 1279 and in power until 1368.

Kublai Khan named his kingdom Yuan in 1271, and the Yuan dynasty is often described as beginning in that year.

1673 [see *Sung *sb.* a]. **1738** tr. *J. B. Du Halde's Descr. of Empire of China & Chinese Tartary* I. 214 (*heading*) The twentieth dynasty, call'd Ywen, which contains nine emperors in the space of eighty nine years. **1788** Gibbon *Decl. & Fall* VI. lxiv. 298 The annals of the Moguls or Yuen. **1836** J. F. Davis *Chinese* I. v. 182 On the accession of Koblai Khan, the first of the *Yuen* dynasty,..an order was promulgated to burn all the books of the Taou sect. **1948** D. Diringer *Alphabet* 355 This character..was only sparsely used but it lingered on at the imperial Chancery under the Yüan dynasty. **1966** F. Schurmann *Ideology & Organization in Communist China* i. 53 The Yüan Mongols relied heavily on a traditional bureaucratic elite to rule the country. Many of these bureaucrats remained loyal to the Yüan after 1368. **1977** *N.Y. Rev. Bks.* 26 May 22/1 The wreckers had found, during their work, the foundations of a gate of the Yuan era.

b. Used *attrib.* and *absol.* of the art and porcelain of the Yuan period.

1888 F. Hirth *Ancient Porc.* 50 Pieces of a surface which bears no resemblance to any of the classical Sung or Yüan monochrome vessels..are very common. **1933** [see *Sung *sb.* b]. **1969** R. Quest *Cerberus Murders* iv. 30 Whistler and his circle went in for blue-and-white..but it was all late stuff... This is all Hsüan-tê or Yüan. **1978** *Nagel's Encycl.-Guide: China* 199 In old Yuan opera, one actor alone, the hero of the play, had the privilege of singing certain parts of his role.

yuan[2] (yu,ā·n). Also **yüan.** Pl. **yuan.** [a. Chinese *yuán* round.] **1.** A Chinese unit of currency introduced in 1914, equal to 10 *jiao*; a coin of this value.

1921 J. V. A. MacMurray *Treaties & Agreements with China 1894–1919* I. 853 The Law for the National Currency.—January 1914... Article 2.—The unit of the national coins shall be called *yuan*, and the *yuan* shall contain..23·97795048 grammes of pure silver. **1927** *Glasgow Herald* 14 Jan. 8 The surtax imposition will yield..an advantage of 12,000,000 yuan of revenue. **1949**, etc. [see *jiao]. **1962** E. Snow *Other Side of River* (1963) ii. 23 At the Hsin Ch'iao a small suite consisting of a sitting room, bedroom and bath cost me 24 yuan..a day. **1976** W. H. Canaway *Willow-Pattern War* xvii. 173 Yang Ma-wei gave me a fistful of yüan and told me to..find Thupten at the tea-house, and pay him. **1982** C. Thomas *Jade Tiger* ii. 51 He offered the stallkeeper one of his own grubby ten yuan notes.

2. *Chinese Archæol.* A flat ring or perforated disc made of jade, widely circulated from the Shang period to the Hang dynasty.

1912 *Field Museum Nat. Hist.* X. v. 154 There are three kinds of annular jade objects, called *pi*..*yüan*..and *huan*... The former is a disk with a round perforation in the centre, the two latter are rings. **1958** W. Willetts *Chinese Art* I. ii. 89 The *yüan* has a perforation twice the width of the body substance, so that its diameter is half that of the whole ring. **1963** K. Chang *Archaeol. Anc. China* ix. 276 A ceremonial pit was uncovered which contained over twenty stone discs of various sizes and a number of jade and stone ceremonial objects (circular *yüan*, square *tsung* tubes, *wuan-kui*, etc.).

yuan[3] (yu,ā·n). Also **Yuan.** Pl. **yuan.** [a. Chinese *yuàn* courtyard, yard.] Each of several government institutions (e.g. *guo wu yuan* the State Council, *waiyuxueyuan* a foreign languages institute) in China.

1928 *China Year Bk.* xxvi. 1234 Administrative Court (*P'ing Cheng Yuan*)... The principal officials of the Administrative Court are the President (Yuan Chang)..and 15 judges. **1938** E. Teichman *Affairs of China* xiv. 205 On the governmental side the Central Executive Committee of the Kuomintang gives birth to the National Government of the Republic of China, composed of a President..and five *Yuan*, or governing committees: the Legislative *Yuan*, charged with the making of laws; the Judicial *Yuan*, charged with the administration of justice..the Examination *Yuan* charged with selection of officials of the public service; the Control *Yuan*, charged with the supervision of the national administration..and the Executive *Yuan*, charged with the actual administration of the government. **1947** *Sun* (Baltimore) 22 Aug. 6/3 They..will be asked to participate in the selection of the National Assembly and the legislative Yuan. **1967** *Sunday Times* 14 May 6/1 Chiang's executive Yuan (Cabinet) has discussed measures. **1979** *China Yearbk.* ix. 98/2 In the event of a dispute among the various Yuan, the President may call a meeting of the Presidents of the Yuan concerned for consultation on a solution.

‖ **yuan hsiao** (yüă·n χʸiau). Also **yüan hsiao** and with hyphen. [Chinese *yuánxiāo* (in Wade-Giles *yüanhsiao*), f. *yuán* first + *xiāo* night.] A sweet rice-flour dumpling made for the Chinese Lantern Festival (15 January in the lunar calendar).

1956 B. Y. Chao *How to cook & eat in Chinese* II. xvii. 210 Orange soup with *yüan-hsiao*... Knead the glutinous rice flour with ½ cup hot..water. Then make into globules of about ½ inch in diameter. These are the *yüan-hsiao*. **1972** K. Lo *Chinese Food* I. 56 *Yuan hsiao* is a form of Chinese festival sweet, served in a rather bland rice soup. *Ibid.* 57 We Chinese love these *yuan hsiaos*, partly, I think, because of happy childhood memories of being allowed to do the rolling.

Yucatec (yū·kătek). Also **Yucate·co** and with small initial. [ad. Sp. *yucateco*, f. *Yucatán*, earlier *Yocotán*, adapted from a Maya name for the language of the Mayan Chontal Indians.] **a.** An American Indian of the Yucatán Peninsula in eastern Mexico; such Indians collectively. **b.** *colloq.* Any present-day inhabitant of the Peninsula or of the Mexican state of Yucatán in its northern part.

1843 J. L. Stephens *Incidents of Travel in Yucatan* I. vi. 139 No native ever calls himself a Yucateco, but always a Macegual or native of the land of Maya. **1845** *Trans. Amer. Ethnol. Soc.* I. 107 The Yucatecs differed materially from the Mexicans with regard to the time of the solar year. **1875** H. H. Bancroft *Native Races Pacific States* II. xxi. 675 So great was the horror in which the Yucatecs held this crime that they did not always wait for conviction,..but sometimes punished a suspected person. **1912** *Contemp. Rev.* Feb. 257 The better Yucatecos do not lean to this profession [of clergyman], which is unendowed. **1966** T. Pynchon *Crying of Lot 49* v. 119 He was part-owner here with a yucateco who still believed in the Revolution. **1974** *Encycl. Brit. Micropædia* X. 841 The Yucatec were the Classical Maya who were conquered by the Spanish and whose calendar, architecture, and hieroglyphic writing marked them as a highly civilized people. **1983** *Word Ways* Aug. 152 Originate in Yucatan and you are a Yucatec.

c. The language of the Yucatán Indians, a Mayan language.

1940 F. Johnson in *Maya & their Neighbors* vi. 107 The divisions of the Yucatec-speaking Maya are relatively indistinct. **1954** J. E. S. Thompson *Rise & Fall Maya Civilization* i. 28 Yucatec is spoken by many whites and mestizos of Yucatán as a second language and is said to be easy to learn. **1977** *Language* LIII. 296 Tall people can be reclassified by one of the long classifiers in Bantu..., Japanese, and Yucatec.

d. *attrib.* or as *adj.*

1875 H. H. Bancroft *Native Races of Pacific States* II. xxi. 665 A Yucatec noble who wedded a woman of inferior degree, descended to her social level. **1934** A. Toynbee *Study of Hist.* I. 123 The Yucatec Society was apparently incorporated into the Mexic Society by conquest at about the turn of the twelfth and thirteenth centuries of the Christian Era. **1956** *Publ. Amer. Dial. Soc.* XXVI. 25 It is their speech which Vasquez..describes as 'yucateco Spanish', characterized by Mayan phonemes and Mayan phrases. **1975** *Sci. Amer.* Oct. 74/3 They spoke a dialect unlike the Yucatec Maya dialect heard generally throughout Yucatán; it was Chontal Maya, one of the dialects of the Cholan Maya group. **1983** *Washington Post* 13 July E-3/1 You can be sure that Yucatec farmers don't waste maize on their cows. **1983** *Times Lit. Suppl.* 7 Oct. 1090/5 The Yucatec city of Chichen Itza.

Hence **Yucate·can** *a.* and *sb.*

1869 *Proc. Amer. Philos. Soc.* XI. 5 The most important dialects of the Maya are the Yucatecan, the Quiche, the Cakcquichel, [etc.]. **1886** *U.S. Cons. Rep.* LXVII. 495 A fair sample of Yucatecan agriculture. **1909** *Athenæum* 4 Dec. 688/1 Of the Yucatecans themselves the authors have nothing good to say. **1931** E. H. Morris *Temple of Warriors* xix. 228 Gold and copper were the only metals known to pre-Columbian Yucatecans. **1950** *Caribbean Q.* II. II. 30 Actual settlement was limited at first to the Belize district, and even that was made precarious by the attack of the Spaniards and Yucatecan Indians.

Yuchi (yū·tʃi). Also 8–9 **Euchee, Euchi(e, Uchee.** [a. Creek, of uncertain origin.] **a.** (A member of) an Indian people formerly inhabiting the region of the Savannah river in Georgia and South Carolina, and now incorporated into the Creek nation in Oklahoma.

1738 W. Stephens *Jrnl.* in *Colonial Rec. Georgia* (1906) IV. 75 He understood they were a Party of the Euchies. **1741** in *South Carolina Hist. Soc. Coll.* (1887) IV. 40 Thomas Jones..was Employed..as a Linguist to the Creeks and Euchees. **1744** in *Georgia Hist. Soc. Coll.* (1840) I. 145 Their cattle..had strayed away and eat the Uckee's corn. **1818** *Lynchburg* (Va.) *Press* 25 Dec. 3/1 The captain..reports to have taken three warriors, a Creek, a Choctaw, and a Uchee. **1893** *Amer. Anthropologist* VI. 280 The Yuchi believe themselves to be the offspring of the sun. **1965** [see *Creek *sb.*[2]]. **1975** W. L. Ballard in J. M. Crawford *Stud. Southeastern Indian Lang.* 163, I spent approximately six weeks in Sapulpa, Oklahoma, interviewing Yuchis.

b. The language of this people.

1836 *Trans. & Coll. Amer. Antiq. Soc.* II. 96 These five languages, the Muskhogee and the Hitchittee, the Uchee, the Natches, and the Alibamon or Coosada are, it is believed, the only ones spoken by the different tribes of the Creek confederacy. **1909** F. S. Speck *Ethnol. Yuchi Indians* 15 It is quite certain now that Yuchi is spoken in only one dialect. **1975** W. L. Ballard in J. M. Crawford *Stud. Southeastern Indian Lang.* 163 A number of tapes of conversations in Yuchi were made.

yuck (yɒk), *sb.*[1] *slang* (orig. *U.S.*). Also **yuk.** [Origin unknown.] A fool; a boor; anyone disliked or despised.

1943 H. A. Smith *Life in Putty Knife Factory* xiv. 239 *Yuck* is a word introduced into the language by Fred Allen. A yuck is a dope who makes a practice of going around appearing on quiz programs. That was its original definition; it now means a dope of any description. **1948** R. Chandler *Let.* 27 Jan. (1981) 105 The public's capacity and adaptability to a quality of entertainment which the yucks seem to be afraid to give them. **1957** M. Shulman *Rally round Flag, Boys!* (1958) vi. 67 The yucks who look at television don't know the difference between Ernest Hemingway and Huntz Hall. **1972** P. Rowlands *Fugitive Mind* xi. 132 'Is your brother a yuk?' Clare asked Sally. 'Oh yes! He's a *terrible* yuk! He shouts a lot, he breaks my toys, he pushes me over.' **1979** J. Wainwright *Duty Elsewhere* xx. 56 Three no-good yucks had felt like playing footsie with the law.

yuck (yɒk), *int.*, *sb.*[2], and *a. slang.* Also **yuk.** [Imitative. Cf. *yech *int.*, *yuck *v.*[1]] **A.** *int.* An expression of strong distaste or disgust.

1966 R. H. Rimmer *Harrad Experiment* (1967) 25 Across the table; Dorothy Stapleton and Valerie Something-or-other belong (yucks, is that the right word?) to Herber Snyder and Peter Longini. **1970** *It* 12–25 Feb. 16/4 The whole tenor of the epistle is that of one elite talking to another without reference to..those who have paid...Yuk. **1976** G. Moffat *Short Time to Live* vii. 58 'Fish pie perhaps, and parsley sauce.' 'Yuk,' said Arabella. **1981** P. Dickinson *Seventh Raven* x. 130 You took a harmless animal and chopped it up..to please your God—yuck, they thought. **1983** D. Simpson *Puppet for Corpse* xx. 172 It was the way he talked about her... 'You know what older women are, wink, wink.'.. Yuk!

B. *sb.*[2] Messy, unpleasant, or distasteful material. *lit.* and *fig.*

1966 *New Statesman* 19 Aug. 258/1 Rotting wodges of chilly yuck which once were apples and pears. **1971** P. Purser *Holy Father's Navy* xxxviii. 187 There was a lot of yuk which I didn't investigate too closely, and a bit of ear definitely in the wrong place. **1977** *Times* 17 Oct. 12/6, I asked Nancy Grimes, a freelance who arranges plants for people... 'The offices now are so ugly and so standard, such standardized urban yuk... They want to see something that can actually survive and grow here. It gives them reassurance.' **1981** M. E. Atkins *Palimpsest* xii. 118 One of those syndicated advice columns... All noble sentiments and romantic yuk.

C. *adj.* **1.** = *yucky *a.* a.

1971 *TV Times* (Austral.) 24 Feb. 39/2 Business was a bit yuk and I was bugged by this lack of confidence. **1973** P. Dickinson *Green Gene* ii. 28 She's got a really yuck family, even worse than mine.

2. Comb. **yuck-making** *ppl. a.* = *sick-making* ppl. adj. s.v. *sick *a.* 11.

1972 *Courier-Mail* (Brisbane) 30 Mar. 1/10 The BBC yesterday described a song about the former Australian Prime Minister..as the biggest 'yuk-making' piece of propaganda in politics. **1975** *Listener* 7 Aug. 168/3 Nasty, yuk-making remarks.

yuck (yɒk), *v.*[1] *Canad. dial.* [Imitative.] *intr.* To vomit.

1963 *Amer. Speech* XXXVIII. 301 [Newfoundland.] *Yuck*, to vomit. **1981** *Publ. Amer. Dial. Soc.* LXVIII. 54 [Newfoundland.] *To vomit*, yuck.

yuck (yɒk), *v.*[2] *slang.* (chiefly *N. Amer.*). Also **yuk.** [Origin unknown. Cf. *yock *sb.* (and *v.*).] *intr.* **a.** To fool around; to act so as to cause laughter. **b.** To laugh. Also *to yuck it up*.

1964 S. Bellow *Herzog* 119 And Gersbach, boisterous, yucking it up, poured whisky, wine, pounded the table. **1967** *Boston Sunday Herald* 2 Apr. (T.V. Mag.) 9/2 Russ Tamblyn and Sidney Poitier yuk it up as the lightly clad barbarians; Rosanna Schiaffino and Beba Loncar play it cool as the lightly clad camp followers. **1969** *Listener* 23 Jan. 98/3 An American watches British television..for visions of America. Stray ones move us: Wally Schirra yukking it up in space. **1974** *Publishers Weekly* 12 Aug. 55/3 Laurel and Hardy fans..should enjoy this semi-biography... Old-timers who yukked when they saw the movies way-back-when should swell the market. **1975** *Time Out* 11 Apr. 36/4 Pryor has them yukking at whitey one moment and at themselves the next.

Hence **yuck** *sb.*[3], a laugh.

1971 *Daily Colonist* (Victoria, B.C.) 10 Feb. 5/2 The biggest yuks, as might be expected, are to be found right here in Canada. **1976** *National Observer* (U.S.) 16 Oct. 10/3 The biggest yuck of the night was when Mr. T. called Mrs. Llewelyn 'Mrs. Rreweryn'. **1977** *Canadian* 2 Apr. 20/1 Humor is his forte. Looking for yuks? Phone Sammy. **1984** *Sun-Times* (Chicago) 17 Feb. 49/2 The movie gets its yuks with slapstick scenes where one guy goes out the window when the other guy comes in the door.

yucky (yɒ·ki), *a. slang.* Also **yukky.** [f. *yuck *a.*] **a.** Nasty, unpleasant; sickly sentimental.

1970 D. Uhnak *Ledger* v. 79 She wanted to go to a lousy, yukky secretarial college. **1977** *Oxford Times* 1 July 15 The sweetness is fused with enough real feeling to avoid being sugary, except for the rather yucky spoken introduction to 'Meadows of Springtime'. **1980** *Sunday Times* 13 Jan. 61/1 To develop an improved instrument for doctors dealing with emergency cases of perforated lungs, the research boys set to work with butchers' skewers and a lot of belly of pork. It may sound yukky to the squeamish. **1981** M. Gordon *Company of Women* III. 240 It's only bats, I say... 'They're weird,' says Linda. 'Yucky.'

b. Messy, 'gooey'.

1975 *Times Lit. Suppl.* 13 June 661/3 Peanut butter, that yucky staple standby of the American snack-eater. **1977** J. Wilson *Making Hate* v. 62 Let's get these yucky things off and get you washed.

Hence **yu·ckiness.**

1982 E. North *Ancient Enemies* ix. 120 The ringing in my ears and general lassitude and yuckiness. **1984** *InfoWorld* 14 May 18/1 So you go for 100 shares... Let's say that by November it soars back up to 68 5/8, where it peaked before all this market yuckiness began.

yud (yud), dial. var. Head *sb. rare.*

1874 [see *innards *sb. pl.*]. **1882** E. L. Chamberlain *Gloss. W. Worcestershire Words* 35 *Yud*,..head. **1974** W. Foley *Child in Forest* I. 100 I'll 'old her yud still, and..you get a good dose down 'er gullet.

Yüeh[1] (yū·ə). Also **Yueh.** [f. the name *Yüeh Chou* of a town (now called Shaoxing) in Zhejiang Province, China.] A type of stoneware distinguished by a celadon glaze, first produced in the Six Dynasties period and perfected during the Tang dynasty. Freq. *attrib.*, esp. as *Yüeh ware*.

[**1887** *Jrnl. R. Asiatic Soc. N. China Branch* XXII. 134 The author of the *T'ao-Shuo* begins his treatise on ancient porcelains with the Yüeh-chou potteries of the T'ang dynasty.] **1910** S. W. Bushell *Descr. Chinese Pott. & Porc.* II. ii. 35 The Hsing-chou porcelain resembles silver, while the Yueh-chou porcelain resembles jade... Hsing is inferior to Yueh. **1915** R. L. Hobson *Chinese Pott. & Porc.* v. 59 Surely this cannot be far removed from the 'secret colour' of the Yueh ware. **1933** *Burlington Mag.* Sept. 122/2 When Yüeh Chou was capital of a principality, Yüeh ware was made exclusively for the princely court and was known as *pi sê yao* or ware of forbidden colour. **1958** W. Willetts *Chinese Art* II. vi. 439 In 1930 a Yüeh Kiln was found by Yonayama at Tê-ch'ing near Hangchow. *Ibid.* vi. 442 A number of whole bowls in the shape of lotus flowers are illustrated, all reputedly being Yüeh *yao*. **1972** *Trans. Oriental Ceramic Soc.* XXXVIII. 23 The fully evolved ware which T'ang poets praised is generally termed *Yüeh ware*..the earlier product being called *Old Yüeh* by the Japanese.

Yüeh[2] (yū·ə). Also **Yueh.** [Chinese.] (A member of) a group of peoples originally living in the coastal provinces of southern China, who expanded into south-east Asia during the third century B.C.

1901 E. A. Parker *China* ii. 23 The..Yüeh tribes..seem to have very soon lost their separate identity, and to have either permanently retired into Annam proper or to have been merged into the Chinese. **1934** K. S. Latourette *Chinese* I. ii 51 Both Wu and Yüeh seem to have depended in part upon boats for their victories, navigating these craft on the sea and on the rivers and lakes in which their possessions abounded. **1966** W. G. Goddard *Formosa* i. 16 The Yueh in coastal China. **1972** M. Sheppard *Taman Indera* 5 At the beginning of the Christian era well-established trade links existed between South China, India and the Red Sea. Many different traders and sailors joined in operating this route,..leaving the Yuehs to control the final sector. **1974** *Encycl. Brit. Macropædia* XIX. 120/1 A long-held notion that identified the Vietnamese with one tribe of the Viets of southern China (Yüeh in Chinese) has been abandoned.

Yüeh[3] (yū·ə). Also **Yue.** [ad. Chinese *Yuè*, a former name for Guangdong province.] A Chinese dialect spoken in parts of the provinces of Guangdong and Guangxi. Freq. *attrib.*

1954 M. A. Pei *Dict. Ling.* 34 *Cantonese*..spoken in Kwang-tung... The indigenous name of this vernacular is *Yüeh*. **1961** Chang-tu Hu et al. *China* v. 101 Yüeh or Cantonese is spoken by some forty million people in China and abroad. **1974** *Encycl. Brit. Macropædia* XVI. 801/2 The most important representative of the Yüeh languages is Standard Cantonese of Canton, Hong Kong, and Macao. **1978** *Nagel's Encycl.-Guide: China* 70 Yue dialect: most of Guang dong and south eastern Guang xi. **1982** C. Thomas *Jade Tiger* ii. 43 The man spoke in northern Min dialect as opposed to his own expatriate Yue dialect.

‖**yüeh**[4] (yū·ə). *Archæol.* Also **yueh.** [Chinese *yuè*.] A bronze battle-axe or halberd, esp. one of the Shang period.
1956 W. C. White *Bronze Culture Ancient China* 4 *Axes* (Yüeh), the most common type of axe is found in a variety of shapes. *Ibid.* 58 (*caption*) Axe-head (Yüeh), with socket for hafting. 1960 Cheng Te-k'un *Archaeol. in China* II. iv. 69 Larger, and consequently richer, tombs would be provided with bronze vessels as well as weapons, such as *ko* dagger axes..*yueh* broad-axes. 1964 M. Medley *Handbk. Chinese Art* 14/2 Axes, called *yüeh*, *ch'i* or *fu*, are either tanged or socketed. 1978 *New Archaeol. Finds in China* II. 29 Also found was a bronze weapon *yueh*.

‖**yüeh ch'in** (yū·ə tʃin, kin). Also **yueh-ch'in, yu-kin, yukin,** 9 **yuĕ kin.** [Chinese (Pinyin *yuè qín*), lit. 'moon guitar'.] A Chinese lute with four strings and a flat, circular body.
1839 *Chinese Repository* VIII. 44 The *pepa* and *yuĕ kin* are of easy purchase. 1909 *Cent. Dict.* Suppl., *Yu-kin*,..a Chinese lute or guitar with a large circular body, a short neck, and four strings. 1954 *Grove's Dict. Mus.* (ed. 5) II. 239/2 *Yüeh ch'in*.., flat lute of four strings. Used..to accompany ballads. Now rare. 1962 E. Snow *Other Side of River* (1963) lxxiii. 566 They may specialize in piano, violin, cello, flute, or one of the standard Chinese strings: *p'i-p'a, yang-ch'in, yueh-ch'in*, and others. 1971 J. R. Bernasconi *Collectors' Gloss.* 396 *Yukin*, a Chinese instrument with four strings played by plucking. 1975 [see *san hsien].

yugawaralite (yugæ·wɑrāləit). *Min.* [f. *Yugawara*, name of a town in Japan near where it was first found + -l- + -ite[1].] A hydrated aluminosilicate of calcium, $CaAl_2Si_6O_{16}.4H_2O$, that is a member of the zeolite group and occurs as colourless or white monoclinic crystals having a vitreous lustre.
1952 Sakurai & Hayashi in *Sci. Rep. Yokohama Nat. Univ.* Section II. 1. 77 We can not identify this mineral with the other zeolites, and we may consider this to be a new zeolite. We wish to [call] it 'Yugawaralite' after its locality. 1969 *Acta Cryst.* B. XXV. 1190/1 It is difficult to place yugawaralite in any of the seven recognized groups of zeolites, although it shows certain similarities to both mordenite and heulandite. 1978 *Mineral. Rec.* IX. 296/1 Yugawaralite, a relatively rare calcium zeolite,..has been found in specimens mined from the Khandivali quarry near Bombay, India.

‖**yugen** (yū·gən). Also **yūgen** and with capital initial. [Jap., f. *yū* dark + *gen* the unfathomable.] In traditional Japanese Court culture, esp. poetry and, later, the Nō play, a hidden quality of graceful beauty or mystery; profound aestheticism.
1921 A. Waley *Nō Plays of Japan* 21 The difficult term yūgen which occurs constantly..is derived from Zen literature. It means 'what lies beneath the surface'; the subtle, as opposed to the obvious; the hint, as opposed to the statement. 1932 *Times Lit. Suppl.* 6 Oct. 715/1 Such a couplet..can hardly be said to capture the spirit of *yugen*. 1959 *Ibid.* 15 May 291/2 This brief basking in the Shōgun's favour led Zeami to stress the importance in performance of Yūgen, or elegant beauty and gentleness. 1970 *Daily Tel.* 16 May 9/4 His smooth curving movements of a fan, together with circling movements across the stage..gave a poetic suggestion of flight, illustrating the *Yugen*—indirection, allusiveness, mystery—which lies at the heart of Nō.

Yugo (yū·go), colloq. abbrev. of *Yugoslavian *a.* and *sb.*
1941 *Daily News* (N.Y.) 20 Mar. 19/1 (*heading*) Yugos and Axis in compromise. *Ibid.* 24 Mar. 3/3 (*heading*) Yugo disorders, Greek warning stay Axis pact. 1963 I. Fleming *On H.M. Secret Service* xi. 117 'Which one was it, anyway?' 'One of the Yugos. Bertil.' 1982 'I. I. Magdalen' *Search for Anderson* i. x. 47 There was someing wrong about that Yugo shoot-out. It stank.

Yugoslav (yū·goslāv, yūgoslā·v), *sb.* and *a.* Also **Jugo-**; 9 **Iugo-Slav, Yougo-Slave, Yugo-Slave.** [ad. G. *Jugoslawe* (F. *Yougoslave*), f. Serbo-Croat *jugo-*, comb. form of *jug* south + G. *Slawe* Slav *sb.*] **A.** *sb.* **a.** (A member of) various groups of southern Slavs, comprising the Serbs, Croats, and Slovenes; also, since 30 Oct. 1918, a native or inhabitant of the State of Yugoslavia. **b.** *rare.* The Slavonic language dominant in Yugoslavia; = *Serbo-Croat* s.v. *Serbo-.
1853 L. H. Kerr tr. *Robert's Slave Prov. Turkey* in *Ranke's Hist. Servia* (ed. 2) 382 Toasts were drunk to..the New Servian Kingdom, which will reunite all the Yugo-Slaves under the eternal patronage of the house of Romanoff. 1867 Mackenzie & Irby *Trav. Slavonic Provinces of Turkey-in-Europe* xxv. 369 The situation was reversed when the Russians had shaken off the Tartar, and the Iugo-Slav fell under the Turk. 1881 Mrs. A. O. Brodie tr. *Tissot's Unknown Hungary* I. 111 The Yougo-Slaves, or Slaves of the south of Austria. 1917 F. S. Copeland tr. *Vosnjak's Bulwark against Germany* xv. 250 The unification of the Jugoslavs. 1922 *Encycl. Brit.* XXX. 372/1 The Yugoslavs..are the most numerous people in the [Balkan] peninsula. 1943 J. B. Priestley *Daylight on Saturday* xxv. 194 If you've got the morale—and even if you've got hardly anything else, like the Jugo-Slavs—you stand up and fight. 1948 R. A. D. Forrest *Chinese Lang.* i. 28 The feature [*sc.* distinctive tone]..survives to this day in one Lithuanian dialect, with traces in Scandinavian and in a variety of Yugoslav. 1973 *Times* 29 Oct. 12/7 The new Electra is Danica Mastilovic, a Yugoslav, making her debut in the house.
B. *adj.* Of, pertaining to, or designating the people or state of Yugoslavia.
1853 L. H. Kerr tr. *Robert's Slave Prov. Turkey* in *Ranke's Hist. Servia* (ed. 2) 378 Gaï's 'Illyrian Journal', the organ of the Serb and Yugo-Slave interests. 1916 B. Vosnjak *Jugoslav Nationalism* 11 There have been..three Jugoslav state creations. 1920 *Edin. Rev.* July 42 Whatever else may be said of the Yugoslav movement..it is at any rate a national movement having its origins within..the peoples whose destiny it affects. 1967 *Listener* 29 June 843/3 Italian, Yugoslav, or Spanish workers have migrated to West Germany..or Switzerland. 1981 L. Deighton *XPD* xxviii. 226 The duty officers could be sure of a bottle of Yugoslav riesling.

Yugoslavian (yūgoslā·viăn), *a.* and *sb.* Also **Jugo-**. [f. prec. + -ian.] **A.** *adj.* = *Yugoslav *a.* **B.** *sb.* **a.** A native or inhabitant of Yugoslavia. **b.** *rare.* The Serbo-Croat language.
1923 W. J. Locke *Moordius & Co.* xiii. 180, I must be back to give dinner to the Jugo-Slavian minister. 1924 *Contemp. Rev.* Apr. 448 They cannot effectively crush the revolutionaries on Yugoslavian soil. 1949 E. Pound *Pisan Cantos* lxxx. 96 White boy says: do you speak Jugoslavian? 1953 A. Smith *Blind White Fish in Persia* x. 204 Firstly, our Yugoslavian visas had not come through. 1962 A. Lurie *Love & Friendship* ii. 31 They didn't have them [*sc.* Boy Scouts] in Yugoslavia. Anyway, only for Yugoslavians. 1977 *New Yorker* 19 Sept. 49/2 There was a Yugoslavian boy who had brought along a portable silent keyboard with a weight attached to each key, to regulate the action. 1983 *Times* 3 Oct. 1/4 The Greek and the Yugoslavian were accustomed to the heat.

yuh (yə), repr. a colloq. (esp. Black English) pronunc. of You.
1906 [see *outa]. 1922 [see *tough nut* s.v. *tough *a.* 10 a]. 1933 *Publ. Texas Folklore Soc.* XI. 101 She tol' me to ax yuh. 1952 [see *macoumère]. 1967 E. Brathwaite in Ramchand & Gray *West Indian Poetry* (1972) 25 Yuh does get up, walk 'bout. 1969 [see *secko]. 1977 *Rolling Stone* 30 June 80/3 So get back to me as soon as you can, will yuh.

yuk, var. *yuck *int.*, *sb.*[2], and *a.*

Yukaghir (yū·kāgiɹ, yukāgi·ɹ), *sb.* (*a.*) Also **Yukaghire, Yukagir(e).** **a.** (A member of) a Mongoloid people of Arctic Siberia. **b.** The Palæo-Siberian language (of unknown affiliation) of this people. Also *attrib.* or as *adj.*
1842 C. H. Cottrell *Recoll. Siberia* iv. 104 The Yukaghires, settled on the banks of the river Anuiy, maintain themselves the whole year on the reindeer they kill in spring and autumn. 1879 C. H. Eden *Frozen Asia* ix. 208 The head-quarters of the Yukagires is on the River Anyui. 1898 [see *Koryak]. 1906 *Daily Chron.* 4 Apr. 7/5 Th[e] sole survivors of a group of ten Yukagirs, are charged with having eaten the man's nephew... After that the Yukagir, although he had caught a swan, continued to eat human flesh. 1908 T. G. Tucker *Introd. Natural Hist. Lang.* 149 The Hyperborean speeches of Asia, some of which may or may not form a family, include..*Yukaghir*. 1932 [see *Chukchee, Chukchi *sb.* and *a.*]. 1948 D. Diringer *Alphabet* 35 Sad love-story of a Yukaghir girl. 1951 W. K. Matthews *Languages U.S.S.R.* ii. 3 Yukagir (Odul), the mother-tongue of fewer than 500 persons in Northern Yakutia. 1964 tr. *Levin & Potapov's Peoples of Siberia* 789 The name 'Yukagir' was borrowed by the Russians from the Yakuts, but its origin is probably Tungusic. 1972 W. B. Lockwood *Panorama Indo-Europ. Lang.* 154 There are isolated languages spoken by diminutive populations. These are Gilyak..and, with a thousand speakers..or less..Yukagir (Yukagir National Area). 1981 M. C. Smith *Gorky Park* i. xiv. 213 He was more Siberian than any of us... The Borodins..lived with the Yukagir, the reindeer herders. 1983 *Word* XXXIV. 217 The focus system of Yukagir, and case and negation in Uralic are treated in syntax sections.

‖**yukata** (yuka·ta). Also **Yukata, yukatta**; 9 **ukata.** [Jap., f. *yu* hot water, bath + *kata* short for *katabira* a light kimono.] A light cotton kimono, freq. with stencil designs, orig. intended to be worn after a bath, but now also used as a housecoat.
1822 F. Shoberl tr. *Titsingh's Illustr. Japan* II. 254 The *ukata*, a robe of fine linen; it is put on in coming out of the bath to dry the body. 1881 Satow & Hawes *Handbk. Japan* p. xvi, Japanese loose cotton gowns (*yukata*). 1886 J. La Farge *Let.* 3 Sept. in *Artist's Lett. from Japan* 229 A—..rode along with only a partial covering of *yukatta*, and attracted no attention. 1936 K. Nohara *True Face of Japan* v. 130 Dons the *yukata*, the light, informal, Japanese house-jacket of coloured cotton. 1960 B. Leach *Potter in Japan* vi. 133 My Yukata (provided cotton kimono) reached to my knees. *Ibid.* ix. 195 Visitors were strolling the streets in 'yukatta' (cotton kimono provided by hotels). 1970 *Guardian* 12 Dec. 6/6 The donning of the cool cotton *yukata* robe and slippers is the first sloughing of Western identity. 1981 J. Melville *Sort of Samurai* ix. 75 All four of them were now wearing cotton *yukatas*.

Yukawa (yukā·wā). *Nucl. Physics.* The name of H. *Yukawa* (b. 1907), Japanese physicist, used *attrib.* with reference to the theory of the strong interaction between nucleons put forward by him, in which it is mediated by the exchange of particles (*Yukawa particles*) subsequently identified with pions; **Yukawa potential,** a potential function of the form $V = V_0(r/r_0)^{-1}\exp(-r/r_0)$, occurring in Yukawa's theory of the nuclear force.
1938 *Nature* 1 Oct. 592/2 The discovery of a new particle, the Yukawa particle or 'heavy electron', has given a new orientation to many of our ideas. 1948 *Proc. Cambr. Philos. Soc.* XLIV. 90 As an illustration of the treatment developed above the nuclear field as described by the Yukawa potential $U(r) = -b\lambda r^{-1}e^{-\lambda r}$ will be treated. 1964 *Listener* 29 Oct. 661/1 Immediately after the war, many physicists..became deeply involved in these problems, especially the nature of protons, neutrons, and the Yukawa mesons. 1968 C. G. Kuper *Introd. Theory Superconductivity* xv. 258 The Yukawa force between nucleons has a pairing part, and the methods of the BCS theory are applicable. 1973 R. J. Blin-Stoyle *Fund. Interactions & Nucleus* i. 3 The earlier Lagrangian formulation of strong interactions based on Yukawa type interactions has continued to play an important role. 1974 Frauenfelder & Henley *Subatomic Physics* viii. 187 If virtual pions are exchanged between nucleons, the basic Yukawa reaction $N \to N' + \pi$ should conserve isospin. 1977 P. D. B. Collins *Introd. Regge Theory & High Energy Physics* i. 42 The simplest form of potential which has the short-range character appropriate to strong interactions is the Yukawa potential $U(r) = g^2e^{-\mu r}/r$, where g^2 is the coupling strength and μ^{-1} is the range.

Yuki (yū·ki), *sb.* (*a.*) Also 9 **Yuka, Yukeh,** etc. [a. Wintu *yu·keh*, lit. 'stranger, enemy'.] **a.** (A member of) a group of linguistically related American Indian peoples, comprising the Yuki, Huchnom, and Wappo tribes, inhabiting the coast of north-western California. **b.** The language spoken by this people. Also *attrib.* or as *adj.*
1858 S. P. Storms *Let.* 14 Aug. in *Ann. Rep. Commissioner of Indian Affairs for 1858* CIX. 307 About three thousand Nome Cults or Yukas make this valley [*sc.* Round Valley in northern California] their headquarters. 1863 *Hist. Mag.* VII. 123/1 The Yukeh, or as the name is variously spelt, Yuka, Yuques, and Uca, are the original inhabitants of the..Round Valley in Tehama County, California. 1875 H. H. Bancroft *Native Races Pacific States* III. ii. iii. 648 In Round Valley, northern California, there is the..Yuka language. *Ibid.* 643 At Humboldt Bay a language called Patawat is mentioned, and in Round Valley the Yuka. 1877 S. Powers *Tribes Calif.* xiv. 129 If a Yuki stumbles and falls on the march..it is a bad omen. *Ibid.*, He has seen Yuki dead left on the field. 1900 J. Frazer *Golden Bough* (ed. 2) I. i. 34 When the men of the Yuki tribe of Indians in California were away fighting, the women at home..danced..in a circle, chanting and waving leafy wands. 1923 A. L. Kroeber *Anthropol.* iii. 90 In the native Californian language known as Yuki, *ko* means *go*. 1939 H. M. Wormington *Ancient Man in North Amer.* vi. 256 Among living Indians in North America the Cahuilla tribes of inland southern California and the Pomo and Yuki of northern coastal California, are thought to show the greatest number of Amurian traits. 1965 *Canad. Jrnl. Linguistics* X. 99 Hokan-Siouan in Sapir's arrangement includes six major constituent units..Hokan-Coahuiltecan, Yuki, [etc.]. 1974 *Encycl. Brit. Micropædia* X. 844/1 Warfare was apparently frequent—between certain communities, between the different Yuki groups, and with other Californian Indians. *Ibid.* 844/2 The Yuki..were organized into communities composed of several scattered settlements. 1981 A. B. Kehoe *North Amer. Indians* vii. 376/2 Yuki and its sister language Wappo, both spoken north of San Francisco Bay.

yukky, var. *yucky *a.*

Yukon (yū·kǫn). [The name of the *Yukon* Territory in north-west Canada.] *Yukon stove*, a lightweight portable stove consisting of a small metal box divided into firebox and oven.
1898 W. B. Haskell *Two Years in Klondyke* 75 The 'Yukon stove'..is a small sheet iron box with an oven at the back and a telescope pipe. 1943 W. Chase *Sourdough Pot* xvi. 97 The stove, a sheet-iron affair, known as a Yukon stove, had a limited capacity. 1974 W. Hunt *North of 53* iv. 15 A wood-fired 'Yukon' stove, either square or round bottomed, and containing a small oven at the back end, provided cooking facilities and kept the cabin warm in the winter.
Also **Yu·koner,** an inhabitant of the Yukon Territory; **Yu·konesque** *a.*
c 1898 Yukoner [see *Hoochinoo 2]. 1924 M. H. Mason *Arctic Forests* 84 These things have made the Yukoners the finest, most generous, and most virile population to be found on the whole continent of America. 1934 A. Huxley *Beyond Mexique Bay* 128 Our Yukonesque stampedes into any business that seems, at any given moment, to be doing well. 1977 Yukoner [see *outsider 1].

yulo (yū·loᵘ). Also **yuloh, yulow,** etc. [Prob. ad. Chinese (Cantonese) *iū-lō* to scull a boat, f. *iū* to shake + *lō* oar.] A Chinese sculling oar (see quot. 1899). So also as *v. intr.* to scull a boat with such an oar. Hence **yu·lohing** *vbl. sb.*
1878 H. A. Giles *Gloss. Reference* 170 Yuloh, to, to scull a boat with an oar at the stern. From the Shanghai pronunciation of..*yao* to work..*lu* an oar. 1888 A. J. Little *Through Yang-tse Gorges* 320 In addition to these, two lateral yuloes (sculls worked by a screw motion),..were attached to the sides. 1899 I. L. Bishop *Yangtze Valley & Beyond* xiii. 145 Others are toiling at *yu-lows*, big broad-bladed sculls, worked over the stern or parallel to the junk's side. 1905 *N. & Q.* 22 Apr. 305/1 The *yuloh* is the single oar used over the stern for the propulsion of sampans and barges, after the manner sometimes called sculling in England. To *yuloh* is to row a boat in that fashion. The meaning is literally 'push and pull wood'. 1911 J. D. Ball *Chinese at Home* xvi. 203 Ferryboats slowly cross the river... The loads of the coolies are put in the bows of the boat, where also occasionally is to be seen a leper, who is not allowed amongst the other passengers. The ferryman *yeeoo-loes* at the stern. 1921 *Outward Bound* June 36/1 The junk is propelled when possible by great oars, called by the Chinese 'yulos', projecting out from either side. 1927 *Chambers's Jrnl.* Mar. 163/1 We pass a fishing boat; a woman with a willowy figure bending to the yulow at one end. 1941 J. Homer *Dawn Watch in China* iv. 93 Junks bound downstream were manned on the foredeck with six or eight or a dozen oarsmen, who stood, each to his oar poling forward into the fast water and singing in unison the minor wordless river chant of the sacred Yangtze. Now and then, the helmsman would *yuloh* in a high-pitched scream were he called forth the wind. 1966 G. R. G. Worcester *Sail & Sweep in China* ii. 11 In rowing, as in so many other arts, the Chinese show their great independence of thought; and in the yuloh or sculling oar, they demonstrate a perfect example of applied mechanics which can, confidently, be dated back to the Later Han Dynasty AD25–220. *Ibid.* 12 When operated by a Chinese, even by a child, yulohing appears to be supremely simple. 1981 *Jrnl. R. Naval Med. Service* LXVII. 46/2 Navigable by one man using a large yulo at the stern.

yum (yʌm), *int.* [Echoic.] An exclamation of pleasurable anticipation, with implication of sensual or gustatory satisfaction; freq. reduplicated as *yum-yum*, etc.
Some of the reduplicated examples are not clearly distinguishable from those given s.v. *yum-yum *sb.* and *a.* below.
1878 *Burlington Hawkeye* in *Irish Monthly* VI. 688 How we would like to get hold of the man... Alone. In the woods, with a revolver in our hip-pocket. Revenge is sweet; yum, yum, yum. 1899 Kipling *Stalky & Co.* 239 Pretty lips..Seem to say—Come away. Kissy! come, come!..Yummy-yum-yum! 1922 Joyce *Ulysses* 173 Kissed her mouth. Yum. 1942 O. Nash *Good Intentions* 143 And fish are only something about which some people say, 'Yum yum, right out of the water and fried to a delicate golden brown.' 1959 'J. Christopher' *Scent of White Poppies* iii. 40 When she had gone, Bella said: 'Yum-yum. Really luscious.' 1962 A. Lurie *Love & Friendship* iii. 45 Miranda cut a piece of warm coffee cake, handed it to Emmy... 'Oh, yum. Thank you.' 1982 S. Paretsky *Indemnity Only* xii. 155 'Lotty talked her into..making homemade enchiladas, yum-yum.' 'Yum-yum,' the two little girls chorused.
Hence **yu·mmy** *a. colloq.*, delicious, delectable; also as *int.*
1899 [see *yum *int.*]. 1934 Webster, Yummy *adj.* 1950 'S. Ransome' *Deadly Miss Ashley* xii. 147 Lora's attractive face or Dorothea's yummy figure. 1955 J. P. Donleavy *Ginger Man* xix. 213 Sitting, facing one another across the white table. Bacon and eggs, tea, bread and butter. Yummy. 1955 H. Kurnitz *Invasion of Privacy* (1956) vi. 47, I adore movie stars. Gregory Peck! Yummy! 1970 P. Zelver *Honey Bunch* xx. 96 Thanks a lot for the lemonade. It was yummy. 1979 *Evening Standard* 18 Sept. 23/2 Mr. Zamoyski is a handsome devil himself—such a yummy photograph on the back of the jacket.

Yuma[1] (yū·mă), *sb.* and *a.* Also 9 **Umea.** [a. Sp., ad. Pima-Papago *yu·mi*.] **A.** *sb.* **a.** (A member of) an Indian people inhabiting south-west Arizona and the adjoining areas of Mexico and California, now officially referred to as the Quechan. **b.** The language of this people. **B.** *adj.* Of, pertaining to, or designating this people or their language.
1831 J. O. Pattie in R. G. Thwaite *Early Western Trav.* (1905) XVIII. 188 Here we found the tribe of Umeas. 1849 A. W. Whipple *Jrnl.* 1 Dec. in H. R. Schoolcraft *Hist. & Stat. Information Indian Tribes* (1852) II. iv. 118 To this day among the Yumas I have never seen anger expressed. *Ibid.*, I will add a vocabulary of the Yuma (or rather Cuchan) language. 1877 H. S. Gatschet *Indian Lang. Pacific States* 156 The term *opa*, composing several of these tribal names, is taken from the Yuma, and means man. 1891 D. G. Brinton *Amer. Race* II. ii. 109 It has also been called..the Apache, that being the Yuma word for 'fighting men'. 1892 *Amer. Anthropologist* Oct. 326 This pamphlet..is the fourth of the author's Yuma series. 1907 [see *G string 2 a]. 1942 *Amer. Anthropologist* July–Sept. 425 Certain structural features of the Yuma language should be outlined... Yuma has three 'parts of speech'. 1965 *Language* XLI. 305 Yuma has /a/. 1973 A. H. Whiteford *North Amer. Indian Arts* 13 Modeling and paddling is a technique..still used by the Papago and Yuma of Arizona. 1979 *Tucson Mag.* Apr. 27/3 In 1780 Teodoro de Croix..ordered that two new settlements of Spaniards be established among the Yuma Indians on the Lower Colorado River.

Yuma[2] (yū·mă). The name of *Yuma* county in north-eastern Colorado, used chiefly *attrib.* to designate the remains of a prehistoric culture discovered there, and applied esp. to a type of projectile point. Cf. *Folsom.
1932 E. B. Renaud in *Proc. Colorado Museum Nat. Hist.* 19 Nov. 5 Yuma and Folsom artifacts. *Ibid.*, The number and variety of artifacts of this splendid collection from the Yuma district contributed greatly to our better knowledge of the points which, in recognition, we named 'Yuma points' to distinguish them from the true Folsom type. 1948 A. L. Kroeber *Anthropol.* (rev. ed.) xvi. 684 Points called Yumas have now and then been found in Folsom association, as at Clovis. 1949 *Time* 12 Sept. 69/1 Not much is known about Yuma Man, for no Yuma skeleton has yet been found. 1960 C. Winick *Dict. Anthropol.* 277/2 Yuma implements are unfluted.

Yuman (yū·măn), *a.* and *sb.* [f. *Yuma[1]: cf. -an.] **A.** *adj.* Of, pertaining to, or designating various related Indian peoples of Arizona, Mexico and California, or the languages spoken by them. **B.** *sb.* **a.** A member of this group of peoples. **b.** A language family of Hokan stock to which the languages of these people belong.
1891 J. W. Powell in *7th Ann. Rep. Bureau Amer. Ethnol.* 137 With the exception of certain small areas possessed by Shoshonean tribes, Indians of the Yuman stock occupied the Colorado River from its mouth as far up as Cataract Creek. 1901 G. W. James *Indian Basketry* x. 161 The carrying frame and net of the Mohave Indians, of the Yuman stock, dwelling about the mouth of the Colorado River. 1920 *Univ. Calif. Publ. Amer. Archaeol. & Ethnol.* XVI. 478 The Yuma, who call themselves Kwichyana..are known to other Yumans by dialect variants of the same name. 1933 L. Spier *Yuman Tribes of Gila River* 151 The Maricopa have transposed the normal Yuman word from south to west. 1950 *Nat. Hist.* Feb. 76/3 They speak essentially the same language—a dialect of the Yuman tongue. 1952 *Amer. Anthropologist* Jan.–Mar. 80 The cultivated plants and the agricultural methods of the Yumans. 1963 [see *Shastan *a.* and *sb.*]. 1965 [see *Mohave]. 1970 *Language* XLVI. 533 The Yuman languages..were early recognized as constituting a linguistic family. 1974 *Encycl. Brit. Micropædia* X. 845/3 The total number of Yuman peoples remaining in the 1970s..was uncertain. 1978 *Language* LIV. 219 It crops up again in Southern California, in one subgroup of the Yuman family. *Ibid.* 505 The present collection indicates that such a tradition has now also been established for Yuman, a Hokan family of Arizona, California.

yump (yʌmp), *v. slang.* [Alteration of Jump *v.*, repr. the supposed pronunciation of it by Swedish speakers or the Norw. *jump* jump (sb.), *jumpe* jump (vb.).] *intr.* Of a rally car or its driver: to leave the ground while taking a crest at speed. So **yu·mping** *vbl. sb.*
1962 *Motoring News* 22 Nov. 7 (*caption*) 'Yumping', a pastime dear to Scandinavian rallyists, was seen practised to the full for the first time in this country during the R.A.C. Rally. 1968 [see *roll *v.*[2] 18 a]. 1968 *Sun* 12 Nov. 8/5 Yumping happens..when cars literally jump into the air. They take off when taking crests at speed.
Also as *sb.*, an instance of 'yumping' (see also quot. 1980).
1975 *Country Life* 4 Dec. 1529/2 Clark had an enormous 'yump' (Swedish rally parlance for a car leaving the ground and hurtling for some distance before landing heavily). 1980 *Sunday Express* 24 Aug. 23 (Advt.), Yump: Rally jargon for a sharp crest in the road, which causes a car to take off at speed.

yumpie (yʌ·mpi). *colloq.* (orig. *U.S.*). Also **yump, Yumpie.** [f. the initial letters of *young upwardly mobile people*: see -ie.] = *yuppie.
1984 *Economist* 17 Mar. 40 Mr. Hart seems to have drawn much of his support from young upwardly mobile people and young urban professionals—yumps and yuppies, as they are called. 1984 *Times* 21 Mar. 14/2 Sometimes 'Yuppies' are referred to as 'Yumpies', meaning Young, upwardly-mobile people. 1984 *N.Y. Times Rev. Bks.* 25 Mar. 20 The yumpies climbing the ladder of success with great agility can be described as upscaling. 1984 *Chicago Sun-Times* 25 Mar. 7 The ultimate Yumpie status symbol is to have a child suffering flash-card burn-out.

yum-yum: see *yum int.

yum-yum (yʌm,yʌm), *sb. slang.* [Redupl. of *yum *int.*] **a.** An action providing a pleasurable or delicious sensation; love-making; also (*Naval slang*), love-letters. Also *yum-yum girl* (euphem.), prostitute.
1885 *Punch* 3 Jan. 4/1 You and me's got our notions of yum-yum, as isn't fur wide o' the mark. 1939 A. Huxley *After Many a Summer* i. iv. 45 Enjoying what she called 'a bit of yum-yum'. 1943 Hunt & Pringle *Service Slang* 71 *Yum yum*, love letters. 1960 'S. Harvester' *Chinese Hammer* xxv. 202 Yum-yum tarts from Hong Kong or ex-temple dancers from Java. 1962 A. Buchwald *How Much is that in Dollars?* 124 One of the chiefs said, 'Don't let her kid you. All her girls are really yum-yum girls from the dance halls'. 1962 *Times* 12 July 13/5 Would the sin bosun (chaplain) frown, one wonders, at hearing a love-letter described as a 'yum-yum'? 1967 S. Beckett *Texts for Nothing* iii. 86 Come, ducky, it's time for yum-yum.
b. *spec.* Tasty food (of various kinds).
1889 'Aunt Babette's' *Cook Book* 330 *Yum-Yums*. Grate one cocoanut, reserving the milk; one pound and a half of pulverized sugar, whites of three eggs. 1932 O. O. McIntyre *Another 'Odd' Book* 5 Whenever I mention a longing for such old-fashioned yum-yums as gingersnaps..the response is immediate. 1941 J. Smiley *Hash House Lingo* 58 *Yum-yum*, sugar.

yum-yum, *a. slang.* [f. as prec.] Excellent, first-class; delectable.
1890 Barrère & Leland *Dict. Slang* II. 428/2 *Yum-yum* (London), first-rate, elegant. 1922 Joyce *Ulysses* 372 Howth settled for slumber tired of long days, of yumyum rhododendrons. 1962 R. Prawer Jhabvala *Get Ready for Battle* iii. 135, I wish my cook could learn..how to make cheese pakoras like these, they are absolutely yum-yum.

Yunani (yunā·ni), *a.* Also **Unani.** [a. Arab. *yūnānī*, lit. 'Greek'.] Designating a Western system of medicine (opp. *Ayurvedic *a.*). Occas. also applied to other disciplines (see quot. 1958).
1922 [see *Ayurvedic *a.*]. 1958 O. Caroe *Pathans* iii. 45 The Yunani or Greek lore which so largely influenced Islamic *literae humaniores*. 1977 *Times Lit. Suppl.* 3 June 684/2 The traditional Hindu (Ayurvedic), Muslim (Unani) and Chinese medicine... Unani is Arabic for 'Ionian' and thus declares its descent from Greek medical thought. 1979 *Social Sci. & Med.* XIII. B. 339 Yūnāni medicine which lasted for a few centuries in the court of medieval Baghdad. 1979 V. Mehta *Mamaji* iii. 39 The Unani system, which is based on ancient Greek medicine and relies on metallic compounds.

Yunca (yu·ŋkă). Also **Yunga, Yunka.** [Amer. Sp., a. Quechua *yunca* plain, valley.] **a.** The Chipayan language formerly spoken by a group of Indian peoples inhabiting the coast of Peru; = *Mochica *sb.* b. **b.** A member of such a people. Also *attrib.* or as *adj.*
1853 F. L. Hawks tr. *Von Tschudi's Peruv. Antiquities* v. 117 The Quichua language has various dialects strongly marked..the *Yunca*, in the bishopric of Truxillo. 1871, etc. [see *Mochica *a.* and *sb.*]. 1908 *Athenæum* 11 Apr. 447/1 Customary with the Natchez, Muizcas, Pokomames, Yuncas, and Incas of the American continent. 1950 J. A. Mason in J. H. Steward *Handbk. S. Amer. Indians* VI. 194 The *Yunca, Mochica*, or *Chimu* language..is practically extinct. 1954 *Internat. Jrnl. Amer. Ling.* Jan. 24 (*heading*) Yunka, language of the Peruvian coastal culture. *Ibid.*, A comparison of the Yunka language with Quechua has been made.

Yung Chêng (yuŋ tʃeŋ). Also **Yung Cheng, Yung-ching,** etc. The name of the reign of the third Chinese Emperor of the Ching dynasty (1723–35), used *attrib.* and *absol.* to denote a kind of porcelain produced during his reign, characterized by its delicate colouring.
1902 W. G. Gulland *Chinese Porcelain* II. 384 It [is]..impossible to tell whether many of these rose pieces were made during this reign or early in the following, but they are all spoken of as Yung-ching pieces. 1906 R. L. Hobson *Porcelain* iv. 34 The Yung-chêng porcelain has a character of its own and includes many innovations worthy of separate notice. 1908 J. F. Blacker *Chats on Oriental China* ix. 115 The contrast between the delicacy of the detail is striking when compared with the broad treatment of the Kang-he period. On the one hand there is fine stipple work. This is Yung-ching. On the other hand there is a broad wash of colour. 1925 R. L. Hobson *Later Ceramic Wares of China* vii. 62 (*heading*) Yung Chêng Porcelain (1723–35). 1949 G. Savage *Ceramics for Collector* ii. 83 Celadons take on a bluish tinge, and obviously have some cobalt mixed with the glaze colour. Colour is applied in washes and fine, detailed drawing usually indicates Yung Chêng. 1971 R. Quest *Death of Sinner* v. 42 'You mean it's not K'ang Hsi?' 'It's later—Yung Cheng.' 1976 Scott & Koski *Walk-In* x. 55 A wealthy importer of usually genuine Chinese *objets d'art*, mostly Yung-chêng porcelain.

Yunnanese (yunănī·z), *sb.* and *a.* Also 9 **Yun-Nese.** [Chinese *Yúnnán* (see below) + -ese.] **A.** *sb.* **a.** A native or inhabitant of Yunnan, a province in S.W. China; also *collect.*, the people of Yunnan. **b.** The dialect of Yunnan. **B.** *adj.* Of or pertaining to Yunnan.
1849 *Ann. Propagation Faith* Mar. 109 A young physician from the province of Yun-nan evinced more generosity... This young Yun-Nese had led so strange a life. 1909 *N. China Herald* 18 Sept. 671/3 With their catch crops the Yunnanese in the eastern part of the province seemed as satisfied as they were at the time before wheat took the place of the now-departed opium. 1927 *Blackw. Mag.* Mar. 308/1 He had a full knowledge of Yunnanese, but these men were Lolos from Szechuan, and understood not a word. 1933 N. Waln *House of Exile* iii. ii. 200 These Yunnanese soldiers were hired henchmen that Sun Yat-sen had brought to Canton to police the city. 1948 G. H. Johnston *Death takes Small Bites* iv. 85 'It's usually thronged with people,' Coates continued boredly. 'Kachins and Shans, Yunnanese, Lisus, Burmese, Miaos.' 1959 C. Ogburn *Marauders* (1960) iii. 92 Cantonese and Yunnanese..slammed their vehicles from one gear to another with a grinding scream. 1959 *Times* 19 Mar. 14/7 It was found expedient not to have Yunnanese mules in Lucifer's section. 1964 M. A. K. Halliday et al. in J. A. Fishman *Readings Sociol. of Lang.* (1968) 145 In China, you speak Cantonese if you come from Canton, Pekinese if you come from Peking and Yunnanese if you come from Yunnan. 1981 P. Barr *Chinese Alice* iv. 94 Golden Persimmon sang..and the Yunnanese watched her.

yup (yʌp), colloq. (orig. U.S.) var. of Yes. Cf. Yep, dial. pron. of Yes.
1906 *Cent. Mag.* Jan. 410/2 'Will you go—if I swear?' 'Yup,' said Pinchas, airing his American. 1919 G. Murray *Aristophanes & War Party* 36 *Demos*. You see those rows and rows of people? *Offal-monger.* Yup. 1923 E. F. Wyatt *Invisible Gods* ii. iv. 78 'Paul as mean as ever?' 'Yup,' replied Hancock..'and a little meaner.' 1947 [see *supercolossal* adj. s.v. *super- 9 a]. 1959 E. H. Clements *High Tension* iv. 63 'You're staying here, then?' 'Yup.' 1968 S. Challis *Death on Quiet Beach* v. 75 'Will you pay the fine?'..'Yup.' 1978 J. Irving *World according to Garp* xii. 232 'Is that you, Roger?' 'Yup.'

Yupik (yū·pik), *a.* and *sb.* [See quot. 1951[1].] **A.** *adj.* Of, pertaining to, or designating an Eskimo-Aleut language spoken in Siberia and Alaska, or the speakers of it. **B.** *sb.* This language.
1951 M. Swadesh *Internat. Jrnl. Amer. Ling.* Apr. 69/1 The phonologic material shows an unmistakable cleavage between two groups of dialects which may be conveniently called the Yupic and Inupik divisions by reference to their treatment of the root for *human being* (juk versus inuk; these are the words for *real person* in Kuskokwim and Barrow respectively). *Ibid.* 70/2 The dichotomy between Yupik and Inupik is clear and geographically abrupt. *Ibid.*, We find nothing in the [Wales] dialect which minimizes the depth of its separation from Yupik Eskimo. 1967 W. H. Oswalt *Alaskan Eskimos* p. xiv, In Alaska two languages, Inupik and Yupik are spoken. *Ibid.* iv. 115 The Yupik speakers..would be expected to have had a greater long-term stability than the Inuit. 1974 *Encycl. Brit. Macropædia* VI. 963/2 In 1961 a program was started..with the active participation of Yupik Eskimos, for working out a systematic Eskimo orthography in the Roman alphabet. *Ibid.*, Greenlandic *ůvoq*..is related to the Asian Yupik form *ugu-*. 1977 *New Yorker* 20 June 55/2 The big river delivers the wood to the Yupik Eskimos of the western coast. 1980 M. E. Krauss *Alaska Native Languages* App. ii. 91 There is..the well known sharp division between Yupik and what we in Alaska usually call Inupiaq. *Ibid.* 103 With a dense and increasing population, almost entirely Yupik, the Yupik language remained..strong.

yuppie (yʌ·pi). *colloq.* (orig. U.S.). Also **Yuppie.** [f. the initial letters of *young urban professional*: see -ie.] A jocular term for a member of a socio-economic group comprising young professional people working in cities. Also *attrib.* Cf. *yumpie.
1984 Piesman & Hartley (*title*) The yuppie handbook. 1984 [see *yumpie]. 1984 *Times* 21 Mar. 14/2 A new term has been introduced into the American political lexicon... It is 'Yuppie', which stands for Young, urban professional people. 1984 *Observer* 8 Apr. 12/1 We have got to break this yuppie image. 1984 *Guardian* 22 Oct. 6/6 The yuppies themselves, in the 25–34 age group, supported Senator Gary Hart in the primaries.

Yurak (yiūˑræk). [Native name.] = *Nenets (see quot. 1972). Also *Comb.*, as *Yurak-Samoyed(e)*.
1882 H. Lansdell *Through Siberia* I. viii. 103 Reclus ..speaks of the Yurak-Samoyedes as still practising their bloody rites. 1911 *Daily News* 3 Nov. 1 Great praise is due to the Russian Red Cross Society for the noble way in which the women of this corps work among the revolting and diseased Yuraks. 1939 L. H. Gray *Foundations of Lang.* 369 The languages of the Uralic family are as follows..*Samoyede* group: Yurak, Yenisei-Samoyede, [etc.]. 1952 *Trans. Philol. Soc.* 118 Yurak-Samoyede *paju, pal'* 'close, thick'. 1954, etc. [see *Nenets]. 1964 tr. *Levin & Potapov's Peoples of Siberia* 547 Before the Revolution the Russians called the Nentsy Samoyeds or Yuraks. 1972 *Language* XLVIII. 208 Calling the language and people 'Yurak' is an anachronism. Since 1917 a native self-designation has come into use in the Soviet Union, which I render as 'Nenets'. 'Yurak' arose from a term applied to the eastern Nenets by their Enets and Nganasan neighbours. 1975 G. F. Cushing tr. *Hajdu's Finno-Ugrian Languages & Peoples* iii. 216 The Nenets are generally termed Yurak-Samoyeds.

Yurok (yiūˑrǫk), *sb.* and *a.* Also 9 **Euroc, Youruk.** [a. Karok *yúruk* (a considerable distance downriver; cf. *yurúkva·rara* Yurok Indian, lit. 'downriver person'.] **A.** *sb.* **a.** (A member of) an Indian people of northern California. **b.** The language of this people, distantly related to Algonquian and Wiyot. **B.** *adj.* Of, pertaining to, or designating this people or their language.
1851 [see *Karok]. 1872 *Overland Monthly* Apr. 326/2, I have seen a half-dozen tatterdemalion Eurocs.. come rushing down through the *chaparral*. 1875 H. H. Bancroft *Native Races Pacific States* III. 641 On the lower Klamath, the Euroc language prevails. 1904 *Univ. Calif. Publ. Amer. Anthrop. & Ethnol.* II. iii. 95 The belief in a previous world..is not uncommon. The Torint of the Eskimo..the Waghe of the Yurok. 1913 *Amer. Anthropologist* XV. 621 Our knowledge of Wiyot and Yurok is still very incomplete. 1921 [see *culturally *adv.*]. 1946 L. Bloomfield in H. Hoijer et al. *Ling. Structures Native America* 85 Two languages of California, Wiyot and Yurok, have been suspected of kinship with Algonquian. 1954 E. A. Hoebel *Law of Primitive Man* ii. 25 An aggrieved Yoruk who felt he had a legitimate claim engaged the legal services of two nonrelatives from a community other than his own. 1962 [see *Karok]. 1973 A. H. Whiteford *N. Amer. Indian Arts* 47 Half-twist overlay was used by the Yurok..for fine flexible mats. 1977 R. Holland *Self & Social Context* ii. 40 Witness the astonishingly beautiful analogy between the oral anxiety of the Yurok people and the uncertainty of the salmon run on which they depend as a culture. 1981 A. B. Kehoe *N. Amer. Indians* vii. 376/2 Macro-Algonkian is represented by two language isolates, Yurok..and Wiyot.

Yurrup (yʌ·rəp). Repr. a supposed U.S. pronunc. of *Europe*.
1883 *National Police Gaz.* 17 Mar. 3/1 The entire court of shyster lawyers, wives and contractors' daughters who have been to Yurrup. 1960 N. Mitford *Don't tell Alfred* iii. 34 There are the [American] business men trying to make a better position for themselves at home as experts on Yurrup. 1980 R. McCrum *In Secret State* xiii. 123 American students enthusing about Yurrup.

Yuruk (yiūˑruk), *sb.* (*a.*). [a. Turk. *yürük* (also used) = nomad.] (A member of) a nomadic people inhabiting Anatolia. Also *attrib.* or as *adj.*
1869 H. F. Tozer *Researches Highlands of Turkey* I. i. 13 In one of these glades we found a tribe of Yuruk with their flocks. 1896 D. G. Hogarth *Wandering Scholar in Levant* iii. 53 The 'Turk', most rightly so called, is the despised *Yuruk*, the 'wanderer', a name applied to the half-settled population, roaming in summer.., collected in the winter into villages. 1907 G. Bell *Let.* 1 May (1927) I. xi. 233 So we rode back along the beautiful grassy shores of the lake, where the Yuruks were watching their flocks and herds. 1959 *Listener* 6 Aug. 221/1 Three weeks among the Yuruks. 1963 *Times* 9 May 14/6 The Yürüks are a tribe of nomads who spend their winter in the equable south of Anatolia and move up to the plateaux and highlands for the summer. 1974 *Encycl. Brit. Micropædia* X. 848/2 *Yürük rugs*, handwoven by nomadic people in various parts of Anatolia... Rugs from Eastern Anatolia, many of them Kurdish rugs..but classed as Yürük, show a wide range of rich and unusual colour shades. 1983 J. Thompson *Carpet Magic* 19/2 (*caption*) One member of a Turkish nomad, or Yuruk, family on migration.

yus, obs. form of Yes. Add: Still current in *dial.* and non-Standard use. Also **yhus** (*once*), **yuss.** (Examples.)
a 1300 [see Yes *adv.* 2 a, 3 a]. 1828 W. Carr *Dial. Craven* (ed. 2) II. 279 *Yus*, Yes. 1888 S. O. Addy *Gloss. Words Sheffield* 292 *Yus*, *adv.*, yes. 1890 Kipling *Many Inventions* (1893) 284 Ho yuss! 'Taint much though, is it? 1898 J. D. Brayshaw *Slum Silhouettes* 15 'Yus!' responded his wife. 1934 J. Franklyn *This Gutter Life* xx. 159 Yhus, movin' aht, are yer? Norra bad flet. 1968 C. M. Vines *Little Nut-Brown Man* i. 32 'Please, may I ask you something?' 'Yus!' said he [*sc.* Lord Beaverbrook], employing, in good humour, this cheerful Canadianism.

yusho (yū·ʃoᵘ). *Path.* [Jap., f. *yu* oil + *sho* disease.] A disease characterized by the development of brown staining of the skin and severe acne, caused by the ingestion of polychlorinated biphenyls.
1969 Goto & Higuchi in *Fukuoka Acta Med.* LX. 409 (*heading*) The symptomatology of yusho (chlorobiphenyls poisoning) in dermatology. *Ibid.*, We thought its cause might be rice oil... The case has been called Yusho. 1973 G. L. Waldbott *Health Effects of Environmental Pollutants* xviii. 226/1 In 1968 about 1000 persons in southern Japan developed darkened skin..and severe acne... The illness was named 'Yusho', or rice oil disease, because the patients had eaten food cooked with contaminated rice oil. 1977 *Lancet* 22 Jan. 176/1 Chisato Hirayama provides a chapter on the clinical aspects of *yusho* (oil disease), the important skin symptoms being reviewed separately.

Yusufzai (yū·sufzai), *sb.* and *a.* Also **Yusafzai**; 9 **Eusofzye, Eusafzai,** etc. [Pers. *yūsuf* Joseph + *-zāī* bringing forth.] **A.** *sb.* (A member of) a Pathan tribal group inhabiting the North-West Frontier Province of India. **B.** *adj.* Of, pertaining to, or designating this people.
1815 M. Elphinstone *Caubul* iii. i. 328 The Eusofzyes are a very numerous tribe. 1838 in *Parl. Papers 1839* XL. 29 Shah Shooja-ool-Moolk disclaims all title to.. Peshaur, with the Eusefzye territory. 1857 S. Cotton *Nine Years on North-West Frontier* (1868) ix. 220 The column placed under his orders on the Euzofzaie Border. 1886 [see *-ology, ology 2]. 1887 Kipling *Plain Tales from Hills* (1888) 24 A Eusufzai horse-thief. 1899 *Dict. Pathan Tribes* 234/1 *Yusufzai* (73,000), a great group of Pathan tribes which includes those of the Black Mountain, the Bunerwals, the Swatis, the people of Dir and the Panjkora valley, and the inhabitants of the Yusufzai plain in British territory. 1900 R. Warburton *Eighteen Years in Khyber* v. 52 Mukarrub Khan of Panjtar..was another pleasing character connected with Yusufzai politics. 1940 P. Sykes *Hist. Afghanistan* I. xix. 290 In the autumn of that same year Babur..decided on an expedition against the Yusufzais. 1953 O. Caroe *Soviet Empire* v. 86 Another Yusufzai *maira*. 1958 —— *Pathans* i. 14 A Yusufzai or a Khalil..if he is asked who he is, will always reply, 'I am an Afghan.' 1978 'M. M. Kaye' *Far Pavilions* xi. 170 The men of his own troop were drawn largely from the Border tribes: Yusafzai, Orakzai and Khattak. 1983 J. C. Griffiths *Queen of Spades* ii. 26 Hussein Rahman, a Yusufzai Pathan.

Yuvaraja (yuvārā·dʒă). Also **Jubaraj, Yuvaraj, Yuveraj, Yuvraj.** [a. Hindi, f. Skr. *yuva-* young + *rājā*: see Raja, rajah.] The male heir to an Indian state or principality; the crown prince.
1893 R. Lethbridge *Golden Bk. India* p. xv, In some of the Orissa Tributary Mahāls..the heir-apparent is styled *Jubaraj* or *Yuvaraj*. 1916 N. Singh *King's Indian Allies* i. xii. 150 To distinguish the heir-apparent from the other sons, it is customary to call the former by some distinctive title, *Yuvaraja* in the case of Hindus. 1931 *Times of India* 6 Nov. 5/4 On May 28, 1930, he was received, with the Yuvraj, at Court at Buckingham Palace. 1937 W. H. Saumarez Smith *Let.* 3 Jan. in *Young Man's Country* (1977) ii. 50 On the 30th, I saw Prince Jaya's father, the Yuvaraja. 1953 E. M. Forster *Hill of Devi* 161 Vikky, henceforward to be referred to as the Yuvraj (Crown Prince), had now developed into a charming and intelligent youth. 1978 'M. M. Kaye' *Far Pavilions* iii. 54 The young crown-prince, Lalji, Yuveraj of Gulkote. 1982 K. Singh *Heir Apparent* i. 1 The birth of a Yuvaraj (heir apparent) was always a matter for rejoicing.

yuzbashi (yuzbɑ·ʃi). Also **Yuzbachi, Yuz-bashi, Yuzbashi.** [Turk. *yüzbaşı*, lit. 'one who is head of a hundred', f. *yüz* hundred + *baş* head.] A captain in the Turkish army; in the Turkish navy, a first lieutenant. Cf. *bimbashi.
1876 [see *bimbashi]. 1907 *Blackw. Mag.* June 806/2 Our procession was headed by a Yuz-bashi. 1908 M. Pickthall *Children of Nile* xxxiii. 268 Word of his coolness during the bombardment having reached the ear of power, he found himself raised to the rank of yuzbashi and honoured with a small command. 1920 *Glasgow Herald* 21 Jan. 4 The author..is an umquile Yuzbashi of the Imperial Ottoman Naval College. 1969 R. Millar *Kut* v. 87 The Turkish *yuzbashi* (equivalent to a captain, literally means the commander of 100 troops) was ushered into the office still blindfolded.

‖**yūzen** (yū·zen). Also **yuzen** and with capital initial. The name of Miyazaki *Yūzen*-sai (fl. mid-18th cent.), Japanese inventor of a technique of dyeing silk fabric in which rice-paste is applied to areas which are not to be dyed, used *attrib.* and *absol.* with reference to this process and the designs produced. Also **yūzen-zome** (see quot. 1983).
1902 [see *habutai]. 1911 *Encycl. Brit.* XV. 183 The difference between the results of the ordinary and the yūzen process of dyeing is, in fact, the difference between a stencilled sketch and a finished picture. 1958 K. Nomachi *Jap. Textiles* 21 Yuzen pattern is the design printed by the Yuzen process, with beautiful and graceful flowers and birds..as the motifs. The Yuzen process is of three kinds: Yuzen proper, hand painting, and stencil printing. 1964 Newman & Ryerson *Jap. Art* x. 77 *Yūzen zome*, a method of producing elaborate multi-colour designs by starch dyeing. Came into vogue in the Edo period. 1968 G. T. Webb tr. *Seiroku's Arts Japan* II. 214/2 The design of this *furisode*..a garment traditionally worn only by young girls, is executed in a variety of techniques, including *yūzen* and *shibori* dyeing and gold appliqué. In *yūzen-zome*, a hand-painted dyeing method, the patterns are first outlined with a color-resist of rice paste applied with a paper funnel or a chopstick-like wooden implement, and later filled in with dyes of any color desired. 1980 *Textile Designs Japan* (rev. ed.) I. 21 The invention of a new starch-resist dyeing technique, known as *yūzen*, during this period..led to undreamed of achievements in the dyeing of free-style graphic designs. 1983 Y. Tazawa *Biogr. Dict. Jap. Art* 652 During and after the Genroku era..stencil dyeing was developed and became popular, for a large number of *yūzen* designs are mentioned... From amongst these dyes, the term *yūzen-zome* came to be used thereafter to refer to *all* such pattern dyeing and it is still used thus to the present day.

Yvorne (ivǭ·ɹn). The name of a village in the Vaud canton of S.W. Switzerland, used *absol.* to designate a white wine produced in the region.
1871 J. Murray *Handbk. for Travellers in Switzerland* (ed. 14) p. xxxii, *Yvorne* is considered the best Swiss white wine. 1888 [see *Cortaillod]. 1935 A. L. Simon *Wines & Liqueurs* 61 *Yvorne*, one of the best Swiss white wines from Aigle, in the Upper Rhône Valley. 1962 *Economist* 29 Dec. 1283/1 The best whites are perhaps Yvorne and Dézaley. 1980 G. Greene *Doctor Fischer of Geneva* ix. 54 Beside every plate was a bottle of good Yvorne.

Z

Z. Add: In the U.S. pronounced (zī). **I. 2. Z-bend,** a series of bends in a road forming a shape like a letter Z; **Z-fold** *a.* (of print-out paper) in a continuous strip that comes folded in alternate directions in a stacked pile; **Z-plan** *Archit.*, the ground-plan of a type of Scottish castle having a central block with a tower placed at each of two diagonally opposite corners; **Z-plastic** *a. Surg.*, involving the use of Z-shaped incisions; also as *sb.*, Z-plastic surgery; so **Z-plasty,** a technique in which one or more Z-shaped incisions is made (the diagonals forming one straight line) and the two triangular flaps of skin so formed are rotated and drawn across the diagonal before being stitched, so as to give a less obvious Z-shaped scar and minimize the effect of contraction; an operation in which this technique is used; also *Comb.*, as **Z-shaped** *a.*, in the shape of a Z; *spec.* in *Archæol.*, designating a rod motif found on Pictish stones.
1887 Macgibbon & Ross *Castellated & Domestic Architect. Scotl.* II. 6 As one form of plan is designated the L plan, it has occurred to us that the form we are now considering might be called the Z plan. We have accordingly adopted this nomenclature. 1901 *Proc. Soc. Antiquaries Scotl.* XI. 91 The double-disc and crescent symbols of the Pictish stones may be connected with the worship of the Blessed Virgin, the Z and V-shaped rods being her floriated sceptre. 1908 F. R. Fraprie *Castles & Keeps of Scotl.* i. 22 A new and entirely Scotch plan is very common. This has been called the zigzag or Z plan. 1913 S. L. McCurdy in *Surg., Gynecol. & Obstetr.* XVI. 209 (*heading*) Z-plastic surgery. *Ibid.* 212/2 (*caption*) Z-plastic operation of the neck for burn scar. 1927 *Sc. N & Q.* 3rd Ser. V. 2/1 Two plates of silver engraved with the double disc and Z-shaped rod symbol. 1934 *Surg., Gynecol. & Obstetr.* LVIII. 178/1 Davis has been interested in tracing the history of the use of Z plastic. 1940 *Ibid.* LXX. 942/1, 2 patients returned for further work after a Z plasty with skin graft. 1958 *New Statesman* 1 Nov. 590/2 The shops were grouped round the angles of a Z-bend in the road. 1964 R. Battle *Plastic Surg.* xii. 316 A Z-plasty should be done only on one aspect of the finger at a time, thereby avoiding complete disruption of the circulation. 1967 *Electronics* 6 Mar. 282/2 The smudge is gone from Z-fold paper. 1967 I. Henderson *Picts* v. 104 A selection of the commoner symbols is illustrated here, the most common of all being the crescent with an applied V-shaped rod, the double disc with an applied Z-shaped rod. 1973 J. Leasor *Host of Extras* v. 69 About twelve miles of diabolical Z-bends. 1977 *Proc. R. Soc. Med.* LXX. 256/2 The Achilles tendon is lengthened by Z-plasty and then the posterior capsules of the ankle and subtalar joints are incised transversely. 1977 *Clinics in Plastic Surg.* IV. 207/1 Z-shaped techniques other than Z-plasties, which are referred to as 'Z-plastics', differ in the movement or lack of movement of the flaps formed by the zigzag incision. 1978 A. & G. Ritchie *Anc. Monuments Orkney* 79 Began 1560, it is an excellent example of a Z-plan castle, comprising towers at diagonally opposite corners of a main block. 1982 *Computerworld* (U.S.) 15 Mar. 66 It [*sc.* a digital plotter] also uses Z-fold paper and disposable fiber-tip pens.
4. a. (Further examples.) Also conventionally representing the sound of snoring. Also as *v. intr.* to make such a noise or noises.
1909 H. G. Wells *Tono-Bungay* i. ii. 67 He had a way of drawing air in at times through his teeth that gave a whispering zest to his speech. It's a sound I can only represent as a soft Zzzz. *Ibid.* iii. ii. 326 He meditated for a time and Zzzzed softly. 1924 *Dialect Notes* V. 259 *Z-z-z* (buzzing, or snoring). 1951 *Blue Book Magazine* Jan. 25/1 A spark danced between two terminals, a filament snake spat an irate, 'Zzzt!' 1966 L. Cohen *Beautiful Losers* i. 16 Hiccup, jerk, zzzzzz, snort. 1967 V. C. Welburn *Johnny so Long* ii. i. 46 Lola: (*makes buzzing noise*) Zzzzzzz. 1975 *New Yorker* 21 Apr. 36/3 David sits in the chair, puts his arms on the armrests, presses his neck against the back of the chair, and moves his feet together. 'Zzzz,' he says, and his head falls forward. 1976 *Cambridge Independent Press* 16 Dec. ii. 3/2 The zzzzz-noise of the electric hare gliding past the opening traps grabs everyone's attention. 1983 *Private Eye* 4 Nov. 6/2 Once you have hit on a commercial product you just go on producing more of the same, over and ..zzzz..over and zzzz..over and..zzzzzzzzzzzzzzzz zzzzzzzzzzzzzzzzzzzzzzzzzz. 1984 *Wall Street Jrnl.* 9 Oct. 28/2 We suspect public interest..more nearly resembles a cartoonist's depiction of a man sawing wood—ZZZZZZZZZ. 1984 *Oxford Star* 29/30 Nov. 19/3 Zzzzing off for forty winks on a regular basis may not sound much like Action Man stuff, but for Alex Gardner it's the most exciting part of the day.
b. In *colloq.* phr. *to catch some z's* and varr., to get some sleep (where z represents the sound of snoring). *U.S.*
Pronounced (zīz) in the U.S.
1963 *Amer. Speech* XXXVIII. 174 An onomatopoetic construction reported four times is *get some Z's*... Variants occurring once were: *bagging Z's, copping some Z's, cutting Z's*, and *knocking out Z's*. 1973 A. Dundes *Mother Wit* 238 Got to go..cop me some z's. 1977 C. McFadden *Serial* (1978) xxxii. 71/1 All Harvey wanted.. was to pop his Sominex and catch a few z's.
II. 5. *axis of z*: now always *z-axis*; also *transf.*
1929, etc. [see *X 3 a]. 1967 *Electronics* 6 Mar. 2 (Advt.), Plug-in markers offer not only variable bandwidth, but also Z-axis or pulse-type marking.
8. Genetics. *Z* is used to designate the male-determining sex chromosome in species in which the female rather than the male is the heterogametic sex.
1917 T. H. Morgan in *Amer. Naturalist* LI. 534 Since the female here [*sc.* in pigeons] is the heterozygotic sex (ZW) the results are such as would follow a direct influence on the sex chromosomes when the polar body is eliminated. 1925 *Ibid.* LIX. 133 The locus of the male tendency gene (M) is in the 'Z-chromosome' of which two are present in the male and one in the female. 1966 *Lancet* 24 Dec. 1397/2 The phenotypic expression of plumage factors on the Z chromosome of birds seems to be a function, principally, of gene dosage. 1971 [see *heterogametic *a.*]. 1976 *Nature* 17 June 598/2 In avian species, the heterogametic (Z W) female sex possesses W-linked histocompatibility antigen.
9. Physics. *Z* is the symbol for the atomic number of an element.
1931 *Proc. R. Soc.* A. CXXXIII. 234 We have taken Z = 8 (oxygen). 1962 F. I. Ordway et al. *Basic Astronautics* xii. 502 Electrons such as those in the outer Van Allen radiation belt are easily stopped by a few millimeters of a low-*Z* material such as aluminum or magnesium. 1978 P. W. Atkins *Physical Chem.* xiv. 438 The next atom to build is lithium, *Z* = 3.
10. *Z* is used to denote one of the two directions of twist in spinning (see quot. 1935); hence *z-spun* adj.
1935 [see *S II. 8]. 1964 H. Hodges *Artifacts* ix. 128 In thigh spinning, for example, a right-handed person will almost always produce Z-spun yarn. 1980 A. Fritz *Fibre of Clothing* iii. 40 There are two types of twist possible in a yarn. One is an S twist, the other a Z twist.
11. *Particle Physics.* *Z* is the symbol of a heavy, uncharged vector boson that forms a triplet with the two Ws.
1967 S. Weinberg in *Physical Rev. Lett.* XIX. 1265/2 The only unequivocal new predictions made by this model have to do with the couplings of the neutral intermediate meson Z_μ. *Ibid.* 1266/1 Our Z_μ and W_μ mesons get their mass from the spontaneous breaking of the symmetry. 1971 —— in *Ibid.* XXVII. 1688 This procedure..resulted in a model involving electrons, electron-type neutrinos, charged intermediate bosons (W_μ), neutral intermediate bosons (Z_μ), [etc.]. 1977 *Dædalus* Fall 32 The family of intermediate vector bosons, of which the photon is a member, is believed to contain one heavy charged particle and its antiparticle, called the W^+ and W^-, and one even heavier neutral particle, called the Z^0. 1982 [see *W 4b]. 1983 *New Scientist* 12 May 355 (*heading*) CERN physicists find the Z particle. 1983 *Nature* 25 Aug. 686/2 This resulted in a total of six examples of the Z particle (four decaying into e^+e^- and two into $\mu^+\mu^-$) and 52 $W^\pm$ particles.
III. Abbreviations. **12. a.** ZANU, Zanu, Zimbabwe African National Union; ZAPU, Zapu, Zimbabwe African People's Union; ZBB (*U.S.*), zero-base(d) budgeting; Z-DNA (*Biochem.*), DNA in which the double helix has a left-handed rather than the usual right-handed twist and the sugar phosphate backbone follows a zigzagged course; ZPG, zero population growth. See also *Z band* (s.v. *Z line), *Z line.
1963 *Times* 10 Aug. 5/2 The split in the Southern Rhodesian African nationalist movement has come to a head. A breakaway group..has formed..the Zimbabwe African National Union. The president is the Rev. Ndabaninge Sithole, formerly one of Mr. Nkomo's staunchest lieutenants, as are all the members of the new Zanu executive. 1964 *Ann. Reg.* 1963 ii. v. 105 On 8 August a new organization, the Zimbabwe African National Union (Z.A.N.U.), was formed with the Rev. Sithole as leader. 1977 *Times* 17 Sept. 15/3 Nobody wants to clear an arena for a final fight between Zapu and Zanu armies for supremacy. 1961 *Guardian* 18 Dec. 1/3 A new African political party, to be known as the Zimbabwe African People's Union, has been launched in Southern Rhodesia..by Mr. J. M. N. Nkomo... Mr. Nkomo said.. ZAPU would press for immediate negotiations for a fresh constitutional arrangement. 1972 J. Biggs-Davison *Africa—Hope Deferred* xi. 100 The rival parties Z.A.P.U. and Z.A.N.U. 1977 *Daily Times* (Lagos) 27 Jan. 3/2 Mr. Moyo..was with Nkomo in ZAPU before the revolt which Nkomo faced prior to 1963. 1976 *N.Y. Times* 27 Aug. D1 Z.B.B., as it is widely known, calls for the justification of all spending in relation to priorities. 1978 *National Civic Review* LXVII. 132 ZBB was formally delineated and put into practice for the first time by Texas Instruments, Inc., 15 years ago. 1979 A. H.-J. Wang et al. in *Nature* 13 Dec. 681/2 In looking at this left-handed helix..it is apparent that the ribose-phosphate backbone follows a zig-zag course resulting from alternating residue conformations. Accordingly, we propose to call this Z-DNA. 1983 *Sci. Amer.* Dec. 92/1 In *Z* DNA..the repeating unit of the helix is not a single base pair, as it is in *A* and *B* DNA, but rather two successive base pairs: *G–C* followed by *C–G*. 1970 *N.Y. Times* 3 June 61/6 Of all the cries that have arisen, perhaps none is quite so superficially appealing, or so profound in its long-range social and economic implications, as zero population growth—'ZPG' in the slogans of the day. 1978 *Nature* 6 Apr. 491/3 This may have reduced the maximum possible number of children per female to around five, and it does not then require harsh assumptions about mortality rates to end up with ZPG.
b. In combinations containing the abbreviation Z followed by a word, as **Z-car,** a police patrol car (after the title of a popular U.K. television series, from the radio call-sign 'zulu' allotted therein to a group of such cars); also used allusively; **Z-day** *Mil.* = *zero day* s.v. *zero *sb.* 7 a.
1961 *Radio Times* 28 Dec. 29/2 The call-sign is ZULU—they call them Z-cars. There are two young constables in each, ready to deal with trouble. 1963 *Daily Mail* 23 Apr. 1/6 (*heading*) Z-car crashes. 1965 *Sunday Mail* (Brisbane) 19 Dec. 32/5 Two black Z-cars—powerful Humbers—keep driving around the half-mile circle every five minutes watching for strangers. 1976 'D. Craig' *Faith Hope & Death* xviii. 127 It's not like Barlow and Z Cars, all that shouting stuff and strong jaws. These boys, they was slimy. 1925 Fraser & Gibbons *Soldier & Sailor Words & Phrases* 312 The opening of the main attack of the battle of the Somme was fixed for June 29th, and notified beforehand as 'Z Day'. 1930 S. Sassoon *Mem. Infantry Officer* iv. 61 Operation Orders..notified us that Thursday was 'Z' (or zero) day. 1938 Blunden *On Several Occasions* In the sour concrete hole the corporal shows his muddy map, his Z Day zone of fire.

za (zā). Also **'za.** U.S. slang abbrev. of *pizza.
1968–70 *Current Slang* (Univ. S. Dakota) III–IV. 140 *Za*.., pizza. 1980 L. Birnbach et al. *Official Preppy Handbk.* 44/2 You can go for 'za, burgers, and ice cream without signing out. 1983 *Verbatim* IX. iv. 1/1 One of the boys called up and asked the parlor to *bag the za* (meaning 'cancel the pizza'). 1984 *Amer. Speech* LIX. 89 In surveying my classes for campus slang I learned that pizza is referred to as *za* and to get or eat a pizza is to *do a za*.

zabaglione (zabalˠo·ne). Also **sambaglione, zabaione,** etc. [a. It., perh. ult. ad. late Lat. *sabaia* an Illyrian drink.] A dessert consisting of egg yolks, sugar, and (usu. Marsala) wine, whipped to a frothy texture over a gentle heat and served either hot or cold. Cf. *sabayon.
1899 M. Ninet *Dainty Meals for Small Households* xiv. 223 *Sambaglione.*—Put into a saucepan about six new-laid eggs, and two sherry-glassfuls of Madeira. 1901 W. G. Waters *Cook's Decameron* 175/1 Zabajone is a kind of syllabub. *Ibid.* 175/2 (*heading*) Iced zabajone. 1914 C. E. Edwards *Bohemian in San Francisco* 104 This was followed by a glass of zabaione for dessert. 1917 *Blackw. Mag.* Jan. 111/1 We dined off onion pilaff, and sponge-cake of maize flour with zabayone. 1924 M. Kennedy *Constant Nymph* xvi. 222 She had instructed Roberto to beat up some eggs..that she might make zabaglione. 1932 *Times Lit. Suppl.* 10 Nov. 836/4 At a single opening of the book we find..zambaglione (the reviewer would prefer to omit the lemon-juice but to add a pinch of yeast). 1958 *Times* 2 Oct. 11/4 A nation that is content at midday with shepherd's pie..must dine out off kebabs and zabagliones. 1960 *House & Garden* May 94/3 A copper zabaglione pan. 1974 *Country Life* 25 Apr. 1006/1 The best-known Italian sweet, *zabaglione*. 1981 R. Airth *Once a Spy* ix. 100 Zac ordered blackberry, peach, pineapple [ices] and *zabaione*. 1984 *Listener* 1 Nov. 42/2 Whatever food is mentioned, from aubergines to zabaglione, they are ready to garnish it with some unsavoury gobbet of introductory junk music.

†**zabernism** (zæ·bəɹniz'm). *Obs.* [f. *Zabern*, German name of the village of Saverne in Alsace + -ism.] The misuse of military power or authority; bullying, aggression (see quot. 1921). Also **za·bernize** *v. intr.*
1914 *Daily News & Leader* 6 Apr. 1/7 A quaint illustration of an attempt to 'zabernise' in business life..is published..to-night. 1916 [see *rounding *vbl. sb.*[1] 1 a]. 1918 *Nineteenth Cent.* Dec. 992 The acquisitive commercialists and financiers of Germany—the Zabernism of industry and *l.s.d.*—were up to the eyes in it [*sc.* the war]. 1918 H. G. Wells *In Fourth Year* iii. 36 Both countries have been slaves to Kruppism and Zabernism. 1920 *Glasgow Herald* 7 Apr. 8 The advance of the Government troops into the Ruhr district, coupled with a marked exhibition of 'Zabernism', has stiffened the Spartacist resistance. 1921 E. Weekley *Etym. Dict. Mod. Eng.* 1658/2 *Zabernism* (*hist.*), military jackbootery. From an incident at *Saverne* (Ger. *Zabern*) in Alsace (1912), when an excited Ger. subaltern cut down a lame cobbler who smiled at him.

‖**zabuton** (zabu·tǫn). [Jap., f. *za* sitting, a seat + *buton* f. *futon* cushion, padded mattress.] A flat floor cushion on which one sits or kneels.

1889 M. B. Huish *Japan & its Art* vii. 85 The Japanese ..is only comfortable when resting on his knees and heels on a cushion (*zabuton*). **1928** J. I. Bryan *Jap. All* vi. 76 In a Japanese house the guest does not sit, but kneels down and rests on the heels with toes out behind. A cushion, called *zabuton*, is given to rest the knees on. **1960** *Sunset* Mar. 137/1 The word *zabuton* applies to a Japanese floor cushion designed for sitting. It comes from the Japanese words *za* meaning to sit, and *futon* meaning bedding. **1973** A. Broinowski *Take One Ambassador* v. 57 There were *shoji* screens and *zabuton* cushions on the *tatami* round the table. *Ibid.* xii. 193 She brought a *zabuton* and sat on the floor on that.

zac (zæk). *Austral. slang.* Also **zack, zak.** [Origin unknown.] A sixpence.

1898 *Bulletin* (Sydney) 1 Oct. 14 6d., a 'zack'. **1924** *Truth* (Sydney) 27 Apr. 6 *Zack*, sixpence. **1941** *Coast to Coast 1941* 213 'Only one, then,' I said. 'We'll only have one more, so we'll make it a sixpence in.' 'Don't worry about the zac,' said Tom. **1952** M. Tripp *Faith is Windsock* ii. 29 Arthur spent the break losing bobs and tanners while the tall Australian reckoned his profits in dieners and zaks. **1962** *John o' London's* 8 Mar. 229/2 And all you get for it is a zack. **1966** *Tel.* (Brisbane) 13 Oct. 13/2 The zack, now the 5c piece, is in such demand that very often its scarcity makes one wonder if it is gradually going out of circulation. **1966** P. Mathers *Trap* ii. 31 Two taps with an improvised hammer and anvil, a two bob and a zac. **1977** *Sunday Sun* (Brisbane) 1 May 16 When it comes to unique competitions the people of outback Winton reckon they're the full quid—and you can bet your last zac or traybit on it.

zacate: see *SACATE, ZACATE; **zacaton**: see *SACATON, ZACATON.

'zackly, 'zactly (zæ·kli, zæ·ktli). Repr. a dial. or colloq. pronunc. of Exactly *adv.*

1886 Baumann *Londinismen* 239/1 *'Zactly*.., exactly. **1893** H. A. Shands *Some Peculiarities of Speech in Mississippi* 69 *Zackly*.., negro for *exactly*. **1929** W. Faulkner *Sartoris* iii. vi. 232 'Yessuh,' Simon agreed readily. 'Hit struck me jes' 'zackly dat way.' **1929** H. Williamson *Beautiful Years* (rev. ed.) vi. 63 'Isn't the sunshine lovely, Mr. Lewis?' 'Eh, ooh? 'Zactly!' **1973** *Amer. Speech* 1970 XLV. 77 Her dress was 'zactly like mine. **1983** R. Sutcliff *Blue Remembered Hills* xii. 91 Some people believed that Mr. Snow was, in West Country parlance 'not quite zackly'... I think he was just simple in the true sense of the word.

zaddik, zadik, varr. *TSADDIK.

Zadokite (zē¹·dŏkait), *sb.* and *a.* [f. the name of *Zadok*, a high priest of Israel in the time of King David + -ite¹.] **A.** *sb.* A member of a Jewish sect which seceded from orthodox Judaism in the second century B.C., and traced its authority back to Zadok.

1910 S. Schechter *Fragments of Zadokite Work* p. xxi, It is the Zadokites from which the Sect derived its spiritual pedigree. **1920** *Encycl. Relig. & Ethics* XI. 43/2 The Sadducees were thus not a religious party at all, but simply a body of people bound together by a common interest to maintain the existing *régime*. The name is explained as meaning 'Zadokites', and was given to them by their opponents, the Pharisees, who borrowed it from an earlier age..when the descendants of Zadok, who then filled the high-priestly office, identified themselves with Hellenism in its most dangerous forms. **1960** tr. *Noth's Hist. Israel* iii. ii. 316 It may also be that..some deported Zadokites returned to Jerusalem. **1974** *Encycl. Brit. Macropædia* XIV. 190/1 The Zadokites may have constituted the survival of an ancient Jebusite (Canaanite) royal priesthood.

B. *adj.* Of, pertaining to, or designating the members of this sect; *spec.* applied to fragments of sectarian texts discovered in Cairo in 1896–7 and later traced back to the Dead Sea Scrolls.

1910 S. Schechter (*title*) Fragments of a Zadokite work. **1912** R. H. Charles *Fragments of Zadokite Work* p. vii, There is no question as to the genuineness of the orders of the Zadokite Priests and Levites. *Ibid.* p. x, The Zadokite Party represents an attempt at reform beginning within the ranks of the priesthood and extending outwards so as to embrace a strong lay element. **1921** J. Moffatt *Approach to New Testament* i. 60 In the so called Zadokite document of Jewish piety, just before the days of Jesus, the idea of a new covenant, a covenant of repentance, began to be linked to the expectation of a messiah. **1954** [see *Qumran]. **1974** *Encycl. Brit. Macropædia* II. 938/2 Another sectarian book of ordinances is the Damascus Document (the Zadokite Fragments). The work was already known from two medieval copies before the discovery of the Dead Sea Scrolls, but fragments of it also were found in Qumrān.

‖ **zadruga** (zadrū·ga). Pl. **zadrugas, zadruge.** Also with capital initial. [Serbo-Croat, = patriarchal commune, association.] A type of patriarchal social unit traditional to (agricultural) Serbians and other southern Slavic peoples, orig. comprising an extended family group which worked the land and lived communally round the main house; the customs and rules associated with this type of unit.

1887 *Encycl. Brit.* XXII. 146/2 The Slavs know nothing of private property,—the land being held in common under the care of the *vladika* or *stareshina*, as in the Servian *zadrugas* at the present day. **1900** 'Odysseus' *Turkey in Europe* viii. 375 The old system of Zadruga, or communal village based upon the family. **1911** Prince Lazarovich-Hrebelianovich *Servian People* i. 39 The basis of the Serb organisation is the family, either in its narrowest sense of blood-relationship, in communistic organisation, or other individuals grouped together for common work and with common possessions. These forms are called 'Zadruga'. **1934** *N.Y. Times* 24 June iv. 3/4 Zadruga in Serbia means a big family where brothers and sisters..lead a community life, under the leadership usually of the eldest member of the family. **1943** L. Adamic *My Native Land* 214 From their Russian homeland the Slavs brought a democratic institution called *zadruga*, a clan or family cooperative, which some of the tribes tried to extend and adjust to the wider forms of government necessary in their new homelands. *Ibid.* 216 They lived in their primitive villages and held onto their Old-Slavic *zadrugé* and 'heart culture'—decency, friendliness, hospitality. **1963** *Times Lit. Suppl.* 25 Jan. 49/3 The *dvor* or peasant household..is the same elastic unit which was familiar in medieval Europe, and survives today in the *zadruge* of the Balkans. **1979** *Internat. Jrnl. Sociol. of Law* VII. 270 The old customary Slavonic 'family' law, the zadruga. **1980** A. Toffler *Third Wave* ii. 44 People tended to live in large, multi-generational households,..from the 'joint family' in India to the 'zadruga' in the Balkans.

‖ **zaftig** (zæ·ftig), *a. U.S. colloq.* Also **zoftig, zofti(c)k.** [Yiddish, a. G. *saftig* juicy.] Of a woman: plump, curvaceous, 'sexy'.

1937 M. Levin *Old Bunch* ii. 394 He could see himself on the road, whizzing by a flaming poster—a beautiful girl, snappy, modern, *zaftig*. **1950** *Commentary* IX. 460/2 The owner of the local barber shop..installed a *zoftig* blond manicurist. **1960** 'E. McBain' *Heckler* x. 91 A *zaftig* redhead..in her early thirties. **1967** P. Welles *Babyhip* (1969) xxiv. 139 'And see how *zoftik* she is.' Mrs. Green felt pleased she had spotted someone over-weight. **1970** S. J. Perelman *Baby, it's Cold Inside* 188 This demure but *zoftick* freshman, with a brain rivalling Spinoza's encased in the body of a Lollobrigida. **1973** R. Hayes *Hungarian Game* xxx. 179 That *zoftig* colonel wasn't a colonel at all. I checked her out in Budapest. **1981** *Gossip* (Holiday Special) 11/2 Zaftig Dolly Parton..once described herself as looking like a 'hooker with a heart of gold'.

Zaghlulist (zaglū·list), *sb.* and *a.* Also **Zaghloulist.** [f. the name of the Egyptian politician *Zaghlūl Sa'd* (1857–1927) + -ist.] **A.** *sb.* An adherent or supporter of the nationalist and separatist principles and policies of Zaghlūl Sa'd. **B.** *adj.* Of, pertaining to, or designating members of this political group.

1921 *Times* 29 Dec. 9/3 The Zaghlulists have raised the familiar Indian cry of non-cooperation. **1924** *Times* 14 Jan. 11/5 The comparative nonentity who was his Zaghlulist opponent. **1924** [see *omdah, omdeh]. **1927** *Daily Express* 9 Sept. 1/1 A congress of Zaghlulist parliamentarians to decide on a policy consequent on the death of their leader will open on September 15 or immediately after, when the Wafdists (champions of Home Rule)..arrive. **1943** Ld. Wavell *Allenby in Egypt* vii. 102 The completeness of the Zaghlulist victory surprised every one..even Zaghlulists themselves. **1955** H. W. Jarvis *Pharaoh to Farouk* xxxii. 267 The Zaghloulists, however, were determined that a 'Tantah' incident *should* be repeated in some part of Egypt. **1971** P. Mansfield *British in Egypt* xxii. 252 Allenby.. departed to genuine expressions of regret from the Egyptian public including the Zaghlulists.

‖ **zaguan** (zagwā·n, sa-). Also 9 **saguan.** [Sp., = vestibule, hallway.] The passage running from the front door to the central patio in houses in South and Central America and in the south-western U.S.

1851 *Harper's Mag.* Sept. 465/2 Don Pedro was heard within, moving toward the 'Saguan'. **1863** *Rio Abajo Press* 28 Apr. 1/2 She had just seen Juanito's ghost in the saguan door. **1880** G. W. Cable *Grandissimes* 131 It was a long, narrowing perspective of arcades, lattices, balconies, *zaguans*, dormer windows, and blue sky. **1921** *Chambers's Jrnl.* Mar. 258/1 There is the *zaguan* or vestibule at the outside, then the *cancel* or grating, and next the *patio* within. *Ibid.* Nov. 821/1 A couple of Moors —in the vestibule or *zaguan*—conduct a peculiar performance. **1927** D. H. Lawrence *Mornings in Mexico* 12 Some stranger comes into the *zaguan*.

Zahal (tsaha·l). [Heb., f. *Ṣĕḇā' Hăgānāh Lĕ-Yiśrā'ēl* Israel Defence Force.] The name applied by the Israelis to their defence forces, formed originally in 1948 by the fusion of pre-independence military organizations.

1959 A. M. Heller *Israel's Odyssey* 61 The Haganah disbanded, the Irgunist and the Sternist forces and their members were fused into the Zahal, made up of the initial Hebrew letters of Zva Haganah L'Israel—Israel's National Army. **1969** A. Perlmutter *Mil. & Politics in Israel* v. 55 Zahal's chief of staff was delegated the operational function of maintaining and training the army. **1971** *Scope* (S. Afr.) 19 Mar. 17/3 Very few top-ranking Israeli officers gave their Zahal (Defence Force) more than one chance in three of pulling it off. **1975** C. A. Haddad *Moroccan* i. 7 The twins did not have to lose two years to the Zahal, as the [Israeli] army does not draft religious girls. **1982** Murray & Viotti *Defense Policies of Nations* ix. 378 The leadership opted for a unified command for *Zahal* in which there would be one general headquarters and a chief of staff.

‖ **zaibatsu** (zaibæ·tsu). Also **Zaibatsu.** [Jap., f. *zai* wealth + *batzu* clique.] In Japan, a large capitalist organization, usu. based on a single family having controlling interests in a variety of companies, of a type that existed before the war of 1939–45; since 1947, a cartel or conglomerate. Also, the members of such an enterprise.

1937 *Econ. Jrnl.* June 272 These concerns are popularly known in Japan as the *Zaibatsu* or money-cliques. **1947** R. Benedict *Chrysanthemum & Sword* iv. 93 A chosen financial oligarchy, the famous Zaibatsu. **1957** *Pacific Affairs* XXX. 219 The *zaibatsu* and the landlords. **1959** R. Kirkbride *Tamiko* (1960) vi. 41 By getting the right people to sit for you. The zaibatsu. The Royal Family. **1964** *Asia Mag.* 26 July 15/2 While the companies being merged presently are old Zaibatsu connected firms, and the groups are sponsoring the moves, the economics of the mergers themselves have little to do with such connections. **1965** *Economist* 11 Dec. 1219/2 The *Zaibatsu* are becoming concerned about the problems of seniority and management. **1970** [see *conglomerate *sb.* 3]. **1974** P. Gore-Booth *With Great Truth & Respect* 45 When a community, historically prone to become the victim of unreason, finds itself in a position of success amid failure, its only recourse is probably to go to ground. The Japanese *Zaibatsu* understood this; a really successful Japanese business man was and is rarely obtrusive. **1982** K. Ohmae *Mind of Strategist* (1983) ix. 109 Japan's prewar *zaibatsu* holding companies.

‖ **zaikai** (zai·kai). [Jap., f. *zai* wealth + *kai* community.] In Japan: financial circles; the business world or power élite who control it.

1968 C. Yanaga *Big Business in Jap. Politics* ii. 32 A new postwar term, zaikai,..practically supersedes zaibatsu... It is frequently interpreted..as a synonym for 'business circles', 'financial circles', and even 'business community'. More inclusive than zaibatu, it is nevertheless restricted to big business... Zaikai also denotes the place where the craving for political power is openly expressed and gratified. **1970** *Times* 19 Aug. 5/2 A Japanese economic magazine recently conducted a survey of opinion among leading representatives of the *zaikai*, the world of business and high finance. **1975** G. L. Curtis in E. F. Vogel *Mod. Jap. Organization & Decision-Making* 38 The men who make up this elite in Japan spend an extraordinary amount of time in so-called *zaikai* activities..that are not directly related to their own companies, but which seek to represent the interests of the business community as a whole. **1981** J. B. Kidd in P. G. O'Neill *Tradition & Mod. Japan* 50 The process of *nemawashi* operating in the senior levels of the *zaikai* world enable the companies to secure long-term goals.

zaire (zai,ɪ°·ɪ). [f. *Zaïre*, local name of the Congo River in Central Africa.] The basic monetary unit of the Republic of Zaïre; a coin of this value, equal to 100 makuta (see *likuta).

1967 *Times* 9 Nov. 6 They tried to recruit former Katangese gendarmes... They offered seven zaires (£5) to every man who enlisted. **1971, 1972** [see *likuta]. **1976** L. Sanders *Hamlet Warning* (1977) xvi. 133 A bribe... Five thousand zaire—ten thousand dollars.

Zairean (zai,ɪ°·riăn), *sb.* and *a.* Also **Zairian.** [f. the name of the Republic of *Zaïre* (cf. prec.) + -an, -ian.] **A.** *sb.* A native or inhabitant of the Republic of Zaïre, formerly the Democratic Republic of the Congo. **B.** *adj.* Of or pertaining to Zaïre.

1972 *Observer* 19 Nov. 2/1 Dead victims..have been found dumped close to the frontier with Zaire... They were taken there by contract drivers, many of them Zaireans. **1973** *Black World* May 80 The dollar value of this manpower is considerable and most of it is defrayed by the Zairian Government. **1973** *Daily Tel.* 1 Oct. 19 Two British geologists yesterday described their fortnight's ordeal under armed Zairean Gendarme guards. **1973** *Times* 11 Dec. (Zaire Suppl.) p. vii/2 None of the young Zairians danced. **1977** *Times* 12 Apr. 13/1 Both the Moroccans and Zaireans are using their own planes to transport the troops. **1977** *Daily Tel.* 14 Apr. 1/8 The spokesman said that on Saturday two Zairean jets violated Zambian airspace. **1982** S. Johnson *Marburg Virus* x. 121 The logistic support of the Zairian army.

Zairese (zai,ɪ°rī·z), *a.* and *sb.* [f. as prec. + -ese.] = *Zairean *sb.* and *a.* Cf. next.

1974 *N.Y. Times* 10 Feb. 11/1 A 700-page, 30,000 word Swahili dictionary..is the work of a Zairese priest. **1975** *Business Week* (Industrial ed.) 27 Oct. 94H On the other side of the lake, the Zairese would like to build a new recovery unit. **1977** *Bangladesh Times* 19 Jan. 11/6 A lack of co-ordination between Belgian and Zairese security officials at Brussels airport, where the Belgian monarch was greeting President Mobutu. **1978** *Washington Post* 18 June A18/4 In the name of 'authenticity', he declared that in French revolution fashion Zairese should call each other 'citizen'.

Zairois (zai,ɪ°rwa·), *sb.* and *a.* Also **Zaïrois.** [a. Fr.] = *Zairean *sb.* and *a.*

1973 J. J. MacKelvey *Man against Tsetse* iv. 228 The state of war against tsetse as the Zairois assess it will dictate the role that this institution may play in further research. **1974** *Observer* 22 Sept. 20/3 The Zairois people are warm, helpful and charming. **1975** R. Barclay *Ernest Bevin & Foreign Office* viii. 159 It is in any case an offence for a Zaïrois to wear such an 'un-authentic' item of dress as a tie. **1977** *New Day* Summer 10 (*caption*) A Zairois student learns how to take a skin smear. **1982** *Financial Times* 26 Jan. 22/1 The Government intends the industry to employ an eventual 700 Zairois.

‖ **zakat** (zăkā·t). Also **zakah, zakkat,** † **zecchat.** [Pers. *zakāt*, Turk. *zekât*, etc., ad. Arab. *zakāh*.] An obligatory tax payable annually under Islamic law on certain kinds of property in order to raise money for charitable and religious objects.

1802 J. Pinkerton *Mod. Geogr.* I. 452 [Turkey.] This revenue is partly derived from..the *zecchat*. **1875** Bellew & Chapman in T. D. Forsyth *Rep. Mission to Yarkund* ii. 103 The *zakāt* is a Mahommedan tax... It is one part in forty of all live-stock, and of merchandise entering the country. **1957** Ld. Hailey *African Survey 1956* x. 671 In Mauretania those who are not liable to *zakkat* pay the personal tax and in addition a capitation fee on every animal. **1960** C. Geertz *Religion of Java* x. 123 The fifth pillar [of Islam] is the *zakah*, the religious tax. **1979** *Observer* 4 Mar. 12/3 Their principal objection was to 'zakat', a partly-compulsory Islamic welfare tax which General Zia proposes to introduce in July. According to Shi-ism zakat has to be an individual act of conscience and can never be levied compulsorily by the State, he said.

‖ **zakuska** (zăku·skă). Also **zakouska.** Pl. **zakuskas, zakuski.** [a. Russ. *zakúska* (usu. as pl. *zakúski*).] An hors d'œuvre. Freq. pl.

The sing. form is occas. erron. const. as pl.

1885 A. J. C. Hare *Stud. in Russia* v. 203 The refreshment-room is enormous,..with a great buffet at one end, whither the Russians resort before dinner for the customary *zakuska* of pickles, sardines, vodki, &c. **1919** H. Walpole *Secret City* i. xii. 79 The 'Zakuska' were on a side-table near the door—herrings and ham and smoked fish and radishes and mushrooms and tongue and caviare. **1920** E. & P. Sykes *Through Deserts & Oases Central Asia* iv. 73 It began with many *zakuskas*, consisting principally of dubious-looking tinned fish. **1922** W. S. Maugham *On Chinese Screen* vi. 27 The emotion..was diverted by the appearance of two Chinese servants in long silk robes and four-sided hats with cocktails and zakouski. **1933** H. Nicolson *Let.* 7 Nov. (1966) 156 There were also little snippets of lobster in tomato cocktails, which were among the least successful of *zakouska* that I have ever known. **1951** V. Nabokov *Speak, Memory* xiv. 213, I detest crowds, harried waiters, Bohemians, vermouth concoctions, coffee, *zakuski*, floor shows and so forth. **1981** T. Keene *Skyshroud* i. 18 The *Zakouski*, the *borshch* and the *blini* of his homeland.

‖ **zamacueca** (zamakwē¹·kă, sa-) Also 9 **-cuca.** [Amer. Sp.] A South American, esp. Chilean, dance in which a couple move around one another, accompanied by chords on the guitar and rhythmical handclapping. Shortened as *cueca.

1855 E. R. Smith *Araucanians* vi. 65 The fashionable dances soon gave way to the more national *Zamacúca*. The Zamacúca has been very much reviled by foreigners, who have seen it only in sea-port towns. **1895** L. Grove et al. *Dancing* xii. 375 The 'Zamacueca' is an inimitable dance, in which the handkerchief plays a prominent part. **1917** [see *cueca]. **1974** *Encycl. Brit. Macropædia* I. 669/1 The Creole people of the coast and plains [of South America] have been replacing the indigenous music of the mountains with their own music, especially with the ubiquitous *cueca*, or *zamacueca*.

zamarra. Add: Also 9 **semara.** (Earlier examples.)

1839 E. E. Napier *Jrnl.* 4 May in *Excursions along Shores of Mediterranean* (1842) II. vi. 81, I observed a tall, gentlemanly-looking man, dressed in a semara. **1841** G. Borrow *Zincali* I. ii. i. 231 Another Gypsy.. dressed in a zamarra of sheep-skin.

Zambian (zæ·mbiăn), *sb.* and *a.* [f. *Zambia* (see def.) + -an.] **A.** *sb.* A native, citizen, or inhabitant of Zambia (formerly Northern Rhodesia), a country in south central Africa.

1964 *Economist* 15 Feb. 607/2 With only 180 'Zambians' (Northern Rhodesia's name when independent) taking school certificate this year. **1976** *Drum* (E. Afr. ed.) Aug. 8/1 Another reason for the shortage here is that the township council has turned down licence applications from non-black Zambians and foreigners. **1977** 'O. Jacks' *Autumn Heroes* xiv. 207 Everyone had learned to freeze into silence the moment the Zambian made pronouncements.

B. *adj.* Of, pertaining to, or characteristic of Zambia or its people.

1964 *Times* 13 Nov. 11/3 A European in Zambia whose father was born in Britain might decide after a year to acquire Zambian citizenship. **1966** *Listener* 15 Dec. 879/2 The whole Rhodesian situation has driven Zambian politicians into a kind of neurosis. **1970** *Cape Times* 28 Oct. 2/5, I was furious when we found ourselves on the Zambian side of the lake. **1977** *Whitaker's Almanack 1978* 588/1 The Zambian Minister of Home Affairs signed deportation orders for 55 aliens who it was said had entered the country illegally.

Hence **Za·mbianize** *v. trans.*, to make Zambian in character, by replacing foreign personnel by native Zambians; **Zambianiza·tion.**

1968 *Times* 18 Nov. (Zambia Suppl.) p. v/2 The Zambian government wants to see the whole industry Zambianized, with no whites and indeed no foreign Africans in any jobs. **1968** *Economist* 14 Dec. 26/2 ANC claims simply that it could have done better, picking out the soaring cost of living, wasteful public expenditure and falling standards in the police force—the last two linked with too-rapid Zambianisation. **1969** *Listener* 14 Aug. 221/3 That may have led to the proposal to 'Zambianise' the judiciary. **1976** G. C. Bond *Politics of Change in Zambian Community* x. 166 Its policy of rapid Zambianization siphoned off the more educated and progressive elements. **1984** *Mining Jrnl.* 26 Oct. 287/2 Zambianization of the mining industry was a natural consequence of independence.

Zamboni (zæmbō̆ᵘ·ni). Chiefly *N. Amer.* Also **zamboni.** [See quot. 1965.] A proprietary name for a machine used to resurface ice rinks.

1965 *Official Gaz.* (U.S. Patent Office) 16 Feb. TM 93 Frank J. Zamboni & Co., Paramount, Calif... *Zamboni* for ice resurfacing machines and dump attachments therefor. First use July 1962. **1968** *Trade Marks Jrnl.* 3 Apr. 517/2 *Zamboni*... Machines for resurfacing ice rinks and parts and fittings therefor. **1974** *Kingston* (Ontario) *Whig-Standard* 11 July 14/2 With 12 scrapers, I could do that ice in nine minutes. With the Zamboni, it took a little longer..but I will have to admit it does do a better job. **1978** *Winnipeg Free Press* 25 Sept. 53/4 A crack in the boards near the door used by the zamboni helped Guindon open the scoring. **1984** *N.Y. Times* 11 Feb. 1. 22/6 Then the ice was glazed by a smooth-nosed French cousin of the Zamboni before another Soviet skater..glided to gold medals.

zambuk (zæ·mbɒk). *Austral.* and *N.Z. colloq.* Also **zambuc, -buck,** and with capital initial. [A proprietary name of a type of antiseptic ointment.] A first-aider, a St. John's ambulance man or woman, esp. at a sporting occasion. Also *attrib.* and *transf.*

[**1904** *Trade Marks Jrnl.* 23 Nov. 1406 *Zam-buk*... Chemical substances prepared for use in medicine and pharmacy, but not including gelatine capsules... The Bile Bean Manufacturing Company,..Leeds; vendors of proprietary medicines.] **1918** [see *gutser]. **1943** *Amer. Speech* XVIII. 95 A few trade names have caught the public fancy, and become generalised. Thus *Zambuk*, a brand of ointment, is regularly used for a 'first-aid man' (usually a member of the St. John's Ambulance Corps), or even as an appeal for first aid. **1950** B. Sutton-Smith *Our Street* iv. 50 Everybody would shout, 'Zambuk! Zambuk!' until the first-aid man came to the rescue. **1965** *N.Z. Listener* 5 Nov. 4/5 Perhaps he might have concussion... A zambuck came with sal volatile. **1968** *Wanganui Photo News* 31 Aug. 13 (*caption*) No injuries to attend to so these 'Zambuk' ladies get on with their knitting at the basketball recently. **1969** *N.Z. Listener* 23 May 12/3 The driver was dead... There was a girl in a bad way mixed up with him. 'Better leave it for the zambucs,' Payne said. **1972** P. Mathers *Wort Papers* 95 They even supplied the zambuck, a priest with brown attaché case containing oils and waters for extreme unction.

zamindari, -y, varr. Zemindary in Dict. and Suppl.

‖ **zampogna** (zæmpǫ·nʸă). *Mus.* [It.:–LL. **sumpōnia, symphōnia* (see Symphony). Cf. Sp. *zampoña*, Pg. *sanfon(h)a*.] A traditional wind-blown bagpipe of southern Italy having two chanters and two drones; also, any wood-wind instrument.

1740 Grassineau *Mus. Dict.* 343 *Zampogna*, sometimes written *Sampogna*..is..any instrument that sounds like a Flute and particularly a Bag-pipe, being an assemblage of divers pipes of different sizes. **1801** T. Busby *Dict. Mus.*, *Zampogna*, or *Sampogna*, the flute-à-bec, or common flute. **1826** M. Kelly *Reminisc* (1975) ii. 30 The peasantry and shepherds from Abruzzo, Calabria, and Apulia,..come from the mountains in their sheepskin dresses, playing on their various instruments, some on the Zampogna, (a kind of bagpipe), others on the Colascione. **1876** Stainer & Barrett *Dict. Mus. Terms* 455/1 A rough-toned reed instrument without a bag is also called Zampogna or Zampugna. **1954** *Grove's Dict. Mus.* (ed. 5) I. 351/1 Italian pipers..make a practice of coming into the towns at the Christmas season to serenade the images of the Infant Christ, which are set up at the roadside. They also formerly visited Britain as strolling musicians, but they appear to have abandoned this habit many years ago; they were all players of the *zampogna*. **1977** *Early Mus.* Oct. 555/2 It was..good to see no less than two sets of zampognas, from the early 19th century. **1983** *N.Y. Times* 26 June xi. (Westchester Weekly) 13/1 Vocalists will be accompanied on such classic Old World instruments as the zampogna.

Zande (zæ·ndi). Also 9 **Zandey**; 9–**Zandeh.** Pl. **Zandes, Azande** († **A-Zande**). [Native name.] **a.** (A member of) a people of mixed ethnic origin inhabiting central Africa. **b.** The language of this people. Also *attrib.* as *adj.*

1873 E. E. Frewer tr. *Schweinfurth's Heart of Afr.* II. xiii. 6 As marks of nationality, all the 'Zandey' score themselves with three or four tattooed squares. **1884** *Encycl. Brit.* XVII. 474/1 The A-Zandey are to be regarded as rather of mixed Negroid than of pure Negro stock. *Ibid.* The Zandey language..appears to be everywhere spoken with considerable uniformity. **1891** A. H. Keane tr. *W. Junker's Trav. in Africa* II. iv. 102 The hair, arranged Zandeh-fashion, though somewhat carelessly, in tresses, projected under a tarbush round the occiput. **1902** *Encycl. Brit.* XXXI. 230/1 The Zandehs proper..are now found to stretch, with interruptions, from the White Nile above the Sobat confluence to the Shari affluent of Lake Chad. **1918** R. G. C. Brock in *Sudan Notes & Records* Oct. 249 The Zande tribe. *Ibid.*, The true Azande live in the Belgian Congo. *Ibid.* 253 The Azande are divided into different clans and most of these believe that they return to the earth in the form of some animal. *Ibid.* 257 The ordinary Zande dance is a very dull affair. **1938** E. M. North *Bk. of Thousand Tongues* 353/2 Zande... Spoken by perhaps 300,000 people in northeastern Belgian Congo, eastern French Equatorial Africa and southern Anglo-Egyptian Sudan. **1955** M. Gluckman *Custom & Conflict in Afr.* iv. 85 If a Zande murdered a fellow-tribesman with his spear he was tried in court by his chief. *Ibid.* 90, I find it difficult to see exactly how the Azande witchcraft charges work. **1978** J. Skorupski in Hookway & Pettit *Action & Interpretation* 85 Winch's account of the meaning of Zande magic is ultimately not at all unlike that which some symbolist writers might give.

zanella. Add: Also 9 **zenilla.** (Earlier example.)

1876 F. Kilvert *Diary* 28 Apr. (1940) III. 272, I found she had taken my umbrella and left me a much better one, a fine silk umbrella in place of my zenilla.

zany, *sb.* Restrict *arch.* and *dial.* to senses in Dict. and add: **2. c, e.** (Further examples.)

1925 *Sunday at Home* Nov. 91/2 He went capering about all his tasks with a zany-like glee. **1929** C. Day Lewis *Transitional Poem* i. 11 A burly wind playing the zany In fields of barleycorn. **1976** G. Langford (*title*) Death of the early morning hero. Episodes of a zany in love.

3. Now apprehended as a simple adj., and the dominant use of the word, with the sense: comically idiotic, crazily ridiculous.

In quot. 1938, 'simple-minded' (cf. sense 2e in Dict.).

1918 G. B. Shaw in *Nation* 22 June 308/2 Before Shakespear touched Hamlet there was a zany Hamlet who mopped and mowed. **1938** L. MacNeice *I crossed Minch* viii. 117 The gardener who was apparently zany, used to level his stick at an obelisk. **1957** Manvell & Huntley *Technique Film Music* v. 204 *Sloppy Jalopy* (UPA), a zany cartoon. **1959** *House & Garden* June 76 Luncheon-mats of the subtlest as well as the zaniest designs. **1978** J. Krantz *Scruples* xii. 349 Television shows that lean heavily on the brand of humor known as 'zany', consisting largely of sight gags and the sight of appealing people making cheerful fools of themselves. **1983** D. Cecil *Portrait of Lamb* ii. ii. 143 A sympathetic spirit able to appreciate his more characteristic and zany vein of humour.

Hence **za·nily** *adv.*; **za·niness.**

1958 S. Ellin *Eighth Circle* (1959) i. i. 11 In Central Park sea lions barked zanily at the sky. **1960** *Sat. Rev.* 6 Feb. 13/2 Ginsberg, for all his carefully cultivated (and natural) zaniness, is a writer far above Kerouac. **1966** M. Laurence *Jest of God* ix. 170 I've..emptied the crucial and precious capsules out of my window, zanily. **1976** *Times Lit. Suppl.* 21 May 602/2 His wider appeal owes much more to his modern zaniness and irreverence. **1980** L. Birnbach et al. *Official Preppy Handbk.* 111/1 Many of these forays into zaniness actually wind up involving mayhem or destruction of property. **1984** *Financial Times* 25 July 13/6 The play is a staple of the repertory and amateur theatre diet, and while it may not inspire to the zanily surreal heights of *See How They Run* or *Madame Louise*, Dighton is certainly a name worth conjuring with. **1985** *Listener* 21 Mar. 28/3 How could anything with Jerry Lewis not fall into a comic range somewhere between surrealistic zaniness and childish destructiveness?

zanza, zanze, varr. *sansa.

Zanzibari (zæ:nzibā·ri), *sb.* and *a.* [f. *Zanzibar*, name of an island off the east coast of Africa, now part of Tanzania + *-i.] **A.** *sb.* A native or inhabitant of Zanzibar. **B.** *adj.* Of, pertaining to, or characteristic of Zanzibar or its people.

1888 G. W. Brooke *Let.* 18 July in M. E. Herbert *Afr. Slave Trade* (1890) 12 The Zanzibaris express horror and disgust at the bare idea of associating with them. *Ibid.* 13 The women and children are hurried off to the Zanzibari camps. **1890** W. Booth *In Darkest Eng.* i. 12 Mr. Stanley's Zanzibaris lost faith. **1917** G. B. Shaw in *Daily Express* 28 Nov. 3/2 When the Sultan of Zanzibar ordered the admiral of his second-hand penny steamboat to go out and sink the British fleet, and the poor devil actually went, we laughed... No doubt Ireland, north and south, teems with Zanzibari courage. **1959** *Daily Tel.* 1 Dec. 8/6 It is uphill work to persuade the Zanzibaris to make a partial switch to other products, which they could fairly easily do in their fertile island. **1966** D. Wilson *Quarter of Mankind* xvi. 205, I have sat on a train in Manchuria and heard Zanzibari visitors ask their guides why colonialism had not yet been expelled from Hongkong. **1975** *Nat. Geographic* Apr. 501/1 Towering above all a tree bearing tiny reddish buds—the famous cloves, basis of the Zanzibari economy. **1985** *Daily Tel.* 20 Aug. 4/7 Zanzibaris with foreign exchange were allowed to import essential goods.

zaouia, var. *zawiya.

zap (zæp), *int. slang* (orig. *U.S.*). [Echoic.] **1.** Used to represent the sound of a ray gun, laser, bullet, etc.; also *fig.*, expressing any sudden or dramatic event.

1929 P. F. Nowlan in *Washington Post* 7 May 16/3 Ahead of me was one of those golden dragon Mongols, with a deadly disintegrator ray... Br-r-rr-r-z-zzz-zap. **1962** *Amer. Speech* XXXVII. 288 The jokester, pretending to be a creature from outer space, pointed his cosmic ray gun (finger) at his friend's genitals and exclaimed, 'Zap! You're sterile.' **1967** L. Deighton *Only when I Larf* (1968) xii. 160 Shouting idiotic things and going, 'Whoop,' 'Zap,' and 'Yap,' all the time. **1968** *Maclean's Mag.* Mar. 77 Bang! Zap! Pow! With laser beams and cracking doomsday machines, the deadly-serious superheroes. **1970** *Daily Tel.* (Colour Suppl.) 15 May 34/4 I'm against the war in Vietnam. But I'm not among the people who say let's stop Vietnam, zap. **1971** *Frendz* 21 May 17/1 Getting down to a blow job, she suddenly produces a razor and zap—the man bleeding and gushing blood, is screaming on the floor. **1974** *Globe & Mail* (Toronto) 12 Sept. 7/2 We have been told..we needn't worry, the sections are not being enforced. Then zap, another homosexual is denied entry for being honest. **1978** *Fortune* 18 Dec. 101 (Advt.), The solution receives a positive charge, the truck a negative charge. Zap! The primer and the GMC are joined with a magnetic-like bond. **1985** *Parade Mag.* 31 Mar. 9 A staff meeting will be Wick just shooting out those things one after another—zap, zap, zap.

2. zap gun, a ray gun or the like.

1969 K. Vonnegut *Slaughterhouse-Five* iv. 65 Billy's will was paralyzed by a zap gun aimed at him from one of the portholes. **1976** *Publishers Weekly* 15 Mar. 59/1 Plot is subordinated to character exploration, but there's more adventure in that than in many a zap-gun epic. **1977** *Sunday Sun* (Brisbane) 30 Jan. 7/1 New York police have confiscated two space-age 'zap' guns from blind singer Stevie Wonder... The guns look like flashlights and fire two darts attached to 20 ft. thin copper wires.

zap (zæp), *v. slang* (orig. *U.S.*). Also (*rare*) **zapp** [Echoic.] **I.** *trans.* **1. a.** To kill, esp. with a gun; to deal a sudden blow to.

1942 Berry & Van den Bark *Amer. Thes. Slang* §118/3 *Kill; murder*,..wipe out, work off, zap. **1965** *Time* 10 Dec. 34 *Zap*..means to clobber. **1966** *Current Slang* (Univ. S. Dakota) Fall 6 *Zap*, v., to slap... I got *zapped* when I tried that. **1969** I. Brown *Rhapsody of Words* 143 In Vietnam a man knocked out was said to be zapped. **1970** *Courier-Mail* (Brisbane) 14 Dec. 3/3 A council workman on-duty during the week zapped any rat foolish enough to wiggle a whisker within a hundred yards of the place. **1971** *Sunday Times* 28 Mar. 9/1 He wants to prove a Hanoi man can zapp a Saigon man. **1971** *Radio Times* 18 Nov. 15/4 This year the system has zapped the counter-culture member in the sneakiest way of all, by robbing him of a decent way of making a living after graduation. **1977** *Time* 6 June 55/2 Proto..fires a beam of electrons at the pellet, zapping it with a jolt equal to 8 trillion watts. **1979** *Mod. Photogr.* Oct. 64 You can't run a darkroom without plug-in power, so you'd better make it electrically safe or you might get zapped! **1981** *Observer* 2 Aug. 10 God is not going to zap women for coming forth. **1982** N. Freeling *Wolfnight* 161 Unbureaucratically, any bugger who shoots, you zap. **1984** *Weekly World News* (U.S.) 25 Dec. 29/1 Teenager Vickie Parker was zapped to death by 640 volts of electricity when she wandered onto the tracks of an elevated train and accidentally touched the 'live' third rail. **1985** *NY. Rev. Bks.* 9 May 23/3 (Advt.), In a New York restaurant, a young man celebrating with friends was zapped in the face by a man with an aerosol spray can.

b. To put an end to, do away with.

1976 *Courier-Mail* (Brisbane) 22 Apr. 21/4 We can zap a headache almost immediately. **1982** *Sunday Sun-Times* (Chicago) 17 Oct. 62 (*heading*) Atari seeks to zap X-rated video games.

2. To fail (someone) in a test, course, etc.; to punish (see also quot. 1969).

1961 *Amer. Speech* XXXVI. 149 The cadet who is *zapped* is the recipient of a large number of demerits or other cadet punishments. The term was probably taken from a favorite cadet newspaper comic strip, 'B.C.', where *zapp* is used as the sound of any blow. **1969** *Current Slang* (Univ. S. Dakota) Winter 12 *Zap*, v., to 'put down'; to put someone in his place. **1976** *National Observer* (U.S.) 17 Jan. 8/5 A graduate student whose 'scholarly potential' is not overwhelmingly lauded 'is going to get zapped'.

3. To overwhelm emotionally.

1967 *Punch* 26 July 123/1 I'll be zapped with love, blow the mind of straight people. **1970** *New Yorker* 22 July 4/3 If the music doesn't zap you..you can contemplate..movies on the wall. **1971** *Daily Colonist* (Victoria, B.C.) 30 May 47/1 Our politicians turn to the architects, presuming them to be the theatrical stage managers of the city to zap the masses with compelling masques and follies. **1974** A. Lurie *War between Tates* (1977) vi. 134 If The Book is published in time, and the right people in Washington read it, it's going to really zap them. **1977** *It* May 31/3 (Advt.), Bring you own sounds!!, and get zapped. **1983** *Theology* Jan. 15 A well-known evangelist invited the undergraduates of Oxford to allow themselves to be 'zapped by the Holy Spirit'.

4. To send, put, or hit in a forceful way.

1967 *Time* 22 Dec. 56 For quick acceleration..the nickel-cadmium batteries would cut in briefly, could zap the car from a standstill to 50 m.p.h. in 20 seconds. **1972** D. Delman *Week to Kill* 139, I nosed the car out of town and on to 118, where I zapped it into high. **1974** *Farm & Country* 9 Apr. 11 (*heading*) Computers zap farmers through space-age door. **1976** *National Observer* (U.S.) 29 May 12/4 Won't they be surprised when Ms. Klutz limberly zaps the ball over the net.

5. To demonstrate against or at.

1972 *Sat. Rev.* 12 Feb. 24/1 Homosexuals..'zap' (confront) politicians until they express themselves..on equal housing..for homosexuals. **1980** *Observer* 3 Aug. 8/5 Columbia and Warner Bros. were 'zapped' last week; this Tuesday it's the turn of 20th Century Fox. An itinerant army of 1,000 striking actors..will parade at the old studio's grimy portals.

6. *Computers.* To erase or change (an item in a program).

1982 *Times* 14 Jan. (Information Technol. Suppl.) p. v/5 When the program is erased, the PROM is said to be 'zapped'. **1983** *80 Microcomputing* Jan. 29/3 On DRS 304, RB 2C you will find the byte to be 20H. Zap this to 18H. **1983** *Your Computer* Sept. 86/1 Since I keep a hard copy listing of the assembly of MODEM7, the easiest thing to do was to zap the offending byte.

II. *intr.* **7.** To move quickly and with vigour.

1968 *Maclean's* Sept. 55 Nothing is quite as sad as watching Lynn watching Lightfoot zap off out of a parking lot. **1972** *Observer* 27 Feb. 33/5 The well-known routine of zapping from studio to studio. **1977** *Mod. Boating* (Austral.) Jan. 30/3 We're zapping over Kogarah Bay with 45 miles an hour on the clock. **1981** *Times* 22 July 12/4 When those self-satisfied pop singers and dizzy girls from Hollywood zap in and out they are not drinking themselves silly at our expense. **1985** *Times* 6 Apr. 11/1 Several smaller craft zap past.

8. To use a fast-forward facility on a video recorder to go quickly *through* the advertisements in a recorded television programme; to switch *through* other channels for the duration of the advertisements when watching programmes off-air.

1983 [implied in *zapping *vbl. sb.*]. **1984** *Broadcast* 7 Dec. 27/2 People are beginning to record the best commercial programmes on their VCRs so they can zap through the commercials. **1985** *Marxism Today* May 34/1 People with the technology use it to avoid commercial breaks, either by zapping through other channels or by fast-forwarding material recorded off-air.

Hence **zapped** *ppl. a.* and *pa. pple.*

1962 *Amer. Speech* XXXVII. 71 'Zapped.'..I first heard it in 1952 while I was an undergraduate at Brown University. The term was in vogue..to designate precisely the process by which a student..had his 'comeuppance' in class or on an examination. **1966** *Punch* 21 Dec. 911/3 Maybe truce negotiations won't be possible until the Viet Cong are zapped to the point of accepting the impossibility of military victory. **1976** *National Observer* (U.S.) 22 May 12/4 Despite his allure, could she really endure him? But the chap's got her zapped. **1980** R. Adams *Girl in Swing* xix. 254 She was pale and sweating; clearly what Mr. Steinberg would call 'zapped'.

zap (zæp), *sb. slang* (orig. *U.S.*). [f. *zap *v.*, *int.*] **1.** Liveliness, energy, power, drive; also, a strong emotional effect.

1968 *N.Y. Times* 2 Aug. 3 When the heat's too much and the gin's lost its zap.., tranquilize your jangled nerves with the Swinging Wonder. **1972** *Publishers' Weekly* 6 Mar. 62/2 As for those lyrics—probably only the over-30s will dig them. Anyone older or younger won't grasp the 'organic' zap of rock's years of innocence. **1975** *Harpers & Queen* May 128/3 The zap of his language drawn from every dialect of the underground. **1979** *Chatelaine* (Canada) Jan. 50 She loves sports, especially skiing, but found she had lost some of her old zap. **1979** *Mod. Photogr.* Oct. 68 Electricity arrives at your neighborhood at a level of 2400 [*read* 240] volts... If the transformers were perfect you would get precisely the 110 volts you wanted and could care less about the big zap lurking outside. **1984** *New Yorker* 16 Apr. 141/1 He gives the film a manic zap.

2. A demonstration (by a group against something).

1972 *Sat. Rev.* 12 Feb. 26 Despite six zaps, New York's Mayor Lindsay has consistently refused to meet with any homosexual delegation. **1974** *Times* 7 Oct. 12/3 A demonstration, alternatively described as a community action or a zap, had been planned..in Brixton's Tesco supermarket.

3. *Computers.* A change in a program.

1983 *80 Microcomputing* Jan. 29/1, I would like to provide the following zaps to TRSDOS 1·3 for the Model III. To provide a 30ms track stepping rate you must change the nine bytes listed below.

4. A short, varying sound of the kind expressed by 'zap!'

1984 *Verbatim* X. iii. 17/2 The whir of the flippers—pin-ball machines—the zaps of the video games.

Zapata (zăpā·tă). The name of Emilio *Zapata* (1879–1919), Mexican revolutionary, used *attrib.* to designate a type of moustache in which the two ends extend downwards to the chin.

Zapata was portrayed with a moustache of this kind by Marlon Brando in the film *Viva Zapata!* in 1952.

1968 *Punch* 25 Dec. 903/1 He stood now, in the centre of the enormous white vault of the school's auditorium, a willowy figure in a green velveteen frock-coat, Zapata moustache and a Donovan cap. **1972** A. Ross *London Assignment* 84 One of the younger men, wearing nothing now but his Zapata moustache.., came up for a fresh joint. **1977** *N.Y. Rev. Bks.* 4 Aug. 11/1 He is a heavy, buccaneering figure with a Zapata mustache. **1983** *Manch. Guardian Weekly* 22 May 20/1, I was delighted to see a few shots of Karajan as stage actor—hiding behind a Zapata moustache as a bit-part cardsharper in the Carmen film he made in 1967.

‖ **zapateado** (þapateā·ðo). Also **Zapateado.** [Sp., f. *zapato* shoe.] **a.** A flamenco dance which involves complex rhythmic syncopated stamping of the heels and toes in imitation of castanets.

1845 R. Ford *Handbk. Spain* I. 190 The chief dances are the *Jota* of Arragon..the *Zapateado* and *Seguidilla* of La Mancha..and the *Zortico* of Biscay. **1902** *Encycl. Brit.* XXVII. 374/2 Other provincial dances now in existence are the *Jaleo de Jerez*,..the *Zapateado*, the *Zorongo*.. and the *Tripola Trapola*. **1952** *Ballet Ann.* VI. 30 Antonio is in certain details of the Spanish dance the greatest exponent I have seen... Both his Bolero and his Zapateado must be seen to be believed. **1958** *Times* 23 Sept. 3/4 There is a good selection of flamenco dances, including solo and *à deux* Zapateados. **1973** *Daily Tel.* 19 Nov. 12/1 They danced without music and—like flamenco dancers performing a Zapateado—they used a very wide variety of sounds produced with their heels and toes so that music was superfluous. **1985** *Listener* 28 Feb. 18/3 Soon the girls are dancing with sensuous frenzy and a young man leaps on a table to execute a fine *zapateado*.

b. Dancing or footwork of this kind.

1959 *Sunday Times* 17 May 20/8 José's imprisonment, his rage expressed in zapateado, and the drama of Carmen's death, are adroitly planned. **1980** 'M. Fonteyn' *Magic of Dance* 56 Antonio, an electrifying dancer of flamenco and particularly of the language of foot and heel beats called zapateado. **1981** *Daily Tel.* 26 Aug. 9/8 Maya..has a superb grasp of subtle and varied zapateado.

zapote (zăpō̆ᵘ·te). [Sp.: see Sapota.] = Sapote.

1842 S. Maverick *Jrnl.* 16 Dec. (1953) ix. 185 Fruit—oranges, Zapotes, etc. **1909** *Chambers's Jrnl.* Aug. 557/2 The Zapote..should secure the appreciation of the English fruitarian. **1926** D. H. Lawrence *Plumed Serpent* xiv. 229 A big basket piled with mangoes, papayas, zapotes—all the tropical fruits one did not want, in hot weather. **1948** G. M. Foster *Empire's Children* 27/2 Fruit trees include the fig,..white and black *zapotes*.

Zapotec (zæ·potek), *sb.* and *a.* Also 8- **Zapoteca, -o,** 9 **Zapoteque.** [ad. Sp. *zapoteco, zapoteca*, ad. Nahuatl *tzapoteca*, pl. of *tzapotecatl*, lit. 'person of the place of the sapodilla'.]

A. *sb.* A member of an American Indian people of southern Mexico.

1797 *Encycl. Brit.* XI. 668/1 To the south-east..were the great provinces of the Mixtecas, the Zapotecas, and the Chiapanecas. **1875** *Ibid.* I. 694/2 In Chiapa were the Zapotecs, in Yucatan the Mayas. **1877** L. H. Morgan *Anc. Soc.* ii. vii. 195 The confederacy was confronted by hostile..tribes..: the tribes of Chiapas, and the Zapotecs. **1953** S. Bedford *Sudden View* iii. v. 231 The ancient Zapotecs..who may have come from Asia across the Behring Straits, and who are believed to have built.. these temples at Mitla and Monte Albàn. **1979** P. Theroux *Old Patagonian Express* v. 75 These Indians—the Zapotecs—were a matrilineal people.

b. Any of several dialects or languages of the Zapotecs.

1881 *Encycl. Brit.* XII. 828/1 The other chief stock or at least not yet classified Mexican tongues are the Miztec and Zapotec of Oajaca, [etc.]. **1936** E. C. Parsons *Mitla* i. 16 These ladies understand Zapoteco..but they do not speak it. **1965** *Language* XLI. 73 In Trique, Isthmus Zapotec, and certain other languages of Mexico, certain clause types have the preferred ordering predicate, subject, object. **1973** *Times* 24 Mar. 11/8 The student of architecture we met in the square of Oaxaca was talking Zapotec to his girl-friend.

B. *adj.* Of or pertaining to the Zapotecs.

1861 [see *expectant *ppl. a.* 1 b]. **1927** D. H. Lawrence *Mornings in Mexico* 45 Four words in the *idioma*, the Zapotec language..'You understand them?' **1934** A. Huxley *Beyond Mexique Bay* 46 A bas-relief of an ithyphallic man—the work..of the pre-Zapotec occupants of the site. **1972** *Bk. Thousand Tongues* (rev. ed.) 464/2 The Zapoteco language is spoken in a number of regional dialects. **1980** *Sci. Amer.* Feb. 46/3 Its hieroglyphs, mostly carved from 500 B.C. to A.D. 700, record the rise and decline of the Zapotec state.

Hence **Zapote·can** *a.* and *sb.*

1922 K. Al-Shimas *Mexican Southland* vii. 122 The Zapotecan tongue is spoken by at least 500,000 souls. *Ibid.* 126 Those accosted made answer in Zapotecan. **1962** E. Birney *Ice Cod Bell or Stone* 54 Rhythmic and Zapotecan-proud the classic women dance. **1978** *Language* 506/2 In the Zapotecan family of Oaxaca, 'Choapan Zapotec phonology', by Larry and Rosemary Lyman.. employs 'a hierarchically oriented framework' of five levels, from phoneme to 'phonological sentence'.

zapper (zæ·pəɪ). [f. *zap *v.* + -er¹.] **a.** A person, technique, etc., that kills or does away with something. **b.** *spec.* Any of various devices for destroying or warding off pests; also (properly with capital initial), a proprietary name in the U.S. for an agricultural machine of this kind.

1969 *Guardian* 27 May 8/5 The United States colonel who formerly commanded them..is now busy evolving a 'zapper' technique for scything down nippa palm groves where Vietcong are reported to be hiding with direct fire artillery. **1970** *Courier-Mail* (Brisbane) 14 Dec. 3/3 The rat zapper does not work on Saturdays or Sundays, but there were still no rats to be seen there yesterday. **1973** *Sci. Amer.* Sept. 74/2 One particularly promising application, making use of small portable zappers, is for greenhouses and other horticultural enterprises where the cost of soil preparation per acre is not an important consideration. For row crops..the self-propelled Zapper is presently economic only when chemical weed-control costs are higher than $15 per acre. **1975** *Official Gaz.* (U.S. Patent Office) 25 Mar. TM 271/2 *Zapper*. For agricultural machine which employs microwaves to control vegetation and vegetation damaging pests. **1976** *Nature* 12 Feb. 441/1 The general message is that although various alternatives are promising, there is no magic 'insect zapper', as Kennedy put it, to replace chemical poisons in the near future. **1977** *Globe & Mail* (Toronto) 30 July 4/1 How many bugs do zappers zap? **1983** *Monitor* (McAllen, Texas) 23 June 10C (Advt.), 1 acre bug zapper. $59·97. 35 watt ultraviolet fluorescent bulb has an effective range up to 1 acre for outdoor comfort.

zapping (zæ·piŋ), *vbl. sb. slang.* [f. *zap *v.* + -ing¹.] The action of *zap *v.*; *spec.* the practice of skipping advertisements when watching television programmes.

1972 *Daily Colonist* (Victoria, B.C.) 8 Jan. 17/7, I wouldn't take the zapping he has taken, to use the contemporary word, for a million dollars. **1983** *Times* 27 Oct. 8/3 The practice is known in the US as 'zapping'. Apparently people who have remote control devices are substantially more likely..to switch over to another channel when ads come on. **1983** *Austral. Personal Computer* Nov. 32/2 You could add suitable sound-effects to an applications program—zapping noises as a word processor deletes characters, maybe? **1984** *Listener* 9 Feb. 14/2 For the ITV companies there is the additional problem of 'zapping' to contend with—the habitual use of the fast-forward button to bypass the commercial breaks in recorded material.

zappy (zæ·pi), *a. slang.* [f. *zap *sb.* + -y¹.] Lively, amusing, energetic; striking.

1969 *Guardian* 1 Mar. 5/2 The Minister wore in his lapel..a zappy coloured badge of the 'Cocoa makes you sterile' type. **1971** 'J. Mayo' *Asking for It* xxiv. 155 They were both giving each other big zappy grins. **1972** *Observer* 16 July 30/2 We badly need some zappy young editor who will start making his more garrulous authors prune their books. **1983** *Times Lit. Suppl.* 30 Dec. 1448/3 The best of contemporary American prose sometimes has a zappy elegance. **1984** *Listener* 3 May 30/4 The company felt the need for a zappier profile.

zarf, zurf. For etym. read [ad. Arab. *ẓarf* vessel.] and add later examples.

1854 R. Curzon *Armenia* 80 One brought coffee on a tray..and then came a man bringing to each of us a cup, well frothed up, and in a zarf, or outer cup. **1978** *Times* 2 May 13/3 (*caption*) A diamond and emerald zarf, mounted in gold and silver, 19th century, 6·5 cm high.

zariba, *sb.* Add: **c.** (Further examples.) In these uses usu. as **zareba.**

1916 'Taffrail' *Pincher Martin* iii. 42 The orchestra.. took their places behind a zareba of bunting-covered biscuit-boxes and hired palms erected in front of the stage. **1940** G. Greene *Power & Glory* ii. ii. 128 The shadows of the palms pointed at him like a zareba of sabres. **1961** Wodehouse *Ice in Bedroom* vii. 52 Owing to his *obiter dicta* having to be filtered through a zareba of white hair, it was not always easy to catch exactly what Mr. Cornelius said. **1965** G. Maxwell *House of Elrig* x. 139 The final approach to any birds' nest was an affair of struggle and achievement. My hand reaches through the twigged zareba of basket and basket handle, touches the soft lining, touches the firm intimacy of the eggs. **1974** *Author* Spring 33 A novelist..has direct access to his readers, with no thorny zariba of a cast of actors to battle through. **1984** *Listener* 3 May 24/2 The two previous ones faced the reader in their opening chapters with an almost impenetrable zareba of thickset descriptions of the committee structure.

‖ **zarzuela** (þaɹþuē·la). [Sp.] A traditional form of popular musical comedy in Spain.

In its theatrical sense the word is said to be derived from the name *Real Sitio de la Zarzuela* in Madrid, where these works were first performed in the early 17th century.

1888 *Lippincott's Mag.* July 51 It had all transpired in a flash, like some of the absurd things he had seen in pantomimes by the *zarzuela* companies at the theatres. **1922** J. Hergesheimer *Bright Shawl* (1923) 54 The zarzuelas, the operettas of one act, largely improvised with local allusions. **1947** A. Einstein *Music in Romantic Era* xvii. 328 The way was being prepared for the Italianization of Spanish music as early as Calderón's period, which brought forth the old *zarzuela*, a mixture of song and dance with spoken dramatic dialogue. **1973** *Oxford Times* 30 Mar. 14/5 The modern zarzuelas are really operettas or lyric dramas, which are a kind of Spanish equivalent of Gilbert and Sullivan and the Viennese operetta, full of tunes which have become part of Spain's popular culture. **1973** R. A. Crampsey *Puerto Rico* viii. 130 Less common are the old *zarzuelas*, the traditional musical dramas. **1983** *Listener* 21 Apr. 31/2 A Catalan born in 1871,..he [*sc.* Amadeo Vives] was evidently best known as a *zarzuela* composer.

zastruga, var. *sastruga.

1982 B. Aldiss *Helliconia Spring* ii. 89 Only Yuli had experience of the tundras and zastrugi, which stretched away to the north of the Quzint.

zat (zæt), repr. a colloq. pronunc. of the cricket appeal 'How's that?' (see How *adv.* 2 b in Dict. and Suppl.).

1934 *Humorist* 27 Jan. 38/3 Not a bad record..for a game that consists chiefly of standing about..and shouting 'Zat?' at intervals. **1966** M. Woodhouse *Tree Frog* xxiii. 167 Outside the window a sharp yelp of 'Zat?' sounded.

zatch (zætʃ). *vulg.* [Perh. corruption of Satchel in similar slang sense.] The buttocks; the female genitals; an act of copulation.

1950 E. B. White *Let.* 6 June (1976) 321 You are just sticking out your zatch, and many a tosspan and strutfart will run you through. **1971** R. Dentry *Encounter at Kharmel* v. 88 Scotsmen playing the bagpipes give me a pain in the prick... Pathan tribesmen playing them is enough to make the harlot of Jerusalem snatch her zatch! **1980** J. Krantz *Princess Daisy* xii. 191 You're going to take her home and give her a zatch.

‖ **zawiya** (zā·wiă). Also **zawia, zawiyah, zawya; zaouia.** [Arab. *zāwiya* (hence F. *zaouia*) corner, prayer room.] In North Africa, a Muslim religious community or its mosque, usually containing the shrine of a holy person.

1836 E. W. Lane *Acct. Manners & Customs Mod. Egyptians* II. xi. 190 These lights were not hung merely in honour of the Prophet: they were near a *zawiyeh* (or small mosque) in which is buried the sheykh Durweesh El-Ashmawee; and this night was his Moolid. **1874** R. L. Playfair *Handbk. Trav. Algeria* ii. 70/2 The *Djamâa*, or rather *Zaouia* of *Abd-er-Rahman eth-Thalebi*..contains the tomb of that saint, who died in 1471. **1911** *Encycl. Brit.* XXVII. 289/1 Kufra..is a centre of the Senūssite brotherhood, whose *zawya* (convent) at Jof, in Kebabo, ranks in importance with that of Jarabub. **1911** D. S. Margoliouth *Mohammedanism* v. 184 A *zawiyah*, or 'hermitage' was offered him by the Pasha. **1924** W. M. Sloane *Greater France in Afr.* viii. 166 The sacred colleges of the Moslem fraternities, zaouias, are the scenes of wild, maniacal, religious orgies. **1958** N. Epton *Saints & Sorcerers* ii. 28 The sheikh (this word signifies either the head of a *zaouia* or of a tribe..) lives close to the *zaouia* with his family and the devotees. **1968** T. Blunsum *Libya* x. 105 The *zawias*, or houses of learning and worship, were documentation centres where copies of the Koran were prepared for distribution. **1977** *Times* 30 Apr. 11/3 Nefta..a village..whose roof lines are broken by the cupolas of mosques and zaouias. **1978** J. Updike *Coup* (1979) iii. 122 Who was it used to tell reporters I was a *zawiya* Berber too pious to be seen?

‖ **zazen** (zāze·n). Also 8 **Sasen**; **Zazen.** [Jap., f. *za* sitting, a seat + *zen* *Zen.] Zen meditation.

1727 J. G. Scheuchzer tr. *Kæmpfer's Hist. Japan* I. iii. i. 242 A posture, which is thought to engage one's mind in so profound a meditation, and to wrap it up so entirely within itself, that the body lies for a while as it were sens less... This profound Enthusiasm is by them call'd *Safen* [sic: read *Sasen*]. **1897** *Princ., Pract. & Enlightenment Soto Sect* 22 How can you think of not-thinking? That is thinking of nothing. This is the most important art of Zazen. **1907** *Jrnl. Pali Text Soc.* 1906–7 36 Zazen is not meant to induce a trance or a state of self-hypnotization. **1960** Koestler *Lotus & Robot* ii. xi. 257 Zazen meditation, unlike Yoga, holds no promise of supernatural rewards. **1973** A. Broinowski *Take One Ambassador* viii. 108 Three hours of daily *zazen* meditation. **1983** M. Furlong *Cousins* xxi. 167 He had taken to practising what he hoped was *zazen* in the yard.

Z band: see *Z line (as a main entry).

Zealander. Add: **b.** A native or inhabitant of New Zealand; orig. and esp. a Maori.

1773 W. Bayly *Jrnl.* 12 Apr. in R. McNab *Hist. Rec. N.Z.* (1914) II. 207 The Zeelanders never eat greens of any kind. **1821** M. Edgeworth *Let.* 7 Dec. (1971) 288 Captain Thompson..went some years ago to New Zealand and when he was taking leave of the Zealanders one of their chiefs consented to accompany him to England. **1949** E. Pound *Pisan Cantos* lxxx. 92 And persuaded an Aussie or Zealander or S. African To kneel with him in prayer.

zearalenone (zī,ăræ·lĕnō̆ᵘn). *Biochem.* [f. Zea + *-ralenone*, f. resorcylic acid *lactone* + -ene, repr. a double bond + -one.] A white crystalline bicyclic latone, $C_{18}H_{22}O_5$, that is a metabolic product of certain cereal fungi and causes disorders of the reproductive system in pigs.

1966 W. H. Urry et al. in *Tetrahedron Lett.* xxvii. 3109 Study of the substance of striking physiological activity isolated from the mycelia of the fungus Gibberella zeae.. has shown that it is one of the enantiomorphs of 6-(10-hydroxy-6-oxo-*trans*-1-undecenyl)-β-resorcylic acid lactone I... Zearalenone I..gives reactions that indicate it has one olefinic and one ketonic group, and two phenolic hydroxyls and an ester group in a β-resorcylate structure. **1977** *Lancet* 26 Mar. 671/2 Analytical tests failed to show the presence of aflatoxin,..although zearalenone was found in maize samples. **1978** [see *trichothecene]. **1981** M. L. Clarke et al. *Veterinary Toxicol.* (ed. 2) viii. 277/1 The phenolic macrolide zearalenone (F-2 toxin) produced by *Fusarium graminearum*..and other *Fusarium* species growing on maize, barley and wheat destined for incorporation into animal feeds, is probably one of the most widespread and economically important mycotoxins.

zearat, var. *ziarat.

zeatin (zī·ătin). *Biochem.* [f. Zea + *-t-* + -in¹.] A purine derivative occurring as a cytokinin in maize kernels and other plants; 6-(4-hydroxy-3-methylbut-2-enyl)-aminopurine, $C_{10}H_{13}N_5O$.

1963 D. S. Letham in *Life Sciences* No. 8. 572 For this factor the name zeatin is proposed since it was first isolated in crystalline form from *Zea mays*. **1973** D. W. Krogmann *Biochem. Green Plants* xi. 196 Coconut milk.. contains a high concentration of zeatin. **1980** *Physiologia Plantarum* XLIX. 304/1 All the cytokinins, at $10^{-4}M$ and above, inhibited both the initiation and the emergence of lateral roots, zeatin being the most powerful inhibitor.

zeaxanthin (zī,ăzæ·nþin). *Biochem.* [a. G. *zeaxanthin* (P. Karrer et al. 1929, in *Helvetica Chim. Acta* XII. 791): see Zea and Xanthin.] A xanthophyll, $C_{40}H_{56}O_4$, originally isolated from Indian corn, *Zea mays*.

1929 *Chem. Abstr.* XXIII. 4480 (*heading*) Zeaxanthin, a new carotinoid pigment in maize. **1934** [see *xanthophyll a]. **1978** [see *violaxanthin].

zebra, *sb.* Add: **2. c.** Also, a striped prison uniform.

1895 *Harper's Weekly* 10 Aug. 753/3 At present I understand that he is in limbo, wearing the famous 'zebra'—the penitentiary dress. **1935** A. J. Pollock *Underworld Speaks* 136/1 *Zebra*, striped prison clothing.

d. A zebra crossing. *colloq.*

1951 O. Lancaster in *Daily Express* 15 Dec. 1/5 If we can only find a Zebra, we can sit down and relax. **1959** *Woman* 24 Jan. 4/3, I often wait ten minutes or more, even at a zebra, while super new cars speed past. **1968** *Listener* 15 Aug. 201/3 Outside, on the road, trucks wait for me to cross a zebra. **1976** *S. Wales Echo* 27 Nov. 9/2 [He] pleaded guilty to stopping a car in a zebra-controlled area.

3. *zebra marking, skin*; *zebra-marked, -striped* (examples) adjs.; **zebra crossing,** a pedestrian crossing marked by broad black and white stripes on the road and Belisha beacons on the kerb; **zebra danio** (dē¹·nio), a small Indian freshwater fish with horizontal dark and light stripes, *Brachydanio rerio* (family Cyprinidæ), which is popular as an aquarium fish; **zebra fish**: also, any of several striped tropical fishes, esp. the zebra danio; (further examples); **zebra spider,** any of several striped spiders of the family Salticidæ; **zebra-wood** (further examples).

[**1949** *Surveyor* 8 July 407/1 Investigations by the Road Research Laboratory..have led to the full-scale trials of the striped (zebra) markings for pedestrian crossings. **1950** *Times* 18 Feb. 2/5 Experiments are now being carried out..to test the efficacy of the zebra striped pedestrian crossing when illuminated at night.] **1950** *Surveyor* 23 June 365/2 The initial values of the percentage of drivers giving way are higher on the 'zebra' crossings than on the plain. **1977** B. Pym *Quartet in Autumn* ii. 23 He..called out angrily after a car which had failed to stop at a zebra crossing. **1917** W. T. Innes *Goldfish Varieties & Tropical Aquarium Fishes* (ed. 2) vii. 86 *Danio rerio*, Zebra Danio. **1962** D. W. Tucker tr. *Sterba's Freshw. Fishes of World* 265 Some species, such as the Zebra Danio, will spawn over gravelly bottoms even when there are no plants present. **1980** *Sci. Amer.* Feb. 127/2 McCutchen has employed his rig to study the motion of a small fish, a zebra danio (*Brachydanio rerio*), which is about three centimeters long. **1895** *Aquarium* Apr. 172/1 (*heading*) Brazilian Zebra Fish (*Heros facetus*). **1925** *Aquatic Life* June 18/1 Danio rerio, those swift, graceful, blue and white and sometimes golden striped zebra fish, have many admirers. **1934** C. W. Coates *Tropical Fishes as Pets* v. 49 The Zebra Fish (*Pterois volitans*), a marine beast from the Indian Ocean, erects all his fins, faces his future meal, and appears to drift toward it. **1962** *Listener* 22 Nov. 852/2 The three black-and-white-striped zebra fish were speed-merchants [in the tank]. **1979** H. F. Axelrod et al. *Exotic Marine Fishes* 121 The scorpion fishes..often called butterfly cod or zebra fishes..are mostly hardy, attractive predators. **1924** J. A. Thomson *Science Old & New* v. 30 Butterflies like the unpalatable zebra-marked Heliconius, which insectivorous birds leave unmolested. **1949** Zebra marking [see *zebra crossing* above]. **1973** G. Beare *Snake on Grave* xiii. 68 Walls, floor, and ceiling were all done in a zebra-skin motif. Everything..was covered in a zebra skin. **1978** S. Naipaul *North of South* ii. vi. 230 Copies of *Playboy* were scattered on a zebra-skin rug. **1866** *Hours at Home* III. 331/1 Next to the garden-spider, the hunting or zebra spider is the most common **1966** C. Sweeney *Scurrying Bush* vi. 87 The commonest of these [jumping spiders] was banded black and white like a zebra... Zebra spiders are able to leap..in an eighteen-inch arc. **1852** Zebra-striped [see *dove orchis* s.v. *dove *sb.* 5 b]. **1950** [see *zebra crossing* above]. **1976** *National Observer* (U.S.) 21 Aug. 1/3 There was Clarabell in his green-and-white zebra-striped clown suit. **1934**, etc. Zebrawood [see *zingana 2].

zebrano (zē-, zĕbrā·no). [Irreg. f. Zebra *sb.*] Striped wood furnished by various species of African trees, esp. ones of the genus *Microberlinia*.

1928 *Sunday Express* 29 July 15/4 We take the most handsome pieces of..sapele mahogany, or of lesser known woods like macassar, ebony, zebrano, amboyna. **1934** *Archit. Rev.* LXXV. 144 (*caption*) The furniture is in Indian rosewood and Japanese chestnut and the panelling in zebrano. **1965** *Wireless World* Sept. 37 (Advt.), Finish zebrano, mahogany, walnut or teak veneers. **1980** E. Scott *Illustr. Encycl. Working in Wood* 248 *Microberlinia brazzavillensis* Zebrano. Normally coarse texture; interlocked grain.

zebrina (zĕbrai·nă). *Bot.* Also **Zebrina.** [mod.L. (A. Schnitzlein 1849, in *Bot. Zeitung* VII. 870/2), f. Zebr(a, in allusion to the striped leaves of some species + -ina².] A creeping recumbent herb of the genus of this name (family Commelinaceæ) that is native to central America, bears ovate, often striped, leaves and clusters of small flowers, and is grown as a greenhouse and indoor plant.

1946 M. Free *All about House Plants* iii. 15 English Ivy, Tradescantia, and Zebrina (to trail over edge [of a window-box]). **1960** *Times* 30 Jan. 11/2 The green or white and green striped zebrinas. **1963** [see *periclinal a. (sb.) 2 b]. **1978** *Homes & Gardens* Oct. 42/2 Among these summer sun-lovers are beloperone..and zebrina.

zebroid (zī-, ze·broid), *a.* and *sb.* [f. Zebr(a + -oid.] **A.** *adj.* (In Dict. s.v. Zebra *sb.* Further example.)

1974 *Nature* 22 Mar. 296/2 G. Nobis's study of the horse teeth recognised three groups: two larger species with primitive 'zebroid' features.

B. *sb.* The offspring of a horse and a zebra.
1899 *Tablet* 25 Nov. 848/2 The zebroid, or hybrid between the horse and the zebra, 'will be the mule of the 20th century'. 1926 *Daily Colonist* (Victoria, B.C.) 18 July 24/1 The first man to have any practical success in crossbreeding the zebra and the horse is Dr. W. E. Hastings of Mt. Vernon, Ind. The result is a new mule, the zebroid. 1973 *Daily Tel.* (Colour Suppl.) 5 Oct. 31/4 Another Kenya 'character'..crossed ponies with zebras, and used the resulting 'zebroids' as pack animals for safaris. 1983 *Listener* 20 Jan. 25/1 Christina Dodwell, in *Travels with Fortune*, paddled a dug-out canoe, and rode a camel and a zebroid, in her journey through Africa.

zebrule (zī·brūl, ze·brūl). [f. ZEBR(A + M)ULE[1].] = next.
1903 in P. Fleming *Bayonets to Lhasa* (1961) viii. 102 In my Section are two Zebrules, half zebra and half donkey. 1978 J. MORRIS *Farewell the Trumpets* vii. 132 By December 13, 1903 every last yak and zebrule had crossed the frontier of Tibet.

zedonk (zī·dǫŋk, ze·dǫŋk). Also **zeedonk.** [f. ZE(BRA + DONK(EY.] The offspring of a male zebra and a female donkey. Cf. *ZONKEY.
1971 *Daily Tel.* 17 Apr. 8/5 The staff at Colchester Zoo have described their newcomer as a 'Zedonk'. 1976 *Observer* (Colour Suppl.) 5 Sept. 42/3 Donkeys crossed with zebras, sometimes called zeedonks, have been used like mules..as pack animals. 1978 *Panorama* (Austral.) Aug. 13/2 Zareeba the zedonk (half zebra, half donkey) is one of only five of his kind known in the world.

Zeeman (zēi·mān). *Physics.* The name of P. *Zeeman* (1865–1943), Dutch physicist, used *attrib.* with reference to the splitting of a spectral line into three or more closely spaced components when the light source is in a magnetic field not strong enough to produce the Paschen–Back effect.
Observed by Zeeman in 1897.
[1897 *Phil. Mag.* XLIV. 503 A theoretical analysis.. can be developed in connexion with Zeeman's phenomenon, which may help to throw light on the nature of the electric vibrations in the molecule.] 1899 *Rep. Brit. Assoc. Adv. Sci. 1898* 789 An effect converse to the Zeeman effect occurs in those cases where the body is absorbing instead of emitting light. 1904 [see *MAGNETO-GYRIC *a.* 1]. 1926 [see *LARMOR]. 1962 W. B. THOMPSON *Introd. Plasma Physics* i. 3 There are several ways of inferring the existence of magnetic fields, of which the most direct is from a measurement of the Zeeman splitting of spectral lines. 1971 *New Scientist* 18 Feb. 381/1 When a nucleus is placed in a high magnetic field, each energy level splits into several sublevels known as magnetic sublevels or Zeeman levels. 1978 PASACHOFF & KUTNER *University Astron.* viii. 219 (*caption*) From studies of the splitting of lines that are sensitive to the Zeeman effect, the magnetic field on the sun can be mapped.

zeep (zīp), *v. rare*[−1]. [? var. ZIP *v.*] *trans.* To elicit a zipping sound from.
1935 S. BECKETT *Echo's Bones*, Tires bleeding voiding zeep the highway.

Zeiss (zəis). The name of Carl *Zeiss* (1816–88), German optical instrument maker, used *attrib.* to designate binoculars manufactured by the firm he founded.
1905 W. JAMES *Let.* 3 Apr. (1920) II. 224 Now I will stop, and use my Zeiss glass on the land, which is getting nearer. 1912 A. CONAN DOYLE *Lost World* viii. 126 He had his Zeiss glasses in his hand. 1938 S. BECKETT *Murphy* iv. 60, I have worshipped her from afar... All last June, through Zeiss glasses. 1971 R. DENTRY *Encounter at Kharmel* vii. 127 There were..Zeiss binoculars..at the back of the glove compartment. 1983 T. POCOCK *1945* vii. 217 The luckiest found Leica cameras and Zeiss binoculars.

‖**zeitgeber** (tsəi·tgēbəɹ). *Physiol.* Pl. same or (anglicized) **zeitgebers.** [Ger. (J. Aschoff 1954, in *Naturwissenschaften* XLI. 49), f. *zeit* time + *geber* giver.] A rhythmically occurring event, esp. in the environment, which acts as a cue in the regulation of certain biological rhythms in an organism.
1964 E. BÜNNING *Physiological Clock* ii. 9 The change of light and dark or alternations of high and low temperatures have a synchronizing effect on the endodiurnal rhythm; they function as 'Zeitgeber' (cues, synchronizers). 1969 *New Scientist* 21 Aug. 369/2 Light is known to be the *zeitgeber* for the ant-lion's solar-day rhythm. 1975 D. VINCE-PRUE *Photoperiodism in Plants* v. 169 The signals responsible for entrainment have been called *zeitgebers*. 1975 *Nature* 27 Nov. 291/2 The various rhythms respond to changes in the phase or the period of the entraining cycle (the Zeitgeber). 1983 *Brit. Med. Jrnl.* 6 Aug. 426/2 Melatonin is secreted by the pineal gland during the night... In rats it will act as a 'zeitgeber' or synchroniser of the rest-activity cycle under conditions of zero environmental input.

Zeitgeist. Add: Also **Zeit Geist, Zeit-Geist** (both *rare*), and with small initial. (Earlier and later examples.)
For 1893 M. ARNOLD in Dict. read 1873 M. ARNOLD. 1848 M. ARNOLD *Let.* Nov. (1932) 95, I..took up Obermann, and refuged myself with him..against your Zeit Geist. 1876 *Mind* I. 369 There is a *Zeitgeist*, he says. 1884 *Macmillan's Mag.* Aug. 254 For realism in one form or another is the zeitgeist which will master us all. 1889 G. B. SHAW *Let.* Aug. (1965) I. 222 My business is to incarnate the Zeitgeist. 1933 A. HUXLEY *Let.* 9 Oct. (1969) 374 The Zeitgeist is a most dismal animal and I wish to heaven one cd escape from its clutches. 1946 AUDEN *Litany & Anthem for S. Matthew's Day*, May we worship neither the flux of chance, nor the wheel of fortune, nor the spiral of the zeit-geist. 1972 *Science* 2 June 991/3 A clear mark of the Zeitgeist of the late 1960's and the 1970's is the increased demand for participation in decision-making by those affected by it. 1982 D. PIPER *Image of Poet* i. 13 Shakespeare becomes in a sense an ever-changing embodiment of the *Zeitgeist*.

zek (zek). [Russ., prob. repr. pronunc. of *z/k*, abbrev. of *zaklyuchënnyĭ* prisoner.] In the U.S.S.R., a person confined in a prison or forced labour camp.
1968 T. P. WHITNEY tr. *Solzhenitsyn's First Circle* p. x, All the zeks at the Mavrino sharashka belonged, though they were not at the time in hard-labor camps, to the realm of GULAG. 1977 *Guardian Weekly* 26 June 22/2 And didn't his author work gratefully too in the same *sharashka*, or Island of Paradise, as the zeks called these 'soft' research camps? 1982 T. BINYON *Swan Song* viii. 58 They got him for parasitism and a few other offences. Now he was presumably..east of the Urals and taking the zeks for their bread ration.

Zeldovich (ze·ldǒvitʃ). *Chem.* The name of Y. B. *Zeldovich* (b. 1914), U.S.S.R. physicist, used *attrib.* to designate a mechanism proposed by him for the oxidation of nitrogen to nitric oxide in flames via a two-stage free-radical reaction.
1973 S. J. WILLIAMSON *Fund. Air Pollution* x. 298 The formation of NO occurs primarily through two simultaneous reactions known as the Zeldovich mechanism: $N + O_2 = NO + O$; $O + N_2 = NO + N$. 1982 *Sci. Amer.* Feb. 94/1 At temperatures low enough to suppress the Zeldovich reactions other reactions still generate nitric oxide.

zelkova (zelko·vă). Also **Zelkova.** [mod.L. (E. Spach 1841, in *Ann. des Sci. Nat.: Bot.* 2nd Ser. XV. 352), f. *zelkoua, tselkwa*, cited by Spach as local names for *Z. carpinifolia* in the Caucasus: cf. mod.Russ. *dzel'kova grabolistnaya*.] A deciduous tree of the genus of this name (family Ulmaceæ), which is native to China, Japan, and the Caucasus, and bears toothed leaves and small green flowers. Also *zelkova tree*.
1893 A. D. WEBSTER *Hardy Oranamental Flowering Trees & Shrubs* 134 Zelkova Tree..is a handsome, large growing tree, with oblong deeply-crenated leaves, and small inconspicuous flowers. 1957 M. HADFIELD *Brit. Trees* 226 The zelkova..deserves more extensive planting. 1976 *Daily Tel.* 5 July 8/6 Three thousand Zelkova trees imported from Germany are being planted in Peterborough to replace elms killed by Dutch Elm disease. 1978 *Vole* Dec. 27/1 Zelkovas are a rather special group of trees. They belong to the Elm family and have delicate, toothed leaves.

zemindary. Add: Also **zamindary.** (Examples of this form and **zamindari.**)
1914 W. G. LAWRENCE *Let.* 30 July in *Home Lett. T. E. Lawrence* (1954) 563 We all ate together of food one could stick one's teeth in, thick Zamindari bread. 1917 *Chambers's Jrnl.* Dec. 781/1 There is no such man within my zamindary as can catch the living snake. 1932 *Ann. Reg. 1931* I. 155 Movements against the payment of rents and land revenue, particularly in the United Provinces, where the *zamindari* (landlord) system is prevalent. 1968 *Times* 6 Apr. (Pakistan Suppl.) p. v/4 After independence, many Hindu zamindars fled, the zamindary system was abolished and the land divided into tiny portions and sold to the Muslims. 1976 M. S. HOQUE *Hunger* II. vi. 41 Parents now think that their son is a Zamindari estate on auction sale.

‖**zemirah** (zəmīə·ra). *Judaism.* Pl. **zemirot(h** (zəmīə·rōᵘt, s-). [Heb.] A religious song sung in Hebrew at Sabbath meals.
1831 [see *MITZVAH]. 1892 I. ZANGWILL *Childr. Ghetto* II. xviii. 82 When supper was over grace was chanted, and then the *Zemiroth* were sung—songs summing up, in light and jingling metre, the very essence of holy joyousness. 1973 *Jewish Chron.* 9 Feb. 21/5 The service was followed by supper and zemirot in the communal hall.

zemstvo. Add: Pl. **zemstvos,** ‖**zemstva.**
1958 *Times Lit. Suppl.* 10 Jan. 14/3 The history of the pressure for political freedom was closely interwoven with the development of the organs of local self-government, the *Zemstva*. 1967 *Listener* 2 Nov. 559/1 At any time in those final decades of the last century even a central assembly of representatives of the *zemstvos* would have constituted an important safety-valve. 1980 *Times. Lit. Suppl.* 14 Nov. 1306/3 Linked closely to the work of the *zemstva*, new local units of administration, medicine became the first true profession in Russia. 1983 P. USTINOV *My Russia* xi. 122 Members of local *zemstvos* might soon enjoy a voice in the internal government of the country.

Zen (zen). Also 8 **Sen.** [a. Jap. *zen*, ad. Chin. *chán* quietude, ad. Skr. *dhyāna* meditation.] A school of Mahayana Buddhism that emphasizes meditation and personal awareness and became influential in Japanese life from the 13th century after being introduced from China.
1727 J. G. SCHEUCHZER tr. *Kæmpfer's Hist. Jap.* I. II. v. 199 In the 1850 streets of this city, there were 1050 [families] of the *Ten Dai's* Religion,..11 016 of *Sen.* 1834 [see *SHINGON]. 1911 *Encycl. Relig. & Ethics* IV. 704/1 Meditation came to have more weight than the other two factors, until in China and Japan there arose a sect, the Zen..in which it is the most essential part of the entire teaching. 1921 *Eastern Buddhist* (Japan) I. 13 Zen in its essence is the art of seeing into the nature of one's own being. 1960 *Spectator* 15 July 101 *What* do they know about Zen—the programme-writer, the film-makers, the beatniks, the lot? 1967 D. & E. T. RIESMAN *Conversations in Japan* 123 Another boy, who..was their top judo athlete, said that he got his values from Zen. 1976 A. DAVIS *Television* 122 By the late sixties, religious programmes were beginning to reflect the interest of young people in Eastern religions such as Buddhism, Hinduism, Zen and the cult of the Maharishi. 1977 J. F. FIXX *Compl. Bk. Running* ii. 14 Our society puts considerable emphasis on personal development and the maximizing of one's potential. Zen, transcendental meditation, assertiveness training..and similar movements are all directed at making us fulfilled human beings.

2. *attrib.* and *Comb.*, esp. in *Zen Buddhism, Buddhist.*
1881 *Trans. Asiatic Soc. Japan* IX. 179 (*heading*) Zen sect. 1894, etc. [see *SOTO]. 1902 *Encycl. Brit.* XXIX. 681/1 The Zen doctrines of Buddhism, which contributed so much to the development of the heroic and the sentimental,..were therefore favourable to the stability of military feudalism. 1921 *Eastern Buddhist* (Japan) I. 13 (*heading*) Zen Buddhism as purifier and liberator of life. *Ibid.* 26 Whether an enlightened Zen master or an ignoramus of the first degree, neither can escape the so-called laws of nature. 1923 *Ibid.* II. 341 It remained for the Chinese Zen Buddhists to invent their own methods according to their own needs and insight. 1947 *Archit. Rev.* CII. 32/2 Taoism and nature mysticism, transmitted through the vehicle of Zen Buddhism. 1950 A. HUXLEY *Themes & Variations* 98 As the Zen Masters like paradoxically to put it, 'Buddha never taught the saving truth.' 1960 KOESTLER *Lotus & Robot* II. x. 234 The monk's rudeness..was in the right tradition of Zen-teasing. 1965 W. SWAAN *Jap. Lantern* v. 58 The canons of Japanese aesthetic appreciations were formulated to a very large degree under the influence of Zen Buddhist philosophy. 1971 'G. BLACK' *Time for Pirates* iii. 45 Mr. Akamoro..contrived..to leak the thought.. that it is time for Japanese initiative to take over with Zen Buddhism for the moral trimmings. 1979 B. MALAMUD *Dubin's Lives* viii. 280 She said she had been talking to a Zen Master in South San Francisco... 'I'm thinking of entering a Zen commune... I expect to become a Zen disciple.'

zendo (ze·ndo). [a. Jap. *zendō*, f. *zen* *ZEN + *dō* hall.] A place for Zen Buddhist meditation and study.
1959 *Encounter* XII. I. 20 The *Zendo* or meditation-hall. 1968 *Time* 18 Oct. 62/3 Students must report to the *zendo* (meditation hall) by 5. 1974 *Country Life* 14 Feb. 326/1, I was taken to the Zendo, the special hall for contemplation, by a young monk. 1981 'E. V. CUNNINGHAM' *Case of Sliding Pool* vii. 77, I thought I would drive down to the Zendo... I feel a need to talk to the Roshi.

Zener (zī·nəɹ). Also **zener. 1.** The name of K. E. *Zener* (1903–61), U.S. psychologist, used *attrib.* to designate a pack of 25 cards that he designed for use in parapsychology experiments, containing five each of five different cards, each showing a simple symbol.
1934 J. B. RHINE *Extra-Sensory Perception* iv. 50 We have once since substituted a 'heart' for the 'waves' figure but later returned to the latter. I shall hereinafter call these cards the 'Zener cards'. 1940 *Proc. Soc. Psychical Res.* XLVI. 153 The guesser at Zener cards all unwittingly was guessing correctly..a card which was one or two places earlier or later in the sequence. 1949 *Mind* LVIII. 390 Extra-sensory perception (ESP) is investigated by means of the card-guessing technique. Zener cards are generally used. 1969 *Listener* 6 Mar. Mar. 301/2 Rhine devised a new kind of test for ESP. He made packs of 25 cards called Zenercards, with geometrical designs: cross, square, circle, waves and star. 1978 D. BLOODWORTH *Crosstalk* xiv. 119 There are a dozen ways of beating the odds with the Zener packs they use for card-guessing tests.

2. *Electronics.* The name of C. M. *Zener* (b. 1905), U.S. physicist. **a.** Used *attrib.* to denote various concepts, etc., connected with or arising from his researches, as **Zener breakdown** = *Zener effect* below; **Zener diode,** a junction diode in which the forward characteristic is like that of an ordinary diode but there is a sudden large increase in reverse current at a certain constant reverse voltage owing to the Zener effect or the avalanche effect, making it useful as a voltage regulator and in switching circuits; **Zener effect,** the increase in reverse current of a Zener diode when attributed to the tunnelling of current-carriers through the transition region rather than to the avalanche effect; **Zener voltage,** the voltage at which Zener breakdown occurs; the reverse breakdown voltage of a Zener diode.

1956 L. P. HUNTER *Handbk. Semiconductor Electronics* III. 17 The two mechanisms which have been observed for semiconductor contacts in the absence of thermal breakdown are Zener breakdown..and avalanche breakdown. 1962 [see *Zener effect* below]. 1981 NASHELSKY & BOYLESTAD *Devices: Discrete & Integrated* ii. 29 As V_Z [*sc.* the Zener voltage] decreases to very low levels, such as −5V, another mechanism, called Zener breakdown, will contribute to the sharp change in the characteristic. 1957 R. F. SHEA *Transistor Circuit Engineering* iii. 62 One of the simplest methods of obtaining such a [stable collector-voltage] supply is with the aid of a diode biased into breakdown in the reverse direction (so-called Zener diode). 1975 *Gramophone* Aug. 384/3 Stabilisation of the feedback input stages is achieved with a zener diode. 1981 J. C. SPROTT *Introd. Mod. Electronics* vi. 136 In the forward direction a Zener diode behaves like any other diode. 1957 W. C. DUNLAP *Introd. Semiconductors* viii. 168 Internal field emission, often called Zener effect because of the early theoretical contributions made to the subject by C. Zener,..is analogous to field emission from metals. 1962 SIMPSON & RICHARDS *Physical Princ. Junction Transistors* iv. 69 Lower voltage units usually have negative temperature coefficients indicating the existence of Zener effect. In such devices the transition region is apparently too thin to allow appreciable carrier multiplication to take place and Zener breakdown occurs before the critical avalanche condition has been reached. 1952 *Proc. IRE* XL. 1349/2 The Zener voltage for this junction was 21·5 volts. 1969 J. J. SPARKES *Transistor Switching* i. 13 At Zener voltages of about 6V the slope resistance of the diode characteristic is minimal. 1980 C. F. G. DELANEY *Electronics for Physicist* ix. 215 The Zener voltage..is a function of the resistivity (that is, of the doping) of the materials from which the diode is constructed.

b. *ellipt.* A Zener diode.
1965 *Wireless World* July 14 (Advt.), Ask for details of zeners from 400 mW to 50W, and 3·6V to 200V. 1976 *Pract. Electronics* Oct. 791/1 To provide any given voltage, it is only necessary to select a Zener having a voltage exceeding the required voltage by 0·7 volts.

Zengakuren (zengăkū·rĕn). [Jap., acronym f. *Zen* Nihon *Gakusei* Jichikai So*ren*go, = All-Japan Federation of Student Self-Government Associations (formed in 1948).] In Japan, an extreme left-wing student movement, noted for its violent interventions in national politics. Also *attrib.* or as *adj.*
1952 E. S. COLBERT *Left Wing in Japanese Politics* v. 299 The Communist-dominated student organization, the *Zengakuren*. *Ibid.* 300 On May 23, [1950] at a *Zengakuren* convention. 1960 *Times* 21 Sept. 16/6 He is rather naturally unloved by the members of Zengakuren, the radical student movement. 1967 D. & E. T. RIESMAN *Conversations in Japan* 121 The large Zengakuren ideas and practical accomplishments. 1970 *Guardian* 1 Apr. 11/1 The students, armed with quarter-staves,..with the colours and battle-cry of the Zengakuren sect to which they belonged. 1975 *New Yorker* 18 Aug. 50/2 Five student groups, including two main ones called Kakumaruha (the Revolutionary Marxists) and Chukakuha (the Middle Core), both composed of self-styled Trotskyites, are offshoots of the original Zengakuren, a radical association that led demonstrations against the security treaty and instigated campus riots that succeeded in closing down a number of universities over a period of months in 1969.

zenick. In def. for *zenik* read *suricatta.* Add: = MEERKAT 2 b, SURICATE. (Earlier example.)
1801 J. BARROW *Acct. Trav. S. Afr.* I. iii. 231 Others of this genus [*sc.* Viverra] are the muskiliatte cat, or zenik,.. the tigrina [etc.].

zenilla, obs. var. ZANELLA in Dict. and Suppl.

zenithal, *a.* (In Dict. s.v. ZENITH.) Add: *zenithal projection*, any of a class of map projections in which a portion of the globe is projected on to a plane tangential to a point on that portion, usu. made the centre of the map.
1882 T. CRAIG *Treat. on Projections* v. 89 The name zenithal projections is..derived from the fact that they can always be considered as the representation of the hemisphere situated above the horizon of the given point, and having the zenith for pole. 1910, etc. [see *PROJECTION *sb.* 7b]. 1961 L. F. BROSNAHAN *Sounds of Language* iii. 74 A continuous distribution of the O blood-group gene frequencies in Europe was..redrawn on a map of zenithal equal-area projection. 1974 *Encycl. Brit. Micropædia* IV. 587/3 *Gnomic map*, type of zenithal projection in which the Earth's grid is projected by radials from a point at the centre of the sphere into a tangent plane so that all great circles are represented by straight lines.

zenithward, *adv.* (In Dict. s.v. ZENITH.) (Earlier example.)
1868 CARLYLE *Jrnl.* 8 June in J. A. Froude *T. Carlyle* (1884) II. xxxi. 371 It was as a ray of everlasting light and insight this, that had shot itself *zenithward* from the soul of a man.

Zenker (ze·ŋkəɹ). **1.** The name of F. A. *Zenker* (1825–98), German pathologist, used in the possessive and occas. *attrib.* to designate (*a*) a hyaline degeneration of striated muscle occurring chiefly in cases of acute infectious disease, esp. typhoid and cholera (described by Zenker in 1864); (*b*) a pathological diverticulum at the junction of the pharynx and œsophagus (described by Zenker & von Ziemssen in *Cycl. Practice of Med.* (1878) VIII. 1–214).
1890 BILLINGS *Med. Dict.* II. 789/2 *Zenker's degeneration*, waxy degeneration as seen in muscles after acute attacks of fever. 1898 W. S. LAZARUS-BARLOW *Man. Gen. Path.* xii. 521 In Zenker's degeneration the muscular fibres themselves look dull and semi-opaque, and microscopically are found to have lost their striation. 1910 *Lippincott's New Med. Dict.* 1103/2 Zenker's pulsion diverticulum. 1932 *Jrnl. Amer. Med. Assoc.* XCVIII. 965/2 The symptoms of a Zenker pulsion diverticulum depend on the size of the pouch. 1970 PASSMORE & ROBSON *Compan. Med. Stud.* II. xxiv. 25/1 In typhoid fever, patches of necrosis develop in striated muscles. Here, the fibres lose all structure and become glassy and eosinophilic (Zenker's hyaline degeneration). 1975 *Year Bk. Ear, Nose & Throat* 279 A variety of neuromuscular disorders, with or without Zenker's diverticulum.

2. *Histology.* [The name of Konrad *Zenker*, German histologist, who described the fluid in 1894 (*Arch. f. path. Anat. und Physiol.* CXXXV. 147).] *Zenker's fluid*: a fixative (see quot. 1902).
1902 E. A. SCHÄFER *Essentials of Histol.* (ed. 6) 397 General methods of preserving and hardening tissues and organs.—The fluids which are more commonly used are.. Müller's fluid (bichromate of potash 2½ parts, sulphate of soda 1 part, water 100 parts); Zenker's fluid (which is Müller's fluid containing 5 parts per cent. of mercuric chloride, to which 5 c.c. of acetic acid is added at the time of use); [etc.]. 1941 [see *FIXER 2]. 1976 *Path. Ann.* XI. 130 Tumors that have been fixed in Zenker's fluid almost invariably show affinity for chrome salts, so that brown granules are easily discerned.

zeno- (zīno), f. Gr. Ζηνο-, comb. form of Ζεύς, used as a word-forming element with the sense 'the planet Jupiter', as **zenoce·ntric** *a.*, measured or expressed with reference to the centre of Jupiter; **zenogra·phic** *a.*, measured or expressed with reference to the surface of Jupiter.
1968 *Dissertation Abstr.* B. XXIX. 858/1 For each year the study has reconfirmed that there are two zenocentric intervals of Io's position in which we observe this influence. 1978 *Nature* 14 Sept. 111/1 The ±3·3° variation of the zenocentric declination of the Earth is sufficient to lead to a 15° variation of the..sub- Io longitude of the emission. 1971 *Icarus* XIV. 343 A bright spot at zenographic latitude 23°.8 N displayed the shortest rotation period ever recorded on Jupiter. 1979 *Nature* 5 July 42/2 Values..synthesised on a 10° by 10° zenographic θ (co-latitude) and λ (longitude) grid.

zenophobia: see *XENOPHOBIA.

zeolite. Add: Hence **zeoli·tically** *adv.*, as in a zeolite.
1951 C. PALACHE et al. *Dana's Syst. Min.* (ed. 7) II. 993 The water content can vary zeolitically over a range from $8H_2O$ to $5H_2O$. 1969 H. T. EVANS tr. *G. Hägg's Gen. & Inorg. Chem.* xxi. 510 Water in a solid phase may.. be zeolitically bound.., so that the proportion of water can vary without breaking down the crystal structure. 1980 *Nuclear Technol.* LI. 143/2 This study shows that similar enrichment of ^{234}U occurs in zeolitically altered volcanic ash and tuff.

Zep, also **Zepp** and with small initial, colloq. abbrev. of ZEPPELIN *sb.*, *v.*
1915, etc. [in Dict. s.v. ZEPPELIN]. 1915 W. OWEN *Let.* 12 June (1967) 338 The spot where a Zep. Bomb fell in a cross-roads. 1915 A. HUXLEY *Let.* Oct. (1969) 79 The dear old Boches and their Zepps, which always get back bomb for bomb. 1916 *Morning Post* 15 Mar. 1 (Advt.), Anti-Zep blinds. 1919 C. ORR *Glorious Thing* ii. 20 We're exceedingly preoccupied with the war—Zepp. raids and things. 1931 *Flight* 28 Aug. 855/1 The most interesting feature of our trip was a personally escorted tour of the entire Zep whilst crossing over France. *a* 1935 T. E. LAWRENCE *Mint* (1955) II. xxii. 159 There were Zepps in a cloud (sausages and mashed) and Adam and Eve on a raft (Hoxtonian for fried eggs on toast). 1974 *Listener* 7 Feb. 177/3 This aeroplane..flew towards the zepp, and he started firing.

Zephiran (ze·fĭrăn). *Pharm.* A proprietary name for an antiseptic preparation of benzalkonium chloride.
1935 *Trade Marks Jrnl.* 17 July 887/1 *Zephiran*... Chemical substances prepared for use in medicine and pharmacy. Bayer Products Limited,.. London, E.C.2; merchants and manufacturers. 1936 *Official Gaz.* (U.S. Patent Office) 28 Jan. 687/2 I. G. Farbenindustrie Aktiengesellschaft, Frankfort-on-the-Main... *Zephiran* for antiseptic and disinfectant. 1951 A. GROLLMAN *Pharmacol. & Therapeutics* xxv. 503 Benzalkonium chloride (Zephiran *) is of comparatively low toxicity. 1974 M. C. GERALD *Pharmacology* xxvi. 448 Iodine tincture.., Zephiran, and hexachlorophene are representative examples of local anti-infective agents.

Zéphirine Drouhin (ze·firĭn dru,æn). A thornless shrub or climbing Bourbon rose of the variety of this name, which bears fragrant pink flowers and was first introduced in France in 1868.
1931 H. H. THOMAS *Rose Bk.* viii. 72 The old thornless Rose (Zéphirine Drouhin) makes an excellent bush, and its lovely, fragrant blooms of soft rose-colour..are freely produced for weeks together. 1940 A. CHRISTIE *Sad Cypress* II. xii. 193 Do you know the name of this rose? It is Zephyrine Droughin [*sic*]. 1983 *Woman's Jrnl.* Jan. 85/2 Zephirine Drouhin is a fragrant climber with cerise-pink semi-double flowers and is thornless.

zephyr (ze·fəɹ), *v.* [f. the sb.] *intr.* To blow like a zephyr. Hence **ze·phyring** *ppl. a.*
1922 HARDY *Late Lyrics & Earlier* 111 An aura zephyring round, That care infected not. 1939 JOYCE *Finnegans Wake* 418 Since longsephyring sighs sought heartseast for their orience? 1973 J. JONES *Touch of Danger* xxiii. 131 A light little breeze zephyred in..from the open water.

Zepp: see *ZEP.

Zeppelin, *sb.* Add: Hence also (both *rare*[−1]) **Ze·ppelinist,** a member of the crew of a Zeppelin; **Zeppelini·stic** *a.*, resembling a Zeppelin in shape.
1930 KIPLING *Limits & Renewals* (1932) 328 He called Saint Jukamus a militarist and an impostor—this defaitist of a Zeppelinistic belly! 1937 F. MORISON *War on Great Cities* iv. 127 The demons who drove and dealt death cared as little for us as the earlier Zeppelinists cared.

zeppole (ze·pole). *U.S.* Pl. **zeppoli.** [It.] A kind of doughnut.
1976 *Monitor* (McAllen, Texas) 14 Oct. c 2/6 The mingled smells of salsiccia, bracciola, zeppole and calzone wafted from the stalls of food vendors around Father Zemo Square. 1979 *New Yorker* 8 Oct. 32/1 There aren't a lot of food venders. We wanted to differentiate this from other street fairs. Also, I didn't want people dropping zeppoli all over our books.

zero, *sb.* **1. a.** Delete 'Now *rare*' and add further examples.
1940 E. T. BELL *Devel. Math.* iii. 48 The introduction of zero as a symbol denoting the absence of units or of certain powers of ten..has been rated as one of the greatest practical inventions of all time. 1959 KOESTLER *Sleepwalkers* II. iii. 105 The Indian system of numerals based on the symbol zero. 1969 P. B. JORDAIN *Condensed Computer Encycl.* 572 Users can be confused and misled by leading zeros: they may hide the true size of the numeral and make it harder to grasp at a glance.

4. c. *Linguistics.* In grammar, the absence of an overt mark, written or spoken, as against its presence in corresponding positions elsewhere (e.g. *cut* pa.t. as against *putted*).
1891 S. C. VASU *Ashṭâdhyâyî of Pâṇini* 56 In Sanskrit Grammar, this 'lopa' is considered as a substitute or âdeṣa, and as such this grammatical *zero* has all the rights and liabilities of the thing it replaces. 1914 BLOOMFIELD *Introd. Study Lang.* v. 154 If..we take into view..*amo* 'I love', *amāvit* 'he loved', *amētur* 'he may be loved', it is possible to call them all related by affixation, the kernel being *am-*... In this instance the group does not contain a word that equals the kernel, or, as we might say, has 'affix zero'. 1933 —— *Language* xiii. 209 In *sheep : sheep* the plural-suffix is replaced by zero. 1946 *Jrnl. Amer. Oriental Soc.* LXVI. 98/1 Zero counts as an ending if it has the same function as an overt ending in another paradigm. 1957 *Eng. Lang. Teaching* (British Council) Oct.–Dec. 11 If the pronoun is object, zero should be used if the subject is a personal pronoun. 1972 R. QUIRK et al. *Gram. Contemporary Eng.* xiii. 866 With time adjuncts, omission of the preposition is usual whether the pronoun is *that* or zero..: That is the time (that) he arrives (at). 1979 *Amer. Speech* LIV. 31 Fifteen items take zero as the plural suffix.

6. a. Also, an absence or lack of anything; nothing. (Further examples.)
1883 *Proc. Inst. Mech. Engineers* Jan. 74 We have again a zero of current. 1915 J. HUNEKER *Ivory, Apes & Peacocks* iii. 35 We should soon reach a zero if we only registered the absence of 'necessary' traits in our poet. 1967 *Boston Globe* 21 May (Confidential Chat) 8/3 True, there is a certain amount of 'freedom' in their use but the wearing of them adds up to absolutely zero! 1977 C. MCCARRY *Secret Lovers* xii. 159 'What about the airlines, the hotel, car rentals.' 'Zero, I'd have used phony paper, wouldn't you?'

b. *spec. Mil.*, the time or the day when an attack or operation is due to begin.
hour of zero: now *obs.* (superseded by *zero hour* or *zero*).
1916 *King's R. Rifle Corps Chron. 1915* 104 Zero, the time the gas and smoke was to start, was 5.50 a.m. 1918 E. A. MACKINTOSH *War, the Liberator* IV. 124 We've got to be at Battalion Headquarters at 5, although I don't suppose zero'll be for a good time after that. 1924 KIPLING *Debits & Credits* (1926) 314 The men's teeth chatterin' behind their masks between rum-issue an' zero. 1942 W. S. CHURCHILL in *Second World War* (1950) IV. I. xiii. 225 It should be assumed..that zero [for an attack on Madagascar] should be about April 30. 1954 W. FAULKNER *Fable* 10 At zero, nobody left the trench except the officers and a few N.C.O.'s. 1983 L. MACDONALD *Somme* v. 41 The main body was ordered to Thiepval..timing their arrival for Zero plus two hours.. to stride on to consolidate the third objective.

7. a. zero-base, -based *adjs.*, applied to a budget and to budgeting in which each item is costed anew, rather than in relation to its size or status in the previous budget; cf. ZBB s.v. *Z 12 a; **zero-crossing,** the crossing of the horizontal axis by a function as it passes through zero and changes sign; a point where this occurs; also *attrib.* with reference to the analysis of complex waveforms through the study of such points; **zero day** *Mil.*, the day on which an attack or operation is scheduled to begin; also *transf.*; **zero hour:** also *transf.*; (*b*) the hour when something is at its lowest ebb (*nonce-use*); (*c*) the hour from which the time of day is measured; **zero-point** (later examples); *spec.* in *Physics*, used *attrib.* with reference to properties and phenomena in quantized systems at absolute zero; esp. *zero-point energy* (see quot. 1935).

1970 P. A. PYHRR in *Harvard Business Rev.* Nov.–Dec. 111 In this speech [Arthur F.] Burns identified the basic need for what we at Texas Instruments have come to call zero-base budgeting... Burns was advocating that government agencies start from ground zero, as it were, with each year's budget and present their requests for appropriations in such a fashion that all funds can be allocated on the basis of cost/benefit or some similar kind of evaluative analysis. 1977 *Wisconsin State Jrnl.* 1 Feb. 1 Zero-base budgeting was used by President Carter when he was governor of Georgia. He asked state agencies to justify their existence and the programs they administer from zero up, rather than merely bringing in requests for more programs, money and staff. 1976 *National Observer* (U.S.) 27 Mar. 3/3 To curb inflation, he advocates a 'zero based' Federal budget, in which each program must be justified every year. 1983 *Financial Times* 10 Nov. 9/1 Sir Douglas Wass..proposed two specific changes... The first would be to introduce 'zero-based budgetting' to make spending departments consider their action if the money available to them were drastically cut. 1950 *Jrnl. Acoustical Soc. Amer.* XXII. 821/2 Vowel sounds were the first to become intelligible, presumably because for them the density of zero crossings in the input to the time quantizer is lower than it is for consonants. 1969 *New Scientist* 1 May 225/1 Equipment capable of analysing and recognizing speech..has achieved considerable success using the so-called zero crossing technique. 1982 *Electronics Today Internat.* Oct. 78/2 Switching the zero-crossing point of the mains cycle eliminates the need for RFI suppression. 1929 *Papers Mich. Acad. Sci., Arts & Lett.* X. 335/2 *Zero day*, the day for an attack. 1938 *Brit. Jrnl. Psychol.* XXVIII. 325 On zero-day (the day preceding the beginning of training with the aid of the electric shock) each rat was immersed six times in the tank. 1947 L. HASTINGS *Dragons are Extra* iv. 80, I..fixed a zero-day three months ahead. 1978 R. V. JONES *Most Secret War* xxxix. 350, 20th October had been fixed as Zero Day for rocket attacks on London to begin. 1930 *Daily Express* 23 May 1/6 Psychologists have fixed 11 a.m. as the zero hour of the worker. 1939 WODEHOUSE *Uncle Fred in Springtime* xviii. 269 Your duties will not begin till after dinner. Zero hour is at nine-thirty sharp. 1939 JOYCE *Finnegans Wake* 403 Methought..I heard at zero hour as 'twere the peal of vixen's laughter among midnight's chimes. 1946 K. TENNANT *Lost Haven* (1947) vii. 98 Desperate over the nearness of zero hour for the punt service, he seized her by the hand and snatched her along. 1953 A. HUXLEY *Let.* 2 Feb. (1969) 664 Incidentally, zero hour for tax-payments is approaching. 1978 A. PRICE *'44 Vintage* viii. 86 He had..the..impression.. that they had been travelling at breakneck speed,..as though the Americans were determined to deliver them on time for some impossible pre-arranged zero hour. 1826 *Mem. Astron. Soc.* II. 469 Every star observed in the course of the sweep which can be identified with a star in any catalogue, and whose place is determined with certainty, is taken as a zero point. 1895 *Physical Rev.* II. 326 The balls were placed horizontally in their neutral position, and the reading of the zero-point of the scale taken. 1935 J. DOUGALL tr. *Born's Atomic Physics* 339 According to the classical theory, the state of least energy of an oscillator is that of zero energy. According to wave mechanics, however, the ground state has a finite energy $E = \frac{1}{2}h\nu_0$...This zero-point energy can be explained by Heisenberg's uncertainty principle. 1938 R. W. LAWSON tr. *Hevesy & Paneth's Man. Radioactivity* (ed. 2) xx. 194 If account be taken of zero-point energy, it is found that at low temperatures the lighter isotope must have the higher vapour pressure, whereas if there is no zero-point energy it will have the lesser vapour pressure. 1955 H. B. G. CASIMIR in W. Pauli *Niels Bohr* 130 The interaction with the zero point vibrations of the crystal lattice leads to an interaction between electrons. 1969 P. ANDERSON in Cockburn & Blackburn *Student Power* 263 When neither society nor man are anywhere put in question, culture stops. In England, it has gradually slowed towards zero point. 1970 G. K. WOODGATE *Elem. Atomic Struct.* iv. 67 A quantized radiation field has a zero-point energy equivalent to a mean-square electric field so that even in a vacuum there are fluctuations in this zero-point radiation field. *Ibid.*, Zero-point fluctuations. 1973 *Sci. Amer.* Jan. 91/1 In quantum physics the ground state is not a state of absolute rest or motionlessness but only a singular 'zero point' form of motion in which there are no quasiparticles.

c. (Further examples.) Also more widely as adj. in the sense 'no, not any'.
1938 *Brit. Jrnl. Psychol.* XXVIII. 329 If each such contact [*sc.* a mild electric shock received by rats] had been counted as an error, the number of cases of zero-error would have been considerably reduced. 1960 [see *social distance* s.v. *SOCIAL *a.* 12]. 1962 *Times* 30 Oct. 4/6 Good design points include 'zero torque'. 1972 *Listener* 22 June 820/3 Why is the notion of zero economic growth so seductive for those who embrace environmental causes? 1976 *New Yorker* 24 May 28/1 Because he had zero toads, Howard had to content himself with the tub of thin green gruel. 1978 *Peace News* 6 Oct. 2/1 We can't call together the group at zero notice. 1981 *TV Picture Life* Mar. 39/3 Jackie claims they now have 'zero communication'.

d. *Linguistics.* In sense *4 c: denoting an absence of a feature (as an inflection, or a phonetic or syntactic element) that is present in other cases, and is often indicated by the following word; **zero-derivation** *Linguistics*, derivation in which the parent word is not altered; the use of a word with a different grammatical function or in a different (though related) sense; hence **zero-derivative** *a.* and *sb.*, **-derived** *a.*; **zero grade** *Philol.*, the absence or extreme reduction of an ablaut vowel from a syllable.
1905 *Amer. Jrnl. Philol.* XXVI. 179 The intrusion of the *ē* grade on the zero grade..is..attested for the nasal verbs of the Indo-Iranian group. 1926 BLOOMFIELD in *Language* II. 160 *Alternation*... Absence of sound may be a phonetic or formal alternant... Such an alternant is a zero element. The postulation of zero elements is necessary for Sanskrit..for Primitive Indo-European..and probably economical for English (singular *book* with affix zero, as opposed to *book-s*, cf *f-oo-t*; *f-ee-t*). 1933 —— *Language* xiii. 215 Another extreme case is that of zero-alternants.., in which a constituent is entirely lacking, as in the plurals *sheep, deer, moose, fish*. *Ibid.* xiv. 236 In English, the nouns *longlegs, bright-eyes, butterfingers* are exocentric, because they occur both as singulars, and, with a zero-affix, as plurals (*that longlegs, those longlegs*). *Ibid.* xv. 252 We have zero-anaphora for participles after forms of *be* and *have*, as in *You were running faster than I was.* 1942 *Language* XVIII. 170 In *He cut it* there is a zero morpheme meaning 'past time' after *cut*. 1947 *Ibid.* XXIII. 340 We may set up the tentative portmanteau.. as an alternant of that constituent morpheme which it resembles phonemically, and set up a zero morph as an alternant of the other constituent morpheme. 1954 M. PEI *Dict. Linguistics* 238 *Zero ending*, in morphology, the bare stem of a word when used as such in discourse is said to have zero ending. 1959 *Brno Studies in English* I. 43 The indefinite article and its zero plural variant. 1960 H. MARCHAND *Categories & Types of Present-Day English Word-Formation* v. 295 (*heading*) Zero-derivation as a 'specifically English' process. *Ibid.* 297 There are quite a few vbs with French roots for which no French verbs are recorded and which may accordingly be treated as zero derivatives: *feeble* vb..., *master* vb, [etc.]. 1963 F. T. VISSER *Hist. Syntax Eng. Lang.* I. iv. 538 The Authorised Version of the Bible clearly fights shy of it: in those places where the Hebrew has a zero-clause, it uses a relative pronoun printed in italics (e.g. Ps. 7, 5, 'he is fallen into the ditch *which* he made'). 1964 *English Studies* XLV. (Suppl.) 63 The *be going + infinitive* group has had a partially independent development. Its opposite in this case is the *will + zero infinitive* group. 1965 *Language* XLI. 519 The unfortunate myth that there is some essential connection between aorist aspect and stems consisting of zero-grade root plus accented thematic vowel in Indo-European. 1971 *Canad. Jrnl. Linguistics* Fall 31 The use of zero morphs would have greatly facilitated the description..of several problems in Japanese. 1976 *Archivum Linguisticum* VII. 129 Zero-derivation..must be regarded an extremely productive word-formative process both in English and German, but also in other languages. *Ibid.* 132 The Direct Object-type based on (17b) has two different surface realizations in English: PAYMENT and also the zero-derived noun *payØ*. *Ibid.* 133, I believe..that *act* (noun) must be regarded as zero-derivative (*actØ*) from the corresponding verb. 1979 *Dictionaries* I. 19 The last point concerns zero-derivatives or homographs. 1981 *Amer. Speech* LVI. 229 Other details on which there have been ethnic speculations are the zero genitive of nouns..and the zero subject of a relative clause.

e. Special Combs. (see also senses 7 a in Dict. and Suppl., *7 d): **zero-balance** *a.*, applied to a bank account operated with no continuing balance, funds being transferred into it when necessary to just the extent required to meet drawings made on it; **zero beat,** a condition existing between two equal frequencies, in which no beats are produced; *spec.* in *Radio*, applied *attrib.* to a method of reception in which the incoming signal is mixed with a receiver-generated oscillation of the same frequency as the carrier wave (see *homodyne* s.v. *HOMO-); **zero-coupon** *a.*, applied to a bond carrying no interest but issued below its redemption price; **zero-energy** *a.* (Nucl. Physics), applied to a small reactor, usu. built for research purposes, that develops so little power that no cooling and little shielding are required; **zero g** or **G** = *zero gravity* below; **zero gravity,** the state or condition in which there is no apparent force of gravity acting on a body, either because the force is locally weak, or because both the body and its surroundings are freely and equally accelerating under the force; = *WEIGHTLESSNESS; **zero grazing** *vbl. sb.* (Agric.) = SOILING *vbl. sb.*[4] 1; so **zero-graze** *v. trans.* = SOIL *v.*[4] 1; **zero-grazed** *ppl. a.* = SOILED *ppl. a.*[2]; **zero growth,** an absence of increase (in population, production, etc.); **zero-length** *a.* (of a rocket launcher) no longer than the rocket it supports; **zero norm,** in a period of pay restraint, a recommended value of zero for the percentage increase in pay; **zero option,** a disarmament proposal that if the Soviet Union would withdraw its SS-20 missiles from Europe the U.S.A. would abandon its plans to deploy Pershing and cruise missiles there; **zero population growth,** an absence of any increase (or decrease) in a population over a period of time; abbrev. ZPG s.v. *Z12 a; **zero-power** *a.* (Nucl. Physics) = *zero-energy* adj. above; **zero rating,** a rating of zero for the purposes of value added tax; also as *vbl. sb.*, the practice of assessing the VAT at zero for a particular item; so **zero-rate** *v. trans.*, to assess at a VAT rate of zero; **zero-rated** *a.*; **zero sound:** see *SOUND *sb.*[3] 1 d; **zero-sum** *a.*, in the theory of games, applied to a game in which the sum of the winnings of all the players is always zero; also *transf.*, denoting any situation in which advantage to one participant necessarily leads to disadvantage to one or more of the others; **zero tillage** (Agric.) = *sod planting* s.v. *SOD *sb.*[1] 5 b; **zero-zero** (Aeronaut.), (*a*) a situation in which both the (horizontal) visibility and the cloud ceiling are technically zero; (*b*) used *attrib.* to designate an ejection seat that works even at rest and at zero altitude.
1974 *U.S. Investor/Eastern Banker* 26 Aug. 42/2 With a Zero Balance account, a customer will pay a small charge for actual activity. 1983 *Fortune* 18 Apr. 76/2 A controlled disbursement account is a type that bankers call a 'zero balance account'—it contains no cash at the end of the day after all checks have been paid. There are no funds left idle. 1927 *Mod. Wireless* Mar. 334/2 The results of fading on an ordinary set using high-frequency amplification were large compared with these effects in zero-beat reception. 1957 *Practical Wireless* XXXIII. 370/1 Tune a signal to zero-beat at the high-wavelength end of the scale. 1982 *Amer. Jrnl. Physics* L. 137/1 With this arrangement, one can explore the zero-beat situation (when frequencies are matched) as well as a variety of non-zero-beat situations. 1979 *Jrnl. Finance* XXXIV. 189 The relationship between yield curves for zero coupon bonds and coupon-bearing bonds is important. 1983 *Chicago Sun-Times* 29 May 5 Following the birth of their first child, the couple purchased $7,000 worth of zero coupon bonds in their daughter's name. 1951 *Canad. Chem. & Process Industries* Jan. 42 (*heading*) Zero Energy Experimental Pile. 1954 R. STEPHENSON *Introd. Nucl. Engin.* iii. 82 The smaller one is known as ZEEP (zero energy experimental pile) and is very similar..except that its power level is limited to a few watts. 1981 *Nucl. Energy* XX. 467 Fast neutron fluence measurements in the core of a zero-energy research reactor. 1952 A. C. CLARKE *Islands in Sky* v. 80 She was escorted by an elderly woman who seemed to be quite at home under zero 'g' and gave Linda a helpful push when she showed signs of being stuck. 1962 F. I. ORDWAY et al. *Basic Astronautics* xii. 477 Walking will be impossible in zero G. 1970 N. ARMSTRONG et al. *First on Moon* vi. 127 Deke Slayton ate the same food we did... In our zero-G situation we were always full. 1978 *Radio Times* 28 Jan. 59/2 *Horizon* explores all these aspects, with footage from Skylab and reminiscences from ex-inhabitants—the astronauts who call this world zero gravity—zero G for short. 1951 A. C. CLARKE *Sands of Mars* i. 5 I'll take you into the zero-gravity section and see how you manage there. 1968 *New Scientist* 12 Sept. 545/1 Astronauts of the future will probably have to build vehicles in space. This will entail joining different metals together under zero-gravity conditions. 1978 [see *zero g* above]. 1979 *United States 1980/81* (Penguin Travel Guides) 553 You'll be offered the opportunity to.. experience weightlessness in a zero-gravity machine. 1984 *News* (Mexico City) 12 Mar. 22/4 Two of the first drugs that will be produced in zero gravity in mass quantities are beta cells..and interferon. 1956 *Britannica Bk. of Year* 316/1 Studies indicated that green-chopping, sometimes called zero-grazing or soiling (bringing the pasture to the livestock), resulted in forage yields about double those from grazing. 1958 *Agriculture* LXV. 129 Zero grazed animals also put on greater weight during the summer months than those out grazing. *Ibid.* 131 The cattle have been zero grazed since May 1956. 1970 R. JEFFRIES *Dead Man's Bluff* vi. 56 Cows were zero-grazed and never stepped off concrete. 1978 *Exper. Husbandry* No. 33, 18 (*heading*) Beef from spring-born zero-grazed Friesians—comparison of bulls, steers and late castrates. 1984 'D. ARCHER' *Ambridge Yrs.* 19 There are other herds that never go out into pasture... They have their grass cut and carted to them, a method called 'zero-grazing'. 1973 *Science* 15 June 1143 The possibility of zero growth in the population of the United States. 1976 *Atlantic Monthly* Jan. 4 Can economies have simultaneously zero growth, rapid inflation, substantial unemployment, and a balance of payments deficit? 1976 *N.Y. Times Mag.* 4 July 73 Power to stop new construction is wielded through the..water board.., a majority of which is dedicated to zero growth. 1954 K. W. GATLAND *Developm. Guided Missile* (ed. 2) iii. 79 (*caption*) A later G.A.P.A. rocket being adjusted on a 'zero-length' launcher. 1966 *Economist* 17 Dec. 1214/1 There is no bogy-man so dismal to those who run Britain's economy as the dreaded Zero Norm, the spectre who is supposed to rule over pay negotiations in the six-month period of 'severe restraint' that replaces the total pay freeze from New Year's Day. 1976 F. ZWEIG *New Acquisitive Society* II. i. 80 A zero norm which equalizes everybody is easier to endure. 1981 *Washington Post* 5 July A20/1 By reviving controversy about a moratorium and the 'zero option', Brandt's trip appears to have realized some of the fears of those in the West German government and opposition party. 1983 *Financial Times* 10 Feb. 11/6 Mr. George Bush, the U.S. Vice-President, last night confirmed Washington's willingness to consider alternatives to its zero option proposals for banning intermediate range nuclear missiles from Europe. 1967 *Science* 10 Nov. 732/2 Most discussions of the population crisis lead logically to zero population growth as the ultimate goal. 1974 *Environmental Conservation* I. 15/1 It is my estimate that zero population growth will be reached..some time during the next decade. 1950 *Nucleonics* Sept. 104/2 *Zero-power reactor*, an experimental nuclear reactor operated at low neutron flux and at a power level so low that not only is no forced cooling required but also fission-product activity in the fuel is sufficiently low to allow the fuel to be handled after use without serious hazard. 1983 *Trans. Amer. Nucl. Soc.* XLIV. 528/2 Plate versus pin geometry continues to be an issue for Zero-Power Plutonium Reactor (ZPPR) analysis. 1972 *Daily Tel.* 22 Mar. 19/3 Zero-rating a transaction, rather than exempting it, is advantageous, because a trader is allowed credit for any VAT paid on his inputs. 1975 *Times* 11 Feb. 6/4 The first thing the Government could do would be to zero-rate the theatre for VAT. 1972 *Daily Tel.* 22 Mar. 17/4 A firm which supplies zero-rated goods or services gets complete relief from Value Added Tax both on its purchases and on its sales. 1976 *Horse & Hound* 3 Dec. 28 (Advt.), Rubber riding boots. Best quality, shiny leather-look finish. Price £3·45 (zero rated). 1971 *Nature* 3 Dec. 310/3 The recently reported claim of the salvage industries to be given zero rating for value-added tax.. can also be extended. 1972 *Daily Tel.* 22 Mar. 17 Mr Barber explained the difference between exemption from VAT and 'zero rating'. 1984 *Bookseller* 3 Nov. 1857/1 Books are the essential tools of society and zero-rating is a very efficient way of encouraging their wide availability. 1944 VON NEUMANN & MORGENSTERN *Theory of Games* ii. 47 An important viewpoint in classifying games is this: Is the sum of all payments received by all players (at the end of the game) always zero; or is this not the case?.. We shall call games of the first mentioned type zero-sum games. 1966 S. BEER *Decision & Control* x. 210 Perhaps the contestants in most important games nowadays (from labour disputes..to international diplomacy) too readily regard their games as zero-sum. 1967 L. B. ARCHER in Wills & Yearsley *Handbk. Management Technol.* vii. 121 Everybody *can* win. Manufacturing is not a zero-sum game. 1971 *Times Lit. Suppl.* 22 Oct. 1335/3 C. Wright Mills..used a zero-sum conception of power (i.e., the more one person had the less was available to others). 1980 *Ibid.* 26 Sept. 1072/2 In Europe [in the 1930s] class conflict was seen as a zero-sum game in which one group could only benefit at the expense of another. 1983 *Listener* 15 Dec. 31/1 We live in a zero-sum world, and it was inconceivable that the strains of setting up a completely new operation within an essentially fixed budget would not eventually begin to tell. 1971 *New Scientist* 25 Mar. 663/1 Even fairly low levels of pesticide destroyed these bacteria... The situation is made even worse by the growing popularity of 'zero-tillage'. 1979 *Austral. Financial Rev.* 16 Aug., Farmers using zero tillage leave straw and stubble on the field and this trash cover helps to eliminate erosion, conserve moisture,.. and reduce salinity problems. 1939 HIXSON & COLODNY *Word Ways* xvi. 141 *Zero-zero* (no visibility in any direction). 1947 *Shell Aviation News* No. 113. 7/2 There are few greater nervous strains than that experienced by a pilot of a transport in deciding to switch over from instruments and manually bring the ship in these last few hundred feet when visibility is 'zero-zero'. 1961 *Aeroplane* C. 593/2 The most interesting item in the Martin-Baker Aircraft Co., Ltd., display..is the prototype rocket-assisted 'zero-zero' ejection seat for VTOL aircraft. 1967 *Times Rev. Industry* June 53/1 Category 3C represents visibility of less than 50 metres and includes 'zero-zero' conditions under which safety experts believe operations will not be possible for a very long time. 1977 P. WAY *Super-Celeste* 215 They had explained to him the controls of his zero-zero ejection seat.

zero (zīə·ro), *v.* [f. the sb.] **I. 1.** *trans.* To set the sights of (a rifle) using targets at known distances.
1913 A. G. FULTON *Notes on Rifle Shooting* 30 (Advt.), A Zero Target, by means of which any novice may correctly zero his own rifle. 1918 H. MCBRIDE *Emma Gees* ix. 119 'Zeroing' a rifle is the process of testing it out on a range of known distances and setting the sights to suit one's individual peculiarities of aiming. 1958 L. VAN DER POST *Lost World Kalahari* vii. 142, I had not yet fired at a live target, though..I had zeroed it on a marked one. 1979 D. LOWDEN *Boudapesti 3* xxxii. 176 The rifle had been zeroed..at an elevation of 200.

2. zero in. a. *trans.* To range guns or missiles on (a target). Usu. *pass.*
1944 *Newsweek* 8 Jan. 45 Don't you know the Jerries have that road zeroed in?—a phrase meaning the Germans had sighted their guns on the road and needed only to pull their triggers. 1945 *Finito! Po Valley Campaign* (15th Army Group) 41 Road junction 711 was not only mined, but zeroed in. 1965 H. KAHN *On Escalation* iv. 74 American bases overseas and American targets on the mainland are at all times zeroed in by Soviet missiles, and vice versa. 1971 *Scope* (S. Afr.) 19 Mar. 20/4 These roads were well covered by strongpoints, and zeroed-in by artillery.

b. *trans.* To aim (a weapon) at a target. Const. *on.*
1944 *Life* 14 Aug. 57/1 Germans who had retreated out of town 'zeroed in' mortar shells among troops and light tanks which tried to follow. 1950 *N.Y. Herald* 26 Aug. 2/1 Guns are 'zeroed in' on the junction. 1961 *Time* (Atlantic ed.) 17 Feb. 16 The Russians now have some 50 ICBMs ready to go, presumably zeroed in on U.S. targets.

c. *intr.* To focus attention *on*, to concentrate *on*. Also *transf.* (const. *on*), to get a closer view of a subject with a camera.
1959 *Guardian* 1 Dec. 15/4 So far..Governor Rockefeller has spoken out on space research, the housing of Puerto Ricans, crop subsidies... Now he is zeroing-in on the social drinkers. 1961 D. HUFF *Score* (1962) p. ix, This book..will narrow the subject down to the specific material you are likely to encounter in a variety of fields...Finally this book will zero-in on the increasingly important psychological tests. 1971 *Daily Tel.* 18 Oct. 14 An excited Taiwanese photographer zeroed in on the frail figure..with a long-range lens. 1972 *Screw* 12 June 21/2 The storyline, such as it is, zeros in on a classy whorehouse in a banana republic. 1974 A. DILLARD *Pilgrim at Tinker Creek* viii. 143 Zero in on a well-watered shore. 1976 *Word 1971* XXVII. 142 Wittgenstein zeroes in upon linguistic constraints and logical conditions as they are made manifest in common, ordinary speech. 1983 *Listener* 14 Apr. 33/1 Its staff were slowly but surely zeroing in on the essential secret of successful breakfast broadcasting in this country. 1985 *Weekly World News* (U.S.) 1 Jan. 11/1 Security cameras were installed to stop vandalism, but they zeroed in on the athletes as they traipsed back and forth across a corridor from the shower to their locker room.

d. *intr.* To move towards, as if to a target. Also *fig.* Const. *on.*
1959 *Guardian* 26 Sept. 5/2 The squadrons of flies that zeroed in on his shiny pate. 1961 J. STEINBECK *Winter of our Discontent* xxi. 300 For twenty years that check has zeroed in on the first of every month. 1965 J. A. MICHENER *Source* (1966) 797 A series of bullets was beginning to zero-in on her, and in a few more steps she was sure to be hit. 1968 K. AMIS *I want it Now* i. 27 By the time he got back to the girl two other men had zeroed in on her. 1972 W. GARNER *Ditto, Brother Rat!* iv. 30, I zeroed in on the downstairs bar. 1974 *Newsweek* 8 Apr. 51/1 In recent months, instrumented spacecraft have zeroed in on Venus, Mars and Jupiter to provide astronomers with a wealth of significant new data on those planets. 1978 G. A. SHEEHAN *Running & Being* xii. 174 A pacemaker..had zeroed in on the perfect pace.

II. 3. To fix the zero hour for (a military operation).
1926 *Blackw. Mag.* Dec. 774/1 Plan no. 7 will be put in operation to-night. It will be zeroed as from 23.00 hours.

4. = *ZEROIZE *v.*
1949 W. F. GOODELL in J. F. Blackburn *Components Handbk.* x. 331 This zero convention makes it possible to 'zero' the dial of a synchro motor on one unit of a system. 1951 C. W. KENNEDY *Inspection & Gaging* ix. 293 An error would result if the air gage were zeroed against a 5-microinch surface and then used to compare the diameter of a bore with a 200-microinch surface roughness. 1969 [see *ODOMETER]. 1974 *Physics Bull.* Mar. 108/2 An offset control allows the pen to be zeroed at any point on the chart. 1982 *Homes & Gardens* Jan. 17/3 One button..can zero the read-out at any point, allowing you to add ingredients without emptying the bowl.

III. 5. zero out. *trans.* To eliminate.
1967 *Word* XXIII. 316 We 'zero out' the voicing of /b/, which is not at all to deny that /b/ is voiced. 1972 W. LABOV *Language in Inner City* ii. 52 The deletion of *are* has reached such a high point that it is effectively zeroed out for many speakers. 1982 *Daily Tel.* 25 Jan. 12/7 Watch..for word of new budget cuts, including Federal programmes that are to be 'zeroed out'.

zeroable (zīə·ro,ăb'l), *a.* [f. ZERO *sb.* + -ABLE.] **1.** *Linguistics.* That may be omitted from a sentence without loss of meaning.
1965 *Language* XLI. 395 Metaphorical and idiomatic objects are not pronounable or zeroable. 1975 *Rev. des Lang. Vivantes* XLI. 239 The relation between 'They described Vesuvius' and its metalinguistic, zeroable, expansion into 'They described Vesuvius *which is a volcano*'.

2. Capable of being set to read zero.
1974 *Physics Bull.* Aug. 349/1 A zeroable offset counter system is provided which makes it easy for the operator to measure the angular error between the ground reference face and the crystallographic planes.

ze·roing, *vbl. sb.* [f. *ZERO *v.* + -ING[1].] **1. a.** The adjustment of an instrument to give a reading of zero.
1949 W. F. GOODELL in J. F. Blackburn *Components Handbk.* x. 331 It is to facilitate the zeroing of synchros that they are provided with standardized flanges that are accurately concentric with the shafts. 1954 *Electronic Engin.* XXVI. 118/1 The connexion of Dekatrons in cascade presents difficulties in that a carry makes necessary a cumbersome circuit for resetting or zeroing. 1975 *Chem. Engin.* 10 Nov. 223/1 A process-control computer..has been extended to control zeroing of selected instrument transmitters to improve the accuracy of the input.

b. The adjustment or setting of the sights of a gun.
1975 V. CANNING *Kingsford Mark* v. 83 On the rifle were telescopic sights... He would..fire a few rounds and check the zeroing. 1979 D. LOWDEN *Boudapesti 3* xxxii. 176 The proper zeroing target had been used.

2. *Linguistics.* The deletion or omission of part of a linguistic form or structure.
1956 *Language* XXXII. 645 In all or nearly all dialects the first of two contiguous unstressed vowels may or may not be modified by shortening and/or raising or zeroing. 1965 *Ibid.* XLI. 393 In *I prefer that I should go first* there is no zeroing, but in the transform of this *I prefer for me to go first→I prefer to go first.* 1970 J. W. GAIR *Colloquial Sinhalese Clause Structures* iv. 54 It is convenient to treat such clauses as derived by context-governed deletion from full clauses like them in every respect save the presence of such constituents. The term 'zeroing' will be reserved for such discourse deletion.

zeroize (zīə·ro,əiz), *v.* [f. ZERO *sb.* + -IZE.] *trans.* To adjust (an instrument or device) to give a zero reading, esp. in order to calibrate it; to assign a value of zero to. So **ze·roizing** *vbl. sb.*
1908 *Brit. Patent 23,895* 3 Such clutches have already been employed for the simultaneous zeroizing of two indicating mechanisms. 1914 E. M. HORSBURGH *Mod. Instruments & Methods Calculation* 87 Mr. Trinks has invented a device which disengages the pawls from the number wheels when the latter are being zeroised. *Ibid.*, The zeroising crank is fixed on the right-hand side of the carriage, and the zeroising is effected by a half revolution of this crank. 1928 [see *REGISTER *sb.*[1] 10 c]. 1945 *Chambers's Jrnl.* Dec. 671/1 It is said to enable operational errors to be rectified in a matter of moments, and the calculations to be continued without the waste of time involved by 'zeroising' the machine and restarting the calculations from the beginning. 1956 G. A. MONTGOMERIE *Digital Calculating Machines* ii. 28 The other major control is the zeroizing or clearance key or handle at the right-hand side of the keyboard. 1974 *Software World* V. v. 12 Integer Function JCV12C converts a positive integer L (L < 256) to a character in A1 format, zeroizing bits 24 to 1.

zeroth (zīə·rōᵘþ). *Math.* and *Sci.* Also (*rare*) **zero'th.** [f. ZERO *sb.* + -TH[2].] Coming next in a series before the one conventionally regarded as the first.
1896 *Electrician* 10 Jan. 350/2 In order to have the zeroth Bessel normal function, we need only let the conductance and permittance per unit length of cable both vary directly as the distance from $x = 0$. 1932 *Physical Rev.* XL. 56 The zeroth order wave functions which they use do indeed indicate that tetrahedral symmetry should give high stability. 1956 E. H. HUTTEN *Lang. Mod. Physics* iv. 143 From this the theorem is derived that one body in contact with another may be used as a thermometer, or that one body 'reads' the temperature of another. This theorem is sometimes referred to..as the zeroth law of thermodynamics. 1969 *Nature* 15 Nov. 642/1 The energy of transition between the zeroth and the first energy levels. 1976 J. H. CONWAY *On Numbers & Games* 0. 3 In this zeroth part, our topic is the notion of number. 1984 *QL User* Dec. 33 LET input. vector = VEC 5 declares a vector of 6 words (BCPL vectors start at their zero'th element) pointed to by INPUT.VECTOR.

zerovalent (zīərovē̆ɪ·lĕnt). *Chem.* [f. ZERO *sb.* + *-VALENT.] Having an actual or formal valency of zero. Hence **zerova·lency.**
1940 *Chambers's Techn. Dict.* 919/2 Zero-valent. 1953 BARNETT & WILSON *Inorg. Chem.* xxii. 282 Since the electrons forming these covalencies are supplied solely by the molecules of carbon monoxide, the metal atom is said to have zero-valency. 1965 *Jrnl. Chem. Soc.* 847 The analogous reaction..of the naphthalene complex in its zerovalent formulation..cannot be distinguished from the possible consecutive reactions. 1977 *Inorg. Chem.* XVI. 1313 (*heading*) Reactions of the zerovalent complex $Ni[HP(C_6H_5)_2]_4$. 1984 *Science* 22 June 1330/2 A greater proportion of nickel in the catalytically active zero-valent state.

zest, *sb.*[1] **1.** Delete 'Now *rare* or *Obs.*' and add later examples. Also in Fr. form *zeste* (see etym. in Dict.).
1958 L. DURRELL *Mountolive* viii. 162 How good the taste of Dubonnet with a *zeste de citron*. 1967 *Guardian* 3 Feb. 8/5 The thin outer skin of oranges, known as the zest. 1974 *Homes & Gardens* Jan. 68/2 Add candied peel, lemon zeste, salt and nutmeg and mix. 1979 N. GORDIMER *Burger's Daughter* II. 269 He had fished the slice of lemon out of the bottom of his glass and was gobbling the skin with a mouth drawn by the zest. 1981 *Family Weekly* (U.S.) 12 July 10/3 Grated zest of 1 lemon. 1982 J. GRIGSON *Fruit Bk.* 256 Orange juice and a little zest can also be added to tomato and carrot soups.

ze·sty (ze·sti), *a.* [f. ZEST *sb.*[1] + -Y[1].] **1.** Of a taste or a food: piquant, agreeably sharp.
1934 in WEBSTER. 1936 [see *PICK-UP *sb.* f]. 1953 'A. BRIDGE' *Place to Stand* v. 67 The meal was plain and good, with the zesty taste of country cooking. 1969 *Daily Tel.* 5 Feb. 15/5 The food is spicy and zesty enough to make you nearly spring out of your seat. 1971 A. G. SEABERG *Menu-Design-Merchandising* 214 The zesty garlic butter brings out the best in this epicurean treat from the sea.

2. *fig.* Energetic, lively, stimulating.
1952 *Time* 15 Dec. 33/1 Zesty as a two-year-old rounding into the stretch, Britain's aged Prime Minister Winston Churchill last week entered his 79th year enthusiastically agallop on all his old hobbies. 1958 *Oxf. Mag.* 22 May 459/1 The little farrago ends with a bound of zesty life-assertion and merriment. 1959 *News Chron.* 23 July 5/5 To provide its zesty performance, Rootes' engineers have developed an aluminium cylinder head. 1972 *Time* 17 Apr. 43/2 Housing starts remain zesty, at an annual rate of 2,500,000 units in recent months. 1981 *N.Y. Times Mag.* 27 Apr. 18/3 The zesty lexicographer from New Zealand neither equivocates nor jazzes around. 1982 *Observer* 10 Oct. 40/8 This column's Third Rule of Television—'If it works, change it'—was given another zesty outing last week.

zeta[2]. Add: **2. zeta potential** *Physical Chem.*, the potential difference that exists across the electrical double layer at the interface of a solid and a liquid.
1939 E. A. HAUSER *Colloidal Phenomena* ix. 91 In simple cases, this double layer can be regarded as an electric condenser, the potential of which is generally termed electrokinetic, or ζ (zeta), potential. 1982 *Nature* 21 Jan. 267/2 They [*sc.* electrokinetic phenomena] arise whenever there is relative motion between a charged interface and a liquid and are usually interpreted in terms of the zeta potential.

zetetic, *a.* and *sb.* Add: Hence **zete·tical** *a.*, only in *Zetetical Society*, a nineteenth-century society with mystical beliefs; also *ellipt.*

1881 (*programme*) The Zetetical Society... Committee. .. J. M. Fells... G. B. Shaw... Sidney Webb. 1882 G. B. Shaw *Let.* 30 Jan. (1965) I. 47, I have overtaxed the patience of the Almighty, and he has smitten me; and, through me, the Zetetical Society. 1905 —— *Let.* 3 Jan. (1972) II. 486 After about a year of the Zetetical I joined the Dialectical. 1921 —— *Immaturity* (1930) p. xl, Not until..1879, did I for the first time rise to my feet in a little debating club called the Zetetical Society. 1962 D. H. Laurence *Bernard Shaw's Platform & Pulpit* p. x, A meeting of the Zetetical Society.

zetetically *adv.* (later example).

1872 *Zetetic* July 8/2 (Advt.), The life and teachings of Jesus Christ zetetically considered.

zeugmatography (ziūgmǽtǫ·grǎfi). *Med.* [f. Gr. ζεῦγμα, ζευγματ- a yoking (in allusion to the coupling of the electromagnetic and magnetic fields) + -o + -graphy.] A form of imaging using the principles of nuclear magnetic resonance to obtain and display the structural details of soft tissue. Hence **zeugma·togram,** a picture produced by zeugmatography; **zeugmatogra·phic** *a.*, involving or produced by zeugmatography.

1973 P. C. Lauterbur in *Nature* 16 Mar. 190/1 Because the interaction may be regarded as a coupling of the two fields by the object, I propose that image formation by this technique be known as zeugmatography. *Ibid.* 191/1 At low radio-frequency power..the two capillaries gave nearly identical images in the zeugmatogram. *Ibid.* 191/2 Zeugmatographic techniques should find many useful applications in studies of the internal structures, states, and composition of microscopic objects. 1981 *Brit. Med. Jrnl.* 7 Nov. 1212/1 Nuclear magnetic resonance (or zeugmatography) is looming on the horizon.

zeze (zē¹·zē¹). [a. Swahili *zeze.*] A zither-like string instrument of eastern and central Africa.

1860 R. F. Burton *Lake Region Central Africa* II. xviii. 291 The zeze, or banjo, resembles in sound the monochord Arabian rubabah, the rude ancestor of the Spanish guitar. 1978 *Times* 19 Aug. 3/3 (*caption*) A member of the Tanzania National Dance Troupe playing a 13-string zeze.

Zhdanovism (ʒdā·nǒviz'm, -fiz'm). [f. the name *Zhdanov* (see below) + -ism.] The policy of rigorous ideological control of literature and cultural life generally that was developed in the post-war period by A. A. Zhdanov (1896–1948), Russian politician. So **Zhda·novist, -ite** *adjs.*, of, pertaining to, or resembling Zhdanov or Zhdanovism.

1957 C. Hunt *Guide to Communist Jargon* xi. 37 The Zhdanovite decrees on literature and the arts of 1946–7. 1958 *Encounter* Nov. 35/1 The ideal of Zhdanovism was, precisely, the reduction of literature to 'a small cog and a small screw' in the mechanism of the totalitarian state. 1962 H. Swayze *Polit. Control Lit. in U.S.S.R., 1946–59* ii. 26 (*heading*) The heyday of Zhdanovism, 1946–1952. 1966 *Listener* 3 Nov. 659/3 The decree..contained the Zhdanovist denunciation of the works of Shostakovich and Khachaturian. 1975 *Times Lit. Suppl.* 31 Oct. 1280/5 Its account of Gramsci's career and the early history of the Communist Party can only be described as Zhdanovite. 1977 *Ibid.* 21 Jan. 76/4 Socialist realism he gives deservedly short shift. But the practice of Zhdanovism does not exhaust the subject.

‖ **zhuyin zimu** (dʒūyin dzimū). Also with accents and (in Wade-Giles transliteration) as **chu-yin tzu-mu.** [Chinese, f. *zhùyīn* phonetic notation (f. *zhù* notes + *yīn* sound) + *zìmǔ* letters of the alphabet (f. *zì* word, character + *mǔ* mother).] The national phonetic alphabet of China made up of symbols based on Chinese characters, first adopted in 1918. Also *ellipt.* as **zhuyin.**

1938 E. M. North *Bk. Thousand Tongues* 89/1 North Mandarin colloquial, Peking dialect, or Kuoyü... Chu Yin phonetics... Wang Chao phonetics... Roman characters. 1960 Chang-tu Hu et al. *China* v. 107 A set of thirty-nine phonetic symbols, *chu-yin tzu-mu*, officially promulgated in 1918 by the government. 1968 P. Kratochvíl *Chinese Lang. Today* v. 168 The purpose of the first official Chinese phonemic transcription called *zhùyīn zìmǔ* 'Pronunciation Alphabet'..was to serve as a stepping stone towards learning the characters, and also as a tool for promulgating the National Language. 1978 *Nagel's Encycl.-Guide: China* 95 The thirty-nine letters of the *zhuyin zimu* alphabet were taken from old, very simple Chinese characters.

‖ **ziarat** (zi,ā·rat). Also 8 **zeearut; zearat, ziarath, ziarut.** [ad. Hindi f. Urdu, f. Arab. *ziyārat* pilgrimage.] A Muslim place of pilgrimage, a shrine; a pilgrimage to such a place.

1776 N. B. Halhed tr. *Code Gentoo Laws* xii. 187 Places of *zeeàrut* (or religious Walks). 1913 *19th Cent.* May 993 This is said to occur frequently in the large shrines or Ziarats, such as Mashad. 1916 M. Diver *Desmond's Daughter* ii. ix. 121 A graphic tale of the manner in which his tribe had come by their first *ziarut*. 1925 *Blackw. Mag.* Dec. 796/2 That is where my brothre lives, Sahib; he likes to be next to his Ziarat. 1934 *N.-W. Frontier Province Gazetteer Peshawar Dist.* (ed. 3) I. i. 153 For the mass of the people the local *ziarats* have to suffice. 1967 F. Rasul *Bengal to Birmingham* vi. 61 At the appointed time, led by our group leader, we went for the *ziarath* (meeting the respected one). 1976 [see *zikr].

zibib (zi·bib, ‖ zəbī·b). Also **zibeeb.** [ad. Arab. *zabīb* dried grapes, (in Egypt) zibib.] A colourless, strongly alcoholic Egyptian drink made from raisins and drunk with added water, which turns it white.

Pronounced (zi·bib) by N.Z. servicemen in Egypt in the 1939–45 war.—Ed.

1836 E. W. Lane *Acct. Manners & Customs Mod. Egyptians* II. i. 19 In the same manner, many *shurbetlees* (or sellers or sherbet) carry about for sale *zebeeb* (or infusion of raisins). 1958 L. Durrell *Mountolive* ix. 181 He drank quite a lot of *zibib* according to the proprietor. 1977 J. Hutchison *Danger has No Face* (1978) iv. 41 Would I have a glass of zibeeb? He asked me. 1980 J. Hove *Flowers of Forest* i. 8, I wandered..sherry glass in hand, imagining it zibib or some other sharp foreign drink.

Ziegfeld (zī·gfeld). The name of Florenz *Ziegfeld* (1869–1932), American theatre manager and producer, used *attrib.* with reference to the follies that he staged annually from 1907 to 1931; so *Ziegfeld girl*, an actress taking part in such a revue.

1913 *Green Bk.* Jan. 72 Never before was so much beauty shown so much as in this latest of the Ziegfeld nonsensicalities. 1915 *New Republic* 31 July 336/1 Ziegfeld Follies 1915, a musical comedy produced by F. Ziegfeld Jr. 1917 *Ibid.* 7 July 278/1 If the Ziegfeld chorus were clothed in brown jaegers.., the paucity of the entertainment..would be shockingly revealed. 1923 G. Ade *Let.* 9 Jan. (1973) 87 Do not permit encores unless actually demanded. This is most important and is the secret of the success of a good Ziegfeld show. 1929 etc. [see *folly *sb.*¹ 5 b]. 1932 G. Greene *Stamboul Train* ii. i. 70 A baronet had married a Ziegfeld girl. 1976 Botham & Donnelly *Valentino* iv. 35 He chatted to the Ziegfeld Follies girls.

Ziegler (zī·gləɹ). *Chem.* The name of Karl *Ziegler* (1898–1973), German chemist, used *attrib.* to designate a trialkyl aluminium–titanium tetrachloride catalyst discovered by him for the synthesis of stereoregular isotactic polymers of high density and crystallinity from an ethylene or propylene monomer; also, *loosely*, = next.

1957 [see *isoprene]. 1961 [see *stereoregular *a.*]. 1966 Phillips & Williams *Inorg. Chem.* II. xxxiii. 560 This reaction is of interest in connexion with the polymerization of olefins by Ziegler catalysts. 1980 *Nature* 20 Mar. 213/3 [They] produced thin films by exposing acetylene gas to concentrated solutions of Ziegler catalysts.

Ziegler–Natta (zī·gləɹ næ·tā). *Chem.* The names of K. *Ziegler* (see prec.) and Giulio *Natta* (1903–79), Italian chemist, used *attrib.* to designate any catalyst of the class including the Ziegler catalyst, consisting in general of a transition metal halide and a non-transition metal organic derivative, and used with any olefin monomer. So *Ziegler–Natta catalysis*.

1965 Phillips & Williams *Inorg. Chem.* I. x. 382 Whereas the ionic catalysts appear to be capable of general use with any unsaturated molecule,..Ziegler–Natta catalysts are usually employed in the polymerization of olefinic hydrocarbons—notably, ethylene and proplyene. 1974 *Encycl. Brit. Micropædia* X. 880/3 The Ziegler–Natta catalysts include many mixtures of halides of transition metals, especially titanium, chromium, vanadium, and zirconium, with organic derivatives of nontransition metals, particularly alkyl aluminium compounds. 1980 M. Orchin et al. *Vocab. Org. Chem.* xiv. 535 It has been suggested that Ziegler–Natta catalysis may involve metal–carbene formation.

Ziehl (tsīl). *Bacteriol.* The name of F. *Ziehl* (1857–1926), German neurologist, used *attrib.* and in the possessive to designate a red stain consisting of an alcoholic solution of fuchsine in an aqueous solution of phenol; so **Ziehl–Neelsen** [F. K. A. *Neelsen* (1854–94), German bacteriologist], applied to a method for identifying acid-fast organisms such as tubercle bacilli by staining with Ziehl's stain, decolorizing with sulphuric acid (and sometimes also alcohol), and counterstaining with methylene blue: acid-fast organisms retain the original red colour.

1892 G. M. Sternberg *Man. Bacteriol.* i. iv. 29 (*heading*) Carbol-fuschsin (Ziehl's solution) *Ibid.* 30 (*heading*) The Ziehl-Neelson [*sic*] method. 1967 K. M. Smith *Insect Virol.* x. 189 Immature polyhedra in the nuclei of the hemolymph cells are easily differentiated by staining the smears with Ziehl fuchsin. 1974 R. M. Kirk et al. *Surgery* ii. 21 *Mycobacterium*..*tuberculosis*..stained by the Ziehl-Neelsen method is acid-fast and alcohol-fast.

ziff (zif). *Austral.* (and *N.Z.*) *slang.* [Origin unknown.] A beard.

1919 W. H. Downing *Digger Dialects* 54 *Ziff*, a beard. 1924 C. J. Dennis *Rose of Spadgers* 137 'E lobbed in on us sudden, ziff an' all. 1934 *Bulletin* (Sydney) 2 May 25/4 All the Druids in that show wore long, white nightgowns and ziffs down to where the tops of their trousers should have been. 1947 J. Morrison *Sailors belong Ships* 97 We all called him The Prophet. He had a long ziff. 1971 *N.Z. Listener* 19 Apr. 56/5 So up he goes and finds he knows one of them, the one with the ziff. 1981 G. Kelly *Always Afternoon* xii. 211 'Better get rid of that ziff,' she said pointing to his embryonic beard.

zig, the first syllable of *zigzag*: (Examples of *zig* sb. and vb. used without *zag*.) Hence **zi·gging** *vbl. sb.* and *ppl. a.*

1969 *Southern Rev.* July 760 After that We drove back, zigging, cautious. 1971 G. Ewart *Gavin Ewart Show* i. 11 An ant zigs quietly over the windowsill. 1977 *Time* 17 Oct. 24/3 The Carter Administration's push for Geneva will surely require much more shoving, not to mention zigging and zagging, in the weeks ahead if it is to succeed. 1978 H. Wouk *War & Remembrance* vii. 72 The transports were on a zig away. 1982 A. Melville-Ross *Trigger* v. 45 What's his course?..Go deep if it zigs towards us. *Ibid.* 53 'They've turned ninety degrees to starboard.'.. 'A zig, sir?' 'Afraid not. They've altered course.' 1983 *Sci. Amer.* July 14/2 Subtler to see is the line just below, whose zigging and zagging is 180 degrees out of phase with the top line. 1985 S. Vanauken *Under the Mercy* x. 246 One word more on change and development,.. the zigs and zags and the farings forward on one's pilgrimage to God.

zigabo, zigaboo, varr. *jigaboo.

‖ **Zigeuner** (tsigoi·nəɹ). fem. **-erin,** pl. **-erinnen.** [Ger., cogn. with Zingano, Zingaro.] A gypsy. **Zigeu:nerbaro·n,** a gypsy baron (in allusion to the operetta *Der Zigeunerbaron* (1885) by Johann Strauss).

[1841 G. Borrow *Zincali* I. i. 2 They are styled in Russia, Zigáni; in Turkey and Persia, Zingarri; and in Germany, Zigeuner.] 1845 Thackeray *Legend of Rhine* x, in *G. Cruikshank's Table-Bk.* I. 198 Here should come the gleemen and jongleurs,..the dark-eyed nut-brown Zigeunerinnen. 1963 *Times* 12 Jan. 9/7 He was a violinist, a travelling musician. His coat had a black fur collar; he looked somewhat like a refined *Zigeunerbaron*. 1964 *Listener* 25 June 1043/3 His [*sc.* Bartók's] passion for the true Hungary and its folk-music as opposed to the *Zigeunerbaron* travesties. 1970 N. Freeling *Kitchen Bk.* xix. 184 There were no musicians, though I recognized Philippa's zigeuner guitarist.

ziggety (zi·geti), *int.*, *adj.*, and *adv.* *slang* (orig. and chiefly *U.S.*). Also **ziggedy, ziggetty, ziggity.** [App. var. of (*hot*) *diggety* s.v. *hot *a.* 12 c.] **1.** *int.* Usu. preceded by *hot* and followed by *dog* or another monosyllable: = *hot diggety* (*dog*) s.v. *hot *a.* 12 c.

1924 *Dialect Notes* V. 265 I'll be damned, hot ziggety damn. 1926 Maines & Grant *Wise-Crack Dict.* 9/2 *Hot ziggetty dog*, expressing unlimited admiration. 1933 R. James *Worth Remembering* iii. 62 Hot ziggity zig! 1942 Berrey & Van den Bark *Amer. Thes. Slang* §277/7 Hot ziggety! hot ziggety damn! *or* darn! hot ziggety dog! 1944 M. Shulman *Feather Merchants* xiii. 78 'Ziggetty!' he said. 1950 A. Buckeridge *Jennings goes to School* viii. 155 Oh, hefty ziggety door knobs! 1961 R. Longrigg *Daughters of Mulberry* viii. 97 'Hot ziggety,' said an old American. 'That's six grand I made.' 1984 *New Yorker* 21 May 46/1 Mr. Deforest entered with his face bright, his hands folded behind him. 'Well, hot ziggetty, a holiday for me. What have we got going here?'

2. *adj.* and *adv.* = Zigzag *a.* and *adv.* Also redupl., as *ziggety-zaggety*.

1935 J. T. Farrell *Judgment Day* ii. 25 He pursued a ziggedy course along the sidewalk. 1956 H. Gold *Man who was not with It* (1965) xi. 86, I had gone backwards to walk sideways, ziggety-zaggety.

zigsaw, zigzaw, obs. varr. *jig-saw *sb.* b.

1912 H. Maxwell *Early Chron. Rel. Scotl.* I. 16 One of those zigzaw puzzles which had a fleeting vogue two or three years ago. 1919 D. Wyllarde *Holiday Husb.* xiii. 167 As neatly as if she had found the right pieces of a zigsaw puzzle.

zigzag, *sb.*, *a.*, *adv.* Add: **A.** *sb.* **5. (Zig(-)Zag.)** A proprietary name for cigarette paper.

1909 *Official Gaz.* (U.S. Patent Office) 14 Dec. 594/1 Braunstein & Cie, Paris... *Zig Zag*... Cigarette-paper. 1927 *Trade Marks Jrnl.* 13 Apr. 675 Zig-zag No. 114... Cigarette papers. Société anonyme des anciens Établissements Braustein Frères.., Paris. 1968 *Current Slang* (Univ. S. Dakota) Fall 52 *Zig-zag*, paper of high quality which is commonly used in rolling marijuana. 1977 C. McFadden *Serial* (1978) xxx. 67/1 She stuffed her..Zig Zags back into her purse.

6. *attrib.*, as **zigzag machine,** a sewing machine with a swing needle that may be used to produce a zigzag stitch and decorative stitches derived from it.

[1950 *Vogue Pattern Book* Apr.–May 81/2 It was Pfaff that developed the famous Zig-Zag Model 130.] 1952 *Consumers' Res. Bull.* Sept. 11/1 All the zig-zag machines but one..were heavy. 1963 *Which?* June 165/2 For plain zig zag machines, the *width* of the stitch limits the range of patterns they can make. 1978 *Detroit Free Press* 5 Mar. D14 (Advt.), Fashionmate zig zag machine featuring our front drop-in bobbin.

B. *adj.* **2*.** *Mil. slang* (chiefly *U.S.*). Drunk.

1918 Hamilton & Corbin *Echoes from over There* (1919) 125 He got a trifle zig-zag. 1919 W. H. Downing *Digger Dialects* 54 *Zig-zag*, drunk. 1923 E. Paul *Impromptu* 149 He groped and floundered..not completely 'zigzag'. 1930 Brophy & Partridge *Songs & Slang Brit. Soldier* 181 *Zig-zag*, drunk. 1961 *Times* 27 Apr. 17/2 What is that to a nation which uses some 400 synonyms for 'drunk'—from 'all geezed up' to 'zig-zag'?

3. zigzag connection *Electr. Engin.*, a form of star connection of three-phase circuits, each branch of which is interconnected and contains portions of two consecutive phases.

1922 P. Kemp *Alternating Current Electr. Engin.* (ed. 2) xiii. 188 This affects the magnetising current and may result in an appreciable increase in iron loss owing to flux distortion, and to minimize this effect zig-zag connections are sometimes adopted. 1947 R. Lee *Electronic Transformers & Circuits* iii. 47 Unbalanced direct current in the half-wave rectifiers requires larger transformers than in the full-wave rectifiers. This is partly overcome in three-phase transformers by the use of zigzag connections.

zigzag, *v.* Add: **1. b.** Of a sewing-machine: to make zigzag stitches.

1950 *Consumer Rep.* May 212/1 The *Necchis* [*sc.* sewing-machines] which zigzagged were inferior to the.. *Necchis* which didn't. 1956 *Sears, Roebuck Catal.* Fall-Winter 1068/1 Whether you want to embroider..zig-zag or sew a straight seam, you'll find the '84' unsurpassed. 1976 *Woman's Weekly* 6 Nov. 68/2 (Advt.), It's a pocket-size sewing machine which you hold in one hand—that sews, bastes, hems, zig-zags, sews on buttons and even zippers.

2. b. To traverse in a zigzag manner.

1930 Bird & Ryan *Recall Public Officers* 42 He literally zigzagged the whole tremendous territory, visiting almost every hamlet. 1978 J. Wainwright *Ripple of Murders* 78 The vans..zig-zagged the near-deserted streets.

zigzagged, *a.* Add: Hence **zi·gzaggedly** *adv.*

1921 W. de la Mare *Mem. Midget* xxxiii. 230 How zigzaggedly you talk. What has poetry to do with Mr Crimble? 1977 R. Katz *Ziggurat* vi. 55 It [*sc.* a ball] rolled zigzaggedly for a while.

zig-zig, U.S. Mil. slang var. *jig-a-jig.

1918 H. V. O'Brien *Diary* 30 Jan. in *Wine, Women & War* (1926) 18 Zig-zig—nothing else but. 1930 in S. Longstreet *Canvas Falcons* (1970) 271 Zig-zig wif me? 1962 W. Robinson *Barbara* (1964) 135 'Allo, baybee! Comment alley vooz—zigzig?

zikkurat, ziggurat. Add: The latter is now the usual spelling. **b.** *transf.* and *fig.*

1959 *Times* 21 Oct. 11/3 The burnished ziggurats of copper saucepans. 1970 *Daily Tel.* (Colour Suppl.) 30 Oct. 26/3 The bags are abandoned, of course, and join the rest of the overkill of trash imagery, now heaped into ziggurats, festooning vegetation, scrawled in livid drifts on the downs. 1979 *Jrnl. R. Soc. Arts* Nov. 761/1 His Dallas Chapel in the form of a spiral ziggurat..borrows quite directly from the ninth-century minaret at Samarra. 1980 *Bee Culture* July 376 The photograph will give an idea of the real engineering performed to make this temple a 'ziggurat' of comb—upward from the base above the regular [honey]combs.

‖ **zikr** (zik'r). Also **zikir.** [ad. Arab *dhikr* remembrance.] A Muslim ritual prayer in which an expression of praise is continually repeated.

1836 E. W. Lane *Acct. Manners & Customs Mod. Egyptians* II. xi. 170 They had not yet begun their performances or *zikrs*, in concert; but one old durweesh, standing between the two rows, was performing a zikr alone; repeating the name of God (Alláh), and bowing his head each time that he uttered the word, alternately to the right and left. 1877 *Encycl. Brit.* VII. 114/1 The Zikr consists mainly in a chant, always becoming louder and more violent, of the first attribute [of God]. 1900 'Odysseus' *Turkey in Europe* v. 193 Dervishes..have also their own characteristic form of worship called Zikr, consisting of the repetition, sometimes continued during several hours, of some religious formula, such as 'There is no God but God', or the ninety and nine names of the Deity. 1923 *Blackw. Mag.* Aug. 251/1 Dervishes shouting themselves into ecstasy at their zikr. 1954 M. Murray in G. B. Gardner *Witchcraft Today* 16 The solemn *zikr* of the Egyptian peasant. 1976 *Bangladesh Times* (Dacca) 12 Jan. 2/5 Monday, January 26: after Esha prayer Khatme Holy Quran, Khatme Gousia and Khatme Khajegan at 2 a.m. Zikir, Zearat and Munazat.

Zilavka (zila·fkā). The name of a white wine of Yugoslavia.

1926 P. M. Shand *Bk. Wine* x. 255 *Bosnia-Herzegovina* ..The principal wines are the red growth of Blatina..and the white Zilavka, a potent, greenish-yellow wine with a Muscatelle bouquet. 1954 M. Krippner *Yugoslavia Invites* 199 Zilavka, a yellow wine, and Blatina, a heavy claret, both come from Herzegovina. 1965 *Sun* 23 Jan. 6/6, I tried Zilavka—a wine..from the Macedonian vineyards near the Greek frontier. 1978 *Chicago* June 220/2 We recommend Zilavka from Mostar vineyard, white but with enough richness and strength for meats.

zilch (ziltʃ), *sb.* (and *a.*) *slang* (orig. and chiefly *U.S.*). [Origin uncertain.] Nothing, nil; also as *adj.*, no; non-existent.

1931 *Ballyhoo* I. i. 1 (*heading*) President Henry P. Zilch. Chairman of the Board Charles D. Zilch. Treasurer Otto Zilch. *Ibid.* ii. 10 (*caption*) 'Mr. Zilch, you don't often stay in so long.' 'No I don't often lose my bathing trunks.' 1940 Berrey & Van den Bark *Amer. Thes. Slang* §184/1 *Dinglegoofer, Mr. Zilch*, indefinite nicknames.] 1966 *Current Slang* (Univ. S. Dakota) Winter 8 *Zilch*, adj. Nothing, zero... What a day—*zilch* from everybody. 1967 P. Welles *Babyhip* ii. 25 Half-starved, no doubt. The old whore probably fed him zilch. 1973 *Daily Colonist* (Victoria, B.C.) 14 Jan. 13/7, I feel that since I was elected to the board of directors last year I have accomplished zilch. 1976 *New Musical Express* 17 Apr. 6/6 My knowledge of classical music is zilch 1976 *Billings* (Montana) *Gaz.* 27 June 2-E/2 In the light of data developed from the city's own 1975 studies, the 17th Street project rates zilch in priorities. 1977 *Tel.* (Brisbane) 3 Feb. 14/3 Georgeous faces but zilch talent. 1977 *Playgirl* May 12/2 Our sex life is practically zilch, and he almost never pays any attention to me. 1984 *Daily Tel.* 8 Mar. 36/5 The power of the legislature over the executive being slightly better than zilch, any MP..who bounces the Home Office deserves a small roll of drums. 1984 *Sounds* 1 Dec. 38/3 Three further 45s ensued in 1979 and '80, plus an album which didn't sell. After that, zilch.

zillah. Add: *zillah parishad*: see *parishad.

zillion (zi·lyən). *slang* (chiefly *U.S.*). [f. Z + M)illion.] A very large but indefinite number.

1944 D. Runyon *Runyon à la Carte* (1946) 165, I love him a zillion dollars' worth. 1947 *Esquire* May 40/2 Faithful to their zillions of fans. 1976 *National Observer* (U.S.) 10 Jan. 10/4 A zillion or so years ago, while I was a student nurse. 1976 J. Crosby *Snake* (1977) xvii. 93 She was going to break the story to her zillion readers. 1983 *Sunday Tel.* 9 Oct. 20/2 Broken Hill Proprietary..is Australia's biggest company and a zillion times bigger than his own. 1984 *Guardian* 29 Oct. 9/2 The whiff of news managers at work, rather than an urge to hear about British Telecom's zillion-pound share sale from the horse's mouth, took me to BT's big press conference on Friday.

Hence **zillionai·re,** a very rich person; **zi·llionth** *a.*, following very many others; 'umpteenth'; *sb.*, a tiny fraction *of* anything.

1946 D. Runyon *Runyon à la Carte* ix. 143 She plays the frost for all who are not well established as practically zillionaires. 1959 I. Fleming *Goldfinger* iv. 50 He's a zillionaire himself. . He's crawling with money. 1969 *Daily Tel.* (Colour Suppl.) 15 Aug. 21/3 A sprinkling of.. Texas zillionaires, film stars, socialites and international playboys. 1972 *Good Housekeeping* Apr. 69/1 'We have strict controls on television,' said Mrs. Hamon [an American]. 'We don't hear a zillionth of what you hear.' 1975 *Listener* 24 Apr. 554/3 Anchorman Peter Snow..said, for the zillionth time, 'Well, there it is.' 1983 'J. Gash' *Sleepers of Erin* xviii. 141 The Heindricks' scam was so big that even zillionaires were keen on its successful execution. 1984 *Listener* 6 Dec. 26/3 For the zillionth time, the hardback edition only is credited in reviews.

Zimba (zi·mbā). (A member of) an African people that was active in the vicinity of the Zambezi in the sixteenth century.

1901 G. M. Theal tr. J. dos Santos in *Rec. South-Eastern Africa* VII. xvii. 291 Facing Tete on the other side of the river..there are two tribes of Kaffirs who eat human flesh, one called the Mumbos and the other the Zimbas or Muzimbas. 1913 C. A. Stigand *Land of Zinj* i. 17 The Zimba, a powerful tribe of barbarians who lived N.E. of Tete on the Zambezi, are now first heard of in these parts. In 1588 they invaded Kilwa, and the next year passed up the coast and invaded Mombasa. 1968 R. Oliver in J. Biggs-Davison *Africa—Hope Deferred* (1972) iii. 26 The Zimbas who swept up the east coast of Africa in the late sixteenth century. 1972 *Stand. Encycl. S. Afr.* VI. 341/1 Mirale occupied Mombasa in 1589, but the Portuguese received unexpected help from the interior when the mysterious cannibal Zimba tribe arrived from the south, destroying Mombasa and most of its occupants. 1974 *Encycl. Brit. Macropædia* XVII. 277/2 The Zimba.. were followers of a Manganja chief living to the south of the Shire River.

Zimbabwe (zimbā·bwe). Also **zimbabwe.** [The name in Bantu of the first such ruin to be discovered, in what is now the State of Zimbabwe.] One of the numerous ruined stone-walled settlements scattered across Zimbabwe and neighbouring countries and dating from medieval times.

1902 Hall & Neal *Anc. Ruins Rhodesia* iii. 36 They introduced fresh features in building, as shown..by new Zimbabwes, which they erected. 1929 J. Buchan *Courts of Morning* ii. 288 The subsidiary towers..recalled in their shape pictures she had seen of the Rhodesian Zimbabwes. 1931 G. Caton-Thompson (*title*) The Zimbabwe culture. 1963 R. Summers in E. Bacon *Vanished Civilisations* iii. 36/2 Other 'Zimbabwes'—over 200 of them..—lie scattered all over Southern Rhodesia. 1976 *Times* 17 July 14/6 Another, early, stone-walled *zimbabwe* of the type best known from the Great Zimbabwe ruins in Rhodesia has been discovered..in Mozambique. 1980 *Nature* 1 May 5/3 In 1975..after independence in Mozambique one of the Zimbabwes—Manyikeni, 50 km west of Vilanculos was excavated.

Zimbabwean (zimbā·bweăn, -bæb-), *sb.* and *a.* [f. *Zimbabwe* (see below) + -an.] **A.** *sb.* **a.** Before 1980, when Rhodesia became Zimbabwe, (*a*) an African nationalist name for a black Rhodesian; (*b*) an inhabitant of a future state of Zimbabwe. **b.** Since 1980, a native or inhabitant of Zimbabwe. **B.** *adj.* Of or pertaining to Zimbabweans or Zimbabwe.

1967 *Times* 30 Dec. 4 Let us work untiringly for the total mobilization of the four million Zimbabweans for Chimurenga (war of liberation). 1973 *Black Panther* 28 Apr. 9/1 (*caption*) Brother Kumbirai Kangai, freedom fighter in the Zimbabwean people's struggle. 1975 *Times* 11 Feb. 12/2 Whatever the future holds in store for an independent Zimbabwe, the white Zimbabwean will continue to play his part alongside his black counterpart. 1976 *Daily Tel.* 19 Aug. 14 Zimbabwean Nationalists who patiently tried a negotiated settlement cannot now be expected to lay down arms. 1979 *African Affairs* LXXVIII. 253 (*heading*) Zimbabwean economic and social historiography since 1970. 1984 *News* (Mexico City) 12 Mar. 7/2 Nkomo..suggested Zimbabweans were no freer than they were under the white-minority colonial government.

zimbalom, -n, var. *cimbalom, cimbelom.

1910 F. W. Galpin *Old Eng. Instrum. Music* iv. 66 The large Dulcimer, used in Hungarian Bands at the present time, is known by the Magyar musicians as the *zimbalom*. 1925 *Glasgow Herald* 17 Oct. 6/2 The zimbalon, which is an improved type of dulcimer, has been honoured with a d.-s. record made by Nitza Godolban. 1975 *New Yorker* 19 May 90/1, I spent two weeks in Shiraz and Persepolis, listening for hours every day to percussion from all over the world:..to varieties of dulcimer (zither, santir, zimbalon).

‖ **zimbel** (tsi·mbəl). *Mus.* [Ger., ad. L. *cymbalum* (see Cymbal).] = Cymbal 3.

1910 F. W. Galpin *Old Eng. Instrum. Music* xiv. 264 We also find an organ-stop called the *Zimbel*..and this.. was a compound stop or mixture intended to represent the sound of bells. 1976 *Gramophone* Oct. 628/3 An exuberant finale contrasts the zimbel with the cornet. 1980 *New Grove Dict. Mus.* XIII. 791/1 'Mixture' was normally used to denote the Principal-scaled chorus Mixture as distinct from the high Zimbeln or the solo Cornets.

zimes, var. *tzimmes. **Zimmenthal,** var. *Simmental.

Zimmer (zi·məɹ). Also **zimmer.** [Maker's name.] A proprietary name for orthopædic appliances, used esp. *attrib.* to designate a kind of walking frame.

1951 *Trade Marks Jrnl.* 11 Apr. 358/1 *Zimmer*... Orthopaedic appliances and instruments. Zimmer Orthopaedic Limited, .London, W.1..; manufacturers. 1957 *Official Gaz.* (U.S. Patent Office) 16 Apr. TM 116/2 *Zimmer*. For orthopedic appliances and fracture equipment. First use in May of 1927. 1974 Passmore & Robson *Compan. Med. Stud.* III. ii. xxxiv. 103/1 For patients with unsteadiness and ataxia, a simple walking frame of 'Zimmer' pattern is valuable. 1981 *Church Times* 10 Apr. 14/4, I had felt too ill to try and use my zimmer-frame and calipers. 1982 E. Dewhurst *Whoever I Am* iii. 44 Incapable of walking without a Zimmer frame.

zinc, *sb.* Add: **1. c.** Galvanized iron. Chiefly *S. Afr.* and *W. Indies.* Cf. *zinc roof* in sense 2 c below.

1873 F. Boyle *To Cape for Diamonds* xix. 252 These hoppers are made of perforated zinc, or wire, the latter being preferable. 1953 R. Mais *Hills were Joyful Together* i. vi. 57 Surjue went through a zinc fence—through a hinged zinc sheet in a zinc fence that looked like a solid stretch of iron sheeting down one side of a lane. 1973 *Eastern Province Herald* (Port Elizabeth) 23 Jan., The brazier, zink bath and cooking utensils. were obtained from Coloureds in the area. 1977 *Daily Express* 29 Jan. 7/3 All six members of the cast take turns to strip off and bathe in a zinc bath.

d. The zinc-covered bar of a café or public house; by metonymy, a café. (A gallicism.)

1914 *19th Cent.* Feb. 286 Elections are made or marred *chez le marchand de vin*, or, as the Parisians familiarly call him, le *mastroquet*... The 'organised democracy marching towards the good of progress and liberty, etc.'..is in reality nothing else than the disorganised and demoralised *plebs* marching towards the *zinc* of the *mastroquet*. *a* 1936 Kipling *Something of Myself* (1937) iv. 81 My barmaid..had watched it [*sc.* evil] across the zinc she was always swabbing off. 1948 W. Fortescue *Beauty for Ashes* xx. 148 We even supplied a little bar with a *zinc* and a motherly Frenchwoman to look after it. 1965 V. Canning *Whip Hand* xii. 138, I..tottered to the nearest zinc, and called for a triple cognac. 1979 A. M. Stein *Rolling Heads* vi. 103 The last of the cafés shut down—it was a laborers' zinc down a back street.

2. a. *zinc-lining.* **b.** *zinc-coated, -covered, -plated, -topped.* **c. zinc-air** *a.*, applied to a type of primary cell employing a zinc anode, a porous carbon cathode able to utilize atmospheric oxygen, and an alkaline electrolyte; **zinc chloride,** a white, crystalline, deliquescent solid, $ZnCl_2$, used as a preservative, a flux, and in Leclanché cells; **zinc chromate,** a toxic, yellow, water-insoluble powder, $ZnCrO_4$, used as a pigment; **zinc chrome** = *zinc yellow* below; **zinc green,** a mixture of zinc and cobalt oxides used as a pigment; **zinc roof** *S. Afr.* and *W. Indies*, a corrugated roof of galvanized iron; **zinc sulphide,** (*U.S.*) **sulfide,** a yellow, water-insoluble powder, ZnS, used as a pigment and as a phosphor; **zinc white** (earlier example); **zinc yellow,** a greenish-yellow pigment consisting principally of zinc chromate.

1970 C. L. Mantell *Batteries & Energy Systems* vi. 63 These portable zinc-air batteries are being used in manpack transceivers, night vision devices, and space satellite communications. 1978 *Jrnl. Electrochem. Soc.* CXXV. 333c/2 The Gould zinc-air button cell is..used in hearing aids and other devices which require a moderately high current. 1851 H. Watts tr. *Gmelin's Hand-bk. Chem.* V. xxviii. 32 ZnCl, 9ZnO..[is prepared] by precipitating aqueous zinc-chloride with an insufficient quantity of ammonia and filtering immediately. 1911 *Encycl. Brit.* XXVIII. 984/1 Zinc chloride, $ZnCl_2$, is produced by heating the metal in dry chlorine gas, when it distils over as a white translucent mass. 1981 Brady & Holum *Fund. Chem.* xx. 700 Zinc chloride, which is exceptionally soluble in water.., has a range of uses that extend from embalming, to fireproofing lumber, to the refining of petroleum. 1851 H. Watts tr. *Gmelin's Hand-bk. Chem.* V. xxviii. 48 The crystals..really consisted of ordinary zinc-sulphate containing small quantities of zinc-chromate intimately bound up with them. 1974 *Encycl. Brit. Macropædia* IV. 571/2 Zinc yellow, a basic zinc chromate, is used as a corrosion-inhibiting primer on aircraft parts fabricated from aluminum or magnesium. 1892 G. H. Hurst *Painters' Colours, Oils, & Varnishes* iv. 132 Zinc chrome is a yellow pigment of good colour and body. 1963 *Times* 22 Apr. 6/5 Zinc chromes, with zinc chromate as the major constituent, form an important class of pigments. 1930 *Jrnl. Iron & Steel Inst.* CXXI. 749 An investigation of the corrosion fatigue of zinc-coated steel specimens. 1981 M. C. Smith *Gorky Park* iii. iv. 361 The sables..climbed zinc-coated mesh walls. 1890 A. Conan Doyle *Firm of Girdlestone* xxx. 238 A very seedy-looking individual..was leaning with his elbows upon the zinc-covered counter. 1847 *Brit. Pat. 11,616* 5 Processes for manufacturing on a large scale zinc yellow..and zinc green. 1973 E. Lalor in T. C. Patton *Pigment Handbk.* I. 850/1 Much of the zinc yellow manufactured was mixed with Prussian blue to make various shades of a permanent zinc green. 1912 C. N. Moody *Saints of Formosa* vi. 132 Boxes had to be made or furnished with zinc-lining. 1859 G. A. Sala *Twice round Clock* 117 There is the rotunda of the Bank of England, with its many-slamming, zinc-plated doors. 1966 *McGraw-Hill Encycl. Sci. & Technol.* IV. 531/1 Zinc-plated coatings are usually purer than hot-dipped coatings. 1883 'R. Iron' *Story Afr. Farm* I. i. i. 11 The zinc roofs of the out-buildings, the stone walls of the 'kraals', all reflected the fierce sunlight. 1946 U. Krige *Way Out* vi. 82 There was a mass of dry maize leaves under a zinc roof jutting out from the wall. 1851 H. Watts tr. *Gmelin's Hand-bk. Chem.* V. xxviii. 19 Nöggerath and Bischof..found in an old mine a quantity of zinc-sulphide mixed with washings of the ore. 1937 *Discovery* Feb. 44/2 The [television] screen is coated with a powder consisting of zinc sulphide and other chemicals. 1974 *Encycl. Brit. Micropædia* X. 883/3 Zinc sulfide is insoluble in water but dissolves in mineral acids. 1938 E. Ambler *Cause for Alarm* (1940) xiii. 203 A zinc-topped bar and four marble-topped tables. 1847 *Brit. Pat. 11,616* 9 The dryer is added to the color prepared with the zinc white. 1847 Zinc yellow [see *zinc green* above]. 1901 A. C. Wright tr. *Bersch's Manuf. Mineral & Lake Pigments* xiv. 152 Zinc yellow may be prepared by the immediate precipitation of a solution of zinc sulphate by a solution of potassium chromate. 1974 *Encycl. Brit. Macropædia* XIII. 889/1 Zinc yellow has a greenish cast because of a high content of chromic oxide.

d. As *adj.*: Having a basically greyish colour.

1922 'K. Mansfield' *Let.* 11 Aug. (1977) 271 A zinc greengage or two. 1960 *New Yorker* 1 Oct. 44/2 Faultless eyes gone blank beneath the immense Zinc-and-gunmetal northern sky.

zincian (zi·ŋkiăn), *a.* *Min.* [f. Zinc *sb.* + *-ian 2.] Of a mineral: having a (small) proportion of a constituent element replaced by zinc.

1930 W. T. Schaller in *Amer. Mineralogist* XV. 571 Zinc—zincian. 1968 I. Kostov *Mineralogy* ix. 498 Pisanite is a cuprian variety..; sommairite zincian melanterite. 1977 *Amer. Mineralogist* LXII. 463/2 The proportion of zinc is probably insufficient to justify a new name... It might, therefore, be called a zincian takovite.

zinckenite. Add: Now the usu. spelling of Zinkenite.

‖ **zindan** (zindā·n). [Pers. *zindān*, Turk. *zindan*.] A prison in Persia or neighbouring parts.

1889 G. N. Curzon *Russia in Central Asia* vii. 217 In another part of the citadel was the Zindan, or prison. 1924 *Glasgow Herald* 29 May 9/7 In Persia..I saw for the first time a Central Asian underground prison called Zindan... Prisoners..cannot possibly escape out of a deep pit with only a circular opening at the top. 1959 *Listener* 21 May 883/1 The Zindan or prison where Stoddart and Conolly spent so many miserable months [in Bokhara].

zine (zīn). *U.S. colloq.* Shortened form of *fanzine.

1965 *New Yorker* 21 Aug. 24 The fanzines, or fan magazines, which are also known as amazines, or amateur magazines, do a great deal of research on the Golden Age, and some of the faneds, or fan-magazine editors, do a remarkably good job. Some zines specialize—like *ERBdom*, which caters to the fans of Edgar Rice Burroughs—but most of them concentrate on the fantasy adventure or superhero comics, on which most fanac, or fan activity, centers. 1975 *Piece of Action* (Houston) Jan. 6 The above zine should not be confused with another proposed zine called *The Atavachron*. 1982 *Amer. Speech* LVII. 20 A publisher of a fanzine would send copies of his zine to a central location for collation with other zines and subsequent distribution. 1985 *Times* 25 Jan. 12/2 A zine is what its addicts call a postal games magazine, of which there are about 50 in the country.

zineb (zi·neb). [f. *zinc ethylene bisdithiocarbamate*, the systematic name: see Ethylene, Bis-².] A white powder used as a fungicide on vegetables and fruit; Zn(·S·CS·NH·CH_2·)$_2$.

1950 *Phytopathology* XL. 118 The Subcommittee on Fungicide Nomenclature of The American Phytopathological Society, cooperating with the Interdepartmental Committee on Pest Control, has selected common names for five commercially-available fungicidal chemicals... *Zineb* for the fungicidal chemical, zinc ethylene bisdithiocarbamate... *Ziram* for the fungicidal chemical, zinc dimethyl dithiocarbamate. 1966 *McGraw-Hill Encycl. Sci. & Technol.* V. 563/2 Organic fungicides have become increasingly important since 1934... Examples are..nabam, zineb, and maneb, the sodium, zinc, and manganese salts, respectively, of ethylenebis(dithiocarbamic acid). 1981 [see *ziram]. 1981 *Bull. Environmental Contamination & Toxicol.* XXVII. 418 Zineb is an important fungicide widely used in agriculture.

Zinfandel. Add: **b.** The grape from which this wine is made. In full, *Zinfandel grape*.

1880 *Californian* II. 219/1 The favorite wine grapes are the Zinfandel, Riesling and Chasselas. 1977 H. Fast *Immigrants* iv. 254 'What kind of grapes do you grow for the market?' 'Zinfandels mostly. We have some Thompons—' 'Ah, well the color of zinfandels is excellent, but we want a sweeter grape.' 1980 *Times* 23 Sept. 10/1 We were..in California. . My companion..grows Chardonnay, Cabernet and Zinfandel grapes.

zing (ziŋ), *sb. colloq.* (orig. *U.S.*). [Echoic.] **1.** A sharp, high-pitched ringing sound; a twang.

1911 D. Runyon *Tents of Trouble* 22, I felt him fall and I sensed the 'zing' of a boob-face Arab's knife. 1922 C. Sandburg *Slabs of Sunburnt West* 63 Then the axmen came and the chips flew to the zing of steel and handle. 1930 E. Ferber *Cimarron* xx. 331 He seized the typewriter by its steel bar and plumped it to the floor with a force that wrung a protesting whine and zing from its startled insides. 1961 E. Williams *George* xxv. 403 There was the confident zing of tennis-rackets in the Parks. 1970 *Washington Post* 30 Sept. B-1/1 The butterfly chair, that zing-zong-zang of tubular metal.

2. Energy, vigour, liveliness; zest; a quality that induces alertness or vitality.

1918 *Independent* 14 Dec. 369/1 They were the picked athletes of the whole English Army and were doing their calisthenics with a precision and spirit I have never seen equaled anywhere. The 'pep', 'zing' and 'vim' were thrilling. 1931 [see *belt *v.* 4 b (*b*)]. 1941 G. Kersh *They die with their Boots Clean* ii. 83 I'll soon get that paleness off your faces and put some zing into those limbs. 1955 'S. Ransome' *Deadly Bedfellows* i. 7 Lake Haven's air.. was full of zing and bounce. 1964 *Punch* 27 May 796/3 The show had zing. 1979 *Brit. Jrnl. Photogr.* 21 Sept. 917/1 In the old days a photographer used to go to work with a holdall and could be on the other side of the world. That doesn't happen very often nowadays in Fleet Street. It's lost its zing. 1983 *Daily Tel.* 21 Dec. 9/2 Freshly-squeezed lemon or lime has a unique impact, adding a zing to almost any combination in flavours. 1985 *Spectator* 28 Sept. 9 While death has not lost its sting, sex has undoubtedly lost its zing.

zing (ziŋ), *int.* Chiefly *U.S.* [Echoic. Cf. *zingo *int.*] Representing the sudden advent of a new situation or emotion.

1919 Wodehouse *Damsel in Distress* vi. 75 The generous blood of the Belphers boiled over, and then—*zing*. They jerked him off to Vine Street. 1948 *Sun* (Baltimore) 3 Dec. 14/2 A method of freezing concentrated orange juice was developed, and zing! the first thing they knew they had a new and depression-beating industry on their hands. 1955 W. Gaddis *Recognitions* i. iv. 161 String a good piece of piano wire across the road..and take a couple of shots at them. They go after you..and zing zing zing there go their heads just like that. 1977 M. French *Women's Room* (1978) iv. 248 So one day you meet this guy, right? And, zing! He is gorgeous!

zing (ziŋ), *v. colloq.* (orig. *U.S.*). [f. the sb. or int.] **1.** *intr.* **a.** To make a sharp, high-pitched ringing or whining sound; to travel rapidly producing such a sound.

1920 S. Lewis *Main Street* ii. 18 Go zinging along on a fast ice-boat. 1949 N. Marsh *Swing, Brother, Swing* v. 88 Lord Pastern banged, and rattled, and zinged... 'Oh,' she thought, 'how vulnerable he is among his tympani!' 1960 I. Cross *Backward Sex* i. 12 It would be better if this old chap did not..hear the bullet zing past. 1962 S. Plath *Johnny Panic & Bible of Dreams* (1977) iii. 247 The bees, now Charlie had lifted the top off the hive, were zinging out and dancing round. 1963 T. Pynchon *V.* xi. 345 Had his coincidence, the accident to shatter the surface of this stagnant pool and send all the mosquitoes of hope zinging away to the exterior night; had it happened? 1977 G. Durrell *Golden Bats & Pink Pigeons* ii. 38 A group of zosterops,..zinging and twittering to each other.

b. *fig.* To move energetically and with ease; to abound *with* energy.

1961 *John o' London's* 29 June 724/2 She zings along the tight-rope which passes for plot. 1969 N. Freeling *Tsing-Boum* xx. 145 They had been busy enough already in Holland! In Marseilles they had been zinging with ambitious energy. 1973 *Playboy* May 44 As pure escapist entertainment..the movie zings right along. 1973 *Daily Tel.* 7 Nov. 15/2 When..she denounces him to her husband for the imaginary seduction..of a 16-year-old girl pupil, matters zing into focus.

2. *trans.* With *up*. To enliven, invigorate. *U.S.*

1970 *New Yorker* 14 Nov. 154 Charles Revson is the philosopher-king of the cosmetic world... He claims to know by instinct how to 'zing up' a face. 1978 *Chicago* June 234/3 Delicious, spicy pickled-pepper relish.. which you can use to zing up the otherwise mild dishes.

3. To abuse; to criticize. *U.S.*

1974 *Evening Herald* (Rock Hill, S. Carolina) 18 Apr. 7/4 In the eighth, Mitterwald doubled and..the rest of the Cubs zinged him for not hitting a fourth homer. 1975 *New Yorker* 21 Apr. 61/3 Brodie told me..of another investigator who, in his haste, 'zinged a dead man' and delivered an investigative report concerning his insurability.

4. To deliver (a witticism, question, etc.) with speed and force. *U.S.*

1975 W. Safire *Before Fall* iii. vi. 179 [Nixon] read what they wrote and surprised them later by remembering it and occasionally zinging it back at them. *Ibid.* vii. i. 474, I zinged out a couple of one-liners. 1977 *Time* 10 Jan. 26/2 Her usual practice of zinging brash, hostile questions at world leaders.

Hence **zi·nging** *ppl. a.*

1954 G. Durrell *Three Singles to Adventure* vi. 144 A group of sandflies discovered us with zinging cries of joy. 1963 *Times* 16 Jan. 6/5 Mr. Richard Daley, the last of the big city political bosses, has promised him a rough ride in a 'swinging, zinging, campaign, the roughest, toughest ever seen in Chicago'. 1970 'D. Halliday' *Dolly & Cookie Bird* xi. 170 There was a sharp pop, a clang, and the zinging noise of a ricocheting bullet. 1972 *Daily Tel.* (Colour Suppl.) 4 Feb. 8/3 She beat me like a rug, pounded me like dough, using karate chops, zinging finger stabs, incredible flicks. 1978 G. Durrell *Garden of Gods* iii. 59 The air was..full of the zinging cries of cicadas.

zingana. [See Zingano.] **1.** Fem. of Zingano.

2. = *zebrano.

1934 A. L. Howard *Man. Timbers of World* (ed. 2) 582 *Zebrano* or *Zingana*. ? *Cynometra* aff. *C. Lujai* Willd... West Coast of Africa... Known as 'zebra' wood in the United Kingdom and America. 1947 J. C. Rich *Materials & Methods of Sculpture* x. 297 Zebrawood, also known as Zingana and Zebrano, is a tropical wood imported from Africa. 1957 N. Clifford *Timber Identification* iv. 134 (*table*) Common name..Zebrano... Alternative names.. Zingana; Zebrawood... Botanical name..*Brachystegia fleuryana* Chev... Habitat: West Africa. 1973 *Observer* (Colour Suppl.) 8 Apr. 13/1 (Advt.), The grapefruit knife's got a stainless steel blade, zingana wood handle. 1974 F. N. Howes *Dict. Useful & Everyday Plants* 286 Zingana, *Microberlinia brazzavillensis*, W. Afr., a commercial timber.

Zingaro. Add: **2.** pl. *I* (or *The*) *Zingari*: the name of an amateur cricket club founded in 1845.

1846 W. Denison *Cricketer's Compan.* 1845 p. xiii, Everybody knows that the 'Zingari' are a tribe of wanderers... Just such a race of individuals is the Club which bears the name and title of 'I Zingari'. As a Club, they have neither habitation nor home. 1867 'Ouida' *Under Two Flags* I. v. 106 The Household [Cavalry] played the Zingari. 1905 H. A. Vachell *Hill* x. 222 After the Zingari Match, Desmond got his Flannels. 1922 Joyce *Ulysses* 731 A new raincoat on him with the muffler in the Zingari colours. 1948 E. Waugh *Loved One* i. 5 Sir Ambrose wore dark grey flannels, an Eton Rambler tie, an I Zingari ribbon in his boater hat. 1978 *Times* 17 June 13/6 The Household Brigade might turn out in jack-boots, gauntlets, [etc.]..to contend with the Zingari.

zinger (zi·ŋəɹ). *U.S. slang.* [f. *zing *sb.* + -er¹.] **1.** Something outstandingly good of its kind.

1955 M. Allingham *Beckoning Lady* vi. 99, I don't know why it was such a zinger, unless it was that it was very big and very cheap. 1968 *Times* 30 May 10, I think every actress needs one zinger of a part early in her career. 1973 *N.Y. Times Bk. Rev.* 4 Nov. 79 (*caption*) A zinger of a novel. 1976 *New Yorker* 9 Feb. 84/2 It's a zinger of a scene: an educated, socially conscious woman dating a lumpen lost soul. 1980 R. Adams *Girl in Swing* v. 63 My private collection was becoming what an American friend..described as a 'zinger'.

2. a. A wisecrack; a punch line.

1970 *Time* 12 Oct. 7 Ann-Margret is giving him a hard time on the home front, too, tossing out little zingers about his advancing age like 'Flab is reality'. 1970 *Life* 30 Oct. 40 In casual chatter the zingers [of Dick Cavett] are just as fast and frequent as they are on the show. 1975 *Homemaker's Mag.* Oct. 28/2 The Vancouver Status of Women group is planning to put out a booklet of useful zingers... Sometimes you just need one sharp line to show you mean business. 1979 *Courier-Mail* (Brisbane) 31 Mar. 76 (*caption*) One more zinger about my height, and I shall be compelled to thrash you mercilessly about the feet and ankles. 1980 *Maledicta* III. 254 Sorry. After having exhibited immense self-control not to quip,.. I just had to shoot off this little zinger.

b. A surprise question; an unexpected turn of events, e.g. in a plot.

1973 R. Thomas *If you can't be Good* (1974) vii. 51, I would drone along..asking tired questions... Then I would throw in the zinger and watch what happened. 1976 *Publishers Weekly* 2 Feb. 91/1 There's a zinger toward the end, in which the nominal hit man gets hit, but it doesn't really compensate for the tedium the reader's gone through. 1983 *Fortune* 11 July 134/2 The supervisor should also encourage discussion..of what may seem like personal issues... In the process, though, be wary of 'zingers', heart-to-heart matters brought up at an awkward time, often near the end of a meeting.

Zingg (ziŋ, tsiŋ). *Petrol.* The name of Theodor *Zingg* (b. 1905), Swiss meteorologist and engineer, used *attrib.* with reference to his system of classification of pebble shapes, in which two ratios formed from three mutually perpendicular diameters are used to assign a pebble to one or other of certain basic shape classes.

1941 *Jrnl. Sedimentary Petrol.* XI. 67/1 (*heading*) Relation between sphericity and Zingg shape classes. 1949 F. J. Pettijohn *Sedimentary Rocks* ii. 49 (*caption*) Chart showing relationship between sphericity and Zingg shape indices. 1979 *Nature* 8 Feb. 496/3 Very simple probabilistic methods were used to estimate the formation of the four basic Zingg shapes (disks, spheres, blades and rods).

zingiber (zi·ndʒibəɹ). [See Ginger.] = Ginger *sb.* 1, 2.

1902 L. H. Bailey *Cycl. Amer. Hort.* IV. 2010/1 Zingibers are occasionally cultivated as stove decorative plants. 1970 *Guardian* 2 June 12/5 Bottles..of everything from Advocaat to Zingiber wine. 1971 *Country Life* 20 May 1252/2 But the particular forte of most [herbs] is as permanent foliage foils for other, brighter things as well as in their own right, from angelica to zingiber.

zingo (zi·ŋgo), *int.* Chiefly *U.S.* [Echoic. Cf. *zing *int.*] = *zing *int.*

1914 *Sat. Even. Post* 17 Jan. 7/1 Just when he was bursting with happiness because he was going to be a real big leaguer and one of us—zingo! he was back where he started. 1941 B. Schulberg *What makes Sammy Run?* ii. 32 It didn't take nearly this long to think. It went zingo, just a look. 1968 *Sunday Times* 25 Aug. 29/5 The stock is 24 asked and I reach for it and I get 200 shares and it moves up 28, zingo. 1970 *New Yorker* 17 Oct. 40/3 Zingo, another pair of jokers..bust in.

zingy (zi·ŋi), *a.* [f. *zing *sb.* + -y¹.] Energetic, exciting, lively. Also *ellipt.*

1948 [see *mambo 1]. 1962 *Guardian* 7 Feb 9/1 A zingy collection..that every with-it girl is going to adore. 1966 *Daily Tel.* 17 Aug. 11/5 With her own fashion taste, a delicate blend of the elegant and the zingy, a model like Fiona Campbell-Walter looks set for a great comeback career. 1968 *Globe & Mail* (Toronto) 3 Feb. 11/1 With some zingy comedy, they're an act that the regular club-haunters around town wait for. 1975 *Observer* 9 Nov. 22/6 A zingy Moselle, from a world famous estate. 1976 *Publishers Weekly* 16 Aug. 116/1 Some zingy, down-to-earth comments... She doesn't mince words.

Zinjanthropus (zindʒæ·nþrǒpǒs). *Palæont.* [mod.L. (L. S. B. Leakey 1959, in *Nature* 15 Aug. 491/2), f. *Zinj*, ancient name for East Africa + Gr. ἄνθρωπος man.] = *Nutcracker Man* s.v. *nut-cracker 5.

1959 *Nature* 15 Aug. 493/2 *Zinjanthropus* comes from Olduvai Gorge. 1961, 1962 [see *Nutcracker Man* s.v. *nut-cracker 5]. 1973 B. J. Williams *Evolution & Human Origins* ix. 143/1 The species *Australopithecus robustus* includes '*Zinjanthropus*' from East Africa. 1977 A. Hallam *Planet Earth* 286 From the early beds.. are recognized two kinds of hominids. One is *A. boisei* (sometimes called Zinjanthropus).

zinnober (zi·nobəɹ). Also **zinnobar.** [a. G. *zinnober* Cinnabar, vermilion.] *zinnober green*: = *chrome green* (*b*) s.v. Chrome 3.

1895 *Montgomery Ward Catal.* Spring & Summer 252/3 Zinnobar Green, Light. Zinnobar Green, medium. Zinnobar Green, deep. 1897 *Sears, Roebuck Catal.* 360/3 Colors for Artists... Zinnober Green, light. 1942 Gettens & Stout *Painting Materials* 178 Zinnober Green is a term ordinarily synonymous with chrome green..which is a processed mixture of chrome yellow and Prussian blue. More specifically, it is given to mixtures that are olive in hue. 1973 F. Taubes *Painter's Dict. Materials & Methods* 253 Zinnober green is a mixture of Prussian blue and cadmium yellow.

Zinoviev (zinǫ·vief). Also **Zinovieff.** [The assumed name of Y. A. Radomyslsky (1883–1936).] *Zinoviev letter*: a letter published in the press in 1924 as having been sent by Zinoviev, a Russian statesman, to British Communists and urging them to commit subversive acts; it was later discovered to be a forgery.

1924 *Times* 28 Oct. 8/1 (*heading*) The Zinovieff letter. *Ibid.*, If the Zinovieff letter is a forgery, it shows the amount of scoundrelliness with which we are surrounded. 1925 (*title*) The 'Zinoviev' letter: report of investigation by British delegation to Russia. 1957 *Encycl. Brit.* XIII. 554/2 In the general election of 1924 the Labour party met with a serious reverse, largely because of the confusion attending the publication of the 'Zinoviev' letter. 1958 C. Cockburn *Crossing Line* iv. 64 The timely forgery by the Intelligence Service of the 'Zinoviev letter'. 1972 'M. Sinclair' *Norslag* xi. 101 The stories in the 'twenties about the Zinovieff letter. 1979 *Guardian* 28 Apr. 32/8 Mr Benn told a Bristol audience that the Tories..were dragging out the old Zinoviev Letter technique.

Hence **Zino·vievite,** in Russia, a supporter of Zinoviev and his anti-Stalin faction in the 1920s and 1930s; also *attrib.* or as *adj.*

1937 *Foreign Affairs* XVI. 30 The Trotskyites and Zinovievites. *Ibid.* 64 Designated by the government of the U.S.S.R. as members of a 'Trotskyite–Zinovievite Center'. 1970 S. Talbott tr. *Khrushchev Remembers* (1971) i. 28 The Fifteenth Party Congress, at which Stalin and his supporters squared off against the Zinovievites, or 'Leningrad opposition', as they were then called. 1974 T. P. Whitney tr. *Solzhenitsyn's Gulag Archipelago* I. i. viii. 329 Centers keep creeping in all the time. . Trotskyite–Zinovievite Centers, Rightist–Bukharinite Centers, but all of them are crushed.

Zionism. Add: (Later examples.) Now concerned chiefly with the development of the State of Israel (see *Israel 3).

1948 *Sunday Pictorial* 18 July 7/6 The Government took office solemnly pledged to support Zionism. 1949 [see *revisionism 2]. 1955 *Ann. Reg. 1954* 241 Social democracy..and Zionism were..repeatedly condemned. 1955 *Times* 9 May 9/3 It happened that Jewish workers in considerable numbers were the first to come to Palestine in answer to the appeal of Zionism. 1975 *Globe & Mail* (Toronto) 14 Nov. 7/6 To [Jews], Zionism is and remains a precious term of honor, tied to the heart of their religious and national dreams.

Zionist. (In Dict. s.v. Zionism.) Add: **1.** (Later examples.)

1906 *Times* 29 Oct. 10/5 The Zionist Congress, which is the nearest approach to a Jewish Parliament, permitting women deputies equally with men. 1923 *Jewish Chron.* 19 Jan. 5/1 It will be seen from this how far right the Zionist leaders, Dr. Weizmann and Mr. Sokolow in particular, were in the estimate they put upon that Statement. 1940 E. Gill *Autobiogr.* vii. 252 The agricultural, arboricultural and horticultural work of the young Zionist colonies..superb in itself and of great educational benefit to the Arabs. 1955 *Times* 9 May 9/3 The private sector—politically represented in the main by the General Zionists. 1966 C. Potok *Chosen* (1967) xiii. 213 He had become involved in Zionist activities and was always attending meetings where he spoke about the importance of Palestine as a Jewish homeland. 1977 *Time* 30 May 10/2 Begin..joined the youth organization of the Zionist-Revisionists, a group of right wing militants who condemned the regular Zionist leadership as misguided and soft.

2. A member of any of a group of independent churches in southern Africa similar to pentecostal churches but containing distinctive African elements of worship and belief. [Named after the first such church, the Christian Catholic Apostolic Church in *Zion*, brought from Chicago to S. Africa in 1904.]

1948 B. G. M. Sundkler *Bantu Prophets in S. Africa* ii. 55 Theologically the Zionists are now a syncretistic Bantu movement with healing, speaking with tongues, purification rites, and taboos as the main expressions of their faith. 1956 H. Bloom *Episode* xvii. 318 Among the crowd were a number of Sunday pilgrims, location Zionists of various sects. 1970 *Standard Encycl. Southern Afr.* II. 55/1 Most 'Zionists' have a characteristic form of dress, worn to all services by all members: long dresses and capes, mostly white, decorated with coloured figures (stars, crosses, rings, angels, etc.), veils for women and special forms of headdress for men. *Ibid.*, The 'Zionist' phase has led to extensive proliferation of small groups. 1977 *Time* 27 June 18/3 Some aspects of the old Soweto still exist: the neatly kept gardens of middle-class black homes;..the Zionists, an Africanized Christian sect, famous for their daylong religious dances that begin at prayer services in backyard tents on Saturday nights.

Zionward(s, *adv.* (Earlier example.)

1705 R. Smyth *Let.* 17 Feb. in *Lett. J. Pinney* (1939) 117, I am sure there are too few yᵗ sett their faces Zionward, here in this Town.

zip, *sb.*¹ Add: **3.** Nothing, nought, zero. Cf. *zilch *sb.* (and *a.*). *colloq.* (orig. and chiefly *U.S.*).

1900 *Dialect Notes* II. 70 *Zip*, *n.*, a zero in marks. 1904 *N.Y. World Mag.* 1 May 6/5 'Zip' is the same as 'zero'. 1927 *Amer. Speech* III. 455 *Swabo*, *zip*, zero. 1972 D. E. Westlake *Cops & Robbers* (1973) viii. 111 Before this, neither one of us had known zip about stocks and bonds. 1976 *Times* 27 Apr. 10/8 The successful team are said to have won one-nothing, one-zero or, more fashionably now, one-zip. 1977 J. Cheever *Falconer* 65 Armed robbery. Zip to ten. Second offense. 1980 J. Krantz *Princess Daisy* xxv. 443 No launch, no commercials, no nothing. Zip! Finished! Over!

4. Also **zipp.** A form of fastener for clothes, luggage, etc., consisting of two flexible strips with interlocking projections closed or opened by a sliding clip pulled along them. In full, *zip-fastener, -fastening* (see sense 5 below). Cf. *zipper.

1928 E. M. Forster *Arthur Snatchfold* in *Life to Come* (1972) 102 He felt the shirt..and he gave the zip at the throat a downward pull. 1940 *Punch* 5 June 612/2 Miss Fisher used to wear some lovely plum-coloured trousers with a zip to match. 1957 J. Braine *Room at Top* ix. 90 There was too much messing about with buttons and zips and straps. 1969 *Homes & Gardens* Nov. 190/1, I lost marks in a dressmaking competition because the zipp was machine-stitched in. 1972 *Lancet* 10 June 1271/1 The plain apron front has an off-centre . fastening by a heavy-duty metal 'Zipp', 30 in. long. 1985 *Vogue* July 77 Over this go..flared shorts..some with zips placed to show not pockets but a sliver of flesh.

5. *attrib.* and *Comb.*, as (sense *4), *zip bag, -bedding, -case, -fastener, -fastening, -front, jacket, pocket, shirt, side, suit, top; zip-fastened, -topped* adjs.; also with advbs. forming adjs., as *zip-down, -in, -off, -on, -out, -over, -up*; **zip gun** *U.S. colloq.*, a cheap home-made or makeshift gun; **zip lock** *U.S.*, used *attrib.* to denote plastic bags with an airtight fastening of two interlocking strips; also (a proprietary name) **Ziploc**; **zip-top** *a.* = *ring-pull* adj. s.v. *ring *sb.*¹ 18 a.

1937 L. MacNeice in Auden & MacNeice *Lett. from Iceland* 133 The permutations..of zip bags, Of compacts ..and coiffures. 1948 W. Fortescue *Beauty for Ashes* xxx. 235 Only at Croydon did I discover that my small zip bags had been stowed away in a locker directly under my legs. 1933 T. E. Lawrence *Home Lett.* (1954) 384 The new zipp-bedding is a great success. 1960 N. Kneale *Quatermass & Pit* i. 22 The Minister's Private Secretary enters quickly, clutching a zip-case. 1971 C. Bonington *Annapurna South Face* 241 [Breeches with] zip-down side to allow putting on over boots and crampons. 1950 J. Cannan *Murder Included* vii. 158 A brown zip-fastened handbag. 1927 *Daily Express* 22 Nov. 3/5 Many of the new sports suits have zip-fasteners. 1973 A. Price *October Men* vii. 95 He stumbled down the nearest alleyway.. fumbling as he went for the zip-fastener on his fly. 1927 *Daily Express* 6 Sept. 3/4 The airwoman's costume of tango suède, complete from the zip fastening to the little hat..is attracting many admirers. 1942 'M. Innes' *Daffodil Affair* i. 24 The pomps of death: dissolution had once been a comfortably solid affair. Now it was papier-mâché coffins and zipp-fastening shrouds. 1965 *N.Y. Herald-Tribune* 11 Apr. 8 Zip-front seersucker 'skimma'. 1974 *Harrods Xmas Catal.* 9/1 Hostess gown with zip-front. 1950 *N.Y. Post* 29 Sept. 2 Three Bronx schoolboys were held by police today after admitting shooting off a home-made 'zip' gun... It was fashioned of a 6-inch stainless steel tube taped to a wooden block with an ordinary closet bolt for a 'trigger'. By means of a rubber band, a long ·22 caliber bullet could be shot from it. 1971 B. Malamud *Tenants* 103, I..had thoughts to kill him off with my zip gun but was afraid to. 1984 *Listener* 7 June 7/2 With the decline of the chicano movement and the increase in sophisticated weapons—zip-guns were replaced with sawn-off shotguns and sometimes automatic weapons—violence rocketed in the Seventies. 1974 *Amer. Speech 1970* XLV. 179 *Chil-dodger*.., zip-in garment to be worn under jackets, coats, or over lounge clothes for extra warmth. 1958 *Spectator* 6 June 729/1 A blue zip jacket and slacks. 1970 *Official Gaz.* (U.S. Patent Office) 17 Feb. TM134 *Ziploc*. The Dow Chemical Company... Filed 9-9-68. 1977 C. McFadden *Serial* (1978) i. 107/1 Spenser rummaged among the Ziploc bags in his briefcase. 1982 *Town Crier* (McAllen, Texas) 31 Mar. 1-A/4 Wrapped in foil in a clear zip lock bag. 1957 *Housewife* Sept. 26 Zip-off covers for easy cleaning. 1974 *Harper's & Queen* Sept. 36/2 Zip-off mink £925, extra maxi piece £300. 1959 *Housewife* June 57 Silk cushions ..with zip-on covers. 1963 *New Yorker* 26 Oct. 6 Enter the Chesterfield in Black Tweed..with warm acrylic pile zip-out liner. 1962 'A. Gilbert' *No Dust in Attic* v. 63 He'd left his luggage..one of the zip-over bags. 1979 *Daily Tel.* 20 Nov. 3/1 When Mrs Robabeh Moheby..was strip-searched, a bundle of banknotes was found in a zip pocket in her knickers. 1976 *Morecambe Guardian* 7 Dec. 32/4 (Advt.), Boys short-sleeved, zip shirts. 1940 *Punch* 4 Dec. p. xvii (Advt.), A Dunlopillo sleeping bag. It has..soft upper lining, zip side,..and makes an ideal holdall. 1978 F. Mullally *Deadly Payoff* ii. 25 He handed Fernandez a zip-suit of grubby white fatigues. 1970 *Times* 16 Feb. p. iii, Ring-pull and zip-top cans are already available. 1974 *Harrods Xmas Catal.* 18/1 Bag..with side pocket, zip top. 1976 *Globe & Mail* (Toronto) 30 Dec. 27/6 Here plastic fish surface and bob for zip-top rings from beer and pop cans. 1979 *This England* Winter 19/3 She folded her cap inside her apron and pushed both into her zip-topped bag. 1959 M. Shadbolt *New Zealanders* 26 He wore an open-neck check shirt and a zip-up jacket. 1973 'D. Rutherford' *Kick Start* i. 12 The zip-up calf-length boots. 1985 *Times* 24 Jan. 3/2 He was described as aged between 18 and 20, of West Indian appearance..wearing a beige zip-up jacket.

zip (zip), *sb.*² *U.S.* Also **Z.I.P., ZIP, Zip.** [f. the initials of *Z*oning *I*mprovement *P*lan.] Used esp. *attrib.* in *zip code*: a series of digits representing a particular area in a city, etc., used in addressing mail (see quots.).

1963 *N.Y. Times* 5 May 86/3 Z.I.P. codes, for the present at least, are for big business, and more particularly big users of the mails such as publishers, banks, insurance companies and mail-order houses. 1964 *N.Y. Times Book Rev.* 5 Apr. 3/2 Among his petty peeves are . Zip codes and automatic telephone dialling. 1969 *Computers & Humanities* IV. 69 The zip code has been another, but less successful, step towards using computer recognition to improve our lives. *Ibid.*, Once a person has miswritten the zip, the computer just misdirects the mail. 1973 *Black Panther* 21 July 16/2 (Please print) Name Address..City..State/Zip. 1975 *New Yorker* 5 May 2 (Advt.), In ordering a change of address, subscribers should give four weeks' notice, providing both old and new addresses, with Zip Codes. 1977 *Chicago Tribune* 2 Oct. vi. 26/4 A 13-cent U.S. commemorative honoring 50 years of talking pictures will go on sale Thursday in Hollywood, Cal., where the postmaster Zip is 90028. 1980 *Christian Sci. Monitor* (Midwestern ed.) 4 Dec. 2/2 The Postal Board of Governors.. withheld final approval of a controversial plan to expand ZIP codes to nine digits.

zip, *v.* (In Dict. s.v. Zip *sb.*) Add: **1.** Also, to move briskly or with speed.

1852 *Knickerbocker* XL. 182 How we did '*z-i-p*!' Seven miles, at one time, in less than seven minutes. 1897 Kipling *Day's Work* (1898) 234, I heard my flanges zippin' along the ties. 1907 A. Bennett *Grim Smile of Five Towns* 222 'Let her zip,' said Mr. Colclough. 1922 S. Lewis *Babbitt* xviii. 221 Everything zips at the Chatham Road Church. 1929 *Evening News* 3 Jan. 4/2 He..hardly saw it [*sc.* the ball] after it pitched, as it zipped off the ground and whipped round his bat to take the off stump. **1958** *Daily Mail* 24 Feb. 12/2, I found the man..zipping round the garden..on a toy trike. **1967** *Electronics* 6 Mar. 46/2 Rail travelers will be able to make telephone calls while zipping along at speeds of up to 160 miles an hour from New York City to Washington, D.C. **1976** *Guardian Weekly* 26 Sept. 9/3 The millions of Orange County commuters, who zip past on the Santa Ana Freeway each morning. **1984** *Times* 30 Nov. 11/2 Even though she is likely to spend as much time immersed to the elbows in washing up as zipping down the slopes, there is no stigma of 'service' attached to the job.

2. *trans.* To close with a zip-fastener. Freq. const. *up* and with a person being or getting dressed as obj. Also *intr.* for *pass.* and *refl.* Also *fig.*

1932 A. Huxley *Brave New World* iii. 50 He zipped up his trousers. **1936** *Sears Catal.* 163/4 Easy to 'Zip' baby in and out! **1939** *Punch* 11 Oct. 412/2 The most marvellous outfits..that you step into and zip up. **1942** *Time* 23 Feb. 78 (Advt.), Zips flawlessly. **1942** in *Amer. Speech* (1943) XVIII. 305/2 Zip your lip. **1944** *Penguin New Writing* xx. 60 Zipping on her enchanted house-coat over her night-dress. **1948** *Chicago Tribune* 28 Mar. (Comics) 4 Stand still, now, while I zip you up! **1956** G. N. Patterson *God's Fool* i. 14 Shivering in the icy atmosphere, [I] zipped myself into my double-layer sleeping bag. **1968** B. Hines *Kestrel for Knave* 23 His mother pulled her skirt on and tried to zip it on the hip. **1973** J. Drummond *Bang! Bang! You're Dead!* viii. 13 Authority, including the B.B.C., did not speculate about guilt. Mouths were carefully zipped. **1980** T. Barling *Goodbye Piccadilly* ii. 36 The cistern flushed and Cave emerged, zipping up.

3. *trans.* To beat (an opposing team) comprehensively by not allowing it to score. Cf. *zip *sb.*[1] 3.

1976 *Daily News* (N.Y.) 16 Jan. 82 The Adams Division leaders zipped Los Angeles, 4–0. **1978** *Time* 16 Jan. 66 The Broncos went out and beat the hell out of them, then the next week, went and zipped Cleveland.

Hence **zipped** *ppl. a.*, fastened or provided with a zip. Also *zipped-in, -up* adjs. Also *fig.*

1944 [see *slide fastener* s.v. *slide- a]. **1946** 'S. Russell' *To Bed with Grand Music* i. 13 She picked up her mother's zipped travelling bag. **1959** *Listener* 2 Apr. 593/2 The chemical configuration of starch..is a long zipped-up chain of glucose molecules. **1959** *Times* 5 Oct. 6/2 A zipped-in detachable lining. **1966** T. Pynchon *Crying of Lot 49* vi. 158 Blobb inquired around about the Trystero organization, running into zipped mouths nearly every way he turned. **1972** D. Haston *In High Places* xi. 117 Down suit and fully zipped windproofs. **1982** M. Kenyon *God Squad Bod* ii. 22 Zipped-up moneybags, satchels and briefcases.

Zipf (zipf). The name of George Kingsley *Zipf* (1902–50), American linguist, used in the possessive in *Zipf's law*, a principle in *Psycholinguistics* (see quots.).

1960 É. Delavenay *Introd. Machine Transl.* v. 68 To lend support to arguments based on Zipf's law, Yngve produced evidence likely to convince unbelievers. **1967** M. Schlauch *Language & Study of Languages Today* vii. 143 Some three decades ago, G. K. Zipf undertook to investigate what relation, if any, exists between the length of words and their frequency of occurrence in sample languages chosen for investigation. He came to the conclusion that the length of a word tends to decrease as its relative frequency of use increases... The formula..has come to be known as 'Zipf's Law', often quoted if also sometimes questioned. **1971** *Jrnl. General Psychol.* Oct. 297 Zipf's law predicts an inverse ratio in the rank order distribution of varieties such that a minimal number of varieties covers a maximal number of responses and a maximal number of varieties covers a minimal number of responses. **1980** *Verbatim* Spring 9/2 A psycholinguist will say that this phenomenon is working proof of Zipf's Law, which states loosely that one can generally determine the relative age of a particular word or phrase by how short it has become.

Ziph (zif). [Origin unknown.] An invented language used at Winchester College.

1853 De Quincey *Autobiogr. Sketches* I. vii. 209 Lord Westport and I communicated our thoughts occasionally by means of a language..which bore the name of *Ziph*. The language and the name were both derived..from Winchester. **1922** O. Jespersen *Lang.* viii. 150 'Ziph' or 'Hypernese' (at Winchester) substitutes *wa* for the first of two initial consonants and inserts *p* or *g*, making 'breeches' into *wareechepes*. **1942** E. Partridge *Usage & Abusage* 160/2 *Gibberish* is applied mostly to Ziph (*shagall wege gogo* = shall we go).

zi·pless, *a.* *coarse slang.* [f. *zip *sb.*[1] 4 + -less.] Denoting a brief and passionate sexual encounter.

1973 E. Jong *Fear of Flying* i. 11 My response..was..to evolve my fantasy of the Zipless Fuck... Zipless because when you came together zippers fell away like petals. **1978** G. Vidal *Kalki* iii. 79 Girls who feared flying tended to race blindly through zipless fucks. **1984** *Times* 2 Nov. 11/1 This small hand I'm shaking..launched a thousand zipless erotic encounters for her heroine.

zipper (zi·pəɹ), *sb.* orig. *U.S.* [f. Zip *sb.* + -er[1].]

1. = *zip *sb.* 4. Also *transf.* and *fig.*

Zipper was registered in the U.S. as a trade mark in April 1925 (with use of the term claimed since June 1923), but in the sense 'boots made of rubber and fabric'. It is no longer a proprietary term in any of its uses. Quot. 1925, which appeared in the first Supplement to the *O.E.D.* (1933), and in the *Dictionary of Americanisms*, does not appear in surviving copies of *Scribner's Mag.*

1925 *Scribner's Mag.* Oct. 22/2 (Advt.), No fastening is so quick, secure, or popular as the 'zipper'. **1928** *Daily Express* 11 Oct. 5/5 Bootees..fastening with 'zippers', press studs, or inset clips. **1933** A. G. Macdonell *England, their England* viii. 130 Brown leather jerkins fastened up the front with that singular arrangement which is called a zipper. **1957** *New Yorker* 29 June 24/2 He hated fiddling with things like zippers caught on tiny strips of cloth. **1959** *Listener* 2 Apr. 593/2 We possess a biological zipper—an enzyme—which enables us to undo the chain [of the glucose molecules of starch]. **1966** F. Shaw et al. *Lern Yerself Scouse* 21 *Purra zipper on it*, please be silent. **1979** R. Jaffe *Class Reunion* i. xi. 105 She reached for the zipper of her skirt.

2. *attrib.* and *Comb.*, as *zipper bag, fastener, fastening*, etc.

1925 *Harper's Bazaar* June 108/3 (Advt.), A 'zipper' closing bag sometimes used to carry champagne. **1925** *Scribner's Mag.* June 31/1 (Advt.), The most convenient and attractive form of the pocket purse is this one of lizard calf, leather lined, with a zipper fastening. **1928** *Daily Express* 15 Aug. 4/3 One 12-inch zipper fastener. *Ibid.* 11 Oct. 5/5 Zipper-fastened. **1937** D. Aldis *Time at her Heels* ix. 196 He was wearing his dark blue zipper jacket. **1939** A. Keith *Land below Wind* xx. 311 There were two zipper bags with dutiable goods. **1941** [see *fly *sb.*[3] 4 a]. **1959** C. Williams *Man in Motion* vii. 81 The briefcase..was a slender one, of the type with no handles, zipper-closed. **1960** M. Spark *Ballad Peckham Rye* ii. 22 He sat down among his belongings, which were partly in and partly out of his zipper bag. **1969** *Sears Catal.* Spring/Summer 6 Center zipper pocket divides bag into two handy sections. **1978** *Lancashire Life* Mar. 70 (Advt.), Ladies zipper coats £11.50. **1982** R. Rendell *Master of Moor* xvi. 175 He put on the zipper jacket and went out.

zi·pper, *v.* orig. *U.S.* [f. the sb.] *trans.* To fasten with a zipper. Also *transf.* and *fig.* Freq. const. *up*.

1930 J. Lait *Put on Spot* 215 Zipper. To shut something up, principally the mouth. **1935** N. Ersine *Underworld & Prison Slang* 80 *Zipper, v.*, to cease talking. (Almost invariably the term is a command.) '*Zipper* that mug!' **1961** *John o' London's* 5 Oct. 398/2 Jean Renoir himself capering about zippered up..in a bear-suit. **1966** D. F. Galouye *Lost Perception* vi. 60 Driving wind..tunnelled through the TUT's open cab and Forsythe zippered up his jacket. **1971** *Nature* 3 Sept. 48/2 It seems to act by zippering together chromosomes which have their homologous telomeres attached next to each other at the nuclear membrane. **1974** H. L. Foster *Ribbin', Jivin', & playin' Dozens* vi. 284 After he tucked in his shirt, he closed his pants, zippered them up, put his belt back, and then left. **1984** *Sunday Tel.* (Colour Suppl.) 19 Feb. 12/3 Much was made of the 'keep it zippered' dressing-down he gave to his fellow-astronauts.

zi·ppered, *ppl. a.* orig. *U.S.* [f. *zipper *sb.* + -ed[2].] Fastened with a zipper; having a zipper.

1939 *Times* 31 Mar. 19/7 A charming wool frock with sun-ray decoration round the neck and neatly zippered might be chosen. **1941** *Time* 25 Aug. 2/3 His zippered ankle-high shoes. **1959** *Harrods News* Summer 15 Pigskin clothes brush with zippered top. **1971** B. Malamud *Tenants* 35 A bulky zippered briefcase was squeezed under his left arm. **1982** S. Radley *Talent for Destruction* vi. 36 A thin man in a zippered fawn cardigan.

Zippo (zi·po). Also **zippo.** The proprietary name of a make of cigarette lighter.

1934 *Official Gaz.* (U.S. Patent Office) 10 July 270 Zippo Manufacturing Company, Bradford, Pa... *Zippo* for pocket lighter of the pyrophoric type. **1938** *Trade Marks Jrnl.* 16 Mar. 320/1 *Zippo*..Pyrophoric lighters. Miriam Barcroft Blaisdell..trading as Zippo Manufacturing Company,..Bradford, State of Pennsylvania, United States of America; manufacturers. **1964** G. McDonald *Running Scared* iii. 35 The Zippo lighter he had given Casey. **1966** G. Lyall *Shooting Script* xxii. 174 Luiz flicked a Zippo under my nose. **1977** McKnight & Tobler *Bob Marley* ix. 108 A novel sleeve made to look like a zippo lighter. *Ibid.* 112 Gone was the zippo, replaced by a picture of Marley smoking a kingsize spliff. **1982** R. Ludlum *Parsifal Mosaic* viii. 111 He extracted one [cigarette] and lit it with an old, tarnished Zippo purchased a quarter of a century ago.

zippy (zi·pi), *a.* *colloq.* (orig. *U.S.*). [f. Zip *sb.* 2 + -y[1].] Bright, lively, energetic; fresh, invigorating; fast, speedy.

1904 G. Ade *True Bills* 108 Vivian, our bright-eyed little Daughter..is the zippiest High-Flyer that speeds the Boulevard. **1917** *National Police Gazette* 25 Aug. 2/4 It bubbles o'er with song and jest, Its fun is keen and 'zippy'. **1926** *Bulletin* (Glasgow) 26 Oct. 18/2 A zippy show is 'Cheerio' at the Glasgow Empire this week. **1939** *Better English* Jan. 29 (*heading*) Your words—are they zippy? **1953** *Amer. Mercury* Jan. 67/2 The March evening was zippy outside but cupped in it the breath of spring. **1959** I. Jefferies *Thirteen Days* xii. 202 My idea..was to get a zippy load into Richon and then move on to T.A. **1966** R. H. Rimmer *Harrad Experiment* (1967) 161 All we have to do now is to put together a zippy script. **1971** *Daily Tel.* 23 June 11/6 The engine warms up rapidly and provides zippy acceleration. **1980** M. Broadbent *Great Vintage Wine Bk.* 134 Ch. Potensac. Zippy little wine. **1984** *Guardian* 5 Nov. 12/1 One advantage to being on Castleton social services committee is that they do have zippy meetings.

Hence **zi·ppily** *adv.*; **zi·ppiness.**

1924 Wodehouse *Bill the Conqueror* viii. 152 This series on Bookmakers' Swindling Methods..had always reached a fair level of zippiness; but never..had it so outzipped itself as in the present instalment. **1983** *Times* 10 June 17/6 The piece is zippily staged..and sung with immense liveliness.

ziram (zəi·ræm). [f. *zi*nc dimethyl dithiocarb*am*ate, the systematic name: see Carbamate.] A white powder used as a fungicide, esp. on vegetables and some fruit crops; $Zn(\cdot S\cdot CS\cdot N(CH_3)_2)_2$.

1950 [see *zineb]. **1960** [see *ferbam]. **1969** *New Scientist* 8 May 299/2 The two main breakdown products of the fungicide, ziram (zinc dimethyldithiocarbamate) were carbon disulphide and dimethylamine. **1981** *Jrnl. Agric. & Food Chem.* XXIX. 729 The limit of detection in water solutions for zineb, ziram, and thiram was 0·05, 0·01, and 0·01 ppm, respectively.

Ziranian, var. *Siryenian *sb.* and *a.*

zircaloy (zɜ̄·ɹkăloi). Also **zircalloy.** [f. Zirc(onium + Alloy *sb.*] Any of several alloys of zirconium, tin, and other metals that are used chiefly as cladding for nuclear reactor fuel.

1956 Thomas & Forscher in *Jrnl. Metals* VIII. 640/1 Zircaloy-2, a low alloy of zirconium, was developed by the Westinghouse Atomic Power Div. during 1952. **1963** H. R. Clauser *Encycl. Engin. Materials & Processes* 769/1 Zircaloy resists corrosion very well in high-temperature water and steam. **1973** *Nature* 2 Feb. 318/1 The rods are zirconium alloy (zircalloy) tubes filled with pellets of uranium dioxide (UO_2) ceramic. **1977** J. H. Schemel *ASTM Manual on Zirconium & Hafnium* (ASTM Special Technical Publ. No. 639) i. 6 Zircaloy is used to designate a series of zirconium, tin, iron, chromium, nickel alloys developed by the U. S. Navy Nuclear Propulsion Program for nuclear service in water-cooled reactors. Zircaloy-1 and Zircaloy-3 are obsolete. Zircaloy-2 and Zircaloy-4 are the most commonly used alloys in nuclear service.

zircon. Add: **b. zircon blue,** a light blue colour.

1949 *Brit. Colour Council Dict. Colours* III. 28/2 *Zircon blue*... Matched to specimens of the precious stone at the B.M. **1972** 'E. Peters' *Death to Landlords!* i. 30 Withdrawing her zircon-blue eyes from the heavens. **1976** *Evening Post* (Nottingham) 13 Dec. 12/3 (Advt.), Rover 2000 1969. H. Zircon blue with black trim.

zirconian, *a.* Add: *spec.* in *Min.*, applied to a mineral in which zirconium replaces a (small) proportion of some constituent element (cf. *-ian 2).

1930 W. T. Schaller in *Amer. Mineralogist* XV. 572. **1963** *Doklady Acad. Sci. USSR: Earth Sci. Sect.* CXLI. 1301/1 Apatite, nepheline, and perovskite were observed, frequently forming intergrowths with the zirconian schorlomite. **1972** *Nature* 31 Mar. 197/2 The appearance of zirconian SiO_2-free minerals in the granitic fraction is rather strange.

zirconolite (zɜ̄ɹkǫ·nŏləit). *Min.* [ad. Russ. *tsirkonolit* (L. S. Borodin et al. 1956, in *Doklady Akad. Nauk SSSR* CX. 845): see Zircon and -lite.] A mixed oxide of (essentially) calcium, zirconium, and titanium, now regarded as identical with zirkelite.

1957 *Chem. Abstr.* LI. 6440 (*heading*) The new mineral zirconolite, a complex oxide of the type AB_3O_7. **1975** *Amer. Mineralogist* LX. 341/1 Zirconolite, $CaZrTi_2O_7$, was described in 1956..as a new mineral and the zirkelite of Blake and Smith was considered to be zirconolite... New analyses and X-ray studies of type material show that all these are a single mineral species of composition $(Ca,Th,RE)Zr(Ti,Nb,Fe)_2O_7$. The name zirkelite has priority. **1980** *Nature* 17 Jan. 282/1 Synthetic zirconolite, $CaZrTi_2O_7$, has attracted interest recently, as it has been proposed as a constituent phase of an artificial rock (SYNROC) which may immobilise, in solid solution, the elements occurring in high-level nuclear reactor wastes. **1981** [see *zirkelite].

zirkelite (zɜ̄·ɹkĕləit). *Min.* [f. the name of Ferdinand *Zirkel* (1838–1912), German mineralogist + -ite[1].] A black monoclinic (pseudocubic) oxide of zirconium, calcium, thorium, titanium, rare earths, and other elements, $(Ca,Th,Ce)Zr(Ti,Nb)_2O_7$.

1895 Hussak & Prior in *Mineral. Mag.* XI. 86 Zirkelite. A new calcium zirconate and titanate. **1962** *Mineral. Abstr.* XV. 538/1 Zirkelite is widely distributed in the Palaeozoic ultrabasic and alkaline complexes of the Kola Peninsula where it occurs as an accessory mineral in calcite-amphibole rocks; [etc.]. **1972** *Nature* 31 Mar. 215/1 We find that terrestrial zirkelite bears the closest comparison with the tranquillityite group, and yet contains sufficient CaO..to provide a possible link with any calcic lunar zirconian phases yet to be established. **1975** [see *zirconolite]. **1977** *Amer. Mineralogist* LXII. 408 Zirconolite and niobozirconolite..are synonymous with zirkelite. **1981** K. Frye *Encycl. Mineral.* 220/1 There is considerable controversy over the exact formulas of zirkelite and zirconolite. Some evidence suggests that there is no difference.

zit (zit). *slang* (chiefly *N. Amer.*). [Origin unknown.] A pimple. Also in extended and *fig.* use. Occas. in *Comb.*

1966 G. C. Sauer *Man. Skin Dis.* (ed. 2) 284 *Zits*, term in teen-agers' vernacular for 'pimples' of acne. **1966** *Current Slang* (Univ. S. Dakota) Summer 5 *Zit*, something bad or unpleasant; crude... What's that *zit* on your coat? **1972** B. Rodgers *Queens' Vernacular* 45 *Vit* (late '60s, fr. sl. *zit* = a pimple: boy with Pepsi consciousness, *i.e.*

nothing going on between the ears). **1975** *Atlantic Monthly* Mar. 51 One splendid effort in 1971 featured ..districts with remarkable pimples in their boundary lines, zits that popped up to include the home of one liberal incumbent in the district of another liberal incumbent. **1975** *Maclean's Mag.* June 59/1 When did you last have a zit on your face? **1977** *Amer. Speech* 1975 L. 69 *Zit n*, reddish mark caused by kissing. 'She has a big zit on her neck.' **1977** J. Wambaugh *Black Marble* (1978) vi. 77 How old is the little zit-faced, coke-snorting, hash-smoking son of a bitch? **1980** *Courier-Mail* (Brisbane) 30 Apr. 46/3 (*caption*) You know playing with teenagers will give you zits. **1984** S. Townsend *Growing Pains A. Mole* 105 Forgot to send you an invite, zit face, but come anyway, dress as a warlock or you won't get in.

zita (zi·ta). Pl. **zite, ziti.** [It.] A tubular variety of pasta resembling large macaroni.

1845 E. Acton *Mod. Cookery* (ed. 4) p. xxx, *Zita*, Naples maccaroni. **1943** A. L. Simon *Conc. Encycl. Gastron.* iv. 141/1 *Zita*, pl. *zite*, one of the fancy Italian pastes... It is made of the gluten of hard wheat and eggs, like Macaroni, in hollow, straight tubes of larger diameter than Macaroni. **1964** *Guardian* 22 May 8/6 *Ziti*, cut macaroni tubes slightly curved. **1978** R. F. Capon *Food for Thought* vii. 106 Why not sauces for ziti, for occhi di lupo, [etc.]? **1979** *Tucson Mag.* Sept. 68/2 Bargain specialties—like gnocchi, zita, linguine, and so-so meat dishes. **1983** *Fortune* 21 Feb. 78/1 For the one-time shopper, goods from auto-parts to ziti.

zither. Add: **b.** *zither-playing, -tinkling* adjs.

1982 C. McIntosh *Swan King* viii. 83 A fun-loving, zither-playing, somewhat bohemian character. **1925** E. Sitwell et al. *Poor Young People* 2 Among the zither-tinkling round green leaves.

zi·ther, *v.* [f. the sb.] *intr.* To play the zither. Also *fig.* (occas. also *trans.*). Hence **zi·thering** *ppl. a.*

Some of the examples could equally well be interpreted as being of echoic origin, with the sense 'to make a sibilant humming sound, buzz lightly'.

1906 W. J. Locke *Beloved Vagabond* (1907) ix. 120 We wandered and fiddled and zithered and tambourined through France. **1930** R. Campbell *Adamastor* 71 The sunlight, zithering their flanks with fire, Flashes between the shadows as they pass. **1958** J. Townsend *Young Devils* iv. 35 His squeaky voice zithered through morning or afternoon. **1973** *Art Internat.* Mar. 57/1 Balla's zithering *Rhythm of the Violinist* is considerably more effective. **1979** H. McLeave *Double Exposure* i. 1 The helicopter..trailing a zithering shadow. **1981** B. Carter *Black Fox Running* xxvii. 176 Insects zithered and chirred.

ziti: see *zita.

zitkamer, zit-kamer, varr. *sitkamer.

zizith, var. *tsitsith.

zizz, *sb.* Add: **1. a.** (Later examples.) Also extended to other whizzing or buzzing noises (see quots.).

1904 G. A. B. Dewar *Glamour of Earth* vi. 131 The zizz of the cricket, or the shrill of the bat. **1955** D. Barton *Glorious Life* xxv. 232 The sustained, high-pitched zizz of a party was audible. **1965** *Listener* 17 June 900/3 The zizz of a trishaw's wheels passing on the road. **1976** *Drive* May–June 53/2 Gear lever zizz is irritating.

b. Gaiety, liveliness, 'sparkle'. *colloq.*

1942 Berrey & Van den Bark *Amer. Thes. Slang* § 240/2 *Animation; spirit; vim;*..zing, zip,..zizz. **1970** *Gourmet* Jan. 18/2 No party got into full swing until Tallulah arrived to put her particular type of zizz into it. **1983** *Times* 22 Feb. 12/6 The Queensgate centre lacks, perhaps, finesse and a touch of zizz.

2. Also **ziz.** A short sleep, a nap. Cf. *Z 4 b. *slang.*

1941 *Tee Emm* Aug. 17 He could not have caught our Pilot Officer Prune at three o'clock one afternoon having a zizz full-length on a mess settee. **1960** 'N. Shute' *Trustee from Toolroom* v. 105 'Captain's having a ziz now,' said the navigator. 'Supper's at eleven o'clock, Greenwich. He's getting up for that.' **1970** P. Dickinson tr. *Aristophanes' Wasps* in *Plays* I. 169 Just what I aim to forget by having A quiet ziz. **1979** M. Tabor *Baker's Daughter* i. 31 Philip's having a zizz He can't stay awake. **1985** *Guardian* 24 Jan. 1/3 They would not film any lord who had drifted off in the warmth of the lights for a refreshing zizz.

zizz, *v.* [f. the sb.] **1.** *intr.* To make a whizzing or buzzing sound. Occas. *trans.* (causally). Also const. *up*, to liven up.

a **1934** in Webster s.v., Trolleys went zizzing along. **1961** E. Williams *George* ix. 112 The pince-nez zizzed back to her lapel. **1963** *Punch* 1 May 621/3 Then she [*sc.* a bee] saw the carpet and zizzed. **1965** *Harper's Bazaar* Feb. 21/2 An adventurous buyer deciding that model 127 is just the thing to zizz up his mid-season collection. **1970** T. Lewis *Jack's Return Home* 89 The banger'd started zizzing furiously. **1970** *Daily Tel.* 16 Dec. 11 Darts and circles about on the floor after its wheels are zizzed smartly. **1978** *Ibid.* 19 July 12/5 The gearchange..is light but slightly notchy, and tended to 'zizz' on the overrun.

2. *slang.* To doze or sleep. Occas. *trans* with *away*.

1942 [see below]. **1961** D. Moore *Highway of Fear* xxxvi. 240 Reckon this sector's safe. Might as well zizz. **1972** K. Bonfiglioli *Don't point that Thing at Me* xii. 94, I zizzed away the worst of the afternoon, awaking some three hours later. **1978** *Sunday Mail Color Mag.* (Brisbane) 30 Apr. 20/6 When everyone inside the building had zizzed off he had sneaked inside.

Hence **zi·zzing** *vbl. sb.* and *ppl. a.*

1942 *Gen* 1 Sept. 13/1 Sleeping is 'zizzing' whether it's on the job or in the hammock. **1951** J. Strachey *Man on Pier* 37 The flies on the window-pane woke up and started to rage together with a venomous zizzing. *Ibid.* 88 Confusion continued, it seemed to Ned, to the old background of zizzing flies, bellowing cows, walks, and hot, cloudy weather. **1961** E. Williams *George* ix. 111 There was a terrified zizzing noise from an obedient coil-spring.

zi·zzy, *a.* *colloq.* [f. Zizz *sb.* + -y[1].] Showy, spectacular; lively, uninhibited.

1966 *Guardian* 5 Feb. 7/6 And who's going to pay for 'Danger Man' with that zizzy Patrick McGoohan, then? **1975** *New Yorker* 21 Apr. 24/3 If you accept the silly, zizzy obviousness, it can make you laugh helplessly. **1976** *Times* 4 Oct. 7 My wife said I should wear a dark suit but I did risk a particularly zizzy tie. **1983** *Guardian Weekly* 6 Feb. 1/2 Zizzy little TV charts.

Z line (zed, *U.S.* zī, ləin). *Histology.* [Partial tr. G. *schicht z* z layer (T. W. Engelmann 1873, in *Arch. f. die ges. Physiol.* VII. 37), f. initial letter of *zwischenscheibe* intervening disc.] A transverse dark line in a fibril of striated muscle formed by Krause's membrane (see *Krause b); the membrane itself.

1916 Jordan & Ferguson *Text-bk. Histol.* iv. 105 This stripe or accessory disk..bisects the portion of the J disk between the Z line and the succeeding Q disk. **1954** *Nature* 22 May 976/1 The series elastic component is provided either by the actin filaments themselves, or, more probably, by their mode of attachment to the Z-line. **1979** *Sci. Amer.* May 94/2 In muscle fibers the actin filaments are anchored to flat protein structures called *Z lines*, which are emplaced between every two contractile units.

Also **Z band.**

1950 A. W. Ham *Histology* xix. 283/1 When a substantial degree of contraction has occurred, an appearance, often referred to as a 'reversal of striations', becomes apparent. This is due to the substance of the myofibril on each side of the Z band, which was formerly light, becoming dark, and the dark material of the Q band becoming light. **1964** G. H. Haggis et al. *Introd. Molecular Biol.* iv. 101 The segment from one Z-band to the next, along a fibril, is termed a sarcomere. **1970** [see *Krause b].

zloty (zlǫ·ti, ‖ zwǫ·ti). Pl. **zloty, zlotys.** [a. Polish *złoty*, f. *złoto* gold, cogn. w. Russ. *zóloto*: see Gold.] **a.** A gold or silver coin of monarchic Poland. **b.** The monetary unit of the Polish republic; a note or coin of the republican currency.

1915 *Publ. Scottish Hist. Soc.* lix. 63 The fourth witness hath borne witness in the following words:—'I..have seen how Jan Furman, a Scot, hath taken..40 Polish zloty, from that Kilian.' **1923** *Times* 3 Mar. 16/6 The Polish Minister of Finance has decided that State loans shall..be effected in Polish zlotys (a zloty is equivalent to a Swiss franc), at the current market rate. **1923** *Times* 13 Aug. 14/5 The zloty, or gold franc, the nominal unit of Poland. **1944** V. G. Garvin tr. *R. Gary's Forest of Anger* xxiii. 90 They've imposed a fine of 100,000 zlotys on Pinski! **1960** S. Becker tr. *Schwarz-Bart's Last of Just* ii. 40 When you have a few zlotys in hand, you can come back here and reimburse me. **1970** *New Yorker* 6 June 33/1 He promised me a room, food, and a small salary in Polish marks. The zloty wasn't yet established as a currency. **1983** *Nature* 28 July 299/2 The Polish Government has raised the price of edible salt to between 11 zloty and 17 zloty per kilo.

Zoar, obs. var. *Zohar.

zob (zǫb). *U.S. slang.* *rare.* [Origin unknown.] A weak or contemptible person; a fool.

1911 W. F. Kirk *Right off Bat* 13 He came here in the early Spring with all the try-out mob Striving to bat like Wagner and to slide (spikes first) like Cobb. Some of the vets cried, 'Bonehead!' Others remarked, 'Poor zob!' **1920** S. Lewis *Main Street* xxxv. 416 And the same thing goes for that crowd of crabs and snobs Down East, and next time you hear some zob from Yahooville-on-the-Hudson chewing the rag..you tell him that no..Westerner would have New York for a gift! **1922** —— *Babbitt* x. 140, I don't know how you fellows feel about prohibition, but the way it strikes me is that it's a mighty beneficial thing for the poor zob that hasn't got any will-power but for fellows like us, it's an infringement of personal liberty. **1942** Berrey & Van den Bark *Amer. Thes. Slang* § 396 Terms of disparagement..yaphead, yazzihamper, zob.

zocalo (zǫ·kālo). Also **zócalo** and with capital initial. [a. Sp.] In Mexico, a public square, a plaza.

1884 J. W. Steele *To Mexico by Palace Car* v. 75 The square in front of the Cathedral, called the Zócalo, is the place of universal resort. **1884** F. H. Ober *Trav. in Mexico* xii. 232 Beneath us is the great square, with the smaller one, the zocalo, or pleasure garden, in its center. **1912** E. H. Blickfeld *Mexican Journey* x. 119 The Zocalo, as it is called, is the real center of Mexico City. **1927** D. H. Lawrence *Mornings in Mexico* 39 And a *plaza* is a *zocalo*, a hub. **1957** M. Lowry *Let.* 17 Mar. (1967) 398, I read of revolutions, counterrevolutions..in the zócalo. **1975** *Times* 24 May 10/3 Oaxaca..its zocalo, with its trees and its arcades.

‖**zoco** (zǫ·ko). Also **Soko.** [Sp., ad. Arab.: see *souk.] = *souk; also *transf.*

1892 M. Thomas *Scamper through Spain & Tangier* xii. 232 The Soko is an unpaved square surrounded with booths, outside the walls, where on Thursdays and Sundays, as in Spain, the market is held. **1903** A. F. Calvert *Impressions of Spain* 96 Toledo—? She is at least faithful to the dead past.. So she retains her old *Soko*, and will have naught to do with the correct *Plaza de la Constitucion.* **1921** *Chambers's Jrnl.* Dec. 817/2 Tangier..cosmopolitan enough as it seems to be to any one who ambles through the little *zoco*, has been administratively mismanaged. **1924** *Ibid.* Sept. 689/2 The minaret of the new mosque overlooking the big zoco cuts through the blue. **1965** C. D. Eby *Siege of Alcázar* (1966) i. 34 The Gobierno resembled the *zoco* of a North African town except that the jewellery boxes, bundles of clothing, ..and whimpering children were not for sale.

zodico, var. *Zydeco. **zoetekoekie,** var. *soetkoekie.

zoetrope. (Earlier example.)

1867 'Aunt Carrie' *Popular Pastimes for Field & Fireside* 229 The Zoetrope is a newly invented toy. It presents a series of striking optical delusions.

zoftick, zoftig, zoftik, varr. *zaftig *a.*

zograscope (zǫ·grăskōᵘp). *Obs.* exc. *Hist.* [Etym. uncertain, perh. f. Zogra(pher, Zogra(phy + -scope.] An optical instrument, consisting of a vertically suspended convex lens in front of a pivotally adjustable mirror mounted on a stand, designed for the viewing of prints in magnified form and with stereoscopic effect.

1753 in G. Adams *Descr. & Use Universal Trigonometrical Octant* 3 (Advt.), Zograscopes for viewing perspective Prints. **1953** *Ann. Sci.* IX. 315 (*heading*) The zograscope or optical diagonal machine. *Ibid.*, No reference to the zograscope has so far been found in the literature of the eighteenth or nineteenth centuries, save in certain..advertisements listing instruments made..by [George Adams]. **1969** E. H. Pinto *Treen* 284 Usually described and sold as Georgian shaving mirrors; this, they are not; they are zograscopes,..probably invented about 1750... Also known, during its long life, as an optical machine..intended for viewing prints, etc., in a magnified form.

zohar (zōᵘ·haɹ). Also 7 **zoar.** [Heb., lit. 'light, splendour'.] The major text of Jewish Cabbalism, in the form of an allegorical interpretation of the Pentateuch.

1682 W. Payne *Learning & Knowl.* 12 A studying Judiciary Schemes, and fanciful Cabalas, and mysterious zoars. **1837** *British & Foreign Rev.* V. 419 The work called Zohar is written in Chaldaic, and develops the mysterious science called *Cabala*. **1843** G. Borrow *Bible in Spain* III. xv. 295 He knew more Zohar and more secrets than the wisest of them. **1888** J. C. Murray tr. *Maimon's Autobiogr.* xiv. 95 The principal work for the study of the Cabbalah is the *Zohar*, which is written in a very lofty style in the Syrian language. **1932** A. Bension *Zohar* i. 11 Although they derived their inspiration from the same source—Zohar and Kabbala—Ashkenazi and Sepharadi mysticism inevitably took different directions. **1941** [see *Tel Avivian *sb.* and *a.*]. **1965** J. A. Michener *Source* (1966) 662 Ximeno had given Dr. Abulafia a manuscript of the Zohar, the arcane book of Kabbalism. **1978** I. B. Singer *Shosha* i. 8 There stood on our shelves volumes of the Zohar.., and other cabalistic works.

zoid *sb.* Delete '(Disused)' and add further examples.

1960 *Bot. Gaz.* CXXII. 33/1 Most algae, excluding the Cyanophyta, reproduce sexually by gametes and asexually by spores of various kinds (collectively called 'zoids', except for aplanospores). **1981** *Austral. Jrnl. Zool.* XXIX. 365 Subterminal ovicell complexes bud distally from part of the complex and are generally composed of female zoid with ovicell, lateral zoid, and ovicell zoid with apical chamber.

zoite (zōᵘ·əit). *Zool.* [The suffix used as an independent word.] (See quots.).

1963 E. N. Kozloff in J. Ludvík et al. *Progr. Protozool.* (Proc. First Internat. Congr. Protozool., Prague, 1961) 78 It appears advisable to reconsider the terminology applied to stages in the life-histories of gregarines and coccidians... *Zoite*, an infective stage, produced by division of a zygote, and usually surrounded by an envelope which is either a zoitocyst or zygocyst. **1969** *New Scientist* 29 May 465/1 Another stage in the development of *Toxoplasma* is the cystic form or zoite. **1977** *Jrnl. Protozool.* XXIV. 36/1 Electronmicroscopic studies revealed that the structure of *T. gondii* zoites (trophozoites, merozoites, 'spores', etc.) is similar to that of the zoites of the coccidia *Eimeria* and *Isospora*. **1979** *Ibid.* XXVI. 437/1 Uninucleate, dinucleate, and multinucleate zoites. **1980** *Jrnl. Parasitol.* LXVI. 67/1 Ultrastructurally, cysts of *T*[*oxoplasma*] *gondii* and *H*[*ammondia*] *hammondi* cannot be distinguished with certainty in skeletal muscle. Classification is based on the typical distribution of cysts and their zoites in the intermediate host.

-zoite (zōᵘ·əit), *suffix.* [f. Gr. ζῷ-ον animal + -ite[1].] A word-forming element used in zoological terms in the sense 'spore', as in *merozoite* s.v. *mero-[1], *tachyzoite* s.v. *tachy-.

Zollinger–Ellison syndrome (zǫ·lindʒəɹ e·lisən). *Path.* [Named after M. *Zollinger* (b. 1903) and E. H. *Ellison* (1918–70), American physicians, who described the syndrome in

1955 (*Ann. Surg.* CXLII. 709).] A syndrome characterized by excessive gastric acid secretion (producing recurrent peptic ulcers) associated with a gastrin-secreting tumour or hyperplasia of the islet cells of the pancreas.

1956 Eiseman & Maynard in *Gastroenterology* XXXI. 302 For the sake of simplicity we propose this clinical entity be called the Zollinger-Ellison syndrome. **1960** *Jrnl. R. Coll. of Surgeons of Edin.* V. 191 Since that time [*sc.* 1955] over 100 instances of co-existent peptic ulcer and islet cell tumours have been described and the association appears to constitute a clinical entity called, by general consent, the 'Zollinger-Ellison syndrome'. **1962** [see *islet 3]. **1974** R. M. Kirk et al. *Surgery* vi. 85/1 In the Zollinger-Ellison syndrome, ulcers may occur at the duodenojejunal junction.

Zöllner (tsö·lnəɹ, z-). *Psychol.* Also **Zoellner.** The name of the German astronomer and physicist, Johann Karl Friedrich *Zöllner* (1834–82), used *attrib.* and in the possessive to designate the optical illusion noted by him of parallel lines which, when marked with short diagonal lines, appear to converge. Now usu. as *Zöllner illusion.*

1890 W. James *Princ. Psychol.* II. xx. 232 In what is known as Zöllner's pattern.., the long parallels tip towards each other the moment we draw the short slanting lines over them. **1911** *Encycl. Brit.* XXVIII. 142/1 (*caption*) Zoellner's Figure showing an illusion of direction. **1922** K. Dunlap *Elem. Sci. Psychol.* xiii. 295 In the Zöllner figure, the long lines, really parallel, seem to converge. **1955** H. E. Garrett *Gen. Psychol.* v. 179 (*caption*) Zoellner illusion. The four horizontal lines are parallel. **1971** [see *Hering]. **1980** *Sci. Amer.* Jan. 91/1 The Zöllner illusion, which exhibits assimilation at extremely small angles and contrast at larger angles.

zolo go, var. *Zydeco.

zombie (zǫ·mbi). Also **zombi** and with capital initial. [Of W. Afr. origin; cf. Kongo *nzambi* god, *zumbi* fetish.] **1.** In the West Indies and southern states of America, a soulless corpse said to have been revived by witchcraft; formerly, the name of a snake-deity in voodoo cults of or deriving from West Africa and Haiti.

1819 R. Southey *Hist. Brazil* III. xxxi. 24 Zombi, the title whereby he [chief of Brazilian natives] was called, is the name for the Deity, in the Angolan tongue... NZambi is the word for Deity. **1872** Schele de Vere *Americanisms* 138 *Zombi*, a phantom or a ghost, not unfrequently heard in the Southern States in nurseries and among the servants. **1886** *Century Mag.* Apr. 815/2 This spiritual influence or potentate is the recognized antagonist and opposite of Obi, the great African manitou or deity, or him whom the Congoes vaguely generalize as Zombi. **1929** W. B. Seabrook *Magic Island* ii. ii. 94 At this very moment, in the moonlight, there are *zombies* working on this island. **1943** R. Ottley *New World* 46 Adding the zombies, jumbies, and obeah men to the gallery of voodoo characters. **1966** G. Greene *Comedians* iv. 104 Luckily no one dared move on the roads at night; it was the hour when only zombies worked or else the Tontons Macoute. **1979** J. Rhys *Smile Please* 30 Zombies were black shapeless things. They could get through a locked door and you heard them walking up to your bed. You didn't see them, you felt their hairy hands round your throat. **1984** *Times* 26 Jan. 12/6 A zombie, as every schoolboy knows, is a person who has been killed and raised from the dead by sinister voodoo priests called bocors.

2. *fig.* A dull, apathetic, or slow-witted person. Also as a general term of disparagement. *colloq.*

1936 H. L. Mencken *Amer. Language* (ed. 4) xi. 587 Any performer [in a film] not a Caucasian is a *zombie*. **1941** H. MacInnes *Above Suspicion* ix. 80 He nodded..in the direction of those concentrating on the mastication of specially chosen vitamins to build a specially chosen race. 'Zombies is, I believe, the technical term,' suggested Richard. **1946** J. B. Priestley *Bright Day* xi. 329 They've spent their lives starving their imagination, just starving it to death. And now they're zombies. **1957** J. Braine *Room at Top* i. 17 To Charles and me it was always Dead Dufton and the councillors and chief officials and anyone we didn't approve of were called zombies. **1961** [see *love *sb.*[1] 4]. **1974** S. Middleton *Holiday* xiii. 233 He had no time for her as a zombie, preferring her moody volatility to this flabby acquiescence. **1981** P. Carey *Bliss* iv. 156 They'll give us electric shocks... They'll give us pills and make us zombies. **1984** *Guardian* 22 Oct. 3/1 Mr. Dawson describes the committee as a parliament of zombies.

3. *Canad. Mil. slang.* In the war of 1939–45, an opprobious nickname applied to men conscripted for home defence.

1943 *Daily Express* 16 Sept. 4/1 The Canadian Government is reducing its 'Home Guard' army.. These troops were jocularly dubbed 'Zombies', after the Voodoo cult which insists that dead men can be made to walk and act as if they were alive. **1946** [see *old man 1 a]. **1953** D. M. Le Bourdais *Nation of North* 245 The first men were drafted for service... Contemptuously referred to as 'zombies', they were never taken seriously by the military authorities. **1963** A. S. Morton *Kingdom of Canada* 481 A nasty distinction arose between the volunteers for service overseas and the conscripts for home defence, who were given the pungent nickname of 'zombie', a West Indian word for impotent spirits. **1978** *Daily Colonist Mag.* (Victoria, B.C.) 1 July 12/1 When the Canadian Army was struggling on the Western Front in the early winter of 1944 and there was an urgent call for reinforcements, yet, in the military camps in Vernon and Terrace the Zombies mutinied when orders came for their movement overseas.

4. A long mixed drink consisting of several kinds of rum, liqueur, and fruit juice (see quot. 1958).

1942 M. K. Rawlings *Cross Creek* xvii. 221 There is a passion fruit liqueur that is the primary ingredient..of that marvelous..drink, the Zombie. **1958** A. L. Simon *Dict. Wines, Spirits & Liqueurs* 167/1 *Zombie*,..lime juice;..pineapple juice;..Falernum, or simple syrup;..White Label Rum;..Gold Label Rum;..Jamaica Rum;..Demerara Rum;..apricot liqueur. Shake well and strain... Garnish with..orange and..mint. **1968** J. M. Ullman *Lady on Fire* vi. 80 The bartender..went off to prepare a zombie. Forbes hated zombies, but it was the longest drink that came to mind. **1977** *Zigzag* Apr. 10/1 It's a Polynesian drink, a rum drink..a killer drink. There was a restaurant I found very close to The Cage that I went to every day... Like I would wake up, go there at noon, have a Zombie, go back and start writing.

5. a. *attrib.*

1956 M. Stearns *Story of Jazz* (1957) xviii. 222 Thelonius Monk, whose weird..and pioneering modulations were referred to as 'Zombie music' by the musicians themselves, more in awe than anger. **1958** A. Wilson *Middle Age of Mrs. Eliot* i. 77 The breathing zombie orchestra around her. **1966** *New Statesman* 14 Jan. 58/3 A dream-sequence in a cemetery with zombie-hands sprouting like crocuses. **1968** P. Ableman *Vac* xxiv. 113 Stop clashing those zombie lips and glide to the bar. **1973** [see *one-nighter* s.v. *one B. 33]. **1976** D. Lodge *Changing Places* v. 162 He could send home, when the time came, some zombie replica of himself.

b. *Comb.*, as **zombie-like** *a.*, characteristic of or resembling a zombie; lifeless, unfeeling.

1957 J. Kerouac *On Road* (1958) 302 His arms hanging zombie-like at his sides. **1962** *Times* 25 July 13/1 Future ..where everybody lives a zombie-like existence. **1975** *Publishers Weekly* 10 Feb. 57/3 White Brian is a zombie-like boy who wanders unseeing and unseen through life. **1983** *Times* 4 Oct. 10/6 On state occasions, a few old men shuffle on to the balcony of the Kremlin and raise their hands in zombie-like salutation.

Hence **zombie·sque, zo·mboid** [-oid] *adjs.*; **zo·mbi,ism.**

1956 M. Stearns *Story of Jazz* (1957) v. 51 The Calinda dance is connected with zombiism in Haiti. **1972** *Vogue* Jan. 7/3 Ponderax..sidetracks the appetite but leaves the character..muted, zombiesque. **1974** *Observer* 3 Feb. 31/5 Zombiesque Security Guard charged with shooting. **1974** *Radio Times* 18 Mar. 22/4 We're not putting up with 'Zombyism'. The aim of our programme is to give hope. **1975** *Sunday Times* (Colour Suppl.) 20 July 11/3 'I'm David Bowie,' he intones with a zomboid air. **1979** *Guardian* 18 Oct. 11/6 Some of the most zomboid heroines in recent fiction. **1983** *Daily Tel.* 15 Oct. 17/1 Zombi-ism exists and is a social phenomenon that can be explored logically.

zona. Add **b. zona fasciculata, glomerulosa, reticularis** [mod.L. (see Fasciculate *a.*, etc.), coined in Ger. by J. Arnold 1866, in *Arch. f. path. Anat. & Physiol.* XXXV. 66], the middle, outer, and inner layers respectively of the cortex of the adrenal gland; **zona radiata,** a radially striated form of the zona pellucida surrounding the ova of certain vertebrates as seen in the light microscope.

1874 R. J. Dunglison *Dict. Med. Sci.* (new ed.) 1130/1 Zona reticularis... Zona glomerulosa... Zona fasciculata. **1881** F. M. Balfour *Treat. Compar. Embryol.* II. iv. 55 The ovum [of Teleosei] when laid is usually invested in the zona radiata only. **1883** E. Klein *Elem. Histol.* xliii. 340 The cortex of the [suprarenal] gland consists of an outer, middle, and inner zone... The outer one is the *zona glomerulosa*. *Ibid.* 341 Next follows the middle zone, or *zona fasciculata*. *Ibid.*, Next follows the inner zone, or *zona reticularis*. **1925** E. B. Wilson *Cell* (ed. 3) iv. 272 The zona radiata of the vertebrate ovum, conspicuously shown in the fishes, is a thick and often double membrane traversed by fine radial canals. **1964** Parker & Haswell *Text-bk. Zool.* (ed. 7) II. 319 The ovum [of *Salmo trutta*, the brown trout] is covered by a thick membrane, the zona radiata. **1965** Lee & Knowles *Animal Hormones* iv. 70 The cells in the innermost layer are in the form of a reticulum—the zona reticularis. *Ibid.*, The zona glomerulosa secretes mineralocorticoid. *Ibid.* 71 Hypophysectomy results in shrinkage mainly of the zona fasciculata and reticularis, and the blood glucose falls.

zonal, *a.* Add: **4.** *Soil Sci.* [a. F. *zonal* (N. Sibirtsev 1897, in *Compt. Rend. de la VII*[e]. *Session, Congr. géol. internat.* (1899) II. iii. v. 80).] Of a soil: regarded as characteristic of a particular climatic or geographic zone and as reflecting the predominant influence of the climate in its formation.

1908 *Jrnl. Agric. Sci.* III. 84 The seven fundamental groups of 'zonal' soils just enumerated are spread over the surface of large continents in zones which coincide with the physico-geographical zones of those continents. **1927**, etc. [see *intrazonal *a.*]. **1952** [see *azonal *a.*]. **1972** J. G. Cruickshank *Soil Geogr.* iv. 110 Apart from being an over-simplification of reality, the zonal system had the unfortunate effect of restricting, in the mind of the student, the distribution of each zonal soil within the limits of its climatic zone.

zonation. a. Delete '(see Zone *sb.* 7)' and add examples. Also *spec.* in *Ecol.*, the distribution of plants into specific zones which are characterized by their dominant species.

1898 *Rep. Brit. Assoc. Adv. Sci.* *1897* 863 (*heading*) The zonal constitution and disposition of plant formations. By Frederic E. Clements. *Ibid.*, The author has here reviewed the phytogeographical contributions bearing upon the subject in hand, with especial reference to the part they have played in the elaboration of the conception of zonation. In addition, he has endeavoured to demonstrate the fundamental universality of zonation in all divisions of the floral covering. **1904** [see *ecotone]. **1926** Tansley & Chipp *Study of Veg.* iv. 53 Adjacent plant communities are often arranged in more or less definite zones, following one another in constant order. This is specially noticeable in ascending a mountain (where it is called altitudinal zonation). **1936** J. Muir *Geol. Tampico Region, Mexico* 121 Reliable local zonation of both the Eocene and the Oligocene has been worked out in the Tampico area. **1969** *Nature* 15 Mar. 1005/2 Marine work during the first eight months concentrated on community structure and zonation on cliffs, reef flats and in mangrove woodland. **1971** *Scot. Jrnl. Geol.* VII. 306 It is first necessary therefore to establish a regional metamorphic zonation. **1981** *Birds* Autumn 63/3, I would have preferred to have seen an accompanying annotated sketch so that the plant zonations could be easily recognised.

b. Substitute for def: Formation of zones in the oocytes of certain plants and animals. (Earlier and later examples.)

1899 *Bot. Gaz.* XXVIII. 237 There is a stage called zonation in which the nuclei, usually in metaphase, are lined up around the ooplasm. **1975** *Acta Anat.* XCIII. 512 The zonation of the ooplasm reveals that a majority of the oocytes [of the teleost *Schizothorax richardsonii*] have a darkly stained inner and a lightly stained outer zone.

zone, *sb.* Add: **2. c.** (*Town-*)*Planning.* A district or an area of land subject to particular restrictions concerning use and development.

1909 H. I. Triggs *Town Planning* iv. 177 The usual method in formulating town building plans on a large scale is to divide the urban area into building zones. **1910** F. Howkins *Housing Acts 1890–1909, & Town Planning* ix. 125 Certain portions of the area would be reserved for the erection of better class residences... This would be similar to the 'zone' system which has been adopted in certain towns on the Continent. **1939** H. M. Lewis *City Planning* xvi. 166 A single map will show the subdivision of the area into zones, which in a typical case might be defined as follows: One family zones. Two family zones. [etc.]. **1953** [see *smokeless *a.* 2]. **1964** J. S. Scott *Dict. Building* 364 Certain areas can be kept for light industry, others for heavy industry, dwellings, offices..and so on, each area being called a zone.

d. *N. Amer. Football* and *Basketball.* A specific area of the court to be defended by a particular player; also, a mode of defensive play employing this system (cf. *zone defence*, sense 9 a below).

1927 G. S. Warner *Football Coaches & Players* 191 In the zone defense the players playing back of the line of scrimmage..are so stationed that they can knock down or intercept any pass that comes into their territory or zone. **1942** C. Bee *Zone Defense & Attack* i. 1 Certain coaches believe only the 'zone' can be called a team defense. **1964** Anderson & Albeck *Coaching Better Basketball* ix. 209 You cannot win consistently..utilizing the zone. **1971** L. Koppett *N.Y. Times Guide to Spectator Sports* iii. 83 Each player is given a specific portion of the floor as 'his territory', and he guards, in turn, any offensive player who enters his 'zone'. **1979** *Farmington* (New Mexico) *Daily Times* 27 May 8A/7 The defense must be a man-to-man with no presses or zones allowed.

e. One of those areas of Germany and Austria occupied by British, American, French, or Russian forces after the war of 1939–45 until 1955. Subsequently occas. applied to East German territory; also *transf.*

1945 *Times* 13 Feb. 4/1 The conference of Mr. Churchill, President Roosevelt, and Marshal Stalin, held at Yalta, in the Crimea, has drawn up military plans for the final defeat of Germany... The forces of the three Powers will each occupy a separate zone of Germany, and a central control commission will have headquarters in Berlin. France will be invited to take a zone of occupation. **1947** *Daily Tel.* 25 Sept. 5/8 German machinery..will be delivered to Hungary in the next three months under a trade agreement signed between Hungary and the Soviet zone of Germany. **1954** W. Faulkner *Fable* 128 Frenchmen had been spending their leaves..among the combat-troop rest-billets not only throughout the entire French Army zone, but the American and the British ones too. **1956** *Ann. Reg.* *1955* iii. v. 223 Some political leaders expressed the view that German interests would be furthered by direct negotiations with the Soviet Union, and that it would ease conditions for the people of the Soviet Zone if there was limited collaboration with the East German authorities. **1963** 'J. le Carré' *Spy who came in from Cold* xiii. 127 The GDR... The Zone if you prefer. **1964** L. Deighton *Funeral in Berlin* vii. 50 Not too near the Sektor boundary and within a mile of the Soviet Zone... Sometimes we prefer to put our cargo on ice in the zone... Anywhere from Lübeck to Leipzig. **1976** W. D. Graf *German Left since 1945* iii. 77 This movement aimed at the restoration of German unity... It was founded on the basis of proposals put forward by the communist leadership in the Soviet Zone.

8*. A hole in certain punched cards that is punched above the column of holes representing non-zero digits and is used in conjunction with these latter holes to represent non-numerical characters. Usu. *attrib.*

1950 W. W. Stifler *High-Speed Computing Devices* 149 For alphabetical representations, two perforations in a single column are used for each letter; one of these is a zone punch (0, 11, or 12) while the other perforation is made in the position identifying one of the digits 1 to 9. **1959** M. H. Wrubel *Primer of Programming for Digital Computers* ii. 33 Alphabetic information is recorded by using two punches in the same column: the upper punch (sometimes called the 'zone') is always a 12, 11, or 0. **1970** O. Dopping *Computers & Data Processing* ii. 44 A letter in the English alphabet is coded with one zone punch (12, 11, or 0) together with one under-punch. **1972** W. R. Price *Introd. Data Processing* vii. 179 The correspondence between the Hollerith zones and the 1401 zones. **1979** Davis & McCormack *Information Age* vi. 98 Later, when the need for alphabetic data arose, the zone positions, rows 12 and 11, were added.

9. a. *zone-mind*; **zone centre,** *spec.* in *Teleph.*, an exchange which acts as a main switching centre in an area containing a number of exchange groups; **zone defence,** (*U.S.* **defense**), *N. Amer. Football* and *Basketball*, a system of defensive play whereby each player guards an allotted portion of the field of play; **zone electrophoresis,** electrophoresis in which a solid but porous medium such as paper is used to ensure that the components remain separated in zones or bands according to their differing electrophoretic mobilities; **zone fossil** *Geol.*, a fossil characteristic of a particular zone or belt of strata; **zone levelling,** a process similar to zone refining in which the molten zone is passed repeatedly to and fro to produce a more homogeneous material; so **zone-level** *v. trans.*, **-levelled** *ppl. a.*; **zone leveller,** an apparatus used for zone levelling; **zone melting** = *zone refining* below; so **zone-melt** *v. trans.*; **zone plate** (examples); **zone refining,** a method of refining used to produce semiconductors and metals of very high purity by causing narrow zones of molten material to travel slowly along an otherwise solid rod or bar, so that impurities become concentrated at one end or the other if their solubilities in the liquid and the solid phases differ; so **zone-refine** *v. trans.*, **-refined** *ppl. a.*; **zone refiner,** an apparatus used for zone refining; **zone therapy,** a technique in which different parts of the feet (or palms) are massaged to relieve conditions in different parts of the body with which they are held to be associated; **zone time,** mean solar time at the standard meridian on which the local time zone is based, taken as the standard time throughout the zone (cf. *Zulu time* s.v. *Zulu *sb.* and *a.* 6).

1934 G. S. Berkeley *Traffic & Trunking Princ. in Automatic Telephony* i. 12 Level 94 is used for trunk calls instead of level 0 in the case of exchanges in Trunk Zone Centres. **1948** [see *route *v.* c]. **1960** R. Syski *Congestion Theory in Telephone Systems* ii. 53 A suitable exchange (zone centre) is established within each zone to handle the long-distance calls to and from that zone. **1971** *Gloss. Electrotechnical, Power Terms (B.S.I.)* iii. ii. 8 *Zone centre*, exchange acting as the main switching centre. **1927** Zone defense [see sense 2d above]. **1929** H. C. Carlson *You & Basket Ball* vi. 128 Zone defense, in which each individual is responsible for a certain zone. **1937** C. Allen *Better Basketball* xviii. 291 It is to be noticed in charting these penetrating offensive plays against a zone defense that the setup of the offense is identical with that used in penetrating the man-for-man defense. **1970** G. Sullivan *Pro Football A to Z* 341 *Zone defense*, a type of pass defense in which each of the three linebackers and four deep backs is assigned to cover a specific area of the field. Zone defense contrasts with man-to-man defense. **1952** *Proc. Soc. Exper. Biol. & Med.* LXXX. 42 (*heading*) Zone electrophoresis in a starch supporting medium. **1964** G. H. Haggis et al. *Introd. Molecular Biol.* ii. 23 (*caption*) The separation of some of the human plasma proteins by zone electrophoresis. **1975** Davis & Simpkins in Williams & Wilson *Biologist's Guide to Princ. & Techniques Pract. Biochem.* iv. 100 A common feature of the use of all supporting media is that the substances migrate as distinct zones which at the end of the analysis can be readily detected by suitable analytical techniques... The term zone electrophoresis has been applied to this method. **1904** Zone-fossil [see Zonal *a.* 3]. **1969** Bennison & Wright *Geol. Hist. Brit. Isles* ii. 23 It may be possible to recognize a zone without actually finding a specimen of the zone fossil itself if other highly characteristic species are found. **1975** J. G. Evans *Environment Early Man Brit. Isles* iii. 69 Evolution needed the temporal and biological continuum which Lyell's theory made possible. And from this came the concept of the 'zone fossil'. **1953** W. G. Pfann in *Trans. Amer. Inst. Mining & Metall. Engineers* CXCIV. 752/1 The discussion of zone-melting has been confined to the categories of zone-leveling and zone-refining. **1956** *Bell Syst. Technical Jrnl.* XXXV. 657, *p* fluctuations in zone leveled material are generally coarse. *Ibid.* 660 A zone leveler has been developed to provide growth conditions suitable for the production of quality germanium single crystals. **1974** *Jrnl. Physics* D. VII. 33 The mixed alloy was then placed in a porcelain boat and zone-levelled by passing through 10–15 zones. **1978** P. W. Atkins *Physical Chem.* x. 301 A modification of the technique, zone levelling, is used to introduce controlled amounts of impurity (for example, indium into germanium). **1952** W. G. Pfann in *Jrnl. Metals* IV. 747/1 A number of procedures will be indicated which have in common the traversal of a relatively long charge of solid alloy by a small molten zone. Such methods will be denoted by the general term *zone-melting*. **1965** Phillips & Williams *Inorg. Chem.* I. viii. 305 All such techniques as precipitation, partition, distillation, crystallization, chromatography, and zone-melting are based on phase equilibria. **1982** *Materials Lett.* I. 33/2 Dielectrically isolated single crystals of large extent may be fabricated by zone melting a thin silicon sheet that is encapsulated between SiO_2 layers. **1983** *Rev. Sci. Instruments* LIV. 385/2 The use of the heater assembly in the specific case of zone melting of mercury cadmium telluride compounds is described. **1932** Blunden *Halfway House* 55 When gong-like struck The violent crisis of zone-minds That chilled us with clouds turned winds. **1890** T. Preston *Theory of Light* ix. 178 A zone plate has therefore the power of a condensing lens. **1937** G. S. Monk *Light* xii. 167 The intensity of the image produced with a zone plate will be greater if alternate zones are not blocked out but are left transmitting, with a phase difference of one half period introduced between them and adjacent zones. **1978** *Sci. Amer.* Nov. 65/1 When the zone plate is illuminated with an X-ray plane wave, a converging spherical wave will come out. **1952** W. G. Pfann in *Jrnl. Metals* IV. 750/2 The particular merit of zone-refining becomes evident when repeated crystallizations are desired. **1956** —— in *Chem. & Engin. News* XXXIV. 1443/3 F. Montariol and coworkers..found that zone-refined aluminum cannot be hardened by cold-working. **1962** *New Scientist* 5 Apr. 813/2 To zone-refine and produce single crystals from such [refractory] materials, N. V. Philips's Gloeilampenfabrieken, Eindhoven, ..has developed a carbon-arc image furnace. **1973** J. G. Tweeddale *Materials Technol.* I. vi. 152 Probably the most effective and generally used method of refining to this kind of standard is zone refining. **1978** P. W. Atkins *Physical Chem.* x. 301 Bismuth is normally regarded as a hard, brittle metal, yet when it has been zone refined it forms rods which can be bent without fracture. **1983** *Metallurgical Trans.* A. XIV. 223/2 Straining electrode experiments were performed in zone refined and vacuum melted nickel alloys. **1959** *Times* 24 Nov. 19/7 It will be possible to obtain high quality single-crystal rods after a few passes in the zone-refiner. **1980** *Analytical Chem.* LII. 1738/1 A novel zone refiner is described, in which a single helical heater rotates in an annular sample space. **1917** W. H. FitzGerald *Zone Therapy* xvi. 157 Dr. Roemer..examined him in a characteristic zone therapy way. He searched the patient's fingers with a metal comb to find out what was the matter with his teeth. **1971** N. Saunders *Alternative London* xv. 123 Zone Therapy is based on the premise that one zone of the body acts as a microcosm of the whole. **1979** D. E. Bayly in A. Hill *Visual Encycl. of Unconventional Med.* 61 The origin of the reflex method is obscure. It is said that it came from China to the West... It is known to have been used by the natives of Kenya, and also by some American Indian tribes. At the beginning of this century it was called zone therapy by one Dr Fitzgerald [*sic*] in America who used it as a form of anaesthesia to render the patient insensible to pain when performing small operations, and to ease childbirth. **1908** H. B. Morse *Trade & Admin. Chinese Empire* viii. 203 The Eighteen Provinces roughly extend from longitude 98° to 122° E., comprising the seventh and eighth hours of the Zone time east of Greenwich. **1930** *Daily Express* 16 Aug. 1/3 The passengers wonder whether they should retire by Greenwich or zone time. **1981** G. Watkins *Exercises in Astro-Navigation* ii. 22 A vessel's chronometer or watch is not generally adjusted or altered at any time while the vessel is at sea, but her clock is altered as she moves through each zone in such a way that it should always be indicating the correct Zone Time for the vessel's position.

zone, *v.* Add: **4.** (*Town-*)*Planning.* To divide (a city, land, etc.) into areas subject to particular planning restrictions; to designate (a specific area) for use or development in this manner. Occas. *intr.* Also (*U.S.*) const. *out*, to forbid (the siting of an enterprise) in a given area. orig. *U.S.*

1916 *N.Y. Times* 4 Feb. 17/2 The plan to zone the city and regulate the height of buildings. **1919** *Melbourne Argus* 28 Aug. 6 The question of 'zoning' the metropolitan area, or separating the city into districts, in order that regulations may be applied to control the erection of shops and factories near residential sites has recently been occupying the attention of the Melbourne City Council. **1934** W. H. Heath in E. Betham *House Building 1934–36* xviii. 180 There is practically no area around London that is zoned in a reasonable manner. **1939** H. M. Lewis *City Planning* xvi. 169 All the frontage of main streets was placed in business zones although..only a small fraction of areas so zoned can ever be used for that purpose. **1967** *Boston Sunday Herald* 26 Mar. 1. 9/4 Planners..are concerned that a community will be thoughtfully zoned overall. **1971** P. Gresswell *Environment* 267 There is no guarantee that land zoned for housing will be released by the landowners. **1975** *N.Y. Times* 16 Oct. 29/1 A law that would 'zone out' massage parlors from the Times Square area on the principle that their proliferation is not sound community planning. **1976** *National Observer* (U.S.) 14 Aug. 7/2 When a municipality zones for industry and commerce for local tax benefits it..must zone to permit adequate housing within the means of the employes involved in such uses. **1977** *Chicago Tribune Mag.* 2 Oct. 8/2 At that time, which was before horse racing was zoned out of the city, the track was on Stony Island Avenue near 63rd Street. **1978** J. Updike *Coup* (1979) vii. 274 The land, they say, is zoned for agribusiness.

5. To restrict the distribution of (a commodity) to a designated area; used *spec.* concerning the allocation of foodstuffs in the war of 1939–45.

1942 *Hansard Lords* 3 June 103 We have arranged that the deliveries of bread shall be zoned. **1945** *Daily Herald* 31 Aug. 2/1 (Advt.), Cyder, like many other things, is now zoned to save transport and labour. **1952** *Ann. Reg.* *1951* 394 The Group scheme of the National Film Finance Corporation has been announced. This 'zoned' a considerable proportion of British production.

Hence **zo·ner,** one employed in the application of planning restrictions to particular areas.

1962 *Punch* 6 June 848/2 Planners and zoners won't apparently make up their minds. **1976** *Daily Tel.* 4 Apr. 3/6 This district contains about 1,000 homes approved by town zoners..after they were assured restrictive deed clauses would make the district a 'permanent adult community'.

zoned, *ppl. a.* Add: **4. b.** Distributed according to zones. Cf. *zone *v.* 5.

1943 *Daily Express* 16 Sept. 3/3 Compelling some grocers to buy their zoned cake through the co-ops.

5. (*Town-*)*Planning.* Designated for a particular type of use or development. orig. *U.S.*

1920 *Michigan Rep.* (Michigan Supreme Court) May 210 A penalty is provided for violation of said ordinance with a mandate prohibiting departments of the city from issuing permits for erection of the forbidden buildings within such zoned district. **1939** *Florida* (Fed. Writers' Project) iii. 396 Zoned residential, business, and industrial districts. **1970** *Cape Times* 28 Oct. 20/3 (Advt.), Paarl shopping centre. On 11000 sq. ft., zoned special business with free car park..in fast developing area.

Zonian (zōᵘ·niăn), *sb.* (*a.*) [f. Panama Canal *Zone* (see below) + -ian.] An American inhabitant of the Panama Canal Zone, a ten mile wide strip of land crossing the Isthmus of Panama on both sides of the Panama Canal, which was granted to the United States as a territory in 1904 and became independent in 1978. Also *attrib.* or as *adj.*

1910 *Everybody's Mag.* Mar. 322/1 Whatever your job, digging a canal, or driving a taxi-cab, or writing stock quotations, don't wear a uniform... The Zonian has a uniform, though, in spite of himself. It is the umbrella. **1950** *Social Forces* Dec. 161/1 The man marrying a girl of mixed blood may be unaware of the general Zonian prejudice that classes all Panamanians as 'colored'. **1951** *Panama Canal Rev.* 3 Aug. 9/3 Zonians..need no longer do without—just because the commissaries are closed. **1964** *Observer* 12 Jan. 1/5 'Zonians', American inhabitants of the [Panama Canal] Zone. **1979** P. Theroux *Old Patagonian Express* xii. 170 Within a very few months the treaty would be ratified. I told this to a Zonian lady... The Zonians, 3,000 workers for the Panama Canal Company..saw the treaty as a sell-out.

zoning, *vbl. sb.* (In Dict. s.v. Zone *v.*) Add: **1.** (Later examples.)

1942 *Times* 5 June 2/3 Further steps, including the zoning of the distribution of manufactured goods..are being investigated by the Ministry. **1945** *Daily Tel.* 17 May 5/7 In the zoning of Germany the United States Navy has been allotted the ports of Bremen and Bremerhaven. **1946** *How Britain was fed in War Time* (H.M.S.O.) iii. 22 The transport of food from production or import points to wholesalers' stores was effected by 'zoning' schemes... The country was divided into..zones, and movement between different zones..was allowed only by permit. **1968** *New York City* (Michelin) 16 After the First World War... A new zoning law regulated the height of buildings in relation to the width of the streets.

2. *spec.* in (*Town-*)*Planning*, the regulation of land use by particular planning restrictions in designated areas. orig. *U.S.*

1912 *Proc. 4th Nat. Conference City Planning* 190 In view of..existing indirect zoning laws, why not add some of the building laws in force in Frankfurt-on-the-Main? **1914** *Proc. 6th Nat. Conference City Planning* 92 The well-established principle of zoning that has been in operation for a generation or more in that country [*sc.* Germany]. **1921** *Glasgow Herald* 25 Mar. 8/3 'Zoning' means the locating of industries in the areas which are best suited for them and the reserving for housing of districts best adapted for this purpose. **1940** Graves & Hodge *Long Week-End* xi. 175 The Town-Planning Act of 1932 perpetuated this cleavage... Now there was 'zoning'..segregating families according to their incomes. **1943** J. S. Huxley *TVA* 31 Some States have introduced compulsory 'agricultural zoning' by which certain types of farming are prohibited in certain areas. **1966** R. F. Babcock (*title*) The zoning game: municipal practices and policies. **1978** J. A. Michener *Chesapeake* xiii. 830 Crossing over to the rivers south of Annapolis,..Chris had a chance to see how lack of zoning and policing had encouraged this shoreline to become a marine slum.

zonk (zǫŋk), *int.* (*sb.*) *slang.* [Echoic.] Representing the sound of a blow or heavy impact, used to indicate finality. Occas. as *sb.* (see quots.).

1949 T. Rattigan *Harlequinade* 38 Just sit there and relax and I'll dash and get you an enormous zonk of whisky. **1958** *Spectator* 15 Aug. 218/2, I..hurl it with a great zonk into the waste-paper basket. **1961** *Radio Times* 16 Nov. 47/2, I never took a note when I was interrogating. The moment you got hold of a piece of paper they'd think 'ah-hah..' and zonk! they'd button up. **1968** L. Deighton *Only when I Larf* i. 12 Silas..closed the safe door a few times. Zonk. It closed with a clang. **1979** R. Blythe *View in Winter* i. 64 He was a man with a catapult. He'd knock a pheasant down—*zonk!*

zonk, *v.* *slang.* [f. prec.]. **1.** *trans.* To hit, strike, or knock. Also *fig.*

1950 A. Melville *Castle in Air* I. in *Plays of Year* III. 338 If the Third Earl found that his wife had nipped off with another man while he was away at the Crusades, he'd have zonked her over the head with his kitbag. 1959 P. Bull *I know Face* xi. 201 We found ourselves back in my flat..zonking down the drink. 1960 I. Cross *Backward Sex* 188 She zonked me again on the head with this hairbrush. 1975 *New Yorker* 21 July 67/1 William Green tried to assure them that care had been taken to put provisions in the bill to see to it that New England 'doesn't get zonked'. 1979 G. Watson *Black Jack* xxii. 178, I felt zonked by this idea. It had never occurred to me. 1982 *Observer* 14 Nov. 15 ICI has invented a new adaptation to ethylene crackers that will zonk the competition and make feedstock costs less critical.

2. *intr.* To fail; to lose consciousness, to die.
1968 *Listener* 14 Mar. 352/3 If Johnny zonked, it would be bad for my book. 1977 *N.Y. Times Mag.* 4 Dec. 142 In a burst of determination, she'd been sitting in the bathtub doing her breathing for five hours straight—in one nostril, out the other—until she zonked and went rigid.

3. const. *out*. **a.** *intr.* To fall heavily asleep. **b.** *trans.* To overcome or knock out (in *fig.* senses).
1970 J. Sangster *Touchfeather, Too* iii. 75 He left me at seven a.m. and I zonked out until after mid-day. 1973 *Austral. Women's Weekly* 26 Dec. 32/5, I sank into my bed.., zonked myself out with sleeping pills, and woke up Friday. 1980 *Telegraph* (Brisbane) 21 Mar. 6/3 It's J.R.'s power that zonks women out. 1984 *N.Y. News Mag.* 18 Mar. 18/2 If mothers zonk out at three in the afternoon every day, they may continue that pattern after it's no longer necessary. 1985 *Sunday Times* 24 Feb. 36/6 No Junoesque oarswomen though... 'I think I row because it zonks me out, then I don't row with anyone.'

Hence **zo·nking** *ppl. a.* (freq. as quasi-*adv.* in *zonking great*).
1958 *Spectator* 25 July 130/2 He would give one a zonking great clip on the ear. 1959 P. Bull *I know Face* vi. 100 She was now technically a 'star' owing to her zonking success as Claudia. *Ibid.* vii. 126 She..is a zonking great film star. 1973 *Daily Tel.* (Colour Suppl.) 9 Feb. 36/4 *Long Day's Journey*..was the first big, zonking part he played after his cancer. 1976 *Times* 21 May 4/7 Rather than play these zonking great parts..I will try to find some dazzling little cameo roles.

zonked (zǫŋkt), *ppl. a.* (chiefly *pred.*). *slang.* [f. *ZONK *v.* + -ED¹.] **1.** Intoxicated by drugs or alcohol; 'stoned'. Freq. const. *out*. Also *transf.* and *fig.*
1959 *Esquire* Nov. 70/1 *Zonked*, one who is stoned, high, drunk. 1967 *New Scientist* 19 Oct. 185/1 Most of the drivers one meets should not be allowed to take charge of a car when sober—let alone when three parts zonked. 1967 P. Welles *Babykip* iv. 53 If only Mr Green weren't Jewish, he could swing around the world on the magic carpet completely zonked out. 1968 T. Wolfe *Electric Kool-Aid Acid Test* vi. 70 Everybody..had taken acid and they were zonked. 1972 J. Wambaugh *Blue Knight* (1973) xiv. 246 We sat..drinking *arak* and wine, and then beer, and we all got pretty zonked. 1973 H. Nielsen *Severed Key* x. 107 I'm serious. Zonked about her. Way out. 1975 *Publishers Weekly* 20 Jan. 78/1 Susan begins an affair with a zonked-out type who calls himself Commander Cloud. 1977 *Rolling Stone* 24 Mar. 84/1 Thousands of young people squeezed themselves into Radio City Music Hall to enjoy, scream at, get zonked to Jethro Tull. 1979 *Daily Tel.* 25 Apr. 21/5 A..Caucasian woman obviously zonked out..and a tracery of leaves resembling cannabis.

2. Exhausted, tired *out*.
1972 *Maclean's Mag.* Oct. 40/1 This portrait of his wife..zonked out on a floating sofa. 1976 J. Farris *Fury* i. 10 You just collapsed and..pulled the covers up around your head... You were completely zonked. 1978 *Washington Post Mag.* 19 Mar. 42/2 Patricia Wells, three hours after providing the high point of 2,300 people's evenings, was 'zonked' and went back to her hotel to bed. 1980 *Daily Tel.* 28 July 8/6 'Fairly zonked' by his non-stop 17 weeks of filming, he is recharging himself for the next stage.

zonkey (zǫ·ŋki). [f. Z(EBRA + D)ONKEY.] The offspring of a zebra and a donkey. Cf. *ZEDONK.
1953 *N.Y. Herald Tribune* 2 Sept. 1/7 Mr. [Gene] Holter explained that a zonkey was half zebra and half donkey, a combination the zoomen had never seen. *Ibid.*, 'Zonkey is not exactly a scientific name and I'm no scientist' Mr. Holter said, 'But I don't know what else you would call it.' 1973 *Indian Express* 29 Oct. 6/8 A zebra and a donkey are expecting an offspring next March, Brooklyn's Prospect Park Zoo authorities say. It will be New York's first zonkey, though. The others were born in western zoos. 1983 *N.Y. Times* 6 Mar. x. 53/4 Melancholy exemplars abound: a male camel who recently injured a foreleg;..and a morose-looking zonkey—the mother a zebra, the father a donkey.

zonky (zǫ·ŋki), *a.* (*sb.*) *slang.* Also **zonkey**. [f. *ZONK(ED *ppl. a.* + -Y¹.] Odd, weird, 'freaky'. Hence as *sb.*, a person in this state.
1972 *Daily Colonist* (Victoria, B.C.) 26 Apr. 22/5 'Wow, is this zonky?' breathed an admiring observer softly, as she gazed incredulously at the newly opened, all-cardboard room. 1975 *Globe & Mail* (Toronto) 25 Oct. 34/1 Unlike the honkey-tonk zonkies who used to hang out at his..workshop, shooting up or having sex in public view...Warhol has rarely, personally, done anything scandalous. 1977 *N.Y. Rev. Bks.* 28 Apr. 12/3 That combination of the new and old seems to have escaped most journalists except—in one of his zonky moments of insight—Hunter Thompson. 1979 B. Malamud *Dubin's Lives* i. 22 All the guy there does is..shakes his head. Makes you feel zonky..twittery. 1980 *Times* 6 Nov. 12/6 His book is really a study in ideas—or to coin an appropriately zonkey term—*weirdology*.

zonule. Add: **2.** *Geol.* (See quot. 1928.)
1928 C. L. & M. A. Fenton in *Amer. Naturalist* XI. 21 We have discussed the need for some term to designate the rocks bearing a faunule with several stratigraphers. In the course of one conversation Dr. Weller suggested 'zonule', and it has been approved by others... We propose the term, therefore, with the following definition: A zonule is the stratum or strata which contain a faunule, its thickness and area being limited by the vertical and horizontal range of that faunule. 1958 *Bull. Geol. Soc. Amer.* LXIX. 113/1 A zonule is a biostratigraphic unit that is recognizable in a sedimentary basin or similar restricted area of sedimentation. 1976 *BMR Jrnl. Austral. Geol. & Geophysics* I. 109 The assemblage can be referred to the middle Eocene *Proteacidites confragosus* Zonule on the basis of the presence of the nominate species.

zoo. Add: **2.** *transf.* A (diverse) collection, esp. of people; the place where they are assembled. (Freq. mildly contemptuous.)
1924 Galsworthy *White Monkey* I. ii. 11 You won't keep me in your Zoo, my dear. I shan't hang around and feed on crumbs. 1935 E. Blunden *Edward Gibbon & his Age* 14 He [sc. Gibbon] passed through Oxford, gathering little but materials for his future monody on a moribund zoo of dons. 1964 in *Current Slang* (Univ. S. Dakota) (1967) Spring 5 *Zoo*, place where students congregate. 1975 J. I. M. Stewart *Young Pattullo* i. 9 The Glencorrys were..the principal figures of fun in Ninian's and my own family zoo.

3. Comb. *zoo-crazy* adj.; **zoo-keeper**, an animal attendant employed in a zoological garden; also, a zoo owner or director; **zooman** *U.S. colloq.* = prec.; similarly **zoowoman** (*rare*).
1938 L. MacNeice *Zoo* iv. 71 A curate who was Anglo-Catholic..and .zoo-crazy. 1936 L. R. Brightwell *Zoo you Knew?* viii. 163 Zoo keepers gave up whistling 'Ta-ra-ra-boom-dehay'. 1960 D. H. S. Risdon *Zoo Keeper* vi. 53 Minor ailments should be curable by the Zoo keeper if he knows his job properly. 1977 *Monitor* (McAllen, Texas) 17 July 2F/4 Zookeepers caring for a two-week-old orphaned hippopotamus have taken to..wading into a pool..to nurse the hippo from a bottle. 1930 *Time* 13 Oct. 32/3 U.S. zoomen, animal catchers and park directors attended the annual meeting of the American Association of Zoological Parks and Aquariums... One zooman had too many elephants. *Ibid.* 33/2 Conspicuous among the zoomen was the only zoowoman in the world. 1942 Berrey & Van den Bark *Amer. Thes. Slang* §458/17 *Zooman*, a zoo director or caretaker. *Ibid.* §624/16 Animal man, zooman. 1973 *Times Lit. Suppl.* 9 Mar. 262/4 The increasing value of the *Yearbook* to zoomen and biologists.

zoo-. Add: **zo:o,archæo·logy**, the study of the animal remains of archæological sites; hence **zo:o,archæo·logist**; **zooce·ntric** *a.*, centred upon the animal world; regarding or treating the animal kingdom as a central fact; **zo·ochore** (-kō³ɪ) [Gr. χωρεῖν to spread], a plant whose seeds are dispersed by animals; hence **zoocho·rous** *a.*; **zoo·chory**, the dissemination of plant seeds by animals; **zoomagnetism** (earlier example); **zoonose**: read **zoono·sis** (pl. **-noses** (-nō³·sīz)); usu. restricted to diseases transmitted naturally to man from animals; (further examples); hence **zoono·tic** *a.*; **zoopla·nkter** [*PLANKTER], an individual organism of the zooplankton; **zooplankto·nic** *a.*, of, pertaining to, or consisting of zooplankton; **zoopraxo·graphy** [Gr. πρᾶξις], the study of animal locomotion; **zo:osemio·tics** *sb. pl.* (const. *sing.*), the study of animal communication through the investigation of signalling behaviour in and between species; **zootheism** (earlier example).
1984 *Nature* 1 Mar. 88/2 John Speth's *Bison Kills and Bone Counts* extends such an invitation to zooarchaeologists and to everyone who uses faunal evidence from archaeological sites to reconstruct past human diet. 1972 *Science* 20 Oct. 297/2 Recently the *Atlas of Animal Bones* by Elisabeth Schmid has become available for research workers in zooarcheology. 1985 *Times Lit. Suppl.* 7 June 646/1 Taphonomy..has..only recently become an integral part of zoo-archaeology (or, as the subject of faunal analysis is more usually called in Europe, archaeozoology). 1882 *Trans. Anthropol. Soc. Washington* I. 93 In later times a few of this school have expanded their scheme to embrace the animal world in general, rendering it zoöcentric instead of anthropocentric. 1977 J. L. Harper *Population Biol. Plants* 433 Virtually all of the work reported is that of zoologists and the research is zoocentric. 1905 F. E. Clements *Res. Methods Ecol.* 216 Migration results when spores, seeds, fruits, offshoots, or plants are moved out of their home by water, wind, animals, man, gravity,..or mechanical propulsion. Corresponding to these agents there may be recognized the following groups:..Animals, zoochores. *Ibid.* 218 Species which grow in exposed grassy or barren habitats are for the most part anemochores, while those that are found in the shelter of forests and thickets are usually zoochorous. 1960 *McGraw-Hill Encycl. Sci. & Technol.* X. 499/2 Animal dispersal, zoochory, is divided into epizoochory (barbed or sticky disseminules, desmochores..) and endozoochory (disseminules eaten and egested by animals). 1969 L. van der Pijl *Princ. Dispersal in Higher Plants* v. 24 We enter here the more general field of zoochory. *Ibid.*, All following zoochorous classes can be subdivided by crosswise partitions as follows. 1974 *Nature* 8 Feb. 407/1 Modes of dispersal, namely aerial (both active and passive), hydrochorous, zoochorous and anthropochorous, are discussed at some length. 1980 *Botanisk Tidsskrift* LXXV. 159 Dispersal is probably mostly by means of water flowing through pores and channels in the soil, but zoochory also plays a role. *a* 1834 S. T. Coleridge *Table Talk* (1884) 73 Nine years has the subject of Zoo-magnetism been before me. 1894 G. M. Gould *Med. Dict.* 1631/1 *Zoonosis*, any disease communicated or communicable from one of the lower animals to man. 1956 *Nature* 3 Mar. 407/2 When a zoonosis gets under way, man-to-man contact may be sufficient to keep the infection spreading. 1972 N. D. Levine in T.-T. Chen *Res. Protozool.* IV. 340 This was the first proof that simian malaria is a true natural zoonosis. 1974 R. Zeledón in K. Elliott et al. *Trypanosomiasis & Leishmaniasis* 51 Chagas' disease..became a zoonosis when the reduviid insect vectors adapted to human dwellings. 1900 Dorland *Med. Dict.* 769/2 Zoonotic. 1956 *Nature* 3 Mar. 407/2 In searching for explanations of..zoonotic outbreaks, there are limitations in the taxonomic approach which should be borne in mind. 1980 *Brit. Med. Jrnl.* 29 Mar. 928/2 Zoonotic pathogens, such as salmonellas.., may be present in any type of slurry. 1963 *Spec. Sci. Rep. U.S. Fish & Wildlife Service* No. 452 (*title*) A towed pump and shipboard filtering system for sampling small zooplankters. 1979 *Nature* 1 Feb. 353/2 Zooplankters must also cope with the consequences of living in a transparent medium. 1911 *Rep. Brit. Assoc.* 422 In the high Alpine lakes there exists an outstanding production of zooplanktonic organisms. 1964 *Oceanogr. & Marine Biol.* II. 152 This technique..is well suited to analysis of zooplanktonic extracts. 1978 *Nature* 20 July 246/1 Determinations of element concentrations in the fecal pellets from a common zooplanktonic species, the euphausiid *Meganyctiphanes norvegica*, are now available for 18 elements. 1891 E. Muybridge (*title*) The science of animal locomotion (zoöpraxography). 1893 —— *Descriptive Zoöpraxography* 2 In the presentation of a lecture on Zoöpraxography the course usually adopted is to project..a series of the most important phases of some act of animal motion. 1947 L. Edwards *Reminisc. Sporting Artist* xv. 151 The science of animal motion (Zoöpraxography) cannot be entirely ignored by artists, more especially since the advent of instantaneous photography has familiarised the public with the camera's version of animal movement. 1974 *Country Life* 2 May 1059/1 His [sc. Lionel Edwards'] preoccupation with zoopraxography is a desire to capture a precise impression of the rhythm of venery and racing. 1963 T. A. Sebeok in *Language* XXXIX. 465 The term *zoosemiotics*—constructed in an exchange between Rulon Wells and me—is proposed for the discipline, within which the science of signs intersects with ethology, devoted to the scientific study of signaling behavior in and across animal species. 1968 *Language* XLIV. 211 The Section of Semiotics and Linguistics of the Laboratoire d'Anthropologie Sociale is assembling a library of offprints on the areas within its purview (viz., linguistics, oral and written literature,..theory of languages, scientific languages, zoösemiotics, and the like). 1978 *New Yorker* 17 Apr. 78 In a collection of papers written by various experts in the field of..'zoosemiotics'—in other words, animal communication—each writer tries valiantly to define what he means by the term. 1881 *Abstr. Trans. Anthropol. Soc. Washington* 128 Let us hope that American students will not fall into this line of error by assuming that zoötheism is the lowest stage, because this is the status of mythology most widely spread on the continent.

zoochlorella (zō³:oklore·lă). *Bot.* Also **Zoo-**. Pl. **-æ**. [mod.L. (coined in Ger. by K. Brandt 1881, in *Arch. für Anat. & Physiol.: Physiol. Abt.* 571), f. Zoo- + CHLOR-¹ + L. *-ella* (see -EL²).] One of the numerous green unicellular organisms, believed to be algæ, that are present as symbionts in the cytoplasm of many freshwater invertebrates.
1889 in *Cent. Dict.* 1899 F. S. Lee tr. *M. Verworn's Gen. Physiol.* ii. 84 Among such symbiotic organisms are especially many algæ, the *Zooxanthellæ* and the *Zoochlorellæ*, the nature of which as independent organisms has been for a long time in dispute. They occur abundantly in the cells of lower animals. 1924 J. A. Thomson *Sci. Old & New* xxvii. 150 The same kind of co-operation is illustrated by a number of green Protozoa, in cases where the green colour has been shown to be due..to a partnership with minute algae (Zoochlorellae and Zooxanthellae). 1972 M. S. Gardiner *Biol. of Invertebrates* i. 34/1 Zooxanthellae are limited to marine species, and zoochlorellae almost entirely to freshwater species. 1979 *Nature* 5 July 58/2 Traces of chlorophyll *b* in extracts of pigments from intact animals suggested that *Amphistegina* spp. and *Amphisorus hemprichii* also contained symbiotic zoochlorellae.

zoogeographic, *a.* (Examples.)
1907 *Jrnl. Geol.* XV. 296 Zoögeographic works in all points confirm the conclusions. 1971 *Nature* 9 July 88/1 The largest zoogeographic group so far recovered..is found in the southern shallow waters of the Florida-Halteras slope.

zooglœa. Add: Hence **zooglœ·al** *a.*
1934 A. T. Henrici *Biol. of Bacteria* xviii. 289 *Nitrosomonas* species may appear in cultures in two forms—as zoögloeal masses which rest at the bottom of liquid cultures, in which the cells are non-motile; and as motile 'swarmers'. 1976 *Ann. Rev. Microbiol.* XXX. 265 Protozoa crawl about the zoogloeal mass and occur in the liquor; most are free-swimming or stalked ciliates.

zoological, *a.* Add: **a.** (Earlier example.)
1807 J. Pinkerton *Mod. Geogr.* (ed. 2) I. p. xxvii, Conceiving that the zoological part might admit of some

improvements..the author applied to Dr. Shaw of the British Museum.

b. Also *zoological park*.
1899 *N.Y. Times* 9 Nov. 14/3 A splendid stretch of 261 acres of country, that is destined to accommodate one of the finest collections of wild animals in the world, and will be known as the New York Zoological Park, was formally opened..yesterday. 1935 *Chambers's Encycl.* X. 811/1 The Zoological Park of Edinburgh,..one of the most beautiful. 1978 Jordan & Ormrod *Last Great Wild Beast Show* i. 48 Stellingen itself remains the finest example of the use of moated enclosures in zoological parks.

zoologizing, *vbl. sb.* (Earlier example.)
1815 [see Entomologize *v.*].

zoology. Add: Also popularly with pronunc. (zū,ǫ·lŏdʒi). (Earlier example.)
Similarly with the derivatives.

zoom, *v.* Add: **1.** Also, to travel or move (as if) with a 'zooming' sound; to move at speed, to hurry. Also *loosely*, to go hastily. Freq. with advbs. *colloq.*
1924 *Brit. Weekly* 18 Dec. 270/1 Trams zoom along and buses rattle past. 1946 Wodehouse *Joy in Morning* xxix. 280 How would it be..to zoom off immediately, without waiting to pack. 1960 T. McLean *Kings of Rugby* 118 Hewitt soon zoomed away on the right. 1976 *National Observer* (U.S.) 14 Aug. 16/1 Every night..a speedboat zooms into the Jones Beach Marine Theater. 1977 G. Durrell *Golden Bats & Pink Pigeons* v. 121 Three cleaner fish..worked assiduously on their three customers, zooming in to suck the parasites off their skin.

2. a. (Later examples.) Also *transf.* In recent use, often not distinguished from sense 1.
1920 *Blackw. Mag.* July 71/2 The bird checked, swerved and dived and zoomed back into level flight again. 1934 [see *DOG-FIGHT *v.*]. 1940 *War Illustr.* 19 Jan. 620/3 The mother-ship would be guarded by 350 m.p.h. fighters that would zoom up into the skies about her on the first hint of danger. 1962 S. Carpenter in *Into Orbit* 75 At 28,000 feet, diving at 900 mph..we suddenly pulled through and started to zoom up again..This manœuvre converted some of our speed into zoom energy. 1980 J. Ditton *Copley's Hunch* II. i. 115 He zoomed up and over to gain height.

b. *trans.* To cause (an aircraft) to zoom; also, to fly over (an obstacle) in this manner.
1918 *Independent* 16 Nov. 208/2, I 'gave 'er all the gun' and 'zoomed' the château—that is, I almost went up the front of the place. 1928 V. Pagé *Mod. Aircraft* 521 A machine should never be 'zoomed' or made to jump into the air by a too-rapid movement of the elevator flaps.

c. *intr.* fig. Of prices, costs, etc.: to rise sharply; to soar or rocket. *colloq.*
1970 *Daily Tel.* (Colour Suppl.) 6 Feb. 17/3 They must double labour and work overtime. Costs zoom. 1976 *National Observer* (U.S.) 17 July 3/1 By March 1978..the dropout total would zoom to 498,300—50 times the total as of March 1972. 1981 *Times* 12 Sept. 2/1 He did not think that the Prime Minister had ever said the economy was going to zoom.

3. *Cinematogr.* and *Photogr.* **a.** *intr.* Of a camera, lens, etc.: to close up on a subject (esp. rapidly) without losing focus; more generally, to alter range by variation of focal length. Freq. const. *in* (*on*). Also *fig.*
1948 *Jrnl. Soc. Motion Picture Engin.* LI. 296 Does the speed change while zooming?..The speed is independent of the zoom. 1959 Halas & Manvell *Technique Film Animation* xix. 237 The scene opens with a full screen live-action back-projection shot which zooms to a miniature. 1962 *Daily Tel.* 8 June 23/7 The lens is capable of 'zooming-in' on a set target up to a mile distant. 1970 *Amateur Photographer* 11 Mar. 13 (Advt.), Needle-sharp f1.8 lens—zooms from telephoto to wide angle. 1973 H. J. Eysenck *Inequality of Man* ii. 84 The computer will continue to select items in such a way as to 'zoom in' on the crucial set of items which will really test the subject's IQ, avoiding all the useless items which are below or above his true level. 1978 *N.Y. Times* 30 Mar. D18/2 The TV camera zoomed in on a triumphant Holmes.

b. *trans.* To cause (a lens fitment, camera, etc.) to alter range in this manner.
1952 *Applied Electronics Ann. 1951* 57/1 The construction makes it possible for the operator to 'zoom' the lenses after the ball or player. 1975 *Physics Bull.* Nov. 481/2 The image magnification, AEI claims, can be zoomed continuously up to × 15 000 000 with no change in image focus or brightness. 1979 *SLR Camera* Mar. 5/3 Have you tried focusing and zooming a lens that operates in the opposite direction to others in your gadget bag?

Hence **zoo·ming** *vbl. sb.* and *ppl. a.*
1917 [see Zoom *v.* 2]. 1923 *Blackw. Mag.* July 7/1 We settled into steady, zooming flight. 1961 G. Millerson *Telev. Production* iii. 34 (*caption*) The zoom angle can be adjusted anywhere in the zooming range. 1982 L. Cook *Under Etna* I. i. 11 'I hate this zooming-in bit.' She gazed hypnotically at the strip of runway coming up to meet them.

zoom, *sb.* (In Dict. s.v. Zoom *v.*) Add: **1.** *Aeronaut.* (Further examples.) Also *attrib.*
1918 *Blackw. Mag.* June 762/1 The Hun's third repetition of the manœuvre was varied by a straight zoom instead of a climbing turn. 1932 [see *BLACKING *vbl. sb.* 1 c]. 1960 *Aeroplane* 2 Dec. 741/1 At the zoom altitudes which the Mirage is capable of reaching, there is insufficient airflow through the engine to maintain afterburner combustion.

2. *Cinematogr.* and *Photogr.* **a.** A camera shot in which the range is (usu. rapidly) shortened to close-up without loss of focus; this process.
1934 [see *FADE-OUT 1]. 1948 [see *ZOOM *v.* 3 a]. 1962 *Movie* Dec. 6/3 The film ends on a zoom into a close up of Tolly's clenched fist. 1967 McLuhan & Fiore *Medium is Massage* 128 The audience had..been preconditioned by television commercials to abrupt zooms.., no story lines, flash cuts. 1977 J. Hedgecoe *Photographer's Handbk.* 125 (*heading*) Zoom and tilt.

b. = *zoom lens*, sense 4 below.
1974 *Some Technical Terms & Slang* (Granada Television), *Zoom*, lens allowing a change of view from long-shot to close-up. 1978 *Amateur Photographer* 29 Nov. 120 He uses a Nikon with 24mm, 43–86mm and 80–200mm zooms, and uses Kodachrome exclusively. 1982 'A. J. Quinnell' *Snap Shot* vi. 99 [He] had taken the telephoto lens off the camera and replaced it with an all-purpose zoom.

3. *fig.* Zest, energy; sparkle, zip. Cf. *ZOOM *int. colloq.*
1964 in Hamblett & Deverson *Generation X* 97 Perhaps also a man has to have a bit of an inferiority complex..to give him that extra zoom as a lover. A smug, inbred type like Roger doesn't care if he's a dead loss to a woman between the sheets. *a* 1974 R. Crossman *Diaries* (1977) III. 297, I am definitely losing political zest, looking and feeling more detached, with less zoom, watching, not believing in things, not as enthusiastic or inspired.

4. Comb. **zoom lens** *Cinematogr.* and *Photogr.*, a camera lens whose focal length (and hence the magnification and the field of view) can be smoothly varied while the image remains in focus; cf.* VARIFOCAL *a.*; **zoom shot**, a camera shot taken with a zoom lens.
1936 *World Film News* Apr. 12 No British manufacturer makes a nice, cheap zoom-lens. 1949 R. H. Alder *Movie Making for Everyone* vi. 67 Most amazing of all is the 'Zoom' lens, in which the angle of view can be changed while the picture is actually being taken. 1962 L. Deighton *Ipcress File* xxiv. 154 A zoom lens, one that would change its focal length. 1978 J. Gardner *Dancing Dodo* xix. 146 The drifting began. Like a huge, long pull-back on a zoom lens so that everything diminished. 1984 A. C. & A. Duxbury *Introd. World's Oceans* iii. 104 The Alvin was equipped with a new color television camera that had a special zoom lens for inspecting the vent animals. 1966 *Listener* 24 Feb. 286/1 A pre-credit titles vertical zoom-shot is served as *hors d'œuvre* for those with appetites for impudently expert camera work. 1977 T. Allbeury *Man with President's Mind* x. 108 A zoom shot on the representatives of OPEC.

zoom (zūm), *int.* [See the vb.] Representing a 'zooming' sound, such as that made by something travelling at speed. Freq. used *fig.* to denote a sudden rise (to success, etc.) or equivalent fall.
1942 D. Powell *Time to be Born* vi. 130 Men..just lucky enough to hold a job a few years and then—zoom! 1942 Berrey & Van den Bark *Amer. Thes. Slang* § 2/11 All of a sudden..socko..whang, zoom. 1976 *National Observer* (U.S.) 25 Dec. 4/2 People began to talk about him. An article appeared in Advertising Age, a trade magazine. Then he wrote a guest column for the Hollywood Reporter. Zoom.

Zoomar (zū·mar). Also **zoomar**. A proprietary name in the U.S. for a make of zoom lens.
1946 *Jrnl. Soc. Motion Picture Engineers* XLVII. 465 (*caption*) Optical principle of Zoomar lens. 1947 *Official Gaz.* (U.S. Patent Office) 10 June 177/2 'Zoomar' for varifocal camera lenses. 1961 A. L. M. Sowerby *Dict. Photogr.* (ed. 19) 712 The first zoom lens for a miniature camera, the Voigtländer Zoomar, gave focal lengths from 36 mm to 82 mm. 1971 D. E. Westlake *I gave at Office* (1972) 68 Rudy had a zoomar lens and got as close as he could for greater clarity. 1973 D. Osmond-Smith tr. *Bettetini's Lang. & Technique of Film* ii. 94 By making use of a complicated system of lenses, called 'pancinor' or 'zoomar', it is in fact possible to vary in a continuous progression the focal length of the shot and, therefore, to move the lens with apparent continuity towards or away from an object or person.

zoomorph. (Earlier example.)
1889 [see *SKEUOMORPH 1].

zoomorphism. 1. (Earlier example.)
1822 tr. *Malte-Brun's Universal Geogr.* I. xxiii. 576 The most gross is the religion of the Egyptians, in which the attributes of the divinity were represented under the figures of animals... This may be termed *zoomorphism*.

zo,omorpho·sed, *ppl. a. rare.* [f. Zoomorph, after Metamorphosed *ppl. a.*] Of a decorative or symbolic design: formed into an animal-like shape; rendered zoomorphic (sense 1).
1955 *Proc. Prehist. Soc.* XXI. 234 The famous sheath with zoomorphosed wave-tendril,..which forms part of his Third La Tène style. 1967 *Antiquaries Jrnl.* XLVII. 211 In two recent papers de Navarro has discussed the most common and indeed interesting occurrence of the horse in Celtic art—as part of a zoomorphosed lyre.

zoon (zūn), *v.* (and *int.*) *U.S. colloq.* [Echoic; cf. Zoom *v.*] **a.** *intr.* To make a humming or buzzing sound; to move quickly. **b.** *trans.* To (cause to) travel with such a sound; to propel. Also as *int.* Hence **zoo·ning** *vbl. sb.* (also applied *spec.* in *Black English* to a style of preaching and response characterized by the repetition of words and phrases with tonal variation).
1883 J. C. Harris *Nights with Uncle Remus* xxxvii. 224 Bimeby Brer Rabbit year de skeeters come zoonin' 'roun', en claimin' kin wid 'im. 1909 *Dialect Notes* III. 391 *Zoon*, *v.i.* and *tr.*, to make a humming or buzzing sound, to cause to make such a sound. 'That rock came *zoonin'* by my head.' 'Watch me *zoon* this rock.' 1911 M. Johnston *Long Roll* xv. 197 Zoon—Zoon—Zoon! O Lord! listen to that shell. 1922 *Outward Bound* Nov. 137/1 The zoonings and ploppings of blundering winged intruders. 1950 W. L. James in *Phylon* XVI. i. 19 The prayer maker is too full for any utterance which is not colored tonally by his emotions. This is called 'Zooning'. 1977 J. L. Dillard *Lexicon Black Eng.* iii. 55 The 'cries'..may come relatively early in the [church] service and show the importance of tonal phenomena in the Black sermon. There is even a term for one such practice: *zooning*, 'crying (a word or phrase) over and over with variations'.

‖ **zoon politikon** (zō³·ǫn poli·tikǫn). [Gr.] = *political animal* s.v. *POLITICAL *a.* 6 (q.v.).
1958 W. Stark *Sociology of Knowledge* 238 Anybody who regards man as necessarily..a *zoon politikon*. 1971 A. Giddens *Capitalism & Mod. Social Theory* 5 In the Greek *polis* every man—that is, every free citizen—was a *zoon politikon*: the social and political were inextricably fused.

zoophagan, *sb.* (Earlier example.)
1840 [see *entomophagan* s.v. *ENTOMO-].

zoophilia (zō³ofi·liă). *Psychol.* Also **zoöphilia**. [f. Gr. ζῷον animal + φιλία affection.] Attraction to animals that acts as an outlet for some form of sexual energy, formerly not implying sexual intercourse or bestiality.
[1899 Rebman tr. *Krafft-Ebing's Psychopathia Sexualis* iii. 267 In close relation to stuff-fetichism, certain cases must be considered in which beasts exercise an aphrodisical influence over human beings. One feels tempted to call it *Zoophilia Erotica*. This perversion seems to be rooted in a fetichism the object of which is the skin of the beast.] 1906 H. Ellis *Stud. Psychol. Sex* V. 71 There is..the more or less sexual pleasure sometimes experienced..in the sight of copulating animals. This I would propose to call Mixoscopic Zoophilia; it falls within the range of normal variations. 1908 M. E. Paul tr. *Bloch's Sexual Life of our Time* xxiii. 641 We will first describe zoophilia, a sexual inclination towards animals without actual sexual intercourse. 1940 Hinsie & Shatzky *Psychiatric Dict.* 558/2 Zoöphilia..is a term coined by Krafft-Ebing to denote sexual excitement caused by the stroking and fondling of animals. It does not refer to sexual intercourse with animals. 1960 *Arch. Gen. Psychiatry* III. 442/1 Zoophilia was known in antiquity. 1966 R. & D. Morris *Men & Apes* iii. 65 The extent to which zoophilia involving monkeys was actually practised is difficult to assess. 1978 *Daily Tel.* 2 Dec. 16 One of your contemporaries referred in its review to the 'zoophilia' in this film. A simpler word is animalism or bestiality.

zo,o-philic, *a.* [f. as Zoophile + -ic.] **a.** Characterized by zoophilism; animal-loving. **b.** *Psychol.* Characterized by zoophilia.
1947 *New Biol.* 152 Our chief concern, however, is with the dog as a beast of burden. In England, presumably owing to the prevalence of zoophilic organisations such as the R.S.P.C.A. such practices are illegalised. 1951 J. Steinbeck *Log* p. xxiv, A sexual, a religious, a zoophilic or a gustatory impulse. 1965 *Movie* Spring 23/2 Although some spectators may derive a premature zoophilic *frisson* from Marnie's words to her horse, 'Oh, Forio, if you want to bite anyone, bite me,' we don't learn that there is anything psychologically wrong with Marnie until the sequence in which we meet Mother.

zoot (zūt). *U.S. slang.* [See next.] **1.** A zoot suit.
1965 'Malcolm X' *Autobiogr.* iv. 59, I saw some of the real Roxbury hipsters eyeing my zoot. 1973 C. Himes *Cotton gonna kill me Yet* in *Black on Black* 196 This George Brown was strictly an icky, drape-shaped in a fine brown zoot with a pancho conk slicker'n mine. 1973 T. Pynchon *Gravity's Rainbow* II. 246 Where'd you get that zoot you're wearing, there?

2. *Comb.* **zoot-shirt**, a (brightly coloured) shirt designed to be worn with a zoot suit.
1959 A. Fullerton *Yellow Ford* viii. 95 He wore a multi-coloured zoot shirt. 1961 *Times* 8 Mar. 14/7 For men students..zoot shirts..are banned.

Hence **zoo·ty** *a.*, in the style of a zoot suit; (strikingly) fashionable, 'sharp'.
1946 Mezzrow & Wolfe *Really the Blues* IV. xvi. 313 Colored kids..work on their dungarees, pegging the legs till they're real sharp and zooty. 1952 *Amer. Speech* XXVII. 20 What the zootie character and the dude have in common is an overfastidious regard for clothing. 1964 S. Bellow *Herzog* 240 Her lover, too, with long jaws and zooty sideburns. 1974 *Listener* 8 Aug. 166/1 The suits were..the zooty type—that's the American style of suiting with a straw hat.

zoot suit (zū·t sⁱūt). orig. *U.S. slang.* [Redupl. rhyming formation on Suit *sb.*] **1.** A type of man's suit of exaggerated style popular in the 1940s (orig. worn by U.S. Blacks), characterized by a long, draped jacket with padded shoulders, and high-waisted tapering trousers.

1942 Gilbert & O'Brien *Zoot Suit* (song) 3, I want a Zoot Suit with a reat pleat, with a drape shape. 1942 [see *REET *a.*]. 1949 R. Chandler *Little Sister* xxix. 218 Taking knives away from grease-balls in zoot suits. 1951 E. Paul *Springtime in Paris* xv. 269, I saw a few Zoot suits, please believe me, on Negroes headed for the bar at No. 12. 1969 *Time* 14 July 16/2 Chavez became a *pachuco*, affecting a zoot suit with pegged pants. [illegible]

2. *transf.* in various *Mil.* uses. Chiefly *U.S.*
1943 *Yank* 15 Oct. 3 Some of the Japs even wear our jungle 'zoot suits'. 1945 Baker *Austral. Lang.* 158 Zoot suit (which, of course, came originally from U.S. jive slang) for the crude civilian clothes given to discharged servicemen. 1952 D. Clarke *Eleventh at War* xvii. 216 The warm 'tank suits'..were soon known to the troops as 'zoot suits', and were aptly described as a 'mass of zip-fasteners joined together by windproof and waterproof material'. An ingenious manipulation of the zips could convert the zoot suit into a sleeping-bag. [illegible]

Hence **zoot-suited** *a.*, dressed in a zoot suit; **zoot-suiter**, a zoot-suited person; *spec.* one of a group or gang of young men wearing zoot suits.
1969 *Zoot-suited* [see *crash-helmeted* adj. s.v. *CRASH *sb.*¹ 2]. 1971 A. Price *Alamut Ambush* ix. 105 [illegible] 1943 *Chicago Daily News* 12 June 6/2 A new human variety has just entered the fauna [in California]—the 'zoot suiters'. 1952 *Zoot-suiter* [see *DADDY-O]. [illegible]

zooxanthella (zō³:ǫzænþe·lă). *Bot.* Also **Zoo-**. Pl. **-æ**. [mod.L. (coined in Ger. by K. Brandt 1881, in *Arch. für Anat. & Physiol.: Physiol. Abt.* 572), f. Zoo- + Xanth(o- + L. *-ella* (see -EL²).] One of the numerous yellow-brown unicellular organisms present in the cytoplasm of many radiolarians, corals, and other marine invertebrates, probably as symbionts.
1889 in *Cent. Dict.* 1899, 1924 [see *ZOOCHLORELLA]. 1957 Nolan & Gojdics in T. T. Chen *Res. Protozool.* II. 235 The zooxanthellae found in Radiolaria and Foraminifera (as well as in some lower invertebrates) are probably specialized dinoflagellates or cryptomonads. Presumably they play the same role that zoochlorellae do in freshwater forms, but a mutually beneficial relationship has not been experimentally proved. 1972 [see *ZOOCHLORELLA]. 1975 *Nature* 6 Nov. 37/1 The only chlorophyll in corals is contained in the brown zooxanthellae. [illegible]

‖ **zoppa** (tsǫ·pā), *a.* *Mus.* [It., fem. of *zoppo* limping (formerly also used).] (See quots.) Freq. in phr. *alla zoppa*.
1740 Grassineau *Mus. Dict.*, *Zoppo*, lame..hopping; [illegible] 1880 Grove *Dict. Mus.* IV. 514/1 *Zoppa, alla*, a term applied to a rhythm in which the second quaver in a bar of 2-4 time is accentuated. [illegible] 1963 *Times* 25 Feb. 5/1 Once heard, its closing gavotte, with the crisp, jaunty, and insinuating appoggiaturas of the kind called *zoppa*, hangs in the mind for ever.

Zoque (sō³·ke), *sb.* (and *a.*) [a. Sp., of uncertain origin.] Any of a group of Central-American Indian languages of the Mixe-Zoquean family; this group of languages collectively. Also *attrib.* or as *adj.*
1891 D. Brinton *Amer. Race* III. vi. 144 (*heading*) Zoque linguistic stock. 1911 Thomas & Swanton *Indian Lang. Mex. & Central Amer.* 60 (*heading*) Zoque. 1940 F. Johnson in *Maya & Neighbors* vi. 209 The territory in which the Zoque language was spoken has scarcely been changed on any map since Thomas. 1953 [see *non-phonemic* s.v. *NON- 3]. 1964 E. A. Nida *Toward Sci. Transl.* vi. 134 As in Zoque, spoken in Mexico, the Biblical expression 'Perfect love casts out fear' becomes 'we do not fear because we love.' 1977 [see *MIXTEC].

zores, -us, varr. *TSORES.

‖ **zori** (zǫ·ri), *sb. pl.* Also 9 **sori**. [Jap., f. *sō* grass, (rice) straw + *ri* footwear, sole.] Japanese thonged sandals with straw (or leather, wood, etc.) soles.
1823 F. Schoberl *Japan* v. 131 The shoes of the Japanese consist of straw soles or slips of wood. Those in common use are called *sori*. 1884 [see *GETA]. 1939 A. Keith *Land below Wind* xvii. 298 Even her *sori* were blue, with sapphire soles and bright blue straps which came between the toes. 1962 *Amer. Speech* XXXVII. 288 Japanese *zori* or the American adaptation, thong sandals. 1970 J. Kirkup *Japan behind Fan* 180 On summer days, *zori* of a specially fine quality, made of bamboo sheaths, may be used for strolling in the garden. 1984 *Coaching Award Scheme* (Brit. Judo Assoc.) 9/1 Zori (flip-flops) are compulsory wear at BJA events and should be worn off the mat in Clubs, Schools, etc.

zorrino: see Zorro.

zos-grass (zǫ·sgrɑs). [Abbrev. of Zos(tera + Grass *sb.*] = Zostera.
[illegible]

Zotzil, var. *Tzotzil.

Zouave. 1. c. (Earlier examples.)
[illegible]

‖ **Zou-Zou** (zū·zū). *Obs. exc. Hist.* Also **Zu-Zu**. [Fr.] Colloq. diminutive of Zouave.
1860 *Leisure Hour* 15 Mar. 290/2 The gamins of Paris, we believe, first applied to the world-renowned Zouaves the pet name of Zou-Zous; and France has confirmed the pleasant diminutive. [illegible]

zowie (zau·i, zau·i), *int.* *U.S. colloq.* An exclamation of astonishment (generally, or as a reaction to a sudden or surprising act), and freq. of admiration.
[illegible]

zoysia (zoi·ziă). [mod.L. (C. L. Willdenow 1801, in *Neue Schriften Gesellsch. Naturforsch. Freunde Berlin* III. 440), f. the name of Carl von Zoys zu Laubach (1756–1800), Austrian botanist + -IA¹.] A perennial grass of the genus of this name, native to eastern Asia, and sometimes used for lawns in subtropical regions. Also *zoysia grass*.
[illegible]

zubr. Add: Also 8 **zuber**. (Earlier example.)
1763 J. Bell *Trav.* I. 294 The stags are of two kinds; one called zuber, the same with the German crownhirsh, but somewhat larger.

‖ **zucca** (zu·kā). *rare.* Pl. **zucche**. [It.: see next.] A gourd, esp. a pumpkin.
Shelley's (plural) form is erroneous.
1818 Shelley *Let.* 9 Nov. (1964) II. 45 Vast heaps of many coloured zucchi or pumpkins..piled as winter food for the hogs. 1946 Blunden *Shelley* xvii. 272 Perhaps..it was Mary who placed a zucca on the grave.

zucchini (zukī·ni), *sb. pl.* [a. It., pl. of *zucchino* (small) marrow, dim. of *zucca* gourd.] **a.** Courgettes. Also const. as *sing.*
The usual word for the vegetable in N. America and Australia.
1929 *Sunset* Feb. 58/2 Wash the zuccini and slice thinly into a baking pan. [illegible]

b. *attrib.*
[illegible]

Zuckerkandl (zu·kəɪkænd'l). *Anat.* The name of E. *Zuckerkandl* (1849–1910), Austrian anatomist, used *attrib.*, in the possessive, and with *of* to designate the para-aortic bodies. [Described by Zuckerkandl in *Verhandl. d. Anat. Ges.* (1901) XIX. 95.]
[illegible]

zufolo (tsū·folo, z-). *Mus.* Also **zuffolo**. [a. It. *zuf(f)olo*.] A flageolet, a small flute or whistle (see quots.).
[illegible]

zug (tsūg). Also **Zug**. The name (formerly proprietary) for a variety of waterproofed leather used esp. for the uppers of climbing boots.
[illegible]

‖ **zugtrompete** (tsū·ktrɔmpe·tə). *Mus.* [Ger., f. *zug* pulling, tugging + *trompete* Trumpet.] A slide trumpet.
[illegible]

‖ **Zugunruhe** (tsū·kunrū,ə). *Ornith.* [Ger.] Migratory restlessness; the migratory drive in birds.
[illegible]

‖ **Zugzwang** (tsū·ktsvaŋ). *Chess.* [Ger., f. *zug* move + *zwang* compulsion, obligation.] A position in which a player is obliged to move but cannot do so without disadvantage; the disagreeable obligation to make such a move. Freq. in *Zugzwang*. Also *transf.*
[illegible]

Zulu, *sb.* and *a.* Add: **1.** (*sb.* or *adj.*) **c.** A derogatory term for a Black person. *U.S.*
[illegible]

2. (Earlier examples.)
1836 W. C. Harris *Wild Sports Southern Africa* 150 Andries..possessed a smattering of zooloo, and we thus

hoped to be able to proceed without the aid of a sworn interpreter. 1849 *Jrnl. Amer. Oriental Soc.* I. 50 The Zulu alphabet..contains the same letters as the English... The English language abounds with short words, but in the Zulu such words are very few.

4. *Zulu hat*, a kind of straw hat with a wide brim. *Obs. exc. Hist.*
[illegible]

5. A kind of fishing-boat formerly used in Scotland. *Obs. exc. Hist.*
[illegible]

6. The radio code word for the letter 'z'; *spec.* (in full *Zulu time*) (Aeronaut. colloq.) = *zone time* s.v. *ZONE *sb.* 9 a.
[illegible]

zumbi, var. *JUMBY.

Zuñi. Delete ‖ and add: Also **Zuni**. Pl. usu. **Zuñi**. **a.** (Earlier and later examples.) Also *Zuni Indian*.
[illegible]

b. The language of this people.
[illegible]

‖ **zuppa** (tsu·pa). [It.] Soup. *Comb.*, esp. as *zuppa di pesce* (pe·ʃe), fish soup. Also *zuppa inglese* and *fig.* in phr. *zuppa inglese* (ingl·ze) [lit. 'English soup'], a rich trifle.
[illegible]

zur. For 'southern dial.' read 'dial. (chiefly south-western)' and *attrib.* So *zure*. (Earlier and later examples.)
[illegible]

Zurich (zū³·rik, ‖ tsü·rix). Also **Zürich**. The name of a city on Lake Zurich in Switzerland, used *attrib.* to designate porcelain manufactured there in the eighteenth century.
[illegible] *European Ceramic Art* 42 Zurich porcelain at its best shows a belated Rococo and a rare beauty of colour. 1981 *Times* 25 May 10/5 Among features in the sale..were..Zurich porcelains. *Ibid.*, £3,721, for an attractive Zurich hunting group of about 1770.

b. *Zurich gnome*: see *GNOME¹ 1 c.
[illegible]

‖ **zurla** (zū³·rlā). *Mus.* [Serbo-Croatian, f. next.] A kind of oriental shawm introduced into Yugoslavia by gypsies.
[illegible]

‖ **zurna** (zʊ·ɹnă). *Mus.* Also 9 **zourna**; 20 **surna**. [Turk.; cf. Pers. *surnā*.] A Turkish pipe resembling a bagpipe or shawm.
[illegible]

zussmanite (zʊ·smənait). *Min.* [f. the name of J. *Zussman* (b. 1924), English mineralogist + -ITE¹.] A rhombohedral aluminosilicate of potassium, ferrous iron, and other metals, K(Fe²⁺,Mg,Mn)₁₃(Si,Al)₁₈O₄₂(OH)₁₄, found as pale green tabular crystals.
[illegible]

‖ **zut** (züt), *int.* [Fr.] An exclamation expressing annoyance, contempt, impatience, etc.
[illegible]

zuur-veld. (Earlier examples.)
[illegible]

Zu-Zu, var. *Zou-Zou. **zwart wit pens**, var. *SWARTWITPENS.

zwieback (tsvī·bak). Also **zwei-**. Pl. **-(s)**. [a. Ger., f. *zwie* (*zwei*) twice + *backen* Bake.] A (sweet) rusk or biscuit made by baking a small loaf, and then toasting slices until they are dry and crisp.
[illegible] give me something' s not necessarily cross-sexed when it is based on bottles and zwieback. 1957 V. Nabokov *Pnin* v. 132 The various biscuits, wafers, pretzels, zwiebacks. 1978 J. Irving *World according to Garp* xv. 287 Hope gave Nicky a zwieback and he stopped crying.

‖ **zwischenzug** (tsvi·ʃəntsūk). *Chess.* [Ger., f. *zwischen* intermediate + *zug* move.] An interim or intermediary move.
[illegible]

zwitterion (tsvi·tər,ai,ən). *Chem.* Formerly also **zwitter-ion** and with capital initial. [a. G. *zwitter-ion* (F. W. Küster 1897, in *Zeitschr. f. anorg. Chem.* XIII. 136), f. *zwitter* hermaphrodite, hybrid.] A molecule or ion having separate positively and negatively charged atoms or groups.
[illegible]

Hence **zwi:tterio·nic** *a.*
[illegible]

Zydeco (zai·dəko). *U.S. Blacks.* Also **zodico**, **zolo go**. [? Creole pronunc. of Fr. *les haricots* from dance-tune title 'Les haricots sont pas salés'.] A kind of Afro-American dance music of southern Louisiana; the dance itself. Also *attrib.*
[illegible]

zygomic (zai·gəmik), *a.* *Law.* [f. Zyg(o- + Gr. *νόμος* law + -ic.] In the terminology of A. Kocourek: a jural relation which involves an act of the evolution of which directly abridges..the freedom of the servus in the enjoyment..of the substance of a legal advantage. Opp. *MESONOMIC *a.*
[illegible]

zygo-. Add: **zygoge·nesis** [-GENESIS], reproduction involving the formation of a zygote; so **zygogene·tic** *a.*; **zy·gomere** *Cytology* [*-MERE], a site on a chromosome thought to be responsible for the initiation of pairing between homologous chromosomes during zygotene in eukaryotes; **zygone·ma** *Cytology* [f. Gr. *νῆμα* thread] [illegible]; **zy·gophore** *Bot.* [-PHORE], a differentiated hypha in Zygomycetes that takes part in conjugation; hence **zygopho·ric** *a.*; **zy·gosome** *Cytology* [ad. G. *zygosom* (E. Strasburger 1904, in *Sitzungsber. d. k. Preuss. Akad. d. Wissensch.* 606): see *-SOME[4]] = *BIVALENT *sb.*

1950 *Adv. Genetics* III. 194 The most common mode of animal reproduction is, however, sexual reproduction or gamogony or zygogenesis. *Ibid.* 198 In other parthenogenetic animals both parthenogenetic and zygogenetic reproduction are present. 1973 B. J. WILLIAMS *Evolution & Human Origins* iii. 37/2 It [*sc.* random genetic drift] includes all events that lead to sampling error in random zygogenesis. 1978 *Biol. Bull* CLV. 273 (*heading*) *Artemia* hemoglobins: genetic variation in parthenogenetic and zygogenetic populations. 1966 J. SYBENGA in *Genetica* XXXVII. 188 General occurrence of localization of the function of initiation of chromosome pairing (long-distance attraction) in discrete units on specific loci is considered a useful working hypothesis. In analogy to 'centromere', 'chromomere' and 'telomere' the term '*zygomere*' is proposed for such units. 1981 *Cytologia* XLVI. 527 Since the bivalent formation has not been disturbed, at least one of two zygomeres seems to be able to have a complete activity. 1911 *Q. Jrnl. Microsc. Sci.* LVII. 32 The debatable stages of the meiotic prophases in which parasyndesis and its associated phenomena occur—leptonema, zygonema..—have been dealt with by many experienced cytologists. *Ibid.* 33 By the time the zygonema is fairly far advanced we do get appearances not unlike what may occasionally..be found in the condensation of a somatic chromosome. 1976 *Nature* 8 Apr. 534/2 It is generally known that during zygonema (stages XII–XIV in rat spermatogenesis) the homologous sets of sister chromatid pairs begin to come together and associate with one another. 1904 A. F. BLAKESLEE in *Science* 3 June 866/1 In all species of both homo- and heterothallic groups..the swollen portions (progametes) from which the gametes are cut off do not grow toward each other..but arise as a result of the stimulus of contact between more or less differentiated hyphæ (zygophores). 1970 J. WEBSTER *Introd. Fungi* II. ii. 116 When two compatible strains approach each other aerial club-shaped branches or zygophores develop which show directional growth towards zygophores of the opposite strain. Zygophores of the same strain repel each other. 1904 Zygophoric [see *HETEROGAMIC *a.*]. 1978 *Canad. Jrnl. Bot.* LVI. 1061 One or more slender, lateral zygophoric filaments proliferate from the subterminal portion of a septate, erect hypha. 1905 *Ann. Bot.* XIX. 249 A similar operation of the law of chance has been suggested by Strasburger ('04) in the separation of the chromatin granules as a result of the division of the 'zygosome'. 1910 [see *parasynaptic* adj. s.v. *PARA-¹ 1]. 1974 *Jrnl. Heredity* LXV. 257/1 The varying amounts of the *q* segments present in the zygosome may account in large measure for the physical and mental deviation of these mongoloid patients from the usual spectrum of characteristics typical of mongoloids bearing three independent chromosomes 21.

zygocactus (zəi·gəkæktəs). *Bot.* Also **Zygo-.** [mod.L. (K. Schumann 1890, in C. F. Martius *Flora Brasiliensis* IV. II. 223): see ZYGO- and CACTUS.] Any cactus of the Brazilian genus *Zygocactus* (sometimes included in *Schlumbergera*), the members of which have branched and jointed stems bearing zygomorphic flowers in various shades of red, and are freq. grown as houseplants.

1950 V. HIGGINS *Cactus Grower's Guide* iv. 54 Two other Epiphyllums which have been much cultivated are now placed in Schlumbergera—*Schlumbergera Gaertneri* and *S. Russelliana*; both are similar in habit to Zygocactus but the flowers are regular. 1962 *Amateur Gardening* 24 Mar. 29/1 Zygocactus should be watered throughout the year. 1980 *Daily Tel.* 24 Sept. 14/5 Among the new plants on show are ..zygocactus in pastel colours with a future as room plants, from Rochford.

zygology (zəigǫ·lŏdʒi). [f. ZYGO- + -OLOGY: coined by Mr. C. G. Hardie, Magdalen College, Oxford.] The branch of technology concerned with joining and fastening. Hence **zygolo·gical** *a.*, **zygo·logist**.

1970 *Assembly & Fastener Engin.* Oct. 48/3 We at Oxford Polytechnic are now offering courses in Zygology. *Ibid.*, I do not wish to suggest that all your readers should be considered as practising zygologists. 1971 *New Electronics* May 56 (Advt.), We are zygologists—experts in fastening techniques. We have specialised in riveting for years. 1973 *Oxford Times* 14 Dec. 40 (Advt.), Oxford Polytechnic... Postgraduate diploma in zygology. 1976 W. C. WAKE *Adhesion* i. 4 Adhesion science should thus include adhesives and joints under its wing and is, if the reader likes classification, a branch of zygology. 1978 *Engin. Materials & Design* Apr. 37/2 Not that adhesion is the only zygological process available for joining one piece of plastics material to another.

Zygomycetes (zəigǫmai·sīːz), *sb. pl. Bot.* Also (*rare*) **zygo-.** [mod.L., ad. G. *Zygomyceten* (O. Brefeld *Bot. Untersuchungen über Schimmelpilze* (1872) I. 53): see ZYGO-, MYCETES *sb. pl.*] A class of saprophytic and parasitic fungi in which sexual reproduction is by fusion of usu. similar gametangia to produce a zygospore and asexual reproduction is by means of non-motile spores; fungi of this class. Occas. in *sing.* **Zygomycete** (-mai·sīt). Hence **zygomyce·tous** *a.*

[1874 *Q. Jrnl. Microsc. Sci.* XIV. 56 Brefeld does not admit that *Chætocladium* and *Piptocephalis* possess sporangia, but only conidia. According to his views, therefore, the term *Zygomycetes* is more expressive than *Mucorini*, which he restricts to the sporangiferous *Zygomycetes*. This, however, appears to us founded on an error.] 1887 H. E. F. GARNSEY tr. *A. de Bary's Compar. Morphol. & Biol. of Fungi* vi. 345 This coincidence with a fixed period of the year is at least not a general rule in the zygospores of the Zygomycetes. 1928 C. W. DODGE tr. *Gäumann's Compar. Morphol. Fungi* xxxvi. 621 A convergent development has apparently occurred in the Zygomycetous sporangia which have become gonotoconts. 1930 H. M. FITZPATRICK *Lower Fungi. Phycomycetes* ii. 34 The origin of the Zygomycete line is somewhat more obscure, though forms possessing one or more undoubted zygomycetous characters exist among the Ancylistales and Chytridiales. 1937 GWYNNE-VAUGHAN & BARNES *Struct. & Developm. Fungi* (ed. 2) 16 The Zygomycetes..are the first fungi to colonise dung. 1952 C. J. ALEXOPOULOS *Introd. Mycol.* vii. 180 Such a theory is based almost entirely on the asexual cycle, the zygomycetous reproduction having no counterpart in the present-day Saprolegniaceae which might give us a clue to its origin. 1978 *Bio Systems* X. 97/2 There are several eukaryote groups where there is, so far, no solidly based evidence for a flagellate ancestry: ..(4) zygomycete fungi. 1979 I. K. ROSS *Biol. of Fungi* xiii. 378 There are three main methods by which spores are actively released: the bursting of a turgid cell (ascomycetes and some zygomycetes), the rounding off of a surface under tension (some zygomycetes, some basidiomycetes) and the so-called ballistospore discharge. 1982 *Phytopathology* LXXII. 1102 (*heading*) Synoptic keys to the genera and species of zygomycetous mycorrhizal fungi.

zygosity (zəigǫ·siti). *Genetics.* [f. ZYGOS(IS + -ITY.] **a.** The genetic relationship of twins, triplets, etc., in respect of their being either monozygotic or dizygotic.

1952 *New Biol.* XII. 42 Instead of exchanging skin between the twin animals whose zygosity it is desired to establish, grafts are being transplanted from both of them to a third, unrelated recipient. 1971 *Nature* 23 July 277/1 Before recent developments with marker genes, it was not possible to assign with certainty the zygosity of a substantial proportion of twin pairs. 1978 *Jrnl. R. Soc. Med.* LXXI. 311/2 The type of twinning and determination of zygosity are given attention.

b. The degree of genetic similarity between alleles which determines whether an individual is homozygotic or heterozygotic for the characteristic expressed.

1967 *Jrnl. Clin. Investigation* XLVI. 681 (*heading*) Relationship between $Rh_0(D)$ zygosity and red cell $Rh_0(D)$ antigen content in family members. *Ibid.*, The members of two families showed a poor correspondence between antibody binding and zygosity. 1972 *Transplantation* XIV. 793/1 Efforts have been made to determine HL-A zygosity of unrelated subjects by use of the gene-dose effect.

zygospore. Add: Hence **zygospo·ric** *a.*, of the nature of or producing zygospores.

1906 *Science* 27 July 122/2 Zygosporic cultures of the 'Harvest Strain' have..been kept running for nearly ten years. 1970 J. WEBSTER *Introd. Fungi* II. ii. 141 In some species [of the genus *Endogone*]..the fruit-body is entirely zygosporic.

zygote. Add: **zygotic** *a.* (examples); **zygo·tically** *adv.*, in the zygote; in terms of the zygote.

1909 W. E. CASTLE *Inheritance in Rabbits* (Carnegie Inst. Publ. No. 114) 58 The enumeration of the conceivable different varieties of gray rabbit, all alike in appearance but all different in breeding capacity, *i.e.*, of different zygotic formula. 1915 *Jrnl. Genetics* V. 45 Zygotically therefore the three forms [of rabbit] may be represented thus: Self-coloured..CCSS, Himalayan..CCss, Albino.. ccss. 1931 *Ibid.* XXIV. 448 It is probable that the new combinations will be less successful than the old, both zygotically and gametically. 1977 J. COHEN *Reproduction* ix. 163 The second phase of embryology begins, controlled by zygotic genes. 1980 *Genetics* XCVI. 187 The possibility that essential loci in the zeste-white region of the *Drosophila melanogaster* X chromosome are expressed both maternally and zygotically has been tested.

zygotene (zəi·gotīn). *Cytology.* [a. F. *zygotène* (V. Grégoire 1907, in *La Cellule* XXIV. 371): see ZYGO- and *-TENE.] The second stage of the prophase of meiosis, following leptotene, during which homologous chromosomes begin to pair.

1911 *Jrnl. Morphol.* XXII. 752 This view..goes on to show that after the last spermatogonial mitosis the chromosomes become very delicate slender threads, the leptotene condition..; these then approximate themselves parallel into pairs making the zygotene condition. 1939 [see *DIAKINESIS]. 1974 *Nature* 9 Aug. 469/2 At leptotene of meiotic prophase in many organisms, all the telomeres become gathered together and attached to a small area of the nuclear envelope, presumably so as to facilitate pairing during zygotene.

Zyklon (zəi·klǫn). Also **Cyclon.** [a. G. *Zyklon*, of unknown etym.] Hydrogen cyanide adsorbed on, or released from, a carrier in the form of small tablets, used as a fumigant and formerly as a poison gas. Usu. as *Zyklon B*.

[1926 *Official Gaz.* (U.S. Patent Office) 9 Nov. 298/1 Deutsche Gesellschaft für Schädlingsbekämpfung, m.b.H., Frankfort-on-the-Main... *Zyklon*... Apparatus for measuring the quantities of substances which generate poisonous gases—for instance, hydrocyanic acid.] 1939 METCALF & FLINT *Destructive & Useful Insects* (ed. 2) ix. 281 The other type of dry cyanides, such as the zyklon products, undergo no chemical change when exposed. 1944 *Chem. Abstr.* XXXVIII. 3416 The application of Cyclon B (0.4 g./cc.) for 24 hrs. destroyed all insects but imparted a peculiar taste to the tobaccos. 1964 L. DEIGHTON *Funeral in Berlin* xxxi. 169 With Cyclon B they killed two and a half million at Auschwitz. 1975 W. CRAIG *Strasbourg Legacy* (1976) i. 9 Former SS soldiers..had functioned so anonymously in the camps that hardly anyone lived who could identify them as guards once manning machine guns or dropping Zyklon B tablets into gas chambers. 1977 *Times* 8 June 9/7 [He] was tried and acquitted at Nürnberg in 1948 for supplying the SS with Zyklon-B gas. 1978 H. WOUK *War & Remembrance* xi. 111 Zyklon B, the powerful insecticide they have been using right along at the camp to fumigate the barracks, may be the surprisingly simple solution.

zymo-. Add: **zy·mogram** *Biochem.* and *Genetics*, a strip of electrophoretic medium showing enzymes separated by a technique such as zone electrophoresis.

1957 HUNTER & MARKERT in *Science* 28 June 1295/2 We propose the term *zymogram* to refer to strips in which the location of enzymes is demonstrated by histochemical methods. 1978 *Nature* 2 Mar. 77/2 The high degree of gene duplication in these species often confounds the genetic interpretation of zymograms. For example, how many loci code for an enzyme represented by a single electrophoretic band? 1981 *Histochemistry* LXXIII. 311 Electrofocused zymograms display species and organ differences.

zymosan (zəi·mosæn). *Biochem.* [f. ZYMO-, after *glucosan*, *hexosan*.] An insoluble polysaccharide of the cell wall of yeast, used in the assay of properdin.

1943 E. E. ECKER et al. in *Jrnl. Immunol.* XLVII. 185 The preparation of human complement lacking in third component. The third component is specifically removed from or inactivated in human serum by the insoluble carbohydrate prepared from fresh yeast... The insoluble carbohydrate is hereafter referred to as 'zymosan'. 1973 *Sci. Amer.* Nov. 60/3 The incubation of normal blood serum..with certain polysaccharides derived from microbial cells (such as zymosan, a carbohydrate of the yeast cell membrane) gives rise to enzymes that activate the complement factors C3 and C5.

Zyrenian, var. *SIRYENIAN *sb.* and *a.*

Zyrian (zi·riăn), *sb.* and *a.* Also **Syrian, Syryane, Syryen,** (and esp.) **Zyryan.** [ad. Russ. *Zyryánin*: see -IAN.] **A.** *sb.* A member of the Komi people of northern central U.S.S.R. **b.** The language of this people; = *KOMI b. **B.** *adj.* Of or pertaining to this people or their language.

1886 *Encycl. Brit.* XXI. 79/2 The Permians,..including ..the Zyrians in Vologda, Archangel, Vyatka, and Perm. 1926 *Chambers's Encycl.* VIII. 101/1 The Syriān is spoken by a large population in the districts of Perm, Viatka, Archangel, and Vologda. 1932, 1933 [see *PERMIAN *a.* (*sb.*) 2]. 1942 [see *KOMI]. 1948 D. DIRINGER *Alphabet* II. viii. 482 Other peoples..such as the Zyryans or Syryans (now called Komi). 1951 W. K. MATTHEWS *Languages U.S.S.R.* iii. 20 The Yuraks have loans from Zyryan (Komi), a Finnic language. *Ibid.* 25 Zyryan resistance to the Russians was less dogged and implacable than Ugrian. 1955 *Trans. Philol. Soc. 1954* 99 The forms are: Norwegian Lappish *miettá*..Syryane *ma* [etc.]. 1972 *Language* XLVIII. 848, 7b is given by Hockett for 'German,..French..., Zyryan'. 1978 K. RÉDEI (*title*) Zyrian folklore texts.

BIBLIOGRAPHY

This is a list of the works most frequently quoted in the four volumes of this *Supplement to the O.E.D.*, other than those already listed in the bibliography (1933) to the *O.E.D.* itself.

No revision of the bibliography to the *O.E.D.* has been attempted.

The information presented in this new bibliography reflects the work of many people. Since it was begun in 1963 those primarily involved have been (in chronological order) N. C. Sainsbury (1963–6), Miss J. M. Hawkins (1966–7), A. J. Augarde (1967–9), M. W. Grose (1969–72), G. D. Hargreaves (1973–5), and J. Paterson (since 1975), as well as members of the Department's permanent library-research team. The bibliography was assembled in its final form during 1982.

Publications of learned institutions are listed here under the name of the institution; translations under the name of the translator. The country or city of publication has not usually been given except for newspapers published outside London, and U.K. editions of books published in an earlier year abroad. Authors' dates of birth and death are usually given when any posthumous works are listed.

November 1982

J. Paterson
Bibliographical Editor

A

'AARONS, E. S.' (Paul Ayres & Edward Ronns) *Assignment treason* 1956
ABBOTT, David *Inorganic chemistry* 1965
ABBOTT, John Henry Macartney *Tommy Cornstalk* 1902
ABBS, Akosua *Ashanti boy* 1959
ABERCROMBIE, David *English phonetic texts* 1964
Problems and principles: studies in the teaching of English as a second language 1956
—— et al. eds. *In honour of Daniel Jones: papers contributed on the occasion of his eightieth birthday* 1964
Aberdeen Press and Journal 1922–39 (continued as *Press and Journal*)
ABRAHAM, George Dixon *Modern mountaineering* 1933
ABRAHAM, Louis Arnold & HAWTREY, Stephen Charles eds. *A parliamentary dictionary* 1956
— (ed. 2) 1964
ABRAHAMS, Beth-Zion tr. *The life of Glückel [Segal] of Hameln 1646–1724, written by herself* 1962
ABRAHAMS, Peter Henry *Dark testament* 1942
The path of thunder 1952
Return to Goli 1953
Wild conquest 1951
ABRAHAMS, Roger D. ed. *Jump-rope rhymes: a dictionary* 1969
Positively black 1970
Académie des Sciences (Paris) *Compte rendu hebdomadaire des séances* 1835–
Academy of Natural Sciences of Philadelphia *Proceedings* 1841–
Academy of Sciences of the U.S.S.R. *Doklady* [French ed.] 1935–47
Doklady: earth science sections 1959–
Soviet physics: Doklady [Eng. trans.] 1956–
Accountant, The 1874–
ACHEBE, Chinua *Girls at war, and other stories* 1972
A man of the people 1966
Things fall apart 1958
ACKERLEY, Joe Randolph (1896–1967) *My father and myself* 1968
ACLAND, Leopold George Dyke (1876–1948) *The early Canterbury runs* (ser. 1) 1930
— (complete ed.) 1951
Acronyms dictionary (Gale Research Co.) 1960 (also later editions used, and supplements with title *New acronyms and initialisms*)
Acta crystallographica 1948–
Acta radiologica 1920–
ACTON, Harold Mario Mitchell *Memoirs of an aesthete* 1948
Actors by daylight; or, Pencillings in the pit 2 vols. 1838–9
ADAMI, John George *The principles of pathology* 2 vols. 1909–10
ADAMS, Andy *The outlet* 1905
ADAMS, Bertram Martin ('Bill') *Ships and women: an autobiography* 1936
ADAMS, Briggs Kilburn *The American spirit: letters* 1918
ADAMS, Carsbie Clifton, et al. *Space flight* 1958
ADAMS, Edwin Plimpton tr. *A. Einstein's The meaning of relativity* 1922
ADAMS, Frank Davis *Aeronautical dictionary* 1959
ADAMS, Ramon Frederick *Western words: a dictionary of the range, cow camp and trail* 1944
— (rev. ed.) 1968
ADAMS, Richard George *Shardik* 1974
Watership Down 1972
ADAMSON, Arthur Wilson *A textbook of physical chemistry* 1973
ADBURGHAM, Alison *Shops and shopping, 1800–1914* 1964
ADBY, Paul Raymond & DEMPSTER, Michael Alan Howarth *Introduction to optimization methods* 1974
ADDYMAN, Frank T. tr. *A. M. Villon's Practical treatise on the leather industry* 1901
ADE, George (1866–1944) *Artie* 1896
Doc' Horne 1899
Fables in slang 1900 (UK 1902)
Forty modern fables 1901
Hand-made fables 1920
In Babel 1903
Knocking the neighbors 1912
Letters ed. T. Tobin 1973
More fables 1900 (UK 1902)
People you know 1903
True bills 1904
Adelphi, The 1923–7; 1930–
Admiralty *See* United Kingdom. Admiralty
Advances in chemistry series 1950–
Advances in genetics 1947–
Adventure (New York) 1910–
Advisory Committee for Aeronautics *See* United Kingdom. Advisory ——
Advocate-News (Barbados) 1895–
Aeronautical journal 1897–1923; 1968–
Aeronautical Research Committee (*later* Council) *See* United Kingdom. Advisory Committee for Aeronautics
Aëronautical Society of Great Britain *Annual report* 1866–95
Aeronautics 1939–
Aeroplane, The 1911–68 (with title *The Aeroplane and astronautics* 1959–62, and *The Aeroplane and commercial aviation news* 1962–8)
Aeroplane spotter, The 1941–8
AFLALO, Frederick George *A sketch of the natural history of Australia* 1896
Africa. Journal of the International Institute of African Languages and Culture 1928–40; 1943–
African encyclopedia 1974
AGAR, Wilfred Eade *Cytology, with special reference to the Metazoan nucleus* 1920
AGATE, James Evershed *More first nights* 1937
Red letter nights 1944
Age, The (Melbourne) 1859–
AGGER, Leo Thomas *Introduction to electricity* 1971
Agricultural and biological chemistry 1955–
Agriculture 1939–
AHARONI, Joseph *The special theory of relativity* 1959
AIKEN, Joan Delano *The butterfly picnic* 1972
Last movement 1977
AIKEN, John *Nightly deadshade* 1971
AINSWORTH, Leopold *Confessions of a planter in Malaya* 1933
'AINSWORTH, Milo' (Peter Fison) *Murder is catching* 1959
Ainsworth's magazine 1842–54
Air Conference [London], 1920 *Proceedings* 1920
Air News (Chicago) 1941–6
Aircraft engineering 1929–
'AIRD, Catherine' (Kinn Hamilton McIntosh) *Henrietta who?* 1968
AIRTH, Rennie *Snatch!* 1969
AITKEN, Adam Jack, MCINTOSH, A., & PÁLSSON, H. eds. *Edinburgh studies in English and Scots* 1971
AITKEN, William *Automatic telephone systems* 3 vols., 1921–4
AITON, William *Hortus Kewensis; or, A catalogue of plants cultivated in the Royal Botanic Garden at Kew* 3 vols. 1789
— (ed. 2, ed. by W. T. Aiton) 5 vols. 1810–13
ALBANESI, Effie Maria *For love of Anne Lambert* 1910
ALBEE, Edward *Who's afraid of Virginia Woolf?* 1962
ALBERT, Abraham Adrian *Modern higher algebra* 1937 (UK 1938)
ALBERT, Arthur Lemuel *Fundamentals of telephony* 1943
Radio fundamentals 1948
Alberta historical review 1957–
ALBERY, James (1838–89) *Dramatic works* ed. W. Albery 2 vols. 1939
ALBRIGHT, William Foxwell *The archaeology of Palestine* 1949
ALCOCK, Leslie *By South Cadbury* 1972
ALCOTT, Louisa May *Little men* 1871
An old-fashioned girl 1870
ALDEN, William Livingston *The adventures of Jimmy Brown* 1885
ALDINGTON Richard *All men are enemies* 1933
The colonel's daughter 1931
Death of a hero 1929
The strange life of Charles Waterton 1782–1865 1949
ALDISS, Brian Wilson *The airs of earth* 1963
The moment of eclipse 1970
A soldier erect 1971
—— & HARRISON, Harry Max eds. *Decade: the 1950's* 1976
Decade: the 1940's 1975
'ALEXANDER, Mrs.' (Mrs. Annie French Hector) *A choice of evils* 3 vols. 1894
ALEXANDER, James Edward *An expedition of discovery into the interior of Africa* 2 vols. 1838
Sketches in Portugal during the Civil War of 1834 1835
ALEXANDER, Jerome *Colloid chemistry, theoretical and applied* 6 vols. 1926–46
ALEXANDER, Samuel *Space, time and deity* 2 vols. 1920
ALEXANDER, William *Sketches of life among my ain folk* 1875
ALEXOPOULOS, Constantine John *Introductory mycology* 1952
— (ed. 2) 1962
ALGREN, Nelson *The man with the golden arm* 1949
A walk on the wild side 1956 (UK 1957)
ALLBEURY, Theo Edward Le Bouthillier ('Ted') *A choice of enemies* 1973
The lantern network 1978
The only good German 1976
Snowball 1974
The special collection 1975
ALLEE, Warder Clyde *Animal aggregations: a study in general sociology* 1931

ALLEN, Arthur Charles *The skin: a clinicopathological treatise* 1954
— (ed. 2) 1967
ALLEN, Clifford *A textbook of psychosexual disorders* 1962
— (ed. 2) 1969
ALLEN, David Elliston *British tastes: an enquiry into the likes and dislikes of the regional consumer* 1968
ALLEN, Frederick Lewis *Only yesterday: an informal history of the nineteen-twenties* 1931
ALLEN, Geoffrey Freeman *British Rail after Beeching* 1966
ALLEN, Herbert Stanley *Photo-electricity: the liberation of electrons by light* 1913
— (ed. 2) 1925
ALLEN, James Lane *The choir invisible* 1897
ALLEN, Jules Verne *Cowboy lore* 1933
ALLEN, Percy Stafford (1869–1933) *Letters* ed. H. M. Allen 1939
ALLEN, Ralph *Home made banners* 1946
ALLEN, Reginald Lancelot Mountford *Colour chemistry* 1971
ALLEN, William Hervey *Anthony Adverse* 1933
ALLEN, William Sidney *Vox Graeca: a guide to the pronunciation of classical Greek* 1968
Vox Latina: a guide to the pronunciation of classical Latin 1965
ALLINGER, Norman Louis, et al. *Organic chemistry* 1971
ALLINGHAM, Margery Louise (1904–66) *The beckoning lady* 1955
Cargo of eagles 1968
Coroner's pidgin 1945
Dancers in mourning 1937
Death of a ghost 1934
The fashion in shrouds 1938
Flowers for the judge 1936
Hide my eyes 1958
Look to the lady 1931
Mr. Campion and others 1939
— (another ed.) 1950
More work for the undertaker 1948 (UK 1949)
Mystery mile 1930
The tiger in the smoke 1952
Traitor's purse 1941
ALLINGHAM, Philip *Cheapjack* 1934
ALLIS, Marguerite *English prelude* 1936
ALLPORT, Gordon Willard *The nature of prejudice* 1954
Personality: a psychological interpretation 1937 (UK 1938)
ALLUM, Peter Antony *Politics and society in post-war Naples 1945–1970* 1971 (UK 1973)
ALMEDINGEN, Edith Martha *Frossia* 1943
ALSTON, Arthur Reginald *Test commentary* 1956
Alta California (San Francisco) 1849–91
Amateur gardening 1884–
Amateur photographer, The 1884–
Amateur radio handbook, The (Radio Society of Great Britain) 1939
— (ed. 2) 1940
— (ed. 3) 1961
— (ed. 4, with title *The radio communication handbook*) 1968
AMBLER, Eric *Cause for alarm* 1938
The dark frontier 1936
Doctor Frigo 1974
The light of day 1962
The mask of Dimitrios 1939
The night-comers 1956
Uncommon danger 1937
AMBROSE, Edmund Jack & EASTY, Dorothy M. *Cell biology* 1970
AMBROSE, Kay *The ballet-lover's pocket-book* 1943
American, The: a national journal 1880–1900
American Academy of Arts and Sciences *Bulletin* 1948–
Memoirs 1785–1818, new ser. 1826–
Proceedings 1846–1958 (continued as *Dædalus*)
American Academy of Political and Social Science *Annals* 1890–
American Association for the Advancement of Science *Proceedings* 1848–
American Association of Petroleum Geologists *Bulletin* 1918–
Memoirs 1962–
American ballads and folk songs ed. J. A. & A. Lomax 1934
American chemical journal 1879–1913
American Chemical Society *Journal* 1879–
American city, The 1909–
American Dialect Society *Publications* 1944–
American dictionary of printing and bookmaking ed. W. W. Pasko 1894
American documentation 1950–69
American dyestuff reporter 1917–
American folk music occasional 1964–
American Geological Institute *Glossary of geology and related sciences* 1957
— (ed. 2), with Supplement, 1960
— (ed. 3) 1972
American heart journal 1925–
American Institute of Electrical Engineers *Transactions* 1884–
American journal of anatomy 1901–
American journal of diseases of children 1911–
American journal of mathematics 1878–
American journal of obstetrics and diseases of women and children 1868– (from 1920 with title *American journal of obstetrics and gynecology*)
American journal of ophthalmology 1884–
American journal of pathology 1925–
American journal of philology 1880–
American journal of physical anthropology 1918–
American journal of physics 1940–
American journal of physiology 1898–
American journal of psychiatry 1921–
American journal of psychology 1887–
American journal of public health 1912–
American journal of roentgenology 1913–
American journal of sociology 1895–
American journal of the medical sciences 1827–
American machinist, The 1877–
American magazine 1905–
American mail order fashions, 1880–1900: pictures and copy reproduced from original catalogs and magazines of the times (Americana Review) 1961
American Mathematical Society *Bulletin* 1894–
Transactions 1900–
American Medical Association *Journal* 1883–
American Mercury, The 1924–
American mineralogist, The 1916–
American museum, The 1787–92; 1798
American Museum of Natural History *Bulletin* 1881–
Memoirs 1893–
American notes and queries 1857; 1888–92; 1941–50; 1962–
American Oriental Society *Journal* 1843–
American Pediatric Society *Transactions* 1888–
American Philological Association *Transactions* 1869–
American philosophical quarterly 1964–
American Philosophical Society *Proceedings* 1838–
Transactions 1769–
American pioneer, The 1842–3
American poetry review 1972–
American political science review 1906–
American railroad journal 1832–86
American review of reviews 1907–28
American Society for Metals *Metals handbook* 1936–
American Society for Testing Materials *Proceedings* 1898–
American speech 1925–
American Veterinary Medical Association *Journal* 1915–
American Water Works Association *Journal* 1914–
Americana annual, The: an encyclopedia of current events 1923–
AMES, Delano L. *Murder, maestro, please* 1952
AMIN, Adibah tr. *S. Ahmad's No harvest but a thorn* 1972
AMIS, Kingsley William *The anti-death league* 1966
Girl, 20 1971
I like it here 1958
Jake's thing 1978
Lucky Jim 1954 [dated 1953]
New maps of hell: a survey of science fiction 1960 (UK 1961)
Take a girl like you 1960
AMIS, Martin Louis *The Rachel papers* 1973
Success 1978
AMOS, Stanley William & BIRKINSHAW, Douglas Crosbie *Television engineering* 4 vols. 1953–8
— Vol. I, 2nd impr., rev. 1957
— Vol. II, 2nd impr., rev. 1958
Amrita Bazar Patrika (Calcutta) 1868–
Analog science fact—science fiction (title varies) 1960–
Analysis 1933–
Analyst, The: a journal of pure and applied mathematics 1874–
Anatomical record 1906–
ANDERSON, Gene *Coring and core analysis handbook* 1975
ANDERSON, J. W. *Fur trader's story* 1961
ANDERSON, James M. *Structural aspects of language change* 1973
ANDERSON, John Neil *Applied dental materials* 1956
— (ed. 2) 1961
ANDERSON, John Richard Lane *Death on the rocks* 1973
ANDERSON, Lewis Flint *The Anglo-Saxon scop* 1903
ANDERSON, Maxwell & STALLINGS, Laurence *What price glory?* in *Three American plays* 1926
ANDERSON, Nels *The hobo* 1923
ANDERSON, Robert Henry *The trees of New South Wales* 1932
— (ed. 2) 1947
ANDERSON, William Arnold D. ed. *Pathology* 1948
Anderson (South Carolina) *Independent* 1924– (1925–49 with title *Anderson Independent-Tribune*)
ANDERSSON, Carl Johan *Lake Ngami; or, Explorations and discoveries, during four years' wanderings in the wilds of South Western Africa* 1856
'ANDOM, R.' (Alfred Walter Barrett) *We three and Troddles* 1894
ANDRADE, Edward Neville da Costa *The structure of the atom* 1923
— (ed. 3) 1927
ANDREE, Richard Vernon *Selections from modern abstract algebra* 1958
ANDREW, Warren tr. *E.D.P. de Robertis's General cytology* 1948
Textbook of comparative histology 1959
ANDREWS, Tom Gaylord ed. *Methods of psychology* 1948
Anglo-American cataloguing rules: British text 1967
Annals of internal medicine 1927–
Annals of mathematics 1884–9; 1900–
Annals of surgery 1885–
Annals of the Congress of the U.S. See: United States. Congress. Debates *Debates and proceedings* 1789–1824
Annals of tropical medicine and parasitology 1907–
Annals See also under the names of particular institutions
ANNAN, Noel Gilroy *Leslie Stephen* 1951
Annual reports on the progress of chemistry (Chemical Society of London) 1904–
Annual review of microbiology 1947–
Annual review of nuclear science 1952–
ANSCOMBE, Gertrude Elizabeth Margaret tr. *L. J. J. Wittgenstein's philosophical investigations* 1953
ANSOFF, Harry Igor *Corporate strategy: an analytic approach to business policy for growth and expansion* 1965 (UK 1968)
ANSON, Peter Frederick *Bishops at large* 1964
Scots fisherfolk 1950
'ANTHONY, Evelyn' (Eve Stephens) *The Malaspiga exit* 1974
Anthropological linguistics 1959–
Antibiotic medicine 1955–61 (from 1956 with title *Antibiotic medicine and clinical therapy*)
Antibiotics and chemotherapy 1951–62
Antiquaries' journal 1921–
Antiquity: a quarterly review of archæology 1927–
Anzac book, The: written and illustrated in Gallipoli by the men of Anzac 1916
APEL, Willi ed. *The Harvard dictionary of music* 1944
— (ed. 2) 1970
Approved names See: United Kingdom. General Medical Council
Arable farmer and vegetable grower 1966–
ARBER, Agnes *The Gramineae: a study of cereal, bamboo, and grass* 1934
Water plants: a study of aquatic angiosperms 1920
ARCHER, William *The theatrical 'world'* 5 vols. 1894–8
Architect, The (title varies) 1869–
Architectural review 1896–
Archives internationales de pharmacodynamie et de thérapie 1898–
Archives of biochemistry 1924– (from 1951 with title *Archives of biochemistry and biophysics*)
Archives of dermatology and syphilology (title varies) 1920–
Archives of general psychiatry 1959–
Archives of internal medicine 1908–
Archives of Maryland 1883–
Archives of neurology (London) 1899–1907
Archives of neurology and psychiatry (London) 1909–50
Archives of neurology and psychiatry (Chicago) 1919–59
Archives of ophthalmology 1879–
Archives of pathology 1928–
Archives of surgery 1920–
Archivum linguisticum 1949–80 (vols. 1970–80 are new series)
Arctic terms 1955 *See* United States. Arctic, Desert and Tropic Information Centre
ARDAGH, John *The new France* 1970
ARDREY, Robert *The territorial imperative* 1966 (UK 1967)
ARENDT, Hannah *Origins of totalitarianism* 1951
— (new ed.) 1967
AREY, Leslie Brainerd *Developmental anatomy: a textbook and laboratory manual of embryology* 1924 (7 later editions to 1974)
ARGENTI, John *Management techniques: a practical guide* 1969
Argus (Melbourne) 1846–1957 (title varies)
ARGYLE, John Michael *Psychology and social problems* 1964
The psychology of interpersonal behaviour 1967
Religious behaviour 1958
The social psychology of work 1972

Aristotelian Society for the Systematic Study of Philosophy *Proceedings* 1888–96; 1900–
Supplementary volumes 1918–
Arizona Daily Star (Tucson, Arizona) 1879–
ARKELL, William Joscelyn *The Jurassic system in Great Britain* 1933
ARLEN, Michael *Man's mortality* 1933
'Piracy' 1922
The romantic lady 1921
'ARMSTRONG, Anthony' (George Anthony Armstrong Willis) *Taxi!* 1930
ARMSTRONG, Charlotte *The black-eyed stranger* 1951 (UK 1952)
Seven seats to the moon 1969
ARMSTRONG, Louis *Satchmo: my life in New Orleans* 1954 (UK 1955)
ARMSTRONG, Neil, et al. *First on the moon* 1970
ARMSTRONG, Terence, ROBERTS, Brian, & SWITHINBANK, Charles *Illustrated glossary of snow and ice* 1966
ARMSTRONG, William *Stocks and stock-jobbing in Wall Street* 1848
Army and Navy Co-operative Society, Ltd. (later Army and Navy Stores, Ltd.) *Price lists* and *Catalogues* 1878–
ARNOLD, James *The Shell book of country crafts* 1968
ARNOLD, Thomas (1823–1900) *New Zealand letters of Thomas Arnold the younger, with further letters from Van Diemen's Land and letters of Arthur Hugh Clough; 1847–1851* ed. J. Bertram 1966
ARONSON, Joseph *The encyclopedia of furniture* 1965 (UK 1966)
ASBURY, Herbert *Sucker's progress: an informal history of gambling in America from the colonies to Canfield* 1938
ASH, Edward Cecil *The practical dog book* 1930
ASHFORD, Daisy *The young visiters; or, Mr. Salteena's plan* 1919
'ASHFORD, Jeffrey' (Roderic Graeme Jeffries) *The double run* 1973
Prisoner at the Bar 1969
Ashmolean Museum *See* Oxford University. Ashmolean Museum
ASHTON-WARNER, Sylvia *Spinster* 1958
Asiatic Society of Bengal (later Royal ——) *Journal* (title varies) 1832–
Asiatic Society of Japan *Transactions* 1874–
ASIMOV, Isaac *Earth is room enough* 1957 (UK 1960)
Fantastic voyage: a novel based on the screenplay by H. Kleiner 1966
Inside the atom 1956
The naked sun 1957 (UK 1958)
Aspects of translation (The Communication Research Centre, University College, London) 1958
Association football ('Know the game' series) 1948
— (rev. ed.) 1953
Association for Computing Machinery *Communications* 1958–
Journal 1954–
Association of American Geographers *Annals* 1911–
ASTON, William George *A history of Japanese literature* 1899
Astounding science fiction 1933–60
Astrophysical journal 1895–
ATHERTON, Gertrude Franklin *Perch of the devil* 1914
ATKINS, William ed. *The art and practice of printing* 6 vols. 1932–3
ATKINSON, James *Herbert & Procter's Telephony* *See* HERBERT, Thomas Ernest
ATKINSON, Richard John Copland *Field archaeology* 1946
— (ed. 2) 1953
ATTENBOROUGH, David Frederick *Life on earth: a natural history* 1979
ATTWATER, Donald ed. *The Catholic encyclopædic dictionary* 1931
— (ed. 2) 1949
The Christian churches of the East 2 vols. 1961
ATWATER, Mary (Meigs) *Crime in corn-weather* (UK ed. with title *Murder in midsummer*) 1935
AUDEN, Wystan Hugh *About the house* 1965 (UK 1966)
The age of anxiety: a baroque eclogue 1947 (UK 1948)
Another time 1940
City without walls, and other poems 1969
Collected poetry 1945
The dance of death 1933
The dyer's hand and other essays 1962 (UK 1963)
The enchafèd flood; or, The romantic iconography of the sea 1950 (UK 1951)
For the time being 1944 (UK 1945)
Homage to Clio 1960
Look, stranger! 1936
New year letter (UK title of US *The double man*) 1941
Nones 1951 (UK 1952)
The orators: an English study 1932
Poems 1930
— (ed. 2) 1933
The shield of Achilles 1955
—— & ISHERWOOD, Christopher William Bradshaw *The ascent of F6* 1936 (US 1937)
The dog beneath the skin; or, Where is Francis? 1935
Journey to a war 1939
On the frontier 1938
—— & KALLMAN, Chester tr. *E. J. Schikaneder & K. L. Giesecke's libretto to Mozart's The magic flute* 1956 (UK 1957)
—— & MACNEICE, Frederick Louis *Letters from Iceland* 1937
—— et al. *I believe* 1939: see *I believe*
AUERBACH, Charlotte *Genetics in the atomic age* 1956
Auk, The 1884–
Aurora (Philadelphia) 1790–1830; 1834–5 (title varies)
AUSTEN, Jane (1775–1817) *Letters* ed. R. W. Chapman 2 vols. or 1 vol. 1923
— (ed. 2) 1952
Minor works ed. R. W. Chapman 1954
AUSTIN, John Langshaw *How to do things with words* (lectures, 1955) 1962
Sense and sensibilia (lectures, last delivered 1959) 1962
AUSTIN, Oliver Luther *Birds of the world* 1961 (UK 1962)
Australasian, The (Melbourne) 1864–1946
Australasian post (Melbourne) 1946–52
Australia. Commonwealth Scientific and Industrial Research Organization. Division of Entomology *The insects of Australia* 1970
Australian, The 1964–
Australian encyclopaedia, The ed. A. H. Chisholm 10 vols. 1965
Australian house and garden 1948–
Australian short stories 1951 *See* MURDOCH, Walter & DRAKE-BROCKMAN, H. F. Y.
— (2nd ser.) 1963 *See* 'JAMES, Brian'
Australian women's weekly 1933–
Australians in England: a complete record of the cricket tour of 1882 1882
Autocar, The 1895–
Autocar handbook, The 1906–
Automobile engineer, The 1910– (with title *Internal combustion engineering* 1912–14)
AVEBURY, John (Lubbock), 1st Baron *The scenery of England and the causes to which it is due* 1902
AVERY, Gillian *The greatest Gresham* 1962
Aviation age 1950–8
Aviation week 1947–
AVIS, Frederick Compton *Boxing reference dictionary* 1954
The sportsman's glossary 1961
AVIS, Walter Spencer, et al. eds. *A dictionary of Canadianisms on historical principles* 1967
Awake! (Watchtower Bible and Tract Soc.) 1919–
AYCKBOURN, Alan *Relatively speaking* 1968
AYER, Alfred Jules *The central questions of philosophy* 1973
The foundations of empirical knowledge 1940
Language, truth and logic 1936
— (ed. 2) 1946
Philosophical essays 1954
The problem of knowledge 1956
AYERST, David *'Guardian': biography of a newspaper* 1971

B

B., W. *See* BAUCKE, William
B.B.C. *See* British Broadcasting Corporation
BP shield international 1961–
B.S.I. *See* British Standards Institution
BABCOCK, Ernest Brown & CLAUSEN, Roy Elwood *Genetics in relation to agriculture* 1918
BABSON, Marian *Cover-up story* 1971
The stalking lamb 1974
BACH, Emmon *An introduction to transformational grammars* 1964
BACHARACH, Alfred Louis ed. *British music of our time* 1946
ed. *The musical companion* 1934
BACHMANN, Lawrence Paul *The ultimate act* 1972
BACON, Alice Mabel *Japanese girls and women* 1891
A Japanese interior 1893
BACON, Gertrude *Balloons, airships and flying machines* 1905
BACON, Thomas *First impressions and studies from nature in Hindostan* 1837
Bacteriological reviews 1937–
BADDELEY, V. C. C. *See* CLINTON-BADDELEY, Victor Clinton
BADEN-POWELL, Robert Stevenson Smyth *Scouting for boys* 1908
BAGBY, George William (1828–83) *The old Virginia gentleman, and other sketches* ed. T. N. Page 1910
BAGEHOT, Walter (1826–77) *Collected works* ed. N. St. John-Stevas 1965–
BAGG, Lyman Hotchkiss *Four years at Yale* 1871
BAGGS, Thomas Alexander *Back from the front: an eye-witness's narrative of the beginnings of the Great War, 1914* 1914
BAGLEY, Desmond *The enemy* 1977
The freedom trap 1971
The snow tiger 1975
The spoilers 1969
The tightrope men 1973
Bahamian review 1952–
BAILAR, John Christian, et al. *Comprehensive inorganic chemistry* 5 vols. 1973
BAILEY, Alan Robert *A text-book of metallurgy* 1954
— (ed. 2) 1960
BAILEY, Hamilton & LOVE, Robert John McNeill *A short practice of surgery* 2 vols. 1932 (and many later editions used)
BAILEY, Henry Christopher *Dead man's shoes* 1942
Mr. Fortune's practice 1923
BAILEY, John (1864–1931) *Letters and diaries* ed. S. Bailey 1935
BAILEY, Liberty Hyde ed. *Cyclopedia of American agriculture* 4 vols. 1907–9
ed. *Cyclopedia of American horticulture* 4 vols. 1900–2
Manual of cultivated plants 1924
— (rev. ed.) 1949
ed. *The standard cyclopedia of horticulture* 6 vols. 1914–17
BAILLIE, Granville Hugh *Watches* 1929
BAILLIE-SCOTT, M. H. *Houses and gardens* 1906
BAINBRIDGE, Beryl *Sweet William* 1975
Young Adolf 1978
BAINES, Anthony ed. *Musical instruments through the ages* 1961
BAINES, John Thomas *Explorations in South-West Africa* 1864
BAINES, Thomas *Greenhouse and stove plants* 1885
BAINES, William Mortimer *The narrative of Edward Crewe; or, Life in New Zealand* 1874
BAIRD, Irene *Waste heritage* 1939
BAIRD, Robert *A view of the valley of the Mississippi; or, The emigrant's and traveller's guide to the West* 1832
BAKER, Cyril Clarence Thomas *Dictionary of mathematics* 1961
BAKER, Dorothy *Young man with a horn* 1938
BAKER, John Randal *Principles of biological microtechnique: a study of fixation and dyeing* 1958
Race 1974
BAKER, Robert Horace *Astronomy* 1930
— (ed. 8) 1964
BAKER, Roger Denio *Essential pathology* 1961
BAKER, Sidney John *Australia speaks: a supplement to 'The Australian language'* 1953
The Australian language 1945
— (ed. 2) 1966
The drum: Australian character and slang 1959 (UK 1960)
New Zealand slang 1941
A popular dictionary of Australian slang 1941
— (ed. 3) 1943
Balance and Columbian Repository (Hudson, NY) 1801–8
BALCHIN, Nigel Marlin *Darkness falls from the air* 1942
Lord, I was afraid 1947
The small back room 1943
A way through the wood 1951
BALDWIN, Faith *Innocent bystander* 1933 (UK 1935)
BALDWIN, James Arthur *Another country* 1962 (UK 1963)
Go tell it on the mountain 1953 (UK 1954)
BALDWIN, James Mark ed. *Dictionary of philosophy and psychology* 3 vols. 1901–5
BALFOUR, Isaac Bayley tr. *C. E. von Goebel's Organography of plants* 2 vols. 1900–5
BALINSKY, Boris Ivan *An introduction to embryology* 1960
BALL, John Dudley *The cool cottontail* 1966 (UK 1967)
Five pieces of jade 1972
BALL, Kenneth *Fiat 600, 600D autobook* 1970
BALLANTYNE, David Watt *The Cunninghams* 1948 (UK 1963)
Ballet annual, The 1947–
BANCROFT, Hubert Howe *The native races of the Pacific states of North America* 5 vols. 1875–6
BAND, George C. *Road to Rakaposhi* 1955
BANFIELD, Edmund James *The confessions of a beachcomber* 1908
Bangladesh Times (Dacca) 1974–
BANIM, John *The smuggler* 3 vols. 1831
BANKS, Michael Edward Borg *Commando climber* 1955
BANNERMAN, David Armitage *The birds of the British Isles* 12 vols. 1953–63
BANNISTER, Roger Gilbert *First four minutes* 1955

Banton, Michael Parker *The policeman in the community* 1964
West African city: a study of tribal life in Freetown 1957
Baptist magazine 1809–1904
'Barbellion, W. N. P.' (Bruce Frederick Cummings) *The journal of a disappointed man* 1919
Barber, Charles Laurence *Linguistic change in present-day English* 1964
Barber, Horatio *The aeroplane speaks* 1916
'Barbican, James' *The confessions of a rum-runner* 1927
Barclay, Florence Louisa *The broken halo* 1913
The rosary 1909
Barclay, Rachel Mary tr. *E. Kraepelin's Dementia praecox and paraphrenia* ed. G. M. Robertson 1919
Barford, John Edward Quintus *Climbing in Britain* 1946
Baring, Maurice *Landmarks in Russian literature* 1910
Barker, George Granville *News of the world* 1950
Two plays 1958
Barker, Lewellys Franklin *Monographic medicine* See under title
The nervous system and its constituent neurones 1899
—, Hoskins, R. G., & Mosenthal, H. O. eds. *Endocrinology and metabolism* 5 vols. 1922
Barlow, James *Term of trial* 1961
Barlow, Peter *A new mathematical and philosophical dictionary* 1814
Barlow, Walter Sydney Lazarus *A manual of general pathology* 1898
Barnard, Howard Clive & Lauwerys, Joseph Albert *A handbook of British educational terms* 1963
Barnard, Keppel Harcourt *A pictorial guide to South African fishes* 1947
Barnard, Robert *Death on the high C's* 1977
Barnes, Dallas *See the woman* 1973 (UK 1974)
Barnes, Harold *Oceanography and marine biology* 1959
Barnett, Denis Oliver *Denis Oliver Barnett: in happy memory . . . His letters from France and Flanders, October 1914–August 1915* 1915
Barnett, Edward de Barry *Mechanism of organic chemical reactions* 1956
Barnett, Henrietta Octavia Weston *Canon Barnett: his life, work and friends* 1918
Barnham, Denis *One man's window: an account of 10 weeks of war, Malta, 1942* 1956
Barnhart, Clarence Lewis, Steinmetz, S., & Barnhart, R. K. *The Barnhart dictionary of new English since 1963* (UK ed. with title *A dictionary of new English 1963–1972*) 1973
The second Barnhart dictionary of new English 1980
Barnum, Phineas Taylor *Life, written by himself* 1855
Struggles and triumphs; or, Forty years' recollections 1869
— (another ed.) 1871
Baron, Joseph Alexander *From the city, from the plough* 1948
The human kind 1953
Rosie Hogarth 1951
There's no home 1950
With hope, farewell 1952
Barr, Ann & York, Peter *The official Sloane Ranger handbook: the first guide to what really matters in life* 1982
Barrett, Charles Leslie *On the wallaby: quest and adventure in many lands* 1942
Barrie, Sir James Matthew *The admirable Crichton* 1914
Plays (Uniform ed.) 12 vols. 1918–38
'Barrington, Pamela' (Muriel Vere Barling) *Accessory to murder* 1968
Barron, Harry *Modern plastics* 1945
Barron, John N. *The language of painting: an informal dictionary* 1967
Barrow, John *An account of travels into the interior of southern Africa, in the years 1797 and 1798* 2 vols. 1801–4
— (ed. 2) 2 vols. 1806
Barrowman, James *A glossary of Scotch mining terms* 1886
'Barry, C.' (Charles Bryson) *The detective's holiday* 1926
Barstow, Stan *A kind of loving* 1960
Barth, John Simmons *The floating opera* 1956
— (rev. ed.) 1967
'Bartimeus' (Lewis Anselm da Costa Ricci) *Naval occasions* 1914
A tall ship and other naval occasions 1915
Bartlett, Adeline Courtney *The larger rhetorical patterns in Anglo-Saxon poetry* 1935
Bartlett, Ellis Ashmead *The battlefields of Thessaly* 1897
Bartlett, Vernon Oldfield *This is my life* 1937
Bartley, Samuel Howard *Vision* 1941
Barton, Frank Townend *The kennel encyclopædia* 1928

Barton, R. tr. *C. Blasis's The code of Terpsichore: the art of dancing* 1830
Bartz, Patricia McBride *South Korea* 1972
Bashforth, George Reginald *The manufacture of iron and steel* 2 vols. 1948–51
— (ed. 2) 4 vols. 1957–62
Baskin, Wade tr. *F. de Saussure's Course in general linguistics* 1959 (UK 1960)
Bates, Herbert Ernest *The day of the tortoise* 1961
The jacaranda tree 1949
Love for Lydia 1952
Bateson, William *Mendel's principles of heredity* 1909
Mendel's principles of heredity: a defence 1902
Problems of genetics 1913
Bathgate, Alexander *Colonial experiences; or, Sketches of people and places in the province of Otago, New Zealand* 1874
Battersby, Albert *A guide to stock control* 1962
Mathematics in management 1966
Network analysis for planning and scheduling 1964
— (ed. 2) 1967
Battey, Maurice Hugh *Mineralogy for students* 1972
Baucke, William *Where the white man treads* 1905
— (ed. 2) 1928
Baumann, Heinrich *Londinismen: Slang und Cant* 1886 [dated 1887]
Baxt, George *A queer kind of death* 1966 (UK 1967)
Bayley, Roger Child *The complete photographer* 1906
Bayliss, William Maddock *Principles of general physiology* 1915
Bayly, Brian *Introduction to petrology* 1968
Baynes, Helton Godwin tr. *C. G. Jung's Psychological types; or, The psychology of individuation* 1923
Bazell, Charles Ernest *Linguistic form* 1953
— et al. eds. *In memory of J. R. Firth* 1966
Beach, Rex Ellingwood *Pardners* 1905 (UK 1912)
Beadle, John Hanson *The undeveloped West; or, Five years in the Territories* 1873
Western wilds, and the men who redeem them 1878
— (another ed.) c 1882
Beadle, Muriel *These ruins are inhabited* 1961 (UK 1963)
Bean, Charles Edwin Woodrow *The 'Dreadnought' of the Darling* 1911
Bean, William Jackson *Trees and shrubs hardy in the British Isles* vols. I–II, 1914; vol. III, 1933
Beardsley, Aubrey (1872–98) *Letters* ed. H. Maas, J. L. Duncan, & W. G. Good 1971
Beaton, Cecil Walter Hardy *The wandering years: diaries 1922–39* 1961
Beattie, William *Life and letters of Thomas Campbell* 3 vols. 1849
Beaty, Arthur David *Cone of silence* 1959
Excellency 1977
Beaumont, Cyril William & Idzikowsky, Stanislas *A manual of the theory and practice of classical theatrical dancing (méthode Cecchetti)* 1922
Beaumont, Reginald Arthur ed. *Aeronautical engineering* 1942
Beautiful British Columbia, land of new horizons 1959–
Beaver (Winnipeg) 1920–
Becker, Stephen D. tr. *A. Schwarz-Bart's The last of the just* 1960 (UK 1961)
Beckett, Samuel Barclay *Eh Joe, and other writings* 1967
More pricks than kicks 1934
Murphy 1938
No's knife: collected shorter prose 1967
Proust 1931
Stories and texts for nothing (tr. by Beckett of his *Nouvelles et textes pour rien* 1955) 1967
Waiting for Godot (tr. by Beckett of his *En attendant Godot* 1953) 1954 (UK 1956)
Whoroscope 1930
Beckson, Karl & Ganz, Arthur *A reader's guide to literary terms* 1960 (UK 1961)
Bedford, Randolph *True Eyes and the whirlwind* 1903
Bedoukian, Paul Z. *Perfumery synthetics and isolates* 1951
Beebe, Charles William *Jungle peace* 1918 (UK 1919)
Beecheno, Henry Oliver *An introduction to business studies* 1962
Beer, Max *A history of British socialism* 2 vols. 1919–20
Beer, Stafford *Cybernetics and management* 1959
Decision and control 1966
Beer, Thomas *The mauve decade: American life at the end of the nineteenth century* 1926
Beerbohm, Henry Maximilian *And even now* 1920
Around theatres 2 vols. 1924 (in *Collected works* Vols. VIII & IX)
— (another ed.) 1 vol. 1953

A Christmas garland 1912
Letters to Reggie Turner ed. R. Hart-Davis 1964
More 1899
Observations 1925
Seven men 1919
A variety of things 1928 (in *Collected works* Vol. X)
Zuleika Dobson; or, An Oxford love story 1911
Beevor, Antony *The violent brink* 1975
Begbie, Edward Harold *Life of William Booth, the founder of the Salvation Army* 2 vols. 1920
Behan, Brendan Francis *Borstal boy* 1958
Behre, Frank ed. *Contributions to English syntax and philology* 1962
Behrend, Arthur *The Samarai affair* 1973
Belfast Telegraph 1870–
Bell, Alexander Melville *Visible speech: the science of universal alphabetics* 1867
Bell, Eric Temple *The development of mathematics* 1940
— (ed. 2) 1945
Bell, Gertrude Margaret Lowthian (1868–1926) *Letters* sel. and ed. F. Bell 2 vols. 1927
Bell, Harold Sill *American petroleum refining* 1923
— (ed. 3) 1945
'Bell, Josephine' (Doris Bell Ball) *Crime in our time* 1962
Murder in hospital 1937
The port of London murders 1938
The seeing eye 1958
Bell, Louis *The telescope* 1922
Bell, Peter & Coombe, David tr. *Strasburger's textbook of botany* (tr. from 28th German ed., rewritten by R. Harder et al.) 1965
— (rev. ed.; tr. from 30th German ed., rewritten by D. von Denffer et al.) 1976
Bell, Robert Charles *Board and table games from many civilizations* 2 vols. 1960–9
Bell system technical journal 1922–
Bell's life in London and sporting chronicle 1822–86
Bellairs, Angus d'Albini *The life of reptiles* 2 vols. 1969
Bellairs, George *Death in High Provence* 1957
Bellanti, Joseph Alphonso *Immunology* 1971
Belloc, Joseph Hilaire Pierre René *Cautionary tales for children* 1907
Charles the First, King of England 1933
Essays of a Catholic layman in England 1931
A history of England 4 vols. 1925–31
Pongo and the bull 1910
Verses and sonnets 1896
Bellow, Saul *The adventures of Augie March* 1953
Dangling man 1944
Herzog 1964 (UK 1965)
Seize the day 1956 (UK 1957)
The victim 1947 (UK 1948)
Benchley, Nathaniel Goddard *Welcome to Xanadu* 1968
Bender, Arnold Eric *Dictionary of nutrition and food technology* 1960
Benedict, Ruth (Fulton) *The chrysanthemum and the sword: patterns of Japanese culture* 1946 (UK 1947)
Patterns of culture 1934 (UK 1935)
Benét, Stephen Vincent *John Brown's body* 1928
Thirteen o'clock 1937 (UK 1938)
Bennett, Arnold *See* Bennett, Enoch Arnold
Bennett, Colin Noël *The guide to kinematography* 1917
— et al. *The handbook of kinematography: the history, theory and practice of motion photography and projection* 1911
Bennett, Enoch Arnold (1867–1931) *Anna of the Five Towns* 1902
The card 1911
Clayhanger 1910
The Grand Babylon Hotel 1902
A great man 1904
The grim smile of the Five Towns 1907
Hilda Lessways 1911
Imperial Palace 1930
Journals, 1896–1928 ed. N. Flower 3 vols. 1932–3
Lemons 1923
Letters ed. J. Hepburn 3 vols. 1966–70
Lilian 1922
The lion's share 1916
Lord Raingo 1926
A man from the North 1898
The matador of the Five Towns, and other stories 1912
The old wives' tale 1908
The pretty lady 1918
The Regent 1913
Riceyman Steps 1923
The roll-call 1918
Tales of the Five Towns 1905
These twain 1915 (UK 1916)
The truth about an author 1903
Bennett, Ivy V. P. *Delinquent and neurotic children* 1960
Bennett, John Godolphin *Witness* 1962
Bennett, Louise *Jamaican humour in dialect* 1943

Bennett, Mary Montgomerie *Christison of Lammermoor* 1927
Bennison, George Mills & Wright, Alan Edward *The geological history of the British Isles* 1969
Benson, Edward Frederic *David Blaize* 1916
The image in the sand 1905
Mapp and Lucia 1931
Thorley Weir 1913
Benson, Eugene Patrick *The bulls of Ronda* 1976
Bentley, Edmund Clerihew *Trent's last case* 1913
Trent's own case 1936
Bentley, Robert & Trimen, Henry *Medicinal plants* 4 vols. 1880
Bentley, William (1759–1819) *Diary* 4 vols. 1905–14
Bentley's miscellany 1837–68
Benton, Kenneth Carter *The red hen conspiracy* 1977
Spy in Chancery 1972
Benton, Thomas Hart *Thirty years' view; or, A history of the working of the American Government for thirty years, from 1820 to 1850* 2 vols. 1854–6
Berckman, Evelyn *The Victorian album* 1973
Bereiter, Carl & Engelmann, Siegfried *Teaching disadvantaged children in the preschool* 1966
Beresford, Maurice *The lost villages of England* 1954
Berg, George Charles *The unconscious significance of hair* 1951
Berg, Leila *Risinghill: death of a comprehensive school* 1968
Berg, Paul C. *A dictionary of new words in English* 1953
Berget, Alphonse *The conquest of the air: aeronautics, aviation; history, theory, practice* 1909
Bergman, Andrew *Hollywood and Le Vine* 1975
Berkeley, Edmund Callis *Giant brains, or machines that think* 1949
— & Wainwright, Lawrence *Computers: their operation and applications* 1956
Berkeley, George Fitz-Hardinge & Berkeley, J. *Italy in the making* 3 vols. 1932–40
Berkman, Al *Singers' glossary of show business jargon* 1961
Berlin, Isaiah *Karl Marx, his life and environment* 1939
Bermant, Chaim Icyk *Coming home* 1976
Bernard, John (1756–1828) *Retrospections of the stage* ed. W. B. Bernard 2 vols. 1830
Berrey, Lester V. & Van den Bark, Melvin *The American thesaurus of slang* 1942
— (ed. 2) as US 1953, UK 1954
Berry, Frederic Arnon, Bollay, E., & Beers, Norman R. eds. *Handbook of meteorology* 1945
Berry, William Turner & Poole, Herbert Edmund *Annals of printing* 1966
Berton, Pierre *Klondike fever* 1958
Bertram, James Munro *The shadow of a war: a New Zealander in the Far East, 1939–1946* 1947
Bessey, Ernst Athearn *Morphology and taxonomy of fungi* 1950
Bessinger, Jess B. & Creed, Robert P. eds. *Medieval and linguistic studies in honor of F. P. Magoun* 1965 (US ed. has title *Franciplegius*)
Best one-act plays 1932–
Betham, Raymond Vernon *Repeat the instructions* 1968
Bethell, Nicholas & Burg, David tr. *A. I. Solzhenitsyn's Cancer ward* 2 vols. 1968–9
Betjeman, John *Collected poems* 1958
Continual dew 1937
A few late chrysanthemums 1954
High and low 1966
Mount Zion; or, In touch with the infinite 1931
New bats in old belfries 1945
Selected poems 1948
Summoned by bells 1960
Bettelheim, Frederick A. *Experimental physical chemistry* 1971
Bevan, John Acton ed. *Essentials of pharmacology* 1969
Bews, John William *The world's grasses* 1929
Bhavnani, Enakshi *The dance in India* 1965
Bible *New English Bible* 3 vols. 1961–70
Bibliographical Society *Transactions* 1892–1919 (then merged with *The Library*)
Bickel, Walter tr. *R. Hering's Dictionary of classical and modern cookery* 1958
Biden, C. Leo *Sea-angling fishes of the Cape* 1930
Bigelow, Jacob *Florula Bostoniensis: a collection of plants of Boston and its environs* 1814
Billings, John Davis *Hardtack and coffee* 1887
Billings, Marland Pratt *Structural geology* 1942
— (ed. 2) 1954
— (ed. 3) 1972
Billings Gazette (Billings, Montana) 1901–
Bingham, John Michael Ward, 7th Baron Clanmorris *God's defector* 1976

The marriage bureau murders 1977
Binyon, Robert Laurence *Collected poems* 2 vols. 1931
Biochemical and biophysical research communications 1959–
Biochemical journal 1906–
Biochemistry (Easton, Pennsylvania) 1962–
Biochimica et biophysica acta 1947–
Biological abstracts 1926–
Biological bulletin 1898–
Biological reviews and biological proceedings 1926– (from 1936 with title *Biological reviews*)
Birch, Anthony & Hutton-Stott, Francis *The veteran motor car pocketbook* 1963
Bird, Kenneth *Smash a glass image* 1968
Bird, Robert Montgomery *The Hawks of Hawk-Hollow* 2 vols. 1835
Bird, William Richard *Off-trail in Nova Scotia* 1956
These are the Maritimes 1959
Birkbeck, Morris *Notes on a journey through France* 1814 (Appendix, 1815)
Birkhoff, Garrett & MacLane, Saunders *A survey of modern algebra* 1941
Birmingham, Maisie Poynter *The heat of the sun* 1976
Birmingham News (Birmingham, Alabama) 1888–
Birmingham Post (title varies) 1857–
Birnbach, Lisa, et al. *The official Preppy handbook* 1980
Birnbaum, Henrik & Puhvel, Jaan eds. *Ancient Indo-European dialects: proceedings of the Conference on Indo-European Linguistics at U.C.L.A. in 1963* 1966
Bissell, Richard Pike *High water* 1954 (UK 1955)
Bithell, Richard *A counting-house dictionary* 1882
'Black, Gavin' (Oswald Wynd) *The bitter tea* 1972 (UK 1973)
The golden cockatrice 1974
A time for pirates 1971
You want to die, Johnny? 1966
Black, Jack *You can't win* 1926
'Black, Lionel' (Dudley Raymond Barker) *Death has green fingers* 1971
A healthy way to die 1976
Outbreak 1968
Black, Max ed. *The importance of language* 1962
The labyrinth of language 1968
The nature of mathematics 1933
ed. *Philosophy in America* 1965
Black, Peter *The biggest aspidistra in the world: a personal celebration of fifty years of the B.B.C.* 1972
Black, Rhona M. *The elements of palaeontology* 1970
Black mask, The 1920–
Black panther 1969–
Black scholar 1969–
Black world 1970–
Blacker, Carlos Paton *Eugenics: Galton and after* 1952
Blackshaw, Alan *Mountaineering. From hill walking to Alpine climbing* 1965
— (rev. ed.) 1970
Blackshaw, Harold & Brightman, Rainald *Dictionary of dyeing and textile printing* 1961
Blackwelder, Richard Eliot *Taxonomy: a text and reference book* 1967
Blaikie, Thomas *Diary of a Scotch gardener at the French court at the end of the 18th century* ed. F. Birrell 1931
Blair & Ketchum's country journal (Brattleboro, Vermont) 1974–
'Blaisdell, Anne' (Barbara Elizabeth Linington) *Practice to deceive* 1971
Blake, Geoffrey Charles & Trott, John Richard *Periodontology* 1962
'Blake, Nicholas' (Cecil Day Lewis) *The case of the abominable snowman* 1941
The deadly joker 1963
The dreadful hollow 1953
Malice in wonderland 1940
A question of proof 1935
The smiler with the knife 1939
There's trouble brewing 1937
Thou shell of death 1936
The whisper in the gloom 1954
See also: Lewis, Cecil Day
Blake, William (1757–1827) *Complete writings* ed. G. Keynes 1966 (Oxford Standard Authors ed.; reprinted with corrections 1972)
Bland, David *The illustration of books* 1951
Blast: review of the great English vortex ed. P. Wyndham Lewis 1914–15
Bleaney, Betty Isabelle & Bleaney, Brebis *Electricity and magnetism* 1957
'Bleeck, Oliver' (Ross Thomas) *The Procane chronicle* 1971 (UK ed. 1972 with title *The thief who painted sunlight*)
Blesh, Rudolph Pickett *Shining trumpets: a history of jazz* 1946 (UK 1949)

They all played ragtime: the true story of an American music 1950 (UK 1958)
Blish, James *A case of conscience* 1958 (UK 1959)
A clash of cymbals 1959
Fallen star 1957
Blishen, Edward *Roaring boys* 1955
Bloch, Bernard & Trager, George Leonard *Outline of linguistic analysis* 1942
Blom, Eric Walter *Music in England* 1942
Blood: the journal of hematology 1946–
Bloodgood, Lida & Santini, Piero eds. *The horseman's dictionary* 1963
Bloodworth, Dennis *Any number can play* 1972
The clients of Omega 1975
Crosstalk 1978
Bloomfield, Leonard *An introduction to the study of language* 1914
Language 1933
Blow, Claude Montague *Rubber technology and manufacture* 1971
Blues unlimited 1963–
Blunden, Edmund Charles *After the bombing, and other short poems* 1949
Choice or chance: new poems 1934
Cricket country 1944
De bello Germanico: a fragment of trench history 1930
An elegy, and other poems 1937
The face of England 1932
The harbingers 1916
The mind's eye 1934
Near and far: new poems 1929
Poems 1930
Poems, 1930–1940 1940
Retreat 1928
Romantic poetry and the fine arts 1942
Shells by a stream 1944
The shepherd, and other poems 1922
A summer's fancy 1930
Undertones of war 1928
Blunt, Wilfrid Jasper W. *Of flowers and a village* 1963
Bodkin, Maud *Archetypal patterns in poetry: psychological studies of imagination* 1934
Bodleian Library record 1938–
Boger, Louise Ade *The dictionary of world pottery and porcelain* 1971
Boltz, Cecil Leonard *Basic radio* 1943
Bond, James *Birds of the West Indies* 1936
— (another ed.) 1960
Bone, John Herbert A. *Petroleum and petroleum wells* 1865
Bonfiglioli, Kyril *Don't point that thing at me* 1972
Something nasty in the woodshed 1976
Bonington, Christian *Annapurna South Face* 1971
The next horizon: autobiography II 1973
Bonner, James Frederick *Plant biochemistry* 1950
Booker, Christopher *The neophiliacs: a study of the revolution in English life in the Fifties and Sixties* 1969
Bookman, The 1891–1934
Bookseller, The (title varies) 1858–
Booth, Andrew Donald & Booth, Kathleen H. V. *Automatic digital calculators* 1953
Booth, Michael R. ed. *English plays of the nineteenth century* 4 vols. 1969–73
Booth, William *In darkest England and the way out* 1890
Born, Max & Wolf, Emil *Principles of optics* 1959
Borradaile, Lancelot Alexander & Potts, Frank Armitage *The Invertebrata* 1932
Borror, Donald Joyce & DeLong, Dwight Moore *An introduction to the study of insects* 1954
— (rev. ed.) 1964
Boston, Massachusetts. Registry Dept. *Records relating to the early history of Boston* 39 vols. 1876–1909
Boston Daily Globe (Boston, Mass.) morning ed. 1872–
— evening ed. 1878–
— Sunday ed., as *Boston Sunday Globe* 1877–
Boston Evening Transcript (title varies) 1830–1941
Boston Herald 1846–
Boston medical and surgical journal 1828–1928
Boston News-Letter (title varies) 1704–76
Boston Sunday Globe See *Boston Daily Globe*
Boston Sunday Herald (title varies) 1861–
Boswell, James *London journal 1762–63* ed. F. A. Pottle 1950
Botanical gazette 1876–
Botkin, Benjamin Albert ed. *Lay my burden down: a folk history of slavery* 1945
ed. *A treasury of American folklore* 1944
ed. *A treasury of Southern folklore* 1949
Bottome, Phyllis *Under the skin* 1950
Boulenger, Edward George *Apes and monkeys* 1936

Boulger, George Simonds *The uses of plants: a manual of economic botany* 1889
Wood: a manual of the natural history and industrial applications of the timber of commerce 1902
Bourne, Geoffrey Howard ed. *Cytology and cell physiology* 1942
— (ed. 2) 1951; corrected impression 1952
Bova, Benjamin *Multiple man* 1976 (UK 1977)
Bowen, Elizabeth Dorothea Cole *Ann Lee's, and other stories* 1926
The cat jumps, and other stories 1934
Collected impressions 1950
The death of the heart 1938
The demon lover, and other stories 1945
Encounters 1923
The heat of the day 1949
The hotel 1927
The house in Paris 1935
The last September 1929
The little girls 1964
Look at all those roses 1941
Seven winters 1942
A time in Rome 1960
To the north 1932
A world of love 1955
Bowen, Frank Charles *Sea slang: a dictionary of the old-timers' expressions and epithets* 1929
Bowen, Godfrey *Wool away! The technique and art of shearing* 1955
— (ed. 2) 1956
'Bower, Bertha Muzzy' (Bertha Muzzy Sinclair) *The Parowan bonanza* 1923
The phantom herd 1916
Bower, Frederic Orpen *Botany of the living plant* 1919
Plant-life on land considered in some of its biological aspects 1911
Bowers, Fredson Thayer *Bibliography and textual criticism* 1964
Principles of bibliographical description 1949
Textual and literary criticism 1959
Bowles, Samuel *Our new West. Records of travel between the Mississippi River and the Pacific Ocean* 1869
Bowley, Arthur Lyon & Stamp, Josiah Charles *The national income, 1924: a comparative study of the income of the United Kingdom in 1911 and 1924* 1927
Bowls ('Know the game' series) 1962
Bowman, Walter Parker & Ball, Robert Hamilton *Theatre language: a dictionary of terms in English from medieval to modern times* 1961
Bowman, William Cameron, Rand, M. J., & West, G. B. *Textbook of pharmacology* 1968
Bowra, Cecil Maurice *The Romantic imagination* 1950
Box, Charles *The English game of cricket* 1877
Boyd, Edward & Parkes, Roger *The dark number* 1973
Boylston, Herbert Melville *An introduction to the metallurgy of iron and steel* 1928
Boys' magazine 1922–
Brace, Charles Loring *The new West; or, California in 1867–1868* 1869
Brace, Gerald Warner *The spire* 1952 (UK 1953)
Bradbury, Malcolm Stanley *Eating people is wrong* 1959
The history man 1975
Stepping westward 1965
Bradbury, Ray Douglas *Fahrenheit 451* 1953 (UK 1954)
The illustrated man 1951 (UK 1952)
Bradby, Mary Katharine *Psycho-analysis and its place in life* 1919
Braddon, Russell Reading *Nancy Wake: the story of a very brave woman* 1956
The year of the angry rabbit 1964
Bradford, Gershom *A glossary of sea terms* 1927
— (new ed.) 1954
Bradley, Francis Herbert *The principles of logic* 1883
Bradshaw's railway guide (title varies) 1841–1961
Bradshaw's railway manual, shareholders' guide, and official directory for 1869 (Vol. XXI) 1869
Bradstreet's: a journal of trade, finance and public economy 1879–1933
Bragg, Melvyn *The hired man* 1969
Without a city wall 1968
Brahmacharini Usha *A Ramakrishna-Vedanta wordbook* 1962
'Brahms, Caryl' (Doris Caroline Abrahams) & 'Simon, S. J.' (Simon Jasha Skidelsky) *Titania has a mother* 1944
Trottie True 1946
Brain, Walter Russell *Diseases of the nervous system* 1933
Speech disorders 1961
Brain: a journal of neurology 1878–
Braine, John *Life at the top* 1962
Room at the top 1957
The Vodi 1959
Braithwaite, David *Fairground architecture* 1968

Bram, Gary & Downs, C. *Manufacturing technology* 1975
'Bramah, Ernest' (Ernest Bramah Smith) *A little flutter* 1930
Bramble, Albert York *Air-plane flight* 1952
Brandon, David Godfrey *Modern techniques in metallography* 1966
Brandon, H. ed. *Poverty, mendicity and crime . . . To which is added a dictionary of the flash or cant language* 1839
Brandon, John Gordon *The pawnshop murder* 1936
Brandt, David Julius O. *The manufacture of iron and steel* 1953
— (ed. 2) 1960
Branford, Jean *A dictionary of South African English* 1978
— (new ed.) 1980
Branston, Frank *An up and coming man* 1977
Braun, Hugh *Parish churches: their architectural development in England* 1970
Braun, Lilian Jackson *The cat who ate Danish Modern* 1967
Brawley, Ernest *The rap* 1974
— (paperback ed.) 1975
Bray, Warwick M. & Trump, David Hilary *A dictionary of archaeology* 1970
Brayshaw, J. Dodsworth *Slum silhouettes: stories of London life* 1898
Brazil, Angela *The new girl at St. Chad's: a story of school life* 1912
Breck, Samuel (1771–1862) *Recollections, with passages from his note-books* ed. H. E. Scudder 1877
Breland, Osmond Philip *Animal friends and foes* 1957
Brenan, Edward Fitz-Gerald (Gerald Brenan) *The face of Spain* 1950
Brennan, Paul, Hesselyn, Ray, & Bateson, Henry *Spitfires over Malta* 1943
'Brent of Bin Bin' (Stella Maria Sarah Miles Franklin) *Up the country: a tale of the early Australian squattocracy* 1928
Brereton, Cloudesley Shovell Henry & Rothwell, Frederick tr. *H. L. Bergson's Laughter* 1911
Brewster, Ben R. tr. *L. Althusser & E. Balibar's Reading Capital* 1970
— (paperback ed.) 1975
Brickhill, Paul & Norton, Conrad *Escape to danger* 1946
'Bridge, Ann' (Mary Dolling, Lady O'Malley) *Four-part setting* 1939
Frontier passage 1942
Peking picnic 1932
The Portuguese escape 1958
Bridges, Robert Seymour *The testament of beauty* 1929
Briggs, Charles Frederick *The adventures of Harry Franco* 2 vols. 1839
Briggs, Fred N. & Knowles, Paulden Ford *Introduction to plant breeding* 1967
Brill, Abraham Arden tr. *S. Freud's The interpretation of dreams* 1913
tr. *S. Freud's Selected papers on hysteria and other psychoneuroses* 1909
tr. *S. Freud's Three contributions to sexual theory* 1910
tr. *S. Freud's Totem and taboo: resemblances between the psychic lives of savages and neurotics* 1918
Brinkley, Frank *Oriental series* 12 vols. 1901–2
Britain's industrial future (Liberal Industrial Inquiry) 1928
Britannica book of the year 1938– (UK eds. 1938–41 with title *Encyclopædia Britannica book of the year*)
British Academy *Proceedings* 1905–
British birds 1907–
British Broadcasting Corporation *B.B.C. handbook* (title varies: also *B.B.C. year-book* and *B.B.C. annual*) 1928–
The B.B.C. quarterly 1946–54
B.B.C. war report ed. D. Hawkins et al. 1946
Glossary of broadcasting terms See: Farquharson, M. G., et al.
British chess magazine 1881–
British commonwealth forest terminology (Empire Forestry Association) 2 parts 1953–7
British dental journal 1903–
British Dominions year book, The (British Dominions Insurance Co.) 1916–24
British Engineering Standards Association *British standard glossary of terms used in electrical engineering* 1926 *See* British Standards Institution. Glossaries. *Electrical engineering*
British Interplanetary Society *Journal* 1934–9; 1946–
British journal of dermatology 1888–
British journal of medical psychology 1923–
British journal of ophthalmology 1917–
British journal of psychology 1904–
British journal of radiology 1924–
British journal of sociology 1950–
British manufacturer Nov.–Dec. 1919

British medical bulletin 1943–
British medical dictionary, The ed. A. S. MacNalty 1961
British pharmaceutical codex 1907 (and several later editions used)
British pharmacopœia 1864 (and many later editions used)
British Pharmacopœia Commission *Approved names* See: United Kingdom. General Medical Council
British printer, The 1888–
British School (of Archaeology) at Athens *Annual* 1895–
British Standards Institution *B.S.I. News* 1956
Nomenclature of commercial timbers, including sources of supply 1946
— (rev. ed.) 1955
Glossaries:
Aeronautical terms (BS 185) 1949, 1950, 1951, and other eds.
Automatic data processing (BS 3527) 1962, and several later parts
Caravan terms (BS AU129) 1967, 1975
Concrete and reinforced concrete (BS 2787) 1956
Dentistry (BS 4492) 1969
Documentation terms (BS 5408) 1976
Electrical engineering (BS 205) 1926, 1936, 1943
Electrotechnical, power, telecommunication, electronics, lighting and colour terms (BS 4727) 1971–
General building terms (BS 3589) 1963
Glass industry (BS 3447) 1962
Highway engineering terms (BS 892) 1940, 1954, 1967
Landscape work (BS 3975) 1966–9
Letterpress rotary printing terms (BS 3814) 1964
Mining terms (BS 3618) 1963–74
Nuclear science (BS 3455) 1962
Offset lithographic printing (BS 4277) 1968
Plastics industry (BS 1755) 1951, 1967–74
Radiology (BS 2957) 1955
Railway signalling (BS 719) 1936
Refrigeration (BS 1584) 1949
Sanitation terms (BS 4118) 1967
Telegraphs and telephones (BS 204) 1924, 1930, 1943 (*Telecommunication*), 1960
Timber preservation (BS 4261) 1968
British weekly, The: a journal of social and Christian progress 1886–
Brittain, Vera Mary *Testament of youth* 1933
Britten, Frederick James *Old clocks and watches* 1899
Britton, Karl *John Stuart Mill* 1953
Britton, Nathaniel Lord *North American trees* 1908
Brno studies in English 1959–
Broad, Charlie Dunbar *The mind and its place in nature* 1925
Perception, physics, and reality 1914
Scientific thought 1923
Brodal, Alf *Neurological anatomy in relation to clinical medicine* 1948
Broinowski, Alison *Take one ambassador* 1973
Bromfield, Louis *The rains came* 1937
Bromige, Iris Amy Edna *The enchanted garden* 1956
Bromley, Gordon *In the absence of the body* 1972
Bronowski, Jacob *The ascent of man* 1973
Bronson-Howard, George *Enemy to society* 1911
God's man 1915
Brontë, Anne *Agnes Grey* 1847
The tenant of Wildfell Hall 3 vols. 1848
Brooke, Rupert Chawner (1887–1915) *Collected poems* 1918
Letters ed. G. Keynes 1968
Brooke-Rose, Christine Frances E. *The languages of love* 1957
Brooks, Frederick Tom *Plant diseases* 1928
Brooks, Henry *Natal: a history and description of the colony* 1876
Brophy, John & Partridge, Eric Honeywood eds. *The long trail: what the British soldier sang and said in the Great War of 1914–18* 1965
Songs and slang of the British soldier, 1914–18 1930
— (ed. 3) 1931
Brose, Henry Herman L. A. tr. *A. J. W. Sommerfeld's Atomic structure and spectral lines* 1923
— (ed. 3) 1934
Brosnahan, Leonard Francis *Some Old English sound changes: an analysis in the light of modern phonetics* 1953
The sounds of language 1961
Brown, Bernard *Talking pictures* 1931
Brown, George Mackay *An Orkney tapestry* 1969
Brown, Ivor John Carnegie *Say the word* 1947
Words in our time 1958
Brown, James Alexander Campbell *The social psychology of industry* 1954
Brown, John (M.D., of Edinburgh; 1810–82) *Letters; with letters from Ruskin, Thackeray, and others* ed. J. Brown & D. W. Forrest 1907
— (ed. 3) 1912

Brown, John *The chancer* 1972
Brown, Michael Barratt *The economics of imperialism* 1974
Brown, Samuel R. *The western gazetteer; or, Emigrant's directory* 1817
Brown, Walter Varian & Bertke, Eldridge Melvin *Textbook of cytology* 1969
Browne, Daniel Jay *The sylva americana; or, A description of the forest trees indigenous to the United States* 1832
Browning, Carl Hamilton *Bacteriology* 1925
Bruner, Jerome Seymour *Beyond the information given: studies in the psychology of knowing* ed. J. M. Anglin 1973 (UK 1974)
Brunt, David *Meteorology* 1928
Bruton, Eric Moore *Dictionary of clocks and watches* 1962
Bryan, William Alanson *Natural history of Hawaii* 1915
Bryant, Edwin *What I saw in California* 1848
— (ed. 5) 1849
'Bryant, Peter' (Peter Bryan George) *Two hours to doom* 1958
Bryden, Henry Anderson *Animals of Africa* 1900
Kloof and karroo: sport, legend, and natural history in Cape Colony 1889
Tales of South Africa 1896
Bryson, Herbert Courtney *The gramophone record* 1935
Buchan, John *The blanket of the dark* 1931
Castle Gay 1930
The courts of the morning 1929
The dancing floor 1926
The gap in the curtain 1932
Greenmantle 1916
The house of the four winds 1935
Huntingtower 1922
The island of sheep 1936
John Macnab 1925
Memory hold-the-door 1940
Mr. Standfast 1919
Nelson's history of the war 24 vols. 1915–19
Prester John 1910
A prince of the captivity 1933
The thirty-nine steps 1915
The three hostages 1924
Buchanan, Robert James McLean *The blood in health and disease* 1909
Buck, Carl Darling *Comparative grammar of Greek and Latin* 1933
Buck, Franklin Augustus (1826–1909) *A Yankee trader in the gold rush* ed. K. A. White 1930
Buck, Peter Henry *The coming of the Maori* 1925
— (rev. ed.) 1949 (corrected impression, 1950)
Buckler, Ernest Redmond *Ox bells and fireflies* 1968
'Buckmaster, Henrietta' (Henrietta Stephens) *The walking trip* 1972
Buenos Aires Herald 1876–
Bull, Peter *I know the face, but . . .* 1959
'Bull-Us, Hector' (James Kirke Paulding) *The diverting history of John Bull and Brother Jonathan* 1812
— (ed. 2) 1813
— (new ed.) 1835
Bullens, Denison Kingsley *Steel and its heat treatment* 1916
Buller, Walter Lawry *A history of the birds of New Zealand* 1873
Manual of the birds of New Zealand 1882
Bulletin (Glasgow) 1915–60 (1924–60 with title *Bulletin and Scots pictorial*)
Bulletin (Sydney) 1880–
Bulletin of entomological research 1910–
Bulletin of the atomic scientists 1945–
Bulletin See also under the names of particular institutions
Bullins, Ed *The theme is blackness: 'The corner' and other plays* 1973
Bullock, Alan Louis Charles ed. *The twentieth century: a promethean age* 1971
Bulwer, Henry Earle *A glossary of technical terms employed in connection with church bells* 1901
Bunn, Alfred *The stage: both before and behind the curtain* 3 vols. 1840
Bunting, Brian Talbot *The geography of soil* 1965
Burbridge, William Frank *From balloon to bomber* 1946
Burchell, William John *Travels in the interior of Southern Africa* 2 vols. 1822–4
Burdett, Geoffrey Arnold T. ed. *Automatic control handbook* 1962
Bureau of American Ethnology *See* United States. Bureau of American Ethnology
Bureau of Standards *See* United States. National Bureau of Standards
Burge, Cyril Gordon ed. *Complete book of aviation* 1935
'Burgess, Anthony' (John Anthony Burgess Wilson) *Beds in the east* 1959

A clockwork orange 1962
The doctor is sick 1960
The enemy in the blanket 1958
MF 1971
The right to an answer 1960
Time for a tiger 1956
Burgess, Francis Henry *A dictionary of sailing* 1961
Burke, Carl F. *God is beautiful, man: interpretations of Bible passages and stories* 1969 (UK 1970)
Burke, Jackson Frederick *Death trick* 1975 (UK 1976)
Burke, John *Till death us do part* (adapted from the television series by J. Speight) 1967
Burke, Thomas *Nights in town* 1915
Burke, Thomas A. ed. *Polly Peablossom's wedding, and other tales* 1851
Burkitt, Miles Crawford *Prehistory: a study of early cultures in Europe and the Mediterranean basin* 1921
Burley, Dan *Original handbook of Harlem Jive* 1944
Burley, William John *Death in a salubrious place* 1973
Guilt edged 1971
To kill a cat 1970
Burlington magazine 1903–
Burn, Joshua Harold *Drugs, medicines and man* 1962
Lecture notes on pharmacology 1948
— (ed. 9) 1968
Burnett, Frances Eliza Hodgson *Little Lord Fauntleroy* 1886
A little princess 1905
The making of a marchioness 1901
The secret garden 1911
The shuttle 1907
T. Tembarom 1913
Burnett, John ed. *Useful toil: autobiographies of working people from the 1820s to the 1920s* 1974
Burnett, William Riley *Little Caesar* 1929
Vanity Row 1952 (UK 1953)
Burney, Frances (Mme D'Arblay) (1752–1840) *Journals and letters* ed. J. Hemlow et al. 1972–
Burnham, George P. *Memoirs of the United States Secret Service* 1872
Burnside, William *Theory of groups of finite order* 1897
Burrard, Gerald, et al. *Big game hunting in the Himalayas and Tibet* 1925
Burroughs, William, *Jr.* *Speed* 1970
Burroughs, William Seward *Junkie* 1953
The naked lunch 1959
Burrow, Thomas *The Sanskrit language* 1955
Burrows, E. Julie *Like an evening gone* 1973
'Burton, Alfred' (John Mitford) *The adventures of Johnny Newcome in the navy* 1818
Burton, Maurice *Wild animals of the British Isles* 1960
Burton, Richard Francis *A plain and literal translation of the Arabian nights' entertainments* 10 vols. 1885–6 (unexpurgated Kamashastra Soc. ed.)
Burton-Brown, Theodore *Early mediterranean migrations. An essay in archaeological interpretation* 1960
Bury, John Bagnell *A history of Greece to the death of Alexander the Great* 1900
Busby, Roger *Deadlock* 1971
Pattern of violence 1973
Bush, Harry Derrick *Atomic and nuclear physics: theoretical principles* 1962
Bushell, Stephen Wootton *Chinese art* 2 vols. 1904–6
— (ed. 2) 1909
Business week 1929–
Busy man's magazine (Toronto) 1896–1911 (continued as *Maclean's magazine*)
Butcher, A. Judith *Copy-editing: the Cambridge handbook* 1975
Butcher, Roger William *A new illustrated British flora* 2 parts 1961
Butler, Gwendoline *Coffin following* 1968
A coffin for Pandora 1973
Coffin in Oxford 1962
Butler, Richard *Where all the girls are sweeter* 1975
Butler, Samuel (1835–1902) *Erewhon; or, Over the range* 1872
Erewhon revisited twenty years later 1901
A first year in Canterbury settlement 1863
Note-books ed. H. F. Jones 1912
The way of all flesh ed. R. A. Streatfeild 1903
Butterworth, Hezekiah *Zigzag journeys in the Western States of America* 1884
Butterworth, Michael *Remains to be seen* 1976
Byatt, Antonia Susan *The virgin in the garden* 1978
Byng, John (5th Viscount Torrington) *The Torrington diaries: containing the tours through England and Wales 1781–94* ed. C. B. Andrews 4 vols. 1934–8
Byrne, Brian Oswald Donn *Destiny Bay* 1928

C

CB magazine (Communications Publication Corporation, Oklahoma City) 1964–
CERN *See* European Organization for Nuclear Research
CRC journal (Community Relations Commission) 1972–
C.S.I.R.O. *See* Australia. Commonwealth Scientific and Industrial Research Organization
Cabinet maker and retail furnisher (title varies) 1880–
'Cable, Boyd' (Ernest Andrew Ewart) *Action Front* 1916
Doing their bit: war work at home 1916
The Old Contemptibles 1919
'Cade, Alexander' (Kenneth Methold) *Turn up a stone* 1969
Cain, James Mallahan *The postman always rings twice* 1934
Caird, Janet *Perturbing spirit* 1966
Calder, Nigel David Ritchie *The restless earth* 1972
Caldwell, Erskine Preston *God's little acre* 1933
Caldwell, William *Schopenhauer's system in its philosophical significance* 1896
Calgary Herald (Calgary, Alberta) 1883–
California. University *Publications in geography* 1913–
California Academy of Natural Sciences *Proceedings* 1854–
California Historical Society *Quarterly* 1922–
Calkins, Gary Nathan *The biology of the protozoa* 1926
—— & Summers, Francis Marion *Protozoa in biological research* 1941
Call-Bulletin (San Francisco) 1856– (title varies)
Callow, Clive *Power from the sea: the search for North Sea oil and gas* 1973
'Calvin, Henry' (Clifford Leonard Clark Hanley) *It's different abroad* 1963
A nice friendly town 1967
The poison chasers 1971
Cambridge ancient history, The ed. J. B. Bury et al. 12 vols. 1923–39
— (ed. 3 of Vols. I & II) ed. I. E. S. Edwards et al. 1970–7
Cambridge and Dublin mathematical journal 1846–55
Cambridge review: a journal of university life and thought 1879–
Cameron, Agnes Deans *The new North: being some account of a woman's journey through Canada to the Arctic* 1909
Cameron, Alexander Thomas & Gilmour, Clifford R. *The biochemistry of medicine* 1933
Cameron, Angus, Frank, R., & Leyerle, J. eds. *Computers and Old English concordances* 1970
Cameron, David Kerr *Willie Gavin, crofter man: portrait of a vanished lifestyle* 1980
Camp, Walter Chauncey *Football without a coach* 1920
Campbell, Alistair *Old English grammar* 1959
Campbell, Allan McCulloch *Episomes* 1969
Campbell, Helen Stuart *Darkness and daylight; or, Lights and shadows of New York life* 1891
Campbell, Ignatius Roy Dunnachie (1901–57) *Adamastor* 1930
Broken record 1934
Collected poems 3 vols. 1949–60
The flaming terrapin 1924
Flowering rifle 1939
The Georgiad 1931
Light on a dark horse 1951
Lorca: an appreciation of his poetry 1952
Mithraic emblems 1936
Poems 1930
tr. *Poems of Baudelaire* 1952
Portugal 1957
Talking Bronco 1946
The wayzgoose 1928
Campbell, Mrs. Patrick *See* Shaw, George Bernard
Campbell, R. W. *Private Spud Tamson* 1915
Campbell, Roy *See* Campbell, Ignatius Roy Dunnachie
Campbell, Thomas (1733–95) *Diary of a visit to England in 1775* ed. J. L. Clifford 1947
Campion, Charles Thomas tr. *A. Schweitzer's On the edge of the primeval forest* 1922
Canada. Department of the Environment. Fisheries and Marine Service *Fisheries fact sheet* 1973–
Canada. Geological Survey *Reports of progress* (later *Annual reports*) 1857–1906
Canada. Senate *Official report of debates* 1870–
Canadian annual review of public affairs 1903–
Canadian antiques collector (title varies) 1966–
Canadian geographical journal 1930–
Canadian historical review 1920–
Canadian journal of linguistics 1961–
Canadian journal of physics 1951–
Canadian journal of research 1929–50 (from 1935 divided into subject series)

Canadian Medical Association *Journal* 1911–
Canadian mineralogist 1957–
Cancer (American Cancer Society) 1948–
Cancer research 1941–
'Candy, Edward' (Alison Neville) *Words for murder perhaps* 1971
Canfield, Dorothea Frances *Fables for parents* 1937 (UK 1938)
Understood Betsy 1917 (UK 1922)
Canfield, Henry Spofford *A maid of the frontier* 1898
Cannan, Joanna Maxwell *All is discovered* 1962
And all I learned 1951
And be a villain 1958
High table 1931
Little I understood 1948
Long shadows 1955
Murder included 1950
No walls of jasper 1930
Canning, Victor *Flight of the grey goose* 1973
The great affair 1970
The melting man 1968
The python project 1967
Queen's pawn 1969
The rainbird pattern 1972
The whip hand 1965
Cantarow, Abraham & Schepartz, Bernard *Biochemistry* 1954
— (ed. 4) 1967
Cape Argus (Cape Town) 1888– (title varies)
Cape monthly magazine 1857–81
Cape Times (Cape Town) 1876–
Capell, Richard *Simiomata: a Greek note book, 1944–1945* 1946
Capote, Truman *The grass harp* 1951 (UK 1952)
In cold blood: a true account of a multiple murder and its consequences 1965 (UK 1966)
Captain, The: a magazine for boys 1899–1924
Cardus, Neville *Autobiography* 1947
Close of play 1956
Days in the sun: a cricketer's journal 1924
Carew, Dorothea Petrie *Many years, many girls* 1967
'Carew, Francis Wylde' (Arthur E. G. Way) *No. 747: being the autobiography of a gipsy* 1891
Carew, Jan Alwin Rynveld *Black Midas* 1958
The wild coast 1958
Caribbean quarterly 1949–
Carlile, M. J. & Skehel, J. J. *Evolution in the microbial world* 1974 (24th Symposium of the Society for General Microbiology)
Carlquist, Sherwin John *Hawaii: a natural history* 1970
Carlson, Elof Axel *The gene: a critical history* 1966
Carman, Philip Crosbie *Chemical constitution and properties of engineering materials* 1949
'Carmichael, Harry' (Leopold Horace Ognall) *Naked to the grave* 1972
The seeds of hate 1960
Too late for tears 1973
Carmichael, Robert Daniel *Introduction to the theory of groups of finite order* 1937
Carnap, Rudolf *Meaning and necessity: a study in semantics and modal logic* 1947
Carnegie Institution of Washington *Publications* 1902–
Year-book 1902–
Carpenter, George Herbert *The biology of insects* 1928
Carpenter, John Richard *An ecological glossary* 1938
Carpenter, Kathleen E. *Life in inland waters* 1928
Carpenter, Rhys *Greek sculpture: a critical review* 1960
Carpenter, Spencer Cecil *Christianity according to S. Luke* 1919
Carr, Edward Hallett *A history of Soviet Russia: The Bolshevik revolution, 1917–1923* 3 vols. 1950–3
The interregnum, 1923–1924 1954
Socialism in one country, 1924–1926 3 vols. 1958–64
Foundations of a planned economy, 1926–1929 by E. H. Carr & R. W. Davies 3 vols. 1969–71
Carr, George Shoobridge *A synopsis of elementary results in pure and applied mathematics* 2 parts 1880–6
Carr, John Dickson *Below suspicion* 1949 (UK 1950)
Carr-Saunders, Alexander Morris & Jones, David Caradog *A survey of the social structure of England and Wales, as illustrated by statistics* 1927
Carrier, Robert *Great dishes of the world* 1963
Carrington, Charles Edmund *Rudyard Kipling* 1955
Carroll, James *Mortal friends* 1978
Carroll, John Bissell *The study of language* 1953
——, Davies, P., & Richman, B. *The American Heritage word frequency book* 1971
Carruthers, William Alexander *See* Caruthers
Carson, Gerald *Cornflake crusade* 1957 (UK 1959)

Carson, Rachel Louise *Silent spring* 1962 (UK 1963)
Carson, Robert Andrew Glendinning *Coins: ancient, mediaeval & modern* 1962
Carstairs, John Paddy *The concrete kimono* 1965
Carter, John Waynflete *ABC for book-collectors* 1952
— (ed. 3) 1961
Carter, Philip Youngman *Mr. Campion's falcon* 1970
Mr. Campion's farthing 1969
Cartwright, George *A journal of transactions and events, during a residence of nearly sixteen years on the coast of Labrador* 3 vols. 1792
Caruthers, William Alexander *The Kentuckian in New York; or, The adventures of three Southerns* 2 vols. 1834
Carvic, Heron *Miss Seeton sings* 1973 (UK 1974)
Cary, Arthur Joyce Lunel (1888–1957) *The African witch* 1936
Aissa saved 1932
An American visitor 1933
The captive and the free 1959
Castle Corner 1938
Except the Lord 1953
Herself surprised 1941
Mister Johnson 1939
Casamassa, Jack V. ed. *Jet aircraft power systems: principles and maintenance* 1950
Cass, James P. *Pulp and paper* 2 vols. 1952
Cassell's cyclopædia of photography ed. B. E. Jones 1911
Cassell's dictionary of abbreviations 1966 *See* Gurnett, John William & Kyte, Colin Henry John
Cassell's family magazine 1874–97
Cassidy, Frederic Gomes *S. Robertson's The development of modern English* (ed. 2) 1954 *See* Robertson, Stuart
Jamaica talk: three hundred years of the English language in Jamaica 1961
—— & Le Page, Robert Brock eds. *Dictionary of Jamaican English* 1967
Castle, Don *Do your own time* 1958
'Castle, John' & 'Hailey, Arthur' *Flight into danger* 1958
Cather, Willa Sibert *Death comes for the Archbishop* 1927
Catholic dictionary of theology, A ed. H. F. Davis et al. 1962–
Catholic encyclopedia, The 17 vols. 1907–17
Catholic Herald 1884–
Catto, Maxwell Joseph *Bird on the wing* 1966
'Caudwell, Christopher' (Christopher St. John Sprigg) *Illusion and reality* 1937
Cavell, Stanley Louis *The world viewed: reflections on the ontology of film* 1971
Caws, Peter *The philosophy of science* 1965
Cecil, Edward Christian David *Two quiet lives: Dorothy Osborne, Thomas Gray* 1948
Cecil, Mirabel *Heroines in love 1750–1974* 1974
Cercle Linguistique de Prague *Travaux* 1929–
Ceylon Daily News (Colombo) 1918–
Ceylon Observer (Colombo) 1834–
Chadwick, Hector Munro *The origin of the English nation* 1907
Chadwick, Henry *The art of pitching and fielding* 1886
Chadwick, Leigh E. tr. *W. Linsenmaier's Insects of the world* 1972
Chaffers, William *Marks and monograms on pottery and porcelain* 1863
— (ed. 2) 1866
Challinor, John *A dictionary of geology* 1961
Chalmers, Simon *Death on a quiet beach* 1968
Chamberlain, Basil Hall *Things Japanese* 1890
Chamberlain, E. *The Indiana gazetteer; or, Topographical dictionary of the State of Indiana* (ed. 3) 1849
Chamberlain, John (1553?–1628) *Letters* ed. N. E. McClure 2 vols. 1939
Chamberlin, John *Hosiery, yarns and fabrics* 1926
Chamberlin, Thomas Chrowder & Salisbury, Rollin Dean *Geology* 3 vols. 1904–6 (UK 1905–6)
Chambers, Edmund Kerchever *The English folk-play* 1933
The mediæval stage 2 vols. 1903
Chambers, Eric *Photolitho-offset* 1967
Chambers, Robert William *The common law* 1911
The firing line 1908
Chambers's technical dictionary ed. C. F. Tweney & L. E. C. Hughes 1940 (and several later editions used)
Chandler, Maurice Henry *Ceramics in the modern world: man's first technology comes of age* 1967
Chandler, Raymond Thornton (1888–1959) *The big sleep* 1939
Farewell, my lovely 1940
The high window 1942 (UK 1943)

The lady in the lake 1943 (UK 1944)
The long good-bye 1953
Raymond Chandler speaking [letters and an unfinished novel] 1962 (paperback ed. 1966)
Trouble is my business, and other stories 1950
Chang, Kwang-chih *See* Kwang-chih Chang
Channing, George G. *Early recollections of Newport, R.I., from the year 1793 to 1811* 1868
Chaplin, Charles Spencer *My autobiography* 1964
Chapman, Abraham ed. *New Black voices: an anthology of contemporary Afro-American literature* 1972
Chapman, James *Travels in the interior of South Africa* 2 vols. 1868
Chapman, Reginald Frederick *The insects: structure and function* 1969
Chapman, Robert McDonald & Bennett, Jonathan eds. *An anthology of New Zealand verse* 1956
Chapman, Ronald George *Father Faber* 1961
Chapman, Royal Norton *Animal ecology: with especial reference to insects* 1931
Chapman, Valentine Jackson *Coastal vegetation* 1964
Charles-Edwards, Thomas Charles & Richardson, Brian *They saw it happen* Vol. III (1689–1897) 1958
Charlesworth, John Kaye *The quaternary era* 2 vols. 1957
Charlton, Lionel Evelyn Oswald *Britain at war: the Royal Air Force from September 1939 to September 1945* 5 vols. 1941–7
Chase, John Centlivres *The Cape of Good Hope and the Eastern Province of Algoa Bay* ed. J. S. Christophers 1843
Chatelaine (Toronto) 1928–
Chatfield, Charles Hugh & Taylor, Charles Fayette *The airplane and its engine* 1928
— (ed. 2) 1932
Chatterton, Edward Keble *The auxiliary patrol* 1923
Cheadle, Walter Butler *Journal of a trip across Canada, 1862–1863* 1931
Cheeseman, Thomas Frederick *Manual of the New Zealand flora* 1906
Chemical abstracts 1907–
Chemical and engineering news 1942–
Chemical gazette 1842–59
Chemical news and journal of physical science (title varies) 1859–1932
Chemical Society of London *Annual reports on the progress of chemistry* See under title
Journal 1862–
Proceedings 1890–1914
Quarterly review 1947–71
Chemistry and industry 1932–
Chemistry in Britain 1965–
Cheng, Thomas Clement *The biology of animal parasites* 1964
Cheshire, Geoffrey Leonard *Bomber pilot* 1943
Chesterfield, 4th Earl of (Philip Dormer Stanhope) (1694–1773) *Letters* ed. B. Dobrée 6 vols. 1932
Chesterton, Gilbert Keith *Autobiography* 1936
The club of queer trades 1905
The flying inn 1914
Four faultless felons 1930
George Bernard Shaw 1910
The incredulity of Father Brown 1926
The innocence of Father Brown 1911
The man who was Thursday: a nightmare 1908
The Napoleon of Notting Hill 1904
Orthodoxy 1909
The thing 1929
The wisdom of Father Brown 1914
Cheyney, Reginald Evelyn Peter S. *I'll say she does!* 1945
You can always duck 1943
Chicago (title varies) 1952–
Chicago Daily News 1876–1960
Chicago Record 1881–1901
Chicago Tribune 1847–
Childe, Vere Gordon *The Danube in prehistory* 1929
The dawn of European civilization 1925
— (ed. 3) 1939
— (ed. 6) 1957
The most ancient East 1928
— (rewritten with title *New light on the most ancient East*) 1934
— (rewritten) 1952
China now 1970–
Chomsky, Avram Noam *Aspects of the theory of syntax* 1965
The logical structure of linguistic theory 1975
—— & Halle, Morris *The sound pattern of English* 1968
Choppin, Gregory Robert *Experimental nuclear chemistry* 1961
Christian 1973–
Christian Century, The 1900–
Christian Science Monitor (Boston, Mass.) 1908–
Christian World, The 1857–

Christie, Agatha Mary Clarissa (1890–1976) *The ABC murders* 1936
An autobiography 1977
The body in the library 1942
Come, tell me how you live 1946
Curtain: Poirot's last case 1975
Elephants can remember 1972
Evil under the sun 1941
Lord Edgware dies 1933
The mirror crack'd from side to side 1962
The moving finger 1943
Murder in the mews 1937
The murder of Roger Ackroyd 1926
Murder on the Orient Express 1934
The mysterious affair at Styles 1921
The pale horse 1961
Peril at End House 1932
Poirot investigates 1924
Postern of fate 1973
Third girl 1966
'Christopher, John' (Christopher Samuel Youd) *The twenty-second century* 1954
Christophersen, Paul *The articles: a study of their theory and use in English* 1939
Chrysler, Charles Byron *White slavery* 1911
Chujoy, Anatole *The dance encyclopedia* 1949
— (rev. ed. by A. Chujoy & P. W. Manchester) 1967
Church, Richard Thomas *The voyage home* 1964
Church Missionary Society *A grammar and vocabulary of the language of New Zealand* ed. S. Lee from materials supplied by T. Kendall 1820
Church quarterly review 1875–1968
Churchill, Randolph Spencer & Gilbert, Martin *Winston S. Churchill* 1966–
Churchill, Winston (1871–1947) *Coniston* 1906
The crisis 1901
Churchill, Winston Leonard Spencer *The end of the beginning: war speeches, 1942* compiled by C. Eade 1943
Into battle: speeches compiled by R. S. Churchill 1941
My early life: a roving commission 1930
The Second World War 6 vols. 1948–53 (UK 1948–54)
Secret session speeches compiled by C. Eade 1946
Victory: war speeches, 1945 compiled by C. Eade 1946
The world crisis 6 vols. 1923–31
'Churton, Henry' (Albion Winegar Tourgée) *Toinette* 1874
Cist, Charles *Cincinnati in 1841: early annals and future prospects* 1841
ed. *The Cincinnati miscellany* 2 vols. 1845–6
Sketches and statistics of Cincinnati in 1851 1851
Clabburn, Pamela *The needleworker's dictionary* 1976
Clacy, Ellen *A lady's visit to the gold diggings of Australia in 1852–53* 1853
Claiborne, John Francis Hamtramck *Life and correspondence of John A. Quitman* 2 vols. 1860
Clancey, Phillip Alexander *The birds of Natal and Zululand* 1964
Clapham, Alfred William *Romanesque architecture in western Europe* 1936
Clapperton, Richard *No news on Monday* 1968
Clarendonian, The 1919–
Clark, Alfred Joseph *Applied pharmacology* 1923
— (ed. 6) 1937
Clark, David *Plane and geodetic surveying for engineers* 2 vols. 1921
Clark, Donald Thomas & Gottfried, Bert A. *Dictionary of business and finance* 1957
—— (Apollo ed., with title *University dictionary of business and finance*) 1967
Clark, Douglas *Death after evensong* 1969
Dread and water 1976
The gimmel flask 1977
Premedicated murder 1975
Sick to death 1971
Clark, George Thomas *Leland Stanford, war governor of California* 1931
Clark, Grahame *See* Clark, John Grahame Douglas
Clark, John Desmond *The prehistory of southern Africa* 1959
Clark, John Grahame Douglas *Archaeology and society* 1939
— (ed. 3) 1957
World prehistory 1961
— (ed. 2) 1977
Clark, Kenneth McKenzie *Another part of the wood: a self-portrait* 1974
Civilisation: a personal view 1969
The nude: a study of ideal art 1956
Clark, Matthew *Home trade* 1930
Clark, Nigel George *Modern organic chemistry: an introduction* 1964
Clark, Percy Missen *The autobiography of an old drifter* 1936
Clark, Ronald William & Pyatt, Edward Charles *Mountaineering in Britain* 1957

Clark, Walter John *International language, past, present & future* 1907
Clark, Wilfrid Edward Le Gros *The tissues of the body* 1939
Clarke, Arthur Charles *The deep range* 1957
The exploration of space 1951
The sands of Mars 1951
Clarke, Austin *Later poems* 1961
Clarke, George Sydenham *Fortification: its past achievements, recent development and future progress* 1890
Clarke, John Henrik ed. *Harlem, U.S.A.* 1964
Clarke, Marcus Andrew Hislop *His natural life* 1874 (UK 3 vols. 1875)
Clarke, William *Every night book; or, Life after dark* 1827
Clarke, William Eagle *Studies in bird migration* 2 vols. 1912
Clarke County Democrat (Grove Hill, Alabama) 1856– (title varies)
Classical Association *Proceedings* 1904–
Classical quarterly 1907–
Classification of occupations 1960 See United Kingdom. General Register Office
Classification of occupations and directory of occupational titles 1972 See United Kingdom. Department of Employment
Clauser, Henry Ray ed. *The encyclopedia of engineering materials and processes* 1963
Clauson, Jon *High Commissioner* 1966
The long pursuit 1967
Peter's Pence 1974
The safe house 1975
The sundowners 1952
Clegg, John *The freshwater life of the British Isles* 1952
Clegg, John Trafford *Works: stories, sketches and rhymes in the Rochdale dialect* 2 vols. 1895–8
Cleife, Kenneth Philip H. *The slick and the dead* 1972
Cleland, John *Memoirs of a woman of pleasure* 2 vols. 1749
Clements, Eileen Helen *High tension* 1959
Honey for the Marshal 1960
Clements, Frederic Edward *Plant indicators: the relation of plant communities to process and practice* 1920
Plant succession: an analysis of the development of vegetation 1916
Research methods in ecology 1905
Clements, Rex *A gipsy of the Horn* 1924
Clerk, Dugald *The gas engine* 1886
Clerke, Agnes Mary *The system of the stars* 1890
Cleugh, James tr. *R. Jungk's Brighter than a thousand suns* 1958
Cleveland Plain Dealer (Cleveland, Ohio) 1845–
Clevely, Hugh Desmond *Public enemy* 1953
Clinical pharmacology and therapeutics 1960–
Clinton-Baddeley, Victor Clinton *No case for the police* 1970
Only a matter of time 1969
To study a long silence 1972
Clowes, Evelyn Mary *On the wallaby through Victoria* 1911
Clune, Francis Patrick (Frank) *The red heart: sagas of Centralia* 1944
Roaming round the Darling 1936
Clute, Willard Nelson *The common names of plants and their meanings* 1931
Coast to coast: Australian stories 1941–
Cobb, Richard Charles *Reactions to the French Revolution* 1972
Cobine, James Dillon *Gaseous conductors* 1941
Cockayne, Leonard *New Zealand plants and their story* 1910
Cockburn, Alexander & Blackburn, Robin eds. *Student power* 1969
Cockerell, Douglas *Bookbinding, and the care of books* 1901
Cockett, Sydney Russell & Hilton, Kenneth Arthur *Dyeing of cellulosic fibres and related processes* 1961
Cockle, G. R. ed. *Car and locomotive cyclopedia of American practice* 1966
— (ed. 3) 1974
Cockton, Henry *The life and adventures of Valentine Vox, the ventriloquist* 1840
Coe, Malcolm James *The ecology of the Alpine zone of Mount Kenya* 1967
'Coe, Tucker' (Donald Edwin Westlake) *Wax apple* 1970 (UK 1973)
Cohen, Antoine *The phonemes of English: a phonemic study of the vowels and consonants of Standard English* 1952
Cohen, Gerda Lesley *What's wrong with hospitals?* 1964
Cohen, Julius Berend *Organic chemistry for advanced students* 2 vols. 1907–13
Cohen, Laurence Jonathan *The diversity of meaning* 1962
— (ed. 2) 1966
Cohen, Leonard *Beautiful losers* 1966 (UK 1970)
Cohen, Morris Raphael & Nagel, Ernest *An introduction to logic and scientific method* 1934

Cohen, Murray *Sensible words: linguistic practice in England 1640–1785* 1977
Cohn, Nik *Pop from the beginning* 1969
— (reissued with title *Awopbopaloobop alopbamboom: pop from the beginning*) 1970
Coke, Desmond Francis Talbot *The bending of a twig* 1906
Sandford of Merton: a story of Oxford life 1903
Cold Spring Harbor symposia on quantitative biology 1933–
Cole, George Douglas Howard *Workshop organisation* 1923
—— & Cole, Margaret Isabel *Murder at the munition works* 1940
Coleman, James Smoot *Nigeria* 1958
Colenso, John William *Ten Weeks in Natal* 1855
Coleridge, Christabel Rose *Charlotte Mary Yonge: her life and letters* 1903
Coleridge, Samuel Taylor (1772–1834) *Philosophical lectures* ed. K. Coburn 1949
Colinvaux, Paul Alfred *Introduction to ecology* 1973
Collector's guide 1942–
College English: an official organ of the National Council of Teachers of English (US) 1939–
College humor 1921–
Collier, John Henry Noyes *His monkey wife; or, Married to a chimp* 1930
Collier, Richard Hughesdon *The city that wouldn't die: London, May 10–11, 1941* 1959
A house called Memory 1960
Collier's: the national weekly (US) 1905–57
Collingwood, Robin George (1889–1943) *An autobiography* 1939
The idea of history 1946
The idea of nature 1945
The new Leviathan; or, Man, society, civilization and barbarism 1942
The principles of art 1938
Roman Britain 1923
Collingwood, William Gershom *Scandinavian Britain* 1908
Collins, Archie Frederick *Manual of wireless telegraphy* 1906
Collins, Gilbert *The valley of eyes unseen* 1923
'Collins, Tom' (Joseph Furphy) *Such is life: being extracts from the diary of Tom Collins* 1903
Collins music encyclopaedia by J. A. Westrup & F. L. Harrison 1959
Collinson, William Edward *Contemporary English: a personal speech record* 1927
Collocott, Thomas Charles ed. *Dictionary of science and technology* 1971
Colman, George ('the younger', 1762–1836) *John Bull; or, The Englishman's fireside* 1803 (pirated ed.)
— (authorized ed.) 1805
Colman, Henry *Report on the agriculture of Massachusetts* See: Massachusetts. Agricultural Survey
Colonial and Indian Exhibition, 1886 *Reports on the colonial sections of the exhibition* ed. H. T. Wood 1887
Colorado magazine 1923–
Colt, Miriam Davis *We went to Kansas* 1862
Columbus Evening Dispatch (Columbus, Ohio) 1871–
Columbus News (Columbus, Montana) 1901–
Colyer, Charles Norman & Hammond, Cyril Oswald *Flies of the British Isles* 1951
Commentary, incorporating Contemporary Jewish record (American Jewish Committee) 1945–
Commission Géologique de la Finlande *Bulletin* 1895–
Committee of Inquiry on Decimal Currency *Report* 1963 See United Kingdom. Parliamentary papers
Committee on Broadcasting *Report* 1960 See United Kingdom. Parliamentary papers
Committee on the Future of Broadcasting (Chairman: Lord Annan) *Report* 1977 (Cmnd. 6753)
Commonwealth Scientific and Industrial Research Organisation *See* Australia. Commonwealth Scientific ...
Compte rendu (Académie des Sciences) *See* Académie des Sciences
Compton, Arthur Holly *Atomic quest* 1956
Compton-Burnett, Ivy *Darkness and day* 1951
A house and its head 1935
More women than men 1933
Computer journal 1958–
Computers and automation 1953–
Computers and the humanities: a newsletter 1966–
Comstock, John Henry & Comstock, Anna *A manual for the study of insects* 1895
Conan Doyle, Arthur *See* Doyle, Arthur Conan
Concise Oxford dictionary of opera by H. Rosenthal & J. Warrack 1964
Condon, Edward Uhler & Odishaw, Hugh *Handbook of physics* 1958
— (ed. 2) 1967
—— & Shortley, G. H. *The theory of atomic spectra* 1935

Condon, Richard *The Manchurian candidate* 1959 (UK 1960)
The whisper of the axe 1976
Congressional globe See: United States. Congress. Debates
Congressional record See: United States. Congress. Debates
Connecticut *Public records of the Colony of Connecticut* 1850–90
Connecticut Academy of Arts and Sciences *Transactions* 1866–
Connecticut Agricultural Experiment Station *Annual report* 1877–
Bulletin 1877–
Connecticut probate records See: Manwaring, Charles William
Conners, Bernard F. *Don't embarrass the Bureau* 1972 (UK 1973)
Connoisseur, The: an illustrated magazine for collectors 1901–
Connolly, Cyril Vernon *A romantic friendship: the letters of Cyril Connolly to Noel Blakiston* 1975
Conquest: a magazine of modern endeavour 1919–26
Conrad, Joseph (Teodor Józef Konrad Korzeniowski) *Almayer's folly: a story of an eastern river* 1895
Lord Jim 1900
Nostromo: a tale of the seaboard 1904
The rover 1923
The shadow-line: a confession 1917
Suspense: a Napoleonic novel 1925
Victory: an island tale 1915
—— & Hueffer, Ford Madox *Romance* 1903
Constable, John (1776–1837) *Correspondence* ed. R. B. Beckett 6 vols. 1962–8
Constellation, The 1829–32
Consumer reports 1942–
'Contact' (Alan John Bott) *An airman's outings* 1917
Conway, Robert Seymour, et al. *The præ-Italic dialects of Italy* 3 vols. 1933
Cook, James Henry *Fifty years on the old frontier, as cowboy, hunter, guide, scout, and ranchman* 1923
Cook, Robin *The crust on its uppers* 1962
Cooke, Deryck *The language of music* 1959
Cooke, John Esten *The Virginia comedians; or, Old days in the Old Dominion* 2 vols. 1854
Cooke, Nelson Magor & Markus, John *Electronics and nucleonics dictionary* 1960
Cooke, Rose Terry *Steadfast: the story of a saint and a sinner* 1889
'Coolidge, Susan' (Sarah Chauncey Woolsey) *What Katy did* 1873
What Katy did at school 1873 (UK 1874)
Coon, Carleton Stevens *The races of Europe* 1939
Cooper, Courtney Ryley *Here's to crime* 1937
Cooper, David Graham *Psychiatry and anti-psychiatry* 1967
Cooper, Evelyn Barbara *Drown him deep* 1966
Cooper, James Fenimore *The deerslayer* 2 vols. 1841 (UK eds. 1 & 2, both 3 vols., both also 1841)
Home as found 2 vols. 1838 (UK ed. as *Eve Effingham; or, Home* 3 vols. 1838)
Homeward bound; or, The chase 2 vols. 1838 (UK ed. 3 vols. 1838)
Ned Myers; or, A life before the mast 1843 (UK ed. 2 vols. 1843)
The oak-openings 2 vols. 1848 (UK ed. as *The bee-hunter* 3 vols. 1848)
The red rover 3 vols. 1827
Cooper, John Butler *Coo-oo-ee! A tale of bushmen from Australia to Anzac* 1916
Cooper, Leonard *The accomplices* 1960
Cooper, Lettice Ulpha *Tea on Sunday* 1973
Cooper, Susan Augusta Fenimore *Rural hours* 1850
'Cooper, William' (Harry Summerfield Hoff) *Memoirs of a new man* 1966
The struggles of Albert Woods 1952
Copi, Irving Marmer *Symbolic logic* 1954
Copleston, Frederick Charles *A history of philosophy* 9 vols. 1946–75
Copson, Edward Thomas *Metric spaces* 1968
Corcoran, Dennis *Pickings from the portfolio of the reporter of the New Orleans 'Picayune'* 1846
'Cordell, Alexander' (George Alexander Graber) *The bright Cantonese* 1967
Corelli, Marie *The secret power* 1921
Thelma 3 vols. 1887
Coriat, Isador H. *Abnormal psychology* 1911
Cork Examiner 1841–
Corkhill, Thomas W. ed. *A concise building encyclopaedia* 1932
Corlett, Peter Norman & Tinsley, John David *Practical programming* 1968
Cormack, William Epps *Narrative of a journey across the island of Newfoundland* 1856
Cornell, Frederick Carruthers *The glamour of prospecting* 1920
Cornell University Agricultural Experiment Station *Bulletin* 1888–
Memoirs 1913–

Corner, Edred John Henry *The life of plants* 1964
The natural history of palms 1966
Wayside trees of Malaya 2 vols. 1940
Cornish, John *The provincials* 1951
Corri, Eugene *Thirty years a boxing referee* 1915
'Corrigan, Mark' (Norman Lee) *Why do women —?* 1963
Corson, Dale R. & Lorrain, Paul *Introduction to electromagnetic fields and waves* 1962
'Corvo, Baron' *See* Rolfe, Frederick William
'Cory, Desmond' (John Lloyd McCarthy) *Bennett* 1977
Sunburst 1971
Cosgrave, Patrick *Cheyney's law* 1977
Cosmopolitan, The 1886–1925
Cotton, Charles A. *Landscape, as developed by processes of normal erosion* 1941
Volcanoes as landscape forms 1944
Cotton, Frank & Wilkinson, Geoffrey *Advanced inorganic chemistry* 1962
— (ed. 2) 1966
— (ed. 3) 1972
Cotton, Harry *Advanced electrical technology* 1967
Cottrell, Alan Howard *An introduction to metallurgy* 1967
Coues, Elliott ed. *New light on the early history of the Greater Northwest: the manuscript journals of Alexander Henry .. and of David Thompson* 3 vols. 1897
Coulter, John Merle, Barnes, C. R., & Cowles, H. C. *A textbook of botany for colleges and universities* 2 vols. 1910–11
Coulter, Stephen *The château* 1974
Embassy 1969
The Soyuz affair 1977
Country gentleman, The 1853–65; 1898–
Country gentlemen's catalogue 1894–
Country gentlemen's magazine 1900–
Country life 1897–
Countryman, The 1927–
Courant, Richard & Robbins, Herbert Ellis *What is mathematics?* 1941
Courier and Advertiser (Dundee) 1926–
Courier-Journal (Louisville, Kentucky) 1868–
Courier-Mail (Brisbane) 1953–
Course, Alfred George *A dictionary of nautical terms* 1962
Coursing calendar and review of the season 1857–1918
Courtenay, Florence *Physical beauty: how to develop and preserve it* 1922
Courtier, Sidney Hobson *Death in dream time* 1959
Cousteau, Jacques Yves & Dumas, Frédéric *The silent world* 1953
Covarrubias, Miguel *Island of Bali* 1937
— (another ed.) 1972
Cowan, James *The Maoris of New Zealand* 1910
Cowan, Lester ed. *Recording sound for motion pictures* 1931
Coward, Noël Pierce *Australia visited, 1940* 1941
Blithe spirit: an improbable farce 1941 (UK 1942)
Design for living: a comedy 1933
Future indefinite 1954
Middle East diary 1944
Play parade 6 vols. 1934–62
Present indicative 1937
Private lives 1930
To-night at 8.30: plays 3 vols. 1936
Cowell, Colin Robert, et al. *Inlays, crowns, and bridges* ed. G. F. Kantorowicz 1963
Cowell, Emilie Marguerite *The Cowells in America: the diary of Mrs. S. Cowell, 1860–1861* ed. M. W. Disher 1934
Cowell, Joseph *Thirty years passed among the players in England and America* 1844
Cowper, William (1731–1800) *Correspondence* ed. T. Wright 4 vols. 1904
Cox, Charles Brian & Dyson, Anthony Edward *The twentieth-century mind: history, ideas and literature in Britain* 3 vols. 1972
Cox, James Stevens *An illustrated dictionary of hairdressing and wigmaking* 1966
Cox, Nigel Shaun Maturin & Grose, Michael William eds. *The organization and handling of bibliographic records by computer* 1967
Cox, Ross *Adventures on the Columbia River* 2 vols. 1831 (US 1 vol. 1832)
Coxe, William *A view of the cultivation of fruit trees, and the management of orchards and cider* 1817
Coxhead, Eileen Elizabeth *The figure in the mist* 1955
One green bottle 1951
The thankless muse 1967
A wind in the west 1949
'Craddock, Charles Egbert' (Mary Noailles Murfree) *In the Tennessee mountains* 1884
Craddock, Harry *The Savoy cocktail book* 1930
'Craig, David' (Allan James Tucker) *A dead liberty* 1974
Young men may die 1970

Craig, Elizabeth *Collins family cookery* 1957
Craigie, William Alexander & Hulbert, James R. eds. *A dictionary of American English on historical principles* See under title
Cramer's Pittsburgh almanac 1812–14; 1816–17
Crandall, Irving Bardshar *Theory of vibrating systems and sound* 1927
Crandall, Lee Saunders *The management of wild mammals in captivity* 1964
Crane, Harold Hart *Letters 1916–32* ed. B. Weber 1952
— (paperback ed.) 1965
Crane, Stephen *Maggie, a girl of the streets* 1893
The red badge of courage. An episode of the American Civil War 1895
Crapsey, Edward *The nether side of New York; or, The vice, crime, and poverty of the great metropolis* 1872
Craske, Margaret & Beaumont, Cyril William *The theory and practice of the allegro in classical ballet* 1930
Crawford, James Coutts *Recollections of travel in New Zealand and Australia* 1880
'Crawford, Robert' (Hugh Crauford Rae) *Whip hand* 1972
Crawford, Samuel *Basic engineering processes* 1964
Crawfurd, John *History of the Indian archipelago* 3 vols. 1820
Creed, David *The trial of Lobo Icheka* 1971
Crescendo 1962–
'Crèvecœur, J. H. St. John de' (Michel Guillaume St. Jean de Crèvecœur) *Letters from an American farmer* 1782
Crew, Peter *Encyclopædic dictionary of mountaineering* 1968
Cricket 1882–1913
Cricketer, The 1921–
'Crispin, Edmund' (Robert Bruce Montgomery) *Buried for pleasure* 1948
Frequent hearses 1950
Glimpses of the moon 1977
The long divorce 1951
Critchley, Macdonald *Developmental dyslexia* 1964
The parietal lobes 1953
Critic, The (New York) 1881–1906
Critical review; or, Annals of literature 1756–1817
Crockett, David *Colonel Crockett's exploits and adventures in Texas .. written by himself* 1836 (attributed to Richard Penn Smith)
Narrative of the life of David Crockett 1834
Sketches and eccentricities of Colonel David Crockett See under title
Crockett, Samuel Rutherford *The dew of their youth* 1910
Croft, John Michael *Spare the rod* 1954
Crofts, Freeman Willis *Inspector French and the Cheyne mystery* 1926
Croker, Bithia Mary *The Company's servant: a romance of Southern India* 1907
Croker, Temple Henry, et al. *The complete dictionary of the arts and sciences* 3 vols. 1764–6
'Crompton, Richmal' (Richmal Crompton Lamburn) *Just — William* 1922
Still — William 1925
William again 1923
William and the evacuees 1940
William and the moon rocket 1954
William carries on 1942
William does his bit 1941
William — the detective 1935
Cronin, Archibald Joseph *The citadel* 1937
Hatter's Castle 1931
The stars look down 1935
Cronin, Bernard *Timber wolves* 1920
'Cronin, Michael' (Brendan Leo Cronin) *Dead . . . and done with* 1959
Cronin, Vincent *The wise man from the west* 1955
Crooke, William *The tribes and castes of the North-Western Provinces and Oudh* 4 vols. 1896
Croome, Honoria Renée Minturn *The faithless mirror* 1946
The forgotten place 1957
You've gone astray 1944
Crosby, Elizabeth Caroline, et al. *Correlative anatomy of the nervous system* 1962
Crosby, John Campbell *Nightfall* 1976 (UK 1977 with title *Snake*)
Cross, Charles Frederick & Bevan, Edward John *A text-book of paper-making* 1888
Cross, Ian Robert *The backward sex* 1960
The God boy 1957 (UK 1958)
'Cross, James' (Hugh Jones Parry) *To hell for half-a-crown* 1967
Crossman, Richard Howard Stafford (1907–74) *The diaries of a Cabinet Minister* 3 vols. 1975–7
Croudace, Glynn *The scarlet bikini* 1970
Crowest, Frederick James *Beethoven* 1899
Crowther, Sir Geoffrey *Outline of money* 1940
— (rev. ed.) 1948
Crowther, James Gerald & Whiddington, Richard *Science at war* 1947

Croy, Homer *How motion pictures are made* 1918
R.F.D. No. 3 1924
Cruickshank, James G. *Soil geography* 1972
Crump, Barry John *A good keen man* 1960
Hang on a minute, mate 1961 (UK 1963)
Crump, James Irving & Maul, Norman *Our airliners* 1940
Crystal, David *Linguistics* 1971
—— & Quirk, Charles Randolph *Systems of prosodic and paralinguistic features in English* 1964
Culbertson, Ely *Contract bridge blue book* 1930
Contract bridge complete 1936
Culley, Richard Spelman *A hand-book of practical telegraphy* 1863
'Culotta, Nino' (John Patrick O'Grady) *They're a weird mob* 1957 (UK 1958)
Cultivator, The: a monthly journal for the farm and the garden 1834–65
Cultural news from India 1960–
Culver, Charles Mortimer tr. *E. Landolt's The refraction and accommodation of the eye and their anomalies* 1886
Cumberland, Marten *Murmurs in the Rue Morgue* 1959
Cumberland News 1910–
Cuming, Fortescue *Sketches of a tour to the western country* 1810
Cummings, Edward Estlin *eimi* 1933
The enormous room 1922
Selected letters ed. F. W. Dupee & G. Stade 1969
Cunningham, Daniel John *Text-book of anatomy* 1902
— (ed. 4) ed. A. Robinson 1913
Cunnington, Cecil Willett *English women's clothing in the present century* 1952
——, Cunnington, P. E., & Beard, C. *A dictionary of English costume* 1960
Curme volume of linguistic studies 1930 See Hatfield, James Taft, et al.
Current researches in anesthesia and analgesia 1922–56
Current slang (University of South Dakota) 1966–71
Current trends in linguistics 1963–
'Curtis, James' *The gilt kid* 1936
They drive by night 1938
Cusack, Norman Edward *The electrical and magnetic properties of solids* 1958
Cushing, George Frederick tr. *P. Hajdú's Finno-Ugrian languages and peoples* 1975
Cushing, Harvey Williams *The pituitary body and its disorders* 1912
Custom car 1970–
Cuthill, Ronald tr. *J. Schober's Silk and the silk industry* 1930
Cyclopedia of American government ed. A. C. McLaughlin & A. B. Hart 3 vols. 1914

D

Dædalus 1955– (continuation of *Proceedings* of the American Academy of Arts and Sciences)
Daglish, Eric Fitch tr. *E. Schneider-Leyer's Dogs of the world* 1964
Daily Ardmoreite (Ardmore, Oklahoma) 1893–
Daily Colonist (Victoria, British Columbia) 1886–
Daily Evening Bulletin (San Francisco) 1855–95
Daily Graphic 1890–1926
Daily Herald 1912–64
Daily Mirror 1903–
Daily Nation (Nairobi, Kenya) 1960–
Daily News (New York) 1919–
Daily News (Perth, Western Australia) 1882–
Daily Oklahoman (Oklahoma City) 1894–
Daily Picayune (New Orleans) 1836–1914
Daily Progress (Charlottesville, Virginia) 1892–
Daily Record (Glasgow) 1895–
Daily Report 1904–9
Daily Sketch 1909–71
Daily Universal Register 1785–7 (continued as *The Times*)
Dakin, Douglas *The unification of Greece, 1770–1923* 1972
Dale, Celia *Other people* 1964
A spring of love 1960
Dales, Rodney Phillips *Annelids* 1963
Dallas, Eneas Sweetland *Kettner's book of the table: a manual of cookery* 1877
Dallimore, William & Jackson, Albert Bruce *A handbook of coniferæ, including ginkgoaceæ* 1923
Dalton, Edward Tuite *Descriptive ethnology of Bengal* 1872
'Daly, Rann' (Edward Vance Palmer) *The outpost* 1924
Daly, Reginald Aldworth *The changing world of the Ice Age* 1934
Igneous rocks and the depths of the earth 1933
Dana, Edmund *Geographical sketches on the western country* 1819
Dancing times 1894–

Dangerfield, Stanley & Howell, Elsworth eds. *The international encyclopaedia of dogs* 1971
Daniel, Glyn Edmund *A hundred years of archaeology* 1950
Darbishire, Otto Bernhard tr. *A. von Buzágh's Colloid systems* 1937
D'Arblay, Mme Frances *See* Burney, Frances
Darling, Malcolm Lyall *At freedom's door* 1949
Wisdom and waste in the Punjab village 1934
Darlington, Cyril Dean *Cytology* 1965
Recent advances in cytology 1932
—— & Mather, Kenneth *The elements of genetics* 1949
Darlington, Philip Jackson *Zoogeography: the geographical distribution of animals* 1957
Daubenmire, Rexford F. *Plants and environment* 1947
Davey, Wheeler Pedlar *A study of crystal structure and its applications* 1934
David, Elizabeth *A book of Mediterranean food* 1950
— (another ed.) 1955
French country cooking 1951
French provincial cooking 1960
Italian food 1954
Davids, Thomas William Rhys *Buddhism: being a sketch of the life and teachings of Gautama the Buddha* 1877
Davidson, Lionel *A long way to Shiloh* 1966
The night of Wenceslas 1960
The Rose of Tibet 1962
Davies, Benjamin Lionel *Technology of plastics: manufacture: structure: design* 1949
Davies, David Pettit *Handling the big jets* 1967
Davies, Edmund Frank *Illyrian venture: the story of the British military mission to enemy-occupied Albania, 1943–44* 1952
Davies, Leslie Purnell *The shadow before* 1971
What did I do tomorrow? 1972
Davies, Rodney Deane & Palmer, Henry Procter *Radio studies of the universe* 1959
Davin, Daniel Marcus *For the rest of our lives* 1947
The gorse blooms pale 1947
ed. *New Zealand short stories* 1953
For 2nd ser. *see* Stead, Christian Karlson
Not here, not now 1970
Roads from home 1949
The sullen bell 1956
'Daviot, Gordon' (Elizabeth Mackintosh) *The man in the queue* 1929
See also: 'Tey, Josephine'
Davis, Alec Edward *Package and print: the development of container and label design* 1967
Davis, Arthur Hoey *See* 'Rudd, Steele'
Davis, Charles Carroll *The marine and fresh-water plankton* 1955
Davis, Frederick Clyde *See* 'Ransome, Stephen'
Davis, John Gilbert *A dictionary of dairying* 1950
— (ed. 2) 1955
Davis, Kingsley *Human society* 1949 (1948 = 'preliminary ed.')
Davis, Norman & Wrenn, Charles Leslie eds. *English and medieval studies presented to J. R. R. Tolkien on the occasion of his 70th birthday* 1962
Davis, Peter Hadland & Heywood, Vernon Hilton *Principles of angiosperm taxonomy* 1963
Davison, Frank Dalby *Children of the dark people: an Australian folk tale* 1936
Dusty: the story of a sheep dog 1946
Davson, Hugh *Textbook of general physiology* 1951
Dawe, Edward Arthur *Paper and its uses: a treatise for printers, stationers and others* 1914
Dawkins, Richard *The selfish gene* 1976
Dawson, Henry Christopher *The age of the gods: a study in the origins of culture in prehistoric Europe and the ancient East* 1928
Day, Charles Russell *The music and musical instruments of Southern India and the Deccan* 1891
Day, Frederick T. *An introduction to paper: its manufacture and use* 1962
Day, Michael Herbert *Guide to fossil man* 1965
Day Lewis, Cecil *See* Lewis, Cecil Day
Dean, Francis Medcalf *Naturally occurring oxygen ring compounds* 1963
Deans, John ed. *Pioneers of Canterbury: Deans letters, 1840–1854* 1937
Dearmer, Percy *The parson's handbook* 1899
— (ed. 13, by C. E. Pocknee) 1965
Debates and proceedings in the Congress of the United States See: United States. Congress. Debates
De Beer, Gavin Rylands *Embryology and evolution* 1930
Vertebrate zoology 1928
— (ed. 2) 1951
De Bres, Joris tr. *E. Mandel's Late capitalism* 1975
Dedham, Massachusetts *The early records of the town of Dedham, Massachusetts* ed. D. G. Hill 5 vols. 1886–99
Dee, Liang-lao tr. *Hsia Chih-yen's The coldest winter in Peking* 1978
Deeping, George Warwick *Kitty* 1927
Roper's Row 1929
Second youth 1919
The secret sanctuary; or, The saving of John Stretton 1923
Sincerity 1912
Sorrell and son 1925
Three rooms 1924
Deer, William Alexander, et al. *An introduction to the rock-forming minerals* 1966
Rock-forming minerals 5 vols. 1962–3
Defendorf, Allen Ross tr. *E. Kraepelin's Clinical psychiatry: a text-book for students and physicians* 1902
De Ferranti, Basil ed. *Living with the computer* 1971
Deighton, Len *Billion-dollar brain* 1966
An expensive place to die 1967
Funeral in Berlin 1964
Horse under water 1963
The Ipcress file 1962
Len Deighton's London dossier with contributions by A. Bailey et al. 1967
Only when I larf 1968
SS-GB 1978
Spy story 1974
Twinkle, twinkle, little spy 1976
XPD 1981
Yesterday's spy 1975
De Kerchove, René *International maritime dictionary* 1948
'Delafield, E. M.' (Edmée Elizabeth Monica Dashwood) *Thank Heaven fasting* 1932
De la Mare, Walter John *The burning-glass, and other poems* 1945
Collected stories for children 1947
The connoisseur, and other stories 1926
Crossings: a fairy play 1921
ed. *Early one morning, in the spring: chapters on children and on childhood* 1935
The fleeting, and other poems 1933
The listeners, and other poems 1912
The Lord Fish 1933
Memoirs of a midget 1921
Memory, and other poems 1938
Private view 1953
The riddle, and other stories 1923
Songs of childhood 1902
This year, next year 1937
The three mulla-mulgars 1910
The traveller 1946
The veil, and other poems 1921
The wind blows over 1936
Winged chariot 1951
Deland, Margaret Wade *Old Chester tales* 1898
De la Roche, Mazo *Finch's fortune* 1931
Jalna 1927
The master of Jalna 1933
Whiteoak harvest 1936
Whiteoaks (US ed. with title *Whiteoaks of Jalna*) 1929
Young Renny 1935
Delavenay, Émile *An introduction to machine translation* tr. by K. M. Delavenay and the author 1960
Delineator 1873–1937
De Lisser, Herbert George *The cup and the lip* 1956
Dell, Jeffrey *Nobody ordered wolves* 1935
Delman, David *Sudden death* 1972 (UK 1973)
Demant, Vigo Auguste *Christian polity* 1936
Religion and the decline of capitalism 1952
De Mauny, Erik *The huntsman in his career* 1949
Democrat See *Clarke County Democrat*
Democratic State Journal (Sacramento, California) 1852–8
De Morgan, William Frend (1839–1917) *Alice-for-short* 1907
Joseph Vance 1906
The old madhouse (completed by M. E. De Morgan) 1919
The old man's youth and the young man's old age (completed by M. E. De Morgan) 1920 (UK 1921)
When ghost meets ghost 1914
Dempsey, Jack (William Harrison Dempsey) *Championship fighting* 1950
Denison, William *Cricket: sketches of the players* 1846
Denlinger, William Watson *The complete Boston [terrier]* 1955
Dennis, Clarence James *The moods of Ginger Mick* 1916
The songs of a sentimental bloke 1915 (1916 impression used)
Denny, Grace Goldena *Fabrics and how to know them* 1923
Dennys, Nicholas Belfield *A descriptive dictionary of British Malaya* 1894
Dent, Colin *Quantity surveying by computer* 1964
Dentry, Robert *Encounter at Kharmel* 1971

Department of Education and Science See United Kingdom
Depuy, Charles H. & Chapman, Orville L. Molecular reactions and photochemistry 1972
Derby, George Horatio See 'Phoenix, John'
Derbyshire Times 1854–
Derrick, June Teaching English to immigrants 1966
Desch, Cecil Henry Metallography 1910
De Sélincourt, Hugh The cricket match 1924
Desha, Lucius Junius Organic chemistry: the chemistry of the compounds of carbon 1936
Design: a monthly journal for manufacturers and designers 1949–
Design and work 1876–81
Design engineering 1971–
De Sitter, Lamoraal Ulbo Structural geology 1956
De Sola, Ralph & De Sola, Dorothy A dictionary of cooking 1969
De Toledano, Ralph ed. Frontiers of jazz 1947
Detroit Free Press (title varies) 1835–
Deuel, Harry James The lipids: their chemistry and biochemistry 3 vols. 1951–7
Deutsch, Sid Theory and design of television receivers 1951
Deutscher, Isaac Marxism in our time ed. T. Deutscher 1971 (UK 1972)
Stalin: a political biography 1949
Devanny, Jean Bushman Burke 1930
The butcher shop 1926
Dawn beloved 1928
Old savage, and other stories 1927
Developmental biology 1959–
De Voto, Bernard Augustine Across the wide Missouri 1947
De Vries, Leonard Victorian advertisements 1968
De Vries, Peter The glory of the hummingbird 1974 (UK 1975)
The Mackerel plaza 1958
Dewees, William B. Letters from an early settler of Texas ed. 'C. Cordelle' (E. C. Kimball) 1852
Dexter, Colin Last seen wearing 1976
Dial, The (Chicago) 1880–1929
Dick, Everett Newfon The Dixie frontier 1948
Dick, William Brisbane The American Hoyle; or, Gentleman's hand-book of games 1864 (and several later editions used)
Dickens, Charles (1812–70) Letters ed. M. House et al. 1965–
Pictures from Italy 1846
Dickens, Monica Enid The happy prisoner 1946
The heart of London 1961
Man overboard 1958
No more meadows 1953
One pair of feet 1942
One pair of hands 1939
Dickinson, Emily Elizabeth (1830–86) Poems ed. T. H. Johnson 3 vols. 1955
Dickinson, Gordon Cawood Maps and air photographs 1969
Dickinson, Peter Malcolm de B. The lizard in the cup 1972
The poison oracle 1974
A pride of heroes 1969
Sleep and his brother 1971
Dickinson, Reginald Ernest Electric trains 1927
Dickinson, Thomas Albert The aeronautical dictionary 1945
Dickson, Mora A world elsewhere 1964
Dickson, William Puride tr. T. Mommsen's The history of Rome 4 vols. 1862–6
Dictionary of American English on historical principles, A ed. W. A. Craigie & J. R. Hulbert 4 vols. 1938–44
Dictionary of Americanisms on historical principles, A ed. M. M. Mathews 2 vols. 1951
Dictionary of Canadianisms on historical principles, A See: Avis, Walter Spencer, et al.
Dictionary of Jamaican English See: Cassidy, Frederic Gomes & Le Page, R. B.
Dictionary of national biography ed. L. Stephen et al. 63 vols. & Supplements 1885–
Dictionary of occupational terms 1921 See United Kingdom. Ministry of Labour
Dictionary of South African English, A See: Branford, Jean
Diehl, Edith Bookbinding: its background and technique 2 vols. 1946
Dillard, Joey Lee Black English: its history and usage in the United States 1972
Dilley, Arthur Urbane Oriental rugs and carpets 1931
Dills, Lanie The 'official' CB slanguage language dictionary and cross-reference ed. D. Gilbertson 1975
— (rev. ed.) 1976
Diment, Adam The bang bang birds 1968
The dolly dolly spy 1967
The great spy race 1968
Think Inc. 1971
Di Mona, Joseph Last man at Arlington 1973 (UK 1974)
Dingwall, William Orr ed. A survey of linguistic science 1971

Dirac, Paul Adrien Maurice The principles of quantum mechanics 1930
— (ed. 3) 1947
Diringer, David The alphabet: a key to the history of mankind 1948
Discovery: a monthly popular journal of knowledge 1920–66
Diseases of the nervous system 1940–
Disraeli, Benjamin The voyage of Captain Popanilla 1828
Dissertation abstracts 1952–
Ditchburn, Robert William Light 1952
— (ed. 2) 1963
Ditmars, Raymond Lee The reptile book 1907
Snakes of the world 1931
Diver, Katherine Helen Maud Candles in the wind 1909
'Divine, David' (Arthur Durham Divine) The King of Fassarai 1950
'Dix, Dorothy' (Elizabeth Meriwether Gilmer) Fables of the elite 1902
Dobie, James Frank ed. Rainbow in the morning 1965 ('reprint ed.' of Vol. V of Publications of the Texas Folklore Society 1926)
Dobson, Gordon Miller Bourne Exploring the atmosphere 1963
— (ed. 2) 1968
Documentary history of American industrial society, A ed. J. R. Commons et al. 11 vols. 1910–11
Documents relative to the colonial history of the State of New-York, procured in Holland, England and France collected by J. R. Brodhead; ed. E. B. O'Callaghan & B. Fernow 15 vols. 1853–87
Dodd, Robert Edward Chemical spectroscopy 1962
'Doesticks, Q. K. Philander' (Mortimer Neal Thomson) Doesticks, what he says 1855
Doklady See: Academy of Sciences of the U.S.S.R.
Dolinsky, Meyer There is no silence 1959
'Dominic, R. B.' (Mary Jane Latsis & Martha Henissart) Epitaph for a lobbyist 1974
See also: 'Lathen, Emma'
Dominy, Eric Norman Teach yourself judo 1954
Dommett, William Erskine A dictionary of aircraft 1918
Doncaster, Leonard An introduction to the study of cytology 1920
Donleavy, James Patrick The ginger man 1955
Donovan, Charles Henry Wynne With Wilson in Matabeleland; or, Sport and war in Zambesia 1894
Doogue, Raymond Brian & Moreland, John Munn New Zealand sea anglers' guide 1960
'Doone, Jice' (Vance Marshall) Timely tips for new Australians 1926
Dopping, Olle Computers and data processing 1970
Dorson, Richard M. ed. Peasant customs and savage myths: selections from the British folklorists 2 vols. 1968
Dos Passos, John Roderigo The best times: an informal memoir 1966 (UK 1968)
The big money 1936
The 42nd parallel 1930
1919 1932
Streets of night 1923
Three soldiers 1921 (UK 1922)
Dougall, John tr. M. Born's Atomic physics 1935
Dougherty, George S. The criminal as a human being 1924
Douglas, George Norman London street games 1916
South wind 1917
Douglas, Jack D. & Johnson, John M. eds. Existential sociology 1977
Douglas, Keith Castellain (1920–44) Alamein to Zem Zem 1946
Douglas, Lloyd Cassel Forgive us our trespasses 1932 (UK 1937)
White banners 1936
Douglas, Norman See Douglas, George Norman
'Douglas, O.' (Anna Buchan) Penny plain 1920
Douglas-Todd, Charles Henry The popular whippet 1961
Douglass, Frederick Narrative of the life of Frederick Douglass, an American slave 1845
Down beat 1934–
Downey, Hal ed. Handbook of hematology 4 vols. 1938
Downing, W. H. Digger dialects 1919
Downside review 1880–
Dowson, Ernest Christopher (1867–1900) Letters ed. D. Flower & H. Maas 1967
Doxiadis, Constantinos Apostolou Between dystopia and utopia 1966 (UK 1968)
Doyle, Arthur Conan The green flag, and other stories of war and sport 1900
His last bow 1917
The history of spiritualism 2 vols. 1926
The lost world 1912
The Maracot Deep, and other stories 1929
The valley of fear 1915

Drabble, Margaret The Garrick year 1964
Jerusalem the golden 1967
The millstone 1965
A summer bird-cage 1962
The waterfall 1969
Drackett, Philip Arthur Motor rallying 1963
Drake, Daniel Natural and statistical view; or, Picture of Cincinnati and the Miami country 1815
Draper, Alfred The death penalty 1972
Swansong for a rare bird 1970
Drayton, John A view of South-Carolina, as respects her natural and civil concerns 1802
Dreiser, Theodore An American tragedy 2 vols. 1925 (UK 1926)
The financier 1912
Driscoll, Peter The white lie assignment 1971
The Wilby conspiracy 1972 (UK 1973)
Drive 1967–
Driver, Christopher The disarmers 1964
Drum (East African ed.) 1965–
Drummond, Charles Death and the leaping ladies 1968
A death at the bar 1972
Death at the furlong post 1967
The odds on death 1969
'Drummond, Ivor' (Roger Erskine Longrigg) The jaws of the watchdog 1973
The man with the tiny head 1969
The power of the bug 1974
Drummond, June Bang! Bang! You're dead! 1973
The black unicorn 1959
The Gantry episode 1968
Drummond, Montagu tr. G. Haberlandt's Physiological plant anatomy 1914
Drury, William D. ed. The book of gardening: a handbook of horticulture 1900
Duane, Alexander tr. E. Fuchs's Text-book of ophthalmology 1892
Dublin University review 1885–7
Duckett's register 1946–
'Dudley-Gordon, Tom' (Dudley Barker et al.) Coastal Command at war 1943
Duff, David Skene Victoria in the Highlands 1968
Duffy, Maureen That's how it was 1962
Duggar, Benjamin Minge Fungous diseases of plants 1909
Duke-Elder, William Stewart Parsons' Diseases of the eye See: Parsons, John Herbert
Text-book of ophthalmology 7 vols. 1932–54
—— et al. System of ophthalmology 1958–
Du Maurier, Daphne The breaking point: eight stories 1959
Rebecca 1938
Du Maurier, George Louis Palmella Busson The Martian 1897
The young George du Maurier: a selection of his letters, 1860–67 ed. D. du Maurier 1951
Dumfries and Galloway Standard 1843–
Dunbar, Carl Owen & Rodgers, John Principles of stratigraphy 1957
Dunbar, Janet Mrs. G. B. S.: a biographical portrait of Charlotte Shaw 1963
Duncan, Lee Over the wall 1936
Duncan, Lew Wallace & Scott, Charles F. History of Allan and Woodson Counties, Kansas 1901
Duncan, William Murdoch The big timer 1973
Dundes, Alan ed. Mother wit from the laughing barrel: readings in the interpretation of Afro-American folklore 1973
Dundy, Elaine The dud avocado 1958
Dunlap, William Memoirs of George Fred. Cooke, Esq., late of the Theatre Royal, Covent Garden 2 vols. 1813
Thirty years ago; or, The memoirs of a water drinker 2 vols. 1836
Dunlop, Derrick Melville ed. Textbook of medical treatment 1939
— (ed. 10) ed. D. M. Dunlop & S. Alstead 1966
Dunn, Robert The shameless diary of an explorer 1907
Dunne, Finley Peter Mr. Dooley in peace and in war 1898
Dunphy, Eamon Only a game? The diary of a professional footballer ed. P. Ball 1976
Durante, James Francis ('Jimmy') & Kefoed, Jack C. Night clubs 1931
Durrant, Philip John & Durrant, Beryl Introduction to advanced inorganic chemistry 1962
Durrell, Gerald Malcolm The Bafut beagles 1954
Catch me a Colobus 1972
The drunken forest 1956
Encounters with animals 1958
Menagerie Manor 1964
My family and other animals 1956
The overloaded ark 1953
Three singles to adventure 1954
Two in the bush 1966
The whispering land 1961
A zoo in my luggage 1960

Durrell, Lawrence George Balthazar 1958
Bitter lemons 1957
Clea 1960
Justine 1957
Mountolive 1958
Nunquam 1970
Spirit of place: letters and essays on travel ed. A. G. Thomas 1969
Tunc 1968
Durst, Paul Badge of infamy 1968
Dury, George Harry ed. Essays in geomorphology 1966
Du Toit, Alexander Logie Our wandering continents 1937
Dwight, Margaret van Horn A journey to Ohio in 1810 ed. M. Farrand 1912
Dyer, Robert Nine years of an actor's life 1833
Dyott, William Diary 1781–1845 ed. R. W. Jeffery 2 vols. 1907
Dyson, Edward George Fact'ry 'ands 1906

E

Eaddy, Percy Allen Hull down 1933
Eagleson, Robert D. & McKie, Ian The terminology of Australian national football 3 parts 1968–9 (Sydney University: Australian Language Research Centre: Occasional Papers Nos. 12–14)
Eagleton, Terence Francis Criticism and ideology: a study in Marxist literary theory 1976
Earl, John How to choose and use tuners and amplifiers 1970
Earle, Augustus A narrative of a nine months' residence in New Zealand, in 1827; together with a journal of a residence in Tristan D'Acunha 1832
— (new ed.) ed. E. H. McCormick 1966
Early music 1973–
Earth-science reviews 1966–
East Africa journal 1964–
East African annual 1936–
East African Standard (Nairobi, Kenya) 1902–
East Hampton, New York Records 5 vols. 1887–1905
East London (Daily) Dispatch (East London, Cape Province) 1872–1925
Eastern Evening News (Norwich) 1882–
Eastern Province Herald (Port Elizabeth, S. Africa) 1845–
Eastlake, Charles Lock Hints on household taste in furniture, upholstery and other details 1868
Eastman, Charles Rochester, et al. tr. K. A. von Zittel's Text-book of palæontology 3 vols. 1900–25
Eastman, Max Forrester tr. L. Trotsky's The history of the Russian Revolution 3 vols. 1932–3
Eaton, Amos A manual of botany for the northern states 1817
— (ed. 2 with title —— for the northern and middle states) 1818
Eaton, Timothy, & Co. Ltd. A shopper's view of Canada's past: pages from Eaton's catalogues 1886–1930 1969
Eberhart, Mignon Good Danger money 1974 (UK 1975)
Ebers, John Seven years of the King's Theatre 1828
Eby, Cecil DeGrotte The siege of the Alcázar 1965 (UK 1966)
Eccles, William Henry Wireless telegraphy and telephony: a handbook of formulae, data and information 1915
Ecclestone, Alan A staircase for silence 1977
Ecological monographs 1931–
Ecology 1920–
Economic geology and Bulletin of the Society of Economic Geologists 1905–
Economic journal 1891–
Economist, The 1843–
Eddington, Arthur Stanley The internal constitution of the stars 1926
The nature of the physical world 1928
New pathways in science 1935
Report on the relativity theory of gravitation 1918
Space, time and gravitation: an outline of the general relativity theory 1920
Eddison, Eric Rücker (1888–1945) A fish dinner in Memison 1941
— (new ed.) 1968
The Mezentian Gate 1958
— (new ed.) 1972
Mistress of mistresses 1935
The worm Ouroboros 1922
Eddy, Arthur Jerome Cubists and Post-Impressionism 1914 (UK 1915)
Eddy, Mary Baker Science and health 1875
Edelman, Jacob Murray Political language 1977
Edgcumbe, Kenelm Industrial electrical measuring instruments 1908
Edgeworth, Maria Letters from England 1813–1844 ed. C. Colvin 1971
Maria Edgeworth in France and Switzerland: selections from the Edgeworth family letters ed. C. Colvin 1979
Edib, Halidé The clown and his daughter 1935
Edinburgh Evening News 1873– (title varies)
Edinburgh medical and surgical journal 1805–55
Edinburgh new philosophical journal 1826–64
Edinburgh Obstetrical Society Transactions 1868–1921
Edington, George Miller & Gilles, Herbert Michel Pathology in the tropics 1969
— (ed. 2) 1976
Edlin, Herbert Leeson Collins guide to tree planting and cultivation 1970
The forester's handbook 1953
Edmonton Journal (Edmonton, Alberta) 1903–
Edward, David B. The history of Texas 1836
Edwardes, Annie A Girton girl 3 vols. 1885
Edwards, Anthony William Fairbank Likelihood: an account of the statistical concept of likelihood and its application to scientific inference 1972
Edwards, Elwyn Hartley Saddlery 1963
Edwards, John Newman Shelby and his men; or, The war in the west 1867
Edwards's botanical register 1815–47
'Egan, Lesley' (Barbara Elizabeth Linington) Blind search 1977
Paper chase 1972 (UK 1973)
Egan, Pierce Anecdotes of the turf, the chase, the ring and the stage 1827
Boxiana; or, Sketches of ancient and modern pugilism 3 vols. 1812–21 (Vol. IV, 1824, by J. Badcock)
— (new ser.) 2 vols. 1828–9
The life of an actor 1825
Eggleston, Edward The circuit rider: a tale of the heroic age 1874
The end of the world: a love story 1872
The Hoosier school-master 1871 (UK 1872)
Egleton, Clive Seven days to a killing 1973
Ehrenberg, Victor L. From Solon to Socrates: Greek history and civilisation during the sixth and fifth centuries B.C. 1968
Einstein, Albert See Adams, Edwin Plimpton
Einstein, Alfred Music in the Romantic era 1947
Eiseley, Loren Corey The immense journey 1957
Ekwall, Bror Oscar Eilert The place-names of Lancashire 1922
Electrical communication 1922–
Electrical world (title varies) 1899–
Electrician, The 1861–1952
Electrochemical industry 1902–9 (1905–9 with title Electrochemical and metallurgical industry)
Electronic engineering 1941–
Electronics (New York) 1930–
Eliason, Norman E. Tarheel talk: an historical study of the English language in North Carolina to 1860 1956
Eliot, Charles Norton Edgcumbe A Finnish grammar 1890
See also: 'Odysseus'
'Eliot, George' (Marian Evans) (1819–80) tr. L. Feuerbach's The essence of Christianity 1854
The George Eliot letters ed. G. S. Haight 7 vols. 1954–6
tr. D. F. Strauss's The life of Jesus, critically examined 3 vols. 1846
Eliot, Thomas Stearns (1888–1965) tr. Anabasis by 'St.-J. Perse' (A. St. Léger Léger) 1930
Ara vos prec 1920
Burnt Norton 1941
The cocktail party: a comedy 1950
Collected poems, 1909–1935 1936
Collected poems, 1909–1962 1963
The confidential clerk 1954
The dry salvages 1941
East Coker 1940
The elder statesman 1959
Elizabethan essays 1934
The family reunion 1939
Little Gidding 1942
Murder in the cathedral 1935
Notes towards the definition of culture 1948
Old Possum's book of practical cats 1939
On poetry and poets 1957
Poems 1919
Prufrock and other observations 1917
The rock: a pageant play 1934
Selected essays 1917–1932 1932
— (ed. 3) 1951; new impression 1953
Sweeney agonistes 1932
The waste land 1922 (UK 1923)
The waste land: a facsimile and transcript of the original drafts including the annotations of Ezra Pound ed. V. Eliot 1971
'Ellan, B. J.' Spitfire!: the experiences of a fighter pilot 1942
Ellery Queen's mystery magazine 1941–
Ellin, Stanley The bind 1970 (UK ed. with title The man from nowhere 1971)
The eighth circle 1958 (UK 1959)
Elliot, John Herbert Duel 1969
Ellis, Carleton The chemistry of synthetic resins 2 vols. 1935
Ellis, Henry Havelock Studies in the psychology of sex 7 vols. 1897–1928
Ellis, William A journal of a tour around Hawaii 1825
Emblen, Donald Lewis Peter Mark Roget, the word and the man 1970
Emerson, Ralph Waldo (1803–82) Journals ed. E. W. Emerson & W. E. Forbes 10 vols. 1909–14
Emory, William Hemsley Notes of a military reconnaissance, from Fort Leavenworth, in Missouri, to San Diego, in California 1848
Empey, Arthur Guy 'Over the top' by an American soldier who went, together with Tommy's dictionary of the trenches 1917 (UK also 1917 with title From the fire step)
Empson, William The gathering storm 1940
Poems 1935
Seven types of ambiguity 1930
Some versions of pastoral 1935
The structure of complex words 1951
Encounter 1953–
Encyclopædia Britannica See Britannica book of the year
Encyclopedia Canadiana ed. J. E. Robbins 10 vols. 1957–8
Encyclopaedia Judaica ed. C. Roth & G. Wigoder 16 vols. 1971–2
Encyclopædia medica ed. C. Watson 15 vols. 1899–1910
Encyclopedia of chemical technology See: Kirk, Raymond E. & Othmer, Donald F.
Encyclopaedia of New Zealand, An ed. A. H. McLintock 3 vols. 1966
Encyclopedia of philosophy, The ed. P. Edwards et al. 8 vols. 1967
Encyclopedia of polymer science and technology ed. H. Mark et al. 1964–
Encyclopedia of psychology ed. H. J. Eysenck et al. 3 vols. 1972
Encyclopædia of religion and ethics ed. J. Hastings 13 vols. 1908–26
Encyclopedia of sports, games and pastimes 1935
Encyclopædia of the social sciences ed. E. R. A. Seligman et al. 15 vols. 1930–5
Encyclopædic dictionary, The by R. Hunter et al. 7 vols. 1879–88
— (reissue with Suppl. 1902–4)
Encyclopædic dictionary of physics ed. J. Thewlis et al. 9 vols. 1961–4; Suppl. 2 vols. 1966–7
Endeavour 1942–
Endocrinology 1917–
Engel, Carl A descriptive catalogue of the musical instruments in the South Kensington Museum 1870
— (ed. 2) 1874
Engineering 1866–
Engineering news-record 1874–
English journal (US National Council of Teachers of English) 1912–
English language teaching 1946– (from 1973 with title English language teaching journal)
English mechanic, The (title varies) 1865–1923
English studies: a journal of English letters and philology 1919–
Englishwoman, The 1909–21
Ennes, Harold Eugene Television broadcasting: equipment, systems and operating fundamentals 1971
Ensor, Robert Charles Kirkwood England, 1870–1914 1936
Entomological Society of London (later Royal ...) Proceedings 1926–76 (from 1936 divided into separate series)
Transactions 1834–
Entwistle, William James Aspects of language 1953
Environmental conservation 1974–
Erdman, Paul The silver bears 1974
Erskine, Noel Underworld and prison slang 1935
Erskine-Murray, James A handbook of wireless telegraphy 1907
Esau, Katherine Plant anatomy 1953
Esquire 1933–
Essays and studies by members of the English Association 1910–
Essays in criticism: a quarterly journal of literary criticism 1951–
Essex Institute [Salem, Massachusetts] historical collections 1859–
Essig, Edward Oliver College entomology 1942
Etc.: a review of general semantics 1943–
Etherington, Harold ed. Nuclear engineering handbook 1958
Eugene Register-Guard (Eugene, Oregon) 1955–
European Organization for Nuclear Research (Conseil Européen pour la Recherche Nucléaire) Proceedings of CERN symposium 2 vols. 1956
Evans, Arthur John The Palace of Minos 7 vols. 1921–36
Evans, Bergen & Evans, Cornelia A dictionary of contemporary American usage 1957

Evans, Charles See Evans, Robert Charles
Evans, Emyr Estyn Irish folk ways 1957
Evans, George Ewart Ask the fellows who cut the hay 1956
The horse in the furrow 1960
The pattern under the plough 1966
Where beards wag all 1970
Evans, Howard T. tr. G. Hägg's General and inorganic chemistry 1969
Evans, John G. The environment of early man in the British Isles 1975
Evans, Philip The bodyguard man 1973
Evans, Robert Charles On climbing 1956
Evans, Robley Dunglison The atomic nucleus 1955
Evans-Pritchard, Edward Evan Social anthropology 1951
—— et al. The institutions of primitive society 1954
Evelyn, John (1620–1706) Diary ed. E. S. de Beer 6 vols. 1955
Evening News (Edinburgh): see Edinburgh Evening News
Evening News (London) 1881–1980
Evening Post (Nottingham) 1878–
Evening Post (Wellington, New Zealand) 1865–
Evening Standard 1860–1980 (continued as New Standard)
Evening Sun (Baltimore, Maryland) 1910–
Evening Telegram (St. John's, Newfoundland) 1879–
Everett, Thomas H. Living trees of the world 1969
Everybody's magazine (US) 1899–1929
Ewart, Alfred James tr. W. Pfeffer's The physiology of plants 3 vols. 1900–6
Ewart, Gavin The Gavin Ewart show: poems 1971
Experiment station record (United States Department of Agriculture) 1889–1946
Experimental wireless and the wireless engineer 1923–31
Eye opener 1902–22
Eyre, Mary A lady's walks in the south of France in 1863 1865

F

Fabian News 1891–
Fairbanks Daily News-Miner (Fairbanks, Alaska) 1903–
Fairbridge, Rhodes Whitmore ed. The encyclopedia of atmospheric sciences and astrogeology 1967
—— ed. The encyclopedia of geomorphology 1968
Fairchild, Henry Pratt ed. Dictionary of sociology 1944
Falla, Robert Alexander, Sibson, R. B., & Turbott, E. G. A field guide to the birds of New Zealand and outlying islands 1966
Famous plays 12 vols. 1931–9
Faraday, Wilfred Barnard ed. A glossary of aeronautical terms 1919
Faraday Society Transactions 1905–
Farber, Marvin The foundation of phenomenology: Edmund Husserl and the quest for a rigorous science of philosophy 1943
Farmer, Fannie Merritt The Boston Cooking-School cook book 1896 (and several later editions used)
Farmer and stockbreeder, The 1889–
Farmers weekly 1934–
Farnie, Henry Brougham The golfer's manual 1857
Farnol, John Jeffery The broad highway 1910
The definite object 1917
Farquharson, M. G., et al. Glossary of broadcasting terms 1942
Farrell, James Thomas Studs Lonigan: a trilogy (Young Lonigan; The young manhood of Studs Lonigan; Judgment day) 1935 (UK 1936)
Young Lonigan: a boyhood in Chicago streets 1932
The young manhood of Studs Lonigan 1934
Farrer, Reginald John My rock-garden 1907
Farrow, Edward Samuel A dictionary of military terms 1918
Fast, Howard Melvin The immigrants 1977
Fastnedge, Ralph English furniture styles, from 1500 to 1830 1955
Faulkner, William Absalom, Absalom! 1936
As I lay dying 1930 (UK 1935)
A fable 1954 (UK 1955)
Go down, Moses, and other stories 1942
The hamlet 1940
Light in August 1932 (UK 1933)
Sanctuary 1931
Sartoris 1929 (UK 1932)
The sound and the fury 1929
The wild palms 1939
Fay, Albert Hill A glossary of the mining and mineral industry 1920
Fearon, Diana Murder-on-Thames 1960
Fearon, William Robert An introduction to biochemistry 1934

Feather, Leonard Geoffrey The encyclopedia of jazz 1955 (UK 1956)
Inside be-bop 1949
Federation proceedings (Federation of American Societies for Experimental Biology) 1942–
Feinsilver, Lillian Mermin The taste of Yiddish 1970
Feldenkrais, Moshé Judo: the art of defence and attack 1941
Femina (Bombay) 1959–
Fennessy, J. C. The sonnet in the bottle 1951
'Fenwick, Elizabeth' (Elizabeth Fenwick Way) Impeccable people 1971
A long way down 1959
Ferber, Edna Dawn O'Hara, the girl who laughed 1911
Ferguson, Charles D. The experiences of a forty-niner during thirty-four years' residence in California and Australia ed. F. T. Wallace 1888
Fergusson, Bernard Edward The watery maze: the story of combined operations 1961
Fergusson, Erna Our Southwest 1940
'Fern, Fanny' (Sara Payson Parton) Ginger-snaps 1870
'Ferrars, Elizabeth' (Morna Doris Brown) Breath of suspicion 1972
Murder in time 1976
Ferris, Paul The detective 1976
Ferris, Richard How it flies; or, The conquest of the air 1910
Fessenden, Thomas G. Pills, poetical, political, and philosophical 1809
Festival of Britain, 1951. South Bank Exhibition Catalogue of exhibits 1951
Ficchi, Rocco F. Electrical interference 1964
Fick, Carl The Danziger transcript 1971 (UK 1973)
Field, Edward Salisbury A six-cylinder courtship 1907
Field, George Chromatography; or, A treatise on colours and pigments 1835
Field, Samuel & Weill, A. Dudley Electroplating: a survey of modern practice 1930
Field archæology (Ordnance Survey Professional Papers) 1932
— (ed. 4) 1963
Fielding, Alexander Wallace ('Xan') The stronghold: the four seasons in the White Mountains of Crete 1953
Fiennes, Celia (1662–1741) Journeys ed. C. Morris 1947
Fiennes, Gerard F. G. Twisleton-Wykeham- I tried to run a railway 1967
Fieser, Louis Frederick & Fieser, Mary A. Advanced organic chemistry 1961
Organic chemistry 1944
Filippini, Alessandro The international cook book 1906
Filson, John The discovery, settlement, and present state of Kentucke 1784
Financial Times 1888–
Finar, Ivor Lionel Organic chemistry 2 vols. 1959–64
Findlay, Alexander tr. W. Ostwald's The principles of inorganic chemistry 1902
Findlay, John Niemeyer Values and intentions: a study in value-theory and philosophy of mind 1961
Finer, Samuel Edward The man on horseback: the role of the military in politics 1962
Fingleton, John Henry Webb The Ashes crown the year: a Coronation cricket diary 1954
Four chukkas to Australia: the 1958–59 M.C.C. tour of Australia 1960
Finito! The Po Valley Campaign 1945 (15th Army Group) 1945
Fink, Donald Glen ed. Electronics engineers' handbook 1975
Principles of television engineering 1940
Finlayson, Hedley Herbert The red centre: man and beast in the heart of Australia 1935
Finlayson, Roderick David Brown man's burden 1938
The schooner came to Atia 1952
Firth, Anthony Tall, balding, thirty-five 1966
Firth, John Rupert Papers in linguistics, 1934–1951 1957
Speech 1930
The tongues of men 1937
Firth, Raymond William Elements of social organization 1951
Fischer, Martin Henry tr. W. Ostwald's Handbook of colloid-chemistry 1915
Fisher, Ernest Arthur An introduction to Anglo-Saxon architecture and sculpture 1959
Fisher, James Maxwell McConnell & Lockley, Ronald Mathias Sea-birds: an introduction to the natural history of the sea-birds of the North-Atlantic 1954
Fisher, Joseph William & Hartree, Douglas Rayner tr. M. Born's The mechanics of the atom 1927
Fisher, Ronald Aylmer The genetical theory of natural selection 1930
Statistical methods for research workers 1925
Fisher, Seymour The female orgasm: psychology, physiology, fantasy 1972 (UK 1973)
Fisher, William Rogers tr. A. F. W. Schimper's Plant-geography upon a physiological basis 1903
See also: Schlich, William A manual of forestry
Fishman, Joshua Aaron ed. Readings in the sociology of language 1968
Fithian, Philip Vickers Journal and letters, 1767–1774 ed. J. R. Williams 1900
Journal, 1775–1776 ed. R. G. Albion & L. Dodson 1934
Fitzgerald, Francis Scott Key (1896–1940) The beautiful and damned 1922
The great Gatsby 1925 (UK 1926)
The last tycoon 1941 (UK 1949)
Letters ed. A. Turnbull 1963 (UK 1964)
Tender is the night 1934
This side of Paradise 1920 (UK 1921)
Fitzgibbon, Theodora The art of British cooking 1965
Fitzsimons, Frederick William The snakes of South Africa 1910
— (new ed.) 1912
Fixx, James F. The complete book of running 1977
Flagg, Edmund The far west; or, A tour beyond the mountains 2 vols. 1838
Flattely, Frederick William & Walton, Charles Livesey The biology of the sea-shore 1922
Flavell, John Hurley The developmental psychology of Jean Piaget 1963
Fleming, Ian Lancaster (1908–64) For your eyes only 1960
Goldfinger 1959
The man with the golden gun 1965
Moonraker 1955
On Her Majesty's Secret Service 1963
Thunderball 1961
You only live twice 1964
Fleming, Joan Margaret Kill or cure 1968
Max Bones 1962
Nothing is the number when you die 1965
You won't let me finish 1973
Young man, I think you're dying 1970
Fleming, John Ambrose The principles of electric wave telegraphy 1906
Short lectures to electrical artisans 1886
Fleming, Robert Peter Brazilian adventure 1933
News from Tartary: a journey from Peking to Kashmir 1936
Flemming, Louis Andrew Practical tanning 1903
Fletcher, Charles Robert Leslie & Kipling, Rudyard A school history of England 1911
'Fletcher, David' (Dulan Friar Barber) Don't whistle 'Macbeth' 1976
Flexner, Stuart Berg Listening to America 1982
Flight 1908– (title varies; from 1962 with title Flight international)
Flitch, John Ernest Crawford Modern dancing and dancers 1911
Florey, Howard Walter, et al. Antibiotics: a survey of penicillin, streptomycin and other antimicrobial substances from fungi, actinomycetes, bacteria, and plants 2 vols. 1949
Florey, Mary Ethel The clinical application of antibiotics 4 vols. 1952–61
Florida FL Reporter 1962–
Florida plantation records See: Jones, George Noble
Florkin, Marcel & Stotz, Elmer Henry eds. Comprehensive biochemistry 1962–
Flügel, John Carl A hundred years of psychology 1833–1933 1933
Flying 1917–19
Flying (New York) 1927–
Flying (title varies) 1924–8
'Flynt, Josiah' (Josiah Flynt Willard) The rise of Ruderick Clowd 1903 (UK 1904)
Tramping with tramps 1899 (UK 1900)
The world of graft 1901
—— & 'Walton, Francis' (Alfred Hodder) The powers that prey 1900
Focal dictionary of photographic technologies, The See: Spencer, Douglas Arthur
Focal encyclopedia of film and television techniques 1969
Focal encyclopedia of photography, The 1956
— (rev. Desk ed.) 1969
Focus 1945–
Foley, Winifred A child in the forest 1974
Folk-lore 1890–
Foot, Michael Richard Daniell SOE in France: an account of the work of the British Special Operations Executive in France 1940–1944 1966
Foote, Henry Stuart Texas and the Texans; or, Advance of the Anglo-Americans to the south-west 2 vols. 1841
Foote, Horace A companion to the theatres, and manual of the British drama 1829
Forbes, Duncan The heart of Malaya 1966
Forbes, Henry Ogg A hand-book to the primates 2 vols. 1894

Ford, Edmund Brisco Mendelism and evolution 1931
Moths 1955
Ford, Ford Madox (until 1919 had name Joseph L. Ford H. Madox Hueffer) The good soldier: a tale of passion 1915
Letters ed. R. M. Ludwig 1965
A man could stand up 1926
No more parades 1925
The panel: a sheer comedy 1912
Ford, Kenneth William The world of elementary particles 1963
Ford, Paul Leicester The Honorable Peter Stirling and what people thought of him 1894 (UK 1898)
Ford, Sewell Inez and Trilby May 1921
Shorty McCabe 1906 (UK 1908)
Side-stepping with Shorty 1908
Ford-Robertson, F. C. ed. Terminology of forest science, technology, practice and products 1971
Foreign review and continental miscellany 1828–30
Forester, Cecil Scott The commodore 1945
The good shepherd 1955
Hornblower and the Atropos 1953
Mr. Midshipman Hornblower 1950
The ship 1943
A ship of the line 1938
Forestry 1927–
Forestry bureau bulletin See: United States. Department of Agriculture. Bureau of Forestry
Forgan, Robert The golfer's handbook 1881
Forrester, Larry A girl called Fathom 1967
Forster, Edward Morgan (1879–1970) Abinger harvest: a miscellany 1936
Aspects of the novel 1927
The hill of Devi: being letters from Dewas State Senior 1953
Howards End 1910
The longest journey 1907
Maurice 1971 (written 1913–14, revised 1959–60)
A passage to India 1924
A room with a view 1908
Two cheers for democracy 1951
Where angels fear to tread 1905
Forster, Harold Flowering lotus: a view of Java 1958
Forsyth, Frederick The day of the jackal 1971
The dogs of war 1974
Fortes, Meyer ed. Social structure: studies presented to A. R. Radcliffe-Brown 1949
Fortnum & Mason Ltd. Price list c1938
Fortune 1930–
Forum, The (New York) 1886–1930
Forum, The (Johannesburg) 1938–
Fosdick, Harry Emerson A pilgrimage to Palestine 1927 (UK 1928)
Foster, Adriance Sherwood & Gifford, Ernest Milton Comparative morphology of vascular plants 1959
Foster, Brian The changing English language 1968
Foster, Frank Pierce ed. An illustrated encyclopædic medical dictionary 4 vols. 1888–94
Foster, Harry La Tourette The adventures of a tropical tramp 1922
A beachcomber in the Orient 1923
A tropical tramp with the tourists 1925
Foster, Herbert L. Ribbin', jivin', and playin' the dozens: the unrecognized dilemma of inner city schools 1974
Foster, Idris Llewellyn & Alcock, Leslie eds. Culture and environment: essays in honour of Sir Cyril Fox 1963
Foster, Robert Frederick Foster's complete Hoyle 1897
Foster-Harris, William The look of the old West 1955
Fowler, Henry Watson A dictionary of modern English usage 1926
— (ed. 2, rev. by E. A. Gowers) 1965
Fowler, Jacob (1765–1850) Journal ed. E. Coues 1898
Fowles, Anthony Dupe negative 1970
Pastime 1974
Fowles, John The collector 1963
The French lieutenant's woman 1969
The magus 1966
Fox, Caroline (1819–71) Journals ed. W. Monk 1972
Fox, Charles Educational psychology: its problems and methods 1925
Fox, Cyril Fred The archaeology of the Cambridge region 1923
Fox, Lawrence Webster Diseases of the eye 1904
Fox, Leslie The numerical solution of two-point boundary problems in ordinary differential equations 1957
—— & Mayers, D. F. Computing methods for scientists and engineers 1968
Fox, Richard Middleton & Fox, Jean Walker Introduction to comparative entomology 1964

Foye, William Owen ed. Principles of medicinal chemistry 1974
Fozard, Basil Instrumentation and control of nuclear reactors 1963
Fraas, Arthur Paul Aircraft power plants 1943
Fraenkel, Gottfried S. & Gunn, D. L. The orientation of animals 1940
Fraenkel-Conrat, Heinz Ludwig & Wagner, R. R. ed. Comprehensive virology 1974–
Frame, Janet The edge of the alphabet 1962
The lagoon: stories 1951
Owls do cry 1957 (UK 1961)
Francis, Dick (Richard Stanley Francis) Blood sport 1967
Bonecrack 1971
Dead cert 1962
Enquiry 1969
Flying finish 1966
For kicks 1965
Forfeit 1968
In the frame 1976
Knock down 1974
Nerve 1964
Odds against 1965
Rat race 1970
Risk 1977
Slay-ride 1973
Francis, Peter W. Volcanoes 1976
Frank, Andre Gunder Latin America: underdevelopment or revolution 1969
Frank, Pat Harry Hart Seven days to never 1957
Frankau, Gilbert More of us, being the present-day adventures of 'One of us': a novel in verse 1937
'Tid'apa': what does it matter? 1914 (UK 1915)
Frankau, Pamela The winged horse 1953
Franklin, Benjamin (1706–90) Writings ed. A. H. Smyth 10 vols. 1905–7
Franklin, John Narrative of a journey to the shores of the Polar Sea, 1819–22 1823
Franklin, Stella Maria Sarah Miles All that swagger 1936 (UK 1947)
My brilliant career 1901
See also: 'Brent of Bin Bin'
Franklyn, Julian E. A dictionary of nicknames 1962
A dictionary of rhyming slang 1960
Fraser, Edward & Gibbons, John Soldier and sailor words and phrases 1925
'Fraser, James' (Alan White) A cock-pit of roses 1969
Heart's ease in death 1977
Who steals my name? 1976
Frayn, Michael The Russian interpreter 1966
The tin men 1965
Frayne, John George & Wolfe, H. Elements of sound recording 1949
Frazee, Irving Augustus & Bedell, Earl L. Automotive fundamentals 1949
Free, Montague All about house plants 1946
Freebody, John William Henry Telegraphy 1959
Freedomways 1961–
Freeling, Nicolas Because of the cats 1963
Criminal conversation 1965
Double-barrel 1964
The Dresden Green 1966
Dressing of diamond 1974
The king of the rainy country 1966
Lake isle 1976
A long silence 1972
Love in Amsterdam 1962
Strike out where not applicable 1967
Tsing-boum 1969
What are the bugles blowing for? 1975
Freeman, E. E. tr. A. Vandel's Biospeleology: the biology of cavernicolous animals 1965
Freeman, Gillian Jack would be a gentleman 1959
The liberty man 1955
Freeman, Mary Eleanor (Wilkins) See Wilkins, Mary Eleanor
Freemantle, Brian The November man 1976
Fremdsprachen 1957–
Fremlin, Celia Margaret Appointment with yesterday 1972
By horror haunted: stories 1974
The jealous one 1965
Frémont, John Charles Report of the exploring expedition to the Rocky Mountains in .. 1842, and to Oregon and North California in .. 1843–44 1845
Frends 1971–2
Freud, Sigmund See Brill, Abraham Arden; Riviere, Joan; Strachey, James
Friedländer, Michael The Jewish religion 1891
Friend, John Albert Newton A text-book of physical chemistry 2 vols. 1932–5
Friend, The 1843–
Fries, Charles Carpenter American English grammar: the grammatical structure of present-day American English with especial reference to social differences or class dialects 1940
Frisch, Otto Robert ed. The nuclear handbook 1958

Fritsch, Felix Eugen The structure and reproduction of the algae 2 vols. 1935–45
Frost, Robert Lee (1874–1963) Collected poems 1930
— (new ed.) 1939
Letters to Louis Untermeyer ed. L. Untermeyer 1964
—— & Frost, Elinor Family letters ed. A. Grade 1972
Froud, Nina, et al. tr. P. Montagné's Larousse gastronomique 1961
Fruton, Joseph Stewart & Simmonds, Sofia General biochemistry 1953
— (ed. 2) 1958
Fry, Christopher The lady's not for burning: a comedy 1949
Venus observed 1950
Fry, Roger Eliot (1866–1934) Cézanne: a study of his development 1927
Letters ed. D. Sutton 2 vols. 1972
Transformations: essays on art 1926
Frye, Northrop Anatomy of criticism 1957
ed. Romanticism reconsidered 1963
ed. Sound and poetry 1957
Fulford, Roger Thomas Baldwin ed. Dearest child: letters between Queen Victoria and the Princess Royal, 1858–1861 1964
ed. Dearest Mama: letters between Queen Victoria and the Crown Princess of Prussia, 1861–1864 1968
ed. Your dear letter: private correspondence of Queen Victoria and the Crown Princess of Prussia, 1865–1871 1971
Fullarton, John H. Troop target 1944
Fuller, George D. & Conard, H. S. tr. J. Braun-Blanquet's Plant sociology: the study of plant communities 1932
Fuller, Jane G. Uncle John's flower-gatherers 1869
Fuller, Richard Buckminster Operating manual for Spaceship Earth 1969
Untitled epic poem on the history of industrialization 1962
Fuller, Roy Broadbent The ruined boys 1959
The second curtain 1953
Funk & Wagnalls Co. A standard dictionary of the English language ed. I. K. Funk et al. 2 vols. 1893–5
— (new ed.) 2 vols. 1928
Standard dictionary of folklore, mythology, and legend ed. M. Leach 2 vols. 1949–50

G

Gaddis, William The recognitions 1955
Gaiger, Sydney Herbert & Davies, Gwilym Owen Veterinary pathology and bacteriology 1932
Galaxy, The 1866–78
Galbraith, John Kenneth The affluent society 1958
Gale, Frederick Echoes from old cricket fields; or, Sketches of cricket and cricketers from the earliest history of the game to the present time 1871
The game of cricket 1887
The life of the Hon. Robert Grimston 1885
The public school matches, and those we meet there 1853
Gallaher, David & Stead, W. J. The complete rugby footballer on the New Zealand system 1906
Gallico, Paul William The foolish immortals 1953
The snow goose 1941
Galouye, Daniel Francis The lost perception 1966
Galsworthy, John Captures 1923
The country house 1907
Five tales 1918
The fugitive 1913
In Chancery 1920
The Inn of Tranquillity: studies and essays 1912
Maid in waiting 1931
The man of property 1906
The roof 1929
A sheaf [essays] 1916
The silver spoon 1926
Swan song 1928
To let 1921
The white monkey 1924
Galton, Francis Natural inheritance 1889
Galwey, Geoffrey Valentine The lift and the drop 1948
Gamble, Charles Frederick Snowden The story of a North Sea air station 1928
Gamble, James Sykes A manual of Indian timbers 1881
Gammond, Peter ed. The Decca book of jazz 1958
ed. Duke Ellington: his life and music 1958
Gandalf's garden 1968–9

GARDINER, Alan Henderson *Egypt of the Pharaohs: an introduction* 1961
The theory of proper names 1940
The theory of speech and language 1932
GARDNER, Erle Stanley *The case of the blonde bonanza* 1962 (UK 1967)
The case of the queenly contestant 1967 (UK 1973)
The case of the stuttering bishop 1936 (UK 1937)
The D.A. draws a circle 1939 (UK 1940)
GARDNER, G. B. *Keris and other Malay weapons* 1936
GARDNER, Helen Louise *The business of criticism* 1959
GARDNER, Hy *So what else is new?* 1959
GARDNER, John Edmund *A complete state of death* 1969
The corner men 1969
Founder member 1969
Madrigal 1967
GARFIELD, Brian Wynne *Hopscotch* 1975
GARNER, Harry Mason *Oriental blue and white* 1954
GARNER, William *A big enough wreath* 1974
The deep, deep freeze 1968
Ditto, Brother Rat! 1972
The us or them war 1969
GARNETT, David *War in the air: September 1939 to May 1941* 1941
GARRARD, Lewis Hector *Wah-to-Yah, and the Taos trail* 1850
GARRATT, George Alfred *The mechanical properties of wood* 1931
GARRETT, Robert *Run down: the world of Alan Brett* 1970
GARROD, Dorothy A. E. & BATE, D. M. A. *The Stone Age of Mount Carmel* Vol. I 1937
GARTNER, Lloyd P. *The Jewish immigrant in England, 1870–1914* 1960
'GARVE, Andrew' (Paul Winterton) *Boomerang* 1969
The golden deed 1960
The late Bill Smith 1971
Murder in Moscow 1951
GARVIN, James Louis *The economic foundations of peace: or, World-partnership as the truer basis of the League of Nations* 1919
GASCOYNE, David Emery *Night thoughts* 1956
Opening day 1933
Poems 1937–1942 1943
A short survey of surrealism 1935
A vagrant, and other poems 1950
'GASKELL, A. P.' (Alexander Gaskell Pickard) *The big game, and other stories* 1947
GASKELL, Elizabeth Cleghorn (1810–65) *A dark night's work* 1863
Letters ed. J. A. V. Chapple & A. Pollard 1966
GASKELL, Philip *A new introduction to bibliography* 1972
GASS, Ian Graham, SMITH, P. J., & WILSON, R. C. L. eds. *Understanding the earth: a reader in the earth sciences* 1971
GASTON, William James ('Bill') *Drifting death* 1964
GATES, Reginald Ruggles *A botanist in the Amazon Valley* 1927
Heredity in man 1929
Human genetics 2 vols. 1946
GATLAND, Kenneth William *Development of the guided missile* 1952
— (ed. 2) 1954
GATSCHET, Albert Samuel *The Klamath Indians of southwestern Oregon* 1890
Gay News 1972–
GAYNOR, Frank ed. *Pocket encyclopedia of atomic energy* 1950
GEAR, Charles William *Introduction to computer science* 1973
GÉBLER, Ernest *Shall I eat you now?* 1969
GEDDES, Patrick *Cities in evolution: an introduction to the town planning movement and to the study of civics* 1915
— & THOMSON, John Arthur *Sex* 1914
GEDDES, Paul *The Ottawa allegation* 1973
GELL, Philip George H. & COOMBS, Robin Royston A. eds. *Clinical aspects of immunology* 1963
Gen: the Services' fortnightly 1942–5
General linguistics 1955–
General systems 1956–
Genetics 1916–
Geo abstracts 1972–
Geochimica et cosmochimica acta 1950–
Geofysiske publikationer (title varies) 1919–
Geographical review 1916–
Geological Society of America *Bulletin* 1890–
Memoirs 1934–
Special papers 1934–
Geological Society of London *Proceedings* 1826–45; 1952–
Quarterly journal 1845–
Transactions 1811–56
Geologiska Föreningens i Stockholm *Förhandlingar* 1872–
Geologists' Association *Proceedings* 1859–
Geomorphological abstracts 1960–5

GEORGE, Charles B. *Forty years on the rail* 1887 (UK 1888)
GEORGE, David Lloyd *Family letters 1885–1936* ed. K. O. Morgan 1973
GEORGE, Russell D. *Minerals and rocks: their nature, occurrence and uses* 1943
Georgia *The colonial records of the State of Georgia* ed. A. D. Candler et al. 1904–
Georgia Historical Society *Collections* 9 vols. 1840–1916
GERALD, Michael C. *Pharmacology: an introduction to drugs* 1974
GÉRIN, Winifred *Charlotte Brontë: the evolution of a genius* 1967
Elizabeth Gaskell: a biography 1976
Germanic review 1926–
GERTH, Hans Heinrich & MARTINDALE, Don Albert tr. *M. Weber's Ancient Judaism* 1952
GIBB, Jocelyn ed. *Light on C. S. Lewis* 1965
'GIBBON, Lewis Grassic' (James Leslie Mitchell) *Sunset song* 1932
GIBBONS, Floyd Phillips *The red Napoleon* 1929 (UK 1930)
GIBBONS, Stella Dorothea *Cold Comfort Farm* 1932
The matchmaker 1949
A pink front door 1959
GIBBS, Philip Hamilton *The battles of the Somme* 1917
Unchanging quest 1925
GIBBS-SMITH, Charles Harvard *The aeroplane: an historical survey of its origins and development* 1960
GIBSON, Guy Penrose *Enemy coast ahead* 1946
GIBSON, William Ralph Boyce tr. *E. Husserl's Ideas: general introduction to pure phenomenology* 1931
GIDDENS, Anthony *Studies in social and political theory* 1977
GIELGUD, Val Henry *The candle-holders* 1970
Conduct of a member 1967
In such a night 1974
A necessary end 1969
GIFFEN, George *With bat and ball: twenty-five years' reminiscences of Australian and Anglo-Australian cricket* 1898
GIFFORD, Thomas *The Cavanaugh quest* 1976 (UK 1977)
'GILBERT, Anthony' (Lucy Beatrice Malleson) *And death came too* 1956
Death against the clock 1958
Death takes a wife 1959
Don't open the door 1945
Knock, knock, who's there? 1964
Missing from her home 1969
No dust in the attic 1962
Ring for a noose 1963
Tenant for the tomb 1971
GILBERT, Martin *Winston S. Churchill* See: CHURCHILL, Randolph Spencer & GILBERT, Martin
GILBERT, Michael Francis *Blood and judgement* 1959
The body of a girl 1972
Close quarters 1947
The dust and the heat 1967
Flash point 1974
Sky high 1955
GILBERT-CARTER, Humphrey tr. *C. Raunkiaer's The life forms of plants* 1934
GILES, Henry Earl *Harbin's Ridge* 1951
GILES, Herbert Allen *A glossary of reference on subjects connected with the Far East* 1878
GILES, Howard ed. *Language, ethnicity and intergroup relations* 1977
GILES, Kenneth *Death and Mr. Prettyman* 1967
Death cracks a bottle 1969
Death in diamonds 1967
A death in the church 1970
A file on death 1973
Some beasts no more 1965
'GILES, Norman' (N. R. McKeown) *The ridge of white waters* 1934
GILES, Peter *A short manual of comparative philology for classical students* 1895
GILL, Edwin Leonard *First guide to South African birds* 1936
GILL, Gerald Byron & WILLIS, Martin Richard *Pericyclic reactions* 1974
GILL, Merle Avery *Underworld slang* 1929
GILLEN, Lucy *Return to Deepwater* 1975
GILLESPIE, Alexander Douglas *Letters from Flanders* 1916
GILLIARD, Ernest Thomas *Living birds of the world* 1958
GILMAN, Caroline *Recollections of a southern matron* 1838
GILMAN, Henry *Organic chemistry: an advanced treatise* 3 vols. 1938–53
— (ed. 2) 2 vols. 1943
GILMOUR, Samuel Carter *Paper: its making, merchandising and usage* 1955
GILPIN, Alan *Dictionary of economic terms* 1966

GIMSON, Alfred Charles *An introduction to the pronunciation of English* 1962
GISH, Lillian & PINCHOT, A. *Lillian Gish: the movies, Mr. Griffith and me* 1969
GISSING, George Robert *New Grub Street* 3 vols. 1891
The odd women 3 vols. 1893
The private papers of Henry Ryecroft 1903
— & WELLS, Herbert George *George Gissing and H. G. Wells: their friendship and correspondence* ed. R. A. Gettman 1961
GIVEN, Meta H. *Modern encyclopedia of cooking* 2 vols. 1947
GLAISTER, Geoffrey Ashall *Glossary of the book* 1960
GLASCOCK, William Nugent *The naval sketch-book; or, The Service afloat and ashore* 2 vols. 1826
— (ed. 2) 2 vols. 1826
— (2nd ser.) 2 vols. 1834
GLASGOW, Ellen Anderson Gholson (1874–1945) *The deliverance: a romance of the Virginia tobacco fields* 1904
Letters ed. B. Rouse 1958
GLASGOW, Roy Stanley *Principles of radio engineering* 1936
Glasgow Herald 1805–
GLASS, David Victor & EVERSLEY, D. E. C. eds. *Population in history* 1965
GLASSER, Otto ed. *The science of radiology* 1933 (UK 1934)
GLASSOP, Lawson *We were the rats* 1944
GLASSTONE, Samuel *An introduction to electrochemistry* 1942
Principles of nuclear reactor engineering 1956
Sourcebook on atomic energy 1950
Text-book of physical chemistry 1940 (UK 1941)
— (ed. 2) 1946 (UK 1948)
GLATT, Max Meier, et al. *The drug scene in Great Britain* 1967
GLAZEBROOK, Richard Tetley ed. *A dictionary of applied physics* 5 vols. 1922–3
GLAZIER, Richard *A manual of historic ornament* 1899
GLEASON, Henry Allan (b. 1882) & CRONQUIST, Arthur *The natural geography of plants* 1964
GLEASON, Henry Allan (b. 1917) *An introduction to descriptive linguistics* 1955
— (ed. 2) 1961
Linguistics and English grammar 1965
GLEMSER, Bernard *A dear Hungarian friend* 1966
GLISAN, Rodney *Journal of army life* 1874
GLOAG, John Edwards *A short dictionary of furniture, containing 1764 terms used in Britain and America* 1952
Globe and Mail (Toronto) 1844– (1844–1936 with title *Globe*)
GLUCKMAN, Herman Max *Custom and conflict in Africa* 1955
GLUECK, Bernard & LIND, John Edward tr. *A. Adler's The neurotic constitution: outlines of a comparative individualistic psychology and psychotherapy* 1917 (UK 1921)
GLYN, Anthony Geoffrey L. S. *The dragon variation* 1969
GLYN, Elinor *'It', and other stories* 1927
GODDEN, Jon & GODDEN, Margaret Rumer *Two under the Indian sun* 1966
GODDEN, Margaret Rumer *Black narcissus* 1939
The greengage summer 1958
In this house of Brede 1969
Kingfishers catch fire 1953
GODFREY, Eve *Retail selling and organization* 1962
GODFREY, James William & AMOS, Stanley William *Sound recording and reproduction* 1952
GODFREY, Philip *Back-stage: a survey of the contemporary English theatre from behind the scenes* 1933
GOLD, Herbert *The man who was not with it* 1956 (UK 1965)
GOLD, Robert S. *A jazz lexicon* 1964
Jazz talk 1975
GOLDBERG, Isaac *The wonder of words* 1938
GOLDIN, Hyman Elias ed. *Dictionary of American underworld lingo* 1950
GOLDING, Louis *Magnolia Street* 1932
GOLDING, William Gerald *Free fall* 1959
Lord of the flies 1954
Pincher Martin 1956
The spire 1964
GOLLANCZ, Victor *My dear Timothy: an autobiographical letter* 1952
GOMBRICH, Ernst Hans Josef *Art and illusion* 1960
The story of art 1950
GOOD, Carter Victor ed. *Dictionary of education* 1945
— (ed. 2) 1959
Good food guide, The 1951–
Good housekeeping (London) 1922–
Good housekeeping (New York) 1885–

Good housekeeping's cookery book: compiled by The Good Housekeeping Institute 1948
— (rev. ed.) 1954
— (rev. ed.) 1960
— (ed. 4) 1956
Good housekeeping's home encyclopaedia 1951
— (ed. 2) 1956
Good motoring 1935–
GOODCHILD, George Frederick & TWENEY, C. F. *A technological and scientific dictionary* 1904–6
GOODFIELD, June *Courier to Peking* 1973
GOODGE, William T. *Hits! skits! and jingles!* 1899
GOODIER, James Hillis *Dictionary of painting and decorating trade terms* 1961
GOODMAN, Clark ed. *The science and engineering of nuclear power* 2 vols. 1947–9
GOODMAN, Richard Merton ed. *Genetic disorders of man* 1970
GOODMAN, William Louis *The history of woodworking tools* 1964
GOODWIN, Derek *Pigeons and doves of the world* 1967
GOOLDEN, Barbara *At the foot of the hills* 1956
For richer, for poorer 1959
GORDIMER, Nadine *Burger's daughter* 1979
The lying days 1953
Six feet of the country: 15 short stories 1956
A world of strangers 1958
GORDON, Eric Valentine *An introduction to Old Norse* 1927
GORDON, Mildred & GORDON, Gordon *The informant* 1973
Ordeal 1976 (UK 1977)
'GORDON, Richard' (Gordon Stanley Ostlere) *Doctor at sea* 1953
Doctor in the house 1952
GORDON, Rupert Montgomery & LAVOIPIERRE, Michel M. J. *Entomology for students of medicine* 1962
GORDON, Taylor *Born to be* 1929
GORER, Geoffrey Edgar Solomon *Africa dances* 1935
GORES, Joseph N. *Dead skip* 1972 (UK 1973)
Hammett 1975 (UK 1976)
GORMAN, James Thomas *Modern weapons of war* 1942
GORTNER, Ross Aiken *Outlines of biochemistry* 1929
GOSSE, Edmund William *Father and son: a study of two temperaments* 1907
GOSVAMI, O. *The story of Indian music* 1957
GOTLIEB, Calvin Carl & HUME, J. N. P. *High-speed data processing* 1958
GOTT, John (1830–1906) *Letters of Bishop Gott, arranged by members of his family* 1918
GOULD, George Milbry *A dictionary of new medical terms* 1905
GOULD, Joseph *The letter-press printer* 1876
— (ed. 3) 1884
GOULD, Julius & KOLB, William L. eds. *A dictionary of the social sciences* 1964
GOWERS, Ernest Arthur *A dictionary of modern English usage* (ed. 2) 1965 See FOWLER, Henry Watson
GOWING, Margaret Mary *Britain and atomic energy, 1939–1945* 1964
GRABAU, Amadeus William *Principles of stratigraphy* 1913
A textbook of geology 2 vols. 1921
GRABBE, Eugene Munter, RAMO, S., & WOOLDRIDGE, D. E. eds. *Handbook of automation, computation and control* 3 vols. 1958–61
GRACE, Alfred Augustus *The tale of a timber town* 1914
GRACE, William Gilbert *Cricket* 1891
'GRAEME, Bruce' (Graham Montague Jeffries) *Tomorrow's yesterday* 1972
Two and two make five 1973
GRAFF, Willem Laurens *Language and languages* 1932
'GRAHAM, James' (Henry Patterson) *Bloody passage* 1974
GRAHAM, Neill *Murder in a dark room: a Solo Malcolm thriller* 1973
GRAHAM, Stephen *Europe—whither bound? being letters of travel from the capitals of Europe* 1921
GRAHAME, Kenneth *The wind in the willows* 1908
GRAHAME-WHITE, Claude & HARPER, Harry *The aeroplane: past, present and future* 1911
Aircraft in the Great War: a record and study 1915
Gramophone, The 1923–
Granada Television *Some technical terms and slang* 1974
GRANT, George Monro *Ocean to ocean: Sandford Fleming's expedition through Canada in 1872* 1873
GRANT, James A. *A walk across Africa* 1864
Granta, The (Cambridge) 1889–
GRANVILLE, Wilfred *A dictionary of sailors' slang* 1962
A dictionary of theatrical terms 1952
GRATTAN, John Henry Grafton & GURREY, Percival *Our living language: a new guide to English grammar* 1925

GRAU, Robert *The theatre of science: a volume of progress and achievement in the motion picture industry* 1914
GRAVES, Charles Patrick Ranke *Life line* 1941
GRAVES, Robert Ranke *Claudius the god and his wife Messalina* 1934
Collected poems 1938
The feather bed 1923
Good-bye to all that: an autobiography 1929
I, Claudius 1934
Mock Beggar Hall 1924
The pier-glass 1921
Poems, 1926–1930 1931
Poems, 1938–1945 1946
Poems, 1953 1953
Seven days in new Crete 1949
Welchman's hose 1925
Whipperginny 1923
— & HODGE, Alan Searle *The long week-end: a social history of Great Britain, 1918–1939* 1940
GRAY, Alexander tr. *R. Grelling's The crime* 3 vols. 1917–18
GRAY, Cecil William Turpie *Contingencies, and other essays* 1947
GRAY, Dulcie Winifred C. *Dead give away* 1974
GRAY, James Henry *The boy from Winnipeg* 1970
GRAY, John *Archaeology and the Old Testament world* 1962
GRAY, Laurence F. & GRAHAM, Richard *Radio transmitters* 1961
GRAY, Louis Herbert *Foundations of language* 1939
GRAYMORE, Clive N. ed. *Biochemistry of the eye* 1970
Great Exhibition *Official descriptive and illustrated catalogue of the Great Exhibition of the Works of Industry of all Nations* 5 parts 1851
GREATOREX, Wilfred *Crossover* 1976
GREEN, Abel & LAURIE, Joe *Show biz, from vaude to video* 1951
GREEN, Bennett Wood *Word-book of Virginia folk-speech* 1899
GREEN, Helen *At the actors' boarding house* 1906
The Maison de Shine: more stories of the actors' boarding house 1908
'GREEN, Henry' (Henry Vincent Yorke) *Living* 1929
GREEN, John Herbert *Basic clinical physiology* 1969
GREEN, Lawrence George *Great African mysteries* 1935
In the land of afternoon 1949
Tavern of the seas 1947
To the river's end 1948
GREEN, Martin Burgess *Children of the sun: a narrative of 'decadence' in England after 1918* 1976
GREEN, Peter tr. *R. Escarpit's The novel computer* 1966
GREENE, Henry Graham *The basement room, and other stories* 1935
Brighton rock 1938
A burnt-out case 1961
The comedians 1966
The confidential agent 1939
The end of the affair 1951
England made me 1935
A gun for sale 1936
The heart of the matter 1948
The honorary consul 1973
The human factor 1978
It's a battlefield 1934
Journey without maps 1936
The lawless roads: a Mexican journey 1939
The Ministry of Fear: an entertainment 1943
Nineteen stories 1947
Our man in Havana 1958
The power and the glory 1940
The quiet American 1955
The third man, and The fallen idol 1950
Travels with my aunt 1969
GREENER, Michael *The Penguin dictionary of commerce* 1970
GREENLEE, Sam *The spook who sat by the door* 1969
GREENOUGH, James Bradstreet & KITTREDGE, George Lyman *Words and their ways in English speech* 1901 (UK 1902)
GREENWOOD, Frederick & GREENWOOD, James *Under a cloud* 3 vols. 1860
GREENWOOD, Peter Humphrey *J. R. Norman's A history of fishes* (ed. 2) 1963 See NORMAN, John Roxborough
GREER, Germaine *The female eunuch* 1970
GREGG, Josiah *Commerce of the prairies; or, The journal of a Santa Fé trader* 2 vols. 1844
GREGORY, Edwin *Metallurgy* 1938
GREGORY, Jackson *The maid of the mountain: a romance of the California wilderness* (UK ed. with title *Bab of the backwoods*) 1925
Man to man 1920 (UK 1921)
GREGORY, John Walter & BARRETT, Benjamin Hilton *General stratigraphy* 1931

GRENFELL, Wilfred Thomason *A Labrador doctor* 1919 (UK 1920)
GRESSWELL, Peter *Environment: an alphabetical handbook* 1971
GREVILLE, Robert Fulke (1751–1824) *Diaries* ed. F. M. Bladon 1930
GREW, Sydney *The art of the player-piano: a text-book for student and teacher* 1922
'GREY OWL' (George Stansfeld Belaney) *The men of the last frontier* 1931
GRIDER, John McGavock *War birds: diary of an unknown aviator* 1926 (UK 1927)
GRIERSON, Edward *A crime of one's own* 1967
Reputation for a song 1952
GRIERSON, George Abraham *Linguistic survey of India* 11 vols. 1903–22
GRIERSON, John *High failure* 1936
GRIEVE, Maud *A modern herbal* 2 vols. 1931
GRIFFIN, Gerald *The collegians* 3 vols. 1829
GRIFFITH, Thomas *The waist-high culture* 1959 (UK 1960)
GRIGSON, Geoffrey Edward Harvey *The Englishman's flora* 1955
GRIMBLE, Arthur Francis *A pattern of islands* 1952
Return to the islands 1957
GROLLMAN, Arthur *Pharmacology and therapeutics* 1951 (and several later editions used)
GROOM, Percy & BALFOUR, Isaac Bayley tr. *J. B. Warming's Oecology of plants* 1909
GROSSINGER, Jennie *The art of Jewish cooking* 1958
GROSSMITH, George & GROSSMITH, Walter Weedon *The diary of a nobody* 1892
GROUT, Frank Fitch *Petrography and petrology* 1932
Growth (Menasha, Wisconsin) 1937–
GRUBER, Jeffrey S. *Lexical structures in syntax and semantics* 1976
GRUNDY, George Beardoe *Fifty-five years at Oxford* 1945
Guardian, The (continuation of *The Manchester Guardian*) 1959–
Guardian Weekly, The (continuation of *Manchester Guardian Weekly*) 1968–
GUENTHER, Ernest *The essential oils* 6 vols. 1949–52
GUGGISBERG, Charles Albert Walter *S.O.S. Rhino* 1966
GUILFOYLE, William Robert *Australian plants suitable for gardens, parks, timber reserves, etc.* 1911
GULBENKIAN, Nubar *Pantaraxia* 1965
GULLAND, William Giuseppi *Chinese porcelain* 1898
GULLICK, John Michael *Malaysia* 1969
GULLIVER, Sam *The Vulcan bulletins* 1974
'GUN BUSTER' (John Austin) *Return via Dunkirk* 1940
GUNN, Mrs. Aeneas (Jeannie) *We of the Never-Never* 1908
GUNN, J. S. *An opal terminology* 1971 (Sydney University: Australian Language Research Centre, Occasional Paper No. 15)
The terminology of the shearing industry 1965 (as above, Occasional Papers Nos. 5 & 6)
GUNGWU, Wang See WANG GUNGWU
GUNN, Thom *Fighting terms: poems* 1954
The sense of movement 1957
GUNTER, Archibald Clavering *Mr. Potter of Texas* 1888
GÜNTHER, Alfred *Microphotography in the library* 1962 (Unesco; reprinted from *Unesco bulletin for libraries* Vol. XVI, No. 1, Jan.–Feb. 1962)
GURNETT, John William & KYTE, Colin Henry John *Cassell's dictionary of abbreviations* 1966
GURNEY, Oliver Robert *The Hittites* 1952
GURR, Edward *The rational use of dyes in biology, and general staining methods* 1965
Synthetic dyes in biology, medicine and chemistry 1971
'GUTHRIE, John' (John Brodie) *The little country* 1935
GUTHRIE, Virgil B. *Petroleum products handbook* 1960
GUTHRIE-SMITH, William Herbert *Tutira: the story of a New Zealand sheep station* 1921
GUY, Albert Glasgow *Physical metallurgy for engineers* 1962
Guy's Hospital reports 1836–
GUYTON, Arthur Clifton *Textbook of medical physiology* 1956
GWYNNE-VAUGHAN, Helen Charlotte I. & BARNES, Bertie Frank *The structure and development of the fungi* 1927
GZOWSKI, Peter *Peter Gzowski's book about this country in the morning* 1974

H

HAAGNER, Alwin Karl & IVY, Robert Henry *Sketches of South African bird life* 1908

HAAS, Paul & HILL, Thomas George *An introduction to the chemistry of plant products* 1913
HABBERTON, John *The Jericho road: a story of western life* 1877
HABER, Heinz *Man in space* 1953
HACKFORTH-JONES, Frank Gilbert *One-one-one: stories of the Navy* 1942
Sixteen bells: stories 1946
'HACKSTON, James' (Harold Frederick N. Gye) *Father clears out* 1966
HADDON, Alfred Cort *The races of man and their distribution* 1909
HADDON, Archibald *Green room gossip* 1922
HADEN, Ernest Faber, HAN, M. S. & HAN, Y. W. *A resonance-theory for linguists* 1962
HADFIELD, Miles *British trees* 1957
HAGGAR, Reginald George *The concise encyclopedia of continental pottery and porcelain* 1960
A dictionary of art terms 1962
See also: MANKOWITZ, W. & HAGGAR, R. G.
HAGGARD, Henry Rider *Benita: an African romance* 1906
'HAGGARD, William' (Richard Henry Michael Clayton) *The arena* 1961
Closed circuit 1960
The hard sell 1965
The hardliners 1970
The high wire 1963
The power house 1966
The protectors 1972
HAGGIS, Geoffrey Harvey ed. *Introduction to molecular biology* 1964
HAGUE, Douglas Chalmers *Managerial economics: analysis for business decisions* 1969
'HAIG, Alec' *Peruvian printout* 1974
HAILEY, William Malcolm (Baron Hailey) *An African survey* 1938
— (rev. ed.) 1957
HALAS, John & MANVELL, Arnold Roger *The technique of film animation* 1959
HALBERSTAM, David *The powers that be* 1979
HALDANE, John Burdon Sanderson & HUXLEY, Julian Sorell *Animal biology* 1927
HALE, Edward Everett *If, yes, and perhaps* 1868
HALE, John *The grudge fight* 1964
HALE, Mason Ellsworth *The biology of lichens* 1967
HALE, Susan (1833–1910) *Letters* ed. C. P. Atkinson 1919
HALIBURTON, Thomas Chandler *The attaché; or, Sam Slick in England* 2 vols. 1843
— (2nd ser.) 2 vols. 1844
Sam Slick's wise saws and modern instances; or, what he said, did, or invented 2 vols. 1853
HALL, Abraham Oakley *The Manhattaner in New Orleans; or, Phases of 'Crescent City' life* 1851
'HALL, Adam' (Trevor Dudley-Smith) *The Kobra manifesto* 1976
The 9th directive 1966
The Striker portfolio 1969
The Tango briefing 1973
See also: 'TREVOR, Elleston'
HALL, Archibald John *The standard handbook of textiles* 1946
— (ed. 7) 1969
A student's textbook of textile science 1963
HALL, Bert & NILES, John Jacob *One man's war: the story of the Lafayette escadrille* 1929
HALL, Frederick Michael *An introduction to abstract algebra* 2 vols. 1966–9
HALL, James *Letters from the West* 1828
Statistics of the West, at the close of the year 1836 1836
HALL, James Norman *Kitchener's mob: the adventures of an American in the British Army* 1916
HALL, John Scoville ed. *Radar aids to navigation* 1947
HALL, John Whitney *Japan from prehistory to modern times* 1970
HALL, Mildred Lillington ed. *Newnes complete amateur photography* See under title
HALL, Richard Pinkham *Protozoology* 1953
HALL, Robert Anderson *External history of the Romance languages* 1974
Introductory linguistics 1964
HALL, Thomas Winthrop *Tales* 1899
HALL, William Thomas *Textbook of quantitative analysis* 1930
HALLAM, Anthony ed. *Planet earth: an encyclopedia of geology* 1977
HALLIDAY, David *Introductory nuclear physics* 1950
'HALLIDAY, Dorothy' (Dorothy Dunnett) *Dolly and the cookie bird* 1970
Dolly and the doctor bird 1971
Dolly and the nanny bird 1976
Dolly and the starry bird 1973
HALLIDAY, Michael Alexander Kirkwood, McINTOSH, A., & STREVENS, P. *The linguistic sciences and language teaching* 1964
HALLOCK, Charles *American club list and sportsman's glossary* 1878
HALPERT, Herbert & STORY, G. M. eds. *Christmas mumming in Newfoundland: essays in anthropology, folklore and history* 1969
HAM, Arthur Worth *Histology* 1950
HAMILTON, Alexander (1712–56) *Hamilton's itinerarium: being a narrative of a journey from Annapolis, Maryland, through Delaware [etc.]...1744* ed. A. B. Hart 1907
HAMILTON, Augustus *Maori art* 1901 (first published in 5 parts 1896–1900)
HAMILTON, Bruce *Too much of water* 1958
HAMILTON, Cosmo *Prisoners of hope* 1924
HAMILTON, Edward Walter *Diary 1880–1885* ed. D. W. R. Bahlman 2 vols. 1972
HAMILTON, Elizabeth Vidal *The Mordaunts: an eighteenth-century family* 1965
HAMILTON, Ferelith ed. *The world encyclopedia of dogs* 1971
HAMILTON, Francis *An account of the kingdom of Nepal* 1819
HAMILTON, Frederic Spencer *P.J.: the Secret Service boy* 1922
'HAMILTON, Gail' (Mary Abigail Dodge) *Gala-days* 1863
HAMILTON, Henrietta *Answer in the negative* 1959
HAMILTON, Ian *The man with the brown paper face* 1967
The thrill machine 1972
HAMILTON, Stanislaus Murray ed. *Letters to Washington, and accompanying papers* 5 vols. 1898–1902
HAMILTON, Walter *The Æsthetic Movement in England* 1882
HAMMERTON, John Alexander ed. *ABC of the RAF* 1941
ed. *Countries of the world* 6 vols. (originally published in 40 parts) 1924–5
HAMMETT, Samuel Dashiell *The Dain curse* 1929 (UK 1930)
The Maltese falcon 1930
Red harvest 1929
HAMMOND, Rolt *Mobile and movable cranes* 1963
HAMP, Eric P., HOUSEHOLDER, F. W., & AUSTERLITZ, R. eds. *Readings in linguistics II* 1966
HAMPTON, Christopher James *The philanthropist: a bourgeois comedy* 1970
Savages 1974
HAMSHER, Donald Henze ed. *Communication system engineering handbook* 1967
HAMSON, Denys Otto Harry *We fell among Greeks* 1946
'HAN SUYIN' (Elizabeth Comber) *A mortal flower, China: autobiography, history* 1966
HANCOCK, Lyn *There's a seal in my sleeping bag* 1972
HANCOCK, William Keith *Australia* 1930
Handbook of hardwoods, A 1956 See United Kingdom. Department of Scientific and Industrial Research. Forest Products Research Laboratory
— (ed. 2) 1972 See United Kingdom. Department of the Environment. Building Research Establishment
Handbook of softwoods, A 1957 See United Kingdom. Department of Scientific and Industrial Research. Forest Products Research Laboratory
HANDY, William Christopher ed. *Blues: an anthology* 1926
Father of the blues: an autobiography ed. A. Bontemps 1941 (UK 1957)
HANLEY, William *Blue dreams; or, The end of romance and the continued pursuit of happiness* 1971
HANNETT, John *Bibliopegia* 1835
HANSEN, Joseph *Fadeout* 1970 (UK 1972)
HANSON, John Lloyd *A dictionary of economics and commerce* 1965
HAPPÉ, Louis Bernard *Basic motion picture technology* 1971
Happy landings (Air Ministry) 1946–8
HARBEN, William Nathaniel *Abner Daniel* 1902
The Georgians 1904
Westerfelt 1901
HARBORD, Frank William & HALL, John William *The metallurgy of steel* 1904
— (ed. 7) 1923
HARCOURT, Palma *At high risk* 1977
A fair exchange 1975
HARDEN, Donald Benjamin *The Phoenicians* 1962
HARDY, Alister Clavering *The open sea, its natural history* Part I *The world of plankton* 1956
— Part II *Fish and fisheries* 1959
HARDY, Arthur Cobb & PERRIN, Fred Hiram *The principles of optics* 1932
HARDY, Thomas (1840–1928) *Collected letters* ed. R. L. Purdy & M. Millgate 1978–
A group of noble dames 1891
Late lyrics and earlier, with many other verses 1922
One rare fair woman: letters to Florence Henniker 1893–1922 ed. E. Hardy & F. B. Pinion 1972
HARDY, William George *Unfulfilled* 1952

HARE, Richard Mervyn *Freedom and reason* 1963
The language of morals 1952
HARGAN, James *Glossary of prison language* 1935 (typescript, referred to on p. 361 of *Journal of abnormal and social psychology*, December 1935)
HARGREAVES, Elisabeth *Fair green weed* 1972
HARGREAVES, Reginald *The enemy at the gate; famous sieges* 1945
HARKER, Alfred *The natural history of igneous rocks* 1909
Petrology for students: an introduction to the study of rocks under the microscope 1895
HARLEY, John Brian *Ordnance Survey maps: a descriptive manual* 1975
HARMAN, Richard Alexander, MILNER, A., & MELLERS, W. *Man and his music* 1962
HARMAR, Hilary *Pet Library's chihuahua guide* 1968
HARMON, Daniel Williams *A journal of voyages and travels in the interiour of North America* 1820
Harmsworth encyclopaedia, The ed. G. Sandeman 8 vols. (originally published in 40 parts) 1905–6
Harmsworth's wireless encyclopedia ed. O. J. Lodge 3 vols. (originally published in 24 parts) 1923–4
HARMSWORTH, Alfred Charles William, et al. *Motors and motor-driving* 1902
HARNEY, William Edward *Taboo* 1943
— (ed. 3) 1944
HARNWELL, Gaylord Probasco *Principles of electricity and electromagnetism* 1938
HARPER, Henry William *Letters from New Zealand, 1857–1911, being some account of life and work in the province of Canterbury, South Island* 1914
HARPER, John L. *Population biology of plants* 1977
Harpers and Queen 1970–
Harper's bazaar (New York) 1867–
— (UK ed.) 1929–70
Harper's weekly 1857–1916
HARRINGTON, Robert Edward *Quintain* 1977
HARRIS, Cora May *Eve's second husband* 1911
HARRIS, Frank See HARRIS, James Thomas Frank
HARRIS, Fred Henry ed. *Dartmouth out o' doors* 1913
HARRIS, Henry *Nucleus and cytoplasm* 1968
HARRIS, James Thomas Frank *My life and loves* 4 vols. 1922–7
HARRIS, Joel Chandler *Mingo, and other sketches in black and white* 1884
Nights with Uncle Remus 1883
Sister Jane, her friends and acquaintances 1896
Uncle Remus and his friends 1892
HARRIS, John Norman *The weird world of Wes Beattie* 1963 (UK 1964)
HARRIS, Leldon M. *An introduction to deepwater floating drilling operations* 1972
HARRIS, Miriam Coles *The tents of wickedness* 1907
HARRIS, Rosemary *The double snare* 1974
The nice girl's story 1968
HARRIS, Stanley *Fundamental principles of contract bridge* 1937
HARRIS, Walter Kilroy *Outback in Australia; or, Three Australian overlanders* 1913
HARRIS, William Cornwallis *The wild sports of southern Africa: being the narrative of an expedition from the Cape of Good Hope..to the Tropic of Capricorn* 1839
HARRIS, Zellig Sabbettai *Methods in structural linguistics* 1951
HARRISON, A. W. C. *The manufacture of lakes and precipitated pigments* 1930
HARRISON, Constance Cary *Woman's handiwork in modern homes* 1881
HARRISON, Harry Max *The Technicolor time machine* 1967 (UK 1968)
HARRISON, Henry Sydnor *Queed* 1911
HARRISON, Michael *Reported safe arrival* 1943
Harrison Mayer. A complete service to the craft potter [Catalogue of Harrison Mayer Ltd., Meir, Stoke-on-Trent] *c* 1977
HARRISSON, Barbara *Orang-utan* 1962
HARRISSON, Thomas Harnett & MADGE, Charles Henry eds. *War begins at home* 1940
HARROD, Leonard Montague *The librarians' glossary* 1938
— (ed. 2) 1959
— (ed. 3) 1971
HARROW, Benjamin & SHERWIN, Carl P. eds. *Textbook of biochemistry* 1935
HART, Basil Henry Liddell *Europe in arms* 1937
HART, Frances Noyes *The Bellamy trial* 1927
HART, Fred H. *The Sazerac lying club: a Nevada book* 1878
HART, Horace *Notes on a century of typography at the University Press, Oxford, 1693–1794* 1900
HART, Jerome *A vigilante girl* 1910

HART, Norman de Villiers *The bridge players' bedside book* 1939
HARTE, Francis Bret *Gabriel Conroy* 1876 (US 1 vol., UK 3 vols.)
The story of a mine 1877
— & 'TWAIN, MARK' (S. L. Clemens) *Sketches of the sixties: being forgotten material now collected..from The Californian* ed. J. Howell 1926
HARTLEY, Leslie Poles *Eustace and Hilda* 1947
The go-between 1953
The hireling 1957
A perfect woman 1955
Two for the river, and other stories 1961
HARTMANN, Reinhard Rudolf K. & STORK, Francis C. *Dictionary of language and linguistics* 1972
HARTREE, Douglas Rayner *Calculating instruments and machines* 1949 (UK 1950)
HARTSHORNE, Albert *Old English glasses* 1897
HARTSHORNE, Charles Henry *English medieval embroidery* 1848
Harvard studies in classical philology 1890–
Harvard University. Computation Laboratory *Annals* 1946–
'HARVESTER, Simon' (Henry Gibbs) *The Chinese hammer* 1960
A corner of the playground 1973
Treacherous road 1966
HARVEY, Peter & BOHLMAN, Kenneth John *Stereo F.M. radio handbook* 1974
HARVEY, Ruth *Curtain time* 1949
HASELGROVE, Maurice Lawrence *Photographers' dictionary* 1962
HASTINGS, James ed. *Encyclopædia of religion and ethics* See under title
HASTINGS, Lewis *Dragons are extra* 1947
HASTINGS, Macdonald *Cork and the serpent* 1955
HASTON, Dougal *In high places* 1972
HATCH, Frederick Henry & RASTALL, Robert Heron *The petrology of the sedimentary rocks* 1913
— (ed. 4, revised by J. T. Greensmith) 1965
HATFIELD, James Taft, LEOPOLD, W., & ZIEGLSCHMID, A. J. F. eds. *Curme volume of linguistic studies* 1930
HAURWITZ, Bernhard *Dynamic meteorology* 1941
HAUSMAN, Louis *Clinical neuroanatomy, neurophysiology and neurology* 1958
HAWKES, Jessie Jacquetta & HAWKES, Charles Francis Christopher *Prehistoric Britain* 1943
— (rev. ed.) 1947
HAWKEY, Raymond & BINGHAM, Roger *Wild card* 1974
HAWKINS, Benjamin (1754–1816) *A sketch of the Creek country, in 1798 and 1799* (Georgia Historical Society) 1848
HAWKINS, Charles Caesar & WALLIS, F. *The dynamo: its theory, design and manufacture* 1893
HAWKINS, John & HAWKINS, Ward *Death watch, and The missing witness* 1958 (UK 1959)
HAWKINS, Nehemiah, et al. *Hawkins' electrical guide* 1914
'HAY, Ian' (John Hay Beith) *The first hundred thousand: being the unofficial history of a unit of 'K(1)'* [Kitchener's First Army] 1915
Housemaster 1936
A knight on wheels 1914
The last million 1919
The lighter side of school life 1914
The poor gentleman 1928
'The right stuff': some episodes in the career of a North Briton 1908
A safety match 1911
— & 'ARMSTRONG, Anthony' (A. A. Willis) *Orders are orders* 1933
HAY, John Milton *The bread-winners: a social study* 1884
HAY, Malcolm Vivian *Foot of pride: the pressure of Christendom on the people of Israel for 1900 years* 1950
HAY, Roy & SYNGE, P. M. *The dictionary of garden plants* 1969
HAY, William Delisle *Brighter Britain! or, Settler and Maori in northern New Zealand* 2 vols. 1882
Three hundred years hence; or, A voice from posterity 1881
HAYASHI, Takashi ed. *Olfaction and taste II* 1967 (Proceedings of the 2nd International Symposium on Olfaction and Taste, Tokyo, 1965)
HAYCRAFT, Howard *Murder for pleasure: the life and times of the detective story* 1942
HAYDEN, Arthur *Chats on old furniture: a practical guide for collectors* 1905
HAYDON, Benjamin Robert (1786–1846) *The life of Haydon, from his autobiography and journals* ed. T. Taylor 3 vols. 1853
HAYES, Arnold Richard W. *Revision physics for sixth forms* 1962
HAYES, Augustus Allen *New Colorado and the Santa Fé trail* 1880 (UK 1881)
HAYES, Roy *The Hungarian game* 1973

HAYES, William *The genetics of bacteria and their viruses* 1964
HAYGARTH, Henry William *Recollections of bush life in Australia during a residence of eight years in the interior* 1848
HAYMAKER, Webb Edward tr. *R. Bing's Textbook of nervous diseases* 1939
HAYS, David G. *Introduction to computational linguistics* 1967
HAYWARD, Charles Brian *Practical aeronautics* 1912
HAYWARD, Harry Maxwell & HARARI, Manya tr. *B. Pasternak's Dr. Zhivago* 1958
HAYWARD, Helena ed. *The Connoisseur's handbook of antique collecting: a dictionary of furniture, silver, ceramics, glass, fine art, etc.* 1960
HAYWARD, John Davey *Prose literature since 1939* 1947
HAZARD, Thomas Benjamin (1756–1845) *Nailer Tom's diary* ed. C. Hazard 1930
Heal and Son *Heal's catalogue 1853–1934* 1972 [facsimile reproductions from catalogues]
HEALD, Frederick Deforest *Introduction to plant pathology* 1937
Manual of plant diseases 1926
HEALD, Timothy Villiers *Deadline* 1975
Just desserts 1977
Let sleeping dogs die 1976
HEALEY, Edna *Lady unknown: the life of Angela Burdett-Coutts* 1978
HEARN, Patricio Lafcadio T. C. *In ghostly Japan* 1899
Japan: an attempt at interpretation 1904
Kokoro: hints and echoes of Japanese inner life 1896
Kottō: being Japanese curios, with sundry cobwebs 1902
HEARNE, John *Stranger at the gate* 1956
Hearst's international 1925–
Heart: a journal for the study of the circulation 1909–33
HEATH, Peter tr. *G. A. Wetter's Soviet ideology today* 1966
HEAVISIDE, Oliver *Electromagnetic theory* 3 vols. 1893–9
HECHT, Ben & MACARTHUR, Charles *The front page* 1928
HEDGECOE, John *The photographer's handbook* 1977
HEFFNER, Roe-Merrill S. *General phonetics* 1949
HEFFRON, Dorris *Crusty crossed* 1976
HEFLIN, Woodford Agee ed. *Aerospace glossary* 1959
ed. *The United States Air Force dictionary* 1956
HEILBRON, Ian Morris ed. *Dictionary of organic compounds* 3 vols. 1934–7
HEIN, Leonard William *An introduction to electronic data processing for business* 1961
HEINLEIN, Robert Anson *The door into summer* 1957 (UK 1960)
HEINRICH, Eberhardt William *Microscopic petrography* 1956
HEITLER, Walter *The quantum theory of radiation* 1936
HELLER, Joseph *Catch-22* 1961 (UK 1962)
Something happened 1974
HELLMAN, Lillian *The little foxes* 1939
Pentimento: a book of portraits 1973 (UK 1974)
An unfinished woman: a memoir 1969
HELLYER, Arthur George Lee *T. W. Sanders' Encyclopaedia of gardening* (ed. 22) 1952
The encyclopaedia of garden work and terms 1954
Practical gardening for amateurs 1935
HELME, Elizabeth tr. *F. Le Vaillant's Travels from the Cape of Good-Hope, into the interior parts of Africa* 2 vols. 1790
HELMER, Olaf tr. *A. Tarski's Introduction to logic and to the methodology of deductive sciences* 1941
HEMINGWAY, Ernest *Across the river and into the trees* 1950
Death in the afternoon 1932
A farewell to arms 1929
The fifth column [a play], *and The first forty-nine stories* 1938 (UK 1939)
For whom the bell tolls 1940
Green hills of Africa 1935 (UK 1936)
In our time 1924
— (enlarged ed.) 1925 (UK 1926)
Men without women 1927 (UK 1928)
The old man and the sea 1952
The sun also rises 1926 (UK ed. with title *Fiesta* 1927)
To have and have not 1937
The torrents of spring: a romantic novel in honour of the passing of a great race 1926 (UK 1933)
HEMPSTEAD, Joshua *Diary, 1711–1758* 1901
HENDERSON, A. G. tr. *V. Cousin's The philosophy of the beautiful: lectures* 1854
HENDERSON, Alexander Morell & PARSONS, Talcott tr. *M. Weber's Theory of social and economic organization* 1947

HENDERSON, David Kennedy & GILLESPIE, Robert Dick *A text-book of psychiatry for students and practitioners* 1927
— (ed. 9) 1962
HENDERSON, George C. *Keys to crookdom* 1924
HENDERSON, Stephen *Understanding the new Black poetry: Black speech and Black music as poetic references* 1973
HENDRICK, Burton Jesse *The life and letters of Walter H. Page* 3 vols. 1922–5
HENISSART, Paul *Winter quarry* 1976
HENNEY, Julian Keith *Principles of radio* 1929
ed. *The radio engineering handbook* 1933
— (ed. 5) 1959
— & DUDLEY, Beverley eds. *Handbook of photography* 1939
HENNING, Rachel (1826–1914) *Letters* ed. D. Adams 1966
HENROT, Thérèse *Belgium* (English version by R. E. Wolf and the author) 1961
HENRY, Frank Souder *Printing for school and shop* 1917
'HENRY, Joan' (Constance Ann Standage) *Who lie in gaol* 1952
'HENRY, O.' (William Sydney Porter, 1862–1910) *Cabbages and kings* 1904
The four million 1906 (UK 1916)
The gentle grafter 1908 (UK 1928)
Heart of the west 1907 (UK 1912)
Options 1909 (UK 1916)
Roads of destiny 1909
Rolling stones 1912 (UK 1916)
Strictly business 1910 (UK 1917)
The trimmed lamp 1907 (UK 1915)
Whirligigs 1910
HENRY, Thomas Anderson *The plant alkaloids* 1913
HENTY, George Alfred *Through Russian snows: a story of Napoleon's retreat from Moscow* 1896
HEPWORTH, Cecil Milton *Animated photography: the ABC of the cinematograph* 1897
HERAK, Milan & STRINGFIELD, V. T. eds. *Karst: important karst regions of the northern hemisphere* 1972
HERBERT, Alan Patrick *Holy deadlock* 1934
Independent Member 1950
Laughing Ann, and other poems 1925
Plain Jane 1927
The water gipsies 1930
What a word! 1935
HERBERT, Alfred Xavier *Capricornia* 1938 (UK 1939)
HERBERT, Thomas Ernest *Telegraphy: a detailed exposition of the telegraph system of the British Post Office* 1906
Telephony: an elementary exposition of the telephone system of the British Post Office 1923
— (new ed., by T. E. Herbert & W. S. Procter) 2 vols. 1932–40
— (new ed., by J. Atkinson) 2 vols. 1948–50
HERBERT, Xavier See HERBERT, Alfred Xavier
Here and now: an independent monthly review 1949–
HEREN, Louis *Growing up on 'The Times'* 1978
HERGESHEIMER, Joseph *The bright shawl* 1922 (UK 1923)
HERON, Alastair ed. *Towards a Quaker view of sex: an essay by a group of Friends* 1963
HERON, James *The Celtic Church in Ireland* 1898
HERON, Patrick *The changing forms of art* 1955
'HERRIOT, James' (J. A. Wight) *It shouldn't happen to a vet* 1972
HERSKOWITZ, Irwin Herman *Genetics* 1962
HERVEY, George F. *A handbook of card games* 1963
HESS, Fr. Cuthbert (Lawrence Anthony Hess) ed. *God and the supernatural: a Catholic statement of the Christian faith* 1920
HEWER, Evelyn Everard *Text-book of histology for medical students* 1937
HEWITT, James *Yoga* 1960
HEWLETT, Richard Tanner *A manual of bacteriology, clinical and applied* 1898
— (ed. 2) 1902
HEYER, Georgette *A blunt instrument* 1938
Death in the stocks 1935
False colours 1963
They found him dead 1937
HEYERDAHL, Thor *Fatu-Hiva: back to nature* 1974
The Kon-Tiki expedition See LYON, Francis Hamilton
HICK, John Harwood ed. *The myth of God incarnate* 1977
HICKEY, William (1749?–1830) *Memoirs* ed. A. Spencer 4 vols. 1913–25
— ed. P. Quennell 1960
Hi-fi answers 1972–
Hi-fi sound 1967–77
HIGGIN, Louis *Handbook of embroidery* ed. M. Alford 1880
HIGGINS, George Vincent *A city on a hill* 1975
Dreamland 1977
The friends of Eddie Coyle 1972
The judgement of Deke Hunter 1976

'HIGH JINKS, JUNIOR' (Harold Poe Swartwood) *Choice slang* 1914
High times (New York) 1974–
HIGHAM, Edward Spencer & HIGHAM, William Richard Joseph *High speed rugby* 1960
HIGHAM, Robert Robin Alger *A handbook of papermaking* 1963
Higher education: report See: United Kingdom. Parliamentary papers. Committee on Higher Education
HIGHSMITH, Patricia *The tremor of forgery* 1969
HILDICK, Edmund Wallace *The boy at the window* 1960
The loop 1977
HILDRETH, James *Dragoon campaigns to the Rocky Mountains* 1836
HILER, Hilaire *Notes on the technique of painting* 1934
HILGARD, Ernest Ropiequet *Theories of learning* 1948
— (ed. 2) 1956 (UK 1958)
HILGARD, Eugene Waldemar *Soils: their formation, properties, composition, and relations to climate and plant growth in the humid and arid regions* 1906
HILGENDORF, Frederick William *Weeds of New Zealand, and how to eradicate them* 1926
HILL, Albert Frederick *Economic botany* 1937
HILL, Archibald Anderson *Introduction to linguistic structures* 1958
HILL, Archie *Summer's end* 1976
HILL, Beth & HILL, Ray *Spirit in stone: petroglyphs of the Northwest Coast Indians* 1974
HILL, Peter *The hunters* 1976
HILL, Reginald *A fairly dangerous thing* 1972
Ruling passion 1973
HILL, Susan *Gentleman and ladies* 1976
Strange meeting 1971
HILLARY, Edmund *High adventure* 1955
HILLARY, Richard Hope *The last enemy* 1942
HILLEBRAND, William *Flora of the Hawaiian Islands* 1888
HILLIARD, Noel Harvey *Maori girl* 1960
A piece of land 1963
HILLIS, Marjorie (Marjorie Hillis Roulston) *Orchids on your budget; or, Live smartly on what have you* 1937 (UK 1938)
HILLYARD, George Whiteside *Forty years of first-class lawn tennis* 1924
HILTON, James *Lost horizon* 1933
See also: 'TREVOR, Glen'
HILTON, John Buxton *Gamekeeper's gallows* 1976
HILTON, Timothy *The Pre-Raphaelites* 1970
HIMES, Chester *Black on black: Baby sister, and selected writings* 1973
Blind man with a pistol 1969
The heat's on 1966
HIND, Herbert Lloyd *Brewing* 2 vols. 1938–40
HINDMARSH, William Russell ed. *Atomic spectra* 1967
Hindu, The (Madras) 1878–
Hindustan Times (New Delhi) 1924–
HINE, Gerald John & BROWNELL, G. L. eds. *Radiation dosimetry* 1956
HINSIE, Leland Earl & SHATZKY, Jacob *Psychiatric dictionary, with encyclopedic treatment of modern terms* 1940
— (ed. 3, by L. E. Hinsie & R. J. Campbell) 1960
HINXMAN, Margaret *End of a good woman* 1976
HISCOCK, Eric Charles *Around the world in Wanderer III* 1956
HOBART, George Vere *Jim Hickey: a story of the one-night stands* 1904
See also: 'MCHUGH, Hugh'
HOBSON, Hank *Mission house murder* 1959
HOBSON, John Atkinson *The problem of the unemployed* 1896
HOBSON, Laura Keane (Zametkin) *The celebrity* 1951 (UK 1953)
'HOBSON, Polly' (Julia Evans) *Titty's dead* 1968
HOBSON, Richmond Pearson *Nothing too good for a cowboy* 1955
HOCKETT, Charles Francis *A course in modern linguistics* 1958
ed. *A Leonard Bloomfield anthology* 1970
A manual of phonology 1955
HODGE, Frederick Webb ed. *Handbook of American Indians north of Mexico* 2 vols. 1907–10
HODGE, Herbert *Cab, sir?* 1939
HODGE, Hiram C. *Arizona as it is* 1877
HODGES, Henry *Artifacts: an introduction to early materials and technology* 1964
HODGINS, Eric *Mr. Blandings builds his dream house* 1946 (UK 1947)
HODGKINSON, Harry *Doubletalk: the language of Communism* 1955
HODKIN, Frederick William & COUSEN, Arnold *A textbook of glass technology* 1925
HODSON, James Lansdale *Our two Englands* 1936
HODZA, Aaron C. & FORTUNE, George *Shona praise poetry* 1979
HOFFMAN, Charles Fenno *A winter in the west* 2 vols. 1835 (UK ed., also 2 vols. 1835, with title —— *in the far west*)
HOGAN, Jeremiah Joseph *The English language in Ireland* 1927
An outline of English philology 1934
HOGARTH, David George *The Nearer East* 1902
HOGG, Alfred George *Redemption from this world; or, The supernatural in Christianity* 1922
HOGGART, Herbert Richard *Auden: an introductory essay* 1951
The uses of literacy: aspects of working-class life, with special reference to publications and entertainments 1957
HOLBROOK, David Kenneth *Flesh wounds* 1966
HOLDEN, Anne *The girl on the beach* 1973
HOLIDAY, Billie *Lady sings the blues* 1956 (UK 1958; new impr. 1973)
Holiday 1946–
HOLLAND, John Henry *The useful plants of Nigeria* 4 vols. 1908–22
HOLLAND, Ray *Self and social context* 1977
HOLLANDER, Zander ed. *The modern encyclopedia of basketball* 1969
HOLLEY, Marietta E. *My opinions and Betsey Bobbet's* 1873
HOLLINSHEAD, William Henry *Textbook of anatomy* 1962
HOLLOWAY, Christopher John *Language and intelligence* 1951
HOLMAN-HUNT, Diana *My grandmothers and I* 1960
HOLMES, Arthur *The nomenclature of petrology, with references to selected literature* 1920
Principles of physical geology 1944
— (ed. 2) 1965
HOLMES, Harry Nicholls *Introductory colloid chemistry* 1934
HOLMES, John Clellon *Nothing more to declare* 1968
HOLMES, Mary Jane *Tempest and sunshine; or, Life in Kentucky* 1854
HOLMES, Oliver Wendell (the younger) *The Pollock–Holmes letters* See: POLLOCK, Frederick
—— & LASKI, Harold Joseph *Holmes–Laski letters: the correspondence of Mr. Justice Holmes and Harold J. Laski, 1916–1935* ed. M. DeW. Howe 2 vols. 1953
HOLMES, Sarah Katherine (Stone) *Brokenburn: the journal of Kate Stone, 1861–1868* ed. J. Q. Anderson 1955
HOLMES, Thomas K. *The man from Tall Timber* 1919
HOLROYDE, Peggy *Indian music* 1972
HOLT, Edwin Bissell *The Freudian wish and its place in ethics* 1915
HOLTBY, Winifred *Poor Caroline* 1931
South Riding 1936
HOME, Lord (Alexander Frederick Douglas-Home) *The way the wind blows* 1976
'HOME, Michael' (Charlie Christopher Bush) *The house of shade* 1942
Home chat 1895–
Homes and gardens 1923–
HONE, Joseph *The flowers of the forest* 1980
The sixth directorate 1975
HOOKER, Joseph Dalton *Handbook of the New Zealand flora* 2 parts 1864–7
See also: HUXLEY, Leonard
HOOL, George Albert & JOHNSON, Nathan Clarke *Concrete engineers' handbook* 1918
HOOPER, Johnson Jones *Some adventures of Captain Simon Suggs* 1845
— (another ed.) 1846
HOOPER, Robert *Lexicon medicum; a new medical dictionary* 1802
— (ed. 7, rev. by K. Grant) 1839
'HOPE, Anthony' (Anthony Hope Hawkins) *The prisoner of Zenda* 1894
Tristram of Blent 1901
HOPE, William Edward Stanton *Diggers' paradise* 1956
HOPKINS, Gerard Manley (1844–89) *Further letters* ed. C. C. Abbott 1938
— (ed. 2, by C. C. Abbott) 1956
Journals and papers ed. G. Storey 1959
Letters to Robert Bridges ed. C. C. Abbott 1935
— (2nd rev. impr.) 1955
Note-books and papers ed. H. House 1937
Poems ed. R. Bridges 1918
— (ed. 3, by W. H. Gardner) 1948
— (ed. 4, by W. H. Gardner & N. H. MacKenzie) 1967
Sermons and devotional writings ed. C. Devlin 1959
A vision of the mermaids (facsimile ed. of manuscript text dated 1862) 1929
HOPKINS, Keith ed. *Hong Kong, the industrial colony: a political, social and economic survey* 1971
HOPKINSON, Diana *The incense-tree* 1968
Horizon: a review of literature and art 1940–50
HORNSBY, Henry *Lonesome valley* 1949
HORNUNG, Ernest William *The amateur cracksman* 1899
The black mask 1901
HOROWITZ, Irving Lewis ed. *Masses in Latin America* 1970
ed. *The new sociology* 1964
HORROBIN, David Frederick *Science is God* 1969
Horse and hound 1884–
HORSLEY, Terence *Find, fix and strike: the work of the Fleet Air Arm* 1943
Hortus third: a concise dictionary of plants cultivated in the United States and Canada (Liberty Hyde Bailey Hortorium) 1976
HORWILL, Herbert William *A dictionary of modern American usage* 1935
— (ed. 2) 1944
HORWOOD, Harold Andrew *Newfoundland* 1969
HOSTETTER, Gordon L. & BEESLEY, Thomas Q. *It's a racket!* 1929
HOSTETTLER, Rudolf, et al. *Technical terms of the printing industry* 1949
— (ed. 3) 1959
Hot car 1968–
Hotel world, The 1875–
House and garden (New York) 1901–
UK ed. 1920–4
— (new series) 1946–
HOUSEHOLD, Geoffrey Edward West *Doom's caravan* 1971
Rogue male 1939
HOUSEHOLDER, Fred Walter *Linguistic speculations* 1971
—— & SAPORTA, Sol eds. *Problems in lexicography* [report of the Conference on Lexicography, Indiana University, 1960] 1962
Housewife 1939–
HOUSMAN, Alfred Edward (1859–1936) *Collected poems* 1939
Letters ed. H. Maas 1971
More poems 1936
HOUSMAN, Laurence *The unexpected years* 1937
HOUSTON, Margaret Bell *The witch man* 1922
Houston Chronicle (Houston, Texas) 1901–
Hovering craft and hydrofoil 1961–
HOVLAND, Carl I., LUMSDAINE, A. A., & SHEFFIELD, F. D. *Experiments on mass communication* 1949
HOWARD, Alexander Liddon *A manual of the timbers of the world* 1920
— (rev. ed.) 1934
HOWARD, Elizabeth Jane *After Julius* 1965
'HOWARD, Hartley' (Leopold Horace Ognall) *Highway to murder* 1973
Nice day for a funeral 1972
HOWARD, John A. *Aerial photo-ecology* 1970
HOWARD, Richard tr. *S. de Beauvoir's Force of circumstance* 1965
Howard journal 1921–
HOWARTH, Thomas Graham *South's British butterflies, based extensively on the classic by Richard South* 1973
HOWAT, Gerald Malcolm D. & TAYLOR, A. J. P. eds. *Dictionary of world history* 1973
HOWE, Ellic ed. *The London compositor: documents relating to wages, working conditions and customs of the London printing trade 1785–1900* 1947
HOWE, Henry *Historical collections of Ohio* 1847
HOWELLS, William Dean *Literature and life: studies* 1902
HOWITT, William *The rural and domestic life of Germany* 1842
Howitt's journal of literature and popular progress 1847–8
HOWSON, Albert Geoffrey *A handbook of terms used in algebra and analysis* 1972
HOYLE, Edmond *Hoyle's games modernized* (new ed., rev. by L. H. Dawson) 1923
— (ed. 20, rev. by L. H. Dawson) 1950
HOYLE, Fred *The black cloud* 1957
Frontiers of astronomy 1955
Galaxies, nuclei and quasars 1965 (UK 1966)
Ice 1981
The nature of the universe: a series of broadcast lectures 1950
HUBBARD, Clifford Lionel Barry *Dogs in Britain* 1948
The observer's book of dogs 1945
HUBBARD, Philip Maitland *The custom of the country* 1967 (UK 1969)
High tide 1970 (UK 1971)
Picture of Millie 1964
HUBBLE, Edwin Powell *The realm of the nebulæ* 1936
HUDSON, Charles Thomas & GOSSE, P. H. *The rotifera* 2 vols. & Suppl. 1886–9
HUDSON, Derek Rommel *Munby: man of two worlds. The life and diaries of Arthur J. Munby 1828–1910* 1972
HUDSON, Kenneth *The jargon of the professions* 1978
HUDSON, William Henry *Far away and long ago: a history of my early life* 1918
Green mansions: a romance of the tropical forest 1904

Hudson's Bay Record Society *Publications* 1938–
HUEFFER, Ford Madox *See* FORD, Ford Madox
HUGHES, Arthur Frederick William *The mitotic cycle: the cytoplasm and nucleus during interphase and mitosis* 1952
HUGHES, Dorothy Belle *The expendable man* 1963 (UK 1964)
HUGHES, George Edward & LONDEY, David George *The elements of formal logic* 1965
HUGHES, Mary Vivian *A London child of the seventies* 1934
HUGHES, Richard Arthur Warren *A high wind in Jamaica* 1929
In hazard 1938
HUGHES, Ted *Crow: from the life and songs of the crow* 1970
The hawk in the rain 1957
Lupercal 1960
HUGHES, Thomas Patrick *A dictionary of Islam* 1885
HUKE, Douglas Wynne *Introduction to natural and synthetic rubbers* 1961
HULL, Clark Leonard *Principles of behavior: an introduction to behavior theory* 1943
HULL, Richard Francis Carrington, et al. tr. *C. G. Jung's Collected works* 20 vols. 1953–79
HULME, Alfred Cresswell *The biochemistry of fruits and their products* 2 vols. 1970–1
HULME, Kathryn Cavarly *The nun's story* 1956
Human world: a quarterly review of English letters 1970–
HUMPHREY, George *Thinking* 1951
HUMPHREY, John Herbert & WHITE, R. G. *Immunology for students of medicine* 1963
HUMPHREYS, David *The Yankey in England* 1815
HUMPHRIES, Derek W. & HUMPHRIES, Evelyn E. tr. *H. & G. Termier's Erosion and sedimentation* 1963
HUNT, John Leslie & PRINGLE, Alan George *Service slang* 1943
HUNT, Robert Nigel Carew *A guide to Communist jargon* 1957
HUNT, Robert William Gainer *The reproduction of colour* 1957
Hunt's merchants' magazine and commercial review See *Merchants' magazine and commercial review*
HUNTER, Alan James Herbert *Gently at a gallop* 1971
Gently coloured 1969
Gently continental 1967
Gently French 1973
HUNTER, John Marvin ed. *The trail drivers of Texas* 1920
'HUNTER, Matthew' *The Cambridgeshire disaster* 1967
HUNTER, Robert, et al. *The encyclopædic dictionary* See under title
HUNTING, Henry Gardner *The vicarion* 1926
HURREN, Bernard John *Eastern Med.* 1943
HURST, Fannie *Humoresque: a laugh on life with a tear behind it* 1919
HURSTON, Zora Neale *Mules and men* 1935
— (another ed.) 1970
HUSCHKE, Ralph E. ed. *Glossary of meteorology* 1959
HUSKEY, Harry Douglas & KORN, G. A. *Computer handbook* 1962
Hutching's illustrated California magazine 1856–61
HUTCHINGS, Ernest A. D. *A survey of printing processes* 1970
HUTCHINSON, Arthur Stuart Menteth *If winter comes* 1921
Once aboard the lugger: the history of George and his Mary 1908
One increasing purpose 1925
This freedom 1922
HUTCHINSON, George Evelyn *A treatise on limnology* 1957–
HUTCHINSON, Horace Gordon *Golfing* 1893
HUTCHINSON, Robert William *A first course in wireless* 1926
Hutchinson's pictorial history of the war ed. W. V. Hutchinson 1939–45
HUTCHISON, Isobel Wylie *North to the rime-ringed sun: being the record of an Alaskan-Canadian journey made in 1933–34* 1934
HUTTEN, Ernest Hirschlaff *The language of modern physics* 1956
HUTTON, John Henry *Caste in India: its nature, function and origins* 1946
HUXLEY, Aldous Leonard (1894–1963) *Adonis and the alphabet* 1956
After many a summer 1939
Along the road: notes and essays 1925
Antic hay 1923
Ape and essence 1948 (UK 1949)
Beyond the Mexique Bay 1934
Brave new world 1932
Brave new world revisited 1958 (UK 1959)
Brief candles 1930
Crome yellow 1921
Do what you will: essays 1929
Ends and means: an enquiry into the nature of ideals and into the methods employed for their realization 1937
Essays new and old 1926
Eyeless in Gaza 1936
The genius and the goddess 1955
Grey eminence: a study in religion and politics 1941
Island 1962
Jesting Pilate: the diary of a journey 1926
Leda 1920
Letters ed. Grover Smith 1969
Music at night, and other essays 1931
The olive tree, and other essays 1936
The perennial philosophy 1945 (UK 1946)
Point counter point 1928
Proper studies 1927
Themes and variations 1950
Those barren leaves 1925
Time must have a stop 1944 (UK 1945)
Vulgarity in literature 1930
HUXLEY, Anthony Julian *An illustrated history of gardening* 1978
Plant and planet 1974
HUXLEY, Elspeth Josceline *Back street new worlds* 1964
HUXLEY, Julian Sorell *Bird-watching and bird behaviour* 1930
Essays in popular science 1926
Essays of a biologist 1923
Evolution in action 1953
Evolution: the modern synthesis 1942
Evolutionary ethics 1943
The individual in the animal kingdom 1912
On living in a revolution 1944
Problems of relative growth 1932
Religion without revelation 1927
— (rev. ed.) 1957
TVA: adventure in planning 1943
Unesco: its purpose and philosophy 1946
The uniqueness of man 1941
What dare I think? 1931
—— & HADDON, Alfred Cort *We Europeans* 1935
See also: HALDANE, John Burdon Sanderson
HUXLEY, Leonard *Life and letters of Sir Joseph Dalton Hooker* (based on materials collected and arranged by Lady Hooker) 2 vols. 1918
HVISTENDAHL, Herbert Selwyn *Engineering units and physical quantities* 1964
HYAMS, Edward Solomon *Gentian violet* 1953
Into the dream 1957
Soil and civilization 1952
Taking it easy 1958
HYAMS, Joseph *The pool* 1978
'HYDE, Robin' (Iris Guiver Wilkinson) *Check to your king: the life history of Charles, Baron de Thierry, King of Nukahiva, Sovereign Chief of New Zealand* 1936
Nor the years condemn 1938
Passport to hell: the story of John Douglas Stark, Bomber, Fifth Regiment, New Zealand Expeditionary Forces 1936
HYLAND, Henry Stanley *Top bloody secret* 1969
Who goes hang? 1958
HYLANDER, Clarence John *The Macmillan wild flower book* 1954
HYMAN, Libbie Henrietta *Comparative vertebrate anatomy* 1922
— (ed. 2) 1942
The invertebrates 5 vols. 1940–59
HYNE, Charles John C. W. *Mr. Horrocks, purser* 1902
HYNES, Martin *Medical bacteriology* (ed. 7) 1961
— (ed. 8) 1964
For earlier eds. see WHITBY, Sir Lionel E. H.
HYNES, Samuel Lynn *The Auden generation: literature and politics in England in the 1930's* 1976

I

IBM journal of research and development 1957–
IEEE *See* Institute of Electrical and Electronics Engineers
IRE *See* Institute of Radio Engineers
I.U.P.A.C. *See* International Union of Pure and Applied Chemistry
I believe: the personal philosophies of certain eminent men and women of our time [by W. H. Auden et al.] 1939 (UK 1940)
ICKES, Harold LeClair (1874–1952) *Secret diary* 3 vols. 1953–4 (UK 1955)
IDDINGS, Joseph Paxson *Igneous rocks: composition, texture and classification, description and occurrence* 2 vols. 1909–13
Rock minerals 1906
Idler magazine (title varies) 1894–1911
IDRIESS, Ion Llewellyn *The cattle king* 1936
The great boomerang 1941
In crocodile land: wandering in Northern Australia 1946
Isles of despair 1947
Lasseter's last ride: an epic of Central Australian gold discovery 1931
Ilkeston Advertiser and Erewash Valley Weekly News 1881–
Illinois. Department of Agriculture *Transactions* 1855–1921 (1855–71 with title *Transactions of the Illinois State Agricultural Society*)
Illinois State Historical Society *Journal* 1908–
Transactions 1899–
Illuminating Engineering Society (U.S.) *Transactions* 1906–40
Illustrated weekly of India (Bombay) 1879–
IMMS, Augustus Daniel *A general textbook of entomology* 1925
— (ed. 9) 1957 *See* RICHARDS, Owain Westmacott & DAVIES, R. G.
Immunology 1958–
Incorporated linguist 1962–
Independent, The (New York; later Boston, Mass.) 1848–1928
Index-Journal (Greenwood, South Carolina) 1897–
Indian journal of medical research 1913–
Indian music journal 1964–
Indiana. House of Representatives *Journal* 1816–46
Indiana Historical Society *Publications* 1895–
Indiana magazine of history 1905– (1905–13 with title *Indiana quarterly magazine of history*)
Industrial and engineering chemistry 1923–46
Industrial chemist and chemical manufacturer, The 1925–
INGE, William Ralph *Lay thoughts of a dean* 1926
INGRAHAM, Joseph Holt *The South-West, by a Yankee* 2 vols. 1835
INGRAM, George *Cockney cavalcade* 1935
'Stir' 1933
'Stir' train 1935
INGRAM, J. S. *The Centennial Exposition, described and illustrated* 1876
Ink 1971–2
Inland printer 1883–
Inlander, The (University of Michigan) 1891–1931 (title varies)
Inner life, The: essays in liberal evangelicalism 2nd ser. 1925
INNES, Hammond *The big footprints* 1977
The blue ice 1948
Campbell's kingdom 1952
'INNES, Michael' (John Innes Mackintosh Stewart) *Appleby plays chicken* 1956
Appleby's answer 1973
Appleby's other story 1974
An awkward lie 1971
Christmas at Candleshoe 1953
The daffodil affair 1942
Death at the President's lodging 1936
Hamlet, revenge! 1937
The journeying boy 1949
Old Hall, New Hall 1956
The open house 1972
Operation Pax 1951
See also under real name
Inside Kenya today 1953–
Institut Pasteur *Annales* 1887–
Institute of Cost and Management Accountants *Terminology of management and financial accountancy* 1974
Institute of Electrical and Electronics Engineers *IEEE transactions on computers* 1968–
IEEE transactions on electronic computers 1963–7
Proceedings 1963–
Institute of Radio Engineers *Proceedings* 1913–62
Transactions on electronic computers 1952–63
Institution of Civil Engineers *Minutes of proceedings* 1837–1937
Institution of Electrical Engineers *Journal* 1889–
Proceedings 1949–
Institution of Mechanical Engineers *Proceedings* 1847–
Institution (later Royal —) of Naval Architects *Transactions* 1860–
International and comparative law quarterly 1952–
International Congress of Genetics, 6th, 1932 *Proceedings* 2 vols. 1932
International Congress of Linguists, 8th, 1957 *Proceedings* 1958
International Congress of Medicine, 17th, 1913 23 sections 1913–14
International Congress of Phonetic Sciences, 2nd, 1935 *Proceedings* 1936
——, 3rd, 1938 *Proceedings* 1939
International Congress of Soil Science, 1st, 1927 *Proceedings and papers* 5 vols. 1928
——, 2nd, 1930 *Proceedings and papers* 7 vols. 1932–5
——, 7th, 1960 *Transactions* 4 vols. 1960
International encyclopaedia of the social sciences 17 vols. 1968
International Herald Tribune See *New York Herald Tribune International*

International journal of American linguistics 1917–
International journal of psycho-analysis 1920–
International Union of Pure and Applied Chemistry *Information bulletin* 1956–77
Nomenclature of inorganic chemistry 1959
— (ed. 2) 1971
Nomenclature of organic chemistry 3 parts 1958–65
— (ed. 2) 3 parts 1966–71
Inter-Ocean (Chicago) 1872–1914
Into orbit, by the seven astronauts of Project Mercury (US ed. with title *We seven*) 1962
Inverness Courier 1817–
IONESCU, Ghiţa & GELLNER, E. *Populism, its meaning and national characteristics* 1969
Iowa State Agricultural Society *Report* 1854–99
Irish Times 1859–
Iron and Steel Institute *Journal* 1871–
IRONSIDE, Janey *A fashion alphabet* 1968
IRVING, John *The 158-pound marriage* 1974
The world according to Garp 1978
IRVING, John James Cawdell *Royal navalese: a glossary* 1946
IRWIN, Godfrey *American tramp and underworld slang* 1931
ISHERWOOD, Christopher William Bradshaw *Goodbye to Berlin* 1939
Mr. Norris changes trains 1935
Prater violet 1945 (UK 1946)
A single man 1964
See also: AUDEN, Wystan Hugh
Isis, The (Oxford) 1892–1914; 1919–36; 1945–
Islander, The (Victoria, British Columbia) 1953–
It 1966–75 (1966–7 with title *International Times*)

J

JACKSON, Emily *A history of hand-made lace* 1900
JACKSON, George *Soledad Brother: the prison letters of George Jackson* 1970 (UK 1971)
JACKSON, Kenneth Hurlstone *Language and history in early Britain* 1953
JACKSON, Louis E. & HELLYER, C. R. *A vocabulary of criminal slang, with some examples of common usages* 1914
JACKSON, Richard A. *Mechanism: an introduction to the study of organic reactions* 1972
JACOB, Gordon Percival Septimus *Orchestral technique* 1931
JACOB, Henry ed. *On the choice of a common language* 1946
A planned auxiliary language 1947
JACOBS, Arthur *A new dictionary of music* 1958
JACOBS, Roderick A. & ROSENBAUM, Peter S. eds. *Readings in English transformational grammar* 1970
JACOBSEN, Charles W. *Oriental rugs: a complete guide* 1962
JACOBSON, Dan *The confessions of Joseph Baisz* 1977
A dance in the sun 1956
The trap 1955
JAFFE, Rona *Class reunion* 1979
— (Dell paperback ed.) 1980
The fame game 1969 (UK 1970)
JAKOBSON, Roman & HALLE, Morris *Fundamentals of language* 1956
JAMES, Brian *England v Scotland* 1969
'JAMES, Brian' (John Lawrence Tierney) ed. *Australian short stories* (2nd ser.) 1963
For 1st ser. *see* MURDOCH, Walter & DRAKE-BROCKMAN, H. F. Y.
JAMES, Edwin *Account of an expedition from Pittsburgh to the Rocky Mountains, performed in the years 1819, 1820. Compiled from the notes of Major Long, Mr. T. Say, and other gentlemen of the party* 3 vols. 1823
JAMES, Glenn & JAMES, Robert Clarke eds. *Mathematics dictionary* (multilingual ed.) 1959
JAMES, Henry (1843–1916) *The ambassadors* 1903
The awkward age 1899
The golden bowl 2 vols. 1904
The ivory tower 1917
The middle years 1917
Notebooks ed. F. O. Matthiessen & K. B. Murdock 1947
The reverberator 2 vols. 1888
The sense of the past 1917
The spoils of Poynton 1897
The wings of the dove 2 vols. 1902 (UK ed. in 1 vol.)
JAMES, Leigh *The chameleon file* 1967 (UK 1968)
JAMES, Phyllis Dorothy *Death of an expert witness* 1977
Shroud for a nightingale 1971
An unsuitable job for a woman 1972
JAMES, Walter *A word-book of wine* 1959
JAMES, William (1842–1910) *Collected essays and reviews* ed. R. B. Perry 1920
Essays in radical empiricism ed. R. B. Perry 1912
Letters ed. H. James 2 vols. 1920
The meaning of truth: a sequel to Pragmatism 1909
Memories and studies ed. H. James 1911
A pluralistic universe 1909
Pragmatism: a new name for some old ways of thinking 1907
Some problems of philosophy: a beginning of an introduction to philosophy 1911
Talks to teachers on psychology, and to students on some of life's ideals 1899
A text-book of psychology 1892
The will to believe, and other essays in popular philosophy 1897
See also: PERRY, Ralph Barton
JAMESON, Margaret Ethel Storm *A richer dust* 1931
Jane's all the world's aircraft (title varies) 1909–
Jane's fighting ships (title varies) 1898–
Jane's freight containers 1968–
Jane's surface skimmer systems (title varies) 1967–
JARVIS, Derek tr. *D. Hess's Plant physiology* 1975
Jazz and blues 1971–3
JEFFERIES, Ian *Dignity and purity* 1960
House-surgeon 1966
It wasn't me! 1961
Thirteen days 1959
JEFFRIES, Roderic Graeme *Dead man's bluff* 1970
Evidence of the accused 1961
Exhibit no. thirteen 1962
A traitor's crime 1968
JEKYLL, Gertrude *Colour in the flower garden* 1908
— (ed. 3, with title *Colour schemes for the flower garden*) 1914
Wood and garden 1899
JENKINS, Daniel Thomas *The educated society* 1966
JENKINS, Elizabeth *The tortoise and the hare* 1954
JENKINS, Geoffrey *A bridge of magpies* 1974
A twist of sand 1959
JENKINS, Herbert George *John Dene of Toronto: a comedy of Whitehall* 1919
JENKINS, James Travis *The fishes of the British Isles* 1925
— (ed. 2) 1936
JENKINS, Vivian Gordon J. *Lions down under: the British Isles Rugby tour of Australia and New Zealand, 1959* 1960
JENKINS, Warren *The Ohio gazetteer, and traveler's guide* 1837
JENNESS, Diamond *The Indians of Canada* 1932
— (ed. 2) 1934
JENNETT, Seán *The making of books* 1951
JENNINGS, Paul *The living village* 1968
JERRARD, Harold George & MCNEILL, Donald Burgess *A dictionary of scientific units, including dimensionless numbers and scales* 1963
JERROLD, William Blanchard *London* (illustr. G. Doré) 1872
Jersey Evening Post (St. Helier, Jersey) 1890–
JESPERSEN, Jens Otto Harry *Essentials of English grammar* 1933
Growth and structure of the English language 1905
An international language 1928
Language: its nature, development and origin 1922
Mankind, nation and individual from a linguistic point of view 1925
A modern English grammar on historical principles 7 parts 1909–49 (parts 6 & 7 by Jespersen & N. Haislund)
The philosophy of grammar 1924
The system of grammar 1933
Jewish Chronicle 1841–
Jewish encyclopedia, The ed. I. Singer 12 vols. 1901–6
Jewish manual, the; or, Practical information in Jewish and modern cookery, with a collection of valuable recipes & hints relating to the toilette, edited by a Lady 1846
JHABVALA, Ruth Prawer *Get ready for battle* 1962
The householder 1960
A new dominion 1972
To whom she will 1955
JOESTEN, Joachim *They call it intelligence* 1963
JOHANNSEN, Albert *A descriptive petrography of the igneous rocks* 4 vols. 1931–8
— (ed. 2) Vol. I 1939
JOHANNSEN, Hano & ROBERTSON, Andrew *Management glossary* ed. E. F. L. Brech 1968
John Edwards Memorial Foundation *Quarterly* (title varies) 1965–
John o' London's weekly 1919–54; revived as *John o' London's* 1959–
Johns Hopkins Hospital *Bulletin* 1889–1966
JOHNSON, Bryan Stanley *Travelling people* 1963
JOHNSON, Lady Bird (Mrs. Lyndon Baines Johnson) *A White House diary* 1970
JOHNSON, Myron ed. *American advertising 1800–1900* 1960
JOHNSON, Pamela Hansford *The humbler creation* 1959
Night and silence, who is here? 1963
The unspeakable Skipton 1959
JOHNSON, Paul Bede *Enemies of society* 1977
JOHNSON, William Ernest *Logic* 3 parts 1921–4
JOHNSTON, Edward *Writing and illuminating, and lettering* 1906
JOHNSTON, George Henry *Death takes small bites* 1948
JOHNSTON, Harry Hamilton *A comparative study of the Bantu and semi-Bantu languages* 2 vols. 1919–22
JOHNSTON, James Finlay Weir *Notes on North America; agricultural, economical and social* 2 vols. 1851
JOHNSTON, Ronald *The black camels of Qashran* 1970
JOLY, John *The surface-history of the earth* 1925
JONES, Bernard Edward ed. *Cassell's cyclopædia of photography* See *Cassell's* ——
The cinematograph book 1915
JONES, Cheslyn Peter M., WAINWRIGHT, G., & YARNOLD, E. eds. *The study of liturgy* 1978
JONES, Daniel *An outline of English phonetics* 1918 (and several later editions used)
The phoneme: its nature and use 1950
JONES, Eli Stanley *The Christ of the Indian road* 1925
JONES, Enid Huws *Margery Fry, the essential amateur* 1966
JONES, Ernest tr. *S. Ferenczi's Contributions to psycho-analysis* 1916
Papers on psycho-analysis 1913
— (rev. ed.) 1918
JONES, Ernest Beachcroft *Instrument technology* 3 vols. 1953–7
JONES, George Noble *Florida plantation records* ed. V. B. Phillips & J. D. Glunt 1927
JONES, Gwyn *A history of the Vikings* 1968
JONES, Henry Festing *Samuel Butler, author of Erewhon, 1835–1902: a memoir* 2 vols. 1919
JONES, John Morris *A Welsh grammar, historical and comparative* 1913
JONES, Mervyn *Potbank* 1961
JONES, Owen *The grammar of ornament* 1856
JONES, Philip Mitchell *The fifth defector* 1967
JONES, Reginald Victor *Most secret war* 1978
JONES, Robert Walter *Dictionary of banking* (ed. 10) 1951 *See* THOMSON, William
JONG, Erica Mann *Fanny: being the true history of the adventures of Fanny Hackabout-Jones* 1980
JORDAIN, Philip Bernard ed. *Condensed computer encyclopedia* 1969
'JORDAN, David' *Black account* 1975
Nile green 1973
JORDAN, David Starr *A guide to the study of fishes* 2 vols. 1905
JOSEPH, Horace William Brindley *An introduction to logic* 1906
— (ed. 2) 1916
JOSEPH, Michael Kennedy *I'll soldier no more* 1958
Journal of abnormal psychology 1906– (1921–64 with title *Journal of abnormal and social psychology*)
Journal of agricultural research 1913–49
Journal of anatomy and physiology 1866– (from 1916 with title *Journal of anatomy*)
Journal of applied physics 1937–
Journal of applied physiology 1948–
Journal of bacteriology 1916–
Journal of biological chemistry 1905–
Journal of biophysical and biochemical cytology 1955–61
Journal of cellular and comparative physiology 1932– (from 1966 with title *Journal of cellular physiology*)
Journal of chemical education 1924–
Journal of chemical physics 1933–
Journal of chromatography 1958–
Journal of clinical endocrinology and metabolism 1941–
Journal of clinical investigation 1924–
Journal of comparative neurology 1891– (1904–10 with title *Journal of comparative neurology and psychology*)
Journal of comparative psychology 1921–46
Journal of ecology 1913–
Journal of economic entomology 1908–
Journal of education 1879–
Journal of endocrinology 1939–
Journal of English and Germanic philology 1903–
Journal of experimental psychology 1916–
Journal of experimental zoology 1904–
Journal of general physiology 1918–
Journal of general psychology 1928–
Journal of genetic psychology 1954–
Journal of genetics 1910–
Journal of geophysical research 1949–
Journal of glaciology 1947–
Journal of heredity 1914–
Journal of hygiene 1901–

Journal of immunology 1916–
Journal of industrial and engineering chemistry 1909–22
Journal of infectious diseases 1904–
Journal of investigative dermatology 1938–
Journal of laboratory and clinical medicine 1915–
Journal of linguistics 1965–
Journal of medical research 1901–24
Journal of mental science 1855–1962
Journal of molecular biology 1959–
Journal of morphology (title varies) 1887–
Journal of natural philosophy, chemistry and the arts 1797–1813
Journal of neurology and psychopathology 1920–37
Journal of neurology, neurosurgery and psychiatry 1944–
Journal of nutrition 1928–
Journal of obstetrics and gynaecology of the British Empire (later —— *Commonwealth*) 1902–
Journal of organic chemistry 1936–
Journal of paleontology 1927–
Journal of pathology and bacteriology 1892–
Journal of pediatrics 1932–
Journal of pharmaceutical sciences 1911–
Journal of pharmacology and experimental therapeutics 1909–
Journal of philosophy 1904– (1904–20 with title *Journal of philosophy, psychology and scientific methods*)
Journal of physiology 1878–
Journal of political economy 1892–
Journal of polymer science 1946–
Journal of protozoology 1954–
Journal of scientific instruments 1922–
Journal of sedimentary petrology 1931–
Journal of social psychology 1930–
Journal of soil science 1949–
Journal of speech and hearing disorders 1936–
Journal of symbolic logic 1936–
Journal of theological studies 1899–
Journal of tropical medicine and hygiene 1898– (1898–1906 with title *Journal of tropical medicine*)
Journal See also under the names of particular institutions
JOWITT, William Allen & WALSH, Clifford eds. *Dictionary of English law* 2 vols. 1959
JOYCE, James Augustine Aloysius (1882–1941) *Chamber music* 1907
Dubliners 1914
Exiles 1918
Finnegans wake 1939 (quotations are mostly taken from ed. 3, 1964, but dated simply 1939)
Giacomo Joyce ed. R. Ellmann 1968
Letters ed. S. Gilbert & R. Ellmann 3 vols. 1957–66
Pomes penyeach 1927
A portrait of the artist as a young man 1916 (quotations are mostly taken from the 1964 Viking Press ed., 15th printing, 1969)
Stephen hero: part of the first draft of 'A portrait of the artist as a young man' ed. T. Spencer 1944
— (ed. J. J. Slocum & H. Cahoon) 1955
Ulysses 1922 (quotations are mostly taken from the Random House ed., 1946, but dated simply 1922)
JOYCE, Thomas Athol *Mexican archaeology* 1914
JUBB, Kenneth Victor F. & KENNEDY, Peter C. *Pathology of domestic animals* 2 vols. 1963
— (ed. 2) 1970
JUDD, Charles Hubbard tr. *W. Wundt's Outlines of psychology* 1897
JUDGE, Arthur William ed. *Modern motor cars* 3 vols. 1924
Modern petrol engines 1946
Stereoscopic photography: its application to science, industry and education 1926
Judge, The 1881–1939
JUKES-BROWNE, Alfred Joseph *The student's handbook of stratigraphical geology* 1902
JUNG, Carl Gustav *See* BAYNES, Helton Godwin; also HULL, Richard F. C., et al.

K

KAHN, David *The codebreakers: the story of secret writing* 1967 (UK 1968)
KAHN, Herman *On escalation: metaphors and scenarios* 1965
Kaleidoscope, The; or, Literary and scientific mirror 1818–31
KANE, Harry Hubbell *Opium-smoking in America and China* 1882
Kansas City Star (Kansas City, Missouri) 1880–
Kansas City Times (Kansas City, Missouri) 1868–
KANTOR, Jacob Robert *An objective psychology of grammar* 1936
KANTOROWICZ, G. F. ed. *Inlays, crowns, and bridges* See: COWELL, Colin Robert, et al.
KAPP, Gilbert *Dynamos, alternators, and transformers* 1893
KARCH, Robert Randolph & BUBER, Edward J. *Graphic arts procedures: the offset processes* 1967
KARP, David *All honorable men* 1956
Leave me alone 1957
KARSNER, Howard Thomas *Human pathology* 1927
KATZ, Jerrold Jacob *The philosophy of language* 1966
KAUFFMAN, Donald T. *Dictionary of religious terms* 1967
KAUFFMANN, Stanley Jules *If it be love* 1960
The tightrope 1952 (UK ed. 1953 with title *The philanderer*)
KAY, Stephen *Travels and researches in Caffraria* 1833
KAYE, Marvin *A lively game of death* 1972
'KAYE, Mary Margaret' (Mary Margaret Kaye Hamilton) *The far pavilions* 1978
House of shade 1959
KAYE-SMITH, Sheila *Mrs. Gailey* 1951
KAZMANN, Raphael G. *Modern hydrology* 1965
KEANE, Augustus Henry *The Boer states: land and people* 1900
Man, past and present 1899
tr. *A. Hovelacque's The science of language* 1877
KEANE, Charles Alexander *Modern organic chemistry* 1909
KEATING, Henry Reymond Fitzwalter *Bats fly up for Inspector Ghote* 1974
Filmi, filmi, Inspector Ghote 1976
Inspector Ghote hunts the peacock 1968
Inspector Ghote plays a joker 1969
KEATON, Buster (Joseph Francis Keaton) *My wonderful world of slapstick* 1960 (UK 1967)
KEATS, John (1795–1821) *Letters* ed. M. B. Forman 2 vols. 1931
Letters, 1814–1821 ed. H. E. Rollins 2 vols. 1958
KEENEY, Arthur Hail *Ocular examination: basis and technique* 1970
KEEPNEWS, Orrin & GRAUER, William *A pictorial history of jazz: people and places from New Orleans to modern jazz* 1955
KEESING, Felix Maxwell *Cultural anthropology, the science of custom* 1958
KEFAUVER, Estes *Crime in America* ed. S. Shalett 1951 (UK 1952)
KEHOE, Vincent J. R. *Aficionado! The pictorial encyclopedia of the Fiesta de Toros of Spain* 1959
The technique of film and television make-up 1957
KEIM, De Benneville Randolph *Sheridan's troopers on the borders: a winter campaign on the plains* 1870
KEITH, Agnes Newton *Land below the wind* 1939
KEITH, Arthur *The antiquity of man* 1915
Human embryology and morphology 1902
KELLER, Helen Adams *The story of my life* ed. J. A. Macy 1903
KELLY, Mary Theresa *The Christmas egg* 1958
The spoilt kill 1961
KELLY, Michael *Spinifex* 1970
KEMBLE, Frances Anne *Records of later life* 3 vols. 1882
KEMELMAN, Harry Gregory *Monday the rabbi took off* 1972
Saturday the rabbi went hungry 1966 (UK 1967)
Wednesday the rabbi got wet 1976
KEMP, Ian *British G.I. in Vietnam* 1969
KEMP, James Furman *A handbook of rocks, for use without the microscope* 1896
KEMP, Peter Kemp *Fleet Air Arm* 1954
KEMP, Peter Mant MacIntyre *Alms for oblivion* 1961
Mine were of trouble 1957
No colours or crest 1958
KENDALL, George Wilkins *Narrative of the Texan Santa Fé expedition* 2 vols. 1844
KENDALL, Maurice George & BUCKLAND, William R. *A dictionary of statistical terms* 1957
KENDREW, Wilfrid George *The climates of the continents* 1922
KENNEDY, Ludovic Henry Coverley *Very lovely people* 1969
KENNEDY, Margaret *The constant nymph* 1924
The heroes of Clone 1957
Lucy Carmichael 1951
KENNEISON, William C. & SPILMAN, Alan J. B. *Dictionary of printing, papermaking and bookbinding* 1963
KENT, Ruth Kimball *The language of journalism: a glossary of print-communications terms* 1970
KENYON, John Samuel *American pronunciation: a text-book of phonetics for students of English* 1924
— (ed. 6) 1935
KENYON, Kathleen Mary *Archaeology in the Holy Land* 1960
Digging up Jericho 1957
KENYON, Michael Forbes *Deep pocket* 1978
The 100,000 welcomes 1970
Mr Big 1975
The whole hog 1967
Kenyon review 1939–70
KEPHART, Horace *Camping and woodcraft* 2 vols. 1916–17
— (new ed.) 1921 (2 vols. in 1)
Our southern highlanders 1913
KER, William Paton *The Dark Ages* 1904
KEROUAC, Jack *On the road* 1957 (UK 1958)
KERSH, Cyril *The aggravations of Minnie Ashe* 1970
KERSH, Gerald *They die with their boots clean* 1941
KERSLEY, Leo & SINCLAIR, Janet *A dictionary of ballet terms* 1952
KETTLE, Edgar Hartley *The pathology of tumours* 1916
Kew bulletin (title varies) 1887–
KEYNES, John Maynard *The economic consequences of the peace* 1919
The general theory of employment, interest and money 1936
A treatise on money 2 vols. 1930
KEYNES, John Neville *Studies and exercises in formal logic* 1884 (and several later editions used)
KICHENSIDE, Geoffrey Michael & WILLIAMS, Alan Reginald *British railway signalling* 1963
KILVERT, Robert Francis (1840–79) *Diary* ed. W. Plomer 3 vols. 1938–40
— (selections, ed. W. Plomer) 1944
KIMENYE, Barbara *Kalasanda revisited* 1966
KINCAID, Dennis *British social life in India, 1608–1937* 1938
KING, Francis Henry *The widow* 1957
KING, Gordon John *The audio handbook* 1975
KING, John Edward & COOKSON, Christopher *The principles of sound and inflexion as illustrated in the Greek and Latin languages* 1888
KING, Martin Luther *The trumpet of conscience* 1968
KING, Woodie ed. *Black short story anthology* 1972
KINGSLEY, John Sterling *Comparative anatomy of vertebrates* 1912
KINGSLEY, Nelson *Diary* (1849–51) ed. F. J. Teggart 1914
Kingston Whig-Standard (Kingston, Ontario) 1926–
KINLOCH, Alexander Angus Airlie *Large game shooting in Thibet and the Northwest* 2 vols. 1869–76
KINSEY, Alfred Charles, et al. *Sexual behavior in the human female* 1953
Sexual behavior in the human male 1948
KIPLING, Rudyard (1865–1936) *Actions and reactions* 1909
A book of words 1928
Debits and credits 1926
A diversity of creatures 1917
From sea to sea 2 vols. 1899 (UK 1900)
Independence 1923
ed. *The Irish Guards in the Great War* 2 vols. 1923
Just so stories 1902
Land and sea tales 1923
Letters of travel 1920
Limits and renewals 1932
The phantom 'rickshaw, and other tales 1888 (UK 1890, revised)
Puck of Pook's Hill 1906
Rewards and fairies 1910
Something of myself for my friends known and unknown 1937
Songs from books 1912 (UK 1913)
Wee Willie Winkie, and other child stories 1888 (UK 1890)
The years between 1919
See also: FLETCHER, Charles Robert Leslie
KIRBY, Thomas Austin & WOOLF, H. B. *Philologica: the [Kemp] Malone anniversary studies* 1949
KIRK, Raymond E. & OTHMER, Donald F. eds. *Encyclopedia of chemical technology* 17 vols. 1947–60
— (ed. 2) 23 vols. 1963–71
KIRK, Raymond Maurice, MAYNARD, J. D., & HENRY, A. N. *Surgery* 1974
KIRK, Thomas *The forest flora of New Zealand* 1889
KIRKALDY, John Francis *General principles of geology* 1954
KIRKBRIDE, Ronald *Tamiko* 1959
'KIRKE, Edmund' (James Roberts Gilmore) *My southern friends* 1863
KIRKLAND, Caroline Matilda *Forest life: a tale* 2 vols. 1842
A new home—who'll follow? Or, Glimpses of western life 1839
Western clearings 1845
KIRKLAND, John *The modern baker, confectioner and caterer* 6 vols. 1907–9
KIRKUP, James Harold *Japan behind the fan* 1970
The only child 1957
Tropic temper: a memoir of Malaya 1963

KIRWAN, Daniel Joseph *Palace and hovel; or, Phases of London life* 1870
KLEENE, Stephen Cole *Introduction to metamathematics* 1952
Mathematical logic 1967
KLIBANSKY, Raymond ed. *Contemporary philosophy: a survey* 4 vols. 1968–71
KNEALE, William Calvert *Probability and induction* 1949
—— & KNEALE, Martha *The development of logic* 1962
KNECHT, Edmund, RAWSON, C., & LOEWENTHAL, R. *A manual of dyeing* 2 vols. 1893
KNEEBONE, Geoffrey Thomas *Mathematical logic and the foundation of mathematics* 1963
KNICKERBOCKER, William Skinkle ed. *Twentieth century English* 1946
KNIGHT, Albert E. *The complete cricketer* 1906
'KNIGHT, Brigid' (Kathleen Henrietta Sinclair) *The covenant* 1943 (UK 1944)
Westward the sun (UK ed. with title *The sun climbs slowly*) 1942
KNIGHT, Damon ed. *One hundred years of science fiction* 1969
KNIGHT, Sarah (Kemble) *The journals of Madam Knight, and Rev. Mr. Buckingham ..written in 1704 & 1710* 1825
KNIGHT, Sophia *Window on Shanghai: letters from China 1965–67* 1967
KNOTT, Frank Alexander *Clinical bacteriology* 1939
KNOWLES, Robert Edward *The attic guest* 1909
KNOX, Alexander *Glossary of geographical and topographical terms* 1904
KNOX, 'Bill' *See* KNOX, William
KNOX, Ronald Arbuthnott *Broadcast minds* 1932
The footsteps at the lock 1928
God and the atom 1945
The hidden stream: a further collection of Oxford conferences 1952
In soft garments: a collection of Oxford conferences 1942
Still dead 1934
KNOX, William ('Bill') *Blacklight* 1967
Little drops of blood 1962
The tallyman 1969
KOBER, Arthur *Parm me* 1945
KOCH, Sigmund ed. *Psychology: a study of a science* 6 vols. 1959–63
KOCHMAN, Thomas ed. *Rappin' and stylin' out: communication in urban black America* 1972
KOEBEL, William Henry *In the Maoriland bush* 1911
The return of Joe, and other New Zealand stories 1907
KOESTLER, Arthur *The age of longing* 1951
Arrow in the blue: an autobiography 1952
Insight and outlook: an inquiry into the common foundations of science, art and social ethics 1949
The invisible writing [2nd vol. of *Arrow in the blue*] 1954
The lotus and the robot 1960
Promise and fulfilment: Palestine 1917–1949 1949
Scum of the earth 1941
Spanish testament 1937
Thieves in the night: chronicle of an experiment 1946
The trail of the dinosaur, and other essays 1955
Twilight bar: an escapade in four acts 1945
The yogi and the commissar, and other essays 1945
—— et al. *The god that failed: six studies in communism* ed. R. H. S. Crossman 1950
Koninklijke Akademie van Wetenschappen te Amsterdam. Section of Sciences *Proceedings* (English ed.) 1899– (1955– with title *Proceedings* of the Koninklijke Nederlandsche Akademie van Wetenschappen)
KOPPETT, Leonard *The New York Times guide to spectator sports* 1971
KOPS, Bernard *The Hamlet of Stepney Green: a sad comedy* 1959
KÖRNER, Stephan *The philosophy of mathematics* 1960
KOSTOV, Ivan *Mineralogy* tr. by the author, ed. P. G. Embrey & J. Phemister 1968
KRACKE, Roy Rachford & GARVER, Hortense Elton *Diseases of the blood and atlas of hematology* 1937
KRAMSCH, Dieter M. tr. *E. Grundmann's General cytology* 1966
KRANTZ, Judith *Scruples* 1978
KRAPP, George Philip *The English language in America* 2 vols. 1925
KRATOCHVÍL, Paul *The Chinese language today: features of an emerging standard* 1968
KROEBER, Alfred Louis *Anthropology* 1923
— (rev. ed.) 1948
KROGH, Schack August S. *Osmotic regulation in aquatic animals* 1939
KROHN, F. tr. *G. Glaser de Cew's Magneto- and dynamo-electric machines* ed. P. Higgs 1884
KRUEGER, Robert G., GILLHAM, N. W., & COGGIN, J. H. *Introduction to microbiology* 1973

KUHN, Irene *Assigned to adventure* 1938
KUPER, Charles Goethe *An introduction to the theory of superconductivity* 1968
KURATH, Hans *Studies in area linguistics* 1972
A word geography of the Eastern United States 1949
KURTZ, Harry *Invasion of privacy* 1955 (UK 1956)
KWANG-CHIH CHANG *The archaeology of ancient China* 1963
— (ed. 3) 1977
KYLE, Robert *Venus examines* (UK ed. with title *The love laboratory*) 1968

L

LABARRE, Émile Joseph *A dictionary of paper and paper-making terms* 1937
LA BERN, Arthur Joseph *Goodbye Piccadilly, farewell Leicester Square* 1966
LABOV, William *Language in the inner city: studies in the Black English vernacular* 1972
'LACKLAND, Thomas' (George Canning Hill) *Homespun; or, Five-and-twenty years ago* 1867
LACY, Ed *Double trouble* 1965
The freeloaders 1962
Ladies' field, The 1896–1928
Ladies' home journal 1883–
Ladies' repository 1841–76
LAFITTE, Paul *The person in psychology: reality or abstraction* 1957
LAGLER, Karl F., BARDACH, J. E., & MILLER, R. R. *Ichthyology* 1962
LAGOWSKI, Joseph John *Modern inorganic chemistry* 1973
LAHEE, Frederic Henry *Field geology* 1916
LAIDLER, F. F. ed. *A glossary in English, French, German, Spanish of terms used in home economics education* 1963
LAKE, Philip & RASTALL, Robert Heron *A textbook of geology* 1910
LAMB, Charles (1775–1834) & LAMB, Mary Ann (1764–1847) *Letters* ed. E. V. Lucas 3 vols. 1935
LAMB, Lynton *Man in a mist* 1974
Picture frame 1972
LAMB, Sydney M. *Outline of stratificational grammar* 1966
LAMB, Warren *Posture and gesture: an introduction to the study of physical behaviour* 1965
LAMBERT, Constant *Music ho!* 1934
LAMBERT, Richard Stanton *Ariel and all his quality* 1940
'LA MERI' (Russell Meriwether Hughes) *Spanish dancing* 1948
— (ed. 2) 1967
Lancashire Dialect Society *Journal* 1951–
Lancashire life 1947–
'LANCASTER, G. B.' (Edith Joan Lyttleton) *Jim of the ranges* 1910
Sons o' men 1904
LANCASTER, Osbert *Homes sweet homes* 1939
Lancaster, Massachusetts *Early records 1643–1725* ed. H. S. Nourse 1884
LANDAU, Thomas ed. *Encyclopaedia of librarianship* 1958
— (ed. 2) 1961
Landfall: a New Zealand quarterly 1947–
LANDY, Eugene E. *The underground dictionary* 1971
LANE, Arthur *Greek pottery* 1948
'LANE, Linacre' *The ABZ of Scouse: how to talk proper in Liverpool, Vol. 2* ed. F. Spiegel 1966 (For Vol. 1 *see* SHAW, Frank, et al.)
LANG, R. F. tr. *F. A. Henglein's Chemical technology* 1969
LANGBRIDGE, Rosamond *The flame and the flood* 1903
LANGFORD, Michael John *Basic photography* 1965
LANGFORD-SMITH, Fritz *Radio designer's handbook* 1940
— (ed. 4) 1953
LANGTON, Mary Beach *How to know oriental rugs: a handbook* 1904
Language: journal of the Linguistic Society of America 1925–
Language learning 1948–
LANHAM, Peter *Blanket boy's moon* (based on an original story by A. S. Mopeli-Paulus) 1953
LANKESTER, Edwin Ray *Diversions of a naturalist* 1915
LANMAN, Charles *Letters from the Alleghany Mountains* 1849
A summer in the wilderness; embracing a canoe voyage up the Mississippi and around Lake Superior 1847
Lantern: politics, amusement, literature (New Orleans) 1886–9
LAQUEUR, Walter ed. *A dictionary of politics* 1971
LARDNER, Ringgold Wilmer *Gullible's travels* 1917 (UK 1926)

LARKIN, Philip Arthur *Jill* 1946
The less deceived 1955
The north ship 1945
LARSELL, Olof *Anatomy of the nervous system* 1942
— (ed. 2) 1951
LASKI, Audrey Louise *The keeper* 1968
Seven other years 1967
LASKI, Harold Joseph *Holmes–Laski letters* See: HOLMES, Oliver Wendell
LASKI, Marghanita *Little boy lost* 1949
Love on the supertax 1944
Tory heaven; or, Thunder on the right 1948
The village 1952
See also: 'RUSSELL, Sarah'
LASKY, Melvin J. *Utopia and revolution* 1976 (UK 1977)
LASLETT, Thomas Peter R. ed. *Household and family in past time* 1972
'LATHEN, Emma' (Mary Jane Latsis & Martha Henissart) *Accounting for murder* 1964 (UK 1965)
Banking on death 1961 (UK 1962)
By hook or by crook 1975
Come to dust 1968 (UK 1969)
The longer the thread 1971 (UK 1972)
Murder without icing 1972 (UK 1973)
Pick up sticks 1970 (UK 1971)
Sweet and low 1974
When in Greece 1969
See also: 'DOMINIC, R. B.'
'LAUCHMONEN' *Old Thom's harvest* 1965
LAURIE, Peter *Scotland Yard: a personal inquiry* 1970
LAVER, James *Taste and fashion, from the French Revolution until to-day* 1937
Victoriana 1966
LAW, James Thomas *Law's Grocer's manual* 1896
LAWRENCE, David Herbert (1885–1930) *Aaron's rod* 1922
Apocalypse 1931 (UK 1932)
Birds, beasts and flowers: poems 1923
Collected letters ed. H. T. Moore 2 vols. 1962
A collier's Friday night 1934
David: a play 1926
England, my England, and other stories 1922 (UK 1924)
Etruscan places 1932
Fantasia of the unconscious 1922
Kangaroo 1923
Lady Chatterley's lover 1928
Last poems ed. R. Aldington & G. Orioli 1932
Letters ed. A. Huxley 1932 (See also *Collected letters*)
Look! We have come through! 1917
The lost girl 1920
Love among the haystacks, and other pieces 1930
Love poems and others 1913
The lovely lady 1933 (dated 1932)
A modern lover 1934
Mornings in Mexico 1927
Nettles 1930
New poems 1918
Pansies: poems 1929
Phoenix: posthumous papers ed. E. D. McDonald 1936
Phoenix II: uncollected, unpublished and other prose works ed. F. W. Roberts & H. T. Moore 1968
The plumed serpent (Quetzalcoatl) 1926
Psychoanalysis and the unconscious 1921
The rainbow 1915
Reflections on the death of a porcupine, and other essays 1925
St. Mawr, together with The princess 1925
Sea and Sardinia 1921
Sons and lovers 1913
Studies in classic American literature 1923
Touch and go: a play 1920
The virgin and the gipsy 1930
The white peacock 1911
The widowing of Mrs. Holroyd: a drama 1914
The woman who rode away, and other stories 1928
Women in love 1920 (UK 1921)
—— & SKINNER, Mary Louisa *The boy in the bush* 1924
LAWRENCE, George H. M. *Taxonomy of vascular plants* 1951
LAWRENCE, Thomas Edward (1888–1935) *The home letters of T. E. Lawrence and his brothers* ed. M. R. Lawrence 1954
Letters ed. D. Garnett 1938
The mint: notes made in the R.A.F. Depot between August and December 1922, and at Cadet College in 1925 by 352087 A/c Ross 1936 (UK, rev. 1955)
Seven pillars of wisdom: a triumph 1926
— (trade ed.) 1935
LAWRENCE, William Witherle *Beowulf and epic tradition* 1928
LAWSON, Henry Archibald Hertzberg (1867–1922) *Stories* ed. C. Mann 3 ser. 1964
Verses, popular and humorous 1900

LAWSON, Robert William tr. *G. von Hevesy & F. A. Paneth's A manual of radioactivity* 1926
— (ed. 2) 1938
tr. *A. Einstein's Relativity: the special and the general theory: a popular exposition* 1920
LAYARD, Edgar Leopold *The birds of South Africa: a descriptive catalogue* 1867
LEACH, Bernard *A potter in Japan 1952–1954* 1960
LEACH, Edmund Ronald *Lévi-Strauss* 1970
LEACH, John Albert *An Australian bird book* 1911
LEACOCK, Stephen Butler *Over the footlights, and other fancies* 1923
LEAKEY, Louis Seymour Bazett *The Stone Age cultures of Kenya Colony* 1931
LEAR, Edward (1812–88) *Indian journal* ed. R. Murphy 1953
Laughable lyrics 1877
LEASOR, Thomas James *A host of extras* 1973
They don't make them like that any more 1969
LEAVIS, Frank Raymond *The great tradition: George Eliot, Henry James, Joseph Conrad* 1948
—— & THOMPSON, Arthur Denys H. *Culture and environment: the training of critical awareness* 1933
LEAVIS, Queenie Dorothy *Fiction and the reading public* 1932
Lebende Sprachen 1956–
'LE CARRÉ, John' (David John M. Cornwell) *The honourable schoolboy* 1977
The looking-glass war 1965
A murder of quality 1962
A small town in Germany 1968
Smiley's people 1979 (UK 1980)
The spy who came in from the cold 1963
Tinker tailor soldier spy 1974
LEE, Harper *To kill a mockingbird* 1960
LEE, John D. *The ninth man* 1976
LEE, John Alexander *Children of the poor* 1934
LEE, Julius & KNOWLES, Francis *Animal hormones* 1965
LEE, Laurie *Cider with Rosie* 1959
LEE, Rawdon B. *A history and description of the modern dogs of Great Britain and Ireland* 3 parts 1893–4
LEE, Reuben *Electronic transformers and circuits* 1947
LEECH, Geoffrey N. *English in advertising* 1966
LEEDS, Herbert Daniel & WEINBERG, Gerald Marvin *Computer programming fundamentals* 1961
LEES, Dan *The rainbow conspiracy* 1971
Rape of a quiet town 1973
Zodiac 1972
LEES-MILNE, James *Ancestral voices* 1975
LEESON, Charles Roland & LEESON, Thomas Sydney *Histology* 1966
— (ed. 2) 1970
LEFF, Gordon A. *History and social theory* 1969
LEFSCHETZ, Solomon *Introduction to topology* 1949
LEGMAN, Gershon *The fake revolt* 1967
LEHMANN, Rosamond Nina *The ballad and the source* 1944
Dusty answer 1927
The echoing grove 1953
The gipsy's baby, and other stories 1946
Invitation to the waltz 1932
A letter to a sister 1931
A note in music 1930
The weather in the streets 1936
LEIGHTON, Robert *The complete book of the dog* 1922
The new book of the dog: a comprehensive natural history of British dogs and their foreign relatives 1907
LEINSTER, Mary *Dew on the leaf* 1926
Leisure hour, The 1852–1905
LEJEUNE, Anthony *Duel in the shadows* 1962
LEMARCHAND, Elizabeth *Death of an old girl* 1967
Death on Doomsday 1971
Let or hindrance 1973
LEONARD, Elmore *Unknown man, no. 89* 1977
LEONARD, Zenas *Narrative of the adventures of Zenas Leonard* 1839
— (ed. W. F. Wagner) 1904
LEPSCHY, Giulio Ciro *A survey of structural linguistics* 1970
LESLIE, John Randolph Shane *The Anglo-Catholic* 1929
The Oppidan 1922
LESSING, Doris May *Five: short novels* 1953
The golden notebook 1962
In pursuit of the English 1960
Shikasta 1979
This was the old chief's country: stories 1951
Letters of Miss Riversdale [anon. novel] 3 vols. 1803
LEVI, Herbert W. & LEVI, Lorna R. tr. *A. Kaestner's Invertebrate zoology* 3 vols. 1967–70
LEVY, Mervyn Montague *The Studio dictionary of art terms* 1961

LEWES, Vivian Byam *Oil fuel* 1913
LEWIS, Alfred Henry *The apaches of New York* 1912
The boss, and how he came to rule New York 1903
Confessions of a detective 1906
Wolfville days 1902
LEWIS, Arthur William *A glossary of woodworking terms* 1966
LEWIS, Cecil Day tr. *Virgil's Aeneid* 1952
Beechen vigil, and other poems 1925
The buried day 1960
Child of misfortune 1939
The friendly tree 1936
tr. *Virgil's Georgics* 1940
A hope for poetry 1934
An Italian visit 1953
The Otterbury incident 1948
Overtures to death, and other poems 1938
Poems, 1943–1947 1948
A time to dance, and other poems 1935
Transitional poem 1929
Word over all 1943
See also: 'BLAKE, Nicholas'
LEWIS, Clarence Irving *A survey of symbolic logic* 1918
Symbolic logic 1932
LEWIS, Clive Staples (1898–1963) *The allegory of love: a study in medieval tradition* 1936
Letters ed. W. H. Lewis 1966
Perelandra 1943
Poems ed. W. Hooper 1964
The problem of pain 1940
The Screwtape letters 1942
Studies in words 1960
Surprised by joy 1955
That hideous strength: a modern fairy-tale for grown-ups 1945
LEWIS, Dominic Bevan Wyndham *On straw, and other conceits* 1927
LEWIS, Ethelreda *Mantis* 1926
LEWIS, Griselda ed. *Handbook of crafts* 1960
LEWIS, Meriwether & CLARK, William *Original journals of the Lewis and Clark expedition, 1804–1806* ed. R. G. Thwaites 8 vols. 1904–5
—— & ORDWAY, John *The journals of Captain Meriwether Lewis and Sergeant John Ordway, kept on the expedition of western exploration, 1803–1806* ed. M. M. Quaife 1916
LEWIS, Percy Wyndham (1882–1957) *The apes of God* 1930
Blasting and bombardiering 1937
Letters ed. W. K. Rose 1963
LEWIS, Roy (John Royston Lewis) *Blood money* 1973
Witness my death 1976
LEWIS, Sinclair (Harry Sinclair Lewis) *Arrowsmith* (UK ed. with title *Martin Arrowsmith*) 1925
Babbitt 1922
Bethel Merriday 1940
Cass Timberlane 1945 (UK 1946)
Free air 1919 (UK 1924)
It can't happen here 1935
Main Street: the story of Carol Kennicott 1920
The man who knew Coolidge 1928
Mantrap 1926
Our Mr. Wrenn: the romantic adventures of a gentle man 1914
LEWIS, Wilfrid James *The language of cricket* 1934
LEWIS, Wyndham *See* LEWIS, Percy Wyndham
LEWY, Willy *The conquest of space* 1949 (UK 1950)
LEYDEN, John tr. *Malay annals* 1821
—— & ERSKINE, William tr. *Memoirs of Zehir-Ed-Din Muhammed Baber, Emperor of Hindustan, written by himself in the Jaghatai Turki* 1826
Liberty 1924–30
Library, The 1889–
Library association record 1899–
Library journal: journal of the American Library Association 1876–
LICHINE, Alexis *Encyclopædia of wines and spirits* 1967
Wines of France 1952
LIDDELL, Billy (William Beveridge Liddell) *My soccer story* 1960
LIDDELL, Robert *The Morea* 1958
LIEBERMAN, Jethro K. & RHODES, Neil S. *The complete CB handbook* (new ed.) 1976
LIEF, Max *Hangover* 1929
Life and work (Church of Scotland) 1879–
LILICO, James Ewart *Sheep dog memoirs* 1934
LILLEY, Arthur Augustus *Twenty-four years of cricket* 1912
LILLYWHITE, Frederick ed. *The guide to cricketers* 1853–66
LILLYWHITE, James *James Lillywhite's cricketers' annual* 1872–7; 1879–93
LILLYWHITE, John *John Lillywhite's cricketers' companion* (title varies) 1865–85
LINCOLN, Abraham (1809–65) *Collected works* ed. R. P. Basler 9 vols. 1953

LINCOLN, Joseph Crosby *Cap'n Warren's wards* 1911
Partners of the tide 1905
LINDGREN, Ernest *The art of the film* 1948
LINDOP, Audrey Erskine *Sight unseen* 1969
LINDSAY, Jack *Life rarely tells* 1958
LINDZEY, Gardner ed. *Handbook of social psychology* 2 vols. 1954
— (ed. 2, by Lindzey & E. Aronson) 5 vols. 1968–9
LINGEMAN, Richard R. *Drugs from A to Z: a dictionary* 1969
Lingua: international review of general linguistics 1947–
LININGTON, Barbara Elizabeth *See* 'BLAISDELL, Anne'; 'EGAN, Lesley'; & 'SHANNON, Dell'
LINKLATER, Eric Robert Russell *Juan in America* 1931
Magnus Merriman 1934
Poet's pub 1929
Private Angelo 1946
Linnean Society of London *Proceedings* 1838–
Linnean Society of New South Wales *Proceedings* 1876–
LINNETT, John Wilfred *The electronic structure of molecules: a new approach* 1964
LINSSEN, Eugene F. *Beetles of the British Isles* 2 ser. 1959
LINTON, David Leslie ed. *Sheffield and its region* 1956
Lippincott's medical dictionary by R. W. Greene et al. 1897
Lippincott's monthly magazine (title varies) 1868–1915
Lippincott's new medical dictionary ed. H. W. Cattell 1910
Listener, The 1929–
LISTER, Raymond George *Decorative cast ironwork in Great Britain* 1960
Decorative wrought ironwork in Great Britain 1957
LITCHFIELD, Frederick *Pottery and porcelain: a guide to collectors* 1900
Literary digest 1890–1938
Literary guide 1894–1954
Little folks: a magazine for children 1871–1933
Liverpool Echo 1879–
Living off the land: a manual of bushcraft, compiled from articles contributed to 'Salt', the army education journal (Australia) 1944
LIVINGSTON, Milton Stanley *Particle physics: the high energy frontier* 1968
LLOYD, Albert Lancaster *Folk song in England* 1967
LLOYD, Christopher *The well-tempered garden* 1970
LLOYD, Hannibal Evans tr. *J. B. von Spix & C. F. P. von Martius's Travels in Brazil, 1817–20* 2 vols. 1824
LLOYD GEORGE, David *See* GEORGE, David Lloyd
LOBECK, Armin Kohl *Geomorphology* 1939
LOCK, Joan *Lady policeman* 1968
LOCKE, William John *Ancestor Jorico* 1929
At the gate of Samaria 1894
The beloved vagabond 1906
The coming of Amos 1924
Derelicts 1897
The glory of Clementina Wing 1911
The house of Baltazar 1920
Idols 1898
Jaffery 1915
Moordius and Co. 1923
The morals of Marcus Ordeyne 1905
The mountebank 1921
The old bridge 1926
The red planet 1917
The rough road 1918
Septimus 1909
Simon the jester 1910
Stella maris 1913
Stories near and far 1926
The tale of Triona 1922
The wonderful year 1916
LOCKET, George Hazlewood & MILLIDGE, Alfred Frank *British spiders* 2 vols. 1951–3
LOCKHART, Laurence William Maxwell *Fair to see* 3 vols. 1871
LOCKRIDGE, Frances Louise & LOCKRIDGE, Richard *Murder has its points* 1961 (UK 1962)
Practise to deceive: a Captain Heimrich mystery 1957 (UK 1959, as by 'Francis Richards')
LOCKWOOD, William Burley *Panorama of Indo-European languages* 1972
LOCKYER, Keith Gerald *An introduction to critical path analysis* 1964
LODGE, David John *The British Museum is falling down* 1965
Changing places 1975
LOGAN, Olive *Before the footlights and behind the scenes* 1870
LOMAX, Alan *Mister Jelly Roll: the fortunes of Jelly Roll Morton, New Orleans Creole and inventor of jazz* 1950 (UK 1952)

LONDON, Jack (John Griffith London) (1876–1916) *The call of the wild* 1903
The iron heel 1907 (UK 1908)
The jacket (US ed. with title *The star rover*) 1915
Letters ed. K. Hendricks & I. Shepard 1965 (UK 1966)
The road 1907 (UK 1914)
A son of the sun 1912 (UK 1913)
The valley of the moon 1913
White Fang 1906 (UK 1907)
London & Globe Telephone & Maintenance Company, Ltd. *List of subscribers* 1884 [reproduced in facsimile in *Three Victorian telephone directories* 1970]
London at table; or, How, when, and where to dine and order a dinner; and where to avoid dining 1851
London magazine: a monthly review of literature 1954–
London Mathematical Society *Proceedings* 1865–
London medical gazette 1827–51
London medical record 1873–87
London mercury 1919–39
London review 1835–6
London review of books 1979–
London society 1862–98
LONG, Constance Ellen tr. *C. G. Jung's Collected papers on analytical psychology* 1916
LONG, Joseph W. *American wild-fowl shooting* 1874
LONG, Ralph Bernard *The sentence and its parts: a grammar of contemporary English* 1961
LONGSTREET, Stephen *The real jazz, old and new* 1956
LONGWELL, Chester Ray, KNOPF, A., & FLINT, R. R. *Outlines of physical geology* 1934
'LORAC, E. C. R.' (Edith Caroline Rivett) *Ask a policeman* 1955
Murder in the mill-race 1952
LORAND, Sándor ed. *Psycho-analysis today: its scope and function* 1933
LORIMER, George Horace *Jack Spurlock—prodigal* 1908
Letters from a self-made merchant to his son 1902
Old Gorgon Graham: more letters from a self-made merchant to his son 1904
LORIMER, Hilda Lockhart *Homer and the monuments* 1950
Los Angeles Times 1881–
LOUNSBURY, Warren C. *Backstage from A to Z: a glossary of technical stage terms* 1959
LOVELOCK, Ian ('Yann') *The vegetable book: an unnatural history* 1972
LOVESEY, Peter Harmer *Invitation to a dynamite party* 1974
LOWELL, Robert Traill Spence *Near the ocean* 1967
Notebook 1970
LOWIE, Robert Heinrich *The history of ethnological theory* 1937 (UK 1938)
LOWNDES, Marie Adelaide Belloc *Diaries and letters 1911–1947* ed. S. Lowndes 1971
The Terriford mystery 1924
LOWRY, Clarence Malcolm (1909–57) *Selected letters* ed. H. Breit & M. B. Lowry 1965 (UK 1967)
Ultramarine 1933
Under the volcano 1947
LOYN, Henry Royston *Anglo-Saxon England and the Norman Conquest* 1962
'LUARD, L.' (William Blaine Luard) *All hands* 1933
Conquering seas 1935
LUARD, Nicholas *The Robespierre serial* 1975
Travelling horseman 1975
LUBBOCK, Adelaide *Australian roundabout* 1963
LUBBOCK, Alfred Basil *Round the Horn before the mast* 1902
LUBBOCK, John *See* AVEBURY, John (Lubbock), 1st Baron
LUCAS, Dione & HUME, Rosemary *'Au petit cordon bleu': an array of recipes* 1936
LUCAS, Edward Verrall *Over Bemerton's: an easy-going chronicle* 1908
Reading, writing and remembering: a literary record 1932
The vermilion box 1916
LUCAS, Norman *The C.I.D.* 1967
LUCIA, Ellis *Klondike Kate: the life and legend of Kitty Rockwell* 1962
LUDLUM, Robert *The Holcroft covenant* 1978
The Matlock paper 1973
LUGARD, Frederick John Dealtry (1858–1945) *Diaries* ed. M. Perham 4 vols. 1959–63
LULL, Richard Swann *Organic evolution* 1917
LUNN, Arnold Henry Moore *Harrovians* 1913
LUPTON, Thomas *Management and the social sciences* 1966
— (ed. 2) 1970
LURIE, Alison *Love and friendship* 1962
The nowhere city 1965
Real people 1969
The war between the Tates 1974

LUSH, Vicesimus *Auckland journals, 1850–63* ed. A. Drummond 1971
Thames journals, 1868–82 ed. A. Drummond 1975
LUTYENS, Mary ed. *Effie in Venice: unpublished letters of Mrs. John Ruskin written from Venice between 1849–1852* 1965
Millais and the Ruskins 1967
The Ruskins and the Grays 1972
LYALL, Gavin Tudor *Blame the dead* 1972
Judas country 1975
The most dangerous game 1964
Shooting script 1966
LYDEKKER, Richard *The deer of all lands* 1898
LYELL, Thomas Reginald G. *Slang, phrase and idiom in colloquial English and their use* 1931
LYLES, Sanders Truman *Biology of microorganisms* 1969
LYNDE, Francis *The grafters* 1904
The quickening 1906
LYON, Francis Hamilton tr. *T. Heyerdahl's The Kon-Tiki Expedition: by raft across the South Seas* (US ed. with title *Kon-Tiki: across the Pacific by raft*) 1950
LYON, Lilian Helen Bowes *Bright feather fading* 1936
The white hare, and other poems 1934
LYONS, Albert Michael Neil *Clara: some scattered chapters in the life of a hussey* 1912
LYONS, John *Introduction to theoretical linguistics* 1968
Semantics 2 vols. 1977
Structural semantics: an analysis of part of the vocabulary of Plato 1963
LYTTLETON, Raymond Arthur *Mysteries of the solar system* 1968
LYTTON, David *The goddam white man* 1960

M

MCADOO, William Gibbs *Crowded years* 1931
MACALISTER, Robert Alexander Steward *A textbook of European archaeology* 1921 (only Vol. I published)
MACARDLE, Dorothy *Children of Europe: a study of the children of liberated countries* 1949
MACAULAY, Emilie Rose (1887–1958) *Crewe train* 1926
Dangerous ages 1921
Going abroad 1934
I would be private 1937
Keeping up appearances 1928
Last letters to a friend, 1952–58 ed. C. B. Smith 1962
Letters to a friend, 1950–52 ed. C. B. Smith 1961
Letters to a sister ed. C. B. Smith 1964
Orphan Island 1924
Personal pleasures 1935
Potterism: a tragi-farcical tract 1920
Staying with relations 1930
Told by an idiot 1923
The towers of Trebizond 1956
'MCBAIN, Ed' (Evan Hunter) *Ax* [UK: *Axe*] 1964
Cop hater 1956 (UK 1958)
Doll 1965 (UK 1966)
Guns 1976 (UK 1977)
Give the boys a great big hand 1960 (UK 1962)
Hail, hail, the gang's all here! 1971
Hail to the chief 1973
Let's hear it for the deaf man 1973
Long time no see 1977
The pusher 1959
Ten plus one 1963 (UK 1964)
MCBAIN, James William *Colloid science* 1950
MACBETH, George Mann *A war quartet* 1969
MCCABE, John *Mr. Laurel and Mr. Hardy* 1961 (UK 1962)
MCCABE, Joseph tr. *E. Haeckel's The wonders of life* 1904
McCall's sewing, in colour 1964
MCCARDELL, Roy Larcom *Conversations of a chorus girl* 1903
MCCARTHY, Ian Ellery *Nuclear reactions* 1970
MCCARTHY, Mary Therese *Birds of America* [a novel] 1971
A charmed life 1955 (UK 1956)
The company she keeps 1942 (UK 1943)
The group 1963
The groves of Academe 1952 (UK 1953)
Memories of a Catholic girlhood 1957
On the contrary 1961 (UK 1962)
MCCARTHY, Wilson *The detail* 1973
MCCARTNEY, William *Olfaction and odours: an osphrésiological essay* 1968
MCCLANE, Albert Jules ed. *McClane's Standard fishing encyclopedia* 1965
MCCLOY, Helen *A change of heart* 1973
Panic 1944 (UK 1972)
The sleepwalker 1974
Through a glass, darkly 1950 (UK 1951)
MCCLUNG, Clarence Erwin ed. *Handbook of microscopical technique* 1929
MCCLUNG, Nellie Letitia *Clearing in the west* 1935
The stream runs fast 1945

MCCLURE, David (1748–1820) *Diary* with notes by F. B. Dexter 1899
MCCLURE, James Howe *The caterpillar cop* 1972
Rogue eagle 1976
Snake 1975
MCCORMICK, John & MASCAREÑAS, Mario Sevilla *The complete aficionado* 1967
MCCOWAN, Daniel *Animals of the Canadian Rockies* 1936
MCCRACKEN, Daniel Delbert *Digital computer programming* 1957
MCCRUM, Robert *In the secret state* 1980
MCCULLERS, Carson *Clock without hands* 1961
The heart is a lonely hunter 1940 (UK 1943)
The member of the wedding 1946
MCCULLOCH, Walter Fraser *Woods words: a comprehensive dictionary of loggers' terms* 1958
MCCULLOUGH, Colleen *The thorn birds* 1977
MCCUTCHAN, Philip Donald *Call for Simon Shard* 1974
Storm south 1959
MCCUTCHEON, George Barr *The rose in the ring* 1910
MCDAVID, Raven I. *See* MENCKEN, Henry Louis
'MACDIARMID, Hugh' (Christopher Murray Grieve) *The company I've kept* 1966
To circumjack Cencrastus; or, The curly snake 1930
MACDONALD, Betty Heskett *The egg and I* 1945 (UK 1946)
MCDONALD, Gregory *Confess, Fletch* 1976 (UK 1977)
MACDONALD, John Dann *The brass cupcake* 1950 (UK 1955)
The executioners 1958 (UK 1959)
The girl, the gold watch, and everything 1962
ed. *The lethal sex: the 1959 anthology of the Mystery Writers of America* 1959 (UK 1962)
Pale gray for guilt 1968 (UK 1969)
'MACDONALD, John Ross' *The Galton case* See: 'MACDONALD, Ross'
MACDONALD, Philip *Mystery of the dead police* 1933
'MACDONALD, Ross' (Kenneth Millar) *The Galton case* 1959 (UK 1960)
The instant enemy 1968
MACDONELL, Archibald Gordon *Autobiography of a cad* 1938
England, their England 1933
MCDOUGALL, William *An outline of abnormal psychology* 1926
MACE, Cecil Alec ed. *British philosophy in the mid-century* 1957
MCEVOY, Joseph Patrick *Hollywood girl* 1929 (UK 1930)
Show girl 1928
MCEWAN, Ian *The comfort of strangers* 1981
MCFADDEN, Cyra *The serial: a year in the life of Marin County* 1977 (UK 1978)
MCGAFFEY, Kenneth *The sorrows of a show girl* 1908
MACGILL, Patrick *The brown brethren* 1917
MCGILVERY, Robert Warren *Biochemistry: a functional approach* 1970
MCGIRR, Edmund *Bardel's murder* 1973
Death pays the wages 1970
An entry of death 1969
The lead-lined coffin 1968
A murderous journey 1974
MCGIVERN, William Peter *The big heat* 1953
Caprifoil 1972 (UK 1973)
M'GOWAN, George tr. *A. Bernthsen's A text-book of organic chemistry* 1889
McGraw-Hill dictionary of modern economics, The 1965
McGraw-Hill encyclopedia of science and technology 15 vols. 1960
— (new ed.) 15 vols. 1966
McGraw-Hill yearbook of science and technology 1963–
MACHLIN, Milton *Pipeline* 1976
'MCHUGH, Hugh' (George Vere Hobart) *Back to the woods: the story of a fall from grace* 1903
Skidoo! 1906
You can search me 1905
MCILVANNEY, William *Laidlaw* 1977
MACINNES, Colin *Absolute beginners* 1959
City of spades 1957
Mr. Love and Justice 1960
MCINNES, Graham *The road to Gundagai* 1965
MACINNES, Hamish *Climb to the lost world* 1974
MACINNES, Helen *Agent in place* 1976
Message from Malaga 1972
The Salzburg connection 1969
MCINTOSH, Louis *Oxford folly* 1956
MACKAIL, Denis George *How amusing! And a lot of other fables* 1929
MACKAY, Kenneth *Out back* 1893
— (ed. 2) 1893
MCKELLAR, Thomas Peter Huntly *Experience and behaviour* 1968
MCKENNEY, Thomas Loraine *Memoirs, official and personal; with sketches of travels among the northern and southern Indians* 2 vols. 1846

MACKENZIE, Donald *Postscript to a dead letter* 1973
Raven and the kamikaze 1977
MACKENZIE, Edward Montague Compton *April fools* 1930
Carnival 1912 (some quotations taken from 1912 US ed.)
The early life and adventures of Sylvia Scarlett 1918
Gallipoli memories 1929
Guy and Pauline 1915
The heavenly ladder 1924
My life and times 10 vols. 1963–71
The old men of the sea 1924
Sinister Street 2 vols. 1913–14
Sylvia and Michael: the later adventures of Sylvia Scarlett 1919
Whisky galore 1947
MACKENZIE, John *Ten years north of the Orange River: a story of every-day life and work among the South African tribes, from 1859 to 1869* 1871
MACKENZIE, Norman Ian *Conviction* 1958
MCKERROW, Ronald Brunlees *An introduction to bibliography for literary students* 1927
MCKIE, Ronald Cecil H. *The company of animals* 1965
MACKINNON, Doris Livingston & HAWES, R. S. J. *An introduction to the study of protozoa* 1961
MACKINTOSH, Ewart Alan (1893–1917) *War, the liberator, and other pieces* 1918
MCKNIGHT, Cathy & TOBLER, John *Bob Marley: the roots of reggae* 1977
MCKNIGHT, George Harley *English words and their background* 1923
MACLAREN-ROSS, Julian *The doomsday book* 1961
Until the day she dies 1960
MACLEAN, Fitzroy Hew *Eastern approaches* 1949
MACLEAN, Magnus ed. *Modern electric practice* 6 vols. 1904–5
MCLEAN, Sarah Pratt *Cape Cod folks* 1881
MCLEAN, Terence Power *Kings of Rugby: the British Lions' 1959 tour of New Zealand* 1960
Maclean's magazine 1912–
MCLEAVE, Hugh *A question of negligence* 1970 (UK 1973)
MACLEOD, Iain *Bridge is an easy game* 1952
'MACLEOD, Robert' (William Knox) *Burial in Portugal* 1973
MCLINTOCK, Alexander Hare ed. *A descriptive atlas of New Zealand* 1959
MCLUHAN, Herbert Marshall *The Gutenberg galaxy* 1962
The mechanical bride: folklore of industrial man 1951 (UK 1967)
Understanding media: the extensions of man 1964
—— & FIORE, Quentin *The medium is the massage* 1967
MCMAHON, Patrick Joseph *Aircraft propulsion* 1971
MCMULLEN, Edwin Wallace *English topographic terms in Florida, 1563–1874* 1953
MACNAB, Angus *The bulls of Iberia: an account of the bullfight* 1957
MACNEICE, Frederick Louis (1907–63) tr. *Aeschylus' Agamemnon* 1936
Astrology ed. D. Hill 1964
Autumn journal: a poem 1939
Autumn sequel: a poem 1954
Christopher Columbus: a radio play 1944
The dark tower, and other radio scripts 1947
The earth compels: poems 1938
tr. *Goethe's Faust* 1951
Holes in the sky: poems 1944–1947 1948
I crossed the Minch 1938
The last ditch 1940
Modern poetry: a personal essay 1938
Poems 1935
The poetry of W. B. Yeats 1941
See also: AUDEN, Wystan Hugh
MCNEILE, Alan Hugh *An introduction to the study of the New Testament* 1927
MACQUARRIE, John ed. *A dictionary of Christian ethics* 1967
MACQUOID, Percy *A history of English furniture* 1904
MCRAE, Archibald Graham *The hill called Grazing: the story of a Transvaal farm* 1956
MCRAE, Thomas Watson *The impact of computers on accounting* 1964
MACREADY, William Charles (1793–1873) *Diaries* ed. W. Toynbee 2 vols. 1912
MACVICAR, Angus *The painted doll affair* 1973
MADDOX, Harry Alfred *A dictionary of stationery* 1923
Mademoiselle 1935–
Madras. Music Academy *Journal* 1930–
MAERZ, Aloys John & PAUL, Morris Rea *A dictionary of color* 1930
MAGUIRE, Michael *Scratchproof* 1976
MAIDEN, Joseph Henry *The forest flora of New South Wales* 8 vols. 1904–25
MAIER, Richard Ali & MAIER, Barbara M. *Comparative animal behavior* 1970

MAILER, Norman *Advertisements for myself* 1959 (UK 1961)
Cannibals and Christians 1966 (UK 1967)
The naked and the dead 1948 (UK 1949)
MAINES, George H. & GRANT, Bruce *Wise-crack dictionary* 1926
Mainichi Daily News (Tokyo) 1922– (title varies)
MAJOR, Clarence *Dictionary of Afro-American slang* 1970
MALAMUD, Bernard *The assistant* 1957
Dubin's lives 1979
The fixer 1966 (UK 1967)
The natural 1952 (UK 1963)
A new life 1961 (UK 1962)
The tenants 1971
Malay Mail (Kuala Lumpur) 1896–
'MALCOLM X' (Malcolm Little) *Autobiography* 1965
Maledicta: the international journal of verbal aggression (Waukesha, Wisconsin) 1977–
'MALET, Lucas' (Mary St. Leger Harrison) *The dogs of want: a modern comedy of errors* 1924
The wages of sin 3 vols. 1891
MALINOWSKI, Bronislaw Kasper (1884–1942) *Coral gardens and their magic* 2 vols. 1935
A scientific theory of culture, and other essays 1944
Sex and repression in savage society 1927
MALLETT, Edward *Telegraphy and telephony, including wireless* 1929
'MALLOCH, Peter' (William Murdoch Duncan) *Kickback* 1973
MALTHUS, Thomas Robert (1766–1834) *Travel diaries* ed. P. James 1966
Man: a monthly record of anthropological science 1901–
Management engineering 1921–3
Manchester Guardian Weekly 1919–68
MANCHON, Joseph *Le slang: lexique de l'anglais familier et vulgaire, précédé d'une étude sur la prononciation et la grammaire populaires* 1923
MANDER, John *Static society: the paradox of Latin America* 1969
MANING, Frederick Edward *Old New Zealand* 1863
MANKOWITZ, Cyril Wolf & HAGGAR, Reginald George *The concise encyclopedia of English pottery and porcelain* 1957
MANN, Anthony *Tiara* 1973
MANN, Charles Riborg & MILLIKAN, Robert Andrews tr. *P. Drude's The theory of optics* 1902
MANN, Jessica *Mrs. Knox's profession* 1972
The sticking place 1974
MANN, Mary E. *In summer shade* 3 vols. 1893
MANNING, Frederic *The middle parts of fortune* 2 vols. 1929
MANNING, Olivia *The great fortune* 1960
The rain forest 1974
'MANSFIELD, Katherine' (Kathleen Middleton Murry) (1888–1923) *Bliss, and other stories* 1920
Collected stories 1945
The doves' nest, and other stories 1923
Journal ed. J. M. Murry 1954
Letters to John Middleton Murry, 1913–1922 ed. J. M. Murry 1951
Something childish, and other stories 1924
MANSFIELD, Walter Kenneth *Elementary nuclear physics* 1958
MANVELL, Arnold Roger & HUNTLEY, John *The technique of film music* 1957
MANWARING, Charles William ed. *A digest of the early Connecticut probate records* 3 vols. 1904–6
MARAIS, Eugène Nielen *My friends the baboons* [anon. tr.] 1939
MARBUT, C. F. tr. *K. D. Glinka's The great soil groups of the world and their development* 1927
MARCHANT, Edgar Walford *Radio telegraphy and telephony* 1923
MARCUS, Stanley *Minding the store: a memoir* 1974 (UK 1975)
MARCUS, Steven *The other Victorians* 1966
MARDER, Irving *The Paris bit* 1967
MARETT, Robert Ranulph ed. *Anthropology and the classics: six lectures* 1908
MARGERISON, Donald & EAST, George C. *An introduction to polymer chemistry* 1967
MARIN, Alfred C. *The clash of distant thunder* 1968 (UK 1969)
Rise with the wind 1969
Marine engineer (title varies) 1879–1972
Mariner's mirror 1911–
MARION, Fulgence *Wonderful balloon ascents* [anon. tr.] 1870
'MARJORIBANKS' *The fluff-hunters* 1903
MARKFIELD, Wallace *To an early grave* 1964 (UK 1965)
MARKS, Percy *Collector's choice* 1972
The plastic age 1924
MARKS, Robert W. *The dymaxion world of Buckminster Fuller* 1960

MARLOTH, Hermann Wilhelm Rudolf (1855–1931) *The flora of South Africa* 4 vols. 1913–32
MARPLES, Morris *Public school slang* 1940
University slang 1950
MARQUAND, Allan *Greek architecture* 1909
MARQUAND, John Phillips *H.M. Pulham, Esquire* 1941 (UK 1942)
Wickford Point 1939
MARQUARD, Leopold *The peoples and policies of South Africa* 1952
MARRIOTT, Harry *Cariboo cowboy* 1966
MARRIOTT, William *Hints to meteorological observers* 1881
— (ed. 6) 1906
MARSH, Edward Howard *Rupert Brooke: a memoir* 1918
MARSH, John Thompson *Self-smoothing fabrics* 1962
MARSH, Ngaio Edith *Artists in crime* 1938
Black as he's painted 1974
Colour scheme 1943
Dead water 1963 (UK 1964)
Death and the dancing footman 1941 (UK 1942)
Death in a white tie 1938
Death in ecstasy 1936
Death of a fool 1956 (UK ed. 1957 with title *Off with his head*)
Death of a peer 1940 (UK ed. 1941 with title *Surfeit of Lampreys*)
False scent 1959 (UK 1960)
Final curtain 1947
Grave mistake 1978
Hand in glove 1962
A man lay dead 1934
Off with his head See above: *Death of a fool*
Opening night 1951
Overture to death 1939
Singing in the shrouds 1958 (UK 1959)
Surfeit of Lampreys See above: *Death of a peer*
Swing, brother, swing 1949
Tied up in tinsel 1972
Vintage murder 1937
MARSHAK, Robert Eugene *Meson physics* 1952 (UK 1953)
MARSHALL, Agnes B. *Mrs. A. B. Marshall's cookery book* 1888
— ('tenth thousand') 1889
Mrs. A. B. Marshall's larger cookery book of extra recipes 1892
MARSHALL, Bruce *George Brown's schooldays* 1946
MARSHALL, Humphry *Arbustrum Americanum: the American grove; or, An alphabetical catalogue of forest trees and shrubs, natives of the American United States* 1785
MARSHALL, Sybil *An experiment in education* 1963
MARSHALL, Thomas Humphrey *Citizenship and social class, and other essays* 1950
MARTIN, Basil Kingsley *Editor: a second volume of autobiography, 1931–45* 1968
Father figures: a first volume of autobiography, 1897–1931 1966
MARTIN, Helen *Tillie, a Mennonite maid* 1904 (UK 1905)
MARTIN, Joseph ed. *A new and comprehensive gazetteer of Virginia, and the District of Columbia* 1835
MARTIN, Laurence Cleveland & HYNES, Martin *Clinical endocrinology for practitioners and students* 1948
— (ed. 3, by L. C. Martin alone) 1961
MARTIN, Mary Ann *Our Maoris* 1884
MARTIN, Robert Denis tr. *W. Wickler's Mimicry in plants and animals* 1968
MARTIN, William *The New Zealand nature book* 1929 (UK 1930)
MARX, Groucho (Julius Henry Marx) (1895–1977) *The Groucho letters: letters from and to Groucho Marx* 1967
MARX, Heinrich Karl *See* MOORE, Samuel & AVELING, E. B.
Maryland historical magazine 1906–28
MASCALL, Eric Lionel *Christian theology and natural science: some questions on their relations* 1956
The recovery of unity: a theological approach 1958
MASEFIELD, John Edward *Ballads* 1903
Ballads and poems 1910
The bird of dawning 1933
Captain Margaret: a romance 1908
The Conway, from her foundation to the present day 1933
Dead Ned: the autobiography of a corpse 1938
The everlasting mercy 1911
Gautama the enlightened, and other verse 1941
The hawbucks 1929
Live and kicking Ned 1939
Lollingdon Downs and other poems, with sonnets 1917
Odtaa 1926
Reynard the fox; or, The ghost heath run 1919
Sard Harker 1924
A tarpaulin muster 1907
The tragedy of Nan, and other plays 1909

Mason, Alfred Edward Woodley The house of the arrow 1924
Miranda of the balcony 1899
The truants 1904
Mason, Frederick A. tr. G. von Georgievics's A text-book of dye chemistry 1920
Mason, Monck Aeronautica; or, Sketches illustrative of the theory and practice of aerostation 1838
Mason, Richard Lakin The world of Suzie Wong 1957
Massachusetts. Agricultural Survey First report on the agriculture of Massachusetts by H. Colman 1838
Second report 1839
Massachusetts (Colony). House of Representatives Journals 1919–
Massingham, Harold John The curious traveller 1950
World without end 1932
Masson, Madeleine Birds of passage 1950
Masterman, John Cecil Fate cannot harm me 1935
Masters, John Bhowani Junction 1954
Bugles and a tiger: a personal adventure 1956
Coromandel! 1955
The deceivers 1952
Far, far the mountain peak 1957
The lotus and the wind 1953
Nightrunners of Bengal 1951
Trial at Monomoy 1964
Masters, Ted Surfing made easy 1962
Masur, Harold Q. ed. Murder most foul 1973
Materials and technology: a systematic encyclopedia of the technology of materials used in industry and commerce 8 vols. 1968–75
Mates, Benson Elementary logic 1965
Mathematical reviews 1940–
Mathematical tables and other aids to computation 1943–59
'Mather, Berkeley' (John Evan W. Davies) The break in the line 1970
Snowline 1973
The springers 1968
Mathews, Albert Prescott Physiological chemistry 1916
Mathews, Anne Memoirs of Charles Mathews, comedian 4 vols. 1838–9
Mathews, Catharine Van Cortlandt Andrew Ellicott: his life and letters 1908
Mathews, Mitford M. ed. A dictionary of Americanisms on historical principles See under title
Mathewson, Christopher Pitching in a pinch; or, Baseball from the inside 1912
Second base Sloan 1917
Matsell, George Washington Vocabulum; or, The rogue's lexicon, compiled from the most authentic sources 1859
Matthews, Leonard Harrison The life of mammals 2 vols. 1969–71
Maugham, Robin (Robert Cecil Romer Maugham) The servant 1972
Maugham, William Somerset The bishop's apron: a study in the origins of a great family 1906
The bread-winner 1930
Cakes and ale; or, The skeleton in the cupboard 1930
The circle 1921
The constant wife 1926
Liza of Lambeth 1897
The moon and sixpence 1919
Of human bondage 1915
On a Chinese screen 1922
Our betters 1923
The painted veil 1925
The razor's edge 1944
Sheppey 1933
The summing up 1938
Theatre 1937
Then and now 1946
A writer's notebook 1949
Mawer, Allen & Stenton, Frank Merry Introduction to the survey of English place-names 1924
Maxwell, Nicole Witch-doctor's apprentice 1962
May, Derwent James Laughter in Djakarta 1973
Maycock, William Perren Electric lighting and power distribution 3 parts 1892–3
Electric wiring, fittings, switches, and lamps 1899
Mayer, Edgar Clinical application of sunlight and artificial radiation 1926
Mayer, Ralph The artist's handbook of materials and techniques 1940
— (UK ed.) 1951
A dictionary of art terms and techniques 1969
Mayer-Gross, Willy, Slater, Eliot T. O., & Roth, Martin Clinical psychiatry 1954
— (ed. 3) 1969
Maylard, Alfred Ernest A treatise on the surgery of the alimentary canal 1896
Mayr, Ernst Animal species and evolution 1963
Principles of systematic zoology 1969
——, Linsley, E. G., & Usinger, R. L. Methods and principles of systematic zoology 1953

Mead, Margaret Growing up in New Guinea: a comparative study of primitive education 1930
Male and female: a study of the sexes in a changing world 1949
Meade, Elizabeth ('Lillie') Thomasina A sweet girl graduate 1891
Meccano magazine 1916–
Medawar, Peter Brian & Medawar, Jean Shinglewood The life science: current ideas of biology 1977
Medbery, James Knowles Men and mysteries of Wall Street 1870
Meddelelser om Grønland (Copenhagen) 1879–
Medical annual, The 1883–
Medical journal of Australia 1914–
Medical news (Philadelphia) 1843–1905
Medical record (New York) 1866–1922
Medical Research Council Mathematics and computer science in biology and medicine 1965
A system of bacteriology in relation to medicine 9 vols. 1929–31
Medium ævum 1932–
Medway, Gathorne (Gathorne-Hardy), Baron The wild mammals of Malaya and offshore islands including Singapore 1969
Meglitsch, Paul Allen Invertebrate zoology 1967
Melbourne Truth (title varies) 1902–
Melchior, Ib Sleeper agent [anon. tr.] 1975 (UK 1976)
Mellor, Joseph William A comprehensive treatise on inorganic and theoretical chemistry 1922–
Modern inorganic chemistry 1912
Melly, Alan George Heywood Owning-up 1965
Melody maker, The 1926–
Melville, Herman Pierre; or, The ambiguities 1852
Redburn: his first voyage (UK ed. in 2 vols.) 1849
Typee: a peep at Polynesian life 1846
'Melville, Jennie' (Gwendoline Butler) Ironwood 1972
Nun's castle 1974
Mencken, Henry Louis The American language: a preliminary inquiry into the development of English in the United States 1919
— (rev. ed.) 1921
— (ed. 3) 1923
— (ed. 4) 1936
— Supplement 2 vols. 1945–8
— (abridged ed. with new material by R. I. McDavid) 1963
Mendelsohn, Oscar Adolf The dictionary of drink and drinking 1965
Men's hockey ('Know the game' series) 1950
— (rev. ed.) 1965
Merchants' magazine and commercial review 1839–70 (1851–60 with title Hunt's merchants' magazine and commercial review)
'Mercury' dictionary of textile terms, The 1950
Meredith, George (1828–1909) Letters ed. C. L. Cline 3 vols. 1970
Mereness, Newton Dennison ed. Travels in the American colonies 1690–1783 1916
Merriman, Arthur Douglas A dictionary of metallurgy 1958
Merton, Robert King Social theory and social structure 1949
— (rev. ed.) 1957
— (rev. ed.) 1968
Merwin, Samuel & Webster, Henry Kitchell Calumet 'K' 1901
Metabolism: clinical and experimental 1952–
Metals handbook See: American Society for Metals
Metcalf, Clell Lee & Flint, Wesley Pillsbury Fundamentals of insect life 1932
Destructive and useful insects 1928
— (ed. 4) 1962
Meteorological Office See United Kingdom
Metronome 1885–
Meyer, Jerome Sydney & Hanlon, Stuart Fun with the new math 1966
Meynell, Laurence Walter Double fault 1965
Hooky gets the wooden spoon 1977
The thirteen trumpeters: a Hooky Hefferman story 1973
Virgin luck 1963
Mezzrow, Milton & Wolfe, Bernard Really the blues 1946 (UK 1957)
Miall, Arthur Bernard tr. C. Guenther's A naturalist in Brazil 1931
Miall, Stephen ed. A new dictionary of chemistry 1940
— (ed. 2, by S. Miall & L. M. Miall) 1949
Michaux, François André Histoire des arbres forestiers de l'Amérique Septentrionale 3 vols. 1810–13
Michelin Tire Corporation New York City 1968
Michelin Tyre Company Michelin guide to Great Britain (ed. 7) 1923
Michell, John Henry & Belz, Maurice Henry The elements of mathematical analysis 2 vols. 1937

Michener, James Albert Chesapeake 1978
The source 1965
Michigan Academy of Science, Arts and Letters Papers Vol. X (for 1928) 1929
Michigan State Agricultural Society Transactions 1849–61
Mikes, H. George Down with everybody! 1951
Milk and honey: Israel explored 1950
Milburn, Clara Emily Mrs. Milburn's diaries: an Englishwoman's day-to-day reflections, 1939–45 ed. P. Donnelly 1979
Miles, Beryl The stars my blanket 1954
Mill, John Stuart (1806–73) Earlier letters ed. F. E. Mineka 2 vols. 1963 (Vols. XII & XIII of Collected works)
The early draft of J. S. Mill's autobiography ed. J. Stillinger 1961
Letters ed. H. S. R. Elliot 2 vols. 1910
Millar, George Reid Horned pigeon 1946
Maquis 1945
Millar, Margaret Ask for me tomorrow 1976 (UK 1977)
The soft talkers 1957
Miller, Arthur All my sons 1947
Collected plays 1957 (UK 1958)
The crucible 1953 (UK 1956)
Death of a salesman 1949
The misfits 1961
Miller, Denis The Chinese jade affair 1973
Miller, Edwin Cyrus Plant physiology, with reference to the green plant 1931
Miller, Henry Valentine Black spring 1936
Nexus 1960 (UK 1964)
Plexus 1953 (UK 1963)
Sexus 2 vols. 1949 (UK 1969)
Tropic of Cancer 1934
Tropic of Capricorn 1939
Miller, Hugh The open city 1973
Miller, Joaquin First fam'lies in the Sierras 1875 (US 1876)
Life amongst the Modocs 1873
Millerson, Gerald The technique of television production 1961
Millett, Kate Flying 1974 (UK 1975)
Sexual politics 1970 (UK 1971)
Milliken, E. J. 'Arry ballads 1892
Millin, Sarah Gertrude The South Africans 1926
Mills, Charles Wright The power élite 1956
White collar: the American middle classes 1951
Mills, James Report to the Commissioner 1972
'Mills, Osmington' (Vivian Collin Brooks) Enemies of the bride 1966
Headlines make murder 1962
Stairway to murder 1959
Mills, P. W. F. The elements of practical flying 1935
Milne, Alan Alexander First plays 1919
Winnie-the-Pooh 1926
Milne, Christopher Robin The enchanted places 1974
Milner, Christina Andrea & Milner, Richard Black players: the secret world of black pimps 1973
Milton, William Fitzwilliam, Viscount & Cheadle, Walter Butler The North-West Passage by land 1865
Milton Keynes Express 1973–
Minchin, Edward Alfred An introduction to the study of the protozoa, with special reference to the parasitic forms 1912
Mineralogical abstracts 1920–
Ministry of . . . See United Kingdom. Ministry of . . .
Minter, Davide Caroline ed. Modern needlecraft 1932
Missingham, Hal A student's guide in commercial art 1948
Mitchell, Geoffrey Duncan Sociology: the study of social systems 1959
Mitchell, Gladys Maude W. Laurels are poison 1942
The murder of Busy Lizzie 1973
The mystery of a butcher's shop 1929
Spotted hemlock 1958
Mitchell, James Smear job 1975
Mitchell, Margaret Gone with the wind 1936
Mitchison, Naomi Margaret We have been warned 1935
Mitford, Jessica Lucy Freeman The American way of death 1963
Mitford, Nancy Freeman The blessing 1951
Christmas pudding 1932
Don't tell Alfred 1960
Love in a cold climate 1949
Pigeon pie 1940
The pursuit of love 1945
Mittelholzer, Edgar Austin Shadows move among them 1951
'Mixer, The' (Henry Kirk) The transport workers' song book c 1926
A modern junior dictionary, specially adapted for use in Australia and New Zealand 1944
Modern Language Association of America Publications 1889–

Modern language notes 1886–
Modern language review 1905–
Modern law review 1937–
Modern philology 1903–
Moffat, Gwen The Corpse road 1974
Deviant death 1973
Miss Pink at the edge of the world 1975
Mogey, John McFarlane Family and neighbourhood: two studies in Oxford 1956
Mohr, Charles Theodore Plant life of Alabama 1901
Molloy, Edward ed. Radio and television engineers' reference book 1954
Moloney, Peter A plea for Mersey; or, The gentle art of insinuendo 1966
Moncrieff, Robert Wighton The chemical senses 1944
Money, Charles L. Knocking about in New Zealand 1871
Monier-Williams, Monier See Williams, Monier
Monist, The: a quarterly magazine devoted to the philosophy of science 1890–
Monographic medicine 7 vols. 1916 (by various authors)
Monsarrat, Nicholas John Turney The cruel sea 1951
This is the schoolroom 1939
Montagu, Mary Wortley (1689–1762) Complete letters ed. R. Halsband 3 vols. 1965–7
Montague, Charles Edward Disenchantment 1922
A hind let loose 1910
Montgomery, Lucy Maud Anne of Avonlea 1909
Anne of Green Gables 1908
Anne of Ingleside 1939
Montgomery Ward & Co. Catalogue and buyers' Guide No. 57, Spring & Summer 1895
Monthly repository of theology (title varies) 1806–38
Monypenny, William Flavelle & Buckle, George Earle The life of Disraeli 6 vols. 1910–20
Moore, Brian Catholics 1972
Moore, Doris Langley E. Nesbit: a biography 1933
Moore, Francis Cruger How to build a home 1897
Moore, John S. ed. The goods and chattels of our forefathers: Frampton Cotterell and district probate inventories 1539–1804 1976
Moore, Preston L., et al. ed. Drilling practices manual 1974
Moore, Raymond Cecil Introduction to historical geology 1949
— ed. Treatise on invertebrate paleontology 1953–
——, Lalicker, C. G., & Fischer, A. G. Invertebrate fossils 1952
Moore, Ruth Candlemas Bay 1950
Moore, Samuel & Aveling, Edward Bibbins tr. K. Marx's Capital: a critical analysis of capitalist production 1887
Moore, Walter John Physical chemistry 1950
Moore, Wilfred George A dictionary of geography 1949
Moorehead, Alan McCrae No room in the ark 1959
Rum jungle 1953
Moorhouse, Walter Wilson The study of rocks in thin section 1959
Moran, James The composition of reading matter 1965
Moran, Patrick Alfred P. An introduction to probability theory 1968
More, Hannah (1745–1833) Letters ed. R. B. Johnson 1925
Morecambe Guardian 1920–
Morehead, Albert Page Human pathology 1965
Morgan, Charles Langbridge The flashing stream 1938
The fountain 1932
The judge's story 1947
Portrait in a mirror 1929
The voyage 1940
Morgan, John Glanfil Involved 1967
Morgan, Mary Ella How to dress a doll 1908
Morgan, Thomas Hunt, et al. The mechanism of Mendelian heredity 1915
Morgan, William Alphonse ed. The 'House' on sport, by members of the London stock exchange 1898
— (suppl. vol.) 1899
Morice, Anne Killing with kindness 1974
The men in her death 1981
Murder in mimicry 1977
Morison, Samuel Eliot The European discovery of America: the northern voyages 1971
The European discovery of America: the southern voyages 1974
Morley, Christopher Darlington Thorofare 1942 (UK 1943)
Kitty Foyle 1939
Morning Herald and Daily Advertiser 1780–1869
Morning Star 1966–
'Morphy, Marcelle, Countess' (Marcelle Azra Forbes) ed. Recipes of all nations 1935

Morrell, Robert Selby, et al. eds. Synthetic resins and allied plastics 1937
Morris, Charles William Signs, language and behavior 1946
Morris, Clara (Clara Morris Harriott) Life on the stage: my personal experiences and recollections 1901
Morris, Desmond John The mammals: a guide to the living species 1965
Manwatching: a field guide to human behaviour 1977
The naked ape 1967
See also: Morris, Ramona
'Morris, John' (John Hearne & Morris Cargill) Fever grass 1969
Morris, Ramona & Morris, Desmond John Men and apes 1966
Men and pandas 1966
Men and snakes 1965
Morris, Terence & Morris, Pauline Pentonville: a sociological study of an English prison 1963
Morris, William News from nowhere; or, An epoch of rest 1890 (UK 1891)
Morris owners' manual 1925
Morse, Eric Wilton Fur trade canoe routes of Canada: then and now 1969
Morse, Mary The unattached: a report of the project carried out by the National Association of Youth Clubs 1965
Morse, Philip McCord Vibration and sound 1936
'Mortimer, Geoffrey' (Walter M. Gallichan) Like stars that fall 1895
Mortimer, John Clifford A voyage round my father 1971
Mortimer, Penelope Daddy's gone a-hunting 1958
Morton, Henry Canova Vollam In search of England 1927
In search of South Africa 1948
Morton, Jack Football 1972
Moss, Stirling Craufurd In the track of speed 1957
Motor, The 1903–
Motor boat 1904–
Motor-car world 1899–1905
Motor-cycle, The 1903–
Motor manual, The 1903–59
Mottram, Ralph Hale The Spanish farm 1924
Movie 1962–
Moyer, James Ambrose & Wostrel, J. F. Radio handbook: including television and sound motion pictures 1931
Moyes, Patricia The curious affair of the third dog 1973
Dead men don't ski 1959
Murder à la mode 1963
Who saw her die? 1970
Muehlenberg, Gotthilf Henry Ernest Catalogus plantarum Americæ Septentrionalis...; or, A catalogue of the...plants of North America, arranged according to the sexual system of Linnæus 1813
Muir, Edwin (1887–1959) Collected poems, 1921–1958 ed. W. Muir & J. C. Hall 1960
Muir, Robert Text-book of pathology 1924
—— & Ritchie, James Manual of bacteriology 1897 (and several later editions used)
Muir, Ward Observations of an orderly: some glimpses of life and work in an English war hospital 1917
Mulford, Clarence Edward Bar-20 1907 (UK 1914)
The Bar-20 three 1921
Black Buttes 1923
The coming of Cassidy—and the others 1913
Cottonwood Gulch 1925
Hopalong Cassidy 1910
Hopalong Cassidy's protégé 1926
Johnny Nelson 1920 (UK 1921)
The man from Bar-20: a story of the cow-country 1918
The orphan 1908
Rustlers' valley 1924
—— & Clay, John Wood Buck Peters, ranchman 1912 (UK 1921)
Mulgan, John Alan Edward Man alone 1939
Mullally, Frederic The Munich involvement 1968
Mullard, Chris Black Britain; with an account of recent events at the Institute of Race Relations by A. Kirby 1973
Mullin, Glen Hawthorne Adventures of a scholar tramp 1925
Mumford, John Kimberley Oriental rugs 1900 (UK 1901)
Mumford, Lewis City development: studies in disintegration and renewal 1945 (UK 1946)
The city in history: its origins, its transformations, and its prospects 1961
Munby, Arthur Joseph See Hudson, Derek Rommel
Munro, George Campbell Birds of Hawaii 1944
Munro, Neil The daft days 1907
Munroe, Kirk The golden days of '49: a tale of the California diggings 1889

Munroe, Ruth Learned Schools of psychoanalytic thought 1955 (UK 1957)
Munrow, David Instruments of the Middle Ages and Renaissance 1976
Munson, Kenneth Pioneer aircraft 1903–1914 1969
Murdoch, Jean Iris The bell 1958
Flight from the enchanter 1956
Henry and Cato 1976
Nuns and soldiers 1980
The sacred and profane love machine 1974
The sandcastle 1957
The sea, the sea 1978
A severed head 1961
The time of the angels 1966
Under the net 1954
An unofficial rose 1962
A word child 1975
Murdoch, Walter & Drake-Brockman, Henrietta F. Y. eds. Australian short stories 1951
For 2nd ser. see 'James, Brian'
Murphy, Gardner Personality 1947
Murray, Amelia Matilda Letters from the United States, Cuba, and Canada 2 vols. 1856
Murray, Henry Alexander Explorations in personality 1938
Murray, Katharine Maud Elisabeth Caught in the web of words: James A. H. Murray and the Oxford English Dictionary 1977
Murray, Peter John & Murray, Linda A dictionary of art and artists 1959
Murray, William Buckley The sweet ride 1967
Murry, John Middleton The necessity of pacifism 1937
Pencillings: little essays on literature 1923
The voyage 1924
Museum of Comparative Zoology, Harvard University Bulletin 1863–
Musical Association, The Proceedings 1875–
Musical quarterly 1915–
Musical times 1844–
Mycologia 1909–
Myers, Arthur Wallis Twenty years of lawn tennis 1921
Myers, Leonard Merrin Electron optics: theoretical and practical 1939

N

NATO Automatic translation of languages: papers presented at NATO Summer School held in Venice, July 1962 1966
N.Z.E.F. See New Zealand Expeditionary Force
Nabokov, Vladimir Vladimirovich Bend sinister 1947
The defence tr. by M. Scammell in collaboration with the author 1964
The gift tr. by M. Scammell and the author 1963
Invitation to a beheading tr. by D. Nabokov and the author 1960
Lolita 2 vols. 1955 (US 1 vol. 1958)
Look at the harlequins! 1974 (UK 1975)
Nabokov's dozen: a collection of thirteen stories 1958 (UK 1959)
Pale fire 1962
Pnin 1957
The real life of Sebastian Knight 1941
Speak, memory (US ed. with title Conclusive evidence) 1951
— (UK ed. 2) 1967
Nadell, Aaron Projecting sound pictures: a practical textbook for projectionists and managers 1931
Nagel, Ernest The structure of science: problems in the logic of scientific explanation 1961
'Na Gopaleen, Myles' (Brian O'Nolan) (1911–66) The best of Myles: a selection from 'Cruiskeen Lawn' ed. K. O'Nolan 1968
See also: 'O'Brien, Flann'
Naipaul, Shivadhar Srinivasa Black and white 1980
Naipaul, Vidiadhar Surajprasad An area of darkness 1964
In a free state 1971
The mimic men 1967
The mystic masseur 1957
Namier, Lewis Bernstein Conflicts: studies in contemporary history 1942
Narlikar, Jayant The structure of the universe 1977
Narragansett historical register 1882–91
Nash, Frederic Ogden The face is familiar 1940 (UK 1942)
Family reunion 1950 (UK 1951)
Good intentions 1942 (UK 1943)
I'm a stranger here myself 1938
Versus 1949
You can't get there from here 1957
Nasmith, Joseph The students' cotton spinning 1892
Nason, Leonard Hastings Chevrons 1926 (UK 1927)
Nation, The (Barbados) 1973–

Nation review (Melbourne) 1958–
National Academy of Sciences (Washington) Proceedings 1863–94; 1915–
National Bureau of Standards See United States. National ——
National observer (US) 1962–77
National police gazette (US) (title varies) 1845–
Natural history 1919–
Navy news 1954–
Nayler, Joseph Lawrence Dictionary of aeronautical engineering 1959
Neale, R. E. ed. Whittaker's electrical engineer's pocket-book 1920 See under title
Neblette, Carroll Bernard Photography: its principles and practice 1927
Nehru, Jawaharlal An autobiography 1936
Neill, Alexander Sutherland A dominie's log 1915 [dated 1916]
Neill! Neill! Orange peel!: a personal view of ninety years 1972
— (rev. ed.) 1973
Summerhill: a radical approach to education 1962
Nelson, George & Wright, Henry Niccolls Tomorrow's house 1945
Nelson, Stanley R. All about jazz 1934
Nesbit, Edith Five children and it 1902
The phoenix and the carpet 1904
The railway children 1906
Neuphilologische Mitteilungen 1899–
Neurology 1951–
'Neville, Margot' Ladies in the dark 1965
Nevin, Charles Merrick Principles of structural geology 1931
New acronyms and initialisms See Acronyms dictionary
New and complete dictionary of arts and sciences, A 4 vols. 1754–5
New biology ed. M. L. Johnson et al. 31 vols. 1945–60
New Castle (Delaware) Records of the court of New Castle on Delaware (Colonial Society of Pennsylvania) 1904
New Catholic encyclopedia ed. W. J. McDonald 15 vols. 1967
New England journal of medicine, The 1928–
New English weekly: a review of public affairs, literature and the arts (title varies) 1932–49
New Gould medical dictionary 1949
— (ed. 2) 1956
New Hampshire (Colony) Probate records of the province of New Hampshire 9 vols. 1907–41
New Jersey Archives of the State of New Jersey 16 vols. 1880–1917
New Jewish encyclopedia, The ed. D. Bridger 1962
New left review 1960–
New musical express 1952–
New Oxford history of music, The ed. J. A. Westrup et al. 1954–
New phytologist 1902–
New republic, The 1914–
New scientist, The 1956–
New society 1962–
New sporting magazine 1831–70
New statesman, The (title varies) 1913–
New York 1968–
New York Academy of Sciences Annals 1877–
New York dramatic news 1894–6
New York Evening Post 1802–1919
New York Herald Tribune (title varies) 1841–
New York Herald Tribune International (title varies) 1887–
New York law journal 1949–
New York medical journal 1865–1923
New York review of books 1963–
New York State Documents relative to the colonial history of the State of New-York See under title
New York State. Department of Correctional Services Guidelines to volunteer services 1974
New York Times, The 1857–
New York Times book review 1896–
New York World 1860–1931
New York World-Telegram 1931–66 (1950–66 with title New York World-Telegram and Sun)
New Yorker, The 1925–
New Zealand. Parliament. House of Representatives Appendix to the journals 1858–
New Zealand Expeditionary Force Chronicles 1916–19
New Zealand Expeditionary Force Times (2nd N.Z.E.F.) 1941–?5
New Zealand illustrated magazine 1901–5
New Zealand Institute (from 1935 Royal ——) Transactions and proceedings 1868–
New Zealand journal 1840–52
New Zealand journal of agriculture (title varies) 1910–
New Zealander listener 1939–
New Zealand News 1927–
New Zealand short stories 1953 See Davin, Daniel Marcus
— (2nd ser.) 1966 See Stead, Christian Karlson

New Zealand timber journal 1954–
New Zealand woman's weekly 1934–
Newbold, Thomas John Political and statistical account of the British settlements in the Straits of Malacca, viz., Pinang, Malacca and Singapore 2 vols. 1839
Newdigate-Newdegate, Anne Emily The Cheverels of Cheverel Manor 1898
Newman, Edwin Sunday punch 1979
Newman, Gordon F. The price 1974
Sir, you bastard 1970
You nice bastard 1972
Newman, Maxwell Herman Alexander Elements of the topology of plane sets of points 1939
Newnes complete amateur photography ed. M. L. Hall 1958
Newnes concise encyclopaedia of electrical engineering ed. M. G. Say 1962
Newnes concise encyclopaedia of nuclear energy 1962
News and Courier (Charleston, South Carolina) 1873–
News and Observer (Raleigh, North Carolina) 1872–
News and Press (Darlington, South Carolina) 1903–
News Chronicle 1930–60
News review 1936–50
Newspaper and general reader's pocket companion 2 parts 1855–6
Newsweek 1933–
Newth, George S. A text-book of inorganic chemistry 1894
— (ed. 12) 1907
'Newton, Francis' (Eric John Ernest Hobsbawm) The jazz scene 1959
Newton, Joseph Introduction to metallurgy 1938
Newton, Kenneth & Steeds, William The motor vehicle 1929 (and several later editions used)
Newton, Peter High country days 1949
Sheep thief 1972
Wayleggo 1947
Nichols, David Echinoderms 1962
Nichols, John Beverley The sweet and twenties 1958
Nicholson, John Kenyon The barker 1927
'Nicholson, Kate' (Judith Fay) Hook, line and sinker 1966
Nicholson, Meredith A Hoosier chronicle 1912
Nickerson, Elinor Barkley Kayaks to the Arctic 1967
Nicol, Abioseh The truly married woman, and other stories 1965
Nicolson, Harold George Curzon: the last phase, 1919–25; a study in post-war diplomacy 1934
Diaries and letters, 1930–1939 ed. N. Nicolson 1966
—— 1939–1945 ed. N. Nicolson 1967
—— 1945–1962 ed. N. Nicolson 1968
King George V: his life and reign 1952
Public faces 1932
Nida, Eugene Albert Morphology: the descriptive analysis of words 2 vols. 1944
— (another ed.) 1946
— ('ed. 2') 1949
Toward a science of translating 1964
Nielsen, Helen After midnight 1966 (UK 1967)
The brink of murder 1976
The severed key 1973
Niesewand, Peter The underground connection 1978
Niland, D'Arcy The big smoke 1959
Call me when the Cross turns over 1957 (UK 1958)
The shiralee 1955
Niles' weekly register 1814–37; continued as Niles' national register 1837–49
Nilson, Arvid The timber trees of New South Wales 1884
Nisbett, Alex The technique of the sound studio 1962
Nitsch, F. A. A general and introductory view of Professor Kant's Principles concerning man, the world and the Deity 1796
Nkosi, Lewis The rhythm of violence 1964
Noble, Montague Alfred Those 'Ashes': the Australian tour of 1926 1927
Nohara, Komakichi The true face of Japan 1936
Nolan, Frederick The Oshawa project 1974
Noller, Carl Robert Chemistry of organic compounds 1951
Textbook of organic chemistry 1951
Nordau, Max Simon Degeneration [anon. tr.] 1895
Norman, Barry The matter of mandrake 1967
Norman, Frank Bang to rights: an account of prison life 1958
Norman, John Roxborough A history of fishes 1931
— (ed. 2, by P. H. Greenwood) 1963
Norman, Richard Oswald Chandler Principles of organic synthesis 1968
Norris, Henry H. An introduction to the study of electrical engineering 1907
Norris, William Fisher & Oliver, Charles A. eds. System of diseases of the eye 4 vols. 1897–9
North, Barbara & North, Robert tr. M. Duverger's Political parties 1954
— (ed. 2) 1959
'North, Gil' (Geoffrey Horne) Sergeant Cluff and the day of reckoning 1967
Sergeant Cluff rings true 1972
North, John David The measure of the universe 1965
North, Lockhart The parasites 1928
North Carolina (Colony) Colonial records ed. W. L. Saunders 10 vols. 1886–90
Northern Territory News (Darwin, Australia) 1952–
Norton, Edward Felix The fight for Everest: 1924 1925
Norton, Olive Marion Dead on prediction 1970
Now lying dead 1967
A school of liars 1966
Norton, Oliver Willcox Army letters 1861–1865: being extracts from private letters to relatives and friends from a soldier in the field during the late civil war 1903
Norwich Mercury, Norfolk News and Journal (title varies) 1714–
Nott, John The cook's and confectioner's dictionary; or, The accomplished housewife's companion 1723
Nourse, Henry Stedman ed. The early records of Lancaster, Massachusetts See: Lancaster, Massachusetts
Novak, Emil Gynecological and obstetric pathology; with clinical and endocrine relations 1940
Nowakowski, T. Z. & Clarke, A. J. tr. V. L. Kretovich's Principles of plant biochemistry 1966
Nowell-Smith, Simon Harcourt ed. Edwardian England 1964
Nowottny, Winifred May T. The language poets use 1962
Noyce, Cuthbert Wilfrid Francis South Col: one man's adventure on the ascent of Everest, 1953 1954
Nuclear instruments and methods 1957– (1957–8 with title Nuclear instruments)
Nucleonics 1947–
'Number 1500' Life in Sing Sing 1904
Numbers: a quarterly review 1954–
Nunn, Thomas Percy Education: its data and first principles 1920
Nursing times and journal of midwifery 1905–
Nuttall, Thomas A journal of travels into the Arkansa territory, during the year 1819 1821
Nye, Edgar Wilson Baled hay 1884
Bill Nye and Boomerang; or, The tale of a meek-eyed mule 1881
Nyren, John The young cricketer's tutor 1833

O

Oakley, Kenneth Page Frameworks for dating fossil man 1964
Oates, Joyce Carol Bellefleur 1981
O'Brien, Edna August is a wicked month 1965
Country girls 1971
The lonely girl 1962
'O'Brien, Flann' (Brian O'Nolan) At Swim-Two-Birds 1939
The hard life: an exegesis of squalor 1961
See also: 'Na Gopaleen, Myles'
O'Brien, S. E. & Stephens, A. G. Material for a dictionary of Australian slang, 1900–1910 (typescript in Mitchell Library, Sydney)
Obstetrics and gynecology 1953–
O'Casey, Sean (1880–1964) Letters ed. D. Krause 1975–
The plough and the stars 1926
The star turns red 1940
Two plays [Juno and the paycock and The shadow of a gunman] 1925
Windfalls: stories, poems and plays 1934
Oceanography and marine biology: an annual review 1963–
O'Connell, Daniel (1775–1847) Correspondence ed. M. R. O'Connell 1972–
'O'Connor, Frank' (Michael Francis O'Donovan) Bones of contention, and other stories 1936
O'Connor, Jimmy The eleventh commandment 1976
Odell, George Clinton Densmore ed. Annals of the New York stage 15 vols. 1927–49
O'Donnell, Edwin P. Great big doorstep 1941
O'Donnell, Lilian The face of the crime 1968 (UK 1969)
O'Donnell, Peter The impossible virgin 1971
Sabre-tooth 1966
The silver mistress 1973
'Odysseus' (Charles Norton Edgcumbe Eliot) Turkey in Europe 1900
O'Faolain, Julia No country for young men 1980
O'Faolain, Sean A nest of simple folk 1933

Official encyclopedia of bridge ed. R. L. Frey & A. F. Truscott 1964
Offshore (US) 1954–
Offshore engineer 1975–
Offshore platforms and pipelining (The Petroleum Publishing Co.) 1976
Ogden, Charles Kay tr. H. Vaihinger's The philosophy of 'as if' 1924
—— & Richards, Ivor Armstrong The meaning of meaning: a study of the influence of language upon thought and of the science of symbolism 1923
Ogden, Robert Morris tr. K. Koffka's The growth of the mind: an introduction to child-psychology 1924
Ogilvy, Charles Stanley & Anderson, John T. Excursions in number theory 1966
Ogilvy, David Confessions of an advertising man 1963 (UK 1964)
O'Hara, John Henry Appointment in Samarra 1934 (UK 1935)
Pal Joey 1940 (UK 1952)
O'Hara, Kenneth The bird-cage 1968
The ghost of Thomas Penry 1977
Ohio archaeological and historical quarterly (title varies) 1887–
Ohwi, Jisaburō Flora of Japan 1965
Oil and gas journal 1910–
Okey, Thomas An introduction to the art of basket-making 1912
Oliver, Paul Savannah syncopators: African retentions in the blues 1970
Screening the blues 1968
Oliver, Walter Reginald Brook New Zealand birds 1930
Olsson, Yngve On the syntax of the English verb 1961
O'Neill, Eugene Gladstone (1888–1953) Ah, wilderness! and Days without end: two plays 1933 (UK 1934)
Anna Christie See below: The hairy ape
Beyond the horizon 1920
Desire under the elms 1925
Dynamo 1929
The Emperor Jones 1921 (UK 1925)
The great god Brown, The fountain, The moon of the Caribbees, and other plays 1926
The hairy ape, and other plays 1922 (UK 1923)
Hughie 1959
The iceman cometh 1946 (UK 1947)
Lazarus laughed 1927
Long day's journey into night 1956
Marco Millions 1927
A moon for the misbegotten 1952 (UK 1953)
The moon of the Caribbees, and six other plays of the sea 1919 (UK 1923)
More stately mansions ed. D. Gallup 1964 (UK 1965)
Mourning becomes Electra: a trilogy 1931 (UK 1932)
Strange interlude 1928
A touch of the poet 1957
Ophthalmic review 1881–1916
Opie, Iona M. B. & Opie, Peter M. Children's games in street and playground 1969
The lore and language of schoolchildren 1959
See also Oxford dictionary of nursery rhymes, The
Optical Society of America Journal 1917–
Optometry today: the vision care profession (American Optometric Association) 1971
Orczy, Emmuska, Baroness Lady Molly of Scotland Yard 1910
The Scarlet Pimpernel 1905
Ordway, Frederick Ira, Gardner, J. P., & Sharpe, M. R. Basic astronautics 1962
Orgill, Douglas William The Jasius pursuit 1973
Oriental Ceramics Society Transactions 1923–
Origo, Iris Images and shadows: part of a life 1970
Orr, John tr. I. Iordan's An introduction to Romance linguistics 1937
Orton, John Kingsley ('Joe') What the butler saw 1969
'Orvis, Kenneth' (Kenneth Lemieux) The damned and the destroyed 1962
'Orwell, George' (Eric Arthur Blair) (1903–50) Animal farm 1945
Burmese days 1934
A clergyman's daughter 1935
Collected essays, journalism and letters ed. 'S. Orwell' & I. Angus 4 vols. 1968
Coming up for air 1939
Critical essays 1946
— (new ed.) 1951
Down and out in Paris and London 1933
England your England, and other essays 1953
The English people 1947
Homage to Catalonia 1938
Inside the whale, and other essays 1940
Keep the aspidistra flying 1936
The lion and the unicorn: socialism and the English genius 1941
Nineteen eighty-four 1949
The road to Wigan Pier: on industrial England and its political future 1937
Shooting an elephant, and other essays 1950

Osborne, John Dental mechanics for students 1940
Osborne, John James The entertainer 1957
Look back in anger 1957
West of Suez 1971
The world of Paul Slickey 1959
—— & Creighton, Anthony Epitaph for George Dillon 1958
Osgood, Charles Egerton Method and theory in experimental psychology 1953
Osler, William The principles and practice of medicine 1892 (also later editions used)
—— & McCrae, Thomas A system of medicine 7 vols. 1907–10
Ottawa Citizen, The (title varies) 1844–
Ottawa Journal 1885–
Ottaway, Andrew Kenneth Cosway Education and society: an introduction to the sociology of education 1953
Our towns: a close-up, 1939–42 (Women's Group on Public Welfare, Hygiene Committee) 1943
Out west 1902–23
Outlook, The 1898–1928
Outlook (New York) 1893–1935 (title varies)
Outward bound 1920–4
Overland monthly 1868–75; 1883–1935
Owen, David Edward English philanthropy, 1660–1960 1964 (UK 1965)
Owen, Francis Diary, 1837–38 ed. G. E. Cory 1926
Owen, Wilfred Edward Salter (1893–1918) Collected letters ed. H. Owen & J. Bell 1967
Collected poems ed. C. Day Lewis 1963
Oxford book of birds, The by B. Campbell 1964
Oxford book of food plants, The by S. G. Harrison, G. B. Masefield, & M. Wallis 1969
Oxford book of invertebrates, The by D. Nicholls & J. A. L. Cooke 1971
Oxford book of wild flowers, The by S. Ary & M. Gregory 1960
Oxford classical dictionary, The ed. M. Cary et al. 1949
— (ed. 2, by N. G. L. Hammond & H. H. Scullard) 1970
Oxford companion to art, The ed. H. Osborne 1970
Oxford companion to French literature, The by P. Harvey & J. E. Heseltine 1959
Oxford companion to music, The by P. A. Scholes 1938 (also later editions used)
Oxford companion to ships and the sea, The ed. P. Kemp 1976
Oxford companion to sports and games, The ed. J. Arlott 1975
Oxford companion to the decorative arts, The ed. H. Osborne 1975
Oxford companion to the theatre, The ed. P. M. Hartnoll 1951
—(ed. 2) 1957
Oxford computer explained, The: data processing at Oxford University Press 1967
Oxford dictionary of nursery rhymes, The ed. I. & P. Opie 1951
Oxford dictionary of the Christian Church, The ed. F. L. Cross 1957
— (ed. 2, by F. L. Cross & E. A. Livingstone) 1974
Oxford diocesan magazine 1902–
Oxford history of South Africa, The ed. M. Wilson & L. Thompson 2 vols. 1969–71
Oxford junior encyclopaedia ed. L. E. Salt et al. 13 vols. 1948–56
Oxford Mail 1928–
Oxford textbook of medicine ed. D. J. Weatherall, J. G. G. Ledingham, & D. A. Warrell 2 vols. 1983
Oxford University Report of Commission of Inquiry 2 vols. 1966
Oxford University. Ashmolean Museum Report of the Visitors 1889–
Oxoniensia (Oxford Architectural and Historical Society) 1936–
Oz 1967–73

P

PMLA See: Modern Language Association of America
Packard, Vance Oakley The hidden persuaders 1957
Packer, John & Vaughan, James A modern approach to organic chemistry 1958
Packer, Joy Apes and ivory 1953
'Packer, Vin' (Marijane Meaker) Don't rely on Gemini 1969 (UK 1970)
Page, Emma A fortnight by the sea 1973
Page, Ewan Stafford & Wilson, Leslie Blackett An introduction to computational combinatorics 1979
Page, Victor Wilfred Modern aircraft 1927 (UK 1928)
Paget, Richard Arthur Surtees Human speech 1930

Paine, Ralph Delahaye Comrades of the rolling ocean 1923
Palmer, Charles Skeele tr. W. Nernst's Theoretical chemistry, from the standpoint of Avogadro's Rule and thermodynamics 1895
Palmer, Edward Vance Golconda 1948
Men are human 1930
The passage 1930
Separate lives 1931
See also: 'Daly, Rann'
Palmer, Elisabeth tr. A. Martinet's Elements of general linguistics 1964
Palmer, Eve The plains of Camdeboo 1966
—— & Pitman, Norah Trees of South Africa 1961
Trees of southern Africa 3 vols. 1972–3
Palmer, Harold Edward A grammar of spoken English on a strictly phonetic basis 1924
Palmer, Lionel Stanley Wireless principles and practice 1928
Pankhurst, Emmeline My own story 1914
Pap, Arthur Elements of analytic philosophy 1949
Pappe, Karl Wilhelm Ludwig Silva Capensis; or, A description of South African forest-trees and arborescent shrubs 1854
Synopsis of the edible fishes at the Cape of Good Hope 1853
Park, William Robb R. ed. Plastics film technology 1969
Parker, Dorothy After such pleasures 1933 (UK 1934)
Parker, Philip Electronics 1950
Parker, Robert Brown God save the child 1974 (UK 1975)
The Godwulf manuscript 1974
Parker, Robert Lueling tr. P. Niggli's Rocks and mineral deposits 1954
Parker, Tony The twisting lane 1969
The unknown citizen 1963
—— & Allerton, Robert The courage of his convictions 1962
Parker, William Hosken Health and disease in farm animals 1970
Parker, William Newton tr. R. E. E. Wiedersheim's Elements of the comparative anatomy of vertebrates 1886
Parkes, Malcolm B. English cursive book hands 1250–1500 1969
Parkes, Roger The guardians 1973
Parkman, Ebenezer (1703–82) The diary of Ebenezer Parkman, of Westborough, Mass. ed. H. M. Forbes 1899
Parlett, David Sidney A short dictionary of languages 1967
Parliamentary papers See United Kingdom. Parliamentary papers
Parrish, Albert Mechanical engineer's reference book 1973
Parry, Dennis Going up—going down 1953
Parsons, John Herbert Diseases of the eye 1907
— (ed. 12, by W. S. Duke-Elder) 1954
Parsons, Nell Wilson Upon a sagebrush harp 1969
Parsons, Talcott The social system 1952
The structure of social action 1937
—— & Shils, Edward Albert eds. Toward a general theory of action 1951
Partington, James Riddick General and inorganic chemistry for university students 1946
A text-book of inorganic chemistry for university students 1921
Partisan review 1934–
Partridge, Eric Honeywood A charm of words: essays and papers on language 1960
A dictionary of abbreviations 1942
A dictionary of catch phrases British and American, from the Sixteenth Century to the present day 1977
A dictionary of clichés 1940
A dictionary of forces' slang, 1939–45 1948
A dictionary of R.A.F. slang 1945
A dictionary of slang and unconventional English 1937 (also later eds. and supplements used)
A dictionary of the underworld, British and American 1950 [dated 1949]
English gone wrong 1957
Here, there and everywhere 1950
Slang to-day and yesterday 1933
— (ed. 3) 1951 [dated 1950]
Usage and abusage: a guide to good English 1942 (UK 1947)
Words, words, words! 1933
The world of words 1938
See also: Brophy, John; also 'Vigilans'
Paryski, Edwin tr. T. Korzybski et al., Antibiotics 2 vols. 1967
Pasachoff, Jay M. & Kutner, Marc Leslie University astronomy 1978
Passmore, John Arthur A hundred years of philosophy 1957
Passmore, Reginald & Robson, James Scott eds. A companion to medical studies 3 vols. 1968–74
Pasternak, Boris See Hayward, Harry M. & Harari, Manya

Patel, H. M., et al. Say not the struggle nought availeth: essays in honour of A. D. Gorwala's seventy-fifth birthday 1975
Patent Office See United Kingdom
'Paterson, Peter' (James Glass Bertram) Glimpses of real life as seen in the theatrical world and in Bohemia 1864
Pathological Society of London Transactions 1853–1907
Pathology annual 1966–
Paton, Alan Cry, the beloved country: a story of comfort in desolation 1948
Hope for South Africa 1958
'Patrick, James' A Glasgow gang observed 1973
Patten, Brian Arthur The irrelevant song 1971
Little Johnny's confession 1967
Patterson, Edward McWilliam Topology 1956
— (ed. 2) 1959
—— & Rutherford, D. E. Elementary abstract algebra 1965
Patterson, Raymond Murray Finlay's river 1968
Paul, Elliot Harold A narrow street 1942
Springtime in Paris 1951
Paul, Maurice Eden & Paul, Cedar tr. J. Stalin's Leninism 2 vols. 1928–33
Paulding, James Kirke John Bull in America; or, The new Munchausen 1825
Westward ho! 2 vols. 1832
See also: 'Bull-Us, Hector'
Pauli, Wolfgang ed. Niels Bohr and the development of physics: essays dedicated to Niels Bohr 1955
Pauling, Linus Carl The nature of the chemical bond and the structure of molecules and crystals 1939
'Paxton, Philip' (S. A. Hammett) A stray Yankee in Texas 1853
Payne, Charles Rockwell tr. O. Pfister's The psychoanalytic method 1917
Peace News 1936–
Peach, Benjamin Neave & Horne, John Chapters on the geology of Scotland 1930
Peacocke, Arthur Robert & Drysdale, Robert Brown The molecular basis of heredity 1965
Peake, Harold John Edward & Fleure, Herbert John Hunters and artists 1927
Peake, Mervyn Laurence Gormenghast 1950
Titus alone 1959
Titus Groan 1946
Pear, Tom Hatherley English social differences 1955
Voice and personality 1931
Pearce, Brian tr. S. Amin's Accumulation on a world scale: a critique of the theory of underdevelopment 2 vols. 1974
tr. E. Preobrazhensky's The new economics 1965
Pearson, T. Gilbert ed. Birds of America 3 vols. 1917
Pearson, William Harrison (Bill) Coal flat 1963
Pearson's magazine 1896–1939
Peattie, Donald Culross The road of a naturalist 1941 (UK 1946)
Peck, George Wilbur Peck's bad boy and his pa 1883
Peck's boss book 1884
Peck's sunshine 1882
Peck, John Mason A gazetteer of Illinois 1834
— (ed. 2) 1837
A guide for emigrants 1831
Pediatrics: the journal of the American Academy of Pediatrics 1948–
Peel, Dorothy Constance E. Life's enchanted cup: an autobiography, 1872–1933 1933
Peel City Guardian and Chronicle (Peel, Isle of Man) 1882–
Peele, Robert ed. Mining engineers' handbook 1918 — (ed. 2) 1927
Peele, Talmadge Lee The neuroanatomical basis for clinical neurology 1954
Pegler, Martin M. The dictionary of interior design 1966 (UK 1967)
Pei, Mario Andrew Glossary of linguistic terminology 1966
The story of language 1949 (UK 1952)
Words in sheep's clothing 1969 (UK 1970)
—— & Gaynor, Frank A dictionary of linguistics 1954
Peirce, Charles Santiago Sanders (1839–1914) Collected papers ed. C. Hartshorne, P. Weiss, & A. W. Burks 8 vols. 1931–58
Pender, Harold ed. American handbook for electrical engineers 1914
Penderel-Brodhurst, James & Layton, Edwin J. A glossary of English furniture of the historic periods 1925
Pendlebury, John Devitt Stringfellow The archaeology of Crete: an introduction 1939
Penfield, Wilder Graves Cytology and cellular pathology of the nervous system 1932
Penguin book of Australian ballads, The ed. R. B. Ward 1964
Penguin music magazine 1946–9
Penguin new writing, The ed. J. F. Lehmann 1940–50
Pennsylvania magazine of history and biography 1877–
Penny magazine 1832–45
Penrose annual (title varies) 1895–
'Pentecost, Hugh' (Judson Pentecost Philips) Girl watcher's funeral 1969 (UK 1970)
See also under real name
People, The 1881–1971
People's Journal (Dundee) 1858–
Pereira, Michael Pigeon's blood 1970
Perelman, Sidney Joseph Baby, it's cold inside 1970
Crazy like a fox 1944 (UK 1945)
Westward ha! or, Around the world in eighty clichés 1947
Perkin, Harold Key profession: the history of the Association of University Teachers 1969
Perkowski, Robert L. & Stral, Lee Philip The joy of CB 1976
Perry, Frank Ernest & Ryder, Frank Raymond Dictionary of banking (ed. 11) 1965 See Thomson, William
Perry, John Howard Chemical engineers' handbook 1934
— (ed. 3) 1950
Perry, Ralph Barton The thought and character of William James as revealed in unpublished correspondence, together with his published writings 1842–1910 2 vols. 1935 (UK 1936)
Perry, Richard The world of the tiger 1964
Perry, Ritchie John A. One good death deserves another 1976
Personal computer world 1978–
Pervin, William Joseph Foundations of general topology 1964
'Peters, Ellis' (Edith Mary Pargeter) Black is the colour of my true-love's heart 1967
The piper on the mountain 1966
Peterson, Harold Leslie ed. Encyclopædia of firearms 1964
Peterson, Roger Tory & Fisher, James Wild America 1956
Petrie, Rhona Despatch of a dove 1969
Murder by precedent 1964
Petroleum refiner 1942–61
Petroleum review 1968–
Petticoat 1966–76
Pettijohn, Francis John Sedimentary rocks 1949
— (ed. 2) 1957
Pettman, Charles Africanderisms: a glossary of South African colloquial words and phrases of place and other names 1913
Pharmacopœia of the United States of America 1820–
Phelps, Elizabeth Stuart See Ward, Elizabeth Stuart
Philadelphia Bulletin (title varies) 1847–
Philadelphia Inquirer (title varies) 1829–
Philip, John Researches in South Africa 2 vols. 1828
Philips, Judson Pentecost A dead ending 1962 (UK 1963)
Escape a killer 1971 (UK 1972)
The twisted people 1965
The vanishing senator 1972 (UK 1973)
See also: 'Pentecost, Hugh'
Philips technical review 1936–
Phillips, Courtenay Stanley Goss & Williams, Robert Joseph Paton Inorganic chemistry 2 vols. 1965–6
Phillips, David Graham (1867–1911) The plum tree 1905
Susan Lenox: her fall and rise 1917
Phillpotts, Eden The portreeve 1906
The secret woman 1905
Philosophical review 1892–
Philosophy: the journal of the British Institute of Philosophy 1931–
Philosophy of science 1934–
'Phoenix, John' (George Horatio Derby) Phoenixiana; or, Sketches and burlesques 1856
Photochemistry and photobiology 1962–
Photo-miniature, The 1899–1932
Physical review 1893– (from 1970 divided into subject series)
Physical review letters 1958–
Physical Society (of London) Proceedings 1874–
Physics bulletin 1950–
Physiological abstracts 1916–37
Physiological reviews 1921–
Phytochemistry 1961–
Phytopathology 1911–
Picart, Bernard The ceremonies and religious customs of the various nations of the known world [anon. tr.] 1733
Picayune (New Orleans): see Daily Picayune
Pick of today's short stories, The ed. J. S. Pudney 1949–63
Pickard, George Lewson Descriptive physical oceanography: an introduction 1963
Picken, Laurence Ernest Rowland The organization of cells and other organisms 1960
Picken, Mary Brooks The fashion dictionary 1957
The language of fashion 1939
Pickering, Robert Easton Himself again 1966
Pierce, Robert Morris Dictionary of aviation 1911
Pieris, Ralph Sinhalese social organization: the Kandyan period 1956
Piggott, Stuart The neolithic cultures of the British Isles 1954
Pigman, William Ward Carbohydrates 1957
Chemistry of the carbohydrates 1948
Pigou, Arthur Cecil The economics of welfare 1920
Pike, Albert Prose sketches and poems, written in the western country 1834
Pike, Kenneth Lee Phonetics: a critical analysis of phonetic theory and a technic for the practical description of sounds 1943
Language in relation to a unified theory of the structure of human behavior 3 parts 1954–60
— (ed. 2) 1967
Pilley, Dorothy Eleanor Climbing days 1935
Pillsbury, Donald Marion, Shelley, W. B., & Kligman, Albert M. Dermatology 1956
Pinero, Arthur Wing The magistrate 1892
The second Mrs. Tanqueray 1895
Pinner, David Ritual 1967
Pinter, Harold The birthday party 1959
The caretaker 1960
The dumb waiter 1960
The homecoming 1965
The room 1960
Pinto, Edward Henry Treen and other wooden bygones: an encyclopædia and social history 1969
Pioneer, The (Lucknow) 1865–
Pirie, Antoinette ed. Lens metabolism in relation to cataract: proceedings of a symposium 1962
Pirie, Norman Wingate Food resources, conventional and novel 1969
Pirsson, Louis Valentine & Schuchert, Charles A text-book of geology 2 parts 1915 (Pt. I by Pirsson, Pt. II by Schuchert)
— (ed. 2, by Pirsson alone) 2 parts 1920–4
Pitts, Denis This city is ours 1975 (UK ed. 1976 with title Target Manhattan)
Pix (Sydney) 1938– (title varies)
Planché, James Robinson Extravaganzas ed. T. F. D. Croker & S. Tucker 5 vols. 1879
Planning: a broadsheet (PEP: Political and Economic Planning) 1933–
Plant physiology 1926–
Plath, Sylvia (1932–63) Ariel 1965
The bell jar 1963
The colossus 1960
— (another ed.) 1967
Crossing the water 1971
Platt, Kim The pushbutton butterfly 1970 (UK 1971)
Player, Robert Let's talk of graves, of worms, and epitaphs 1975
Plays of the year ed. J. C. Trewin 1949–
Plomer, William Charles Franklyn I speak of Africa 1927
Museum pieces 1952
Plumptre, Anne tr. H. Lichtenstein's Travels in southern Africa 2 vols. 1812–15
Pocknee, C. E. See Dearmer, Percy
Poetry (Chicago) 1912–
Pohl, Frederik & Kornbluth, Cyril M. The space merchants 1953 (UK 1955)
Polack, Joel Samuel Manners and customs of the New Zealanders 2 vols. 1840
New Zealand: being a narrative of travels and adventures 2 vols. 1838
Police journal 1928–
Police review (title varies) 1893–
Political quarterly 1930–
Political science quarterly 1886–
Pollitt, John Depression and its treatment 1965
Pollock, Albin Jay The underworld speaks 1935
Pollock, Frederick & Holmes, Oliver Wendell (the younger) The Pollock–Holmes letters: correspondence of Sir Frederick Pollock and Mr. Justice Holmes, 1874–1932 ed. M. de W. Howe 2 vols. 1942
Polunin, Nicholas Vladimir Introduction to plant geography and some related sciences 1960
Ponting, Herbert George The great white south: being an account of experiences with Captain Scott's South Pole Expedition 1921
Poole, A. L. & Adams, Nancy M. Trees and shrubs of New Zealand 1963
Pope-Hennessy, James Robert Louis Stevenson 1974
Popenoe, Wilson Manual of tropical and subtropical fruits 1920
Popley, Herbert Arthur The music of India 1921
Popper, Karl Raimund The logic of scientific discovery 1959
The open society and its enemies 2 vols. 1945
The poverty of historicism 1957
Porcupine, The 1860–1915; 1923
'Porte Crayon' (David Hunter Strother) Virginia illustrated 1857

Porter, Mrs. G. S. See Stratton-Porter, Mrs. Gene
Porter, Horace C. tr. E. Strasburger's A textbook of botany 1898
Porter, Joyce The chinks in the curtain 1967
Dover two 1965
It's murder with Dover 1973
A meddler and her murder 1972
Rather a common sort of crime 1970
Sour cream with everything 1966
Porter, Katherine Anne Ship of fools 1962
Porter's Spirit of the times 1856–9
Portola Institute The last whole earth catalog 1972
Portsmouth, Rhode Island Early records ed. A. Perry & C. S. Brigham 1901
Post, Waldron Kintzing Harvard stories: sketches of the undergraduate 1893
Post-Herald (Birmingham, Alabama) 1951–
Post Office electrical engineers' journal 1908–
Postal, Paul Martin Aspects of phonological theory 1968
Postgate, Raymond William The plain man's guide to wine 1951
— (ed. 2) 1965
Postman, Neil Crazy talk, stupid talk 1976
Potok, Chaim The chosen 1967
In the beginning 1975
My name is Asher Lev 1972
Potter, David British Elizabethan stamps: the story of the postage stamps of the United Kingdom 1971
Potter, Dennis Christopher G. The glittering coffin 1960
Potter, Helen Beatrix Journal, 1881–1897 transcribed from code by L. Linder 1966
Potter, Jeremy Foul play 1967
Potter, Simeon Language in the modern world 1960
Modern linguistics 1957
Our language 1950
Pottle, Frederick Albert Stretchers: the story of a hospital unit on the Western Front 1929 (UK 1930)
Potts, Jean Death of a stray cat 1955 (UK 1956)
The evil wish 1962
Potts, Thomas Henry Out in the open: a budget of scraps of natural history, gathered in New Zealand 1882
Poultry chronicle 1854–5
Pound, Ezra Loomis Cantos LII–LXXI 1940
Canzoni 1911
Cathay: translations for the most part from the Chinese of Rihaku 1915
The classic anthology defined by Confucius 1955
A draft of XXX cantos 1930
Hugh Selwyn Mauberley: life and contacts 1920
Letters See Selected letters
Lustra 1916
— (another ed., with earlier poems) 1917
Pavannes and divisions 1918
Personae: the collected poems 1909
The Pisan cantos 1948 (UK 1949)
Section: rock-drill; 85–95 de los cantares 1955
— (another ed.) 1957
Selected letters, 1907–1941 ed. D. D. Paige 1971
Thrones: 96–109 de los cantares 1959
tr. Sophocles' Women of Trachis 1956
Poutsma, Hendrik A grammar of late modern English 2 parts in 4 vols. 1904–26
— (ed. 2 of Part I) 2 vols. 1928–9
Powdrill, Ernest Arthur Vocabulary of land planning 1961
Powell, Anthony Dymoke The acceptance world 1955
Afternoon men 1931
A buyer's market 1952
Casanova's Chinese restaurant 1960
A question of upbringing 1951
The valley of bones 1964
Venusberg 1932
What's become of Waring? 1939
Powell, Dawn A time to be born 1942 (UK 1943)
Turn, magic wheel 1936
Powell, John Leonard & Crasemann, Bernd Quantum mechanics 1961
Powell, Martin Beynon & Higman, Graham eds. Finite simple groups 1971
Power, Tyrone Impressions of America, 1833–1835 2 vols. 1836
Power, William James Tyrone Sketches in New Zealand, with pen and pencil 1849
Power farming 1957–
Powys, John Cowper A Glastonbury romance 1932 (UK 1933)
Maiden Castle 1936 (UK 1937)
Visions and revisions: a book of literary devotions 1915
Powys, Marian Lace and lace-making 1953
Practical motorist 1954–
Practical wireless 1932–
Practitioner, The: a monthly journal of therapeutics and public health 1868–
Practitioners library of medicine and surgery, The ed. G. Blumer 16 vols. 1932–41
Pratt, Lyde S. The chemistry and physics of organic pigments 1947
Prehistoric Society (of East Anglia) Proceedings 1908–
Press, The (Christchurch, N.Z.) 1861–
Press and Journal (Aberdeen) 1939– (continuation of Aberdeen Press and Journal)
Price, Anthony The Alamut ambush 1971
Colonel Butler's wolf 1972
The '44 vintage 1978
October men 1973
Other paths to glory 1974
Our man in Camelot 1975
War game 1976
Price, Henry Habberley Perception 1932
Thinking and experience 1953
Price, Pamela Vandyke France: a food and wine guide 1966
Price, Stanley Just for the record 1961
Prichard, Katharine Susannah Coonardoo: the well in the shadow 1929
Winged seeds 1950
Priebsch, Robert Charles & Collinson, William Edward The German language 1934
Priestley, John Boynton Angel Pavement 1930
Bright day 1946
Daylight on Saturday 1943
Delight 1949
English journey 1934
Festival at Farbridge 1951
The good companions 1929
Home is tomorrow 1949
It's an old country 1967
Let the people sing 1939
The magicians 1954
Margin released 1962
Salt is leaving 1966
Saturn over the water 1961
They walk in the city 1936
Three men in new suits 1945
—— & Hawkes, Jessie Jacquetta Journey down a rainbow 1955
Priestley, Ronald & Wisdom, Thomas Henry Good driving—the B.S.M. way 1965
Pringle-Pattison, Andrew Seth The idea of God in the light of recent philosophy 1917
Printers' ink 1888–1967
Prior, Allan The interrogators 1965
The operators 1966
Prior, Arthur Norman Formal logic 1955
Time and modality 1957
Pritchett, Victor Sawdon A cab at the door, an autobiography: early years 1968
The gentle barbarian: the life and work of Turgenev 1977
Private eye 1962–
Proceedings See under the names of particular institutions
Procter, Henry Richardson The principles of leather manufacture 1903
Procter, Maurice Exercise Hoodwink 1967
Man in ambush 1958
Proctor, Michael Charles Faraday & Yeo, Peter The pollination of flowers 1973
Product engineering 1930–
Providence, Rhode Island Early records 21 vols. 1892–1915
Pryce-Jones, Alan ed. The new outline of modern knowledge 1956
Psyche (subtitle varies) 1921–37
Psychiatry 1938–
Psychoanalytic review 1913–
Psychological bulletin 1904–
Psychological review 1894–
Public Ledger (Philadelphia) 1836–1934
Publications See under the names of particular institutions
Publishers' circular (title varies) 1837–1959
Publishers' weekly 1873–
Puck (New York) 1877–1918
Pugh, Marshall Last place left 1969
A wilderness of monkeys 1958
Pulgram, Ernst Introduction to the spectography of speech 1959
Punnett, Reginald Crundall Mendelism 1905 (and several later editions used)
Purseglove, John William Tropical crops: dicotyledons 2 vols. 1968
Purser, Philip The Holy Father's navy 1971
The Twentymen 1967
Putnam, Donald Fulton ed. Canadian regions 1952
Putnam, Samuel Whitehall tr. E. da Cunha's Rebellion in the backlands 1944
Puzo, Mario Fools die 1978
Pycraft, William Plane ed. The standard natural history 1931
Pyke, Magnus Alfred Food and society 1968
Pym, Barbara Mary Crampton No fond return of love 1961
Quartet in autumn 1977
Some tame gazelle 1950
Pynchon, Thomas The crying of Lot 49 1966
Gravity's rainbow 1973
V.: a novel 1963

Q

'Q' (Arthur Thomas Quiller-Couch) The mayor of Troy 1906
See also under real name
Quarterly cumulative index medicus 1927–56
Quarterly journal of medicine 1907–
Quarterly register 1886–1934
Quarterly review of biology 1926–
'Quartermain, James' (James Broom Lynne) Rock of diamond 1972
'Queen, Ellery' (Frederic Dannay & Manfred Bennington Lee) The four of hearts 1938 (UK 1939)
The fourth side of the triangle 1965
The French powder mystery 1930
The Roman hat mystery 1929
Queen, The: an illustrated journal and review (title varies) 1861–1970
Quennell, Peter Courtney The marble foot: an autobiography, 1905–1938 1976
'Quentin, Patrick' (Richard Wilson Webb & Hugh Callingham Wheeler) The follower 1950
Puzzle for fiends 1946 (UK 1947)
Shadow of guilt 1959
Suspicious circumstances 1957
Quick, Herbert Yellowstone nights 1911
Quiller-Couch, Arthur Thomas Foe-Farrell 1918
Major Vigoureux 1907
On the art of writing 1916
Studies in literature 3 ser. 1918–29
See also: 'Q'
Quincy Whig (Quincy, Illinois) 1838–1915 (title varies)
Quine, Willard Van Orman From a logical point of view 1953
Mathematical logic 1940
Methods of logic 1950 (UK 1952)
Set theory and its logic 1963
Word and object 1960
Quinland; or, Varieties in American life 2 vols. 1857
Quinn, Arthur Hobson Pennsylvania stories 1899
Quinton, Anthony Meredith The nature of things 1973
Quirk, Charles Randolph The use of English 1962
—— & Smith, Albert Hugh eds. The teaching of English 1959
—— et al. A grammar of contemporary English 1972
See also: Crystal, David

R

R.A.F. journal See Royal Air Force journal
R.A.F. news See Royal Air Force news
Rabinowitch, Eugene Isakovich Photosynthesis and related processes 2 vols. (Vol. II in 2 parts) 1945–56
Rackham, Bernard tr. E. Hannover's Pottery and porcelain Vol. I 1925 (Vols. II & III, also 1925, tr. by W. W. Worster; whole work ed. by B. Rackham)
Raddall, Thomas Head Hangman's beach 1966
Radin, Paul tr. J. Vendryès's Language 1925
Radio communication handbook, The See Amateur radio handbook
Radio review: a monthly record of scientific progress in radiotelegraphy and telephony 1919–22
Radio times, The 1923–
Radiology 1923–
Rae, Hugh Crauford A few small bones 1968
The marksman 1971
The shooting gallery 1972
Rae, John The custard boys 1960
Raffé, Walter George Dictionary of the dance 1964
Railroad and engineering journal 1887–92
Railway magazine 1897–
'Raine, Richard' (Raymond Harold Sawkins) Night of the hawk 1968
Raine, William Macleod Bucky O'Connor: a tale of the unfenced border 1910 (UK 1920)
Troubled waters 1924
Raleigh, Walter Alexander R. (1861–1922) Letters ed. Lady Raleigh 2 vols. 1926
— (new ed.) 2 vols. 1928
Shakespeare 1907
Rampini, Charles J. G. Letters from Jamaica 1873
Ramsay, Diana Descent into the dark 1975
A little murder music 1972
No cause to kill 1974
Ramsbottom, John Mushrooms and toadstools 1953
Ramsden, Evelyn tr. E. Gram & H. Weber's Plant diseases in orchard, nursery and garden crops 1952

Ramsey, Frederic & Smith, Charles Edward Jazzmen 1939 (UK 1940)
Ramsey, Leonard Gerald Gwynne ed. The Connoisseur new guide to antique English pottery, porcelain and glass 1961
Ramson, William Stanley Australian English: an historical study of the vocabulary, 1788–1898 1966
Rand, Austin Loomer Mammals of the eastern Rockies and western plains of Canada 1948
Rand Daily Mail (Johannesburg) 1902–
Randall-Diehl, Anna Two thousand words and their definitions; not in Webster's Dictionary 1888
Randolph Enterprise (Elkins, West Virginia) 1874–
Ranjitsinhji, Kumar Shri The Jubilee book of cricket 1897
Rankine, William John Macquorn A manual of applied mechanics 1858
Ransome, Arthur Michell (1884–1967) Autobiography ed. R. Hart-Davis 1976
Great Northern? 1947
Secret water 1939
'Ransome, Stephen' (Frederick Clyde Davis) The deadly Miss Ashley 1950 (US ed. published under author's real name)
Without a trace 1962
Rao, G. Subba Indian words in English: a study in Indo-British cultural and linguistic relations 1954
Raper, John Robert Genetics of sexuality in higher fungi 1966
Raphael, Chaim A feast of history: the drama of Passover through the ages 1972
Raphael, Frederic Michael The glittering prizes 1976
The limits of love 1960
Raphael, John E. Modern Rugby football 1918
Ratcliffe, John Ashworth The physical principles of wireless 1929
Rathbone, Julian Diamonds bid 1967
Joseph 1979
Kill cure 1975
Rattigan, Terence Mervyn The deep blue sea 1952
Flare path 1942
French without tears: a play 1937
Ross: a dramatic portrait 1960
While the sun shines 1944
Who is Sylvia? 1951
The Winslow boy 1946
Rattray, Robert Sutherland Ashanti 1923
Raven, Simon Arthur Noël The survivors 1976
Raverat, Gwendolyn Mary Period piece: a Cambridge childhood 1952
Rawlings, Marjorie Kinnan The yearling 1938
Rawlinson, William Modern foundry operations and equipment 1928
Ray, Cyril ed. The compleat imbiber 1956–71
Merry England 1960
Rayleigh, John William (Strutt), 3rd Baron The theory of sound 2 vols. 1877–8
Raymond, Ernest The jesting army 1930
Rayner, Dorothy Helen The stratigraphy of the British Isles 1967
RCA review 1936–
Read, Herbert Edward Annals of innocence and experience 1940
Art and industry 1934
Art and society 1937
Collected poems, 1913–25 1926
A concise history of modern painting 1959
The contrary experience: autobiographies 1963
Education through art 1943
Icon and idea: the function of art in the development of human consciousness 1955
The meaning of art 1931
The tenth muse: essays in criticism 1957
Read, Herbert Harold & Watson, Janet V. Introduction to geology 2 vols. (Vol. II in 2 parts) 1962–75
Read, John A text-book of organic chemistry 1926
Read, Piers Paul The Villa Golitsyn 1981
Reader's digest 1922–
Reader's digest family guide to the law 1971
Real estate review 1923–
Reaney, Percy Hide The origin of English place-names 1960
Recent progress in hormone research 1947–
Record, Samuel James & Hess, Robert William Timbers of the New World 1943
—— & Mell, Clayton Dissinger Timbers of tropical America 1924
Recruiter's bulletin (U.S. Marine Corps) 1915–21
Red Cross magazine 1916–20
Redbook (title varies) 1903–
Rede, Leman Thomas The road to the stage 1827
Redfield, Robert Peasant society and culture: an anthropological approach to civilization 1956
Reece, Gordon tr. F. Hund's The history of quantum theory 1974
Reece, Robert H. Night bombing with the Bedouins 1919
Reese, John Terence & Dormer, Albert The bridge player's dictionary 1959
Referee, The 1877–1928
Register of debates in Congress See: United States. Congress. Debates
Reichenbach, Hans Elements of symbolic logic 1947
Reid, John Cowie ed. The Kiwi laughs: an anthology of New Zealand prose humour 1961
Reid, Thomas Mayne The headless horseman: a strange tale of Texas 2 vols. & 1 vol. 1866
The young voyageurs 1854
Reifer, Mary A dictionary of new words 1955
Rein, Johann Justus Japan: travels and researches [anon. tr.] 1884
Reisner, Robert George The jazz titans; including 'The parlance of hip' 1960
Reisz, Karel The technique of film editing 1953
— (ed. 2) 1968
Reith, John Charles Walsham (1889–1971) The Reith diaries ed. C. Stuart 1975
Release and resettlement: an explanation of your position and rights Issued by H.M. Government to all serving members of H.M. Forces 1945
Rendell, Ruth A guilty thing surprised 1970
Make death love me 1979
A new lease of death 1967
One across, two down 1971
Some lie and some die 1973
Report. Indian affairs See: United States Report of the Commissioner of Indian Affairs
Report of the Committee of Inquiry on Decimal Currency See: United Kingdom. Parliamentary papers
Report of the Committee on Broadcasting 1960 See: United Kingdom. Parliamentary papers
Reporter: the magazine of facts and ideas (New York) 1949–
Republican Review (Albuquerque, New Mexico) 1870–
Return to the attack: with the New Zealand Division in North Africa (Army Board, N.Z.) 1944
Review of English studies 1925–
Review of scientific instruments 1930–
Reynolds, John The pioneer history of Illinois, 1673–1818 1852
Reynolds, Mack (Dallas McCord Reynolds) After some tomorrow 1967
Rhode Island Historical Society Collections 1827–1941
Rhys, Jean Voyage in the dark 1934
Wide Sargasso Sea 1966
Rice, Clara Mabel Dictionary of geological terms 1940
Rice, David Talbot English art, 871–1100 1952
Rice, Elmer Leopold A voyage to Purilia 1930
Rice, Jack C. The materials and methods of sculpture 1947
Richards, Frank James Old-soldier sahib 1936
Richards, Ivor Armstrong Practical criticism: a study of literary judgement 1929
Principles of literary criticism 1925
Richards, Ivor ed.: see Charles Kay
Richards, Owain Westmacott & Davies, R. G. eds. A. D. Imms's A general textbook of entomology (ed. 9) 1957
Richards, Paul Westmacott The tropical rain forest: an ecological study 1952
Richards, Richard Kohler Arithmetic operations in digital computers 1955
Richardson, Alan ed. A dictionary of Christian theology 1969
Richardson, Edward Gick Sound: a physical text-book 1927
'Richardson, Henry Handel' (Edith Florence Lindesay Robertson) The fortunes of Richard Mahony 1917
Richardson, John A dictionary, Persian, Arabic, and English 2 vols. 1777–80
Richler, Mordecai Cocksure 1968
Richmond Enquirer (Richmond, Virginia) 1804–77
Richmond News-Leader (Richmond, Virginia) 1903–
Richmond Times-Dispatch (Richmond, Virginia) 1903–
Richmond Whig (Richmond, Virginia) 1824–88 (title varies)
Richmond–Atkinson papers, The ed. G. H. Scholefield 2 vols. 1960
Richter, George Holmes Textbook of organic chemistry 1938
Richter, Gisela M. A. & Milne, Marjorie J. Shapes and names of Athenian vases 1935
Rickert, Margaret Painting in Britain: the Middle Ages 1954
Rickman, Eric Come racing with me 1951
Riddell, Charlotte Eliza Lawson (Mrs. J. H. Riddell) The senior partner 3 vols. 1881
Ridge, William Pett Affectionate regards 1929
Ridley, Cecilia Anne Cecilia: the life and letters of Cecilia Ridley, 1819–1845 ed. Viscountess Ridley 1958
Rieger, Rigomar, Michaelis, A., & Green, M. M. A glossary of genetics and cytogenetics, classical and molecular 1968
Riesenberg, Felix Golden Gate: the story of San Francisco harbour 1940
Riesman, David Individualism reconsidered, and other essays 1954
—— Denney, R. N., & Glazer, N. The lonely crowd: a study of the changing American character 1950
Riis, Jacob August How the other half lives: studies among the tenements of New York 1890 (UK 1891)
Rimmer, Robert Henry The Premar experiments 1975 (UK 1976)
Riordan, John Stochastic service systems 1962
Ritchie, Anna & Ritchie, Graham The ancient monuments of Orkney 1978
Ritchie, Anne (Thackeray) (1837–1923) Letters, with forty-two additional letters from W. M. Thackeray ed. H. Ritchie 1924
Rittenhouse, Isabella Maud Maud ed. R. L. Strout 1939
Rivers, William Halse R. Kinship and social organisation 1914
Riviere, Joan tr. S. Freud's Introductory lectures on psycho-analysis 1922
— ed. tr. S. Freud's Collected papers 5 vols. 1924–50
Robb, Frank Sea hunters 1953
Roberts, E. M. A flying fighter: an American above the lines in France 1918
Roberts, Elizabeth Madox The time of man 1926 (UK 1927)
Roberts, Geoffrey Keith A dictionary of political analysis 1971
Roberts, Hermese E. The third ear 1970
Roberts, John D., Stewart, R., & Caserio, M. C. Organic chemistry 1971
Roberts, Nesta The face of France 1976
Roberts, Paul Ernest Historical geography of India 1916
Robertson, A. Trevor (Eileen Arbuthnot Robertson) Ordinary families 1933
Robertson, Edith Tennent & Goodding, Evelyn Lindsay Botany for the Caribbean 1963
Robertson, Stuart The development of modern English 1934 (UK 1936)
— (ed. 2, rev. by F. G. Cassidy) 1954
Robins, Elizabeth The magnetic north 1904
Robins, Robert Henry General linguistics: an introductory survey 1964
Robinson, Gilbert Wooding Soils: their origin, constitution and classification 1932
Robinson, Henry Crabb (1775–1867) Diary ed. D. Hudson 1967
Robinson, Paul A. Freudian Left 1969
Robinson, William The English flower garden 1883 (and several later editions used)
The wild garden 1870 (and several later editions used)
Roche, Harriet A. On trek in the Transvaal; or, Over berg and veldt in South Africa 1878
Rodd, Ernest Harry ed. Chemistry of carbon compounds 5 vols. in 10 parts 1951–62
Rodgers, Bruce The queens' vernacular: a gay lexicon 1972
Roe, Thomas The embassy of Sir Thomas Roe to the court of the Great Mogul, 1615–1619, as narrated in his journal and correspondence ed. W. Foster 2 vols. 1899
Rogers, Walter Thomas (d. 1912) Dictionary of abbreviations 1913
Roget, Samuel Romilly A dictionary of electrical terms 1924 (and later editions used)
Rohan, Criena The delinquents 1962
Rolfe, Frederick William S. A. L. M. ('Baron Corvo') (1860–1913) The desire and pursuit of the whole: a romance of modern Venice 1934
Hadrian the Seventh 1904
Nicholas Crabbe; or, The one and the many ed. C. Woolf 1958
Rolling stone 1967–
'Rolph, C. H.' (Cecil Rolph Hewitt) ed. The human sum 1957
ed. Women of the streets 1955
Romer, Alfred Sherwood The procession of life 1968
The vertebrate body 1949
Vertebrate paleontology 1933
Romilly, Joseph Romilly's Cambridge diary 1832–42 ed. J. P. T. Bury 1967
Rook, Arthur James, Wilkinson, D. S., & Ebling, F. J. G. eds. Textbook of dermatology 2 vols. 1968
Roosenburg, Henriette The walls came tumbling down 1957
Rose, Geoffrey Keith The story of the 2/4th Oxfordshire and Buckinghamshire Light Infantry 1920
Rose, Howard N. A thesaurus of slang 1934
Rose, William ed. An outline of modern knowledge 1931
Rosen, Ephraim & Gregory, Ian Abnormal psychology 1965

Ross, Alan John Australia 55: a journal of the M.C.C. tour 1955
Australia 63 1963
ed. The cricketer's companion 1960
Ross, Alan Strode Campbell Etymology, with especial reference to English 1958
Ross, Angus The Bradfords business 1974
The Dunfermline affair 1973
The Manchester thing 1970
Ross, David Alexander Introduction to oceanography 1970
Ross, Edward Alsworth The Russian Soviet republic 1923
Ross, James Handbook of the diseases of the nervous system 1885
'Ross, Jonathan' (John Rossiter) The burning of Billy Toober 1974
Dead at first hand 1969
I know what it's like to die 1976
Ross, Rodger James Television film engineering 1966
Rossetti, Dante Gabriel (1828–82) Letters ed. O. Doughty & J. R. Wahl 4 vols. 1965–7
Rossiter, John The manipulators 1973
A rope for General Dietz 1972
Rosten, Leo Calvin The joys of Yiddish 1968
Roth, Cecil A history of the Marranos 1932
Roth, Philip Goodbye, Columbus 1959
Portnoy's complaint 1969
Rotman, Jonah Joseph The theory of groups 1965
Round, Frank Eric Introduction to the lower plants 1969
Rowlands, John James Spindrift from a house by the sea 1960
Rowley, Jennifer E. Mechanised in-house information systems 1979
Rowse, Alfred Leslie A Cornish childhood 1942
The early Churchills: an English family 1956
The England of Elizabeth 1950
The English spirit: essays 1944
The expansion of Elizabethan England 1955
Tudor Cornwall 1941
The use of history 1946
Royal Aeronautical Society Journal 1923–67
Handbook of aeronautics 1931
Royal Air Force journal ?1942–6
Royal Air Force news 1961–
Royal Asiatic Society: Straits Branch (later Malayan Branch) Journal 1878–1920 (later 1923– Malayan Branch)
Royal Asiatic Society of Bengal See Asiatic —
Royal Entomological Society of London See Entomological —
Royal Horticultural Society Dictionary of gardening ed. F. J. Chittenden et al. 4 vols. 1951
Journal 1846–55; 1866–
Royal Institution of Great Britain Proceedings 1851– (1851–1928 with title Notices of the proceedings)
Royal Society of Edinburgh Proceedings 1832–(from 1941 divided into subject sections)
Royal Society of Medicine Proceedings 1907–
Royall, Anne The black book; or, A continuation of travels in the United States 3 vols. 1828–9
Royce, Kenneth Spider underground 1973
Trap Spider 1974
Royde-Smith, Naomi Gwladys Incredible tale 1932
Ruark, Arthur Edward & Urey, Harold Clayton Atoms, molecules and quanta 1930
Rubinstein, Helena The art of feminine beauty 1930
Ruck, Amy Roberta (Berta) Disturbing charm 1919
'Rudd, Steele' (Arthur Hoey Davis) From selection to city 1909
Our new selection 1903
Rudder, The 1890–
Rudinger, Edith ed. The consumer's car glossary (Consumers' Association) 1966
— (ed. 2) 1967
ed. Wills and probate (Consumers' Association) 1967
Ruell, Patrick Red Christmas 1972
Rules of the game: the complete illustrated encyclopedia of all the sports of the world (by the Diagram Group) 1974
Rumbaugh, Duane M. ed. Language learning by a chimpanzee: the Lana project 1977
Rumney, George R. Climatology and the world's climates 1968
Runes, Dagobert David ed. The dictionary of philosophy 1942
Runyon, Alfred Damon Furthermore ed. E. C. Bentley 1938
Guys and dolls 1931 (UK 1932)
More than somewhat ed. E. C. Bentley 1937
Runyon à la carte 1944 (UK 1946)
Take it easy 1938
Rushdie, Salman Midnight's children 1981
Ruskin, Effie See Lutyens, Mary
Russell, Alexander A treatise on the theory of alternating currents 2 vols. 1904–6
Russell, Archer Bush ways: a Bush-lover's wanderings on plain and range in Central and Eastern Australia 1944
Gone nomad 1936
A tramp-royal in wild Australia, 1928–1929 1934
Russell, Bertrand Arthur William The analysis of mind 1921
Autobiography 3 vols. 1967–9
A history of Western philosophy 1945 (UK 1946)
An inquiry into meaning and truth 1940
Marriage and morals 1929
An outline of philosophy 1927
Principia mathematica See: Whitehead, Alfred North & Russell, B. A. W.
The principles of mathematics 1903
Religion and science 1935
Russell, Charles Marion Trails plowed under 1927
Russell, Frederick Stratten & Yonge, Charles Maurice The seas: our knowledge of life in the sea and how it is gained 1928
— (rev. ed.) 1936
Russell, George Oscar Speech and voice 1931
Russell, Loris Shano Everyday life in colonial Canada 1973
Russell, Martin James Deadline 1971
Double hit 1973
Murder by the mile 1975
Russell, Ralph tr. Aziz Ahmad's The shore and the wave 1971
'Russell, Sarah' (Marghanita Laski) To bed with grand music 1946
Russenholt, Edgar Stanford The heart of the continent: being the history of Assiniboia—the truly typical Canadian community 1968
'Rutherford, Douglas' (James Douglas Rutherford McConnell) The creeping flesh 1963
The gilt-edged cockpit 1969
Kick start 1973
The long echo 1957
Rutherford, Ernest Radio-activity 1904
—— Chadwick, J., & Ellis, C. D. Radiations from radioactive substances 1930
Rutten, Martin Gerard The geology of western Europe 1969
Rycroft, Charles A critical dictionary of psychoanalysis 1968
Ryder, Jonathan Trevayne 1973 (UK 1974)
Ryle, Gilbert The concept of mind 1949

S

S.L.R. camera 1964
S.P.E. See Society for Pure English
Sachs, Albert Louis The jail diary of Albie Sachs 1966
Sachs, Curt The history of musical instruments 1940 (UK 1942)
Sackville-West, Victoria Mary The Edwardians 1930
Sadeek, Sheik M. Windswept, and other stories 1969
Sadleir, Michael Excursions in Victorian bibliography 1922
Trollope: a commentary 1927
Sadler, William Samuel Theory and practice of psychiatry 1936
Safire, William L. The new language of politics: an anecdotal dictionary of catchwords, slogans, and political usage 1968
Sagan, Carl The cosmic connection: an extraterrestrial perspective 1973 (UK 1974)
Sagan, Edward Cosmetics: science and technology 1957
Sage, Rufus B. Scenes in the Rocky Mountains, and in Oregon, California, New Mexico, Texas, and the grand prairies 1846
St. Clair, Leonard A fortune in death 1976
St. John, John Henry Herbert Pakeha rambles through Maori lands 1873
St. Louis Globe-Democrat (St. Louis, Missouri) 1875–
Saintsbury, George Edward Bateman Notes on a cellar-book 1920
The peace of the Augustans 1916
A short history of French literature 1882
'Saki' (Hector Hugh Munro) Beasts and super-beasts 1914
The chronicles of Clovis 1912
Reginald 1904
Reginald in Russia, and other sketches 1910
Salaman, John Stephen ed. Dictionary of American sports 1961
Salinger, Jerome David The catcher in the rye 1951
Franny and Zooey 1962
Salisbury, Harrison Evans The shook-up generation 1958 (UK 1959)
Salter, Charles tr. G. von Georgievics's The chemistry of dyestuffs 1903
tr. G. von Georgievics's The chemical technology of textile fibres 1902
Salter, William Henry Zoar; or, The evidence of psychical research concerning survival 1961
Sampson, Anthony Terrell S. Anatomy of Britain 1962
Sampson, Edmund Tales of the fancy 1889
Samuels, Michael Louis Linguistic evolution: with special reference to English 1972
San Francisco Examiner (title varies) 1865–
Sandburg, Carl Chicago poems 1916
Cornhuskers 1918
Slabs of the sunburnt west 1922
Smoke and steel 1920
Sanders, George Memoirs of a professional cad 1960
Sanders, Lawrence The Anderson tapes 1970
Sanders, Leonard The Hamlet warning 1976 (UK 1977)
Sanderson, C. C. ed. Pedigree dogs as recognised by the Kennel Club 1927
Sandilands, John Western Canadian dictionary and phrase-book 1912
Sangster, Jimmy Touchfeather 1968
Your friendly neighbourhood death peddler 1971
Sansom, George Bailey Japan: a short cultural history 1931
Sansom, William The face of innocence 1951
Santa Fé Republican (Santa Fé, New Mexico) 1847–
Sapir, Edward Language: an introduction to the study of speech 1921
Saporta, Sol & Bastian, Jarvis R. eds. Psycholinguistics: a book of readings 1961
'Sapper' (Herman Cyril McNeile) The black gang 1922
Bull-dog Drummond: the adventures of a demobilised officer who found peace dull 1920
The female of the species 1928
The finger of fate 1930
The third round 1924
Sargent, Charles Sprague Manual of the trees of North America 1905
Sargent, Epes Winthrop The technique of the photoplay 1912
— (ed. 2) 1913
— (ed. 3) 1916
Sargeson, Frank I saw in my dream 1949
A man and his wife 1940
— (new ed.) 1944
Memoirs of a peon 1965
That summer, and other stories 1946
Saroyan, William The daring young man on the flying trapeze, and other stories 1934 (UK 1935)
Sarton, George Alfred Leon A history of science 2 vols. 1952–9 (UK 1953–9)
Sasieni, Lewis Sidney The principles and practice of optical dispensing and fitting 1962
Sassoon, Siegfried Loraine Collected poems, 1908–56 1961
Memoirs of a fox-hunting man 1928
Memoirs of an infantry officer 1930
Sherston's progress 1936
Satchell, William The land of the lost: a tale of the New Zealand gum-country 1902
Saturday evening post: an illustrated weekly magazine (title varies) 1821–1969
Saturday night (Toronto) 1887–
Saturday review (US) 1924– (title varies: 1924–51 with title Saturday review of literature; 1972–3 divided into four: Saturday review of education; society; the arts; the sciences; 1973–5 recombined as Saturday review world; 1975– with title Saturday review)
Saturday Westminster gazette ?1899–1922
Saucier, Walter J. Principles of meteorological analysis 1955
Saucy stories 1916–25
Saunders, Margaret Baillie Litany Lane 1909
Saunders, Ripley Dunlap Colonel Todhunter of Missouri 1911
Savage, George The antique collector's handbook 1959
Dictionary of antiques 1970
Porcelain through the ages 1954
—— & Newman, Harold An illustrated dictionary of ceramics 1974
Sawyer, Frederic Henry Read The inhabitants of the Philippines 1900
Saxon, Lyle, Dreyer, E., & Tallant, R. Gumbo ya-ya: a collection of Louisiana folk tales 1945
Say, Maurice George ed. Newnes concise encyclopaedia of electrical engineering See under title
Sayers, Dorothy Leigh Busman's honeymoon 1937
Clouds of witness 1926
The five red herrings 1931
Gaudy night 1935
Hangman's holiday 1933
Have his carcase 1932
Lord Peter views the body 1928
Murder must advertise 1933
The nine tailors 1934
Strong poison 1930
Unnatural death 1927
The unpleasantness at the Bellona Club 1928

Sayers, Dorothy Leigh & 'Eustace, Robert' (Eustace Robert Barton) *The documents in the case* 1930
Schaller, George B. *The year of the gorilla* 1964 (UK 1965)
Schapera, Isaac *The Bantu-speaking tribes of South Africa* 1937
The Khoisan peoples of South Africa: Bushmen and Hottentots 1930
Scharf, Bedrich *Engineering and its language* 1971
Schilpp, Paul Arthur ed. *The philosophy of Bertrand Russell* 1944
ed. *The philosophy of Rudolf Carnap* 1963
Schlauch, Margaret *The English language in modern times, since 1400* 1959
Schlich, William *A manual of forestry* 5 vols. (Vols. IV & V by W. R. Fisher) 1889–96
Schneider, Frank *Qualitative organic microanalysis* 1946
Scholarly publishing 1969–
Scholem, Gershom Gerhard *Major trends in Jewish mysticism* 1941
Schoonmaker, Frank & Marvel, Tom *The complete wine book* 1935
Schreiber, Charlotte Elizabeth *Journal, 1833–1891* extracts ed. Earl of Bessborough 2 vols. 1950–2
Journals, 1869–85 ed. M. J. Guest 2 vols. 1911
Schreiner, Olive *From man to man* 1926
Schuchert, Charles See Pirsson, Louis Valentine
Schuell, Hildred *Aphasia theory and therapy: selected lectures and papers* ed. L. F. Sies 1974
Schulberg, Budd Wilson *The disenchanted* 1950 (UK 1951)
What makes Sammy run? 1941
Schumpeter, Joseph Alois *Business cycles: a theoretical, historical, and statistical analysis of the capitalist process* 2 vols. 1939
Schwartz, Robert J. *A complete dictionary of abbreviations* 1955
Schwarz, Urs & Hadik, Laszlo *Strategic terminology* 1966
Sciama, Dennis William *Modern cosmology* 1971
Science abstracts 1898–1940 (from 1903 divided into subject sections)
Science journal 1965–71
Science news 1946–60
Science news letter 1921– (from 1966 with title *Science news*)
Science progress (title varies) 1894–8; 1906–40; 1946–
Science survey 1960–
Scientific monthly 1915–57
Sclater, Philip Lutley & Thomas, M. R. O. *The book of antelopes* 17 parts (4 vols.) 1894–1900
Sclater, William Lutley See Stark, Arthur Cowell
Scollins, Richard & Titford, John *Ey up, mi duck!* 3 parts 1976–7
Scope 1966–
Scots observer: a weekly journal of religious and national interest 1926–34
Scotsman, The 1817– (1855–9 with title *The Daily Scotsman*)
Scott, Clement William & Howard, C. *The life and reminiscences of E. L. Blanchard* 2 vols. 1891
Scott, Edward *All about the latest dances* 1919
Scott, Gavin *Hot pursuit* 1977
Scott, Jack S. *A clutch of vipers* 1979
Scott, John Somerville *A dictionary of building* 1964
A dictionary of civil engineering 1958
Scott, Paul Mark *Staying on* 1977
Scott, Richard John Ernst ed. *Gould's medical dictionary* 1926
Scott, Robert Falcon (1868–1912) *Scott's last expedition* ed. L. Huxley 2 vols. 1913
The voyage of the 'Discovery' 2 vols. 1905
Scott, Virgil Joseph & Koski, Dominic *Walk-in* 1976 (UK 1977)
Scott, Walter (1771–1832) *Letters* ed. H. J. C. Grierson et al. 12 vols. 1932–7
Scott, William Berryman *An introduction to geology* 1897
Scottish field 1903–
Scottish journal of theology 1948–
Scottish national dictionary ed. W. Grant & D. D. Murison 10 vols. 1931–76
Scottish review 1975–
Screen 1959–
Scripture, Edward Wheeler *The elements of experimental phonetics* 1902
Scrutiny: a quarterly review 1932–53
Sculthorpe, Cyril Duncan *The biology of aquatic vascular plants* 1967
Scupham, Peter *The hinterland* 1977
Sea breezes 1919–39; 1946–
Seafarers' log 1939–
Seaman, Donald *The bomb that could lip-read* 1974
Sears, Roebuck & Co. *Catalogs* c 1876–

Seaton, Albert Edward *A manual of marine engineering* 1883
Sedgwick, Anne Douglas *The little French girl* 1924
Segrè, Emilio Gino *Experimental nuclear physics* 3 vols. 1953–9
Sela, Owen *The Bengali inheritance* 1975
Selbie, William Boothby *The psychology of religion* 1924
Seldon, Arthur & Pennance, F. C. *Everyman's dictionary of economics* 1965
Self, Margaret Cabell *The horseman's encyclopedia* 1946
Selvon, Samuel *A brighter sun* 1952
Senhouse, Roger tr. *Colette's Chéri, and The last of Chéri* 1951
Senn, Charles Herman *The new century cookery book: practical gastronomy and recherché cookery* 1901
Serling, Robert J. *The president's plane is missing* 1967 (UK 1968)
Serra, José Autures *Modern genetics* 3 vols. 1965–8
Service, Robert William *Ballads of a cheechako* 1909 (UK 1910)
Ploughman of the moon: an adventure into memory 1945 (UK 1946)
Rhymes of a rolling stone 1912 (UK 1913)
Songs of a sourdough 1907 (UK 1908)
The trail of '98: a northland romance 1910 (UK 1911)
Sessions papers [series of reports of criminal cases heard in London]:
1. *The whole proceedings of the Sessions of the Peace, and Oyer and Terminer for the City of London and County of Middlesex* 1730–1824
2. *Sessions' paper . . held at the Justice Hall, in the Old Bailey* 1825–33
3. Central Criminal Court *Minutes* 1835–1913
Seton, Ernest Evan Thompson *Trail of an artist-naturalist* 1940
Seton-Watson, Christopher *Italy from liberalism to fascism, 1870–1925* 1967
Severson, John Hugh ed. *Great surfing* 1967
Sewall, William *Diary, 1797–1846* ed. J. Goodell 1930
Sewanee review 1892–
Sewell, Anna *Black Beauty: the autobiography of a horse* 1877
Sewell, Edward Humphrey D. *Rugby football up to date* 1921
Sewell, George Harold *Amateur film-making* 1938
— (ed. 2) 1951
Seymour, Gerald *The glory boys* 1976
Shackleton, Ernest Henry *South: the story of Shackleton's last expedition, 1914–1917* 1919
Shadbolt, Maurice Francis Richard *Among the cinders* 1965
The New Zealanders: a sequence of stories 1959
Strangers and journeys 1972
Shand, Philip Morton *A book of wine* 1926
Shands, Harley Cecil *Semiotic approaches to psychiatry* 1970
Shands, Hubert Anthony *Some peculiarities of speech in Mississippi* 1893
'Shannon, Dell' (Barbara Elizabeth Linington) *Murder with love* 1971 (UK 1972)
No holiday for crime 1973 (UK 1974)
Rain with violence 1967 (UK 1969)
The ringer 1971 (UK 1972)
Unexpected death 1970 (UK 1971)
With a vengeance 1966 (UK 1968)
Shapiro, Nat & Hentoff, Nat eds. *Hear me talkin' to ya: the story of jazz by the men who made it* 1955
Sharcott, Margaret *A place of many winds* 1960
Sharp, Lester Whyland *Introduction to cytology* 1921
— (ed. 3) 1934
Sharp, Margery *Cluny Brown* 1944
The gipsy in the parlour 1954
The nutmeg tree 1937
Something light 1960
Sharpe, Frederick Dew *Sharpe of the Flying Squad* 1938
Sharpe, May Churchill *Chicago May: her story* 1928
Sharpe, Thomas Ridley *Porterhouse blue* 1974
Wilt 1976
Sharps, Wallace Samuel *Dictionary of cinematography and sound recording* 1959
Shaw, Dennis Frederick *An introduction to electronics* 1962
— (ed. 2) 1970
Shaw, Frank, Spiegl, Fritz, & Kelly, Stan *Lern yerself Scouse: how to talk proper in Liverpool* 1966 (For Vol. 2 see 'Lane, Linacre')
Shaw, George Bernard (1856–1950) *The admirable Bashville* (in *Cashel Byron's profession*, rev. ed.) 1901
Androcles and the lion: a fable play 1914
The apple cart: a political extravaganza 1930

Back to Methuselah: a metabiological Pentateuch 1921
Bernard Shaw and Mrs. Patrick Campbell: their correspondence ed. A. Dent 1952
Buoyant billions 1949
Caesar and Cleopatra See *Three plays for Puritans*
Captain Brassbound's conversion See *Three plays for Puritans*
Collected letters ed. D. H. Laurence 1965–
The common sense of municipal trading 1904
The Dark Lady of the sonnets See *Misalliance*
The devil's disciple See *Three plays for Puritans*
The doctor's dilemma, Getting married, and The shewing-up of Blanco Posnet 1911
Ellen Terry and Bernard Shaw: a correspondence See: Terry, Ellen & Shaw, George Bernard
Everybody's political what's what? 1944
Fanny's first play See *Misalliance*
Geneva 1939
Getting married See *The doctor's dilemma*
Heartbreak House, Great Catherine, and Playlets of the War 1919
How to become a musical critic ed. D. H. Laurence 1960
In good King Charles's golden days 1939
The intelligent woman's guide to socialism and capitalism 1928
Letters See *Collected letters*
Letters to Granville Barker ed. C. B. Purdom 1956
London music in 1888–89 as heard by Corno di Bassetto 1937
The millionairess See *The simpleton of the Unexpected Isles*
Misalliance, The Dark Lady of the sonnets, and Fanny's first play. With a treatise on parents and children 1914
On the rocks See *Too true to be good*
Our theatres in the Nineties: criticisms contributed week by week to The Saturday Review 3 vols. 1932
Peace Conference hints 1919
Pen portraits and reviews 1931
— (Standard ed.) 1932
Platform and pulpit ed. D. H. Laurence 1962
Pygmalion in *Nash's magazine* Nov.–Dec. 1914
— (combined ed. with *Androcles and the lion* and *Overruled*) 1916
The quintessence of Ibsenism 1891
— *now completed to the death of Ibsen* 1913
The shewing-up of Blanco Posnet See *The doctor's dilemma*
The simpleton of the Unexpected Isles, The six of Calais, and The millionairess 1936
Three plays for Puritans (The devil's disciple, Caesar and Cleopatra, Captain Brassbound's conversion) 1901
To a young actress: letters to Molly Tompkins ed. P. Tompkins 1960
Too true to be good, Village wooing, and On the rocks 1934
Translations and tomfooleries 1926
What I really wrote about the War 1930
— (Standard ed.) 1931
Widowers' houses: a comedy 1893
Shaw, Herman *A text-book of aeronautics* 1919
Shaw, Irwin *Beggarman, thief* 1977
Rich man, poor man 1970
The troubled air 1951
Shaw, William Napier *Forecasting weather* 1911
— (ed. 2) 1923
— (ed. 3) 1940
—— & Austin, Elaine *Manual of meteorology* 3 vols. 1926–30
She 1955–
Shea, Timothy Edward *Transmission networks and wave filters* 1930
Shedlock, John South tr. *K. W. J. H. Riemann's Dictionary of music* 1893
Sheehan, George A. *Running and being: the total experience* 1978
'Shelby, Brit' (James Grady) *The great pebble affair* 1976
Sheldon, Sidney *Bloodline* 1978
Shell aviation news 1931–
Shell International Petroleum Company *Bitumen* (Shell Briefing Service) 1977
Information handbook, 1974–5 1974
Protecting the world's crops (Shell Briefing Service) 1977
Shelley, Percy Bysshe (1792–1822) *Letters* ed. F. L. Jones 2 vols. 1964
Shelly, Lou ed. *Hepcats jive talk dictionary* 1945
Shelton, Jane de Forest *The salt-box house: eighteenth century life in a New England hill town* 1900
Shepardson, George Defrees *Telephone apparatus* 1917
Shepherd, Edwin Colston *The military aeroplane* 1941
Sheppard, Mubin *Taman indera: a royal pleasure ground. Malay decorative arts and pastimes* 1972
Sheppard, Samuel Edward *Photo-chemistry* 1914

—— & Mees, Charles Edward Kenneth *Investigations on the theory of the photographic process* 1907
Sheppard, Thomas Harvey *Dictionary of railway slang* 1965
— (ed. 2) 1966
Sheridan, Alan tr. *J. Lacan's Écrits* 1977
Sheridan, Elizabeth *Betsy Sheridan's journal: letters from Sheridan's sister, 1784–1786 and 1788–1790* ed. W. LeFanu 1960
Sherif, Muzafer *The psychology of social norms* 1936
Sherrard, Owen Aubrey *Two Victorian girls: with extracts from the Hall diaries* ed. A. R. Mills 1966
Sherrington, Charles Scott *The integrative action of the nervous system* 1906
Sherwood, Adiel *A gazetteer of the State of Georgia* 1827
Sherwood, Robert Emmet *Idiot's delight* 1936
Shevelov, George Yury *A prehistory of Slavic: the historical phonology of common Slavic* 1965
Shillaber, Benjamin Penhallow *The life and sayings of Mrs. Partington* 1854
Shoberl, Frederic tr. *I. Titsingh's Illustrations of Japan* 1822
Shooting times and country magazine 1882–
Short, Eirian *Embroidery and fabric collage* 1967
Shrader, Robert Louis *Electronic communication* 1959
Shulman, Max *Barefoot boy with cheek* 1943
Shurr, Gertrude & Yocom, Rachael Dunaven *Modern dance: techniques and teaching* 1949
'Shute, Nevil' (Nevil Shute Norway) *Beyond the black stump* 1956
The chequer board 1947
The far country 1952
In the wet 1953
Landfall 1940
Lonely road 1932
No highway 1948
On the beach 1957
Pastoral 1944
Pied piper 1942
The rainbow and the rose 1958
Requiem for a Wren 1955
Round the bend 1951
Ruined city 1938
A town like Alice 1950
Trustee from the toolroom 1960
Shuttleworth, Charles *Malayan safari* 1965
Sichel, Allan *The Penguin book of wines* 1965
Sidgwick, Arthur & Sidgwick, Eleanor Mildred *Henry Sidgwick: a memoir* 1906
Sidgwick, Cecily *Sack and sugar* 1926
Victorian 1922
Sidgwick, Nevil Vincent *The chemical elements and their compounds* 2 vols. 1950
The electronic theory of valency 1927
Sidney, Richard John Hamilton *In British Malaya to-day* 1927
Sieff, Israel *Memoirs* 1970
Sierra Club bulletin (San Francisco) 1893–
Silberrad, Una Lucy *The letters of Jean Armiter: a novel* 1923
Silberstein, Ludwik *The theory of relativity* 1914
Silliman, Benjamin *Manual on the cultivation of the sugar cane* 1833
Sillitoe, Alan *The loneliness of the long-distance runner* 1959
Saturday night and Sunday morning 1958
Sim, Thomas Robertson *The forests and forest flora of the colony of the Cape of Good Hope* 1907
Simak, Clifford Donald *They walked like men* 1963
Simms, William Gilmore *Guy Rivers: a tale of Georgia* 2 vols. 1834
The partisan: a tale of the revolution 2 vols. 1835
Southward ho! A spell of sunshine 1854
The wigwam and the cabin; or, Tales of the South 2 ser. 1845
Simon, André Louis ed. *A concise encyclopædia of gastronomy* 9 vols. 1939–46
A dictionary of wines, spirits and liqueurs 1958
Guide to good food and wines (rev. ed.) 1963
—— & Howe, Robin *A dictionary of gastronomy* 1970
Simon, Edith *The past masters* 1953
Simon, Roger Lichtenberg *The big fix* 1973 (UK 1974)
Wild turkey 1975 (UK 1976)
Simonds, Herbert Rumsey & Ellis, Carleton *Handbook of plastics* 1943
'Simons, Roger' (Margaret & Ivor Punnett) *A frame for murder* 1960
Simonsen, John Lionel *The terpenes* 2 vols. 1931–2
Simpson, George Gaylord *Principles of animal taxonomy* 1961
Simpson, John Hamilton & Richards, Roger Smith *Physical principles and applications of junction transistors* 1962

Sims, George Frederick *Hunters Point* 1973
The last best friend 1967
The sand dollar 1969
Sinclair, Michael *A long time sleeping* 1975
Norslag 1972
Sinclair, Upton Beall *The brass check: a study of American journalism* 1919
The jungle 1906
Oil! 1927
The presidential agent 1944 (UK 1945)
Singer, Isaac Bashevis *Shosha* 1978
Singha, Rina & Massey, R. *Indian dances, their history and growth* 1967
Sisam, Kenneth *Studies in the history of Old English literature* 1953
Sissons, Thomas Michael B. & French, P. eds. *Age of austerity* 1963
Sitwell, Edith Louisa *Bath* 1932
Bucolic comedies 1923
Gardeners and astronomers 1953
I live under a black sun 1937
The sleeping beauty 1924
Troy Park 1925
Victoria of England 1936
The wooden Pegasus 1920
Sitwell, Francis Osbert S. *Miracle on Sinai* 1933
Sitwell, Sacheverell *Golden wall and mirador* 1961
6,000 words: a supplement to Webster's Third New International Dictionary 1976
Skeat, Walter William (1866–1953) *Malay magic* 1900
Skelton, John (? 1460–1529) tr. *Diodorus Siculus' Bibliotheca historica* ed. F. M. Salter & H. L. R. Edwards 2 vols. 1956–7
Skerl, John George Anthony tr. *A. L. Wegener's The origin of continents and oceans* 1924
Sketch, The: a journal of art and actuality 1893–1959
Sketches and eccentricities of Colonel David Crockett, of West Tennessee 1833
Skilbeck, Oswald *ABC of film and TV working terms* 1960
Skinner, Cornelia Otis *Madame Sarah* 1967
Slater, Mary *Caribbean cooking for pleasure* 1970
Slatter, Gordon Cyril *A gun in my hand* 1959 (UK 1960)
Slim, William Joseph *Defeat into victory* 1956
Slimming, John *Temiar jungle* 1958
Small, Austin J. *Frozen gold* 1924
Small, Ronald John *The study of landforms: a textbook of geomorphology* 1970
Smart, John Jamieson Carswell *Between science and philosophy: an introduction to the philosophy of science* 1968
Smart set, The: a magazine of cleverness 1900–30
Smeaton, Amethe tr. *R. Carnap's The logical syntax of language* 1937
Smiley, Jack *Hash house lingo* 1941
Smith, Alexander *Introduction to general inorganic chemistry* 1906
Smith, Andrew *The diary of Dr. Andrew Smith, director of the 'Expedition for exploring Central Africa', 1834–1836* ed. P. R. Kirby 2 vols. 1939–40
Smith, Annie Lorrain *A handbook of British lichens* 1921
Lichens 1921
Smith, Anthony John Francis *Blind white fish in Persia* 1953
Throw out two hands 1963
Smith, Ashley *The East-Enders* 1961
Smith, Charles Alphonso *New words self-defined* 1919
Smith, Charles Henry *Bill Arp, so called: a side show of the Southern side of the war* 1866
Smith, Dennis John *Discovering railwayana* 1977
Smith, Dorothy Gladys ('Dodie') *Dear octopus: a comedy* 1938
I capture the castle 1949
Smith, Edgar Fahs tr. *V. von Richter's Chemistry of the carbon compounds; or, Organic chemistry* 1886
— (ed. 2) 1892
— (ed. 3) 2 vols. 1899–1900
Smith, Egerton *A guide to English traditions and public life* 1953
Smith, Gilbert Morgan *Cryptogamic botany* 1938
Smith, Godfrey *The business of loving* 1961
The flaw in the crystal 1954
Smith, John B. *Explanation of terms used in entomology* 1906
Smith, John Hugh *Digital logic: basic theory and practice* 1973
Smith, Julie P. *The Widow Goldsmith's daughter* 1870
Smith, Kenneth M. *Insect virology* 1967
Smith, Laura Alexandrine *The music of the waters: a collection of the sailors' chanties, or working songs of the sea, of all maritime nations* 1888

Smith, Logan Pearsall *Words and idioms: studies in the English language* 1925 (UK 1928)
Smith, Martin Cruz *Gorky Park* 1981
Smith, Robert Allan *Semiconductors* 1959
Smith, Robert Kimmel *Ransom* 1971 (UK 1972)
Smith, Seba *The select letters of Major Jack Downing* 1834
Smith, Solomon Franklin *The theatrical apprenticeship and anecdotical recollections of Sol. Smith* 1846
Smith, W. H. Saumarez *A young man's country: letters of a subdivisional officer of the Indian Civil Service 1936–1937* 1977
Smith, Walter George *Allergy and tissue metabolism* 1964
Smith, Wilbur *Gold mine* 1970
Smith, William tr. *J. G. Fichte's The characteristics of the present age* 1847
Smith, William Hawley *The promoters: a novel without a woman* 1904
Smyth, Henry Dewolf *A general account of the development of methods of using atomic energy for military purposes under the auspices of the United States government, 1940–1945* 1945
Smyth, James Desmond *An introduction to animal parasitology* 1962
Smyth, John Ferdinand Dalziel *A tour in the United States of America* 2 vols. 1784
Smythies, John Raymond & Corbett, Lionel *Psychiatry for students of medicine* 1976
Snell, Walter Henry & Dick, Esther Amelia *A glossary of mycology* 1957
Sniečkus, Antanas *Soviet Lithuania on the road of prosperity* [anon. tr.] 1974
Snook, Barbara Lilian *English historical embroidery* 1960
Snow, Charles Percy *The affair* 1960
The conscience of the rich 1958
Corridors of power 1964
Homecomings 1956
The masters 1951
The new men 1954
Strangers and brothers 1940
Snow, Edgar *The other side of the river: Red China today* 1962 (UK 1963)
Red star over China 1937
Snow, John Augustine *Cricket rebel: an autobiography* 1976
Social science abstracts 1929–33
Society for Experimental Biology and Medicine *Proceedings* 1903–
Society for Pure English *Tracts* Nos. 1–66, 1919–48
Society of Biblical Archaeology *Proceedings* 1878–1918
Transactions 1872–93
Society of Chemical Industry *Journal* 1882–1950 (from 1917 divided into sections)
Society of Dyers and Colourists *Journal* 1884–
Society of Telegraph Engineers (and of Electricians) *Journal* 1872–88
Soil science 1916–
Sollas, Hertha Beatrice Coryn & Sollas, William Johnson tr. *E. Suess's The face of the earth* 5 vols. 1904–24
Solzhenitsyn, Alexander Isayevitch See Bethell, Nicholas & Burg, David; Whitney, Thomas Porter
Somerville, Edith Œnone & 'Ross, Martin' (Violet Florence Martin) *All on the Irish shore: Irish sketches* 1903
The real Charlotte 1894
Somerville-Large, Peter *Couch of earth* 1975
Eagles near the carcase 1977
Sommerstein, Alan H. *The sound pattern of ancient Greek* 1973
Sorokin, Pitrim Aleksandrovich *Sociological theories of today* 1966
Sounds 1970–
South, Richard *The butterflies of the British Isles* 1906
The moths of the British Isles 2 ser. 1907–8
South Africa. Senate *Debates* 1911–
South African Philosophical Society *Transactions* 1877–1909
South Carolina Gazette (Charleston, S. Carolina) 1732–?1775 (title varies)
South Carolina Historical Society *Collections* 5 vols. 1857–97
South China Morning Post (Hong Kong) 1904–
South Wales Echo (Cardiff) 1884–
Southall, James Powell Cocke ed. & tr. *H. L. F. von Helmholtz et al.'s Treatise on physiological optics* 3 vols. 1924
Southampton, New York *Records* 6 vols. 1874–1915
Southerly: the magazine of the Australian English Association 1939–
Southern Evening Echo (Southampton) 1958–
Southern folklore quarterly 1937–
Southern literary messenger 1834–64
Southward, John *Modern printing: a handbook of the principles and practice of typography and the auxiliary arts* 4 vols. 1898–1900 (and several later editions used)

Southwestern historical quarterly 1912–
Southwood, Thomas Richard Edmund & Leston, Dennis *Land and water bugs of the British Isles* 1959
Southworth, Louis *Felon in disguise* 1966
Soviet physics: Doklady See: Academy of Sciences of the U.S.S.R.
Soyinka, Wole (Akinwande Oluwole Soyinka) *Kongi's harvest* 1967
The lion and the jewel 1963
The road 1965
Spaceflight 1956–
Spackman, William Mode *An armful of warm girl* 1978
Spanner, Edward Frank *The naviators* 1926
Spare rib 1972–
Spark, Muriel Sarah *The bachelors* 1960
The ballad of Peckham Rye 1960
The comforters 1957
The Mandelbaum gate 1965
The prime of Miss Jean Brodie 1961
Sparkes, John J. *Transistor switching and sequential circuits* 1969
Sparks, Bruce Wilfred *Geomorphology* 1960
Spartanburg Herald (Spartanburg, S. Carolina) 1890–
Spencer, Douglas Arthur *The Focal dictionary of photographic technologies* 1973
Spencer, John ed. *The English language in West Africa* 1971
Spender, Stephen Harold *Collected poems, 1928–1953* 1955
The edge of being 1949
European witness: impressions of Germany in 1945 1946
Learning laughter 1952
The making of a poem 1955
tr. *E. H. Toller's Pastor Hall* 1939
Poems of dedication 1947
Ruins and visions 1942
The still centre 1939
World within world: autobiography 1951
—— & Gili, J. L. tr. *F. García Lorca's Poems* 1939
Sphere, The 1900–64
'Spillane, Mickey' (Frank Morrison Spillane) *The big kill* 1951
Spinks, John William Trainer tr. *G. Herzberg's Atomic spectra and atomic structure* 1937
Spirit of the times: a chronicle of the turf, agriculture, field sports, literature and the stage 1831–61
Spitzer, Leo *Linguistics and literary history* 1948
Spock, Benjamin McLane *Decent and indecent: our personal and political behavior* 1970
Spons' Mechanics' own book: a manual for handicraftsmen and amateurs 1885
Sport (title varies) 1946–57
Spring, Robert Howard *These lovers fled away* 1955
Spring Rice, Margaret Lois *Working-class wives: their health and conditions* 1939
Springer, Clifford Harry, Herlihy, R. E., & Beggs, R. I. *Advanced methods and models* 1965
Springfield Republican (Springfield, Massachusetts) (weekly ed.) 1824–
Sprott, Walter John Herbert *Social psychology* 1952
Spruce, Richard *Notes of a botanist on the Amazon and Andes..1849–64* ed. A. R. Wallace 2 vols. 1908
Squire, Amos Osborne *Sing Sing doctor* 1935
Srb, Adrian Morris & Owen, Roy D. *General genetics* 1952
— (ed. 2, by A. M. Srb et al.) 1965
Stage year book, The 1908–28; 1949–
Stamp, Laurence Dudley *Britain's structure and scenery* 1946
— (ed. 5) 1960
ed. *A glossary of geographical terms* 1961
An introduction to stratigraphy: British Isles 1923
Standard, The 1827–1916
Standard encyclopaedia of southern Africa 12 vols. 1970–6
Stanford, Charles Villiers & Forsyth, Cecil *A history of music* 1916
Stanley, Rupert *Text-book on wireless telegraphy* 1914
— (new ed.) 2 vols. 1919
Stanton, Isabel Alice *A dictionary for medical secretaries* 1960
Stanwell-Fletcher, Theodora Morris *Driftwood Valley* 1946
Staples, Frank Alston *Water-colour painting is fun* 1948 (UK ed. 1951 with title *Water-colour painting*)
Star, The (Sheffield) 1938–
Stark, Arthur Cowell & Sclater, William Lutley *The birds of South Africa* 4 vols. (I–III by Stark, IV by Sclater) 1900–6
Stark, Freya Madeline (1893–) *Baghdad sketches* 1937

Letters ed. L. Moorehead 1974–
Letters from Syria 1942
Riding to the Tigris 1959
The southern gates of Arabia: a journey in the Hadhramaut 1936
The valleys of the Assassins, and other Persian travels 1934
Stark, Werner *The sociology of knowledge* 1958
State (Columbia, S. Carolina) 1891–
Staunton, Harold *The chess-player's handbook* 1847
Stead, Christian Karlson ed. *New Zealand short stories* (2nd ser.) 1966
For 1st ser. see Davin, Daniel Marcus
Stead, Christina Ellen *The man who loved children* 1940
Stearns, Marshall Winslow *The story of jazz* 1956 (UK 1957)
Stedman, Thomas Lathrop *A practical medical dictionary* 1911 (and many later editions used)
Steele, John (1764–1815) *Papers* ed. H. M. Wagstaff 2 vols. 1924
Steen, Marguerite *Phoenix rising* 1952
The tower 1959
Steffens, Joseph Lincoln (1866–1936) *Autobiography* 2 vols. 1931
Stegner, Wallace Earle *Wolf Willow* 1962
Stein, Gertrude *The autobiography of Alice B. Toklas* 1933
Steinbeck, John Ernst *Cannery Row* 1945
East of Eden 1952
The grapes of wrath 1939
In dubious battle 1936
The long valley 1938
Of mice and men 1937
The pearl 1947 (UK 1948)
A Russian journal 1948
Sweet Thursday 1954
Tortilla flat 1935
The wayward bus 1947
Steiner, George *In Bluebeard's castle: some notes towards the re-definition of culture* 1971
Language and silence: essays 1958–1966 1967
Step, Edward *Bees, wasps, ants and allied insects of the British Isles* 1932
Shell life: an introduction to the British Mollusca 1901
Stephenson, George E. & Kruse, R. S. *The radio manual* 1928
Sterling, Thomas L. *The evil of the day* 1955
Stern, Nils Gustav *Meaning and change of meaning, with special reference to the English language* 1931
Stevens, Stanley Smith ed. *Handbook of experimental psychology* 1951
Stevens, Wallace (1879–1955) *Collected poems* 1954
Letters ed. H. Stevens 1967
Stevenson, Robert Louis & Osbourne, Lloyd *The ebb-tide: a trio and a quartette* 1894
Stevenson-Hamilton, James *Wild life in South Africa* 1947
Steward, Julian Haynes ed. *Handbook of South American Indians* 7 vols. 1946–59
Stewart, Ian Nicholas *Concepts of modern mathematics* 1975
Stewart, John Innes Mackintosh *Eight modern writers* 1963
The gaudy 1974
The guardians 1955
The Madonna of the astrolabe 1977
The man who won the pools 1961
A memorial service 1976
A use of riches 1957
Young Pattullo 1975
See also: Innes, Michael
Stewart, Mary *Madam, will you talk?* 1954
My brother Michael 1960
Nine coaches waiting 1958
'Stiff, Dean' (Nels Anderson) *The milk and honey route* 1931
Stifler, William Warren ed. *High-speed computing devices* 1950
Stimpson, George William *A book about a thousand things* 1946
Stinton, Darrol *The anatomy of the aeroplane* 1966
Stivens, Dallas George ('Dal') *The courtship of Uncle Henry: a collection of tales and stories* 1946
Stocqueler, Joachim Hayward *The oriental interpreter and treasury of East India knowledge* 1848
Stokoe, William John *The caterpillars of the British butterflies* 1944
The caterpillars of British moths 2 ser. 1948
— (new ed.) 2 ser. 1958
Stone, Hannah M. & Stone, Abraham *A marriage manual* 1935 (UK 1936)
Stone, Kate See Holmes, Sarah Katherine (Stone)
Stone, Louis *Jonah* 1911
'Stone, Zachary' *The Modigliani scandal* 1979
Stopes, Marie Charlotte Carmichael *Married love: a new contribution to the solution of sex difficulties* 1918
Stoppard, Tom *Dirty linen, and New-found-land* 1976
Jumpers 1972
The real Inspector Hound 1968
Rosencrantz and Guildenstern are dead 1967
Travesties 1975
Storey, David Malcolm *Saville* 1976
This sporting life 1960
Stornoway Gazette and West Coast Advertiser 1917–
Storr, Catherine *The Chinese egg* 1975
Marianne and Mark 1960
Tales from a psychiatrist's couch 1977
Story, Jack Trevor *Dishonourable member* 1969
Something for nothing 1963
Stouffer, Samuel Andrew *Measurement and prediction* 1950
Stout, George Frederick *Analytic psychology* 1896
Stout, Rex Todhunter *Fer-de-lance* 1934
If death ever slept 1957 (UK 1958)
Over my dead body 1940
Please pass the guilt 1973 (UK 1974)
The red box 1936 (UK 1937)
Red threads 1939 (UK 1941)
Stowe, Harriet Elizabeth (Beecher) *House and home papers* 1865
We and our neighbors 1875
Strachey, Alix & Strachey, James tr. *S. Freud's Collected papers* Vol. III See Riviere, Joan, et al.
Strachey, Giles Lytton *Characters and commentaries* 1933
Eminent Victorians 1918
Queen Victoria 1921
Strachey, James tr. *S. Freud's Group psychology and the analysis of the ego* 1922
tr. *S. Freud's Totem and taboo* 1950
—— et al. tr. *S. Freud's Complete psychological works* 24 vols. 1953–74
See also: Strachey, Alix
Straits Times (Malaysia) 1845– (title varies)
Stranathan, James Docking *The 'particles' of modern physics* 1942
Strand, Stanley *Marketing dictionary* 1962
Strand Electrical and Engineering Company, Ltd. *Glossary of technical theatrical terms* 1947
Strang, Barbara Mary Hope *A history of English* 1970
Metaphors and models: an inaugural lecture 1965
Modern English structure 1962
Strangways, Arthur Henry Fox *The music of Hindostan* 1914
Stratton-Porter, Gene *Freckles* 1904
A girl of the Limberlost 1909
The harvester 1911
Laddie 1913
Straumann, Heinrich *Newspaper headlines: a study in linguistic method* 1935
Strauss, Victor *The printing industry* 1967
Strawson, Peter Frederick *Individuals: an essay in descriptive metaphysics* 1959
Streatfeild, Noel *Aunt Clara* 1952
Ballet shoes 1936
I ordered a table for six 1942
Streetwalker [anon.] 1959
Strevens, Peter Derek *New orientations in the teaching of English* 1977
Papers in language and language teaching 1965
Strickberger, Monroe Wolf *Genetics* 1968
Strong, Charles Augustus *Why the mind has a body* 1903
Strong, Leonard Alfred George *All fall down* 1944
The bay 1941
The director 1944
Othello's occupation 1945
Sea wall 1933
Sun on the water, and other stories 1940
Trevannion 1948
Which I never 1950
Stroud, John Anthony *The shorn lamb* 1960
Touch and go 1961
Stuart, Granville (1834–1918) *Forty years on the frontier* ed. P. C. Phillips 2 vols. 1925
Stuart, Jesse *Men of the mountains* 1941
Stuart, Ruth MacEnery *In Simpkinsville* 1897
Stubbs, Jean *The painted face* 1974
Studholme, Edgar Channon *Te Waimate: early station life in New Zealand* 1940
— (ed. 2) 1954
Studies in English (University of Cape Town) 1970– (title varies)
Studies in English literature (University of Tokyo) 1919–
Sturgis, Russell *A dictionary of architecture and building: biographical, historical and descriptive* 3 vols. 1901–2
Sturtevant, Edgar Howard *Comparative grammar of the Hittite language* 1933
Styron, William *Confessions of Nat Turner* 1967
Sophie's choice 1979

Sudworth, George Bishop *Nomenclature of the arborescent flora of the United States* 1897
Sullivan, Joseph M. *Criminal slang* 1908
Summers, Montague (Alphonsus J.-M. A. M. Summers) *The supernatural omnibus* 1931
Summerton, Margaret *A memory of darkness* 1967
A small wilderness 1959
The sunset hour 1957
Sumner, James Batcheller & Myrbäck, Karl eds. *The enzymes* 2 vols. in 4 parts 1950–2
—— & Somers, George Frederick *Chemistry and methods of enzymes* 1943
Sumner, William Graham *Folkways: a study of the sociological importance of usages, manners, customs, mores, and morals* 1907
Sun, The (Baltimore, Maryland) 1837–
Sun, The (London) 1964–
Sun (New York) 1833– (title varies)
Sun-Herald (Sydney) 1953–
Sunday Advocate-News (Barbados) 1968–
Sunday Bulletin (Philadelphia): see *Philadelphia Bulletin*
Sunday Dispatch 1928–61
Sunday Express (London) 1918–
Sunday Express and Home Journal (Johannesburg) 1937–
Sunday Mail (Glasgow) 1919–
Sunday Nation (Nairobi) 1960–
Sunday Pictorial 1915–63
Sunday Post (Glasgow) 1920–
Sunday Times (Johannesburg) 1906–
Sunday Times (Lagos) 1953–
Sunday Times (London) 1822–
Sunday Truth (Brisbane) 1960–71
Sunset (title varies) 1898–
Suppes, Patrick *Introduction to logic* 1957
Surfer: the international surfing magazine 1960–
Surgery 1937–
Surgery, gynecology and obstetrics 1905–
Survey: a journal of East and West studies 1955–
Sutcliff, Robert *Travels in some parts of North America in the years 1804, 1805, and 1806* 1811
'Sutherland, Joan' (Joan Collings) *The circle of the stars* 1924
Suttie, Jane Isabel tr. *S. Ferenczi's Further contributions to the theory and technique of psycho-analysis* 1926
Sutton, Oliver Graham *The science of flight* 1949
'Suyin, Han' See 'Han Suyin'
Suzuki, Daisetsu Teitarō *Zen Buddhism and its influence on Japanese culture* 1938
Svensk botanisk tidskrift 1907–
Sverdrup, Harald Ulrik, Johnson, M. W., & Fleming, R. H. *The oceans: their physics, chemistry, and general biology* 1942
Swaan, Wim *Japanese lantern* 1965
Swan, Lester A. & Papp, Charles Steven *The common insects of North America* 1972
Swan, Michael Lancelot *British Guiana* 1957
Swann, Harry Kirke *A dictionary of English and folk-names of British birds* 1913
Swanson, Carl Pontius *Cytology and cytogenetics* 1957 (UK 1958)
Swanton, Michael J. ed. *The dream of the Rood* 1970
Swayne, James Colin *A concise glossary of geographical terms* 1956
Sweeney, Charles *The scurrying bush* 1966
Sweet, Alexander Edwin & Knox, John Armoy *On a Mexican mustang through Texas, from the Gulf to the Rio Grande* 1883 (UK 1884)
Sweet-Escott, Bickham *Baker Street irregular* 1965
Swell's night guide, The 1841
— (another ed.) 1846
— (another ed.) 1847
Swettenham, Frank Athelstane *British Malaya: an account of the origin and progress of British influence in Malaya* 1907
Swinburne, Algernon Charles (1837–1909) *Letters* ed. C. Y. Lang 6 vols. 1959–62
Swing music 1935–6
Swinnerton, Frank Arthur *The doctor's wife comes to stay* 1949
Swinson, Arthur Horace *Six minutes to sunset: the story of General Dyer and the Amritsar affair* 1964
Swyer, Gerald Isaac M. *Reproduction and sex* 1954
Sydney, University of. Australian Language Research Centre *Occasional papers* 1964–
Sydney Morning Herald 1842–
Sydney slang dictionary, The ? 1882
Symonds, John Addington (1840–93) *Letters* ed. H. M. Schueller & R. L. Peters 3 vols. 1967–9
Symons, Julian Gustave *The end of Solomon Grundy* 1964
The narrowing circle 1954
A three pipe problem 1975
Symons, R. D. *Many trails* 1963
Where the wagon led 1973
Symposia in applied mathematics *Proceedings* 1947–

Synge, John Millington *Letters to Molly: J. M. Synge to Maire O'Neill 1906–1909* ed. A. Saddlemyer 1971
The playboy of the western world 1907

T

TV times 1955–
Tablet, The 1840–
'Tackline' *Holiday sailor* 1945
'Taffrail' (Henry Taprell Dorling) *Carry on!: naval sketches and stories* 1916
Pincher Martin, O.D.: a story of the inner life of the Royal Navy 1916
Tak, Montie *Truck talk: the language of the open road* 1971
Takagaki, Shinzō & Sharp, Harold E. *The techniques of judo* 1957
Talbot, Frederick Arthur Ambrose *Moving pictures: how they are made and worked* 1912
— (rewritten) 1923
Practical cinematography and its applications 1913
Talbot, Patrick Henry Brabazon *Principles of fungal taxonomy* 1971
Talbot, Percy Amaury *The peoples of southern Nigeria* 4 vols. 1926
Tamarack review 1956–
Tansley, Arthur George *The British islands and their vegetation* 1939
—— & Chipp, Thomas Ford eds. *Aims and methods in the study of vegetation* 1926
Tararua (Tararua Tramping Club, NZ) 1947–
Tarkington, Newton Booth *The gentleman from Indiana* 1899
In the arena 1905
Penrod 1914
Tassie, L. J. *The physics of elementary particles* 1973
'Tate, Richard' (Anthony Masters) *Birds of a bloodied feather* 1974
Taubes, Frederic *The painter's dictionary of materials and methods* 1973
Tawney, Richard Henry *Religion and the rise of capitalism* 1926
Taylor, Alan John Percivale *The course of German history* 1945
English history, 1914–45 1965
See also: Howat, G. M. D. & Taylor, A. J. P.
Taylor, Anne Marjorie *The language of World War II* 1944
— (rev. ed.) 1948
Taylor, Benjamin Franklin *Between the gates: summer rambles in California* 1878
January and June: being out-door thinkings and fireside musings 1854
Summer-savory, gleaned from rural nooks in pleasant weather 1879
The world on wheels, and other sketches 1874
Taylor, Charles Fayette *The internal-combustion engine in theory and practice* 2 vols. 1960–8
Taylor, Edith *The serpent under it* 1973 (UK 1974)
Taylor, Elizabeth *At Mrs Lippincote's* 1945
The sleeping beauty 1953
Taylor, Frederick *A manual of the practice of medicine* 1890
Taylor, Hugh Stott ed. *A treatise on physical chemistry* 2 vols. 1924
Taylor, John (1753–1824) *Arator: being a series of agricultural essays, practical and political* 1813
— (ed. 2) 1814
Taylor, Michael tr. *C. Metz's Film language: a semiotics of the cinema* 1974
Taylor, Nancy Margaret ed. *Early travellers in New Zealand* 1959
Taylor, Richard *Te ika a Maui; or, New Zealand and its inhabitants* 1855
— (ed. 2) 1870
Taylor, Selwyn Francis, Cotton, L. T., & Murray, J. G. *A short textbook of surgery* 1967
— (ed. 2) 1968
Taylor, Tom *The ticket-of-leave man* c 1863
Technology: the monthly review of training and education for industry 1957–
Technology week 1966–
Tee emm [training memoranda] (Air Ministry) 1941–6
Teilhet, Darwin & Teilhet, Hildegard *Feather cloak murders* 1937
Telegraph, The (Brisbane) 1872– (title varies)
Telegraph and telephone journal 1914–33
Tempest, Paul *Lag's lexicon: a comprehensive dictionary and encyclopædia of the English prison to-day* 1950
'Templar' (Arthur Bassett Hopkins) *The poker manual* 1895
Temple, William (1881–1944) *Citizen and churchman* 1941
Thoughts in war-time 1940
Tennant, Kylie *The battlers* 1941
Foveaux 1939

The joyful condemned 1953
Lost haven 1946 (UK 1947)
Ride on stranger 1943
'Tensas, Madison' (Henry Clay Lewis) *Odd leaves from the life of a Louisiana 'swamp doctor'* 1850
Terkel, Louis ('Studs') ed. *American dreams, lost and found* 1980
Terman, Frederick Emmons *Radio engineering* 1932
— (ed. 4, with title *Electronic and radio engineering*) 1955
Radio engineers' handbook 1943
Terms used in forestry and logging 1905 See United States. Department of Agriculture. Bureau of Forestry
Terry, Ellen (1847–1928) & Shaw, George Bernard (1856–1950) *Ellen Terry and Bernard Shaw: a correspondence* ed. C. St. John 1931
Tetrahedron letters 1959–
Textile Institute *Textile terms and definitions* 1954
— (ed. 3) 1957
'Tey, Josephine' (Elizabeth Mackintosh) *Brat Farrar* 1949
The daughter of time 1951
The Franchise affair 1948
A shilling for candles 1936
The singing sands 1952
To love and be wise 1950
See also: 'Daviot, Gordon'
Thackeray, Christopher ('Kit') *Crownbird* 1979
Theimer, Walter & Campbell, Peter *Encyclopaedia of world politics* 1950
Theodorson, George A. & Theodorson, Achilles G. *A modern dictionary of sociology* 1970
Theology: a monthly journal of historic Christianity 1920–
Theroux, Paul *The consul's file* 1977
The mosquito coast 1981
The Old Patagonian Express: by train through the Americas 1979
Thewlis, J., et al. eds. *Encyclopaedic dictionary of physics* See under title
Thicknesse, Philip *A year's journey through France and part of Spain* 2 vols. 1777
— (ed. 2) 2 vols. 1778
— (ed. 3) 2 vols. 1789
Thilly, Frank tr. *F. Paulsen's Introduction to philosophy* 1895
Thirkell, Angela Margaret *August folly* 1936
Before lunch 1939
The headmistress 1944
This week magazine 1935–
Thomas, Donald Michael *The white hotel* 1981
Thomas, David *Travels through the Western country in the summer of 1816* 1819
Thomas, Dylan Marlais (1914–53) *Collected poems, 1934–52* 1952
Deaths and entrances: poems 1946
18 poems 1934
The map of love: verse and prose 1939
Portrait of the artist as a young dog 1940
A prospect of the sea, and other stories and prose writings ed. D. Jones 1955
Quite early one morning: broadcasts ed. A. T. Davies 1954
Selected letters ed. C. Fitzgibbon 1966
Under Milk Wood: a play for voices 1954
Thomas, Graham Stuart *The old shrub roses* 1955
Thomas, Jack *No banners: the story of A. and H. Newton* 1955
Thomas, Mary *Dictionary of embroidery stitches* 1934
Thomas, Northcote Whitridge *Kinship organisations and group marriage in Australia* 1906
Thomas, Ronald Stuart *An acre of land* 1952
The minster 1953
The stones of the field 1946
Tares 1961
Thomas, Ross *The backup men* 1971
If you can't be good 1973 (UK 1974)
The porkchoppers 1972 (UK 1974)
The Singapore wink 1969
Spy in the vodka 1966 (UK 1967)
See also: 'Bleeck, Oliver'
Thomas, Samuel Evelyn *Elements of economics* 1925
Thompson, Arthur Beeby *Oil-field exploration and development* 2 vols. 1925
Thompson, C. Patrick *Cocktails* 1919
Thompson, Flora *Candleford Green* 1943
Lark Rise 1939
Over to Candleford 1941
Still glides the stream 1948
Thompson, George *Travels and adventures in southern Africa* 1827
Thompson, William Bell *An introduction to plasma physics* 1962
Thompson, William Tappan *Major Jones's sketches of travel* 1848
Thomson, Arthur Landsborough ed. *A new dictionary of birds* 1964

Thomson, John Arthur *The biology of the seasons* 1911
Science old and new 1924
Thomson, June *Death cap* 1973
The long revenge 1974
A question of identity 1978
Thomson, Mortimer Neal *See* 'Doesticks, Q. K. Philander'
Thomson, William *Dictionary of banking: a concise encyclopædia of banking law and practice* 1911
— (ed. 10, by R. W. Jones) 1951
— (ed. 11, by F. E. Perry & F. R. Ryder) 1965
Thoreau, Henry David (1817–62) *Correspondence* ed. W. Harding & C. Bode 1958
Writings ed. B. Torrey & F. B. Sanborn 20 vols. 1906
Thornburg, Newton *Cutter and Bone* 1976
Thornbury, William David *Principles of geomorphology* 1954
Thorstensen, Thomas Clayton *Practical leather technology* 1969
Thurber, James Grover *My life and hard times* 1933
The years with Ross 1959
Thwaites, Reuben Gold ed. *Early western travels, 1748–1846: a series of annotated reprints* 32 vols. 1904–7
Tidyman, Ernest *Dummy* 1974
Tiemann, Hugh Philip *Iron and steel: a pocket encyclopedia, including allied industries and sciences* 1910
— (ed. 2) 1919
Tilden, Josephine Elizabeth *Algae and their life relations: fundamentals of phycology* 1935
Tillyard, Robert John *The insects of Australia and New Zealand* 1926
Tilman, Harold William *The ascent of Nanda Devi* 1937
Nepal Himalaya 1952
Timber trades journal 1873–
Time: the weekly news-magazine 1923– (several eds. used)
Time and tide 1920–
Time out 1968–
Times educational supplement, The 1910–
Times literary supplement, The 1902–
Times of India, The 1838–
Times review of industry, The 1947–
Times-Picayune (New Orleans) 1914–
Tinbergen, Nikolaas *The herring gull's world* 1953
Tindale, Norman Bartlett & Lindsay, Harold Arthur *Rangatira, the high-born: a Polynesian saga* 1959
Titchener, Edward Bradford tr. *W.M. Wundt's Principles of physiological psychology* 1904
Titford, William Jowit *Sketches towards a Hortus Botanicus Americanus* 1811
Titus, Harold *'Timber'* 1922
To-day 1893–1905; 1916–23
Toffler, Alvin *Future shock* 1970
— (Bantam ed.) 1971
Tolkien, John Ronald Reuel (1892–1973) *The fellowship of the ring (The lord of the rings,* part 1) 1954
The hobbit; or, There and back again 1937
The return of the king (The lord of the rings, part 3) 1955
The Silmarillion ed. C. Tolkien 1977
The two towers (The lord of the rings, part 2) 1954
Tomlinson, Alfred Charles *The shaft* 1978
Topeka Daily Capital (Topeka, Kansas) 1879–
Topley, William Whiteman C. & Wilson, Graham Selby *The principles of bacteriology and immunity* 2 vols. 1929
Topsfield, Massachusetts *Town records* 2 vols. 1917–20
Toronto Daily Star 1892–
'Torr, Dominic' (John B. S. Pedler) *The treason line* 1968
Torrey, Bradford *The foot-path way* 1892
Tottle, Charles Ronald *The science of engineering materials* 1966
Tourgée, Albion Winegar *Button's Inn* 1887
See also: 'Churton, Henry'
Townsend, Albert Alan *The structure of turbulent shear flow* 1956
Towster, Julian *Political power in the U.S.S.R., 1917–47: the theory and structure of government in the Soviet state* 1948
Toynbee, Arnold Joseph *Civilization on trial* 1948
East to west: a journey round the world 1958
An historian's approach to religion 1956
A study of history 12 vols. 1934–61
Trade marks journal (Patent Office) 1876–
Trager, George Leonard & Smith, Henry Lee *An outline of English structure* 1951
Traill, Sinclair ed. *Concerning jazz* 1957
ed. *Play that music: a guide to playing jazz* 1956

Trans-action: social science and the community (Washington Univ.) 1963–
Transactions See under the names of particular institutions
Transatlantic review 1959–77
Travers, Ben *Mischief* 1925
Treadgold, Mary *We couldn't leave Dinah* 1941
Treadwell, Bill ed. *Big book of swing* 1946
Tregonning, Kennedy Gordon Phillip *A history of modern Malaya* 1964
'Trevanian' *The Loo sanction* 1973 (UK 1974)
The main 1976 (UK 1977)
'Trevor, Elleston' (Trevor Dudley-Smith) *A place for the wicked* 1968
See also: 'Hall, Adam'
'Trevor, Glen' (James Hilton) *Murder at school* 1931
'Trevor, William' (William Trevor Cox) *The boarding-house* 1965
The children of Dynmouth 1976
Trevor-Roper, Hugh Redwald (Baron Dacre of Glanton) *The last days of Hitler* 1947
Trevor-Roper, Patrick Dacre *Ophthalmology* 1955
Trewin, John Courtenay ed. *Plays of the year* See under title
Tribune 1937–
Trimmer, Eric J., et al. *The visual dictionary of sex* 1977 (UK 1978)
Tripp, Miles Barton *Faith is a windsock* 1952
Five minutes with a stranger 1971
One is one 1968
Tritton, Harold Percy *Time means tucker* 1959
Trollope, Anthony (1815–82) *The American Senator* 3 vols. 1877
The Duke's children 3 vols. 1880
Is he Popenjoy? 3 vols. 1878
John Caldigate 3 vols. 1879
The Kellys and the O'Kellys 3 vols. 1848
Letters ed. B. A. Booth 1951
The Macdermots of Ballycloran 3 vols. 1847
Phineas Finn, the Irish Member 2 vols. 1869
Phineas Redux 2 vols. 1874
The Prime Minister 4 vols. 1876
Rachel Ray 2 vols. 1863
Tales of all countries 1861 — (2nd ser.) 1863
Troubridge, Laura (1853–1929) *Life amongst the Troubridges* ed. J. Hope-Nicholson 1966
Troup, Gordon John Fordyce *Masers* 1959
— (ed. 2, with title *Masers and lasers*) 1963
Troup, Robert Scott *Silvicultural systems* 1928
The silviculture of Indian trees 3 vols. 1921
Truman, Margaret *Harry S. Truman* 1973
Truth (Melbourne): see *Melbourne Truth*
Truth (Sydney) 1890–1958
Truth: a weekly journal (London) 1877–1957
Tucker, Arthur Wilson *Wild talent* 1954 (UK 1955)
Tucker, Denys William tr. *G. Sterba's Freshwater fishes of the world* 1962
Tucker, Thomas George *Introduction to the natural history of language* 1908
Tucker, Wilson *See* Tucker, Arthur Wilson
Tuckett, Mariana *Mariana's diary: a record of a holiday at Falmouth* ed. H. Fox (Royal Cornwall Polytechnic Society) *c* 1975
Tucson Daily Citizen (Tucson, Arizona) 1879–
Tucson magazine (Tucson, Arizona) 1975–
Tunis, Edwin *Indians* 1959
Turner, Bill *Circle of squares* 1969
Sex trap 1968
Turner, Charles Cyril *Aerial navigation of to-day: a popular account of the evolution of aeronautics* 1910
Aircraft of to-day 1917
Turner, Charles Godfrey *The happy wanderer* ed. E. M. R. Rice 1922
Turner, Ernest Sackville *Taking the cure* 1967
Turner, Ethel Sybil *The little larrikin* 1896
Turner, Francis John & Verhoogen, Jean *Igneous and metamorphic petrology* 1951
— (ed. 2) 1960
Turner, George W. *The English language in Australia and New Zealand* 1966
ed. *Good Australian English and good New Zealand English* 1972
Turner, John Evan *A theory of direct realism, and the relation of realism to idealism* 1925
Turner, Victor Witter *Schism and continuity in an African society: a study in Ndembu village life* 1957
Turner, William *An introduction to human anatomy, including the anatomy of the tissues* 2 vols. 1875–7
Tuscaloosa News (Tuscaloosa, Alabama) 1910–
Tuve, Rosemund *Seasons and months: studies in a tradition of Middle English poetry* 1933
'Twain, Mark' (Samuel Langhorne Clemens, 1835–1910) *The American claimant* 1892
Autobiography ed. A. B. Paine 2 vols. 1924
The celebrated jumping frog of Calaveras County, and other sketches ed. J. Paul 1867
Christian Science 1907
A Connecticut Yankee in King Arthur's court 1889

Following the equator: a journey around the world 1897
Letters ed. A. B. Paine 2 vols. 1917 (UK, abridged, 1 vol. 1920)
Letters from Hawaii ed. A. G. Day 1966 (UK 1967)
Letters to his publishers, 1867–1894 ed. H. Hill 1967
The man that corrupted Hadleyburg, and other stories and essays 1900
Old times on the Mississippi 1876
The prince and the pauper 1881 [dated 1882]
Screamers: a gathering of scraps of humour, delicious bits, and short stories 1871
Sketches new and old 1875
Speeches with introd. by W. D. Howells 1910
Speeches with introd. by A. B. Paine 1923
The stolen white elephant 1882
See also: Harte, Francis Bret & 'Twain', Mark
—— & Warner, Charles Dudley *The gilded age* 1873 [dated 1874]
Tweeddale, James George *Materials technology* 2 vols. 1973
Tweney, Charles Frederick *Dictionary of naval and military terms* 1914
Twenhofel, William Henry *Principles of sedimentation* 1939
—— & Shrock, Robert Rakes *Invertebrate paleontology* 1935
Twentieth century, The 1951–
Twisleton, Ellen *Letters, 1852–1862* ed. E. T. Vaughan 1928
Tyrrell, George Walter *The principles of petrology* 1926
Tyrrell, James Williams *Across the Sub-Arctics of Canada* 1897

U

UCT studies in English See *Studies in English* (University of Cape Town)
Udden, A. D. tr. *N. Bohr's The theory of spectra and atomic constitution* 1922
Ulanov, Barry *A history of jazz in America* 1952 (UK 1958)
Ullman, Stephen *The principles of semantics* 1951
Ulverston News (Ulverston, Cumbria) 1883–
Undercurrents 1972–
Underhill, Evelyn *Mysticism: a study in the nature and development of man's spiritual consciousness* 1911
Underwood, Benton J. *Experimental psychology: an introduction* 1949
'Underwood, Michael' (John Michael Evelyn) *Murder with malice* 1977
A pinch of snuff 1974
Reward for a defector 1973
United Free Church of Scotland *Missionary record* (title varies) 1901–29
United Kingdom. Admiralty *Manual of seamanship* 2 vols. 1908–9
— Vol. II (rev. ed.) 1923
Manual of seamanship for boys' training ships of the Royal Navy 1883
United Kingdom. Advisory Committee for Aeronautics (*later* Aeronautical Research Committee: *later* Aeronautical Research Council) *Technical reports* 1910–
United Kingdom. Department of Education and Science *A Language for life. Report of the Committee of Inquiry..under the chairmanship of Sir Alan Bullock* 1975
United Kingdom. Department of Employment *Classification of occupations and directory of occupational titles* 3 vols. 1972
United Kingdom. Department of Scientific and Industrial Research. Forest Products Research Laboratory *A handbook of hardwoods* 1956 (for ed. 2, see below)
A handbook of softwoods 1957
United Kingdom. Department of the Environment. Building Research Establishment *Handbook of hardwoods* (ed. 2, rev. by R. H. Farmer) 1972 (for ed. 1, see above)
United Kingdom. General Medical Council (*later* British Pharmacopœia Commission) *Approved names* 1957–66; 1970–
United Kingdom. General Register Office *Classification of occupations* 1960
United Kingdom. Meteorological Office *Meteorological glossary* 1918 (and several later editions used)
United Kingdom. Ministry of Agriculture *Journal* 1919–
United Kingdom. Ministry of Information *Manpower: the story of Britain's mobilization for war* 1944
United Kingdom. Ministry of Labour *A dictionary of occupational terms, based on the classification of occupations used in the census of population, 1921* 1927

United Kingdom. Parliamentary papers [*Note* Volume numbers in citations refer mostly to the bound collections in the Bodleian Library. The second page number in citations is the starting-page of the particular report, etc., in the overall numbering within the volume.]
Committee on Broadcasting, 1960 (Chairman: H. Pilkington) *Report* 1962 in *Parl. Papers 1961–2* (Cmnd. 1753) IX. 259
Committee of Inquiry on Decimal Currency (Chairman: the Earl of Halsbury) *Report* 1963 in *Parl. Papers 1962–3* (Cmnd. 2145) XI. 195
Committee on Higher Education, 1961–3 (Chairman: Lord Robbins) *Higher education: report* 1963 in *Parl. Papers 1962–3* (Cmnd. 2154) XI. 639
United Kingdom. Patent Office *Official journal* (title varies) 1884–
See also *Trade marks journal*
United Service journal and naval and military magazine (title varies) 1829–43
United States. Arctic, Desert and Tropic Information Center *Glossary of arctic and subarctic terms* 1955
United States. Bureau of American Ethnology *Annual report* 1881–1933
Bulletin 1886–
United States. Bureau of Fisheries *Report* See: United States. Commissioner of Fisheries
United States. Bureau of Forestry *See* —— Department of Agriculture
United States. Bureau of Standards *See* —— National Bureau of Standards
United States. Commissioner of Fisheries *Report* (title varies) 1873–1941
United States. Commissioner of Patents *Annual report* 1839–
United States. Congress. Debates *Congressional globe, The: containing the debates and proceedings..of Congress* 1834–73
Congressional record, The: containing the proceedings and debates..of the Congress 1873–
Debates and proceedings in the Congress of the United States, 1789–1824 42 vols. 1834–56
Register of debates in Congress, 1824–37 14 vols. 1825–37
Senate hearings *Illicit narcotics traffic: hearings before the Subcommittee..of the Committee on the Judiciary* (84th Congress, 1st Session) 1956
United States. Department of Agriculture *Bulletin* 1913–27
Farmers' bulletin 1889–
Yearbook 1894– (1923–5 with title *Agricultural yearbook*; 1926– with title *Yearbook of agriculture*)
Bureau of Animal Industry *Annual report* 1884–1911
Bureau of Forestry *Bulletin* 1887–1913
Terms used in forestry and logging 1905 (*Bulletin* No. 61)
United States. Geological Survey *Annual report* 1880–1932
Bulletin 1883–
Monographs 1882–1929
Professional papers 1902–
Water-supply papers (title varies) 1896–
United States. Laws and statutes *The statutes at large of the United States of America* 1845–
United States. National Bureau of Standards *Journal of research* 1928–
United States. National Cancer Institute *See* United States. Public Health Service
United States. National Museum *Bulletin* 1875–
Proceedings 1878–
Reports 1884–
United States. Patent Office *Official gazette* 1872–
United States. Public Health Service. National Cancer Institute *Journal* 1940–
United States *Report of the Commissioner of Indian Affairs to the Secretary of the Interior* (title varies) 1825–
United States *Senate hearings* VIII *See* United States. Congress. Senate hearings
United States. Statutes *See* United States. Laws and statutes
United States. War Department *Reports of explorations and surveys to ascertain the most practicable and economical route for a railroad from the Mississippi River to the Pacific Ocean made in 1853–54* 12 vols. 1855–9
Short guide to Great Britain 1942
United States magazine and democratic review 1837–51
United States 1980/81 (Penguin Travel Guides) 1979
United Telephone Company, Ltd. *List of subscribers..classified into professions and trades* (ed. 6) 1885 [reproduced in facsimile in *Three Victorian telephone directories* 1970]
List of subscribers: instructions 1885 [*ibid.*]
Universe, The 1860–

Updike, Daniel Berkeley *Printing types: their history, forms and use* 2 vols. 1922
Updike, John Hoyer *The coup* 1978 (UK 1979)
Couples 1968
Rabbit, run 1960 (UK 1961)
Upfield, Arthur William *Bony and the white savage* 1961
Bushranger of the skies 1940
Murder must wait 1953
The will of the tribe 1962
Urban, Wilbur Marshall *Valuation: its nature and laws; being an introduction to the general theory of value* 1909
Urmson, James Opie ed. *The concise encyclopædia of western philosophy and philosophers* 1960
Philosophical analysis: its development between the two world wars 1956
Urquhart, John W. *Dynamo construction: a practical handbook for the use of engineer-constructors and electricians-in-charge* 1881
Uttley, Alison (Alice Jane Uttley) *The country child* 1931

V

Vachell, Horace Annesley *Dew of the sea, and other stories* 1927
Virgin 1929
Vaile, Pembroke Arnold *Modern golf* 1909
Vaizey, George De Horne *Tangled web* 1974
Vaizey, Jessie *A college girl* 1913
More about Pixie 1904
Vance, John Holbrook *The deadly isles* 1969 (UK 1970)
Vance, Louis Joseph *Baroque* 1923
Cynthia-of-the-minute 1911
The destroying angel 1912 (UK 1913)
Vancouver Province 1898–
Vancouver Sun (title varies) 1888–
'Vandercook' (William Cooper) *Yachts and yachting* 1873
Van der Post, Laurens Jan *The heart of the hunter* 1961
The lost world of the Kalahari 1958
A story like the wind 1972
Van Deventer, C. N. *An introduction to general aeronautics* 1965
Van de Wetering, Janwillem *The corpse on the dike* 1976 (UK 1977)
Tumbleweed 1976
Vanity fair (New York) 1914–36
Van Nostrand's scientific encyclopedia 1938
— (ed. 3) 1958
Van Oosten, Henrik *The Dutch gardener; or, The compleat florist* [anon. tr.] 1703
Van Sterkenburg, P. G. J., et al. eds. *Lexicologie: een bundel opstellen voor F. de Tollenaere* 1977
Van Thal, Herbert Maurice ed. *Fanfare for Ernest Newman* 1955
Van Tyne, Josselyn & Berger, Andrew John *Fundamentals of ornithology* 1959
Van Vechten, Carl *Nigger heaven* 1926
Varaday, Desmond *Gara-Yaka: the story of a cheetah* 1964
Gara-Yaka's domain 1966
Variety 1905–
Vaughan-Thomas, Wynford *Anzio* 1961
Vawter, Bruce *The conscience of Israel: pre-exile prophets and prophecy* 1961
Venkatachalam, Govindraj *Dance in India* 1948
Venkataraman, Krishnasami ed. *The chemistry of synthetic dyes* 8 vols. 1952–78
Venner, Norman *The imperfect impostor* 1925
Verbatim: the language quarterly 1974–
Vermandel, Janet Gregory *Dine with the devil* 1970
Vermont. Board of Agriculture *Annual report* (title varies) 1872–1908
Verne, Jules *The clipper of the clouds* [anon. tr.] 1887
Verney, John *Friday's tunnel* 1959
Verney, Maria Margaret ed. *Verney letters of the eighteenth century, from the manuscripts at Claydon House* 2 vols. 1930
Verschoyle, Terence Trevor Hamilton tr. *A. E. Haas's Atomic theory* 1927
Vesey-Fitzgerald, Brian Seymour ed. *The book of the dog* 1948
Veterinary journal 1875–95; 1900–48
Veterinary record 1888–
Victoria, Queen See Fulford, Roger T. B.
Victoria (Australia). Mining Department *Mining surveyors' reports* 1859–63
Vidal, Gore *The city and the pillar* 1948 (UK 1949)
In a yellow wood 1947
Kalki 1978
'Vigilans' (Eric Honeywood Partridge) *Chamber of horrors: a glossary of official jargon both English and American* 1952

Village voice (New York) 1955–
Villiers, Douglas ed. *Next year in Jerusalem* 1976
Vince-Prue, Daphne *Photoperiodism in plants* 1975
Vincent, E. A. tr. *A. Rittmann's Volcanoes and their activity* 1962
Vincent, William Thomas *Recollections of Fred Leslie* 2 vols. 1894
Vines, Walter Sherrard *Humours unreconciled* 1928
Virginia *Calendar of Virginia state papers and other manuscripts* ed. W. P. Palmer et al. 11 vols. 1875–93
Virology 1955–
Visser, Fredericus Theodorus *An historical syntax of the English language* 3 parts in 4 vols. 1963–73
Vitamins and hormones 1943–
Viteles, Morris Simon *Industrial psychology* 1933
Voegelin, Charles Frederick & Voegelin, Florence Marie R. *Classification and index of the world's languages* 1977
Vogue 1892–
Voice (New York) 1884–95
Vole 1977–81
Von Arnim, Mary Annette *The enchanted April* 1922
Fräulein Schmidt and Mr. Anstruther 1907
Introduction to Sally 1926
Vonnegut, Kurt *Cat's cradle* 1963
Slaughterhouse-five; or, The children's crusade 1969
Von Engeln, Oscar Diedrich *Geomorphology* 1942
Von Neumann, John (1903–57) *Collected works* ed. A. H. Taub 6 vols. 1961–3
Von Sternberg, Joseph *Fun in a Chinese laundry* 1965 (UK 1966)
Voorloper: an interim presentation of materials for a dictionary of South African English on historical principles ed. W. Branford et al. 1976

W

Waddington, Conrad Hal *Principles of embryology* 1956
Wadman, Howard Douglas *Life sentence: a New Zealand play in three acts* 1949
Wahlstrom, Ernest Eugene *Introduction to theoretical igneous petrology* 1950
Wain, John Barrington *The contenders* 1958
Hurry on down 1953
ed. *Interpretations: essays on twelve English poems* 1955
Strike the father dead 1962
A travelling woman 1959
Wainwright, John William *The big tickle* 1969
Death in a sleeping city 1965
The devil you don't 1973
Dig the grave and let him lie 1971
Do nothin' till you hear from me 1977
Edge of extinction 1968
The evidence I shall give 1974
The hard hit 1974
High-class kill 1973
The last buccaneer 1971
A nest of rats 1977
Night is a time to die 1972
A pride of pigs 1973
Requiem for a loser 1972
Square dance 1975
The take-over men 1969
Talent for murder 1967
A touch of malice 1973
Who goes next? 1976
Wakeman, Reginald Leslie *The chemistry of commercial plastics* 1947
Waldron, Ronald Alan *Sense and sense development* 1967
Walker, Andrew Jamieson & Mott, Owen E. tr. *A. F. Holleman's A text-book of organic chemistry* 2 parts 1903
— (ed. 2) 1907
Walker, Ernest Pillsbury, et al. *Mammals of the world* 3 vols. 1964–7
Walker, James *Introduction to physical chemistry* 1899
Walker, Martin *The National Front* 1977
Wall, Bernard, et al. tr. *P. Teilhard de Chardin's The phenomenon of man* 1959
Wall Street Journal 1889–
Wallace, Richard Horatio Edgar *The double* 1928
The feathered serpent 1927
The flying Fifty-Five 1922
The Flying Squad 1928
The gunner 1928
The lady of Ascot 1930
The mind of Mr. J. G. Reeder 1925
The missing million 1923

Wallace (*cont.*)
More educated Evans 1926
Room 13 1924
When the gangs came to London 1932
Wallace, Robert *The rural economy and agriculture of Australia and New Zealand* 1891
Wallis, Arthur James & Blair, Charles F. *Thunder above* 1956 (UK 1959)
Waln, Nora *The house of exile* 1933
Walpole, Horace (1717–97) *Correspondence* ed. W. S. Lewis et al. (Yale ed.) 40 vols. 1937–80
Letters ed. Mrs P. Toynbee 16 vols. 1903–5
— Suppl. ed. P. Toynbee 3 vols. 1918–25
Walpole, Hugh Seymour *The cathedral* 1922
The dark forest 1916
Fortitude 1913
The fortress 1932
Hans Frost 1929
Judith Paris 1931
Rogue Herries 1930
The secret city 1919
Vanessa 1933
The young enchanted 1921
Walsh, John James *Understanding paraplegia* 1964
Walsh, John William Tudor *Photometry* 1926
Wambaugh, Joseph *The black marble* 1978
The blue knight 1972 (UK 1973)
Wandsworth Borough News 1884–
Wang Gungwu ed. *Malaysia: a survey* 1964
Wanganui Chronicle (Wanganui, NZ) 1856–
War illustrated, The 1914–19
War illustrated, The 1939–47
War weekly 1939–41
Ward, Edmund *The hanged man* 1976
Ward, Elizabeth Stuart (Phelps) *The story of Avis* 1877
Trotty's wedding tour 1874
Ward, Emma Georgina Elizabeth *Outside Paris during the two sieges* 1871
Ward, Mrs. Humphry *See* Ward, Mary Augusta
Ward, James *The realm of ends; or, Pluralism and theism* 1911
Ward, Mary Augusta (Mrs. Humphry Ward) *Diana Mallory* 1908
The marriage of William Ashe 1905
Ward, Russel Braddock ed. *The Penguin book of Australian ballads* See under title
Ward-Jackson, Cyril Henry *It's a piece of cake; or, R.A.F. slang made easy* 1943
— (rev. ed.) 1945
Warder, John A. *Hedges and evergreens* 1858
Warming, Johannes Eugenius Bülow *Oecology of plants* 1909
Warner, Francis R. Le P. *Early poems* 1964
Killing time 1976
Lying figures 1972
Maquettes 1972
Meeting ends 1974
Warner, James Franklin *A universal dictionary of musical terms* 1842
Warner, Pelham Francis *How we recovered the Ashes* 1904
My cricketing life 1921
Warner, Sylvia Townsend *Lolly Willowes; or, The loving huntsman* 1926
Warren, Howard Crosby ed. *Dictionary of psychology* 1934
Warren, Robert Penn *Incarnations: poems 1966–68* 1968 (UK 1970)
Night rider 1939
Warwick, Wayne *Surfriding in New Zealand* 1968
Washington, George *Diaries, 1748–1799* ed. J. C. Fitzpatrick 4 vols. 1925
Washington Academy of Sciences *Journal* 1911–
Washington Post 1877–
Wason, Robert Alexander *Happy Hawkins* 1909
Waterhouse, Keith Spencer *Billy Liar* 1959
Waterman, Talbot Howe *The physiology of Crustacea* 2 vols. 1960–1
Waters, Stanley *Indentures indorsed* 1967
Watkins, Alfred *Photography: the Watkins manual of exposure and development* 1902
— (ed. 2) 1903
Watson, Alfred Edward Thomas *The turf* 1898
ed. *The young sportsman* 1900
Watson, Colin *Bump in the night* 1960
Coffin scarcely used 1958
Hopjoy was here 1962
Watson, George Marr *The Roxburghshire wordbook* 1923
Watson, Henry Clay *Camp fires of the revolution; or, The war of independence* 1850
Watson, Hewett Cottrell *Cybele Britannica; or, British plants and their geographical relations* 4 vols. 1847–59
Watson, James Dewey *The double helix: a personal account of the discovery of DNA* 1968
Watson, John Fanning *Annals of Philadelphia* 1830
Watson, William *A text-book of physics* 1899
Watson-Watt, Robert Alexander *Three steps to victory: a personal account* 1957

Watsonia: journal of the Botanical Society of the British Isles 1949–
Watt, John Mitchell & Breyer-Brandwijk, Maria Gerdina *The medicinal and poisonous plants of southern Africa* 1932
— (ed. 2, with title —— *of southern and eastern Africa*) 1962
Waugh, Alexander Raban *The last chukka* 1928
The loom of youth 1917
Waugh, Auberon Alexander *The foxglove saga* 1960
Waugh, Evelyn Arthur St. John (1903–66) *Black mischief* 1932
Brideshead revisited 1945
Decline and fall 1928
Diaries ed. M. Davie 1976
A handful of dust 1934
Labels: a Mediterranean journal 1930
Letters ed. M. Amory 1980
The loved one 1948
Men at arms 1952
Ninety-two days 1934
Officers and gentlemen 1955
Put out more flags 1942
Remote people 1931
Scoop 1938
Unconditional surrender 1961
Vile bodies 1930
Waugh, Hillary Baldwin *The con game* 1968
Finish me off 1970 (UK 1971)
Last seen wearing — 1952 (UK 1953)
Parrish for the defense 1974 (UK 1975)
Pure poison 1966 (UK 1967)
Way, Peter *Super-Celeste* 1977
Way to play, The: the illustrated encyclopedia of the games of the world 1975
Wayre, Philip *Wind in the reeds* 1965
Weale, Robert Alexander *The eye and its function* 1960
Weatherly, Keith *The roo shooter* 1968
Weaver, John Ernest & Clements, Frederic Edward *Plant ecology* 1929
— (ed. 2) 1938
Weaver, Robert ed. *Canadian short stories* (2nd ser.) 1968
Webb, Martha Beatrice *My apprenticeship* 1926
Webb, Robert F. *Motorists' dictionary* 1963
Webster, Arthur Gordon *The theory of electricity and magnetism* 1897
Webster, Douglas & Webster, Molly *Comparative vertebrate morphology* 1974
Webster, Jean (Alice Jean Chandler Webster) *Dear enemy* 1915 (UK 1916)
Webster, John *Introduction to fungi* 1970
Webster, Noah *New international dictionary of the English language* ed. W. T. Harris & F. S. Allen 1909
— (ed. 2, ed. W. A. Neilson, T. A. Knott, & P. W. Carhart) 2 vols. 1934
— (ed. 3, ed. P. B. Gove) 2 vols. 1961 (see also *6,000 words*)
Webster, Robert *Gems: their sources, description and identification* 2 vols. 1962
Webster's sports dictionary 1976
Wedgwood, Cicely Veronica *The trial of Charles I* 1964
Wedgwood, Josiah (1730–95) *Selected letters* ed. A. Finer & G. Savage 1965
Week-end review 1930–4
Weekend world (Johannesburg) 1968–
Weekley, Ernest *Something about words* 1935
Words and names 1932
Weekly dispatch 1914–28
Weekly News (Auckland) 1863–
Weekly register 1811–14
Weems, Mason Locke *The life of General F. Marion* 1809
Weeton, Ellen *Journal of a governess, 1807–1825* 2 vols. 1969 (reprint of 1936–9 ed. with new editorial matter by J. J. Bagley & E. Hall)
Weichert, Charles Kipp *Anatomy of the chordates* 1951
Weidman, Jerome *I can get it for you wholesale* 1937 (UK 1939)
Weingarten, Joseph A. *An American dictionary of slang and colloquial speech* 1954
'Welcome, John' (John Brennan) *Hell is where you find it* 1968
Stop at nothing 1959
Wellek, René & Warren, Edward Austin *Theory of literature* 1949
Wellington, John Harold *Southern Africa: a geographical study* 1955
Wells, Alfred Kingsley *Outline of historical geology* 1938
Wells, Herbert George *All aboard for Ararat* 1940
The anatomy of frustration: a modern synthesis 1936
Ann Veronica: a modern love story 1909
Apropos of Dolores 1938
The autocracy of Mr. Parham 1930
Babes in the darkling wood 1940
Brynhild; or, The show of things 1937

Experiment in autobiography 2 vols. 1934
The first men in the moon 1901
The history of Mr. Polly 1910
The holy terror 1939
The invisible man: a grotesque romance 1897
Joan and Peter: the story of an education 1918
Love and Mr. Lewisham 1900
Men like gods 1923
The new Machiavelli 1910 (UK 1911)
The new world order 1940
The shape of things to come 1933
Things to come: a film story based on 'The shape of things to come' 1935
Tono-Bungay 1909
The war of the worlds 1898
The wheels of chance 1896
The work, wealth and happiness of mankind 1931 (UK 1932)
You can't be too careful: a sample of life 1941
See also: Gissing, George Robert & Wells, H. G.
Wells, Kenneth McNeill *The Owl Pen reader* 1969
Wells, Robert *The pastrycook and confectioner's guide* 1889
'Wells, Tobias' (Deloris Florine Forbes) *Dead by the light of the moon* 1967 (UK 1968)
Welty, Eudora *The optimist's daughter* 1973
Wentworth, Harold *American dialect dictionary* 1944
—— & Flexner, Stuart Berg *Dictionary of American slang* 1960
— (new ed.) 1967
— (new ed.) 1975
'Wentworth, Patricia' (Dora Amy Turnbull) *In the balance* 1941 (UK ed. 1942 with title *Danger point*)
Wenyon, Charles Morley *Protozoology* 2 vols. 1926
Wertheim, Willem Frederik *Evolution and revolution: the rising waves of emancipation* [anon. tr.] 1974
Weseen, Maurice Harley *Crowell's dictionary of English grammar and handbook of American usage* 1928
A dictionary of American slang 1934 (UK 1935)
Wesker, Arnold *Chicken soup with barley* (in *New English dramatists* I) 1959
Chips with everything 1962
I'm talking about Jerusalem 1960
The kitchen (in *New English dramatists* II) 1960
Roots 1959
Wesley, John (1703–91) *Journal* ed. N. Curnock 8 vols. 1909–16
Wesleyan-Methodist magazine 1822–1913
West, Anthony *The trend is up* 1960 (UK 1961)
'West, Rebecca' (Cicily Isabel Andrews) *Black lamb and grey falcon: the record of a journey through Yugoslavia in 1937* 2 vols. 1941 (UK 2 vols. 1942)
The fountain overflows 1957
The harsh voice: four short novels 1935
The modern 'Rake's progress' 1934
St. Augustine 1933
The strange necessity: essays and reviews 1928
The thinking reed 1936
West, Richard Gilbert *Pleistocene geology and biology, with especial reference to the British Isles* 1968
West, Richard Leaf *Sketches from Vietnam* 1968
West Lancashire Evening Gazette (Blackpool) 1929–
Western Mail (Cardiff) 1869–
Western Morning News (Plymouth) 1860–
Westheimer, David *Going public* 1973
Westlake, Donald Edwin *Bank shot* 1972
Cops and robbers 1972 (UK 1973)
I gave at the office 1971 (UK 1972)
Up your banners 1969 (UK 1970)
See also: 'Coe, Tucker'
Weston, Carolyn *Poor, poor Ophelia* 1972 (UK 1973)
Rouse the demon 1976 (UK 1977)
Weston, Hubert Claude *Sight, light and efficiency* 1949
— (ed. 2, with title *Sight, light and work*) 1962
Weston, Richard *The universal botanist and nurseryman* 4 vols. 1770–7
Westwater, Frank Lorimer *Teach yourself electronic computers* 1962
— (rev. ed.) 1964
Weyl, Alfred Richard *Guided missiles* 1949
Weyl, Claus Hugo Hermann *The classical groups: their invariants and representations* 1939
Weyman, Stanley John *The new rector* 2 vols. 1891
Shrewsbury: a romance 1898
Wharton, Edith Newbold *The custom of the country* 1913
The fruit of the tree 1907
The greater inclination 1899
The house of mirth 1905
Human nature 1933
Summer 1917

Whatmough, Joshua *Language: a modern synthesis* 1956
Wheatley, Dennis Yates *The man who missed the war* 1945
Wheeler, Alwynne Cooper *The fishes of the British Isles and north-west Europe* 1969
Wheeler, Keith *Epitaph for Mr. Wynn* 1971 (UK 1972)
Wheeler, Robert Eric Mortimer *Archaeology from the earth* 1954
Wheeler, William Morton *Social life among the insects* 1922
Where: information on education (Advisory Centre for Education) 1960–
Whetton, Harry ed. *Practical printing and binding* 1946
Which? (Consumers' Association Ltd.) 1957–
Whig-Standard (Kingston, Ontario): see *Kingston Whig-Standard*
Whishaw, Lorna *As far as you'll take me* 1958
Whitby, Lionel Ernest Howard *Medical bacteriology* 1928
— (ed. 4) 1944
— (ed. 5, by Whitby & M. Hynes) 1951
For later eds. *see* Hynes, Martin
—— & Britton, Cedric John C. *Disorders of the blood* 1935
White, Abraham, Handler, P., & Smith, E. L. ed. *Principles of biochemistry* 1954
— (ed. 4) 1968
White, Alan *The long silence* 1976
White, Elwyn Brooks *Letters* [1908–76] ed. D. L. Guth 1976
White, Florence *Good English food, local and regional* 1952
White, Harvey Elliott *Introduction to atomic spectra* 1934
White, Howard ed. *Raincoast chronicles, first five: stories and history of the B.C. coast* 1975
White, Jon Ewbank Manchip *The garden game* 1973
Nightclimber 1968
White, Judith tr. *N. Poulantzas's Fascism and dictatorship: the Third International and the problem of fascism* 1974
White, Osmar Egmont D. *Silent Reach* 1978
White, Patrick Victor Martindale *The eye of the storm* 1973
Riders in the chariot 1961
The tree of man 1955 (UK 1956)
Voss 1957
White, Peter Royden *Planning for public transport* 1976
White, Stewart Edward *Arizona nights* 1907
Blazed trail stories, and Stories of the wild life 1904
The claim jumpers 1901
Conjuror's house: a romance of the free forest 1903
The forest 1903
The riverman 1908
The rules of the game 1910 (UK 1913)
The silent places 1904
The westerners 1901
White, Terence Hanbury *The elephant and the kangaroo* 1947 (UK 1948)
The once and future king 1958
The sword in the stone 1938
White, Theodore Harold *The making of the President 1960* 1961 (UK 1962)
— *1964* 1965
— *1972* 1973 (UK 1974)
Whiteford, Andrew Hunter *North American Indian arts* 1970
— (new ed.) 1973
Whitehead, Alfred North *Adventures of ideas* 1933
The concept of nature 1920
An enquiry concerning the principles of natural knowledge 1919
Process and reality: an essay in cosmology 1929
Science and the modern world 1925 (UK 1926)
—— & Russell, Bertrand Arthur William *Principia mathematica* 3 vols. 1910–13
Whiteman, Paul & McBride, Mary Margaret *Jazz* 1926
Whitman, Charles *Death out of focus* 1970
Whitman, Walt (1819–92) *Daybooks and notebooks* ed. W. White 3 vols. 1978
Drum-taps 1865
November boughs 1888
Passage to India 1871
Specimen days and Collect 1882–3 (1 vol.)
Whitmore, Frank Clifford *Organic chemistry* 1937
Whitney, Thomas Porter tr. *A. I. Solzhenitsyn's The Gulag archipelago* 3 vols. 1974–8
Whittaker's electrical engineer's pocket-book ed. K. Edgcumbe 1903
— (ed. 4, ed. R. E. Neale) 1920
Whitworth, George Clifford *An Anglo-Indian dictionary: a glossary of Indian terms used in English* 1885
Whyte, William Hollingsworth *The organization man* 1956 (UK 1957)

Wiesengrund-Adorno, Theodor, et al. *The authoritarian personality* 1950
Wiggin, Kate Douglas *Polly Oliver's problem* 1893
Rebecca of Sunnybrook Farm 1903 (UK 1904)
Wigginton, Brooks Eliot ed. *The foxfire book: log dressing; log cabin building; mountain crafts and foods;..and other affairs of plain living* 1972
ed. *Foxfire 3* 1975
Wigglesworth, Vincent Brian *The life of insects* 1964
Wight, John *Mornings at Bow Street* 1824
Wilberforce, Herbert William Wrangham *Lawn tennis* 1889
Wilcox, Ruth Turner *The dictionary of costume* 1969 (UK 1970)
Wilde, Oscar Fingall O'Flahertie Wills (1854–1900) *An ideal husband* 1899
The importance of being earnest: a trivial comedy for serious people 1899
Lady Windermere's fan: a play about a good woman 1893
Letters ed. R. Hart-Davis 1962
The picture of Dorian Gray 1891 (first published in a shorter version in *Lippincott's monthly magazine*, July 1890)
Poems 1881
Ravenna 1878 (Newdigate Prize poem)
A woman of no importance 1894
Wildeblood, Peter *Against the law* 1955
The main chance 1957
Wilder, Harris Hawthorne *A laboratory manual of anthropometry* 1920
Wilder, Thornton Niven *The bridge of San Luis Rey* 1927
Heaven's my destination 1934
Our town 1938
Wilenski, Reginald Howard *The modern movement in art* 1927
Wilkerson, Albert E. ed. *The rights of children: emergent concepts in law and society* 1973
Wilkes' Spirit of the times (title varies) 1859–1902
Wilkins, Mary Eleanor *By the light of the soul* 1906
A New England nun, and other stories 1891
Williams, Bobby Joe *Evolution and origins: an introduction to physical anthropology* 1973
Williams, Bryan L. & Wilson, Keith eds. *A biologist's guide to principles and techniques of practical biochemistry* 1975
Williams, Charles *Man in motion* 1959
Man on a leash 1973 (UK 1974)
Williams, Charles Walter Stansby *All Hallows' Eve* 1945
The greater trumps 1932
The place of the lion 1931
War in heaven 1930
Williams, David *Murder for Treasure* 1980
Treasure by degrees 1977
Williams, Gordon John *Economic geology of New Zealand* 1965
Williams, Howel, Turner, F. J., & Gilbert, C. M. *Petrography* 1954
Williams, John Lee *The territory of Florida* 1837
Williams, Laurence Frederic Rushbrook *The State of Israel* 1957
Williams, Martin T. ed. *The art of jazz: essays on the nature and development of jazz* 1959 (UK 1960)
Williams, Monier (*later* Monier Monier-Williams) *An elementary grammar of the Sanscrit language* 1846
Hinduism 1877
Indian wisdom 1875
Williams, Raymond Henry *Culture and society, 1780–1950* 1958
Keywords 1976
The long revolution 1961
Williams, Samuel Robinson *Magnetic phenomena* 1931
Williams, Tennessee (Thomas Lanier Williams) *Cat on a hot tin roof* 1955
The glass menagerie 1945
Orpheus descending 1958
A streetcar named Desire 1947
Williams, William Carlos *Collected earlier poems* 1951
Collected later poems 1950
Selected essays 1954
Williamson, Henry *The beautiful years* 1921
— (rev. ed.) 1929
Tarka the otter 1927
Williamson, Hugh *Methods of book design* 1956
— (ed. 2) 1966
Williamson, William Durkee *The history of the state of Maine* 2 vols. 1832
Willich, Anthony Florian Madinger *The domestic encyclopædia* 4 vols. 1802
— (US ed., with additions by J. Mease) 5 vols. 1803–4
Elements of the critical philosophy 1798
Willis, Rupert Allan *Pathology of tumours* 1948

Willmott, Peter *Adolescent boys of East London* 1966
The evolution of a community: a study of Dagenham after forty years 1963
Willock, Colin Dennistoun *Death at flight* 1956
Death in covert 1961
The enormous zoo: a profile of the Uganda national parks 1964
Willoughby, Florance Barrett *Alaskans all* 1933
Wills, Gordon & Yearsley, Ronald eds. *Handbook of management technology* 1967
Wilmot, Catherine *An Irish peer on the continent (1801–1803): being a narrative of the tour of Stephen, 2nd Earl Mount Cashell, through France, Italy, etc., as related by Catherine Wilmot* ed. T. U. Sadleir 1920
Wilmot, Martha *More letters from Martha Wilmot: impressions of Vienna, 1819–1829* ed. by the Marchioness of Londonderry & H. M. Hyde 1935
—— & Wilmot, Catherine *The Russian journals of Martha and Catherine Wilmot, 1803–1808* ed. by the Marchioness of Londonderry & H. M. Hyde 1934
Wilson, Angus Frank Johnstone *Anglo-Saxon attitudes* 1956
As if by magic 1973
A bit off the map, and other stories 1957
Hemlock and after 1952
The middle age of Mrs Eliot 1958
No laughing matter 1967
The old men at the zoo 1961
The strange ride of Rudyard Kipling: his life and works 1977
Such darling dodos, and other stories 1950
The wrong set, and other stories 1949
Wilson, Bryan R. *Sects and society: a sociological study of three religious groups in Britain* 1961
Wilson, Charles William (1836–1905) *Mapping the frontier* ed. G. F. G. Stanley 1970
Wilson, Colin Henry *The outsider* 1956
Wilson, Edmund *Axel's castle: a study in the imaginative literature of 1870–1930* 1931
Devil take the hindmost: a year of the slump (US ed. with title *The American jitters*) 1932
I thought of Daisy 1929
The wound and the bow: seven studies in literature 1941
Wilson, Edmund Beecher *The cell in development and inheritance* 1896
— (ed. 2) 1900 (UK 1902)
— (ed. 3 with title —— *and heredity*) 1925
Wilson, George Buckley L. *A dictionary of ballet* 1957
— (rev. ed.) 1961
Wilson, Harold *See* Wilson, James Harold
Wilson, Harriette *Memoirs written by herself* 4 vols. 1825
Wilson, Harry Leon *Ma Pettengill* 1919
Ruggles of Red Gap 1915
— (new ed.) 1917
Somewhere in Red Gap 1916
The spenders 1902
Wilson, Jacqueline *Hide and seek* 1972
Wilson, James Harold *The governance of Britain* 1976
Wilson, John Alexander *Missionary life and work in New Zealand, 1833 to 1862: being the private journal of the late Rev. J. A. Wilson* ed. C. J. Wilson 1889
Wilson, John Boyd *Reason and morals* 1961
Wilson, John Fleming *The land claimers* 1911
Wilson, Marjorie Kerr tr. *K. Z. Lorenz's Man meets dog* 1954
Wilson, Percy & Webb, G. W. *Modern gramophones and electrical reproducers* 1929
Wilson, Richard Albert *The birth of language* 1937
Wilson, Robin J. *Introduction to graph theory* 1972
Wilton, Mary M. (Egerton), Countess of ed. *The art of needlework* [by E. Stone] *c* 1840
Wine, women and war [1918 journal of Howard Vincent O'Brien, published anonymously] ed. K. E. Harriman 1926
Winearls, Jane *Modern dance: the Jooss-Leeder method* 1958
— (ed. 2) 1968
Winick, Charles *Dictionary of anthropology* 1956 (UK 1957, new impr. 1960)
Winslow, Pauline Glen *The counsellor heart* 1980
Death of an angel 1975
The Witch Hill murder 1977
Winstedt, Richard Olof *The Malays: a cultural history* 1947
Winter's crimes ed. G. Hardinge et al. 1969–
Winthrop, Theodore (1828–61) *John Brent* 1862 (UK 1883)
Life in the open air 1863
Winton, John Richard ed. *A dictionary of economic terms* (ed. 3) 1951
Wintrobe, Maxwell Myer *Clinical hematology* 1942

Wireless world 1913–
Wirth, Louis & Shils, Edward Albert tr. *K. Mannheim's Ideology and Utopia: an introduction to the sociology of knowledge* 1936
Wisconsin. State Historical Society *Reports and collections* (title varies) 1855–
Wisden cricketers' almanack (title varies) 1864–
Wisdom, Thomas Henry *High-performance driving for you* 1966
Wister, Owen *The Virginian: a horseman of the plains* 1902
Wittfogel, Karl August *Oriental despotism: a comparative study of total power* 1957
Wittgenstein, Ludwig Josef Johann *Tractatus logico-philosophicus* [anon. tr.] 1922
See also: Anscombe, Gertrude E. M.
Witts, Francis Edward (1783–1854) *The diary of a Cotswold parson* ed. D. Verey 1978
Witwer, Harry Charles *Fighting blood* 1923
From baseball to Boches 1918
The leather pushers 1921
Wodehouse, Pelham Grenville *The adventures of Sally* 1923
Aunts aren't gentlemen: a Jeeves and Bertie story 1974
Bill the conqueror 1924
Blandings Castle and elsewhere 1935
Carry on, Jeeves! 1925
The clicking of Cuthbert 1922
The code of the Woosters 1938
The coming of Bill 1920 (US ed. 1919 with title *Their mutual child*)
Company for Henry 1967
A damsel in distress 1919
Eggs, beans and crumpets 1940
Frozen assets 1964
The heart of a goof 1926
Hot water 1932
The inimitable Jeeves 1923
Jill the reckless 1921 (US ed. 1920 with title *The little warrior*)
Laughing gas 1936
Leave it to Psmith 1924
Lord Emsworth and others 1937
Louder and funnier 1932
The luck of the Bodkins 1935
Meet Mr. Mulliner 1927
Money in the bank 1942 (UK 1946)
Mr. Mulliner speaking 1929
Much obliged, Jeeves 1971
Pearls, girls and Monty Bodkin 1972
Performing flea: a self-portrait in letters 1953
Plum pie 1966
Right ho, Jeeves 1934
Sam the sudden 1925
Service with a smile 1961 (UK 1962)
Something fresh 1915
Tales of St. Austin's 1903
Uncle Fred in the springtime 1939
Very good, Jeeves! 1930
Wolf, Frederick Adolph & Wolf, Frederick Taylor *The fungi* 2 vols. 1947
Wolfe, Bernard *Limbo* 1952 (UK ed. 1953 with title *Limbo '90*)
See also: Mezzrow, Milton
Wolfe, Thomas Clayton (1900–38) *Look homeward, angel: a story of the buried life* 1929 (UK 1930)
The web and the rock 1939
You can't go home again 1940
Wolfe, Tom *The electric kool-aid acid test* 1968
The kandy-colored tangerine-flake streamline baby 1965 (UK 1966)
Woman 1937–
Woman's day (New York) 1937–
Woman's journal 1927–
Woman's own 1932–
Women's Group on Public Welfare. Hygiene Committee *Our towns: a close-up, 1939–42* 1943
Wood, Christopher *'Terrible hard', says Alice* 1970
Wood, Clement & Goddard, Gloria *Dictionary of American slang* 1926
Wood, Eric Stuart *Collins field guide to archaeology* 1963
Wood, Ernest Egerton *Yoga dictionary* 1956
Wood, James Alex F. *North beat* 1973
Wood, Thomas *Cobbers: a personal record of a journey..to Australia* 1934
Woodford, Gwendolyn Green tr. *M. Gignoux's Stratigraphic geology* 1955
Woodforde, James (1740–1803) *The diary of a country parson* ed. J. Beresford 5 vols. 1924–31
Woodgate, Gordon Kemble *Elementary atomic structure* 1970
Woodham-Smith, Cecil Blanche *Florence Nightingale, 1820–1910* 1950
Woodhouse, Martin C. *Blue bone* 1973
Mama doll 1972
Rock baby 1968
Tree frog 1966

Woodhouse, Thomas *Artificial silk* 1927
Woodruff, Hiram Washington *The trotting horse of America: how to train and drive him* 1868
'Woodruff, Philip' (Philip Mason) *The island of Chamba* 1950
Whatever dies 1948
Woods, Katharine Seymour *Rural crafts of England* 1949
'Woods, Sara' (Sara Bowen-Judd) *Done to death* 1974
Enter certain murderers 1966
Enter the corpse 1973
Woodworth, Robert Sessions *Experimental psychology* 1938
— (rev. ed., by Woodworth & H. Schlosberg) 1954
Psychology: a study of mental life 1921 (UK 1922) (Several later editions also used)
Wooldridge, Sidney William & Morgan, Ralph Sisk *The physical basis of geography: an outline of geomorphology* 1937
— (ed. 2, with title *An outline of geomorphology*) 1959
Woolf, Leonard Sidney *The village in the jungle* 1913
Woolf, Virginia (1882–1941) *Between the acts* 1941
The captain's death bed, and other essays ed. L. Woolf 1950
The common reader 1925
— (2nd ser.) 1932
Jacob's room 1922
Letters ed. N. Nicolson & J. Trautmann 6 vols. 1975–80
Mrs. Dalloway 1925
Night and day 1919
Orlando: a biography 1928
Roger Fry: a biography 1940
A room of one's own 1929
To the lighthouse 1927
The voyage out 1915
The waves 1931
A writer's diary ed. L. Woolf 1953
The years 1937
Woollcott, Alexander *While Rome burns* 1934
Woolley, Charles Leonard *Alalakh: an account of the excavations at Tell Atchana, in the Hatay, 1937–1949* 1955
Wootton, Barbara Frances *Social science and social pathology* 1959
Testament for social science 1950
Word 1945–
Word study 1925–69
Words 1934–41
Wordsworth, Dorothy (1771–1855) *Journals* ed. E. de Sélincourt 2 vols. 1941
Work, Milton Cooper *Complete contract bridge* 1929
Work 1889–1924
World archaeology 1969–
World of cricket monthly 1976–
World review 1936–53
World's work (New York) 1900–32
Worrall, Arthur James *English idioms for foreign students* 1932
Worsley, Peter Maurice *The trumpet shall sound: a study of 'cargo' cults in Melanesia* 1957
Worster, William W. tr. *E. Hannover's Pottery and porcelain* See: Rackham, Bernard
Worsthorne, Peregrine Gerard *The socialist myth* 1971
Wrenn, Charles Leslie *Word and symbol* 1967
Wright, Barbara tr. *R. Queneau's Between blue and blue* 1967
Wright, David McKee *Station ballads and other verses* 1897
Wright, Frank Lloyd *Autobiography* 1932
Wright, George Payling & Symmers, William St. Clair eds. *Systemic pathology* 2 vols. 1966
Wright, Harold Bell *The winning of Barbara Worth* 1911
Wright, Peter *The language of British industry* 1974
Wright, Thomas *The romance of the lace pillow* 1919
Wyburn, George McReath, Warbrick, J. G., & Todd, M. E. *Concise anatomy* 1967
Wykes, Alan *Gambling* 1964
Wyld, Henry Cecil *A short history of English* 1914
Wyllie, Peter J. ed. *Ultramafic and related rocks* 1967
Wymondham and Attleborough Express 1975–
Wynd, Oswald *Sumatra seven zero* 1968
Walk softly, men praying 1967
'Wyndham, John' (John Wyndham P. L. B. Harris) *The day of the triffids* 1951
The Midwich cuckoos 1957
The seeds of time 1956
Trouble with lichen 1960

Wynne family *The Wynne diaries, 1789–1820* ed. A. Fremantle 3 vols. 1935–40
Wynter, Sylvia *The hills of Hebron* 1962

X

'X, Malcolm' *See* 'Malcolm X'
Xanthakis, John N. ed. *Solar physics* 1967

Y

Yachting (US) 1907–
Yachting monthly 1906–
Yadin, Yigael *Hazor* 1972
Yale alumni magazine 1937–
Yale literary magazine 1836–
Yankee: a monthly magazine for Yankees everywhere 1935–42; 1945–
Yapp, William Brunsdon *An introduction to animal physiology* 1939
Yearbook of astronomy 1961–
Year-book of wireless telegraphy and telephony, The 1913–25
Year's work in English studies, The 1921–
Yeates, Victor M. *Winged victory* 1934
Yeats, William Butler (1865–1939) *The Countess Kathleen, and various legends and lyrics* 1892
In the seven woods: being poems chiefly of the Irish heroic age 1903
The king of the great clock tower: commentaries and poems 1934
Last poems and two plays 1939
Letters ed. A. Wade 1954
October blast 1927
Per amica silentia lunae 1918
Poems 1895
Poems (variorum ed. by P. Allt & R. K. Alspach) 1957
Responsibilities: poems and a play 1914
Reveries over childhood and youth 1915 (UK 1916)
The trembling of the veil 1922
The wanderings of Oisin, and other poems 1889
The wind among the reeds 1899
Words for music perhaps, and other poems 1932
Yesterday's shopping: the Army & Navy Stores catalogue 1907: a facsimile of the Army & Navy Co-operative Society's 1907 issue of 'Rules of the Society and price list of articles sold at the Stores' introduced by A. Adburgham 1969
Yonge, Charlotte Mary *Abbeychurch; or, Self-control and self-conceit* 1844
Burnt out: a story for mothers' meetings 1879
Chantry House 2 vols. 1886
The chaplet of pearls; or, The white and black Ribaumont 1868
The Christmas mummers 1858
Heartsease; or, The brother's wife 2 vols. 1854
Lady Hester; or, Ursula's narrative 1874
Magnum bonum; or, Mother Carey's brood 3 vols. 1879
My young Alcides: a faded photograph 2 vols. 1875
Nuttie's father 1885
An old woman's outlook in a Hampshire village 1892
Our new mistress; or, Changes at Brookfield Earl 1888
The pillars of the house; or, Under wode, under rode 4 vols. 1873
Pioneers and founders; or, Recent workers in the mission field 1871
The pupils of St. John the Divine 1868 (dated 1867)
The release; or, Caroline's French kindred 1896
That stick 2 vols. 1892
The three brides 2 vols. 1876
The two guardians; or, Home in this world 1852
The two sides of the shield 2 vols. 1885
Womankind 1876
'York, Andrew' (Christopher Nicole) *The eliminator* 1966
The expurgator 1972
'Yorke, Margaret' (Margaret Beda Nicholson) *Grave matters* 1973
Mortal remains 1974
The silent witness 1972
Yorkshire Post 1866–
Young, Alexander Bell Filson *The complete motorist: being an account of the evolution and construction of the modern motor car* 1904
Young, Charles Augustus *A text-book of general astronomy for colleges and scientific schools* 1888 (UK 1889)
Young, Emily Hilda *William* 1925
Young, Francis Brett *Portrait of Clare* 1927
Young, Geoffrey Winthrop ed. *Mountain craft* 1920
Young, John Zachary *An introduction to the study of man* 1971

Young (*cont.*)
The life of vertebrates 1950
A model of the brain 1964
Young, Wayland Hilton *Eros denied* 1965
Young Englishwoman, The 1865–77
Yoxall, Harold Waldo *A fashion of life* 1966
'Yuill, P. B.' (Gordon Maclean Williams) *The bornless keeper* 1974
Yule, George Udny *An introduction to the theory of statistics* 1911
— (ed. 11, by Yule & M. G. Kendall) 1937

Z

Zandvoort, Reinard Willem *A handbook of English grammar* 1957
—— et al. *Wartime English: materials for a linguistic history of World War II* 1957
Zangwill, Israel *Ghetto comedies* 1907
Ghetto tragedies 1893
Zeuner, Friedrich Eberhard *Dating the past: an introduction to geochronology* 1946
— (ed. 4) 1958

Zigrosser, Carl & Gaehde, Christa M. *A guide to the collecting and care of original prints* 1965
Zigzag 1969–
Zipf, George Kingsley *The psycho-biology of language: an introduction to dynamic philology* 1935 (UK 1936)
Zuckerman, Solly *The social life of monkeys and apes* 1932
Zweig, Ferdynand *The new acquisitive society* 1976